CRC Handbook
of
Chemistry and Physics

79th Edition

EDITORIAL ADVISORY BOARD

CRC Handbook
of
Chemistry and Physics

A Ready-Reference Book of Chemical and Physical Data

HANDBOOK OF CHEMISTRY AND PHYSICS

1998-1999

79th

EDITION

CRC PRESS LLC

Editor-in-Chief

David R. Lide, Ph.D.

Former Director, Standard Reference Data
National Institute of Standards and Technology

CRC Press
Boca Raton   Boston   London   New York   Washington, D.C.

PREFACE
79th Edition

The 79th Edition of the *Handbook of Chemistry and Physics* features updates and expansions of a number of tables, as well as the introduction of several new topics. Among the new tables are:

- Diamagnetic Susceptibility of Selected Organic Compounds
- Magnetic Susceptibility of the Elements and Inorganic Compounds
- Spectroscopic Constants of Diatomic Molecules
- Dielectric Strength of Insulating Materials
- Carbohydrate Names and Symbols
- Global Temperature Trend, 1866-1996

Many additional compounds have been incorporated in the table Concentrative Properties of Aqueous Solutions: Density, Refractive Index, Freezing Point Lowering, and Viscosity. The section on Scientific Abbreviations and Symbols has been expanded significantly, as has the table Threshold Limits for Airborne Contaminants. Two previous tables on dielectric constants of liquids have been consolidated into a single table with a more convenient format, and the table on Acid-Base Indicators has been recast to make it easier to locate indicators of interest. The long table on Line Spectra of the Elements, which appeared in the *Handbook* in very small type for a great many years, has been revised and reformatted in a more legible fashion. Finally, the following tables have been completely updated with the most recent data available:

- Atomic Masses and Abundances
- Critical Constants
- Enthaply of Vaporization
- Table of the Isotopes
- Electron Affinities
- Atmospheric Concentration of Carbon Dioxide,1958-1996

The Editor will appreciate suggestions on new topics for the *Handbook* (and on older tables whose usefulness is questionable). Address all comments to Editor, *Handbook of Chemistry and Physics*, CRC Press LLC, 2000 Corporate Blvd. N. W., Boca Raton, FL 33431. Comments may also be sent by electronic mail to dlide@earthlink.net.

The *Handbook of Chemistry and Physics* is dependent on the efforts of many contributors throughout the world. The list of current contributors follows this Preface. Many valuable suggestions have come from the Editorial Advisory Board. Special thanks are due to H.P.R. Frederikse, who made major improvements in the section on Properties of Solids during his tenure as Associate Editor. Finally, the production staff at CRC Press is instrumental to its success. I wish particularly to thank Paul Gottehrer for the meticulous care and dedication he showed over many years as production manager, and to commend Susan Fox for an excellent production job on this edition.

David R. Lide
February 26, 1998

This Edition is Dedicated to my Uncle, William Murrell Lide

Note on the Ordering of Chemical Compounds: The decision on the order in which to list chemical compounds in a table is always difficult. An alphabetical list by name has the disadvantage that several different synonyms are often in common use, with the result that a reader may conclude incorrectly that a compound is not present if he looks it up under the wrong name. Listing by line formula is satisfactory for simple inorganic compounds, but is cumbersome for organics. A listing by molecular formula is attractive because clear rules can be given for locating a compound, but the user may have to go to some effort to determine the molecular formula. In this book the choice is made on pragmatic grounds. The long tables, Physical Constants of Organic Compounds and Physical Constants of Inorganic Compounds, are ordered by systematic name, but indexes to synonyms, formulas, and CAS Registry Numbers are provided. If the table is very short and includes only common substances, the listing is usually alphabetical by common name or formula. The remaining tables are ordered by molecular formula using a modification of the Hill convention. In this convention the molecular formula is written with C first, H second, and then all other elements in alphabetical order of their chemical symbols. For tables with organic compounds only, the sequence of entries is determined by the alphabetical order of elements in the molecular formula and the number of atoms of each element, in ascending order, e.g., C_3H_7Cl, C_3H_7N, C_3H_7NO, $C_3H_7NO_2$, etc. (For organic compounds, a quick way to determine the molecular formula is to use the Physical Constants of Organic Compounds table, which starts on Page 3-1, and its synonym index on Page 3-586.) In tables containing non-carbon compounds, those are listed first, followed by a separate listing of compounds that do contain carbon. This is in contrast to the strict Hill convention as followed by Chemical Abstracts Service, where the molecular formulas beginning with A and B precede the formulas for carbon-containing compounds, while those beginning with D... Z follow. For tabular displays, as opposed to an index, it appears more convenient to the user if the non-carbon compounds are listed as a block, rather than being split by the longer list of carbon compounds.

CURRENT CONTRIBUTORS

Douglas Ambrose
9 The Crest
Surbiton
Surrey KT5 8JZ
England

Lev I. Berger
California Institute of Electronics
and Materials Science
2115 Flame Tree Way
Hemet, California 92545

R.W. Clay
Department of Physics and
Mathematical Physics
The University of Adelaide
Adelaide, South Australia 5005
Australia

A. K. Covington
Department of Chemistry
University of Newcastle
Newcastle upon Tyne NE1 7RU
England

Hans Dolezalek
1812 Drury Lane
Alexandria, Virginia 22307

K.F. Eckerman
Oak Ridge National Laboratory
Oak Ridge, Tennessee 37831

Robert B. Fox
6115 Wiscassett Rd.
Bethesda, Maryland 20816

H. P. R. Frederikse
9625 Dewmar Lane
Kensington, Maryland 20895

J.R. Fuhr
Atomic Physics Division
National Institute of Standards and
Technology
Gaithersburg, Maryland 20899

A.C. Gregory
Department of Physics and
Mathematical Physics
The University of Adelaide
Adelaide, South Australia 5005
Australia

Karl A. Gschneidner
Ames Laboratory
Energy and Mineral Resources
Research Institute
Iowa State University
Ames, Iowa 50011

C. R. Hammond
17 Greystone Rd.
West Hartford, Connecticut 06107

Robert F. Hampson
Chemical Kinetics Division
National Institute of Standards and
Technology
Gaithersburg, Maryland 20899

Norman E. Holden
National Nuclear Data Center
Brookhaven National Laboratory
Upton, New York 11973

Henry V. Kehiaian
ITODYS
1 rue Guy de la Brosse
75005 Paris
France

J. A. Kerr
Atmospheric Chemistry Group
EAWAG
CH-8600 Dubendorf-Zurich
Switzerland

William F. Krupke
Lawrence Livermore Laboratory
Livermore, California 94550

Sharon G. Lias
National Institute of Standards and
Technology
Gaithersburg, Maryland 20899

Arthur Maki
15012 24 Ave., S. E.
Mill Creek, Washington 98012

William C. Martin
Atomic Physics Division
National Institute of Standards and
Technology
Gaithersburg, Maryland 20899

Thomas M. Miller
Air Force Research Laboratory/
VSBP
29 Randolph Rd.
Hanscom AFB, Massachusetts
01731-3010

Dale L. Perry
Lawrence Berkeley Laboratory
1 Cyclotron Road
Berkeley, California 94720

Sidney L. Phillips
Camatx/Basic Chemistry
171 El Toyonal
Orinda, California 94563

Joseph Reader
Atomic Physics Division
National Institute of Standards and
Technology
Gaithersburg, Maryland 20899

B. N. Taylor
Physics Laboratory
National Institute of Standards and
Technology
Gaithersburg, Maryland 20899

Thomas G. Trippe
Particle Data Group
Lawrence Berkeley Laboratory
1 Cyclotron Road
Berkeley, California 94720

Petr Vanýsek
Department of Chemistry
Northern Illinois University
DeKalb, Illinois 60115

Wolfgang L. Wiese
Atomic Physics Division
National Institute of Standards and
Technology
Gaithersburg, Maryland 20899

Christian Wohlfarth
Institut für Physikalische Chemie
Martin Luther University
D-06217 Merseburg
Germany

TABLE OF CONTENTS

SECTION 5: THERMOCHEMISTRY, ELECTROCHEMISTRY, AND KINETICS

SECTION 6: FLUID PROPERTIES

SECTION 7: BIOCHEMISTRY

SECTION 8: ANALYTICAL CHEMISTRY

SECTION 9: MOLECULAR STRUCTURE AND SPECTROSCOPY

SECTION 10: ATOMIC, MOLECULAR, AND OPTICAL PHYSICS

SECTION 11: NUCLEAR AND PARTICLE PHYSICS

SECTION 12: PROPERTIES OF SOLIDS

SECTION 13: POLYMER PROPERTIES

SECTION 14: GEOPHYSICS, ASTRONOMY, AND ACOUSTICS

Section 1
Basic Constants, Units, and Conversion Factors

Section 1
Basic Constants, Units, and Conversion Factors

FUNDAMENTAL PHYSICAL CONSTANTS

The 1986 CODATA Recommended Values

By E. Richard Cohen and Barry N. Taylor as published in the *Journal of Research of the National Bureau of Standards*, 92, 85, 1987. Discussions of the background, data selection and evaluation procedures are presented in CODATA Bulletin Number 63, November 1986, ''The 1986 Adjustment of the Fundamental Physical Constants'', a Report of the CODATA Task Group on Fundamental Physical Constants (36 pages) published by Pergamon Press.

The 1986 recommended values of the fundamental physical constants are given in five tables. Table 1 is an abbreviated list containing the quantities which should be of greatest interest to most users. Table 2 is a more complete compilation. Table 3 is a list of related ''maintained units and standard values.'' Table 4 contains a number of scientifically, technologically, and metrologically useful energy conversion factors. Table 5 is an extended covariance matrix containing the variances, covariances, and correlation coefficients of the unknowns and a number of different constants (included for convenience) from which the like quantities of other constants may be calculated. (B. N. Taylor, W. H. Parker, and D. N. Langenberg, *Rev. Mod. Phys.*, 41, 375, 1969. Such a matrix is necessary because the variables in a least-square adjustment are correlated.

Table 1
SUMMARY OF THE 1986 RECOMMENDED VALUES OF THE FUNDAMENTAL PHYSICAL CONSTANTS

Quantity	Symbol	Value	Units	Relative Uncertainty (ppm)
speed of light in vacuum	c	299 792 458	$m\,s^{-1}$	(exact)
permeability of vacuum	μ_o	$4\pi \times 10^{-7}$	$N\,A^{-2}$	
		$=12.566\,370\,614\ldots$	$10^{-7}\,N\,A^{-2}$	(exact)
permittivity of vacuum	ϵ_o	$1/\mu_o c^2$		
		$=8.854\,187\,817\ldots$	$10^{-12}\,F\,m^{-1}$	(exact)
Newtonian constant of gravitation	G	6.672 59(85)	$10^{-11}\,m^3\,kg^{-1}\,s^{-2}$	128
Planck constant	h	6.626 0755(40)	$10^{-34}\,J\,s$	0.60
$h/2\pi$	$\hbar$	1.054 572 66(63)	$10^{-34}\,J\,s$	0.60
elementary charge	e	1.602 177 33(49)	$10^{-19}\,C$	0.30
magnetic flux quantum, $h/2e$	Φ_o	2.067 834 61(61)	$10^{-15}\,Wb$	0.30
electron mass	m_e	9.109 3897(54)	$10^{-31}\,kg$	0.59
proton mass	m_p	1.672 6231(10)	$10^{-27}\,kg$	0.59
proton-electron mass ratio	m_p/m_e	1836.152 701(37)		0.020
fine-structure constant, $\mu_o ce^2/2h$	α	7.297 353 08(33)	10^{-3}	0.045
inverse fine-structure constant	α^{-1}	137.035 9895(61)		0.045
Rydberg constant, $m_e c\alpha^2/2h$	R_∞	10 973 731.534(13)	m^{-1}	0.0012
Avogadro constant	N_A, L	6.022 1367(36)	$10^{23}\,mol^{-1}$	0.59
Faraday constant, $N_A e$	F	96 485.309(29)	$C\,mol^{-1}$	0.30
molar gas constant	R	8.314 510(70)	$J\,mol^{-1}\,K^{-1}$	8.4
Boltzmann constant, R/N_A	k	1.380 658(12)	$10^{-23}\,J\,K^{-1}$	8.5
Stefan–Boltzmann constant, $(\pi^2/60)k^4/\hbar^3 c^2$	σ	5.670 51(19)	$10^{-8}\,W\,m^{-2}\,K^{-4}$	34
Non-SI units used with SI				
electron volt, $(e/C)J = \{e\}\,J$	eV	1.602 177 33(49)	$10^{-19}\,J$	0.30
(unified) atomic mass unit, $1\,u = m_u = \frac{1}{12}m(^{12}C)$	u	1.660 5402(10)	$10^{-27}\,kg$	0.59

NOTE: An abbreviated list of the fundamental constants of physics and chemistry based on a least-squares adjustment with 17 degrees of freedom. The digits in parentheses are the one-standard-deviation uncertainty in the last digits of the given value. Since the uncertainties of many entries are correlated, the full covariance matrix must be used in evaluating the uncertainties of quantities computed from them.

Table 2
THE 1986 RECOMMENDED VALUES OF THE FUNDAMENTAL PHYSICAL CONSTANTS

Quantity	Symbol	Value	Units	Relative Uncertainty (ppm)
GENERAL CONSTANTS				
Universal Constants				
speed of light in vacuum	c	299 792 458	$m\,s^{-1}$	(exact)
permeability of vacuum	μ_o	$4\pi \times 10^{-7}$	$N\,A^{-2}$	
		$=12.566\,370\,614\ldots$	$10^{-7}\,N\,A^{-2}$	(exact)
permittivity of vacuum	ϵ_o	$1/\mu_o c^2$		
		$=8.854\,187\,817\ldots$	$10^{-12}\,F\,m^{-1}$	(exact)
Newtonian constant of gravitation	G	6.672 59(85)	$10^{-11}\,m^3\,kg^{-1}\,s^{-2}$	128
Planck constant	h	6.626 0755(40)	$10^{-34}\,J\,s$	0.60
in electron volts, $h/\{e\}$		4.135 6692(12)	$10^{-15}\,eV\,s$	0.30
$\hbar/2\pi$	$\hbar$	1.054 572 66(63)	$10^{-34}\,J\,s$	0.60
in electron volts, $\hbar/\{e\}$		6.582 1220(20)	$10^{-16}\,eV\,s$	0.30

Quantity	Symbol	Value	Units	Relative Uncertainty (ppm)
Planck mass, $(\hbar c/G)^{\frac{1}{2}}$	m_P	2.17671(14)	10^{-8} kg	64
Planck length, $\hbar/m_P c = (\hbar G/c^3)^{\frac{1}{2}}$	l_P	1.61605(10)	10^{-35} m	64
Planck time, $l_P/c = (\hbar G/c^5)^{\frac{1}{2}}$	t_P	5.39056(34)	10^{-44} s	64

Electromagnetic Constants

Quantity	Symbol	Value	Units	Relative Uncertainty (ppm)
elementary charge	e	1.60217733(49)	10^{-19} C	0.30
	e/h	2.41798836(72)	10^{14} A J^{-1}	0.30
magnetic flux quantum, $h/2e$	Φ_o	2.06783461(61)	10^{-15} Wb	0.30
Josephson frequency-voltage ratio	$2e/h$	4.8359767(14)	10^{14} Hz V^{-1}	0.30
quantized Hall conductance	e^2/h	3.87404614(17)	10^{-5} S	0.045
quantized Hall resistance,	R_H	25812.8056(12)	Ω	0.045
$\quad h/e^2 = \mu_o c/2\alpha$				
Bohr magneton, $e\hbar/2m_e$	μ_B	9.2740154(31)	10^{-24} J T^{-1}	0.34
$\quad$ in electron volts, $\mu_B/\{e\}$		5.78838263(52)	10^{-5} eV T^{-1}	0.089
$\quad$ in hertz, μ_B/h		1.39962418(42)	10^{10} Hz T^{-1}	0.30
$\quad$ in wavenumbers, μ_B/hc		46.686437(14)	m^{-1} T^{-1}	0.30
$\quad$ in kelvins, μ_B/k		0.6711099(57)	K T^{-1}	8.5
nuclear magneton, $e\hbar/2m_p$	μ_N	5.0507866(17)	10^{-27} J T^{-1}	0.34
$\quad$ in electron volts, $\mu_N/\{e\}$		3.15245166(28)	10^{-8} eV T^{-1}	0.089
$\quad$ in hertz, μ_N/h		7.6225914(23)	MHz T^{-1}	0.30
$\quad$ in wavenumbers, μ_N/hc		2.54262281(77)	10^{-2} m^{-1} T^{-1}	0.30
$\quad$ in kelvins, μ_N/k		3.658246(31)	10^{-4} K T^{-1}	8.5

ATOMIC CONSTANTS

Quantity	Symbol	Value	Units	Relative Uncertainty (ppm)
fine-structure constant, $\mu_o c e^2/2h$	α	7.29735308(33)	10^{-3}	0.045
inverse fine-structure constant	α^{-1}	137.0359895(61)		0.045
Rydberg constant, $m_e c\alpha^2/2h$	R_∞	10973731.534(13)	m^{-1}	0.0012
$\quad$ in hertz, $R_\infty c$		3.2898419499(39)	10^{15} Hz	0.0012
$\quad$ in joules, $R_\infty hc$		2.1798741(13)	10^{-18} J	0.60
$\quad$ in eV, $R_\infty hc/\{e\}$		13.6056981(40)	eV	0.30
Bohr radius, $\alpha/4\pi R_\infty$	a_o	0.529177249(24)	10^{-10} m	0.045
Hartree energy, $e^2/4\pi\epsilon_o a_o = 2R_\infty hc$	E_h	4.3597482(26)	10^{-18} J	0.60
$\quad$ in eV, $E_h/\{e\}$		27.2113961(81)	eV	0.30
quantum of circulation	$h/2m_e$	3.636948 07(33)	10^{-4} m^2 s^{-1}	0.089
	h/m_e	7.273896 14(65)	10^{-4} m^2 s^{-1}	0.089

Electron

Quantity	Symbol	Value	Units	Relative Uncertainty (ppm)
electron mass	m_e	9.1093897(54)	10^{-31} kg	0.59
		5.48579903(13)	10^{-4} u	0.023
$\quad$ in electron volts, $m_e c^2/\{e\}$		0.51099906(15)	MeV	0.30
electron-muon mass ratio	m_e/m_μ	4.83633218(71)	10^{-3}	0.15
electron-proton mass ratio	m_e/m_p	5.44617013(11)	10^{-4}	0.020
electron-deuteron mass ratio	m_e/m_d	2.72443707(6)	10^{-4}	0.020
electron-α-particle mass ratio	m_e/m_α	1.37093354(3)	10^{-4}	0.021
electron specific charge	$-e/m_e$	$-1.75881962(53)$	10^{11} C kg^{-1}	0.30
electron molar mass	$M(e), M_e$	5.48579903(13)	10^{-7} kg/mol	0.023
Compton wavelength, $h/m_e c$	λ_C	2.42631058(22)	10^{-12} m	0.089
$\quad \lambda_C/2\pi = \alpha a_o = \alpha^2/4\pi R_\infty$	λ_C	3.86159323(35)	10^{-13} m	0.089
classical electron radius, $\alpha^2 a_o$	r_e	2.81794092(38)	10^{-15} m	0.13
Thomson cross section, $(8\pi/3)r_e^2$	σ_e	0.66524616(18)	10^{-28} m^2	0.27
electron magnetic moment	μ_e	928.47701(31)	10^{-26} J T^{-1}	0.34
$\quad$ in Bohr magnetons	μ_e/μ_B	1.001159652193(10)		1×10^{-5}
$\quad$ in nuclear magnetons	μ_e/μ_N	1838.282000(37)		0.020
electron magnetic moment anomaly, $\mu_e/\mu_B - 1$	a_e	1.159652193(10)	10^{-3}	0.0086
electron g-factor, $2(1 + a_e)$	g_e	2.002319304386(20)		1×10^{-5}
electron-muon magnetic moment ratio	μ_e/μ_μ	206.766967(30)		0.15
electron-proton magnetic moment ratio	μ_e/μ_p	658.2106881(66)		0.010

Muon

Quantity	Symbol	Value	Units	Relative Uncertainty (ppm)
muon mass	m_μ	1.8835327(11)	10^{-28} kg	0.61
		0.113428913(17)	u	0.15
$\quad$ in electron volts, $m_\mu c^2/\{e\}$		105.658389(34)	MeV	0.32
muon-electron mass ratio	m_μ/m_e	206.768262(30)		0.15
muon molar mass	$M(\mu), M_\mu$	1.13428913(17)	10^{-4} kg/mol	0.15
muon magnetic moment	μ_μ	4.4904514(15)	10^{-26} J T^{-1}	0.33
$\quad$ in Bohr magnetons,	μ_μ/μ_B	4.84197097(71)	10^{-3}	0.15
$\quad$ in nuclear magnetons,	μ_μ/μ_N	8.8905981(13)		0.15

Quantity	Symbol	Value	Units	Relative Uncertainty (ppm)
muon magnetic moment anomaly,				
$[\mu_\mu/(e\hbar/2m_\mu)] - 1$	a_μ	1.1659230(84)	10^{-3}	7.2
muon g-factor, $2(1 + a_\mu)$	g_μ	2.002331846(17)		0.0084
muon–proton				
magnetic moment ratio	μ_μ/μ_p	3.18334547(47)		0.15
		Proton		
proton mass	m_p	1.6726231(10)	10^{-27} kg	0.59
		1.007276470(12)	u	0.012
in electron volts, $m_p c^2/\{e\}$		938.27231(28)	MeV	0.30
proton–electron mass ratio	m_p/m_e	1836.152701(37)		0.020
proton–muon mass ratio	m_p/m_μ	8.8802444(13)		0.15
proton specific charge	e/m_p	9.5788309(29)	10^7 C kg^{-1}	0.30
proton molar mass	$M(p), M_p$	1.007276470(12)	10^{-3} kg/mol	0.012
proton Compton wavelength, $h/m_p c$	$\lambda_{C,p}$	1.32141002(12)	10^{-15} m	0.089
$\lambda_{C,p}/2\pi$	$\lambda_{C,p}$	2.10308937(19)	10^{-16} m	0.089
proton magnetic moment	μ_p	1.41060761(47)	10^{-26} J T^{-1}	0.34
in Bohr magnetons	μ_p/μ_B	1.521032202(15)	10^{-3}	0.010
in nuclear magnetons	μ_p/μ_N	2.792847386(63)		0.023
diamagnetic shielding correction				
for protons in pure water,				
spherical sample, 25 °C, $1 - \mu_p'/\mu_p$	σ_{H_2O}	25.689(15)	10^{-6}	–
shielded proton moment	μ_p'	1.41057138(47)	10^{-26} J T^{-1}	0.34
(H_2O, sph., 25 °C)				
in Bohr magnetons	μ_p'/μ_B	1.520993129(17)	10^{-3}	0.011
in nuclear magnetons	μ_p'/μ_N	2.792775642(64)		0.023
proton gyromagnetic ratio	γ_p	26752.2128(81)	10^4 s^{-1} T^{-1}	0.30
	$\gamma_p/2\pi$	42.577469(13)	MHz T^{-1}	0.30
uncorrected (H_2O, sph., 25 °C)	γ_p'	26751.5255(81)	10^4 s^{-1} T^{-1}	0.30
	$\gamma_p'/2\pi$	42.576375(13)	MHz T^{-1}	0.30
		Neutron		
neutron mass	m_n	1.6749286(10)	10^{-27} kg	0.59
		1.008664904(14)	u	0.014
in electron volts, $m_n c^2/\{e\}$		939.56563(28)	Mev	0.30
neutron–electron mass ratio	m_n/m_e	1838.683662(40)		0.022
neutron–proton mass ratio	m_n/m_p	1.001378404(9)		0.009
neutron molar mass	$M(n), M_n$	1.008664904(14)	10^{-3} kg/mol	0.014
neutron Compton wavelength, $h/m_n c$	$\lambda_{C,n}$	1.31959110(12)	10^{-15} m	0.089
$\lambda_{C,n}/2\pi$	$\lambda_{C,n}$	2.10019445(19)	10^{-16} m	0.089
neutron magnetic moment *	μ_n	0.96623707(40)	10^{-26} J T^{-1}	0.41
in Bohr magnetons	μ_n/μ_B	1.04187563(25)	10^{-3}	0.24
in nuclear magnetons	μ_n/μ_N	1.91304275(45)		0.24
neutron–electron				
magnetic moment ratio	μ_n/μ_e	1.04066882(25)	10^{-3}	0.24
neutron–proton				
magnetic moment ratio	μ_n/μ_p	0.68497934(16)		0.24
		Deuteron		
deuteron mass	m_d	3.3435860(20)	10^{-27} kg	0.59
		2.013553214(24)	u	0.012
in electron volts, $m_d c^2/\{e\}$		1875.61339(57)	MeV	0.30
deuteron–electron mass ratio	m_d/m_e	3670.483014(75)		0.020
deuteron–proton mass ratio	m_d/m_p	1.999007496(6)		0.003
deuteron molar mass	$M(d), M_d$	2.013553214(24)	10^{-3} kg/mol	0.012
deuteron magnetic moment *	μ_d	0.43307375(15)	10^{-26} J T^{-1}	0.34
in Bohr magnetons,	μ_d/μ_B	0.4669754479(91)	10^{-3}	0.019
in nuclear magnetons,	μ_d/μ_N	0.857438230(24)		0.028
deuteron–electron				
magnetic moment ratio	μ_d/μ_e	0.4664345460(91)	10^{-3}	0.019
deuteron–proton				
magnetic moment ratio	μ_d/μ_p	0.3070122035(51)		0.017

PHYSICO-CHEMICAL CONSTANTS

Quantity	Symbol	Value	Units	Relative Uncertainty (ppm)
Avogadro constant	N_A, L	6.022 1367(36)	10^{23} mol^{-1}	0.59
atomic mass constant, $\frac{1}{12} m(^{12}C)$	m_u	1.6605402(10)	10^{-27} kg	0.59
in electron volts, $m_u c^2/\{e\}$		931.49432(28)	MeV	0.30
Faraday constant	F	96485.309(29)	C mol^{-1}	0.30
molar Planck constant	$N_A h$	3.99031323(36)	10^{-10} J s mol^{-1}	0.089
	$N_A hc$	0.11962658(11)	J m mol^{-1}	0.089
molar gas constant	R	8.314510(70)	J mol^{-1} K^{-1}	8.4

Table 2
THE 1986 RECOMMENDED VALUES OF THE
FUNDAMENTAL PHYSICAL CONSTANTS (continued)

Quantity	Symbol	Value	Units	Relative Uncertainty (ppm)
Boltzmann constant, R/N_A	k	1.380 658(12)	10^{-23} J K^{-1}	8.5
in electron volts, $k/\{e\}$		8.617 385(73)	10^{-5} eV K^{-1}	8.4
in hertz, k/h		2.083 674(18)	10^{10} Hz K^{-1}	8.4
in wavenumbers, k/hc		69.503 87(59)	m^{-1} K^{-1}	8.4
molar volume (ideal gas), RT/p				
$T = 273.15$ K, $p = 101\,325$ Pa	V_m	22.414 10(19)	L/mol	8.4
Loschmidt constant, N_A/V_m	n_0	2.686 763(23)	10^{25} m^{-3}	8.5
$T = 273.15$ K, $p = 100$ kPa	V_m	22.711 08(19)	L/mol	8.4
Sackur-Tetrode constant				
(absolute entropy constant), **				
$\frac{5}{2} + \ln\{(2\pi m_u kT_1/h^2)^{\frac{3}{2}} kT_1/p_0\}$				
$T_1 = 1$ K, $p_0 = 100$ kPa	S_0/R	$-1.151\,693(21)$		18
$p_0 = 101\,325$ Pa		$-1.164\,856(21)$		18
Stefan–Boltzmann constant,				
$(\pi^2/60)k^4/\hbar^3c^2$	σ	5.670 51(19)	10^{-8} W m^{-2} K^{-4}	34
first radiation constant, $2\pi hc^2$	c_1	3.741 7749(22)	10^{-16} W m^2	0.60
second radiation constant, hc/k	c_2	0.014 387 69(12)	m K	8.4
Wien displacement law constant,				
$b = \lambda_{max}T = c_2/4.965\,114\,23\ldots$	b	2.897 756(24)	10^{-3} m K	8.4

NOTE: This list of the fundamental constants of physics and chemistry is based on a least-squares adjustment with 17 degrees of freedom. The digits in parentheses are the one-standard-deviation uncertainty in the last digits of the given value. Since the uncertainties of many of these entries are correlated, the full covariance matrix must be used in evaluating the uncertainties of quantities computed from them.

*   The scalar magnitude of the neutron moment is listed here. The neutron magnetic dipole is directed oppositely to that of the proton, and corresponds to the dipole associated with a spinning negative charge distribution. The vector sum, $\mu_d = \mu_p + \mu_n$, is approximately satisfied.

**   The entropy of an ideal monatomic gas of relative atomic weight A_r is given by $S = S_0 + \frac{3}{2}R$ in $A_r - R$ in $(p/p_0) + \frac{5}{2}R$ in(T/K).

Table 3
MAINTAINED UNITS AND STANDARD VALUES

Quantity	Symbol	Value	Units	Relative Uncertainty (ppm)
electron volt, (e/C) J $= \{e\}$ J	eV	1.602 177 33(49)	10^{-19} J	0.30
(unified) atomic mass unit,	u	1.660 5402(10)	10^{-27} kg	0.59
$1\ u = m_u = \frac{1}{12}m(^{12}C)$				
standard atmosphere	atm	101 325	Pa	(exact)
standard acceleration of gravity	g_n	9.806 65	m s^{-2}	(exact)
'As–Maintained' Electrical Units				
BIPM maintained ohm, Ω_{69-BI}				
$\Omega_{BI85} \equiv \Omega_{69-BI}(1\ \text{Jan}\ 1985)$	Ω_{BI85}	$1 - 1.563(50)\times10^{-6}$	Ω	
		$= 0.999\,998\,437(50)$	Ω	0.050
Drift rate of Ω_{69-BI}	$\dfrac{d\Omega_{69-BI}}{dt}$	$-0.0566(15)$	$\mu\Omega$/a	—
BIPM maintained volt,	V_{76-BI}	$1 - 7.59(30)\times10^{-6}$	V	
$V_{76-BI} \equiv 483\,594\,\text{GHz}(h/2e)$		$= 0.999\,992\,41(30)$	V	0.30
BIPM maintained ampere,	A_{BI85}	$1 - 6.03(30)\times10^{-6}$	A	
$A_{BIPM} = V_{76-BI}/\Omega_{69-BI}$		$= 0.999\,993\,97(30)$	A	0.30
X-Ray Standards				
Cu x-unit :	xu(CuKα_1)	1.002 077 89(70)	10^{-13} m	0.70
$\lambda(\text{CuK}\alpha_1) \equiv 1537.400$ xu				
Mo x-unit :	xu(MoKα_1)	1.002 099 38(45)	10^{-13} m	0.45
$\lambda(\text{MoK}\alpha_1) \equiv 707.831$ xu				
Å* :	Å*	1.000 014 81(92)	10^{-10} m	0.92
$\lambda(\text{WK}\alpha_1) \equiv 0.209\,100$ Å*				

Table 3
MAINTAINED UNITS AND STANDARD VALUES
(continued)

Quantity	Symbol	Value	Units	Relative Uncertainty (ppm)
lattice spacing of Si (in vacuum, 22.5 °C), + $d_{220} = a/\sqrt{8}$	a	0.543 101 96(11)	nm	0.21
$d_{220} = a/\sqrt{8}$	d_{220}	0.192 015 540(40)	nm	0.21
molar volume of Si, $M(\text{Si})/\rho(\text{Si}) = N_A a^3/8$	$V_m(\text{Si})$	12.058 8179(89)	cm^3/mol	0.74

NOTE: A summary of "maintained" units and "standard" values and their relationship to SI units, based on a least-squares adjustment with 17 degrees of freedom. The digits in parentheses are the one-standard-deviation uncertainty in the last digits of the given value. Since the uncertainties of many of these entries are correlated, the full covariance matrix must be used in evaluating the uncertainties of quantities computed from them.

+ The lattice spacing of single-crystal Si can vary by parts in 10^7 depending on the preparation process. Measurements at PTB indicate also the possibility of distortions from exact cubic symmetry of the order of 0.2 ppm.

Table 4
ENERGY CONVERSION FACTORS

	J	kg	m^{-1}	Hz
1 J =	1	$1/\{c^2\}$ 1.112 650 06×10^{-17}	$1/\{hc\}$ 5.034 1125(30)×10^{24}	$1/\{h\}$ 1.509 188 97(90)×10^{33}
1 kg =	$\{c^2\}$ 8.987 551 787×10^{16}	1	$\{c/h\}$ 4.524 4347(27)×10^{41}	$\{c^2/h\}$ 1.356 391 40(81)×10^{50}
1 m^{-1} =	$\{hc\}$ 1.986 4475(12)×10^{-25}	$\{h/c\}$ 2.210 2209(13)×10^{-42}	1	$\{c\}$ 299 792 458
1 Hz =	$\{h\}$ 6.626 0755(40)×10^{-34}	$\{h/c^2\}$ 7.372 5032(44)×10^{-51}	$1/\{c\}$ 3.335 640 952×10^{-9}	1
1 K =	$\{k\}$ 1.380 658(12)×10^{-23}	$\{k/c^2\}$ 1.536 189(13)×10^{-40}	$\{k/hc\}$ 69.503 87(59)	$\{k/h\}$ 2.083 674(18)×10^{10}
1 eV =	$\{e\}$ 1.602 177 33(49)×10^{-19}	$\{e/c^2\}$ 1.782 662 70(54)×10^{-36}	$\{e/hc\}$ 806 554.10(24)	$\{e/h\}$ 2.417 988 36(72)×10^{14}
1 u =	$\{m_u c^2\}$ 1.492 419 09(88)×10^{-10}	$\{m_u\}$ 1.660 5402(10)×10^{-27}	$\{m_u c/h\}$ 7.513 005 63(67)×10^{14}	$\{m_u c^2/h\}$ 2.252 342 42(20)×10^{23}
1 hartree =	$\{2R_\infty hc\}$ 4.359 7482(26)×10^{-18}	$\{2R_\infty h/c\}$ 4.850 8741(29)×10^{-35}	$\{2R_\infty\}$ 21 947 463.067(26)	$\{2R_\infty c\}$ 6.579 683 8999(78)×10^{15}

	K	eV	u	hartree
1 J =	$1/\{k\}$ 7.242 924(61)×10^{22}	$1/\{e\}$ 6.241 5064(19)×10^{18}	$1/\{m_u c^2\}$ 6.700 5308(40)×10^9	$1/\{2R_\infty hc\}$ 2.293 7104(14)×10^{17}
1 kg =	$\{c^2/k\}$ 6.509 616(55)×10^{39}	$\{c^2/e\}$ 5.609 5862(17)×10^{35}	$1/\{m_u\}$ 6.022 1367(36)×10^{26}	$\{c/2R_\infty h\}$ 2.061 4841(12)×10^{34}
1 m^{-1} =	$\{hc/k\}$ 0.014 387 69(12)	$\{hc/e\}$ 1.239 842 44(37)×10^{-6}	$\{h/m_u c\}$ 1.331 025 22(12)×10^{-15}	$1/\{2R_\infty\}$ 4.556 335 2672(54)×10^{-8}
1 Hz =	$\{h/k\}$ 4.799 216(41)×10^{-11}	$\{h/e\}$ 4.135 6692(12)×10^{-15}	$\{h/m_u c^2\}$ 4.439 822 24(40)×10^{-24}	$1/\{2R_\infty c\}$ 1.519 829 8508(18)×10^{-16}
1 K =	1	$\{k/e\}$ 8.617 385(73)×10^{-5}	$\{k/m_u c^2\}$ 9.251 140(78)×10^{-14}	$\{k/2R_\infty hc\}$ 3.166 829(27)×10^{-6}
1 eV =	$\{e/k\}$ 11 604.45(10)	1	$\{e/m_u c^2\}$ 1.073 543 85(33)×10^{-9}	$\{e/2R_\infty hc\}$ 0.036 749 309(11)
1 u =	$\{m_u c^2/k\}$ 1.080 9478(91)×10^{13}	$\{m_u c^2/e\}$ 931.494 32(28)×10^6	1	$\{m_u c/2R_\infty h\}$ 3.423 177 25(31)×10^7
1 hartree =	$\{2R_\infty hc/k\}$ 3.157 733(27)×10^5	$\{2R_\infty hc/e\}$ 27.211 3961(81)	$\{2R_\infty h/m_u c\}$ 2.921 262 69(26)×10^{-8}	1

NOTE: To use this table note that all entries on the same line are equal; the unit at the top of a column applies to all of the values beneath it. Example: 1 eV = 806 544.10 m^{-1}.

Table 5
EXPANDED COVARIANCE AND CORRELATION COEFFICIENT MATRIX FOR THE 1986 RECOMMENDED SET OF FUNDAMENTAL PHYSICAL CONSTANTS

	α^{-1}	K_V	K_Ω	μ_μ/μ_p	e	$\overline{h}$	m_e	$\overline{N_A}$	F
α^{-1}	1997	−1062	925	3267	−3059	−4121	−127	127	−2932
K_V	−0.080	87988	90	−1737	89050	177038	174914	−174914	−85864
K_Ω	0.416	0.006	2477	1513	−835	−744	1105	−1105	−1939
μ_μ/μ_p	0.498	−0.040	0.207	21523	−5004	−6742	−208	208	−4796
e	−0.226	0.989	−0.055	−0.112	92109	181159	175042	−175042	−82933
h	−0.154	0.997	−0.025	−0.077	0.997	358197	349956	−349956	−168797
m_e	−0.005	0.997	0.038	−0.002	0.975	0.989	349702	−349702	−174660
N_A	0.005	−0.997	−0.038	0.002	−0.975	−0.989	−1.000	349702	174660
F	−0.217	−0.956	−0.129	−0.108	−0.902	−0.931	−0.975	0.975	91727

The elements of the covariance matrix appear on and above the major diagonal in (parts in 10^9)^2; correlation coefficients appear in *italics* below the diagonal. The values are given to as many as six digits only as a matter of consistency. The correlation coefficient between m_e and N_A appears as −1.000 in this table because the auxiliary constants were considered to be exact in carrying out the least-squares adjustment. When the uncertainties of m_p/m_e and M_p are properly taken into account, the correlation coefficient is −0.999 and the variances of m_e and N_A are slightly increased.

STANDARD ATOMIC WEIGHTS (1995)

This table of atomic weights is reprinted from the 1995 report of the IUPAC Commission on Atomic Weights and Isotopic Abundances. The Standard Atomic Weights apply to the elements as they exist naturally on Earth, and the uncertainties take into account the isotopic variation found in most laboratory samples. Further comments on the variability are given in the footnotes.

The number in parentheses following the atomic weight value gives the uncertainty in the last digit. An entry in brackets indicates the mass number of the longest-lived isotope of an element that has no stable isotopes and for which a Standard Atomic Weight cannot be defined because of wide variability in isotopic composition (or complete absence) in nature.

REFERENCE

IUPAC Commission on Atomic Weights and Isotopic Abundances, Atomic Weights of the Elements, 1995, *Pure Appl. Chem.*, 68, 2339, 1996.

Name	Symbol	At. no.	Atomic Weight	Footnotes		
Actinium	Ac	89	[227]			
Aluminum	Al	13	26.981538(2)			
Americium	Am	95	[243]			
Antimony	Sb	51	121.760(1)	g		
Argon	Ar	18	39.948(1)	g		r
Arsenic	As	33	74.92160(2)			
Astatine	At	85	[210]			
Barium	Ba	56	137.327(7)			
Berkelium	Bk	97	[247]			
Beryllium	Be	4	9.012182(3)			
Bismuth	Bi	83	208.98038(2)			
Bohrium	Bh	107	[262]			
Boron	B	5	10.811(7)	g	m	r
Bromine	Br	35	79.904(1)			
Cadmium	Cd	48	112.411(8)	g		
Calcium	Ca	20	40.078(4)	g		
Californium	Cf	98	[251]			
Carbon	C	6	12.0107(8)	g		r
Cerium	Ce	58	140.116(1)	g		
Cesium	Cs	55	132.90545(2)			
Chlorine	Cl	17	35.4527(9)		m	
Chromium	Cr	24	51.9961(6)			
Cobalt	Co	27	58.933200(9)			
Copper	Cu	29	63.546(3)			r
Curium	Cm	96	[247]			
Dubnium	Db	105	[262]			
Dysprosium	Dy	66	162.50(3)	g		
Einsteinium	Es	99	[252]			
Erbium	Er	68	167.26(3)	g		
Europium	Eu	63	151.964(1)	g		
Fermium	Fm	100	[257]			
Fluorine	F	9	18.9984032(5)			
Francium	Fr	87	[223]			
Gadolinium	Gd	64	157.25(3)	g		
Gallium	Ga	31	69.723(1)			
Germanium	Ge	32	72.61(2)			
Gold	Au	79	196.96655(2)			
Hafnium	Hf	72	178.49(2)			
Hassium	Hs	108	[265]			
Helium	He	2	4.002602(2)	g		r
Holmium	Ho	67	164.93032(2)			
Hydrogen	H	1	1.00794(7)	g	m	r
Indium	In	49	114.818(3)			
Iodine	I	53	126.90447(3)			

Name	Symbol	At. no.	Atomic Weight	Footnotes		
Iridium	Ir	77	192.217(3)			
Iron	Fe	26	55.845(2)			
Krypton	Kr	36	83.80(1)	g	m	
Lanthanum	La	57	138.9055(2)	g		
Lawrencium	Lr	103	[262]			
Lead	Pb	82	207.2(1)	g		r
Lithium	Li	3	6.941(2)	g	m	r
Lutetium	Lu	71	174.967(1)	g		
Magnesium	Mg	12	24.3050(6)			
Manganese	Mn	25	54.938049(9)			
Meitnerium	Mt	109	[266]			
Mendelevium	Md	101	[258]			
Mercury	Hg	80	200.59(2)			
Molybdenum	Mo	42	95.94(1)	g		
Neodymium	Nd	60	144.24(3)	g		
Neon	Ne	10	20.1797(6)	g	m	
Neptunium	Np	93	[237]			
Nickel	Ni	28	58.6934(2)			
Niobium	Nb	41	92.90638(2)			
Nitrogen	N	7	14.00674(7)	g		r
Nobelium	No	102	[259]			
Osmium	Os	76	190.23(3)	g		
Oxygen	O	8	15.9994(3)	g		r
Palladium	Pd	46	106.42(1)	g		
Phosphorus	P	15	30.973761(2)			
Platinum	Pt	78	195.078(2)			
Plutonium	Pu	94	[244]			
Polonium	Po	84	[209]			
Potassium	K	19	39.0983(1)	g		
Praseodymium	Pr	59	140.90765(2)			
Promethium	Pm	61	[145]			
Protactinium	Pa	91	231.03588(2)			
Radium	Ra	88	[226]			
Radon	Rn	86	[222]			
Rhenium	Re	75	186.207(1)			
Rhodium	Rh	45	102.90550(2)			
Rubidium	Rb	37	85.4678(3)	g		
Ruthenium	Ru	44	101.07(2)	g		
Rutherfordium	Rf	104	[261]			
Samarium	Sm	62	150.36(3)	g		
Scandium	Sc	21	44.955910(8)			
Seaborgium	Sg	106	[263]			
Selenium	Se	34	78.96(3)			
Silicon	Si	14	28.0855(3)			r
Silver	Ag	47	107.8682(2)	g		
Sodium	Na	11	22.989770(2)			
Strontium	Sr	38	87.62(1)	g		r
Sulfur	S	16	32.066(6)	g		r
Tantalum	Ta	73	180.9479(1)			
Technetium	Tc	43	[98]			
Tellurium	Te	52	127.60(3)	g		
Terbium	Tb	65	158.92534(2)			
Thallium	Tl	81	204.3833(2)			
Thorium	Th	90	232.0381(1)	g		
Thulium	Tm	69	168.93421(2)			
Tin	Sn	50	118.710(7)	g		
Titanium	Ti	22	47.867(1)			
Tungsten	W	74	183.84(1)			

Name	Symbol	At. no.	Atomic Weight	Footnotes	
Uranium	U	92	238.0289(1)	g	m
Vanadium	V	23	50.9415(1)		
Xenon	Xe	54	131.29(2)	g	m
Ytterbium	Yb	70	173.04(3)	g	
Yttrium	Y	39	88.90585(2)		
Zinc	Zn	30	65.39(2)		
Zirconium	Zr	40	91.224(2)	g	

g   geological specimens are known in which the element has an isotopic composition outside the limits for normal material. The difference between the atomic weight of the element in such specimens and that given in the table may exceed the stated uncertainty.

m   modified isotopic compositions may be found in commercially available material because it has been subjected to an undisclosed or inadvertent isotopic fractionation. Substantial deviations in atomic weight of the element from that given the table can occur.

r   range in isotopic composition of normal terrestrial material prevents a more precise atomic weight being given; the tabulated atomic weight value should be applicable to any normal material.

ATOMIC MASSES AND ABUNDANCES

This table lists the mass (in atomic mass units, symbol u) and the natural abundance (in percent) of the stable nuclides and a few important radioactive nuclides. A complete table of all nuclides may be found in Section 10 ("Table of the Isotopes").

The atomic masses were taken from the 1995 evaluation of Audi and Wapstra (References 2, 3). Mass values were rounded in accordance with the stated uncertainty. The number in parentheses following the mass value is the uncertainty (specifically, the rounded value of the estimated standard deviation) in the last digit given.

Natural abundance values are also followed by uncertainties in the last digit(s) of the stated values. This uncertainty includes both the estimated measurement uncertainty and the reported range of variation in different terrestrial sources of the element (see Reference 4 for more details). A * in the Abundance column indicates a radioactive nuclide not present in nature or an element whose isotopic composition varies so widely that a meaningful natural abundance cannot be defined.

REFERENCES

1. Holden, N. E., "Table of the Isotopes", in Lide, D. R., Ed., *CRC Handbook of Chemistry and Physics, 79th Ed.*, CRC Press, Boca Raton, FL, 1998.
2. Audi, G. and Wapstra, A. H., *Nucl. Phys.*, A595, 409, 1995.
3. Audi, G. and Wapstra, A. H., Atomic Mass Data Center, World Wide Web site, <http://csnwww.in2p3.fr>. Note that the print version in Reference 2 gives mass excess in energy units but not the actual atomic mass. The latter is included in the electronic version.
4. IUPAC Commission on Atomic Weights and Isotopic Abundances, *Pure Appl. Chem.*, 63, 991, 1991.

Z	Isotope	Mass in u	Abundance in %	Z	Isotope	Mass in u	Abundance in %
1	^{1}H	1.007 825 032(1)	99.985(1)	18	^{38}Ar	37.962 732 2(5)	0.063(1)
1	^{2}H	2.014 101 778(1)	0.015(1)	18	^{39}Ar	38.964 313(5)	*
1	^{3}H	3.016 049 268(1)	*	18	^{40}Ar	39.962 383 123(3)	99.600(3)
2	^{3}He	3.016 029 310(1)	0.000137(3)	19	^{39}K	38.963 706 8(3)	93.2581(44)
2	^{4}He	4.002 603 250(1)	99.999863(3)	19	^{40}K	39.963 998 7(3)	0.0117(1)
3	^{6}Li	6.015 122 3(5)	7.5(2)	19	^{41}K	40.961 826 0(3)	6.7302(44)
3	^{7}Li	7.016 004 0(5)	92.5(2)	20	^{40}Ca	39.962 591 1(3)	96.941(18)
4	^{9}Be	9.012 182 1(4)	100	20	^{42}Ca	41.958 618 3(4)	0.647(9)
5	^{10}B	10.012 937 0(3)	19.9(2)	20	^{43}Ca	42.958 766 8(5)	0.135(6)
5	^{11}B	11.009 305 5(5)	80.1(2)	20	^{44}Ca	43.955 481 1(9)	2.086(12)
6	^{12}C	12 by definition	98.90(3)	20	^{46}Ca	45.953 693(3)	0.004(3)
6	^{13}C	13.003 354 838(1)	1.10(3)	20	^{48}Ca	47.952 534(4)	0.187(4)
6	^{14}C	14.003 241 988(4)	*	21	^{45}Sc	44.955 910(1)	100
7	^{14}N	14.003 074 005(1)	99.634(9)	22	^{46}Ti	45.952 629(1)	8.0(1)
7	^{15}N	15.000 108 898(1)	0.366(9)	22	^{47}Ti	46.951 764(1)	7.3(1)
8	^{16}O	15.994 914 622(2)	99.762(15)	22	^{48}Ti	47.947 947(1)	73.8(1)
8	^{17}O	16.999 131 5(2)	0.038(3)	22	^{49}Ti	48.947 871(1)	5.5(1)
8	^{18}O	17.999 160 4(9)	0.200(12)	22	^{50}Ti	49.944 792(1)	5.4(1)
9	^{19}F	18.998 403 21(8)	100	23	^{50}V	49.947 163(1)	0.250(2)
10	^{20}Ne	19.992 440 176(2)	90.48(3)	23	^{51}V	50.943 964(1)	99.750(2)
10	^{21}Ne	20.993 846 74(4)	0.27(1)	24	^{50}Cr	49.946 050(1)	4.345(13)
10	^{22}Ne	21.991 385 5(2)	9.25(3)	24	^{52}Cr	51.940 512(2)	83.789(18)
11	^{23}Na	22.989 769 7(2)	100	24	^{53}Cr	52.940 654(2)	9.501(17)
12	^{24}Mg	23.985 041 9(2)	78.99(3)	24	^{54}Cr	53.938 885(1)	2.365(7)
12	^{25}Mg	24.985 837 0(2)	10.00(1)	25	^{55}Mn	54.938 050(1)	100
12	^{26}Mg	25.982 593 0(2)	11.01(2)	26	^{54}Fe	53.939 615(1)	5.8(1)
13	^{27}Al	26.981 538 4(1)	100	26	^{56}Fe	55.934 942(1)	91.72(30)
14	^{28}Si	27.976 926 533(2)	92.23(1)	26	^{57}Fe	56.935 399(1)	2.1(1)
14	^{29}Si	28.976 494 72(3)	4.67(1)	26	^{58}Fe	57.933 280(1)	0.28(1)
14	^{30}Si	29.973 770 22(5)	3.10(1)	27	^{59}Co	58.933 200(2)	100
15	^{31}P	30.973 761 5(2)	100	28	^{58}Ni	57.935 348(2)	68.077(9)
16	^{32}S	31.972 070 7(1)	95.02(9)	28	^{60}Ni	59.930 791(2)	26.223(8)
16	^{33}S	32.971 458 5(1)	0.75(4)	28	^{61}Ni	60.931 060(2)	1.140(1)
16	^{34}S	33.967 866 8(1)	4.21(8)	28	^{62}Ni	61.928 349(2)	3.634(2)
16	^{36}S	35.967 080 9(3)	0.02(1)	28	^{64}Ni	63.927 970(2)	0.926(1)
17	^{35}Cl	34.968 852 71(4)	75.77(7)	29	^{63}Cu	62.929 601(2)	69.17(3)
17	^{37}Cl	36.965 902 60(5)	24.23(7)	29	^{65}Cu	64.927 794(2)	30.83(3)
18	^{36}Ar	35.967 546 3(3)	0.337(3)	30	^{64}Zn	63.929 147(2)	48.6(3)
18	^{37}Ar	36.966 775 9(3)	*	30	^{66}Zn	65.926 037(2)	27.9(2)

Z	Isotope	Mass in u	Abundance in %	Z	Isotope	Mass in u	Abundance in %
30	^{67}Zn	66.927 131(2)	4.1(1)	46	^{108}Pd	107.903 894(4)	26.46(9)
30	^{68}Zn	67.924 848(2)	18.8(4)	46	^{110}Pd	109.905 15(1)	11.72(9)
30	^{70}Zn	69.925 325(4)	0.6(1)	47	^{107}Ag	106.905 093(6)	51.839(7)
31	^{69}Ga	68.925 581(3)	60.108(9)	47	^{109}Ag	108.904 756(3)	48.161(7)
31	^{71}Ga	70.924 705(2)	39.892(9)	48	^{106}Cd	105.906 458(6)	1.25(4)
32	^{70}Ge	69.924 250(2)	21.23(4)	48	^{108}Cd	107.904 183(6)	0.89(2)
32	^{72}Ge	71.922 076(2)	27.66(3)	48	^{110}Cd	109.903 006(3)	12.49(12)
32	^{73}Ge	72.923 459(2)	7.73(1)	48	^{111}Cd	110.904 182(3)	12.80(8)
32	^{74}Ge	73.921 178(2)	35.94(2)	48	^{112}Cd	111.902 757(3)	24.13(14)
32	^{76}Ge	75.921 403(2)	7.44(2)	48	^{113}Cd	112.904 401(3)	12.22(8)
33	^{75}As	74.921 596(2)	100	48	^{114}Cd	113.903 358(3)	28.73(28)
34	^{74}Se	73.922 477(2)	0.89(2)	48	^{116}Cd	115.904 755(3)	7.49(12)
34	^{76}Se	75.919 214(2)	9.36(11)	49	^{113}In	112.904 061(4)	4.3(2)
34	^{77}Se	76.919 915(2)	7.63(6)	49	^{115}In	114.903 878(5)	95.7(2)
34	^{78}Se	77.917 310(2)	23.78(9)	50	^{112}Sn	111.904 821(5)	0.97(1)
34	^{80}Se	79.916 522(2)	49.61(10)	50	^{114}Sn	113.902 782(3)	0.65(1)
34	^{82}Se	81.916 700(2)	8.73(6)	50	^{115}Sn	114.903 346(3)	0.34(1)
35	^{79}Br	78.918 338(2)	50.69(7)	50	^{116}Sn	115.901 744(3)	14.53(1)
35	^{81}Br	80.916 291(3)	49.31(7)	50	^{117}Sn	116.902 954(3)	7.68(7)
36	^{78}Kr	77.920 386(7)	0.35(2)	50	^{118}Sn	117.901 606(3)	24.23(11)
36	^{80}Kr	79.916 378(4)	2.25(2)	50	^{119}Sn	118.903 309(3)	8.59(4)
36	^{82}Kr	81.913 485(3)	11.6(1)	50	^{120}Sn	119.902 197(3)	32.59(10)
36	^{83}Kr	82.914 136(3)	11.5(1)	50	^{122}Sn	121.903 440(3)	4.63(3)
36	^{84}Kr	83.911 507(3)	57.0(3)	50	^{124}Sn	123.905 275(1)	5.79(5)
36	^{86}Kr	85.910 610(1)	17.3(2)	51	^{121}Sb	120.903 818(2)	57.36(8)
37	^{85}Rb	84.911 789(3)	72.165(20)	51	^{123}Sb	122.904 216(2)	42.64(8)
37	^{87}Rb	86.909 183(3)	27.835(20)	52	^{120}Te	119.904 02(1)	0.096(2)
38	^{84}Sr	83.913 425(4)	0.56(1)	52	^{122}Te	121.903 047(2)	2.603(4)
38	^{86}Sr	85.909 262(2)	9.86(1)	52	^{123}Te	122.904 273(2)	0.908(2)
38	^{87}Sr	86.908 879(2)	7.00(1)	52	^{124}Te	123.902 819(2)	4.816(6)
38	^{88}Sr	87.905 614(2)	82.58(1)	52	^{125}Te	124.904 425(2)	7.139(6)
39	^{89}Y	88.905 848(3)	100	52	^{126}Te	125.903 306(2)	18.95(1)
40	^{90}Zr	89.904 704(2)	51.45(3)	52	^{128}Te	127.904 461(2)	31.69(1)
40	^{91}Zr	90.905 645(2)	11.22(4)	52	^{130}Te	129.906 223(2)	33.80(1)
40	^{92}Zr	91.905 040(2)	17.15(2)	53	^{127}I	126.904 468(4)	100
40	^{94}Zr	93.906 316(3)	17.38(4)	54	^{124}Xe	123.905 896(2)	0.10(1)
40	^{96}Zr	95.908 276(3)	2.80(2)	54	^{126}Xe	125.904 269(7)	0.09(1)
41	^{93}Nb	92.906 378(2)	100	54	^{128}Xe	127.903 530(2)	1.91(3)
42	^{92}Mo	91.906 810(4)	14.84(4)	54	^{129}Xe	128.904 779 4(9)	26.4(6)
42	^{94}Mo	93.905 088(2)	9.25(3)	54	^{130}Xe	129.903 508(1)	4.1(1)
42	^{95}Mo	94.905 841(2)	15.92(5)	54	^{131}Xe	130.905 082(1)	21.2(4)
42	^{96}Mo	95.904 679(2)	16.68(5)	54	^{132}Xe	131.904 154(1)	26.9(5)
42	^{97}Mo	96.906 021(2)	9.55(3)	54	^{134}Xe	133.905 394 5(9)	10.4(2)
42	^{98}Mo	97.905 408(2)	24.13(7)	54	^{136}Xe	135.907 220(8)	8.9(1)
42	^{100}Mo	99.907 477(6)	9.63(3)	55	^{133}Cs	132.905 447(3)	100
43	^{98}Tc	97.907 216(4)	*	56	^{130}Ba	129.906 310(7)	0.106(2)
43	^{99}Tc	98.906 255(2)	*	56	^{132}Ba	131.905 056(3)	0.101(2)
44	^{96}Ru	95.907 598(8)	5.52(6)	56	^{134}Ba	133.904 503(3)	2.417(27)
44	^{98}Ru	97.905 287(7)	1.88(6)	56	^{135}Ba	134.905 683(3)	6.592(18)
44	^{99}Ru	98.905 939(2)	12.7(1)	56	^{136}Ba	135.904 570(3)	7.854(36)
44	^{100}Ru	99.904 220(2)	12.6(1)	56	^{137}Ba	136.905 821(3)	11.23(4)
44	^{101}Ru	100.905 582(2)	17.0(1)	56	^{138}Ba	137.905 241(3)	71.70(7)
44	^{102}Ru	101.904 350(2)	31.6(2)	57	^{138}La	137.907 107(4)	0.0902(2)
44	^{104}Ru	103.905 430(4)	18.7(2)	57	^{139}La	138.906 348(3)	99.9098(2)
45	^{103}Rh	102.905 504(3)	100	58	^{136}Ce	135.907 14(5)	0.19(1)
46	^{102}Pd	101.905 608(3)	1.02(1)	58	^{137}Ce	136.907 78(5)	*
46	^{104}Pd	103.904 035(5)	11.14(8)	58	^{138}Ce	137.905 99(1)	0.25(1)
46	^{105}Pd	104.905 084(5)	22.33(8)	58	^{139}Ce	138.906 647(8)	*
46	^{106}Pd	105.903 483(5)	27.33(3)	58	^{140}Ce	139.905 434(3)	88.48(10)

Z	Isotope	Mass in u	Abundance in %	Z	Isotope	Mass in u	Abundance in %
58	^{141}Ce	140.908 271(3)	*	72	^{180}Hf	179.946 549(3)	35.100(7)
58	^{142}Ce	141.909 240(4)	11.08(10)	73	^{180}Ta	179.947 466(3)	0.012(2)
59	^{141}Pr	140.907 648(3)	100	73	^{181}Ta	180.947 996(3)	99.988(2)
60	^{142}Nd	141.907 719(3)	27.13(12)	74	^{180}W	179.946 706(5)	0.13(4)
60	^{143}Nd	142.909 810(3)	12.18(6)	74	^{182}W	181.948 206(3)	26.3(2)
60	^{144}Nd	143.910 083(3)	23.80(12)	74	^{183}W	182.950 224(3)	14.3(1)
60	^{145}Nd	144.912 569(3)	8.30(6)	74	^{184}W	183.950 933(3)	30.67(15)
60	^{146}Nd	145.913 112(3)	17.19(9)	74	^{186}W	185.954 362(3)	28.6(2)
60	^{148}Nd	147.916 889(3)	5.76(3)	75	^{185}Re	184.952 956(3)	37.40(2)
60	^{150}Nd	149.920 887(4)	5.64(3)	75	^{187}Re	186.955 751(3)	62.60(2)
61	^{143}Pm	142.910 928(4)	*	76	^{184}Os	183.952 491(3)	0.02(1)
61	^{145}Pm	144.912 744(4)	*	76	^{186}Os	185.953 838(3)	1.58(30)
61	^{147}Pm	146.915 134(3)	*	76	^{187}Os	186.955 748(3)	1.6(3)
62	^{144}Sm	143.911 995(4)	3.1(1)	76	^{188}Os	187.955 836(3)	13.3(7)
62	^{147}Sm	146.914 893(3)	15.0(2)	76	^{189}Os	188.958 145(3)	16.1(8)
62	^{148}Sm	147.914 818(3)	11.3(1)	76	^{190}Os	189.958 445(3)	26.4(12)
62	^{149}Sm	148.917 180(3)	13.8(1)	76	^{192}Os	191.961 479(4)	41.0(8)
62	^{150}Sm	149.917 271(3)	7.4(1)	77	^{191}Ir	190.960 591(3)	37.3(5)
62	^{152}Sm	151.919 728(3)	26.7(2)	77	^{193}Ir	192.962 924(3)	62.7(5)
62	^{154}Sm	153.922 205(3)	22.7(2)	78	^{190}Pt	189.959 930(7)	0.01(1)
63	^{151}Eu	150.919 846(3)	47.8(15)	78	^{192}Pt	191.961 035(4)	0.79(6)
63	^{153}Eu	152.921 226(3)	52.2(15)	78	^{194}Pt	193.962 664(3)	32.9(6)
64	^{152}Gd	151.919 788(3)	0.20(1)	78	^{195}Pt	194.964 774(3)	33.8(6)
64	^{154}Gd	153.920 862(3)	2.18(3)	78	^{196}Pt	195.964 935(3)	25.3(6)
64	^{155}Gd	154.922 619(3)	14.80(5)	78	^{198}Pt	197.967 876(5)	7.2(2)
64	^{156}Gd	155.922 120(3)	20.47(4)	79	^{197}Au	196.966 552(3)	100
64	^{157}Gd	156.923 957(3)	15.65(3)	80	^{196}Hg	195.965 815(4)	0.15(1)
64	^{158}Gd	157.924 101(3)	24.84(12)	80	^{198}Hg	197.966 752(3)	9.97(8)
64	^{160}Gd	159.927 051(3)	21.86(4)	80	^{199}Hg	198.968 262(3)	16.87(10)
65	^{159}Tb	158.925 343(3)	100	80	^{200}Hg	199.968 309(3)	23.10(16)
66	^{156}Dy	155.924 278(7)	0.06(1)	80	^{201}Hg	200.970 285(3)	13.18(8)
66	^{158}Dy	157.924 405(4)	0.10(1)	80	^{202}Hg	201.970 626(3)	29.86(20)
66	^{160}Dy	159.925 194(3)	2.34(6)	80	^{204}Hg	203.973 476(3)	6.87(4)
66	^{161}Dy	160.926 930(3)	18.9(2)	81	^{203}Tl	202.972 329(3)	29.524(14)
66	^{162}Dy	161.926 795(3)	25.5(2)	81	^{205}Tl	204.974 412(3)	70.476(14)
66	^{163}Dy	162.928 728(3)	24.9(2)	82	^{204}Pb	203.973 029(3)	1.4(1)
66	^{164}Dy	163.929 171(3)	28.2(2)	82	^{206}Pb	205.974 449(3)	24.1(1)
67	^{165}Ho	164.930 319(3)	100	82	^{207}Pb	206.975 881(3)	22.1(1)
68	^{162}Er	161.928 775(4)	0.14(1)	82	^{208}Pb	207.976 636(3)	52.4(1)
68	^{164}Er	163.929 197(4)	1.61(2)	83	^{209}Bi	208.980 383(3)	100
68	^{166}Er	165.930 290(3)	33.6(2)	84	^{209}Po	208.982 416(3)	*
68	^{167}Er	166.932 045(3)	22.95(15)	85	^{210}At	209.987 131(9)	*
68	^{168}Er	167.932 368(3)	26.8(2)	86	^{211}Rn	210.990 585(8)	*
68	^{170}Er	169.935 460(3)	14.9(2)	86	^{222}Rn	222.017 570(3)	*
69	^{169}Tm	168.934 211(3)	100	87	^{223}Fr	223.019 731(3)	*
70	^{168}Yb	167.933 894(5)	0.13(1)	88	^{223}Ra	223.018 497(3)	*
70	^{170}Yb	169.934 759(3)	3.05(6)	88	^{225}Ra	225.023 604(3)	*
70	^{171}Yb	170.936 322(3)	14.3(2)	88	^{226}Ra	226.025 403(3)	*
70	^{172}Yb	171.936 378(3)	21.9(3)	89	^{227}Ac	227.027 747(3)	*
70	^{173}Yb	172.938 207(3)	16.12(21)	90	^{229}Th	229.031 755(3)	*
70	^{174}Yb	173.938 858(3)	31.8(4)	90	^{232}Th	232.038 050(2)	100
70	^{176}Yb	175.942 568(3)	12.7(2)	91	^{231}Pa	231.035 879(3)	*
71	^{175}Lu	174.940 768(3)	97.41(2)	92	^{234}U	234.040 946(2)	0.0055(5)
71	^{176}Lu	175.942 682(3)	2.59(2)	92	^{235}U	235.043 923(2)	0.7200(12)
72	^{174}Hf	173.940 040(3)	0.162(3)	92	^{238}U	238.050 783(2)	99.2745(60)
72	^{176}Hf	175.941 402(3)	5.206(5)	93	^{237}Np	237.048 167(2)	*
72	^{177}Hf	176.943 220(3)	18.606(4)	94	^{239}Pu	239.052 157(2)	*
72	^{178}Hf	177.943 698(3)	27.297(4)	94	^{244}Pu	244.064 198(5)	*
72	^{179}Hf	178.945 815(3)	13.629(6)				

ELECTRON CONFIGURATION OF NEUTRAL ATOMS IN THE GROUND STATE

Atomic no.	n = Element	K 1 s	L 2 s	L 2 p	M 3 s	M 3 p	M 3 d	N 4 s	N 4 p	N 4 d	N 4 f	O 5 s	O 5 p	O 5 d	O 5 f	P 6 s	P 6 p	P 6 d	Q 7 s
1	H	1																	
2	He	2																	
3	Li	2	1																
4	Be	2	2																
5	B	2	2	1															
6	C	2	2	2															
7	N	2	2	3															
8	O	2	2	4															
9	F	2	2	5															
10	Ne	2	2	6															
11	Na	2	2	6	1														
12	Mg	2	2	6	2														
13	Al	2	2	6	2	1													
14	Si	2	2	6	2	2													
15	P	2	2	6	2	3													
16	S	2	2	6	2	4													
17	Cl	2	2	6	2	5													
18	Ar	2	2	6	2	6													
19	K	2	2	6	2	6		1											
20	Ca	2	2	6	2	6		2											
21	Sc	2	2	6	2	6	1	2											
22	Ti	2	2	6	2	6	2	2											
23	V	2	2	6	2	6	3	2											
24	Cr	2	2	6	2	6	5*	1											
25	Mn	2	2	6	2	6	5	2											
26	Fe	2	2	6	2	6	6	2											
27	Co	2	2	6	2	6	7	2											
28	Ni	2	2	6	2	6	8	2											
29	Cu	2	2	6	2	6	10*	1											
30	Zn	2	2	6	2	6	10	2											
31	Ga	2	2	6	2	6	10	2	1										
32	Ge	2	2	6	2	6	10	2	2										
33	As	2	2	6	2	6	10	2	3										
34	Se	2	2	6	2	6	10	2	4										
35	Br	2	2	6	2	6	10	2	5										
36	Kr	2	2	6	2	6	10	2	6										
37	Rb	2	2	6	2	6	10	2	6			1							
38	Sr	2	2	6	2	6	10	2	6			2							
39	Y	2	2	6	2	6	10	2	6	1		2							
40	Zr	2	2	6	2	6	10	2	6	2		2							
41	Nb	2	2	6	2	6	10	2	6	4*		1							
42	Mo	2	2	6	2	6	10	2	6	5		1							
43	Tc	2	2	6	2	6	10	2	6	5		2							
44	Ru	2	2	6	2	6	10	2	6	7		1							
45	Rh	2	2	6	2	6	10	2	6	8		1							
46	Pd	2	2	6	2	6	10	2	6	10*									
47	Ag	2	2	6	2	6	10	2	6	10		1							
48	Cd	2	2	6	2	6	10	2	6	10		2							
49	In	2	2	6	2	6	10	2	6	10		2	1						
50	Sn	2	2	6	2	6	10	2	6	10		2	2						
51	Sb	2	2	6	2	6	10	2	6	10		2	3						
52	Te	2	2	6	2	6	10	2	6	10		2	4						
53	I	2	2	6	2	6	10	2	6	10		2	5						
54	Xe	2	2	6	2	6	10	2	6	10		2	6						
55	Cs	2	2	6	2	6	10	2	6	10		2	6			1			
56	Ba	2	2	6	2	6	10	2	6	10		2	6			2			

Atomic no.	$n=$ Element	K 1 s	L 2 s	p	M 3 s	p	d	N 4 s	p	d	f	O 5 s	p	d	f	P 6 s	p	d	Q 7 s
57	La	2	2	6	2	6	10	2	6	10		2	6	1		2			
58	Ce	2	2	6	2	6	10	2	6	10	1*	2	6	1		2			
59	Pr	2	2	6	2	6	10	2	6	10	3	2	6			2			
60	Nd	2	2	6	2	6	10	2	6	10	4	2	6			2			
61	Pm	2	2	6	2	6	10	2	6	10	5	2	6			2			
62	Sm	2	2	6	2	6	10	2	6	10	6	2	6			2			
63	Eu	2	2	6	2	6	10	2	6	10	7	2	6			2			
64	Gd	2	2	6	2	6	10	2	6	10	7	2	6	1		2			
65	Tb	2	2	6	2	6	10	2	6	10	9*	2	6			2			
66	Dy	2	2	6	2	6	10	2	6	10	10	2	6			2			
67	Ho	2	2	6	2	6	10	2	6	10	11	2	6			2			
68	Er	2	2	6	2	6	10	2	6	10	12	2	6			2			
69	Tm	2	2	6	2	6	10	2	6	10	13	2	6			2			
70	Yb	2	2	6	2	6	10	2	6	10	14	2	6			2			
71	Lu	2	2	6	2	6	10	2	6	10	14	2	6	1		2			
72	Hf	2	2	6	2	6	10	2	6	10	14	2	6	2		2			
73	Ta	2	2	6	2	6	10	2	6	10	14	2	6	3		2			
74	W	2	2	6	2	6	10	2	6	10	14	2	6	4		2			
75	Re	2	2	6	2	6	10	2	6	10	14	2	6	5		2			
76	Os	2	2	6	2	6	10	2	6	10	14	2	6	6		2			
77	Ir	2	2	6	2	6	10	2	6	10	14	2	6	7		2			
78	Pt	2	2	6	2	6	10	2	6	10	14	2	6	9		1			
79	Au	2	2	6	2	6	10	2	6	10	14	2	6	10		1			
80	Hg	2	2	6	2	6	10	2	6	10	14	2	6	10		2			
81	Tl	2	2	6	2	6	10	2	6	10	14	2	6	10		2	1		
82	Pb	2	2	6	2	6	10	2	6	10	14	2	6	10		2	2		
83	Bi	2	2	6	2	6	10	2	6	10	14	2	6	10		2	3		
84	Po	2	2	6	2	6	10	2	6	10	14	2	6	10		2	4		
85	At	2	2	6	2	6	10	2	6	10	14	2	6	10		2	5		
86	Rn	2	2	6	2	6	10	2	6	10	14	2	6	10		2	6		
87	Fr	2	2	6	2	6	10	2	6	10	14	2	6	10		2	6		1
88	Ra	2	2	6	2	6	10	2	6	10	14	2	6	10		2	6		2
89	Ac	2	2	6	2	6	10	2	6	10	14	2	6	10		2	6	1	2
90	Th	2	2	6	2	6	10	2	6	10	14	2	6	10		2	6	2	2
91	Pa	2	2	6	2	6	10	2	6	10	14	2	6	10	2*	2	6	1	2
92	U	2	2	6	2	6	10	2	6	10	14	2	6	10	3	2	6	1	2
93	Np	2	2	6	2	6	10	2	6	10	14	2	6	10	4	2	6	1	2
94	Pu	2	2	6	2	6	10	2	6	10	14	2	6	10	6*	2	6		2
95	Am	2	2	6	2	6	10	2	6	10	14	2	6	10	7	2	6		2
96	Cm	2	2	6	2	6	10	2	6	10	14	2	6	10	7*	2	6	1	2
97	Bk	2	2	6	2	6	10	2	6	10	14	2	6	10	9	2	6		2
98	Cf	2	2	6	2	6	10	2	6	10	14	2	6	10	10	2	6		2
99	Es	2	2	6	2	6	10	2	6	10	14	2	6	10	11	2	6		2
100	Fm	2	2	6	2	6	10	2	6	10	14	2	6	10	12	2	6		2
101	Md	2	2	6	2	6	10	2	6	10	14	2	6	10	13	2	6		2
102	No	2	2	6	2	6	10	2	6	10	14	2	6	10	14	2	6		2
103	Lr	2	2	6	2	6	10	2	6	10	14	2	6	10	14	2	6	1	2
104	Rf	2	2	6	2	6	10	2	6	10	14	2	6	10	14	2	6	2	2

* Note irregularity.

REFERENCE

W. L. Wiese and G. A. Martin, in *A Physicist's Desk Reference*, American Institute of Physics, New York, 1989, 94.

INTERNATIONAL TEMPERATURE SCALE OF 1990 (ITS-90)

B. W. Mangum

A new temperature scale, the International Temperature Scale of 1990 (ITS-90), was officially adopted by the Comité International des Poids et Mesures (CIPM), meeting 26—28 September 1989 at the Bureau International des Poids et Mesures (BIPM). The ITS-90 was recommended to the CIPM for its adoption following the completion of the final details of the new scale by the Comité Consultatif de Thermométrie (CCT), meeting 12—14 September 1989 at the BIPM in its 17th Session. The ITS-90 became the official international temperature scale on 1 January 1990. The ITS-90 supersedes the present scales, the International Practical Temperature Scale of 1968 (IPTS-68) and the 1976 Provisional 0.5 to 30 K Temperature Scale (EPT-76).

The ITS-90 extends upward from 0.65 K, and temperatures on this scale are in much better agreement with thermodynamic values that are those on the IPTS-68 and the EPT-76. The new scale has subranges and alternative definitions in certain ranges that greatly facilitate its use. Furthermore, its continuity, precision, and reproducibility throughout its ranges are much improved over that of the present scales. The replacement of the thermocouple with the platinum resistance thermometer at temperatures below 961.78°C resulted in the biggest improvement in reproducibility.

The ITS-90 is divided into four primary ranges:

1. Between 0.65 and 3.2 K, the ITS-90 is defined by the vapor pressure-temperature relation of ^{3}He, and between 1.25 and 2.1768 K (the λ point) and between 2.1768 and 5.0 K by the vapor pressure-temperature relations of ^{4}He. T_{90} is defined by the vapor pressure equations of the form:

$$T_{90}/K = A_0 + \sum_{i=1}^{9} A_i \left[\left(\ln(p/\mathrm{Pa}) - B \right)/C \right]^i$$

   The values of the coefficients A_i, and of the constants A_0, B, and C of the equations are given below.

2. Between 3.0 and 24.5561 K, the ITS-90 is defined in terms of a ^{3}He or ^{4}He constant volume gas thermometer (CVGT). The thermometer is calibrated at three temperatures — at the triple point of neon (24.5561 K), at the triple point of equilibrium hydrogen (13.8033 K), and at a temperature between 3.0 and 5.0 K, the value of which is determined by using either ^{3}He or ^{4}He vapor pressure thermometry.

3. Between 13.8033 K (−259.3467°C) and 1234.93 K (961.78°C), the ITS-90 is defined in terms of the specified fixed points given below, by resistance ratios of platinum resistance thermometers obtained by calibration at specified sets of the fixed points, and by reference functions and deviation functions of resistance ratios which relate to T_{90} between the fixed points.

4. Above 1234.93 K, the ITS-90 is defined in terms of Planck's radiation law, using the freezing-point temperature of either silver, gold, or copper as the reference temperature.

Full details of the calibration procedures and reference functions for various subranges are given in:

The International Temperature Scale of 1990, *Metrologia*, 27, 3, 1990; errata in *Metrologia*, 27, 107, 1990.

Defining Fixed Points of the ITS-90

Material[a]	Equilibrium state[b]	Temperature	
		T_{90} (K)	t_{90} (°C)
He	VP	3 to 5	−270.15 to −268.15
e-H$_2$	TP	13.8033	−259.3467
e-H$_2$ (or He)	VP (or CVGT)	≈17	≈ −256.15
e-H$_2$ (or He)	VP (or CVGT)	≈20.3	≈ −252.85
Ne[c]	TP	24.5561	−248.5939
O$_2$	TP	54.3584	−218.7916
Ar	TP	83.8058	−189.3442
Hg[c]	TP	234.3156	−38.8344
H$_2$O	TP	273.16	0.01
Ga[c]	MP	302.9146	29.7646
In[c]	FP	429.7485	156.5985
Sn	FP	505.078	231.928
Zn	FP	692.677	419.527
Al[c]	FP	933.473	660.323
Ag	FP	1234.93	961.78
Au	FP	1337.33	1064.18
Cu[c]	FP	1357.77	1084.62

Defining Fixed Points of the ITS-90 (continued)

[a] e-H_2 indicates equilibrium hydrogen, that is, hydrogen with the equilibrium distribution of its ortho and para states. Normal hydrogen at room temperature contains 25% para hydrogen and 75% ortho hydrogen.

[b] VP indicates vapor pressure point; CVGT indicates constant volume gas thermometer point; TP indicates triple point (equilibrium temperature at which the solid, liquid, and vapor phases coexist); FP indicates freezing point, and MP indicates melting point (the equilibrium temperatures at which the solid and liquid phases coexist under a pressure of 101 325 Pa, one standard atmosphere). The isotopic composition is that naturally occurring.

[c] Previously, these were secondary fixed points.

Values of Coefficients in the Vapor Pressure Equations for Helium

Coef.or constant	^{3}He 0.65—3.2 K	^{4}He 1.25—2.1768 K	^{4}He 2.1768—5.0 K
A_0	1.053 447	1.392 408	3.146 631
A_1	0.980 106	0.527 153	1.357 655
A_2	0.676 380	0.166 756	0.413 923
A_3	0.372 692	0.050 988	0.091 159
A_4	0.151 656	0.026 514	0.016 349
A_5	−0.002 263	0.001 975	0.001 826
A_6	0.006 596	−0.017 976	−0.004 325
A_7	0.088 966	0.005 409	−0.004 973
A_8	−0.004 770	0.013 259	0
A_9	−0.054 943	0	0
B	7.3	5.6	10.3
C	4.3	2.9	1.9

CONVERSION OF TEMPERATURES FROM THE 1948 AND 1968 SCALES TO ITS-90

This table gives temperature corrections from older scales to the current International Temperature Scale of 1990 (see the preceding table for details on ITS-90). The first part of the table may be used for converting Celsius temperatures in the range -180 to 4000°C from IPTS-68 or IPTS-48 to ITS-90. Within the accuracy of the corrections, the temperature in the first column may be identified with either t_{68}, t_{48}, or t_{90}. The second part of the table is designed for use at lower temperatures to convert values expressed in kelvins from EPT-76 or IPTS-68 to ITS-90.

The references give analytical equations for expressing these relations. Note that Reference 1 supersedes Reference 2 with respect to corrections in the 630 to 1064°C range.

REFERENCES

1. Burns, G. W. et al., in *Temperature: Its Measurement and Control in Science and Industry,* Vol. 6, Schooley, J. F., Ed., American Institute of Physics, New York, 1993.
2. Goldberg, R. N. and Weir, R. D., *Pure and Appl. Chem.*, 1545, 1992.

t/°C	t_{90}-t_{68}	t_{90}-t_{48}	t/°C	t_{90}-t_{68}	t_{90}-t_{48}	t/°C	t_{90}-t_{68}	t_{90}-t_{48}
-180	0.008	0.020	270	-0.039	0.028	720	0.00	0.45
-170	0.010	0.017	280	-0.039	0.030	730	0.02	0.49
-160	0.012	0.007	290	-0.039	0.032	740	0.03	0.53
-150	0.013	0.000	300	-0.039	0.034	750	0.03	0.56
-140	0.014	0.001	310	-0.039	0.035	760	0.04	0.60
-130	0.014	0.008	320	-0.039	0.036	770	0.05	0.63
-120	0.014	0.017	330	-0.040	0.036	780	0.05	0.66
-110	0.013	0.026	340	-0.040	0.037	790	0.05	0.69
-100	0.013	0.035	350	-0.041	0.036	800	0.05	0.72
-90	0.012	0.041	360	-0.042	0.035	810	0.05	0.75
-80	0.012	0.045	370	-0.043	0.034	820	0.04	0.76
-70	0.011	0.045	380	-0.045	0.032	830	0.04	0.79
-60	0.010	0.042	390	-0.046	0.030	840	0.03	0.81
-50	0.009	0.038	400	-0.048	0.028	850	0.02	0.83
-40	0.008	0.032	410	-0.051	0.024	860	0.01	0.85
-30	0.006	0.024	420	-0.053	0.022	870	0.00	0.87
-20	0.004	0.016	430	-0.056	0.019	880	-0.02	0.87
-10	0.002	0.008	440	-0.059	0.015	890	-0.03	0.89
0	0.000	0.000	450	-0.062	0.012	900	-0.05	0.90
10	-0.002	-0.006	460	-0.065	0.009	910	-0.06	0.92
20	-0.005	-0.012	470	-0.068	0.007	920	-0.08	0.93
30	-0.007	-0.016	480	-0.072	0.004	930	-0.10	0.94
40	-0.010	-0.020	490	-0.075	0.002	940	-0.11	0.96
50	-0.013	-0.023	500	-0.079	0.000	950	-0.13	0.97
60	-0.016	-0.026	510	-0.083	-0.001	960	-0.15	0.97
70	-0.018	-0.026	520	-0.087	-0.002	970	-0.16	0.99
80	-0.021	-0.027	530	-0.090	-0.001	980	-0.18	1.00
90	-0.024	-0.027	540	-0.094	0.000	990	-0.19	1.02
100	-0.026	-0.026	550	-0.098	0.002	1000	-0.20	1.04
110	-0.028	-0.024	560	-0.101	0.007	1010	-0.22	1.05
120	-0.030	-0.023	570	-0.105	0.011	1020	-0.23	1.07
130	-0.032	-0.020	580	-0.108	0.018	1030	-0.23	1.10
140	-0.034	-0.018	590	-0.112	0.025	1040	-0.24	1.12
150	-0.036	-0.016	600	-0.115	0.035	1050	-0.25	1.14
160	-0.037	-0.012	610	-0.118	0.047	1060	-0.25	1.17
170	-0.038	-0.009	620	-0.122	0.060	1070	-0.25	1.19
180	-0.039	-0.005	630	-0.125	0.075	1080	-0.26	1.20
190	-0.039	-0.001	640	-0.11	0.12	1090	-0.26	1.20
200	-0.040	0.003	650	-0.10	0.15	1100	-0.26	1.2
210	-0.040	0.007	660	-0.09	0.19	1200	-0.30	1.4
220	-0.040	0.011	670	-0.07	0.24	1300	-0.35	1.5
230	-0.040	0.014	680	-0.05	0.29	1400	-0.39	1.6
240	-0.040	0.018	690	-0.04	0.32	1500	-0.44	1.8
250	-0.040	0.021	700	-0.02	0.37	1600	-0.49	1.9
260	-0.040	0.024	710	-0.01	0.41	1700	-0.54	2.1

$t/°C$	$t_{90}\text{-}t_{68}$	$t_{90}\text{-}t_{48}$	T/K	$T_{90}\text{-}T_{76}$	$T_{90}\text{-}T_{68}$	T/K	$T_{90}\text{-}T_{76}$	$T_{90}\text{-}T_{68}$
1800	-0.60	2.2	28		-0.005	77		0.008
1900	-0.66	2.3	29		-0.006	78		0.008
2000	-0.72	2.5	30		-0.006	79		0.008
2100	-0.79	2.7	31		-0.007	80		0.008
2200	-0.85	2.9	32		-0.008	81		0.008
2300	-0.93	3.1	33		-0.008	82		0.008
2400	-1.00	3.2	34		-0.008	83		0.008
2500	-1.07	3.4	35		-0.007	84		0.008
2600	-1.15	3.7	36		-0.007	85		0.008
2700	-1.24	3.8	37		-0.007	86		0.008
2800	-1.32	4.0	38		-0.006	87		0.008
2900	-1.41	4.2	39		-0.006	88		0.008
3000	-1.50	4.4	40		-0.006	89		0.008
3100	-1.59	4.6	41		-0.006	90		0.008
3200	-1.69	4.8	42		-0.006	91		0.008
3300	-1.78	5.1	43		-0.006	92		0.008
3400	-1.89	5.3	44		-0.006	93		0.008
3500	-1.99	5.5	45		-0.007	94		0.008
3600	-2.10	5.8	46		-0.007	95		0.008
3700	-2.21	6.0	47		-0.007	96		0.008
3800	-2.32	6.3	48		-0.006	97		0.009
3900	-2.43	6.6	49		-0.006	98		0.009
4000	-2.55	6.8	50		-0.006	99		0.009
			51		-0.005	100		0.009
T/K	$T_{90}\text{-}T_{76}$	$T_{90}\text{-}T_{68}$	52		-0.005	110		0.011
			53		-0.004	120		0.013
5	-0.0001		54		-0.003	130		0.014
6	-0.0002		55		-0.002	140		0.014
7	-0.0003		56		-0.001	150		0.014
8	-0.0004		57		0.000	160		0.014
9	-0.0005		58		0.001	170		0.013
10	-0.0006		59		0.002	180		0.012
11	-0.0007		60		0.003	190		0.012
12	-0.0008		61		0.003	200		0.011
13	-0.0010		62		0.004	210		0.010
14	-0.0011	-0.006	63		0.004	220		0.009
15	-0.0013	-0.003	64		0.005	230		0.008
16	-0.0014	-0.004	65		0.005	240		0.007
17	-0.0016	-0.006	66		0.006	250		0.005
18	-0.0018	-0.008	67		0.006	260		0.003
19	-0.0020	-0.009	68		0.007	270		0.001
20	-0.0022	-0.009	69		0.007	273.16		0.000
21	-0.0025	-0.008	70		0.007	300		-0.006
22	-0.0027	-0.007	71		0.007	400		-0.031
23	-0.0030	-0.007	72		0.007	500		-0.040
24	-0.0032	-0.006	73		0.007	600		-0.040
25	-0.0035	-0.005	74		0.007	700		-0.055
26	-0.0038	-0.004	75		0.008	800		-0.089
27	-0.0041	-0.004	76		0.008	900		-0.124

INTERNATIONAL SYSTEM OF UNITS (SI)

The International System of units (SI) was adopted by the 11th General Conference on Weights and Measures (CGPM) in 1960. It is a coherent system of units built from seven *SI base units*, one for each of the seven dimensionally independent base quantities: they are the meter, kilogram, second, ampere, kelvin, mole, and candela, for the dimensions length, mass, time, electric current, thermodynamic temperature, amount of substance, and luminous intensity, respectively. The definitions of the SI base units are given below. The *SI derived units* are expressed as products of powers of the base units, analogous to the corresponding relations between physical quantities but with numerical factors equal to unity.

In the International System there is only one SI unit for each physical quanitity. This is either the appropriate SI base unit itself or the appropriate SI derived unit. However, any of the approved decimal prefixes, called *SI prefixes*, may be used to construct decimal multiples or submultiples of SI units.

It is recommended that only SI units be used in science and technology (with SI prefixes where appropriate). Where there are special reasons for making an exception to this rule, it is recommended always to define the units used in terms of SI units. This section was reprinted with the permission of IUPAC.

Definitions of SI Base Units

Meter — the meter is the length of path travelled by light in vacuum during a time interval of 1/299 792 458 of a second (17th CGPM, 1983).

Kilogram — The kilogram is the unit of mass; it is equal to the mass of the international prototype of the kilogram (3rd CGPM, 1901).

Second — The second is the duration of 9 192 631 770 periods of the radiation corresponding to the transition between the two hyperfine levels of the ground state of the cesium-133 atom (13th CGPM, 1967).

Ampere — The ampere is that constant current which, if maintained in two straight parallel conductors of infinite length, of negligible circular cross-section, and placed 1 meter apart in vacuum, would produce between these conductors a force equal to 2×10^{-7} newton per meter of length (9th CGPM, 1948).

Kelvin — The kelvin, unit of thermodynamic temperature, is the fraction 1/273.16 of the thermodynamic temperature of the triple point of water (13th CGPM, 1967)

Mole — The mole is the amount of substance of a system which contains as many elementary entities as there are atoms in 0.012 kilogram of carbon-12. When the mole is used, the elementary entities must be specified and may be atoms, molecules, ions, electrons, other particles, or specified groups of such particles (14th CGPM, 1971).

Examples of the use of the mole:

1 mol of H_2 contains about 6.022×10^{23} H_2 molecules, or 12.044×10^{23} H atoms

1 mol of HgCl has a mass of 236.04 g

1 mol of Hg_2Cl_2 has a mass of 472.08 g

1 mol of Hg_2^{2+} has a mass of 401.18 g and a charge of 192.97 kC

1 mol of $Fe_{0.91}S$ has a mass of 82.88 g

1 mol of e^- has a mass of 548.60 μg and a charge of –96.49 kC

1 mol of photons whose frequency is 10^{14} Hz has energy of about 39.90 kJ

Candela — The candela is the luminous intensity, in a given direction, of a source that emits monochromatic radiation of frequency 540×10^{12} hertz and that has a radiant intensity in that direction of (1/683) watt per steradian (16th CGPM, 1979).

Names and Symbols for the SI Base Units

Physical quantity	Name of SI unit	Symbol for SI unit
length	meter	m
mass	kilogram	kg
time	second	s
electric current	ampere	A
thermodynamic temperature	kelvin	K
amount of substance	mole	mol
luminous intensity	candela	cd

SI Derived Units with Special Names and Symbols

Physical quantity	Name of SI unit	Symbol for SI unit	Expression in terms of SI base units	
frequency[1]	hertz	Hz	s^{-1}	
force	newton	N	$m\ kg\ s^{-2}$	
pressure, stress	pascal	Pa	$N\ m^{-2}$	$= m^{-1}\ kg\ s^{-2}$
energy, work, heat	joule	J	$N\ m$	$= m^2\ kg\ s^{-2}$
power, radiant flux	watt	W	$J\ s^{-1}$	$= m^2\ kg\ s^{-3}$
electric charge	coulomb	C	$A\ s$	
electric potential, electromotive force	volt	V	$J\ C^{-1}$	$= m^2\ kg\ s^{-3}\ A^{-1}$
electric resistance	ohm	Ω	$V\ A^{-1}$	$= m^2\ kg\ s^{-3}\ A^{-2}$
electric conductance	siemens	S	Ω^{-1}	$= m^{-2}\ kg^{-1}\ s^3\ A^2$
electric capacitance	farad	F	$C\ V^{-1}$	$= m^{-2}\ kg^{-1}\ s^4\ A^2$
magnetic flux density	tesla	T	$V\ s\ m^{-2}$	$= kg\ s^{-2}\ A^{-1}$
magnetic flux	weber	Wb	$V\ s$	$= m^2\ kg\ s^{-2}\ A^{-1}$
inductance	henry	H	$V\ A^{-1}\ s$	$= m^2\ kg\ s^{-2}\ A^{-2}$
Celsius temperature[2]	degree Celsius	°C	K	
luminous flux	lumen	lm	cd sr	
illuminance	lux	lx	$cd\ sr\ m^{-2}$	
activity[3] (radioactive)	becquerel	Bq	s^{-1}	
absorbed dose[3] (of radiation)	gray	Gy	$J\ kg^{-1}$	$= m^2\ s^{-2}$
dose equivalent[3] (dose equivalent index)	sievert	Sv	$J\ kg^{-1}$	$= m^2\ s^{-2}$
plane angle[4]	radian	rad	1	$= m\ m^{-1}$
solid angle[4]	steradian	sr	1	$= m^2\ m^{-2}$

(1) For radial (circular) frequency and for angular velocity the unit rad s^{-1}, or simply s^{-1}, should be used, and this may not be simplified to Hz. The unit Hz should be used only for frequency in the sense of cycles per second.

(2) The Celsius temperature θ is defined by the equation:

$$\theta/°C = T/K - 273.15$$

The SI unit of Celsius temperature interval is the degree Celsius, °C, which is equal to the kelvin, K. °C should be treated as a single symbol, with no space between the ° sign and the letter C. (The symbol °K, and the symbol °, should no longer be used.)

(3) The units gray and sievert are admitted for reasons of safeguarding human health.

(4) The units radian and steradian are described as 'SI supplementary units'. However, in chemistry, as well as in physics, they are usually treated as dimensionless derived units, and this was recognized by CIPM in 1980. Since they are then of dimension 1, this leaves open the possibility of including them or omitting them in expressions of SI derived units, In practice this means that rad and sr may be used when appropriate and may be omitted if clarity is not lost thereby.

SI Prefixes

To signify decimal multiples and submultiples of SI units the following prefixes may be used.

Factor	Prefix	Symbol	Factor	Prefix	Symbol
10^{24}	yotta	Y	10^{-1}	deci	d
10^{21}	zetta	Z	10^{-2}	centi	c
10^{18}	exa	E	10^{-3}	milli	m
10^{15}	peta	P	10^{-6}	micro	μ
10^{12}	tera	T	10^{-9}	nano	n
10^{9}	giga	G	10^{-12}	pico	p
10^{6}	mega	M	10^{-15}	femto	f
10^{3}	kilo	k	10^{-18}	atto	a
10^{2}	hecto	h	10^{-21}	zepto	z
10^{1}	deka	da	10^{-24}	yocto	y

Prefix symbols should be printed in roman (upright) type with no space between the prefix and the unit symbol.

Example kilometer, km

When a prefix is used with a unit symbol, the combination is taken as a new symbol that can be raised to any power without the use of parentheses.

Examples

$$1 \text{ cm}^3 = (0.01 \text{ m})^3 = 10^{-6} \text{ m}^3$$
$$1 \text{ } \mu s^{-1} = (10^{-6} \text{ s})^{-1} = 10^6 \text{ s}^{-1}$$
$$1 \text{ V/cm} = 100 \text{ V/m}$$
$$1 \text{ mmol/dm}^3 = \text{mol m}^{-3}$$

A prefix should never be used on its own, and prefixes are not to be combined into compound prefixes.

Example   pm, not $\mu\mu$m

The names and symbols of decimal multiples and sub-multiples of the SI base unit of mass, the kg, which already contains a prefix, are constructed by adding the appropriate prefix to the word gram and symbol g.

Examples   mg, not μkg; Mg, not kkg

The SI prefixes are not to be used with °C.

Units in Use Together with the SI

These units are not part of the SI, but it is recognized that they will continue to be used in appropriate contexts. SI prefixes may be attached to some of these units, such as milliliter, ml; millibar, mbar; megaelectronvolt, MeV; kilotonne, ktonne.

Physical quantity	Name of unit	Symbol for unit	Value in SI units
time	minute	min	60 s
time	hour	h	3600 s
time	day	d	86 400 s
plane angle	degree	°	$(\pi/180)$ rad
plane angle	minute	′	$(\pi/10\ 800)$ rad
plane angle	second	″	$(\pi/648\ 000)$ rad
length	ångström[1]	Å	10^{-10} m
area	barn	b	10^{-28} m^2
volume	litre	l, L	dm^3 $\quad = 10^{-3}$ m^3
mass	tonne	t	Mg $\quad = 10^3$ kg
pressure	bar[1]	bar	10^5 Pa $= 10^5$ N m^{-2}
energy	electronvolt[2]	eV $(= e \times V)$	$\approx 1.60218 \times 10^{-19}$ J
mass	unified atomic mass unit[2,3]	u $(= m_a(^{12}C)/12)$	$\approx 1.66054 \times 10^{-27}$ kg

(1) The ångström and the bar are approved by CIPM for 'temporary use with SI units', until CIPM makes a further recommendation, However, they should not be introduced where they are not used at present.

(2) The values of these units in terms of the corresponding SI units are not exact, since they depend on the values of the physical constants e (for the electronvolt) and N_A (for the unified atomic mass unit), which are determined by experiment.

(3) The unified atomic mass unit is also sometimes called the dalton, with symbol Da, although the name and symbol have not been approved by CGPM.

Atomic Units

For the purposes of quantum mechanical calculations of electronic wavefunctions, it is convenient to regard certain fundamental constants (and combinations of such constants) as though they were units. They are customarily called *atomic units* (abbreviated: au), and they may be regarded as forming a coherent system of units for the calculation of electronic properties in theoretical chemistry, although there is no authority from CGPM for treating them as units. The first five atomic units in the table below have special names and symbols. Only four of these are independent; all others may be derived by multiplication and division in the usual way, and the table includes a number of examples.

The relation of atomic units to the corresponding SI units involves the values of the fundamental physical constants, and is therefore not exact. The numerical values in the table are based on the 1986 CODATA values of the fundamental constants. The numerical results of calculations in theoretical chemistry are frequently quoted in atomic units, or as numerical values in the form *(physical quantity)/(atomic unit)*, so that the reader may make the conversion using the current best estimates of the physical constants.

Physical quantity	Name of unit	Symbol for unit	Definition and value of unit in SI
mass	electron rest mass	m_e	$m_e \approx 9.1095 \times 10^{-31}$ kg
charge	elementary charge	e	$e \approx 1.6022 \times 10^{-19}$ C
action	Planck constant/2π	$\hbar$	$\hbar = h/2\pi \approx 1.0546 \times 10^{-34}$ J s
length	bohr	a_0	$4\pi\varepsilon_0\hbar^2/m_e e^2 \approx 5.2918 \times 10^{-11}$ m
energy	hartree	E_h	$\hbar^2/m_e a_0^2 \approx 4.3598 \times 10^{-18}$ J
time	au of time	$\hbar/E_h$	$\approx 2.4189 \times 10^{-17}$ s
velocity[1]	au of velocity	$a_0 E_h/\hbar$	$\approx 2.1877 \times 10^6$ m s^{-1}
force	au of force	E_h/a_0	$\approx 8.2389 \times 10^{-8}$ N
momentum, linear	au of momentum	$\hbar/a_0$	$\approx 1.9929 \times 10^{-24}$ N s
electric current	au of current	$eE_h/\hbar$	$\approx 6.6236 \times 10^{-3}$ A
electric field	au of electric field	E_h/ea_0	$\approx 5.1422 \times 10^{11}$ V m^{-1}
electric dipole moment	au of electric dipole	ea_0	$\approx 8.4784 \times 10^{-30}$ C m
magnetic flux density	au of magnetic flux density	$\hbar/ea_0^2$	$\approx 2.3505 \times 10^5$ T
magnetic dipole moment[2]	au of magnetic dipole	$e\hbar/m_e$	$= 2\mu_B \approx 1.8548 \times 10^{-23}$ J T^{-1}

(1) The numerical value of the speed of light, when expressed in atomic units, is equal to the reciprocal of the fine structure constant α; $c/(\text{au of velocity}) = c\hbar/a_0 E_h = \alpha^{-1}$ 137.04.
(2) The atomic unit of magnetic dipole moment is twice the Bohr magneton, μ_B.

REFERENCES

Quantities, Units and Symbols in Physical Chemistry, Ian Mills, Editor, Blackwell Scientific Publications, Oxford, 1987.
Symbols, Units, Nomenclature and Fundamental Constants in Physics, E. R. Cohen and P. Giacomo, Editors, International Union of Pure and Applied Physics Document 25, 1987; also appears in Physica 146A, 1—8, 1987.
ISO Standards Handbook 2, Units of Measurement, International Organization for Standardization, Geneva, 1982.

CONVERSION FACTORS

The following table gives conversion factors from various units of measure to SI units. It is reproduced from NIST Special Publication 811, *Guide for the Use of the International System of Units* (Superintendent of Documents, U.S. Government Printing Office, 1991), which in turn was derived from IEEE Std 268—1982, *IEEE Standard Metric Practice* (© 1982 by the Institute of Electrical and Electronics Engineers, Inc.).

The SI values are expressed in terms of the base, supplementary, and derived units of SI in order to provide a coherent presentation of the conversion factors and facilitate computations (see the table "International System of Units" in this Section). Powers of ten can be avoided by using SI prefixes and shifting the decimal point if necessary. Conversion from a non-SI unit to a different non-SI unit may be carried out by using this table in two stages, e.g.,

1 cal (thermochemical) = 4.184 J
1 Btu (mean) = 1.05587 E+03 J

Thus, 1 Btu (mean) = (1.05587 E+03/4.184) cal (thermochemical)
= 252.359 cal (thermochemical).

Conversion factors are presented for ready adaptation to computer readout and electronic data transmission. The factors are written as a number equal to or greater than one and less than ten with six or less decimal places. This number is followed by the letter E (for exponent), plus or minus symbol, and two digits which indicate the power of 10 by which the number must be multiplied to obtain the correct value. For example:

$$3.523\,907\ E{-}02 \text{ is } 3.523\,907 \times 10^{-2}$$

or

$$0.035\,239\,07$$

Similarly:

$$3.386\,389\ E{+}03 \text{ is } 3.386\,389 \times 10^{3}$$

or

$$3\,386.389$$

An asterisk (*) after the sixth decimal place indicates that the conversion factor is exact and that all subsequent digits are zero. All other conversion factors have been rounded to the figures given in accordance with accepted practice. Where less than six decimal places are shown, more precision is not warranted.

To convert from	to	Multiply by
abampere	ampere (A)	1.000 000*E + 01
abcoulomb	coulomb (C)	1.000 000*E + 01
abfarad	farad (F)	1.000 000*E + 09
abhenry	henry (H)	1.000 000*E − 09
abmho	siemens (S)	1.000 000*E + 09
abohm	ohm (Ω)	1.000 000*E − 09
abvolt	volt (V)	1.000 000*E − 08
acre foot	meter³ (m³)	1.233 5 E + 03
acre	meter² (m²)	4.046 873 E + 03
ampere hour	coulomb (C)	3.600 000*E + 03
ångström	meter (m)	1.000 000*E − 10
are	meter² (m²)	1.000 000*E + 02
astronomical unit	meter (m)	1.495 979 E + 11
atmosphere (standard)	pascal (Pa)	1.013 250*E + 05
atmosphere (technical = 1 kgf/cm²)	pascal (Pa)	9.806 650*E + 04
bar	pascal (Pa)	1.000 000*E + 05
barn	meter² (m²)	1.000 000*E − 28
barrel (for petroleum, 42 gal)	meter³ (m³)	1.589 873 E − 01
board foot	meter³ (m³)	2.359 737 E − 03
British thermal unit (International Table)	joule (J)	1.055 056 E + 03
British thermal unit (mean)	joule (J)	1.055 87 E + 03

To convert from	to	Multiply by
British thermal unit (thermochemical)	joule (J)	1.054 350 E+03
British thermal unit (39°F)	joule (J)	1.059 67 E+03
British thermal unit (59°)	joule (J)	1.054 80 E+03
British thermal unit (60°)	joule (J)	1.054 68 E+03
Btu (International Table) · ft/(h · ft² · °F) (thermal conductivity)	watt per meter kelvin [W/(m · K)]	1.730 735 E+00
Btu (thermochemical) · ft/(h · ft² · °F) (thermal conductivity)	watt per meter kelvin [W/(m · K)]	1.729 577 E+00
Btu (International Table) · in/(h · ft² · °F) (thermal conductivity)	watt per meter kelvin [W/(m · K)]	1.442 279 E−01
Btu (thermochemical) · in/(h · ft² · °F) (thermal conductivity)	watt per meter kelvin [W/(m · K)]	1.441 314 E−01
Btu (International Table) · in/(s · ft² · °F) (thermal conductivity)	watt per meter kelvin [W/(m · K)]	5.192 204 E+02
Btu (thermochemical) · in/(s · ft² · °F) (thermal conductivity)	watt per meter kelvin [W/(m · K)]	5.188 732 E+02
Btu (International Table)/h	watt (W)	2.930 711 E−01
Btu (International Table)/s	watt (W)	1.055 056 E+03
Btu (thermochemical)/h	watt (W)	2.928 751 E−01
Btu (thermochemical)/min	watt (W)	1.757 250 E+01
Btu (thermochemical)/s	watt (W)	1.054 350 E+03
Btu (International Table)/ft²	joule per meter² (J/m²)	1.135 653 E+04
Btu (thermochemical)/ft²	joule per meter² (J/m²)	1.134 893 E+04
Btu (International Table)/(ft² · h)	watt per meter² (W/m²)	3.154 591 E+00
Btu (International Table)/(ft² · s)	watt per meter² (W/m²)	1.135 653 E+04
Btu (thermochemical)/(ft² · h)	watt per meter² (W/m²)	3.152 481 E+00
Btu (thermochemical)/(ft² · min)	watt per meter² (W/m²)	1.891 489 E+02
Btu (thermochemical)/(ft² · s)	watt per meter² (W/m²)	1.134 893 E+04
Btu (thermochemical)/(in² · s)	watt per meter² (W/m²)	1.634 246 E+06
Btu (International Table)/(h · ft² · °F)	watt per meter² kelvin [W/(m² · K)]	5.678 263 E+00
Btu (thermochemical)/(h · ft² · °F)	watt per meter² kelvin [W/(m² · K)]	5.674 466 E+00
Btu (International Table)/(s · ft² · °F)	watt per meter² kelvin [W/(m² · K)]	2.044 175 E+04
Btu (thermochemical)/(s · ft² · °F)	watt per meter² kelvin [W/(m² · K)]	2.042 808 E+04
Btu (International Table)/lb	joule per kilogram (J/kg)	2.326 000*E+03
Btu (thermochemical)/lb	joule per kilogram (J/kg)	2.324 444 E+03
Btu (International Table)/(lb · °F) (specific heat capacity)	joule per kilogram kelvin (J/(kg · K)]	4.186 800*E+03
Btu (thermochemical)/(lb · °F) (specific heat capacity)	joule per kilogram kelvin [J/(kg · K)]	4.184 000*E+03
Btu (International Table)/ft³	joule per meter³ (J/m³)	3.725 895 E+04
Btu (thermochemical)/ft³	joule per meter³ (J/m³)	3.723 402 E+04
bushel	meter³ (m³)	3.523 907 E−02
calorie (International Table)	joule (J)	4.186 800*E+00
calorie (mean)	joule (J)	4.190 02 E+00
calorie (thermochemical)	joule (J)	4.184 000*E+00
calorie (15 °C)	joule (J)	4.185 80 E+00
calorie (20 °C)	joule (J)	4.181 90 E+00
calorie (kilogram, International Table)	joule (J)	4.186 800*E+03
calorie (kilogram, mean)	joule (J)	4.190 02 E+03
calorie (kilogram, thermochemical)	joule (J)	4.184 000*E+03
cal (thermochemical)/cm²	joule per meter² (J/m²)	4.184 000*E+04
cal (International Table)/g	joule per kilogram (J/kg)	4.186 800*E+03
cal (thermochemical)/g	joule per kilogram (J/kg)	4.184 000*E+03
cal (International Table)/(g · °C)	joule per kilogram kelvin [J/(kg · K)]	4.186 800*E+03
cal (thermochemical)/(g · °C)	joule per kilogram kelvin [J/(kg · K)]	4.184 000*E+03

To convert from	to	Multiply by
cal (thermochemical)/min	watt (W)	6.973 333 E−02
cal (thermochemical)/s	watt (W)	4.184 000*E+00
cal (thermochemical)/(cm² · min)	watt per meter² (W/m²)	6.973 333 E−02
cal (thermochemical)/(cm² · s)	watt per meter² (W/m²)	4.184 000 E+04
cal (thermochemical)/(cm · s · °C)	watt per meter kelvin [W/(m·K)]	4.184 000 E+02
cd/in²	candela per meter² (cd/m²)	1.550 003 E+03
carat (metric)	kilogram (kg)	2.000 000*E−04
centimeter of mercury (0 °C)	pascal (Pa)	1.333 22 E+03
centimeter of water (4 °C)	pascal (Pa)	9.806 38 E+01
centipoise	pascal second (Pa . s)	1.000 000*E−03
centistokes	meter² per second (m²/s)	1.000 000*E−06
chain	meter² (m²)	5.067 075 E+01
circular mil	meter² (m²)	5.067 075 E−10
clo	kelvin meter² per watt (K·m²/W)	2.003 712 E−01
cup	milliliter (mL)	2.366 E+02
curie	becquerel (Bq)	3.700 000*E+10
darcy[a]	meter² (m²)	9.869 233 E−13
day	second (s)	8.640 000*E+04
day (sidereal)	second (s)	8.616 409 E+04
degree (angle)	radian (rad)	1.745 329 E−02
degree Celsius	kelvin (K)	$T_K = t_{°C} + 273.15$
degree centigrade	[see note below]	
degree Fahrenheit	degree Celsius (°C)	$t_{°C} = (t_{°F}-32)/1.8$
degree Fahrenheit	kelvin (K)	$T_K = (t_{°F}+459.67)/1.8$
degree Rankine	kelvin (K)	$T_K = T_{°R}/1.8$
°F · h · ft²/Btu (International Table)	kelvin meter² per watt (K·m²/W)	1.761 102 E−01
°F · h · ft²/Btu (thermochemical)	kelvin meter² per watt (K·m²/W)	1.762 280 E−01
°F · h · ft²/[Btu (International Table) · in] (thermal resistivity)	kelvin meter² per watt (K·m²/W)	6.933 472 E+00
°F · h · ft²/[Btu (thermochemical) · in] (thermal resistivity)	kelvin meter² per watt (K·m²/W)	6.938 112 E+00
denier	kilogram per meter (kg/m)	1.111 111 E−07
dyne	newton (N)	1.000 000*E−05
dyne · cm	newton meter (N · m)	1.000 000*E−07
dyne/cm²	pascal (Pa)	1.000 000*E−01
electronvolt	joule (J)	1.602 19 E−19
EMU of capacitance	farad (F)	1.000 000*E+09
EMU of current	ampere (A)	1.000 000*E+01
EMU of electric potential	volt (V)	1.000 000*E−08
EMU of inductance	henry (H)	1.000 000*E−09
EMU of resistance	ohm (Ω)	1.000 000*E−09
ESU of capacitance	farad (F)	1.112 650 E−12
ESU of current	ampere (A)	3.335 641 E−10
ESU of electric potential	volt (V)	2.997 925 E+02
ESU of inductance	henry (H)	8.987 552 E+11
ESU of resistance	ohm (Ω)	8.987 552 E+11
erg	joule (J)	1.000 000*E−07
erg/cm² · s	watt per meter² (W/m²)	1.000 000*E−03
erg/s	watt (W)	1.000 000*E−07
faraday (based on carbon-12)	coulomb (C)	9.648 70 E+04
faraday (chemical)	coulomb (C)	9.649 57 E+04
faraday (physical)	coulomb (C)	9.652 19 E+04

To convert from	to	Multiply by
fathom	meter (m)	1.828 8 E+00
fermi (femtometer)	meter (m)	1.000 000*E−15
fluid ounce (US)	meter³ (m³)	2.957 353 E−05
foot	meter (m)	3.048 000*E−01
foot (US survey)	meter (m)	3.048 006 E−01
foot of water (39.2 °F)	pascal (Pa)	2.988 98 E+03
ft²	meter² (m²)	9.290 304*E−02
ft²/h (thermal diffusivity)	meter² per second (m²/s)	2.580 640*E−05
ft²/s	meter² per second (m²/s)	9.290304*E−02
ft³ (volume; section modulus)	meter³ (m³)	2.831 685 E−02
ft³/min	meter³ per second (m³/s)	4.719 474 E−04
ft³/s	meter³ per second (m³/s)	2.831 685 E−02
ft⁴ (second moment of area)[b]	meter⁴ (m⁴)	8.630 975 E−03
ft/h	meter per second (m/s)	8.466 667 E−05
ft/min	meter per second (m/s)	5.080 000*E−03
ft/s	meter per second (m/s)	3.048 000*E−01
ft/s²	meter per second² (m/s²)	3.048 000*E−01
footcandle	lux (lx)	1.076 391 E+01
footlambert	candela per meter² (cd/m²)	3.426 259 E+00
ft · lbf	joule (J)	1.355 818 E+00
ft · lbf/h	watt (W)	3.766 161 E−04
ft · lbf/min	watt (W)	2.259 697 E−02
ft · lbf/s	watt (W)	1.355 818 E+00
ft · poundal	joule (J)	4.214 011 E−02
g, standard acceleration of free fall	meter per second² (m/s²)	9.806 650*E+00
gal	meter per second² (m/s²)	1.000 000*E−02
gallon (Canadian liquid)	meter³ (m³)	4.546 090*E−03
gallon (UK liquid)	meter³ (m³)	4.546 090*E−03
gallon (US liquid)	meter³ (m³)	3.785 412 E−03
gallon (US liquid) per day	meter³ per second (m³/s)	4.381 264 E−08
gallon (US liquid) per minute	meter³ per second (m³/s)	6.309 020 E−05
gallon (US liquid) per (hp · h) (SFC, specific fuel consumption)	meter³ per joule (m³/J)	1.410 089 E−09
gamma	tesla (T)	1.000 000*E−09
gauss	tesla (T)	1.000 000*E−04
gilbert	ampere (A)	7.957 747 E−01
gill (UK)	meter³ (m³)	1.420 654 E−04
gill (US)	meter³ (m³)	1.182 941 E−04
grade	degree (angular)	9.000 000*E−01
grade	radian (rad)	1.570 796 E−02
grain	kilogram (kg)	6.479 891*E−05
grain/gal (US liquid)	kilogram per meter³ (kg/m³)	1.711 806 E−02
gram	kilogram (kg)	1.000 000*E−03
g/cm³	kilogram per meter³ (kg/m³)	1.000 000*E+03
gram-force/cm²	pascal (PA)	9.806 650*E+01
hectare	meter² (m²)	1.000 000*E+04
horsepower (550 ft · lbf/s)	watt (W)	7.456 999 E+02
horsepower (boiler)	watt (W)	9.809 50 E+03
horsepower (electric)	watt (W)	7.460 000*E+02
horsepower (metric)	watt (W)	7.354 99 E+02
horsepower (water)	watt (W)	7.460 43 E+02
horsepower (UK)	watt (W)	7.457 0 E+02
hour	second (s)	3.600 000*E+03
hour (sidereal)	second (s)	3.590 170 E+03
hundredweight (long)	kilogram (kg)	5.080 235 E+01
hundredweight (short)	kilogram (kg)	4.535 924 E+01

To convert from	to	Multiply by
inch..........	meter (m)	2.540 000*E − 02
inch of mercury (32 °F) [c]..........	pascal (Pa)	3.386 38 E + 03
inch of mercury (60 °F) [c]..........	pascal (Pa)	3.376 85 E + 03
inch of water (39.2 °F)	pascal (Pa)	2.490 82 E + 02
inch of water (60 °F)..........	pascal (Pa)	2.488 4 E + 02
in^2..........	$meter^2$ (m^2)..........	6.451 600*E − 04
in^3 (volume; section modulus) [d]	$meter^3$ (m^3)..........	1.638 706 E − 05
in^3/min..........	$meter^3$ per second (m^3/s)..........	2.731 177 E − 07
in^4 (second moment of area) [b]	$meter^4$ (m^4)..........	4.162 314 E − 07
in/s	meter per second (m/s)..........	2.540 000*E − 02
in/s^2..........	meter per $second^2$ (m/s^2)..........	2.540 000*E − 02
kayser..........	1 per meter (1/m)..........	1.000 000*E + 02
kelvin..........	degree Celsius	$t_{°C} = T_K − 273.15$
kilocalorie (International Table)..........	joule (J)	4.186 800*E + 03
kilocalorie (mean)	joule (J)	4.190 02 E + 03
kilocalorie (thermochemical)	joule (J)	4.184 000*E + 03
kilocalorie (thermochemical)/min	watt (W)	6.973 333 E + 01
kilocalorie (thermochemical)/s..........	watt (W)	4.184 000*E + 03
kilogram-force (kgf)	newton (N)	9.806 650*E + 00
kgf · m..........	newton meter (N · m)..........	9.806 650*E + 00
kgf · s^2/m (mass)	kilogram (kg)	9.806 650*E + 00
kgf/cm^2..........	pascal (Pa)	9.806 650*E + 04
kgf/m^2..........	pascal (Pa)	9.806 650*E + 00
kgf/mm^2..........	pascal (Pa)	9.806 650*E + 06
km/h..........	meter per second (m/s)	2.777 778 E − 01
kilopond (1 kp = 1 kgf)	newton (N)	9.806 650*E + 00
kW · h..........	joule (J)	3.600 000*E + 06
kip (1000 lbf)	newton (N)	4.448 222 E + 03
kip/in^2 (ksi)..........	pascal (Pa)	6.894 757 E + 06
knot (international)..........	meter per second (m/s)..........	5.144 444 E − 01
lambert	candela per $meter^2$ (cd/m^2)..........	1/π *E + 04
lambert	candela per $meter^2$ (cd/m^2)..........	3.183 099 E + 04
langley	joule per $meter^2$ (J/m^2)..........	4.184 000*E + 04
light year [e]	meter (m)	9.460 73 E + 15
liter [f]	$meter^3$ (m^3)..........	1.000 000*E − 03
lumen per ft^2..........	lumen per $meter^2$ (lm/m^2)..........	1.076 391 E + 01
maxwell	weber (Wb)..........	1.000 000*E − 08
mho	siemens (S)	1.000 000*E + 00
microinch	meter (m)	2.540 000*E − 08
micron	meter (m)	1.000 000*E − 06
mil..........	meter (m)	2.540 000*E − 05
mile (international)..........	meter (m)	1.609 344*E + 03
mile (US statute)	meter (m)	1.609 3 E + 03
mile (international nautical)..........	meter (m)	1.852 000*E + 03
mile (US nautical)..........	meter (m)	1.852 000*E + 03
mi^2 (international)..........	$meter^2$ (m^2)..........	2.589 988 E + 06
mi^2 (US statute)	$meter^2$ (m^2)..........	2.589 998 E + 06
mi/h (international)..........	meter per second (m/s)..........	4.470 400*E − 01
mi/h (international)..........	kilometer per hour (km/h)..........	1.609 344*E + 00
mi/min (international)	meter per second (m/s)..........	2.682 240*E + 01
mi/s (international)..........	meter per second (m/s)..........	1.609 344*E + 03
millibar	pascal (Pa)	1.000 000*E + 02
millimeter of mercury (0 °C) [c]	pascal (Pa)	1.333 22 E + 02
minute (angle)	radian (rad)	2.908 882 E − 04
minute..........	second (s)	6.000 000*E + 01
minute (sidereal)	second (s)	5.983 617 E + 01
oersted..........	ampere per meter (A/m)	7.957 747 E + 01
ohm centimeter..........	ohm meter (Ω · m)	1.000 000*E − 02

To convert from	to	Multiply by
ohm circular-mil per ft	ohm meter ($\Omega \cdot$ m)	1.662 426 E − 09
ounce (avoirdupois)	kilogram (kg)	2.834 952 E − 02
ounce (troy or apothecary)	kilogram (kg)	3.110 348 E − 02
ounce (UK fluid)	meter^3 (m^3)	2.841 307 E − 05
ounce (US fluid)	meter^3 (m^3)	2.957 353 E − 05
ounce-force	newton (N)	2.780 139 E − 01
ozf $\cdot$ in	newton meter (N $\cdot$ m)	7.061 552 E − 03
oz (avoirdupois)/gal (UK liquid)	kilogram per meter^3 (kg/m^3)	6.236 023 E + 00
oz (avoirdupois)/gal (US liquid)	kilogram per meter^3 (kg/m^3)	7.489 152 E + 00
oz (avoirdupois)/in^3	kilogram per meter^3 (kg/m^3)	1.729 994 E + 03
oz (avoirdupois)/ft^2	kilogram per meter^2 (kg/m^2)	3.051 517 E − 01
oz (avoirdupois)/yd^2	kilogram per meter^2 (kg/m^2)	3.390 575 E − 02
parsec	meter (m)	3.085 678 E + 16
peck (US)	meter^3 (m^3)	8.809 768 E − 03
pennyweight	kilogram (kg)	1.555 174 E − 03
perm (0 °C)	kilogram per pascal second meter^2 [kg/Pa $\cdot$ s $\cdot$ m^2)]	5.721 35 E − 11
perm (23 °C)	kilogram per pascal second meter^2 [kg/(Pa $\cdot$ s $\cdot$ m^2)]	5.745 25 E − 11
perm $\cdot$ in (0 °C)	kilogram per pascal second meter [kg/(Pa $\cdot$ s $\cdot$ m)]	1.453 22 E − 12
perm $\cdot$ in (23 °C)	kilogram per pascal second meter [kg/(Pa $\cdot$ s $\cdot$ m)]	1.459 29 E − 12
phot	lumen per meter^2 (lm/m^2)	1.000 000*E + 04
pica (printer's)	meter (m)	4.217 518 E − 03
pint (US dry)	meter^3 (m^3)	5.506 105 E − 04
pint (US liquid)	meter^3 (m^3)	4.731 765 E − 04
point (printer's)	meter (m)	3.514 598*E − 04
poise (absolute viscosity)	pascal second (Pa $\cdot$ s)	1.000 000*E − 01
pound (avoirdupois)[g]	kilogram (kg)	4.535 924 E − 01
pound (troy or apothecary)	kilogram (kg)	3.732 417 E − 01
lb/ft	kilogram per meter (kg/m)	1.488 164 E + 00
lb $\cdot$ ft^2 (moment of inertia)	kilogram meter^2 (kg $\cdot$ m^2)	4.214 011 E − 02
lb $\cdot$ in^2 (moment of inertia)	kilogram meter^2 (kg $\cdot$ m^2)	2.926 397 E − 04
lb/ft $\cdot$ h	pascal second (Pa $\cdot$ s)	4.133 789 E − 04
lb/ft $\cdot$ s	pascal second (Pa $\cdot$ s)	1.488 164 E + 00
lb/ft^2	kilogram per meter^2 (kg/m^2)	4.882 428 E + 00
lb/ft^3	kilogram per meter^3 (kg/m^3)	1.601 846 E + 01
lb/gal (UK liquid)	kilogram per meter^3 (kg/m^3)	9.977 633 E + 01
lb/gal (US liquid)	kilogram per meter^3 (kg/m^3)	1.198 264 E + 02
lb/h	kilogram per second (kg/s)	1.259 979 E − 04
lb/hp $\cdot$ h (SFC, specific fuel consumption	kilogram per joule (kg/J)	1.689 659 E − 07
lb/in	kilogram per meter (kg/m)	1.785 797 E + 01
lb/in^3	kilogram per meter^3 (kg/m^3)	2.767 990 E + 04
lb/min	kilogram per second (kg/s)	7.559 873 E − 03
lb/s	kilogram per second (kg/s)	4.535 924 E − 01
lb/yd^3	kilogram per meter^3 (kg/m^3)	5.932 764 E − 01
poundal	newton (N)	1.382 550 E − 01
poundal/ft^2	pascal (Pa)	1.488 164 E + 00
poundal $\cdot$ s/ft^2	pascal second (Pa $\cdot$ s)	1.488 164 E + 00
pound-force (lbf)[h]	newton (N)	4.448 222 E + 00
lbf $\cdot$ ft	newton meter (N $\cdot$ m)	1.355 818 E + 00
lbf $\cdot$ ft/in	newton meter per meter (N $\cdot$ m/m)	5.337 866 E + 01
lbf $\cdot$ in	newton meter (N $\cdot$ m)	1.129 848 E − 01
lbf $\cdot$ in/in	newton meter per meter (N $\cdot$ m/m)	4.448 222 E + 00
lbf $\cdot$ s/ft^2	pascal second (Pa $\cdot$ s)	4.788 026 E + 01

To convert from	to	Multiply by
lbf · s/in².	pascal second (Pa · s)	6.894 757 E+03
lbf/ft	newton per meter (N/m)	1.459 390 E+01
lbf/ft²	pascal (Pa)	4.788 026 E+01
lbf/in.	newton per meter (N/m)	1.751 268 E+02
lbf/in² (psi)	pascal (Pa)	6.894 757 E+03
lbf/lb (thrust/weight [mass] ratio)	newton per kilogram (N/kg)	9.806 650 E+00
quad	joule (J)	1.055 E+18
quart (US dry)	meter³ (m³)	1.101 221 E−03
quart (US liquid)	meter³ (m³)	9.463 529 E−04
rad (absorbed dose)	gray (Gy)	1.000 000*E−02
rem (dose equivalent)	sievert (Sv)	1.000 000*E−02
rhe	1 per pascal second [1/Pa · s]	1.000 000*E+01
rod	meter (m)	5.029 210 E+00
roentgen	coulomb per kilogram (C/kg)	2.58 E−04
second (angle)	radian (rad)	4.848 137 E−06
second (sidereal)	second (s)	9.972 696 E−01
shake	second (s)	1.000 000*E−08
slug	kilogram (kg)	1.459 390 E+01
slug/ft · s	pascal second (Pa · s)	4.788 026 E+01
slug/ft³	kilogram per meter³ (kg/m³)	5.153 788 E+02
statampere	ampere (A)	3.335 641 E−10
statcoulomb	coulomb (C)	3.335 641 E−10
statfarad	farad (F)	1.112 650 E−12
stathenry	henry (H)	8.987 552 E+11
statmho	siemens (S)	1.112 650 E−12
statohm	ohm (Ω)	8.987 552 E+11
statvolt	volt (V)	2.997 925 E+02
stere	meter³ (m³)	1.000 000*E+00
stilb	candela per meter² (cd/m²)	1.000 000*E+04
stokes (kinematic viscosity)	meter² per second (m²/s)	1.000 000*E−04
tablespoon	milliliter (mL)	1.479 E+01
teaspoon	meter³ (m³) milliliter (mL)	4.929 E+00
tex	kilogram per meter (kg/m)	1.000 000*E−06
therm (EEG) [i]	joule (J)	1.055 060*E+08
therm (US) [i]	joule (J)	1.054 804*E+08
ton (assay)	kilogram (kg)	2.916 667 E−02
ton (long, 2240 lb)	kilogram (kg)	1.016 047 E+03
ton (metric)	kilogram (kg)	1.000 000*E+03
ton (explosive energy of one ton of TNT)	joule (J)	4.184 E+09 [j]
ton of refrigeration (12 000 Btu/h)	watt (W)	3.517 E+03
ton (register)	meter³ (m³)	2.831 685 E+00
ton (short, 2000 lb)	kilogram (kg)	9.071 847 E+02
ton (long)/yd³	kilogram per meter³ (kg/m³)	1.328 939 E+03
ton (short)/yd³	kilogram per meter³ (kg/m³)	1.186 553 E+03
ton (short)/h	kilogram per second (kg/s)	2.519 958 E−01
ton-force (2000 lbf)	newton (N)	8.896 443 E+03
tonne	kilogram (kg)	1.000 000*E+03
torr (mmHg, 0 °C) [c]	pascal (Pa)	1.333 22 E+02
unit pole	weber (Wb)	1.256 637 E−07
W · h	joule (J)	3.600 000*E+03
W · s	joule (J)	1.000 000*E+00
W/cm²	watt per meter² (W/m²)	1.000 000*E+04
W/in²	watt per meter² (W/m²)	1.550 003 E+03

To convert from	to	Multiply by
yard	meter (m)	9.144 000*E−01
yd^2	meter^2 (m^2)	8.361 274 E−01
yd^3	meter^3 (m^3)	7.645 549 E−01
yd^3/min.	meter^3 per second (m^3/s)	1.274 258 E−02
year (365 days)	second (s)	3.153 600*E+07
year (sidereal)	second (s)	3.155 815 E+07
year (tropical)	second (s)	3.155 693 E+07

Note: The centigrade temperature scale is obsolete. The unit, degree centigrade, is only approximately equal to the degree Celsius.

[a] The darcy is a unit for measuring permeability of porous solids.

[b] This is sometimes called the moment of section or area moment of ineria of a plane section about a specified axis.

[c] Conversion factors for mercury manometer pressure units are calculated using the standard value for the acceleration of gravity and the density of mercury at the stated temperature. Higher levels of precision are not justified because the definitions of the units do not take into account the compressibility of mercury or the density value change caused by the revised practical temperature scale, ITS-90.

[d] The exact conversion factor is 1.638 706 4*E–05.

[e] This conversion factor is based on the astronomical unit of time of one day (86 400 seconds); an interval of 36 525 days is one Julian century. (See the Astronomical Almanac for the Year 1991, page K6, U.S. Government Printing Office, Washington, D.C.)

[f] In 1964 the General Conference on Weights and Measures reestablished the name liter as a special name for the cubic decimeter. Between 1901 and 1964, the liter was slightly larger (1.000 028 dm^3); in the use of high-accuracy volume data of that time interval, this fact must be kept in mind.

[g] The exact conversion factor is 4.535 923 7*E–01.

[h] The exact conversion factor is 4.448 221 615 260 5*E+00.

[i] The therm (EEC) is legally defined in the Council Directive of 20 December 1979, Council of the European Communities. The therm (US) is legally defined in the Federal Register of July 27, 1968. Although the therm (EEC), which is based on the International Table Btu, is frequently used by engineers in the US, the therm (US) is the legal unit used by the US natural gas industry.

[j] Defined (not measured) value.

Note concerning the foot:

The U.S. Metric Law of 1866 gave the relationship, 1 meter equals 39.37 inches. Since 1893 the U.S. yard has been derived from the meter. In 1959 a refinement was made in the definition of the yard to bring the U.S. yard and the yard used in other countries into agreement. The U.S. yard was changed from 3600/3937 m to 0.9144 m exactly. The new length is shorter by exactly two parts in a million. At the same time it was decided that any data in feet derived from and published as a result of geodetic surveys within the United States would remain with the old standard (1 ft = 1200/3937 m) until further decision This foot is named the U.S. survey foot and has the following relationships:

1 rod (pole or perch) = 16 $\frac{1}{2}$ feet
1 chain = 66 feet
1 mile (U.S. statute) = 5280 feet

CONVERSION OF TEMPERATURES

From	To	
°Celsius	°Fahrenheit	$t_F = (t_C \times 1.8) + 32$
	Kelvin	$T_K = t_C + 273.15$
	°Rankine	$T_R = (t_C + 273.15) \times 1.8$
°Fahrenheit	°Celsius	$t_C = \dfrac{t_F - 32}{1.8}$
	Kelvin	$T_k = \dfrac{t_F - 32}{1.8} + 273.15$
	°Rankine	$T_R = t_F + 459.67$
Kelvin	°Celsius	$t_C = T_K - 273.15$
	°Rankine	$T_R = T_K \times 1.8$
°Rankine	°Fahrenheit	$t_F = T_R - 459.67$
	Kelvin	$T_K = \dfrac{T_R}{1.8}$

DESIGNATION OF LARGE NUMBERS

	U.S.A.	Other countries
10^6	million	million
10^9	billion	milliard
10^{12}	trillion	billion
10^{15}	quadrillion	billiard
10^{18}	quintillion	trillion

CONVERSION FACTORS FOR ENERGY UNITS

$E = h\nu = hc\tilde{\nu} = kT; \; E_m = LE$

| | wavenumber $\tilde{\nu}$ | frequency ν | energy E | | | molar energy E_m | | temperature T |
	cm^{-1}	MHz	aJ	eV	E_h	kJ/mol	kcal/mol	K
$\tilde{\nu}$: 1 cm^{-1}	$\hat{=}\,1$	2.997925×10^4	1.986447×10^{-5}	1.239842×10^{-4}	4.556335×10^{-6}	11.96266×10^{-3}	2.85914×10^{-3}	1.438769
ν: 1 MHz	$\hat{=}\,3.33564 \times 10^{-5}$	1	6.626076×10^{-10}	4.135669×10^{-9}	1.519830×10^{-10}	3.990313×10^{-7}	9.53708×10^{-8}	4.79922×10^{-5}
E: 1 aJ	$\hat{=}\,50341.1$	1.509189×10^9	1	6.241506	0.2293710	602.2137	143.9325	7.24292×10^4
1 eV	$\hat{=}\,8065.54$	2.417988×10^8	0.1602177	1	3.674931×10^{-2}	96.4853	23.0605	1.16045×10^4
1 E_h	$\hat{=}\,219474.63$	6.579684×10^9	4.359748	27.2114	1	2625.500	627.510	3.15773×10^5
E_m: 1 kJ/mol	$\hat{=}\,83.5935$	2.506069×10^6	1.660540×10^{-3}	1.036427×10^{-2}	3.808798×10^{-4}	1	0.239006	120.272
1 kcal/mol	$\hat{=}\,349.755$	1.048539×10^7	6.947700×10^{-3}	4.336411×10^{-2}	1.593601×10^{-3}	4.184	1	503.217
T: 1 K	$\hat{=}\,0.695039$	$2.083\,67 \times 10^4$	1.380658×10^{-5}	8.61738×10^{-5}	3.16683×10^{-6}	8.31451×10^{-3}	1.98722×10^{-3}	1

Examples of the use of this table: 1 aJ $\hat{=}$ 50341 cm^{-1}

1 eV $\hat{=}$ 96.4853 kJ mol^{-1}

The symbol $\hat{=}$ should be read as meaning "corresponds to" or "is equivalent to".

CONVERSION FACTORS FOR PRESSURE UNITS

	Pa	kPa	MPa	bar	atmos	Torr	μmHg	psi
Pa	1	0.001	0.000001	0.00001	9.8692×10^{-6}	0.0075006	7.5006	0.0001450377
kPa	1000	1	0.001	0.01	0.0098692	7.5006	7500.6	0.1450377
MPa	1000000	1000	1	10	9.8692	7500.6	7500600	145.0377
bar	100000	100	0.1	1	0.98692	750.06	750060	14.50377
atmos	101325	101.325	0.101325	1.01325	1	760	760000	14.69594
Torr	133.322	0.133322	0.000133322	0.00133322	0.00131579	1	1000	0.01933672
μmHg	0.133322	0.00013322	1.33322×10^{-7}	1.33322×10^{-6}	1.31579×10^{-6}	0.001	1	1.933672×10^{-5}
psi	6894.757	6.894757	0.006894757	0.06894757	0.068046	51.7151	51715.1	1

To convert a pressure value from a unit in the left hand column to a new unit, multiply the value by the factor appearing in the column for the new unit. For example:

$1\ kPa = 9.8692 \times 10^{-3}$ atmos

$1\ Torr = 1.33322 \times 10^{-4}\ MPa$

Notes: μmHg is often referred to as "micron"

Torr is essentially identical to mmHg

psi is an abbreviation for the unit pound–force per square inch

psia (as a term for a physical quantity) implies the true (absolute) pressure

psig implies the true pressure minus the local atmospheric pressure

CONVERSION FACTORS FOR THERMAL CONDUCTIVITY UNITS

MULTIPLY by appropriate factor to OBTAIN →	Btu_{IT} h^{-1} ft^{-1} F^{-1}	Btu_{IT} in. h^{-1} ft^{-2} F^{-1}	Btu_{th} h^{-1} ft^{-1} F^{-1}	Btu_{th} in. h^{-1} ft^{-2} F^{-1}	cal_{IT} s^{-1} cm^{-1} C^{-1}	cal_{th} s^{-1} cm^{-1} C^{-1}	$kcal_{th}$ h^{-1} m^{-1} C^{-1}	J s^{-1} cm^{-1} K^{-1}	W cm^{-1} K^{-1}	W m^{-1} K^{-1}	mW cm^{-1} K^{-1}
Btu_{IT} h^{-1} ft^{-1} F^{-1}	1	12	0.999331	8.32776×10^{-2}	2.41909×10^{2}	2.41747×10^{2}	0.671520	57.7789	57.7789	0.577789	5.77789×10^{-2}
Btu_{IT} in. h^{-1} ft^{-2} F^{-1}	12	1	11.9920	0.999331	2.90291×10^{2}	2.90096×10^{2}	8.05824	6.93347×10^{2}	6.93347×10^{2}	6.93347	0.693347
Btu_{th} h^{-1} ft^{-1} F^{-1}	1.00067	8.33891×10^{-2}	1	8.33333×10^{-2}	2.42071×10^{2}	2.41909×10^{2}	0.671969	57.8176	57.8176	0.578176	5.78176×10^{-2}
Btu_{th} in. h^{-1} ft^{-2} F^{-1}	12.0080	1.00067	12	1	2.90485×10^{3}	2.90291×10^{3}	8.06363	6.93811×10^{2}	6.93811×10^{2}	6.93811	0.693811
cal_{IT} s^{-1} cm^{-1} C^{-1}	4.13379×10^{-3}	3.44482×10^{-4}	4.13102×10^{-3}	3.44252×10^{-4}	1	1.00067	3.60241×10^{2}	4.1868	4.1868	4.1868×10^{2}	4.1868×10^{3}
cal_{th} s^{-1} cm^{-1} C^{-1}	4.13656×10^{-3}	3.44713×10^{-4}	4.13379×10^{-3}	3.44482×10^{-4}	1.00067	1	3.6×10^{2}	4.184	4.184	4.184×10^{2}	4.184×10^{3}
$kcal_{th}$ h^{-1} m^{-1} C^{-1}	1.48916	0.124097	1.48816	0.124014	2.77592×10^{-3}	2.77778×10^{-3}	1	1.16222×10^{-2}	1.16222×10^{-2}	1.16222	11.6222
J s^{-1} cm^{-1} K^{-1}	1.73073×10^{-2}	1.44228×10^{-3}	1.72958×10^{-2}	1.44131×10^{-3}	4.1868	4.184	1.16222×10^{-2}	1	1	1×10^{2}	1×10^{3}
W cm^{-1} K^{-1}	1.73073×10^{-2}	1.44228×10^{-3}	1.72958×10^{-2}	1.44131×10^{-3}	4.1868	4.184	1.16222×10^{-2}	1	1	1×10^{2}	1×10^{3}
W m^{-1} K^{-1}	1.73073	0.144228	1.72958	0.144131	4.1868×10^{2}	4.184×10^{2}	1.16222	1×10^{2}	1×10^{2}	1	10
mW cm^{-1} K^{-1}	17.3073	1.44228	17.2958	1.44131	4.1868×10^{3}	4.184×10^{3}	11.6222	1×10^{3}	1×10^{3}	10	1

CONVERSION FACTORS FOR ELECTRICAL RESISTIVITY UNITS

To convert from multiply by appropriate factor to Obtain →

	abΩ cm	μΩ cm	Ω cm	StatΩ cm	Ω m	Ω cir. mil ft⁻¹	Ω in.	Ω ft
abohm centimeter	1	1×10^{-3}	10^{-9}	1.113×10^{-21}	10^{-11}	6.015×10^{-3}	3.937×10^{-10}	3.281×10^{-11}
microohm centimeter	10^3	1	10^{-6}	1.113×10^{-18}	10^{-6}	6.015	3.937×10^{-7}	3.281×10^{-6}
ohm centimeter	10^8	10^6	1	1.113×10^{-12}	1×10^{-2}	6.015×10^6	3.937×10^{-1}	3.281×10^{-2}
statohm centimeter (esu)	8.987×10^{20}	8.987×10^{17}	8.987×10^{11}	1	8.987×10^9	5.406×10^{18}	3.538×10^{11}	2.949×10^{10}
ohm meter	10^{11}	10^8	10^2	1.113×10^{-10}	1	6.015×10^8	3.937×10^1	3.281
ohm circular mil per foot	1.662×10^2	1.662×10^{-1}	1.662×10^{-7}	1.850×10^{-19}	1.662×10^{-9}	1	6.54×10^{-6}	5.45×10^{-9}
ohm inch	2.54×10^9	2.54×10^6	2.54	2.827×10^{-12}	2.54×10^{-2}	1.528×10^7	1	8.3×10^{-2}
ohm foot	3.048×10^{10}	3.048×10^7	3.048×10^{-1}	3.3924×10^{-11}	3.048×10^{-1}	1.833×10^8	12	1

CONVERSION FACTORS FOR CHEMICAL KINETICS

Equivalent second order rate constants

A \ B	cm³ mol⁻¹ s⁻¹	dm³ mol⁻¹ s⁻¹	m³ mol⁻¹ s⁻¹	cm³ molecule⁻¹ s⁻¹	(mm Hg)⁻¹ s⁻¹	atm⁻¹ s⁻¹	ppm⁻¹ min⁻¹	m² kN⁻¹ s⁻¹
1 cm³ mol⁻¹ s⁻¹ =	1	10^{-3}	10^{-6}	1.66×10^{-24}	$1.604 \times 10^{-5} T^{-1}$	$1.219 \times 10^{-2} T^{-1}$	2.453×10^{-9}	$1.203 \times 10^{-4} T^{-1}$
1 dm³ mol⁻¹ s⁻¹ =	10^{3}	1	10^{-3}	1.66×10^{-21}	$1.604 \times 10^{-2} T^{-1}$	$12.19 \, T^{-1}$	2.453×10^{-6}	$1.203 \times 10^{-1} T^{-1}$
1 m³ mol⁻¹ s⁻¹ =	10^{6}	10^{3}	1	1.66×10^{-18}	$16.04 \, T^{-1}$	$1.219 \times 10^{4} T^{-1}$	2.453×10^{-3}	$120.3 \, T^{-1}$
1 cm³ molecule⁻¹ s⁻¹ =	6.023×10^{23}	6.023×10^{20}	6.023×10^{17}	1	$9.658 \times 10^{18} T^{-1}$	$7.34 \times 10^{21} T^{-1}$	1.478×10^{15}	$7.244 \times 10^{19} T^{-1}$
1 (mm Hg)⁻¹ s⁻¹ =	$6.236 \times 10^{4} T$	$62.36 \, T$	$6.236 \times 10^{-2} T$	$1.035 \times 10^{-19} T$	1	760	4.56×10^{-2}	7.500
1 atm⁻¹ s⁻¹	$82.06 \, T$	$8.206 \times 10^{-2} T$	$8.206 \times 10^{-5} T$	$1.362 \times 10^{-22} T$	1.316×10^{-3}	1	6×10^{-5}	9.869×10^{-3}
1 ppm⁻¹ min⁻¹ = at 298 K, 1 atm total pressure	4.077×10^{8}	4.077×10^{5}	407.7	6.76×10^{-16}	21.93	1.667×10^{4}	1	164.5
1 m² kN⁻¹ s⁻¹ =	$8314 \, T$	$8.314 \, T$	$8.314 \times 10^{-3} T$	$1.38 \times 10^{-20} T$	0.1333	101.325	6.079×10^{-3}	1

To convert a rate constant from one set of units A to a new set B find the conversion factor for the row A under column B and multiply the old value by it, e.g. to convert cm³ molecule⁻¹ s⁻¹ to m³ mol⁻¹ s⁻¹ multiply by 6.023×10^{17}.

Table adapted from High Temperature Reaction Rate Data No. 5, The University, Leeds (1970).

Equivalent third order rate constants

A \ B	cm⁶ mol⁻² s⁻¹	dm⁶ mol⁻² s⁻¹	m⁶ mol⁻² s⁻¹	cm⁶ molecule⁻² s⁻¹	(mm Hg)⁻² s⁻¹	atm⁻² s⁻¹	ppm⁻² min⁻¹	m⁴ kN⁻² s⁻¹
1 cm⁶ mol⁻² s⁻¹ =	1	10^{-6}	10^{-12}	2.76×10^{-48}	$2.57 \times 10^{-10} T^{-2}$	$1.48 \times 10^{-4} T^{-2}$	1.003×10^{-19}	$1.447 \times 10^{-8} T^{-2}$
1 dm⁶ mol⁻² s⁻¹ =	10^{6}	1	10^{-6}	2.76×10^{-42}	$2.57 \times 10^{-4} T^{-2}$	$148 \, T^{-2}$	1.003×10^{-13}	$1.447 \times 10^{-2} T^{-2}$
1 m⁶ mol⁻² s⁻¹ =	10^{12}	10^{6}	1	2.76×10^{-36}	$257 \, T^{-2}$	$1.48 \times 10^{8} T^{-2}$	1.003×10^{-7}	$1.447 \times 10^{4} T^{-2}$
1 cm⁶ molecule⁻² s⁻¹ =	3.628×10^{47}	3.628×10^{41}	3.628×10^{35}	1	$9.328 \times 10^{37} T^{-2}$	$5.388 \times 10^{43} T^{-2}$	3.64×10^{28}	$5.248 \times 10^{39} T^{-2}$
1 (mm Hg)⁻² s⁻¹ =	$3.89 \times 10^{9} T^{2}$	$3.89 \times 10^{3} T^{2}$	$3.89 \times 10^{-3} T^{2}$	$1.07 \times 10^{-38} T^{2}$	1	5.776×10^{5}	3.46×10^{-5}	56.25
1 atm⁻² s⁻¹ =	$6.733 \times 10^{3} T^{2}$	$6.733 \times 10^{-3} T^{2}$	$6.733 \times 10^{-9} T^{2}$	$1.86 \times 10^{-44} T^{2}$	1.73×10^{-6}	1	6×10^{-11}	9.74×10^{-5}
1 ppm⁻² min⁻¹ = at 298 K, 1 atm total pressure	9.97×10^{18}	9.97×10^{12}	9.97×10^{6}	2.75×10^{-29}	2.89×10^{4}	1.667×10^{10}	1	1.623×10^{6}
1 m⁴ kN⁻² s⁻¹ =	$6.91 \times 10^{7} T^{2}$	$6.91 \, T^{2}$	$69.1 \times 10^{-5} T^{2}$	$1.904 \times 10^{-40} T^{2}$	0.0178	1.027×10^{4}	6.16×10^{-7}	1

From *J. Phys. Chem. Ref. Data*, 9, 470, 1980, by permission of the authors and the copyright owner, the American Institute of Physics.

CONVERSION FACTORS FOR IONIZING RADIATION

CONVERSION BETWEEN SI AND OTHER UNITS

Quantity	Symbol for quantity	Expression in SI units	Special name for SI units	Symbols using special names	Conventional units	Symbol for conventional unit	Value of conventional unit in SI units
Activity	A	1 per second / s^{-1}	becquerel	Bq	curie	Ci	3.7×10^{10} Bq
Absorbed dose	D	joule per kilogram / $J\,kg^{-1}$	gray	Gy	rad	rad	0.01 Gy
Absorbed dose rate	$\dot{D}$	joule per kilogram second / $J\,kg^{-1}\,s^{-1}$		$Gy\,s^{-1}$	rad	$rad\,s^{-1}$	$0.01\ Gy\,s^{-1}$
Average energy per ion pair	W	joule / J			electronvolt	eV	1.602×10^{-19} J
Dose equivalent	H	joule per kilogram / $J\,kg^{-1}$	sievert	Sv	rem	rem	0.01 Sv
Dose equivalent rate	$\dot{H}$	joule per kilogram second / $J\,kg^{-1}\,s^{-1}$		$Sv\,s^{-1}$	rem per second	$rem\,s^{-1}$	$0.01\ Sv\,s^{-1}$
Electric current	I	ampere / A			ampere	A	1.0 A
Electric potential difference	U, V	watt per ampere / $W\,A^{-1}$	volt	V	volt	V	1.0 A
Exposure	X	coulomb per kilogram / $C\,kg^{-1}$			roentgen	R	$2.58 \times 10^{-4}\ C\,kg^{-1}$
Exposure rate	$\dot{X}$	coulomb per kilogram second / $C\,kg^{-1}\,s^{-1}$			roentgen	$R\,s^{-1}$	$2.58 \times 10^{-4}\ C\,kg^{-1}\,s^{-1}$
Fluence	Φ	1 per meter squared / m^{-2}			1 per centimeter squared	cm^{-2}	$1.0 \times 10^{4}\ m^{-2}$
Fluence rate	$\dot{\Phi}$	1 per meter squared second / $m^{-2}\,s^{-1}$			1 per centimeter squared second	$cm^{-2}\,s^{-1}$	$1.0 \times 10^{4}\ m^{-2}\,s^{-1}$
Kerma	K	joule per kilogram / $J\,kg^{-1}$	gray	Gy	rad	rad	0.01 Gy
Kerma rate	$\dot{K}$	joule per kilogram second / $J\,kg^{-1}\,s^{-1}$		$Gy\,s^{-1}$	rad per second	$rad\,s^{-1}$	$0.01\ Gy\,s^{-1}$
Lineal energy	y	joule per meter / $J\,m^{-1}$			kiloelectron volt per micrometer	$keV\,\mu m^{-1}$	$1.602 \times 10^{-10}\ J\,m^{-1}$
Linear energy transfer	L	joule per meter / $J\,m^{-1}$			kiloelectron volt per micrometer	$keV\,\mu m^{-1}$	$1.602 \times 10^{-10}\ J\,m^{-1}$
Mass attenuation coefficient	μ/ρ	meter squared per kilogram / $m^2\,kg^{-1}$			centimeter squared per gram	$cm^2\,g^{-1}$	$0.1\ m^2\,kg^{-1}$
Mass energy transfer coefficient	μ_{tr}/ρ	meter squared per kilogram / $m^2\,kg^{-1}$			centimeter squared per gram	$cm^2\,g^{-1}$	$0.1\ m^2\,kg^{-1}$
Mass energy absorption coefficient	μ_{en}/ρ	meter squared per kilogram / $m^2\,kg^{-1}$			centimeter squared per gram	$cm^2\,g^{-1}$	$0.1\ m^2\,kg^{-1}$
Mass stopping power	S/ρ	joule meter squared per kilogram / $J\,m^2\,kg^{-1}$			MeV centimeter squared per gram	$MeV\,cm^2\,g^{-1}$	$1.602 \times 10^{-14}\ J\,m^2\,kg^{-1}$
Power	P	joule per second / $J\,s^{-1}$	watt	W	watt	W	1.0 W
Pressure	P	newton per meter squared / $N\,m^{-2}$	pascal	Pa	torr	torr	(101325/760)Pa
Radiation chemical yield	G	mole per joule / $mol\,J^{-1}$			molecules per 100 electron volts	molecules $(100\ eV)^{-1}$	$1.04 \times 10^{-7}\ mol\,J^{-1}$
Specific energy	z	joule per kilogram / $J\,kg^{-1}$	gray	Gy	rad	rad	0.01 Gy

CONVERSION OF RADIOACTIVITY UNITS FROM MBq TO mCi AND μCi

MBq	mCi	MBq	mCi	MBq	μCi	MBq	μCi
7000	189.	500	13.5	30	810	1	27
6000	162.	400	10.8	20	540	0.9	24
5000	135.	300	8.1	10	270	0.8	21.6
4000	108.	200	5.4	9	240	0.7	18.9
3000	81.	100	2.7	8	220	0.6	16.2
2000	54.	90	2.4	7	189	0.5	13.5
1000	27.	80	2.16	6	162	0.4	10.8
900	24.	70	1.89	5	135	0.3	8.1
800	21.6	60	1.62	4	108	0.2	5.4
700	18.9	50	1.35	3	81	0.1	2.7
600	16.2	40	1.08	2	54		

CONVERSION OF RADIOACTIVITY UNITS FROM mCi AND μCi TO MBq

mCi	MBq	mCi	MBq	μCi	MBq	μCi	MBq
200	7400	10	370	1000	37.0	80	2.96
150	5550	9	333	900	33.3	70	2.59
100	3700	8	296	800	29.6	60	2.22
90	3330	7	259	700	25.9	50	1.85
80	2960	6	222	600	22.2	40	1.48
70	2590	5	185	500	18.5	30	1.11
60	2220	4	148	400	14.8	20	0.74
50	1850	3	111	300	11.1	10	0.37
40	1480	2	74.0	200	7.4	5	0.185
30	1110	1	37.0	100	3.7	2	0.074
20	740			90	3.33	1	0.037

CONVERSION OF RADIOACTIVITY UNITS

100 TBq (10^{14} Bq)	=	2.7 kCi (2.7×10^3 Ci)	100 kBq (10^5 Bq)	=	2.7 μCi (2.7×10^{-6} Ci)
10 TBq (10^{13} Bq)	=	270 Ci (2.7×10^2 Ci)	10 kBq (10^4 Bq)	=	270 nCi (2.7×10^{-7} Ci)
1 TBq (10^{12} Bq)	=	27 Ci (2.7×10^1 Ci)	1 kBq (10^3 Bq)	=	27 nCi (2.7×10^{-8} Ci)
100 GBq (10^{11} Bq)	=	2.7 Ci (2.7×10^0 Ci)	100 Bq (10^2 Bq)	=	2.7 nCi (2.7×10^{-9} Ci)
10 GBq (10^{10} Bq)	=	270 mCi (2.7×10^{-1} Ci)	10 Bq (10^1 Bq)	=	270 pCi (2.7×10^{-10} Ci)
1 GBq (10^9 Bq)	=	27 mCi (2.7×10^{-2} Ci)	1 Bq (10^0 Bq)	=	27 pCi (2.7×10^{-11} Ci)
100 MBq (10^8 Bq)	=	2.7 mCi (2.7×10^{-3} Ci)	100 mBq (10^{-1} Bq)	=	2.7 pCi (2.7×10^{-12} Ci)
10 MBq (10^7 Bq)	=	270 μCi (2.7×10^{-4} Ci)	10 mBq (10^{-2} Bq)	=	270 fCi (2.7×10^{-13} Ci)
1 MBq (10^6 Bq)	=	27 μCi (2.7×10^{-5} Ci)	1 mBq (10^{-3} Bq)	=	27 fCi (2.7×10^{-14} Ci)

CONVERSION OF ABSORBED DOSE UNITS

SI Units		Conventional
100 Gy (10^2 Gy)	=	10,000 rad (10^4 rad)
10 Gy (10^1 Gy)	=	1,000 rad (10^3 rad)
1 Gy (10^0 Gy)	=	100 rad (10^2 rad)
100 mGy (10^{-1} Gy)	=	10 rad (10^1 rad)
10 mGy (10^{-2} Gy)	=	1 rad (10^0 rad)
1 mGy (10^{-3} Gy)	=	100 mrad (10^{-1} rad)
100 μGy (10^{-4} Gy)	=	10 mrad (10^{-2} rad)
10 μGy (10^{-5} Gy)	=	1 mrad (10^{-3} rad)
1 μGy (10^{-6} Gy)	=	100 μrad (10^{-4} rad)
100 nGy (10^{-7} Gy)	=	10 μrad (10^{-5} rad)
10 nGy (10^{-8} Gy)	=	1 μrad (10^{-6} rad)
1 nGy (10^{-9} Gy)	=	100 nrad (10^{-7} rad)

CONVERSION OF DOSE EQUIVALENT UNITS

100 Sv (10^2 Sv)	=	10,000 rem (10^4 rem)
10 Sv (10^1 Sv)	=	1,000 rem (10^3 rem)
1 Sv (10^0 Sv)	=	100 rem (10^2 rem)
100 mSv (10^{-1} Sv)	=	10 rem (10^1 rem)
10 mSv (10^{-2} Sv)	=	1 rem (10^0 rem)
1 mSv (10^{-3} Sv)	=	100 mrem (10^{-1} rem)
100 μSv (10^{-4} Sv)	=	10 mrem (10^{-2} rem)
10 μSv (10^{-5} Sv)	=	1 mrem (10^{-3} rem)
1 μSv (10^{-6} Sv)	=	100 μrem (10^{-4} rem)
100 nSv (10^{-7} Sv)	=	10 μrem (10^{-5} rem)
10 nSv (10^{-8} Sv)	=	1 μrem (10^{-6} rem)
1 nSv (10^{-9} Sv)	=	100 nrem (10^{-7} rem)

VALUES OF THE GAS CONSTANT IN DIFFERENT UNIT SYSTEMS

In SI units the value of the gas constant, R, is:

$R = 8.314510$ Pa m^3 K^{-1} mol^{-1}

$\quad = 8314.510$ Pa L K^{-1} mol^{-1}

$\quad = 0.08314510$ bar L K^{-1} mol^{-1}

This table gives the appropriate value of R for use in the ideal gas equation, $PV = nRT$, when the variables are expressed in other units. The following conversion factors for pressure units were used in generating the table:

1 atm = 101325 Pa

1 psi = 6894.757 Pa

1 torr (mmHg) = 133.322 Pa [at 0°C]

1 in Hg = 3386.38 Pa [at 0°C]

1 in H_2O = 249.082 Pa [at 4°C]

1 ft H_2O = 2988.98 Pa [at 4°C]

The advice of Prabir K. Chandra is appreciated.

Units of V, T, n			Units of P						
V	T	n	kPa	atm	psi	mmHg	in Hg	in H_2O	ft H_2O
ft^3	K	mol	0.2936241	0.00289785	0.0425866	2.20237	0.0867074	1.17882	0.0982355
		lb·mol	133.1857	1.31444	19.3169	998.978	39.3298	534.706	44.5589
	°R	mol	0.1631245	0.00160991	0.0236592	1.22354	0.0481708	0.654903	0.0545753
		lb·mol	73.99204	0.730245	10.7316	554.987	21.8499	297.059	24.7549
cm^3	K	mol	8314.510	82.0578	1205.92	62364.1	2455.28	33380.6	2781.72
		lb·mol	3771398	37220.8	546995	28287900	1113700	15141200	1261770
	°R	mol	4619.172	45.5877	669.954	34646.7	1364.04	18544.8	1545.40
		lb·mol	2095221	20678.2	303886	15715500	618720	8411770	700982
L	K	mol	8.314510	0.0820578	1.20592	62.3641	2.45528	33.3806	2.78172
		lb·mol	3771.398	37.2208	546.995	28287.9	1113.70	15141.2	1261.77
	°R	mol	4.619172	0.0455877	0.669954	34.6467	1.36404	18.5448	1.54540
		lb·mol	2095.221	20.6782	303.886	15715.5	618.720	8411.77	700.982
m^3	K	mol	0.008314510	0.0000820578	0.00120592	0.0623641	0.00245528	0.0333806	0.00278172
		lb·mol	3.771398	0.0372208	0.546995	28.2879	1.11370	15.1412	1.26177
	°R	mol	0.004619172	0.0000455877	0.000669954	0.0346467	0.00136404	0.0185448	0.00154540
		lb·mol	2.095221	0.0206782	0.303886	15.7155	0.618720	8.41177	0.700982

Section 2
Symbols, Terminology, and Nomenclature

SYMBOLS AND TERMINOLOGY FOR PHYSICAL AND CHEMICAL QUANTITIES

The International Organization for Standardization (ISO), International Union of Pure and Applied Chemistry (IUPAC), and the International Union of Pure and Applied Physics (IUPAP) have jointly developed a set of recommended symbols for physical and chemical quantities. Consistent use of these recommended symbols helps assure unambiguous scientific communication. The list below is reprinted from Reference 1 with permission from IUPAC. Full details may be found in the following references:

1. Ian Mills, Ed., *Quantities, Units, and Symbols in Physical Chemistry*, Blackwell Scientific Publications, Oxford, 1988.
2. E. R. Cohen and P. Giacomo, *Symbols, Units, Nomenclature, and Fundamental Constants in Physics*, Document IUPAP-25, 1987; also published in *Physica*, 146A, 1—68, 1987.
3. *ISO Standards Handbook 2: Units of Measurement*, International Organization of Standardization, Geneva, 1982.

GENERAL RULES

The value of a physical quantity is expressed as the product of a numerical value and a unit, e.g.:

$T = 300$ K
$V = 26.2$ cm^3
$C_p = 45.3$ J mol^{-1} K^{-1}

The symbol for a physical quantity is always given in italic (sloping) type, while symbols for units are given in roman type. Column headings in tables and axis labels on graphs may conveniently be written as the physical quantity symbol divided by the unit symbol, e.g.:

T/K
V/cm^3
C_p/J mol K^{-1}

The values in the table or graph axis are then pure numbers.

Subscripts to symbols for physical quantities should be italic if the subscript refers to another physical quantity or to a number, e.g.:

C_p — heat capacity at constant pressure
B_n — nth virial coefficient

Subscripts which have other meanings should be in roman type:

m_p — mass of the proton
E_k — kinetic energy

The following tables give the recommended symbols for the major classes of physical and chemical quantities. The expression in the Definition column is given as an aid in identifying the quantity but is not necessarily the complete or unique definition. The SI Unit gives one (not necessarily unique) expression for the coherent SI unit for the quantity. Other equivalent unit expressions, including those which involve SI prefixes, may be used.

Name	Symbol	Definition	SI unit

SPACE AND TIME

Name	Symbol	Definition	SI unit
cartesian space coordinates	x, y, z		m
spherical polar coordinates	r, θ, ϕ		m, 1, 1
generalized coordinate	q, q_i		(varies)
position vector	r	$r = xi + yj + zk$	m
length	l		m
special symbols:			
height	h		
breadth	b		
thickness	d, δ		
distance	d		

Name	Symbol	Definition	SI unit
radius	r		
diameter	d		
path length	s		
length of arc	s		
area	A, A_s, S		m^2
volume	$V, (v)$		m^3
plane angle	$\alpha, \beta, \gamma, \theta, \phi \ldots$	$\alpha = s/r$	rad, 1
solid angle	ω, Ω	$\omega = A/r^2$	sr, 1
time	t		s
period	T	$T = t/N$	s
frequency	v, f	$v = 1/T$	Hz
circular frequency, angular frequency	ω	$\omega = 2\pi v$	rad s^{-1}, s^{-1}
characteristic time interval, relaxation time, time constant	τ, T	$\tau = \lvert dt/d\ln x \rvert$	s
angular velocity	ω	$\omega = d\phi/dt$	rad s^{-1}, s^{-1}
velocity	$v, u, w, c, \dot{r}$	$v = dr/dt$	m s^{-1}
speed	v, u, w, c	$v = \lvert v \rvert$	m s^{-1}
acceleration	$a, (g)$	$a = dv/dt$	m s^{-2}

CLASSICAL MECHANICS

Name	Symbol	Definition	SI unit
mass	m		kg
reduced mass	μ	$\mu = m_1 m_2/(m_1 + m_2)$	kg
density, mass density	ρ	$\rho = m/V$	kg m^{-3}
relative density	d	$d = \rho/\rho^{\bullet}$	1
surface density	ρ_A, ρ_S	$\rho_A = m/A$	kg m^{-2}
specific volume	v	$v = V/m = 1/\rho$	m^3 kg^{-1}
momentum	p	$p = mv$	kg m s^{-1}
angular momentum, action	L	$L = r \times p$	J s
moment of inertia	I, J	$I = \sum m_i r_i^2$	kg m^2
force	F	$F = dp/dt = ma$	N
torque, moment of a force	$T, (M)$	$T = r \times F$	N m
energy	E		J
potential energy	E_p, V, Φ	$E_p = -\int F \cdot ds$	J
kinetic energy	E_k, T, K	$E_k = \frac{1}{2}mv^2$	J
work	W, w	$W = \int F \cdot ds$	J
Hamilton function	H	$H(q, p)$ $= T(q, p) + V(q)$	J
Lagrange function	L	$L(q, \dot{q})$ $= T(q, \dot{q}) - V(q)$	J

Name	Symbol	Definition	SI unit
pressure	p, P	$p = F/A$	Pa, N m^{-2}
surface tension	γ, σ	$\gamma = dW/dA$	N m^{-1}, J m^{-2}
weight	$G, (W, P)$	$G = mg$	N
gravitational constant	G	$F = Gm_1 m_2/r^2$	N m^2 kg^{-2}
normal stress	σ	$\sigma = F/A$	Pa
shear stress	τ	$\tau = F/A$	Pa
linear strain, relative elongation	ε, e	$\varepsilon = \Delta l/l$	1
modulus of elasticity, Young's modulus	E	$E = \sigma/\varepsilon$	Pa
shear strain	γ	$\gamma = \Delta x/d$	1
shear modulus	G	$G = \tau/\gamma$	Pa
volume strain, bulk strain	θ	$\theta = \Delta V/V_0$	1
bulk modulus, compression modulus	K	$K = -V_0(dp/dV)$	Pa
viscosity, dynamic viscosity	η, μ	$\tau_{x,z} = \eta(dv_x/dz)$	Pa s
fluidity	ϕ	$\phi = 1/\eta$	m kg^{-1} s
kinematic viscosity	v	$v = \eta/\rho$	m^2 s^{-1}
friction coefficient	$\mu, (f)$	$F_{frict} = \mu F_{norm}$	1
power	P	$P = dW/dt$	W
sound energy flux	P, P_a	$P = dE/dt$	W
acoustic factors, reflection factor	ρ	$\rho = P_r/P_0$	1
acoustic absorption factor	$\alpha_a, (\alpha)$	$\alpha_a = 1 - \rho$	1
transmission factor	τ	$\tau = P_{tr}/P_0$	1
dissipation factor	δ	$\delta = \alpha_a - \tau$	1

ELECTRICITY AND MAGNETISM

Name	Symbol	Definition	SI unit
quantity of electricity, electric charge	Q		C
charge density	ρ	$\rho = Q/V$	C m^{-3}
surface charge density	σ	$\sigma = Q/A$	C m^{-2}
electric potential	V, ϕ	$V = dW/dQ$	V, J C^{-1}
electric potential difference	$U, \Delta V, \Delta\phi$	$U = V_2 - V_1$	V
electromotive force	E	$E = \int (F/Q) \cdot ds$	V
electric field strength	E	$E = F/Q = -\text{grad } V$	V m^{-1}
electric flux	Ψ	$\Psi = \int D \cdot dA$	C
electric displacement	D	$D = \varepsilon E$	C m^{-2}
capacitance	C	$C = Q/U$	F, C V^{-1}
permittivity	ε	$D = \varepsilon E$	F m^{-1}

Name	Symbol	Definition	SI unit
permittivity of vacuum	ε_0	$\varepsilon_0 = \mu_0^{-1} c_0^{-2}$	$F\,m^{-1}$
relative permittivity	ε_r	$\varepsilon_r = \varepsilon/\varepsilon_0$	1
dielectric polarization (dipole moment per volume)	P	$P = D - \varepsilon_0 E$	$C\,m^{-2}$
electric susceptibility	χ_e	$\chi_e = \varepsilon_r - 1$	1
electric dipole moment	p, μ	$p = Qr$	$C\,m$
electric current	I	$I = dQ/dt$	A
electric current density	j, J	$I = \int j \cdot dA$	$A\,m^{-2}$
magnetic flux density, magnetic induction	B	$F = Qv \times B$	T
magnetic flux	Φ	$\Phi = \int B \cdot dA$	Wb
magnetic field strength	H	$B = \mu H$	$A\,m^{-1}$
permeability	μ	$B = \mu H$	$N\,A^{-2}$, $H\,m^{-1}$
permeability of vacuum	μ_0		$H\,m^{-1}$
relative permeability	μ_r	$\mu_r = \mu/\mu_0$	1
magnetization (magnetic dipole moment per volume)	M	$M = B/\mu_0 - H$	$A\,m^{-1}$
magnetic susceptibility	$\chi, \kappa, (\chi_m)$	$\chi = \mu_r - 1$	1
molar magnetic susceptibility	χ_m	$\chi_m = V_m \chi$	$m^3\,mol^{-1}$
magnetic dipole moment	m, μ	$E_p = -m \cdot B$	$A\,m^2$, $J\,T^{-1}$
electrical resistance	R	$R = U/I$	Ω
conductance	G	$G = 1/R$	S
loss angle	δ	$\delta = (\pi/2) + \phi_I - \phi_U$	1, rad
reactance	X	$X = (U/I)\sin\delta$	Ω
impedance (complex impedance)	Z	$Z = R + iX$	Ω
admittance (complex admittance)	Y	$Y = 1/Z$	S
susceptance	B	$Y = G + iB$	S
resistivity	ρ	$\rho = E/j$	$\Omega\,m$
conductivity	κ, γ, σ	$\kappa = 1/\rho$	$S\,m^{-1}$
self-inductance	L	$E = -L(dI/dt)$	H
mutual inductance	M, L_{12}	$E_1 = L_{12}(dI_2/dt)$	H
magnetic vector potential	A	$B = \nabla \times A$	$Wb\,m^{-1}$
Poynting vector	S	$S = E \times H$	$W\,m^{-2}$

QUANTUM MECHANICS

Name	Symbol	Definition	SI unit
momentum operator	$\hat{p}$	$\hat{p} = -i\hbar\nabla$	$m^{-1}\,J\,s$
kinetic energy operator	$\hat{T}$	$\hat{T} = -(\hbar^2/2m)\nabla^2$	J

Name	Symbol	Definition	SI unit
hamiltonian operator	$\hat{H}$	$\hat{H} = \hat{T} + V$	J
wavefunction, state function	Ψ, ψ, ϕ	$\hat{H}\psi = E\psi$	$(\mathrm{m}^{-3/2})$
probability density	P	$P = \psi^{*}\psi$	(m^{-3})
charge density of electrons	ρ	$\rho = -eP$	$(\mathrm{C\ m}^{-3})$
probability current density	S	$S = -i\hbar(\psi^{*}\nabla\psi - \psi\nabla\psi^{*})/2m_{e}$	$(\mathrm{m}^{-2}\,\mathrm{s}^{-1})$
electric current density of electrons	j	$j = -eS$	$(\mathrm{A\ m}^{-2})$
matrix element of operator $\hat{A}$	$A_{ij}, \langle i\vert\hat{A}\vert j\rangle$	$A_{ij} = \int\psi_{i}^{*}\hat{A}\psi_{j}\,\mathrm{d}\tau$	(varies)
expectation value of operator $\hat{A}$	$\langle A\rangle, \bar{A}$	$\langle A\rangle = \int\psi^{*}\hat{A}\psi\,\mathrm{d}\tau$	(varies)
hermitian conjugate of $\hat{A}$	$\hat{A}^{\dagger}$	$(\hat{A}^{\dagger})_{ij} = (A_{ji})^{*}$	(varies)
commutator of $\hat{A}$ and $\hat{B}$	$[\hat{A}, \hat{B}], [\hat{A}, \hat{B}]_{-}$	$[\hat{A}, \hat{B}] = \hat{A}\hat{B} - \hat{B}\hat{A}$	(varies)
anticommutator	$[\hat{A}, \hat{B}]_{+}$	$[\hat{A}, \hat{B}]_{+} = \hat{A}\hat{B} + \hat{B}\hat{A}$	(varies)
spin wavefunction	$\alpha; \beta$		1
coulomb integral	H_{AA}	$H_{AA} = \int\psi_{A}^{*}\hat{H}\psi_{A}\,\mathrm{d}\tau$	J
resonance integral	H_{AB}	$H_{AB} = \int\psi_{A}^{*}\hat{H}\psi_{B}\,\mathrm{d}\tau$	J
overlap integral	S_{AB}	$S_{AB} = \int\psi_{A}^{*}\psi_{B}\,\mathrm{d}\tau$	1

ATOMS AND MOLECULES

Name	Symbol	Definition	SI unit
nucleon number, mass number	A		1
proton number, atomic number	Z		1
neutron number	N	$N = A - Z$	1
electron rest mass	m_{e}		kg
mass of atom, atomic mass	m_{a}, m		kg
atomic mass constant	m_{u}	$m_{u} = m_{a}(^{12}C)/12$	kg
mass excess	Δ	$\Delta = m_{a} - Am_{u}$	kg
elementary charge, proton charge	e		C
Planck constant	h		J s
Planck constant/2π	$\hbar$	$\hbar = h/2\pi$	J s
Bohr radius	a_{0}	$a_{0} = 4\pi\varepsilon_{0}\hbar^{2}/m_{e}e^{2}$	m
Hartree energy	E_{h}	$E_{h} = \hbar^{2}/m_{e}a_{0}^{2}$	J
Rydberg constant	R_{∞}	$R_{\infty} = E_{h}/2hc$	m^{-1}
fine structure constant	α	$\alpha = e^{2}/4\pi\varepsilon_{0}\hbar c$	1
ionization energy	E_{i}		J

Name	Symbol	Definition	SI unit
electron affinity	E_{ea}		J
dissociation energy	E_d, D		J
from the ground state	D_0		J
from the potential minimum	D_e		J
principal quantum number (H atom)	n	$E = -hcR/n^2$	1
angular momentum quantum numbers	see under Spectroscopy		
magnetic dipole moment of a molecule	m, μ	$E_p = -m \cdot B$	J T^{-1}
magnetizability of a molecule	ξ	$m = \xi B$	J T^{-2}
Bohr magneton	μ_B	$\mu_B = e\hbar/2m_e$	J T^{-1}
nuclear magneton	μ_N	$\mu_N = (m_e/m_p)\mu_B$	J T^{-1}
magnetogyric ratio (gyromagnetic ratio)	γ	$\gamma = \mu/L$	C kg^{-1}
g factor	g		1
Larmor circular frequency	ω_L	$\omega_L = (e/2m)B$	s^{-1}
Larmor frequency	ν_L	$\nu_L = \omega_L/2\pi$	Hz
longitudinal relaxation time	T_1		s
transverse relaxation time	T_2		s
electric dipole moment of a molecule	p, μ	$E_p = -p \cdot E$	C m
quadrupole moment of a molecule	$Q; \Theta$	$E_p = \frac{1}{2}Q : V'' = \frac{1}{3}\Theta : V''$	C m^2
quadrupole moment of a nucleus	eQ	$eQ = 2\langle \Theta_{zz} \rangle$	C m^2
electric field gradient tensor	q	$q_{\alpha\beta} = -\partial^2 V/\partial\alpha\partial\beta$	V m^{-2}
quadrupole interaction energy tensor	χ	$\chi_{\alpha\beta} = eQq_{\alpha\beta}$	J
electric polarizability of a molecule	α	$p \text{ (induced)} = \alpha E$	$\text{C m}^2 \text{ V}^{-1}$
activity (of a radioactive substance)	A	$A = -dN_B/dt$	Bq
decay (rate) constant, disintegration (rate) constant	λ	$A = \lambda N_B$	s^{-1}
half life	$t_{\frac{1}{2}}, T_{\frac{1}{2}}$		s
mean life	τ		s
level width	Γ	$\Gamma = \hbar/\tau$	J

Name	Symbol	Definition	SI unit
disintegration energy	Q		J
cross section (of a nuclear reaction)	σ		m^2

SPECTROSCOPY

Name	Symbol	Definition	SI unit
total term	T	$T = E_{tot}/hc$	m^{-1}
transition wavenumber	$\tilde{\nu}, (\nu)$	$\tilde{\nu} = T' - T''$	m^{-1}
transition frequency	ν	$\nu = (E' - E'')/h$	Hz
electronic term	T_e	$T_e = E_e/hc$	m^{-1}
vibrational term	G	$G = E_{vib}/hc$	m^{-1}
rotational term	F	$F = E_{rot}/hc$	m^{-1}
spin orbit coupling constant	A	$T_{s.o.} = A\langle \hat{L} \cdot \hat{S} \rangle$	m^{-1}
principal moments of inertia	$I_A; I_B; I_C$	$I_A \leqslant I_B \leqslant I_C$	kg m^2
rotational constants,			
in wavenumber	$\tilde{A}; \tilde{B}; \tilde{C}$	$\tilde{A} = h/8\pi^2 c I_A$	m^{-1}
in frequency	$A; B; C$	$A = h/8\pi^2 I_A$	Hz
inertial defect	Δ	$\Delta = I_C - I_A - I_B$	kg m^2
asymmetry parameter	κ	$\kappa = \dfrac{(2B - A - C)}{(A - C)}$	1
centrifugal distortion constants,			
S reduction	$D_J; D_{JK}; D_K; d_1; d_2$		m^{-1}
A reduction	$\Delta_J; \Delta_{JK}; \Delta_K; \delta_J; \delta_K$		m^{-1}
harmonic vibration wavenumber	$\omega_e; \omega_r$		m^{-1}
vibrational anharmonicity constant	$\omega_e x_e; x_{rs}; g_{tt'}$		m^{-1}
vibrational quantum numbers	$v_r; l_t$		1
Coriolis zeta constant	ζ_{rs}^{α}		1
angular momentum quantum numbers	see additional information below		
degeneracy, statistical weight	g, d, β		1
electric dipole moment of a molecule	p, μ	$E_p = -p \cdot E$	C m
transition dipole moment of a molecule	M, R	$M = \int \psi' p \psi'' d\tau$	C m
molecular geometry, interatomic distances,			
equilibrium distance	r_e		m
zero-point average distance	r_z		m
ground state distance	r_0		m

Name	Symbol	Definition	SI unit
substitution structure distance	r_s		m
vibrational coordinates,			
internal coordinates	R_i, r_i, θ_j, etc.		(varies)
symmetry coordinates	S_i		(varies)
normal coordinates			
mass adjusted	Q_r		$kg^{\frac{1}{2}}\,\dot{m}$
dimensionless	q_r		1
vibrational force constants,			
diatomic	$f, (k)$	$f = \partial^2 V / \partial r^2$	$J\,m^{-2}$
polyatomic,			
internal coordinates	f_{ij}	$f_{ij} = \partial^2 V / \partial r_i \partial r_j$	(varies)
symmetry coordinates	F_{ij}	$F_{ij} = \partial^2 V / \partial S_i \partial S_j$	(varies)
dimensionless normal coordinates	$\phi_{rst\ldots},$ $k_{rst\ldots}$		m^{-1}
nuclear magnetic resonance (NMR),			
magnetogyric ratio	γ	$\gamma = \mu / I\hbar$	$C\,kg^{-1}$
shielding constant	σ_A	$B_A = (1 - \sigma_A)B$	1
chemical shift, δ scale	δ	$\delta = 10^6 (v - v_0)/v_0$	1
(indirect) spin–spin coupling constant	J_{AB}	$\hat{H}/h = J_{AB}\boldsymbol{I}_A \cdot \boldsymbol{I}_B$	Hz
direct (dipolar) coupling constant	D_{AB}		Hz
longitudinal relaxation time	T_1		s
transverse relaxation time	T_2		s
electron spin resonance, electron paramagnetic resonance (ESR, EPR),			
magnetogyric ratio	γ	$\gamma = \mu / s\hbar$	$C\,kg^{-1}$
g factor	g	$hv = g\mu_B B$	1
hyperfine coupling constant,			
in liquids	a, A	$\hat{H}_{hfs}/h = a\boldsymbol{S} \cdot \boldsymbol{I}$	Hz
in solids	T	$\hat{H}_{hfs}/h = \boldsymbol{S} \cdot \boldsymbol{T} \cdot \boldsymbol{I}$	Hz

Angular momentum	Operator symbol	Quantum number symbol		
		Total	Z-axis	z-axis
electron orbital	$\hat{L}$	L	M_L	Λ
one electron only	$\hat{l}$	l	m_l	λ

Angular momentum	Operator symbol	Quantum number symbol		
		Total	Z-axis	z-axis
electron spin	$\hat{S}$	S	M_S	Σ
one electron only	$\hat{s}$	s	m_s	σ
electron orbital + spin	$\hat{L} + \hat{S}$			$\Omega = \Lambda + \Sigma$
nuclear orbital (rotational)	$\hat{R}$	R		K_R, k_R
nuclear spin	$\hat{I}$	I	M_I	
internal vibrational				
spherical top	$\hat{l}$	$l(l\zeta)$		K_l
other	$\hat{j}, \hat{\pi}$			$l(l\zeta)$
sum of $R + L(+j)$	$\hat{N}$	N		K, k
sum of $N + S$	$\hat{J}$	J	M_J	K, k
sum of $J + I$	$\hat{F}$	F	M_F	

ELECTROMAGNETIC RADIATION

Name	Symbol	Definition	SI unit
wavelength	λ		m
speed of light			
in vacuum	c_0		m s^{-1}
in a medium	c	$c = c_0/n$	m s^{-1}
wavenumber in vacuum	$\tilde{v}$	$\tilde{v} = v/c_0 = 1/n\lambda$	m^{-1}
wavenumber (in a medium)	σ	$\sigma = 1/\lambda$	m^{-1}
frequency	v	$v = c/\lambda$	Hz
circular frequency, pulsatance	ω	$\omega = 2\pi v$	s^{-1}, rad s^{-1}
refractive index	n	$n = c_0/c$	1
Planck constant	h		J s
Planck constant/2π	$\hbar$	$\hbar = h/2\pi$	J s
radiant energy	Q, W		J
radiant energy density	ρ, w	$\rho = Q/V$	J m^{-3}
spectral radiant energy density			
in terms of frequency	ρ_v, w_v	$\rho_v = d\rho/dv$	J m^{-3} Hz^{-1}
in terms of wavenumber	$\rho_{\tilde{v}}, w_{\tilde{v}}$	$\rho_{\tilde{v}} = d\rho/d\tilde{v}$	J m^{-2}
in terms of wavelenglth	ρ_λ, w_λ	$\rho_\lambda = d\rho/d\lambda$	J m^{-4}
Einstein transition probabilities			
spontaneous emission	A_{nm}	$dN_n/dt = -A_{nm}N_n$	s^{-1}
stimulated emission	B_{nm}	$dN_n/dt = -\rho_{\tilde{v}}(\tilde{v}_{nm}) \times B_{nm}N_n$	s kg^{-1}
stimulated absorption	B_{mn}	$dN_n/dt = \rho_{\tilde{v}}(\tilde{v}_{nm})B_{mn}N_m$	s kg^{-1}
radiant power, radiant energy per time	Φ, P	$\Phi = dQ/dt$	W

Name	Symbol	Definition	SI unit
radiant intensity	I	$I = d\Phi/d\Omega$	$W\ sr^{-1}$
radiant exitance, (emitted radiant flux)	M	$M = d\Phi/dA_{source}$	$W\ m^{-2}$
irradiance, (radiant flux received)	$E, (I)$	$E = d\Phi/dA$	$W\ m^{-2}$
emittance	ε	$\varepsilon = M/M_{bb}$	1
Stefan–Boltzmann constant	σ	$M_{bb} = \sigma T^4$	$W\ m^{-2}\ K^{-4}$
first radiation constant	c_1	$c_1 = 2\pi h c_0^2$	$W\ m^2$
second radiation constant	c_2	$c_2 = h c_0/k$	$K\ m$
transmittance, transmission factor	τ, T	$\tau = \Phi_{tr}/\Phi_0$	1
absorptance, absorption factor	α	$\alpha = \Phi_{abs}/\Phi_0$	1
reflectance, reflection factor	ρ	$\rho = \Phi_{refl}/\Phi_0$	1
(decadic) absorbance	A	$A = -\lg(1-\alpha_i)$	1
napierian absorbance	B	$B = -\ln(1-\alpha_i)$	1
absorption coefficient			
(linear) decadic	a, K	$a = A/l$	m^{-1}
(linear) napierian	α	$\alpha = B/l$	m^{-1}
molar (decadic)	ε	$\varepsilon = a/c = A/cl$	$m^2\ mol^{-1}$
molar napierian	κ	$\kappa = \alpha/c = B/cl$	$m^2\ mol^{-1}$
absorption index	k	$k = \alpha/4\pi\tilde{\nu}$	1
complex refractive index	$\hat{n}$	$\hat{n} = n + ik$	1
molar refraction	R, R_m	$R = \dfrac{(n^2-1)}{(n^2+2)} V_m$	$m^3\ mol^{-1}$
angle of optical rotation	α		1, rad

SOLID STATE

Name	Symbol	Definition	SI unit
lattice vector	R, R_0		m
fundamental translation vectors for the crystal lattice	$a_1; a_2; a_3,$ $a; b; c$	$R = n_1 a_1 + n_2 a_2 + n_3 a_3$	m
(circular) reciprocal lattice vector	G	$G \cdot R = 2\pi m$	m^{-1}
(circular) fundamental translation vectors for the reciprocal lattice	$b_1; b_2; b_3,$ $a^*; b^*; c^*$	$a_i \cdot b_k = 2\pi\delta_{ik}$	m^{-1}
lattice plane spacing	d		m
Bragg angle	θ	$n\lambda = 2d\sin\theta$	1, rad

Name	Symbol	Definition	SI unit
order of reflection	n		1
order parameters			
short range	σ		1
long range	s		1
Burgers vector	b		m
particle position vector	r, R_j		m
equilibrium position vector of an ion	R_0		m
displacement vector of an ion	u	$u = R - R_0$	m
Debye–Waller factor	B, D		1
Debye circular wavenumber	q_D		m^{-1}
Debye circular frequency	ω_D		s^{-1}
Grüneisen parameter	γ, Γ	$\gamma = \alpha V / \kappa C_V$	1
Madelung constant	$\alpha, \mathcal{M}$	$E_{coul} = \dfrac{\alpha N_A z_+ z_- e^2}{4\pi\varepsilon_0 R_0}$	1
density of states	N_E	$N_E = dN(E)/dE$	$J^{-1}\,m^{-3}$
(spectral) density of vibrational modes	N_ω, g	$N_\omega = dN(\omega)/d\omega$	$s\,m^{-3}$
resistivity tensor	ρ_{ik}	$E = \rho \cdot j$	$\Omega\,m$
conductivity tensor	σ_{ik}	$\sigma = \rho^{-1}$	$S\,m^{-1}$
thermal conductivity tensor	λ_{ik}	$J_q = -\lambda \cdot \mathrm{grad}\,T$	$W\,m^{-1}\,K^{-1}$
residual resistivity	ρ_R		$\Omega\,m$
relaxation time	τ	$\tau = l/v_F$	s
Lorenz coefficient	L	$L = \lambda/\sigma T$	$V^2\,K^{-2}$
Hall coefficient	A_H, R_H	$E = \rho \cdot j + R_H(B \times j)$	$m^3\,C^{-1}$
thermoelectric force	E		V
Peltier coefficient	Π		V
Thomson coefficient	$\mu, (\tau)$		$V\,K^{-1}$
work function	Φ	$\Phi = E_\infty - E_F$	J
number density, number concentration	$n, (p)$		m^{-3}
gap energy	E_g		J
donor ionization energy	E_d		J
acceptor ionization energy	E_a		J
Fermi energy	E_F, ε_F		J
circular wave vector, propagation vector	k, q	$k = 2\pi/\lambda$	m^{-1}
Bloch function	$u_k(r)$	$\psi(r) = u_k(r)\exp(ik \cdot r)$	$m^{-3/2}$
charge density of electrons	ρ	$\rho(r) = -e\psi^*(r)\psi(r)$	$C\,m^{-3}$

Name	Symbol	Definition	SI unit
effective mass	m^*		kg
mobility	μ	$\mu = v_{drift}/E$	$m^2\,V^{-1}\,s^{-1}$
mobility ratio	b	$b = \mu_n/\mu_p$	1
diffusion coefficient	D	$dN/dt = -DA(dn/dx)$	$m^2\,s^{-1}$
diffusion length	L	$L = \sqrt{D\tau}$	m
characteristic (Weiss) temperature	$\theta,\ \theta_W$		K
Curie temperature	T_C		K
Néel temperature	T_N		K

STATISTICAL THERMODYNAMICS

Name	Symbol	Definition	SI unit
number of entities	N		1
number density of entities, number concentration	$n,\ C$	$n = N/V$	m^{-3}
Avogadro constant	$L,\ N_A$		mol^{-1}
Boltzmann constant	$k,\ k_B$		$J\,K^{-1}$
gas constant (molar)	R	$R = Lk$	$J\,K^{-1}\,mol^{-1}$
molecular position vector	$r\,(x, y, z)$		m
molecular velocity vector	$c(c_x, c_y, c_z),$ $u(u_x, u_y, u_z)$	$c = dr/dt$	$m\,s^{-1}$
molecular momentum vector	$p(p_x, p_y, p_z)$	$p = mc$	$kg\,m\,s^{-1}$
velocity distribution function (Maxwell)	$f(c_x)$	$f(c_x) = (m/2\pi kT)^{\frac{1}{2}}$ $\times \exp(-mc_x^2/2kT)$	$m^{-1}\,s$
speed distribution function (Maxwell–Boltzmann)	$F(c)$	$F(c) = (m/2\pi kT)^{3/2}$ $\times 4\pi c^2 \exp(-mc^2/2kT)$	$m^{-1}\,s$
average speed	$\bar{c},\ \bar{u},$ $\langle c \rangle,\ \langle u \rangle$	$\bar{c} = \int cF(c)dc$	$m\,s^{-1}$
generalized coordinate	q		(m)
generalized momentum	p	$p = \partial L/\partial \dot{q}$	$(kg\,m\,s^{-1})$
volume in phase space	Ω	$\Omega = (1/h)\int p\,dq$	1
probability	P		1
statistical weight, degeneracy	$g,\ d,\ W,\ \omega,\ \beta$		1
density of states	$\rho(E)$	$\rho(E) = dN/dE$	J^{-1}
partition function, sum over states, for a single molecule	$q,\ z$	$q = \sum_i g_i \exp(-\varepsilon_i/kT)$	1
for a canonical ensemble (system, or assembly)	$Q,\ Z$		1

Name	Symbol	Definition	SI unit
microcanonical ensemble	Ω		1
grand (canonical ensemble)	Ξ		1
symmetry number	σ, s		1
reciprocal temperature parameter	β	$\beta = 1/kT$	J^{-1}
characteristic temperature	Θ		K

GENERAL CHEMISTRY

Name	Symbol	Definition	SI unit
number of entities (e.g. molecules, atoms, ions, formula units)	N		1
amount (of substance)	n	$n_B = N_B/L$	mol
Avogadro constant	L, N_A		mol^{-1}
mass of atom, atomic mass	m_a, m		kg
mass of entity (molecule, or formula unit)	m_f, m		kg
atomic mass constant	m_u	$m_u = m_a(^{12}C)/12$	kg
molar mass	M	$M_B = m/n_B$	$kg\,mol^{-1}$
relative molecular mass (relative molar mass, molecular weight)	M_r	$M_{r,B} = m_B/m_u$	1
molar volume	V_m	$V_{m,B} = V/n_B$	$m^3\,mol^{-1}$
mass fraction	w	$w_B = m_B/\Sigma m_i$	1
volume fraction	ϕ	$\phi_B = V_B/\Sigma V_i$	1
mole fraction, amount fraction, number fraction	x, y	$x_B = n_B/\Sigma n_i$	1
(total) pressure	p, P		Pa
partial pressure	p_B	$p_B = y_B p$	Pa
mass concentration (mass density)	γ, ρ	$\gamma_B = m_B/V$	$kg\,m^{-3}$
number concentration, number density of entities	C, n	$C_B = N_B/V$	m^{-3}
amount concentration, concentration	c	$c_B = n_B/V$	$mol\,m^{-3}$
solubility	s	$s_B = c_B$ (saturated solution)	$mol\,m^{-3}$
molality (of a solute)	$m, (b)$	$m_B = n_B/m_A$	$mol\,kg^{-1}$

Name	Symbol	Definition	SI unit
surface concentration	Γ	$\Gamma_B = n_B/A$	$mol\ m^{-2}$
stoichiometric number	ν		1
extent of reaction, advancement	ξ	$\Delta\xi = \Delta n_B/\nu_B$	mol
degree of dissociation	α		1

CHEMICAL THERMODYNAMICS

Name	Symbol	Definition	SI unit
heat	q, Q		J
work	w, W		J
internal energy	U	$\Delta U = q + w$	J
enthalpy	H	$H = U + pV$	J
thermodynamic temperature	T		K
Celsius temperature	θ, t	$\theta/°C = T/K - 273.15$	°C
entropy	S	$dS \geq dq/T$	$J\,K^{-1}$
Helmholtz energy, (Helmholtz function)	A	$A = U - TS$	J
Gibbs energy, (Gibbs function)	G	$G = H - TS$	J
Massieu function	J	$J = -A/T$	$J\,K^{-1}$
Planck function	Y	$Y = -G/T$	$J\,K^{-1}$
surface tension	γ, σ	$\gamma = (\partial G/\partial A_s)_{T,p}$	$J\,m^{-2}, N\,m^{-1}$
molar quantity X	X_m	$X_m = X/n$	(varies)
specific quantity X	x	$x = X/m$	(varies)
pressure coefficient	β	$\beta = (\partial p/\partial T)_V$	$Pa\,K^{-1}$
relative pressure coefficient	α_p	$\alpha_p = (1/p)(\partial p/\partial T)_V$	K^{-1}
compressibility,			
isothermal	κ_T	$\kappa_T = -(1/V)(\partial V/\partial p)_T$	Pa^{-1}
isentropic	κ_S	$\kappa_S = -(1/V)(\partial V/\partial p)_S$	Pa^{-1}
linear expansion coefficient	α_l	$\alpha_l = (1/l)(\partial l/\partial T)$	K^{-1}
cubic expansion coefficient	α, α_V, γ	$\alpha = (1/V)(\partial V/\partial T)_p$	K^{-1}
heat capacity,			
at constant pressure	C_p	$C_p = (\partial H/\partial T)_p$	$J\,K^{-1}$
at constant volume	C_V	$C_V = (\partial U/\partial T)_V$	$J\,K^{-1}$
ratio of heat capacities	$\gamma, (\kappa)$	$\gamma = C_p/C_V$	1
Joule–Thomson coefficient	μ, μ_{JT}	$\mu = (\partial T/\partial p)_H$	$K\,Pa^{-1}$
second virial coefficient	B	$pV_m = RT(1 + B/V_m + \cdots)$	$m^3\,mol^{-1}$
compression factor (compressibility factor)	Z	$Z = pV_m/RT$	1

Name	Symbol	Definition	SI unit
partial molar quantity X	X_B, (X'_B)	$X_B = (\partial X/\partial n_B)_{T, p, n_{j \neq B}}$	(varies)
chemical potential (partial molar Gibbs energy)	μ	$\mu_B = (\partial G/\partial n_B)_{T, p, n_{j \neq B}}$	$\mathrm{J\,mol^{-1}}$
absolute activity	λ	$\lambda_B = \exp(\mu_B/RT)$	1
standard chemical potential	$\mu^\bullet$, μ°		$\mathrm{J\,mol^{-1}}$
standard partial molar enthalpy	$H_B{}^\bullet$	$H_B{}^\bullet = \mu_B{}^\bullet + TS_B{}^\bullet$	$\mathrm{J\,mol^{-1}}$
standard partial molar entropy	$S_B{}^\bullet$	$S_B{}^\bullet = -(\partial \mu_B{}^\bullet/\partial T)_p$	$\mathrm{J\,mol^{-1}\,K^{-1}}$
standard reaction Gibbs energy (function)	$\Delta_r G^\bullet$	$\Delta_r G^\bullet = \sum_B \nu_B \mu_B{}^\bullet$	$\mathrm{J\,mol^{-1}}$
affinity of reaction	A, $(\mathscr{A})$	$A = -(\partial G/\partial \xi)_{p, T}$ $= -\sum_B \nu_B \mu_B$	$\mathrm{J\,mol^{-1}}$
standard reaction enthalpy	$\Delta_r H^\bullet$	$\Delta_r H^\bullet = \sum_B \nu_B H_B{}^\bullet$	$\mathrm{J\,mol^{-1}}$
standard reaction entropy	$\Delta_r S^\bullet$	$\Delta_r S^\bullet = \sum_B \nu_B S_B{}^\bullet$	$\mathrm{J\,mol^{-1}\,K^{-1}}$
equilibrium constant	$K^\bullet$, K	$K^\bullet = \exp(-\Delta_r G^\bullet/RT)$	1
equilibrium constant, pressure basis	K_p	$K_p = \prod_B p_B{}^{\nu_B}$	$\mathrm{Pa}^{\Sigma\nu}$
concentration basis	K_c	$K_c = \prod_B c_B{}^{\nu_B}$	$\mathrm{(mol\,m^{-3})}^{\Sigma\nu}$
molality basis	K_m	$K_m = \prod_B m_B{}^{\nu_B}$	$\mathrm{(mol\,kg^{-1})}^{\Sigma\nu}$
fugacity	f, $\tilde{p}$	$f_B = \lambda_B \lim_{p \to 0}(p_B/\lambda_B)_T$	Pa
fugacity coefficient	ϕ	$\phi_B = f_B/p_B$	1
activity and activity coefficient referenced to Raoult's law, (relative) activity	a	$a_B = \exp\left[\dfrac{\mu_B - \mu_B{}^*}{RT}\right]$	1
activity coefficient	f	$f_B = a_B/x_B$	1
activities and activity coefficients referenced to Henry's law, (relative) activity, molality basis	a_m	$a_{m, B} = \exp\left[\dfrac{\mu_B - \mu_B{}^\bullet}{RT}\right]$	1
concentration basis	a_c	$a_{c, B} = \exp\left[\dfrac{\mu_B - \mu_B{}^\bullet}{RT}\right]$	1

Name	Symbol	Definition	SI Unit
mole fraction basis	a_x	$a_{x,B} = \exp\left[\dfrac{\mu_B - \mu_B^\bullet}{RT}\right]$	1
activity coefficient,			
molality basis	γ_m	$a_{m,B} = \gamma_{m,B}\, m_B / m^\bullet$	1
concentration basis	γ_c	$a_{c,B} = \gamma_{c,B}\, c_B / c^\bullet$	1
mole fraction basis	γ_x	$a_{x,B} = \gamma_{x,B}\, x_B$	1
ionic strength,			
molality basis	I_m, I	$I_m = \tfrac{1}{2}\Sigma m_B z_B^2$	mol kg^{-1}
concentration basis	I_c, I	$I_c = \tfrac{1}{2}\Sigma c_B z_B^2$	mol m^{-3}
osmotic coefficient,			
molality basis	ϕ_m	$\phi_m = (\mu_A^* - \mu_A)/(RTM_A\Sigma m_B)$	1
mole fraction basis	ϕ_x	$\phi_x = (\mu_A - \mu_A^*)/(RT\ln x_A)$	1
osmotic pressure	Π	$\Pi = c_B RT$ (ideal dilute solution)	Pa

(i) *Symbols used as subscripts to denote a chemical process or reaction*
These symbols should be printed in roman (upright) type, without a full stop (period).

vaporization, evaporation (liquid→gas)	vap
sublimation (solid→gas)	sub
melting, fusion (solid→liquid)	fus
transition (between two phases)	trs
mixing of fluids	mix
solution (of solute in solvent)	sol
dilution (of a solution)	dil
adsorption	ads
displacement	dpl
immersion	imm
reaction in general	r
atomization	at
combustion reaction	c
formation reaction	f

(ii) *Recommended superscripts*

standard	$\ominus$, o
pure substance	*
infinite dilution	∞
ideal	id
activated complex, transition state	‡
excess quantity	E

Name	Symbol	Definition	SI unit
CHEMICAL KINETICS			
rate of change of quantity X	$\dot{X}$	$\dot{X} = dX/dt$	(varies)
rate of conversion	$\dot{\xi}$	$\dot{\xi} = d\xi/dt$	mol s^{-1}
rate of concentration change (due to chemical reaction)	r_B, v_B	$r_B = dc_B/dt$	$\text{mol m}^{-3}\,\text{s}^{-1}$
rate of reaction (based on amount concentration)	v	$v = \dot{\xi}/V$ $= v_B^{-1}dc_B/dt$	$\text{mol m}^{-3}\,\text{s}^{-1}$
partial order of reaction	n_B	$v = k\Pi c_B^{n_B}$	1
overall order of reaction	n	$n = \Sigma n_B$	1
rate constant, rate coefficient	k	$v = k\Pi c_B^{n_B}$	$(\text{mol}^{-1}\,\text{m}^3)^{n-1}\text{s}^{-1}$
Boltzmann constant	k, k_B		J K^{-1}
half life	$t_{\frac{1}{2}}$	$c(t_{\frac{1}{2}}) = c_0/2$	s
relaxation time	τ	$\tau = 1/(k_1 + k_{-1})$	s
energy of activation, activation energy	E_a, E	$E_a = RT^2\, d\ln k/dT$	J mol^{-1}
pre-exponential factor	A	$k = A\exp(-E_a/RT)$	$(\text{mol}^{-1}\,\text{m}^3)^{n-1}\text{s}^{-1}$
volume of activation	$\Delta^{\ddagger}V$	$\Delta^{\ddagger}V = -RT\times$ $(\partial\ln k/\partial p)_T$	$\text{m}^3\,\text{mol}^{-1}$
collision diameter	d	$d_{AB} = r_A + r_B$	m
collision cross-section	σ	$\sigma_{AB} = \pi d_{AB}^2$	m^2
collision frequency	Z_A		s^{-1}
collision number	Z_{AB}, Z_{AA}		$\text{m}^{-3}\text{s}^{-1}$
collision frequency factor	z_{AB}, z_{AA}	$z_{AB} = Z_{AB}/Lc_Ac_B$	$\text{m}^3\,\text{mol}^{-1}\text{s}^{-1}$
standard enthalpy of activation	$\Delta^{\ddagger}H^{\circ}, \Delta H^{\ddagger}$		J mol^{-1}
standard entropy of activation	$\Delta^{\ddagger}S^{\circ}, \Delta S^{\ddagger}$		$\text{J mol}^{-1}\,\text{K}^{-1}$
standard Gibbs energy of activation	$\Delta^{\ddagger}G^{\circ}, \Delta G^{\ddagger}$		J mol^{-1}
quantum yield, photochemical yield	ϕ		1
ELECTROCHEMISTRY			
elementary charge (proton charge)	e		C
Faraday constant	F	$F = eL$	C mol^{-1}

Name	Symbol	Definition	SI unit
charge number of an ion	z	$z_B = Q_B/e$	1
ionic strength	I_c, I	$I_c = \frac{1}{2}\sum c_i z_i^2$	$mol\,m^{-3}$
mean ionic activity	$a_\pm$	$a_\pm = m_\pm \gamma_\pm / m^\circ$	1
mean ionic molality	$m_\pm$	$m_\pm^{(\nu_+ + \nu_-)} = m_+^{\nu_+} m_-^{\nu_-}$	$mol\,kg^{-1}$
mean ionic activity coefficient	$\gamma_\pm$	$\gamma_\pm^{(\nu_+ + \nu_-)} = \gamma_+^{\nu_+} \gamma_-^{\nu_-}$	1
charge number of electrochemical cell reaction	$n, (z)$		1
electric potential difference (of a galvanic cell)	$\Delta V, E, U$	$\Delta V = V_R - V_L$	V
emf, electromotive force	E	$E = \lim_{I \to 0} \Delta V$	V
standard emf, standard potential of the electrochemical cell reaction	E°	$E^\circ = -\Delta_r G^\circ / nF$ $= (RT/nF) \ln K^\circ$	V
standard electrode potential	E°		V
emf of the cell, potential of the electrochemical cell reaction	E	$E = E^\circ - (RT/nF)$ $\times \sum \nu_i \ln a_i$	V
pH	pH	$pH \approx -lg\left[\dfrac{c(H^+)}{mol\,dm^{-3}}\right]$	1
inner electric potential	ϕ	$\nabla\phi = -E$	V
outer electric potential	ψ	$\psi = Q/4\pi\varepsilon_0 r$	V
surface electric potential	χ	$\chi = \phi - \psi$	V
Galvani potential difference	$\Delta\phi$	$\Delta_\alpha^\beta \phi = \phi^\beta - \phi^\alpha$	V
volta potential difference	$\Delta\psi$	$\Delta_\alpha^\beta \psi = \psi^\beta - \psi^\alpha$	V
electrochemical potential	$\tilde{\mu}$	$\tilde{\mu}_B^\alpha = (\partial G/\partial n_B^\alpha)$	$J\,mol^{-1}$
electric current	I	$I = dQ/dt$	A
(electric) current density	j	$j = I/A$	$A\,m^{-2}$
(surface) charge density	σ	$\sigma = Q/A$	$C\,m^{-2}$
electrode reaction rate constant	k	$k_{ox} = I_a/(nFA \prod_i c_i^{n_i})$	(varies)
mass transfer coefficient, diffusion rate constant	k_d	$k_{d,B} = |\nu_B| I_{l,B}/nFcA$	$m\,s^{-1}$
thickness of diffusion layer	δ	$\delta_B = D_B/k_{d,B}$	m

Name	Symbol	Definition	SI unit
transfer coefficient (electrochemical)	α	$\alpha_c = \dfrac{-|v|RT}{nF}\dfrac{\partial \ln|I_c|}{\partial E}$	1
overpotential	η	$\eta = E_I - E_{I=0} - IR_u$	V
electrokinetic potential (zeta potential)	ζ		V
conductivity	$\kappa, (\sigma)$	$\kappa = j/E$	$S\,m^{-1}$
conductivity cell constant	K_{cell}	$K_{cell} = \kappa R$	m^{-1}
molar conductivity (of an electrolyte)	Λ	$\Lambda_B = \kappa/c_B$	$S\,m^2\,mol^{-1}$
ionic conductivity, molar conductivity of an ion	λ	$\lambda_B = |z_B|Fu_B$	$S\,m^2\,mol^{-1}$
electric mobility	$u, (\mu)$	$u_B = v_B/E$	$m^2\,V^{-1}\,s^{-1}$
transport number	t	$t_B = j_B/\Sigma j_i$	1
reciprocal radius of ionic atmosphere	κ	$\kappa = (2F^2I/\varepsilon RT)^{\frac{1}{2}}$	m^{-1}

COLLOID AND SURFACE CHEMISTRY

Name	Symbol	Definition	SI unit
specific surface area	a, a_s, s	$a = A/m$	$m^2\,kg^{-1}$
surface amount of B, adsorbed amount of B	n_B^s, n_B^a		mol
surface excess of B	n_B^σ		mol
surface excess concentration of B	$\Gamma_B, (\Gamma_B^\sigma)$	$\Gamma_B = n_B^\sigma/A$	$mol\,m^{-2}$
total surface excess concentration	$\Gamma, (\Gamma^\sigma)$	$\Gamma = \sum_i \Gamma_i$	$mol\,m^{-2}$
area per molecule	a, σ	$a_B = A/N_B^\sigma$	m^2
area per molecule in a filled monolayer	a_m, σ_m	$a_{m,B} = A/N_{m,B}$	m^2
surface coverage	θ	$\theta = N_B^\sigma/N_{m,B}$	1
contact angle	θ		1, rad
film thickness	t, h, δ		m
thickness of (surface or interfacial) layer	τ, δ, t		m
surface tension, interfacial tension	γ, σ	$\gamma = (\partial G/\partial A_s)_{T,p}$	$N\,m^{-1}, J\,m^{-2}$
film tension	Σ_f	$\Sigma_f = 2\gamma_f$	$N\,m^{-1}$
reciprocal thickness of the double layer	κ	$\kappa = [2F^2I_c/\varepsilon RT]^{\frac{1}{2}}$	m^{-1}
average molar masses			
number-average	M_n	$M_n = \Sigma n_i M_i/\Sigma n_i$	$kg\,mol^{-1}$
mass-average	M_m	$M_m = \Sigma n_i M_i^2/\Sigma n_i M_i$	$kg\,mol^{-1}$

Name	Symbol	Definition	SI unit
Z-average	M_Z	$M_Z = \Sigma n_i M_i^3 / \Sigma n_i M_i^2$	kg mol^{-1}
sedimentation coefficient	s	$s = v/a$	s
van der Waals constant	λ		J
retarded van der Waals constant	β, B		J
van der Waals–Hamaker constant	A_H		J
surface pressure	π^s, π	$\pi^s = \gamma^0 - \gamma$	N m^{-1}

TRANSPORT PROPERTIES

Name	Symbol	Definition	SI unit
flux (of a quantity X)	J_X, J	$J_X = A^{-1} dX/dt$	(varies)
volume flow rate	$q_V, \dot{V}$	$q_v = dV/dt$	$\text{m}^3 \text{ s}^{-1}$
mass flow rate	$q_m, \dot{m}$	$q_m = dm/dt$	kg s^{-1}
mass transfer coefficient	k_d		m s^{-1}
heat flow rate	ϕ	$\phi = dq/dt$	W
heat flux	J_q	$J_q = \phi/A$	W m^{-2}
thermal conductance	G	$G = \phi/\Delta T$	W K^{-1}
thermal resistance	R	$R = 1/G$	K W^{-1}
thermal conductivity	λ, k	$\lambda = J_q/(dT/dl)$	$\text{W m}^{-1} \text{ K}^{-1}$
coefficient of heat transfer	$h, (k, K, \alpha)$	$h = J_q/\Delta T$	$\text{W m}^{-2} \text{ K}^{-1}$
thermal diffusivity	a	$a = \lambda/\rho c_p$	$\text{m}^2 \text{ s}^{-1}$
diffusion coefficient	D	$D = J_n/(dc/dl)$	$\text{m}^2 \text{ s}^{-1}$

The following symbols are used in the definitions of the dimensionless quantities: mass (m), time (t), volume (V), area (A), density (ρ), speed (v), length (l), viscosity (η), pressure (p), acceleration of free fall (g), cubic expansion coefficient (α), temperature (T), surface tension (γ), speed of sound (c), mean free path (λ), frequency (f), thermal diffusivity (a), coefficient of heat transfer (h), thermal conductivity (k), specific heat capacity at constant pressure (c_p), diffusion coefficient (D), mole fraction (x), mass transfer coefficient (k_d), permeability (μ), electric conductivity (κ), and magnetic flux density (B).

Name	Symbol	Definition	SI unit
Reynolds number	Re	$Re = \rho v l/\eta$	1
Euler number	Eu	$Eu = \Delta p/\rho v^2$	1
Froude number	Fr	$Fr = v/(lg)^{\frac{1}{2}}$	1
Grashof number	Gr	$Gr = l^3 g\alpha\Delta T\rho^2/\eta^2$	1
Weber number	We	$We = \rho v^2 l/\gamma$	1
Mach number	Ma	$Ma = v/c$	1
Knudsen number	Kn	$Kn = \lambda/l$	1
Strouhal number	Sr	$Sr = lf/v$	1
Fourier number	Fo	$Fo = at/l^2$	1
Péclet number	Pe	$Pe = vl/a$	1
Rayleigh number	Ra	$Ra = l^3 g\alpha\Delta T\rho/\eta a$	1
Nusselt number	Nu	$Nu = hl/k$	1
Stanton number	St	$St = h/\rho v c_p$	1

Name	Symbol	Definition	SI units
Fourier number for mass transfer	Fo^*	$Fo^* = Dt/l^2$	1
Péclet number for mass transfer	Pe^*	$Pe^* = vl/D$	1
Grashof number for mass transfer	Gr^*	$Gr^* = l^3 g \left(\dfrac{\partial \rho}{\partial x}\right)_{T,P} \left(\dfrac{\Delta x \rho}{\eta}\right)$	1
Nusselt number for mass transfer	Nu^*	$Nu^* = k_d l/D$	1
Stanton number for mass transfer	St^*	$St^* = k_d/v$	1
Prandtl number	Pr	$Pr = \eta/\rho a$	1
Schmidt number	Sc	$Sc = \eta/\rho D$	1
Lewis number	Le	$Le = a/D$	1
magnetic Reynolds number	Rm, Re_m	$Rm = v\mu\kappa l$	1
Alfvén number	Al	$Al = v(\rho\mu)^{\frac{1}{2}}/B$	1
Hartmann number	Ha	$Ha = Bl(\kappa/\eta)^{\frac{1}{2}}$	1
Cowling number	Co	$Co = B^2/\mu\rho v^2$	1

NOMENCLATURE OF CHEMICAL COMPOUNDS

The International Union of Pure and Applied Chemistry (IUPAC) maintains several commissions that deal with the naming of chemical substances. In general, the approach of IUPAC is to present rules for arriving at names in a systematic manner, rather than recommending a unique name for each compound. Thus there are often several alternative "IUPAC names", depending on which nomenclature system is used, each of which may have advantages in specific applications. However, each of these names will be unambiguous.

Organizations such as the Chemical Abstacts Service and the Beilstein Institute that prepare indexes to the chemical literature must adopt a system for selecting unique names in order to avoid excessive cross referencing. Chemical Abstracts Service uses a system which groups together compounds derived from a single parent compound. Thus most index names are inverted (e.g., Benzene, bromo rather than bromobenzene; Acetic acid, sodium salt rather than sodium acetate). In this *Handbook* the CAS Index Names are used only in the table "Physical Constants of Organic Compounds". Other tables use more familiar names which, with a few possible exceptions, conform to one of the IUPAC naming systems.

Recommended names for the most common substituent groups, ligands, ions, and organic rings are given in the two following tables, "Nomenclature for Inorganic Ions and Ligands" and "Organic Substituent Groups and Ring Systems". For the basics of macromolecular nomenclature, see "Naming Organic Polymers" in Section 13.

Some of the most useful recent guides to chemical nomenclature, prepared by IUPAC and other organizations such as the International Union of Biochemistry and Molecular Biology (IUBMB) and the American Chemical Society are listed below . These books contain citations to the more detailed nomenclature documents in each area.

Inorganic Chemistry

International Union of Pure and Applied Chemistry, *Nomenclature of Inorganic Chemistry, Recommendations 1990,* edited by Leigh, G.J., Blackwell Scientific Publications, Oxford, 1990.

Block, B.P., Powell, W.H., and Fernelius, W.C., *Inorganic Chemical Nomenclature, Principles and Practice*, American Chemical Society, Washington, 1990.

Organic Chemistry

International Union of Pure and Applied Chemistry, *A Guide to IUPAC Nomenclature of Organic Compounds, Recommendations 1993,* edited by Panico, R., Powell, W.H., and Richer, J.-C., Blackwell Scientific Publications, Oxford, 1993.

International Union of Pure and Applied Chemistry, *Glossary of Class Names of Organic Compounds and Reactive Intermediates Based on Structure*, edited by Moss, G.P., Smith, P.A.S., and Tavernier, D., *Pure & Appl. Chem*, 67, 1307, 1995.

Rhodes, P.H., *The Organic Chemist's Desk Reference*, Chapman & Hall, London, 1995.

International Union of Pure and Applied Chemistry, *Basic Terminology of Stereochemistry*, edited by Moss, G.P., *Pure & Applied Chemistry*, 68, 2193, 1966.

Macromolecular Chemistry

International Union of Pure and Applied Chemistry, *Compendium of Macromolecular Nomenclature*, edited by, Metanomski, W.V., Blackwell Scientific Publications, Oxford, 1991.

International Union of Pure and Applied Chemistry, *Glossary of Basic Terms in Polymer Science*, edited by Jenkins, A.D., Kratochvil, P., Stepto, R.F.T., and Suter, U.W., *Pure & Appl. Chem*, 68, in press.

Biochemistry

International Union of Biochemistry and Molecular Biology, *Biochemical Nomenclature and Related Documents*, *2nd Edition, 1992*, Portland Press, London, 1993; includes recommendations of the IUPAC-IUBMB Joint Commission on Biochemical Nomenclature.

International Union of Biochemistry and Molecular Biology, *Enzyme Nomenclature, 1992*, Academic Press, Orlando, FL, 1992.

IUPAC-IUBMB Joint Commission on Biochemical Nomenclature, *Nomenclature of Carbohydrates*, *Recommendations 1996*, edited by McNaught, A.D., *Pure & Appl. Chem.*, 68, 1919, 1996.

General

Chemical Abstracts Service, *Naming and Indexing Chemical Substances for Chemical Abstracts, Appendix IV, Chemical Abstracts 1994 Index Guide.*

NOMENCLATURE FOR INORGANIC IONS AND LIGANDS

See the table *Nomenclature of Chemical Compounds* for references. The assistance of Warren H. Powell in preparing this list is gratefully acknowledged.

Group	As cation	As anion	As ligand	As prefix in organic compounds
H	hydrogen	hydride	hydrido	
F	fluorine	fluoride	fluoro	fluoro
Cl	chlorine	chloride	chloro	chloro
Br	bromine	bromide	bromo	bromo
I	iodine	iodide	iodo	iodo
ClO	chlorosyl	hypochlorite	hypochlorito	chlorosyl
ClO_2	chloryl	chlorite	chlorito	chloryl
ClO_3	perchloryl	chlorate	chlorato	perchloryl
ClO_4		perchlorate		
IO	iodosyl	hypoiodite		iodoso
IO_2	iodyl			iodyl; iodoxy
O		oxide	oxo	oxo
O_2		peroxide (O_2^{2-})	peroxo	peroxy
		hyperoxide (O_2^-)		
HO		hydroxide	hydroxo	hydroxy
HO_2		hydrogen peroxide	hydrogen peroxo	hydroperoxy
S		sulfide	thio; sulfido	thio; thioxo
HS		hydrogen sulfide	thiolo	mercapto
S_2		disulfide	disulfido	
SO	sulfinyl; thionyl			sulfinyl
SO_2	sulfonyl; sulfuryl	sulfoxylate		sulfonyl
SO_3		sulfite	sulfito	
HSO_3		hydrogen sulfite	hydrogen sulfito	
S_2O_3		thiosulfate	thiosulfato	
SO_4		sulfate	sulfato	
Se		selenide	seleno	seleno; selenoxo
SeO	seleninyl			seleninyl
SeO_2	selenonyl			selenonyl
SeO_3		selenite	selenito	
SeO_4		selenate	selenato	
Te		telluride	telluro	telluro; telluroxo
CrO_2	chromyl			
UO_2	uranyl			
NpO_2	neptunyl			
PuO_2	plutoryl			
AmO_2	americyl			
N		nitride	nitrido	
N_3		azide	azido	
NH		imide	imido	imino
NH_2		amide	amido	amino
NHOH		hydroxylamide	hydroxylamido	hydroxyamino
N_2H_3		hydrazide	hydrazido	hydrazino; diazanyl
NO	nitrosyl		nitrosyl	nitroso
NO_2	nitryl		nitro	nitro
ONO		nitrite	nitrito	
NS	thionitrosyl			
NO_3		nitrate	nitrato	
N_2O_3		hyponitrite	hyponitrito	
P		phosphide	phosphido	phosphinidyne
PO	phosphoryl			phosphoroso; phosphinylidyne
PO_2	phospho			
PS	thiophosphoryl			phosphinothioylidyne; thiophosphorozo
PH_2O_3		hypophosphite	hypophosphito	
PHO_3		phosphite	phosphito	

Group	As cation	As anion	As ligand	As prefix in organic compounds
PO_4		phosphate	phosphato	
AsO_4		arsenate	arsenato	
VO	vanadyl			
CO	carbonyl		carbonyl	carbonyl
CS	thiocarbonyl			thiocarbonyl
CH_3O		methanolate	methoxo	methoxy
C_2H_5O		ethanolate	ethoxo	ethoxy
CH_3S		methanethiolate	methanethiolato	methylthio
C_2H_5S		ethanethiolate	ethanethiolato	ethylthio
CN	cyanogen	cyanide	cyano	cyano
OCN		cyanate	cyanato	cyanato
SCN		thiocyanate	thiocyanato	thiocyanato
$SeCN$		selenocyanate	selenocyanato	selenocyanato
$TeCN$		tellurocyanate	tellurocyanato	tellurocyanato
CO_3		carbonate	carbonato	
HCO_3		hydrogen carbonate	hydrogen carbonato	carboxycarbonyl
C_2O_4		oxalate	oxalato	

ORGANIC SUBSTITUENT GROUPS AND RING SYSTEMS

The first part of this table lists substituent groups and their line formulas. A substituent group is defined by IUPAC as a group that replaces one or more hydrogen atoms attached to a parent structure. Such groups are sometimes called radicals, but IUPAC now reserves the term radical for a free molecular species with unpaired electrons. IUPAC does not recommend some of these names, which are marked here with asterisks (e.g., amyl*), but they are included in this list because they are often encountered in the older literature. Substituent group names which are formed by systematic rules (e.g., methyl from methane, ethyl from ethane, etc.) are included here only for the first few members of a homologous series.

In the second part of the table a number of common organic ring compounds are shown, with the conventional numbering of the ring positions indicated.

The help of Warren H. Powell in preparing this table is greatly appreciated. Pertinent references may be found in the table *Nomenclature of Chemical Compounds*.

SUBSTITUENT GROUPS

acetamido (acetylamino)	CH_3CONH-
acetoacetyl	CH_3COCH_2CO-
acetonyl	CH_3COCH_2-
acetyl	CH_3CO-
acryloyl* (1-oxo-2-propenyl)	$CH_2=CHCO-$
alanyl (from alanine)	$CH_3CH(NH_2)CO-$
β-alanyl	$H_2N(CH_2)_2CO-$
allyl (2-propenyl)	$CH_2=CHCH_2-$
allylidene (2-propenylidene)	$CH_2=CHCH=$
amidino (aminoiminomethyl)	$H_2NC(=NH)-$
amino	H_2N-
amyl* (pentyl)	$CH_3(CH_2)_4-$
anilino (phenylamino)	C_6H_5NH-
anisidino	$CH_3OC_6H_4NH-$
anthranoyl (2-aminobenzoyl)	$2-H_2NC_6H_4CO-$
arsino	AsH_2-
azelaoyl (from azelaic acid)	$-OC(CH_2)_7CO-$
azido	N_3-
azino	$=N-N=$
azo	$-N=N-$
azoxy	$-N(O)=N-$
benzal* (benzylidene)	$C_6H_5CH=$
benzamido (benzoylamino)	C_6H_5CONH-
benzhydryl (diphenylmethyl)	$(C_6H_5)_2CH-$
benzoxy* (benzoyloxy)	C_6H_5COO-
benzoyl	C_6H_5CO-
benzyl	$C_6H_5CH_2-$
benzylidene	$C_6H_5CH=$
benzylidyne	$C_6H_5C=$
biphenylyl	$C_6H_5C_6H_5-$
biphenylene	$-C_6H_4-C_6H_4-$
butoxy	C_4H_9O-
sec-butoxy (1-methylpropoxy)	$C_2H_5CH(CH_3)O-$
tert-butoxy (1,1-dimethylethoxy)	$(CH_3)_3CO-$
butyl	$CH_3(CH_2)_3-$
sec-butyl (1-methylpropyl)	$CH_3CH_2CH(CH_3)-$
tert-butyl (1,1-dimethylethyl)	$(CH_3)_3C-$
butyryl (1-oxobutyl)	$CH_3(CH_2)_2CO-$
caproyl* (hexanoyl)	$CH_3(CH_2)_4CO-$
capryl* (decanoyl)	$CH_3(CH_2)_8CO-$
capryloyl* (octanoyl)	$CH_3(CH_2)_6CO-$
carbamido (carbamoylamino)	$H_2NCONH-$
carbamoyl (aminocarbonyl)	H_2NCO-
carbamyl (aminocarbonyl)	H_2NCO-
carbazoyl (hydrazinocarbonyl)	$H_2NNHCO-$
carbethoxy (ethoxycarbonyl)	C_2H_5OCO-
carbonyl	$=C=O$
carboxy	$HOOC-$
cetyl* (hexadecyl)	$CH_3(CH_2)_{15}-$
chloroformyl (chlorcarbonyl)	$ClCO-$
cinnamoyl	$C_6H_5CH=CHCO-$
cinnamyl (3-phenyl-2-propenyl)	$C_6H_5CH=CHCH_2-$
cinnamylidene	$C_6H_5CH=CHCH=$
cresyl* (hydroxymethylphenyl)	$HO(CH_3)C_6H_4-$
crotonoyl	$CH_3CH=CHCO-$
crotyl (2-butenyl)	$CH_3CH=CHCH_2-$
cyanamido (cyanoamino)	$NCNH-$
cyanato	$NCO-$
cyano	$NC-$
decanedioyl	$-OC(CH_2)_8CO-$
decanoyl	$CH_3(CH_2)_8CO-$
diazo	$N_2=$
diazoamino	$-NHN=N-$
disilanyl	H_3SiSiH_2-
disiloxanyloxy	$H_3SiOSiH_2O-$
disulfinyl	$-S(O)S(O)-$
dithio	$-SS-$
enanthoyl* (heptanoyl)	$CH_3(CH_2)_5CO-$
epoxy	$-O-$
ethenyl (vinyl)	$CH_2=CH-$
ethynyl	$HC≡C-$
ethoxy	C_2H_5O-
ethyl	CH_3CH_2-
ethylene	$-CH_2CH_2-$
ethylidene	$CH_3CH=$
ethylthio	C_2H_5S-
formamido (formylamino)	$HCONH-$
formyl	$HCO-$
furmaroyl (from fumaric acid)	$-OCCH=CHCO-$
furfuryl (2-furanylmethyl)	$OC_4H_3CH_2-$
furfurylidene (2-furanylmethylene)	$OC_4H_3CH=$
glutamoyl (from glutamic acid)	$-OC(CH_2)_2CH(NH_2)CO-$
glutaryl (from glutaric acid)	$-OC(CH_2)_3CO-$
glycylamino	H_2NCH_2CONH-
glycoloyl; glycolyl (hydroxyacetyl)	$HOCH_2CO-$
glycyl (aminoacetyl)	H_2NCH_2CO-
glyoxyloyl; glyoxylyl (oxoacetyl)	$HCOCO-$
guanidino	$H_2NC(=NH)NH-$
guanyl (aminoiminomethyl)	$H_2NC(=NH)-$
heptadecanoyl	$CH_3(CH_2)_{15}CO-$
heptanamido	$CH_3(CH_2)_5CONH-$
heptanedioyl	$-OC(CH_2)_5CO-$
heptanoyl	$CH_3(CH_2)_5CO-$
hexadecanoyl	$CH_3(CH_2)_{14}CO-$
hexamethylene (1,6-hexanediyl)	$-(CH_2)_6-$
hexanedioyl	$-OC(CH_2)_4CO-$
hippuryl (N-benzoylglycyl)	$C_6H_5CONHCH_2CO-$
hydrazino	H_2NNH-
hydrazo	$-HNNH-$
hydrocinnamoyl	$C_6H_5(CH_2)_2CO-$
hydroperoxy	$HOO-$
hydroxyamino	$HONH-$
hydroxy	$HO-$
imino	$HN=$
iodoso* (iodosyl)	$OI-$
iodyl	O_2I-
isoamyl* (isopentyl; 3-methylbutyl)	$(CH_3)_2CH(CH_2)_2-$
isobutenyl (2-methyl-1-propenyl)	$(CH_3)_2C=CH-$
isobutoxy (2-methylpropoxy)	$(CH_3)_2CHCH_2O-$
isobutyl (2-methylpropyl)	$(CH_3)_2CHCH_2-$
isobutylidene (3-methylpropylidene)	$(CH_3)_2CHCH=$
isobutyryl (2-methyl-1-oxopropyl)	$(CH_3)_2CHCO-$

isocyanato	OCN-
isocyano	CN-
isohexyl (4-methylpentyl)	$(CH_3)_2CH(CH_2)_3-$
isoleucyl (from isoleucine)	$C_2H_5CH(CH_3)CH(NH_2)CO-$
isonitroso* (hydroxyamino)	HON=
isopentyl (3-methylbutyl)	$(CH_3)_2CH(CH_2)_2-$
isopentylidene (3-methylbutylidene)	$(CH_3)_2CHCH_2CH=$
isopropenyl (1-methylethenyl)	$CH_2=C(CH_3)-$
isopropoxy (1-methylethoxy)	$(CH_3)_2CHO-$
isopropyl (1-methylethyl)	$(CH_3)_2CH-$
isopropylidene (1-methylethylidene)	$(CH_3)_2C=$
isothiocyanato (isothiocyano)	SCN-
isovaleryl* (3-methyl-1-oxobutyl)	$(CH_3)_2CHCH_2CO-$
lactoyl (from lactic acid)	$CH_3CH(OH)CO-$
lauroyl (from lauric acid)	$CH_3(CH_2)_{10}CO-$
lauryl (dodecyl)	$CH_3(CH_2)_{11}-$
leucyl (from leucine)	$(CH_3)_2CHCH_2CH(NH_2)CO-$
levulinoyl (from levulinic acid)	$CH_3CO(CH_2)_2CO-$
malonyl (from malonic acid)	$-OCCH_2CO-$
mandeloyl (from mandelic acid)	$C_6H_5CH(OH)CO-$
mercapto	HS-
mesityl	$2,4,6-(CH_3)_3C_6H_2-$
methacryloyl (from methacrylic acid)	$CH_2=C(CH_3)CO-$
methallyl (2-methyl-2-propenyl)	$CH_2=C(CH_3)CH_2-$
methionyl (from methionine)	$CH_3SCH_2CH_2CH(NH_2)CO-$
methoxy	CH_3O-
methyl	H_3C-
methylene	$H_2C=$
methylthio	CH_3S-
myristoyl (from myristic acid)	$CH_3(CH_2)_{12}CO-$
myristyl (tetradecyl)	$CH_3(CH_2)_{13}-$
naphthyl	$(C_{10}H_7)-$
naphthylene	$-(C_{10}H_6)-$
neopentyl (2,2-dimethylpropyl)	$(CH_3)_3CCH_2-$
nitramino (nitroamino)	O_2NNH-
nitro	O_2N-
nitrosamino (nitrosoamino)	ONNH-
nitrosimino (nitrosoimino)	ONN=
nitroso	ON-
nonanoyl (from nonanoic acid)	$CH_3(CH_2)_7CO-$
oleoyl (from oleic acid)	$CH_3(CH_2)_7CH=CH(CH_2)_7CO-$
oxalyl (from oxalic acid)	-OCCO-
oxo	O=
palmitoyl (from palmitic acid)	$CH_3(CH_2)_{14}CO-$
pentamethylene (1,5-pentanediyl)	$-(CH_2)_5-$
pentyl	$CH_3(CH_2)_4-$
tert-pentyl	$CH_3CH_2C(CH_3)_2-$
phenacyl	$C_6H_5COCH_2-$
phenacylidene	$C_6H_5COCH=$
phenethyl (2-phenylethyl)	$C_6H_5CH_2CH_2-$
phenoxy	C_6H_5O-
phenyl	C_6H_5-
phenylene (benzenediyl)	$-C_6H_4-$
phosphino* (phosphanyl)	H_2P-
phosphinyl* (phosphinoyl)	$H_2P(O)-$
phospho	O_2P-
phosphono	$(HO)_2P(O)-$
phthaloyl (from phthalic acid)	$1,2-C_6H_4(CO-)_2$

picryl (2,4,6-trinitrophenyl)	$2,4,6-(NO_2)_3C_6H_2-$
pimeloyl (from pimelic acid)	$-OC(CH_2)_5CO-$
piperidino (1-piperidinyl)	$C_5H_{10}N-$
pivaloyl (from pivalic acid)	$(CH_3)_3CCO-$
prenyl (3-methyl-2-butenyl)	$(CH_3)_2C=CHCH_2-$
propargyl (2-propynyl)	$HC'CCH_2-$
1-propenyl	$-CH=CHCH_2$
2-propenyl (allyl)	$CH_2=CHCH_2-$
propionyl* (propanyl)	CH_3CH_2CO-
propoxy	$CH_3CH_2CH_2O-$
propyl	$CH_3CH_2CH_2-$
propylidene	$CH_3CH_2CH=$
pyrryl (pyrrolyl)	C_3H_4N-
salicyloyl (2-hydroxybenzoyl)	$2-HOC_6H_4CO-$
selenyl* (selanyl; hydroseleno)	HSe-
seryl (from serine)	$HOCH_2CH(NH_2)CO-$
siloxy	H_3SiO-
silyl	H_3Si-
silylene	$H_2Si=$
sorboyl (from sorbic acid)	$CH_3CH=CHCH=CHCO-$
stearoyl (from stearic acid)	$CH_3(CH_2)_{14}CO-$
stearyl (octadecyl)	$CH_3(CH_2)_{17}-$
styryl (2-phenylethenyl)	$C_6H_5CH=CH-$
suberoyl (from suberic acid)	$-OC(CH_2)_6CO-$
succinyl (from succinic acid)	$-OCCH_2CH_2CO-$
sulfamino (sulfoamino)	$HOSO_2NH-$
sulfamoyl (sulfamyl)	H_2NSO_2-
sulfanilyl [(4-aminophenyl)sulfonyl]	$4-H_2NC_6H_4SO_2-$
sulfeno	HOS-
sulfhydryl (mercapto)	HS-
sulfinyl	OS=
sulfo	HO_3S-
sulfonyl (sulfuryl)	$-SO_2-$
terephthaloyl	$1,4-C_6H_4(CO-)_2$
tetramethylene	$-(CH_2)_4-$
thienyl (from thiophene)	$(C_4H_3S)-$
thiocarbonyl (carbothionyl)	=CS
thiocarboxy	HOSC-
thiocyanato (thiocyano)	NCS-
thionyl* (sulfinyl)	-SO-
threonyl (from threonine)	$CH_3CH(OH)CH(NH_2)CO-$
toluidino [(methylphenyl)amino]	$CH_3C_6H_4NH-$
toluoyl (methylbenzoyl)	$CH_3C_6H_4CO-$
tolyl (methylphenyl)	$CH_3C_6H_4-$
α-tolyl (benzyl)	$C_6H_5CH_2-$
tolylene (methylphenylene)	$-(CH_3C_6H_3)-$
tosyl [(4-methylphenyl) sulfonyl)]	$4-CH_3C_6H_4SO_2-$
triazano	$H_2NNHNH-$
trimethylene (1,3-propanediyl)	$-(CH_2)_3-$
trityl (triphenylmethyl)	$(C_6H_5)_3C-$
valeryl* (pentanoyl)	$CH_3(CH_2)_3CO-$
valyl (from valine)	$(CH_3)_2CHCH(NH_2)CO-$
vinyl (ethenyl)	$CH_2=CH-$
vinylidene (ethenylidene)	$CH_2=C=$
xylidino [(dimethylphenyl)amino]	$(CH_3)_2C_6H_3NH-$
xylyl (dimethylphenyl)	$(CH_3)_2C_6H_3-$
xylylene [phenelenebis(methylene)]	$-CH_2C_6H_4CH_2-$

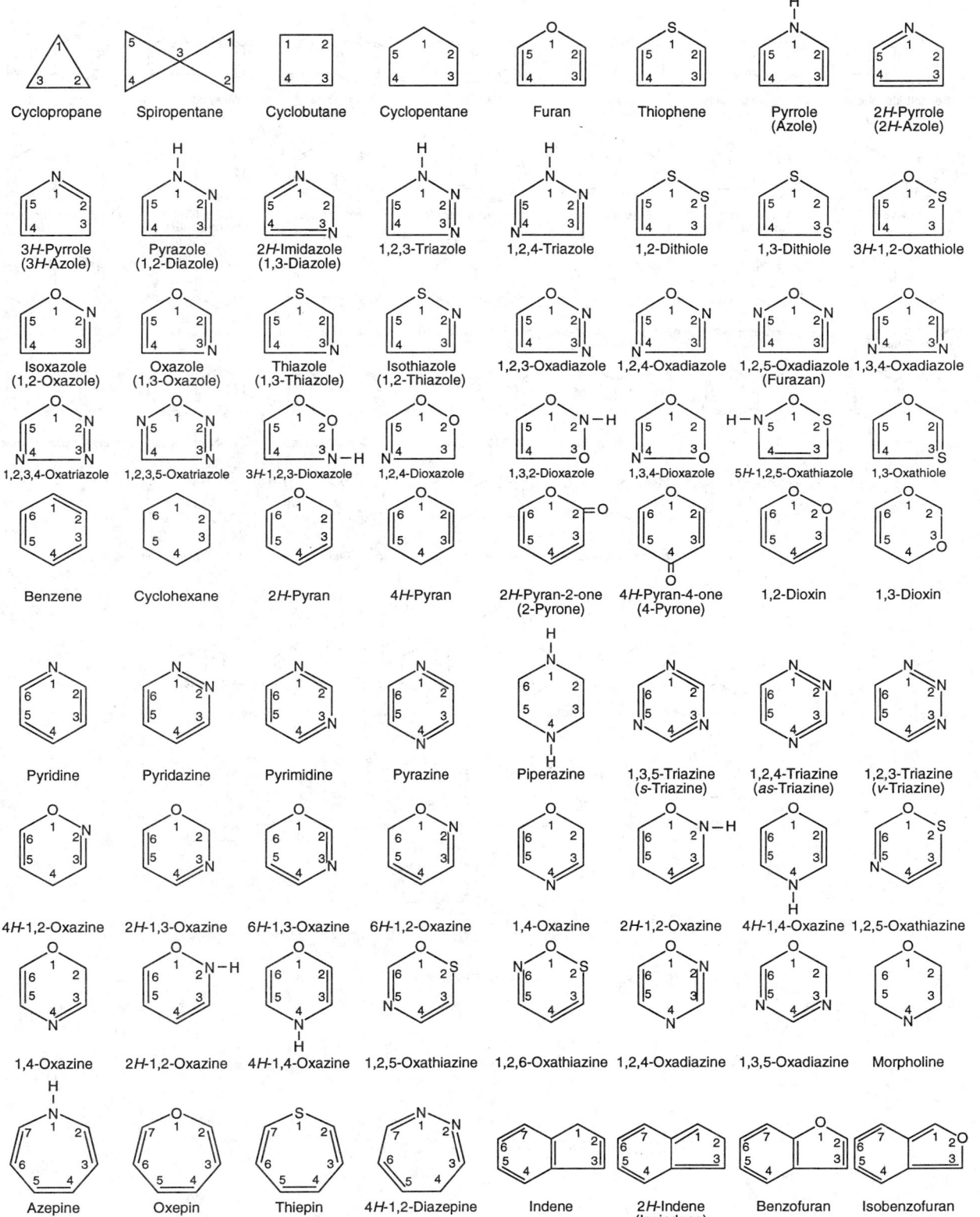

Cyclopropane   Spiropentane   Cyclobutane   Cyclopentane   Furan   Thiophene   Pyrrole (Azole)   2H-Pyrrole (2H-Azole)

3H-Pyrrole (3H-Azole)   Pyrazole (1,2-Diazole)   2H-Imidazole (1,3-Diazole)   1,2,3-Triazole   1,2,4-Triazole   1,2-Dithiole   1,3-Dithiole   3H-1,2-Oxathiole

Isoxazole (1,2-Oxazole)   Oxazole (1,3-Oxazole)   Thiazole (1,3-Thiazole)   Isothiazole (1,2-Thiazole)   1,2,3-Oxadiazole   1,2,4-Oxadiazole   1,2,5-Oxadiazole (Furazan)   1,3,4-Oxadiazole

1,2,3,4-Oxatriazole   1,2,3,5-Oxatriazole   3H-1,2,3-Dioxazole   1,2,4-Dioxazole   1,3,2-Dioxazole   1,3,4-Dioxazole   5H-1,2,5-Oxathiazole   1,3-Oxathiole

Benzene   Cyclohexane   2H-Pyran   4H-Pyran   2H-Pyran-2-one (2-Pyrone)   4H-Pyran-4-one (4-Pyrone)   1,2-Dioxin   1,3-Dioxin

Pyridine   Pyridazine   Pyrimidine   Pyrazine   Piperazine   1,3,5-Triazine (s-Triazine)   1,2,4-Triazine (as-Triazine)   1,2,3-Triazine (v-Triazine)

4H-1,2-Oxazine   2H-1,3-Oxazine   6H-1,3-Oxazine   6H-1,2-Oxazine   1,4-Oxazine   2H-1,2-Oxazine   4H-1,4-Oxazine   1,2,5-Oxathiazine

1,4-Oxazine   2H-1,2-Oxazine   4H-1,4-Oxazine   1,2,5-Oxathiazine   1,2,6-Oxathiazine   1,2,4-Oxadiazine   1,3,5-Oxadiazine   Morpholine

Azepine   Oxepin   Thiepin   4H-1,2-Diazepine   Indene   2H-Indene (Isoindene)   Benzofuran   Isobenzofuran

Benzo[b]thiophene

Benzo[c]thiophene

Indole

3H-Indole

1H-Indole

Cyclopenta[b]pyridine

Pyrano[3,4-b]-pyrrole

Indazole

Benzisoxazole (Indoxazine)

Benzoxazole

2,1-Benzisoxazole

Naphthalene

1,2,3,4-Tetra-hydronaphthalene (Tetralin)

Octahydronaphthalene (Decalin)

2H-1-Benzopyran (2H-Chromene)

2H-1-Benzopyran-2-one (Coumarin)

4H-1-Benzopyran-4-one (Chromen-4-one)

1H-2-Benzopyran-1-one (Isocoumarin)

3H-2-Benzopyran-1-one (Isochromen-3-one)

Quinoline

Isoquinoline

Cinnoline

Quinazoline

1,8-Naphtyridine

1,7-Naphtyridine

1,5-Naphtyridine

1,6-Naphtyridine

2H-1,3-Benzoxazine

2H-1,4-Benzoxazine

1H-2,3-Benzoxazine

4H-3,1-Benzoxazine

2H-1,2-Benzoxazine

4H-1,4-Benzoxazine

Anthracene

Phenanthrene

Phenalene

Fluorene

Carbazole

Xanthene

Acridine

Norpinane (Bicyclo[3.1.1]heptane)

7H-Purine

Steroid ring system

R = Nearly always methyl
R' = Usually methyl
R'' = Various groups

SCIENTIFIC ABBREVIATIONS AND SYMBOLS

This table lists some symbols, abbreviations, and acronyms encountered in the physical sciences. Most entries in italic type are symbols for physical quantities; for more details on these, see the table "Symbols and Terminology for Physical and Chemical Quantities" in this section. Additional information on units may be found in the table "International System of Units" in Section 1. Many of the terms to which these abbreviations refer are included in the tables "Definitions of Scientific Terms" in Section 2 and "Techniques for Materials Characterization" in Section 12.

Publication practices vary with regard to the use of capital or lower case letters for many abbreviations. An effort has been made to follow the most common practices in this table, but much variation is found in the literature. Likewise, policies on the use of periods in an abbreviation vary considerably. Periods are generally omitted in this table unless they are necessary for clarity. Periods should never appear in SI units. The SI prefixes (m, k ,M, etc.) are not listed here, since they should never be used alone, but selected combinations with SI units (e.g., mg, kV, MW) are included.

Abbreviations are listed in alphabetical order without regard to case. Entries beginning with Greek letters fall at the end of the table.

a	absorption coefficient, acceleration, activity	$as, asym$	asymmetrical (as chemical descriptor)
a_0	Bohr radius	ASCII	American National Standard Code for Information Interchange
A	ampere, adenine (in genetic code)		
Å	ångstrom	ASE	aromatic stabilization energy
A	absorbance, area, Helmholtz energy, mass number	Asn	asparagine
		Asp	aspartic acid
A_H	Hall coefficient	at	atomization
A_r	atomic weight (relative atomic mass)	atm	standard atmosphere
AAS	atomic absorption spectroscopy	ATP	adenosine 5'-triphosphate
Abe	abequose	ATR	attenuated total internal reflection
abs	absolute	at.wt.	atomic weight
ac	alternating current	AU	astronomical unit
Ac	acetyl	av	average
AcOH	acetic acid	avdp	avoirdupois
ACT	activated complex theory	b	barn
ACTH	adrenocorticotropic hormone	B	magnetic flux density, second virial coefficient, susceptance
Ade	adenine		
ADP	adenosine diphosphate	bar	bar (pressure unit)
ads	adsorption	bbl	barrel
ae	eon (10^9 years)	bcc	body centered cubic
AES	atomic emission spectroscopy, Auger electron spectroscopy	BCS	Bardeen-Cooper-Schrieffer (theory)
		BDE	bond dissociation energy
AF	audio frequency	Bé	Baumé
AFM	atomic force microscopy	BET	Brunauer-Emmett-Teller (method)
AI	artificial intelligence	BeV	billion electronvolt
AIM	atoms in molecules	Bhn	Brinell hardness number
Al	Alfen number	Bi	biot
Ala	alanine	BN	bond number
alc	alcohol	BNS	nuclear backscattering spectroscopy
aliph.	aliphatic	BO	bond order, Born-Oppenheimer (approximation)
alk.	alkaline	BOD	biochemical oxygen demand
All	allose	bp	boiling point
Alt	altrose	bpy	2,2'-bipyridine
am	amorphous solid	Bq	becquerel
Am	amyl	BRE	bond resonance energy
AM	amplitude modulation	BSSE	basis set superposition error
AMP	adenosine 5'-monophosphate	Btu	British thermal unit
amu	atomic mass unit (recommended symbol is u)	bu	bushel
anh, anhyd	anhydrous	Bu	butyl
antilog	antilogarithm	Bz	benzoyl
AO	atomic orbital	Bzl	benzyl
AOM	angular overlap model	c	combustion reaction
Api	apiose	c	amount concentration, specific heat, velocity
APS	appearance potential spectroscopy	c_0	speed of light in vacuum
APW	augmented plane wave	C	coulomb, cytosine (in genetic code)
aq	aqueous	°C	degree Celsius
Ar	aryl	C	capacitance, heat capacity, number concentration
Ara	arabinose	ca.	approximately
Ara-ol	arabinitol	cal	calorie
Arg	arginine		

calc	calculated		cwt	hundredweight (112 pounds)
CARS	coherent anti-Stokes Raman spectroscopy		Cy	cyclohexyl
CAS RN	Chemical Abstracts Service Registry Number		cyl	cylinder
CAT	clear-air turbulence, computerized axial tomography		Cys	cysteine
CBS	complete basis set		d	day, deuteron
cc	cubic centimeter		d	distance, density, dextrorotatory
CCD	charge-coupled device		D	debye unit
cd	candela, condensed phase		D	diffusion coefficient, dissociation energy, electric displacement
c.d.	current density			
CD	circular dichroism		Da	dalton
CDP	cytidine 5'-diphosphate		DA	donor-acceptor (complex)
CEPA	couplet electron pair approximation		dB	decibel
cf.	compare		dc	direct current
cfm	cubic feet per minute		DE	delocalization energy
cgs	centimeter-gram-second system		dec	decomposes
CHF	coupled Hartree-Fock (method)		deg	degree
Ci	curie		den	density
CI	configuration interaction, chemical ionization		det	determinant
CIDEP	chemically induced dynamic electron polarization		dev	deviation
CIDNP	chemically induced dynamic nuclear polarization		DFT	density functional theory
cir	circular		diam	diameter
CKFF	Cotton-Kraihanzel force field		dil	dilute, dilution
CL	cathode luminescence		DIM	diatomics in molecules
cm	centimeter		dm	decimeter
c.m.	center of mass		dmf, DMF	N,N-dimethylformamide
c.m.c.	critical micelle concentration		dmso, DMSO	dimethylsulfoxide
CMO	canonical molecular orbital		DNA	deoxyribonucleic acid
CMP	cytidine 5'-monophosphate		DNase	deoxyribonuclease
CN	coordination number		DNMR	dynamic nuclear magnetic resonance
CNDO	complete neglect of differential overlap		DOS	density of states
Co	Cowling number		doz	dozen
COD	chemical oxygen demand		d.p.	degree of polymerization
conc	concentrated, concentration		dpl	displacement
const	constant		dpm	disintegrations per minute
cos	cosine		dps	disintegrations per second
cosh	hyperbolic cosine		dr	dram
COSY	correlation spectroscopy (in NMR)		dRib	2-deoxyribose
cot	cotangent		DRIFT	diffuse reflectance infrared Fourier transform
coth	hyperbolic cotangent		DRS	diffuse reflectance spectroscopy
cp	candle power		DSC	differential scanning calorimetry
cP	centipoise		DTA	differential thermal analysis
Cp	cyclopentadienyl		dyn	dyne
CP	chemically pure		e	electron, base of natural logarithms
CPA	coherent potential approximation		e	elementary charge, linear strain
cpd	contact potential difference		E	electric field strength, electromotive force, energy, modulus of elasticity, entgegen (*trans* configuration)
cps	cycles per second			
CPT	charge conjugation-space inversion-time reversal (theorem)		E_h	Hartree energy
			EA	electron affinity
CPU	central processing unit		EAN	effective atomic number
cr, cryst	crystalline (phase)		ECP	effective core potential
CRU	constitutional repeating unit		ECR	electron cyclotron resonance
csc	cosecant		ED	electron diffraction, effective dose
ct	carat		EDS	energy dispersive X-ray spectroscopy
CT	charge transfer		EDTA	ethylenediaminetetraacetic acid
CTEM	conventional transmission electron microscopy		EELS	electron energy loss spectroscopy
CTP	cytidine 5'-triphosphate		EFFF	energy factored force field
CTR	controlled thermonuclear reaction		EHMO	extended Hückel molecular orbital
cu	cubic		EHT	extended Hückel theory
CV	cyclic voltammetry		emf	electromotive force
CVD	chemical vapor deposition		emu	electromagnetic unit system
cw	continuous wave		en	ethylenediamine

ENDOR	electron-nuclear double resonance
EOS	equation of state
EPMA	electron probe microanalysis
EPR	electron paramagnetic (spin) resonance
eq, eqn	equation
eqQ	quadrupole coupling constant
erf	error function
erg	erg
ESCA	electron spectroscopy for chemical analysis
e.s.d.	estimated standard deviation
ESD	electron stimulated desorption
ESR	electron spin resonance
est	estimate, estimated
esu	electrostatic unit system
Et	ethyl
ET	electron transfer, ephemeris time
Et_2O	diethyl ether
e.u.	entropy unit
Eu	Euler number
eV	electronvolt
EWG	electron withdrawing group
EXAFS	extended x-ray absorption fine structure
EXELFS	extended energy loss fine structure
exp	exponential function
expt	experimental
ext	external
f	formation reaction
f	activity coefficient, aperture ratio, focal length, force constant, frequency, fugacity
F	farad
°F	degree Fahrenheit
F	Faraday constant, force, angular momentum
FAD	flavin adenine dinucleotide
fcc	face centered cubic
FEL	free electron laser
FEM	field emission microscopy
FEMO	free electron molecular orbital
FET	field effect transistor
fid	free induction decay
FIM	field ion microscopy
FIR	far infrared
fl	fluid (phase)
FM	frequency modulation
Fo	Fourier number
fp	freezing point
fpm	feet per minute
fps	feet per second, foot-pound-second system
Fr	franklin
Fr	Froude number
Fru	fructose
FSGO	floating spherical Gaussian orbital
ft	foot
ft-lb	foot pound
FT	Fourier transform
FTIR	Fourier transform infrared spectroscopy
Fuc	fucose
Fuc-ol	fucitol
fus	fusion (melting)
g	gram, gas
g	acceleration due to gravity, degeneracy, statistical weight, Landé g-factor

G	gauss, guanine (in genetic code)
G	electrical conductance, Gibbs energy, gravitational constant, sheer modulus
gal	gallon
Gal	gal, galileo, galactose
GalN	galactosamine
GC	gas chromatography
GC-MS	gas chromatography-mass spectrometry
GDMS	glow discharge mass spectroscopy
gem	geminal (on the same carbon atom)
GeV	gigaelectronvolt
GIAO	gauge invariant atomic orbital
gl	glacial
GLC	gas-liquid chromatography
Glc	glucose
GlcN	glucosamine
Glc-ol	glucitol
Gln	glutamine
Glu	glutamic acid
Gly	glycine
GMP	guanosine 5'-triphosphate
GMT	Greenwich mean time
gpm	gallons per minute
gps	gallons per second
gr	grain
Gr	Grashof number
GTO	gaussian type atomic orbital
Gua	guanine
Gul	gulose
GUT	grand unified theory
GVB	generalized valence bond
GWS	Glashow-Weinberg-Salam (theory)
Gy	gray, gigayear
h	helion, hour
h	Planck constant
H	henry
H	enthalpy, Hamiltonian function, magnetic field
H_0	Hubble constant
ha	hectare
Ha	Hartmann number
Hacac	acetylacetone
HAM	hydrogenic atoms in molecules
hav	haversine
Hb	hemoglobin
hcp	hexagonal closed packed
Hea	ethanolamine
HEIS	high energy ion scattering
HEP	high energy physics
HF	high frequency
hfs	hyperfine structure
Him	imidazole
His	histidine
HMO	Hückel molecular orbital
HOMO	highest occupied molecular orbital
hp	horsepower
HPLC	high-performance liquid chromatography
Hpz	pyrazole
hr	hour
HREELS	high resolution electron energy loss spectroscopy
HREM	high resolution electron microscopy
HSAB	hard-soft acid-base (theory)

HSE	homodesmotic stabilization energy
Hz	hertz
i	square root of minus one
I	electric current, ionic strength, moment of inertia, nuclear spin angular momentum, radiant intensity
IAT	international atomic time
i-Bu	isobutyl
IC	integrated circuit
ICP	inductively coupled plasma
ICR	ion cyclotron resonance
id	ideal (solution)
ID	inside diameter
Ido	idose
IDP	inosine 5'-diphosphate
IE	ionization energy
i.e.p.	isoelectric point
IEPA	independent electron pair approximation
IF	intermediate frequency
IGLO	individual gauge for localized orbitals
Ile	isoleucine
Im	imaginary part
imm	immersion
IMPATT	impact ionization avalanche transit time
in.	inch
INDO	intermediate neglect of differential overlap
INS	inelastic neutron scattering, ion neutralization spectroscopy
int	internal
I/O	input/output
IP	ionization potential
IPN	interpenetrating polymer network
i-Pr	isopropyl
IPR	isotopic perturbation of resonance
IPTS	International Practical Temperature Scale
IR	infrared
IRAS	reflection-absorption infrared spectroscopy
IRC	intrinsic reaction coordinate
isc	intersystem crossing
ISE	isodesmic stabilization energy
ISS	ion scattering spectroscopy
ITP	inosine 5'-triphosphate
ITS	International Temperature Scale (1990)
IU	international unit
j	angular momentum, electric current density
J	joule
J	angular momentum, electric current density, flux, Massieu function
k	absorption index, Boltzmann constant, rate constant, thermal conductivity, wave vector
K	kelvin
K	absorption coefficient, bulk modulus, equilibrium constant, kinetic energy
kb	kilobar, kilobase (DNA or RNA)
kcal	kilocalorie
KE	kinetic energy
keV	kiloelectronvolt
kg	kilogram
kgf	kilogram force
kJ	kilojoule
km	kilometer
Kn	Knudsen number

kPa	kilopascal
kt	karat
kV	kilovolt
kva	kilovolt ampere
kW	kilowatt
kwh	kilowatt hour
l	liquid, liter
l	angular momentum, length, levorotatory
L	liter, lambert
L	Avogadro constant, inductance, Lagrange function
lat.	latitude
lb	pound
lbf	pound force
lc	liquid crystal
LC	liquid chromatography
LCAO	linear combination of atomic orbitals
LD	lethal dose
Le	Lewis function
LED	light emitting diode
LEED	low-energy electron diffraction
LEIS	low energy ion scattering
Leu	leucine
LFER	linear free energy relationship
lim	limit
LIMS	laser ionization mass spectroscopy, laboratory information management system
liq	liquid
lm	lumen
ln	logarithm (natural)
LNDO	local neglect of differential overlap
log	logarithm (common)
long.	longitude
LST	local sidereal time
LT	local time
LTE	local thermodynamic equilibrium
LUMO	lowest unoccupied molecular orbital
lut	lutidine
lx	lux
ly	langley
l.y.	light year
Lys	lysine
Lyx	lyxose
m	meter, molal (as in 0.1 m solution), metastable (isotope)
m	magnetic dipole moment, mass, molality, angular momentum component, *meta* (as chemical descriptor)
M	molar (as in 0.1 M solution), metal (in chemical formulas)
M	magnetization, molar mass, mutual inductance, torque, angular momentum component
M_r	molecular weight (relative molar mass)
Ma	Mach number
Man	mannose
MASNMR	magic angle spinning nuclear magnetic resonance
max	maximum
MBE	molecular beam epitaxy
MBPT	many body perturbation theory
MC	Monte Carlo (method)
MCD	magnetic circular dichroism
MCPF	modified couple pair functional
MCSCF	multi-configurational self-consistent field

MD	molecular dynamics		NEXAFS	near-edge x-ray absorption fine structure
Me	methyl		ng	nanogram
MEP	molecular electrostatic potential		NIR	near infrared
MERP	minimum energy reaction path		nm	nanometer
Mes	mesityl		NMR	nuclear magnetic resonance
MESFET	metal-semiconductor field-effect transistor		NNDO	neglect of nonbonded differential overlap
Met	methionine		NO	natural orbital
meV	millielectronvolt		NOE	nuclear Overhauser effect
MeV	megaelectronvolt		NPA	natural population analysis
MF	molecular formula		NQR	nuclear quadrupole resonance
mg	milligram		NRA	prompt nuclear reaction analysis
MHD	magnetohydrodynamics		ns	nanosecond
mi	mile		NTP	normal temperature and pressure
MIM	molecules-in-molecules		Nu	nucleophile
min	minimum, minute		Nu	Nusselt number
MINDO	modified intermediate neglect of differential overlap		o	ortho (as chemical descriptor)
MIR	mid-infrared		obs, obsd	observed
misc	miscible		OD	optical density, outside diameter
MKS	meter-kilogram-second system		Oe	oersted
MKSA	meter-kilogram-second-ampere system		ORD	optical rotatory dispersion
mL, ml	milliliter		oz	ounce
mm	millimeter		p	proton
MM	molecular mechanics		p	dielectric polarization, electric dipole moment, momentum, pressure, para (as chemical descriptor)
mmf	magnetomotive force			
mmHg	millimeter of mercury		P	poise
MO	molecular orbital		P	power, pressure, probability, sound energy flux
mol	mole		Pa	pascal
mol.wt.	molecular weight		PA	proton affinity
mon	monomeric form		PAS	photoacoustic spectroscopy
MOS	metal-oxide semiconductor		pc	parsec
MOSFET	metal-oxide semiconductor field-effect transistor		PCR	polymerase chain reaction
mp	melting point		PD	potential difference
MPa	megapascal		pdl	poundal
MPA	Mulliken population analysis		pe	probable error
Mpc	megaparsec		Pe	Péclet number
MRI	magnetic resonance imaging		PES	photoelectron spectroscopy
mRNA	messenger RNA		PET	positron emission tomography
ms	millisecond		peth	petroleum ether
MS	mass spectroscopy		pf	power factor
MSL	mean sea level		pg	picogram
Mur	muramic acid		pH	negative log of hydrogen ion concentration
mV	millivolt		Ph	phenyl
mW	milliwatt		Phe	phenylalanine
MW	megawatt, microwave, molecular weight		pI	isoelectric point
Mx	maxwell		pip	piperidine
n	neutron		pK	negative log of ionization constant
n	amount of substance, number density, principal quantum number, refractive index, normal (in chemical formulas)		pm	picometer
			PMO	perturbational molecular orbital
			PNDO	partial neglect of differential overlap
N	newton		PNRA	prompt nuclear reaction analysis
N	angular momentum, neutron number		pol	polymeric form
N_A	Avogadro constant		ppb	parts per billion
N_E	density of states		ppm	parts per million
NAA	neutron activation analysis		PPP	Pariser-Parr-Pople (method)
NAD	nicotinamide adenine dinucleotide		ppt	parts per thousand, precipitate
NADH	reduced NAD		Pr	propyl
NADP	nicotinamide adenine dinucleotide phosphate		Pr	Prandtl number
NAO	natural atomic orbital		PRDDO	partial retention of diatomic differential overlap
NBO	natural bond order		Pro	proline
nbp	normal boiling point		ps	picosecond
Neu	neuraminic acid		PS	photoelectron spectroscopy

PSD	photon stimulated desorption
Psi	psicose
psi	pounds per square inch
psia	pounds per square inch absolute
psig	pounds per square inch gage
pt	pint
PVT	pressure-volume-temperature
py	pyridine
q	electric field gradient, flow rate, heat, wave vector (phonons)
Q	electric charge, heat, partition function, quadrupole moment, radiant energy, vibrational normal coordinate
QCD	quantum chromodynamics
QED	quantum electrodynamics
Q.E.D.	quod erat demonstrandum (which was to be proved)
QSAR	quantitative structure-activity relationship
QSO	quasi-stellar object (quasar)
qt	quart
quad	quadrillion Btu (= 1.05510^{18} J)
Qui	quinovose
q.v.	quod vide (which you should see)
r	reaction
r	position vector, radius
R	roentgen, alkyl radical (in chemical formulas)
°R	degree Rankine
R	electrical resistance, gas constant, molar refraction, Rydberg constant
RA	right ascension
rad	radian
RAIRS	reflection-absorption infrared spectroscopy
RAM	random access memory
RBS	Rutherford backscattering spectroscopy
RE	resonance energy
Re	real part
RED	radial electron distribution
REM	reflection electron microscopy
rem	roentgen equivalent man
RF	radiofrequency
Rha	rhamnose
RHEED	reflection high-energy electron diffraction
RHF	restricted Hartree-Fock (theory)
RIA	radioimmunoassay
Rib	ribose
Ribulo	ribulose
rms	root mean square
RNA	ribonucleic acid
RNase	ribonuclease
rRNA	ribosomal RNA
ROHF	restricted open shell Hartree-Fock
ROM	read only memory
RPA	random phase approximation
rpm	revolutions per minute
rps	revolutions per second
RRK	Rice-Ramsperger-Kassel (theory)
RRKM	Rice-Ramsperger-Kassel-Marcus (theory)
RRS	resonance Raman spectroscopy
RS	Raman spectroscopy
Ry	rydberg
s	second, solid
s	path length, solubility, spin angular momentum, symmetry number, symmetrical (as stereochemical descriptor)

S	siemens
S	area, entropy, probability current density, Poynting vector, symmetry coordinate, spin angular momentum
SALC	symmetry adapted linear combinations
SALI	surface analysis by laser ionization
SAM	scanning Auger microscopy
SANS	small angle neutron scattering
Sar	sarcosine
sat, satd	saturated
SAXS	small angle x-ray scattering
s-Bu	sec-butyl
Sc	Schmidt number
SCE	saturated calomel electrode
SCF	self-consistent field
SCR	silicon-controlled rectifier
sd	standard deviation
sec	secant, second
sec	secondary (in chemical name)
SEELFS	surface sensitive energy loss fine structure
SEM	scanning electron microscope
sepn	separation
Ser	serine
SERS	surface-enhanced Raman spectroscopy
SET	single electron transfer
SEXAFS	surface extended x-ray absorption fine structure
Sh	Sherwood number
SI	International System of Units
SIMS	secondary ion mass spectroscopy
sin	sine
SINDO	symmetrically orthogonalized INDO method
sinh	hyperbolic sine
SIPN	semi-interpenetrating polymer network
SLAM	scanning laser acoustic microscopy
sln	solution
SMO	semiempirical molecular orbital
SMOW	Standard Mean Ocean Water
SNMS	sputtered neutral mass spectroscopy
SNU	solar neutrino unit
SO	spin orbital
sol	soluble, solution
soln	solution
SOMO	singly occupied molecular orbital
Sor	sorbose
sp gr	specific gravity
SPM	scanned probe microscopy
sq	square
sr	steradian
Sr	Strouhal number
SSMS	spark source mass spectroscopy
St	stoke
St	Stanton number
std, stnd	standard (state)
STEM	scanning transmission electron microscope
STM	scanning tunneling microscopy
STO	Slater type orbital
STP	standard temperature and pressure
sub	sublimation, sublimes
Sv	sievert
t	metric tonne, triton
t	Celsius temperature, thickness, time, transport number
T	tesla

T	kinetic energy, period, term value, temperature (thermodynamic), torque, transmittance
Tag	tagatose
Tal	talose
tan	tangent
tanh	hyperbolic tangent
t-Bu	$tert$-butyl
TCA	trichloroacetic acid
TCE	trichloroethylene
tcne	tetracyanoethylene
TCSCF	two configuration self-consistent field
TE	transverse electric
TED	transmission electron diffraction, transferred electron device
TEM	transmission electron microscopy, transverse electromagnetic
temp	temperature
$tert$	tertiary (in chemical name)
TFD	Thomas-Fermi-Dirac (method)
TGA	thermo-gravimetric analysis
theor	theoretical
thf, THF	tetrahydrofuran
Thr	threonine
Thy	thymine
TL	thermoluminescence
TLC	thin-layer chromatography
TM	transverse magnetic
Tol	tolyl
Torr	torr
tRNA	transfer RNA
Trp	tryptophan
trs	transition
TS	transition state
tsp	teaspoon
Tyr	tyrosine
u	unified atomic mass unit
u	Bloch function, electric mobility, velocity
U	uracil (in genetic code)
U	electric potential difference, internal energy
UDP	uridine 5'-diphosphate
UHF	ultrahigh frequency, unrestricted Hartree-Fock (theory)
UMP	uridine 5'-monophosphate
$uns, unsym$	unsymmetrical (as chemical descriptor)
UPES	ultraviolet photoelectron spectroscopy
UPS	ultraviolet photoelectron spectroscopy
ur	urea
Ura	uracil
USP	United States Pharmacopeia
UT	universal time
UTP	uridine 5'-triphosphate
UV	ultraviolet
v	reaction rate, specific volume, velocity, vibrational quantum number, vicinal (as chemical descriptor)
V	volt
V	electric potential, potential energy, volume
Val	valine
vap	vaporization
VB	valence band, valence bond
VCD	vibrational circular dichroism
VHF	very high frequency

vic	vicinal (on adjacent carbon atoms)
VIS	visible region of the spectrum
vit	vitreous
VSEPR	valence shell electron pair repulsion
VSLI	very large scale integrated (circuit)
VUV	vacuum ultraviolet
v/v	volume per volume (volume of solute divided by volume of solution, expressed as percent)
w	energy density, mass fraction, velocity, work
W	watt
W	radiant energy, statistical weight, work
WAXS	wide angle x-ray scattering
Wb	weber
We	Weber number
WKB	Wentzel-Kramers-Brillouin (method)
wt	weight
w/v	weight per volume (mass of solute divided by volume of solution, generally expressed as g/100 mL)
w/w	weight per weight (mass of solute divided by mass of solution, expressed as percent)
x	mole fraction
X	X unit, halogen (in chemical formula)
X	reactance
XAFS	x-ray absorption fine structure
XANES	x-ray absorption near-edge structure
XPES	x-ray photoelectron spectroscopy
XPS	x-ray photoelectron spectroscopy
XRD	x-ray diffraction
XRF	x-ray fluorescence
XRS	x-ray spectroscopy
Xyl	xylose
y, yr	year
Y	admittance, Planck function, Young's modulus
yd	yard
z	charge number (of an ion), collision frequency factor
Z	atomic number, compression factor, collision number, impedance, partition function, zusammen (cis configuration)
ZDO	zero differential overlap
ZPE, ZPVE	zero point vibrational energy
ZULU	Greenwich mean time
α	alpha particle
α	absorption coefficient, degree of dissociation, electric polarizability, expansion coefficient, fine structure constant
β	beta particle
γ	photon
γ	activity coefficient, conductivity, magnetogyric ratio, mass concentration, ratio of heat capacities, surface tension
Γ	Gruneisin parameter, level width, surface concentration
δ	chemical shift, Dirac delta function, Kronecker delta, loss angle
Δ	inertia defect, mass defect
ε	emittance, Levi-Civita symbol, linear strain, molar absorption coefficient, permittivity
η	overpotential, viscosity
θ	Bragg angle, temperature, scattering angle, surface coverage
Θ	quadrupole moment

κ	compressibility, conductivity, magnetic susceptibility, molar absorption coefficient, transmission coefficient
λ	absolute activity, radioactive decay constant, thermal conductivity, wavelength
Λ	angular momentum, ionic conductivity
μ	muon
μ	chemical potential, electric dipole moment, electric mobility, friction coefficient, Joule-Thompson coefficient, magnetic dipole moment, mobility, permeability
μF	microfarad
μg	microgram
μm	micrometer
μs	microsecond
ν	frequency, kinematic velocity, stoichiometric number, wavenumber
ν_e	neutrino
π	pion
Π	osmotic pressure, Peltier coefficient
ρ	density, reflectance, resistivity
σ	electrical conductivity, cross section, normal stress, shielding constant (NMR), Stefan-Boltzmann constant, surface tension
τ	transmittance, chemical shift, shear stress, relaxation time
ϕ	electrical potential, fugacity coefficient, osmotic coefficient, quantum yield, volume fraction, wavefunction
Φ	magnetic flux, potential energy, radiant power, work function
χ	magnetic susceptibility, electronegativity
χ_e	electric susceptibility
ψ	wavefunction
ω	circular frequency, angular velocity, harmonic vibration wavenumber, statistical weight
Ω	ohm
Ω	axial angular momentum, solid angle

GREEK, RUSSIAN, AND HEBREW ALPHABETS

The following table presents the Hebrew, Greek, and Russian alphabets, their letters, the names of the letters, and the English equivalents.

HEBREW[1,3]				GREEK[4]			RUSSIAN		
א		aleph	' [2]	Α α	alpha	a	А а		a
ב		beth	b, bh	Β β	beta	b	Б б		b
							В в		v
ג		gimel	g, gh	Γ γ	gamma	g, n	Г г		g
ד		daleth	d, dh	Δ δ	delta	d	Д д		d
ה		he	h	Ε ε	epsilon	e	Е е		e
ו		waw	w	Ζ ζ	zeta	z	Ж ж		zh
ז		zayin	z	Η η	eta	ē	З з		z
ח		heth	ḥ	Θ θ	theta	th	И и Й й		i, ĭ
ט		teth	ṭ	Ι ι	iota	i	К к		k
י		yodh	y	Κ κ	kappa	k	Л л		l
כ	ך	kaph	k, kh	Λ λ	lambda	l	М м		m
ל		lamedh	l	Μ μ	mu	m	Н н		n
מ	ם	mem	m	Ξ ξ	xi	x	О о		o
נ	ן	nun	n	Ο ο	omicron	o	П п		p
ס		samekh	s	Π π	pi	p	Р р		r
ע		ayin	'	Ρ ρ	rho	r, rh	С с		s
פ	ף	pe	p, ph	Σ σ ς	sigma	s	Т т		t
צ	ץ	sadhe	ṣ	Τ τ	tau	t	У у		u
ק		qoph	q	Υ υ	upsilon	y, u	Ф ф		f
ר		resh	r	Φ φ	phi	ph	Х х		kh
שׂ		sin	ś	Χ χ	chi	ch	Ц ц		ts
שׁ		shin	sh	Ψ ψ	psi	ps	Ч ч		ch
ת		taw	t, th	Ω ω	omega	ō	Ш ш		sh
							Щ щ		shch
							Ъ ъ [5]		"
							Ы ы		y
							Ь ь [6]		'
							Э э		e
							Ю ю		yu
							Я я		ya

[1] Where two forms of a letter are given, the second one is the form used at the end of a word.

[2] Not represented in transliteration when initial.

[3] The Hebrew letters are primarily consonants; a few of them are also used secondarily to represent certain vowels, when provided at all, is by means of a system of dots or strokes adjacent to the consonated characters.

[4] The letter gamma is transliterated "n" only before velars; the letter upsilon is transliterated "u" only as the final element in diphthongs.

[5] This sign indicates that the immediately preceding consonant is not palatized even though immediately followed by a palatized vowel.

[6] This sign indicates that the immediately preceding consonant is palatized even though not immediately followed by a palatized vowel.

DEFINITIONS OF SCIENTIFIC TERMS

Brief definitions of selected terms of importance in chemistry, physics, and related fields of science are given in this section. The selection process emphasizes the following types of terms:
- Physical quantities
- Units of measure
- Classes of chemical compounds and materials
- Important theories, laws, and basic concepts.

Individual chemical compounds are not included.

Definitions have taken wherever possible from the recommendations of international or national bodies, especially the International Union of Pure and Applied Chemistry (IUPAC) and International Organization for Standardization (ISO). For physical quantities and units, the recommended symbol is also given. The source of such definitions is indicated by the reference number in brackets following the definition. In many cases these official definitions have been edited in the interest of stylistic consistency and economy of space. The user is referred to the original source for further details.

* An asterisk following a term indicates that further information can be found by consulting the index of this handbook under the entry for that term.

REFERENCES

1. *ISO Standards Handbook 2, Units of Measurement*, International Organization for Standardization, Geneva, 1992.
2. *Quantities, Units, and Symbols in Physical Chemistry, Second Edition*, International Union of Pure and Applied Chemistry, Blackwell Scientific Publications, Oxford, 1993.
3. *Compendium of Chemical Terminology*, International Union of Pure and Applied Chemistry, Blackwell Scientific Publications, Oxford, 1987.
4. *A Guide to IUPAC Nomenclature of Organic Compounds*, International Union of Pure and Applied Chemistry, Blackwell Scientific Publications, Oxford, 1993.
5. *Glossary of Class Names of Organic Compounds and Reactive Intermediates Based on Structure, Pure and Applied Chemistry*, 67, 1307, 1995.
6. *Compendium of Analytical Nomenclature*, International Union of Pure and Applied Chemistry, Blackwell Scientific Publications, Oxford, 1987.
7. *Nomenclature of Inorganic Chemistry*, International Union of Pure and Applied Chemistry, Blackwell Scientific Publications, Oxford, 1990.
8. *Glossary of Basic Terms in Polymer Science, Pure and Applied Chemistry*, 68, 2287, 1996.
9. *The International Temperature Scale of 1990, Metrologia*, 27, 107, 1990.
10. *Compilation of ASTM Standard Definitions*, American Society of Testing and Materials, Philadelphia, 1990.
11. *ASM Metals Reference Book*, American Society for Metals, Metals Park, OH, 1983.

Ab initio **method** - An approach to quantum-mechanical calculations on molecules which starts with the Schrödinger equation and carries out a complete integration, without introducing empirical factors derived from experimental measurement.

Absorbance (A) - Defined as $-\log(1-\alpha) = \log(1/\tau)$, where α is the absorptance and τ the transmittance of a medium through which a light beam passes. [2]

Absorbed dose (D) - For any ionizing radiation, the mean energy imparted to an element of irradiated matter divided by the mass of that element. [1]

Absorptance (α) - Ratio of the radiant or luminous flux in a given spectral interval absorbed in a medium to that of the incident radiation. Also called absorption factor. [1]

Absorption coefficient (a) - The relative decrease in the intensity of a collimated beam of electromagnetic radiation, as a result of absorption by a medium, during traversal of an infinitesimal layer of the medium, divided by the length traversed. [1]

Absorption coefficient, molar (ε) - Absorption coefficient divided by amount-of-substance concentration of the absorbing material in the sample solution $(\varepsilon = a/c)$. The SI unit is m^2/mol. Also called extinction coefficient, but usually in units of $mol^{-1}dm^3cm^{-1}$. [2]

Acceleration - Rate of change of velocity with respect to time.

Acceleration due to gravity (g)* - The standard value (9.80665 m/s^2) of the acceleration experienced by a body in the earth's gravitational field. [1]

Acenes - Polycyclic aromatic hydrocarbons consisting of fused benzene rings in a rectilinear arrangement. [5]

Acid - Historically, a substance that yields an H^+ ion when it dissociates in solution, resulting in a pH<7. In the Brönsted definition, an acid is a substance that donates a proton in any type of reaction. The most general definition, due to G.N. Lewis, classifies any chemical species capable of accepting an electron pair as an acid.

Acid dissociation constant (K_a)* - The equilibrium constant for the dissociation of an acid HA through the reaction $HA + H_2O \rightleftharpoons A^- + H_3O^+$. The quantity $pK_a = -\log K_a$ is often used to express the acid dissociation constant.

Actinides - The elements of atomic number 89 through 103, e.g., Ac, Th, Pa, U, Np, Pu, Am, Cm, Bk, Cf, Es, Fm, Md, No, Lr. [7]

Activation energy* - In general, the energy that must be added to a system in order for a process to occur, even though the process may already be thermodynamically possible. In chemical kinetics, the activation energy is the height of the potential barrier separating the products and reactants. It determines the temperature dependence of the reaction rate.

Activity - For a mixture of substances, the absolute activity λ of substance B is defined as $\lambda_B = \exp(\mu_B/RT)$, where μ_B is the chemical potential of substance B, R the gas constant, and T the thermodynamic temperature. The relative activity a is defined as $a_B = \exp[(\mu_B-\mu_B°)/RT]$, where $\mu_B°$ designates the chemical potential in the standard state. [2]

Activity coefficient (γ)* - Ratio of the activity a_B of component B of a mixture to the concentration of that component. The value of γ depends on the method of stating the composition. For mole fraction x_B, the relation is $a_B = \gamma_B x_B$; for molarity c_B, it is $a_B = \gamma_B c_B/c°$, where $c°$ is the standard state composition (typically chosen as 1 mol/L); for molality

m_B, it is $a_B = \gamma_B m_B/m°$, where $m°$ is the standard state molality (typically 1 mol/kg). [2]

Activity, of radioactive substance (A) - The average number of spontaneous nuclear transitions from a particular energy state occurring in an amount of a radionuclide in a small time interval divided by that interval. [1]

Acyl groups - Groups formed by removing the hydroxy groups from oxoacids that have the general structure $RC(=O)(OH)$ and replacement analogues of such acyl groups. [5]

Adiabatic process - A thermodynamic process in which no heat enters or leaves the system.

Admittance (Y) - Reciprocal of impedance. $Y = G + iB$, where G is conductance and B is susceptance. [1]

Adsorption - A process in which molecules of gas, of dissolved substances in liquids, or of liquids adhere in an extremely thin layer to surfaces of solid bodies with which they are in contact. [10]

Albedo* - The ratio of the light reflected or scattered from a surface to the intensity of incident light. The term is often used in reference to specific types of terrain or to entire planets.

Alcohols - Compounds in which a hydroxy group, -OH, is attached to a saturated carbon atom. [5]

Aldehydes - Compounds $RC(=O)H$, in which a carbonyl group is bonded to one hydrogen atom and to one R group. [5]

Aldoses - Aldehydic parent sugars (polyhydroxyaldehydes $H[CH(OH)]_nC(=O)H$, $n>1$) and their intramolecular hemiacetals. [5]

Aldoximes - Oximes of aldehydes: RCH=NOH. [5]

Alfvén number (Al) - A dimensionless quantity used in plasma physics, defined by $Al = v(\rho\mu)^{1/2}/B$, where ρ is density, v is velocity, μ is permeability, and B is magnetic flux density. [2]

Alfven waves - Very low frequency waves which can exist in a plasma in the presence of a uniform magnetic field. Also called magnetohydrodynamic waves.

Alicyclic compounds - Aliphatic compounds having a carbocyclic ring structure which may be saturated or unsaturated, but may not be a benzenoid or other aromatic system. [5]

Aliphatic compounds - Acyclic or cyclic, saturated or unsaturated carbon compounds, excluding aromatic compounds. [5]

Alkali metals - The elements lithium, sodium, potassium, rubidium, cesium, and francium.

Alkaline earth metals - The elements calcium, strontium, barium, and radium. [7]

Alkaloids - Basic nitrogen compounds (mostly heterocyclic) occurring mostly in the plant kingdom (but not excluding those of animal origin). Amino acids, peptides, proteins, nucleotides, nucleic acids, and amino sugars are not normally regarded as alkaloids. [5]

Alkanes - Acyclic branched or unbranched hydrocarbons having the general formula C_nH_{2n+2}, and therefore consisting entirely of hydrogen atoms and saturated carbon atoms. [5]

Alkenes - Acyclic branched or unbranched hydrocarbons having one carbon-carbon double bond and the general formula C_nH_{2n}. Acyclic branched or unbranched hydrocarbons having more than one double bond are alkadienes, alkatrienes, etc. [5]

Alkoxides - Compounds, ROM, derivatives of alcohols, ROH, in which R is saturated at the site of its attachment to oxygen and M is a metal or other cationic species. [5]

Alkyl groups - Univalent groups derived from alkanes by removal of a hydrogen atom from any carbon atom: C_nH_{2n+1}-. The groups derived by removal of a hydrogen atom from a terminal carbon atom of unbranched alkanes form a subclass of normal alkyl (n-alkyl) groups. The groups RCH_2-, R_2CH-, and R_3C- (R not equal to H) are primary, secondary, and tertiary alkyl groups, respectively. [5]

Alkynes - Acyclic branched or unbranched hydrocarbons having a carbon-carbon triple bond and the general formula C_nH_{2n-2}, RC≡CR'.

Acyclic branched or unbranched hydrocarbons having more than one triple bond are known as alkadiynes, alkatriynes, etc. [5]

Allotropy - The occurrence of an element in two or more crystalline forms.

Allylic groups - The group $CH_2=CHCH_2$- (allyl) and derivatives formed by substitution. The term 'allylic position' or 'allylic site' refers to the saturated carbon atom. A group, such as -OH, attached at an allylic site is sometimes described as "allylic". [5]

Amagat volume unit - A non-SI unit previously used in high pressure science. It is defined as the molar volume of a real gas at one atmosphere pressure and 273.15 K. The approximate value is 22.4 L/mol.

Amides - Derivatives of oxoacids $RC(=O)(OH)$ in which the hydroxy group has been replaced by an amino or substituted amino group. [5]

Amine oxides - Compounds derived from tertiary amines by the attachment of one oxygen atom to the nitrogen atom: R_3N^+-O^-. By extension the term includes the analogous derivatives of primary and secondary amines. [5]

Amines - Compounds formally derived from ammonia by replacing one, two, or three hydrogen atoms by hydrocarbyl groups, and having the general structures RNH_2 (primary amines), R_2NH (secondary amines), R_3N (tertiary amines). [5]

Amino acids* - Compounds containing both a carboxylic acid group (-COOH) and an amino group (-NH_2). The most important are the α-amino acids, in which the -NH_2 group in attached to the C atom adjacent to the -COOH group. In the ß-amino acids, there is an intervening carbon atom. [4]

Ampere (A)* - The SI base unit of electric current. [1]

Ampere's law - The defining equation for the magnetic induction B, viz., $dF = Idl \times B$, where dF is the force produced by a current I flowing in an element of the conductor dl pointing in the direction of the current.

Ångström ($Å$) - A unit of length used in spectroscopy, crystallography, and molecular structure, equal to 10^{-10} m.

Angular momentum (L) - The angular momentum of a particle about a point is the vector product of the radius vector from this point to the particle and the momentum of the particle; i.e., $L = r \times p$. [1]

Angular velocity (ω) - The angle through which a body rotates per unit time.

Anilides - Compounds derived from oxoacids $RC(=O)(OH)$ by replacing the -OH group by the -NHPh group or derivative formed by ring substitution. Also used for salts formed by replacement of a nitrogen-bound hydrogen of aniline by a metal. [5]

Anion - A negatively charged atomic or molecular particle.

Antiferroelectricity* - An effect analogous to antiferromagnetism in which electric dipoles in a crystal are ordered in two sublattices that are polarized in opposite directions, leading to zero net polarization. The effect vanishes above a critical temperature.

Antiferromagnetism* - A type of magnetism in which the magnetic moments of atoms in a solid are ordered into two antiparallel aligned sublattices. Antiferromagnets are characterized by a zero or small positive magnetic susceptibility. The susceptibility increases with temperature up to a critical value, the Néel temperature, above which the material becomes paramagnetic.

Antiparticle - A particle having the same mass as a given elementary particle and a charge equal in magnitude but opposite in sign.

Appearance potential* - The lowest energy which must be imparted to the parent molecule to cause it to produce a particular specified parent ion. This energy, usually stated in eV, may be imparted by electron impact, photon impact, or in other ways. More properly called appearance energy. [3]

Appearance potential spectroscopy (APS) - See Techniques for Materials Characterization, page 12-1.

Are (a) - A unit of area equal to 100 m^2. [1]

Arenes - Monocyclic and polycyclic aromatic hydrocarbons. See aromatic compounds. [5]

Aromatic compounds - Compounds whose structure includes a cyclic delocalized π-electron system. Historical use of the term implies a ring containing only carbon (e.g., benzene, naphthalene), but it is often generalized to include heterocyclic structures such as pyridine and thiophene. [5]

Arrhenius equation - A key equation in chemical kinetics which expresses the rate constant k as $k = A\exp(-E_a/RT)$, where E_a is the activation energy, R the molar gas constant, and T the temperature. A is called the preexponential factor and, for simple gas phase reactions, may be identified with the collision frequency.

Arsines - AsH_3 and compounds derived from it by substituting one, two or three hydrogen atoms by hydrocarbyl groups. $RAsH_2$, R_2AsH, R_3As (R not equal to H) are called primary, secondary and tertiary arsines, respectively. [5]

Aryl groups - Groups derived from arenes by removal of a hydrogen atom from a ring carbon atom. Groups similarly derived from heteroarenes are sometimes subsumed in this definition. [5]

Astronomical unit (AU)* - The mean distance of the earth from the sun, equal to $1.49597870 \times 10^{11}$ m.

Atomic absorption spectroscopy (AAS) - See Techniques for Materials Characterization, page 12-1.

Atomic emission spectroscopy (AES) - See Techniques for Materials Characterization, page 12-1.

Atomic force microscopy (AFM) - See Techniques for Materials Characterization, page 12-1.

Atomic mass* - The mass of a nuclide, normally expressed in unified atomic mass units (u).

Atomic mass unit (u)* - A unit of mass used in atomic, molecular, and nuclear science, defined as the mass of one atom of ^{12}C divided by 12. Its approximate value is 1.66054×10^{-27} kg. Also called the unified atomic mass unit. [1]

Atomic number (Z) - A characteristic property of an element, equal to the number of protons in the nucleus.

Atomic weight (A_r)* - The ratio of the average mass per atom of an element to 1/12 of the mass of nuclide ^{12}C. An atomic weight can be defined for a sample of any given isotopic composition. The standard atomic weight refers to a sample of normal terrestrial isotopic composition. The term relative atomic mass is synonymous with atomic weight. [2]

Attenuated total reflection (ATR) - See Techniques for Materials Characterization, page 12-1.

Auger effect - An atomic process in which an electron from a higher energy level fills a vacancy in an inner shell, transferring the released energy to another electron which is ejected.

Aurora - An atmospheric phenomenon in which streamers of light are produced when electrons from the sun are guided into the thermosphere by the earth's magnetic field. It occurs in the polar regions at altitudes of 95—300 km.

Avogadro constant (N_A)* - The number of elementary entities in one mole of a substance.

Azeotrope - A liquid mixture in a state where the variation of vapor pressure with composition at constant temperature (or, alternatively, the variation of normal boiling point with composition) shows either a maximum or a minimum. Thus when an azeotrope boils the vapor has the same composition as the liquid.

Azides - Compounds bearing the group $-N_3$, viz. $-N=N^+=N^-$; usually attached to carbon, e.g. PhN_3, phenyl azide or azidobenzene. Also used for salts of hydrazoic acid, HN_3, e.g. NaN_3, sodium azide. [5]

Azines - Condensation products, $R_2C=NN=CR_2$, of two moles of a carbonyl compound with one mole of hydrazine. [5]

Azo compounds - Derivatives of diazene (diimide), $HN=NH$, wherein both hydrogens are substituted by hydrocarbyl groups, e.g., $PhN=NPh$, azobenzene or diphenyldiazene. [5]

Balmer series - The series of lines in the spectrum of the hydrogen atom which corresponds to transitions between the state with principal quantum number $n = 2$ and successive higher states. The wavelengths are given by $1/\lambda = R_H(1/4 - 1/n^2)$, where $n = 3, 4, \ldots$ and R_H is the Rydberg constant for hydrogen. The first member of the series ($n = 2 \rightleftharpoons 3$), which is often called the H_α line, falls at a wavelength of 6563 Å.

Bar (bar) - A unit of pressure equal to 10^5 Pa.

Bardeen-Cooper-Schrieffer (BCS) theory - A theory of superconductivity which is based upon the formation of electron pairs as a result of an electron-lattice interaction. The theory relates the superconducting transition temperature to the density of states and the Debye temperature.

Barn (b) - A unit used for expressing cross sections of nuclear processes, equal to 10^{-28} m^2.

Barrel - A unit of volume equal to 158.9873 L.

Baryon - Any elementary particle built up from three quarks. Examples are the proton, neutron, and various short-lived hyperons. Baryons have odd half-integer spins.

Base - Historically, a substance that yields an OH⁻ ion when it dissociates in solution, resulting in a pH>7. In the Brönsted definition, a base is a substance capable of accepting a proton in any type of reaction. The more general definition, due to G.N. Lewis, classifies any chemical species capable of donating an electron pair as a base.

Becquerel (Bq)* - The SI unit of radioactivity (disintegrations per unit time), equal to s^{-1}. [1]

Beer's law - An approximate expression for the change in intensity of a light beam that passes through an absorbing medium, viz., $\log(I/I_0) = -\varepsilon c l$, where I_0 is the incident intensity, I is the final intensity, ε is the molar (decadic) absorption coefficient, c is the molar concentration of the absorbing substance, and l is the path length. Also called the Beer-Lambert law

Binding energy* - A generic term for the energy required to decompose a system into two or more of its constituent parts. In nuclear physics, the binding energy is the energy difference between a nucleus and the separated nucleons of which it is composed (the mass equivalent of the mass defect). In atomic physics, it is the energy required to remove an electron from an atom.

Biot (Bi) - A name sometimes used for the unit of current in the emu system.

Birefringence - A property of certain crystals in which two refracted rays result from a single incident light ray. One, the ordinary ray, follows the normal laws of refraction, while the other, the extraordinary ray, exhibits a variable refractive index which depends on the direction in the crystal.

Black body radiation* - The radiation emitted by a perfect black body, i.e., a body which absorbs all radiation incident on it and reflects none. The wavelength dependence of the radiated energy density ρ (energy per unit volume per unit wavelength range) is given by the Planck formula

$$\rho = \frac{8\pi hc}{\lambda^5 \left(e^{hc/\lambda kT} - 1\right)}$$

where λ is the wavelength, h is Planck's constant, c is the speed of light, k is the Boltzmann constant, and T is the temperature.

Black hole - A very dense object, formed in a supernova explosion, whose gravitational field is so large that no matter or radiation can escape from the object.

Bloch wave function - A solution of the Schrödinger equation for an electron moving in a spatially periodic potential; used in the band theory of solids.

Bohr magneton (μ_B)* - The atomic unit of magnetic moment, defined as

$eh/4\pi m_e$, where h is Planck's constant, m_e the electron mass, and e the elementary charge. It is the moment associated with a single electron spin.

Bohr, bohr radius (a_0)* - The radius of the lowest orbit in the Bohr model of the hydrogen atom, defined as $\varepsilon_o h^2/\pi m_e e^2$, where ε_o is the permittivity of a vacuum, h is Planck's constant, m_e the electron mass, and e the elementary charge. It is customarily taken as the unit of length when using atomic units.

Boiling point - The temperature at which the liquid and gas phases of a substance are in equilibrium at a specified pressure. The normal boiling point is the boiling point at normal atmospheric pressure (101.325 kPa).

Boltzmann constant (k)* - The molar gas constant R divided by Avogadro's constant.

Boltzmann distribution - An expression for the equilibrium distribution of molecules as a function of their energy, in which the number of molecules in a state of energy E is proportional to $\exp(-E/kT)$, where k is the Boltzmann constant and T is the temperature.

Bond strength - See Dissociation energy.

Born-Haber cycle* - A thermodynamic cycle in which a crystalline solid is converted to gaseous ions and then reconverted to the solid. The cycle permits calculation of the lattice energy of the crystal.

Bose-Einstein distribution - A modification of the Boltzmann distribution which applies to a system of particles that are bosons. The number of particles of energy E is proportional to $[e^{(E-\mu)/kT}-1]^{-1}$, where μ is a normalization constant, k is the Boltzmann constant, and T is the temperature.

Boson - A particle that obeys Bose-Einstein Statistics; specifically, any particle with spin equal to zero or an integer. This includes the photon, pion, deuteron, and all nuclei of even mass number.

Boyle's law - The empirical law, exact only for an ideal gas, which states that the volume of a gas is inversely proportional to its pressure at constant temperature.

Bragg angle (θ) - Defined by the equation $n\lambda = 2d\sin\theta$, which relates the angle θ between a crystal plane and the diffracted x-ray beam, the wavelength λ of the x-rays, the crystal plane spacing d, and the diffraction order n (any integer).

Bravais lattices* - The 14 distinct crystal lattices that can exist in three dimensions. They include three in the cubic crystal system, two in the tetragonal, four in the orthorhombic, two in the monoclinic, and one each in the triclinic, hexagonal, and trigonal systems.

Breakdown voltage - The potential difference at which an insulating substance undergoes a physical or chemical change that causes it to become a conductor, thus allowing current to flow through the sample.

Bremsstrahlung - Electromagnetic radiation generated when the velocity of a charged particle is reduced (literally, "braking radiation"). An example is the x-ray continuum resulting from collisions of electrons with the target in an x-ray tube.

Brewster angle - The angle of incidence for which the maximum degree of plane polarization occurs when a beam of unpolarized light is incident on the surface of a medium of refractive index n. At this angle, the angle between the reflected and refracted beams is 90°. The value of the Brewster angle is $\tan^{-1}n$.

Brillouin scattering - The scattering of light by acoustic phonons in a solid or liquid.

Brillouin zone - A region of allowed wave vectors and energy levels in a crystalline solid, which plays a part in the propagation of waves through the lattice.

British thermal unit (Btu) - A non-SI unit of energy, equal to approximately 1055 J. Several values of the Btu, defined in slightly different ways, have been used.

Brownian motion - The random movements of small particles suspended in a fluid, which arise from collisions with the fluid molecules.

Brunauer-Emmett-Teller method (BET) - See Techniques for Materials Characterization, page 12-1.

Buffer* - A solution designed to maintain a constant pH when small amounts of a strong acid or base are added. Buffers usually consist of a fairly weak acid and its salt with a strong base. Suitable concentrations are chosen so that the pH of the solution remains close to the pK_a of the weak acid.

Calorie (cal) - A non-SI unit of energy, originally defined as the heat required to raise the temperature of 1 g of water by 1°C. Several calories of slightly different values have been used. The thermochemical calorie is now defined as 4.184 J.

Candela (cd)* - The SI base unit of luminous intensity. [1]

Capacitance (C) - Ratio of the charge acquired by a body to the change in potential. [1]

Carbamates - Salts or esters of carbamic acid, $H_2NC(=O)OH$, or of N-substituted carbamic acids: $R_2NC(=O)OR'$, (R' = hydrocarbyl or a cation). The esters are often called urethanes or urethans, a usage that is strictly correct only for the ethyl esters. [5]

Carbenes - The electrically neutral species H_2C: and its derivatives, in which the carbon is covalently bonded to two univalent groups of any kind or a divalent group and bears two nonbonding electrons, which may be spin-paired (singlet state) or spin-non-paired (triplet state). [5]

Carbinols - An obsolete term for substituted methanols, in which the name carbinol is synonymous with methanol. [5]

Carbohydrates - Originally, compounds such as aldoses and ketoses, having the stoichiometric formula $C_n(H_2O)_n$ (hence "hydrates of carbon"). The generic term carbohydrate now includes mono-, oligo-, and polysaccharides, as well as their reaction products and derivatives. [5]

Carboranes - A contraction of carbaboranes. Compounds in which a boron atom in a polyboron hydride is replaced by a carbon atom with maintenance of the skeletal structure. [5]

Carboxylic acids - Oxoacids having the structure $RC(=O)OH$. The term is used as a suffix in systematic name formation to denote the $-C(=O)OH$ group including its carbon atom. [5]

Carnot cycle - A sequence of reversible changes in a heat engine using a perfect gas as the working substance, which is used to demonstrate that entropy is a state function. The Carnot cycle also provides a means to calculate the efficiency of a heat engine.

Catalyst - A substance that participates in a particular chemical reaction and thereby increases its rate but without a net change in the amount of that substance in the system. [3]

Catenanes, catena compounds - Hydrocarbons having two or more rings connected in the manner of links of a chain, without a covalent bond. More generally, the class catena compounds embraces functional derivatives and hetero analogues. [5]

Cation - A positively charged atomic or molecular particle.

Centipoise (cP) - A common non-SI unit of viscosity, equal to mPa s.

Centrifugal distortion - An effect in molecular spectroscopy in which rotational levels are lowered in energy, relative to the values of a rigid rotor, as the rotational angular momentum increases. The effect may be understood classically as a stretching of the bonds in the molecule as it rotates faster, thus increasing the moment of inertia.

Ceramic - A nonmetallic material of very high melting point.

Cerenkov radiation - Light emitted when a beam of charged particles travels through a medium at a speed greater than the speed of light in the medium. It is typically blue in color.

Cgs system of units - A system of units based upon the centimeter, gram, and second. The cgs system has been supplanted by the International System (SI).

Chalcogens - The Group VIA elements (oxygen, sulfur, selenium, tellurium, and polonium). Compounds of these elements are called chalcogenides. [7]

Chaotic system - A complex system whose behavior is governed by deterministic laws but whose evolution can vary drastically when small changes are made in the initial conditions.

Charge - See Electric charge.

Charles' law - The empirical law, exact only for an ideal gas, which states that the volume of a gas is directly proportional to its temperature at constant pressure.

Charm - A quantum number introduced in particle physics to account for certain properties of elementary particles and their reactions.

Chelate - A compound characterized by the presence of bonds from two or more bonding sites within the same ligand to a central metal atom. [3]

Chemical potential - For a mixture of substances, the chemical potential of constituent B is defined as the partial derivative of the Gibbs energy G with respect to the amount (number of moles) of B, with temperature, pressure, and amounts of all other constituents held constant. Also called partial molar Gibbs energy. [2]

Chemical shift* - A small change in the energy levels (and hence in the spectra associated with these levels) resulting from the effects of chemical binding in a molecule. The term is used in fields such as NMR, Mössbauer, and photoelectron spectroscopy, where the energy levels are determined primarily by nuclear or atomic effects.

Chiral molecule - A molecule which cannot be superimposed on its mirror image. A common example is an organic molecule containing a carbon atom to which four different atoms or groups are attached. Such molecules exhibit optical activity, i.e., they rotate the plane of a polarized light beam.

Chlorocarbons - Compounds consisting solely of chlorine and carbon. [5]

Chromatography* - A method for separation of the components of a sample in which the components are distributed between two phases, one of which is stationary while the other moves. In gas chromatography the gas moves over a liquid or solid stationary phase. In liquid chromatography the liquid mixture moves through another liquid, a solid, or a gel. The mechanism of separation of components may be adsorption, differential solubility, ion-exchange, permeation, or other mechanisms. [6]

Clapeyron equation - A relation between pressure and temperature of two phases of a pure substance that are in equilibrium, viz., $dp/dT = \Delta_{trs}S/\Delta_{trs}V$, where $\Delta_{trs}S$ is the difference in entropy between the phases and $\Delta_{trs}V$ the corresponding difference in volume.

Clathrates - Inclusion compounds in which the guest molecule is in a cage formed by the host molecule or by a lattice of host molecules. [5]

Clausius (Cl) - A non-SI unit of entropy or heat capacity defined as cal/K = 4.184 J/K. [2]

Clausius-Clapeyron equation - An approximation to the Clapeyron equation applicable to liquid-gas and solid-gas equilibrium, in which one assumes an ideal gas with volume much greater than the condensed phase volume. For the liquid-gas case, it takes the form $d(\ln p)/dT = \Delta_{vap}H/RT^2$, where R is the molar gas constant and $\Delta_{vap}H$ is the molar enthalpy of vaporization. For the solid-gas case, $\Delta_{vap}H$ is replaced by the molar enthalpy of sublimation, $\Delta_{sub}H$.

Clausius-Mosotti equation - A relation between the dielectric constant ε_r at optical frequencies and the polarizability α:

$$\frac{\varepsilon_r - 1}{\varepsilon_r + 2} = \frac{\rho N_A \alpha}{3M\varepsilon_0}$$

where ρ is density, N_A is Avogadro's number, M is molar mass, and ε_0 is the permittivity of a vacuum.

Clebsch-Gordon coefficients - A set of coefficients used to describe the vector coupling of angular momenta in atomic and nuclear physics.

Codon - A set of three bases, chosen from the four primary bases found in the DNA molecule (uracil, cytosine, adenine, and guanine), which specifies the production of a particular amino acid or carries some other genetic instruction. For example, the codon UCA specifies the amino acid serine, CAG specifies glutamine, etc. There are a total of 64 codons.

Coercive force - The magnetizing force at which the magnetic flux density is equal to zero. [10]

Coercivity* - The maximum value of coercive force that can be attained when a magnetic material is symmetrically magnetized to saturation induction. [10]

Coherent anti-Stokes Raman spectroscopy (CARS) - See Techniques for Materials Characterization, page **12**-1.

Colloid - Molecules or polymolecular particles dispersed in a medium that have, at least in one direction, a dimension roughly between 1 nm and 1 μm. [3]

Color center - A defect in a crystal that gives rise to optical absorption, thus changing the color of the material. A common type is the F-center, which results when an electron occupies the site of a negative ion.

Compressibility (κ)* - The fractional change of volume as pressure is increased, viz., $\kappa = -(1/V)(dV/dp)$. [1]

Compton wavelength (λ_C)* - In the scattering of electromagnetic radiation by a free particle (e.g., electron, proton), $\lambda_C = h/mc$ is the increase in wavelength, at a 90° scattering angle, corresponding to the transfer of energy from radiation to particle. Here h is Planck's constant, c the speed of light, and m the mass of the particle.

Conductance (G)* - For direct current, the reciprocal of resistance. More generally, the real part of admittance. [1]

Conductivity, electrical (σ)* - The reciprocal of the resistivity. [1]

Conductivity, thermal - See Thermal conductivity.

Congruent transformation - A phase transition (melting, vaporization, etc.) in which the substance preserves its exact chemical composition.

Constitutional repeating unit (CRU) - In polymer science, the smallest constitutional unit, the repetition of which constitutes a regular macromolecule, i.e., a macromolecule with all units connected identically with respect to directional sense. [8]

Copolymer - A polymer derived from more than one species of monomer. [8]

Coriolis effect - The deviation from simple trajectories when a mechanical system is described in a rotating coordinate system. It affects the motion of projectiles on the earth and in molecular spectroscopy leads to an important interaction between the rotational and vibrational motions. The effect may be described by an additional term in the equations of motion, called the Coriolis force.

Cosmic rays* - High energy nuclear particles, electrons, and photons, originating mostly outside the solar system, which continually bombard the earth's atmosphere.

Coulomb (C)* - The SI unit of electric charge, equal to A s. [1]

Coulomb's law - The statement that the force F between two electrical charges q_1 and q_2 separated by a distance r is $F = (4\pi\varepsilon_0)^{-1}q_1q_2/r^2$, where ε_0 is the permittivity of a vacuum.

Covalent bond - A chemical bond between two atoms whose stability results from the sharing of two electrons, one from each atom.

Cowling number (Co) - A dimensionless quantity used in plasma physics, defined by $Co = B^2/\mu\rho v^2$, where ρ is density, v is velocity, μ is permeability, and B is magnetic flux density. [2]

CPT theorem - A theorem in particle physics which states that any local Lagrangian theory that is invariant under proper Lorentz transformations is also invariant under the combined operations of charge conjugation, C, space inversion, P, and time reversal, T, taken in any order.

Critical point* - In general, the point on the phase diagram of a two-phase system at which the two coexisting phases have identical properties and therefore represent a single phase. At the liquid-gas critical point of a pure substance, the distinction between liquid and gas vanishes, and the vapor pressure curve ends. The coordinates of this point are called the critical temperature and critical pressure. Above the critical temperature, it is not possible to liquefy the substance.

Cross section (σ)* - A measure of the probability of collision (or other interaction) between a beam of particles and a target which it encounters. In rough terms it is the effective area the target particles present to the incident ones; however, the precise definition depends on the nature of the interaction. A general definition of σ is the number of encounters per unit time divided by nv, where n is the concentration of incident particles and v their velocity.

Crosslink - In polymer science, a small region in a macromolecule from which at least four chains emanate, and formed by reactions involving sites or groups on existing macromolecules or by interactions between existing macromolecules. [8]

Crown compounds - Macrocyclic polydentate compounds, usually uncharged, in which three or more coordinating ring atoms (usually oxygen or nitrogen) are or may become suitably close for easy formation of chelate complexes with metal ions or other cationic species. [5]

Crust* - The outer layer of the solid earth, above the Mohorovicic discontinuity. Its thickness averages about 35 km on the continents and about 7 km below the ocean floor.

Cryoscopic constant (E_f)* - The constant that expresses the amount by which the freezing point T_f of a solvent is lowered by a non-dissociating solute, through the relation $\Delta T_f = E_f m$, where m is the molality of the solute.

Curie (Ci) - A non-SI unit of radioactivity (disintegrations per unit time), equal to 3.7×10^{10} s^{-1}.

Curie temperature (T_C)* - For a ferromagnetic material, the critical temperature above which the material becomes paramagnetic. Also applied to the temperature at which the spontaneous polarization disappears in a ferroelectric solid. [1]

Cyanohydrins - Alcohols substituted by a cyano group, most commonly, but not limited to, examples having a CN and an OH group attached to the same carbon atom. They are formally derived from aldehydes or ketones by the addition of hydrogen cyanide. [5]

Cycloalkanes - Saturated monocyclic hydrocarbons (with or without side chains). See alicyclic compounds. Unsaturated monocyclic hydrocarbons having one endocyclic double or one triple bond are called cycloalkenes and cycloalkynes, respectively. [5]

Cyclotron resonance - The resonant absorption of energy from a system in which electrons or ions that are orbiting in a uniform magnetic field are subjected to radiofrequency or microwave radiation. The resonance frequency is given by $v = eH/2\pi m^*c$, where e is the elementary charge, H is the magnetic field strength, m^* is the effective mass of the charged particle, and c is the speed of light. The effect occurs in both solids (involving electrons or holes) and in low pressure gasses (involving ions)

Dalton (Da) - A name sometimes used in biochemistry for the unified atomic mass unit (u).

De Broglie wavelength - The wavelength associated with the wave representation of a moving particle, given by h/mv, where h is Planck's constant, m the particle mass, and v the velocity.

De Haas-Van Alphen effect - An effect observed in certain metals and semiconductors at low temperatures and high magnetic fields, characterized by a periodic variation of magnetic susceptibility with field strength.

Debye equation* - The relation between the relative permittivity (dielectric constant) ε_r, polarizability α, and permanent dipole moment μ in a dielectric material whose molecules are free to rotate. It takes the form

$$\frac{\varepsilon_r - 1}{\varepsilon_r + 2} = \frac{\rho N_A}{3M\varepsilon_0}\left(\alpha + \frac{\mu^2}{3kT}\right)$$

where ρ is density, N_A is Avogadro's number, M is molar mass, and ε_0 is the permittivity of a vacuum.

Debye length - In the Debye-Hückel theory of ionic solutions, the effective thickness of the cloud of ions of opposite charge which surrounds each given ion and shields the Coulomb potential produced by that ion.

Debye temperature (θ_D)* - In the Debye model of the heat capacity of a crystalline solid, $\theta_D = hv_D/k$, where h is Planck's constant, k is the Boltzmann constant, and v_D is the maximum vibrational frequency the crystal can support. For $T \ll \theta_D$, the heat capacity is proportional to T^3.

Debye unit (D) - A non-SI unit of electric dipole moment used in molecular physics, equal to 3.335641×10^{-30} C m.

Debye-Waller factor (D) - The factor by which the intensity of a diffraction line is reduced because of lattice vibrations. [1]

Defect - Any departure from the regular structure of a crystal lattice. A Frenkel defect results when an atom or ion moves to an interstitial position and leaves behind a vacancy. A Schottky defect involves either a vacancy where the atom has moved to the surface or a structure where a surface atom has moved to an interstitial position.

Degree of polymerization - The number of monomeric units in a macromolecule or an oligomer molecule. [8]

Dendrite - A tree-like crystalline pattern often observed, for example, in ice crystals and alloys in which the crystal growth branches repeatedly.

Density (ρ)* - In the most common usage, mass density or mass per unit volume. More generally, the amount of some quantity (mass, charge, energy, etc.) divided by a length, area, or volume.

Density of states (N_E, ρ) - The number of one-electron states in an infinitesimal interval of energy, divided by the range of that interval and by volume. [1]

Dew point* - The temperature at which liquid begins to condense as the temperature of a gas mixture is lowered. In meteorology, it is the temperature at which moisture begins to condense on a surface in contact with the air.

Diamagnetism - A type of magnetism characterized by a negative magnetic susceptibility, so that the material, when placed in an external magnetic field, becomes weakly magnetized in the direction opposite to the field. This magnetization is independent of temperature.

Diazo compounds - Compounds having the divalent diazo group, $=N^+=N^-$, attached to a carbon atom, e.g., $CH_2=N_2$ diazomethane. [5]

Dielectric constant (ε)* - Ratio of the electric displacement in a medium to the electric field strength. Also called permittivity. [1]

Dienes - Compounds that contain two fixed double bonds (usually assumed to be between carbon atoms). Dienes in which the two double-bond units are linked by one single bond are termed conjugated. [5]

Differential scanning calorimetry (DSC) - See Techniques for Materials Characterization, page 12-1.

Differential thermal analysis (DTA) - See Techniques for Materials Characterization, page 12-1.

Diffusion* - The migration of atoms, molecules, ions, or other particles as a result of some type of gradient (concentration, temperature, etc.).

Diopter - A unit used in optics, formally equal to m^{-1}. It is used in expressing dioptic power, which is the reciprocal of the focal length of a lens.

Dipole moment, electric (p,μ)* - For a distribution of equal positive and negative charge, the magnitude of the dipole moment vector is the positive charge multiplied by the distance between the centers of positive and negative charge distribution. The direction is given by the line from the center of negative charge to the center of positive charge.

Dipole moment, magnetic (m, μ) - Formally defined in electromagnetic theory as a vector quantity whose vector product with the magnetic flux density equals the torque. The magnetic dipole generated by a current I flowing in a small loop of area A has a magnetic moment of magnitude IA. In atomic and nuclear physics, a magnetic moment is associated with the angular momentum of a particle; e.g., an electron with orbital angular momentum l exhibits a magnetic moment of $-el/2m_e$ where e is the elementary charge and m_e the mass of the electron. [1]

Disaccharides - Compounds in which two monosaccharides are joined by a glycosidic bond. [5]

Dislocation - An extended displacement of a crystal from a regular lattice. An edge dislocation results when one portion of the crystal has partially slipped with respect to the other, resulting in an extra plane of atoms extending through part of the crystal. A screw dislocation transforms successive atomic planes into the surface of a helix.

Dispersion - Splitting of a beam of light (or other electromagnetic radiation) of mixed wavelengths into the constituent wavelengths as a result of the variation of refractive index of the medium with wavelength.

Dissociation constant* - The equilibrium constant for a chemical reaction in which a compound dissociates into its constituent parts.

Dissociation energy (D_e)* - For a diatomic molecule, the difference between the energies of the free atoms at rest and the minimum in the potential energy curve. The term bond dissociation energy (D_0), which can be applied to polyatomic molecules as well, is used for the difference between the energies of the fragments resulting when a bond is broken and the energy of the original molecule in its lowest energy state. The term bond strength implies differences in enthalpy rather than energy.

Domain - A small region of a solid in which the magnetic or electric moments of the individual units (atoms, molecules, or ions) are aligned in the same direction.

Domain wall - The transition region between adjacent ferromagnetic domains, generally a layer with a thickness of a few hundred ångström units. Also called Bloch wall.

Doppler effect - The change in the apparent frequency of a wave (sound, light, or other) when the source of the wave is moving relative to the observer.

Dose equivalent (H) - The product of the absorbed dose of radiation at a point of interest in tissue and various modifying factors which depend on the type of tissue and radiation. [1]

Drift velocity - The velocity of charge carriers (electrons, ions, etc.) moving under the influence of an electric field in a medium which subjects the carriers to some frictional force.

Dyne (dyn) - A non-SI (cgs) unit of force, equal to 10^{-5} N.

Ebullioscopic constant (E_b)* - The constant that expresses the amount by which the boiling point T_b of a solvent is raised by a non-dissociating solute, through the relation $\Delta T_b = E_b\, m$, where m is the molality of the solute.

Eddy currents - Circulating currents set up in conducting bulk materials or sheets by varying magnetic fields.

Effinghausen effect - The appearance of a temperature gradient in a current carrying conductor that is placed in a transverse magnetic field. The direction of the gradient is perpendicular to the current and the field.

Eigenvalue - An allowed value of the constant a in the equation $Au = au$, where A is an operator acting on a function u (which is called an eigenfunction). In quantum mechanics, the outcome of any observation is an eigenvalue of the corresponding operator. Also called characteristic value.

Einstein - A non-SI unit used in photochemistry, equal to one mole of photons.

Einstein temperature (θ_V) - In the Einstein theory of the heat capacity of a crystalline solid, $\theta_V = h\nu/k$, where h is Planck's constant, k is the Boltzmann constant, and ν is the vibrational frequency of the crystal.

Einstein transition probability - A constant in the Einstein relation $A_{ij} + B_{ij}\rho$ for the probability of a transition between two energy levels i and j in a radiation field of energy density ρ. The A_{ij} coefficient describes the probability of spontaneous emission, while B_{ij} and B_{ji} govern the probability of stimulated emission and absorption, respectively ($B_{ij} = B_{ji}$).

Elastic limit - The greatest stress which a material is capable of sustaining without any permanent strain remaining after complete release of the stress. [10]

Elastic modulus - See Young's modulus.

Electric charge (Q) - The quantity of electricity; i.e., the property that controls interactions between bodies through electrical forces.

Electric current (I) - The charge passing through a circuit per unit time. [1]

Electric displacement (D) - A vector quantity whose magnitude equals the electric field strength multiplied by the permittivity of the medium and whose direction is the same as that of the field strength.

Electric field strength (E) - The force exerted by an electric field on a point charge divided by the electric charge. [1]

Electric potential (V) - A scalar quantity whose gradient is equal to the negative of the electric field strength.

Electrical conductance - See Conductance

Electrical resistance - See Resistance

Electrical resistivity - See Resistivity.

Electrochemical series* - An arrangement of reactions which produce or consume electrons in an order based on standard electrode potentials. A common arrangement places metals in decreasing order of their tendency to give up electrons.

Electrode potential* - The electromotive force of a cell in which the electrode on the left is the standard hydrogen electrode and that on the right is the electrode in question. [2]

Electrolysis - The decomposition of a substance as a result of passing an electric current between two electrodes immersed in the sample.

Electromotive force (emf) - The energy supplied by a source divided by the charge transported through the source. [1]

Electron* - An elementary particle in the family of leptons, with negative charge and spin of 1/2.

Electron affinity* - The energy difference between the ground state of a gas-phase atom or molecule and the lowest state of the corresponding negative ion.

Electron cyclotron resonance (ECR) - See Techniques for Materials Characterization, page **12**-1.

Electron energy loss spectroscopy (EELS) - See Techniques for Materials Characterization, page **12**-1.

Electron nuclear double resonance (ENDOR) - See Techniques for Materials Characterization, page **12**-1.

Electron paramagnetic resonance (EPR) - See Techniques for Materials Characterization, page **12**-1.

Electron probe microanalysis (EPMA) - See Techniques for Materials Characterization, page **12**-1.

Electron spectroscopy for chemical analysis (ESCA) - See Techniques for Materials Characterization, page **12**-1.

Electron spin (s) - The quantum number, equal to 1/2, that specifies the intrinsic angular momentum of the electron.

Electron stimulated desorption (ESD) - See Techniques for Materials Characterization, page **12-1**.

Electron volt (eV)* - A non-SI unit of energy used in atomic and nuclear physics, equal to approximately 1.602177×10^{-19} J. The electron volt is defined as the kinetic energy acquired by an electron upon acceleration through a potential difference of 1 V. [1]

Electronegativity* - A parameter originally introduced by Pauling which describes, on a relative basis, the power of an atom or group of atoms to attract electrons from the same molecular entity. [3]

Electrophoresis - The motion of macromolecules or colloidal particles in an electric field. [3]

Emissivity (ϵ)* - Ratio of the radiant flux emitted per unit area to that of an ideal black body at the same temperature. Also called emittance. [1]

Emu - The electromagnetic system of units, based upon the cm, g, and s plus the emu of current (sometimes called the abampere).

Enantiomers - A chiral molecule and its non-superposable mirror image. The two forms rotate the plane of polarized light by equal amounts in opposite directions. Also called optical isomers.

Energy (E, U)* - The characteristic of a system that enables it to do work.

Energy gap* - In the theory of solids, the region between two energy bands, in which no bound states can occur.

Enols, alkenols - The term refers specifically to vinylic alcohols, which have the structure $HOCR'=CR_2$. Enols are tautomeric with aldehydes ($R' = H$) or ketones (R' not equal to H). [5]

Enthalpy (H)* - A thermodynamic function, especially useful when dealing with constant-pressure processes, defined by $H = E + PV$, where E is energy, P pressure, and V volume. [1]

Enthalpy of combustion* - The enthalpy change in a combustion reaction. Its negative is the heat released in combustion.

Enthalpy of formation, standard* - The enthalpy change for the reaction in which a substance is formed from its constituent elements, each in its standard reference state (normally refers to 1 mol, sometimes to 1 g, of the substance).

Enthalpy of fusion* - The enthalpy change in the transition from solid to liquid state.

Enthalpy of sublimation - The enthalpy change in the transition from solid to gas state.

Enthalpy of vaporization* - The enthalpy change in the transition from liquid to gas state.

Entropy (S)* - A thermodynamic function defined such that when a small quantity of heat dQ is received by a system at temperature T, the entropy of the system is increased by dQ/T, provided that no irreversible change takes place in the system. [1]

Entropy unit (e.u.) - A non-SI unit of entropy, equal to 4.184 J/K mol.

Ephemeris time - Time measured in tropical years from January 1, 1900.

Epoxy compounds - Compounds in which an oxygen atom is directly attached to two adjacent or non-adjacent carbon atoms of a carbon chain or ring system; thus cyclic ethers. [5]

Equation of continuity - Any of a class of equations that express the fact that some quantity (mass, charge, energy, etc.) cannot be created or destroyed. Such equations typically specify that the rate of increase of the quantity in a given region of space equals the net current of the quantity flowing into the region.

Equation of state* - An equation relating the pressure, volume, and temperature of a substance or system.

Equilibrium constant (K)* - For a chemical reaction aA + bB $\rightleftharpoons$ cC + dD, the equilibrium constant is defined by:

$$K = \frac{a_C{}^c \cdot a_D{}^d}{a_A{}^a \cdot a_B{}^b}$$

where a_i is the activity of component i. To a certain approximation, the activities can be replaced by concentrations. The equilibrium constant is related to $\Delta_r G°$, the standard Gibbs energy change in the reaction, by $RT \ln K = -\Delta_r G°$.

Equivalent conductance - See Conductivity, electrical

Erg (erg) - A non-SI (cgs) unit of energy, equal to 10^{-7} J.

Esters - Compounds formally derived from an oxoacid RC(=O)(OH) and an alcohol, phenol, heteroarenol, or enol by linking, with formal loss of water from an acidic hydroxy group of the former and a hydroxy group of the latter. [5]

Esu - The electrostatic system of units, based upon the cm, g, and s plus the esu of charge (sometimes called the statcoulomb or franklin).

Ethers - Compounds with formula ROR, where R is not equal to H. [5]

Euler number (Eu) - A dimensionless quantity used in fluid mechanics, defined by $Eu = \Delta p / \rho v^2$, where p is pressure, ρ is density, and v is velocity. [2]

Eutectic - The point on a two-component solid-liquid phase diagram which represents the lowest melting point of any possible mixture. A liquid having the eutectic composition will freeze at a single temperature without change of composition.

Excitance (M) - Radiant energy flux leaving an element of a surface divided by the area of that element. [1]

Exciton - A localized excited state consisting of a bound electron-hole pair in a molecular or ionic crystal. The exciton can propagate through the crystal.

Exosphere - The outermost part of the earth's atmosphere, beginning at about 500 to 1000 km above the surface. It is characterized by densities so low that air molecules can escape into outer space.

Expansion coefficient - See thermal expansion coefficient.

Extended electron energy loss fine structure (EXELFS) - See Techniques for Materials Characterization, page **12-1**.

Extended x-ray absorption fine structure (EXAFS) - See Techniques for Materials Characterization, page **12-1**.

Extinction coefficient - See Absorption coefficient, molar

F-Center - See Color center

Fahrenheit temperature (°F) - The temperature scale based on the assignment of 32°F = 0°C and a temperature interval of °F =(5/9)°C; i.e., $t/°F = (9/5)t/°C + 32$.

Farad (F)* - The SI unit of electric capacitance, equal to C/V. [1]

Faraday constant (F)* - The electric charge of 1 mol of singly charged positive ions; i.e., $F = N_A e$, where N_A is Avogadro's constant and e is the elementary charge. [1]

Faraday effect* - The rotation of the plane of plane-polarized light by a medium placed in a magnetic field parallel to the direction of the light beam. The effect can be observed in solids, liquids, and gasses.

Fatty acids - Aliphatic monocarboxylic acids derived from or contained in esterified form in an animal or vegetable fat, oil, or wax. Natural fatty acids commonly have a chain of 4 to 28 carbons (usually unbranched and even-numbered), which may be saturated or unsaturated. By extension, the term is sometimes used to embrace all acyclic aliphatic carboxylic acids. [5]

Fermat's principle - The law that a ray of light traversing one or more media will follow a path which minimizes the time required to pass between two given points.

Fermi (f) - Name sometimes used in nuclear physics for the femtometer.

Fermi level - The highest energy of occupied states in a solid at zero temperature. Sometimes called Fermi energy. The Fermi surface is the surface in momentum space formed by electrons occupying the Fermi level.

Fermi resonance - An effect observed in vibrational spectroscopy when an overtone of one fundamental vibration closely coincides in energy with another fundamental of the same symmetry species. It leads to a splitting of vibrational bands.

Fermi-Dirac distribution - A modification of the Boltzmann distribution which takes into account the Pauli exclusion principle. The number of particles of energy E is proportional to $[e^{(E-\mu)/kT}+1]^{-1}$, where μ is a normalization constant, k the Boltzmann constant, and T the temperature. The distribution is applicable to a system of fermions.

Fermion - A particle that obeys Fermi-Dirac statistics. Specifically, any particle with spin equal to an odd multiple of 1/2. Examples are the electron, proton, neutron, muon, etc.

Ferrimagnetism* - A type of magnetism in which the magnetic moments of atoms in a solid are ordered into two nonequivalent sublattices with unequal magnetic moments, leading to a nonzero magnetic susceptibility.

Ferrite - A ferrimagnetic material of nominal formula MFe_2O_4, where M is a divalent metal; widely used in microwave switches and other solid state devices.

Ferroelectricity* - The retention of electric polarization by certain materials after the external field that produced the polarization has been removed.

Ferromagnetism* - A type of magnetism in which the magnetic moments of atoms in a solid are aligned within domains which can in turn be aligned with each other by a weak magnetic field. Some ferromagnetic materials can retain their magnetization when the external field is removed, as long as the temperature is below a critical value, the Curie temperature. They are characterized by a large positive magnetic susceptibility.

Fick's law - The statement that the flux J of a diffusing substance is proportional to the concentration gradient, i.e., $J = -D(dc/dx)$, where D is called the diffusion coefficient.

Field - A mathematical construct which describes the interaction between particles resulting from gravity, electromagnetism, or other physical phenomena. In classical physics a field is described by equations. Quantum field theory introduces operators to represent the physical observables.

Field emission microscopy (FEM) - See Techniques for Materials Characterization, page **12**-1.

Field ion microscopy (FIM) - See Techniques for Materials Characterization, page **12**-1.

Fine structure - The splitting in spectral lines that results from interactions of the electron spin with the orbital angular momentum.

Fine structure constant (α)* - Defined as $e^2/2hc\epsilon_0$, where e is the elementary charge, h Planck's constant, c the speed of light, and ϵ_0 the permittivity of a vacuum. It is a measure of the strength of the electromagnetic interaction between particles.

First radiation constant (c_1)* - Constant ($=2\pi hc^2$) in the equation for the radiant excitance M_λ of a black body:

$$M_\lambda = \frac{c_1 \lambda^{-5} \Delta\lambda}{e^{c_2/\lambda T} - 1}$$

where λ is the wavelength, T is the temperature, and $c_2 = hc/k$ is the second radiation constant.

Flash point - The lowest temperature at which vapors above a volatile combustible substance will ignite in air when exposed to a flame. [10]

Fluence (F) - Term used in photochemistry to specify the energy per unit area delivered in a given time interval, for example by a laser pulse. [2]

Fluorocarbons - Compounds consisting solely of fluorine and carbon. [5]

Fluxoid - The quantum of magnetic flux in superconductivity theory, equal to $hc/2e$, where h is Planck's constant, c the velocity of light, and e the elementary charge.

Force (F) - The rate of change of momentum with time. [1]

Force constants (f, k)* - In molecular vibrations, the coefficients in the expression of the potential energy in terms of atom displacements from their equilibrium positions. In a diatomic molecule, $f = d^2V/dr^2$, where $V(r)$ is the potential energy and r is the interatomic distance. [2]

Fourier number (Fo) - A dimensionless quantity used in fluid mechanics, defined by $Fo = at/l^2$, where a is thermal diffusivity, t is time, and l is length. [2]

Fourier transform infrared spectroscopy (FTIR) - A technique for obtaining an infrared spectrum by use of an interferometer in which the path length of one of the beams is varied. A Fourier transformation of the resulting interferogram yields the actual spectrum. The technique is also used for NMR and other types of spectroscopy.

Fractals - Geometrical objects that are self-similar under a change of scale; i.e., they appear similar at all levels of magnification. They can be considered to have fractional dimensionality. Examples occur in diverse fields such as geography (rivers and shorelines), biology (trees), and solid state physics (amorphous materials).

Franck-Condon principle - An important principle in molecular spectroscopy which states that the nuclei in a molecule remain essentially stationary while an electronic transition is taking place. The physical interpretation rests on the fact that the electrons move much more rapidly than the nuclei because of their much smaller mass.

Franklin (Fr) - Name sometimes given to the unit of charge in the esu system.

Fraunhofer diffraction - Diffraction of light in situations where the source and observation point are so far removed that the wave surfaces may be considered planar.

Fraunhofer lines - Sharp absorption lines in the spectrum of sunlight, caused by absorption of the solar blackbody radiation by atoms near the sun's surface.

Free radical - See Radicals. The term "free radical" is often used more broadly for molecules that have a paramagnetic ground state (e.g., O_2) and sometimes for any transient or highly reactive molecular species.

Freezing point - See Melting point

Frequency (ν)* - Number of cycles of a periodic phenomenon divided by time. [1]

Fresnel diffraction - Diffraction of light in a situation where the source and observation point are sufficiently close together that the curvature of the wave surfaces must be taken into account.

Froude number (Fr) - A dimensionless quantity used in fluid mechanics, defined by $Fr = v/(lg)^{1/2}$, where v is velocity, l is length, and g is acceleration due to gravity. [2]

Fugacity (f_B) - For a gas mixture, the fugacity of component B is defined as the absolute activity λ_B times the limit, as the pressure p approaches zero at constant temperature, of p_B/λ_B. [2]

Fullerenes - Compounds composed solely of an even number of carbon atoms, which form a cage-like fused-ring polycyclic system with twelve five-membered rings and the rest six-membered rings. The archetypal example is [60]fullerene, where the atoms and bonds delineate a truncated icosahedron. The term has been broadened to include any closed cage structure consisting entirely of three-coordinate carbon atoms. [5]

Fulvalenes - The hydrocarbon fulvalene and its derivatives formed by substitution (and by extension, analogues formed by replacement of one or more carbon atoms of the fulvalene skeleton by a heteroatom). [5]

Fulvenes - The hydrocarbon fulvene and its derivatives formed by substitution (and by extension, analogues formed by replacement of one or more carbon atoms of the fulvene skeleton by a heteroatom). [5]

Fundamental vibrational frequencies* - In molecular spectroscopy, the characteristic vibrational frequencies obtained when the vibrational energy is expressed in normal coordinates. They determine the primary features of the infrared and Raman spectra of the molecule.

γ - Name sometimes used for microgram.

γ-rays* - Electromagnetic radiation (photons) with energy greater than about 0.1 MeV (wavelength less than about 1 pm).

g-Factor of the electron* - The proportionality factor in the equation relating the magnetic moment μ of an electron to its total angular momentum quantum number J, i.e., $\mu = -g\mu_B J$, where μ_B is the Bohr magneton. Also called Landé factor.

Gal - A non-SI unit of acceleration, equal to 0.01 m/s. Also called galileo.

Gallon (US) - A unit of volume equal to 3.785412 L.

Gallon (UK, Imperial) - A unit of volume equal to 4.546090 L.

Gauss (G) - A non-SI unit of magnetic flux density (B) equal to 10^{-4} T.

Gaussian system of units - A hybrid system used in electromagnetic theory, which combines features of both the esu and emu systems.

Gel - A colloidal system with a finite, but usually rather small, yield stress (the sheer stress at which yielding starts abruptly). [3]

Genetic code* - The set of relations between each of the 64 codons of DNA and a specific amino acid (or other genetic instruction).

Gibbs energy (G)* - An important function in chemical thermodynamics, defined by $G = H - TS$, where H is the enthalpy, S the entropy, and T the thermodynamic temperature. Sometimes called Gibbs free energy and, in older literature, simply "free energy". [2]

Gibbs phase rule - The relation $F = C - P + 2$, where C is the number of components in a mixture, P is the number of phases, and F is the degrees of freedom, i.e., the number of intensive variables that can be changed independently without affecting the number of phases.

Glass transition temperature* - The temperature at which an amorphous polymer is transformed, in a reversible way, from a viscous or rubbery condition to a hard and relatively brittle one. [10]

Glow discharge mass spectroscopy (GDMS) - See Techniques for Materials Characterization, page **12**-1.

Gluon - A hypothetical particle postulated to take part in the binding of quarks, in analogy to the role of the photon in electromagnetic interactions.

Glycerides - Esters of glycerol (propane-1,2,3-triol) with fatty acids, widely distributed in nature. They are by long-established custom subdivided into triglycerides, 1,2- or 1,3-diglycerides, and 1- or 2-monoglycerides, according to the number and positions of acyl groups. [5]

Glycols - Dihydric alcohols in which two hydroxy groups are on different carbon atoms, usually but not necessarily adjacent. Also called diols. [5]

Grain (gr) - A non-SI unit of mass, equal to 64.79891 mg.

Grain boundary - The interface between two regions of different crystal orientation.

Grashof number (Gr) - A dimensionless quantity used in fluid mechanics, defined by $Gr = l^3 g \alpha \Delta T \rho^2 / \eta^2$, where T is temperature, ρ is density, l is length, η is viscosity, α is cubic expansion coefficient, and g is acceleration of gravity. [2]

Gravitational constant (G)* - The universal constant in the equation for the gravitational force between two particles, $F = Gm_1m_2/r^2$, where r is the distance between the particles and m_1 and m_2 are their masses. [1]

Gray (Gy)* - The SI unit of absorbed dose of radiation, equal to J/kg. [1]

Gregorian calendar - The modification of the Julian calendar introduced in 1582 by Pope Gregory XII which specified that a year divisible by 100 is a leap year only if divisible by 400.

Grignard reagents - Organomagnesium halides, RMgX, having a carbon-magnesium bond (or their equilibrium mixtures in solution with $R_2Mg + MgX_2$). [5]

Gruneisen parameter (γ) - Defined by $\gamma = \alpha_V / \kappa \, c_V \rho$, where α_V is the cubic thermal expansion coefficient, κ is the isothermal compressibility, c_V is the specific heat capacity at constant volume, and ρ is the mass density. γ is independent of temperature for most crystalline solids. [1]

Gyromagnetic ratio (γ) - Ratio of the magnetic moment of a particle to its angular momentum. Also called magnetogyric ratio.

Hadron - Any elementary particle that can take part in the strong interaction. Hadrons are subdivided into baryons, with odd half integer spins, and mesons, which have zero or integral spin.

Hall effect* - The development of a transverse potential difference V in a conducting material when subjected to a magnetic field H perpendicular to the direction of the current. The potential difference is given by $V = R_H BJt$, where B is the magnetic induction, J the current density, t the thickness of the specimen in the direction of the potential difference, and R_H is called the Hall coefficient.

Halocarbon - A compound containing no elements other than carbon, hydrogen, and one or more halogens. In common practice, the term is used mainly for compounds of no more than four or five carbon atoms.

Halogens - The elements F, Cl, Br, I, and At. Compounds of these elements are called halogenides or halides. [7]

Hamiltonian (H) - An expression for the total energy of a mechanical system in terms of the momenta and positions of constituent particles. In quantum mechanics, the Hamiltonian operator appears in the eigenvalue equation $H\psi = E\psi$, where E is an energy eigenvalue and ψ the corresponding eigenfunction.

Hardness* - The resistance of a material to deformation, indentation, or scratching. Hardness is measured on various scales, such as Mohs, Brinell, Knoop, Rockwell, and Vickers. [10]

Hartmann number (Ha) - A dimensionless quantity used in plasma physics, defined by $Ha = Bl(\kappa/\eta)^{1/2}$, where B is magnetic flux density, l is length, κ is electric conductivity, and η is viscosity. [2]

Hartree (E_h)* - An energy unit used in atomic and molecular science, equal to approximately $4.3597482 \times 10^{-18}$ J.

Hartree-Fock method - A iterative procedure for solving the Schrödinger equation for an atom or molecule in which the equation is solved for each electron in an initial assumed potential from all the other electrons. The new potential that results is used to repeat the calculation and the procedure continued until convergence is reached. Also called self-consistent field (SCF) method.

Heat capacity* - Defined in general as dQ/dT, where dQ is the amount of heat that must be added to a system to increase its temperature by a small amount dT. The heat capacity at constant pressure is $C_p = (\partial H/\partial T)_p$; that at constant volume is $C_V = (\partial E/\partial T)_V$, where H is enthalpy, E is internal energy, p is pressure, V is volume, and T is temperature. An upper case C normally indicates the molar heat capacity, while a lower case c is used for the specific (per unit mass) heat capacity. [1]

Heat of formation, vaporization, etc. - See corresponding terms under Enthalpy.

Hectare (ha) - A unit of area equal to 10^4 m^2. [1]

Heisenberg uncertainty principle - The statement that two observable properties of a system that are complementary, in the sense that their quantum-mechanical operators do not commute, cannot be specified simultaneously with absolute precision. An example is the position and momentum of a particle; according to this principle, the uncertainties in position Δq and momentum Δp must satisfy the relation $\Delta p \Delta q \geq h/4\pi$, where h is Planck's constant.

Heitler-London model - An early quantum-mechanical model of the hydrogen atom which introduced the concept of the exchange interaction between electrons as the primary reason for stability of the chemical bond.

Helicon - A low-frequency wave generated when a metal at low temperature is exposed to a uniform magnetic field and a circularly polarized electric field.

Helmholz energy (A) - A thermodynamic function defined by $A = E - TS$, where E is the energy, S the entropy, and T the thermodynamic temperature. [2]

Hemiacetals - Compounds having the general formula $R_2C(OH)OR'$ (R' not equal to H). [5]

Henry (H)* - The SI unit of inductance, equal to Wb/A. [1]

Henry's law * - An expression which applies to an ideal dilute solution in which one or more gasses are dissolved, viz., $p_i = H_i x_i$, where p_i is the partial pressure of component i above the solution, x_i is its mole fraction in the solution, and H_i is the Henry's law constant (a characteristic of the given gas and solvent, as well as the temperature).

Hermitian operator - An operator A that satisfies the relation $\int u_m * A u_n dx = (\int u_n * A u_m \, dx)*$, where * indicates the complex conjugate. The eigenvalues of Hermitian operators are real, and eigenfunctions belonging to different eigenvalues are orthogonal.

Hertz (Hz) - The SI unit of frequency, equal to s^{-1}. [1]

Heterocyclic compounds - Cyclic compounds having as ring members atoms of at least two different elements, e.g., quinoline, 1,2-thiazole, bicyclo[3.3.1]tetrasiloxane. [5]

Heusler alloys - Alloys of manganese, copper, aluminum, nickel, and sometimes other metals which find important uses as permanent magnets.

Holography - A technique for creating a three-dimensional image of a object by recording the interference pattern between a light beam diffracted from the object and a reference beam. The image can be reconstructed from this pattern by a suitable optical system.

Homopolymer - A polymer derived from one species of (real, implicit, or hypothetical) monomer. [8]

Hooke's law - The statement that the ratio of stress to strain is a constant in a totally elastic medium.

Horse power - A non-SI unit of energy, equal to approximately 746 W.

Hubble constant - The ratio of the recessional velocity of an extragalactic object to the distance of that object. Its value is about $2 \times 10^{-18} \, s^{-1}$.

Huckel theory - A simple approximation for calculating the energy of conjugated molecules in which only the resonance integrals between neighboring bonds are considered. Also called CNDO method (complete neglect of differential overlap).

Hume-Rothery rules - A set of empirical rules for predicting the occurrence of solid solutions in metallic systems. The rules involve size, crystal structure, and electronegativity.

Hund's rules - A series of rules for predicting the sequence of energy states in atoms and molecules. One of the important results is that when two electrons exist in different orbitals, the state with their spins parallel (triplet state) lies at lower energy than the state with antiparallel spins (singlet).

Hydrazines - Hydrazine (diazane), H_2NNH_2, and its hydrocarbyl derivatives. When one or more substituents are acyl groups, the compound is a hydrazide. [5]

Hydrocarbon - A compound containing only carbon and hydrogen. [5]

Hydrolysis - A reaction occurring in water in which a chemical bond is cleaved and a new bond formed with the oxygen atom of water.

Hyperfine structure - Splitting of energy levels and spectral lines into several closely spaced components as a result of interaction of nuclear spin angular momentum with other angular momenta in the atom or molecule.

Hysteresis* - An irreversible response of a system (parameter A) as a function of an external force (parameter F), usually symmetric with respect to the origin of the A vs. F graph after the initial application of the force. A common example is magnetic induction vs. magnetic field strength in a ferromagnet.

Ideal gas law - The equation of state $pV = RT$, which defines an ideal gas, where p is pressure, V molar volume, T temperature, and R the molar gas constant.

Ideal solution - A solution in which solvent-solvent and solvent-solute interactions are identical, so that properties such as volume and enthalpy are exactly additive. Ideal solutions follow Raoult's law, which states that the vapor pressure p_i of component i is $p_i = x_i p_i^*$, where x_i is the mole fraction of component i and p_i^* the vapor pressure of the pure substance i.

Ignition temperature* - The lowest temperature at which combustion of a material will occur spontaneously under specified conditions. Sometimes called autoignition temperature, kindling point. [10]

Imides - Diacyl derivatives of ammonia or primary amines, especially those cyclic compounds derived from diacids. Also used for salts having the anion RN_2^-. [5]

Impedence (Z) - The complex representation of potential difference divided by the complex representation of current. In terms of reactance X and resistance R, the impedance is given by $Z = R + iX$. [1]

Index of refraction (n)* - For a non-absorbing medium, the ratio of the velocity of electromagnetic radiation *in vacuo* to the phase velocity of radiation of a specified frequency in the medium. [1]

Inductance - The ratio of the electromagnetic force induced in a coil by a current to the rate of change of the current.

Inductive coupled plasma mass spectroscopy (ICPMS) - See Techniques for Materials Characterization, page **12-1**.

Inertial defect - In molecular spectroscopy, the quantity $I_c - I_a - I_b$ for a molecule whose equilibrium configuration is planar, where I_a, I_b, and I_c are the effective principal moments of inertia. The inertial defect for a rigid planar molecule would be zero, but vibration-rotation interactions in a real molecule lead to a positive inertial defect.

Insulator - A material in which the highest occupied energy band (valence band) is completely filled with electrons, while the next higher band (conduction band) is empty. Solids with an energy gap of 5 eV or more are generally considered as insulators at room temperature. Their conductivity is less than 10^{-6} S/m and increases with temperature.

Intercalation compounds - Compounds resulting from reversible inclusion, without covalent bonding, of one kind of molecule in a solid matrix of another compound, which has a laminar structure. The host compound, a solid, may be macromolecular, crystalline, or amorphous. [5]

International System of Units (SI)* - The unit system adopted by the General Conference on Weights and Measures in 1960. It consists of seven base units (meter, kilogram, second, ampere, kelvin, mole, candela), plus derived units and prefixes. [1]

International Temperature Scale (ITS-90)* - The official international temperature scale adopted in 1990. It consists of a set of fixed points and equations which enable the thermodynamic temperature to be determined from operational measurements. [9]

Ion - An atomic or molecular particle having a net electric charge. [3]

Ion exchange - A process involving the adsorption of one or several ionic species accompanied by the simultaneous desorption (displacement) of one or more other ionic species. [3]

Ion neutralization spectroscopy (INS) - See Techniques for Materials Characterization, page **12-1**.

Ionic strength (I) - A measure of the total concentration of ions in a solution, defined by $I = 1/2 \Sigma_i z_i^2 m_i$, where z_i is the charge of ionic species i and m_i is its molality. For a 1-1 electrolyte at molality m, $I = m$.

Ionization constant* - The equilibrium constant for a reaction in which a substance in solution dissociates into ions.

Ionization potential* - The minimum energy required to remove an electron from an isolated atom or molecule (in its vibrational ground state) in the gaseous phase. More properly called ionization energy. [3]

Irradiance (E) - The radiant energy flux incident on an element of a surface, divided by the area of that element. [1]

Isentropic process - A thermodynamic process in which the entropy of the system does not change.

Ising model - A model describing the coupling between two atoms in a ferromagnetic lattice, in which the interaction energy is proportional to the negative of the product of the spin components along a specified axis.

Isobar - A line connecting points of equal pressure on a graphical representation of a physical system.

Isochore - A line or surface of constant volume on a graphical representation of a physical system.

Isoelectric point* - The pH of a solution or dispersion at which the net charge on the macromolecules or colloidal particles is zero. In electrophoresis there is no motion of the particles in an electric field at the isoelectric point.

Isomers - In chemistry, compounds that have identical molecular formulas but differ in the nature or sequence of bonding of their atoms or in the arrangement of their atoms in space. In physics, nuclei of the same atomic number Z and mass number A but in different energy states. [3]

Isomorphs - Substances of different chemical nature but having the same crystal structure.

Isotactic macromolecule - A tactic macromolecule, essentially comprising only one species of repeating unit which has chiral or prochiral atoms in the main chain in a unique arrangement with respect to its adjacent constitutional units. [8]

Isotherm - A line connecting points of equal temperature on a graphical representation of a physical system.

Isothermal process - A thermodynamic process in which the temperature of the system does not change.

Isotones - Nuclides having the same neutron number N but different atomic number Z. [3]

Isotopes - Two or more nuclides with the same atomic number Z but different mass number A. The term is sometimes used synonymously with nuclide, but it is preferable to reserve the word nuclide for a species of specific Z and A. [3]

Jahn-Teller effect - An interaction of vibrational and electronic motions in a nonlinear molecule which removes the degeneracy of certain electronic energy levels. It can influence the spectrum, crystal structure, and magnetic properties of the substance.

Johnson noise - Electrical noise generated by random thermal motion of electrons in a conductor or semiconductor. Also called thermal noise.

Josephson effect - The tunneling of electron pairs through a thin insulating layer which separates two superconductors. When a potential difference is applied to the superconductors, an alternating current is generated whose frequency is precisely proportional to the potential difference. This effect has important applications in metrology and determination of fundamental physical constants.

Joule (J)* - The SI unit of energy, equal to N m. [1]

Joule-Thomson coefficient (μ) - A parameter which describes the temperature change when a gas expands adiabatically through a nozzle from a high pressure to a low pressure region. It is defined by $\mu = (\partial T / \partial p)_H$, where H is enthalpy.

Julian calendar - The calendar introduced by Julius Caeser in 46 B.C. which divided the year into 365 days with a leap year of 366 days every fourth year.

Julian date (JD) - The number of days elapsed since noon Greenwich Mean Time on January 1, 4713 B.C. Thus January 1, 2000, 0h (midnight) will be JD 2,451,543.5. This dating system was introduced by Joseph Scaliger in 1582.

Kaon - One of the elementary particles in the family of mesons. Kaons have a spin of zero and may be neutral or charged.

Kelvin (K)* - The SI base unit of thermodynamic temperature. [1]

Kepler's laws - The three laws of planetary motion, which established the elliptical shape of planetary orbits and the relation between orbital dimensions and the period of rotation.

Kerr effect* - An electrooptical effect in which birefringence is induced in a liquid or gas when a strong electric field is applied perpendicular to the direction of an incident light beam. The Kerr constant k is given by $n_1 - n_2 = k\lambda E^2$, where λ is the wavelength, E is the electric field strength, and n_1 and n_2 are the indices of refraction of the ordinary and extraordinary rays, respectively.

Ketenes - Compounds in which a carbonyl group is connected by a double bond to an alkylidene group: $R_2C=C=O$. [5]

Ketones - Compounds in which a carbonyl group is bonded to two carbon atoms: $R_1R_2C=O$ (neither R may be H). [5]

Kilogram (kg)* - The SI base unit of mass. [1]

Kinetic energy (E_k, T) - The energy associated with the motion of a system of particles in a specified reference frame. For a single particle of mass m moving at velocity v, $E_k = 1/2 mv^2$.

Kirchhoff's laws - Basic rules for electric circuits, which state (a) the algebraic sum of the currents at a network node is zero and (b) the algebraic sum of the voltage drops around a closed path is zero.

Klein-Gordon equation - A relativistic extension of the Schrödinger equation.

Klein-Nishima formula - An expression for the scattering cross section of a photon by an unbound electron, based upon the Dirac electron theory.

Knight shift - The change in magnetic resonance frequency of a nucleus in a metal relative to the same nucleus in a diamagnetic solid. The effect is due to the polarization of the conduction electrons in the metal.

Knudsen number (Kn) - A dimensionless quantity used in fluid mechanics, defined by $Kn = \lambda / l$, where λ is mean free path and l is length. [2]

Kondo effect - A large increase in electrical resistance observed at low temperatures in certain dilute alloys of a magnetic metal in a nonmagnetic material.

Kramers-Kronig relation - A set of equations relating the real and imaginary parts of the index of refraction of a medium

Lactams - Cyclic amides of amino carboxylic acids, having a 1-azacycloalkan-2-one structure, or analogues having unsaturation or heteroatoms replacing one or more carbon atoms of the ring. [5]

Lactones - Cyclic esters of hydroxy carboxylic acids, containing a 1-oxacycloalkan-2-one structure, or analogues having unsaturation or heteroatoms replacing one or more carbon atoms of the ring. [5]

Lagrangian function (L) - A function used in classical mechanics, defined as the kinetic energy minus the potential energy for a system of particles.

Lamb shift - The small energy difference between the $^2S_{1/2}$ and $^2P_{1/2}$ levels in the hydrogen atom, which results from interactions between the electron and the radiation field.

Laminar flow - Smooth, uniform, non-turbulent flow of a gas or liquid in parallel layers, with little mixing between layers. It is characterized by small values of the Reynolds number.

Landé g-factor - See g-Factor of the electron

Langevin function - The mathematical function $L(x) = (e^x + e^{-x})/(e^x - e^{-x}) - 1/x$, which occurs in the expression for the average dipole moment of a group of rotating polar molecules in an electric field: $\mu_{av} = \mu L(\mu E / kT)$, where μ is the electric dipole moment of a single molecule, E is the electric field strength, k is the Boltzmann constant, and T is the temperature.

Lanthanides - The elements of atomic number 57 through 71, which share common chemical properties: La, Ce, Pr, Nd, Pm, Sm, Eu, Gd, Tb, Dy, Ho, Er, Tm, Yb, Lu. [7]

Larmor frequency (v_L) - The precession frequency of a magnetic dipole in an applied magnetic field. In particular, a nucleus in a magnetic field of strength B has a Larmor frequency of $\gamma B / 2\pi$, where γ is the magnetogyric ratio of the nucleus.

Laser* - A device in which an optical cavity is filled with a medium where a population inversion can be produced by some means. When the resonant frequency of the cavity bears the proper relation to the separation of the inverted energy levels, stimulated emission occurs, producing a highly monochromatic, coherent beam of light.

Laser ionization mass spectroscopy (LIMS) - See Techniques for Materials Characterization, page **12**-1.

Lattice constants* - Parameters specifying the dimensions of a unit cell in a crystal lattice, specifically the lengths of the cell edges and the angles between them.

Lattice energy* - The energy per ion pair required to separate completely the ions in a crystal lattice at a temperature of absolute zero.

Laue diagram - A diffraction pattern produced when an x-ray beam passes through a thin slice of a crystal and impinges on a detector behind the crystal.

Lenz's law - The statement that the current induced in a circuit by a change in magnetic flux is so directed as to oppose the change in flux

Leonard-Jones potential - A simple but useful function for approximating the interaction between two neutral atoms or molecules separated by a distance r by writing the potential energy as $U(r) = 4\varepsilon\{(r_0/r)^{12} - (r_0/r)^6\}$, where ε and r_0 are adjustable parameters. In this form the depth of the potential well is ε and the minimum occurs at $2^{1/6}r_0$. The $(1/r)^{12}$ term is often replaced by other powers of $1/r$.

Lepton - One of the class of elementary particles that do not take part in the strong interaction. Included are the electron, muon, and neutrino. All leptons have a spin of 1/2.

Lewis number (Le) - A dimensionless quantity used in fluid mechanics, defined by $Le = a/D$, where a is thermal diffusivity and D is diffusion coefficient. [2]

Ligand field theory - A description of the structure of crystals containing a transition metal ion surrounded by nonmetallic ions (ligands). It is based on construction of molecular orbitals involving the d-orbitals of the central metal ion and combinations of atomic orbitals of the ligands.

Light year (l.y.) - A unit of distance used in astronomy, defined as the distance light travels in one year in a vacuum. Its approximate value is 9.46073×10^{15} m.

Lignins - Macromolecular constituents of wood related to lignans, composed of phenolic propylbenzene skeletal units, linked at various sites and apparently randomly. [5]

Ligroin - The petroleum fraction consisting mostly of C_7 and C_8 hydrocarbons and boiling in the range 90-140°C; commonly used as a laboratory solvent.

Lipids - A loosely defined term for substances of biological origin that are soluble in nonpolar solvents. They consist of saponifiable lipids, such as glycerides (fats and oils) and phospholipids, as well as nonsaponifiable lipids, principally steroids. [5]

Lipoproteins - Clathrate complexes consisting of a lipid enwrapped in a protein host without covalent binding, in such a way that the complex has a hydrophilic outer surface consisting of all the protein and the polar ends of any phospholipids. [5]

Liter (L)* - A synonym for cubic decimeter. [1]

Lithosphere* - The outer layer of the solid earth, extending from the base of the mantle to the surface of the crust.

Lorentz contraction - The reduction in length of a moving body in the direction of motion, given by the factor $(1-v^2/c^2)^{1/2}$, where v is the velocity of the body and c the velocity of light. Also known as the FitzGerald-Lorentz contraction.

Lorentz force - The force exerted on a point charge Q moving at velocity v in the presence of external fields E and B. It is given (in SI units) by $F = Q(E + v \times B)$.

Loss angle (δ) - For a dielectric material in an alternating electromagnetic field, δ is the phase difference between the current and the potential difference. The function $\tan \delta$ is a measure of the ratio of the power dissipated in the dielectric to the power stored.

Low energy electron diffraction (LEED) - See Techniques for Materials Characterization, page **12**-1.

Lumen (lm)* - The SI unit of luminous flux, equal to cd sr. [1]

Luminous flux (Φ) - The intensity of light from a source multiplied by the solid angle. The SI unit is lumen. [1]

Lux (lx)* - The SI unit of illuminance, equal to cd sr m^{-2}. [1]

Lyddane-Sachs-Teller relation - A relation between the phonon frequencies and dielectric constants of an ionic crystal which states that $(\omega_T/\omega_L)^2 = \varepsilon(\infty)/\varepsilon(0)$, where ω_T is the angular frequency of transverse optical phonons, ω_L that of longitudinal optical phonons, $\varepsilon(0)$ is the static dielectric constant, and $\varepsilon(\infty)$ the dielectric constant at optical frequencies.

Lyman series - The series of lines in the spectrum of the hydrogen atom which corresponds to transitions between the ground state (principal quantum number $n = 1$) and successive excited states. The wavelengths are given by $1/\lambda = R_H(1-1/n^2)$, where $n = 2,3,4,\ldots$ and R_H is the Rydberg constant for hydrogen. The first member of the series ($n = 1 \leftrightarrow 2$), which is often called the Lyman-α line, falls at a wavelength of 1216 Å, and the series converges at 912 Å, the ionization limit of hydrogen.

Mach number (Ma) - A dimensionless quantity used in fluid mechanics, defined by $Ma = v/c$, where v is velocity and c is the speed of sound. [2]

Macromolecule - A molecule of high relative molecular mass (molecular weight), the structure of which essentially comprises the multiple repetition of units derived, actually or conceptually, from molecules of low relative molecular mass [8]

Madelung constant* - A constant characteristic of a particular crystalline material which gives a measure of the electrostatic energy binding the ions in the crystal.

Magnetic field strength (H) - An axial vector quantity, the curl of which is equal to the current density, including the displacement current. [1]

Magnetic induction (B) - An axial vector quantity such that the force exerted on an element of current is equal to the vector product of this element and the magnetic induction. [1]

Magnetic moment - See Dipole moment, magnetic.

Magnetic susceptibility (χ_m, κ)* - Defined by $\chi_m = (\mu-\mu_0)/\mu_0$, where μ is the permeability of the medium and μ_0 the permeability of a vacuum. [1]

Magnetization (M) - Defined by $M = (B/\mu_0)-H$, where B is magnetic induction, H magnetic field strength, and μ_0 the permeability of a vacuum. [1]

Magnetogyric ratio (γ) - Ratio of the magnetic moment of a particle to its angular momentum. Also called gyromagnetic ratio.

Magneton - See Bohr magneton, Nuclear magneton.

Magnetostriction* - The change in dimensions of a solid sample when it is placed in a magnetic field.

Magnon - A quantum of magnetic energy associated with a spin wave in a ferromagnetic or antiferromagnetic crystal.

Mantle - The layer of the earth between the crust and the liquid outer core, which begins about 2900 km below the earth's surface.

Maser - A device in which a microwave cavity is filled with a medium where a population inversion can be produced by some means. When the resonant frequency of the cavity bears the proper relation to the separation of the inverted energy levels, the device can serve as an amplifier or oscillator at that frequency.

Mass (m)* - Quantity of matter. Mass can also be defined as "resistance to acceleration".

Mass defect (B) - Defined by $B = Zm(^1H) + Nm_n - m_a$, where Z is the atomic number, $m(^1H)$ is the mass of the hydrogen atom, N is the neutron number, m_n is the rest mass of the neutron, and m_a is the mass of the atom in question. Thus Bc^2 can be equated to the binding energy of the nucleus if the binding energy of atomic electrons is neglected. [1]

Mass excess (Δ) - Defined by $\Delta = m_a - Am_u$, where m_a is the mass of the atom, A the number of nucleons, and m_u the unified atomic mass constant ($m_u = 1$ u). [1]

Mass fraction (w_B) - The ratio of the mass of substance B to the total mass of a mixture. [1]

Mass number (A) - A characteristic property of a specific isotope of an element, equal to the sum of the number of protons and neutrons in the nucleus.

Mass spectrometry - An analytical technique in which ions are separated according to the mass/charge ratio and detected by a suitable detector. The ions may be produced by electron impact on a gas, a chemical reaction, energetic vaporization of a solid, etc. [6]

Massieu function - A thermodynamic function defined by $J = -A/T$, where A is the Helmholz energy and T the thermodynamic temperature. [2]

Matthiessen's rule - The statement that the electrical resistivity ρ of a metal can be written as $\rho = \rho_L + \rho_i$, where ρ_L is due to scattering of conduction electrons by lattice vibrations and ρ_i to scattering by impurities and imperfections. If the impurity concentration is small, ρ_i is temperature independent.

Maxwell (Mx)* - A non-SI unit of magnetic field strength (H) equal to 10^{-8} Wb. [1]

Maxwell's equations - The fundamental equations of electromagnetism. In a form appropriate to SI units, they are:

curl $H = \partial D/\partial t + j$

div $B = 0$

curl $E = -\partial B/\partial t$

div $D = \rho$

where H is the magnetic field strength, B the magnetic induction, E the electric field strength, D the electric displacement, j the current density, ρ the charge density, and t is time.

Maxwell-Boltzmann distribution - An expression for the fraction of molecules $f(v)$ in a gas that have velocity v within a specified interval. It takes the form

$$f(v) = 4\pi (M/2\pi RT)^{3/2} v^2 e^{-Mv^2/2RT}$$

where M is the molar mass, R the molar gas constant, and T the temperature.

Mean free path* - The average distance a gas molecule travels between collisions.

Meissner effect - The complete exclusion of magnetic induction from the interior of a superconductor.

Melting point* - The temperature at which the solid and liquid phases of a substance are in equilibrium at a specified pressure (normally taken to be atmospheric unless stated otherwise).

Mercaptans - A traditional term abandoned by IUPAC, synonymous with thiols. This term is still widely used. [5]

Meson - Any elementary particle that has zero or integral spin. Mesons are responsible for the forces between protons and neutrons in the nucleus.

Mesosphere - The part of the earth's atmosphere extending from the top of the stratosphere (about 50 km above the surface) to 80-90 km. It is characterized by a decrease in temperature with increasing altitude.

Metal - A material in which the highest occupied energy band (conduction band) is only partially filled with electrons. The electrical conductivity of metals generally decreases with temperature.

Metallocenes - Organometallic coordination compounds in which one atom of a transition metal such as iron, ruthenium or osmium is bonded to and only to the face of two cyclopentadienyl ligands which lie in parallel planes. [5]

Meter (m)* - The SI base unit of length. [1]

Methine group - In organic compounds, the -C= group. [5]

Mho - An archaic name for the SI unit siemens (reciprocal ohm).

Micelle - A particle formed by the aggregation of surfactant molecules (typically, 10 to 100 molecules) in solution. For aqueous solutions, the hydrophilic end of the molecule is on the surface of the micelle, while the hydrophobic end (often a hydrocarbon chain) points toward the center. At the critical micelle concentration (cmc) the previously dissolved molecules aggregate into a micelle.

Micron (μ) - An obsolete name for micrometer.

Mie scattering - The scattering of light by spherical dielectric particles whose diameter is comparable to the wavelength of the light.

Milky way - The band of light in the night sky resulting from the stars in the galactic plane. The term is also used to denote the galaxy in which the sun is located.

Miller indices (hkl) - A set of indices used to label planes in a crystal lattice. [2]

Millimeter of mercury (mmHg) - A non-SI unit of pressure, equal to 133.322 Pa. The name is generally considered interchangeable with torr.

Mobility (μ)* - In solid state physics, the drift velocity of electrons or holes in a solid divided by the applied electric field strength. The term is used in a similar sense in other fields.

Molality (m) - A measure of concentration of a solution in which one states the amount of substance (i.e., number of moles) of solute per kilogram of solvent. Thus a 0.1 molal solution (often written as 0.1 m) has $m = 0.1$ mol/kg.

Molar mass - The mass of one mole of a substance. It is normally expressed in units of g/mol, in which case its numerical value is identical with the molecular weight (relative molecular mass). [1]

Molar quantity - It is often convenient to express an extensive quantity (e.g., volume, enthalpy, heat capacity, etc.) as the actual value divided by amount of substance (number of moles). The resulting quantity is called molar volume, molar enthalpy, etc

Molar refraction (R) - A property of a dielectric defined by the equation $R = V_m[(n^2-1)/(n^2+2)]$, where n is the index of refraction of the medium (at optical wavelengths) and V_m the molar volume. It is related to the polarizability α of the molecules that make up the medium by the Lorenz-Lorentz equation, $R = N_A \alpha/3\varepsilon_0$, where N_A is Avogadro's constant and ε_0 is the permittivity of a vacuum.

Molarity (c) - A measure of concentration of a solution in which one states the amount of substance (i.e., number of moles) of solute per liter of solution. Thus a 0.1 molar solution (often referred to as 0.1 M) has a concentration $c = 0.1$ mol/L.

Mole (mol)* - The SI base unit of amount of substance. [1]

Mole fraction (x_B) - The ratio of the amount of substance (number of moles) of substance B to the total amount of substance in a mixture. [1]

Molecular orbital - See Orbital.

Molecular weight (M_r)* - The ratio of the average mass per molecule or specified entity of a substance to 1/12 of the mass of nuclide ^{12}C. Also called relative molar (or molecular) mass. [1]

Moment of inertia (I) - The moment of inertia of a body about an axis is the sum (or integral) of the products of its elements of mass and the squares of their distances from the axis. [1]

Momentum (p) - The product of mass and velocity. [1]

Monomer - A substance consisting of molecules which can undergo polymerization, thereby contributing constitutional units to the essential structure of a macromolecule. [8]

Monosaccharides - A term which includes aldoses, ketoses, and a wide variety of derivatives. [5]

Mössbauer effect - The recoilless emission of γ-rays from nuclei bound in a crystal under conditions where the recoil energy associated with the γ emission is taken up by the crystal as a whole. This results in a very narrow line width, which can be exploited in various types of precise measurements.

Muon* - An unstable elementary particle of spin 1/2 and mass about 200 times that of the electron.

Naphtha - The petroleum fraction consisting mostly of C_6 to C_8 hydrocarbons and boiling in the range 80-120°C. Solvents derived from this fraction include ligroin and petroleum ether.

Nautical mile - A non-SI unit of length, equal to exactly 1852 m.

Navier-Stokes equations - A set of complex equations for the motion of a viscous fluid subject to external forces.

Néel temperature (T_N)* - The critical temperature above which an antiferromagnetic substance becomes paramagnetic. [1]

Nernst effect - The production of an electric field in a conductor subject to an applied magnetic field and containing a transverse temperature gradient. The electric field is perpendicular to the magnetic field and the temperature gradient.

Network - In polymer science, a highly ramified macromolecule in which essentially each constitutional unit is connected to each other constitutional unit and to the macroscopic phase boundary by many permanent paths through the macromolecule, the number of such paths increasing with the number of intervening bonds. The paths must on the average be coextensive with the macromolecule. [8]

Neutrino - A stable elementary particle in the lepton family. Neutrinos have zero (or at least near-zero) rest mass and spin 1/2.

Neutron* - An elementary particle on spin 1/2 and zero charge. The free neutron has a mean lifetime of 887 seconds. Neutrons and protons, which are collectively called nucleons, are the constituents of the nucleus.

Neutron activation analysis (NAA) - See Techniques for Materials Characterization, page 12-1.

Neutron number (N) - A characteristic property of a specific isotope of an element, equal to the number of neutrons in the nucleus.

Newton (N)* - The SI unit of force, equal to m kg s^{-2}. [1]

Nitriles - Compounds having the structure RC≡N; thus C-substituted derivatives of hydrocyanic acid, HC≡N. [5]

Nitrosamines - N-Nitroso amines: compounds of the structure R_2NNO. Compounds RNHNO are not ordinarily isolatable, but they, too, are nitrosamines. The name is a contraction of N-nitrosoamine and, as such, does not require the N locant. [5]

Nuclear magnetic resonance (NMR)* - A widely used technique in which the resonant absorption of radiofrequency radiation by magnetic nuclei in a magnetic field is measured. The results give important information on the local environment of each nucleus.

Nuclear magneton (μ_N)* - The unit of nuclear magnetic moment, defined as $eh/4\pi m_p$, where h is Planck's constant, m_p the proton mass, and e the elementary charge.

Nuclear quadrupole resonance (NQR) - See Techniques for Materials Characterization, page 12-1.

Nuclear reaction analysis (NRA) - See Techniques for Materials Characterization, page 12-1.

Nuclear spin (I) - The quantum number that specifies the intrinsic angular momentum of a particular nucleus. The magnitude of the angular momentum is given by $[I(I+1)]^{1/2} h/2\pi$, where h is Planck's constant.

Nucleic acids* - Macromolecules, the major organic matter of the nuclei of biological cells, made up of nucleotide units, and hydrolyzable into certain pyrimidine or purine bases (usually adenine, cytosine, guanine, thymine, uracil), D-ribose or 2-deoxy-D-ribose. [5]

Nucleon - A collective term for the proton and neutron.

Nucleosides - Ribosyl or deoxyribosyl derivatives (rarely, other glycosyl derivatives) of certain pyrimidine or purine bases. They are thus glycosylamines or N-glycosides related to nucleotides by the lack of phosphorylation. [5]

Nucleotides - Compounds formally obtained by esterification of the 3′ or 5′ hydroxy group of nucleosides with phosphoric acid. They are the monomers of nucleic acids and are formed from them by hydrolytic cleavage. [5]

Nuclide - A species of atoms in which each atom has identical atomic number Z and identical mass number A. [3]

Nusselt number (Nu) - A dimensionless quantity used in fluid mechanics, defined by $Nu = hl/k$, where h is coefficient of heat transfer, l is length, and k is thermal conductivity. [2]

Nyquist theorem - An expression for the mean square thermal noise voltage across a resistor, given by $4RkT\Delta f$ where R is the resistance, k the Boltzmann constant, T the temperature, and Δf the frequency band within which the voltage is measured.

Octanol-water partition coefficient (P)* - A measure of the way in which a compound will partition itself between the octanol and water phases in the two-phase octanol-water system, and thus an indicator of certain types of biological activity. Specifically, P is the ratio of the concentration (in moles per liter) of the compound in the octanol phase to that in the water phase at infinite dilution. The quantity normally reported is log P.

Oersted (Oe) - A non-SI unit of magnetic field (H), equal to 79.57747 A/m.

Ohm (Ω)* - The SI unit of electric resistance, equal to V/A. [1]

Ohm's law - A relation among electric current I, potential difference V, and resistance R, viz., $I = V/R$. The resistance is constant at constant temperature to high precision for many materials.

Olefins - Acyclic and cyclic hydrocarbons having one or more carbon-carbon double bonds, apart from the formal ones in aromatic compounds. The class olefins subsumes alkenes and cycloalkenes and the corresponding polyenes. [5]

Oligomer - A substance consisting of molecules of intermediate relative molecular mass (molecular weight), the structure of which essentially comprises the multiple repetition of units derived, actually or conceptually, from molecules of low relative molecular mass. In contrast to a polymer, the properties of an oligomer can vary significantly with the removal of one or a few of its units. [8]

Oligopeptides - Peptides containing from three to nine amino groups. [5]

Onsager relations - An important set of equations in the thermodynamics of irreversible processes. They express the symmetry between the transport coefficients describing reciprocal processes in systems with a linear dependence of flux on driving forces.

Optical rotary power - Angle by which the plane of polarization of a light beam is rotated by an optically active medium, divided by path length and by concentration of the active constituent. Depending on whether mass or molar concentration is used, the modifier "specific" or "molar" is attached. [2]

Orbital - A one-electron wavefunction. Atomic orbitals are classified as s-, p-, d,- or f-orbitals according to whether the angular momentum quantum number $l = 0, 1, 2,$ or 3. Molecular orbitals, which are usually constructed as linear combinations of atomic orbitals, describe the distribution of electrons over the entire molecule.

Oscillator strength (f) - A measure of the intensity of a spectroscopic transition, defined by

$$f = \frac{8\pi^2 M e \nu}{3he^2}\left|\mu_{ij}\right|^2$$

where ν is the frequency, μ_{ij} the transition dipole moment, m_e the mass of the electron, e the elementary charge, and h Planck's constant.

Osmosis - The flow of a solvent in a system in which two solutions of different concentration are separated by a semipermeable membrane which cannot pass solute molecules. The solvent will flow from the side of lower concentration to that of higher concentration, thus tending to equalize the concentrations. The pressure that must be applied to the more concentrated side to stop the flow is called the osmotic pressure.

Osmotic coefficient (ϕ) - Defined by $\phi = \ln a_A/(M_A \Sigma m_B)$, where M_A is the molar mass of substance A (normally the solvent), a_A is its activity, and the m_B are molalities of the solutes. [1]

Osmotic pressure (Π) - The excess pressure necessary to maintain osmotic equilibrium between a solution and the pure solvent separated by a membrane permeable only to the solvent. In an ideal dilute solution $\Pi = c_B RT$, where c_B is the amount-of-substance concentration of the solute, R is the molar gas constant, and T the temperature. [1,2]

Ostwald dilution law - A relation for the concentration dependence of the molar conductivity Λ of an electrolyte solution, viz.,

$$\frac{1}{\Lambda} = \frac{1}{\Lambda^\circ} + \frac{\Lambda c}{K(\Lambda^\circ)^2}$$

where c is the solute concentration, K is the equilibrium constant for dissociation of the solute, and Λ° is the conductivity at $c\Lambda = 0$.

Ounce (oz) - A non-SI unit of mass. The avoirdupois ounce equals 28.34952 g, while the troy ounce equals 31.10348 g.

Overpotential (η) - In an electrochemical cell, the difference between the potential of an electrode and its zero-current value.

Oximes - Compounds of structure $R_2C=NOH$ derived from condensation of aldehydes or ketones with hydroxylamine. Oximes from aldehydes may be called aldoximes; those from ketones may be called ketoximes. [5]

Oxo compounds - Compounds containing an oxygen atom, =O, doubly bonded to carbon or another element. The term thus embraces aldehydes, carboxylic acids, ketones, sulfonic acids, amides and esters. [5]

Ozonides - The 1,2,4-trioxolanes formed by the reaction of ozone at a carbon-carbon double bond, or the analogous compounds derived from acetylenic compounds. [5]

Pair production - A process in which a photon is converted into a particle and its antiparticle (e.g., an electron and positron) in the electromagnetic field of a nucleus.

Paraffins - Obsolescent term for saturated hydrocarbons, commonly but not necessarily acyclic. Still widely used in the petrochemical industry, where the term designates acyclic saturated hydrocarbons, and stands in contradistinction to naphthenes. [5]

Paramagnetism* - A type of magnetism characterized by a positive magnetic susceptibility, so that the material becomes weakly magnetized in the direction of an external field. The magnetization disappears when the field in removed. In the simplest approximation (Curie's law) the susceptibility is inversely proportional to temperature.

Parity - The property of a quantum-mechanical wave function that describes its behavior under the symmetry operation of coordinate inversion. A parity of +1 (or even) is assigned if the wave function does not change sign when the signs of all the coordinates are changed; the parity is -1 (or odd) if the wave function changes sign under this operation.

Parsec (pc) - A unit of distance defined as the distance at which 1 astronomical unit (AU) subtends an angle of 1 second of arc. It is equal to 206264.806 AU or 3.085678×10^{16} m.

Particle induced x-ray emission (PIXE) - See Techniques for Materials Characterization, page **12-1**.

Partition function (q,z) - For a single molecule, $q = \Sigma_i g_i \exp(\varepsilon_i/kT)$, where ε_i is an energy level of degeneracy g_i, k the Boltzmann constant, and T the absolute temperature; the summation extends over all energy states. For a system of N non-interacting molecules which are indistinguishable, as in an ideal gas, the canonical partition function $Q = qN/N!$.

Pascal (Pa)* - The SI unit of pressure, equal to N/m². [1]

Paschen series - The series of lines in the spectrum of the hydrogen atom which corresponds to transitions between the state with principal quantum number $n = 3$ and successive higher states. The wavelengths are given by $1/\lambda = R_H(1/9 - 1/n^2)$, where $n = 4,5,6,\ldots$ and R_H is the Rydberg constant. The first member of the series ($n = 3 \leftrightarrow 4$), which is often called the P_α line, falls in the infrared at a wavelength of 1.875 μm.

Paschen-Back effect - In atomic spectroscopy, the decoupling of electron spin from orbital angular momentum as the strength of an external magnetic field is increased.

Pauli exclusion principle - The statement that two electrons in an atom cannot have identical quantum numbers; thus if there are two electrons in the same orbital, their spin quantum numbers must be of opposite sign.

Pearson symbol - A code for designating crystallographic information, including the crystal system, the lattice type, and the number of atoms per unit cell.

Péclet number (Pe) - A dimensionless quantity used in fluid mechanics, defined by $Pe = vl/a$, where v is velocity, l is length, and a is thermal diffusivity. [2]

Peltier effect - The absorption or generation of heat (depending on the current direction) which occurs when an electric current is passed through a junction between two materials.

Peptides - Amides derived from two or more amino carboxylic acid molecules (the same or different) by formation of a covalent bond from the carbonyl carbon of one to the nitrogen atom of another with formal loss of water. [5]

Permeability (μ) - Magnetic induction divided by magnetic field strength; i.e. $\mu = B/H$. The relative permeability $\mu_r = \mu/\mu_0$, where μ_0 is the permeability of a vacuum. [1]

Permittivity (ε) - Ratio of the electric displacement in a medium to the electric field strength. Also called dielectric constant. [1]

Peroxides - Compounds of structure ROOR in which R may be any organic group. In inorganic chemistry, salts of the anion O_2^{-2} [5]

Peroxy acids - Acids in which an acidic -OH group has been replaced by an -OOH group; e.g. $CH_3C(=O)OOH$ peroxyacetic acid, $PhS(=O)_2OOH$ benzeneperoxysulfonic acid. [5]

Petroleum ether - The petroleum fraction consisting of C_5 and C_6 hydrocarbons and boiling in the range 35-60°C; commonly used as a laboratory solvent.

pH* - A convenient measure of the acid-base character of a solution, usually defined by pH = $-\log [c(H^+)/\text{mol L}^{-1})]$, where $c(H^+)$ is the concentration of hydrogen ions. The more precise definition is in terms of activity rather than concentration. [2]

Phenols - Compounds having one or more hydroxy groups attached to a benzene or other arene ring. [5]

Phonon - A quantum of energy associated with a vibrational mode of a crystal lattice.

Phosphines - PH_3 and compounds derived from it by substituting one, two or three hydrogen atoms by hydrocarbyl groups. RPH_2, R_2PH and R_3P (R not equal to H) are called primary, secondary and tertiary phosphines, respectively. [5]

Phosphonium compounds - Salts (and hydroxides) $[R_4P]^+X^-$ containing tetracoordinate phosphonium ion and the associated anion. [5]

Phosphonium ylides - Compounds having the structure $R_3P^+\text{-}C^-R_2 \rightleftharpoons R_3P=CR_2$. Also known as Wittig reagents. [5]

Phosphoresence - The process by which a molecule is excited by light to a higher electronic state and then undergoes a radiationless transition to a state of different multiplicity from which it decays, after some delay, to the ground state. The emitted light is normally of longer wavelength than the exciting light because vibrational energy has been dissipated.

Photoelectric effect - The complete absorption of a photon by a solid with the emission of an electron.

Photon - An elementary particle of zero mass and spin 1/2. The photon is involved in electromagnetic interactions and is the quantum of electromagnetic radiation.

Photon stimulated desorption (PSD) - See Techniques for Materials Characterization, page **12**-1.

Pinacols - Tetra(hydrocarbyl)ethane-1,2-diols, $R_2C(OH)C(OH)R_2$, of which the tetramethyl example is the simplest one and is itself commonly known as pinacol. [5]

Pion - An elementary particle in the family of mesons. Pions have zero spin and may be neutral or charged. They participate in the strong interaction which holds the nucleus together.

pK* - The negative logarithm (base 10) of an equilibrium constant K. For pK_a, see Acid dissociation constant.

Planck constant (h)* - The elementary quantum of action, which relates energy to frequency through the equation $E = h\nu$.

Planck distribution - See Black body radiation

Planck function - A thermodynamic function defined by $Y = -G/T$, where G is Gibbs energy and T thermodynamic temperature. [2]

Plasma - A highly ionized gas in which the charge of the electrons is balanced by the charge of the positive ions, so that the system as a whole is electrically neutral.

Plasmon - A quantum associated with a plasma oscillation in the electron gas of a solid.

Point group* - A group of symmetry operations (rotations, reflections, etc.) that leave a molecule invariant. Every molecular conformation can be assigned to a specific point group, which plays a major role in determining the spectrum of the molecule.

Poise (P) - A non-SI unit of viscosity, equal to 0.1 Pa s.

Poiseuille's equation - A formula for the rate of flow of a viscous fluid through a tube:

$$\frac{dV}{dt} = \frac{\left(p_1{}^2 - p_2{}^2\right)\pi r^4}{16l\eta p_0}$$

where V is the volume as measured at pressure p_0; p_1 and p_2 are the pressures at each end of the tube; r is the radius and l the length of the tube; and η is the viscosity.

Poisson ratio (μ) - The absolute value of the ratio of the transverse strain to the corresponding axial strain resulting from uniformly distributed axial stress below the proportional limit (i.e., where Hooke's law is valid). [10]

Polariton - A quantum associated with the coupled modes of photons and optical phonons in an ionic crystal.

Polarizability (α)* - The change in dipole moment of a molecule produced by an external electric field; specifically, $\alpha_{ab} = \partial p_a/\partial E_b$, where p_a is the dipole moment component on the a axis and E_b is the component of the electric field strength along the b axis. [2]

Polymer - A substance composed of molecules of high relative molecular mass (molecular weight), the structure of which essentially comprises the multiple repetition of units derived, actually or conceptually, from molecules of low relative molecular mass. A single molecule of a polymer is called a macromolecule. [8]

Polypeptides - Peptides containing 10 or more amino acid residues. See also Peptides. [5]

Polysaccharides - Compounds consisting of a large number of monosaccharides linked glycosidically. This term is commonly used only for those containing more than ten monosaccharide residues. Also called glycans. [5]

Porphyrins - Natural pigments containing a fundamental skeleton of four pyrrole nuclei united through the α-positions by four methine groups to form a macrocyclic structure (porphyrin is designated porphine in Chemical Abstracts indexes). [5]

Positron - The antiparticle of the electron. It has the same mass and spin as an electron, and an equal but opposite charge.

Positronium - The hydrogen-like "atom" formed from a positron nucleus and an electron. Its lifetime is very short because of annihilation of the positron and electron.

Potential - See Electric potential

Potential energy (E_p, V, U) - The portion of the energy of a system that is associated with its position in a force field.

Pound (lb) - A non-SI unit of mass, equal to 0.4535924 kg.

Power (P) - Rate of energy transfer. For electrical circuits, this is equal to the product of current and potential difference, $P = IV$. [1]

Poynting vector (S) - For electromagnetic radiation, the vector product of the electric field strength and the magnetic field strength. [1]

Prandtl number (Pr) - A dimensionless quantity used in fluid mechanics, defined by $Pr = \eta/\rho a$, where η is viscosity, ρ is density, and a is thermal diffusivity. [2]

Pressure* - Force divided by area. [1]

Proteins - Naturally occurring and synthetic polypeptides having molecular weights greater than about 10,000 (the limit is not precise). See also Peptides. [5]

Proton* - A stable elementary particle of unit positive charge and spin 1/2. Protons and neutrons, which are collectively called nucleons, are the constituents of the nucleus.

Pulsar - A neutron star which rotates rapidly and emits electromagnetic radiation in regular pulses at a frequency related to the rotation period.

Purine bases* - Purine and its substitution derivatives, especially naturally occurring examples. [5]

Pyrimidine bases* - Pyrimidine and its substitution derivatives, especially naturally occurring examples. [5]

Q-switching - A technique for obtaining very high power from a laser by keeping the Q factor of the laser cavity low while the population inversion builds up, then suddenly increasing the Q to initiate the stimulated emission.

Quad - A unit of energy defined as 10^{15} Btu, equal to approximately 1.055056×10^{18} J.

Quadrupole moment - A coefficient of the third term (after monopole and dipole) in the power series expansion of the electric potential of an array of charges. A nucleus of spin greater than 1/2 has a nonvanishing nuclear quadrupole moment which can interact with the electric field gradient of the surrounding electrons. Molecular quadrupole moments have an influence on intermolecular forces.

Quality factor (Q) - The ratio of the absolute value of the reactance of an electrical system to the resistance; thus a measure of the energy stored per cycle relative to the energy dissipated.

Quantum yield - In photochemistry, the number of moles transformed in a specific process, either physically (e.g., by emission of photons) or chemically, per mole of photons absorbed by the system. [3]

Quark - An elementary entity which has not been directly observed but is considered a constituent of protons, neutrons, and other hadrons.

Quasar - An extragalactic object emitting electromagnetic radiation at a very high power level and showing a very large red shift, thus indicating that the object is receding at a speed approaching the speed of light.

Quasicrystal - A solid having conventional crystalline properties but whose lattice does not display translational periodicity.

Quaternary ammonium compounds - Derivatives of ammonium compounds, $NH_4{}^+ Y^-$, in which all four of the hydrogens bonded to nitrogen have been replaced with hydrocarbyl groups. Compounds having a carbon-nitrogen double bond (i.e. $R_2C=N^+R_2Y^-$) are more accurately called iminium compounds. [5]

Quinones - Compounds having a fully conjugated cyclic dione structure, such as that of benzoquinones, derived from aromatic compounds by conversion of an even number of -CH= groups into -C(=O)- groups with any necessary rearrangement of double bonds. [5]

Racemic mixture - A mixture of equal amounts of a pair of enantiomers (optical isomers); such a mixture is not optically active.

Rad - A non-SI unit of absorbed dose of radiation, equal to 0.01 Gy.

Radiance (L) - The radiant intensity in a given direction from an element of a surface, divided by the area of the orthogonal projection of this element on a plane perpendicular to the given direction. [1]

Radiant intensity (I) - The radiant energy flux leaving an element of a source within an element of solid angle, divided by that element of solid angle. [1]

Radicals - Molecular entities possessing an unpaired electron, such as ·CH_3, ·SnH_3, ·Cl. (In these formulas the dot, symbolizing the unpaired electron, should be placed so as to indicate the atom of highest spin density, if this is possible). [5]

Raman effect - The inelastic scattering of light by a molecule, in which the incident photon either gives up to, or receives energy from, one of the internal vibrational modes of the molecule. The scattered light thus has either a lower frequency (Stokes radiation) or higher frequency (anti-Stokes radiation) than the incident light. These shifts provide a measure of the normal vibrational frequencies of the molecule.

Rankine cycle - A thermodynamic cycle which can be used to calculate the ideal performance of a heat engine that uses a condensable vapor as the working fluid (e.g., a steam engine or a heat pump).

Rankine temperature - A thermodynamic temperature scale based on a temperature interval °R = (5/9) K; i.e., T/°R = (9/5)T/K = t/°F + 459.67.

Raoult's law - The expression for the vapor pressure p_i of component i in an ideal solution, viz., $p_i = x_i p_{i0}$, where x_i is the mole fraction of component i and p_{i0} the vapor pressure of the pure substance i.

Rare earth elements - The elements Sc, Y, and the lanthanides (La, Ce, Pr, Nd, Pm, Sm, Eu, Gd, Tb, Dy, Ho, Er, Tm, Yb, Lu). [7]

Rayleigh number (Ra) - A dimensionless quantity used in fluid mechanics, defined by $Ra = l^3 g\alpha\Delta T\rho/\eta a$, where l is length, g is acceleration of gravity, α is cubic expansion coefficient, T is temperature, ρ is density, η is viscosity, and a is thermal diffusivity. [2]

Rayleigh scattering - The scattering of light by particles which are much smaller than the wavelength of the light. It is characterized by a scattered intensity which varies as the inverse fourth power of the wavelength.

Rayleigh wave - A guided elastic wave along the surface of a solid; also called surface acoustic wave.

Reactance (X) - The imaginary part of impedance. For an inductive reactance L and a capacitive reactance C in series, the reactance is $X = L\omega - 1/(C\omega)$, where ω is 2π times the frequency of the current. [1]

Red shift - A displacement of a spectral line toward longer wavelengths. This can occur through the Doppler effect (e.g., in the light from receding galaxies) or, in the general theory of relativity, from the effects of a star's gravitational field.

Reflectance (ρ) - Ratio of the radiant or luminous flux at a given wavelength that is reflected to that of the incident radiation. Also called reflection factor. [1]

Reflection high energy electron diffraction (RHEED) - See Techniques for Materials Characterization, page **12**-1.

Relative humidity* - The ratio of the partial pressure of water vapor in air to the saturation vapor pressure of water at the same temperature, expressed as a percentage. [10]

Relative molar mass - See Molecular weight.

Rem - A non-SI unit of dose equivalent, equal to 0.01 Sv.

Resistance (R) - Electric potential difference divided by current when there is no electromotive force in the conductor. This definition applies to direct current. More generally, resistance is defined as the real part of impedance. [1]

Resistivity (ρ) - Electric field strength divided by current density when there is no electromotive force in the conductor. Resistivity is an intrinsic property of a material. For a conductor of uniform cross section with area A and length L, and whose resistance is R, the resistivity is given by $\rho = RA/L$. [1]

Reynolds number (Re) - A dimensionless quantity used in fluid mechanics, defined by $Re = \rho v l/\eta$, where ρ is density, v is velocity, l is length, and η is viscosity. [2]

Rheology - The study of the flow of liquids and deformation of solids. Rheology addresses such phenomena as creep, stress relaxation, anelasticity, nonlinear stress deformation, and viscosity.

Ribonucleic acids (RNA) - Naturally occurring polyribonucleotides. See also nucleic acids, nucleosides, nucleotides, ribonucleotides. [5]

Ribonucleotides - Nucleotides in which the glycosyl group is a ribosyl group. See also nucleotides. [5]

Roentgen (R) - A unit used for expressing the charge (positive or negative) liberated by x-ray or γ radiation in air, divided by the mass of air. A roentgen is defined as 2.58×10^{-4} C/kg.

Rotational constants - In molecular spectroscopy, the constants appearing in the expression for the rotational energy levels as a function of the angular momentum quantum numbers. These constants are proportional to the reciprocals of the principal moments of inertia, averaged over the vibrational motion.

Rutherford back scattering (RBS) - See Techniques for Materials Characterization, page **12**-1.

Rydberg constant (R_∞)* - The fundamental constant which appears in the equation for the energy levels of hydrogen-like atoms; i.e., $E_n = hcR_\infty Z^2\mu/n^2$, where h is Planck's constant, c the speed of light, Z the atomic number, μ the reduced mass of nucleus and electron, and n the principal quantum number ($n = 1, 2, \ldots$).

Rydberg series - A regular series of lines in the spectrum of an atom or molecule, with the spacing between successive lines becoming smaller as the frequency increases (wavelength decreases). The series eventually converges to a limit which usually corresponds to the complete removal of an electron from the atom or molecule.

Sackur-Tetrode equation* - An equation for the molar entropy S_m of an ideal monatomic gas: $S_m = R\ln(e^{5/2} V/N_A\Lambda^3)$, where R is the molar gas constant, V is the volume, and N_A is Avogadro's number. The constant Λ is given by $\Lambda = h/(2\pi mkT)^{1/2}$, where h is Planck's constant, m the atomic mass, k the Boltzmann constant, and T the temperature.

Salinity (S)* - A parameter used in oceanography to describe the concentration of dissolved salts in seawater. It is defined in terms of electrical conductivity relative to a standard solution of KCl. When expressed in units of parts per thousand, S may be roughly equated to the concentration of dissolved material in grams per kilogram of seawater.

Salt - An ionic compound formed by the reaction of an acid and a base.

Scanned probe microscopy (SPM) - See Techniques for Materials Characterization, page **12**-1.

Scanning electron microscopy (SEM) - See Techniques for Materials Characterization, page **12**-1.

Scanning laser acoustic microscopy (SLAM) - See Techniques for Materials Characterization, page **12**-1.

Scanning transmission electron microscopy (STEM) - See Techniques for Materials Characterization, page **12**-1.

Scanning tunneling microscopy (STM) - See Techniques for Materials Characterization, page **12**-1.

Schiff bases - Imines bearing a hydrocarbyl group on the nitrogen atom: $R_2C=NR'$ (R' not equal to H). Considered by many to be synonymous with azomethines. [5]

Schmidt number (Sc) - A dimensionless quantity used in fluid mechanics, defined by $Sc = \eta/\rho D$, where η is viscosity, ρ is density, and D is diffusion coefficient. [2]

Schottky barrier - A potential barrier associated with a metal-semiconductor contact. It forms the basis for the rectifying device known as the Schottky diode.

Schrödinger equation - The basic equation of wave mechanics which, for systems not dependent on time, takes the form:

$$-(\hbar/2m)\nabla^2\psi + V\psi = E\psi$$

where ψ is the wavefunction, V is the potential energy expressed as a function of the spatial coordinates, E is an energy eigenvalue, ∇^2 is the Laplacian operator, $\hbar$ is Planck's constant divided by 2π, and m is the mass.

Second (s)* - The SI base unit of time. [1]

Second radiation constant (c₂)* - See First radiation constant.

Secondary ion mass spectroscopy (SIMS) - See Techniques for Materials Characterization, page **12**-1.

Seebeck effect - The development of a potential difference in a circuit where two different metals or semiconductors are joined and their junctions maintained at different temperatures. It is the basis of the thermocouple.

Selenides - Compounds having the structure RSeR (R not equal to H). They are thus selenium analogues of ethers. Also used for metal salts of H_2Se. [5]

Semicarbazones - Compounds having the structure $R_2C=NNHC(=O)NH_2$, formally derived by condensation of aldehydes or ketones with semicarbazide [$NH_2NHC(=O)NH_2$]. [5]

Semiconductor - A material in which the highest occupied energy band (valence band) is completely filled with electrons at $T = 0$ K, and the energy gap to the next highest band (conduction band) ranges from 0 to 4 or 5 eV. With increasing temperature electrons are excited into the conduction band, leading to an increase in the electrical conductivity.

Semiquinones - Radical anions having the structure -O-Z-O· where Z is an ortho- or para-arylene group or analogous heteroarylene group; they are formally generated by the addition of an electron to a quinone. [5]

SI units* - The International System of Units adopted in 1960 and recommended for use in all scientific and technical fields. [1]

Siemens (S)* - The SI unit of electric conductance, equal to Ω^{-1}. [1]

Sievert (Sv)* - The SI unit of dose equivalent (of radiation), equal to J/kg. [1]

Silanes - Saturated silicon hydrides, analogues of the alkanes; i.e. compounds of the general formula Si_nH_{2n+2}. Silanes may be subdivided into silane, oligosilanes, and polysilanes. Hydrocarbyl derivatives are often referred to loosely as silanes. [5]

Silicones - Polymeric or oligomeric siloxanes, usually considered unbranched, of general formula [-OSiR₂-]ₙ (R not equal to H). [5]

Siloxanes - Saturated silicon-oxygen hydrides with unbranched or branched chains of alternating silicon and oxygen atoms (each silicon atom is separated from its nearest silicon neighbors by single oxygen atoms). [5]

Skin effect - The concentration of high frequency alternating currents near the surface of a conductor.

Slater orbital - A particular mathematical expression for the radial part of the wave function of a single electron, which is used in quantum-mechanical calculations of the energy and other properties of atoms and molecules.

Small angle neutron scattering (SANS) - See Techniques for Materials Characterization, page **12**-1.

Snell's law - The relation between the angle of incidence i and the angle of refraction r of a light beam which passes from a medium of refractive index n_0 to a medium of index n_1, viz., $\sin i/\sin r = n_1/n_0$.

Solar constant* - The mean radiant energy flux from the sun on a unit surface normal to the direction of the rays at the mean distance of the earth from the sun. The value is approximately 1373 W/m².

Solar wind - The stream of high velocity hydrogen and helium ions emitted by the sun which flows through the solar system and beyond.

Soliton - A spatially localized wave in a solid or liquid that can interact strongly with other solitons but will afterwards regain its original form.

Solubility* - A quantity expressing the maximum concentration of some material (the solute) that can exist in another liquid or solid material (the solvent) at thermodynamic equilibrium at specified temperature and pressure. Common measures of solubility include the mass of solute per unit mass of solution (mass fraction), mole fraction of solute, molality, molarity, and others.

Solubility product constant (K_{sp})* - The equilibrium constant for the dissolution of a sparsely soluble salt into its constituent ions.

Space group* - A group of symmetry operations (reflections, rotations, etc.) that leave a crystal invariant. A total of 230 space groups have been identified.

Spark source mass spectroscopy (SSMS) - See Techniques for Materials Characterization, page **12**-1.

Specific gravity - Ratio of the mass density of a material to that of water. Since one must specify the temperature of both the sample and the water to have a precisely defined quantity, the use of this term is now discouraged.

Specific heat - Heat capacity divided by mass. See Heat capacity.

Specific quantity - It is often convenient to express an extensive quantity (e.g., volume, enthalpy, heat capacity, etc.) as the actual value divided by mass. The resulting quantity is called specific volume, specific enthalpy, etc.

Specific rotation [α]θλ- For an optically active substance, defined by $[\alpha]^\theta_\lambda = \alpha/\gamma l$, where α is the angle through which plane polarized light is rotated by a solution of mass concentration γ and path length l. Here θ is the Celsius temperature and λ the wavelength of the light at which the measurement is carried out. Also called specific optical rotatory power. [2]

Spin (s, I)* - A measure of the intrinsic angular momentum of a particle, which it possesses independent of its orbital motion. The symbol s is used for the spin quantum number of an electron, while I is generally used for nuclear spin.

Spiro compounds - Compounds having one atom (usually a quaternary carbon) as the only common member of two rings. [5]

Stacking fault - An error in the normal sequence of layer growth in a crystal.

Standard mean ocean water (SMOW) - A standard sample of pure water of accurately known isotopic composition which is maintained by the International Atomic Energy Agency. It is used for precise calibration of density and isotopic composition measurements.

Standard reduction potential ($E°$) - The zero-current potential of a cell in which the specified reduction reaction occurs at the right-hand electrode and the left-hand electrode is the standard hydrogen electrode. Also called Standard electrode potential.

Standard state - A defined state (specified temperature, pressure, concentration, etc.) for tabulating thermodynamic functions and carrying out thermodynamic calculations. The standard state pressure is usually taken as 100,000 Pa (1 bar), but various standard state temperatures are used. [2]

Stanton number (St) - A dimensionless quantity used in fluid mechanics, defined by $St = h/\rho v c_p$, where h is coefficient of heat transfer, ρ is density, v is velocity, and c_p is specific heat capacity at constant pressure. [2]

Stark effect - The splitting of an energy level of an atom or molecule, and hence a splitting of spectral lines arising from that level, as a result of the application of an external electric field.

Statistical weight (g) - The number of distinct states corresponding to the same energy level. Also called degeneracy.

Stefan-Boltzmann constant (σ)* - Constant in the equation for the radiant exitance M (radiant energy flux per unit area) from a black body at thermodynamic temperature T, viz. $M = \sigma T^4$. [1]

Stibines - SbH_3 and compounds derived from it by substituting one, two or three hydrogen atoms by hydrocarbyl groups: R_3Sb. $RSbH_2$, R_2SbH, and R_3Sb (R not equal to H) are called primary, secondary and tertiary stibines, respectively. [5]

Stochastic process - A process which involves random variables and whose outcome can thus be described only in terms of probabilities.

Stoichiometric number (ν) - The number appearing before the symbol for each compound in the equation for a chemical reaction. By convention, it is negative for reactants and positive for products. [2]

Stokes (St) - A non-SI unit of kinematic viscosity, equal to 10^{-4} m^2/s.

Stokes' law - The statement, valid under certain conditions, that the viscous force F experienced by a sphere of radius a moving at velocity v in a medium of viscosity η is given by $F = -6\pi\eta av$.

Strain - The deformation of a body that results from an applied stress.

Stratosphere - The part of the earth's atmosphere extending from the top of the troposphere (typically 10 to 15 km above the surface) to about 50 km. It is characterized by an increase in temperature with increasing altitude.

Stress - Force per unit area (pressure) applied to a body. Tensile stress tends to stretch or compress the body in the direction of the applied force. Sheer stress results from a tangential force which tends to twist the body.

Strong interaction - The short range (order of 1 fm) attractive forces between protons, neutrons, and other hadrons which are responsible for the stability of the nucleus.

Strouhal number (Sr) - A dimensionless quantity used in fluid mechanics, defined by $Sr = lf/v$, where l is length, f is frequency, and v is velocity. [2]

Structure factor - In x-ray crystallography, the sum of the scattering factors of all the atoms in a unit cell, weighted by an appropriate phase factor. The intensity of a given reflection is proportional to the square of the structure factor.

Sublimation pressure - The pressure of a gas in equilibrium with a solid at a specified temperature.

Sulfides - Compounds having the structure RSR (R not equal to H). Such compounds were once called thioethers. In an inorganic sense, salts or other derivatives of hydrogen sulfide. [5]

Sulfones - Compounds having the structure, $RS(=O)_2R$ (R not equal to H), e.g. $C_2H_5S(=O)_2CH_3$, ethyl methyl sulfone. [5]

Sulfonic acids - $HS(=O)_2OH$, sulfonic acid, and its S-hydrocarbyl derivatives. [5]

Sulfoxides - Compounds having the structure $R_2S=O$ (R not equal to H), e.g. $Ph_2S=O$, diphenyl sulfoxide. [5]

Superconductor - A material that experiences a nearly total loss of electrical resistivity below a critical temperature T_c. The effect can occur in pure metals, alloys, semiconductors, organic compounds, and certain inorganic solids.

Superfluid - A fluid with near-zero viscosity and extremely high thermal conductivity. Liquid helium exhibits these properties below 2.186 K (the λ point).

Supernova - A star in the process of exploding because of instabilities which follow the exhaustion of its nuclear fuel.

Surface analysis by laser ionization (SALI) - See Techniques for Materials Characterization, page **12-1**.

Surface tension (γ,σ)* - The force per unit length in the plane of the interface between a liquid and a gas, which resists an increase in the area of that surface. It can also be equated to the surface Gibbs energy per unit area.

Surfactant - A substance which lowers the surface tension of the medium in which it is dissolved, and/or the interfacial tension with other phases, and accordingly is positively adsorbed at the liquid-vapor or other interfaces. [3]

Susceptance (B) - Imaginary part of admittance. [1]

Svedberg - A non-SI unit of time, used to express sedimentation coefficients, equal to 10^{-13} s.

Syndiotactic macromolecule - A tactic macromolecule, essentially comprising alternating enantiomeric configurational base units which have chiral or prochiral atoms in the main chain in a unique arrangement with respect to their adjacent constitutional units. In this case the repeating unit consists of two configurational base units that are enantiomeric. [8]

Tacticity - The orderliness of the succession of configurational repeating units of a macromolecule or oligomer molecule. In a tactic macromolecule essentially all the configurational repeating units are identical with respect to directional sense. See Configurational repeating unit, Isotactic, Syndiotactic. [8]

Tautomerism - Isomerism of the general form G-X-Y=Z $\rightleftharpoons$ X=Y-Z-G, where the isomers (called tautomers) are readily interconvertible; the atoms connecting the groups X, Y, Z are typically any of C, H, O, or S, and G is a group which becomes an electrofuge (i.e., a group that does not carry away the bonding electron pair when it leaves its position in the molecule) or nucleofuge (a group that does carry away the bonding electrons when leaving) during isomerization. The commonest case, when the electrofuge is H^+, is also known as prototropy. A common example, written so as to illustrate the general pattern given above, is keto-enol tautomerism, such as

$$H-O-C(CH_3)=CH-CO_2Et \text{ (enol)} \rightleftharpoons (CH_3)C(=O)-CH_2-CO_2Et \text{ (keto)}$$

In some cases the interconversion rate between tautomers is slow enough to permit isolation of the separate keto and enol forms. [5]

Tensile strength* - In tensile testing, the ratio of maximum load a body can bear before breaking to original cross-sectional area. Also called ultimate strength. [11]

Terpenes - Hydrocarbons of biological origin having carbon skeletons formally derived from isoprene $[CH_2=C(CH_3)CH=CH_2]$. [5]

Terpenoids - Natural products and related compounds formally derived from isoprene units. They contain oxygen in various functional groups. The skeleton of terpenoids may differ from strict additivity of isoprene units by the loss or shift of a methyl (or other) group. [5]

Tesla (T)* - The SI unit of magnetic flux density (B), equal to V s/m². [1]

Thermal conductivity* - Rate of heat flow divided by area and by temperature gradient. [1]

Thermal diffusivity - Thermal conductivity divided by density and by specific heat capacity at constant pressure. [1]

Thermal expansion coefficient (α)* - The linear expansion coefficient is defined by $\alpha_l = (1/l)(dl/dT)$; the volume expansion coefficient by $\alpha_V = (1/V)(dV/dT)$. [1]

Thermionic emission - The emission of electrons from a solid as a result of heat. The effect requires a high enough temperature to impart sufficient kinetic energy to the electrons to exceed the work function of the solid.

Thermodynamic laws - The foundation of the science of thermodynamics:

First law: The internal energy of an isolated system is constant; if energy is supplied to the system in the form of heat dq and work dw, then the change in energy $dU = dq + dw$.

Second law: No process is possible in which the only result is the transfer of heat from a reservoir and its complete conversion to work.

Third law: The entropy of a perfect crystal approaches zero as the thermodynamic temperature approaches zero.

Thermoelectric power - For a bar of a pure material whose ends are at different temperatures, the potential difference divided by the difference in temperature of the ends. See also Seeback effect.

Thermogravimetric analysis (TGA) - See Techniques for Materials Characterization, page **12**-1.

Thermosphere - The layer of the earth's atmosphere extending from the top of the mesosphere (typically 80-90 km above the surface) to about 500 km. It is characterized by a rapid increase in temperature with increasing altitude up to about 200 km, followed by a leveling off in the 300-500 km region.

Thiols - Compounds having the structure RSH (R not equal to H). Also known by the term mercaptans (abandoned by IUPAC); e.g. CH_3CH_2SH, ethanethiol. [5]

Thomson coefficient (μ, τ) - The heat power developed in the Thomson effect (whereby heat is evolved in a conductor when a current is flowing in the presence of a temperature gradient), divided by the current and the temperature difference. [1]

Tonne (t) - An alternative name for megagram (1000 kg). [1]

Torque (T) - For a force F that produces a torsional motion, $T = r \times F$, where r is a vector from some reference point to the point of application of the force.

Torr - A non-SI unit of pressure, equal to 133.322 Pa. The name is generally considered interchangeable with millimeter of mercury.

Townsend coefficient - In a radiation counter, the number of ionizing collisions by an electron per unit path length in the direction of an applied electric field.

Transducer - Any device that converts a signal from acoustical, optical, or some other form of energy into an electrical signal (or vice versa) while preserving the information content of the original signal.

Transistor - A voltage amplifier using controlled electron currents inside a semiconductor.

Transition metals - Elements characterized by a partially filled d subshell. The First Transition Series comprises Sc, Ti, V, Cr, Mn, Fe, Co, Ni, Cu. The Second and Third Transition Series include the lanthanides and actinides, respectively. [7]

Transition probability* - See Einstein transition probability.

Transmittance (τ) - Ratio of the radiant or luminous flux at a given wavelength that is transmitted to that of the incident radiation. Also called transmission factor. [1]

Tribology - The study of frictional forces between solid surfaces.

Triple point* - The point in p,T space where the solid, liquid, and gas phases of a substance are in thermodynamic equilibrium. The corresponding temperature and pressure are called the triple point temperature and triple point pressure.

Troposphere - The lowest part of the earth's atmosphere, extending to 10-15 km above the surface. It is characterized by a decrease in temperature with increasing altitude. The exact height varies with latitude and season.

Tunnel diode - A device involving a p-n junction in which both sides are so heavily doped that the Fermi level on the p-side lies in the valence band and on the n-side in the conduction band. This leads to a current-voltage curve with a maximum, so that the device exhibits a negative resistance in some regions.

Ultraviolet photoelectron spectroscopy (UPS) - See Techniques for Materials Characterization, page **12**-1.

Umklapp process - A process involving the interaction of three or more waves (lattice or electron) in a solid in which the sum of the wave vectors does not equal zero.

Unified atomic mass unit (u)* - A unit of mass used in atomic, molecular, and nuclear science, defined as the mass of one atom of ^{12}C divided by 12. Its approximate value is 1.66054×10^{-27} kg. [1]

Universal time (t_U, UT) - Mean solar time counted from midnight at the Greenwich meridian. Also called Greenwich mean time (GMT). The interval of mean solar time is based on the average, over one year, of the time between successive transits of the sun across the observer's meridian.

Vacancy - A missing atom or ion in a crystal lattice.

Van Allen belts - Two toroidal regions above the earth's atmosphere containing protons and electrons. The outer belt at about 25,000 km above the surface is probably of solar origin. The inner belt at about 3000 km contains more energetic particles from outside the solar system.

Van der Waals' equation* - An equation of state for fluids which takes the form:

$$pV_m = RT\left(\frac{1}{V_m - b} - \frac{a}{V_m^2}\right)$$

where p is pressure, V_m is molar volume, T is temperature, R is the molar gas constant, and a and b are characteristic parameters of the substance which describe the effect of attractive and repulsive intermolecular forces, respectively.

Van der Waals' force - The weak attractive force between two molecules which arises from electric dipole interactions. It can lead to the formation of stable but weakly bound dimer molecules or clusters.

Van't Hoff equation - The equation expressing the temperature dependence of the equilibrium constant K of a chemical reaction:

$$\frac{d\ln k}{dT} = \frac{\Delta_r H^\circ}{RT^2}$$

where $\Delta_r H^\circ$ is the standard enthalpy of reaction, R the molar gas constant, and T the temperature. Also called van't Hoff isochore.

Vapor pressure* - The pressure of a gas in equilibrium with a liquid (or, in some usage, a solid) at a specified temperature.

Varistor - A device that utilizes the properties of certain metal oxides with small amounts of impurities, which show abrupt nonlinearities at specific voltages where the material changes from a semiconductor to an insulator.

Velocity (v) - Rate of change of distance with time.

Verdet constants (V)* - Angle of rotation of a plane polarized light beam passing through a medium in a magnetic field, divided by the field strength and by the path length.

Virial equation of state* - An equation relating the pressure p, molar volume V_m, and temperature T of a real gas in the form of an expansion in powers of the molar volume, viz., $pV_m = RT(1 + BV_m^{-1} + CV_m^{-2} + \ldots)$, where R is the molar gas constant. B is called the second virial coefficient, C the third virial coefficient, etc. The virial coefficients are functions of temperature.

Viscosity (η)* - The proportionality factor between sheer rate and sheer stress, defined through the equation $F = \eta A(dv/dx)$, where F is the tangential force required to move a planar surface of area A at velocity v relative to a parallel surface separated from the first by a distance x. Sometimes called dynamic or absolute viscosity. The term kinematic viscosity (symbol ν) is defined as η divided by the mass density.

Volt (V)* - The SI unit of electric potential, equal to W/A. [1]

Volume fraction (ϕ_j) - Defined as $V_j / \Sigma_i V_i$, where V_j is the volume of the specified component and the V_i are the volumes of all the components of a mixture prior to mixing. [2]

Watt (W)* - The SI unit of power, equal to J/s. [1]

Wave function - A function of the coordinates of all the particles in a quantum mechanical system (and, in general, of time) which fully describes the state of the system. The product of the wave function and its complex conjugate is proportional to the probability of finding a particle at a particular point in space.

Weak interaction - The weak forces (order of 10^{-12} of the strong interaction) between elementary particles which are responsible for beta decay and other nuclear effects.

Weber (Wb)* - The SI unit of magnetic flux, equal to V s. [1]

Weber number (We) - A dimensionless quantity used in fluid mechanics, defined by $We = \rho v^2 l/\gamma$, where ρ is density, v is velocity, l is length, and γ is surface tension. [2]

Weight - That force which, when applied to a body, would give it an acceleration equal to the local acceleration of gravity. [1]

Wiedeman-Franz law - The law stating that the thermal conductivity k and electrical conductivity σ of a pure metal are related by $k = L\sigma T$, where T is the temperature and L (called the Lorenz ratio) has the approximate value 2.45×10^{-8} V^2/K^2.

Wien displacement law - The relation, which can be derived from the Planck formula for black body radiation, that $\lambda_{max}T = 0.0028978$ m K, where λ_{max} is the wavelength of maximum radiance at temperature T.

Wigner-Seitz method - A method of calculating electron energy levels in a solid using a model in which each electron is subject to a spherically symmetric potential.

Wittig reagents - See phosphonium ylides.

Work (W) - Force multiplied by the displacement in the direction of the force. [1]

Work function (Φ)* - The energy difference between an electron at rest at infinity and an electron at the Fermi level in the interior of a substance. It is thus the minimum energy required to remove an electron from the interior of a solid to a point just outside the surface. [1]

X unit (X) - A unit of length used in x-ray crystallography, equal to approximately 1.002×10^{-13} m.

X-ray photoelectron spectroscopy (XPS) - See Techniques for Materials Characterization, page **12-1**.

Yield strength - The stress at which a material exhibits a specified deviation (often chosen as 0.2% for metals) from proportionality of stress and strain. [11]

Young's modulus (E) - In tension or compression of a body below its elastic limit, the ratio of stress to corresponding strain. Since strain is normally expressed on a fractional basis, Young's modulus has dimensions of pressure. Also called elastic modulus. [11]

Zeeman effect - The splitting of an energy level of an atom or molecule, and hence a splitting of spectral lines arising from that level, as a result of the application of an external magnetic field.

Zener diode - A control device utilizing a p-n junction with a well defined reverse-bias avalanche breakdown voltage.

Zeotrope - A liquid mixture that shows no maximum or minimum when vapor pressure is plotted against composition at constant temperature. See Azeotrope.

Zero-point energy - The energy possessed by a quantum mechanical system as a result of the uncertainty principle even when it is in its lowest energy state; e.g., the difference between the lowest energy level of a harmonic oscillator and the minimum in the potential well.

Zeta potential (ζ) - The electric potential at the surface of a colloidal particle relative to the potential in the bulk medium at a long distance. Also called electrokinetic potential.

Zwitterions - Neutral compounds having formal unit electrical charges of opposite sign. Some chemists restrict the term to compounds with the charges on non-adjacent atoms. Sometimes referred to as inner salts, dipolar ions (a misnomer). [5]

Section 3
Physical Constants of Organic Compounds

PHYSICAL CONSTANTS OF ORGANIC COMPOUNDS

The basic physical constants for over 12,000 organic compounds are presented in this table, along with structures and references to other sources of information. An effort has been made to include the compounds most frequently encountered in the laboratory, the workplace, and the environment. The selection was based mainly on the appearance of the compounds in various specialized tables in this *Handbook* and in other widely used reference sources, such as the *Merck Index* and the *DIPPR Database of Pure Compound Properties*. The occurrence of a compound on regulatory lists of hazardous chemicals was also taken into consideration, as was the availability of reliable physical constant data. Clearly, criteria of this type are somewhat subjective, and compounds considered important by some users have undoubtedly been omitted. Suggestions for additional compounds or other improvements are welcomed.

The data in the table have been taken from many sources, including both compilations and the primary literature. Where conflicts were found, the value deemed most reliable was chosen. Some of the useful compilations of physical property data are listed at the end of this introduction.

The table is arranged alphabetically by the primary name, which is generally the Index Name from the 8th or 9th Collective Index of Chemical Abstracts Service (CAS). In a few cases, especially pesticides and pharmaceuticals, the common name is used rather than the more complex systematic name. By convention, CAS Index Names are written in inverted order, e.g., chloromethane is listed as methane, chloro and ethyl acetate as acetic acid, ethyl ester. Furthermore, certain important compounds are listed under Index Names which differ from the names by which they are commonly known (e.g. aniline appears as benzenamine and acetone as 2-propanone). In order to facilitate the location of compounds in the table, three indexes are provided:

- **Synonym Index:** Includes common synonyms, but not the primary name by which the table is arranged.
- **Molecular Formula Index:** Lists compounds by molecular formula in the Hill order (see Preface to this *Handbook*).
- **CAS Registry Number Index:** Lists compounds by Chemical Abstracts Service Registry Number.

Two lines of data appear for each compound. The explanation of the data fields follows.

Top Line:
- **No.:** An identification number used in the indexes and to identify the structure diagrams.
- **Name:** Primary name, generally the CAS Index Name.
- **Mol. Form.:** The molecular formula written in the Hill convention.
- **CAS RN:** The Chemical Abstracts Service Registry Number assigned by CAS as a unique identifier for the compound.
- **Merck No:** Monograph Number in *The Merck Index, Eleventh Edition*. It should be noted that this is not a unique identifier for a single compound, since several derivatives or isomers of a compound may be included in the same Monograph.
- **Beil. Ref:** Citation to the *Beilstein Handbook of Organic Chemistry*. An entry of 5-18-11-01234, for example, indicates that the compound may be found in the 5th Series, Volume 18, Subvolume 11, page 1234.
- **Solubility:** Solubility in common solvents on a relative scale: 1 = insoluble; 2 = slightly soluble; 3 = soluble; 4 = very soluble; 5 = miscible; 6 = decomposes. See List of Abbreviations for the solvent abbreviations.

Bottom line:
- **Synonym:** A synonym in common use. When the primary name is non-systematic, the systematic name appears here.
- **Mol. Wt.:** Molecular weight (relative molar mass) as calculated with the 1991 IUPAC Standard Atomic Weights.
- **mp/°C:** Normal melting point in °C. Although some values are quoted to 0.1°C, uncertainties are typically several degrees Celsius. A value is sometimes followed by "dec", indicating decomposition is observed at the stated temperature (so that it is probably not a true melting point). See the List of Abbreviations for other abbreviations.
- **bp/°C:** Boiling point in °C. When available, the normal boiling point is given first, without a superscript. This is the temperature at which the liquid phase is in equilibrium with the vapor at a pressure of 760 mmHg (101.325 kPa). Boiling point values at reduced pressure are also given in many cases; here the superscript indicates the pressure in mmHg. A "dec" or "exp" following the value indicates decomposition or explosion has been observed at the boiling point. A simple entry of "exp" (sometimes followed by a temperature) indicates explosion may occur on heating, even below the boiling point. An entry of "sub" indicates that no boiling point is available, but measurable vapor (sublimation) pressure has been observed upon heating the solid. A temperature may be given, but no precise meaning can be attached because the pressure is not specified.
- **den/g cm^{-3}:** density (mass per unit volume) in g/cm^{-3}. The superscript indicates the temperature in °C. Values are given only for the liquid and solid phases, and all values are true densities, not specific gravities. The number of decimal places gives a rough estimate of the accuracy of the value.
- **n_D:** Refractive index, at the temperature indicated by the superscript. Unless otherwise indicated, all values refer to a wavelength of 589 nm (sodium D line). Values are given only for liquids and solids.

Structures are given, when available, in the section following the main table, using the **No.** in the first column as the linking identifier.

LIST OF ABBREVIATIONS

Ac	Acetyl	EtOH	Ethanol
ace	Acetone	exp	Explodes
acid	Acids	gl	Glacial
AcOEt	Ethyl acetate	HOAc	Acetic acid
alk	Alkalis	hp	Heptane
anh	Anhydrous	hx	Hexane
aq	Aqueous	lig	Ligroin
bz	Benzene	MeOH	Methanol
chl	Chloroform	os	Organic solvents
con	Concentrated	peth	Petroleum ether
ctc	Carbon tetrachloride	$PhNO_2$	Nitrobenzene
cyhex	Cyclohexane	PrOH	1-Propanol
dec	Decomposes	py	Pyridine
dil	Dilute	sub	Sublimes
diox	Dioxane	sulf	Sulfuric acid
DMF	Dimethyl formamide	temp.	Temperature
DMSO	Dimethyl sulfoxide	tfa	Trifluoroacetic acid
DMSO-d_6	Deuterated dimethyl sulfoxide	tol	Toluene
eth	Ethyl ether	xyl	Xylene

REFERENCES

1. Lide, D. R., and Milne, G. W. A., Editors, *Handbook of Data on Organic Compounds, Third Edition*, CRC Press, Boca Raton, FL, 1993.
2. Luckenbach, R., Editor, *Beilstein Handbook of Organic Chemistry*, Springer-Verlag, Heidelberg. Data can also be accessed through the on-line Beilstein Database.
3. Budavari, S., Editor, *The Merck Index, Eleventh Edition*, Merck & Co., Rahway, NJ, 1989.
4. Daubert, T. E., Danner, R. P., Sibul, H. M., and Stebbins, C. C., *Physical and Thermodynamic Properties of Pure Compounds: Data Compilation*, extant 1994 (core with 4 supplements), Taylor & Francis, Bristol, PA (also available as database).
5. *TRC Thermodynamic Tables*, Thermodynamic Research Center, Texas A & M University, College Station, TX.
6. Riddick, J. A., Bunger, W. B., and Sakano, T. K., *Organic Solvents, Fourth Edition*, John Wiley & Sons, New York, 1986.
7. *Physical Constants of Hydrocarbon and Non-Hydrocarbon Compounds*, ASTM Data Series DS 4B, ASTM, Philadelphia, 1988.
8. Stevenson, R. M., and Malanowski, S., *Handbook of the Thermodynamics of Organic Compounds*, Elsevier, New York, 1987.
9. Buckingham, J., Editor, *Dictionary of Organic Compounds, Fifth Edition* (and Supplements), Chapman and Hall, New York, 1982-1993.
10. Lide, D. R., and Kehiaian, H. V., *Handbook of Thermophysical and Thermochemical Data*, CRC Press, Boca Raton, FL, 1994.

PHYSICAL CONSTANTS OF ORGANIC COMPOUNDS (continued)

No.	Name / Synonym	Mol. Form. / Mol. Wt.	CAS RN / mp/°C	Merck No. / bp/°C	Beil. Ref. / den/g cm⁻³	Solubility / n_D
1	Abate / Phosphorothioic acid, O,O'-(thiodi-4,1-phenylene) O,O,O',O'-tetramethyl ester	$C_{16}H_{20}O_6P_2S_3$ / 466.48	3383-96-8 / 30	9075	/ 1.32	
2	1-Acenaphthylenamine, 1,2-dihydro- / 1-Acenaphthenamine	$C_{12}H_{11}N$ / 169.23	40745-44-6 / 135	sub	2-12-00-00764	bz 4; EtOH 4
3	Acenaphthylene / Acenaphthalene	$C_{12}H_8$ / 152.20	208-96-8 / 92.5	280; 150^{28}	4-05-00-02138 / 0.8987^{16}	H_2O 1; EtOH 4; eth 4; bz 4
4	Acenaphthylene, 5-bromo-1,2-dihydro- / 5-Bromoacenaphthene	$C_{12}H_9Br$ / 233.11	2051-98-1 / 52	335	4-05-00-01839 / 1.4392^{52}	H_2O 1; EtOH 2; chl 2 / 1.6565^{54}
5	Acenaphthylene, 5-chloro-1,2-dihydro- / 5-Chloroacenaphthene	$C_{12}H_9Cl$ / 188.66	5209-33-6 / 70.5	319; 163^{13}	4-05-00-01837 / 1.1954^{70}	1.6169^{100}
6	Acenaphthylene, 1,2-dihydro- / Acenaphthene	$C_{12}H_{10}$ / 154.21	83-32-9 / 93.4	23 / 279	4-05-00-01834 / 1.222^{20}	H_2O 1; EtOH 2; bz 4; chl 2 / 1.6048^{95}
7	Acenaphthylene, 1,2-dihydro-5-iodo- / 5-Iodoacenaphthene	$C_{12}H_9I$ / 280.11	6861-64-9 / 64.3		2-05-00-00497 / 1.5000^{20}	bz 4; EtOH 4 / 1.6909^{65}
8	1,2-Acenaphthylenedione	$C_{12}H_6O_2$ / 182.18	82-86-0 / 261	sub	4-07-00-02498 / 1.4800^{20}	H_2O 1; EtOH 2; bz 2; lig 3
9	Acenaphthylene, 1,2,2a,3,4,5-hexahydro-	$C_{12}H_{14}$ / 158.24	480-72-8 / 12.8	249	4-05-00-01566 / 1.0290^{20}	chl 3
10	Acephate / Phosphoramidothioic acid, acetyl-, O,S-dimethyl ester	$C_4H_{10}NO_3PS$ / 183.17	30560-19-1 / 88		/ 1.35^{20}	
11	Acetaldehyde / Ethanal	C_2H_4O / 44.05	75-07-0 / -123	32 / 20.1	4-01-00-03094 / 0.7834^{18}	H_2O 5; EtOH 5; eth 5; bz 5 / 1.3316^{20}
12	Acetaldehyde, chloro- / Chloroacetaldehyde	C_2H_3ClO / 78.50	107-20-0 / -16.3	2108 / 85.5	4-01-00-03134	eth 3
13	Acetaldehyde, chloro-, trimer / Chloroacetaldehyde, trimer	$C_6H_9Cl_3O_3$ / 235.49	65438-35-9 / 87	142^{10}	4-19-00-04718	eth 4
14	Acetaldehyde, ethoxy- / α-Ethoxyacetaldehyde	$C_4H_8O_2$ / 88.11	22056-82-2		3-01-00-03181 / 0.942^{20}	H_2O 4; ace 4; EtOH 4 / 1.3956^{20}
15	Acetaldehyde, hydroxy- / Glycolaldehyde	$C_2H_4O_2$ / 60.05	141-46-8 / 97		4-01-00-03955 / 1.366^{100}	chl 3 / 1.4772^{19}
16	Acetaldehyde, methoxy- / Methoxyacetaldehyde	$C_3H_6O_2$ / 74.08	10312-83-1	92	4-01-00-03955 / 1.005^{25}	H_2O 4; ace 4; eth 4; EtOH 4 / 1.3950^{20}
17	Acetaldehyde, oxime / Acetaldoxime	C_2H_5NO / 59.07	107-29-9 / 45	35 / 115	4-01-00-03121 / 0.9656^{20}	H_2O 3; EtOH 5; eth 5; chl 3 / 1.4264^{20}
18	Acetaldehyde, phenylhydrazone	$C_8H_{10}N_2$ / 134.18	935-07-9 / 99.5	150^{40}; 135^{21}	3-15-00-00079	EtOH 4
19	Acetaldehyde, tetramer / Metaldehyde II (tetramer)	$C_4H_{12}O_4$ / 124.14	66056-06-2 / 47	110	4-19-00-05643	ace 4; bz 4; eth 4; EtOH 4
20	Acetaldehyde, tribromo- / Bromal	C_2HBr_3O / 280.74	115-17-3	1372 / 174	4-01-00-03155 / 2.6649^{25}	ace 4; eth 4; EtOH 4 / 1.5939^{20}
21	Acetaldehyde, trichloro- / Chloral	C_2HCl_3O / 147.39	75-87-6 / -57.5	9538 / 97.8	4-01-00-03142 / 1.512^{20}	H_2O 4; EtOH 3; eth 3 / 1.4580^{20}
22	Acetamide / Ethanamide	C_2H_5NO / 59.07	60-35-5 / 81	36 / 222.0	4-02-00-00399 / 0.9986^{85}	H_2O 4; EtOH 4 / 1.4278
23	Acetamide, N-acetyl-	$C_4H_7NO_2$ / 101.11	625-77-4 / 79	223.5	4-02-00-00416	H_2O 3; EtOH 3; eth 3; chl 3
24	Acetamide, N-acetyl-N-ethyl-	$C_6H_{11}NO_2$ / 129.16	1563-83-3	197	4-04-00-00352 / 1.0092^{20}	EtOH 4 / 1.4513^{20}
25	Acetamide, N-acetyl-N-methyl-	$C_5H_9NO_2$ / 115.13	1113-68-4 / -25	195; 114.5^{61}	4-04-00-00183 / 1.0663^{25}	H_2O 5; eth 1 / 1.4502^{25}
26	Acetamide, N-acetyl-N-phenyl-	$C_{10}H_{11}NO_2$ / 177.20	1563-87-7 / 37.5	200^{100}; 142^{11}	4-12-00-00385	H_2O 2; EtOH 3; bz 3; tol 3
27	Acetamide, N-(aminocarbonyl)-	$C_3H_6N_2O_2$ / 102.09	591-07-1 / 218	sub 180	3-03-00-00119	H_2O 2; EtOH 3; eth 2
28	Acetamide, N-(2-aminoethyl)-	$C_4H_{10}N_2O$ / 102.14	1001-53-2 / 51	136^{27}	4-04-00-01193	H_2O 3; EtOH 3; eth 1; bz 3
29	Acetamide, N-(4-aminophenyl)- / p-Aminoacetanilide	$C_8H_{10}N_2O$ / 150.18	122-80-5 / 166.5	422 / 267	4-13-00-00137	H_2O 3; EtOH 4; eth 4
30	Acetamide, N-[(4-aminophenyl)sulfonyl]- / Sulfacetamide	$C_8H_{10}N_2O_3S$ / 214.25	144-80-9 / 183	8870	4-14-00-02662	H_2O 2; EtOH 3; eth 1; ace 4
31	Acetamide, N-[4-(aminosulfonyl)phenyl]- / Acetylsulfanilamide	$C_8H_{10}N_2O_3S$ / 214.25	121-61-9 / 219.5	97	4-14-00-02704	H_2O 3; EtOH 3; ace 3
32	Acetamide, N-(5-(aminosulfonyl)-1,3,4-thiadiazol-2-yl]- / Acetazolamide	$C_4H_6N_4O_3S_2$ / 222.25	59-66-5 / 260.5	45	4-27-00-08219	H_2O 2
33	Acetamide, N-(aminothioxomethyl)- / Acetylthiourea	$C_3H_6N_2OS$ / 118.16	591-08-2 / 165		4-03-00-00354	H_2O 2; EtOH 3; eth 2
34	Acetamide, N-[4-[[(aminothioxomethyl)hydrazono]methyl]phenyl]- / Thiacetazone	$C_{10}H_{12}N_4OS$ / 236.30	104-06-3 / 225 dec	9218	4-14-00-00075	H_2O 1; os 1; CS_2 1
35	Acetamide, N-[1,1'-biphenyl]-2-yl-	$C_{14}H_{13}NO$ / 211.26	2113-47-5 / 121	355	4-12-00-03224	H_2O 1; EtOH 4; eth 4
36	Acetamide, N-bromo- / N-Bromoacetamide	C_2H_4BrNO / 137.96	79-15-2 / 103.5	1386	4-02-00-00417	eth 4

No.	Name / Synonym	Mol. Form. / Mol. Wt.	CAS RN / mp/°C	Merck No. / bp/°C	Beil. Ref. / den/g cm^{-3}	Solubility / n_D
37	Acetamide, N-(4-bromophenyl)- / p-Bromoacetanilide	C_8H_8BrNO / 214.06	103-88-8 / 168	1387	4-12-00-01504 / 1.717[25]	H_2O 1; EtOH 3; eth 2; bz 2
38	Acetamide, N-butyl-	$C_6H_{13}NO$ / 115.18	1119-49-9 /	229	4-04-00-00565 / 0.896[25]	1.4388[25]
39	Acetamide, N-butyl-N-phenyl-	$C_{12}H_{17}NO$ / 191.27	91-49-6 / 24.5	281	4-12-00-00379 / 0.9912[20]	chl 2 / 1.5146[20]
40	Acetamide, N,N'-carbonylbis-	$C_5H_8N_2O_3$ / 144.13	638-20-0 / 154.5	sub	3-03-00-00121	H_2O 2; EtOH 2; eth 3; ace 3
41	Acetamide, 2-chloro- / Chloroacetamide	C_2H_4ClNO / 93.51	79-07-2 / 121	2109 / 225	4-02-00-00490	H_2O 3; EtOH 4; eth 2
42	Acetamide, 2-chloro-N,N-bis(1-methylethyl)-	$C_8H_{16}ClNO$ / 177.67	7403-66-9 / 48.5	163[55]; 86[2.7]	4-04-00-00516	1.4619[25]
43	Acetamide, 2-chloro-N-3-chloroallyl- / Chloroacetamide, N-3-chloroallyl	$C_5H_7Cl_2NO$ / 168.02	100130-25-4 / 52.5	112[0.5]	4-04-00-01087	peth 4
44	Acetamide, 2-chloro-N,N-di-2-propenyl- / Allidochlor	$C_8H_{12}ClNO$ / 173.64	93-71-0 / 250	116[1]; 90[0.7]	4-04-00-01064	H_2O 2; EtOH 3 / 1.4932[25]
45	Acetamide, 2-chloro-N-hexyl- / α-Chloro-N-hexylacetamide	$C_8H_{16}ClNO$ / 177.67	5326-81-8 / 108.5	100[0.2]	4-04-00-00713	peth 4
46	Acetamide, 2-chloro-N-(3-methoxypropyl)- / α-Chloro-N-(3-methoxypropyl)acetamide	$C_6H_{12}ClNO_2$ / 165.62	1709-03-1 / 30	88[0.5]	4-04-00-01645	1.4712[25]
47	Acetamide, 2-chloro-N-(1-methylpropyl)-	$C_6H_{12}ClNO$ / 149.62	32322-73-9 / 45.5	68[0.7]	4-04-00-00621	peth 4
48	Acetamide, N-(4-chlorophenyl)-	C_8H_8ClNO / 169.61	539-03-7 / 179	333	4-12-00-01178 / 1.385[22]	H_2O 1; EtOH 3; eth 4; ctc 2
49	Acetamide, 2-cyano- / Cyanoacetamide	$C_3H_4N_2O$ / 84.08	107-91-5 / 121.5	2695	4-02-00-01891	H_2O 4
50	Acetamide, N,N-diacetyl- / Triacetamide	$C_6H_9NO_3$ / 143.14	641-06-5 / 79		4-02-00-00416	eth 4
51	Acetamide, 2,2-dichloro-	$C_2H_3Cl_2NO$ / 127.96	683-72-7 / 99.4	234	4-02-00-00505	H_2O 3; EtOH 3; eth 3; ace 2
52	Acetamide, 2,2-dichloro-N-[2-hydroxy-1-(hydroxymethyl)-2-(4-nitrophenyl)ethyl]-, [R*,(R*,R*)]- / Chloramphenicol	$C_{11}H_{12}Cl_2N_2O_5$ / 323.13	56-75-7 / 150.5	2068 / sub	3-13-00-02268	ace 4; EtOH 4; chl 4
53	Acetamide, N,N-diethyl-	$C_6H_{13}NO$ / 115.18	685-91-6 /	185.5	4-04-00-00349 / 0.9130[17]	H_2O 3; EtOH 3; eth 5; ace 5 / 1.4374[17]
54	Acetamide, 2-(diethylamino)-N-(2,6-dimethylphenyl)- / Lidocaine	$C_{14}H_{22}N_2O$ / 234.34	137-58-6 / 68.5	5359 / 180-2[4]	4-12-00-02538	bz 4; eth 4; EtOH 4; chl 4
55	Acetamide, N,N-dimethyl- / N,N-Dimethylethanamide	C_4H_9NO / 87.12	127-19-5 / -18.59	3216 / 165	4-04-00-00180 / 0.9366[25]	H_2O 5; EtOH 5; eth 5; ace 5 / 1.4380[20]
56	Acetamide, N-(2,4-dimethylphenyl)-	$C_{10}H_{13}NO$ / 163.22	2050-43-3 / 129.3	170[10]	4-12-00-02548	EtOH 4; chl 4
57	Acetamide, N,N-diphenyl- / Diphenylacetamide	$C_{14}H_{13}NO$ / 211.26	519-87-9 / 103	3315 / sub	4-12-00-00381	H_2O 2; EtOH 3; eth 2; chl 2
58	Acetamide, N,N-dipropyl-	$C_8H_{17}NO$ / 143.23	1116-24-1 /	209.5	4-04-00-00476 / 0.8992[17]	EtOH 4 / 1.4419[17]
59	Acetamide, N-(2-ethoxyphenyl)-	$C_{10}H_{13}NO_2$ / 179.22	581-08-8 / 79	>240	3-13-00-00779	H_2O 1; EtOH 3; eth 3; chl 3
60	Acetamide, N-(4-ethoxyphenyl)- / Phenacetin	$C_{10}H_{13}NO_2$ / 179.22	62-44-2 / 137.5	7155	4-13-00-01092 / 1.571	H_2O 2; EtOH 3; eth 2; ace 3
61	Acetamide, N-ethyl- / N-Ethylacetamide	C_4H_9NO / 87.12	625-50-3 /	205; 104[18]	4-04-00-00347 / 0.942[4]	H_2O 5; EtOH 5; chl 3; HOAc 3 / 1.4338[20]
62	Acetamide, N-ethyl-N-phenyl-	$C_{10}H_{13}NO$ / 163.22	529-65-7 / 55	260	4-12-00-00379 / 0.9938[60]	H_2O 3; eth 3; ctc 3
63	Acetamide, 2-fluoro- / Fluoroacetic acid amide	C_2H_4FNO / 77.06	640-19-7 / 108	4095 / sub	4-02-00-00454	H_2O 3; ace 3; chl 2
64	Acetamide, N-9H-fluoren-2-yl- / 2-(Acetylamino)fluorene	$C_{15}H_{13}NO$ / 223.27	53-96-3 / 193	4083	4-12-00-03373	H_2O 1; EtOH 3; eth 3; HOAc 3
65	Acetamide, N-(2-hydroxyethyl)-	$C_4H_9NO_2$ / 103.12	142-26-7 / 63.5	166-7[8]	4-04-00-01535 / 1.1079[25]	H_2O 5; ace 3; bz 2; lig 2 / 1.4674[20]
66	Acetamide, N-(2-hydroxy-1-naphthalenyl)-	$C_{12}H_{11}NO_2$ / 201.22	117-93-1 / 235 dec	sub	4-13-00-02095	ace 4; bz 4; eth 4; EtOH 4
67	Acetamide, N-(4-hydroxyphenyl)- / Acetaminophen	$C_8H_9NO_2$ / 151.17	103-90-2 / 170	40	4-13-00-01091 / 1.293[21]	H_2O 1; EtOH 4
68	Acetamide, 2-mercapto-N-2-naphthalenyl- / Thionalide	$C_{12}H_{11}NOS$ / 217.29	93-42-5 / 111.5	9273	3-12-00-03052	H_2O 1; EtOH 4; os 4
69	Acetamide, N-[2-(5-methoxy-1H-indol-3-yl)ethyl]- / Melatonin	$C_{13}H_{16}N_2O_2$ / 232.28	73-31-4 / 117	5695	5-22-12-00042	
70	Acetamide, N-(2-methoxyphenyl)-	$C_9H_{11}NO_2$ / 165.19	93-26-5 / 87.5	304	4-13-00-00833	H_2O 4; EtOH 4; eth 3; ace 3
71	Acetamide, N-(4-methoxyphenyl)- / p-Acetanisidine	$C_9H_{11}NO_2$ / 165.19	51-66-1 / 131	43	4-13-00-01092	ace 4; EtOH 4; chl 4
72	Acetamide, N-methyl-	C_3H_7NO / 73.09	79-16-3 / 28	205	4-04-00-00176 / 0.9371[25]	ace 4; bz 4; eth 4; EtOH 4 / 1.4301[20]
73	Acetamide, N-[(methylamino)carbonyl]-	$C_4H_8N_2O_2$ / 116.12	623-59-6 / 180.5	dec	4-04-00-00207	H_2O 3; EtOH 2; eth 2; chl 3

No.	Name Synonym	Mol. Form. Mol. Wt.	CAS RN mp/°C	Merck No. bp/°C	Beil. Ref. den/g cm^{-3}	Solubility n_D
74	Acetamide, N-(1-methylethyl)-N-phenyl- N-Isopropylacetanilide	C$_{11}$H$_{15}$NO 177.25	5461-51-8 40.5	264	3-12-00-00466	lig 4
75	Acetamide, N-methyl-N-(2-methylphenyl)-	C$_{10}$H$_{13}$NO 163.22	573-26-2 55.5	260	3-12-00-01854	EtOH 3; chl 3
76	Acetamide, N-(2-methylphenyl)-	C$_9$H$_{11}$NO 149.19	120-66-1 110	296	4-12-00-01755 1.168^{15}	H$_2$O 2; EtOH 3; eth 3; ace 3
77	Acetamide, N-(3-methylphenyl)-	C$_9$H$_{11}$NO 149.19	537-92-8 65.5	303	4-12-00-01823 1.141^{15}	H$_2$O 2; EtOH 4; eth 4; chl 3
78	Acetamide, N-(4-methylphenyl)-	C$_9$H$_{11}$NO 149.19	103-89-9 152	307	4-12-00-01902 1.212^{15}	eth 4; EtOH 4
79	Acetamide, N-methyl-N-phenyl- N-Methylacetanilide	C$_9$H$_{11}$NO 149.19	579-10-2 103	5931 256	4-12-00-00378 1.0036^{105}	H$_2$O 3; EtOH 3; eth 3; chl 3 1.576
80	Acetamide, 2-nitro-N-phenyl- α-Nitroacetanilide	C$_8$H$_8$N$_2$O$_3$ 180.16	10151-95-8 94	100$^{0.1}$	1-12-00-00193 1.419^{15}	bz 4; EtOH 4; lig 4
81	Acetamide, N-(2-nitrophenyl)-	C$_8$H$_8$N$_2$O$_3$ 180.16	552-32-9 94	100$^{0.1}$	4-12-00-01574 1.419^{15}	H$_2$O 3; EtOH 3; eth 4; bz 3
82	Acetamide, N-(3-nitrophenyl)-	C$_8$H$_8$N$_2$O$_3$ 180.16	122-28-1 155	100$^{0.0074}$	4-12-00-01595	H$_2$O 3; EtOH 3; eth 1; chl 3
83	Acetamide, N-(4-nitrophenyl)-	C$_8$H$_8$N$_2$O$_3$ 180.16	104-04-1 216	100$^{0.008}$	4-12-00-01632	H$_2$O 2; EtOH 3; eth 2; chl 2
84	Acetamide, N-(5-nitro-2-propoxyphenyl)- 5'-Nitro-2'-propoxyacetanilide	C$_{11}$H$_{14}$N$_2$O$_4$ 238.24	553-20-8 102.5	6550	3-13-00-00881	
85	Acetamide, N-(5-nitro-2-thiazolyl)- Aminitrozole	C$_5$H$_5$N$_3$O$_3$S 187.18	140-40-9 264.5	421	4-27-00-04676	alk 3
86	Acetamide, N-phenyl- Acetanilide	C$_8$H$_9$NO 135.17	103-84-4 114.3	42 304	4-12-00-00373 1.2190^{15}	H$_2$O 2; EtOH 4; eth 3; ace 4
87	Acetamide, N-(phenylmethyl)-	C$_9$H$_{11}$NO 149.19	588-46-5 61	157^2	4-12-00-02230	EtOH 4; eth 4
88	Acetamide, N-phenyl-N-propyl- N-Propylacetanilide	C$_{11}$H$_{15}$NO 177.25	2437-98-1 49	268	4-12-00-00379	eth 4; EtOH 4; lig 4
89	Acetamide, N-3-pyridinyl-	C$_7$H$_8$N$_2$O 136.15	5867-45-8 133	326.5	5-22-09-00017	H$_2$O 3; EtOH 3; eth 2; bz 2
90	Acetamide, N-4-quinolinyl-	C$_{11}$H$_{10}$N$_2$O 186.21	32433-28-6 176	sub	5-22-10-00250	EtOH 4
91	Acetamide, N,N'-(sulfonyldi-4,1- phenylene)bis- Acedapsone	C$_{16}$H$_{16}$N$_2$O$_4$S 332.38	77-46-3 290	16	4-13-00-01361	EtOH 4
92	Acetamide, N-(5,6,7,9-tetrahydro-10-hydroxy- 1,2,3-trimethoxy-9-oxobenzo[a]heptalen-7-yl)- , (S)- Colchiceine	C$_{21}$H$_{23}$NO$_6$ 385.42	477-27-0 178.5	2469	4-14-00-00943 1.24^{25}	H$_2$O 2; EtOH 4; eth 1; bz 1
93	Acetamide, N-(5,6,7,9-tetrahydro-1,2,3,10- tetramethoxy-9-oxobenzo[a]heptalen-7-yl)-, (S)- Colchicine	C$_{22}$H$_{25}$NO$_6$ 399.44	64-86-8 156	2470	4-14-00-00946	H$_2$O 4; EtOH 4
94	Acetamide, 2,2,2-tribromo- Tribromoacetamide	C$_2$H$_2$Br$_3$NO 295.76	594-47-8 121.5	sub	3-02-00-00485	H$_2$O 4; eth 4; EtOH 4
95	Acetamide, 2,2,2-trichloro-	C$_2$H$_2$Cl$_3$NO 162.40	594-65-0 142	240	4-02-00-00520	H$_2$O 2; EtOH 4; eth 4
96	Acetamide, 2,2,2-trichloro-N,N-diethyl-	C$_6$H$_{10}$Cl$_3$NO 218.51	2430-00-4 27	112^{12}	4-04-00-00351	1.4900^{24}
97	Acetamide, 2,2,2-trichloro-N,N-dimethyl-	C$_4$H$_6$Cl$_3$NO 190.46	7291-33-0 12	231 dec	4-04-00-00182 1.390^{20}	bz 4; chl 4 1.5017^{25}
98	Acetamide, 2,2,2-trichloro-N-phenyl-	C$_8$H$_6$Cl$_3$NO 238.50	2563-97-5 96	169^{19}	4-12-00-00377	EtOH 4
99	Acetamide, 2,2,2-trifluoro-N,N-dimethyl-	C$_4$H$_6$F$_3$NO 141.09	1547-87-1	133	4-04-00-00182 1.232^{21}	1.3611^{25}
100	Acetic acid Ethanoic acid	C$_2$H$_4$O$_2$ 60.05	64-19-7 16.6	47 117.9	4-02-00-00094 1.0446^{25}	H$_2$O 5; EtOH 5; eth 5; ace 5 1.3720^{20}
101	Acetic acid, (acetyloxy)-	C$_4$H$_6$O$_4$ 118.09	13831-30-6 67.5	145^{12}	4-03-00-00576	H$_2$O 4; EtOH 4; eth 4; ace 4
102	Acetic acid, (acetyloxy)-, ethyl ester	C$_6$H$_{10}$O$_4$ 146.14	623-86-9	179	4-03-00-00584 1.0880^{20}	eth 4; EtOH 4; HOAc 4 1.4112^{20}
103	Acetic acid, [2-(aminocarbonyl)phenoxy]- Salicylamide O-acetic acid	C$_9$H$_9$NO$_4$ 195.17	25395-22-6	8298	4-10-00-00184	
104	Acetic acid, aminooxo- Oxamic acid	C$_2$H$_3$NO$_3$ 89.05	471-47-6 210 dec	6870	4-02-00-01857	H$_2$O 2; EtOH 1; eth 1
105	Acetic acid, aminooxo-, hydrazide Semioxamazide	C$_2$H$_5$N$_3$O$_2$ 103.08	515-96-8 221 dec	8397	4-02-00-01868	H$_2$O 2; EtOH 1; eth 1; alk 4
106	Acetic acid, ammonium salt Ammonium, acetate	C$_2$H$_7$NO$_2$ 77.08	631-61-8 114	514 dec	1.17^{20}	H$_2$O 4; EtOH 3; ace 2
107	Acetic acid, anhydride Acetic anhydride	C$_4$H$_6$O$_3$ 102.09	108-24-7 -73	48 139.5	4-02-00-00386 1.082^{20}	H$_2$O 4; EtOH 3; eth 5; bz 3 1.3901^{20}
108	Acetic acid, anhydride with nitric acid Acetyl nitrate	C$_2$H$_3$NO$_4$ 105.05	591-09-3	92 22^{70}; exp 60	4-02-00-00435 1.24^{15}	
109	Acetic acid, azido-	C$_2$H$_3$N$_3$O$_2$ 101.06	18523-48-3 16	116^{12}; 92^3	4-02-00-00541 1.354^{33}	
110	Acetic acid, barium salt	C$_4$H$_6$BaO$_4$ 255.42	543-80-6	975	4-02-00-00094 2.468^{25}	H$_2$O 3

No.	Name Synonym	Mol. Form. Mol. Wt.	CAS RN mp/°C	Merck No. bp/°C	Bell. Ref. den/g cm⁻³	Solubility n_D
111	Acetic acid, bromo- Bromoacetic acid	$C_2H_3BrO_2$ 138.95	79-08-3 50	1388 208	4-02-00-00526 1.9335[50]	H_2O 5; EtOH 5; eth 5; ace 3 1.4804[50]
112	Acetic acid, bromochloro- Bromochloroacetic acid	$C_2H_2BrClO_2$ 173.39	5589-96-8 31.5	215; 104[11]	4-02-00-00532 1.9848[31]	
113	Acetic acid, bromochloro-, ethyl ester Ethyl bromochloroacetate	$C_4H_6BrClO_2$ 201.45	22524-32-9	174 dec	4-02-00-00532 1.5890[22]	eth 4; EtOH 4 1.4639[24]
114	Acetic acid, bromo-, 1,1-dimethylethyl ester	$C_6H_{11}BrO_2$ 195.06	5292-43-3	73-4[25]	3-02-00-00482	eth 4; EtOH 4 1.4430[20]
115	Acetic acid, bromo-, ethyl ester Ethyl bromacetate	$C_4H_7BrO_2$ 167.00	105-36-2	168.5	4-02-00-00527 1.5032[20]	H_2O 1; EtOH 5; eth 5; ace 3 1.4489[20]
116	Acetic acid, bromofluoro- Bromofluoroacetic acid	$C_2H_2BrFO_2$ 156.94	359-25-1 50	183	4-02-00-00531	H_2O 4; EtOH 4; chl 4
117	Acetic acid, bromo-, methyl ester Methyl 2-bromoacetate	$C_3H_5BrO_2$ 152.98	96-32-2	132	4-02-00-00527 1.6350[20]	H_2O 1; EtOH 3; eth 3; ace 3 1.4520[20]
118	Acetic acid, bromo-, 2-methylpropyl ester	$C_6H_{11}BrO_2$ 195.06	59956-48-8	188; 74.5[10]	1-02-00-00096 1.3269[20]	ace 4; eth 4; EtOH 4
119	Acetic acid, bromo-, phenyl ester	$C_8H_7BrO_2$ 215.05	620-72-4 32	140[20]	3-06-00-00599	eth 4; EtOH 4
120	Acetic acid, bromo-, propyl ester Propyl bromoacetate	$C_5H_9BrO_2$ 181.03	35223-80-4	176	4-02-00-00528 1.4099[20]	ace 4; eth 4; EtOH 4 1.4518[20]
121	Acetic acid, butyl ester Butyl acetate	$C_6H_{12}O_2$ 116.16	123-86-4 -78	1535 126.1	4-02-00-00143 0.8825[20]	H_2O 2; EtOH 5; eth 5; ace 3 1.3941[20]
122	Acetic acid, calcium salt	$C_4H_6CaO_4$ 158.17	62-54-4 dec	1643	4-02-00-00094	H_2O 3; EtOH 2 1.55
123	Acetic acid, chloro- Chloroacetic acid (β)	$C_2H_3ClO_2$ 94.50	79-11-8 63	2111 189.3	4-02-00-00474 1.4043[40]	H_2O 4; EtOH 3; eth 3; bz 3 1.4351[55]
124	Acetic acid, chloro-, anhydride Chloroacetic anhydride	$C_4H_4Cl_2O_3$ 170.98	541-88-8 46	2112 203	4-02-00-00487 1.5497[20]	
125	Acetic acid, chloro-, butyl ester	$C_6H_{11}ClO_2$ 150.60	590-02-3	183	4-02-00-00482 1.0704[20]	eth 4; EtOH 4 1.4297[20]
126	Acetic acid, chloro-, 2-chloroethyl ester	$C_4H_6Cl_2O_2$ 157.00	3848-12-2	203; 102[20]	4-02-00-00481 1.3600[25]	eth 4 1.4619[25]
127	Acetic acid, chlorodifluoro- Chlorodifluoroacetic acid	$C_2HClF_2O_2$ 130.48	76-04-0 25	122	4-02-00-00497	chl 3 1.3559[20]
128	Acetic acid, chloro-, 1,1-dimethylethyl ester tert-Butyl chloroacetate	$C_6H_{11}ClO_2$ 150.60	107-59-5	1563 150; 50[10]	3-02-00-00444	H_2O 6 1.4260[20]
129	Acetic acid, chloro-, 1,2-ethanediyl ester	$C_6H_8Cl_2O_4$ 215.03	6941-69-1 45.5	157[8]	3-02-00-00448	H_2O 1; eth 4
130	Acetic acid, chloro-, ethyl ester Ethyl chloroacetate	$C_4H_7ClO_2$ 122.55	105-39-5 -21	3741 144.3	4-02-00-00481 1.1585[20]	H_2O 1; EtOH 5; eth 5; ace 5 1.4215[20]
131	Acetic acid, chlorofluoro- Chlorofluoroacetic acid	$C_2H_2ClFO_2$ 112.49	471-44-3	162	4-02-00-00493 1.532[25]	1.4085[25]
132	Acetic acid, chlorofluoro-, butyl ester Chlorofluoroacetic acid, butyl ester	$C_6H_{10}ClFO_2$ 168.60	368-34-3	165.5	3-02-00-00453 1.124[25]	1.4067[25]
133	Acetic acid, chlorofluoro-, ethyl ester	$C_4H_6ClFO_2$ 140.54	401-56-9	129	4-02-00-00493 1.225[20]	1.3927[20]
134	Acetic acid, chlorofluoro-, methyl ester Methyl chlorofluoroacetate	$C_3H_4ClFO_2$ 126.51	433-52-3	116	4-02-00-00493 1.323[25]	1.3903[25]
135	Acetic acid, chlorofluoro-, propyl ester Chlorofluoroacetic acid, propyl ester	$C_5H_8ClFO_2$ 154.57	348-70-9	147	3-02-00-00453 1.170[25]	1.3994[25]
136	Acetic acid, chloro-, 2-hydroxyethyl ester	$C_4H_7ClO_3$ 138.55	35280-53-6	240 dec; 86[1.5]	4-02-00-00447 1.3300[20]	H_2O 4; EtOH 4 1.4609[20]
137	Acetic acid, chloro-, 2-methoxyethyl ester	$C_5H_9ClO_3$ 152.58	13361-36-9	86-6[9]	3-02-00-00447 1.2015[20]	eth 4 1.4382[20]
138	Acetic acid, chloro-, methyl ester Chloroacetic acid, methyl ester	$C_3H_5ClO_2$ 108.52	96-34-4 -32.1	5965 129.5	4-02-00-00480 1.236[20]	ace 4; bz 4; eth 4; EtOH 4 1.4218[20]
139	Acetic acid, chloro-, 1-methylethyl ester Isopropyl chloroacetate	$C_5H_9ClO_2$ 136.58	105-48-6	150.5	4-02-00-00482 1.0888[20]	eth 4 1.4382[20]
140	Acetic acid, chloro-, 4-methylphenyl ester	$C_9H_9ClO_2$ 184.62	15150-39-7 32	254; 162[41]	4-06-00-02112 1.1840[35]	eth 4; EtOH 4
141	Acetic acid, (4-chloro-2-methylphenoxy)- MCPA	$C_9H_9ClO_3$ 200.62	94-74-6 120	5645	4-06-00-01991	H_2O 2; EtOH 4; eth 4; bz 3
142	Acetic acid, chloro-, 1-methylpropyl ester	$C_6H_{11}ClO_2$ 150.60	17696-64-9	163.5	3-02-00-00443 1.060[20]	eth 4; EtOH 4 1.425[19]
143	Acetic acid, chloro-, 2-methylpropyl ester	$C_6H_{11}ClO_2$ 150.60	13361-35-8	170	4-02-00-00483 1.0612[20]	ace 4; eth 4 1.4255[20]
144	Acetic acid, chlorooxo-, ethyl ester	$C_4H_5ClO_3$ 136.53	4755-77-5	137	4-02-00-01853 1.2226[20]	bz 4; eth 4
145	Acetic acid, chlorooxo-, methyl ester	$C_3H_3ClO_3$ 122.51	5781-53-3	119	3-02-00-01583 1.3316[20]	1.4189[20]
146	Acetic acid, (4-chlorophenoxy)- (p-Chlorophenoxy)acetic acid	$C_8H_7ClO_3$ 186.59	122-88-3 156.5		4-06-00-00845	H_2O 4; chl 2
147	Acetic acid, chloro-, phenyl ester Phenyl chloroacetate	$C_8H_7ClO_2$ 170.60	620-73-5 44.5	232.5	3-06-00-00598 1.2202[44]	H_2O 1; EtOH 4; eth 4 1.5146[44]
148	Acetic acid, chloro-, phenylmethyl ester	$C_9H_9ClO_2$ 184.62	140-18-1	147.5[9]; 85[0.4]	4-06-00-02264 1.1223[4]	eth 4; EtOH 4 1.5426[18]
149	Acetic acid, chloro-, propyl ester Propyl chloroacetate	$C_5H_9ClO_2$ 136.58	5396-24-7	161	4-02-00-00482 1.1022[20]	eth 4 1.4261[20]

No.	Name Synonym	Mol. Form. Mol. Wt.	CAS RN mp/°C	Merck No. bp/°C	Beil. Ref. den/g cm^{-3}	Solubility n_D
150	Acetic acid, cyano- Cyanoacetic acid	$C_3H_3NO_2$ 85.06	372-09-8 66	2696 108[0.15]; 160 dec	4-02-00-01888	H_2O 3; EtOH 3; eth 3; chl 2
151	Acetic acid, cyano-, butyl ester	$C_7H_{11}NO_2$ 141.17	5459-58-5	231; 115[15]	4-02-00-01891 1.0010[20]	
152	Acetic acid, cyano-, ethyl ester Ethyl cyanoacetate	$C_5H_7NO_2$ 113.12	105-56-6 -22.5	3744 205	4-02-00-01889 1.0654[20]	eth 4; EtOH 4 1.4175[20]
153	Acetic acid, cyano-, hydrazide Cyacetacide	$C_3H_5N_3O$ 99.09	140-87-4 114.5	2688	4-02-00-01895	H_2O 4; EtOH 4
154	Acetic acid, cyano-, methyl ester Methyl cyanoacetate	$C_4H_5NO_2$ 99.09	105-34-0 -22.5	200.5	4-02-00-01889 1.1225[25]	eth 4; EtOH 4 1.4176[20]
155	Acetic acid, cyclohexyl ester Cyclohexyl acetate	$C_8H_{14}O_2$ 142.20	622-45-7	173; 96[75]	4-06-00-00036 0.968[20]	eth 4; EtOH 4 1.442[20]
156	Acetic acid, cyclohexylidene-, methyl ester Methyl cyclohexylideneacetate	$C_9H_{14}O_2$ 154.21	40203-74-5	99[20]	4-09-00-00124 1.0021[19]	1.4838[19]
157	Acetic acid, decyl ester Decyl acetate	$C_{12}H_{24}O_2$ 200.32	112-17-4 -15	244	4-02-00-00168 0.8671[20]	H_2O 1; EtOH 3; eth 3; bz 3 1.4273[20]
158	Acetic acid, diazo-, ethyl ester Diazoacetic ester	$C_4H_6N_2O_2$ 114.10	623-73-4 -22	2980 140 dec	4-03-00-01495 1.0852[18]	H_2O 2; EtOH 5; eth 5; bz 5 1.4605[20]
159	Acetic acid, dibromo- Dibromoacetic acid	$C_2H_2Br_2O_2$ 217.84	631-64-1 49	195[250]; 130[16]	4-02-00-00533	H_2O 4; EtOH 4; eth 3
160	Acetic acid, dibromo-, ethyl ester Ethyl dibromoacetate	$C_4H_6Br_2O_2$ 245.90	617-33-4	194	4-02-00-00533 1.8991[20]	H_2O 1; EtOH 5; eth 5 1.5017[13]
161	Acetic acid, dibromofluoro-, ethyl ester Ethyl dibromofluoroacetate	$C_4H_5Br_2FO_2$ 263.89	565-53-7	83[33]	4-02-00-00534 1.894[25]	1.4633[25]
162	Acetic acid, dichloro- Dichloroacetic acid	$C_2H_2Cl_2O_2$ 128.94	79-43-6 13.5	3037 194; 102[20]	4-02-00-00498 1.5634[20]	H_2O 5; EtOH 5; eth 5; ace 3 1.4658[20]
163	Acetic acid, dichloro-, butyl ester Butyl dichloroacetate	$C_6H_{10}Cl_2O_2$ 185.05	29003-73-4	193.5	4-02-00-00502 1.1820[20]	eth 4; EtOH 4 1.4420[20]
164	Acetic acid, dichloro-, ethyl ester Ethyl dichloroacetate	$C_4H_6Cl_2O_2$ 157.00	535-15-9	155; 56[10]	4-02-00-00501 1.2827[20]	H_2O 2; EtOH 5; eth 5; ace 3 1.4386[20]
165	Acetic acid, dichloro-, 2-hydroxyethyl ester	$C_4H_6Cl_2O_3$ 173.00	17704-30-2	81-2[0.5]	3-02-00-00460 1.438[20]	EtOH 4 1.4735[20]
166	Acetic acid, dichloro-, methyl ester Methyl dichloroacetate	$C_3H_4Cl_2O_2$ 142.97	116-54-1 -51.9	142.9	4-02-00-00501 1.3774[20]	H_2O 1; EtOH 3; ctc 3 1.4429[20]
167	Acetic acid, dichloro-, 1-methylethyl ester	$C_5H_8Cl_2O_2$ 171.02	25006-60-4	163.5	4-02-00-00502 1.2053[20]	eth 4; EtOH 4 1.4328[20]
168	Acetic acid, [2,3-dichloro-4-(2-methylene-1-oxobutyl)phenoxy]- Ethacrynic acid	$C_{13}H_{12}Cl_2O_4$ 303.14	58-54-8 122.5	3669		
169	Acetic acid, (2,4-dichlorophenoxy)- 2,4-D	$C_8H_6Cl_2O_3$ 221.04	94-75-7 140.5	2802 160[0.4]	4-06-00-00908	H_2O 1; EtOH 3; bz 2
170	Acetic acid, (2,4-dichlorophenoxy)-, 2-butoxyethyl ester 2,4-D 2-Butoxyethyl ester	$C_{14}H_{18}Cl_2O_4$ 321.20	1929-73-3	156-62[1]	4-06-00-00911 1.232[20]	
171	Acetic acid, (2,4-dichlorophenoxy)-, butyl ester 2,4-D Butyl ester	$C_{12}H_{14}Cl_2O_3$ 277.15	94-80-4	2802	3-06-00-00706	
172	Acetic acid, (2,4-dichlorophenoxy)-, methyl ester 2,4-D Methyl ester	$C_9H_8Cl_2O_3$ 235.07	1928-38-7 119	141[18]	4-06-00-00909	
173	Acetic acid, (2,4-dichlorophenoxy)-, 1-methylethyl ester 2,4-D Isopropyl ester	$C_{11}H_{12}Cl_2O_3$ 263.12	94-11-1 5	140[1]	4-06-00-00910 1.26[25]	1.5209[25]
174	Acetic acid, diethoxy-, ethyl ester	$C_8H_{16}O_4$ 176.21	6065-82-3	199	4-03-00-01494 0.985[25]	1.4100[20]
175	Acetic acid, difluoro- Difluoroacetic acid	$C_2H_2F_2O_2$ 96.03	381-73-7 -1	133	4-02-00-00455 1.526[25]	1.3470[20]
176	Acetic acid, difluoro-, ethyl ester Ethyl difluoroacetate	$C_4H_6F_2O_2$ 124.09	454-31-9	100	4-02-00-00455 1.1893[9]	H_2O 1
177	Acetic acid, 1,1-dimethylethyl ester tert-Butyl acetate	$C_6H_{12}O_2$ 116.16	540-88-5	1537 95.1	4-02-00-00151 0.8665[20]	EtOH 3; eth 3; chl 3; HOAc 3 1.3855[20]
178	Acetic acid, dodecyl ester	$C_{14}H_{28}O_2$ 228.38	112-66-3 1.3	265; 180[40]	4-02-00-00170 0.8652[22]	1.4439[20]
179	Acetic acid ethenyl ester Vinyl acetate	$C_4H_6O_2$ 86.09	108-05-4 -93.2	9896 72.5	4-02-00-00176 0.9317[20]	H_2O 1; EtOH 5; eth 3; ace 3 1.3959[20]
180	Acetic acid, ethoxy- 2-Ethoxyacetic acid	$C_4H_8O_3$ 104.11	627-03-2	206.5	4-03-00-00574 1.1021[20]	H_2O 4; EtOH 4; eth 4; chl 3 1.4194[20]
181	Acetic acid, ethoxy-, ethyl ester	$C_6H_{12}O_3$ 132.16	817-95-8	158	4-03-00-00581 0.9702[20]	EtOH 3; eth 3; ace 3 1.4039[20]
182	Acetic acid, ethoxy-, methyl ester Methyl ethoxyacetate	$C_5H_{10}O_3$ 118.13	17640-26-5	147	4-03-00-00578 1.0112[15]	ace 4; eth 4; EtOH 4
183	Acetic acid, ethoxy-, 5-methyl-2-(1-methylethyl)cyclohexyl ester, (1α,2β,5α)-	$C_{14}H_{26}O_3$ 242.36	579-94-2	155[20]	2-06-00-00047 0.9545[20]	eth 4; EtOH 4; chl 4
184	Acetic acid, (ethoxysulfonyl)-, ethyl ester Acetic acid, sulfo, diethyl ester	$C_6H_{12}O_5S$ 196.22	59376-54-4	116[0.5]	3-04-00-00058 1.232[16]	1.436[21]
185	Acetic acid, 2-ethylbutyl ester 2-Ethylbutyl acetate	$C_8H_{16}O_2$ 144.21	10031-87-5 <-100	162.5	4-02-00-00161 0.8790[20]	H_2O 1; EtOH 3; eth 3; ctc 3 1.4109[20]
186	Acetic acid ethyl ester Ethyl acetate	$C_4H_8O_2$ 88.11	141-78-6 -83.6	3713 77.11	4-02-00-00127 0.9003[20]	H_2O 3; EtOH 5; eth 5; ace 4 1.3723[20]

No.	Name / Synonym	Mol. Form. / Mol. Wt.	CAS RN / mp/°C	Merck No. / bp/°C	Beil. Ref. / den/g cm^{-3}	Solubility / n_D
187	Acetic acid, 2-ethylhexyl ester / 2-Ethylhexyl acetate	$C_{10}H_{20}O_2$ / 172.27	103-09-3 / -80	6683 / 199	4-02-00-00166 / 0.8718[20]	H_2O 1; EtOH 3; eth 3 / 1.4204[20]
188	Acetic acid, (ethylthio)-	$C_4H_8O_2S$ / 120.17	627-04-3 / -8.5	/ 164[83]; 109[5]	4-03-00-00603 / 1.1497[20]	H_2O 4; EtOH 4; eth 4
189	Acetic acid, fluoro- / Fluoroethanoic acid	$C_2H_3FO_2$ / 78.04	144-49-0 / 35.2	4096 / 168	4-02-00-00446 / 1.3693[36]	H_2O 3; EtOH 3
190	Acetic acid, fluoro-, 1,1-dimethylethyl ester	$C_6H_{11}FO_2$ / 134.15	406-74-6 /	/ 132	4-02-00-00450 / 0.9904[20]	1.386[20]
191	Acetic acid, fluoro-, ethyl ester	$C_4H_7FO_2$ / 106.10	459-72-3 /	/ 120	4-02-00-00447 / 1.098[25]	H_2O 4 / 1.3755[20]
192	Acetic acid, fluoro-, methyl ester / Methyl fluoroacetate	$C_3H_5FO_2$ / 92.07	453-18-9 / -35	/ 104.5	4-02-00-00447 / 1.1613[15]	
193	Acetic acid, heptyl ester / Heptyl acetate	$C_9H_{18}O_2$ / 158.24	112-06-1 / -50.2	/ 193	4-02-00-00162 / 0.8750[15]	H_2O 1; EtOH 3; eth 3; ctc 3 / 1.4150[20]
194	Acetic acid, hexyl ester	$C_8H_{16}O_2$ / 144.21	142-92-7 / -80.9	/ 171.5	4-02-00-00159 / 0.8779[15]	eth 4; EtOH 4 / 1.4092[20]
195	Acetic acid, hydrazide	$C_2H_6N_2O$ / 74.08	1068-57-1 / 67	/ 137[25]	4-02-00-00435	H_2O 3; EtOH 3; eth 2
196	Acetic acid, hydroxy- / Glycolic acid	$C_2H_4O_3$ / 76.05	79-14-1 / 79.5	4394 / 100	4-03-00-00571	H_2O 3; EtOH 3; eth 3
197	Acetic acid, hydroxy-, ethyl ester	$C_4H_8O_3$ / 104.11	623-50-7 /	/ 160	4-03-00-00580 / 1.0826[23]	eth 4; EtOH 4 / 1.4180[20]
198	Acetic acid, hydroxy-, methyl ester	$C_3H_6O_3$ / 90.08	96-35-5 /	/ 149; 52[17]	4-03-00-00578 / 1.1677[18]	H_2O 3; EtOH 5; eth 5
199	Acetic acid, hydroxy-, propyl ester / Hydroxyacetic acid, propyl ester	$C_5H_{10}O_3$ / 118.13	90357-58-7 /	/ 170.5	4-03-00-00588 / 1.0631[18]	1.4231[18]
200	Acetic acid, iodo- / Iodoacetic acid	$C_2H_3IO_2$ / 185.95	64-69-7 / 82.5	4918 / dec	4-02-00-00534	H_2O 3; EtOH 3; eth 2; chl 2
201	Acetic acid, iodo-, ethyl ester	$C_4H_7IO_2$ / 214.00	623-48-3 /	/ 179	4-02-00-00535 / 1.8173[13]	EtOH 3; eth 3 / 1.5079[13]
202	Acetic acid, mercapto- / Thioglycolic acid	$C_2H_4O_2S$ / 92.12	68-11-1 / -16.5	9265 / 120[20]	4-03-00-00600 / 1.3253[20]	H_2O 5; EtOH 5; eth 5; chl 2 / 1.5080[20]
203	Acetic acid, mercapto-, ethyl ester	$C_4H_8O_2S$ / 120.17	623-51-8 /	/ 157	4-03-00-00616 / 1.0964[15]	EtOH 3; eth 3; ctc 2 / 1.4582[20]
204	Acetic acid, mercapto-, methyl ester	$C_3H_6O_2S$ / 106.15	2365-48-2 /	/ 42-3[10]	4-03-00-00614	eth 4; EtOH 4 / 1.4657[20]
205	Acetic acid, methoxy-	$C_3H_6O_3$ / 90.08	625-45-6 /	/ 203.5	4-03-00-00574 / 1.1768[20]	H_2O 3; EtOH 3; eth 3 / 1.4168[20]
206	Acetic acid, methoxy-, ethyl ester / Ethyl methoxyacetate	$C_5H_{10}O_3$ / 118.13	3938-96-3 /	/ 142	3-03-00-00381 / 1.0118[15]	H_2O 2; EtOH 4; eth 4; ctc 3 / 1.4050[20]
207	Acetic acid, methoxy-, methyl ester	$C_4H_8O_3$ / 104.11	6290-49-9 /	/ 131	4-03-00-00578 / 1.0511[20]	H_2O 2; EtOH 4; eth 4; ace 3 / 1.3962[20]
208	Acetic acid, methyl ester / Methyl acetate	$C_3H_6O_2$ / 74.08	79-20-9 / -98	5932 / 56.87	4-02-00-00122 / 0.9342[20]	H_2O 4; eth 4; EtOH 4 / 1.3614[20]
209	Acetic acid, 1-methylethyl ester / Isopropyl acetate	$C_5H_{10}O_2$ / 102.13	108-21-4 / -73.4	5093 / 88.6	4-02-00-00141 / 0.8718[20]	H_2O 3; EtOH 3; eth 5; ace 3 / 1.3773[20]
210	Acetic acid, [[5-methyl-2-(1-methylethyl)cyclohexyl]oxy]-, [1R-(1α,2β,5α)]- / l-Methoxyacetic acid	$C_{12}H_{22}O_3$ / 214.30	40248-63-3 / 54	/ 171[11]	4-06-00-00157	H_2O 2; EtOH 4; eth 3; ace 4
211	Acetic acid, 2-methylphenyl ester / o-Cresyl acetate	$C_9H_{10}O_2$ / 150.18	533-18-6 /	2588 / 208	4-06-00-01960 / 1.0533[15]	eth 4; EtOH 4 / 1.5002[20]
212	Acetic acid, 3-methylphenyl ester / m-Cresyl acetate	$C_9H_{10}O_2$ / 150.18	122-46-3 / 12	2587 / 212	4-06-00-02047 / 1.043[20]	bz 4; eth 4; EtOH 4 / 1.4978[20]
213	Acetic acid, 4-methylphenyl ester	$C_9H_{10}O_2$ / 150.18	140-39-6 /	/ 212.5	4-06-00-02112 / 1.0512[17]	H_2O 2; EtOH 3; eth 3; ctc 2 / 1.5163[22]
214	Acetic acid, 1-methylpropyl ester / sec-Butyl acetate	$C_6H_{12}O_2$ / 116.16	105-46-4 / -98.9	1536 / 112	4-02-00-00148 / 0.8748[20]	H_2O 1; EtOH 3; eth 3; ctc 2 / 1.3888[20]
215	Acetic acid, 2-methylpropyl ester / Isobutyl acetate	$C_6H_{12}O_2$ / 116.16	110-19-0 / -98.8	5014 / 116.5	4-02-00-00149 / 0.8712[20]	H_2O 2; EtOH 5; eth 5; ace 3 / 1.3902[20]
216	Acetic acid, (methylthio)-	$C_3H_6O_2S$ / 106.15	2444-37-3 /	/ 130-1[27]	4-03-00-00603 / 1.221[20]	1.495[20]
217	Acetic acid, (2-naphthalenyloxy)- / 2-Naphthoxyacetic acid	$C_{12}H_{10}O_3$ / 202.21	120-23-0 / 156	6317 /	4-06-00-04274	H_2O 3; EtOH 3; eth 3; HOAc 3
218	Acetic acid, nitro-, ethyl ester / Ethyl nitroacetate	$C_4H_7NO_4$ / 133.10	626-35-7 /	/ 106[25]; 83[6]	4-02-00-00537 / 1.1953[20]	H_2O 2; EtOH 5; eth 4; dil alk 3 / 1.4250[20]
219	Acetic acid, 2-nitrophenyl ester	$C_8H_7NO_4$ / 181.15	610-69-5 / 40.5	/ 253 dec; 141[11]	4-06-00-01256	H_2O 3; EtOH 4; eth 4; ace 4
220	Acetic acid, nonyl ester / Nonyl acetate	$C_{11}H_{22}O_2$ / 186.29	143-13-5 / -26	6597 / 210	4-02-00-00167 / 0.8785[15]	1.426[20]
221	Acetic acid, octadecyl ester	$C_{20}H_{40}O_2$ / 312.54	822-23-1 / 34.5	/ 208[9]	4-02-00-00171 / 0.8510[30]	EtOH 4
222	Acetic acid, octyl ester	$C_{10}H_{20}O_2$ / 172.27	112-14-1 / -38.5	/ 210	4-02-00-00165 / 0.8705[20]	H_2O 1; EtOH 3; eth 3; ctc 2 / 1.4150[20]
223	Acetic acid, oxo- / Glyoxylic acid	$C_2H_2O_3$ / 74.04	298-12-4 / 98	4407 /	4-03-00-01489	H_2O 4; EtOH 2; eth 2; bz 2
224	Acetic acid, oxo(phenylamino)-, ethyl ester	$C_{10}H_{11}NO_3$ / 193.20	1457-85-8 / 65.5	/ 280	3-12-00-00550	ace 4; bz 4; eth 4; EtOH 4

No.	Name / Synonym	Mol. Form. / Mol. Wt.	CAS RN / mp/°C	Merck No. / bp/°C	Beil. Ref. / den/g cm⁻³	Solubility / n_D
225	Acetic acid, [(4-oxo-2-phenyl-4H-1-benzopyran-7-yl)oxy]-, ethyl ester Efloxate	$C_{19}H_{16}O_5$ 324.33	119-41-5 123.7	3490	4-18-00-00695	chl 3
226	Acetic acid, 2,2'-oxybis-	$C_4H_6O_5$ 134.09	110-99-6 148 dec		4-03-00-00577	H_2O 4; eth 4; EtOH 4
227	Acetic acid, pentyl ester Pentyl acetate	$C_7H_{14}O_2$ 130.19	628-63-7 -70.8	149.2	4-02-00-00152 0.8756²⁰	H_2O 2; EtOH 5; eth 5; ctc 3 1.4023²⁰
228	Acetic acid, phenoxy- Phenoxyacetic acid	$C_8H_8O_3$ 152.15	122-59-8 98.5	7223 285 dec	4-06-00-00634	H_2O 3; EtOH 4; eth 4; bz 4
229	Acetic acid, phenoxy-, ethyl ester Ethyl phenoxyacetate	$C_{10}H_{12}O_3$ 180.20	2555-49-9	109-12³	4-06-00-00635 1.1082²⁰	1.5080²⁰
230	Acetic acid, phenoxy-, methyl ester Methyl phenoxyacetate	$C_9H_{10}O_3$ 166.18	2065-23-8	245	4-06-00-00635 1.1493²⁰	eth 4; EtOH 4 1.5155²⁰
231	Acetic acid, phenyl ester Phenyl acetate	$C_8H_8O_2$ 136.15	122-79-2	7238 196; 75⁸	4-06-00-00611 1.0780²⁰	H_2O 2; EtOH 5; eth 5; ctc 3 1.5035²⁰
232	Acetic acid, 2-phenylethyl ester	$C_{10}H_{12}O_2$ 164.20	103-45-7 -31.1	232.6	4-06-00-03073 1.0883²⁰	eth 4; EtOH 4 1.5171²⁰
233	Acetic acid, phenylmethyl ester Benzyl acetate	$C_9H_{10}O_2$ 150.18	140-11-4 -51.3	1137 213	4-06-00-02262 1.0550²⁰	H_2O 2; EtOH 5; eth 3; ace 3 1.5232²⁰
234	Acetic acid, potassium salt Potassium acetate	$C_2H_3KO_2$ 98.14	127-08-2 292	7580	4-02-00-00094 1.57²⁵	H_2O 4; EtOH 3; eth 1; ace 1
235	Acetic acid, 2-propenyl ester Allyl acetate	$C_5H_8O_2$ 100.12	591-87-7	103.5	4-02-00-00180 0.9275²⁰	H_2O 2; EtOH 5; eth 5; ace 3 1.4049²⁰
236	Acetic acid, propyl ester Propyl acetate	$C_5H_{10}O_2$ 102.13	109-60-4 -93	7853 101.54	4-02-00-00138 0.8878²⁰	H_2O 2; EtOH 5; eth 5; ctc 3 1.3842²⁰
237	Acetic acid, sodium salt Sodium acetate	$C_2H_3NaO_2$ 82.03	127-09-3 324	8513	4-02-00-00094 1.528²⁵	H_2O 4; EtOH 2 1.464
238	Acetic acid, sulfo- Sulfoacetic acid	$C_2H_4O_5S$ 140.12	123-43-3 85	8932 245 dec	4-04-00-00102	H_2O 4; ace 4; EtOH 4
239	Acetic acid, 2,2'-sulfonylbis- Sulfonyldiacetic acid	$C_4H_6O_6S$ 182.15	123-45-5 187	8940	3-03-00-00424	H_2O 4; EtOH 4; eth 3; sulf 3
240	Acetic acid, 2,2'-thiobis- Thiodiglycolic acid	$C_4H_6O_4S$ 150.16	123-93-3 129	9260	4-03-00-00612	H_2O 2; EtOH 4; bz 3
241	Acetic acid, tribromo- Tribromoacetic acid	$C_2HBr_3O_2$ 296.74	75-96-7 132	9521 245 dec	4-02-00-00534	H_2O 3; EtOH 3; eth 3
242	Acetic acid, tribromo-, ethyl ester Ethyl tribromoacetate	$C_4H_5Br_3O_2$ 324.79	599-99-5	225	3-02-00-00485 2.2260²⁰	EtOH 3; eth 3 1.5438¹³
243	Acetic acid, trichloro- Trichloroacetic acid	$C_2HCl_3O_2$ 163.39	76-03-9 57.5	9539 196.5	4-02-00-00508 1.6126⁶⁴	H_2O 4; EtOH 3; eth 3; ctc 2 1.4603⁶¹
244	Acetic acid, trichloro-, anhydride Trichloroacetic anhydride	$C_4Cl_6O_3$ 308.76	4124-31-6	223 dec	4-02-00-00518 1.6908²⁰	eth 4; HOAc 4
245	Acetic acid, trichloro-, butyl ester Butyl trichloroacetate	$C_6H_9Cl_3O_2$ 219.49	3657-07-6	204	4-02-00-00515 1.2778²⁰	ctc 3 1.4525²⁵
246	Acetic acid, trichloro-, 2-chloroethyl ester	$C_4H_4Cl_4O_2$ 225.89	4974-21-4	217	4-02-00-00515 1.5357²⁰	eth 4 1.4813²⁰
247	Acetic acid, trichloro-, 1,1-dimethylethyl ester	$C_6H_9Cl_3O_2$ 219.49	1860-21-5 25.5	54-5⁷	4-02-00-00516 1.2363²⁵	eth 4; EtOH 4 1.4398²⁵
248	Acetic acid, trichloro-, ethyl ester Ethyl trichloroacetate	$C_4H_5Cl_3O_2$ 191.44	515-84-4 167.5	9539	4-02-00-00514 1.3836²⁰	H_2O 1; EtOH 3; eth 3; bz 3 1.4505²⁰
249	Acetic acid, trichloro-, 2-hydroxyethyl ester	$C_4H_5Cl_3O_3$ 207.44	33560-17-7	130-4¹²	4-02-00-00474 1.532²⁰	EtOH 4 1.4775²⁰
250	Acetic acid, trichloro-, 2-methoxyethyl ester 2-Methoxyethyl trichloroacetate	$C_5H_7Cl_3O_3$ 221.47	35449-34-4 14.5	98-9¹⁷	3-02-00-00474 1.3826²⁰	bz 4; eth 4; EtOH 4 1.4563²⁰
251	Acetic acid, trichloro-, 1-methylpropyl ester	$C_6H_9Cl_3O_2$ 219.49	4484-80-4	93-4²⁴	3-02-00-00472 1.2636²⁰	bz 4; eth 4; EtOH 4 1.4483²⁰
252	Acetic acid, trichloro-, 2-methylpropyl ester Isobutyl trichloroacetate	$C_6H_9Cl_3O_2$ 219.49	33560-15-5	188	4-02-00-00515 1.2636²⁰	bz 4; eth 4; EtOH 4 1.4483²⁰
253	Acetic acid, trichloro-, methyl ester Methyl trichloroacetate	$C_3H_3Cl_3O_2$ 177.41	598-99-2 -17.5	153.8	4-02-00-00513 1.4874²⁰	H_2O 1; EtOH 4; eth 4; ctc 3 1.4572²⁰
254	Acetic acid, trichloro-, 1-methylethyl ester Isopropyl trichloroacetate	$C_5H_7Cl_3O_2$ 205.47	3974-99-0	175; 66¹⁵	4-02-00-00515 1.3034²⁰	bz 4; eth 4; EtOH 4 1.4428²⁰
255	Acetic acid, trichloro-, 3-methylbutyl ester	$C_7H_{11}Cl_3O_2$ 233.52	57392-55-9	217	4-02-00-00516 1.2314²⁰	eth 4; EtOH 4 1.4521²⁰
256	Acetic acid, (2,4,5-trichlorophenoxy)- 2,4,5-T	$C_8H_5Cl_3O_3$ 255.48	93-76-5 153	8999	4-06-00-00973	H_2O 1; EtOH 3; bz 4
257	Acetic acid, (2,4,5-trichlorophenoxy)-, 2-butoxyethyl ester 2,4,5-T Butoxyethyl ester	$C_{14}H_{17}Cl_3O_4$ 355.64	2545-59-7	163-6¹	4-06-00-00976 1.280²⁰	ctc 3
258	Acetic acid, (2,4,5-trichlorophenoxy)-, butyl ester 2,4,5-T Butyl ester	$C_{12}H_{13}Cl_3O_3$ 311.59	93-79-8 28.5		4-06-00-00975	
259	Acetic acid, (2,4,5-trichlorophenoxy)-, 1-methylethyl ester 2,4,5-T Isopropyl ester	$C_{11}H_{11}Cl_3O_3$ 297.56	93-78-7 45		4-06-00-00974	chl 4
260	Acetic acid, (2,4,5-trichlorophenoxy)-, 1-methylpropyl ester 2,4,5-T sec-Butyl ester	$C_{12}H_{13}Cl_3O_3$ 311.59	61792-07-2 39		4-06-00-00975	
261	Acetic acid, trichloro-, propyl ester	$C_5H_7Cl_3O_2$ 205.47	13313-91-2	187	4-02-00-00515 1.3221²⁰	eth 4; EtOH 4 1.4501²⁰
262	Acetic acid, trichloro-, trichloromethyl ester	$C_3Cl_6O_2$ 280.75	6135-29-1 34	191.5	2-03-00-00015 1.6733³⁵	bz 4; eth 4; chl 4; lig 4

No.	Name / Synonym	Mol. Form. / Mol. Wt.	CAS RN / mp/°C	Merck No. / bp/°C	Beil. Ref. / den/g cm⁻³	Solubility / n_D
263	Acetic acid, trifluoro- / Trifluoroacetic acid	$C_2HF_3O_2$ / 114.02	76-05-1 / -15.2	9595 / 73	4-02-00-00458 / 1.5351[25]	H_2O 3; EtOH 3; eth 3; ace 3
264	Acetic acid, trifluoro-, anhydride	$C_4F_6O_3$ / 210.03	407-25-0 / -65	/ 39.5	4-02-00-00469 / 1.490[25]	1.269[25]
265	Acetic acid, trifluoro-, butyl ester / Butyl trifluoroacetate	$C_6H_9F_3O_2$ / 170.13	367-64-6 /	/ 102	4-02-00-00464 / 1.0268[22]	chl 3 / 1.353[22]
266	Acetic acid, trifluoro-, 1,1-dimethylethyl ester	$C_6H_9F_3O_2$ / 170.13	400-52-2 /	/ 83	4-02-00-00465 /	chl 3 / 1.3300[25]
267	Acetic acid, trifluoro-, ethyl ester	$C_4H_5F_3O_2$ / 142.08	383-63-1 /	/ 61	4-02-00-00463 / 1.194[20]	1.308[20]
268	Acetic acid, trifluoro-, 2-methylpentyl ester / 2-Methylpentyl trifluoroacetate	$C_8H_{13}F_3O_2$ / 198.19	10042-29-2 /	/ 140	4-02-00-00466 / 1.0504[20]	chl 2
269	Acetic acid, trifluoro-, propyl ester / Propyl trifluoroacetate	$C_5H_7F_3O_2$ / 156.10	383-66-4 /	/ 82.5	4-02-00-00464 / 1.1285[25]	chl 3 / 1.3233[22]
270	Acetochlor / Acetamide, 2-chloro-N-(ethoxymethyl)-N-(2-ethyl-6-methylphenyl)-	$C_{14}H_{20}ClNO_2$ / 269.77	34256-82-1 /			
271	Acetonitrile / Methyl cyanide	C_2H_3N / 41.05	75-05-8 / -43.8	62 / 81.6	4-02-00-00419 / 0.7857[20]	H_2O 5; EtOH eth 5; ace 5 / 1.3442[30]
272	Acetonitrile, amino- / Aminoacetonitrile	$C_2H_4N_2$ / 56.07	540-61-4 /	423 / 58[15]	4-04-00-02363 /	EtOH 4
273	Acetonitrile, (butylamino)-	$C_6H_{12}N_2$ / 112.17	3010-04-6 /	/ 85[9]	4-04-00-02382 / 0.891[25]	1.4337[20]
274	Acetonitrile, chloro- / Chloromethyl cyanide	C_2H_2ClN / 75.50	107-14-2 /	/ 126.5	4-02-00-00492 / 1.1930[20]	eth 4; EtOH 4 / 1.4202[25]
275	Acetonitrile, (cyclohexylamino)-	$C_8H_{14}N_2$ / 138.21	1074-58-4 / 18	/ 76[1]	4-12-00-00073 / 0.9657[25]	
276	Acetonitrile, cyclohexylidene- / Cyclohexylideneacetonitrile	$C_8H_{11}N$ / 121.18	4435-18-1 /	/ 107-8[22]	4-09-00-00124 / 0.9483[15]	eth 4; EtOH 4 / 1.4382[25]
277	Acetonitrile, dibromo- / Dibromoacetonitrile	C_2HBr_2N / 198.84	3252-43-5 /	/ 67-9[24]	4-02-00-00533 / 2.296[25]	1.5393[20]
278	Acetonitrile, dichloro- / Dichloroacetonitrile	C_2HCl_2N / 109.94	3018-12-0 /	/ 112.5	4-02-00-00506 / 1.369[20]	MeOH 3 / 1.4391[25]
279	Acetonitrile, dichlorofluoro- / Dichlorofluoroacetonitrile	C_2Cl_2FN / 127.93	353-82-2 / -110	/ 33	4-02-00-00508 / 1.3909[25]	1.3682[20]
280	Acetonitrile, (diethylamino)-	$C_6H_{12}N_2$ / 112.17	3010-02-4 /	/ 170	4-04-00-02378 / 0.8660[20]	H_2O 3 / 1.4260[20]
281	Acetonitrile, (dimethylamino)- / Glycinonitrile, N,N-dimethyl-	$C_4H_8N_2$ / 84.12	926-64-7 /	/ 137.5	4-04-00-02368 / 0.8649[20]	H_2O 4; EtOH 4 / 1.4095[20]
282	Acetonitrile, hydroxy- / Glyconitrile	C_2H_3NO / 57.05	107-16-4 / <-72	/ 183 dec; 119[24]	4-03-00-00598 /	H_2O 4; EtOH 4; eth 4; bz 1 / 1.4117[19]
283	Acetonitrile, iodo- / Iodoacetonitrile	C_2H_2IN / 166.95	624-75-9 /	/ 185	2-02-00-00206 / 2.307[25]	1.5744[20]
284	Acetonitrile, methoxy-	C_3H_5NO / 71.08	1738-36-9 /	/ 119	3-03-00-00399 / 0.9492[20]	H_2O 2; EtOH 3; eth 3; ace 3 / 1.3831[20]
285	Acetonitrile, (methyleneamino)- / Methyleneaminoacetonitrile	$C_3H_4N_2$ / 68.08	109-82-0 / 129	5976 /		
286	Acetonitrile, (methylthio)- / Cyanomethyl methyl sulfide	C_3H_5NS / 87.15	35120-10-6 /	/ 61-3[15]	4-03-00-00630 / 1.039[25]	1.4826[20]
287	Acetonitrile, phenoxy- / Phenoxyacetonitrile	C_8H_7NO / 133.15	3598-14-9 /	/ 239.5	4-06-00-00640 / 1.0991[20]	eth 4; EtOH 4 / 1.5246[20]
288	Acetonitrile, trichloro- / Trichloroacetonitrile	C_2Cl_3N / 144.39	545-06-2 / -42	9540 / 85.7	4-02-00-00524 / 1.4403[25]	H_2O 1 / 1.4409[20]
289	Acetonitrile, trifluoro- / Trifluoroacetonitrile	C_2F_3N / 95.02	353-85-5 /	/ -68.8	4-02-00-00472 /	
290	Acetonitrile, trinitro- / Trinitroacetonitrile	$C_2N_4O_6$ / 176.05	630-72-8 / 41.5	/ exp 220	0-02-00-00229 /	eth 4
291	Acetyl bromide / Acetic acid, bromide	C_2H_3BrO / 122.95	506-96-7 / -96	76 / 76	4-02-00-00398 / 1.6625[18]	eth 5; ace 3; bz 5; chl 5 / 1.4486[20]
292	Acetyl bromide, bromo- / Bromoacetyl bromide	$C_2H_2Br_2O$ / 201.85	598-21-0 /	/ 148.5	4-02-00-00530 / 2.312[22]	ace 3; ctc 3 / 1.5449[20]
293	Acetyl bromide, trichloro- / Trichloroacetyl bromide	C_2BrCl_3O / 226.28	34069-94-8 /	/ 143	3-02-00-00476 / 1.898[15]	bz 4; eth 4
294	Acetyl chloride / Ethanoyl chloride	C_2H_3ClO / 78.50	75-36-5 / -112.8	79 / 50.7	4-02-00-00395 / 1.1051[20]	eth 5; ace 5; bz 5; ctc 3 / 1.3886[20]
295	Acetyl chloride, chloro- / Chloroacetyl chloride	$C_2H_2Cl_2O$ / 112.94	79-04-9 / -22	2054 / 106	4-02-00-00488 / 1.4202[20]	eth 5; ace 3; ctc 3 / 1.4530[20]
296	Acetyl chloride, chlorofluoro- / Chlorofluoroacetyl chloride	C_2HCl_2FO / 130.93	359-32-0 /	/ 69.5	3-02-00-00453 / 1.468[25]	1.3992[25]
297	Acetyl chloride, dichloro- / Dichloroacetyl chloride	C_2HCl_3O / 147.39	79-36-7 /	3040 / 108	4-02-00-00504 / 1.5315[16]	H_2O 6; EtOH 6; eth 5 / 1.4591[20]
298	Acetyl chloride, ethoxy-	$C_4H_7ClO_2$ / 122.55	14077-58-8 /	/ 125	3-03-00-00396 / 1.1170[25]	ace 4; eth 4 / 1.4204[20]
299	Acetyl chloride, methoxy-	$C_3H_5ClO_2$ / 108.52	38870-89-2 /	/ 112.5	3-03-00-00396 / 1.1871[20]	eth 3; ace 3; ctc 3; chl 4 / 1.4199[20]
300	Acetyl chloride, trichloro- / Trichloroacetyl chloride	C_2Cl_4O / 181.83	76-02-8 /	/ 117.9	4-02-00-00519 / 1.6202[20]	eth 5 / 1.4695[20]
301	Acetyl fluoride / Acetic acid, fluoride	C_2H_3FO / 62.04	557-99-3 / -84	/ 20.8	4-02-00-00393 / 1.032[25]	EtOH 5; eth 5; bz 3; chl 3

No.	Name Synonym	Mol. Form. Mol. Wt.	CAS RN mp/°C	Merck No. bp/°C	Beil. Ref. den/g cm^{-3}	Solubility n_D
302	Acetyl iodide	C_2H_3IO 169.95	507-02-8 88	108	4-02-00-00399 2.0673[20]	eth 4 1.5491[20]
303	Acetyl isothiocyanate Isothiocyanic acid, acetyl	C_3H_3NOS 101.13	13250-46-9	132.5	3-03-00-00278 1.1523[13]	eth 3; CS_2 3 1.5231[18]
304	Aciifluorfen Benzoic acid, 5-[2-chloro-4-(trifluoromethyl)phenoxy]-2-nitro-	$C_{14}H_7ClF_3NO_5$ 361.66	50594-66-6 150	105		
305	Aconitane-3,8,13,14,15-pentol, 20-ethyl-1,6,16-trimethoxy-4-(methoxymethyl)-, Aconine	$C_{25}H_{41}NO_9$ 499.60	509-20-6 132	110	5-21-06-00310	H_2O 3; EtOH 3; eth 2; chl 3
306	Aconitane-3,8,13,14,15-pentol, 20-ethyl-1,6,16-trimethoxy-4-(methoxymethyl)-, 8-acetate 14-benzoate, Aconitine	$C_{34}H_{47}NO_{11}$ 645.75	302-27-2 204	113	5-21-06-00310	bz 4; EtOH 4; chl 4
307	Aconitane-3,8,13,14-tetrol, 20-ethyl-1,6,16-trimethoxy-4-(methoxymethyl)-, 8-acetate 14-benzoate, (1α,3alp Indaconitine	$C_{34}H_{47}NO_{10}$ 629.75	4491-19-4 202-3 dec	4842	5-21-06-00308	eth 4; EtOH 4; chl 4
308	Aconitane-3,8,13,14-tetrol, 20-ethyl-1,6,16-trimethoxy-4-(methoxymethyl)-, 8-acetate 14-(3,4-dimethoxybenzoate),(1α,3α,6α,14α,16β)- Pseudoaconitine	$C_{36}H_{51}NO_{12}$ 689.80	127-29-7 214	7924	5-21-06-00309	eth 4; EtOH 4
309	Aconitane-4,8,9-triol, 20-ethyl-1,14,16-trimethoxy-, 4-[2-(acetylamino)benzoate], (1α,14α,16β)- Lappaconitine	$C_{32}H_{44}N_2O_8$ 584.71	32854-75-4 217.5	5237	5-21-06-00291	H_2O 1; EtOH 2; eth 2; bz 3
310	Aconitane-8,13,14-triol, 20-ethyl-1,6,16-trimethoxy-4-(methoxymethyl)-, 8-acetate 14-(3,4-dimethoxybenzoate), Bikhaconitine	$C_{36}H_{51}NO_{11}$ 673.80	6078-26-8 164	1234	5-21-06-00303	eth 4; EtOH 4; chl 4
311	Aconitane-8,13,14-triol, 1,6,16-trimethoxy-4-(methoxymethyl)-20-methyl-, 8-acetate 14-benzoate, Delphinine	$C_{33}H_{45}NO_9$ 599.72	561-07-9 199	2867	5-21-06-00302	H_2O 1; chl 3; ace 3; eth 3
312	4-Acridinamine	$C_{13}H_{10}N_2$ 194.24	578-07-4 108	183.5	5-22-11-00008	ace 4; bz 4; eth 4; EtOH 4
313	9-Acridinamine Aminacrine	$C_{13}H_{10}N_2$ 194.24	90-45-9 241	418	5-22-11-00008	EtOH 3; ace 3; dil HCl 4
314	Acridine Dibenzo[b,e]pyridine	$C_{13}H_9N$ 179.22	260-94-6 106(form a); 110(form b)	117 344.86	5-20-08-00199 1.005[20]	H_2O 2; EtOH 4; eth 4; bz 4
315	Acridine, 9-chloro- 9-Chloroacridine	$C_{13}H_8ClN$ 213.67	1207-69-8 121	sub	5-20-08-00210	H_2O 4; EtOH 4
316	3,6-Acridinediamine Proflavine	$C_{13}H_{11}N_3$ 209.25	92-62-6 285	7780	5-22-11-00322	H_2O 3; EtOH 4; eth 2; bz 2
317	3,9-Acridinediamine, 7-ethoxy- Ethacridine	$C_{15}H_{15}N_3O$ 253.30	442-16-0 226	3668	5-22-12-00243	
318	Acridine, 9-phenyl- 9-Phenylacridine	$C_{19}H_{13}N$ 255.32	602-56-2 184	404	5-20-08-00564	H_2O 1; EtOH 2; eth 3; bz 4
319	Adenosine β-D-Ribofuranoside, adenine-9	$C_{10}H_{13}N_5O_4$ 267.24	58-61-7 235.5	143	4-26-00-03598	H_2O 2; EtOH 1
320	5'-Adenylic acid Adenosine 5'-monophosphate	$C_{10}H_{14}N_5O_7P$ 347.22	61-19-8 195 dec	148	4-26-00-03615	H_2O 4; EtOH 3
321	Ajmalan-17,21-diol, (17R,21α)- Ajmaline	$C_{20}H_{26}N_2O_2$ 326.44	4360-12-7 206	184	4-23-00-03212	H_2O 1; EtOH 3; eth 2; bz 2
322	Alachlor Acetamide, 2-chloro-N-(2,6-diethylphenyl)-N-(methoxymethyl)-	$C_{14}H_{20}ClNO_2$ 269.77	15972-60-8 40	193 100[0.02]	1.133[25]	
323	β-Alanine 3-Aminopropanoic acid	$C_3H_7NO_2$ 89.09	107-95-9 200 dec	196	4-04-00-02526 1.437[19]	H_2O 3; EtOH 2; eth 1; ace 1
324	DL-Alanine DL-2-Aminopropanoic acid	$C_3H_7NO_2$ 89.09	302-72-7 300 dec	195 sub 250	4-04-00-02481 1.424[25]	H_2O 3; EtOH 4
325	L-Alanine (S)-2-Aminopropanoic acid	$C_3H_7NO_2$ 89.09	56-41-7 297 dec	195 sub 250	4-04-00-02480 1.432[22]	H_2O 3; EtOH 2; eth 1; ace 1
326	Alanine, 3-amino- 2,3-Diaminopropionic acid	$C_3H_8N_2O_2$ 104.11	515-94-6 110	2962	4-04-00-02581	H_2O 4
327	DL-Alanine, N-benzoyl-	$C_{10}H_{11}NO_3$ 193.20	1205-02-3 165.5	dec	4-09-00-00794	H_2O 3; EtOH 3; eth 2; DMSO 2
328	β-Alanine, N,N-diethyl-, ethyl ester Ethyl 3-(diethylamino)propionate	$C_9H_{19}NO_2$ 173.26	5515-83-3 83-4[12]		4-04-00-02538 0.881[25]	1.4253[20]
329	β-Alanine, N-(2,4-dihydroxy-3,3-dimethyl-1-oxobutyl)-, (R)- Pantothenic acid	$C_9H_{17}NO_5$ 219.24	79-83-4	6964	4-04-00-02569	H_2O 4; bz 4; eth 4
330	β-Alanine, N-(3-ethoxy-3-oxopropyl)-N-methyl-, ethyl ester Propionic acid, 3,3'-iminodi-N-methyl, diethyl ester	$C_{11}H_{21}NO_4$ 231.29	6315-60-2 136-8[4]		3-04-00-01278 1.0172[20]	eth 4; EtOH 4 1.4421[20]
331	Alanine, 2-methyl- α-Aminoisobutyric acid	$C_4H_9NO_2$ 103.12	62-57-7 335	457 sub 280	4-04-00-02616	H_2O 4; EtOH 2; eth 1

No.	Name / Synonym	Mol. Form. / Mol. Wt.	CAS RN / mp/°C	Merck No. / bp/°C	Beil. Ref. / den/g cm^{-3}	Solubility / n_D
332	β-Alanine, N-methyl-, ethyl ester	$C_6H_{13}NO_2$ 131.17	2213-08-3	80^{21}	3-04-00-01264 1.0064^{20}	eth 4; EtOH 4 1.4443^{20}
333	L-Alanine, 3-(2-propenylsulfinyl)-, (S)- Alliin	$C_6H_{11}NO_3S$ 177.22	556-27-4 165	251	4-04-00-03147	H_2O 4
334	Aldicarb Propanal, 2-methyl-2-(methylthio)-, O-[(methylamino)carbonyl]oxime	$C_7H_{14}N_2O_2S$ 190.27	116-06-3 99	216	1.195^{25}	
335	Aldoxycarb Propanal, 2-methyl-2-(methylsulfonyl)-, O-[(methylamino)carbonyl]oxime	$C_7H_{14}N_2O_4S$ 222.27	1646-88-4			
336	Allethrin 2-Methyl-4-oxo-3-(2-propenyl)-2-cyclopenten-1-yl 2,2-dimethyl-3-(2-methyl-1-propenyl) cyclopropane carboxylate	$C_{19}H_{26}O_3$ 302.41	584-79-2		1.010^{20}	
337	D-Allose	$C_6H_{12}O_6$ 180.16	2595-97-3 128	279	4-01-00-04299	H_2O 4
338	D-Altrose	$C_6H_{12}O_6$ 180.16	1990-29-0 103.5	319	4-01-00-04300	H_2O 4
339	Aluminum, trimethyl Trimethyl aluminum	C_3H_9Al 72.09	75-24-1 15.4	130; 20^8	0.752^{20}	
340	Aluminum, tris(2-hydroxypropanoato-O1,O2)- Aluminum lactate	$C_9H_{15}AlO_9$ 294.19	18917-91-4 350			H_2O 3
341	Ametryn N-Ethyl-N'-(1-methylethyl)-6-(methylthio)1,3,5-triazine-2,4-diamine	$C_9H_{17}N_5S$ 227.33	834-12-8 88	402		
342	Amitraz N-Methylbis(2,4-xylyliminomethyl)amine	$C_{19}H_{23}N_3$ 293.41	33089-61-1 86	503	1.128^{20}	
343	Androsta-1,4-dien-3-one, 17-hydroxy-, (17β)- Boldenone	$C_{19}H_{26}O_2$ 286.41	846-48-0 165	1327		
344	Androsta-1,4-dien-3-one, 17-hydroxy-17-methyl-, (17β)- Methandrostenolone	$C_{20}H_{28}O_2$ 300.44	72-63-9 166	5862	4-08-00-01119	
345	Androstane	$C_{19}H_{32}$ 260.46	24887-75-0 50	668 60$^{0.003}$	4-05-00-01211	ace 4; eth 4; EtOH 4; peth 4
346	Androstane-17-carboxylic acid, (5β,17β)- Etiocholanic acid	$C_{20}H_{32}O_2$ 304.47	438-08-4 228.5	3824 sub 160	3-09-00-02644	
347	Androstan-17-one, 3-hydroxy-, (3β,5α)- Epiandrosterone	$C_{19}H_{30}O_2$ 290.45	481-29-8 178	3562	4-08-00-00641	
348	Androstan-17-one, 3-hydroxy-, (3α,5α)- Androsterone	$C_{19}H_{30}O_2$ 290.45	53-41-8 185	673	4-08-00-00642	H_2O 2; EtOH 3; eth 3; ace 3
349	Androstan-3-one, 17-hydroxy-, (5α,17β)- Stanolone	$C_{19}H_{30}O_2$ 290.45	521-18-6 181	8753 sub 135	4-08-00-00634	
350	Androstan-3-one, 17-hydroxy-17-methyl-, (5α,17β)- Mestanolone	$C_{20}H_{32}O_2$ 304.47	521-11-9 192.5	5816	4-08-00-00656	AcOEt 2
351	Androstan-3-one, 2-methyl-17-(1-oxopropoxy)-, (2α,5α,17β)- Dromostanolone propionate	$C_{23}H_{36}O_3$ 360.54	521-12-0 128	3436	4-08-00-00654	
352	Androst-4-ene-3,17-dione 4-Androstene-3,17-dione	$C_{19}H_{26}O_2$ 286.41	63-05-8 143(form a); 173(form b)	671	4-07-00-02381	
353	Androst-4-ene-3,11,17-trione Adrenosterone	$C_{19}H_{24}O_3$ 300.40	382-45-6 222	165 sub	4-07-00-02796	H_2O 2; EtOH 3; eth 3; ace 3
354	Androst-4-en-3-one, 4-chloro-17-hydroxy-, (17β)- Clostebol	$C_{19}H_{27}ClO_2$ 322.87	1093-58-9 189	2409	4-08-00-00989	
355	Androst-4-en-3-one, 9-fluoro-11,17-dihydroxy-17-methyl-, (11β,17β)- Fluoxymesterone	$C_{20}H_{29}FO_3$ 336.45	76-43-7 270	4113	4-08-00-02057	
356	Androst-4-en-3-one, 17-hydroxy-, (17β)- Testosterone	$C_{19}H_{28}O_2$ 288.43	58-22-0 155	9109	4-08-00-00974	H_2O 1; EtOH 3; eth 3; ace 3
357	Androst-4-en-3-one, 17-hydroxy-7,17-dimethyl-, (7β,17β)- Calusterone	$C_{28}H_{48}O$ 400.69	17021-26-0 157.5	1730	4-08-00-01029	
358	Androst-4-en-3-one, 17-hydroxy-17-methyl-, (17β)- 17-Methyltestosterone	$C_{20}H_{30}O_2$ 302.46	58-18-4 163.5	6044	4-08-00-01010	eth 4; EtOH 4
359	Androst-4-en-3-one, 17-(1-oxopropoxy)-, (17β)- Testosterone-17-propionate	$C_{22}H_{32}O_3$ 344.49	57-85-2 120	9115	3-08-00-00897	eth 4; py 4; EtOH 4
360	Anilazine 2,4-Dichloro-6-(o-chloroanilino)-s-triazine	$C_9H_5Cl_3N_4$ 275.52	101-05-3 160	685	1.8^{20}	
361	Aniline, 3,3'-azobis[N,N-dimethyl- Azobenzene, 3,3'-bis(dimethylamino)	$C_{16}H_{20}N_4$ 268.36	21232-53-1 118	sub	0-16-00-00305	bz 4; EtOH 4; chl 4
362	Aniline, p-isopentyl-	$C_{11}H_{17}N$ 163.26	104177-72-2	257	4-12-00-02858 0.928^{15}	1.5305^{20}
363	Anisole, 3-isopropyl-2-methyl- o-Cresol, 3-isopropyl, methyl ether	$C_{11}H_{16}O$ 164.25	31202-12-7 -0.5	67^3	4-06-00-03328 0.9540^{25}	1.5148^{20}
364	Anisole, 2,3,5-trinitro- 2,3,5-Trinitroanisole	$C_7H_5N_3O_7$ 243.13	7539-25-5 107.4		2-06-00-00253 1.618^{15}	ace 4; bz 4; EtOH 4

No.	Name / Synonym	Mol. Form. / Mol. Wt.	CAS RN / mp/°C	Merck No. / bp/°C	Beil. Ref. / den/g cm^{-3}	Solubility / n_D
365	2-Anthracenamine	$C_{14}H_{11}N$ 193.25	613-13-8 238.8	sub	4-12-00-03435	H_2O 1; EtOH 3; os 2; con sulf 1
366	Anthracene	$C_{14}H_{10}$ 178.23	120-12-7 215.0	712 339.9	4-05-00-02281 1.28[25]	H_2O 1; EtOH 3; eth 2; ace 2
367	Anthracene, 9,10-bis(phenylmethyl)- 9,10-Dibenzylanthracene	$C_{28}H_{22}$ 358.48	3613-42-1 246.5		3-05-00-02640 1.1787[16]	H_2O 1; eth 2; bz 2; lig 1
368	1-Anthracenecarboxylic acid 1-Anthroic acid	$C_{15}H_{10}O_2$ 222.24	607-42-1 251.5	sub	4-09-00-02670	H_2O 1; EtOH 3; eth 3; bz 2
369	2-Anthracenecarboxylic acid 2-Anthroic acid	$C_{15}H_{10}O_2$ 222.24	613-08-1 281	sub	4-09-00-02671	HOAc 4
370	9-Anthracenecarboxylic acid 9-Anthroic acid	$C_{15}H_{10}O_2$ 222.24	723-62-6 217 dec	sub	4-09-00-02671	H_2O 1; EtOH 3
371	2-Anthracenecarboxylic acid, 9,10-dihydro-4,5-dihydroxy-9,10-dioxo- Rhein	$C_{15}H_8O_6$ 284.22	478-43-3 321	8175 sub	4-10-00-04088	H_2O 2; EtOH 2; eth 2; ace 2
372	2-Anthracenecarboxylic acid, 9,10-dihydro-9,10-dioxo-	$C_{15}H_8O_4$ 252.23	117-78-2 291	sub	4-10-00-03200	EtOH 2; eth 1; ace 3; bz 1
373	2-Anthracenecarboxylic acid, 7-β-D-glucopyranosyl-9,10-dihydro-3,5,6,8-tetrahydroxy-1-methyl-9,10-dioxo- Carminic acid	$C_{22}H_{20}O_{13}$ 492.39	1260-17-9 136 dec	1850	3-10-00-04874	H_2O 3; EtOH 3; eth 2; bz 1
374	Anthracene, 1-chloro- 1-Chloroanthracene	$C_{14}H_9Cl$ 212.68	4985-70-0 83.5		4-05-00-02292 1.1707[100]	H_2O 1; EtOH 3; eth 3; bz 3 1.6959[100]
375	Anthracene, 9,10-dibromo- 9,10-Dibromoanthracene	$C_{14}H_8Br_2$ 336.03	523-27-3 226	3000 sub	4-05-00-02295	H_2O 1; EtOH 2; eth 2; bz 2
376	Anthracene, 9,10-dihydro- 9,10-Dihydroanthracene	$C_{14}H_{12}$ 180.25	613-31-0 111	305	4-05-00-02182 1.215[20]	H_2O 1; EtOH 3; eth 3; bz 3
377	Anthracene, 1,3-dimethyl- 1,3-Dimethylanthracene	$C_{16}H_{14}$ 206.29	610-46-8 83	143[2]	4-05-00-02327	eth 4; EtOH 4
378	Anthracene, 2,10-dimethyl- Anthracene, 9,3-dimethyl	$C_{16}H_{14}$ 206.29	27532-76-9 84	140-5[2]	4-05-00-02328	H_2O 1; EtOH 4; eth 4
379	9,10-Anthracenedione Anthraquinone	$C_{14}H_8O_2$ 208.22	84-65-1 286	717 377	4-07-00-02556 1.438[20]	H_2O 1; EtOH 2; eth 2; bz 2
380	9,10-Anthracenedione, 1-amino- 1-Aminoanthraquinone	$C_{14}H_9NO_2$ 223.23	82-45-1 253.5	428 sub	4-14-00-00429	ace 4; bz 4; EtOH 4; chl 4
381	9,10-Anthracenedione, 2-amino- 2-Aminoanthraquinone	$C_{14}H_9NO_2$ 223.23	117-79-3 304.5	sub	4-14-00-00447	H_2O 1; EtOH 2; eth 1; ace 3
382	9,10-Anthracenedione, 3-amino-1,2-dihydroxy-	$C_{14}H_9NO_4$ 255.23	3963-78-8 >300	sub	2-14-00-00185	aq NH_3 4
383	9,10-Anthracenedione, 2-amino-1-hydroxy- Anthraquinone, 2-amino-1-hydroxy	$C_{14}H_9NO_3$ 239.23	568-99-0 258	sub	4-14-00-00891	bz 4; eth 4; EtOH 4
384	9,10-Anthracenedione, 1-amino-2-methyl- C.I. Solvent Orange 35	$C_{15}H_{11}NO_2$ 237.26	82-28-0 205.5		4-14-00-00496	H_2O 1; EtOH 3; eth 2; bz 3
385	9,10-Anthracenedione, 1-bromo-	$C_{14}H_7BrO_2$ 287.11	632-83-7 186.5	sub	3-07-00-04076	bz 4; EtOH 4
386	9,10-Anthracenedione, 2-bromo-	$C_{14}H_7BrO_2$ 287.11	572-83-8 205.8	sub	3-07-00-04076	bz 4
387	9,10-Anthracenedione, 1-chloro-	$C_{14}H_7ClO_2$ 242.66	82-44-0 163	sub	4-07-00-02559	H_2O 1; EtOH 2; eth 5; bz 3
388	9,10-Anthracenedione, 2-chloro-	$C_{14}H_7ClO_2$ 242.66	131-09-9 211	sub	4-07-00-02559	H_2O 1; EtOH 2; eth 1; bz 2
389	9,10-Anthracenedione, 1,5-diamino-	$C_{14}H_{10}N_2O_2$ 238.25	129-44-2 319	sub	4-14-00-00479	H_2O 1; EtOH 2; eth 2; ace 2
390	9,10-Anthracenedione, 2,7-dibromo- 2,7-Dibromoanthraquinone	$C_{14}H_8Br_2O_2$ 366.01	605-42-5 249	sub	2-07-00-00718	bz 4; HOAc 4
391	9,10-Anthracenedione, 1,2-dihydroxy- Alizarin	$C_{14}H_8O_4$ 240.22	72-48-0 289.5	237	4-08-00-03256	H_2O 2; EtOH 3; eth 3; ace 3
392	9,10-Anthracenedione, 1,4-dihydroxy- Quinizarin	$C_{14}H_8O_4$ 240.22	81-64-1 200	8094	4-08-00-03260	H_2O 3; EtOH 3; eth 3; bz 3
393	9,10-Anthracenedione, 1,5-dihydroxy- Anthrarufin	$C_{14}H_8O_4$ 240.22	117-12-4 280	719 sub	4-08-00-03268	H_2O 1; EtOH 2; eth 2; ace 2
394	9,10-Anthracenedione, 1,7-dihydroxy- 9,10-Anthraquinone, 1,7-dihydroxy	$C_{14}H_8O_4$ 240.22	569-08-4 293.3	sub	4-08-00-03270	H_2O 1; EtOH 3; eth 3; bz 3
395	9,10-Anthracenedione, 1,8-dihydroxy- Danthron	$C_{14}H_8O_4$ 240.22	117-10-2 193	2813 sub	4-08-00-03271	H_2O 1; EtOH 3; eth 3; ace 3
396	9,10-Anthracenedione, 2,7-dihydroxy-	$C_{14}H_8O_4$ 240.22	572-93-0 353.8	sub	4-08-00-03273	H_2O 1; EtOH 3; eth 2; bz 2
397	9,10-Anthracenedione, 1,8-dihydroxy-3-(hydroxymethyl)- Aloe-emodol	$C_{15}H_{10}O_5$ 270.24	481-72-1 223.5	303 sub	4-08-00-03578	bz 4; eth 4; EtOH 4
398	9,10-Anthracenedione, 1,8-dihydroxy-3-methyl- Chrysophanic acid	$C_{15}H_{10}O_4$ 254.24	481-74-3 196	2263 sub	4-08-00-03277 0.92[25]	bz 4; HOAc 4
399	9,10-Anthracenedione, 1,2-dihydroxy-3-nitro- Alizarine Orange	$C_{14}H_7NO_6$ 285.21	568-93-4 244 dec	240 sub	4-08-00-03259	H_2O 2; EtOH 3; bz 3; chl 3
400	9,10-Anthracenedione, 1,2-dihydroxy-4-nitro-	$C_{14}H_7NO_6$ 285.21	2243-71-2 289 dec	sub	4-08-00-03259	bz 4; EtOH 4; chl 4; HOAc 4

No.	Name Synonym	Mol. Form. Mol. Wt.	CAS RN mp/°C	Merck No. bp/°C	Beil. Ref. den/g cm^{-3}	Solubility n_D
401	9,10-Anthracenedione, 1,8-dihydroxy-2,4,5,7-tetranitro- Chrysamminic acid	C$_{14}$H$_4$N$_4$O$_{12}$ 420.21	517-92-0 exp	2253 dec	1-08-00-00723	eth 4; EtOH 4
402	9,10-Anthracenedione, 1,4-dimethyl-	C$_{16}$H$_{12}$O$_2$ 236.27	1519-36-4 140.5	sub	4-07-00-02583	H$_2$O 1; EtOH 2; bz 3; xyl 3
403	9,10-Anthracenedione, 2,3-dimethyl-	C$_{16}$H$_{12}$O$_2$ 236.27	6531-35-7 210.8	sub	3-07-00-04131	EtOH 3; bz 3; chl 2; xyl 3
404	9,10-Anthracenedione, 2,6-dimethyl-	C$_{16}$H$_{12}$O$_2$ 236.27	3837-38-5 242	sub	3-07-00-04132	EtOH 4; HOAc 4; tol 4
405	9,10-Anthracenedione, 1,5-dinitro-	C$_{14}$H$_8$N$_2$O$_6$ 298.21	82-35-9 385	sub	4-07-00-02562	H$_2$O 1; EtOH 2; eth 2; bz 2
406	9,10-Anthracenedione, 1,2,3,5,6,7-hexahydroxy- Rufigallol	C$_{14}$H$_8$O$_8$ 304.21	82-12-2	8267 sub	4-08-00-03750	H$_2$O 1; EtOH 2; eth 2; ace 3
407	9,10-Anthracenedione, 1-hydroxy-	C$_{14}$H$_8$O$_3$ 224.22	129-43-1 193.8	sub	4-08-00-02586	H$_2$O 1; EtOH 3; eth 3; bz 3
408	9,10-Anthracenedione, 2-hydroxy-	C$_{14}$H$_8$O$_3$ 224.22	605-32-3 306	sub	4-08-00-02598	H$_2$O 1; EtOH 3; eth 3; KOH 3
409	9,10-Anthracenedione, 2-methyl- 2-Methylanthraquinone	C$_{15}$H$_{10}$O$_2$ 222.24	84-54-8 177	5943 sub	4-07-00-02574	bz 4; EtOH 4; HOAc 4
410	9,10-Anthracenedione, 1-nitro-	C$_{14}$H$_7$NO$_4$ 253.21	82-34-8 231.5	270[7]	4-07-00-02561	H$_2$O 1; EtOH 2; eth 2; ace 3
411	9,10-Anthracenedione, 2-nitro-	C$_{14}$H$_7$NO$_4$ 253.21	605-27-6 184.5	sub	4-07-00-02561	H$_2$O 1; EtOH 2; eth 2; ace 3
412	9,10-Anthracenedione, 1,2,5,6-tetrahydroxy-	C$_{14}$H$_8$O$_6$ 272.21	632-77-9 340	sub	3-08-00-04288	py 4; EtOH 4; HOAc 4
413	9,10-Anthracenedione, 1,2,5,8-tetrahydroxy- Quinalizarin	C$_{14}$H$_8$O$_6$ 272.21	81-61-8 >275	8059	4-08-00-03683	H$_2$O 2; EtOH 2; eth 2; ace 2
414	9,10-Anthracenedione, 1,3,5,7-tetrahydroxy-	C$_{14}$H$_8$O$_6$ 272.21	632-82-6 360	sub	4-08-00-03683	ace 4; bz 4; EtOH 4; HOAc 4
415	9,10-Anthracenedione, 1,2,3-trihydroxy- Anthragallol	C$_{14}$H$_8$O$_5$ 256.21	602-64-2 313	713 sub 290	4-08-00-03567	H$_2$O 2; EtOH 3; eth 3; CS$_2$ 3
416	9,10-Anthracenedione, 1,2,4-trihydroxy- Purpurin	C$_{14}$H$_8$O$_5$ 256.21	81-54-9 259	7962 sub	4-08-00-03568	H$_2$O 2; EtOH 4; eth 3; bz 4
417	9,10-Anthracenedione, 1,2,6-trihydroxy-	C$_{14}$H$_8$O$_5$ 256.21	82-29-1 330	459	4-08-00-03568	H$_2$O 3; EtOH 3; eth 2; bz 3
418	9,10-Anthracenedione, 1,2,7-trihydroxy-	C$_{14}$H$_8$O$_5$ 256.21	602-65-3 374	462	4-08-00-03569	bz 4; EtOH 4; HOAc 4
419	9,10-Anthracenedione, 1,3,8-trihydroxy-6-methyl- Emodin	C$_{15}$H$_{10}$O$_5$ 270.24	518-82-1 257	3518 sub	4-08-00-03575	eth 4; EtOH 4
420	Anthracene, 9-ethyl- 9-Ethylanthracene	C$_{16}$H$_{14}$ 206.29	605-83-4 60		4-05-00-02326 1.0413^{99}	H$_2$O 1; EtOH 3; eth 3 1.6767^{99}
421	Anthracene, 9-ethyl-9,10-dihydro- 9-Ethyl-9,10-dihydroanthracene	C$_{16}$H$_{16}$ 208.30	605-82-3	321.5	0-05-00-00649 1.048^{18}	bz 4; eth 4; EtOH 4
422	Anthracene, 1,2,3,4,5,6-hexahydro- 1,2,3,4,5,6-Hexahydroanthracene	C$_{14}$H$_{16}$ 184.28	6109-22-4 68.5	160[15]	3-05-00-01676	bz 4
423	Anthracene, 1-methyl- 1-Methylanthracene	C$_{15}$H$_{12}$ 192.26	610-48-0 85.5	199.5	4-05-00-02310 1.0471^{99}	H$_2$O 1; EtOH 3; eth 3; bz 3 1.6802^{99}
424	Anthracene, 2-methyl- 2-Methylanthracene	C$_{15}$H$_{12}$ 192.26	613-12-7 209	sub	4-05-00-02311 1.80^0	H$_2$O 1; EtOH 2; eth 2; ace 1
425	Anthracene, 9-methyl- 9-Methylanthracene	C$_{15}$H$_{12}$ 192.26	779-02-2 81.5	196-7[12]	4-05-00-02312 1.065^{99}	EtOH 3; eth 3; ace 3; bz 3 1.6959^{99}
426	Anthracene, 9-nitro- 9-Nitroanthracene	C$_{14}$H$_9$NO$_2$ 223.23	602-60-8 146	275[17]	4-05-00-02296	H$_2$O 1; EtOH 2; ace 4; chl 2
427	Anthracene, 1,2,3,4,5,6,7,8-octahydro-	C$_{14}$H$_{18}$ 186.30	1079-71-6 78	294	4-05-00-01584 0.9703^{80}	H$_2$O 1; EtOH 3; bz 4; ctc 2 1.5372^{80}
428	Anthracene, 9-phenyl- 9-Phenylanthracene	C$_{20}$H$_{14}$ 254.33	602-55-1 156	417	4-05-00-02628	H$_2$O 1; EtOH 3; eth 3; bz 3
429	1,2,10-Anthracenetriol Anthrarobin	C$_{14}$H$_{10}$O$_3$ 226.23	577-33-3 208	718	4-06-00-07601	H$_2$O 2; EtOH 4; eth 4; ace 4
430	1,8,9-Anthracenetriol Anthralin	C$_{14}$H$_{10}$O$_3$ 226.23	1143-38-0 179	714	4-06-00-07602	H$_2$O 1; EtOH 3; eth 2; ace 3
431	1-Anthracenol	C$_{14}$H$_{10}$O 194.23	610-50-4 158	234[13]	4-06-00-04928	H$_2$O 1; EtOH 4; eth 4; NaOH 3
432	9-Anthracenol Anthranol	C$_{14}$H$_{10}$O 194.23	529-86-2 152	716	4-06-00-04930	
433	9(10H)-Anthracenone Anthrone	C$_{14}$H$_{10}$O 194.23	90-44-8 155	721	4-06-00-04930	ace 3; bz 3; con sulf 3; dil alk 3
434	5,9,14,18-Anthrazinetetrone, 6,15-dihydro- Indanthrene	C$_{28}$H$_{14}$N$_2$O$_4$ 442.43	81-77-6 470-500 dec	4846	5-24-09-00504	H$_2$O 1; EtOH 1; eth 1; ace 1
435	8'-Apo-β,ψ-carotenal, 3-hydroxy-, (3R)- β-Citraurin	C$_{30}$H$_{40}$O$_2$ 432.65	650-69-1 147	2326	3-08-00-01599	H$_2$O 1; EtOH 4; eth 4; ace 4
436	β-D-Arabinopyranose Arabinose (D)(β)	C$_5$H$_{10}$O$_5$ 150.13	6748-95-4 155.5		4-01-00-04215 1.625^{25}	H$_2$O 4
437	DL-Arginine	C$_6$H$_{14}$N$_4$O$_2$ 174.20	7200-25-1 217 dec	805	3-04-00-01359	H$_2$O 1; EtOH 1; eth 1; bz 1
438	L-Arginine	C$_6$H$_{14}$N$_4$O$_2$ 174.20	74-79-3 244 dec	805	4-04-00-02648	H$_2$O 3; EtOH 2; eth 1

No.	Name / Synonym	Mol. Form. / Mol. Wt.	CAS RN / mp/°C	Merck No. / bp/°C	Beil. Ref. / den/g cm⁻³	Solubility / n_D
439	Arsenic acid, triethyl ester / Ethyl arsenate	$C_6H_{15}AsO_4$ / 226.10	15606-95-8	/ 236.5	4-01-00-01359 / 1.3021[20]	1.4343[20]
440	Arsenous acid, triethyl ester / Ethyl arsenite	$C_6H_{15}AsO_3$ / 210.10	3141-12-6	/ 165.5	4-01-00-01358 / 1.2239[20]	1.4369[13]
441	Arsine, diethyl- / Diethylarsine	$C_4H_{11}As$ / 134.05	692-42-2	/ 105	3-04-00-01797 / 1.1338[24]	ace 4; bz 4; eth 4; EtOH 4 1.4709
442	Arsine, dimethyl- / Dimethylarsine	C_2H_7As / 106.00	593-57-7 / -136.1	/ 36	4-04-00-03665 / 1.208[29]	ace 4; bz 4; eth 4; EtOH 4
443	Arsine, diphenyl- / Diphenylarsine	$C_{12}H_{11}As$ / 230.14	829-83-4	/ 174[25]	4-16-00-01138 / 1.30[25]	eth 4; EtOH 4
444	Arsine, ethyl- / Ethylarsine	C_2H_7As / 106.00	593-59-9	/ 36	3-04-00-01797 / 1.214[22]	eth 4; EtOH 4
445	Arsine, ethyldifluoro-	$C_2H_5AsF_2$ / 141.98	430-40-0 / -38.7	/ 94.3	3-04-00-01799 / 1.708[17]	
446	Arsine, methyl- / Methylarsine	CH_5As / 91.97	593-52-2 / -143	/ 2	3-04-00-01795	ace 4; eth 4; EtOH 4
447	Arsine, oxophenyl- / Oxophenylarsine	C_6H_5AsO / 168.03	637-03-6 / 145	6903	4-16-00-01160	H_2O 1; EtOH 2; eth 1; bz 4
448	Arsine, triethyl- / Triethylarsine	$C_6H_{15}As$ / 162.11	617-75-4	/ 138.5	4-04-00-03666 / 1.150[20]	ace 4; eth 4; EtOH 4 1.467[20]
449	Arsine, trimethyl- / Trimethylarsine	C_3H_9As / 120.03	593-88-4 / -87.3	/ 52	4-04-00-03665 / 1.144[15]	bz 4; eth 4; EtOH 4
450	Arsine, triphenyl- / Triphenylarsine	$C_{18}H_{15}As$ / 306.24	603-32-7 / 61	/ 360	4-16-00-01139 / 1.2634[18]	H_2O 1; EtOH 2; eth 4; bz 4 1.6888[21]
451	Arsinic acid, dimethyl- / Cacodylic acid	$C_2H_7AsO_2$ / 138.00	75-60-5 / 195	1603	4-04-00-03681	H_2O 4; EtOH 3; eth 1
452	Arsinous acid, dimethyl-, anhydride / Cacodyl oxide	$C_4H_{12}As_2O$ / 225.98	503-80-0 / -25	/ 150	3-04-00-01814 / 1.4816[15]	eth 4; EtOH 4 1.5225[9]
453	Arsinous chloride, dimethyl- / Dimethylchloroarsine	C_2H_6AsCl / 140.44	557-89-1 / <-45	/ 109; 77[37]	4-04-00-03670 / 1.505[12]	EtOH 4 1.5203[12]
454	Arsinous chloride, diphenyl- / Chlorodiphenylarsine	$C_{12}H_{10}AsCl$ / 264.59	712-48-1 / 44	/ 337	3-16-00-00944 / 1.4820[16]	ace 4; bz 4; eth 4; EtOH 4 1.6332[56]
455	Arsonic acid, [4-[(aminocarbonyl)amino]phenyl]- / Carbarsone	$C_7H_9AsN_2O_4$ / 260.08	121-59-5 / 174	1788	3-16-00-01112	H_2O 2; EtOH 2; eth 1; chl 1
456	Arsonic acid, (4-aminophenyl)- / Arsanilic acid	$C_6H_8AsNO_3$ / 217.06	98-50-0 / 232	818	4-16-00-01190 / 1.9571[10]	H_2O 3; EtOH 2; eth 3; ace 1
457	Arsonic acid, ethyl- / Ethanearsonic acid	$C_2H_7AsO_3$ / 154.00	507-32-4 / 99.5	3677 / 209-11[12]	4-04-00-03682	H_2O 4; EtOH 4
458	Arsonic acid, methyl- / Methanearsonic acid	CH_5AsO_3 / 139.97	124-58-3 / 160.5	5864	4-04-00-03682	H_2O 3; EtOH 3
459	Arsonic acid, (4-nitrophenyl)- / Nitarsone	$C_6H_8AsNO_5$ / 247.04	98-72-6 / >310 dec	6483	4-16-00-01185	H_2O 2; EtOH 2; DMSO-d_6 2
460	Arsonic acid, phenyl- / Benzenearsonic acid	$C_6H_7AsO_3$ / 202.04	98-05-5 / 158- dec	1075	4-16-00-01183	H_2O 4; EtOH 4
461	Arsonic acid, propyl- / 1-Propanearsonic acid	$C_3H_9AsO_3$ / 168.02	107-34-6 / 134.5	7810	4-04-00-03682	H_2O 4; EtOH 4; eth 1
462	Arsonous dichloride, phenyl- / Dichlorophenylarsine	$C_6H_5AsCl_2$ / 222.93	696-28-6 / -19	/ 255	4-16-00-01161 / 1.6516[20]	bz 4; eth 4; EtOH 4 1.6386[15]
463	Arsonous difluoride, methyl- / Difluoromethylarsine	CH_3AsF_2 / 127.95	420-24-6 / -29.7	/ 76.5	3-04-00-01796 / 1.924[18]	
464	Arsonous diiodide, phenyl-	$C_6H_5AsI_2$ / 405.84	6380-34-3 / 15	205[14] 185[10]	3-16-00-00958 / 1.626[15]	
465	DL-Ascorbic acid	$C_6H_8O_6$ / 176.13	62624-30-0 / 168.5		5-18-05-00026	H_2O 4; EtOH 3; eth 1; bz 1
466	L-Ascorbic acid / Vitamin C	$C_6H_8O_6$ / 176.13	50-81-7 / 190-2 dec	855	5-18-05-00026 / 1.65[25]	H_2O 4; EtOH 3; eth 1; bz 1
467	L-Ascorbic acid, 6-deoxy- / 6-Desoxy-L-ascorbic acid	$C_6H_8O_5$ / 160.13	528-81-4 / 168	2880 / sub 160	4-18-00-02301	H_2O 4; ace 4; EtOH 4
468	L-Asparagine / α-Aminosuccinamic acid	$C_4H_8N_2O_3$ / 132.12	70-47-3 / 235	859	4-04-00-03004 / 1.543[15]	H_2O 3; EtOH 1; eth 1; MeOH 1
469	DL-Asparagine, monohydrate	$C_4H_{10}N_2O_4$ / 150.13	69833-18-7 / 182.5	859 / 214 dec	2-04-00-00900 / 1.4540[15]	H_2O 2; EtOH 1; eth 1; bz 1
470	DL-Aspartic acid	$C_4H_7NO_4$ / 133.10	617-45-8 / 275		2-04-00-00900 / 1.6622[13]	H_2O 2; EtOH 1; eth 1; bz 1
471	L-Aspartic acid / L-Aminosuccinic acid	$C_4H_7NO_4$ / 133.10	56-84-8 / 270	862	4-04-00-02998 / 1.6603[13]	H_2O 2; EtOH 1; eth 1; bz 1
472	L-Aspartic acid, N-[2-[(2-amino-2-oxoethyl)amino]-2-carboxyethyl]- / Lycomarasmine	$C_9H_{15}N_3O_7$ / 277.23	7611-43-0 / 227-9 dec	5491	3-04-00-01521	
473	Aspidospermidine, 1-acetyl-17-methoxy- / Aspidospermine	$C_{22}H_{30}N_2O_2$ / 354.49	466-49-9 / 208	872 / 220[2]	5-23-12-00210	H_2O 2; EtOH 3; eth 2; bz 3
474	Asulam / Methyl [(4-aminophenyl)sulfonyl]carbamate	$C_8H_{10}N_2O_4S$ / 230.24	3337-71-1 / 144			
475	Atisine / Anthorine	$C_{22}H_{33}NO_2$ / 343.51	466-43-3 / 58.5	881		eth 4; EtOH 4; chl 4
476	Atrazine / 6-Chloro-N-ethyl-N '-(1-methylethyl)-1,3,5-triazine-2,4-diamine	$C_8H_{14}ClN_5$ / 215.69	1912-24-9 / 173	886		

No.	Name / Synonym	Mol. Form. / Mol. Wt.	CAS RN / mp/°C	Merck No. / bp/°C	Beil. Ref. / den/g cm^{-3}	Solubility / n_D
477	Avermectin B1a / Abamectin	$C_{48}H_{72}O_{14}$ / 873.09	71751-41-2 / 152	1		
478	7-Azabicyclo[4.1.0]heptane / 2,3-Tetramethyleneaziridine	$C_6H_{11}N$ / 97.16	286-18-0 / 21	150; 50[22]	5-20-04-00304 / 0.9480[27]	bz 4; eth 4; EtOH 4
479	9-Azabicyclo[3.3.1]nonan-3-one, 9-methyl- / Pseudopelletierine	$C_9H_{15}NO$ / 153.22	552-70-5 / 54	7936 / 246	5-21-07-00059 / 1.001[100]	H_2O 4; eth 4; EtOH 4 / 1.4760[100]
480	1-Azabicyclo[2.2.2]octane / Quinuclidine	$C_7H_{13}N$ / 111.19	100-76-5 / 158	8109	5-20-04-00335	H_2O 4; ace 4; eth 4; EtOH 4
481	8-Azabicyclo[3.2.1]octane-2-carboxylic acid, 3-(benzoyloxy)-8-methyl-, [1R-(exo,exo)]- / Benzoylecgonine	$C_{16}H_{19}NO_4$ / 289.33	519-09-5 / 195	1125	5-22-05-00053	bz 4; EtOH 4
482	8-Azabicyclo[3.2.1]octane-2-carboxylic acid, 3-(benzoyloxy)-8-methyl-, methyl ester, [1R-(exo,exo)]- / Cocaine	$C_{17}H_{21}NO_4$ / 303.36	50-36-2 / 98	2450 / 187[0.1]	5-22-05-00054	H_2O 2; EtOH 4; eth 4; ace 3 / 1.5022[98]
483	8-Azabicyclo[3.2.1]octane-2-carboxylic acid, 3-hydroxy-8-methyl-, [1R-(exo,exo)]- / Ecgonine	$C_9H_{15}NO_3$ / 185.22	481-37-8 / 205	3467	5-22-05-00053	H_2O 4; EtOH 4
484	8-Azabicyclo[3.2.1]octane, 8-methyl- / Tropane	$C_8H_{15}N$ / 125.21	529-17-9	9689 / 166	5-20-04-00331 / 0.925[15]	
485	1-Azabicyclo[2.2.2]octan-3-ol / 3-Quinuclidinol	$C_7H_{13}NO$ / 127.19	1619-34-7 / 221	8110 / sub 120	5-21-01-00271	ace 3
486	8-Azabicyclo[3.2.1]octan-3-ol, 8-methyl-, exo- / Pseudotropine	$C_8H_{15}NO$ / 141.21	135-97-7 / 109	7937 / 241	5-21-01-00219	H_2O 4; EtOH 4; eth 2; bz 3
487	8-Azabicyclo[3.2.1]octan-3-ol, 8-methyl-, benzoate (ester), endo-	$C_{15}H_{19}NO_2$ / 245.32	19145-60-9 / 41.5	/ 177.5	5-21-01-00226	H_2O 4; eth 4; EtOH 4
488	8-Azabicyclo[3.2.1]octan-3-ol, 8-methyl-, benzoate (ester), exo- / Tropacocaine	$C_{15}H_{19}NO_2$ / 245.32	537-26-8 / 49	9686 / dec	5-21-01-00226 / 1.0426[100]	bz 4; eth 4; EtOH 4; peth 4 / 1.5080[100]
489	8-Azabicyclo[3.2.1]octan-3-ol, 8-methyl-, endo- / Tropine	$C_8H_{15}NO$ / 141.21	120-29-6 / 64	9693 / 233	5-21-01-00219 / 1.016[100]	H_2O 4; eth 4; EtOH 4 / 1.4811[100]
490	8-Azabicyclo[3.2.1]octan-3-one, 8-methyl-	$C_8H_{13}NO$ / 139.20	532-24-1 / 43	/ 227; 113[25]	5-21-07-00032 / 1.9872[100]	EtOH 3; eth 3; ace 3; bz 3 / 1.4598[100]
491	8-Azabicyclo[3.2.1]oct-2-ene-2-carboxylic acid, 8-methyl-, (1R)- / Ecgonidine	$C_9H_{13}NO_2$ / 167.21	484-93-5 / 226-30 dec	3466	5-22-01-00446	H_2O 4
492	1H-Azepine, hexahydro- / Hexamethyleneimine	$C_6H_{13}N$ / 99.18	111-49-9	/ 138	5-20-04-00003 / 0.8643[22]	H_2O 3; EtOH 4; eth 4 / 1.4631[20]
493	2H-Azepin-2-one, hexahydro- / Caprolactam	$C_6H_{11}NO$ / 113.16	105-60-2 / 69.3	1762 / 270	5-21-06-00444	H_2O 4; bz 4; EtOH 4; chl 4
494	Azetidine	C_3H_7N / 57.10	503-29-7 /	/ 63	5-20-01-00136 / 0.8436[20]	ace 4; bz 4; eth 4; EtOH 4 / 1.4287[25]
495	2-Azetidinecarboxylic acid	$C_4H_7NO_2$ / 101.11	2517-04-6 / 217 dec	923	5-22-01-00015	
496	2-Azetidinone	C_3H_5NO / 71.08	930-21-2 / 73.5	/ 106[15]	5-21-06-00312	eth 4; EtOH 4; chl 4
497	Azinphos-methyl / O,O-Dimethyl S-[(4-oxo-1,2,3-benzotriazin-3(4H)-yl)methyl]phosphorodithioate	$C_{10}H_{12}N_3O_3PS_2$ / 317.33	86-50-0 / 73	926	/ 1.44[20]	
498	Aziridine / Ethyleneimine	C_2H_5N / 43.07	151-56-4 / -77.9	3760 / 56	5-20-01-00003 / 0.832[25]	H_2O 5; EtOH 3; eth 4; chl 2
499	1-Aziridineethanol	C_4H_9NO / 87.12	1072-52-2 /	/ 168	5-20-01-00014 / 1.088[25]	/ 1.4560[20]
500	Aziridine, 1-methyl-	C_3H_7N / 57.10	1072-44-2 /	/ 27.5	5-20-01-00007 / 0.7572[10]	H_2O 5 / 1.3885[19]
501	Aziridine, 2-methyl- / 1,2-Propylenimine	C_3H_7N / 57.10	75-55-8 /	/ 67	5-20-01-00150 / 0.812[16]	
502	Azocine, octahydro-	$C_7H_{15}N$ / 113.20	1121-92-2 / 29	/ 51-3[15]	5-20-04-00156 / 0.896[25]	/ 1.4720[20]
503	1H-Azonine, octahydro-	$C_8H_{17}N$ / 127.23	5661-71-2 / -19	/ 91[24]	5-20-04-00203 / 0.83[25]	/ 1.4760[20]
504	Azulene / Bicyclo[5.3.0]decapentaene	$C_{10}H_8$ / 128.17	275-51-4 / 99	939 / 270 dec; 125[10]	4-05-00-01636	H_2O 1; EtOH 3; eth 3; ace 3
505	Azulene, 1,4-dimethyl-7-(1-methylethyl)- / Guaiazulene	$C_{15}H_{18}$ / 198.31	489-84-9 / 31.5	4464 / 167-8[12]	4-05-00-01751 / 0.973[20]	EtOH 3; eth 3; AcOEt 3
506	Azulene, 4,8-dimethyl-2-(1-methylethyl)-	$C_{15}H_{18}$ / 198.31	529-08-8 / 32.3	/ 122[1.0]	4-05-00-01754 / 0.9735[19]	eth 4; EtOH 4
507	5-Azulenemethanol, 1,2,3,4,5,6,7,8-octahydro-α,α,3,8-tetramethyl-, / Guaiol	$C_{15}H_{26}O$ / 222.37	489-86-1 / 91	4466 / 288 dec; 165[17]	3-06-00-00412 / 0.9074[100]	H_2O 1; EtOH 3; eth 3 / 1.4716[100]
508	Bacitracin	$C_{66}H_{103}N_{17}O_{16}S$ / 1422.71	1405-87-4 / 223	948		H_2O 4; EtOH 4
509	Balan / Benzenamine, N-butyl-N-ethyl-2,6-dinitro-4-(trifluoromethyl)-	$C_{13}H_{16}F_3N_3O_4$ / 335.28	1861-40-1 / 66	1048 / 121[0.5]; 148[7]		

No.	Name / Synonym	Mol. Form. / Mol. Wt.	CAS RN / mp/°C	Merck No. / bp/°C	Beil. Ref. / den/g cm^{-3}	Solubility / n_D
510	Barban / Carbamic acid, (3-chlorophenyl)-, 4-chloro-2-butynyl ester	$C_{11}H_9Cl_2NO_2$ / 258.10	101-27-9 / 75	969		
511	Basic copper carbonate / Copper, [μ-[carbonato(2-)-O:O']]dihydroxydi-	$CH_2Cu_2O_5$ / 221.12	12069-69-1 / 200 dec	2634	3.8[20]	
512	Bayleton / 2-Butanone, 1-(4-chlorophenoxy)-3,3-dimethyl-1-(1H-1,2,4-triazol-1-yl)-	$C_{14}H_{16}ClN_3O_2$ / 293.75	43121-43-3 / 82	9507	1.22[20]	
513	Benalaxyl / Methyl N-(2,6-dimethylphenyl)-N-(phenylacetyl)-DL-alaninate	$C_{20}H_{23}NO_3$ / 325.41	71626-11-4 / 79		1.27[25]	
514	Bendiocarb / 1,3-Benzodioxol-4-ol, 2,2-dimethyl-, methylcarbamate	$C_{11}H_{13}NO_4$ / 223.23	22781-23-3 / 130	1044	1.25[20]	
515	Benomyl / Carbamic acid, [1-[(butylamino)carbonyl]-1H-benzimidazol-2-yl]-, methyl ester	$C_{14}H_{18}N_4O_3$ / 290.32	17804-35-2 / dec	1053		
516	Bensulfuron-methyl / 2-[[[[[(4,6-Dimethoxypyrimidin-2-yl)amino]carbonyl]amino]sulfonyl]methyl]benzoic acid, methyl ester	$C_{16}H_{18}N_4O_7S$ / 410.41	83055-99-6 / 187			
517	Bensulide / Phosphorodithioic acid, O,O-bis(1-methylethyl) S-[2-[(phenylsulfonyl)amino]ethyl] ester	$C_{14}H_{244}NO_4PS_3$ / 397.52	741-58-2 / 34.4		1.224 [20]	
518	Bentazon / 1H-2,1,3-Benzothiadiazin-4(3H)-one, 3-(1-methylethyl)-, 2,2-dioxide	$C_{10}H_{12}N_2O_3S$ / 240.28	25057-89-0 / 138	1060		
519	Benz[a]anthracene / 1,2-Benzanthracene	$C_{18}H_{12}$ / 228.29	56-55-3 / 84	1069	4-05-00-02549	EtOH 4
520	Benz[a]anthracene, 7,12-dimethyl- / 9,10-Dimethyl-1,2-benzanthracene	$C_{20}H_{16}$ / 256.35	57-97-6 / 122.5	3224	4-05-00-02587	ace 4; bz 4
521	Benz[a]anthracene, 11-methyl-	$C_{19}H_{14}$ / 242.32	6111-78-0 / 118	360 dec	4-05-00-02568	H_2O 1; EtOH 3; HOAc 3
522	Benz[a]anthracene, 3-methyl- / 3-Methylbenz[a]anthracene	$C_{19}H_{14}$ / 242.32	2498-75-1 / 156.5	160[0.1]	4-05-00-02562	H_2O 1; EtOH 3; ace 3; bz 3
523	Benz[a]anthracene, 8-methyl-	$C_{19}H_{14}$ / 242.32	2381-31-9 / 156.5	272[3]; 160[0.1]	4-05-00-02566 / 1.2310[0]	H_2O 1; EtOH 3; eth 3; bz 3
524	Benzadox / Acetic acid, [(benzoylamino)oxy]-	$C_9H_9NO_4$ / 195.17	5251-93-4 / 140			
525	Benzaldehyde / Benzenecarboxaldehyde	C_7H_6O / 106.12	100-52-7 / -26	1065 / 179.0	4-07-00-00505 / 1.0415[10]	H_2O 2; EtOH 5; eth 5; ace 4 / 1.5463[20]
526	Benzaldehyde, 2-bromo-	C_7H_5BrO / 185.02	6630-33-7 / 21.5	230	4-07-00-00578	H_2O 1; EtOH 4; bz 4; ctc 2 / 1.5925[20]
527	Benzaldehyde, 3-bromo-	C_7H_5BrO / 185.02	3132-99-8 /	234.5	4-07-00-00579	H_2O 1; EtOH 4; eth 4; ctc 2 / 1.5935[20]
528	Benzaldehyde, 4-bromo-	C_7H_5BrO / 185.02	1122-91-4 / 67	66-8[2]	4-07-00-00579	H_2O 1; EtOH 4; bz 4; chl 2
529	Benzaldehyde, 2-chloro-	C_7H_5ClO / 140.57	89-98-5 / 12.4	211.9	4-07-00-00561 / 1.2483[20]	H_2O 2; EtOH 3; eth 3; ace 3 / 1.5662[20]
530	Benzaldehyde, 3-chloro-	C_7H_5ClO / 140.57	587-04-2 / 17.5	213.5	4-07-00-00566 / 1.2410[20]	H_2O 2; EtOH 3; eth 3; ace 3 / 1.5650[20]
531	Benzaldehyde, 4-chloro-	C_7H_5ClO / 140.57	104-88-1 / 47.5	213.5	4-07-00-00568 / 1.196[61]	H_2O 3; EtOH 4; eth 4; ace 3 / 1.555[61]
532	Benzaldehyde, 3-chloro-4-hydroxy-	$C_7H_5ClO_2$ / 156.57	2420-16-8 / 139	150[14]	4-08-00-00296	eth 4; EtOH 4
533	Benzaldehyde, 5-chloro-2-hydroxy-	$C_7H_5ClO_2$ / 156.57	635-93-8 / 100.3	105[12]	4-08-00-00224	H_2O 1; EtOH 4; eth 3; alk 3
534	Benzaldehyde, 3,5-dibromo-2-hydroxy- / 3,5-Dibromosalicylaldehyde	$C_7H_4Br_2O_2$ / 279.92	90-59-5 / 86	3011 / sub	3-08-00-00188	bz 4; eth 4; chl 4
535	Benzaldehyde, 2,5-dichloro- / 2,5-Dichlorobenzaldehyde	$C_7H_4Cl_2O$ / 175.01	6361-23-5 / 58	232	4-07-00-00576	bz 4; eth 4; EtOH 4; chl 4
536	Benzaldehyde, 3,4-dichloro- / 3,4-Dichlorobenzaldehyde	$C_7H_4Cl_2O$ / 175.01	6287-38-3 / 44	247.5	4-07-00-00576	H_2O 1; EtOH 3; eth 3; ctc 2
537	Benzaldehyde, 3,5-dichloro- / 3,5-Dichlorobenzaldehyde	$C_7H_4Cl_2O$ / 175.01	10203-08-4 / 65	240	4-07-00-00577	ace 4; bz 4; eth 4; EtOH 4
538	Benzaldehyde, 3,4-diethoxy- / 3,4-Diethoxybenzaldehyde	$C_{11}H_{14}O_3$ / 194.23	2029-94-9 / 22	279; 200[50]	4-08-00-01766 / 1.0100[22]	EtOH 4
539	Benzaldehyde, 4-(diethylamino)-	$C_{11}H_{15}NO$ / 177.25	120-21-8 / 41	172[10]	4-14-00-00066	H_2O 4; EtOH 3; eth 3; bz 3
540	Benzaldehyde, 2,3-dihydroxy-	$C_7H_6O_3$ / 138.12	24677-78-9 / 108	235; 120[16]	3-08-00-01979	ace 4; EtOH 4; HOAc 4
541	Benzaldehyde, 2,4-dihydroxy- / β-Resorcylaldehyde	$C_7H_6O_3$ / 138.12	95-01-2 / 135	8159 / 226[22]	4-08-00-01753	H_2O 3; EtOH 4; eth 4; bz 2
542	Benzaldehyde, 3,4-dihydroxy- / Protocatechualdehyde	$C_7H_6O_3$ / 138.12	139-85-5 / 153 dec	7908	4-08-00-01762	H_2O 3; EtOH 4; eth 4
543	Benzaldehyde, 2,4-dimethoxy- / 2,4-Dimethoxybenzaldehyde	$C_9H_{10}O_3$ / 166.18	613-45-6 / 72	290; 165[10]	4-08-00-01754	H_2O 1; EtOH 3; eth 3; bz 3
544	Benzaldehyde, 2,5-dimethoxy- / 2,5-Dimethoxybenzaldehyde	$C_9H_{10}O_3$ / 166.18	93-02-7 / 52	270; 146[10]	4-08-00-01759	H_2O 2; EtOH 3; eth 3

No.	Name Synonym	Mol. Form. Mol. Wt.	CAS RN mp/°C	Merck No. bp/°C	Beil. Ref. den/g cm^{-3}	Solubility n_D
545	Benzaldehyde, 3,4-dimethoxy- Veratraldehyde	$C_9H_{10}O_3$ 166.18	120-14-9 43	9852 281; 155[10]	4-08-00-01765	H_2O 2; EtOH 4; eth 4; chl 2
546	Benzaldehyde, 3,5-dimethoxy- 3,5-Dimethoxybenzaldehyde	$C_9H_{10}O_3$ 166.18	7311-34-4 46.3	151[16]	4-08-00-01786	H_2O 2; EtOH 3; bz 3; peth 2
547	Benzaldehyde, 2,4-dimethyl- 2,4-Dimethylbenzaldehyde	$C_9H_{10}O$ 134.18	15764-16-6 -9	218	4-07-00-00706	EtOH 3; eth 3; ace 3; bz 3
548	Benzaldehyde, 2,5-dimethyl- Isoxylaldehyde	$C_9H_{10}O$ 134.18	5779-94-2	220	4-07-00-00707 0.9500[20]	EtOH 4; eth 3; ace 3; bz 3
549	Benzaldehyde, 3,5-dimethyl-	$C_9H_{10}O$ 134.18	5779-95-3 9	221	4-07-00-00707	ace 4; bz 4; eth 4; EtOH 4
550	Benzaldehyde, 4-(dimethylamino)- Ehrlich's reagent	$C_9H_{11}NO$ 149.19	100-10-7 74.5	3219 176-7[17]	4-14-00-00051 1.0254[100]	H_2O 2; EtOH 3; eth 3; ace 3
551	Benzaldehyde, 2,4-dinitro- 2,4-Dinitrobenzaldehyde	$C_7H_4N_2O_5$ 196.12	528-75-6 72	3265 200[15]	3-07-00-00923	H_2O 2; EtOH 3; eth 3; bz 3
552	Benzaldehyde, 2-ethoxy-	$C_9H_{10}O_2$ 150.18	613-69-4 21	248	4-08-00-00180	EtOH 5; eth 5; chl 2
553	Benzaldehyde, 3-ethoxy-	$C_9H_{10}O_2$ 150.18	22924-15-8	245.5	0-08-00-00060 1.0768[20]	EtOH 3; eth 3; bz 3 1.5408[20]
554	Benzaldehyde, 4-ethoxy-	$C_9H_{10}O_2$ 150.18	10031-82-0 13.5	249	4-08-00-00254 1.08[21]	EtOH 4; eth 4; bz 4
555	Benzaldehyde, 3-ethoxy-4-hydroxy- Ethylvanillin	$C_9H_{10}O_3$ 166.18	121-32-4 77.5	3815	4-08-00-01765	H_2O 2; EtOH 3; eth 3; bz 3
556	Benzaldehyde, 5-ethoxy-2-hydroxy- 5-Ethoxy-2-hydroxybenzaldehyde	$C_9H_{10}O_3$ 166.18	80832-54-8 51.5	230	3-08-00-02004	eth 4; EtOH 4; chl 4
557	Benzaldehyde, 4-ethoxy-3-methoxy-	$C_{10}H_{12}O_3$ 180.20	120-25-2 64.5	168[13]	4-08-00-01766	H_2O 2; EtOH 3; eth 3; bz 3
558	Benzaldehyde, 2-ethyl-	$C_9H_{10}O$ 134.18	22927-13-5	209	4-07-00-00696 1.0216[17]	1.5425[17]
559	Benzaldehyde, 2-fluoro-	C_7H_5FO 124.11	446-52-6 -44.5	175	4-07-00-00559 1.178[25]	1.5234[20]
560	Benzaldehyde, 3-fluoro-	C_7H_5FO 124.11	456-48-4	173	4-07-00-00559 1:17[25]	1.5206[20]
561	Benzaldehyde, 4-fluoro-	C_7H_5FO 124.11	459-57-4 -10	181.5	4-07-00-00559 1.1810[19]	
562	Benzaldehyde, 2-(β-D-glucopyranosyloxy)- Helicin	$C_{13}H_{16}O_7$ 284.27	618-65-5 175	4544	5-17-07-00150	H_2O 4; EtOH 4
563	Benzaldehyde, hydrazone Benzylidene hydrazine	$C_7H_8N_2$ 120.15	5281-18-5 16	140[14]	4-07-00-00530	EtOH 3
564	Benzaldehyde, 2-hydroxy- Salicylaldehyde	$C_7H_6O_2$ 122.12	90-02-8 -7	8295 197	4-08-00-00176 1.1674[20]	H_2O 2; EtOH 5; eth 5; ace 4 1.5740[20]
565	Benzaldehyde, 3-hydroxy-	$C_7H_6O_2$ 122.12	100-83-4 108	240	4-08-00-00240 1.1179[130]	H_2O 2; EtOH 3; eth 3; ace 3
566	Benzaldehyde, 4-hydroxy- p-Hydroxybenzaldehyde	$C_7H_6O_2$ 122.12	123-08-0 117	4741	4-08-00-00251 1.129[130]	H_2O 2; EtOH 4; eth 4; ace 2 1.5705[130]
567	Benzaldehyde, 2-hydroxy-4,6-dimethoxy-	$C_9H_{10}O_4$ 182.18	708-76-9 70	193[25]; 165[10]	4-08-00-02717	H_2O 1; EtOH 4; eth 4; bz 4
568	Benzaldehyde, 3-hydroxy-4,5-dimethoxy-	$C_9H_{10}O_4$ 182.18	29865-90-5 62.5	179[12]	4-08-00-02718	bz 4; EtOH 4; lig 4
569	Benzaldehyde, 4-hydroxy-3,5-dimethoxy- Syringaldehyde	$C_9H_{10}O_4$ 182.18	134-96-3 113	8996 192-3[14]	4-08-00-02718	H_2O 2; EtOH 4; eth 4; bz 4
570	Benzaldehyde, 2-hydroxy-3-methoxy-	$C_8H_8O_3$ 152.15	148-53-8 44.5	265.5	4-08-00-01747	H_2O 2; EtOH 4; eth 4; ctc 4
571	Benzaldehyde, 2-hydroxy-5-methoxy-	$C_8H_8O_3$ 152.15	672-13-9 4	247.5	4-08-00-01759	eth 4; EtOH 4
572	Benzaldehyde, 3-hydroxy-4-methoxy-	$C_8H_8O_3$ 152.15	621-59-0 114	179[15]	4-08-00-01764 1.196[25]	H_2O 2; EtOH 3; eth 3; bz 3
573	Benzaldehyde, 3-hydroxy-5-methoxy- 3-Hydroxy-5-methoxybenzaldehyde	$C_8H_8O_3$ 152.15	57179-35-8 130.5	sub 110	3-08-00-02073	bz 4; EtOH 4; HOAc 4
574	Benzaldehyde, 4-hydroxy-3-methoxy- Vanillin	$C_8H_8O_3$ 152.15	121-33-5 81.5	9839 285	4-08-00-01763 1.056[25]	H_2O 2; EtOH 4; eth 4; ace 4
575	Benzaldehyde, 2-hydroxy-3-methyl- m-Tolualdehyde, 2-hydroxy	$C_8H_8O_2$ 136.15	824-42-0 17	208.5	4-08-00-00430	eth 4; EtOH 4; chl 4
576	Benzaldehyde, 2-hydroxy-4-methyl-	$C_8H_8O_2$ 136.15	698-27-1 60.5	223	4-08-00-00434	eth 4; EtOH 4; chl 4
577	Benzaldehyde, 2-hydroxy-5-methyl-	$C_8H_8O_2$ 136.15	613-84-3 56	217.5	4-08-00-00432 1.091[59]	eth 4; EtOH 4; chl 4 1.547[59]
578	Benzaldehyde, 2-hydroxy-6-methyl-	$C_8H_8O_2$ 136.15	18362-36-2 32.3	230	4-08-00-00430	bz 4
579	Benzaldehyde, 2-hydroxy-, oxime Salicylaldoxime	$C_7H_7NO_2$ 137.14	94-67-7 57	8296	1-08-00-00520	H_2O 2; EtOH 4; eth 4; bz 4
580	Benzaldehyde, 2-iodo-	C_7H_5IO 232.02	26260-02-6 37	129[14]	4-07-00-00582	H_2O 2; ace 3
581	Benzaldehyde, 3-iodo-	C_7H_5IO 232.02	696-41-3 57	125[13]	4-07-00-00583	EtOH 3; ace 3; bz 3
582	Benzaldehyde, 4-iodo-	C_7H_5IO 232.02	15164-44-0 77.5	265	4-07-00-00584	H_2O 2; EtOH 3; bz 3
583	Benzaldehyde, 2-methoxy-	$C_8H_8O_2$ 136.15	135-02-4 37.5	243.5	4-08-00-00180 1.1326[20]	H_2O 1; EtOH 3; eth 4; ace 4 1.5600[20]
584	Benzaldehyde, 3-methoxy-	$C_8H_8O_2$ 136.15	591-31-1	231	4-08-00-00241 1.1187[20]	H_2O 1; EtOH 3; eth 4; ace 4 1.5530[20]

No.	Name / Synonym	Mol. Form. / Mol. Wt.	CAS RN / mp/°C	Merck No. / bp/°C	Beil. Ref. / den/g cm^{-3}	Solubility / n_D
585	Benzaldehyde, 4-methoxy- / p-Anisaldehyde	$C_8H_8O_2$ / 136.15	123-11-5 / 0	693 / 248; 134[12]	4-08-00-00252 / 1.119[15]	H_2O 1; EtOH 5; eth 5; ace 4 / 1.5730[20]
586	Benzaldehyde, 4-methoxy-3-methyl-	$C_9H_{10}O_2$ / 150.18	32723-67-4	80-5[1]	4-08-00-00432 / 1.025[25]	1.5670[20]
587	Benzaldehyde, 2-methyl- / o-Tolualdehyde	C_8H_8O / 120.15	529-20-4	9453 / 200; 94[10]	4-07-00-00667 / 1.0328[20]	H_2O 2; EtOH 3; eth 3; ace 4 / 1.5462[20]
588	Benzaldehyde, 3-methyl-	C_8H_8O / 120.15	620-23-5	199	4-07-00-00669 / 1.0189[21]	H_2O 2; EtOH 5; eth 5; ace 4 / 1.5413[21]
589	Benzaldehyde, 4-methyl- / p-Tolualdehyde	C_8H_8O / 120.15	104-87-0	204-5.106[10]	4-07-00-00672 / 1.019[17]	H_2O 2; EtOH 5; eth 5; ace 5 / 1.5454[20]
590	Benzaldehyde, 4-(1-methylethyl)- / Cuminaldehyde	$C_{10}H_{12}O$ / 148.20	122-03-2	2623 / 235.5	4-07-00-00723 / 0.9755[20]	H_2O I; EtOH 3; eth 3; ctc 2 / 1.5301[20]
591	Benzaldehyde, 2-nitro-	$C_7H_5NO_3$ / 151.12	552-89-6 / 43.5	153[23]	4-07-00-00584 / 1.2844[20]	H_2O 2; EtOH 4; eth 4; ace 4
592	Benzaldehyde, 3-nitro-	$C_7H_5NO_3$ / 151.12	99-61-6 / 58.5	164[23]	4-07-00-00591 / 1.2792[20]	H_2O 2; EtOH 3; eth 3; ace 4
593	Benzaldehyde, 4-nitro-	$C_7H_5NO_3$ / 151.12	555-16-8 / 107	sub	4-07-00-00598 / 1.496[25]	H_2O 2; EtOH 4; eth 2; bz 3
594	Benzaldehyde, oxime, (E)-	C_7H_7NO / 121.14	622-31-1 / 35	119[10]	4-07-00-00527 / 1.145[20]	H_2O 3; EtOH 4; eth 4
595	Benzaldehyde, oxime, (Z)-	C_7H_7NO / 121.14	622-32-2 / 36.5	200	4-07-00-00527 / 1.1111[20]	bz 4; eth 4; EtOH 4 / 1.5908[20]
596	Benzaldehyde, pentafluoro-	C_7HF_5O / 196.08	653-37-2 / 20	167	4-07-00-00561	1.4506[20]
597	Benzaldehyde, 3-phenoxy-	$C_{13}H_{10}O_2$ / 198.22	39515-51-0	140[0.1]	4-08-00-00242 / 1.147[25]	1.5954[20]
598	Benzaldehyde, 2,3,5,6-tetramethyl- / Durene aldehyde	$C_{11}H_{14}O$ / 162.23	17432-37-0 / 20	135[11]	3-07-00-01148	1.5560[30]
599	Benzaldehyde, 2,4,6-trimethyl-	$C_{10}H_{12}O$ / 148.20	487-68-3 / 14	238.5	4-07-00-00730 / 1.0154[25]	H_2O 1; EtOH 3; eth 3; ace 3
600	Benzamide / Benzoic acid amide	C_7H_7NO / 121.14	55-21-0 / 129.1	1067 / 290	4-09-00-00725 / 1.0792[130]	H_2O 2; EtOH 4; eth 2; bz 2
601	Benzamide, 2-amino-	$C_7H_8N_2O$ / 136.15	88-68-6 / 110.5 dec	300	4-14-00-01010	H_2O 3; EtOH 3; eth 2; bz 2
602	Benzamide, 4-amino-N-[2-(diethylamino)ethyl]-, monohydrochloride / Procainamide hydrochloride	$C_{13}H_{22}ClN_3O$ / 271.79	614-39-1 / 167	7762	4-14-00-01154	H_2O 4; EtOH 3; eth 1; bz 1
603	Benzamide, N-[(4-aminophenyl)sulfonyl]- / Sulfabenzamide	$C_{13}H_{12}N_2O_3S$ / 276.32	127-71-9 / 181.5	8868	4-14-00-02664	
604	Benzamide, 2-bromo-	C_7H_6BrNO / 200.03	4001-73-4 / 160.5	sub	4-09-00-01013	H_2O 2; EtOH 3; eth 2; tfa 2
605	Benzamide, 3-bromo-	C_7H_6BrNO / 200.03	22726-00-7 / 155.3	sub	4-09-00-01015	H_2O 2; EtOH 3
606	Benzamide, 5-bromo-N,2-dihydroxy- / 5-Bromosalicylhydroxamic acid	$C_7H_6BrNO_3$ / 232.03	5798-94-7 / 232 dec	1424	4-10-00-00221	
607	Benzamide, N-butyl-4-chloro-2-hydroxy- / Buclosamide	$C_{11}H_{14}ClNO_2$ / 227.69	575-74-6 / 91.5	1450		
608	Benzamide, 5-chloro-N-(3,4-dichlorophenyl)-2-hydroxy- / 3',4',5-Trichlorosalicylanilide	$C_{13}H_8Cl_3NO_2$ / 316.57	642-84-2 / 247	9558	4-12-00-01273	
609	Benzamide, 3,5-dibromo-N-(4-bromophenyl)-2-hydroxy- / Tribromsalan	$C_{13}H_8Br_3NO_2$ / 449.92	87-10-5 / 227	9529		
610	Benzamide, 3,5-dichloro-N-(3,4-dichlorophenyl)-2-hydroxy- / 3,3',4',5-Tetrachlorosalicylanilide	$C_{13}H_7Cl_4NO_2$ / 351.01	1154-59-2 / 161	9127		
611	Benzamide, N,N-diethyl-4-hydroxy-3-methoxy- / Ethamivan	$C_{12}H_{17}NO_3$ / 223.27	304-84-7 / 95	3673	4-10-00-01481	chl 3
612	Benzamide, N,N-diethyl-3-methyl- / DEET	$C_{12}H_{17}NO$ / 191.27	134-62-3	2848 / 160[19]; 111[1]	4-09-00-01716 / 0.996[20]	H_2O 4; bz 4; eth 4; EtOH 4 / 1.5212[20]
613	Benzamide, N,2-dihydroxy- / Salicylhydroxamic acid	$C_7H_7NO_3$ / 153.14	89-73-6 / 168	8300 / sub	4-10-00-00192	H_2O 2; EtOH 4; eth 4; HOAc 3
614	Benzamide, 3,5-dinitro- / Nitromide	$C_7H_5N_3O_5$ / 211.13	121-81-3 / 184	6533	4-09-00-01351	H_2O 4
615	Benzamide, 2-ethoxy- / Ethenzamide	$C_9H_{11}NO_2$ / 165.19	938-73-8 / 133	3685	4-10-00-00175	H_2O 2; EtOH 4; eth 4; chl 2
616	Benzamide, 2-hydroxy- / Salicylamide	$C_7H_7NO_2$ / 137.14	65-45-2 / 142	8297 / 181.5[14]	4-10-00-00169 / 1.175[140]	H_2O 2; EtOH 3; eth 2
617	Benzamide, 2-hydroxy-N-phenyl- / Salicylanilide	$C_{13}H_{11}NO_2$ / 213.24	87-17-2 / 136.5	8299	4-12-00-00906	H_2O 3; EtOH 2; eth 2; bz 2
618	Benzamide, 4-methoxy-	$C_8H_9NO_2$ / 151.17	3424-93-9 / 166.5	295	4-10-00-00433	H_2O 4; EtOH 4
619	Benzamide, 2-methyl- / o-Toluamide	C_8H_9NO / 135.17	527-85-5 / 147	9454	4-09-00-01701	H_2O 2; EtOH 4; eth 2; bz 2
620	Benzamide, N-methyl-	C_8H_9NO / 135.17	613-93-4 / 82	291; 167[12]	4-09-00-00727	EtOH 3; ace 3
621	Benzamide, N-(2-methylphenyl)-	$C_{14}H_{13}NO$ / 211.26	584-70-3 / 145.5		4-12-00-01757 / 1.205[15]	H_2O 2; EtOH 3; eth 3

No.	Name / Synonym	Mol. Form. / Mol. Wt.	CAS RN / mp/°C	Merck No. / bp/°C	Beil. Ref. / den/g cm^{-3}	Solubility / n_D
622	Benzamide, N-(3-methylphenyl)- / m-Benzotoluidide	$C_{14}H_{13}NO$ / 211.26	582-77-4 / 125		4-12-00-01825 / 1.170[15]	EtOH 4
623	Benzamide, N-(4-methylphenyl)-	$C_{14}H_{13}NO$ / 211.26	582-78-5 / 158		4-12-00-01910 / 1.202[15]	eth 4; EtOH 4
624	Benzamide, 2-nitro-	$C_7H_6N_2O_3$ / 166.14	610-15-1 / 176.6	317	4-09-00-01049	H_2O 3; EtOH 3; eth 3
625	Benzamide, 3-nitro-	$C_7H_6N_2O_3$ / 166.14	645-09-0 / 142.7	312.5	4-09-00-01061	H_2O 3; EtOH 3; eth 3
626	Benzamide, N-phenyl- / Benzanilide	$C_{13}H_{11}NO$ / 197.24	93-98-1 / 163	1068 / sub 117	4-12-00-00417 / 1.315[25]	H_2O 1; EtOH 2; eth 2; HOAc 2
627	Benzamide, N,N,4-trimethyl-	$C_{10}H_{13}NO$ / 163.22	14062-78-3 / 41	156[10]	0-09-00-00486	H_2O 2; EtOH 2
628	Benz[b]indeno[1,2-d]pyran-9(6H)-one, 6a,7-dihydro-3,4,6a,10-tetrahydroxy- / Hematein	$C_{16}H_{12}O_6$ / 300.27	475-25-2 / 250 dec	4553	4-18-00-03343	H_2O 1; EtOH 2; eth 1; bz 1
629	Benz[b]indeno[1,2-d]pyran-3,4,6a,9,10(6H)-pentol, 7,11b-dihydro-, cis-(+)- / Hematoxylin	$C_{16}H_{14}O_6$ / 302.28	517-28-2 / 140	4556	2-17-00-00273	H_2O 2; EtOH 3; eth 2; alk 3
630	Benz[c]acridine / 12-Azabenz[a]anthracene	$C_{17}H_{11}N$ / 229.28	225-51-4 / 132		5-20-08-00519	bz 4; eth 4; EtOH 4
631	Benz[e]acephenanthrylene / Benzo[b]fluoranthene	$C_{20}H_{12}$ / 252.32	205-99-2 / 168		4-05-00-02686	H_2O 1; bz 5
632	1H-Benz[e]indene, 2,3-dihydro-	$C_{13}H_{12}$ / 168.24	4944-94-9	294.5	4-05-00-01865 / 1.066[20]	H_2O 1; HOAc 3 / 1.6290[20]
633	Benzenamine / Aniline	C_6H_7N / 93.13	62-53-3 / -6.0	687 / 184.1	4-12-00-00223 / 1.0217[20]	H_2O 3; EtOH 5; eth 5; ace 5 / 1.5863[20]
634	Benzenamine, 4-[(4-aminophenyl)(4-imino-2,5-cyclohexadien-1-ylidene)methyl]-2-methyl-, monohydrochloride / Magenta I	$C_{20}H_{20}ClN_3$ / 337.85	632-99-5 / 200 dec	5528		H_2O 2; EtOH 2; eth 1
635	Benzenamine, 4,4'-azobis- / p-Diaminoazobenzene	$C_{12}H_{12}N_4$ / 212.25	538-41-0 / 250.5	2954	4-16-00-00515	H_2O 2; EtOH 3; bz 4; chl 4
636	Benzenamine, 2,6-bis(1-methylethyl)-	$C_{12}H_{19}N$ / 177.29	24544-04-5 / -45	257	3-12-00-02764 / 0.94[25]	1.5332[20]
637	Benzenamine, 3,5-bis(trifluoromethyl)-	$C_8H_5F_6N$ / 229.12	328-74-5	85[15]	3-12-00-02498 / 1.487[25]	1.4335[20]
638	Benzenamine, 2-bromo- / o-Bromoaniline	C_6H_6BrN / 172.02	615-36-1 / 32	229	4-12-00-01487 / 1.578[20]	H_2O 1; EtOH 3; eth 3 / 1.6113[20]
639	Benzenamine, 3-bromo- / m-Bromoaniline	C_6H_6BrN / 172.02	591-19-5 / 18.5	251	4-12-00-01491 / 1.5793[20]	H_2O 2; EtOH 3; eth 3 / 1.6260[20]
640	Benzenamine, 4-bromo- / p-Bromoaniline	C_6H_6BrN / 172.02	106-40-1 / 66.4	1392 / dec	4-12-00-01497 / 1.4970[100]	H_2O 1; EtOH 3; eth 3; chl 2
641	Benzenamine, 2-bromo-4,6-dichloro-	$C_6H_4BrCl_2N$ / 240.91	697-86-9 / 83.5	273	2-12-00-00355	bz 4; EtOH 4; chl 4
642	Benzenamine, 4-bromo-N,N-diethyl-	$C_{10}H_{14}BrN$ / 228.13	2052-06-4 / 38	270	4-12-00-01500	H_2O 1; EtOH 4; eth 4
643	Benzenamine, 2-bromo-N,N-dimethyl-	$C_8H_{10}BrN$ / 200.08	698-00-0	107-8[14]	4-12-00-01487 / 1.3839[25]	EtOH 4 / 1.5748[25]
644	Benzenamine, 3-bromo-N,N-dimethyl-	$C_8H_{10}BrN$ / 200.08	16518-62-0 / 11	259	4-12-00-01492 / 1.3651[20]	EtOH 4; HOAc 4; con acid 3
645	Benzenamine, 4-bromo-N,N-dimethyl-	$C_8H_{10}BrN$ / 200.08	586-77-6 / 55	264	4-12-00-01499 / 1.3220[100]	H_2O 1; EtOH 3; eth 4; os 3
646	Benzenamine, 2-bromo-4,6-dinitro-	$C_8H_4BrN_3O_4$ / 262.02	1817-73-8 / 153.5	sub	4-12-00-01734	EtOH 4; ace 4; HOAc 3
647	Benzenamine, 2-bromo-4-methyl-	C_7H_8BrN / 186.05	583-68-6 / 26	240	4-12-00-01992 / 1.510[20]	H_2O 1; EtOH 3; eth 3 / 1.5999[20]
648	Benzenamine, 2-bromo-5-methyl- / 2-Bromo-5-methylaniline	C_7H_8BrN / 186.05	53078-85-6 / 46	129[15]	3-12-00-01998 / 1.470[25]	eth 4; EtOH 4
649	Benzenamine, 3-bromo-4-methyl-	C_7H_8BrN / 186.05	7745-91-7 / 26	255.5	4-12-00-01991	eth 4
650	Benzenamine, 4-bromo-2-methyl-	C_7H_8BrN / 186.05	583-75-5 / 59.5	240	4-12-00-01804	H_2O 2; EtOH 3; eth 4; chl 2
651	Benzenamine, 4-bromo-3-methyl-	C_7H_8BrN / 186.05	6933-10-4 / 81	240	4-12-00-01856	H_2O 1; EtOH 4; chl 2
652	Benzenamine, 5-bromo-2-methyl-	C_7H_8BrN / 186.05	39478-78-9 / 33	255 dec; 165[25]	3-12-00-01923	eth 4; EtOH 4
653	Benzenamine, 2-bromo-6-nitro- / Aniline, 2-bromo-6-nitro-	$C_6H_5BrN_2O_2$ / 217.02	59255-95-7 / 74.5		2-12-00-00402 / 1.988[25]	EtOH 4
654	Benzenamine, 4-bromo-2-nitro-	$C_6H_5BrN_2O_2$ / 217.02	875-51-4 / 111.5	sub	3-12-00-01670	EtOH 4
655	Benzenamine, 4-butyl-	$C_{10}H_{15}N$ / 149.24	104-13-2	261	4-12-00-02807 / 0.945[20]	ctc 2
656	Benzenamine, N-butyl- / N-Butylaniline	$C_{10}H_{15}N$ / 149.24	1126-78-9 / -14.4	243.5	4-12-00-00256 / 0.9323[20]	eth 4; EtOH 4 / 1.5341[20]
657	Benzenamine, 4,4'-carbonimidoylbis[N,N-dimethyl- / 4,4'-Dimethylaminobenzophenonimide	$C_{17}H_{21}N_3$ / 267.37	492-80-8 / 136		4-14-00-00256	H_2O 1; EtOH 3; eth 2

No.	Name Synonym	Mol. Form. Mol. Wt.	CAS RN mp/°C	Merck No. bp/°C	Beil. Ref. den/g cm⁻³	Solubility n_D
658	Benzenamine, 2-chloro- o-Chloroaniline	C₆H₈ClN 127.57	95-51-2 -14	2118 208.8	4-12-00-01115	H₂O 1; EtOH 5; eth 3; ace 3 1.5895[20]
659	Benzenamine, 3-chloro- m-Chloroaniline	C₆H₈ClN 127.57	108-42-9 -10.4	2118 230.5	4-12-00-01137 1.2161[20]	H₂O 1; EtOH 5; eth 5; ace 5 1.5941[20]
660	Benzenamine, 4-chloro- p-Chloroaniline	C₆H₈ClN 127.57	106-47-8 72.5	2118 232	4-12-00-01166 1.429[19]	H₂O 3; EtOH 3; eth 3; chl 3 1.5546[87]
661	Benzenamine, 4-chloro-N,N-diethyl-	C₁₀H₁₄ClN 183.68	2873-89-4 45.5	252; 113[6]	4-12-00-01170	EtOH 3
662	Benzenamine, 2-chloro-N,N-dimethyl-	C₈H₁₀ClN 155.63	698-01-1	205	4-12-00-01116 1.1067[20]	bz 4; EtOH 4 1.5578[20]
663	Benzenamine, 4-chloro-N,N-dimethyl-	C₈H₁₀ClN 155.63	698-69-1 35.5	231	4-12-00-01169 1.0480[100]	EtOH 3
664	Benzenamine, 2-chloro-6-fluoro- Aniline, 2-chloro-6-fluoro	C₆H₅ClFN 145.56	363-51-9	91[30]	4-12-00-01238 1.316[23]	1.5511[23]
665	Benzenamine, 2-chloro-, hydrochloride Aniline, 2-chloro, hydrochloride	C₆H₇Cl₂N 164.03	137-04-2 235		3-12-00-01281 1.505[18]	H₂O 4
666	Benzenamine, 2-chloro-5-methoxy-	C₇H₈ClNO 157.60	2401-24-3 27	142[18]	4-13-00-00999	
667	Benzenamine, 4-chloro-2-methoxy- 4-Chloro-o-anisidine	C₇H₈ClNO 157.60	93-50-5 52	260	4-13-00-00885	1.5848[25] EtOH 3; eth 3; bz 3; chl 3
668	Benzenamine, 2-chloro-4-methyl- 2-Chloro-p-toluidine	C₇H₈ClN 141.60	615-65-6 7	220	3-12-00-02152 1.151[20]	EtOH 2; bz 2 1.5748[22]
669	Benzenamine, 2-chloro-5-methyl- 6-Chloro-m-toluidine	C₇H₈ClN 141.60	95-81-8 29.5	229	4-12-00-01849	EtOH 4
670	Benzenamine, 2-chloro-N-methyl-	C₇H₈ClN 141.60	932-32-1	218	4-12-00-01116 1.1735[11]	EtOH 3; ace 3; bz 3 1.5780[25]
671	Benzenamine, 3-chloro-2-methyl- 3-Chloro-o-toluidine	C₇H₈ClN 141.60	87-60-5 1	245	4-12-00-01800	H₂O 3; EtOH 3; eth 1; bz 1 1.5880[20]
672	Benzenamine, 3-chloro-4-methyl- 3-Chloro-p-toluidine	C₇H₈ClN 141.60	95-74-9 26	243	4-12-00-01985	EtOH 3; ctc 2
673	Benzenamine, 4-chloro-2-methyl-	C₇H₈ClN 141.60	95-69-2 30.3	244	4-12-00-01796	EtOH 3; ctc 2
674	Benzenamine, 4-chloro-3-methyl- 4-Chloro-m-toluidine	C₇H₈ClN 141.60	7149-75-9 83.5	241	4-12-00-01850	ace 4; bz 4; EtOH 4
675	Benzenamine, 4-chloro-N-methyl-	C₇H₈ClN 141.60	932-96-7	240	4-12-00-01168 1.169[11]	EtOH 3; ace 3; bz 3 1.5835[20]
676	Benzenamine, 5-chloro-2-methyl- 5-Chloro-o-toluidine	C₇H₈ClN 141.60	95-79-4 26	239; 140[38]	4-12-00-01794	EtOH 4
677	Benzenamine, 5-chloro-2-nitro-	C₆H₅ClN₂O₂ 172.57	1635-61-6 127.8	sub	4-12-00-01673	eth 4; EtOH 4
678	Benzenamine, 3-chloro-N-phenyl-	C₁₂H₁₀ClN 203.67	101-17-7	338	4-12-00-01141 1.2[25]	H₂O 1; EtOH 3; eth 3; bz 3 1.6513[20]
679	Benzenamine, 4-chloro-N-phenyl-	C₁₂H₁₀ClN 203.67	1205-71-6 74	335	4-12-00-01170	bz 4; eth 4; EtOH 4; lig 4
680	Benzenamine, N-cyclohexyl-	C₁₂H₁₇N 175.27	1821-36-9 16	279; 192[73]	4-12-00-00268 1.0155[20]	H₂O 1; EtOH 3; eth 3; bz 3 1.5610[20]
681	Benzenamine, 2,4-dibromo-	C₆H₅Br₂N 250.92	615-57-6 79.5	156[74]	4-12-00-01532 2.260[20]	EtOH 3; eth 3; chl 3; HOAc 3
682	Benzenamine, 2,6-dibromo-	C₆H₅Br₂N 250.92	608-30-0 87.5	263	4-12-00-01535	EtOH 4; eth 4; bz 4; chl 4
683	Benzenamine, 3,4-dibromo- 3,4-Dibromoaniline	C₆H₅Br₂N 250.92	615-55-4 80.5	100	4-12-00-01535	EtOH 4
684	Benzenamine, N,N-dibutyl-	C₁₄H₂₃N 205.34	613-29-6 -32.2	274.8	4-12-00-00256 0.9037[20]	H₂O 1; EtOH 5; eth 5; ace 4 1.5186[20]
685	Benzenamine, 2,3-dichloro- 2,3-Dichloroaniline	C₆H₅Cl₂N 162.02	608-27-5 24	252	0-12-00-00621	EtOH 3; eth 4; ace 3; bz 2
686	Benzenamine, 2,4-dichloro- 2,4-Dichloroaniline	C₆H₅Cl₂N 162.02	554-00-7 63.5	245	4-12-00-01241 1.567[20]	H₂O 2; EtOH 3; eth 3; chl 2
687	Benzenamine, 2,5-dichloro- 2,5-Dichloroaniline	C₆H₅Cl₂N 162.02	95-82-9 50	251	4-12-00-01250	H₂O 2; EtOH 3; eth 3; bz 3
688	Benzenamine, 2,6-dichloro- 2,6-Dichloroaniline	C₆H₅Cl₂N 162.02	608-31-1 39		2-12-00-00337	H₂O 2; EtOH 3; eth 3
689	Benzenamine, 3,4-dichloro- 3,4-Dichloroaniline	C₆H₅Cl₂N 162.02	95-76-1 72	3041 272	4-12-00-01257	EtOH 3; eth 3; bz 2; chl 2
690	Benzenamine, 3,5-dichloro- 3,5-Dichloroaniline	C₆H₅Cl₂N 162.02	626-43-7 52	261	4-12-00-01274	H₂O 1; EtOH 3; eth 3; ctc 3
691	Benzenamine, 2,6-dichloro-4-ethoxy- Aniline, 2,6-dichloro-4-ethoxy	C₈H₉Cl₂NO 206.07	51225-20-8 46	275	2-13-00-00276	bz 4; eth 4; EtOH 4
692	Benzenamine, 2,6-dichloro-4-nitro- 2,6-Dichloro-4-nitroaniline	C₆H₄Cl₂N₂O₂ 207.02	99-30-9 191		4-12-00-01681	EtOH 3; DMSO-d₈ 2; acid 3
693	Benzenamine, 3,4-diethoxy- 3,4-Diethoxyaniline	C₁₀H₁₅NO₂ 181.23	39052-12-5 48	132-5[0.45]	4-13-00-02507	bz 4; eth 4; EtOH 4
694	Benzenamine, 2,6-diethyl-	C₁₀H₁₅N 149.24	579-66-8 3	243	4-12-00-02841 0.906[25]	1.5452[20]
695	Benzenamine, N,N-diethyl- N,N-Diethylaniline	C₁₀H₁₅N 149.24	91-66-7 -38.8	3102 216.3	4-12-00-00252 0.9307[20]	H₂O 2; EtOH 3; eth 4; ace 3 1.5409[20]
696	Benzenamine, N,N-diethyl-2-methyl-	C₁₁H₁₇N 163.26	606-46-2 -60	209	4-12-00-01748 0.9286[20]	H₂O 2; EtOH 5; eth 5; ctc 3 1.5153[20]
697	Benzenamine, N,N-diethyl-4-methyl-	C₁₁H₁₇N 163.26	613-48-9	229	4-12-00-01875 0.9242[16]	H₂O 2; EtOH 5; eth 5

No.	Name / Synonym	Mol. Form. / Mol. Wt.	CAS RN / mp/°C	Merck No. / bp/°C	Beil. Ref. / den/g cm⁻³	Solubility / n_D
698	Benzenamine, N,N-diethyl-4-nitro-	$C_{10}H_{14}N_2O_2$ 194.23	2216-15-1 77.5		4-12-00-01618 1.225[25]	EtOH 3; lig 2
699	Benzenamine, N,N-diethyl-4-nitroso- Aniline, N,N-diethyl-4-nitroso	$C_{10}H_{14}N_2O$ 178.23	120-22-9 87.5		4-12-00-01559 1.24[15]	H_2O 2; EtOH 3; eth 3; ace 3
700	Benzenamine, 2,4-difluoro- 2,4-Difluoroaniline	$C_6H_5F_2N$ 129.11	367-25-9 -7.5	3131 170	4-12-00-01112 1.268[25]	1.5063[20]
701	Benzenamine, 2,4-diiodo- 2,4-Diiodoaniline	$C_6H_5I_2N$ 344.92	533-70-0 95.5	3173	3-12-00-01508 2.748[25]	ace 4; bz 4; eth 4; EtOH 4
702	Benzenamine, 2,5-dimethoxy-	$C_8H_{11}NO_2$ 153.18	102-56-7 82.5	270	4-13-00-02548	H_2O 3; EtOH 3; chl 3; lig 3
703	Benzenamine, 2,6-dimethoxy- 2,6-Dimethoxyaniline	$C_8H_{11}NO_2$ 153.18	2734-70-5 76.5	146[23]	4-13-00-02526	bz 4; eth 4; EtOH 4; lig 4
704	Benzenamine, 2,3-dimethyl- 2,3-Xylidine	$C_8H_{11}N$ 121.18	87-59-2 <-15	221.5	4-12-00-02497 0.9931[20]	H_2O 2; EtOH 4; eth 4; ctc 3 1.5684[20]
705	Benzenamine, 2,4-dimethyl- 2,4-Xylidine	$C_8H_{11}N$ 121.18	95-68-1 -14.3	214	4-12-00-02545 0.9723[20]	H_2O 2; EtOH 3; eth 3; bz 3 1.5569[20]
706	Benzenamine, 2,5-dimethyl- 2,5-Xylidine	$C_8H_{11}N$ 121.18	95-78-3 15.5	214	4-12-00-02567 0.9790[21]	H_2O 2; eth 3; ctc 3 1.5591[21]
707	Benzenamine, 2,6-dimethyl- 2,6-Xylidine	$C_8H_{11}N$ 121.18	87-62-7 11.2	215	4-12-00-02521 0.9842[20]	eth 4; EtOH 4 1.5610[20]
708	Benzenamine, 3,4-dimethyl- 3,4-Xylidine	$C_8H_{11}N$ 121.18	95-64-7 51	228	4-12-00-02502 1.076[18]	H_2O 2; eth 3; chl 2; lig 4
709	Benzenamine, 3,5-dimethyl- 3,5-Xylidine	$C_8H_{11}N$ 121.18	108-69-0 9.8	220.5	4-12-00-02561 0.9706[20]	H_2O 2; eth 3; ctc 3 1.5581[20]
710	Benzenamine, N,2-dimethyl-	$C_8H_{11}N$ 121.18	611-21-2	207.5	4-12-00-01746 0.9709[20]	H_2O 1; EtOH 5; eth 5; ace 3 1.5649[20]
711	Benzenamine, N,3-dimethyl-	$C_8H_{11}N$ 121.18	696-44-6	206.5	4-12-00-01815	H_2O 1; EtOH 5; eth 5; ace 3 1.5557[25]
712	Benzenamine, N,4-dimethyl-	$C_8H_{11}N$ 121.18	623-08-5	210	4-12-00-01874 0.9348[55]	H_2O 1; EtOH 5; eth 5; ace 3 1.5568[20]
713	Benzenamine, N,N-dimethyl- N,N-Dimethylaniline	$C_8H_{11}N$ 121.18	121-69-7 2.4	3223 194.1	4-12-00-00243 0.9557[20]	H_2O 2; EtOH 3; eth 3; ace 3 1.5582[20]
714	Benzenamine, 2-(1,1-dimethylethyl)-	$C_{10}H_{15}N$ 149.24	6310-21-0	234	4-12-00-02823 0.977[15]	EtOH 4; eth 4; bz 4; acid 3 1.5453[20]
715	Benzenamine, 4-(1,1-dimethylethyl)-	$C_{10}H_{15}N$ 149.24	769-92-6 17	241	4-12-00-02828 0.952[15]	H_2O 2; EtOH 5; eth 5; bz 4 1.5380[20]
716	Benzenamine, N-(1,1-dimethylethyl)- N-tert-Butylaniline	$C_{10}H_{15}N$ 149.24	937-33-7	215; 95[19]	4-12-00-00257	EtOH 3; ace 4; bz 4; chl 4 1.5270[20]
717	Benzenamine, N,N-dimethyl-, hydrochloride	$C_8H_{12}ClN$ 157.64	5882-44-0 90		4-12-00-00243 1.1156[19]	H_2O 4; EtOH 4; chl 4
718	Benzenamine, N,N-dimethyl-2-nitro-	$C_8H_{10}N_2O_2$ 166.18	610-17-3 -20	146[20]	4-12-00-01564 1.1794[20]	H_2O 3; EtOH 4; eth 3; chl 4 1.6102[20]
719	Benzenamine, N,N-dimethyl-3-nitro-	$C_8H_{10}N_2O_2$ 166.18	619-31-8 60.5	282.5	4-12-00-01591 1.313[17]	H_2O 1; EtOH 3; eth 3
720	Benzenamine, N,N-dimethyl-4-nitroso- p-Nitroso-N,N-dimethylaniline	$C_8H_{10}N_2O$ 150.18	138-89-6 92.5	6559	4-12-00-01558 1.145[20]	H_2O 2; EtOH 3; eth 3; chl 3
721	Benzenamine, N,N-dimethyl-4-(phenylazo)- 4-(Dimethylamino)azobenzene	$C_{14}H_{15}N_3$ 225.29	60-11-7 117	3218 dec	3-16-00-00340	H_2O 1; EtOH 4; eth 3; chl 2
722	Benzenamine, N,N-dimethyl-4-(2-phenylethenyl)-, (E)-	$C_{16}H_{17}N$ 223.32	838-95-9 150	145[0.1]	4-12-00-03400	H_2O 1; EtOH 3; chl 3
723	Benzenamine, N,N-dimethyl-4-[[3-(trifluoromethyl)phenyl]azo]- Azobenzene, 4-dimethylamino-3'-trifluoromethyl	$C_{15}H_{14}F_3N_3$ 293.29	328-96-1 80	76[8]	4-16-00-00487	chl 3
724	Benzenamine, 2,3-dinitro- 2,3-Dinitroaniline	$C_6H_5N_3O_4$ 183.12	602-03-9 128		4-12-00-01689 1.646[50]	H_2O 1; EtOH 3; eth 2
725	Benzenamine, 2,4-dinitro- 2,4-Dinitroaniline	$C_6H_5N_3O_4$ 183.12	97-02-9 180	3263	4-12-00-01689 1.615[14]	H_2O 1; EtOH 2; ace 2; HCl 2
726	Benzenamine, 2,6-dinitro- 2,6-Dinitroaniline	$C_6H_5N_3O_4$ 183.12	606-22-4 141.5	3264	4-12-00-01729	H_2O 1; EtOH 2; eth 3; bz 3
727	Benzenamine, 3,5-dinitro- 3,5-Dinitroaniline	$C_6H_5N_3O_4$ 183.12	618-87-1 163		4-12-00-01729 1.601[50]	H_2O 1; EtOH 3; eth 3; ace 2
728	Benzenamine, N,N-diphenyl- Triphenylamine	$C_{18}H_{15}N$ 245.32	603-34-9 127	365	4-12-00-00276 0.774[0]	H_2O 1; EtOH 2; eth 3; bz 3 1.353[16]
729	Benzenamine, 4-(diphenylmethyl)-	$C_{19}H_{17}N$ 259.35	603-38-3 84.5	248[12]	4-12-00-03470	bz 4; eth 4; lig 4
730	Benzenamine, N,N-dipropyl-	$C_{12}H_{19}N$ 177.29	2217-07-4	242	4-12-00-00255 0.9104[20]	H_2O 1; EtOH 3; eth 3; ace 3 1.5271[20]
731	Benzenamine, 4,4'-(1,2-ethanediyl)bis-	$C_{14}H_{16}N_2$ 212.29	621-95-4 137	sub	4-13-00-00401	H_2O 1; EtOH 4
732	Benzenamine, 4,4'-(1,2-ethenediyl)bis-, (E)-	$C_{14}H_{14}N_2$ 210.28	7314-06-9 231	sub	4-13-00-00453	H_2O 2; bz 2; MeOH 3; CS_2 2
733	Benzenamine, 2-ethenyl-	C_8H_9N 119.17	3867-18-3	112[20]; 97[8]	4-12-00-02914 1.0181[20]	ace 3; bz 3 1.6124[20]
734	Benzenamine, 3-ethenyl-	C_8H_9N 119.17	15411-43-5	112-5[12]	4-12-00-02915 1.0198[20]	ace 3; bz 3 1.6069[26]
735	Benzenamine, 4-ethenyl-	C_8H_9N 119.17	1520-21-4 23.5	116[9]	4-12-00-02915 1.010[20]	ace 3; bz 3 1.6250[22]

No.	Name / Synonym	Mol. Form. / Mol. Wt.	CAS RN / mp/°C	Merck No. / bp/°C	Beil. Ref. / den/g cm⁻³	Solubility / n_D
736	Benzenamine, 2-ethoxy- / o-Phenetidine	$C_8H_{11}NO$ / 137.18	94-70-2 / <-21	7187 / 232.5	4-13-00-00807	H_2O 2; EtOH 3; eth 3; ctc 2 / 1.5560[20]
737	Benzenamine, 4-ethoxy- / p-Phenetidine	$C_8H_{11}NO$ / 137.18	156-43-4 / 2.4	7188 / 254	4-13-00-01017 / 1.0652[16]	H_2O 2; EtOH 3; eth 3; chl 3 / 1.5528[20]
738	Benzenamine, 3-ethoxy-N,N-diethyl-	$C_{12}H_{19}NO$ / 193.29	1864-92-2	/ 286; 97[0.6]	4-13-00-00970	EtOH 3; bz 3; HOAc 3 / 1.5325[25]
739	Benzenamine, 2-ethoxy-5-nitro- / o-Phenetidine, 5-nitro	$C_8H_{10}N_2O_3$ / 182.18	136-79-8 / 96.5	6538 / 205-6[14]	4-13-00-00896	eth 4; EtOH 4
740	Benzenamine, 2-ethyl- / 2-Ethylaniline	$C_8H_{11}N$ / 121.18	578-54-1 / -43	/ 209.5	4-12-00-02411 / 0.983[22]	H_2O 2; EtOH 4; eth 4; chl 2 / 1.5584[22]
741	Benzenamine, 3-ethyl-	$C_8H_{11}N$ / 121.18	587-02-0 / -64	/ 214; 94[6]	4-12-00-02417 / 0.9896[25]	eth 4; EtOH 4
742	Benzenamine, 4-ethyl- / p-Ethylaniline	$C_8H_{11}N$ / 121.18	589-16-2 / -4.8	/ 217.5	4-12-00-02419 / 0.9679[20]	H_2O 2; EtOH 4; eth 4; ctc 2 / 1.5554[20]
743	Benzenamine, N-ethyl- / N-Ethylaniline	$C_8H_{11}N$ / 121.18	103-69-5 / -63.5	3722 / 203.0	4-12-00-00250 / 0.9625[20]	H_2O 1; EtOH 5; eth 5; ace 4 / 1.5559[20]
744	Benzenamine, 2-ethyl-N-(2-ethylphenyl)-	$C_{16}H_{19}N$ / 225.33	64653-59-4 / 29	/ 336[603;] / 173[10]	4-12-00-02413	H_2O 1; EtOH 4; eth 4; chl 2 / 1.5550[25]
745	Benzenamine, 2-ethyl-6-methyl-	$C_9H_{13}N$ / 135.21	24549-06-2 / -33	/ 231	4-12-00-02638 / 0.968[25]	1.5525[20]
746	Benzenamine, N-ethyl-2-methyl-	$C_9H_{13}N$ / 135.21	94-68-8 / <-15	/ 216	4-12-00-01747 / 0.948[25]	EtOH 3; eth 3 / 1.5456[20]
747	Benzenamine, N-ethyl-3-methyl-	$C_9H_{13}N$ / 135.21	102-27-2	/ 221	4-12-00-01816	EtOH 3; eth 3 / 1.5451[20]
748	Benzenamine, N-ethyl-4-methyl- / N-Ethyl-4-toluidine	$C_9H_{13}N$ / 135.21	622-57-1	/ 217	4-12-00-01875 / 0.9391[16]	EtOH 3; eth 3
749	Benzenamine, N-ethyl-N-methyl-	$C_9H_{13}N$ / 135.21	613-97-8	/ 204	4-12-00-00252 / 0.92[55]	H_2O 1; EtOH 5; eth 5; ctc 3
750	Benzenamine, N-ethyl-N-nitroso-	$C_8H_{10}N_2O$ / 150.18	612-64-6	/ 119-20[15]	4-16-00-00865 / 1.0874[20]	H_2O 1; HOAc 3
751	Benzenamine, N-ethyl-N-phenyl-	$C_{14}H_{15}N$ / 197.28	606-99-5	/ 295.5; 147[10]	4-12-00-00276 / 1.0377[20]	eth 4; EtOH 4 / 1.6095[20]
752	Benzenamine, 2-fluoro- / o-Fluoroaniline	C_6H_6FN / 111.12	348-54-9 / -34.6	/ 175; 55[12]	4-12-00-01099 / 1.1513[21]	H_2O 1; EtOH 3; eth 3; ctc 2 / 1.5421[20]
753	Benzenamine, 3-fluoro- / m-Fluoroaniline	C_6H_6FN / 111.12	372-19-0	/ 188	4-12-00-01101 / 1.1561[19]	H_2O 2; EtOH 3; eth 3; chl 2 / 1.5436[20]
754	Benzenamine, 4-fluoro- / p-Fluoroaniline	C_6H_6FN / 111.12	371-40-4 / -0.8	4098 / 182; 85[19]	4-12-00-01104 / 1.1725[20]	H_2O 2; EtOH 3; eth 3; ctc 2 / 1.5195[20]
755	Benzenamine, 4-fluoro-2-methyl-	C_7H_8FN / 125.15	452-71-1 / 14.2	/ 94[16]	4-12-00-01786 / 1.1263[18]	eth 3; ace 3; bz 3; ctc 3 / 1.5363[18]
756	Benzenamine, hydrochloride / Aniline, hydrochloride	C_6H_8ClN / 129.59	142-04-1 / 198	687	4-12-00-00223 / 1.2215[4]	H_2O 4; EtOH 4; eth 1; chl 1
757	Benzenamine, N-hydroxy- / Phenylhydroxylamine	C_6H_7NO / 109.13	100-65-2 / 83.5	7266	4-15-00-00004	bz 4; eth 4; EtOH 4; chl 4
758	Benzenamine, N-hydroxy-4-methyl-	C_7H_9NO / 123.15	623-10-9 / 96	/ 117 dec	4-15-00-00019	eth 4; EtOH 4; chl 4
759	Benzenamine, N-hydroxy-N-nitroso-, ammonium salt / Cupferron	$C_6H_9N_3O_2$ / 155.16	135-20-6 / 163.5	2625	4-16-00-00889	DMSO 2
760	Benzenamine, 3-iodo-	C_6H_6IN / 219.02	626-01-7 / 33	/ 145[15]	4-12-00-01543	H_2O 1; EtOH 3; chl 3 / 1.6811[20]
761	Benzenamine, 4-iodo- / p-Iodoaniline	C_6H_6IN / 219.02	540-37-4 / 67.5	4920	4-12-00-01544	H_2O 2; EtOH 3; eth 3; os 3
762	Benzenamine, 2-iodo-4-methyl- / 2-Iodo-4-methylaniline	C_7H_8IN / 233.05	29289-13-2 / 40	/ dec	4-12-00-01996	ace 4; bz 4; eth 4; EtOH 4
763	Benzenamine, 5-iodo-2-methyl- / o-Toluidine, 5-iodo	C_7H_8IN / 233.05	83863-33-6 / 48.5	/ 273 dec	1-12-00-00391	EtOH 4
764	Benzenamine, N,N'-methanetetraylbis-	$C_{13}H_{10}N_2$ / 194.24	622-16-2 / 169	/ 331; 175[20]	4-12-00-00866	H_2O 2; EtOH 2; eth 2; bz 3
765	Benzenamine, N,N'-methanetetraylbis[4-methyl-	$C_{15}H_{14}N_2$ / 222.29	726-42-1 / 58.5	/ 221[20]	4-12-00-01938 / 1.1500[20]	
766	Benzenamine, 2-methoxy- / o-Anisidine	C_7H_9NO / 123.15	90-04-0 / 6.2	/ 224	4-13-00-00806 / 1.0923[20]	H_2O 2; EtOH 3; eth 3; ace 3 / 1.5715[10]
767	Benzenamine, 3-methoxy- / m-Anisidine	C_7H_9NO / 123.15	536-90-3 / -1	/ 251	4-13-00-00953 / 1.096[20]	H_2O 2; EtOH 3; eth 3; ace 3 / 1.5794[20]
768	Benzenamine, 4-methoxy- / p-Anisidine	C_7H_9NO / 123.15	104-94-9 / 57.2	/ 243	4-13-00-01015 / 1.071[57]	H_2O 3; EtOH 4; eth 4; ace 3 / 1.5559[60]
769	Benzenamine, 2-methoxy-N,N-dimethyl-	$C_9H_{13}NO$ / 151.21	700-75-4	/ 210	4-13-00-00816 / 1.0160[23]	chl 4
770	Benzenamine, 2-methoxy-5-methyl- / 5-Methyl-o-anisidine	$C_8H_{11}NO$ / 137.18	120-71-8 / 53	/ 235	3-13-00-01577	H_2O 2; EtOH 3; eth 3; bz 3
771	Benzenamine, 3-methoxy-4-methyl-	$C_8H_{11}NO$ / 137.18	16452-01-0 / 56.5	/ 251	4-13-00-01666	bz 4; eth 4; EtOH 4; lig 4
772	Benzenamine, 4-methoxy-2-methyl-	$C_8H_{11}NO$ / 137.18	102-50-1 / 29.5	/ 248.5	4-13-00-01698 / 1.065[25]	EtOH 4 / 1.5647[20]
773	Benzenamine, 5-methoxy-2-methyl-	$C_8H_{11}NO$ / 137.18	50868-72-9 / 47	/ 253; 140[20]	4-13-00-01711	eth 4

No.	Name / Synonym	Mol. Form. / Mol. Wt.	CAS RN / mp/°C	Merck No. / bp/°C	Beil. Ref. / den/g cm^{-3}	Solubility / n_D
774	Benzenamine, 2-methoxy-6-methyl-	$C_8H_{11}NO$ / 137.18	50868-73-0 / 31	119[16]	3-13-00-01549	EtOH 4
775	Benzenamine, 2-methoxy-5-nitro- / 5-Nitro-o-anisidine	$C_7H_8N_2O_3$ / 168.15	99-59-2 / 118		4-13-00-00896 / 1.2068[15]	H$_2$O 3; EtOH 4; eth 3; ace 4
776	Benzenamine, 3-methoxy-4-nitro- / m-Anisidine, 4-nitro-	$C_7H_8N_2O_3$ / 168.15	16292-88-9 / 169	sub	4-13-00-01005	ace 4; EtOH 4
777	Benzenamine, 5-methoxy-2-nitro-	$C_7H_8N_2O_3$ / 168.15	16133-49-6 / 131	sub	4-13-00-01003	EtOH 4
778	Benzenamine, 2-methyl- / o-Toluidine	C_7H_9N / 107.16	95-53-4 / -16.3	9462 / 200.3	4-12-00-01744 / 0.9984[20]	H$_2$O 2; EtOH 5; eth 5; ctc 5 / 1.5725[20]
779	Benzenamine, 3-methyl- / m-Toluidine	C_7H_9N / 107.16	108-44-1 / -31.2	9462 / 203.3	4-12-00-01813 / 0.9889[20]	ace 4; bz 4; eth 4; EtOH 4 / 1.5681[20]
780	Benzenamine, 4-methyl- / p-Toluidine	C_7H_9N / 107.16	106-49-0 / 43.7	9462 / 200.4	4-12-00-01866 / 0.9619[20]	H$_2$O 2; EtOH 4; eth 3; ace 3 / 1.5534[45]
781	Benzenamine, N-methyl- / N-Methylaniline	C_7H_9N / 107.16	100-61-8 / -57	5941 / 196.2	4-12-00-00241 / 0.9891[20]	H$_2$O 1; EtOH 3; eth 3; ctc 3 / 1.5684[20]
782	Benzenamine, 4-(6-methyl-2-benzothiazolyl)-	$C_{14}H_{12}N_2S$ / 240.33	92-36-4 / 194.8	434	4-27-00-05052	EtOH 2; eth 2; bz 2; HOAc 2
783	Benzenamine, N-(3-methylbutyl)-	$C_{11}H_{17}N$ / 163.26	2051-84-5 /	254.5	4-12-00-00259 / 0.8912[55]	eth 4; EtOH 4 / 1.5305[20]
784	Benzenamine, 4,4'-methylenebis- / p,p'-Diaminodiphenylmethane	$C_{13}H_{14}N_2$ / 198.27	101-77-9 / 92.5	2958 / 398; 257[18]	4-13-00-00390	H$_2$O 2; EtOH 4; eth 4; bz 4
785	Benzenamine, 4,4'-methylenebis[2-chloro- / 4,4'-Methylenebis[2-chloroaniline]	$C_{13}H_{12}Cl_2N_2$ / 267.16	101-14-4 /	5978	1-13-00-00074	ctc 3
786	Benzenamine, 4,4'-methylenebis[N,N-dimethyl- / Michler's Base	$C_{17}H_{22}N_2$ / 254.38	101-61-1 / 91.5	6099 / 390 dec; 183[3]	4-13-00-00390	H$_2$O 1; EtOH 2; eth 4; bz 4
787	Benzenamine, 2-(1-methylethenyl)-	$C_9H_{11}N$ / 133.19	52562-19-3 /	95[13]	4-12-00-02925 / 0.976[25]	1.5722[20]
788	Benzenamine, 2-(1-methylethyl)- / 2-Isopropylaniline	$C_9H_{13}N$ / 135.21	643-28-7 /	221; 95[13]	4-12-00-02630 / 0.9760[12]	H$_2$O 1; eth 3; bz 3; ctc 3
789	Benzenamine, 4-(1-methylethyl)- / Cumidine	$C_9H_{13}N$ / 135.21	99-88-7 /	2622 / 225	4-12-00-02632 / 0.9530	
790	Benzenamine, N-(1-methylethyl)- / N-Isopropylaniline	$C_9H_{13}N$ / 135.21	768-52-5 /	203	4-12-00-00255 / 0.9526[25]	EtOH 3; eth 3; ace 3; bz 3 / 1.5380[20]
791	Benzenamine, 2-methyl-, hydrochloride / o-Toluidine hydrochloride	$C_7H_{10}ClN$ / 143.62	636-21-5 / 215		4-12-00-01744	H$_2$O 4; EtOH 4
792	Benzenamine, N-methyl-, hydrochloride	$C_7H_{10}ClN$ / 143.62	2739-12-0 / 122.7		4-12-00-00241 / 1.0660[131]	H$_2$O 4; EtOH 3; eth 1; bz 1
793	Benzenamine, 2-methyl-5-(1-methylethyl)-	$C_{10}H_{15}N$ / 149.24	2051-53-8 / -16	241	3-12-00-02733 / 0.9942[20]	ctc 3; CS$_2$ 3 / 1.5387[20]
794	Benzenamine, 4-methyl-N-(1-methylethyl)-	$C_{10}H_{15}N$ / 149.24	10436-75-6 /	220	4-12-00-01876 / 0.9226[20]	1.5332
795	Benzenamine, 2-methyl-N-(2-methylphenyl)-	$C_{14}H_{15}N$ / 197.28	617-00-5 / 52.5	316; 192[23]	4-12-00-01750	H$_2$O 2
796	Benzenamine, 3-methyl-N-(3-methylphenyl)-	$C_{14}H_{15}N$ / 197.28	626-13-1 / -12-	319.5	2-12-00-00467	H$_2$O 2; EtOH 3; eth 3; peth 3
797	Benzenamine, 4-methyl-N-(4-methylphenyl)-	$C_{14}H_{15}N$ / 197.28	620-93-9 / 79.8	330.5	4-12-00-01879	eth 4; peth 4
798	Benzenamine, 2-methyl-4-[(2-methylphenyl)azo]- / 2',3-Dimethyl-4-aminoazobenzene	$C_{14}H_{15}N_3$ / 225.29	97-56-3 / 102		4-16-00-00528	eth 4; EtOH 4
799	Benzenamine, 2-methyl-3-nitro-	$C_7H_8N_2O_2$ / 152.15	603-83-8 / 92	305	4-12-00-01811 / 1.3780[15]	H$_2$O 2; EtOH 3; eth 3; bz 3
800	Benzenamine, 2-methyl-4-nitro-	$C_7H_8N_2O_2$ / 152.15	99-52-5 / 133.5		4-12-00-01809 / 1.1586[140]	H$_2$O 2; EtOH 3; bz 3; HOAc 3
801	Benzenamine, 2-methyl-5-nitro- / 5-Nitro-o-toluidine	$C_7H_8N_2O_2$ / 152.15	99-55-8 / 105.5		4-12-00-01807	H$_2$O 2; EtOH 3; eth 3; ace 3
802	Benzenamine, 2-methyl-6-nitro-	$C_7H_8N_2O_2$ / 152.15	570-24-1 / 96		4-12-00-01807 / 1.1900[100]	H$_2$O 2; EtOH 3; eth 3; bz 3
803	Benzenamine, 4-methyl-2-nitro-	$C_7H_8N_2O_2$ / 152.15	89-62-3 / 116.3		4-12-00-02000 / 1.16[121]	H$_2$O 2; EtOH 3; chl 3
804	Benzenamine, N-methyl-2-nitro-	$C_7H_8N_2O_2$ / 152.15	612-28-2 / 38	158[18]	4-12-00-01564	H$_2$O 2; EtOH 3; eth 3; ace 3
805	Benzenamine, N-methyl-4-nitro-	$C_7H_8N_2O_2$ / 152.15	100-15-2 / 152	dec	4-12-00-01616 / 1.201[155]	H$_2$O 1; EtOH 3; eth 2; bz 3
806	Benzenamine, N-methyl-N-nitroso-	$C_7H_8N_2O$ / 136.15	614-00-6 / 14.7	225 dec; 121[13]	4-16-00-00864 / 1.1240[20]	H$_2$O 1; EtOH 3; eth 3 / 1.5769[20]
807	Benzenamine, N-methyl-N-phenyl- / Methyldiphenylamine	$C_{13}H_{13}N$ / 183.25	552-82-9 / -7.5	5973 / 293.5	4-12-00-00275 / 1.0476[20]	H$_2$O 1; EtOH 2; ctc 3; MeOH 2 / 1.6193[20]
808	Benzenamine, 3-methyl-N-phenyl-	$C_{13}H_{13}N$ / 183.25	1205-64-7 / 30	316; 183[17]	4-12-00-01818	bz 4; eth 4; EtOH 4 / 1.6350[20]
809	Benzenamine, 2-methyl-N-(phenylmethylene)-	$C_{14}H_{13}N$ / 195.26	5877-55-4 /	314	4-12-00-01751 / 1.041[20]	H$_2$O 1; ace 3 / 1.6310[25]
810	Benzenamine, 3-methyl-N-(phenylmethylene)-	$C_{14}H_{13}N$ / 195.26	5877-58-7 / 31	315	3-12-00-01958 / 1.0390[25]	ace 3; os 3 / 1.6353[25]

No.	Name / Synonym	Mol. Form. / Mol. Wt.	CAS RN / mp/°C	Merck No. / bp/°C	Beil. Ref. / den/g cm^{-3}	Solubility / n_D
811	Benzenamine, 4-methyl-N-(phenylmethylene)-	$C_{14}H_{13}N$ / 195.26	2272-45-9 / 35	318; 178[11]	2-12-00-00496	ace 4
812	Benzenamine, 2-(1-methylpropyl)-	$C_{10}H_{15}N$ / 149.24	55751-54-7 /	120-2[16]	4-12-00-02814 / 0.9574[20]	EtOH 3; ace 3; bz 3; ctc 2
813	Benzenamine, 4-(1-methylpropyl)-	$C_{10}H_{15}N$ / 149.24	30273-11-1 /	238; 118[15]	4-12-00-02814 / 0.949[15]	bz 4; eth 4 / 1.5360[29]
814	Benzenamine, 4-(2-methylpropyl)-	$C_{10}H_{15}N$ / 149.24	30090-17-6 /	238; 112[11]	4-12-00-02818 / 0.949[15]	ace 4; bz 4; EtOH 4
815	Benzenamine, 4-methyl-N-propyl- / N-Propyl-p-toluidine	$C_{10}H_{15}N$ / 149.24	54837-90-0 /	235	4-12-00-01876 / 0.9243[20]	1.5367
816	Benzenamine, N-(1-methylpropyl)- / N-sec-Butylaniline	$C_{10}H_{15}N$ / 149.24	6068-69-5 /	225	4-12-00-00257	chl 2; os 3 / 1.5333[20]
817	Benzenamine, N-(2-methylpropyl)-	$C_{10}H_{15}N$ / 149.24	588-47-6 /	232; 113[11]	4-12-00-00257 / 0.940[18]	H_2O 1; eth 4; bz 4 / 1.5328[20]
818	Benzenamine, N-methyl-N,2,4,6-tetranitro- / Nitramine	$C_7H_5N_5O_8$ / 287.15	479-45-8 / 131.5	6488 / exp 180	4-16-00-00895 / 1.57[10]	H_2O 1; EtOH 2; eth 2; ace 3
819	Benzenamine, 2-(methylthio)- / Aniline, 2-methylthio	C_7H_9NS / 139.22	2987-53-3 /	234	4-13-00-00910 / 1.111[25]	1.6239[20]
820	Benzenamine, 4-(methylthio)-	C_7H_9NS / 139.22	104-96-1 /	272.5	4-13-00-01289 / 1.1379[20]	EtOH 3; eth 3; ace 3; bz 3 / 1.6395[20]
821	Benzenamine, nitrate / Aniline nitrate	$C_6H_8N_2O_3$ / 156.14	542-15-4 / 190 dec	687	4-12-00-00223 / 1.358[4]	H_2O 4; eth 4; EtOH 4
822	Benzenamine, 2-nitro- / o-Nitroaniline	$C_6H_6N_2O_2$ / 138.13	88-74-4 / 71.2	6504 / 284	4-12-00-01563 / 1.442[15]	H_2O 2; EtOH 3; eth 4; ace 4
823	Benzenamine, 3-nitro- / m-Nitroaniline	$C_6H_6N_2O_2$ / 138.13	99-09-2 / 114	6503 / 306 dec	4-12-00-01589 / 0.991[25]	H_2O 2; EtOH 3; eth 3; ace 3
824	Benzenamine, 4-nitro- / p-Nitroaniline	$C_6H_6N_2O_2$ / 138.13	100-01-6 / 147	6505 / 332	4-12-00-01613 / 1.424[20]	H_2O 1; EtOH 3; eth 3; ace 3
825	Benzenamine, N-nitro-	$C_6H_6N_2O_2$ / 138.13	645-55-6 / 47.3	exp	4-16-00-00889	H_2O 4; bz 4; eth 4; EtOH 4
826	Benzenamine, 4-nitro-N-phenyl-	$C_{12}H_{10}N_2O_2$ / 214.22	836-30-6 / 135.3	211[30]	4-12-00-01619	H_2O 1; EtOH 4; ace 2; HOAc 4
827	Benzenamine, 5-nitro-2-propoxy- / Aniline, 5-nitro-2-propoxy	$C_9H_{12}N_2O_3$ / 196.21	553-79-7 / 49	6551	4-13-00-00897	EtOH 4
828	Benzenamine, 4-nitroso-	$C_6H_6N_2O$ / 122.13	659-49-4 / 173.5	dec	4-12-00-01558	H_2O 3; EtOH 3; bz 2
829	Benzenamine, 4-nitroso-N-phenyl- / p-Nitrosodiphenylamine	$C_{12}H_{10}N_2O$ / 198.22	156-10-5 / 143	6560	4-12-00-01560	H_2O 2; EtOH 4; eth 4; bz 4
830	Benzenamine, N-nitroso-N-phenyl- / N-Nitroso-N-diphenylamine	$C_{12}H_{10}N_2O$ / 198.22	86-30-6 / 66.5		4-16-00-00865	EtOH 2; bz 3; chl 2
831	Benzenamine, 4-octyl-	$C_{14}H_{23}N$ / 205.34	16245-79-7 / 20	310; 138[5]	4-12-00-02887 / 0.9128[20]	eth 4
832	Benzenamine, 4,4'-oxybis- / Bis(p-aminophenyl) ether	$C_{12}H_{12}N_2O$ / 200.24	101-80-4 / 189 dec		4-13-00-01038	
833	Benzenamine, 2,3,4,5,6-pentamethyl-	$C_{11}H_{17}N$ / 163.26	2243-30-3 / 152.5	277.5	3-12-00-02758	eth 4; EtOH 4
834	Benzenamine, 2-phenoxy-	$C_{12}H_{11}NO$ / 185.23	2688-84-8 / 45.8	308; 172[14]	4-13-00-00809	EtOH 3; eth 3; ace 3; bz 3
835	Benzenamine, 3-phenoxy-	$C_{12}H_{11}NO$ / 185.23	3586-12-7 / 37	315; 180[10]	4-13-00-00955 / 1.1583[25]	EtOH 3; eth 3; ace 3; bz 3
836	Benzenamine, N-phenyl- / Diphenylamine	$C_{12}H_{11}N$ / 169.23	122-39-4 / 52.9	3317 / 302	4-12-00-00271 / 1.158[22]	H_2O 1; EtOH 4; eth 3; ace 4
837	Benzenamine, 4-(phenylazo)- / p-Aminoazobenzene	$C_{12}H_{11}N_3$ / 197.24	60-09-3 / 127	430 / >360	2-16-00-00149	H_2O 2; EtOH 3; eth 3; bz 3
838	Benzenamine, 2-(2-phenylethyl)-	$C_{14}H_{15}N$ / 197.28	5697-85-8 / 33	193[19]	4-12-00-03311 / 1.0430[20]	
839	Benzenamine, 4-(phenylmethyl)-	$C_{13}H_{13}N$ / 183.25	1135-12-2 / 34.5	300	4-12-00-03281 / 1.038[25]	eth 4; EtOH 4; lig 4
840	Benzenamine, N-(phenylmethylene)- / Benzylideneaniline	$C_{13}H_{11}N$ / 181.24	538-51-2 / 54	1152 / 310	2-12-00-00113 / 1.038[55]	H_2O 1; EtOH 3; eth 3; chl 2 / 1.600[100]
841	Benzenamine, N-2-propenyl-	$C_9H_{11}N$ / 133.19	589-09-3 /	219; 106[12]	4-12-00-00265 / 0.982[25]	H_2O 2; EtOH 3; eth 5; ace 3 / 1.563[20]
842	Benzenamine, 2-propyl-	$C_9H_{13}N$ / 135.21	1821-39-2 /	226; 116[15]	4-12-00-02580 / 0.9602[20]	H_2O 1; EtOH 3; eth 3 / 1.5427[20]
843	Benzenamine, N-propyl-	$C_9H_{13}N$ / 135.21	622-80-0 /	222; 98[11]	4-12-00-00254 / 0.9443[20]	eth 4; EtOH 4 / 1.5428[20]
844	Benzenamine, N-sulfinyl-	C_6H_5NOS / 139.18	1122-83-4 /	200	4-12-00-01075 / 1.236[25]	1.6270[20]
845	Benzenamine, 4,4'-sulfinylbis- / 4,4'-Sulfinyldianiline	$C_{12}H_{12}N_2OS$ / 232.31	119-59-5 / 175 dec	8927	4-13-00-01306	H_2O 3; EtOH 3
846	Benzenamine, 4,4'-sulfonylbis- / Dapsone	$C_{12}H_{12}N_2O_2S$ / 248.31	80-08-0 / 175.5	2820	4-13-00-01306	EtOH 3; DMSO-d_6 2
847	Benzenamine, 2,3,4,5-tetramethyl- / Aniline, 2,3,4,5-tetramethyl	$C_{10}H_{15}N$ / 149.24	2217-45-0 / 69	259.5	3-12-00-02743	eth 4; EtOH 4; lig 4
848	Benzenamine, 2,3,4,6-tetramethyl-	$C_{10}H_{15}N$ / 149.24	488-71-1 / 23.5	255	4-12-00-02849 / 0.978[24]	EtOH 4
849	Benzenamine, N,N,2,6-tetramethyl-	$C_{10}H_{15}N$ / 149.24	769-06-2 / -36	196; 88[20]	4-12-00-02522 / 0.9147[20]	

No.	Name Synonym	Mol. Form. Mol. Wt.	CAS RN mp/°C	Merck No. bp/°C	Beil. Ref. den/g cm^{-3}	Solubility n_D
850	Benzenamine, 4,4'-thiobis- Bis(4-aminophenyl) sulfide	$C_{12}H_{12}N_2S$ 216.31	139-65-1 108.5		4-13-00-01306	H_2O 2; EtOH 4; eth 4; bz 4
851	Benzenamine, 2,4,6-tribromo- 2,4,6-Tribromoaniline	$C_6H_4Br_3N$ 329.82	147-82-0 122	9522 300	4-12-00-01538 2.35[20]	H_2O 1; EtOH 2; eth 3; chl 3
852	Benzenamine, 2,3,4-trichloro- 2,3,4-Trichloroaniline	$C_6H_4Cl_3N$ 196.46	634-67-3 73	292	0-12-00-00626	EtOH 4
853	Benzenamine, 2,4,5-trichloro-	$C_6H_4Cl_3N$ 196.46	636-30-6 96.5	270	4-12-00-01277	EtOH 3; eth 3; CS_2 4; lig 2
854	Benzenamine, 2,4,6-trichloro-	$C_6H_4Cl_3N$ 196.46	634-93-5 78.5	262	4-12-00-01281	H_2O 1; EtOH 3; eth 3; chl 3
855	Benzenamine, 2-(trifluoromethyl)- o-(Trifluoromethyl)aniline	$C_7H_6F_3N$ 161.13	88-17-5 35.5	68[15]	4-12-00-01786 1.282[25]	1.4810[20]
856	Benzenamine, 3-(trifluoromethyl)- m-(Trifluoromethyl)aniline	$C_7H_6F_3N$ 161.13	98-16-8 5.5	187; 74-5[10]	4-12-00-01843 1.3047[12]	H_2O 2; EtOH 3; eth 3 1.4787[20]
857	Benzenamine, 4-(trifluoromethyl)-	$C_7H_6F_3N$ 161.13	455-14-1 38	117.5[60]	4-12-00-01982 1.283[27]	1.4815[25]
858	Benzenamine, 2,4,5-trimethyl-	$C_9H_{13}N$ 135.21	137-17-7 68	234.5	4-12-00-02645 0.957[25]	EtOH 4
859	Benzenamine, 2,4,6-trimethyl- Mesitylamine	$C_9H_{13}N$ 135.21	88-05-1 -5	232.5	4-12-00-02648 0.9633[25]	ctc 2 1.5495[20]
860	Benzenamine, N,N,2-trimethyl- N,N-Dimethyl-o-toluidine	$C_9H_{13}N$ 135.21	609-72-3 -60	194.1	4-12-00-01747 0.9286[20]	eth 4; EtOH 4 1.5152[20]
861	Benzenamine, N,N,3-trimethyl-	$C_9H_{13}N$ 135.21	121-72-2	212	4-12-00-01815 0.9410[20]	EtOH 5; eth 5 1.5492[20]
862	Benzenamine, N,N,4-trimethyl-	$C_9H_{13}N$ 135.21	99-97-8	211	4-12-00-01874 0.9366[20]	H_2O 1; EtOH 5; eth 5; ctc 3 1.5366[20]
863	Benzenamine, 2,4,6-trinitro-	$C_6H_4N_4O_6$ 228.12	489-98-5 193.5	exp	4-12-00-01735 1.762[10]	H_2O 1; EtOH 2; eth 2; ace 3
864	Benzenamine, 2,4,6-trinitro-N-(2,4,6-trinitrophenyl)- Dipicrylamine	$C_{12}H_5N_7O_{12}$ 439.21	131-73-7 244 dec	3340	4-12-00-01737	H_2O 1; EtOH 1; eth 2; ace 2
865	Benzenaminium, N-ethyl-3-hydroxy-N,N-dimethyl-, chloride Ammonium, dimethyl ethyl 3-hydroxyphenyl,chloride	$C_{10}H_{16}ClNO$ 201.70	116-38-1	3487	4-13-00-00969	
866	Benzenaminium, N,N,N-trimethyl-, iodide Phenyl trimethyl ammonium iodide	$C_9H_{14}IN$ 263.12	98-04-4 224	7289	4-12-00-00249	H_2O 4; EtOH 3; ace 2; chl 1
867	Benzene [6]Annulene	C_6H_6 78.11	71-43-2 5.5	1074 80.0	4-05-00-00583 0.8765[20]	H_2O 2; EtOH 5; eth 5; ace 5 1.5011[20]
868	Benzeneacetaldehyde Phenylacetaldehyde	C_8H_8O 120.15	122-78-1 33.5	7236 195	4-07-00-00664 1.0272[20]	H_2O 2; EtOH 5; eth 5; ace 3 1.5255[20]
869	Benzeneacetaldehyde, 3,4-dimethoxy-	$C_{10}H_{12}O_3$ 180.20	5703-21-9	125[1.5]	4-08-00-01828 1.148[20]	1.5426[20]
870	Benzeneacetaldehyde, 2-methoxy-	$C_9H_{10}O_2$ 150.18	33567-59-8	245; 117[17]	1-08-00-00543 1.0897[20]	ace 4; EtOH 4 1.5393[20]
871	Benzeneacetaldehyde, 4-methoxy-	$C_9H_{10}O_2$ 150.18	5703-26-4	255.5	4-08-00-00426 1.096[20]	1.5359[20]
872	Benzeneacetaldehyde, α-methyl-	$C_9H_{10}O$ 134.18	93-53-8	203.5	1.0089[20]	EtOH 4 1.5176[20]
873	Benzeneacetaldehyde, 2-methyl-	$C_9H_{10}O$ 134.18	10166-08-2	221; 92[10]	1-07-00-00163 1.0241[10]	eth 4; EtOH 4; chl 4
874	Benzeneacetaldehyde, 4-methyl-	$C_9H_{10}O$ 134.18	104-09-6 40	221.5	4-07-00-00705 1.0052[20]	eth 4; EtOH 4; chl 4 1.5255[20]
875	Benzeneacetaldehyde, 2-nitro-	$C_8H_7NO_3$ 165.15	1969-73-9 22.5	133[5]	2-07-00-00229	bz 4; eth 4; EtOH 4
876	Benzeneacetaldehyde, α-oxo-, aldoxime Isonitrosoacetophenone	$C_8H_7NO_2$ 149.15	532-54-7 129	5077	4-07-00-02132	H_2O 2; chl 3
877	Benzeneacetaldehyde, α-phenyl-	$C_{14}H_{12}O$ 196.25	947-91-1	315 dec; 157[7]	4-07-00-01400 1.1061[21]	H_2O 1; EtOH 4; eth 4; bz 4 1.5920[21]
878	Benzeneacetamide α-Phenylacetamide	C_8H_9NO 135.17	103-81-1 157	7237	4-09-00-01632	H_2O 2; EtOH 3; eth 2; bz 2
879	Benzeneacetamide, N-(aminocarbonyl)- Phenacemide	$C_9H_{10}N_2O_2$ 178.19	63-98-9 215	7154	4-09-00-01636	bz 4; eth 4; EtOH 4
880	Benzeneacetamide, 4-butoxy-N-hydroxy- Bufexamac	$C_{12}H_{17}NO_3$ 223.27	2438-72-4 154	1462		
881	Benzeneacetamide, α-ethyl- α-Phenylbutyramide	$C_{10}H_{13}NO$ 163.22	90-26-6 86	7249 185[16]	4-09-00-01818	H_2O 3; ace 2; ctc 3
882	Benzeneacetamide, α-(1-methylethyl)-	$C_{11}H_{15}NO$ 177.25	5470-47-3 111.5	180[14]	3-09-00-02518	eth 4; EtOH 4
883	Benzeneacetic acid Phenylacetic acid	$C_8H_8O_2$ 136.15	103-82-2 76.7	7239 265.5	4-09-00-01614 1.228[6]	H_2O 2; EtOH 4; eth 4; ace 3
884	Benzeneacetic acid, α-acetyl-, ethyl ester	$C_{12}H_{14}O_3$ 206.24	5413-05-8	156[22]	4-10-00-02776 1.0855[20]	eth 4; EtOH 4 1.5176[20]
885	Benzeneacetic acid, α-amino- α-Phenylglycine	$C_8H_9NO_2$ 151.17	69-91-0 292 dec	7263 sub 255	4-14-00-01317	alk 3; os 2
886	Benzeneacetic acid, 4-amino- p-Aminophenylacetic acid	$C_8H_9NO_2$ 151.17	1197-55-3 200 dec	475	3-14-00-01182	H_2O 1; EtOH 2; DMSO-d_6 2
887	Benzeneacetic acid, anhydride	$C_{16}H_{14}O_3$ 254.29	1555-80-2 73.3	195[12]	4-09-00-01629	eth 4; chl 4

No.	Name Synonym	Mol. Form. Mol. Wt.	CAS RN mp/°C	Merck No. bp/°C	Beil. Ref. den/g cm^{-3}	Solubility n_D
888	Benzeneacetic acid, 4-bromo-	$C_8H_7BrO_2$ 215.05	1878-68-8 116	sub	3-09-00-02275	H_2O 2; EtOH 4; eth 4; CS_2 4
889	Benzeneacetic acid, 4-bromo-α-hydroxy-, (±)- Acetic acid, (4-bromophenyl)-hydroxy(DL)	$C_8H_7BrO_3$ 231.05	7021-04-7 119	1412	3-10-00-00480	H_2O 4; EtOH 4; eth 4; bz 4
890	Benzeneacetic acid, α-butyl-	$C_{12}H_{16}O_2$ 192.26	24716-09-4	182[20]; 118[0.1]	4-09-00-01897 1.0225[19]	1.5071[19]
891	Benzeneacetic acid, α-chloro-, ethyl ester, (S)- Acetic acid, chlorophenyl-, ethyl ester, L-(+)-	$C_{10}H_{11}ClO_2$ 198.65	10606-73-2	162[45]; 133[12]	3-09-00-02266 1.1594[20]	1.5152[20]
892	Benzeneacetic acid, α-chloro-α-phenyl-, ethyl ester	$C_{16}H_{15}ClO_2$ 274.75	52460-86-3 43.5	185[14]	4-09-00-02508	eth 4; EtOH 4
893	Benzeneacetic acid, α-cyano-, ethyl ester	$C_{11}H_{11}NO_2$ 189.21	4553-07-5	275 dec; 165[20]	3-09-00-04262 1.091[20]	ace 4; bz 4; eth 4; EtOH 4 1.5012[25]
894	Benzeneacetic acid, 4-[2-(diethylamino)-2-oxoethoxy]-3-methoxy-, propyl ester Acetic acid, amide,(4(carboxy(propyl ester)methyl)-2-methoxyphenoxy)-N,N-diethyl	$C_{18}H_{27}NO_5$ 337.42	1421-14-3	7813 210-2[0.7]		H_2O 1; EtOH 3; chl 3
895	Benzeneacetic acid, 2,5-dihydroxy- Homogentisic acid	$C_8H_8O_4$ 168.15	451-13-8 153	4658	4-10-00-01506	H_2O 4; EtOH 4; eth 4; bz 1
896	Benzeneacetic acid, α,4-dihydroxy-3-methoxy- Vanilmandelic acid	$C_9H_{10}O_5$ 198.18	55-10-7 131-3 dec	9840	3-10-00-02100	H_2O 4; ace 4; eth 4
897	Benzeneacetic acid, α-ethyl-	$C_{10}H_{12}O_2$ 164.20	90-27-7 47.5	271	2-09-00-00356	eth 3; bz 3; ctc 3
898	Benzeneacetic acid, α-ethyl-, 2-(diethylamino)ethyl ester Butethamate	$C_{16}H_{25}NO_2$ 263.38	14007-64-8	1516 167-9[11]	3-09-00-02463	1.4909[20]
899	Benzeneacetic acid, ethyl ester Ethyl phenylacetate	$C_{10}H_{12}O_2$ 164.20	101-97-3 -29.4	3794 227	4-09-00-01618 1.0333[20]	eth 4; EtOH 4 1.4980[20]
900	Benzeneacetic acid, α-ethyl-, methyl ester	$C_{11}H_{14}O_2$ 178.23	2294-71-5 77.5	228	3-09-00-02462	eth 4; EtOH 4
901	Benzeneacetic acid, 4-fluoro- (p-Fluorophenyl)acetic acid	$C_8H_7FO_2$ 154.14	405-50-5 86	4105 164[2]	4-09-00-01672	
902	Benzeneacetic acid, α-(hydroxymethyl)- 8-methyl-8-azabicyclo[3.2.1]oct-3-yl ester endo-(±)- Atropine	$C_{17}H_{23}NO_3$ 289.37	51-55-8 118.5	891 sub 95	5-21-01-00235	H_2O 4; EtOH 4; eth 1; chl 2
903	Benzeneacetic acid, α-hydroxy-, (S)-	$C_8H_8O_3$ 152.15	17199-29-0 134		4-10-00-00564 1.341[25]	H_2O 4; eth 4; EtOH 4; chl 4
904	Benzeneacetic acid, 2-hydroxy-	$C_8H_8O_3$ 152.15	614-75-5 148	240	4-10-00-00536	H_2O 2; eth 3; chl 2
905	Benzeneacetic acid, 3-hydroxy-	$C_8H_8O_3$ 152.15	621-37-4 132	190[11]	4-10-00-00541	H_2O 4; EtOH 4; eth 4; bz 3
906	Benzeneacetic acid, 4-hydroxy-	$C_8H_8O_3$ 152.15	156-38-7 152	sub	4-10-00-00543	H_2O 2; EtOH 4; eth 4
907	Benzeneacetic acid, α-hydroxy-, ethyl ester, (±)-	$C_{10}H_{12}O_3$ 180.20	4358-88-7 35.5	254	4-10-00-00569 1.1258[20]	EtOH 3; eth 3; ctc 2; CS_2 3
908	Benzeneacetic acid, α-(hydroxymethyl)-, 9-methyl-3-oxa-9-azatricyclo[3.3.1.02,4]non-7-yl ester, [(7S)-(1α,2β,4β,5α,7β)]- Scopolamine	$C_{17}H_{21}NO_4$ 303.36	51-34-3	8361	2-27-00-00064	
909	Benzeneacetic acid, α-hydroxy-, ethyl ester	$C_{10}H_{12}O_3$ 180.20	774-40-3 35	150[21]	2-10-00-00121 1.1270[20]	eth 4; EtOH 4
910	Benzeneacetic acid, α-hydroxy-, methyl ester, (±)-	$C_9H_{10}O_3$ 166.18	4358-87-6 58	250 dec; 144[20]	2-10-00-00120 1.1756[20]	EtOH 4; chl 4
911	Benzeneacetic acid, α-hydroxy-, 8-methyl-8-azabicyclo[3.2.1]oct-3-yl ester, endo-(±)- Homatropine	$C_{16}H_{21}NO_3$ 275.35	87-00-3 99.5	4649	5-21-01-00234	H_2O 2; EtOH 3; eth 3; ace 3
912	Benzeneacetic acid, α-hydroxy-, 3-methylbutyl ester Mandelic acid isoamyl ester	$C_{13}H_{18}O_3$ 222.28	5421-04-5	5600 172[11]	4-10-00-00571	
913	Benzeneacetic acid, 4-hydroxy-3-methoxy- Homovanillic acid	$C_9H_{10}O_4$ 182.18	306-08-1 143.5	4662	4-10-00-01509	
914	Benzeneacetic acid, α-hydroxy-α-methyl- Atrolactic acid	$C_9H_{10}O_3$ 166.18	515-30-0 116.5	889	2-10-00-00157	H_2O 4; ace 4; bz 4; EtOH 4
915	Benzeneacetic acid, α-(hydroxymethyl)- Tropic acid	$C_9H_{10}O_3$ 166.18	529-64-6 118	9691	3-10-00-00564	eth 4; EtOH 4
916	Benzeneacetic acid, α-(hydroxymethyl)-, 8-azabicyclo[3.2.1]oct-3-yl ester, [3(S)-endo]- Norhyoscyamine	$C_{16}H_{21}NO_3$ 275.35	537-29-1 140.5	6623	5-21-01-00219	EtOH 4; chl 4
917	Benzeneacetic acid, α-(hydroxymethyl)-, 8-methyl-8-azabicyclo[3.2.1]oct-3-yl ester, [3(S)-endo]- Hyoscyamine	$C_{17}H_{23}NO_3$ 289.37	101-31-5 108.5	4795	5-21-01-00235	H_2O 2; EtOH 4; eth 2; bz 2

No.	Name / Synonym	Mol. Form. / Mol. Wt.	CAS RN / mp/°C	Merck No. / bp/°C	Beil. Ref. / den/g cm^{-3}	Solubility / n_D
918	Benzeneacetic acid, α-hydroxy-α-phenyl- Benzilic acid	$C_{14}H_{12}O_3$ 228.25	76-93-7 151	1089 180 dec	4-10-00-01256	H_2O 2; EtOH 4; eth 4; ace 2
919	Benzeneacetic acid, α-hydroxy-α-phenyl-, ethyl ester	$C_{16}H_{16}O_3$ 256.30	52182-15-7 34	201[21]	4-10-00-01258 1.5620[20]	eth 4; EtOH 4
920	Benzeneacetic acid, α-hydroxy-α-phenyl-, methyl ester	$C_{15}H_{14}O_3$ 242.27	76-89-1 75.8	187[13]	4-10-00-01258	eth 4; EtOH 4
921	Benzeneacetic acid, 2-methoxy-	$C_9H_{10}O_3$ 166.18	93-25-4 124	100-1[2]	4-10-00-00536	H_2O 3; EtOH 4; eth 4; ace 4
922	Benzeneacetic acid, 4-methoxy-	$C_9H_{10}O_3$ 166.18	104-01-8 87	138[2]	4-10-00-00544	H_2O 1; EtOH 4; eth 3; bz 3
923	Benzeneacetic acid, 4-methoxy-, ethyl ester	$C_{11}H_{14}O_3$ 194.23	14062-18-1	138-40[70]	4-10-00-00549 1.097[25]	1.5075[20]
924	Benzeneacetic acid, α-methyl-, (±)-	$C_9H_{10}O_2$ 150.18	2328-24-7 <-20	263	4-09-00-01779 1.1[0]	1.5237[20]
925	Benzeneacetic acid, 3-methyl-	$C_9H_{10}O_2$ 150.18	621-36-3 62	120-3[26]	4-09-00-01792	H_2O 3; chl 3
926	Benzeneacetic acid, 4-methyl-	$C_9H_{10}O_2$ 150.18	622-47-9 93	265	4-09-00-01795	bz 4; eth 4; EtOH 4
927	Benzeneacetic acid, α-methylene- Atropic acid	$C_9H_8O_2$ 148.16	492-38-6 106.5	890 267 dec	4-09-00-02050	H_2O 2; EtOH 3; eth 3; bz 3
928	Benzeneacetic acid, α-methylene-, 8-methyl- 8-azabicyclo[3.2.1]oct-3-yl ester, endo- Apoatropine	$C_{17}H_{21}NO_2$ 271.36	500-55-0 62	770	5-21-01-00228	H_2O 2; EtOH 4; eth 4; ace 4
929	Benzeneacetic acid, methyl ester	$C_9H_{10}O_2$ 150.18	101-41-7	216.5	4-09-00-01617 1.0622[16]	H_2O 1; EtOH 5; eth 5; ace 3 1.5075[20]
930	Benzeneacetic acid, α-(1-methylethyl)-	$C_{11}H_{14}O_2$ 178.23	3508-94-9 63	159[14]	3-09-00-02518	eth 4; EtOH 4
931	Benzeneacetic acid, α-methyl-α-phenyl-	$C_{15}H_{14}O_2$ 226.27	5558-66-7 176	sub	4-09-00-02535	H_2O 2; EtOH 3; eth 4; bz 3
932	Benzeneacetic acid, 2-methylpropyl ester	$C_{12}H_{16}O_2$ 192.26	102-13-6	247	4-09-00-01619 0.999[18]	H_2O 1; EtOH 3; eth 3
933	Benzeneacetic acid, 4-nitro- p-Nitrophenylacetic acid	$C_8H_7NO_4$ 181.15	104-03-0 154	6543	4-09-00-01687	H_2O 2; EtOH 3; eth 3; bz 3
934	Benzeneacetic acid, α-oxo-	$C_8H_6O_3$ 150.13	611-73-4 66	163[15]	4-10-00-02737	H_2O 4; EtOH 3; eth 3; ctc 2
935	Benzeneacetic acid, α-oxo-, ethyl ester	$C_{10}H_{10}O_3$ 178.19	1603-79-8	256.5	4-10-00-02738 1.1222[25]	1.5190[25]
936	Benzeneacetic acid, α-phenyl- Diphenylacetic acid	$C_{14}H_{12}O_2$ 212.25	117-34-0 147	3316 194[25]	4-09-00-02492 1.257[15]	H_2O 2; EtOH 4; eth 3; chl 3
937	Benzeneacetic acid, α-phenyl-, anhydride	$C_{28}H_{22}O_3$ 406.48	1760-46-9 98	222[15]	4-09-00-02497	bz 4; chl 4
938	Benzeneacetic acid, α-phenyl-, 2- (diethylamino)ethyl ester, hydrochloride Adiphenine hydrochloride	$C_{20}H_{26}ClNO_2$ 347.88	50-42-0 113.5	151	3-09-00-03297	H_2O 4; EtOH 2; eth 2
939	Benzeneacetic acid, α-phenyl-, ethyl ester	$C_{16}H_{16}O_2$ 240.30	3468-99-3 60	195[25]	4-09-00-02493 1.1860[20]	H_2O 1; EtOH 4; eth 4; CS_2 4
940	Benzeneacetic acid, 2-phenylethyl ester	$C_{16}H_{16}O_2$ 240.30	102-20-5 26.5	177-8[4.5]	3-09-00-02183 1.077[25]	EtOH 4
941	Benzeneacetic acid, α-(phenylmethylene)- α-Phenylcinnamic acid	$C_{15}H_{12}O_2$ 224.26	3368-16-9 172.5	7253 sub	3-09-00-03438	eth 4; EtOH 4
942	Benzeneacetic acid, α-(phenylmethylene)-, ethyl ester	$C_{17}H_{16}O_2$ 252.31	24446-63-7 33.5	214-5[28]	3-09-00-03416 1.0971[18]	1.5972[18]
943	Benzeneacetic acid, α-propyl-, (±)- Valeric acid, 2-phenyl-, (±)-	$C_{11}H_{14}O_2$ 178.23	7782-21-0 58	280	4-09-00-01868	eth 4; EtOH 4
944	Benzeneacetic acid, 2,3,6-trichloro- Chlorfenac	$C_8H_5Cl_3O_2$ 239.48	85-34-7 161	2085	4-09-00-01681	
945	Benzeneacetonitrile Benzyl cyanide	C_8H_7N 117.15	140-29-4 -23.8	1145 233.5	4-09-00-01663 1.0205[15]	1.5211[25]
946	Benzeneacetonitrile, 4-amino-	$C_8H_8N_2$ 132.17	3544-25-0 46	312	4-14-00-01312	H_2O 2; EtOH 3; CS_2 2; os 3
947	Benzeneacetonitrile, α-bromo- α-Bromobenzyl cyanide	C_8H_6BrN 196.05	5798-79-8 29	1400 242 dec; 133[12]	3-09-00-02278 1.539[29]	H_2O 1; EtOH 4; eth 4; ace 4
948	Benzeneacetonitrile, 2-bromo-	C_8H_6BrN 196.05	19472-74-3 1	140[13]	4-09-00-01682	H_2O 1; EtOH 4
949	Benzeneacetonitrile, 2-chloro-	C_8H_6ClN 151.60	2856-63-5 24	251	4-09-00-01674 1.1737[18]	
950	Benzeneacetonitrile, 3-chloro-	C_8H_6ClN 151.60	1529-41-5	134-6[10]	4-09-00-01675 1.283[25]	1.5437[20]
951	Benzeneacetonitrile, α-ethyl-	$C_{10}H_{11}N$ 145.20	769-68-6	241	4-09-00-01822 0.977[14]	H_2O 1; EtOH 3; eth 3; bz 3
952	Benzeneacetonitrile, 2-fluoro-	C_8H_6FN 135.14	326-62-5	232; 102[10]	4-09-00-01671 1.059[25]	1.5009[20]
953	Benzeneacetonitrile, 4-fluoro-	C_8H_6FN 135.14	459-22-3	119-20[18]	4-09-00-01673 1.126[25]	1.5002[20]

No.	Name Synonym	Mol. Form. Mol. Wt.	CAS RN mp/°C	Merck No. bp/°C	Beil. Ref. den/g cm^{-3}	Solubility n_D
954	Benzeneacetonitrile, α-[(6-O-β-D-glucopyranosyl-β-D-glucopyranosyl)oxy]-, (R)- Amygdalin	$C_{20}H_{27}NO_{11}$ 457.43	29883-15-6 224.5	629	5-17-08-00118	H_2O 4; EtOH 2; eth 1; chl 1
955	Benzeneacetonitrile, α-(β-D-glucopyranosyloxy)- Mandelonitrile glucoside	$C_{14}H_{17}NO_6$ 295.29	138-53-4 122	5602		H_2O 3; EtOH 4
956	Benzeneacetonitrile, α-hydroxy-, (±)-	C_8H_7NO 133.15	613-88-7 22	170	3-10-00-00473 1.1165[20]	eth 4; EtOH 4; chl 4
957	Benzeneacetonitrile, 2-methoxy-	C_9H_9NO 147.18	7035-03-2 69.8	141[15]	4-10-00-00538	bz 4
958	Benzeneacetonitrile, 4-methoxy-	C_9H_9NO 147.18	104-47-2 286.5		4-10-00-00555 1.0845[20]	EtOH 3; eth 3; chl 3 1.5309[20]
959	Benzeneacetonitrile, α-methyl-	C_9H_9N 131.18	1823-91-2 231		4-09-00-01783 0.9854[20]	eth 4; EtOH 4 1.5095[25]
960	Benzeneacetonitrile, 2-methyl-	C_9H_9N 131.18	22364-68-7 244		4-09-00-01791 1.0156[22]	bz 4; eth 4; EtOH 4 1.5252[20]
961	Benzeneacetonitrile, 3-methyl-	C_9H_9N 131.18	2947-60-6 248; 133[15]		4-09-00-01792 1.0022[22]	H_2O 1; EtOH 3; eth 3; bz 3 1.5233[20]
962	Benzeneacetonitrile, 4-methyl-	C_9H_9N 131.18	2947-61-7 18	242.5	4-09-00-01796 0.992[25]	H_2O 1; EtOH 3; eth 3; bz 3 1.5190[20]
963	Benzeneacetonitrile, α-(1-methylethyl)-	$C_{11}H_{13}N$ 159.23	5558-29-2 247		4-09-00-01879 0.967[15]	bz 4; EtOH 4 1.5038[25]
964	Benzeneacetonitrile, α-(2-methylpropyl)- Valeronitrile, 4-methyl-2-phenyl-	$C_{12}H_{15}N$ 173.26	5558-31-6 264; 137[15]		4-09-00-01903 0.942[16]	bz 4; eth 4; EtOH 4
965	Benzeneacetonitrile, 2-nitro-	$C_8H_6N_2O_2$ 162.15	610-66-2 84	178[12]; 138[1]	3-09-00-02283	ace 4; bz 4; eth 4; EtOH 4
966	Benzeneacetonitrile, 3-nitro-	$C_8H_6N_2O_2$ 162.15	621-50-1 63	180[15]	4-09-00-01687	bz 4; eth 4; EtOH 4; chl 4
967	Benzeneacetonitrile, 4-nitro- p-Nitrobenzyl cyanide	$C_8H_6N_2O_2$ 162.15	555-21-5 117	6512 195-7[12]	4-09-00-01689	H_2O 2; EtOH 3; eth 3; bz 3
968	Benzeneacetonitrile, α-oxo-	C_8H_5NO 131.13	613-90-1 32.5	206	4-10-00-02743	H_2O 1; EtOH 4; eth 4; chl 2
969	Benzeneacetonitrile, α-phenyl-	$C_{14}H_{11}N$ 193.25	86-29-3 74.3	184[16]	4-09-00-02505	EtOH 3; eth 4; chl 3; lig 2
970	Benzeneacetonitrile, α-propyl-	$C_{11}H_{13}N$ 159.23	5558-78-1 255; 126[13]		4-09-00-01869 0.9425[20]	bz 4; EtOH 4 1.5000[20]
971	Benzeneacetyl chloride Phenylacetyl chloride	C_8H_7ClO 154.60	103-80-0 170[250]; 105[24]		4-09-00-01631 1.1682[20]	eth 4 1.5325[20]
972	Benzeneacetyl chloride, α-(acetyloxy)-	$C_{10}H_9ClO_3$ 212.63	1638-63-7 36	152[33]; 125[10]	1-10-00-00089	eth 4; bz 4; chl 4; lig 4
973	Benzeneacetyl chloride, α-chloro-	$C_8H_6Cl_2O$ 189.04	2912-62-1 120[23]		2-09-00-00308 1.196[25]	1.5440[20]
974	Benzeneacetyl chloride, α-ethyl-	$C_{10}H_{11}ClO$ 182.65	36854-57-6 110-2[10]		4-09-00-01817 1.5169[20]	ctc 3
975	Benzeneacetyl chloride, α-phenyl-	$C_{14}H_{11}ClO$ 230.69	1871-76-7 56.5	170[16]	3-09-00-03300	lig 3
976	Benzene, azido-	$C_6H_5N_3$ 119.13	622-37-7 -27.5	70[11]	4-05-00-00759 1.0860[20]	H_2O 1; EtOH 2; eth 2 1.5589[25]
977	Benzene, 1-azido-4-bromo-	$C_6H_4BrN_3$ 198.02	2101-88-4 20	105[10]; 69[2]	4-05-00-00761 1.5827[25]	H_2O 1; EtOH 2; eth 4; bz 4
978	Benzene, 1-azido-4-chloro-	$C_6H_4ClN_3$ 153.57	3296-05-7 20	96[20]	4-05-00-00760 1.2634[25]	H_2O 1; eth 3
979	Benzene, 1-azido-2-iodo-	$C_6H_4IN_3$ 245.02	54467-95-7 90-100[0.9]		0-05-00-00278 1.8893[25]	ace 4; EtOH 4 1.6631[25]
980	Benzene, (azidomethyl)-	$C_7H_7N_3$ 133.15	622-79-7 108[23]; 74[11]		4-05-00-00877 1.0655[25]	H_2O 1; EtOH 5; eth 5 1.5341[25]
981	Benzene, 1-azido-2-methyl-	$C_7H_7N_3$ 133.15	31656-92-5 <-10	90.5[30]; 76[12]	4-05-00-00876 1.0648[25]	eth 4
982	Benzene, 1-azido-4-methyl-	$C_7H_7N_3$ 133.15	2101-86-2 180 dec; 80[10]		4-05-00-00876 1.0527[23]	eth 4; EtOH 4
983	Benzene, 1,2-bis(bromomethyl)-	$C_8H_8Br_2$ 263.96	91-13-4 95	128-30[4.5]	4-05-00-00929 1.988[25]	H_2O 1; EtOH 3; eth 3; ctc 3
984	Benzene, 1,3-bis(bromomethyl)-	$C_8H_8Br_2$ 263.96	626-15-3 77	135-40[20]	4-05-00-00946 1.959[25]	H_2O 1; EtOH 3; eth 3; chl 3
985	Benzene, 1,4-bis(bromomethyl)-	$C_8H_8Br_2$ 263.96	623-24-5 144.5	245	4-05-00-00970 2.012[25]	H_2O 1; EtOH 4; eth 2; bz 3
986	Benzene, 1,2-bis(chloromethyl)-	$C_8H_8Cl_2$ 175.06	612-12-4 55	239.5	4-05-00-00927 1.393[25]	H_2O 1; EtOH 4; eth 4; ctc 3
987	Benzene, 1,3-bis(chloromethyl)-	$C_8H_8Cl_2$ 175.06	626-16-4 34.2	251.5	4-05-00-00944 1.302[20]	H_2O 1; EtOH 4; eth 4; chl 2
988	Benzene, 1,4-bis(chloromethyl)-	$C_8H_8Cl_2$ 175.06	623-25-6 100	245 dec; 135[16]	4-05-00-00967 1.417[25]	H_2O 1; EtOH 4; eth 4; ace 4
989	Benzene, 1,4-bis(dichloromethyl)-	$C_8H_8Cl_4$ 243.95	7398-82-5 9.5		3-05-00-00857 1.606[25]	H_2O 1; EtOH 3; eth 3; bz 3
990	Benzene, 1,4-bis(1,1-dimethylethyl)- Benzene, 1,4-di-tert-butyl	$C_{14}H_{22}$ 190.33	1012-72-2 79.5	238; 109[15]	4-05-00-01163 0.9850[20]	H_2O 1; EtOH 3; eth 3

No.	Name / Synonym	Mol. Form. / Mol. Wt.	CAS RN / mp/°C	Merck No. / bp/°C	Beil. Ref. / den/g cm^{-3}	Solubility / n_D
991	Benzene, 1,4-bis(2-hydroxybenzylideneamino)-2-nitro Ethionine	$C_{20}H_{15}N_3O_4$ 361.36	13073-35-3 195	3693		
992	Benzene, 1,2-bis(1-methylethyl)-	$C_{12}H_{18}$ 162.27	577-55-9 -57	204	4-05-00-01125 0.8701[20]	H_2O 1; EtOH 5; eth 5; ace 5 1.4960[20]
993	Benzene, 1,3-bis(1-methylethyl)-	$C_{12}H_{18}$ 162.27	99-62-7 -63.1	203.2	4-05-00-01125 0.8559[20]	H_2O 1; EtOH 5; eth 5; ace 5 1.4883[20]
994	Benzene, 1,4-bis(1-methylethyl)-	$C_{12}H_{18}$ 162.27	100-18-5 -17	210.3	4-05-00-01126 0.8568[20]	H_2O 1; EtOH 5; eth 5; ace 5 1.4898[20]
995	Benzene, 1,4-bis(phenylmethyl)-	$C_{20}H_{18}$ 258.36	793-23-7 87.5	225[18]	4-05-00-02512	EtOH 3; eth 2; bz 3; chl 3
996	Benzene, 1,4-bis(trichloromethyl)-	$C_8H_4Cl_6$ 312.84	68-36-0 109	1313	4-05-00-00968	chl 3
997	Benzene, 1,3-bis(trifluoromethyl)-	$C_8H_4F_6$ 214.11	402-31-3	116	4-05-00-00943 1.3790[25]	1.3916[25]
998	Benzene, bromo- Bromobenzene	C_6H_5Br 157.01	108-86-1 -30.6	1394 156.0	4-05-00-00670 1.4950[20]	H_2O 1; EtOH 4; eth 4; bz 4 1.5597[20]
999	Benzene, 1-bromo-2-(bromomethyl)-	$C_7H_6Br_2$ 249.93	3433-80-5 31	129[19]	4-05-00-00836	eth 4; EtOH 4; HOAc 4
1000	Benzene, 1-bromo-4-(bromomethyl)- p-Bromobenzyl bromide	$C_7H_6Br_2$ 249.93	589-15-1 63	1397	4-05-00-00836	H_2O 2; EtOH 3; eth 4; bz 3
1001	Benzene, (3-bromo-3-butenyl)- 1-Butene, 2-bromo-4-phenyl	$C_{10}H_{11}Br$ 211.10	62692-40-4	118[21]; 90[5]	3-05-00-01209 1.2907[20]	ace 4 1.5450[20]
1002	Benzene, (4-bromo-2-butenyl)- 2-Butene, 1-bromo-4-phenyl-	$C_{10}H_{11}Br$ 211.10	40734-75-6	126-30[10.5]	4-05-00-01377 1.2660[20]	eth 4 1.5678[20]
1003	Benzene, (4-bromobutoxy)-	$C_{10}H_{13}BrO$ 229.12	1200-03-9 41	154[18]	4-06-00-00558	EtOH 2; ctc 2
1004	Benzene, 1-bromo-2-chloro-	C_6H_4BrCl 191.45	694-80-4 -12.3	204	4-05-00-00680 1.637[25]	H_2O 1; bz 4; ctc 2 1.5809[20]
1005	Benzene, 1-bromo-3-chloro-	C_6H_4BrCl 191.45	108-37-2 -21.5	196	4-05-00-00680 1.6302[20]	H_2O 1; EtOH 4; eth 4 1.5771[20]
1006	Benzene, 1-bromo-4-chloro-	C_6H_4BrCl 191.45	106-39-8 68	196	4-05-00-00681 1.576[71]	H_2O 1; EtOH 2; eth 3; bz 3 1.5531[70]
1007	Benzene, 1-bromo-3-(chloromethyl)-	C_7H_6BrCl 205.48	932-77-4 32.5	119[18]	4-05-00-00832	EtOH 4
1008	Benzene, 1-bromo-4-(chloromethyl)- p-Bromobenzyl chloride	C_7H_6BrCl 205.48	589-17-3 50	1398 236	4-05-00-00832	H_2O 1; EtOH 4; eth 4; peth 3
1009	Benzene, 1-bromo-4-cyclohexyl-	$C_{12}H_{15}Br$ 239.16	25109-28-8	160[25]	3-05-00-01258 1.283[25]	bz 4; chl 4 1.5584[20]
1010	Benzene, 1-bromo-2,3-dichloro-	$C_6H_3BrCl_2$ 225.90	56961-77-4 60	245	0-05-00-00209	bz 4; eth 4; chl 4
1011	Benzene, 1-bromo-2,4-dichloro-	$C_6H_3BrCl_2$ 225.90	1193-72-2 26	235	3-05-00-00564	bz 4; eth 4; chl 4
1012	Benzene, 1-bromo-3,5-dichloro-	$C_6H_3BrCl_2$ 225.90	19752-55-7 83	232	2-05-00-00162	H_2O 1; EtOH 3; eth 3; bz 4
1013	Benzene, 2-bromo-1,3-dichloro-	$C_6H_3BrCl_2$ 225.90	19393-92-1 65	242	0-05-00-00210	bz 4; eth 4; chl 4
1014	Benzene, 1-bromo-1,4-dichloro-	$C_6H_3BrCl_2$ 225.90	1435-50-3 35	226; 119[20]	3-05-00-00564	bz 4; eth 4; EtOH 4; lig 4
1015	Benzene, 4-bromo-1,2-dichloro-	$C_6H_3BrCl_2$ 225.90	18282-59-2 25	237	3-05-00-00564	H_2O 1; EtOH 2; eth 4; bz 4
1016	Benzene, 4-bromo-1,2-dimethoxy-	$C_8H_9BrO_2$ 217.06	2859-78-1	254.5	4-06-00-05621 1.702[25]	1.5743[20]
1017	Benzene, 1-bromo-2,3-dimethyl-	C_8H_9Br 185.06	576-23-8	214	4-05-00-00928 1.365[25]	ace 4; bz 4; eth 4 1.5587[20]
1018	Benzene, 1-bromo-2,4-dimethyl-	C_8H_9Br 185.06	583-70-0 -17	205	4-05-00-00945 1.3419[20]	H_2O 1; EtOH 4; eth 4; ace 4 1.5501[20]
1019	Benzene, 1-bromo-3,5-dimethyl-	C_8H_9Br 185.06	556-96-7	204	4-05-00-00945 1.362[20]	eth 4; ace 3; bz 3 1.5462[22]
1020	Benzene, 2-bromo-1,3-dimethyl-	C_8H_9Br 185.06	576-22-7	203.5; 100[20]	4-05-00-00945	eth 4; ace 3; bz 3 1.5552[20]
1021	Benzene, 2-bromo-1,4-dimethyl-	C_8H_9Br 185.06	553-94-6 9	199; 88-9[13]	4-05-00-00969 1.3582[18]	H_2O 1; EtOH 4; bz 3 1.5514[18]
1022	Benzene, 4-bromo-1,2-dimethyl-	C_8H_9Br 185.06	583-71-1 -0.2	214.5	4-05-00-00928 1.3708[20]	H_2O 1; EtOH 4; eth 4 1.5530[20]
1023	Benzene, 1-bromo-4-(1,1-dimethylethyl)-	$C_{10}H_{13}Br$ 213.12	3972-65-4 19	231.5	4-05-00-01050 1.2286[20]	H_2O 1; eth 3; bz 3; chl 3 1.5436[20]
1024	Benzene, 1-bromo-2,3-dinitro- 1-Bromo-2,3-dinitrobenzene	$C_6H_3BrN_2O_4$ 247.00	19613-76-4 101.5	320	4-05-00-00749	EtOH 4
1025	Benzene, (1-bromoethenyl)-	C_8H_7Br 183.05	98-81-7 -44	86[14]; 71[3]	4-05-00-01349 1.4025[23]	1.5881[20]
1026	Benzene, 1-bromo-2-ethenyl-	C_8H_7Br 183.05	2039-88-5 -52.8	209.2; 98[20]	4-05-00-01349 1.4160[20]	1.5927[20]
1027	Benzene, 1-bromo-3-ethenyl-	C_8H_7Br 183.05	2039-86-3	90-4[20]	3-05-00-01176 1.4059[20]	1.5933[20]
1028	Benzene, 1-bromo-4-ethenyl-	C_8H_7Br 183.05	2039-82-9 7.7	212; 103[20]	4-05-00-01349 1.3984[20]	H_2O 1; chl 4; HOAc 3 1.5947[20]
1029	Benzene, (2-bromoethenyl)-, (E)-	C_8H_7Br 183.05	588-72-7 7	219 dec; 108[20]	4-05-00-01349 1.4269[16]	H_2O 1; EtOH 5; eth 5; chl 3 1.6093[20]

No.	Name Synonym	Mol. Form. Mol. Wt.	CAS RN mp/°C	Merck No. bp/°C	Beil. Ref. den/g cm^{-3}	Solubility n_D
1030	Benzene, 1,1'-(bromoethenylidene)bis-	$C_{14}H_{11}Br$ 259.15	13249-58-6 41.5	175[11.5]	4-05-00-02177	EtOH 2; eth 3; ace 3; CS_2 3
1031	Benzene, (2-bromoethenyl)-, (Z)-	C_8H_7Br 183.05	588-73-8 -7	55-6[2]	4-05-00-01349 1.4322[10]	1.5990[22]
1032	Benzene, 1-bromo-4-ethoxy-	C_8H_9BrO 201.06	588-96-5 4	231	4-06-00-01045 1.4071[25]	H_2O 1; EtOH 4; eth 4; chl 3 1.5517[20]
1033	Benzene, (2-bromoethoxy)-	C_8H_9BrO 201.06	589-10-6 39	240 dec; 128[20]	4-06-00-00556 1.3555[20]	H_2O 1; EtOH 4; eth 4
1034	Benzene, (1-bromoethyl)-, (±)-	C_8H_9Br 185.06	38661-81-3	202-3. 85[13]	4-05-00-00906 1.3605[20]	EtOH 3; eth 3; bz 3; ctc 2 1.5612[20]
1035	Benzene, 1-bromo-2-ethyl-	C_8H_9Br 185.06	1973-22-4 -67.9	199.3	4-05-00-00905 1.3548[20]	ace 4; bz 4; eth 4; EtOH 4 1.5472[20]
1036	Benzene, 1-bromo-3-ethyl-	C_8H_9Br 185.06	2725-82-8	202	4-05-00-00906 1.3493[20]	1.5465[20]
1037	Benzene, 1-bromo-4-ethyl-	C_8H_9Br 185.06	1585-07-5 -43.5	204	4-05-00-00906 1.3423[20]	ace 4; bz 4; eth 4; EtOH 4 1.5445[20]
1038	Benzene, (2-bromoethyl)-	C_8H_9Br 185.06	103-63-9	219	4-05-00-00907 1.3587[20]	H_2O 1; eth 3; bz 3; ctc 2 1.5372[20]
1039	Benzene, (1-bromoethyl)-, (R)-	C_8H_9Br 185.06	1459-14-9	203	4-05-00-00906 1.3108[23]	H_2O 1; EtOH 3; eth 3 1.5612[20]
1040	Benzene, 1-bromo-2-ethynyl- o-Bromophenylacetylene	C_8H_5Br 181.03	766-46-1	92-3[20]	3-05-00-01351 1.4434[25]	1.5962[25]
1041	Benzene, 1-bromo-3-ethynyl- m-Bromophenylacetylene	C_8H_5Br 181.03	766-81-4	85-6[16]	3-05-00-01351 1.4466[25]	1.5896[25]
1042	Benzene, 1-bromo-4-ethynyl-	C_8H_5Br 181.03	766-96-1 64.5	89[16]	4-05-00-01529	chl 3
1043	Benzene, 1-bromo-3-fluoro-	C_6H_4BrF 175.00	1073-06-9	150	4-05-00-00678 1.7081[20]	ctc 3 1.5257[20]
1044	Benzene, 1-bromo-4-fluoro-	C_6H_4BrF 175.00	460-00-4 -17.4	151.5	4-05-00-00678 1.593[15]	H_2O 1; EtOH 3; eth 3; chl 3 1.5310[15]
1045	Benzene, 1-bromo-4-(hexyloxy)-	$C_{12}H_{17}BrO$ 257.17	30752-19-3	156-7[13]	4-06-00-01047 1.2306[20]	1.5262[20]
1046	Benzene, 1-bromo-2-iodo-	C_6H_4BrI 282.91	583-55-1 9.5	257; 120[15]	4-05-00-00698 2.2570[25]	H_2O 1; EtOH 2; ace 3; HOAc 2 1.6618[25]
1047	Benzene, 1-bromo-3-iodo- m-Bromoiodobenzene	C_6H_4BrI 282.91	591-18-4 -9.3	252; 120[18]	4-05-00-00698	H_2O 1; EtOH 2; HOAc 2
1048	Benzene, 1-bromo-4-iodo-	C_6H_4BrI 282.91	589-87-7 92	252	4-05-00-00698	H_2O 1; EtOH 2; eth 3; chl 2
1049	Benzene, 1-bromo-4-isocyanato- p-Bromophenyl isocyanate	C_7H_4BrNO 198.02	2493-02-9 1418	226	1-12-00-00321	eth 4
1050	Benzene, 1-bromo-2-methoxy-	C_7H_7BrO 187.04	578-57-4 2.5	216	4-06-00-01037 1.5018[20]	H_2O 1; EtOH 4; eth 4 1.5727[20]
1051	Benzene, 1-bromo-3-methoxy-	C_7H_7BrO 187.04	2398-37-0	211; 105[16]	4-06-00-01043 1.5635[20]	H_2O 1; EtOH 3; eth 3; bz 3
1052	Benzene, 1-bromo-4-methoxy-	C_7H_7BrO 187.04	104-92-7 13.5	215	4-06-00-01044 1.4564[20]	H_2O 2; EtOH 4; eth 4; ctc 3 1.5642[20]
1053	Benzene, (bromomethyl)- Benzyl bromide	C_7H_7Br 171.04	100-39-0 -3	1142 201	4-05-00-00829 1.4380[25]	H_2O 1; EtOH 5; eth 5; ctc 3 1.5752[20]
1054	Benzene, 1-bromo-2-methyl- o-Bromotoluene	C_7H_7Br 171.04	95-46-5 -27.8	181.7	4-05-00-00825 1.4232[20]	H_2O 1; EtOH 4; eth 4; bz 4 1.5565[20]
1055	Benzene, 1-bromo-3-methyl- m-Bromotoluene	C_7H_7Br 171.04	591-17-3 -39.8	183.7	4-05-00-00827 1.4099[20]	H_2O 1; EtOH 3; eth 5; ace 3 1.5510[20]
1056	Benzene, 1-bromo-4-methyl- p-Bromotoluene	C_7H_7Br 171.04	106-38-7 28.5	1429 184.3	4-05-00-00827 1.3959[35]	H_2O 1; EtOH 3; eth 3; ace 3 1.5477[20]
1057	Benzene, 1-(bromomethyl)-3,5-dimethyl- Mesitylene, α-bromo-	$C_9H_{11}Br$ 199.09	27129-86-8 40	230; 118[22]	4-05-00-01027	bz 4; eth 4; EtOH 4; chl 4
1058	Benzene, 1,1'-(bromomethylene)bis- Bromodiphenylmethane	$C_{13}H_{11}Br$ 247.13	776-74-9 45	184[20]; 152[2]	4-05-00-01850	EtOH 3; bz 4; chl 3
1059	Benzene, 1-bromo-4-(1-methylethyl)-	$C_9H_{11}Br$ 199.09	586-61-8 -22.5	218.7	4-05-00-00994 1.3145[20]	H_2O 1; eth 3; bz 3; ctc 2 1.5569[20]
1060	Benzene, (2-bromo-1-methylethyl)-	$C_9H_{11}Br$ 199.09	1459-00-3 -59.3	188.5	4-05-00-00995 1.3020[20]	
1061	Benzene, 1,1',1''-(bromomethylidyne)tris-	$C_{19}H_{15}Br$ 323.23	596-43-0 153	230[15]	4-05-00-02500 1.5500[20]	
1062	Benzene, 1-(bromomethyl)-2-methyl-	C_8H_9Br 185.06	89-92-9 21	217; 108[16]	4-05-00-00928 1.3811[23]	H_2O 1; EtOH 3; eth 3; ace 3 1.5730[20]
1063	Benzene, 1-(bromomethyl)-3-methyl-	C_8H_9Br 185.06	620-13-3	212.5	4-05-00-00945 1.3711[23]	H_2O 1; EtOH 4; eth 4; os 3 1.5660[20]
1064	Benzene, 1-(bromomethyl)-4-methyl-	C_8H_9Br 185.06	104-81-4 35	220	4-05-00-00969 1.324[25]	H_2O 1; EtOH 3; eth 4; chl 4
1065	Benzene, 2-bromo-1-methyl-4-(1-methylethyl)-	$C_{10}H_{13}Br$ 213.12	2437-76-5	234.3	4-05-00-01063 1.2689[18]	eth 4; EtOH 4; chl 4 1.5360[20]
1066	Benzene, 1-bromo-2-methyl-4-nitro-	$C_7H_6BrNO_2$ 216.03	7149-70-4 78	141[17]	4-05-00-00860	eth 4
1067	Benzene, 1-(bromomethyl)-3-nitro-	$C_7H_6BrNO_2$ 216.03	3958-57-4 59.3	162[13]	4-05-00-00860	H_2O 1; EtOH 3

No.	Name Synonym	Mol. Form. Mol. Wt.	CAS RN mp/°C	Merck No. bp/°C	Beil. Ref. den/g cm^{-3}	Solubility n_D
1068	Benzene, 1-bromo-3-methyl-2-nitro- 3-Bromo-2-nitrotoluene	$C_7H_6BrNO_2$ 216.03	52414-97-8 28.5	153[30], 129[10]	4-05-00-00860	eth 4; EtOH 4
1069	Benzene, 1-bromo-3-methyl-5-nitro- Toluene, 3-bromo-5-nitro-	$C_7H_6BrNO_2$ 216.03	52488-28-5 83	269.5	3-05-00-00752	eth 4; EtOH 4
1070	Benzene, 1-bromo-2-methyl-3-nitro- Toluene, 2-bromo-6-nitro	$C_7H_6BrNO_2$ 216.03	55289-35-5 42	143[22]	3-05-00-00751	EtOH 4
1071	Benzene, 2-bromo-1-methyl-3-nitro-	$C_7H_6BrNO_2$ 216.03	41085-43-2 41.5	161[30]	4-05-00-00860	EtOH 4
1072	Benzene, 2-bromo-1-methyl-4-nitro-	$C_7H_6BrNO_2$ 216.03	7745-93-9 78	175[38], 150[20]	4-05-00-00861	eth 4
1073	Benzene, 2-bromo-4-methyl-1-nitro- 2-Bromo-4-methyl-1-nitrobenzene	$C_7H_6BrNO_2$ 216.03	40385-54-4 37.8	269; 154[20]	4-05-00-00861	eth 4; EtOH 4
1074	Benzene, 4-bromo-1-methyl-2-nitro- 4-Bromo-2-nitrotoluene	$C_7H_6BrNO_2$ 216.03	60956-26-5 47	256.5	4-05-00-00860	eth 4; EtOH 4
1075	Benzene, 1-bromo-2-nitro-	$C_6H_4BrNO_2$ 202.01	577-19-5 43	258	4-05-00-00728 1.6245[80]	H_2O 1; EtOH 4; eth 3; ace 3
1076	Benzene, 1-bromo-3-nitro-	$C_6H_4BrNO_2$ 202.01	585-79-5 56	265	4-05-00-00729 1.7036[20]	H_2O 2; EtOH 3; eth 3; bz 3 1.5979[20]
1077	Benzene, 1-bromo-4-nitro- p-Nitrobromobenzene	$C_6H_4BrNO_2$ 202.01	586-78-7 127	256	4-05-00-00729 1.948[25]	H_2O 1; EtOH 3; eth 3; bz 3
1078	Benzene, bromopentafluoro- Bromopentafluorobenzene	C_6BrF_5 246.96	344-04-7 -31	137	4-05-00-00680 1.981[25]	1.4490[20]
1079	Benzene, 1-bromo-4-phenoxy- 4-Bromophenyl phenyl ether	$C_{12}H_9BrO$ 249.11	101-55-3 18.72	124-8[3-4]	4-06-00-01047 1.6088[20]	H_2O 1; eth 3; ctc 3 1.6084[20]
1080	Benzene, 1-bromo-2-(2-phenylethenyl)-, (E)- trans-2-Bromostilbene	$C_{14}H_{11}Br$ 259.15	54737-45-0 34	143[0.2]	4-05-00-02164	H_2O I; EtOH 3; peth 3 1.6822[25]
1081	Benzene, 1-bromo-2-(2-phenylethenyl)-, (Z)- Stilbene, 2-bromo (cis)	$C_{14}H_{11}Br$ 259.15	4877-77-4	121[0.5]	4-05-00-02164	H_2O 1; EtOH 3; peth 3 1.6404[25]
1082	Benzene, 1-bromo-4-(1-propenyl)-	C_9H_9Br 197.07	4489-23-0 35	238.5	2-05-00-00372 1.332[36]	1.590[36]
1083	Benzene, 1-bromo-4-(2-propenyl)-	C_9H_9Br 197.07	2294-43-1	224; 96[12]	4-05-00-01363 1.324[15]	1.559[15]
1084	Benzene, (3-bromo-1-propenyl)-	C_9H_9Br 197.07	4392-24-9 34	130[10]	3-05-00-01188 1.3428[30]	EtOH 4 1.613[20]
1085	Benzene, (3-bromopropoxy)-	$C_9H_{11}BrO$ 215.09	588-63-6 10.7	127[18]	4-06-00-00557 1.364[16]	eth 4
1086	Benzene, (1-bromopropyl)-, (±)- Benzene, (1-bromopropyl) (DL)	$C_9H_{11}Br$ 199.09	63790-14-7	105[17]	2-05-00-00305 1.309[19]	1.5517[19]
1087	Benzene, (2-bromopropyl)-	$C_9H_{11}Br$ 199.09	2114-39-8	107-9[16]	4-05-00-00981 1.2908[16]	chl 4 1.5450[20]
1088	Benzene, (3-bromopropyl)-	$C_9H_{11}Br$ 199.09	637-59-2	219.5; 117[25]	4-05-00-00982 1.3106[25]	H_2O 1; eth 4 1.5440[25]
1089	Benzene, 1-bromo-2-(trifluoromethyl)-	$C_7H_4BrF_3$ 225.01	392-83-6	167.5	4-05-00-00831 1.652[25]	1.4817[20]
1090	Benzene, 1-bromo-3-(trifluoromethyl)-	$C_7H_4BrF_3$ 225.01	401-78-5 1	151.5	4-05-00-00831 1.613[25]	1.4716[20]
1091	Benzene, 1-bromo-4-(trifluoromethyl)-	$C_7H_4BrF_3$ 225.01	402-43-7	160	4-05-00-00831 1.607[25]	1.4705[25]
1092	Benzene, 1-bromo-2,3,5-trimethyl- 1-Bromo-2,3,5-trimethylbenzene	$C_9H_{11}Br$ 199.09	31053-99-3 <-15	238	4-05-00-01014	bz 4 1.5516[20]
1093	Benzene, 1-bromo-2,4,5-trimethyl-	$C_9H_{11}Br$ 199.09	5469-19-2 73	234	4-05-00-01014	H_2O 1; EtOH 4
1094	Benzene, 2-bromo-1,3,5-trimethyl-	$C_9H_{11}Br$ 199.09	576-83-0 -1	225	4-05-00-01027 1.3191[10]	H_2O 1; eth 4; bz 3; ctc 2 1.5510[20]
1095	Benzene, 1,1'-(1,3-butadiene-1,4-diyl)bis-, (E,E)-	$C_{16}H_{14}$ 206.29	538-81-8 154.3	352	4-05-00-02319	bz 4; eth 4; EtOH 4; peth 4
1096	Benzene, 1,1'-(1,3-butadiene-1,4-diyl)bis-, (Z,Z)-	$C_{16}H_{14}$ 206.29	5807-76-1 70.5		4-05-00-02318 0.9697[100]	H_2O 1; EtOH 2; eth 4; bz 4 1.6183[100]
1097	Benzene, 1,3-butadienyl-, (E)-	$C_{10}H_{10}$ 130.19	16939-57-4 4.5	76[11]	4-05-00-01536 0.9286[20]	H_2O I; EtOH 3; eth 3; ace 3 1.6089[25]
1098	Benzene, 1,3-butadienyl-, (Z)-	$C_{10}H_{10}$ 130.19	31915-94-3 -57	86[11]	0.9334[20]	1.6095[20]
1099	Benzene, 1,1'-(1,4-butanediyl)bis- 1,4-Diphenylbutane	$C_{16}H_{18}$ 210.32	1083-56-3 52.5	317	4-05-00-01937 0.9880[20]	H_2O 1; EtOH 3; eth 3; chl 3
1100	Benzenebutanoic acid	$C_{10}H_{12}O_2$ 164.20	1821-12-1 52	290	4-09-00-01811	H_2O 3; EtOH 3; eth 3
1101	Benzenebutanoic acid, 4-[bis(2-chloroethyl)amino]- Chlorambucil	$C_{14}H_{19}Cl_2NO_2$ 304.22	305-03-3 65	2064	4-14-00-01715	
1102	Benzenebutanoic acid, α,2-diamino-γ-oxo- Kynurenine	$C_{10}H_{12}N_2O_3$ 208.22	343-65-7 191 dec	5207	4-14-00-02562	
1103	Benzenebutanoic acid, 4-hydroxy-γ-(4-hydroxyphenyl)-γ-methyl- Diphenolic acid	$C_{17}H_{18}O_4$ 286.33	126-00-1 171.5	3312	4-10-00-01890	H_2O 4; ace 4; EtOH 4

No.	Name / Synonym	Mol. Form. / Mol. Wt.	CAS RN / mp/°C	Merck No. / bp/°C	Beil. Ref. / den/g cm^{-3}	Solubility / n_D
1104	Benzenebutanoic acid, γ-methyl-	$C_{11}H_{14}O_2$ 178.23	16433-43-5 13	210[85]; 165[12]	4-09-00-01866 1.0554[15]	eth 4; EtOH 4 1.5167[20]
1105	Benzenebutanol, 4-methoxy-	$C_{11}H_{16}O_2$ 180.25	52244-70-9 3.5	160[8]; 125[1.5]	4-06-00-06005	1.5267[20]
1106	Benzenebutanol, α-methyl-	$C_{11}H_{16}O$ 164.25	2344-71-0	134-5[16]	3-06-00-01954 0.9643[25]	1.518[19]
1107	Benzenebutanol, β-methyl-, (±)- 1-Butanol, 2-methyl-4-phenyl (DL)	$C_{11}H_{16}O$ 164.25	116783-11-0	135[11]	4-06-00-03380 0.9719[20]	ace 4; bz 4; eth 4; EtOH 4 1.5173[16]
1108	Benzenebutanol, β-phenyl-, (R)- 1-Butanol, 2,4-diphenyl-, (-)-	$C_{16}H_{18}O$ 226.32	17297-04-0 51.5	178[1]; 145[0.1]		chl 3 1.5686[25]
1109	Benzene, 1-butenyl-, (E)-	$C_{10}H_{12}$ 132.21	1005-64-7 -43.1	198.7	4-05-00-01374 0.9019[20]	H_2O 1; EtOH 3; eth 3; bz 3 1.5420[20]
1110	Benzene, 1-butenyl-, (Z)-	$C_{10}H_{12}$ 132.21	1560-09-4	189	3-05-00-01205 0.9065[25]	H_2O 1; EtOH 3; eth 3; bz 3 1.5390[20]
1111	Benzene, 2-butenyl-	$C_{10}H_{12}$ 132.21	1560-06-1	176	3-05-00-01207 0.8831[20]	1.5101[20]
1112	Benzene, 3-butenyl-	$C_{10}H_{12}$ 132.21	768-56-9 -70	177	4-05-00-01378 0.8831[20]	H_2O 1; eth 3; bz 3 1.5059[20]
1113	Benzene, 1,1'-(1-butenylidene)bis-	$C_{16}H_{16}$ 208.30	1726-14-3 -1.2	296	4-05-00-02209 0.9934[20]	1.5604[20]
1114	Benzene, butoxy-	$C_{10}H_{14}O$ 150.22	1126-79-0 -19.4	210	4-06-00-00558 0.9351[20]	eth 3; ace 3 1.4969[20]
1115	Benzene, (butoxymethyl)-	$C_{11}H_{16}O$ 164.25	588-67-0	223	4-06-00-02231 0.9227[20]	ace 4; eth 4; EtOH 4 1.4833[20]
1116	Benzene, 1-butoxy-3-methyl-	$C_{11}H_{16}O$ 164.25	23079-65-4	229.2	4-06-00-02040 0.9406[70]	eth 4 1.4970[20]
1117	Benzene, 1-butoxy-4-methyl-	$C_{11}H_{16}O$ 164.25	10519-06-9	229.5	3-06-00-01354 0.9205[25]	eth 3 1.4970[20]
1118	Benzene, butyl- Butylbenzene	$C_{10}H_{14}$ 134.22	104-51-8 -87.9	1549 183.3	4-05-00-01033 0.8601[20]	H_2O 1; EtOH 5; eth 5; ace 5 1.4898[20]
1119	Benzene, (1-butylhexadecyl)- Eicosane, 5-phenyl-	$C_{26}H_{46}$ 358.65	2400-04-6 30.2	197[1]	4-05-00-01223 0.8549[20]	1.4796[20]
1120	Benzene, 1,1'-butylidenebis- 1,1-Diphenylbutane	$C_{16}H_{18}$ 210.32	719-79-9 27	287	4-05-00-01944 0.9928[20]	H_2O 1; EtOH 3; eth 3; bz 3 1.5664[20]
1121	Benzene, 1-butyl-2-methyl-	$C_{11}H_{16}$ 148.25	1595-11-5	208	3-05-00-00999 0.8710[20]	ace 4; bz 4; eth 4 1.4960[20]
1122	Benzene, 1-butyl-3-methyl-	$C_{11}H_{16}$ 148.25	1595-04-6	205	3-05-00-00999 0.8590[20]	ace 4; bz 4; eth 4 1.4910[20]
1123	Benzene, 1-butyl-4-methyl-	$C_{11}H_{16}$ 148.25	1595-05-7 -85	207	4-05-00-01094 0.8586[20]	ace 4; bz 4; eth 4 1.4916[20]
1124	Benzene, 3-butynyl-	$C_{10}H_{10}$ 130.19	16520-62-0	190	4-05-00-01535 0.9258[20]	1.5208[20]
1125	Benzenecarbodithioic acid, 2-hydroxy- Dithiosalicylic acid	$C_7H_6OS_2$ 170.26	527-89-9 49	3381	2-10-00-00078	bz 4; eth 4; EtOH 4
1126	Benzenecarboperoxoic acid Perbenzoic acid	$C_7H_6O_3$ 138.12	93-59-4 42	7109 100[14]	4-09-00-00715	ace 4; bz 4; eth 4; EtOH 4
1127	Benzenecarboperoxoic acid, 3-chloro- 3-Chloroperoxybenzoic acid	$C_7H_5ClO_3$ 172.57	937-14-4 92-4 dec		4-09-00-00972	
1128	Benzenecarboperoxoic acid, 1,1-dimethylethyl ester Benzoyl tert-butyl peroxide	$C_{11}H_{14}O_3$ 194.23	614-45-9	75-6[0.2]	4-09-00-00715 1.021[25]	1.4990[20]
1129	Benzenecarbothioamide, N-phenyl-	$C_{13}H_{11}NS$ 213.30	636-04-4 102	dec	4-12-00-00423	H_2O 1; EtOH 4; eth 3; bz 3
1130	Benzenecarbothioic acid	C_7H_6OS 138.19	98-91-9 24	85-7[10]	4-09-00-01364 1.28[20]	ace 4; bz 4; eth 4; EtOH 4 1.6040[20]
1131	Benzene, chloro- Chlorobenzene	C_6H_5Cl 112.56	108-90-7 -45.2	2121 131.7	4-05-00-00640 1.1058[20]	H_2O 1; EtOH 5; eth 5; bz 4 1.5241[20]
1132	Benzene, 1-chloro-2-(chloromethyl)-	$C_7H_6Cl_2$ 161.03	611-19-8 -17	217	4-05-00-00816 1.2699[0]	H_2O 1; EtOH 2; eth 4; bz 4 1.5530[20]
1133	Benzene, 1-chloro-3-(chloromethyl)-	$C_7H_6Cl_2$ 161.03	620-20-2	216; 110[25]	4-05-00-00816 1.2695[15]	EtOH 4 1.5554[20]
1134	Benzene, 1-chloro-4-(chloromethyl)- 4-Chlorobenzyl chloride	$C_7H_6Cl_2$ 161.03	104-83-6 31	223	4-05-00-00816	ctc 2
1135	Benzene, 1-chloro-4-(chloromethyl)-2-nitro-	$C_7H_5Cl_2NO_2$ 206.03	57403-35-7	122[10]	4-05-00-00858 1.32[25]	1.5855[20]
1136	Benzene, 1-chloro-4-[(chloromethyl)thio]-	$C_7H_6Cl_2S$ 193.10	7205-90-5	128-9[12]	4-06-00-01593 1.346[25]	1.6055[20]
1137	Benzene, 1-chloro-2,3-dimethyl-	C_8H_9Cl 140.61	608-23-1	188	4-05-00-00926 1.053[20]	bz 4; chl 4 1.526[20]
1138	Benzene, 1-chloro-2,4-dimethyl-	C_8H_9Cl 140.61	95-66-9 -32.5	184	4-05-00-00943 1.0579[20]	H_2O 1; ace 3; bz 4 1.5230[25]
1139	Benzene, 2-chloro-1,3-dimethyl-	C_8H_9Cl 140.61	6781-98-2 -35	186	4-05-00-00943 1.053[20]	H_2O 1; ace 3; bz 4; chl 3 1.526[20]
1140	Benzene, 2-chloro-1,4-dimethyl-	C_8H_9Cl 140.61	95-72-7 1.6	187	4-05-00-00965 1.0589[15]	H_2O 1; ace 3; bz 4; ctc 3
1141	Benzene, 4-chloro-1,2-dimethyl-	C_8H_9Cl 140.61	615-60-1 -6	194	3-05-00-00816 1.0682[15]	H_2O 1; ace 3; bz 4; ctc 3
1142	Benzene, 1-chloro-4-(1,1-dimethylethyl)-	$C_{10}H_{13}Cl$ 168.67	3972-56-3	213	4-05-00-01048 1.0075[18]	1.5123[20]

No.	Name Synonym	Mol. Form. Mol. Wt.	CAS RN mp/°C	Merck No. bp/°C	Beil. Ref. den/g cm^{-3}	Solubility n_D
1143	Benzene, (2-chloro-1,1-dimethylethyl)- Neophyl chloride	$C_{10}H_{13}Cl$ 168.67	515-40-2	6375 223; 105[18]	4-05-00-01048 1.047[20]	ace 4; bz 4; eth 4; EtOH 4 1.5247[20]
1144	Benzene, 1-chloro-2,4-dinitro- 1-Chloro-2,4-dinitrobenzene	$C_6H_3ClN_2O_4$ 202.55	97-00-7 53	2135 315	4-05-00-00744 1.4982[75]	H_2O 1; EtOH 2; eth 3; bz 3 1.5857[60]
1145	Benzene, 2-chloro-1,3-dinitro- 2-Chloro-1,3-dinitrobenzene	$C_6H_3ClN_2O_4$ 202.55	606-21-3 88	2136 315	4-05-00-00744 1.6887[16]	H_2O 1; EtOH 3; eth 3; chl 2
1146	Benzene, 1,1'-(1-chloro-1,2-ethenediyl)bis-, (E)-	$C_{14}H_{11}Cl$ 214.69	948-98-1 53.5	322; 128[2]	4-05-00-02161	H_2O 1; EtOH 3; eth 3; bz 3
1147	Benzene, 1,1'-(1-chloro-1,2-ethenediyl)bis-, (Z)-	$C_{14}H_{11}Cl$ 214.69	948-99-2	160-2[12]	4-05-00-02162	H_2O 1; EtOH 3; eth 3; bz 3 1.6281[19]
1148	Benzene, (1-chloroethenyl)-	C_8H_7Cl 138.60	618-34-8 -23	199	4-05-00-01345 1.1016[18]	eth 4; EtOH 4 1.5612[20]
1149	Benzene, 1-chloro-2-ethenyl- o-Chlorostyrene	C_8H_7Cl 138.60	2039-87-4 -63.1	188.7	4-05-00-01345 1.1000[20]	EtOH 3; eth 3; ace 3; ctc 3 1.5649[20]
1150	Benzene, 1-chloro-3-ethenyl-	C_8H_7Cl 138.60	2039-85-2	62-3[6]	4-05-00-01345 1.1168[20]	H_2O 1; EtOH 3; eth 3 1.5625[20]
1151	Benzene, 1-chloro-4-ethenyl-	C_8H_7Cl 138.60	1073-67-2 15.9	192	4-05-00-01345 1.0868[20]	H_2O 1; EtOH 3; eth 3; ace 5 1.5660[20]
1152	Benzene, 1-(1-chloroethenyl)-2,4-dimethyl- Styrene, α-chloro-2,4-dimethyl	$C_{10}H_{11}Cl$ 166.65	74346-30-8	104-5[19]	2-05-00-00381 1.044[13]	1.5446[13]
1153	Benzene, (2-chloroethenyl)-, (E)-	C_8H_7Cl 138.60	4110-77-4	199	4-05-00-01346 1.1095[18]	H_2O 1; EtOH 3; eth 3; ace 3 1.5648[20]
1154	Benzene, 1,1',1''-(1-chloro-1-ethenyl-2-ylidene)tris[4-methoxy- Chlorotrianisene	$C_{23}H_{21}ClO_3$ 380.87	569-57-3 115	2173	4-06-00-07650	
1155	Benzene, (2-chloroethenyl)-, (Z)-	C_8H_7Cl 138.60	4604-28-8	59[3.5]	4-05-00-01346 1.1046[25]	1.5762[25]
1156	Benzene, 1-chloro-2-ethoxy-	C_8H_9ClO 156.61	614-72-2	210	4-06-00-00785 1.1288[15]	bz 4; eth 4; EtOH 4 1.5284[25]
1157	Benzene, 1-chloro-3-ethoxy- m-Chlorophenyl ethyl ether	C_8H_9ClO 156.61	2655-83-6	207	4-06-00-00811 1.1712[20]	H_2O 1; EtOH 4; eth 4; bz 4
1158	Benzene, 1-chloro-4-ethoxy-	C_8H_9ClO 156.61	622-61-7 21	213	4-06-00-00823 1.1254[20]	EtOH 3; eth 3; bz 4; ctc 2 1.5252[20]
1159	Benzene, (2-chloroethoxy)-	C_8H_9ClO 156.61	622-86-6 28	218.5	4-06-00-00556	H_2O 1; EtOH 4; eth 4; ace 4
1160	Benzene, (1-chloroethyl)-, (±)-	C_8H_9Cl 140.61	38661-82-4	81[17]	4-05-00-00898 1.062[20]	EtOH 3; eth 4; bz 3 1.5276[20]
1161	Benzene, 1-chloro-2-ethyl-	C_8H_9Cl 140.61	89-96-3 -82.7	178.4	4-05-00-00897 1.0569[20]	H_2O 1; ace 3; bz 3; ctc 3 1.5218[20]
1162	Benzene, 1-chloro-3-ethyl-	C_8H_9Cl 140.61	620-16-6 -55	183.8	4-05-00-00897 1.0529[20]	ace 4; bz 4; eth 4; EtOH 4 1.5195[20]
1163	Benzene, 1-chloro-4-ethyl-	C_8H_9Cl 140.61	622-98-0 -62.6	184.4	4-05-00-00897 1.0455[20]	H_2O 1; EtOH 5; eth 5; ace 5 1.5175[20]
1164	Benzene, (2-chloroethyl)-	C_8H_9Cl 140.61	622-24-2	197.5; 92[20]	4-05-00-00899 1.069[25]	H_2O 1; EtOH 3; eth 3; ace 3 1.5276[20]
1165	Benzene, [(2-chloroethyl)thio]-	C_8H_9ClS 172.68	5535-49-9	117-8[11]	4-06-00-01469 1.1769[25]	eth 4; chl 4 1.5828[20]
1166	Benzene, 1-chloro-2-fluoro-	C_6H_4ClF 130.55	348-51-6 -43	137.6	4-05-00-00652 1.2233[30]	H_2O 1; ace 3; bz 3 1.4918[30]
1167	Benzene, 1-chloro-3-fluoro-	C_6H_4ClF 130.55	625-98-9	127.6	4-05-00-00652 1.221[25]	
1168	Benzene, 1-chloro-4-fluoro-	C_6H_4ClF 130.55	352-33-0 -26.8	130	4-05-00-00652 1.4990[15]	H_2O 1; EtOH 3; eth 3; bz 3 1.4990[15]
1169	Benzene, 1-chloro-3-fluoro-2-methyl-	C_7H_6ClF 144.58	443-83-4	154	3-05-00-00691 1.191[25]	
1170	Benzene, 2-chloro-4-fluoro-1-methyl-	C_7H_6ClF 144.58	452-73-3	152.5	3-05-00-00691 1.1972[20]	1.5026[20]
1171	Benzene, 1-chloro-2-iodo-	C_6H_4ClI 238.45	615-41-8 0.7	234.5	4-05-00-00695 1.9515[25]	H_2O 1; ace 3; ctc 2 1.6331[25]
1172	Benzene, 1-chloro-4-iodo-	C_6H_4ClI 238.45	637-87-6 57	227	4-05-00-00696 1.886[27]	H_2O 1; EtOH 3; chl 2
1173	Benzene, 1-chloro-2-iodo-4-methyl- Toluene, 4-chloro-3-iodo-	C_7H_6ClI 252.48	2401-22-1 -10	249	3-05-00-00726 1.7940[15]	1.614[23]
1174	Benzene, 4-chloro-1-iodo-2-methyl- Toluene, 5-chloro-2-iodo	C_7H_6ClI 252.48	23399-70-4 -21	240	3-05-00-00726 1.702[17]	1.616[23]
1175	Benzene, 4-chloro-2-iodo-1-methyl-	C_7H_6ClI 252.48	33184-48-4 -25	242.5	3-05-00-00726 1.8358[15]	1.620[23]
1176	Benzene, 1-chloro-2-isocyanato-	C_7H_4ClNO 153.57	3320-83-0 30.5	200; 115[43]	4-12-00-01129	ctc 2
1177	Benzene, 1-chloro-4-isothiocyanato-	C_7H_4ClNS 169.63	2131-55-7 46	249.5	4-12-00-01214	H_2O 1; EtOH 3
1178	Benzene, 1-chloro-2-methoxy-	C_7H_7ClO 142.58	766-51-8 -26.8	198.5	4-06-00-00785 1.1911[20]	H_2O 1; EtOH 3; eth 3; chl 2 1.5480[20]
1179	Benzene, 1-chloro-3-methoxy-	C_7H_7ClO 142.58	2845-89-8	193.5	4-06-00-00811 1.1759[12]	H_2O 1; EtOH 3; eth 3 1.5365[20]
1180	Benzene, 1-chloro-4-methoxy-	C_7H_7ClO 142.58	623-12-1 <-18	197.5	4-06-00-00822 1.201[20]	H_2O 1; EtOH 4; eth 4; ctc 3 1.5390[20]
1181	Benzene, [(chloromethoxy)methyl]-	C_8H_9ClO 156.61	3587-60-8	103[13]	4-06-00-02252 1.1350[20]	1.5192[20]

No.	Name Synonym	Mol. Form. Mol. Wt.	CAS RN mp/°C	Merck No. bp/°C	Beil. Ref. den/g cm⁻³	Solubility n_D
1182	Benzene, (chloromethyl)- Benzyl chloride	C_7H_7Cl 126.59	100-44-7 -45	1143 179	4-05-00-00809 1.1004[20]	H_2O 1; EtOH 5; eth 5; ctc 2 1.5391[20]
1183	Benzene, 1-chloro-2-methyl- o-Chlorotoluene	C_7H_7Cl 126.59	95-49-8 -35.6	2172 159.0	4-05-00-00805 1.0825[20]	H_2O 1; EtOH 3; eth 5; ace 5 1.5268[20]
1184	Benzene, 1-chloro-3-methyl- m-Chlorotoluene	C_7H_7Cl 126.59	108-41-8 -47.8	2172 161.8	4-05-00-00806 1.075[20]	H_2O 1; EtOH 3; eth 5; bz 3 1.5214[19]
1185	Benzene, 1-chloro-4-methyl- p-Chlorotoluene	C_7H_7Cl 126.59	106-43-4 7.5	2172 162.4	4-05-00-00806 1.0697[20]	H_2O 1; EtOH 3; eth 5; ctc 3 1.5150[20]
1186	Benzene, 1-(chloromethyl)-2,4-dimethyl-	$C_9H_{11}Cl$ 154.64	824-55-5 215.5; 110[20]		4-05-00-01012 1.0580[19]	bz 4; eth 4; EtOH 4
1187	Benzene, 1,1'-(chloromethylene)bis-	$C_{13}H_{11}Cl$ 202.68	90-99-3 16	140[3]	4-05-00-01847 1.140[25]	chl 3 1.5951[20]
1188	Benzene, 1-(2-chloro-1-methylethenyl)-4-methyl- Styrene, β-chloro, α-4-dimethyl	$C_{10}H_{11}Cl$ 166.65	30926-60-4 107-8[10]		2-05-00-00381 1.058[20]	1.5549[23]
1189	Benzene, (1-chloro-1-methylethyl)-	$C_9H_{11}Cl$ 154.64	934-53-2 98[1]		4-05-00-00992 1.192[25]	1.5290[25]
1190	Benzene, 1-chloro-2-(1-methylethyl)-	$C_9H_{11}Cl$ 154.64	2077-13-6 -74.4	191.1	4-05-00-00992 1.0341[20]	ace 4; bz 4; eth 4; EtOH 4 1.5168[20]
1191	Benzene, 1-(chloromethyl)-4-ethyl-	$C_9H_{11}Cl$ 154.64	1467-05-6 95-6[15]		4-05-00-01004 1.5290[25]	bz 4; EtOH 4; chl 4
1192	Benzene, 1-chloro-4-(1-methylethyl)-	$C_9H_{11}Cl$ 154.64	2621-46-7 -12.3	198.3	4-05-00-00992 1.0208[20]	H_2O 1; EtOH 5; eth 5; ace 5 1.5117[20]
1193	Benzene, (2-chloro-1-methylethyl)-	$C_9H_{11}Cl$ 154.64	824-47-5 85[13]		4-05-00-00992 1.047[19]	1.5245[19]
1194	Benzene, 1-(chloromethyl)-2-fluoro-	C_7H_6ClF 144.58	345-35-7 86[40]		4-05-00-00813 1.216[25]	1.5150[20]
1195	Benzene, 1-(chloromethyl)-4-fluoro-	C_7H_6ClF 144.58	352-11-4 82[26]		4-05-00-00813 1.207[25]	1.5130
1196	Benzene, 1,1',1''-(chloromethylidyne)tris-	$C_{19}H_{15}Cl$ 278.78	76-83-5 113.5	310	4-05-00-02497	H_2O 1; EtOH 2; eth 4; ace 3
1197	Benzene, 1-(chloromethyl)-4-methoxy-	C_8H_9ClO 156.61	824-94-2 24.5	262.5	4-06-00-02137 1.261[20]	ace 4; bz 4; eth 4 1.580[20]
1198	Benzene, 1-(chloromethyl)-2-methyl-	C_8H_9Cl 140.61	552-45-4 198; 90[20]		4-05-00-00926 1.063[25]	eth 4; EtOH 4 1.5410[25]
1199	Benzene, 1-(chloromethyl)-3-methyl-	C_8H_9Cl 140.61	620-19-9 195.5		4-05-00-00943 1.064[20]	H_2O 1; EtOH 3; eth 3 1.5345[20]
1200	Benzene, 1-(chloromethyl)-4-methyl-	C_8H_9Cl 140.61	104-82-5 201; 90[20]		4-05-00-00966 1.0512[20]	H_2O 1; EtOH 3; eth 5 1.5380
1201	Benzene, 2-chloro-1-methyl-4-(1-methylethyl)-	$C_{10}H_{13}Cl$ 168.67	4395-79-3 216.5		4-05-00-01062 1.0104[25]	H_2O 1; ace 3; bz 3 1.5078[20]
1202	Benzene, 2-chloro-4-methyl-1-(1-methylethyl)-	$C_{10}H_{13}Cl$ 168.67	4395-80-6 215; 113[35]		3-05-00-00956 1.018[15]	1.5180[20]
1203	Benzene, 1-(chloromethyl)-2-nitro-	$C_7H_6ClNO_2$ 171.58	612-23-7 51		4-05-00-00854	H_2O 1; EtOH 3; eth 3; ace 4 1.5557[62]
1204	Benzene, 1-chloro-2-methyl-3-nitro-	$C_7H_6ClNO_2$ 171.58	83-42-1 37.8	238	4-05-00-00854	H_2O 1; EtOH 3 1.5377[69]
1205	Benzene, 1-chloro-2-methyl-4-nitro-	$C_7H_6ClNO_2$ 171.58	13290-74-9 42.5	249	4-05-00-00855	eth 4
1206	Benzene, 1-(chloromethyl)-3-nitro-	$C_7H_6ClNO_2$ 171.58	619-23-8 46	173[34]	4-05-00-00855	ace 4; bz 4; eth 4; EtOH 4 1.5577[62]
1207	Benzene, 1-chloro-3-methyl-5-nitro- 3-Chloro-5-nitrotoluene	$C_7H_6ClNO_2$ 171.58	16582-38-0 61		4-05-00-00855	EtOH 4 1.5404[69]
1208	Benzene, 1-(chloromethyl)-4-nitro- 4-Nitrobenzyl chloride	$C_7H_6ClNO_2$ 171.58	100-14-1 71		4-05-00-00856	H_2O 1; EtOH 3; eth 3; ace 4 1.5647[62]
1209	Benzene, 1-chloro-4-methyl-2-nitro-	$C_7H_6ClNO_2$ 171.58	89-60-1 7	261; 118[11]	4-05-00-00855	H_2O 1; ctc 3 1.5572[20]
1210	Benzene, 2-chloro-1-methyl-4-nitro-	$C_7H_6ClNO_2$ 171.58	121-86-8 66.5	260	4-05-00-00855	H_2O 2; EtOH 3; eth 3; chl 2 1.5470[69]
1211	Benzene, 2-chloro-4-methyl-1-nitro-	$C_7H_6ClNO_2$ 171.58	38939-88-7 24	146[15]	4-05-00-00856	H_2O 1; ctc 3
1212	Benzene, 4-chloro-1-methyl-2-nitro-	$C_7H_6ClNO_2$ 171.58	89-59-8 38	242; 115.5[11]	4-05-00-00853 1.2559[80]	H_2O 1; EtOH 3; eth 3; chl 2
1213	Benzene, 4-chloro-2-methyl-1-nitro-	$C_7H_6ClNO_2$ 171.58	5367-28-2 25.5		4-05-00-00854	H_2O 1; ctc 3 1.5496[65]
1214	Benzene, 1-chloro-2-nitro- o-Chloronitrobenzene	$C_6H_4ClNO_2$ 157.56	88-73-3 32.5	2151 245.5	4-05-00-00721 1.368[242]	H_2O 1; EtOH 3; eth 3; ace 4
1215	Benzene, 1-chloro-3-nitro- m-Chloronitrobenzene	$C_6H_4ClNO_2$ 157.56	121-73-3 44.4	2151 235.5	4-05-00-00722 1.343[50]	H_2O 1; EtOH 3; eth 3; bz 3 1.5374[80]
1216	Benzene, 1-chloro-4-nitro- p-Chloronitrobenzene	$C_6H_4ClNO_2$ 157.56	100-00-5 83.5	2151 242	4-05-00-00723 1.2979[90]	H_2O l; EtOH 2; eth 3; chl 3 1.5376[100]
1217	Benzene, 1-chloro-2-nitro-4-(trifluoromethyl)- 2-Nitro-4-(trifluoromethyl)-1-chlorobenzene	$C_7H_3ClF_3NO_2$ 225.55	121-17-5 -2.5	222; 95[10]	4-05-00-00857 1.511[25]	1.4893[20]
1218	Benzene, 1-chloro-4-nitro-2-(trifluoromethyl)- 4-Chloro-3-(trifluoromethyl)nitrobenzene	$C_7H_3ClF_3NO_2$ 225.55	777-37-7 22	232	4-05-00-00857 1.527[25]	1.5083[26]
1219	Benzene, 4-chloro-1-nitro-2-(trifluoromethyl)- 4-Chloro-2-(trifluoromethyl)nitrobenzene	$C_7H_3ClF_3NO_2$ 225.55	118-83-2		3-05-00-00748 1.526[25]	1.4980[20]
1220	Benzene, chloropentafluoro- Chloropentafluorobenzene	C_6ClF_5 202.51	344-07-0 117.9		4-05-00-00654 1.568[25]	1.4256[20]

No.	Name / Synonym	Mol. Form. / Mol. Wt.	CAS RN / mp/°C	Merck No. / bp/°C	Beil. Ref. / den/g cm^{-3}	Solubility / n_D
1221	Benzene, 1-chloro-4-phenoxy- 4-Chlorophenyl phenyl ether	$C_{12}H_9ClO$ 204.66	7005-72-3	284.5	4-06-00-00826 1.2026[15]	1.599
1222	Benzene, 1-chloro-2-(2-phenylethenyl)-, (E)-	$C_{14}H_{11}Cl$ 214.69	1657-52-9 39.5	209[30]; 139[2]	4-05-00-02161	H_2O 1; EtOH 3; chl 3
1223	Benzene, 1-chloro-3-(2-phenylethenyl)-, (E)-	$C_{14}H_{11}Cl$ 214.69	14064-43-8 73.5	177[0.2]	4-05-00-02161	H_2O 1; EtOH 3; chl 3
1224	Benzene, 1-chloro-4-(phenylmethyl)-	$C_{13}H_{11}Cl$ 202.68	831-81-2 7.5	299; 147-8[8]	4-05-00-01847 1.1247[20]	ace 4
1225	Benzene, 1-chloro-4-(phenylsulfonyl)- Sulphenone	$C_{12}H_9ClO_2S$ 252.72	80-00-2 94	8969	4-06-00-01587	H_2O 1; EtOH 2; eth 3; ace 4
1226	Benzene, (1-chloro-1-propenyl)-	C_9H_9Cl 152.62	35673-03-1	90.5[0.9]	2-05-00-00372 1.085[20]	ace 4; bz 4 1.5635[15]
1227	Benzene, (2-chloro-1-propenyl)-	C_9H_9Cl 152.62	13099-50-8	120[20]; 61[1]	4-05-00-01186 1.0738[19]	ace 4; bz 4; chl 4 1.5565[19]
1228	Benzene, (3-chloro-2-propenyl)-	C_9H_9Cl 152.62	6268-37-7	211	4-05-00-01363 1.073[13]	H_2O 1; eth 3; ace 3; bz 3 1.545[14]
1229	Benzene, 1-chloro-3-(1-propenyl)-, (E)-	C_9H_9Cl 152.62	23204-80-0 8.5	106.7[13]	1.0926[20]	H_2O 1; EtOH 5; eth 5; ace 4 1.5851[20]
1230	Benzene, (3-chloro-1-propenyl)-, (E)-	C_9H_9Cl 152.62	21087-29-6 8.5	106.7[13]	4-05-00-01360 1.0926[20]	ace 4; bz 4; eth 4; EtOH 4 1.5851[20]
1231	Benzene, 1-(3-chloro-2-propenyl)-4-methoxy- Benzene, 1-(3-chloroallyl)-4-methoxy	$C_{10}H_{11}ClO$ 182.65	54644-23-4	120[11]	3-06-00-02416 1.155[8]	1.553[8]
1232	Benzene, 1-chloro-4-(2-propenyloxy)-	C_9H_9ClO 168.62	13997-70-1	106.7[12]	4-06-00-00825 1.131[15]	bz 4; eth 4; EtOH 4 1.5348[25]
1233	Benzene, (3-chloropropoxy)-	$C_9H_{11}ClO$ 170.64	3384-04-1 12	250	4-06-00-00557 1.1167[20]	ace 4; eth 4 1.5235[25]
1234	Benzene, (2-chloropropyl)-, (R)- R-(-)-2-Chloro-1-phenylpropane	$C_9H_{11}Cl$ 154.64	55449-46-2	94[17]	4-05-00-00980 1.038[19]	H_2O 1; ace 3; bz 3; chl 3 1.5198[22]
1235	Benzene, (3-chloropropyl)-	$C_9H_{11}Cl$ 154.64	104-52-9	219.5	4-05-00-00980 1.056[21]	ctc 2 1.5160[25]
1236	Benzene, [(3-chloropropyl)thio]-	$C_9H_{11}ClS$ 186.71	4911-65-3	116-7[4]	3-06-00-00981 1.1536[20]	ace 4; py 4 1.5752[20]
1237	Benzene, 1-chloro-2-(trichloromethyl)- 1-Chloro-2-(trichloromethyl)benzene	$C_7H_4Cl_4$ 229.92	2136-89-2 29.4	264.3	4-05-00-00823 1.5187[20]	H_2O 1; eth 3; ace 3; ctc 2 1.5836[20]
1238	Benzene, 1-chloro-3-(trichloromethyl)- 1-Chloro-3-(trichloromethyl)benzene	$C_7H_4Cl_4$ 229.92	2136-81-4 0.6	255	4-05-00-00823 1.495[14]	H_2O 1; eth 3; ace 3; ctc 2 1.4461[20]
1239	Benzene, 1-chloro-4-(trichloromethyl)- 1-Chloro-4-(trichloromethyl)benzene	$C_7H_4Cl_4$ 229.92	5216-25-1	245	4-05-00-00823 1.4463[20]	ace 4; eth 4
1240	Benzene, 1-chloro-2-(trifluoromethyl)-	$C_7H_4ClF_3$ 180.56	88-16-4 -6	152.2	4-05-00-00814 1.2540[30]	chl 3 1.4513[25]
1241	Benzene, 1-chloro-3-(trifluoromethyl)- Benzotrifluoride, 3-chloro	$C_7H_4ClF_3$ 180.56	98-15-7 -56	137.5	4-05-00-00814 1.3311[25]	1.4438[25]
1242	Benzene, 1-chloro-4-(trifluoromethyl)-	$C_7H_4ClF_3$ 180.56	98-56-6 -33	138.5	4-05-00-00815 1.3340[25]	1.4431[30]
1243	Benzene, 2-chloro-1,3,5-trimethyl-	$C_9H_{11}Cl$ 154.64	1667-04-5 <-20	205	4-05-00-01026 1.0337[30]	eth 4; EtOH 4 1.5212[30]
1244	Benzene, 2-chloro-1,3,5-trinitro- Picryl chloride	$C_6H_2ClN_3O_6$ 247.55	88-88-0 83	7390	4-05-00-00757 1.797[20]	H_2O 1; EtOH 3; eth 2; ace 4
1245	Benzene, cyclobutyl-	$C_{10}H_{12}$ 132.21	4392-30-7	190[155]; 101[41]	4-05-00-01387 0.9378[20]	1.5277[20]
1246	Benzene, 1-cyclohexen-1-yl-	$C_{12}H_{14}$ 158.24	771-98-2 -11	252	4-05-00-01557 0.9939[20]	MeOH 4 1.5718[20]
1247	Benzene, 2-cyclohexen-1-yl-	$C_{12}H_{14}$ 158.24	15232-96-9	115-7[16]	4-05-00-01558 0.9800[20]	1.5530[20]
1248	Benzene, cyclohexyl- Cyclohexylbenzene	$C_{12}H_{16}$ 160.26	827-52-1 7.3	240.1	4-05-00-01424 0.9427[20]	H_2O 1; EtOH 4; eth 3; ctc 2 1.5329[20]
1249	Benzene, (1-cyclohexylethyl)-	$C_{14}H_{20}$ 188.31	4413-16-5	265	4-05-00-01454 0.9773[17]	1.549[17]
1250	Benzene, 1-cyclohexyl-2-methoxy-	$C_{13}H_{18}O$ 190.29	2206-48-6	268	4-06-00-03899 1.007[18]	1.5365[18]
1251	Benzene, 1-cyclohexyl-2-nitro- Cyclohexane, 2-nitrophenyl	$C_{12}H_{15}NO_2$ 205.26	7137-56-6 45	174[16]; 113[0.5]	3-05-00-01259 1.111[23]	1.5472[25]
1252	Benzene, 1-cyclohexyl-4-nitro-	$C_{12}H_{15}NO_2$ 205.26	5458-48-0 58.5	210[25]	4-05-00-01426	bz 4; eth 4
1253	Benzene, (cyclohexyloxy)-	$C_{12}H_{16}O$ 176.26	2206-38-4	261	4-06-00-00565 1.0077[20]	H_2O 1; ace 3; bz 3 1.520[22]
1254	Benzene, cyclopentyl-	$C_{11}H_{14}$ 146.23	700-88-9	219	4-05-00-01409 0.9462[20]	eth 4 1.5280[20]
1255	Benzene, (3-cyclopentylpropyl)-	$C_{14}H_{20}$ 188.31	2883-12-7	272	4-05-00-01455 0.9233[19]	1.5130[19]
1256	Benzene, cyclopropyl- Cyclopropylbenzene	C_9H_{10} 118.18	873-49-4 -31	173.6; 80[37]	4-05-00-01370 0.9317[20]	H_2O 1; eth 3; ace 3; chl 3 1.5285[20]
1257	Benzenedecanoic acid, ι-octyl-	$C_{24}H_{40}O_2$ 360.58	1938-17-6 40.5	199-205[0.09]	2-09-00-00376 0.9310[25]	1.4894[20]
1258	Benzene, decyl- Decylbenzene	$C_{16}H_{26}$ 218.38	104-72-3 -14.4	298	4-05-00-01193 0.8555[20]	ace 4; bz 4; eth 4; EtOH 4 1.4832[20]
1259	1,4-Benzenediacetic acid, dimethyl ester	$C_{12}H_{14}O_4$ 222.24	36076-25-2 51.5	189[15]	2-09-00-00624	EtOH 3; eth 3

No.	Name / Synonym	Mol. Form. / Mol. Wt.	CAS RN / mp/°C	Merck No. / bp/°C	Beil. Ref. / den/g cm^{-3}	Solubility / n_D
1260	1,2-Benzenediamine / o-Phenylenediamine	C$_6$H$_8$N$_2$ / 108.14	95-54-5 / 102.5	7255 / 257	4-13-00-00038	H$_2$O 3; EtOH 4; eth 3; bz 3
1261	1,3-Benzenediamine / m-Phenylenediamine	C$_6$H$_8$N$_2$ / 108.14	108-45-2 / 63.5	7254 / 285	4-13-00-00079 / 1.0096[58]	H$_2$O 4; EtOH 3; eth 3; bz 3 / 1.6339[58]
1262	1,4-Benzenediamine / p-Phenylenediamine	C$_6$H$_8$N$_2$ / 108.14	106-50-3 / 146	7256 / 267	4-13-00-00104	H$_2$O 2; EtOH 3; eth 3; bz 3
1263	1,4-Benzenediamine, N-(4-aminophenyl)- / 4,4'-Diaminodiphenylamine	C$_{12}$H$_{13}$N$_3$ / 199.26	537-65-5 / 158	2957 / dec	4-13-00-00203	eth 4; EtOH 4
1264	1,2-Benzenediamine, 4-chloro-	C$_6$H$_7$ClN$_2$ / 142.59	95-83-0 / 76		4-13-00-00068	H$_2$O 2; EtOH 4; eth 4; bz 3
1265	1,3-Benzenediamine, N,N-diethyl-	C$_{10}$H$_{16}$N$_2$ / 164.25	26513-20-2 / 152	277	2-13-00-00026	EtOH 4
1266	1,4-Benzenediamine, N,N-diethyl- / N,N-Diethyl-p-phenylenediamine	C$_{10}$H$_{16}$N$_2$ / 164.25	93-05-0	261	4-13-00-00109	bz 4
1267	1,2-Benzenediamine, N,N-dimethyl-	C$_8$H$_{12}$N$_2$ / 136.20	2836-03-5	218; 90[22]	4-13-00-00042 / 0.995[22]	H$_2$O 2; EtOH 4; eth; ace 4
1268	1,3-Benzenediamine, N,N-dimethyl-	C$_8$H$_{12}$N$_2$ / 136.20	2836-04-6 / <-20	270; 138[10]	4-13-00-00080 / 0.995[25]	H$_2$O 2; EtOH 4; eth 4
1269	1,2-Benzenediamine, N,N'-dimethyl- / o-Phenylenediamine, N,N'-dimethyl	C$_8$H$_{12}$N$_2$ / 136.20	3213-79-4 / 34.5	247	4-13-00-00042	1.5914[25]
1270	1,4-Benzenediamine, N,N-dimethyl- / Dimethyl-p-phenylenediamine	C$_8$H$_{12}$N$_2$ / 136.20	99-98-9 / 53	3242 / 263	4-13-00-00106 / 1.036[20]	H$_2$O 3; EtOH 4; eth 4; bz 4
1271	1,4-Benzenediamine, N,N'-diphenyl- / N,N'-Diphenyl-p-phenylenediamine	C$_{18}$H$_{16}$N$_2$ / 260.34	74-31-7 / 144	3331	4-13-00-00116	EtOH 2; eth 2; bz 2; chl 2
1272	1,2-Benzenediamine, 4-ethoxy-	C$_8$H$_{12}$N$_2$O / 152.20	1197-37-1 / 71.5	295	3-13-00-01362	H$_2$O 4; EtOH 3; eth; chl 3
1273	1,2-Benzenediamine, 4-methoxy-	C$_7$H$_{10}$N$_2$O / 138.17	102-51-2 / 51	200[21]; 168[11]	4-13-00-01448	eth 4
1274	1,3-Benzenediamine, 4-methoxy- / 4-Methoxy-m-phenylenediamine	C$_7$H$_{10}$N$_2$O / 138.17	615-05-4 / 67.5	5920	4-13-00-01425	EtOH 3; eth 3; DMSO-d$_6$ 2
1275	1,2-Benzenediamine, 3-methyl-	C$_7$H$_{10}$N$_2$ / 122.17	2687-25-4 / 63.5	255	3-13-00-00277	ace 4; bz 4; EtOH 4
1276	1,2-Benzenediamine, 4-methyl- / Toluene-3,4-diamine	C$_7$H$_{10}$N$_2$ / 122.17	496-72-0 / 89.5	265	4-13-00-00260	H$_2$O 4; lig 3
1277	1,3-Benzenediamine, 2-methyl- / 2,6-Diaminotoluene	C$_7$H$_{10}$N$_2$ / 122.17	823-40-5 / 106		3-13-00-00291	H$_2$O 3; EtOH 3; bz 3
1278	1,3-Benzenediamine, 4-methyl- / Toluene-2,4-diamine	C$_7$H$_{10}$N$_2$ / 122.17	95-80-7 / 99	292	4-13-00-00235	H$_2$O 4; EtOH 4; eth 4; bz 4
1279	1,4-Benzenediamine, 2-methyl-	C$_7$H$_{10}$N$_2$ / 122.17	95-70-5 / 64	273.5	4-13-00-00246	H$_2$O 3; EtOH 3; eth 3; bz 2
1280	1,4-Benzenediamine, N-methyl-	C$_7$H$_{10}$N$_2$ / 122.17	623-09-6 / 36	258	4-13-00-00106	H$_2$O 4; EtOH 4; eth; ace 4
1281	1,2-Benzenediamine, 4-nitro- / 4-Nitro-o-phenylenediamine	C$_6$H$_7$N$_3$O$_2$ / 153.14	99-56-9 / 199.5	6544	4-13-00-00075	acid 3
1282	1,2-Benzenediamine, N-phenyl-	C$_{12}$H$_{12}$N$_2$ / 184.24	534-85-0 / 79.5	313	4-13-00-00043	H$_2$O 2; ace 3; bz 3; chl 3
1283	1,4-Benzenediamine, N-phenyl- / p-Aminodiphenylamine	C$_{12}$H$_{12}$N$_2$ / 184.24	101-54-2 / 66	354	4-13-00-00113	H$_2$O 2; EtOH 4; eth 3; chl 2
1284	1,3-Benzenediamine, 4-(phenylazo)-, monohydrochloride / Chrysoidine hydrochloride	C$_{12}$H$_{13}$ClN$_4$ / 248.71	532-82-1 / 118.5	2262	2-16-00-00203	ace 4
1285	1,3-Benzenediamine, N,N,N',N'-tetraethyl- / N,N,N',N'-Tetraethyl-m-phenylenediamine	C$_{14}$H$_{24}$N$_2$ / 220.36	64287-26-9	148[9]	2-13-00-00026 / 0.9522[12]	eth 4; EtOH 4 / 1.5537[12]
1286	1,2-Benzenediamine, N,N,N',N'-tetramethyl-	C$_{10}$H$_{16}$N$_2$ / 164.25	704-01-8 / 8.9	215.5	4-13-00-00042 / 0.9560[20]	
1287	1,4-Benzenediamine, N,N,N',N'-tetramethyl- / Tetramethyl-p-phenylenediamine	C$_{10}$H$_{16}$N$_2$ / 164.25	100-22-1 / 51	9159 / 260	4-13-00-00107	H$_2$O 2; EtOH 4; eth 4; bz 4
1288	Benzene, 1,2-dibromo- / o-Dibromobenzene	C$_6$H$_4$Br$_2$ / 235.91	583-53-9 / 7.1	225	4-05-00-00682 / 1.9843[20]	H$_2$O 1; EtOH 3; eth 5; ace 5 / 1.6155[20]
1289	Benzene, 1,3-dibromo- / m-Dibromobenzene	C$_6$H$_4$Br$_2$ / 235.91	108-36-1 / -7	218	4-05-00-00682 / 1.9523[20]	H$_2$O 1; EtOH 3; eth 5 / 1.6083[17]
1290	Benzene, 1,4-dibromo- / p-Dibromobenzene	C$_6$H$_4$Br$_2$ / 235.91	106-37-6 / 87.3	3001 / 218.5	4-05-00-00683 / 2.261[17]	H$_2$O 1; EtOH 3; eth 4; ace 4 / 1.5742
1291	Benzene, 1,3-dibromo-5-(bromomethyl)- / 3,5-Dibromobenzyl bromide	C$_7$H$_5$Br$_3$ / 328.83	56908-88-4 / 96	173[19]	3-05-00-00719	EtOH 4
1292	Benzene, 1,2-dibromo-3-chloro- / 1,2-Dibromo-3-chlorobenzene	C$_6$H$_3$Br$_2$Cl / 270.35	104514-49-0 / 73.5	264	1-05-00-00117	bz 4; eth 4; chl 4
1293	Benzene, 1,2-dibromo-4-chloro- / 1,2-Dibromo-4-chlorobenzene	C$_6$H$_3$Br$_2$Cl / 270.35	60956-24-3 / 35.5	256	0-05-00-00212	bz 4; eth 4; chl 4
1294	Benzene, 1,3-dibromo-2-chloro- / 1,3-Dibromo-2-chlorobenzene	C$_6$H$_3$Br$_2$Cl / 270.35	19230-27-4 / 72	265	3-05-00-00569	bz 4; eth 4; chl 4
1295	Benzene, 1,4-dibromo-2-chloro- / 1,4-Dibromo-2-chlorobenzene	C$_6$H$_3$Br$_2$Cl / 270.35	3460-24-0 / 40.5	259	0-05-00-00212	bz 4; eth 4; chl 4
1296	Benzene, 2,4-dibromo-1-chloro-	C$_6$H$_3$Br$_2$Cl / 270.35	29604-75-9 / 27	258	0-05-00-00212	bz 4; eth 4; chl 4
1297	Benzene, 1,2-dibromo-3,4-dimethyl- / 3,4-Dibromo-o-xylene	C$_8$H$_8$Br$_2$ / 263.96	24932-49-8 / 7	277	2-05-00-00285 / 1.782[15]	EtOH 4
1298	Benzene, 1,5-dibromo-2,4-dimethyl-	C$_8$H$_8$Br$_2$ / 263.96	615-87-2 / 70	255.5	3-05-00-00839	EtOH 4

No.	Name / Synonym	Mol. Form. / Mol. Wt.	CAS RN / mp/°C	Merck No. / bp/°C	Beil. Ref. / den/g cm⁻³	Solubility / n_D
1299	Benzene, (1,2-dibromoethyl)-	$C_8H_8Br_2$ 263.96	93-52-7 75	133[19]	4-05-00-00909	EtOH 3; eth 3; bz 3; chl 3
1300	Benzene, 2,4-dibromo-1-fluoro-	$C_6H_3Br_2F$ 253.90	1435-53-6 69	103-5[32]	3-05-00-00569 2.047[20]	1.5840[20]
1301	Benzene, (dibromomethyl)-	$C_7H_6Br_2$ 249.93	618-31-5	156[23]	4-05-00-00836 1.8365[28]	H_2O 1; EtOH 5; eth 5 1.6147[20]
1302	Benzene, 1,2-dibromo-3-methyl- 2,3-Dibromotoluene	$C_7H_6Br_2$ 249.93	61563-25-5 30.5	126-8[19]	4-05-00-00835 1.823[15]	1.5984[25]
1303	Benzene, 1,2-dibromo-4-methyl- 3,4-Dibromotoluene	$C_7H_6Br_2$ 249.93	60956-23-2 -10	240; 123[19]	4-05-00-00836 1.819[25]	1.5979[25]
1304	Benzene, 1,3-dibromo-2-methyl- 2,6-Dibromotoluene	$C_7H_6Br_2$ 249.93	69321-60-4 5.5	246	4-05-00-00836 1.812[22]	
1305	Benzene, 1,4-dibromo-2-methyl- 2,5-Dibromotoluene	$C_7H_6Br_2$ 249.93	615-59-8 5.6	236	4-05-00-00835 1.8127[17]	H_2O 1 1.5982[18]
1306	Benzene, 2,4-dibromo-1-methyl-	$C_7H_6Br_2$ 249.93	31543-75-6 -9.7	103-4[11]	4-05-00-00835 1.8176[25]	1.5964[25]
1307	Benzene, 1,2-dibromo-4-nitro- 3,4-Dibromonitrobenzene	$C_6H_3Br_2NO_2$ 280.90	5411-50-7 58.5	296	4-05-00-00732 2.354[8]	bz 4; EtOH 4; HOAc 4 1.9835[111]
1308	Benzene, 1,2-dibromo-3-nitro- 2,6-Dibromonitrobenzene	$C_6H_3Br_2NO_2$ 280.90	13402-32-9 84	sub	3-05-00-00621 2.211[8]	ace 4; bz 4; EtOH 4
1309	Benzene, 1,3-dibromo-5-nitro-	$C_6H_3Br_2NO_2$ 280.90	6311-60-0 106		3-05-00-00621 2.36[8]	bz 4; eth 4; EtOH 4
1310	Benzene, 1,4-dibromo-2-nitro-	$C_6H_3Br_2NO_2$ 280.90	3460-18-2 85.5		4-05-00-00732 2.368[8]	H_2O 1; ace 3; bz 3; chl 2
1311	Benzene, 2,4-dibromo-1-nitro-	$C_6H_3Br_2NO_2$ 280.90	51686-78-3 62		3-05-00-00620 2.356[8]	ace 4; bz 4
1312	Benzene, 1,2-dibromo-3,4,5,6-tetrafluoro-	$C_6Br_2F_4$ 307.87	827-08-7 12	198	2.260[20]	1.5151[25]
1313	Benzene, 2,4-dibromo-1,3,5-trimethyl- Mesitylene, 2,4-dibromo-	$C_9H_{10}Br_2$ 277.99	6942-99-0 65.5	285	3-05-00-00921	ctc 2
1314	Benzene, 1,4-dibutoxy-	$C_{14}H_{22}O_2$ 222.33	104-36-9 45.5	158[15]	4-06-00-05721	ctc 3
1315	1,3-Benzenedicarbonitrile 1,3-Dicyanobenzene	$C_8H_4N_2$ 128.13	626-17-5 162	sub	4-09-00-03297 0.992[40]	H_2O 2; EtOH 4; eth 3; bz 3
1316	1,4-Benzenedicarbonitrile	$C_8H_4N_2$ 128.13	623-26-7 224	sub	4-09-00-03328	H_2O 1; EtOH 2; eth 2; bz 3
1317	1,3-Benzenedicarbonitrile, 2,4,5,6-tetrachloro- Chlorothalonil	$C_8Cl_4N_2$ 265.91	1897-45-6 250	2167 350	1.7[25]	H_2O 1; ace 2; cyhex 2
1318	1,2-Benzenedicarbonyl dichloride Phthaloyl chloride	$C_8H_4Cl_2O_2$ 203.02	88-95-9 15.5	7349 281.1	4-09-00-03261 1.4089[20]	1.5684[20]
1319	1,3-Benzenedicarbonyl dichloride	$C_8H_4Cl_2O_2$ 203.02	99-63-8 43.5	276	4-09-00-03295 1.3880[17]	H_2O 2; EtOH 2; eth 3 1.570[47]
1320	1,4-Benzenedicarbonyl dichloride	$C_8H_4Cl_2O_2$ 203.02	100-20-9 83.5	258; 125[9]	4-09-00-03318	eth 3
1321	1,2-Benzenedicarbonyl difluoride	$C_8H_4F_2O_2$ 170.12	445-69-2 42.5	227	4-09-00-03260 1.3066[50]	H_2O 1; peth 3
1322	1,3-Benzenedicarboxaldehyde	$C_8H_6O_2$ 134.13	626-19-7 89.5	246; 136[13]	4-07-00-02139	H_2O 2; EtOH 4; eth 2; ace 3
1323	1,4-Benzenedicarboxaldehyde	$C_8H_6O_2$ 134.13	623-27-8 117	246	4-07-00-02140	H_2O 2; EtOH 4; eth 3; chl 3
1324	1,2-Benzenedicarboxamide Phthalamide	$C_8H_8N_2O_2$ 164.16	88-96-0 222	7343 dec	4-09-00-03265	H_2O 2; EtOH 2; eth 1
1325	1,2-Benzenedicarboxamide, N,N,N',N'- tetraethyl- N,N,N',N'-Tetraethylphthalamide	$C_{16}H_{24}N_2O_2$ 276.38	83-81-8 36	9137 204[16]	4-09-00-03266	
1326	1,3-Benzenedicarboxamide, N,N,N',N'- tetraethyl- Isophthalamide, N,N,N',N'-tetraethyl	$C_{16}H_{24}N_2O_2$ 276.38	13698-87-8 86	242[12]	4-09-00-03296	ace 4; bz 4; eth 4; EtOH 4
1327	1,2-Benzenedicarboxylic acid Phthalic acid	$C_8H_6O_4$ 166.13	88-99-3 230 dec	7345 dec	4-09-00-03167 2.18[191]	H_2O 2; EtOH 3; eth 2; chl 1
1328	1,3-Benzenedicarboxylic acid Isophthalic acid	$C_8H_6O_4$ 166.13	121-91-5 347	5083 sub	4-09-00-03292	H_2O 2; EtOH 3; eth 1; bz 1
1329	1,4-Benzenedicarboxylic acid Terephthalic acid	$C_8H_6O_4$ 166.13	100-21-0	9093 sub 300	4-09-00-03301	H_2O 1; EtOH 1; eth 1; ctc 2
1330	1,3-Benzenedicarboxylic acid, 2-amino-	$C_8H_7NO_4$ 181.15	39622-79-2 340	sub 267	4-14-00-01899	eth 4; EtOH 4
1331	1,3-Benzenedicarboxylic acid, 5-amino-	$C_8H_7NO_4$ 181.15	99-31-0 360	sub	1-14-00-00636	H_2O 1; EtOH 2
1332	1,2-Benzenedicarboxylic acid, 3-benzoyl-	$C_{15}H_{10}O_5$ 270.24	602-82-4 140.5	dec	3-10-00-04007	bz 4; EtOH 4
1333	1,2-Benzenedicarboxylic acid, bis(2- ethoxyethyl) ester	$C_{16}H_{22}O_6$ 310.35	605-54-9 34	345	2-09-00-00597 1.1229[21]	
1334	1,2-Benzenedicarboxylic acid, bis(2- ethylhexyl) ester Bis(2-ethylhexyl) phthalate	$C_{24}H_{38}O_4$ 390.56	117-81-7 -55	1262 384	4-09-00-03181 0.981[25]	ctc 2 1.4853[20]
1335	1,2-Benzenedicarboxylic acid, bis(3- methylbutyl) ester Diisoamyl phthalate	$C_{18}H_{26}O_4$ 306.40	605-50-5	5006 334 dec	4-09-00-03179 1.0209[16]	EtOH 4 1.4871[20]

No.	Name Synonym	Mol. Form. Mol. Wt.	CAS RN mp/°C	Merck No. bp/°C	Beil. Ref. den/g cm^{-3}	Solubility n_D
1336	1,2-Benzenedicarboxylic acid, bis(2-methylpropyl) ester	C$_{16}$H$_{22}$O$_4$ 278.35	84-69-5	296.5; 159[4]	4-09-00-03177 1.0490[15]	ctc 3
1337	1,2-Benzenedicarboxylic acid, bis(phenylmethyl) ester	C$_{22}$H$_{18}$O$_4$ 346.38	523-31-9 44.0	277[15]	4-09-00-03219	H$_2$O 1; EtOH 4; eth 4; lig 1
1338	1,4-Benzenedicarboxylic acid, bis(2-methylpropyl) ester	C$_{16}$H$_{22}$O$_4$ 278.35	18699-48-4 55	180[6]	4-09-00-03305	eth 3
1339	1,2-Benzenedicarboxylic acid, dibutyl ester Dibutyl phthalate	C$_{16}$H$_{22}$O$_4$ 278.35	84-74-2 -35	1586 340	4-09-00-03175 1.0465[20]	H$_2$O 1; EtOH 5; eth 5; bz 5 1.4911[20]
1340	1,2-Benzenedicarboxylic acid, 3,5-dichloro- Phthalic acid, 3,5-dichloro-	C$_8$H$_4$Cl$_2$O$_4$ 235.02	25641-98-9 164 dec	sub	4-09-00-03270	ace 4; eth 4; EtOH 4
1341	1,2-Benzenedicarboxylic acid, dicyclohexyl ester	C$_{20}$H$_{26}$O$_4$ 330.42	84-61-7 66		4-09-00-03189 1.383[20]	H$_2$O 1; EtOH 3; eth 3; chl 2 1.431[20]
1342	1,2-Benzenedicarboxylic acid, diethyl ester Diethyl phthalate	C$_{12}$H$_{14}$O$_4$ 222.24	84-66-2 -40.5	7345 295	4-09-00-03172 1.232[14]	H$_2$O I; EtOH 5; eth 5; ace 3 1.5000[21]
1343	1,2-Benzenedicarboxylic acid, dimethyl ester Dimethyl phthalate	C$_{10}$H$_{10}$O$_4$ 194.19	131-11-3 5.5	3243 283.7	4-09-00-03170 1.1905[20]	H$_2$O 1; EtOH 5; eth 5; bz 3 1.5138[20]
1344	1,2-Benzenedicarboxylic acid, dioctyl ester Dioctyl phthalate	C$_{24}$H$_{38}$O$_4$ 390.56	117-84-0 25		4-09-00-03180	
1345	1,2-Benzenedicarboxylic acid, diphenyl ester Phenyl phthalate	C$_{20}$H$_{14}$O$_4$ 318.33	84-62-8 73	7278 253[14]	4-09-00-03217	H$_2$O 1; EtOH 2; eth 2; ctc 2
1346	1,3-Benzenedicarboxylic acid, diethyl ester Diethyl isophthalate	C$_{12}$H$_{14}$O$_4$ 222.24	636-53-3 11.5	302	4-09-00-03294 1.1239[17]	H$_2$O 1 1.508[18]
1347	1,3-Benzenedicarboxylic acid, dimethyl ester Dimethyl isophthalate	C$_{10}$H$_{10}$O$_4$ 194.19	1459-93-4 67.5	282	4-09-00-03293 1.194[20]	H$_2$O 2 1.5168[20]
1348	1,3-Benzenedicarboxylic acid, 4,6-dimethyl-	C$_{10}$H$_{10}$O$_4$ 194.19	2790-09-2 266	sub	3-09-00-04298	EtOH 4
1349	1,4-Benzenedicarboxylic acid, 2,5-dichloro- 2,5-Dichloroterephthalic acid	C$_8$H$_4$Cl$_2$O$_4$ 235.02	13799-90-1 306	sub	4-09-00-03335	H$_2$O 2; EtOH 3; eth 4
1350	1,4-Benzenedicarboxylic acid, diethyl ester Diethyl terephthalate	C$_{12}$H$_{14}$O$_4$ 222.24	636-09-9 44	302	4-09-00-03304 1.0898[45]	H$_2$O 1; EtOH 4; eth 3
1351	1,4-Benzenedicarboxylic acid, dimethyl ester Dimethyl terephthalate	C$_{10}$H$_{10}$O$_4$ 194.19	120-61-6 141	288	4-09-00-03303 1.075[141]	H$_2$O 2; EtOH 2; eth 3; chl 3
1352	1,3-Benzenedicarboxylic acid, 4-hydroxy- 4-Hydroxyisophthalic acid	C$_8$H$_6$O$_5$ 182.13	636-46-4 310	4758	4-10-00-02091	H$_2$O 1; EtOH 4; eth 4; chl 1
1353	1,2-Benzenedicarboxylic acid, 3,4,5,6-tetrachloro-, monoethyl ester Phthalic acid, tetrachloro-, monoethyl ester	C$_{10}$H$_6$Cl$_4$O$_4$ 331.97	602-21-1 94.5	250 dec	3-09-00-04206	eth 4; EtOH 4
1354	Benzene, 1,2-dichloro- o-Dichlorobenzene	C$_6$H$_4$Cl$_2$ 147.00	95-50-1 -16.7	3044 180	4-05-00-00654 1.3059[20]	H$_2$O 1; EtOH 3; eth 3; ace 5 1.5515[20]
1355	Benzene, 1,3-dichloro- m-Dichlorobenzene	C$_6$H$_4$Cl$_2$ 147.00	541-73-1 -24.8	3043 173	4-05-00-00657 1.2884[20]	H$_2$O 1; EtOH 3; eth 3; ace 5 1.5459[20]
1356	Benzene, 1,4-dichloro- p-Dichlorobenzene	C$_6$H$_4$Cl$_2$ 147.00	106-46-7 52.7	3045 174	4-05-00-00658 1.2475[55]	H$_2$O 1; EtOH 5; eth 3; ace 5 1.5285[20]
1357	Benzene, 1,2-dichloro-4-(chloromethyl)-	C$_7$H$_5$Cl$_3$ 195.47	102-47-6 37.5	241	4-05-00-00820	H$_2$O 1; EtOH 3; ctc 3; os 3
1358	Benzene, 1,3-dichloro-5-(chloromethyl)-	C$_7$H$_5$Cl$_3$ 195.47	3290-06-0 36	60[0.35]	3-05-00-00699	EtOH 4
1359	Benzene, 1,2-dichloro-4-(dichloromethyl)- Benzylidene chloride, 3,4-dichloro	C$_7$H$_4$Cl$_4$ 229.92	56961-84-3	257	3-05-00-00702 1.515[22]	bz 4; eth 4; EtOH 4
1360	Benzene, 1,2-dichloro-3,4-dimethyl- o-Xylene, 3,4-dichloro	C$_8$H$_8$Cl$_2$ 175.06	68266-67-1 9	231	4-05-00-00926 1.2240[20]	
1361	Benzene, 1,3-dichloro-2,5-dimethyl-	C$_8$H$_8$Cl$_2$ 175.06	38204-89-6 15	222	3-05-00-00855 1.2156[20]	
1362	Benzene, 1,4-dichloro-2,5-dimethyl-	C$_8$H$_8$Cl$_2$ 175.06	1124-05-6 71	222	3-05-00-00855	chl 3
1363	Benzene, 1,5-dichloro-2,4-dimethyl-	C$_8$H$_8$Cl$_2$ 175.06	2084-45-9 68.5	223	3-05-00-00835	bz 4; eth 4; chl 4
1364	Benzene, 2,3-dichloro-1,4-dimethyl-	C$_8$H$_8$Cl$_2$ 175.06	34840-79-4 -2	230	3-05-00-00855 1.2310[20]	
1365	Benzene, 1,1'-(1,2-dichloro-1,2-ethenediyl) bis-, (E)-	C$_{14}$H$_{10}$Cl$_2$ 249.14	951-86-0 143.5	183[18]	1-05-00-00304	H$_2$O 1; EtOH 4; eth 4
1366	Benzene, 1,1'-(1,2-dichloro-1,2-ethenediyl) bis-, (Z)-	C$_{14}$H$_{10}$Cl$_2$ 249.14	5216-32-0 67.5	179[18]	1-05-00-00304	eth 4; EtOH 4
1367	Benzene, 1,2-dichloro-3-ethenyl- 2,3-Dichlorostyrene	C$_8$H$_6$Cl$_2$ 173.04	2123-28-6	92-4[4.5]	3-05-00-01174 1.2826[20]	1.5848[20]
1368	Benzene, 1,2-dichloro-4-ethenyl-	C$_8$H$_6$Cl$_2$ 173.04	2039-83-0	95[5]	4-05-00-01347 1.256[20]	1.5857[20]
1369	Benzene, 1,3-dichloro-2-ethenyl- 2,6-Dichlorostyrene	C$_8$H$_6$Cl$_2$ 173.04	28469-92-3 89	64-5[3]	3-05-00-01174 1.2631[20]	1.5752[20]
1370	Benzene, 1,3-dichloro-5-ethenyl- 3,5-Dichlorostyrene	C$_8$H$_6$Cl$_2$ 173.04	2155-42-2	59[1]	3-05-00-01175 1.225[25]	1.5745[25]
1371	Benzene, 1,4-dichloro-2-ethenyl- 2,5-Dichlorostyrene	C$_8$H$_6$Cl$_2$ 173.04	1123-84-8	72-3[2]	4-05-00-01347 1.4045[20]	1.5798[20]
1372	Benzene, (2,2-dichloroethenyl)-	C$_8$H$_6$Cl$_2$ 173.04	698-88-4	225	4-05-00-01348 1.2531[20]	ace 3; chl 3 1.5852[20]

No.	Name Synonym	Mol. Form. Mol. Wt.	CAS RN mp/°C	Merck No. bp/°C	Bell. Ref. den/g cm^{-3}	Solubility n_D
1373	Benzene, 2,4-dichloro-1-ethenyl- 2,4-Dichlorostyrene	$C_8H_6Cl_2$ 173.04	2123-27-5	81[6]	4-05-00-01347 1.243[25]	1.5828[20]
1374	Benzene, 1,1'-(dichloroethenylidene)bis[4-chloro- 2,2-Dichloro-1,1-bis(4-chlorophenyl)ethylene	$C_{14}H_8Cl_4$ 318.03	72-55-9 89		4-05-00-02177	
1375	Benzene, 2,4-dichloro-1-ethoxy-	$C_8H_8Cl_2O$ 191.06	5392-86-9 32	237	4-06-00-00885	H_2O 1; EtOH 3; eth 3; bz 3
1376	Benzene, (1,2-dichloroethyl)-	$C_8H_8Cl_2$ 175.06	1074-11-9	233.5	4-05-00-00901 1.240[15]	H_2O 1; ace 3; bz 3 1.5544[15]
1377	Benzene, 1,4-dichloro-2-ethyl- 2,5-Dichloroethylbenzene	$C_8H_8Cl_2$ 175.06	54484-63-8	213.5	4-05-00-00901 1.239[25]	eth 2; bz 3; chl 2; peth 2
1378	Benzene, 1,1'-(2,2-dichloroethylidene)bis[4-chloro- 1,1-Dichloro-2,2-bis(p-chlorophenyl)ethane	$C_{14}H_{10}Cl_4$ 320.04	72-54-8 109.5	3049	4-05-00-01884	chl 2
1379	Benzene, (dichlorofluoromethyl)-	$C_7H_5Cl_2F$ 179.02	498-67-9 -26.8	179	4-05-00-00818 1.3138[11]	EtOH 4 1.5180[11]
1380	Benzene, 1,4-dichloro-2-iodo-	$C_6H_3Cl_2I$ 272.90	29682-41-5 22	253	3-05-00-00580	bz 4; eth 4; EtOH 4; chl 4
1381	Benzene, 2,4-dichloro-1-methoxy-	$C_7H_6Cl_2O$ 177.03	553-82-2 28.5	232; 125[10]	4-06-00-00885	chl 2
1382	Benzene, (dichloromethyl)- Benzal chloride	$C_7H_6Cl_2$ 161.03	98-87-3 -17	1064 205	4-05-00-00817 1.26[25]	eth 4; EtOH 4 1.5502[20]
1383	Benzene, 1,2-dichloro-3-methyl- 2,3-Dichlorotoluene	$C_7H_6Cl_2$ 161.03	32768-54-0 6	207.5	4-05-00-00815 1.2458[20]	bz 4 1.5511[20]
1384	Benzene, 1,2-dichloro-4-methyl- 3,4-Dichlorotoluene	$C_7H_6Cl_2$ 161.03	95-75-0 -15.2	208.9	4-05-00-00815 1.2564[20]	H_2O 1; EtOH 5; eth 5; ace 5 1.5471[20]
1385	Benzene, 1,3-dichloro-2-methyl- 2,6-Dichlorotoluene	$C_7H_6Cl_2$ 161.03	118-69-4 25.8	198	4-05-00-00815 1.2686[20]	H_2O 1; chl 3 1.5507[20]
1386	Benzene, 1,3-dichloro-5-methyl- 3,5-Dichlorotoluene	$C_7H_6Cl_2$ 161.03	25186-47-4 26	201.5	4-05-00-00816	H_2O 1
1387	Benzene, 1,4-dichloro-2-methyl- 2,5-Dichlorotoluene	$C_7H_6Cl_2$ 161.03	19398-61-9 5	200	3-05-00-00694 1.2535[20]	H_2O 1; bz 3 1.5449[20]
1388	Benzene, 2,4-dichloro-1-methyl- 2,4-Dichlorotoluene	$C_7H_6Cl_2$ 161.03	95-73-8 -13.5	201	4-05-00-00815 1.2476[20]	H_2O 1; ctc 3 1.5511[20]
1389	Benzene, 1,1'-(dichloromethylene)bis-	$C_{13}H_{10}Cl_2$ 237.13	2051-90-3	305 dec; 190[21]	4-05-00-01848 1.235[18]	eth 3; bz 3; ctc 3
1390	Benzene, 1-(dichloromethyl)-4-methyl-	$C_8H_8Cl_2$ 175.06	23063-36-7 48.5	105[18]	3-05-00-00855	EtOH 2; bz 2; chl 2
1391	Benzene, 1,2-dichloro-3-nitro-	$C_6H_3Cl_2NO_2$ 192.00	3209-22-1 61.5	257.5	4-05-00-00725 1.721[14]	H_2O 1; EtOH 3; eth 3; ace 3
1392	Benzene, 1,2-dichloro-4-nitro-	$C_6H_3Cl_2NO_2$ 192.00	99-54-7 43	255.5	4-05-00-00726 1.4558[75]	H_2O 1; EtOH 3; eth 3; ctc 2
1393	Benzene, 1,3-dichloro-2-nitro- 2,6-Dichloronitrobenzene	$C_6H_3Cl_2NO_2$ 192.00	601-88-7 72.5	130[8]	4-05-00-00726 1.603[17]	eth 4; EtOH 4
1394	Benzene, 1,3-dichloro-5-nitro-	$C_6H_3Cl_2NO_2$ 192.00	618-62-2 65.4		4-05-00-00727 1.4000[100]	H_2O 1; EtOH 3; eth 3 1.4000[100]
1395	Benzene, 1,4-dichloro-2-nitro-	$C_6H_3Cl_2NO_2$ 192.00	89-61-2 56	267	4-05-00-00726 1.439[75]	H_2O 1; EtOH 3; eth 3; bz 3 1.4390[75]
1396	Benzene, 2,4-dichloro-1-nitro-	$C_6H_3Cl_2NO_2$ 192.00	611-06-3 34	258.5	4-05-00-00726 1.4790[80]	H_2O 1; EtOH 3; eth 3; chl 2 1.5512[70]
1397	Benzene, (3,3-dichloro-2-propenyl)-	$C_9H_8Cl_2$ 187.07	38862-78-1 59	242; 142[30]	4-05-00-01363 1.2028[20]	eth 3; bz 3; chl 3; peth 3
1398	Benzene, 1,2-dichloro-3,4,5,6-tetrafluoro- 1,2-Dichlorotetrafluorobenzene	$C_6Cl_2F_4$ 218.97	1198-59-0			
1399	Benzene, 1,3-dichloro-2,4,5,6-tetrafluoro- 1,3-Dichlorotetrafluorobenzene	$C_6Cl_2F_4$ 218.97	1198-61-4			
1400	Benzene, 1,4-dichloro-2,3,5,6-tetrafluoro- 1,4-Dichlorotetrafluorobenzene	$C_6Cl_2F_4$ 218.97	1198-62-5			
1401	Benzene, 1,2-dichloro-4-(trichloromethyl)-	$C_7H_3Cl_5$ 264.36	13014-24-9 25.8	283.1	4-05-00-00824 1.5913[20]	1.5886[20]
1402	Benzene, 2,4-dichloro-1,3,5-trimethyl- Mesitylene, 2,4-dichloro-	$C_9H_{10}Cl_2$ 189.08	57386-83-1 59	243.5	3-05-00-00919	EtOH 3; eth 2; bz 2
1403	Benzene, 1,2-diethenyl- o-Divinylbenzene	$C_{10}H_{10}$ 130.19	91-14-5	82[14]	4-05-00-01540 0.9325[22]	ace 3; bz 3 1.5767[20]
1404	Benzene, 1,3-diethenyl- m-Divinylbenzene	$C_{10}H_{10}$ 130.19	108-57-6 -52.3	121[76]; 52[3]	3-05-00-01367 0.9294[20]	ace 3; bz 3 1.5760[20]
1405	Benzene, 1,4-diethenyl- p-Divinylbenzene	$C_{10}H_{10}$ 130.19	105-06-6 31	95[18]; 34[0.2]	4-05-00-01541 0.913[40]	ace 3; bz 3 1.5835[25]
1406	Benzene, 1,2-diethoxy-	$C_{10}H_{14}O_2$ 166.22	2050-46-6 44	219	4-06-00-05565 1.0075[20]	EtOH 3; eth 4; ctc 3 1.5083[25]
1407	Benzene, 1,3-diethoxy-	$C_{10}H_{14}O_2$ 166.22	2049-73-2 12.4	235	4-06-00-05664 1.0170[15]	H_2O 1; EtOH 3; eth 3
1408	Benzene, 1,4-diethoxy-	$C_{10}H_{14}O_2$ 166.22	122-95-2 72	246	4-06-00-05720	EtOH 4; eth 3; bz 3; ctc 3
1409	Benzene, (diethoxymethoxy)-	$C_{11}H_{16}O_3$ 196.25	14444-77-0	103-4[10]	4-06-00-00610 1.014[25]	1.4822[20]
1410	Benzene, 1,4-diethoxy-2-methyl- 2,5-Diethoxytoluene	$C_{11}H_{18}O_2$ 180.25	41901-72-8 24.5	248	3-06-00-04499 1.0134[15]	bz 4; eth 4; EtOH 4

No.	Name / Synonym	Mol. Form. / Mol. Wt.	CAS RN / mp/°C	Merck No. / bp/°C	Beil. Ref. / den/g cm^{-3}	Solubility / n_D
1411	Benzene, 1-(diethoxymethyl)-4-(1-methylethyl)-	$C_{14}H_{22}O_2$ 222.33	35364-90-0	265; 142[18.5]	2-07-00-00247 0.944[19]	1.484[19]
1412	Benzene, (3,3-diethoxy-1-propynyl)-	$C_{13}H_{16}O_2$ 204.27	6142-95-6	99-100[2]	4-07-00-01173 0.991[25]	1.5170[20]
1413	Benzene, 1,2-diethyl- o-Diethylbenzene	$C_{10}H_{14}$ 134.22	135-01-3 -31.2	184	4-05-00-01065 0.8800[20]	H_2O 1; EtOH 5; eth 5; ace 5 1.5035[20]
1414	Benzene, 1,3-diethyl- m-Diethylbenzene	$C_{10}H_{14}$ 134.22	141-93-5 -83.9	181.1	4-05-00-01066 0.8602[20]	H_2O 1; EtOH 5; eth 5; ace 5 1.4955[20]
1415	Benzene, 1,4-diethyl- p-Diethylbenzene	$C_{10}H_{14}$ 134.22	105-05-5 -42.83	183.7	4-05-00-01067 0.8620[20]	H_2O 1; EtOH 5; eth 5; ace 5 1.4967[20]
1416	Benzene, 1,2-diethyl-3-methyl-	$C_{11}H_{16}$ 148.25	13632-93-4	206.6	4-05-00-01105 0.8910[20]	ace 4; bz 4; eth 4; EtOH 4 1.5105[20]
1417	Benzene, 1,2-diethyl-4-methyl- Toluene, 3,4-diethyl-	$C_{11}H_{16}$ 148.25	13732-80-4	203.6	4-05-00-01106 0.8762[20]	ace 4; bz 4; eth 4; EtOH 4 1.5039[20]
1418	Benzene, 1,3-diethyl-2-methyl- Toluene, 2,6-diethyl-	$C_{11}H_{16}$ 148.25	13632-95-6	208.8	4-05-00-01105 0.8907[20]	ace 4; bz 4; eth 4; EtOH 4 1.5106[20]
1419	Benzene, 1,3-diethyl-5-methyl-	$C_{11}H_{16}$ 148.25	2050-24-0 -74.1	205	4-05-00-01106 0.8748[20]	H_2O 1; EtOH 5; eth 5; ace 5 1.5027[20]
1420	Benzene, 1,4-diethyl-2-methyl-	$C_{11}H_{16}$ 148.25	13632-94-5	207.1	4-05-00-01106 0.8758[20]	ace 4; bz 4; eth 4; EtOH 4 1.5034[20]
1421	Benzene, 2,4-diethyl-1-methyl-	$C_{11}H_{16}$ 148.25	1758-85-6	205	4-05-00-01105 0.8748[20]	ace 4; bz 4; eth 4; EtOH 4 1.5027[20]
1422	Benzene, 1,2-difluoro- o-Difluorobenzene	$C_6H_4F_2$ 114.09	367-11-3 -34	94	4-05-00-00637 1.1599[18]	H_2O 1; ace 3; bz 3; chl 3 1.4451[18]
1423	Benzene, 1,3-difluoro- m-Difluorobenzene	$C_6H_4F_2$ 114.09	372-18-9 -69.0	82.6	4-05-00-00637 1.1572[20]	H_2O 1; ace 3; bz 3 1.4374[20]
1424	Benzene, 1,4-difluoro- p-Difluorobenzene	$C_6H_4F_2$ 114.09	540-36-3 -13	3132 89	4-05-00-00637 1.1701[20]	H_2O 1; ace 3; bz 3; ctc 2 1.4422[20]
1425	Benzene, (difluoromethyl)-	$C_7H_6F_2$ 128.12	455-31-2	140	4-05-00-00801 1.1370[19]	H_2O 1; EtOH 3 1.4577[20]
1426	Benzene, 2,4-difluoro-1-nitro-	$C_6H_3F_2NO_2$ 159.09	446-35-5 9.8	207	4-05-00-00720 1.4571[14]	chl 2 1.5149[14]
1427	Benzene, 1,2-diiodo-	$C_6H_4I_2$ 329.91	615-42-9 27	287; 100[3]	4-05-00-00700 2.54[20]	H_2O 2; EtOH 2 1.7179[20]
1428	Benzene, 1,3-diiodo-	$C_6H_4I_2$ 329.91	626-00-6 40.4	285	4-05-00-00700 2.47[25]	eth 4; EtOH 4; chl 4
1429	Benzene, 1,4-diiodo-	$C_6H_4I_2$ 329.91	624-38-4 131.5	285	4-05-00-00700	H_2O 1; EtOH 3; eth 4; chl 2
1430	Benzene, 1,3-diisocyanato-2-methyl- Toluene-2,6-diisocyanate	$C_9H_6N_2O_2$ 174.16	91-08-7 18.3			H_2O 6; ace 3; bz 3
1431	Benzene, 2,4-diisocyanato-1-methyl- Toluene 2,4-diisocyanate	$C_9H_6N_2O_2$ 174.16	584-84-9 20.5	9456 251	4-13-00-00243 1.2244[20]	ace 4; bz 4; eth 4
1432	Benzene, 1,4-diisothiocyanato- Bitoscanate	$C_8H_4N_2S_2$ 192.27	4044-65-9 132	1318	4-13-00-00174	
1433	1,3-Benzenedimethanamine m-Xylenediamine	$C_8H_{12}N_2$ 136.20	1477-55-0	247	4-13-00-00293 1.052[20]	H_2O 4; eth 4; EtOH 4
1434	1,3-Benzenedimethanol	$C_8H_{10}O_2$ 138.17	626-18-6 57	156[13]	4-06-00-05966 1.1610[18]	H_2O 4; eth 4; EtOH 4
1435	1,4-Benzenedimethanol	$C_8H_{10}O_2$ 138.17	589-29-7 117.5	138-43[1]	4-06-00-05971	H_2O 4; ace 4; eth 4; EtOH 4
1436	Benzene, 1,2-dimethoxy- Veratrole	$C_8H_{10}O_2$ 138.17	91-16-7 22.5	9857 206	4-06-00-05564 1.0810[25]	H_2O 2; EtOH 3; eth 3; ctc 3 1.5827[21]
1437	Benzene, 1,3-dimethoxy-	$C_8H_{10}O_2$ 138.17	151-10-0 -52	217.5	4-06-00-05663 1.0521[25]	H_2O 2; EtOH 3; eth 3; bz 3 1.5231[20]
1438	Benzene, 1,4-dimethoxy-	$C_8H_{10}O_2$ 138.17	150-78-7 59	212.6	4-06-00-05718 1.0375[55]	H_2O 2; EtOH 3; eth 4; bz 4
1439	Benzene, 1,2-dimethoxy-3-methyl-	$C_9H_{12}O_2$ 152.19	4463-33-6	204	4-06-00-05860 1.0335[20]	eth 4; EtOH 4 1.5121[25]
1440	Benzene, 1,2-dimethoxy-4-methyl-	$C_9H_{12}O_2$ 152.19	494-99-5 24	220	4-06-00-05879 1.0509[25]	H_2O 1; ctc 2; os 4 1.5257[25]
1441	Benzene, 1,3-dimethoxy-5-methyl-	$C_9H_{12}O_2$ 152.19	4179-19-5	244	3-06-00-04533 1.0478[15]	bz 4; eth 4; EtOH 4 1.5234[20]
1442	Benzene, 1,4-dimethoxy-2-(1-methylethyl)-	$C_{11}H_{16}O_2$ 180.25	4132-71-2	114-6[15]	4-06-00-05990 1.0129[17]	bz 4; eth 4 1.5103[17]
1443	Benzene, 1-(dimethoxymethyl)-3-nitro-	$C_9H_{11}NO_4$ 197.19	3395-79-7	162-4[19]	4-07-00-00592 1.209[15]	bz 4
1444	Benzene, 1-(dimethoxymethyl)-4-nitro-	$C_9H_{11}NO_4$ 197.19	881-67-4 24	294	4-07-00-00600	bz 4
1445	Benzene, 1,2-dimethoxy-4-nitro-	$C_8H_9NO_4$ 183.16	709-09-1 98	230[15]	4-06-00-05627 1.1888[133]	H_2O 1; EtOH 4; eth 4; chl 3
1446	Benzene, 1,3-dimethoxy-5-nitro-	$C_8H_9NO_4$ 183.16	16147-07-2 89		3-06-00-04347 1.1693[133]	bz 4; EtOH 4
1447	Benzene, 1,4-dimethoxy-2-nitro-	$C_8H_9NO_4$ 183.16	89-39-4 72.5		4-06-00-05786 1.1666[132]	H_2O 1; EtOH 3; bz 3; chl 3
1448	Benzene, 2,4-dimethoxy-1-nitro- 1,3-Dimethoxy-4-nitrobenzene	$C_8H_9NO_4$ 183.16	4920-84-7 76.5		4-06-00-05691 1.1876[132]	H_2O 1; EtOH 3; os 3; lig 1
1449	Benzene, 1,2-dimethoxy-4-(1-propenyl)-	$C_{11}H_{14}O_2$ 178.23	93-16-3 18	270.5	2-06-00-00918 1.0521[20]	1.5616[20]

No.	Name Synonym	Mol. Form. Mol. Wt.	CAS RN mp/°C	Merck No. bp/°C	Beil. Ref. den/g cm⁻³	Solubility n_D
1450	Benzene, 1,2-dimethoxy-4-(2-propenyl)-	$C_{11}H_{14}O_2$ 178.23	93-15-2 -4	254.7	4-06-00-06337 1.0396^{20}	H_2O 1; EtOH 3; eth 3 1.5340^{20}
1451	Benzene, 1,2-dimethoxy-4-(1-propenyl)-, (Z)-	$C_{11}H_{14}O_2$ 178.23	6380-24-1 70	270.5	4-06-00-06325 1.0521^{20}	H_2O 1; ace 3; bz 3 1.5616^{20}
1452	Benzene, 1,2-dimethyl- o-Xylene	C_8H_{10} 106.17	95-47-6 -25.2	9988 144.5	4-05-00-00917 0.8802^{10}	H_2O 1; EtOH 5; eth 5; ace 5 1.5055^{20}
1453	Benzene, 1,3-dimethyl- m-Xylene	C_8H_{10} 106.17	108-38-3 -47.8	9988 139.1	4-05-00-00932 0.8642^{20}	H_2O 1; EtOH 5; eth 5; ace 5 1.4972^{10}
1454	Benzene, 1,4-dimethyl- p-Xylene	C_8H_{10} 106.17	106-42-3 13.2	9988 138.3	4-05-00-00951 0.8611^{20}	H_2O 1; EtOH 5; eth 5; ace 5 1.4958^{20}
1455	Benzene, (1,1-dimethylbutyl)- Pentane, 2-methyl-2-phenyl	$C_{12}H_{18}$ 162.27	1985-57-5	205.5	4-05-00-01118 0.8796^{10}	1.4955^{16}
1456	Benzene, 1,1'-(1,2-dimethyl-1,2-ethenediyl)bis-, (E)-	$C_{16}H_{16}$ 208.30	782-06-9 107	156[16]	4-05-00-02207 0.987^{20}	H_2O 1; EtOH 2; eth 3; MeOH 2
1457	Benzene, 1,1'-(1,2-dimethyl-1,2-ethenediyl)bis-, (Z)-	$C_{16}H_{16}$ 208.30	782-05-8 67.5		4-05-00-02207 1.004^{20}	H_2O 1; EtOH 3; HOAc 3; MeOH 3 1.5612^{78}
1458	Benzene, (1,1-dimethylethoxy)-	$C_{10}H_{14}O$ 150.22	6669-13-2 -24	185.5	4-06-00-00559 0.9214^{20}	
1459	Benzene, (1,1-dimethylethyl)- tert-Butylbenzene	$C_{10}H_{14}$ 134.22	98-06-6 -57.8	1551 169.1	4-05-00-01045 0.8665^{20}	H_2O 1; EtOH 4; eth 4; ace 5 1.4927^{20}
1460	Benzene, 1-(1,1-dimethylethyl)-3,5-dimethyl-	$C_{12}H_{18}$ 162.27	98-19-1 -18	207	4-05-00-01130 0.8668^{20}	ctc 3
1461	Benzene, 1-(1,1-dimethylethyl)-3,5-dimethyl-2,4,6-trinitro- 2,4,6-Trinitro-3,5-dimethyl-tert-butylbenzene	$C_{12}H_{15}N_3O_6$ 297.27	81-15-2 110		4-05-00-01132	H_2O 1; EtOH 2; eth 3; chl 3
1462	Benzene, 1-(1,1-dimethylethyl)-4-methoxy-	$C_{11}H_{16}O$ 164.25	5396-38-3	238	4-06-00-03297 0.9436^{20}	1.5039^{20}
1463	Benzene, 1-(1,1-dimethylethyl)-2-methoxy-4-methyl-3,5-dinitro-	$C_{12}H_{16}N_2O_5$ 268.27	83-66-9 85	185[16]	4-06-00-03402	H_2O 1; EtOH 2; eth 3; chl 3
1464	Benzene, 1-(1,1-dimethylethyl)-2-methyl-	$C_{11}H_{16}$ 148.25	1074-92-6 -50.3	200.4	4-05-00-01096 0.8897^{20}	ace 4; bz 4; eth 4; EtOH 4 1.5076^{20}
1465	Benzene, 1-(1,1-dimethylethyl)-3-methyl-	$C_{11}H_{16}$ 148.25	1075-38-3 -41.4	189.3	4-05-00-01096 0.8657^{20}	ace 4; bz 4; eth 4; EtOH 4 1.4944^{20}
1466	Benzene, 1-(1,1-dimethylethyl)-4-methyl- p-tert-Butyltoluene	$C_{11}H_{16}$ 148.25	98-51-1 -52	190	4-05-00-01097 0.8612^{20}	H_2O 1; EtOH 2; eth 4; ace 3 1.4918^{20}
1467	Benzene, 1-(1,1-dimethylethyl)-3-nitro-	$C_{10}H_{13}NO_2$ 179.22	23132-52-7 2	254; 136[16]	4-05-00-01052 1.0643^{25}	
1468	Benzene, 1-(1,5-dimethyl-4-hexenyl)-4-methyl- ar-Curcumene	$C_{15}H_{22}$ 202.34	644-30-4	140[19]	4-05-00-01465 0.8805^{20}	H_2O 1; bz 3 1.4989^{20}
1469	Benzene, 1,2-dimethyl-3-(1-methylethyl)-	$C_{11}H_{16}$ 148.25	22539-65-7	202.6	4-05-00-01103 0.888^{20}	ace 4; bz 4; eth 4; EtOH 4 1.508^{20}
1470	Benzene, 1,2-dimethyl-4-(1-methylethyl)-	$C_{11}H_{16}$ 148.25	4132-77-8	200	4-05-00-01103 0.8710^{20}	1.4951^{20}
1471	Benzene, 1,3-dimethyl-2-(1-methylethyl)-	$C_{11}H_{16}$ 148.25	14411-75-7	199	4-05-00-01103 0.890^{20}	ace 4; bz 4; eth 4; EtOH 4 1.509^{20}
1472	Benzene, 2,4-dimethyl-1-(1-methylethyl)-	$C_{11}H_{16}$ 148.25	4706-89-2 -82	197	4-05-00-01104 0.8734^{20}	1.4998^{25}
1473	Benzene, 2,4-dimethyl-1-(1-methylpropyl)-	$C_{12}H_{18}$ 162.27	1483-60-9	84[8]	4-05-00-01129 0.8664^{20}	1.4939^{25}
1474	Benzene, 1,2-dimethyl-3-nitro-	$C_8H_9NO_2$ 151.17	83-41-0 15	240	4-05-00-00930 1.1402^{20}	H_2O 1; EtOH 3; ctc 3 1.5441^{20}
1475	Benzene, 1,2-dimethyl-4-nitro- 4-Nitro-o-xylene	$C_8H_9NO_2$ 151.17	99-51-4 30.5	251; 143[21]	4-05-00-00930 1.112^{15}	H_2O 1; EtOH 5 1.5202^{20}
1476	Benzene, 1,3-dimethyl-2-nitro-	$C_8H_9NO_2$ 151.17	81-20-9 15	226	4-05-00-00948 1.112^{15}	H_2O 1; EtOH 4; ctc 3 1.5202^{20}
1477	Benzene, 1,3-dimethyl-5-nitro-	$C_8H_9NO_2$ 151.17	99-12-7 75	274	4-05-00-00948	H_2O 1; EtOH 4; eth 4
1478	Benzene, 1,4-dimethyl-2-nitro-	$C_8H_9NO_2$ 151.17	89-58-7 -25	240.5	4-05-00-00971 1.132^{15}	H_2O 1; EtOH 3 1.5413^{20}
1479	Benzene, 2,4-dimethyl-1-nitro-	$C_8H_9NO_2$ 151.17	89-87-2 9	247; 122[18]	4-05-00-00948 1.135^{15}	H_2O 1; eth 3; ace 3; bz 3 1.5473^{25}
1480	Benzene, (1,1-dimethylpropyl)-	$C_{11}H_{16}$ 148.25	2049-95-8	192.4	4-05-00-01090 0.8748^{20}	1.4958^{20}
1481	Benzene, (1,2-dimethylpropyl)-	$C_{11}H_{16}$ 148.25	4481-30-5	187	4-05-00-01092 0.8672^{16}	1.4972^{16}
1482	Benzene, 1,3-dimethyl-2-propyl-	$C_{11}H_{16}$ 148.25	17059-45-9	207.6	4-05-00-01102 0.8856^{20}	ace 4; bz 4; eth 4; EtOH 4 1.5063^{20}
1483	Benzene, 1,3-dimethyl-5-propyl-	$C_{11}H_{16}$ 148.25	3982-64-7 -59.1	202.2	4-05-00-01102 0.8607^{20}	ace 4; bz 4; eth 4; EtOH 4 1.4952^{20}
1484	Benzene, (2,2-dimethylpropyl)-	$C_{11}H_{16}$ 148.25	1007-26-7	186	4-05-00-01092 0.8581^{18}	1.4884^{18}
1485	Benzene, 1,2-dimethyl-3-propyl- o-Xylene, 3-propyl	$C_{11}H_{16}$ 148.25	17059-44-8	210.7	4-05-00-01101 0.8864^{20}	ace 4; bz 4; eth 4; EtOH 4 1.5075^{20}
1486	Benzene, 1,4-dimethyl-2-propyl-	$C_{11}H_{16}$ 148.25	3042-50-0	204.3	4-05-00-01102 0.8717^{20}	ace 4; bz 4; eth 4; EtOH 4 1.4999^{20}
1487	Benzene, 2,4-dimethyl-1-propyl- m-Xylene, 4-propyl-	$C_{11}H_{16}$ 148.25	61827-85-8	206.6	4-05-00-01102 0.8723^{20}	ace 4; bz 4; eth 4; EtOH 4 1.4998^{20}

No.	Name Synonym	Mol. Form. Mol. Wt.	CAS RN mp/°C	Merck No. bp/°C	Beil. Ref. den/g cm^{-3}	Solubility n_D
1488	Benzene, 1-(1,1-dimethylpropyl)-4-methoxy-	$C_{12}H_{18}O$ 178.27	2050-03-5	238	4-06-00-03383 0.9436[20]	1.5039[20]
1489	Benzene, 1,4-dimethyl-2,3,5-trinitro- p-Xylene, 2,3,5-trinitro-	$C_8H_7N_3O_6$ 241.16	602-27-7 139.5	exp 410	3-05-00-00864 1.59[19]	EtOH 4
1490	Benzene, 2,4-dimethyl-1,3,5-trinitro-	$C_8H_7N_3O_6$ 241.16	632-92-8 184		4-05-00-00950 1.604[19]	H_2O 1; EtOH 2; eth 2; ace 3
1491	Benzene, 1,2-dinitro- o-Dinitrobenzene	$C_6H_4N_2O_4$ 168.11	528-29-0 118.5	318; 194[30]	4-05-00-00738 1.3119[120]	H_2O 1; EtOH 3; bz 3; chl 3 1.565[17]
1492	Benzene, 1,3-dinitro- m-Dinitrobenzene	$C_6H_4N_2O_4$ 168.11	99-65-0 90	291; 167[14]	4-05-00-00739 1.575[118]	H_2O 2; EtOH 4; eth 3; ace 4
1493	Benzene, 1,4-dinitro- p-Dinitrobenzene	$C_6H_4N_2O_4$ 168.11	100-25-4 174	297; 183[34]	4-05-00-00741 1.625[18]	H_2O 1; EtOH 2; ace 3; bz 3
1494	Benzene, 1,1'-(1,2-dinitro-1,2-ethenediyl)bis-, (Z)-	$C_{14}H_{10}N_2O_4$ 270.24	1796-05-0 108.5	dec	4-05-00-02171	H_2O 1; EtOH 3; eth 3; ace 4
1495	Benzene, 2,4-dinitro-1-phenoxy-	$C_{12}H_8N_2O_5$ 260.21	2486-07-9 71	230[27]	4-06-00-01375	eth 4; EtOH 4
1496	Benzene, 2,4-dinitro-1-(2-phenylethenyl)-, (E)- Stilbene, 2,4-dinitro (trans)	$C_{14}H_{10}N_2O_4$ 270.24	56456-42-9 144	exp 412	4-05-00-02170	EtOH 1; bz 2; chl 3; CS_2 3
1497	1,2-Benzenediol Pyrocatechol	$C_6H_6O_2$ 110.11	120-80-9 105	8009 245	4-06-00-05557 1.1493[22]	H_2O 4; bz 4; eth 4; EtOH 4 1.604
1498	1,3-Benzenediol Resorcinol	$C_6H_6O_2$ 110.11	108-46-3 111	8158 178[16]	4-06-00-05658 1.2717[25]	H_2O 3; EtOH 3; eth 3; bz 2
1499	1,4-Benzenediol p-Hydroquinone	$C_6H_6O_2$ 110.11	123-31-9 172.3	4738 287	4-06-00-05712 1.328[15]	H_2O 3; EtOH 4; eth 3; ace 4
1500	1,2-Benzenediol, 4-(2-amino-1-hydroxyethyl)-, (R)- Norepinephrine	$C_8H_{11}NO_3$ 169.18	51-41-2 217 dec	6612		H_2O 2; EtOH 2; eth 2; alk 4
1501	1,4-Benzenediol, 2,5-bis(1,1-dimethylpropyl)- 2,5-Di-tert-pentylhydroquinone	$C_{16}H_{26}O_2$ 250.38	79-74-3 180	3300	3-06-00-04748	
1502	1,3-Benzenediol, 4-bromo-	$C_6H_5BrO_2$ 189.01	6626-15-9 103	140[12]	4-06-00-05687	H_2O 4; EtOH 2; eth 4; bz 2
1503	1,4-Benzenediol, 2-bromo-	$C_6H_5BrO_2$ 189.01	583-69-7 111.5	sub	4-06-00-05780	H_2O 4; EtOH 4; eth 4; bz 4
1504	1,2-Benzenediol, 3-chloro-	$C_6H_5ClO_2$ 144.56	4018-65-9 48.5	110[11]	4-06-00-05613	lig 4
1505	1,2-Benzenediol, 4-chloro-	$C_6H_5ClO_2$ 144.56	2138-22-9 90.5	139[10.5]	4-06-00-05614	H_2O 4; ace 4; eth 4; EtOH 4
1506	1,3-Benzenediol, 5-chloro-	$C_6H_5ClO_2$ 144.56	52780-23-1 118	sub	4-06-00-05684	H_2O 3; eth 3; ace 3; os 3
1507	1,4-Benzenediol, 2-chloro-	$C_6H_5ClO_2$ 144.56	615-67-8 108	263	4-06-00-05767	H_2O 3; EtOH 3; eth 3; bz 4
1508	1,2-Benzenediol, diacetate	$C_{10}H_{10}O_4$ 194.19	635-67-6 64.5	142-3[9]	4-06-00-05582	H_2O 1; EtOH 4; eth 4; chl 4
1509	1,4-Benzenediol, diacetate	$C_{10}H_{10}O_4$ 194.19	1205-91-0 123.5		4-06-00-05741 0.8731[25]	H_2O 3; EtOH 4; eth 4; chl 4
1510	1,3-Benzenediol, 4,6-dichloro-	$C_6H_4Cl_2O_2$ 179.00	137-19-9 113	254	4-06-00-05685	H_2O 4; EtOH 4; eth 4; ace 4
1511	1,2-Benzenediol, 4,5-dimethyl-	$C_8H_{10}O_2$ 138.17	2785-74-2 87.5	160[26]	4-06-00-05950	H_2O 4; EtOH 4; eth 4; bz 2
1512	1,3-Benzenediol, 2,5-dimethyl-	$C_8H_{10}O_2$ 138.17	488-87-9 163	278.5	4-06-00-05970	H_2O 3; EtOH 3; eth 3
1513	1,3-Benzenediol, 4,5-dimethyl-	$C_8H_{10}O_2$ 138.17	527-55-9 136	284	4-06-00-05948 1.1994[20]	H_2O 3; EtOH 4; eth 3; bz 2
1514	1,3-Benzenediol, 4,6-dimethyl-	$C_8H_{10}O_2$ 138.17	615-89-4 126.5	276	4-06-00-05963	H_2O 3; EtOH 3; eth 3
1515	1,2-Benzenediol, 4,4'-(2,3-dimethyl-1,4-butanediyl)bis- Nordihydroguaiaretic acid	$C_{18}H_{22}O_4$ 302.37	500-38-9 185.5	6610	4-06-00-07771	H_2O 2; EtOH 3; eth 3; ace 3
1516	1,2-Benzenediol, 4-(1,1-dimethylethyl)-	$C_{10}H_{14}O_2$ 166.22	98-29-3 54.3	285; 160[22]	4-06-00-06014	tfa 3
1517	1,3-Benzenediol, 2,4-dinitro- 2,4-Dinitroresorcinol	$C_6H_4N_2O_6$ 200.11	519-44-8 147.5	3278	4-06-00-05696	H_2O 2; EtOH 2
1518	1,3-Benzenediol, 4-ethyl-	$C_8H_{10}O_2$ 138.17	2896-60-8 98.5	160[24]; 131[15]	4-06-00-05924	H_2O 2; EtOH 2; eth 2
1519	1,3-Benzenediol, 4-hexyl- 4-Hexylresorcinol	$C_{12}H_{18}O_2$ 194.27	136-77-6 68	4633 334	4-06-00-06048	ace 4; eth 4; EtOH 4; chl 4
1520	1,4-Benzenediol, 2-(hydroxymethyl)- Gentisyl alcohol	$C_7H_8O_3$ 140.14	495-08-9 100	4292 sub 75	4-06-00-07380	H_2O 4; EtOH 4; chl 4
1521	1,2-Benzenediol, 4-[1-hydroxy-2-(methylamino)ethyl]-, (R)- Epinephrine	$C_9H_{13}NO_3$ 183.21	51-43-4 211.5	3569	4-13-00-02927	H_2O 2; EtOH 1; HOAc 3; min acid 3
1522	1,2-Benzenediol, 4-[1-hydroxy-2-[(1-methylethyl)amino]ethyl]- Ethanol, 1(3,4-dihydroxyphenyl)-2(isopropylamino)	$C_{11}H_{17}NO_3$ 211.26	7683-59-2 170.5	5105	3-13-00-02387	
1523	1,2-Benzenediol, 4-iodo- 4-Iodocatechol	$C_6H_5IO_2$ 236.01	76149-14-9 92	sub	4-06-00-05625	ace 4; bz 4; eth 4; EtOH 4
1524	1,3-Benzenediol, 5-iodo- Resorcinol, 5-iodo	$C_6H_5IO_2$ 236.01	64339-43-1 92.5	sub	2-06-00-00821	EtOH 4

No.	Name / Synonym	Mol. Form. / Mol. Wt.	CAS RN / mp/°C	Merck No. / bp/°C	Beil. Ref. / den/g cm^{-3}	Solubility / n_D
1525	1,2-Benzenediol, 3-methoxy-	$C_7H_8O_3$ 140.14	934-00-9 42.8	163[48,] 129[10,]	4-06-00-07329	chl 3
1526	1,3-Benzenediol, 5-methoxy-	$C_7H_8O_3$ 140.14	2174-64-3 80.3	213[16]	4-06-00-07362	H_2O 2; EtOH 4; eth 4; bz 2
1527	1,2-Benzenediol, 3-methyl-	$C_7H_8O_2$ 124.14	488-17-5 68	241	4-06-00-05860	H_2O 3; EtOH 3; bz 3; chl 3
1528	1,2-Benzenediol, 4-methyl-	$C_7H_8O_2$ 124.14	452-86-8 65	251	4-06-00-05878 1.1287[74]	H_2O 3; EtOH 3; eth 3; ace 3 1.5425[74]
1529	1,3-Benzenediol, 2-methyl-	$C_7H_8O_2$ 124.14	608-25-3 120	264	4-06-00-05877	H_2O 4; bz 4; eth 4; EtOH 4
1530	1,3-Benzenediol, 4-methyl-	$C_7H_8O_2$ 124.14	496-73-1 106	268.5	4-06-00-05864	H_2O 3; EtOH 3; eth 3; bz 2
1531	1,3-Benzenediol, 5-methyl- Orcinol	$C_7H_8O_2$ 124.14	504-15-4 107.5	6819 289.5	4-06-00-05892 1.290[4]	H_2O 3; EtOH 3; eth 3; bz 3
1532	1,4-Benzenediol, 2-methyl-	$C_7H_8O_2$ 124.14	95-71-6 128	163[11]	4-06-00-05866	H_2O 4; EtOH 4; eth 4; ace 3
1533	1,2-Benzenediol, 4-[2-(methylamino)ethyl]- Deoxyepinephrine	$C_9H_{13}NO_2$ 167.21	501-15-5 188.5	2885	4-13-00-02606	
1534	1,3-Benzenediol, 4-(3-methylbutyl)-	$C_{11}H_{16}O_2$ 180.25	15116-17-3 67	177[7]	3-06-00-04700	bz 4; eth 4; EtOH 4
1535	1,2-Benzenediol, 4-(1-methylethyl)-	$C_9H_{12}O_2$ 152.19	2138-43-4 78	271	3-06-00-04632	peth 3
1536	1,2-Benzenediol, 3-methyl-6-(1-methylethyl)-	$C_{10}H_{14}O_2$ 166.22	490-06-2 48	267	4-06-00-06019	ace 4; eth 4
1537	1,3-Benzenediol, 2-[3-methyl-6-(1-methylethenyl)-2-cyclohexen-1-yl]-5-pentyl-, (1R-trans)- Cannabidiol	$C_{21}H_{30}O_2$ 314.47	13956-29-1 67	1750 187-90[2]	3-06-00-05362 1.040[40]	H_2O 1; EtOH 3; eth 3; bz 3 1.5404[20]
1538	1,4-Benzenediol, 2-methyl-5-(1-methylethyl)-	$C_{10}H_{14}O_2$ 166.22	2217-60-9 148	290	4-06-00-06019	eth 4; EtOH 4
1539	1,3-Benzenediol, 4-(2-methylpropyl)- Resorcinol, 4-isobutyl-	$C_{10}H_{14}O_2$ 166.22	18979-62-9 62.5	166[6]	3-06-00-04667	eth 4; EtOH 4
1540	1,2-Benzenediol, monoacetate	$C_8H_8O_3$ 152.15	2848-25-1 57.5	189[102,] 148[25]	4-06-00-05581	H_2O 4; ace 4; EtOH 4; chl 4
1541	1,3-Benzenediol, 4-nitro-	$C_6H_5NO_4$ 155.11	3163-07-3 122	178[11]	4-06-00-05691	EtOH 3; eth 3; bz 3; ctc 2
1542	1,3-Benzenediol, 4-[(4-nitrophenyl)azo]- Magneson	$C_{12}H_9N_3O_4$ 259.22	74-39-5 200	5578	4-16-00-00266	H_2O 1; EtOH 2; bz 2; HOAc 2
1543	1,2-Benzenediol, 3-pentadecyl- 3-Pentadecylcatechol	$C_{21}H_{36}O_2$ 320.52	492-89-7 59.5	7061	4-06-00-06109	bz 4; eth 4; EtOH 4
1544	1,3-Benzenediol, 4-pentyl-	$C_{11}H_{16}O_2$ 180.25	533-24-4 72.5	168[6]	3-06-00-04693	bz 4; eth 4; EtOH 4
1545	1,3-Benzenediol, 5-(2-phenylethenyl)-, (E)- Pinosylvin	$C_{14}H_{12}O_2$ 212.25	22139-77-1 156	7419	3-06-00-05577	ace 4; bz 4; chl 4; HOAc 4
1546	1,4-Benzenediol, 2-(phenylmethyl)-	$C_{13}H_{12}O_2$ 200.24	1706-73-6 105.8	230[13]	3-06-00-05404	eth 4; EtOH 4
1547	1,2-Benzenediol, 3-(2-propenyl)-	$C_9H_{10}O_2$ 150.18	1125-74-2	141-5[16]	4-06-00-06334 1.1241[20]	1.5656[20]
1548	1,2-Benzenediol, 4-propyl-	$C_9H_{12}O_2$ 152.19	2525-02-2 60	177[30,] 111[0.2]	3-06-00-04613 1.100[18]	1.4440[18]
1549	1,3-Benzenediol, 4-propyl-	$C_9H_{12}O_2$ 152.19	18979-60-7 82.5	172[14]	4-06-00-05975	H_2O 3; EtOH 2; eth 3
1550	1,3-Benzenediol, 5-propyl-	$C_9H_{12}O_2$ 152.19	500-49-2 82.8	169[8;] 148[3]	3-06-00-04622	ace 4; bz 4; eth 4; EtOH 4
1551	1,4-Benzenediol, 2,3,5,6-tetrabromo-	$C_6H_2Br_4O_2$ 425.70	2641-89-6 244		3-06-00-04440 3.023[21]	eth 4; EtOH 4
1552	1,4-Benzenediol, 2,3,5,6-tetramethyl- Durohydroquinone	$C_{10}H_{14}O_2$ 166.22	527-18-4 233	3451	4-06-00-06028	EtOH 3; eth 2
1553	1,3-Benzenediol, 2,4,6-trimethyl-	$C_9H_{12}O_2$ 152.19	608-98-0 150.8	274.5	4-06-00-06000	bz 4; eth 4; EtOH 4
1554	1,3-Benzenediol, 2,4,6-trinitro- Styphnic acid	$C_6H_3N_3O_8$ 245.11	82-71-3 175.5	8828 sub	4-06-00-05699	eth 4; EtOH 4
1555	Benzene, 1,2-dipropoxy-	$C_{12}H_{18}O_2$ 194.27	6280-98-4	235.5	4-06-00-05566 0.9554[33]	chl 2 1.4950[27]
1556	Benzene, 1,3-dipropoxy- Resorcinol dipropyl ether	$C_{12}H_{18}O_2$ 194.27	56106-37-7	253	3-06-00-04308 1.033[20]	1.5138[83]
1557	Benzene, 1,4-dipropyl-	$C_{12}H_{18}$ 162.27	4815-57-0	109[23]	4-05-00-01124 0.8563[19]	1.4917[19]
1558	1,3-Benzenedisulfonamide, 4-amino-6-chloro- Chloraminophenamide	$C_6H_8ClN_3O_4S_2$ 285.73	121-30-2 254.5	2067	4-14-00-02810	
1559	1,3-Benzenedisulfonamide, 4,5-dichloro- 1,3-Benzenedisulfonic acid, diamide,4,5-dichloro	$C_6H_6ClN_2O_4$ 205.58	120-97-8 228.7	3067	4-11-00-00555	
1560	1,3-Benzenedisulfonic acid, 4-hydroxy- Phenoldisulfonic acid	$C_6H_6O_7S_2$ 254.24	96-77-5 >100 dec	7207	2-11-00-00139	H_2O 4; EtOH 4
1561	1,2-Benzenedithiol	$C_6H_6S_2$ 142.25	17534-15-5 28.5	238.5	4-06-00-05651	EtOH 4; eth 4; bz 4; AcOEt 3

No.	Name / Synonym	Mol. Form. / Mol. Wt.	CAS RN / mp/°C	Merck No. / bp/°C	Beil. Ref. / den/g cm^{-3}	Solubility / n_D
1562	1,3-Benzenedithiol	$C_6H_6S_2$ 142.25	626-04-0 27	245	4-06-00-05705	bz 4; eth 4; EtOH 4
1563	1,3-Benzenedithiol, 4-chloro-	$C_6H_5ClS_2$ 176.69	58593-78-5	145-6[13]	4-06-00-05708 1.393[25]	1.6704[20]
1564	1,2-Benzenedithiol, 4-methyl- Toluene-3,4-dithiol	$C_7H_8S_2$ 156.27	496-74-2 29	9457	4-06-00-05890	chl 3
1565	Benzene, docosyl- Docosane, 1-phenyl-	$C_{28}H_{50}$ 386.71	5634-22-0 51	448	3-05-00-01142 0.8544[20]	
1566	Benzene, dodecyl- 1-Phenyldodecane	$C_{18}H_{30}$ 246.44	123-01-3 3	328	4-05-00-01200 0.8551[20]	H_2O 1 1.4824[20]
1567	Benzene, 1,1'-dodecylidenebis[4-methyl- Dodecane, 1,1-di-(4-tolyl)	$C_{26}H_{38}$ 350.59	55268-62-7	197[1.0]	3-05-00-01925 0.9117[20]	1.5001[20]
1568	Benzene, eicosyl-	$C_{26}H_{46}$ 358.65	2398-68-7 44	429	4-05-00-01222 0.8545[20]	1.4805[20]
1569	Benzeneethanamine Phenylethylamine	$C_8H_{11}N$ 121.18	64-04-0	7186 197.5	4-12-00-02453 0.9580[24]	H_2O 3; eth 4; eth 4; ctc 3 1.5290[25]
1570	Benzeneethanamine, 4-bromo- Phenethylamine, 4-bromo	$C_8H_{10}BrN$ 200.08	73918-56-6	63-72[0.2]	4-12-00-02492 1.29[25]	1.5750[20]
1571	Benzeneethanamine, 3,4-dimethoxy-	$C_{10}H_{15}NO_2$ 181.23	120-20-7	163-5[14]	4-13-00-02604	ctc 3 1.5464[20]
1572	Benzeneethanamine, α,α-dimethyl- Phentermine	$C_{10}H_{15}N$ 149.24	122-09-8	7232 205; 100[21]	4-12-00-02820	
1573	Benzeneethanamine, N,α-dimethyl-, (S)- Methamphetamine	$C_{10}H_{15}N$ 149.24	537-46-2	5859	4-12-00-02589	
1574	Benzeneethanamine, N,β-dimethyl- Phenylpropylmethylamine	$C_{10}H_{15}N$ 149.24	93-88-9	7280 207.5	3-12-00-02689 0.915[25]	bz 4; eth 4; EtOH 4
1575	Benzeneethanamine, N,α-dimethyl-N-(phenylmethyl)-, (+)- Benzphetamine	$C_{17}H_{21}N$ 239.36	156-08-1	1130 127[0.02]	4-12-00-02594	eth 4; EtOH 4; MeOH 4; chl 4 1.5515[19]
1576	Benzeneethanamine, N-ethyl-α-methyl- N-Ethylamphetamine	$C_{11}H_{17}N$ 163.26	457-87-4	3720 105-6[14]	3-12-00-02668	1.4986[25]
1577	Benzeneethanamine, 2-methoxy- Phenethylamine, 2-methoxy	$C_9H_{13}NO$ 151.21	2045-79-6	236.5	3-13-00-01625 1.0400[25]	1.5422[20]
1578	Benzeneethanamine, α-methyl-, (±)- Amphetamine	$C_9H_{13}N$ 135.21	300-62-9	616 203	4-12-00-02587 0.9306[25]	H_2O 2; EtOH 3; eth 2; chl 3 1.518[26]
1579	Benzeneethanamine, β-methyl-	$C_9H_{13}N$ 135.21	582-22-9	210	4-12-00-02634 0.9433[4]	bz 4; eth 4; EtOH 4 1.5255[20]
1580	Benzeneethanamine, 4-methyl-	$C_9H_{13}N$ 135.21	3261-62-9	214	4-12-00-02641 0.93[25]	1.5257[20]
1581	Benzeneethanamine, N-methyl-	$C_9H_{13}N$ 135.21	589-08-2	206	4-12-00-02453 0.93[25]	1.5162[20]
1582	Benzeneethanamine, α-methyl-, (S)-, sulfate (2:1) Dextroamphetamine sulfate	$C_{18}H_{28}N_2O_4S$ 368.50	51-63-8 >300	2932	3-12-00-02665 1.15[25]	H_2O 4
1583	Benzeneethanamine, α-methyl-N-(2,2,2-trichloroethylidene)- Amphecloral	$C_{11}H_{12}Cl_3N$ 264.58	5581-35-1	613 96.[0.5]		1.530
1584	Benzeneethanamine, α-phenyl-	$C_{14}H_{15}N$ 197.28	25611-78-3	311; 175[15]	4-12-00-03313 1.031[15]	eth 4; EtOH 4
1585	Benzeneethanamine, N-(2-phenylethyl)-	$C_{16}H_{19}N$ 225.33	6308-98-1 29	360; 190[15]	4-12-00-02459	eth 4; EtOH 4 1.5550[25]
1586	Benzeneethanamine, 3,4,5-trimethoxy- Mescaline	$C_{11}H_{17}NO_3$ 211.26	54-04-6 35.5	5808 180[12]	4-13-00-02919	H_2O 3; EtOH 3; eth 1; bz 3
1587	Benzene, 1,1'-(1,2-ethanediyl)bis- Dibenzyl	$C_{14}H_{14}$ 182.27	103-29-7 52.5	1219 284	4-05-00-01868 0.9780[25]	H_2O 1; EtOH 3; eth 3; CS_2 3 1.5476[60]
1588	Benzene, 1,1'-[1,2-ethanediylbis(oxy)]bis-	$C_{14}H_{14}O_2$ 214.26	104-66-5 98	182[12]	4-06-00-00573	H_2O 1; EtOH 2; eth 3; chl 3
1589	Benzene, 1,1'-(1,2-ethanediyl)bis[3-methyl-	$C_{16}H_{18}$ 210.32	4662-96-8	298	4-05-00-01943 0.9703[22]	1.5566[22]
1590	Benzene, 1,1'-(1,2-ethanediyl)bis[4-methyl- 1,2-Di-p-tolylethane	$C_{16}H_{18}$ 210.32	538-39-6 85	3386 178[18]	4-05-00-01943	H_2O 1; EtOH 2; bz 3; peth 3
1591	Benzene, 1,1',1'',1'''-(1,2-ethanediylidene)tetrakis-	$C_{26}H_{22}$ 334.46	632-50-8 214.5	360	4-05-00-02746	EtOH 2; bz 3; HOAc 3
1592	Benzeneethanol Phenethyl alcohol	$C_8H_{10}O$ 122.17	60-12-8 -27	7185 218.2	4-06-00-03067 1.0202[20]	H_2O 2; EtOH 5; eth 5 1.5325[20]
1593	Benzeneethanol, β-amino-	$C_8H_{11}NO$ 137.18	7568-92-5 58.3	261; 160[17]	4-13-00-01835	H_2O 4; EtOH 4
1594	Benzeneethanol, 2-amino-	$C_8H_{11}NO$ 137.18	5339-85-5	147-8[3.5]	3-13-00-01679 1.045[25]	H_2O 4 1.5849[20]
1595	Benzeneethanol, 4-bromo-	C_8H_9BrO 201.06	4654-39-1	138[9]	4-06-00-03082 1.436[25]	1.5735[20]
1596	Benzeneethanol, 3-chloro-	C_8H_9ClO 156.61	5182-44-5	133-7[13]	4-06-00-03079 1.181[25]	1.5491[20]
1597	Benzeneethanol, 4-chloro-	C_8H_9ClO 156.61	1875-88-3	259	4-06-00-03079 1.1804[20]	1.5487[20]
1598	Benzeneethanol, 4-chloro-α,α-dimethyl-	$C_{10}H_{13}ClO$ 184.67	5468-97-3 34	126[12]	4-06-00-03291	1.5310[20]
1599	Benzeneethanol, 2,5-dimethoxy-	$C_{10}H_{14}O_3$ 182.22	7417-19-8	176[18]	4-06-00-07393 1.055[25]	1.5395[20]

No.	Name Synonym	Mol. Form. Mol. Wt.	CAS RN mp/°C	Merck No. bp/°C	Beil. Ref. den/g cm^{-3}	Solubility n_D
1600	Benzeneethanol, α,α-dimethyl-	C$_{10}$H$_{14}$O 150.22	100-86-7 24	215	4-06-00-03290 0.9787[16]	1.5173[16]
1601	Benzeneethanol, α,α-diphenyl-	C$_{20}$H$_{18}$O 274.36	4428-13-1 89.5	222[11]	4-06-00-05057	H$_2$O 1; EtOH 4; eth 2; chl 2
1602	Benzeneethanol, β-ethyl-	C$_{10}$H$_{14}$O 150.22	2035-94-1	125[25]; 81[2]	2-06-00-00488 0.989[16]	1.43[16]
1603	Benzeneethanol, 2-ethyl- Phenethyl alcohol, 2-ethyl	C$_{10}$H$_{14}$O 150.22	22545-12-6	249.7; 127[20]	4-06-00-03351 0.9972[20]	1.5304[20]
1604	Benzeneethanol, 4-ethyl- Phenethyl alcohol, 4-ethyl	C$_{10}$H$_{14}$O 150.22	22545-13-7 7.9	250	4-06-00-03355 0.9907[20]	1.5229[20]
1605	Benzeneethanol, α-ethyl-α-methyl-, (±)- 2-Butanol, 3-chloro (erythro, DL)	C$_4$H$_9$ClO 108.57	116783-12-1	235; 56.1[30]	3-06-00-01958 1.0610[25]	eth 4; EtOH 4; chl 4 1.4397[25]
1606	Benzeneethanol, 3-hydroxy-	C$_8$H$_{10}$O$_2$ 138.17	13398-94-2	168-73[4]	4-06-00-05936 1.082[25]	1.5643[20]
1607	Benzeneethanol, α-methyl-	C$_9$H$_{12}$O 136.19	698-87-3	125[25]	4-06-00-03192 0.991[20]	1.5190[20]
1608	Benzeneethanol, β-methyl-	C$_9$H$_{12}$O 136.19	1123-85-9	105-6[11]	2-06-00-00477 0.975[25]	H$_2$O 1; EtOH 3 1.5582[2]
1609	Benzeneethanol, 2-methyl-	C$_9$H$_{12}$O 136.19	19819-98-8 2	243.5	4-06-00-03234 1.016[25]	1.5355[20]
1610	Benzeneethanol, 4-methyl-	C$_9$H$_{12}$O 136.19	699-02-5	244.5; 94[6]	4-06-00-03244 1.0028[20]	1.5267[20]
1611	Benzeneethanol, β-(1-methylethyl)- 1-Butanol, 3-methyl-2-phenyl	C$_{11}$H$_{18}$O 164.25	90499-41-5	130[15]	4-06-00-03390 0.9694[25]	ace 4; bz 4; eth 4; EtOH 4 1.5137[20]
1612	Benzeneethanol, 2-nitro-	C$_8$H$_9$NO$_3$ 167.16	15121-84-3 2	267	3-06-00-01714 1.19[25]	1.5637[20]
1613	Benzeneethanol, α-phenyl-	C$_{14}$H$_{14}$O 198.26	614-29-9 67	177[15]	2-06-00-00637 1.0360[70]	
1614	Benzeneethanol, α-phenyl-, (S)-	C$_{14}$H$_{14}$O 198.26	5773-56-8 67.5	168[10]	4-06-00-04701 1.0358[70]	H$_2$O 1; EtOH 4; eth 4
1615	Benzeneethanol, β-phenyl- 2,2-Diphenylethanol	C$_{14}$H$_{14}$O 198.26	1883-32-5 64.5	193[20]; 144[1]	4-06-00-04724	EtOH 2; eth 3; ace 3; chl 3
1616	Benzeneethanol, α,α,β-trimethyl- 2-Butanol, 2-methyl-3-phenyl	C$_{11}$H$_{18}$O 164.25	3280-08-8 46	197; 106[12]	4-06-00-03389 0.9794[20]	ace 4; bz 4; eth 4; EtOH 4 1.5193[20]
1617	Benzeneethanol, α,β,β-trimethyl-	C$_{11}$H$_{18}$O 164.25	2977-31-3	197	4-06-00-03387	ace 4; bz 4; eth 4; EtOH 4 1.5161[13]
1618	Benzene, 1,1',1'',1'''-(1-ethanyl-2-ylidyne)tetrakis- 1,1,1,2-Tetraphenylethane	C$_{26}$H$_{22}$ 334.46	2294-94-2 144.8	278[21]	4-05-00-02747	bz 4
1619	Benzene, 1,1'-(1,2-ethenediyl)bis-, (E)- trans-Stilbene	C$_{14}$H$_{12}$ 180.25	103-30-0 123	8774 307; 166[12]	4-05-00-02156 0.9707[20]	H$_2$O 1; EtOH 2; eth 4; bz 4 1.6264[17]
1620	Benzene, 1,1'-(1,2-ethenediyl)bis-, (Z)- cis-Stilbene	C$_{14}$H$_{12}$ 180.25	645-49-8 -5	8774 141[12]	4-05-00-02155 1.0143[20]	H$_2$O 1; EtOH 3; eth 3; ace 3 1.6130[20]
1621	Benzene, 1,1'-(1,2-ethenediyl)bis[4-methoxy-	C$_{16}$H$_{16}$O$_2$ 240.30	4705-34-4 214.5	sub	4-06-00-06823	H$_2$O 1; ace 3; bz 3; HOAc 3
1622	Benzene, 1,1'-(1,2-ethenediyl)bis[2-methyl- 2,2'-Dimethylstilbene	C$_{16}$H$_{16}$ 208.30	10311-74-7 83	176-80[10]	4-05-00-02208	eth 4; EtOH 4; MeOH 4
1623	Benzene, 1,1',1'',1'''-(1,2-ethenediylidene)tetrakis-	C$_{26}$H$_{20}$ 332.44	632-51-9 225	420	4-05-00-02780 1.155[0]	H$_2$O 1; EtOH 2; eth 2; bz 4
1624	Benzene, ethenyl- Styrene	C$_8$H$_8$ 104.15	100-42-5 -31	8830 145	4-05-00-01334 0.9060[20]	H$_2$O 1; EtOH 3; eth 3; ace 3 1.5468[20]
1625	Benzene, (1-ethenyl-3-butenyl)- 1,5-Hexadiene, 3-phenyl	C$_{12}$H$_{14}$ 158.24	1076-66-0	78[8]	4-05-00-01555 0.8911[25]	1.5141[25]
1626	Benzene, 1-ethenyl-2,4-dimethyl-	C$_{10}$H$_{12}$ 132.21	2234-20-0	79-80[12]	4-05-00-01386 0.905[20]	1.539[20]
1627	Benzene, 1-ethenyl-3,5-dimethyl-	C$_{10}$H$_{12}$ 132.21	5379-20-4	57-8[4]	4-05-00-01386 0.894[25]	1.5382[20]
1628	Benzene, 2-ethenyl-1,4-dimethyl-	C$_{10}$H$_{12}$ 132.21	2039-89-6	69[10]	4-05-00-01385 0.9072[17]	1.5236[17]
1629	Benzene, 4-ethenyl-1,2-dimethyl-	C$_{10}$H$_{12}$ 132.21	27831-13-6	94-6[20]	4-05-00-01385 0.906[25]	1.5463[20]
1630	Benzene, 1-ethenyl-2-ethyl-	C$_{10}$H$_{12}$ 132.21	7564-63-8 -75.5	187.3; 68[12]	3-05-00-01216 0.9017[25]	1.5380[20]
1631	Benzene, 1-ethenyl-3-ethyl-	C$_{10}$H$_{12}$ 132.21	7525-62-4 -101	190.0	3-05-00-01217 0.8945[20]	1.5351[20]
1632	Benzene, 1-ethenyl-4-ethyl-	C$_{10}$H$_{12}$ 132.21	3454-07-7 -49.7	192.3; 86[20]	4-05-00-01384 0.8884[25]	1.5376[20]
1633	Benzene, 1-ethenyl-2-fluoro-	C$_8$H$_7$F 122.14	394-46-7	46[32]	4-05-00-01342 1.0282[20]	bz 4; eth 4; EtOH 4 1.5200[20]
1634	Benzene, 1-ethenyl-3-fluoro-	C$_8$H$_7$F 122.14	350-51-6	30-1[4]	4-05-00-01343 1.0177[20]	H$_2$O 1; EtOH 3; eth 3; bz 3 1.5170[20]
1635	Benzene, 1-ethenyl-4-fluoro-	C$_8$H$_7$F 122.14	405-99-2 -34.5	67.4[50]; 30[4]	4-05-00-01343 1.0220[20]	H$_2$O 1; EtOH 3; eth 3; bz 3 1.5150[20]
1636	Benzene, 1,1'-ethenylidenebis- 1,1-Diphenylethene	C$_{14}$H$_{12}$ 180.25	530-48-3 8.2	3323 277	4-05-00-02173 1.0232[20]	H$_2$O 1; eth 3; chl 3 1.6085[20]
1637	Benzene, 1-ethenyl-2-methoxy-	C$_9$H$_{10}$O 134.18	612-15-7 29	197; 83[12]	4-06-00-03771 1.0049[17]	ace 4; bz 4; eth 4; EtOH 4 1.5388[20]
1638	Benzene, 1-ethenyl-3-methoxy-	C$_9$H$_{10}$O 134.18	626-20-0	90-3[15]	4-06-00-03773 0.999[16]	H$_2$O 1; EtOH 3; eth 3; bz 3 1.5586[23]

No.	Name Synonym	Mol. Form. Mol. Wt.	CAS RN mp/°C	Merck No. bp/°C	Beil. Ref. den/g cm^{-3}	Solubility n_D
1639	Benzene, 1-ethenyl-4-methoxy-	$C_9H_{10}O$ 134.18	637-69-4	205; 91[13]	4-06-00-03775 1.0001[13]	H_2O 1; EtOH 3; eth 3; bz 3 1.5642[13]
1640	Benzene, 1-ethenyl-2-methyl- o-Methylstyrene	C_9H_{10} 118.18	611-15-4 -68.5	169.8	4-05-00-01367 0.9077[25]	H_2O 1; bz 3; chl 3 1.5437[20]
1641	Benzene, 1-ethenyl-3-methyl- m-Methylstyrene	C_9H_{10} 118.18	100-80-1 -86.3	164	4-05-00-01367 0.9076[25]	H_2O 1; EtOH 3; eth 3; bz 3 1.5411[20]
1642	Benzene, 1-ethenyl-4-methyl- p-Methylstyrene	C_9H_{10} 118.18	622-97-9 -34.1	172.8	4-05-00-01369 0.9173[25]	H_2O 1; bz 3 1.5420
1643	Benzene, 1-ethenyl-4-(1-methylethyl)-	$C_{11}H_{14}$ 146.23	2055-40-5 -44.7	204.1	4-05-00-01408 0.8850[20]	ace 4; bz 4; eth 4; EtOH 4 1.5289[20]
1644	Benzene, 1-ethenyl-3-nitro-	$C_8H_7NO_2$ 149.15	586-39-0 -10	120-1[11]	4-05-00-01351 1.1552[32]	H_2O 1; EtOH 3; eth 3; bz 3 1.5836[20]
1645	Benzene, 1-ethenyl-4-nitro-	$C_8H_7NO_2$ 149.15	100-13-0 29	dec	4-05-00-01351	EtOH 4; eth 4; chl 3; HOAc 3
1646	Benzene, (ethenyloxy)-	C_8H_8O 120.15	766-94-9	155.5	4-06-00-00561 0.9770[20]	H_2O 1; eth 4 1.5224[20]
1647	Benzene, 2-ethenyl-1,3,5-trimethyl-	$C_{11}H_{14}$ 146.23	769-25-5 -37	209	4-05-00-01408 0.9057[20]	ctc 3 1.5296[20]
1648	Benzene, 1,1',1''-(1-ethenyl-2-ylidene)tris-	$C_{20}H_{16}$ 256.35	58-72-0 72.5	220-1[14]	4-05-00-02575 1.0373[78]	H_2O 1; EtOH 3; eth 4; chl 3 1.6292[78]
1649	Benzene, ethoxy- Phenetole	$C_8H_{10}O$ 122.17	103-73-1 -29.5	7189 169.8	4-06-00-00554 0.9651[20]	H_2O 1; EtOH 3; eth 3; ctc 3 1.5076[20]
1650	Benzene, 2-ethoxy-1,3-dinitro- Phenetole, 2,6-dinitro-	$C_8H_8N_2O_5$ 212.16	13027-43-5 60.5	137[3]	3-06-00-00868	eth 4
1651	Benzene, (2-ethoxyethenyl)-	$C_{10}H_{12}O$ 148.20	17655-74-2 0	224.5	0.979[20]	1.5496[20]
1652	Benzene, (2-ethoxyethoxy)- Ethane, 1-ethoxy-2-phenoxy-	$C_{10}H_{14}O_2$ 166.22	19594-02-6	230	4-06-00-00572 1.018[11]	H_2O 1
1653	Benzene, 1-ethoxy-4-ethyl-	$C_{10}H_{14}O$ 150.22	1585-06-4	210	3-06-00-01665 0.9385[17]	ace 4; bz 4; EtOH 4
1654	Benzene, 1-ethoxy-2-fluoro-	C_8H_9FO 140.16	451-80-9 -16.7	171.4	4-06-00-00771 1.0874[17]	H_2O 1; ace 3; bz 3; ctc 2 1.4932[17]
1655	Benzene, 1-ethoxy-3-fluoro-	C_8H_9FO 140.16	458-03-7 -27.5	172; 65.2[15]	3-06-00-00669 1.0716[16]	H_2O 1; bz 3 1.4847[17]
1656	Benzene, 1-ethoxy-4-fluoro-	C_8H_9FO 140.16	459-26-7 -8.5	173; 54[7]	4-06-00-00774 1.0715[18]	H_2O 1; bz 3; ctc 2; chl 3 1.4826[18]
1657	Benzene, 1-ethoxy-4-iodo-	C_8H_9IO 248.06	699-08-1 29	250	4-06-00-01077	bz 4; eth 4; EtOH 4; chl 4
1658	Benzene, (ethoxymethyl)- Benzyl ethyl ether	$C_9H_{12}O$ 136.19	539-30-0	1147 186	4-06-00-02229 0.9478[20]	H_2O 1; EtOH 5; eth 5 1.4955[20]
1659	Benzene, 1-ethoxy-2-methyl-	$C_9H_{12}O$ 136.19	614-71-1	184	4-06-00-01944 0.9592[13]	eth 4; EtOH 4 1.508[13]
1660	Benzene, 1-ethoxy-3-methyl-	$C_9H_{12}O$ 136.19	621-32-9	192	4-06-00-02039 0.949[20]	H_2O 1; EtOH 3; eth 3 1.513[20]
1661	Benzene, 1-ethoxy-4-methyl-	$C_9H_{12}O$ 136.19	622-60-6	188.5	4-06-00-02099 0.9509[18]	H_2O 1; EtOH 3; eth 3; ctc 2 1.5058[18]
1662	Benzene, 1-ethoxy-2-nitro-	$C_8H_9NO_3$ 167.16	610-67-3 2.1	267	4-06-00-01250 1.1903[15]	eth 4; EtOH 4 1.5425[20]
1663	Benzene, 1-ethoxy-3-nitro-	$C_8H_9NO_3$ 167.16	621-52-3 36	264; 169[70]	4-06-00-01271	H_2O 1; EtOH 3; eth 3
1664	Benzene, 1-ethoxy-4-nitro-	$C_8H_9NO_3$ 167.16	100-29-8 60	283	4-06-00-01283 1.1176[100]	H_2O 2; EtOH 2; eth 4; ace 5
1665	Benzene, 1-ethoxy-2-(1-propenyl)- Phenetole, 2-propenyl	$C_{11}H_{14}O$ 162.23	67191-37-1	231; 120[17]	0-06-00-00565 0.9731[24]	1.544[24]
1666	Benzene, (3-ethoxy-1-propenyl)-	$C_{11}H_{14}O$ 162.23	1476-07-9	127-8[22]	3-06-00-02404 0.970[15]	1.547[15]
1667	Benzene, 1-ethoxy-2-propyl- Phenetole, 2-propyl	$C_{11}H_{16}O$ 164.25	101144-90-5	213; 100[16]	0-06-00-00499 0.9240[20]	1.494[20]
1668	Benzene, ethyl- Ethylbenzene	C_8H_{10} 106.17	100-41-4 -94.9	3723 136.1	4-05-00-00885 0.8670[20]	H_2O 1; EtOH 5; eth 5; chl 2 1.4959[20]
1669	Benzene, (1-ethylbutyl)-	$C_{12}H_{18}$ 162.27	4468-42-2 -55.4	209	4-05-00-01117 0.8239[25]	1.4859[20]
1670	Benzene, 1-ethyl-2,3-dimethyl- 3-Ethyl-o-xylene	$C_{10}H_{14}$ 134.22	933-98-2 -49.5	194	4-05-00-01069 0.8881[25]	1.5117[20]
1671	Benzene, 1-ethyl-2,4-dimethyl-	$C_{10}H_{14}$ 134.22	874-41-9 -62.9	188.4	4-05-00-01070 0.8763[20]	ace 4; bz 4; eth 4; EtOH 4 1.5038[20]
1672	Benzene, 1-ethyl-3,5-dimethyl-	$C_{10}H_{14}$ 134.22	934-74-7 -84.3	183.6	4-05-00-01071 0.8608[25]	H_2O 1; EtOH 5; eth 5; ace 5 1.4981[20]
1673	Benzene, 2-ethyl-1,3-dimethyl-	$C_{10}H_{14}$ 134.22	2870-04-4 -16.2	190	4-05-00-01069 0.8864[25]	1.5107[20]
1674	Benzene, 2-ethyl-1,4-dimethyl-	$C_{10}H_{14}$ 134.22	1758-88-9 -53.7	186.9	4-05-00-01070 0.8732[25]	H_2O 1; EtOH 5; eth 5; ace 5 1.5043[20]
1675	Benzene, 4-ethyl-1,2-dimethyl-	$C_{10}H_{14}$ 134.22	934-80-5 -66.9	189.5	4-05-00-01069 0.8706[25]	H_2O 1; EtOH 5; eth 5; ace 5 1.5031[20]
1676	Benzene, 1,1'-(1-ethyl-1,2-ethanediyl)bis-	$C_{16}H_{18}$ 210.32	5223-59-6	290; 152[11]	4-05-00-01939 0.9777[20]	H_2O 1; EtOH 3; eth 3; bz 3 1.5554[20]
1677	Benzene, 1,1'-ethylidenebis- 1,1-Diphenylethane	$C_{14}H_{14}$ 182.27	612-00-0 -17.9	272.6	4-05-00-01880 0.9997[20]	H_2O 1; EtOH 5; eth 5; bz 3 1.5756[20]
1678	Benzene, 1,1'-ethylidenebis[4-chloro- 1,1-Bis(4-chlorophenyl)ethane	$C_{14}H_{12}Cl_2$ 251.15	3547-04-4 56	320	4-05-00-01881	

No.	Name / Synonym	Mol. Form. / Mol. Wt.	CAS RN / mp/°C	Merck No. / bp/°C	Beil. Ref. / den/g cm^{-3}	Solubility / n_D
1679	Benzene, 1,1',1"-ethylidynetris-	$C_{20}H_{18}$ 258.36	5271-39-6 95	207[18]	4-05-00-02510	H_2O 1; EtOH 2; eth 3; bz 3
1680	Benzene, 1-ethyl-2-iodo- o-Ethyliodobenzene	C_8H_9I 232.06	18282-40-1 226		4-05-00-00910 1.6189[10]	ace 4; bz 4 1.5941[22]
1681	Benzene, 1-ethyl-4-iodo-	C_8H_9I 232.06	25309-64-2 -17	209	3-05-00-00801 1.6095[10]	ace 4; bz 4 1.5909[22]
1682	Benzene, 1-ethyl-2-methoxy-	$C_9H_{12}O$ 136.19	14804-32-1	187; 80[14]	3-06-00-01656 0.9636[19]	bz 4; eth 4 1.5142[20]
1683	Benzene, 1-ethyl-3-methoxy-	$C_9H_{12}O$ 136.19	10568-38-4	197; 74[10]	3-06-00-01662 0.9575[18]	H_2O 1; eth 4; bz 3; ctc 3 1.5102
1684	Benzene, 1-ethyl-4-methoxy-	$C_9H_{12}O$ 136.19	1515-95-3	198	4-06-00-03021 0.9624[15]	bz 4; eth 4 1.5120[20]
1685	Benzene, 1-ethyl-2-methyl- o-Ethyltoluene	C_9H_{12} 120.19	611-14-3 -80.8	165.2	4-05-00-00999 0.8807[20]	H_2O 1; EtOH 5; eth 5; ace 5 1.5046[20]
1686	Benzene, 1-ethyl-3-methyl- m-Ethyltoluene	C_9H_{12} 120.19	620-14-4 -95.5	161.3	4-05-00-01001 0.8645[20]	H_2O 1; EtOH 4; eth 4; ace 5 1.4966[20]
1687	Benzene, 1-ethyl-4-methyl- p-Ethyltoluene	C_9H_{12} 120.19	622-96-8 -62.3	162	4-05-00-01003 0.8614[20]	H_2O 1; EtOH 4; eth 4; ace 5 1.4959[20]
1688	Benzene, 1-ethyl-2-(1-methylethyl)- Benzene, 1-ethyl-2-isopropyl	$C_{11}H_{16}$ 148.25	18970-44-0 193		4-05-00-01100 0.888[20]	ace 4; bz 4; eth 4; EtOH 4 1.508[20]
1689	Benzene, 1-ethyl-3-(1-methylethyl)- Benzene, 1-ethyl-3-isopropyl	$C_{11}H_{16}$ 148.25	4920-99-4 <-20	192	3-05-00-01005 0.859[20]	H_2O 1; EtOH 3; eth 3; bz 4 1.4921[20]
1690	Benzene, 1-ethyl-4-(1-methylethyl)- Benzene, 1-ethyl-4-isopropyl	$C_{11}H_{16}$ 148.25	4218-48-8 <-20	196.6	4-05-00-01100 0.8585[20]	H_2O 1; eth 3; bz 4 1.4923[20]
1691	Benzene, (1-ethyl-1-methylpropyl)- Pentane, 3-methyl-3-phenyl	$C_{12}H_{18}$ 162.27	1985-97-3 205		4-05-00-01120 0.8773[15]	1.4972[16]
1692	Benzene, 1-ethyl-2-nitro-	$C_8H_9NO_2$ 151.17	612-22-6 -12.3	232.5	4-05-00-00911 1.1207[20]	H_2O 1; EtOH 4; eth 4; ace 3 1.5356[20]
1693	Benzene, 1-ethyl-3-nitro-	$C_8H_9NO_2$ 151.17	7369-50-8 -37.9	242.5	4-05-00-00911 1.1345[25]	ace 4; eth 4; EtOH 4
1694	Benzene, 1-ethyl-4-nitro-	$C_8H_9NO_2$ 151.17	100-12-9 -12.3	245.5	4-05-00-00912 1.1192[20]	H_2O 1; EtOH 4; eth 4; ace 3 1.5455[20]
1695	Benzene, (1-ethyloctadecyl)- 3-Phenyleicosane	$C_{26}H_{46}$ 358.65	2400-02-4 29.3	202[1]	4-05-00-01222 0.8546[20]	1.4796[20]
1696	Benzene, 1,1'-4-(phenylmethyl)-	$C_{15}H_{16}$ 196.29	620-85-9 -24	297	4-05-00-01926 0.9777[20]	EtOH 4; eth 3; chl 3 1.5616[20]
1697	Benzene, 1,1'-(1-ethyl-1,3-propanediyl)bis- Pentane, 1,3-diphenyl	$C_{17}H_{20}$ 224.35	838-45-9	304; 113[0.8]	4-05-00-01963 0.9734[21]	1.553[21]
1698	Benzene, 1,1'-(2-ethyl-1,3-propanediyl)bis- Propane, 1,1-dibenzyl	$C_{17}H_{20}$ 224.35	1520-45-2	304.5	4-05-00-01964 0.9734[21]	1.553[21]
1699	Benzene, (1-ethyl-1-propenyl)- 3-Phenyl-2-pentene	$C_{11}H_{14}$ 146.23	4701-36-4	198; 92[18]	4-05-00-01404 0.9173[14]	1.5266[15]
1700	Benzene, (1-ethyl-2-propenyl)-	$C_{11}H_{14}$ 146.23	19947-22-9	191.5; 71[12]	3-05-00-01240 0.8458[23]	1.5030[21]
1701	Benzene, (1-ethylpropyl)-	$C_{11}H_{16}$ 148.25	1196-58-3	187.5	4-05-00-01090 0.8649[20]	1.4880[20]
1702	Benzene, 1-ethyl-2-propyl-	$C_{11}H_{16}$ 148.25	16021-20-8	203	4-05-00-01099 0.8744[20]	1.4992[20]
1703	Benzene, 1-ethyl-4-propyl- 1-Ethyl-4-propylbenzene	$C_{11}H_{16}$ 148.25	20024-90-2	205	4-05-00-01100 0.8594[20]	ace 4; bz 4; eth 4; EtOH 4 1.4921[21]
1704	Benzene, (ethylsulfonyl)-	$C_8H_{10}O_2S$ 170.23	599-70-2 42	160[12]	4-06-00-01469 1.1410[20]	bz 4; eth 4; EtOH 4; chl 4
1705	Benzene, (ethylthio)-	$C_8H_{10}S$ 138.23	622-38-8	205	4-06-00-01468 1.0211[20]	EtOH 3 1.5670[20]
1706	Benzene, 1-(ethylthio)-3-methyl- Ethyl m-tolyl sulfide	$C_9H_{12}S$ 152.26	34786-24-8	219	3-06-00-01332 0.9987[20]	1.5590[20]
1707	Benzene, 1-(ethylthio)-4-methyl-	$C_9H_{12}S$ 152.26	622-63-9	220	4-06-00-02156 0.9996[20]	1.555[20]
1708	Benzene, 1-ethyl-2,3,4-trimethyl- Benzene, 4-ethyl-1,2,3-trimethyl	$C_{11}H_{16}$ 148.25	61827-86-9	220.4	4-05-00-01107 0.9019[20]	ace 4; bz 4; eth 4; EtOH 4 1.5180[20]
1709	Benzene, 1-ethyl-2,3,5-trimethyl- 1-Ethyl-2,3,5-trimethylbenzene	$C_{11}H_{16}$ 148.25	18262-85-6	213	4-05-00-01107 0.8897[20]	ace 4; bz 4; eth 4; EtOH 4 1.5118[20]
1710	Benzene, 1-ethyl-2,4,5-trimethyl- 1-Ethyl-2,4,5-trimethylbenzene	$C_{11}H_{16}$ 148.25	17851-27-3 -13.5	213	4-05-00-01108 0.883[20]	ace 4; bz 4; eth 4; EtOH 4 1.5075[20]
1711	Benzene, 2-ethyl-1,3,4-trimethyl- Benzene, 3-ethyl-1,2,4-trimethyl	$C_{11}H_{16}$ 148.25	61827-87-0	216.6	4-05-00-01107 0.895[20]	ace 4; bz 4; eth 4; EtOH 4 1.5133[20]
1712	Benzene, 2-ethyl-1,3,5-trimethyl-	$C_{11}H_{16}$ 148.25	3982-67-0 -15.5	212.4	4-05-00-01107 0.883[20]	ace 4; bz 4; eth 4; EtOH 4 1.5074[20]
1713	Benzene, 5-ethyl-1,2,3-trimethyl- 5-Ethyl-1,2,3-trimethylbenzene	$C_{11}H_{16}$ 148.25	31366-00-4	215.8	4-05-00-01107 0.8863[20]	ace 4; eth 4; EtOH 4 1.5101[20]
1714	Benzene, 1,1'-(1,2-ethynediyl)bis- Diphenylacetylene	$C_{14}H_{10}$ 178.23	501-65-5 62.5	9428 300	4-05-00-02276 0.9657[100]	H_2O 1; EtOH 2; eth 4; ctc 2
1715	Benzene, ethynyl- Ethynylbenzene	C_8H_6 102.14	536-74-3 -44.8	3817 143	4-05-00-01525 0.9300[20]	H_2O 1; EtOH 5; eth 5; ace 3 1.5470[20]
1716	Benzene, (ethynyloxy)-	C_8H_6O 118.14	4279-76-9 -36	61-2[25]	4-06-00-00565 1.0614[20]	eth 4; EtOH 4 1.5125[20]
1717	Benzene, fluoro- Fluorobenzene	C_6H_5F 96.10	462-06-6 -42.2	4099 84.7	4-05-00-00632 1.0225[20]	bz 4; eth 4; EtOH 4; lig 4 1.4684[30]
1718	Benzene, 1-fluoro-2,4-dinitro- 2,4-Dinitrophenyl fluoride	$C_6H_3FN_2O_4$ 186.10	70-34-8 25.8	4101 296	4-05-00-00742 1.4718[54]	EtOH 3; chl 2 1.5690[20]

No.	Name / Synonym	Mol. Form. / Mol. Wt.	CAS RN / mp/°C	Merck No. / bp/°C	Beil. Ref. / den/g cm^{-3}	Solubility / n_D
1719	Benzene, 1-fluoro-2-iodo-	C$_6$H$_4$FI / 222.00	348-52-7 / -41.5	188.6	4-05-00-00693 /	ace 3; bz 3; chl 3 / 1.5910[20]
1720	Benzene, 1-fluoro-4-iodo-	C$_6$H$_4$FI / 222.00	352-34-1 / -27	183	4-05-00-00694 / 1.9523[15]	H$_2$O 3; EtOH 3; eth 3; ace 3 / 1.5270[22]
1721	Benzene, 1-fluoro-3-isothiocyanato-	C$_7$H$_4$FNS / 153.18	404-72-8 /	227	3-12-00-01276 / 1.27[25]	1.6186[20]
1722	Benzene, 1-fluoro-2-methoxy-	C$_7$H$_7$FO / 126.13	321-28-8 / -39	154.5	4-06-00-00771 / 1.5489[17]	H$_2$O 1; eth 3; ctc 3 / 1.4969[17]
1723	Benzene, 1-fluoro-3-methoxy-	C$_7$H$_7$FO / 126.13	456-49-5 / -35	159; 51[14]	4-06-00-00772 / 1.104[25]	1.4876[20]
1724	Benzene, 1-fluoro-4-methoxy-	C$_7$H$_7$FO / 126.13	459-60-9 / -45	157	4-06-00-00773 / 1.1781[18]	eth 3 / 1.4886[18]
1725	Benzene, (fluoromethyl)-	C$_7$H$_7$F / 110.13	350-50-5 / -35	140; 40[14]	4-05-00-00800 / 1.0228[25]	ctc 3 / 1.4892[25]
1726	Benzene, 1-fluoro-2-methyl- / o-Fluorotoluene	C$_7$H$_7$F / 110.13	95-52-3 / -62	4108 / 115	4-05-00-00799 / 1.0041[13]	H$_2$O 1; EtOH 4; eth 4 / 1.4704[20]
1727	Benzene, 1-fluoro-3-methyl- / m-Fluorotoluene	C$_7$H$_7$F / 110.13	352-70-5 / -87	4108 / 115	4-05-00-00799 / 0.9974[20]	H$_2$O 1; EtOH 4; eth 4 / 1.4691[20]
1728	Benzene, 1-fluoro-4-methyl- / p-Fluorotoluene	C$_7$H$_7$F / 110.13	352-32-9 / -56	4108 / 116.6	4-05-00-00800 / 0.9975[20]	H$_2$O 1; EtOH 4; eth 4 / 1.4699[20]
1729	Benzene, 4-fluoro-4-methyl-2-nitro- / Toluene, 4-fluoro-3-nitro-	C$_7$H$_6$FNO$_2$ / 155.13	446-11-7 / 26.5	241	3-05-00-00743 / 1.2619[28]	1.5237[28]
1730	Benzene, 2-fluoro-4-methyl-1-nitro- / 3-Fluoro-4-nitrotoluene	C$_7$H$_6$FNO$_2$ / 155.13	446-34-4 / 53.2	97[3]	4-05-00-00852 / 1.4380[25]	
1731	Benzene, 4-fluoro-1-methyl-2-nitro-	C$_7$H$_6$FNO$_2$ / 155.13	446-10-6 / 27	213; 138[83]	4-05-00-00851 / 1.2686[20]	1.5218[20]
1732	Benzene, 4-fluoro-2-methyl-1-nitro-	C$_7$H$_6$FNO$_2$ / 155.13	446-33-3 / 27.5	217; 97-8[10]	4-05-00-00851 / 1.272[28]	1.5271[20]
1733	Benzene, 1-fluoro-2-nitro- / o-Fluoronitrobenzene	C$_6$H$_4$FNO$_2$ / 141.10	1493-27-2 / -6	215 dec	4-05-00-00718 / 1.3285[18]	eth 4; EtOH 4 / 1.5489[17]
1734	Benzene, 1-fluoro-3-nitro- / m-Fluoronitrobenzene	C$_6$H$_4$FNO$_2$ / 141.10	402-67-5 / 41	199; 86[19]	4-05-00-00719 / 1.3254[19]	H$_2$O 1; EtOH 3; eth 3; bz 2 / 1.5262[15]
1735	Benzene, 1-fluoro-4-nitro- / p-Fluoronitrobenzene	C$_6$H$_4$FNO$_2$ / 141.10	350-46-9 / 21	205	4-05-00-00719 / 1.3300[20]	H$_2$O 1; EtOH 3; eth 3; ctc 2 / 1.5316[20]
1736	Benzene, 1-fluoro-2-(trichloromethyl)-	C$_7$H$_4$Cl$_3$F / 213.47	488-98-2 /	75[5]	3-05-00-00701 / 1.453[25]	1.5432[20]
1737	Benzene, 1-fluoro-2-(trifluoromethyl)-	C$_7$H$_4$F$_4$ / 164.10	392-85-8 /	114.5	4-05-00-00804 / 1.293[25]	1.4040[25]
1738	Benzene, 1-fluoro-3-(trifluoromethyl)-	C$_7$H$_4$F$_4$ / 164.10	401-80-9 / -81.5	101.5	4-05-00-00804 / 1.3021[17]	
1739	Benzene, 1-fluoro-4-(trifluoromethyl)-	C$_7$H$_4$F$_4$ / 164.10	402-44-8 / -41.7	103.5	4-05-00-00804 / 1.293[25]	1.4025[20]
1740	Benzene, 2-fluoro-1,3,5-trimethyl-	C$_9$H$_{11}$F / 138.18	392-69-8 / -36.7	171.5	4-05-00-01025 / 0.9745[25]	1.4809[25]
1741	Benzene, heneicosyl- / Heneicosylbenzene	C$_{27}$H$_{48}$ / 372.68	40775-09-5 / 48	439	4-05-00-01226 / 0.8545[20]	1.4804[20]
1742	Benzene, heptacosyl- / Heptacosylbenzene	C$_{33}$H$_{60}$ / 456.84	61828-25-9 / 64	397[100.] 262[1]	4-05-00-01243 / 0.8543[20]	1.4797[20]
1743	Benzene, heptadecyl- / Heptadecane, 1-phenyl-	C$_{23}$H$_{40}$ / 316.57	14752-75-1 / 32	397	3-05-00-01123 / 0.8546[20]	1.4810[20]
1744	Benzene, 1,1'-(1-heptenylidene)bis- / 1,1-Diphenyl-1-heptene	C$_{19}$H$_{22}$ / 250.38	1530-20-7 /		4-05-00-02252 / 0.9673[18]	EtOH 2 / 1.5648[18]
1745	Benzene, heptyl-	C$_{13}$H$_{20}$ / 176.30	1078-71-3 / -48	240; 109[10]	4-05-00-01143 / 0.8567[20]	H$_2$O 1; bz 3; chl 3 / 1.4865[20]
1746	Benzene, 1,1'-heptylidenebis- / 1,1-Diphenylheptane	C$_{19}$H$_{24}$ / 252.40	1530-05-8 / 14	333.5; 137[1]	4-05-00-01992 / 0.9497[20]	
1747	Benzene, (heptyloxy)-	C$_{13}$H$_{20}$O / 192.30	32395-96-3 / -33.5	267	4-06-00-00560 / 0.9170[15]	ace 4; eth 4; EtOH 4 / 1.4912[20]
1748	Benzenehexacarboxylic acid / Mellitic acid	C$_{12}$H$_6$O$_{12}$ / 342.17	517-60-2 / 286-8 dec	5706	4-09-00-03825 /	H$_2$O 4; EtOH 3; sulf 3
1749	Benzene, hexachloro- / Hexachlorobenzene	C$_6$Cl$_6$ / 284.78	118-74-1 / 231.8	4600 / 325	4-05-00-00670 / 2.044[23]	H$_2$O 1; EtOH 2; eth 3; bz 4 / 1.5691[23]
1750	Benzene, hexacosyl- / Hexacosane, 1-phenyl-	C$_{32}$H$_{58}$ / 442.81	13024-80-1 / 62	388[100.] 250[1]	4-05-00-01242 / 0.8543[20]	1.4798[20]
1751	Benzene, hexadecyl-	C$_{22}$H$_{38}$ / 302.54	1459-09-2 / 27	385	4-05-00-01216 / 0.8547[20]	H$_2$O 1; EtOH 2; eth 4; bz 4 / 1.4813[20]
1752	Benzene, (hexadecyloxy)-	C$_{22}$H$_{38}$O / 318.54	35021-70-6 / 41.8	200[1]	3-06-00-00555 / 0.8434[82]	1.4556[82]
1753	Benzene, 1,3-hexadienyl- / 1-Phenyl-1,3-hexadiene	C$_{12}$H$_{14}$ / 158.24	41635-77-2 /	128[16]	3-05-00-01378 / 0.9253[12]	1.6025[12]
1754	Benzene, 1,5-hexadienyl- / 1,5-Hexadiene, 1-phenyl-	C$_{12}$H$_{14}$ / 158.24	1009-81-0 /	102-3[8]	4-05-00-01554 / 0.9005[25]	1.5421[25]
1755	Benzene, 3,5-hexadienyl- / 6-Phenyl-1,3-hexadiene	C$_{12}$H$_{14}$ / 158.24	39669-95-9 /	100.5[11]	3-05-00-01378 / 0.9304[13]	1.5446[13]
1756	Benzene, hexaethyl- / Hexaethylbenzene	C$_{18}$H$_{30}$ / 246.44	604-88-6 / 129	298	4-05-00-01208 / 0.8305[130]	H$_2$O 1; EtOH 3; eth 4; bz 4 / 1.4736[130]
1757	Benzene, hexafluoro- / Perfluorobenzene	C$_6$F$_6$ / 186.06	392-56-3 / 5.3	80.2	4-05-00-00640 / 1.6184[20]	1.3777[20]

No.	Name / Synonym	Mol. Form. / Mol. Wt.	CAS RN / mp/°C	Merck No. / bp/°C	Beil. Ref. / den/g cm^{-3}	Solubility / n_D
1758	Benzene, hexamethyl- / Mellitene	$C_{12}H_{18}$ / 162.27	87-85-4 / 166.5	263.4	4-05-00-01137 / 1.0630[25]	H_2O 1; EtOH 3; eth 3; ace 3
1759	Benzenehexanoic acid	$C_{12}H_{16}O_2$ / 192.26	5581-75-9 / 23	206[30]	4-09-00-01895	1.5164[21]
1760	Benzene, 2-hexenyl- / 2-Hexene, 1-phenyl	$C_{12}H_{16}$ / 160.26	67590-77-6	108[16]	0-05-00-00501 / 0.8898[16]	1.5058[16]
1761	Benzene, 3-hexenyl-	$C_{12}H_{16}$ / 160.26	35008-86-7	222	4-05-00-01418 / 0.8790[20]	1.5039[20]
1762	Benzene, 5-hexenyl-	$C_{12}H_{16}$ / 160.26	1588-44-9	94-5[10]	3-05-00-01251 / 0.8839[20]	1.5033[20]
1763	Benzene, hexyl-	$C_{12}H_{18}$ / 162.27	1077-16-3 / -61	226.1	4-05-00-01115 / 0.8575[20]	H_2O 1; eth 5; bz 3; peth 3 / 1.4864[20]
1764	Benzene, (1-hexylheptyl)- / 7-Phenyltridecane	$C_{19}H_{32}$ / 260.46	2400-01-3	183-4[20]	4-05-00-01210 / 0.8723[20]	1.4931[18]
1765	Benzene, (hexyloxy)-	$C_{12}H_{18}O$ / 178.27	1132-66-7 / -19	240	4-06-00-00560 / 0.9174[20]	H_2O 1; eth 3 / 1.4921[20]
1766	Benzene, iodo- / Iodobenzene	C_6H_5I / 204.01	591-50-4 / -31.3	4922 / 188.4	4-05-00-00688 / 1.8308[20]	H_2O 1; EtOH 3; eth 5; ace 5 / 1.6200[20]
1767	Benzene, 1-iodo-2,3-dimethoxy- / 3-Iodoveratrole	$C_8H_9IO_2$ / 264.06	25245-33-4 / 45.5	144-5[14]	3-06-00-04262 / 1.7799[20]	1.6127[20]
1768	Benzene, 4-iodo-1,2-dimethoxy-	$C_8H_9IO_2$ / 264.06	5460-32-2 / 35	174[30]	4-06-00-05625	bz 4; EtOH 4
1769	Benzene, 1-iodo-2,3-dimethyl-	C_8H_9I / 232.06	31599-60-7	229	3-05-00-00820 / 1.6395[20]	ace 4 / 1.6074[20]
1770	Benzene, 1-iodo-2,4-dimethyl-	C_8H_9I / 232.06	4214-28-2	231 dec; 111[14]	4-05-00-00947 / 1.6282[16]	H_2O 1; ace 3; bz 3 / 1.6008[16]
1771	Benzene, 1-iodo-3,5-dimethyl- / m-Xylene, 5-iodo-	C_8H_9I / 232.06	22445-41-6	230.5	3-05-00-00840 / 1.6085[18]	ace 4 / 1.5967[18]
1772	Benzene, 2-iodo-1,3-dimethyl-	C_8H_9I / 232.06	608-28-6 / 11.2	229.5	4-05-00-00947 / 1.6158[20]	H_2O 1; ace 3; bz 3 / 1.6035[20]
1773	Benzene, 2-iodo-1,4-dimethyl-	C_8H_9I / 232.06	1122-42-5	227 dec	4-05-00-00970 / 1.6168[17]	H_2O 1; ace 3; bz 3 / 1.5992[17]
1774	Benzene, 4-iodo-1,2-dimethyl- / 4-Iodo-o-xylene	C_8H_9I / 232.06	31599-61-8	231.5	3-05-00-00821 / 1.6334[18]	ace 4 / 1.6049[18]
1775	Benzene, 1-iodo-2-methoxy- / o-Iodoanisole	C_7H_7IO / 234.04	529-28-2	4921 / 241; 91[2]	4-06-00-01070 / 1.8[20]	EtOH 4; eth 4; ace 4; bz 4
1776	Benzene, 1-iodo-4-methoxy-	C_7H_7IO / 234.04	696-62-8 / 53	238; 138[25]	4-06-00-01075	EtOH 3; eth 3; chl 3
1777	Benzene, (iodomethyl)-	C_7H_7I / 218.04	620-05-3 / 24.5	93[10]	4-05-00-00842 / 1.7335[25]	bz 4; eth 4; EtOH 4 / 1.6334[25]
1778	Benzene, 1-iodo-2-methyl-	C_7H_7I / 218.04	615-37-2	211.5	4-05-00-00838 / 1.713[20]	H_2O 1; EtOH 5; eth 5 / 1.6079[20]
1779	Benzene, 1-iodo-3-methyl-	C_7H_7I / 218.04	625-95-6 / -27.2	213	4-05-00-00839 / 1.705[20]	H_2O 1; EtOH 5; eth 5 / 1.6053[20]
1780	Benzene, 1-iodo-4-methyl-	C_7H_7I / 218.04	624-31-7 / 36.5	211	4-05-00-00840 / 1.678[20]	H_2O 1; EtOH 3; eth 3; chl 2
1781	Benzene, 2-iodo-1-methyl-4-(1-methylethyl)- / 2-Iodo-p-cymene	$C_{10}H_{13}I$ / 260.12	56739-95-8	139[23]	3-05-00-00960 / 1.4205[17]	1.5800[17]
1782	Benzene, 2-iodo-4-methyl-1-(1-methylethyl)- / p-Cymene, 3-iodo-	$C_{10}H_{13}I$ / 260.12	4395-81-7	123[13]; 80[5]	3-05-00-00960 / 1.4113[17]	1.5690[17]
1783	Benzene, 1-iodo-2-nitro-	$C_6H_4INO_2$ / 249.01	609-73-4 / 54	290; 162[18]	4-05-00-00732 / 1.9186[75]	H_2O 1; EtOH 3; eth 3
1784	Benzene, 1-iodo-3-nitro-	$C_6H_4INO_2$ / 249.01	645-00-1 / 38.5	280	4-05-00-00733 / 1.9477[50]	H_2O 1; EtOH 3; eth 3
1785	Benzene, 1-iodo-4-nitro-	$C_6H_4INO_2$ / 249.01	636-98-6 / 174.7	288	4-05-00-00734 / 1.8090[155]	H_2O 1; EtOH 3; HOAc 3
1786	Benzene, isocyanato- / Phenyl isocyanate	C_7H_5NO / 119.12	103-71-9	7267 / 163; 55[13]	4-12-00-00864 / 1.0956[20]	eth 4; chl 2 / 1.5368[20]
1787	Benzene, 1-isocyanato-2-methyl-	C_8H_7NO / 133.15	614-68-6	185	4-12-00-01768	H_2O 1; eth 3 / 1.5282[20]
1788	Benzene, 1-isocyanato-2-nitro-	$C_7H_4N_2O_3$ / 164.12	3320-86-3 / 41	137[18]	3-12-00-01535	bz 4; eth 4; chl 4
1789	Benzene, 1-isocyanato-3-nitro-	$C_7H_4N_2O_3$ / 164.12	3320-87-4 / 51	130[11]	4-12-00-01604	bz 4; eth 4; chl 4
1790	Benzene, 1-isocyanato-4-nitro-	$C_7H_4N_2O_3$ / 164.12	100-28-7 / 57	162[20]; 137[11]	4-12-00-01653	bz 4; eth 4; chl 4
1791	Benzene, (isocyanomethyl)-	C_8H_7N / 117.15	10340-91-7	199 dec; 93[55]	4-12-00-02219 / 0.972[15]	1.5193[20]
1792	Benzene, isothiocyanato- / Phenyl isothiocyanate	C_7H_5NS / 135.19	103-72-0 / -21	7268 / 221	4-12-00-00867 / 1.1303[20]	H_2O 1; EtOH 3; eth 3; ctc 3 / 1.6492[23]
1793	Benzene, 1-isothiocyanato-2-methoxy-	C_8H_7NOS / 165.22	3288-04-8	131-2[11]	4-13-00-00855 / 1.1878[20]	1.6458[20]
1794	Benzene, (isothiocyanatomethyl)-	C_8H_7NS / 149.22	622-78-6	243	4-12-00-02276 / 1.1246[16]	H_2O 1; EtOH 5; eth 3 / 1.6049[15]
1795	Benzenemethanamine / Benzylamine	C_7H_9N / 107.16	100-46-9 /	1139 / 185; 90[12]	4-12-00-02155 / 0.9813[20]	H_2O 5; EtOH 5; eth 5; ace 4 / 1.5401[20]
1796	Benzenemethanamine, 2-amino-	$C_7H_{10}N_2$ / 122.17	4403-69-4 / 61	269	4-13-00-00265	EtOH 4

No.	Name / Synonym	Mol. Form. / Mol. Wt.	CAS RN / mp/°C	Merck No. / bp/°C	Beil. Ref. / den/g cm^{-3}	Solubility / n_D
1797	Benzenemethanamine, N-[bis(4-methoxyphenyl)methylene]- / Benzylimidobis(p-methoxyphenyl)methane	$C_{22}H_{21}NO_2$ / 331.41	524-96-9 / 90	1153	3-12-00-02254	eth 4; chl 4
1798	Benzenemethanamine, N,N-bis(phenylmethyl)-	$C_{21}H_{21}N$ / 287.40	620-40-6 / 91.5	385	4-12-00-02183 / 0.9912[95]	H_2O 2; EtOH 2; eth 3; ctc 3
1799	Benzenemethanamine, 4-bromo- / Benzyl amine, 4-bromo	C_7H_8BrN / 186.05	3959-07-7 / 20	250	3-12-00-02350	eth 4
1800	Benzenemethanamine, N-3-butynyl-N-methyl-	$C_{12}H_{15}N$ / 173.26	15240-91-2 /	127[16]	4-12-00-02171 / 0.9372[20]	1.5202[20]
1801	Benzenemethanamine, N-(2-chloroethyl)-N-(phenylmethyl)-, hydrochloride / N-(2-Chloroethyl)dibenzylamine hydrochloride	$C_{16}H_{19}Cl_2N$ / 296.24	55-43-6 /	2138	4-12-00-02180	
1802	Benzenemethanamine, 3-chloro-N-methyl-	$C_8H_{10}ClN$ / 155.63	39191-07-6 /	88[4]	3-12-00-02342	chl 3 / 1.5350[25]
1803	Benzenemethanamine, 2,4-dichloro-	$C_7H_7Cl_2N$ / 176.04	95-00-1 /	124-6[13]	4-12-00-02376	chl 3 / 1.5762[25]
1804	Benzenemethanamine, N,N-diethyl-α-phenyl- / N,N-Diethylbenzhydrylamine	$C_{17}H_{21}N$ / 239.36	519-72-2 / 58.5	3103 / 170[17]	4-12-00-03283	
1805	Benzenemethanamine, 3,4-dimethoxy-	$C_9H_{13}NO_2$ / 167.21	5763-61-1 /	156[12]; 120[3]	4-13-00-02582 / 1.143[25]	chl 3
1806	Benzenemethanamine, α,α-dimethyl-	$C_9H_{13}N$ / 135.21	585-32-0 /	196.5	4-12-00-02633 / 0.9423[20]	1.5181[25]
1807	Benzenemethanamine, N,N-dimethyl- / Dimethylbenzylamine	$C_9H_{13}N$ / 135.21	103-83-3 /	181	4-12-00-02161 / 0.915[0]	H_2O 2; EtOH 5; eth 5 / 1.5011[20]
1808	Benzenemethanamine, 2-ethoxy- / 2-Ethoxybenzylamine	$C_9H_{13}NO$ / 151.21	37806-29-4 /	69-74[0.2]	4-13-00-01676 / 1.015[25]	1.5326[20]
1809	Benzenemethanamine, α-ethyl-	$C_9H_{13}N$ / 135.21	2941-20-0 /	206; 100[16]	4-12-00-02582 / 0.9346[25]	1.5173[23]
1810	Benzenemethanamine, N-ethyl-	$C_9H_{13}N$ / 135.21	14321-27-8 /	194	4-12-00-02163 / 0.9342[17]	H_2O 2; EtOH 3; eth 3; bz 3 / 1.5117[20]
1811	Benzenemethanamine, N-ethyl-N-phenyl- / Ethylbenzylaniline	$C_{15}H_{17}N$ / 211.31	92-59-1 / 35	3728 / 288; 185[22]	4-12-00-02176 / 1.001[55]	H_2O 1; EtOH 3; eth 3; chl 3 / 1.5943[23]
1812	Benzenemethanamine, 2-methoxy-	$C_8H_{11}NO$ / 137.18	6850-57-3 /	229	4-13-00-01675 / 1.051[25]	1.5475[20]
1813	Benzenemethanamine, 4-methoxy-	$C_8H_{11}NO$ / 137.18	2393-23-9 /	236.5	4-13-00-01719 / 1.050[15]	H_2O 2; EtOH 2; eth 2 / 1.5462[20]
1814	Benzenemethanamine, α-methyl-, (±)-	$C_8H_{11}N$ / 121.18	618-36-0 / 32	187	4-12-00-02424 / 0.9395[15]	H_2O 3; EtOH 5; eth 5; chl 3 / 1.5238[25]
1815	Benzenemethanamine, 2-methyl-	$C_8H_{11}N$ / 121.18	89-93-0 / -30	206; 81[15]	4-12-00-02518 / 0.9766[19]	1.5436[19]
1816	Benzenemethanamine, 3-methyl-	$C_8H_{11}N$ / 121.18	100-81-2 /	203.5	4-12-00-02565 / 0.966[25]	1.5360[20]
1817	Benzenemethanamine, 4-methyl-	$C_8H_{11}N$ / 121.18	104-84-7 / 12.5	195	4-12-00-02574 / 0.952[20]	1.5340[20]
1818	Benzenemethanamine, N-methyl-	$C_8H_{11}N$ / 121.18	103-67-3 /	180.5	4-12-00-02161 / 0.9442[18]	H_2O 4
1819	Benzenemethanamine, 4-(1-methylethyl)-	$C_{10}H_{15}N$ / 149.24	4395-73-7 /	227	4-12-00-02836	eth 4; EtOH 4 / 1.5182[17]
1820	Benzenemethanamine, N-(1-methylethyl)-	$C_{10}H_{15}N$ / 149.24	102-97-6 /	200; 93[10]	4-12-00-02165 / 0.892[25]	1.5025[20]
1821	Benzenemethanamine, N-(2-methylphenyl)-	$C_{14}H_{15}N$ / 197.28	5405-13-0 / 60	302.5	4-12-00-02177 / 1.014[65]	ace 4; EtOH 4; chl 4 / 1.5861[65]
1822	Benzenemethanamine, α-methyl-N-(1-phenylethyl)-	$C_{16}H_{19}N$ / 225.33	10024-74-5 /	296.5	4-12-00-02428 / 1.018[15]	1.573
1823	Benzenemethanamine, N-methyl-N-2-propynyl- / Pargyline	$C_{11}H_{13}N$ / 159.23	555-57-7 /	6988 / 96-7[11]	2-12-00-00548 / 0.944[25]	1.5213[20]
1824	Benzenemethanamine, N-(3-methylphenyl)- / m-Toluidine, N-benzyl	$C_{14}H_{15}N$ / 197.28	5405-17-4 /	312	4-12-00-02178 / 1.0083[65]	ace 4; bz 4; eth 4 / 1.5845[65]
1825	Benzenemethanamine, N-(4-methylphenyl)- / p-Toluidine, N-benzyl	$C_{14}H_{15}N$ / 197.28	5405-15-2 / 19.5	319; 181[10]	4-12-00-02178 / 1.0064[65]	bz 4; eth 4; EtOH 4; chl 4 / 1.5832[65]
1826	Benzenemethanamine, α-phenyl- / Benzhydrylamine	$C_{13}H_{13}N$ / 183.25	91-00-9 / 34	1085 / 304; 176[23]	4-12-00-03282 / 1.0633[20]	H_2O 2; ace 3 / 1.5963
1827	Benzenemethanamine, N-phenyl- / Benzylaniline	$C_{13}H_{13}N$ / 183.25	103-32-2 / 37.5	1140 / 306.5	4-12-00-02172 / 1.0298[65]	eth 4; EtOH 4 / 1.6118[25]
1828	Benzenemethanamine, N-(phenylmethyl)- / Dibenzylamine	$C_{14}H_{15}N$ / 197.28	103-49-1 / -26	2993 / 300 dec; 270[250]	4-12-00-02179 / 1.0256[22]	H_2O 1; EtOH 4; eth 4; ctc 3 / 1.5781[20]
1829	Benzenemethanamine, N-phenyl-N-(phenylmethyl)-	$C_{20}H_{19}N$ / 273.38	91-73-6 / 69	226[10]	3-12-00-02225 / 1.0444[80]	H_2O 1; EtOH 2; eth 3; bz 3 / 1.6065[80]
1830	Benzenemethanamine, α-propyl-	$C_{10}H_{15}N$ / 149.24	2941-19-7 /	221; 108[16]	3-12-00-02716 / 0.9366[20]	EtOH 4
1831	Benzene, 1,1',1'',1'''-methanetetraryltetrakis- / Tetraphenylmethane	$C_{25}H_{20}$ / 320.43	630-76-2 / 282	431	4-05-00-02741	H_2O 1; EtOH 1; eth 1; bz 3
1832	Benzenemethanethiol / Thiobenzyl alcohol	C_7H_8S / 124.21	100-53-8 / -30	9249 / 194.5	4-06-00-02632 / 1.058[20]	H_2O 1; EtOH 4; eth 4; ctc 2 / 1.5151[20]

No.	Name Synonym	Mol. Form. Mol. Wt.	CAS RN mp/°C	Merck No. bp/°C	Beil. Ref. den/g cm^{-3}	Solubility n_D
1833	Benzenemethanethiol, 4-chloro-	C_7H_7ClS 158.65	6258-66-8 19.5	113[17]	4-06-00-02770 1.202[25]	1.5893[20]
1834	Benzenemethanethiol, α-methyl-, (S)- Ethanethiol, 1-phenyl (1)	$C_8H_{10}S$ 138.23	33877-11-1	199; 83[10]	4-06-00-03062 1.022[20]	bz 4; eth 4; EtOH 4 1.5593[20]
1835	Benzenemethanimine, α-phenyl-	$C_{13}H_{11}N$ 181.24	1013-88-3	282	4-07-00-01365 1.0847[19]	eth 4 1.6191[19]
1836	Benzenemethanol Benzyl alcohol	C_7H_8O 108.14	100-51-6 -15.2	1138 205.3	4-06-00-02222 1.0419[24]	H_2O 3; EtOH 3; eth 3; ace 3 1.5396[20]
1837	Benzenemethanol, 2-amino-	C_7H_9NO 123.15	5344-90-1 83.5	273	3-13-00-01615	H_2O 3; EtOH 3; eth 3; bz 4
1838	Benzenemethanol, α-(1-aminoethyl)-, hydrochloride, (R*,S*)-(±)- Phenylpropanolamine hydrochloride	$C_9H_{14}ClNO$ 187.67	154-41-6 194	7279	2-13-00-00371	H_2O 4; EtOH 3; eth 1; bz 1
1839	Benzenemethanol, α-(aminomethyl)- Phenylethanolamine	$C_8H_{11}NO$ 137.18	7568-93-6 56.5	7258 160[17]	4-13-00-01801	H_2O 4; EtOH 3
1840	Benzenemethanol, α-(1-aminopropyl)- α-(α-Aminopropyl)benzyl alcohol	$C_{10}H_{15}NO$ 165.24	5897-76-7 79.5	483	3-13-00-01791	
1841	Benzenemethanol, 2-[bis(4- hydroxyphenyl)methyl]- Phenolphthalol	$C_{20}H_{18}O_3$ 306.36	81-92-5 201.5	7211	4-06-00-07623	
1842	Benzenemethanol, α,α-bis(1-methylethyl)-	$C_{13}H_{20}O$ 192.30	4397-05-1	229; 158[60]	4-06-00-03462 0.9755[20]	eth 4 1.5239[20]
1843	Benzenemethanol, 5-bromo-2-hydroxy- Bromosaligenin	$C_7H_7BrO_2$ 203.04	2316-64-5 113	1426	3-06-00-04541	bz 4; eth 4; EtOH 4; chl 4
1844	Benzenemethanol, α-(bromomethyl)-	C_8H_9BrO 201.06	2425-28-7	109-10[2]	4-06-00-03053 1.4994[20]	H_2O 1 1.5800[17]
1845	Benzenemethanol, 2-bromo-α-methyl-	C_8H_9BrO 201.06	5411-56-3	109-10[2]	4-06-00-03053 1.4994[20]	1.5800[17]
1846	Benzenemethanol, α-butyl- 1-Phenyl-1-pentanol	$C_{11}H_{16}O$ 164.25	583-03-9	3921 141[25]; 102[3]	2-06-00-00503 0.9655[20]	ace 4; eth 4; EtOH 4 1.4086[25]
1847	Benzenemethanol, α-butyl-α-methyl- 2-Hexanol, 2-phenyl-	$C_{12}H_{18}O$ 178.27	4396-98-9	129-30[4]	3-06-00-01995 0.9522[20]	1.5091[20]
1848	Benzenemethanol, 2-chloro-	C_7H_7ClO 142.58	17849-38-6 73	230	4-06-00-02589	H_2O 2; EtOH 4; eth 4; lig 4
1849	Benzenemethanol, 4-chloro-	C_7H_7ClO 142.58	873-76-7 75	235	4-06-00-02593	bz 4; eth 4; EtOH 4
1850	Benzenemethanol, 4-chloro-α-(4- chlorophenyl)-α-methyl- 1,1-Bis(4-chlorophenyl)ethanol	$C_{14}H_{12}Cl_2O$ 267.15	80-06-8 70	2086	4-06-00-04718	H_2O 1; EtOH 1; eth 3; bz 3
1851	Benzenemethanol, 4-chloro-α-(4- chlorophenyl)-α-(trichloromethyl)- 1,1-Bis(p-chlorophenyl)-2,2,2-trichloroethanol	$C_{14}H_9Cl_5O$ 370.49	115-32-2 77.5	3075	4-06-00-04722	H_2O 1; os 1
1852	Benzenemethanol, α-(chloromethyl)-	C_8H_9ClO 156.61	1674-30-2	128[17]	4-06-00-03045 1.1926[20]	EtOH 3; eth 4 1.5523[20]
1853	Benzenemethanol, 2,4-dichloro- 2,4-Dichlorobenzyl alcohol	$C_7H_8Cl_2O$ 177.03	1777-82-8 59.5	3048 150[25]	4-06-00-02597	chl 3
1854	Benzenemethanol, 2,4-dichloro-α-methyl-	$C_8H_8Cl_2O$ 191.06	1475-13-4	125-6[7]	4-06-00-03048 1.293[25]	1.5605[20]
1855	Benzenemethanol, α,α-diethyl-	$C_{11}H_{16}O$ 164.25	1565-71-5 <-17	223.5	4-06-00-03381 0.9831[20]	EtOH 4 1.5165[20]
1856	Benzenemethanol, 3,4-dimethoxy-	$C_9H_{12}O_3$ 168.19	93-03-8	298; 172[12]	4-06-00-07381 1.178[17]	H_2O 3; EtOH 3 1.555[17]
1857	Benzenemethanol, α,α-dimethyl- α-Cumyl alcohol	$C_9H_{12}O$ 136.19	617-94-7 36	202	4-06-00-03219 0.9735[20]	H_2O 1; EtOH 3; eth 3; bz 3 1.5325[20]
1858	Benzenemethanol, α,3-dimethyl- 1-(3-Methylphenyl)ethyl alcohol	$C_9H_{12}O$ 136.19	7287-81-2	112[12]	3-06-00-01823 0.9974[15]	eth 4; EtOH 4 1.5240[20]
1859	Benzenemethanol, α,4-dimethyl- 1-(4-Methylphenyl)ethanol	$C_9H_{12}O$ 136.19	536-50-5	3226 219	4-06-00-03242 0.9668[25]	H_2O 1; EtOH 4; eth 4 1.5246[20]
1860	Benzenemethanol, 2,4-dimethyl-	$C_9H_{12}O$ 136.19	16308-92-2 22	232; 150[44]	4-06-00-03250 1.0310[20]	1.5339[20]
1861	Benzenemethanol, 3,5-dimethyl-	$C_9H_{12}O$ 136.19	27129-87-9	219.5	0-06-00-00521 0.927[25]	1.5312[20]
1862	Benzenemethanol, 4-(dimethylamino)-	$C_9H_{13}NO$ 151.21	1703-46-4 69	123[1]	4-13-00-01771 1.059[14]	1.5727[14]
1863	Benzenemethanol, α-[1- (dimethylamino)ethyl]-, [R-(R*,S*)]- (1R,2S)-N-Methylephedrine	$C_{11}H_{17}NO$ 179.26	552-79-4 87.5	5987		H_2O 1; EtOH 3; eth 3; MeOH 3
1864	Benzenemethanol, α-(1,1-dimethylethyl)-	$C_{11}H_{16}O$ 164.25	3835-64-1 45	111[15]	1-06-00-00270	eth 4; EtOH 4
1865	Benzenemethanol, 4-(1,1-dimethylethyl)-	$C_{11}H_{16}O$ 164.25	877-65-6	140[20]	4-06-00-03402 0.928[25]	1.5179[20]
1866	Benzenemethanol, α,α-diphenyl- Triphenylmethanol	$C_{19}H_{16}O$ 260.34	76-84-6 164.2	9653 380	4-06-00-05014 1.199[0]	H_2O 1; EtOH 4; eth 4; ace 3
1867	Benzenemethanol, α,α-dipropyl- 4-Heptanol, 4-phenyl-	$C_{13}H_{20}O$ 192.30	4436-96-8	134[20]	2-06-00-00513 0.9470[15]	1.516[10]
1868	Benzenemethanol, α-ethenyl- 1-Phenylallyl alcohol	$C_9H_{10}O$ 134.18	4393-06-0		2-06-00-00529 1.0249[21]	H_2O 2; EtOH 3; eth 3; bz 3 1.5406[20]
1869	Benzenemethanol, α-ethenyl-α-methyl- 3-Buten-2-ol, 2-phenyl-	$C_{10}H_{12}O$ 148.20	6051-52-1	92-5[7]	4-06-00-03839 0.9965[21]	1.5277[21]

No.	Name / Synonym	Mol. Form. / Mol. Wt.	CAS RN mp/°C	Merck No. bp/°C	Beil. Ref. den/g cm^{-3}	Solubility n_D
1870	Benzenemethanol, α-ethyl- / α-Ethylbenzyl alcohol	$C_9H_{12}O$ 136.19	93-54-9	3727 219	4-06-00-03183 0.9915[25]	bz 4; eth 4; EtOH 4; MeOH 4 1.5169[23]
1871	Benzenemethanol, α-[1-(ethylamino)ethyl]-, (R*,S*)- / N-Ethylnorephedrine	$C_{11}H_{17}NO$ 179.26	37025-57-3 51.5	143[18]	4-13-00-01887	bz 4
1872	Benzenemethanol, α-ethyl-3-hydroxy- / 1-(m-Hydroxyphenyl)-1-propanol	$C_9H_{12}O_2$ 152.19	55789-02-1 107	177[13]	3-06-00-04623	eth 4; EtOH 4
1873	Benzenemethanol, α-ethyl-4-methoxy-	$C_{10}H_{14}O_2$ 166.22	5349-60-0	143[20]	3-06-00-04624	ctc 3 1.5277[20]
1874	Benzenemethanol, α-ethyl-α-methyl-, (R)-	$C_{10}H_{14}O$ 150.22	1006-06-0 -13	211; 113[23]	0.984[25]	H$_2$O 1; EtOH 3; eth 3 1.5185[20]
1875	Benzenemethanol, α-ethyl-α-propyl- / 3-Hexanol, 3-phenyl	$C_{12}H_{18}O$ 178.27	20731-93-5	134[27]	3-06-00-02000 0.964[20]	1.5100[20]
1876	Benzenemethanol, α-ethynyl- / 1-Phenylpropargyl alcohol	C_9H_8O 132.16	4187-87-5 22	114[12]	4-06-00-04066 1.0655[20]	1.5508[20]
1877	Benzenemethanol, α-ethynyl-, carbamate / Carfimate	$C_{10}H_9NO_2$ 175.19	3567-38-2 86.5	1845	4-06-00-04067	
1878	Benzenemethanol, α-ethynyl-α-methyl-	$C_{10}H_{10}O$ 146.19	127-66-2 52.3	217.5; 102[12]	4-06-00-04073 1.0314[20]	
1879	Benzenemethanol, 3-fluoro-	C_7H_7FO 126.13	456-47-3	201; 104[22]	4-06-00-02589 1.164[25]	1.5095[20]
1880	Benzenemethanol, 4-fluoro-	C_7H_7FO 126.13	459-56-3 23	210	4-06-00-02589	1.5080[20]
1881	Benzenemethanol, α-hexyl- / 1-Phenyl-1-heptanol	$C_{13}H_{20}O$ 192.30	614-54-0	275	4-06-00-03454 0.946[25]	1.5024[20]
1882	Benzenemethanol, 2-hydroxy- / Salicyl alcohol	$C_7H_8O_2$ 124.14	90-01-7 87	8294 sub	4-06-00-05896 1.1613[25]	H$_2$O 3; EtOH 3; eth 3; bz 3
1883	Benzenemethanol, 3-hydroxy-	$C_7H_8O_2$ 124.14	620-24-6 73	300 dec	4-06-00-05907 1.161[25]	H$_2$O 4; EtOH 4; eth 4; chl 2
1884	Benzenemethanol, 4-hydroxy- / p-Hydroxybenzyl alcohol	$C_7H_8O_2$ 124.14	623-05-2 124.5	252	4-06-00-05909	H$_2$O 4; EtOH 4; eth 3; bz 4
1885	Benzenemethanol, 2-hydroxy-3-methoxy-	$C_8H_{10}O_3$ 154.17	4383-05-5 61.5	162[12]	4-06-00-07378	chl 2
1886	Benzenemethanol, 4-hydroxy-3-methoxy-	$C_8H_{10}O_3$ 154.17	498-00-0 115	dec	4-06-00-07381	H$_2$O 3; EtOH 3; eth 3; bz 3
1887	Benzenemethanol, 4-hydroxy-α-[(methylamino)methyl]- / Synephrine	$C_9H_{13}NO_2$ 167.21	94-07-5 184.5	8994	2-13-00-00491	
1888	Benzenemethanol, 2-iodo-	C_7H_7IO 234.04	5159-41-1 92	147-9[32]	4-06-00-02605	1.6349[20]
1889	Benzenemethanol, 2-methoxy-	$C_8H_{10}O_2$ 138.17	612-16-8	249	4-06-00-05896 1.0386[25]	H$_2$O 1; EtOH 3; eth 5 1.5455[20]
1890	Benzenemethanol, 3-methoxy-	$C_8H_{10}O_2$ 138.17	6971-51-3	252	4-06-00-05907 1.112[25]	1.5440[20]
1891	Benzenemethanol, 4-methoxy- / Anise alcohol	$C_8H_{10}O_2$ 138.17	105-13-5 25	695 259.1	4-06-00-05909 1.109[26]	H$_2$O 3; EtOH 4; eth 4; ctc 3 1.5420[25]
1892	Benzenemethanol, 4-methoxy-, acetate	$C_{10}H_{12}O_3$ 180.20	104-21-2 84	270; 150[23]	4-06-00-05914 1.105[25]	ctc 3
1893	Benzenemethanol, 2-methoxy-α-methyl-	$C_9H_{12}O_2$ 152.19	13513-82-1 -85.1	125	3-06-00-04563 0.9647[20]	ace 4; bz 4; eth 4; EtOH 4 1.4024[20]
1894	Benzenemethanol, 3-methoxy-α-methyl-	$C_9H_{12}O_2$ 152.19	23308-82-9	133[15]	4-06-00-05929 1.0781[19]	eth 4; EtOH 4 1.5325[20]
1895	Benzenemethanol, 4-methoxy-α-methyl-	$C_9H_{12}O_2$ 152.19	3319-15-1	310 dec; 140[17]	4-06-00-05930 1.0794[20]	ctc 3 1.5310[25]
1896	Benzenemethanol, α-methyl- / α-Methylbenzyl alcohol	$C_8H_{10}O$ 122.17	98-85-1 20	205	2-06-00-00444 1.013[25]	H$_2$O 1; EtOH 4; eth 4 1.5265[20]
1897	Benzenemethanol, 2-methyl- / o-Methylbenzyl alcohol	$C_8H_{10}O$ 122.17	89-95-2 38	224; 118[20]	4-06-00-03109 1.023[40]	eth 4; EtOH 4; chl 4
1898	Benzenemethanol, 3-methyl- / m-Methylbenzyl alcohol	$C_8H_{10}O$ 122.17	587-03-1 <-20	215.5	4-06-00-03162 0.9157[17]	H$_2$O 2; EtOH 4; eth 4; chl 3
1899	Benzenemethanol, 4-methyl- / p-Methylbenzyl alcohol	$C_8H_{10}O$ 122.17	589-18-4 61.5	217	4-06-00-03171 0.978[22]	eth 4; EtOH 4
1900	Benzenemethanol, α-[1-(methylamino)ethyl]-, (R*,S*)-(±)- / (±)-Ephedrine	$C_{10}H_{15}NO$ 165.24	90-81-3 76.5	3561 135[12]	4-13-00-01881 1.1220[20]	H$_2$O 3; EtOH 3; eth 3; bz 3
1901	Benzenemethanol, α-(1-methylethyl)- / 1-Phenyl-2-methylpropyl alcohol	$C_{10}H_{14}O$ 150.22	611-69-8	223	4-06-00-03289 0.9869[14]	H$_2$O 1; EtOH 3; ace 3 1.5193[14]
1902	Benzenemethanol, 4-(1-methylethyl)- / Cumic alcohol	$C_{10}H_{14}O$ 150.22	536-60-7 28	2621 249	4-06-00-03348 0.9818[20]	H$_2$O 1; EtOH 5; eth 5; bz 4 1.5210[20]
1903	Benzenemethanol, α-[1-[methyl(3-phenyl-2-propenyl)amino]ethyl]- / Cinnamedrine	$C_{19}H_{23}NO$ 281.40	90-86-8 75	2299	3-13-00-01734	
1904	Benzenemethanol, α-methyl-α-propyl-	$C_{11}H_{16}O$ 164.25	4383-18-0	216	2-06-00-00505 0.9723[22]	EtOH 4
1905	Benzenemethanol, α-(2-methylpropyl)-	$C_{11}H_{16}O$ 164.25	1565-86-2	236; 112[9]	4-06-00-03380 0.9537[19]	ace 4; bz 4; eth 4; EtOH 4 1.5080[18]
1906	Benzenemethanol, 2-nitro-	$C_7H_7NO_3$ 153.14	612-25-9 74	270; 138[20]	4-06-00-02608	H$_2$O 2; EtOH 3; eth 3

No.	Name / Synonym	Mol. Form. / Mol. Wt.	CAS RN / mp/°C	Merck No. / bp/°C	Beil. Ref. / den/g cm^{-3}	Solubility / n_D
1907	Benzenemethanol, 3-nitro-	C$_7$H$_7$NO$_3$ 153.14	619-25-0 30.5	175-80[3]	4-06-00-02609 1.296[19]	H$_2$O 3; EtOH 3; eth 3; chl 2
1908	Benzenemethanol, 4-nitro- p-Nitrobenzyl alcohol	C$_7$H$_7$NO$_3$ 153.14	619-73-8 96.5	255 dec; 185[12]	4-06-00-02611	H$_2$O 2; EtOH 3; eth 3; ace 2
1909	Benzenemethanol, α-pentyl-	C$_{12}$H$_{18}$O 178.27	4471-05-0	170[50]; 128[5]	4-06-00-03417 0.9497[25]	EtOH 3; eth 3 1.5105[20]
1910	Benzenemethanol, α-phenyl- Benzohydrol	C$_{13}$H$_{12}$O 184.24	91-01-0 69	1100 298; 180[20]	4-06-00-04648	H$_2$O 2; EtOH 4; eth 4; ctc 4
1911	Benzenemethanol, α-2-propenyl-	C$_{10}$H$_{12}$O 148.20	936-58-3	228.5	4-06-00-03838 1.004[18]	1.5289[21]
1912	Benzenemethanol, α-propyl-, (R)-	C$_{10}$H$_{14}$O 150.22	22144-60-1 16	232	3-06-00-01845 0.9740[20]	eth 4; EtOH 4 1.5139[20]
1913	Benzenemethanol, α-[(2-pyridinylamino)methyl]- Phenyramidol	C$_{13}$H$_{14}$N$_2$O 214.27	553-69-5 83.5	7292	5-22-08-00301	
1914	Benzenemethanol, 3,4,5-trimethoxy- 3,4,5-Trimethoxybenzyl alcohol	C$_{10}$H$_{14}$O$_4$ 198.22	3840-31-1	228[25]	4-06-00-07695 1.232[25]	1.5439[20]
1915	Benzene, methoxy- Anisole	C$_7$H$_8$O 108.14	100-66-3 -37.5	699 153.7	4-06-00-00548 0.9940[20]	H$_2$O 1; EtOH 3; eth 3; ace 4 1.5174[20]
1916	Benzene, 1-methoxy-2,3-dimethyl-	C$_9$H$_{12}$O 136.19	2944-49-2 29	199	4-06-00-03097 0.9596[40]	H$_2$O 1; EtOH 4; eth 4; ace 3 1.5120[40]
1917	Benzene, 1-methoxy-2,4-dimethyl-	C$_9$H$_{12}$O 136.19	6738-23-4	192	4-06-00-03127 0.9740[16]	H$_2$O 1; EtOH 3; eth 3; bz 3 1.5190[16]
1918	Benzene, 1-methoxy-3,5-dimethyl-	C$_9$H$_{12}$O 136.19	874-63-5	194; 89[15]	4-06-00-03142 0.9627[15]	H$_2$O 1; EtOH 3; eth 3; bz 3 1.5110[20]
1919	Benzene, 2-methoxy-1,3-dimethyl-	C$_9$H$_{12}$O 136.19	1004-66-6	182.5	4-06-00-03114 0.9619[14]	H$_2$O 1; EtOH 3; eth 3; bz 3 1.5053[14]
1920	Benzene, 2-methoxy-1,4-dimethyl-	C$_9$H$_{12}$O 136.19	1706-11-2	194	4-06-00-03165 0.9693[13]	H$_2$O 1; EtOH 3; eth 3; bz 3 1.5182[15]
1921	Benzene, 4-methoxy-1,2-dimethyl-	C$_9$H$_{12}$O 136.19	4685-47-6	202	4-06-00-03100 0.9744[14]	H$_2$O 1; EtOH 3; eth 3; bz 3 1.5198[14]
1922	Benzene, 1-methoxy-2,3-dinitro- Anisole, 2,3-dinitro-	C$_7$H$_8$N$_2$O$_5$ 198.14	16315-07-4 119		4-06-00-01369 1.23[137]	EtOH 4; lig 4
1923	Benzene, 1-methoxy-2,4-dinitro-	C$_7$H$_8$N$_2$O$_5$ 198.14	119-27-7 94.5	206[12]	4-06-00-01372 1.3364[131]	H$_2$O 2; EtOH 3; eth 3; ace 3 1.546[15]
1924	Benzene, 1-methoxy-3,5-dinitro- 3,5-Dinitroanisole	C$_7$H$_6$N$_2$O$_5$ 198.14	5327-44-6 105.3		4-06-00-01385 1.558[12]	ace 4; bz 4; MeOH 4
1925	Benzene, 2-methoxy-1,3-dinitro-	C$_7$H$_6$N$_2$O$_5$ 198.14	3535-67-9 118		3-06-00-00868 1.3000[128]	EtOH 4
1926	Benzene, 2-methoxy-1,4-dinitro- Anisole, 2,5-dinitro-	C$_7$H$_6$N$_2$O$_5$ 198.14	3962-77-4 97	136-8[2]	4-06-00-01383 1.476[18]	H$_2$O 2; EtOH 3; ace 3; bz 3
1927	Benzene, 4-methoxy-1,2-dinitro-	C$_7$H$_6$N$_2$O$_5$ 198.14	4280-28-8 71	168[3]	4-06-00-01384 1.3332[110]	EtOH 4
1928	Benzene, (2-methoxyethenyl)-	C$_9$H$_{10}$O 134.18	4747-15-3	211.5	2-06-00-00521 0.9894[23]	1.5620[24]
1929	Benzene, 1-methoxy-2-(3-methoxyphenoxy)-	C$_{14}$H$_{14}$O$_3$ 230.26	1655-71-6 54.8	327.5	4-06-00-05670	bz 4; eth 4; EtOH 4
1930	Benzene, (methoxymethyl)- Benzyl methyl ether	C$_8$H$_{10}$O 122.17	538-86-3 -52.6	1154 170	4-06-00-02229 0.9634[20]	H$_2$O 1; EtOH 4; eth 4; bz 3 1.5008[20]
1931	Benzene, 1-methoxy-2-methyl-	C$_8$H$_{10}$O 122.17	578-58-5 -34.1	171	4-06-00-01943 0.985[25]	H$_2$O 1; EtOH 3; eth 3; ace 3 1.5161[20]
1932	Benzene, 1-methoxy-3-methyl-	C$_8$H$_{10}$O 122.17	100-84-5 -47	175.5	4-06-00-02039 0.969[25]	H$_2$O 1; EtOH 3; eth 3; ace 3 1.5130[20]
1933	Benzene, 1-methoxy-4-methyl-	C$_8$H$_{10}$O 122.17	104-93-8 -32	175.5	4-06-00-02098 0.969[25]	H$_2$O 1; EtOH 3; eth 3; chl 3 1.5112[20]
1934	Benzene, 4-methoxy-1-methyl-2-nitro-	C$_8$H$_9$NO$_3$ 167.16	17484-36-5 17	266.5	3-06-00-01384 1.207[25]	1.5525[20]
1935	Benzene, 1-methoxy-2-nitro- 2-Nitroanisole	C$_7$H$_7$NO$_3$ 153.14	91-23-6 10.5	277; 144[4]	4-06-00-01249 1.2540[20]	H$_2$O 1; EtOH 5; eth 5; ctc 3 1.5161[20]
1936	Benzene, 1-methoxy-3-nitro-	C$_7$H$_7$NO$_3$ 153.14	555-03-3 38.5	258	4-06-00-01270 1.373[18]	H$_2$O 1; EtOH 3; eth 4
1937	Benzene, 1-methoxy-4-nitro- Anisole, 4-nitro	C$_7$H$_7$NO$_3$ 153.14	100-17-4 54	274	4-06-00-01282 1.2192[60]	H$_2$O 1; EtOH 4; eth 4; ctc 3 1.5070[60]
1938	Benzene, 1-methoxy-2-phenoxy-	C$_{13}$H$_{12}$O$_2$ 200.24	1695-04-1 79	288; 91[7]	4-06-00-05571	bz 4; eth 4; EtOH 4
1939	Benzene, 1-methoxy-4-(2-phenylethenyl)-, (E)-	C$_{15}$H$_{14}$O 210.28	1694-19-5 136.5	142.5[15]	4-06-00-04855	H$_2$O 1; EtOH 4; eth 4; ace 4
1940	Benzene, 1-methoxy-4-(1-phenylethyl)-	C$_{15}$H$_{16}$O 212.29	2605-18-7	180-2[19]	4-06-00-04711 1.0473[20]	bz 4; eth 4; HOAc 4 1.5725[20]
1941	Benzene, 1-methoxy-2-(1-propenyl)-	C$_{10}$H$_{12}$O 148.20	10577-44-3	224; 104[13]	4-06-00-03793 0.9962[15]	1.560[15]
1942	Benzene, 1-methoxy-4-(1-propenyl)- Anethole	C$_{10}$H$_{12}$O 148.20	104-46-1 21.3	675 234; 115[12]	2-06-00-00523 0.9882[20]	ace 4; bz 4; eth 4; EtOH 4 1.5615[20]
1943	Benzene, 1-methoxy-4-(2-propenyl)- Estragole	C$_{10}$H$_{12}$O 148.20	140-67-0	3657 215.5	4-06-00-03817 0.965[25]	EtOH 4; chl 4 1.5195[20]
1944	Benzene, 1-methoxy-4-(1-propenyl)-, (E)-	C$_{10}$H$_{12}$O 148.20	4180-23-8 21.35	234; 115[12]	4-06-00-03796 0.9882[20]	H$_2$O 2; EtOH 5; eth 5; ace 3 1.5615[20]
1945	Benzene, 1-methoxy-4-propyl-	C$_{10}$H$_{14}$O 150.22	104-45-0	211.5	4-06-00-03181 0.9472[20]	H$_2$O 2; EtOH 3; eth 4; ace 3 1.5045[20]

No.	Name Synonym	Mol. Form. Mol. Wt.	CAS RN mp/°C	Merck No. bp/°C	Beil. Ref. den/g cm⁻³	Solubility n_D
1946	Benzene, 2-methoxy-1,3,5-trinitro- Methyl picrate	$C_7H_5N_3O_7$ 243.13	606-35-9 69		4-06-00-01456 1.4947[80]	H_2O 1; EtOH 4; eth 3; bz 4
1947	Benzene, methyl- Toluene	C_7H_8 92.14	108-88-3 -94.9	9455 110.6	4-05-00-00766 0.8669[20]	H_2O 1; EtOH 5; eth 5; ace 3 1.4961[20]
1948	Benzene, 1-methyl-2,4-bis(1-methylethyl)- Toluene, 2,4-diisopropyl-	$C_{13}H_{20}$ 176.30	1460-98-6	82[7]	4-05-00-01152 0.8636[20]	ctc 3; CS_2 3 1.4912[20]
1949	Benzene, 1-methyl-3,5-bis(1-methylethyl)-	$C_{13}H_{20}$ 176.30	3055-14-9 -60.8	216.5	4-05-00-01152 0.8668[20]	1.4950[20]
1950	Benzene, (1-methyl-1-butenyl)- 2-Phenyl-2-pentene	$C_{11}H_{14}$ 146.23	53172-84-2	202	4-05-00-01402 0.8950[26]	1.5196[26]
1951	Benzene, [(3-methylbutoxy)methyl]-	$C_{12}H_{18}O$ 178.27	122-73-6	237; 118[19]	3-06-00-01457 0.9098[20]	eth 4; EtOH 4 1.4792[20]
1952	Benzene, (1-methylbutyl)-	$C_{11}H_{16}$ 148.25	2719-52-0	199	4-05-00-01088 0.8594[21]	H_2O 1; EtOH 3; eth 3 1.4875[25]
1953	Benzene, (2-methylbutyl)-	$C_{11}H_{16}$ 148.25	3968-85-2	195.4	2-05-00-00332 0.8584[20]	1.4873[20]
1954	Benzene, (3-methylbutyl)-	$C_{11}H_{16}$ 148.25	2049-94-7	195	4-05-00-01089 0.856[10]	H_2O 1; EtOH 3; eth 3; bz 4 1.4867[10]
1955	Benzene, (1-methylcyclopropyl)-	$C_{10}H_{12}$ 132.21	2214-14-4	91[50]	4-05-00-01388 	ace 3; bz 3; chl 3 1.5160[20]
1956	Benzene, 1-methyl-2,4-dinitro- 2,4-Dinitrotoluene	$C_7H_6N_2O_4$ 182.14	121-14-2 71	300 dec	4-05-00-00865 1.3208[71]	H_2O 1; EtOH 3; eth 3; ace 4 1.442
1957	Benzene, 1-methyl-3,5-dinitro-	$C_7H_6N_2O_4$ 182.14	618-85-9 93	sub	4-05-00-00867 1.2772[111]	H_2O 2; EtOH 3; eth 3; bz 3
1958	Benzene, 2-methyl-1,3-dinitro- 2,6-Dinitrotoluene	$C_7H_6N_2O_4$ 182.14	606-20-2 66		4-05-00-00866 1.2833[111]	EtOH 3; chl 3 1.479
1959	Benzene, 2-methyl-1,4-dinitro-	$C_7H_6N_2O_4$ 182.14	619-15-8 52.5		4-05-00-00866 1.282[111]	EtOH 3; bz 3; CS_2 4
1960	Benzene, 4-methyl-1,2-dinitro- 3,4-Dinitrotoluene	$C_7H_6N_2O_4$ 182.14	610-39-9 58.3		4-05-00-00866 1.2594[111]	H_2O 1; EtOH 3; chl 2; CS_2 3
1961	Benzene, 1,1'-methylenebis- Diphenylmethane	$C_{13}H_{12}$ 168.24	101-81-5 25.2	3329 265.0	4-05-00-01841 1.001[26]	H_2O 1; EtOH 3; eth 3; chl 3 1.5753[20]
1962	Benzene, 1,1'-methylenebis[2-bromo- Diphenylmethane, 2,2'-dibromo	$C_{13}H_{10}Br_2$ 326.03	61592-89-0	234-5[40]	3-05-00-01795 1.6197[20]	1.6300[20]
1963	Benzene, 1,1'-methylenebis[4-chloro-	$C_{13}H_{10}Cl_2$ 237.13	101-76-8 55.5	186-90[18]	4-05-00-01848 1.365[17]	EtOH 3
1964	Benzene, 1,1'-methylenebis[4-methyl-	$C_{15}H_{16}$ 196.29	4957-14-6 28	286	4-05-00-01929 0.9800[20]	eth 4; EtOH 4
1965	Benzene, 1,1'-[methylenebis(oxy)]bis[4-chloro- Bis(p-chlorophenoxy)methane	$C_{13}H_{10}Cl_2O_2$ 269.13	555-89-5 70.5	1258 189-94[6]	4-06-00-00833	ace 4; bz 4
1966	Benzene, 1,1'-(1-methylene-1,2-ethanediyl)bis-	$C_{15}H_{14}$ 194.28	948-97-0 48	289	4-05-00-02191 1.1014[20]	1.5903[20]
1967	Benzene, (1-methylene-2-propenyl)-	$C_{10}H_{10}$ 130.19	2288-18-8	60-1[17]	4-05-00-01539 0.9266[20]	H_2O 1; eth 3; bz 3; chl 3 1.5489[20]
1968	Benzene, (1-methylenepropyl)-	$C_{10}H_{12}$ 132.21	2039-93-2	182	4-05-00-01380 0.887[25]	1.5288[20]
1969	Benzene, 1,1'-(1-methyl-1,2-ethanediyl)bis-	$C_{15}H_{16}$ 196.29	5814-85-7 52	280.5	4-05-00-01915 0.9807[20]	1.5700[20]
1970	Benzene, 1,1'-[(1-methyl-1,2-ethanediyl)bis(oxy)]bis- Propane, 1,2-diphenoxy-	$C_{15}H_{16}O_2$ 228.29	69813-63-4 32	175-8[12]	2-06-00-00151 1.0748[33]	ace 4; bz 4; eth 4; EtOH 4 1.5542[33]
1971	Benzene, 1,1'-(1-methyl-1,2-ethenediyl)bis-	$C_{15}H_{14}$ 194.28	779-51-1 82	285.5	4-05-00-02189 0.9857[17]	EtOH 3; eth 3; bz 3; lig 2 1.5635[17]
1972	Benzene, (1-methylethenyl)- α-Methylstyrene	C_9H_{10} 118.18	98-83-9 -23.2	165.4	4-05-00-01364 0.9106[20]	H_2O 1; EtOH 3; eth 3; ace 5 1.5386[20]
1973	Benzene, 1-(1-methylethenyl)-4-(1-methylethyl)-	$C_{12}H_{16}$ 160.26	2388-14-9 -30.6	220.8	4-05-00-01422 0.8936[20]	ace 4; bz 4; eth 4; EtOH 4 1.5238[20]
1974	Benzene, (1-methylethoxy)-	$C_9H_{12}O$ 136.19	2741-16-4 -33	176.8	4-06-00-00557 0.9408[25]	H_2O 3; EtOH 3; ace 3; bz 3 1.4975[20]
1975	Benzene, (1-methylethyl)- Cumene	C_9H_{12} 120.19	98-82-8 -96.0	2619 152.4	4-05-00-00985 0.8618[20]	H_2O 1; EtOH 5; eth 5; ace 5 1.4915[20]
1976	Benzene, 1,1'-(1-methylethylidene)bis-	$C_{15}H_{16}$ 196.29	778-22-3 282.5		4-05-00-01925 0.9980[20]	
1977	Benzene, 1-(1-methylethyl)-2-nitro-	$C_9H_{11}NO_2$ 165.19	6526-72-3	103[9]	4-05-00-00997 1.101[12]	H_2O 1; ace 3; bz 3; HOAc 3 1.5259[20]
1978	Benzene, 1-(1-methylethyl)-4-nitro-	$C_9H_{11}NO_2$ 165.19	1817-47-6	122[9]	4-05-00-00997 1.0830[20]	H_2O 1; ace 3; bz 3; lig 3 1.5367[20]
1979	Benzene, 1-(1-methylethyl)-4-(phenylmethyl)-	$C_{16}H_{18}$ 210.32	886-58-8 13.4	310	4-05-00-01951 1.007[18]	
1980	Benzene, [(1-methylethyl)thio]-	$C_9H_{12}S$ 152.26	3019-20-3	208	4-06-00-01471 0.9852[20]	1.5464[20]
1981	Benzene, (1-methylheptyl)- 2-Phenyloctane	$C_{14}H_{22}$ 190.33	777-22-0	123-5[20]	4-05-00-01157 0.8611[20]	1.4837[20]
1982	Benzene, 1,1',1''-methylidynetris- Triphenylmethane	$C_{19}H_{16}$ 244.34	519-73-3 94	9655 359; 200[10]	4-05-00-02495 1.014[99]	H_2O 1; EtOH 2; eth 4; bz 3 1.5839[99]
1983	Benzene, 1,1',1''-[methylidynetris(oxy)]tris-	$C_{19}H_{16}O_3$ 292.33	16737-44-3 76.5	269[50] dec	4-06-00-00611	eth 4; EtOH 4
1984	Benzene, 1-methyl-2-(1-methylethenyl)-	$C_{10}H_{12}$ 132.21	7399-49-7	175	3-05-00-01214 0.9180[15]	1.5112[30]

No.	Name Synonym	Mol. Form. Mol. Wt.	CAS RN mp/°C	Merck No. bp/°C	Beil. Ref. den/g cm^{-3}	Solubility n_D
1985	Benzene, 1-methyl-3-(1-methylethenyl)-	C$_{10}$H$_{12}$ 132.21	1124-20-5	185	4-05-00-01383 0.9010[25]	1.5335[20]
1986	Benzene, 1-methyl-4-(1-methylethenyl)-	C$_{10}$H$_{12}$ 132.21	1195-32-0 -20	185.3	4-05-00-01383 0.8936[23]	1.5283[23]
1987	Benzene, 1-methyl-3-(1-methylethoxy)-	C$_{10}$H$_{14}$O 150.22	19177-04-9	195	4-06-00-02040 0.931[20]	1.4959[20]
1988	Benzene, 1-methyl-2-(1-methylethyl)- o-Cymene	C$_{10}$H$_{14}$ 134.22	527-84-4 -71.5	2770 178.1	4-05-00-01057 0.8766[20]	H$_2$O 1; EtOH 5; eth 5; ace 5 1.5006[20]
1989	Benzene, 1-methyl-3-(1-methylethyl)- m-Cymene	C$_{10}$H$_{14}$ 134.22	535-77-3 -63.7	2770 175.1	4-05-00-01058 0.8610[20]	H$_2$O 1; EtOH 5; eth 5; ace 5 1.4930[20]
1990	Benzene, 1-methyl-4-(1-methylethyl)- p-Cymene	C$_{10}$H$_{14}$ 134.22	99-87-6 -68.9	2770 177.1	4-05-00-01060 0.8573[20]	H$_2$O 1; EtOH 5; eth 5; ace 5 1.4909[20]
1991	Benzene, 1-methyl-4-(1-methylethyl)-2-nitro-	C$_{10}$H$_{13}$NO$_2$ 179.22	943-15-7	126[10]	4-05-00-01064 1.0744[20]	eth 4; EtOH 4 1.5301[20]
1992	Benzene, [2-methyl-1-(1-methylethyl)propyl]-	C$_{13}$H$_{20}$ 176.30	21777-84-4 -24.5	220.7	4-05-00-01146 0.8821[10]	1.512
1993	Benzene, 1-methyl-2-[(4-methylphenyl)thio]- 2,4'-Ditolyl sulfide	C$_{14}$H$_{14}$S 214.33	4279-70-3	173[11]	4-06-00-02172 1.0774[15]	eth 4; EtOH 4
1994	Benzene, 1-methyl-2-(1-methylpropyl)-	C$_{11}$H$_{16}$ 148.25	1595-06-8	196	4-05-00-01094 0.873[20]	ace 4; bz 4; eth 4; EtOH 4 1.497[20]
1995	Benzene, 1-methyl-2-(2-methylpropyl)-	C$_{11}$H$_{16}$ 148.25	36301-29-8 -73.3	196	4-05-00-01095 0.8649[20]	ace 4; bz 4; eth 4; EtOH 4 1.4935[20]
1996	Benzene, 1-methyl-3-(1-methylpropyl)-	C$_{11}$H$_{16}$ 148.25	1772-10-7	194	4-05-00-01095 0.858[20]	ace 4; bz 4; eth 4; EtOH 4 1.490[20]
1997	Benzene, 1-methyl-3-(2-methylpropyl)-	C$_{11}$H$_{16}$ 148.25	5160-99-6	194	3-05-00-01001 0.8536[20]	ace 4; bz 4; eth 4; EtOH 4 1.4888[20]
1998	Benzene, 1-methyl-4-(1-methylpropyl)-	C$_{11}$H$_{16}$ 148.25	1595-16-0	197	4-05-00-01095 0.8640[19]	bz 4; eth 4; chl 4 1.493[20]
1999	Benzene, 1-methyl-4-(2-methylpropyl)-	C$_{11}$H$_{16}$ 148.25	5161-04-6	196	4-05-00-01096 0.8517[20]	ace 4; bz 4; eth 4; EtOH 4 1.4874[20]
2000	Benzene, 1-methyl-2-nitro- o-Nitrotoluene	C$_7$H$_7$NO$_2$ 137.14	88-72-2 -10	6572 222	4-05-00-00845 1.1611[19]	H$_2$O 1; EtOH 5; eth 5; ctc 3 1.5450[20]
2001	Benzene, 1-methyl-3-nitro- m-Nitrotoluene	C$_7$H$_7$NO$_2$ 137.14	99-08-1 15.5	6572 232	4-05-00-00847 1.1581[20]	H$_2$O 1; EtOH 3; eth 5; bz 3 1.5466[20]
2002	Benzene, 1-methyl-4-nitro- p-Nitrotoluene	C$_7$H$_7$NO$_2$ 137.14	99-99-0 51.6	6572 238.3	4-05-00-00848 1.1038[75]	H$_2$O 1; EtOH 3; eth 4; ace 4
2003	Benzene, (1-methyl-1-nitroethyl)-	C$_9$H$_{11}$NO$_2$ 165.19	3457-58-7	224 dec; 126[15]	4-05-00-00998 1.1025[20]	ace 4; bz 4 1.5209[20]
2004	Benzene, 1-methyl-4-(nitromethyl)-	C$_8$H$_9$NO$_2$ 151.17	29559-27-1 11.5	150-1[35]	4-05-00-00972 1.1234[20]	H$_2$O 1; bz 3 1.5278[20]
2005	Benzene, 1-methyl-2-(2-nitrophenoxy)- 2-Nitrophenyl 2-tolyl ether	C$_{13}$H$_{11}$NO$_3$ 229.24	54106-40-0 39.5	194-6[14]	4-06-00-01946 1.195[20]	
2006	Benzene, 1-methyl-4-(4-nitrophenoxy)-	C$_{13}$H$_{11}$NO$_3$ 229.24	3402-74-2 69	224[25]	4-06-00-02102	bz 4; HOAc 4
2007	Benzene, (1-methylnonadecyl)- Eicosane, 2-phenyl-	C$_{26}$H$_{46}$ 358.65	2398-66-5 29.0	204.5[1]	4-05-00-01222 0.8547[20]	1.4795[20]
2008	Benzene, (1-methyl-1-pentenyl)- 2-Hexene, 2-phenyl	C$_{12}$H$_{16}$ 160.26	20247-89-6	66-8[1]	4-05-00-01419 0.910[20]	1.5200[20]
2009	Benzene, (1-methylpentyl)-	C$_{12}$H$_{18}$ 162.27	6031-02-3	208	4-05-00-01116 0.869[15]	1.492[15]
2010	Benzene, (3-methylpentyl)- (3-Methylpentyl)benzene	C$_{12}$H$_{18}$ 162.27	54410-69-4	220	0-05-00-00444 0.864[14]	1.4896
2011	Benzene, (4-methylpentyl)-	C$_{12}$H$_{18}$ 162.27	4215-86-5	215	4-05-00-01116 0.8568[16]	bz 4; eth 4; EtOH 4
2012	Benzene, 1-methyl-2-phenoxy-	C$_{13}$H$_{12}$O 184.24	3991-61-5 21.7	263.5	4-06-00-01946 1.0480[25]	
2013	Benzene, 1-methyl-3-phenoxy-	C$_{13}$H$_{12}$O 184.24	3586-14-9	272	4-06-00-02041 1.051[25]	1.5727[20]
2014	Benzene, 1-methyl-2-(phenylmethyl)-	C$_{14}$H$_{14}$ 182.27	713-36-0 6.6	280.5	4-05-00-01892 1.0020[20]	1.5763[20]
2015	Benzene, 1-methyl-3-(phenylmethyl)-	C$_{14}$H$_{14}$ 182.27	620-47-3 -28	279	4-05-00-01893 0.9913[20]	H$_2$O 1; EtOH 3; eth 3; bz 3 1.5712[20]
2016	Benzene, 1-methyl-4-(phenylmethyl)-	C$_{14}$H$_{14}$ 182.27	620-83-7 -30	286	4-05-00-01894 0.9976[20]	eth 4; bz 4; EtOH 4; chl 4 1.5712[20]
2017	Benzene, 1-methyl-2-(phenylthio)-	C$_{13}$H$_{12}$S 200.30	13963-35-4	308; 164[12]	4-06-00-02018 1.0893[20]	H$_2$O 1; ace 3; bz 3
2018	Benzene, 1-methyl-3-(phenylthio)-	C$_{13}$H$_{12}$S 200.30	13865-48-0 -6.5	309.5	4-06-00-02080 1.0937[15]	H$_2$O 1; ace 3; bz 3
2019	Benzene, 1-methyl-4-(phenylthio)-	C$_{13}$H$_{12}$S 200.30	3699-01-2 15.7	317	4-06-00-02169 1.0986[25]	H$_2$O 1; ace 3; bz 3 1.6225[25]
2020	Benzene, 1,1'-(1-methyl-1,3-propanediyl)bis-, (±)- Butane, 1,3-diphenyl (DL)	C$_{16}$H$_{18}$ 210.32	116783-21-2	293	4-05-00-01938 0.9722[20]	1.5525[20]
2021	Benzene, 1,1'-(2-methyl-1,3-propanediyl)bis-	C$_{16}$H$_{18}$ 210.32	1520-46-3 -33.7	303	4-05-00-01939 0.9669[20]	
2022	Benzene, 1,1'-(3-methyl-1-propene-1,3-diyl)bis- 1,3-Diphenyl-1-butene	C$_{16}$H$_{16}$ 208.30	7614-93-9 47.5	311	4-05-00-02205 0.9996[20]	1.590[15]
2023	Benzene, 1-methyl-2-(2-propenyl)-	C$_{10}$H$_{12}$ 132.21	1587-04-8	182.5	3-05-00-01212 0.9005[20]	1.5187[20]

No.	Name Synonym	Mol. Form. Mol. Wt.	CAS RN mp/°C	Merck No. bp/°C	Beil. Ref. den/g cm^{-3}	Solubility n_D
2024	Benzene, (2-methyl-1-propenyl)-	$C_{10}H_{12}$ 132.21	768-49-0	$99^{43.5}$	4-05-00-01380 0.9029^{20}	1.5388^{20}
2025	Benzene, (1-methyl-1-propenyl)-, (E)-	$C_{10}H_{12}$ 132.21	768-00-3 -23.5	194.7	4-05-00-01378 0.9138^{25}	H_2O 1; bz 3; chl 3 1.5425^{20}
2026	Benzene, 1-methyl-2-(2-propenyloxy)-	$C_{10}H_{12}O$ 148.20	936-72-1	206.5	4-06-00-01946 0.9698^{15}	H_2O 1; ctc 2 1.5188^{15}
2027	Benzene, 1-methyl-3-(2-propenyloxy)-	$C_{10}H_{12}O$ 148.20	1758-10-7	212.5	4-06-00-02041 0.9564^{20}	bz 4 1.5179^{20}
2028	Benzene, 1-methyl-4-(2-propenyloxy)-	$C_{10}H_{12}O$ 148.20	23431-48-3	214.5	4-06-00-02101 0.9719^{15}	H_2O 1; ace 3; ctc 2 1.5157^{24}
2029	Benzene, (1-methyl-1-propenyl)-, (Z)-	$C_{10}H_{12}$ 132.21	767-99-7	194.7; 177^{500}	3-05-00-01209 0.9191^{25}	H_2O 1; bz 3; chl 3 1.5402^{25}
2030	Benzene, (1-methylpropoxy)-	$C_{10}H_{14}O$ 150.22	10574-17-1	194.5	4-06-00-00559 0.9415^{20}	H_2O 1; eth 3; ctc 2 1.4926^{25}
2031	Benzene, (2-methylpropoxy)-	$C_{10}H_{14}O$ 150.22	1126-75-6	197.5	4-06-00-00559 0.9232^{24}	1.4932^{14}
2032	Benzene, 1-methyl-4-propoxy- Toluene, 4-propoxy	$C_{10}H_{14}O$ 150.22	5349-18-8 0	210.4	3-06-00-01354 0.9496^{0}	
2033	Benzene, [(2-methylpropoxy)methyl]- Benzyl isobutyl ether	$C_{11}H_{16}O$ 164.25	940-49-8	212	3-06-00-01457 0.9233^{20}	H_2O 1; eth 4; chl 4 1.4826^{20}
2034	Benzene, 1-(2-methylpropoxy)-2-nitro- 1-Isobutoxy-2-nitrobenzene	$C_{10}H_{13}NO_3$ 195.22	56245-02-4	277.5	0-06-00-00218 1.1361^{20}	eth 4
2035	Benzene, 1-methylpropyl sec-Butylbenzene	$C_{10}H_{14}$ 134.22	135-98-8 -82.7	1550 173.5	0.8580^{25}	H_2O 1; EtOH 5; eth 5; ace 5 1.4895^{20}
2036	Benzene, 1-methyl-2-propyl-	$C_{10}H_{14}$ 134.22	1074-17-5 -60.3	185	4-05-00-01056 0.8697^{25}	1.4996^{20}
2037	Benzene, 1-methyl-3-propyl-	$C_{10}H_{14}$ 134.22	1074-43-7 -82.5	182	4-05-00-01056 0.8569^{25}	1.4935^{20}
2038	Benzene, 1-methyl-4-propyl-	$C_{10}H_{14}$ 134.22	1074-55-1 -63.6	183.4	4-05-00-01057 0.8544^{25}	H_2O 1; EtOH 3; eth 3 1.4922^{20}
2039	Benzene, (2-methylpropyl)- Isobutylbenzene	$C_{10}H_{14}$ 134.22	538-93-2 -51.4	5018 172.7	4-05-00-01042 0.8532^{20}	H_2O 1; EtOH 5; eth 5; ace 5 1.4866^{20}
2040	Benzene, (methylsulfinyl)-	C_7H_8OS 140.21	1193-82-4 32.0	263.5; 140^{13}	4-06-00-01467	1.5885^{20}
2041	Benzene, (methylthio)-	C_7H_8S 124.21	100-68-5	193	4-06-00-01466 1.0579^{20}	H_2O 1; EtOH 3; ace 4; os 3 1.5868^{20}
2042	Benzene, [(methylthio)methyl]-	$C_8H_{10}S$ 138.23	766-92-7 -30	210; 120^{48}	4-06-00-02633 1.0274^{20}	1.5620^{20}
2043	Benzene, 1-(methylthio)-2-nitro-	$C_7H_7NO_2S$ 169.20	3058-47-7 64.5		4-06-00-01661 1.2628^{78}	H_2O 3; EtOH 3; bz 3; chl 3 1.6246^{78}
2044	Benzene, 1-(methylthio)-4-nitro-	$C_7H_7NO_2S$ 169.20	701-57-5 72	$135-40^{2}$	4-06-00-01687 1.2391^{80}	H_2O 1; ace 3; bz 3 1.6401^{20}
2045	Benzene, 1-methyl-4-(trifluoromethyl)-	$C_8H_7F_3$ 160.14	6140-17-6	129	1.4320^{26}	eth 3; ace 3; bz 3; chl 3 1.4320^{26}
2046	Benzene, 1-methyl-2,3,4-trinitro- 2,3,4-Trinitrotoluene	$C_7H_5N_3O_6$ 227.13	602-29-9 112	301	4-05-00-00872 1.62^{25}	H_2O 1; EtOH 2; eth 3; ace 3
2047	Benzene, 1-methyl-2,4,5-trinitro- 2,4,5-Trinitrotoluene	$C_7H_5N_3O_6$ 227.13	610-25-3 104	exp 300	4-05-00-00872	H_2O 1; EtOH 2; eth 3; ace 3
2048	Benzene, 2-methyl-1,3,5-trinitro- 2,4,6-Trinitrotoluene	$C_7H_5N_3O_6$ 227.13	118-96-7 80.1	9643 exp 240	4-05-00-00873 1.654^{25}	H_2O 1; EtOH 2; eth 3; ace 4
2049	Benzene, nitro- Nitrobenzene	$C_6H_5NO_2$ 123.11	98-95-3 5.7	6509 210.8	4-05-00-00708 1.2037^{20}	H_2O 2; EtOH 4; eth 4; ace 4 1.5562^{20}
2050	Benzene, (2-nitroethenyl)-, (E)-	$C_8H_7NO_2$ 149.15	5153-67-3 60	255	4-05-00-01352	H_2O 1; EtOH 3; eth 4; ace 3
2051	Benzene, (2-nitroethyl)-	$C_8H_9NO_2$ 151.17	6125-24-2 -23	250; 137^{16}	4-05-00-00912 1.126^{24}	1.5407^{19}
2052	Benzene, (nitromethyl)-	$C_7H_7NO_2$ 137.14	622-42-4	226; 135^{25}	4-05-00-00850 1.1596^{20}	ace 4; eth 4 1.5323^{20}
2053	Benzene, 1-nitro-2-phenoxy-	$C_{12}H_9NO_3$ 215.21	2216-12-8 <-20	235^{60}; 184^{8}	4-06-00-01252 1.2539^{22}	bz 4; eth 4; EtOH 4; chl 4 1.575^{20}
2054	Benzene, 1-nitro-4-phenoxy-	$C_{12}H_9NO_3$ 215.21	620-88-2 61	320; 225^{30}	4-06-00-01287	H_2O 1; EtOH 2; eth 3; bz 3
2055	Benzene, 1-nitro-2-(2-phenylethenyl)-, (E)-	$C_{14}H_{11}NO_2$ 225.25	4264-29-3 73	209^{11}	4-05-00-02166	H_2O 1; EtOH 3; eth 4
2056	Benzene, 1-nitro-2-(2-phenylethenyl)-, (Z)-	$C_{14}H_{11}NO_2$ 225.25	52208-62-5 64.8	187^{11}	4-05-00-02166	EtOH 3; eth 4; AcOEt 4; HOAc 4
2057	Benzene, 1-nitro-2-(phenylthio)-	$C_{12}H_9NO_2S$ 231.28	4171-83-9 82	210^{15}	4-06-00-01663	eth 4; EtOH 4
2058	Benzene, 1-nitro-4-(phenylthio)-	$C_{12}H_9NO_2S$ 231.28	952-97-6 56	288^{100}; 240^{25}	4-06-00-01694	eth 4; EtOH 4
2059	Benzene, nitroso- Nitrosobenzene	C_6H_5NO 107.11	586-96-9 68.5	58^{18}	4-05-00-00702	H_2O 1; EtOH 3; eth 3; bz 3
2060	Benzene, 1-nitro-2-(trifluoromethyl)-	$C_7H_4F_3NO_2$ 191.11	384-22-5 32.5	217; 105^{20}	4-05-00-00852	H_2O 1; EtOH 4; bz 4; ctc 2
2061	Benzene, 1-nitro-3-(trifluoromethyl)- 3-(Trifluoromethyl)-1-nitrobenzene	$C_7H_4F_3NO_2$ 191.11	98-46-4 -2.4	202.8; 103^{40}	4-05-00-00853 1.4357^{15}	H_2O 1; EtOH 3; eth 3; ctc 2 1.4719^{20}
2062	Benzene, nonadecyl-	$C_{25}H_{44}$ 344.62	29136-19-4 40	419	3-05-00-01126 0.8545^{20}	1.4807^{20}

No.	Name Synonym	Mol. Form. Mol. Wt.	CAS RN mp/°C	Merck No. bp/°C	Beil. Ref. den/g cm^{-3}	Solubility n_D
2063	Benzenenonanoic acid, θ-nonyl-	$C_{24}H_{40}O_2$ 360.58	1938-22-3 37	200-4[0.09]	2-09-00-00376 0.9312[25]	1.4891[20]
2064	Benzene, nonyl-	$C_{15}H_{24}$ 204.36	1081-77-2 -24	280.5	3-05-00-01075 0.8584[20]	1.4816[20]
2065	Benzene, octacosyl- Octacosylbenzene	$C_{34}H_{62}$ 470.87	61828-26-0 66	405[100-] 268[1]	4-05-00-01243 0.8543[20]	1.4796[20]
2066	Benzene, octadecyl- Octadecylbenzene	$C_{24}H_{42}$ 330.60	4445-07-2 29	408	4-05-00-01219 0.8546[20]	1.4809[20]
2067	Benzene, octyl-	$C_{14}H_{22}$ 190.33	2189-60-8 -36	264.5	4-05-00-01157 0.8562[20]	H$_2$O 1; eth 5; bz 5 1.4845[20]
2068	Benzene, (1-octyldodecyl)- Eicosane, 9-phenyl-	$C_{28}H_{46}$ 358.65	2398-65-4 17.9	196[1.0]	4-05-00-01223 0.8534[20]	1.4790[20]
2069	Benzene, (octyloxy)-	$C_{14}H_{22}O$ 206.33	1818-07-1 8	285	4-06-00-00560 0.9131[15]	H$_2$O 1; EtOH 3; eth 3 1.4875[20]
2070	Benzene, (3-octylundecyl)- 9-(2-Phenylethyl)heptadecane	$C_{25}H_{44}$ 344.62	5637-96-7 -26.7	189[1.0]	4-05-00-01221 0.8560[20]	1.4806[20]
2071	Benzene, 1,1'-oxybis- Diphenyl ether	$C_{12}H_{10}O$ 170.21	101-84-8 26.8	7259 258.0	4-06-00-00568 1.0661[30]	H$_2$O 1; EtOH 3; eth 3; bz 3 1.5787[25]
2072	Benzene, 1,1'-oxybis[4-bromo-	$C_{12}H_8Br_2O$ 328.00	2050-47-7 60.5	339	4-06-00-01048 1.8[25]	H$_2$O 1; EtOH 3; eth 4; bz 3
2073	Benzene, 1,1'-oxybis[4-chloro-	$C_{12}H_8Cl_2O$ 239.10	2444-89-5 30	313	4-06-00-00826 1.1231[20]	H$_2$O 1 1.611[20]
2074	Benzene, 1,1'-oxybis[2-ethyl-	$C_{16}H_{18}O$ 226.32	56911-77-4	318.5	1.0141[18]	H$_2$O 1; eth 3; chl 3 1.5488[18]
2075	Benzene, 1,1'-oxybis[2-methoxy-	$C_{14}H_{14}O_3$ 230.26	1655-70-5 79.5	330.5	3-06-00-04223	H$_2$O 1; EtOH 4; eth 4; lig 1
2076	Benzene, 1,1'-oxybis[2-methyl-	$C_{14}H_{14}O$ 198.26	4731-34-4	271	4-06-00-01947 1.047[24]	bz 4; eth 4; EtOH 4
2077	Benzene, 1,1'-oxybis[3-methyl-	$C_{14}H_{14}O$ 198.26	19814-71-2	284; 136[14]	4-06-00-02043 1.0323[21]	H$_2$O 1; EtOH 3; eth 3; bz 3
2078	Benzene, 1,1'-oxybis[4-methyl-	$C_{14}H_{14}O$ 198.26	1579-40-4 51	285	4-06-00-02103	bz 4; eth 4; EtOH 4
2079	Benzene, 1,1'-[oxybis(methylene)]bis- Benzyl ether	$C_{14}H_{14}O$ 198.26	103-50-4 3.6	1146 298	4-06-00-02240 1.0428[20]	H$_2$O 1; EtOH 5; eth 5; ctc 3 1.5168[20]
2080	Benzene, 1,1'-(oxydi-2,1-ethanediyl)bis-	$C_{16}H_{18}O$ 226.32	2396-53-4	318.5; 86[0.5]	4-06-00-03071 1.0141[18]	eth 4; chl 4 1.5488[18]
2081	Benzene, 1,1'-(oxydiethylidene)bis-, (R*,R*)- (±)- Bis-(1-phenylethyl)ether (DL)	$C_{16}H_{18}O$ 226.32	53776-69-5	280.2	4-06-00-03034 1.0058[15]	H$_2$O 1; eth 3; chl 3 1.5454[21]
2082	Benzene, pentabromo- Pentabromobenzene	C_6HBr_5 472.59	608-90-2 160.5	sub	4-05-00-00687	bz 4; chl 4
2083	Benzene, pentabromomethyl-	$C_7H_3Br_5$ 486.62	87-83-2 288		3-05-00-00720 2.97[17]	H$_2$O 1; EtOH 2; bz 3; HOAc 2
2084	Benzene, pentachloro- Pentachlorobenzene	C_6HCl_5 250.34	608-93-5 86	277	4-05-00-00669 1.834[16]	H$_2$O 1; EtOH 1; eth 2; bz 2
2085	Benzene, pentachloro(dichloromethyl)-	C_7HCl_7 333.25	2136-95-0 119.5	334	4-05-00-00824	EtOH 2
2086	Benzene, pentachloromethyl-	$C_7H_3Cl_5$ 264.36	877-11-2 224.8	301	4-05-00-00823	EtOH 2; eth 2; bz 3; tol 3
2087	Benzene, pentachloronitro- Quintozene	$C_6Cl_5NO_2$ 295.34	82-68-8 144	8108 328 dec	4-05-00-00728 1.718[25]	H$_2$O 1; EtOH 2; bz 3; chl 3
2088	Benzene, pentacosyl- Pentacosylbenzene	$C_{31}H_{56}$ 428.79	61828-06-6 59	380[100-] 244[1]	4-05-00-01241 0.8544[20]	1.4799[20]
2089	Benzene, pentadecyl-	$C_{21}H_{36}$ 288.52	2131-18-2 22	373	3-05-00-01119 0.8548[20]	1.4815[20]
2090	Benzene, pentaethyl- Pentaethylbenzene	$C_{16}H_{26}$ 218.38	605-01-6 <-20	277	4-05-00-01197 0.8971[19]	1.5127[20]
2091	Benzene, pentafluoro- Pentafluorobenzene	C_6HF_5 168.07	363-72-4 -47.3	85.7	4-05-00-00639 1.514[25]	1.3905[20]
2092	Benzene, pentafluoroiodo-	C_6F_5I 293.96	827-15-6 -29	166	4-05-00-00695 2.212[20]	1.4950[25]
2093	Benzene, pentafluoromethoxy-	$C_7H_3F_5O$ 198.09	389-40-2 -37	138.5	4-06-00-00782 1.493[20]	1.4087[20]
2094	Benzene, pentafluoromethyl- 2,3,4,5,6-Pentafluorotoluene	$C_7H_3F_5$ 182.09	771-56-2 -29.8	117.5	1.440[20]	1.4016[25]
2095	Benzene, pentafluoro(trifluoromethyl)- Perfluorotoluene	C_7F_8 236.06	434-64-0 -65.6	104.5	4-05-00-00804	1.3670[20]
2096	Benzene, pentaiodo- Pentaiodobenzene	C_6HI_5 707.60	608-96-8 172	sub	4-05-00-00702	bz 4; chl 4
2097	Benzene, pentamethyl- Pentamethylbenzene	$C_{11}H_{16}$ 148.25	700-12-9 54.5	232	4-05-00-01109 0.917[20]	H$_2$O 1; EtOH 4; bz 4; chl 3 1.527[20]
2098	Benzene, 1,1'-(1,5-pentanediyl)bis- 1,5-Diphenylpentane	$C_{17}H_{20}$ 224.35	1718-50-9 -8	327	4-05-00-01963 0.9812[19]	H$_2$O 1 1.559[19]
2099	Benzenepentanoic acid 5-Phenylvaleric acid	$C_{11}H_{14}O_2$ 178.23	2270-20-4 57.5	190[30]	4-09-00-01864	H$_2$O 2; EtOH 4; os 3
2100	Benzenepentanoic acid, δ-oxo-, methyl ester	$C_{12}H_{14}O_3$ 206.24	1501-04-8 -2	186[19]; 147[8]	4-10-00-02783	H$_2$O 4
2101	Benzenepentanol	$C_{11}H_{18}O$ 164.25	10521-91-2	155[20]	4-06-00-03373 0.9725[20]	eth 4; EtOH 4 1.5156[20]

No.	Name / Synonym	Mol. Form. / Mol. Wt.	CAS RN / mp/°C	Merck No. / bp/°C	Beil. Ref. / den/g cm^{-3}	Solubility / n_D
2102	Benzenepentanol, α-ethyl- / 7-Phenylheptan-3-ol	$C_{13}H_{20}O$ / 192.30	60012-63-7	277.5; 150[20]	3-06-00-02027 / 0.968[20]	1.5115[20]
2103	Benzenepentanol, α-methyl- / 6-Phenyl-2-hexanol	$C_{12}H_{18}O$ / 178.27	38487-94-4	148[18]	3-06-00-01995 / 0.9567[20]	1.5079[20]
2104	Benzene, 1-pentenyl-	$C_{11}H_{14}$ / 146.23	826-18-6	217	4-05-00-01401 / 0.8782[22]	1.5158[20]
2105	Benzene, 3-pentenyl-	$C_{11}H_{14}$ / 146.23	1745-16-0	201	4-05-00-01401 / 0.8884[16]	1.5089[16]
2106	Benzene, 4-pentenyl- / 1-Pentene, 5-phenyl	$C_{11}H_{14}$ / 146.23	1075-74-7	201; 77[10]	3-05-00-01238 / 0.8889[20]	1.5065[20]
2107	Benzene, pentyl- / Amylbenzene	$C_{11}H_{16}$ / 148.25	538-68-1 / -75	635 / 205.4	4-05-00-01085 / 0.8585[20]	H$_2$O 1; EtOH 5; eth 5; ace 5 / 1.4878[20]
2108	Benzene, 1,1'-pentylidenebis- / Pentane, 1,1-diphenyl-	$C_{17}H_{20}$ / 224.35	1726-12-1 / -12	308	4-05-00-01966 / 0.9659[20]	H$_2$O 1 / 1.5511[20]
2109	Benzene, (pentyloxy)- / Pentyl phenyl ether	$C_{11}H_{16}O$ / 164.25	2050-04-6	200; 111[17]	3-06-00-00552 / 0.9270[20]	ace 4; EtOH 4 / 1.4947[20]
2110	Benzene, 1,1'-(phenylmethylene)bis[4-methyl-	$C_{21}H_{20}$ / 272.39	603-39-4 / 56	218[12]	4-05-00-02520 /	H$_2$O 1; EtOH 3; eth 4; bz 4
2111	Benzene, 1-(phenylmethyl)-4-propyl-	$C_{16}H_{18}$ / 210.32	62155-41-3	152-5[10]	4-05-00-01950 / 0.9739[18]	1.3552[20]
2112	Benzene, [(phenylmethyl)sulfonyl]-	$C_{13}H_{12}O_2S$ / 232.30	3112-88-7 / 146		4-06-00-02647 / 1.1261[153]	H$_2$O 1; EtOH 2; eth 2; bz 2
2113	Benzene, [(phenylmethyl)thio]-	$C_{13}H_{12}S$ / 200.30	831-91-4 / 43.5	197[27]	4-06-00-02644 /	H$_2$O 1; EtOH 3; eth 3; con sulf 3
2114	Benzenepropanal / Hydrocinnamic aldehyde	$C_9H_{10}O$ / 134.18	104-53-0 / 47	224; 117[28]	4-07-00-00692 / 1.0190[20]	H$_2$O 1; EtOH 4; eth 5
2115	Benzenepropanal, 2-methyl- / Hydrocinnamaldehyde, o-methyl-	$C_{10}H_{12}O$ / 148.20	19564-40-0	120[13]	2-07-00-00246 / 0.998[19]	1.522[19]
2116	Benzenepropanal, 4-methyl-	$C_{10}H_{12}O$ / 148.20	5406-12-2	223	2-07-00-00247 / 0.999[14]	1.525[14]
2117	Benzenepropanal, α-methyl-4-(1-methylethyl)- / 3-p-Cumenyl-2-methylpropionaldehyde	$C_{13}H_{18}O$ / 190.29	103-95-7	135[99]; 115[9]	4-07-00-00788 / 0.951[15]	bz 4; eth 4; EtOH 4 / 1.5068[20]
2118	Benzenepropanal, α-methyl-β-oxo-	$C_{10}H_{10}O_2$ / 162.19	16837-43-7 / 119.3	155[125]	3-07-00-03497 /	bz 4; EtOH 4; MeOH 4
2119	Benzenepropanamine, α-methyl- / 1-Methyl-3-phenylpropylamine	$C_{10}H_{15}N$ / 149.24	22374-89-6 / 143	223; 101[14]	4-12-00-02811 / 0.9289[15]	EtOH 4 / 1.5152[20]
2120	Benzenepropanaminium, γ-(aminocarbonyl)-N-methyl-N,N-bis(1-methylethyl)-γ-phenyl-, iod / Ammonium, (3-carbamoyl-3,3-diphenylpropyl) diisopropyl methyl,iodide	$C_{23}H_{33}IN_2O$ / 480.43	71-81-8	5090	4-14-00-01853 /	
2121	Benzene, 1,1'-(1,3-propanediyl)bis- / 1,3-Diphenylpropane	$C_{15}H_{16}$ / 196.29	1081-75-0 / 6	300.3	4-05-00-01913 / 1.007[20]	1.5760[20]
2122	Benzene, 1,1'-[1,3-propanediylbis(oxy)]bis-	$C_{15}H_{16}O_2$ / 228.29	726-44-3 / 62	339	4-06-00-00585 /	H$_2$O 1; EtOH 3; eth 3
2123	Benzenepropanenitrile / Hydrocinnamonitrile	C_9H_9N / 131.18	645-59-0 / -1	261; 141[25]	4-09-00-01764 / 1.0016[20]	EtOH 3; eth 3; chl 2 / 1.5266[28]
2124	Benzenepropanenitrile, β-oxo- / Benzoylacetonitrile	C_9H_7NO / 145.16	614-16-4 / 80.5	160[10]	4-10-00-02757 /	H$_2$O 2; EtOH 3; eth 3; bz 3
2125	Benzenepropanethiol	$C_9H_{12}S$ / 152.26	24734-68-7	109[10]	4-06-00-03206 / 1.01[25]	1.5494[20]
2126	Benzenepropanoic acid / Hydrocinnamic acid	$C_9H_{10}O_2$ / 150.18	501-52-0 / 107	4707 / 279.8	4-09-00-01752 / 1.0712[49]	H$_2$O 3; EtOH 3; eth 3; bz 4
2127	Benzenepropanoic acid, α-acetyl-α-(phenylmethyl)-, ethyl ester	$C_{20}H_{22}O_3$ / 310.39	42597-26-2 / 57	233	3-10-00-03350 /	EtOH 2; eth 2
2128	Benzenepropanoic acid, β-(aminomethyl)- / 4-Amino-3-phenylbutyric acid	$C_{10}H_{13}NO_2$ / 179.22	1078-21-3 / 250-5 dec	476	4-14-00-01723 /	
2129	Benzenepropanoic acid, α,α-dimethyl-, methyl ester / Methyl α,α-dimethylhydrocinnamate	$C_{12}H_{16}O_2$ / 192.26	14248-22-7	121-2[23]	4-09-00-01879 / 1.0043[25]	1.4945[25]
2130	Benzenepropanoic acid, β,β-dimethyl-	$C_{11}H_{14}O_2$ / 178.23	1010-48-6 / 60	167[10]	4-09-00-01874 /	H$_2$O 3; ace 3; peth 3 / 1.5182[20]
2131	Benzenepropanoic acid, ethyl ester	$C_{11}H_{14}O_2$ / 178.23	2021-28-5	247.2	4-09-00-01753 / 1.0147[20]	eth 4; EtOH 4 / 1.4954[20]
2132	Benzenepropanoic acid, α-ethyl-β-oxo-, ethyl ester	$C_{13}H_{16}O_3$ / 220.27	24346-56-3	152[7]	3-10-00-03065 / 1.0706[15]	eth 4 / 1.509[15]
2133	Benzenepropanoic acid, α-hydroxy-, (±)- / (±)-3-Phenyllactic acid	$C_9H_{10}O_3$ / 166.18	828-01-3 / 98	149[15]	2-10-00-00154 /	H$_2$O 4; ace 4; eth 4; EtOH 4
2134	Benzenepropanoic acid, 4-hydroxy- / p-Hydroxyhydrocinnamic acid	$C_9H_{10}O_3$ / 166.18	501-97-3 / 130.8	208-10[14]	4-10-00-00631 /	H$_2$O 3; EtOH 3; eth 3; bz 3
2135	Benzenepropanoic acid, 4-hydroxy-3,5-diiodo-α-phenyl- / Iodoalphionic acid	$C_{15}H_{12}I_2O_3$ / 494.07	577-91-3 / 164	4919	4-10-00-01291 /	H$_2$O 1; EtOH 3; eth 3; bz 2
2136	Benzenepropanoic acid, 4-(4-hydroxy-3-iodophenoxy)-3,5-diiodo- / 3,3',5-Triiodothyropropionic acid	$C_{15}H_{11}I_3O_4$ / 635.96	51-26-3	9346	4-10-00-00641 /	
2137	Benzenepropanoic acid, α-methyl- / α-Methyldihydrocinnamic acid	$C_{10}H_{12}O_2$ / 164.20	1009-67-2 / 36.5	272; 152[12]	2-09-00-00357 /	eth 4; EtOH 4

No.	Name Synonym	Mol. Form. Mol. Wt.	CAS RN mp/°C	Merck No. bp/°C	Beil. Ref. den/g cm^{-3}	Solubility n_D
2138	Benzenepropanoic acid, β-methyl-, (±)-	$C_{10}H_{12}O_2$ 164.20	772-17-8 46.5	140-5[3]	4-09-00-01814 1.0701[20]	H_2O 2; peth 3 1.5155[20]
2139	Benzenepropanoic acid, methyl ester Methyl dihydrocinnamate	$C_{10}H_{12}O_2$ 164.20	103-25-3	238.5; 91[4]	4-09-00-01753 1.0455[25]	H_2O 1; EtOH 3; eth 3; bz 3
2140	Benzenepropanoic acid, 1-methylethyl ester	$C_{12}H_{16}O_2$ 192.26	22767-95-9	126[11]	2-09-00-00339 0.9860[25]	eth 4; EtOH 4
2141	Benzenepropanoic acid, β-oxo-, ethyl ester Ethyl benzoylacetate	$C_{11}H_{12}O_3$ 192.21	94-02-0 <0	3726 267 dec; 167[20]	4-10-00-02756 1.1202[15]	H_2O 2; EtOH 3; eth 3 1.531[715]
2142	Benzenepropanoic acid, β-oxo-, methyl ester	$C_{10}H_{10}O_3$ 178.19	614-27-7	265 dec; 151[12]	4-10-00-02755 1.158[29]	ace 4; eth 4; EtOH 4 1.537[20]
2143	Benzenepropanoic acid, α-phenyl-, (±)- Propionic acid, 2,3-diphenyl (DL)	$C_{15}H_{14}O_2$ 226.27	94942-89-9 88(form a); 95(form b)	337.5	4-09-00-02528 1.1495[96]	bz 4; eth 4; EtOH 4
2144	Benzenepropanoic acid, α-(phenylmethyl)-	$C_{16}H_{16}O_2$ 240.30	618-68-8 90	235[18]	4-09-00-02546	bz 4; eth 4; EtOH 4
2145	Benzenepropanoic acid, phenylmethyl ester	$C_{16}H_{16}O_2$ 240.30	22767-96-0	325; 198[20]	4-09-00-01755 1.090[15]	eth 4
2146	Benzenepropanoic acid, β-propyl-	$C_{12}H_{16}O_2$ 192.26	5703-52-6	152[4]	4-09-00-01900 1.025[25]	1.5078
2147	Benzenepropanoic-β-14C acid, α-bromo- Hydrocinnamic acid, α-bromo-	$C_9H_9BrO_2$ 229.07	16503-53-0 52	150[13]	4-09-00-01771 1.4790[25]	ace 4; bz 4; EtOH 4
2148	Benzenepropanol Hydrocinnamyl alcohol	$C_9H_{12}O$ 136.19	122-97-4 <-18	235	4-06-00-03198 0.995[25]	H_2O 3; EtOH 5; eth 5; ctc 3 1.5357[18]
2149	Benzenepropanol, carbamate Phenprobamate	$C_{10}H_{13}NO_2$ 179.22	673-31-4 102	7229		H_2O 1; EtOH 3; chl 3
2150	Benzenepropanol, α,α-dimethyl- Benzyl-tert-butanol	$C_{11}H_{16}O$ 164.25	103-05-9 24.5	121[13]	4-06-00-03380 0.9626[21]	H_2O 1; EtOH 4; eth 4; ace 4 1.5077[21]
2151	Benzenepropanol, β,β-dimethyl-	$C_{11}H_{16}O$ 164.25	13351-61-6 34.5	125[14.5]	2-06-00-00507	eth 4; EtOH 4
2152	Benzenepropanol, α-ethyl-, (S)- 3-Pentanol, 1-phenyl (d)	$C_{11}H_{16}O$ 164.25	71747-37-0 38	143[19]	3-06-00-01953 0.9687[20]	EtOH 4
2153	Benzenepropanol, 4-hydroxy-3-methoxy-	$C_{10}H_{14}O_3$ 182.22	2305-13-7 65	197[15]	4-06-00-07405 1.5545[25]	eth 4; EtOH 4
2154	Benzenepropanol, 4-methoxy-	$C_{10}H_{14}O_2$ 166.22	5406-18-8 26	166[18]	4-06-00-05982	1.5315[20]
2155	Benzenepropanol, α-methyl-	$C_{10}H_{14}O$ 150.22	2344-70-9	239; 123[15]	4-06-00-03275 0.9899[16]	1.517[16]
2156	Benzenepropanol, γ-methyl-	$C_{10}H_{14}O$ 150.22	2722-36-3	139[33]; 110[6]	4-06-00-03286 0.9832[20]	1.4965[20]
2157	Benzenepropanol, γ-methylene- 3-Buten-1-ol, 3-phenyl-	$C_{10}H_{12}O$ 148.20	3174-83-2	123[10]	4-06-00-03840 1.0272[20]	1.5577[20]
2158	Benzenepropanol, β-phenyl-	$C_{15}H_{16}O$ 212.29	3536-29-6 51	310; 190[15]	4-06-00-04754 1.0585[24]	1.5742[19]
2159	Benzenepropanol, γ-propyl- 1-Hexanol, 3-phenyl	$C_{12}H_{18}O$ 178.27	67700-22-5	127[5]	3-06-00-02001 0.955[25]	1.5101[25]
2160	Benzenepropanoyl chloride	C_9H_9ClO 168.62	645-45-4	225 dec; 105[10]	4-09-00-01762 1.135[21]	eth 3; CS_2 3
2161	Benzene, 1,1'-(1-propene-1,3-diyl)bis-	$C_{15}H_{14}$ 194.28	5209-18-7 16	302; 178[15]	4-05-00-02187 1.0120[20]	
2162	Benzene, 1-propenyl-	C_9H_{10} 118.18	637-50-3 -27.1	175.5	3-05-00-01185 0.9019[25]	ace 4; bz 4; eth 4; EtOH 4 1.5508[20]
2163	Benzene, 1-propenyl-, (E)-	C_9H_{10} 118.18	873-66-5 -29.3	178.3	4-05-00-01359 0.9023[25]	H_2O 1; EtOH 5; eth 5; ace 5 1.5506[20]
2164	Benzene, 1-propenyl-, (Z)-	C_9H_{10} 118.18	766-90-5 -61.6	167.5	4-05-00-01359 0.9088[20]	H_2O 1; EtOH 5; eth 5; ace 5 1.5420[20]
2165	Benzene, 2-propenyl-	C_9H_{10} 118.18	300-57-2 -40	156	4-05-00-01362 0.8920[20]	H_2O 1; EtOH 3; eth 3; bz 3 1.5131[20]
2166	Benzene, 1,1'-(1-propenylidene)bis-	$C_{15}H_{14}$ 194.28	778-66-5 52	280-1 149[11]	4-05-00-02192 1.0250[20]	H_2O 1; EtOH 3; bz 3 1.5880[20]
2167	Benzene, (2-propenyloxy)-	$C_9H_{10}O$ 134.18	1746-13-0	191.7	4-06-00-00562 0.9811[20]	H_2O 1; EtOH 3; eth 5; ctc 2 1.5223[20]
2168	Benzene, propoxy-	$C_9H_{12}O$ 136.19	622-85-5 -27	189.9	4-06-00-00556 0.9474[20]	EtOH 3; eth 3 1.5014[20]
2169	Benzene, propyl- Propylbenzene	C_9H_{12} 120.19	103-65-1 -99.5	7856 159.2	4-05-00-00977 0.8620[20]	H_2O 1; EtOH 5; eth 5; ace 5 1.4920[20]
2170	Benzene, (1-propylheptadecyl)- Eicosane, 4-phenyl-	$C_{26}H_{46}$ 358.65	2400-03-5 31.4	199[1]	4-05-00-01223 0.8546[20]	1.4794[20]
2171	Benzene, 1,1'-propylidenebis- 1,1-Diphenylpropane	$C_{15}H_{18}$ 196.29	1530-03-6 13.3	281.6	4-05-00-01920 0.9951[14]	1.5681[14]
2172	Benzene, (propylthio)-	$C_9H_{12}S$ 152.26	874-79-3 -45	220	4-06-00-01470 0.9995[20]	1.5571[20]
2173	Benzene, 1-propynyl-	C_9H_8 116.16	673-32-5	183	4-05-00-01530 0.942[15]	1.563[15]
2174	Benzeneseleninic acid Phenylseleninic acid	$C_6H_6O_2Se$ 189.07	6996-92-5 124.5		4-11-00-00708 1.93[20]	H_2O 2; bz 1; alk 4
2175	Benzene, 1,1'-selenobis-	$C_{12}H_{10}Se$ 233.17	1132-39-4 2.5	301.5	4-06-00-01779 1.351[20]	H_2O 1; EtOH 5; eth 5; bz 3 1.5500[20]

No.	Name / Synonym	Mol. Form. / Mol. Wt.	CAS RN / mp/°C	Merck No. / bp/°C	Beil. Ref. / den/g cm^{-3}	Solubility / n_D
2176	Benzene, 1,1'-selenobis[4-methyl-	$C_{14}H_{14}Se$ 261.23	22077-55-0 69.5	196^{16}	4-06-00-02218	H_2O 1; EtOH 3; eth 2; chl 2
2177	Benzeneselenol	C_6H_6Se 157.07	645-96-5	$183.6; 84^{25}$	4-06-00-01777 1.4865^{15}	H_2O 1; EtOH 3; eth 4; ctc 4
2178	Benzenesulfenyl chloride, 2,4-dinitro- 2,4-Dinitrobenzenesulfenyl chloride	$C_6H_3ClN_2O_4S$ 234.62	528-76-7 99	3267	4-06-00-01772	bz 4; chl 4; HOAc 4; peth 2
2179	Benzenesulfenyl chloride, 4-nitro-	$C_6H_4ClNO_2S$ 189.62	937-32-6 52	$125^{0.1}$	4-06-00-01717	bz 4
2180	Benzenesulfinic acid	$C_6H_6O_2S$ 142.18	618-41-7 84	dec	4-11-00-00003	H_2O 2; EtOH 3; eth 3; bz 3
2181	Benzenesulfinic acid, 2-methyl- o-Toluenesulfinic acid	$C_7H_8O_2S$ 156.21	13165-77-0 80	dec	4-11-00-00008	H_2O 4; ace 4; eth 4; EtOH 4
2182	Benzenesulfinic acid, 4-methyl- p-Toluenesulfinic acid	$C_7H_8O_2S$ 156.21	536-57-2 86.5	9458	4-11-00-00009	H_2O 3; EtOH 4; eth 4; bz 2
2183	Benzene, 1,1'-sulfinylbis-	$C_{12}H_{10}OS$ 202.28	945-51-7 71.2	340^{16}	4-06-00-01489	EtOH 4; eth 4; bz 4; chl 2
2184	Benzene, 1,1'-[sulfinylbis(methylene)]bis-	$C_{14}H_{14}OS$ 230.33	621-08-9 134	210 dec	4-06-00-02651	H_2O 1; EtOH 4; eth 3
2185	Benzenesulfinyl chloride	C_6H_5ClOS 160.62	4972-29-6 38	$71-2^{1.5}$	4-11-00-00005 1.3469^{25}	eth 3; chl 3 1.3470^{25}
2186	Benzenesulfinyl chloride, 4-methyl-	C_7H_7ClOS 174.65	10439-23-3 57	$113^{3.5}$	4-11-00-00010	chl 4
2187	Benzenesulfonamide, 4-amino- Sulfanilamide	$C_6H_8N_2O_2S$ 172.21	63-74-1 165.5	8898	4-14-00-02658 1.08^{25}	H_2O 3; EtOH 3; eth 3; ace 3
2188	Benzenesulfonamide, 4-amino-N-[4-(aminosulfonyl)phenyl]- N4-Sulfanilylsulfanilamide	$C_{12}H_{13}N_3O_4S_2$ 327.39	547-52-4 137	8905	4-14-00-02760	H_2O 2; EtOH 3; eth 3; ace 3
2189	Benzenesulfonamide, 4-amino-N-(aminothioxomethyl)- Sulfathiourea	$C_7H_9N_3O_2S_2$ 231.30	515-49-1 182	8921	4-14-00-02671	H_2O 1; EtOH 2
2190	Benzenesulfonamide, 4-amino-N-(aminocarbonyl)- Sulfanilylurea	$C_7H_9N_3O_3S$ 215.23	547-44-4 147 dec	8906	4-14-00-02667	
2191	Benzenesulfonamide, 4-amino-N-(aminoiminomethyl)- Sulfaguanidine	$C_7H_{10}N_4O_2S$ 214.25	57-67-0 191.5	8879	4-14-00-02668	
2192	Benzenesulfonamide, 4-amino-N-[(butylamino)carbonyl]- Carbutamide	$C_{11}H_{17}N_3O_3S$ 271.34	339-43-5 144.5	1839	4-14-00-02667	
2193	Benzenesulfonamide, 4-amino-N-(6-chloro-3-pyridazinyl)- Sulfachlorpyridazine	$C_{10}H_9ClN_4O_2S$ 284.73	80-32-0 187	8871	4-25-00-02069	
2194	Benzenesulfonamide, 4-amino-N-(2,6-dimethoxy-4-pyrimidinyl)- Sulfadimethoxine	$C_{12}H_{14}N_4O_4S$ 310.33	122-11-2 203.5	8876	4-25-00-03483	
2195	Benzenesulfonamide, 4-amino-N-(3,4-dimethyl-5-isoxazolyl)- Sulfisoxazole	$C_{11}H_{13}N_3O_3S$ 267.31	127-69-5 191	8930	4-27-00-04747	
2196	Benzenesulfonamide, 4-amino-N-(4,6-dimethyl-2-pyrimidinyl)- Sulfamethazine	$C_{12}H_{14}N_4O_2S$ 278.33	57-68-1 198.5	8886	4-25-00-02215	H_2O 3; acid 3; alk 3
2197	Benzenesulfonamide, 4-amino-N-(6-methoxy-3-pyridazinyl)- Sulfamethoxypyridazine	$C_{11}H_{12}N_4O_3S$ 280.31	80-35-3 182.5	8890	4-25-00-03332	
2198	Benzenesulfonamide, 4-amino-N-(5-methyl-3-isoxazolyl)- Sulfamethoxazole	$C_{10}H_{11}N_3O_3S$ 253.28	723-46-6	8889	4-27-00-04685	
2199	Benzenesulfonamide, 4-amino-N-(4-methyl-2-pyrimidinyl)- Sulfamerazine	$C_{11}H_{12}N_4O_2S$ 264.31	127-79-7 236	8884	4-25-00-02159	H_2O 2; EtOH 2; eth 1; ace 2
2200	Benzenesulfonamide, 4-amino-N-(4-methyl-2-thiazolyl)- Sulfamethylthiazole	$C_{10}H_{11}N_3O_2S_2$ 269.35	515-59-3 237	8891	4-27-00-04715	EtOH 4
2201	Benzenesulfonamide, 4-amino-N-(1-phenyl-1H-pyrazol-5-yl)- Sulfaphenazole	$C_{15}H_{14}N_4O_2S$ 314.37	526-08-9	8910	4-25-00-02029	
2202	Benzenesulfonamide, 4-amino-N-pyrazinyl- Sulfapyrazine	$C_{10}H_{10}N_4O_2S$ 250.28	116-44-9 251	8912	4-25-00-02143	H_2O 1; EtOH 1; eth 1; ace 2
2203	Benzenesulfonamide, 4-amino-N-2-pyridinyl- Sulfapyridine	$C_{11}H_{11}N_3O_2S$ 249.29	144-83-2 192	8913	5-22-08-00424	H_2O 1; EtOH 3; bz 1; ctc 1
2204	Benzenesulfonamide, 4-amino-N-2-pyrimidinyl- Sulfadiazine	$C_{10}H_{10}N_4O_2S$ 250.28	68-35-9 255-6 dec	8874	4-25-00-02097	H_2O 2; EtOH 2; ace 2
2205	Benzenesulfonamide, 4-amino-N-2-quinoxalinyl- Sulfaquinoxaline	$C_{14}H_{12}N_4O_2S$ 300.34	59-40-5 247.5	8914	4-25-00-02606	H_2O 2; EtOH 2; ace 2; aq alk 3
2206	Benzenesulfonamide, 4-amino-N-2-thiazolyl- Sulfathiazole	$C_9H_9N_3O_2S_2$ 255.32	72-14-0 175(form a); 202(form b)	8920	4-27-00-04623	H_2O 2; EtOH 2; DMSO-d_6 2

No.	Name / Synonym	Mol. Form. / Mol. Wt.	CAS RN / mp/°C	Merck No. / bp/°C	Beil. Ref. / den/g cm^{-3}	Solubility / n_D
2207	Benzenesulfonamide, N-[(butylamino)carbonyl]-4-methyl- Tolbutamide	C$_{12}$H$_{18}$N$_2$O$_3$S 270.35	64-77-7 128.5	9432	4-11-00-00396 1.245^{25}	H$_2$O 2; EtOH 3; eth 3; chl 3
2208	Benzenesulfonamide, 4-chloro-N-[(propylamino)carbonyl]- Chloropropamide	C$_{10}$H$_{13}$ClN$_2$O$_3$S 276.74	94-20-2	2187	4-11-00-00119	
2209	Benzenesulfonamide, 4-[(2,4-diaminophenyl)azo]- Prontosil	C$_{12}$H$_{14}$ClN$_5$O$_2$S 327.79	103-12-8 249.5	8895	4-16-00-00563	H$_2$O 2; EtOH 3; ace 3; oils 3
2210	Benzenesulfonamide, N,N-dichloro-4-methyl- Dichloramine-T	C$_7$H$_7$Cl$_2$NO$_2$S 240.11	473-34-7 83	3034	3-11-00-00301	H$_2$O 1; EtOH 3; eth 3; bz 3
2211	Benzenesulfonamide, N,4-dimethyl-	C$_8$H$_{11}$NO$_2$S 185.25	640-61-9 78.5		4-11-00-00377 1.340^{25}	eth 4; EtOH 4
2212	Benzenesulfonamide, N,4-dimethyl-N-nitroso- p-Tolylsulfonylmethylnitrosamide	C$_8$H$_{10}$N$_2$O$_3$S 214.25	80-11-5 60	9469	4-11-00-00478	H$_2$O 1; EtOH 4; eth 4
2213	Benzenesulfonamide, 4-[(phenylmethyl)amino]- N4-Benzylsulfanilamide	C$_{13}$H$_{14}$N$_2$O$_2$S 262.33	104-22-3 171	1161	3-14-00-02026	
2214	Benzenesulfonic acid Besylic acid	C$_6$H$_6$O$_3$S 158.18	98-11-3 65	1078	4-11-00-00027	H$_2$O 4; EtOH 4; eth 1; bz 2
2215	Benzenesulfonic acid, 2-amino- Orthanilic acid	C$_6$H$_7$NO$_3$S 173.19	88-21-1 >320 dec	6835	4-14-00-02638	H$_2$O 2; EtOH 1; eth 1
2216	Benzenesulfonic acid, 3-amino- Metanilic acid	C$_6$H$_7$NO$_3$S 173.19	121-47-1 dec	5832	4-14-00-02647	H$_2$O 2; EtOH 2; eth 1
2217	Benzenesulfonic acid, 4-amino- Sulfanilic acid	C$_6$H$_7$NO$_3$S 173.19	121-57-3 288	8901	4-14-00-02655 1.485^{25}	H$_2$O 2; EtOH 1; eth 1
2218	Benzenesulfonic acid, 2-amino-5-[(4-amino-3-sulfophenyl)(4-imino-3-sulfo-2,5-cyclohexadien-1-ylidene)methyl]-3 Fuchsin, acid	C$_{20}$H$_{17}$N$_3$Na$_2$O$_9$S$_3$ 585.55	3244-88-0	103		H$_2$O 2; EtOH 2
2219	Benzenesulfonic acid, sec-butyl ester Benzenesulfonic acid, chloride	C$_{10}$H$_{14}$O$_3$S 214.29	103275-78-1 14.5	251 dec; 120^{10}	4-11-00-00032 1.3830^{15}	H$_2$O 1; EtOH 4; eth 3; ctc 3
2220	Benzenesulfonic acid, 4-chloro- p-Chlorobenzenesulfonic acid	C$_6$H$_5$ClO$_3$S 192.62	98-66-8 67	2122	147-8^{25}	4-11-00-00107	H$_2$O 3; EtOH 3; eth 1; bz 1
2221	Benzenesulfonic acid, 4-chloro-, 4-chlorophenyl ester Ovex	C$_{12}$H$_8$Cl$_2$O$_3$S 303.17	80-33-1 86.5	6856	4-11-00-00109	H$_2$O 1; EtOH 2; ace 3
2222	Benzenesulfonic acid, 4-[[4-(dimethylamino)phenyl]azo]-, sodium salt Methyl orange	C$_{14}$H$_{14}$N$_3$NaO$_3$S 327.34	547-58-0 dec	6019		H$_2$O 2; EtOH 2; eth 1; py 2
2223	Benzenesulfonic acid, ethyl ester Ethyl benzenesulfonate	C$_8$H$_{10}$O$_3$S 186.23	515-46-8 156^{15}	1078	4-11-00-00030 1.2167^{20}	H$_2$O 2; EtOH 3; eth 4; chl 4 1.5081^{20}
2224	Benzenesulfonic acid, 4-hydrazino- p-Sulfophenylhydrazine	C$_6$H$_8$N$_2$O$_3$S 188.21	98-71-5 286	4695	3-15-00-00865	H$_2$O 2; EtOH 2
2225	Benzenesulfonic acid, 2-hydroxy- o-Phenolsulfonic acid	C$_6$H$_6$O$_4$S 174.18	609-46-1 145 dec		4-11-00-00574	H$_2$O 4; EtOH 4
2226	Benzenesulfonic acid, 4-hydroxy- p-Phenolsulfonic acid	C$_6$H$_6$O$_4$S 174.18	98-67-9	7212	4-11-00-00582	H$_2$O 4; EtOH 4
2227	Benzenesulfonic acid, methyl ester Methyl benzenesulfonate	C$_7$H$_8$O$_3$S 172.20	80-18-2 154^{20}		4-11-00-00029 1.2730^{17}	H$_2$O 2; EtOH 4; eth 4; chl 4 1.5151^{20}
2228	Benzenesulfonic acid, 2-methyl-	C$_7$H$_8$O$_3$S 172.20	88-20-0 67.5	128.8^{25}	4-11-00-00228	H$_2$O 4; EtOH 3; eth 1
2229	Benzenesulfonic acid, 4-methyl- p-Toluenesulfonic acid	C$_7$H$_8$O$_3$S 172.20	104-15-4 104.5	9459 140^{20}	4-11-00-00241	H$_2$O 4; EtOH 3; eth 3
2230	Benzenesulfonic acid, 4-methyl-, butyl ester	C$_{11}$H$_{16}$O$_3$S 228.31	778-28-9 164-6^6		4-11-00-00250 1.1319^{20}	H$_2$O 1; eth 3; ctc 2 1.5050^{20}
2231	Benzenesulfonic acid, 4-methyl-, ethyl ester Ethyl p-toluenesulfonate	C$_9$H$_{12}$O$_3$S 200.26	80-40-0 34.5	3814 173^{15}	4-11-00-00248 1.166^{48}	H$_2$O 1; EtOH 3; eth 3; ctc 2
2232	Benzenesulfonic acid, 4-methyl-, methyl ester	C$_8$H$_{10}$O$_3$S 186.23	80-48-8 28.5	292; 186^{22}	4-11-00-00247 1.2087^{40}	H$_2$O 1; EtOH 4; eth 3; bz 4
2233	Benzenesulfonic acid, 4-methyl-, propyl ester	C$_{10}$H$_{14}$O$_3$S 214.29	599-91-7 <-20	189^9	4-11-00-00249 1.144^{20}	1.4998^{20}
2234	Benzenesulfonic acid, 4-(phenylamino)- N-Phenylsulfanilic acid	C$_{12}$H$_{11}$NO$_3$S 249.29	101-57-5 206	7284	3-14-00-02025	H$_2$O 4; EtOH 4
2235	Benzenesulfonic acid, propyl ester Propyl benzenesulfonate	C$_9$H$_{12}$O$_3$S 200.26	80-42-2 162-3^{15}		4-11-00-00031 1.1804^{17}	H$_2$O 2; EtOH 3; eth 4; chl 4 1.5035^{25}
2236	Benzenesulfonothioic acid, 4-chloro-, S-phenyl ester Benzenesulfonic acid, p-chlorothio-, S-phenyl ester	C$_{12}$H$_9$ClO$_2$S$_2$ 284.79	1142-97-8 -86.8	118.5	4-11-00-00223 0.8341^{20}	1.4087^{14}
2237	Benzene, 1,1'-sulfonylbis- Diphenyl sulfone	C$_{12}$H$_{10}$O$_2$S 218.28	127-63-9 128.5	3336 379	4-06-00-01490 1.252^{20}	H$_2$O 1; EtOH 3; eth 3; bz 3
2238	Benzene, 1,1'-sulfonylbis[4-bromo-	C$_{12}$H$_8$Br$_2$O$_2$S 376.07	2050-48-8 172		4-06-00-01652 1.88^{25}	H$_2$O 1; EtOH 2; HOAc 3
2239	Benzene, 1,1'-sulfonylbis[4-chloro-	C$_{12}$H$_8$Cl$_2$O$_2$S 287.17	80-07-9 147.9	250^{10}	4-06-00-01587	H$_2$O 2; EtOH 3; chl 3
2240	Benzene, 1,1'-sulfonylbis[4-methyl-	C$_{14}$H$_{14}$O$_2$S 246.33	599-66-6 159	406	4-06-00-02174	H$_2$O 2; EtOH 3; eth 2; bz 3
2241	Benzene, 1,1'-[sulfonylbis(methylene)]bis-	C$_{14}$H$_{14}$O$_2$S 246.33	620-32-6 152	290 dec	4-06-00-02651	H$_2$O 1; EtOH 2; ace 4; bz 3

No.	Name Synonym	Mol. Form. Mol. Wt.	CAS RN mp/°C	Merck No. bp/°C	Beil. Ref. den/g cm^{-3}	Solubility n_D
2242	Benzenesulfonyl chloride Phenylsulfonyl chloride	C$_6$H$_5$ClO$_2$S 176.62	98-09-9 14.5	1080 251 dec	4-11-00-00049 1.3470[15]	H$_2$O 1; EtOH 4; eth 3; ctc 3
2243	Benzenesulfonyl chloride, 4-(acetylamino)- Acetylsulfanilyl chloride	C$_8$H$_8$ClNO$_3$S 233.68	121-60-8 149	99	4-14-00-02703	EtOH 4; eth 4; bz 3; chl 3
2244	Benzenesulfonyl chloride, 4-bromo- p-Brosyl chloride	C$_6$H$_4$BrClO$_2$S 255.52	98-58-8 76	1395 153[15]	4-11-00-00162	H$_2$O 1; eth 4; chl 3
2245	Benzenesulfonyl chloride, 4-chloro- 	C$_6$H$_4$Cl$_2$O$_2$S 211.07	98-60-2 51	141[15]	4-11-00-00114	eth 4; bz 4
2246	Benzenesulfonyl chloride, 4-fluoro- 	C$_6$H$_4$ClFO$_2$S 194.61	349-88-2 30	106[9]	4-11-00-00102	bz 4; eth 4; chl 4
2247	Benzenesulfonyl chloride, 4-iodo- Pipsyl chloride	C$_6$H$_4$ClIO$_2$S 302.52	98-61-3 85	7458	4-11-00-00169	
2248	Benzenesulfonyl chloride, 2-methoxy- 	C$_7$H$_7$ClO$_3$S 206.65	10130-87-7 56	126[0.3]	4-11-00-00575	eth 3; peth 3
2249	Benzenesulfonyl chloride, 2-methyl- o-Toluenesulfonyl chloride	C$_7$H$_7$ClO$_2$S 190.65	133-59-5 10.2	154[36]	4-11-00-00229 1.3383[20]	H$_2$O 1; EtOH 3; eth 3; bz 3 1.5565[20]
2250	Benzenesulfonyl chloride, 4-methyl- p-Toluenesulfonyl chloride	C$_7$H$_7$ClO$_2$S 190.65	98-59-9 71	9460 145-6[15]	4-11-00-00375	H$_2$O 1; EtOH 3; eth 3; bz 4
2251	Benzenesulfonyl fluoride Phenylsulfonyl fluoride	C$_6$H$_5$FO$_2$S 160.17	368-43-4 	203.5	4-11-00-00049 1.3286[20]	EtOH 3; eth 3 1.4932[18]
2252	Benzenesulfonyl fluoride, 4-amino- p-Sulfanilyl fluoride	C$_6$H$_6$FNO$_2$S 175.18	98-62-4 68.5	8904	4-14-00-02658	
2253	Benzene, 1,2,3,5-tetrabromo- 1,2,3,5-Tetrabromobenzene	C$_6$H$_2$Br$_4$ 393.70	634-89-9 99.5	329	4-05-00-00686	bz 4; eth 4; EtOH 4
2254	Benzene, 1,2,4,5-tetrabromo- 2,3,5,6-Tetrabromobenzene	C$_6$H$_2$Br$_4$ 393.70	636-28-2 182	3.1[20]	4-05-00-00687 3.072[20]	eth 4
2255	Benzene, 1,2,3,4-tetrabromo-5,6-dimethyl- 	C$_8$H$_6$Br$_4$ 421.75	2810-69-7 262	374.5	3-05-00-00820	bz 4
2256	Benzene, 1,2,3,5-tetrabromo-4-methoxy- 2,3,4,6-Tetrabromoanisole	C$_7$H$_4$Br$_4$O 423.72	95970-07-3 113.5	340	3-06-00-00766	EtOH 4
2257	1,2,4,5-Benzenetetracarboxylic acid Pyromellitic acid	C$_{10}$H$_6$O$_8$ 254.15	89-05-4 276	8015	4-09-00-03804	H$_2$O 2; EtOH 3
2258	1,2,4,5-Benzenetetracarboxylic acid, tetramethyl ester	C$_{14}$H$_{14}$O$_8$ 310.26	635-10-9 144	sub	4-09-00-03804	EtOH 4
2259	Benzene, 1,2,3,4-tetrachloro- 1,2,3,4-Tetrachlorobenzene	C$_6$H$_2$Cl$_4$ 215.89	634-66-2 47.5	254	4-05-00-00667	H$_2$O 1; EtOH 2; eth 4; CS$_2$ 4
2260	Benzene, 1,2,4,5-tetrachloro- s-Tetrachlorobenzene	C$_6$H$_2$Cl$_4$ 215.89	95-94-3 139.5	244.5	4-05-00-00668 1.858[22]	H$_2$O 1; EtOH 2; eth 3; bz 3
2261	Benzene, 1,2,3,5-tetrachloro-4,6-dimethyl- 	C$_8$H$_6$Cl$_4$ 243.95	877-09-8 223		4-05-00-00944 1.703[25]	H$_2$O 1; EtOH 1; eth 3; bz 3
2262	Benzene, 1,2,3,5-tetrachloro-4-methyl- 2,3,4,6-Tetrachlorotoluene	C$_7$H$_4$Cl$_4$ 229.92	875-40-1 94	276.5	3-05-00-00702	bz 4; EtOH 4
2263	Benzene, 1,2,4,5-tetrachloro-3-methyl- 2,3,5,6-Tetrachlorotoluene	C$_7$H$_4$Cl$_4$ 229.92	1006-31-1 93.5	sub	3-05-00-00702	eth 4; EtOH 4
2264	Benzene, 1,2,4,5-tetrachloro-3-nitro- 	C$_6$HCl$_4$NO$_2$ 260.89	117-18-0 99.5	304	4-05-00-00728 1.744[25]	H$_2$O 1; EtOH 3; bz 3; chl 3
2265	Benzene, tetracosyl- Tetracosylbenzene	C$_{30}$H$_{54}$ 414.76	61828-05-5 57	373[100.] 238[1]	4-05-00-01239 0.8544[20]	1.4800[20]
2266	Benzene, tetradecyl- 	C$_{20}$H$_{34}$ 274.49	1459-10-5 16	359	4-05-00-01212 0.8549[20]	1.4818[20]
2267	Benzene, 1,1'-tetradecylidenebis- 	C$_{28}$H$_{38}$ 350.59	55268-63-8 17.9	207[1]	4-05-00-02017 0.9187[20]	1.5202[20]
2268	Benzene, 1,2,3,4-tetraethyl- 1,2,3,4-Tetraethylbenzene	C$_{14}$H$_{22}$ 190.33	642-32-0 11.8	252; 121.7[14]	4-05-00-01168 0.8875[20]	eth 4; EtOH 4 1.5125[20]
2269	Benzene, 1,2,4,5-tetraethyl- 1,2,4,5-Tetraethylbenzene	C$_{14}$H$_{22}$ 190.33	635-81-4 10	250	4-05-00-01168 0.8788[20]	eth 4; EtOH 4 1.5054[20]
2270	Benzene, 1,2,3,4-tetrafluoro- 1,2,3,4-Tetrafluorobenzene	C$_6$H$_2$F$_4$ 150.08	551-62-2 	94.3	4-05-00-00639	1.4054[20]
2271	Benzene, 1,2,3,5-tetrafluoro- 1,2,3,5-Tetrafluorobenzene	C$_6$H$_2$F$_4$ 150.08	2367-82-0 -48	84.4	4-05-00-00639 1.319[25]	1.4035[20]
2272	Benzene, 1,2,4,5-tetrafluoro- 1,2,4,5-Tetrafluorobenzene	C$_6$H$_2$F$_4$ 150.08	327-54-8 4.5	90.2	4-05-00-00639 1.4255[20]	1.4075[20]
2273	Benzene, 1,2,3,4-tetraiodo- 1,2,3,4-Tetraiodobenzene	C$_6$H$_2$I$_4$ 581.70	634-68-4 136	sub	0-05-00-00229	eth 4; EtOH 4
2274	Benzene, 1,2,3,5-tetraiodo- 1,2,3,5-Tetraiodobenzene	C$_6$H$_2$I$_4$ 581.70	634-92-4 148	sub	0-05-00-00229	HOAc 4
2275	Benzene, 1,2,4,5-tetraiodo- 1,2,4,5-Tetraiodobenzene	C$_6$H$_2$I$_4$ 581.70	636-31-7 254	sub	0-05-00-00229	HOAc 4
2276	Benzene, 1,2,4,5-tetrakis(1-methylethyl)- 1,2,4,5-Tetraisopropylbenzene	C$_{18}$H$_{30}$ 246.44	635-11-0 118.4	259; 133[17]	4-05-00-01207 0.758[150]	
2277	Benzene, 1,2,3,4-tetramethoxy-5-(2-propenyl)- 	C$_{13}$H$_{18}$O$_4$ 238.28	15361-99-6 25	145[12]	3-06-00-06691 1.087[25]	1.5146[25]
2278	Benzene, 1,2,3,4-tetramethyl- 	C$_{10}$H$_{14}$ 134.22	488-23-3 -6.2	205	4-05-00-01072 0.9052[20]	H$_2$O 1; EtOH 5; eth 5; ace 5 1.5203[20]
2279	Benzene, 1,2,3,5-tetramethyl- Isodurene	C$_{10}$H$_{14}$ 134.22	527-53-7 -23.7	5050 198	4-05-00-01073 0.8903[20]	H$_2$O 1; EtOH 5; eth 5; ace 5 1.5130[20]
2280	Benzene, 1,2,4,5-tetramethyl- Durene	C$_{10}$H$_{14}$ 134.22	95-93-2 79.3	3450 196.8	4-05-00-01076 0.8380[81]	H$_2$O 1; EtOH 5; eth 5; ace 5 1.4790[81]

No.	Name / Synonym	Mol. Form. / Mol. Wt.	CAS RN / mp/°C	Merck No. / bp/°C	Beil. Ref. / den/g cm⁻³	Solubility / n_D
2281	Benzene, 1,1'-(1,1,4,4-tetramethyl-1,4-butanediyl)bis- / Bineophyl	$C_{20}H_{26}$ / 266.43	17648-05-4 / 61.8	146-50[0.5]	4-05-00-01997	chl 3
2282	Benzene, 1,2,4,5-tetramethyl-3,6-dinitro-	$C_{10}H_{12}N_2O_4$ / 224.22	5465-13-4 / 211.5	sub	4-05-00-01080	H_2O 1; EtOH 2; eth 4; bz 3
2283	Benzene, 1,1'-thiobis- / Phenyl sulfide	$C_{12}H_{10}S$ / 186.28	139-66-2 / -25.9	7285 / 296	4-06-00-01488 / 1.1136[20]	H_2O 1; EtOH 3; eth 5; bz 5 / 1.6334[20]
2284	Benzene, 1,1'-thiobis[4-bromo-	$C_{12}H_8Br_2S$ / 344.07	3393-78-0 / 115	268.5[40]	4-06-00-01651 / 1.84[25]	H_2O 1; EtOH 2; eth 3; ctc 3
2285	Benzene, 1,1'-thiobis[2-methyl-	$C_{14}H_{14}S$ / 214.33	4537-05-7 / 64	285	4-06-00-02019	eth 4; EtOH 4; chl 4
2286	Benzene, 1,1'-thiobis[4-methyl-	$C_{14}H_{14}S$ / 214.33	620-94-0 / 57.3	>300; 175[16]	4-06-00-02173	H_2O 1; EtOH 3; eth 4; ace 3
2287	Benzene, 1,1'-[thiobis(methylene)]bis- / Benzyl sulfide	$C_{14}H_{14}S$ / 214.33	538-74-9 / 49.5	1162 / dec	4-06-00-02649 / 1.0583[50]	H_2O 1; EtOH 3; eth 3; CS_2 3
2288	Benzene, 1,1'-thiobis[2,4,6-trinitro- / Picryl sulfide	$C_{12}H_4N_6O_{12}S$ / 456.26	2217-06-3		4-06-00-01775	
2289	Benzenethiol / Phenyl mercaptan	C_6H_6S / 110.18	108-98-5 / -14.9	9285 / 169.1	4-06-00-01463 / 1.0775[20]	H_2O 1; EtOH 3; eth 3; bz 3 / 1.5893[20]
2290	Benzenethiol, 2-amino-	C_6H_7NS / 125.19	137-07-5 / 26	234	4-13-00-00910	EtOH 3; eth 3 / 1.4606[20]
2291	Benzenethiol, 4-amino-	C_6H_7NS / 125.19	1193-02-8 / 46	143[17]	4-13-00-01289	H_2O 3; EtOH 3
2292	Benzenethiol, 4-bromo-	C_6H_5BrS / 189.08	106-53-6 / 73	230.5	4-06-00-01650 / 1.5260[83]	H_2O 2; EtOH 2; eth 4; ctc 4
2293	Benzenethiol, 2-chloro-	C_6H_5ClS / 144.62	6320-03-2	205.5	4-06-00-01570 / 1.2752[10]	H_2O 2; EtOH 2
2294	Benzenethiol, 3-chloro-	C_6H_5ClS / 144.62	2037-31-2	206	4-06-00-01576 / 1.2637[13]	H_2O 1; EtOH 3; eth 3; chl 3
2295	Benzenethiol, 4-chloro-	C_6H_5ClS / 144.62	106-54-7 / 61	206	4-06-00-01581 / 1.1911[20]	H_2O 1; EtOH 4; eth 4; bz 4 / 1.5480[20]
2296	Benzenethiol, 4-ethoxy-	$C_8H_{10}OS$ / 154.23	699-09-2 / 1.6	238	4-06-00-05792 / 1.1070[17]	ace 4; bz 4; eth 4; EtOH 4
2297	Benzenethiol, 3-ethyl- / 3-Ethylthiophenol	$C_8H_{10}S$ / 138.23	62154-77-2	211	3-06-00-01663 / 1.038[20]	1.572[30]
2298	Benzenethiol, 4-ethyl-	$C_8H_{10}S$ / 138.23	4946-13-8	211	4-06-00-03027 / 1.038[20]	1.572[20]
2299	Benzenethiol, 3-methoxy-	C_7H_8OS / 140.21	15570-12-4	224.5; 114[20]	4-06-00-05701 / 1.1313[25]	chl 3 / 1.5874[20]
2300	Benzenethiol, 4-methoxy-	C_7H_8OS / 140.21	696-63-9	228	4-06-00-05790 / 1.1313[25]	EtOH 3; eth 3; bz 3; chl 2 / 1.5801[25]
2301	Benzenethiol, 2-methyl-	C_7H_8S / 124.21	137-06-4 / 15	195	4-06-00-02014 / 1.041[20]	H_2O 1; EtOH 3; eth 4 / 1.570[20]
2302	Benzenethiol, 3-methyl-	C_7H_8S / 124.21	108-40-7 / -20	195	4-06-00-02079 / 1.044[20]	H_2O 1; EtOH 3; eth 5 / 1.572[20]
2303	Benzenethiol, 4-methyl-	C_7H_8S / 124.21	106-45-6 / 43	195	4-06-00-02153 / 1.0220[51]	H_2O 1; EtOH 3; eth 4; chl 3
2304	Benzenethiol, pentafluoro-	C_6HF_5S / 200.13	771-62-0 / -24	143	1.501[25]	1.4645[20]
2305	1,2,3-Benzenetriamine / 1,2,3-Triaminobenzene	$C_6H_9N_3$ / 123.16	608-32-2 / 103	336	3-13-00-00551	H_2O 4; eth 4; EtOH 4
2306	1,2,4-Benzenetriamine	$C_6H_9N_3$ / 123.16	615-71-4 / 97.3	340	4-13-00-00506	H_2O 4; EtOH 4; chl 4
2307	Benzene, 1,2,3-tribromo- / 1,2,3-Tribromobenzene	$C_6H_3Br_3$ / 314.80	608-21-9 / 87.5		3-05-00-00569 / 2.658[25]	eth 4
2308	Benzene, 1,2,4-tribromo- / 1,2,4-Tribromobenzene	$C_6H_3Br_3$ / 314.80	615-54-3 / 44.5	275	4-05-00-00685	H_2O 1; EtOH 3; eth 4; ace 4
2309	Benzene, 1,3,5-tribromo- / 1,3,5-Tribromobenzene	$C_6H_3Br_3$ / 314.80	626-39-1 / 122.8	271	4-05-00-00685	H_2O 1; EtOH 2; eth 3; bz 3
2310	Benzene, 1,2,3-tribromo-5-methoxy- / Anisole, 3,4,5-tribromo	$C_7H_5Br_3O$ / 344.83	73557-60-5 / 93.3	303	2-06-00-00195	ace 4; bz 4; EtOH 4
2311	Benzene, 1,2,4-tribromo-5-methoxy- / 2,4,5-Tribromoanisole	$C_7H_5Br_3O$ / 344.83	95970-10-8 / 105	307	4-06-00-01067	ace 4; bz 4; EtOH 4
2312	Benzene, 1,2,4-tribromo-3-methoxy- / Anisole, 2,3,5-tribromo	$C_7H_5Br_3O$ / 344.83	73931-44-9 / 82	308.5	2-06-00-00192	ace 4; bz 4; EtOH 4
2313	Benzene, 1,3,5-tribromo-2-methoxy-	$C_7H_5Br_3O$ / 344.83	607-99-8 / 88	298	4-06-00-01067 / 2.491[25]	H_2O 2; EtOH 2; ace 4; bz 4
2314	Benzene, 1,3,5-tribromo-2-methyl- / 2,4,6-Tribromotoluene	$C_7H_5Br_3$ / 328.83	6320-40-7 / 70	290	3-05-00-00719 / 2.479[17]	
2315	Benzene, 1,2,3-tribromo-5-nitro- / 3,4,5-Tribromonitrobenzene	$C_6H_2Br_3NO_2$ / 359.80	3460-20-6 / 112	sub	4-05-00-00732 / 2.645[25]	
2316	Benzene, 1,3,5-tribromo-2-nitro-	$C_6H_2Br_3NO_2$ / 359.80	3463-40-9 / 125	177[11]	4-05-00-00732	eth 4; chl 4
2317	1,3,5-Benzenetricarbonyl trichloride	$C_9H_3Cl_3O_3$ / 265.48	4422-95-1 / 36.3	180[16]	4-09-00-03748	chl 3
2318	1,2,3-Benzenetricarboxylic acid / Hemimellitic acid	$C_9H_6O_6$ / 210.14	569-51-7 / 200		4-09-00-03745 / 1.546[20]	eth 4; EtOH 4
2319	1,2,4-Benzenetricarboxylic acid / Trimellitic acid	$C_9H_6O_6$ / 210.14	528-44-9 / 219	9616	4-09-00-03746	H_2O 4; eth 4; EtOH 4
2320	1,3,5-Benzenetricarboxylic acid, 2-chloro- / 2-Chloro-1,3,5-benzenetricarboxylic acid	$C_9H_5ClO_6$ / 244.59	56961-24-1 / 285	sub	4-09-00-03749	H_2O 4; eth 4; EtOH 4

No.	Name Synonym	Mol. Form. Mol. Wt.	CAS RN mp/°C	Merck No. bp/°C	Beil. Ref. den/g cm^{-3}	Solubility n_D
2321	Benzene, 1,2,3-trichloro- 1,2,3-Trichlorobenzene	C$_6$H$_3$Cl$_3$ 181.45	87-61-6 53.5	9542 218.5	4-05-00-00664 1.4533[25]	H$_2$O 1; EtOH 2; eth 4; bz 4
2322	Benzene, 1,2,4-trichloro- 1,2,4-Trichlorobenzene	C$_6$H$_3$Cl$_3$ 181.45	120-82-1 17	9543 213.5	4-05-00-00664 1.459[25]	H$_2$O 1; EtOH 2; eth 4; ctc 2 1.5717[20]
2323	Benzene, 1,3,5-trichloro- 1,3,5-Trichlorobenzene	C$_6$H$_3$Cl$_3$ 181.45	108-70-3 63.5	9544 208	4-05-00-00666	H$_2$O 1; EtOH 2; eth 4; bz 4
2324	Benzene, 1,2,4-trichloro-5-(chloromethyl)-	C$_7$H$_4$Cl$_4$ 229.92	3955-26-8	273	4-05-00-00822 1.547[20]	ace 4; eth 4; EtOH 4
2325	Benzene, 1,2,4-trichloro-5-[(4-chlorophenyl)sulfonyl]- Tetradifon	C$_{12}$H$_6$Cl$_4$O$_2$S 356.06	116-29-0 146	9132	4-06-00-01636 1.151[20]	
2326	Benzene, 1,2,3-trichloro-4-(dichloromethyl)- Benzylidene chloride, 2,3,4-trichloro	C$_7$H$_3$Cl$_5$ 264.36	56961-82-1 84	278	1-05-00-00153	bz 4
2327	Benzene, 1,2,4-trichloro-5-(dichloromethyl)- Benzylidene chloride, 2,4,5-trichloro	C$_7$H$_3$Cl$_5$ 264.36	33429-70-8	280.5	4-05-00-00824 1.5956[20]	bz 4 1.5992[20]
2328	Benzene, 1,1'-(2,2,2-trichloroethylidene)bis[4-chloro- DDT	C$_{14}$H$_9$Cl$_5$ 354.49	50-29-3 108.5	2832 260; 186[0.05]	4-05-00-01885	H$_2$O 1; EtOH 2; eth 4; ace 4
2329	Benzene, 1,1'-(2,2,2-trichloroethylidene)bis[4-methoxy- Methoxychlor	C$_{16}$H$_{15}$Cl$_3$O$_2$ 345.65	72-43-5 87	5913	4-06-00-06691 1.41[25]	H$_2$O 1; EtOH 3; eth 4; bz 4
2330	Benzene, 1,2,4-trichloro-3-methoxy-	C$_7$H$_5$Cl$_3$O 211.47	50375-10-5 45	227	3-06-00-00717	ace 4; EtOH 4
2331	Benzene, 1,2,4-trichloro-5-methoxy-	C$_7$H$_5$Cl$_3$O 211.47	6130-75-2 77.5	254	4-06-00-00963	EtOH 4; ace 4; os 3
2332	Benzene, 1,3,5-trichloro-2-methoxy- 2,4,6-Trichloroanisole	C$_7$H$_5$Cl$_3$O 211.47	87-40-1 61.5	9541 241	4-06-00-01005 1.640[25]	EtOH 3; ace 4; bz 3; chl 3
2333	Benzene, (trichloromethyl)- (Trichloromethyl)benzene	C$_7$H$_5$Cl$_3$ 195.47	98-07-7 -5	1120 221	4-05-00-00820 1.3723[20]	H$_2$O 1; EtOH 3; eth 3; bz 3 1.5580[20]
2334	Benzene, 1,2,3-trichloro-4-methyl- 2,3,4-Trichlorotoluene	C$_7$H$_5$Cl$_3$ 195.47	7359-72-0 43.5	244	4-05-00-00819	H$_2$O 1; EtOH 3; eth 3; ace 3
2335	Benzene, 1,2,3-trichloro-5-methyl- Toluene, 3,4,5-trichloro-	C$_7$H$_5$Cl$_3$ 195.47	21472-86-6 45.5	246	0-05-00-00299	H$_2$O 1; EtOH 3; ace 3; os 3
2336	Benzene, 1,2,4-trichloro-5-methyl- 2,4,5-Trichlorotoluene	C$_7$H$_5$Cl$_3$ 195.47	6639-30-1 82.4	231	4-05-00-00819	H$_2$O 1; EtOH 3; ace 3; os 3
2337	Benzene, 1,2,5-trichloro-3-methyl- 2,3,5-Trichlorotoluene	C$_7$H$_5$Cl$_3$ 195.47	56961-86-5 45.5	230	0-05-00-00299	H$_2$O 1; EtOH 3; ace 3; os 3
2338	Benzene, 1,2,3-trichloro-5-nitro-	C$_6$H$_2$Cl$_3$NO$_2$ 226.45	20098-48-0 72.5		2-05-00-00187 1.807[25]	EtOH 4
2339	Benzene, 1,2,4-trichloro-5-nitro-	C$_6$H$_2$Cl$_3$NO$_2$ 226.45	89-69-0 57.5	288	4-05-00-00728 1.790[23]	H$_2$O 1; EtOH 2; eth 3; bz 3
2340	Benzene, 1,3,5-trichloro-2,4,6-trifluoro- 1,3,5-Trichloro-2,4,6-trifluorobenzene	C$_6$Cl$_3$F$_3$ 235.42	319-88-0	198.4		
2341	Benzene, tricosyl- Tricosylbenzene	C$_{29}$H$_{52}$ 400.73	61828-04-4 54	365[100]; 231[1]	4-05-00-01237 0.8544[20]	1.4801[20]
2342	Benzene, tridecyl- 1-Phenyltridecane	C$_{19}$H$_{32}$ 260.46	123-02-4 10	9575 346	3-05-00-01108 0.855[20]	1.4821[20]
2343	Benzene, 1,3,5-triethoxy- 1,3,5-Triethoxybenzene	C$_{12}$H$_{18}$O$_3$ 210.27	2437-88-9 43.5	170[24]	4-06-00-07362	eth 4; EtOH 4
2344	Benzene, 1,2,4-triethyl- 1,2,4-Triethylbenzene	C$_{12}$H$_{18}$ 162.27	877-44-1	218; 99[15]	4-05-00-01133 0.8738[20]	H$_2$O 1; EtOH 3; eth 3 1.5024[20]
2345	Benzene, 1,3,5-triethyl- 1,3,5-Triethylbenzene	C$_{12}$H$_{18}$ 162.27	102-25-0 -66.5	215.9	4-05-00-01133 0.8631[20]	H$_2$O 1; EtOH 4; eth 4 1.4969[20]
2346	Benzene, 1,2,4-trifluoro-	C$_6$H$_3$F$_3$ 132.09	367-23-7	90	4-05-00-00638 1.264[25]	1.4171[20]
2347	Benzene, 1,3,5-trifluoro- 1,3,5-Trifluorobenzene	C$_6$H$_3$F$_3$ 132.09	372-38-3 -5.5	75.5	4-05-00-00638 1.277[25]	1.4140[20]
2348	Benzene, (trifluoromethyl)- (Trifluoromethyl)benzene	C$_7$H$_5$F$_3$ 146.11	98-08-8 -29.1	1121 102.1	4-05-00-00802 1.1884[20]	EtOH 5; eth 5; ace 5; bz 5 1.4146[20]
2349	Benzene, 1,2,3-triiodo- 1,2,3-Triiodobenzene	C$_6$H$_3$I$_3$ 455.80	608-29-7 116	sub	1-05-00-00122	eth 4; EtOH 4; chl 4
2350	Benzene, 1,2,4-triiodo- 1,2,4-Triiodobenzene	C$_6$H$_3$I$_3$ 455.80	615-68-9 91.5	sub	3-05-00-00585	EtOH 4; chl 4
2351	Benzene, 1,3,5-triiodo- 1,3,5-Triiodobenzene	C$_6$H$_3$I$_3$ 455.80	626-44-8 184.2	sub	4-05-00-00702	H$_2$O 1; EtOH 2; eth 2; bz 2
2352	Benzene, 1,3,5-triiodo-2-methyl- 2,4,6-Triiodotoluene	C$_7$H$_5$I$_3$ 469.83	36994-79-3 118.5	300 dec	1-05-00-00158	EtOH 4
2353	Benzene, 1,2,3-triiodo-5-nitro-	C$_6$H$_2$I$_3$NO$_2$ 500.80	53663-23-3 167		3-05-00-00626 3.256[25]	bz 4; eth 4; EtOH 4; chl 4
2354	Benzene, 1,2,3-trimethoxy-	C$_9$H$_{12}$O$_3$ 168.19	634-36-6 48.5	235	4-06-00-07329 1.1009[45]	H$_2$O 1; EtOH 3; eth 3; bz 3
2355	Benzene, 1,3,5-trimethoxy-	C$_9$H$_{12}$O$_3$ 168.19	621-23-8 54.5	255.5	4-06-00-07362	H$_2$O 1; EtOH 3; eth 3; bz 3
2356	Benzene, 1,2,4-trimethoxy-5-(1-propenyl)-	C$_{12}$H$_{16}$O$_3$ 208.26	494-40-6 67	296	4-06-00-07476 1.165[20]	eth 4; EtOH 4; peth 4 1.5683[20]
2357	Benzene, 1,2,3-trimethyl- 1,2,3-Trimethylbenzene	C$_9$H$_{12}$ 120.19	526-73-8 -25.4	176.1	4-05-00-01007 0.8944[20]	H$_2$O 1; EtOH 5; eth 5; ace 5 1.5139[20]
2358	Benzene, 1,2,4-trimethyl- Pseudocumene	C$_9$H$_{12}$ 120.19	95-63-6 -43.8	7929 169.3	4-05-00-01010 0.8758[20]	H$_2$O 1; EtOH 5; eth 5; ace 5 1.5048[20]

No.	Name Synonym	Mol. Form. Mol. Wt.	CAS RN mp/°C	Merck No. bp/°C	Beil. Ref. den/g cm^{-3}	Solubility n_D
2359	Benzene, 1,3,5-trimethyl- Mesitylene	C_9H_{12} 120.19	108-67-8 -44.7	5810 164.7	4-05-00-01016 0.8652[20]	H$_2$O 1; EtOH 5; eth 5; ace 5 1.4994[20]
2360	Benzene, 1,3,5-trimethyl-2,4-dinitro-	$C_9H_{10}N_2O_4$ 210.19	608-50-4 85.3	exp 418	4-05-00-01029	H$_2$O 1; EtOH 3
2361	Benzene, 1,2,4-trimethyl-5-(1-methylethyl)-	$C_{12}H_{18}$ 162.27	10222-95-4 21	221	3-05-00-01038 0.8795[21]	1.5065[21]
2362	Benzene, 1,2,4-trimethyl-5-nitro-	$C_9H_{11}NO_2$ 165.19	610-91-3 71	265	4-05-00-01015	EtOH 3; chl 3; peth 3
2363	Benzene, 1,3,5-trimethyl-2-nitro-	$C_9H_{11}NO_2$ 165.19	603-71-4 44	255	4-05-00-01028 1.51[25]	EtOH 4
2364	Benzene, 1,3,5-trimethyl-2,4,6-trinitro-	$C_9H_9N_3O_6$ 255.19	602-96-0 238.2	exp 415	4-05-00-01029	H$_2$O 1; EtOH 2; eth 2; ace 3
2365	Benzene, 1,3,5-trinitro- sym-Trinitrobenzene	$C_6H_3N_3O_6$ 213.11	99-35-4 121.5	9639 315	4-05-00-00755 1.4775[152]	H$_2$O 2; EtOH 2; eth 2; ace 4
2366	1,2,3-Benzenetriol Pyrogallol	$C_6H_6O_3$ 126.11	87-66-1 133	8010 309	4-06-00-07327 1.453[4]	H$_2$O 4; EtOH 4; eth 4; ace 3 1.561[134]
2367	1,2,4-Benzenetriol Hydroxyhydroquinone	$C_6H_6O_3$ 126.11	533-73-3 140.5	1081	4-06-00-07338	H$_2$O 4; EtOH 4; eth 4; bz 1
2368	1,3,5-Benzenetriol Phloroglucinol	$C_6H_6O_3$ 126.11	108-73-6 218.5	7301 sub	4-06-00-07361 1.46[25]	H$_2$O 2; EtOH 4; eth 4; ace 3
2369	1,2,3-Benzenetriol, 5-methyl-	$C_7H_8O_3$ 140.14	609-25-6 129	sub	4-06-00-07376	bz 4
2370	1,2,4-Benzenetriol, triacetate	$C_{12}H_{12}O_6$ 252.22	613-03-6 99	300	4-06-00-07342	EtOH 3; chl 3; MeOH 3
2371	1,3,5-Benzenetriol, 2,4,6-trichloro- Trichloro-1,3,5-trihydroxybenzene	$C_6H_3Cl_3O_3$ 229.45	56961-23-0 136	sub	0-06-00-01104	EtOH 4
2372	Benzene, 1,2,4-tris(1-methylethyl)- 1,2,4-Triisopropylbenzene	$C_{15}H_{24}$ 204.36	948-32-3	244	4-05-00-01178 0.8574[25]	1.4896[25]
2373	Benzene, 1,3,5-tris(1-methylethyl)-	$C_{15}H_{24}$ 204.36	717-74-8 -7.4	238	4-05-00-01178 0.8545[20]	ace 3; bz 3; chl 3 1.4882[20]
2374	Benzene, undecyl-	$C_{17}H_{28}$ 232.41	6742-54-7 -5	316	3-05-00-01100 0.8553[20]	1.4828[20]
2375	9,10[1',2']-Benzenoanthracene, 9,10-dihydro- Triptycene	$C_{20}H_{14}$ 254.33	477-75-8	9663	4-05-00-02639	
2376	1H-Benzimidazole N,N'-Methenyl-o-phenylenediamine	$C_7H_6N_2$ 118.14	51-17-2 170.5	1091 >360	5-23-06-00196	H$_2$O 2; EtOH 4; eth 2; bz 1
2377	1H-Benzimidazole, 4,5-dichloro-2-(trifluoromethyl)- Chlorflurazole	$C_8H_3Cl_2F_3N_2$ 255.03	3615-21-2 213.5			
2378	1H-Benzimidazole, 1,5-dimethyl- 1,5-Dimethylbenzimidazole	$C_9H_{10}N_2$ 146.19	10394-35-1 95	300	5-23-06-00380	bz 4; eth 4; EtOH 4
2379	1H-Benzimidazole, 2,5-dimethyl-	$C_9H_{10}N_2$ 146.19	1792-41-2 203	350	5-23-06-00448	eth 4; EtOH 4
2380	1H-Benzimidazole, 5,6-dimethyl- Dimedazole	$C_9H_{10}N_2$ 146.19	582-60-5 205.5	3225 sub	5-23-06-00454	H$_2$O 3; EtOH 3; eth 3; chl 3
2381	1H-Benzimidazole, 1-ethyl-2-methyl- 1-Ethyl-2-methylbenzimidazole	$C_{10}H_{12}N_2$ 160.22	5805-76-5 51	296	5-23-06-00326 1.073[25]	
2382	1H-Benzimidazole, 1-methyl-	$C_8H_8N_2$ 132.17	1632-83-3 66	286	5-23-06-00215 1.1254[20]	peth 3 1.6013[7]
2383	1H-Benzimidazole, 1-phenyl-	$C_{13}H_{10}N_2$ 194.24	2622-60-8 97	221[25]	5-23-06-00222	H$_2$O 2; EtOH 3; MeOH 3
2384	1H-Benzimidazole, 2-phenyl- Phenzidole	$C_{13}H_{10}N_2$ 194.24	716-79-0 293	7245	5-23-08-00437	H$_2$O 2; EtOH 3; bz 2; chl 2
2385	1H-Benzimidazole, 2-(phenylmethyl)- Bendazol	$C_{14}H_{12}N_2$ 208.26	621-72-7 187	1043	5-23-08-00498	bz 4; EtOH 4; gl HOAc 4
2386	2H-Benzimidazole-2-thione, 1,3-dihydro- 2-Benzimidazolethiol	$C_7H_6N_2S$ 150.20	583-39-1 298	1092	5-24-02-00413	EtOH 4
2387	1,2-Benzisothiazole 1-Thia-2-azaindene	C_7H_5NS 135.19	272-16-2 38	220	2-27-00-00016	ctc 3
2388	1,2-Benzisothiazol-3(2H)-one, 1,1-dioxide Saccharin	$C_7H_5NO_3S$ 183.19	81-07-2 228 dec	8282 sub	4-27-00-02649 0.828[25]	H$_2$O 2; EtOH 3; eth 2; ace 3
2389	1,2-Benzisoxazole Indoxazene	C_7H_5NO 119.12	271-95-4	100[26]	4-27-00-01067 1.1727[21]	eth 4 1.5570[20]
2390	2,1-Benzisoxazole Anthranil	C_7H_5NO 119.12	271-58-9 <-18	215	4-27-00-01068 1.1827[20]	ace 4; EtOH 4 1.5845[20]
2391	Benz[j]aceanthrylene, 1,2-dihydro- Cholanthrene	$C_{20}H_{14}$ 254.33	479-23-2 174	2200	4-05-00-02638	H$_2$O 1; EtOH 3; bz 3; HOAc 3
2392	Benz[j]aceanthrylene, 1,2-dihydro-3-methyl- 3-Methylcholanthrene	$C_{21}H_{16}$ 268.36	56-49-5 180	5967 280[80]	4-05-00-02648 1.28[20]	H$_2$O 1
2393	11H-Benzo[a]fluorene	$C_{17}H_{12}$ 216.28	238-84-6 189.5	405	4-05-00-02473	EtOH 2; eth 3; bz 3; chl 3
2394	Benzo[a]phenazine 7,12-Diazabenz[a]anthracene	$C_{18}H_{10}N_2$ 230.27	225-61-6 142.5	>360	5-23-09-00494	H$_2$O 1; EtOH 3; eth 3; ace 3
2395	Benzo[a]pyrene 2,3-Benzopyrene	$C_{20}H_{12}$ 252.32	50-32-8 176.5	1113	4-05-00-02687	chl 4
2396	2H-Benzo[a]quinolizin-2-one, 1,3,4,6,7,11b-hexahydro-9,10-dimethoxy-3-(2-methylpropyl)- 1,2-Benzoquinolizine-2-one, 9,10-dimethoxy-1,2,3,4,6,7,11b-heptahydro-3-isobutyl	$C_{19}H_{27}NO_3$ 317.43	58-46-8 128	9118	5-21-13-00178	chl 3

No.	Name Synonym	Mol. Form. Mol. Wt.	CAS RN mp/°C	Merck No. bp/°C	Beil. Ref. den/g cm^{-3}	Solubility n_D
2397	2H,8H-Benzo[1,2-b:5,4-b']dipyran-2-one, 8,8-dimethyl- Xanthyletin	$C_{14}H_{12}O_3$ 228.25	553-19-5 131.5	9978 140-5[0.1]	5-19-04-00470	EtOH 3; peth 3
2398	2H,8H-Benzo[1,2-b:5,4-b']dipyran-2-one, 5-methoxy-8,8-dimethyl- Xanthoxyletin	$C_{15}H_{14}O_4$ 258.27	84-99-1 133	9975	5-19-06-00039	H_2O 1; EtOH 3; eth 2; ace 3
2399	Benzo[b]thiophene Thianaphthene	C_8H_6S 134.20	95-15-8 32	9232 221	5-17-02-00006 1.1484[32]	H_2O 1; EtOH 4; eth 3; ace 3 1.6374[37]
2400	Benzo[b]thiophene-2-carboxylic acid Thionaphthene-2-carboxylic acid	$C_9H_6O_2S$ 178.21	6314-28-9 240.5	9274	5-18-06-00427	eth 4
2401	Benzo[b]thiophene-2,3-dione	$C_8H_4O_2S$ 164.18	493-57-2 121	247	5-17-11-00251	H_2O 1; EtOH 3; bz 3; HOAc 3
2402	Benzo[b]thiophene, 5-methyl-	C_9H_8S 148.23	14315-14-1 20.5	111-5[12]	5-17-02-00031 1.111[22]	1.615[22]
2403	Benzo[b]thiophene-4-ol	C_8H_6OS 150.20	3610-02-4 79.8	sub	5-17-04-00199	EtOH 4
2404	Benzo[b]thiophen-3(2H)-one, 2-(3-oxobenzo[b]thien-2(3H)-ylidene)-	$C_{16}H_8O_2S_2$ 296.37	522-75-8 359	sub	5-19-05-00281	H_2O 1; EtOH 1; bz 3; chl 2
2405	7H-Benzo[c]carbazole 7-Aza-7H-benzo[c]fluorene	$C_{16}H_{11}N$ 217.27	205-25-4 134	448	5-20-08-00455	bz 4; EtOH 4; HOAc 4
2406	Benzo[c]phenanthrene Tetrahelicene	$C_{18}H_{12}$ 228.29	195-19-7 68	1129	4-05-00-02552	H_2O 1; EtOH 2; lig 2
2407	5H-Benzocyclohepten-5-one, 6,7,8,9-tetrahydro-	$C_{11}H_{12}O$ 160.22	826-73-3 124[7]; 108[1]		4-07-00-01029 1.0780[20]	EtOH 3 1.5698[20]
2408	5H-Benzocyclohepten-5-one, 2,3,4,6-tetrahydroxy- Purpurogallin	$C_{11}H_8O_5$ 220.18	569-77-7 274-5 dec	7963	4-08-00-03456	
2409	3H-1,4-Benzodiazepin-2-amine, 7-chloro-N-methyl-5-phenyl-, 4-oxide Chlorodiazepoxide	$C_{16}H_{14}ClN_3O$ 299.76	58-25-3 236.2	2082		
2410	2H-1,4-Benzodiazepin-2-one, 7-chloro-1,3-dihydro-1-methyl-5-phenyl- Diazepam	$C_{16}H_{13}ClN_2O$ 284.74	439-14-5 132	2977	5-24-04-00300	
2411	2H-1,4-Benzodiazepin-2-one, 7-chloro-1,3-dihydro-5-phenyl- Nordazepam	$C_{15}H_{11}ClN_2O$ 270.72	1088-11-5 216.5	6608	5-24-04-00291	
2412	1,3,2-Benzodioxaborole	$C_6H_5BO_2$ 119.92	274-07-7 12	88[156]; 50[50]	1.2700[20]	1.5070[20]
2413	1,3,2-Benzodioxaphosphole, 2-chloro-	$C_6H_4ClO_2P$ 174.52	1641-40-3 30	80[20]	4-06-00-05598 1.4650[20]	1.5712[20]
2414	1,4-Benzodioxin, 2,3-dihydro-	$C_8H_8O_2$ 136.15	493-09-4 103[6]		5-19-01-00439 1.142[25]	1.5485[20]
2415	1,3-Benzodioxole	$C_7H_6O_2$ 122.12	274-09-9 172.5; 77[27]		5-19-01-00434 1.064[25]	1.5398[20]
2416	1,3-Benzodioxole-5-acetic acid, methyl ester Phenylacetic acid, 3,4-methylenedioxy, methyl ester	$C_{10}H_{10}O_4$ 194.19	326-59-0 153-5[10]		2-19-00-00295 1.246[20]	1.534
2417	1,3-Benzodioxole-5-carboxaldehyde Piperonal	$C_8H_6O_3$ 150.13	120-57-0 37	7445 263	5-19-04-00225	H_2O 2; EtOH 4; eth 5; ace 3
2418	1,3-Benzodioxole-5-carboxylic acid Piperonylic acid	$C_8H_6O_4$ 166.13	94-53-1 229	7447	5-19-07-00300	
2419	1,3-Benzodioxole-5-carboxylic acid, ethyl ester	$C_{10}H_{10}O_4$ 194.19	6951-08-2 18.5	285.5; 135[6]	5-19-07-00300	eth 4; EtOH 4; peth 4
2420	1,3-Benzodioxole-5-carboxylic acid, methyl ester	$C_9H_8O_4$ 180.16	326-56-7 53	273 dec	5-19-07-00300	eth 4; EtOH 4
2421	1,3-Benzodioxole, 5-(chloromethyl)-	$C_8H_7ClO_2$ 170.60	20850-43-5 20.5	134-5[14]	5-19-01-00443 1.312[25]	1.5660[20]
2422	1,3-Benzodioxole, 4,7-dimethoxy-5-(1-propenyl)-, (E)-	$C_{12}H_{14}O_4$ 222.24	17672-88-7 56	303.5	4-19-00-01033	ace 4; bz 4; eth 4; EtOH 4
2423	1,3-Benzodioxole, 4,7-dimethoxy-5-(2-propenyl)- Apiole	$C_{12}H_{14}O_4$ 222.24	523-80-8 29.5	767 294. 179[35]	5-19-03-00307 1.015[20]	ace 4; bz 4; EtOH 4; lig 4 1.5360[20]
2424	1,3-Benzodioxole-5-ethanamine	$C_9H_{11}NO_2$ 165.19	1484-85-1 166[20]		5-19-08-00407 1.225[20]	1.5620[20]
2425	1,3-Benzodioxole, 5-ethenyl-	$C_9H_8O_2$ 148.16	7315-32-4 224		5-19-01-00544 1.1488[18]	eth 4; EtOH 4; chl 4 1.5802
2426	1,3-Benzodioxole-5-methanamine	$C_8H_9NO_2$ 151.17	2620-50-0 138-9[13]		5-19-08-00395 1.214[25]	1.5635[20]
2427	1,3-Benzodioxole-5-methanol	$C_8H_8O_3$ 152.15	495-76-1 58	157[16]	5-19-02-00543	H_2O 2; EtOH 4; eth 3; bz 3
2428	1,3-Benzodioxole, 4-methoxy-6-(2-propenyl)- Myristicin	$C_{11}H_{12}O_3$ 192.21	607-91-0 <-20	6247 276.5	5-19-02-00631 1.1416[20]	H_2O 1; EtOH 2; eth 3; bz 3 1.5403[20]
2429	1,3-Benzodioxole, 5-(2-propenyl)- Safrole	$C_{10}H_{10}O_2$ 162.19	94-59-7 11.2	8287 234.5	5-19-01-00553 1.1000[20]	H_2O 1; EtOH 4; eth 5; chl 5 1.5381[20]
2430	1,3-Benzodioxole, 5-(1-propenyl)-, (E)-	$C_{10}H_{10}O_2$ 162.19	4043-71-4 6.8	253	5-19-01-00552 1.1224[20]	H_2O 1; EtOH 5; eth 5; ace 4 1.5782[20]

No.	Name / Synonym	Mol. Form. / Mol. Wt.	CAS RN / mp/°C	Merck No. / bp/°C	Beil. Ref. / den/g cm^{-3}	Solubility / n_D
2431	1,3-Benzodioxole, 5-propyl- / Dihydrosafrole	$C_{10}H_{12}O_2$ / 164.20	94-58-6		5-19-01-00473	ctc 3
2432	[1,3]Benzodioxolo[5,6-c]-1,3-dioxolo[4,5-i]phenanthridinium, 13-methyl- / Sanguinarine	$C_{20}H_{15}NO_5$ / 349.34	2447-54-3 / 266	8320		ace 4; bz 4; eth 4; EtOH 4
2433	[1,3]Benzodioxolo[5,6-c]phenanthridinium, 1,2-dimethoxy-12-methyl- / Chelerythrine	$C_{21}H_{19}NO_5$ / 365.39	34316-15-9 / 207	2041		chl 4
2434	1,4-Benzodithiin	$C_8H_6S_2$ / 166.27	255-50-5 / 67-70[0.1]		5-19-01-00536 / 1.2799[20]	1.6754[25]
2435	Benzo[e]-1,3-dioxolo[4,5-l][2]benzazecin-12(5H)-one, 4,6,7,13-tetrahydro-9,10-dimethoxy-5-methyl- / Cryptopine	$C_{21}H_{23}NO_5$ / 369.42	482-74-6 / 223	2611	4-27-00-06652 / 1.315[20]	H$_2$O 1; EtOH 2; eth 2; bz 2
2436	Benzo[e]pyrene / 1,2-Benzpyrene	$C_{20}H_{12}$ / 252.32	192-97-2 / 177.5	1114 / 311	4-05-00-02689	
2437	Benzo[f]quinoline / β-Naphthoquinoline	$C_{13}H_9N$ / 179.22	85-02-9 / 94	1115 / 352; 202-5[8]	5-20-08-00220	H$_2$O 2; EtOH 4; eth 4; ace 3
2438	Benzofuran / Coumarone	C_8H_6O / 118.14	271-89-6 / <-18	1098 / 174	5-17-02-00003 / 1.0913[25]	H$_2$O 1; EtOH 3; eth 3 / 1.5615[17]
2439	2-Benzofurancarboxylic acid / Coumarilic acid	$C_9H_6O_3$ / 162.14	496-41-3 / 192.5	2562 / 312.5	5-18-06-00419	EtOH 4
2440	2-Benzofurancarboxylic acid, ethyl ester	$C_{11}H_{10}O_3$ / 190.20	3199-61-9 / 30.5	/ 276; 161[15]	5-18-06-00420 / 1.1656[28]	1.564[27]
2441	Benzofuran, 2-(chloromethyl)-2,3-dihydro- / 2-(Chloromethyl)-2,3-dihydrobenzofuran	C_9H_9ClO / 168.62	53491-32-0 / 41.5	/ 118-9[11]	4-17-00-00421 / 1.2183[7]	ace 4; eth 4 / 1.5620[7]
2442	Benzofuran, 2,3-dihydro- / Coumaran	C_8H_8O / 120.15	496-16-2 / -21.5	2560 / 188.5	5-17-01-00581 / 1.058[25]	eth 4; EtOH 4; chl 4 / 1.5497[20]
2443	Benzofuran, 2,3-dihydro-2-methyl-	$C_9H_{10}O$ / 134.18	1746-11-8 /	/ 197.5	5-17-01-00597 / 1.061[25]	1.5308
2444	Benzofuran, 2,3-dihydro-5-methyl- / 2,3-Dihydro-5-methylbenzofuran	$C_9H_{10}O$ / 134.18	76429-68-0 /	/ 210.5	4-17-00-00423 / 1.0463[19]	1.54[20]
2445	Benzofuran, 2,5-dimethyl- / 2,5-Dimethylbenzofuran	$C_{10}H_{10}O$ / 146.19	29040-46-8 /	/ 220	5-17-02-00045 / 1.031[20]	eth 4 / 1.5534[20]
2446	Benzofuran, 2,6-dimethyl- / 2,6-Dimethylbenzofuran	$C_{10}H_{10}O$ / 146.19	24410-51-3 /	/ 217.5	5-17-02-00046 / 1.051[12]	eth 4 / 1.554[15]
2447	Benzofuran, 2,7-dimethyl- / 2,7-Dimethylfuran	$C_{10}H_{10}O$ / 146.19	59020-74-5 /	/ 216	4-17-00-00512 / 1.036[20]	eth 4 / 1.5546[20]
2448	Benzofuran, 3,5-dimethyl- / 3,5-Dimethylfuran	$C_{10}H_{10}O$ / 146.19	10410-35-2 /	/ 220.5	5-17-02-00047 / 1.036[20]	eth 4 / 1.550[20]
2449	Benzofuran, 3,6-dimethyl- / 3,6-DimethylfuranBenzofuran	$C_{10}H_{10}O$ / 146.19	24410-50-2 /	/ 222	4-17-00-00512 / 1.0456[20]	eth 4 / 1.5505[20]
2450	Benzofuran, 4,6-dimethyl-	$C_{10}H_{10}O$ / 146.19	116668-34-9 /	/ 219	0-17-00-00063 / 1.037[20]	eth 4 / 1.5485[21]
2451	Benzofuran, 4,7-dimethyl- / 4,7-Dimethylbenzofuran	$C_{10}H_{10}O$ / 146.19	28715-26-6 /	/ 216	0-17-00-00063 / 1.041[17]	eth 4 / 1.549[17]
2452	Benzofuran, 5,6-dimethyl- / 5,6-Dimethylbenzofuran	$C_{10}H_{10}O$ / 146.19	24410-52-4 /	/ 221	5-17-02-00048 / 1.060[15]	eth 4 / 1.5516[15]
2453	Benzofuran, 5,7-dimethyl- / 5,7-Dimethylbenzofuran	$C_{10}H_{10}O$ / 146.19	64965-91-9 /	/ 222	0-17-00-00063 / 1.0262[18]	eth 4 / 1.5358[20]
2454	Benzofuran, 6,7-dimethyl- / 6,7-Dimethylfuran	$C_{10}H_{10}O$ / 146.19	35355-36-3 /	/ 218	0-17-00-00063 / 1.038[20]	eth 4 / 1.5478[20]
2455	2-Benzofuranmethanol / 2-Hydroxymethylbenzofuran	$C_9H_8O_2$ / 148.16	55038-01-2 /	/ 149-50[14]	5-17-04-00224 / 1.177[13]	1.5550[11]
2456	Benzofuran, 2-methyl-	C_9H_8O / 132.16	4265-25-2 /	/ 197.5	5-17-02-00022 / 1.0540[20]	eth 4; EtOH 4 / 1.5495[22]
2457	Benzofuran, 3-methyl- / 3-Methylbenzo[b]furan	C_9H_8O / 132.16	21535-97-7 /	/ 197; 86[20]	5-17-02-00025 / 1.0540[25]	H$_2$O 1; EtOH 3; eth 3 / 1.5536[16]
2458	Benzofuran, 5-methyl- / 5-Methylbenzo[b]furan	C_9H_8O / 132.16	18441-43-5 /	/ 198	5-17-02-00030 / 1.0603[19]	H$_2$O 1; EtOH 3; eth 3; sulf 3 / 1.5570[19]
2459	Benzofuran, 7-methyl- / 7-Methylbenzo[b]furan	C_9H_8O / 132.16	17059-52-8 / 109	/ 190.5	5-17-02-00032 / 1.0490[19]	H$_2$O 1; EtOH 3; eth 3; sulf 3 / 1.5525[19]
2460	Benzofuran, 4-methyl-7-(1-methylethyl)- / Benzofuran, 7-isopropyl-4-methyl	$C_{12}H_{14}O$ / 174.24	95835-77-1 /	/ 241.5	0-17-00-00065 / 1.0145[18]	1.5363[16]
2461	2(3H)-Benzofuranone	$C_8H_6O_2$ / 134.13	553-86-6 / 50	/ 249	5-17-10-00006 / 1.2236[14]	
2462	3(2H)-Benzofuranone	$C_8H_6O_2$ / 134.13	7169-34-8 / 102.5	/ 152[15]	5-17-04-00185	bz 4
2463	2(3H)-Benzofuranone, 3-[2-(diethylamino)ethyl]-3-phenyl- / Amolanone	$C_{20}H_{23}NO_2$ / 309.41	76-65-3 / 43.4	603 / 192-4[2.0]	4-18-00-07995	1.5614[25]
2464	5-Benzofuranpropanol, 2-(1,3-benzodioxol-5-yl)-7-methoxy- / Egonol	$C_{19}H_{18}O_5$ / 326.35	530-22-3 / 118	/ 228-30[15]	5-19-09-00583	chl 4
2465	Benzofuran, 4,5,6,7-tetrahydro-3,6-dimethyl-	$C_{10}H_{14}O$ / 150.22	494-90-6 / 86	/ 80[18]	5-17-01-00518 / 0.972[15]	
2466	6H-Benzofuro[3,2-c][1]benzopyran-6-one, 3,9-dihydroxy- / Coumestrol	$C_{15}H_8O_5$ / 268.23	479-13-0 / 385 dec	2565	5-19-06-00405	H$_2$O 1; EtOH 2; eth 1; ace 2

No.	Name / Synonym	Mol. Form. / Mol. Wt.	CAS RN / mp/°C	Merck No. / bp/°C	Beil. Ref. / den/g cm^{-3}	Solubility / n_D
2467	Benzo[g]-1,3-benzodioxolo[5,6-a]quinolizinium, 5,6-dihydro-9,10-dimethoxy- / Berberine	$C_{20}H_{19}NO_5$ / 353.37	2086-83-1 / 145	1169		eth 4; EtOH 4
2468	5H-Benzo[g]-1,3-benzodioxolo[6,5,4-de]quinoline, 6,7,7a,8-tetrahydro-10,11-dimethoxy-7-methyl-, (S)- / Dicentrine	$C_{20}H_{21}NO_4$ / 339.39	517-66-8	3028	4-27-00-06486	chl 3
2469	5H-Benzo[g]-1,3-benzodioxolo[6,5,4-de]quinoline, 6,7,7a,8-tetrahydro-11-methoxy-7-methyl-, (R)- / Laureline	$C_{19}H_{19}NO_3$ / 309.36	81-38-9 / 114	5251	1-27-00-00461	H_2O 1; EtOH 3; eth 3; dil acid 3
2470	5H-Benzo[g]-1,3-benzodioxolo[6,5,4-de]quinolin-12-ol, 6,7,7a,8-tetrahydro-11-methoxy-7-methyl-, (S)- / Bulbocapnine	$C_{19}H_{19}NO_4$ / 325.36	298-45-3 / 199.5	1471	2-27-00-00554	H_2O 1; EtOH 3; chl 4; os 3
2471	Benzo[ghi]perylene / 1,12-Benzperylene	$C_{22}H_{12}$ / 276.34	191-24-2		4-05-00-02766	
2472	Benzo[g]pteridine-2,4(1H,3H)-dione, 7,8-dimethyl- / Lumichrome	$C_{12}H_{10}N_4O_2$ / 242.24	1086-80-2	5468	4-26-00-02538	
2473	Benzo[g]quinoline / 6,7-Benzoquinoline	$C_{13}H_9N$ / 179.22	260-36-6 / 114	202[14]	5-20-08-00195	eth 4; EtOH 4
2474	Benzo[h]quinoline	$C_{13}H_9N$ / 179.22	230-27-3 / 52	339; 233[47]	5-20-08-00215 / 1.2340[20]	H_2O 2; EtOH 3; eth 3; ace 3
2475	Benzoic acid / Benzenecarboxylic acid	$C_7H_6O_2$ / 122.12	65-85-0 / 122.4	1101 / 249.2	4-09-00-00273 / 1.2659[15]	H_2O 2; EtOH 4; eth 4; ace 3 / 1.504[132]
2476	Benzoic acid, 2-acetyl-	$C_9H_8O_3$ / 164.16	577-56-0 / 114.5		110-2[2]	4-10-00-02766	H_2O 4; eth 4; EtOH 4
2477	Benzoic acid, 3-acetyl-	$C_9H_8O_3$ / 164.16	586-42-5 / 172	110-2[2]	4-10-00-02769	H_2O 3; EtOH 5
2478	Benzoic acid, 4-acetyl-	$C_9H_8O_3$ / 164.16	586-89-0 / 208	sub	4-10-00-02769	H_2O 4
2479	Benzoic acid, 4-acetyl-, methyl ester	$C_{10}H_{10}O_3$ / 178.19	3609-53-8 / 95	142[4]	3-10-00-03030	H_2O 4
2480	Benzoic acid, 2-(acetyloxy)- / Aspirin	$C_9H_8O_4$ / 180.16	50-78-2 / 135	873	4-10-00-00138	H_2O 3; EtOH 4; eth 3; bz 2
2481	Benzoic acid, 2-(acetyloxy)-5-bromo- / 5-Bromoacetylsalicylic acid	$C_9H_7BrO_4$ / 259.06	1503-53-3 / 60	1425	2-10-00-00064	H_2O 1; EtOH 4; eth 3
2482	Benzoic acid, 2-(acetyloxy)-, methyl ester / Methyl acetylsalicylate	$C_{10}H_{10}O_4$ / 194.19	580-02-9 / 51.5	5934 / 134-6[9]	4-10-00-00147	eth 4; EtOH 4; chl 4
2483	Benzoic acid, 2-(acetyloxy)-, phenyl ester / Phenyl acetylsalicylate	$C_{15}H_{12}O_4$ / 256.26	134-55-4 / 96	7241	4-10-00-00156	
2484	Benzoic acid, 2-amino- / o-Anthranilic acid	$C_7H_7NO_2$ / 137.14	118-92-3 / 146.5	433 / sub	4-14-00-01004 / 1.412[20]	H_2O 3; EtOH 3; eth 3; bz 2
2485	Benzoic acid, 3-amino- / Aniline-3-carboxylic acid	$C_7H_7NO_2$ / 137.14	99-05-8 / 173	432	4-14-00-01092 / 1.51[25]	H_2O 2; EtOH 2; eth 3; ace 4
2486	Benzoic acid, 4-amino- / Aniline-4-carboxylic acid	$C_7H_7NO_2$ / 137.14	150-13-0 / 188.5	434	4-14-00-01126 / 1.374[20]	H_2O 3; EtOH 3; eth 3; ace 2
2487	Benzoic acid, 2-amino-5-bromo- / 5-Bromoanthranilic acid	$C_7H_6BrNO_2$ / 216.03	5794-88-7 / 219.5	1393	4-14-00-01081	DMSO 3
2488	Benzoic acid, 4-amino-, butyl ester / Butamben	$C_{11}H_{15}NO_2$ / 193.25	94-25-7 / 58	1504 / 173-4[8]	4-14-00-01130	H_2O 1; EtOH 3; eth 3; bz 3
2489	Benzoic acid, 4-(aminocarbonyl)-	$C_8H_7NO_3$ / 165.15	6051-43-0 / 300	sub	4-09-00-03318	H_2O 1; EtOH 1; eth 1; bz 1
2490	Benzoic acid, 5-amino-2-chloro- / 5-Amino-2-chlorobenzoic acid	$C_7H_6ClNO_2$ / 171.58	89-54-3 / 188		4-14-00-01116 / 1.519[15]	EtOH 4
2491	Benzoic acid, 2-amino-3,6-dichloro-	$C_7H_5Cl_2NO_2$ / 206.03	3032-32-4 / 155	sub	4-14-00-01078	ace 4; bz 4; eth 4; EtOH 4
2492	Benzoic acid, 3-amino-2,5-dichloro- / Chloramben	$C_7H_5Cl_2NO_2$ / 206.03	133-90-4 / 200	2063		DMSO 2
2493	Benzoic acid, 4-amino-, 2-(diethylamino)ethyl ester / Procaine	$C_{13}H_{20}N_2O_2$ / 236.31	59-46-1 / 61	7763	4-14-00-01138	H_2O 2; EtOH 3; eth 3; bz 3
2494	Benzoic acid, 4-[[(2-amino-1,4-dihydro-4-oxo-6-pteridinyl)methyl]formylamino]- / Rhizopterin	$C_{15}H_{12}N_6O_4$ / 340.30	119-20-0 / >300	8181	4-26-00-03949	H_2O 1; EtOH 1; eth 1; aq alk 3
2495	Benzoic acid, 2-amino-, ethyl ester	$C_9H_{11}NO_2$ / 165.19	87-25-2 / 13	/ 268	4-14-00-01008 / 1.1174[20]	eth 4; EtOH 4 / 1.5646[20]
2496	Benzoic acid, 3-amino-, ethyl ester	$C_9H_{11}NO_2$ / 165.19	582-33-2	/ 294; 160[5]	4-14-00-01093 / 1.1248[22]	H_2O 2; EtOH 4; eth 4; ctc 3 / 1.5600[22]
2497	Benzoic acid, 4-amino-, ethyl ester / Ethyl aminobenzoate	$C_9H_{11}NO_2$ / 165.19	94-09-7 / 92	3719 / 310	4-14-00-01129	H_2O 1; EtOH 4; eth 4; chl 3
2498	Benzoic acid, 4-amino-2-hydroxy- / p-Aminosalicylic acid	$C_7H_7NO_3$ / 153.14	65-49-6 / 150-1 dec	491	4-14-00-01967	H_2O 3; EtOH 3; eth 3; ace 3
2499	Benzoic acid, 5-amino-2-hydroxy- / Mesalamine	$C_7H_7NO_3$ / 153.14	89-57-6 / 283	5807	4-14-00-02058	H_2O 2; EtOH 1
2500	Benzoic acid, 4-amino-2-hydroxy-, hydrazide / p-Aminosalicylic acid hydrazide	$C_7H_9N_3O_2$ / 167.17	6946-29-8 / 195	492	4-14-00-01995	EtOH 4
2501	Benzoic acid, 3-amino-4-hydroxy-, methyl ester / Orthocaine	$C_8H_9NO_3$ / 167.16	536-25-4 / 143	6836	3-14-00-01477	H_2O 1; EtOH 4; eth 3; bz 2

No.	Name / Synonym	Mol. Form. / Mol. Wt.	CAS RN / mp/°C	Merck No. / bp/°C	Beil. Ref. / den/g cm^{-3}	Solubility / n_D
2502	Benzoic acid, 4-amino-2-hydroxy-, phenyl ester / Phenyl p-aminosalicylate	$C_{13}H_{11}NO_3$ / 229.24	133-11-9 / 153	7243	4-14-00-01979	
2503	Benzoic acid, 2-amino-, methyl ester / Methyl anthranilate	$C_8H_9NO_2$ / 151.17	134-20-3 / 24.5	5942 / 256	4-14-00-01008 / 1.1682[10]	H_2O 2; EtOH 4; eth 4 / 1.5810
2504	Benzoic acid, 3-amino-, methyl ester	$C_8H_9NO_2$ / 151.17	4518-10-9 / 39	152-3[11]	4-14-00-01093 / 1.232[20]	EtOH 4; eth 4; bz 4; chl 4
2505	Benzoic acid, 4-amino-, 2-methylpropyl ester / Isobutyl p-aminobenzoate	$C_{11}H_{15}NO_2$ / 193.25	94-14-4 / 64.5	5017	4-14-00-01130	
2506	Benzoic acid, 2-amino-3-nitro- / 3-Nitroanthranilic acid	$C_7H_6N_2O_4$ / 182.14	606-18-8 / 208.5		4-14-00-01087 / 1.558[15]	eth 4; EtOH 4
2507	Benzoic acid, 4-amino-, propyl ester / Risocaine	$C_{10}H_{13}NO_2$ / 179.22	94-12-2 / 75	8226	4-14-00-01130	bz 4; eth 4; EtOH 4; chl 4
2508	Benzoic acid, 4-(aminosulfonyl)- / Carzenide	$C_7H_7NO_4S$ / 201.20	138-41-0 / 290-2 dec	1885	4-11-00-00690	H_2O 1; EtOH 4; eth 2; bz 1
2509	Benzoic acid, ammonium salt	$C_7H_9NO_2$ / 139.15	1863-63-4 / 194.5	515	4-09-00-00273	H_2O 3
2510	Benzoic acid, anhydride / Benzoic anhydride	$C_{14}H_{10}O_3$ / 226.23	93-97-0 / 42.5	1102 / 360	4-09-00-00550 / 1.989[15]	H_2O 1; EtOH 3; eth 3; chl 2 / 1.5767[15]
2511	Benzoic acid, 3-benzoyl-	$C_{14}H_{10}O_3$ / 226.23	579-18-0 / 163.3	sub	4-10-00-02982	eth 4; EtOH 4
2512	Benzoic acid, 4-benzoyl-	$C_{14}H_{10}O_3$ / 226.23	611-95-0 / 199	sub	3-10-00-03305	H_2O 2; EtOH 3; eth 3; bz 2
2513	Benzoic acid, 3-(benzoylamino)-	$C_{14}H_{11}NO_3$ / 241.25	587-54-2 / 252.5		4-14-00-01105 / 1.510[4]	EtOH 4
2514	Benzoic acid, 4-(benzoylamino)-2-hydroxy- / Benzoyl-PAS	$C_{14}H_{11}NO_4$ / 257.25	13898-58-3 / 260.5	1127	4-14-00-02029	
2515	Benzoic acid, 2-benzoyl-, ethyl ester	$C_{16}H_{14}O_3$ / 254.29	604-61-5 / 59.5		3-10-00-03291 / 1.221[64]	H_2O 1; EtOH 4; eth 4; chl 2 / 1.560[64]
2516	Benzoic acid, 2-benzoyl-, methyl ester	$C_{15}H_{12}O_3$ / 240.26	606-28-0 / 52	351	4-10-00-02977 / 1.1903[19]	H_2O 1; EtOH 4; eth 4; sulf 3 / 1.591[20]
2517	Benzoic acid, 2-[bis(4-hydroxyphenyl)methyl]- / Phenolphthalin	$C_{20}H_{16}O_4$ / 320.34	81-90-3 / 230.5	7210	4-10-00-01932	eth 4; EtOH 4
2518	Benzoic acid, 2-bromo-	$C_7H_5BrO_2$ / 201.02	88-65-3 / 150	sub	4-09-00-01011 / 1.929[25]	H_2O 2; EtOH 3; eth 3; ace 3
2519	Benzoic acid, 3-bromo-	$C_7H_5BrO_2$ / 201.02	585-76-2 / 155		4-09-00-01013 / 1.845[20]	H_2O 1; EtOH 3; eth 3
2520	Benzoic acid, 4-bromo- / p-Bromobenzoic acid	$C_7H_5BrO_2$ / 201.02	586-76-5 / 254.5	>280 / 1396	4-09-00-01017 / 1.894[20]	H_2O 2; EtOH 3; eth 3
2521	Benzoic acid, 2-bromo-, ethyl ester	$C_9H_9BrO_2$ / 229.07	6091-64-1 /	254.5	4-09-00-01012 / 1.4438[15]	ace 4; bz 4; eth 4; EtOH 4 / 1.5455[15]
2522	Benzoic acid, 3-bromo-, ethyl ester	$C_9H_9BrO_2$ / 229.07	24398-88-7 /	261	4-09-00-01014 / 1.4308[19]	ace 4; bz 4; eth 4; EtOH 4 / 1.5430[19]
2523	Benzoic acid, 4-bromo-, ethyl ester	$C_9H_9BrO_2$ / 229.07	5798-75-4 /	263; 125[15]	3-09-00-01405 / 1.4332[17]	H_2O 2; EtOH 3; eth 3; ace 3 / 1.5438[17]
2524	Benzoic acid, 5-bromo-2-hydroxy-	$C_7H_5BrO_3$ / 217.02	89-55-4 / 169.8	sub 100	4-10-00-00217	H_2O 2; EtOH 4; eth 4; ace 2
2525	Benzoic acid, 3-bromo-, methyl ester	$C_8H_7BrO_2$ / 215.05	618-89-3 / 32	125[15]	4-09-00-01014	H_2O 2; EtOH 3; eth 3
2526	Benzoic acid, 4-bromo-, methyl ester	$C_8H_7BrO_2$ / 215.05	619-42-1 / 81		4-09-00-01017 / 1.689[25]	EtOH 3; eth 3; ace 3; bz 4
2527	Benzoic acid, 2-bromo-4-nitro- / 2-Bromo-4-nitrobenzoic acid	$C_7H_4BrNO_4$ / 246.02	16426-64-5 / 166.5	sub 150	3-09-00-01771	eth 4; EtOH 4
2528	Benzoic acid, 2-bromo-5-nitro- / 2-Bromo-5-nitrobenzoic acid	$C_7H_4BrNO_4$ / 246.02	943-14-6 / 180.5	sub	3-09-00-01771	eth 4; EtOH 4
2529	Benzoic acid, 4-bromo-3-nitro- / 4-Bromo-3-nitrobenzoic acid	$C_7H_4BrNO_4$ / 246.02	6319-40-0 / 203.5	sub	4-09-00-01236	EtOH 4
2530	Benzoic acid, 5-bromo-2-nitro- / 2-Nitro-5-bromobenzoic acid	$C_7H_4BrNO_4$ / 246.02	6950-43-2 / 140		1-09-00-00165 / 1.920[18]	bz 4; EtOH 4
2531	Benzoic acid, 4-(butylamino)-, 2-(dimethylamino)ethyl ester, monohydrochloride / Tetracaine hydrochloride	$C_{15}H_{25}ClN_2O_2$ / 300.83	136-47-0 / 147	9123	4-14-00-01172	
2532	Benzoic acid, butyl ester / Butyl benzoate	$C_{11}H_{14}O_2$ / 178.23	136-60-7 / -22.4	1552 / 250.3	4-09-00-00290 / 1.000[20]	H_2O 1; EtOH 5; eth 5; ace 3 / 1.4940[25]
2533	Benzoic acid, 4,4'-carbonylbis-	$C_{15}H_{10}O_5$ / 270.24	964-68-1 / 365	sub	4-10-00-03497	HOAc 4
2534	Benzoic acid, 2-chloro- / o-Chlorobenzoic acid	$C_7H_5ClO_2$ / 156.57	118-91-2 / 140.2	2125 / sub	4-09-00-00956 / 1.544[20]	H_2O 3; EtOH 4; eth 4; ace 4
2535	Benzoic acid, 3-chloro- / m-Chlorobenzoic acid	$C_7H_5ClO_2$ / 156.57	535-80-8 / 158	2124 / sub	4-09-00-00969 / 1.496[25]	H_2O 4; EtOH 3; eth 4; bz 2
2536	Benzoic acid, 4-chloro- / p-Chlorobenzoic acid	$C_7H_5ClO_2$ / 156.57	74-11-3 / 243	2126	4-09-00-00973	H_2O 1; EtOH 4; eth 2; ace 2
2537	Benzoic acid, 2-chloro-, ethyl ester	$C_9H_9ClO_2$ / 184.62	7335-25-3 /	243	4-09-00-00957 / 1.1942[15]	eth 4; EtOH 4 / 1.5242[15]
2538	Benzoic acid, 3-chloro-, ethyl ester	$C_9H_9ClO_2$ / 184.62	1128-76-3 /	243	4-09-00-00970 / 1.1859[15]	H_2O 1; EtOH 3; eth 3 / 1.5223[20]
2539	Benzoic acid, 4-chloro-, ethyl ester	$C_9H_9ClO_2$ / 184.62	7335-27-5 /	237.5	4-09-00-00975 / 1.1873[14]	EtOH 4
2540	Benzoic acid, 3-chloro-4-hydroxy-	$C_7H_5ClO_3$ / 172.57	3964-58-7 / 171	sub	4-10-00-00463	H_2O 2; EtOH 4; eth 4; ace 4

No.	Name / Synonym	Mol. Form. / Mol. Wt.	CAS RN / mp/°C	Merck No. / bp/°C	Beil. Ref. / den/g cm^{-3}	Solubility / n_D
2541	Benzoic acid, 3-chloro-2-hydroxy-, ethyl ester / Salicylic acid, 3-chloro, ethyl ester	$C_9H_9ClO_3$ / 200.62	56961-32-1 / 21	/ 269.5	0-10-00-00101	EtOH 4
2542	Benzoic acid, 5-chloro-2-hydroxy-, methyl ester	$C_8H_7ClO_3$ / 186.59	4068-78-4 / 50	/ 249 dec; 120[12]	4-10-00-00209	EtOH 4
2543	Benzoic acid, 4-chloro-, methyl ester	$C_8H_7ClO_2$ / 170.60	1126-46-1 / 43.5		4-09-00-00974 / 1.382[20]	EtOH 4
2544	Benzoic acid, 2-chloro-3-nitro- / 2-Chloro-3-nitrobenzoic acid	$C_7H_4ClNO_4$ / 201.57	3970-35-2 / 183.5		4-09-00-01225 / 1.662[18]	H_2O 3; EtOH 3
2545	Benzoic acid, 2-chloro-5-nitro-	$C_7H_4ClNO_4$ / 201.57	2516-96-3 / 166.5		4-09-00-01228 / 1.608[18]	H_2O 2; EtOH 3; eth 3; ace 2
2546	Benzoic acid, 3-chloro-2-nitro- / 2-Nitro-3-chlorobenzoic acid	$C_7H_4ClNO_4$ / 201.57	4771-47-5 / 238		4-09-00-01225 / 1.566[18]	eth 4; EtOH 4
2547	Benzoic acid, 4-chloro-3-nitro-	$C_7H_4ClNO_4$ / 201.57	96-99-1 / 182.8		4-09-00-01226 / 1.645[18]	H_2O 1; EtOH 2; ace 2
2548	Benzoic acid, 4-chloro-3-nitro-, methyl ester / Methyl 4-chloro-3-nitrobenzoate	$C_8H_6ClNO_4$ / 215.59	14719-83-6 / 83		4-09-00-01226 / 1.522[18]	EtOH 4
2549	Benzoic acid, 5-chloro-2-nitro-, methyl ester / Methyl 5-chloro-2-nitrobenzoate	$C_8H_6ClNO_4$ / 215.59	51282-49-6 / 48.5		0-09-00-00401 / 1.453[18]	MeOH 4
2550	Benzoic acid, 3-cyano-	$C_8H_5NO_2$ / 147.13	1877-72-1 / 219	/ sub	4-09-00-03296	H_2O 2; EtOH 3; eth 3
2551	Benzoic acid, cyclohexyl ester	$C_{13}H_{16}O_2$ / 204.27	2412-73-9 / <-10	/ 285	4-09-00-00296 / 1.0429[20]	H_2O 1; EtOH 3; eth 3 / 1.5200[20]
2552	Benzoic acid, 2,3-diamino- / 2,3-Diaminobenzoic acid	$C_7H_8N_2O_2$ / 152.15	603-81-6 / 201	/ dec	3-14-00-01172	EtOH 4; HOAc 4
2553	Benzoic acid, 2,4-diamino- / 4-Aminoanthranilic acid	$C_7H_8N_2O_2$ / 152.15	611-03-0 / 140	/ 200 dec	4-14-00-01299	EtOH 4; HOAc 4
2554	Benzoic acid, 3,5-diamino- / 3,5-Diaminobenzoic acid	$C_7H_8N_2O_2$ / 152.15	535-87-5 / 228	2955	4-14-00-01304	H_2O 2; EtOH 3; eth 4; tfa 2
2555	Benzoic acid, 2,4-dibromo- / 2,4-Dibromobenzoic acid	$C_7H_4Br_2O_2$ / 279.92	611-00-7 / 174	/ sub	4-09-00-01027	eth 4; EtOH 4
2556	Benzoic acid, 2,5-dibromo- / 2,5-Dibromobenzoic acid	$C_7H_4Br_2O_2$ / 279.92	610-71-9 / 157	/ sub	4-09-00-01027	eth 4; EtOH 4; chl 4
2557	Benzoic acid, 2,6-dibromo- / 2,6-Dibromobenzoic acid	$C_7H_4Br_2O_2$ / 279.92	601-84-3 / 150.5	/ 209[16]	4-09-00-01028	H_2O 3; EtOH 4; eth 4; ace 3
2558	Benzoic acid, 3,5-dibromo-2-hydroxy- / 3,5-Dibromosalicylic acid	$C_7H_4Br_2O_3$ / 295.91	3147-55-5 / 228	3012	4-10-00-00222	ace 3
2559	Benzoic acid, 3,5-dibromo-2-hydroxy-, methyl ester / Methyl 3,5-dibromosalicylate	$C_8H_6Br_2O_3$ / 309.94	21702-79-4 / 149	/ 181[12]	4-10-00-00223	eth 4
2560	Benzoic acid, 2,6-dibromo-3,4,5-trihydroxy- / Dibromogallic acid	$C_7H_4Br_2O_5$ / 327.91	602-92-6 / 150	3006	2-10-00-00347	H_2O 4; eth 4; EtOH 4
2561	Benzoic acid, 2,4-dichloro- / 2,4-Dichlorobenzoic acid	$C_7H_4Cl_2O_2$ / 191.01	50-84-0 / 164.2	/ sub	4-09-00-00998	H_2O 3; EtOH 3; eth 3; ace 2
2562	Benzoic acid, 2,5-dichloro- / 2,5-Dichlorobenzoic acid	$C_7H_4Cl_2O_2$ / 191.01	50-79-3 / 154.4	/ 301	4-09-00-01005	H_2O 2; EtOH 3; eth 3
2563	Benzoic acid, 2,6-dichloro- / 2,6-Dichlorobenzoic acid	$C_7H_4Cl_2O_2$ / 191.01	50-30-6 / 144	/ sub	4-09-00-01005	H_2O 3; EtOH 3; eth 3; bz 3
2564	Benzoic acid, 3,5-dichloro- / 3,5-Dichlorobenzoic acid	$C_7H_4Cl_2O_2$ / 191.01	51-36-5 / 188	/ sub	4-09-00-01008	H_2O 2; EtOH 3; eth 3; lig2
2565	Benzoic acid, 4-[(dichloroamino)sulfonyl]- / Halazone	$C_7H_5Cl_2NO_4S$ / 270.09	80-13-7 / 195 dec	4503	2-11-00-00220	H_2O 2; chl 2; HOAc 4; peth 1
2566	Benzoic acid, 3,5-dichloro-2-hydroxy-	$C_7H_4Cl_2O_3$ / 207.01	320-72-9 / 220.5	/ sub	4-10-00-00213	H_2O 2; EtOH 4; eth 3
2567	Benzoic acid, 3,6-dichloro-2-methoxy- / Dicamba	$C_8H_6Cl_2O_3$ / 221.04	1918-00-9 / 115	3026	/ 1.57[25]	
2568	Benzoic acid, 2,3-dihydroxy-	$C_7H_6O_4$ / 154.12	303-38-8 / 204		4-10-00-01414 / 1.542[20]	H_2O 3; EtOH 3; eth 3; ace 2
2569	Benzoic acid, 2,4-dihydroxy- / β-Resorcylic acid	$C_7H_6O_4$ / 154.12	89-86-1 / 225-7 dec	8160	4-10-00-01420	H_2O 3; EtOH 3; eth 3; bz 3
2570	Benzoic acid, 2,5-dihydroxy- / Gentisic acid	$C_7H_6O_4$ / 154.12	490-79-9 / 199.5	4290	4-10-00-01441	H_2O 4; EtOH 4; eth 4; ace 3
2571	Benzoic acid, 3,4-dihydroxy- / Protocatechuic acid	$C_7H_6O_4$ / 154.12	99-50-3 / 200-2 dec	7909	4-10-00-01459 / 1.524[4]	H_2O 2; EtOH 4; eth 3; bz 1
2572	Benzoic acid, 2-(3,6-dihydroxy-9H-xanthen-9-yl)- / Fluorescin	$C_{20}H_{14}O_5$ / 334.33	518-44-5 / 126	4087	2-18-00-00307	H_2O 1; EtOH 3; eth 3; ace 3
2573	Benzoic acid, 2,4-dihydroxy-6-methyl- / o-Orsellinic acid	$C_8H_8O_4$ / 168.15	480-64-8 / 176 dec	6834	4-10-00-01526	EtOH 3; eth 3
2574	Benzoic acid, 2,4-dihydroxy-6-methyl-, ethyl ester	$C_{10}H_{12}O_4$ / 196.20	2524-37-0 / 132	/ sub	4-10-00-01526	eth 4; EtOH 4
2575	Benzoic acid, 3,4-dihydroxy-5-[(3,4,5-trihydroxybenzoyl)oxy]- / Digallic acid	$C_{14}H_{10}O_9$ / 322.23	536-08-3 / 268-70 dec	3135	3-10-00-02086	ace 4; EtOH 4
2576	Benzoic acid, 3,4-dimethoxy- / Veratric acid	$C_9H_{10}O_4$ / 182.18	93-07-2 / 181	/ sub	9854	4-10-00-01460	H_2O 1; EtOH 4; eth 4; chl 2
2577	Benzoic acid, 3,5-dimethoxy- / 3,5-Dimethoxybenzoic acid	$C_9H_{10}O_4$ / 182.18	1132-21-4 / 185.5	/ sub	4-10-00-01501	eth 4; EtOH 4

No.	Name / Synonym	Mol. Form. / Mol. Wt.	CAS RN / mp/°C	Merck No. / bp/°C	Beil. Ref. / den/g cm^{-3}	Solubility / n_D
2578	Benzoic acid, 3,4-dimethoxy-, ethyl ester	$C_{11}H_{14}O_4$ / 210.23	3943-77-9 / 43.5	295.5	4-10-00-01469	eth 4; EtOH 4
2579	Benzoic acid, 3,4-dimethoxy-, methyl ester	$C_{10}H_{12}O_4$ / 196.20	2150-38-1 / 60.8	283	4-10-00-01467	bz 4; eth 4; EtOH 4
2580	Benzoic acid, 2,4-dimethyl- / 4-Carboxy-1,3-dimethylbenzene	$C_9H_{10}O_2$ / 150.18	611-01-8 / 90	268	4-09-00-01801	H_2O 2; EtOH 3; ace 3; bz 3
2581	Benzoic acid, 2,5-dimethyl-	$C_9H_{10}O_2$ / 150.18	610-72-0 / 132	sub	4-09-00-01802 / 1.069[21]	H_2O 1; EtOH 3; eth 3; ace 3
2582	Benzoic acid, 2,6-dimethyl- / 2,6-Dimethylbenzoic acid	$C_9H_{10}O_2$ / 150.18	632-46-2 / 116	274.5; 155[17]	4-09-00-01798	H_2O 2; EtOH 3; eth 3; lig 2
2583	Benzoic acid, 3,5-dimethyl- / Mesitylenic acid	$C_9H_{10}O_2$ / 150.18	499-06-9 / 171.1	sub	4-09-00-01806	H_2O 2; EtOH 4; eth 4
2584	Benzoic acid, 2-(dimethylamino)-	$C_9H_{11}NO_2$ / 165.19	610-16-2 / 72	sub	4-14-00-01017	H_2O 4; eth 4; EtOH 4
2585	Benzoic acid, 4-(dimethylamino)- / p-(Dimethylamino)benzoic acid	$C_9H_{11}NO_2$ / 165.19	619-84-1 / 242.5	3221	4-14-00-01164	EtOH 3; eth 2
2586	Benzoic acid, 6-[[6-[2-(dimethylamino)ethyl]-4-methoxy-1,3-benzodioxol-5-yl]acetyl]-2,3-dimethoxy- / Narceine	$C_{23}H_{27}NO_8$ / 445.47	131-28-2 / 138	6341	5-19-09-00051	
2587	Benzoic acid, 2-[[4-(dimethylamino)phenyl]azo]- / Methyl red	$C_{15}H_{15}N_3O_2$ / 269.30	493-52-7 / 183	6037	4-16-00-00504	H_2O 2; EtOH 3; ace 4; bz 4
2588	Benzoic acid, 4-(1,1-dimethylethyl- / p-tert-Butylbenzoic acid	$C_{11}H_{14}O_2$ / 178.23	98-73-7 / 164.5		4-09-00-01884	H_2O 1; EtOH 4; bz 4; chl 3
2589	Benzoic acid, 2,4-dimethyl-, methyl ester / Methyl 2,4-dimethylbenzoate	$C_{10}H_{12}O_2$ / 164.20	23617-71-2 / -2	232.5	2-09-00-00350	1.5052[20]
2590	Benzoic acid, 3,4-dinitro- / 3,4-Dinitrobenzoic acid	$C_7H_4N_2O_6$ / 212.12	528-45-0 / 166	3268	4-09-00-01242	H_2O 2; EtOH 4; eth 4
2591	Benzoic acid, 3,5-dinitro- / 3,5-Dinitrobenzoic acid	$C_7H_4N_2O_6$ / 212.12	99-34-3 / 205	3269	4-09-00-01242	EtOH 4; HOAc 4
2592	Benzoic acid, 3,5-dinitro-, butyl ester / Butyl 3,5-dinitrobenzoate	$C_{11}H_{12}N_2O_6$ / 268.23	10478-02-1 / 62.5		4-09-00-01246 / 1.488	EtOH 3; DMSO-d$_6$ 2
2593	Benzoic acid, 3,5-dinitro-, ethyl ester / Ethyl 3,5-dinitrobenzoate	$C_9H_8N_2O_6$ / 240.17	618-71-3 / 94		4-09-00-01243 / 1.295[111]	EtOH 3 / 1.560
2594	Benzoic acid, 3,3'-[(1,6-dioxo-1,6-hexanediyl)diimino]bis[2,4,6-triiodo- / Iodipamide	$C_{20}H_{14}I_6N_2O_6$ / 1139.77	606-17-7 / 307 dec	4916	4-14-00-01122	H_2O 1; EtOH 2; eth 2; ace 2
2595	Benzoic acid, 4-[(dipropylamino)sulfonyl]- / Probenecid	$C_{13}H_{19}NO_4S$ / 285.36	57-66-9 / 195	7760	4-11-00-00691	
2596	Benzoic acid, 2-ethoxy-	$C_9H_{10}O_3$ / 166.18	134-11-2 / 20.7	211-2[39]	3-10-00-00098	H_2O 2; EtOH 2; ctc 2
2597	Benzoic acid, 3-ethoxy-	$C_9H_{10}O_3$ / 166.18	621-51-2 / 137	sub	3-10-00-00245	H_2O 3; EtOH 3; eth 3; bz 3
2598	Benzoic acid, 2-ethoxy-, ethyl ester / Ethyl 2-ethoxybenzoate	$C_{11}H_{14}O_3$ / 194.23	6290-24-0	251	3-10-00-00117 / 1.1005[20]	eth 4; EtOH 4
2599	Benzoic acid, 4-ethoxy-, ethyl ester / Ethyl 4-ethoxybenzoate	$C_{11}H_{14}O_3$ / 194.23	23676-09-7	275	4-10-00-00368 / 1.076[12]	H_2O 1; EtOH 4; eth 4; ctc 2
2600	Benzoic acid, 2-ethyl-	$C_9H_{10}O_2$ / 150.18	612-19-1 / 68	259	4-09-00-01789 / 1.0431[100]	eth 4; EtOH 4 / 1.5099[100]
2601	Benzoic acid, 3-ethyl-	$C_9H_{10}O_2$ / 150.18	619-20-5 / 47		4-09-00-01791 / 1.042[100]	eth 4; EtOH 4 / 1.5345[100]
2602	Benzoic acid, ethyl ester / Ethyl benzoate	$C_9H_{10}O_2$ / 150.18	93-89-0 / -34	3725 / 212	4-09-00-00285 / 1.051[15]	H_2O 1; EtOH 3; eth 5; ace 3 / 1.5007[20]
2603	Benzoic acid, 2-fluoro-	$C_7H_5FO_2$ / 140.11	445-29-4 / 126.5		4-09-00-00950 / 1.460[25]	H_2O 2; EtOH 4; eth 4; bz 1
2604	Benzoic acid, 3-fluoro-	$C_7H_5FO_2$ / 140.11	455-38-9 / 124		4-09-00-00952 / 1.474[25]	H_2O 2; eth 3
2605	Benzoic acid, 4-fluoro- / p-Fluorobenzoic acid	$C_7H_5FO_2$ / 140.11	456-22-4 / 185	4100	4-09-00-00953 / 1.479[25]	H_2O 2; EtOH 3; eth 3; ace 2
2606	Benzoic acid, 4-fluoro-, ethyl ester	$C_9H_9FO_2$ / 168.17	451-46-7 / 26	210	4-09-00-00953 / 1.146[25]	eth 4; EtOH 4 / 1.4864[20]
2607	Benzoic acid, 2-formyl-	$C_8H_6O_3$ / 150.13	119-67-5 / 98		4-10-00-02748 / 1.404[25]	H_2O 3; EtOH 4; eth 4
2608	Benzoic acid, 6-formyl-2,3-dimethoxy- / Opianic acid	$C_{10}H_{10}O_5$ / 210.19	519-05-1 / 150	6806	4-10-00-03863	EtOH 3; eth 3
2609	Benzoic acid, 3-formyl-4-hydroxy-	$C_8H_6O_4$ / 166.13	584-87-2 / 244	sub	4-10-00-03622	eth 4; EtOH 4
2610	Benzoic acid, hexyl ester	$C_{13}H_{18}O_2$ / 206.28	6789-88-4	272; 139[8]	4-09-00-00293 / 0.9793[20]	H_2O 1; EtOH 3; ace 3
2611	Benzoic acid, hydrazide	$C_7H_8N_2O$ / 136.15	613-94-5 / 115	267 dec	4-09-00-00922	H_2O 3; EtOH 3; eth 2; ace 2
2612	Benzoic acid, 2-hydroxy- / Salicylic acid	$C_7H_6O_3$ / 138.12	69-72-7 / 158	8301 / 211[20]	4-10-00-00125 / 1.443[20]	H_2O 2; EtOH 4; eth 4; ace 4 / 1.565
2613	Benzoic acid, 4-hydroxy- / p-Salicylic acid	$C_7H_6O_3$ / 138.12	99-96-7 / 214.5	4742	4-10-00-00345 / 1.46[25]	H_2O 2; EtOH 4; eth 3; ace 3
2614	Benzoic acid, 2-hydroxy-, butyl ester	$C_{11}H_{14}O_3$ / 194.23	2052-14-4 / -5.9	271	4-10-00-00153 / 1.0728[20]	ctc 2 / 1.5115[20]
2615	Benzoic acid, 4-hydroxy-, butyl ester / Butylparaben	$C_{11}H_{14}O_3$ / 194.23	94-26-8 / 68.5	1583	4-10-00-00375	H_2O 2; EtOH 3; ctc 2

No.	Name Synonym	Mol. Form. Mol. Wt.	CAS RN mp/°C	Merck No. bp/°C	Beil. Ref. den/g cm^{-3}	Solubility n_D
2616	Benzoic acid, 2-hydroxy-, 2-carboxyphenyl ester Salsalate	C$_{14}$H$_{10}$O$_5$ 258.23	552-94-3 147	8307	4-10-00-00165	ace 2
2617	Benzoic acid, 2-hydroxy-3,5-diiodo- 3,5-Diiodosalicylic acid	C$_7$H$_4$I$_2$O$_3$ 389.92	133-91-5 235.5	3175	4-10-00-00226	H$_2$O 2; EtOH 4; eth 4; bz 1
2618	Benzoic acid, 4-hydroxy-3,5-diiodo-	C$_7$H$_4$I$_2$O$_3$ 389.92	618-76-8 237	260 dec	4-10-00-00479	H$_2$O 1; EtOH 4; eth 4; bz 2
2619	Benzoic acid, 2-hydroxy-, ethyl ester Ethyl salicylate	C$_9$H$_{10}$O$_3$ 166.18	118-61-6 45	3804 150-1[10]	4-10-00-00149 1.1326[20]	H$_2$O 1; EtOH 5; eth 4; ctc 3 1.5296[20]
2620	Benzoic acid, 3-hydroxy-, ethyl ester	C$_9$H$_{10}$O$_3$ 166.18	7781-98-8 74		4-10-00-00321 1.0680[131]	H$_2$O 2; EtOH 3; eth; chl 2
2621	Benzoic acid, 4-hydroxy-, ethyl ester Ethylparaben	C$_9$H$_{10}$O$_3$ 166.18	120-47-8 117	3792 297.5	4-10-00-00367	H$_2$O 2; EtOH 4; eth 4; chl 2
2622	Benzoic acid, 2-hydroxy-, 2-hydroxyethyl ester Glycol salicylate	C$_9$H$_{10}$O$_4$ 182.18	87-28-5 37	4395 173[15]	3-10-00-00138 1.2526[15]	H$_2$O 2; EtOH 4; eth 4; bz 4
2623	Benzoic acid, 4-hydroxy-3-iodo- 4-Hydroxybenzoic acid, 3-iodo	C$_7$H$_5$IO$_3$ 264.02	37470-46-5 174.3	sub	4-10-00-00478	eth 4; EtOH 4
2624	Benzoic acid, 5-hydroxy-2-iodo- 3-Hydroxybenzoic acid, 6-iodo	C$_7$H$_5$IO$_3$ 264.02	57772-57-3 198	sub 160	3-10-00-00261	eth 4; EtOH 4
2625	Benzoic acid, 4-hydroxy-3-methoxy- Vanillic acid	C$_8$H$_8$O$_4$ 168.15	121-34-6 211.5	9838 sub	4-10-00-01459	H$_2$O 2; EtOH 4; eth 3; DMSO 3
2626	Benzoic acid, 4-hydroxy-3-methoxy-, ethyl ester	C$_{10}$H$_{12}$O$_4$ 196.20	617-05-0 44	292	3-10-00-01413	H$_2$O 1; EtOH 4; eth 4; chl 3
2627	Benzoic acid, 4-hydroxy-3-methoxy-, methyl ester	C$_9$H$_{10}$O$_4$ 182.18	3943-74-6 64	286	3-10-00-01410	EtOH 3; chl 2; peth 3
2628	Benzoic acid, 4-hydroxy-3-methoxy-, 2-methylpropyl ester	C$_{12}$H$_{16}$O$_4$ 224.26	7152-88-7 56.5	125[2]	3-10-00-01415	chl 2
2629	Benzoic acid, 2-hydroxy-3-methyl- o-Cresotic acid	C$_8$H$_8$O$_3$ 152.15	83-40-9 165.5	2585	4-10-00-00601	H$_2$O 2; EtOH 3; eth 3; bz 3
2630	Benzoic acid, 2-hydroxy-4-methyl- m-Cresotic acid	C$_8$H$_8$O$_3$ 152.15	50-85-1 177	2584	4-10-00-00617	H$_2$O 2; EtOH 3; eth 4; bz 3
2631	Benzoic acid, 2-hydroxy-5-methyl- p-Cresotic acid	C$_8$H$_8$O$_3$ 152.15	89-56-5 151	2586	4-10-00-00610	H$_2$O 2; EtOH 3; eth 3; bz 3
2632	Benzoic acid, 3-hydroxy-5-methyl-	C$_8$H$_8$O$_3$ 152.15	585-81-9 210	sub	4-10-00-00609	H$_2$O 3; EtOH 4; eth 4; chl 2
2633	Benzoic acid, 2-hydroxy-, 3-methylbutyl ester Isopentyl salicylate	C$_{12}$H$_{16}$O$_3$ 208.26	87-20-7 278; 151[15]	5007	4-10-00-00153 1.0535[20]	H$_2$O 1; EtOH 4; eth 3; ctc 2 1.5080[20]
2634	Benzoic acid, 2-hydroxy-, methyl ester Methyl salicylate	C$_8$H$_8$O$_3$ 152.15	119-36-8 -8	6038 222.9	4-10-00-00143 1.181[25]	eth 4; EtOH 4; chl 4 1.535[20]
2635	Benzoic acid, 2-hydroxy-, 1-methylethyl ester Isopropyl salicylate	C$_{10}$H$_{12}$O$_3$ 180.20	607-85-2 238		4-10-00-00152 1.0729[20]	H$_2$O 1; EtOH 5; eth 5 1.5065[20]
2636	Benzoic acid, 3-hydroxy-, methyl ester	C$_8$H$_8$O$_3$ 152.15	19438-10-9 73	281; 178[17]	4-10-00-00319 1.1528[100]	EtOH 3; bz 3; chl 2; peth 3
2637	Benzoic acid, 4-hydroxy-, methyl ester Methylparaben	C$_8$H$_8$O$_3$ 152.15	99-76-3 131	6021 275 dec	4-10-00-00360	H$_2$O 2; EtOH 4; eth 4; ace 4
2638	Benzoic acid, 2-hydroxy-3-methyl-, methyl ester	C$_9$H$_{10}$O$_3$ 166.18	23287-26-5 29	235	4-10-00-00602 1.1683[25]	1.5354[16]
2639	Benzoic acid, 2-hydroxy-4-methyl-, methyl ester	C$_9$H$_{10}$O$_3$ 166.18	4670-56-8 27.5	243	2-10-00-00139 1.1483[15]	1.5378[15]
2640	Benzoic acid, 2-hydroxy-5-methyl-, methyl ester	C$_9$H$_{10}$O$_3$ 166.18	22717-57-3 -1	244.5	4-10-00-00611 1.1673[25]	1.5351[15]
2641	Benzoic acid, 2-hydroxy-3-methyl-6-(1-methylethyl)-	C$_{11}$H$_{14}$O$_3$ 194.23	4389-53-1 136	sub	4-10-00-00738	eth 4; EtOH 4
2642	Benzoic acid, 2-hydroxy-6-methyl-3-(1-methylethyl)- o-Thymotic acid	C$_{11}$H$_{14}$O$_3$ 194.23	548-51-6 127	9342 sub	3-10-00-00629	bz 4; eth 4; EtOH 4
2643	Benzoic acid, 2-hydroxy-, 2-methylpropyl ester Isobutyl salicylate	C$_{11}$H$_{14}$O$_3$ 194.23	87-19-4 5.9	261	3-10-00-00121 1.0639[20]	H$_2$O 1; EtOH 3; eth 3; ctc 3 1.5087[20]
2644	Benzoic acid, 2-hydroxy-, 1-naphthalenyl ester 1-Naphthyl salicylate	C$_{17}$H$_{12}$O$_3$ 264.28	550-97-0 83	6334	2-10-00-00052	eth 4
2645	Benzoic acid, 2-hydroxy-, 2-naphthalenyl ester 2-Naphthyl salicylate	C$_{17}$H$_{12}$O$_3$ 264.28	613-78-5 95.5	6335	4-10-00-00158 1.11[116]	H$_2$O 1; EtOH 2; eth 3; bz 3
2646	Benzoic acid, 2-hydroxy-3-nitro- 3-Nitrosalicylic acid	C$_7$H$_5$NO$_5$ 183.12	85-38-1 148	6553	4-10-00-00228	H$_2$O 2; EtOH 4; eth 4; ace 3
2647	Benzoic acid, 2-hydroxy-5-nitro- 5-Nitrosalicylic acid	C$_7$H$_5$NO$_5$ 183.12	96-97-9 229.5	6554	4-10-00-00255 1.650[20]	H$_2$O 2; EtOH 4; eth 4; ace 4
2648	Benzoic acid, 2-hydroxy-5-[(4-nitrophenyl)azo]- Alizarine Yellow R	C$_{13}$H$_9$N$_3$O$_5$ 287.23	2243-76-7 253-4 dec	241	4-16-00-00372	H$_2$O 4; EtOH 4
2649	Benzoic acid, 2-hydroxy-, pentyl ester	C$_{12}$H$_{16}$O$_3$ 208.26	2050-08-0 270		4-10-00-00153 1.064[15]	H$_2$O 2; EtOH 5; eth 5 1.506[20]
2650	Benzoic acid, 2-hydroxy-, phenyl ester Phenyl salicylate	C$_{13}$H$_{10}$O$_3$ 214.22	118-55-8 130.5	7282 173[12]	4-10-00-00154 1.2614[30]	H$_2$O 1; EtOH 4; eth 3; ace 4

No.	Name Synonym	Mol. Form. Mol. Wt.	CAS RN mp/°C	Merck No. bp/°C	Beil. Ref. den/g cm^{-3}	Solubility n_D
2651	Benzoic acid, 2-hydroxy-, phenylmethyl ester Benzyl salicylate	$C_{14}H_{12}O_3$ 228.25	118-58-1	1160 320	4-10-00-00157 1.1799[20]	H_2O 2; EtOH 3; eth 3; ctc 3 1.5805[20]
2652	Benzoic acid, 2-hydroxy-, propyl ester	$C_{10}H_{12}O_3$ 180.20	607-90-9 97	239	4-10-00-00152 1.0979[20]	ctc 3; CS_2 3 1.5161[20]
2653	Benzoic acid, 4-hydroxy-, propyl ester Propylparaben	$C_{10}H_{12}O_3$ 180.20	94-13-3 97	7879	4-10-00-00374 1.0630[102]	H_2O 1; EtOH 3; eth 3; chl 2 1.5050[102]
2654	Benzoic acid, 2-hydroxy-5-[[4-[(2-pyridinylamino)sulfonyl]phenyl]azo]- Sulfasalazine	$C_{18}H_{14}N_4O_5S$ 398.40	599-79-1 220 dec	8917	5-22-08-00433	
2655	Benzoic acid, 2-hydroxy-5-sulfo- Sulfosalicylic acid	$C_7H_6O_6S$ 218.19	97-05-2 120	8944	4-11-00-00702	H_2O 5; EtOH 5; eth 4
2656	Benzoic acid, 2,2'-iminobis- Diphenylamine-2,2'-dicarboxylic acid	$C_{14}H_{11}NO_4$ 257.25	579-92-0 296-7 dec	3318	4-14-00-01058	
2657	Benzoic acid, 2-iodo- o-Iodobenzoic acid	$C_7H_5IO_2$ 248.02	88-67-5 163	4923 exp 233	4-09-00-01030 2.25[25]	H_2O 2; EtOH 4; eth 4; ace 2
2658	Benzoic acid, 3-iodo-	$C_7H_5IO_2$ 248.02	618-51-9 188.3	sub	4-09-00-01033	H_2O 2; EtOH 4; eth 2
2659	Benzoic acid, 4-iodo-	$C_7H_5IO_2$ 248.02	619-58-9 270	sub	4-09-00-01035 2.184[20]	H_2O 2; EtOH 2; eth 3
2660	Benzoic acid, 2-iodo-, methyl ester	$C_8H_7IO_2$ 262.05	610-97-9	280; 146[16]	4-09-00-01031	EtOH 3 1.6052[20]
2661	Benzoic acid, 3-iodo-, methyl ester	$C_8H_7IO_2$ 262.05	618-91-7 54.5	277; 150[18]	4-09-00-01034	H_2O 1; EtOH 3; eth 4; ace 4
2662	Benzoic acid, 4-iodo-, methyl ester	$C_8H_7IO_2$ 262.05	619-44-3 114.8	sub	4-09-00-01036 2.020[10]	EtOH 3; eth 3
2663	Benzoic acid, 2-mercapto- o-Thiosalicylic acid	$C_7H_6O_2S$ 154.19	147-93-3 168.5	9291 sub	4-10-00-00272	H_2O 3; EtOH 3; eth 3; lig 2
2664	Benzoic acid, 2-methoxy-	$C_8H_8O_3$ 152.15	579-75-9 101	200	4-10-00-00130	H_2O 2; EtOH 4; eth 4; bz 3
2665	Benzoic acid, 3-methoxy-	$C_8H_8O_3$ 152.15	586-38-9 107	170[10]	4-10-00-00316	H_2O 3; EtOH 3; eth 3; bz 3
2666	Benzoic acid, 4-methoxy- p-Anisic acid	$C_8H_8O_3$ 152.15	100-09-4 185	696 276.5	4-10-00-00346	H_2O 1; EtOH 4; eth 4; chl 3
2667	Benzoic acid, 2-(4-methoxybenzoyl)- o-(p-Anisoyl)benzoic acid	$C_{15}H_{12}O_4$ 256.26	1151-15-1 146	702	3-10-00-04429	eth 4; EtOH 4; tol 4
2668	Benzoic acid, 4-methoxy-, butyl ester	$C_{12}H_{16}O_3$ 208.26	6946-35-6	183[40]	3-10-00-00307 1.054[16]	1.5141[16]
2669	Benzoic acid, 2-methoxy-, ethyl ester	$C_{10}H_{12}O_3$ 180.20	7335-26-4	261	4-10-00-00150 1.1124[20]	eth 4; EtOH 4 1.5224[20]
2670	Benzoic acid, 3-methoxy-, ethyl ester	$C_{10}H_{12}O_3$ 180.20	10259-22-0	260.5	4-10-00-00321 1.0993[20]	eth 4; EtOH 4 1.5161[20]
2671	Benzoic acid, 4-methoxy-, ethyl ester	$C_{10}H_{12}O_3$ 180.20	94-30-4 7.5	269.5	4-10-00-00368 1.1038[20]	H_2O 1; EtOH 3; eth 3 1.5254[20]
2672	Benzoic acid, 3-methoxy-2-(methylamino)-, methyl ester Damascenine	$C_{10}H_{13}NO_3$ 195.22	483-64-7 28	2809 271; 147[10]	4-14-00-02073	bz 4; eth 4; EtOH 4; lig 4
2673	Benzoic acid, 2-methoxy-, methyl ester	$C_9H_{10}O_3$ 166.18	606-45-1	246.5	4-10-00-00144 1.1571[19]	H_2O 1; EtOH 3 1.534[19]
2674	Benzoic acid, 3-methoxy-, methyl ester	$C_9H_{10}O_3$ 166.18	5368-81-0	248	4-10-00-00319 1.1310[20]	H_2O 1; EtOH 3 1.5224[20]
2675	Benzoic acid, 4-methoxy-, methyl ester	$C_9H_{10}O_3$ 166.18	121-98-2 49	244	4-10-00-00360	H_2O 1; EtOH 3; eth 3; chl 3
2676	Benzoic acid, 2-methyl- o-Toluic acid	$C_8H_8O_2$ 136.15	118-90-1 103.7	9461 259	4-09-00-01697 1.062[115]	H_2O 1; EtOH 4; eth 4; chl 3 1.512[115]
2677	Benzoic acid, 3-methyl- m-Toluic acid	$C_8H_8O_2$ 136.15	99-04-7 108.7	9461	4-09-00-01712 1.054[112]	H_2O 2; EtOH 4; eth 4; chl 2 1.509
2678	Benzoic acid, 4-methyl- p-Toluic acid	$C_8H_8O_2$ 136.15	99-94-5 179.6	9461	4-09-00-01724	H_2O 1; EtOH 4; eth 4; tfa 2
2679	Benzoic acid, 2-(methylamino)-	$C_8H_9NO_2$ 151.17	119-68-6 180.5	80[0.01]	4-14-00-01015	H_2O 2; EtOH 4; eth 4; bz 4
2680	Benzoic acid, 2-(methylamino)-, ethyl ester Anthranilic acid, N-methyl-, ethyl ester	$C_{10}H_{13}NO_2$ 179.22	35472-56-1 39	266; 172[45]	2-14-00-00213	eth 4
2681	Benzoic acid, 2-(methylamino)-, methyl ester	$C_9H_{11}NO_2$ 165.19	85-91-6 19	255	4-14-00-01016 1.120[15]	H_2O 1; EtOH 3; eth 3 1.5839[15]
2682	Benzoic acid, 2-methyl-, anhydride	$C_{16}H_{14}O_3$ 254.29	607-86-3 38.5	>325; 220[11]	4-09-00-01700	eth 4
2683	Benzoic acid, 3-methyl-, anhydride	$C_{16}H_{14}O_3$ 254.29	21436-44-2 71	230[17]	4-09-00-01715	ace 4; bz 4; eth 4; EtOH 4
2684	Benzoic acid, 2-(4-methylbenzoyl)- 2-(p-Toluyl)benzoic acid	$C_{15}H_{12}O_3$ 240.26	85-55-2 146	9465	4-10-00-02991	H_2O 2; EtOH 4; eth 4; ace 4
2685	Benzoic acid, 3-methyl-, butyl ester Butyl m-toluate	$C_{12}H_{16}O_2$ 192.26	6640-77-3	262.3	3-09-00-00397 1.0040[20]	H_2O 1; EtOH 3; eth 5 1.4950[20]
2686	Benzoic acid, 3,3'-methylenebis[6-hydroxy- 5,5'-Methylenedisalicylic acid	$C_{15}H_{12}O_6$ 288.25	122-25-8 243.5	5984	4-10-00-02337	ace 4; eth 4; EtOH 4
2687	Benzoic acid, methyl ester Methyl benzoate	$C_8H_8O_2$ 136.15	93-58-3 -15	5947 199	4-09-00-00283 1.0933[15]	H_2O 1; EtOH 3; eth 5; ctc 3 1.5164[20]
2688	Benzoic acid, 2-(1-methylethyl)-	$C_{10}H_{12}O_2$ 164.20	2438-04-2 64	160-1[25]	4-09-00-01839	H_2O 2; EtOH 4; eth 4; bz 3
2689	Benzoic acid, 4-(1-methylethyl)- Cumic acid	$C_{10}H_{12}O_2$ 164.20	536-66-3 117.5	2620 sub	4-09-00-01843 1.162[4]	H_2O 2; EtOH 4; eth 4; peth 3

No.	Name / Synonym	Mol. Form. / Mol. Wt.	CAS RN / mp/°C	Merck No. / bp/°C	Beil. Ref. / den/g cm^{-3}	Solubility / n_D
2690	Benzoic acid, 2-methyl-, ethyl ester	$C_{10}H_{12}O_2$ 164.20	87-24-1 <-10	227; 113[18]	4-09-00-01699 1.0325[21]	H_2O 1; EtOH 5; eth 5 1.507[22]
2691	Benzoic acid, 3-methyl-, ethyl ester	$C_{10}H_{12}O_2$ 164.20	120-33-2	234	4-09-00-01713 1.0265[21]	eth 4; EtOH 4 1.5052[22]
2692	Benzoic acid, 4-methyl-, ethyl ester	$C_{10}H_{12}O_2$ 164.20	94-08-6	232	4-09-00-01726 1.0269[18]	H_2O 1; EtOH 5; eth 5 1.5089[18]
2693	Benzoic acid, 1-methylethyl ester	$C_{10}H_{12}O_2$ 164.20	939-48-0	216	4-09-00-00289 1.0163[15]	H_2O 1; EtOH 3; eth 3; ace 3 1.4890[20]
2694	Benzoic acid, 4-(1-methylethyl)-, methyl ester	$C_{11}H_{14}O_2$ 178.23	20185-55-1	126[14]	4-09-00-01843 1.018[19]	1.515[19]
2695	Benzoic acid, 4-(1-methylethyl)-, 2-methylpropyl ester / Benzoic acid, 4-isopropyl, isobutyl ester	$C_{14}H_{20}O_2$ 220.31	6315-03-3	155[14]	2-09-00-00359 0.966[19]	1.497[19]
2696	Benzoic acid, 2-methyl-, methyl ester	$C_9H_{10}O_2$ 150.18	89-71-4 <-50	215	4-09-00-01699 1.068[20]	H_2O 1; EtOH 5; eth 5
2697	Benzoic acid, 3-methyl-, methyl ester	$C_9H_{10}O_2$ 150.18	99-36-5	221	4-09-00-01713 1.061[20]	H_2O 1; EtOH 3; ctc 2
2698	Benzoic acid, 4-methyl-, methyl ester	$C_9H_{10}O_2$ 150.18	99-75-2 33.2	220	4-09-00-01726	H_2O 1; EtOH 4; eth 4
2699	Benzoic acid, 2-methylphenyl ester	$C_{14}H_{12}O_2$ 212.25	617-02-7	309; 155[8.5]	4-09-00-00305 1.114[19]	eth 4; EtOH 4
2700	Benzoic acid, 4-methylphenyl ester	$C_{14}H_{12}O_2$ 212.25	614-34-6 71.5	316	4-09-00-00306	eth 4; EtOH 4
2701	Benzoic acid, 2-methylpropyl ester	$C_{11}H_{14}O_2$ 178.23	120-50-3	242	4-09-00-00291 0.9990[20]	H_2O 1; EtOH 5; eth 5; ace 3
2702	Benzoic acid, 2-[(1-naphthalenylamino)carbonyl]- / Naptalam	$C_{18}H_{13}NO_3$ 291.31	132-66-1 185	6338	3-12-00-02876 1.4[20]	H_2O 1; EtOH 2; ace 2; bz 2
2703	Benzoic acid, 2-nitro- / o-Nitrobenzoic acid	$C_7H_5NO_4$ 167.12	552-16-9 147.5		4-09-00-01046 1.575[20]	H_2O 3; EtOH 4; eth 3; ace 4
2704	Benzoic acid, 3-nitro- / m-Nitrobenzoic acid	$C_7H_5NO_4$ 167.12	121-92-6 141		4-09-00-01055 1.494[20]	H_2O 2; EtOH 4; eth 4; ace 4
2705	Benzoic acid, 4-nitro- / p-Nitrobenzoic acid	$C_7H_5NO_4$ 167.12	62-23-7 242	sub	4-09-00-01072 1.610[20]	ace 4; eth 4; EtOH 4; chl 4
2706	Benzoic acid, 4-nitro-, butyl ester	$C_{11}H_{13}NO_4$ 223.23	120-48-9 35.3	160[8]	3-09-00-01544	bz 4; eth 4
2707	Benzoic acid, 2-nitro-, ethyl ester	$C_9H_9NO_4$ 195.17	610-34-4 30	275; 178[23]	4-09-00-01047	eth 4; EtOH 4
2708	Benzoic acid, 3-nitro-, ethyl ester	$C_9H_9NO_4$ 195.17	618-98-4 47	297	4-09-00-01056	H_2O 1; EtOH 4; eth 4
2709	Benzoic acid, 4-nitro-, ethyl ester	$C_9H_9NO_4$ 195.17	99-77-4 57	186.3; 153[8]	4-09-00-01074	H_2O 1; EtOH 3; eth 3
2710	Benzoic acid, 2-nitro-, methyl ester	$C_8H_7NO_4$ 181.15	606-27-9 -13	275	4-09-00-01047 1.2855[20]	H_2O 1; EtOH 3; eth 3; bz 3
2711	Benzoic acid, 3-nitro-, methyl ester	$C_8H_7NO_4$ 181.15	618-95-1 78	279[60]	4-09-00-01056	H_2O 1; EtOH 2; eth 2; MeOH 2
2712	Benzoic acid, pentachloro- / Perchlorobenzoic acid	$C_7HCl_5O_2$ 294.35	1012-84-6 209.3	sub	4-09-00-01011	EtOH 4; tol 4; HOAc 4
2713	Benzoic acid, pentamethyl- / Permethylbenzoic acid	$C_{12}H_{16}O_2$ 192.26	2243-32-5 210.5	sub	4-09-00-01920	H_2O 1; EtOH 4
2714	Benzoic acid, 2-phenoxy-	$C_{13}H_{10}O_3$ 214.22	2243-42-7 113	355	4-10-00-00132 1.1553[50]	H_2O 1; EtOH 4; eth 4; chl 3
2715	Benzoic acid, 2-(phenylamino)- / N-Phenylanthranilic acid	$C_{13}H_{11}NO_2$ 213.24	91-40-7 183.5	7244	4-14-00-01019	H_2O 1; EtOH 4; eth 2; bz 2
2716	Benzoic acid, phenyl ester / Phenyl benzoate	$C_{13}H_{10}O_2$ 198.22	93-99-2 71	7246 314	4-09-00-00303 1.235[20]	H_2O 1; EtOH 3; eth 3; chl 3
2717	Benzoic acid, 2-phenylhydrazide	$C_{13}H_{12}N_2O$ 212.25	532-96-7 168	314	4-15-00-00169	H_2O 2; EtOH 3; eth 2; bz 3
2718	Benzoic acid, 2-(phenylmethyl)-	$C_{14}H_{12}O_2$ 212.25	612-35-1 118	sub	4-09-00-02516	H_2O 2; EtOH 3; eth 3; bz 3
2719	Benzoic acid, 4-(phenylmethyl)-	$C_{14}H_{12}O_2$ 212.25	620-86-0 158.8	sub	4-09-00-02518	bz 4; eth 4; EtOH 4; chl 4
2720	Benzoic acid, phenylmethyl ester / Benzyl benzoate	$C_{14}H_{12}O_2$ 212.25	120-51-4 21	1141 323.5	4-09-00-00307 1.1121[25]	H_2O 1; EtOH 3; eth 3; ace 3 1.5680[20]
2721	Benzoic acid, 4-[[(phenylmethyl)sulfonyl]amino]- / Carinamide	$C_{14}H_{13}NO_4S$ 291.33	536-95-8 229.5	1163	4-14-00-01264	EtOH 4
2722	Benzoic acid, 2-propenyl ester	$C_{10}H_{10}O_2$ 162.19	583-04-0		4-09-00-00296 1.0569[15]	H_2O 1; EtOH 3; eth 3; ace 3 1.5178[20]
2723	Benzoic acid, o-propyl- / Benzoic acid, 2-propyl	$C_{10}H_{12}O_2$ 164.20	2438-03-1 58	273; 164[20]	4-09-00-01836 1.0020[15]	eth 4; EtOH 4
2724	Benzoic acid, propyl ester	$C_{10}H_{12}O_2$ 164.20	2315-68-6 -51.6	211	4-09-00-00288 1.0230[20]	H_2O 1; EtOH 5; eth 5 1.5000[20]
2725	Benzoic acid, sodium salt	$C_7H_5NaO_2$ 144.11	532-32-1 >300	8527	4-09-00-00273	H_2O 3
2726	Benzoic acid, 2-[[[4-[(2-thiazolylamino)sulfonyl]phenyl]amino]carbonyl]- / Phthalylsulphathiazole	$C_{17}H_{13}N_3O_5S_2$ 403.44	85-73-4 273	7351	4-27-00-04639	H_2O 1; EtOH 2; eth 1; chl 1

No.	Name / Synonym	Mol. Form. / Mol. Wt.	CAS RN / mp/°C	Merck No. / bp/°C	Beil. Ref. / den/g cm^{-3}	Solubility / n_D
2727	Benzoic acid, 2,3,6-trichloro- / 2,3,6-Trichlorobenzoic acid	$C_7H_3Cl_3O_2$ / 225.46	50-31-7 / 124.5		4-09-00-01009	H_2O 2; eth 3
2728	Benzoic acid, 2,4,5-trichloro- / 2,4,5-Trichlorobenzoic acid	$C_7H_3Cl_3O_2$ / 225.46	50-82-8 / 166.5	sub	4-09-00-01010	H_2O 1; EtOH 3; eth 3
2729	Benzoic acid, 2-[[3-(trifluoromethyl)phenyl]amino]- / Flufenamic acid	$C_{14}H_{10}F_3NO_2$ / 281.23	530-78-9 / 133.5	4060	3-14-00-00905	DMSO 3
2730	Benzoic acid, 2,3,4-trihydroxy-	$C_7H_6O_5$ / 170.12	610-02-6 / 221	sub	4-10-00-01971	H_2O 2; EtOH 3; eth 3; ace 3
2731	Benzoic acid, 3,4,5-trihydroxy- / Gallic acid	$C_7H_6O_5$ / 170.12	149-91-7 / 253 dec	4251	3-10-00-02070 / 1.694[6]	H_2O 2; EtOH 4; eth 2; ace 3
2732	Benzoic acid, 3,4,5-trihydroxy-, propyl ester / Propyl gallate	$C_{10}H_{12}O_5$ / 212.20	121-79-9 / 130	7872	4-10-00-02003	H_2O 2
2733	Benzoic acid, 2,4,5-trimethoxy-	$C_{10}H_{12}O_5$ / 212.20	490-64-2 / 145	300	4-10-00-01980	H_2O 4; bz 4; EtOH 4; peth 4
2734	Benzoic acid, 3,4,5-trimethoxy-	$C_{10}H_{12}O_5$ / 212.20	118-41-2 / 172.3	225-7[10]	4-10-00-01996	H_2O 2; EtOH 4; eth 4; chl 4
2735	Benzoic acid, 2,4,6-trinitro- / 2,4,6-Trinitrobenzoic acid	$C_7H_3N_3O_8$ / 257.12	129-66-8 / 228 dec	9640	4-09-00-01362	H_2O 4; EtOH 4; eth 3; ace 3
2736	1H,5H-Benzo[ij]quinolizine-1,6(7H)-dione, 2,3-dihydro- / Julolidine, 1,6-dioxo	$C_{12}H_{11}NO_2$ / 201.22	39052-57-8 / 145.5	190-210[0.3]	4-21-00-05525	EtOH 4
2737	1H,5H-Benzo[ij]quinolizine, 2,3,6,7-tetrahydro- / Julolidine	$C_{12}H_{15}N$ / 173.26	479-59-4 / 40	280 dec; 155[17]	5-20-07-00189 / 1.003[20]	1.568[25]
2738	Benzo[j]fluoranthene / Dibenzo[a,jk]fluorene	$C_{20}H_{12}$ / 252.32	205-82-3 / 166		4-05-00-02687	H_2O 1; EtOH 2; HOAc 2
2739	Benzo[k]fluoranthene / 2,3,1',8'-Binaphthylene	$C_{20}H_{12}$ / 252.32	207-08-9 / 217	480	4-05-00-02686	H_2O 1; EtOH 3; bz 3; HOAc 3
2740	Benzonitrile / Phenyl cyanide	C_7H_5N / 103.12	100-47-0 / -12.7	1107 / 191.1	4-09-00-00892 / 1.0093[15]	H_2O 2; EtOH 5; eth 5; ace 4 / 1.5289[20]
2741	Benzonitrile, 2-amino-	$C_7H_6N_2$ / 118.14	1885-29-6 / 51	263	4-14-00-01013	H_2O 2; EtOH 4; eth 4; ace 4
2742	Benzonitrile, 3-amino-	$C_7H_6N_2$ / 118.14	2237-30-1 / 54.3	289	4-14-00-01095	H_2O 2; EtOH 4; eth 4; ace 4
2743	Benzonitrile, 5-amino-2-methyl-	$C_8H_8N_2$ / 132.17	50670-64-9 / 88	100-10[22]	3-14-00-01201	EtOH 4
2744	Benzonitrile, 2-bromo-	C_7H_4BrN / 182.02	2042-37-7 / 55.5	252	4-09-00-01013	H_2O 3; EtOH 4; chl 2
2745	Benzonitrile, 3-bromo-	C_7H_4BrN / 182.02	6952-59-6 / 39.5	225	4-09-00-01016	EtOH 4; eth 4; chl 2
2746	Benzonitrile, 4-bromo-	C_7H_4BrN / 182.02	623-00-7 / 114	236	4-09-00-01025	H_2O 3; EtOH 3; eth 3; chl 3
2747	Benzonitrile, 2-chloro-	C_7H_4ClN / 137.57	873-32-5 / 46.3	232	4-09-00-00965	H_2O 2; EtOH 3; eth 3; chl 3
2748	Benzonitrile, 3-chloro-	C_7H_4ClN / 137.57	766-84-7 / 41	100[15]	4-09-00-00973	H_2O 1; EtOH 3; eth 3
2749	Benzonitrile, 4-chloro-	C_7H_4ClN / 137.57	623-03-0 / 95	223; 95[5]	4-09-00-00992 / 1.1133[17]	H_2O 2; EtOH 3; eth 3; bz 3
2750	Benzonitrile, 2-(chloromethyl)-	C_8H_6ClN / 151.60	612-13-5 / 61	252	4-09-00-01707	EtOH 4
2751	Benzonitrile, 3,5-dibromo-4-hydroxy- / Bromoxynil	$C_7H_3Br_2NO$ / 276.91	1689-84-5 / 190	1431	4-10-00-00475	
2752	Benzonitrile, 2,6-dichloro- / Dichlobenil	$C_7H_3Cl_2N$ / 172.01	1194-65-6 / 144.5	3029 / 270	4-09-00-01006	
2753	Benzonitrile, 2,6-dimethoxy- / 2,6-Dimethoxybenzonitrile	$C_9H_9NO_2$ / 163.18	16932-49-3 / 119	310	2-10-00-00260	H_2O 2; EtOH 4; eth 2; ace 3
2754	Benzonitrile, 2-ethoxy- / o-Ethoxybenzonitrile	C_9H_9NO / 147.18	6609-57-0 / 5	260.5	0-10-00-00097 / 1.0650[15]	eth 4; EtOH 4; lig 4
2755	Benzonitrile, 4-ethoxy-	C_9H_9NO / 147.18	25117-74-2 / 61.5	258	3-10-00-00344	eth 4; EtOH 4; lig 4
2756	Benzonitrile, 4-fluoro-	C_7H_4FN / 121.11	1194-02-1 / 34.8	188.8	4-09-00-00954 / 1.1070[55]	chl 2; peth 3 / 1.4925[55]
2757	Benzonitrile, 3-formyl-	C_8H_5NO / 131.13	24964-64-5 / 76.5	210	4-10-00-02752	H_2O 4; EtOH 4; eth 4; chl 4
2758	Benzonitrile, 4-formyl-	C_8H_5NO / 131.13	105-07-7 / 100.5	133[12]	4-10-00-02754	H_2O 3; EtOH 4; eth 4; chl 4
2759	Benzonitrile, 2-hydroxy-	C_7H_5NO / 119.12	611-20-1 / 98	149[14]	4-10-00-00191 / 1.1052[100]	H_2O 2; EtOH 4; eth 4; bz 4 / 1.5372[100]
2760	Benzonitrile, 2-methoxy-	C_8H_7NO / 133.15	6609-56-9 / 24.5	255.5	4-10-00-00191 / 1.1063[20]	EtOH 3; eth 4
2761	Benzonitrile, 3-methoxy-	C_8H_7NO / 133.15	1527-89-5 /	111-2[13]	4-10-00-00329 / 1.089[25]	1.5402[20]
2762	Benzonitrile, 4-methoxy-	C_8H_7NO / 133.15	874-90-8 / 61.5	256.5	4-10-00-00441	H_2O 1; EtOH 4; eth 4; bz 3
2763	Benzonitrile, 2-methyl- / o-Tolunitrile	C_8H_7N / 117.15	529-19-1 / -13.5	9463 / 205	4-09-00-01703 / 0.9955[20]	H_2O 1; EtOH 5; eth 5; ctc 2 / 1.5279[20]
2764	Benzonitrile, 3-methyl- / m-Tolunitrile	C_8H_7N / 117.15	620-22-4 / -23	213	4-09-00-01717 / 1.0316[20]	H_2O 1; EtOH 5; eth 5; ctc 2 / 1.5252[20]
2765	Benzonitrile, 4-methyl- / p-Tolunitrile	C_8H_7N / 117.15	104-85-8 / 29.5	9464 / 217.0	4-09-00-01738 / 0.9762[30]	H_2O 1; EtOH 4; eth 4; ctc 2

No.	Name Synonym	Mol. Form. Mol. Wt.	CAS RN mp/°C	Merck No. bp/°C	Beil. Ref. den/g cm^{-3}	Solubility n_D
2766	Benzonitrile, 2-methyl-5-nitro-	$C_8H_6N_2O_2$ 162.15	939-83-3 106	174[18]	3-09-00-02314	ace 4; bz 4; eth 4; EtOH 4
2767	Benzonitrile, pentafluoro-	C_7F_5N 193.08	773-82-0 2.4	162	4-09-00-00956 1.563[20]	1.4402[25]
2768	Benzonitrile, 4-(2-phenylethenyl)-, (Z)- 4-Stilbenecarbonitrile (cis)	$C_{15}H_{11}N$ 205.26	14064-68-7 44.5	130[0.1]	4-09-00-02615	H_2O 2; EtOH 3; chl 3
2769	Benzonitrile, 2-(trifluoromethyl)-	$C_8H_4F_3N$ 171.12	447-60-9 18	205	4-09-00-01705	ctc 3
2770	Benzonitrile, 3-(trifluoromethyl)-	$C_8H_4F_3N$ 171.12	368-77-4 14.5	189	3-09-00-02327 1.2813[20]	1.4508[20]
2771	Benzophenone, 3-(dimethylamino)- 3-(Dimethylamino)benzophenone	$C_{15}H_{15}NO$ 225.29	31766-07-1 47	216[15]	4-14-00-00247	EtOH 3
2772	2H-1-Benzopyran 1,2-Chromene	C_9H_8O 132.16	254-04-6	1112 132[102]; 91[13]	5-17-02-00019 1.0993[16]	H_2O 1 1.5869[24]
2773	1H-2-Benzopyran-3-carboxylic acid, 6-[2-(4,9-dihydro-8-hydroxy-5,7-dimethoxy-4,9-dioxonaphtho[2,3-b]furan-2-yl Collinomycin	$C_{27}H_{20}O_{12}$ 536.45	27267-69-2 281	2478	5-19-08-00331	ace 4; diox 4; chl 4
2774	3H-2-Benzopyran-7-carboxylic acid, 4,6-dihydro-8-hydroxy-3,4,5-trimethyl-6-oxo-, (3R-trans)- Citrinin	$C_{13}H_{14}O_5$ 250.25	518-75-2 178-9 dec	2329	5-18-09-00061	H_2O 1; EtOH 2; eth 2; ace 3
2775	2H-1-Benzopyran-3-carboxylic acid, 2-oxo- Coumarin-3-carboxylic acid	$C_9H_6O_3$ 162.14	531-81-7 190 dec	2564	5-18-08-00323	EtOH 4
2776	1H-2-Benzopyran, 3,4-dihydro- Isochroman	$C_9H_{10}O$ 134.18	493-05-0	90[12]	5-17-01-00595 1.067[25]	1.5444[20]
2777	2H-1-Benzopyran, 3,4-dihydro-	$C_9H_{10}O$ 134.18	493-08-3	215; 98[18]	5-17-01-00593 1.0610[20]	H_2O 3; os 5 1.5444[20]
2778	2H-1-Benzopyran, 3,4-dihydro-2-methyl- 2-Methylchroman	$C_{10}H_{12}O$ 148.20	13030-26-7	224.5	5-17-01-00609 1.0340[20]	bz 4; eth 4; EtOH 4 1.532[13]
2779	2H-1-Benzopyran, 3,4-dihydro-3-methyl- Chroman, 3-methyl	$C_{10}H_{12}O$ 148.20	70401-56-8	102-4[15]	4-17-00-00430 1.002[22]	1.5335[20]
2780	2H-1-Benzopyran, 3,4-dihydro-6-methyl- Chroman, 6-methyl-	$C_{10}H_{12}O$ 148.20	3722-74-5	234	4-17-00-00431 1.0374[14]	1.5392[25]
2781	2H-1-Benzopyran, 3,4-dihydro-7-methyl- Chroman, 7-methyl-	$C_{10}H_{12}O$ 148.20	3722-73-4	141-3[60]	4-17-00-00431 1.028[26]	1.5380[20]
2782	2H-1-Benzopyran, 3,4-dihydro-2-phenyl- Flavan	$C_{15}H_{14}O$ 210.28	494-12-2 46.3	332; 190[16]	5-17-02-00283	bz 4; eth 4; EtOH 4
2783	[1]Benzopyrano[3,4-b]furo[2,3-h][1]benzopyran-6(6aH)-one, 1,2,12,12a-tetrahydro-8,9-dimethoxy-2-(1-methylethen Rotenone	$C_{23}H_{22}O_6$ 394.42	83-79-4 176	8248 210-20[0.5]	5-19-10-00581	H_2O 1; EtOH 3; eth 2; ace 3
2784	[2]Benzopyrano[6,5,4-def][2]benzopyran-1,3,6,8-tetrone	$C_{14}H_4O_6$ 268.18	81-30-1 450	sub 320	5-19-05-00423	H_2O 1; Na_2CO_3 3; HOAc 3
2785	2H-1-Benzopyran-6-ol, 3,4-dihydro-2,8-dimethyl-2-(4,8,12-trimethyltridecyl)-, [2R-[2R*(4R*,8R*)]]- δ-Tocopherol	$C_{27}H_{46}O_2$ 402.66	119-13-1 150[0.001]	9419		H_2O 1; EtOH 4; eth 4; ace 4
2786	2H-1-Benzopyran-7-ol, 3,4-dihydro-3-(4-hydroxyphenyl)-, (S)- Equol	$C_{14}H_{14}O_3$ 230.26	531-95-3 189.5	3583	5-17-05-00413	
2787	2H-1-Benzopyran-6-ol, 3,4-dihydro-2,5,7,8-tetramethyl-2-(4,8,12-trimethyltridecyl)-, [2R-[2R*(4R*,8R*)]]- Vitamin E	$C_{29}H_{50}O_2$ 430.71	59-02-9 3.0	9931 210[0.1]	5-17-04-00168 0.950[25]	H_2O 1; EtOH 3; eth 3; ace 3 1.5045[25]
2788	2H-1-Benzopyran-6-ol, 3,4-dihydro-2,5,7,8-tetramethyl-2-(4,8,12-trimethyltridecyl)-, acetate, [2R-[2R*(4R*,8R* Vitamin E acetate	$C_{31}H_{52}O_3$ 472.75	58-95-7 -27.5	9932 184[0.01]	5-17-04-00168 0.9533[21]	H_2O 1; EtOH 2; eth 3; ace 3 1.497[20]
2789	2H-1-Benzopyran-6-ol, 3,4-dihydro-2,5,8-trimethyl-2-(4,8,12-trimethyltridecyl)- β-Tocopherol	$C_{28}H_{48}O_2$ 416.69	148-03-8	9417 200-10[0.1]	5-17-04-	ace 4; eth 4; EtOH 4; chl 4
2790	2H-1-Benzopyran-6-ol, 3,4-dihydro-2,7,8-trimethyl-2-(4,8,12-trimethyltridecyl)- γ-Tocopherol	$C_{28}H_{48}O_2$ 416.69	7616-22-0 -3	9418 200-10[0.1]		H_2O 1; EtOH 5; eth 5; ace 5
2791	1H-2-Benzopyran-1-one Isocoumarin	$C_9H_6O_2$ 146.15	491-31-6 47	286	5-17-10-00149	H_2O 1; EtOH 4; eth 4; bz 4
2792	2H-1-Benzopyran-2-one Coumarin	$C_9H_6O_2$ 146.15	91-64-5 71	2563 301.7	5-17-10-00143 0.935[20]	H_2O 3; EtOH 3; eth 4; chl 4
2793	4H-1-Benzopyran-4-one	$C_9H_6O_2$ 146.15	491-38-3 59	sub	5-17-10-00139 1.2900[20]	H_2O 2; EtOH 3; eth 3; bz 3
2794	4H-1-Benzopyran-4-one, 7-[[6-O-(6-deoxy-α-L-mannopyranosyl)-β-D-glucopyranosyl]oxy]-2,3-dih Hesperidin	$C_{28}H_{34}O_{15}$ 610.57	520-26-3 262	4591	5-18-05-00218	py 4; EtOH 4; HOAc 4
2795	4H-1-Benzopyran-4-one, 7-[[2-O-(6-deoxy-α-L-mannopyranosyl)-β-D-glucopyranosyl]oxy]-2,3-dihydro-5-hy Naringin	$C_{27}H_{32}O_{14}$ 580.54	10236-47-2	6345	5-18-04-00528	H_2O 2; EtOH 2; eth 1; bz 1

No.	Name Synonym	Mol. Form. Mol. Wt.	CAS RN mp/°C	Merck No. bp/°C	Beil. Ref. den/g cm^{-3}	Solubility n_D
2796	4H-1-Benzopyran-4-one, 3-[(6-deoxy-α-L-mannopyranosyl)oxy]-2-(3,4-dihydroxyphenyl)-5,7-dihydroxy- Quercitrin	$C_{21}H_{20}O_{11}$ 448.38	522-12-3 170	8047	5-18-05-00514	H_2O 1; EtOH 3; eth 1; HOAc 3
2797	1H-2-Benzopyran-1-one, 3,4-dihydro-	$C_9H_8O_2$ 148.16	4702-34-5 176[20]; 112[0.4]		5-17-10-00018 1.197[25]	1.5629[25]
2798	2H-1-Benzopyran-2-one, 3,4-dihydro-	$C_9H_8O_2$ 148.16	119-84-6 25	272	5-17-10-00013 1.169[18]	H_2O 1; EtOH 2; eth 2; ctc 2 1.5563[20]
2799	4H-1-Benzopyran-4-one, 2,3-dihydro- 4-Chromanone	$C_9H_8O_2$ 148.16	491-37-2 36.5	160[50]; 127[13]	5-17-10-00014 1.1291[100]	EtOH 3; eth 4; ace 4; bz 4 1.5750
2800	4H-1-Benzopyran-4-one, 2,3-dihydro-5,7-dihydroxy-2-(3-hydroxy-4-methoxyphenyl)-, (S)- Hesperetin	$C_{16}H_{14}O_6$ 302.28	520-33-2 227.5	4590 sub 205	4-18-00-03215	eth 4; EtOH 4
2801	4H-1-Benzopyran-4-one, 2,3-dihydro-5,7-dihydroxy-3-(3-hydroxy-4-methoxyphenyl)- Flavanone, 4'-methoxy-3',5,7-trihydroxy	$C_{16}H_{14}O_6$ 302.28	99365-26-1 227.5	sub 205	2-18-00-00204	H_2O 1; EtOH 4; eth 3; bz 2
2802	4H-1-Benzopyran-4-one, 2,3-dihydro-5,7-dihydroxy-2-(4-hydroxyphenyl)-, (S)- Naringenin	$C_{15}H_{12}O_5$ 272.26	480-41-1 251	6344	5-18-04-00524	bz 4; eth 4; EtOH 4
2803	4H-1-Benzopyran-4-one, 2,3-dihydro-6-methyl-	$C_{10}H_{10}O_2$ 162.19	39513-75-2 35	141-3[13.5]	5-17-10-00036 1.1245[57]	1.555[57]
2804	2H-1-Benzopyran-2-one, 6,7-dihydroxy- Esculetin	$C_9H_6O_4$ 178.14	305-01-1 276	3645 sub	5-18-03-00202	H_2O 2; EtOH 3; eth 2; ace 3
2805	2H-1-Benzopyran-2-one, 7,8-dihydroxy- Daphnetin	$C_9H_6O_4$ 178.14	486-35-1 262	2816 sub	5-18-03-00211	H_2O 3; EtOH 3; eth 2; bz 2
2806	4H-1-Benzopyran-4-one, 5,7-dihydroxy-2-(4-hydroxyphenyl)- Apigenin	$C_{15}H_{10}O_5$ 270.24	520-36-5 347.5	763	5-18-04-00574	H_2O 1; EtOH 3; py 3; dil alk 4
2807	4H-1-Benzopyran-4-one, 5,7-dihydroxy-3-(4-hydroxyphenyl)- Genistein	$C_{15}H_{10}O_5$ 270.24	446-72-0 301-2 dec	4281	5-18-04-00594	
2808	2H-1-Benzopyran-2-one, 7,8-dihydroxy-6-methoxy- Fraxetin	$C_{10}H_8O_5$ 208.17	574-84-5 231	4178	5-18-04-00332	EtOH 4
2809	4H-1-Benzopyran-4-one, 5,7-dihydroxy-2-(4-methoxyphenyl)- Acacetin	$C_{16}H_{12}O_5$ 284.27	480-44-4 263	9	5-18-04-00575	EtOH 4
2810	4H-1-Benzopyran-4-one, 5,7-dihydroxy-6-methoxy-2-phenyl- Oroxylin A	$C_{16}H_{12}O_5$ 284.27	480-11-5 231.5	6830	5-18-04-00569	ace 4; eth 4; EtOH 4
2811	4H-1-Benzopyran-4-one, 5,7-dihydroxy-2-phenyl- Chrysin	$C_{15}H_{10}O_4$ 254.24	480-40-0 285.5	2261	5-18-04-00076	H_2O 1; EtOH 3; eth 2; ace 3
2812	4H-1-Benzopyran-4-one, 2-(3,4-dihydroxyphenyl)-2,3-dihydro-5,7-dihydroxy-, (S)- Eriodictyol	$C_{15}H_{12}O_6$ 288.26	552-58-9 267 dec	3616	5-18-05-00214	EtOH 4; HOAc 4
2813	4H-1-Benzopyran-4-one, 2-(3,4-dihydroxyphenyl)-3,7-dihydroxy- Fisetin	$C_{15}H_{10}O_6$ 286.24	528-48-3 330	4026	5-18-05-00291	H_2O 1; EtOH 3; eth 2; ace 3
2814	4H-1-Benzopyran-4-one, 2-(3,4-dihydroxyphenyl)-5,7-dihydroxy- Luteolin	$C_{15}H_{10}O_6$ 286.24	491-70-3 329-30 dec	5483	5-18-05-00296	H_2O 2; EtOH 3; eth 3; alk 3
2815	4H-1-Benzopyran-4-one, 2-(3,4-dihydroxyphenyl)-3,5-dihydroxy-7-methoxy- Rhamnetin	$C_{16}H_{12}O_7$ 316.27	90-19-7 295	8170	5-18-05-00495	H_2O 2; EtOH 3; ace 3; dil alk 4
2816	4H-1-Benzopyran-4-one, 2-(2,4-dihydroxyphenyl)-3,5,7-trihydroxy- Morin	$C_{15}H_{10}O_7$ 302.24	480-16-0 303.5	6179	5-18-05-00492	H_2O 2; EtOH 4; eth 2; bz 3
2817	4H-1-Benzopyran-4-one, 2-(3,4-dihydroxyphenyl)-3,5,7-trihydroxy- Quercetin	$C_{15}H_{10}O_7$ 302.24	117-39-5 316.5	8044 sub	5-18-05-00494	H_2O 2; EtOH 3; eth 2; ace 3
2818	2H-1-Benzopyran-2-one, 5,7-dimethoxy- Limettin	$C_{11}H_{10}O_4$ 206.20	487-06-9 149	5370 200 dec	5-18-03-00199	H_2O 2; EtOH 4; eth 1; ace 4
2819	4H-1-Benzopyran-4-one, 8-[(dimethylamino)methyl]-7-methoxy-3-methyl-2-phenyl- Dimefline	$C_{20}H_{21}NO_3$ 323.39	1165-48-6 109.5	3190	5-18-12-00083	chl 3
2820	2H-1-Benzopyran-2-one, 3-ethyl- Coumarin, 3-ethyl-	$C_{11}H_{10}O_2$ 174.20	66898-39-3 72.3	299	5-17-10-00185	eth 4; EtOH 4
2821	2H-1-Benzopyran-2-one, 7-(β-D-glucopyranosyloxy)- Skimmin	$C_{15}H_{16}O_8$ 324.29	93-39-0 220	8506	5-18-01-00409	H_2O 3; EtOH 3; eth 1; chl 1
2822	2H-1-Benzopyran-2-one, 6-(β-D-glucopyranosyloxy)-7-hydroxy- Esculin	$C_{15}H_{16}O_9$ 340.29	531-75-9 205 dec	3646 230 dec	5-18-03-00208	H_2O 2; EtOH 2; eth 2; chl 3

No.	Name / Synonym	Mol. Form. / Mol. Wt.	CAS RN / mp/°C	Merck No. / bp/°C	Beil. Ref. / den/g cm⁻³	Solubility / n_D
2823	2H-1-Benzopyran-2-one, 8-(β-D-glucopyranosyloxy)-7-hydroxy-6-methoxy- / Fraxin	$C_{16}H_{18}O_{10}$ / 370.31	524-30-1 / 205	4179	5-18-04-00334	
2824	4H-1-Benzopyran-4-one, 3-[4-(β-D-glucopyranosyloxy)phenyl]-5,7-dihydroxy- / Sophoricoside	$C_{21}H_{20}O_{10}$ / 432.38	152-95-4 / 274	8675	5-18-04-00597	
2825	2H-1-Benzopyran-2-one, 6-hydroxy-	$C_9H_6O_3$ / 162.14	6093-68-1 / 250		5-18-01-00385 / 1.25²⁵	EtOH 4
2826	2H-1-Benzopyran-2-one, 7-hydroxy- / Umbelliferone	$C_9H_6O_3$ / 162.14	93-35-6 / 230.5 sub	9758	5-18-01-00386	EtOH 4; HOAc 4; chl 4
2827	4H-1-Benzopyran-4-one, 7-hydroxy-3-(4-hydroxyphenyl)- / Daidzein	$C_{15}H_{10}O_4$ / 254.24	486-66-8 / 323 dec	2805 sub	5-18-04-00089	EtOH 3; eth 3
2828	4H-1-Benzopyran-4-one, 5-hydroxy-3-(4-hydroxyphenyl)-7-methoxy- / Prunetin	$C_{16}H_{12}O_5$ / 284.27	552-59-0 / 239.5	7923	5-18-04-00595	
2829	2H-1-Benzopyran-2-one, 7-hydroxy-6-methoxy- / Scopoletin	$C_{10}H_8O_4$ / 192.17	92-61-5 / 204	8363	5-18-03-00203	H_2O 2; EtOH 2; eth 2; bz 1
2830	4H-1-Benzopyran-4-one, 7-hydroxy-3-(4-methoxyphenyl)- / Formononetin	$C_{16}H_{12}O_4$ / 268.27	485-72-3 / 256.5	4157	5-18-04-00089	
2831	2H-1-Benzopyran-2-one, 7-hydroxy-4-methyl- / Hymecromone	$C_{10}H_8O_3$ / 176.17	90-33-5 / 194.5	4792	5-18-01-00439	H_2O 2; EtOH 3; eth 2; chl 2
2832	2H-1-Benzopyran-2-one, 4-hydroxy-3-(3-oxo-1-phenylbutyl)- / Warfarin	$C_{19}H_{16}O_4$ / 308.33	81-81-2 / 161	9950	5-18-04-00162	H_2O 1; EtOH 3; ace 3; diox 3
2833	2H-1-Benzopyran-2-one, 3-methyl- / 3-Methylcoumarin	$C_{10}H_8O_2$ / 160.17	2445-82-1 / 91	292.5	5-17-10-00165	H_2O 2; EtOH 3; chl 3; KOH 1
2834	2H-1-Benzopyran-2-one, 5-methyl- / 5-Methylcoumarin	$C_{10}H_8O_2$ / 160.17	42286-84-0 / 65.8	173¹²	5-17-10-00167	H_2O 2; EtOH 4; eth 4; bz 4
2835	2H-1-Benzopyran-2-one, 6-methyl-	$C_{10}H_8O_2$ / 160.17	92-48-8 / 76.5	304; 174¹⁴	5-17-10-00168	EtOH 4; eth 4; bz 4; chl 2
2836	2H-1-Benzopyran-2-one, 7-methyl- / 7-Methylcoumarin	$C_{10}H_8O_2$ / 160.17	2445-83-2 / 128	171.5¹¹	5-17-10-00169	H_2O 2; EtOH 4; eth 3; HOAc 4
2837	2H-1-Benzopyran-2-one, 8-methyl- / 8-Methylcoumarin	$C_{10}H_8O_2$ / 160.17	1807-36-9 / 110.5	178²⁰	5-17-10-00170	EtOH 4; eth 4; bz 4; ctc 4
2838	4H-1-Benzopyran-4-one, 3-methyl- / Tricromyl	$C_{10}H_8O_2$ / 160.17	85-90-5 /	9574	5-17-10-00164	chl 3
2839	2H-1-Benzopyran-2-one, 3,3'-methylenebis[4-hydroxy- / Dicumarol	$C_{19}H_{12}O_6$ / 336.30	66-76-2 / 290	3080	5-19-06-00682	
2840	4H-1-Benzopyran-4-one, 2-phenyl- / Flavone	$C_{15}H_{10}O_2$ / 222.24	525-82-6 / 100	4030	5-17-10-00552	H_2O 1; EtOH 3; eth 3; ace 3
2841	4H-1-Benzopyran-4-one, 3-phenyl- / Isoflavone	$C_{15}H_{10}O_2$ / 222.24	574-12-9 / 148	5057	5-17-10-00562	
2842	4H-1-Benzopyran-4-one, 3,5,7-trihydroxy-2-(2-hydroxyphenyl)- / Datiscetin	$C_{15}H_{10}O_6$ / 286.24	480-15-9 / 277.5	2822	4-18-00-03281	ace 4; eth 4; EtOH 4
2843	4H-1-Benzopyran-4-one, 3,5,7-trihydroxy-2-(4-hydroxyphenyl)- / Kaempferol	$C_{15}H_{10}O_6$ / 286.24	520-18-3 / 277	5156	5-18-05-00251	H_2O 2; EtOH 4; eth 4; ace 4
2844	4H-1-Benzopyran-4-one, 5,6,7-trihydroxy-2-phenyl- / Baicalein	$C_{15}H_{10}O_5$ / 270.24	491-67-8 / 264-5 dec	954	5-18-04-00569	H_2O 2; EtOH 3; eth 3; ace 3
2845	2H-1-Benzopyran-2-one, 6,7,8-trimethoxy-	$C_{12}H_{12}O_5$ / 236.22	6035-49-0 / 104.2	90-100⁰·²	5-18-04-00333	EtOH 3; eth 3
2846	2H-1-Benzopyran-3,5,7-triol, 2-(3,4-dihydroxyphenyl)-3,4-dihydro-, (2S-cis)-	$C_{15}H_{14}O_6$ / 290.27	35323-91-2 / 242		5-17-08-00447 / 1.344⁴	ace 4; EtOH 4
2847	1-Benzopyrylium, 3,5,7-trihydroxy-2-(4-hydroxy-3,5-dimethoxyphenyl)-, chloride / Malvidin chloride	$C_{17}H_{15}ClO_7$ / 366.75	643-84-5 / >300	5596	5-17-08-00514	H_2O 2; EtOH 3; MeOH 3
2848	1-Benzopyrylium, 3,5,7-trihydroxy-2-(4-hydroxyphenyl)-, chloride / Pelargonidin chloride	$C_{15}H_{11}ClO_5$ / 306.70	134-04-3 / >350	7014	1-18-00-00410	H_2O 3; EtOH 4; chl 2; con sulf 3
2849	1-Benzopyrylium, 3,5,7-trihydroxy-2-(3,4,5-trihydroxyphenyl)-, chloride / Delphinidin	$C_{15}H_{11}ClO_7$ / 338.70	528-53-0 / >350	2866	5-17-08-00514	H_2O 4; EtOH 4; MeOH 4; AcOEt 3
2850	Benzo[rst]pentaphene / 1,2:7,8-Dibenzopyrene	$C_{24}H_{14}$ / 302.38	189-55-9 / 281.5		4-05-00-02803	
2851	2H-1,2,4-Benzothiadiazine-7-sulfonamide, 3-bicyclo[2.2.1]hept-5-en-2-yl-6-chloro-3,4-dihydro-, 1,1-dioxide / Cyclothiazide	$C_{14}H_{16}ClN_3O_4S_2$ / 389.88	2259-96-3 / 234	2760		
2852	2H-1,2,4-Benzothiadiazine-7-sulfonamide, 6-chloro-, 1,1-dioxide / Chlorothiazide	$C_7H_6ClN_3O_4S_2$ / 295.73	58-94-6 / 350 dec	2169	4-27-00-08036	

No.	Name / Synonym	Mol. Form. / Mol. Wt.	CAS RN / mp/°C	Merck No. / bp/°C	Bell. Ref. / den/g cm^{-3}	Solubility / n_D
2853	2H-1,2,4-Benzothiadiazine-7-sulfonamide, 6-chloro-3-(cyclopentylmethyl)-3,4-dihydro-, 1,1-dioxide / Cyclopenthiazide	$C_{13}H_{18}ClN_3O_4S_2$ / 379.89	742-20-1 / 238	2750		
2854	2H-1,2,4-Benzothiadiazine-7-sulfonamide, 6-chloro-3-(dichloromethyl)-3,4-dihydro-, 1,1-dioxide / Trichloromethiazide	$C_8H_8Cl_3N_3O_4S_2$ / 380.66	133-67-5 / 270 dec	9537		H_2O 2; EtOH 3
2855	2H-1,2,4-Benzothiadiazine-7-sulfonamide, 6-chloro-3,4-dihydro-, 1,1-dioxide / Hydrochlorothiazide	$C_7H_8ClN_3O_4S_2$ / 297.74	58-93-5 / 274	4704		
2856	2H-1,2,4-Benzothiadiazine-7-sulfonamide, 6-chloro-3,4-dihydro-3-(2-methylpropyl)-, 1,1-dioxide / Buthiazide	$C_{11}H_{16}ClN_3O_4S_2$ / 353.85	2043-38-1 / 221.5	1519		
2857	2H-1,2,4-Benzothiadiazine-7-sulfonamide, 6-chloro-3,4-dihydro-2-methyl-3-[[(2,2,2-trifluoroethyl)thio]methyl]- / Polythiazide	$C_{11}H_{13}ClF_3N_3O_4S_3$ / 439.89	346-18-9 / 214	7561		
2858	2H-1,2,4-Benzothiadiazine-7-sulfonamide, 3,4-dihydro-6-(trifluoromethyl)-, 1,1-dioxide / Hydroflumethiazide	$C_8H_8F_3N_3O_4S_2$ / 331.30	135-09-1 / 270.5	4716	4-27-00-08035	
2859	2-Benzothiazolamine / 2-Aminobenzothiazole	$C_7H_6N_2S$ / 150.20	136-95-8 / 132	435	4-27-00-04824	H_2O 2; EtOH 3; eth 3; chl 3
2860	6-Benzothiazolamine / 6-Aminobenzothiazole	$C_7H_6N_2S$ / 150.20	533-30-2 / 87	436	4-27-00-04884	H_2O 1; EtOH 3; eth 1
2861	Benzothiazole / Benzosulfonazole	C_7H_5NS / 135.19	95-16-9 / 2	1118 / 231	4-27-00-01069 / 1.2460[20]	H_2O 2; EtOH 4; eth 4; ace 3 / 1.6379[20]
2862	Benzothiazole, 2-chloro- / 2-Chlorobenzothiazole	C_7H_4ClNS / 169.63	615-20-3 / 24	/ 248	4-27-00-01072 / 1.3715[10]	ace 4; eth 4; EtOH 4 / 1.6338[10]
2863	Benzothiazole, 2-methyl- / 2-Methylbenzothiazole	C_8H_7NS / 149.22	120-75-2 / 14	/ 238	4-27-00-01080 / 1.1763[19]	H_2O 1; EtOH 3; chl 3 / 1.6092[19]
2864	Benzothiazole, 2-phenyl- / 2-Phenylbenzothiazole	$C_{13}H_9NS$ / 211.29	883-93-2 / 115	/ 371	4-27-00-01385	H_2O 1; EtOH 3; eth 3; CS_2 3
2865	2-Benzothiazolesulfonamide, 6-ethoxy- / Ethoxzolamide	$C_9H_{10}N_2O_3S_2$ / 258.32	452-35-7 / 189	3711	4-27-00-04404	
2866	2(3H)-Benzothiazolethione / 2-Mercaptobenzothiazole	$C_7H_5NS_2$ / 167.26	149-30-4 / 181	5759	4-27-00-02709 / 1.42[20]	H_2O 1; EtOH 3; eth 2; bz 2
2867	2(3H)-Benzothiazolethione, 3-methyl-	$C_8H_7NS_2$ / 181.28	2254-94-6 / 90	/ 335	4-27-00-02714	H_2O 1; EtOH 2; eth 2; bz 4
2868	2(3H)-Benzothiazolone	C_7H_5NOS / 151.19	934-34-9 / 139	/ 360	4-27-00-02693	H_2O 1; EtOH 4; eth 4
2869	4H-1-Benzothiopyran-4-one, 2,3-dihydro-	C_9H_8OS / 164.23	3528-17-4 / 29	154[12]	5-17-10-00016 / 1.2487[14]	1.6395[20]
2870	1,2,4-Benzotriazine / 1,2,4-Triazanaphthalene	$C_7H_5N_3$ / 131.14	254-87-5 / 75.8	/ 237.5	4-26-00-00166	H_2O 4; eth 4; EtOH 4
2871	1H-Benzotriazole / 1,2,3-Triaza-1H-indene	$C_6H_5N_3$ / 119.13	95-14-7 / 100	1119 / 204[15]	4-26-00-00093	H_2O 2; EtOH 3; bz 3; chl 3
2872	1H-Benzotriazole, 1-methyl-	$C_7H_7N_3$ / 133.15	13351-73-0 / 64.5	/ 270.5	4-26-00-00095	bz 4; EtOH 4; HOAc 4
2873	1H-Benzotriazole, 5-nitro- / 5-Nitrobenzotriazole	$C_6H_4N_4O_2$ / 164.12	2338-12-7 / 217		4-26-00-00129	
2874	1,2,3-Benzoxadiazole, 5,7-dinitro- / 2,3-Benzodiazole, 5,7-dinitro	$C_6H_2N_4O_5$ / 210.11	87-31-0 / 158			EtOH 4
2875	3H-2,1-Benzoxathiol-3-one, 1,1-dioxide	$C_7H_4O_4S$ / 184.17	81-08-3 / 129.5	184[18]	5-19-04-00215	bz 4; chl 4
2876	2-Benzoxazolamine, 5-chloro- / Zoxazolamine	$C_7H_5ClN_2O$ / 168.58	61-80-3 / 184.5	10098	4-27-00-04819	EtOH 4
2877	Benzoxazole / 1-Oxa-3-azaindene	C_7H_5NO / 119.12	273-53-0 / 31	/ 182.5	4-27-00-01069 / 1.1754[20]	H_2O 1; EtOH 3; sulf 3 / 1.5594[20]
2878	Benzoxazole, 2-chloro- / 2-Chlorobenzoxazole	C_7H_4ClNO / 153.57	615-18-9 / 7	/ 201.5	2-27-00-00017 / 1.3453[18]	1.5678[20]
2879	Benzoxazole, 2,5-dimethyl- / 2,5-Dimethylbenzoxazole	C_9H_9NO / 147.18	5676-58-4 /	/ 218.5	4-27-00-01101 / 1.0880[18]	ctc 3 / 1.5412[20]
2880	Benzoxazole, 2-methyl- / 2-Methylbenzoxazole	C_8H_7NO / 133.15	95-21-6 / 9.5	/ 200.5	4-27-00-01078 / 1.1211[20]	H_2O 1; EtOH 4; eth 5 / 1.5497[20]
2881	2-Benzoxazolol / 2-Hydroxybenzoxazole	$C_7H_5NO_2$ / 135.12	69564-68-7 / 141	230[30]	2-27-00-00233	eth 4; EtOH 4
2882	2(3H)-Benzoxazolone	$C_7H_5NO_2$ / 135.12	59-49-4 / 138	335; 230[30]	4-27-00-02677	H_2O 2; EtOH 3; eth 3; tfa 3
2883	2(3H)-Benzoxazolone, 5-chloro- / Chlorzoxazone	$C_7H_4ClNO_2$ / 169.57	95-25-0 / 191.5	2197	4-27-00-02683	EtOH 4; MeOH 4
2884	Benzoyl azide / Benzazide	$C_7H_5N_3O$ / 147.14	582-61-6 / 32	/ exp	4-09-00-00945 / 1.1680[35]	eth 4; EtOH 4
2885	Benzoyl bromide / Benzoic acid, bromide	C_7H_5BrO / 185.02	618-32-6 / -24	/ 218.5	4-09-00-00724 / 1.570[15]	eth 5 / 1.5868[25]
2886	Benzoyl chloride / Benzenecarbonyl chloride	C_7H_5ClO / 140.57	98-88-4 / -1.0	1124 / 197.2; 71[9]	4-09-00-00721 / 1.2120[20]	eth 5; bz 3; ctc 3; CS_2 3 / 1.5537[20]
2887	Benzoyl chloride, 2-bromo-	C_7H_4BrClO / 219.46	7154-66-7 / 11	/ 243	4-09-00-01013	ctc 2 / 1.5963[20]
2888	Benzoyl chloride, 4-bromo-	C_7H_4BrClO / 219.46	586-75-4 / 42	/ 246; 181[125]	4-09-00-01023	EtOH 4; eth 4; bz 4; lig 4

No.	Name / Synonym	Mol. Form. / Mol. Wt.	CAS RN / mp/°C	Merck No. / bp/°C	Beil. Ref. / den/g cm⁻³	Solubility / n_D
2889	Benzoyl chloride, 4-butyl-	$C_{11}H_{13}ClO$ / 196.68	28788-62-7 /	155-6[22]	3-09-00-02520 / 1.051[25]	1.5351[20]
2890	Benzoyl chloride, 2-chloro-	$C_7H_4Cl_2O$ / 175.01	609-65-4 / -4	238	4-09-00-00962 /	ctc 3 / 1.5726[16]
2891	Benzoyl chloride, 3-chloro- / m-Chlorobenzoyl chloride	$C_7H_4Cl_2O$ / 175.01	618-46-2 /	225	4-09-00-00972 /	1.5677[20]
2892	Benzoyl chloride, 4-chloro-	$C_7H_4Cl_2O$ / 175.01	122-01-0 / 16	222	4-09-00-00984 / 1.3770[20]	chl 2 / 1.5756[20]
2893	Benzoyl chloride, 2,4-dichloro- / 2,4-Dichlorobenzoyl chloride	$C_7H_3Cl_3O$ / 209.46	89-75-8 / 16.5	150[34]; 111[7.5]	4-09-00-01002 /	ctc 3 / 1.5895[20]
2894	Benzoyl chloride, 3,4-dichloro- / 3,4-Dichlorobenzoyl chloride	$C_7H_3Cl_3O$ / 209.46	3024-72-4 / 25	242	3-09-00-01379 /	ctc 2
2895	Benzoyl chloride, 4-(1,1-dimethylethyl)-	$C_{11}H_{13}ClO$ / 196.68	1710-98-1 /	135[20]	4-09-00-01887 / 1.007[25]	1.5364[20]
2896	Benzoyl chloride, 3,5-dinitro- / 3,5-Dinitrobenzoyl chloride	$C_7H_3ClN_2O_5$ / 230.56	99-33-2 / 74	3270 / 196[12]	4-09-00-01350 /	eth 3; chl 3
2897	Benzoyl chloride, 2-fluoro-	C_7H_4ClFO / 158.56	393-52-2 / 4	90-2[15]	4-09-00-00951 / 1.328[25]	1.5365[20]
2898	Benzoyl chloride, 3-fluoro-	C_7H_4ClFO / 158.56	1711-07-5 / -30	189	4-09-00-00953 / 1.304[25]	1.5285[20]
2899	Benzoyl chloride, 4-fluoro-	C_7H_4ClFO / 158.56	403-43-0 / 9	82[20]	4-09-00-00954 / 1.342[25]	1.5296[20]
2900	Benzoyl chloride, 2-hydroxy-	$C_7H_5ClO_2$ / 156.57	1441-87-8 / 19	92[15]	3-10-00-00150 / 1.3112[20]	eth 4 / 1.5812[20]
2901	Benzoyl chloride, 4-methoxy- / p-Anisoyl chloride	$C_8H_7ClO_2$ / 170.60	100-07-2 / 24.5	703 / 262.5	4-10-00-00430 / 1.261[20]	eth 3; ace 3; bz 4; ctc 2 / 1.580[20]
2902	Benzoyl chloride, 2-methyl-	C_8H_7ClO / 154.60	933-88-0 /	213.5	3-09-00-02304 /	eth 4; EtOH 4 / 1.5549[20]
2903	Benzoyl chloride, 3-methyl-	C_8H_7ClO / 154.60	1711-06-4 / -23	219.5	4-09-00-01716 / 1.0265[21]	eth 4; EtOH 4 / 1.505[22]
2904	Benzoyl chloride, 4-methyl-	C_8H_7ClO / 154.60	874-60-2 / -3	226	4-09-00-01733 / 1.1686[20]	ctc 3 / 1.5547[20]
2905	Benzoyl chloride, 3-nitro-	$C_7H_4ClNO_3$ / 185.57	121-90-4 / 36	276.5	4-09-00-01061 /	eth 4
2906	Benzoyl chloride, 4-nitro- / p-Nitrobenzoyl chloride	$C_7H_4ClNO_3$ / 185.57	122-04-3 / 75	6511 / 203[105]; 151[15]	4-09-00-01191 /	eth 3
2907	Benzoyl chloride, 4-pentyl-	$C_{12}H_{15}ClO$ / 210.70	49763-65-7 /	121[8]	4-09-00-01910 / 1.036[25]	1.5300[20]
2908	Benzoyl chloride, 4-(pentyloxy)-	$C_{12}H_{15}ClO_2$ / 226.70	36823-84-4 /	196-200[30]	4-10-00-00431 / 1.087[25]	1.5434[20]
2909	Benzoyl fluoride / Benzoic acid, fluoride	C_7H_5FO / 124.11	455-32-3 / -28	154.5	4-09-00-00721 / 1.1400[20]	EtOH 4; eth 4; ctc 3
2910	Benzoyl iodide	C_7H_5IO / 232.02	618-38-2 / 3	128[20]	4-09-00-00725 / 1.746[18]	eth 4; EtOH 4
2911	Berbaman-12-ol, 6,6',7-trimethoxy-2,2'-dimethyl- / Berbamine	$C_{37}H_{40}N_2O_6$ / 608.73	478-61-5 / 198.5	1168	2-27-00-00891 /	H_2O 2; EtOH 3; eth 3; chl 3
2912	Beryllium, bis(2,4-pentanedionato-O,O')-, (T-4)- / Beryllium acetylacetonate	$C_{10}H_{14}BeO_4$ / 207.23	10210-64-7 / 108	1180 / 270		
2913	Bicyclo[2.2.1]hepta-2,5-diene	C_7H_8 / 92.14	121-46-0 / -19.1	89.5	4-05-00-00879 / 0.9064[20]	H_2O 1; EtOH 3; eth 3; ace 3 / 1.4702[20]
2914	Bicyclo[2.2.1]heptane	C_7H_{12} / 96.17	279-23-2 / 87.5	105.3	4-05-00-00258 /	ace 4; bz 4; eth 4; EtOH 4
2915	Bicyclo[4.1.0]heptane / Norcarane	C_7H_{12} / 96.17	286-08-8 /	6605 / 116.5	4-05-00-00257 / 0.853[25]	1.4564[20]
2916	Bicyclo[2.2.1]heptane-2-carboxaldehyde / 2-Norbornanecarboxaldehyde	$C_8H_{10}O$ / 122.17	19396-83-9 /	70-2[22]	4-07-00-00157 / 1.0227[19]	eth 4 / 1.4760[25]
2917	Bicyclo[2.2.1]heptane-2-carboxylic acid, 4,7,7-trimethyl-3-oxo- / d-Camphocarboxylic acid	$C_{11}H_{16}O_3$ / 196.25	18530-30-8 / 127.5	1737	3-10-00-02925 /	bz 4; eth 4; EtOH 4
2918	Bicyclo[2.2.1]heptane, 2-chloro-	$C_7H_{11}Cl$ / 130.62	29342-53-8 / -5	162; 52[11]	4-05-00-00259 / 1.0603[20]	chl 4 / 1.4849[20]
2919	Bicyclo[2.2.1]heptane, 2-chloro-1,7,7-trimethyl- / Bornane, 2-chloro-	$C_{10}H_{17}Cl$ / 172.70	6120-13-4 / 132.8	207	4-05-00-00319 /	eth 4; EtOH 4
2920	Bicyclo[2.2.1]heptane, 2-chloro-1,7,7-trimethyl-, endo- / Bornyl chloride	$C_{10}H_{17}Cl$ / 172.70	464-41-5 / 132	1341 / 207.5	4-05-00-00319 /	bz 4; eth 4; EtOH 4; peth 4
2921	Bicyclo[2.2.1]heptane, 2,2-dimethyl-3-methylene-, (±)-	$C_{10}H_{16}$ / 136.24	565-00-4 / 51.5	158.5	4-05-00-00462 / 0.879[20]	H_2O 1; EtOH 4; eth 4 / 1.4551[54]
2922	Bicyclo[2.2.1]heptane, 2,2-dimethyl-5-methylene-	$C_{10}H_{16}$ / 136.24	497-32-5 /	152	4-05-00-00465 / 0.8591[20]	1.4645[25]
2923	Bicyclo[2.2.1]heptane, 7,7-dimethyl-2-methylene-, (±)- / α-Fenchene (DL)	$C_{10}H_{16}$ / 136.24	2623-54-3 /	155	3-05-00-00389 / 0.8660[20]	ace 4; eth 4; EtOH 4 / 1.4705[20]

No.	Name / Synonym	Mol. Form. / Mol. Wt.	CAS RN / mp/°C	Merck No. / bp/°C	Beil. Ref. / den/g cm^{-3}	Solubility / n_D
2924	Bicyclo[3.1.1]heptane, 6,6-dimethyl-2-methylene-, (1R)-	C$_{10}$H$_{16}$	19902-08-0	7415	4-05-00-00456	bz 4; eth 4; EtOH 4; chl 4
		136.24	-50	164.6	0.8654[20]	1.4789[20]
2925	Bicyclo[3.1.1]heptane, 6,6-dimethyl-2-methylene- β-Pinene	C$_{10}$H$_{16}$	127-91-3	7415		H$_2$O 1; bz 3; EtOH 3; eth 3
		136.24	-61.5	166	0.860[25]	1.4768[25]
2926	Bicyclo[2.2.1]heptane-2,3-dione, 1,7,7-trimethyl-, (1R)-	C$_{10}$H$_{14}$O$_2$	10334-26-6		4-07-00-02039	H$_2$O 3; EtOH 4; eth 4; bz 3
		166.22	199	sub		
2927	Bicyclo[2.2.1]heptane-1-methanamine, 7,7-dimethyl- α-Camphylamine	C$_{10}$H$_{19}$N	54131-44-1		1-12-00-00129	
		153.27		195; 95[12]	0.8688[20]	1.4728[18]
2928	Bicyclo[3.1.1]heptane-2-methanamine, 6,6-dimethyl-, [1S-(1α,2β,5α)]-	C$_{10}$H$_{19}$N	73522-42-6		4-06-00-00384	
		153.27		94-9[27]	0.915[25]	1.4877[20]
2929	Bicyclo[2.2.1]heptane-1-methanesulfonic acid, 7,7-dimethyl-2-oxo-, (1S)- D-Camphorsulfonic acid	C$_{10}$H$_{16}$O$_4$S	3144-16-9	1740	4-11-00-00642	H$_2$O 4; eth 1; HOAc 2
		232.30	195 dec			
2930	Bicyclo[3.1.1]heptane-2-methanol, 6,6-dimethyl-, [1S-(1α,2β,5α)]-	C$_{10}$H$_{18}$O	51152-12-6		4-06-00-00278	
		154.25		70-2[1]	0.986[25]	1.4890[20]
2931	Bicyclo[2.2.1]heptane, 1,3,3-trimethyl-, (+)- Fenchane(d)	C$_{10}$H$_{18}$	116435-30-4		3-05-00-00256	H$_2$O 1; EtOH 3; eth 3; HOAc 2
		138.25	<-15	151	0.8345[20]	1.44714[20]
2932	Bicyclo[2.2.1]heptane, 1,7,7-trimethyl-	C$_{10}$H$_{18}$	464-15-3		4-05-00-00319	H$_2$O 1; EtOH 3; eth 3; AcOEt 3
		138.25		161		
2933	Bicyclo[2.2.1]heptane, 2,2,3-trimethyl-	C$_{10}$H$_{18}$ 138.25	473-19-8 62.5	166	3-05-00-00263 0.8276[67]	ace 4; EtOH 4
2934	Bicyclo[3.1.1]heptane, 2,6,6-trimethyl-, (1α,2α,5α)-(±)- Pinane (DL)	C$_{10}$H$_{18}$	112456-46-9		1-05-00-00048	bz 4; eth 4
		138.25		164.5	0.8551[20]	1.4609[20]
2935	Bicyclo[3.1.1]heptane, 2,6,6-trimethyl-, [1R-(1α,2β,5α)]-	C$_{10}$H$_{18}$	4795-86-2		4-05-00-00318	bz 4; eth 4
		138.25	-53	169	0.8560[20]	1.4629[20]
2936	Bicyclo[4.1.0]heptane, 3,7,7-trimethyl-, [1S-(1α,3β,6α)]-	C$_{10}$H$_{18}$	2778-68-9		4-05-00-00316	
		138.25		169	0.8410[20]	1.4569
2937	Bicyclo[4.1.0]heptane, 3,7,7-trimethyl-, [1S-(1α,3α,6α)]-	C$_{10}$H$_{18}$	6069-97-2		4-05-00-00316	
		138.25		169	0.841[20]	1.456
2938	Bicyclo[2.2.1]heptan-2-ol, 1,7-dimethyl-, (exo,syn)- β-Santenol	C$_9$H$_{16}$O	509-12-6		4-06-00-00242	eth 4
		140.23	101.5	192; 60[15]		
2939	Bicyclo[2.2.1]heptan-2-ol, 1-phenyl-, acetate	C$_{15}$H$_{18}$O$_2$ 230.31	71173-15-4 44	159[13]; 128[1.0]	3-06-00-02755	ctc 2
2940	Bicyclo[2.2.1]heptan-2-ol, 1,3,3-trimethyl-, endo-(±)-	C$_{10}$H$_{18}$O	36386-49-9		3-06-00-00288	eth 4; EtOH 4
		154.25	39	199.5	0.9420[40]	
2941	Bicyclo[2.2.1]heptan-2-ol, 1,3,3-trimethyl-, (1S-endo)- l-α-Fenchyl alcohol	C$_{10}$H$_{18}$O	512-13-0		4-06-00-00278	
		154.25	48	94[20]	0.9034[84]	
2942	Bicyclo[2.2.1]heptan-2-ol, 1,7,7-trimethyl-, endo-(±)-	C$_{10}$H$_{18}$O	6627-72-1		4-06-00-00281	H$_2$O 1; EtOH 4; eth 4; bz 4
		154.25	210.5	sub	1.011[20]	
2943	Bicyclo[2.2.1]heptan-2-ol, 1,7,7-trimethyl-, exo-(±)-	C$_{10}$H$_{18}$O	24393-70-2		4-06-00-00282	H$_2$O 1; EtOH 4; eth 4; bz 2
		154.25	212	sub	1.10[20]	
2944	Bicyclo[3.1.1]heptan-2-ol, 4,6,6-trimethyl-, [1R-(1α,2α,4α,5α)]-	C$_{10}$H$_{18}$O	515-88-8		3-06-00-00284	
		154.25	58	218	0.940[20]	1.4702[20]
2945	Bicyclo[2.2.1]heptan-2-ol, 1,7,7-trimethyl-, acetate, endo-(±)-	C$_{12}$H$_{20}$O$_2$	36386-52-4		4-06-00-00283	ctc 3
		196.29	<-17	223.5	1.4630[20]	1.4630[20]
2946	Bicyclo[2.2.1]heptan-2-ol, 1,7,7-trimethyl-, acetate, exo-(±)-	C$_{12}$H$_{20}$O$_2$	17283-45-3		4-06-00-00283	ace 4; EtOH 4
		196.29		116[21]	0.9841[20]	1.4640[20]
2947	Bicyclo[2.2.1]heptan-2-ol, 4,7,7-trimethyl-, acetate, (1R-endo)- Epiborneol, acetate	C$_{12}$H$_{20}$O$_2$	22621-74-5		2-06-00-00092	
		196.29	<-15	101[11]	0.9872[14]	1.4651[14]
2948	Bicyclo[2.2.1]heptan-2-ol, 1,7,7-trimethyl-, acetate, endo- Borneol, acetate	C$_{12}$H$_{22}$O$_2$	76-49-3	1339	3-06-00-00303	
		198.31	29	221		
2949	Bicyclo[2.2.1]heptan-2-ol, 1,7,7-trimethyl-, formate, (1R-endo)- Bornyl formate (d)	C$_{11}$H$_{18}$O$_2$	74219-20-8		3-06-00-00301	
		182.26		90[10]	1.009[22]	1.4708[15]
2950	Bicyclo[2.2.1]heptan-2-one, 3-(acetyl-17O)-1,7,7-trimethyl- Camphor, 3-acetyl	C$_{12}$H$_{18}$O$_2$	15068-90-3		2-07-00-00563	
		194.27		127-8[15]	1.0314[19]	1.4949[17]

No.	Name Synonym	Mol. Form. Mol. Wt.	CAS RN mp/°C	Merck No. bp/°C	Beil. Ref. den/g cm^{-3}	Solubility n_D
2951	Bicyclo[2.2.1]heptan-2-one, 1-(bromomethyl)-7,7-dimethyl-, (1S)- d-Camphor, 10-bromo-	C$_{10}$H$_{15}$BrO 231.13	64161-50-8 78	265 dec	4-07-00-00218	ace 4; bz 4; eth 4; EtOH 4
2952	Bicyclo[2.2.1]heptan-2-one, 3-bromo-1,7,7-trimethyl-, (1R-endo)-	C$_{10}$H$_{15}$BrO 231.13	10293-06-8 76	274 dec	1.449^{25}	bz 4; eth 4; EtOH 4
2953	Bicyclo[2.2.1]heptan-2-one, 3,3-dibromo-1,7,7-trimethyl-, (1R)- α,α'-Dibromo-d-camphor	C$_{10}$H$_{14}$Br$_2$O 310.03	514-12-5 61	3002	4-07-00-00220 1.854^{21}	H$_2$O 1; EtOH 4; eth 4; bz 4
2954	Bicyclo[2.2.1]heptan-2-one, 3,3-dimethyl-, (1S)-	C$_9$H$_{14}$O 138.21	6069-71-2 41.9	193; 76^{12}	4-07-00-00177 0.807^{20}	eth 4
2955	Bicyclo[3.1.1]heptan-2-one, 6,6-dimethyl-, (1R)-	C$_9$H$_{14}$O 138.21	38651-65-9 -1	209	4-07-00-00176 0.9807^{20}	eth 4; EtOH 4 1.4787^{20}
2956	Bicyclo[3.1.1]heptan-3-one, 6,6-dimethyl-2-methylene-, (1S)-	C$_{10}$H$_{14}$O 150.22	19890-00-7 -1.8	67-9^4	0.9875^{15}	1.4950^{20}
2957	Bicyclo[2.2.1]heptan-2-one, 3-hydroxy-1,7,7-trimethyl- 3-Hydroxycamphor	C$_{10}$H$_{16}$O$_2$ 168.24	10373-81-6 205.5	4747	4-08-00-00081	eth 4; EtOH 4; chl 4
2958	Bicyclo[2.2.1]heptan-2-one, 3-methylene- 3-Methylene-2-norbornanone	C$_8$H$_{10}$O 122.17	5597-27-3	69-71^{11}	4-07-00-00301 0.997^{25}	1.4891^{20}
2959	Bicyclo[2.2.1]heptan-2-one, 1,4,7,7-tetramethyl-	C$_{11}$H$_{18}$O 166.26	10309-50-9 168	213	2-07-00-00113	eth 4; EtOH 4
2960	Bicyclo[2.2.1]heptan-2-one, 1,3,3-trimethyl-, (±)-	C$_{10}$H$_{16}$O 152.24	18492-37-0 -18	3911 193.5	3-07-00-00393 0.9492^{15}	H$_2$O 1; EtOH 4; eth 3; ace 3 1.4702^{20}
2961	Bicyclo[2.2.1]heptan-2-one, 1,7,7-trimethyl-, (1R)- (+)-Camphor	C$_{10}$H$_{16}$O 152.24	464-49-3 178.8	1738 207.4	4-07-00-00213 0.990^{25}	H$_2$O 1; EtOH 4; eth 4; ace 3 1.5462
2962	Bicyclo[2.2.1]heptan-2-one, 1,7,7-trimethyl-, oxime, (1S)- (-)-Camphor oxime	C$_{10}$H$_{17}$NO 167.25	36065-15-3 118	252 dec	4-07-00-00215 1.01^{116}	eth 4; EtOH 4
2963	Bicyclo[2.2.1]heptan-2-one, 4,7,7-trimethyl-, (1S)-	C$_{10}$H$_{16}$O 152.24	10292-98-5 186.5	213	3-07-00-00421	H$_2$O 2; EtOH 4; eth 4; peth 4
2964	Bicyclo[3.1.1]heptan-2-one, 4,6,6-trimethyl-, [1R-(1α,4α,5α)]-	C$_{10}$H$_{16}$O 152.24	515-90-2	222	4-07-00-00210 0.961^{20}	1.4752^{20}
2965	Bicyclo[3.1.1]heptan-3-one, 2,6,6-trimethyl-, [1S-(1α,2β,5α)]-	C$_{10}$H$_{16}$O 152.24	14575-93-0	212	4-07-00-00209 0.964^{25}	1.4735^{25}
2966	Bicyclo[2.2.1]hept-5-ene-2-carbonitrile	C$_8$H$_9$N 119.17	95-11-4 13	82-6^{10}	0.999^{25}	1.4885^{20}
2967	Bicyclo[2.2.1]hept-5-ene-2-carboxaldehyde	C$_8$H$_{10}$O 122.17	5453-80-5	67-70^{12}	2-07-00-00125 1.018^{25}	1.4893^{20}
2968	Bicyclo[2.2.1]hept-2-ene, 2,3-dimethyl- 2-Norbornene, 2,3-dimethyl-	C$_9$H$_{14}$ 122.21	529-16-8	140.5	4-05-00-00427 0.8698^{17}	eth 3; ace 3; bz 3 1.4688^{17}
2969	Bicyclo[3.1.1]hept-2-ene-2-ethanol, 6,6-dimethyl-	C$_{11}$H$_{18}$O 166.26	128-50-7	235; 110^{10}	4-06-00-00398 0.973^{25}	chl 3 1.4930^{20}
2970	Bicyclo[2.2.1]hept-2-ene, 1,7,7-trimethyl-	C$_{10}$H$_{16}$ 136.24	464-17-5 113	146	2-05-00-00105	bz 4; eth 4; EtOH 4
2971	Bicyclo[3.1.1]hept-2-ene, 2,6,6-trimethyl-, (1S)- (-)-α-Pinene	C$_{10}$H$_{16}$ 136.24	7785-26-4 -64	7414 155	4-05-00-00452 0.859^{20}	1.4660^{20}
2972	Bicyclo[3.1.1]hept-2-ene, 2,6,6-trimethyl-, (±)- (±)-α-Pinene	C$_{10}$H$_{16}$ 136.24	2437-95-8 -55	7414 156.2	2-05-00-00097 0.8582^{20}	H$_2$O 1; EtOH 5; eth 5; chl 5 1.4658^{20}
2973	Bicyclo[3.1.1]hept-2-ene, 2,6,6-trimethyl 2-Pinene	C$_{10}$H$_{16}$ 136.24	80-56-8 -64	155.9	0.8539^{25}	1.4632^{25}
2974	Bicyclo[4.1.0]hept-3-ene, 3,7,7-trimethyl-, (1S)-	C$_{10}$H$_{16}$ 136.24	498-15-7	171; 123^{200}	4-05-00-00450 0.8549^{30}	ace 4; bz 4; eth 4 1.469^3
2975	Bicyclo[3.1.1]hept-3-en-2-ol, 4,6,6-trimethyl-, (1α,2β,5α)-	C$_{10}$H$_{16}$O 152.24	1845-30-3 15.5	90^{10}	0.9684^{25}	1.4912^{25}
2976	Bicyclo[3.1.1]hept-3-en-2-ol, 4,6,6-trimethyl-, (1α,2α,5α)-	C$_{10}$H$_{16}$O 152.24	1820-09-3 24	92^{10}	4-06-00-00383 0.9657^{25}	1.4908^{25}
2977	Bicyclo[3.1.1]hept-2-en-6-one, 2,7,7-trimethyl- Chrysanthenone	C$_{10}$H$_{14}$O 150.22	473-06-3	2256 88-9^{12}	4-07-00-00327	EtOH 4 1.4720^{22}
2978	Bicyclo[3.1.1]hept-3-en-2-one, 4,6,6-trimethyl-, (1R)- d-Verbenone	C$_{10}$H$_{14}$O 150.22	18309-32-5 9.8	9862 227.5	4-07-00-00327 0.9978^{20}	H$_2$O 3; EtOH 3; ace 3; bz 3 1.4993^{18}
2979	[Bi-1,4-cyclohexadien-1-yl]-3,3',6,6'-tetrone, 2,2'-dihydroxy-4,4'-dimethyl- Phenicin	C$_{14}$H$_{10}$O$_6$ 274.23	128-68-7 230.5	7194	4-08-00-03660	H$_2$O 2; EtOH 4; chl 4; HOAc 4
2980	Bicyclo[3,1,0]hexane, 1-methyl- 1-Methylbicyclo[3,1,0]hexane	C$_7$H$_{12}$ 96.17	4625-24-5	93.1		

No.	Name / Synonym	Mol. Form. / Mol. Wt.	CAS RN / mp/°C	Merck No. / bp/°C	Beil. Ref. / den/g cm^{-3}	Solubility / n_D
2981	Bicyclo[3.1.0]hexane, 4-methylene-1-(1-methylethyl)-, (±)- / 4(10)-Thujene, (±)-	$C_{10}H_{16}$ / 136.24	15826-80-9	164; 69[30]	3-05-00-00365 / 0.8437[20]	H_2O 1; EtOH 4; eth 3; ace 3 / 1.4676[20]
2982	Bicyclo[3.1.0]hexane, 4-methyl-1-(1-methylethyl)-	$C_{10}H_{18}$ / 138.25	471-12-5	157	4-05-00-00317 / 0.8139[20]	1.4376[20]
2983	Bicyclo[3.1.0]hexan-3-ol, 4-methylene-1-(1-methylethyl)- / 4(10)-Thujene-3-ol	$C_{10}H_{16}O$ / 152.24	59905-55-4	208	4-06-00-00382 / 0.9461[25]	1.4871[25]
2984	Bicyclo[3.1.0]hexan-3-ol, 4-methylene-1-(1-methylethyl)-, [1S-(1α,3β,5α)]-	$C_{10}H_{18}O$ / 152.24	471-16-9	208	4-06-00-00382 / 0.9488[19]	eth 3 / 1.4871[25]
2985	Bicyclo[3.1.0]hexan-3-ol, 4-methyl-1-(1-methylethyl)-	$C_{10}H_{18}O$ / 154.25	513-23-5	209	3-06-00-00278 / 0.9210[20]	1.4621[25]
2986	Bicyclo[3.1.0]hex-2-ene, 2-methyl-5-(1-methylethyl)-	$C_{10}H_{18}$ / 136.24	2867-05-2	151	4-05-00-00451 / 0.8301[20]	1.4515[20]
2987	Bicyclo[3.1.0]hex-3-en-2-one, 4-methyl-1-(1-methylethyl)-	$C_{10}H_{14}O$ / 150.22	24545-81-1	220	0-07-00-00159 / 0.9572[15]	1.48325
2988	1,1'-Bicyclohexyl / Cyclohexylcyclohexane	$C_{12}H_{22}$ / 166.31	92-51-3 / 4	238		H_2O 2; EtOH 3; eth 3
2989	[1,1'-Bicyclohexyl]-2-one	$C_{12}H_{20}O$ / 180.29	90-42-6 / -32	264	4-07-00-00248 / 0.9696[25]	1.4877[25]
2990	Bicyclo[3.3.1]nonane	C_9H_{16} / 124.23	280-65-9 / 145.5	169.5	4-05-00-00293	EtOH 4
2991	Bicyclo[4.2.0]octane	C_8H_{14} / 110.20	278-30-8	136	4-05-00-00276 / 0.8573[20]	1.4613[20]
2992	Bicyclo[2.2.2]octane, 2-methyl- / 2-Methylbicyclo[2.2.2]octane	C_9H_{16} / 124.23	766-53-0 / 33.5	159	4-05-00-00295 / 0.8664[40]	1.4608[40]
2993	Bicyclo[4.2.0]oct-2-ene	C_8H_{12} / 108.18	13367-29-8	138	4-05-00-00413 / 0.8948[20]	1.4810[20]
2994	[1,1'-Bicyclopentyl]-2-ol / Bicyclopentyl, 2-hydroxy	$C_{10}H_{18}O$ / 154.25	4884-25-7 / 20	235.5	3-06-00-00261 / 0.9785[15]	1.4884[17]
2995	[1,1'-Bicyclopentyl]-2-one	$C_{10}H_{16}O$ / 152.24	4884-24-6 / -13	232.5	4-07-00-00198 / 0.9745[21]	1.4763
2996	Bicyclo[7.2.0]undec-4-ene, 4,11,11-trimethyl-8-methylene-, [1R-(1R*,4E,9S*)]- / Caryophyllene	$C_{15}H_{24}$ / 204.36	87-44-5	122[13.5]	1884 / 4-05-00-01182 / 0.9075[20]	bz 4 / 1.4986[20]
2997	Bifenox / Benzoic acid, 5-(2,4-dichlorophenoxy)-2-nitro-, methyl ester	$C_{14}H_9Cl_2NO_5$ / 342.13	42576-02-3 / 85	1228		
2998	Bifenthrin / 2-Methylbiphenyl-3-ylmethyl (Z)-(1RS,3RS)-3-(2-chloro-3,3,3-trifluoroprop-1-enyl)- 2,2-dimethylcyclopropanecarboxylate	$C_{23}H_{22}ClF_3O_2$ / 422.87	82657-04-3 / 69	1229	1.2[125]	
2999	[2,2'-Bi-1H-indene]-1,1',3,3'(2H,2'H)-tetrone, 2,2'-dihydroxy- / Hydrindantin	$C_{18}H_{10}O_8$ / 322.27	5103-42-4 / 250 dec	4698	4-08-00-03701	
3000	21H-Biline-8,12-dipropanoic acid, 2,17-diethenyl-1,10,19,22,23,24-hexahydro-3,7,13,18-tetramethyl-1,19-dioxo- / Bilirubin	$C_{33}H_{36}N_4O_6$ / 584.67	635-65-4	1235	4-26-00-03268	H_2O 1; EtOH 2; eth 2; bz 3
3001	21H-Biline-8,12-dipropanoic acid, 3,18-diethenyl-1,19,22,24-tetrahydro-2,7,13,17-tetramethyl-1,19-dioxo- / Biliverdine	$C_{33}H_{34}N_4O_6$ / 582.66	114-25-0 / >300	1236	4-26-00-03272	H_2O 1; EtOH 3; eth 2; bz 3
3002	[2,2'-Bimorphinan]-3,3',6,6'-tetrol, 7,7',8,8'-tetradehydro-4,5:4',5'-diepoxy-17,17'-dimethyl- / Pseudomorphine	$C_{34}H_{36}N_2O_6$ / 568.67	125-24-6 / 282.5	7934	4-27-00-08724	H_2O 1; EtOH 1; eth 1; chl 1
3003	Binapacryl / 2-Butenoic acid, 3-methyl-, 2-(1-methylpropyl)-4,6-dinitrophenyl ester	$C_{15}H_{18}N_2O_6$ / 322.32	485-31-4 / 70	1237	1.27[20]	
3004	1,1'-Binaphthalene / 1,1'-Binaphthyl	$C_{20}H_{14}$ / 254.33	604-53-5 / 160	>360; 240[12]	4-05-00-02634 / 1.300[20]	H_2O 1; EtOH 2; eth 3; ace 3
3005	2,2'-Binaphthalene	$C_{20}H_{14}$ / 254.33	612-78-2 / 188.3	452	4-05-00-02636	H_2O 1; EtOH 2; eth 3; bz 3
3006	[1,1'-Binaphthalene]-4,4'-diol	$C_{20}H_{14}O_2$ / 286.33	1446-34-0 / 300	sub	4-06-00-07021	eth 4; EtOH 4
3007	2,2'-Bioxirane, (R*,S*)- / Erythritol anhydride	$C_4H_8O_2$ / 86.09	564-00-1 / -19	3621 / 138	5-19-01-00184 / 1.113[18]	H_2O 5; EtOH 4 / 1.4330[20]
3008	1,1'-Biphenyl / Diphenyl	$C_{12}H_{10}$ / 154.21	92-52-4 / 69	3314 / 256.1	4-05-00-01807 / 1.04[20]	H_2O 1; EtOH 3; eth 3; bz 4 / 1.588[77]
3009	[1,1'-Biphenyl]-4-acetic acid / Felbinac	$C_{14}H_{12}O_2$ / 212.25	5728-52-9 / 160.5	3893	4-09-00-02520	
3010	[1,1'-Biphenyl]-2-amine	$C_{12}H_{11}N$ / 169.23	90-41-5 / 51	299	4-12-00-03223	H_2O 1; EtOH 3; eth 3; bz 3
3011	[1,1'-Biphenyl]-4-amine / p-Biphenylamine	$C_{12}H_{11}N$ / 169.23	92-67-1 / 53.5	1248 / 302	4-12-00-03241	H_2O 2; EtOH 3; eth 3; ace 3
3012	[1,1'-Biphenyl]-4-amine, 4'-chloro- / 4-Amino-4'-chlorodiphenyl	$C_{12}H_{10}ClN$ / 203.67	135-68-2 / 134	445	4-12-00-03269	ace 4; bz 4; eth 4

No.	Name Synonym	Mol. Form. Mol. Wt.	CAS RN mp/°C	Merck No. bp/°C	Beil. Ref. den/g cm^{-3}	Solubility n_D
3013	[1,1'-Biphenyl]-2-amine, 6-methyl- Biphenyl, 2-amino-6-methyl	$C_{13}H_{13}N$ 183.25	76472-83-8 43.5	355; 144[2]	3-12-00-03234 1.9060[0]	bz 4; eth 4; EtOH 4
3014	[1,1'-Biphenyl]-4-amine, 3-methyl-	$C_{13}H_{13}N$ 183.25	63019-98-7 43	190[15]	3-12-00-03238	bz 4; eth 4; EtOH 4
3015	[1,1'-Biphenyl]-4-amine, 4'-methyl- Biphenyl, 4-amino-4'-methyl	$C_{13}H_{13}N$ 183.25	1204-78-0 99	190[18]	2-12-00-00771	ace 4; bz 4; eth 4; EtOH 4
3016	1,1'-Biphenyl, 2-bromo-	$C_{12}H_9Br$ 233.11	2052-07-5 1.5	297	4-05-00-01818 1.2175[26]	eth 4; EtOH 4 1.6248[25]
3017	1,1'-Biphenyl, 3-bromo-	$C_{12}H_9Br$ 233.11	2113-57-7	300; 171[17]	4-05-00-01818	H_2O 1 1.6411[20]
3018	1,1'-Biphenyl, 4-bromo-	$C_{12}H_9Br$ 233.11	92-66-0 91.5	310	4-05-00-01819 0.9327[25]	H_2O 1; EtOH 3; eth 3; bz 3
3019	[1,1'-Biphenyl]-4-butanoic acid	$C_{16}H_{16}O_2$ 240.30	6057-60-9 119.5	190[0.3]	4-09-00-02558	eth 4; EtOH 4
3020	[1,1'-Biphenyl]-2-carbonitrile	$C_{13}H_9N$ 179.22	24973-49-7 41	175[13]	3-09-00-03269	eth 4; EtOH 4
3021	[1,1'-Biphenyl]-2-carboxylic acid	$C_{13}H_{10}O_2$ 198.22	947-84-2 114.3	343.5	4-09-00-02472	H_2O 1; EtOH 4; bz 4; HOAc 4
3022	[1,1'-Biphenyl]-4-carboxylic acid	$C_{13}H_{10}O_2$ 198.22	92-92-2 228	sub	4-09-00-02479	H_2O 1; EtOH 3; eth 3; bz 3
3023	1,1'-Biphenyl, 2-chloro-	$C_{12}H_9Cl$ 188.66	2051-60-7 34	274	4-05-00-01816 1.1499[32]	eth 4; EtOH 4; lig 4
3024	1,1'-Biphenyl, 3-chloro-	$C_{12}H_9Cl$ 188.66	2051-61-8 16	284.5	4-05-00-01816 1.1579[25]	ace 4; eth 4; EtOH 4 1.6181[25]
3025	1,1'-Biphenyl, 4-chloro-	$C_{12}H_9Cl$ 188.66	2051-62-9 78.8	292.9; 146[10]	4-05-00-01816	H_2O 1; EtOH 3; eth 3; lig 3
3026	Biphenyl, 2-chloro-2'-methyl- 2-Chloro-2'-methylbiphenyl	$C_{13}H_{11}Cl$ 202.68	19493-31-3 17	276; 174[50]	3-05-00-01799	1.588[25]
3027	1,1'-Biphenyl, 2,2',3,3',4,4',5,5',6,6'- decafluoro-	$C_{12}F_{10}$ 334.12	434-90-2 67.5	206	4-05-00-01816 1.785[20]	
3028	[1,1'-Biphenyl]-2,2'-diamine	$C_{12}H_{12}N_2$ 184.24	1454-80-4 81	162[4]	4-13-00-00356 1.3090[20]	H_2O 3; ace 3; bz 3
3029	[1,1'-Biphenyl]-2,4'-diamine 2,4'-Biphenyldiamine	$C_{12}H_{12}N_2$ 184.24	492-17-1 54.5	1249 363	4-13-00-00360	H_2O 1; EtOH 3; eth 3
3030	[1,1'-Biphenyl]-3,3'-diamine	$C_{12}H_{12}N_2$ 184.24	2050-89-7 93.5	205[0.001]	4-13-00-00363	H_2O 2; eth 4; bz 3
3031	[1,1'-Biphenyl]-4,4'-diamine Benzidine	$C_{12}H_{12}N_2$ 184.24	92-87-5 120	1086 401	4-13-00-00364	H_2O 2; EtOH 3; eth 2; DMSO 2
3032	[1,1'-Biphenyl]-4,4'-diamine, 2,2'-dichloro- o-Dichlorobenzidine	$C_{12}H_{10}Cl_2N_2$ 253.13	84-68-4 165	3046	4-13-00-00384	eth 4; EtOH 4
3033	[1,1'-Biphenyl]-4,4'-diamine, 3,3'-dimethoxy- o-Dianisidine	$C_{14}H_{16}N_2O_2$ 244.29	119-90-4 137	2970	4-13-00-02834	H_2O 1; EtOH 3; eth 3; ace 3
3034	[1,1'-Biphenyl]-4,4'-diamine, 3,3'-dimethyl- o-Tolidine	$C_{14}H_{16}N_2$ 212.29	119-93-7 131.5	9437	4-13-00-00419	H_2O 2; EtOH 4; eth 4; chl 2
3035	[1,1'-Biphenyl]-4,4'-diamine, N,N'-diphenyl- N,N'-Diphenylbenzidine	$C_{24}H_{20}N_2$ 336.44	531-91-9 247	3319	4-13-00-00368	H_2O 1; EtOH 2; eth 2; bz 2
3036	[1,1'-Biphenyl]-4,4'-diamine, tetramethyl-	$C_{16}H_{20}N_2$ 240.35	34314-06-2 198	>360	3-13-00-00429	bz 4; chl 4
3037	1,1'-Biphenyl, 4,4'-dibromo-	$C_{12}H_8Br_2$ 312.00	92-86-4 164	357.5	4-05-00-01820	H_2O 1; EtOH 2; bz 3
3038	[1,1'-Biphenyl]-2,2'-dicarboxylic acid o,o'-Diphenic acid	$C_{14}H_{10}O_4$ 242.23	482-05-3 233.5	3309 sub	4-09-00-03552	H_2O 1; EtOH 3; eth 3; os 3
3039	[1,1'-Biphenyl]-2,2'-dicarboxylic acid, dimethyl ester	$C_{16}H_{14}O_4$ 270.28	5807-64-7 74	204-6[14]	4-09-00-03553	H_2O 1; EtOH 4; eth 3; bz 3
3040	1,1'-Biphenyl, 3,3'-dichloro-	$C_{12}H_8Cl_2$ 223.10	2050-67-1 29	320	3-05-00-01739	bz 4; eth 4; EtOH 4
3041	1,1'-Biphenyl, 4,4'-dichloro-	$C_{12}H_8Cl_2$ 223.10	2050-68-2 149.3	317	4-05-00-01817 1.4420[0]	H_2O 1; EtOH 2; bz 3; chl 2
3042	1,1'-Biphenyl, 2,2'-difluoro-	$C_{12}H_8F_2$ 190.19	388-82-9 118.5		4-05-00-01815 1.393[20]	H_2O 1; EtOH 3; xyl 3
3043	1,1'-Biphenyl, 3,3'-difluoro-	$C_{12}H_8F_2$ 190.19	396-64-5 8	130[14]	4-05-00-01815 1.192[25]	ctc 2 1.5678[20]
3044	1,1'-Biphenyl, 4,4'-difluoro- 4,4'-Difluorodiphenyl	$C_{12}H_8F_2$ 190.19	398-23-2 94.5	3133 254.5	4-05-00-01816	H_2O 1; EtOH 4; eth 3; ace 3
3045	1,1'-Biphenyl, 2,2'-dimethoxy- 2,2'-Dimethoxybiphenyl	$C_{14}H_{14}O_2$ 214.26	4877-93-4 155	307.5	4-06-00-06645 1.268[25]	H_2O 1; EtOH 4; eth 2; bz 4
3046	1,1'-Biphenyl, 3,3'-dimethoxy- 3,3'-Dimethoxybiphenyl	$C_{14}H_{14}O_2$ 214.26	6161-50-8 36	328	4-06-00-06650	H_2O 1; EtOH 4; eth 4; ace 3
3047	1,1'-Biphenyl, 4,4'-dimethoxy-	$C_{14}H_{14}O_2$ 214.26	2132-80-1 175	sub	4-06-00-06651	H_2O 1; EtOH 4; eth 2; bz 4
3048	1,1'-Biphenyl, 2,2'-dimethoxy-5,5'-dimethyl- Biphenyl, 2,2'-dimethoxy-5,5'-dimethyl	$C_{16}H_{18}O_2$ 242.32	7168-55-0 71	188[12]	3-06-00-05447	ace 4; bz 4; eth 4; EtOH 4
3049	1,1'-Biphenyl, 2,5'-dimethoxy-2',5-dimethyl- Biphenyl, 2,5'-dimethoxy-2',5-dimethyl	$C_{16}H_{18}O_2$ 242.32	72935-12-7 66	168[4]	4-06-00-06704	bz 4; peth 4
3050	1,1'-Biphenyl, 2,2'-dimethyl-	$C_{14}H_{14}$ 182.27	605-39-0 19.5	256	4-05-00-01897 0.9906[20]	H_2O 1; EtOH 4; eth 4; ace 3 1.5752[20]
3051	1,1'-Biphenyl, 2,3'-dimethyl-	$C_{14}H_{14}$ 182.27	611-43-8	270	4-05-00-01902 0.9924[20]	H_2O 1; EtOH 4; eth 4; ace 3 1.5810[20]

No.	Name Synonym	Mol. Form. Mol. Wt.	CAS RN mp/°C	Merck No. bp/°C	Beil. Ref. den/g cm⁻³	Solubility n_D
3052	1,1'-Biphenyl, 2,3-dimethyl- 2,3-Dimethylbiphenyl	$C_{14}H_{14}$ 182.27	3864-18-4 42	141[14]; 98[2]	4-05-00-01897	H_2O 1; eth 3; peth 3 1.5845[23]
3053	1,1'-Biphenyl, 2,4'-dimethyl-	$C_{14}H_{14}$ 182.27	611-61-0 	277.2	4-05-00-01903 0.9924[20]	H_2O 1; EtOH 4; eth 4; ace 3 1.5826[20]
3054	1,1'-Biphenyl, 2,4-dimethyl- 2,4-Dimethylbiphenyl	$C_{14}H_{14}$ 182.27	4433-10-7 	273; 144[20]	4-05-00-01897 0.9947[20]	1.5844[20]
3055	1,1'-Biphenyl, 2,5-dimethyl-	$C_{14}H_{14}$ 182.27	7372-85-2 	140[14.5]	4-05-00-01897 0.9931[20]	1.5819[20]
3056	1,1'-Biphenyl, 2,6-dimethyl-	$C_{14}H_{14}$ 182.27	3976-34-9 -5	262.5	4-05-00-01897 0.9907[20]	1.5745[20]
3057	1,1'-Biphenyl, 3,3'-dimethyl-	$C_{14}H_{14}$ 182.27	612-75-9 9	280	4-05-00-01903 0.9995[20]	H_2O 1; EtOH 4; eth 4; ace 3 1.5946[20]
3058	1,1'-Biphenyl, 3,4-dimethyl-	$C_{14}H_{14}$ 182.27	7383-90-6 14.5	289; 153[15]	4-05-00-01905 0.9978[20]	H_2O 1; bz 3 1.5968[20]
3059	1,1'-Biphenyl, 3,4-dimethyl- 3,4-Dimethylbiphenyl	$C_{14}H_{14}$ 182.27	4433-11-8 29.5	282	4-05-00-01903 1.0087[20]	H_2O 1; bz 3; chl 3 1.6036[20]
3060	1,1'-Biphenyl, 3,5-dimethyl- 3,5-Dimethylbiphenyl	$C_{14}H_{14}$ 182.27	17057-88-4 22.5	274.5	4-05-00-01903 0.9990[20]	1.5952[20]
3061	1,1'-Biphenyl, 4,4'-dimethyl-	$C_{14}H_{14}$ 182.27	613-33-2 125	295	4-05-00-01906 0.9171[21]	H_2O 1; EtOH 2; eth 3; ace 3
3062	1,1'-Biphenyl, 2,2'-dinitro-	$C_{12}H_8N_2O_4$ 244.21	2436-96-6 126	305	4-05-00-01826 1.45[25]	H_2O 1; EtOH 4; eth 3; ace 2
3063	1,1'-Biphenyl, 2,4'-dinitro-	$C_{12}H_8N_2O_4$ 244.21	606-81-5 93.5		4-05-00-01827 1.474[25]	H_2O 1; EtOH 4; eth 3; bz 3
3064	[1,1'-Biphenyl]-2,2'-diol	$C_{12}H_{10}O_2$ 186.21	1806-29-7 109	320	4-06-00-06645 1.3420[20]	H_2O 3; EtOH 3; eth 3; ace 3
3065	[1,1'-Biphenyl]-2,4-diol	$C_{12}H_{10}O_2$ 186.21	611-62-1 162.5	342	3-06-00-05387	eth 4
3066	[1,1'-Biphenyl]-3,3'-diol	$C_{12}H_{10}O_2$ 186.21	612-76-0 124.8	247[18]	3-06-00-05388	H_2O 2; EtOH 3; eth 3; bz 3
3067	[1,1'-Biphenyl]-2,2'-diol, 3,3'-dimethyl-	$C_{14}H_{14}O_2$ 214.26	32750-14-4 113	sub	3-06-00-05445	bz 4; eth 4; EtOH 4
3068	[1,1'-Biphenyl]-2,2'-diol, 5,5'-dimethyl-	$C_{14}H_{14}O_2$ 214.26	15519-73-0 153.5	sub	4-06-00-06705	H_2O 2; EtOH 3; eth 4; ace 3
3069	[1,1'-Biphenyl]-4,4'-diol, 3,3',5,5'-tetramethyl-	$C_{16}H_{18}O_2$ 242.32	2417-04-1 221.8	sub	4-06-00-06744	EtOH 2; bz 2; gl HOAc 2; tol 2
3070	[1,1'-Biphenyl]-4,4'-disulfonic acid 4,4'-Biphenyldisulfonic acid	$C_{12}H_{10}O_6S_2$ 314.34	5314-37-4 72.5	>200	3-11-00-00472	H_2O 4
3071	1,1'-Biphenyl, 3-ethoxy- 3-Ethoxybiphenyl	$C_{14}H_{14}O$ 198.26	54852-73-2 35	305	3-06-00-03313	EtOH 3; eth 3; ace 3; bz 3
3072	1,1'-Biphenyl, 2-ethyl-	$C_{14}H_{14}$ 182.27	1812-51-7 -6	266; 128[11]	4-05-00-01896 0.9966[20]	1.5758[25]
3073	1,1'-Biphenyl, 3-ethyl-	$C_{14}H_{14}$ 182.27	5668-93-9 -27.5	283.5	4-05-00-01896 1.048[25]	1.5859[25]
3074	1,1'-Biphenyl, 2-fluoro-	$C_{12}H_9F$ 172.20	321-60-8 73.5	248	4-05-00-01815 1.2452[25]	EtOH 3; eth 3; chl 3; lig 2
3075	1,1'-Biphenyl, 4-fluoro-	$C_{12}H_9F$ 172.20	324-74-3 74.2	253	4-05-00-01815 1.247[25]	EtOH 2; eth 3; gl HOAc 3
3076	1,1'-Biphenyl, 2,2',4,4',6,6'-hexamethyl-	$C_{18}H_{22}$ 238.37	4482-03-5 103.5	297	4-05-00-01989 1.023[50]	H_2O 1; EtOH 2; eth 3; bz 3
3077	1,1'-Biphenyl, 2-iodo-	$C_{12}H_9I$ 280.11	2113-51-1 	190[36]; 158[6]	4-05-00-01820 1.5991[25]	H_2O 1; EtOH 3; eth 3; bz 3 1.6620[20]
3078	1,1'-Biphenyl, 3-iodo-	$C_{12}H_9I$ 280.11	20442-79-9 26.5	188[16]	4-05-00-01821 1.5967[25]	
3079	1,1'-Biphenyl, 4-iodo-	$C_{12}H_9I$ 280.11	1591-31-7 113.5	320; 183[11]	4-05-00-01821	H_2O 1; EtOH 3; eth 3; bz 3
3080	1,1'-Biphenyl, 2-methoxy-	$C_{13}H_{12}O$ 184.24	86-26-0 29	274	4-06-00-04580 1.0233[99]	H_2O 1; EtOH 3; ctc 2; peth 3 1.5641[99]
3081	1,1'-Biphenyl, 4-methoxy-	$C_{13}H_{12}O$ 184.24	613-37-6 90	157[10]	4-06-00-04600 1.0278[100]	H_2O 1; EtOH 3; eth 3 1.5744[100]
3082	1,1'-Biphenyl, 2-methyl-	$C_{13}H_{12}$ 168.24	643-58-3 -0.2	255.5	4-05-00-01855 1.010[22]	H_2O 1; EtOH 3; eth 3 1.5914[20]
3083	1,1'-Biphenyl, 3-methyl-	$C_{13}H_{12}$ 168.24	643-93-6 4.5	272.7	4-05-00-01858 1.0182[17]	H_2O 1; EtOH 3; eth 3; ctc 3 1.5972[20]
3084	1,1'-Biphenyl, 4-methyl-	$C_{13}H_{12}$ 168.24	644-08-6 49.5	267.5	4-05-00-01860 1.015[27]	H_2O 1; EtOH 3; eth 3; ctc 2
3085	1,1'-Biphenyl, 2-nitro- o-Nitrobiphenyl	$C_{12}H_9NO_2$ 199.21	86-00-0 37.2	6513 320	4-05-00-01823 1.44[25]	H_2O 1; EtOH 3; eth 3; chl 3
3086	1,1'-Biphenyl, 3-nitro-	$C_{12}H_9NO_2$ 199.21	2113-58-8 62	227[35]; 143[9]	4-05-00-01823	H_2O 1; EtOH 3; eth 3; HOAc 3
3087	1,1'-Biphenyl, 4-nitro- p-Nitrobiphenyl	$C_{12}H_9NO_2$ 199.21	92-93-3 114	6514 340	4-05-00-01823	H_2O 1; EtOH 2; eth 3; bz 3
3088	[1,1'-Biphenyl]-2-ol o-Phenylphenol	$C_{12}H_{10}O$ 170.21	90-43-7 59	7276 286	4-06-00-04579 1.213[25]	H_2O 1; EtOH 3; eth 4; ace 3
3089	[1,1'-Biphenyl]-3-ol m-Phenylphenol	$C_{12}H_{10}O$ 170.21	580-51-8 78	>300	4-06-00-04597	H_2O 2; EtOH 4; eth 4; bz 4
3090	[1,1'-Biphenyl]-4-ol p-Phenylphenol	$C_{12}H_{10}O$ 170.21	92-69-3 166	7277 305	4-06-00-04600	H_2O 2; EtOH 4; eth 4; chl 4

No.	Name Synonym	Mol. Form. Mol. Wt.	CAS RN mp/°C	Merck No. bp/°C	Beil. Ref. den/g cm^{-3}	Solubility n_D
3091	[1,1'-Biphenyl]-4-ol, acetate	$C_{14}H_{12}O_2$ 212.25	148-86-7 89	196[13]	3-06-00-03326	H_2O 2; EtOH 2; eth 3; bz 3
3092	[1,1'-Biphenyl]-2-ol, 3-chloro- 2-Phenyl-6-chlorophenol	$C_{12}H_9ClO$ 204.66	85-97-2 6	7251 317 dec	4-06-00-04586 1.24[25]	H_2O 1; EtOH 3; eth 3; ace 3 1.6237[30]
3093	1,1'-Biphenyl, 2-(phenylmethyl)-	$C_{19}H_{16}$ 244.34	606-97-3 55	283[110]	3-05-00-02323	bz 4; eth 4; EtOH 4
3094	1,1'-Biphenyl, 4-(phenylmethyl)-	$C_{19}H_{16}$ 244.34	613-42-3 85	285-6[110]	4-05-00-02502 1.171[0]	H_2O 1; EtOH 3; eth 4; bz 4
3095	1,1'-Biphenyl, 2,2',4,4'-tetramethyl-	$C_{16}H_{18}$ 210.32	3976-36-1 41	287	3-05-00-01891	EtOH 3; eth 2; bz 2; chl 3
3096	1,1'-Biphenyl, 2,2',5,5'-tetramethyl-	$C_{16}H_{18}$ 210.32	3075-84-1 51.5	285	3-05-00-01892	eth 3; bz 3
3097	[1,1'-Biphenyl]-3,3',5,5'-tetrol Diresorcinol	$C_{12}H_{10}O_4$ 218.21	531-02-2 310	3361	2-06-00-01129	H_2O 4; eth 4; EtOH 4
3098	2,3'-Bipiperidine 2,3-Bipiperidyl	$C_{10}H_{20}N_2$ 168.28	2467-09-6 68.5	269.5	4-23-00-00512	H_2O 4; EtOH 4
3099	2,2'-Bipyridine α,α'-Dipyridyl	$C_{10}H_8N_2$ 156.19	366-18-7 72	3355 273.5	5-23-08-00016	H_2O 2; EtOH 4; eth 4; bz 4
3100	2,3'-Bipyridine 2,3'-Bipyridyl	$C_{10}H_8N_2$ 156.19	581-50-0 	295.5	5-23-08-00026 1.140[20]	H_2O 1; EtOH 4; eth 4; bz 4 1.6223[20]
3101	2,4'-Bipyridine	$C_{10}H_8N_2$ 156.19	581-47-5 61.5	281	5-23-08-00027	H_2O 4; EtOH 4; eth 4; chl 4
3102	3,3'-Bipyridine 3,3'-Dipyridyl	$C_{10}H_8N_2$ 156.19	581-46-4 68	291.5	5-23-08-00027 1.1614[20]	H_2O 4; EtOH 4; eth 2
3103	3,4'-Bipyridine 3,4'-Bipyridyl	$C_{10}H_8N_2$ 156.19	4394-11-0 62	297; 144[15]	5-23-08-00028	EtOH 4; peth 4
3104	4,4'-Bipyridine γ,γ'-Dipyridyl	$C_{10}H_8N_2$ 156.19	553-26-4 111	3356 305	5-23-08-00028	H_2O 2; EtOH 4; eth 3; bz 4
3105	2,3'-Bipyridine, 1,2,3,6-tetrahydro-, (S)- Anatabine	$C_{10}H_{12}N_2$ 160.22	581-49-7 	661 145-6[10]	5-23-06-00507 1.091[19]	H_2O 5; EtOH 3; eth 3; bz 3 1.5676[20]
3106	[5,5'-Bipyrimidine]- 2,2',4,4',6,6'(1H,1'H,3H,3'H,5H,5'H)-hexone- 5,5'-dihydroxy Alloxantin	$C_8H_6N_4O_8$ 286.16	76-24-4 253-5 dec	282	4-26-00-02782	H_2O 2; EtOH 2; eth 2
3107	2,3'-Biquinoline 2,3'-Biquinolyl	$C_{18}H_{12}N_2$ 256.31	612-81-7 175.8	>400	5-23-10-00009	bz 4; eth 4; EtOH 4; chl 4
3108	Bis[1,3]benzodioxolo[4,5-c:5',6'-g]azecin- 13(5H)-one, 4,6,7,14-tetrahydro-5,14- dimethyl-, (R)- Corycavamine	$C_{21}H_{21}NO_5$ 367.40	521-85-7 149	2540	2-27-00-00621	EtOH 4; chl 4
3109	Bis[1,3]benzodioxolo[4,5-c:5',6'-g]azecin- 13(5H)-one, 4,6,7,14-tetrahydro-5-methyl- Protopine	$C_{20}H_{19}NO_5$ 353.37	130-86-9 208	7912	4-27-00-06881	H_2O 1; EtOH 2; eth 2; bz 2
3110	3H-Bis[1]benzopyrano[3,4-b:6',5'-e]pyran- 7(7aH)-one, 13,13a-dihydro-7a-hydroxy-9,10- dimethoxy-3,3-dimethyl Tephrosin	$C_{23}H_{22}O_7$ 410.42	76-80-2 198	9082	4-19-00-05271	ace 4; eth 4; chl 4
3111	Bismuthine, triphenyl- Triphenylbismuth	$C_{18}H_{15}Bi$ 440.30	603-33-8 77.6	242[14]	4-16-00-01226 1.715[75]	EtOH 2; eth 3; ace 3; bz 3 1.7040[75]
3112	2,2'-Bithiophene	$C_8H_6S_2$ 166.27	492-97-7 33	260	5-19-01-00537	H_2O 1; EtOH 4; eth 3; ctc 3
3113	Boranamine, N,N,1,1-tetramethyl-	$C_4H_{12}BN$ 84.96	1113-30-0 -92	65	4-04-00-04367	eth 4; ace 4; os 5
3114	Boranediamine, 1-fluoro-N,N,N',N'- tetramethyl-	$C_4H_{12}BFN_2$ 117.96	383-90-4 -44.3	106	4-04-00-00303	ace 4; eth 4
3115	Borane, difluoro(4-methylphenyl)-	$C_7H_7BF_2$ 139.94	768-39-8 	128	4-16-00-01674 1.055[25]	bz 4; eth 4 1.4535[25]
3116	Borane, difluorophenyl-	$C_6H_5BF_2$ 125.91	368-98-9 -36.2	98	4-16-00-01661 1.087[25]	bz 4; eth 4 1.4441[25]
3117	Borane, dimethoxy- Dimethoxyborane	$C_2H_7BO_2$ 73.89	4542-61-4 -130.6	25.9	4-01-00-01269	H_2O 6
3118	Borane, triethyl Triethylborane	$C_6H_{15}B$ 98.00	97-94-9 -93	95	0.70[23]	EtOH 3; eth 3 1.3971
3119	Borane, trimethyl- Trimethylborane	C_3H_9B 55.92	593-90-8 -161.5	-20.2		
3120	Borane, trimethyl-, monoammoniate Aminetrimethylboron	$C_3H_{12}BN$ 72.95	1830-95-1 73.5			
3121	Borane, tripropyl-	$C_9H_{21}B$ 140.08	1116-61-6 -56	159	4-04-00-04360 0.7204[25]	
3122	Borane, tris(3-methylbutyl)- Borane, triisopentyl-	$C_{15}H_{33}B$ 224.24	3062-81-5 	119[14]	2-04-00-01023 0.7600[25]	ace 4; eth 4; EtOH 4 1.4321
3123	Borane, tris(2-methylpropyl)-	$C_{12}H_{27}B$ 182.16	1116-39-8 	188; 86[20]	4-04-00-04361 0.7380[25]	bz 4; eth 4; EtOH 4 1.4188[23]
3124	Boric acid, tributyl ester	$C_{12}H_{27}BO_3$ 230.16	688-74-4 	234	4-01-00-01544 0.8567[20]	EtOH 3; eth 4; bz 3; MeOH 4 1.4106[18]
3125	Boric acid, triethyl ester	$C_6H_{15}BO_3$ 145.99	150-46-9 -84.8	120	4-01-00-01365 0.8546[20]	EtOH 5; eth 5 1.3749[20]
3126	Boric acid, trimethyl ester Trimethyl borate	$C_3H_9BO_3$ 103.91	121-43-7 -29.3	9626 67.5	4-01-00-01269 0.915[25]	eth 4; MeOH 4 1.3568[20]

No.	Name Synonym	Mol. Form. Mol. Wt.	CAS RN mp/°C	Merck No. bp/°C	Bell. Ref. den/g cm^{-3}	Solubility n_D
3127	Boric acid, tripropyl ester	$C_9H_{21}BO_3$ 188.07	688-71-1	179.5	4-01-00-01436 0.8576[20]	EtOH 4; eth 5; PrOH 3 1.3948[20]
3128	Boric acid, tris(1-methylethyl) ester Isopropyl borate	$C_9H_{21}BO_3$ 188.07	5419-55-6	140; 75[76]	4-01-00-01488 0.8251[20]	EtOH 4; eth 4; bz 4; PrOH 4 1.3777[20]
3129	Borinic acid, dimethylthio-, methyl ester	C_3H_9BS 87.98	19163-05-4 -84	71	4-04-00-04368	ace 4; eth 4
3130	Borinic acid, diphenyl-	$C_{12}H_{11}BO$ 182.03	2622-89-1 57.5	150-5[20]	4-16-00-01639 1.0740[20]	EtOH 3; bz 3; peth 3
3131	3-Bornanecarboxylic acid Camphane-3-carboxylic acid	$C_{11}H_{18}O_2$ 182.26	91965-23-0 90.5	153[13]	3-09-00-00244	EtOH 4; HOAc 4
3132	Boron, (N,N-dimethylmethanamine)trihydro-, (T-4)- Trimethylamineborane	$C_3H_{12}BN$ 72.95	75-22-9 94	172	4-04-00-00134 0.792[25]	eth 4; EtOH 4
3133	Boronic acid, ethyl-	$C_2H_7BO_2$ 73.89	4433-63-0 40	dec	4-04-00-04379	H_2O 4; eth 4; EtOH 4
3134	Boronic acid, (2-methylphenyl)-	$C_7H_9BO_2$ 135.96	16419-60-6 166.5	dec	3-16-00-01277	bz 4; eth 4; EtOH 4
3135	Boronic acid, phenyl- Benzeneboronic acid	$C_6H_7BO_2$ 121.93	98-80-6 219	1076	4-16-00-01654	H_2O 2; EtOH 3; eth 3; bz 3
3136	Boronic acid, propyl-	$C_3H_9BO_2$ 87.91	17745-45-8 107	dec	3-04-00-01964	H_2O 4; eth 4; EtOH 4
3137	Boron, trifluoro[1,1'-oxybis[ethane]]-, (T-4)- Boron trifluoride etherate	$C_4H_{10}BF_3O$ 141.93	109-63-7 -60.4	1352 125.5	4-01-00-01314 1.3572[20]	eth 4; EtOH 4 1.4447[20]
3138	Boron, trifluoro[oxybis[methane]]-, (T-4)- Boron trifluoride compd. with methyl ether	$C_2H_6BF_3O$ 113.88	353-42-4 -14	127 dec	4-01-00-01245 1.2410[20]	1.302[20]
3139	Bromacil 2,4(1H,3H)-Pyrimidinedione, 5-bromo-6- methyl-3-(1-methylpropyl)-	$C_9H_{13}BrN_2O_2$ 261.12	314-40-9 158	1370	1.55[25]	
3140	Bromopropylate Benzeneacetic acid, 4-bromo-α-(4- bromophenyl)-α-hydroxy-, 1-methylethyl ester	$C_{17}H_{16}Br_2O_3$ 428.12	18181-80-1 77	1422	1.59[20]	
3141	Bufa-20,22-dienolide, 16-(acetyloxy)-3,14- dihydroxy-, (3β,5β,16β)- Bufotalin	$C_{26}H_{36}O_6$ 444.57	471-95-4 223 dec	1466	5-18-04-00443	H_2O 1; EtOH 3; chl 3
3142	Bufa-4,20,22-trienolide, 6-(acetyloxy)-3-(β-D- glucopyranosyloxy)-8,14-dihydroxy-, (3β,6β)- Scilliroside	$C_{32}H_{44}O_{12}$ 620.69	507-60-8 169	8356 dec	4-18-00-03178	H_2O 2; EtOH 4; eth 1; ace 2
3143	sec-Bumeton 1,3,5-Triazine-2,4-diamine, N-ethyl-6- methoxy-N'-(1-methylpropyl)-	$C_{10}H_{19}N_5O$ 225.29	26259-45-0 87			
3144	Butachlor Acetamide, N-(butoxymethyl)-2-chloro-N-(2,6- diethylphenyl)-	$C_{17}H_{26}ClNO_2$ 311.85	23184-66-9 <-5	1498 156.[5]	1.070[25]	
3145	1,2-Butadiene Methylallene	C_4H_6 54.09	590-19-2 -136.2	10.9	4-01-00-00975 0.676[0]	H_2O 1; EtOH 5; eth 5; bz 4 1.4205[1]
3146	1,3-Butadiene Divinyl	C_4H_6 54.09	106-99-0 -108.9	1500 -4.4	4-01-00-00976 *0.6149[25]	H_2O 1; EtOH 3; eth 3; ace 4 1.4292[-25]
3147	1,2-Butadiene, 4-bromo- 4-Bromo-1,2-butadiene	C_4H_5Br 132.99	18668-68-3	110	3-01-00-00929	ace 4 1.5248[20]
3148	1,3-Butadiene, 1-bromo- 1-Bromo-1,3-butadiene	C_4H_5Br 132.99	21890-35-7	93	4-01-00-00989 1.416[0]	eth 4; EtOH 4; peth 4
3149	1,3-Butadiene, 2-bromo-	C_4H_5Br 132.99	1822-86-2	42-3[165]	4-01-00-00989 1.397[20]	eth 4; EtOH 4 1.4988[20]
3150	1,2-Butadiene, 4-chloro-	C_4H_5Cl 88.54	25790-55-0	88	4-01-00-00975 0.9891[20]	ace 4; bz 4; eth 4 1.4775[20]
3151	1,3-Butadiene, 1-chloro-	C_4H_5Cl 88.54	627-22-5	68	3-01-00-00949 0.9606[20]	eth 4; EtOH 4; chl 4 1.4712[20]
3152	1,3-Butadiene, 2-chloro- 2-Chloro-1,3-butadiene	C_4H_5Cl 88.54	126-99-8 -130	59.4	4-01-00-00984 0.956[20]	H_2O 2; eth 5; ace 5; bz 5 1.4583[20]
3153	1,3-Butadiene, 1-chloro-2-methyl- Chloromethylbutadiene	C_5H_7Cl 102.56	35383-51-8	107	3-01-00-00974 0.9710[20]	ace 4; EtOH 4 1.4702[20]
3154	1,3-Butadiene, 1-chloro-3-methyl- 4-Chloroisoprene	C_5H_7Cl 102.56	51034-46-9	100.7	4-01-00-01005 0.9543[20]	ace 4; eth 4; EtOH 4; chl 4 1.4719[20]
3155	1,3-Butadiene, 2-chloro-3-methyl-	C_5H_7Cl 102.56	1809-02-5	93	4-01-00-01004 0.9593[20]	H_2O 1; EtOH 3; eth 4; ace 3 1.4686[20]
3156	1,3-Butadiene, 1,1-dichloro- 1,1-Dichloro-1,3-butadiene	$C_4H_4Cl_2$ 122.98	6061-06-9	42-3[90]	4-01-00-00985 1.1831[20]	H_2O 1; eth 4; bz 4; chl 4 1.5022[20]
3157	1,3-Butadiene, 1,2-dichloro- 1,2-Dichloro-1,3-butadiene	$C_4H_4Cl_2$ 122.98	3574-40-1	62[105]; 35[40]	4-01-00-00985 1.1991[20]	ctc 4 1.4960[20]
3158	1,3-Butadiene, 2,3-dichloro-	$C_4H_4Cl_2$ 122.98	1653-19-6	98	4-01-00-00986 1.1829[20]	chl 4 1.4890[20]
3159	1,3-Butadiene, 2,3-dimethyl- Diisopropenyl	C_6H_{10} 82.15	513-81-5 -76	3228 68.8	4-01-00-01023 0.7222[25]	ctc 3 1.4394[20]
3160	1,3-Butadiene, 1-ethoxy-	$C_6H_{10}O$ 98.14	5614-32-4	110.5	4-01-00-02222 0.8154[20]	ace 4; bz 4; eth 4; EtOH 4 1.4529[20]
3161	1,3-Butadiene, 2-ethoxy-	$C_6H_{10}O$ 98.14	4747-05-1	94.5	4-01-00-02223 0.8177[20]	ace 4; bz 4; eth 4; EtOH 4 1.4400[20]
3162	1,3-Butadiene, 2-fluoro- Fluoroprene	C_4H_5F 72.08	381-61-3	12	4-01-00-00982 0.843[4]	1.400[4]
3163	1,3-Butadiene, 1,1,2,3,4,4-hexachloro- Hexachloro-1,3-butadiene	C_4Cl_6 260.76	87-68-3 -21	215	4-01-00-00988 1.556[25]	H_2O 1; EtOH 3; eth 3 1.5542[20]

No.	Name / Synonym	Mol. Form. / Mol. Wt.	CAS RN / mp/°C	Merck No. / bp/°C	Beil. Ref. / den/g cm^{-3}	Solubility / n_D
3164	1,3-Butadiene, 1,1,2,3,4,4-hexafluoro- 1,3-Butadiene, hexafluoro	C$_4$F$_6$ 162.03	685-63-2 -132	6	4-01-00-00983 1.553[20]	1.378[20]
3165	1,3-Butadiene, 2-iodo-	C$_4$H$_5$I 179.99	19221-28-4	112	3-01-00-00956 1.7278[20]	1.5616
3166	1,2-Butadiene, 4-iodo- 4-Iodo-1,2-butadiene	C$_4$H$_5$I 179.99	67885-08-9	130	3-01-00-00929 1.7129[20]	1.5709[20]
3167	1,2-Butadiene, 4-methoxy-	C$_5$H$_8$O 84.12	36678-06-5	88	4-01-00-02221 0.8286[20]	EtOH 4 1.435[20]
3168	1,3-Butadiene, 1-methoxy-	C$_5$H$_8$O 84.12	3036-66-6	91.5	3-01-00-01975 0.8296[20]	H$_2$O 3; EtOH 3 1.4594[20]
3169	1,3-Butadiene, 2-methoxy-	C$_5$H$_8$O 84.12	3588-30-5	75	4-01-00-02223 0.8272[20]	ace 4; bz 4; eth 4; EtOH 4 1.4442[20]
3170	1,2-Butadiene, 3-methyl-	C$_5$H$_8$ 68.12	598-25-4 -113.6	40.83	4-01-00-01006 0.6806[25]	ace 4; bz 4; eth 4; EtOH 4 1.4203[20]
3171	1,3-Butadiene, 2-methyl- Isoprene	C$_5$H$_8$ 68.12	78-79-5 -145.9	5087 34.0	4-01-00-01001 0.679[20]	H$_2$O 1; EtOH 5; eth 5; ace 5 1.4219[20]
3172	1,3-Butadiene, 1,1,2,4,4-pentafluoro-3-(trifluoromethyl)- 1,3-Butadiene, pentafluoro-2-(trifluoromethyl)-	C$_5$F$_8$ 212.04	384-04-3	39	1.527[0]	ace 4; bz 4; eth 4 1.3000[0]
3173	1,3-Butadiene, 1,2,3,4-tetrachloro-	C$_4$H$_2$Cl$_4$ 191.87	1637-31-6 -4	188	4-01-00-00987 1.515[15]	H$_2$O 1; EtOH 3; eth 3; ace 3 1.5455[20]
3174	1,3-Butadiene, 1,2,3-trichloro- 1,2,3-Trichloro-1,3-butadiene	C$_4$H$_3$Cl$_3$ 157.43	1573-58-6	33-4[7]	3-01-00-00954 1.4060[20]	eth 4; chl 4 1.5262[20]
3175	2,3-Butadien-1-ol	C$_4$H$_6$O 70.09	18913-31-0	128.5	4-01-00-02221 0.9164[20]	ace 4; bz 4; eth 4; EtOH 4 1.4759[20]
3176	1,3-Butadien-1-ol, acetate	C$_8$H$_8$O$_2$ 112.13	1515-76-0	60-1[40]	3-02-00-00295 0.945[25]	1.4690[20]
3177	1,3-Butadiyne	C$_4$H$_2$ 50.06	460-12-8 -36.4	10.3	4-01-00-01116 0.7364[0]	H$_2$O 4; EtOH 3; eth 4; ace 4 1.4189[5]
3178	Butanal Butyraldehyde	C$_4$H$_8$O 72.11	123-72-8 -99	1591 74.8	4-01-00-03229 0.8016[20]	H$_2$O 3; EtOH 5; eth 5; ace 4 1.3843[20]
3179	Butanal, 2-bromo-	C$_4$H$_7$BrO 151.00	24764-97-4	33[17]	4-01-00-03241 1.469[20]	eth 3; ace 3; bz 3; chl 3 1.4683[20]
3180	Butanal, 2-chloro-	C$_4$H$_7$ClO 106.55	28832-55-5	107	4-01-00-03238 1.1072[17]	1.4262[17]
3181	Butanal, 4-chloro-	C$_4$H$_7$ClO 106.55	6139-84-0	50-1[13]	4-01-00-03240 1.108[8]	ace 4; eth 4; EtOH 4 1.4466[8]
3182	Butanal, 2,3-dichloro-	C$_4$H$_6$Cl$_2$O 141.00	55775-41-2	58-60[20]	4-01-00-03240 1.2666[21]	ace 4; eth 4; EtOH 4; chl 4 1.4618[21]
3183	Butanal, 2-ethyl- Diethylacetaldehyde	C$_6$H$_{12}$O 100.16	97-96-1	117-9[160]	4-01-00-03310 0.8110[20]	H$_2$O 2; EtOH 5; eth 5; ctc 2 1.4025[20]
3184	Butanal, 3-hydroxy- Aldol	C$_4$H$_8$O$_2$ 88.11	107-89-1	217 83[20]	4-01-00-03984 1.103[20]	H$_2$O 5; EtOH 5; eth 3; ace 4 1.4238[20]
3185	Butanal, 2-methyl-, (±)-	C$_5$H$_{10}$O 86.13	57456-98-1	92.5	4-01-00-03285 0.8029[20]	ace 4; eth 4; EtOH 4 1.3869[20]
3186	Butanal, 3-methyl- Isovaleraldehyde	C$_5$H$_{10}$O 86.13	590-86-3 -51	5118 92.5	4-01-00-03291 0.7977[20]	H$_2$O 2; EtOH 3; eth 3 1.3902[20]
3187	Butanal, 3-methyl-, oxime Butyraldehyde, 3-methyl, oxime	C$_5$H$_{11}$NO 101.15	626-90-4 48.5	161.3	4-01-00-03293 0.8934[20]	ace 4; eth 4; EtOH 4 1.4367[20]
3188	Butanal, oxime Butyraldehyde, oxime	C$_4$H$_9$NO 87.12	110-69-0 -29.5	154	4-01-00-03234 0.923[20]	H$_2$O 4; EtOH 5; eth 4; ace 4
3189	Butanal, phenylhydrazone Butyraldehyde, phenylhydrazone	C$_{10}$H$_{14}$N$_2$ 162.23	940-54-5 94	192[80]; 152[14]	3-15-00-00080	ctc 2
3190	Butanal, 2-(phenylmethylene)-	C$_{11}$H$_{12}$O 160.22	28467-92-7	157-8[5]	4-07-00-01024 1.0201[22]	1.578[20]
3191	Butanal, 2,2,3-trichloro- 2,2,3-Trichlorobutyraldehyde	C$_4$H$_5$Cl$_3$O 175.44	76-36-8	9545 164	4-01-00-03241 1.3956[20]	H$_2$O 4; eth 4; EtOH 4 1.4755[20]
3192	Butanal, 2,3,4-trihydroxy-, [R-(R*,R*)]- D-Erythrose	C$_4$H$_8$O$_4$ 120.11	583-50-6	3637	4-01-00-04172	H$_2$O 3; EtOH 4
3193	Butanal, 2,3,4-trihydroxy-, [S-(R*,S*)]- Threose (D-)	C$_4$H$_8$O$_4$ 120.11	95-43-2 129	9317	4-01-00-04173	
3194	Butanamide Butyramide	C$_4$H$_9$NO 87.12	541-35-5 114.8	1592 216	4-02-00-00804 0.8850[120]	H$_2$O 2; EtOH 3; eth 2; bz 1 1.4087[130]
3195	Butanamide, N-(aminocarbonyl)-2-bromo-2-ethyl- Carbromal	C$_7$H$_{13}$BrN$_2$O$_2$ 237.10	77-65-6 118	1837	4-03-00-00117 1.544[25]	H$_2$O 2; ace 3; bz 3; chl 2
3196	Butanamide, N-(aminocarbonyl)-2-bromo-3-methyl- Bromisovalum	C$_6$H$_{11}$BrN$_2$O$_2$ 223.07	496-67-3 154	1385 sub	3-03-00-00123 1.56[15]	ace 4; bz 4; eth 4; EtOH 4
3197	Butanamide, N-[5-(aminosulfonyl)-1,3,4-thiadiazol-2-yl]- Butazolamide	C$_6$H$_{10}$N$_4$O$_3$S$_2$ 250.30	16790-49-1	1511	4-27-00-08223	
3198	Butanamide, 2-bromo-2-ethyl- Diethylbromoacetamide	C$_6$H$_{12}$BrNO 194.07	511-70-6 67	3105	3-02-00-00755	H$_2$O 2; EtOH 4; eth 4; bz 4
3199	Butanamide, 2-bromo-2-ethyl-3-methyl- Ibrotamide	C$_7$H$_{14}$BrNO 208.10	466-14-8 50.5	4809 sub	4-02-00-00979	ace 4; bz 4; eth 4; EtOH 4
3200	Butanamide, N,N-diethyl-	C$_8$H$_{17}$NO 143.23	1114-76-7	206	4-04-00-00354	H$_2$O 4; EtOH 4 1.4403[25]
3201	Butanamide, N,N-diethyl-3-methyl- Isovaleryl diethylamide	C$_9$H$_{19}$NO 157.26	533-32-4 211	5122	4-04-00-00356 0.8764[20]	eth 4; EtOH 4 1.4422[20]

No.	Name / Synonym	Mol. Form. / Mol. Wt.	CAS RN / mp/°C	Merck No. / bp/°C	Beil. Ref. / den/g cm⁻³	Solubility / n_D
3202	Butanamide, 2,4-dihydroxy-N-(3-hydroxypropyl)-3,3-dimethyl-, (R)- Dexpanthenol	$C_9H_{19}NO_4$ 205.25	81-13-0	2924 dec	4-04-00-01652 1.20[20]	H_2O 4; EtOH 4; eth 2; MeOH 4 1.497[20]
3203	Butanamide, N,N-dimethyl-	$C_6H_{13}NO$ 115.18	760-79-2 -40	186; 125[100]	4-04-00-00185 0.9064[25]	ace 4; bz 4; eth 4; EtOH 4 1.4391[25]
3204	Butanamide, N,N'-(3,3'-dimethyl[1,1'-biphenyl]-4,4'-diyl)bis[3-oxo- N,N'-Bis(acetoacetyl)-3,3'-dimethylbenzidine	$C_{22}H_{24}N_2O_4$ 380.44	91-96-3 212		3-13-00-00490	DMSO 2
3205	Butanamide, N-(4-hydroxyphenyl)- 4'-Hydroxybutyranilide	$C_{10}H_{13}NO_2$ 179.22	101-91-7 139.5	4745	4-13-00-01109	H_2O 4; EtOH 4
3206	Butanamide, 3-methyl- Isovaleramide	$C_5H_{11}NO$ 101.15	541-46-8 137	5119 226	4-02-00-00902	H_2O 3; EtOH 3; eth 3; peth 4
3207	Butanamide, 3-oxo-N-phenyl- Acetoacetanilide	$C_{10}H_{11}NO_2$ 177.20	102-01-2 86	50	4-12-00-00955	H_2O 2; EtOH 3; eth 3; bz 3
3208	Butanamide, N-phenyl-	$C_{10}H_{13}NO$ 163.22	1129-50-6 97	189[15]	4-12-00-00387 1.134[25]	H_2O 1; EtOH 4; eth 4; chl 2
3209	1-Butanamine Butylamine	$C_4H_{11}N$ 73.14	109-73-9 -49.1	1543 77.0	4-04-00-00540 0.7414[20]	H_2O 5; EtOH 3; eth 3 1.4031[20]
3210	2-Butanamine, (±)- sec-Butylamine, (±)-	$C_4H_{11}N$ 73.14	33966-50-6 <-72	1544 63.5	4-04-00-00618 0.7246[20]	H_2O 3; EtOH 5; eth 5; ace 4 1.3932[20]
3211	1-Butanamine, N-butyl- Dibutylamine	$C_8H_{19}N$ 129.25	111-92-2 -62	3019 159.6	4-04-00-00550 0.7670[20]	H_2O 1; EtOH 4; eth 4; ace 3 1.4177[20]
3212	1-Butanamine, N-butyl-N-nitroso- Dibutylnitrosamine	$C_8H_{18}N_2O$ 158.24	924-16-3 105[8]		4-04-00-03389	
3213	1-Butanamine, N,N-dibutyl- Tributylamine	$C_{12}H_{27}N$ 185.35	102-82-9 -70	9530 216.5	4-04-00-00554 0.7770[20]	H_2O 2; EtOH 4; eth 4; ace 3 1.4299[20]
3214	1-Butanamine, 4,4-diethoxy-	$C_8H_{19}NO_2$ 161.24	6346-09-4 196		4-04-00-01928 0.933[25]	1.4275[20]
3215	1-Butanamine, N,N-dimethyl-	$C_6H_{15}N$ 101.19	927-62-8 95		4-04-00-00546 0.7206[20]	H_2O 5; EtOH 5; eth 5; ace 5 1.3970[20]
3216	2-Butanamine, 2,3-dimethyl-	$C_6H_{15}N$ 101.19	4358-75-2 -70.5	104.5	4-04-00-00733 0.7683[0]	1.4096[17]
3217	2-Butanamine, 3,3-dimethyl-	$C_6H_{15}N$ 101.19	3850-30-4 -20	102	4-04-00-00730 0.7668[20]	H_2O 4 1.4105[25]
3218	1-Butanamine, N-ethyl- Butylethylamine	$C_6H_{15}N$ 101.19	13360-63-9	107.5	4-04-00-00547 0.7398[20]	EtOH 5; eth 5; ace 5; bz 5 1.4040[20]
3219	2-Butanamine, N-ethyl-, (±)- sec-Butyl ethyl amine (DL)	$C_6H_{15}N$ 101.19	116724-10-8 -104.3	98	2-04-00-00636 0.7358[20]	ace 4; bz 4; eth 4; EtOH 4
3220	1-Butanamine, 3-methyl- Isoamylamine	$C_5H_{13}N$ 87.16	107-85-7 96	4994	4-04-00-00696 0.7505[20]	H_2O 5; EtOH 5; eth 5; ace 3 1.4083[20]
3221	1-Butanamine, N-methyl- Butylmethylamine	$C_5H_{13}N$ 87.16	110-68-9 91		4-04-00-00546 0.7637[15]	
3222	2-Butanamine, 2-methyl-	$C_5H_{13}N$ 87.16	594-39-8 -105	77	4-04-00-00694 0.731[25]	H_2O 4; ace 4; eth 4; EtOH 4 1.3954[25]
3223	2-Butanamine, 3-methyl-	$C_5H_{13}N$ 87.16	598-74-3 -50	85.5	2-04-00-00644 0.7574[19]	H_2O 4; EtOH 3 1.4096[18]
3224	1-Butanamine, 2-methyl-N,N-bis(2-methylbutyl)-	$C_{15}H_{33}N$ 227.43	620-43-9 232		0-04-00-00179 0.9[13]	ace 4; bz 4; eth 4; EtOH 4 1.4330[20]
3225	1-Butanamine, 3-methyl-N,N-bis(3-methylbutyl)-	$C_{15}H_{33}N$ 227.43	645-41-0 235		4-04-00-00700 0.7848[20]	H_2O 1; EtOH 4; eth 5; bz 5 1.4331[20]
3226	1-Butanamine, 3-methyl-N-(3-methylbutyl)- Diisoamylamine	$C_{10}H_{23}N$ 157.30	544-00-3 -44	3178 188	4-04-00-00699 0.7672[21]	H_2O 1; EtOH 3; eth 5 1.4235[20]
3227	2-Butanamine, N-(1-methylpropyl)-	$C_8H_{19}N$ 129.25	626-23-3 135		4-04-00-00620 0.7534[20]	H_2O 4; EtOH 3; os 3 1.4162[20]
3228	1-Butanamine, 1,1,2,2,3,3,4,4,4-nonafluoro-N,N-bis(nonafluorobutyl)-	$C_{12}F_{27}N$ 671.10	311-89-7 178		4-02-00-00819 1.884[25]	ace 3 1.291[25]
3229	Butane	C_4H_{10} 58.12	106-97-8 -138.2	1507 -0.5	4-01-00-00236 *0.573[25]	H_2O 3; EtOH 4; eth 4; chl 4 1.3326[20]
3230	Butane, 2,2-bis(ethylsulfonyl)- Sulfonethylmethane	$C_8H_{18}O_4S_2$ 242.36	76-20-0 76	8937 dec	3-01-00-02790 1.199[85]	chl 3
3231	Butane, 1-bromo- Butyl bromide	C_4H_9Br 137.02	109-65-9 -112.4	1553 101.6	4-01-00-00258 1.2758[20]	H_2O 1; EtOH 5; eth 5; ace 5 1.4401[20]
3232	Butane, 2-bromo-, (±)- (±)-sec-Butyl bromide	C_4H_9Br 137.02	5787-31-5 -111.9	1554 91.2	4-01-00-00261 1.2585[20]	ace 4; eth 4; chl 4 1.4366[20]
3233	Butane, 1-bromo-4-chloro-	C_4H_8BrCl 171.46	6940-78-9	175; 63[10]	4-01-00-00264 1.488[20]	H_2O 1; EtOH 3; eth 3; ctc 2 1.4885[20]
3234	Butane, 2-bromo-1-chloro- 2-Bromo-1-chlorobutane	C_4H_8BrCl 171.46	79504-01-1 146.5		4-01-00-00264 1.468[20]	bz 4; eth 4; EtOH 4; chl 4 1.4880[20]
3235	Butane, 1-bromo-3,3-dimethyl-	$C_6H_{13}Br$ 165.07	1647-23-0 138		3-01-00-00409 1.1556[20]	eth 4; EtOH 4; chl 4 1.4440[20]
3236	Butane, 2-bromo-2,3-dimethyl- 2-Bromo-2,3-dimethylbutane	$C_6H_{13}Br$ 165.07	594-52-5 24.5	133; 87[180]	4-01-00-00374 1.1772[10]	eth 4; chl 4 1.4517
3237	Butane, 1-bromo-4-fluoro- 1-Bromo-4-fluorobutane	C_4H_8BrF 155.01	462-72-6 135		4-01-00-00263	eth 4; EtOH 4 1.4370[25]
3238	Butane, 1-bromo-2-methyl-, (S)- d-Amyl bromide	$C_5H_{11}Br$ 151.05	534-00-9 121.6	636	4-01-00-00327 1.2234[20]	H_2O 1; EtOH 3; eth 3; chl 4 1.4451[20]
3239	Butane, 1-bromo-3-methyl- Isoamyl bromide	$C_5H_{11}Br$ 151.05	107-82-4 -112	4996 120.4	4-01-00-00328 1.2071[20]	H_2O 1; EtOH 3; eth 3; ctc 2 1.4420[20]

No.	Name / Synonym	Mol. Form. / Mol. Wt.	CAS RN / mp/°C	Merck No. / bp/°C	Beil. Ref. / den/g cm^{-3}	Solubility / n_D
3240	Butane, 2-bromo-2-methyl- / tert-Amyl bromide	$C_5H_{11}Br$ / 151.05	507-36-8	638 / 108	4-01-00-00327 / 1.197[18]	-1.4421
3241	Butane, 1-chloro- / Butyl chloride	C_4H_9Cl / 92.57	109-69-3 / -123.1	1560 / 78.6	4-01-00-00246 / 0.8862[20]	H_2O 1; EtOH 5; eth 5; ctc 2 / 1.4021[20]
3242	Butane, 2-chloro-, (±)- / sec-Butyl chloride (DL)	C_4H_9Cl / 92.57	53178-20-4 / -131.3	1561 / 68.2	4-01-00-00248 / 0.8732[20]	bz 4; eth 4; EtOH 4; chl 4 / 1.3971[20]
3243	Butane, 3-chloro-1,1-diethoxy- / Butyraldehyde, 3-chloro-, diethyl acetal	$C_8H_{17}ClO_2$ / 180.67	51786-70-0	/ 70-1[12]	4-01-00-03239 / 0.9709[20]	bz 4; eth 4; EtOH 4 / 1.4195[20]
3244	Butane, 3-chloro-1,1-dimethoxy- / Butyraldehyde, 3-chloro-, dimethyl acetal	$C_6H_{13}ClO_2$ / 152.62	50710-40-2	/ 40-4[7]	4-01-00-03239 / 1.0024[20]	1.4160[25]
3245	Butane, 1-chloro-2,3-dimethyl- / 1-Chloro-2,3-dimethylbutane	$C_6H_{13}Cl$ / 120.62	600-06-6	/ 124	4-01-00-00373 / 0.887[22]	eth 4; EtOH 4; chl 4
3246	Butane, 1-chloro-3,3-dimethyl-	$C_6H_{13}Cl$ / 120.62	2855-08-5	/ 115	4-01-00-00369 / 0.8670[20]	eth 4; EtOH 4; chl 4 / 1.4161[20]
3247	Butane, 2-chloro-2,3-dimethyl-	$C_6H_{13}Cl$ / 120.62	594-57-0 / -10.4	/ 112	4-01-00-00373 / 0.8780[20]	ace 4; EtOH 4 / 1.4191[20]
3248	Butane, 3-chloro-2,2-dimethyl-	$C_6H_{13}Cl$ / 120.62	5750-00-5 / 0.9	/ 111	4-01-00-00369 / 0.8767[20]	eth 4 / 1.4182[20]
3249	Butane, 1-(2-chloroethoxy)-	$C_6H_{13}ClO$ / 136.62	10503-96-5	154 dec; / 50[11]	4-01-00-01519 / 0.9335[20]	eth 4 / 1.4155[20]
3250	Butane, 1-chloro-4-fluoro- / 1-Chloro-4-fluorobutane	C_4H_8ClF / 110.56	462-73-7	/ 114.7	4-01-00-00249 / 1.0627[25]	eth 4; EtOH 4 / 1.4020[25]
3251	Butane, 1-chloro-2-methyl-, (±)- / Butane, 1-chloro-2-methyl (DL)	$C_5H_{11}Cl$ / 106.60	114180-21-1	/ 99.9	4-01-00-00324 / 0.8810[15]	EtOH 3; eth 3 / 1.4102[25]
3252	Butane, 1-chloro-3-methyl- / Isoamyl chloride	$C_5H_{11}Cl$ / 106.60	107-84-6 / -104.4	4998 / 98.9	4-01-00-00325 / 0.8750[20]	H_2O 2; EtOH 5; eth 5; chl 4 / 1.4084[20]
3253	Butane, 2-chloro-2-methyl-	$C_5H_{11}Cl$ / 106.60	594-36-5 / -73.5	/ 85.6	4-01-00-00324 / 0.8653[20]	H_2O 2; EtOH 3; eth 3; ctc 3 / 1.4055[20]
3254	Butane, 2-chloro-2,3,3-trimethyl-	$C_7H_{15}Cl$ / 134.65	918-07-0 / 136	/ sub	4-01-00-00411	eth 4
3255	Butane, decafluoro- / Perfluorobutane	C_4F_{10} / 238.03	355-25-9 / -128.2	/ -1.9	4-01-00-00245 / 1.6484[25]	bz 3; chl 3
3256	Butanedial	$C_4H_6O_2$ / 86.09	638-37-9	170 dec; / 58[9]	4-01-00-03642 / 1.064[20]	H_2O 4; ace 4; eth 4; EtOH 4 / 1.4262[18]
3257	Butanediamide / Succinamide	$C_4H_8N_2O_2$ / 116.12	110-14-5 / 268 dec	8837 / sub 125	4-02-00-01922	
3258	1,4-Butanediamine / Putrescine	$C_4H_{12}N_2$ / 88.15	110-60-1 / 27.5	7964 / 158.5	4-04-00-01283 / 0.877[25]	H_2O 3 / 1.4969[20]
3259	1,4-Butanediamine, N,N'-bis(3-aminopropyl)- / Spermine	$C_{10}H_{26}N_4$ / 202.34	71-44-3 / 29	8699 / 150[5]	4-04-00-01301	
3260	1,4-Butanediamine, dihydrochloride / Butane, 1,4-diamino, dihydrochloride	$C_4H_{14}Cl_2N_2$ / 161.07	333-93-7 / 280 dec	/ sub	4-04-00-01283	H_2O 4; EtOH 4; eth 1; bz 1
3261	2,3-Butanediamine, (R*,R*)-(±)-	$C_4H_{12}N_2$ / 88.15	20699-48-3 / -22	57-8[60]	4-04-00-01302 / 0.8499[25]	1.4408[25]
3262	1,4-Butanediamine, N,N,N',N'-tetraethyl- / N,N,N',N'-Tetraethyl-1,4-butanediamine	$C_{12}H_{28}N_2$ / 200.37	69704-44-5	/ 85-6[5]	4-04-00-01286	H_2O 4; eth 4; EtOH 4 / 1.4383[25]
3263	1,4-Butanediamine, N,N,N',N'-tetramethyl- / N,N,N',N'-Tetramethyl-1,4-diaminobutane	$C_8H_{20}N_2$ / 144.26	111-51-3	9157 / 168	4-04-00-01284 / 0.7942[15]	H_2O 5; EtOH 3; eth 3 / 1.4621[25]
3264	Butane, 1,1-dibromo- / 1,1-Dibromobutane	$C_4H_8Br_2$ / 215.92	62168-25-6	91[101]; 72[43]	4-01-00-00266 / 1.80[25]	1.4980[25]
3265	Butane, 1,2-dibromo- / α-Butylene dibromide	$C_4H_8Br_2$ / 215.92	533-98-2 / -65.4	1565 / 166.3	4-01-00-00266 / 1.7915[20]	H_2O 1; eth 3; chl 3 / 1.4025[20]
3266	Butane, 1,3-dibromo- / 1,3-Dibromobutane	$C_4H_8Br_2$ / 215.92	107-80-2	/ 174	4-01-00-00266 / 1.800[20]	H_2O 1; eth 3; ctc 2; chl 3 / 1.507[20]
3267	Butane, 1,4-dibromo- / 1,4-Dibromobutane	$C_4H_8Br_2$ / 215.92	110-52-1 / -16.5	/ 197	4-01-00-00267 / 1.8270[20]	H_2O 1; ctc 2; chl 3 / 1.5190[20]
3268	Butane, 2,3-dibromo- / 2,3-Dibromobutane	$C_4H_8Br_2$ / 215.92	5408-86-6 / -24	/ 161	3-01-00-00296 / 1.7893[22]	H_2O 1; eth 3 / 1.5133[22]
3269	Butane, 1,2-dibromo-2,3-dimethyl- / 1,2-Dibromo-2,3-dimethylbutane	$C_6H_{12}Br_2$ / 243.97	29916-45-8	/ 80[17]	3-01-00-00414 / 1.6033[20]	1.5105[20]
3270	Butane, 1,4-dibromo-2,3-dimethyl- / 1,4-Dibromo-2,3-dimethylbutane	$C_6H_{12}Br_2$ / 243.97	54462-70-3	/ 106-8[22]	4-01-00-00375 / 1.620[20]	1.5128[20]
3271	Butane, 2,3-dibromo-2,3-dimethyl- / 1,2-Dibromo-1,1,2,2-tetramethylethane	$C_6H_{12}Br_2$ / 243.97	594-81-0 / 171 dec	/ 78[10]	4-01-00-00375 / 1.8110[21]	
3272	Butane, 1,2-dibromo-2-methyl- / 1,2-Dibromo-2-methylbutane	$C_5H_{10}Br_2$ / 229.94	10428-64-5	/ 47-8[9]	4-01-00-00330 / 1.6711[20]	1.5088[20]
3273	Butane, 1,4-dibromo-2-methyl-, (R)- / Butane, 1,4-dibromo-2-methyl (d)	$C_5H_{10}Br_2$ / 229.94	69498-28-8	/ 78-9[12]	4-01-00-00330 / 1.695[17]	1.5128[20]
3274	Butane, 2,3-dibromo-2-methyl- / 1,2-Dibromo-1,1,2-trimethylethane	$C_5H_{10}Br_2$ / 229.94	594-51-4 / 7	/ 61-4[17]	4-01-00-00330 / 1.6717[20]	1.5729[25]
3275	Butane, 1,4-dibromo-1,1,2,2,3,3,4,4-octafluoro- / 1,4-Dibromooctafluorobutane	$C_4Br_2F_8$ / 359.84	335-48-8	/ 97		
3276	Butane, 1,1-dichloro- / Butylidene chloride	$C_4H_8Cl_2$ / 127.01	541-33-3	1571 / 113.8	4-01-00-00250 / 1.0863[20]	H_2O 1; chl 3 / 1.4355[20]
3277	Butane, 1,2-dichloro- / 1,2-Dichlorobutane	$C_4H_8Cl_2$ / 127.01	616-21-7	/ 124.1	4-01-00-00250 / 1.1116[25]	H_2O 1; eth 3; ctc 2; chl 3 / 1.4450[20]
3278	Butane, 1,3-dichloro- / 1,3-Dichlorobutane	$C_4H_8Cl_2$ / 127.01	1190-22-3	/ 134	4-01-00-00250 / 1.1158[20]	H_2O 1; eth 3; ctc 2; chl 3 / 1.4445[20]

No.	Name / Synonym	Mol. Form. / Mol. Wt.	CAS RN / mp/°C	Merck No. / bp/°C	Beil. Ref. / den/g cm⁻³	Solubility / n_D
3279	Butane, 1,4-dichloro- / 1,4-Dichlorobutane	$C_4H_8Cl_2$ / 127.01	110-56-5 / -37.3	161	4-01-00-00250 / 1.1408^{20}	chl 4 / 1.4542^{20}
3280	Butane, 2,2-dichloro- / 2,2-Dichlorobutane	$C_4H_8Cl_2$ / 127.01	4279-22-5 / -74	104	4-01-00-00251 / 1.1048^{25}	H_2O 1; chl 3 / 1.4295
3281	Butane, 2,3-dichloro-, (R*,R*)-(±)-	$C_4H_8Cl_2$ / 127.01	2211-67-8 / -80	118; 53.2^{80}	4-01-00-00251 / 1.063^{25}	1.4409^{25}
3282	Butane, 1,1-dichloro-3-methyl- / 1,1-Dichloro-3-methylbutane	$C_5H_{10}Cl_2$ / 141.04	625-66-1	130	4-01-00-00326 / 1.0473^{20}	eth 4; EtOH 4 / 1.4344^{20}
3283	Butane, 1,2-dichloro-2-methyl- / 1,2-Dichloro-2-methylbutane	$C_5H_{10}Cl_2$ / 141.04	23010-04-0	134	3-01-00-00360 / 1.0785^{20}	eth 4; EtOH 4; chl 4 / 1.4432^{21}
3284	Butane, 1,3-dichloro-3-methyl-	$C_5H_{10}Cl_2$ / 141.04	624-96-4	146.7	4-01-00-00326 / 1.0654^{20}	eth 4; chl 4 / 1.4455^{20}
3285	Butane, 1,4-dichloro-2-methyl- / 1,4-Dichloro-2-methylbutane	$C_5H_{10}Cl_2$ / 141.04	623-34-7	168.5	3-01-00-00360 / 1.1003^{25}	chl 4 / 1.4562^{21}
3286	Butane, 2,3-dichloro-2-methyl- / Amylene dichloride	$C_5H_{10}Cl_2$ / 141.04	507-45-9 / 129	645	4-01-00-00325 / 1.0696^{15}	eth 4; EtOH 4 / 1.4450^{18}
3287	Butane, 2,3-dichloro-1,1,1,2,3,4,4,4-octafluoro-	$C_4Cl_2F_8$ / 270.94	355-20-4 / -68	63	4-01-00-00252 / 1.6801^{20}	1.3100^{20}
3288	Butane, 1,4-diiodo-	$C_4H_8I_2$ / 309.92	628-21-7 / 5.8	125^{15} dec	4-01-00-00276 / 2.365^{915}	H_2O 1; ctc 2; os 3 / 1.6239^{15}
3289	Butane, 2,2-dimethyl- / Neohexane	C_6H_{14} / 86.18	75-83-2 / -99	49.7	4-01-00-00367 / 0.6444^{25}	H_2O 1; EtOH 3; eth 3; ace 4 / 1.3688^{20}
3290	Butane, 2,3-dimethyl- / 2,3-Dimethylbutane	C_6H_{14} / 86.18	79-29-8 / -128.8	57.9	4-01-00-00371 / 0.6616^{20}	H_2O 1; EtOH 3; eth 3; ace 4 / 1.3750^{20}
3291	Butanedinitrile / Succinonitrile	$C_4H_4N_2$ / 80.09	110-61-2 / 54.5	8843 / 266	4-02-00-01923 / 0.9867^{60}	H_2O 4; EtOH 3; eth 2; ace 3 / 1.4173^{60}
3292	Butanedinitrile, tetramethyl- / Tetramethylsuccinonitrile	$C_8H_{12}N_2$ / 136.20	3333-52-6 / 170.5		4-02-00-02054 / 1.070^{25}	EtOH 3
3293	Butane, 1,4-dinitro- / 1,4-Dinitrobutane	$C_4H_8N_2O_4$ / 148.12	4286-49-1 / 33.5	176^{13}	4-01-00-00280	H_2O 1; EtOH 2; eth 3; bz 3
3294	Butanedioic acid / Succinic acid	$C_4H_6O_4$ / 118.09	110-15-6 / 188	8840 / 235 dec	4-02-00-01908 / 1.572^{25}	H_2O 2; EtOH 3; eth 3; ace 3 / 1.450
3295	Butanedioic acid, acetyl-, diethyl ester	$C_{10}H_{16}O_5$ / 216.23	1115-30-6	255; 133^{17}	4-03-00-01826 / 1.081^{20}	H_2O 1; EtOH 3; eth 3; bz 3 / 1.4346^{20}
3296	Butanedioic acid, bis(1,1-dimethylethyl) ester / Di-tert-butyl succinate	$C_{12}H_{22}O_4$ / 230.30	926-26-1 / 36.5	3024 / 109-10⁹	4-02-00-01916	
3297	Butanedioic acid, bis(1-methylpropyl) ester / Di-sec-butyl succinate	$C_{12}H_{22}O_4$ / 230.30	626-31-3	256	3-02-00-01665 / 0.9735^{20}	bz 4; eth 4; EtOH 4 / 1.4238^{25}
3298	Butanedioic acid, bis(phenylmethyl) ester	$C_{18}H_{18}O_4$ / 298.34	103-43-5 / 49.5	245^{15}	4-06-00-02271 / 1.256^{25}	H_2O 1; EtOH 3; eth 3; bz 3 / 1.596
3299	Butanedioic acid, bromo-, (±)-	$C_4H_5BrO_4$ / 196.99	584-98-5 / 161		4-02-00-01929 / 2.073^{25}	H_2O 3; EtOH 3; HOAc 2
3300	Butanedioic acid, bromo- / Bromosuccinic acid	$C_4H_5BrO_4$ / 196.99	923-06-8 / 161	1427	1-02-00-00268 / 2.073^{25}	EtOH 4
3301	Butanedioic acid, dibutyl ester / Dibutyl succinate	$C_{12}H_{22}O_4$ / 230.30	141-03-7 / -29.2	3023 / 274.5	4-02-00-01916 / 0.9752^{20}	H_2O 1; EtOH 3; eth 3; bz 3 / 1.4299^{20}
3302	Butanedioic acid, 2,3-dichloro-, (±)- / Succinic acid, 2,3-dichloro-, (±)-	$C_4H_4Cl_2O_4$ / 186.98	1114-09-6 / 175 dec		4-02-00-01928 / 1.844^{15}	ace 4; eth 4; chl 4
3303	Butanedioic acid, diethyl ester / Diethyl succinate	$C_8H_{14}O_4$ / 174.20	123-25-1 / -21	8840 / 217.7	4-02-00-01914 / 1.0402^{20}	H_2O 1; EtOH 5; eth 5; ace 3 / 1.4201^{20}
3304	Butanedioic acid, 2,3-dihydroxy-, (R*,S*)- / meso-Tartaric acid	$C_4H_6O_6$ / 150.09	147-73-9 / 147	9040	4-03-00-01218 / 1.666^{20}	H_2O 4; EtOH 4
3305	Butanedioic acid, 2,3-dihydroxy-, bis(1-methylethyl) ester, (R*,R*)-(±)- / Tartaric acid, diisopropyl ester (DL)	$C_{10}H_{18}O_6$ / 234.25	58167-01-4 / 34	275; 154^{12}	2-03-00-00337 / 1.1166^{20}	ace 4; eth 4; EtOH 4
3306	Butanedioic acid, 2,3-dihydroxy- [R-(R*,R*)]-, bis(phenylmethyl) ester	$C_{18}H_{18}O_6$ / 330.34	622-00-4 / 50	250-70⁴	3-06-00-01537 / 1.2036^{72}	py 4; EtOH 4
3307	Butanedioic acid, 2,3-dihydroxy-, diethyl ester, (R*,R*)-(±)- / Diethyl DL-tartrate	$C_8H_{14}O_6$ / 206.20	57968-71-5 / 18.7	281; 158^{14}	4-03-00-01232 / 1.2046^{20}	H_2O 2; EtOH 5; eth 5; ace 3 / 1.4438^{20}
3308	Butanedioic acid, 2,3-dihydroxy-, dimethyl ester, (R*,R*)-(±)-	$C_6H_{10}O_6$ / 178.14	608-69-5 / 90	280	4-03-00-01232 / 1.2604^{90}	H_2O 4; ace 4; eth 4; EtOH 4
3309	Butanedioic acid, 2,3-dihydroxy- [R-(R*,R*)]- / L-Tartaric acid	$C_4H_8O_6$ / 150.09	87-69-4 / 169	9039	4-03-00-01219	
3310	Butanedioic acid, 2,3-dihydroxy- [R-(R*,R*)]-, dipropyl ester / Tartaric acid, dipropyl ester, (+)-	$C_{10}H_{18}O_6$ / 234.25	2217-14-3 /	303	2-03-00-00337 / 1.1390^{20}	H_2O 4; ace 4; eth 4; EtOH 4
3311	Butanedioic acid, 2,3-dihydroxy-, (R*,R*)-(±)- / DL-Tartaric acid	$C_4H_6O_8$ / 150.09	133-37-9 / 206	9038	4-03-00-01229 / 1.788^{25}	H_2O 3; EtOH 3; eth 2; bz 1
3312	Butanedioic acid, 2,3-dihydroxy-, [S-(R*,R*)]- / D-Tartaric acid	$C_4H_8O_6$ / 150.09	147-71-7 / 172.5	9037	4-03-00-01229 / 1.7598^{20}	DMSO 2 / 1.4955^{20}
3313	Butanedioic acid, 2,3-dihydroxy- [R-(R*,R*)]-, dibutyl ester	$C_{12}H_{22}O_6$ / 262.30	87-92-3 / 22	320	4-03-00-01232 / 1.0909^{20}	H_2O 4; ace 4; EtOH 4 / 1.4451^{20}
3314	Butanedioic acid, 2,3-dihydroxy- [R-(R*,R*)]-, disodium salt / Tartaric acid, disodium salt	$C_4H_4Na_2O_6$ / 194.05	868-18-8	8640		H_2O 3

No.	Name / Synonym	Mol. Form. / Mol. Wt.	CAS RN / mp/°C	Merck No. / bp/°C	Beil. Ref. / den/g cm⁻³	Solubility / n_D
3315	Butanedioic acid, 2,3-dihydroxy- [R-(R*,R*)]-, monoethyl ester Tartaric acid, ethyl ester	$C_6H_{10}O_6$ 178.14	608-89-9 90	3812	3-03-00-01028	H_2O 4; EtOH 4
3316	Butanedioic acid, [(dimethoxyphosphinothioyl)thio]-, diethyl ester Malathion	$C_{10}H_{19}O_6PS_2$ 330.36	121-75-5 2.8	5582 $156^{0.7}$ dec	4-03-00-01136 1.2076^{20}	H_2O 2; EtOH 3; eth 3; bz 3 1.4960^{20}
3317	Butanedioic acid, 2,3-dimethyl-	$C_6H_{10}O_4$ 146.14	13545-04-5 132 dec	$87-90^2$	4-02-00-01998	ace 4; eth 4
3318	Butanedioic acid, dimethyl ester	$C_6H_{10}O_4$ 146.14	106-65-0 19	196.4	4-02-00-01913 1.1198^{20}	H_2O 2; EtOH 3; eth 4; ace 3 1.4197^{20}
3319	Butanedioic acid, dipentyl ester	$C_{14}H_{26}O_4$ 258.36	645-69-2 -10.8	171.5^{18}	3-02-00-01665 0.9616^{20}	
3320	Butanedioic acid, diphenyl ester Diphenyl succinate	$C_{16}H_{14}O_4$ 270.28	621-14-7 121	330; 222.5^{15}	3-06-00-00605	H_2O 1; EtOH 3; eth 3; ace 3
3321	Butanedioic acid, dipropyl ester	$C_{10}H_{18}O_4$ 202.25	925-15-5 -5.9	250.8	4-02-00-01916 1.0020^{20}	ace 4; bz 4; eth 4 1.4250^{20}
3322	Butanedioic acid, ethyl-, (R)-	$C_6H_{10}O_4$ 146.14	4074-24-2 96	181.5	4-02-00-01995 1.0017^{20}	H_2O 4; eth 4; EtOH 4
3323	Butanedioic acid, ethyl methyl ester	$C_7H_{12}O_4$ 160.17	627-73-6 <-20	208.2	0-02-00-00609 1.076^{20}	eth 4; EtOH 4
3324	Butanedioic acid, formyl-, diethyl ester	$C_9H_{14}O_5$ 202.21	5472-38-8	$130-4^{15}$	4-03-00-01819	eth 4; EtOH 4 1.4486^{25}
3325	Butanedioic acid, methyl-, (±)-	$C_5H_8O_4$ 132.12	636-60-2 115	dec	4-02-00-01948 1.4200^0	H_2O 4; EtOH 4; eth 3; chl 2 1.4303
3326	Butanedioic acid, methyl-, diethyl ester	$C_9H_{16}O_4$ 188.22	4676-51-1	217.5	3-02-00-01696 1.012^{25}	1.4199^{20}
3327	Butanedioic acid, methyl-, dimethyl ester	$C_7H_{12}O_4$ 160.17	1604-11-1	196	3-02-00-01696 1.076^{25}	1.4200^{20}
3328	Butanedioic acid, methylene- Itaconic acid	$C_5H_8O_4$ 130.10	97-65-4 175	5130 dec	4-02-00-02228 1.632^{25}	H_2O 3; EtOH 3; eth 2; ace 3
3329	Butanedioic acid, methylene-, diethyl ester	$C_9H_{14}O_4$ 186.21	2409-52-1 58.5	228	4-02-00-02230 1.0467^{20}	EtOH 5; eth 3; ace 4; bz 3 1.4377^{20}
3330	Butanedioic acid, methylene-, dimethyl ester	$C_7H_{10}O_4$ 158.15	617-52-7 38	208	4-02-00-02229 1.1241^{18}	EtOH 3; eth 3; ace 4; MeOH 3 1.4457^{20}
3331	Butanedioic acid, (1-methylethylidene)-	$C_7H_{10}O_4$ 158.15	584-27-0 160.5	174 dec	4-02-00-02251	eth 4; EtOH 4
3332	Butanedioic acid, (1-methylpropyl)-, diethyl ester Succinic acid, sec-butyl-, diethyl ester	$C_{12}H_{22}O_4$ 230.30	69248-35-7 $75-6^{0.2}$		3-02-00-01779 0.9745^{25}	1.4293^{25}
3333	Butanedioic acid, monobutyl ester	$C_8H_{14}O_4$ 174.20	5150-93-6 8.6	136.5^3	3-02-00-01665 1.0732^{20}	1.4360^{20}
3334	Butanedioic acid, mono(2,2-dimethylhydrazide) Daminozide	$C_6H_{12}N_2O_3$ 160.17	1596-84-5 154.5	2810		
3335	Butanedioic acid, monoethyl ester Succinic acid, ethyl ester	$C_6H_{10}O_4$ 146.14	1070-34-4 8	172^{42}; 119^3	4-02-00-01914 1.1466^{20}	H_2O 4; eth 4; EtOH 4 1.4327^{20}
3336	Butanedioic acid, monomethyl ester	$C_5H_8O_4$ 132.12	3878-55-5 58	151^{20}; 122^4	4-02-00-01913	H_2O 3
3337	Butanedioic acid, monopentyl ester Succinic acid, pentyl ester	$C_9H_{16}O_4$ 188.22	97479-78-2 17.2	147^3	2-02-00-00551 1.0460^{20}	1.4378^{20}
3338	Butanedioic acid, monopropyl ester Succinic acid, monopropyl ester	$C_7H_{12}O_4$ 160.17	6946-88-9 15	126^3	3-02-00-01664 1.1071^{20}	1.4343^{20}
3339	Butanedioic acid, oxo- Oxalacetic acid	$C_4H_4O_5$ 132.07	328-42-7 161 dec	6863	4-03-00-01808	
3340	Butanedioic acid, oxo-, diethyl ester Diethyl oxalacetate	$C_8H_{12}O_5$ 188.18	108-56-5 $131-2^{24}$	3791	4-03-00-01809 1.131^{20}	H_2O 1; EtOH 5; eth 5; ace 4 1.4561^{17}
3341	Butanedioic acid, phenyl-, (±)-	$C_{10}H_{10}O_4$ 194.19	10424-29-0 168	dec	4-09-00-03352	H_2O 2; EtOH 4; eth 4; ace 4
3342	Butanedioic acid, tetrahydroxy- Dihydroxytartaric acid	$C_4H_6O_8$ 182.09	76-30-2 114.5	3172	3-03-00-01419	
3343	Butanedioic acid, tetramethyl-	$C_8H_{14}O_4$ 174.20	630-51-3 200	sub	4-02-00-02054 1.30^{25}	bz 4; EtOH 4; chl 4
3344	1,2-Butanediol, (±)-	$C_4H_{10}O_2$ 90.12	26171-83-5	190.5	4-01-00-02507 1.0024^{20}	H_2O 3; EtOH 3; ace 3 1.4378^{20}
3345	1,3-Butanediol 1,3-Butylene glycol	$C_4H_{10}O_2$ 90.12	107-88-0 <-50	1566 207.5	0-01-00-00477 1.0053^{20}	1.4401^{20}
3346	1,4-Butanediol Tetramethylene glycol	$C_4H_{10}O_2$ 90.12	110-63-4 20.1	235	4-01-00-02515 1.0171^{20}	H_2O 5; EtOH 3; eth 2; DMSO 2 1.4460^{20}
3347	2,3-Butanediol, (R*,R*)-(±)-	$C_4H_{10}O_2$ 90.12	6982-25-8 7.6	182.5	4-01-00-02525 1.0033^{20}	H_2O 5; EtOH 3; ace 3 1.4310^{25}
3348	1,4-Butanediol, diacetate	$C_8H_{14}O_4$ 174.20	628-67-1 12	229	4-02-00-00224 1.0479^{15}	1.4251^{15}
3349	2,3-Butanediol, 1,4-dichloro- 1,4-Dichloro-2,3-butanediol	$C_4H_8Cl_2O_2$ 159.01	2419-73-0 126.5	150^{30}	4-01-00-02531	EtOH 4
3350	2,3-Butanediol, 2,3-dimethyl- Pinacol	$C_6H_{14}O_2$ 118.18	76-09-5 43	7408 174.4	4-01-00-02575	H_2O 2; EtOH 4; eth 4; CS_2 2
3351	2,3-Butanediol, 2,3-dimethyl-, hexahydrate	$C_6H_{26}O_8$ 226.27	6091-58-3 45.4	172	2-01-00-00553 0.967^{15}	eth 4; EtOH 4

No.	Name / Synonym	Mol. Form. / Mol. Wt.	CAS RN / mp/°C	Merck No. / bp/°C	Beil. Ref. / den/g cm^{-3}	Solubility / n_D
3352	1,2-Butanediol, 3-methyl- / 3-Methyl-1,2-butanediol	$C_5H_{12}O_2$ / 104.15	50468-22-9 /	/ 206	4-01-00-02549 / 0.9987[0]	eth 4; EtOH 4
3353	1,3-Butanediol, 2-methyl- / 2-Methyl-1,3-butanediol	$C_5H_{12}O_2$ / 104.15	684-84-4 /	/ 200; 115[20]	4-01-00-02546 / 0.9912[20]	H_2O 3; EtOH 3; eth 3
3354	1,3-Butanediol, 3-methyl-	$C_5H_{12}O_2$ / 104.15	2568-33-4 /	/ 202.5	4-01-00-02549 / 0.9448[20]	H_2O 3; EtOH 3 / 1.4452[20]
3355	2,3-Butanediol, 2-methyl- / 2,3-Dihydroxy-2-methylbutane	$C_5H_{12}O_2$ / 104.15	5396-58-7 /	/ 175	4-01-00-02547 / 0.9920[25]	H_2O 4; eth 4; EtOH 4 / 1.4375[20]
3356	1,2-Butanediol, 2-phenyl- / 2-Phenyl-1,2-butanediol	$C_{10}H_{14}O_2$ / 166.22	90925-48-7 / 56	165[23]	3-06-00-04667 /	H_2O 4; eth 4; EtOH 4
3357	1,3-Butanediol, 1-phenyl- / 1-Phenylbutane-1,3-diol	$C_{10}H_{14}O_2$ / 166.22	65469-88-7 / 73.5	175[21]; 130[2]	4-06-00-06006 / 1.0720[25]	bz 4; EtOH 4; lig 4
3358	1,3-Butanediol, 3-phenyl-	$C_{10}H_{14}O_2$ / 166.22	7133-68-8 /	150[2]; 109[0.5]	4-06-00-06009 / 1.0865[20]	1.5341[20]
3359	2,3-Butanedione / Diacetyl	$C_4H_6O_2$ / 86.09	431-03-8 / -2.4	2946 / 88	4-01-00-03644 / 0.9808[18]	H_2O 4; EtOH 5; eth 5; ace 4 / 1.3951[20]
3360	2,3-Butanedione, dioxime / Dimethylglyoxime	$C_4H_8N_2O_2$ / 116.12	95-45-4 / 245.5	3235 / sub 234	3-01-00-03105 /	H_2O 1; EtOH 4; eth 4; bz 2
3361	2,3-Butanedione, monooxime	$C_4H_7NO_2$ / 101.11	57-71-6 / 76.8	/ 185.5	4-01-00-03646 /	H_2O 2; EtOH 4; eth 4; chl 4
3362	1,3-Butanedione, 1-phenyl-	$C_{10}H_{10}O_2$ / 162.19	93-91-4 / 56	/ 261.5	4-07-00-02151 / 1.0599[74]	H_2O 1; eth 3; chl 2 / 1.5678[78]
3363	1,3-Butanedione, 1-(2-pyridinyl)-	$C_9H_9NO_2$ / 163.18	40614-52-6 / 50	140[15]	4-21-00-04750 /	bz 4; eth 4; EtOH 4; chl 4
3364	1,3-Butanedione, 1-(3-pyridinyl)-	$C_9H_9NO_2$ / 163.18	3594-37-4 / 85	171[15]	5-21-10-00180 /	H_2O 2; EtOH 3; eth 3; ace 3
3365	1,3-Butanedione, 1-(4-pyridinyl)-	$C_9H_9NO_2$ / 163.18	75055-73-1 / 62	145[18]	5-21-10-00180 /	bz 4; eth 4; EtOH 4
3366	Butanedioyl dichloride / Succinyl chloride	$C_4H_4Cl_2O_2$ / 154.98	543-20-4 / 20	8844 / 193.3	4-02-00-01921 / 1.3748[20]	eth 3; ace 3; bz 3 / 1.4683[20]
3367	Butanedioyl dichloride, methylene-	$C_5H_4Cl_2O_2$ / 166.99	1931-60-8 /	89[17]	3-02-00-01934 /	ace 3 / 1.4919[20]
3368	1,4-Butanedithiol / Tetramethylenedithiol	$C_4H_{10}S_2$ / 122.26	1191-08-8 / -53.9	/ 195.5	4-01-00-02523 / 1.0021[0]	H_2O 1; EtOH 4; ctc 2 / 1.5290[20]
3369	Butane, 1-(ethenyloxy)- / Butyl vinyl ether	$C_6H_{12}O$ / 100.16	111-34-2 / -92	/ 94	4-01-00-02052 / 0.7888[20]	H_2O 1; EtOH 4; eth 5; ace 4 / 1.4026[20]
3370	Butane, 2-(ethenyloxy)-	$C_6H_{12}O$ / 100.16	1888-85-3 /	/ 81	4-01-00-02054 / 0.7715[20]	1.3970[20]
3371	Butane, 1-(ethenyloxy)-3-methyl- / Isopentyl vinyl ether	$C_7H_{14}O$ / 114.19	39782-38-2 /	/ 112.5	4-01-00-02055 / 0.7826[20]	eth 4; EtOH 4 / 1.4072[20]
3372	Butane, 1-ethoxy- / Butyl ethyl ether	$C_6H_{14}O$ / 102.18	628-81-9 / -124	/ 92.3	4-01-00-01518 / 0.7495[20]	H_2O 1; EtOH 5; eth 5; ace 4 / 1.3818[20]
3373	Butane, 2-ethoxy- / sec-Butyl ethyl ether	$C_6H_{14}O$ / 102.18	2679-87-0 /	/ 81	4-01-00-01572 / 0.7503[20]	H_2O 1; EtOH 4; eth 4 / 1.3802[20]
3374	Butane, 1-ethoxy-3-methyl- / Ethyl isopentyl ether	$C_7H_{16}O$ / 116.20	628-04-6 /	/ 112.5	4-01-00-01681 / 0.7688[21]	eth 4; EtOH 4
3375	Butane, 2-ethoxy-2-methyl- / Ethyl tert-amyl ether	$C_7H_{16}O$ / 116.20	919-94-8 /	/ 102	3-01-00-01626 / 0.751[18]	eth 4; EtOH 4
3376	Butane, 1-(ethylthio)- / Butyl ethyl sulfide	$C_6H_{14}S$ / 118.24	638-46-0 / -95.1	/ 144.3	4-01-00-01558 / 0.8376[20]	EtOH 4; chl 3 / 1.4492[10]
3377	Butane, 1-(ethynyloxy)-	$C_6H_{10}O$ / 98.14	3329-56-4 /	/ 104	4-01-00-02213 / 0.8200[20]	eth 4; EtOH 4 / 1.4067
3378	Butane, 1-fluoro- / Butyl fluoride	C_4H_9F / 76.11	2366-52-1 / -134	/ 32.5	4-01-00-00244 / 0.7789[20]	EtOH 4 / 1.3396[20]
3379	Butane, 2-fluoro- / sec-Butyl fluoride	C_4H_9F / 76.11	359-01-3 / -121.4	/ 25.1	4-01-00-00244 / 0.7559[25]	
3380	Butane, 1,1,2,2,3,4,4-heptachloro-	$C_4H_3Cl_7$ / 299.24	34973-41-6 /	137.5[13.5]	4-01-00-00257 / 1.739[20]	ace 4; ctc 4 / 1.5407[20]
3381	Butane, 1-iodo- / Butyl iodide	C_4H_9I / 184.02	542-69-8 / -103	1572 / 130.6	4-01-00-00271 / 1.6154[20]	H_2O 1; EtOH 5; eth 5; chl 4 / 1.5001[20]
3382	Butane, 2-iodo-, (±)- / Butane, 2-iodo (DL)	C_4H_9I / 184.02	52152-71-3 / -104.2	1573 / 120	4-01-00-00272 / 1.5920[20]	H_2O 1; EtOH 5; eth 5; chl 4 / 1.4991[20]
3383	Butane, 1-iodo-2-methyl-, (S)- / Butane, 1-iodo-2-methyl (d)	$C_5H_{11}I$ / 198.05	29394-58-9 /	148; / 132.6[500]	3-01-00-00366 / 1.5253[20]	eth 4; EtOH 4 / 1.4977[20]
3384	Butane, 1-iodo-3-methyl- / Isoamyl iodide	$C_5H_{11}I$ / 198.05	541-28-6 /	5002 / 147	4-01-00-00331 / 1.5118[20]	H_2O 2; EtOH 5; eth 5; ctc 2 / 1.4939[20]
3385	Butane, 2-iodo-2-methyl-	$C_5H_{11}I$ / 198.05	594-38-7 /	/ 124.5	4-01-00-00331 / 1.4937[20]	H_2O 1; EtOH 5; eth 5 / 1.4981[20]
3386	Butane, 1-isocyanato-3-methyl-	$C_6H_{11}NO$ / 113.16	1611-65-0 /	/ 137	0-04-00-00186 / 0.806[20]	H_2O 1; EtOH 5; eth 5 / 1.406[20]
3387	Butane, 1-isocyano- / Butyl isocyanide	C_5H_9N / 83.13	2769-64-4 /	/ 120	4-04-00-00562 / 0.78[20]	eth 4; EtOH 4
3388	Butane, 1-isothiocyanato-	C_5H_9NS / 115.20	592-82-5 /	/ 168	4-04-00-00596 / 0.9546[20]	eth 4; EtOH 4 / 1.501[20]
3389	Butane, 2-isothiocyanato-, (±)- / sec-Butyl isothiocyanate (DL)	C_5H_9NS / 115.20	116724-11-9 /	/ 159.5	4-04-00-00624 / 0.944[12]	eth 4; EtOH 4
3390	Butane, 1-isothiocyanato-3-methyl-	$C_6H_{11}NS$ / 129.23	628-03-5 /	/ 183	4-04-00-00707 / 0.9419[17]	eth 4; EtOH 4
3391	Butane, 1-methoxy- / Butyl methyl ether	$C_5H_{12}O$ / 88.15	628-28-4 / -115.5	/ 70.1	4-01-00-01518 / 0.7443[20]	H_2O 1; EtOH 5; eth 5; ace 3 / 1.3736[20]

No.	Name Synonym	Mol. Form. Mol. Wt.	CAS RN mp/°C	Merck No. bp/°C	Beil. Ref. den/g cm^{-3}	Solubility n_D
3392	Butane, 2-methoxy-, (S)- sec-Butyl methyl ether (d)	$C_5H_{12}O$ 88.15	66610-39-7	60.7	4-01-00-01571 0.7400[25]	
3393	Butane, 2-methoxy-, (±)- sec-Butyl methyl ether (DL)	$C_5H_{12}O$ 88.15	116783-23-4	60	4-01-00-01571 0.7415[20]	ace 4; eth 4; EtOH 4 1.3680[25]
3394	Butane, 1-methoxy-2-methyl- Methyl (2-methylbutyl) ether	$C_6H_{14}O$ 102.18	62016-48-2	90	4-01-00-01667 0.754[18]	1.3849[20]
3395	Butane, 1-methoxy-3-methyl-	$C_6H_{14}O$ 102.18	626-91-5	91	4-01-00-01681 0.7517[20]	eth 4; EtOH 4 1.3830[20]
3396	Butane, 2-methoxy-2-methyl-	$C_6H_{14}O$ 102.18	994-05-8	86.3	4-01-00-01671 0.7703[20]	eth 4; EtOH 4 1.3855[20]
3397	Butane, 2-methyl- Isopentane	C_5H_{12} 72.15	78-78-4 -159.9	27.8	4-01-00-00320 0.6201[20]	H_2O 1; EtOH 5; eth 5 1.3537[20]
3398	Butane, 1,1'-[methylenebis(oxy)]bis-	$C_9H_{20}O_2$ 160.26	2568-90-3 -58.1	179.2	4-01-00-03029 0.8339[20]	1.4072[17]
3399	Butane, 1-(1-methylethoxy)- Butyl isopropyl ether	$C_7H_{16}O$ 116.20	1860-27-1	109	4-01-00-01519 0.7594[15]	H_2O 1; EtOH 3; eth 3; ace 3 1.3870[15]
3400	Butane, 3-methyl-1-nitro- 3-Methyl-1-nitrobutane	$C_5H_{11}NO_2$ 117.15	627-67-8 97	163.5; 72[27]	4-01-00-00332 0.9458[25]	
3401	Butane, 1-(2-methylpropoxy)- Butyl isobutyl ether	$C_8H_{18}O$ 130.23	17071-47-5	151	4-01-00-01594 0.7980[22]	ace 4; eth 4; EtOH 4 1.4077[21]
3402	Butane, 1-(methylthio)- Butyl methyl sulfide	$C_5H_{12}S$ 104.22	628-29-5 -97.8	123.5	4-01-00-01557 0.8426[20]	EtOH 4; MeOH 4 1.4477[20]
3403	Butanenitrile Propyl cyanide	C_4H_7N 69.11	109-74-0 -111.9	1597 117.6	4-02-00-00806 0.7936[20]	H_2O 2; EtOH 5; eth 5; bz 3 1.3842[20]
3404	Butanenitrile, 4-bromo-	C_4H_6BrN 148.00	5332-06-9	206	4-02-00-00836 1.4967[20]	EtOH 3; eth 3; chl 3 1.4818[20]
3405	Butanenitrile, 4-chloro-	C_4H_6ClN 103.55	628-20-6	192	4-02-00-00827 1.0934[15]	H_2O 1; EtOH 3; eth 3; ctc 2 1.4413[20]
3406	Butanenitrile, 4,4-diethoxy-	$C_8H_{15}NO_2$ 157.21	18381-45-8	104-6[10]	3-03-00-01210 0.937[25]	H_2O 2; EtOH 5; eth 5 1.4186[20]
3407	Butanenitrile, 2-ethyl-	$C_6H_{11}N$ 97.16	617-80-1	145.5	4-02-00-00953	eth 4; EtOH 4 1.3891[24]
3408	Butanenitrile, 2-hydroxy-3-methyl-2-(1-methylethyl)- Diisopropyl ketone, cyanohydrin	$C_8H_{15}NO$ 141.21	4390-75-4 59	111[18]	4-03-00-00884	ace 4; bz 4; eth 4; EtOH 4
3409	Butanenitrile, 2-methyl- 2-Methylbutanenitrile	C_5H_9N 83.13	18936-17-9	125	4-02-00-00892 0.7913[15]	eth 4; EtOH 4 1.3933[20]
3410	Butanenitrile, 3-methyl- Isobutyl cyanide	C_5H_9N 83.13	625-28-5 -101	127.5	4-02-00-00902 0.7914[20]	H_2O 2; EtOH 5; eth 5; ace 4 1.3927[20]
3411	Butanenitrile, 4-phenoxy-	$C_{10}H_{11}NO$ 161.20	2243-43-8 45.5	288	4-06-00-00646	eth 3
3412	Butane, 1-nitro- 1-Nitrobutane	$C_4H_9NO_2$ 103.12	627-05-4	153	4-01-00-00277 0.970[25]	H_2O 2; EtOH 5; eth 5; alk 3 1.4303[20]
3413	Butane, 2-nitro-, (±)-	$C_4H_9NO_2$ 103.12	116781-85-2 -132	140	4-01-00-00278 0.9854[17]	chl 3 1.4044[20]
3414	Butane, 1,1,1,2,3,4,4,4-octafluoro-2,3-bis(trifluoromethyl)- Perfluoro-2,3-dimethylbutane	C_6F_{14} 338.04	354-96-1 -15	59.8		
3415	Butane, 1,1'-oxybis- Dibutyl ether	$C_8H_{18}O$ 130.23	142-96-1 -95.2	1568 140.2	4-01-00-01520 0.7684[20]	H_2O 1; EtOH 5; eth 5; ace 4 1.3992[20]
3416	Butane, 2,2'-oxybis-, (±)- sec-Butyl ether, (±)-	$C_8H_{18}O$ 130.23	17226-28-7	120.5	3-01-00-01533 0.756[25]	ace 4; eth 4; EtOH 4 1.393[25]
3417	Butane, 1,1'-[oxybis(2,1-ethanediyloxy)]bis-	$C_{12}H_{26}O_3$ 218.34	112-73-2 -60	256	4-01-00-02395 0.885[25]	1.4235[20]
3418	Butane, 1,1'-oxybis[3-methyl- Diisoamyl ether	$C_{10}H_{22}O$ 158.28	544-01-4	5000 172.5	4-01-00-01682 0.7777[20]	ace 4; EtOH 4; chl 4 1.4085[20]
3419	Butane, 1,1,2,3,4-pentachloro- Butane, 1,1,2,3,4-pentachloro (liquid)	$C_4H_5Cl_5$ 230.35	77753-24-3	95.5[11]	4-01-00-00256 1.561[18]	EtOH 4; ctc 4 1.5141[18]
3420	Butane, 1,2,2,3,4-pentachloro- 1,2,2,3,4-Pentachlorobutane	$C_4H_5Cl_5$ 230.35	2431-52-9	85[10]	4-01-00-00256 1.5543[20]	ace 4; chl 4 1.5157[20]
3421	Butane, 1-propoxy- Butyl propyl ether	$C_7H_{16}O$ 116.20	3073-92-5	118.1	4-01-00-01519 0.7772[20]	H_2O 1; EtOH 4; eth 4
3422	Butane, 1,1'-sulfinylbis-	$C_8H_{18}OS$ 162.30	2168-93-6 32.6	dec	4-01-00-01561 0.8317[23]	H_2O 1; EtOH 3; eth 3 1.4669[20]
3423	Butane, 1,1'-sulfonylbis-	$C_8H_{18}O_2S$ 178.30	598-04-9 45	291	4-01-00-01561 0.9885[47]	H_2O 1; EtOH 3; eth 3
3424	Butane, 1,1,4,4-tetrabromo-	$C_4H_6Br_4$ 373.71	116779-77-2	138-45[10]	3-01-00-00298 2.529[20]	bz 4; eth 4; chl 4 1.6077[20]
3425	Butane, 1,2,2,3-tetrabromo-	$C_4H_6Br_4$ 373.71	116779-78-3 -2	97.5[7]	3-01-00-00298 2.510[20]	
3426	Butane, 1,2,3,4-tetrabromo-, DL- Butane, 1,2,3,4-tetrabromo (DL)	$C_4H_6Br_4$ 373.71	2657-65-0 118	260; 180[60]	4-01-00-00271	ace 4; eth 4; EtOH 4; lig 4
3427	Butane, 1,1,1,2-tetrachloro- 1,1,1,2-Tetrachlorobutane	$C_4H_6Cl_4$ 195.90	39966-95-5	135	4-01-00-00253 1.3907[20]	1.4920[25]
3428	Butane, 1,2,2,3-tetrachloro- 1,2,2,3-Tetrachlorobutane	$C_4H_6Cl_4$ 195.90	79630-70-9 -48	182	3-01-00-00286 1.4276[18]	ace 4; chl 4 1.491[20]
3429	Butane, 1,2,3,3-tetrachloro- 1,2,3,3-Tetrachlorobutane	$C_4H_6Cl_4$ 195.90	13138-51-7	90[32]; 56[10]	4-01-00-00254 1.4204[20]	ace 4; eth 4; EtOH 4 1.4958[20]
3430	Butane, 2,2,3,3-tetramethyl- 2,2,3,3-Tetramethylbutane	C_8H_{18} 114.23	594-82-1 100.7	106.4	4-01-00-00447 0.8242[20]	H_2O 1; eth 3; chl 3 1.4695[20]

No.	Name / Synonym	Mol. Form. / Mol. Wt.	CAS RN / mp/°C	Merck No. / bp/°C	Beil. Ref. / den/g cm^{-3}	Solubility / n_D
3431	1,2,3,4-Butanetetrol, (R*,S*)- / Erythritol	$C_4H_{10}O_4$ / 122.12	149-32-6 / 121.5	3620 / 330.5	4-01-00-02807 / 1.451[20]	H_2O 3; eth 1; bz 1
3432	1,2,3,4-Butanetetrol, tetranitrate, (R*,S*)- / Erythrityl tetranitrate	$C_4H_6N_4O_{12}$ / 302.11	7297-25-8 / 61	3622	4-01-00-02809	EtOH 4
3433	Butane, 1,1'-thiobis- / Dibutyl sulfide	$C_8H_{18}S$ / 146.30	544-40-1 / -79.7	1590 / 185	4-01-00-01559 / 0.8386[20]	eth 4; EtOH 4; chl 4 / 1.4530[20]
3434	Butane, 2,2'-thiobis-	$C_8H_{18}S$ / 146.30	626-26-6 /	/ 165	4-01-00-01586 / 0.8348[20]	H_2O 1; EtOH 4; eth 4 / 1.4506[20]
3435	Butane, 1,1'-thiobis[2-methyl- / Bis(2-methylbutyl) sulfide	$C_{10}H_{22}S$ / 174.35	96034-00-3 /	165; 95[13]	0-01-00-00387 / 0.8348[20]	eth 4; EtOH 4 / 1.4506[20]
3436	Butane, 1,1'-thiobis[3-methyl- / Isoamyl sulfide	$C_{10}H_{22}S$ / 174.35	544-02-5 / -74.6	/ 211	4-01-00-01689 / 0.8323[20]	H_2O 1; EtOH 5; eth 4 / 1.4520[20]
3437	1-Butanethiol / Butyl mercaptan	$C_4H_{10}S$ / 90.19	109-79-5 / -115.7	1575 / 98.5	4-01-00-01555 / 0.8416[20]	H_2O 2; EtOH 4; eth 4; chl 2 / 1.4440[20]
3438	2-Butanethiol, (±)- / 2-Butanethiol (DL)	$C_4H_{10}S$ / 90.19	91840-99-2 / -165	1576 / 85	4-01-00-01584 / 0.8295[20]	EtOH 4; eth 3; bz 3; ctc 2 / 1.4366[20]
3439	1-Butanethiol, 2-methyl-, (S)- / (+)-2-Methyl-1-butanethiol	$C_5H_{12}S$ / 104.22	20089-07-0 /	/ 118.2	4-01-00-01668 / 0.8420[20]	1.4440[20]
3440	1-Butanethiol, 3-methyl- / Isopentyl mercaptan	$C_5H_{12}S$ / 104.22	541-31-1 /	/ 120	4-01-00-01688 / 0.8350[20]	H_2O 1; EtOH 5; eth 5; ctc 3 / 1.4418[20]
3441	2-Butanethiol, 2-methyl- / 2-Methyl-2-butanethiol	$C_5H_{12}S$ / 104.22	1679-09-0 /	/ 99.1	4-01-00-01674 / 0.8120[20]	1.4385[20]
3442	Butane, 1,1,2-tribromo- / 1,1,2-Tribromocyclobutane	$C_4H_7Br_3$ / 294.81	3675-68-1 /	/ 216.2	2-01-00-00084 / 2.1835[20]	eth 4; EtOH 4; chl 4 / 1.5626[17]
3443	Butane, 1,2,2-tribromo- / 1,2,2-Tribromobutane	$C_4H_7Br_3$ / 294.81	3675-69-2 /	/ 213.8	2-01-00-00085 / 2.1691[20]	eth 4; EtOH 4; chl 4 / 1.5624[10]
3444	Butane, 1,2,3-tribromo- / 1,2,3-Tribromobutane	$C_4H_7Br_3$ / 294.81	632-05-3 / -19	/ 220	4-01-00-00271 / 2.1907[20]	eth 4; EtOH 4; chl 4 / 1.5680[20]
3445	Butane, 1,2,4-tribromo- / 1,2,4-Tribromobutane	$C_4H_7Br_3$ / 294.81	38300-67-3 / -18	/ 215	4-01-00-00271 / 2.170[20]	eth 4; EtOH 4; chl 4 / 1.5608[20]
3446	Butane, 1,3,3-tribromo- / 1,3,3-Tribromobutane	$C_4H_7Br_3$ / 294.81	62127-46-2 /	/ 202.5	3-01-00-00298 / 2.1445[20]	eth 4; EtOH 4; chl 4 / 1.5564[20]
3447	Butane, 2,2,3-tribromo- / 2,2,3-Tribromobutane	$C_4H_7Br_3$ / 294.81	62127-47-3 / 1.8	/ 206	3-01-00-00298 / 2.1723[20]	H_2O 1; EtOH 3; eth 3; ctc 2 / 1.5602[20]
3448	Butane, 1,1,3-trichloro- / 1,1,3-Trichlorobutane	$C_4H_7Cl_3$ / 161.46	13279-87-3 /	/ 152	4-01-00-00252 / 1.317[15]	H_2O 1; EtOH 3; eth 3; chl 4 / 1.4600[15]
3449	Butane, 1,2,3-trichloro- / 1,2,3-Trichlorobutane	$C_4H_7Cl_3$ / 161.46	18338-40-4 /	168; 63[20]	3-01-00-00285 / 1.3164[20]	H_2O 1; EtOH 3; eth 3; chl 4 / 1.4790[20]
3450	Butane, 1,2,4-trichloro- / 1,2,4-Trichlorobutane	$C_4H_7Cl_3$ / 161.46	1790-22-3 /	61-3[10]	4-01-00-00253 / 1.3175[20]	H_2O 1; bz 3; ctc 2; chl 4 / 1.4820[20]
3451	Butane, 2,2,3-trichloro- / 2,2,3-Trichlorobutane	$C_4H_7Cl_3$ / 161.46	10403-60-8 /	/ 144	3-01-00-00285 / 1.2699[20]	H_2O 1; chl 4 / 1.4645[20]
3452	Butane, 2,2,3-trichloro-1,1,1,3,4,4,4-heptafluoro-	$C_4Cl_3F_7$ / 287.39	335-44-4 / 4	/ 98	4-01-00-00253 / 1.7484[20]	1.3530[20]
3453	Butane, 1,2,3-trichloro-2-methyl- / Butane, 2-methyl-1,2,3-trichloro	$C_5H_9Cl_3$ / 175.48	62521-69-1 /	/ 184	3-01-00-00361 / 1.2527[20]	chl 4
3454	Butane, 2,2,3-trichloro-3-methyl- / Butane, 2-methyl-2,3,3-trichloro	$C_5H_9Cl_3$ / 175.48	98070-91-8 /	/ 182.5	3-01-00-00361 / 1.215[15]	chl 4; HOAc 4 / 1.473[21]
3455	Butane, 1,1,3-trimethoxy-	$C_7H_{16}O_3$ / 148.20	10138-89-3 /	/ 157	4-01-00-03985 / 0.935[25]	1.4032[20]
3456	Butane, 1,3,3-trimethoxy-	$C_7H_{16}O_3$ / 148.20	6607-66-5 /	61-3[20]	4-01-00-03995 / 0.94[25]	1.4096[20]
3457	Butane, 2,2,3-trimethyl- / 2,2,3-Trimethylbutane	C_7H_{16} / 100.20	464-06-2 / -25	/ 80.8	4-01-00-00410 / 0.6901[20]	H_2O 1; EtOH 3; eth 3; ace 4 / 1.3864[20]
3458	1,2,3-Butanetriol / 1-Methylglycerol	$C_4H_{10}O_3$ / 106.12	4435-50-1 /	170[20]	4-01-00-02774 /	H_2O 4; EtOH 4 / 1.4462[20]
3459	1,2,4-Butanetriol	$C_4H_{10}O_3$ / 106.12	3068-00-6 /	172-4[12]	2-01-00-00596 / 1.018[20]	H_2O 4; EtOH 4 / 1.4688[20]
3460	Butanoic acid / Butyric acid	$C_4H_8O_2$ / 88.11	107-92-6 / -5.7	1593 / 163.7	4-02-00-00779 / 0.9577[20]	H_2O 5; EtOH 5; eth 5; ctc 2 / 1.3980[20]
3461	Butanoic acid, 2-acetyl-3-methyl-, ethyl ester	$C_9H_{16}O_3$ / 172.22	1522-46-9 /	201; 97-8[20]	4-03-00-01608 / 0.9648[18]	H_2O 1; EtOH 5; eth 5 / 1.4256[18]
3462	Butanoic acid, 2-acetyl-3-oxo-, ethyl ester	$C_8H_{12}O_4$ / 172.18	603-69-0 /	/ 210	4-03-00-01781 / 1.1045[20]	bz 4; eth 4; EtOH 4 / 1.4690[20]
3463	Butanoic acid, 2-amino-, (±)-	$C_4H_9NO_2$ / 103.12	2835-81-6 / 305 dec	/ sub	3-04-00-01296 / 1.2300[20]	H_2O 4; EtOH 2; eth 1; bz 1
3464	Butanoic acid, 4-amino- / γ-Aminobutyric acid	$C_4H_9NO_2$ / 103.12	56-12-2 / 203 dec	441	4-04-00-02600	H_2O 4; EtOH 2; eth 1; ace 2
3465	Butanoic acid, 3-amino-4-hydroxy- / γ-Hydroxy-β-aminobutyric acid	$C_4H_9NO_3$ / 119.12	589-44-6 / 216	455	4-04-00-03216	
3466	Butanoic acid, 4-amino-2-hydroxy-4-oxo-	$C_4H_7NO_4$ / 133.10	66398-52-5 / 147.5		3-03-00-00918 / 1.577[18]	H_2O 3; EtOH 2; eth 1; MeOH 2
3467	Butanoic acid, anhydride / Butyric anhydride	$C_8H_{14}O_3$ / 158.20	106-31-0 / -75	1594 / 200	4-02-00-00802 / 0.9668[20]	eth 3; ctc 2 / 1.4070[20]
3468	Butanoic acid, 2-bromo-, (±)- / DL-α-Bromobutyric acid	$C_4H_7BrO_2$ / 167.00	2385-70-8 / -4	1401 / 217 dec; 181[250]	4-02-00-00833 / 1.5641[20]	H_2O 3; EtOH 3; eth 3
3469	Butanoic acid, 2-bromo-, ethyl ester	$C_8H_{11}BrO_2$ / 195.06	533-68-6 /	177; 43[5]	4-02-00-00834 / 1.3273[20]	H_2O 1; EtOH 5; eth 5; chl 3 / 1.4475[20]

No.	Name / Synonym	Mol. Form. / Mol. Wt.	CAS RN / mp/°C	Merck No. / bp/°C	Beil. Ref. / den/g cm⁻³	Solubility / n_D
3470	Butanoic acid, 4-bromo-, ethyl ester	$C_6H_{11}BrO_2$ 195.06	2969-81-5	82[10]	4-02-00-00836 1.363[25]	1.4559[20]
3471	Butanoic acid, 2-bromo-3-methyl-, (±)-	$C_5H_9BrO_2$ 181.03	10323-40-7 44	230 dec; 138[25]	3-02-00-00706 1.459[20]	H_2O 2; EtOH 4; eth 3; ace 3
3472	Butanoic acid, 3-bromo-3-methyl- β-Bromoisovaleric acid	$C_5H_9BrO_2$ 181.03	5798-88-9 74	1410	4-02-00-00905	bz 4; eth 4; EtOH 4
3473	Butanoic acid, 2-bromo-, methyl ester	$C_5H_9BrO_2$ 181.03	3196-15-4	168	4-02-00-00834 1.4528[20]	EtOH 4 1.4029[25]
3474	Butanoic acid, 4-bromo-, methyl ester	$C_5H_9BrO_2$ 181.03	4897-84-1	186.5	4-02-00-00835 1.42[25]	EtOH 4 1.4567[25]
3475	Butanoic acid, 2-bromo-3-methyl-, ethyl ester	$C_7H_{13}BrO_2$ 209.08	609-12-1	186	4-02-00-00906 1.2780[20]	eth 4; EtOH 4 1.4496[20]
3476	Butanoic acid, 2-bromo-3-methyl-, methyl ester	$C_6H_{11}BrO_2$ 195.06	26330-51-8	177	3-02-00-00706 1.352[13]	eth 4; EtOH 4 1.4530[20]
3477	Butanoic acid, 2-bromo-3-oxo-, ethyl ester	$C_8H_9BrO_3$ 209.04	609-13-2	210 dec; 107[15]	4-03-00-01551 1.4294[16]	eth 4; EtOH 4 1.463[14]
3478	Butanoic acid, 4-bromo-3-oxo-, ethyl ester	$C_6H_9BrO_3$ 209.04	13176-46-0	115[14]	4-03-00-01551 1.4840[20]	eth 4; EtOH 4 1.5281[20]
3479	Butanoic acid, butyl ester Butyl butanoate	$C_8H_{16}O_2$ 144.21	109-21-7 -91.5	1556 166	4-02-00-00789 0.8700[20]	H_2O 1; EtOH 5; eth 5; ctc 3 1.4075[20]
3480	Butanoic acid, 2-chloro-	$C_4H_7ClO_2$ 122.55	4170-24-5	189[627]; 101[15]	4-02-00-00821 1.1796[20]	H_2O 2; EtOH 4; eth 4 1.441[20]
3481	Butanoic acid, 3-chloro-, (±)-	$C_4H_7ClO_2$ 122.55	625-68-3 16	116[22]	4-02-00-00823 1.1898[20]	EtOH 3; eth 4; ctc 2 1.4221[20]
3482	Butanoic acid, 4-chloro-	$C_4H_7ClO_2$ 122.55	627-00-9 16	196[22]; 68[0.2]	4-02-00-00825 1.2236[20]	EtOH 4 1.4642[20]
3483	Butanoic acid, 2-chloro-, ethyl ester	$C_6H_{11}ClO_2$ 150.60	7425-45-8	163.5	4-02-00-00822 1.056[18]	eth 4; EtOH 4 1.4180[25]
3484	Butanoic acid, 3-chloro-, ethyl ester	$C_6H_{11}ClO_2$ 150.60	7425-48-1		4-02-00-00824 1.0517[20]	EtOH 4 1.4246[20]
3485	Butanoic acid, 4-chloro-, ethyl ester	$C_6H_{11}ClO_2$ 150.60	3153-36-4	184	4-02-00-00825 1.0756[20]	ace 4; eth 4; EtOH 4 1.4311[20]
3486	Butanoic acid, 2-chloro-2-methyl- Butyric acid, 2-chloro-2-methyl-	$C_5H_9ClO_2$ 136.58	73758-54-0	200 dec	3-02-00-00688 1.1204[20]	eth 4; EtOH 4 1.4445[20]
3487	Butanoic acid, 2-chloro-3-methyl-	$C_5H_9ClO_2$ 136.58	921-08-4 21	210; 126[32]	4-02-00-00904 1.135[13]	H_2O 1; EtOH 3; eth 3 1.4450[11]
3488	Butanoic acid, 2-chloro-, methyl ester	$C_5H_9ClO_2$ 136.58	26464-32-4	146	3-02-00-00622 1.0979[14]	eth 4; EtOH 4 1.4247[20]
3489	Butanoic acid, 4-chloro-, methyl ester	$C_5H_9ClO_2$ 136.58	3153-37-5	175; 55[4]	4-02-00-00825 1.1201[20]	H_2O 1; EtOH 4; eth 4; ace 3 1.4321[20]
3490	Butanoic acid, 2-chloro-3-methyl-, ethyl ester Ethyl 2-chloro-3-methylbutanoate	$C_7H_{13}ClO_2$ 164.63	91913-99-4	179	0-02-00-00316 1.021[13]	eth 4; EtOH 4
3491	Butanoic acid, 2-chloro-2-methyl-, ethyl ester Butyric acid, 2-chloro-2-methyl, ethyl ester	$C_7H_{13}ClO_2$ 164.63	58190-94-6	176	3-02-00-00688 1.069[14]	eth 4; EtOH 4 1.4388[11]
3492	Butanoic acid, 2-chloro-2-methyl-3-oxo-, ethyl ester Butyric acid, 2-chloro-2-methyl-3-oxo, ethyl ester	$C_7H_{11}ClO_3$ 178.62	37935-39-0	116-7[75]	2-03-00-00433 1.157[18]	EtOH 4 1.4382[18]
3493	Butanoic acid, 4-(4-chloro-2-methylphenoxy)- 4-(4-Chloro-2-methylphenoxy)butyric acid	$C_{11}H_{13}ClO_3$ 228.68	94-81-5 100		4-06-00-01996	
3494	Butanoic acid, 4-chloro-3-oxo-, ethyl ester	$C_6H_9ClO_3$ 164.59	638-07-3 -8	220 dec; 115[14]	4-03-00-01550 1.218[25]	1.4520[20]
3495	Butanoic acid, 2-cyano-2-ethyl-, ethyl ester Cyanoacetic acid, diethyl, ethyl ester	$C_9H_{15}NO_2$ 169.22	1619-56-3	214.5	3-02-00-01761	eth 4; EtOH 4 1.4200[27]
3496	Butanoic acid, cyclohexyl ester	$C_{10}H_{18}O_2$ 170.25	1551-44-6	213	4-06-00-00037 0.9572[0]	H_2O 1; EtOH 3; ctc 2; os 3
3497	Butanoic acid, 2,3-dibromo-	$C_4H_6Br_2O_2$ 245.90	600-30-6 87	100-10[20]	4-02-00-00837	H_2O 2; EtOH 4; eth 4; bz 4
3498	Butanoic acid, 2,3-dibromo-, ethyl ester	$C_6H_{10}Br_2O_2$ 273.95	609-11-0 58.5	113[30]	3-02-00-00634 1.680[20]	H_2O 2; EtOH 3; eth 3; ctc 2
3499	Butanoic acid, 2,4-dibromo-, ethyl ester	$C_6H_{10}Br_2O_2$ 273.95	36847-51-5	149-50[52]	4-02-00-00838 1.6987[20]	H_2O 1; EtOH 3; eth 3 1.4960[20]
3500	Butanoic acid, 2,2-dichloro- Butyric acid, 2,2-dichloro-	$C_4H_6Cl_2O_2$ 157.00	13023-00-2	107-10[14]	2-02-00-00254 1.389[20]	H_2O 2; EtOH 4; eth 4
3501	Butanoic acid, 3,4-dichloro-, butyl ester Butyric acid, 3,4-dichloro, butyl ester	$C_8H_{14}Cl_2O_2$ 213.10	116723-94-5	127-8[15]	4-02-00-00829 1.143[25]	1.454[24]
3502	Butanoic acid, 3,4-dichloro-, methyl ester	$C_5H_8Cl_2O_2$ 171.02	819-93-2	90[15]	4-02-00-00829 1.278[23]	1.459[23]
3503	Butanoic acid, 2,2-dichloro-3-oxo-, ethyl ester	$C_6H_8Cl_2O_3$ 199.03	6134-66-3	206; 91[12]	2-03-00-00427 1.291[16]	EtOH 4 1.4492[17]
3504	Butanoic acid, 2,2-diethyl-3-oxo-, ethyl ester	$C_{10}H_{18}O_3$ 186.25	1619-57-4	216; 64[3]	4-03-00-01625 0.9717[18]	H_2O 1; EtOH 5; eth 5; ctc 3 1.4326[17]
3505	Butanoic acid, 2,3-dihydroxypropyl ester	$C_7H_{14}O_4$ 162.19	557-25-5	280; 117[10]	4-02-00-00798 1.129[18]	H_2O 4; EtOH 4 1.4531[20]
3506	Butanoic acid, 2,2-dimethyl-	$C_6H_{12}O_2$ 116.16	595-37-9 -14	186	4-02-00-00954 0.9276[20]	H_2O 2; EtOH 3; eth 3 1.4145[20]

No.	Name Synonym	Mol. Form. Mol. Wt.	CAS RN mp/°C	Merck No. bp/°C	Beil. Ref. den/g cm^{-3}	Solubility n_D
3507	Butanoic acid, 2,3-dimethyl- Butyric acid, 2,3-dimethyl	$C_6H_{12}O_2$ 116.16	14287-61-7 -1.5	 191.7	4-02-00-00958 0.9275[20]	eth 4; EtOH 4 1.4146[20]
3508	Butanoic acid, 3,3-dimethyl- tert-Butylacetic acid	$C_6H_{12}O_2$ 116.16	1070-83-3 6.5	1538 190	4-02-00-00955 0.9124[20]	EtOH 3; eth 3 1.4096[20]
3509	Butanoic acid, 1,1-dimethylethyl ester	$C_8H_{16}O_2$ 144.21	2308-38-5 	 146	4-02-00-00790 	ace 4; eth 4; EtOH 4 1.4007[17]
3510	Butanoic acid, 3,3-dimethyl-2-oxo-	$C_6H_{10}O_3$ 130.14	815-17-8 90.5	189; 80[15]	4-03-00-01595 	H_2O 2; eth 3; bz 3; chl 3
3511	Butanoic acid, 2,2-dimethyl-3-oxo-, ethyl ester	$C_8H_{14}O_3$ 158.20	597-04-6 	 182	4-03-00-01594 0.9759[20]	eth 4; EtOH 4 1.4180[20]
3512	Butanoic acid, 1,2-ethanediyl ester	$C_{10}H_{18}O_4$ 202.25	105-72-6 -80	 240	4-02-00-00796 1.0005[20]	eth 4; EtOH 4 1.4262[20]
3513	Butanoic acid, ethenyl ester Vinyl butyrate	$C_6H_{10}O_2$ 114.14	123-20-6 	116.7; 64[130]	4-02-00-00792 0.9006[20]	
3514	Butanoic acid, 2-ethyl- Diethylacetic acid	$C_6H_{12}O_2$ 116.16	88-09-5 -31.8	3099 194	4-02-00-00950 0.9239[20]	H_2O 2; EtOH 5; eth 5; ctc 2 1.4132[20]
3515	Butanoic acid, ethyl ester Ethyl butanoate	$C_6H_{12}O_2$ 116.16	105-54-4 -98	3733 121.5	4-02-00-00787 0.8844[15]	H_2O 2; EtOH 3; eth 3; ctc 2 1.4000[20]
3516	Butanoic acid, 2-ethyl-2-methyl-	$C_7H_{14}O_2$ 130.19	19889-37-3 <-20	 207	4-02-00-00981 	EtOH 4 1.4250[20]
3517	Butanoic acid, 2-ethyl-, methyl ester	$C_7H_{14}O_2$ 130.19	816-11-5 	 137	4-02-00-00950 0.8886[12]	 1.4067[12]
3518	Butanoic acid, 2-ethyl-3-oxo-, ethyl ester	$C_8H_{14}O_3$ 158.20	607-97-6 	198.0; 80[10]	4-03-00-01592 0.9847[16]	EtOH 5; eth 5 1.4214[25]
3519	Butanoic acid, 2-ethyl-3-oxo-, methyl ester	$C_7H_{12}O_3$ 144.17	51756-08-2 	 182	4-03-00-01592 0.995[14]	ace 4; eth 4; EtOH 4
3520	Butanoic acid, 2-furanylmethyl ester	$C_9H_{12}O_3$ 168.19	623-21-2 	 212.5	4-17-00-01247 1.0530[20]	H_2O 2; EtOH 3; eth 5
3521	Butanoic acid, heptafluoro-	$C_4HF_7O_2$ 214.04	375-22-4 -17.5	 121	4-02-00-00810 1.651[20]	H_2O 3; eth 3; tol 3; peth 1 1.295[25]
3522	Butanoic acid, heptafluoro-, anhydride	$C_8F_{14}O_3$ 410.06	336-59-4 -43	 106.5	4-02-00-00817 1.665[20]	 1.285[20]
3523	Butanoic acid, heptafluoro-, ethenyl ester	$C_6H_3F_7O_2$ 240.08	356-28-5 	 79	4-02-00-00815 1.418[20]	 1.309
3524	Butanoic acid, heptafluoro-, ethyl ester	$C_8H_5F_7O_2$ 242.09	356-27-4 	 95	4-02-00-00813 1.394[20]	H_2O 2; eth 3; ace 3 1.3011[20]
3525	Butanoic acid, heptafluoro-, methyl ester	$C_5H_3F_7O_2$ 228.07	356-24-1 -86	 80	4-02-00-00812 1.483[20]	H_2O 2; eth 3; ace 3 1.295[20]
3526	Butanoic acid, heptafluoro-, 1-methylethyl ester Isopropyl heptafluorobutyrate	$C_7H_7F_7O_2$ 256.12	425-23-0 	 106	4-02-00-00813 1.324[20]	 1.310[20]
3527	Butanoic acid, heptyl ester	$C_{11}H_{22}O_2$ 186.29	5870-93-9 -57.5	 225.8	4-02-00-00791 0.8637[20]	EtOH 4 1.4231[20]
3528	Butanoic acid, hexyl ester	$C_{10}H_{20}O_2$ 172.27	2639-63-6 -78	 208	4-02-00-00791 0.8652[20]	H_2O 1; EtOH 3; chl 2; os 3 1.4160[15]
3529	Butanoic acid, hydrazide Butyrhydrazide	$C_4H_{10}N_2O$ 102.14	3538-65-6 45.5	138[20]; 120[10]	4-02-00-00807 	H_2O 4; eth 4; EtOH 4
3530	Butanoic acid, 2-hydroxy-, (±)-	$C_4H_8O_3$ 104.11	600-15-7 44.2	260 dec; 140[14]	3-03-00-00561 1.125[20]	H_2O 3; EtOH 3; eth 3
3531	Butanoic acid, 3-hydroxy-, (±)-	$C_4H_8O_3$ 104.11	625-71-8 49	130[12]; 94[0.1]	4-03-00-00761 	H_2O 4; EtOH 4; eth 4; bz 1 1.4424[20]
3532	Butanoic acid, 2-hydroxy-, ethyl ester, (±)- Butyric acid, 2-hydroxy, ethyl ester (DL)	$C_6H_{12}O_3$ 132.16	68057-83-0 	 167	4-03-00-00756 1.0069[20]	EtOH 4 1.4179[20]
3533	Butanoic acid, 3-hydroxy-, ethyl ester, (±)-	$C_6H_{12}O_3$ 132.16	35608-64-1 	185; 76[15]	4-03-00-00762 1.017[20]	H_2O 3; EtOH 3; ctc 2 1.4182[20]
3534	Butanoic acid, 3-hydroxy-, ethyl ester	$C_6H_{12}O_3$ 132.16	5405-41-4 	185; 76[15]	4-03-00-00762 1.017[20]	H_2O 4; EtOH 4 1.4182[20]
3535	Butanoic acid, 3-hydroxy-3-methyl-	$C_5H_{10}O_3$ 118.13	625-08-1 <-32	162[12]	4-03-00-00827 0.9384[20]	H_2O 4; eth 4; EtOH 4 1.5081[20]
3536	Butanoic acid, 2-hydroxy-, 2-methylpropyl ester Butyric acid, 2-hydroxy, isobutyl ester	$C_8H_{16}O_3$ 160.21	116723-95-6 	 196	0-03-00-00302 0.944[15]	 1.4182
3537	Butanoic acid, 3-mercapto-	$C_4H_8O_2S$ 120.17	26473-49-4 	110[10]; 88[2.5]	4-03-00-00769 1.1371[20]	 1.4782[20]
3538	Butanoic acid, 2-methyl-, (±)- (±)-2-Methylbutyric acid	$C_5H_{10}O_2$ 102.13	600-07-7 <-80	 177	4-02-00-00889 0.934[20]	H_2O 2; EtOH 5; eth 5; chl 3 1.4051[20]
3539	Butanoic acid, 3-methyl- Isovaleric acid	$C_5H_{10}O_2$ 102.13	503-74-2 -29.3	5120 176.5	4-02-00-00895 0.931[20]	H_2O 3; EtOH 5; eth 5; chl 5 1.4033[20]
3540	Butanoic acid, 2-methyl-, 10-(acetyloxy)-9,10-dihydro-8,8-dimethyl-2-oxo-2H,8H-benzo[1,2-b:3,4-b']dipyran-9-yl Visnadine	$C_{21}H_{24}O_7$ 388.42	477-32-7 85.5	9915	5-19-06-00272 	H_2O 1; EtOH 3; eth 3
3541	Butanoic acid, 3-(methylamino)-, ethyl ester Butyric acid, 3-methylamino-, ethyl ester	$C_7H_{15}NO_2$ 145.20	68384-70-3 	72[12.5]	3-04-00-01313 0.9282[20]	 1.4250[20]
3542	Butanoic acid, 3-methyl-, anhydride	$C_{10}H_{18}O_3$ 186.25	1468-39-9 	 215	4-02-00-00901 0.9327[20]	eth 4 1.4043[20]
3543	Butanoic acid, 3-methyl-, butyl ester Butyl isovalerate	$C_9H_{18}O_2$ 158.24	109-19-3 		3-02-00-00698 	 1.4058[25]

No.	Name / Synonym	Mol. Form. / Mol. Wt.	CAS RN / mp/°C	Merck No. / bp/°C	Beil. Ref. / den/g cm^{-3}	Solubility / n_D
3544	Butanoic acid, 3-methylbutyl ester / Isopentyl butanoate	$C_9H_{18}O_2$ / 158.24	106-27-4	4997 / 179	4-02-00-00791 / 0.865[19]	H_2O 1; EtOH 4; eth 4 / 1.4110[20]
3545	Butanoic acid, 2-methyl-, butyl ester / Butyric acid, (2-methylbutyl) ester	$C_9H_{18}O_2$ / 158.24	15706-73-7	/ 179	0-02-00-00304 / 0.8620[20]	1.4135[20]
3546	Butanoic acid, 3-methyl-, 1,1-dimethylpropyl ester / tert-Amyl isovalerate	$C_{10}H_{20}O_2$ / 172.27	542-37-0	647 / 173.5	0-02-00-00312 / 0.8729[0]	EtOH 4
3547	Butanoic acid, methyl ester / Methyl butanoate	$C_5H_{10}O_2$ / 102.13	623-42-7 / -85.8	5957 / 102.8	4-02-00-00786 / 0.8984[20]	H_2O 2; EtOH 5; eth 5; ctc 2 / 1.3878[20]
3548	Butanoic acid, 3-methyl-, ethyl ester / Ethyl isovalerate	$C_7H_{14}O_2$ / 130.19	108-64-5 / -99.3	3772 / 135.0	4-02-00-00898 / 0.8656[20]	EtOH 4; eth 4 / 1.3962[20]
3549	Butanoic acid, 1-methylethyl ester / Isopropyl butyrate	$C_7H_{14}O_2$ / 130.19	638-11-9	/ 130.5	4-02-00-00789 / 0.8588[20]	H_2O 1; EtOH 3 / 1.3936[20]
3550	Butanoic acid, 2-methyl-, ethyl ester, (S)-	$C_7H_{14}O_2$ / 130.19	10307-61-6	/ 133; 35[16]	4-02-00-00890 / 0.8689[25]	H_2O 1; EtOH 3; bz 3 / 1.3964[20]
3551	Butanoic acid, 3-methyl-, 3-methylbutyl ester / Isoamyl isovalerate	$C_{10}H_{20}O_2$ / 172.27	659-70-1	5003 / 190.4	4-02-00-00899 / 0.8583[19]	1.4130[19]
3552	Butanoic acid, 3-methyl-, methyl ester / Methyl isovalerate	$C_6H_{12}O_2$ / 116.16	556-24-1	6006 / 116.5	4-02-00-00897 / 0.8808[20]	H_2O 1; EtOH 4; eth 4; ace 4 / 1.3927[20]
3553	Butanoic acid, 3-methyl-, 1-methylethyl ester / Butyric acid, 3-methyl, isopropyl ester	$C_8H_{16}O_2$ / 144.21	32665-23-9	/ 142; 70[55]	3-02-00-00698 / 0.8538[17]	ace 4; eth 4; EtOH 4 / 1.3960[20]
3554	Butanoic acid, 3-methyl-, 5-methyl-2-(1-methylethyl)cyclohexyl ester	$C_{15}H_{28}O_2$ / 240.39	16409-46-4	/ 129[9]	2-06-00-00043 / 0.9089[15]	H_2O 1; EtOH 3; ace 3 / 1.4486[20]
3555	Butanoic acid, 3-methyl-, 2-methylpropyl ester / Isobutyl isovalerate	$C_9H_{18}O_2$ / 158.24	589-59-3	5029 / 168.5	4-02-00-00898 / 0.853[20]	H_2O 1; EtOH 5; eth 5; ace 4 / 1.4057[20]
3556	Butanoic acid, 3-methyl-2-oxo-	$C_5H_8O_3$ / 116.12	759-05-7 / 31.5	/ 170.5	4-03-00-01577 / 0.9968[20]	H_2O 3; EtOH 3; eth 3 / 1.3850[16]
3557	Butanoic acid, 2-methyl-3-oxo-, ethyl ester	$C_7H_{12}O_3$ / 144.17	609-14-3	/ 187	4-03-00-01573 / 0.9941[20]	H_2O 2; EtOH 3; eth 3; ace 4 / 1.4185[20]
3558	Butanoic acid, 2-methyl-3-oxo-, methyl ester	$C_6H_{10}O_3$ / 130.14	17094-21-2	/ 177.4	3-03-00-01225 / 1.0217[25]	eth 4; EtOH 4 / 1.416[24]
3559	Butanoic acid, 1-methylpentyl ester, (S)- / Butyric acid (1-methylpentyl) ester (d)	$C_{10}H_{20}O_2$ / 172.27	116723-96-7	/ 85[20]	1-02-00-00120 / 0.8744[21]	eth 4; EtOH 4
3560	Butanoic acid, 3-methyl-, phenylmethyl ester	$C_{12}H_{16}O_2$ / 192.26	103-38-8	/ 245; 136[25]	4-06-00-02266 / 0.9983[15]	1.4884[20]
3561	Butanoic acid, 3-methyl-, 1,2,3-propanetriyl ester	$C_{18}H_{32}O_6$ / 344.45	620-63-3	/ 332.5	4-02-00-00900 / 0.9984[20]	eth 4; EtOH 4 / 1.4354[20]
3562	Butanoic acid, 3-methyl-, propyl ester / Propyl isovalerate	$C_8H_{16}O_2$ / 144.21	557-00-6	/ 155.9	4-02-00-00898 / 0.8617[20]	eth 4; EtOH 4 / 1.4031[20]
3563	Butanoic acid, 2-methylpropyl ester / Isobutyl butanoate	$C_8H_{16}O_2$ / 144.21	539-90-2	5020 / 156.9	4-02-00-00790 / 0.8364[18]	H_2O 2; EtOH 5; eth 5 / 1.4032[20]
3564	Butanoic acid, 1-methylpropyl ester, (S)- / Butyric acid, sec-butyl ester (d)	$C_8H_{16}O_2$ / 144.21	116836-55-4 / -91.5	/ 152; 54[18]	3-02-00-00600 / 0.8737[13]	EtOH 3; bz 3; CS_2 3; py 3 / 1.4011[20]
3565	Butanoic acid, 3-methyl-, 1,7,7-trimethylbicyclo[2.2.1]hept-2-yl ester, (1R-endo)- / d-Bornyl isovalerate	$C_{15}H_{26}O_2$ / 238.37	53022-14-3	1342 / 257.5	2-06-00-08488 / 0.955[25]	eth 4; EtOH 4
3566	Butanoic acid, octyl ester	$C_{12}H_{24}O_2$ / 200.32	110-39-4 / -55.6	4-02-00-00791 / 244.1	0.8629[20]	EtOH 4 / 1.4267[15]
3567	Butanoic acid, 2-oxo-	$C_4H_8O_3$ / 102.09	600-18-0 / 33	/ 80-2[16]	4-03-00-01524 / 1.200[17]	H_2O 4; EtOH 4; eth 2 / 1.3972[20]
3568	Butanoic acid, 3-oxo- / Acetoacetic acid	$C_4H_8O_3$ / 102.09	541-50-4 / 36.5	51 / 100 dec	4-03-00-01527	H_2O 4; eth 4; EtOH 4
3569	Butanoic acid, 3-oxo-, butyl ester	$C_8H_{14}O_3$ / 158.20	591-60-6 / -35.6	/ 127[50]; 85[8]	4-03-00-01536 / 0.9671[25]	H_2O 2; EtOH 5; bz 5; lig 5 / 1.4137[20]
3570	Butanoic acid, 3-oxo-, 2-chloroethyl ester	$C_6H_9ClO_3$ / 164.59	54527-68-3	/ 198; 107[14]	4-03-00-01534 / 1.190[20]	bz 4; eth 4; EtOH 4 / 1.4430[20]
3571	Butanoic acid, 3-oxo-, ethyl ester / Ethyl acetoacetate	$C_8H_{10}O_3$ / 130.14	141-97-9 / -45	3714 / 180.8	4-03-00-01528 / 1.0368[10]	H_2O 4; EtOH 5; eth 5; bz 3 / 1.4171[20]
3572	Butanoic acid, 4-oxo-, ethyl ester	$C_6H_{10}O_3$ / 130.14	10138-10-0	/ 162; 85[12]	4-03-00-01555 / 1.0490[22]	chl 2 / 1.4287[20]
3573	Butanoic acid, 3-oxo-, methyl ester / Methyl acetoacetate	$C_5H_8O_3$ / 116.12	105-45-3 / 27.5	5933 / 171.7	4-03-00-01527 / 1.0762[20]	H_2O 4; EtOH 5; eth 5; ctc 3 / 1.4184[20]
3574	Butanoic acid, 3-oxo-, 1-methylethyl ester / Isopropyl acetoacetate	$C_7H_{12}O_3$ / 144.17	542-08-5 / -27.3	5094 / 186	4-03-00-01535 / 0.9835[20]	eth 4; EtOH 4; lig 4 / 1.4173[20]
3575	Butanoic acid, 4-oxo-4-(phenylamino)- / Succinanilic acid	$C_{10}H_{11}NO_3$ / 193.20	102-14-7 / 148.5	8839	4-12-00-00468	H_2O 2; EtOH 3; eth 4
3576	Butanoic acid, 3-oxo-2-(phenylmethylene)-, ethyl ester	$C_{13}H_{14}O_3$ / 218.25	620-80-4 / 60.5	/ 296; 180[17]	3-10-00-03158	H_2O 1; EtOH 2; eth 2; bz 2
3577	Butanoic acid, 3-oxo-, 2-propenyl ester	$C_7H_{10}O_3$ / 142.15	1118-84-9 / -85	/ 196; 66.5[14]	4-03-00-01538 / 1.0366[20]	H_2O 3; EtOH 5; bz 5; lig 3 / 1.4398[20]
3578	Butanoic acid, 4-oxo-4-[[4-[(2-thiazolylamino)sulfonyl]phenyl]amino]- / Succinylsulphathiazole	$C_{13}H_{13}N_3O_5S_2$ / 355.40	116-43-8 / 193.5	8850	4-27-00-04637	H_2O 1; EtOH 2; eth 1; ace 2
3579	Butanoic acid, pentyl ester / Amyl butyrate	$C_9H_{18}O_2$ / 158.24	540-18-1 / -73.2	639 / 186.4	4-02-00-00790 / 0.8713[15]	H_2O 1; EtOH 4; eth 4 / 1.4123[20]
3580	Butanoic acid, 2-phenoxy-	$C_{10}H_{12}O_3$ / 180.20	13794-14-4 / 98	/ 258	3-06-00-00616	H_2O 2; EtOH 3; eth 3; ace 3

No.	Name / Synonym	Mol. Form. / Mol. Wt.	CAS RN / mp/°C	Merck No. / bp/°C	Beil. Ref. / den/g cm⁻³	Solubility / n_D
3581	Butanoic acid, 4-phenoxy-, ethyl ester / Ethyl 4-phenoxybutyrate	$C_{12}H_{16}O_3$ / 208.26	2364-59-2 /	170-3[25]	2-06-00-00159 / 1.042[35]	eth 4; EtOH 4 / 1.491[35]
3582	Butanoic acid, 2-phenoxyethyl ester	$C_{12}H_{16}O_3$ / 208.26	23511-70-8 /	251; 88[2]	3-06-00-00571 / 1.0388[21]	ace 4; eth 4; EtOH 4
3583	Butanoic acid, 2-phenoxy-, phenyl ester / 2-Phenoxybutyric acid, phenyl ester	$C_{16}H_{16}O_3$ / 256.30	116836-19-2 / 48.5	202-3[25]	4-06-00-00645 / 1.135[15]	
3584	Butanoic acid, phenyl ester / Phenyl butyrate	$C_{10}H_{12}O_2$ / 164.20	4346-18-3 /	225	4-06-00-00615 / 1.0382[15]	H_2O 1; EtOH 3; eth 3
3585	Butanoic acid, phenylmethyl ester	$C_{11}H_{14}O_2$ / 178.23	103-37-7 /	239	4-06-00-02266 / 1.0111[20]	H_2O 1; EtOH 4; eth 4; ctc 3 / 1.4920[20]
3586	Butanoic acid, 1,2,3-propanetriyl ester / Tributyrin	$C_{15}H_{26}O_6$ / 302.37	60-01-5 / -75	9532 / 307.5	4-02-00-00799 / 1.0350[20]	H_2O 1; EtOH 3; eth 4; ace 3 / 1.4359[20]
3587	Butanoic acid, 2-propenyl ester	$C_7H_{12}O_2$ / 128.17	2051-78-7 /	142; 44.5[15]	4-02-00-00793 / 0.9017[20]	H_2O 1; EtOH 5; eth 5; ctc 2 / 1.4158[20]
3588	Butanoic acid, propyl ester / Propyl butanoate	$C_7H_{14}O_2$ / 130.19	105-66-8 / -95.2	7858 / 143.0	4-02-00-00788 / 0.8730[20]	H_2O 2; EtOH 5; eth 5 / 1.4001[20]
3589	Butanoic acid, 3-thioxo-, ethyl ester / Butyric acid, 3-thioxo, ethyl ester	$C_6H_{10}O_2S$ / 146.21	7740-33-2 /	75[15]	4-03-00-01553 / 1.0554[31]	eth 4; EtOH 4 / 1.4712[26]
3590	Butanoic acid, 2,2,3-trichloro- / 2,2,3-Trichlorobutyric acid	$C_4H_5Cl_3O_2$ / 191.44	5344-55-8 / 60	237	3-02-00-00629 /	eth 4
3591	Butanoic acid, 2,2,3-trichloro-, ethyl ester / Butyric acid, 2,2,3-trichloro, ethyl ester	$C_6H_9Cl_3O_2$ / 219.49	116723-97-8 /	212	0-02-00-00281 / 1.3114[20]	eth 4; EtOH 4
3592	Butanoic acid, 4,4,4-trifluoro-3-oxo-, ethyl ester	$C_6H_7F_3O_3$ / 184.11	372-31-6 /	133	4-03-00-01548 / 1.2586[15]	EtOH 3; eth 3 / 1.3783[15]
3593	1-Butanol / Butyl alcohol	$C_4H_{10}O$ / 74.12	71-36-3 / -89.8	1540 / 117.7	4-01-00-01506 / 0.8098[20]	H_2O 3; EtOH 5; eth 5; ace 4 / 1.3993[20]
3594	2-Butanol / sec-Butyl alcohol	$C_4H_{10}O$ / 74.12	78-92-2 / -114.7	1541 / 99.5	2-01-00-00400 / 0.8063[20]	H_2O 4; EtOH 5; eth 5; ace 4 / 1.3978[20]
3595	1-Butanol, 2-amino-, (±)-	$C_4H_{11}NO$ / 89.14	13054-87-0 / -2	178	4-04-00-01705 / 0.9162[20]	H_2O 5; EtOH 5; eth 5; chl 2 / 1.4489[25]
3596	1-Butanol, 3-amino-	$C_4H_{11}NO$ / 89.14	2867-59-6 /	82-5[19]	4-04-00-01710 /	H_2O 4; EtOH 4 / 1.4534[25]
3597	1-Butanol, 4-amino-	$C_4H_{11}NO$ / 89.14	13325-10-5 /	205; 125[34]	4-04-00-01711 / 0.967[12]	H_2O 3; EtOH 3; eth 1 / 1.4625[20]
3598	2-Butanol, 3-amino-	$C_4H_{11}NO$ / 89.14	42551-55-3 / 19	160; 79[20]	3-04-00-00779 / 0.9399[20]	H_2O 4; eth 4; EtOH 4 / 1.4445[25]
3599	1-Butanol, 2-amino-3-methyl-, (±)-	$C_5H_{13}NO$ / 103.16	16369-05-4 /	75-7[8]	4-04-00-01777 / 0.93[25]	1.4543[20]
3600	1-Butanol, 3-bromo-	C_4H_9BrO / 153.02	6089-12-9 /	69-71[20]	4-01-00-01553 / 1.7107[18]	1.5010[21]
3601	2-Butanol, 3-bromo-, (R*,S*)-(±)- / 2-Butanol, 3-bromo-, erythro-(±)-	C_4H_9BrO / 153.02	19246-39-0 /	154; 53[13]	4-01-00-01580 / 1.4550[20]	eth 4; EtOH 4 / 1.4780[20]
3602	1-Butanol, 2-chloro-	C_4H_9ClO / 108.57	26106-95-6 /	74-6[25]	4-01-00-01549 / 1.062[25]	eth 4; EtOH 4 / 1.4438[20]
3603	1-Butanol, 3-chloro-	C_5H_9ClO / 120.58	2203-35-2 /	175; 66[15]	3-01-00-01517 / 1.0883[20]	EtOH 3; eth 3; ace 3; ctc 2 / 1.4518[20]
3604	1-Butanol, 4-chloro-	C_4H_9ClO / 108.57	928-51-8 /	84-5[16]	4-01-00-01550 / 1.0883[20]	eth 4; EtOH 4 / 1.4518[20]
3605	2-Butanol, 1-chloro- / α-Butylene chlorohydrin	C_4H_9ClO / 108.57	1873-25-2 /	141	4-01-00-01578 / 1.068[25]	EtOH 3; eth 3 / 1.4400[20]
3606	2-Butanol, 3-chloro- / β-Butylene chlorohydrin	C_4H_9ClO / 108.57	563-84-8 /	139	3-01-00-01538 / 1.0669[20]	EtOH 4; eth 4 / 1.4432[20]
3607	2-Butanol, 4-chloro-	C_4H_9ClO / 108.57	2203-34-1 /	67[20]	4-01-00-01578 /	eth 4; EtOH 4 / 1.4408[20]
3608	2-Butanol, 1-chloro-2,3-dimethyl- / 1-Chloro-2,3-dimethyl-2-butanol	$C_6H_{13}ClO$ / 136.62	66235-62-9 /	163	4-01-00-01730 / 1.049[20]	1.4459[20]
3609	2-Butanol, 1-chloro-3,3-dimethyl- / 1-Chloro-3,3-dimethyl-2-butanol	$C_6H_{13}ClO$ / 136.62	36402-31-0 / 13	157.5	4-01-00-01728 / 1.0063[25]	1.4432[25]
3610	2-Butanol, 1-chloro-2-methyl-	$C_5H_{11}ClO$ / 122.59	74283-48-0 /	151	4-01-00-01673 / 1.0161[20]	EtOH 4 / 1.4469[20]
3611	2-Butanol, 3-chloro-2-methyl-	$C_5H_{11}ClO$ / 122.59	21326-62-5 /	141.5	3-01-00-01627 / 1.0295[20]	eth 4; EtOH 4 / 1.4436[20]
3612	1-Butanol, 2,3-dibromo- / 2,3-Dibromo-1-butanol	$C_4H_8Br_2O$ / 231.91	4021-75-4 / 32	99.5[10]	4-01-00-01554 / 1.9475[20]	1.5442[20]
3613	1-Butanol, 3,4-dibromo- / 3,4-Dibromobutanol	$C_4H_8Br_2O$ / 231.91	87018-30-2 /	114[11]	3-01-00-01518 / 1.98[15]	1.548[15]
3614	2-Butanol, 1,4-dibromo- / 2-Hydroxy-1,4-dibromobutane	$C_4H_8Br_2O$ / 231.91	19398-47-1 /	114-5[15]	4-01-00-01581 / 1.953[20]	1.544[20]
3615	1-Butanol, 2,2-dimethyl- / 2,2-Dimethyl-1-butanol	$C_6H_{14}O$ / 102.18	1185-33-7 / <-15	136.5	4-01-00-01726 / 0.8283[20]	H_2O 2; EtOH 3; eth 3 / 1.4208[20]
3616	1-Butanol, 2,3-dimethyl-, (±)-	$C_6H_{14}O$ / 102.18	20281-85-0 /	144.5	4-01-00-01729 / 0.8297[20]	EtOH 3; eth 3; ace 3 / 1.4195[20]
3617	1-Butanol, 3,3-dimethyl-	$C_6H_{14}O$ / 102.18	624-95-3 / -60	143	4-01-00-01729 / 0.844[15]	H_2O 2; EtOH 3; eth 3; ace 3 / 1.4323[15]
3618	2-Butanol, 2,3-dimethyl-	$C_6H_{14}O$ / 102.18	594-60-5 / -14	118.4	4-01-00-01729 / 0.8236[20]	H_2O 3; EtOH 5; eth 5 / 1.4176[20]
3619	2-Butanol, 3,3-dimethyl-, (±)- / 3,3-Dimethylbutan-2-ol (DL-)	$C_6H_{14}O$ / 102.18	20281-91-8 / 5.6	120.4	4-01-00-01727 / 0.8122[25]	H_2O 2; EtOH 3; eth 5 / 1.4148[20]
3620	1-Butanol, 2-ethyl- / 2-Ethyl-1-butanol	$C_6H_{14}O$ / 102.18	97-95-0 / <-15	147	4-01-00-01725 / 0.8326[20]	H_2O 2; EtOH 3; eth 3; chl 3 / 1.4220[20]

No.	Name Synonym	Mol. Form. Mol. Wt.	CAS RN mp/°C	Merck No. bp/°C	Beil. Ref. den/g cm⁻³	Solubility n_D
3621	1-Butanol, 4-fluoro- 4-Fluoro-1-butanol	C_4H_9FO 92.11	372-93-0	58[15]	4-01-00-01546	ace 4; eth 4; EtOH 4 1.3942[15]
3622	1-Butanol, 2,2,3,3,4,4,4-heptafluoro-	$C_4H_3F_7O$ 200.06	375-01-9	95	4-01-00-01547 1.600[20]	EtOH 3; ace 3 1.294[20]
3623	2-Butanol, 1-[(2-hydroxyethyl)amino]- 2-Hydroxyethyl 2-hydroxybutyl amine	$C_6H_{15}NO_2$ 133.19	6967-43-7	137[9]	4-04-00-01725 1.0310[20]	H₂O 4; ace 4; EtOH 4 1.4690[30]
3624	2-Butanol, 3,3'-iminobis- Bis-(3-hydroxy-2-butyl)amine	$C_8H_{19}NO_2$ 161.24	6959-06-4	112-5[3]	3-04-00-00781 0.9775[20]	H₂O 4; ace 4; EtOH 4 1.4162[20]
3625	1-Butanol, 3-methoxy- 3-Methoxy-1-butanol	$C_5H_{12}O_2$ 104.15	2517-43-3	157	4-01-00-02509 0.923[23]	EtOH 4; eth 3; ace 4; chl 2 1.4148[25]
3626	1-Butanol, 2-methyl- 2-Methyl-1-butanol	$C_5H_{12}O$ 88.15	137-32-6	5952 128	4-01-00-01666 0.8150[25]	H₂O 2; EtOH 5; eth 5; ace 4 1.4092[20]
3627	1-Butanol, 3-methyl- Isopentyl alcohol	$C_5H_{12}O$ 88.15	123-51-3 -117.2	5081 131.1	4-01-00-01677 0.8104[20]	ace 4; eth 4; EtOH 4 1.4053[20]
3628	2-Butanol, 2-methyl- tert-Amyl alcohol	$C_5H_{12}O$ 88.15	75-85-4 -8.8	7096 102.4	4-01-00-01668 0.8096[20]	H₂O 3; EtOH 5; eth 5; ace 4 1.4052[20]
3629	2-Butanol, 3-methyl-, (±)- (±)-3-Methyl-2-butanol	$C_5H_{12}O$ 88.15	70116-68-6	5953 112.9	4-01-00-01675 0.8180[20]	H₂O 2; EtOH 5; eth 5; ace 4 1.4089[20]
3630	1-Butanol, 2-methyl-, acetate	$C_7H_{14}O_2$ 130.19	624-41-9	140	4-02-00-00156 0.8740[20]	ace 4; eth 4; EtOH 4 1.4040[20]
3631	1-Butanol, 3-methyl-, acetate Isopentyl acetate	$C_7H_{14}O_2$ 130.19	123-92-2 -78.5	4993 142.5	4-02-00-00157 0.876[15]	H₂O 2; EtOH 5; eth 5; ace 3 1.4000[20]
3632	1-Butanol, 3-methyl-, benzoate Isoamyl benzoate	$C_{12}H_{16}O_2$ 192.26	94-46-2	4995 261	4-09-00-00292 0.993[15]	EtOH 4
3633	1-Butanol, 3-methyl-, carbamate	$C_6H_{13}NO_2$ 131.17	543-86-2 59	220	4-03-00-00058 0.9438[71]	eth 4; EtOH 4 1.4175[20]
3634	2-Butanol, 2-methyl-, carbamate tert-Amyl carbamate	$C_6H_{13}NO_2$ 131.17	590-60-3 86	641	1-03-00-00014	ace 4; bz 4
3635	1-Butanol, 3-methyl-, carbonate (2:1)	$C_{11}H_{22}O_3$ 202.29	2050-95-5	234; 122[16]	3-03-00-00011 0.9067[20]	H₂O 1 1.4174[20]
3636	1-Butanol, 3-methyl-, formate Isopentyl formate	$C_6H_{12}O_2$ 116.16	110-45-2 -93.5	5001 123.5	4-02-00-00030 0.877[20]	H₂O 2; EtOH 3; eth 5; ctc 2 1.3967[20]
3637	1-Butanol, 3-methyl-, nitrate Isoamyl nitrate	$C_5H_{11}NO_3$ 133.15	543-87-3	5004 148	4-01-00-01683 0.996[22]	1.4122[21]
3638	1-Butanol, 3-methyl-2-nitro-	$C_5H_{11}NO_3$ 133.15	77392-54-2	111[10]	3-01-00-01645 1.0889[25]	1.4430[20]
3639	1-Butanol, 3-methyl-, propanoate Isopentyl propanoate	$C_8H_{16}O_2$ 144.21	105-68-0	160.2	4-02-00-00709 0.8697[20]	eth 4; EtOH 4 1.4069[20]
3640	1-Butanol, 2-nitro- 2-Nitro-1-butanol	$C_4H_9NO_3$ 119.12	609-31-4 -47	105[10]	4-01-00-01555 1.1332[25]	H₂O 3; EtOH 5; eth 5; ace 3 1.4390[20]
3641	2-Butanol, 1-nitro- 2-Hydroxy-1-nitrobutane	$C_4H_9NO_3$ 119.12	3156-74-9	204; 75[2]	4-01-00-01582 1.1353[20]	H₂O 3; EtOH 4; eth 4; ace 3 1.4435[20]
3642	2-Butanol, 3-nitro- 3-Nitro-2-butanol	$C_4H_9NO_3$ 119.12	6270-16-2	112[38]; 55[0.5]	4-01-00-01583 1.1166[7]	1.4414[20]
3643	2-Butanol, 1,1,1-trichloro- Ethyl(trichloromethyl)carbinol	$C_4H_7Cl_3O$ 177.46	6111-61-1	171; 83[22]	4-01-00-01579 1.3630[25]	ace 4; bz 4; eth 4; EtOH 4 1.4800[20]
3644	2-Butanol, 2,3,3-trimethyl-	$C_7H_{16}O$ 116.20	594-83-2 17	131	4-01-00-01755 0.8380[25]	ace 4; eth 4; EtOH 4 1.4233[22]
3645	2-Butanone Methyl ethyl ketone	C_4H_8O 72.11	78-93-3 -86.65	5991 79.59	4-01-00-03243 0.7999[25]	H₂O 4; EtOH 5; eth 5; ace 5 1.3788[20]
3646	2-Butanone, 1-chloro- Chloromethyl ethyl ketone	C_4H_7ClO 106.55	616-27-3	137.5	4-01-00-03255 1.0850[20]	MeOH 4 1.4372[20]
3647	2-Butanone, 3-chloro-	C_4H_7ClO 106.55	4091-39-8	115	4-01-00-03256 1.0554[25]	1.4219[20]
3648	2-Butanone, 4-chloro- 1-Chloro-3-butanone	C_4H_7ClO 106.55	6322-49-2	120 dec	4-01-00-03256 1.0680[23]	eth 4; EtOH 4 1.4284[23]
3649	2-Butanone, 3-chloro-3-methyl- 3-Chloro-3-methyl-2-butanone	C_5H_9ClO 120.58	5950-19-6	117.2	3-01-00-02818 1.0083[20]	eth 4; EtOH 4 1.4204[20]
3650	1-Butanone, 4-chloro-1-phenyl-	$C_{10}H_{11}ClO$ 182.65	939-52-6 19.5	130-3[4]	4-07-00-00711 1.137[25]	1.5459[20]
3651	1-Butanone, 4-[4-(4-chlorophenyl)-4-hydroxy- 1-piperidinyl]-1-(4-fluorophenyl)- Haloperidol	$C_{21}H_{23}ClFNO_2$ 375.87	52-86-8 151.5	4511	5-21-02-00377	
3652	1-Butanone, 1-cyclohexyl-	$C_{10}H_{18}O$ 154.25	1462-27-7	94[13]	4-07-00-00080 0.903[20]	1.4537[20]
3653	2-Butanone, 3,4-dibromo- 1,2-Dibromoethyl methyl ketone	$C_4H_6Br_2O$ 229.90	25109-57-3	80-1[10]	3-01-00-02788 1.9662[15]	1.5314[15]
3654	2-Butanone, 1,3-dichloro- 1,3-Dichloro-2-butanone	$C_4H_6Cl_2O$ 141.00	16714-77-5	166.5	3-01-00-02786 1.3116[20]	ace 4; bz 4; eth 4; EtOH 4 1.4686[20]
3655	2-Butanone, 3,4-dichloro-	$C_4H_6Cl_2O$ 141.00	58625-77-7	65-70[16]	3-01-00-02787 1.293[18]	1.4628[18]
3656	2-Butanone, 4-(diethylamino)-	$C_8H_{17}NO$ 143.23	3299-38-5	84[30]	4-04-00-01930 0.8630[20]	ace 4; bz 4; eth 4; EtOH 4 1.4333[24]
3657	2-Butanone, 3,3-dimethyl- Pinacolone	$C_6H_{12}O$ 100.16	75-97-8 -52.5	7409 106.1	4-01-00-03310 0.7229[25]	H₂O 2; EtOH 3; eth 3; ace 3 1.3952[20]
3658	2-Butanone, 3,3-diphenyl- 2,2-Diphenyl-3-butanone	$C_{16}H_{16}O$ 224.30	2575-20-4 41	310.5	4-07-00-01451 1.069[20]	H₂O 1; EtOH 3; eth 4; chl 4 1.5748[20]
3659	2-Butanone, 4-(2-furanyl)-	$C_8H_{10}O_2$ 138.17	699-17-2	203	4-17-00-04559 1.0361[19]	1.4696[17]
3660	2-Butanone, 1-hydroxy-	$C_4H_8O_2$ 88.11	5077-67-8	160; 78[60]	4-01-00-03989 1.0272[20]	H₂O 4; EtOH 4; eth 4 1.4189[20]

No.	Name / Synonym	Mol. Form. / Mol. Wt.	CAS RN / mp/°C	Merck No. / bp/°C	Beil. Ref. / den/g cm⁻³	Solubility / n_D
3661	2-Butanone, 3-hydroxy- / Acetoin	$C_4H_8O_2$ / 88.11	513-86-0 / 15	55 / 143	2-01-00-00870 / 1.0044[20]	H$_2$O 4; ace 4; EtOH 4 / 1.4171[20]
3662	2-Butanone, 4-hydroxy-	$C_4H_8O_2$ / 88.11	590-90-9 /	109[30]; 90[11]	4-01-00-03994 / 1.1089[24]	H$_2$O 5; EtOH 5; eth 5; ace 4 / 1.4585[14]
3663	2-Butanone, 4-(4-hydroxy-3-methoxyphenyl)- / Zingerone	$C_{11}H_{14}O_3$ / 194.23	122-48-5 / 40.5	10072 / 187-8[14]	4-08-00-01866 /	eth 4
3664	2-Butanone, 4-hydroxy-3-methyl-	$C_5H_{10}O_2$ / 102.13	3393-64-4 /	90-5[15]	4-01-00-04009 / 0.993[25]	1.4340[20]
3665	1-Butanone, 1-(2-hydroxyphenyl)-	$C_{10}H_{12}O_2$ / 164.20	2887-61-8 / 10	124-6[14]	4-08-00-00497 / 1.0683[24]	1.5379[25]
3666	1-Butanone, 2-hydroxy-1-phenyl- / Butyrophenone, 2-hydroxy-	$C_{10}H_{12}O_2$ / 164.20	16183-46-3 /	132-3[12]	4-08-00-00502 / 1.077[19]	1.529[19]
3667	1-Butanone, 4-hydroxy-1-phenyl-	$C_{10}H_{12}O_2$ / 164.20	39755-03-8 / 30	187-8[9]	3-08-00-00449 /	EtOH 4
3668	2-Butanone, 3-methyl- / Methyl isopropyl ketone	$C_5H_{10}O$ / 86.13	563-80-4 / -92	94.3	4-01-00-03287 / 0.8051[20]	H$_2$O 2; EtOH 5; eth 5; ace 4 / 1.3880[20]
3669	1-Butanone, 3-methyl-1-(2-methylphenyl)- / Isobutyl 2-tolyl ketone	$C_{12}H_{16}O$ / 176.26	58138-81-1 /	248; 85-6[3]	4-07-00-00765 / 0.9578[20]	1.5104[20]
3670	1-Butanone, 3-methyl-1-(4-methylphenyl)- / Isobutyl 4-tolyl ketone	$C_{12}H_{16}O$ / 176.26	61971-91-3 /	259; 137[18]	4-07-00-00765 / 0.9557[20]	1.5081[20]
3671	1-Butanone, 3-methyl-1-phenyl-	$C_{11}H_{14}O$ / 162.23	582-62-7 /	236.5	4-07-00-00739 / 0.9701[16]	H$_2$O 1; EtOH 5; eth 5; ace 4 / 1.5139[15]
3672	1-Butanone, 1-(4-methylphenyl)- / Butyrophenone, 4-methyl	$C_{11}H_{14}O$ / 162.23	4160-52-5 / 12	251.5	4-07-00-00743 / 0.9745[20]	eth 4; EtOH 4 / 1.5232[20]
3673	2-Butanone, (1-methylpropylidene)hydrazone	$C_8H_{16}N_2$ / 140.23	5921-54-0 /	171.5	4-01-00-03252 / 0.8404[20]	1.4511[20]
3674	1-Butanone, 1-(1-naphthalenyl)- / 1-Butyronaphthone	$C_{14}H_{14}O$ / 198.26	2876-62-2 /	317	4-07-00-01305 / 1.086[25]	bz 4; eth 4; EtOH 4 / 1.596[27]
3675	2-Butanone, oxime	C_4H_9NO / 87.12	96-29-7 / -29.5	152.5	4-01-00-03250 / 0.9232[20]	H$_2$O 3; EtOH 5; eth 5; chl 3 / 1.4410[20]
3676	1-Butanone, 1-phenyl-	$C_{10}H_{12}O$ / 148.20	495-40-9 / 12	228.5	4-07-00-00709 / 0.988[20]	H$_2$O 1; EtOH 5; eth 5; ace 4 / 1.5203[20]
3677	2-Butanone, 1-phenyl-	$C_{10}H_{12}O$ / 148.20	1007-32-5 /	230; 111[16]	4-07-00-00712 / 1.002[20]	H$_2$O 1; EtOH 3; eth 5; ace 4
3678	2-Butanone, 4-phenyl-	$C_{10}H_{12}O$ / 148.20	2550-26-7 / -13	233.5	4-07-00-00713 / 0.9849[22]	H$_2$O 1; EtOH 3; eth 3; ace 4 / 1.511[22]
3679	2-Butanone, 1,3,4-trihydroxy-, (S)- / L-Erythrulose	$C_4H_8O_4$ / 120.11	533-50-6 / dec	3641 / dec	4-01-00-04176 /	H$_2$O 4; EtOH 4
3680	Butanoyl bromide / Butyryl bromide	C_4H_7BrO / 151.00	5856-82-6 /	129	4-02-00-00804 / 1.4162[17]	1.596[17]
3681	Butanoyl chloride / Butyryl chloride	C_4H_7ClO / 106.55	141-75-3 / -89	1598 / 102	4-02-00-00803 / 1.0277[20]	eth 5 / 1.4121[20]
3682	Butanoyl chloride, 2-bromo-	C_4H_6BrClO / 185.45	22118-12-3 /	151	4-02-00-00834 / 1.5320[20]	eth 4 / 1.5320[20]
3683	Butanoyl chloride, 2-chloro-	$C_4H_6Cl_2O$ / 141.00	7623-11-2 /	130.5	3-02-00-00623 / 1.2360[17]	eth 4 / 1.4475[20]
3684	Butanoyl chloride, 3-chloro- / 3-Chlorobutyryl chloride	$C_4H_6Cl_2O$ / 141.00	1951-11-7 /	40-1[12]	4-02-00-00824 / 1.2163[20]	1.4509[20]
3685	Butanoyl chloride, 4-chloro-	$C_4H_6Cl_2O$ / 141.00	4635-59-0 /	173.5	4-02-00-00826 / 1.2581[20]	eth 3 / 1.4616[20]
3686	Butanoyl chloride, 2-chloro-3-methyl- / Butyryl chloride, 2-chloro-3-methyl	$C_5H_8Cl_2O$ / 155.02	35383-59-6 /	148.5; 56[20]	4-02-00-00904 / 1.135[13]	eth 4
3687	Butanoyl chloride, 4-chloro-3-oxo- / Butyryl chloride, 4-chloro-3-oxo	$C_4H_4Cl_2O_2$ / 154.98	41295-64-1 /	117-9[17]	3-03-01207 / 1.4397[20]	bz 4 / 1.4860[20]
3688	Butanoyl chloride, 2,2-dimethyl-	$C_6H_{11}ClO$ / 134.61	5856-77-9 /	132	4-02-00-00955 / 0.9810[20]	eth 4 / 1.4245[20]
3689	Butanoyl chloride, 2,3-dimethyl- / Butyryl chloride, 2,3-dimethyl	$C_6H_{11}ClO$ / 134.61	51760-90-8 /	136; 38[18]	3-02-00-00761 / 0.9795[20]	eth 4
3690	Butanoyl chloride, 3,3-dimethyl-	$C_6H_{11}ClO$ / 134.61	7065-46-5 /	130; 68[100]	4-02-00-00956 / 0.9696[20]	eth 4 / 1.4210[20]
3691	Butanoyl chloride, 2-ethyl-	$C_6H_{11}ClO$ / 134.61	2736-40-5 /	140	4-02-00-00952 / 0.9825[20]	eth 4 / 1.4234[20]
3692	Butanoyl chloride, heptafluoro-	C_4ClF_7O / 232.48	375-16-6 /	38.5	4-02-00-00818 / 1.55[20]	1.288[20]
3693	Butanoyl chloride, 3-methyl- / Isovaleryl chloride	C_5H_9ClO / 120.58	108-12-3 /	5121 / 114	4-02-00-00901 / 0.9844[20]	eth 3 / 1.4149[20]
3694	Butanoyl chloride, 2-methyl-, (±)- / Butyryl chloride, 2-methyl (DL)	C_5H_9ClO / 120.58	57526-28-0 /	116	4-02-00-00891 / 0.9917[20]	1.4170[20]
3695	2-Butenal, (E)- / trans-Crotonaldehyde	C_4H_6O / 70.09	123-73-9 / -76	2599 / 102.2	4-01-00-03447 / 0.8516[20]	H$_2$O 3; EtOH 4; eth 4; ace 4 / 1.4366[20]
3696	2-Butenal, 2-bromo-	C_4H_5BrO / 148.99	24247-53-8 /	63[15]	3-01-00-02981 / 1.5797[13]	1.5184[20]
3697	2-Butenal, 2-chloro- / α-Chlorocrotonaldehyde	C_4H_5ClO / 104.54	53175-28-3 /	147.5	3-01-00-02981 / 1.1404[23]	H$_2$O 2; EtOH 3; eth 3; ctc 3 / 1.4780[25]
3698	2-Butenal, 2-methyl-	C_5H_8O / 84.12	1115-11-3 /	117; 64[110]	3-01-00-02990 / 0.8710[20]	H$_2$O 3; EtOH 5; eth 5 / 1.4475[20]
3699	2-Butenal, 2-methyl-, (E)-	C_5H_8O / 84.12	497-03-0 /	117; 64[119]	4-01-00-03464 / 0.8710[20]	H$_2$O 4; EtOH 4 / 1.4475[20]
3700	2-Butenal, 3-methyl- / Senecialdehyde	C_5H_8O / 84.12	107-86-8 /	8399 / 134	4-01-00-03464 / 0.8722[20]	H$_2$O 3; EtOH 3; eth 3 / 1.4528[20]

No.	Name Synonym	Mol. Form. Mol. Wt.	CAS RN mp/°C	Merck No. bp/°C	Beil. Ref. den/g cm^{-3}	Solubility n_D
3701	2-Butenal, 3-phenyl-	C$_{10}$H$_{10}$O 146.19	1196-67-4 95	124[12]; 109[4]	4-07-00-01011	1.5876[23]
3702	2-Butenamide, (E)-	C$_4$H$_7$NO 85.11	625-37-6 161.5	sub 140	2-02-00-00392 0.9461[120]	H$_2$O 2; EtOH 3; eth 2; bz 3 1.4420[165]
3703	3-Butenamide, 3-methyl-4-phenyl- β-Benzalbutyramide	C$_{11}$H$_{13}$NO 175.23	7236-47-7 133	1063		
3704	1-Butene 1-Butylene	C$_4$H$_8$ 56.11	106-98-9 -185.3	1513 -6.2	4-01-00-00765 *0.588[25]	H$_2$O 1; EtOH 4; eth 4; bz 3 1.3962[20]
3705	2-Butene, (E)- trans-2-Butene	C$_4$H$_8$ 56.11	624-64-6 -105.5	1514 0.8	4-01-00-00781 *0.599[25]	bz 3 1.3848[-25]
3706	2-Butene, (Z)- cis-2-Butene	C$_4$H$_8$ 56.11	590-18-1 -138.9	1514 3.7	4-01-00-00778 *0.616[25]	H$_2$O 1; EtOH 4; eth 4; bz 3 1.3931[-25]
3707	1-Butene, 1-bromo-, (E)- trans-1-Butenyl bromide	C$_4$H$_7$Br 135.00	32620-08-9 -100.3	94.7	1.3209[15]	H$_2$O 1; eth 3; ace 3; bz 3 1.4527[20]
3708	1-Butene, 1-bromo-, (Z)-	C$_4$H$_7$Br 135.00	31849-78-2	86.1	1.3265[15]	H$_2$O 1; eth 3; ace 3; bz 3 1.4536[20]
3709	1-Butene, 2-bromo- 2-Bromo-1-butene	C$_4$H$_7$Br 135.00	23074-36-4 -133.4	88	4-01-00-00775 1.3209[15]	H$_2$O 1; eth 3; ace 3; bz 3 1.4527[20]
3710	1-Butene, 3-bromo- α-Methallyl bromide	C$_4$H$_7$Br 135.00	22037-73-6	7[10]	4-01-00-00775 1.2998[25]	1.4618[20]
3711	1-Butene, 4-bromo-	C$_4$H$_7$Br 135.00	5162-44-7	98.5	4-01-00-00775 1.3230[20]	bz 4; eth 4; EtOH 4 1.4622[20]
3712	2-Butene, 1-bromo-	C$_4$H$_7$Br 135.00	4784-77-4	104.5	4-01-00-00789 1.3371[25]	H$_2$O 1; EtOH 3; eth 3; bz 4 1.4822[20]
3713	2-Butene, 2-bromo-, (E)-	C$_4$H$_7$Br 135.00	3017-71-8 -114.6	85.6	4-01-00-00790 1.3323[15]	H$_2$O 1; EtOH 3; eth 3; bz 4 1.4602[16]
3714	2-Butene, 2-bromo-, (Z)-	C$_4$H$_7$Br 135.00	3017-68-3 -111.5	93.9	4-01-00-00790 1.3416[15]	H$_2$O 1; EtOH 3; eth 3; bz 4 1.4631[19]
3715	1-Butene, 4-bromo-3-chloro-3,4,4-trifluoro- 4-Bromo-3-chloro-3,4,4-trifluoro-1-butene	C$_4$H$_3$BrClF$_3$ 223.42	374-25-4	99.5	4-01-00-00776 1.678[25]	1.4092[25]
3716	1-Butene, 2-bromo-3-methyl- 2-Bromo-3-methyl-1-butene	C$_5$H$_9$Br 149.03	31844-96-9	105	3-01-00-00800 1.2328[20]	bz 4; eth 4; chl 4 1.4504[20]
3717	2-Butene, 1-bromo-3-methyl-	C$_5$H$_9$Br 149.03	870-63-3	131 dec; 50[40]	4-01-00-00824 1.2817[20]	ace 4; bz 4; eth 4; EtOH 4 1.4930[15]
3718	2-Butene, 2-bromo-3-methyl-	C$_5$H$_9$Br 149.03	3017-70-7	119.5	4-01-00-00824 1.2773[20]	ace 4; eth 4; chl 4 1.4738[20]
3719	2-Butene, 2-bromo-3-phenyl- 2-Bromo-3-phenyl-2-butene	C$_{10}$H$_{11}$Br 211.10	90841-14-8	120-30[11]	4-05-00-01379 1.3348[20]	bz 4; eth 4; chl 4 1.5811[20]
3720	1-Butene, 1-chloro-, (E)- trans-1-Chloro-1-butene	C$_4$H$_7$Cl 90.55	7611-87-2	68	3-01-00-00723 0.9205[15]	H$_2$O 1; EtOH 5; eth 4; ace 3 1.4223[15]
3721	1-Butene, 1-chloro-, (Z)- cis-1-Chloro-1-butene	C$_4$H$_7$Cl 90.55	7611-86-1	63.5	3-01-00-00723 0.9153[15]	ace 4; eth 4; EtOH 4; chl 4 1.4194[15]
3722	1-Butene, 2-chloro- 2-Chloro-1-butene	C$_4$H$_7$Cl 90.55	2211-70-3	58.5	4-01-00-00769 0.9107[15]	ace 4; bz 4; eth 4; EtOH 4 1.4165[21]
3723	1-Butene, 3-chloro- 3-Chloro-1-butene	C$_4$H$_7$Cl 90.55	563-52-0	2131 64.5	4-01-00-00769 0.8978[20]	eth 4; ace 4; chl 3 1.4149[20]
3724	1-Butene, 4-chloro-	C$_4$H$_7$Cl 90.55	927-73-1	75	4-01-00-00771 0.9211[20]	ace 4; eth 4; chl 4 1.4233[20]
3725	2-Butene, 1-chloro-, (E)-	C$_4$H$_7$Cl 90.55	4894-61-5	85	4-01-00-00784 0.9295[20]	H$_2$O 1; ace 3; chl 3 1.4350[20]
3726	2-Butene, 1-chloro-, (Z)-	C$_4$H$_7$Cl 90.55	4628-21-1	84.1	4-01-00-00783 0.9426[20]	H$_2$O 1; EtOH 3; ace 3; chl 3 1.4390[20]
3727	2-Butene, 2-chloro-, (E)-	C$_4$H$_7$Cl 90.55	2211-68-9 -105.8	62.8	4-01-00-00785 0.9138[20]	H$_2$O 1; EtOH 5; ace 3; chl 3 1.4190[20]
3728	2-Butene, 2-chloro-, (Z)-	C$_4$H$_7$Cl 90.55	2211-69-0 -117.3	70.6	4-01-00-00785 0.9239[20]	H$_2$O 1; EtOH 5; ace 3; chl 3 1.4240[20]
3729	1-Butene, 3-chloro-2-(chloromethyl)- 3-Chloro-2-(chloromethyl)-1-butene	C$_5$H$_8$Cl$_2$ 139.02	69295-21-2	155	3-01-00-00787 1.1233[20]	ace 4; chl 4 1.4724[20]
3730	2-Butene, 2-chloro-1,1-dimethoxy- Crotonaldehyde-2-chloro, dimethylacetal	C$_6$H$_{11}$ClO$_2$ 150.60	108365-83-9	58[13]	3-01-00-02981 1.074[18]	1.4466[18]
3731	2-Butene, 1-chloro-2,3-dimethyl- 1-Chloro-2,3-dimethyl-2-butene	C$_6$H$_{11}$Cl 118.61	37866-06-1	112	4-01-00-00856 0.9355[20]	ace 4; eth 4; chl 4 1.4605[20]
3732	1-Butene, 1-chloro-2-methyl- 1-Chloro-2-methyl-1-butene	C$_5$H$_9$Cl 104.58	23378-11-2	96.5	3-01-00-00787 0.9170[20]	ace 4; eth 4 1.4141[20]
3733	1-Butene, 1-chloro-3-methyl- 1-Chloro-3-methyl-1-butene	C$_5$H$_9$Cl 104.58	23010-00-6	87	3-01-00-00799	ace 4; eth 4; chl 4 1.4229[20]
3734	1-Butene, 3-chloro-2-methyl- 3-Chloro-2-methyl-1-butene	C$_5$H$_9$Cl 104.58	5166-35-8	94	4-01-00-00819 0.9088[20]	ace 4; eth 4; chl 4 1.4304[20]
3735	2-Butene, 1-chloro-2-methyl- 1-Chloro-2-methyl-2-butene	C$_5$H$_9$Cl 104.58	13417-43-1	110	4-01-00-00822 0.9327[20]	ace 4; eth 4; EtOH 4 1.4481[20]
3736	2-Butene, 1-chloro-3-methyl-	C$_5$H$_9$Cl 104.58	503-60-6	109	4-01-00-00823 0.9273[20]	ace 4; eth 4; EtOH 4; chl 4 1.4485[20]
3737	2-Butene, 2-chloro-3-methyl- 2-Chloro-3-methyl-2-butene	C$_5$H$_9$Cl 104.58	17773-65-8	96	3-01-00-00794 0.9324[20]	ace 4; eth 4; EtOH 4; chl 4 1.4320[20]
3738	1-Butene, 1,1-dibromo- 1,1-Dibromo-1-butene	C$_4$H$_6$Br$_2$ 213.90	73383-24-1	53-5[22]	3-01-00-00727 1.8315[20]	1.5168[20]
3739	1-Butene, 1,2-dibromo- 1,2-dibromo-1-butene	C$_4$H$_6$Br$_2$ 213.90	55030-56-3 -49.5	150	0-01-00-00204 1.887[25]	eth 4; EtOH 4
3740	1-Butene, 3,4-dibromo- 1,2-Dibromo-3-butene	C$_4$H$_6$Br$_2$ 213.90	10463-48-6	52[10]	4-01-00-00776 1.865[24]	1.541[21]

No.	Name / Synonym	Mol. Form. / Mol. Wt.	CAS RN / mp/°C	Merck No. / bp/°C	Beil. Ref. / den/g cm⁻³	Solubility / n_D
3741	2-Butene, 1,3-dibromo- / 2,4-Dibromo-2-butene	$C_4H_6Br_2$ / 213.90	64930-16-1	168 dec	3-01-00-00746 / 1.8768[20]	1.5485[20]
3742	2-Butene, 1,4-dibromo-, (E)-	$C_4H_6Br_2$ / 213.90	821-06-7 / 53.4	203; 74[14]	4-01-00-00791	H_2O 2; EtOH 4; ace 3; chl 2
3743	1-Butene, 1,3-dichloro- / 1,3-Dichloro-1-butene	$C_4H_6Cl_2$ / 125.00	52497-07-1	125	4-01-00-00772 / 1.1341[24]	ace 4; eth 4; EtOH 4; chl 4 / 1.4647[20]
3744	1-Butene, 2,3-dichloro- / 2,3-Dichloro-1-butene	$C_4H_6Cl_2$ / 125.00	7013-11-8	112	4-01-00-00772 / 1.1340[20]	ace 4; eth 4; chl 4 / 1.4580[20]
3745	1-Butene, 3,4-dichloro- / 1,2-Dichloro-3-butene	$C_4H_6Cl_2$ / 125.00	760-23-6 / -61	116	4-01-00-00772 / 1.1170[20]	H_2O 1; EtOH 3; eth 3; bz 4 / 1.4641[20]
3746	2-Butene, 1,1-dichloro- / 1,1-Dichloro-2-butene	$C_4H_6Cl_2$ / 125.00	56800-09-0	126	3-01-00-00741 / 1.1310[20]	1.466[18]
3747	2-Butene, 1,2-dichloro- / 1,2-Dichloro-2-butene	$C_4H_6Cl_2$ / 125.00	13602-13-6	130.5	3-01-00-00741 / 1.1601[20]	ace 4; eth 4; chl 4 / 1.4734[20]
3748	2-Butene, 1,3-dichloro-, (E)-	$C_4H_6Cl_2$ / 125.00	7415-31-8	131; 53[50]	4-01-00-00786 / 1.1585[20]	ace 4; bz 4; eth 4; EtOH 4 / 1.4719[20]
3749	2-Butene, 1,3-dichloro-, (Z)-	$C_4H_6Cl_2$ / 125.00	10075-38-4	131; 34[20]	4-01-00-00786 / 1.1605[20]	ace 4; bz 4; eth 4; EtOH 4 / 1.4735[20]
3750	2-Butene, 1,4-dichloro-, (E)- / 1,4-Dichloro-trans-2-butene	$C_4H_6Cl_2$ / 125.00	110-57-6 / 2.0	155.4	4-01-00-00787 / 1.183[25]	ace 4; bz 4; eth 4; EtOH 4 / 1.4871[25]
3751	2-Butene, 1,4-dichloro-, (Z)-	$C_4H_6Cl_2$ / 125.00	1476-11-5 / -48	152.5	4-01-00-00787 / 1.188[25]	ace 4; bz 4; eth 4; EtOH 4 / 1.4887[25]
3752	2-Butene, 2,3-dichloro-, (E)-	$C_4H_6Cl_2$ / 125.00	1587-29-7	102	4-01-00-00787 / 1.1416[20]	H_2O 1; EtOH 3; eth 3; ace 3 / 1.4582[20]
3753	2-Butene, 2,3-dichloro-, (Z)-	$C_4H_6Cl_2$ / 125.00	1587-26-4	125.5	4-01-00-00787 / 1.1618[20]	H_2O 1; EtOH 3; eth 3; ace 3 / 1.4590[20]
3754	2-Butene, 2,3-dichloro-1,1,1,4,4,4-hexafluoro- / 2,3-Dichlorohexafluoro-2-butene	$C_4Cl_2F_6$ / 232.94	303-04-8 / -67.3	68.5	4-01-00-00787 / 1.6233[20]	1.3459[20]
3755	1-Butene, 3,3-dichloro-2-methyl- / 3,3-Dichloro-2-methyl-1-butene	$C_5H_8Cl_2$ / 139.02	42101-38-2		3-01-00-00787 / 1.1276[18]	ace 4; eth 4; chl 4 / 1.4737[18]
3756	2-Butene, 1,3-dichloro-2-methyl- / 1,3-Dichloro-2-methyl-2-butene	$C_5H_8Cl_2$ / 139.02	25148-87-2	150	3-01-00-00795 / 1.1293[20]	ace 4; chl 4
3757	2-Butene, 1,4-dichloro-2-methyl- / 1,4-Dichloro-2-methyl-2-butene	$C_5H_8Cl_2$ / 139.02	29843-58-1	93[50]; 56[10]	3-01-00-00795 / 1.1526[20]	ace 4; chl 4 / 1.4932[20]
3758	2-Butene, 1,1-diethoxy- / Crotonaldehyde, diethyl acetal	$C_8H_{16}O_2$ / 144.21	10602-34-3	147.5	3-01-00-02980 / 0.8473[18]	H_2O 1; EtOH 5; eth 5; ace 5 / 1.4097[20]
3759	1-Butene, 2,3-dimethyl- / 2,3-Dimethyl-1-butene	C_6H_{12} / 84.16	563-78-0 / -157.3	55.6	4-01-00-00852 / 0.6803[20]	H_2O 1; EtOH 3; eth 3; ace 3 / 1.3995[20]
3760	1-Butene, 3,3-dimethyl-	C_6H_{12} / 84.16	558-37-2 / -115.2	41.2	4-01-00-00850 / 0.6529[20]	H_2O 1; EtOH 3; eth 3; ctc 3 / 1.3763[20]
3761	2-Butene, 2,3-dimethyl- / 2,3-Dimethyl-2-butene	C_6H_{12} / 84.16	563-79-1 / -74.6	73.3	4-01-00-00853 / 0.7080[20]	H_2O 1; EtOH 3; eth 3; ace 3 / 1.4122[20]
3762	2-Butenedinitrile, (E)-	$C_4H_2N_2$ / 78.07	764-42-1 / 96.8	186	4-02-00-02219 / 0.9416[111]	H_2O 3; EtOH 3; eth 3; ace 3 / 1.4349[111]
3763	2-Butenedinitrile, 2-chloro-, (Z)- / Fumaronitrile, chloro-	C_4HClN_2 / 112.52	71200-79-8 / 60.5	172; 68[25]	3-02-00-01909 / 1.2499[20]	1.4957[20]
3764	2-Butenedioic acid (E)- / Fumaric acid	$C_4H_4O_4$ / 116.07	110-17-8 / 287 dec	4200 / sub 165	3-02-00-01890 / 1.635[20]	H_2O 2; EtOH 3; eth 2; ace 2
3765	2-Butenedioic acid (Z)- / Maleic acid	$C_4H_4O_4$ / 116.07	110-16-7 / 130.5	5585	4-02-00-02199 / 1.590[20]	H_2O 4; EtOH 4; eth 3; ace 4
3766	2-Butenedioic acid (E)-, bis(1-methylethyl) ester / Diisopropyl fumarate	$C_{10}H_{16}O_4$ / 200.23	7283-70-7 / 2.1	280	4-02-00-02209	ace 4; eth 4; EtOH 4
3767	2-Butenedioic acid (E)-, bis(2-methylpropyl) ester / Diisobutyl fumarate	$C_{12}H_{20}O_4$ / 228.29	7283-69-4	170[160]; 122[5]	4-02-00-02211 / 0.9760[20]	H_2O 1; EtOH 3; eth 3; ace 3 / 1.4432[20]
3768	2-Butenedioic acid (E)-, bis(phenylmethyl) ester / Benzyl fumarate	$C_{18}H_{18}O_4$ / 296.32	538-64-7 / 59	1149 / 210-11[5]	4-06-00-02275	eth 4; EtOH 4; chl 4
3769	2-Butenedioic acid, 2-bromo-, (E)-	$C_4H_3BrO_4$ / 194.97	584-99-6 / 141	dec	4-02-00-02224	H_2O 4; EtOH 4; eth 4
3770	2-Butenedioic acid, 2-bromo-, (Z)- / 2-Bromofumaric acid	$C_4H_3BrO_4$ / 194.97	644-80-4 / 184.3	200 dec	4-02-00-02224	H_2O 4; EtOH 4
3771	2-Butenedioic acid, 2-chloro-, diethyl ester, (E)- / Diethyl chloromaleate	$C_8H_{11}ClO_4$ / 206.63	626-10-8	235 dec; 125[19]	3-02-00-01928 / 1.1741[20]	ace 4; eth 4; EtOH 4
3772	2-Butenedioic acid, 2-chloro-, diethyl ester, (Z)-	$C_8H_{11}ClO_4$ / 206.63	10302-94-0	250 dec; 127[10]	3-02-00-01909 / 1.1880[20]	eth 4; EtOH 4 / 1.4571[20]
3773	2-Butenedioic acid, 2-chloro-, dimethyl ester, (E)- / Dimethyl chloromaleate	$C_6H_7ClO_4$ / 178.57	19393-45-4	224; 100[17]	3-02-00-01928 / 1.2899[25]	eth 4; EtOH 4 / 1.472[18]
3774	2-Butenedioic acid (E)-, dibutyl ester	$C_{12}H_{20}O_4$ / 228.29	105-75-9	150[4]	4-02-00-02210 / 0.9869[20]	H_2O 1; ace 3; chl 3 / 1.4469[20]
3775	2-Butenedioic acid (E)-, diethyl ester	$C_8H_{12}O_4$ / 172.18	623-91-6 / 1.5	214	4-02-00-02207 / 1.0452[20]	H_2O 1; ace 3; chl 3 / 1.4412[20]
3776	2-Butenedioic acid (Z)-, diethyl ester / Diethyl maleate	$C_8H_{12}O_4$ / 172.18	141-05-9 / -8.8	3113 / 223	4-02-00-02207 / 1.0662[20]	H_2O 1; EtOH 3; eth 3; chl 2 / 1.4416[20]

No.	Name / Synonym	Mol. Form. / Mol. Wt.	CAS RN / mp/°C	Merck No. / bp/°C	Beil. Ref. / den/g cm⁻³	Solubility / n_D
3777	2-Butenedioic acid, 2,3-dihydroxy-, (Z)- / Dihydroxymaleic acid	$C_4H_4O_6$ / 148.07	526-84-1 / 155 dec	3170	4-03-00-01975	H_2O 2; EtOH 3; eth 2; MeOH 2
3778	2-Butenedioic acid (E)-, dimethyl ester	$C_6H_8O_4$ / 144.13	624-49-7 / 103.5	193	4-02-00-02205 / 1.37[20]	H_2O 1; ace 3; chl 3 / 1.4062[111]
3779	2-Butenedioic acid (Z)-, dimethyl ester	$C_6H_8O_4$ / 144.13	624-48-6 / -19	202	4-02-00-02204 / 1.1606[20]	H_2O 1; eth 3; ctc 3; lig 1 / 1.4416[20]
3780	2-Butenedioic acid (Z)-, diphenyl ester	$C_{16}H_{12}O_4$ / 268.27	7242-17-3 / 73	226[15]	4-06-00-00628	ace 4; bz 4; eth 4; EtOH 4
3781	2-Butenedioic acid (E)-, di-2-propenyl ester / Diallyl fumarate	$C_{10}H_{12}O_4$ / 196.20	2807-54-7	140[3]	3-02-00-01906 / 1.060[25]	ace 4; bz 4; eth 4; EtOH 4 / 1.4670[25]
3782	2-Butenedioic acid (Z)-, di-2-propenyl ester / Diallyl maleate	$C_{10}H_{12}O_4$ / 196.20	999-21-3	108-10[3]	4-02-00-02214 / 1.0773[20]	chl 3 / 1.4699[20]
3783	2-Butenedioic acid (E)-, dipropyl ester / Dipropyl fumarate	$C_{10}H_{16}O_4$ / 200.23	14595-35-8	110[5]	4-02-00-02209 / 1.0129[20]	EtOH 3; eth 3 / 1.4435[20]
3784	2-Butenedioic acid (Z)-, dipropyl ester	$C_{10}H_{16}O_4$ / 200.23	2432-63-5	126[12]	4-02-00-02209 / 1.0245[20]	H_2O 1; EtOH 3; eth 3; ace 3 / 1.4434[20]
3785	2-Butenedioic acid, 2-methyl-, (E)- / Mesaconic acid	$C_5H_6O_4$ / 130.10	498-24-8 / 204.5	5806 sub	4-02-00-02231 / 1.466[20]	H_2O 2; EtOH 4; eth 3; bz 2
3786	2-Butenedioic acid, 2-methyl-, (Z)- / Citraconic acid	$C_5H_6O_4$ / 130.10	498-23-7 / 93.5	2323	4-02-00-02230 / 1.617[25]	H_2O 4; eth 2; bz 1; chl 2
3787	2-Butenedioic acid, 2-methyl-, diethyl ester, (E)- / Diethyl mesaconate	$C_9H_{14}O_4$ / 186.21	2418-31-7	229	4-02-00-02232 / 1.0434[20]	ace 4; bz 4; eth 4; EtOH 4 / 1.4488[20]
3788	2-Butenedioic acid, 2-methyl-, diethyl ester, (Z)-	$C_9H_{14}O_4$ / 186.21	691-83-8	228; 120[20]	4-02-00-02232 / 1.0491[20]	eth 4; EtOH 4; HOAc 4 / 1.4467[20]
3789	2-Butenedioic acid, 2-methyl-, dimethyl ester, (E)-	$C_7H_{10}O_4$ / 158.15	617-53-8	203.5	4-02-00-02232 / 1.0914[20]	ace 4; eth 4; EtOH 4 / 1.4512[20]
3790	2-Butenedioic acid, 2-methyl-, dimethyl ester, (Z)-	$C_7H_{10}O_4$ / 158.15	617-54-9	210.5	4-02-00-02232 / 1.1153[20]	ace 4; eth 4; EtOH 4 / 1.4473[20]
3791	2-Butenedioic acid (E)-, monoethyl ester	$C_6H_8O_4$ / 144.13	2459-05-4 / 70	147[16]	4-02-00-02206 / 1.1109[87]	EtOH 3; ace 3; chl 2
3792	2-Butene-1,4-diol, (E)- / trans-2-Butene-1,4-diol	$C_4H_8O_2$ / 88.11	821-11-4 / 25	131[13]	4-01-00-02660 / 1.0700[20]	H_2O 4; EtOH 4 / 1.4755[20]
3793	2-Butene-1,4-diol, (Z)- / cis-2-Butene-1,4-diol	$C_4H_8O_2$ / 88.11	6117-80-2 / 4	235	4-01-00-02660 / 1.0698[20]	H_2O 3; EtOH 4 / 1.4782[20]
3794	3-Butene-1,2-diol	$C_4H_8O_2$ / 88.11	497-06-3	196.5	4-01-00-02658 / 1.0470[20]	H_2O 3; EtOH 3 / 1.4628[21]
3795	2-Butenedioyl dichloride, (E)- / Fumaric acid dichloride	$C_4H_2Cl_2O_2$ / 152.96	627-63-4 / 159		4-02-00-02217 / 1.408[20]	1.5004[18]
3796	2-Butenedioyl dichloride, 2-chloro-, (Z)-	$C_4HCl_3O_2$ / 187.41	17096-37-6 / 185 dec		3-02-00-01909 / 1.564[20]	eth 4; HOAc 4 / 1.5206[20]
3797	1-Butene, 2-ethyl-3-methyl- / 2-Ethyl-3-methylbut-1-ene	C_7H_{14} / 98.19	7357-93-9 / 89		4-01-00-00871 / 0.7150[20]	H_2O 1; eth 3; ace 3; bz 3 / 1.410[20]
3798	2-Butene, 1,1,2,3,4,4-hexachloro- / 2-Butene, 1,1,2,3,4,4-hexachloro (liquid)	$C_4H_2Cl_6$ / 262.78	13045-99-3 / -19	97-8[10]	4-01-00-00789 / 1.650[15]	H_2O 1; chl 3 / 1.5331
3799	1-Butene, 2-methyl- / 2-Methyl-1-butene	C_5H_{10} / 70.13	563-46-2 / -137.5	31.2	4-01-00-00818 / 0.6504[20]	H_2O 1; EtOH 3; eth 3; bz 3 / 1.3378[20]
3800	1-Butene, 3-methyl- / 3-Methyl-1-butene	C_5H_{10} / 70.13	563-45-1 / -168.5	20.1	4-01-00-00825 / 0.6213[25]	H_2O 1; EtOH 5; eth 5; bz 3 / 1.3643[20]
3801	2-Butene, 2-methyl- / 2-Methyl-2-butene	C_5H_{10} / 70.13	513-35-9 / -133.7	644 / 38.5	4-01-00-00820 / 0.6623[20]	H_2O 1; EtOH 3; eth 3; bz 3 / 1.3874[20]
3802	2-Butenenitrile, (E)-	C_4H_5N / 67.09	627-26-9 / -51.5	120	4-02-00-01507 / 0.8239[20]	eth 3; ace 3 / 1.4225[20]
3803	3-Butenenitrile / Allyl cyanide	C_4H_5N / 67.09	109-75-1 / -87	288 / 119	4-02-00-01491 / 0.8341[20]	H_2O 2; EtOH 5; eth 5 / 1.4060[20]
3804	2-Butene, 1,1,1,2,3,4,4,4-octafluoro- / Perfluoro-2-butene	C_4F_8 / 200.03	360-89-4 / -129	1.5	4-01-00-00783 / 1.5297[25]	
3805	2-Butene, 1,1,1,4-pentachloro- / 1,1,1,4,4-Pentachloro-2-butene	$C_4H_3Cl_5$ / 228.33	77753-21-0	78-80[11]	3-01-00-00744 / 1.609[21]	EtOH 4; chl 4 / 1.5538[21]
3806	1-Butene, 1,3,4,4-tetrachloro- / 1,3,4,4-Tetrachloro-1-butene	$C_4H_4Cl_4$ / 193.89	2984-40-9	88[20]	4-01-00-00774 / 1.476[20]	chl 4 / 1.4773[20]
3807	1-Butene, 2,3,3,4-tetrachloro- / 2,3,3,4-Tetrachloro-1-butene	$C_4H_4Cl_4$ / 193.89	64346-45-8	41-2[7]	3-01-00-00726 / 1.4602[20]	ace 4; chl 4 / 1.5135[20]
3808	1-Butene, 1,3,4,4-tetrachloro-1,2,3,4-tetrafluoro- / 1-Butene, 1,3,4,4-tetrachloro-tetrafluoro	$C_4Cl_4F_4$ / 265.85	357-20-0	140.8	4-01-00-00774 / 1.6902[25]	1.429[20]
3809	1-Butene, 2,3,4-trichloro- / 2,3,4-Trichloro-1-butene	$C_4H_5Cl_3$ / 159.44	2431-50-7	60[20]	3-01-00-00725 / 1.3430[20]	ace 4; chl 4 / 1.4944[20]
3810	2-Butene, 1,2,4-trichloro- / 1,2,4-Trichloro-2-butene	$C_4H_5Cl_3$ / 159.44	2431-54-1	67-9[10]	3-01-00-00744 / 1.3843[20]	bz 4; eth 4; EtOH 4; chl 4 / 1.5175[20]
3811	2-Butene, 1,1,3-triethoxy- / 2-Butenal, 3-ethoxy, diethyl acetal	$C_{10}H_{20}O_3$ / 188.27	69190-65-4	192.5	4-01-00-04082 / 0.908[21]	EtOH 4 / 1.430[21]
3812	1-Butene, 2,3,3-trimethyl-	C_7H_{14} / 98.19	594-56-9 / -109.9	77.9	4-01-00-00873 / 0.7050[20]	H_2O 1; eth 3; bz 3; chl 3 / 1.4025[20]
3813	3-Butenoic acid	$C_4H_8O_2$ / 86.09	625-38-7 / -35	169	4-02-00-01491 / 1.0091[20]	H_2O 3; EtOH 5; eth 5 / 1.4239[20]

No.	Name / Synonym	Mol. Form. / Mol. Wt.	CAS RN / mp/°C	Merck No. / bp/°C	Beil. Ref. / den/g cm^{-3}	Solubility / n_D
3814	2-Butenoic acid, (E)- / trans-Crotonic acid	$C_4H_8O_2$ / 86.09	107-93-7 / 72	2600 / 184.7	4-02-00-01498 / 0.9604[77]	H_2O 4; EtOH 4; eth 3; ace 3 / 1.4249[77]
3815	2-Butenoic acid, (Z)- / cis-Crotonic acid	$C_4H_8O_2$ / 86.09	503-64-0 / 15	5048 / 169	4-02-00-01497 / 1.0267[20]	H_2O 4; EtOH 3 / 1.4483[14]
3816	2-Butenoic acid, 3-amino-, ethyl ester, (E)-	$C_6H_{11}NO_2$ / 129.16	41867-20-3 / 34	/ 212 dec	3-03-00-01199 / 1.0219[19]	H_2O 1; EtOH 3; eth 3; bz 3 / 1.4988[22]
3817	2-Butenoic acid, 4-amino-4-oxo-, (Z)- / Maleamic acid	$C_4H_5NO_3$ / 115.09	557-24-4 / 172.5	5583	4-02-00-02218	H_2O 4; EtOH 4
3818	2-Butenoic acid, anhydride / Crotonic acid anhydride	$C_8H_{10}O_3$ / 154.17	623-68-7	/ 247; 129[19]	4-02-00-01505 / 1.0397[20]	eth 4 / 1.4745[20]
3819	2-Butenoic acid, 4-bromo-, ethyl ester, (E)-	$C_6H_9BrO_2$ / 193.04	37746-78-4	/ 97-8[15]	4-02-00-01517 / 1.402[16]	EtOH 4 / 1.4925[20]
3820	2-Butenoic acid, 3-chloro-, (E)-	$C_4H_5ClO_2$ / 120.54	6214-28-4 / 61	/ 195	4-02-00-01510 / 1.1969[69]	H_2O 2; EtOH 4; chl 3; CS_2 4
3821	2-Butenoic acid, 3-chloro-, (Z)-	$C_4H_5ClO_2$ / 120.54	6213-90-7 / 61		4-02-00-01510 / 1.1905[66]	H_2O 2; EtOH 4; peth 4 / 1.4704[66]
3822	2-Butenoic acid, 4-chloro- / γ-Chlorocrotonic acid	$C_4H_5ClO_2$ / 120.54	16197-90-3 / 83	/ 117[13]	4-02-00-01511	eth 4; HOAc 4
3823	2-Butenoic acid, 2-chloro-, ethyl ester / Ethyl β-chlorocrotonate	$C_6H_9ClO_2$ / 148.59	56216-12-7 /	/ 176	4-02-00-01510 / 1.1133[20]	1.453[20]
3824	2-Butenoic acid, 2-chloro-, ethyl ester, (E)- / 2-Butenoic acid, 2-chloro, ethyl ester (trans)	$C_6H_9ClO_2$ / 148.59	77825-53-7	/ 177; 75[30]	4-02-00-01510 / 1.1135[20]	eth 4; EtOH 4 / 1.4538[20]
3825	2-Butenoic acid, 2-chloro-, ethyl ester, (Z)- / 2-Butenoic acid, 2-chloro, ethyl ester (cis)	$C_6H_9ClO_2$ / 148.59	77825-54-8	/ 75[30]	4-02-00-01510 / 1.1021[18]	eth 4; EtOH 4
3826	2-Butenoic acid, 3-chloro-, ethyl ester / 3-Chlorocrotonic acid ethyl ester	$C_6H_9ClO_2$ / 148.59	38624-62-3	/ 180; 75[14]	4-02-00-01511 / 1.1062[20]	1.459[20]
3827	2-Butenoic acid, 3-chloro-, ethyl ester, (E)-	$C_6H_9ClO_2$ / 148.59	6127-92-0	/ 184	4-02-00-01511 / 1.1062[20]	eth 4; EtOH 4 / 1.4592[20]
3828	2-Butenoic acid, 3-chloro-, ethyl ester, (Z)-	$C_6H_9ClO_2$ / 148.59	6127-93-1	/ 161.4	4-02-00-01511 / 1.0860[20]	eth 4; EtOH 4 / 1.4542[19]
3829	2-Butenoic acid, 4-chloro-, ethyl ester / Crotonic acid, 4-chloro-, ethyl ester	$C_6H_9ClO_2$ / 148.59	15333-22-9	/ 93[10]	4-02-00-01511 / 1.090[18]	eth 4 / 1.462[18]
3830	2-Butenoic acid, 2-chloro-, methyl ester, (E)-	$C_5H_7ClO_2$ / 134.56	22038-57-9	/ 161.5; 63[17]	4-02-00-01510 / 1.160[20]	eth 4 / 1.4569[23]
3831	2-Butenoic acid, 3-chloro-, methyl ester, (E)-	$C_5H_7ClO_2$ / 134.56	6372-01-6	/ 64-7[14]	4-02-00-01510 / 1.157[20]	H_2O 2; EtOH 4; eth 4 / 1.4630[20]
3832	2-Butenoic acid, 3-chloro-, methyl ester, (Z)-	$C_5H_7ClO_2$ / 134.56	6214-25-1	/ 142.4; 70[24]	4-02-00-01500 / 1.138[20]	H_2O 2; eth 4; MeOH 4 / 1.4573[19]
3833	2-Butenoic acid, 4-chloro-, methyl ester / Methyl γ-chlorocrotonate	$C_5H_7ClO_2$ / 134.56	15320-72-6	/ 68[11]	4-02-00-01511 / 1.169[18]	1.467[18]
3834	2-Butenoic acid, 3-cyclohexyl- / Cicrotoic acid	$C_{10}H_{16}O_2$ / 168.24	25229-42-9 / 85.5	2272		
3835	3-Butenoic acid, 2,2-dimethyl- / 2,2-Dimethyl-3-butenoic acid	$C_6H_{10}O_2$ / 114.14	10276-09-2 / -6	/ 185	4-02-00-01574 / 0.9567[15]	1.4305[15]
3836	2-Butenoic acid, 2,3-dimethyl-, ethyl ester / Ethyl 2,3-dimethyl-2-butenoate	$C_8H_{14}O_2$ / 142.20	13979-28-7	/ 155	4-02-00-01576 / 0.9072[19]	1.430[19]
3837	2-Butenoic acid, 2-ethyl-, (E)- / 2-Ethylcrotonic acid	$C_6H_{10}O_2$ / 114.14	1187-13-9 / 45.5	/ 209	3-02-00-01329 / 0.9578[50]	eth 4; EtOH 4 / 1.4475[50]
3838	2-Butenoic acid, ethyl ester	$C_6H_{10}O_2$ / 114.14	10544-63-5	/ 136.5	2-02-00-00394 / 0.9175[20]	eth 4; EtOH 4 / 1.4243[20]
3839	2-Butenoic acid, ethyl ester, (E)- / Ethyl trans-crotonate	$C_6H_{10}O_2$ / 114.14	623-70-1	2600 / 138	4-02-00-01500 / 0.9175[20]	H_2O 1; EtOH 3; eth 3 / 1.4243[20]
3840	2-Butenoic acid, ethyl ester, (Z)- / cis-Ethyl crotonate	$C_6H_{10}O_2$ / 114.14	6776-19-8	2600 / 136	4-02-00-01497 / 0.9182[20]	ace 4; eth 4; EtOH 4 / 1.4242[20]
3841	3-Butenoic acid, ethyl ester / Ethyl 3-butenoate	$C_6H_{10}O_2$ / 114.14	1617-18-1	/ 119	4-02-00-01491 / 0.9122[20]	EtOH 3 / 1.4105[20]
3842	3-Butenoic acid, 2-hydroxy- / Vinylglycolic acid	$C_4H_6O_3$ / 102.09	600-17-9 / 43	/ 124[9]	3-03-00-00685 / 1.2020[21]	H_2O 4; eth 4; EtOH 4; chl 4
3843	3-Butenoic acid, 2-hydroxy-, ethyl ester / Ethyl 2-hydroxy-3-butenoate	$C_6H_{10}O_3$ / 130.14	91890-87-8	/ 173 dec; 68[15]	3-03-00-00685 / 1.0470[15]	H_2O 4; eth 4; EtOH 4 / 1.436[13]
3844	2-Butenoic acid, 2-methyl-, (E)- / Tiglic acid	$C_5H_8O_2$ / 100.12	80-59-1 / 64.5	9365 / 198.5	4-02-00-01552 / 0.9641[78]	eth 4; EtOH 4 / 1.4330[76]
3845	2-Butenoic acid, 2-methyl-, (Z)- / Angelic acid	$C_5H_8O_2$ / 100.12	565-63-9 / 45.5	678 / 185	4-02-00-01551 / 0.9834[49]	H_2O 2; EtOH 3; eth 4 / 1.4434[47]
3846	2-Butenoic acid, 3-methyl-	$C_5H_8O_2$ / 100.12	541-47-9 / 69.5	/ 197	4-02-00-01555 / 1.0062[24]	
3847	2-Butenoic acid, methyl ester, (E)- / Methyl trans-crotonate	$C_5H_8O_2$ / 100.12	623-43-8 / -42	2600 / 121	4-02-00-01500 / 0.9444[20]	H_2O 1; EtOH 4; eth 4 / 1.4242[20]
3848	2-Butenoic acid, methyl ester, (Z)- / Methyl cis-crotonate	$C_5H_8O_2$ / 100.12	4358-59-2	5048 / 118	4-02-00-01500	1.4175[20]
3849	2-Butenoic acid, 2-methyl-, ethyl ester, (E)-	$C_7H_{12}O_2$ / 128.17	5837-78-5	/ 156	4-02-00-01553 / 0.9200[20]	1.4340[20]
3850	2-Butenoic acid, 3-methyl-, ethyl ester	$C_7H_{12}O_2$ / 128.17	638-10-8	/ 153.5	4-02-00-01556 / 0.9199[21]	1.4345[20]
3851	2-Butenoic acid, 2-methyl-, methyl ester, (E)-	$C_6H_{10}O_2$ / 114.14	6622-76-0	/ 139	4-02-00-01553 / 0.9498[20]	1.4370[20]
3852	2-Butenoic acid, 3-methyl-, methyl ester	$C_6H_{10}O_2$ / 114.14	924-50-5 / 114	/ 136.5	3-02-00-01313 / 0.9337[20]	1.432[20]

No.	Name / Synonym	Mol. Form. / Mol. Wt.	CAS RN / mp/°C	Merck No. / bp/°C	Beil. Ref. / den/g cm^{-3}	Solubility / n_D
3853	2-Butenoic acid, 4-oxo-4-(phenylamino)-, (Z)- / Maleanilic acid	$C_{10}H_9NO_3$ / 191.19	555-59-9 / 192 dec	5584	3-12-00-00571 / 1.418[30]	
3854	2-Butenoic acid, 3-phenyl-, (Z)-	$C_{10}H_{10}O_2$ / 162.19	704-79-0 / 131.5	170[14]	4-09-00-02057	bz 3; CS$_2$ 3; peth 2
3855	3-Butenoic acid, 4-phenyl- / Styrylacetic acid	$C_{10}H_{10}O_2$ / 162.19	2243-53-0 / 87	302	4-09-00-02054	H$_2$O 1; EtOH 4; eth 4; CS$_2$ 4
3856	2-Butenoic acid, 2-propenyl ester / Allyl crotonate	$C_7H_{10}O_2$ / 126.16	20474-93-5 / 88-9[70]		4-02-00-01503 / 0.9440[20]	1.4465[20]
3857	2-Butenoic acid, 2-propenyl ester, (E)-	$C_7H_{10}O_2$ / 126.16	5453-44-1 / 88-9[70]		4-02-00-01503 / 0.9440[20]	1.4465[20]
3858	3-Buten-2-ol, (±)-	C_4H_8O / 72.11	6118-14-5 / <-100	97.3	4-01-00-02102 / 0.8318[20]	1.4137[20]
3859	2-Buten-1-ol, (E)- / trans-Crotyl alcohol	C_4H_8O / 72.11	504-61-0 / <-30	121.2	4-01-00-02107 / 0.8521[20]	H$_2$O 4; EtOH 5; eth 5; chl 3 / 1.4288[20]
3860	2-Buten-1-ol, (Z)- / cis-Crotyl alcohol	C_4H_8O / 72.11	4088-60-2 /	123	4-01-00-02107 / 0.8662[20]	1.4342[25]
3861	3-Buten-1-ol	C_4H_8O / 72.11	627-27-0 /	113.5	4-01-00-02105 / 0.8424[20]	H$_2$O 3; EtOH 5; eth 5; ace 3 / 1.4224[20]
3862	2-Buten-1-ol, acetate	$C_6H_{10}O_2$ / 114.14	628-08-0 /	132	3-02-00-00283 / 0.9192[20]	bz 4; eth 4; EtOH 4 / 1.4181[20]
3863	2-Buten-1-ol, 2-chloro-	C_4H_7ClO / 106.55	116723-93-4 /	159	3-01-00-01900 / 1.1180[20]	H$_2$O 4; EtOH 4 / 1.4682[20]
3864	2-Buten-1-ol, 3-chloro- / γ-Chlorocrotyl alcohol	C_4H_7ClO / 106.55	40605-42-3 / -42.9	161.5	4-01-00-02108 / 1.095[23]	1.4652[20]
3865	2-Buten-1-ol, 4-chloro- / 4-Chloro-2-buten-1-ol	C_4H_7ClO / 106.55	7523-44-6 /	64-5[2]	4-01-00-02110	eth 4; EtOH 4 / 1.4845[20]
3866	3-Buten-2-ol, 1-chloro- / 1-Chloro-3-buten-2-ol	C_4H_7ClO / 106.55	671-56-7 /	145.5	3-01-00-01893 / 1.111[20]	chl 4 / 1.4643[20]
3867	3-Buten-1-ol, 2-chloro-	C_4H_7ClO / 106.55	75455-41-3 /	66-7[30]	4-01-00-02106 / 1.1044[20]	eth 4; EtOH 4 / 1.4665[20]
3868	3-Buten-2-ol, 3-chloro- / 3-Chloro-3-buten-2-ol	C_4H_7ClO / 106.55	6498-47-1 /	53-7[19]	3-01-00-01893 / 1.1138[23]	eth 4; EtOH 4
3869	2-Buten-1-ol, 3-methyl-	$C_5H_{10}O$ / 86.13	556-82-1 /	140	4-01-00-02129 / 0.848[25]	1.4412[20]
3870	3-Buten-2-ol, 2-methyl-	$C_5H_{10}O$ / 86.13	115-18-4 / -28	97	4-01-00-02132 / 0.82[20]	
3871	3-Buten-2-ol, 3-methyl- / Methyl isopropenyl carbinol	$C_5H_{10}O$ / 86.13	10473-14-0 /	114	4-01-00-02127 / 0.8531[17]	1.4288[17]
3872	2-Buten-1-ol, 3-phenyl- / 2-Butene-1-ol, 3-phenyl	$C_{10}H_{12}O$ / 148.20	1504-54-7 /	145[15]	4-06-00-03839 / 1.035[20]	ctc 3 / 1.5678[20]
3873	3-Buten-2-ol, 4-(2,6,6-trimethyl-1-cyclohexen-1-yl)- / β-Ionol	$C_{13}H_{22}O$ / 194.32	22029-76-1 /	131[15]; 80[0.7]	3-06-00-00401 / 0.9243[20]	EtOH 3; eth 3; ace 3 / 1.4969[20]
3874	3-Buten-2-ol, 4-(2,6,6-trimethyl-2-cyclohexen-1-yl)- / α-Ionol	$C_{13}H_{22}O$ / 194.32	25312-34-9 /	127[15]	3-06-00-00402 / 0.9474[20]	1.4735[20]
3875	3-Buten-2-one / Methyl vinyl ketone	C_4H_6O / 70.09	78-94-4 / 81.4	6052	4-01-00-03444 / 0.864[20]	H$_2$O 3; EtOH 3; eth 4; ace 4 / 1.4081[20]
3876	2-Buten-1-one, 1,3-diphenyl- / Dypnone	$C_{16}H_{14}O$ / 222.29	495-45-4 / 342.5	3460	4-07-00-01682 / 1.1080[15]	eth 4; EtOH 4 / 1.6343[20]
3877	1-Buten-1-one, 2-ethyl- / Diethylketene	$C_6H_{10}O$ / 98.14	24264-08-2 / 88.5		4-01-00-03477 / 0.831[20]	1.4135[11]
3878	3-Buten-2-one, 4-(2-furanyl)-	$C_8H_8O_2$ / 136.15	623-15-4 / 39.5	229 dec; 113[10]	5-17-09-00578 / 1.0496[57]	H$_2$O 1; EtOH 4; eth 4; chl 4 / 1.5788[45]
3879	3-Buten-2-one, 3-methyl- / Isopropenyl methyl ketone	C_5H_8O / 84.12	814-78-8 / -54	98	4-01-00-03462 / 0.8527[20]	EtOH 4 / 1.4220[20]
3880	2-Buten-1-one, 1-phenyl-	$C_{10}H_{10}O$ / 146.19	495-41-0 / 20.5	111-2[9]	3-07-00-01410 / 1.025[15]	1.5626[18]
3881	3-Buten-2-one, 4-phenyl- / Benzylideneacetone	$C_{10}H_{10}O$ / 146.19	122-57-6 / 41.5	1151 / 261	2-07-00-00287 / 1.0368[15]	bz 4; eth 4; EtOH 4; chl 4
3882	3-Buten-2-one, 4-phenyl-, (E)-	$C_{10}H_{10}O$ / 146.19	1896-62-4 / 42	262	4-07-00-01003 / 1.0097[45]	H$_2$O 1; EtOH 4; eth 3; ace 3 / 1.5836[45]
3883	3-Buten-2-one, 4-(2,5,6,6-tetramethyl-1-cyclohexen-1-yl)- / β-Irone	$C_{14}H_{22}O$ / 206.33	79-70-9 /	4978 / 85-90[0.1]	4-07-00-00377 / 0.9434[21]	H$_2$O 2; EtOH 4; eth 4; bz 4 / 1.5017[20]
3884	3-Buten-2-one, 4-(2,5,6,6-tetramethyl-2-cyclohexen-1-yl)- / α-Irone	$C_{14}H_{22}O$ / 206.33	79-69-6 /	4977 / 87-92[0.4]	4-07-00-00361 / 0.9362[20]	1.5002[20]
3885	3-Buten-2-one, 4-(2,6,6-trimethyl-1-cyclohexen-1-yl)-, (E)-	$C_{13}H_{20}O$ / 192.30	79-77-6 /	140[18]; 73[0.1]	4-07-00-00361 / 0.9462[20]	H$_2$O 2; EtOH 5; eth 5; chl 3 / 1.5198[20]
3886	3-Buten-2-one, 4-(2,6,6-trimethyl-2-cyclohexen-1-yl)-, (E)-(±)-	$C_{13}H_{20}O$ / 192.30	30685-95-1 /	146[28]; 75[0.15]	4-07-00-00363 / 0.9298[21]	ace 4; eth 4; EtOH 4 / 1.5041[20]
3887	2-Butenoyl chloride	C_4H_5ClO / 104.54	10487-71-5 /	124.5	2-02-00-00394 / 1.0905[20]	ace 4 / 1.460[18]
3888	2-Butenoyl chloride, 3-methyl-	C_5H_7ClO / 118.56	3350-78-5 /	146	4-02-00-01558 / 1.065[25]	1.4770[20]

No.	Name Synonym	Mol. Form. Mol. Wt.	CAS RN mp/°C	Merck No. bp/°C	Beil. Ref. den/g cm^{-3}	Solubility n_D
3889	1-Buten-3-yne Vinylacetylene	C_4H_4 52.08	689-97-4	5.1	4-01-00-01083 0.7094[0]	H_2O 1; bz 3 1.4161[1]
3890	1-Buten-3-yne, 4-chloro-	C_4H_3Cl 86.52	40589-38-6	56	4-01-00-01085 1.0022[20]	chl 4 1.4656[20]
3891	1-Buten-3-yne, 1-methoxy-	C_5H_6O 82.10	2798-73-4	123 dec	4-01-00-02301 0.9062[20]	H_2O 1; chl 3; os 4 1.4818[20]
3892	1-Buten-3-yne, 2-methyl- Isopropenylacetylene	C_5H_6 66.10	78-80-8 -113	32	4-01-00-01089 0.6801[11]	chl 3 1.4140[20]
3893	Butralin Benzenamine, 4-(1,1-dimethylethyl)-N-(1-methylpropyl)-2,6-dinitro-	$C_{14}H_{21}N_3O_4$ 295.34	33629-47-9 60	1532 135[0.5]		
3894	N-sec-Butyl-2,6-dinitrobenzene Benzenamine, N-(1-methylpropyl)-2,6-dinitro	$C_{10}H_{13}N_3O_4$ 239.23	55702-49-3			
3895	2-Butynal Tetrolaldehyde	C_4H_4O 68.08	1119-19-3 -26	106.5	4-01-00-03540 0.9264[17]	eth 3; ace 3; chl 3 1.446[19]
3896	3-Butyn-2-amine, 2-methyl-	C_5H_9N 83.13	2978-58-7 18	79.5	4-04-00-01145 0.79[25]	1.4235[20]
3897	1-Butyne Ethylacetylene	C_4H_6 54.09	107-00-6 -125.7	8.0	4-01-00-00969 0.6783[0]	H_2O 1; EtOH 3; eth 3 1.3962[20]
3898	2-Butyne Dimethylacetylene	C_4H_6 54.09	503-17-3 -32.3	26.9	4-01-00-00971 0.6910[20]	H_2O 1; EtOH 3; eth 3; ctc 3 1.3921[20]
3899	1-Butyne, 3-chloro-	C_4H_5Cl 88.54	21020-24-6	68.5	4-01-00-00970 1.4218[25]	1.4218[25]
3900	2-Butyne, 1-chloro-	C_4H_5Cl 88.54	3355-17-7	103	4-01-00-00973 1.0152[20]	ace 4; eth 4; EtOH 4 1.4581[20]
3901	1-Butyne, 3-chloro-3-methyl-	C_5H_7Cl 102.56	1111-97-3 -61	76	4-01-00-01000 0.9061[20]	
3902	2-Butynediamide Cellocidin	$C_4H_4N_2O_2$ 112.09	543-21-5 217 dec	1958	4-02-00-02295	H_2O 2; EtOH 2; eth 2; chl 2
3903	2-Butyne-1,4-diamine	$C_4H_8N_2$ 84.12	53878-96-9 44	96[16]	4-04-00-01392	H_2O 4; EtOH 4; MeOH 4
3904	2-Butyne, 1,4-dibromo- 1,4-Dibromo-2-butyne	$C_4H_4Br_2$ 211.88	2219-66-1	92[15]	4-01-00-00974 2.014[18]	eth 3; ace 3; chl 4 1.588[18]
3905	2-Butyne, 1,4-dichloro-	$C_4H_4Cl_2$ 122.98	821-10-3	165.5	4-01-00-00973 1.258[20]	eth 3; ace 3; ctc 2; chl 4 1.5058[20]
3906	2-Butyne, 1,4-diiodo-	$C_4H_4I_2$ 305.88	116529-73-8 54.5	70[0.1]	4-01-00-00975	H_2O 2; EtOH 3; eth 3; ace 3
3907	1-Butyne, 3,3-dimethyl- tert-Butylacetylene	C_6H_{10} 82.15	917-92-0 -78.2	37.7	4-01-00-01022 0.6623[25]	1.3736[20]
3908	2-Butynedinitrile	C_4N_2 76.06	1071-98-3 20.5	76.5	4-02-00-02295 0.9708[25]	1.4647[25]
3909	2-Butynedioic acid, diethyl ester	$C_8H_{10}O_4$ 170.17	762-21-0 1.5	184[200]	4-02-00-02294 1.0075[20]	EtOH 3; eth 3; ctc 3 1.4425[20]
3910	2-Butynedioic acid, dimethyl ester	$C_6H_8O_4$ 142.11	762-42-5	197 dec; 98[20]	4-02-00-02291 1.1564[20]	EtOH 3; eth 3; ctc 3 1.4434[20]
3911	2-Butyne-1,4-diol Bis(hydroxymethyl)acetylene	$C_4H_6O_2$ 86.09	110-65-6 50	238	4-01-00-02687	H_2O 4; EtOH 4; eth 2; ace 4 1.4804[20]
3912	3-Butyne-1,2-diol 1-Butyne-3,4-diol	$C_4H_6O_2$ 86.09	616-28-4 40	64[0.2]	4-01-00-02689	H_2O 4; EtOH 4
3913	2-Butyne-1,4-diol, diacetate 1,4-Diacetoxy-2-butyne	$C_8H_{10}O_4$ 170.17	1573-17-7	122-3[10]	4-02-00-00244	ctc 3 1.4611[20]
3914	2-Butyne, 1,1,1,4,4,4-hexafluoro-	C_4F_6 162.03	692-50-2 -117.4	-24.6	4-01-00-00972	EtOH 3; eth 3; ace 3; ctc 3
3915	2-Butyne, 1-methoxy-	C_5H_8O 84.12	2768-41-4	99.5	3-01-00-01973 0.8496[20]	bz 4; eth 4; EtOH 4 1.4262[20]
3916	1-Butyne, 3-methyl-	C_5H_8 68.12	598-23-2 -89.7	26.3	4-01-00-00999 0.6660[20]	H_2O 1; EtOH 5; eth 5 1.3723[20]
3917	2-Butynoic acid	$C_4H_4O_2$ 84.07	590-93-2 78	203	4-02-00-01690 0.9641[20]	H_2O 4; eth 4; EtOH 4; chl 4
3918	2-Butynoic acid, ethyl ester	$C_6H_8O_2$ 112.13	4341-76-8	163	4-02-00-01691 0.9641[20]	1.4372[20]
3919	2-Butyn-1-ol	C_4H_6O 70.09	764-01-2 -2.2	148	4-01-00-02220 0.9370[20]	eth 4; EtOH 4 1.4530[20]
3920	3-Butyn-1-ol	C_4H_6O 70.09	927-74-2 -63.6	129	4-01-00-02219 0.9257[20]	H_2O 4; EtOH 4 1.4409[20]
3921	3-Butyn-2-ol	C_4H_6O 70.09	2028-63-9 -3	106.5	4-01-00-02218 0.8618[20]	H_2O 4; eth 4; EtOH 4 1.4207[20]
3922	3-Butyn-2-ol, 2-methyl- 1,1-Dimethylpropargyl alcohol	C_5H_8O 84.12	115-19-5 3	5956 104	4-01-00-02229 0.8618[20]	1.4207[20]
3923	3-Butyn-2-one Ethynyl methyl ketone	C_4H_4O 68.08	1423-60-5	84	4-01-00-03539 0.8860[15]	1.4070[20]
3924	3-Butyn-2-one, 4-phenyl-	$C_{10}H_8O$ 144.17	1817-57-8	75-6[0.8]	4-07-00-01175 0.99[25]	1.5762[20]
3925	Butyrac 118 Butanoic acid, 4-(2,4-dichlorophenoxy)-	$C_{10}H_{10}Cl_2O_3$ 249.09	94-82-6 118	2828		
3926	Butyric acid, 3,4-dichloro- 3,4-Dichlorobutyric acid	$C_4H_6Cl_2O_2$ 157.00	29653-38-1 49	123[8]	4-02-00-00828	eth 4; chl 4; HOAc 4
3927	Butyric acid, 4-fluoro- 4-Fluorobutyric acid	$C_4H_7FO_2$ 106.10	462-23-7	76-8[5]	4-02-00-00809	eth 4; EtOH 4 1.3993[25]

No.	Name / Synonym	Mol. Form. / Mol. Wt.	CAS RN / mp/°C	Merck No. / bp/°C	Beil. Ref. / den/g cm⁻³	Solubility / n_D
3928	Butyric acid, 2-iodo-, ethyl ester	$C_6H_{11}IO_2$ 242.06	7425-47-0	100^{21}; 46^1	4-02-00-00839 1.570^{17}	1.4923^{20}
3929	Cadmium, dimethyl- Dimethyl cadmium	C_2H_6Cd 142.48	506-82-1 -4.5	3229 105.5; exp 150	1.9846^{18}	peth 3 1.5488
3930	Calcium cyanide Calcyan	C_2CaN_2 92.11	592-01-8	1664		
3931	Camphoric acid, trans-(±)- Isocamphoric acid, (DL)	$C_{10}H_{16}O_4$ 200.23	560-08-7 195		3-09-00-03879 1.249^{25}	eth 4
3932	Carbamic acid, (aminocarbonyl)-, ethyl ester	$C_4H_8N_2O_3$ 132.12	626-36-8 196.5	dec	4-03-00-00127	H_2O 1; EtOH 2; eth 1; bz 2
3933	Carbamic acid, butyl-, 2-[[(aminocarbonyl)oxy]methyl]-2-methylpentyl ester 2-Methyl-2-propyltrimethylene butylcarbamate carbamate	$C_{13}H_{26}N_2O_4$ 274.36	4268-36-4	9737		
3934	Carbamic acid, butyl-, butyl ester	$C_9H_{19}NO_2$ 173.26	13105-52-7	88^3	4-04-00-00577 0.9221^{20}	eth 4; EtOH 4 1.4359^{20}
3935	Carbamic acid, butyl ester Butyl carbamate	$C_5H_{11}NO_2$ 117.15	592-35-8 53	204 dec; 108^{14}	4-03-00-00054	EtOH 4; chl 2
3936	Carbamic acid, butyl-, ethyl ester	$C_7H_{15}NO_2$ 145.20	591-62-8 -22	202; 100^{15}	4-04-00-00577 0.9434^{26}	1.4278^{26}
3937	Carbamic acid, butyl-, methyl ester	$C_6H_{13}NO_2$ 131.17	2594-21-0 -18	92^{15}	0-04-00-00158 0.9689^{23}	1.4289^{23}
3938	Carbamic acid, N-butyl-N-nitro,butyl ester	$C_9H_{18}N_2O_4$ 218.25	110795-27-2	98^3	1.046^{20}	EtOH 5; eth 5 1.448^{20}
3939	Carbamic acid, (3-chlorophenyl)-, 1-methylethyl ester Chloropropham	$C_{10}H_{12}ClNO_2$ 213.66	101-21-3 41	2188 149^2	4-12-00-01149 1.18^{30}	1.5388^{20}
3940	Carbamic acid, dichloro-, ethyl ester	$C_3H_5Cl_2NO_2$ 157.98	13698-16-3	55-6^{15}	3-03-00-00050 1.349^{25}	1.4595^{20}
3941	Carbamic acid, diethyl- Carbamic acid, N,N-diethyl	$C_5H_{11}NO_2$ 117.15	24579-70-2 -15	171 dec	3-04-00-00222 0.9276^{20}	H_2O 4; eth 4; EtOH 4 1.4206^{20}
3942	Carbamic acid, (1,1-dimethylethyl)-, ethyl ester	$C_7H_{15}NO_2$ 145.20	1611-50-3 21.5	72^{16}	4-04-00-00665 0.943^{15}	
3943	Carbamic acid, (1,1-dimethylethyl)-, methyl ester	$C_6H_{13}NO_2$ 131.17	27701-01-5 27	63.3^{17}	3-04-00-00325 0.966^{15}	
3944	Carbamic acid, (1,1-dimethylethyl)nitro-, ethyl ester	$C_7H_{14}N_2O_4$ 190.20	55696-02-1	56^2 dec	4-04-00-03409 1.049^{20}	eth 4; EtOH 4 1.4331^{20}
3945	Carbamic acid, dimethyl-, 3-methyl-1-phenyl-1H-pyrazol-5-yl ester Pyrolan	$C_{13}H_{15}N_3O_2$ 245.28	87-47-8 50	8013 $160-2^{0.2}$	5-23-10-00530	ctc 3; CS_2 3
3946	Carbamic acid, diphenyl-, ethyl ester	$C_{15}H_{15}NO_2$ 241.29	603-52-1 72	360	4-12-00-00850	H_2O 4; eth 4; bz 4
3947	Carbamic acid, ethyl-, butyl ester Carbamic acid, N-ethyl, butyl ester	$C_7H_{15}NO_2$ 145.20	16246-07-4	66^3	0.9396^{20}	eth 4; EtOH 4 1.4301^{20}
3948	Carbamic acid, ethyl ester Urethan	$C_3H_7NO_2$ 89.09	51-79-6 49	9789 185	4-03-00-00040 0.9862^{21}	H_2O 4; EtOH 4; eth 4; bz 4 1.4144^{51}
3949	Carbamic acid, ethyl-, ethyl ester	$C_5H_{11}NO_2$ 117.15	623-78-9	176	4-04-00-00365 0.9813^{20}	H_2O 4; eth 4; EtOH 4 1.4215^{20}
3950	Carbamic acid, ethylnitro-, ethyl ester Carbamic acid, N-ethyl-N-nitro, ethyl ester	$C_5H_{10}N_2O_4$ 162.15	6274-16-4	107^{31}	4-04-00-03404 1.161^{20}	eth 4; EtOH 4 1.4432^{20}
3951	Carbamic acid, ethylnitro-, methyl ester Carbamic acid, N-ethyl-N-nitro, butyl ester	$C_7H_{14}N_2O_4$ 190.20	6162-79-4	79^3	4-04-00-03403 1.089^{20}	H_2O 4; eth 4; EtOH 4 1.4455^{20}
3952	Carbamic acid, (3-methylbutyl)-, ethyl ester Urethane, isopentyl	$C_8H_{17}NO_2$ 159.23	1611-52-5	218	2-04-00-00648 0.9322^{20}	1.4326^{20}
3953	Carbamic acid, methyl ester Methyl carbamate	$C_2H_5NO_2$ 75.07	598-55-0 54	5958 177	4-03-00-00037 1.1361^{56}	H_2O 4; EtOH 4; eth 3 1.4125^{56}
3954	Carbamic acid, methyl-, ethyl ester	$C_4H_9NO_2$ 103.12	105-40-8	170	4-04-00-00200 1.0115^{20}	H_2O 4; EtOH 4 1.4183^{20}
3955	Carbamic acid, 1-methylethyl ester Isopropyl carbamate	$C_4H_9NO_2$ 103.12	1746-77-6 93	183	4-03-00-00053 0.9951^{66}	
3956	Carbamic acid, (1-methylethyl)-, ethyl ester	$C_6H_{13}NO_2$ 131.17	2594-20-9	79^{15}	4-04-00-00520 0.9531^{20}	eth 4; EtOH 4 1.4229^{20}
3957	Carbamic acid, (1-methylethyl)nitro-, ethyl ester Carbamic acid, N-isopropyl-N-nitro, ethyl ester	$C_6H_{12}N_2O_4$ 176.17	62261-05-6	72^7	4-04-00-03407 1.110^{20}	eth 4; EtOH 4 1.4381^{20}
3958	Carbamic acid, methylphenyl-, methyl ester	$C_9H_{11}NO_2$ 165.19	28685-60-1 44	235	2-12-00-00235 1.2960^{19}	eth 4; lig 4
3959	Carbamic acid, 2-methylpropyl ester Isobutyl carbamate	$C_5H_{11}NO_2$ 117.15	543-28-2 67	5021 207	4-03-00-00056	eth 4; EtOH 4 1.4098^{78}
3960	Carbamic acid, (1-methylpropyl)-, ethyl ester Urethane, sec-butyl	$C_7H_{15}NO_2$ 145.20	10212-74-5 -14	194	4-04-00-00622 0.9404^{26}	1.4267^{26}
3961	Carbamic acid, (2-methylpropyl)-, ethyl ester Isobutyl urethane	$C_7H_{15}NO_2$ 145.20	539-89-9 <-65	110^{30}	4-04-00-00647 0.9432^{20}	eth 4; EtOH 4 1.4288^{20}

No.	Name Synonym	Mol. Form. Mol. Wt.	CAS RN mp/°C	Merck No. bp/°C	Beil. Ref. den/g cm^{-3}	Solubility n_D
3962	Carbamic acid, (1-methylpropyl)-, methyl ester Carbamic acid, N-sec-butyl, methyl ester	$C_6H_{13}NO_2$ 131.17	39076-02-3 	83[16]	0-04-00-00162 0.9651[25]	1.4263[25]
3963	Carbamic acid, nitro-, ethyl ester	$C_3H_6N_2O_4$ 134.09	626-37-9 64	140 dec	4-03-00-00247 1.0074[20]	H_2O 4; EtOH 4; eth 4; lig 2
3964	Carbamic acid, nitropropyl-, ethyl ester Carbamic acid, N-propyl-N-nitro, ethyl ester	$C_6H_{12}N_2O_4$ 176.17	13855-77-1 	66[3]	4-04-00-03406 1.121[20]	eth 4; EtOH 4 1.4431[20]
3965	Carbamic acid, phenyl-, ethyl ester Phenylurethane	$C_9H_{11}NO_2$ 165.19	101-99-5 53	7291 237 dec	4-12-00-00619 1.1064[30]	H_2O 4; EtOH 4; eth 4; bz 3 1.5376[30]
3966	Carbamic acid, phenyl-, 1-methylethyl ester Propham	$C_{10}H_{13}NO_2$ 179.22	122-42-9 90	7828	4-12-00-00620 1.09[20]	bz 4; EtOH 4 1.4989[91]
3967	Carbamic acid, (phenylmethyl)-, ethyl ester	$C_{10}H_{13}NO_2$ 179.22	2621-78-5 49	230	3-12-00-02250	bz 4; eth 4; EtOH 4; chl 4
3968	Carbamic acid, phenyl-, 2-methylpropyl ester	$C_{11}H_{15}NO_2$ 193.25	2291-80-7 86	216	3-12-00-00614	bz 4; eth 4; EtOH 4
3969	Carbamic acid, propyl ester	$C_4H_9NO_2$ 103.12	627-12-3 60	196	4-03-00-00052	ace 4; eth 4; EtOH 4
3970	Carbamic acid, propyl-, ethyl ester	$C_6H_{13}NO_2$ 131.17	623-85-8 	192	4-04-00-00480 0.9921[15]	EtOH 4
3971	Carbamic chloride Carbamyl chloride	CH_2ClNO 79.49	463-72-9 	1785 62 dec	3-03-00-00065	
3972	Carbamic chloride, diethyl- Carbamoyl chloride, N,N-diethyl	$C_5H_{10}ClNO$ 135.59	88-10-8 	186	4-04-00-00379	
3973	Carbamic chloride, dimethyl- Dimethylcarbamoyl chloride	C_3H_6ClNO 107.54	79-44-7 -33	167	4-04-00-00224 1.168[25]	1.4540[20]
3974	Carbamic chloride, methylphenyl-	C_8H_8ClNO 169.61	4285-42-1 88.5	280	4-12-00-00839	eth 4; EtOH 4
3975	Carbamodithioic acid, dimethyl-, 2,4-dinitrophenyl ester	$C_9H_9N_3O_4S_2$ 287.32	89-37-2 152.5		4-06-00-01760 1.54[20]	H_2O 1; EtOH 3; ace 3; bz 3
3976	Carbamothioic acid, butylethyl-, S-propyl ester Pebulate	$C_{10}H_{21}NOS$ 203.35	1114-71-2 	7007 142[20]	0.9458[20]	ace 4; bz 4; MeOH 4 1.4752[20]
3977	Carbamothioic acid, diphenyl-, S-[2-(diethylamino)ethyl] ester Phencarbamide	$C_{19}H_{24}N_2OS$ 328.48	3735-90-8 48.5	7178 120-6[0.01]		eth 4; chl 4; MeOH 4; peth 4
3978	Carbamothioic acid, O-ethyl ester	C_3H_7NOS 105.16	625-57-0 41	dec	4-03-00-00294 1.069[20]	eth 4; EtOH 4; chl 4 1.520[20]
3979	Carbamothioic acid, S-ethyl ester	C_3H_7NOS 105.16	637-98-9 109	dec	4-03-00-00294	eth 4; EtOH 4
3980	Carbamothioic chloride, dimethyl-	C_3H_6ClNS 123.61	16420-13-6 42.5	98[10]	3-04-00-00147	eth 4; chl 3; peth 3
3981	9H-Carbazole Dibenzo[b,d]pyrrole	$C_{12}H_9N$ 167.21	86-74-8 246.2	1792 354.69	5-20-08-00009	H_2O 1; EtOH 2; eth 2; ace 3
3982	9H-Carbazole-9-acetic acid (N-Carbazolyl)acetic acid	$C_{14}H_{11}NO_2$ 225.25	524-80-1 215	1793	5-20-08-00033	eth 4; EtOH 4; chl 4; HOAc 4
3983	9H-Carbazole, 9-acetyl-	$C_{14}H_{11}NO$ 209.25	574-39-0 69	190[6]	5-20-08-00030 1.158[100]	1.640[100]
3984	9H-Carbazole, 9-butyl-	$C_{16}H_{17}N$ 223.32	1484-08-8 58	218-9[19]	5-20-08-00018	eth 4
3985	9H-Carbazole, 9-ethyl-	$C_{14}H_{13}N$ 195.26	86-28-2 68	190[10]	5-20-08-00016 1.059[80]	H_2O 1; EtOH 4; eth 4 1.6394[80]
3986	9H-Carbazole, 3-methyl-	$C_{13}H_{11}N$ 181.24	4630-20-0 208.5	365	5-20-08-00076	bz 4; eth 4
3987	9H-Carbazole, 9-methyl-	$C_{13}H_{11}N$ 181.24	1484-12-4 89.34	343.64; 195[12]	5-20-08-00015	eth 4
3988	9H-Carbazole, 9-(phenylmethyl)-	$C_{19}H_{15}N$ 257.33	19402-87-0 119	267[24]	5-20-08-00023	bz 4
3989	1H-Carbazole, 2,3,4,9-tetrahydro-	$C_{12}H_{13}N$ 171.24	942-01-8 120	327.5	5-20-07-00468	H_2O 1; EtOH 3; eth 4; bz 4
3990	Carbendazim Carbamic acid, 1H-benzimidazol-2-yl-, methyl ester	$C_9H_9N_3O_2$ 191.19	10605-21-7 300 dec	1794	1.45	
3991	Carbofuran 7-Benzofuranol, 2,3-dihydro-2,2-dimethyl-, methylcarbamate	$C_{12}H_{15}NO_3$ 221.26	1563-66-2 151	1810	1.18	
3992	Carbon dioxide Carbonic anhydride	CO_2 44.01	124-38-9 -56.57 (5.11 atm)	1816 sub -78.5 (1 atm)	0-03-00-00004 *0.720[25]	
3993	Carbon disulfide Carbon bisulfide	CS_2 76.14	75-15-0 -111.5	1818 46	4-03-00-00395 1.2632[20]	H_2O 3; EtOH 5; eth 5; chl 3 1.6319[20]
3994	Carbonic acid, bis(1,1-dimethylethyl) ester	$C_9H_{18}O_3$ 174.24	34619-03-9 40	158	4-03-00-00009	EtOH 4
3995	Carbonic acid, bis(1-methylethyl) ester	$C_7H_{14}O_3$ 146.19	6482-34-4 	147	3-03-00-00009 0.9162[20]	EtOH 4 1.3932[20]
3996	Carbonic acid, bis(2-methylphenyl) ester	$C_{15}H_{14}O_3$ 242.27	617-09-4 60	144[0.5]	3-06-00-01256	HOAc 4
3997	Carbonic acid, bis(2-methylpropyl) ester	$C_9H_{18}O_3$ 174.24	539-92-4 	190	4-03-00-00009 0.9138[20]	H_2O 1; EtOH 5; eth 5 1.4072[20]
3998	Carbonic acid, dibutyl ester Dibutyl carbonate	$C_9H_{18}O_3$ 174.24	542-52-9 	1558 207	4-03-00-00008 0.9251[20]	H_2O 1; EtOH 3; eth 3 1.4117[20]

No.	Name / Synonym	Mol. Form. / Mol. Wt.	CAS RN / mp/°C	Merck No. / bp/°C	Beil. Ref. / den/g cm^{-3}	Solubility / n_D
3999	Carbonic acid, diethyl ester / Diethyl carbonate	$C_5H_{10}O_3$ / 118.13	105-58-8 / -43	3738 / 126	4-03-00-00005 / 0.9692[25]	H_2O 1; EtOH 3; eth 3; chl 3 / 1.3845[20]
4000	Carbonic acid, dimethyl ester / Methyl carbonate	$C_3H_6O_3$ / 90.08	616-38-6 / 0.5	5960 / 90.5	4-03-00-00003 / 1.0636[25]	H_2O 1; EtOH 3; eth 3; ctc 2 / 1.3687[20]
4001	Carbonic acid, diphenyl ester / Phenyl carbonate	$C_{13}H_{10}O_3$ / 214.22	102-09-0 / 83	7250 / 306	4-06-00-00629 / 1.1215[87]	H_2O 1; EtOH 3; eth 3; ctc 3
4002	Carbonic acid, dipropyl ester / Dipropyl carbonate˙	$C_7H_{14}O_3$ / 146.19	623-96-1 /	/ 168	4-03-00-00006 / 0.9435[20]	H_2O 2; EtOH 5; eth 5 / 1.4008[20]
4003	Carbonic acid, ethyl methyl ester / Ethyl methyl carbonate	$C_4H_8O_3$ / 104.11	623-53-0 / -14	/ 107.5	4-03-00-00004 / 1.012[20]	eth 4; EtOH 4 / 1.3778[20]
4004	Carbonic acid monosodium salt	$CHNaO_3$ / 84.01	144-55-8 /	8528 /		H_2O 3; EtOH 1
4005	Carbonic chloride fluoride / Carbonyl chloride fluoride	CClFO / 82.46	353-49-1 /		4-03-00-00031 /	
4006	Carbonic dibromide / Carbonyl bromide	CBr_2O / 187.82	593-95-3 /	/ 64.5	2.52[15]	
4007	Carbonic dichloride / Phosgene	CCl_2O / 98.92	75-44-5 / -127.9	7310 / 8	4-03-00-00031 / *1.3719[25]	bz 3; ctc 3; chl 3; tol 3
4008	Carbonic difluoride / Carbonyl fluoride	CF_2O / 66.01	353-50-4 / -111.2	1826 / -84.5	4-03-00-00021 / 1.139[25]	
4009	Carbonic dihydrazide / Carbohydrazide	CH_6N_4O / 90.09	497-18-7 / 154	1811 /	4-03-00-00240 / 1.616[20]	H_2O 4; EtOH 4
4010	Carbonic dihydrazide, 2,2'-diphenyl- / sym-Diphenylcarbazide	$C_{13}H_{14}N_4O$ / 242.28	140-22-7 / 170	3321 / dec	4-15-00-00182 /	H_2O 2; EtOH 3; eth 2; ace 3
4011	Carbon monoxide / Carbon oxide	CO / 28.01	630-08-0 / -205	1820 / -191.5	0-01-00-00720 / 0.7909[-19]	H_2O 2; bz 3; HOAc 3
4012	Carbonochloridic acid, butyl ester / Butyl chloroformate	$C_5H_9ClO_2$ / 136.58	592-34-7 /	/ 142	4-03-00-00025 / 1.074[25]	eth 5; ace 3; ctc 2 / 1.4114[20]
4013	Carbonochloridic acid, 2-chloroethyl ester / 2-Chloroethyl chloroformate	$C_3H_4Cl_2O_2$ / 142.97	627-11-2 /	/ 155	4-03-00-00024 / 1.3847[20]	H_2O 1; EtOH 3; eth 3; ace 3 / 1.4483[20]
4014	Carbonochloridic acid, chloromethyl ester / Chloromethyl chloroformate	$C_2H_2Cl_2O_2$ / 128.94	22128-62-7 /	/ 107	4-03-00-00030 / 1.465[15]	ace 4; eth 4 / 1.4286[22]
4015	Carbonochloridic acid, 3-chloropropyl ester	$C_4H_6Cl_2O_2$ / 157.00	628-11-5 /	/ 177	3-03-00-00025 / 1.2926[25]	H_2O 1 / 1.4456[20]
4016	Carbonochloridic acid, dichloromethyl ester	$C_2HCl_3O_2$ / 163.39	22128-63-8 /	/ 110.5	4-03-00-00031 / 1.5600[15]	ace 4
4017	Carbonochloridic acid, 2-ethoxyethyl ester	$C_5H_9ClO_3$ / 152.58	628-64-8 /	67.2[14] /	3-03-00-00029 / 1.1341[25]	EtOH 4 / 1.4169[25]
4018	Carbonochloridic acid, ethyl ester / Ethyl chloroformate	$C_3H_5ClO_2$ / 108.52	541-41-3 / -80.6	3742 / 95	4-03-00-00023 / 1.1352[20]	bz 4; eth 4; chl 4 / 1.3974[20]
4019	Carbonochloridic acid, 2-methoxyethyl ester	$C_4H_7ClO_3$ / 138.55	628-12-6 /	58.7[13] /	3-03-00-00029 / 1.1905[25]	eth 4 / 1.4163[20]
4020	Carbonochloridic acid, methyl ester / Methyl chlorocarbonate	$C_2H_3ClO_2$ / 94.50	79-22-1 /	5966 / 70.5	4-03-00-00023 / 1.2231[20]	EtOH 5; eth 5; bz 3; ctc 3 / 1.3868[20]
4021	Carbonochloridic acid, 3-methylbutyl ester	$C_6H_{11}ClO_2$ / 150.60	628-50-2 /	/ 154.3	4-03-00-00026 / 1.0288[17]	eth 4; EtOH 4 / 1.4176[20]
4022	Carbonochloridic acid, 1-methylethenyl ester / Isopropenyl chloroformate	$C_4H_5ClO_2$ / 120.54	57933-83-2 /	/ 102	3-03-00-00028 / 1.101[20]	eth 4
4023	Carbonochloridic acid, 1-methylethyl ester / Isopropyl chloroformate	$C_4H_7ClO_2$ / 122.55	108-23-6 /	/ 105	4-03-00-00024 /	eth 4 / 1.4013[20]
4024	Carbonochloridic acid, 2-methylpropyl ester / Isobutyl chlorocarbonate	$C_5H_9ClO_2$ / 136.58	543-27-1 /	5023 / 128.8	4-03-00-00026 / 1.0426[18]	EtOH 3; eth 5; bz 3; chl 3 / 1.4071[18]
4025	Carbonochloridic acid, pentyl ester	$C_6H_{11}ClO_2$ / 150.60	638-41-5 /	60-2[15] /	4-03-00-00026 /	eth 3 / 1.4181[18]
4026	Carbonochloridic acid, phenyl ester / Phenyl chloroformate	$C_7H_5ClO_2$ / 156.57	1885-14-9 /	71[9] /	4-06-00-00629 /	
4027	Carbonochloridic acid, phenylmethyl ester / Carbobenzoxy chloride	$C_8H_7ClO_2$ / 170.60	501-53-1 /	1807 / 103[20]	4-06-00-02278 / 1.195[25]	eth 3; ace 3; bz 3 / 1.5190[20]
4028	Carbonochloridic acid, propyl ester / Propyl chlorocarbonate	$C_4H_7ClO_2$ / 122.55	109-61-5 /	7860 / 115.2	4-03-00-00024 / 1.0901[20]	EtOH 5; eth 5 / 1.4035[20]
4029	Carbonochloridic acid, trichloromethyl ester / Diphosgene	$C_2Cl_4O_2$ / 197.83	503-38-8 / -57	3339 / 128	4-03-00-00033 / 1.6525[14]	eth 4; EtOH 4 / 1.4566[22]
4030	Carbonocyanidic acid, ethyl ester	$C_4H_5NO_2$ / 99.09	623-49-4 /	/ 115.5	4-02-00-01862 / 1.003[25]	H_2O 1; EtOH 3; eth 3; ctc 3 / 1.3820[20]
4031	Carbonodithioic acid, S,S-diethyl ester	$C_5H_{10}OS_2$ / 150.27	623-80-3 /	/ 197	3-03-00-00339 / 1.085[20]	eth 4; EtOH 4 / 1.5237[18]
4032	Carbonothioic dichloride / Thiophosgene	CCl_2S / 114.98	463-71-8 /	/ 73	4-03-00-00281 / 1.508[15]	H_2O 1; EtOH 3; eth 3 / 1.5442[20]
4033	Carbonothioic dihydrazide / 1,3-Diamino-2-thiourea	CH_6N_4S / 106.15	2231-57-4 / 170 dec		4-03-00-00388 /	H_2O 4
4034	Carbonotrithioic acid / Trithiocarbonic acid	CH_2S_3 / 110.22	594-08-1 / -30	9671 / 57 dec	4-03-00-00428 / 1.47[17]	EtOH 4; tol 4; chl 4
4035	Carbon oxide sulfide / Carbonyl sulfide	COS / 60.08	463-58-1 / -138.8	/ -50	4-03-00-00271 / 1.028[17]	H_2O 2; EtOH 3; KOH 4 / 1.24[-87]
4036	Carbon selenide / Carbon diselenide	CSe_2 / 169.93	506-80-9 / -45.5	1817 / 125.5	4-03-00-00436 / 2.6823[20]	1.8454[20]
4037	Carbophenothion / Phosphorodithioic acid, S-[[(4-chlorophenyl)thio]methyl] O,O-diethyl ester	$C_{11}H_{16}ClO_2PS_3$ / 342.87	786-19-6 /	1827 / 82[0.01]	1.271[20]	

No.	Name / Synonym	Mol. Form. / Mol. Wt.	CAS RN / mp/°C	Merck No. / bp/°C	Bell. Ref. / den/g cm^{-3}	Solubility / n_D
4038	Carbosulfan 2,3-Dihydro-2,2-dimethyl-7-benzofuranyl (dibutylamino)thio methylcarbamate	$C_{20}H_{32}N_2O_3S$ 380.55	55285-14-8	126	1.056^{20}	
4039	Carboxin 1,4-Oxathii$N$-3-carboxamide, 5,6-dihydro-2-methyl-N-phenyl-	$C_{12}H_{13}NO_2S$ 235.31	5234-68-4 94	1832		
4040	Card-20(22)-enolide, 3-[(6-deoxy-α-L-mannopyranosyl)oxy]-1,5,11,14,19-pentahydroxy-, (1β,3β)- Ouabain	$C_{29}H_{44}O_{12}$ 584.66	630-60-4 200	6854	5-18-05-00625	H_2O 2; EtOH 4
4041	Card-20(22)-enolide, 3-[(O-2,6-dideoxy-β-D-ribo-hexopyranosyl-(1-4)-O-2,6-dideoxy-β-D-ribo-hexopyranosyl-(1-4)-2,6-dideoxy-β-D, 12,14-dihydroxy Digoxin	$C_{41}H_{64}O_{14}$ 780.95	20830-75-5 248-50 dec	3150	5-18-04-00381	EtOH 4
4042	Card-20(22)-enolide, 3-[(O-2,6-dideoxy-β-D-ribo-hexopyranosyl-(1-4)-O-2,6-dideoxy-β-D-ribo-hexopyranosyl-(1-4)-2,6-dideoxy-β-D, 14,16-dihydroxy Gitoxin	$C_{41}H_{64}O_{14}$ 780.95	4562-36-1 285 dec	4328	5-18-04-00388	
4043	Card-20(22)-enolide, 3-[(O-2,6-dideoxy-β-D-ribo-hexopyranosyl-(1-4)-O-2,6-dideoxy-β-D-ribo-hexopyranosyl-(1-4)-2,6-dideoxy-β-D, 14-dihydroxy Digitoxin	$C_{41}H_{64}O_{13}$ 764.95	71-63-6 255.5	3146	5-18-03-00354	H_2O 2; EtOH 4; eth 3; chl 3
4044	Card-20(22)-enolide, 3,14-dihydroxy-, (3β,5β)- Digitoxigenin	$C_{23}H_{34}O_4$ 374.52	143-62-4 253	3145	5-18-03-00348	EtOH 3; MeOH 4
4045	Card-20(22)-enolide, 3-[(6-O-β-D-glucopyranosyl-β-D-glucopyranosyl)oxy]-14-hydroxy-, Uzarin	$C_{35}H_{54}O_{14}$ 698.81	20231-81-6 269	9809	4-18-00-01497	
4046	Card-20(22)-enolide, 3,11,14-trihydroxy-, (3β,5β,11α)- Sarmentogenin	$C_{23}H_{34}O_5$ 390.52	76-28-8 280	8334	5-18-04-00378	H_2O 1; EtOH 3; eth 1; ace 2
4047	Card-20(22)-enolide, 3,12,14-trihydroxy-, (3β,5β,12β)- Digoxigenin	$C_{23}H_{34}O_5$ 390.52	1672-46-4 222	3149	5-18-04-00380	EtOH 4; chl 2; MeOH 4
4048	Card-20(22)-enolide, 3,14,16-trihydroxy-, (3β,5β,16β)- Gitoxigenin	$C_{23}H_{34}O_5$ 390.52	545-26-6 234	4327	5-18-04-00385	H_2O 1; eth 2; chl 3
4049	Card-20(22)-enolide, 3,5,14-trihydroxy-19-oxo-, (3β,5β)- Strophanthidin	$C_{23}H_{32}O_6$ 404.50	66-28-4 173 dec	8818	5-18-05-00126	H_2O 1; EtOH 3; eth 1; ace 3
4050	4-Carene, (1S,3R,6R)-(-)- Δ4-Carene (l)	$C_{10}H_{16}$ 136.24	5208-49-1	169; 64^{20}	4-05-00-00451 0.8441^{30}	ace 4; bz 4; eth 4 1.4740^{30}
4051	β,β-Carotene β-Carotene, all-$trans$-	$C_{40}H_{56}$ 536.88	7235-40-7 183	1860	4-05-00-02617 1.00^{20}	H_2O 1; EtOH 2; eth 3; ace 3
4052	β,ψ-Carotene γ-Carotene	$C_{40}H_{56}$ 536.88	472-93-5 178	1861	4-05-00-02616	H_2O 1; EtOH 1; eth 2; bz 3
4053	ψ,ψ-Carotene $trans$-Lycopene	$C_{40}H_{56}$ 536.88	502-65-8 175	5492	4-01-00-01165	EtOH 2; eth 3; bz 4; chl 4
4054	β,ε-Carotene, (6'R)- α-Carotene	$C_{40}H_{56}$ 536.88	7488-99-5 187.5	1859	4-05-00-02620 1.00^{20}	bz 4; eth 4; chl 4
4055	β,β-Carotene, 3'-(acetyloxy)-6',7'-didehydro-5,6-epoxy-5,5',6,6',7,8-hexahydro-3,5'-dihydroxy-8-oxo-, Fucoxanthin	$C_{40}H_{60}O_6$ 636.91	3351-86-8 168	4195	5-18-04-00673	eth 4; EtOH 4
4056	β,β-Carotene-3,3'-diol, 5,6:5',6'-diepoxy-5,5',6,6'-tetrahydro- Violaxanthin	$C_{40}H_{56}O_4$ 600.88	126-29-4 208	9902	5-19-03-00464	EtOH 3; eth 3; CS_2 3; peth 1
4057	β,β-Carotene-3,3'-diol, (3R,3'R)- Zeaxanthin	$C_{40}H_{56}O_2$ 568.88	144-68-3 215.5	10019 226-9$^{0.06}$	4-06-00-07017	H_2O 1; EtOH 2; eth 3; ace 3
4058	β,ε-Carotene-3,3'-diol, (3R,3'R,6'R)- Xanthophyll	$C_{40}H_{56}O_2$ 568.88	127-40-2 196	9972	3-06-00-05871	bz 4; eth 4; EtOH 4; peth 4
4059	β,ψ-Caroten-16'-oic acid, 3',4'-didehydro- Torularhodin	$C_{40}H_{52}O_2$ 564.85	514-92-1 211	9476	4-09-00-02767	py 4; chl 4; CS_2 4
4060	ψ,ψ-Caroten-16-ol Lycoxanthin	$C_{40}H_{56}O$ 552.88	19891-74-8 168	5499	4-01-00-02368	H_2O 1; EtOH 2; bz 3; CS_2 3
4061	β,β-Caroten-3-ol, (3R)- Cryptoxanthin	$C_{40}H_{56}O$ 552.88	472-70-8 160	2612	4-06-00-05111	bz 4; chl 4
4062	β,ψ-Caroten-3-ol, (3R)- Rubixanthin	$C_{40}H_{56}O$ 552.88	3763-55-1 160	8265	4-06-00-05110	EtOH 2; bz 3; chl 3; peth 2
4063	β,κ-Caroten-6'-one, 3,3'-dihydroxy-, (3R,3'S,5'R)- Capsanthin	$C_{40}H_{56}O_3$ 584.88	465-42-9 176	1768	4-08-00-02657	
4064	Carpaine 2,15-Dioxa-12,25-diazatricyclo[22.2.2.211,14]triacontane, carpaine deriv.	$C_{28}H_{50}N_2O_4$ 478.72	3463-92-1 121	1866	2-27-00-00209	ace 4; bz 4; eth 4; EtOH 4

No.	Name Synonym	Mol. Form. Mol. Wt.	CAS RN mp/°C	Merck No. bp/°C	Beil. Ref. den/g cm^{-3}	Solubility n_D
4065	CDEC Carbamodithioic acid, diethyl-, 2-chloro-2-propenyl ester	$C_8H_{14}ClNS_2$ 223.79	95-06-7	8882 129[1]	1.088	
4066	Cedrene 1H-3a,7-Methanoazulene, cedrene deriv.	$C_{15}H_{24}$ 204.36	11028-42-5	262.5	3-05-00-01095 0.9342[20]	H_2O 1; bz 3; lig 3 1.5034[20]
4067	Cevane-3,4,12,14,16,17,20-heptol, 3-acetate, (3β,4α,16β)- Sabadine	$C_{29}H_{47}NO_8$ 537.69	124-80-1 258	8281	4-21-00-02883	ace 4; EtOH 4
4068	Cevane-3,4,7,14,15,16,20-heptol, 4,9-epoxy-, (3β,4α,7α,15α,16β)- Germine	$C_{27}H_{43}NO_8$ 509.64	508-65-6 220	4307	5-21-13-00706	bz 3; MeOH 3; alk 3; acid 3
4069	Cevane-3,4,12,14,16,17,20-heptol, 4,9-epoxy-, 3-(2-methyl-2-butenoate), [3β(Z),4α,16β]- Cevadine	$C_{32}H_{49}NO_9$ 591.74	62-59-9 213-4 dec	2025	4-21-00-06820	
4070	Cevane-3,4,6,7,14,15,16,20-octol, 4,9-epoxy-, (3β,4α,6α,7α,15α,16β)- Protoverine	$C_{27}H_{43}NO_9$ 525.64	76-45-9 221	7916	4-21-00-06841	H_2O 1; EtOH 3; bz 3; aq acid 3
4071	Chelidonine Stylophorine	$C_{20}H_{19}NO_5$ 353.37	476-32-4 135.5	2043 220[0.002]	2-27-00-00615	H_2O 1; EtOH 3; eth 3; chl 3
4072	Chinomethionat Carbonic acid, dithio-, cyclic S,S-(6-methyl-2,3-quinoxalinediyl) ester	$C_{10}H_6N_2OS_2$ 234.30	2439-01-2 170	6933		
4073	DL-chiro-Inositol	$C_6H_{12}O_6$ 180.16	18685-70-6 253		4-06-00-07920 1.752[15]	H_2O 4; EtOH 2; eth 1; bz 1
4074	D-chiro-Inositol, 2-deoxy- D-Quercitol	$C_6H_{12}O_5$ 164.16	488-73-3 236	8046	4-06-00-07883 1.584[13]	H_2O 4
4075	Chlorbenside Benzene, 1-chloro-4-[[(4-chlorophenyl)methyl]thio]-	$C_{13}H_{10}Cl_2S$ 269.19	103-17-3 75	2074	1.4210[20]	
4076	Chlorbromuron Urea, N'-(4-bromo-3-chlorophenyl)-N-methoxy-N-methyl-	$C_9H_{10}BrClN_2O_2$ 293.55	13360-45-7 96		1.69[20]	
4077	Chlordane 4,7-Methano-1H-indene, 1,2,4,5,6,7,8,8-octachloro-2,3,3a,4,7,7a-hexahydro-	$C_{10}H_6Cl_8$ 409.78	57-74-9 106	2079 175[1]	1.60[25]	
4078	Chlorfenvinphos Phosphoric acid, 2-chloro-1-(2,4-dichlorophenyl)ethenyl diethyl ester	$C_{12}H_{14}Cl_3O_4P$ 359.57	470-90-6 170[0.05]	2087		
4079	Chlorflurecol 9H-Fluorene-9-carboxylic acid, 2-chloro-9-hydroxy-	$C_{14}H_9ClO_3$ 260.68	2464-37-1		1.496[20]	
4080	Chloridazon 3(2H)-Pyridazinone, 5-amino-4-chloro-2-phenyl-	$C_{10}H_8ClN_3O$ 221.65	1698-60-8 205			
4081	Chlorimuron 2-(4-Chloro-6-methoxypyrimidin-2-ylcarbamoylsulfamoyl) benzoic acid, ethyl ester	$C_{15}H_{15}ClN_4O_6S$ 414.83	90982-32-4 186	2092		
4082	Chlorobenzilate Benzeneacetic acid, 4-chloro-α-(4-chlorophenyl)-α-hydroxy-, ethyl ester	$C_{16}H_{14}Cl_2O_3$ 325.19	510-15-6 37	157[0.07]	1.2816 [20]	
4083	5-Chloro-3-methyl-4-nitro-1H-pyrazole 1H-Pyrazole, 3-chloro-5-methyl-4-nitro-	$C_4H_4ClN_3O_2$ 161.55	6814-58-0 111			
4084	Chloroneb Benzene, 1,4-dichloro-2,5-dimethoxy-	$C_8H_8Cl_2O_2$ 207.06	2675-77-6 134	268		
4085	Chlorosulfuric acid, ethyl ester	$C_2H_5ClO_3S$ 144.58	625-01-4 152.5; 93[100]		4-01-00-01326 1.3502[25]	eth 4; chl 4; lig 4 1.416[20]
4086	Chlorosulfuric acid, methyl ester	CH_3ClO_3S 130.55	812-01-1 -70 134		4-01-00-01252 1.4805[25]	ace 4; bz 4; eth 4 1.4138[18]
4087	Chlorosulfurous acid, ethyl ester	$C_2H_5ClO_2S$ 128.58	6378-11-6 52.5[44]; 32[16]		3-01-00-01316 1.2766[25]	eth 4 1.4550[25]
4088	Chloroxuron Urea, N'-[4-(4-chlorophenoxy)phenyl]-N,N-dimethyl-	$C_{15}H_{15}ClN_2O_2$ 290.75	1982-47-4 151			
4089	Chlorpyrifos Phosphorothioic acid, O,O-diethyl O-(3,5,6-trichloro-2-pyridinyl) ester	$C_9H_{11}Cl_3NO_3PS$ 350.59	2921-88-2 42	2190		
4090	Chlorpyrifos-methyl Phosphorothioic acid, O,O-dimethyl O-(3,5,6-trichloro-2-pyridinyl) ester	$C_7H_7Cl_3NO_3PS$ 322.54	5598-13-0 43			
4091	Chlorsulfuron 2-Chloro-N-[[(4-methoxy-6-methyl-1,3,5-triaziN-2-yl)amino]carbonyl]benzenesulfonamide	$C_{12}H_{12}ClN_5O_4S$ 357.78	64902-72-3 176	2192		
4092	Chlorthiophos Phosphorothioic acid, O-[2,5-dichloro-4-(methylthio)phenyl] O,O-diethyl ester	$C_{11}H_{15}Cl_2O_3PS_2$ 361.25	21923-23-9 150[0.001]			

No.	Name Synonym	Mol. Form. Mol. Wt.	CAS RN mp/°C	Merck No. bp/°C	Beil. Ref. den/g cm^{-3}	Solubility n_D
4093	Cholane 1H-Cyclopenta[a]phenanthrene, hexadecahydro-10,13-dimethyl-17-(1-methylbutyl)-, [8R-[8α,9β,10α,1	$C_{24}H_{42}$ 330.60	548-98-1 90	2198 190[0.001]	4-05-00-01220	
4094	Cholan-24-oic acid Cholanic acid	$C_{24}H_{40}O_2$ 360.58	25312-65-6 163.5	2199	4-09-00-01992	EtOH 3; chl 3; HOAc 3
4095	Cholan-24-oic acid, 3,12-dihydroxy-, (3α,5β,12α)- Deoxycholic acid	$C_{24}H_{40}O_4$ 392.58	83-44-3 177	2881	4-10-00-01608	
4096	Cholan-24-oic acid, 3,6-dihydroxy-, (3α,5β,6α)- Hyodeoxycholic acid	$C_{24}H_{40}O_4$ 392.58	83-49-8 198.5	4794	4-10-00-01600	H_2O 2; EtOH 3; eth 2; ace 2
4097	Cholan-24-oic acid, 3,7-dihydroxy-, (3α,5β,7α)- Chenodiol	$C_{24}H_{40}O_4$ 392.58	474-25-9 119	2044	4-10-00-01604	H_2O 1; EtOH 4; eth 3; ace 4
4098	Cholan-24-oic acid, 3,7-dihydroxy-, (3α,5β,7β)- Ursodiol	$C_{24}H_{40}O_4$ 392.58	128-13-2 203	9801	4-10-00-01604	EtOH 4; eth 2
4099	Cholan-24-oic acid, 3-hydroxy-, (3α,5β)- Lithocholic acid	$C_{24}H_{40}O_3$ 376.58	434-13-9 186	5423	4-10-00-00785	H_2O 1; EtOH 3; eth 2; chl 3
4100	Cholan-24-oic acid, 3,7,12-trihydroxy-, (3α,5β,7α,12α)- Cholic acid	$C_{24}H_{40}O_5$ 408.58	81-25-4 198	2206	4-10-00-02072	H_2O 2; EtOH 3; eth 4; ace 3
4101	Cholan-24-oic acid, 3,7,12-trioxo-, (5β)- Dehydrocholic acid	$C_{24}H_{34}O_5$ 402.53	81-23-2 237	2858	4-10-00-03478	H_2O 1; EtOH 2; eth 1; ace 3
4102	Cholesta-3,5-diene	$C_{27}H_{44}$ 368.65	747-90-0 80	260[13]	4-05-00-01620 0.925[100]	H_2O 1; EtOH 3; eth 5; bz 5
4103	Cholesta-5,7-dien-3-ol, (3β)- 7-Dehydrocholesterol	$C_{27}H_{44}O$ 384.65	434-16-2 150.5	2857	4-06-00-04153	H_2O 1; EtOH 2; eth 3; ace 3
4104	Cholesta-8,24-dien-3-ol, (3β,5α)-	$C_{27}H_{44}O$ 384.65	128-33-6 110	160[0.001]	3-06-00-02828	ace 3; chl 3; MeOH 3
4105	Cholestane, (5α)- 28,29,30-Trinorlanostane	$C_{27}H_{48}$ 372.68	481-21-0 80	2202 250[1]	4-05-00-01227 0.909[88]	H_2O 1; EtOH 4; eth 4; bz 4 1.4887[88]
4106	Cholestane, (5β)- Coprostane	$C_{27}H_{48}$ 372.68	481-20-9 72	2518	4-05-00-01226 0.9119[87]	eth 4; chl 4 1.4884[88]
4107	Cholestanol Dihydrocholesterol	$C_{27}H_{48}O$ 388.68	80-97-7 141.5	2203	4-06-00-03577	eth 4; chl 4
4108	Cholestan-3-ol, (3α,5α)- Epicholestanol	$C_{27}H_{48}O$ 388.68	516-95-0 185.5	3564	4-06-00-03577	chl 3
4109	Cholestan-3-ol, (3β,5β)- Coprosterol	$C_{27}H_{48}O$ 388.68	360-68-9 102	2519	4-06-00-03576	H_2O 1; EtOH 4; eth 4; bz 4
4110	Cholest-5-en-3-ol (3β)- Cholesterol	$C_{27}H_{46}O$ 386.66	57-88-5 148.5	2204 360 dec; 233[0.5]	4-06-00-04000 1.067[20]	H_2O 1; EtOH 2; eth 4; ace 2
4111	Cholest-4-en-3-ol, (3β)- Allocholesterol	$C_{27}H_{46}O$ 386.66	517-10-2 132	253	3-06-00-02604	H_2O 1; EtOH 3; eth 4; ace 4
4112	Cholest-5-en-3-ol, (3α)- Epicholesterol	$C_{27}H_{46}O$ 386.66	474-77-1 141.5	3565	4-06-00-04001	EtOH 2
4113	6-Chrysenamine 6-Aminochrysene	$C_{18}H_{13}N$ 243.31	2642-98-0 210.5	2258	4-12-00-03492	
4114	Chrysene 1,2-Benzophenanthrene	$C_{18}H_{12}$ 228.29	218-01-9 258.2	2259 448	4-05-00-02554 1.274[20]	H_2O 1; EtOH 2; eth 2; ace 2
4115	Chrysene, 5,6-dimethyl- 5,6-Dimethylchrysene	$C_{20}H_{16}$ 256.35	3697-27-6 129.3	200[0.5]	4-05-00-02590	H_2O 1; EtOH 3; CS_2 3; HOAc 3
4116	5,6-Chrysenedione	$C_{18}H_{10}O_2$ 258.28	2051-10-7 239.5	sub	4-07-00-02646	H_2O 1; EtOH 3; eth 2; bz 3
4117	Chrysene, 1-methyl- 1-Methylchrysene	$C_{19}H_{14}$ 242.32	3351-28-8 256.5	sub 140	4-05-00-02572	H_2O 1; EtOH 3
4118	Chrysene, octadecahydro-	$C_{18}H_{30}$ 246.44	2090-14-4 115	353	3-05-00-01107	EtOH 4
4119	Cinchonan-6',9-diol, (8α,9R)- Cupreine	$C_{19}H_{22}N_2O_2$ 310.40	524-63-0 202	2626	4-23-00-03260	EtOH 4
4120	Cinchonan-9-ol, (9S)- Cinchonine	$C_{19}H_{22}N_2O$ 294.40	118-10-5 265	2289	5-23-12-00406	
4121	Cinchonan-9-ol, 10,11-dihydro-, (8α,9R)- Hydrocinchonidine	$C_{19}H_{24}N_2O$ 296.41	485-64-3 229	4705		EtOH 4
4122	Cinchonan-9-ol, 10,11-dihydro-, (9S)- Hydrocinchonine	$C_{19}H_{24}N_2O$ 296.41	485-65-4 268.5	4706		H_2O 3; EtOH 2; eth 1
4123	Cinchonan-9-ol, 10,11-dihydro-6'-methoxy-, (8α,9R)- Hydroquinine	$C_{20}H_{28}N_2O_2$ 326.44	522-66-7 172.5	4737	5-23-13-00340	ace 4; eth 4; EtOH 4; chl 4
4124	Cinchonan-9-ol, 10,11-dihydro-6'-methoxy-, (9S)- Hydroquinidine	$C_{20}H_{28}N_2O_2$ 326.44	1435-55-8 168.5	4736	5-23-13-00340	EtOH 3; eth 3; ace 3; chl 3
4125	Cinchonan-9-ol, 6'-methoxy-, (9R)- Epiquinidine	$C_{20}H_{24}N_2O_2$ 324.42	572-59-8 113	3570	0-23-00-00505	EtOH 4; eth 3
4126	Cinchonan-9-ol, 6'-methoxy-, (9S)- Quinidine	$C_{20}H_{24}N_2O_2$ 324.42	56-54-2 174	8072	5-23-13-00395	H_2O 2; EtOH 3; eth 2; bz 3

No.	Name / Synonym	Mol. Form. / Mol. Wt.	CAS RN / mp/°C	Merck No. / bp/°C	Beil. Ref. / den/g cm^{-3}	Solubility / n_D
4127	Cinchonan-9-ol, 6'-methoxy-, (8α,9R)- / Quinine	$C_{20}H_{24}N_2O_2$ / 324.42	130-95-0 / 57	8075	5-23-13-00395	H_2O 2; EtOH 4; eth 3; ace 2 / 1.625[15]
4128	Cinchonan-9-ol, 6'-methoxy-, carbonate (2:1) (ester), (8α,9R)- / Quinine carbonate	$C_{41}H_{46}N_4O_5$ / 674.84	146-06-5 / 189	8077	4-23-00-03282	
4129	Cinchonan-9-ol, 6'-methoxy-, (8α,9R)-, monoformate (salt) / Quinine, formate	$C_{21}H_{26}N_2O_4$ / 370.45	130-90-5 / 149.5	8081	4-23-00-03274	H_2O 4; EtOH 4; chl 4
4130	Cinchonan-9-ol, 6'-methoxy-, monohydrochloride, (8α,9R)- / Quinine hydrochloride	$C_{20}H_{25}ClN_2O_2$ / 360.88	130-89-2 / 159	8085		H_2O 4; EtOH 4; chl 4
4131	Cinchonan-9-ol, 6'-methoxy-, (8α,9R)-, mono(2-hydroxybenzoate) (salt) / Quinine salicylate	$C_{27}H_{32}N_2O_6$ / 480.56	750-90-3 / 195	8088	4-23-00-03277	EtOH 4; chl 4
4132	Cinchonan-9-ol, 6'-methoxy-, (8α,9R)-, sulfate (2:1) (salt) / Quinine sulfate	$C_{40}H_{50}N_4O_8S$ / 746.93	804-63-7 / 235.2	8089		EtOH 4
4133	Cinchonan-9-ol, (8α,9R)- / Cinchonidine	$C_{19}H_{22}N_2O$ / 294.40	485-71-2 / 210.5 sub	2288	5-23-12-00406	H_2O 1; EtOH 3; eth 2; bz 1
4134	Cinchonan-9-one, 6'-methoxy-, (8α)- / Quininone	$C_{20}H_{22}N_2O_2$ / 322.41	84-31-1 / 108	8093	4-25-00-00204	bz 4; eth 4; EtOH 4
4135	Cinnamic acid, β-propyl- / 2-Hexenoic acid, 3-phenyl-	$C_{12}H_{14}O_2$ / 190.24	4362-01-0 / 94	183-4[14]	3-09-00-02811	bz 4
4136	Cinnoline / 1,2-Benzodiazine	$C_8H_6N_2$ / 130.15	253-66-7 / 38	2309 / 114[0.3]	5-23-07-00124	eth 4; EtOH 4
4137	Ciodrin / 2-Butenoic acid, 3-[(dimethoxyphosphinyl)oxy]-, 1-phenylethyl ester, (E)-	$C_{14}H_{19}O_6P$ / 314.28	7700-17-6 /	2603 / 135[0.03]	1.19[25]	
4138	Clethodim	$C_{17}H_{26}ClNO_3S$ / 359.92	99129-21-2		1.14[20]	
4139	Clofentezine / 3,6-Bis(2-chlorophenyl)-1,2,4,5-tetrazine	$C_{14}H_8Cl_2N_4$ / 303.15	74115-24-5 / 182	2373		
4140	Clomazone / 2-(2-Chlorobenzyl)-4,4-dimethyl-1,2-oxazolidin-3-one	$C_{12}H_{14}ClNO_2$ / 239.70	81777-89-1	3203	1.192[20]	
4141	Cloprop / Propanoic acid, 2-(3-chlorophenoxy)-	$C_9H_9ClO_3$ / 200.62	101-10-0			
4142	Clopyralid / 3,6-Dichloro-2-pyridinecarboxylic acid	$C_6H_3Cl_2NO_2$ / 192.00	1702-17-6 / 151	2398		
4143	Con-5-enin-3-amine, N,N-dimethyl-, (3β)- / Conessine	$C_{24}H_{40}N_2$ / 356.60	546-06-5 / 125.5	2492 / 165-7[0.1]		H_2O 2; chl 3; HOAc 3
4144	Conhydrine, N-methyl- / N-Methylconhydrine	$C_9H_{19}NO$ / 157.26	94826-08-1 /	94-5[11]	2-21-00-00011 / 0.9400[20]	1.4708[19]
4145	Coronene	$C_{24}H_{12}$ / 300.36	191-07-1 / 437.3	525	4-05-00-02830 / 1.371[25]	H_2O 1; bz 2; con sulf 1
4146	Corynan-16-carboxylic acid, 16,17,18,19-tetradehydro-17-methoxy-, methyl ester, (16E)- / Corynantheine	$C_{22}H_{26}N_2O_3$ / 366.46	18904-54-6 / 165.5	2546	4-25-00-01255	EtOH 4
4147	Coumaphos / Coumarin, 3-chloro-7-hydroxy-4-methyl, O-ester with O,O-diethylphosphorothioate	$C_{14}H_{16}ClO_5PS$ / 362.77	56-72-4 / 93	2559	1.474	
4148	o-Cresol, 4-(3-methylpentyl)- / Phenol, 2-methyl-4-(3-methyl pentyl)	$C_{13}H_{20}O$ / 192.30	882-25-7 / 35	145-6[11]		ace 4 / 1.5200[25]
4149	Crotonaldehyde, 2-bromo-, diethyl acetal / 2-Butenal, 2-bromo, diethyl acetal	$C_8H_{15}BrO_2$ / 223.11	98551-14-5 /	86[15]	1-01-00-00380 / 1.2255[21]	1.4565[21]
4150	Crotonic acid, 2-chloro- / 2-Chlorocrotonic acid	$C_4H_5ClO_2$ / 120.54	600-13-5 / 99.8	212; 111[14]	4-02-00-01509	eth 4; EtOH 4
4151	Crotonic acid, 4-chloro-, (E)- / 2-Butenoic acid, 4-chloro ($trans$)	$C_4H_5ClO_2$ / 120.54	26340-58-9 / 83	117-8[13]	4-02-00-01511	eth 4
4152	Crotonic acid, 4,4-dichloro- / 4,4-Dichlorocrotonic acid	$C_4H_4Cl_2O_2$ / 154.98	99980-00-4 / 100.5	130[18]	4-02-00-01512 / 1.3331[99]	eth 4; EtOH 4; chl 4 / 1.4597[99]
4153	Crufomate / Phosphoramidic acid, methyl-, 2-chloro-4-(1,1-dimethylethyl)phenyl methyl ester	$C_{12}H_{19}ClNO_3P$ / 291.71	299-86-5 / 60	2607 / 118[0.01]		
4154	Curan-17-ol, (16α)- / Geissoschizoline	$C_{19}H_{26}N_2O$ / 298.43	18397-07-4 / 135 dec	4274	5-23-12-00211	H_2O 1; EtOH 4; eth 4; chl 4
4155	Cyanamide / Cyanogenamide	CH_2N_2 / 42.04	420-04-2 / 44	2691 / 140[19]	4-03-00-00145 / 1.282[20]	H_2O 4; EtOH 4; eth 3; ace 3 / 1.4418[48]
4156	Cyanamide, diethyl-	$C_5H_{10}N_2$ / 98.15	617-83-4 / -80.6	188	4-04-00-00381 / 0.854[20]	H_2O 1; EtOH 3; eth 3 / 1.4126[25]
4157	Cyanamide, dimethyl-	$C_3H_6N_2$ / 70.09	1467-79-4 /	163.5	4-04-00-00226	ace 4; eth 4; EtOH 4 / 1.4089[19]
4158	Cyanamide, diphenyl- / Diphenylcyanamide	$C_{13}H_{10}N_2$ / 194.24	27779-01-7 / 73.5	237[60]	4-12-00-00857	EtOH 4; lig 4
4159	Cyanamide, di-2-propenyl- / Diallylcyanamide	$C_7H_{10}N_2$ / 122.17	538-08-9 /	2952 / 142[90]; 95[9]	4-04-00-01078	EtOH 3; eth 2; ctc 2
4160	Cyanamide, methylphenyl-	$C_8H_8N_2$ / 132.17	18773-77-8 / 32	137[12]	4-12-00-00840 / 1.040[55]	

No.	Name / Synonym	Mol. Form. / Mol. Wt.	CAS RN / mp/°C	Merck No. / bp/°C	Beil. Ref. / den/g cm⁻³	Solubility / n_D
4161	Cyanazine / Propanenitrile, 2-[[4-chloro-6-(ethylamino)-1,3,5-triazin-2-yl]amino]-2-methyl-	$C_9H_{13}ClN_6$ / 240.70	21725-46-2 / 168	2692		
4162	Cyanic acid / Hydrogen cyanate (HOCN)	CHNO / 43.03	420-05-3 / -86	2693 / 23	4-03-00-00080 / 1.140^{20}	H_2O 4; bz 4; eth 4; chl 4
4163	Cyanic acid, ethyl ester / Ethyl cyanate	C_3H_5NO / 71.08	627-48-5	162 dec; 30^{12}	0.89^{20}	EtOH 4 / 1.3788^{25}
4164	Cyanic acid, sodium salt / Sodium cyanate	CNNaO / 65.01	917-61-3 / 550	8552	1.893^{20}	H_2O 3; EtOH 2; eth 1
4165	Cyanogen bromide / Bromine cyanide	CBrN / 105.92	506-68-3 / 52	2700 / 61.5	4-03-00-00092 / 2.015^{20}	H_2O 3; EtOH 3; eth 3
4166	Cyanogen chloride / Chlorine cyanide	CClN / 61.47	506-77-4 / -6.5	2701 / 13	4-03-00-00090 / 1.186^{20}	H_2O 3; EtOH 3; eth 4
4167	Cyanogen fluoride / Fluorine cyanide	CFN / 45.02	1495-50-7 / -82	-46		
4168	Cycloate / Carbamothioic acid, cyclohexylethyl-, S-ethyl ester	$C_{11}H_{21}NOS$ / 215.36	1134-23-2 / 11.5	145^{10}	1.0156^{30}	
4169	Cyclobutanamine / Aminocyclobutane	C_4H_9N / 71.12	2516-34-9 /	82	4-12-00-00003 / 0.8328^{20}	1.4363^{19}
4170	Cyclobutane / Tetramethylene	C_4H_8 / 56.11	287-23-0 / -90.6	2720 / 12.6	4-05-00-00006 / 0.6890^{25}	H_2O 1; EtOH 4; eth 5; ace 4 / 1.365^{290}
4171	Cyclobutaneacetic acid, 3-(ethoxycarbonyl)-2,2-dimethyl-, ethyl ester / Pinic acid diethyl ester	$C_{13}H_{22}O_4$ / 242.32	28664-03-1 /	$142-6^{10}$	2-09-00-00529 / 1.0104^{20}	1.4496^{70}
4172	Cyclobutane, 1,2-bis(methylene)-	C_6H_8 / 80.13	14296-80-1 /	64	4-05-00-00389 / 0.7698^{20}	1.4232^{20}
4173	Cyclobutanecarbonitrile / Cyanocyclobutane	C_5H_7N / 81.12	4426-11-3 /	149.6		
4174	Cyclobutanecarboxylic acid / Cyclobutylcarboxylic acid	$C_5H_8O_2$ / 100.12	3721-95-7 / -2	190; $74-5^2$	4-09-00-00006 / 1.0599^{20}	H_2O 2; EtOH 5; eth 5 / 1.4400^{20}
4175	Cyclobutanecarboxylic acid, ethyl ester / Ethyl cyclobutanecarboxylate	$C_7H_{12}O_2$ / 128.17	14924-53-9 /	159	3-09-00-00007 / 0.928^{25}	ctc 2 / 1.4261^{20}
4176	Cyclobutanecarboxylic acid, methyl ester / Methyl cyclobutanecarboxylate	$C_6H_{10}O_2$ / 114.14	765-85-5 /	135.5		
4177	Cyclobutane, 1-chloro-1,2,2,3,3,4,4-heptafluoro- / Chloroheptafluorocyclobutane	C_4ClF_7 / 216.49	377-41-3			
4178	Cyclobutane, 1,1-dibromo- / 1,1-Dibromocyclobutane	$C_4H_6Br_2$ / 213.90	33742-81-3 /	160	0-05-00-00017 / 1.930^{20}	bz 4; EtOH 4; lig 4 / 1.5362^{20}
4179	Cyclobutane, 1,2-dibromo- / 1,2-Dibromocyclobutane	$C_4H_6Br_2$ / 213.90	89033-70-5 / 2.5	172.5	0-05-00-00017 / 1.972^{20}	
4180	1,1-Cyclobutanedicarboxylic acid, diethyl ester / Diethyl 1,1-cyclobutanedicarboxylate	$C_{10}H_{16}O_4$ / 200.23	3779-29-1 /	224	4-09-00-02788 / 1.0456^{20}	EtOH 4; ctc 2 / 1.4330^{26}
4181	1,2-Cyclobutanedicarboxylic acid, dimethyl ester, trans- / Dimethyl trans-1,2-cyclobutanedicarboxylate	$C_8H_{12}O_4$ / 172.18	7371-67-7 /	225; 118^{24}	3-09-00-03799 / 1.1276^{20}	1.4450^{20}
4182	1,3-Cyclobutanedicarboxylic acid, cis-	$C_6H_8O_4$ / 144.13	2398-16-5 / 143.5	252	3-09-00-03801	H_2O 4; EtOH 4
4183	1,3-Cyclobutanedicarboxylic acid, 2,2-dimethyl-, cis- / cis-Norpinic acid	$C_8H_{12}O_4$ / 172.18	3211-48-1 / 175	sub 100	4-09-00-02813	eth 4; chl 4
4184	1,3-Cyclobutanedicarboxylic acid, 2,2-dimethyl-, dimethyl ester / Norpinic acid, dimethyl ester	$C_{10}H_{16}O_4$ / 200.23	91057-79-3 /	229; 113^{11}	3-09-00-03829 / 1.070^{14}	1.4459^{17}
4185	1,3-Cyclobutanedicarboxylic acid, trans-	$C_6H_8O_4$ / 144.13	7439-33-0 / 171	sub	3-09-00-03802	H_2O 4; EtOH 4
4186	Cyclobutane, 1,2-diethenyl-, cis-	C_8H_{12} / 108.18	16177-46-1 /	38^{38}	4-05-00-00412 / 0.8010^{20}	1.4563^{20}
4187	Cyclobutane, 1,2-diethenyl-, trans- / trans-1,2-Divinylcyclobutane	C_8H_{12} / 108.18	6553-48-6 /	112.5	4-05-00-00412 / 0.7817^{20}	1.4451^{20}
4188	Cyclobutane, 1,2-dimethyl-, trans- / trans-1,2-Dimethylcyclobutane	C_6H_{12} / 84.16	15679-02-4 / -122.5	58	4-05-00-00088 / 0.703^{20}	1.395^{20}
4189	Cyclobutane, 1,3-dimethyl-, cis- / cis-1,3-Dimethylcyclobutane	C_6H_{12} / 84.16	2398-09-6 /	60.5	4-05-00-00089 / 0.7060^{25}	1.3933^{20}
4190	Cyclobutane, 1,3-dimethyl-, trans- / trans-1,3-Dimethylcyclobutane	C_6H_{12} / 84.16	2398-10-9 /	57.5	4-05-00-00089 / 0.6970^{25}	1.3896^{20}
4191	Cyclobutane, ethyl- / Ethylcyclobutane	C_6H_{12} / 84.16	4806-61-5 / -142.9	70.8	4-05-00-00087 / 0.7284^{20}	H_2O 1; EtOH 5; eth 5; ace 3 / 1.4020^{20}
4192	Cyclobutane, methyl- / Methylcyclobutane	C_5H_{10} / 70.13	598-61-8 / -161.5	36.3	4-05-00-00021 / 0.6884^{20}	H_2O 1; EtOH 5; eth 5; ace 3 / 1.3866^{20}
4193	Cyclobutane, methylene- / exo-Methylenecyclobutane	C_5H_8 / 68.12	1120-56-5 / -134.7	42.2	4-05-00-00216 / 0.7401^{20}	1.4210^{20}
4194	Cyclobutane, octafluoro- / Perfluorocyclobutane	C_4F_8 / 200.03	115-25-3 / -40.1	6666 / -5.9	4-05-00-00008 / $*1.500^{25}$	eth 3
4195	Cyclobutane, 1,2,3,4-tetrabromo- / 1,2,3,4-Tetrabromocyclobutane	$C_4H_4Br_4$ / 371.69	101257-79-8 /	$110^{1.5}$	4-05-00-00012 / 2.5672^{20}	1.6303^{20}
4196	Cyclobutanol / Hydroxycyclobutane	C_4H_8O / 72.11	2919-23-5 /	124	4-06-00-00003 / 0.9218^{15}	1.4371^{20}

No.	Name Synonym	Mol. Form. Mol. Wt.	CAS RN mp/°C	Merck No. bp/°C	Beil. Ref. den/g cm^{-3}	Solubility n_D
4197	Cyclobutanone	C_4H_6O 70.09	1191-95-3 -50.9	99	4-07-00-00003 0.9547^0	H_2O 3; EtOH 4; eth 3; bz 3 1.4215^{20}
4198	Cyclobutene	C_4H_6 54.09	822-35-5	2	4-05-00-00207 0.733^0	ace 4; bz 3; peth 3
4199	Cyclobutene, 1,2-dichloro-3,3,4,4-tetrafluoro-	$C_4Cl_2F_4$ 194.94	377-93-5 -43.4	67	4-05-00-00209 1.534^{25}	
4200	Cyclobutene, hexafluoro-	C_4F_6 162.03	697-11-0 -60	5.5	4-05-00-00208 1.602^{-20}	1.3699^{25}
4201	Cyclobutene, 1-methyl- 1-Methylcyclobutene	C_5H_8 68.12	1489-60-7	37	4-05-00-00216 0.7244^{20}	1.298^{20} 1.4088^{18}
4202	Cyclodecane	$C_{10}H_{20}$ 140.27	293-96-9 10	202	4-05-00-00144 0.8538^{25}	1.4716^{20}
4203	1,2-Cyclodecanedione Sebacil	$C_{10}H_{16}O_2$ 168.24	96-01-5 40.5	8370 104-5^{10}	4-07-00-02009	
4204	Cyclodecanol	$C_{10}H_{20}O$ 156.27	1502-05-2 40.5	125^{12}	4-06-00-00138 0.9606^{20}	EtOH 3 1.4926^{20}
4205	Cyclodecanone	$C_{10}H_{18}O$ 154.25	1502-06-3 28	106-7^{13}	4-07-00-00076 0.9654^{20}	bz 4; eth 4; chl 4 1.4806^{20}
4206	Cyclodecanone, 2-hydroxy- Sebacoin	$C_{10}H_{18}O_2$ 170.25	96-00-4 38.5	8371 134-8^{14}	4-08-00-00037	
4207	Cyclodecene, (E)-	$C_{10}H_{18}$ 138.25	2198-20-1	68-70^{10}	4-05-00-00296 0.8672^{20}	H_2O 1; bz 3; chl 3 1.4822^{20}
4208	Cyclodecene, (Z)-	$C_{10}H_{18}$ 138.25	935-31-9	195	4-05-00-00295 0.8770^{20}	1.4854^{20}
4209	Cyclodecyne	$C_{10}H_{16}$ 136.24	3022-41-1	204	4-05-00-00430 0.8975^{20}	1.4903^{20}
4210	Cyclododecane	$C_{12}H_{24}$ 168.32	294-62-2 60.4	247	4-05-00-00169 0.82^{80}	
4211	Cyclododecanone Cyclodoecanone	$C_{12}H_{22}O$ 182.31	830-13-7 59	126-8^{12}	4-07-00-00100 0.9059^{66}	1.4571^{60}
4212	1,5,9-Cyclododecatriene CDT	$C_{12}H_{18}$ 162.27	4904-61-4 -17	240	4-05-00-01114 0.84^{100}	
4213	1,5,9-Cyclododecatriene, (Z,E,E)- 1,5,9-Cyclododecatriene, (cis, trans, trans)	$C_{12}H_{18}$ 162.27	53859-78-2 -18	100-1^{11}	4-05-00-01115 0.8910^{20}	bz 4; chl 4 1.5058^{20}
4214	1,5,9-Cyclododecatriene, (Z,Z,Z)-	$C_{12}H_{18}$ 162.27	4736-48-5 -1.5	108^{17}		bz 4; chl 4 1.5100^{25}
4215	Cyclododecene, (E)-	$C_{12}H_{22}$ 166.31	1486-75-5	112-5^{17}	4-05-00-00332	bz 4; chl 4 1.4850^{20}
4216	Cyclododecene, (Z)-	$C_{12}H_{22}$ 166.31	1129-89-1	133^{35}; 71^2	4-05-00-00332	bz 4; chl 4 1.4840^{20}
4217	9-Cycloheptadecen-1-one 9-Cycloheptadecene-1-one	$C_{17}H_{30}O$ 250.42	74244-64-7 32.5	343; 159^2	3-07-00-00524 0.9170^{33}	H_2O 2; EtOH 3; bz 3 1.4830^{33}
4218	9-Cycloheptadecen-1-one, (Z)- Civetone	$C_{17}H_{30}O$ 250.42	542-46-1 32.5	2337 343; 159^2	3-07-00-00524	
4219	1,3-Cycloheptadiene	C_7H_{10} 94.16	4054-38-0 -110.4	120.5	4-05-00-00390 0.868^{25}	1.4978^{20}
4220	2,4-Cycloheptadien-1-one, 2,6,6-trimethyl- Eucarvone	$C_{10}H_{14}O$ 150.22	503-93-5	105^{22}; 88^{16}	4-07-00-00311 0.9490^{20}	eth 3; ace 3 1.5087^{20}
4221	Cycloheptane	C_7H_{14} 98.19	291-64-5 -8.0	118.4	4-05-00-00092 0.8098^{20}	H_2O 1; EtOH 4; eth 4; bz 3 1.4436^{20}
4222	Cycloheptane, bromo- Cycloheptyl bromide	$C_7H_{13}Br$ 177.08	2404-35-5	101.5^{40}; 75^{12}	4-05-00-00093 1.2887^{22}	H_2O 1; eth 4; chl 4 1.4996^{20}
4223	Cycloheptanecarboxylic acid Cycloheptylcarboxylic acid	$C_8H_{14}O_2$ 142.20	1460-16-8	259; 130^8	4-09-00-00038 1.0423^{20}	ace 3 1.4753^{20}
4224	Cycloheptanecarboxylic acid, ethyl ester Ethyl cycloheptanecarboxylate	$C_{10}H_{18}O_2$ 170.25	32777-26-7	74^3	4-09-00-00038 0.9515^{20}	1.4482^{20}
4225	1,2-Cycloheptanedione	$C_7H_{10}O_2$ 126.16	3008-39-7 -40	107-9^{17}	4-07-00-01989 1.0583^{22}	EtOH 3 1.4689^{22}
4226	Cycloheptane, methyl- Methylcycloheptane	C_8H_{16} 112.22	4126-78-7	134	4-05-00-00114 0.800^{20}	bz 4; eth 4; EtOH 4; peth 4 1.439^{18}
4227	Cycloheptane, 1,1,2-trimethyl- 1,1,2-Trimethylcycloheptane	$C_{10}H_{20}$ 140.27	35099-89-9	104-5^{100}	3-05-00-00127 0.8243^{20}	1.4527^{20}
4228	Cycloheptane, 1,1,4-trimethyl- 1,1,4-Trimethylcycloheptane	$C_{10}H_{20}$ 140.27	2158-55-6	164	4-05-00-00146 0.801^{20}	1.4420^{20}
4229	Cycloheptanol	$C_7H_{14}O$ 114.19	502-41-0 2	185	4-06-00-00094 0.9554^{20}	H_2O 2; EtOH 4; eth 4 1.40705^{20}
4230	Cycloheptanol, 1-methyl- 1-Methyl-1-cycloheptanol	$C_8H_{16}O$ 128.21	3761-94-2	183.5	4-06-00-00114 0.9392^{21}	1.4677^{22}
4231	Cycloheptanol, 2-methyl- 2-Methylcycloheptanol	$C_8H_{16}O$ 128.21	59777-92-3	191	3-06-00-00083 0.9492^{15}	1.4762^{15}
4232	Cycloheptanol, 4-methyl-	$C_8H_{16}O$ 128.21	90200-61-6	105-6^{39}	3-06-00-00084 0.9134^{31}	1.4574
4233	Cycloheptanol, 2,6,6-trimethyl- 3,3,7-Trimethylcycloheptanol	$C_{10}H_{20}O$ 156.27	33515-82-1	216	1-06-00-00018 0.9096^{24}	1.4369
4234	Cycloheptanone Suberone	$C_7H_{12}O$ 112.17	502-42-1	2728 178.5	4-07-00-00039 0.9508^{20}	H_2O 1; EtOH 4; eth 3 1.4608^{20}
4235	Cycloheptanone, 2-methyl- 2-Methylcycloheptanone	$C_8H_{14}O$ 126.20	932-56-9	185	4-07-00-00051 0.9395^{18}	1.461^{18}
4236	Cycloheptanone, 2-phenyl- 2-Phenylcycloheptanone	$C_{13}H_{16}O$ 188.27	14996-78-2 22.2	141^5; 125^2	4-07-00-01061	1.5395^{20}

No.	Name / Synonym	Mol. Form. / Mol. Wt.	CAS RN / mp/°C	Merck No. / bp/°C	Beil. Ref. / den/g cm^{-3}	Solubility / n_D
4237	Cycloheptanone, 2,2,6-trimethyl-	$C_{10}H_{18}O$ / 154.25	1686-41-5	191	3-07-00-00138 / 0.878[14]	1.4421[16]
4238	Cycloheptanone, 2,6,6-trimethyl- / Tetrahydroeucarvone	$C_{10}H_{18}O$ / 154.25	4436-59-3	208.5	4-07-00-00079 / 0.9095[18]	1.4568[18]
4239	Cycloheptasiloxane, tetradecamethyl- / Tetradecamethylcycloheptasiloxane	$C_{14}H_{42}O_7Si_7$ / 519.08	107-50-6 / -26	154[20]	4-04-00-04130 / 0.9703[20]	1.4040[20]
4240	1,3,5-Cycloheptatriene / Tropilidene	C_7H_8 / 92.14	544-25-2 / -79.5	117; 60.5[122]	4-05-00-00765 / 0.8875[19]	H_2O 1; EtOH 3; eth 3; bz 4 / 1.5343[20]
4241	1,3,6-Cycloheptatriene-1-carboxylic acid, 5,5-dimethyl- / Thujic acid	$C_{10}H_{12}O_2$ / 164.20	499-89-8	9325	4-09-00-01810	
4242	2,4,6-Cycloheptatrien-1-one	C_7H_6O / 106.12	539-80-0 / -7	113[15]; 84[6]	4-07-00-00501 / 1.095[22]	bz 4; chl 4 / 1.6172[22]
4243	2,4,6-Cycloheptatrien-1-one, 2-hydroxy-	$C_7H_6O_2$ / 122.12	533-75-5 / 50.8	sub 40	4-08-00-00159	H_2O 3; eth 3; ace 3; os 3
4244	Cycloheptene	C_7H_{12} / 96.17	628-92-2 / -56	115	4-05-00-00244 / 0.8228[20]	H_2O 3; EtOH 3; eth 3; bz 3 / 1.4552[20]
4245	Cycloheptene, 1-methyl- / 1-Methylcycloheptene	C_8H_{14} / 110.20	1453-25-4	136	4-05-00-00265 / 0.8243[22]	1.4581[20]
4246	Cycloheptene, 5-methyl- / 5-Methylcycloheptene	C_8H_{14} / 110.20	2505-06-8	69-70[38]	3-05-00-00211 / 0.7606[31]	1.4202[31]
4247	Cycloheptene, 1,4,4-trimethyl- / 1,4,4-Trimethyl-1-cycloheptene	$C_{10}H_{18}$ / 138.25	4755-36-6	165	3-05-00-00232 / 0.8185[20]	1.4561[20]
4248	1,3-Cyclohexadiene	C_6H_8 / 80.13	592-57-4 / -89	80.5	4-05-00-00382 / 0.8405[20]	H_2O 1; EtOH 3; eth 4; bz 3 / 1.4755[20]
4249	1,4-Cyclohexadiene / 1,4-Dihydrobenzene	C_6H_8 / 80.13	628-41-1 / -49.2	85.5	4-05-00-00385 / 0.8471[20]	H_2O 1; EtOH 5; eth 5; bz 3 / 1.4725[20]
4250	1,3-Cyclohexadiene-1-carboxaldehyde, 2,6,6-trimethyl- / Safranal	$C_{10}H_{14}O$ / 150.22	116-26-7 / 70[1]	8286	4-07-00-00318 / 0.9734[19]	EtOH 4; peth 4 / 1.5281[19]
4251	1,4-Cyclohexadiene-1,2-dicarbonitrile, 4,5-dichloro-3,6-dioxo- / 2,3-Dichloro-5,6-dicyanobenzoquinone	$C_8Cl_2N_2O_2$ / 227.01	84-58-2 / 214.5	3052	0-10-00-00902	bz 4; HOAc 4; diox 4
4252	1,3-Cyclohexadiene, 1,3-dimethyl- / 1,3-Dimethyl-1,3-cyclohexadiene	C_8H_{12} / 108.18	4573-05-1	135.5	3-05-00-00325 / 0.8373[20]	1.4776[20]
4253	1,3-Cyclohexadiene, 1,4-dimethyl- / 1,4-Dimethyl-1,3-cyclohexadiene	C_8H_{12} / 108.18	26120-52-5	136.5	4-05-00-00409 / 0.830[20]	1.4792[19]
4254	1,3-Cyclohexadiene, 1,5-dimethyl- / 1,5-Dimethyl-1,3-cyclohexadiene	C_8H_{12} / 108.18	1453-17-4	130	1-05-00-00063 / 0.821[20]	1.471[20]
4255	1,3-Cyclohexadiene, 2,5-dimethyl- / 2,5-Dimethyl-1,3-cyclohexadiene	C_8H_{12} / 108.18	2050-33-1	127	3-05-00-00325 / 0.8245[18]	1.4631[18]
4256	1,3-Cyclohexadiene, 2,6-dimethyl- / 1,3-Cyclohexadiene, 3,5-dimethyl	C_8H_{12} / 108.18	2050-32-0	131	0-05-00-00119 / 0.8225[20]	1.4675[20]
4257	2,5-Cyclohexadiene-1,4-dione / p-Benzoquinone	$C_6H_4O_2$ / 108.10	106-51-4 / 115.7	8103 / sub	4-07-00-02065 / 1.318[20]	H_2O 2; EtOH 3; eth 3; chl 3
4258	2,5-Cyclohexadiene-1,4-dione, 2,5-dichloro-3,6-dihydroxy- / Chloranilic acid	$C_6H_2Cl_2O_4$ / 208.98	87-88-7 / 283.5	2072	4-08-00-02707	H_2O 3
4259	2,5-Cyclohexadiene-1,4-dione, 2,5-dihydroxy-3,6-dinitro- / Nitranilic acid	$C_6H_2N_2O_8$ / 230.09	479-22-1 / 100 dec	6489 / exp 170	3-08-00-03351	H_2O 4; EtOH 4; eth 1
4260	2,5-Cyclohexadiene-1,4-dione, 2,5-dihydroxy-3-methoxy-6-methyl- / Spinulosin	$C_8H_8O_5$ / 184.15	85-23-4 / 202.5	8706 / sub 120	3-08-00-03983	H_2O 2; alk 3
4261	2,5-Cyclohexadiene-1,4-dione, 2,5-dihydroxy-3-undecyl- / Embelin	$C_{17}H_{26}O_4$ / 294.39	550-24-3 / 142.5	3513	4-08-00-02769	bz 4; eth 4; EtOH 4
4262	2,5-Cyclohexadiene-1,4-dione, 2,6-dimethoxy- / 2,6-Dimethoxy-p-quinone	$C_8H_8O_4$ / 168.15	530-55-2 / 256	3214 / sub	4-08-00-02710	H_2O 2; EtOH 2; eth 2; tfa 3
4263	2,5-Cyclohexadiene-1,4-dione, 2,3-dimethyl-	$C_8H_8O_2$ / 136.15	526-86-3 / 55	sub	4-07-00-02093	H_2O 2; EtOH 3; eth 3; chl 3
4264	2,5-Cyclohexadiene-1,4-dione, 2,6-dimethyl-	$C_8H_8O_2$ / 136.15	527-61-7 / 72.5	sub	4-07-00-02095 / 1.0479[28]	chl 3
4265	2,5-Cyclohexadiene-1,4-dione, 3-hydroxy-2-methoxy-5-methyl- / Fumigatin	$C_8H_8O_4$ / 168.15	484-89-9 / 116	4201	3-08-00-03374	ace 4; bz 4; eth 4; EtOH 4
4266	2,5-Cyclohexadiene-1,4-dione, 2-methyl-	$C_7H_6O_2$ / 122.12	553-97-9 / 69	sub	4-07-00-02088 / 1.08[75]	H_2O 2; EtOH 3; eth 3
4267	2,5-Cyclohexadiene-1,4-dione, 2-methyl-5-(1-methylethyl)-	$C_{10}H_{12}O_2$ / 164.20	490-91-5 / 45.5	232	4-07-00-02100	chl 3
4268	2,5-Cyclohexadiene-1,4-dione, 2,3,5,6-tetrachloro- / Chloranil	$C_6Cl_4O_2$ / 245.88	118-75-2 / 290	2071 / sub	4-07-00-02083	H_2O 1; EtOH 2; eth 3; chl 2
4269	2,5-Cyclohexadiene-1,4-dione, 2,3,5,6-tetrahydroxy- / Tetroquinone	$C_6H_4O_6$ / 172.09	319-89-1	9177	4-08-00-03604	H_2O 2; EtOH 4; eth 2; ctc 2
4270	2,5-Cyclohexadiene-1,4-dione, 2,3,5,6-tetramethyl- / Duroquinone	$C_{10}H_{12}O_2$ / 164.20	527-17-3 / 111.5	3452	4-07-00-02102	H_2O 1; EtOH 3; eth 3; ace 3
4271	1,3-Cyclohexadiene, 2-methyl- / 4,5-Dihydrotoluene	C_7H_{10} / 94.16	1489-57-2	107.5	3-06-00-00211 / 0.8260[18]	1.4662[18]

No.	Name / Synonym	Mol. Form. / Mol. Wt.	CAS RN / mp/°C	Merck No. / bp/°C	Beil. Ref. / den/g cm^{-3}	Solubility / n_D
4272	1,3-Cyclohexadiene, 5-methyl-, (±)- / 1,3-Cyclohexadiene, 5-methyl (DL)	C_7H_{10} / 94.16	116781-86-3	101.5	3-05-00-00318 / 0.8354[20]	bz 4; eth 4; EtOH 4; lig 4 / 1.4763[20]
4273	1,3-Cyclohexadiene, 1-methyl-4-(1-methylethyl)-	$C_{10}H_{16}$ / 136.24	99-86-5	175	4-05-00-00435 / 0.8375[19]	H_2O 1; EtOH 5; eth 5 / 1.477[19]
4274	1,3-Cyclohexadiene, 2-methyl-5-(1-methylethyl)-, (±)-	$C_{10}H_{16}$ / 136.24	13811-01-3	175.5	3-05-00-00342 / 0.8410[22]	H_2O 1; EtOH 1; eth 3 / 1.4772[19]
4275	1,4-Cyclohexadiene, 1-methyl-4-(1-methylethyl)-	$C_{10}H_{16}$ / 136.24	99-85-4	183	4-05-00-00436 / 0.849[20]	1.4765[14]
4276	1,3-Cyclohexadiene, 1,2,3,4,5,5,6,6-octafluoro-	C_6F_8 / 224.05	377-70-8	63.5	4-05-00-00384 / 1.601[20]	1.3149[20]
4277	1,4-Cyclohexadiene, 1,2,3,3,4,5,6,6-octafluoro-	C_6F_8 / 224.05	775-51-9	57.5	4-05-00-00386 / 1.6010[25]	1.318[18]
4278	2,5-Cyclohexadien-1-one, 4-[bis(4-hydroxyphenyl)methylene]- / Aurin	$C_{19}H_{14}O_3$ / 290.32	603-45-2 / 308-10 dec	899	4-08-00-02646	H_2O 1; EtOH 3; eth 2; bz 1
4279	2,5-Cyclohexadien-1-one, 2,6-dibromo-4-(chloroimino)- / 2,6-Dibromoquinone-4-chlorimide	$C_6H_2Br_2ClNO$ / 299.35	537-45-1 / 83	3010	4-07-00-02086	EtOH 4
4280	2,5-Cyclohexadien-1-one, 2,6-dichloro-4-(chloroimino)- / Gibbs' reagent	$C_6H_2Cl_3NO$ / 210.45	101-38-2 / 66	4315	3-07-00-03377	
4281	2,4-Cyclohexadien-1-one, 3,5-dihydroxy-2,6,6-tris(3-methyl-2-butenyl)-4-(3-methyl-1-oxobutyl)- / Lupulon	$C_{26}H_{38}O_4$ / 414.59	468-28-0 / 93	5482	3-07-00-04753	H_2O 1; EtOH 3; peth 3; hx 3
4282	2,4-Cyclohexadien-1-one, 3,5,6-trihydroxy-4,6-bis(3-methyl-2-butenyl)-2-(3-methyl-1-oxobutyl)-, (R)- / Humulon	$C_{21}H_{30}O_5$ / 362.47	26472-41-3 / 66.5	4673	4-08-00-03410	H_2O 2; EtOH 3; eth 3; ace 3
4283	Cyclohexanamine / Cyclohexylamine	$C_6H_{13}N$ / 99.18	108-91-8 / -17.7	2735 / 134	4-12-00-00008 / 0.8191[20]	H_2O 3; EtOH 4; eth 5; ace 5 / 1.4372[20]
4284	Cyclohexanamine, N-cyclohexyl- / Dicyclohexylamine	$C_{12}H_{23}N$ / 181.32	101-83-7 / -0.1	3085 / 256 dec; 114[9]	4-12-00-00022 / 0.9123[20]	H_2O 2; EtOH 3; eth 3; bz 3 / 1.4842[20]
4285	Cyclohexanamine, N,N-diethyl-	$C_{10}H_{21}N$ / 155.28	91-65-6	193; 85[20]	4-12-00-00019 / 0.872[0]	EtOH 3; ctc 2
4286	Cyclohexanamine, N,N-dimethyl- / Cyclohexyldimethylamine	$C_8H_{17}N$ / 127.23	98-94-2	162	4-12-00-00018	
4287	Cyclohexanamine, 2-ethyl- / 2-Ethylcyclohexylamine	$C_8H_{17}N$ / 127.23	6850-36-8	170.5	3-12-00-00093 / 0.8744[20]	1.4682[20]
4288	Cyclohexanamine, N-ethyl-	$C_8H_{17}N$ / 127.23	5459-93-8	164	4-12-00-00018 / 0.868[0]	H_2O 2; EtOH 5; eth 5; ctc 2
4289	Cyclohexanamine, N-(2-ethylhexyl)-	$C_{14}H_{29}N$ / 211.39	5432-61-1	270; 140[14]	3-12-00-00018 / 0.8473[20]	ctc 2
4290	Cyclohexanamine, 1-ethynyl-	$C_8H_{13}N$ / 123.20	30389-18-5	65-6[20]	4-12-00-00214 / 0.913[25]	1.4817[20]
4291	Cyclohexanamine, N,N'-methanetetraylbis- / Dicyclohexylcarbodiimide	$C_{13}H_{22}N_2$ / 206.33	538-75-0 / 34.5	3086 / 123[6]; 99[0.5]	4-12-00-00072	
4292	Cyclohexanamine, 2-methyl-, cis-	$C_7H_{15}N$ / 113.20	2164-19-4	153.5	4-12-00-00118 / 0.8778[20]	1.4688[20]
4293	Cyclohexanamine, 2-methyl-, trans- / trans-2-Methylcyclohexylamine	$C_7H_{15}N$ / 113.20	931-10-2	150	4-12-00-00118 / 0.8685[20]	1.4650[20]
4294	Cyclohexanamine, 3-methyl-, cis-	$C_7H_{15}N$ / 113.20	1193-16-4	153	4-12-00-00121 / 0.8552[20]	1.4538[20]
4295	Cyclohexanamine, 3-methyl-, trans-	$C_7H_{15}N$ / 113.20	1193-17-5	151.5	4-12-00-00121 / 0.8572[20]	1.4547[20]
4296	Cyclohexanamine, 4-methyl-, cis-	$C_7H_{15}N$ / 113.20	2523-56-0	153.5	4-12-00-00122 / 0.8567[20]	1.4559[20]
4297	Cyclohexanamine, 4-methyl-, trans-	$C_7H_{15}N$ / 113.20	2523-55-9	152	4-12-00-00122 / 0.8543[20]	1.4550[20]
4298	Cyclohexanamine, N-methyl-	$C_7H_{15}N$ / 113.20	100-60-7	147	4-12-00-00017 / 0.8660[23]	H_2O 2; EtOH 4; eth 5; chl 3 / 1.4560[20]
4299	Cyclohexanamine, 4,4'-methylenebis-	$C_{13}H_{26}N_2$ / 210.36	1761-71-3 / 15	320	0.92[75]	
4300	Cyclohexanamine, 4,4'-methylenebis-, [cis(cis)]-	$C_{13}H_{26}N_2$ / 210.36	6693-31-8 / 61	141[2]	4-13-00-00031	H_2O 1 / 1.5014[27]
4301	Cyclohexanamine, 4,4'-methylenebis-, [trans(cis)]-	$C_{13}H_{26}N_2$ / 210.36	6693-30-7 / 36.5	127-8[1.2]	4-13-00-00031 / 0.9608[25]	H_2O 1 / 1.5046[25]
4302	Cyclohexanamine, 4,4'-methylenebis-, [trans(trans)]-	$C_{13}H_{26}N_2$ / 210.36	6693-29-4 / 64.7	131[0.8]	4-13-00-00031	H_2O 1 / 1.5032[25]
4303	Cyclohexanamine, N-(1-methylethyl)-	$C_9H_{19}N$ / 141.26	1195-42-2	60-5[12]	4-12-00-00019 / 0.859[25]	1.4480[20]

No.	Name / Synonym	Mol. Form. / Mol. Wt.	CAS RN / mp/°C	Merck No. / bp/°C	Beil. Ref. / den/g cm⁻³	Solubility / n_D
4304	Cyclohexanamine, N-2-propenyl-	$C_9H_{17}N$ 139.24	6628-00-8	65-6[12]	4-12-00-00021 0.962[25]	1.4664[20]
4305	Cyclohexane Hexahydrobenzene	C_6H_{12} 84.16	110-82-7 6.59	2729 80.73	4-05-00-00027 0.7739[25]	H_2O 1; EtOH 5; eth 5; ace 5 1.4266[20]
4306	Cyclohexaneacetic acid Cyclohexylacetic acid	$C_8H_{14}O_2$ 142.20	5292-21-7 33	245	4-09-00-00038 1.0423[18]	H_2O 2; eth 3; ace 3; os 3 1.4775[20]
4307	Δ1,α-Cyclohexaneacetic acid, 3,3-dimethyl- Allocyclogeranic acid	$C_{10}H_{16}O_2$ 168.24	6790-47-2	145[14]	4-09-00-00161 0.9921[17]	1.4765[17]
4308	Cyclohexaneacetic acid, ethyl ester	$C_{10}H_{18}O_2$ 170.25	5452-75-5	211	4-09-00-00038 0.9537[14]	1.451[14]
4309	Cyclohexaneacetic acid, α-ethyl-1-hydroxy- Cyclobutyrol	$C_{10}H_{18}O_3$ 186.25	512-16-3 81.5	2721 164[24]	4-10-00-00049 1.0010[18]	ace 4; eth 4; EtOH 4; chl 4 1.4680[18]
4310	Cyclohexaneacetic acid, methyl ester Methyl cyclohexaneacetate	$C_9H_{16}O_2$ 156.22	14352-61-5	201	4-09-00-00038 0.9896[24]	1.459[14]
4311	Cyclohexaneacetic acid, 2-methyl- Cyclohexaneacetic acid, 2-methyl	$C_9H_{16}O_2$ 156.22	6617-04-5	145-7[13]	2-09-00-00013 1.012[25]	1.4656[25]
4312	Cyclohexaneacetic acid, 3-methyl- Cyclohexaneacetic acid, 3-methyl	$C_9H_{16}O_2$ 156.22	67451-76-7	148[18]	2-09-00-00013 0.9847[20]	1.495[20]
4313	Cyclohexaneacetonitrile	$C_8H_{13}N$ 123.20	4435-14-7	105[22]	4-09-00-00040 0.9473[21]	eth 4; EtOH 4 1.4787[21]
4314	Cyclohexane, 1,2-bis(methylene)-	C_8H_{12} 108.18	2819-48-9	125; 60-1[90]	4-05-00-00409 0.8229[25]	H_2O 1; EtOH 3; eth 3; ace 4 1.4718[25]
4315	Cyclohexane, bromo- Cyclohexyl bromide	$C_6H_{11}Br$ 163.06	108-85-0 -56.5	2736 166.2	4-05-00-00067 1.3359[20]	H_2O 1; EtOH 5; eth 5; ace 5 1.4957[20]
4316	Cyclohexane, (2-bromoethyl)-	$C_8H_{15}Br$ 191.11	1647-26-3 -57	212	4-05-00-00116 1.2357[20]	1.4899[20]
4317	Cyclohexane, 1-bromo-2-methoxy-	$C_7H_{13}BrO$ 193.08	24618-31-3	72-4[8]	3-06-00-00043 1.3257[20]	1.4871[20]
4318	Cyclohexane, (bromomethyl)-	$C_7H_{13}Br$ 177.08	2550-36-9	76-7[26]	4-05-00-00100 1.2763[25]	bz 4; eth 4; chl 4 1.4907[30]
4319	Cyclohexane, 1-bromo-1-methyl-	$C_7H_{13}Br$ 177.08	931-77-1	158	4-05-00-00100 1.2510[20]	H_2O 1; EtOH 3; chl 3 1.4866[20]
4320	Cyclohexane, 1-bromo-3-methyl- 3-Methylcyclohexyl bromide	$C_7H_{13}Br$ 177.08	13905-48-1	181; 60[11]	3-05-00-00076 1.275[25]	H_2O 1; eth 4; bz 3 1.4979[20]
4321	Cyclohexane, 1-bromo-4-methyl-1-(1-methylethyl)- p-Menthane, 4-bromo	$C_{10}H_{19}Br$ 219.16	116836-10-3	98-9[11]	0-05-00-00051 1.165[20]	1.4872[20]
4322	Cyclohexane, (2-bromo-2-propenyl)- Propylene, 2-bromo-3-cyclohexyl	$C_9H_{15}Br$ 203.12	53608-85-8	90[15]	3-05-00-00224 1.215[17]	eth 4 1.495[17]
4323	Cyclohexane, 1,1'-(1,4-butanediyl)bis- 1,4-Dicyclohexylbutane	$C_{16}H_{30}$ 222.41	6165-44-2 12	305	4-05-00-00357 0.8771[21]	EtOH 2; eth 2; bz 2; chl 2 1.475[21]
4324	Cyclohexanebutanol	$C_{10}H_{20}O$ 156.27	4441-57-0	103-4[4]	4-06-00-00139 0.902[25]	1.4660[20]
4325	Cyclohexane, butyl- Butylcyclohexane	$C_{10}H_{20}$ 140.27	1678-93-9 -74.7	180.9	4-05-00-00146 0.7902[20]	H_2O 1 1.4408[20]
4326	Cyclohexane, 1,1'-butylidenebis- Butane, 1,1-dicyclohexyl-	$C_{16}H_{30}$ 222.41	54890-00-5 -10.5	293	4-05-00-00359 0.8841[16]	1.485[16]
4327	Cyclohexane, 2-butyl-1,1,3-trimethyl- 2-Butyl-1,1,3-trimethylcyclohexane	$C_{13}H_{26}$ 182.35	54676-39-0	94-5[10]	4-05-00-00176 0.8292[19]	1.4563[19]
4328	Cyclohexanecarbonitrile	$C_7H_{11}N$ 109.17	766-05-2			
4329	Cyclohexanecarbonitrile, 1-hydroxy-	$C_7H_{11}NO$ 125.17	931-97-5 35	132[20]	4-10-00-00021 1.0172[20]	H_2O 4; eth 4 1.4693[20]
4330	Cyclohexanecarbonyl chloride	$C_7H_{11}ClO$ 146.62	2719-27-9	180	4-09-00-00026 1.0962[15]	1.4711[29]
4331	Cyclohexanecarboxaldehyde	$C_7H_{12}O$ 112.17	2043-61-0	159.3	4-07-00-00045 0.9035[20]	H_2O 3; eth 3 1.4496[20]
4332	Cyclohexanecarboxylic acid Hexahydrobenzoic acid	$C_7H_{12}O_2$ 128.17	98-89-5 31.5	2730 232.5	4-09-00-00016 1.0334[22]	H_2O 2; EtOH 4; bz 4; ctc 2 1.4530[20]
4333	Cyclohexanecarboxylic acid, 4-(aminomethyl)-, trans- Tranexamic acid	$C_8H_{15}NO_2$ 157.21	1197-18-8 >300	9487	3-14-00-00868	H_2O 4
4334	Cyclohexanecarboxylic acid, ethyl ester	$C_9H_{16}O_2$ 156.22	3289-28-9	196	4-09-00-00018 0.9362[20]	ace 4; eth 4; EtOH 4; chl 4 1.4501[15]
4335	Cyclohexanecarboxylic acid, methyl ester	$C_8H_{14}O_2$ 142.20	4630-82-4	183	4-09-00-00017 0.9954[15]	H_2O 1; EtOH 3; eth 3; ace 3 1.4433[20]
4336	Cyclohexanecarboxylic acid, 3-methyl-4-oxo- Cyclohexanone-4-carboxylic acid, 2-methyl	$C_8H_{12}O_3$ 156.18	101567-36-6 93.5	195[20]	4-10-00-02617	bz 4; EtOH 4; chl 4
4337	Cyclohexanecarboxylic acid, propyl ester Propyl cyclohexanecarboxylate	$C_{10}H_{18}O_2$ 170.25	6739-34-0	215.5	4-09-00-00018 0.9530[15]	ace 4; eth 4; EtOH 4; chl 4 1.4486[15]
4338	Cyclohexanecarboxylic acid, 1,3,4,5-tetrahydroxy-, [1R-(1α,3α,4α,5β)]- Quinic acid	$C_7H_{12}O_6$ 192.17	77-95-2 162.5	8071	4-10-00-02257 1.64[25]	H_2O 4; EtOH 4; HOAc 4
4339	Cyclohexane, chloro- Cyclohexyl chloride	$C_6H_{11}Cl$ 118.61	542-18-7 -44	2738 142	4-05-00-00048 1.000[20]	H_2O 1; EtOH 5; eth 5; ace 5 1.4626[20]
4340	Cyclohexane, cyclopentyl-	$C_{11}H_{20}$ 152.28	1606-08-2	215.1	4-05-00-00328 0.8758[20]	1.4725[20]
4341	Cyclohexane, (3-cyclopentylpropyl)- Propane, 1-cyclohexyl-3-cyclopentyl-	$C_{14}H_{26}$ 194.36	2883-07-0	270	4-05-00-00347 0.8751[20]	1.4765[20]

No.	Name / Synonym	Mol. Form. / Mol. Wt.	CAS RN / mp/°C	Merck No. / bp/°C	Beil. Ref. / den/g cm⁻³	Solubility / n_D
4342	Cyclohexane, 1,1,2,2,3,3,4,4,5,6-decafluoro-5,6-bis(trifluoromethyl)- Cyclohexane, 1,2-dimethyl, perfluoro	C_8F_{16} 400.06	306-98-9 -56	101.5	3-05-00-00098 1.829[25]	1.283[25]
4343	Cyclohexane, 1,1,2,2,3,4,4,5,5,6-decafluoro-3,6-bis(trifluoromethyl)-	C_8F_{16} 400.06	374-77-6	100.5	4-05-00-00124 1.8503[20]	1.2897[20]
4344	Cyclohexane, decyl-	$C_{16}H_{32}$ 224.43	1795-16-0 -1.7	299	4-05-00-00187 0.8186[20]	1.4534[20]
4345	Cyclohexane, 1,1-dibromo- 1,1-Dibromocyclohexane	$C_6H_{10}Br_2$ 241.95	10489-97-1	87[14]	4-05-00-00070 1.7349[20]	1.5392[25]
4346	Cyclohexane, 1,2-dibromo-, cis- cis-1,2-Dibromocyclohexane	$C_6H_{10}Br_2$ 241.95	19246-38-9 9.7	115[14]	4-05-00-00070 1.798[25]	ace 4; bz 4; eth 4; lig 4 1.5514[25]
4347	Cyclohexane, 1,2-dibromo-, trans-(±)-	$C_6H_{10}Br_2$ 241.95	5183-77-7 -4	145[100]; 105[20]	4-05-00-00071 1.7759[20]	ace 4; bz 4; eth 4; EtOH 4 1.5445[19]
4348	Cyclohexane, 1,3-dibromo-, trans- trans-1,3-Dibromocyclohexane	$C_6H_{10}Br_2$ 241.95	29624-17-7 1	116[16]	4-05-00-00072 1.5480[20]	bz 4; EtOH 4
4349	Cyclohexane, 1,4-dibromo-, cis- cis-1,4-Dibromocyclohexane	$C_6H_{10}Br_2$ 241.95	16661-99-7	137-8[25]	4-05-00-00072 1.7834[20]	eth 4 1.5531[20]
4350	1,2-Cyclohexanedicarboxylic acid, diethyl ester, trans-(±)- Cyclohexane-1,2-dicarboxylic acid, diethyl ester (trans, DL)	$C_{12}H_{20}O_4$ 228.29	96836-97-4 133[11]		4-09-00-02803 1.040[20]	eth 4 1.4522[13]
4351	1,3-Cyclohexanedicarboxylic acid, diethyl ester, cis- Cyclohexane-1,3-dicarboxylic acid, diethyl ester (cis)	$C_{12}H_{20}O_4$ 228.29	62059-56-7	288; 151[15]	3-09-00-03817 1.045[20]	1.452[20]
4352	1,3-Cyclohexanedicarboxylic acid, dimethyl ester, cis- Dimethyl cis-1,3-cyclohexanedicarboxylate	$C_{10}H_{16}O_4$ 200.23	6998-82-9	263	4-09-00-02806 1.0997[20]	1.4568[20]
4353	1,3-Cyclohexanedicarboxylic acid, dimethyl ester, trans- Cyclohexane-1,3-dicarboxylic acid, dimethyl ester (trans)	$C_{10}H_{16}O_4$ 200.23	10021-92-8 139[20]		4-09-00-02806 1.1095[20]	1.4577[25]
4354	1,4-Cyclohexanedicarboxylic acid, trans-	$C_8H_{12}O_4$ 172.18	619-82-9 312.5	sub 300	4-09-00-02807	H_2O 2; EtOH 4; eth 2; ace 3
4355	1,4-Cyclohexanedicarboxylic acid, diethyl ester, cis- Cyclohexane-1,4-dicarboxylic acid, diethyl ester (cis)	$C_{12}H_{20}O_4$ 228.29	116724-15-3	151[13]	2-09-00-00524 1.015[20]	eth 4
4356	1,4-Cyclohexanedicarboxylic acid, diethyl ester, trans- Cyclohexane-1,4-dicarboxylic acid, diethyl ester (trans)	$C_{12}H_{20}O_4$ 228.29	19145-96-1 43.5		2-09-00-00524 1.0110[20]	eth 4 1.4337[64]
4357	1,4-Cyclohexanedicarboxylic acid, dimethyl ester, cis- cis-Dimethyl 1,4-cyclohexanedicarboxylate	$C_{10}H_{16}O_4$ 200.23	3399-21-1 14	132.5[10]	4-09-00-02808 1.1112[20]	
4358	Cyclohexane, 1,1-dichloro- 1,1-Dichlorocyclohexane	$C_6H_{10}Cl_2$ 153.05	2108-92-1 -47	171	4-05-00-00050 1.1559[20]	1.4803[20]
4359	Cyclohexane, 1,2-dichloro-, cis- cis-1,2-Dichlorocyclohexane	$C_6H_{10}Cl_2$ 153.05	10498-35-8 -1.5	206.9	4-05-00-00050 1.2021[20]	bz 4 1.4967[20]
4360	Cyclohexane, 1,2-dichloro-, trans-(±)-	$C_6H_{10}Cl_2$ 153.05	5183-79-9 -6.3	189	4-05-00-00051 1.1839[20]	1.4902[20]
4361	Cyclohexane, 1,4-dichloro-, cis- cis-1,4-Dichlorocyclohexane	$C_6H_{10}Cl_2$ 153.05	16749-11-4 18	80.3[25]	4-05-00-00051 1.1900[20]	1.4942[20]
4362	1,3-Cyclohexanedimethanamine Cyclohexane, 1,3-bis(aminomethyl)	$C_8H_{18}N_2$ 142.24	2579-20-6 <-70	220	0.945[20]	H_2O 4; eth 4; EtOH 4
4363	1,3-Cyclohexanedimethanol Cyclohexane, 1,3-bis(hydroxymethyl)	$C_8H_{16}O_2$ 144.21	3971-28-6	105.5	4-06-00-05236 1.036[25]	1.4912[20]
4364	Cyclohexane, 1,1-dimethyl- 1,1-Dimethylcyclohexane	C_8H_{16} 112.22	590-66-9 -33.3	119.6	4-05-00-00117 0.7809[20]	H_2O 1; EtOH 3; eth 3; ace 3 1.4290[20]
4365	Cyclohexane, 1,2-dimethyl-, cis- cis-1,2-Dimethylcyclohexane	C_8H_{16} 112.22	2207-01-4 -49.9	129.8	4-05-00-00118 0.7963[20]	H_2O 1; EtOH 3; eth 5; ace 5 1.4360[20]
4366	Cyclohexane, 1,2-dimethyl-, trans- trans-1,2-Dimethylcyclohexane	C_8H_{16} 112.22	6876-23-9 -90	123.5	4-05-00-00118 0.7760[20]	H_2O 1; EtOH 3; eth 3; ace 3 1.4270[20]
4367	Cyclohexane, 1,3-dimethyl-, cis- cis-1,3-Dimethylcyclohexane	C_8H_{16} 112.22	638-04-0 -75.6	120.1	4-05-00-00121 0.7660[20]	H_2O 1; EtOH 5; eth 5; ace 5 1.4229[20]
4368	Cyclohexane, 1,4-dimethyl-, cis- cis-1,4-Dimethylcyclohexane	C_8H_{16} 112.22	624-29-3 -87.4	124.4	4-05-00-00122 0.7829[20]	H_2O 1; EtOH 5; eth 5; ace 5 1.4230[20]
4369	Cyclohexane, (1,1-dimethylethyl)-	$C_{10}H_{20}$ 140.27	3178-22-1 -41.2	171.5	4-05-00-00147 0.8127[20]	H_2O 1 1.4469[20]
4370	Cyclohexane, 1,3-dimethyl-, trans- trans-1,3-Dimethylcyclohexane	C_8H_{16} 112.22	2207-03-6 -90.1	124.5	0.79[15]	1.4284[25]
4371	Cyclohexane, 1,4-dimethyl-, trans- trans-1,4-Dimethylcyclohexane	C_8H_{16} 112.22	2207-04-7 -36.9	119.4	0.77[15]	1.4185[25]
4372	1,2-Cyclohexanediol, cis-	$C_6H_{12}O_2$ 116.16	1792-81-0 100	120[15]	4-06-00-05193 1.0297[101]	EtOH 3; ace 3; bz 3; chl 2
4373	1,2-Cyclohexanediol, trans-(±)-	$C_6H_{12}O_2$ 116.16	54383-22-1 105	117[13]	4-06-00-05194 1.147[24]	H_2O 4; bz 4; eth 4; EtOH 4
4374	1,4-Cyclohexanediol, trans- trans-1,4-Dihydroxycyclohexane	$C_6H_{12}O_2$ 116.16	6995-79-5 143		4-06-00-05209 1.18[20]	H_2O 3; EtOH 3; eth 1; ace 2

No.	Name / Synonym	Mol. Form. / Mol. Wt.	CAS RN / mp/°C	Merck No. / bp/°C	Bell. Ref. / den/g cm^{-3}	Solubility / n_D
4375	1,2-Cyclohexanedione / 1,2-Dioxocyclohexane	$C_8H_8O_2$ / 112.13	765-87-7 / 40	/ 194	4-07-00-01982 / 1.1187[21]	H_2O 3; EtOH 3; eth 3; bz 3 / 1.4995[20]
4376	1,3-Cyclohexanedione	$C_8H_8O_2$ / 112.13	504-02-9 / 105.5	3161 /	4-07-00-01985 / 1.0861[91]	H_2O 3; EtOH 3; eth 2; ace 3 / 1.4576[102]
4377	1,4-Cyclohexanedione / Tetrahydroquinone	$C_8H_8O_2$ / 112.13	637-88-7 / 78	132[20] /	4-07-00-01986 / 1.0861[91]	H_2O 3; EtOH 3; eth 3; ace 3
4378	1,3-Cyclohexanedione, 5,5-dimethyl- / 5,5-Dimethyldihydroresorcinol	$C_8H_{12}O_2$ / 140.18	126-81-8 / 149-51 dec	3231 /	4-07-00-01999 /	H_2O 2; eth 2; ace 3; ctc 3
4379	1,2-Cyclohexanedione, dioxime / Nioxime	$C_8H_{10}N_2O_2$ / 142.16	492-99-9 / 192	6477 /	4-07-00-01982 /	H_2O 3; ace 3; chl 3; tfa 2
4380	Cyclohexane, docosyl- / Docosane, 1-cyclohexyl-	$C_{28}H_{56}$ / 392.75	61828-07-7 / 54.4	/ 449	3-05-00-00167 / 0.8334[20]	1.4632[20]
4381	Cyclohexane, dodecafluoro- / Perfluorocyclohexane	C_6F_{12} / 300.05	355-68-0 / 62.5 (trip. pt.)	/ 52.8 (subl. pt.)	4-05-00-00048 /	
4382	Cyclohexane, dodecyl-	$C_{18}H_{36}$ / 252.48	1795-17-1 / 12.5	/ 331	4-05-00-00191 / 0.8223[20]	1.4559[20]
4383	Cyclohexane, dotriacontyl- / Dotriacontylcyclohexane	$C_{38}H_{76}$ / 533.02	61828-17-9 / 74.8	/ 527	4-05-00-00204 / 0.8388[20]	1.4668[20]
4384	Cyclohexane, eicosyl- / 1-Cyclohexyleicosane	$C_{26}H_{52}$ / 364.70	4443-55-4 / 48.5	/ 430	4-05-00-00198 / 0.8318[20]	1.4622[20]
4385	Cyclohexaneethanamine, N,α-dimethyl- / Propylhexedrine	$C_{10}H_{21}N$ / 155.28	101-40-6 /	7873 / 205; 82[10]	3-12-00-00108 / 0.8501[20]	EtOH 4 / 1.4600[20]
4386	Cyclohexane, 1,1'-(1,2-ethanediyl)bis- / 1,2-Dicyclohexylethane	$C_{14}H_{26}$ / 194.36	3321-50-4 / 11.5	/ 272.5	4-05-00-00344 / 0.8728[24]	1.4745[24]
4387	Cyclohexaneethanol	$C_8H_{16}O$ / 128.21	4442-79-9 /	/ 208	4-06-00-00119 / 0.9229[20]	EtOH 3; eth 3; bz 3 / 1.4641[20]
4388	Cyclohexane, ethenyl-	C_8H_{14} / 110.20	695-12-5 /	/ 128	4-05-00-00268 / 0.8166[19]	1.455[19]
4389	Cyclohexane, 1-ethenyl-1-methyl-2,4-bis(1-methylethyl)-, (1α,2β,4β)- / β-Elemene	$C_{15}H_{24}$ / 204.36	33880-83-0 /	/ 117-24[15.5]	4-05-00-01180 / 0.8749[20]	1.4935[20]
4390	Cyclohexane, ethyl- / Ethylcyclohexane	C_8H_{16} / 112.22	1678-91-7 / -111.3	/ 131.9	4-05-00-00115 / 0.7880[20]	H_2O 1; EtOH 3; eth 3; ace 3 / 1.4330[20]
4391	Cyclohexane, ethylidene- / Ethylidenecyclohexane	C_8H_{14} / 110.20	1003-64-1 /	/ 136	4-05-00-00267 / 0.822[25]	1.4618[20]
4392	Cyclohexane, 1,1'-ethylidenebis- / 1,1-Dicyclohexylethane	$C_{14}H_{26}$ / 194.36	2319-61-1 / -20.9	/ 256; 112[7]	4-05-00-00345 / 0.9070[25]	1.4887[25]
4393	Cyclohexane, 1-ethylidene-2-methyl- / Cyclohexane, 2-methyl-1-ethylidene	C_9H_{16} / 124.23	40514-70-3 /	/ 158	0-05-00-00078 / 0.823[25]	1.47
4394	Cyclohexane, 1-ethyl-1-methyl-	C_9H_{18} / 126.24	4926-90-3 /	/ 152.2	4-05-00-00135 / 0.8025[25]	1.4419[20]
4395	Cyclohexane, 1-ethyl-2-methyl-, cis-	C_9H_{18} / 126.24	4923-77-7 /	/ 156	4-05-00-00136 / 0.8059[25]	1.4456[20]
4396	Cyclohexane, 1-ethyl-2-methyl-, trans- / Cyclohexane, 1-methyl-2-ethyl (trans)	C_9H_{18} / 126.24	4923-78-8 /	/ 151.7	4-05-00-00136 / 0.790[25]	1.4381[20]
4397	Cyclohexane, 1-ethyl-3-methyl-, cis- / cis-1-Ethyl-3-methylcyclohexane	C_9H_{18} / 126.24	19489-10-2 /	/ 156	4-05-00-00136 / 0.8094[20]	1.4432[20]
4398	Cyclohexane, 1-ethyl-3-methyl-, trans-	C_9H_{18} / 126.24	4926-76-5 /	/ 150	4-05-00-00136 / 0.7972[20]	1.4382[20]
4399	Cyclohexane, 1-ethyl-4-methyl-, cis- / cis-1-Ethyl-4-methylcyclohexane	C_9H_{18} / 126.24	4926-78-7 /	/ 152.6	4-05-00-00136 / 0.7969[20]	1.4374[20]
4400	Cyclohexane, 1-ethyl-4-methyl-, trans- / trans-1-Methyl-4-ethylcyclohexane	C_9H_{18} / 126.24	6236-88-0 / -80.8	/ 149	4-05-00-00136 / 0.7798[20]	1.4304[20]
4401	Cyclohexane, 1-ethyl-1-methyl-2,4-bis(1-methylethyl)-, [1R-(1α,2β,4β)]-	$C_{15}H_{30}$ / 210.40	515-12-8 /	/ 115-9[10]	4-05-00-00185 / 0.8509[20]	bz 4; eth 4; peth 4 / 1.4640[20]
4402	Cyclohexane, 1,1'-(2-ethyl-1,3-propanediyl)bis- / Propane, 1,3-dicyclohexyl-2-ethyl	$C_{17}H_{32}$ / 236.44	54833-34-0 /	/ 306	4-05-00-00363 / 0.8845[20]	1.483[21]
4403	Cyclohexane, fluoro- / Cyclohexyl fluoride	$C_6H_{11}F$ / 102.15	372-46-3 / 13	/ 101	4-05-00-00044 / 0.9279[20]	H_2O 1; py 3 / 1.4146[20]
4404	Cyclohexane, heptadecyl- / Heptadecane, 1-cyclohexyl-	$C_{23}H_{46}$ / 322.62	19781-73-8 / 38	/ 397	3-05-00-00161 / 0.8290[20]	1.4603[20]
4405	Cyclohexane, heptyl-	$C_{13}H_{26}$ / 182.35	5617-41-4 / -30	/ 244	4-05-00-00174 / 0.8109[20]	1.4484[20]
4406	Cyclohexane, 1,2,3,4,5,6-hexachloro-, (1α,2β,3α,4β,5α,6β)- / β-Lindane	$C_6H_6Cl_6$ / 290.83	319-85-7 /	/ 60[0.50]	4-05-00-00061 / 1.89[19]	H_2O 1; EtOH 2; bz 2; chl 2
4407	Cyclohexane, 1,2,3,4,5,6-hexachloro-, (1α,2α,3α,4β,5α,6β)- / δ-Lindane	$C_6H_6Cl_6$ / 290.83	319-86-8 / 141.5	/ 60[0.36]	4-05-00-00056 /	
4408	Cyclohexane, 1,2,3,4,5,6-hexachloro-, (1α,2α,3β,4α,5β,6β)-(±)- / Cyclohexane, 1,2,3,4,5,6-hexachloro, (α, DL)	$C_6H_6Cl_6$ / 290.83	60291-32-9 / 159.5	/ 288	4-05-00-00060 / 1.87[20]	H_2O 1; EtOH 4; bz 3; chl 3
4409	Cyclohexane, 1,2,3,4,5,6-hexachloro-, (1α,2α,3β,4α,5α,6β)- / Lindane	$C_6H_6Cl_6$ / 290.83	58-89-9 / 112.5	5379 / 323.4	4-05-00-00058 /	ace 4; bz 4
4410	Cyclohexane, hexacosyl- / Hexacosylcyclohexane	$C_{32}H_{64}$ / 448.86	61828-11-3 / 64	/ 484	4-05-00-00203 / 0.8359[20]	1.4649[20]

No.	Name Synonym	Mol. Form. Mol. Wt.	CAS RN mp/°C	Merck No. bp/°C	Beil. Ref. den/g cm⁻³	Solubility n_D
4411	Cyclohexane, hexadecyl- Hexadecane, 1-cyclohexyl-	$C_{22}H_{44}$ 308.59	6812-38-0 33.6	385	4-05-00-00196 0.8279[20]	1.4596[20]
4412	Cyclohexanehexanoic acid Caproic acid, 6-cyclohexyl	$C_{12}H_{22}O_2$ 198.31	4354-56-7 34	180[11]	4-09-00-00095 0.9626[20]	eth 3; chl 2 1.4750[20]
4413	Cyclohexane, hexatriacontyl- Hexatriacontylcyclohexane	$C_{42}H_{84}$ 589.13	61828-21-5 80	551	4-05-00-00205 0.8402[20]	1.4678[20]
4414	Cyclohexane, hexyl-	$C_{12}H_{24}$ 168.32	4292-75-5 -43	224	4-05-00-00170 0.8076[20]	1.4462[20]
4415	Cyclohexane, iodo- Cyclohexyl iodide	$C_6H_{11}I$ 210.06	626-62-0	180 dec; 81[20]	4-05-00-00078 1.6244[20]	H_2O 1; EtOH 3; eth 3; ace 3 1.5477[20]
4416	Cyclohexane, isocyanato- Cyclohexyl isocyanate	$C_7H_{11}NO$ 125.17	3173-53-3	172	4-12-00-00071 0.98[25]	1.4551[20]
4417	Cyclohexane, isothiocyanato-	$C_7H_{11}NS$ 141.24	1122-82-3	220	4-12-00-00073 0.996[25]	H_2O 1; EtOH 3; eth 3; ctc 2 1.5375[20]
4418	Cyclohexanemethanamine	$C_7H_{15}N$ 113.20	3218-02-8	160	4-12-00-00123 0.87[25]	1.4630[20]
4419	Cyclohexanemethanol Cyclohexylcarbinol	$C_7H_{14}O$ 114.19	100-49-2 -43	2737 183	4-06-00-00106 0.9297[20]	eth 4; EtOH 4 1.4644[20]
4420	Cyclohexanemethanol, α-methyl-	$C_8H_{16}O$ 128.21	1193-81-3	189	1-06-00-00012 0.928[25]	EtOH 4; eth 4; ctc 2 1.4656[20]
4421	Cyclohexanemethanol, α-methyl-, (S)-	$C_8H_{16}O$ 128.21	3113-98-2	105[35]; 82[12]	4-06-00-00117 0.9254[20]	1.4635[25]
4422	Cyclohexanemethanol, 2-methyl-, cis- Cyclohexane, 1-hydroxymethyl-2-methyl (cis)	$C_8H_{16}O$ 128.21	3937-45-9	188.5	4-06-00-00122 0.9342[20]	1.4689[20]
4423	Cyclohexanemethanol, 2-methyl-, trans- Cyclohexane, 1-hydroxymethyl-2-methyl (trans)	$C_8H_{16}O$ 128.21	3937-46-0	192	4-06-00-00122 0.9224[20]	1.4665[20]
4424	Cyclohexanemethanol, 4-methyl-	$C_8H_{16}O$ 128.21	34885-03-5	75[2.5]	2-06-00-00030 0.9074[20]	1.4617[20]
4425	Cyclohexane, methoxy-	$C_7H_{14}O$ 114.19	931-56-6 -74.4	133	4-06-00-00026 0.8756[20]	eth 4; EtOH 4 1.4355[20]
4426	Cyclohexane, methyl- Methylcyclohexane	C_7H_{14} 98.19	108-87-2 -126.6	100.9	4-05-00-00094 0.7694[20]	H_2O 1; EtOH 3; eth 3; ace 5 1.4231[20]
4427	Cyclohexane, (3-methylbutyl)-	$C_{11}H_{22}$ 154.30	54105-76-9	196.5	3-05-00-00143 0.8023[20]	bz 4; lig 4 1.4420[20]
4428	Cyclohexane, methylene- Methylenecyclohexane	C_7H_{12} 96.17	1192-37-6 -106.7	102.5	4-05-00-00250 0.8074[20]	H_2O 1; eth 3; bz 3; chl 3 1.4523[20]
4429	Cyclohexane, 1,1'-methylenebis- Dicyclohexylmethane	$C_{13}H_{24}$ 180.33	3178-23-2	110-5[18]	4-05-00-00340 0.8750[20]	1.4752[20]
4430	Cyclohexane, 1,1'-(1-methyl-1,2-ethanediyl)bis- 1,2-Dicyclohexylpropane	$C_{15}H_{28}$ 208.39	41851-34-7 -22	284.5	4-05-00-00350 0.8724[21]	1.479[21]
4431	Cyclohexane, (1-methylethyl)- Isopropylcyclohexane	C_9H_{18} 126.24	696-29-7 -89.4	154.8	4-05-00-00134 0.8023[20]	H_2O 1; EtOH 4; eth 4; ace 5 1.4410[20]
4432	Cyclohexane, 1,1'-(1-methylethylidene)bis- Propane, 2,2-dicyclohexyl	$C_{15}H_{28}$ 208.39	54934-90-6 15.6	286	4-05-00-00350 0.9001[23]	1.490[23]
4433	Cyclohexane, 1,1',1''-methylidynetris- Methane, tricyclohexyl-	$C_{19}H_{34}$ 262.48	1610-24-8 48	325.5	4-05-00-00504 0.9273[50]	H_2O 1; EtOH 2; eth 3; bz 3 1.4986[40]
4434	Cyclohexane, 1-methyl-4-methylene-	C_8H_{14} 110.20	2808-80-2	122	4-05-00-00271 0.7910[19]	1.4465[18]
4435	Cyclohexane, 1-methyl-3-(1-methylethenyl)- m-Menth-8-ene	$C_{10}H_{18}$ 138.25	16605-36-0	170	3-05-00-00235 0.8178[20]	1.4546
4436	Cyclohexane, 1-methyl-2-(1-methylethyl)-	$C_{10}H_{20}$ 140.27	16580-23-7	171	4-05-00-00150 0.8134[22]	1.447[21]
4437	Cyclohexane, 1-methyl-3-(1-methylethyl)-	$C_{10}H_{20}$ 140.27	16580-24-8	166.5	3-05-00-00132 0.7963[20]	1.44[24]
4438	Cyclohexane, 1-methyl-4-(1-methylethyl)-	$C_{10}H_{20}$ 140.27	99-82-1 -87.6	170.7	4-05-00-00151 0.797[25]	1.4373[20]
4439	Cyclohexane, 1-methyl-4-(1-methylethyl)-, cis-	$C_{10}H_{20}$ 140.27	6069-98-3 -89.9	172	4-05-00-00151 0.8039[20]	H_2O 1; EtOH 4; eth 4; bz 3 1.4431[20]
4440	Cyclohexane, 1-methyl-4-(1-methylethyl)-, trans-	$C_{10}H_{20}$ 140.27	1678-82-6 -86.3	170.6	4-05-00-00151 0.7928[20]	bz 4; eth 4; EtOH 4; lig 4 1.4366[20]
4441	Cyclohexane, 1-methyl-3-(1-methylethylidene)-	$C_{10}H_{18}$ 138.25	13828-34-7	174	1-05-00-00043 0.8214[20]	1.4670[20]
4442	Cyclohexane, 1-methyl-4-(1-methylethylidene)-	$C_{10}H_{18}$ 138.25	1124-27-2	173	4-05-00-00302 0.819[21]	1.4568[21]
4443	Cyclohexane, 1-methyl-1-nitro- 1-Methyl-1-nitrocyclohexane	$C_7H_{13}NO_2$ 143.19	59368-15-9 -71	118-20[50]	4-05-00-00103 1.0384[20]	1.4598[20]
4444	Cyclohexane, 1-methyl-2-nitro- 2-Methylnitrocyclohexane	$C_7H_{13}NO_2$ 143.19	74221-86-6	100[20]	4-05-00-00103 1.046[25]	1.4608[25]
4445	Cyclohexane, 1-methyl-4-nitro- 1-Methyl-4-nitrocyclohexane	$C_7H_{13}NO_2$ 143.19	89895-45-4	102[20]	4-05-00-00103 1.037[25]	1.4567[23]
4446	Cyclohexane, 1-methyl-2-pentyl- 1-Methyl-2-pentylcyclohexane	$C_{12}H_{24}$ 168.32	54411-01-7	217.5	4-05-00-00171 0.815[20]	1.4487[20]
4447	Cyclohexane, 1,1'-(2-methyl-1,3-propanediyl)bis- 1,3-Dicyclohexyl-2-methylpropane	$C_{16}H_{30}$ 222.41	2883-08-1 0.6	295.2	4-05-00-00358 0.8715[20]	1.4756[20]

No.	Name Synonym	Mol. Form. Mol. Wt.	CAS RN mp/°C	Merck No. bp/°C	Beil. Ref. den/g cm⁻³	Solubility n_D
4448	Cyclohexane, (1-methylpropyl)- sec-Butylcyclohexane	$C_{10}H_{20}$ 140.27	7058-01-7	179.3	4-05-00-00146 0.8131[20]	H_2O 1; ace 3 1.4467[20]
4449	Cyclohexane, 1-methyl-1-propyl- 1-Methyl-1-propylcyclohexane	$C_{10}H_{20}$ 140.27	4258-93-9	172	4-05-00-00148 0.8101[20]	1.4440[20]
4450	Cyclohexane, 1-methyl-2-propyl- 1-Methyl-2-propylcyclohexane	$C_{10}H_{20}$ 140.27	4291-79-6	176; 56[13]	4-05-00-00148 0.8130[19]	1.4468[19]
4451	Cyclohexane, 1-methyl-3-propyl- 1-Methyl-3-propylcyclohexane	$C_{10}H_{20}$ 140.27	4291-80-9	164.5	4-05-00-00149 0.7895[21]	1.4377[20]
4452	Cyclohexane, 1-methyl-4-propyl- 1-Methyl-4-propylcyclohexane	$C_{10}H_{20}$ 140.27	4291-81-0	175.5	4-05-00-00149 0.797[20]	1.4393[20]
4453	Cyclohexane, (2-methylpropyl)-	$C_{10}H_{20}$ 140.27	1678-98-4 -95	171.3	4-05-00-00147 0.7952[20]	H_2O 1; EtOH 3; eth 4; ace 3 1.4386[20]
4454	Cyclohexane, nitro- Nitrocyclohexane	$C_6H_{11}NO_2$ 129.16	1122-60-7 -34	205; 95[22]	4-05-00-00081 1.0610[20]	H_2O 1; EtOH 3; lig 3 1.4612[19]
4455	Cyclohexane, (nitromethyl)- (Nitromethyl)cyclohexane	$C_7H_{13}NO_2$ 143.19	2625-30-1	123[40]; 98[10]	4-05-00-00103 1.0482[20]	1.4705[20]
4456	Cyclohexane, nonadecyl-	$C_{25}H_{50}$ 350.67	22349-03-7 45	420	4-05-00-00197 0.8310[20]	1.4616[20]
4457	Cyclohexane, nonyl-	$C_{15}H_{30}$ 210.40	2883-02-5 -10	282	4-05-00-00183 0.8163[20]	1.4519[20]
4458	Cyclohexane, octacosyl- Octacosylcyclohexane	$C_{34}H_{68}$ 476.91	61828-13-5 68	499	4-05-00-00203 0.8370[20]	1.4656[20]
4459	Cyclohexane, octadecyl- Octadecane, 1-cyclohexyl-	$C_{24}H_{48}$ 336.65	4445-06-1 41.6	409; 175[1]	4-05-00-00197 0.8300[20]	1.4610[20]
4460	Cyclohexane, octyl-	$C_{14}H_{28}$ 196.38	1795-15-9 -20	264	4-05-00-00178 0.8138[20]	1.4503[20]
4461	Cyclohexane, 1,1'-oxybis-	$C_{12}H_{22}O$ 182.31	4645-15-2 -36	242.5	3-06-00-00019 0.9227[20]	1.4741[20]
4462	Cyclohexane, pentacosyl- Pentacosylcyclohexane	$C_{31}H_{62}$ 434.83	61828-10-2 62	476	4-05-00-00202 0.8353[20]	1.4645[20]
4463	Cyclohexane, pentadecyl-	$C_{21}H_{42}$ 294.56	6006-95-7 29	373	4-05-00-00195 0.8267[20]	1.4588[20]
4464	Cyclohexane, 1,1'-(1,5-pentanediyl)bis- Pentane, 1,5-dicyclohexyl-	$C_{17}H_{32}$ 236.44	54833-31-7 -13.6	325	4-05-00-00362 0.8718[21]	1.479[21]
4465	Cyclohexane, pentyl- Pentylcyclohexane	$C_{11}H_{22}$ 154.30	4292-92-6 -57.5	203.7	4-05-00-00164 0.8037[20]	ace 4; bz 4; eth 4; EtOH 4 1.4437[20]
4466	Cyclohexane, 1,1'-(1,3-propanediyl)bis- 1,3-Dicyclohexylpropane	$C_{15}H_{28}$ 208.39	3178-24-3 -17	291.5	4-05-00-00349 0.8728[24]	1.4736[24]
4467	Cyclohexanepropanenitrile, 2-oxo-	$C_9H_{13}NO$ 151.21	4594-78-9	138-42[10]	4-10-00-02623 1.0181[20]	ctc 2 1.4755[20]
4468	Cyclohexanepropanoic acid	$C_9H_{16}O_2$ 156.22	701-97-3 16	276.5	4-09-00-00056 0.912[25]	H_2O 3; eth 3; ctc 2 1.4638[20]
4469	Cyclohexanepropanol, α-methyl- 3-Cyclohexyl-1-methylpropanol	$C_{10}H_{20}O$ 156.27	10528-67-3	112[16]	3-06-00-00123 0.903[21]	1.464[21]
4470	Cyclohexane, 2-propenyl-	C_9H_{16} 124.23	2114-42-3	132	4-05-00-00283 0.8135[20]	EtOH 4; eth 3; ace 3; bz 3 1.4500[20]
4471	Cyclohexane, propyl- Propylcyclohexane	C_9H_{18} 126.24	1678-92-8 -94.9	156.7	4-05-00-00134 0.7936[20]	H_2O 1; EtOH 5; eth 3; ace 5 1.4370[20]
4472	Cyclohexane, 1,1'-propylidenebis- Propane, 1,1-dicyclohexyl-	$C_{15}H_{28}$ 208.39	54934-91-7 -23.5	283.6	4-05-00-00350 0.8886[23]	1.485[23]
4473	Cyclohexane, 2-propynyl-	C_9H_{14} 122.21	17715-00-3	157.5	4-05-00-00419 0.8449[20]	eth 3 1.4605[20]
4474	Cyclohexane, tetracosyl- Tetracosylcyclohexane	$C_{30}H_{60}$ 420.81	61828-09-9 59.5	467	4-05-00-00202 0.8347[20]	1.4641[20]
4475	Cyclohexane, tetradecyl-	$C_{20}H_{40}$ 280.54	1795-18-2 24	360	4-05-00-00194 0.8254[20]	1.4579[20]
4476	Cyclohexane, 1,1'-tetradecylidenebis- 1,1-Dicyclohexyltetradecane	$C_{26}H_{50}$ 362.68	55334-08-2 37.6	406	4-05-00-00374 0.8735[20]	1.4799[20]
4477	Cyclohexane, 1,1,3,4-tetramethyl- 1,1,3,4-Tetramethylcyclohexane	$C_{10}H_{20}$ 140.27	24612-75-7	160	4-05-00-00159 0.7976[20]	1.4380[20]
4478	Cyclohexane, 1,1,3,5-tetramethyl-, cis- cis-1,1,3,5-Tetramethylcyclohexane	$C_{10}H_{20}$ 140.27	50876-32-9	152.5	4-05-00-00159 0.7813[20]	1.4319[20]
4479	Cyclohexane, tetratriacontyl- Tetratriacontylcyclohexane	$C_{40}H_{80}$ 561.08	61828-19-1 78	540	4-05-00-00205 0.8395[20]	1.4673[20]
4480	Cyclohexanethiol Cyclohexyl mercaptan	$C_6H_{12}S$ 116.23	1569-69-3	158.9	4-06-00-00072 0.9782[20]	ace 4; bz 4; eth 4; EtOH 4 1.4921[20]
4481	Cyclohexanethione Thiocyclohexanone	$C_6H_{10}S$ 114.21	2720-41-4	82[17]; 74[11]	3-07-00-00039	EtOH 3; eth 3; ace 3 1.5375[20]
4482	Cyclohexane, triacontyl- Triacontylcyclohexane	$C_{36}H_{72}$ 504.97	61828-15-7 71.6	514	4-05-00-00204 0.8379[20]	1.4662[20]
4483	Cyclohexane, tricosyl- Tricosylcyclohexane	$C_{29}H_{58}$ 406.78	61828-08-8 57	459	4-05-00-00202 0.8341[20]	1.4637[20]
4484	Cyclohexane, tridecyl- Tridecane, 1-cyclohexyl-	$C_{19}H_{38}$ 266.51	6006-33-3 18.5	346	4-05-00-00192 0.8239[20]	1.4570[20]
4485	Cyclohexane, 1,1,2-trimethyl- 1,1,2-Trimethylcyclohexane	C_9H_{18} 126.24	7094-26-0 -29	145.2	4-05-00-00137 0.7963[25]	1.4382[20]
4486	Cyclohexane, 1,1,3-trimethyl- Cyclogeraniolane	C_9H_{18} 126.24	3073-66-3 -65.7	136.6	4-05-00-00137 0.7749[25]	1.4295[20]
4487	Cyclohexane, 1,1,4-trimethyl- 1,1,4-Trimethylcyclohexane	C_9H_{18} 126.24	7094-27-1	135	4-05-00-00137 0.7685[25]	1.4251[20]

No.	Name / Synonym	Mol. Form. / Mol. Wt.	CAS RN / mp/°C	Merck No. / bp/°C	Beil. Ref. / den/g cm^{-3}	Solubility / n_D
4488	Cyclohexane, 1,2,3-trimethyl- / 1,2,3-Trimethylcyclohexane	C_9H_{18} / 126.24	1678-97-3 /	/ 142	4-05-00-00137 / 0.7898[20]	/ 1.4346[20]
4489	Cyclohexane, 1,2,3-trimethyl-, (1α,2β,3α)-	C_9H_{18} / 126.24	1678-81-5 / -66.9	/ 144	4-05-00-00137 / 0.777[25]	/ 1.430[20]
4490	Cyclohexane, 1,2,3-trimethyl-, (1α,2α,3α)-	C_9H_{18} / 126.24	1839-88-9 / -85	/ 151.7	4-05-00-00137 / 0.7989[25]	/ 1.4405[20]
4491	Cyclohexane, 1,2,3-trimethyl-, (1α,2α,3β)-	C_9H_{18} / 126.24	7667-55-2 / -85.7	/ 151.2	4-05-00-00137 / 0.7995[25]	/ 1.4399[20]
4492	Cyclohexane, 1,2,4-trimethyl-, (1α,2α,4α)-	C_9H_{18} / 126.24	1678-80-4 / -77.5	/ 146.6	4-05-00-00138 / 0.783[25]	/ 1.4340[20]
4493	Cyclohexane, 1,2,4-trimethyl-, (1α,2β,4α)-	C_9H_{18} / 126.24	7667-59-6 / -86	/ 142.9	4-05-00-00138 / 0.7683[25]	/ 1.4266[20]
4494	Cyclohexane, 1,2,4-trimethyl-, (1α,2α,4β)-	C_9H_{18} / 126.24	7667-58-5 / -91.8	/ 146.7	4-05-00-00138 / 0.7870[25]	/ 1.4345[20]
4495	Cyclohexane, 1,2,4-trimethyl-, (1α,2β,4β)-	C_9H_{18} / 126.24	7667-60-9 / -83.5	/ 142.9	4-05-00-00138 / 0.7870[25]	/ 1.4341[20]
4496	Cyclohexane, 1,3,5-trimethyl-, (1α,3α,5α)-	C_9H_{18} / 126.24	1795-27-3 / -49.7	/ 138.5	4-05-00-00138 / 0.7708[20]	H$_2$O 1; eth 3; bz 3; ctc 3 / 1.4269[20]
4497	Cyclohexane, 1,3,5-trimethyl-, (1α,3α,5β)- / trans-1,3,5-Trimethylcyclohexane	C_9H_{18} / 126.24	1795-26-2 / -107.4	/ 140.5	4-05-00-00138 / 0.7794[20]	bz 4; eth 4; lig 4 / 1.4307[20]
4498	Cyclohexane, undecafluoro- / Undecafluorocyclohexane	C_6HF_{11} / 282.06	308-24-7 /	/ 62.0		
4499	Cyclohexane, undecafluoro(trifluoromethyl)- / Perfluoro(methylcyclohexane)	C_7F_{14} / 350.05	355-02-2 / -44.7	/ 76.3	4-05-00-00097 / 1.7878[25]	ace 3; bz 3; ctc 3; tol 3 / 1.285[17]
4500	Cyclohexanol / Cyclohexyl alcohol	$C_6H_{12}O$ / 100.16	108-93-0 / 25.4	2731 / 160.8	4-06-00-00020 / 0.9624[20]	H$_2$O 3; EtOH 3; eth 3; ace 3 / 1.4641[20]
4501	Cyclohexanol, 2-amino-, trans-(±)- / Cyclohexanol, 2-amino (trans, DL)	$C_6H_{13}NO$ / 115.18	33092-82-9 / 68	219; 105[10]	4-13-00-00540 /	bz 4; chl 4; HOAc 4
4502	Cyclohexanol, 3-(aminomethyl)-3,5,5-trimethyl- / 1-Hydroxy-3-aminomethyl-3,5,5-trimethylcyclohexane	$C_{10}H_{21}NO$ / 171.28	15647-11-7 / 45.5	/ 265	/ 0.969[25]	/ 1.4904[20]
4503	Cyclohexanol, 2-bromo- / 2-Bromo-1-cyclohexanol	$C_6H_{11}BrO$ / 179.06	24796-87-0 /	85[10]; 60[1]	4-06-00-00063 / 1.4604[20]	/ 1.5169[20]
4504	Cyclohexanol, 2-butyl-, trans- / trans-2-Butylcyclohexanol	$C_{10}H_{20}O$ / 156.27	35242-05-8 /	111-2[16]	3-06-00-00122 / 0.9020[20]	H$_2$O 1; eth 3; ace 3; bz 3 / 1.4641[20]
4505	Cyclohexanol, 2-chloro-, trans- / trans-2-Chlorocyclohexanol	$C_6H_{11}ClO$ / 134.61	6628-80-4 / 29	93[20]	4-06-00-00064 / 1.146[16]	bz 4; eth 4; EtOH 4; chl 4 / 1.4899[20]
4506	Cyclohexanol, 2-chloro-, cis-(±)-	$C_6H_{11}ClO$ / 134.61	116783-28-9 / 36.5	93-4[26]	4-06-00-00064 / 1.1261[25]	bz 4; EtOH 4; chl 4 / 1.4894[25]
4507	Cyclohexanol, 4-chloro-, trans- / trans-4-Chlorocyclohexanol	$C_6H_{11}ClO$ / 134.61	29538-77-0 / 82.5	106[14]	4-06-00-00068 / 1.1435[13]	EtOH 3; eth 3; bz 3; chl 3 / 1.4930[17]
4508	Cyclohexanol, 1,2-dimethyl-, cis- / Cyclohexanol, 1,2-dimethyl-, stereoisomer	$C_8H_{16}O$ / 128.21	19879-11-9 / 23.2	82.8[25]	4-06-00-00121 / 0.9250[20]	/ 1.4625[20]
4509	Cyclohexanol, 1,2-dimethyl-, trans- / Cyclohexanol, 1,2-dimethyl-, stereoisomer	$C_8H_{16}O$ / 128.21	19879-12-0 / 13.2	74[25]	4-06-00-00121 / 0.9187[20]	/ 1.4590[20]
4510	Cyclohexanol, 1,3-dimethyl-, cis- / cis-1,3-Dimethylcyclohexanol	$C_8H_{16}O$ / 128.21	15466-94-1 / 27.5	84[25]	3-06-00-00092 / 0.9022[20]	/ 1.4575[20]
4511	Cyclohexanol, 1,3-dimethyl-, trans- / trans-1,3-Dimethylcyclohexanol	$C_8H_{16}O$ / 128.21	15466-93-0 / 14.5	77.6[25]	3-06-00-00093 / 0.8894[30]	/ 1.4507[20]
4512	Cyclohexanol, 1,4-dimethyl-, cis-	$C_8H_{16}O$ / 128.21	16980-61-3 / 24	83.5[25]	3-06-00-00098 / 0.9011[30]	/ 1.4564[20]
4513	Cyclohexanol, 2,2-dimethyl- / 2,2-Dimethylcyclohexanol	$C_8H_{16}O$ / 128.21	1193-46-0 / 8	/ 177	/ 0.9225[20]	/ 1.4648[20]
4514	Cyclohexanol, 2,4-dimethyl- / 2,4-Dimethylcyclohexanol	$C_8H_{16}O$ / 128.21	69542-91-2 /	/ 177.5	4-06-00-00124 / 0.900[20]	/ 1.4500[20]
4515	Cyclohexanol, 2,6-dimethyl- / 2,6-Dimethyl-1-cyclohexanol	$C_8H_{16}O$ / 128.21	5337-72-4 / 32.5	/ 176	2-06-00-00028 / 0.944[25]	/ 1.4600[20]
4516	Cyclohexanol, 3,3-dimethyl- / 3,3-Dimethylcyclohexanol	$C_8H_{16}O$ / 128.21	767-12-4 / 11.5	185; 99.5[35]	3-06-00-00089 / 0.9128[14]	/ 1.4606[15]
4517	Cyclohexanol, 3,4-dimethyl- / 3,4-Dimethylcyclohexyl alcohol	$C_8H_{16}O$ / 128.21	5715-23-1 /	/ 189	3-06-00-00091 / 0.9073[16]	/ 1.458[16]
4518	Cyclohexanol, 3,5-dimethyl- / 3,5-Dimethylcyclohexanol	$C_8H_{16}O$ / 128.21	5441-52-1 / 11.6	/ 187	2-06-00-00029 / 0.898[20]	/ 1.4550[20]
4519	Cyclohexanol, 4,4-dimethyl- / 4,4-Dimethylcyclohexanol	$C_8H_{16}O$ / 128.21	932-01-4 / 16	/ 186	4-06-00-00120 / 0.9250[20]	/ 1.4613[20]
4520	Cyclohexanol, 3-(dimethylamino)-	$C_8H_{17}NO$ / 143.23	6890-03-5 / 73	/ 231	3-13-00-00719 / 0.9737[25]	EtOH 4 / 1.4852[20]
4521	Cyclohexanol, 2-(1,1-dimethylethyl)-	$C_{10}H_{20}O$ / 156.27	13491-79-7 / 45	139[95]	3-06-00-00125 / 0.902[25]	
4522	Cyclohexanol, 1-ethyl- / 1-Ethylcyclohexanol	$C_8H_{16}O$ / 128.21	1940-18-7 / 34.5	/ 166	4-06-00-00115 / 0.9227[25]	H$_2$O 2; bz 3; ctc 2; peth 3 / 1.4633[20]
4523	Cyclohexanol, 2-ethyl-, trans-(±)- / Cyclohexanol, 2-ethyl (trans, DL)	$C_8H_{16}O$ / 128.21	89886-25-9 /	79[12]	4-06-00-00117 / 0.9193[21]	ace 4; bz 4; eth 4; peth 4 / 1.4640[21]
4524	Cyclohexanol, 2-ethyl-, cis-(±)-	$C_8H_{16}O$ / 128.21	116697-35-9 /	/ 181	3-06-00-00085 / 0.9274[21]	ace 4; bz 4; eth 4; peth 4 / 1.4655[21]
4525	Cyclohexanol, 4-ethyl- / 4-Ethylcyclohexanol	$C_8H_{16}O$ / 128.21	4534-74-1 /	84[10]	2-06-00-00026 / 0.889[25]	/ 1.4625[20]
4526	Cyclohexanol, 1-ethyl-2-methyl- / 1-Ethyl-2-methylcyclohexanol	$C_9H_{18}O$ / 142.24	102370-18-3 /	182; 82[73]	4-06-00-00133 / 0.9235[20]	/ 1.458[20]

No.	Name / Synonym	Mol. Form. / Mol. Wt.	CAS RN	mp/°C	bp/°C	Merck No.	Beil. Ref.	den/g cm⁻³	Solubility	n_D
4527	Cyclohexanol, 1-ethyl-3-methyl- / 1-Ethyl-3-methylcyclohexanol	C₉H₁₈O 142.24	62067-44-1		88^{20}		2-06-00-00035	0.9013^{20}		1.459
4528	Cyclohexanol, 1,1'-(1,2-ethenediyl)bis-	C₁₄H₂₂O₂ 222.33	78-54-6	112.5	182^{13}		4-06-00-06076		H₂O 1; EtOH 4; eth 4; ace 4	
4529	Cyclohexanol, 1-ethynyl-	C₈H₁₂O 124.18	78-27-3	31.5	174		4-06-00-00348	0.9873^{20}	H₂O 1; EtOH 3; bz 3; chl 2	1.4822^{20}
4530	Cyclohexanol, 1-methyl- / 1-Methylcyclohexanol	C₇H₁₄O 114.19	590-67-0	25	155; 70^{25}		4-06-00-00095	0.9194^{20}	H₂O 1; EtOH 3; bz 3; chl 3	1.4595^{20}
4531	Cyclohexanol, 2-methyl-, *cis*-(±)- / (±)-*cis*-2-Methylcyclohexanol	C₇H₁₄O 114.19	615-38-3	7	165		4-06-00-00100	0.9360^{20}	EtOH 4	1.4640^{20}
4532	Cyclohexanol, 2-methyl-, *trans*-(±)- / (±)-*trans*-2-Methylcyclohexanol	C₇H₁₄O 114.19	615-39-4	-4	167.5		4-06-00-00100	0.9247^{20}	eth 4; EtOH 4	1.4616^{20}
4533	Cyclohexanol, 3-methyl-, *cis*-(±)- / *cis*-3-Methylcyclohexanol	C₇H₁₄O 114.19	5454-79-5	-5.5	168; 94^{12}			0.9155^{20}	eth 4; EtOH 4	1.4752^{20}
4534	Cyclohexanol, 3-methyl-, *trans*-(±)- / *trans*-3-Methylcyclohexanol	C₇H₁₄O 114.19	7443-55-2	-0.5	167; 84^{13}			0.9214^{30}	eth 4; EtOH 4	1.4580^{20}
4535	Cyclohexanol, 4-methyl-, *cis*- / *cis*-4-Methylcyclohexanol	C₇H₁₄O 114.19	7731-28-4	-9.2	173		4-06-00-00105	0.9170^{20}	eth 4; EtOH 4	1.4614^{20}
4536	Cyclohexanol, 4-methyl-, *trans*- / *trans*-4-Methylcyclohexanol	C₇H₁₄O 114.19	7731-29-5		174		4-06-00-00105	0.9118^{21}	H₂O 2; EtOH 5; eth 3	1.4561^{20}
4537	Cyclohexanol, 2-methylene-	C₇H₁₂O 112.17	4065-80-9		$82\text{-}4^{13}$		4-06-00-00208	0.955^{20}		1.4843^{20}
4538	Cyclohexanol, 1-(1-methylethyl)-	C₉H₁₈O 142.24	3552-01-0		$76\text{-}9^{10}$		4-06-00-00131	0.928^{19}		1.4683^{19}
4539	Cyclohexanol, 2-(1-methylethyl)-, *cis*- / *cis*-2-Isopropylcyclohexanol	C₉H₁₈O 142.24	10488-25-2	52.5	77^{13}		4-06-00-00131	0.9223^{25}		1.4665^{25}
4540	Cyclohexanol, 4-(1-methylethyl)-, *cis*- / *cis*-4-Isopropylcyclohexanol	C₉H₁₈O 142.24	22900-08-9		$72^{1.9}$		3-06-00-00111	0.9195^{20}		1.4671^{20}
4541	Cyclohexanol, 4-(1-methylethyl)-, *trans*- / *trans*-4-Isopropylcyclohexanol	C₉H₁₈O 142.24	15890-36-5		$79.5^{1.8}$		3-06-00-00112	0.9140^{20}		1.4658^{20}
4542	Cyclohexanol, 1-methyl-4-(1-methylethenyl)- / β-Terpineol	C₁₀H₁₈O 154.25	138-87-4	32.5	210; 90^{10}		4-06-00-00254	0.917^{20}		1.4747^{20}
4543	Cyclohexanol, 2-methyl-5-(1-methylethenyl)-, [1*S*-(1α,2β,5α)]-	C₁₀H₁₈O 154.25	22567-21-1		225; 107^{14}		3-06-00-00256	0.9274^{20}		1.4780^{20}
4544	Cyclohexanol, 5-methyl-2-(1-methylethenyl)-, [1*R*-(1α,2β,5α)]-	C₁₀H₁₈O 154.25	89-79-2	78	93^{14}		3-06-00-00257	0.911^{20}	H₂O 2; EtOH 3; eth 3	1.4723^{20}
4545	Cyclohexanol, 5-methyl-2-(1-methylethenyl)-, acetate, [1*R*-(1α,2β,5α)]-	C₁₂H₂₀O₂ 196.29	57576-09-7	85	$112\text{-}4^{8}$		4-06-00-00255	0.925^{25}		1.4566^{20}
4546	Cyclohexanol, 1-methyl-4-(1-methylethyl)-	C₁₀H₂₀O 156.27	21129-27-1		208.5		4-06-00-00147	0.90^{20}		1.4619^{20}
4547	Cyclohexanol, 2-methyl-5-(1-methylethyl)-, [1*R*-(1α,2α,5β)]-	C₁₀H₂₀O 156.27	5563-78-0		217.5; 102^{18}		4-06-00-00148	0.9012^{20}	eth 4; EtOH 4	1.4632^{20}
4548	Cyclohexanol, 5-methyl-2-(1-methylethyl)-, (1α,2β,5α)-(±)- / (±)-Menthol	C₁₀H₂₀O 156.27	15356-70-4	38	216	5723	4-06-00-00151	0.903^{15}	H₂O 1; EtOH 4; eth 4; ace 4	1.4615^{20}
4549	Cyclohexanol, 5-methyl-2-(1-methylethyl)-, acetate, [1*R*-(1α,2β,5α)]-	C₁₂H₂₂O₂ 198.31	2623-23-6		109^{10}		4-06-00-00153	0.9185^{20}		1.4469^{20}
4550	Cyclohexanol, 2-methyl-1-propyl- / 2-Methyl-1-propylcyclohexanol	C₁₀H₂₀O 156.27	24580-48-1		$97\text{-}8^{34}$		3-06-00-00127	0.919^{20}		1.48^{20}
4551	Cyclohexanol, 3-methyl-1-propyl- / 3-Methyl-1-propylcyclohexanol	C₁₀H₂₀O 156.27	24580-52-7		199; $96\text{-}8^{20}$		1-06-00-00018	0.8903^{21}		1.4566^{24}
4552	Cyclohexanol, 1-phenyl- / 1-Phenylcyclohexan-1-ol	C₁₂H₁₆O 176.26	1589-60-2	63.5	157^{28}; 112^{5}		4-06-00-03908	1.035^{16}	bz 4; eth 4; EtOH 4; lig 4	1.5415^{16}
4553	Cyclohexanol, 2-phenyl-, *cis*-(±)- / Cyclohexanol, 2-phenyl (*cis, DL*)	C₁₂H₁₆O 176.26	40960-73-4	41.5	$140\text{-}1^{16}$		4-06-00-03909	1.035^{16}		1.5415^{16}
4554	Cyclohexanol, 2-phenyl-, *trans*-(±)- / Cyclohexanol, 2-phenyl (*trans, DL*)	C₁₂H₁₆O 176.26	40960-69-8	56.5	153^{16}		4-06-00-03909		EtOH 4; chl 4	
4555	Cyclohexanol, 1-(2-propenyl)-	C₉H₁₆O 140.23	1123-34-8		190		4-06-00-00235	0.9341^{22}		1.4756^{22}
4556	Cyclohexanol, 2-(2-propenyl)-, *trans*- / Cyclohexanol, 2-allyl-, *trans*-	C₉H₁₆O 140.23	24844-28-8		$94\text{-}6^{15}$		4-06-00-00235	0.947^{20}	HOAc 4	1.4778^{20}
4557	Cyclohexanol, 1-propyl- / 1-Hydroxy-1-propylcyclohexane	C₉H₁₈O 142.24	5445-24-9		180 dec; 85^{20}		4-06-00-00129	0.934^{12}		1.468^{12}
4558	Cyclohexanol, 2-propyl-, *cis*- / *cis*-2-Propylcyclohexanol	C₉H₁₈O 142.24	5857-86-3		84^{10}		4-06-00-00130	0.9247^{11}		1.4688^{11}
4559	Cyclohexanol, 2-propyl-, *trans*- / *trans*-2-Propylcyclohexanol	C₉H₁₈O 142.24	5846-43-5		90^{14}		4-06-00-00130	0.9162^{11}		1.4668^{11}
4560	Cyclohexanol, 4-propyl-, *trans*- / *trans*-4-Propylcyclohexanol	C₉H₁₈O 142.24	77866-58-1		208; 103^{14}		3-06-00-00106	0.9072^{17}		1.4613^{17}
4561	Cyclohexanol, 4-propyl-, *cis*-	C₉H₁₈O 142.24	98790-22-8		99^{17}		3-06-00-00105	0.9102^{16}		1.4652^{16}
4562	Cyclohexanol, 1,2,2-trimethyl- / 1,2,2-Trimethylcyclohexanol	C₉H₁₈O 142.24	31720-69-1		81^{20}; 76^{16}		3-06-00-00114	0.9230^{20}	bz 4; eth 4; EtOH 4	1.4680^{20}

No.	Name / Synonym	Mol. Form. / Mol. Wt.	CAS RN / mp/°C	Merck No. / bp/°C	Beil. Ref. / den/g cm^{-3}	Solubility / n_D
4563	Cyclohexanol, 1,2,6-trimethyl- / 1,2,6-Trimethylcyclohexanol	$C_9H_{18}O$ / 142.24	96244-13-2 /	/ 78[22]	1-06-00-00017 / 0.9126[15]	ace 4; bz 4; eth 4; EtOH 4 / 1.4598[15]
4564	Cyclohexanol, 1,3,5-trimethyl- / 1,3,5-Trimethylcyclohexanol	$C_9H_{18}O$ / 142.24	90760-75-1 /	/ 181	3-06-00-00117 / 0.8876[17]	eth 4; EtOH 4; chl 4 / 1.454[16]
4565	Cyclohexanol, 2,2,5-trimethyl- / 2,2,5-Trimethylcyclohexanol	$C_9H_{18}O$ / 142.24	73210-25-0 /	/ 188	0-06-00-00022 / 0.8955[22]	EtOH 4 / 1.4569[20]
4566	Cyclohexanol, 2,2,6-trimethyl- / 1,1,3-Trimethyl-2-cyclohexanol	$C_9H_{18}O$ / 142.24	10130-91-3 / 51	187; 87[28]	4-06-00-00135 / 0.9128[20]	eth 4; EtOH 4; chl 4 / 1.4600[20]
4567	Cyclohexanol, 2,3,3-trimethyl-	$C_9H_{18}O$ / 142.24	116724-17-5 / 28	/ 197	1-06-00-00016 / 0.9002[25]	ace 4; EtOH 4 / 1.4572[25]
4568	Cyclohexanol, 2,3,6-trimethyl- / 2,3,6-Trimethylcyclohexanol	$C_9H_{18}O$ / 142.24	58210-03-0 /	/ 195	0-06-00-00022 / 0.9117[17]	EtOH 4; chl 4
4569	Cyclohexanol, 3,3,5-trimethyl-, cis- / cis-3,3,5-Trimethylcyclohexanol	$C_9H_{18}O$ / 142.24	933-48-2 / 37.3	202; 92[12]	4-06-00-00135 / 0.9006[16]	H_2O 1; EtOH 3; eth 3; chl 3 / 1.4550[16]
4570	Cyclohexanol, 3,3,5-trimethyl-, trans- / trans-3,3,5-Trimethylcyclohexanol	$C_9H_{18}O$ / 142.24	767-54-4 / 55.8	189.2	4-06-00-00135 / 0.8631[60]	H_2O 1; EtOH 3; eth 3; chl 3
4571	Cyclohexanone / Pimelic ketone	$C_8H_{10}O$ / 98.14	108-94-1 / -31	2732 / 155.4	4-07-00-00015 / 0.9478[20]	H_2O 3; EtOH 3; eth 3; ace 3 / 1.4507[20]
4572	Cyclohexanone, 2-acetyl-	$C_8H_{12}O_2$ / 140.18	874-23-7 /	/ 111-2[18]	4-07-00-01997 / 1.078[25]	ctc 3 / 1.5138[20]
4573	Cyclohexanone, 2,6-bis(phenylmethylene)-	$C_{20}H_{18}O$ / 274.36	897-78-9 / 117.5	185-95[20]	3-07-00-02661 /	EtOH 2; bz 3; HOAc 3
4574	Cyclohexanone, 2-bromo-	C_6H_9BrO / 177.04	822-85-5 /	89-90[14]	4-07-00-00034 / 1.340[25]	1.5085[25]
4575	Cyclohexanone, 2-butyl- / 2-Butylcyclohexanone	$C_{10}H_{18}O$ / 154.25	1126-18-7 /	70[2]	4-07-00-00079 / 0.905[20]	H_2O 1; ctc 2; os 4 / 1.4545[20]
4576	Cyclohexanone, 2-butylidene- / 2-Butylidenecyclohexanone	$C_{10}H_{16}O$ / 152.24	7153-14-2 /	98-100[10]	4-07-00-00181 / 0.935[20]	ace 4; bz 4; eth 4; EtOH 4 / 1.4800[20]
4577	Cyclohexanone, 2-chloro- / 2-Chlorocyclohexanone	C_6H_9ClO / 132.59	822-87-7 / 23	82[15]	4-07-00-00032 / 1.160[20]	eth 3; bz 3; ctc 2; diox 3 / 1.4825[20]
4578	Cyclohexanone, 4-chloro- / 4-Chlorocyclohexanone	C_6H_9ClO / 132.59	21299-26-3 /	95[17]	4-07-00-00032 /	eth 4 / 1.4867[20]
4579	Cyclohexanone, 2-(2-chlorophenyl)-2-(methylamino)-, (±)- / Ketamine	$C_{13}H_{16}ClNO$ / 237.73	6740-88-1 / 92.5	5174		
4580	Cyclohexanone, 2,2-dimethyl- / 2,2-Dimethyl-1-cyclohexanone	$C_8H_{14}O$ / 126.20	1193-47-1 / -20.5	172	4-07-00-00055 / 0.9145[20]	1.4486[20]
4581	Cyclohexanone, 2,3-dimethyl-, cis- / cis-2,3-Dimethylcyclohexanone	$C_8H_{14}O$ / 126.20	1551-88-8 /	178.5; 99[57]	3-07-00-00088 / 0.9159[20]	1.4505[20]
4582	Cyclohexanone, 2,3-dimethyl-, trans- / trans-2,3-Dimethylcyclohexanone	$C_8H_{14}O$ / 126.20	1551-89-9 /	96[57]	4-07-00-00057 / 0.908[22]	1.4511[25]
4583	Cyclohexanone, 2,4-dimethyl-, (2S-trans)- / Cyclohexanone, 2,4-dimethyl (trans, d)	$C_8H_{14}O$ / 126.20	93921-42-7 /	179; 69[17]	3-07-00-00093 / 0.9004[16]	ace 4; bz 4; eth 4 / 1.4488[22]
4584	Cyclohexanone, 2,4-dimethyl-, cis-	$C_8H_{14}O$ / 126.20	116783-29-0 /	177	3-07-00-00093 / 0.910[15]	1.4493[15]
4585	Cyclohexanone, 2,5-dimethyl-, trans-(±)- / trans-(±)-2,5-Dimethylcyclohexanone	$C_8H_{14}O$ / 126.20	66395-23-1 /	172	4-07-00-00059 / 0.9025[20]	eth 4; EtOH 4 / 1.4446[20]
4586	Cyclohexanone, 2,6-dimethyl- / 2,6-Dimethylcyclohexanone	$C_8H_{14}O$ / 126.20	2816-57-1 /	175	2-07-00-00025 / 0.925[25]	1.4460[20]
4587	Cyclohexanone, 3,3-dimethyl- / 3,3-Dimethylcyclohexanone	$C_8H_{14}O$ / 126.20	2979-19-3 /	180; 72-3[25]	4-07-00-00056 / 0.909[15]	1.4482[17]
4588	Cyclohexanone, 3,4-dimethyl- / 3,4-Dimethylcyclohexanone	$C_8H_{14}O$ / 126.20	5465-09-8 /	187	3-07-00-00089 / 0.906[20]	1.4507[20]
4589	Cyclohexanone, 3,5-dimethyl- / 3,5-Dimethylcyclohexanone	$C_8H_{14}O$ / 126.20	2320-30-1 /	182.5	4-07-00-00058 / 0.896[20]	1.4407[20]
4590	Cyclohexanone, 4,4-dimethyl- / 4,4-Dimethylcyclohexanone	$C_8H_{14}O$ / 126.20	4255-62-3 / 39	73[14]	4-07-00-00056 / 0.932[20]	1.4537[24]
4591	Cyclohexanone, 2-[(dimethylamino)methyl]- / 2-[(Dimethylamino)methyl]cyclohexanone	$C_9H_{17}NO$ / 155.24	15409-60-6 /	92[10.5]	4-14-00-00008 / 0.9504[20]	H_2O 2; EtOH 3; eth 3 / 1.4672[20]
4592	Cyclohexanone, 4-(1,1-dimethylpropyl)-	$C_{11}H_{20}O$ / 168.28	16587-71-6 /	124-5[16]	4-07-00-00096 / 0.920[25]	1.4677[20]
4593	Cyclohexanone, 4-ethyl- / 4-Ethylcyclohexanone	$C_8H_{14}O$ / 126.20	5441-51-0 /	193	3-07-00-00083 / 0.895[20]	1.4515[20]
4594	Cyclohexanone, 2-ethylidene- / 2-Ethylidenecyclohexanone	$C_8H_{12}O$ / 124.18	1122-25-4 /	92[20]	4-07-00-00146 / 0.962[20]	H_2O 1; os 4 / 1.4882[20]
4595	Cyclohexanone, 2-ethyl-2-methyl- / 2-Ethyl-2-methylcyclohexanone	$C_9H_{16}O$ / 140.23	17206-52-9 /	195	2-07-00-00031 / 0.9037[17]	1.4515[17]
4596	Cyclohexanone, 2-methyl-, (±)- / (±)-2-Methylcyclohexanone	$C_7H_{12}O$ / 112.17	24965-84-2 / -13.9	165	4-07-00-00041 / 0.9250[20]	H_2O 1; EtOH 3; eth 3 / 1.4483[25]
4597	Cyclohexanone, 3-methyl-, (±)-	$C_7H_{12}O$ / 112.17	625-96-7 / -73.5	169; 65[15]	4-07-00-00043 / 0.9136[20]	H_2O 1; EtOH 3; eth 3 / 1.4456[20]
4598	Cyclohexanone, 4-methyl- / 4-Methylcyclohexanone	$C_7H_{12}O$ / 112.17	589-92-4 / -40.6	170	4-07-00-00044 / 0.9138[20]	H_2O 1; EtOH 3; eth 3; ctc 2 / 1.4451[20]
4599	Cyclohexanone, 2-(1-methylethyl)- / 2-Isopropylcyclohexanone	$C_9H_{16}O$ / 140.23	1004-77-9 / 72.5	83[17]	4-07-00-00066 / 0.922[16]	H_2O 1; EtOH 3; eth 3; ace 3 / 1.4564[15]
4600	Cyclohexanone, 4-(1-methylethyl)- / 4-Isopropylcyclohexanone	$C_9H_{16}O$ / 140.23	5432-85-9 /	139-40[100]	4-07-00-00066 / 0.9158[25]	1.4552[25]
4601	Cyclohexanone, 2-methyl-5-(1-methylethenyl)-, (2S-trans)-	$C_{10}H_{16}O$ / 152.24	619-02-3 / -11	221.5	0.9253[20]	1.4717[20]

No.	Name Synonym	Mol. Form. Mol. Wt.	CAS RN mp/°C	Merck No. bp/°C	Beil. Ref. den/g cm^{-3}	Solubility n_D
4602	Cyclohexanone, 5-methyl-2-(1-methylethenyl)-	C$_{10}$H$_{16}$O 152.24	529-00-0	100[18]	4-07-00-00190 0.9198[20]	1.4675[20]
4603	Cyclohexanone, 5-methyl-2-(1-methylethenyl)-, *trans*-	C$_{10}$H$_{16}$O 152.24	29606-79-9	100[18]	4-07-00-00190 0.9198[20]	1.4675[20]
4604	Cyclohexanone, 2-methyl-5-(1-methylethyl)-, (2S-*trans*)-	C$_{10}$H$_{18}$O 154.25	13163-73-0	220; 96[15]	4-07-00-00085 0.904[20]	ace 4; EtOH 4; chl 4 1.4553[20]
4605	Cyclohexanone, 5-methyl-2-(1-methylethyl)-, (2R-*cis*)- Menthone(d)	C$_{10}$H$_{18}$O 154.25	1196-31-2	204; 85[14]	4-07-00-00087 0.8947[20]	H$_2$O 2; EtOH 3; eth 3; ace 3 1.4503[20]
4606	Cyclohexanone, 5-methyl-2-(1-methylethyl)-, (2S-*trans*)- (-)-Menthone	C$_{10}$H$_{18}$O 154.25	14073-97-3 -6	5724 207	4-07-00-00188 0.8954[20]	H$_2$O 2; EtOH 5; eth 5; ace 3 1.4505[20]
4607	Cyclohexanone, 5-methyl-2-(1-methylethylidene)-	C$_{10}$H$_{16}$O 152.24	15932-80-6	95-6[11]	4-07-00-00188 0.9367[20]	1.4869[20]
4608	Cyclohexanone, 5-methyl-2-(1-methylethylidene)-, (R)- Pulegone	C$_{10}$H$_{16}$O 152.24	89-82-7	7955 224	4-07-00-00188 0.9346[45]	H$_2$O 1; EtOH 5; eth 5; ctc 3 1.4894[20]
4609	Cyclohexanone, 4-(2-methylpropyl)-	C$_{10}$H$_{18}$O 154.25	42061-72-3	218; 98[12]	4-07-00-00082 0.8914[24]	1.4502[24]
4610	Cyclohexanone, oxime (Hydroxyimino)cyclohexane	C$_6$H$_{11}$NO 113.16	100-64-1 90	206	4-07-00-00021	H$_2$O 3; EtOH 3; eth 3; chl 2
4611	Cyclohexanone, 2-(1-piperidinylmethyl)- Pimeclone	C$_{12}$H$_{21}$NO 195.30	534-84-9	7399 118-20[14]	5-20-02-00330	
4612	Cyclohexanone, 2-propyl- 2-Propylcyclohexanone	C$_9$H$_{16}$O 140.23	94-65-5	197	4-07-00-00064 0.927[20]	H$_2$O 1; EtOH 3; eth 4; ace 3 1.4538[20]
4613	Cyclohexanone, 4-propyl-	C$_9$H$_{16}$O 140.23	40649-36-3	213; 97.5[15]	3-07-00-00116 0.9072[19]	1.4530[19]
4614	Cyclohexanone, 2,2,3-trimethyl- 2,2,3-Trimethylcyclohexanone	C$_9$H$_{16}$O 140.23	39257-08-4 66-7[10]		4-07-00-00069 0.9234[15]	1.4569[15]
4615	Cyclohexanone, 2,2,4-trimethyl- 2,2,4-Trimethylcyclohexanone	C$_9$H$_{16}$O 140.23	35413-38-8	86-7[12]	3-07-00-00125 0.8914[20]	1.4520[20]
4616	Cyclohexanone, 2,2,5-trimethyl- 2,2,5-Trimethylcyclohexanone	C$_9$H$_{16}$O 140.23	933-36-8	183	3-07-00-00126 0.8871[24]	1.4432[24]
4617	Cyclohexanone, 2,2,6-trimethyl-	C$_9$H$_{16}$O 140.23	2408-37-9 -31.8	178.5	4-07-00-00069 0.9043[18]	1.4470[20]
4618	Cyclohexanone, 2,3,6-trimethyl- 2,3,6-Trimethylcyclohexanone	C$_9$H$_{16}$O 140.23	42185-47-7	190	3-07-00-00126 0.9129[18]	1.4464[21]
4619	Cyclohexanone, 2,4,4-trimethyl- 2,4,4-Trimethylcyclohexanone	C$_9$H$_{16}$O 140.23	2230-70-8	191	4-07-00-00070 0.902[20]	1.4493[20]
4620	Cyclohexanone, 2,4,5-trimethyl- 2,4,5-Trimethylcyclohexanone	C$_9$H$_{16}$O 140.23	51299-55-9	193	4-07-00-00072 0.897[20]	1.4479[20]
4621	Cyclohexanone, 2,4,6-trimethyl- 2,4,6-Trimethylcyclohexanone	C$_9$H$_{16}$O 140.23	90645-54-8	185	3-07-00-00127 0.8992[20]	1.4458[20]
4622	Cyclohexanone, 3,3,5-trimethyl- Dihydroisophorone	C$_9$H$_{16}$O 140.23	873-94-9	189	4-07-00-00070 0.8919[19]	1.4454[15]
4623	Cyclohexanone, 3,4,4-trimethyl- 3,4,4-Trimethylcyclohexanone	C$_9$H$_{16}$O 140.23	40441-35-8 80-1[13]		3-07-00-00122 0.911[25]	1.4552[20]
4624	Cyclohexasiloxane, dodecamethyl- Dodecamethylcyclohexasiloxane	C$_{12}$H$_{36}$O$_6$Si$_6$ 444.93	540-97-6 -3	3400 245	4-04-00-04129 0.9672[25]	H$_2$O 1 1.4015[20]
4625	Cyclohexasiloxane, 2,4,6,8,10,12-hexamethyl- Hexamethylcyclohexasiloxane	C$_6$H$_{24}$O$_6$Si$_6$ 360.77	6166-87-6 -79		4-04-00-04100 1.006[20]	H$_2$O 1 1.3944[20]
4626	Cyclohexene Tetrahydrobenzene	C$_6$H$_{10}$ 82.15	110-83-8 -103.5	2733 82.9	4-05-00-00218 0.8110[20]	H$_2$O 1; EtOH 5; eth 5; ace 5 1.4465[20]
4627	Cyclohexene, 1-bromo-	C$_6$H$_9$Br 161.04	2044-08-8	165	4-05-00-00236 1.3901[20]	eth 3; ace 4; bz 3; chl 4 1.5134[20]
4628	Cyclohexene, 3-bromo-	C$_6$H$_9$Br 161.04	1521-51-3	80-2[40]	4-05-00-00237 1.3890[2c]	H$_2$O 1; eth 3; bz 3; chl 3 1.5320[20]
4629	3-Cyclohexene-1-carbonitrile	C$_7$H$_9$N 107.16	100-45-8	39[1.5]	4-09-00-00116	ctc 3 1.4716[20]
4630	1-Cyclohexenecarbonitrile 1-Cyanocyclohexene	C$_7$H$_9$N 107.16	1855-63-6	81[12]		
4631	1-Cyclohexene-1-carboxaldehyde	C$_7$H$_{10}$O 110.16	1192-88-7	72[15]	4-07-00-00132 0.9694[20]	EtOH 3; eth 3 1.5005[20]
4632	3-Cyclohexene-1-carboxaldehyde	C$_7$H$_{10}$O 110.16	100-50-5 2	105	4-07-00-00134 0.9692[20]	ace 3; ctc 2; MeOH 3
4633	1-Cyclohexene-1-carboxaldehyde, 4-(1-methylethenyl)- Perillaldehyde	C$_{10}$H$_{14}$O 150.22	2111-75-3	7119	2-07-00-00130	
4634	1-Cyclohexene-1-carboxaldehyde, 4-(1-methylethenyl)-, (R)-	C$_{10}$H$_{14}$O 150.22	5503-12-8	238; 99[7]	4-07-00-00316 0.953[20]	1.5058[20]
4635	1-Cyclohexene-1-carboxaldehyde, 4-(1-methylethyl)- Phellandral	C$_{10}$H$_{16}$O 152.24	21391-98-0	223	4-07-00-00187 0.93[20]	1.4911[20]
4636	1-Cyclohexene-1-carboxaldehyde, 2,6,6-trimethyl- β-Cyclocitral	C$_{10}$H$_{16}$O 152.24	432-25-7	95-100[15]	4-07-00-00192 0.959[15]	1.4971[15]

No.	Name Synonym	Mol. Form. Mol. Wt.	CAS RN mp/°C	Merck No. bp/°C	Beil. Ref. den/g cm⁻³	Solubility n_D
4637	1-Cyclohexene-1-carboxylic acid Cyclohexene-1-carboxylic acid	$C_7H_{10}O_2$ 126.16	636-82-8 38	241	4-09-00-00111 1.109^{20}	H_2O 2; EtOH 3; ace 3 1.4902^{20}
4638	3-Cyclohexene-1-carboxylic acid	$C_7H_{10}O_2$ 126.16	4771-80-6 17	234.5	4-09-00-00114 1.0820^{20}	H_2O 4; EtOH 3; ace 3 1.4814^{20}
4639	3-Cyclohexene-1-carboxylic acid, ethyl ester	$C_9H_{14}O_2$ 154.21	15111-56-5	194.5	4-09-00-00114 0.9688^{20}	1.4578^{20}
4640	3-Cyclohexene-1-carboxylic acid, methyl ester	$C_8H_{12}O_2$ 140.18	6493-77-2	80^{20}	4-09-00-00114 0.995^{25}	1.4610^{20}
4641	2-Cyclohexene-1-carboxylic acid, 2,6-dimethyl-4-oxo, ethyl ester	$C_{11}H_{16}O_3$ 196.25	6102-15-4	$157-8^{18}$	4-10-00-02674 1.0493^{20}	ace 3 1.4773^{20}
4642	Cyclohexene, 1-chloro- 1-Chlorocyclohexene	C_6H_9Cl 116.59	930-66-5	142.5	4-05-00-00230 1.0361^{19}	eth 3; ace 3; ctc 3; chl 3 1.4797^{20}
4643	Cyclohexene, decafluoro- Perfluorocyclohexene	C_6F_{10} 262.05	355-75-9	52.0	4-05-00-00229 1.6650^{25}	1.293^{20}
4644	1-Cyclohexene-1,2-dicarboxylic acid, diethyl ester Cyclohexene-1,2-dicarboxylic acid, diethyl ester	$C_{12}H_{18}O_4$ 226.27	92687-41-7 160^{14}		3-09-00-03940 1.0803^{19}	 1.4747^{19}
4645	1-Cyclohexene-1,3-dicarboxylic acid, diethyl ester Diethyl 1-cyclohexene-1,3-dicarboxylate	$C_{12}H_{18}O_4$ 226.27	38511-07-8 150^{12}		3-09-00-03950 1.0772^{20}	 1.4722^{20}
4646	1-Cyclohexene-1,4-dicarboxylic acid, dimethyl ester Cyclohexene-1,4-dicarboxylic acid, dimethyl ester	$C_{10}H_{14}O_4$ 198.22	22646-79-3 39	153^{20}	4-09-00-02897	ace 4; eth 4; EtOH 4
4647	Cyclohexene, 1,2-dichloro-3,3,4,4,5,5,6,6-octafluoro-	$C_6Cl_2F_8$ 294.96	336-19-6 -70	113	4-05-00-00231 1.719^{20}	 1.375^{20}
4648	Cyclohexene, 1,2-dimethyl-	C_8H_{14} 110.20	1674-10-8 -84.1	138	4-05-00-00268 0.8220^{25}	1.4620^{20}
4649	Cyclohexene, 1,3-dimethyl- 1,3-Dimethyl-1-cyclohexene	C_8H_{14} 110.20	2808-76-6	127	4-05-00-00270 0.799^{25}	1.449^{20}
4650	Cyclohexene, 1,4-dimethyl- 1,4-Dimethyl-1-cyclohexene	C_8H_{14} 110.20	2808-79-9 -59.4	128.5	4-05-00-00270 0.8005^{20}	1.4457^{20}
4651	Cyclohexene, 1,5-dimethyl- Cyclohexene, 2,4-dimethyl-	C_8H_{14} 110.20	2808-77-7	126	3-05-00-00216 0.8009^{25}	1.450^{20}
4652	Cyclohexene, 1,6-dimethyl-	C_8H_{14} 110.20	1759-64-4	131	4-05-00-00269 0.815^{20}	1.454^{20}
4653	Cyclohexene, 3,3-dimethyl- 3,3-Dimethylcyclohexene	C_8H_{14} 110.20	695-28-3	119	4-05-00-00268 0.804^{20}	1.445^{20}
4654	Cyclohexene, 3,5-dimethyl- 3,5-Dimethylcyclohexene	C_8H_{14} 110.20	823-17-6	124.5	3-05-00-00216 0.8005^{18}	1.443^{18}
4655	Cyclohexene, 3,6-dimethyl- 3,6-Dimethylcyclohexene	C_8H_{14} 110.20	19550-40-4	125	4-05-00-00271 0.802^{25}	1.4433^{25}
4656	Cyclohexene, 4,4-dimethyl-	C_8H_{14} 110.20	14072-86-7 -74.4	117	4-05-00-00268 0.7996^{20}	1.4420^{20}
4657	Cyclohexene, 1-ethenyl-	C_8H_{12} 108.18	2622-21-1	145	4-05-00-00405 0.8623^{15}	H_2O 1; eth 3; bz 3; MeOH 4 1.4915^{20}
4658	Cyclohexene, 4-ethenyl-	C_8H_{12} 108.18	100-40-3 -108.9	128	4-05-00-00406 0.8299^{20}	H_2O 1; eth 3; bz 3; peth 3 1.4639^{20}
4659	Cyclohexene, 6-ethenyl-6-methyl-1-(1-methylethyl)-3-(1-methylethylidene)-, (S)-	$C_{15}H_{24}$ 204.36	5951-67-7 $120-30^8$		3-05-00-01083 0.8782^{20}	ace 4; bz 4 1.5130^{26}
4660	Cyclohexene, 1-ethyl-	C_8H_{14} 110.20	1453-24-3 -109.9	137	4-05-00-00266 0.8176^{25}	1.4567^{20}
4661	Cyclohexene, 3-ethyl- 3-Ethyl-1-cyclohexene	C_8H_{14} 110.20	2808-71-1	134	4-05-00-00266 0.814^{20}	1.451^{20}
4662	Cyclohexene, 4-ethyl- 4-Ethyl-1-cyclohexene	C_8H_{14} 110.20	3742-42-5	133	4-05-00-00267 0.810^{20}	1.449^{20}
4663	Cyclohexene, 1-ethyl-4-methyl- 1-Ethyl-4-methylcyclohexene	C_9H_{16} 124.23	62088-36-2	149	4-05-00-00286 0.8169^{16}	1.453^{16}
4664	1-Cyclohexene-1-methanol, 4-(1-methylethenyl)-	$C_{10}H_{16}O$ 152.24	536-59-4 244; 12.5^{12}		2-06-00-00102 0.9690^{20}	 1.5005^{20}
4665	1-Cyclohexene-1-methanol, 2,4,4-trimethyl- β-Cyclolavandulol	$C_{10}H_{18}O$ 154.25	103985-40-6 $105-6^{15}$		4-06-00-00259 0.9153^{20}	 1.4814^{20}
4666	1-Cyclohexene-1-methanol, 2,6,6-trimethyl- Cyclogeranyl alcohol	$C_{10}H_{18}O$ 154.25	472-20-8 43.5		3-06-00-00259 0.945^{19}	1.487^{19}
4667	2-Cyclohexene-1-methanol, 2,6,6-trimethyl- α-Cyclogeraniol	$C_{10}H_{18}O$ 154.25	6627-74-3	215	4-06-00-00257 0.9382^{20}	1.4843^{20}
4668	3-Cyclohexene-1-methanol, α,α,4-trimethyl-, (±)-	$C_{10}H_{18}O$ 154.25	2438-12-2 40.5	220	4-06-00-00252 0.9337^{20}	ace 4; bz 4; eth 4; EtOH 4 1.4831^{20}
4669	3-Cyclohexene-1-methanol, α,α,4-trimethyl-, acetate	$C_{12}H_{20}O_2$ 196.29	80-26-2 140^{40}; 105^{11}		4-06-00-00253 0.9659^{21}	H_2O 1; EtOH 3; eth 3; bz 3 1.4689^{21}
4670	Cyclohexene, 1-methyl-	C_7H_{12} 96.17	591-49-1 -120.4	110.3	4-05-00-00245 0.8102^{20}	H_2O 1; eth 3; bz 3; ctc 3 1.4503^{20}

No.	Name / Synonym	Mol. Form. / Mol. Wt.	CAS RN / mp/°C	Merck No. / bp/°C	Beil. Ref. / den/g cm^{-3}	Solubility / n_D
4671	Cyclohexene, 3-methyl-, (±)- / (±)-3-Methylcyclohexene	C_7H_{12} / 96.17	56688-75-6 / -115.5	/ 104	4-05-00-00247 / 0.7990[20]	bz 4; eth 4; chl 4; peth 4 / 1.4414[20]
4672	Cyclohexene, 4-methyl-	C_7H_{12} / 96.17	591-47-9 / -115.5	/ 102.7	4-05-00-00248 / 0.7991[20]	H_2O 1; EtOH 3; eth 3 / 1.4414[20]
4673	Cyclohexene, 3-methylene-6-(1-methylethyl)- / β-Phellandrene	$C_{10}H_{16}$ / 136.24	555-10-2 /	7152 / 171.5	2-05-00-00087 / 0.8520[20]	H_2O 1; EtOH 1; eth 3 / 1.4788[20]
4674	Cyclohexene, 3-methylene-6-(1-methylethyl)-, (+)-	$C_{10}H_{16}$ / 136.24	6153-16-8 /	/ 171.5	4-05-00-00436 / 0.8520[20]	eth 4 / 1.4788[20]
4675	Cyclohexene, 4-methylene-1-(1-methylethyl)-	$C_{10}H_{16}$ / 136.24	99-84-3 /	/ 173.5	4-05-00-00437 / 0.838[22]	/ 1.4754[22]
4676	Cyclohexene, 1-methyl-3-(1-methylethenyl)-	$C_{10}H_{16}$ / 136.24	38738-60-2 /	/ 175	4-05-00-00434 / 0.8479[18]	/ 1.4760[18]
4677	Cyclohexene, 1-methyl-4-(1-methylethenyl)- / Limonene	$C_{10}H_{16}$ / 136.24	138-86-3 / -95	5371 / 178	4-05-00-00440 / 0.8402[21]	H_2O 1; EtOH 5; eth 5; ctc 3 / 1.4727[19]
4678	Cyclohexene, 1-methyl-5-(1-methylethenyl)-	$C_{10}H_{16}$ / 136.24	13898-73-2 /	/ 176; 95[20]	4-05-00-00435 / 0.8481[20]	eth 4; EtOH 4 / 1.4804[20]
4679	Cyclohexene, 5-methyl-1-(1-methylethenyl)-, (±)- / Cyclohexene, 1-isopropenyl-5-methyl (DL)	$C_{10}H_{16}$ / 136.24	105065-24-5 /	/ 184	3-05-00-00336 / 0.8594[20]	/ 1.4975
4680	Cyclohexene, 1-methyl-4-(1-methylethyl)-	$C_{10}H_{18}$ / 138.25	5502-88-5 /	/ 174.5	4-05-00-00300 / 0.8457[15]	/ 1.4735[20]
4681	Cyclohexene, 1-methyl-4-(1-methylethyl)-, (R)-	$C_{10}H_{18}$ / 138.25	1195-31-9 /	/ 173	4-05-00-00299 / 0.8246[18]	bz 4; EtOH 4; peth 4 / 1.4563[18]
4682	Cyclohexene, 3-methyl-6-(1-methylethyl)-, (3R-trans)-	$C_{10}H_{18}$ / 138.25	5113-93-9 /	/ 167; 55-6[12]	4-05-00-00300 / 0.824[20]	/ 1.461[20]
4683	Cyclohexene, 4-methyl-1-(1-methylethyl)-, (R)-	$C_{10}H_{18}$ / 138.25	619-52-3 /	/ 167.5	4-05-00-00301 / 0.8078[20]	H_2O 1; EtOH 3; eth 3; bz 3 / 1.4503[15]
4684	Cyclohexene, 1-methyl-4-(1-methylethylidene)- / Terpinolene	$C_{10}H_{16}$ / 136.24	586-62-9 /	/ 186	4-05-00-00437 / 0.8632[15]	H_2O 1; EtOH 5; eth 5; bz 3 / 1.4883[20]
4685	Cyclohexene, 1-methyl-4-(5-methyl-1-methylene-4-hexenyl)-, (S)-	$C_{15}H_{24}$ / 204.36	495-61-4 /	/ 129-30[10.5]	4-05-00-01174 / 0.8673[20]	/ 1.4880[20]
4686	5-Cyclohexene-1,2,3,4-tetrone, 5,6-dihydroxy- / Rhodizonic acid	$C_6H_2O_6$ / 170.08	118-76-3 / 248 dec	sub	4-08-00-03609 /	EtOH 3
4687	Cyclohexene, 1,2,3-trimethyl- / 1,2,3-Trimethylcyclohexene	C_9H_{16} / 124.23	72312-48-2 /	/ 150	1-05-00-00040 / 0.8347[11]	/ 1.463[11]
4688	Cyclohexene, 1,3,5-trimethyl- / 1,3,5-Trimethylcyclohexene	C_9H_{16} / 124.23	3643-64-9 /	/ 141	3-05-00-00226 / 0.8025[11]	/ 1.449[13]
4689	Cyclohexene, 1,4,4-trimethyl- / 1,4,4-Trimethylcyclohexene	C_9H_{16} / 124.23	3419-71-4 /	/ 140	1-05-00-00039 / 0.8032[19]	/ 1.444[23]
4690	Cyclohexene, 1,4,5-trimethyl- / 1,4,5-Trimethylcyclohexene	C_9H_{16} / 124.23	20030-32-4 /	/ 145	2-05-00-00050 / 0.805[20]	/ 1.4482[20]
4691	Cyclohexene, 1,5,5-trimethyl- / α-Cyclogeraniolene	C_9H_{16} / 124.23	503-46-8 /	/ 140	4-05-00-00286 / 0.7981[23]	/ 1.4461[21]
4692	Cyclohexene, 1,6,6-trimethyl- / 1,6,6-Trimethylcyclohexene	C_9H_{16} / 124.23	69745-49-9 /	/ 147	3-05-00-00225 / 0.8217[20]	/ 1.456[20]
4693	Cyclohexene, 1,5,6-trimethyl-	C_9H_{16} / 124.23	116724-18-6 /	/ 140	3-05-00-00225 / 0.829[25]	/ 1.4572[25]
4694	2-Cyclohexen-1-ol	$C_6H_{10}O$ / 98.14	822-67-3 /	/ 164	4-06-00-00196 / 0.9923[15]	EtOH 3; ace 3 / 1.4790[25]
4695	3-Cyclohexen-1-ol	$C_6H_{10}O$ / 98.14	822-66-2 /	/ 164; 68[16]	4-06-00-00200 / 0.9845[23]	H_2O 2; eth 3; ace 3 / 1.4851[22]
4696	2-Cyclohexen-1-ol, 2-methyl-5-(1-methylethenyl)-	$C_{10}H_{16}O$ / 152.24	99-48-9 /	/ 228	2-06-00-00102 / 0.9484[25]	/ 1.4942[25]
4697	2-Cyclohexen-1-ol, 3-methyl-6-(1-methylethyl)-	$C_{10}H_{18}O$ / 154.25	491-04-3 /	/ 97-8[15.5]	2-06-00-00065 / 0.9119[25]	/ 1.4729[25]
4698	2-Cyclohexen-1-ol, 6-methyl-3-(1-methylethyl)- / Carvenol	$C_{10}H_{18}O$ / 154.25	586-27-6 /	/ 220	4-06-00-00253 / 0.925[22]	/ 1.479
4699	3-Cyclohexen-1-ol, 4-methyl-1-(1-methylethyl)-	$C_{10}H_{18}O$ / 154.25	562-74-3 /	/ 209	4-06-00-00250 / 0.926[20]	/ 1.4785[19]
4700	2-Cyclohexen-1-ol, 1,3,5-trimethyl- / Cyclohexen-3-ol, 1,3,5-trimethyl	$C_9H_{16}O$ / 140.23	33843-55-9 / 46	/ 87-90[17]	1-06-00-00036 / 0.9132[20]	/ 1.4735[19]
4701	2-Cyclohexen-1-ol, 2,4,4-trimethyl- / 2,4,4-Trimethyl-2-cyclohexen-1-ol	$C_9H_{16}O$ / 140.23	73741-61-4 /	/ 193	1-06-00-00036 / 0.9310[25]	bz 4; EtOH 4
4702	2-Cyclohexen-1-ol, 3,5,5-trimethyl- / Isophorol	$C_9H_{16}O$ / 140.23	470-99-5 /	/ 69[5]	3-06-00-00230 / 0.9128[20]	/ 1.4717[20]
4703	2-Cyclohexen-1-one	C_6H_8O / 96.13	930-68-7 / -53	/ 170	4-07-00-00124 / 0.9620[25]	EtOH 4; ace 3 / 1.4883[20]
4704	3-Cyclohexen-1-one	C_6H_8O / 96.13	4096-34-8 /	/ 170	4-07-00-00126 / 0.9620[25]	ace 4; EtOH 4 / 1.4842[18]
4705	2-Cyclohexen-1-one, 2,3-dimethyl-	$C_8H_{12}O$ / 124.18	1122-20-9 /	/ 93-6[20]	4-07-00-00148 / 0.9695[20]	EtOH 3; eth 3 / 1.4995[20]
4706	2-Cyclohexen-1-one, 2,5-dimethyl- / 2,5-Dimethyl-2-cyclohexenone	$C_8H_{12}O$ / 124.18	14845-35-3 /	/ 189.5	1-07-00-00051 / 0.938[22]	eth 4; EtOH 4 / 1.4753[22]
4707	2-Cyclohexen-1-one, 3,5-dimethyl-	$C_8H_{12}O$ / 124.18	1123-09-7 /	/ 208.5	4-07-00-00149 / 0.9400[20]	EtOH 3; eth 3 / 1.4812[20]

No.	Name Synonym	Mol. Form. Mol. Wt.	CAS RN mp/°C	Merck No. bp/°C	Beil. Ref. den/g cm^{-3}	Solubility n_D
4708	2-Cyclohexen-1-one, 3,6-dimethyl-	$C_8H_{12}O$ 124.18	15329-10-9	75[19]	4-07-00-00150 1.007[18]	H_2O 1; EtOH 3; eth 3 1.4805[18]
4709	2-Cyclohexen-1-one, 2-hydroxy-3-methyl-6-(1-methylethyl)- Diosphenol	$C_{10}H_{16}O_2$ 168.24	490-03-9 83	3292 109[10]		
4710	2-Cyclohexen-1-one, 2-methyl-	$C_7H_{10}O$ 110.16	1121-18-2	178.5	4-07-00-00131 0.966[20]	bz 3 1.4833[20]
4711	2-Cyclohexen-1-one, 3-methyl-	$C_7H_{10}O$ 110.16	1193-18-6 -21	201	4-07-00-00130 0.9693[20]	H_2O 5; bz 3 1.49475[20]
4712	2-Cyclohexen-1-one, 5-methyl- 5-Methyl-2-cyclohexenone	$C_7H_{10}O$ 110.16	7214-50-8	185	4-07-00-00133 0.947[25]	ace 4; bz 4; EtOH 4 1.4739[25]
4713	3-Cyclohexen-1-one, 2-methyl-	$C_7H_{10}O$ 110.16	31883-98-4 -21	201	4-07-00-00131 0.9693[20]	bz 4 1.4947[20]
4714	3-Cyclohexen-1-one, 4-methyl-	$C_7H_{10}O$ 110.16	5259-65-4	171; 74[17]	4-07-00-00131 0.9551[20]	EtOH 3; ace 3; bz 3 1.4652[20]
4715	2-Cyclohexen-1-one, 2-methyl-5-(1-methylethenyl)-, (S)- (+)-Carvone	$C_{10}H_{14}O$ 150.22	2244-16-8 <15	1883 231	4-07-00-00315 0.965[20]	H_2O 2; EtOH 4; eth 3; chl 3 1.4989[20]
4716	2-Cyclohexen-1-one, 3-methyl-6-(1-methylethyl)-, (±)- (±)-Piperitone	$C_{10}H_{16}O$ 152.24	6091-52-7 -19	7443 232.5	4-07-00-00186 0.9331[20]	ace 4; EtOH 4 1.4845[20]
4717	2-Cyclohexen-1-one, 6-methyl-3-(1-methylethyl)-, (±)- Carvenone (DL)	$C_{10}H_{16}O$ 152.24	23733-68-8	235.5	4-07-00-00187 0.9263[20]	ace 4 1.4826[20]
4718	2-Cyclohexen-1-one, 3-methyl-6-(1-methylethylidene)- Piperitenone	$C_{10}H_{14}O$ 150.22	491-09-8	120-2[14]	4-07-00-00314 0.9774[20]	EtOH 4; eth 4 1.5294[20]
4719	2-Cyclohexen-1-one, 3,4,4-trimethyl- Cyclohexen-3-one, 1,6,6-trimethyl	$C_9H_{14}O$ 138.21	17299-41-1	217; 98[13]	4-07-00-00164 0.944[25]	1.4889[25]
4720	2-Cyclohexen-1-one, 3,4,6-trimethyl- Cyclohexen-3-one, 1,4,6-trimethyl	$C_9H_{14}O$ 138.21	7474-10-4	213	4-07-00-00170 0.943[17]	1.484[17]
4721	2-Cyclohexen-1-one, 3,5,5-trimethyl- Isophorone	$C_9H_{14}O$ 138.21	78-59-1 -8.1	215.2	4-07-00-00165 0.9255[20]	1.4766[18]
4722	2-Cyclohexen-1-one, 3,6,6-trimethyl- 3,6,6-Trimethyl-2-cyclohexenone	$C_9H_{14}O$ 138.21	23438-77-9	208; 86-8[15]	3-07-00-00286 0.927[24]	1.4798[20]
4723	Cyclohexylamine, N,4-dimethyl- Cyclohexane, 1-methylamino-4-methyl	$C_8H_{17}N$ 127.23	90226-23-6 -15	196	0.8188[22]	1.4175[22]
4724	Cyclononane	C_9H_{18} 126.24	293-55-0 11	178.4	4-05-00-00133 0.8463[25]	1.4666[20]
4725	Cyclononanone	$C_9H_{16}O$ 140.23	3350-30-9 34	148[24]; 94[12]	4-07-00-00062 0.9560[20]	EtOH 3 1.4729[20]
4726	Cyclononasiloxane, octadecamethyl- Octadecamethylcyclononasiloxane	$C_{18}H_{54}O_9Si_9$ 667.39	556-71-8	188[20]	3-04-00-01886	bz 4; lig 4 1.4070[20]
4727	Cyclononene, (E)-	C_9H_{16} 124.23	3958-38-1	94-6[30]	4-05-00-00280 0.8615[20]	bz 4 1.4799[20]
4728	Cyclononene, (Z)-	C_9H_{16} 124.23	933-21-1	168	4-05-00-00280 0.8671[20]	bz 4 1.4805[20]
4729	Cyclononyne	C_9H_{14} 122.21	6573-52-0	178	4-05-00-00419 0.8972[20]	1.4890[20]
4730	2,4(1H)-Cyclo-3,4-secoakuammilanium, 3,17-dihydroxy-16-(methoxycarbonyl)-4-methyl-, (3β,16R)- Echitamine	$C_{22}H_{30}N_2O_5$ 402.49	6871-44-9 206	3474		H_2O 3; EtOH 3; eth 3; chl 3
4731	1,3-Cyclooctadecadiene	$C_{18}H_{32}$ 248.45	6568-58-7	115[3]	4-05-00-00503 0.8814[25]	1.4899[20]
4732	1,4-Cyclooctadiene	C_8H_{12} 108.18	1073-07-0 -53	145	4-05-00-00321 0.8754[20]	
4733	1,5-Cyclooctadiene	C_8H_{12} 108.18	111-78-4 -56.4	150.8	3-05-00-00321 0.8818[25]	bz 4 1.4905[25]
4734	1,5-Cyclooctadiene, (Z,Z)-	C_8H_{12} 108.18	1552-12-1 -70	151; 51-2[25]	4-05-00-00403 0.8818[25]	bz 4 1.4905[25]
4735	Cyclooctanamine Aminocyclooctane	$C_8H_{17}N$ 127.23	5452-37-9 -48	190	4-12-00-00128 0.928[25]	1.4804[20]
4736	Cyclooctane	C_8H_{16} 112.22	292-64-8 14.8	149	4-05-00-00111 0.8349[20]	H_2O 1; bz 3; lig 3 1.4586[20]
4737	Cyclooctane, butyl- Butylcyclooctane	$C_{12}H_{24}$ 168.32	16538-93-5	90-1[10]	4-05-00-00170 0.8260[25]	1.4585[25]
4738	Cyclooctanecarboxaldehyde Formylcyclooctane	$C_9H_{16}O$ 140.23	6688-11-5	96[15]	3-07-00-00112 0.94[25]	1.4748[20]
4739	Cyclooctanol	$C_8H_{16}O$ 128.21	696-71-9 25.1	99[16]	4-06-00-00113 0.9740[20]	EtOH 3 1.4871[20]
4740	Cyclooctanone	$C_8H_{14}O$ 126.20	502-49-8	196	4-07-00-00049 0.9581[20]	H_2O 1; EtOH 3; ace 3; bz 3 1.4694[20]
4741	Cyclooctasiloxane, hexadecamethyl- Hexadecamethylcyclooctasiloxane	$C_{16}H_{48}O_8Si_8$ 593.24	556-68-3 31.5	290	3-04-00-01886 1.177[25]	bz 4; lig 4 1.4060[20]
4742	1,3,5,7-Cyclooctatetraene [8]Annulene	C_8H_8 104.15	629-20-9 -4.7	140.5	4-05-00-01331 0.9206[20]	EtOH 3; eth 3; ace 3; bz 3 1.5381[20]
4743	1,3,5,7-Cyclooctatetraene, 1-bromo-	C_8H_7Br 183.05	7567-22-8	52-3[18]	4-05-00-01334 1.4206[25]	ace 3 1.5870[25]
4744	1,3,5,7-Cyclooctatetraene, 1-butyl-	$C_{12}H_{16}$ 160.26	13402-37-4	98[20]	4-05-00-01417 0.8876[25]	1.5083[25]

No.	Name Synonym	Mol. Form. Mol. Wt.	CAS RN mp/°C	Merck No. bp/°C	Beil. Ref. den/g cm⁻³	Solubility n_D
4745	1,3,5,7-Cyclooctatetraene, 1-chloro- Chlorocyclooctatetraene	C_8H_7Cl 138.60	29554-49-2	50-1[5.5]	4-05-00-01334 1.1199[25]	H_2O 1; ace 3; bz 3; chl 3 1.5542[25]
4746	1,3,5,7-Cyclooctatetraene, 1-ethyl-	$C_{10}H_{12}$ 132.21	13402-35-2	81[37]	4-05-00-01373 0.8996[25]	1.5187[25]
4747	1,3,5,7-Cyclooctatetraene, 1-methyl-	C_9H_{10} 118.18	2570-12-9	84.5[67]	4-05-00-01358 0.8978[25]	H_2O 1; EtOH 3; eth 3; bz 3 1.5249[25]
4748	1,3,5,7-Cyclooctatetraene, propyl- Cyclooctatetraene, propyl	$C_{11}H_{14}$ 146.23	13402-36-3	73[9]	4-05-00-01400 0.8870[25]	1.5131[25]
4749	1,3,5-Cyclooctatriene	C_8H_{10} 106.17	1871-52-9 -83	145.5	4-05-00-00882 0.8971[25]	1.5035[25]
4750	1,3,6-Cyclooctatriene	C_8H_{10} 106.17	3725-30-2 -62	68[60]	4-05-00-00883 0.8940[25]	
4751	Cyclooctene, (E)-	C_8H_{14} 110.20	931-89-5 -59	143	4-05-00-00263 0.8483[20]	EtOH 3; ctc 2; chl 3 1.4741[25]
4752	Cyclooctene, (Z)-	C_8H_{14} 110.20	931-87-3 -12	138	3-05-00-00209 0.8472[20]	EtOH 3; eth 3; ctc 3 1.4698[20]
4753	2-Cycloocten-1-ol	$C_8H_{14}O$ 126.20	3212-75-7	74[2]	4-06-00-00217 0.9756[25]	1.4959[25]
4754	3-Cycloocten-1-one	$C_8H_{12}O$ 124.18	4734-90-1	200.5	4-07-00-00142 0.990[20]	ace 4; eth 4; chl 4 1.4953[25]
4755	Cyclooctyne	C_8H_{12} 108.18	1781-78-8	158	4-05-00-00401 0.868[20]	1.4850[20]
4756	15H-Cyclopenta[a]phenanthrene, 16,17-dihydro- 1,2-Cyclopentenophenanthrene	$C_{17}H_{14}$ 218.30	482-66-6 135.5	2749	4-05-00-02423	H_2O 1; EtOH 3; peth 3
4757	Cyclopenta[c]furo[3',2':4,5]furo[2,3-h][1]benzopyran-1,11-dione, 2,3,6a,8,9,9a-hexahydro-4-methoxy-, (6aR- Aflatoxin B2	$C_{17}H_{14}O_6$ 314.29	7220-81-7 287.5	168	5-19-10-00552	
4758	Cyclopenta[c]pyran-1(4aH)-one, 5,6,7,7a-tetrahydro-4,7-dimethyl- Nepetalactone	$C_{10}H_{14}O_2$ 166.22	490-10-8	6383 71-2[0.05]	4-17-00-04622 1.0663[25]	1.4859[25]
4759	Cyclopentadecane	$C_{15}H_{30}$ 210.40	295-48-7 61.3		4-05-00-00181 0.8364[61]	1.4592[61]
4760	Cyclopentadecanol Exaltol	$C_{15}H_{30}O$ 226.40	4727-17-7 80.5	177[11]; 145[0.3]	4-06-00-00185 0.930[20]	1.4555[98]
4761	Cyclopentadecanone	$C_{15}H_{28}O$ 224.39	502-72-7 63	120[0.3]	4-07-00-00113 0.8895[25]	H_2O 2; EtOH 3; ace 3 1.4637[60]
4762	Cyclopentadecanone, 2-methyl- 2-Methylcyclopentadecanone	$C_{16}H_{30}O$ 238.41	52914-66-6	171-3[12]	3-07-00-00208 0.9213[16]	1.4812[16]
4763	Cyclopentadecanone, 3-methyl- Muscone	$C_{16}H_{30}O$ 238.41	541-91-3	6222 329; 130[0.5]	4-07-00-00118 0.922[17]	ace 4; eth 4; EtOH 4 1.4802[17]
4764	1,3-Cyclopentadiene Pyropentylene	C_5H_6 66.10	542-92-7 -85	2744 41	4-05-00-00377 0.8021[20]	H_2O 1; EtOH 5; eth 5; ace 3 1.4440[20]
4765	1,3-Cyclopentadiene, 5-ethylidene-	C_7H_8 92.14	3839-50-7	32[17]	4-05-00-00878 0.865[18]	1.526[20]
4766	1,3-Cyclopentadiene, 1,2,3,4,5,5-hexachloro- Perchlorocyclopentadiene	C_5Cl_6 272.77	77-47-4 -9	239; 48[0.3]	4-05-00-00381 1.7019[25]	1.5658[20]
4767	1,3-Cyclopentadiene, 5-methylene- Fulvene	C_6H_6 78.11	497-20-1	7-8[56]	4-05-00-00764 0.8241[20]	H_2O 1; bz 3; chl 3 1.4920[20]
4768	1,3-Cyclopentadiene, 5-(1-methylethylidene)-	C_8H_{10} 106.17	2175-91-9	49-50[11]	4-05-00-00974 0.881[20]	1.5474[20]
4769	1,3-Cyclopentadiene, 5-(2-propenylidene)-, (E)- Fulvene, 6-vinyl(trans)	C_8H_8 104.15	116862-65-8 -35	45[12] 20[0.15]	0.898[20]	H_2O 1; bz 3; chl 3
4770	1,3-Cyclopentadiene, 1,2,3,4-tetrachloro-5,5-dimethoxy-	$C_7H_6Cl_4O_2$ 263.93	2207-27-4	109	4-07-00-00286 1.501[25]	1.5282[20]
4771	Cyclopentanamine Cyclopentylamine	$C_5H_{11}N$ 85.15	1003-03-8 -82.7	108	4-12-00-00004 0.8689[20]	ace 3; bz 3; chl 3 1.4728[25]
4772	Cyclopentanamine, 1-methyl-	$C_6H_{13}N$ 99.18	40571-45-7	121	4-12-00-00113 0.8025[25]	H_2O 4 1.4408[25]
4773	Cyclopentanamine, 2-methyl-	$C_6H_{13}N$ 99.18	41223-14-7	123	0-12-00-00007 0.801[20]	H_2O 4
4774	Cyclopentane Pentamethylene	C_5H_{10} 70.13	287-92-3 -93.8	2746 49.3	4-05-00-00014 0.7457[20]	H_2O I; EtOH 5; eth 5; ace 5 1.4065[20]
4775	Cyclopentaneacetic acid 2-Cyclopentylacetic acid	$C_7H_{12}O_2$ 128.17	1123-00-8 13.5	228	4-09-00-00033 1.0216[18]	1.4523[18]
4776	Cyclopentaneacetic acid, 3-methyl- 3-Methylcyclopentaneacetic acid	$C_8H_{14}O_2$ 142.20	63370-69-4	237	3-09-00-00060 0.9818[20]	1.4504[20]
4777	Cyclopentane, bromo- Cyclopentyl bromide	C_5H_9Br 149.03	137-43-9	137.5	4-05-00-00019 1.3873[20]	ctc 2 1.4886[20]
4778	Cyclopentane, butyl- Butylcyclopentane	C_9H_{18} 126.24	2040-95-1 -108	156.6	4-05-00-00139 0.7846[20]	ace 4; bz 4; eth 4; EtOH 4 1.4316[20]
4779	Cyclopentanecarbonitrile Cyanocyclopentane	C_6H_9N 95.14	4254-02-8			
4780	Cyclopentanecarboxaldehyde	$C_6H_{10}O$ 98.14	872-53-7	133.5	4-07-00-00037 0.9371[20]	H_2O 4; eth 4; EtOH 4 1.4432[20]
4781	Cyclopentanecarboxylic acid Cyclopentanoic acid	$C_6H_{10}O_2$ 114.14	3400-45-1 -7	212; 104[11]	4-09-00-00011 1.0527[20]	H_2O 2; ctc 2; MeOH 3 1.4532[20]

No.	Name / Synonym	Mol. Form. / Mol. Wt.	CAS RN / mp/°C	Merck No. / bp/°C	Beil. Ref. / den/g cm⁻³	Solubility / n_D
4782	Cyclopentanecarboxylic acid, 1-amino- Cycloleucine	$C_6H_{11}NO_2$ 129.16	52-52-8 330 dec	2740	4-14-00-00974	
4783	Cyclopentanecarboxylic acid, 1-methyl- 1-Methylcyclopentanecarboxylic acid	$C_7H_{12}O_2$ 128.17	5217-05-0	219	4-09-00-00035 1.0218[20]	eth 4; EtOH 4 1.4529[20]
4784	Cyclopentanecarboxylic acid, 2-methyl-5-(1-methylethyl)- Puleganic acid	$C_{10}H_{18}O_2$ 170.25	528-23-4 -18	152[25]; 139[12]	4-09-00-00080 0.9642[20]	1.4524[24]
4785	Cyclopentanecarboxylic acid, 2-oxo-, ethyl ester	$C_8H_{12}O_3$ 156.18	611-10-9	221; 110[16]	4-10-00-02602 1.0781[21]	eth 3; bz 3 1.4519[20]
4786	Cyclopentanecarboxylic acid, 1,2,2,3-tetramethyl- Campholic acid	$C_{10}H_{18}O_2$ 170.25	464-88-0 106	254; 146[12]	2-09-00-00019	H_2O 2; EtOH 3; eth 3
4787	Cyclopentanecarboxylic acid, 3-isopropyl-1-methyl- 	$C_{12}H_{22}O_2$ 198.31	104068-42-0 18.5	257	4-09-00-00079 0.9698[19]	1.4563[20]
4788	Cyclopentanecarboxylic acid, 2-methyl-5-(1-methylethyl)-, ethyl ester Cyclopentanecarboxylic acid, 2-isopropyl-5-methyl, ethyl ester	$C_{12}H_{22}O_2$ 198.31	116530-99-5	145[4]	1-09-00-00017 0.9178[12]	1.4405[12]
4789	Cyclopentane, chloro- Cyclopentyl chloride	C_5H_9Cl 104.58	930-28-9	114	4-05-00-00018 1.0051[20]	H_2O 1; eth 3; ace 3; bz 3 1.4510[20]
4790	Cyclopentane, decyl- Decane, 1-cyclopentyl-	$C_{15}H_{30}$ 210.40	1795-21-7 -22	279	4-05-00-00186 0.8110[20]	ace 4; bz 4; eth 4; EtOH 4 1.4486[20]
4791	1,1-Cyclopentanedicarboxylic acid, diethyl ester	$C_{11}H_{18}O_4$ 214.26	4167-77-5	210	4-09-00-02793 1.0186[20]	1.4370[25]
4792	1,2-Cyclopentanedicarboxylic acid, dimethyl ester, trans-(±)- Cyclopentane-1,2-dicarboxylic acid, dimethyl ester (DL, trans)	$C_9H_{14}O_4$ 186.21	80656-12-8	118-9[17]	4-09-00-02794 1.1130[19]	1.4498[17]
4793	1,3-Cyclopentanedicarboxylic acid, cis-	$C_7H_{10}O_4$ 158.15	876-05-1 121	300 dec	4-09-00-02795	H_2O 3; EtOH 3; eth 3; ace 3
4794	1,3-Cyclopentanedicarboxylic acid, 1,2,2-trimethyl-, (1R-cis)-	$C_{10}H_{16}O_4$ 200.23	124-83-4 189.1		4-09-00-02851 1.186[20]	H_2O 2; EtOH 4; eth 4; ace 3
4795	1,3-Cyclopentanedicarboxylic acid, 1,2,2-trimethyl-, dimethyl ester, (1R-cis)- Camphoric acid, dimethyl ester, (+)-	$C_{12}H_{20}O_4$ 228.29	15797-21-4 <-16	264	4-09-00-02851 1.015[20]	eth 4; EtOH 4 1.4627[19]
4796	Cyclopentane, 1,1-diethyl- 1,1-Diethylcyclopentane	C_9H_{18} 126.24	2721-38-2 -95.6	150.5	4-05-00-00141 0.7988[25]	1.4388[20]
4797	Cyclopentane, 1,2-diethyl-, cis- cis-1,2-Diethylcyclopentane	C_9H_{18} 126.24	932-39-8 -118	153.6	4-05-00-00141 0.7920[25]	1.4355[20]
4798	Cyclopentane, 1,2-diethyl-, trans- trans-1,2-Diethylcyclopentane	C_9H_{18} 126.24	932-40-1 -95	147.5	4-05-00-00141 0.7792[25]	bz 4; eth 4; peth 4 1.4295[20]
4799	Cyclopentane, 1,3-diethyl- 1,3-Diethylcyclopentane	C_9H_{18} 126.24	19398-75-5	150	0-05-00-00045 0.783[20]	1.430[20]
4800	Cyclopentane, 1,1-dimethyl- 1,1-Dimethylcyclopentane	C_7H_{14} 98.19	1638-26-2 -69.8	87.5	4-05-00-00105 0.7499[25]	1.4136[20]
4801	Cyclopentane, 1,2-dimethyl-, cis- cis-1,2-Dimethylcyclopentane	C_7H_{14} 98.19	1192-18-3 -54	99.5	4-05-00-00106 0.7680[25]	1.4222[20]
4802	Cyclopentane, 1,2-dimethyl-, trans- trans-1,2-Dimethylcyclopentane	C_7H_{14} 98.19	822-50-4 -117.6	91.9	4-05-00-00107 0.7468[25]	1.4120[20]
4803	Cyclopentane, 1,3-dimethyl-, cis- cis-1,3-Dimethylcyclopentane	C_7H_{14} 98.19	2532-58-3 -133.7	90.8	4-05-00-00107 0.7402[25]	1.4089[20]
4804	Cyclopentane, 1,3-dimethyl-, trans- trans-1,3-Dimethylcyclopentane	C_7H_{14} 98.19	1759-58-6 -134	91.7	4-05-00-00108 0.7443[25]	1.4107[20]
4805	Cyclopentane, (1,1-dimethylethyl)-	C_9H_{18} 126.24	3875-52-3 -95.8	144.9	3-05-00-00124 0.7870[25]	1.4338[20]
4806	1,3-Cyclopentanediol, cis-	$C_5H_{10}O_2$ 102.13	16326-97-9 31	105[5]; 90[1]	3-06-00-04057 1.099[15]	1.4792[20]
4807	1,3-Cyclopentanediol, trans- Cyclopentane-1,3-diol (trans)	$C_5H_{10}O_2$ 102.13	16326-98-0 40	80-5[0.1]	3-06-00-04057 1.094[25]	1.4830[20]
4808	Cyclopentane, docosyl- Docosylcyclopentane	$C_{27}H_{54}$ 378.73	62016-55-1 45	433	4-05-00-00201 0.8295[20]	bz 4; eth 4 1.4608[20]
4809	Cyclopentane, dodecyl-	$C_{17}H_{34}$ 238.46	5634-30-0 -5	312	3-05-00-00156 0.8158[20]	ace 4; bz 4; eth 4; EtOH 4 1.4518[20]
4810	Cyclopentaneethanamine, N,α-dimethyl- Cyclopentamine	$C_9H_{19}N$ 141.26	102-45-4 	2745 171	4-12-00-00137	1.4500[20]
4811	Cyclopentaneethanol	$C_7H_{14}O$ 114.19	766-00-7	183; 96[24]	4-06-00-00110 0.9180[20]	eth 4 1.4577[20]
4812	Cyclopentane, ethyl- Ethylcyclopentane	C_7H_{14} 98.19	1640-89-7 -138.4	103.5	4-05-00-00104 0.7665[20]	H_2O 1; EtOH 5; eth 5; ace 5 1.4198[20]
4813	Cyclopentane, ethylidene- Ethylidenecyclopentane	C_7H_{12} 96.17	2146-37-4 -129.5	115	4-05-00-00253 0.8020[20]	1.4481[20]
4814	Cyclopentane, 1-ethyl-1-methyl- 1-Ethyl-1-methylcyclopentane	C_8H_{16} 112.22	16747-50-5 -143.8	121.6	4-05-00-00126 0.7767[25]	ace 4; bz 4; eth 4; EtOH 4 1.4272[20]
4815	Cyclopentane, 1-ethyl-2-methyl-, cis- cis-1-Methyl-2-ethylcyclopentane	C_8H_{16} 112.22	930-89-2 -106	128	4-05-00-00126 0.7852[20]	1.4293[20]

No.	Name Synonym	Mol. Form. Mol. Wt.	CAS RN mp/°C	Merck No. bp/°C	Beil. Ref. den/g cm^{-3}	Solubility n_D
4816	Cyclopentane, 1-ethyl-2-methyl-, *trans*-	C$_8$H$_{16}$ 112.22	930-90-5 -105.9	121.2	4-05-00-00126 0.7649[25]	1.4219[20]
4817	Cyclopentane, 1-ethyl-3-methyl-, *cis*- 1-Methyl-*cis*-3-ethylcyclopentane	C$_8$H$_{16}$ 112.22	2613-66-3 	121	4-05-00-00127 0.7724[20]	ace 4; bz 4; eth 4; EtOH 4 1.4203[20]
4818	Cyclopentane, 1-ethyl-3-methyl-, *trans*- 1-Methyl-*trans*-3-ethylcyclopentane	C$_8$H$_{16}$ 112.22	2613-65-2 -108	121	4-05-00-00127 0.7619[20]	1.4186[20]
4819	Cyclopentane, heneicosyl-	C$_{26}$H$_{52}$ 364.70	6703-82-8 42	420	4-05-00-00200 0.8286[20]	1.4602[20]
4820	Cyclopentane, heptyl- Heptane, 1-cyclopentyl-	C$_{12}$H$_{24}$ 168.32	5617-42-5 -53	224	4-05-00-00173 0.8010[20]	ace 4; bz 4; eth 4; EtOH 4 1.4421[20]
4821	Cyclopentane, hexacosyl- Hexacosylcyclopentane	C$_{31}$H$_{62}$ 434.83	62016-58-4 56	468	4-05-00-00203 0.8326[20]	1.4628[20]
4822	Cyclopentane, hexadecyl- Hexadecane, 1-cyclopentyl-	C$_{21}$H$_{42}$ 294.56	6812-39-1 21	368	4-05-00-00195 0.8228[20]	ace 4; bz 4; eth 4; EtOH 4 1.4564[20]
4823	Cyclopentane, hexyl-	C$_{11}$H$_{22}$ 154.30	4457-00-5 -73	203	4-05-00-00168 0.7965[20]	ace 4; bz 4; eth 4; EtOH 4 1.4392[20]
4824	Cyclopentane, iodo- Cyclopentyl iodide	C$_5$H$_9$I 196.03	1556-18-9 	166.5	4-05-00-00020 1.7096[20]	H$_2$O 1; eth 3; bz 3; ctc 2 1.5447[20]
4825	Cyclopentanemethanol	C$_6$H$_{12}$O 100.16	3637-61-4 	163	4-06-00-00092 0.9332[20]	1.4579[20]
4826	Cyclopentane, methyl- Methylcyclopentane	C$_6$H$_{12}$ 84.16	96-37-7 -142.5	71.8	4-05-00-00084 0.7486[20]	H$_2$O 1; EtOH 5; eth 5; ace 5 1.4097[20]
4827	Cyclopentane, methylene- Methylenecyclopentane	C$_6$H$_{10}$ 82.15	1528-30-9 	75.5	4-05-00-00241 0.7787[20]	bz 3; chl 3 1.4355[20]
4828	Cyclopentane, (1-methylethyl)- Isopropylcyclopentane	C$_8$H$_{16}$ 112.22	3875-51-2 -111.4	126.5	4-05-00-00125 0.7765[20]	H$_2$O 1; EtOH 5; eth 3; ace 5 1.4258[20]
4829	Cyclopentane, 1-methyl-2-(1-methylethyl)- Cyclopentane, 1-isopropyl-2-methyl-	C$_9$H$_{18}$ 126.24	89223-57-4 	143	4-05-00-00141 0.7832[15]	1.4279[20]
4830	Cyclopentane, 1-methyl-1-nitro- 1-Methyl-1-nitrocyclopentane	C$_6$H$_{11}$NO$_2$ 129.16	30168-50-4 	181; 92[40]	4-05-00-00087 1.0395[20]	1.4504[20]
4831	Cyclopentane, 1-methyl-2-nitro- 1-Methyl-2-nitrocyclopentane	C$_6$H$_{11}$NO$_2$ 129.16	102153-88-8 	185 dec; 98[40]	1-05-00-00011 1.0381[22]	1.4488[22]
4832	Cyclopentane, (1-methylpropyl)-	C$_9$H$_{18}$ 126.24	4850-32-2 	154.4	3-05-00-00124 0.7905[25]	1.4357[20]
4833	Cyclopentane, 1-methyl-1-propyl- 1-Methyl-1-propylcyclopentane	C$_9$H$_{18}$ 126.24	16631-63-3 	146	4-05-00-00140 0.795[25]	1.437[20]
4834	Cyclopentane, 1-methyl-2-propyl-, *cis*- *cis*-1-Methyl-2-propylcyclopentane	C$_9$H$_{18}$ 126.24	932-43-4 -104	152.6	4-05-00-00140 0.7881[25]	1.4343[20]
4835	Cyclopentane, 1-methyl-2-propyl-, *trans*- *trans*-1-Methyl-2-propylcyclopentane	C$_9$H$_{18}$ 126.24	932-44-5 -123	146.4	4-05-00-00140 0.7735[25]	1.4274[20]
4836	Cyclopentane, 1-methyl-3-propyl-, *cis*- *cis*-1-Methyl-3-propylcyclopentane	C$_9$H$_{18}$ 126.24	2443-04-1 	148	3-05-00-00125 0.776[25]	1.426[20]
4837	Cyclopentane, 1-methyl-3-propyl-, *trans*- *trans*-1-Methyl-3-propylcyclopentane	C$_9$H$_{18}$ 126.24	2443-03-0 	148	3-05-00-00125 0.776[25]	1.426[20]
4838	Cyclopentane, (2-methylpropyl)- Isobutylcyclopentane	C$_9$H$_{18}$ 126.24	3788-32-7 -115.2	148	4-05-00-00140 0.7769[25]	1.4298[20]
4839	Cyclopentane, nitro- Nitrocyclopentane	C$_5$H$_9$NO$_2$ 115.13	2562-38-1 	90-1[40]	4-05-00-00020 1.0776[23]	H$_2$O l; bz 3; chl 3 1.4538[20]
4840	Cyclopentane, (nitromethyl)- (Nitromethyl)cyclopentane	C$_6$H$_{11}$NO$_2$ 129.16	2625-31-2 	110[35]	1-05-00-00011 1.0713[20]	1.4587[20]
4841	Cyclopentane, nonyl-	C$_{14}$H$_{28}$ 196.38	2882-98-6 -29	262	4-05-00-00181 0.8081[20]	ace 4; bz 4; eth 4; EtOH 4 1.4467[20]
4842	Cyclopentane, octadecyl- Octadecane, 1-cyclopentyl-	C$_{23}$H$_{46}$ 322.62	62016-53-9 30	391	4-05-00-00196 0.8254[20]	ace 4; bz 4; eth 4; EtOH 4 1.4581[20]
4843	Cyclopentane, octyl- Octane, 1-cyclopentyl-	C$_{13}$H$_{26}$ 182.35	1795-20-6 -44	243	4-05-00-00176 0.8048[20]	1.4446[20]
4844	Cyclopentane, pentadecyl- Pentadecane, 1-cyclopentyl-	C$_{20}$H$_{40}$ 280.54	4669-01-6 17	355	4-05-00-00195 0.8213[20]	ace 4; bz 4; eth 4; EtOH 4 1.4554[20]
4845	Cyclopentane, pentatriacontyl- Pentatriacontylcyclopentane	C$_{40}$H$_{80}$ 561.08	61827-98-3 74	531	4-05-00-00205 0.8374[20]	1.4660[20]
4846	Cyclopentane, pentyl-	C$_{10}$H$_{20}$ 140.27	3741-00-2 -83	180	4-05-00-00160 0.7912[20]	ace 4; bz 4; eth 4; EtOH 4 1.4356[20]
4847	Cyclopentanepropanoic acid	C$_8$H$_{14}$O$_2$ 142.20	140-77-2 	130-2[12]	4-09-00-00046 0.996[25]	1.4570[20]
4848	Cyclopentane, 2-propenyl-	C$_8$H$_{14}$ 110.20	3524-75-2 -110.7	125	4-05-00-00272 0.793[25]	chl 3 1.4412[20]
4849	Cyclopentane, propyl- Propylcyclopentane	C$_8$H$_{16}$ 112.22	2040-96-2 -117.3	131	4-05-00-00125 0.7763[20]	H$_2$O 1; EtOH 5; eth 5; ace 5 1.4266[20]
4850	Cyclopentane, tetradecyl- Tetradecane, 1-cyclopentyl-	C$_{19}$H$_{38}$ 266.51	1795-22-8 9	341	3-05-00-00158 0.8196[20]	1.4543[20]
4851	Cyclopentane, 1,1,3,3-tetramethyl- 1,1,3,3-Tetramethylcyclopentane	C$_9$H$_{18}$ 126.24	50876-33-0 -88.4	118	4-05-00-00143 0.7469[25]	1.4125[20]
4852	Cyclopentanethiol Cyclopentyl mercaptan	C$_5$H$_{10}$S 102.20	1679-07-8 	132.1	4-06-00-00015 0.9550[20]	
4853	Cyclopentane, triacontyl- Triacontylcyclopentane	C$_{35}$H$_{70}$ 490.94	61827-93-8 65	498	4-05-00-00204 0.8350[20]	1.4644[20]
4854	Cyclopentane, tridecyl- Tridecane, 1-cyclopentyl-	C$_{18}$H$_{36}$ 252.48	6006-34-4 5	327	4-05-00-00192 0.8178[20]	ace 4; bz 4; eth 4; EtOH 4 1.4531[20]
4855	Cyclopentane, 1,1,2-trimethyl- Camphocean	C$_8$H$_{16}$ 112.22	4259-00-1 -21.6	114; 53[100]	4-05-00-00127 0.7660[20]	1.4199[20]

No.	Name Synonym	Mol. Form. Mol. Wt.	CAS RN mp/°C	Merck No. bp/°C	Beil. Ref. den/g cm⁻³	Solubility n_D
4856	Cyclopentane, 1,1,3-trimethyl- 1,1,3-Trimethylcyclopentane	C₈H₁₆ 112.22	4516-69-2 -142.4	104.9	3-05-00-00111 0.7439[25]	1.4112[20]
4857	Cyclopentane, 1,2,3-trimethyl-, (1α,2β,3α)- Cyclopentane, 1,2,3-trimethyl-, trans-1,2,cis-1,3-	C₈H₁₆ 112.22	19374-46-0 -112.7	110.4	4-05-00-00128 0.7492[25]	1.4138[20]
4858	Cyclopentane, 1,2,3-trimethyl-, (1α,2α,3α)- Cyclopentane, 1,2,3-trimethyl-, cis-1,2,cis-1,3-	C₈H₁₆ 112.22	2613-69-6 -116.4	123	4-05-00-00128 0.7751[25]	1.4262[20]
4859	Cyclopentane, 1,2,3-trimethyl-, (1α,2α,3β)-	C₈H₁₆ 112.22	15890-40-1 -112	117.5	4-05-00-00128 0.7661[25]	1.4218[20]
4860	Cyclopentane, 1,2,4-trimethyl-, (1α,2β,4α)-	C₈H₁₆ 112.22	16883-48-0 -130.8	109.3	4-05-00-00129 0.7430[25]	1.4106[20]
4861	Cyclopentane, 1,2,4-trimethyl-, (1α,2α,4α)- Cyclopentane, 1,2,4-trimethyl-, cis-1,2,cis-1,4-	C₈H₁₆ 112.22	2613-72-1 -132.3	116.8	4-05-00-00129 0.758[25]	1.4186[20]
4862	Cyclopentane, 1,2,4-trimethyl-, (1α,2α,4β)-	C₈H₁₆ 112.22	4850-28-6 -132.6	116.7	4-05-00-00129 0.7592[25]	1.4186[20]
4863	Cyclopentane, undecyl-	C₁₆H₃₂ 224.43	6785-23-5 -10	296	4-05-00-00189 0.8135[20]	ace 4; bz 4; eth 4; EtOH 4 1.4503[20]
4864	Cyclopentanol Cyclopentyl alcohol	C₅H₁₀O 86.13	96-41-3 -17.5	2747 140.4	4-06-00-00005 0.9488[20]	H₂O 2; EtOH 3; eth 3; ace 3 1.4530[20]
4865	Cyclopentanol, acetate Cyclopentyl acetate	C₇H₁₂O₂ 128.17	933-05-1	51-2[12]	4-06-00-00007 0.9522[16]	
4866	Cyclopentanol, 1-ethynyl- 1-Ethynyl-1-cyclopentanol	C₇H₁₀O 110.16	17356-19-3 27	157.5	4-06-00-00339 0.962[25]	1.4751[20]
4867	Cyclopentanol, 1-methyl-	C₆H₁₂O 100.16	1462-03-9 36	136; 53[30]	4-06-00-00086 0.9044[23]	1.4429[23]
4868	Cyclopentanol, 2-methyl-, cis- cis-2-Methylcyclopentanol	C₆H₁₂O 100.16	25144-05-2	148.5	3-06-00-00055 0.9379[16]	1.4504[16]
4869	Cyclopentanol, 2-methyl-, trans- trans-2-Methylcyclopentanol	C₆H₁₂O 100.16	25144-04-1	150.5	4-06-00-00091 0.9248[16]	1.4499[16]
4870	Cyclopentanol, 3-methyl- 3-Methylcyclopentanol	C₆H₁₂O 100.16	18729-48-1	148.5	2-06-00-00015 0.9158[16]	eth 4; EtOH 4 1.4487[16]
4871	Cyclopentanol, 1-phenyl-	C₁₁H₁₄O 162.23	10487-96-4	132-3[18]	4-06-00-03884 1.0530[20]	1.5479[20]
4872	Cyclopentanol, 1-propyl-	C₈H₁₆O 128.21	1604-02-0 -37.5	173.5	3-06-00-00101 0.9040[25]	1.4502[25]
4873	Cyclopentanol, 2-propyl-, cis- cis-2-Propylcyclopentanol	C₈H₁₆O 128.21	25172-42-3	79-80[12]	3-06-00-00101 0.9163[9]	1.4600[9]
4874	Cyclopentanol, 2-propyl-, trans- trans-2-Propylcyclopentanol	C₈H₁₆O 128.21	25172-43-4	78-9[10]	3-06-00-00101 0.9016[9]	1.4565[9]
4875	Cyclopentanol, 1,2,2-trimethyl- 1,2,2-Trimethylcyclopentanol	C₈H₁₆O 128.21	1121-95-5	156; 60[15]	1-06-00-00014 0.9101[20]	1.4513[20]
4876	Cyclopentanol, 1,2,4-trimethyl- 1,2,4-Trimethylcyclopentanol	C₈H₁₆O 128.21	34103-97-4	158	1-06-00-00014 0.8850[21]	1.4424[21]
4877	Cyclopentanol, 1,2,5-trimethyl- 1,2,5-Trimethylcyclopentanol	C₈H₁₆O 128.21	33840-38-9	56-60[8]	1-06-00-00014 0.9121[15]	1.4554[16]
4878	Cyclopentanone Adipic ketone	C₅H₈O 84.12	120-92-3 -51.3	2748 130.5	4-07-00-00005 0.9487[20]	H₂O 1; EtOH 3; eth 5; ace 3 1.4366[20]
4879	Cyclopentanone, 2-acetyl- 2-Acetylcyclopentanone	C₇H₁₀O₂ 126.16	1670-46-8	72-5[20]	4-07-00-01993 1.0431[25]	1.4906[20]
4880	Cyclopentanone, 2-chloro-	C₅H₇ClO 118.56	694-28-0	72-4[12]	4-07-00-00012 1.185[25]	1.4750[20]
4881	Cyclopentanone, 2-cyclopentylidene-	C₁₀H₁₄O 150.22	825-25-2	135[25]	4-07-00-00320 1.0179[18]	1.5215[18]
4882	Cyclopentanone, 2,5-dimethyl- 2,5-Dimethylcyclopentanone	C₇H₁₂O 112.17	4041-09-2	146.5	4-07-00-00048 0.882[25]	1.4310[20]
4883	Cyclopentanone, 2-ethyl-5-methyl- 2-Ethyl-5-methylcyclopentanone	C₈H₁₄O 126.20	51686-60-3	164.5	3-06-00-02552 0.900[12]	1.4360[22]
4884	Cyclopentanone, 4-isopropyl-2-methyl- Cyclopentanone, 3-isopropyl-5-methyl	C₉H₁₆O 140.23	90645-61-7	191.5	2-07-00-00034 0.8862[20]	1.4413[19]
4885	Cyclopentanone, 2-methyl-	C₆H₁₀O 98.14	1120-72-5 -75	139.5	4-07-00-00036 0.9139[20]	H₂O 3; EtOH 4; eth 4; ace 4 1.4364[20]
4886	Cyclopentanone, 3-methyl-, (±)- (±)-3-Methylcyclopentanone	C₆H₁₀O 98.14	6195-92-2 -58.4	144	4-07-00-00037 0.913[22]	H₂O 3; EtOH 4; eth 4; ace 4 1.4329[20]
4887	Cyclopentanone, 2-(1-methylethyl)- 2-Isopropylcyclopentanone	C₈H₁₄O 126.20	14845-55-7	175	4-07-00-00060 0.9105[20]	1.4425[15]
4888	Cyclopentanone, 3-methyl-2-(3-methylbutyl)- Cyclopentanone, 2-isopentyl-3-methyl	C₁₁H₂₀O 168.28	52033-97-3	98-9[8]	3-07-00-00181 0.8938[20]	1.4537[20]
4889	Cyclopentanone, 2-methyl-2-(1-methylethyl)- Cyclopentanone, 2-isopropyl-2-methyl-	C₉H₁₆O 140.23	32116-65-7	97.5[45]	3-07-00-00130 0.9067[16]	1.4495[16]
4890	Cyclopentanone, 2-methyl-5-(1-methylethyl)- Cyclopentanone, 2-isopropyl-5-methyl	C₉H₁₆O 140.23	6784-18-5	182	3-07-00-00130 0.889[20]	1.4402[20]
4891	Cyclopentanone, 4-methyl-2-(1-methylethyl)- Cyclopentanone, 2-isopropyl-4-methyl	C₉H₁₆O 140.23	69770-98-5	183	1-07-00-00026 0.8850[20]	1.4392[20]
4892	Cyclopentanone, oxime	C₅H₉NO 99.13	1192-28-5 57.8	196	4-07-00-00008	H₂O 4; bz 4
4893	Cyclopentanone, 2-(phenylmethyl)- 2-Benzylcyclopentanone	C₁₂H₁₄O 174.24	2867-63-2	152[18]; 84[0.3]	4-07-00-01043 1.038[20]	1.534[13]
4894	Cyclopentanone, 2-propyl- 2-Propylcyclopentanone	C₈H₁₄O 126.20	1193-70-0 -68.3	183	4-07-00-00060 0.9017[20]	1.4429[20]

No.	Name / Synonym	Mol. Form. / Mol. Wt.	CAS RN / mp/°C	Merck No. / bp/°C	Beil. Ref. / den/g cm^{-3}	Solubility / n_D
4895	Cyclopentanone, 3-propyl- / 3-Propylcyclopentanone	$C_8H_{14}O$ / 126.20	82322-93-8	190.5	3-07-00-00099 / 0.9041[20]	1.4456[12]
4896	Cyclopentanone, 2,2,4-trimethyl- / 2,2,4-Trimethylcyclopentanone	$C_8H_{14}O$ / 126.20	28056-54-4 / -40.6	158	4-07-00-00061 / 0.877[25]	1.4300[20]
4897	Cyclopentanone, 2,2,5-trimethyl- / 2,2,5-Trimethylcyclopentanone	$C_8H_{14}O$ / 126.20	4573-09-5	152	3-07-00-00107 / 0.8781[20]	1.4306[20]
4898	Cyclopentanone, 2,3,5-trimethyl- / 2,3,5-Trimethylcyclopentanol	$C_8H_{14}O$ / 126.20	90112-26-8	158	1-07-00-00021 / 0.8778[19]	1.4316[19]
4899	Cyclopentanone, 2,4,4-trimethyl- / 2,4,4-Trimethylcyclopentanone	$C_8H_{14}O$ / 126.20	4694-12-6 / -25.6	160.5	4-07-00-00061 / 0.8785[18]	1.433[18]
4900	1H-Cyclopentapyrimidine-2,4(3H,5H)-dione, 3-cyclohexyl-6,7-dihydro- / Lenacil	$C_{13}H_{18}N_2O_2$ / 234.30	2164-08-1 / 290	5318	5-24-07-00375 / 1.32[25]	py 4
4901	Cyclopentasiloxane, decamethyl- / Decamethylcyclopentasiloxane	$C_{10}H_{30}O_5Si_5$ / 370.77	541-02-6 / -38	2841 / 210	4-04-00-04128 / 0.9593[20]	H_2O 1 / 1.3982[20]
4902	Cyclopentasiloxane, 2,4,6,8,10-pentamethyl-	$C_5H_{20}O_5Si_5$ / 300.64	6166-86-5 / -108	169	4-04-00-04099 / 0.9985[20]	1.3912[20]
4903	Cyclopentene	C_5H_8 / 68.12	142-29-0 / -135.1	44.2	4-05-00-00209 / 0.7720[20]	H_2O 1; EtOH 3; eth 3; bz 3 / 1.4225[20]
4904	3-Cyclopentene-1-acetic acid	$C_7H_{10}O_2$ / 126.16	767-03-3 / 19	93-4[2.5]	1.047[25]	1.4676[20]
4905	1-Cyclopentene-1-acetonitrile	C_7H_9N / 107.16	22734-04-9	124[100]	3-09-00-00151 / 0.951[25]	1.4670[20]
4906	Cyclopentene, 1-butyl- / 1-Butyl-1-cyclopentene	C_9H_{16} / 124.23	2423-01-0 / -88.4	157	4-05-00-00289 / 0.8035[25]	1.4486[20]
4907	1-Cyclopentenecarbonitrile / 1-Cyanocyclopentene	C_6H_7N / 93.13	3047-38-9			
4908	1-Cyclopentene-1-carboxaldehyde	C_6H_8O / 96.13	6140-65-4 / -32	146	4-07-00-00127 / 0.970[21]	1.4872[17]
4909	1-Cyclopentene-1-carboxylic acid	$C_6H_8O_2$ / 112.13	1560-11-8 / 123.5	210	4-09-00-00109	H_2O 3; EtOH 3; eth 3; peth 3
4910	1-Cyclopentene-1-carboxylic acid, 2,3,3-trimethyl-	$C_9H_{14}O_2$ / 154.21	5587-63-3 / 135	255.5	4-09-00-00150	eth 4; EtOH 4
4911	1-Cyclopentene-1-carboxylic acid, 2,3,3-trimethyl-, methyl ester	$C_{10}H_{16}O_2$ / 168.24	1460-88-4	203.5	0-09-00-00058 / 0.9656[20]	1.4571[20]
4912	2-Cyclopentene-1-carboxylic acid, 1,2,3-trimethyl- / Laurolenic acid	$C_9H_{14}O_2$ / 154.21	6894-69-5 / 6.5	192[100]	2-09-00-00040 / 1.0318[25]	1.4766
4913	3-Cyclopentene-1-carboxylic acid, 2,2,3-trimethyl- / α-Campholytic acid	$C_9H_{14}O_2$ / 154.21	6709-22-4 / 40.5	241.5; 162[45]	4-09-00-00150 / 1.0145[18]	1.4712[17]
4914	Cyclopentene, 3-chloro-	C_5H_7Cl / 102.56	96-40-2 / 25-31[30]		4-05-00-00212 / 1.0577[15]	eth 4; EtOH 4; chl 4 / 1.4708[26]
4915	1-Cyclopentene-1,2-dicarboxylic acid, diethyl ester / Diethyl 1-cyclopentene-1,2-dicarboxylate	$C_{11}H_{16}O_4$ / 212.25	70202-92-5	140[16]	3-09-00-03938 / 1.0805[20]	1.4652[20]
4916	1-Cyclopentene-1,3-dicarboxylic acid, diethyl ester / Cyclopentene-1,3-dicarboxylic acid, diethyl ester	$C_{11}H_{16}O_4$ / 212.25	30689-41-9	168[21]	3-09-00-03939 / 1.1121[24]	1.4564
4917	Cyclopentene, 1,2-dichloro-3,3,4,4,5,5-hexafluoro-	$C_5Cl_2F_6$ / 244.95	706-79-6 / -105.8	90.7	4-05-00-00213 / 1.6546[20]	1.3676[20]
4918	Cyclopentene, 1,2-dimethyl-	C_7H_{12} / 96.17	765-47-9 / -90.4	105.8	4-05-00-00254 / 0.7928[25]	1.4448[20]
4919	Cyclopentene, 1,3-dimethyl- / 1,3-Dimethylcyclopentene	C_7H_{12} / 96.17	62184-82-1	92	4-05-00-00255 / 0.766[20]	1.428[20]
4920	Cyclopentene, 1,4-dimethyl- / 1,4-Dimethylcyclopentene	C_7H_{12} / 96.17	19550-48-2	93.2	4-05-00-00255 / 0.779[20]	1.4283[20]
4921	Cyclopentene, 1,5-dimethyl-	C_7H_{12} / 96.17	16491-15-9 / -118	99	4-05-00-00254 / 0.780[20]	1.4331[20]
4922	Cyclopentene, 4,4-dimethyl-	C_7H_{12} / 96.17	19037-72-0	88	4-05-00-00254 / 0.766[25]	1.423[20]
4923	Cyclopentene, 1,5-dimethyl-4-(1-methylethyl)- / Cyclopentene, 2,3-dimethyl-4-isopropyl	$C_{10}H_{18}$ / 138.25	6912-05-6	163	1-05-00-00046 / 0.8085[22]	1.4503[22]
4924	Cyclopentene, 1-ethyl-	C_7H_{12} / 96.17	2146-38-5 / -118.5	106.3	4-05-00-00252 / 0.7936[25]	1.4412[20]
4925	Cyclopentene, 3-ethyl-	C_7H_{12} / 96.17	694-35-9	97.8	4-05-00-00252 / 0.7784[25]	1.4315[20]
4926	Cyclopentene, 4-ethyl- / 4-Ethylcyclopentene	C_7H_{12} / 96.17	3742-38-9	98.3	4-05-00-00253 / 0.778[25]	1.431[20]
4927	Cyclopentene, 4-isopropyl-1-methyl- / 4-Isopropyl-1-methylcyclopentene	C_9H_{16} / 124.23	90769-70-3	142.5	1-05-00-00041 / 0.7945[21]	1.4403[21]
4928	Cyclopentene, 1-methyl-	C_6H_{10} / 82.15	693-89-0 / -126.5	75.5	4-05-00-00239 / 0.7748[25]	1.4322[20]
4929	Cyclopentene, 3-methyl- / 3-Methylcyclopentene	C_6H_{10} / 82.15	1120-62-3	64.9	4-05-00-00240 / 0.7572[25]	1.4216[20]
4930	Cyclopentene, 4-methyl-	C_6H_{10} / 82.15	1759-81-5 / -160.8	65.7	4-05-00-00240 / 0.7634[25]	1.4209[20]

No.	Name Synonym	Mol. Form. Mol. Wt.	CAS RN mp/°C	Merck No. bp/°C	Beil. Ref. den/g cm^{-3}	Solubility n_D
4931	Cyclopentene, 3-methyl-1-(1-methylethyl)- Pulegene	C_9H_{16} 124.23	51115-02-7	138.5	1-05-00-00040 0.7912[22]	1.4380[23]
4932	Cyclopentene, octachloro- Perchlorocyclopentene	C_5Cl_8 343.68	706-78-5 40	283	4-05-00-00214 1.8200[50]	H_2O 1; EtOH 4 1.5660[50]
4933	1-Cyclopentene-1-pentanoic acid, δ-hydroxy-3,5-bis(1-methylpropyl)-β-oxo-	$C_{18}H_{30}O_4$ 310.43	53109-18-5 183		3-10-00-04194 1.269[20]	eth 4; EtOH 4
4934	Cyclopentene, 1-pentyl-	$C_{10}H_{18}$ 138.25	4291-98-9 -83	179	4-05-00-00306 0.8085[25]	1.4516[20]
4935	Cyclopentene, 1-propyl- 1-Propyl-1-cyclopentene	C_8H_{14} 110.20	3074-61-1 -100.3	131.2	4-05-00-00271 0.7978[25]	1.4452[20]
4936	2-Cyclopentene-1-tridecanoic acid, (S)- Chaulmoogric acid	$C_{18}H_{32}O_2$ 280.45	29106-32-9 68.5	2037 247-8[20]	4-09-00-00216	eth 4; chl 4
4937	Cyclopentene, 1,2,3-trimethyl-	C_8H_{14} 110.20	473-91-6	121.6	3-05-00-00219 0.8039[15]	1.4464[16]
4938	Cyclopentene, 1,5,5-trimethyl- 1,5,5-Trimethylcyclopentene	C_8H_{14} 110.20	62184-83-2	109	3-05-00-00219 0.7824[20]	1.4324[20]
4939	4-Cyclopentene-1,2,3-trione, 4,5-dihydroxy- Croconic acid	$C_5H_2O_5$ 142.07	488-86-8 150	dec	4-08-00-03342	H_2O 4; EtOH 3
4940	2-Cyclopentene-1-undecanoic acid, (R)- Hydnocarpic acid	$C_{16}H_{28}O_2$ 252.40	459-67-6 60.5	4679	4-09-00-00215	EtOH 4; chl 4; peth 4
4941	2-Cyclopenten-1-one Cyclopenten-3-one	C_5H_6O 82.10	930-30-3	136; 40[12]	4-07-00-00122 0.989[15]	eth 4; EtOH 4 1.4629[15]
4942	2-Cyclopenten-1-one, 2,3-dihydroxy- Reductic acid	$C_5H_6O_3$ 114.10	80-72-8 212	8134	4-08-00-01714	H_2O 3; EtOH 3; eth 2; ace 2
4943	2-Cyclopenten-1-one, 2,4-dimethyl-	$C_7H_{10}O$ 110.16	23048-13-7	165	4-07-00-00138 0.9423[20]	1.4655[20]
4944	2-Cyclopenten-1-one, 2-methyl-	C_6H_8O 96.13	1120-73-6	157	4-07-00-00127 0.9808[16]	1.4762[15]
4945	2-Cyclopenten-1-one, 3-methyl- 3-Methyl-2-cyclopentenone	C_6H_8O 96.13	2758-18-1	157.5	4-07-00-00126 0.9712[20]	1.4714[20]
4946	2-Cyclopenten-1-one, 5-methyl- Cyclopenten-3-one, 4-methyl	C_6H_8O 96.13	14963-40-7	140	3-07-00-00229 0.942[18]	1.4460[18]
4947	2-Cyclopenten-1-one, 3-(1-methylethyl)-	$C_8H_{12}O$ 124.18	1619-28-9	214	2-07-00-00061 0.9378[20]	1.4788[20]
4948	2-Cyclopenten-1-one, 5-methyl-2-(1-methylethyl)-	$C_9H_{14}O$ 138.21	5587-79-1	189	3-07-00-00291 0.9143[20]	1.4660[20]
4949	2-Cyclopenten-1-one, 3-methyl-2-(2-pentenyl)-, (Z)- Jasmone	$C_{11}H_{16}O$ 164.25	488-10-8	5141 258; 134[12]	4-07-00-00337 0.9437[22]	H_2O 2; EtOH 3; eth 3; ctc 3 1.4979[22]
4950	2-Cyclopenten-1-one, 3-methyl-2-pentyl-	$C_{11}H_{18}O$ 166.26	1128-08-1	143[22]; 116[12]	4-07-00-00230 0.9165[18]	1.4767[20]
4951	2-Cyclopenten-1-one, 3-phenyl-	$C_{11}H_{10}O$ 158.20	3810-26-2 -23	234.2	4-07-00-01180 0.9711[20]	EtOH 3; eth 2; ace 3; chl 3 1.5440[20]
4952	Cycloprate Cyclopropanecarboxylic acid, hexadecyl ester	$C_{20}H_{38}O_2$ 310.52	54460-46-7	154[0.05]		
4953	Cyclopropanamine Cyclopropylamine	C_3H_7N 57.10	765-30-0 -35.4	50.5	4-12-00-00003 0.8240[20]	H_2O 5; EtOH 3; eth 3; chl 3 1.4210[20]
4954	Cyclopropa[d]naphthalene, 1,1a,4,4a,5,6,7,8-octahydro-2,4a,8,8-tetramethyl-, [1aS-(1aα,4aβ,8aR*)]- Thujopsene	$C_{15}H_{24}$ 204.36	470-40-6	9327	4-05-00-01189	
4955	Cyclopropane Trimethylene	C_3H_6 42.08	75-19-4 -127.4	2755 -32.8	4-05-00-00003 *0.617[25]	H_2O 3; EtOH 4; eth 4; bz 3 1.3799[-42]
4956	Cyclopropanecarbonitrile Cyclopropyl cyanide	C_4H_5N 67.09	5500-21-0	135.1	4-09-00-00005 0.8946[20]	eth 3; ctc 2; hx 3 1.4229[20]
4957	Cyclopropanecarbonyl chloride, 2-phenyl-, trans- trans-2-Phenylcyclopropanecarbonyl chloride	$C_{10}H_9ClO$ 180.63	939-87-7	108-10[2]	3-09-00-02774 1.163[25]	1.5560[20]
4958	Cyclopropanecarboxylic acid	$C_4H_6O_2$ 86.09	1759-53-1 18.5	183	4-09-00-00003 1.0885[20]	H_2O 3; EtOH 3; eth 3; ctc 2 1.4390[20]
4959	Cyclopropanecarboxylic acid, 2,2-dimethyl-3-(2-methyl-1-propenyl)-, (1R-cis)-	$C_{10}H_{16}O_2$ 168.24	26771-11-9 41	135[12]; 95[0.1]	4-09-00-00168	eth 4; EtOH 4; chl 4
4960	Cyclopropanecarboxylic acid, 2,2-dimethyl-3-(2-methyl-1-propenyl)-, 2-methyl-4-oxo-3-(2,4-pentadienyl)-2-cyclopenten-1-yl ester, [1R-[1α[S*(Z)],3β]]- Pyrethrin 1	$C_{21}H_{28}O_3$ 328.45	121-21-1	7978 170[0.1] dec	3-09-00-00215 1.5192[18]	H_2O 1; EtOH 3; eth 3; ctc 3 1.5192[18]
4961	Cyclopropanecarboxylic acid, ethyl ester Ethyl cyclopropylcarboxylate	$C_6H_{10}O_2$ 114.14	4606-07-9	134	4-09-00-00003 0.9608[15]	1.4190[20]
4962	Cyclopropanecarboxylic acid, 3-(3-methoxy-2-methyl-3-oxo-1-propenyl)-2,2-dimethyl-, 2-methyl-4-oxo-3-(2,4-pentadienyl)-2-cyclopenten-1-yl ester [1R-[1α[S*(Z)],3β(E)]]- Pyrethrin 2	$C_{22}H_{28}O_5$ 372.46	121-29-9	7978 200[0.1] dec	3-09-00-03988	H_2O 1; EtOH 3; eth 3; ctc 3 1.5258[20]
4963	Cyclopropanecarboxylic acid, methyl ester Methyl cyclopropanecarboxylate	$C_5H_8O_2$ 100.12	2868-37-3	114.9	4-09-00-00003 0.9848[20]	ace 3; chl 3 1.4144[19]

No.	Name Synonym	Mol. Form. Mol. Wt.	CAS RN mp/°C	Merck No. bp/°C	Beil. Ref. den/g cm^{-3}	Solubility n_D
4964	Cyclopropanecarboxylic acid, 2-methyl- 2-Methylcyclopropanecarboxylic acid	$C_5H_8O_2$ 100.12	29555-02-0	191; 96.5[14]	2-09-00-00006 1.0267[20]	1.4441[20]
4965	Cyclopropanecarboxylic acid, 2,2-dimethyl-3-(2-methyl-1-propenyl)-, methyl ester, trans-(±)- Chrysanthemum monocarboxylic acid, methyl ester (DL, trans)	$C_{11}H_{18}O_2$ 182.26	15543-70-1	108[20]; 95[13]	4-09-00-00169 0.9274[25]	1.4614[25]
4966	Cyclopropane, (chloromethyl)-	C_4H_7Cl 90.55	5911-08-0 -90.9	88	4-05-00-00013 0.98[15]	1.4350[20]
4967	1,1-Cyclopropanedicarboxylic acid, diethyl ester	$C_9H_{14}O_4$ 186.21	1559-02-0	216; 100[12]	4-09-00-02786 1.0535[25]	EtOH 4; eth 4 1.4345[18]
4968	1,2-Cyclopropanedicarboxylic acid, diethyl ester, cis- Cyclopropane-1,2-dicarboxylic acid, diethyl ester (cis)	$C_9H_{14}O_4$ 186.21	710-43-0		2-09-00-00513 1.062[12]	eth 4; EtOH 4 1.4450[20]
4969	1,2-Cyclopropanedicarboxylic acid, dimethyl ester Cyclopropane-1,2-dicarboxylic acid, dimethyl ester	$C_7H_{10}O_4$ 158.15	702-28-3	219; 110[3]	1.1584[16]	eth 4; EtOH 4 1.4472[14]
4970	1,2-Cyclopropanedicarboxylic acid, trans-(±)- Cyclopropane-1,2-dicarboxylic acid (trans, DL)	$C_5H_6O_4$ 130.10	58616-95-8 175	210[30]	4-09-00-02786 1.4600[20]	H_2O 4; eth 4; EtOH 4
4971	Cyclopropane, 1,1-dimethyl- gem-Dimethylcyclopropane	C_5H_{10} 70.13	1630-94-0 -109	20.6	4-05-00-00024 0.6604[20]	H_2O 1; EtOH 3; eth 4; sulf 4 1.3668[20]
4972	Cyclopropane, 1,2-dimethyl-, cis- cis-1,2-Dimethylcyclopropane	C_5H_{10} 70.13	930-18-7 -140.9	37.0	4-05-00-00025 0.6889[25]	H_2O 1; EtOH 3; eth 4; ctc 2 1.3829[20]
4973	Cyclopropane, 1,2-dimethyl-, trans- trans-1,2-Dimethylcyclopropane	C_5H_{10} 70.13	2402-06-4 -149.6	28.2	4-05-00-00026 0.6648[25]	eth 4; EtOH 4 1.3713[20]
4974	Cyclopropane, ethyl- Ethylcyclopropane	C_5H_{10} 70.13	1191-96-4 -149.2	35.9	4-05-00-00023 0.6790[25]	1.3786[20]
4975	Cyclopropane, 1-ethyl-1-methyl- Cyclopropane, 1-methyl-1-ethyl	C_6H_{12} 84.16	53778-43-1 -130.2	56.8	4-05-00-00091 0.6968[25]	1.3887[20]
4976	Cyclopropane, 1-ethyl-2-methyl-, cis- cis-1-Methyl-2-ethylcyclopropane	C_6H_{12} 84.16	19781-68-1	58.7	3-05-00-00063 0.6935[20]	1.3846[20]
4977	Cyclopropane, 1-ethyl-2-methyl-, trans- trans-1-Methyl-2-ethylcyclopropane	C_6H_{12} 84.16	19781-69-2	59	3-05-00-00063 0.7146[20]	ctc 3 1.3953[20]
4978	Cyclopropanemethanol, 2,2-dimethyl-3-(2-methyl-1-propenyl)-	$C_{10}H_{18}O$ 154.25	5617-92-5	66-9[0.07]	4-06-00-00264 0.888[25]	1.4757[20]
4979	Cyclopropanemethanol, α-methyl-	$C_5H_{10}O$ 86.13	765-42-4 -32.1	123.5	4-06-00-00020 0.8805[20]	1.4316[20]
4980	Cyclopropane, methoxy- Cyclopropyl methyl ether	C_4H_8O 72.11	540-47-6 -119	2756 44.7	3-06-00-00003 0.8100[20]	H_2O 4; bz 4; eth 4; EtOH 4 1.3802[20]
4981	Cyclopropane, methyl- Methylcyclopropane	C_4H_8 56.11	594-11-6 -177.3	0.7	4-05-00-00013 0.6912[20]	eth 4; EtOH 4
4982	Cyclopropane, (1-methylethenyl)-	C_6H_{10} 82.15	4663-22-3	70	4-05-00-00243 0.7500[20]	1.4252[20]
4983	Cyclopropane, (1-methylethyl)-	C_6H_{12} 84.16	3638-35-5 -112.9	58.3	4-05-00-00090 0.6936[25]	1.3865[20]
4984	Cyclopropane, propyl- Propane, 1-cyclopropyl-	C_6H_{12} 84.16	2415-72-7	69.1	4-05-00-00090 0.7062[25]	1.3930[20]
4985	Cyclopropane, 1,1,2-trimethyl- 1,1,2-Trimethylcyclopropane	C_6H_{12} 84.16	4127-45-1 -138.2	54	4-05-00-00091 0.6897[25]	1.3864[20]
4986	Cyclopropane, 1,2,3-trimethyl-, (1α,2α,3α)-	C_6H_{12} 84.16	4806-58-0	66	3-05-00-00063 0.7130[25]	1.3970[20]
4987	Cyclopropane, 1,2,3-trimethyl-, (1α,2α,3β)-	C_6H_{12} 84.16	4806-59-1	59.7	3-05-00-00063 0.6929[25]	1.3873[20]
4988	Cyclopropanone	C_3H_4O 56.06	5009-27-8 stable only at low temp.			
4989	1H-Cycloprop[e]azulen-4-ol, decahydro-1,1,4,7-tetramethyl-, (1aR,4R,4aS,7R,7aS,7bS)- Ledol	$C_{15}H_{26}O$ 222.37	577-27-5 105	5313 292	0.9078[100]	ace 4; eth 4; EtOH 4 1.4667[110]
4990	Cyclopropene	C_3H_4 40.06	2781-85-3	-36 dec		
4991	Cyclotetrasiloxane, ethenylheptamethyl-	$C_9H_{24}O_4Si_4$ 308.63	3763-39-1	84[20]	0.9505[20]	ctc 3; CS_2 3 1.4034[20]
4992	Cyclotetrasiloxane, heptamethyl- Heptamethylcyclotetrasiloxane	$C_7H_{22}O_4Si_4$ 282.59	15721-05-8 -27	165	4-04-00-04125 0.9583[20]	H_2O I 1.3965[20]
4993	Cyclotetrasiloxane, octamethyl- Octamethylcyclotetrasiloxane	$C_8H_{24}O_4Si_4$ 296.62	556-67-2 17.5	6668 175.8	4-04-00-04125 0.9561[20]	H_2O I; ctc 3 1.3968[20]
4994	Cyclotetrasiloxane, octaphenyl-	$C_{48}H_{40}O_4Si_4$ 793.19	546-56-5 200.5	330[1]	4-16-00-01530	H_2O I; EtOH 2; bz 3; chl 3
4995	Cyclotetrasiloxane, 2,4,6,8-tetrabutyl-2,4,6,8-tetramethyl- Siloxane, butylmethyl(cyclic tetramer)	$C_{20}H_{48}O_4Si_4$ 464.94	14685-29-1	291	4-04-00-04173 0.9230[20]	1.4300[20]
4996	Cyclotetrasiloxane, 2,4,6,8-tetraethenyl-2,4,6,8-tetramethyl-	$C_{12}H_{24}O_4Si_4$ 344.66	2554-06-5 -43.5	224; 111[12]	4-04-00-04184 0.9875[20]	ctc 3; CS_2 3

No.	Name Synonym	Mol. Form. Mol. Wt.	CAS RN mp/°C	Merck No. bp/°C	Beil. Ref. den/g cm^{-3}	Solubility n_D
4997	Cyclotetrasiloxane, 2,4,6,8-tetraethyl-2,4,6,8-tetramethyl- Cyclotetrasiloxane, 2,4,6,8-tetramethyl-2,4,6,8-tetraethyl	$C_{12}H_{32}O_4Si_4$ 352.73	7623-01-0 -43.5	 245; 111[12]	4-04-00-04154 0.9600[20]	ctc 3; CS_2 3
4998	Cyclotetrasiloxane, 2,4,6,8-tetraethyl-2,4,6,8-tetraphenyl- Cyclotetrasiloxane, 1,3,5,7-tetraethyl-1,3,5,7-tetraphenyl	$C_{32}H_{40}O_4Si_4$ 601.01	18758-34-4 106	 212[0.1]	3-16-00-01211 1.1100[20]	1.5430[25]
4999	Cyclotetrasiloxane, 2,4,6,8-tetramethyl-	$C_4H_{16}O_4Si_4$ 240.51	2370-88-9 -65	 134.5	4-04-00-04099 0.9912[20]	H_2O I 1.3870[20]
5000	Cyclotetrasiloxane, 2,4,6,8-tetramethyl-2,4,6,8-tetraphenyl-	$C_{28}H_{32}O_4Si_4$ 544.90	77-63-4 99	 237[115]	3-16-00-01211 1.1183[20]	H_2O 1; ace 5; hp 5 1.5461[20]
5001	Cyclotrisilazane, 2,2,4,4,6,6-hexamethyl- 1,1,3,3,5,5-Hexamethylcyclotrisilazane	$C_6H_{21}N_3Si_3$ 219.51	1009-93-4 -10	 188	4-04-00-04114 0.9196[20]	1.448[20]
5002	Cyclotrisiloxane, hexamethyl- Dimethylsiloxane cyclic trimer	$C_6H_{18}O_3Si_3$ 222.46	541-05-9 64.5	 134	4-04-00-04123 1.1200[20]	H_2O 1
5003	Cyclotrisiloxane, 2,4,6-triethyl-2,4,6-trimethyl- 2,4,6-Triethyl-2,4,6-trimethylcyclotrisiloxane	$C_9H_{24}O_3Si_3$ 264.54	15901-49-2 -3	 199	4-04-00-04154	ctc 3; CS_2 3
5004	Cyclotrisiloxane, 2,4,6-triethyl-2,4,6-triphenyl- 2,4,6-Triethyl-2,4,6-triphenylcyclotrisiloxane	$C_{24}H_{30}O_3Si_3$ 450.76	546-33-8 177.5	 166[0.025]	3-16-00-01211 1.0952[25]	H_2O 1 1.5402[25]
5005	Cyclotrisiloxane, 2,4,6-trimethyl-2,4,6-triphenyl-	$C_{21}H_{24}O_3Si_3$ 408.68	546-45-2 100	 190[1.5]	3-16-00-01211 1.1062[20]	1.5397[20]
5006	1,4,8-Cycloundecatriene, 2,6,6,9-tetramethyl-, (E,E,E)- Humulene	$C_{15}H_{24}$ 204.36	6753-98-6 123[10]	4672 	4-05-00-01171 0.8905[20]	1.5038[20]
5007	7,20-Cycloveatchane-1,12,15-triol, 21-ethyl-4-methyl-16-methylene-, (1α,12α,15β)- Luciculine	$C_{22}H_{35}NO_3$ 361.52	5008-52-6 149	6286 165[0.02]	4-21-00-02584	EtOH 4
5008	Cyfluthrin Cyano(4-fluoro-3-phenoxyphenyl)methyl 3-(2,2-dichloroethenyl)-2,2-dimethylcyclopropanecarboxylate	$C_{22}H_{18}Cl_2FNO_3$ 434.29	68359-37-5 60	2764		
5009	Cygon Phosphorodithioic acid, O,O-dimethyl S-[2-(methylamino)-2-oxoethyl] ester	$C_5H_{12}NO_3PS_2$ 229.26	60-51-5 52	3209 117[0.1]	1.277[65]	
5010	Cyhalothrin	$C_{23}H_{19}ClF_3NO_3$ 449.86	91465-08-6 49.2			
5011	Cyhexatin Stannane, tricyclohexylhydroxy-	$C_{18}H_{34}OSn$ 385.18	13121-70-5 196	2767		
5012	Cypermethrin Cyclopropanecarboxylic acid, 3-(2,2-dichloroethenyl)-2,2-dimethyl-, cyano(3-phenoxy-phenyl)methyl ester	$C_{22}H_{19}Cl_2NO_3$ 416.30	52315-07-8 70	2775	1.25[20]	
5013	Cyprazine 1,3,5-Triazine-2,4-diamine, 6-chloro-N-cyclopropyl-N'-(1-methylethyl)-	$C_9H_{14}ClN_5$ 227.70	22936-86-3 167			
5014	L-Cysteine Propanoic acid, 2-amino-3-mercapto-, (R)-	$C_3H_7NO_2S$ 121.16	52-90-4 240 dec	2787	4-04-00-03144	H_2O 4; ace 4; EtOH 4
5015	L-Cysteine, N-acetyl- Acetylcysteine	$C_5H_9NO_3S$ 163.20	616-91-1 109.5	82	4-04-00-03160	
5016	L-Cysteine, S-(2-amino-2-carboxyethyl)-, (R)- L-Lanthionine	$C_6H_{12}N_2O_4S$ 208.24	922-55-4 293-5 dec	5234	3-04-00-01593	
5017	L-Cysteine, S-(carboxymethyl)- Carbocysteine	$C_5H_9NO_4S$ 179.20	638-23-3 206	1809	4-04-00-03150	
5018	L-Cysteine, S,S'-methylenebis- Djenkolic acid	$C_7H_{14}N_2O_4S_2$ 254.33	498-59-9 300-50 dec	3393	3-04-00-01591	
5019	L-Cysteine, methyl ester, hydrochloride Methylcysteine hydrochloride	$C_4H_{10}ClNO_2S$ 171.65	18598-63-5 140.5	5668	3-04-00-01601	
5020	L-Cystine 3,3'-Dithiobis(2-aminopropanoic acid)	$C_6H_{12}N_2O_4S_2$ 240.30	56-89-3 260-1 dec	2788	4-04-00-03155 1.677[25]	H_2O 2; EtOH 1; eth 1; bz 1
5021	Cytidine 4-Amino-1-β-D-ribofuranosyl-2(1H)-pyrimidinone	$C_9H_{13}N_3O_5$ 243.22	65-46-3 230-1 dec	2792	4-25-00-03667	H_2O 4; EtOH 2
5022	3'-Cytidylic acid Cytidine 3'-monophosphate	$C_9H_{14}N_3O_8P$ 323.20	84-52-6 233 dec	2794	4-25-00-03669	H_2O 3; EtOH 3
5023	1,3-Decadiene 1-Hexyl-1,3-butadiene	$C_{10}H_{18}$ 138.25	2051-25-4 169		4-01-00-01056 0.752[30]	bz 4
5024	1,9-Decadiene α,ω-Decadiene	$C_{10}H_{18}$ 138.25	1647-16-1 167		4-01-00-01056 0.75[25]	1.4325[20]
5025	2,4-Decadienoic acid, (E,E)- 2,4-Decadienoic acid (trans-2, trans-4)	$C_{10}H_{16}O_2$ 168.24	30361-33-2 49.5	 121[0.3]	4-02-00-01730	1.5058[31]
5026	2,4-Decadienoic acid, methyl ester, (E,E)-	$C_{11}H_{18}O_2$ 182.26	7328-33-8 70[0.2]		4-02-00-01731 0.982[22]	1.4918[22]
5027	2,4-Decadienoic acid, methyl ester, (E,Z)-	$C_{11}H_{18}O_2$ 182.26	4493-42-9 71[0.15]		4-02-00-01731 0.9128[22]	1.4874[22]

No.	Name / Synonym	Mol. Form. / Mol. Wt.	CAS RN / mp/°C	Merck No. / bp/°C	Beil. Ref. / den/g cm^{-3}	Solubility / n_D
5028	2,4-Decadienoic acid, methyl ester, (Z,E)- 2,4-Decadienoic acid, methyl ester (cis-2, trans-4)	$C_{11}H_{18}O_2$ 182.26	108965-84-0	76-80[0.5]	4-02-00-01731 0.9131[23]	1.4876[23]
5029	2,4-Decadienoic acid, methyl ester, (Z,Z)- 2,4-Decadienoic acid, methyl ester (cis-2, cis-4)	$C_{11}H_{18}O_2$ 182.26	108965-86-2	78[0.6]	4-02-00-01731 0.9095[23]	1.4830[23]
5030	Decanal Capraldehyde	$C_{10}H_{20}O$ 156.27	112-31-2 -5	208.5	4-01-00-03366 0.830[15]	H_2O 1; EtOH 3; eth 3; ace 3 1.4287[20]
5031	Decanamide Capramide	$C_{10}H_{21}NO$ 171.28	2319-29-1 108		4-02-00-01050 0.822[130]	ace 4; eth 4; EtOH 4 1.4621[110]
5032	1-Decanamine	$C_{10}H_{23}N$ 157.30	2016-57-1 17	220.5	4-04-00-00783 0.7936[20]	H_2O 2; EtOH 5; eth 5; ace 5 1.4369[20]
5033	Decane n-Decane	$C_{10}H_{22}$ 142.28	124-18-5 -29.7	174.1	4-01-00-00464 0.7300[20]	H_2O 1; EtOH 5; eth 3; ctc 2 1.4102[20]
5034	Decane, 1-bromo-	$C_{10}H_{21}Br$ 221.18	112-29-8 -29.2	240.6	4-01-00-00470 1.0702[20]	H_2O 1; eth 4; ctc 3; chl 4 1.4557[20]
5035	Decane, 1-bromo-10-fluoro- 1-Bromo-10-fluorodecane	$C_{10}H_{20}BrF$ 239.17	334-61-2	131-2[11]	4-01-00-00471 1.152[20]	eth 4; EtOH 4 1.4512[25]
5036	Decane, 1-chloro-	$C_{10}H_{21}Cl$ 176.73	1002-69-3 -31.3	223.4	4-01-00-00469 0.8705[20]	H_2O 1; eth 4; ctc 3; chl 4 1.4379[20]
5037	Decane, 1-chloro-10-fluoro- 1-Chloro-10-fluorodecane	$C_{10}H_{20}ClF$ 194.72	334-62-3	115[9]	4-01-00-00469 0.957[20]	eth 4; EtOH 4 1.4333[25]
5038	Decane, 1,10-dibromo- Decamethylene dibromide	$C_{10}H_{20}Br_2$ 300.08	4101-68-2 28	161[9]; 128[4]	4-01-00-00471 1.335[30]	H_2O 1; EtOH 2; eth 3 1.4927[25]
5039	Decane, 5,6-dibromo- 5,6-Dibromodecane	$C_{10}H_{20}Br_2$ 300.08	77928-86-0	119[9]	4-01-00-00472 1.3484[20]	1.4912[20]
5040	Decane, 1,10-dichloro- Decamethylene dichloride	$C_{10}H_{20}Cl_2$ 211.17	2162-98-3 15.6	167-8[28]	4-01-00-00469 0.9945[25]	1.4586[25]
5041	Decanedinitrile	$C_{10}H_{16}N_2$ 164.25	1871-96-1	204[16]	4-02-00-02089 0.9313[20]	H_2O 1; chl 3 1.4474[20]
5042	Decanedioic acid Sebacic acid	$C_{10}H_{18}O_4$ 202.25	111-20-6 130.8	8369 295[100]; 232[10]	4-02-00-02078 1.2705[20]	H_2O 2; EtOH 3; eth 3; bz 1 1.422[133]
5043	Decanedioic acid, bis(2-ethylhexyl) ester Bis(2-ethylhexyl) sebacate	$C_{26}H_{50}O_4$ 426.68	122-62-3 -48	1263 256[5]	4-02-00-02083 0.912[25]	ace 4; bz 4; EtOH 4 1.451[25]
5044	Decanedioic acid, dibutyl ester Dibutyl sebacate	$C_{18}H_{34}O_4$ 314.47	109-43-3 -10	344.5	4-02-00-02081 0.9405[15]	H_2O 1; eth 3; ctc 3 1.4433[15]
5045	Decanedioic acid, diethyl ester	$C_{14}H_{26}O_4$ 258.36	110-40-7 5	305; 188[19]	4-02-00-02080 0.9646[20]	H_2O 2; EtOH 3; ace 3; bz 1 1.4306[20]
5046	Decanedioic acid, dimethyl ester	$C_{12}H_{22}O_4$ 230.30	106-79-6 38	175[20]; 144[5]	4-02-00-02080 0.9882[28]	H_2O 1; EtOH 3; eth 3; ace 3 1.4355[28]
5047	1,10-Decanediol Decamethylene glycol	$C_{10}H_{22}O_2$ 174.28	112-47-0 74	2842 192[20]	4-01-00-02613	H_2O 2; EtOH 4; eth 2; DMSO 3
5048	2,9-Decanediol	$C_{10}H_{22}O_2$ 174.28	14021-92-2 33	114[5]	3-01-00-02232	1.4505[21]
5049	Decanedioyl dichloride	$C_{10}H_{16}Cl_2O_2$ 239.14	111-19-3 -2.5	220[75]; 165[11]	4-02-00-02088 1.1212[20]	1.4684[18]
5050	Decane, docosafluoro- Perfluorodecane	$C_{10}F_{22}$ 538.07	307-45-9	144.2		
5051	Decane, 1-(ethenyloxy)- Decyl vinyl ether	$C_{12}H_{24}O$ 184.32	765-05-9 -41	101[10]	4-01-00-02057 0.812[20]	1.4346[20]
5052	Decane, 1-fluoro- Decyl fluoride	$C_{10}H_{21}F$ 160.28	334-56-5 -35	186.2	4-01-00-00468 0.8194[20]	eth 4 1.4085
5053	Decane, 1-iodo-	$C_{10}H_{21}I$ 268.18	2050-77-3 -16.3	132[15]	4-01-00-00472 1.2546[20]	H_2O 1; EtOH 3; eth 3; ctc 3 1.4858[20]
5054	Decane, 2-methyl- 2-Methyldecane	$C_{11}H_{24}$ 156.31	6975-98-0 -48.9	189.3	4-01-00-00491 0.7368[20]	1.4154[20]
5055	Decane, 3-methyl- 3-Methyldecane	$C_{11}H_{24}$ 156.31	13151-34-3 -92.9	188.1	4-01-00-00491 0.7422[20]	1.4177[20]
5056	Decane, 4-methyl- 4-Methyldecane	$C_{11}H_{24}$ 156.31	2847-72-5 -77.5	187		1.4352[20]
5057	Decanenitrile Caprinitrile	$C_{10}H_{19}N$ 153.27	1975-78-6 -17.9	243. 106[10]	4-02-00-01051 0.8199[20]	ace 4; eth 4; EtOH 4; chl 4 1.4296[20]
5058	1-Decanethiol Decyl mercaptan	$C_{10}H_{22}S$ 174.35	143-10-2 -26	240.6	4-01-00-01821 0.8443[20]	H_2O 1; EtOH 3; eth 3 1.4509[20]
5059	Decanoic acid Capric acid	$C_{10}H_{20}O_2$ 172.27	334-48-5 31.9	1759 268.7	4-02-00-01041 0.8858[40]	ace 4; bz 4; eth 4; EtOH 4 1.4288[40]
5060	Decanoic acid, anhydride	$C_{20}H_{38}O_3$ 326.52	2082-76-0 24.7		4-02-00-01049 0.8865[25]	eth 4; EtOH 4 1.400[25]
5061	Decanoic acid, 2-bromo-	$C_{10}H_{19}BrO_2$ 251.16	2623-95-2 4	140-1[2]	4-02-00-01054 1.1912[24]	eth 4 1.4595[24]
5062	Decanoic acid, decyl ester	$C_{20}H_{40}O_2$ 312.54	1654-86-0 9.7	219[15]	4-02-00-01045 0.8586[20]	eth 4 1.4423[20]
5063	Decanoic acid, ethyl ester Ethyl caprate	$C_{12}H_{24}O_2$ 200.32	110-38-3 -20	3734 241.5	4-02-00-01044 0.8650[20]	eth 4; EtOH 4; chl 4 1.4256[20]
5064	Decanoic acid, 10-fluoro- Capric acid, 10-fluoro	$C_{10}H_{19}FO_2$ 190.26	334-59-8 49	135-8[10]	4-02-00-01051	eth 4; EtOH 4; lig 4
5065	Decanoic acid, 2-hexyl- 2-Hexyldecanoic acid	$C_{16}H_{32}O_2$ 256.43	25354-97-6	4630 140-50[0.02]	4-02-00-01189	1.4432[24]

No.	Name Synonym	Mol. Form. Mol. Wt.	CAS RN mp/°C	Merck No. bp/°C	Beil. Ref. den/g cm^{-3}	Solubility n_D
5066	Decanoic acid, methyl ester	$C_{11}H_{22}O_2$ 186.29	110-42-9 -18	224	4-02-00-01044 0.8730[20]	H_2O 1; EtOH 4; eth 4; ctc 2 1.4259[20]
5067	Decanoic acid, 1-methylethyl ester	$C_{13}H_{26}O_2$ 214.35	2311-59-3	121[10]	4-02-00-01045 0.8543[20]	1.4221[25]
5068	Decanoic acid, 2-octyl-	$C_{18}H_{36}O_2$ 284.48	619-39-6 38.5	215[13]	4-02-00-01254 0.8447[70]	eth 4; EtOH 4
5069	Decanoic acid, propyl ester	$C_{13}H_{26}O_2$ 214.35	30673-60-0	128.5[10]	4-02-00-01045 0.8623[20]	1.4280[20]
5070	1-Decanol Capric alcohol	$C_{10}H_{22}O$ 158.28	112-30-1 6.9	2847 231.1	4-01-00-01815 0.8297[20]	H_2O 1; EtOH 5; eth 5; ace 5 1.4372[20]
5071	2-Decanol, (±)- 2-Decanol (DL)	$C_{10}H_{22}O$ 158.28	74742-10-2 -2.4	211	4-01-00-01823 0.8250[20]	EtOH 3; eth 5; ace 5; bz 3 1.4326[25]
5072	4-Decanol 1-Propylheptyl alcohol	$C_{10}H_{22}O$ 158.28	2051-31-2 -11	210.5	4-01-00-01824 0.8261[20]	H_2O 1; EtOH 3; ctc 3 1.4320[20]
5073	1-Decanol, 10-chloro-	$C_{10}H_{21}ClO$ 192.73	51309-10-5 12.5	185-9[15]	4-01-00-01821 0.9630[25]	eth 4; EtOH 4 1.4578[20]
5074	1-Decanol, 10-fluoro- 10-Fluoro-1-decanol	$C_{10}H_{21}FO$ 176.27	334-64-5 22	136-7[15]	4-01-00-01821 0.919[20]	eth 4; EtOH 4 1.4322[25]
5075	2-Decanone Methyl octyl ketone	$C_{10}H_{20}O$ 156.27	693-54-9 14	210; 96[12]	4-01-00-03367 0.8248[20]	H_2O 1; EtOH 3; eth 3; ctc 2 1.4255[20]
5076	3-Decanone Ethyl heptyl ketone	$C_{10}H_{20}O$ 156.27	928-80-3 2.5	203	4-01-00-03368 0.8251[20]	EtOH 3; eth 3; ctc 3 1.4252[20]
5077	4-Decanone Hexyl propyl ketone	$C_{10}H_{20}O$ 156.27	624-16-8 -9	206.5	4-01-00-03368 0.824[20]	H_2O 1; EtOH 5; eth 5 1.4240[21]
5078	Decanoyl chloride Caprinoyl chloride	$C_{10}H_{19}ClO$ 190.71	112-13-0 -34.5	95	4-02-00-01050 0.919[25]	eth 3; ctc 3 1.4410[20]
5079	Decasiloxane, docosamethyl- Decasiloxane, dicosamethyl	$C_{22}H_{66}O_9Si_{10}$ 755.62	556-70-7	183.[4]	3-04-00-01881 0.925[20]	bz 4; lig 4 1.3988[20]
5080	2-Decenal	$C_{10}H_{18}O$ 154.25	3913-71-1	230	4-01-00-03511 0.845[17]	1.4533[17]
5081	3-Decenal	$C_{10}H_{18}O$ 154.25	58474-80-9	93-4[14]	4-01-00-03512 0.850[15]	1.4462[15]
5082	1-Decene	$C_{10}H_{20}$ 140.27	872-05-9 -66.3	170.5	3-01-00-00858 0.7408[20]	H_2O 1; EtOH 5; eth 5 1.4215[20]
5083	4-Decene	$C_{10}H_{20}$ 140.27	19689-18-0	170.6	4-01-00-00902 0.7404[20]	1.4243[20]
5084	5-Decene, (E)-	$C_{10}H_{20}$ 140.27	7433-56-9 -73	171	4-01-00-00902 0.7401[20]	H_2O 1; EtOH 5; eth 5; ctc 2 1.4243[20]
5085	5-Decene, (Z)-	$C_{10}H_{20}$ 140.27	7433-78-5 -112	171; 73[20]	3-01-00-00859 0.7445[20]	H_2O 1; EtOH 5; eth 5; ctc 2 1.4258[20]
5086	1-Decene, 2-bromo- 2-Bromo-1-decene	$C_{10}H_{19}Br$ 219.16	3017-67-2	115-6[22]	3-01-00-00859 1.0842[20]	1.4629[20]
5087	2-Decene, 1-bromo- 1-Bromo-2-decene	$C_{10}H_{19}Br$ 219.16	14304-30-4	121[17]	4-01-00-00902 1.074[18]	lig 4 1.4716[18]
5088	2-Decenoic acid Δ 2-Decenoic acid	$C_{10}H_{18}O_2$ 170.25	3913-85-7 12	165[15]	4-02-00-01606 0.9280[18]	1.4616[20]
5089	3-Decenoic acid	$C_{10}H_{18}O_2$ 170.25	15469-77-9 18	154-63[11]	4-02-00-01606 0.914[15]	1.4510[18]
5090	4-Decenoic acid Deconic acid telomer	$C_{10}H_{18}O_2$ 170.25	26303-90-2	149[13]	4-02-00-01607 0.9197[20]	bz 4; eth 4 1.4497[20]
5091	9-Decenoic acid Caproleic acid	$C_{10}H_{18}O_2$ 170.25	14436-32-9	158[21]; 142[4]	4-02-00-01605 0.9238[15]	eth 4; EtOH 4 1.4507[15]
5092	9-Decen-1-ol Decylenic alcohol	$C_{10}H_{20}O$ 156.27	13019-22-2	236	4-01-00-02184 0.876[25]	1.4480[20]
5093	3-Decen-2-one Heptylidene acetone	$C_{10}H_{18}O$ 154.25	10519-33-2	102-3[15.3]	4-01-00-03512 0.8473[20]	1.4480[20]
5094	1-Decen-3-yne	$C_{10}H_{16}$ 136.24	33622-26-3	76[20]	4-01-00-01105 0.7873[20]	1.4620[20]
5095	1-Decen-4-yne	$C_{10}H_{16}$ 136.24	24948-66-1	73-4[22]	3-01-00-01049 0.7880[20]	1.445[20]
5096	2-Decen-4-yne	$C_{10}H_{16}$ 136.24	116668-40-7	55[5]	3-01-00-01049 0.7850[25]	1.4609[25]
5097	1-Decyne Octylacetylene	$C_{10}H_{18}$ 138.25	764-93-2 -44	174	4-01-00-01054 0.7655[20]	H_2O 1; EtOH 3; eth 3; os 3 1.4265[20]
5098	3-Decyne	$C_{10}H_{18}$ 138.25	2384-85-2	177	4-01-00-01055 0.7619[25]	1.4315[20]
5099	4-Decyne	$C_{10}H_{18}$ 138.25	2384-86-3	74.5[19]	3-01-00-01017 0.772[17]	1.436[17]
5100	5-Decyne Dibutylacetylene	$C_{10}H_{18}$ 138.25	1942-46-7 -73	177; 78.8[25]	4-01-00-01055 0.7690[20]	H_2O 1; EtOH 3; eth 3 1.4331[20]
5101	4-Decyne, 3,3-dimethyl- 3,3-Dimethyl-4-decyne	$C_{12}H_{22}$ 166.31	70732-45-5	86[20]	3-01-00-01026 0.7731[20]	1.4399[20]
5102	Deltamethrin Cyano(3-phenoxyphenyl)methyl-3-(2,2-dibromoethenyl)-2,2-dimethylcyclopropanecarboxylate	$C_{22}H_{19}Br_2NO_3$ 505.21	52918-63-5 99	2869		
5103	Demeton S methyl Phosphorothioic acid, S-[2-(ethylthio)ethyl] O,O-dimethyl ester	$C_6H_{15}O_3PS_2$ 230.29	919-86-8	89[0.15]; 118[1]	1.207[20]	

No.	Name Synonym	Mol. Form. Mol. Wt.	CAS RN mp/°C	Merck No. bp/°C	Beil. Ref. den/g cm^{-3}	Solubility n_D
5104	Desmedipham Carbamic acid, [3-[[(phenylamino)carbonyl]oxy]phenyl]-, ethyl ester	$C_{16}H_{16}N_2O_4$ 300.31	13684-56-5 120			
5105	Dialifor Phosphorodithioic acid, S-[2-chloro-1-(1,3-dihydro-1,3-dioxo-2H-isoindol-2-yl)ethyl]	$C_{14}H_{17}ClNO_4PS_2$ 393.85	10311-84-9 68	2949		
5106	Diallate Carbamothioic acid, bis(1-methylethyl)-, S-(2,3-dichloro-2-propenyl) ester	$C_{10}H_{17}Cl_2NOS$ 270.22	2303-16-4	2950 150[9]		
5107	8,8'-Diapo-ψ,ψ-carotenedioic acid Crocetin (trans)	$C_{20}H_{24}O_4$ 328.41	27876-94-4 286	2592	3-02-00-02018	H_2O 2; EtOH 2; eth 1; bz 1
5108	6,6'-Diapo-ψ,ψ-carotenedioic acid, monomethyl ester, 9-cis- Bixin	$C_{25}H_{30}O_4$ 394.51	6983-79-5 198	1320	3-02-00-02021	H_2O 1; EtOH 3; eth 2; ace 3
5109	Diarsine, tetraethyl- Tetraethyldiarsenic	$C_8H_{20}As_2$ 266.09	612-08-8 186		3-04-00-01832 1.1388[24]	eth 4; EtOH 4 1.4709
5110	Diarsine, tetramethyl- Cacodyl	$C_4H_{12}As_2$ 209.98	471-35-2 -6	1602 165	4-04-00-03683 1.447[15]	eth 4; EtOH 4
5111	1,4-Diazabicyclo[2.2.2]octane Triethylenediamine	$C_6H_{12}N_2$ 112.17	280-57-9 159	9584	5-23-03-00487	chl 3
5112	Diazene, bis(2-ethoxyphenyl)- Azobenzene, 2,2'-diethoxy-	$C_{16}H_{18}N_2O_2$ 270.33	613-43-4 131	240 dec	4-16-00-00138	H_2O 1; EtOH 3; HCl 3
5113	Diazene, bis(4-ethoxyphenyl)-	$C_{16}H_{18}N_2O_2$ 270.33	588-52-3 162	dec	4-16-00-00174	H_2O 1; EtOH 2; eth 3; bz 3
5114	Diazene, bis(2-methylphenyl)-	$C_{14}H_{14}N_2$ 210.28	584-90-7 55.5		2-16-00-00019 1.0215[65]	H_2O 1; EtOH 3; eth 4; bz 4 1.6180[65]
5115	Diazene, bis(3-methylphenyl)-, (E)-	$C_{14}H_{14}N_2$ 210.28	51437-67-3 54		4-16-00-00070 1.0123[66]	EtOH 4; eth 4; bz 4 1.6152[66]
5116	Diazene, bis(2-methylphenyl)-, 1-oxide, (Z)- Azoxybenzene, 2,2'-dimethyl (trans)	$C_{14}H_{14}N_2O$ 226.28	51284-68-5 60		4-16-00-00064 1.0215[65]	bz 4; eth 4; EtOH 4 1.6180[65]
5117	Diazene, bis(2-methylphenyl)-, 1-oxide, (E)- Azoxybenzene, 2,2'-dimethyl(trans)	$C_{14}H_{14}N_2O$ 226.28	116723-89-8		2-16-00-00318 1.0215[65]	H_2O 1; EtOH 3; eth 3; bz 3 1.6180[65]
5118	Diazene, bis(3-methylphenyl)-, 1-oxide, (Z)- Azoxybenzene, 3,3'-dimethyl (trans)	$C_{14}H_{14}N_2O$ 226.28	71297-97-7 39		4-16-00-00071 1.0123[66]	bz 4; eth 4; EtOH 4; lig 4 1.6152[66]
5119	Diazene, bis(3-methylphenyl)-, 1-oxide, (E)- Azoxybenzene, 3,3'-dimethyl(trans)	$C_{14}H_{14}N_2O$ 226.28	116723-90-1 39		2-16-00-00319 1.0123[66]	H_2O 1; EtOH 3; eth 3; bz 3 1.6152[66]
5120	Diazene, (4-bromophenyl)phenyl-, 1-oxide Azoxybenzene, 4-bromo-	$C_{12}H_9BrN_2O$ 277.12	16109-68-5 93.5		3-16-00-00584 1.4138[100]	H_2O 1; EtOH 3; bz 3; MeOH 3
5121	Diazenecarbothioic acid, phenyl-, 2-phenylhydrazide Dithizone	$C_{13}H_{12}N_4S$ 256.33	60-10-6 165-9 dec	3383	4-16-00-00018	H_2O 1; EtOH 2; eth 2; chl 3
5122	Diazenecarboxylic acid, phenyl-, 2-phenylhydrazide Diphenylcarbazone	$C_{13}H_{12}N_4O$ 240.26	538-62-5 157 dec	3322	4-16-00-00017	H_2O 1; EtOH 4; bz 4; chl 4
5123	Diazene, dibutyl Azobutane	$C_8H_{18}N_2$ 142.24	2159-75-3 60[18]			
5124	Diazenedicarboxamide Azodicarbonamide	$C_2H_4N_4O_2$ 116.08	123-77-3 212 dec	932	3-03-00-00234	
5125	Diazene, dimethyl- Azomethane	$C_2H_6N_2$ 58.08	503-28-6 -78	1.5	3-04-00-01747 0.743[0]	ace 4; eth 4; EtOH 4 1.4199[19]
5126	Diazene, dimethyl-, (E)-	$C_2H_6N_2$ 58.08	4143-41-3 -78	1.5	4-04-00-03366 0.743[0]	ctc 3; hp 3
5127	Diazene, di-1-naphthalenyl- 1,1'-Azobisnaphthalene	$C_{20}H_{14}N_2$ 282.34	487-10-5 190	sub	2-16-00-00026	H_2O 1; EtOH 2; ace 3; bz 4
5128	Diazene, di-2-naphthalenyl- 2,2'-Azonaphthalene	$C_{20}H_{14}N_2$ 282.34	582-08-1 208	sub	2-16-00-00026	H_2O 1; EtOH 2; eth 2; bz 3
5129	Diazene, diphenyl-, (E)- trans-Azobenzene	$C_{12}H_{10}N_2$ 182.22	17082-12-1 68.5	930 293	4-16-00-00007 1.203[20]	H_2O 2; EtOH 3; eth 3; bz 3 1.6266[78]
5130	Diazene, diphenyl-, (Z)- cis-Azobenzene	$C_{12}H_{10}N_2$ 182.22	1080-16-6 71	930	4-16-00-00007	H_2O 2; EtOH 3; eth 3; bz 3
5131	Diazene, diphenyl-, 1-oxide, (Z)- cis-Azoxybenzene	$C_{12}H_{10}N_2O$ 198.22	20972-43-4 87	937	4-16-00-00010 1.166[20]	1.633[20]
5132	Diazene, dipropyl Azopropane	$C_6H_{14}N_2$ 114.19	821-67-0 114			
5133	Diazene, (4-ethoxyphenyl)phenyl-	$C_{14}H_{14}N_2O$ 226.28	7466-38-8 85	339.5	2-16-00-00040 1.0400[100]	EtOH 4; eth 4; ace 3; bz 4 1.6419[100]
5134	Diazene, ethylphenyl-	$C_8H_{10}N_2$ 134.18	935-08-0 177		4-16-00-00007 0.9628[22]	bz 4; eth 4; EtOH 4 1.5579
5135	Diazene, (2-methoxyphenyl)phenyl-	$C_{13}H_{12}N_2O$ 212.25	6319-21-7 41	195-7[14]	3-16-00-00081 1.0728[100]	H_2O 1; EtOH 3; ace 3; os 4
5136	Diazene, (3-methoxyphenyl)phenyl-	$C_{13}H_{12}N_2O$ 212.25	34238-81-8 33.3	193[15]	4-16-00-00155 1.1023[53]	H_2O 1; EtOH 3; ace 3; os 4
5137	Diazene, (4-methoxyphenyl)phenyl-	$C_{13}H_{12}N_2O$ 212.25	2396-60-3 56	340	2-16-00-00040 1.12[75]	H_2O 1; EtOH 3; eth 3; ace 3
5138	Diazene, (3-methylphenyl)phenyl- Azobenzene, 3-methyl	$C_{13}H_{12}N_2$ 196.25	17478-66-9 18.5	175[19]	3-16-00-00047 1.065[20]	
5139	Diazene, (4-methylphenyl)phenyl- p-Methylazobenzene	$C_{13}H_{12}N_2$ 196.25	949-87-1 71.5	312	3-16-00-00049	bz 4; eth 4; chl 4

No.	Name / Synonym	Mol. Form. / Mol. Wt.	CAS RN / mp/°C	Merck No. / bp/°C	Beil. Ref. / den/g cm^{-3}	Solubility / n_D
5140	1H-1,4-Diazepine, 1-[(4-chlorophenyl)phenylmethyl]hexahydro-4-methyl- Homochlorocyclizine	$C_{19}H_{23}ClN_2$ 314.86	848-53-3	4653	5-23-03-00248	
5141	1H-1,4-Diazepine, hexahydro-1-methyl-	$C_6H_{14}N_2$ 114.19	4318-37-0	154	5-23-03-00241 0.9111[20]	1.4769[20]
5142	Dibenz[a,h]anthracene 1,2:5,6-Dibenzanthracene	$C_{22}H_{14}$ 278.35	53-70-3 269.5	2989	4-05-00-02722	H_2O 1; EtOH 2; ace 3; bz 3
5143	Dibenz[a,j]acridine 7-Azadibenz[a,j]anthracene	$C_{21}H_{13}N$ 279.34	224-42-0 216		5-20-08-00656	
5144	5H-Dibenz[b,f]azepine-5-carboxamide Carbamazepine	$C_{15}H_{12}N_2O$ 236.27	298-46-4 190.2	1783	5-20-08-00247	
5145	Dibenz[c,e]oxepin-5,7-dione	$C_{14}H_8O_3$ 224.22	6050-13-1 217	sub	5-17-11-00525	H_2O 1; eth 2
5146	10H-Dibenzo[a,d]cyclohepten-10-one, 5,11-dihydro-	$C_{15}H_{12}O$ 208.26	6374-70-5 30	203-4[7]	4-07-00-01675 1.1635[20]	1.6324[20]
5147	5H-Dibenzo[a,d]cyclohepten-5-one, 10,11-dihydro-	$C_{15}H_{12}O$ 208.26	1210-35-1 30	203-4[7]	4-07-00-01674 1.1635[20]	1.6324[20]
5148	13H-Dibenzo[a,g]fluorene	$C_{21}H_{14}$ 266.34	207-83-0 174.5	195-200[0.5]	4-05-00-02693	H_2O 1; EtOH 2; eth 2; bz 3
5149	6H-Dibenzo[a,g]quinolizine, 5,8,13,13a-tetrahydro-2,3,9,10-tetramethoxy-13-methyl-, (13S-trans)- Corydaline (d)	$C_{22}H_{27}NO_4$ 369.46	518-69-4 136	2543	5-21-06-00173	bz 4; eth 4; EtOH 4; chl 4
5150	6H-Dibenzo[a,g]quinolizin-2-ol, 5,8,13,13a-tetrahydro-3,9,10-trimethoxy-13-methyl-, (13S-trans)- Isocorybulbine	$C_{21}H_{25}NO_4$ 355.43	22672-74-8 187.5	5045	5-21-06-00173 1.045[20]	H_2O 1; EtOH 3; chl 3; acid 3
5151	6H-Dibenzo[a,g]quinolizin-3-ol, 5,8,13,13a-tetrahydro-2,9,10-trimethoxy-13-methyl-, (13S-trans)- Corybulbine	$C_{21}H_{25}NO_4$ 355.43	518-77-4 237.5	2539	5-21-06-00173	H_2O 1; EtOH 2; eth 2; ace 3
5152	13H-Dibenzo[a,h]fluoren-13-one 1,2:6,7-Dibenzo-9-fluorenone	$C_{21}H_{12}O$ 280.33	4599-94-4 214	sub 200	3-07-00-02897	
5153	11H-Dibenzo[b,e][1,4]dioxepin-11-one, 2,4,7,9-tetrachloro-3-hydroxy-8-methoxy-1,6-dimethyl- Diploicin	$C_{16}H_{10}Cl_4O_5$ 424.06	527-93-5 232	3345	5-19-06-00364	
5154	6H-Dibenzo[b,d]pyran-1-ol, 6,6,9-trimethyl-3-pentyl- Cannabinol	$C_{21}H_{26}O_2$ 310.44	521-35-7 77	1751 185[0.05]	5-17-04-00567	H_2O 1; EtOH 3; eth 3; ace 3
5155	4H-Dibenzo[de,g]quinoline-10,11-diol, 5,6,6a,7-tetrahydro-6-methyl-, (R)- Apomorphine	$C_{17}H_{17}NO_2$ 267.33	58-00-4 195 dec	776	5-21-05-00333	H_2O 2; EtOH 3; eth 3; ace 3
5156	4H-Dibenzo[de,g]quinoline-2,9-diol, 5,6,6a,7-tetrahydro-1,10-dimethoxy-6-methyl-, (S)- Boldine	$C_{19}H_{21}NO_4$ 327.38	476-70-0 163	1328	5-21-06-00118	EtOH 4; chl 4
5157	4H-Dibenzo[de,g]quinoline, 5,6,6a,7-tetrahydro-1,2,9,10-tetramethoxy-6-methyl-, (S)- d-Glaucine	$C_{21}H_{25}NO_4$ 355.43	475-81-0 120	4332	5-21-06-00122	ace 4; EtOH 4; chl 4
5158	4H-Dibenzo[de,g]quinolin-1-ol, 5,6,6a,7-tetrahydro-2,11-dimethoxy-6-methyl-, (S)- Isothebaine	$C_{19}H_{21}NO_3$ 311.38	568-21-8 203.5	5116	5-21-05-00658	H_2O 1; EtOH 5; eth 2; chl 5
5159	4H-Dibenzo[de,g]quinolin-10-ol, 5,6,6a,7-tetrahydro-1,2-dimethoxy-, (R)- Tsuduranine	$C_{18}H_{19}NO_3$ 297.35	517-97-5 204	9710	5-21-05-00655	ace 4; eth 4; EtOH 4
5160	4H-Dibenzo[de,g]quinolin-11-ol, 5,6,6a,7-tetrahydro-10-methoxy-6-methyl-, (R)- Apocodeine	$C_{18}H_{19}NO_2$ 281.35	641-36-1 123.5	771	5-21-05-00334	EtOH 2; eth 3; ace 3; bz 3
5161	4H-Dibenzo[de,g]quinolin-11-ol, 5,6,6a,7-tetrahydro-1,2,10-trimethoxy-6-methyl-, (S)- Isocorydine	$C_{20}H_{23}NO_4$ 341.41	475-67-2 185	5046	5-21-06-00132	chl 4
5162	4H-Dibenzo[de,g]quinolin-1-ol, 5,6,6a,7-tetrahydro-2,10,11-trimethoxy-6-methyl-, (S)- Corydine	$C_{20}H_{23}NO_4$ 341.41	476-69-7 149	2545	5-21-06-00134	eth 4; EtOH 4; chl 4
5163	Dibenzofuran 2,2'-Biphenylene oxide	$C_{12}H_8O$ 168.19	132-64-9 86.5	287	5-17-02-00234 1.0886[99]	H_2O 2; EtOH 3; eth 4; ace 3 1.6079[99]
5164	Dibenzofuran, 2-bromo- 2-Bromodibenzofuran	$C_{12}H_7BrO$ 247.09	86-76-0 110	220[40]	5-17-02-00238	EtOH 3; HOAc 2
5165	Dibenzofuran, 3-bromo- 3-Bromodibenzofuran	$C_{12}H_7BrO$ 247.09	26608-06-0 121	220[40]	4-17-00-00588	H_2O 1; EtOH 3; eth 3
5166	Dibenzofuran, 3-nitro- 3-Nitrodibenzofuran	$C_{12}H_7NO_3$ 213.19	5410-97-9 181.5	180-5[3]	5-17-02-00239	H_2O 1; EtOH 2; eth 2; HOAc 3
5167	Dibenzothiophene	$C_{12}H_8S$ 184.26	132-65-0 99.5	332.5	5-17-02-00239	H_2O 3; EtOH 4; bz 4; chl 3
5168	Diborane(6), mu-(methylthio)- Diborane, methylthio	CH_8B_2S 73.76	91572-15-5 -101.5	53	4-01-00-01287	ace 4; bz 4

No.	Name / Synonym	Mol. Form. / Mol. Wt.	CAS RN mp/°C	Merck No. bp/°C	Beil. Ref. den/g cm^{-3}	Solubility n_D
5169	Dicarbonic acid, diethyl ester / Pyrocarbonic acid diethyl ester	$C_6H_{10}O_5$ / 162.14	1609-47-8 /	8008 / 93-4[18]	4-03-00-00018 / 1.120[20]	ace 4; EtOH 4; lig 4 / 1.3960[20]
5170	Dichlorvos / Phosphoric acid, 2,2-dichloroethenyl dimethyl ester	$C_4H_7Cl_2O_4P$ / 220.98	62-73-7 /	3069 / 140[20]; 84[1]	1.415[25]	
5171	Diclofop-methyl / Propanoic acid, 2-[4-(2,4-dichlorophenoxy)phenoxy]-, methyl ester	$C_{16}H_{14}Cl_2O_4$ / 341.19	51338-27-3 / 40	3072 / 176[0.1]		
5172	Dicrotophos / Phosphoric acid, 3-(dimethylamino)-1-methyl-3-oxo-1-propenyl dimethyl ester, (E)-	$C_8H_{16}NO_5P$ / 237.19	141-66-2 /	3077 / 400; 130[0.1]	1.216[15]	
5173	Diethatyl-ethyl / Glycine, N-(chloroacetyl)-N-(2,6-diethylphenyl)-, ethyl ester	$C_{16}H_{22}ClNO_3$ / 311.81	38727-55-8 /			
5174	Difenoconazole / cis,trans-3-Chloro-4-[4-methyl-2-(1H-1,2,4-triazol-1-ylmethyl)-1,3-dioxolan-2-yl]phenyl 4-chlorophenyl ether	$C_{19}H_{17}Cl_2N_3O_3$ / 406.27	119446-68-3 / 76	220[0.03]		
5175	Difenzoquat methyl sulfate / 1H-Pyrazolium, 1,2-dimethyl-3,5-diphenyl-, methyl sulfate	$C_{17}H_{17}N_2 \cdot CH_3O_4S$ / 360.43	43222-48-6 / 157			
5176	Diflubenzuron / Benzamide, N-[[(4-chlorophenyl)amino]carbonyl]-2,6-difluoro-	$C_{14}H_9ClF_2N_2O_2$ / 310.69	35367-38-5 / 239	3128		
5177	1,4:5,8-Dimethanonaphthalene, 1,2,3,4,10,10-hexachloro-1,4,4a,5,8,8a-hexahydro-, / Aldrin	$C_{12}H_8Cl_6$ / 364.91	309-00-2 / 104	219	3-05-00-01385	H_2O 1; EtOH 3; eth 3; ace 3
5178	2,7:3,6-Dimethanonaphth[2,3-b]oxirene, 3,4,5,6,9,9-hexachloro-1a,2,2a,3,6,6a,7,7a-octahydro-,(1aα,2β,2aα,3β,6β,6aα,7β,7aα)- / Dieldrin	$C_{12}H_8Cl_6O$ / 380.91	60-57-1 / 175.5	3093 / 1.75[25]	5-17-02-00087	H_2O 1; EtOH 2; ace 3; bz 3
5179	Dimethipin / 2,3-Dihydro-5,6-dimethyl-1,4-dithiin, 1,1,4,4-tetraoxide	$C_6H_{10}O_4S_2$ / 210.28	55290-64-7 / 165			
5180	Dimethyl tetrachloroterephthalate / 1,4-Benzenedicarboxylic acid, 2,3,5,6-tetrachloro-, dimethyl ester	$C_{10}H_6Cl_4O_4$ / 331.97	1861-32-1 / 155	2830		
5181	Dinitramine / 1,3-Benzenediamine, N3,N3-diethyl-2,4-dinitro-6-(trifluoromethyl)-	$C_{11}H_{13}F_3N_4O_4$ / 322.24	29091-05-2 / 98			
5182	Dinocap / 2-Butenoic acid, 2-(1-methylheptyl)-4,6-dinitrophenyl ester	$C_{18}H_{24}N_2O_6$ / 364.40	6119-92-2 /	136[0.01]		
5183	Dinoseb / Phenol, 2-(1-methylpropyl)-4,6-dinitro-	$C_{10}H_{12}N_2O_5$ / 240.22	88-85-7 / 40	3282	1.265[45]	
5184	2,3-Dioxabicyclo[2.2.2]oct-5-ene, 1-methyl-4-(1-methylethyl)- / Ascaridole	$C_{10}H_{16}O_2$ / 168.24	512-85-6 / 3.3	852 / 115[15]; 39[0.2]	5-19-01-00319 / 1.0103[20]	H_2O 1; EtOH 3; ace 3; bz 3 / 1.4769[20]
5185	1,7-Dioxadispiro[4.0.4.2]dodeca-3,9-diene-2,8-dione, trans- / Anemonin	$C_{10}H_8O_4$ / 192.17	508-44-1 / 158	674	5-19-05-00101	chl 4
5186	1,3-Dioxane / 1,3-Dioxacyclohexane	$C_4H_8O_2$ / 88.11	505-22-6 / -45	/ 106.1	5-19-01-00011 / 1.0342[20]	H_2O 5; EtOH 5; eth 5; ace 5 / 1.4165[20]
5187	1,4-Dioxane / 1,4-Dioxacyclohexane	$C_4H_8O_2$ / 88.11	123-91-1 / 11.8	3294 / 101.5	5-19-01-00016 / 1.0337[20]	H_2O 5; EtOH 5; eth 5; ace 5 / 1.4224[20]
5188	1,4-Dioxane, 2,3-dichloro-	$C_4H_8Cl_2O_2$ / 157.00	95-59-0 / 30	/ 80-2[10]	4-19-00-00030 / 1.468[20]	H_2O 1; eth 4; ace 4; bz 4 / 1.4928[20]
5189	1,3-Dioxane, 2,4-dimethyl-	$C_6H_{12}O_2$ / 116.16	766-20-1 /	/ 116.5	5-19-01-00072 / 0.9392[20]	H_2O 2; acid 4 / 1.4136[20]
5190	1,3-Dioxane-4,6-dione, 2,2-dimethyl- / Meldrum's acid	$C_6H_8O_4$ / 144.13	2033-24-1 / 94	5696	5-19-05-00008	
5191	1,3-Dioxane-2-ethanol, 4-hydroxy-α,6-dimethyl- / Paraldol	$C_8H_{11}O_4$ / 171.17	19404-07-0 / 90	/ 90[15]	4-01-00-03987 / 1.116[20]	H_2O 4; eth 4; EtOH 4 / 1.4610[20]
5192	1,4-Dioxane, 2,2,3,3,5,5,6-heptachloro- / 1,4-Dioxane, heptachloro	$C_4HCl_7O_2$ / 329.22	6629-96-5 / 55	/ 123-8[8]	4-19-00-00032	H_2O 1; eth 4; ace 4; bz 4
5193	1,3-Dioxane, 4-methyl-	$C_5H_{10}O_2$ / 102.13	1120-97-4 / -44.5	/ 114	5-19-01-00054 / 0.9758[20]	H_2O 2; os 4 / 1.4159[20]
5194	1,3-Dioxane, 4-methyl-4-phenyl- / m-Dioxane, 4-methyl-4-phenyl-	$C_{11}H_{14}O_2$ / 178.23	1200-73-3 / 37.5	/ 256	5-19-01-00481 / 1.0864[20]	H_2O 1; chl 2; os 3 / 1.5240[20]
5195	1,3-Dioxane, 2-phenyl-	$C_{10}H_{12}O_2$ / 164.20	772-01-0 / 41	/ 253	5-19-01-00459 / 1.6053[60]	EtOH 4; eth 4
5196	1,3-Dioxane, 4-phenyl-	$C_{10}H_{12}O_2$ / 164.20	772-00-9 /	/ 247	5-19-01-00464 / 1.1038[20]	H_2O 1; os 3 / 1.5306[18]
5197	1,3-Dioxan-5-ol, 2-methyl-	$C_5H_{10}O_3$ / 118.13	3774-03-6 /	/ 176; 79[18]	4-19-00-00624 / 1.0705[17]	H_2O 4 / 1.4375[17]
5198	1,3,2-Dioxathiane, 4-methyl-, 2-oxide / 1,3-Butanediol, sulfite	$C_4H_8O_3S$ / 136.17	4426-51-1 / 5	/ 185	4-01-00-02510 / 1.2352[20]	EtOH 4; eth 4; ace 4; bz 4 / 1.4661[20]

No.	Name / Synonym	Mol. Form. / Mol. Wt.	CAS RN / mp/°C	Merck No. / bp/°C	Beil. Ref. / den/g cm⁻³	Solubility / n_D
5199	1,3,2-Dioxathiolane, 2,2-dioxide	$C_2H_4O_4S$ / 124.12	1072-53-3 / 99	sub	3-01-00-02110	ace 4; bz 4; eth 4; EtOH 4
5200	1,3,2-Dioxathiolane, 2-oxide	$C_2H_4O_3S$ / 108.12	3741-38-6 / -11	173	4-01-00-02409 / 1.4402[20]	H_2O 4; EtOH 4; eth 4; ace 4 / 1.4463[20]
5201	Dioxathion / Phosphorodithioic acid, S,S'-1,4-dioxane-2,3-diyl O,O,O',O'-tetraethyl ester	$C_{12}H_{26}O_6P_2S_4$ / 456.55	78-34-2 / -20	3296	1.257[26]	
5202	1,4-Dioxin	$C_4H_4O_2$ / 84.07	290-67-5	76	5-19-01-00275 / 1.115[20]	eth 3; ace 3; bz 3 / 1.4350[20]
5203	1,4-Dioxin, 2,3-dihydro-	$C_4H_6O_2$ / 86.09	543-75-9	94.1	5-19-01-00181 / 1.0836[20]	ctc 3 / 1.4372[20]
5204	1,3-Dioxolane / 1,3-Dioxacyclopentane	$C_3H_6O_2$ / 74.08	646-06-0 / -95	78	5-19-01-00006 / 1.060[20]	H_2O 3; EtOH 3; eth 3; ace 3 / 1.3974[20]
5205	1,3-Dioxolane, 2-(3-bromophenyl)-	$C_9H_9BrO_2$ / 229.07	17789-14-9	132.5	5-19-01-00448 / 1.514[25]	1.5627[20]
5206	1,3-Dioxolane-4-carboxaldehyde, 2,2-dimethyl-	$C_6H_{10}O_3$ / 130.14	5736-03-8 /	74[50]	5-19-04-00041	H_2O 4 / 1.4189[25]
5207	1,3-Dioxolane-4-methanol, 2,2-dimethyl- / Isopropylidene glycerol	$C_6H_{12}O_3$ / 132.16	100-79-8	82[10]	5-19-02-00362 / 1.064[20]	1.4383[20]
5208	1,3-Dioxolane-4-methanol, 2-methyl-	$C_5H_{10}O_3$ / 118.13	3773-93-1	187	5-19-02-00350 / 1.1243[17]	EtOH 4 / 1.4413[17]
5209	1,3-Dioxolane-4-methanol, 2-phenyl-	$C_{10}H_{12}O_3$ / 180.21	1708-39-0	143-4[2]	5-19-02-00562 / 1.1916[17]	1.5389[17]
5210	1,3-Dioxolane, 2-methyl-	$C_4H_8O_2$ / 88.11	497-26-7	81.5	5-19-01-00042 / 0.9811[20]	H_2O 4; EtOH 5; eth 5 / 1.4035[17]
5211	1,3-Dioxolan-2-one / Ethylene glycol carbonate	$C_3H_4O_3$ / 88.06	96-49-1 / 36.4	248	5-19-04-00006 / 1.3214[39]	H_2O 5; EtOH 5; eth 5; bz 5 / 1.4148[50]
5212	1,3-Dioxolan-4-one, 2,5-bis(trichloromethyl)- / Chloralide	$C_5H_2Cl_6O_3$ / 322.79	554-21-2 / 116	272.5	4-19-00-01571	eth 4; HOAc 4
5213	1,3-Dioxolan-2-one, 4-methyl-	$C_4H_6O_3$ / 102.09	108-32-7 / -48.8	242	5-19-04-00021 / 1.2047[20]	H_2O 4; EtOH 4; eth 4; ace 4 / 1.4189[20]
5214	1,3-Dioxolo[4,5-g]isoquinoline, 5,6,7,8-tetrahydro-4-methoxy-6-methyl- / Hydrocotarnine	$C_{12}H_{15}NO_3$ / 221.26	550-10-7 / 56	4715	4-27-00-06395	H_2O 1; EtOH 3; eth 3; ace 3
5215	1,3-Dioxolo[4,5-g]isoquinoline, 5,6,7,8-tetrahydro-6-methyl- / Hydrohydrastinine	$C_{11}H_{13}NO_2$ / 191.23	494-55-3 / 66	303	4732	4-27-00-06306	ace 4; bz 4; eth 4; EtOH 4
5216	1,3-Dioxolo[4,5-g]isoquinolin-5-ol, 5,6,7,8-tetrahydro-4-methoxy-6-methyl- / Cotarnine	$C_{12}H_{15}NO_4$ / 237.26	82-54-2 / 132-3 dec	2551	4-27-00-06448	H_2O 2; EtOH 3; eth 3; bz 3
5217	1,3-Dioxolo[4,5-g]isoquinolin-5-ol, 5,6,7,8-tetrahydro-6-methyl- / Hydrastinine	$C_{11}H_{13}NO_3$ / 207.23	6592-85-4 / 116.5	4689	4-27-00-06396	H_2O 3; EtOH 4; eth 4; chl 4
5218	1,3-Dioxolo[4,5-h]quinolin-8(9H)-one, 6-methoxy-9-methyl- / Casimiroin	$C_{12}H_{11}NO_4$ / 233.22	477-89-4	1894	4-27-00-06618	chl 2
5219	Diphenamid / Benzeneacetamide, N,N-dimethyl-α-phenyl-	$C_{16}H_{17}NO$ / 239.32	957-51-7 / 135	3305	1.17[23.3]	
5220	Diphosphoramide, octamethyl- / Schradan	$C_8H_{24}N_4O_3P_2$ / 286.25	152-16-9 / 17	188-22[0.3]	8351	4-04-00-00288 / 1.1343[25]	H_2O 4; EtOH 4; chl 4 / 1.462[25]
5221	Diphosphoric acid, tetraethyl ester / Tetraethyl pyrophosphate	$C_8H_{20}O_7P_2$ / 290.19	107-49-3 / 170 dec	155[3]	9138	4-01-00-01342 / 1.1847[20]	H_2O 5; EtOH 5; eth 5; ace 5 / 1.4180[20]
5222	Dipropetryn / 1,3,5-Triazine-2,4-diamine, 6-(ethylthio)-N,N'-bis(1-methylethyl)-	$C_{11}H_{21}N_5S$ / 255.39	4147-51-7 / 105	3349		
5223	5H,10H-Dipyrrolo[1,2-a:1',2'-d]pyrazine-5,10-dione / Pyrocoll	$C_{10}H_6N_2O_2$ / 186.17	484-73-1 / 268	sub	5-24-08-00263	eth 4; EtOH 4; chl 4
5224	Diquat dibromide / Dipyrido[1,2-a:2',1'-c]pyrazinediium, 6,7-dihydro-, dibromide	$C_{12}H_{12}N_2 \cdot 2Br$ / 344.05	85-00-7 / 337	3359	1.24[20]	
5225	Diselenide, diphenyl / Diphenyl diselenide	$C_{12}H_{10}Se_2$ / 312.13	1666-13-3 / 63.5	202[11]	4-06-00-01781 / 1.557[80]	EtOH 3; eth 3; xyl 3; MeOH 3 / 1.743[20]
5226	Disilane, 1,2-dichloro-1,1,2,2-tetramethyl-	$C_4H_{12}Cl_2Si_2$ / 187.22	4342-61-4	49-50[18]	4-04-00-04280 / 1.010[20]	1.4548[20]
5227	Disilane, 1,2-difluoro-1,1,2,2-tetramethyl- / 1,2-Difluorotetramethyldisilane	$C_4H_{12}F_2Si_2$ / 154.31	661-68-7	94	4-04-00-04279 / 0.9120[20]	1.3837[20]
5228	Disilane, hexamethyl- / Permethyldisilane	$C_6H_{18}Si_2$ / 146.38	1450-14-2 / 13.5	113.5	4-04-00-04277 / 0.7247[22]	H_2O 1; eth 3; ace 3; bz 3 / 1.4229[20]
5229	Disiloxane, 1,3-bis(bromomethyl)-1,1,3,3-tetramethyl- / 1,3-Bis(bromomethyl)tetramethyldisiloxane	$C_6H_{16}Br_2OSi_2$ / 320.17	2351-13-5 /	102-4[15]	4-04-00-04027 / 1.3918[25]	1.4719[25]
5230	Disiloxane, 1,3-bis(chloromethyl)-1,1,3,3-tetramethyl-	$C_6H_{16}Cl_2OSi_2$ / 231.27	2362-10-9 / -90	204; 92[21]	4-04-00-04026 / 1.0381[20]	1.4398[20]
5231	Disiloxane, 1,3-bis(dichloromethyl)-1,1,3,3-tetramethyl- / 1,3-Bis-(dichloromethyl)tetramethyl disiloxane	$C_6H_{14}Cl_4OSi_2$ / 300.16	2943-70-6 /	149-50[50]	4-04-00-04031 / 1.2213[20]	1.4660[20]
5232	Disiloxane, 1,3-dichloro-1,1,3,3-tetramethyl-	$C_4H_{12}Cl_2OSi_2$ / 203.22	2401-73-2 / -37.5	138	4-04-00-04119 / 1.038[20]	

No.	Name / Synonym	Mol. Form. / Mol. Wt.	CAS RN / mp/°C	Merck No. / bp/°C	Beil. Ref. / den/g cm^{-3}	Solubility / n_D
5233	Disiloxane, 1,3-diethenyl-1,1,3,3-tetramethyl-	$C_8H_{18}OSi_2$ 186.40	2627-95-4 -99.7	39	4-04-00-04080 0.811[20]	1.4123[20]
5234	Disiloxane, 1,3-dimethoxy-1,1,3,3-tetramethyl- Bis(methoxydimethylsilyl) oxide	$C_6H_{18}O_3Si_2$ 194.38	18187-24-1	139	4-04-00-04117 0.9048[20]	1.3835[20]
5235	Disiloxane, hexaethyl- 1,1,1,3,3,3-Hexaethyldisiloxane	$C_{12}H_{30}OSi_2$ 246.54	994-49-0	233; 129[30]	4-04-00-04055 0.8589[0]	1.4340[20]
5236	Disiloxane, hexamethyl- Hexamethyldisiloxane	$C_6H_{18}OSi_2$ 162.38	107-46-0 -66	99	4-04-00-04018 0.7638[20]	H_2O 1 1.3774[20]
5237	Disiloxane, 1,1,3,3-tetramethyl-	$C_4H_{14}OSi_2$ 134.33	3277-26-7	72	4-04-00-03991 0.7572[20]	1.3700[20]
5238	Disiloxane, 1,1,3,3-tetramethyl-1,3-diphenyl-	$C_{16}H_{22}OSi_2$ 286.52	56-33-7	155-8[13]	4-16-00-01475 0.9763[20]	ctc 3 1.5176[20]
5239	Disiloxane, 1,1,1-trimethyl-3,3,3-triphenyl- 1,1,1-Trimethyl-3,3,3-triphenyldisiloxane	$C_{21}H_{24}OSi_2$ 348.59	799-53-1 51	349	4-16-00-01486 1.0320[25]	ctc 3; CS_2 3
5240	Dispiro[5.1.5.1]tetradecane Dispiro[5,1,5,1]tetradecane	$C_{14}H_{24}$ 192.34	184-97-4 10	131[10]	4-05-00-00491 0.872[26]	1.4735[27]
5241	Distannoxane, hexabutyl- Bis(tributyltin) oxide	$C_{24}H_{54}OSn_2$ 596.11	56-35-9 45	180[2]		
5242	Disulfide, bis(1,1-dimethylethyl)	$C_8H_{18}S_2$ 178.36	110-06-5 -5	88[21]	4-01-00-01638 0.9226[20]	1.4899[20]
5243	Disulfide, bis(3-methylbutyl)	$C_{10}H_{22}S_2$ 206.42	2051-04-9	247	4-01-00-01689 0.9192[20]	1.4864[20]
5244	Disulfide, bis(1-methylethyl)	$C_6H_{14}S_2$ 150.31	4253-89-8 -69	177	4-01-00-01503 0.9435[20]	1.4916[20]
5245	Disulfide, bis(3-methylphenyl)	$C_{14}H_{14}S_2$ 246.40	20333-41-9 -21	177[3]	4-06-00-02085	EtOH 3; eth; ace 3
5246	Disulfide, bis(4-methylphenyl)	$C_{14}H_{14}S_2$ 246.40	103-19-5 47.5	210-15[20]	4-06-00-02206 1.114[51]	H_2O 1; EtOH 3; eth 4; ace 3
5247	Disulfide, bis(3-nitrophenyl) Nitrophenide	$C_{12}H_8N_2O_4S_2$ 308.34	537-91-7 84	6539	4-06-00-01686	EtOH 2; eth 3; chl 2
5248	Disulfide, bis(phenylmethyl) Dibenzyl disulfide	$C_{14}H_{14}S_2$ 246.40	150-60-7 71.5	2995	4-06-00-02760	H_2O 2; EtOH 3; eth 3; bz 3
5249	Disulfide, bis(trifluoromethyl)	$C_2F_6S_2$ 202.14	372-64-5	34.6	4-03-00-00278	EtOH 4; peth 4
5250	Disulfide, diacetyl Acetyl disulfide	$C_4H_6O_2S_2$ 150.22	592-22-3 20	105[18]	4-02-00-00564	eth 4; EtOH 4
5251	Disulfide, dibenzoyl	$C_{14}H_{10}O_2S_2$ 274.36	644-32-6 134.5	dec	3-09-00-01977	H_2O 1; EtOH 2; eth 2; CS_2 3
5252	Disulfide, dibutyl	$C_8H_{18}S_2$ 178.36	629-45-8	117[20]; 85[3]	4-01-00-01562 0.9371[20]	H_2O 1; EtOH 5; eth 5 1.4923[20]
5253	Disulfide, diethyl Diethyl disulfide	$C_4H_{10}S_2$ 122.26	110-81-6 -101.5	154.1	4-01-00-01397 0.9931[20]	H_2O 2; EtOH 5; eth 5 1.5073[20]
5254	Disulfide, dimethyl Dimethyl disulfide	$C_2H_6S_2$ 94.20	624-92-0 -85	109.8	4-01-00-01281 1.0625[20]	H_2O 1; EtOH 5; eth 5 1.5289[20]
5255	Disulfide, di-1-naphthalenyl	$C_{20}H_{14}S_2$ 318.46	39178-11-5 91		4-06-00-04245 1.144[20]	H_2O 1; EtOH 2; eth 4; chl 2
5256	Disulfide, di-2-naphthalenyl	$C_{20}H_{14}S_2$ 318.46	5586-15-2 139.5		4-06-00-04320 1.144[145]	H_2O 1; EtOH 4; eth 4; lig 1 1.4555[20]
5257	Disulfide, dipentyl	$C_{10}H_{22}S_2$ 206.42	112-51-6	119[7]	4-01-00-01654 0.9221[20]	H_2O 1; eth 3 1.4889[20]
5258	Disulfide, diphenyl	$C_{12}H_{10}S_2$ 218.34	882-33-7 62	310	4-06-00-01560 1.353[20]	H_2O 1; EtOH 3; eth 3; bz 3
5259	Disulfide, dipropyl	$C_6H_{14}S_2$ 150.31	629-19-6 -85.6	193.5	4-01-00-01454 0.9599[20]	1.4981[20]
5260	Disulfoton Phosphorodithioic acid, O,O-diethyl S-[2-(ethylthio)ethyl] ester	$C_8H_{19}O_2PS_3$ 274.41	298-04-4 -25	3371 108[0.01]; 128[1]	1.144[20]	
5261	1,2-Dithiane	$C_4H_8S_2$ 120.24	505-20-4 32.5	80[14]; 60[5]	5-19-01-00010	eth 3; bz 3; chl 3 1.5981[25]
5262	1,3-Dithiane	$C_4H_8S_2$ 120.24	505-23-7 54	89[14]	5-19-01-00013	bz 4; eth 4; chl 4 1.5981[25]
5263	1,4-Dithiane	$C_4H_8S_2$ 120.24	505-29-3 112.3	199.5	5-19-01-00036	H_2O 2; EtOH 3; eth 3; ctc 3
5264	4H-1,3,5-Dithiazine, dihydro-5-methyl- 1,3,5-Dithiazine, 4,5-dihydro-5-methyl	$C_4H_9NS_2$ 135.25	6302-94-9 65	185	4-27-00-06285	eth 4; EtOH 4; HOAc 4
5265	4H-1,3,5-Dithiazine, dihydro-2,4,6-trimethyl-, (2α,4α,6α)-	$C_6H_{13}NS_2$ 163.31	638-17-5 46	dec	2-27-00-00525 1.0613[50]	eth 4; EtOH 4
5266	1,3-Dithiolane 1,3-Dithiacyclopentane	$C_3H_6S_2$ 106.21	4829-04-3 -50	175	5-19-01-00008 1.259[17]	EtOH 3; eth 3; xyl 3 1.5975[15]
5267	Dithiopyr S,S'-Dimethyl-2-difluoromethyl-4-isobutyl-6-trifluoromethylpyridine-3,5-dicarbothioate	$C_{15}H_{16}F_5NO_2S_2$ 401.42	97886-45-8 65			
5268	Docosane n-Docosane	$C_{22}H_{46}$ 310.61	629-97-0 44.4	368.6	4-01-00-00572 0.7944[20]	H_2O 1; EtOH 3; eth 4; chl 3 1.4455[20]
5269	Docosane, 5-butyl- 5-n-Butyldocosane	$C_{26}H_{54}$ 366.71	55282-16-1 208	244[10]	4-01-00-00584 0.8058[20]	1.4503[20]
5270	Docosane, 7-butyl- 7-Butyldocosane	$C_{26}H_{54}$ 366.71	55282-15-0 3.2	241[10]	4-01-00-00584 0.8046[20]	1.4499[20]

No.	Name / Synonym	Mol. Form. / Mol. Wt.	CAS RN / mp/°C	Merck No. / bp/°C	Beil. Ref. / den/g cm⁻³	Solubility / n_D
5271	Docosane, 9-butyl- / 9-Butyldocosane	$C_{26}H_{54}$ / 366.71	55282-14-9 / 1.3	242.5^{10}	4-01-00-00584 / 0.8044^{20}	1.4498^{20}
5272	Docosane, 11-butyl- / 11-Butyldocosane	$C_{26}H_{54}$ / 366.71	13475-76-8 /	242.5^{10}	4-01-00-00584 / 0.8041^{20}	1.4499^{20}
5273	Docosane, decyl- / Docosane, 11-decyl	$C_{32}H_{66}$ / 450.88	28261-97-4 / 1.0	286^{10}	4-01-00-00597 / 0.8131^{20}	1.4546^{20}
5274	Docosane, 7-hexyl- / 7-Hexyldocosane	$C_{28}H_{58}$ / 394.77	55373-86-9 / 19.3	258^{10}	4-01-00-00590 / 0.8080^{20}	1.4517^{20}
5275	Docosane, 9-octyl- / 9-Octyldocosane	$C_{30}H_{62}$ / 422.82	55319-83-0 / 8.6	272.5^{10}	4-01-00-00594 / 0.8114^{20}	1.4537^{20}
5276	Docosanoic acid / Behenic acid	$C_{22}H_{44}O_2$ / 340.59	112-85-6 / 81	1033 / 306^{60}	4-02-00-01290 / 0.8223^{90}	H_2O 2; EtOH 2; eth 2 / 1.4270^{100}
5277	Docosanoic acid, ethyl ester	$C_{24}H_{48}O_2$ / 368.64	5908-87-2 / 50	240^{10}	4-02-00-01292	eth 4; EtOH 4
5278	Docosanoic acid, methyl ester	$C_{23}H_{46}O_2$ / 354.62	929-77-1 / 54		4-02-00-01291	eth 4; EtOH 4 / 1.4339^{60}
5279	1-Docosanol	$C_{22}H_{46}O$ / 326.61	661-19-8 / 72.5	$180^{0.22}$	4-01-00-01906	H_2O 2; EtOH 4; eth 2; chl 3
5280	1-Docosene	$C_{22}H_{44}$ / 308.59	1599-67-3 / 38	367	3-01-00-00883 / 0.794^{25}	
5281	13-Docosenoic acid	$C_{22}H_{42}O_2$ / 338.57	1072-39-5 / 34.7	242^5	4-02-00-01676 / 0.8532^{20}	1.4444^{70}
5282	13-Docosenoic acid, (E)- / Brassidic acid	$C_{22}H_{42}O_2$ / 338.57	506-33-2 / 61.9	1359 / 282^{30}, 256^{10}	4-02-00-01677 / 0.8585^{57}	1.4347^{100}
5283	13-Docosenoic acid, (Z)- / Erucic acid	$C_{22}H_{42}O_2$ / 338.57	112-86-7 / 33.5	3619 / 265^{15}	4-02-00-01676 / 0.860^{55}	H_2O 1; EtOH 3; eth 4; ctc 3 / 1.4758^{20}
5284	13-Docosen-1-ol, (Z)- / Erucyl alcohol	$C_{22}H_{44}O$ / 324.59	629-98-1 / 34.7	240^{10}	4-01-00-02209 / 0.8416^{33}	bz 4; EtOH 4; peth 4
5285	1,11-Dodecadiene	$C_{12}H_{22}$ / 166.31	5876-87-9 /	208.5	4-01-00-01067 / 0.7702^{20}	1.4400^{20}
5286	3,9-Dodecadiyne	$C_{12}H_{18}$ / 162.27	61827-89-2 /			
5287	5,7-Dodecadiyne / Dibutylbutadiyne	$C_{12}H_{18}$ / 162.27	1120-29-2 /	103^8	4-01-00-01130	
5288	Dodecanal / Lauraldehyde	$C_{12}H_{24}O$ / 184.32	112-54-9 / 44.5	185^{100}, $100^{2.5}$	4-01-00-03380 / 0.8352^{15}	H_2O 1; EtOH 2; eth 3 / 1.435^{22}
5289	Dodecanamide	$C_{12}H_{25}NO$ / 199.34	1120-16-7 / 110	199^{12}	4-02-00-01103	H_2O 1; EtOH 3; eth 2; ace 3 / 1.4287^{110}
5290	Dodecanamide, N,N-diethyl-	$C_{16}H_{33}NO$ / 255.44	3352-87-2 /	$166-7^2$	4-04-00-00357 / 0.847^{25}	chl 3 / 1.4545^{20}
5291	1-Dodecanamine	$C_{12}H_{27}N$ / 185.35	124-22-1 / 28.3	259	4-04-00-00794 / 0.8015^{20}	H_2O 2; EtOH 5; eth 5; bz 5 / 1.4421^{20}
5292	Dodecane	$C_{12}H_{26}$ / 170.34	112-40-3 / -9.6	216.3	4-01-00-00498 / 0.7487^{20}	H_2O 1; EtOH 4; eth 4; ace 4 / 1.4216^{20}
5293	Dodecane, 1-bromo- / Lauryl bromide	$C_{12}H_{25}Br$ / 249.23	143-15-7 / -9.5	5260 / 276	4-01-00-00502 / 1.0399^{20}	H_2O 1; EtOH 3; eth 3; ace 5 / 1.4583^{20}
5294	Dodecane, 1-bromo-12-fluoro- / 1-Bromo-12-fluorododecane	$C_{12}H_{24}BrF$ / 267.22	353-29-7 /	$85.8^{0.15}$	4-01-00-00502	ace 4; eth EtOH 4 / 1.4524^{20}
5295	Dodecane, 1-chloro-	$C_{12}H_{25}Cl$ / 204.78	112-52-7 / -9.3	260	4-01-00-00501 / 0.8687^{20}	H_2O 1; EtOH 4; ace 5; bz 3 / 1.4433^{20}
5296	Dodecane, 1,12-dibromo-	$C_{12}H_{24}Br_2$ / 328.13	3344-70-5 / 41	215^{15}	4-01-00-00503	H_2O 1; EtOH 4; eth 3; chl 4
5297	Dodecane, 1,1-dimethoxy- / Lauraldehyde, dimethyl acetal	$C_{14}H_{30}O_2$ / 230.39	14620-52-1 /	$132-4^5$	4-01-00-03381	eth 4; EtOH 4 / 1.4310^{25}
5298	Dodecane, 1,12-dimethoxy- / 1,12-Dodecanediol, dimethyl ether	$C_{14}H_{30}O_2$ / 230.39	73120-52-2 / 11.5	$155-6^{15}$	4-01-00-02627 / 0.8563^{22}	1.436^{22}
5299	Dodecanedioic acid	$C_{12}H_{22}O_4$ / 230.30	693-23-2 / 128	222^{25}	4-02-00-02126 / 1.15^{25}	tfa 3
5300	Dodecanedioic acid, dimethyl ester	$C_{14}H_{26}O_4$ / 258.36	1731-79-9 / 31.9	170^{10}	4-02-00-02126	H_2O 1; AcOEt 3
5301	1,12-Dodecanediol / 1,12-Dihydroxydodecane	$C_{12}H_{26}O_2$ / 202.34	5675-51-4 / 81.3	189^{12}	4-01-00-02627	tfa 3
5302	Dodecane, 1-iodo-	$C_{12}H_{25}I$ / 296.23	4292-19-7 / 0.3	298.2	4-01-00-00503 / 1.1999^{20}	H_2O 1; EtOH 3; eth 5; ace 5 / 1.4840^{20}
5303	Dodecanenitrile / Lauronitrile	$C_{12}H_{23}N$ / 181.32	2437-25-4 / 4	277; 198^{100}	4-02-00-01104 / 0.8240^{20}	H_2O 1; EtOH 5; eth 5; ace 5 / 1.4361^{20}
5304	1-Dodecanethiol	$C_{12}H_{26}S$ / 202.40	112-55-0 /	$142-5^{15}$	4-01-00-01851 / 0.8435^{20}	H_2O 1; EtOH 3; eth 3; chl 3 / 1.4589^{20}
5305	Dodecanoic acid / Lauric acid	$C_{12}H_{24}O_2$ / 200.32	143-07-7 / 43.2	5254 / 91.4	4-02-00-01082 / 0.8679^{50}	H_2O 1; EtOH 4; eth 4; ace 3 / 1.4183^{82}
5306	Dodecanoic acid, anhydride	$C_{24}H_{46}O_3$ / 382.63	645-66-9 / 41.8		4-02-00-01101 / 0.8533^{70}	EtOH 4 / 1.4292^{70}
5307	Dodecanoic acid, 2-bromo-	$C_{12}H_{23}BrO_2$ / 279.22	111-56-8 / 32	$157-9^2$	4-02-00-01106 / 1.1474^{74}	bz 4; eth 4; EtOH 4; lig 4 / 1.4585^{24}
5308	Dodecanoic acid, 2-bromoethyl ester	$C_{14}H_{27}BrO_2$ / 307.27	6309-50-8 /	176^{12}; 145^3	4-02-00-01092 / 1.088^{25}	1.4547^{25}
5309	Dodecanoic acid, 2-bromo-, methyl ester	$C_{13}H_{25}BrO_2$ / 293.24	617-60-7 /	$169-71^{12}$	4-02-00-01107 / 1.113^{15}	1.4572^{25}

No.	Name Synonym	Mol. Form. Mol. Wt.	CAS RN mp/°C	Merck No. bp/°C	Beil. Ref. den/g cm^{-3}	Solubility n_D
5310	Dodecanoic acid, 2,3-dihydroxypropyl ester, (±)- Glycerol, 1-laurate (DL)	$C_{15}H_{30}O_4$ 274.40	40738-26-9 63	186[2]	4-02-00-01096 0.9248[97]	EtOH 2; eth 4; ace 4; bz 3 1.4350[86]
5311	Dodecanoic acid, 1,2-ethanediyl ester Ethylene glycol dilaurate	$C_{26}H_{50}O_4$ 426.68	624-04-4 56.6	4393 188[20]	4-02-00-01094	eth 4; EtOH 4
5312	Dodecanoic acid, ethyl ester Ethyl laurate	$C_{14}H_{28}O_2$ 228.38	106-33-2 -10	3774 271; 154[15]	4-02-00-01092 0.8618[20]	H_2O 1; EtOH 4; eth 5; ctc 2 1.4311[20]
5313	Dodecanoic acid, 2-(2-hydroxyethoxy)ethyl ester Diethylene glycol monolaurate	$C_{16}H_{32}O_4$ 288.43	141-20-8 17.5	3110 >270	0.96[25]	EtOH 5; eth 5; ace 5; bz 3
5314	Dodecanoic acid, 2-methyl-	$C_{13}H_{26}O_2$ 214.35	2874-74-0 22	153[1]	4-02-00-01121 0.890[18]	
5315	Dodecanoic acid, methyl ester Methyl laurate	$C_{13}H_{26}O_2$ 214.35	111-82-0 5.2	267	4-02-00-01090 0.8702[20]	H_2O 1; EtOH 5; eth 5; ace 5 1.4319[20]
5316	Dodecanoic acid, 1-methylethyl ester	$C_{15}H_{30}O_2$ 242.40	10233-13-3	196[60]; 117[2]	4-02-00-01092 0.8536[20]	eth 4; EtOH 4 1.4280[25]
5317	Dodecanoic acid, phenyl ester Phenyl laurate	$C_{18}H_{28}O_2$ 276.42	4228-00-6 24.5	210[15]	4-06-00-00618 0.9354[30]	ace 4; eth 4; EtOH 4
5318	Dodecanoic acid, phenylmethyl ester	$C_{19}H_{30}O_2$ 290.45	140-25-0 8.5	209-11[12]	4-06-00-02267 0.9429[25]	bz 4; eth 4; EtOH 4; peth 4 1.4812[24]
5319	Dodecanoic acid, 1,2,3-propanetriyl ester	$C_{39}H_{74}O_6$ 639.01	538-24-9		4-02-00-01098 0.8986[55]	H_2O 1; EtOH 3; eth 3; ace 4 1.4404[60]
5320	Dodecanoic acid, propyl ester Propyl laurate	$C_{15}H_{30}O_2$ 242.40	3681-78-5	205[60]; 124[2]	4-02-00-01092 0.8600[20]	1.4335[20]
5321	1-Dodecanol Lauryl alcohol	$C_{12}H_{26}O$ 186.34	112-53-8 24	3402 259	4-01-00-01844 0.8309[24]	H_2O 1; EtOH 3; eth 3; bz 2
5322	2-Dodecanol	$C_{12}H_{26}O$ 186.34	10203-28-8 19	252	3-01-00-01793 0.8286[20]	1.4400[20]
5323	3-Dodecanol	$C_{12}H_{26}O$ 186.34	10203-30-2 25	130[15]	4-01-00-01854 0.8223[32]	
5324	6-Dodecanol	$C_{12}H_{26}O$ 186.34	6836-38-0 30	225; 119[9]	3-01-00-01794 0.820[40]	eth 4; EtOH 4
5325	2-Dodecanone Decyl methyl ketone	$C_{12}H_{24}O$ 184.32	6175-49-1 21	246.5	4-01-00-03382 0.8198[20]	H_2O 1; EtOH 3; eth 3; ace 3 1.4330[20]
5326	6-Dodecanone Amyl hexyl ketone	$C_{12}H_{24}O$ 184.32	6064-27-3 10	112[9]	4-01-00-03383	1.4302[20]
5327	1-Dodecanone, 1-phenyl-	$C_{18}H_{28}O$ 260.42	1674-38-0 47	201[9]; 181[5]	4-07-00-00847 0.8794[18]	H_2O 1; ace 3; ctc 2 1.4700[18]
5328	Dodecanoyl chloride	$C_{12}H_{23}ClO$ 218.77	112-16-3 -17	145[18]	4-02-00-01103 0.9169[25]	eth 4 1.4458[20]
5329	1,3,6,10-Dodecatetraene, 3,7,11-trimethyl-, (E,E)- α-Farnesene	$C_{15}H_{24}$ 204.36	502-61-4 129-32[12]	3883	3-01-00-01067 0.8410[20]	H_2O 1; eth 3; ace 3; peth 5 1.4836[20]
5330	2,6,10-Dodecatrienal, 3,7,11-trimethyl-	$C_{15}H_{24}O$ 220.35	19317-11-4 172-4[14]		4-01-00-03603 0.893[18]	1.4995
5331	1,6,10-Dodecatriene, 7,11-dimethyl-3-methylene-, (E)- β-Farnesene	$C_{15}H_{24}$ 204.36	18794-84-8 121-2[9]	3884	4-01-00-01133 0.8363[20]	ace 4; eth 4; chl 4 1.4899[20]
5332	1,6,10-Dodecatrien-3-ol, 3,7,11-trimethyl-, [S-(Z)]-	$C_{15}H_{26}O$ 222.37	142-50-7 276; 70[0.1]	6388	4-01-00-02336 0.8778[20]	EtOH 4; eth 3; ace 3; os 3 1.4898[20]
5333	2,6,10-Dodecatrien-1-ol, 3,7,11-trimethyl-, (E,E)-	$C_{15}H_{26}O$ 222.37	106-28-5 160[10]		4-01-00-02335 0.8846[20]	H_2O 1; EtOH 4; eth 3; ace 3 1.4877[20]
5334	2,6,10-Dodecatrien-1-ol, 3,7,11-trimethyl-, (Z,E)-	$C_{15}H_{26}O$ 222.37	3790-71-4 156[12]; 120[0.3]		4-01-00-02335 0.8908[20]	ace 4; eth 4; EtOH 4 1.4877[20]
5335	1-Dodecene	$C_{12}H_{24}$ 168.32	112-41-4 -35.2	213.8	4-01-00-00914 0.7584[20]	H_2O 1; EtOH 3; eth 3; ace 3 1.4300[20]
5336	2-Dodecenedioic acid, (E)- Traumatic acid	$C_{12}H_{20}O_4$ 228.29	6402-36-4 165.5	9493	4-02-00-02279	eth 4; EtOH 4; chl 4
5337	2-Dodecenoic acid	$C_{12}H_{22}O_2$ 198.31	4412-16-2 17.1	155[3]; 127[0.15]	4-02-00-01619 0.9265[20]	1.4629[25]
5338	4-Dodecenoic acid Linderic acid	$C_{12}H_{22}O_2$ 198.31	505-92-0 1.3	171[13]	4-02-00-01619 0.9081[15]	bz 4; eth 4; chl 4 1.4529[20]
5339	5-Dodecenoic acid	$C_{12}H_{22}O_2$ 198.31	2761-84-4 1.3	170-2[13]	4-02-00-01619 0.9081[20]	
5340	11-Dodecenoic acid	$C_{12}H_{22}O_2$ 198.31	65423-25-8 20	171[13]; 144[3]	4-02-00-01618 0.9014[20]	1.4510[20]
5341	11-Dodecenoic acid, methyl ester Methyl 11-dodecenoate	$C_{13}H_{24}O_2$ 212.33	29972-79-0 138-9[13]		4-02-00-01618 0.8789[22]	1.4414[20]
5342	1-Dodecen-3-yne	$C_{12}H_{20}$ 164.29	74744-36-8 78[4]		4-01-00-01112 0.7858[25]	1.4510[25]
5343	1-Dodecyne Decylacetylene	$C_{12}H_{22}$ 166.31	765-03-7 -19	215	4-01-00-01066 0.7788[20]	1.4340[20]
5344	2-Dodecyne	$C_{12}H_{22}$ 166.31	629-49-2 -9	105[15]	0-01-00-00261 0.7917[15]	1.4828[20]
5345	3-Dodecyne	$C_{12}H_{22}$ 166.31	6790-27-8 95[12]		3-01-00-01025 0.7871[20]	ace 4; eth 4 1.4442[20]

No.	Name / Synonym	Mol. Form. / Mol. Wt.	CAS RN / mp/°C	Merck No. / bp/°C	Beil. Ref. / den/g cm^{-3}	Solubility / n_D
5346	6-Dodecyne	$C_{12}H_{22}$ / 166.31	6975-99-1	/ 210; 100[14]	4-01-00-01067 / 0.7871[20]	ace 4; eth 4; EtOH 4 / 1.4442[20]
5347	Dodine / Guanidine, dodecyl-, monoacetate	$C_{13}H_{29}N_3 \cdot C_2H_4O_2$ / 287.45	2439-10-3 / 136	3406		
5348	Dotriacontane / Bicetyl	$C_{32}H_{66}$ / 450.88	544-85-4 / 69.7	/ 467	4-01-00-00595 / 0.8124[20]	H_2O 1; EtOH 2; eth 3; bz 4 / 1.4550[20]
5349	Eburnamenine-14-carboxylic acid, 14,15-dihydro-14-hydroxy-, methyl ester, $(3\alpha,14\beta,16\alpha)$- / Vincamine	$C_{21}H_{26}N_2O_3$ / 354.45	1617-90-9 / 231.5	9888	4-25-00-01252	
5350	Eicosane	$C_{20}H_{42}$ / 282.55	112-95-8 / 36.8	/ 343	4-01-00-00563 / 0.7886[20]	H_2O 1; eth 3; ace 4; bz 3 / 1.4425[20]
5351	Eicosane, 1-cyclopentyl- / Cyclopentane, eicosyl	$C_{25}H_{50}$ / 350.67	22331-38-0 / 38	/ 413	4-05-00-00198 / 0.8276[20]	ace 4; bz 4; eth 4; EtOH 4 / 1.4595[20]
5352	Eicosanedioic acid / 1,18-Octadecanedicarboxylic acid	$C_{20}H_{38}O_4$ / 342.52	2424-92-2 / 125.5	/ 233-4[2]	4-02-00-02185	eth 3
5353	Eicosanedioic acid, diethyl ester	$C_{24}H_{46}O_4$ / 398.63	42235-39-2 / 54.5	/ 240[12]; 230[2]	3-02-00-01881	eth 4; EtOH 4
5354	Eicosane, 9-octyl- / Eicosane, 9-ocytl	$C_{28}H_{58}$ / 394.77	13475-77-9 / 0.5	/ 257[10]; 199[0.5]	4-01-00-00590 / 0.8075[20]	1.4515[20]
5355	Eicosanoic acid / Arachidic acid	$C_{20}H_{40}O_2$ / 312.54	506-30-9 / 76.5	791 / 328 dec; 204[1]	4-02-00-01275 / 0.8240[100]	H_2O 1; EtOH 2; eth 4; bz 3 / 1.425[100]
5356	Eicosanoic acid, ethyl ester	$C_{22}H_{44}O_2$ / 340.59	18281-05-5 / 50	/ 295[100]; 186[2]	3-02-00-01067	H_2O 1; EtOH 3; eth 3; bz 1
5357	Eicosanoic acid, methyl ester	$C_{21}H_{41}O_2$ / 325.56	1120-28-1 / 54.5	/ 215[10]	4-02-00-01276	bz 4; eth 4; EtOH 4; chl 4 / 1.4317[60]
5358	1-Eicosanol / Arachic alcohol	$C_{20}H_{42}O$ / 298.55	629-96-9 / 66.1	/ 309; 222[3]	4-01-00-01900 / 0.8405[20]	H_2O 1; EtOH 2; ace 4; bz 3 / 1.4350[20]
5359	2-Eicosanol / 1-Methyl-1-nonadecanol	$C_{20}H_{42}O$ / 298.55	4340-76-5 / 63.5	/ 357	4-01-00-01902 / 0.8378[20]	ace 4; bz 4 / 1.4912[80]
5360	5,8,11,14-Eicosatetraenoic acid / 5,8,11,14-Eicosotetraenoic acid	$C_{20}H_{32}O_2$ / 304.47	7771-44-0 / -49.5	/ 169[0.15]	4-02-00-01802 / 0.9219[20]	H_2O 1; EtOH 3; eth 3; ace 3 / 1.4824[20]
5361	5,8,11,14-Eicosatetraenoic acid, (all-Z)- / Arachidonic acid	$C_{20}H_{32}O_2$ / 304.47	506-32-1 / -49.5	792 / 163[1]	4-02-00-01802 / 0.9082[20]	ace 4; eth 4; EtOH 4; peth 4 / 1.4824[20]
5362	1-Eicosene	$C_{20}H_{40}$ / 280.54	3452-07-1 / 28.5	/ 341; 151[2]	4-01-00-00934 / 0.7882[30]	H_2O 1; bz 3; peth 3 / 1.4440[30]
5363	11-Eicosenoic acid / Δ 11-Eicosenoic acid	$C_{20}H_{38}O_2$ / 310.52	2462-94-4 / 24	/ 267[15]	4-02-00-01673 / 0.8826[25]	EtOH 4; MeOH 4
5364	9-Eicosenoic acid	$C_{20}H_{38}O_2$ / 310.52	506-31-0 / 23	/ 220[6]; 170[0.1]	4-02-00-01672 / 0.8882[25]	1.4597[25]
5365	9-Eicosenoic acid, (Z)-	$C_{20}H_{38}O_2$ / 310.52	29204-02-2 / 24.5	/ 220[6]	4-02-00-01672 / 0.8882[25]	
5366	1-Eicosyne	$C_{20}H_{38}$ / 278.52	765-27-5 / 36	/ 340	4-01-00-01077 / 0.8073[20]	bz 4; peth 4 / 1.4501[20]
5367	Emetan-6'-ol, 7',10,11-trimethoxy- / Cephaeline	$C_{28}H_{38}N_2O_4$ / 466.62	483-17-0 / 115.5	1970	5-23-13-00611	ace 4; EtOH 4; MeOH 4; chl 4
5368	Emetan, 6',7',10,11-tetramethoxy- / Emetine	$C_{29}H_{40}N_2O_4$ / 480.65	483-18-1 / 74	3517	4-23-00-03422	H_2O 1; EtOH 3; eth 3; ace 3
5369	Endosulfan / 5,9-Methano-2,4,3-benzodioxathiepin, 6,7,8,9,10,10-hexachloro-1,5,5a,6,9,9a-hexahydro-, 3-oxide	$C_9H_6Cl_6O_3S$ / 406.93	115-29-7 / 106	3529 / 106[0.7]	1.745[20]	
5370	Endothall, disodium salt / 7-Oxabicyclo[2.2.1]heptane-2,3-dicarboxylic acid, disodium salt	$C_8H_{10}Na_2O_5$ / 232.14	145-73-3 / 144	3530	1.431[20]	
5371	4,7-Epoxyisobenzofuran-1,3-dione, hexahydro-3a,7a-dimethyl-, $(3a\alpha,4\beta,7\beta,7a\alpha)$- / Cantharidin	$C_{10}H_{12}O_4$ / 196.20	56-25-7 / 218	1755 / sub 84	5-19-05-00051	H_2O 1; EtOH 2; eth 2; ace 2
5372	EPTC / Carbamothioic acid, dipropyl-, S-ethyl ester	$C_9H_{19}NOS$ / 189.32	759-94-4 /	3580 / 127[20]	0.9546[30]	
5373	Ergoline-8-carboxamide, 9,10-didehydro-N-(2-hydroxy-1-methylethyl)-6-methyl-, $[8\alpha(S)]$- / Ergometrinine	$C_{19}H_{23}N_3O_2$ / 325.41	479-00-5 / 195-7 dec	3599	4-25-00-00950	chl 4
5374	Ergoline-8-carboxamide, 9,10-didehydro-6-methyl-, (8β)- / Lysergamide	$C_{16}H_{17}N_3O$ / 267.33	478-94-4 / 137.5	5505	4-25-00-00937	EtOH 2; ace 2; os 2
5375	Ergoline-8-carboxylic acid, 9,10-didehydro-6-methyl-, (8α)- / Isolysergic acid	$C_{16}H_{16}N_2O_2$ / 268.32	478-95-5 / 218 dec	5065	4-25-00-00935	H_2O 2; EtOH 2; py 3
5376	Ergoline-8-carboxylic acid, 9,10-didehydro-6-methyl-, (8β)- / Lysergic acid	$C_{16}H_{16}N_2O_2$ / 268.32	82-58-6 / 240 dec	5506	4-25-00-00934	H_2O 2; EtOH 3; eth 2; bz 2
5377	Ergostane, (5α)-	$C_{28}H_{50}$ / 386.71	511-20-6 / 85	3602	4-05-00-01234	ace 4; eth 4; chl 4
5378	Ergostane, (5β)- / Coproergostane	$C_{28}H_{50}$ / 386.71	511-21-7 / 64	2516	3-05-00-01143	eth 4; chl 4

No.	Name Synonym	Mol. Form. Mol. Wt.	CAS RN mp/°C	Merck No. bp/°C	Beil. Ref. den/g cm^{-3}	Solubility n_D
5379	Ergostan-3-ol, (3β,5α)- Ergostanol	$C_{28}H_{50}O$ 402.70	6538-02-9 144.5	3603	4-06-00-03602	H_2O 1; eth 3; chl 3
5380	Ergosta-5,7,9(11),22-tetraen-3-ol, (3β,22E)- Dehydroergosterol	$C_{28}H_{42}O$ 394.64	516-85-8 146	2861 230[0.5]	4-06-00-04838	ace 4; bz 4; eth 4; EtOH 4
5381	Ergosta-5,7,22-trien-3-ol, (3β,10α,22E)- Pyrocalciferol	$C_{28}H_{44}O$ 396.66	128-27-8 94	8007	3-06-00-03098	H_2O 1; EtOH 3; chl 3; MeOH 3
5382	Ergosta-5,7,22-trien-3-ol, (3β,9β,10α,22E)- Lumisterol	$C_{28}H_{44}O$ 396.66	474-69-1 118	5471	4-06-00-04408	H_2O 1; EtOH 3; eth 4; ace 4
5383	Ergosta-5,7,22-trien-3-ol, (3β,22E)- Ergosterol	$C_{28}H_{44}O$ 396.66	57-87-4 170	3607 250[0.01]	4-06-00-04407	H_2O 1; EtOH 2; eth 2; bz 3
5384	Ergost-7-en-3-ol, (3β,5α)- γ-Ergostenol	$C_{28}H_{48}O$ 400.69	516-78-9 146	3606	4-06-00-04029	eth 3
5385	Ergost-8(14)-en-3-ol, (3β,5α)- α-Ergostenol	$C_{28}H_{48}O$ 400.69	632-32-6 131	3604	4-06-00-04031	EtOH 3; eth 3; bz 3; chl 3
5386	Ergost-5-en-3-ol, (3β,24R)- Campesterol	$C_{28}H_{48}O$ 400.69	474-62-4 157.5	1735	3-06-00-02680	
5387	Ergotaman-3',6',18-trione, 12'-hydroxy-2',5'- bis(1-methylethyl)-, Ergocornine	$C_{31}H_{39}N_5O_5$ 561.68	564-36-3 182-4 dec	3591	4-25-00-00963	H_2O 1; EtOH 3; ace 3; bz 3
5388	Ergotaman-3',6',18-trione, 12'-hydroxy-2',5'- bis(1-methylethyl)-, Ergocorninine	$C_{31}H_{39}N_5O_5$ 561.68	564-37-4 228 dec	3592	4-25-00-00963	ace 4; bz 4; EtOH 4; chl 4
5389	Ergotaman-3',6',18-trione, 12'-hydroxy-2'-(1- methylethyl)-5'-(2-methylpropyl)- Ergocryptine	$C_{32}H_{41}N_5O_5$ 575.71	511-09-1 212-4 dec	3595	4-25-00-00964	H_2O 1; EtOH 3; chl 3
5390	Ergotaman-3',6',18-trione, 12'-hydroxy-2'-(1- methylethyl)-5'-(2-methylpropyl)-, (5'α,8α)- Ergocryptinine	$C_{33}H_{41}N_5O_5$ 587.72	511-10-4 245 dec	3596	4-25-00-00964	ace 4; chl 4
5391	Ergotaman-3',6',18-trione, 12'-hydroxy-2'-(1- methylethyl)-5'-(phenylmethyl)- Ergocristine	$C_{35}H_{39}N_5O_5$ 609.73	511-08-0 175 dec	3593	4-25-00-00966	H_2O 1; EtOH 3; ace 3; chl 3
5392	Ergotaman-3',6',18-trione, 12'-hydroxy-2'-(1- methylethyl)-5'-(phenylmethyl)- Ergocristinine	$C_{35}H_{39}N_5O_5$ 609.73	511-07-9 237-8 dec	3594	4-25-00-00967	H_2O 1; EtOH 2; ace 2; chl 2
5393	Ergotaman-3',6',18-trione, 12'-hydroxy-2'- methyl-5'-(2-methylpropyl)-, Ergosine	$C_{30}H_{37}N_5O_5$ 547.65	561-94-4 228 dec	3601	4-25-00-00961	ace 3; chl 3; MeOH 2
5394	Ergotaman-3',6',18-trione, 12'-hydroxy-2'- methyl-5'-(phenylmethyl)-, (5'α)- Ergotamine	$C_{33}H_{55}N_5O_5$ 601.83	113-15-5 213-4 dec	3609	4-25-00-00964	bz 4; eth 4; chl 4
5395	Ergotaman-3',6',18-trione, 12'-hydroxy-2'- methyl-5'-(phenylmethyl)-, Ergotaminine	$C_{33}H_{35}N_5O_5$ 581.67	639-81-6 252 dec	3610	4-25-00-00966	H_2O 1; EtOH 2; ace 2; bz 2
5396	Erythromycin Propiocine	$C_{37}H_{67}NO_{13}$ 733.94	114-07-8 191	3626	5-18-10-00398	ace 4; eth 4; EtOH 4; chl 4
5397	Estra-1,3,5,7,9-pentaen-17-one, 3-hydroxy- Equilenin	$C_{18}H_{18}O_2$ 266.34	517-09-9 258.5	3581 sub 170	4-08-00-01420	EtOH 2; ace 2; chl 2
5398	Estra-1,3,5(10),7-tetraen-17-one, 3-hydroxy- Equilin	$C_{18}H_{20}O_2$ 268.36	474-86-2 239	3582 sub 170	4-08-00-01366	H_2O 2; EtOH 3; ace 3; diox 3
5399	Estra-1,3,5(10)-triene-3,17-diol (17β)- Estradiol	$C_{18}H_{24}O_2$ 272.39	50-28-2 178.5	3653	4-06-00-06611	ace 4; EtOH 4; Diox 4
5400	Estra-1,3,5(10)-triene-3,17-diol, (17α)- α-Estradiol	$C_{18}H_{24}O_2$ 272.39	57-91-0 221.5	3654	4-06-00-06611	H_2O 1; EtOH 3; eth 2; ace 3
5401	Estra-1,3,5(10)-triene-3,17-diol, (8α,17β)- Isoestradiol	$C_{18}H_{24}O_2$ 272.39	517-04-4 181	5051	4-06-00-06611	EtOH 3; diox 3
5402	Estra-5,7,9-triene-3,17-diol, (3β,17β)- Hexahydroequilenin	$C_{18}H_{24}O_2$ 272.39	517-07-7 168.5	4607 sub 150	4-06-00-06618	
5403	Estra-1,3,5(10)-triene-3,17-diol (17β)-, 3- benzoate Estradiol benzoate	$C_{25}H_{28}O_3$ 376.50	50-50-0 196	3655	4-09-00-00406	
5404	Estra-1,3,5(10)-triene-3,16,17-triol, (16α,17β)- Estriol	$C_{18}H_{24}O_3$ 288.39	50-27-1 288 dec	3659	4-06-00-07550 1.27[25]	EtOH 3; eth 2; bz 2; tfa 2
5405	Estra-1,3,5(10)-triene-3,16,17-triol, (16β,17β)- 16-Epiestriol	$C_{18}H_{24}O_3$ 288.39	547-81-9 290	3567	4-06-00-07549	
5406	Estra-1,3,5(10)-trien-17-one, 3-hydroxy- Estrone	$C_{18}H_{22}O_2$ 270.37	53-16-7 260.2	3660	3-08-00-01171 1.236[25]	H_2O 1; EtOH 2; eth 2; ace 3
5407	Estra-1,3,5(10)-trien-17-one, 3-hydroxy-, (8α)- 8-Isoestrone	$C_{18}H_{22}O_2$ 270.37	517-06-6 247	5052	3-08-00-01170	eth 4; Diox 4
5408	Estr-4-en-3-one, 17-(1-oxo-3-phenylpropoxy)-, (17β)- Nandrolone phenpropionate	$C_{27}H_{34}O_3$ 406.57	62-90-8	6283	4-09-00-01758	
5409	Ethalfluralin Benzenamine, N-ethyl-N-(2-methyl-2- propenyl)-2,6-dinitro-4-(trifluoromethyl)-	$C_{13}H_{14}F_3N_3O_4$ 333.27	55283-68-6 57	3671 256 dec		
5410	Ethanamine Ethylamine	C_2H_7N 45.08	75-04-7 -80.5	3718 16.5	4-04-00-00307 *0.677[25]	H_2O 5; EtOH 5; eth 5 1.3663[20]
5411	Ethanamine, 2-[(3-butyl-1-isoquinolinyl)oxy]- N,N-dimethyl- Dimethisoquin	$C_{17}H_{24}N_2O$ 272.39	86-80-6 146	3207 155-7[3]	5-21-03-00414	H_2O 3; EtOH 3 1.5486[20]

No.	Name Synonym	Mol. Form. Mol. Wt.	CAS RN	mp/°C	Merck No. bp/°C	Beil. Ref. den/g cm⁻³	Solubility n_D
5412	Ethanamine, 2-chloro-N,N-bis(2-chloroethyl)- 2,2',2"-Trichlorotriethylamine	$C_8H_{12}Cl_3N$ 204.53	555-77-1 -4	9560 143-4[15]	4-04-00-00447	bz 4; eth 4; EtOH 4
5413	Ethanamine, 2-chloro-N-(2-chloroethyl)-N-methyl- Mechlorethamine	$C_5H_{11}Cl_2N$ 156.05	51-75-2 -60	5655 87[18]; 64[5]		H_2O 2; ctc 5; DMF 5
5414	Ethanamine, 2,2-diethoxy-	$C_6H_{15}NO_2$ 133.19	645-36-3 -78	163	4-04-00-01918 0.9159[25]	H_2O 4; eth 4; EtOH 4; chl 4 1.4123[25]
5415	Ethanamine, 2,2-diethoxy-N,N-diethyl-	$C_{10}H_{23}NO_2$ 189.30	3616-57-7	194.5	4-04-00-01919 0.863[25]	H_2O 3; EtOH 3; eth 3; ctc 2 1.4189[25]
5416	Ethanamine, 2,2-diethoxy-N,N-dimethyl-	$C_8H_{19}NO_2$ 161.24	3616-56-6	170.5	3-04-00-00870 0.885[7]	H_2O 4; EtOH 3; eth 3; ace 3 1.4129[20]
5417	Ethanamine, N,N-diethyl- Triethylamine	$C_6H_{15}N$ 101.19	121-44-8 -114.7	9582 89	4-04-00-00322 0.7275[20]	H_2O 3; EtOH 3; eth 3; ace 4 1.4010[20]
5418	Ethanamine, N,N-diethyl-, hydrochloride	$C_6H_{16}ClN$ 137.65	554-68-7 260 dec	sub 245	4-04-00-00322 1.0689[21]	H_2O 4; EtOH 4; chl 4
5419	Ethanamine, 2,2-dimethoxy-	$C_4H_{11}NO_2$ 105.14	22483-09-6	135-9[95]	4-04-00-01918 0.965[25]	1.4170[20]
5420	Ethanamine, 2,2-dimethoxy-N-methyl-	$C_5H_{13}NO_2$ 119.16	122-07-6	140	2-04-00-00759 0.928[25]	1.4115[20]
5421	Ethanamine, N,N-dimethyl-	$C_4H_{11}N$ 73.14	598-56-1 -140	36.5	4-04-00-00312 0.675[20]	1.3705[25]
5422	Ethanamine, 2-(diphenylmethoxy)-N,N-dimethyl- Diphenhydramine	$C_{17}H_{21}NO$ 255.36	58-73-1	3308	4-06-00-04658	
5423	Ethanamine, 2-ethoxy-	$C_4H_{11}NO$ 89.14	110-76-9	108	4-04-00-01411 0.8512[20]	H_2O 5; EtOH 5; eth 5; ace 3 1.4101[20]
5424	Ethanamine, 2-ethoxy-N,N-dimethyl-	$C_6H_{15}NO$ 117.19	26311-17-1	121	4-04-00-01425 0.806[20]	1.406[20]
5425	Ethanamine, 2-ethoxy-N-(2-ethoxyethyl)- Bis-(2-ethoxyethyl) amine	$C_8H_{19}NO_2$ 161.24	124-21-0 -50	194	4-04-00-01516 0.883[25]	1.4213[20]
5426	Ethanamine, 2-ethoxy-N-methyl- Methyl(2-ethoxyethyl)amine	$C_5H_{13}NO$ 103.16	38256-94-9	115	3-04-00-00647 0.8363[20]	ace 4; bz 4; eth 4; EtOH 4 1.4147[20]
5427	Ethanamine, N-ethyl- Diethylamine	$C_4H_{11}N$ 73.14	109-89-7 -49.8	3100 55.5	4-04-00-00313 0.7056[20]	H_2O 4; EtOH 5; eth 3; ctc 3 1.3864[20]
5428	Ethanamine, N-ethyl-, hydrochloride	$C_4H_{12}ClN$ 109.60	660-68-4 228.5		4-04-00-00313 1.0477[22]	H_2O 4; EtOH 4
5429	Ethanamine, N-ethyl-N-hydroxy-	$C_4H_{11}NO$ 89.14	3710-84-7 10	133	4-04-00-03304 0.8669[20]	1.4195[20]
5430	Ethanamine, N-ethyl-N-methyl-	$C_5H_{13}N$ 87.16	616-39-7 -196	66	4-04-00-00321 0.703[25]	H_2O 4; EtOH 4; eth 4 1.3879[25]
5431	Ethanamine, N-ethyl-N-nitro-	$C_4H_{10}N_2O_2$ 118.14	7119-92-8	206.5	4-04-00-03403 1.057[15]	eth 4; EtOH 4
5432	Ethanamine, N-ethyl-N-nitroso- N-Nitrosodiethylamine	$C_4H_{10}N_2O$ 102.14	55-18-5	6557 176.9	4-04-00-03386 0.9422[20]	H_2O 3; EtOH 3; eth 3; chl 2 1.4386[20]
5433	Ethanamine, hydrochloride	C_2H_8ClN 81.54	557-66-4 109.5	3718	4-04-00-00307 1.2160[20]	H_2O 4; EtOH 4
5434	Ethanamine, N-hydroxy- N-Ethylhydroxylamine	C_2H_7NO 61.08	624-81-7 59-60 dec		4-04-00-03304 0.9079[20]	H_2O 4; EtOH 4 1.4152[66]
5435	Ethanamine, N-methyl-, hydrochloride Ethyl methyl amine, hydrochloride	$C_3H_{10}ClN$ 95.57	624-60-2 128		2-04-00-00589 1.0874[20]	H_2O 4; EtOH 4; eth 1; chl 3
5436	Ethanamine, 1,1,2,2,2-pentafluoro-N,N-bis(pentafluoroethyl)-	$C_6F_{15}N$ 371.05	359-70-6	70.3	4-02-00-00471 1.736[20]	1.262[25]
5437	Ethanamine, N-(phenylmethylene)-	$C_9H_{11}N$ 133.19	6852-54-6	196	4-07-00-00518 0.937[20]	H_2O 1; EtOH 3; eth 3 1.5378[15]
5438	Ethanaminium, 2-(acetyloxy)-N,N,N-trimethyl-, bromide Acetyl choline bromide	$C_7H_{16}BrNO_2$ 226.11	66-23-9 146	80	4-04-00-01446	H_2O 4
5439	Ethanaminium, 2-(acetyloxy)-N,N,N-trimethyl-, chloride Acetyl choline chloride	$C_7H_{16}ClNO_2$ 181.66	60-31-1 150	81	4-04-00-01446	H_2O 3; EtOH 3; eth 1
5440	Ethanaminium, 2-[(aminocarbonyl)oxy]-N,N,N-trimethyl-, chloride Carbamoylcholine chloride	$C_6H_{15}ClN_2O_2$ 182.65	51-83-2 210 dec	1780	4-04-00-01455	H_2O 4; EtOH 2; eth 1; chl 1
5441	Ethanaminium, 2,2',2"-[1,2,3-benzenetriyltris(oxy)]tris[N,N,N-triethyl-, tr Gallamine triethiodide	$C_{30}H_{60}I_3N_3O_3$ 891.54	65-29-2 147.5	4249	4-06-00-07333	H_2O 4; EtOH 4; eth 2; ace 2
5442	Ethanaminium, 2-chloro-N,N,N-trimethyl-, chloride	$C_5H_{13}Cl_2N$ 158.07	999-81-5 239 dec	2103	4-04-00-00445	
5443	Ethanaminium, 2-[(cyclohexylhydroxyphenylacetyl)oxy]-N,N-diethyl-N-methyl-, bromide Oxyphenonium bromide	$C_{21}H_{34}BrNO_3$ 428.41	50-10-2 191.5	6928	3-10-00-00911	H_2O 4; EtOH 2
5444	Ethanaminium, N-[4-[[4-(diethylamino)phenyl]phenylmethylene]-2,5-cyclohexadien-1-ylidene]-N-ethyl-, sulfate (1 Brilliant green	$C_{27}H_{34}N_2O_4S$ 482.64	633-03-4	1367	4-13-00-02281	H_2O 4; EtOH 4
5445	Ethanaminium, N,N-diethyl-N-methyl-2-[(9H-xanthen-9-ylcarbonyl)oxy]-, bromide Methantheline bromide	$C_{21}H_{26}BrNO_3$ 420.35	53-46-3 174.5	5869	5-18-06-00590	H_2O 3; EtOH 3; eth 1; chl 3

No.	Name / Synonym	Mol. Form. / Mol. Wt.	CAS RN / mp/°C	Merck No. / bp/°C	Beil. Ref. / den/g cm^{-3}	Solubility / n_D
5446	Ethanaminium, 2-hydroxy-N,N,N-trimethyl-, chloride Choline chloride	$C_5H_{14}ClNO$ 139.63	67-48-1 305-5 dec	2208	4-04-00-01443	H_2O 4; EtOH 4
5447	Ethanaminium, N,N,N-triethyl-, bromide Tetraethylammonium bromide	$C_8H_{20}BrN$ 210.16	71-91-0 285-7 dec	9133	4-04-00-00331 1.3970[20]	H_2O 4; EtOH 4; chl 4; MeOH 4
5448	Ethane	C_2H_6 30.07	74-84-0 -182.8	3676 -88.6	4-01-00-00108 0.5446[-89]	bz 4
5449	Ethane, 1,2-bis(2-chloroethoxy)-	$C_6H_{12}Cl_2O_2$ 187.07	112-26-5 	232	4-01-00-02379 1.195[20]	ctc 3 1.4592[25]
5450	Ethane, 1,1-bis(ethylthio)-	$C_6H_{14}S_2$ 150.31	14252-42-7 	184	4-01-00-03162 0.9706[20]	eth 3; ace 3; chl 3 1.5025[20]
5451	Ethane, 1,2-bis(ethylthio)-	$C_6H_{14}S_2$ 150.31	5395-75-5 180	217; 108[27]	4-01-00-02452 0.9815[20]	EtOH 3; eth 3 1.5118[20]
5452	Ethane, 1,2-bis(methylthio)-	$C_4H_{10}S_2$ 122.26	6628-18-8 	183; 79[11]	4-01-00-02451 1.0371[20]	H_2O 4; ace 4; eth 4; EtOH 4 1.5292[20]
5453	Ethanebis(thioic) acid, S,S-diethyl ester	$C_6H_{10}O_2S_2$ 178.28	615-85-0 27	235	1-02-00-00244 1.0565[21]	eth 4
5454	Ethane, bromo- Ethyl bromide	C_2H_5Br 108.97	74-96-4 -118.6	3730 38.5	4-01-00-00150 1.4604[20]	H_2O 2; EtOH 5; eth 5; chl 5 1.4239[20]
5455	Ethane, 1-bromo-1-chloro- 1-Bromo-1-chloroethane	C_2H_4BrCl 143.41	593-96-4 	83	4-01-00-00155 1.667[10]	1.4660[2-]
5456	Ethane, 1-bromo-2-chloro-	C_2H_4BrCl 143.41	107-04-0 -16.7	107	4-01-00-00155 1.7392[20]	H_2O 2; EtOH 3; eth 3; chl 3 1.4908[20]
5457	Ethane, 1-bromo-2-chloro-1,1,2-trifluoro- 1-Bromo-2-chloro-1,1,2-trifluoroethane	$C_2HBrClF_3$ 197.38	354-06-3 	52.5	4-01-00-00156 1.8636[25]	1.3738[20]
5458	Ethane, 2-bromo-2-chloro-1,1,1-trifluoro- Halothane	$C_2HBrClF_3$ 197.38	151-67-7 	4517 50.2; 20[243]	1.871[20]	H_2O 2; peth 3 1.3697[0]
5459	Ethane, 2-bromo-1,1-diethoxy-	$C_6H_{13}BrO_2$ 197.07	2032-35-1 	66-7[18]	4-01-00-03151 1.310[25]	EtOH 3; eth 3 1.4387[20]
5460	Ethane, 2-bromo-1,1-dimethoxy-	$C_4H_9BrO_2$ 169.02	7252-83-7 	149	3-01-00-02672 1.430[20]	eth 3; ace 3; chl 3 1.4450[20]
5461	Ethane, 1-bromo-2-ethoxy- 2-Bromoethyl ethyl ether	C_4H_9BrO 153.02	592-55-2 	127.5	4-01-00-01386 1.3852[0]	H_2O 2; EtOH 5; eth 5 1.4447[20]
5462	Ethane, 1-bromo-2-fluoro-	C_2H_4BrF 126.96	762-49-2 	71.5	4-01-00-00154 1.7044[25]	eth 4; EtOH 4 1.4236[20]
5463	Ethane, 1-bromo-2-iodo- 1-Bromo-2-iodoethane	C_2H_4BrI 234.86	590-16-9 28	163	4-01-00-00168 2.516[29]	
5464	Ethane, (bromomethoxy)-	C_3H_7BrO 138.99	53588-92-4 	110	3-01-00-02594 1.4402[20]	eth 4 1.4515[20]
5465	Ethane, 1-bromo-2-methoxy-	C_3H_7BrO 138.99	6482-24-2 	110 dec	4-01-00-01386 1.4623[20]	1.44753[20]
5466	Ethane, 2-bromo-1,1,1-trifluoro-	$C_2H_2BrF_3$ 162.94	421-06-7 -93.9	26	4-01-00-00154 1.7881[20]	1.3331[20]
5467	Ethane, chloro- Ethyl chloride	C_2H_5Cl 64.51	75-00-3 -138.7	3740 12.3	4-01-00-00124 *0.8902[25]	H_2O 2; EtOH 4; eth 5; chl 2 1.3676[20]
5468	Ethane, 1-chloro-1-(2-chloroethoxy)- 1-Chloroethyl 2-chloroethyl ether	$C_4H_8Cl_2O$ 143.01	1462-34-6 	56[17]	3-01-00-02655 1.1867[20]	eth 4; EtOH 4; chl 4 1.4473[20]
5469	Ethane, 2-chloro-1,1-diethoxy-	$C_6H_{13}ClO_2$ 152.62	621-62-5 	157.4	4-01-00-03134 1.0180[20]	H_2O 2; EtOH 5; eth 5; ctc 2 1.4170[20]
5470	Ethane, 1-chloro-1,1-difluoro- 1-Chloro-1,1-difluoroethane	$C_2H_3ClF_2$ 100.50	75-68-3 -130.8	-9.7	1.107[25]	H_2O 1; bz 3
5471	Ethane, 2-chloro-1,1-difluoro- 1-Chloro-2,2-difluoroethane	$C_2H_3ClF_2$ 100.50	338-65-8 	35.1		
5472	Ethane, 2-chloro-1,1-dimethoxy-	$C_4H_9ClO_2$ 124.57	97-97-2 	127.5	4-01-00-03134 1.068[20]	EtOH 2; eth 2; bz 2; ctc 2 1.4150[20]
5473	Ethane, 1-chloro-1-ethoxy-	C_4H_9ClO 108.57	7081-78-9 	93.5	4-01-00-03119 0.9655[20]	1.4053[20]
5474	Ethane, 1-chloro-2-ethoxy-	C_4H_9ClO 108.57	628-34-2 	107.5	4-01-00-01375 0.9895[20]	H_2O 2; eth 5; chl 3 1.4113[20]
5475	Ethane, 1-chloro-1-fluoro- 1-Chloro-1-fluoroethane	C_2H_4ClF 82.50	1615-75-4 	16.2	4-01-00-00127	
5476	Ethane, 1-chloro-2-fluoro-	C_2H_4ClF 82.50	762-50-5 	59	4-01-00-00127 1.1747[20]	eth 4; EtOH 4 1.3775[20]
5477	Ethane, 1-chloro-2-iodo-	C_2H_4ClI 190.41	624-70-4 -15.6	140	4-01-00-00167 2.1644[25]	
5478	Ethane, (chloromethoxy)- Chloromethyl ethyl ether	C_3H_7ClO 94.54	3188-13-4 	83	4-01-00-03047 1.0188[15]	1.4040[20]
5479	Ethane, 1-chloro-1-methoxy-	C_3H_7ClO 94.54	1538-87-0 	73	4-01-00-03119 0.9902[20]	eth 4 1.4004[20]
5480	Ethane, 1-chloro-2-methoxy-	C_3H_7ClO 94.54	627-42-9 	92.5	4-01-00-01375 1.0345[20]	H_2O 4; eth 4 1.4111[20]
5481	Ethane, 1-chloro-2-(methylthio)-	C_3H_7ClS 110.61	542-81-4 	60[30]; 44[20]	4-01-00-01406 1.1225[20]	EtOH 3; eth 3; ace 3 1.4902[20]
5482	Ethane, 1-chloro-1-nitro- 1-Chloro-1-nitroethane	$C_2H_4ClNO_2$ 109.51	598-92-5 	124.5	4-01-00-00172 1.2837[20]	H_2O 1; EtOH 3; ctc 3; alk 3 1.4224[20]
5483	Ethane, 1-chloro-2-nitro- 2-Chloro-1-nitroethane	$C_2H_4ClNO_2$ 109.51	625-47-8 	173.5	4-01-00-00173 1.3550[20]	1.407[7]
5484	Ethane, chloropentafluoro- Chloropentafluoroethane	C_2ClF_5 154.47	76-15-3 -99.4	-37.9	4-01-00-00129 1.5678[-42]	H_2O 1; EtOH 3; eth 3 1.2678[-42]

No.	Name / Synonym	Mol. Form. / Mol. Wt.	CAS RN / mp/°C	Merck No. / bp/°C	Beil. Ref. / den/g cm⁻³	Solubility / n_D
5485	Ethane, 1-chloro-1,1,2,2-tetrafluoro- / 1-Chloro-1,1,2,2-tetrafluoroethane	C_2HClF_4 / 136.48	354-25-6 / -117	/ -10	4-01-00-00129	
5486	Ethane, 1-chloro-1,2,2,2-tetrafluoro- / 1-Chloro-1,2,2,2-tetrafluoroethane	C_2HClF_4 / 136.48	2837-89-0 /	/ -12		
5487	Ethane, 1-chloro-1,1,2-trifluoro / 1-Chloro-1,1,2-trifluoroethane	$C_2H_2ClF_3$ / 118.49	421-04-5 /	/ 12		
5488	Ethane, 1-chloro-1,2,2-trifluoro / 1-Chloro-1,2,2-trifluoroethane	$C_2H_2ClF_3$ / 118.49	431-07-2 /	/ 17		
5489	Ethane, 2-chloro-1,1,1-trifluoro- / 2-Chloro-1,1,1-trifluoroethane	$C_2H_2ClF_3$ / 118.49	75-88-7 / -105.5	/ 6.1	4-01-00-00128 / 1.389[0]	1.3090[0]
5490	Ethane, 1-chloro-1,1,2-trifluoro-2-iodo- / Ethane, 1-chloro-2-iodo-1,1,2-trifluoro	C_2HClF_3I / 244.38	354-26-7 /	/ 77[630]	4-01-00-00167 / 2.181[20]	1.4320[20]
5491	Ethanedial / Glyoxal	$C_2H_2O_2$ / 58.04	107-22-2 / 15	4405 / 50.4	4-01-00-03625 / 1.14[20]	H_2O 4; EtOH 3; eth 3 / 1.3826[20]
5492	Ethanedial, dioxime	$C_2H_4N_2O_2$ / 88.07	557-30-2 / 178 dec	/ sub	3-01-00-03084	H_2O 4; EtOH 4; eth 4
5493	Ethanediamide / Oxamide	$C_2H_4N_2O_2$ / 88.07	471-46-5 / 350 dec	6871 /	4-02-00-01860 / 1.667[20]	H_2O 2; EtOH 2; eth 1
5494	Ethanediamide, N,N'-diethyl-	$C_6H_{12}N_2O_2$ / 144.17	615-84-9 / 175	/ 169[4]	4-04-00-00359 / 1.159[4]	EtOH 4
5495	Ethanediamide, N,N'-diphenyl-	$C_{14}H_{12}N_2O_2$ / 240.26	620-81-5 / 254	/ >360	4-12-00-00463	bz 4
5496	1,2-Ethanediamine / Ethylenediamine	$C_2H_8N_2$ / 60.10	107-15-3 / 11.1	3752 / 117	4-04-00-01166 / 0.8979[20]	H_2O 4; EtOH 5; eth 1; bz 1 / 1.4565[20]
5497	1,2-Ethanediamine, N-(2-aminoethyl)- / Diethylenetriamine	$C_4H_{13}N_3$ / 103.17	111-40-0 / -39	/ 207	4-04-00-01238 / 0.9569[20]	H_2O 5; EtOH 5; eth 1; lig 3 / 1.4810[25]
5498	1,2-Ethanediamine, N-(2-aminoethyl)-N'-[2-[(2-aminoethyl)amino]ethyl]- / Tetraethylenepentamine	$C_8H_{23}N_5$ / 189.30	112-57-2 /	/ 341.5	4-04-00-01244 /	H_2O 3 / 1.5042[20]
5499	1,2-Ethanediamine, N,N'-bis(2-aminoethyl)- / Trientine	$C_6H_{18}N_4$ / 146.24	112-24-3 / 12	9579 / 266.5	4-04-00-01242 /	H_2O 3; EtOH 3; acid 3 / 1.4971[20]
5500	1,2-Ethanediamine, N,N'-bis(phenylmethyl)- / Benzathine	$C_{16}H_{20}N_2$ / 240.35	140-28-3 / 26	1072 / 195[4]	4-12-00-02321 / 1.024[20]	bz 4; eth 4; EtOH 4 / 1.5635[20]
5501	1,2-Ethanediamine, N,N-diethyl-	$C_6H_{16}N_2$ / 116.21	111-74-0 /	/ 146	4-04-00-01174 / 0.8280[20]	H_2O 4; eth 4; EtOH 4; tol 4 / 1.4340[20]
5502	1,2-Ethanediamine, N,N-diethyl- / N,N-Diethylethylenediamine	$C_6H_{16}N_2$ / 116.21	100-36-7 /	/ 144	4-04-00-01175 / 0.8280[20]	H_2O 5; EtOH 3; eth 3; ctc 3 / 1.4340[20]
5503	1,2-Ethanediamine, N,N-dimethyl-	$C_4H_{12}N_2$ / 88.15	110-70-3 /	/ 120	4-04-00-01171 / 0.828[15]	EtOH 3; eth 3; dil HCl 3
5504	1,2-Ethanediamine, N,N-dimethyl-N'-(phenylmethyl)-	$C_{11}H_{18}N_2$ / 178.28	103-55-9 /	/ 122-4[11]	4-12-00-02320 / 0.922[25]	1.5089[20]
5505	1,2-Ethanediamine, N,N-dimethyl-N'-(phenylmethyl)-N'-2-pyridinyl- / Tripelennamine	$C_{16}H_{21}N_3$ / 255.36	91-81-6 /	9651 / 167-72[0.1]	5-22-08-00378	
5506	1,2-Ethanediamine, N,N-dimethyl-N'-phenyl-N'-(phenylmethyl)- / Phenbenzamine	$C_{17}H_{22}N_2$ / 254.38	961-71-7 /	7176 / 179-80[7]	4-12-00-02324 / 1.016[25]	1.5794[25]
5507	1,2-Ethanediamine, N,N-dimethyl-N'-2-pyridinyl-N'-(2-thienylmethyl)- / Methapyrilene	$C_{14}H_{19}N_3S$ / 261.39	91-80-5 /	5871 / 173-5[3]	5-22-08-00401	1.5915[20]
5508	1,2-Ethanediamine, N,N'-diphenyl- / 1,2-Dianilinoethane	$C_{14}H_{16}N_2$ / 212.29	150-61-8 / 74	2969 / 229[12]; 178[2]	4-12-00-00986	H_2O 1; EtOH 3; eth 3; tfa 2
5509	1,2-Ethanediamine, N-ethyl-	$C_4H_{12}N_2$ / 88.15	110-72-5 /	/ 129	4-04-00-01174 / 0.837[25]	1.4385[20]
5510	1,2-Ethanediamine, N'-ethyl-N,N-dimethyl- / Ethylenediamine, N,N-dimethyl-N'-ethyl	$C_6H_{15}N_2$ / 115.20	123-83-1 /	/ 134.5	4-04-00-01174 / 0.738[25]	1.4222[20]
5511	1,2-Ethanediamine, N-[(4-methoxyphenyl)methyl]-N',N'-dimethyl-N-2-pyridinyl- / Pyrilamine	$C_{17}H_{23}N_3O$ / 285.39	91-84-9 /	7996 / 201[5]	5-22-08-00381	
5512	1,2-Ethanediamine, N-methyl-	$C_3H_{10}N_2$ / 74.13	109-81-9 /	/ 115	4-04-00-01170 / 0.841[25]	1.4395[20]
5513	1,2-Ethanediamine, monohydrate	$C_2H_{10}N_2O$ / 78.11	6780-13-8 / 10	/ 118	4-04-00-01166 / 0.964[20]	H_2O 4 / 1.4500[20]
5514	1,2-Ethanediamine, N-1-naphthalenyl- / N-(1-Naphthyl)ethylenediamine	$C_{12}H_{14}N_2$ / 186.26	551-09-7 /	6329 / 204[9]	4-12-00-03101 / 1.114[25]	1.6648[25]
5515	1,2-Ethanediamine, N,N,N',N'-tetraethyl-	$C_{10}H_{24}N_2$ / 172.31	150-77-6 /	/ 192	4-04-00-01177 / 0.808[25]	1.4343[20]
5516	1,2-Ethanediamine, N,N,N',N'-tetramethyl- / N,N,N',N'-Tetramethyldiaminoethane	$C_6H_{16}N_2$ / 116.21	110-18-9 / -55	/ 121	4-04-00-01172 / 0.77[25]	1.4179[20]
5517	Ethane, 1,1-dibromo- / Ethylidene dibromide	$C_2H_4Br_2$ / 187.86	557-91-5 / -63	/ 109	4-01-00-00157 / 2.0554[20]	H_2O 1; EtOH 3; eth 4; ace 3 / 1.5128[20]
5518	Ethane, 1,2-dibromo- / Ethylene dibromide	$C_2H_4Br_2$ / 187.86	106-93-4 / 9.9	3753 / 131.6	4-01-00-00158 / 2.1791[20]	ace 4; bz 4; eth 4; EtOH 4 / 1.5387[20]
5519	Ethane, 1,2-dibromo-1-chloro-1,2,2-trifluoro- / 1,2-Dibromo-1-chloro-1,2,2-trifluoroethane	$C_2Br_2ClF_3$ / 276.28	354-51-8 / 50	/ 93		
5520	Ethane, 1,2-dibromo-1,1-dichloro-	$C_2H_2Br_2Cl_2$ / 256.75	75-81-0 / -26	/ 195	4-01-00-00161 / 2.135[20]	ace 4; bz 4; eth 4; EtOH 4 / 1.5662[20]
5521	Ethane, 1,2-dibromo-1,2-dichloro-	$C_2H_2Br_2Cl_2$ / 256.75	683-68-1 / -26	/ 195	4-01-00-00161 / 2.135[20]	H_2O 1; EtOH 3; eth 3; ace 3 / 1.5662[20]

No.	Name / Synonym	Mol. Form. / Mol. Wt.	CAS RN / mp/°C	Merck No. / bp/°C	Beil. Ref. / den/g cm^{-3}	Solubility / n_D
5522	Ethane, 1,2-dibromo-1,1-difluoro- / Genetron 132b-B2	$C_2H_2Br_2F_2$ / 223.84	75-82-1 / -61.3	92.5	4-01-00-00160 / 2.2238[20]	1.4456[20]
5523	Ethane, 1,2-dibromo-1-ethoxy-	$C_4H_8Br_2O$ / 231.91	2983-26-8 /	80[20]	4-01-00-03152 / 1.7320[20]	EtOH 4; chl 4 / 1.5044[20]
5524	Ethane, 1,2-dibromo-1,1,2,2-tetrachloro-	$C_2Br_2Cl_4$ / 325.64	630-25-1 /	202.5	3-01-00-00190 / 2.713[25]	
5525	Ethane, 1,2-dibromo-1,1,2,2-tetrafluoro- / 1,2-Dibromotetrafluoroethane	$C_2Br_2F_4$ / 259.82	124-73-2 / -110.4	47.3	4-01-00-00160 / 2.149[25]	1.361[25]
5526	Ethane, 1,2-dibromo-1,1,2-trifluoro- / Halon 2302	$C_2HBr_2F_3$ / 241.83	354-04-1 /	76	4-01-00-00160 / 2.274[27]	1.4191[24]
5527	Ethane, 1,2-dibutoxy / Ethylene glycol dibutyl ether	$C_{10}H_{22}O_2$ / 174.28	112-48-1 / -69.1	203.3	/ 0.8319[25]	
5528	Ethane, 1,1-dichloro- / Ethylidene dichloride	$C_2H_4Cl_2$ / 98.96	75-34-3 / -96.9	3766 / 57.4	4-01-00-00130 / 1.1757[20]	H_2O 2; EtOH 4; eth 4; ace 3 / 1.4164[20]
5529	Ethane, 1,2-dichloro- / Ethylene dichloride	$C_2H_4Cl_2$ / 98.96	107-06-2 / -35.5	3754 / 83.5	4-01-00-00131 / 1.2351[20]	H_2O 2; EtOH 4; eth 5; ace 3 / 1.4448[20]
5530	Ethane, 1,1-dichloro-2,2-diethoxy-	$C_6H_{12}Cl_2O_2$ / 187.07	619-33-0 /	183.5	4-01-00-03140 / 1.1383[14]	EtOH 3; eth 3; ctc 2
5531	Ethane, 1,2-dichloro-1,1-difluoro-	$C_2H_2Cl_2F_2$ / 134.94	1649-08-7 / -101.2	46.8	4-01-00-00135 / 1.4163[20]	1.36193[20]
5532	Ethane, 1,2-dichloro-1,2-difluoro-	$C_2H_2Cl_2F_2$ / 134.94	431-06-1 / -101.2	46.8	4-01-00-00135 / 1.4163[20]	1.3619[20]
5533	Ethane, 1,2-dichloro-1-ethoxy-	$C_4H_8Cl_2O$ / 143.01	623-46-1 /	145	4-01-00-03136 / 1.1370[20]	chl 2 / 1.4435[20]
5534	Ethane, 1,1-dichloro-1-fluoro- / 1,1-Dichloro-1-fluoroethane	$C_2H_3Cl_2F$ / 116.95	1717-00-6 / -103.5	32	4-01-00-00134 / 1.250[10]	1.3600[10]
5535	Ethane, 1,1-dichloro-1-fluoro-	$C_2H_3Cl_2F$ / 116.95	430-57-9 / -60	73.7	4-01-00-00134 / 1.3814[20]	1.4132[20]
5536	Ethane, 1,1-dichloro-1-nitro- / Ethide	$C_2H_3Cl_2NO_2$ / 143.96	594-72-9 /	123.5	4-01-00-00173	ctc 3
5537	Ethane, 1,1-dichloro-1,2,2,2-tetrafluoro- / 1,1-Dichlorotetrafluoroethane	$C_2Cl_2F_4$ / 170.92	374-07-2 / -56.6	4	4-01-00-00136 / *1.455[25]	bz 4; eth 4; EtOH 4 / 1.3092[0]
5538	Ethane, 1,2-dichloro-1,1,2,2-tetrafluoro- / Refrigerant 114	$C_2Cl_2F_4$ / 170.92	76-14-2 / -94	2608 / 3.8	4-01-00-00137 / *1.455[25]	eth 4; EtOH 4 / 1.3092[0]
5539	Ethane, 1,2-dichloro-1,1,2-trifluoro-	$C_2HCl_2F_3$ / 152.93	354-23-4 / -78	28	4-01-00-00136 / 1.50[25]	
5540	Ethane, 2,2-dichloro-1,1,1-trifluoro- / 2,2-Dichloro-1,1,1-trifluoroethane	$C_2HCl_2F_3$ / 152.93	306-83-2 / -107	27.82	4-01-00-00135 / 1.4638[25]	
5541	Ethane, 1,1-diethoxy- / Acetal	$C_6H_{14}O_2$ / 118.18	105-57-7 / -100	31 / 102.2	4-01-00-03103 / 0.8254[20]	H_2O 3; EtOH 5; eth 5; ace 4 / 1.3834[20]
5542	Ethane, 1,2-diethoxy- / Ethylene glycol diethyl ether	$C_6H_{14}O_2$ / 118.18	629-14-1 /	119.4	4-01-00-02379 / 0.8484[20]	ace 4; bz 4; eth 4; EtOH 4 / 1.3860[20]
5543	Ethane, 1,1-diethoxy-2-iodo-	$C_6H_{13}IO_2$ / 244.07	51806-20-3 /	132[90]; 100[10]	4-01-00-03156 / 1.489[22]	1.4734[22]
5544	Ethane, 1,1-difluoro-	$C_2H_4F_2$ / 66.05	75-37-6 / -117	-24.9	4-01-00-00120 / *0.896[20]	1.3011[-72]
5545	Ethane, 1,2-difluoro- / Freon 152	$C_2H_4F_2$ / 66.05	624-72-6 /	30.7	4-01-00-00121	bz 4; eth 4; chl 4
5546	Ethane, 1,1-diiodo- / 1,1-Diiodoethane	$C_2H_4I_2$ / 281.86	594-02-5 /	179.5	4-01-00-00169 / 2.84[25]	ace 4; eth 4; EtOH 4; chl 4 / 1.673[20]
5547	Ethane, 1,2-diiodo-	$C_2H_4I_2$ / 281.86	624-73-7 / 83	200	4-01-00-00169 / 3.325[20]	H_2O 2; EtOH 3; eth 3; ace 3 / 1.871[20]
5548	Ethane, 1,1-dimethoxy- / Dimethyl acetal	$C_4H_{10}O_2$ / 90.12	534-15-6 / -113.2	3215 / 64.5	4-01-00-03103 / 0.8501[20]	H_2O 3; EtOH 3; eth 3; ace 4 / 1.3668[20]
5549	Ethane, 1,2-dimethoxy- / Ethylene glycol dimethyl ether	$C_4H_{10}O_2$ / 90.12	110-71-4 / -69.20	3213 / 85	4-01-00-02376 / 0.8691[20]	H_2O 3; EtOH 3; eth 3; ace 3 / 1.3796[20]
5550	Ethanedinitrile / Cyanogen	C_2N_2 / 52.04	460-19-5 / -27.9	2698 / -21.1	4-02-00-01863 / 0.9537[-21]	H_2O 3; EtOH 3; eth 3
5551	Ethane, 1,1-dinitro- / 1,1-Dinitroethane	$C_2H_4N_2O_4$ / 120.06	600-40-8 /	185.5	4-01-00-00174 / 1.349[24]	H_2O 2; EtOH 3; eth 3
5552	Ethane, 1,2-dinitro- / 1,2-Dinitroethane	$C_2H_4N_2O_4$ / 120.06	7570-26-5 / 39.5	94-6[5]	4-01-00-00175 / 1.4597[20]	eth 4; EtOH 4 / 1.4468[20]
5553	Ethanedioic acid / Oxalic acid	$C_2H_2O_4$ / 90.04	144-62-7 / 189.5 dec	6865 / sub 157	4-02-00-01819 / 1.900[17]	H_2O 3; EtOH 4; eth 2; bz 1
5554	Ethanedioic acid, bis(2-chloroethyl) ester	$C_6H_8Cl_2O_4$ / 215.03	7208-92-6 / 45	150[4]	4-02-00-01849	bz 4; EtOH 4
5555	Ethanedioic acid, bis(3-methylbutyl) ester	$C_{12}H_{22}O_4$ / 230.30	2051-00-5 / -9	267.5	4-02-00-01851 / 0.968[11]	eth 4; EtOH 4
5556	Ethanedioic acid, bis(1-methylethyl) ester	$C_8H_{14}O_4$ / 174.20	615-81-6 /	191	4-02-00-01850 / 1.001[20]	eth 4; EtOH 4 / 1.4100[20]
5557	Ethanedioic acid, bis(2-methylpropyl) ester / Diisobutyl oxalate	$C_{10}H_{18}O_4$ / 202.25	2050-61-5 /	229	4-02-00-01850 / 0.9737[20]	ace 4; eth 4; EtOH 4 / 1.4180[20]
5558	Ethanedioic acid, dibutyl ester	$C_{10}H_{18}O_4$ / 202.25	2050-60-4 / -30.5	241; 96[2]	4-02-00-01850 / 0.9873[20]	H_2O 1; EtOH 3; eth 3 / 1.4234[20]
5559	Ethanedioic acid, dicyclohexyl ester / Dicyclohexyl oxalate	$C_{14}H_{22}O_4$ / 254.33	620-82-6 / 43.5	190[73]	3-06-00-00026	H_2O 1; EtOH 4; eth 4; MeOH 3
5560	Ethanedioic acid, diethyl ester / Diethyl oxalate	$C_6H_{10}O_4$ / 146.14	95-92-1 / -40.6	3115 / 185.7	4-02-00-01848 / 1.0785[20]	H_2O 2; EtOH 5; eth 5; ace 5 / 1.4101[20]

No.	Name / Synonym	Mol. Form. / Mol. Wt.	CAS RN / mp/°C	Merck No. / bp/°C	Beil. Ref. / den/g cm^{-3}	Solubility / n_D
5561	Ethanedioic acid, dihydrate / Oxalic acid dihydrate	$C_2H_6O_6$ / 126.07	6153-56-6 / 101.5		4-02-00-01819 / 1.653[18]	H_2O 3; EtOH 3; eth 2
5562	Ethanedioic acid, dihydrazide	$C_2H_6N_4O_2$ / 118.10	996-98-5 / 243 dec		4-02-00-01868 / 1.458[22]	H_2O 3; EtOH 2; eth 2; bz 2
5563	Ethanedioic acid, dimethyl ester / Dimethyl oxalate	$C_4H_6O_4$ / 118.09	553-90-2 / 54.3	6020 / 163.5	4-02-00-01847 / 1.1716[60]	H_2O 2; EtOH 3; eth 3; ace 3 / 1.379[82]
5564	Ethanedioic acid, di-2-propenyl ester / Diallyl oxalate	$C_8H_{10}O_4$ / 170.17	615-99-6 /	/ 217	4-02-00-01851 / 1.1582[20]	H_2O 1; EtOH 3; ace 3; bz 3 / 1.4481[20]
5565	Ethanedioic acid, dipropyl ester	$C_8H_{14}O_4$ / 174.20	615-98-5 / -44.3	/ 211	4-02-00-01849 / 1.0188[20]	H_2O 2; EtOH 5; eth 3 / 1.4158[20]
5566	Ethanedioic acid, disodium salt / Oxalic acid, disodium salt	$C_2Na_2O_4$ / 134.00	62-76-0 / 260 dec	8601	4-02-00-01819 / 2.34[25]	H_2O 2; EtOH 1; eth 1
5567	Ethanedioic acid, ethyl methyl ester	$C_5H_8O_4$ / 132.12	615-52-1 /	/ 173.7	0-02-00-00535 / 1.1555[0]	eth 4; EtOH 4
5568	Ethanedioic acid, lead (II) salt / Lead oxalate	C_2O_4Pb / 295.22	814-93-7 / 300 dec	5294	4-02-00-01819 / 5.28	H_2O I; EtOH 1; HNO_3 3
5569	1,2-Ethanediol / Ethylene glycol	$C_2H_6O_2$ / 62.07	107-21-1 / -13	3755 / 197.3	4-01-00-02369 / 1.1088[20]	H_2O 5; EtOH 3; ace 5 / 1.4318[20]
5570	1,1-Ethanediol, diacetate / Ethylidene diacetate	$C_6H_{10}O_4$ / 146.14	542-10-9 / 18.9	3767 / 169	4-02-00-00282 / 1.070[25]	eth 4; EtOH 4 / 1.3985[25]
5571	1,2-Ethanediol, diacetate / Ethylene glycol diacetate	$C_6H_{10}O_4$ / 146.14	111-55-7 / -31	3756 / 190	4-02-00-00217 / 1.1043[20]	H_2O 4; EtOH 5; eth 5; ace 5 / 1.4159[20]
5572	1,2-Ethanediol, dibenzoate	$C_{16}H_{14}O_4$ / 270.28	94-49-5 / 73.5	/ 360 dec	4-09-00-00356	H_2O 1; eth 3; chl 3
5573	1,2-Ethanediol, diformate	$C_4H_6O_4$ / 118.09	629-15-2 /	/ 174	4-02-00-00037 / 1.193[0]	H_2O 2; EtOH 3; eth 3 / 1.3580
5574	1,2-Ethanediol, dinitrate / Ethylene glycol dinitrate	$C_2H_4N_2O_6$ / 152.06	628-96-6 / -22.3	/ 198.5	4-01-00-02413 / 1.4918[20]	eth 4; EtOH 4
5575	1,2-Ethanediol, 1,2-diphenyl-, (R^*,R^*)-(±)-	$C_{14}H_{14}O_2$ / 214.26	655-48-1 / 122.5	/ >300	4-06-00-06682	H_2O 1; EtOH 4; eth 4; ace 3
5576	1,2-Ethanediol, dipropanoate	$C_8H_{14}O_4$ / 174.20	123-80-8 /	/ 211	4-02-00-00715 / 1.020[15]	H_2O 2; EtOH 5; eth 5; ace 4
5577	1,2-Ethanediol, monoacetate / Ethylene glycol, monoacetate	$C_4H_8O_3$ / 104.11	542-59-6 /	3757 / 188	4-02-00-00214 / 1.108[15]	H_2O 5; EtOH 5; eth 5
5578	1,2-Ethanediol, monobenzoate	$C_9H_{10}O_3$ / 166.18	94-33-7 / 45	/ 150[10]	4-09-00-00354 / 1.110[30]	EtOH 4
5579	1,2-Ethanediol, 1-phenyl- / Styrene glycol	$C_8H_{10}O_2$ / 138.17	93-56-1 / 67.5	8831 / 273	4-06-00-05939	H_2O 4; eth 4; bz 4; EtOH 4
5580	1,2-Ethanediol, 1,1,2,2-tetraphenyl- / Benzopinacol	$C_{26}H_{22}O_2$ / 366.46	464-72-2 / 182	1110	4-06-00-07053	H_2O 1; EtOH 2; eth 3; ace 3
5581	1,1-Ethanediol, 2,2,2-tribromo- / Acetaldehyde, tribromo, hydrate	$C_2H_3Br_3O_2$ / 298.76	507-42-6 / 53.5	1373 / dec	4-01-00-03155 / 2.5661[40]	eth 4; EtOH 4
5582	1,1-Ethanediol, 2,2,2-trichloro- / Chloral hydrate	$C_2H_3Cl_3O_2$ / 165.40	302-17-0 / 57	2061 / 96 dec	4-01-00-03143 / 1.9081[20]	H_2O 4; bz 4; eth 4; EtOH 4
5583	Ethanedione, di-2-furanyl-, dioxime / α-Furildioxime	$C_{10}H_8N_2O_4$ / 220.18	522-27-0 / 167	4218	4-19-00-02008	EtOH 2; eth 2; bz 2; lig 2
5584	Ethanedione, diphenyl- / Benzil	$C_{14}H_{10}O_2$ / 210.23	134-81-6 / 94.8	1087 / 347	4-07-00-02502 / 1.084[102]	H_2O 1; EtOH 4; eth 4; ace 3
5585	Ethanedioyl dichloride / Oxalyl chloride	$C_2Cl_2O_2$ / 126.93	79-37-8 / -16	6867 / 63.5	4-02-00-01853 / 1.4785[20]	eth 3 / 1.4316[20]
5586	Ethane, 1,1-diproxy- / 1,2-Dipropoxyethane	$C_8H_{18}O_2$ / 146.23	18854-56-3 /			
5587	1,2-Ethanedisulfonic acid / Ethylenedisulfonic acid	$C_2H_6O_6S_2$ / 190.20	110-04-3 / 173	3678	4-04-00-00078	Diox 4
5588	Ethanedithioamide / Rubeanic acid	$C_2H_4N_2S_2$ / 120.20	79-40-3 / 170 dec	8256	4-02-00-01871	H_2O 2; EtOH 2; con sulf 3
5589	Ethane(dithioic) acid	$C_2H_4S_2$ / 92.19	594-03-6 /	/ 66[85]; 37[15]	4-02-00-00572 / 1.24[20]	H_2O 3; EtOH 4; eth 4; ace 4
5590	1,2-Ethanedithiol / Ethylenedimercaptan	$C_2H_6S_2$ / 94.20	540-63-6 / -41.2	3679 / 146	4-01-00-02450 / 1.234[20]	H_2O 1; EtOH 3; eth 3; ace 3 / 1.5590[20]
5591	Ethane, 1-ethoxy-2-methoxy- / 1-Ethoxy-2-methoxyethane	$C_5H_{12}O_2$ / 104.15	5137-45-1 /	/ 102.1	3-01-00-02078 / 0.8529[20]	1.3868[20]
5592	Ethane, 1,1'-[ethylidenebis(oxy)]bis[2-chloro-	$C_6H_{12}Cl_2O_2$ / 187.07	14689-97-5 /	/ 195	4-01-00-03104 / 1.1737[20]	eth 4; EtOH 4 / 1.4527[20]
5593	Ethane, fluoro- / Ethyl fluoride	C_2H_5F / 48.06	353-36-6 / -143.2	/ -37.6	4-01-00-00120 / 0.7182[20]	H_2O 3; EtOH 4; eth 4 / 1.2656[20]
5594	Ethane, hexabromo- / Hexabromoethane	C_2Br_6 / 503.45	594-73-0 /	/ 200 dec	3-01-00-00193 / 3.823[20]	EtOH 2; eth 2; CS_2 2 / 1.863
5595	Ethane, hexachloro- / Hexachloroethane	C_2Cl_6 / 236.74	67-72-1 / 187 (triple point)	4601	4-01-00-00148 / 2.091[20]	H_2O 1; EtOH 4; eth 4; bz 3
5596	Ethane, hexafluoro- / Perfluoroethane	C_2F_6 / 138.01	76-16-4 / -100.7	/ -78.1	4-01-00-00123 / 1.590[-78]	H_2O 1; EtOH 2; eth 2
5597	Ethane, iodo- / Ethyl iodide	C_2H_5I / 155.97	75-03-6 / -111.1	3769 / 72.5	4-01-00-00163 / 1.9358[20]	H_2O 2; EtOH 5; eth 3; chl 3 / 1.5133[20]
5598	Ethane, isocyanato- / Ethyl isocyanate	C_3H_5NO / 71.08	109-90-0 /	/ 60	4-04-00-00402 / 0.9031[20]	H_2O 5; EtOH 5; eth 5 / 1.3808[20]
5599	Ethane, isocyano-	C_3H_5N / 55.08	624-79-3 / <-66	/ 79	4-04-00-00342 / 0.7402[20]	H_2O 4; EtOH 5; eth 5; ace 3 / 1.3622[20]
5600	Ethane, isothiocyanato- / Ethyl isothiocyanate	C_3H_5NS / 87.15	542-85-8 / -5.9	3771 / 131.5	4-04-00-00403 / 0.9990[20]	H_2O 1; EtOH 5; eth 5 / 1.5130[20]

No.	Name / Synonym	Mol. Form. / Mol. Wt.	CAS RN	mp/°C	Merck No. bp/°C	Beil. Ref. den/g cm⁻³	Solubility n_D
5601	Ethane, 1,1',1'',1'''-[methanetetrayltetrakis(oxy)]tetrakis-	$C_9H_{20}O_4$ 192.26	78-09-1		159.5	4-03-00-00006 0.9186^{20}	EtOH 5; eth 5; ctc 3 1.3905^{25}
5602	Ethane, methoxy- Ethyl methyl ether	C_3H_8O 60.10	540-67-0 -113	3783 7.4		4-01-00-01314 0.7251^0	H_2O 3; EtOH 5; eth 5; ace 3 1.3420^4
5603	Ethane, 1,1'-[methylenebis(oxy)]bis- Diethoxymethane	$C_5H_{12}O_2$ 104.15	462-95-3 -66.5	88		4-01-00-03027 0.8319^{20}	H_2O 4; EtOH 5; eth 5; ace 4 1.3748^{18}
5604	Ethane, 1,1',1''-[methylidynetris(oxy)]tris- Ethyl orthoformate	$C_7H_{16}O_3$ 148.20	122-51-0	143; 60²⁰		4-02-00-00025 0.8909^{20}	EtOH 3; eth 3 1.3922^{20}
5605	Ethane, 1,1',1''-[methylidynetris(thio)]tris-	$C_7H_{16}S_3$ 196.40	6267-24-9	235 dec; 127¹²		4-02-00-00093 1.053^{20}	eth 4; EtOH 4 1.5410^{15}
5606	Ethane, (methylthio)- Ethyl methyl sulfide	C_3H_8S 76.16	624-89-5 -105.9	66.7		4-01-00-01392 0.8422^{20}	H_2O 1; EtOH 5; eth 3; chl 3 1.4404^{20}
5607	Ethane, nitro- Nitroethane	$C_2H_5NO_2$ 75.07	79-24-3 -89.5	6518 114.0		4-01-00-00170 1.0448^{25}	H_2O 2; EtOH 5; eth 5; ace 3 1.3917^{20}
5608	Ethane, 1,1'-oxybis- Diethyl ether	$C_4H_{10}O$ 74.12	60-29-7 -116.3	3762 34.5		4-01-00-01314 0.7138^{20}	H_2O 2; EtOH 5; eth 5; ace 4 1.3526^{20}
5609	Ethane, 1,1'-oxybis[2-bromo-	$C_4H_8Br_2O$ 231.91	5414-19-7	115³²		4-01-00-01386 1.8222^{27}	1.5131^{27}
5610	Ethane, 1,1'-oxybis[1-chloro-	$C_4H_8Cl_2O$ 143.01	6986-48-7	116.5		4-01-00-03120 1.1060^{25}	eth 4; EtOH 4; chl 4 1.4186^{25}
5611	Ethane, 1,1'-oxybis[2-chloro- Bis(2-chloroethyl) ether	$C_4H_8Cl_2O$ 143.01	111-44-4 -51.9	3055 178.5		4-01-00-01375 1.22^{20}	H_2O 1; EtOH 3; eth 3; ace 3 1.451^{20}
5612	Ethane, 1,1'-oxybis[2-ethoxy- Diethylene glycol diethyl ether	$C_8H_{18}O_3$ 162.23	112-36-7 -45	3108 188		4-01-00-02394 0.9063^{20}	H_2O 4; EtOH 4; eth 3; os 4 1.4115^{20}
5613	Ethane, 1,1'-oxybis[2-methoxy- Diethylene glycol dimethyl ether	$C_6H_{14}O_3$ 134.18	111-96-6 -68	3148 162		4-01-00-02393 0.9434^{20}	H_2O 5; EtOH 5; eth 5 1.4097^{20}
5614	Ethane, 1,1'-[oxybis(methyleneoxy)]bis- Bis(ethoxymethyl)ether	$C_6H_{14}O_3$ 134.18	5648-29-3	140.6			
5615	Ethane, 1,1'-oxybis[1,2,2,2-tetrachloro- Bis(1,2,2,2-tetrachloroethyl) ether	$C_4H_2Cl_8O$ 349.68	41284-12-2 40.2	188; 130¹¹		2-01-00-00681	bz 4; peth 4
5616	Ethane, pentabromo- 1,1,1,2,2-Pentabromoethane	C_2HBr_5 424.55	75-95-6 56.5	210³⁰⁰		3-01-00-00193 3.312^{20}	H_2O 1; EtOH 3; eth 4; chl 3
5617	Ethane, pentachloro- Pentachloroethane	C_2HCl_5 202.29	76-01-7 -29	7058 159.8		4-01-00-00147 1.6796^{20}	H_2O 1; EtOH 5; eth 5 1.5025^{20}
5618	Ethane, pentachlorofluoro-	C_2Cl_5F 220.28	354-56-3 101.3	135		4-01-00-00148 1.74^{25}	H_2O 1; EtOH 3; eth 3
5619	Ethaneperoxoic acid Peroxyacetic acid	$C_2H_4O_3$ 76.05	79-21-0 -0.2	7107 110		4-02-00-00390 1.226^{15}	H_2O 4; EtOH 3; eth 4; sulf 4 1.3974^{20}
5620	Ethane, 1,1'-selenobis-	$C_4H_{10}Se$ 137.08	627-53-2 55	108		4-01-00-01411 1.2300^{20}	1.4768^{20}
5621	Ethane, 1,1'-sulfinylbis-	$C_4H_{10}OS$ 106.19	70-29-1 14	104²⁵; 90¹⁵		4-01-00-01395 1.0092^{22}	H_2O 4; eth 4; EtOH 4
5622	Ethanesulfonic acid Ethylsulfonic acid	$C_2H_6O_3S$ 110.13	594-45-6 -17	123¹		4-04-00-00033 1.3341^{25}	H_2O 4; EtOH 4 1.4335^{20}
5623	Ethanesulfonic acid, 2-amino- Taurine	$C_2H_7NO_3S$ 125.15	107-35-7 328	9043		4-04-00-03289	H_2O 4
5624	Ethanesulfonic acid, 2-(methylamino)- N-Methyltaurine	$C_3H_9NO_3S$ 139.18	107-68-6 241.5	6043		3-04-00-01699	H_2O 4; EtOH 1; eth 1
5625	Ethanesulfonic acid, 2-[[(3α,5β,7α,12α)-3,7,12-trihydroxy-24-oxocholan-24-yl]amino]- Taurocholic acid	$C_{28}H_{45}NO_7S$ 515.71	81-24-3 125 dec	9044		4-10-00-02078	H_2O 4; EtOH 4; eth 2; AcOEt 2
5626	Ethane, 1,1'-sulfonylbis-	$C_4H_{10}O_2S$ 122.19	597-35-3 73.5	248		4-01-00-01396 1.357^{20}	H_2O 3; eth 3; bz 4; peth 1
5627	Ethanesulfonyl chloride	$C_2H_5ClO_2S$ 128.58	594-44-5	174		4-04-00-00034 1.357^{22}	eth 4; CS_2 3 1.4531^{20}
5628	Ethanesulfonyl chloride, 2-bromo- 2-Bromoethanesulfonyl chloride	$C_2H_4BrClO_2S$ 207.48	54429-56-0	102¹³		4-04-00-00037 1.921^{20}	1.5242^{20}
5629	Ethanesulfonyl chloride, 2-chloro-	$C_2H_4Cl_2O_2S$ 163.02	1622-32-8	201.5		4-04-00-00036 1.555^{20}	1.4920^{20}
5630	Ethane, 1,1'-tellurobis-	$C_4H_{10}Te$ 185.72	627-54-3	137.5		4-01-00-01412 1.599^{15}	EtOH 4 1.5182^{15}
5631	Ethane, 1,1,1,2-tetrabromo- 1,2,2,2-Tetrabromoethane	$C_2H_2Br_4$ 345.65	630-16-0 0	112¹⁸ dec		4-01-00-00162 2.8747^{20}	ace 4; bz 4; eth 4; EtOH 4 1.6277^{20}
5632	Ethane, 1,1,2,2-tetrabromo- Acetylene tetrabromide	$C_2H_2Br_4$ 345.65	79-27-6 0	9121 243.5; 151⁵⁴		4-01-00-00162 2.9655^{20}	H_2O 1; EtOH 5; eth 5; ace 3 1.6353^{20}
5633	1,1,2,2-Ethanetetracarboxylic acid, tetraethyl ester	$C_{14}H_{22}O_8$ 318.32	632-56-4 77	305 dec		4-02-00-02415 1.064^{80}	EtOH 3 1.4105^{80}
5634	Ethane, 1,1,1,2-tetrachloro- 1,1,1,2-Tetrachloroethane	$C_2H_2Cl_4$ 167.85	630-20-6 -70.2	130.5		4-01-00-00143 1.5406^{20}	H_2O 2; EtOH 5; eth 5; ace 3 1.4821^{20}
5635	Ethane, 1,1,2,2-tetrachloro- Acetylene tetrachloride	$C_2H_2Cl_4$ 167.85	79-34-5 -43.8	9125 146.5		4-01-00-00144 1.5953^{20}	H_2O 2; EtOH 5; eth 5; ace 3 1.4940^{20}
5636	Ethane, 1,1,2,2-tetrachloro-2,2-difluoro- 1,1-Difluoroperchloroethane	$C_2Cl_4F_2$ 203.83	76-11-9 40.6	91.5		4-01-00-00146 1.649^{25}	H_2O 1; EtOH 3; eth 3; chl 3
5637	Ethane, 1,1,2,2-tetrachloro-1,2-difluoro- Tetrachloro-1,2-difluoroethane	$C_2Cl_4F_2$ 203.83	76-12-0 26	93		4-01-00-00146 1.6447^{25}	H_2O 1; EtOH 3; eth 3; chl 3 1.4130^{25}
5638	Ethane, 1,1,2,2-tetrachloro-1-fluoro-	C_2HCl_4F 185.84	354-14-3 -82.6	116		4-01-00-00145 1.5497^{17}	1.4390^{20}

No.	Name Synonym	Mol. Form. Mol. Wt.	CAS RN mp/°C	Merck No. bp/°C	Beil. Ref. den/g cm⁻³	Solubility n_D
5639	Ethane, 1,1,1,2-tetrachloro-2-fluoro- 1,1,1,2-Tetrachloro-2-fluoroethane	C_2HCl_4F 185.84	354-11-0 -95.3	116.5		
5640	Ethane, 1,1,1,2-tetrafluoro- 1,1,1,2-Tetrafluoroethane	$C_2H_2F_4$ 102.03	811-97-2 -103.3	-26.08	4-01-00-00123 1.2072[25]	H_2O 1; eth 3
5641	Ethane, 1,1,2,2-tetrafluoro- 1,1,2,2-Tetrafluoroethane	$C_2H_2F_4$ 102.03	359-35-3 -89	-19.9		
5642	Ethane, 1,1,2,2-tetrafluoro-1,2-dinitro- Ethane, 1,2-dinitro-1,1,2,2-tetrafluoro	$C_2F_4N_2O_4$ 192.03	356-16-1 -41.5	58.5	3-01-00-00203 1.6024[25]	H_2O 1; ace 3 1.3265[25]
5643	Ethanethioamide Thioacetamide	C_2H_5NS 75.13	62-55-5 115.5	9246	4-02-00-00565	H_2O 4; EtOH 4; eth 2; bz 2
5644	Ethane, 1,1'-thiobis- Diethyl sulfide	$C_4H_{10}S$ 90.19	352-93-2 -103.9	3809 92.1	4-01-00-01394 0.8362[20]	H_2O 2; EtOH 3; eth 3; ctc 2 1.4430[20]
5645	Ethane, 1,1'-thiobis[2-chloro- Mustard gas	$C_4H_8Cl_2S$ 159.08	505-60-2 13.5	6225 216	4-01-00-01407 1.2741[20]	1.5313[20]
5646	Ethanethioic acid Thioacetic acid	C_2H_4OS 76.12	507-09-5 <-17	9247 93; 26[35]	4-02-00-00542 1.064[20]	H_2O 3; EtOH 4; eth 5; ace 4 1.4648[20]
5647	Ethanethioic acid, S-ethyl ester	C_4H_8OS 104.17	625-60-5 	116.4	4-02-00-00543 0.9792[20]	H_2O 1; EtOH 4; eth 4 1.4583[21]
5648	Ethanethiol Ethyl mercaptan	C_2H_6S 62.14	75-08-1 -147.8	3680 35.1	4-01-00-01390 0.8315[25]	H_2O 2; EtOH 3; eth 3; ace 3 1.4310[20]
5649	Ethanethiol, 2-amino- Cysteamine	C_2H_7NS 77.15	60-23-1 99.5	2785 dec	4-04-00-01570	H_2O 4; EtOH 4
5650	Ethanethiol, 2-(butylamino)- 2-(Butylamino)ethanethiol	$C_6H_{15}NS$ 133.26	5842-00-2 	112-5[10]	4-04-00-01602 0.901[25]	1.4711[20]
5651	Ethanethiol, 2-chloro- 2-Chloroethanethiol	C_2H_5ClS 96.58	4325-97-7 	113	4-01-00-01406 1.1826[20]	eth 4; EtOH 4; Diox 4 1.4929[20]
5652	Ethanethiol, 2,2'-thiobis-	$C_4H_{10}S_3$ 154.32	3570-55-6 	135-6[10]	4-01-00-02455 1.183[25]	1.5982[20]
5653	Ethane, 1,1,2-tribromo- 1,1,2-Tribromoethane	$C_2H_3Br_3$ 266.76	78-74-0 -29.3	188.93	4-01-00-00161 2.6210[20]	H_2O 3; EtOH 3; eth 3; bz 3 1.5933[20]
5654	1,1,2-Ethanetricarboxylic acid, triethyl ester	$C_{11}H_{18}O_6$ 246.26	7459-46-3 	99[05]	4-02-00-02364 1.074[25]	1.4290[20]
5655	Ethane, 1,1,1-trichloro- Methylchloroform	$C_2H_3Cl_3$ 133.40	71-55-6 -30.4	9549 74.0	4-01-00-00138 1.3390[20]	H_2O 2; EtOH 3; eth 5; chl 3 1.4379[20]
5656	Ethane, 1,1,2-trichloro- 1,1,2-Trichloroethane	$C_2H_3Cl_3$ 133.40	79-00-5 -36.6	9550 113.8	4-01-00-00139 1.4397[20]	H_2O 1; EtOH 3; eth 3; chl 3 1.4714[20]
5657	Ethane, 1,1,1-trichloro-2,2-diethoxy- Chloral, diethyl acetal	$C_6H_{11}Cl_3O_2$ 221.51	599-97-3 197		4-01-00-03144 1.266[20]	H_2O 2; EtOH 5; eth 5; glycerol 5 1.4586[25]
5658	Ethane, 1,1,1-trichloro-2,2-difluoro 1,1,1-Trichloro-2,2-difluoroethane	$C_2HCl_3F_2$ 169.38	354-12-1 	73		
5659	Ethane, 1,2,2-trichloro-1,1-difluoro-	$C_2HCl_3F_2$ 169.38	354-21-2 	71.9	4-01-00-00141 1.5447[20]	1.3889[20]
5660	Ethane, 1,2,2-trichloro-1,2-difluoro- 1,2,2-Trichloro-1,2-difluoroethane	$C_2HCl_3F_2$ 169.38	354-15-4 	72.5		
5661	Ethane, 1,1,2-trichloro-2-fluoro-	$C_2H_2Cl_3F$ 151.39	359-28-4 	102	4-01-00-00141 1.5393[20]	H_2O 1 1.4390[20]
5662	Ethane, 1,1,1-trichloro-2,2,2-trifluoro- 1,1,1-Trichlorotrifluoroethane	$C_2Cl_3F_3$ 187.38	354-58-5 14.2	46.1	4-01-00-00142 1.5790[20]	H_2O 1; EtOH 3; eth 3; chl 3 1.3610[35]
5663	Ethane, 1,1,2-trichloro-1,2,2-trifluoro- 1,1,2-Trichlorotrifluoroethane	$C_2Cl_3F_3$ 187.38	76-13-1 -35	47.7	4-01-00-00142 1.5635[25]	H_2O 1; EtOH 3; eth 5; bz 5 1.3557[25]
5664	Ethane, 1,1,1-triethoxy-	$C_8H_{18}O_3$ 162.23	78-39-7 	145	4-02-00-00137 0.8847[25]	H_2O 1; EtOH 5; eth 5; ctc 5 1.3980[20]
5665	Ethane, 1,1,1-trifluoro- Methylfluoroform	$C_2H_3F_3$ 84.04	420-46-2 -111.3	-47.5	4-01-00-00122	eth 3; chl 3
5666	Ethane, 1,1,2-trifluoro- 1,1,2-Trifluoroethane	$C_2H_3F_3$ 84.04	430-66-0 -84	5	4-01-00-00123	
5667	Ethane, 1,1,1-trifluoro-2-iodo-	$C_2H_2F_3I$ 209.94	353-83-3 	54.5	4-01-00-00166 2.13[25]	1.4009[20]
5668	Ethane, 1,1,1-trifluoro-2-nitro-	$C_2H_2F_3NO_2$ 129.04	819-07-8 	96	4-01-00-00172 1.3914[20]	eth 4 1.3394[20]
5669	Ethane, 1,1,1-trimethoxy-	$C_5H_{12}O_3$ 120.15	1445-45-0 	108	4-02-00-00127 0.9438[25]	eth 4; EtOH 4 1.3859[25]
5670	Ethanimidamide, N,N'-bis(4-ethoxyphenyl)-, monohydrochloride Holocaine, hydrochloride	$C_{18}H_{23}ClN_2O_2$ 334.85	620-99-5 191	7153	3-13-00-01069	H_2O 4; EtOH 4; chl 4
5671	Ethanimidamide, monohydrochloride Acetamidine, hydrochloride	$C_2H_7ClN_2$ 94.54	124-42-5 177.5	37	4-02-00-00428	H_2O 4; EtOH 4
5672	Ethanol Ethyl alcohol	C_2H_6O 46.07	64-17-5 -114.1	3716 78.2	4-01-00-01289 0.7893[20]	H_2O 5; EtOH 5; eth 5; ace 5 1.3611[20]
5673	Ethanol, 2-[2-(acetyloxy)ethoxy]-	$C_6H_{12}O_4$ 148.16	2093-20-1 	114-5[7]	2-02-00-00155 1.1208[20]	1.4320[20]
5674	Ethanol, 1-amino- Acetaldehyde ammonia	C_2H_7NO 61.08	75-39-8 97	33 110 dec		H_2O 3; eth 2
5675	Ethanol, 2-amino- Ethanolamine	C_2H_7NO 61.08	141-43-5 10.5	3681 171	4-04-00-01406 1.0180[20]	H_2O 5; EtOH 5; eth 2; bz 2 1.4541[20]
5676	Ethanol, 2-[(2-aminoethyl)amino]- N-(Aminoethyl)ethanolamine	$C_4H_{12}N_2O$ 104.15	111-41-1 	239; 105[10]	4-04-00-01558 1.0286[20]	H_2O 5; EtOH 5; ace 3; bz 2 1.4863[20]
5677	Ethanol, 2-[(2-amino-2-methylpropyl)amino]-	$C_6H_{16}N_2O$ 132.21	68750-16-3 5	236		chl 3 1.4698[25]

No.	Name Synonym	Mol. Form. Mol. Wt.	CAS RN mp/°C	Merck No. bp/°C	Beil. Ref. den/g cm^{-3}	Solubility n_D
5678	Ethanol, 2-[2-[2-(3-aminopropoxy)ethoxy]ethoxy]-	C$_9$H$_{21}$NO$_4$ 207.27	49542-66-7 -50	184[10]	1.0663[20]	H$_2$O 4; EtOH 4 1.4668[20]
5679	Ethanol, 2,2'-[(3-aminopropyl)imino]bis- N-(3-Aminopropyl)diethanolamine	C$_7$H$_{18}$N$_2$O$_2$ 162.23	4985-85-7	160[1]	3-04-00-00720	
5680	Ethanol, 1-amino-2,2,2-trichloro- Chloral ammonia	C$_2$H$_4$Cl$_3$NO 164.42	507-47-1 73	2057 100 dec	4-01-00-03147	bz 4; eth 4; EtOH 4
5681	Ethanol, 2-azido-	C$_2$H$_5$N$_3$O 87.08	1517-05-1	75[40]	4-01-00-01389 1.146[24]	H$_2$O 4
5682	Ethanol, 2-[bis(3-methylbutyl)amino]- Diisopentyl 2-hydroxyethylamine	C$_{12}$H$_{27}$NO 201.35	3574-43-4	248	3-04-00-00685 0.8492[20]	eth 4; EtOH 4; chl 4 1.4435[20]
5683	Ethanol, 2-[bis(1-methylethyl)amino]- N,N-Diisopropyl-2-aminoethanol	C$_8$H$_{19}$NO 145.25	96-80-0	190	4-04-00-01504 0.826[25]	1.4417[20]
5684	Ethanol, 2-[bis(2-methylpropyl)amino]- Ethanolamine, N,N-diisobutyl	C$_{10}$H$_{23}$NO 173.30	4535-66-4	214	4-04-00-01509 0.8407[20]	eth 4; EtOH 4; lig 4; chl 4 1.4355[20]
5685	Ethanol, 2-bromo- Ethylene bromohydrin	C$_2$H$_5$BrO 124.97	540-51-2	3749 150; 51[4]	4-01-00-01385 1.7629[20]	H$_2$O 5; EtOH 5; eth 5; os 4 1.4915[20]
5686	Ethanol, 2-bromo-, acetate	C$_4$H$_7$BrO$_2$ 167.00	927-68-4 -13.8	162.5	4-02-00-00136 1.514[20]	H$_2$O 5; EtOH 5; eth 5; chl 4 1.457[23]
5687	Ethanol, 2-butoxy- Ethylene glycol monobutyl ether	C$_6$H$_{14}$O$_2$ 118.18	111-76-2 -74.8	1559 168.4	4-01-00-02380 0.9015[20]	H$_2$O 5; EtOH 5; eth 5; ctc 2 1.4198[20]
5688	Ethanol, 2-(2-butoxyethoxy)- Diethylene glycol monobutyl ether	C$_8$H$_{18}$O$_3$ 162.23	112-34-5 -68	1557 231	4-01-00-02394 0.9553[20]	H$_2$O 5; EtOH 4; eth 4; ace 4 1.4306[20]
5689	Ethanol, 2-(2-butoxyethoxy)-, acetate	C$_{10}$H$_{20}$O$_4$ 204.27	124-17-4 -32	245	3-02-00-00308 0.985[20]	ace 4; eth 4; EtOH 4 1.4262[20]
5690	Ethanol, 2-[2-(2-butoxyethoxy)ethoxy]-	C$_{10}$H$_{22}$O$_4$ 206.28	143-22-6	278	4-01-00-02402 0.9890[20]	EtOH 4; MeOH 4 1.4389[20]
5691	Ethanol, 2-(butylamino)-	C$_6$H$_{15}$NO 117.19	111-75-1	199; 91-2[11]	3-04-00-00682 0.8907[20]	H$_2$O 4; EtOH 4; eth 4 1.4437[20]
5692	Ethanol, 2,2'-(butylimino)bis-	C$_8$H$_{19}$NO$_2$ 161.24	102-79-4	275; 80[35]	4-04-00-01520 0.9592[20]	chl 3 1.4625[20]
5693	Ethanol, 2-chloro- Ethylene chlorohydrin	C$_2$H$_5$ClO 80.51	107-07-3 -67.5	3750 128.6	4-01-00-01372 1.2019[20]	H$_2$O 5; EtOH 5; eth 2; chl 3 1.4419[20]
5694	Ethanol, 1-chloro-, acetate	C$_4$H$_7$ClO$_2$ 122.55	5912-58-3	123	4-02-00-00282 1.110[20]	eth 4 1.409[20]
5695	Ethanol, 2-chloro-, acetate β-Chloroethyl acetate	C$_4$H$_7$ClO$_2$ 122.55	542-58-5	2137 145	4-02-00-00136 1.178[20]	H$_2$O 1; EtOH 5; eth 5; ctc 3 1.4234[20]
5696	Ethanol, 2-chloro-, benzoate	C$_9$H$_9$ClO$_2$ 184.62	939-55-9 118	255	4-09-00-00288 1.1789[20]	eth 4; EtOH 4
5697	Ethanol, 2-chloro-, carbonate (2:1) Bis(2-chloroethyl) carbonate	C$_5$H$_8$Cl$_2$O$_3$ 187.02	623-97-2 8	241	4-03-00-00006 1.3506[20]	H$_2$O 1 1.461[20]
5698	Ethanol, 2-(2-chloroethoxy)-	C$_4$H$_9$ClO$_2$ 124.57	628-89-7	79-81[5]	3-01-00-02078 1.18[25]	H$_2$O 4; EtOH 5; eth 5 1.4529[20]
5699	Ethanol, 2-chloro-, methanesulfonate	C$_3$H$_7$ClO$_3$S 158.61	3570-58-9	125-6[9]	4-04-00-00012 1.39[25]	chl 3 1.4562[20]
5700	Ethanol, 2-[2-[4-[(4-chlorophenyl)phenylmethyl]-1-piperazinyl]ethoxy]- Hydroxyzine	C$_{21}$H$_{27}$ClN$_2$O$_2$ 374.91	68-88-2	4786	5-23-01-00462	
5701	Ethanol, 2-chloro-, phosphate (3:1)	C$_6$H$_{12}$Cl$_3$O$_4$P 285.49	115-96-8	330; 194[10]	4-01-00-01379 1.39[25]	ctc 3 1.4721[20]
5702	Ethanol, 2-chloro-, phosphite (3:1)	C$_6$H$_{12}$Cl$_3$O$_3$P 269.49	140-08-9	112-5[2]	4-01-00-01378 1.328[25]	1.4868[20]
5703	Ethanol, 2-(dibutylamino)-	C$_{10}$H$_{23}$NO 173.30	102-81-8	114[18]	4-04-00-01506	
5704	Ethanol, 2,2-dichloro- 2,2-Dichloroethanol	C$_2$H$_4$Cl$_2$O 114.96	598-38-9	146	4-01-00-01383 1.4040[25]	H$_2$O 2; EtOH 3; eth 3; ctc 2 1.4626[25]
5705	Ethanol, 2,2-diethoxy-	C$_6$H$_{14}$O$_3$ 134.18	621-63-6	167	4-01-00-03958 0.9[25]	eth 4; EtOH 4 1.4073[20]
5706	Ethanol, 2-(diethylamino)- 2-(Diethylamino)ethanol	C$_6$H$_{15}$NO 117.19	100-37-8	3101 163	4-04-00-01471 0.8921[20]	H$_2$O 5; EtOH 3; eth 3; ace 3 1.4412[20]
5707	Ethanol, 2-[2-(diethylamino)ethoxy]-	C$_8$H$_{19}$NO$_2$ 161.24	140-82-9	221.5; 92[7]	4-04-00-01474 0.9421[25]	1.4480[20]
5708	Ethanol, 2,2-difluoro-	C$_2$H$_4$F$_2$O 82.05	359-13-7 -28.2	95.5	4-01-00-01369 1.3084[17]	ace 4; eth 4; EtOH 4 1.3345[11]
5709	Ethanol, 2-(dimethylamino)- Deanol	C$_4$H$_{11}$NO 89.14	108-01-0 -59	2834 134	4-04-00-01424 0.8866[20]	H$_2$O 5; EtOH 5; eth 5; chl 3 1.4300[20]
5710	Ethanol, 2-[(1,1-dimethylethyl)amino]-	C$_6$H$_{15}$NO 117.19	4620-70-6 44	176.5; 72[14]	4-04-00-01509 0.8818[20]	
5711	Ethanol, 2,2',2'',2'''-[(4,8-di-1-piperidinylpyrimido[5,4-d]pyrimidine- Dipyridamole	C$_{24}$H$_{40}$N$_8$O$_4$ 504.63	58-32-2 163	3354	4-26-00-03840	
5712	Ethanol, 2-(dipropylamino)-	C$_8$H$_{19}$NO 145.25	3238-75-3	196	4-04-00-01502 0.8576[20]	eth 4; EtOH 4 1.4402[20]
5713	Ethanol, 2,2'-[1,2-ethanediylbis(oxy)]bis- Triethylene glycol	C$_6$H$_{14}$O$_4$ 150.17	112-27-6 -7	9585 285	4-01-00-02400 1.1274[15]	H$_2$O 5; EtOH 5; eth 2; bz 5 1.4531[20]
5714	Ethanol, 2,2'-[1,2-ethanediylbis(oxy)]bis-, diacetate Triethylene glycol, diacetate	C$_{10}$H$_{18}$O$_6$ 234.25	111-21-7 -50	286	4-02-00-00215 1.1153[20]	H$_2$O 4; eth 4; EtOH 4
5715	Ethanol, 2,2'-(1,2-ethanediyldiimino)bis-	C$_6$H$_{16}$N$_2$O$_2$ 148.21	4439-20-7 97.5	135-7[1]	4-04-00-01559	H$_2$O 3

No.	Name / Synonym	Mol. Form. / Mol. Wt.	CAS RN / mp/°C	Merck No. / bp/°C	Beil. Ref. / den/g cm^{-3}	Solubility / n_D
5716	Ethanol, 2-(ethenyloxy)-	$C_4H_8O_2$ 88.11	764-48-7	141.6	4-01-00-02387 0.9821[20]	H_2O 3; EtOH 3; eth 3; bz 3 1.4564[17]
5717	Ethanol, 2-ethoxy- Ethylene glycol monoethyl ether	$C_4H_{10}O_2$ 90.12	110-80-5 -70	3707 135	4-01-00-02377 0.9297[20]	H_2O 4; ace 4; eth 4; EtOH 4 1.4080[20]
5718	Ethanol, 2-ethoxy-, acetate Ethylene glycol monoethyl ether acetate	$C_6H_{12}O_3$ 132.16	111-15-9 -61.7	3708 156.4	4-02-00-00214 0.9740[20]	H_2O 4; ace 4; eth 4; EtOH 4 1.4054[20]
5719	Ethanol, 2-ethoxy-, carbonate (2:1)	$C_9H_{18}O_5$ 206.24	2049-74-3	245.5	3-03-00-00017 1.0439[20]	ace 4; eth 4; EtOH 4 1.4227[20]
5720	Ethanol, 2-(2-ethoxyethoxy)- Diethylene glycol monoethyl ether	$C_6H_{14}O_3$ 134.18	111-90-0	1806 196	4-01-00-02393 0.9885[20]	H_2O 5; EtOH 5; eth 4; ace 5 1.4300[20]
5721	Ethanol, 2-(2-ethoxyethoxy)-, acetate Diethylene glycol monoethyl ether acetate	$C_8H_{16}O_4$ 176.21	112-15-2 -25	218.5	3-02-00-00308 1.0096[20]	H_2O 4; ace 4; eth 4; EtOH 4 1.4213[20]
5722	Ethanol, 2-(ethylamino)-	$C_4H_{11}NO$ 89.14	110-73-6	169.5	4-04-00-01465 0.914[20]	H_2O 4; EtOH 4; eth 4; chl 3 1.444[20]
5723	Ethanol, 2,2'-(ethylimino)bis-	$C_6H_{15}NO_2$ 133.19	139-87-7 -50	247	3-04-00-00693 1.0135[20]	H_2O 4; EtOH 4; eth 2 1.4663[20]
5724	Ethanol, 2-[ethyl(3-methylphenyl)amino]-	$C_{11}H_{17}NO$ 179.26	91-88-3	118[1.5]	4-12-00-01819	ctc 3 1.5540[20]
5725	Ethanol, 2-(ethylsulfonyl)- Ethylsulfonylethyl alcohol	$C_4H_{10}O_3S$ 138.19	513-12-2	3810	4-01-00-02430	chl 2
5726	Ethanol, 2-(ethylthio)- 2-(Ethylthio)ethanol	$C_4H_{10}OS$ 106.19	110-77-0 -100	3813 184	4-01-00-02430 1.0166[20]	H_2O 2; EtOH 3; ace 4 1.4867[20]
5727	Ethanol, 2-fluoro- Ethylene fluorohydrin	C_2H_5FO 64.06	371-62-0 -26.4	103.5	4-01-00-01366 1.1040[20]	H_2O 5; EtOH 5; eth 5; ace 4 1.3647[18]
5728	Ethanol, 2-(hexyloxy)-	$C_8H_{18}O_2$ 146.23	112-25-4 -45.1	208	4-01-00-02383 0.8878[20]	H_2O 2; EtOH 4; eth 4 1.4291[20]
5729	Ethanol, 2-hydrazino- 2-Hydrazinoethanol	$C_2H_8N_2O$ 76.10	109-84-2 -70	4696 219; 120[17.5]	4-04-00-03348 1.119[25]	H_2O 4; EtOH 4; MeOH 4
5730	Ethanol, 2,2'-iminobis- Diethanolamine	$C_4H_{11}NO_2$ 105.14	111-42-2 28	3097 268.8	4-04-00-01514 1.0966[20]	H_2O 4; EtOH 4; eth 2; bz 2 1.4776[20]
5731	Ethanol, 2-iodo-	C_2H_5IO 171.97	624-76-0	176 dec	4-01-00-01387 2.1967[20]	H_2O 4; eth 4; EtOH 4 1.5713[20]
5732	Ethanol, 2-mercapto- 2-Mercaptoethanol	C_2H_6OS 78.14	60-24-2	5760 158; 55[13]	4-01-00-02428 1.1143[20]	H_2O 3; EtOH 3; eth 3; bz 3 1.4996[20]
5733	Ethanol, 2-methoxy- Ethylene glycol monomethyl ether	$C_3H_8O_2$ 76.10	109-86-4 -85.1	5961 124.1	4-01-00-02375 0.9647[20]	H_2O 5; EtOH 5; eth 5; ace 3 1.4024[20]
5734	Ethanol, methoxy-, acetate Acetic acid, 2-methoxyethyl ester	$C_5H_{10}O_3$ 118.13	32718-56-2	145; 40[12]	2-02-00-00154 1.0074[19]	H_2O 3; EtOH 3; eth 3; ctc 2 1.4002[20]
5735	Ethanol, 2-methoxy-, acetate 2-Methoxyethyl acetate	$C_5H_{10}O_3$ 118.13	110-49-6 -70	5962 143	4-02-00-00214 1.0074[19]	H_2O 4; EtOH 4 1.4002[20]
5736	Ethanol, 2-methoxy-, carbonate (2:1) Bis(2-methoxyethyl) carbonate	$C_7H_{14}O_5$ 178.19	626-84-6	231	3-03-00-00017 1.0988[20]	EtOH 3; eth 3; ace 3 1.4204[20]
5737	Ethanol, 2-(2-methoxyethoxy)- 2-(2-Methoxyethoxy)ethanol	$C_5H_{12}O_3$ 120.15	111-77-3	5959 193	4-01-00-02392 1.035[20]	H_2O 5; EtOH 4; eth 4; ace 5 1.4264[20]
5738	Ethanol, 2-(methylamino)- N-Methyl-2-ethanolamine	C_3H_9NO 75.11	109-83-1	5939 158	4-04-00-01422 0.937[20]	H_2O 5; EtOH 5; eth 5 1.4385[20]
5739	Ethanol, 2-[(3-methylbutyl)amino]- N-Isopentylethanolamine	$C_7H_{17}NO$ 131.22	34240-76-1	210; 82[5.5]	2-04-00-00728 0.8822[20]	H_2O 4; eth 4; EtOH 4 1.4447[20]
5740	Ethanol, 2-(1-methylethoxy)- 2-Isopropoxyethanol	$C_5H_{12}O_2$ 104.15	109-59-1	145	4-01-00-02380 0.9030[20]	H_2O 5; EtOH 5; eth 5; ace 3 1.4095[20]
5741	Ethanol, 2-[(1-methylethyl)amino]-	$C_5H_{13}NO$ 103.16	109-56-8 128.5	173	3-04-00-00681 0.8970[20]	H_2O 5; EtOH 5; eth 5 1.4395[20]
5742	Ethanol, 2,2'-(methylimino)bis-	$C_5H_{13}NO_2$ 119.16	105-59-9 -21	247	4-04-00-01517 1.038[25]	H_2O 4 1.4685[20]
5743	Ethanol, 2-(methylphenylamino)-	$C_9H_{13}NO$ 151.21	93-90-3	150[14]	4-12-00-00280 0.9993[15]	H_2O 3; EtOH 4; eth 4; ace 4
5744	Ethanol, 2-[(2-methylphenyl)amino]-	$C_9H_{13}NO$ 151.21	136-80-1	285.5	4-12-00-01750 1.0794[20]	eth 4; EtOH 4 1.5675[20]
5745	Ethanol, 2-[(4-methylphenyl)amino]-	$C_9H_{13}NO$ 151.21	2933-74-6 42.5	287	4-12-00-01880	bz 4; eth 4; EtOH 4; chl 4
5746	Ethanol, 2-(2-methylpropoxy)-	$C_6H_{14}O_2$ 118.18	4439-24-1	160	4-01-00-02382 0.8900[20]	1.4143[20]
5747	Ethanol, 2-[(2-methylpropyl)amino]- Ethanolamine, N-isobutyl	$C_6H_{15}NO$ 117.19	17091-40-6	199; 90[16]	3-04-00-00683 0.8818[20]	H_2O 4; EtOH 4; eth 4 1.4402[20]
5748	Ethanol, 2-[(2-methylpropyl)amino]-, 4-aminobenzoate (ester) Butethamine	$C_{13}H_{20}N_2O_2$ 236.31	2090-89-3	1517		
5749	Ethanol, 2-(methylthio)-	C_3H_8OS 92.16	5271-38-5	68-70[20]	4-01-00-02429 1.063[20]	H_2O 4; eth 4; EtOH 4 1.4861[30]
5750	Ethanol, 2-(2-naphthalenylamino)- Ethanolamine, N-(β-naphthyl)	$C_{12}H_{13}NO$ 187.24	36190-77-9 52	197-8[3]	4-12-00-03131	eth 4
5751	Ethanol, 2,2',2''-nitrilotris- Triethanolamine	$C_6H_{15}NO_3$ 149.19	102-71-6 20.5	9581 335.4	4-04-00-01524 1.1242[20]	H_2O 5; EtOH 2; bz 2 1.4852[20]
5752	Ethanol, 2-nitro- 2-Nitroethanol	$C_2H_5NO_3$ 91.07	625-48-9 -80	194; 102[10]	4-01-00-01388 1.270[15]	H_2O 5; EtOH 5; eth 5; bz 1 1.4438[19]
5753	Ethanol, 2,2'-oxybis- Diethylene glycol	$C_4H_{10}O_3$ 106.12	111-46-6 -10.4	3109 245.8	4-01-00-02390 1.1197[15]	H_2O 3; EtOH 3; eth 3; chl 3 1.4472[20]
5754	Ethanol, 2,2'-oxybis-, diacetate Diethylene glycol, diacetate	$C_8H_{14}O_5$ 190.20	628-68-2 18	200	4-02-00-00216 1.1068[15]	EtOH 4 1.4348[20]
5755	Ethanol, 2,2'-oxybis-, dibenzoate	$C_{18}H_{18}O_5$ 314.34	120-55-8 33.5	280[24]; 250[1]	4-09-00-00356 1.1690[15]	H_2O 4; EtOH 4

No.	Name / Synonym	Mol. Form. / Mol. Wt.	CAS RN / mp/°C	Merck No. / bp/°C	Beil. Ref. / den/g cm^{-3}	Solubility / n_D
5756	Ethanol, 2,2'-[oxybis(2,1-ethanediyloxy)]bis- / Tetraethylene glycol	$C_8H_{18}O_5$ / 194.23	112-60-7 / -6.2	/ 328	4-01-00-02403 / 1.1285[15]	H_2O 4; EtOH 3; eth 3; ctc 3 / 1.4577[20]
5757	Ethanol, 2-(pentyloxy)-	$C_7H_{16}O_2$ / 132.20	6196-58-3 /	/ 182	4-01-00-02383 / 0.8918[15]	1.4213[25]
5758	Ethanol, 2-phenoxy- / 2-Phenoxyethanol	$C_8H_{10}O_2$ / 138.17	122-99-6 / 14	7226 / 245	4-06-00-00571 / 1.102[22]	H_2O 1; EtOH 3; eth 3; chl 3 / 1.534[20]
5759	Ethanol, 2-(phenylamino)-	$C_8H_{11}NO$ / 137.18	122-98-5 /	/ 279.5; 150[10]	4-12-00-00277 / 1.0945[20]	H_2O 2; EtOH 4; eth 4; chl 4 / 1.5760[20]
5760	Ethanol, 2,2'-(phenylimino)bis-	$C_{10}H_{15}NO_2$ / 181.23	120-07-0 / 57	/ 200[10]	4-12-00-00285 / 1.201[60]	ace 4; bz 4; eth 4; EtOH 4
5761	Ethanol, 2-(phenylmethoxy)-	$C_9H_{12}O_2$ / 152.19	622-08-2 / <-75	/ 256	4-06-00-02241 / 1.0640[20]	H_2O 4; eth 4; EtOH 4 / 1.5233[20]
5762	Ethanol, 2-[(phenylmethyl)amino]-	$C_9H_{13}NO$ / 151.21	104-63-2 /	/ 153-6[12]	4-12-00-02184 / 1.065[25]	1.5430[20]
5763	Ethanol, 2-(2-propenyloxy)-	$C_5H_{10}O_2$ / 102.13	111-45-5 /	/ 158.5	4-01-00-02388 / 0.9580[20]	H_2O 5; EtOH 4; bz 3; ctc 3 / 1.4358[20]
5764	Ethanol, 2-propoxy- / 2-Propoxyethanol	$C_5H_{12}O_2$ / 104.15	2807-30-9 /	/ 149.8	4-01-00-02379 / 0.9112[20]	H_2O 3; EtOH 4; eth 4 / 1.4133[20]
5765	Ethanol, 2-(propylamino)-	$C_5H_{13}NO$ / 103.16	16369-21-4 /	/ 183	3-04-00-00681 / 0.9005[20]	1.4428[20]
5766	Ethanol, 2,2'-thiobis- / 2,2'-Thiodiethanol	$C_4H_{10}O_2S$ / 122.19	111-48-8 / -10.2	9259 / 282	4-01-00-02437 / 1.1793[25]	H_2O 5; EtOH 5; eth 3; bz 2 / 1.5211[20]
5767	Ethanol, 2,2,2-tribromo- / Tribromoethanol	$C_2H_3Br_3O$ / 282.76	75-80-9 / 81	9525 / 92-3[10]	3-01-00-01362 /	bz 4; eth 4; EtOH 4
5768	Ethanol, 2,2,2-trichloro- / 2,2,2-Trichloroethanol	$C_2H_3Cl_3O$ / 149.40	115-20-8 / 19	9551 / 152; 52[11]	4-01-00-01383 /	H_2O 2; EtOH 5; eth 5; ctc 2 / 1.4861[20]
5769	Ethanol, 2,2,2-trichloro-1-ethoxy- / Chloral alcoholate	$C_4H_7Cl_3O_2$ / 193.46	515-83-3 / 56.5	2056 / 115.5	4-01-00-03144 / 1.143[40]	H_2O 3; EtOH 3; eth 3
5770	Ethanol, 2,2,2-trifluoro-	$C_2H_3F_3O$ / 100.04	75-89-8 / -43.5	/ 74	4-01-00-01370 / 1.3842[20]	EtOH 4; eth 3; ace 3; bz 3 / 1.2907[22]
5771	Ethanone, 2-(acetyloxy)-1-phenyl-	$C_{10}H_{10}O_3$ / 178.19	2243-35-8 / 49	/ 270	4-08-00-00368 / 1.1169[65]	H_2O 1; EtOH 4; eth 4; bz 2 / 1.5036[65]
5772	Ethanone, 1-(3-aminophenyl)-	C_8H_9NO / 135.17	99-03-6 / 98.5	/ 289.5	4-14-00-00096 /	H_2O 2; EtOH 3
5773	Ethanone, 1-(4-aminophenyl)-	C_8H_9NO / 135.17	99-92-3 / 106	/ 294; 195[15]	4-14-00-00100 /	eth 4; EtOH 4
5774	Ethanone, 2-amino-1-phenyl- / Phenacylamine	C_8H_9NO / 135.17	613-89-8 / 20	7159 / 251	4-14-00-00114 /	H_2O 1; eth 3; ctc 2 / 1.6160[20]
5775	Ethanone, 1-(2-benzofuranyl)-	$C_{10}H_8O_2$ / 160.17	1646-26-0 / 76	/ 126[11]	5-17-10-00172 /	H_2O 3
5776	Ethanone, 1-[1,1'-biphenyl]-4-yl-	$C_{14}H_{12}O$ / 196.25	92-91-1 / 121	/ 326	4-07-00-01407 / 1.2510[0]	H_2O 1; EtOH 4; ace 4; chl 2
5777	Ethanone, 2-bromo-1-(4-bromophenyl)- / p-Bromophenacyl bromide	$C_8H_8Br_2O$ / 277.94	99-73-0 / 111	1415 /	4-07-00-00652 /	H_2O 1; EtOH 3; eth 3; chl 3
5778	Ethanone, 2-bromo-1-(4-chlorophenyl)- / p-Chlorophenacyl bromide	C_8H_8BrClO / 233.49	536-38-9 / 96.5	2153 /	4-07-00-00651 /	
5779	Ethanone, 2-bromo-1-(4-methylphenyl)-	C_9H_9BrO / 213.07	619-41-0 / 51	/ 157[14]	4-07-00-00705 /	eth 4; EtOH 4
5780	Ethanone, 1-(3-bromophenyl)-	C_8H_7BrO / 199.05	2142-63-4 / 7.5	/ 133[19]	4-07-00-00646 /	H_2O 1; ace 3; bz 3 / 1.5755[20]
5781	Ethanone, 1-(4-bromophenyl)- / p-Bromoacetophenone	C_8H_7BrO / 199.05	99-90-1 / 50.5	1390 / 257; 130[11]	4-07-00-00647 / 1.647[25]	H_2O 1; EtOH 3; eth 3; bz 3 / 1.647
5782	Ethanone, 2-bromo-1-phenyl- / ω-Bromoacetophenone	C_8H_7BrO / 199.05	70-11-1 / 50.5	1391 / 135[18]	4-07-00-00649 / 1.647[20]	H_2O 1; EtOH 3; eth 4; bz 4
5783	Ethanone, 1-(5-bromo-2-thienyl)-	C_8H_5BrOS / 205.08	5370-25-2 / 94.5	/ 103[4]	5-17-09-00392 /	EtOH 2; ctc 3
5784	Ethanone, 2-chloro-1-(4-chlorophenyl)-	$C_8H_8Cl_2O$ / 189.04	937-20-2 / 101.5	/ 270	4-07-00-00644 /	EtOH 3; bz 3; MeOH 3
5785	Ethanone, 2-chloro-1,2-diphenyl-	$C_{14}H_{11}ClO$ / 230.69	447-31-4 / 68.5	/ dec	3-07-00-02106 /	EtOH 3; chl 2; alk 1
5786	Ethanone, 2-chloro-1-(4-hydroxy-3-methoxyphenyl)-	$C_9H_9ClO_3$ / 200.62	6344-28-1 / 102	/ 208[22]	3-08-00-02114 /	H_2O 2; EtOH 3
5787	Ethanone, 2-chloro-1-(4-methylphenyl)-	C_9H_9ClO / 168.62	4209-24-9 / 57.5	/ 261.5	4-07-00-00704 /	eth 4; EtOH 4
5788	Ethanone, 1-(3-chlorophenyl)-	C_8H_7ClO / 154.60	99-02-5 /	/ 244; 129[30]	4-07-00-00638 / 1.2130[40]	EtOH 3; eth 3; ace 3 / 1.5494[20]
5789	Ethanone, 1-(4-chlorophenyl)- / p-Chloroacetophenone	C_8H_7ClO / 154.60	99-91-2 / 20	2114 / 232	4-07-00-00639 / 1.1922[20]	H_2O 1; EtOH 5; eth 5; chl 3 / 1.5550[20]
5790	Ethanone, 2-chloro-1-phenyl- / ω-Chloroacetophenone	C_8H_7ClO / 154.60	532-27-4 / 56.5	2115 / 247	4-07-00-00641 / 1.324[15]	H_2O 1; EtOH 4; eth 4; ace 3
5791	Ethanone, 1-(5-chloro-2-thienyl)-	C_8H_5ClOS / 160.62	6310-09-4 / 52	/ 117[17]; 88[4.5]	5-17-09-00391 /	H_2O 4; eth 4; EtOH 4
5792	Ethanone, 1-cyclobutyl-	$C_8H_{10}O$ / 98.14	3019-25-8 /	/ 138	4-07-00-00037 / 0.9020[20]	H_2O 2 / 1.4322[19]
5793	Ethanone, 1-(1-cyclohexen-1-yl)-	$C_8H_{12}O$ / 124.18	932-66-1 / 73	/ 201.5	4-07-00-00143 / 0.9655[20]	EtOH 3; eth 3 / 1.4881[20]
5794	Ethanone, 1-cyclohexyl-	$C_8H_{14}O$ / 126.20	823-76-7 /	/ 180.5	4-07-00-00053 / 0.9176[20]	H_2O 1; eth 3 / 1.4565[16]
5795	Ethanone, 1-cyclopentyl-	$C_7H_{12}O$ / 112.17	6004-60-0 /	/ 158.5	4-07-00-00047 / 0.916[20]	eth 3 / 1.4409[20]

No.	Name Synonym	Mol. Form. Mol. Wt.	CAS RN mp/°C	Merck No. bp/°C	Beil. Ref. den/g cm^{-3}	Solubility n_D
5796	Ethanone, 1-cyclopropyl- Cyclopropyl methyl ketone	C_5H_8O 84.12	765-43-5 -68.3	111.3	4-07-00-00014 0.8984[20]	H_2O 4; eth 4; EtOH 4 1.4251[20]
5797	Ethanone, 1-(3,5-dibromophenyl)-	$C_8H_6Br_2O$ 277.94	14401-73-1 68	198[15]	3-07-00-00984	bz 4; eth 4; EtOH 4
5798	Ethanone, 2,2-dibromo-1-phenyl-	$C_8H_6Br_2O$ 277.94	13665-04-8 36.5	159[13]	4-07-00-00653	EtOH 3; eth 3; chl 3
5799	Ethanone, 1-(2,4-dichlorophenyl)-	$C_8H_6Cl_2O$ 189.04	2234-16-4 33.5		4-07-00-00643	H_2O 1 1.5640[20]
5800	Ethanone, 1-(2,5-dichlorophenyl)-	$C_8H_6Cl_2O$ 189.04	2476-37-1 12	118[12]	4-07-00-00643 1.321[30]	1.5595[30]
5801	Ethanone, 1-(3,4-dichlorophenyl)-	$C_8H_6Cl_2O$ 189.04	2642-63-9 76	135[12]	4-07-00-00643	H_2O 1; ctc 3; lig 3
5802	Ethanone, 2,2-dichloro-1-phenyl-	$C_8H_6Cl_2O$ 189.04	2648-61-5 20.5	249	4-07-00-00644 1.340[16]	EtOH 3; bz 3; ctc 3 1.5686[20]
5803	Ethanone, 1-(2,4-dihydroxyphenyl)- Resacetophenone	$C_8H_8O_3$ 152.15	89-84-9 146	8143	4-08-00-01792 1.18[141]	H_2O 1; EtOH 3; eth 2; bz 2
5804	Ethanone, 1-(3,4-dihydroxyphenyl)-2-(methylamino)- Adrenalone	$C_9H_{11}NO_3$ 181.19	99-45-6 235-6 dec	161	4-14-00-00832	H_2O 2; EtOH 2; eth 2
5805	Ethanone, 1-(3,4-dimethoxyphenyl)-	$C_{10}H_{12}O_3$ 180.20	1131-62-0 51	287	4-08-00-01815	H_2O 4; bz 4; EtOH 4; chl 4
5806	Ethanone, 1-[3-(dimethylamino)phenyl]-	$C_{10}H_{13}NO$ 163.22	18992-80-8 43	148[13]	4-14-00-00096	ace 4; eth 4
5807	Ethanone, 1-[10-[3-(dimethylamino)propyl]-10H-phenothiazin-2-yl]- Acepromazine	$C_{19}H_{22}N_2OS$ 326.46	61-00-7 27	230[0.5]	4-27-00-02911	
5808	Ethanone, 1-[4-(1,1-dimethylethyl)phenyl]-	$C_{12}H_{16}O$ 176.26	943-27-1	136-8[20]	4-07-00-00769 0.9705[25]	1.518[15]
5809	Ethanone, 1-(2,4-dimethylphenyl)-	$C_{10}H_{12}O$ 148.20	89-74-7	228	4-07-00-00729 1.0121[15]	eth 4; EtOH 4 1.5340[20]
5810	Ethanone, 1-(2,5-dimethylphenyl)-	$C_{10}H_{12}O$ 148.20	2142-73-6 -18.1	232.5	4-07-00-00728 0.9963[19]	H_2O 1; EtOH 4; eth 4; bz 4 1.5291[20]
5811	Ethanone, 1-(3,4-dimethylphenyl)-	$C_{10}H_{12}O$ 148.20	3637-01-2 -3	246.5	4-07-00-00728 1.0090[14]	H_2O 1; EtOH 4; eth 4; bz 4 1.5413[15]
5812	Ethanone, 1,2-diphenyl-	$C_{14}H_{12}O$ 196.25	451-40-1 60	320	4-07-00-01393 1.201[0]	H_2O 2; EtOH 3; eth 3; ctc 3
5813	Ethanone, 2-ethoxy-1,2-diphenyl-	$C_{16}H_{16}O_2$ 240.30	574-09-4 62	194-5[20]	3-08-00-01279 1.1016[17]	bz 4; eth 4; EtOH 4; lig 4 1.5727[17]
5814	Ethanone, 2-ethoxy-1-phenyl- 2-Ethoxyacetophenone	$C_{10}H_{12}O_2$ 164.20	14869-39-7 43	243.5	4-08-00-00365 1.0036[78]	eth 4; EtOH 4; lig 4
5815	Ethanone, 1-(4-fluorophenyl)-	C_8H_7FO 138.14	403-42-9 -45	196	4-07-00-00635 1.1382[25]	H_2O 1; bz 3; chl 3 1.5081[25]
5816	Ethanone, 2-fluoro-1-phenyl-	C_8H_7FO 138.14	450-95-3 29	90-1[12]	4-07-00-00636 1.152[20]	1.5200[20]
5817	Ethanone, 1-(2-furanyl)-	$C_6H_6O_2$ 110.11	1192-62-7 33	175	5-17-09-00381 1.098[20]	H_2O 1; EtOH 3; eth 3 1.5017[20]
5818	Ethanone, 1-[4-(β-D-glucopyranosyloxy)phenyl]- Picein	$C_{14}H_{18}O_7$ 298.29	530-14-3 195.5	7367	5-17-07-00151	H_2O 2; EtOH 3; eth 3; chl 1
5819	Ethanone, 1-(1H-indol-3-yl)-	$C_{10}H_9NO$ 159.19	703-80-0 192.3	144[10]	5-21-08-00297	EtOH 4
5820	Ethanone, 1-(1H-pyrrol-2-yl)-	C_6H_7NO 109.13	1072-83-9 90	220	5-21-07-00204	H_2O 3; EtOH 3; eth 3
5821	Ethanone, 1-(1-hydroxycyclohexyl)-	$C_8H_{14}O_2$ 142.20	1123-27-9	125.5; 91[11]	4-08-00-00023 1.0248[25]	eth 4; EtOH 4 1.4670[25]
5822	Ethanone, 2-hydroxy-1,2-diphenyl-, (±)- (±)-Benzoin	$C_{14}H_{12}O_2$ 212.25	579-44-2 137	1103 344; 194[12]	4-08-00-01279 1.310[20]	EtOH 4; chl 4
5823	Ethanone, 1-(2-hydroxy-4-methoxyphenyl)-	$C_9H_{10}O_3$ 166.18	552-41-0 52.5	158[20]	4-08-00-01793 1.3102[81]	bz 4; eth 4; EtOH 4; chl 4 1.5452[81]
5824	Ethanone, 1-(4-hydroxy-3-methoxyphenyl)- Apocynin	$C_9H_{10}O_3$ 166.18	498-02-2 115	772 297; 234[15]	4-08-00-01814	H_2O 2; EtOH 3; eth 4; ace 3
5825	Ethanone, 1-[6-hydroxy-2-(1-methylethenyl)-5-benzofuranyl]- Euparin	$C_{13}H_{12}O_3$ 216.24	532-48-9 121.5	3857	5-18-02-00029	eth 3; bz 3; chl 3; NaOH 2
5826	Ethanone, 1-(2-hydroxy-4-methylphenyl)-	$C_9H_{10}O_2$ 150.18	6921-64-8 21	245	4-08-00-00480 1.1012[10]	1.5527[13]
5827	Ethanone, 1-(2-hydroxy-5-methylphenyl)-	$C_9H_{10}O_2$ 150.18	1450-72-2 50	210; 120[20]	4-08-00-00472 1.079[53]	bz 4; eth 4; EtOH 4; chl 4
5828	Ethanone, 1-(4-hydroxy-2-methylphenyl)-	$C_9H_{10}O_2$ 150.18	875-59-2 128	313	4-08-00-00469 1.0592[135]	eth 4; EtOH 4 1.5369[135]
5829	Ethanone, 1-(1-hydroxy-2-naphthalenyl)-	$C_{12}H_{10}O_2$ 186.21	711-79-5 101	325 dec	4-08-00-01175	bz 4; HOAc 4
5830	Ethanone, 1-(2-hydroxyphenyl)-	$C_8H_8O_2$ 136.15	118-93-4 5	218	4-08-00-00320 1.1307[20]	eth 4; EtOH 4; HOAc 4 1.5584[20]
5831	Ethanone, 1-(3-hydroxyphenyl)-	$C_8H_8O_2$ 136.15	121-71-1 96	296; 153[5]	4-08-00-00334 1.0992[109]	H_2O 2; EtOH 4; eth 4; bz 4 1.5348[109]
5832	Ethanone, 1-(4-hydroxyphenyl)-	$C_8H_8O_2$ 136.15	99-93-4 109.5	147-8[3]	4-08-00-00339 1.1090[109]	H_2O 2; EtOH 4; eth 4 1.5577[109]
5833	Ethanone, 2-hydroxy-1-phenyl-	$C_8H_8O_2$ 136.15	582-24-1 90	125[12]; 56[1]	4-08-00-00365 1.0963[99]	H_2O 3; EtOH 3; eth 3; chl 3

No.	Name / Synonym	Mol. Form. / Mol. Wt.	CAS RN / mp/°C	Merck No. / bp/°C	Beil. Ref. / den/g cm⁻³	Solubility / n_D
5834	Ethanone, 2-[6-(2-hydroxy-2-phenylethyl)-1-methyl-2-piperidinyl]-1-phenyl-, [2R-[2α,6α(S*)]]- Lobeline	$C_{22}H_{27}NO_2$ 337.46	90-69-7 130.5	5432	5-21-12-00627	H_2O 2; EtOH 3; eth 3; ace 4
5835	Ethanone, 1-(3-hydroxy-2-thienyl)-	$C_8H_6O_2S$ 142.18	5556-07-0 51.5	47[0.2]	5-18-01-00121 1.5000[20]	EtOH 3 1.5795[20]
5836	Ethanone, 1-(3-iodophenyl)-	C_8H_7IO 246.05	14452-30-3	129[8]; 117[4]	4-07-00-00654	bz 3 1.622[20]
5837	Ethanone, 1-(4-iodophenyl)-	C_8H_7IO 246.05	13329-40-3 86	153[18]	4-07-00-00654	EtOH 3; eth 2; bz 3; HOAc 3
5838	Ethanone, 2-iodo-1-phenyl-	C_8H_7IO 246.05	4636-16-2 34.4	154[15]	4-07-00-00654	ace 3
5839	Ethanone, 2-methoxy-1,2-diphenyl-	$C_{15}H_{14}O_2$ 226.27	3524-62-7 49.5	188-9[15]	4-08-00-01280 1.1278[14]	bz 4; eth 4; EtOH 4
5840	Ethanone, 1-(3-methoxyphenyl)-	$C_9H_{10}O_2$ 150.18	586-37-8 95.5	240	4-08-00-00334 1.0343[19]	H_2O 3; EtOH 3; ace 3; ctc 3 1.5410[20]
5841	Ethanone, 1-(4-methoxyphenyl)-	$C_9H_{10}O_2$ 150.18	100-06-1 38.5	258	4-08-00-00340 1.0818[41]	H_2O 2; EtOH 3; eth 3; ace 3 1.547[41]
5842	Ethanone, 2-methoxy-1-phenyl-	$C_9H_{10}O_2$ 150.18	4079-52-1 8	245; 125[19]	4-08-00-00365 1.0897[20]	H_2O 2; EtOH 3; ace 3 1.5393[20]
5843	Ethanone, 1-(1-methylcyclohexyl)-	$C_9H_{16}O$ 140.23	2890-62-2	186.2	4-07-00-00067 0.9504[20]	1.4543[20]
5844	Ethanone, 1-(3-methylcyclohexyl)-	$C_9H_{16}O$ 140.23	7193-78-4	197	3-07-00-00120 0.912[19]	1.4517[19]
5845	Ethanone, 1-(4-methylcyclohexyl)-	$C_9H_{16}O$ 140.23	1879-06-7	196	3-07-00-00121 0.9055[18]	1.4509[18]
5846	Ethanone, 1-(1-methylcyclopentyl)-	$C_8H_{14}O$ 126.20	13388-93-7	48.4[10]	3-07-00-00102 0.9104[20]	1.4430[20]
5847	Ethanone, 1-(2-methylcyclopentyl)-	$C_8H_{14}O$ 126.20	1601-00-9	170.5; 59.6[15]	3-07-00-00103 0.9222[4]	1.4434[16]
5848	Ethanone, 1-[4-(1-methylethyl)phenyl]-	$C_{11}H_{14}O$ 162.23	645-13-6	254	4-07-00-00747 0.9753[15]	1.5235[20]
5849	Ethanone, 1-(1-methyl-1H-pyrrol-2-yl)-	C_7H_9NO 123.15	932-16-1	201[252]; 93[22]	5-21-07-00205 1.0445[15]	EtOH 3; bz 3; chl 3 1.5403[15]
5850	Ethanone, 1-[2-methyl-5-(1-methylethyl)phenyl]-	$C_{12}H_{16}O$ 176.26	1202-08-0 <-20	249.5	4-07-00-00772 0.956[20]	1.5181[20]
5851	Ethanone, 1-(2-methylphenyl)-	$C_9H_{10}O$ 134.18	577-16-2	214	4-07-00-00697 1.026[20]	1.5276[20]
5852	Ethanone, 1-(3-methylphenyl)-	$C_9H_{10}O$ 134.18	585-74-0	220	4-07-00-00699 1.0165[0]	EtOH 3; eth 3; ace 3; ctc 2 1.533[15]
5853	Ethanone, 1-(4-methylphenyl)-	$C_9H_{10}O$ 134.18	122-00-9 28	226; 93.5[7]	4-07-00-00701 1.0051[20]	bz 4; eth 4; EtOH 4; chl 4 1.5335[20]
5854	Ethanone, 2,2'-(1-methyl-2,6-piperidinediyl)bis[1-phenyl-, cis- Lobelanine	$C_{22}H_{25}NO_2$ 335.45	579-21-5 99	5430	5-21-11-00432	ace 4; bz 4; EtOH 4; chl 4
5855	Ethanone, 1-(6-methyl-3-pyridinyl)-	C_8H_9NO 135.17	36357-38-7 17.6	144[50]	5-21-07-00425 1.0168[25]	H_2O 4 1.5302[25]
5856	Ethanone, 1-(5-methyl-2-thienyl)-	C_7H_8OS 140.21	13679-74-8 27.5	232.5	5-17-09-00425 1.1185[25]	eth 3; ace 3; bz 3; chl 2 1.5604
5857	Ethanone, 1-(1-naphthalenyl)-	$C_{12}H_{10}O$ 170.21	941-98-0 34	297	4-07-00-01292 1.1171[21]	H_2O 1; EtOH 3; eth 3; ace 3 1.6280[22]
5858	Ethanone, 1-(2-naphthalenyl)-	$C_{12}H_{10}O$ 170.21	93-08-3 56	302	4-07-00-01294	EtOH 2; ctc 2
5859	Ethanone, 1-(1-naphthalenyl)-2-phenyl- Benzyl 1-naphthyl ketone	$C_{18}H_{14}O$ 246.31	605-85-6 66.5	194-6[0.05]	4-07-00-01790	eth 4
5860	Ethanone, 1-(3-nitrophenyl)-	$C_8H_7NO_3$ 165.15	121-89-1 81	202; 167[18]	4-07-00-00656	H_2O 4; EtOH 4; chl 2
5861	Ethanone, 1-(4-nitrophenyl)-	$C_8H_7NO_3$ 165.15	100-19-6 81.8	165[5]	4-07-00-00657	eth 4; EtOH 4
5862	Ethanone, 2-nitro-1-phenyl-	$C_8H_7NO_3$ 165.15	614-21-1 106	158[16]; 142[10]	4-07-00-00659	eth 4; EtOH 4 1.5468[30]
5863	Ethanone, 1-(2-nitrophenyl)- Acetophene,2-nitro	$C_8H_7N_3O$ 161.16	577-59-3 28.5	178[32]; 158[18]	4-07-00-00655 1.2370[25]	H_2O 1; EtOH 4; eth 4; chl 4 1.5468[20]
5864	Ethanone, 1-phenyl- Acetophenone	C_8H_8O 120.15	98-86-2 20	65 202	4-07-00-00619 1.0281[20]	H_2O 1; EtOH 3; eth 3; ace 3 1.5372[20]
5865	Ethanone, 1,1'-(1,3-phenylene)bis-	$C_{10}H_{10}O_2$ 162.19	6781-42-6 32	152[15]	4-07-00-02156	H_2O 2; EtOH 3; bz 3; chl 3
5866	Ethanone, 1-phenyl-, oxime	C_8H_9NO 135.17	613-91-2 60	245	3-07-00-00954 1.0515[78]	H_2O 2; EtOH 4; eth 4; ace 4
5867	Ethanone, 1-(2-pyridinyl)-	C_7H_7NO 121.14	1122-62-9	192	5-21-07-00385 1.077[25]	EtOH 3; eth 3; ctc 2; HOAc 3 1.5203[20]
5868	Ethanone, 1-(3-pyridinyl)- Methyl 3-pyridyl ketone	C_7H_7NO 121.14	350-03-8 13.5	6034 220	5-21-07-00394	H_2O 3; EtOH 3; eth 3; acid 3 1.5341[20]
5869	Ethanone, 1-(4-pyridinyl)-	C_7H_7NO 121.14	1122-54-9 16	212	5-21-07-00400 1.097[25]	EtOH 2; eth 2; acid 2 1.5282[25]

No.	Name Synonym	Mol. Form. Mol. Wt.	CAS RN mp/°C	Merck No. bp/°C	Beil. Ref. den/g cm^{-3}	Solubility n_D
5870	Ethanone, 1-(2-thienyl)-	C_8H_6OS 126.18	88-15-3 10.5	213.5	5-17-09-00387 1.1679[20]	H_2O 2; EtOH 5; eth 5; ctc 3 1.5667[20]
5871	Ethanone, 2,2,2-trichloro-1-phenyl-	$C_8H_5Cl_3O$ 223.49	2902-69-4	256.5	4-07-00-00645 1.425[18]	eth 4; EtOH 4
5872	Ethanone, 2,2,2-trifluoro-1-phenyl-	$C_8H_5F_3O$ 174.12	434-45-7 -40	153	4-07-00-00637 1.279[20]	1.4583[20]
5873	Ethanone, 1-(2,3,4-trihydroxyphenyl)- Gallacetophenone	$C_8H_8O_4$ 168.15	528-21-2 173	4248	4-08-00-02721	H_2O 3; EtOH 4; eth 3; ace 4
5874	Ethanone, 1-(2,3,4-trimethoxyphenyl)-	$C_{11}H_{14}O_4$ 210.23	13909-73-4 15.8	296	4-08-00-02721	1.5384[20]
5875	Ethanone, 1-(2,4,5-trimethylphenyl)-	$C_{11}H_{14}O$ 162.23	2040-07-5 10.5	246.5	4-07-00-00751 1.0039[15]	H_2O 1; EtOH 4; eth 4; bz 4 1.541[15]
5876	Ethanone, 1-(2,4,6-trimethylphenyl)-	$C_{11}H_{14}O$ 162.23	1667-01-2	241; 120[12]	4-07-00-00749 0.9754[20]	H_2O 1; EtOH 3; eth 3; ace 3 1.5175[20]
5877	Ethanone, 1-(3,4,5-trimethylphenyl)-	$C_{11}H_{14}O$ 162.23	2047-21-4 4.7	101.5[3]	4-07-00-00749 1.0037[25]	1.5420[25]
5878	Ethenaminium, N,N,N-trimethyl-, hydroxide Neurine	$C_5H_{13}NO$ 103.16	463-88-7	6393	3-04-00-00442	H_2O 4; eth 4; EtOH 4
5879	Ethene Ethylene	C_2H_4 28.05	74-85-1 -169	3748 -103.7	4-01-00-00677 0.5678[-104]	H_2O 1; EtOH 3; eth 3; ace 2 1.363[100]
5880	Ethene, bromo- Vinyl bromide	C_2H_3Br 106.95	593-60-2 -137.8	15.8	4-01-00-00718 1.4933[20]	H_2O 1; EtOH 3; eth 3; ace 3 1.4380[20]
5881	Ethene, 1-bromo-2-chloro- Ethylene, 1-bromo-2-chloro-	C_2H_2BrCl 141.39	3018-09-5 -86.7	84.6	3-01-00-00671 1.7972[15]	1.4982
5882	Ethene, 2-bromo-1,1-dichloro-	C_2HBrCl_2 175.84	5870-61-1 -88.5	107.5	4-01-00-00720 1.9053[15]	
5883	Ethene, 1-bromo-1,2-difluoro- 1-Bromo-1,2-difluoroethylene	C_2HBrF_2 142.93	358-99-6	19	0-01-00-00189 1.8434[25]	1.3846[0]
5884	Ethene, chloro- Vinyl chloride	C_2H_3Cl 62.50	75-01-4 -153.7	9898 -13.3	4-01-00-00700 0.9106[20]	H_2O 2; EtOH 3; eth 4 1.3700[20]
5885	Ethene, 2-chloro-1,1-difluoro- 1,1-Difluoro-2-chloroethylene	C_2HClF_2 98.48	359-10-4 -138.5	-18.5	4-01-00-00703	
5886	Ethene, 1-chloro-2-ethoxy- 2-Chlorovinyl ethyl ether	C_4H_7ClO 106.55	928-56-3	120	4-01-00-02061 1.0386[20]	1.4385[20]
5887	Ethene, (2-chloroethoxy)- 2-Chloroethyl vinyl ether	C_4H_7ClO 106.55	110-75-8 -70	2139 108	4-01-00-02051 1.0495[20]	EtOH 4; eth 4; chl 2 1.4378[20]
5888	Ethene, 1-chloro-2-iodo-	C_2H_2ClI 188.40	20244-71-7 -38.2	119	3-01-00-00674 2.2298[25]	
5889	Ethene, chlorotrifluoro- Chlorotrifluoroethylene	C_2ClF_3 116.47	79-38-9 -158	-27.8	4-01-00-00704 1.54[-60]	bz 3; chl 3 1.38[0]
5890	Ethene, 1,1-dibromo-	$C_2H_2Br_2$ 185.85	593-92-0	92	4-01-00-00720 2.1179[21]	EtOH 3; eth 3; ace 3; bz 3
5891	Ethene, 1,2-dibromo-, (E)- trans-1,2-Dibromoethylene	$C_2H_2Br_2$ 185.85	590-12-5 -6.5	108	4-01-00-00721 2.2308[20]	H_2O 1; EtOH 4; eth 4; ace 3 1.5505[18]
5892	Ethene, 1,2-dibromo-, (Z)- cis-1,2-Dibromoethylene	$C_2H_2Br_2$ 185.85	590-11-4 -53	112.5	4-01-00-00720 2.2464[20]	H_2O 1; EtOH 4; eth 4; ace 3 1.5428[20]
5893	Ethene, 1,1-dibromo-2-ethoxy- Ethylene, 1,1-dibromo-2-ethoxy-	$C_4H_6Br_2O$ 229.90	77295-79-5	172; 74[15]	2-01-00-00473 1.769[18]	eth 4
5894	Ethene, 1,1-dichloro- Vinylidene chloride	$C_2H_2Cl_2$ 96.94	75-35-4 -122.5	9900 31.6	4-01-00-00706 1.213[20]	H_2O 1; EtOH 3; eth 4; ace 3 1.4249[20]
5895	Ethene, 1,2-dichloro-, (E)- trans-1,2-Dichloroethylene	$C_2H_2Cl_2$ 96.94	156-60-5 -49.8	86 48.7	4-01-00-00709 1.2565[20]	H_2O 2; EtOH 5; eth 5; ace 5 1.4454[20]
5896	Ethene, 1,2-dichloro-, (Z)- cis-1,2-Dichloroethylene	$C_2H_2Cl_2$ 96.94	156-59-2 -80	86 60.1	4-01-00-00707 1.2837[20]	H_2O 2; EtOH 5; eth 5; ace 5 1.4490[20]
5897	Ethene, 1,1-dichloro-2,2-difluoro-	$C_2Cl_2F_2$ 132.92	79-35-6 -116	19	4-01-00-00711 1.555[-20]	1.383[-20]
5898	Ethene, 1,2-dichloro-1,2-difluoro-	$C_2Cl_2F_2$ 132.92	598-88-9 -130.5	21.1	1.4950[0]	1.3777[0]
5899	Ethene, 1,2-dichloro-1-ethoxy-	$C_4H_6Cl_2O$ 141.00	42345-82-4	128.2	3-01-00-02950 1.1972[25]	1.4558[17]
5900	Ethene, 1,1-dichloro-2-fluoro- 1,1-Dichloro-2-fluoroethylene	C_2HCl_2F 114.93	359-02-4 -108.8	37.5	4-01-00-00711 1.3732[16]	1.4031[16]
5901	Ethene, 1,1-diethoxy-	$C_6H_{12}O_2$ 116.16	2678-54-8	68[100]	4-01-00-03420 0.7932[20]	1.3643[21]
5902	Ethene, 1,1-difluoro- Vinylidene fluoride	$C_2H_2F_2$ 64.03	75-38-7 -144	-85.7	4-01-00-00696	eth 4; EtOH 4
5903	Ethene, 1,2-difluoro-, (Z)- cis-1,2-Difluoroethylene	$C_2H_2F_2$ 64.03	1630-77-9			
5904	Ethene, 1,2-diiodo-, (Z)-	$C_2H_2I_2$ 279.85	590-26-1 -14	72.5[16]	4-01-00-00724 3.0625[20]	eth 3; chl 3
5905	Ethene, ethoxy- Ethyl vinyl ether	C_4H_8O 72.11	109-92-2 -115.8	35.5	4-01-00-02049 0.7589[20]	H_2O 2; EtOH 3; eth 5; ctc 2 1.3767[20]
5906	Ethene, fluoro- Vinyl fluoride	C_2H_3F 46.04	75-02-5 -160.5	-72	4-01-00-00694	H_2O 1; EtOH 3; ace 3
5907	Ethene, iodo- Iodoethylene	C_2H_3I 153.95	593-66-8	56	4-01-00-00722 2.037[20]	eth 4; EtOH 4 1.5385[20]
5908	Ethene, methoxy- Methyl vinyl ether	C_3H_6O 58.08	107-25-5 -122	5.5	4-01-00-02049 0.7725[0]	H_2O 2; EtOH 4; eth 4; ace 4 1.3730[0]
5909	Ethene, (methylsulfonyl)-	$C_3H_6O_2S$ 106.15	3680-02-2	122-4[24]	4-01-00-02065 1.2117[20]	eth 3; ace 3 1.4636[20]

No.	Name / Synonym	Mol. Form. / Mol. Wt.	CAS RN / mp/°C	Merck No. / bp/°C	Beil. Ref. / den/g cm^{-3}	Solubility / n_D
5910	Ethene, (methylthio)-	C_3H_6S / 74.15	1822-74-8 /	/ 69.5	4-01-00-02065 / 0.9026[20]	eth 3; ace 3; chl 3 / 1.4837[20]
5911	Ethene, nitro-	$C_2H_3NO_2$ / 73.05	3638-64-0 / -55.5	/ 98.5	4-01-00-00725 / 1.2212[14]	EtOH 4; eth 4; ace 4; bz 4 / 1.4282[20]
5912	Ethene, 1,1'-oxybis- / Divinyl ether	C_4H_6O / 70.09	109-93-3 / -101	9899 / 28.3	4-01-00-02058 / 0.7733[20]	H_2O 1; EtOH 5; eth 5; ace 5 / 1.3989[20]
5913	Ethene, 1,1'-sulfonylbis-	$C_4H_6O_2S$ / 118.16	77-77-0 / -26	/ 234.5	4-01-00-02068 / 1.177[25]	/ 1.4765[20]
5914	Ethene, tetrabromo- / Tetrabromoethylene	C_2Br_4 / 343.64	79-28-7 / 56.5	/ 226	4-01-00-00722 /	H_2O 1; EtOH 3; eth 3; ace 3
5915	Ethenetetracarbonitrile / Tetracyanoethylene	C_6N_4 / 128.09	670-54-2 / 199	9129 / 223	4-02-00-02450 / 1.348[25]	eth 2; ace 3; bz 2; ctc 2 / 1.560[25]
5916	Ethenetetracarboxylic acid, tetraethyl ester	$C_{14}H_{20}O_8$ / 316.31	6174-95-4 / 58	/ 328 dec; 203[13]	4-02-00-02450 /	H_2O 1; EtOH 4; eth 4
5917	Ethene, tetrachloro- / Tetrachloroethylene	C_2Cl_4 / 165.83	127-18-4 / -22.3	9126 / 121.3	4-01-00-00715 / 1.6227[20]	H_2O 1; EtOH 5; eth 5; bz 5 / 1.5053[20]
5918	Ethene, tetrafluoro- / Tetrafluoroethylene	C_2F_4 / 100.02	116-14-3 / -142.5	/ -75.9	4-01-00-00698 / 1.519[-76]	H_2O 1
5919	Ethene, tetraiodo- / Tetraiodoethylene	C_2I_4 / 531.64	513-92-8 / 187	9151 / sub	4-01-00-00724 / 2.983[20]	bz 4; chl 4
5920	Ethene, 1,1'-thiobis-	C_4H_6S / 86.16	627-51-0 / 20	/ 84	4-01-00-02068 / 0.9174[15]	H_2O 2; EtOH 5; eth 5; ace 3
5921	Ethene, tribromo-	C_2HBr_3 / 264.74	598-16-3 /	/ 164	4-01-00-00722 / 2.7082[0]	H_2O 2; EtOH 4; eth 3; ace 3 / 1.6045[16]
5922	Ethene, trichloro- / Trichloroethylene	C_2HCl_3 / 131.39	79-01-6 / -84.7	9552 / 87.2	4-01-00-00712 / 1.4642[20]	H_2O 2; EtOH 5; eth 5; ace 3 / 1.4773[20]
5923	Ethene, trichlorofluoro-	C_2Cl_3F / 149.38	359-29-5 / -108.9	/ 71	4-01-00-00715 / 1.5460[20]	chl 3 / 1.4379[20]
5924	Ethene, trifluoro- / Trifluoroethylene	C_2HF_3 / 82.03	359-11-5 /	/ -51	4-01-00-00697 / 1.26[-70]	H_2O 1; EtOH 2; eth 3
5925	Ethene, trifluoroiodo-	C_2F_3I / 207.92	359-37-5 /	/ 30	4-01-00-00723 / 2.284[25]	/ 1.4143[0]
5926	1,4-Etheno-3H,7H-benzo[1,2-c:3,4-c']dipyran-3,7-dione, 9-(3-furanyl)-1,4,4a,5,6,6a,9,10,10a,10b-decahydro-4-hydroxy-4a,10a-dimethyl-, [1R-(1α,4β,4aα,6aβ,9β,10aα,10b)] / Columbin	$C_{20}H_{22}O_8$ / 358.39	546-97-4 /	2489 /	5-19-10-00546 /	
5927	Ethenone / Ketene	C_2H_2O / 42.04	463-51-4 / -151	5177 / -49.8	4-01-00-03418 /	eth 2; ace 2
5928	Ethenone, diphenyl- / Diphenylketene	$C_{14}H_{10}O$ / 194.23	525-06-4 /	3327 / 267.5	4-07-00-01653 / 1.1107[13]	/ 1.615[14]
5929	Ethephon / Phosphonic acid, (2-chloroethyl)-	$C_2H_6ClO_3P$ / 144.49	16672-87-0 / 74	3686 /	/ 1.2	
5930	Ethiolate / Carbamothioic acid, diethyl-, S-ethyl ester	$C_7H_{15}NOS$ / 161.27	2941-55-1 /	/ 52[0.5]; 142[87]	/ 0.9791[30]	
5931	Ethion / Phosphorodithioic acid, S,S'-methylene O,O,O',O'-tetraethyl ester	$C_9H_{22}O_4P_2S_4$ / 384.48	563-12-2 / -13	3691 / 165[0.3]	/ 1.22[20]	
5932	Ethirimol / 4(1H)-Pyrimidinone, 5-butyl-2-(ethylamino)-6-methyl-	$C_{11}H_{19}N_3O$ / 209.29	23947-60-6 / 160	3695 /	/ 1.21[25]	
5933	Ethofumesate / 2-Ethoxy-2,3-dihydro-3,3-dimethyl-5-benzofuranyl methanesulfonate	$C_{13}H_{18}O_5S$ / 286.35	26225-79-6 / 71	3697 /	/ 1.14	
5934	Ethoprop / Phosphorodithioic acid, O-ethyl S,S-dipropyl ester	$C_8H_{19}O_2PS_2$ / 242.34	13194-48-4 /	3702 / 88[0.2]	/ 1.094[20]	
5935	Ethylene, 1-bromo-1,2-dichloro-, trans- / Ethylene, 1-bromo-1,2-dichloro (dis)	C_2HBrCl_2 / 175.84	6795-75-1 / -83.5	/ 113.8	3-01-00-00671 / 1.9133[15]	/ 1.5218[15]
5936	Ethyne / Acetylene	C_2H_2 / 26.04	74-86-2 / -80.7 (trip. pt.)	84 / -84.7 (subl. point)	4-01-00-00939 / *0.377[25]	H_2O 2; EtOH 2; ace 3; bz 3
5937	Ethyne, bromo- / Bromoacetylene	C_2HBr / 104.93	593-61-3 /	/ 4.7	3-01-00-00919 /	eth 4
5938	Ethyne, chloro- / Chloroacetylene	C_2HCl / 60.48	593-63-5 / -126	/ -30	4-01-00-00957 /	EtOH 2
5939	Ethyne, dibromo-	C_2Br_2 / 183.83	624-61-3 / -24	/ exp 76	3-01-00-00919 /	ace 4; bz 4; eth 4; EtOH 4
5940	Ethyne, dichloro- / Dichloroacetylene	C_2Cl_2 / 94.93	7572-29-4 / -66	/ 33	4-01-00-00957 / 1.261[20]	EtOH 3; eth 3; ace 3 / 1.42790[20]
5941	Ethyne, diiodo- / Diiodoacetylene	C_2I_2 / 277.83	624-74-8 / 81.5	/ exp	4-01-00-00958 /	ace 4; bz 4; eth 4; EtOH 4
5942	Ethyne, ethoxy-	C_4H_6O / 70.09	927-80-0 /	/ 50	4-01-00-02211 / 0.8000[20]	/ 1.3796[20]
5943	Ethyne, fluoro- / Fluoroacetylene	C_2HF / 44.03	2713-09-9 /	/	4-01-00-00957 /	
5944	Ethyne, methoxy-	C_3H_4O / 56.06	6443-91-0 /	/	4-01-00-02211 / 0.8001[20]	eth 4; EtOH 4 / 1.3812[20]

No.	Name / Synonym	Mol. Form. / Mol. Wt.	CAS RN / mp/°C	Merck No. / bp/°C	Beil. Ref. / den/g cm^{-3}	Solubility / n_D
5945	Etrimfos / Phosphorothioic acid, O-(6-ethoxy-2-ethyl-4-pyrimidinyl) O,O-dimethyl ester	$C_{10}H_{17}N_2O_4PS$ / 292.30	38260-54-7 / -3.35	3847	1.195[20]	
5946	Famphur / Phosphorothioic acid, O-[4-[(dimethylamino)sulfonyl]phenyl] O,O-dimethyl ester	$C_{10}H_{16}NO_5PS_2$ / 325.35	52-85-7 / 53	3882		
5947	Fenamiphos / Phosphoramidic acid, (1-methylethyl)-, ethyl 3-methyl-4-(methylthio)phenyl ester	$C_{13}H_{22}NO_3PS$ / 303.36	22224-92-6 / 49	3901	1.15[20]	
5948	Fenarimol / 5-Pyrimidinemethanol, α-(2-chlorophenyl)-α-(4-chlorophenyl)-	$C_{17}H_{12}Cl_2N_2O$ / 331.20	60168-88-9 / 118	3903		
5949	Fenbuconazole / 4-(4-Chlorophenyl)-2-phenyl-2-(1H-1,2,4-triazol-1-ylmethyl)butyronitrile	$C_{20}H_{19}ClN_4$ / 350.85	114369-43-6 / 125			
5950	Fenbutatin oxide / Distannoxane, hexakis(2-methyl-2-phenylpropyl)-	$C_{60}H_{78}OSn_2$ / 1052.70	13356-08-6 / 138	3907		
5951	Fenitrothion / Phosphorothioic acid, O,O-dimethyl O-(3-methyl-4-nitrophenyl) ester	$C_9H_{12}NO_5PS$ / 277.24	122-14-5	3922 / 118[0.05]; 164[1]	1.3227[25]	
5952	Fenoxycarb / Ethyl 2-(4-phenoxyphenoxy)ethylcarbamate	$C_{17}H_{19}NO_4$ / 301.34	79127-80-3 / 53			
5953	Fenpropathrin / Cyano(3-phenoxyphenyl)methyl 2,2,3,3-tetramethylcyclopropanecarboxylate	$C_{22}H_{23}NO_3$ / 349.43	64257-84-7 / 47		1.15[25]	
5954	Fensulfothion / Phosphorothioic acid, O,O-diethyl O-[4-(methylsulfinyl)phenyl] ester	$C_{11}H_{17}O_4PS_2$ / 308.36	115-90-2	3943 / 140[0.01]	1.202[20]	
5955	Fenthion / Phosphorothioic acid, O,O-dimethyl O-[3-methyl-4-(methylthio)phenyl] ester	$C_{10}H_{15}O_3PS_2$ / 278.33	55-38-9 / 7.5	3945 / 87[0.01]	1.246[20]	
5956	Fenvalerate / Benzeneacetic acid, 4-chloro-α-(1-methylethyl)-, cyano(3-phenoxyphenyl)methyl ester	$C_{25}H_{22}ClNO_3$ / 419.91	51630-58-1 / dec	3952	1.15[25]	
5957	Ferbam / Iron, tris(dimethylcarbamodithioato-S,S')-, (OC-6-11)-	$C_9H_{18}FeN_3S_6$ / 416.51	14484-64-1 / 180 dec	3954		
5958	Ferrate(2-), chloro[7,12-diethenyl-3,8,13,17-tetramethyl-21H,23H-porphine-2,18-dipropanoato(4-)-N21,N22,N23, Hemin	$C_{34}H_{32}ClFeN_4O_4$ / 651.95	16009-13-5 / >300	4563		
5959	Ferrate(2-), [7,12-diethenyl-3,8,13,17-tetramethyl-21H,23H-porphine-2,18-dipropanoato(4-)-N21,N22,N23,N24 Hematin	$C_{34}H_{33}FeN_4O_5$ / 633.51	15489-90-4 / >200	4554	4-26-00-03047	H_2O 1; EtOH 3; eth 1; alk 3
5960	Ferrocene / Bis(cyclopentadienyl)iron	$C_{10}H_{10}Fe$ / 186.04	102-54-5 / 172.5	3985 / 249		H_2O 1
5961	Ferrocene, 1,1'-bis(1,1,3,3-tetramethyl-3-phenyldisiloxanyl)- / Disiloxane, 1-(cyclopentadienyl)-1,1,3,3-tetramethyl-3-phenyl-, Fe deriv.	$C_{30}H_{42}FeO_2Si_4$ / 602.85	12321-04-9 / 18.7	200-5[0.03]	1.1063[25]	1.5473[25]
5962	Fluazifop-butyl / (±)-2-[4-[[5-(Trifluoromethyl)-2-pyridinyl]oxy]phenoxy]propanoic acid, butyl ester	$C_{19}H_{20}F_3NO_4$ / 383.37	69806-50-4 / 13	4049 / 170[0.5]	1.21[20]	
5963	Fluazifop-P-butyl / (R)-2-[4-(5-Trifluoromethyl-2-pyridyloxy)phenoxy)propionic acid, butyl ester	$C_{19}H_{20}F_3NO_4$ / 383.37	79241-46-6 / 5			
5964	Fluchloralin / Benzenamine, N-(2-chloroethyl)-2,6-dinitro-N-propyl-4-(trifluoromethyl)-	$C_{12}H_{13}ClF_3N_3O_4$ / 355.70	33245-39-5 / 42	4052		
5965	Flucythrinate / Cythrin	$C_{26}H_{23}F_2NO_4$ / 451.47	70124-77-5 / 108[0.35]	4055	1.189[22]	
5966	Fluoranthene / 1,2-(1,8-Naphthylene)benzene	$C_{16}H_{10}$ / 202.26	206-44-0 / 107.8	384	4-05-00-02463 / 1.252[0]	H_2O 1; EtOH 3; eth 3; bz 3
5967	9H-Fluorene / 2,2'-Methylenebiphenyl	$C_{13}H_{10}$ / 166.22	86-73-7 / 114.8	4081 / 295	4-05-00-02142 / 1.203[0]	H_2O 1; EtOH 2; eth 3; ace 3
5968	9H-Fluorene, 2-bromo- / 2-Bromofluorene	$C_{13}H_9Br$ / 245.12	1133-80-8 / 113.5	185[135]	4-05-00-02147	H_2O 1; EtOH 3; chl 4; HOAc 3
5969	9H-Fluorene-2-carboxylic acid, 9-oxo-	$C_{14}H_8O_3$ / 224.22	784-50-9 / 341	sub 340	4-10-00-03040	H_2O 1; EtOH 3; HOAc 3
5970	9H-Fluorene, 9-[(2-chlorophenyl)methylene]- / Fluorene, 9-(2-chlorobenzylidene)-	$C_{20}H_{13}Cl$ / 288.78	1643-49-8 / 69.5	180[0.7]	4-05-00-02631	EtOH 4; HOAc 4
5971	9H-Fluorene-2,7-diamine / 2,7-Diaminofluorene	$C_{13}H_{12}N_2$ / 196.25	525-64-4 / 166	4082	4-13-00-00449	H_2O 1; EtOH 3; chl 3
5972	9H-Fluorene, 2,7-dichloro- / 2,7-Dichlorofluorene	$C_{13}H_8Cl_2$ / 235.11	7012-16-0 / 128	sub	4-05-00-02147	bz 4; chl 4

No.	Name Synonym	Mol. Form. Mol. Wt.	CAS RN mp/°C	Merck No. bp/°C	Beil. Ref. den/g cm^{-3}	Solubility n_D
5973	1H-Fluorene, dodecahydro-	$C_{13}H_{22}$ 178.32	5744-03-6 	 260	4-05-00-00489 0.92[25]	 1.5012[20]
5974	9H-Fluorene, 9-methyl- 9-Methylfluorene	$C_{14}H_{12}$ 180.25	2523-37-7 46.5	 154-6[15]	4-05-00-02185 1.0263[66]	H_2O 1; EtOH 3; eth 3; ace 3 1.610[66]
5975	9H-Fluoren-9-one	$C_{13}H_8O$ 180.21	486-25-9 84	 341.5	4-07-00-01629 1.1300[99]	H_2O 1; EtOH 3; eth 4; ace 3 1.6309[99]
5976	9H-Fluoren-9-one, 2-chloro-	$C_{13}H_7ClO$ 214.65	3096-47-7 125.5	 sub	4-07-00-01634 	EtOH 4
5977	9H-Fluoren-9-one, 2-nitro-	$C_{13}H_7NO_3$ 225.20	3096-52-4 224.3	 sub	4-07-00-01636 	EtOH 2; ace 3; sulf 3; HOAc 3
5978	9H-Fluoren-9-one, 1,2,3,4-tetrahydro- Phentydrone	$C_{13}H_{12}O$ 184.24	634-19-5 81.5	7235 139-40[0.05]	4-07-00-01302 	
5979	9H-Fluoren-9-one, 2,4,7-trinitro- 2,4,7-Trinitrofluorenone	$C_{13}H_5N_3O_7$ 315.20	129-79-3 176	9641 	4-07-00-01638 	H_2O 2; ace 4; bz 4; chl 4
5980	Fluorodifen Benzene, 2-nitro-1-(4-nitrophenoxy)-4-(trifluoromethyl)-	$C_{13}H_7F_3N_2O_5$ 328.20	15457-05-3 94	 	 	
5981	Fluridone 4(1H)-Pyridinone, 1-methyl-3-phenyl-5-[3-(trifluoromethyl)phenyl]-	$C_{19}H_{14}F_3NO$ 329.32	59756-60-4 155	 	 	
5982	Fluroxypyr [(4-Amino-3,5-dichloro-6-fluoro-2-pyridyl)oxy]acetic acid	$C_7H_5Cl_2FN_2O_3$ 255.03	69377-81-7 232	4128 	 	
5983	Fluvalinate 2-[[2-Chloro-4-(trifluoromethyl)phenyl]amino]-3-methylbutanoic acid, (±)-cyano(3-phenoxy phenyl)methyl ester	$C_{26}H_{22}ClF_3N_2O_3$ 502.92	102851-06-9 	 >450	 1.29[25]	
5984	Folpet 1H-Isoindole-1,3(2H)-dione, 2-[(trichloromethyl)thio]-	$C_9H_4Cl_3NO_2S$ 296.56	133-07-3 177	4142 	 	
5985	Fomesafen 5-[2-Chloro-4-(trifluoromethyl)phenoxy]-N-(methylsulfonyl)-2-nitrobenzamide	$C_{15}H_{10}ClF_3N_2O_6S$ 438.77	72178-02-0 220	 	 1.28[20]	
5986	Fonofos Phosphonodithioic acid, ethyl-, O-ethyl S-phenyl ester	$C_{10}H_{15}OPS_2$ 246.33	944-22-9 	4147 130[0.1]	 1.16[25]	
5987	Formaldehyde Methanal	CH_2O 30.03	50-00-0 -92	4148 -19.1	4-01-00-03017 0.815[-20]	H_2O 3; EtOH 3; eth 5; ace 5
5988	Formaldehyde, oxime	CH_3NO 45.04	75-17-2 2.5	4-01-00-03055 109[15]	 1.133[25]	H_2O 3; EtOH 4; eth 4
5989	Formamide Methanamide	CH_3NO 45.04	75-12-7 2.55	4151 220	4-02-00-00045 1.1334[20]	H_2O 5; EtOH 5; eth 2; ace 3 1.4472[20]
5990	Formamide, N-cyclohexyl-	$C_7H_{13}NO$ 127.19	766-93-8 39.5	 260; 113[0.7]	4-12-00-00038 1.0123[17]	
5991	Formamide, N,N-diethyl-	$C_5H_{11}NO$ 101.15	617-84-5 	 177.5	4-04-00-00346 0.9080[19]	H_2O 5; EtOH 4; eth 4; ace 5 1.4321[25]
5992	Formamide, N,N-dimethyl- N,N-Dimethylformamide	C_3H_7NO 73.09	68-12-2 -60.4	3232 153	4-04-00-00171 0.944[25]	H_2O 5; EtOH 5; eth 5; ace 5 1.4305[20]
5993	Formamide, N,N-diphenyl-	$C_{13}H_{11}NO$ 197.24	607-00-1 73.5	 337.5; 189[13]	4-12-00-00370 	H_2O 1; EtOH 3; eth 3; bz 3
5994	Formamide, N-ethyl- N-Ethylformamide	C_3H_7NO 73.09	627-45-2 	 198	4-04-00-00346 0.9552[20]	H_2O 5; EtOH 5; eth 5 1.4320[20]
5995	Formamide, N-methyl- N-Methylformamide	C_2H_5NO 59.07	123-39-7 -3.8	4-04-00-00170 199.5	 1.011[19]	H_2O 4; ace 4; EtOH 4 1.4319[20]
5996	Formamide, N-(2-methylphenyl)-	C_8H_9NO 135.17	94-69-9 62	 288	3-12-00-01852 1.086[55]	H_2O 3; EtOH 4
5997	Formamide, N-(3-methylphenyl)-	C_8H_9NO 135.17	3085-53-8 <-18	 290 dec; 176[17]	3-12-00-01962 	H_2O 3; NaOH 3
5998	Formamide, N-methyl-N-phenyl-	C_8H_9NO 135.17	93-61-8 14.5	 243	4-12-00-00370 1.0948[20]	H_2O 2; EtOH 3; ace 3; ctc 2 1.5589[20]
5999	Formamide, N-phenyl- Formanilide	C_7H_7NO 121.14	103-70-8 46	4152 271	4-12-00-00368 1.1186[50]	H_2O 3; EtOH 4; eth 3; bz 3
6000	Formamide, N-(2,2,2-trichloro-1-hydroxyethyl)- Chloral formamide	$C_3H_4Cl_3NO_2$ 192.43	515-82-2 120	2060 	3-02-00-00059 	ace 4; eth 4; EtOH 4
6001	Formetanate hydrochloride Methanimidamide, N,N-dimethyl-N'-[3-[[(methylamino)carbonyl]oxy]phenyl]-, monohydrochloride	$C_{11}H_{16}ClN_3O_2$ 257.72	23422-53-9 	 	 	
6002	Formic acid Methanoic acid	CH_2O_2 46.03	64-18-6 8.3	4153 101	4-02-00-00003 1.220[20]	H_2O 5; EtOH 5; eth 5; ace 4 1.3714[20]
6003	Formic acid, ammonium salt Ammonium formate	CH_5NO_2 63.06	540-69-2 	548 	4-02-00-00003 	
6004	Formic acid, butyl ester Butyl formate	$C_5H_{10}O_2$ 102.13	592-84-7 -91.5	 106.1	4-02-00-00028 0.8885[20]	H_2O 2; EtOH 5; eth 5; ace 3 1.3912[20]
6005	Formic acid, cyclohexyl ester	$C_7H_{12}O_2$ 128.17	4351-54-6 163	 	4-06-00-00035 1.0057[0]	H_2O 1; EtOH 3; eth 4; HCOOH 3 1.4430[20]

No.	Name Synonym	Mol. Form. Mol. Wt.	CAS RN mp/°C	Merck No. bp/°C	Beil. Ref. den/g cm^{-3}	Solubility n_D
6006	Formic acid, ethenyl ester Vinyl formate	$C_3H_4O_2$ 72.06	692-45-5		4-02-00-00033	
6007	Formic acid, ethyl ester Ethyl formate	$C_3H_6O_2$ 74.08	109-94-4 -79.6	3763 54.4	4-02-00-00023 0.9168[20]	H_2O 3; EtOH 5; eth 5; ace 4 1.3598[10]
6008	Formic acid, heptyl ester Heptyl formate	$C_8H_{16}O_2$ 144.21	112-23-2	178.1	4-02-00-00031 0.8784[20]	H_2O 1; EtOH 5; eth 5 1.4140[20]
6009	Formic acid, hexyl ester Hexyl formate	$C_7H_{14}O_2$ 130.19	629-33-4 -62.6	155.5	4-02-00-00031 0.8813[20]	H_2O 1; EtOH 5; eth 5 1.4071[20]
6010	Formic acid, methyl ester Methyl formate	$C_2H_4O_2$ 60.05	107-31-3 -99	5994 31.7	4-02-00-00020 0.9742[20]	H_2O 4; EtOH 5; eth 3; chl 3 1.3433[20]
6011	Formic acid, 1-methylethyl ester Isopropyl formate	$C_4H_8O_2$ 88.11	625-55-8	68.2	4-02-00-00027 0.8728[20]	H_2O 2; EtOH 5; eth 5; ace 4 1.3678[20]
6012	Formic acid, 1-methylpropyl ester	$C_5H_{10}O_2$ 102.13	589-40-2	97	4-02-00-00029 0.8846[20]	H_2O 2; EtOH 5; eth 5; ace 3 1.3865[20]
6013	Formic acid, 2-methylpropyl ester Isobutyl formate	$C_5H_{10}O_2$ 102.13	542-55-2 -95.8	5026 98.2	4-02-00-00029 0.8776[20]	H_2O 2; EtOH 5; eth 5; ace 4 1.3857[20]
6014	Formic acid, octyl ester	$C_9H_{18}O_2$ 158.24	112-32-3 -39.1	198.8	4-02-00-00031 0.8744[20]	H_2O 1; EtOH 3; eth 5; ctc 2 1.4208[15]
6015	Formic acid, pentyl ester Pentyl formate	$C_6H_{12}O_2$ 116.16	638-49-3 -73.5	130.4	4-02-00-00030 0.8853[20]	H_2O 1; EtOH 5; eth 5 1.3992[20]
6016	Formic acid, phenylmethyl ester Benzyl formate	$C_8H_8O_2$ 136.15	104-57-4	1148 203; 84[10]	4-06-00-02262 1.081[20]	H_2O 1; EtOH 3; eth 5; ace 3 1.5154[20]
6017	Formic acid, 2-propenyl ester Allyl formate	$C_4H_6O_2$ 86.09	1838-59-1	83.6	3-02-00-00046 0.9460[20]	H_2O 2; EtOH 3; eth 5
6018	Formic acid, propyl ester Propyl formate	$C_4H_8O_2$ 88.11	110-74-7 -92.9	7871 80.9	4-02-00-00026 0.9058[20]	H_2O 2; EtOH 5; eth 5; ctc 2 1.3779[20]
6019	Formyl fluoride Fluoroformaldehyde	CHFO 48.02	1493-02-3 -142.2	4162 -26.5	4-02-00-00042 1.1950[-30]	
6020	Fosetyl-Al Ethyl hydrogen phosphonate, aluminum salt	$C_6H_{18}AlO_9P_3$ 354.11	39148-24-8 >300	4167		
6021	Fosthiazate (RS)-S-sec-Butyl O-ethyl 2-oxo-1,3-thiazolidin-3-ylphosphonothioate	$C_9H_{18}NO_3PS_2$ 283.35	98886-44-3			
6022	D-Fructopyranose, 3-O-α-D-glucopyranosyl- D-Turanose	$C_{12}H_{22}O_{11}$ 342.30	547-25-1 168	9728	5-17-07-00213	H_2O 4; EtOH 3; MeOH 3
6023	β-D-Fructose β-Levulose	$C_6H_{12}O_6$ 180.16	53188-23-1 103 dec	4185	4-01-00-04401 1.60[20]	H_2O 4; EtOH 3; ace 4; MeOH 3
6024	Fulminic acid Carbyloxime	CHNO 43.03	506-85-4 unstable in pure form			eth 3
6025	Furan Oxacyclopentadiene	C_4H_4O 68.08	110-00-9 -85.6	4206 31.5	5-17-01-00291 0.9514[20]	H_2O 1; EtOH 4; eth 4; ace 3 1.4214[20]
6026	2-Furanacetic acid	$C_6H_6O_3$ 126.11	2745-26-8 68.5	102[0.4]	5-18-06-00206	H_2O 3; bz 3; MeOH 3; peth 3
6027	2-Furanacetic acid, butyl ester 2-Furylacetic acid, butyl ester	$C_{10}H_{14}O_3$ 182.22	4915-23-5 110-11[13]		4-18-00-04062 1.0232[25]	1.4558[25]
6028	2-Furanacetic acid, ethyl ester	$C_8H_{10}O_3$ 154.17	4915-21-3 88[15]		5-18-06-00206 1.0763[25]	1.4571[25]
6029	2-Furanacetic acid, methyl ester	$C_7H_8O_3$ 140.14	4915-22-4 87-8[21]		5-18-06-00206 1.1250[20]	1.4638[25]
6030	3-Furanacetic acid, tetrahydro-2,2-dimethyl-5-oxo- Terpenylic acid	$C_8H_{12}O_4$ 172.18	26754-48-3 90	9100	5-18-08-00032	H_2O 4
6031	2-Furanacetonitrile 2-Furylacetonitrile	C_6H_5NO 107.11	2745-25-7 75-80[20]		5-18-06-00207 1.0854[25]	eth 4; EtOH 4 1.4693[20]
6032	Furan, 2,5-bis(1,1-dimethylethyl)-	$C_{12}H_{20}O$ 180.29	4789-40-6 210		5-17-01-00439 0.8372[0]	H_2O 1; EtOH 3; eth 3 1.4369[20]
6033	Furan, 2-bromo-	C_4H_3BrO 146.97	584-12-3 103		5-17-01-00295 1.6500[20]	H_2O 2; EtOH 3; eth 3; ace 3 1.4980[20]
6034	Furan, 3-bromo- 3-Bromofuran	C_4H_3BrO 146.97	22037-28-1 103		5-17-01-00295 1.6606[20]	ace 4; bz 4; eth 4; EtOH 4 1.4958[20]
6035	Furan, 2-(bromomethyl)-	C_5H_5BrO 161.00	4437-18-7 32-4[2]		4-17-00-00268 1.557[20]	1.5380[20]
6036	Furan, 2-(bromomethyl)tetrahydro-	C_5H_9BrO 165.03	1192-30-9 170; 70[22]		5-17-01-00080 1.3653[20]	EtOH 3; eth 3 1.4850[20]
6037	Furan, 2-(butoxymethyl)-	$C_9H_{14}O_2$ 154.21	56920-82-2 190		2-17-00-00114 0.9516[20]	eth 4; EtOH 4 1.4522[20]
6038	Furan, 2,2'-sec-butylidenedi- Butane, 2,2-di(2-furyl)	$C_{12}H_{14}O_2$ 190.24	100121-73-1 64-6[1]		4-19-00-00287 1.0330[20]	eth 4; EtOH 4 1.4970[20]
6039	2-Furancarbonitrile	C_5H_3NO 93.08	617-90-3 147		5-18-06-00122 1.0822[20]	EtOH 3; eth 3 1.4798[20]
6040	2-Furancarbonyl chloride	$C_5H_3ClO_2$ 130.53	527-69-5 -2 173		5-18-06-00110 1.324[25]	H_2O 1; eth 3; ctc 2; chl 3 1.5310[20]
6041	2-Furancarboxaldehyde Furfural	$C_5H_4O_2$ 96.09	98-01-1 -36.5 161.7	4214	5-17-09-00292 1.1594[20]	H_2O 3; EtOH 4; eth 5; ace 4 1.5261[20]
6042	3-Furancarboxaldehyde	$C_5H_4O_2$ 96.09	498-60-2 145; 70-2[43]		5-17-09-00372 1.109[20]	1.4945[20]
6043	2-Furancarboxaldehyde, 5-bromo-	$C_5H_3BrO_2$ 174.98	1899-24-7 83.5 201; 112[16]		5-17-09-00320	eth 4; EtOH 4

No.	Name / Synonym	Mol. Form. / Mol. Wt.	CAS RN / mp/°C	Merck No. / bp/°C	Beil. Ref. / den/g cm^{-3}	Solubility / n_D
6044	2-Furancarboxaldehyde, 5-(hydroxymethyl)- / 5-(Hydroxymethyl)-2-furaldehyde	$C_6H_6O_3$ / 126.11	67-47-0 / 31.5	4764 / 114-6[1]	5-18-01-00130 / 1.2062[25]	H_2O 3; EtOH 3; eth 2; bz 3 / 1.5627[18]
6045	2-Furancarboxaldehyde, 5-methyl-	$C_6H_6O_2$ / 110.11	620-02-0	/ 187; 89[28]	5-17-09-00404 / 1.1072[18]	H_2O 3; EtOH 4; eth 5; ctc 2 / 1.5264[20]
6046	2-Furancarboxaldehyde, 5-nitro-	$C_5H_3NO_4$ / 141.08	698-63-5 / 35.5	/ 128-32[10]	5-17-09-00323	H_2O 2; peth 3
6047	2-Furancarboxaldehyde, oxime, (E)-	$C_5H_5NO_2$ / 111.10	620-03-1	/ 205 dec; 98[9]	5-17-09-00306 / 1.1550[80]	H_2O 3; EtOH 4; eth 4; bz 4
6048	2-Furancarboxaldehyde, oxime, (Z)-	$C_5H_5NO_2$ / 111.10	1450-58-4 / 91	/ 100[12]	5-17-09-00306 / 1.3800[20]	H_2O 2; EtOH 4; eth 4; bz 4
6049	2-Furancarboxaldehyde, tetrahydro-	$C_5H_8O_2$ / 100.12	7681-84-7	/ 142; 46[29]	5-17-09-00027 / 1.0727[20]	H_2O 4; eth 4 / 1.4366[20]
6050	2-Furancarboxylic acid / 2-Furoic acid	$C_5H_4O_3$ / 112.08	88-14-2 / 133.5	4219 / 231	5-18-06-00102	H_2O 3; EtOH 3; eth 4; ace 2
6051	3-Furancarboxylic acid	$C_5H_4O_3$ / 112.08	488-93-7 / 122.5	/ sub 105	5-18-06-00196	H_2O 2; EtOH 3; eth 4; AcOEt 3
6052	2-Furancarboxylic acid, butyl ester	$C_9H_{12}O_3$ / 168.19	583-33-5	/ 233	4-18-00-03918 / 1.0555[20]	H_2O 1; EtOH 3; eth 3; bz 3 / 1.4740
6053	2-Furancarboxylic acid, 5-chloro-, ethyl ester / Ethyl 5-chloro-2-furancarboxylate	$C_7H_7ClO_3$ / 174.58	4301-39-7 2	/ 217; 108[23]	5-18-06-00131 / 1.2418[25]	
6054	3-Furancarboxylic acid, 4,5-dihydro-5-oxo-	$C_5H_4O_4$ / 128.08	585-68-2 / 164	/ dec	4-18-00-05333	H_2O 4; EtOH 3; MeOH 3
6055	3-Furancarboxylic acid, 2,5-dimethyl-, ethyl ester	$C_9H_{12}O_3$ / 168.19	29113-63-1	/ 208.5; 96[19]	5-18-06-00233 / 1.0478[23]	/ 1.4686[20]
6056	2-Furancarboxylic acid, ethyl ester	$C_7H_8O_3$ / 140.14	614-99-3 / 34.5	/ 196.8	5-18-06-00104 / 1.1174[21]	H_2O 1; EtOH 5; eth 5; ace 5 / 1.4797[21]
6057	2-Furancarboxylic acid, 2-furanylmethyl ester	$C_{10}H_8O_4$ / 192.17	615-11-2 / 27.5	/ 122[2]	4-18-00-03936 / 1.2347[25]	ace 4; bz 4; eth 4; EtOH 4 / 1.5280[20]
6058	2-Furancarboxylic acid, hexyl ester / α-Furoic acid, hexyl ester	$C_{11}H_{16}O_3$ / 196.25	39251-86-0	/ 105-7[1]	2-18-00-00267 / 1.0170[20]	H_2O 1; EtOH 3; eth 3; bz 5
6059	2-Furancarboxylic acid, methyl ester	$C_6H_6O_3$ / 126.11	611-13-2	/ 181.3	5-18-06-00103 / 1.1786[21]	H_2O 1; EtOH 3; eth 3; bz 3 / 1.4860[20]
6060	2-Furancarboxylic acid, 5-methyl-	$C_6H_6O_3$ / 126.11	1917-15-3 / 109.5	/ 105[1]	5-18-06-00217	H_2O 2; EtOH 3; eth 3; bz 2
6061	2-Furancarboxylic acid, 1-methylethyl ester	$C_8H_{10}O_3$ / 154.17	6270-34-4	/ 198.5	0-18-00-00275 / 1.0655[24]	H_2O 1; EtOH 3; eth 3; ace 3 / 1.4682[24]
6062	2-Furancarboxylic acid, 2-methylpropyl ester	$C_9H_{12}O_3$ / 168.19	20279-53-2	/ 222	2-18-00-00266 / 1.0388[20]	ace 4; bz 4; eth 4; EtOH 4 / 1.4676[20]
6063	3-Furancarboxylic acid, methyl ester	$C_6H_6O_3$ / 126.11	13129-23-2	/ 160	5-18-06-00196 / 1.1173[15]	H_2O 1; EtOH 3; ace 3 / 1.4676[20]
6064	3-Furancarboxylic acid, 2-methyl-, ethyl ester	$C_8H_{10}O_3$ / 154.17	28921-35-9	/ 85-7[20]	5-18-06-00214 / 1.0102[25]	eth 4 / 1.4620[25]
6065	3-Furancarboxylic acid, 5-methyl- / 3-Furoic acid, 5-methyl-	$C_6H_6O_3$ / 126.11	21984-93-0 / 119	/ sub	5-18-06-00216	H_2O 4; eth 4; EtOH 4
6066	2-Furancarboxylic acid, 5-nitro-	$C_5H_3NO_5$ / 157.08	645-12-5 / 186	/ sub	5-18-06-00139	H_2O 3; EtOH 3; eth 3; ace 2
6067	2-Furancarboxylic acid, octyl ester / α-Furoic acid, octyl ester	$C_{13}H_{20}O_3$ / 224.30	39251-88-2	/ 126-7[1]	2-18-00-00267 / 0.9885[20]	H_2O 1; EtOH 3
6068	2-Furancarboxylic acid, pentyl ester / 2-Furoic acid, pentyl ester	$C_{10}H_{14}O_3$ / 182.22	4996-48-9	/ 95-7[1]	2-18-00-00266 / 1.0335[20]	H_2O 1; EtOH 5
6069	2-Furancarboxylic acid, phenylmethyl ester	$C_{12}H_{10}O_3$ / 202.21	5380-40-5	/ 179-81[18]	4-18-00-03921 / 1.1623[22]	ace 4; eth 4 / 1.5550[20]
6070	2-Furancarboxylic acid, 2-propenyl ester	$C_8H_8O_3$ / 152.15	4208-49-5	/ 207.5	5-18-06-00105 / 1.115[25]	eth 3; ace 3; ctc 2 / 1.4945[20]
6071	2-Furancarboxylic acid, propyl ester	$C_8H_{10}O_3$ / 154.17	615-10-1	/ 210.9; 112[20]	5-18-06-00104 / 1.0745[20]	H_2O 1; EtOH 3; eth 3; ace 3 / 1.4787[20]
6072	2-Furancarboxylic acid, tetrahydro-	$C_5H_8O_3$ / 116.12	16874-33-2 / 21	/ 145[25]	5-18-06-00010 / 1.1912[20]	H_2O 4 / 1.4612[20]
6073	Furan, 2-chloro- / 2-Chlorofuran	C_4H_3ClO / 102.52	3187-94-8	/ 78	5-17-01-00294 / 1.1923[20]	ace 4; eth 4; EtOH 4 / 1.4569[20]
6074	Furan, 3-chloro- / 3-Chlorofuran	C_4H_3ClO / 102.52	50689-17-3	/ 80	5-17-01-00295 / 1.2094[20]	ace 4; eth 4 / 1.4601[20]
6075	Furan, 2-(chloromethyl)-	C_5H_5ClO / 116.55	617-88-9	/ 49[26]	5-17-01-00323 / 1.1783[20]	bz 4; eth 4; EtOH 4 / 1.4941[20]
6076	Furan, 2-(cyclohexyloxy)- / Cyclohexyl 2-furyl ether	$C_{10}H_{14}O_2$ / 166.22	99172-57-3	/ 118-9[28]	4-17-00-01219 / 1.0200[28]	ace 4; eth 4; EtOH 4 / 1.4861[28]
6077	Furan, 2,3-dibromo- / 2,3-Dibromofuran	$C_4H_2Br_2O$ / 225.87	30544-34-4	/ 167	5-17-01-00295 / 1.9938[25]	/ 1.5458[25]
6078	Furan, 2,5-dibromo- / 2,5-Dibromofuran	$C_4H_2Br_2O$ / 225.87	32460-00-7 / 9.5	/ 164.5	5-17-01-00295 / 2.27[20]	/ 1.5455[20]
6079	2,3-Furandicarboxylic acid / Isocitrate lactone	$C_6H_4O_5$ / 156.09	4282-24-0 / 226	/ sub	5-18-07-00043	H_2O 4; EtOH 4; eth 2; AcOEt 4
6080	2,4-Furandicarboxylic acid	$C_6H_4O_5$ / 156.09	4282-28-4 / 266	/ sub	5-18-07-00045	H_2O 2; EtOH 4; ace 4; chl 2
6081	2,5-Furandicarboxylic acid / Dehydromucic acid	$C_6H_4O_5$ / 156.09	3238-40-2 / 342	/ sub	5-18-07-00047 / 1.7400[20]	H_2O 2; EtOH 2

No.	Name / Synonym	Mol. Form. / Mol. Wt.	CAS RN / mp/°C	Merck No. / bp/°C	Beil. Ref. / den/g cm⁻³	Solubility / n_D
6082	3,4-Furandicarboxylic acid, diethyl ester / 2,3-Bis(ethoxycarbonyl)furan	$C_{10}H_{12}O_5$ / 212.20	30614-77-8 /	/ 155[13]	5-18-07-00053 / 1.14[25]	/ 1.4717[20]
6083	2,5-Furandicarboxylic acid, dimethyl ester / Dimethyl 2,5-furandicarboxylate	$C_8H_8O_5$ / 184.15	4282-32-0 / 112	/ 154[15]	5-18-07-00048 /	eth 4; EtOH 4; chl 4
6084	Furan, 2-(diethoxymethyl)-	$C_9H_{14}O_3$ / 170.21	13529-27-6 /	/ 191.5	5-17-09-00299 / 0.9976[20]	EtOH 4 / 1.4451[20]
6085	Furan, 2,2-diethyltetrahydro- / 2,2-Diethyltetrahydrofuran	$C_8H_{16}O$ / 128.21	1193-35-7 /	/ 146	4-17-00-00107 / 0.8703[20]	ace 4; bz 4; eth 4; EtOH 4 / 1.4317[20]
6086	Furan, 2,3-dihydro- / 2,3-Dihydrofuran	C_4H_6O / 70.09	1191-99-7 /	/ 54.5	5-17-01-00172 / 0.927[25]	/ 1.4239[20]
6087	Furan, 2,5-dihydro-2,5-dimethoxy- / 2,5-Dimethoxy-2,5-dihydrofuran	$C_6H_{10}O_3$ / 130.14	332-77-4 /	/ 161	5-17-05-00073 / 1.073[25]	/ 1.4339[20]
6088	Furan, 2,4-dimethyl- / 3,5-Dimethylfuran	C_6H_8O / 96.13	3710-43-8 /	/ 94	5-17-01-00343 / 0.8993[20]	/ 1.4371[20]
6089	Furan, 2,5-dimethyl- / 2,5-Dimethylfuran	C_6H_8O / 96.13	625-86-5 / -62.8	/ 93.5	5-17-01-00344 / 0.8883[20]	H_2O 1; EtOH 3; eth 3; ace 3 / 1.4363[20]
6090	Furan, 2-(1,1-dimethylethyl)- / 2-tert-Butylfuran	$C_8H_{12}O$ / 124.18	7040-43-9 /	/ 119.5	5-17-01-00371 / 0.869[20]	ace 4; eth 4; EtOH 4 / 1.4373[20]
6091	2,5-Furandione / Maleic anhydride	$C_4H_2O_3$ / 98.06	108-31-6 / 52.8	5586 / 202	5-17-11-00055 / 1.314[60]	H_2O 3; eth 3; ace 3; chl 3
6092	2,5-Furandione, 3-chloro-	C_4HClO_3 / 132.50	96-02-6 / 33	/ 196	5-17-11-00063 / 1.5375[25]	/ 1.4980[20]
6093	2,5-Furandione, dihydro- / Succinic anhydride	$C_4H_4O_3$ / 100.07	108-30-5 / 119	8841 / 261	5-17-11-00006 / 1.2[20]	H_2O 1; EtOH 3; eth 2; chl 3
6094	2,5-Furandione, dihydro-3-methyl-	$C_5H_6O_3$ / 114.10	4100-80-5 / 34	/ 239	5-17-11-00011 / 1.22[25]	
6095	2,5-Furandione, dihydro-3-methylene-	$C_5H_4O_3$ / 112.08	2170-03-8 / 69	/ 139[30]; 114[18]	5-17-11-00066 /	eth 2; chl 4
6096	2,5-Furandione, dihydro-3-phenyl-, (±)- / Succinic anhydride, phenyl (DL)	$C_{10}H_8O_3$ / 176.17	112489-85-7 / 54	/ 204[22]	5-17-11-00275 /	ace 4; bz 4; eth 4; EtOH 4
6097	2,5-Furandione, 3,4-dimethyl-	$C_6H_6O_3$ / 126.11	766-39-2 / 96	/ 223	5-17-11-00069 / 1.107[100]	H_2O 2; EtOH 4; eth 4; bz 4
6098	2,5-Furandione, 3-methyl-	$C_5H_4O_3$ / 112.08	616-02-4 / 7.5	/ 213.5	5-17-11-00065 / 1.2469[16]	ace 4; eth 4; EtOH 4 / 1.4710[21]
6099	2,5-Furandione, 3,3,4,4-tetrafluorodihydro-	$C_4F_4O_3$ / 172.04	699-30-9 /	/ 54.5	5-17-11-00007 / 1.6209[20]	/ 1.3240[20]
6100	Furan, 2,5-diphenyl-	$C_{16}H_{12}O$ / 220.27	955-83-9 / 91	/ 344	5-17-02-00398 /	H_2O 1; EtOH 4; eth 4; ace 3
6101	Furan, 2,2'-[dithiobis(methylene)]bis-	$C_{10}H_{10}O_2S_2$ / 226.32	4437-20-1 / 10	/ 167[13]; 112[0.5]	4-17-00-01258 /	EtOH 4
6102	Furan, 2-ethenyl-	C_6H_6O / 94.11	1487-18-9 / -94	/ 99.5	5-17-01-00477 / 0.9445[19]	/ 1.4992[19]
6103	Furan, 2-ethoxy- / Ethyl 2-furyl ether	$C_6H_8O_2$ / 112.13	5809-07-4 /	/ 125.5	4-17-00-01219 / 0.9849[23]	/ 1.4500[23]
6104	Furan, 2-(ethoxymethyl)-	$C_7H_{10}O_2$ / 126.16	6270-56-0 /	/ 149	5-17-03-00340 / 0.9844[20]	H_2O 1; EtOH 3; eth 3 / 1.4523[20]
6105	Furan, 2-ethyl- / 2-Ethylfuran	C_6H_8O / 96.13	3208-16-0 /	/ 92	5-17-01-00338 / 0.911[15]	EtOH 3; eth 3; ace 3; bz 3 / 1.4466[23]
6106	Furan, 2-ethyl-5-methyl- / 2-Methyl-5-ethylfuran	$C_7H_{10}O$ / 110.16	1703-52-2 /	/ 118.5	5-17-01-00357 / 0.8883[20]	bz 4; eth 4; EtOH 4 / 1.4473[20]
6107	Furan, 2-ethyltetrahydro- / 2-Ethyltetrahydrofuran	$C_6H_{12}O$ / 100.16	1003-30-1 /	/ 109	5-17-01-00095 / 0.8570[19]	ace 4; bz 4; eth 4; EtOH 4 / 1.4147[19]
6108	Furan, 2-ethynyl-	C_6H_4O / 92.10	18649-64-4 /	/ 105.5	4-17-00-00396 / 0.9919[20]	/ 1.5055[20]
6109	Furan, 2,2,3,3,4,4,5-heptafluorotetrahydro-5-(nonafluorobutyl)- / Perfluoro-2-butyltetrahydrofuran	$C_8F_{16}O$ / 416.06	335-36-4 /	/ 102.6		
6110	Furan, 2-iodo- / 2-Furyl iodide	C_4H_3IO / 193.97	54829-48-0 /	/ 43-5[15]	5-17-01-00295 / 2.024[20]	eth 4 / 1.5661[20]
6111	Furan, 3-iodo- / 3-Iodofuran	C_4H_3IO / 193.97	29172-20-1 /	/ 133; 37-8[22]	5-17-01-00296 / 2.045[20]	H_2O 1; eth 3 / 1.5610[20]
6112	Furan, 2-isopropoxy- / 2-Furyl isopropyl ether	$C_7H_{10}O_2$ / 126.16	98272-34-5 /	/ 135.5	4-17-00-01219 / 0.9689[20]	/ 1.4419[20]
6113	2-Furanmethanamine / Furfurylamine	C_5H_7NO / 97.12	617-89-0 /	/ 145.5	5-18-09-00541 / 1.0995[20]	H_2O 5; EtOH 5; eth 3; chl 3 / 1.4908[20]
6114	2-Furanmethanamine, N-(2-furanylmethyl)- / Difurfurylamine	$C_{10}H_{11}NO_2$ / 177.20	18240-50-1 /	/ 135-42[15]	4-18-00-07092 / 1.1045[20]	eth 4 / 1.5168[20]
6115	2-Furanmethanamine, N-methyl-	C_6H_9NO / 111.14	4753-75-7 /	/ 149	5-18-09-00541 / 0.989[25]	/ 1.4729[20]
6116	2-Furanmethanamine, tetrahydro- / Tetrahydrofurfurylamine	$C_5H_{11}NO$ / 101.15	4795-29-3 /	/ 153	5-18-09-00488 / 0.9752[20]	H_2O 4; eth 4; EtOH 4 / 1.4551[20]
6117	2-Furanmethanethiol	C_5H_6OS / 114.17	98-02-2 /	/ 157	5-17-03-00351 / 1.1319[20]	H_2O 1; chl 2 / 1.5329[20]
6118	2-Furanmethanol / Furfuryl alcohol	$C_5H_6O_2$ / 98.10	98-00-0 / -31	4215 / 171	5-17-03-00338 / 1.1296[20]	H_2O 5; EtOH 4; eth 4; chl 4 / 1.4869[20]
6119	2-Furanmethanol, acetate	$C_7H_8O_3$ / 140.14	623-17-6 /	/ 179	5-17-03-00344 / 1.1175[20]	H_2O 1; EtOH 3; eth 3 / 1.4327[20]
6120	2-Furanmethanol, 5-(aminomethyl)tetrahydro-	$C_6H_{13}NO_2$ / 131.17	589-14-0 / <-70	/ 103[0.3]	1.1021[25]	/ 1.4870[25]

No.	Name Synonym	Mol. Form. Mol. Wt.	CAS RN mp/°C	Merck No. bp/°C	Beil. Ref. den/g cm^{-3}	Solubility n_D
6121	2-Furanmethanol, α-methyl-	$C_6H_8O_2$ 112.13	4208-64-4	162.5	5-17-03-00374 1.0739[25]	1.4827[15]
6122	2-Furanmethanol, 5-methyl-	$C_6H_8O_2$ 112.13	3857-25-8	195 dec; 81[23]	5-17-03-00387 1.0769[20]	eth 4; EtOH 4 1.4853[20]
6123	2-Furanmethanol, propanoate	$C_8H_{10}O_3$ 154.17	623-19-8	195	2-17-00-00115 1.1085[20]	H_2O 2; EtOH 3; eth 5; ace 3
6124	2-Furanmethanol, tetrahydro- Tetrahydrofurfuryl alcohol	$C_5H_{10}O_2$ 102.13	97-99-4 <-80	9146 178	5-17-03-00115 1.0524[20]	ace 4; eth 4 1.4520[20]
6125	2-Furanmethanol, tetrahydro-, acetate	$C_7H_{12}O_3$ 144.17	637-64-9	193; 89[18]	5-17-03-00119 1.0624[20]	H_2O 4; eth 4; EtOH 4; chl 4 1.4350[25]
6126	2-Furanmethanol, tetrahydro-, propanoate	$C_8H_{14}O_3$ 158.20	637-65-0	205.5	5-17-03-00119 1.044[20]	eth 4; EtOH 4; chl 4
6127	Furan, 2-methoxy- 2-Methoxyfuran	$C_5H_6O_2$ 98.10	25414-22-6	110.5	5-17-03-00283 1.0646[25]	1.4468[25]
6128	Furan, 2-(methoxymethyl)-	$C_6H_8O_2$ 112.13	13679-46-4	132	5-17-03-00340 1.0163[20]	H_2O 1; EtOH 3; eth 4 1.4570[20]
6129	Furan, 2-(methoxymethyl)-5-nitro- 2-(Methoxymethyl)-5-nitrofuran	$C_6H_7NO_4$ 157.13	586-84-5	5917 104-5[3]	4-17-00-01254 1.281[20]	EtOH 4 1.5325[20]
6130	Furan, 2-methyl- 2-Methylfuran	C_5H_6O 82.10	534-22-5 -87.5	65	5-17-01-00322 0.9132[20]	H_2O 2; EtOH 3; eth 3; ctc 2 1.4342[20]
6131	Furan, 3-methyl- 3-Methylfuran	C_5H_6O 82.10	930-27-8	66	5-17-01-00330 0.923[18]	H_2O 1; EtOH 3; eth 3 1.4330[19]
6132	Furan, 2,2'-methylenebis-	$C_9H_8O_2$ 148.16	1197-40-6	94[22]	5-19-01-00545 1.102[20]	H_2O 1; EtOH 4; eth 4; ace 4 1.5049[20]
6133	Furan, 3-(4-methyl-3-pentenyl)-	$C_{10}H_{14}O$ 150.22	539-52-6	185.5	5-17-01-00515 0.9017[20]	1.4705[21]
6134	Furan, 2-nitro-	$C_4H_3NO_3$ 113.07	609-39-2 30	134[123]; 84[13]	5-17-01-00296	H_2O 3; EtOH 3; eth 3
6135	Furan, 2-(octyloxy)- 2-Furyl octyl ether	$C_{12}H_{20}O_2$ 196.29	100314-90-7	129-30[18]	4-17-00-01219 0.9214[28]	eth 4 1.4520[20]
6136	2-Furanol, 3,4-bis(1,3-benzodioxol-5-ylmethyl)tetrahydro- Cubebin	$C_{20}H_{20}O_6$ 356.38	18423-69-3 131.5	2616	5-19-12-00048	eth 4; EtOH 4; chl 4
6137	3-Furanol, tetrahydro-	$C_4H_8O_2$ 88.11	453-20-3	181	5-17-03-00064 1.09[25]	1.4500[20]
6138	2-Furanol, tetrahydro-, benzoate	$C_{11}H_{12}O_3$ 192.21	3333-44-6	302; 139[2]	5-17-03-00060 1.137[20]	eth 4; EtOH 4
6139	2(3H)-Furanone, 3-acetyldihydro- α-Acetylbutyrolactone	$C_6H_8O_3$ 128.13	517-23-7	77 107-8[5]	5-17-11-00016 1.1846[20]	H_2O 4 1.4585[20]
6140	2(3H)-Furanone, 5-butyldihydro-	$C_8H_{14}O_2$ 142.20	104-50-7	132-3[20]	5-17-09-00073 0.9796[19]	EtOH 3; ctc 2 1.4451[19]
6141	2(3H)-Furanone, 5-(chloromethyl)dihydro- Valeric acid, 5-chloro-4-hydroxy-, γ-lactone	$C_5H_7ClO_2$ 134.56	39928-72-8	132-5[12]	5-17-09-00025 1.625[18]	EtOH 2; eth 2; bz 2
6142	2(3H)-Furanone, dihydro- γ-Butyrolactone	$C_4H_6O_2$ 86.09	96-48-0 -43.3	1596 204	5-17-09-00007 1.1284[16]	ace 4; bz 4; eth 4; EtOH 4 1.4341[20]
6143	2(3H)-Furanone, dihydro-3-hydroxy-4,4-dimethyl-, (R)- Pantolactone	$C_6H_{10}O_3$ 130.14	599-04-2 92	6963	5-18-01-00022	
6144	2(3H)-Furanone, dihydro-3-(hydroxyphenylmethyl)-4-[(1-methyl-1H-imidazol-5-yl)methyl]-, Isopilosine	$C_{16}H_{18}N_2O_3$ 286.33	491-88-3 187	5085	1-27-00-00612	EtOH 4
6145	2(3H)-Furanone, dihydro-3-methyl-	$C_5H_8O_2$ 100.12	1679-47-6	78-81[10]	5-17-09-00028 1.063[25]	1.4325[20]
6146	2(3H)-Furanone, dihydro-5-methyl-, (±)-	$C_5H_8O_2$ 100.12	57129-69-8 -31	206	5-17-09-00024 1.0465[25]	H_2O 5; EtOH 3; ace 3; ctc 2 1.4328[20]
6147	2(3H)-Furanone, dihydro-3-methylene- α-Methylene butyrolactone	$C_5H_6O_2$ 98.10	547-65-9	5981 85-6[10]	5-17-09-00127 1.1206[20]	H_2O 3; EtOH 4; eth 3; ace 3 1.4650[20]
6148	2(3H)-Furanone, dihydro-5-methylene-	$C_5H_6O_2$ 98.10	10008-73-8	85-6[10]	5-17-09-00124 1.1206[20]	ace 4; bz 4; eth 4; EtOH 4 1.4650[20]
6149	2(3H)-Furanone, dihydro-5-pentyl- 4-Hydroxynonanoic acid lactone	$C_9H_{16}O_2$ 156.22	104-61-0		5-17-09-00084	
6150	2(3H)-Furanone, 5-ethyldihydro-	$C_6H_{10}O_2$ 114.14	695-06-7 -18	215.5	5-17-09-00040 1.0261[20]	H_2O 4; EtOH 4 1.4495[20]
6151	2(3H)-Furanone, 3-ethyldihydro-4-[(1-methyl-1H-imidazol-5-yl)methyl]-, (3S-cis)- Pilocarpine	$C_{11}H_{16}N_2O_2$ 208.26	92-13-7 34	7395 260[5]	2-27-00-00694	H_2O 3; EtOH 3; eth 2; bz 2
6152	2(3H)-Furanone, 5-heptyldihydro- 4-Hydroxyundecanoic acid lactone	$C_{11}H_{20}O_2$ 184.28	104-67-6	286	5-17-09-00098 0.9494[20]	EtOH 4 1.4512[20]
6153	2(3H)-Furanone, 5-methyl-	$C_5H_6O_2$ 98.10	591-12-8 18	56[12]	5-17-09-00120 1.084[20]	H_2O 3; EtOH 3; eth 3; ctc 2 1.4476[20]
6154	2(5H)-Furanone, 5-methyl-	$C_5H_6O_2$ 98.10	591-11-7 <-17	209; 98[15]	5-17-09-00121 1.0810[20]	H_2O 5; EtOH 3; eth 3 1.4454[20]
6155	2(5H)-Furanone, 5-methylene- Protoanemonin	$C_5H_4O_2$ 96.09	108-28-1	7907 73[11]	5-17-09-00371	H_2O 2; chl 3
6156	Furan, 2,2'-[oxybis(methylene)]bis-	$C_{10}H_{10}O_3$ 178.19	4437-22-3	101[2]	5-17-03-00347 1.1405[20]	H_2O 1 1.5088[20]
6157	Furan, 2-phenoxy- 2-Furyl phenyl ether	$C_{10}H_8O_2$ 160.17	60698-31-9	105-6[18]	4-17-00-01219 1.1010[23]	eth 3 1.5418[23]

No.	Name / Synonym	Mol. Form. / Mol. Wt.	CAS RN	mp/°C / bp/°C	Merck No.	Beil. Ref. / den/g cm^{-3}	Solubility / n_D
6158	Furan, 2-(phenoxymethyl)- / Furfuryl phenyl ether	$C_{11}H_{10}O_2$ / 174.20	4437-23-4	/ 139-40[20]		5-17-03-00340 / 1.123[17]	/ 1.5535[17]
6159	Furan, 2-phenyl- / 2-Phenylfuran	$C_{10}H_8O$ / 144.17	17113-33-6	/ 107-8[18]		5-17-02-00139 / 1.083[20]	ace 4; bz 4 / 1.5920[20]
6160	Furan, 2-(phenylthio)- / 2-Furyl phenyl sulfide	$C_{10}H_8OS$ / 176.24	16003-14-8	/ 119-20[8]		5-17-03-00285 / 1.1341[26]	eth 4; EtOH 4 / 1.5976[20]
6161	2-Furanpropanoic acid	$C_7H_8O_3$ / 140.14	935-13-7	58 / 229		5-18-06-00224	H_2O 4; eth 4; chl 4
6162	2-Furanpropanoic acid, methyl ester	$C_8H_{10}O_3$ / 154.17	37493-31-5	/ 89[15]		5-18-06-00224 / 1.088[20]	
6163	2-Furanpropanoic acid, β-oxo-, ethyl ester	$C_9H_{10}O_4$ / 182.18	615-09-8	/ 142-3[10]		5-18-08-00143 / 1.165[17]	H_2O 1; EtOH 3; eth 3 / 1.4662[20]
6164	2-Furanpropanoic acid, tetrahydro- / Propionic acid, 3-(2-tetrahydrofuryl)	$C_7H_{12}O_3$ / 144.17	935-12-6	/ 263		5-18-06-00033 / 1.1135[20]	ace 4; EtOH 4 / 1.4578[25]
6165	2-Furanpropanoic acid, tetrahydro-, ethyl ester	$C_9H_{16}O_3$ / 172.22	4525-36-4	/ 222; 73[2]		5-18-06-00034 / 1.023[7]	ace 4; EtOH 4 / 1.440[20]
6166	2-Furanpropanol, tetrahydro-α-[2-(tetrahydro-2-furanyl)ethyl]- / 1,5-Bis(2-tetrahydrofuryl)-3-pentanol	$C_{13}H_{24}O_3$ / 228.33	6265-26-5	/ 116[0.05]		4-19-00-00704	chl 2 / 1.4541[25]
6167	Furan, 2-propyl- / 2-Propylfuran	$C_7H_{10}O$ / 110.16	4229-91-8	/ 115		5-17-01-00354 / 0.8876[20]	EtOH 3; eth 3; ace 3 / 1.4549[20]
6168	Furan, tetrahydro- / Tetrahydrofuran	C_4H_8O / 72.11	109-99-9	109-99-9 / -108.3	9144 / 65	5-17-01-00027 / 0.8892[20]	H_2O 3; EtOH 4; eth 4; ace 4 / 1.4050[20]
6169	Furan, tetrahydro-2,5-dimethoxy-	$C_8H_{12}O_3$ / 132.16	696-59-3	/ 145.7		5-17-05-00004 / 1.02[25]	
6170	Furan, tetrahydro-2-(methoxymethyl)-	$C_6H_{12}O_2$ / 116.16	19354-27-9	/ 140		5-17-03-00116 / 0.952[25]	/ 1.4270[20]
6171	Furan, tetrahydro-2-methyl- / 2-Methyltetrahydrofuran	$C_5H_{10}O$ / 86.13	96-47-9	/ 78		5-17-01-00078 / 0.8552[20]	H_2O 3; EtOH 4; eth 4; ace 4 / 1.4059[21]
6172	Furan, tetrahydro-3-methyl- / 3-Methyltetrahydrofuran	$C_5H_{10}O$ / 86.13	13423-15-9	/ 86.5		5-17-01-00081 / 0.8642[20]	ace 4; bz 4; eth 4; EtOH 4 / 1.4122[20]
6173	Furan, tetrahydro-2-propyl-	$C_7H_{14}O$ / 114.19	3208-22-8	/ 136		5-17-01-00116 / 0.8547[20]	/ 1.4242[20]
6174	Furan, tetrahydro-2,2,4,4-tetramethyl- / 2,2,4,4-Tetramethyltetrahydrofuran	$C_8H_{16}O$ / 128.21	3358-28-9	/ 121		4-17-00-00109	H_2O 1; EtOH 5; eth 5 / 1.4074[20]
6175	Furan, tetraphenyl- / 2,3,4,5-Tetraphenylfuran	$C_{28}H_{20}O$ / 372.47	1056-77-5	175 / 220		4-17-00-00810	H_2O 1; EtOH 3; eth 3; ace 3
6176	Furazan, dimethyl-	$C_4H_8N_2O$ / 98.10	4975-21-7	-7 / 156		4-27-00-07096 / 1.0528[14]	eth 4; EtOH 4 / 1.4327[20]
6177	3aH-Furo[2,3-b]indol-3a-ol, 8a-(5-ethenyl-1-azabicyclo[2.2.2]oct-2-yl)-2,3,8,8a-tetrahydro- / Conquinamine	$C_{19}H_{24}N_2O_2$ / 312.41	464-86-8	2505 / 123		2-27-00-00667	H_2O 2; EtOH 3; eth 3; chl 3
6178	3aH-Furo[2,3-b]indol-3a-ol, 8a-(5-ethenyl-1-azabicyclo[2.2.2]oct-2-yl)-2,3,8,8a-tetrahydro- / Quinamine	$C_{19}H_{24}N_2O_2$ / 312.41	464-85-7	8060 / 185.5		2-27-00-00667	H_2O 1; EtOH 4; eth 3; ace 3
6179	Furo[2,3-b]quinoline, 4,8-dimethoxy- / Fagarine	$C_{13}H_{11}NO_3$ / 229.24	524-15-2	3879 / 142		4-27-00-02211	H_2O 2; EtOH 3; eth 3; bz 3
6180	Furo[2,3-b]quinoline, 4-methoxy- / Dictamnine	$C_{12}H_9NO_2$ / 199.21	484-29-7	3079 / 133.5		4-27-00-02030	H_2O 2; EtOH 4; eth 3; chl 3
6181	Furo[2,3-b]quinoline, 4,7,8-trimethoxy- / Skimmianine	$C_{14}H_{13}NO_4$ / 259.26	83-95-4	8505 / 177		4-27-00-02296	H_2O 1; EtOH 3; eth 2; chl 3
6182	Furo[2,3-b]quinolin-4(2H)-one, 3,9-dihydro-8-methoxy-9-methyl-2-(1-methylethyl)-, (R)- / Lunacrine	$C_{16}H_{19}NO_3$ / 273.33	82-40-6	5473		4-27-00-03566	chl 3
6183	4H-Furo[3,2-c]pyran-2(6H)-one, 4-hydroxy- / Patulin	$C_7H_6O_4$ / 154.12	149-29-1	7002 / 111		5-18-03-00005	H_2O 3; EtOH 3; eth 3; ace 3
6184	1H,12H-Furo[3',2':4,5]furo[2,3-h]pyrano[3,4-c][1]benzopyran-1,12-dione, 3,4,7a,10a-tetrahydro-5-meth / Aflatoxin G1	$C_{17}H_{12}O_7$ / 328.28	1165-39-5	/ 245		5-19-12-00081	
6185	5H-Furo[3,2-g][1]benzopyran-5-one, 4,9-dimethoxy-7-methyl- / Khellin	$C_{14}H_{12}O_5$ / 260.25	82-02-0	5189 / 154-5 dec	180-200[0.05]	5-19-06-00320	H_2O 1; EtOH 3; eth 2; ace 3
6186	5H-Furo[3,2-g][1]benzopyran-5-one, 4-methoxy-7-methyl- / Visnagin	$C_{13}H_{10}O_4$ / 230.22	82-57-5	9916 / 144.5		5-19-06-00030	H_2O 2; EtOH 2; chl 4
6187	7H-Furo[3,2-g][1]benzopyran-7-one, 9-methoxy- / Methoxsalen	$C_{12}H_8O_4$ / 216.19	298-81-7	5911 / 148		5-19-06-00015	H_2O 2; EtOH 4; eth 2; ace 2
6188	7H-Furo[3,2-g][1]benzopyran-7-one, 3-methoxy-2-(1-methylethyl)- / Peucedanin	$C_{15}H_{14}O_4$ / 258.27	133-26-6	7144 / 85	276-81[17]	5-19-06-00042	H_2O 2; EtOH 3; eth 3; bz 2
6189	7H-Furo[3,2-g][1]benzopyran-7-one, 9-[(3-methyl-2-butenyl)oxy]- / Imperatorin	$C_{16}H_{14}O_4$ / 270.28	482-44-0	4839 / 102		5-19-06-00016	H_2O 2; EtOH 3; eth 3; bz 3
6190	3-Furoic acid, 5-bromo-, ethyl ester / β-Furoic acid, 5-bromo, ethyl ester	$C_7H_7BrO_3$ / 219.03	32460-20-1	17 / 235; 135[34]		5-18-06-00198 / 1.528[20]	eth 4; EtOH 4
6191	Furo[3',4':6,7]naphtho[2,3-d]-1,3-dioxol-6(5aH)-one, 5,8,8a,9-tetrahydro-9-hydroxy-5-(3,4,5-trimethoxyphenyl)- / Picropodophyllin	$C_{22}H_{22}O_8$ / 414.41	477-47-4	7385 / 228		5-19-10-00665	ace 4; bz 4; eth 4; EtOH 4

No.	Name Synonym	Mol. Form. Mol. Wt.	CAS RN mp/°C	Merck No. bp/°C	Beil. Ref. den/g cm^{-3}	Solubility n_D
6192	Furo[3',4':6,7]naphtho[2,3-d]-1,3-dioxol-6(5aH)-one, 5,8,8a,9-tetrahydro-9-hydroxy-5-(3,4,5-trimetho Podophyllotoxin	C$_{22}$H$_{22}$O$_8$ 414.41	518-28-5 183	7520	5-19-10-00666	H$_2$O 2; EtOH 4; eth 1; ace 3
6193	Galactaric acid Mucic acid	C$_6$H$_{10}$O$_8$ 210.14	526-99-8 255 dec	4237	4-03-00-01292	
6194	Galactitol Dulcose	C$_6$H$_{14}$O$_6$ 182.17	608-66-2 189.5	4238 275-80[1]	4-01-00-02844 1.47[20]	H$_2$O 3; EtOH 2; eth 1; bz 1
6195	β-D-Galactopyranoside, (2α,3β,5α,15β,25R)-2,15-dihydroxyspirostan-3-yl Digitonin	C$_{56}$H$_{92}$O$_{29}$ 1229.33	11024-24-1 237.5	3144	5-19-03-00603	
6196	β-D-Galactopyranoside, (3β,5α,22β,25S)-spirosolan-3-yl O-β-Tomatine	C$_{50}$H$_{83}$NO$_{21}$ 1034.20	17406-45-0 270	9471	4-27-00-01954	EtOH 4; diox 4
6197	D-Galactose	C$_6$H$_{12}$O$_6$ 180.16	59-23-4 170	4241	4-01-00-04336	H$_2$O 4; EtOH 2; eth 1; bz 1
6198	Galactose, 6-deoxy-3-O-methyl- Digitalose	C$_7$H$_{14}$O$_5$ 178.19	4481-08-7 119	3142	4-01-00-04270	H$_2$O 4
6199	D-Galacturonic acid	C$_6$H$_{10}$O$_7$ 194.14	685-73-4 166	4242	1-03-00-00306	H$_2$O 3; EtOH 3; eth 1
6200	Galanthamine Lycoremine	C$_{17}$H$_{21}$NO$_3$ 287.36	357-70-0 126.5	4245	4-27-00-02184	ace 4; EtOH 4; chl 4
6201	Galanthan-1,2-diol, 3,12-didehydro-9,10-[methylenebis(oxy)]-, (1α,2β)- Lycorine	C$_{16}$H$_{17}$NO$_4$ 287.32	476-28-8 280 sub	5498	2-27-00-00547	H$_2$O 1; EtOH 2; eth 2; chl 2
6202	Gardona Phosphoric acid, 2-chloro-1-(2,4,5-trichlorophenyl)ethenyl dimethyl ester	C$_{10}$H$_9$Cl$_4$O$_4$P 365.96	961-11-5			
6203	Gelsemine Spiro[3,5,8-ethanylylidene-1H-pyrano[3,4-c]pyridine-10,3'-[3H]indol]-2'(1'H)-one, 5-ethenyl	C$_{20}$H$_{22}$N$_2$O$_2$ 322.41	509-15-9 178	4277	4-27-00-07526	ace 4; bz 4; eth 4; EtOH 4
6204	Gibb-3-ene-1,10-dicarboxylic acid, 2,4a,7-trihydroxy-1-methyl-8-methylene-, 1,4a-lactone, Gibberellic acid	C$_{19}$H$_{22}$O$_6$ 346.38	77-06-5 234	4313	5-18-09-00269	ace 4; EtOH 4; MeOH 4
6205	D-Glucaric acid D-Tetrahydroxyadipic acid	C$_6$H$_{10}$O$_8$ 210.14	87-73-0 125.5	4344	4-03-00-01291	H$_2$O 4; EtOH 4; eth 2; chl 2
6206	D-Glucitol Sorbitol	C$_6$H$_{14}$O$_6$ 182.17	50-70-4 111	8680 295[3.5]	4-01-00-02839 1.489[20]	H$_2$O 4; ace 4 1.3330[20]
6207	D-Glucitol, 1-amino-1-deoxy- Glucamine	C$_6$H$_{15}$NO$_5$ 181.19	488-43-7 127	4343	4-04-00-01913	H$_2$O 4; EtOH 4
6208	D-Glucitol, 1-deoxy-1-(methylamino)- N-Methylglucamine	C$_7$H$_{17}$NO$_5$ 195.22	6284-40-8 128.5	5995	4-04-00-01914	H$_2$O 3
6209	D-Gluconic acid Gluconic acid	C$_6$H$_{12}$O$_7$ 196.16	526-95-4 131	4350	4-03-00-01255	H$_2$O 3; EtOH 2; eth 1; bz 1
6210	D-Gluconic acid, 4-O-β-D-galactopyranosyl- Lactobionic acid	C$_{12}$H$_{22}$O$_{12}$ 358.30	96-82-2	5219	5-17-07-00436	H$_2$O 4; EtOH 2; eth 1; HOAc 2
6211	Gluconic acid, β-lactone Oxetane, gluconic acid deriv.	C$_6$H$_{10}$O$_6$ 178.14	3087-62-5 153		2-18-00-00190 1.610[-5]	H$_2$O 2; EtOH 1; eth 1; ace 1
6212	D-Gluconic acid, δ-lactone δ-D-Gluconolactone	C$_6$H$_{10}$O$_6$ 178.14	90-80-2	4351	5-18-05-00011	
6213	D-Gluconic acid, monosodium salt Gluconic acid, sodium salt	C$_6$H$_{11}$NaO$_7$ 218.14	527-07-1	8569	4-03-00-01255	H$_2$O 3
6214	β-D-Glucopyranose, 4-O-β-D-galactopyranosyl-	C$_{12}$H$_{22}$O$_{11}$ 342.30	5965-66-2 254		5-17-07-00196 1.59[20]	H$_2$O 4; EtOH 2; eth 1; chl 1
6215	α-L-Glucopyranose, 2-(methylamino)-2-deoxy- N-Methyl-α-L-glucosamine	C$_7$H$_{15}$NO$_5$ 193.20	42852-95-9	5996		
6216	α-D-Glucopyranose, pentaacetate	C$_{16}$H$_{22}$O$_{11}$ 390.34	604-68-2 113.3 sub		5-17-07-00318	H$_2$O 2; EtOH 2; eth 3; chl 3
6217	β-D-Glucopyranose, pentaacetate	C$_{16}$H$_{22}$O$_{11}$ 390.34	604-69-3 134 sub		5-17-07-00319 1.2740[20]	H$_2$O 1; EtOH 2; eth 2; bz 3
6218	α-D-Glucopyranoside, β-D-fructofuranosyl Sucrose	C$_{12}$H$_{22}$O$_{11}$ 342.30	57-50-1 185.5	8855	5-17-08-00399 1.5805[17]	H$_2$O 3; EtOH 2; eth 1; py 3 1.5376
6219	α-D-Glucopyranoside, β-D-fructofuranosyl O-α-D-galactopyranosyl-(1 Raffinose	C$_{18}$H$_{32}$O$_{16}$ 504.44	512-69-6 80	8120	4-17-00-03801 1.465[25]	H$_2$O 4; EtOH 1; eth 1; MeOH 4
6220	α-D-Glucopyranoside, α-D-glucopyranosyl Trehalose	C$_{12}$H$_{22}$O$_{11}$ 342.30	99-20-7 203	9496	5-17-08-00414 1.58[24]	H$_2$O 4; EtOH 3; eth 1; bz 1
6221	α-D-Glucopyranoside, O-α-D-glucopyranosyl-(13)-β-D-fructofuranosyl Melezitose	C$_{18}$H$_{32}$O$_{16}$ 504.44	597-12-6 153	5698	5-17-08-00414 1.5565[25]	H$_2$O 4
6222	β-D-Glucopyranoside, 2-(hydroxymethyl)phenyl Salicin	C$_{13}$H$_{18}$O$_7$ 286.28	138-52-3 207	8293 240 dec	5-17-07-00113 1.434[20]	H$_2$O 4; EtOH 4; HOAc 4
6223	β-D-Glucopyranoside, 2-(hydroxymethyl)phenyl, 6-benzoate Populin	C$_{20}$H$_{22}$O$_8$ 390.39	99-17-2 180	7571	5-17-07-00343	

No.	Name / Synonym	Mol. Form. / Mol. Wt.	CAS RN / mp/°C	Merck No. / bp/°C	Beil. Ref. / den/g cm^{-3}	Solubility / n_D
6224	β-D-Glucopyranoside, 4-hydroxyphenyl / Arbutin	$C_{12}H_{16}O_7$ / 272.25	497-76-7 / 199.5	799	5-17-07-00110	H_2O 4; EtOH 3; eth 2; bz 1
6225	β-D-Glucopyranoside, 4-(3-hydroxy-1-propenyl)-2,6-dimethoxyphenyl / Syringin	$C_{17}H_{24}O_9$ / 372.37	118-34-3 / 192	8997	4-17-00-03002	EtOH 4
6226	β-D-Glucopyranoside, 4-(3-hydroxy-1-propenyl)-2-methoxyphenyl / Coniferin	$C_{16}H_{22}O_8$ / 342.35	531-29-3 / 186	2498	4-17-00-02999	H_2O 3; EtOH 2; eth 1; py 3
6227	α-D-Glucopyranoside, methyl / α-Methylglucoside	$C_7H_{14}O_6$ / 194.18	97-30-3 / 168	5997 / 200$^{0.2}$	5-17-07-00010 / 1.46^{30}	H_2O 4
6228	α-D-Glucopyranoside, methyl 2,3,4,6-tetra-O-methyl- / α-D-Glucose, pentamethyl ether	$C_{11}H_{22}O_6$ / 250.29	605-81-2 / 180$^{0.4}$		5-17-07-00029 / 1.0944^{20}	H_2O 4; ace 4; eth 4; EtOH 4 / 1.4466^{20}
6229	α-D-Glucopyranoside, 1,3,4,6-tetra-O-acetyl-β-D-fructofuranosyl, tetraacetate / Octaacetyl sucrose	$C_{28}H_{38}O_{19}$ / 678.60	126-14-7 / 86.5	8856 / 250^1	5-17-08-00410 / 1.27^{16}	H_2O 2; EtOH 3; eth 3; ace 3 / 1.4660
6230	α-D-Glucopyranosiduronic acid, (3β,20β)-20-carboxy-11-oxo-30-norolean-12-en-3-yl 2-O- / Glycyrrhizic acid	$C_{42}H_{62}O_{16}$ / 822.94	1405-86-3 / 220 dec	4401	4-18-00-05156	H_2O 4
6231	β-D-Glucopyranosiduronic acid, 2,2,2-trichloroethyl / Urochloralic acid	$C_8H_{11}Cl_3O_7$ / 325.53	97-25-6 / 142	9798	4-18-00-05113	H_2O 4; EtOH 4
6232	β-D-Glucose	$C_6H_{12}O_6$ / 180.16	28905-12-6 / 150		4-01-00-04306 / 1.5620^{18}	H_2O 4; EtOH 2; eth 1; py 3
6233	D-Glucose, 2-amino-2-deoxy- / D-Glucosamine	$C_6H_{13}NO_5$ / 179.17	3416-24-8	4352	4-04-00-02017	H_2O 4
6234	D-Glucose, 6-O-α-L-arabinopyranosyl- / Vicianose	$C_{11}H_{20}O_{10}$ / 312.27	14116-69-9 / 210 dec	9879	5-17-06-00112	H_2O 4
6235	D-Glucose, 6-deoxy- / Quinovose	$C_6H_{12}O_5$ / 164.16	7658-08-4 / 139.5	8106	4-01-00-04260	H_2O 4; EtOH 4
6236	D-Glucose, 6-O-(6-deoxy-α-L-mannopyranosyl)- / Rutinose	$C_{12}H_{22}O_{10}$ / 326.30	90-74-4 / 189-92 dec	8277	5-17-06-00269	H_2O 4; EtOH 4
6237	D-Glucose, 6-O-β-D-xylopyranosyl- / Primeverose	$C_{11}H_{20}O_{10}$ / 312.27	26531-85-1 / 210	7752	4-17-00-02447	H_2O 4; MeOH 4
6238	D-Glucuronic acid	$C_6H_{10}O_7$ / 194.14	6556-12-3 / 165	4360	4-03-00-01996	H_2O 4; EtOH 4
6239	D-Glucuronic acid, γ-lactone / D-Glucuronolactone	$C_6H_8O_6$ / 176.13	32449-92-6 / 177.5	4362	5-18-05-00033 / 1.76^{20}	H_2O 3; EtOH 2; bz 1; MeOH 2
6240	DL-Glutamic acid	$C_5H_9NO_4$ / 147.13	617-65-2 / 199 dec		4-04-00-03028 / 1.4601^{20}	H_2O 2; EtOH 1; eth 2; CS_2 1
6241	L-Glutamic acid / (S)-2-Aminopentanedioic acid	$C_5H_9NO_4$ / 147.13	56-86-0 / 160 dec	4363 / sub 175	4-04-00-03028 / 1.538^{20}	
6242	L-Glutamic acid, N-(4-aminobenzoyl)- / N-(p-Aminobenzoyl)-L-(+)-glutamic acid	$C_{12}H_{14}N_2O_5$ / 266.25	4271-30-1 / 173	437	4-14-00-01153	
6243	L-Glutamic acid, N-[4-[[(2-amino-1,4-dihydro-4-oxo-6-pteridinyl)methyl]amino]benzoyl]- / Folic acid	$C_{19}H_{19}N_7O_6$ / 441.40	59-30-3 / 250 dec	4140	4-26-00-03944	py 4; EtOH 4; HOAc 4
6244	L-Glutamic acid, N-[4-[[(2-amino-5-formyl-1,4,5,6,7,8-hexahydro-4-oxo-6-pteridinyl)methyl]amino]benzoyl]- / Folinic acid	$C_{20}H_{23}N_7O_7$ / 473.45	58-05-9 / 245 dec	4141	4-26-00-03881	H_2O 2
6245	L-Glutamic acid, 5-ethyl ester / γ-Ethyl L-glutamate	$C_7H_{13}NO_4$ / 175.18	1119-33-1 / 191	4364	4-04-00-03033	H_2O 2
6246	DL-Glutamic acid, 3-hydroxy-	$C_5H_9NO_5$ / 163.13	5985-23-9 / 209 dec		4-04-00-03253	H_2O 4; EtOH 1; eth 1; bz 1
6247	L-Glutamic acid, 5-[2-[4-(hydroxymethyl)phenyl]hydrazide] / Agaritine	$C_{12}H_{17}N_3O_4$ / 267.28	2757-90-6 / 205-9 dec	175		H_2O 4
6248	L-Glutamic acid, monosodium salt / Glutamic acid, monosodium salt	$C_5H_8NNaO_4$ / 169.11	142-47-2	6165		H_2O 3
6249	L-Glutamine / 2-Aminoglutaramic acid	$C_5H_{10}N_2O_3$ / 146.15	56-85-9 / 185-6 dec	4365	4-04-00-03038	H_2O 3; EtOH 1; eth 1; bz 1
6250	Glycine / Aminoacetic acid	$C_2H_5NO_2$ / 75.07	56-40-6 / 290 dec	4386	4-04-00-02349 / 1.161^{20}	H_2O 4; EtOH 1; eth 1; ace 2
6251	Glycine, N-acetyl- / Aceturic acid	$C_4H_7NO_3$ / 117.10	543-24-8 / 206	74	4-04-00-02399	H_2O 4; ace 4; EtOH 4
6252	Glycine, N-(4-aminobenzoyl)- / p-Aminohippuric acid	$C_9H_{10}N_2O_3$ / 194.19	61-78-9 / 198.5	454	4-14-00-01152	ace 4; bz 4; EtOH 4
6253	Glycine, N-(aminoiminomethyl)- / Glycocyamine	$C_3H_7N_3O_2$ / 117.11	352-97-6 / 282	4391	4-04-00-02414	H_2O 2; EtOH 2; eth 2
6254	Glycine, N-(aminoiminomethyl)-N-methyl- / Creatine	$C_4H_9N_3O_2$ / 131.13	57-00-1 / 303 dec	2570	4-04-00-02426 / 1.33^{25}	H_2O 3; EtOH 2; eth 1
6255	Glycine, N-benzoyl- / Hippuric acid	$C_9H_9NO_3$ / 179.18	495-69-2 / 191.5	4636	4-09-00-00778 / 1.371^{20}	H_2O 3; EtOH 3; eth 2; bz 2
6256	Glycine, N,N-bis(carboxymethyl)- / Nitrilotriacetic acid	$C_6H_9NO_6$ / 191.14	139-13-9 / 242 dec	6499	4-04-00-02441	H_2O 2; EtOH 3; DMSO-d_6 2

PHYSICAL CONSTANTS OF ORGANIC COMPOUNDS (continued)

No.	Name / Synonym	Mol. Form. / Mol. Wt.	CAS RN / mp/°C	Merck No. / bp/°C	Beil. Ref. / den/g cm^{-3}	Solubility / n_D
6257	Glycine, N,N-bis[2-[bis(carboxymethyl)amino]ethyl]- / Pentetic acid	$C_{14}H_{23}N_3O_{10}$ / 393.35	67-43-6	7083	4-04-00-02454	
6258	Glycine, N,N-bis(ethoxycarbonyl)-, ethyl ester / Aminoacetic acid, N,N-(dicarboethoxy), ethyl ester	$C_{10}H_{17}NO_6$ / 247.25	127665-32-1 / 36.5	152^{10}	0-04-00-00365	bz 4; eth 4; EtOH 4
6259	Glycine, N,N-bis(2-hydroxyethyl)- / Bicine	$C_6H_{13}NO_4$ / 163.17	150-25-4 / 193-5 dec	1221	4-04-00-02390	H_2O 4; EtOH 1
6260	Glycine, N-(carboxymethyl)- / Iminodiacetic acid	$C_4H_7NO_4$ / 133.10	142-73-4 / 247.5	4832	4-04-00-02428	H_2O 2; EtOH 1; eth 1
6261	Glycine, N,N-dimethyl- / N,N-Dimethylglycine	$C_4H_9NO_2$ / 103.12	1118-68-9 / 185.5	3233	4-04-00-02365	H_2O 4; EtOH 3; eth 3; ace 3
6262	Glycine, N,N-1,2-ethanediylbis[N-(carboxymethyl)- / Edetic acid	$C_{10}H_{16}N_2O_8$ / 292.25	60-00-4 / 245 dec	3484	4-04-00-02449	
6263	Glycine, ethyl ester	$C_4H_9NO_2$ / 103.12	459-73-4	149; 58^{18}	4-04-00-02355 / 1.0275^{10}	H_2O 5; EtOH 5; eth 5; ace 5 / 1.4242^{10}
6264	Glycine, N-(N-L-γ-glutamyl-L-cysteinyl)- / Glutatione	$C_{10}H_{17}N_3O_6S$ / 307.33	70-18-8 / 195	4369	4-04-00-03165	H_2O 4; EtOH 1; eth 1; DMF 3
6265	Glycine, N-glycyl- / 2-(Aminoacetamido)acetic acid	$C_4H_8N_2O_3$ / 132.12	556-50-3 / 215 dec	4399	4-04-00-02459	H_2O 3
6266	Glycine, hydrazide / Aminoacetic acid hydrazide	$C_2H_7N_3O$ / 89.10	14379-80-7 / 80.5	150 dec	3-04-00-01121	EtOH 2; eth 1; chl 3
6267	Glycine, N-(4-hydroxyphenyl)- / N-(4-Hydroxyphenyl)glycine	$C_8H_9NO_3$ / 167.16	122-87-2 / 245-7 dec	4771	4-13-00-01210	H_2O 2; EtOH 2; eth 1; ace 2
6268	Glycine, N-methyl- / Sarcosine	$C_3H_7NO_2$ / 89.09	107-97-1 / 212-3 dec	8331	4-04-00-02363	
6269	Glycine, N-methyl-N-(1-oxododecyl)-, sodium salt / Gardol	$C_{15}H_{28}NNaO_3$ / 293.38	137-16-6	4267		H_2O 2
6270	Glycine, N-phenyl- / Phenylaminoacetic acid	$C_8H_9NO_2$ / 151.17	103-01-5 / 127.5	7262	4-12-00-00872	H_2O 4; EtOH 4
6271	Glycine, N-phenyl-, ethyl ester	$C_{10}H_{13}NO_2$ / 179.22	2216-92-4 / 58	273.5	4-12-00-00873	eth 4; EtOH 4
6272	Glycine, N-(phenylmethyl)-, ethyl ester	$C_{11}H_{15}NO_2$ / 193.25	6436-90-4	$175-9^{50}$	4-12-00-02277	EtOH 4; eth 4; bz 4 / 1.5041^{20}
6273	Glycine, N-[(3α,5β,7α,12α)-3,7,12-trihydroxy-24-oxocholan-24-yl]- / Glycocholic acid	$C_{26}H_{43}NO_6$ / 465.63	475-31-0 / 166.5	4390	4-10-00-02077	H_2O 2; EtOH 4; eth 2
6274	Glycogen	$(C_6H_{10}O_5)_x$ / 162.14	9005-79-2	4392		H_2O 4; EtOH 1; eth 1
6275	Glyodin / 1H-Imidazole, 2-heptadecyl-4,5-dihydro-, monoacetate	$C_{20}H_{40}N_2 \cdot C_2H_4O_2$ / 368.60	556-22-9	4404	1.035^{20}	
6276	Glyphosate / Glycine, N-(phosphonomethyl)-	$C_3H_8NO_5P$ / 169.07	1071-83-6 / 230 dec	4408		
6277	Glyphosine / Glycine, N,N-bis(phosphonomethyl)-	$C_4H_{11}NO_8P_2$ / 263.08	2439-99-8	4409		
6278	Guanidine / Aminomethanamidine	CH_5N_3 / 59.07	113-00-8 / 50	4475	4-03-00-00148	H_2O 4; EtOH 4
6279	Guanidine, N,N'-bis(2-methylphenyl)-	$C_{15}H_{17}N_3$ / 239.32	97-39-2 / 179		4-12-00-01764 / 1.10^{20}	H_2O 2; EtOH 2; eth 4; chl 3
6280	Guanidine, cyano- / Dicyanodiamide	$C_2H_4N_4$ / 84.08	461-58-5 / 211	3082	4-03-00-00160 / 1.404^{14}	H_2O 3; EtOH 3; eth 1; ace 3
6281	Guanidine, N,N'-dimethyl-N''-(phenylmethyl)- / Bethanidine	$C_{10}H_{15}N_3$ / 177.25	55-73-2 / 196	1208		
6282	Guanidine, N,N'-diphenyl- / 1,3-Diphenylguanidine	$C_{13}H_{13}N_3$ / 211.27	102-06-7 / 150	3325 / 170 dec	4-12-00-00769 / 1.13^{20}	H_2O 2; EtOH 3; eth 4; ctc 3
6283	Guanidine, (3-methyl-2-butenyl)- / Galegine	$C_6H_{13}N_3$ / 127.19	543-83-9 / 62.5	4246 / dec	3-04-00-00465	H_2O 4; EtOH 4
6284	Guanidine, N-methyl-N'-nitro-N-nitroso- / N-Methyl-N'-nitro-N-nitrosoguanidine	$C_2H_5N_5O_3$ / 147.09	70-25-7	6017	4-04-00-03386	DMSO 3
6285	Guanidine, monohydrochloride	CH_6ClN_3 / 95.53	50-01-1 / 182.3		4-03-00-00148 / 1.354^{20}	H_2O 4; EtOH 4
6286	Guanidine, mononitrate	$CH_6N_4O_3$ / 122.08	506-93-4 / 217	dec	4-03-00-00148	H_2O 4; EtOH 4
6287	Guanidine, nitro- / Nitroguanidine	$CH_4N_4O_2$ / 104.07	556-88-7 / 239 dec	6529	4-03-00-00249	H_2O 2; EtOH 2; eth 1; alk 4
6288	Guanidine, N,N',N''-triphenyl-	$C_{19}H_{17}N_3$ / 287.36	101-01-9 / 146.5	dec	4-12-00-00866 / 1.163^{20}	H_2O 2; EtOH 3
6289	Guanosine / 2-Amino-1,9-dihydro-9-β-D-ribofuranosyl-6H-purin-6-one	$C_{10}H_{13}N_5O_5$ / 283.24	118-00-3 / 239 dec	4480	4-26-00-03901	H_2O 2; EtOH 1; eth 1; HOAc 4
6290	D-Gulose / Gulose, D-	$C_6H_{12}O_6$ / 180.16	4205-23-6 / dec	4490	4-01-00-04333	H_2O 4
6291	Haloxyfop (2-ethoxyethyl) / (RS)-2-[4-(3-Chloro-5-trifluoromethyl-2-pyridyloxy)phenoxy]propionic acid, ethyl ester	$C_{17}H_{15}ClF_3NO_4$ / 389.76	87237-48-7 / 60		1.34	

No.	Name Synonym	Mol. Form. Mol. Wt.	CAS RN mp/°C	Merck No. bp/°C	Beil. Ref. den/g cm^{-3}	Solubility n_D
6292	7H-Benz[de]anthracen-7-one Benzanthrone	$C_{17}H_{10}O$ 230.27	82-05-3 170	1070		bz 2
6293	Helenine	$C_{15}H_{20}O_2$ 232.32	1407-14-3 78	275; 192[10]	4-17-00-05030	bz 4; eth 4; EtOH 4
6294	Heneicosane	$C_{21}H_{44}$ 296.58	629-94-7 40.5	356.5	4-01-00-00569 0.7919[20]	H_2O 1; EtOH 2; peth 3 1.4441[20]
6295	Heneicosane, 11-decyl-	$C_{31}H_{64}$ 436.85	55320-06-4 10.0	282.0[10]	4-01-00-00595 0.8116[20]	1.4540[20]
6296	Heneicosane, 11-(2,2-dimethylpropyl)-	$C_{26}H_{54}$ 366.71	55282-10-5 -21	238.0[10]	4-01-00-00585 0.8031[20]	1.4491[20]
6297	Heneicosane, 11-phenyl- Benzene, (1-decylundecyl)-	$C_{27}H_{48}$ 372.68	6703-80-6 20.8	205[1.0]	4-05-00-01226 0.8531[20]	1.4788[20]
6298	10-Heneicosene, 11-phenyl-	$C_{27}H_{46}$ 370.66	6703-78-2 48.2	203[1.0] 190[0.5]	4-05-00-01506 0.8638[20]	1.4922[20]
6299	Hentriacontane Untriacontane	$C_{31}H_{64}$ 436.85	630-04-6 67.9	458	4-01-00-00594 0.781[68]	EtOH 2; eth 2; bz 2; chl 2 1.4278[90]
6300	16-Hentriacontanone	$C_{31}H_{82}O$ 450.83	502-73-8 83.8	250[0.1]	4-01-00-03413 0.7947[91]	H_2O 1; EtOH 2; eth 3; ace 2 1.4297[94]
6301	Heptachlor epoxide 2,5-Methano-2H-indeno[1,2-b]oxirene, 2,3,4,5,6,7,7-heptachloro-1a,1b,5,5a,6,6a-hexahydro-, (1aα,1bβ,2α,5α,5aβ,6β,6aα)-	$C_{10}H_5Cl_7O$ 389.32	1024-57-3 160			
6302	Heptacosane	$C_{27}H_{56}$ 380.74	593-49-7 59.5	442	4-01-00-00586 0.7796[60]	H_2O 1; EtOH 1; eth 2 1.4345[65]
6303	14-Heptacosanone	$C_{27}H_{54}O$ 394.73	542-50-7 77.5	238[0.1]	4-01-00-03411 0.7986[81]	
6304	Heptadecanal Margaric aldehyde	$C_{17}H_{34}O$ 254.46	629-90-3 36	204[26]	4-01-00-03395	bz 4; eth 4
6305	1-Heptadecanamine	$C_{17}H_{37}N$ 255.49	4200-95-7 49	336	4-04-00-00824 0.8510[20]	H_2O 1; EtOH 3; eth 3 1.4510[20]
6306	Heptadecane	$C_{17}H_{36}$ 240.47	629-78-7 22	302.0	4-01-00-00548 0.7780[20]	H_2O 1; EtOH 2; eth 3; ctc 2 1.4369[20]
6307	Heptadecane, 1-bromo- 1-Bromoheptadecane	$C_{17}H_{35}Br$ 319.37	3508-00-7 32	349	4-01-00-00549 0.9916[20]	H_2O 1; chl 4 1.4625[20]
6308	Heptadecane, 1,17-dibromo-	$C_{17}H_{34}Br_2$ 398.26	81726-82-1 38	208[3]	3-01-00-00564	chl 4
6309	Heptadecane, 9-hexyl- 9-Hexylheptadecane	$C_{23}H_{48}$ 324.63	55124-79-3 -19.4	213[10.] 157[0.5]	4-01-00-00578 0.7976[20]	1.4465[20]
6310	Heptadecane, 2-methyl- 2-Methylheptadecane	$C_{18}H_{38}$ 254.50	1560-89-0 5.7	311	4-01-00-00556 0.7838[15]	1.4394[15]
6311	Heptadecane, 9-methyl- 9-Methylheptadecane	$C_{18}H_{38}$ 254.50	18869-72-2	172-4[10]	3-01-00-00568 0.7810[20]	1.4388[20]
6312	Heptadecanenitrile	$C_{17}H_{33}N$ 251.46	5399-02-0 34	349	4-02-00-01195 0.8315[20]	H_2O 1; EtOH 2; eth 4; chl 2 1.4467[20]
6313	Heptadecane, 9-octyl-	$C_{25}H_{52}$ 352.69	7225-64-1 -13.8	231.5[10]	4-01-00-00583 0.8020[20]	1.4487[20]
6314	Heptadecanoic acid Margaric acid	$C_{17}H_{34}O_2$ 270.46	506-12-7 61.3	5634 227[100]	4-02-00-01193 0.8532[60]	EtOH 2; eth 3; ace 3; bz 3 1.4342[60]
6315	Heptadecanoic acid, ethyl ester	$C_{19}H_{38}O_2$ 298.51	14010-23-2 28	185[5]	4-02-00-01194 0.8517[30]	ace 4; bz 4; eth 4; EtOH 4
6316	Heptadecanoic acid, methyl ester	$C_{18}H_{36}O_2$ 284.48	1731-92-6 30	185[9]; 152[0.05]	4-02-00-01194	H_2O 1; EtOH 3; eth 4; ace 3
6317	1-Heptadecanol Heptadecyl alcohol	$C_{17}H_{36}O$ 256.47	1454-85-9 53.8	333	4-01-00-01884 0.8475[20]	H_2O 1; EtOH 3; eth 3
6318	2-Heptadecanol	$C_{17}H_{36}O$ 256.47	16813-18-6 54	155[1]	4-01-00-01885	eth 4; EtOH 4 1.4407[37]
6319	9-Heptadecanol	$C_{17}H_{36}O$ 256.47	624-08-8 61	174[9]	4-01-00-01886	H_2O 1; EtOH 3; eth 3; ace 3 1.4262[80]
6320	2-Heptadecanone Pentadecyl methyl ketone	$C_{17}H_{34}O$ 254.46	2922-51-2 48	320	4-01-00-03395 0.8049[48]	H_2O 1; EtOH 2; eth 4; ace 3
6321	9-Heptadecanone	$C_{17}H_{34}O$ 254.46	540-08-9 53	251.5; 142[1.5]	4-01-00-03396 0.8140[48]	EtOH 2; MeOH 3
6322	1-Heptadecene Hexahydroaplotaxene	$C_{17}H_{34}$ 238.46	6765-39-5 11.5	300	4-01-00-00929 0.7852[20]	H_2O 1; eth 4; bz 3; lig 5 1.4432[20]
6323	8-Heptadecene, 9-octyl-	$C_{25}H_{50}$ 350.67	24306-18-1	227[10]	4-01-00-00936 0.8086[20]	1.4554[20]
6324	9-Heptadecenoic acid	$C_{17}H_{32}O_2$ 268.44	10136-52-4 12.5	175[0.5]	4-02-00-01632 0.8942[20]	1.4598[20]
6325	2,4-Heptadienal, (E,E)-	$C_7H_{10}O$ 110.16	4313-03-5	84.5	4-01-00-03549 0.881[25]	1.5315[20]
6326	1,2-Heptadiene Butylallene	C_7H_{12} 96.17	2384-90-9	105.5	4-01-00-01027 0.7306[18]	1.432[18]
6327	1,4-Heptadiene	C_7H_{12} 96.17	5675-22-9	93	3-01-00-00999 0.7270[20]	bz 4; eth 4; peth 4 1.4370[20]
6328	1,5-Heptadiene 1-Methyldiallyl	C_7H_{12} 96.17	1541-23-7	94	3-01-00-00999 0.7186[20]	ace 4; bz 4; eth 4; EtOH 4 1.4200[20]

No.	Name Synonym	Mol. Form. Mol. Wt.	CAS RN mp/°C	Merck No. bp/°C	Beil. Ref. den/g cm⁻³	Solubility n_D
6329	2,4-Heptadiene	C_7H_{12} 96.17	628-72-8	108	4-01-00-01029 0.7384^{20}	H_2O 1; EtOH 3; eth 3; ace 3 1.4578^{20}
6330	1,3-Heptadiene, 2,6-dimethyl-	C_9H_{16} 124.23	2436-84-2	141	4-01-00-01051 0.7648^{10}	1.4606^{22}
6331	1,5-Heptadiene, 2,6-dimethyl-	C_9H_{16} 124.23	6709-39-3 -70	143	4-01-00-01051 0.7648^{25}	
6332	2,4-Heptadiene, 2,4-dimethyl-	C_9H_{16} 124.23	74421-05-9	138	4-01-00-01051 0.7750^4	1.4587^4
6333	2,4-Heptadiene, 3,5-dimethyl-	C_9H_{16} 124.23	101935-28-8	142	4-01-00-01052 0.7728^{20}	1.4487^{20}
6334	1,6-Heptadiene-3,5-dione, 1,7-bis(4-hydroxy-3-methoxyphenyl)-, (E,E)- Curcumin	$C_{21}H_{20}O_6$ 368.39	458-37-7 183	2681	4-08-00-03697	EtOH 4; HOAc 4
6335	1,5-Heptadiene, 3-methyl- 3-Methyl-1,5-heptadiene	C_8H_{14} 110.20	4894-62-6 -57	111.	4-01-00-01040 0.7250^{25}	
6336	1,5-Heptadien-4-ol	$C_7H_{12}O$ 112.17	5638-26-6	156; 68^{24}	4-01-00-02248 0.8598^{20}	ace 4 1.4510^{25}
6337	1,6-Heptadien-4-ol 4-Hydroxy-1,6-heptadiene	$C_7H_{12}O$ 112.17	2883-45-6	151	4-01-00-02249 0.864^{25}	1.4505^{20}
6338	1,6-Heptadien-4-ol, 4-methyl- 1,6-Heptadiene-4-ol, 4-methyl	$C_8H_{14}O$ 126.20	25201-40-5	158.4	4-01-00-02260 0.8611^{20}	ace 4; EtOH 4 1.4500^{23}
6339	3,5-Heptadien-2-one	$C_7H_{10}O$ 110.16	3916-64-1	88^{28}	4-01-00-03549 0.8946^{19}	eth 4 1.5177^{19}
6340	2,5-Heptadien-4-one, 2,6-dimethyl- Phorone	$C_9H_{14}O$ 138.21	504-20-1 28	7307 197.5	4-01-00-03564 0.8850^{20}	H_2O 2; EtOH 3; eth 3; ace 3 1.4998^{20}
6341	1,6-Heptadien-3-yne 1,6-Heptadiene-3-yne	C_7H_8 92.14	5150-80-1	110	3-01-00-01061 0.787^{25}	bz 3; peth 3 1.4694^{25}
6342	1,5-Heptadiyne	C_7H_8 92.14	764-56-7	26^{30}	2-01-00-00247 0.8100^{21}	bz 4; peth 4 1.4521^{21}
6343	1,6-Heptadiyne	C_7H_8 92.14	2396-63-6 -85	112	4-01-00-01121 0.8164^{17}	H_2O 1; bz 3; HOAc 3 1.451^{17}
6344	Heptanal Heptaldehyde	$C_7H_{14}O$ 114.19	111-71-7 -43.3	4578 152.8	4-01-00-03314 0.8132^{25}	H_2O 1; EtOH 5; eth 5; ctc 2 1.4113^{20}
6345	Heptanal, 5-methyl-	$C_8H_{16}O$ 128.21	75579-88-3	72^{25}	4-01-00-03345 0.8164^{25}	1.4144^{25}
6346	Heptanal, oxime Enanthaldoxime	$C_7H_{15}NO$ 129.20	629-31-2 57.5	195	4-01-00-03316 0.8583^{55}	H_2O 2; EtOH 3; eth 3 1.4210^{20}
6347	Heptanal, 2-(phenylmethylene)-	$C_{14}H_{18}O$ 202.30	122-40-7 80	$174-5^{20}$	2-07-00-00310 0.9711^{20}	H_2O 1; ace 3; ctc 3 1.5381^{20}
6348	Heptanamide	$C_7H_{15}NO$ 129.20	628-62-6 96	254	4-02-00-00963 0.8521^{110}	H_2O 3; EtOH 3; eth 3; chl 2 1.4217^{110}
6349	1-Heptanamine Heptylamine	$C_7H_{17}N$ 115.22	111-68-2 -18	156	4-04-00-00734 0.7754^{20}	H_2O 2; EtOH 5; eth 5; chl 2 1.4251^{20}
6350	2-Heptanamine Tuaminoheptane	$C_7H_{17}N$ 115.22	123-82-0	9712 142	3-04-00-00374 0.7665^{19}	H_2O 2; EtOH 3; eth 3; chl 2 1.4199^{19}
6351	4-Heptanamine	$C_7H_{17}N$ 115.22	16751-59-0	139.5	4-04-00-00745 0.767^{20}	1.4172^{20}
6352	1-Heptanamine, N-heptyl- Diheptylamine	$C_{14}H_{31}N$ 213.41	2470-68-0 31.5	271; 135^9	4-04-00-00736 0.7956^{21}	H_2O 2; EtOH 3; eth 4
6353	2-Heptanamine, 6-methyl-, (±)-	$C_8H_{19}N$ 129.25	5984-58-7	6678 155	4-04-00-00764 0.767^{25}	1.4209^{20}
6354	2-Heptanamine, 6-methyl-N-(3-methylbutyl)- Octamylamine	$C_{13}H_{29}N$ 199.38	502-59-0	6671 $100-1^7$	3-04-00-00387	
6355	Heptane	C_7H_{16} 100.20	142-82-5 -90.6	4580 98.5	4-01-00-00376 0.6837^{20}	H_2O 1; EtOH 4; eth 5; ace 5 1.3878^{20}
6356	Heptane, 1-bromo- Heptyl bromide	$C_7H_{15}Br$ 179.10	629-04-9 -58	179	4-01-00-00391 1.1400^{20}	H_2O 1; EtOH 4; eth 4; ctc 2 1.4502^{20}
6357	Heptane, 2-bromo- 2-Heptyl bromide	$C_7H_{15}Br$ 179.10	1974-04-5 47	166	3-01-00-00431 1.1277^{20}	H_2O 1; bz 4; ctc 3; chl 3 1.4503^{20}
6358	Heptane, 4-bromo- 4-Heptyl bromide	$C_7H_{15}Br$ 179.10	998-93-6	84.6^{72}; 60^{18}	4-01-00-00392 1.1351^{20}	H_2O 1; bz 3; ctc 3; chl 3 1.4495^{20}
6359	Heptane, 1-bromo-7-fluoro-	$C_7H_{14}BrF$ 197.09	334-42-9	85^{11}	4-01-00-00392	eth 4; EtOH 4 1.4463^{20}
6360	Heptane, 1-chloro- Heptyl chloride	$C_7H_{15}Cl$ 134.65	629-06-1 -69.5	159	4-01-00-00389 0.8758^{20}	H_2O 1; EtOH 5; eth 5; ctc 2 1.4256^{20}
6361	Heptane, 2-chloro- 2-Chloroheptane	$C_7H_{15}Cl$ 134.65	1001-89-4	46^{19}	4-01-00-00390 0.8672^{20}	H_2O 1; eth 4; bz 3; chl 3 1.4221^{20}
6362	Heptane, 3-chloro- 3-Chloroheptane	$C_7H_{15}Cl$ 134.65	999-52-0	144; 48.3^{20}	4-01-00-00390 0.8960^{20}	bz 4; eth 4 1.4228^{20}
6363	Heptane, 4-chloro- 4-Chloroheptane	$C_7H_{15}Cl$ 134.65	998-95-8	144	4-01-00-00390 0.8710^{20}	bz 4; eth 4 1.4237^{20}
6364	Heptane, 3-chloro-3-ethyl-	$C_9H_{19}Cl$ 162.70	28320-89-0	46^3	3-01-00-00510 0.8856^{20}	EtOH 4; chl 4 1.4400^{20}
6365	Heptane, 1-chloro-7-fluoro-	$C_7H_{14}ClF$ 152.64	334-43-0	70^{10}	4-01-00-00390 0.993^{20}	bz 4; eth 4; EtOH 4; chl 4 1.4222^{25}
6366	Heptane, 2-chloro-2-methyl- 2-Chloro-2-methylheptane	$C_8H_{17}Cl$ 148.68	4325-49-9	50^{15}	4-01-00-00428 0.8568^{25}	bz 4; eth 4; EtOH 4; chl 4 1.4240^{25}
6367	Heptane, 2-chloro-6-methyl- 2-Chloro-6-methylheptane	$C_8H_{17}Cl$ 148.68	2350-19-8	74^{35}	3-01-00-00472	bz 4; eth 4; EtOH 4; chl 4 1.4260^{15}
6368	Heptane, 3-(chloromethyl)-	$C_8H_{17}Cl$ 148.68	123-04-6	172	4-01-00-00430 0.8769^{20}	H_2O 1; EtOH 3; eth 3; ace 3 1.4319^{20}

No.	Name / Synonym	Mol. Form. / Mol. Wt.	CAS RN / mp/°C	Merck No. / bp/°C	Beil. Ref. / den/g cm⁻³	Solubility / n_D
6369	Heptane, 3-chloro-3-methyl- / 3-Methyl-3-chloroheptane	$C_8H_{17}Cl$ / 148.68	5272-02-6	64[27]	4-01-00-00430 / 0.8764[20]	bz 4; eth 4; EtOH 4; chl 4 / 1.4317[20]
6370	Heptane, 4-chloro-4-methyl- / 2-Chloro-2-propylpentane	$C_8H_{17}Cl$ / 148.68	61764-94-1	50[12]	3-01-00-00477 / 0.8690[20]	bz 4; eth 4; EtOH 4; chl 4 / 1.4310[15]
6371	1,7-Heptanediamine	$C_7H_{18}N_2$ / 130.23	646-19-5 / 28.9	224	4-04-00-01354	EtOH 3; eth 3; ace 3
6372	Heptane, 1,2-dibromo- / 1,2-Dibromoheptane	$C_7H_{14}Br_2$ / 258.00	42474-21-5	228	3-01-00-00432 / 1.5086[20]	1.4986[20]
6373	Heptane, 1,5-dibromo- / 1,5-Dibromoheptane	$C_7H_{14}Br_2$ / 258.00	1622-10-2	113-5[11]	4-01-00-00393 / 1.536[15]	1.5041[15]
6374	Heptane, 1,7-dibromo- / Heptamethylene dibromide	$C_7H_{14}Br_2$ / 258.00	4549-31-9 / 41.7	263	4-01-00-00393 / 1.5306[20]	H_2O 1; eth 3; ace 3; bz 3 / 1.5034[20]
6375	Heptane, 2,3-dibromo- / 2,3-Dibromoheptane	$C_7H_{14}Br_2$ / 258.00	21266-88-6	101[17]	3-01-00-00432 / 1.5139[20]	1.4992[20]
6376	Heptane, 3,4-dibromo- / 3,4-Dibromoheptane	$C_7H_{14}Br_2$ / 258.00	21266-90-0	107[24]	3-01-00-00433 / 1.5182[20]	1.5010[20]
6377	Heptane, 1,1-dichloro- / 1,1-Dichlorohexane	$C_7H_{14}Cl_2$ / 169.09	821-25-0	191	4-01-00-00390 / 1.011[20]	H_2O 1; eth 3; bz 3; ctc 3 / 1.4440[20]
6378	Heptane, 1,2-dichloro- / 1,2-Dichloroheptane	$C_7H_{14}Cl_2$ / 169.09	10575-87-8	68-72[7]	3-01-00-00430 / 1.064[20]	bz 4; eth 4; chl 4 / 1.4490[20]
6379	Heptane, 1,7-dichloro- / 1,7-Dichloroheptane	$C_7H_{14}Cl_2$ / 169.09	821-76-1	124-5[35]	4-01-00-00390 / 1.0408[25]	1.4565[25]
6380	Heptane, 2,2-dichloro- / 2,2-Dichloroheptane	$C_7H_{14}Cl_2$ / 169.09	65786-09-6	77[25]	3-01-00-00430 / 1.012[20]	bz 4; eth 4; chl 4 / 1.4440[20]
6381	Heptane, 4,4-dichloro-	$C_7H_{14}Cl_2$ / 169.09	89796-76-9	86[27]	2-01-00-00117 / 1.008[17]	eth 4; chl 4 / 1.448[17]
6382	Heptane, 2,6-dichloro-2,6-dimethyl-	$C_9H_{18}Cl_2$ / 197.15	35951-36-1 / 43	93[16]	3-01-00-00513	bz 4; chl 4
6383	Heptane, 2,2-dimethyl- / 2,2-Dimethylheptane	C_9H_{20} / 128.26	1071-26-7 / -113	132.7	4-01-00-00457 / 0.7105[20]	H_2O 1; eth 3; ace 4; bz 5 / 1.4016[20]
6384	Heptane, 2,3-dimethyl- / 2,3-Dimethylheptane	C_9H_{20} / 128.26	3074-71-3 / -116	140.5	4-01-00-00457 / 0.7260[20]	H_2O 1; EtOH 5; eth 5; ace 5 / 1.4088[20]
6385	Heptane, 2,4-dimethyl- / 2,4-Dimethylheptane	C_9H_{20} / 128.26	2213-23-2	132.9	4-01-00-00457 / 0.7115[25]	H_2O 1; EtOH 5; eth 5; ace 5 / 1.4034[20]
6386	Heptane, 2,5-dimethyl- / 3,6-Dimethylheptane	C_9H_{20} / 128.26	2216-30-0	136	4-01-00-00457 / 0.7198[20]	ace 4; bz 4; eth 4; EtOH 4 / 1.4033[20]
6387	Heptane, 2,6-dimethyl- / 2,6-Dimethylheptane	C_9H_{20} / 128.26	1072-05-5 / -102.9	135.2	4-01-00-00458 / 0.7089[20]	chl 2 / 1.4011[20]
6388	Heptane, 3,3-dimethyl- / 3,3-Dimethylheptane	C_9H_{20} / 128.26	4032-86-4	137.3	4-01-00-00458 / 0.7254[20]	H_2O 1; EtOH 5; eth 3; ace 4 / 1.4087[20]
6389	Heptane, 3,4-dimethyl- / 3,4-Dimethylheptane	C_9H_{20} / 128.26	922-28-1	140.6	3-01-00-00514 / 0.7314[20]	H_2O 1; eth 3; ace 4; bz 5 / 1.4108[20]
6390	Heptane, 3,5-dimethyl- / 3,5-Dimethylheptane	C_9H_{20} / 128.26	926-82-9	136	4-01-00-00458 / 0.7225[20]	H_2O 1; eth 3; ace 4; bz 5 / 1.4083[20]
6391	Heptane, 4,4-dimethyl- / 4,4-Dimethylheptane	C_9H_{20} / 128.26	1068-19-5	135.2	4-01-00-00458 / 0.7221[20]	H_2O 1; eth 3; ace 4; bz 5 / 1.4076[20]
6392	Heptanedinitrile	$C_7H_{10}N_2$ / 122.17	646-20-8 / -31.4	155[14]	4-02-00-02006 / 0.949[18]	H_2O 1; EtOH 5; eth 5; chl 5 / 1.4472[20]
6393	Heptanedioic acid / Pimelic acid	$C_7H_{12}O_4$ / 160.17	111-16-0 / 106	7401 / 272[100]; 212[10]	4-02-00-02003 / 1.329[15]	H_2O 3; EtOH 3; eth 3; bz 1
6394	Heptanedioic acid, diethyl ester / Diethyl pimelate	$C_{11}H_{20}O_4$ / 216.28	2050-20-6 / -24	254; 140[15]	4-02-00-02004 / 0.9945[20]	eth 4; EtOH 4 / 1.4305[20]
6395	Heptanedioic acid, dimethyl ester	$C_9H_{16}O_4$ / 188.22	1732-08-7 / -21	120[10]; 80[1]	4-02-00-02004 / 1.0625[20]	H_2O 2; EtOH 3; eth 3; bz 3 / 1.4309[20]
6396	Heptanedioic acid, monoethyl ester	$C_9H_{16}O_4$ / 188.22	33018-91-6 / 10	182[18]; 160[4]	3-02-00-01742	1.4415[20]
6397	1,4-Heptanediol	$C_7H_{16}O_2$ / 132.20	40646-07-9	243; 127[4]	4-01-00-02579 / 0.9542[20]	1.4510[25]
6398	1,5-Heptanediol	$C_7H_{16}O_2$ / 132.20	60096-09-5	135-6[11]	3-01-00-02213 / 0.9705[20]	eth 4; EtOH 4; chl 4 / 1.4571[22]
6399	1,7-Heptanediol	$C_7H_{16}O_2$ / 132.20	629-30-1 / 22.5	262	4-01-00-02580 / 0.9569[25]	eth 4; EtOH 4 / 1.4520[25]
6400	2,4-Heptanediol	$C_7H_{16}O_2$ / 132.20	20748-86-1	107-8[8]	4-01-00-02580 / 0.926[20]	1.4386[25]
6401	2,4-Heptanediol, 3-methyl- / 3-Methylheptan-2,4-diol	$C_8H_{18}O_2$ / 146.23	6964-04-1	115[3]	0-01-00-00491 / 0.928[20]	H_2O 1; EtOH 3; ctc 2 / 1.4459[20]
6402	2,3-Heptanedione / Acetyl valeryl	$C_7H_{12}O_2$ / 128.17	96-04-8	46[13]	4-01-00-03697 / 0.919[18]	1.4150[18]
6403	2,6-Heptanedione	$C_7H_{12}O_2$ / 128.17	13505-34-5 / 33.5	222; 97[11]	4-01-00-03699 / 0.9399[37]	bz 4; eth 4; EtOH 4 / 1.4277[37]
6404	3,5-Heptanedione, 2,2,6,6-tetramethyl- / Dipivaloylmethane	$C_{11}H_{20}O_2$ / 184.28	1118-71-4	72-3[6]	4-01-00-03729 / 0.883[25]	ctc 2 / 1.4589[20]
6405	Heptane, 3-[(ethenyloxy)methyl]-	$C_{10}H_{20}O$ / 156.27	103-44-6 / -100	76[32]	4-01-00-02056 / 0.8108[20]	1.4247[25]
6406	Heptane, 1-ethoxy-	$C_9H_{20}O$ / 144.26	1969-43-3 / -68.3	166	4-01-00-01733 / 0.790[16]	
6407	Heptane, 3-ethyl- / 3-Ethylheptane	C_9H_{20} / 128.26	15869-80-4 / -114.9	143.0	4-01-00-00456 / 0.7225[20]	1.4093[20]
6408	Heptane, 4-ethyl- / 4-Ethylheptane	C_9H_{20} / 128.26	2216-32-2	141.2	4-01-00-00457 / 0.7241[25]	H_2O 1; EtOH 5; eth 3; ace 5 / 1.4096[20]

No.	Name / Synonym	Mol. Form. / Mol. Wt.	CAS RN / mp/°C	Merck No. / bp/°C	Beil. Ref. / den/g cm⁻³	Solubility / n_D
6409	Heptane, 3-ethyl-2-methyl- / Heptane, 2-methyl-3-ethyl	$C_{10}H_{22}$ / 142.28	14676-29-0 /	/ 163	4-01-00-00480 / 0.7398^{25}	/ 1.4174^{20}
6410	Heptane, 3-ethyl-3-methyl-	$C_{10}H_{22}$ / 142.28	17302-01-1 /	/ 163.8	4-01-00-00481 / 0.7463^{25}	/ 1.4208^{20}
6411	Heptane, 3-ethyl-4-methyl- / 3-Ethyl-4-methylheptane	$C_{10}H_{22}$ / 142.28	52896-91-0 /	/ 165	4-01-00-00481 / 0.7468^{25}	/ 1.4207^{20}
6412	Heptane, 3-ethyl-5-methyl- / 3-Ethyl-5-methylheptane	$C_{10}H_{22}$ / 142.28	52896-90-9 /	/ 158.2	4-01-00-00481 / 0.7368^{25}	/ 1.4164^{20}
6413	Heptane, 4-ethyl-2-methyl- / 4-Ethyl-2-methylheptane	$C_{10}H_{22}$ / 142.28	52896-88-5 /	/ 158	4-01-00-00481 / 0.7322^{25}	/ 1.4137^{20}
6414	Heptane, 4-ethyl-3-methyl- / 4-Ethyl-3-methylheptane	$C_{10}H_{22}$ / 142.28	52896-89-6 /	/ 165	4-01-00-00481 / 0.7466^{25}	/ 1.4206^{20}
6415	Heptane, 4-ethyl-4-methyl- / 4-Ethyl-4-methylheptane	$C_{10}H_{22}$ / 142.28	17302-04-4 /	/ 160.8	4-01-00-00481 / 0.7472^{25}	/ 1.4210^{25}
6416	Heptane, 5-ethyl-2-methyl-	$C_{10}H_{22}$ / 142.28	13475-78-0 /	/ 159.7	4-01-00-00481 / 0.7318^{25}	/ 1.4134^{20}
6417	Heptane, 1-fluoro-	$C_7H_{15}F$ / 118.19	661-11-0 / -73	/ 117.9	4-01-00-00387 / 0.8062^{20}	H_2O 1; eth 3; ace 3; bz 3 / 1.3854^{20}
6418	Heptane, hexadecafluoro- / Perfluoroheptane	C_7F_{16} / 388.05	335-57-9 / -78	/ 82.5	4-01-00-00388 / 1.7333^{20}	ace 4; eth 4; EtOH 4; chl 4 / 1.2618^{20}
6419	Heptane, 1-iodo-	$C_7H_{15}I$ / 226.10	4282-40-0 / -48.2	/ 204	4-01-00-00393 / 1.3791^{20}	H_2O 1; EtOH 3; eth 3; ace 3 / 1.4904^{20}
6420	Heptane, 2-iodo- / 2-Iodoheptane	$C_7H_{15}I$ / 226.10	18589-29-2 /	/ 98^{50}	4-01-00-00393 / 1.304^{20}	H_2O 1; ace 3; bz 3; ctc 3 / 1.4826
6421	Heptane, 1-methoxy-	$C_8H_{18}O$ / 130.23	629-32-3 /	/ 151	3-01-00-01682 / 0.7862^{15}	ace 4; eth 4; EtOH 4 / 1.4073^{20}
6422	Heptane, 2-methyl- / 2-Methylheptane	C_8H_{18} / 114.23	592-27-8 / -108.9	/ 117.6	4-01-00-00428 / 0.6980^{20}	H_2O 1; EtOH 5; eth 3; ace 5 / 1.3949^{20}
6423	Heptane, 3-methyl-, (S)-	C_8H_{18} / 114.23	6131-25-5 / -120	/ 118.9	4-01-00-00429 / 0.7051^{16}	H_2O 1; EtOH 3; eth 3; ace 5 / 1.4002^{18}
6424	Heptane, 4-methyl- / 4-Methylheptane	C_8H_{18} / 114.23	589-53-7 / -121	/ 117.7	4-01-00-00431 / 0.7046^{20}	H_2O 1; EtOH 5; eth 3; ace 5 / 1.3979^{20}
6425	Heptane, 3-methylene-	C_8H_{16} / 112.22	1632-16-2 /	/ 120	4-01-00-00884 / 0.7270^{20}	H_2O 1; eth 4; bz 4; peth 4 / 1.4157^{20}
6426	Heptane, 4-methylene- / 2-Propyl-1-pentene	C_8H_{16} / 112.22	15918-08-8 /	/ 117.7	4-01-00-00885 / 0.7198^{25}	/ 1.4136^{20}
6427	Heptane, 4-(1-methylethyl)- / 4-Isopropylheptane	$C_{10}H_{22}$ / 142.28	52896-87-4 /	/ 158.9	4-01-00-00481 / 0.7354^{25}	/ 1.4153^{20}
6428	Heptanenitrile	$C_7H_{13}N$ / 111.19	629-08-3 /	/ 183; $70-2^{10}$	4-02-00-00963 / 0.8106^{20}	H_2O 1; eth 3; ace 3; bz 3 / 1.4104^{30}
6429	Heptane, 1-nitro- / 1-Nitroheptane	$C_7H_{15}NO_2$ / 145.20	693-39-0 /	/ 194	4-01-00-00395 / 0.9476^{17}	eth 4; EtOH 4
6430	Heptane, 1,1'-oxybis-	$C_{14}H_{30}O$ / 214.39	629-64-1 /	/ 258.5	4-01-00-01733 / 0.8008^{20}	eth 4; EtOH 4 / 1.4275^{20}
6431	Heptane, 3,3'-[oxybis(methylene)]bis- / Ether, dihexyl,2,2'-diethyl	$C_{16}H_{34}O$ / 242.45	10143-60-9 /	/ 269.4	4-01-00-01785 /	ctc 2 / 1.4325^{20}
6432	Heptane, 1,1,1,2,2,3,3,4,4,5,5,6,6,7,7-pentadecafluoro- / 1H-Pentadecafluoroheptane	C_7HF_{15} / 370.06	375-83-7 /	/ 96.0	4-01-00-00388 / 1.725^{25}	/ 1.2690^{25}
6433	Heptane, 2,2,4,6,6-pentamethyl- / 2,2,4,6,6-Pentamethylheptane	$C_{12}H_{26}$ / 170.34	13475-82-6 / -67	/ 177.8	4-01-00-00510 / 0.7463^{20}	/ 1.4440^{20}
6434	Heptane, 4-propyl- / 4-Propylheptane	$C_{10}H_{22}$ / 142.28	3178-29-8 /	/ 157.5	4-01-00-00480 / 0.7321^{25}	/ 1.4135^{20}
6435	Heptane, 1,1'-thiobis-	$C_{14}H_{30}S$ / 230.46	629-65-2 / 70	/ 298	4-01-00-01739 / 0.8416^{20}	H_2O 1; eth 3 / 1.4606^{20}
6436	1-Heptanethiol / Heptyl mercaptan	$C_7H_{16}S$ / 132.27	1639-09-4 / -43	/ 177	4-01-00-01738 / 0.8427^{20}	H_2O 1; EtOH 5; eth 5; chl 3 / 1.4521^{20}
6437	Heptane, 2,2,3-trimethyl- / 2,2,3-Trimethylheptane	$C_{10}H_{22}$ / 142.28	52896-92-1 /	/ 157.8	4-01-00-00481 / 0.7385^{25}	/ 1.4168^{20}
6438	Heptane, 2,2,4-trimethyl- / 2,2,4-Trimethylheptane	$C_{10}H_{22}$ / 142.28	14720-74-2 /	/ 148.3	4-01-00-00481 / 0.7237^{25}	H_2O 1; bz 3; ctc 3; chl 3 / 1.4092^{20}
6439	Heptane, 2,2,5-trimethyl- / 2,2,5-Trimethylheptane	$C_{10}H_{22}$ / 142.28	20291-95-6 /	/ 150.8	4-01-00-00482 / 0.7243^{25}	/ 1.4101^{20}
6440	Heptane, 2,2,6-trimethyl- / 2,2,6-Trimethylheptane	$C_{10}H_{22}$ / 142.28	1190-83-6 / -105	/ 148.9	4-01-00-00482 / 0.7200^{25}	/ 1.4078^{20}
6441	Heptane, 2,3,3-trimethyl- / 2,3,3-Trimethylheptane	$C_{10}H_{22}$ / 142.28	52896-93-2 /	/ 160.2	4-01-00-00482 / 0.7450^{25}	/ 1.4202^{20}
6442	Heptane, 2,3,4-trimethyl- / 2,3,4-Trimethylheptane	$C_{10}H_{22}$ / 142.28	52896-95-4 /	/ 161	4-01-00-00482 / 0.7447^{25}	/ 1.4195^{20}
6443	Heptane, 2,3,5-trimethyl- / 2,3,5-Trimethylheptane	$C_{10}H_{22}$ / 142.28	20278-85-7 /	/ 160.7	4-01-00-00482 / 0.7413^{25}	/ 1.4169^{20}
6444	Heptane, 2,3,6-trimethyl- / 2,3,6-Trimethylheptane	$C_{10}H_{22}$ / 142.28	4032-93-3 /	/ 156.0	4-01-00-00482 / 0.7305^{25}	/ 1.4131^{20}
6445	Heptane, 2,4,4-trimethyl- / 2,4,4-Trimethylheptane	$C_{10}H_{22}$ / 142.28	4032-92-2 /	/ 151.0	4-01-00-00482 / 0.7308^{25}	/ 1.4142^{20}
6446	Heptane, 2,4,5-trimethyl- / 2,4,5-Trimethylheptane	$C_{10}H_{22}$ / 142.28	20278-84-6 /	/ 156.5	4-01-00-00483 / 0.7373^{25}	/ 1.4160^{20}
6447	Heptane, 2,4,6-trimethyl- / 2,4,6-Trimethylheptane	$C_{10}H_{22}$ / 142.28	2613-61-8 /	/ 147.6	4-01-00-00483 / 0.7190^{25}	/ 1.4071^{20}
6448	Heptane, 2,5,5-trimethyl- / 2,5,5-Trimethylheptane	$C_{10}H_{22}$ / 142.28	1189-99-7 /	/ 152.8	4-01-00-00483 / 0.7362^{25}	/ 1.4149^{20}

No.	Name / Synonym	Mol. Form. / Mol. Wt.	CAS RN / mp/°C	Merck No. / bp/°C	Beil. Ref. / den/g cm⁻³	Solubility / n_D
6449	Heptane, 3,3,4-trimethyl- / 3,3,4-Trimethylheptane	$C_{10}H_{22}$ / 142.28	20278-87-9	/ 161.9	4-01-00-00483 / 0.7527[25]	/ 1.4236[20]
6450	Heptane, 3,3,5-trimethyl- / 3,3,5-Trimethylheptane	$C_{10}H_{22}$ / 142.28	7154-80-5	/ 155.7	4-01-00-00483 / 0.7248[20]	H_2O 1; bz 3; ctc 3; chl 3 / 1.4170[20]
6451	Heptane, 3,4,4-trimethyl- / 3,4,4-Trimethylheptane	$C_{10}H_{22}$ / 142.28	20278-88-0	/ 161.1	4-01-00-00484 / 0.7535[25]	/ 1.4235[20]
6452	Heptane, 3,4,5-trimethyl- / 3,4,5-Trimethylheptane	$C_{10}H_{22}$ / 142.28	20278-89-1	/ 162.5	4-01-00-00484 / 0.7519[25]	/ 1.4229[20]
6453	1,4,7-Heptanetriol	$C_7H_{16}O_3$ / 148.20	3920-53-4 / -35	231[25]; 146[1]	4-01-00-02787 / 1.075[18]	H_2O 3; EtOH 3; ace 3 / 1.4725[20]
6454	2,4,6-Heptanetrione	$C_7H_{10}O_3$ / 142.15	626-53-9 / 49	121[10]	4-01-00-03783 / 1.0599[40]	H_2O 4; eth 4; EtOH 4 / 1.4930[20]
6455	Heptanoic acid / Enanthylic acid	$C_7H_{14}O_2$ / 130.19	111-14-8 / -7.5	4581 / 222.2	4-02-00-00958 / 0.9181[20]	H_2O 2; EtOH 3; eth 3; ace 3 / 1.4170[20]
6456	Heptanoic acid, anhydride	$C_{14}H_{26}O_3$ / 242.36	626-27-7 / -12.4	/ 269.5	4-02-00-00962 / 0.9321[20]	H_2O 1; EtOH 3; eth 3 / 1.4335[15]
6457	Heptanoic acid, 2-bromo- / 2-Bromoheptanoic acid	$C_7H_{13}BrO_2$ / 209.08	2624-01-3	250 dec; 147[12]	4-02-00-00967 / 1.319[15]	ace 4; eth 4 / 1.471[18]
6458	Heptanoic acid, 7-bromo-	$C_7H_{13}BrO_2$ / 209.08	30515-28-7 / 31	280; 165[12]	4-02-00-00968	ace 4; bz 4; eth 4; EtOH 4
6459	Heptanoic acid, butyl ester / Butyl enanthate	$C_{11}H_{22}O_2$ / 186.29	5454-28-4 / -67.5	/ 226.2	3-02-00-00768 / 0.8638[20]	ace 4; bz 4; eth 4; EtOH 4 / 1.4204[20]
6460	Heptanoic acid, 2,2-dimethyl-6-oxo- / Geronic acid	$C_9H_{16}O_3$ / 172.22	461-11-0	278; 169[12]	3-03-00-01264 / 1.0211[20]	eth 4; EtOH 4 / 1.4488[20]
6461	Heptanoic acid, 4,4-dimethyl-6-oxo- / Isogeronic acid	$C_9H_{16}O_3$ / 172.22	471-04-5 / 50	178[20]; 162[10]	3-03-00-01266	H_2O 4; bz 4; eth 4; EtOH 4
6462	Heptanoic acid, ethyl ester / Ethyl oenanthate	$C_9H_{18}O_2$ / 158.24	106-30-9 / -66.1	3790 / 187	4-02-00-00960 / 0.8817[20]	H_2O 2; EtOH 3; eth 3; ctc 2 / 1.4100[20]
6463	Heptanoic acid, 7-fluoro-	$C_7H_{13}FO_2$ / 148.18	334-28-1	133[10]	4-02-00-00964 / 1.039[20]	/ 1.4207[25]
6464	Heptanoic acid, heptyl ester / Heptyl heptanoate	$C_{14}H_{28}O_2$ / 228.38	624-09-9 / -33	/ 277	4-02-00-00961 / 0.8649[20]	H_2O 1; EtOH 3; eth 3 / 1.4320[20]
6465	Heptanoic acid, hexyl ester / Hexyl heptanoate	$C_{13}H_{26}O_2$ / 214.35	1119-06-8 / -48	/ 261	3-02-00-00768 / 0.8611[20]	ace 4; bz 4; eth 4; EtOH 4 / 1.429[15]
6466	Heptanoic acid, methyl ester / Methyl heptanoate	$C_8H_{16}O_2$ / 144.21	106-73-0 / -56	4581 / 174	4-02-00-00960 / 0.8815[20]	H_2O 2; EtOH 3; eth 3; ace 2 / 1.4152[20]
6467	Heptanoic acid, 2-methylpropyl ester	$C_{11}H_{22}O_2$ / 186.29	7779-80-8	/ 208	1-02-00-00145 / 0.8593[20]	ace 4; bz 4; eth 4; EtOH 4
6468	Heptanoic acid, octyl ester / Octyl enanthate	$C_{15}H_{30}O_2$ / 242.40	5132-75-2 / -22.5	/ 290	3-02-00-00768 / 0.8596[20]	ace 4; bz 4; eth 4; EtOH 4 / 1.4349[15]
6469	Heptanoic acid, 6-oxo-	$C_7H_{12}O_3$ / 144.17	3128-07-2 / 40.2	251[280]; 135[1]	4-03-00-01598	H_2O 4; ace 4; eth 4; EtOH 4 / 1.4306[25]
6470	Heptanoic acid, pentyl ester / Amyl heptanoate	$C_{12}H_{24}O_2$ / 200.32	7493-82-5 / -50	/ 245.4	4-02-00-00960 / 0.8623[20]	ace 4; bz 4; eth 4; EtOH 4 / 1.4263[15]
6471	Heptanoic acid, propyl ester / Propyl heptanoate	$C_{10}H_{20}O_2$ / 172.27	7778-87-2 / -63.5	/ 208	3-02-00-00767 / 0.8641[15]	ace 4; bz 4; eth 4; EtOH 4 / 1.4183[15]
6472	1-Heptanol / Heptyl alcohol	$C_7H_{16}O$ / 116.20	111-70-6 / -34	4582 / 176.4	4-01-00-01731 / 0.8219[20]	H_2O 2; EtOH 5; eth 5; ctc 2 / 1.4249[20]
6473	2-Heptanol, (±)-	$C_7H_{16}O$ / 116.20	52390-72-4	4583 / 159	4-01-00-01740 / 0.8167[20]	H_2O 2; EtOH 3; eth 3; ctc 2 / 1.4210[20]
6474	3-Heptanol, (S)-	$C_7H_{16}O$ / 116.20	26549-25-7 / -70	157; 66[18]	2-01-00-00444 / 0.8227[20]	H_2O 2; EtOH 3; eth 3; ctc 2 / 1.4201[20]
6475	4-Heptanol / Dipropylcarbinol	$C_7H_{16}O$ / 116.20	589-55-9 / -41.2	/ 156	4-01-00-01743 / 0.8183[20]	H_2O 1; EtOH 3; eth 3 / 1.4205[20]
6476	4-Heptanol, acetate	$C_9H_{18}O_2$ / 158.24	5921-84-6	/ 171	4-02-00-00163 / 0.8741[0]	eth 4 / 1.4105[19]
6477	1-Heptanol, 7-chloro- / Heptamethylene chlorohydrin	$C_7H_{15}ClO$ / 150.65	55944-70-2 / 11	150[20]	4-01-00-01738 / 0.9998[15]	EtOH 4; peth 4 / 1.4537[25]
6478	2-Heptanol, 1-chloro- / 1-Chloro-2-heptanol	$C_7H_{15}ClO$ / 150.65	53660-21-2	93[13]	4-01-00-01741 / 0.9885[20]	EtOH 3; eth 3; ace 3 / 1.4499[20]
6479	2-Heptanol, 2,6-dimethyl- / 2,6-Dimethyl-2-heptanol	$C_9H_{20}O$ / 144.26	13254-34-7	/ 173	2-01-00-00457 / 0.8186[20]	/ 1.4242[20]
6480	3-Heptanol, 2,3-dimethyl- / Butylisopropylmethylcarbinol	$C_9H_{20}O$ / 144.26	19549-71-4	173; 75[16]	3-01-00-01752 / 0.8395[20]	/ 1.4355[20]
6481	3-Heptanol, 2,6-dimethyl- / 2,6-Dimethyl-3-heptanol	$C_9H_{20}O$ / 144.26	19549-73-6	/ 175	3-01-00-01753 / 0.8212[20]	/ 1.4246[20]
6482	3-Heptanol, 3,5-dimethyl-	$C_9H_{20}O$ / 144.26	19549-74-7	95-7[50]	4-01-00-01811 / 0.8177[25]	/ 1.4251[20]
6483	3-Heptanol, 3,6-dimethyl- / 3,6-Dimethylheptan-3-ol	$C_9H_{20}O$ / 144.26	1573-28-0	/ 173	4-01-00-01810 / 0.8285[16]	eth 4; EtOH 4; lig 4 / 1.4326[16]
6484	4-Heptanol, 2,4-dimethyl- / 2,4-Dimethyl-4-heptanol	$C_9H_{20}O$ / 144.26	19549-77-0	171; 70[26]	4-01-00-01810 / 0.8215[20]	/ 1.4292[20]
6485	4-Heptanol, 2,6-dimethyl- / Diisobutylcarbinol	$C_9H_{20}O$ / 144.26	108-82-7	/ 174.5	4-01-00-01810 / 0.8114[20]	H_2O 1; EtOH 3; eth 3; ctc 2 / 1.4242[20]
6486	4-Heptanol, 3,5-dimethyl- / 3,5-Dimethyl-4-heptanol	$C_9H_{20}O$ / 144.26	19549-79-2	/ 186	4-01-00-01811 / 0.836[18]	/ 1.4283[20]
6487	4-Heptanol, 4-ethyl-	$C_9H_{20}O$ / 144.26	597-90-0	/ 182	4-01-00-01809 / 0.8350[20]	eth 4; EtOH 4 / 1.4332[20]

No.	Name / Synonym	Mol. Form. / Mol. Wt.	CAS RN / mp/°C	Merck No. / bp/°C	Beil. Ref. / den/g cm⁻³	Solubility / n_D
6488	3-Heptanol, 3-ethyl-2-methyl-	$C_{10}H_{22}O$ 158.28	66719-37-7	193	1-01-00-00214 0.8455[20]	1.4378
6489	3-Heptanol, 3-ethyl-6-methyl-	$C_{10}H_{22}O$ 158.28	66719-40-2 83-6[15]		0-01-00-00427 0.852[25]	1.4409
6490	3-Heptanol, 5-ethyl-4-methyl-	$C_{10}H_{22}O$ 158.28	66731-94-0 86-8[14]		3-01-00-01770 0.865[17]	1.4565[17]
6491	1-Heptanol, 7-fluoro-	$C_7H_{15}FO$ 134.19	408-16-2 98-9[12]		4-01-00-01736 0.956[20]	eth 4; EtOH 4 1.4197[25]
6492	1-Heptanol, 3-methyl- 3-Methyl-1-heptanol	$C_8H_{18}O$ 130.23	1070-32-2 101[20]		4-01-00-01782 0.824[24]	1.4295[25]
6493	1-Heptanol, 4-methyl- 4-Methyl-1-heptanol	$C_8H_{18}O$ 130.23	817-91-4 188		4-01-00-01789 0.8065[25]	EtOH 4 1.4253[25]
6494	1-Heptanol, 6-methyl- 6-Methyl-1-heptanol	$C_8H_{18}O$ 130.23	1653-40-3 -106	188; 95.8[20]	4-01-00-01782 0.8176[25]	H_2O 1; EtOH 3; eth 3 1.4251[25]
6495	2-Heptanol, 2-methyl- 2-Methyl-2-heptanol	$C_8H_{18}O$ 130.23	625-25-2 -50.4	156	4-01-00-01780 0.8142[20]	H_2O 1; EtOH 3; eth 3 1.4250[20]
6496	2-Heptanol, 3-methyl- 3-Methyl-2-heptanol	$C_8H_{18}O$ 130.23	31367-46-1 166.1		0.8177[25]	H_2O 1; EtOH 3; eth 3; ctc 3 1.4199[25]
6497	2-Heptanol, 5-methyl- 5-Methyl-2-heptanol	$C_8H_{18}O$ 130.23	54630-50-1 -61	170	4-01-00-01783 0.8174[21]	
6498	2-Heptanol, 6-methyl-	$C_8H_{18}O$ 130.23	4730-22-7 -105	174	3-01-00-01728 0.8218[20]	1.4238[10]
6499	3-Heptanol, 2-methyl-, (±)-	$C_8H_{18}O$ 130.23	100296-26-2 -85	167.5	4-01-00-01781 0.8235[20]	H_2O 2; EtOH 3; eth 3; ctc 3 1.4265[20]
6500	3-Heptanol, 3-methyl- 2-Ethyl-2-hexanol	$C_8H_{18}O$ 130.23	5582-82-1 -83	163	4-01-00-01783 0.8282[20]	H_2O 1; EtOH 3; eth 3; ctc 3 1.4279[20]
6501	3-Heptanol, 4-methyl- 4-Methyl-3-heptanol	$C_8H_{18}O$ 130.23	14979-39-6 -123	170	4-01-00-01789 0.827[25]	1.4300[20]
6502	3-Heptanol, 5-methyl-	$C_8H_{18}O$ 130.23	18720-65-5 -91.2	172	4-01-00-01783 0.8425[25]	1.433[24]
6503	4-Heptanol, 2-methyl- 6-Methyl-4-heptanol	$C_8H_{18}O$ 130.23	21570-35-4 -81	164	4-01-00-01781 0.8207[20]	eth 4; EtOH 4 1.4203
6504	4-Heptanol, 3-methyl-, (R*,S*)-(±)- 4-Heptanol, 3-methyl(DL)	$C_8H_{18}O$ 130.23	92737-91-2 162		3-01-00-01731 0.8335[25]	H_2O 2; EtOH 3; eth 3; ctc 3 1.4211[25]
6505	4-Heptanol, 4-methyl- 4-Hydroxy-4-methylheptane	$C_8H_{18}O$ 130.23	598-01-6 -82	161	4-01-00-01789 0.8248[20]	H_2O 1; EtOH 3; eth 3; ctc 3 1.4258[20]
6506	2-Heptanol, 6-methyl-, acetate Acetic acid, (6-methyl-2-heptyl) ester	$C_{10}H_{20}O_2$ 172.27	67952-57-2 187		0-02-00-00135 0.8474[20]	EtOH 4 1.413[20]
6507	3-Heptanol, 6-methyl-, acetate Acetic acid, (6-methyl-3-heptyl) ester	$C_{10}H_{20}O_2$ 172.27	32764-34-4 184.5		0-02-00-00135 0.8554[20]	EtOH 4 1.4160[20]
6508	4-Heptanol, 4-propyl-	$C_{10}H_{22}O$ 158.28	2198-72-3 191		4-01-00-01831 0.8337[21]	bz 4; eth 4; EtOH 4 1.4355[21]
6509	4-Heptanol, 2,4,6-trimethyl-	$C_{10}H_{22}O$ 158.28	60836-07-9 181; 79[12]		4-01-00-01833 0.823[21]	1.4334[18]
6510	2-Heptanone Methyl pentyl ketone	$C_7H_{14}O$ 114.19	110-43-0 -35	4584 151.0	4-01-00-03318 0.8111[20]	H_2O 4; EtOH 3; eth 3 1.4088[20]
6511	3-Heptanone Ethyl butyl ketone	$C_7H_{14}O$ 114.19	106-35-4 -39	147	4-01-00-03321 0.8183[20]	H_2O 1; EtOH 5; eth 5; ctc 2 1.4057[20]
6512	4-Heptanone Dipropyl ketone	$C_7H_{14}O$ 114.19	123-19-3 -33	3351 144	4-01-00-03323 0.8174[20]	H_2O 1; EtOH 5; eth 5; ctc 3 1.4069[20]
6513	2-Heptanone, 1-chloro- 1-Chloro-2-heptanone	$C_7H_{13}ClO$ 148.63	41055-92-9 83[16]		4-01-00-03320 0.802[20]	eth 4; EtOH 4 1.4371[20]
6514	4-Heptanone, 2,6-dimethyl- Diisobutyl ketone	$C_9H_{18}O$ 142.24	108-83-8 -41.5	169.4	4-01-00-03360 0.8062[20]	H_2O 1; EtOH 5; eth 5; ctc 3 1.412[21]
6515	4-Heptanone, 3,5-dimethyl- Di-sec-butyl ketone	$C_9H_{18}O$ 142.24	19549-84-9 162		3-01-00-02893 0.826[14]	1.4193[14]
6516	3-Heptanone, 6-(dimethylamino)-4,4-diphenyl-, hydrochloride Methadone hydrochloride	$C_{21}H_{28}ClNO$ 345.91	1095-90-5 235	5852	3-14-00-00279	H_2O 4; EtOH 4
6517	4-Heptanone, 2-hydroxy-	$C_7H_{14}O_2$ 130.19	54862-92-9 95[12]		4-01-00-04034 0.9296[19]	1.4357[19]
6518	2-Heptanone, 4-hydroxy-3-methyl- 4-Heptanol-2-one, 3-methyl	$C_8H_{18}O_2$ 144.21	56072-27-6 110[16]		3-01-00-03249 0.9238[20]	1.442[20]
6519	3-Heptanone, 5-hydroxy-5-methyl- 3-Heptanol-5-one, 3-methyl	$C_8H_{18}O_2$ 144.21	39121-37-4 85[15]		4-01-00-04045 0.9315[15]	1.4367[15]
6520	2-Heptanone, 3-methyl- 3-Methyl-2-heptanone	$C_8H_{16}O$ 128.21	2371-19-9 164		3-01-00-02878 0.8218[20]	H_2O 2; EtOH 3; eth 3; ace 3 1.4172[20]
6521	2-Heptanone, 6-methyl-	$C_8H_{16}O$ 128.21	928-68-7 167		4-01-00-03344 0.8151[20]	H_2O 2; EtOH 4; eth 4; ace 5 1.4162[20]
6522	3-Heptanone, 2-methyl- Butyl isopropyl ketone	$C_8H_{16}O$ 128.21	13019-20-0 158		4-01-00-03343 0.8163[20]	H_2O 1; EtOH 3; eth 3; ace 4 1.4115[20]
6523	3-Heptanone, 6-methyl-	$C_8H_{16}O$ 128.21	624-42-0 164		4-01-00-03344 0.8304[20]	H_2O 1; EtOH 3; eth 3; bz 3 1.4209[20]
6524	4-Heptanone, 2-methyl- Isobutyl propyl ketone	$C_8H_{16}O$ 128.21	626-33-5 155		4-01-00-03343 0.813[22]	H_2O 1; EtOH 3; eth 3
6525	4-Heptanone, 3-methyl- 3-Methyl-4-heptanone	$C_8H_{16}O$ 128.21	15726-15-5 154		4-01-00-03344 0.817[20]	1.4103[25]
6526	1-Heptanone, 1-phenyl-	$C_{13}H_{18}O$ 190.29	1671-75-6 16.4	283.3	4-07-00-00780 0.9516[20]	ace 4; eth 4; EtOH 4 1.5060[20]
6527	Heptanoyl chloride	$C_7H_{13}ClO$ 148.63	2528-61-2 -83.8	125.2	4-02-00-00963 0.9590[20]	eth 3; ctc 2; lig 4 1.4345[18]

No.	Name / Synonym	Mol. Form. / Mol. Wt.	CAS RN / mp/°C	Merck No. / bp/°C	Beil. Ref. / den/g cm⁻³	Solubility / n_D
6528	Heptasiloxane, hexadecamethyl- / Hexadecamethylheptasiloxane	$C_{16}H_{48}O_6Si_7$ / 533.15	541-01-5 / -78	270	4-04-00-04129 / 0.9012[20]	bz 4; lig 4 / 1.3965[20]
6529	1,3,5-Heptatriene	C_7H_{10} / 94.16	2196-23-8 / 23	113.5; 31[25]	4-01-00-01097 / 0.764[20]	1.5079
6530	2-Heptenal / Butylacrolein	$C_7H_{12}O$ / 112.17	2463-63-0	166	4-01-00-03478 / 0.864[17]	1.4468[17]
6531	3-Heptenal	$C_7H_{12}O$ / 112.17	89896-73-1	151	4-01-00-03480 / 0.851[15]	eth 3; ace 3; bz 3 / 1.4348[15]
6532	5-Hepten-2-amine, N,6-dimethyl- / Isometheptene	$C_8H_{19}N$ / 141.26	503-01-5	5070 / 177	3-04-00-00467	eth 4; EtOH 4
6533	1-Heptene	C_7H_{14} / 98.19	592-76-7 / -119.7	93.6	4-01-00-00857 / 0.6970[20]	H_2O 1; EtOH 3; eth 3; ctc 2 / 1.3998[20]
6534	2-Heptene, (E)-	C_7H_{14} / 98.19	14686-13-6 / -109.5	98	4-01-00-00860 / 0.7012[20]	H_2O 1; EtOH 3; eth 3; ace 3 / 1.4045[20]
6535	2-Heptene, (Z)- / cis-2-Heptene	C_7H_{14} / 98.19	6443-92-1	98.4	3-01-00-00824 / 0.708[20]	H_2O 1; EtOH 3; eth 3; ace 3 / 1.406[20]
6536	3-Heptene, (E)- / trans-3-Heptene	C_7H_{14} / 98.19	14686-14-7 / -136.6	95.7	4-01-00-00861 / 0.6981[20]	H_2O 1; EtOH 3; eth 3; ace 3 / 1.4043[20]
6537	3-Heptene, (Z)- / cis-3-Heptene	C_7H_{14} / 98.19	7642-10-6 / -136.6	95.8	4-01-00-00861 / 0.7030[20]	H_2O 1; EtOH 3; eth 3; ace 3 / 1.4059[20]
6538	1-Heptene, 1-chloro- / 1-Chloro-1-heptene	$C_7H_{13}Cl$ / 132.63	2384-75-0	155	3-01-00-00823 / 0.8948[20]	ace 4; bz 4; eth 4; chl 4 / 1.4380[20]
6539	1-Heptene, 2-chloro-	$C_7H_{13}Cl$ / 132.63	65786-11-0	139	3-01-00-00823 / 0.8895[20]	eth 4; EtOH 4; chl 4 / 1.4349[20]
6540	3-Heptene, 4-chloro-	$C_7H_{13}Cl$ / 132.63	2431-24-5	139	3-01-00-00827 / 0.883[14]	eth 4; EtOH 4; chl 4 / 1.437[14]
6541	2-Heptene, 6-chloro-2-methyl-	$C_8H_{15}Cl$ / 146.66	80325-37-7	60-1[15]	2-01-00-00200 / 0.8931[18]	ace 4; bz 4; eth 4; EtOH 4 / 1.4458[18]
6542	2-Heptene, 2,3-dimethyl- / 2,3-Dimethyl-2-heptene	C_9H_{18} / 126.24	3074-64-4 / -108.5	145.2	4-01-00-00897 / 0.731[25]	1.4519[20]
6543	1-Heptene, 2-methyl- / 2-Methyl-1-heptene	C_8H_{16} / 112.22	15870-10-7 / -90	119.3	4-01-00-00881 / 0.7104[25]	1.4123[20]
6544	1-Heptene, 3-methyl- / 3-Methyl-1-heptene	C_8H_{16} / 112.22	4810-09-7	111	4-01-00-00883 / 0.707[25]	1.406[20]
6545	1-Heptene, 4-methyl- / 4-Methyl-1-heptene	C_8H_{16} / 112.22	13151-05-8	112.8	4-01-00-00884 / 0.713[25]	1.410[20]
6546	1-Heptene, 5-methyl- / 5-Methyl-1-heptene	C_8H_{16} / 112.22	13151-04-7	113.3	4-01-00-00884 / 0.7125[25]	1.4094[20]
6547	1-Heptene, 6-methyl- / 6-Methyl-1-heptene	C_8H_{16} / 112.22	5026-76-6	113.2	4-01-00-00883 / 0.7079[25]	1.4070[20]
6548	2-Heptene, 2-methyl- / 2-Methyl-2-heptene	C_8H_{16} / 112.22	627-97-4	122.6	4-01-00-00882 / 0.7200[25]	H_2O 1; eth 3; bz 3; ctc 3 / 1.4170[20]
6549	2-Heptene, 3-methyl-, (E)- / 3-Methyl-trans-2-heptene	C_8H_{16} / 112.22	22768-20-3	122	4-01-00-00883 / 1.419[20]	
6550	2-Heptene, 3-methyl-, (Z)- / 2-Heptene, 3-methyl (cis)	C_8H_{16} / 112.22	22768-19-0	122	3-01-00-00843 / 0.725[25]	1.419[20]
6551	2-Heptene, 4-methyl-, (E)- / 4-Methyl-trans-2-heptene	C_8H_{16} / 112.22	66225-17-0	114	4-01-00-00885 / 0.712[25]	1.410[20]
6552	2-Heptene, 4-methyl-, (Z)- / 2-Heptene, 4-methyl (cis)	C_8H_{16} / 112.22	66225-16-9	114	4-01-00-00885 / 0.712[25]	1.408[20]
6553	2-Heptene, 5-methyl- / 2-Methyl-5-heptene	C_8H_{16} / 112.22	73548-72-8	117	3-01-00-00843 / 0.714[25]	1.410[20]
6554	3-Heptene, 2-methyl- / 2-Methyl-3-heptene	C_8H_{16} / 112.22	17618-76-7 / -107.5	112	4-01-00-00882 / 0.702[25]	1.407[20]
6555	3-Heptene, 3-methyl- / 2-Ethyl-2-hexene	C_8H_{16} / 112.22	7300-03-0	121	3-01-00-00843 / 0.724[25]	1.418[20]
6556	3-Heptene, 4-methyl- / 4-Methyl-3-heptene	C_8H_{16} / 112.22	4485-16-9	121	1-01-00-00094 / 0.721[25]	1.417[20]
6557	3-Heptene, 5-methyl-	C_8H_{16} / 112.22	13172-91-3	112	4-01-00-00884 / 0.709[25]	1.410[20]
6558	3-Heptene, 6-methyl- / 2-Methyl-4-heptene	C_8H_{16} / 112.22	3404-57-7	115	4-01-00-00882 / 0.709[25]	1.410[20]
6559	3-Heptene, 4-propyl- / 4-Propyl-3-heptene	$C_{10}H_{20}$ / 140.27	4485-13-6	160.5	4-01-00-00906 / 0.7518[17]	1.4302[18]
6560	1-Heptene, 1,1,2,3,3,4,4,5,5,6,6,7,7,7-tetradecafluoro- / Perfluoro-1-heptene	C_7F_{14} / 350.05	355-63-5	81.0	4-01-00-00859	
6561	2-Heptenoic acid	$C_7H_{12}O_2$ / 128.17	18999-28-5 / -19	226.5	4-02-00-01577 / 0.9575[20]	1.4488[20]
6562	4-Heptenoic acid	$C_7H_{12}O_2$ / 128.17	35194-37-7	98-100[5]	4-02-00-01578 / 0.949[15]	1.4418[15]
6563	5-Heptenoic acid	$C_7H_{12}O_2$ / 128.17	3593-00-8	223	4-02-00-01578 / 0.9496[20]	1.4444[20]
6564	6-Heptenoic acid	$C_7H_{12}O_2$ / 128.17	1119-60-4 / -6.5	226	4-02-00-01576 / 0.9515[14]	1.4404[14]
6565	2-Heptenoic acid, 6-methyl-	$C_8H_{14}O_2$ / 142.20	90112-75-7 / 3	239.5	1-02-00-00194 / 0.938[20]	1.4511[10]
6566	1-Hepten-4-ol	$C_7H_{14}O$ / 114.19	3521-91-3	152.1	3-01-00-01936 / 0.8384[22]	1.4347[20]
6567	2-Hepten-1-ol, (E)-	$C_7H_{14}O$ / 114.19	33467-76-4	178; 75[10]	3-01-00-01936 / 0.8516[20]	EtOH 3; ace 3 / 1.4460[20]

No.	Name / Synonym	Mol. Form. / Mol. Wt.	CAS RN / mp/°C	Merck No. / bp/°C	Beil. Ref. / den/g cm⁻³	Solubility / n_D
6568	2-Hepten-1-ol, (Z)- / cis-2-Hepten-1-ol	$C_7H_{14}O$ / 114.19	55454-22-3	178; 82[20]	4-01-00-02155 / 0.8421[20]	1.4410[20]
6569	2-Hepten-4-ol, (±)-	$C_7H_{14}O$ / 114.19	115113-98-9	153	4-01-00-02155 / 0.8445[20]	EtOH 3; eth 3 / 1.4373[20]
6570	3-Hepten-2-ol	$C_7H_{14}O$ / 114.19	98991-54-9	66-7[16]	4-01-00-02156 / 0.834[17]	1.4391[18]
6571	6-Hepten-2-ol	$C_7H_{14}O$ / 114.19	24395-10-6	64-5[13]	2-01-00-00488 / 0.8484[18]	1.4387[20]
6572	6-Hepten-3-ol / 1-Heptene-4-ol	$C_7H_{14}O$ / 114.19	19781-77-2	60-2[11]	4-01-00-02154 / 0.8447[18]	1.4369[18]
6573	1-Hepten-4-ol, 4-methyl- / 1-Heptene-4-ol, 4-methyl	$C_8H_{16}O$ / 128.21	1186-31-8	159.5	3-01-00-01945 / 0.8345[20]	eth 4; EtOH 4 / 1.4479[18]
6574	2-Hepten-4-ol, 5-methyl-	$C_8H_{16}O$ / 128.21	4048-31-1	69-70[18]	2-01-00-00490 / 0.8473[20]	1.4411[20]
6575	2-Hepten-4-ol, 6-methyl- / 6-Methyl-2-hepten-4-ol	$C_8H_{16}O$ / 128.21	4798-62-3	75[21]; 67[11]	2-01-00-00490 / 0.8354[15]	1.4378[20]
6576	3-Hepten-2-ol, 2-methyl-	$C_8H_{16}O$ / 128.21	116668-45-2	62-3[14]	2-01-00-00490 / 0.8398[10]	1.4416[10]
6577	4-Hepten-2-ol, 2-methyl-	$C_8H_{16}O$ / 128.21	116668-44-1 / -28	63-4[20]	3-01-00-01944 / 0.8424[14]	1.4407[14]
6578	4-Hepten-3-ol, 3-methyl-	$C_8H_{16}O$ / 128.21	4048-30-0	103[75]; 62[15]	2-01-00-00490 / 0.8477[17]	1.4465[20]
6579	4-Hepten-3-ol, 4-methyl-	$C_8H_{16}O$ / 128.21	81280-12-8	66[17]	4-01-00-02170 / 0.8525[18]	1.4479[18]
6580	5-Hepten-2-ol, 6-methyl-	$C_8H_{16}O$ / 128.21	1569-60-4	175	1-01-00-00230 / 0.8545[20]	1.4505[20]
6581	6-Hepten-1-ol, 3-methyl-	$C_8H_{16}O$ / 128.21	4048-32-2	97-9[22]	2-01-00-00491 / 0.8562[19]	1.4470[14]
6582	6-Hepten-2-ol, 2-methyl- / 1,1-Dimethyl-5-hexen-1-ol	$C_8H_{16}O$ / 128.21	77437-98-0	65-6[16]	3-01-00-01945 / 0.8393[16]	1.4393[16]
6583	1-Hepten-3-one / Butyl vinyl ketone	$C_7H_{12}O$ / 112.17	2918-13-0	42-3[15]	4-01-00-03477 / 0.8434[20]	1.4305[19]
6584	4-Hepten-2-one	$C_7H_{12}O$ / 112.17	24332-22-7	61-2[20]	3-01-00-03004 / 0.8618[21]	1.4290[21]
6585	6-Hepten-2-one / Methyl 4-pentenyl ketone	$C_7H_{12}O$ / 112.17	21889-88-3	41-3[10]	4-01-00-03478 / 0.8673[18]	1.4350[18]
6586	6-Hepten-3-one / 1-Hepten-5-one	$C_7H_{12}O$ / 112.17	2565-39-1	142	4-01-00-03478 / 0.8487[18]	1.4254[18]
6587	3-Hepten-2-one, (E)-	$C_7H_{12}O$ / 112.17	5609-09-6	62[15]	4-01-00-03481 / 0.8496[20]	H_2O 1; EtOH 4; eth 4 / 1.4436[20]
6588	3-Hepten-2-one, (Z)- / 3-Hepten-2-one (cis)	$C_7H_{12}O$ / 112.17	69668-88-8	34-9[11]	4-01-00-03481 / 0.8440[20]	1.4325[20]
6589	5-Hepten-2-one, (E)-	$C_7H_{12}O$ / 112.17	1071-94-9	153	4-01-00-03480 / 0.8445[20]	1.4309[20]
6590	3-Hepten-2-one, 3-ethyl-4-methyl-	$C_{10}H_{18}O$ / 154.25	54244-90-5	190; 83[14]	3-01-00-03028 / 0.8559[20]	1.4535[20]
6591	4-Hepten-2-one, 3-ethyl-4-methyl-	$C_{10}H_{18}O$ / 154.25	53252-08-7	69[11]	3-01-00-03028 / 0.8450[20]	1.4405[20]
6592	4-Hepten-3-one, 5-ethyl-4-methyl-	$C_{10}H_{18}O$ / 154.25	22319-28-4	80[10]	4-01-00-03517 / 0.8622[10]	1.4545[19]
6593	5-Hepten-2-one, 4-ethyl-6-methyl- / 2-Hepten-6-one, 4-ethyl-2-methyl	$C_{10}H_{18}O$ / 154.25	104035-66-7	74-5[13]	4-01-00-03517 / 0.8430[20]	1.4546[20]
6594	5-Hepten-3-one, 5-ethyl-4-methyl-	$C_{10}H_{18}O$ / 154.25	74764-56-0 / 74.5	135-9[3]	4-01-00-03517 / 0.8564[21]	1.4452[21]
6595	1-Hepten-3-one, 5-methyl-	$C_8H_{14}O$ / 126.20	44829-76-7	40-1[7]	3-01-00-03014 / 0.8479[19]	1.4360[19]
6596	2-Hepten-4-one, 5-methyl- / 5-Methyl-2-hepten-4-one	$C_8H_{14}O$ / 126.20	81925-81-7	171	4-01-00-03496 / 0.848[25]	1.4400[25]
6597	2-Hepten-4-one, 6-methyl-	$C_8H_{14}O$ / 126.20	49852-35-9	170	4-01-00-03494 / 0.841[25]	1.4388[25]
6598	3-Hepten-2-one, 6-methyl-	$C_8H_{14}O$ / 126.20	2009-74-7	179	3-01-00-03011 / 0.8443[17]	1.4427[10]
6599	4-Hepten-2-one, 6-methyl- / 2-Methyl-3-hepten-6-one	$C_8H_{14}O$ / 126.20	39273-81-9	163	1-01-00-00385 / 0.8345[20]	1.4315
6600	4-Hepten-3-one, 5-methyl- / Homomesityl oxide	$C_8H_{14}O$ / 126.20	1447-26-3	166.5	3-01-00-03012 / 0.8591[20]	1.4476[20]
6601	5-Hepten-2-one, 3-methyl- / 3-Methyl-5-hepten-2-one	$C_8H_{14}O$ / 126.20	38552-72-6	62-4[20]	4-01-00-03496 / 0.8463[18]	1.4345[18]
6602	5-Hepten-2-one, 5-methyl-	$C_8H_{14}O$ / 126.20	10339-67-0	84[45]	4-01-00-03494 / 0.8647[16]	1.4434[16]
6603	5-Hepten-2-one, 6-methyl-	$C_8H_{14}O$ / 126.20	110-93-0	173.5	4-01-00-03493 / 0.8546[16]	eth 4; EtOH 4 / 1.4445[20]
6604	5-Hepten-3-one, 2-methyl-	$C_8H_{14}O$ / 126.20	77958-21-5	161.5	1-01-00-00385 / 0.842[20]	1.4310[20]
6605	5-Hepten-3-one, 5-methyl- / 5-Methylhept-5-en-3-one	$C_8H_{14}O$ / 126.20	1190-34-7	63[19]	3-01-00-03011 / 0.8524[21]	1.4367[21]
6606	1-Hepten-3-yne	C_7H_{10} / 94.16	2384-73-8	110	4-01-00-01097 / 0.7603[20]	ace 4; bz 4; EtOH 4; peth 4 / 1.4520[25]
6607	1-Hepten-4-yne, 6,6-dimethyl-	C_9H_{14} / 122.21	31508-08-4	125	0.758[20]	ace 4; bz 4; EtOH 4; peth 4 / 1.4312[20]

No.	Name / Synonym	Mol. Form. / Mol. Wt.	CAS RN / mp/°C	Merck No. / bp/°C	Beil. Ref. / den/g cm^{-3}	Solubility / n_D
6608	6-Hepten-4-yn-3-ol, 3-ethyl- / Diethylvinylethynylcarbinol	$C_9H_{14}O$ / 138.21	3142-84-5 /	/ 62[4]	3-01-00-02034 / 0.8875[20]	eth 4; EtOH 4 / 1.4800[20]
6609	1-Heptyne	C_7H_{12} / 96.17	628-71-7 / -81	/ 99.7	4-01-00-01025 / 0.7328[20]	H_2O 2; EtOH 5; eth 5; bz 3 / 1.4087[20]
6610	2-Heptyne / 1-Methyl-2-butylacetylene	C_7H_{12} / 96.17	1119-65-9 /	/ 112	4-01-00-01026 / 0.744[25]	H_2O 1; EtOH 5; eth 5; bz 3 / 1.4230[20]
6611	3-Heptyne / 1-Ethyl-2-propylacetylene	C_7H_{12} / 96.17	2586-89-2 / -130.5	/ 107.2	4-01-00-01027 / 0.7336[25]	H_2O 1; EtOH 5; eth 5; bz 3 / 1.4189[20]
6612	1-Heptyne, 1-bromo- / 1-Bromo-1-heptyne	$C_7H_{11}Br$ / 175.07	19821-84-2 /	164; 69[25]	3-01-00-00998 / 1.2120[22]	ace 4; eth 4; EtOH 4; chl 4 / 1.4678[22]
6613	2-Heptyne, 1-bromo- / 2-Heptynyl bromide	$C_7H_{11}Br$ / 175.07	18495-26-6 /	104[55]; 84[20]	4-01-00-01026 /	ace 4; eth 4; EtOH 4 / 1.4878[25]
6614	1-Heptyne, 1-chloro- / 1-Chloro-1-heptyne	$C_7H_{11}Cl$ / 130.62	51556-10-6 /	/ 141	3-01-00-00997 / 0.9250[24]	eth 4; EtOH 4 / 1.4411[24]
6615	2-Heptyne, 7-chloro- / 1-Chloro-5-heptyne	$C_7H_{11}Cl$ / 130.62	70396-13-3 /	/ 166	3-01-00-00998 /	eth 4; EtOH 4 / 1.4507[25]
6616	3-Heptyne, 7-chloro- / 4-Heptynyl chloride	$C_7H_{11}Cl$ / 130.62	51575-85-0 /	/ 164	3-01-00-00999 /	eth 4; EtOH 4 / 1.4517[20]
6617	3-Heptyne, 2,6-dimethyl-	C_9H_{16} / 124.23	19549-97-4 /	/ 130.6	3-01-00-01015 / 0.785[20]	ace 4; eth 4
6618	3-Heptyne, 5,5-dimethyl- / 5,5-Dimethyl-3-heptyne	C_9H_{16} / 124.23	23097-98-5 /	69[100]	3-01-00-01015 / 0.7610[20]	ace 4; eth 4 / 1.4360[20]
6619	3-Heptyne, 5-ethyl-5-methyl- / 5-Ethyl-5-methyl-3-heptyne	$C_{10}H_{18}$ / 138.25	61228-10-2 /	88[100]	3-01-00-01021 / 0.7714[20]	ace 4; eth 4 / 1.4386[20]
6620	1-Heptyne, 1-iodo- / 1-Iodo-1-heptyne	$C_7H_{11}I$ / 222.07	54573-13-6 /	90-2[17]	3-01-00-00998 / 1.4701[19]	1.5123[19]
6621	Hexacene	$C_{26}H_{16}$ / 328.41	258-31-1 / 380	sub	4-05-00-02834 /	H_2O 1; EtOH 1
6622	Hexacosane	$C_{26}H_{54}$ / 366.71	630-01-3 / 56.4	/ 412.2	4-01-00-00583 / 0.7783[60]	bz 4; lig 4; chl 4 / 1.4357[60]
6623	Hexacosane, 13-dodecyl-	$C_{38}H_{78}$ / 535.04	55517-73-2 / 13.7	/ 272[1]	4-01-00-00600 / 0.8188[20]	1.4577[20]
6624	Hexacosanoic acid / Cerotic acid	$C_{26}H_{52}O_2$ / 396.70	506-46-7 / 88.5		4-02-00-01310 / 0.8198[100]	H_2O 1; EtOH 4; eth 4 / 1.4301[100]
6625	Hexacosanoic acid, methyl ester / Methyl hexacosanoate	$C_{27}H_{54}O_2$ / 410.72	5802-82-4 / 63.8	286[15]; 237[2]	4-02-00-01311 /	chl 3 / 1.4301[80]
6626	1-Hexacosanol	$C_{26}H_{54}O$ / 382.71	506-52-5 / 80	305 [20] dec	4-01-00-01912 /	H_2O 1; EtOH 3; eth 3
6627	6,10-Hexadecadiyne	$C_{16}H_{26}$ / 218.38	10160-99-3 /	157[10]	3-01-00-01067 / 0.7907[20]	1.4523[20]
6628	Hexadecanal	$C_{16}H_{32}O$ / 240.43	629-80-1 / 35	200[29]	4-01-00-03393 /	H_2O 1; EtOH 3; eth 3; ace 3
6629	Hexadecanamide	$C_{16}H_{33}NO$ / 255.44	629-54-9 / 107	236[12]	4-02-00-01182 / 1.0000[20]	H_2O 1; EtOH 2; eth 2; ace 2
6630	Hexadecanamide, N-phenyl-	$C_{22}H_{37}NO$ / 331.54	6832-98-0 / 91.8	283[17]; 132[10]	4-12-00-00400 /	ace 4; bz 4; EtOH 4; chl 4
6631	1-Hexadecanamine	$C_{16}H_{35}N$ / 241.46	143-27-1 / 46.8	/ 322.5	4-04-00-00818 / 0.8129[20]	H_2O 1; EtOH 4; eth 4; ace 3 / 1.4496[20]
6632	Hexadecane / Cetane	$C_{16}H_{34}$ / 226.45	544-76-3 / 18.1	/ 286.8	4-01-00-00537 / 0.7733[20]	H_2O 1; EtOH 2; eth 5; ctc 3 / 1.4345
6633	Hexadecane, 1-bromo-	$C_{16}H_{33}Br$ / 305.34	112-82-3 / 18	/ 336	4-01-00-00542 / 0.9991[20]	H_2O 1; eth 3 / 1.4618[25]
6634	Hexadecane, 1-chloro-	$C_{16}H_{33}Cl$ / 260.89	4860-03-1 / 17.9	/ 322	4-01-00-00542 / 0.8652[20]	H_2O 1 / 1.4505[20]
6635	Hexadecane, 1,16-dibromo- / Hexadecamethylene dibromide	$C_{16}H_{32}Br_2$ / 384.24	45223-18-5 / 56	204[4]	2-01-00-00138 /	chl 4
6636	Hexadecane, 1,1-dimethoxy- / Palmitaldehyde, dimethyl acetal	$C_{18}H_{38}O_2$ / 286.50	2791-29-9 / 10	144[2]	4-01-00-03393 / 0.8542[20]	ace 4; eth 4; EtOH 4 / 1.4382[25]
6637	Hexadecane, 6,11-dipentyl-	$C_{26}H_{54}$ / 366.71	15874-03-0 / -16.2	231[10]	4-01-00-00586 / 0.8072[20]	1.4502[20]
6638	Hexadecane, 1-(ethenyloxy)-	$C_{18}H_{36}O$ / 268.48	822-28-6 / 16	160[2]	4-01-00-02057 / 0.821[27]	1.4444[25]
6639	Hexadecane, 1-fluoro- / 1-Fluorohexadecane	$C_{16}H_{33}F$ / 244.44	408-38-8 / 18	/ 289	4-01-00-00542 / 0.8321[20]	eth 4; lig 4 / 1.4317[20]
6640	Hexadecane, 1-iodo-	$C_{16}H_{33}I$ / 352.34	544-77-4 / 22	212[15]	4-01-00-00543 / 1.123[25]	H_2O 1; EtOH 2; eth 3; ace 3 / 1.4806[20]
6641	Hexadecanenitrile	$C_{16}H_{31}N$ / 237.43	629-79-8 / 31	/ 333	4-02-00-01183 / 0.8303[20]	H_2O 1; EtOH 4; eth 4; ace 4 / 1.4450[20]
6642	Hexadecane, 1,1'-oxybis-	$C_{32}H_{66}O$ / 466.88	4113-12-6 / 55	270 dec	4-01-00-01878 / 0.978[19]	H_2O 1; EtOH 3; eth 3; chl 3
6643	1-Hexadecanethiol / Cetyl mercaptan	$C_{16}H_{34}S$ / 258.51	2917-26-2 / 19	123-8[0.5]	4-01-00-01881 /	H_2O 1; EtOH 2; eth 3; ctc 2
6644	Hexadecanoic acid / Palmitic acid	$C_{16}H_{32}O_2$ / 256.43	57-10-3 / 63.1	6947 / 351.5	4-02-00-01157 / 0.8527[62]	H_2O 1; EtOH 3; eth 5; ace 3 / 1.43345[60]
6645	Hexadecanoic acid, aluminum salt	$C_{48}H_{93}AlO_6$ / 793.24	555-35-1 /	/ 361	4-02-00-01157 /	H_2O 1; EtOH 3; peth 3; bz 3
6646	Hexadecanoic acid, anhydride	$C_{32}H_{62}O_3$ / 494.84	623-65-4 / 64		4-02-00-01181 / 0.8388[83]	eth 4 / 1.4364[68]
6647	Hexadecanoic acid, butyl ester	$C_{20}H_{40}O_2$ / 312.54	111-06-8 / 16.9		4-02-00-01167 /	H_2O 1; EtOH 3; eth 3 / 1.4312[50]

No.	Name Synonym	Mol. Form. Mol. Wt.	CAS RN mp/°C	Merck No. bp/°C	Beil. Ref. den/g cm^{-3}	Solubility n_D
6648	Hexadecanoic acid, 2-[(dichloroacetyl)amino]-3-hydroxy-3-(4-nitrophenyl)propyl ester, [R-(R*,R*)]- Chloramphenicol palmitate	$C_{27}H_{42}Cl_2N_2O_6$ 561.55	530-43-8 90	2069	4-13-00-02753	bz 4; eth 4; EtOH 4
6649	Hexadecanoic acid, 1,2-ethanediyl ester	$C_{34}H_{66}O_4$ 538.90	624-03-3 72		4-02-00-01169 0.8594[78]	H_2O 1; EtOH 1; eth 3; ace 4
6650	Hexadecanoic acid, ethyl ester	$C_{18}H_{36}O_2$ 284.48	628-97-7 24	191[10]	4-02-00-01165 0.8577[25]	H_2O 1; EtOH 3; eth 3; ace 3 1.4347[34]
6651	Hexadecanoic acid, hexadecyl ester Cetyl palmitate	$C_{32}H_{64}O_2$ 480.86	540-10-3 54	2023	4-02-00-01168 0.989[20]	eth 4; EtOH 4 1.4398[70]
6652	Hexadecanoic acid, 2-hydroxyethyl ester	$C_{18}H_{36}O_3$ 300.48	4219-49-2 51	173-4[3]	4-02-00-01168 0.8768[60]	EtOH 4
6653	Hexadecanoic acid, methyl ester	$C_{17}H_{34}O_2$ 270.46	112-39-0 30	417; 148[2]	4-02-00-01165 0.8247[75]	H_2O 1; EtOH 4; eth 3; ace 4
6654	Hexadecanoic acid, 1-methylethyl ester	$C_{19}H_{38}O_2$ 298.51	142-91-6 13.5	160[2]	4-02-00-01167 0.8404[38]	ace 4; bz 4; eth 4; EtOH 4 1.4364[25]
6655	Hexadecanoic acid, phenylmethyl ester Benzyl palmitate	$C_{23}H_{38}O_2$ 346.55	41755-60-6 36		3-06-00-01481 0.9109[35]	bz 4; eth 4; EtOH 4; chl 4 1.4689[50]
6656	Hexadecanoic acid, 1,2,3-propanetriyl ester Tripalmitin	$C_{51}H_{98}O_6$ 807.34	555-44-2 66.5	9648 315	4-02-00-01176 0.8752[70]	H_2O 1; EtOH 2; eth 4; bz 3 1.4381[80]
6657	Hexadecanoic acid, propyl ester	$C_{19}H_{38}O_2$ 298.51	2239-78-3 20.4	190[12]	4-02-00-01167 0.8455[33]	1.4392[25]
6658	Hexadecanoic acid, 3,7,11,15-tetramethyl- Phytanic acid	$C_{20}H_{40}O_2$ 312.54	14721-66-5 -65			
6659	1-Hexadecanol Cetyl alcohol	$C_{16}H_{34}O$ 242.45	36653-82-4 49.3	2020 334	4-01-00-01876 0.8187[50]	H_2O 1; EtOH 2; eth 4; ace 3 1.4283[79]
6660	2-Hexadecanol	$C_{16}H_{34}O$ 242.45	14852-31-4 44	314	4-01-00-01882 0.8338[20]	1.4479[20]
6661	3-Hexadecanol	$C_{16}H_{34}O$ 242.45	593-03-3 50	152[4]	4-01-00-01882 0.8000[67]	
6662	1-Hexadecanol, acetate	$C_{18}H_{36}O_2$ 284.48	629-70-9 -18.5	220-5[205]	4-02-00-00171 0.8574[25]	H_2O 1; EtOH 2; ctc 3 1.4438[20]
6663	3-Hexadecanone	$C_{16}H_{32}O$ 240.43	18787-64-9 43	184[17]; 140[2]	4-01-00-03394	chl 3
6664	1-Hexadecanone, 1-phenyl-	$C_{22}H_{36}O$ 316.53	6697-12-7 59	251[15]	4-07-00-00860 0.8692[76]	1.4675[76]
6665	Hexadecanoyl chloride	$C_{16}H_{31}ClO$ 274.87	112-67-4 12	199[20]	4-02-00-01182 0.9016[25]	eth 4 1.4514[20]
6666	6,10,14-Hexadecatrienoic acid Hiragonic acid	$C_{16}H_{26}O_2$ 250.38	4444-12-6	180-90[15]	3-02-00-01505 0.9296[20]	eth 4; EtOH 4 1.4850[50]
6667	1-Hexadecene 1-Cetene	$C_{16}H_{32}$ 224.43	629-73-2 4.1	284.9	4-01-00-00927 0.7811[20]	H_2O 1; EtOH 3; eth 3; ctc 3 1.4412[20]
6668	7-Hexadecenoic acid Dinoroleic acid	$C_{16}H_{30}O_2$ 254.41	1191-75-9 33	230[10]	0-02-00-00460	EtOH 4; eth 4
6669	9-Hexadecenoic acid	$C_{16}H_{30}O_2$ 254.41	2091-29-4 0.5	181[1]	4-02-00-01629	bz 4; eth 4; peth 4
6670	9-Hexadecenoic acid, (Z)- Palmitoleic acid	$C_{16}H_{30}O_2$ 254.41	373-49-9 -0.1	131[0.06]	4-02-00-01629	
6671	1-Hexadecen-3-ol, 3,7,11,15-tetramethyl- Isophytol	$C_{20}H_{40}O$ 296.54	505-32-8	5084 107-10[0.01]	2-01-00-00504 0.8519[20]	bz 4; eth 4; EtOH 4 1.4571[20]
6672	2-Hexadecen-1-ol, 3,7,11,15-tetramethyl-, [R-[R*,R*-(E)]]- Phytol	$C_{20}H_{40}O$ 296.54	150-86-7	7362 203[10]	4-01-00-02208 0.8497[25]	1.4595[25]
6673	1-Hexadecyne	$C_{16}H_{30}$ 222.41	629-74-3 15	284	4-01-00-01073 0.7965[20]	bz 4 1.4440[20]
6674	2-Hexadecyne	$C_{16}H_{30}$ 222.41	629-75-4 20	160[15]	3-01-00-01028 0.8039[20]	
6675	2,4-Hexadienal, (E,E)-	C_6H_8O 96.13	142-83-6	174; 76[30]	4-01-00-03545 0.898[20]	1.5384[20]
6676	1,2-Hexadiene Propylallene	C_6H_{10} 82.15	592-44-9	76	4-01-00-01011 0.7149[20]	eth 4; chl 4 1.4282[20]
6677	1,4-Hexadiene	C_6H_{10} 82.15	592-45-0	65	4-01-00-01013 0.7000[20]	H_2O 1; eth 4 1.4150[20]
6678	1,5-Hexadiene Biallyl	C_6H_{10} 82.15	592-42-7 -140.7	59.4	4-01-00-01013 0.6878[25]	H_2O 1; EtOH 3; eth 3; bz 3 1.4042[20]
6679	2,3-Hexadiene Methylethylallene	C_6H_{10} 82.15	592-49-4	78	4-01-00-01016 0.684[15]	1.395[20]
6680	2,4-Hexadiene	C_6H_{10} 82.15	592-46-1 -79	80	3-01-00-00985 0.7196[20]	H_2O 1; EtOH 3; eth 3; chl 3 1.4500[20]
6681	1,3-Hexadiene, (E)-	C_6H_{10} 82.15	20237-34-7 -102.4	73.2	4-01-00-01011 0.6995[25]	1.4406[20]
6682	1,3-Hexadiene, (Z)-	C_6H_{10} 82.15	14596-92-0	73.1	4-01-00-01011 0.7033[25]	1.4379[20]
6683	1,4-Hexadiene, (E)-	C_6H_{10} 82.15	7319-00-8 -138.7	65.0	0.695[25]	1.4104[20]
6684	1,4-Hexadiene, (Z)-	C_6H_{10} 82.15	7318-67-4	66.3	0.695[25]	eth 4 1.4049[20]
6685	2,4-Hexadiene, (E,E)-	C_6H_{10} 82.15	5194-51-4 -44.9	82.2	4-01-00-01016 0.7101[25]	1.4510[20]
6686	2,4-Hexadiene, (E,Z)-	C_6H_{10} 82.15	5194-50-3 -96.1	83.5	4-01-00-01016 0.7185[25]	1.4560[20]

No.	Name Synonym	Mol. Form. Mol. Wt.	CAS RN mp/°C	Merck No. bp/°C	Beil. Ref. den/g cm⁻³	Solubility n_D
6687	2,4-Hexadiene, (Z,Z)-	C_6H_{10} 82.15	6108-61-8	85	0.7298[25]	1.4606[20]
6688	1,3-Hexadiene, 3-chloro-	C_6H_9Cl 116.59	101870-06-8	68[117]	4-01-00-01012 0.9390[20]	eth 4; chl 4 1.4770[20]
6689	1,5-Hexadiene, 1,1,2,3,3,4,4,5,6,6-decachloro-	C_6Cl_{10} 426.59	29030-84-0 49	121[0.03]	4-01-00-01016 1.905[52]	ace 4; eth 4; chl 4 1.6012[51]
6690	2,4-Hexadiene, 1,3-dichloro-	$C_6H_8Cl_2$ 151.03	73454-83-8	80-2[17]	4-01-00-01017 1.1456[20]	bz 4; chl 4 1.5271[20]
6691	1,5-Hexadiene, 2,5-dimethyl-	C_8H_{14} 110.20	627-58-7 -75.6	114.3	4-01-00-01042 0.743[20]	H_2O 1; ace 3; chl 3 1.43995[21]
6692	2,4-Hexadiene, 2,5-dimethyl-	C_8H_{14} 110.20	764-13-6 14	134.5	4-01-00-01043 0.7577[25]	H_2O 1; EtOH 3; eth 3; bz 3 1.4785[20]
6693	1,5-Hexadiene-3,4-diol, (R*,R*)-(±)-	$C_6H_{10}O_2$ 114.14	19700-97-1 21.7	90-1[8]	1.017[19]	H_2O 3; EtOH 3; eth 3; chl 3 1.4790[19]
6694	1,5-Hexadiene, 3-ethyl-2-methyl- 1,5-Hexadiene, 2-methyl-3-ethyl	C_9H_{16} 124.23	73398-15-9 -70	145	0-01-00-00260 0.7594[20]	
6695	1,5-Hexadiene, 2-methyl-	C_7H_{12} 96.17	4049-81-4 -128.8	88.1	4-01-00-01030 0.7153[25]	1.4183[20]
6696	2,4-Hexadiene, 2-methyl-	C_7H_{12} 96.17	28823-41-8 -74.2	111.5	4-01-00-01030 0.7411[25]	
6697	2,4-Hexadiene, 1,3,4,6-tetrachloro-	$C_6H_6Cl_4$ 219.92	100367-45-1	84-9[2]	3-01-00-00987 1.4013[20]	MeOH 4; chl 4 1.5465[20]
6698	2,4-Hexadienoic acid Sorbic acid	$C_6H_8O_2$ 112.13	110-44-1 134.5	8677 228 dec; 153[50]	2-02-00-00452 1.204[19]	H_2O 3; EtOH 3; eth 4; chl 3
6699	2,4-Hexadienoic acid, ethyl ester	$C_8H_{12}O_2$ 140.18	110318-09-7	195.5 85[20]	2-02-00-00452 0.9506[20]	H_2O 1; EtOH 3; eth 3; chl 3 1.4951[20]
6700	2,4-Hexadienoic acid, ethyl ester, (E,E)-	$C_8H_{12}O_2$ 140.18	2396-84-1	195.5	4-02-00-01703 0.9506[20]	eth 4; EtOH 4; chl 4 1.4951[20]
6701	2,5-Hexadienoic acid, 3-methoxy-5-methyl-4-oxo- Penicillic acid	$C_8H_{10}O_4$ 170.17	90-65-3 83	7033	3-03-00-01467	H_2O 3; EtOH 4; eth 4; ace 3
6702	2,4-Hexadienoic acid, methyl ester Methyl 2,4-hexadienoate	$C_7H_{10}O_2$ 126.16	1515-80-6 15	180; 70[20]	2-02-00-00452 0.9777[20]	H_2O 1; EtOH 3; eth 3 1.5025[22]
6703	2,4-Hexadienoic acid, methyl ester, (E,E)-	$C_7H_{10}O_2$ 126.16	689-89-4 15	180	4-02-00-01703 0.9777[20]	eth 4; EtOH 4 1.5025[22]
6704	2,4-Hexadienoic acid, potassium salt, (E,E)-	$C_6H_7KO_2$ 150.22	24634-61-5 >270 dec	7661	1.361[25]	H_2O 4; EtOH 3
6705	1,4-Hexadien-3-ol	$C_6H_{10}O$ 98.14	1070-14-0	50-2[15]	4-01-00-02238 0.859[23]	
6706	1,5-Hexadien-3-ol 3-Hydroxy-1,5-hexadiene	$C_6H_{10}O$ 98.14	924-41-4	133.5	4-01-00-02239 0.8596[25]	1.4464[25]
6707	2,4-Hexadien-1-ol 1-Hydroxy-2,4-hexadiene	$C_6H_{10}O$ 98.14	111-28-4 30.5	8678 76[12]	4-01-00-02239 0.8967[23]	H_2O 1; EtOH 3; eth 3 1.4981[20]
6708	3,5-Hexadien-2-ol 1,3-Hexadien-5-ol	$C_6H_{10}O$ 98.14	3280-51-1	77-8[26]	4-01-00-02237 0.8678[20]	EtOH 4 1.4896[20]
6709	2,4-Hexadien-1-ol, (E,E)-	$C_6H_{10}O$ 98.14	17102-64-6 30.5	76[12]	4-01-00-02239 0.8967[20]	eth 4; EtOH 4 1.4981[20]
6710	3,5-Hexadien-2-one, 6-phenyl-	$C_{12}H_{12}O$ 172.23	4173-44-8 68	171[15]; 155[12]	4-07-00-01183	bz 4; eth 4; EtOH 4; chl 4
6711	2,4-Hexadienoyl chloride, (E,E)-	C_6H_7ClO 130.57	2614-88-2	78[15]	3-02-00-01460 1.0666[19]	ace 4 1.5545[20]
6712	1,3-Hexadien-5-yne	C_6H_6 78.11	10420-90-3 -81	83.5	4-01-00-01120 0.7806[20]	bz 4 1.5095[20]
6713	1,5-Hexadien-3-yne Divinylacetylene	C_6H_6 78.11	821-08-9 -88	85	4-01-00-01120 0.7851[20]	H_2O 1; bz 3 1.5035[20]
6714	1,5-Hexadien-3-yne, 2,5-dimethyl- Diisopropenylacetylene	C_8H_{10} 106.17	3725-05-1	123	4-01-00-01124 0.7863[25]	bz 4; chl 4 1.4845[20]
6715	1,4-Hexadiyne	C_6H_6 78.11	10420-91-4 <-80	81.7	3-01-00-01057 0.825[0]	bz 4; chl 4
6716	1,5-Hexadiyne Bipropargyl	C_6H_6 78.11	628-16-0 -6	86	4-01-00-01118 0.8049[20]	H_2O 1; EtOH 3; eth 3; ace 3 1.4380[23]
6717	2,4-Hexadiyne Dimethyldiacetylene	C_6H_6 78.11	2809-69-0 67.8	129.5	4-01-00-01119	EtOH 4; eth 4
6718	Hexanal Caproaldehyde	$C_6H_{12}O$ 100.16	66-25-1 -56	1761 131	4-01-00-03296 0.8335[20]	H_2O 2; EtOH 4; eth 4; ace 3 1.4039[20]
6719	Hexanal, 2-ethyl- 2-Ethylhexanal	$C_8H_{16}O$ 128.21	123-05-7 <-100	163	4-01-00-03345 0.8540[20]	H_2O 1; EtOH 3; eth 3; ctc 2 1.4142[20]
6720	Hexanal, 3-methyl- Caproaldehyde, 3-methyl	$C_7H_{14}O$ 114.19	19269-28-4	143	2-01-00-00756 0.8203[20]	H_2O 1; EtOH 3; eth 3 1.4122[20]
6721	Hexanamide	$C_6H_{13}NO$ 115.18	628-02-4 101	255	4-02-00-00929 0.999[20]	bz 4; eth 4; EtOH 4; chl 4 1.4200[110]
6722	Hexanamide, N-phenyl-	$C_{12}H_{17}NO$ 191.27	621-15-8 95		4-12-00-00392 1.112[25]	eth 4; EtOH 4
6723	1-Hexanamine Hexylamine	$C_6H_{15}N$ 101.19	111-26-2 -22.9	132.8	4-04-00-00709 0.7660[20]	H_2O 2; EtOH 5; eth 5; chl 3 1.4180[20]
6724	2-Hexanamine, (±)-	$C_6H_{15}N$ 101.19	68107-05-1 -19	117.5	4-04-00-00721 0.7533[20]	EtOH 3; eth 3 1.4080[25]

No.	Name / Synonym	Mol. Form. / Mol. Wt.	CAS RN / mp/°C	Merck No. / bp/°C	Bell. Ref. / den/g cm^{-3}	Solubility / n_D
6725	1-Hexanamine, N,N-dihexyl-	$C_{18}H_{39}N$ 269.51	102-86-3	261.7	4-04-00-00711 0.7976[21]	H_2O 1; EtOH 4; eth 4; acid 3
6726	1-Hexanamine, 2-ethyl- 2-Ethyl-1-hexylamine	$C_8H_{19}N$ 129.25	104-75-6		4-04-00-00766	
6727	1-Hexanamine, N-hexyl-	$C_{12}H_{27}N$ 185.35	143-16-8 -13.1	236; 75[1]	4-04-00-00711 0.7889[20]	EtOH 3; eth 3 1.4339[20]
6728	1-Hexanamine, 3-methyl- Hexylamine, 3-methyl-	$C_7H_{17}N$ 115.22	65530-93-0	149; 67[45]	2-04-00-00653 0.772[26]	1.4249[25]
6729	1-Hexanamine, 4-methyl- Hexylamine, 4-methyl-	$C_7H_{17}N$ 115.22	34263-68-8	153	2-04-00-00654 0.7802[20]	1.4238[20]
6730	2-Hexanamine, 4-methyl- Methylhexaneamine	$C_7H_{17}N$ 115.22	105-41-9 132.5	6000	4-04-00-00747 0.7655[20]	H_2O 2; EtOH 4; eth 4; chl 4 1.4150[25]
6731	Hexane	C_6H_{14} 86.18	110-54-3 -95.3	4613 68.7	4-01-00-00338 0.6548[25]	H_2O 1; EtOH 4; eth 3; chl 3 1.3749[20]
6732	Hexane, 1-bromo- Hexyl bromide	$C_6H_{13}Br$ 165.07	111-25-1 -84.7	155.3	4-01-00-00352 1.1744[20]	H_2O 1; EtOH 5; eth 5; ace 3 1.4478[20]
6733	Hexane, 2-bromo- 2-Bromohexane	$C_6H_{13}Br$ 165.07	3377-86-4	144; 78[90]	4-01-00-00353 1.1658[20]	H_2O 1; EtOH 4; eth 3; ace 3 1.4832[25]
6734	Hexane, 3-bromo- 3-Bromohexane	$C_6H_{13}Br$ 165.07	3377-87-5	142	4-01-00-00353 1.1799[20]	ace 4; eth 4; EtOH 4; chl 4 1.4472[20]
6735	Hexane, 1-bromo-6-fluoro- 1-Bromo-6-fluorohexane	$C_6H_{12}BrF$ 183.06	373-28-4	67-8[11]	4-01-00-00353 1.293[20]	ace 4; eth 4; EtOH 4; chl 4 1.4436[25]
6736	Hexane, 1-bromo-6-methoxy-	$C_7H_{15}BrO$ 195.10	50592-87-5	112[35]; 98[19]	4-01-00-01704 1.1887[25]	1.4469[25]
6737	Hexane, 1-chloro- Hexyl chloride	$C_6H_{13}Cl$ 120.62	544-10-5 -94	2144 135	4-01-00-00349 0.8785[20]	H_2O 1; EtOH 3; eth 3; ace 3 1.4199[20]
6738	Hexane, 2-chloro- 2-Hexyl chloride	$C_6H_{13}Cl$ 120.62	638-28-8	122.5	4-01-00-00349 0.8694[21]	ace 4; bz 4; eth 4; EtOH 4 1.4142[22]
6739	Hexane, 3-chloro- 3-Chlorohexane	$C_6H_{13}Cl$ 120.62	2346-81-8	123	4-01-00-00349 0.8684[20]	ace 4; bz 4; eth 4; EtOH 4 1.4163[20]
6740	Hexane, 2-chloro-2,5-dimethyl- 2-Chloro-2,5-dimethylhexane	$C_8H_{17}Cl$ 148.68	29342-44-7	86[100]	4-01-00-00434 0.8476[18]	ace 4; bz 4; eth 4; EtOH 4 1.4232[20]
6741	Hexane, 3-chloro-2,3-dimethyl-	$C_8H_{17}Cl$ 148.68	101654-30-2	41-3[12]	3-01-00-00481 0.8869[20]	EtOH 4 1.4333[25]
6742	Hexane, 1-chloro-6-fluoro- 1-Chloro-6-fluorohexane	$C_6H_{12}ClF$ 138.61	1550-09-0	168; 62[15]	4-01-00-00350 1.015[20]	eth 4; EtOH 4; chl 4 1.4168[25]
6743	Hexane, 1-chloro-3-methyl-	$C_7H_{15}Cl$ 134.65	101257-63-0	151	2-01-00-00119 0.8766[20]	eth 4; EtOH 4; chl 4 1.4274[207]
6744	Hexane, 2-chloro-2-methyl-	$C_7H_{15}Cl$ 134.65	4398-65-6	135 dec; 60[52]	4-01-00-00398 0.8635[20]	eth 4; EtOH 4; chl 4 1.4200[20]
6745	Hexane, 2-chloro-5-methyl- 2-Chloro-5-methylhexane	$C_7H_{15}Cl$ 134.65	58766-17-9	138 dec	3-01-00-00436 0.863[20]	eth 4; EtOH 4; chl 4
6746	Hexane, 3-chloro-3-methyl-	$C_7H_{15}Cl$ 134.65	43197-78-0	135	3-01-00-00438 0.8787[20]	eth 4; EtOH 4; chl 4 1.4250[20]
6747	Hexane, 3-chloro-2,2,3-trimethyl-	$C_9H_{19}Cl$ 162.70	102449-95-6	64-5[13]	3-01-00-00515 0.9010[20]	eth 4; EtOH 4; chl 4 1.4465[20]
6748	Hexanedial	$C_6H_{10}O_2$ 114.14	1072-21-5 -8	92-4[9]	4-01-00-03686 1.003[19]	bz 4; eth 4; EtOH 4 1.4350[20]
6749	1,6-Hexanediamine Hexamethylenediamine	$C_6H_{16}N_2$ 116.21	124-09-4 41.5	4614 205	4-04-00-01320	H_2O 4; EtOH 3; bz 3
6750	1,6-Hexanediamine, N,N'-dibutyl- N,N'-Dibutylhexamethylenediamine	$C_{14}H_{32}N_2$ 228.42	4835-11-4	137-40[3-4]	4-04-00-01327	1.4470[25]
6751	2,5-Hexanediamine, 2,5-dimethyl-	$C_8H_{20}N_2$ 144.26	23578-35-0	63-4[8]	4-04-00-01363 0.832[25]	1.4459[20]
6752	1,6-Hexanediamine, N,N,N',N'-tetramethyl-	$C_{10}H_{24}N_2$ 172.31	111-18-2	209.5	4-04-00-01322 0.806[25]	1.4359[20]
6753	1,6-Hexanediaminium, N,N'-di-9H-fluoren-9-yl-N,N,N',N'-tetramethyl-dibromide Hexafluorenium bromide	$C_{36}H_{42}Br_2N_2$ 662.55	317-52-2 153.5	4606	4-12-00-03397	
6754	Hexane, 1,2-dibromo- 1,2-Dibromohexane	$C_6H_{12}Br_2$ 243.97	624-20-4	104[36]; 82[12]	3-01-00-00392 1.5774[20]	bz 4; eth 4; chl 4 1.5024[20]
6755	Hexane, 1,4-dibromo- 1,4-Dibromohexane	$C_6H_{12}Br_2$ 243.97	25118-28-9	106-8[15]	3-01-00-00392 1.602[15]	1.5084[15]
6756	Hexane, 1,5-dibromo- 1,5-Dibromohexane	$C_6H_{12}Br_2$ 243.97	627-96-3	153[100]; 103[9]	4-01-00-00354 1.5989[20]	1.5072[15]
6757	Hexane, 1,6-dibromo-	$C_6H_{12}Br_2$ 243.97	629-03-8 -2.3	245.5	4-01-00-00354 1.5948[15]	H_2O 1; eth 3; ace 3; ctc 2 1.5037[20]
6758	Hexane, 2,3-dibromo- 2,3-Dibromohexane	$C_6H_{12}Br_2$ 243.97	6423-02-5	90[16]	3-01-00-00393 1.5812[20]	1.5025[20]
6759	Hexane, 2,5-dibromo-, (R*,R*)-(±)- Hexane, 2,5-dibromo(DL)	$C_6H_{12}Br_2$ 243.97	54462-68-9 -44.64	108[30] 94[13]	3-01-00-00393 1.5788[20]	H_2O 1; eth 3; ace 3; ctc 3 1.5007[20]
6760	Hexane, 3,4-dibromo-	$C_6H_{12}Br_2$ 243.97	89583-12-0	80-1[13]	3-01-00-00394 1.6027[20]	1.5043[20]
6761	Hexane, 1,2-dichloro- 1,2-Dichlorohexane	$C_6H_{12}Cl_2$ 155.07	2162-92-7	173	4-01-00-00350 1.085[15]	eth 4; chl 4
6762	Hexane, 1,6-dichloro- 1,6-Dichlorohexane	$C_6H_{12}Cl_2$ 155.07	2163-00-0	204	4-01-00-00350 1.0677[20]	H_2O 1; eth 3; ctc 3; chl 3 1.4572[20]
6763	Hexane, 2,2-dichloro- 2,2-Dichlorohexane	$C_6H_{12}Cl_2$ 155.07	42131-89-5	68[49]	3-01-00-00390 1.0150[25]	eth 4; chl 4 1.4353[25]

No.	Name / Synonym	Mol. Form. / Mol. Wt.	CAS RN / mp/°C	Merck No. / bp/°C	Beil. Ref. / den/g cm^{-3}	Solubility / n_D
6764	Hexane, 2,3-dichloro- / 2,3-Dichlorohexane	$C_6H_{12}Cl_2$ / 155.07	54305-87-2	163.5	4-01-00-00350 / 1.0527[11]	eth 4; chl 4
6765	Hexane, 2,5-dichloro-2,5-dimethyl- / 2,5-Dimethyl-2,5-dichlorohexane	$C_8H_{16}Cl_2$ / 183.12	6223-78-5 / 67.5		4-01-00-00435 / 0.9543[20]	bz 4; eth 4; EtOH 4; chl 4
6766	Hexane, 2,5-dichloro-, (R^*,R^*)-(±)- / (±)-2,5-Dichlorohexane	$C_6H_{12}Cl_2$ / 155.07	41761-12-0 / -38	177; 106[91]	4-01-00-00350 / 1.0474[20]	chl 4 / 1.4491[20]
6767	Hexane, 3,3-diethyl-	$C_{10}H_{22}$ / 142.28	17302-02-2	166.3	4-01-00-00484 / 0.7575[25]	1.4258[20]
6768	Hexane, 3,4-diethyl- / 3,4-Diethylhexane	$C_{10}H_{22}$ / 142.28	19398-77-7	163.9	4-01-00-00484 / 0.7472[25]	1.4190[20]
6769	Hexane, 1,6-diiodo- / Hexamethylene diiodide	$C_6H_{12}I_2$ / 337.97	629-09-4 / 9.5	163[17]; 141[10]	4-01-00-00356 / 2.0416[15]	H_2O 1; EtOH 4; eth 4 / 1.5899[15]
6770	Hexane, 1,6-diisocyanato- / Hexamethylene diisocyanate	$C_8H_{12}N_2O_2$ / 168.20	822-06-0	122[10]; 94[1]	4-04-00-01349 / 1.0528[20]	1.4585[20]
6771	Hexane, 2,2-dimethyl- / 2,2-Dimethylhexane	C_8H_{18} / 114.23	590-73-8 / -121.1	106.8	4-01-00-00432 / 0.6953[20]	ace 4; bz 4; eth 4; EtOH 4 / 1.3935[20]
6772	Hexane, 2,3-dimethyl- / 2,3-Dimethylhexane	C_8H_{18} / 114.23	584-94-1	115.6	4-01-00-00432 / 0.6912[25]	ace 4; bz 4; EtOH 4; lig 4 / 1.4011[20]
6773	Hexane, 2,4-dimethyl- / 2,4-Dimethylhexane	C_8H_{18} / 114.23	589-43-5	109.5	0.6962[25]	1.3929[25]
6774	Hexane, 2,5-dimethyl- / 2,5-Dimethylhexane	C_8H_{18} / 114.23	592-13-2 / -91	109.1	4-01-00-00434 / 0.6901[25]	H_2O 1; EtOH 5; eth 3; ace 5 / 1.3925[20]
6775	Hexane, 3,3-dimethyl- / 3,3-Dimethylhexane	C_8H_{18} / 114.23	563-16-6 / -126.1	111.9	4-01-00-00435 / 0.7100[20]	H_2O 1; EtOH 5; eth 4; ace 4 / 1.4001[20]
6776	Hexane, 3,4-dimethyl- / 3,4-Dimethylhexane	C_8H_{18} / 114.23	583-48-2	117.7	4-01-00-00436 / 0.7151[25]	H_2O 1; EtOH 5; eth 3; ace 5 / 1.4041[20]
6777	Hexanedinitrile / Adiponitrile	$C_6H_8N_2$ / 108.14	111-69-3	295	4-02-00-01975 / 0.9676[20]	H_2O 2; EtOH 3; eth 2; chl 3 / 1.4380[20]
6778	Hexane, 1,6-dinitro- / 1,6-Dinitrohexane	$C_6H_{12}N_2O_4$ / 176.17	4286-47-9 / 37.5	100-3[0.3]	4-01-00-00357	H_2O 1; HOAc 3
6779	Hexanedioic acid / Adipic acid	$C_6H_{10}O_4$ / 146.14	124-04-9 / 153.2	152 / 337.5	4-02-00-01956 / 1.360[25]	H_2O 2; EtOH 4; eth 3; HOAc 1
6780	Hexanedioic acid, bis(2-ethylbutyl) ester / Adipic acid, di-2-ethylbutyl ester	$C_{18}H_{34}O_4$ / 314.47	10022-60-3 / -15	200[19]	4-02-00-01964 / 0.934[25]	ace 4; EtOH 4 / 1.4434[20]
6781	Hexanedioic acid, bis(2-ethylhexyl) ester / Bis(2-ethylhexyl) adipate	$C_{22}H_{42}O_4$ / 370.57	103-23-1 / -67.8	214[5]	4-02-00-01964 / 0.922[25]	ace 4; eth 4; EtOH 4 / 1.4474[20]
6782	Hexanedioic acid, bis(1-methylethyl) ester	$C_{12}H_{22}O_4$ / 230.30	6938-94-9 / -1.1	120[6.5]	4-02-00-01961 / 0.9569[20]	ace 4; eth 4; EtOH 4 / 1.4247[20]
6783	Hexanedioic acid, bis(2-methylpropyl) ester	$C_{14}H_{26}O_4$ / 258.36	141-04-8	186-8[15]	4-02-00-01962 / 0.9543[19]	1.4301[20]
6784	Hexanedioic acid, dibutyl ester	$C_{14}H_{26}O_4$ / 258.36	105-99-7 / -32.4	165[10]	4-02-00-01961 / 0.9613[20]	H_2O 1; EtOH 5; eth 5 / 1.4369[20]
6785	Hexanedioic acid, diethyl ester	$C_{10}H_{18}O_4$ / 202.25	141-28-6 / -19.8	245	4-02-00-01960 / 1.0076[20]	H_2O 1; EtOH 3; eth 3 / 1.4272[20]
6786	Hexanedioic acid, dimethyl ester	$C_8H_{14}O_4$ / 174.20	627-93-0 / 10.3	115[13]	4-02-00-01959 / 1.0600[20]	H_2O 1; EtOH 3; eth 3; ctc 3 / 1.4283[20]
6787	Hexanedioic acid, dipropyl ester	$C_{12}H_{22}O_4$ / 230.30	106-19-4 / -15.7	151[11]	4-02-00-01961 / 0.9790[20]	eth 4; EtOH 4; chl 4 / 1.4314[20]
6788	Hexanedioic acid, 2-methyl-	$C_7H_{12}O_4$ / 160.17	626-70-0 / 64	209[13]	4-02-00-02010	H_2O 4; eth 4; EtOH 4; chl 4
6789	Hexanedioic acid, 3-methyl-, (±)- / Adipic acid, 3-methyl (DL)	$C_7H_{12}O_4$ / 160.17	81177-02-8 / 97	230[30]	4-02-00-02014	H_2O 4; eth 4; EtOH 4; ace 4
6790	Hexanedioic acid, 3-methyl-, diethyl ester / Diethyl 3-methyladipate	$C_{11}H_{20}O_4$ / 216.28	55877-01-5	258; 126.5[10]	4-02-00-02015 / 0.9948[20]	1.4335
6791	Hexanedioic acid, monoethyl ester	$C_8H_{14}O_4$ / 174.20	626-86-8 / 29	285	4-02-00-01960 / 0.9796[20]	EtOH 3; eth 3; peth 3 / 1.4311[20]
6792	Hexanedioic acid, monomethyl ester	$C_7H_{12}O_4$ / 160.17	627-91-8 / 9	158[10]	3-02-00-01711 / 1.0623[20]	EtOH 3 / 1.4283[20]
6793	1,3-Hexanediol	$C_6H_{14}O_2$ / 118.18	21531-91-9	123[13]	4-01-00-02555 / 0.958[22]	1.4461[22]
6794	1,4-Hexanediol	$C_6H_{14}O_2$ / 118.18	16432-53-4	134-5[18.5]	4-01-00-02556 / 0.982[16]	eth 4; EtOH 4 / 1.4530[16]
6795	1,5-Hexanediol	$C_6H_{14}O_2$ / 118.18	928-40-5	237; 90[0.5]	4-01-00-02556 / 0.981[25]	1.4511[20]
6796	1,6-Hexanediol / Hexamethylene glycol	$C_6H_{14}O_2$ / 118.18	629-11-8 / 42.8	4610 / 208	4-01-00-02556	H_2O 3; EtOH 3; eth 2; ace 3 / 1.4579[25]
6797	2,3-Hexanediol	$C_6H_{14}O_2$ / 118.18	617-30-1 / 60	205	4-01-00-02561 / 0.990[15]	H_2O 4; eth 4; EtOH 4 / 1.4510[15]
6798	2,4-Hexanediol	$C_6H_{14}O_2$ / 118.18	19780-90-6	211; 105[9]	4-01-00-02562 / 0.9516[21]	1.4418[21]
6799	2,5-Hexanediol / Diisopropanol	$C_6H_{14}O_2$ / 118.18	2935-44-6 / 43	218; 86[1]	4-01-00-02562 / 0.9610[20]	H_2O 3; EtOH 3; eth 3; ctc 2 / 1.4475[20]
6800	3,4-Hexanediol, 3,4-diethyl- / 3,4-Diethyl-3,4-hexanediol	$C_{10}H_{22}O_2$ / 174.28	6931-71-1 / 28	230	4-01-00-02621 / 0.960[13]	eth 4; EtOH 4 / 1.467[13]
6801	2,5-Hexanediol, 2,5-dimethyl- / 1,1,4,4-Tetramethyl-1,4-butanediol	$C_8H_{18}O_2$ / 146.23	110-03-2 / 92	214	4-01-00-02600. / 0.898[20]	H_2O 3; EtOH 4; bz 4; chl 4
6802	1,3-Hexanediol, 2-ethyl- / Ethohexadiol	$C_8H_{18}O_2$ / 146.23	94-96-2 / -40	3699 / 244	4-01-00-02597 / 0.9325[22]	H_2O 2; EtOH 3; eth 3 / 1.4497[20]

No.	Name / Synonym	Mol. Form. / Mol. Wt.	CAS RN / mp/°C	Merck No. / bp/°C	Beil. Ref. / den/g cm^{-3}	Solubility / n_D
6803	2,4-Hexanediol, 2-methyl-	$C_7H_{16}O_2$ 132.20	66225-35-2	121[30]	2-01-00-00554 0.9321[18]	1.4407[18]
6804	2,4-Hexanedione Propionylacetone	$C_6H_{10}O_2$ 114.14	3002-24-2	160	4-01-00-03687 0.959[20]	1.4516[20]
6805	2,5-Hexanedione Acetonylacetone	$C_6H_{10}O_2$ 114.14	110-13-4 -5.5	63 194	4-01-00-03688 0.7370[20]	H_2O 4; bz 4; eth 4; EtOH 4 1.4232[20]
6806	3,4-Hexanedione Bipropionyl	$C_6H_{10}O_2$ 114.14	4437-51-8 -10	130	4-01-00-03690 0.941[21]	1.4130[21]
6807	3,4-Hexanedione, 2,5-dimethyl- Biisobutyryl	$C_8H_{14}O_2$ 142.20	4388-87-8	144.5	4-01-00-03713 0.9232[20]	1.4206[20]
6808	2,5-Hexanedione, 3-hydroxy- 3-Hydroxy-2,5-hexanedione	$C_6H_{10}O_3$ 130.14	61892-85-1	62-7[0.5]	3-01-00-03317	bz 4; eth 4 1.4497[25]
6809	2,3-Hexanedione, 5-methyl- 2-Methylhexa-4,5-dione	$C_7H_{12}O_2$ 128.17	13706-86-0	138	4-01-00-03701 0.908[22]	1.4119[20]
6810	2,5-Hexanedione, 3-methyl- 3-Methyl-2,5-hexanedione	$C_7H_{12}O_2$ 128.17	4437-50-7	196; 71[10]	3-01-00-03137 0.9527[20]	eth 4; EtOH 4 1.4260[20]
6811	3,4-Hexanedione, 2,2,5,5-tetramethyl- Pivalil	$C_{10}H_{18}O_2$ 170.25	4388-88-9 -2	169	4-01-00-03726 0.8776[20]	H_2O 1; eth 5 1.4157[20]
6812	1,6-Hexanedithiol	$C_6H_{14}S_2$ 150.31	1191-43-1	118-9[15]	4-01-00-02559 0.983[25]	1.5110[20]
6813	Hexane, 1-(ethenyloxy)- Hexyl vinyl ether	$C_8H_{16}O$ 128.21	5363-64-4	143	4-01-00-02055 0.7966[20]	1.4171[20]
6814	Hexane, 1-ethoxy-	$C_8H_{18}O$ 130.23	5756-43-4	143	4-01-00-01697 0.7722[20]	eth 4; EtOH 4 1.4008[20]
6815	Hexane, 3-ethyl- 3-Ethylhexane	C_8H_{18} 114.23	619-99-8	118.6	4-01-00-00431 0.7136[20]	H_2O 1; EtOH 5; eth 5; ace 5 1.4018[20]
6816	Hexane, 3-ethyl-2,2-dimethyl- 3-Ethyl-2,2-dimethylhexane	$C_{10}H_{22}$ 142.28	20291-91-2	156.1	4-01-00-00484 0.7447[25]	1.4197[20]
6817	Hexane, 3-ethyl-2,3-dimethyl- 3-Ethyl-2,3-dimethylhexane	$C_{10}H_{22}$ 142.28	52897-00-4	166	4-01-00-00484 0.7599[25]	1.4270[207]
6818	Hexane, 3-ethyl-2,4-dimethyl- 3-Ethyl-2,4-dimethylhexane	$C_{10}H_{22}$ 142.28	7220-26-0	162	4-01-00-00484 0.7514[25]	1.4225[20]
6819	Hexane, 3-ethyl-2,5-dimethyl- 3-Ethyl-2,5-dimethylhexane	$C_{10}H_{22}$ 142.28	52897-04-8	154.1	4-01-00-00484 0.7368[25]	1.4157[20]
6820	Hexane, 3-ethyl-3,4-dimethyl- 3-Ethyl-3,4-dimethylhexane	$C_{10}H_{22}$ 142.28	52897-06-0	162.1	4-01-00-00485 0.7596[25]	1.4267[207]
6821	Hexane, 4-ethyl-2,2-dimethyl- Hexane, 2,2-dimethyl-4-ethyl	$C_{10}H_{22}$ 142.28	52896-99-8	147	4-01-00-00485 0.7330[20]	1.4131[20]
6822	Hexane, 4-ethyl-2,3-dimethyl- 4-Ethyl-2,3-dimethylhexane	$C_{10}H_{22}$ 142.28	52897-01-5	162	4-01-00-00485 0.7516[25]	1.4226[20]
6823	Hexane, 4-ethyl-2,4-dimethyl- 4-Ethyl-2,4-dimethylhexane	$C_{10}H_{22}$ 142.28	52897-03-7	161.1	4-01-00-00485 0.7525[25]	1.4235[20]
6824	Hexane, 4-ethyl-3,3-dimethyl- 4-Ethyl-3,3-dimethylhexane	$C_{10}H_{22}$ 142.28	52897-05-9	162.9	4-01-00-00485 0.7598[25]	1.4269[20]
6825	Hexane, 3-ethyl-2-methyl- 3-Ethyl-2-methylhexane	C_9H_{20} 128.26	16789-46-1	138	4-01-00-00459 0.7310[20]	1.4106[20]
6826	Hexane, 3-ethyl-3-methyl- 3-Ethyl-3-methylhexane	C_9H_{20} 128.26	3074-76-8	140.6	4-01-00-00459 0.7371[25]	1.4140[20]
6827	Hexane, 3-ethyl-4-methyl- 4-Ethyl-3-methylhexane	C_9H_{20} 128.26	3074-77-9	140	3-01-00-00514 0.7420[20]	1.4134[20]
6828	Hexane, 4-ethyl-2-methyl- 4-Ethyl-2-methylhexane	C_9H_{20} 128.26	3074-75-7	133.8	4-01-00-00459 0.7195[25]	1.4063[20]
6829	Hexane, 1-fluoro- Hexyl fluoride	$C_6H_{13}F$ 104.17	373-14-8 -103	91.5	4-01-00-00348 0.7995[20]	eth 3; bz 3 1.3738[20]
6830	Hexane, 2-fluoro- 2-Fluorohexane	$C_6H_{13}F$ 104.17	372-54-3	86	4-01-00-00348 0.7916[20]	1.3693[20]
6831	Hexane, 1-iodo- Hexyl iodide	$C_6H_{13}I$ 212.07	638-45-9 -74.2	181	4-01-00-00355 1.4397[20]	H_2O 1 1.4929[20]
6832	Hexane, 2-iodo-, (R)- Hexane, 2-iodo (I)	$C_6H_{13}I$ 212.07	59654-13-6	90[70]; 45[9]	4-01-00-00355 1.4354[17]	H_2O 1; ace 3; chl 3 1.4878[25]
6833	Hexane, 1-methoxy- Hexyl methyl ether	$C_7H_{16}O$ 116.20	4747-07-3	126.1		
6834	Hexane, 2-methyl- 2-Methylhexane	C_7H_{16} 100.20	591-76-4 -118.2	90.0	4-01-00-00397 0.6787[20]	H_2O 1; EtOH 3; eth 5; ace 5 1.3848[20]
6835	Hexane, 3-methyl-, (S)- (+)-3-Methylhexane	C_7H_{16} 100.20	6131-24-4 -119	92	4-01-00-00399 0.6860[20]	H_2O 1; EtOH 3; eth 5; ace 5 1.3887[20]
6836	Hexane, 2-methyl-3-methylene- 2-Isopropyl-1-pentene	C_8H_{16} 112.22	16746-02-4	113	4-01-00-00887 0.721[25]	1.414[20]
6837	Hexane, 2-methyl-4-methylene- 1-Pentene, 2-ethyl-4-methyl-	C_8H_{16} 112.22	3404-80-6	110.3	4-01-00-00888 0.7152[25]	1.4105[20]
6838	Hexane, 3-methyl-4-methylene-	C_8H_{16} 112.22	3404-67-9	112.5	4-01-00-00890 0.725[25]	1.4142[20]
6839	Hexane, 2-methyl-3-(1-methylethyl)- Hexane, 3-isopropyl-2-methyl-	$C_{10}H_{22}$ 142.28	62016-13-1	165	4-01-00-00484 0.7436[25]	1.4195[20]
6840	Hexanenitrile Capronitrile	$C_6H_{11}N$ 97.16	628-73-9 -80.3	163.6	4-02-00-00930 0.8051[20]	H_2O 1; EtOH 3; eth 3; chl 2 1.4068[20]
6841	Hexane, 1-nitro- 1-Nitrohexane	$C_6H_{13}NO_2$ 131.17	646-14-0	193; 84[21]	4-01-00-00356 0.9396[20]	H_2O 1; EtOH 3; eth 3; ace 3 1.4270[20]
6842	Hexane, 1,1'-oxybis-	$C_{12}H_{26}O$ 186.34	112-58-3	223	4-01-00-01698 0.7936[20]	H_2O 1; eth 3; ctc 2 1.4204[20]

No.	Name Synonym	Mol. Form. Mol. Wt.	CAS RN mp/°C	Merck No. bp/°C	Beil. Ref. den/g cm⁻³	Solubility n_D
6843	Hexane, tetradecafluoro- Perfluorohexane	C_6F_{14} 338.04	355-42-0 -87.1	56.6	4-01-00-00348 1.6995[20]	H₂O 1; eth 3; bz 3; chl 3 1.2515[20]
6844	Hexane, 2,2,3,3-tetramethyl- 2,2,3,3-Tetramethylhexane	$C_{10}H_{22}$ 142.28	13475-81-5 -54	160.3	4-01-00-00485 0.7609[25]	1.4282[20]
6845	Hexane, 2,2,3,4-tetramethyl- 2,2,3,4-Tetramethylhexane	$C_{10}H_{22}$ 142.28	52897-08-2	157	4-01-00-00485 0.7513[25]	1.4216[20]
6846	Hexane, 2,2,3,5-tetramethyl- 2,2,3,5-Tetramethylhexane	$C_{10}H_{22}$ 142.28	52897-09-3	148.4	4-01-00-00485 0.7336[25]	1.4142[20]
6847	Hexane, 2,2,4,4-tetramethyl- 2,2,4,4-Tetramethylhexane	$C_{10}H_{22}$ 142.28	51750-65-3	153.8	4-01-00-00486 0.7424[25]	1.4208[20]
6848	Hexane, 2,2,4,5-tetramethyl- 2,2,4,5-Tetramethylhexane	$C_{10}H_{22}$ 142.28	16747-42-5	147.9	4-01-00-00486 0.7316[25]	1.4132[20]
6849	Hexane, 2,2,5,5-tetramethyl- 2,2,5,5-Tetramethylhexane	$C_{10}H_{22}$ 142.28	1071-81-4 -12.6	137.4	4-01-00-00486 0.7148[25]	1.4055[20]
6850	Hexane, 2,3,3,4-tetramethyl- 2,3,3,4-Tetramethylhexane	$C_{10}H_{22}$ 142.28	52897-10-6	164.6	4-01-00-00486 0.7656[25]	1.4298[20]
6851	Hexane, 2,3,3,5-tetramethyl- 2,3,3,5-Tetramethylhexane	$C_{10}H_{22}$ 142.28	52897-11-7	153.1	4-01-00-00486 0.7449[25]	1.4196[20]
6852	Hexane, 2,3,4,4-tetramethyl- 2,3,4,4-Tetramethylhexane	$C_{10}H_{22}$ 142.28	52897-12-8	161.6	4-01-00-00487 0.7586[25]	1.4267[20]
6853	Hexane, 2,3,4,5-tetramethyl- 2,3,4,5-Tetramethylhexane	$C_{10}H_{22}$ 142.28	52897-15-1	156.2	4-01-00-00487 0.7456[25]	1.4204[20]
6854	Hexane, 3,3,4,4-tetramethyl- 3,3,4,4-Tetramethylhexane	$C_{10}H_{22}$ 142.28	5171-84-6	170.0	3-01-00-00534 0.7789[25]	1.4368[20]
6855	Hexane, 1,1'-thiobis-	$C_{12}H_{26}S$ 202.40	6294-31-1	230; 136[20]	4-01-00-01706 0.8411[20]	1.4586[20]
6856	1-Hexanethiol Hexyl mercaptan	$C_6H_{14}S$ 118.24	111-31-9 -81	151	4-01-00-01705 0.8424[20]	H₂O 1; EtOH 4; eth 4 1.4496[20]
6857	2-Hexanethiol	$C_6H_{14}S$ 118.24	1679-06-7 -147	142	4-01-00-01711 0.8345[20]	H₂O 1; EtOH 3; eth 3; bz 3 1.4451[20]
6858	3-Hexanethiol 3-Mercaptohexane	$C_6H_{14}S$ 118.24	1633-90-5	57[25]	4-01-00-01713 0.9206[20]	EtOH 3 1.4496[20]
6859	Hexane, 2,2,3-trimethyl- 2,2,3-Trimethylhexane	C_9H_{20} 128.26	16747-25-4	133.6	4-01-00-00459 0.7257[25]	1.4106[20]
6860	Hexane, 2,2,4-trimethyl- 2,2,4-Trimethylhexane	C_9H_{20} 128.26	16747-26-5 -120	126.5	4-01-00-00459 0.711[20]	1.4033[20]
6861	Hexane, 2,2,5-trimethyl- 2,2,5-Trimethylhexane	C_9H_{20} 128.26	3522-94-9 -105.7	124.0	4-01-00-00460 0.7072[20]	H₂O 1; EtOH 4; eth 4; ace 4 1.3997[20]
6862	Hexane, 2,3,3-trimethyl- 2,3,3-Trimethylhexane	C_9H_{20} 128.26	16747-28-7 -116.8	137.7	4-01-00-00461 0.7345[25]	1.4141[20]
6863	Hexane, 2,3,4-trimethyl- 2,3,4-Trimethylhexane	C_9H_{20} 128.26	921-47-1	139.1	4-01-00-00461 0.7354[25]	1.4144[20]
6864	Hexane, 2,3,5-trimethyl- 2,3,5-Trimethylhexane	C_9H_{20} 128.26	1069-53-0 -127.9	131.4	4-01-00-00461 0.7218[20]	1.4051[20]
6865	Hexane, 2,4,4-trimethyl- 2,4,4-Trimethylhexane	C_9H_{20} 128.26	16747-30-1 -113.4	130.7	4-01-00-00461 0.7201[25]	1.4074[20]
6866	Hexane, 3,3,4-trimethyl- 3,3,4-Trimethylhexane	C_9H_{20} 128.26	16747-31-2 -101.2	140.5	4-01-00-00462 0.7414[25]	1.4178[20]
6867	1,2,5-Hexanetriol	$C_6H_{14}O_3$ 134.18	10299-30-6	181[10]; 149[3]	4-01-00-02784 1.1012[20]	H₂O 4; EtOH 4
6868	1,2,6-Hexanetriol 1,2,6-Trihydroxyhexane	$C_6H_{14}O_3$ 134.18	106-69-4	170[3]	4-01-00-02784 1.03[20]	1.58[20]
6869	2,3,4-Hexanetriol	$C_6H_{14}O_3$ 134.18	93972-93-1 47	256.5	3-01-00-02349	H₂O 4; EtOH 4
6870	Hexanoic acid Caproic acid	$C_6H_{12}O_2$ 116.16	142-62-1 -3	1760 205.2	4-02-00-00917 0.9274[20]	H₂O 1; EtOH 3; eth 3; chl 3 1.4163[20]
6871	Hexanoic acid, 6-(acetylamino)- ε-Acetamidocaproic acid	$C_8H_{15}NO_3$ 173.21	57-08-9 104.5	38	4-04-00-02700	
6872	Hexanoic acid, 2-acetyl-, ethyl ester	$C_{10}H_{18}O_3$ 186.25	1540-29-0	221.5	4-03-00-01616 0.9523[20]	ace 4; eth 4 1.4301[20]
6873	Hexanoic acid, 6-amino- ε-Aminocaproic acid	$C_6H_{13}NO_2$ 131.17	60-32-2 205	442	4-04-00-02695	H₂O 4; EtOH 1; MeOH 2
6874	Hexanoic acid, anhydride	$C_{12}H_{22}O_3$ 214.30	2051-49-2 -41	255 dec	4-02-00-00928 0.9240[15]	eth 4; EtOH 4 1.4297[20]
6875	Hexanoic acid, 2-bromo-, (±)-	$C_6H_{11}BrO_2$ 195.06	2681-83-6 4	240; 140[23]	3-02-00-00736 1.2810[33]	EtOH 3; eth 3
6876	Hexanoic acid, 6-bromo-	$C_6H_{11}BrO_2$ 195.06	4224-70-8 35	165-70[20]	4-02-00-00940	peth 4
6877	Hexanoic acid, 2-bromo-, ethyl ester, (±)-	$C_8H_{15}BrO_2$ 223.11	63927-44-6	208.7	4-02-00-00938 1.2210[25]	EtOH 3
6878	Hexanoic acid, 5-bromo-, ethyl ester Caproic acid, 5-bromo,ethyl ester	$C_8H_{15}BrO_2$ 223.11	90202-08-7	121[21]; 70[3]	3-02-00-00737 1.1908[25]	1.4525[25]
6879	Hexanoic acid, 6-bromo-, ethyl ester	$C_8H_{15}BrO_2$ 223.11	25542-62-5 33	126-7[21]	4-02-00-00940 1.238[23]	1.4566[21]
6880	Hexanoic acid, butyl ester	$C_{10}H_{20}O_2$ 172.27	626-82-4 -64.3	208	4-02-00-00922 0.8653[20]	H₂O 1; EtOH 3; eth 5 1.4152[20]
6881	Hexanoic acid, 2-cyano-, ethyl ester 2-Cyanocaproic acid, ethyl ester	$C_9H_{15}NO_2$ 169.22	7391-39-1	105[9]	3-02-00-01746 0.988[15]	1.4248[20]
6882	Hexanoic acid, 2-ethyl- 2-Ethylhexanoic acid	$C_8H_{16}O_2$ 144.21	149-57-5	228; 120[13]	3-02-00-00804 0.9031[25]	H₂O 3; EtOH 2; eth 3; ctc 3 1.4241[20]

No.	Name / Synonym	Mol. Form. / Mol. Wt.	CAS RN / mp/°C	Merck No. / bp/°C	Beil. Ref. / den/g cm^{-3}	Solubility / n_D
6883	Hexanoic acid, 3-ethyl- / 3-Ethylhexanoic acid	$C_8H_{16}O_2$ / 144.21	41065-91-2 /	/ 106[5]	4-02-00-01008 / 0.911[30]	/ 1.4287[25]
6884	Hexanoic acid, ethyl ester / Ethyl hexanoate	$C_8H_{16}O_2$ / 144.21	123-66-0 / -67	3735 / 167	4-02-00-00921 / 0.873[20]	eth 4; EtOH 4 / 1.4073[20]
6885	Hexanoic acid, 2-ethyl-, ethyl ester	$C_{10}H_{20}O_2$ / 172.27	2983-37-1 /	/ 90[28]	4-02-00-01004 / 0.8586[25]	/ 1.4123[25]
6886	Hexanoic acid, 2-ethyl-3-methyl- / 2-Ethyl-3-methylhexanoic acid	$C_9H_{18}O_2$ / 158.24	74581-94-5 /	/ 231	3-02-00-00832 / 0.9060[25]	/ 1.4302[25]
6887	Hexanoic acid, 3-ethyl-3-methyl- / 3-Ethyl-3-methylhexanoic acid	$C_9H_{18}O_2$ / 158.24	50902-82-4 /	/ 139-40[22]	4-02-00-01036 / 0.9205[25]	/ 1.4375[25]
6888	Hexanoic acid, 6-fluoro- / Caproic acid, 6-fluoro	$C_6H_{11}FO_2$ / 134.15	373-05-7 /	/ 138[28]; 67[0.6]	4-02-00-00932 /	eth 4; EtOH 4 / 1.4166[25]
6889	Hexanoic acid, heptyl ester	$C_{13}H_{28}O_2$ / 214.35	6976-72-3 / -34.4	/ 261	4-02-00-00923 / 0.8612[0]	ace 4; bz 4; eth 4; EtOH 4 / 1.4293[15]
6890	Hexanoic acid, hexyl ester	$C_{12}H_{24}O_2$ / 200.32	6378-65-0 / -55	/ 246	4-02-00-00922 / 0.865[18]	ace 4; bz 4; eth 4; EtOH 4 / 1.4264[15]
6891	Hexanoic acid, 2-hydroxy-, (±)-	$C_6H_{12}O_3$ / 132.16	636-36-2 / 61.3	/ 270	4-03-00-00838 /	H_2O 4; EtOH 4; eth 3; chl 4
6892	Hexanoic acid, 3-hydroxy-2,2-dimethyl-, β-lactone / Caproic acid, 2,2-dimethyl-3-hydroxy, β-lactone	$C_8H_{14}O_2$ / 142.20	90112-82-6 / 143.5	/ 93-100[14]	DMSO 2 / 1.4260[20]
6893	Hexanoic acid, 2-methyl-, (±)- / Caproic acid, 2-methyl (DL)	$C_7H_{14}O_2$ / 130.19	22160-12-9 /	/ 215.5	4-02-00-00969 / 0.9612[20]	H_2O 5; EtOH 5; eth 5; ace 5 / 1.4195[20]
6894	Hexanoic acid, 3-methyl-	$C_7H_{14}O_2$ / 130.19	3780-58-3 /	/ 213; 116[15]	2-02-00-00298 / 0.9187[20]	/ 1.4222[20]
6895	Hexanoic acid, 4-methyl-, (±)-	$C_7H_{14}O_2$ / 130.19	22160-41-4 / -80	/ 217.5	4-02-00-00974 / 0.9215[20]	H_2O 2; EtOH 3; eth 3; ace 3 / 1.4211[20]
6896	Hexanoic acid, 5-methyl-	$C_7H_{14}O_2$ / 130.19	628-46-6 / <-25	/ 216	4-02-00-00970 / 0.9138[21]	ace 4; bz 4; eth 4; EtOH 4 / 1.4220[20]
6897	Hexanoic acid, 3-methylbutyl ester	$C_{11}H_{22}O_2$ / 186.29	2198-61-0 /	/ 225.5	3-02-00-00727 / 0.861[20]	H_2O 1; EtOH 3; eth 3
6898	Hexanoic acid, methyl ester / Methyl caproate	$C_7H_{14}O_2$ / 130.19	106-70-7 / -71	/ 149.5	4-02-00-00921 / 0.8846[20]	H_2O 1; EtOH 4; eth 4; ace 3 / 1.4049[20]
6899	Hexanoic acid, 3-methyl-, ethyl ester / Ethyl 3-methylhexanoate	$C_9H_{18}O_2$ / 158.24	41692-47-1 /	/ 177	2-02-00-00298 / 0.8679[20]	/ 1.4119[20]
6900	Hexanoic acid, 4-methyl-, ethyl ester / Ethyl 4-methylhexanoate	$C_9H_{18}O_2$ / 158.24	1561-10-0 /	/ 180	4-02-00-00974 / 0.8708[20]	/ 1.4051[20]
6901	Hexanoic acid, 4-methyl-5-oxo-	$C_7H_{12}O_3$ / 144.17	6818-07-1 /	/ 160[15]; 140[5]	4-03-00-01601 / 0.9803[20]	/ 1.4320[20]
6902	Hexanoic acid, 4-methyl-5-oxo-, ethyl ester / Valeric acid, 4-acetyl, ethyl ester	$C_9H_{16}O_3$ / 172.22	53068-88-5 /	/ 114[15]	4-03-00-01601 / 0.9803[20]	/ 1.4282[25]
6903	Hexanoic acid, 4-methyl-5-oxo-, methyl ester / Valeric acid, 4-acetyl, methyl ester	$C_8H_{14}O_3$ / 158.20	36045-56-4 /	/ 113-7[16]	4-03-00-01601 / 1.0988[20]	/ 1.4288[20]
6904	Hexanoic acid, octyl ester	$C_{14}H_{28}O_2$ / 228.38	4887-30-3 / -28	/ 275	3-02-00-00727 / 0.8603[20]	ace 4; bz 4; eth 4; EtOH 4 / 1.4326[15]
6905	Hexanoic acid, 5-oxo-	$C_6H_{10}O_3$ / 130.14	3128-06-1 / 13.5	/ 274.5	4-03-00-01583 / 1.09[25]	H_2O 3; EtOH 3; eth 3; ctc 2 / 1.4451[20]
6906	Hexanoic acid, 5-oxo-, ethyl ester / Butyric acid, 4-acetyl, ethyl ester	$C_8H_{14}O_3$ / 158.20	13984-57-1 /	/ 221.5	4-03-00-01583 / 0.989[25]	/ 1.4277[20]
6907	Hexanoic acid, pentyl ester / Amyl caproate	$C_{11}H_{22}O_2$ / 186.29	540-07-8 / -47	640 / 226	4-02-00-00922 / 0.8612[25]	EtOH 3; eth 3; ace 3; ctc 2 / 1.4202[25]
6908	Hexanoic acid, 1,2,3-propanetriyl ester	$C_{21}H_{38}O_6$ / 386.53	621-70-5 / -60	/ >200	4-02-00-00926 / 0.9867[20]	H_2O 1; EtOH 5; eth 5; ace 4 / 1.4427[20]
6909	Hexanoic acid, propyl ester	$C_9H_{18}O_2$ / 158.24	626-77-7 / -68.7	/ 187	4-02-00-00922 / 0.8672[20]	eth 4; EtOH 4 / 1.4170[20]
6910	Hexanoic acid, 3,5,6-trichloro-2,2,3,4,4,5,6,6-octafluoro- / Hexanoic acid, 2,2,3,4,4,5,6,6-octafluoro-3,5,6-trichloro	$C_6HCl_3F_8O_2$ / 363.42	2106-54-9 /	/ 105-6[5]	4-02-00-00937 / 1.860[20]	/ 1.3903[20]
6911	1-Hexanol / Caproyl alcohol	$C_6H_{14}O$ / 102.18	111-27-3 / -44.6	4615 / 157.6	4-01-00-01694 / 0.8136[20]	H_2O 2; EtOH 3; eth 5; ace 3 / 1.4178[20]
6912	2-Hexanol, (±)-	$C_6H_{14}O$ / 102.18	20281-86-1 /	/ 140	4-01-00-01708 / 0.8159[20]	H_2O 2; EtOH 3; eth 3; ctc 2 / 1.4144[20]
6913	2-Hexanol, (R)-	$C_6H_{14}O$ / 102.18	26549-24-6 /	/ 137	3-01-00-01663 / 0.8178[18]	H_2O 2; EtOH 3; eth 3
6914	3-Hexanol, (±)-	$C_6H_{14}O$ / 102.18	17015-11-1 /	/ 135	4-01-00-01711 / 0.8182[20]	H_2O 2; EtOH 3; eth 5; ace 3 / 1.4167[20]
6915	3-Hexanol, (R)- / 3-Hexanol (l)	$C_6H_{14}O$ / 102.18	13471-42-6 /	/ 135	/ 0.8213[20]	H_2O 2; EtOH 3; eth 5; ace 3 / 1.4140[20]
6916	1-Hexanol, 6-chloro-	$C_6H_{13}ClO$ / 136.62	2009-83-8 /	/ 107[12]	4-01-00-01704 / 1.0241[20]	H_2O 2; EtOH 4; eth 4 / 1.4550[20]
6917	2-Hexanol, 1-chloro- / 1-Chloro-2-hexanol	$C_6H_{13}ClO$ / 136.62	52802-07-0 /	/ 73-5[12]	4-01-00-01710 / 1.0139[20]	ace 4; eth 4; EtOH 4 / 1.4478[20]
6918	3-Hexanol, 1-chloro- / 1-Chloro-3-hexanol	$C_6H_{13}ClO$ / 136.62	52418-81-2 /	/ 120[35]; 78[6]	4-01-00-01712 / 1.003[25]	ace 4; bz 4; eth 4; EtOH 4 / 1.446[25]
6919	3-Hexanol, 5-chloro-	$C_6H_{13}ClO$ / 136.62	58588-28-6 /	/ 78-9[13]	4-01-00-01712 / 1.0012[15]	ace 4; bz 4; eth 4; EtOH 4 / 1.4433[19]
6920	1-Hexanol, 3,5-dimethoxy-	$C_8H_{18}O_3$ / 162.23	90952-10-6 /	/ 114[13]	4-01-00-02785 / 0.9631[25]	/ 1.4329[25]

No.	Name / Synonym	Mol. Form. / Mol. Wt.	CAS RN / mp/°C	Merck No. / bp/°C	Beil. Ref. / den/g cm^{-3}	Solubility / n_D
6921	1-Hexanol, 2,5-dimethyl-	$C_8H_{18}O$ 130.23	6886-16-4	179.5	0-01-00-00422 0.8280[20]	1.5095[17]
6922	2-Hexanol, 2,3-dimethyl- 2,3-Dimethyl-2-hexanol	$C_8H_{18}O$ 130.23	19550-03-9	160; 62[10]	3-01-00-01737 0.8365[20]	1.4335[20]
6923	2-Hexanol, 2,5-dimethyl- 2,5-Dimethyl-2-hexanol	$C_8H_{18}O$ 130.23	3730-60-7	154.5	4-01-00-01792 0.8227[20]	1.4208[20]
6924	2-Hexanol, 3,4-dimethyl-	$C_8H_{18}O$ 130.23	19550-05-1	165.5	3-01-00-01740 0.840[15]	1.4325[15]
6925	3-Hexanol, 2,2-dimethyl- Propyl-tert-butylcarbinol	$C_8H_{18}O$ 130.23	4209-90-9	156	3-01-00-01737 0.8342[20]	1.4275[20]
6926	3-Hexanol, 2,3-dimethyl-	$C_8H_{18}O$ 130.23	4166-46-5	61-2[18]	4-01-00-01791 0.8371[20]	1.4309[20]
6927	3-Hexanol, 2,5-dimethyl- 2,5-Dimethyl-3-hexanol	$C_8H_{18}O$ 130.23	19550-07-3	157.5	4-01-00-01793 0.8212[20]	1.4246[20]
6928	3-Hexanol, 3,5-dimethyl-, (±)- 3-Hexanol, 3,5-dimethyl (DL)	$C_8H_{18}O$ 130.23	19113-78-1	152	4-01-00-01792 0.8373[20]	bz 4; eth 4; EtOH 4 1.4300[20]
6929	3-Hexanol, 4,4-dimethyl-	$C_8H_{18}O$ 130.23	19550-09-5	160.4	4-01-00-01793 0.8341[20]	1.4345[20]
6930	1-Hexanol, 2-ethyl- 2-Ethyl-1-hexanol	$C_8H_{18}O$ 130.23	104-76-7 -70	3764 184.6	4-01-00-01783 0.8319[25]	H_2O 1; EtOH 3; eth 3; ace 3 1.4300[20]
6931	3-Hexanol, 3-ethyl-	$C_8H_{18}O$ 130.23	597-76-2	160	4-01-00-01790 0.8373[20]	H_2O 1; EtOH 3; eth 3; ace 3 1.4300[20]
6932	1-Hexanol, 2-ethyl-3-methyl-	$C_9H_{20}O$ 144.26	66794-04-5	193; 83[10]	4-02-00-02107 0.8486[20]	1.4455[20]
6933	2-Hexanol, 3-ethyl-2-methyl-	$C_9H_{20}O$ 144.26	66794-02-3	71.5[14]	4-01-00-01812 0.8334[25]	1.4331[25]
6934	3-Hexanol, 3-ethyl-2-methyl-	$C_9H_{20}O$ 144.26	66794-03-4	177.5	4-01-00-01811 0.8445[25]	1.4369[25]
6935	3-Hexanol, 3-ethyl-5-methyl-	$C_9H_{20}O$ 144.26	597-77-3	172; 68[18]	3-01-00-01755 0.8396[22]	eth 4; EtOH 4 1.4346[13]
6936	1-Hexanol, 6-fluoro-	$C_6H_{13}FO$ 120.17	373-32-0	85-6[14]	4-01-00-01703 0.975[20]	eth 4; EtOH 4 1.4141[25]
6937	1-Hexanol, 2-methyl-, (±)- 1-Hexanol, 2-methyl (DL)	$C_7H_{16}O$ 116.20	111768-04-8	164; 71[15]	4-01-00-01745 0.8270[20]	eth 4; EtOH 4 1.4226[20]
6938	1-Hexanol, 2-methyl-, (R)- (R)-2-Methyl-1-hexanol	$C_7H_{16}O$ 116.20	66050-98-4	164; 71[15]	0.8313[13]	eth 4; EtOH 4 1.4245[17]
6939	1-Hexanol, 3-methyl-, (±)-	$C_7H_{16}O$ 116.20	111768-08-2	169; 91[13]	4-01-00-01748 0.8258[20]	H_2O 1; EtOH 4; eth 4; ace 3 1.4245[20]
6940	1-Hexanol, 4-methyl-, (±)- 1-Hexanol, 4-methyl (DL)	$C_7H_{16}O$ 116.20	111768-05-9	173	4-01-00-01749 0.8239[20]	ace 4; bz 4; eth 4; EtOH 4 1.4219[20]
6941	1-Hexanol, 5-methyl- 5-Methyl-1-hexanol	C_7H_{16} 100.20	627-98-5	170; 53-5[15]	4-01-00-01748 0.8119[20]	eth 4; EtOH 4 1.4175[20]
6942	2-Hexanol, 2-methyl-	$C_7H_{16}O$ 116.20	625-23-0	143	4-01-00-01745 0.8119[20]	H_2O 2; EtOH 5; eth 5 1.4175[20]
6943	2-Hexanol, 3-methyl- 3-Methyl-2-hexanol	$C_7H_{16}O$ 116.20	2313-65-7	79-81[52]	0.8820[25]	H_2O 1; EtOH 4; eth 4; ace 3 1.4198[18]
6944	2-Hexanol, 5-methyl- 5-Methyl-2-hexanol	$C_7H_{16}O$ 116.20	627-59-8	151; 78[28]	4-01-00-01747 0.814[20]	H_2O 1; EtOH 3; eth 3 1.4180[20]
6945	3-Hexanol, 2-methyl-, (±)- (±)-2-Methyl-3-hexanol	$C_7H_{16}O$ 116.20	100295-82-7	143	4-01-00-01746 0.8407[20]	H_2O 1; EtOH 3; eth 3; ctc 3 1.4149[20]
6946	3-Hexanol, 3-methyl- 2-Ethyl-2-pentanol	$C_7H_{16}O$ 116.20	597-96-6	143	4-01-00-01749 0.8233[20]	H_2O 2; EtOH 3; eth 3; ctc 3 1.4231[20]
6947	3-Hexanol, 5-methyl-, (±)- (±)-5-Methyl-3-hexanol	$C_7H_{16}O$ 116.20	100295-83-8	148	4-01-00-01747 0.827[20]	H_2O 1; EtOH 3; eth 3; ctc 2 1.4128[20]
6948	1-Hexanol, 5-methyl-2-(1-methylethyl)- 2-Isopropyl-5-methyl-1-hexanol	$C_{10}H_{22}O$ 158.28	2051-33-4	211	4-01-00-01833 0.8345[20]	H_2O 1; EtOH 4 1.4369[20]
6949	3-Hexanol, 1-phenyl-	$C_{12}H_{18}O$ 178.27	2180-43-0 34	170[50]; 146[16]	4-06-00-03418 0.9525[20]	
6950	3-Hexanol, 2,2,5,5-tetramethyl-	$C_{10}H_{22}O$ 158.28	55073-86-4 52.5	170	3-01-00-01772	ace 4; eth 4; EtOH 4; peth 4
6951	1-Hexanol, 3,5,5-trimethyl-	$C_9H_{20}O$ 144.26	3452-97-9	194	4-01-00-01813 0.8236[25]	1.4300[25]
6952	2-Hexanol, 2,3,4-trimethyl-	$C_9H_{20}O$ 144.26	21102-13-6	57[5]	3-01-00-01756 0.853[15]	ace 4; eth 4; EtOH 4; peth 4 1.4415[15]
6953	3-Hexanol, 2,2,3-trimethyl-	$C_9H_{20}O$ 144.26	5340-41-0	170	3-01-00-01755 0.8474[20]	EtOH 3; eth 3; ace 3 1.4402[20]
6954	3-Hexanol, 2,3,5-trimethyl-	$C_9H_{20}O$ 144.26	65927-60-8	72[21]	4-01-00-01813 0.8256[20]	ace 4; eth 4; EtOH 4 1.4321[20]
6955	3-Hexanol, 2,4,4-trimethyl-	$C_9H_{20}O$ 144.26	66793-92-8	170	3-01-00-01756 0.8489[20]	EtOH 3; eth 3; ace 3 1.4395[20]
6956	3-Hexanol, 2,5,5-trimethyl-	$C_9H_{20}O$ 144.26	66793-72-4	77[32]	3-01-00-01756 0.8250[20]	ace 4; eth 4; EtOH 4 1.4286[20]
6957	3-Hexanol, 3,4,4-trimethyl-	$C_9H_{20}O$ 144.26	66793-74-6	165.5	0-01-00-00425 0.8322[21]	EtOH 3; eth 3; ace 3 1.4341[21]
6958	3-Hexanol, 3,5,5-trimethyl- 3,5,5-Trimethylhexan-3-ol	$C_9H_{20}O$ 144.26	66810-87-5	62[14]	3-01-00-01755 0.8350[20]	ace 4; eth 4; EtOH 4 1.4352[20]
6959	2-Hexanone Butyl methyl ketone	$C_6H_{12}O$ 100.16	591-78-6 -55.5	5955 127.6	4-01-00-03298 0.8113[20]	H_2O 2; EtOH 5; eth 5; ace 3 1.4007[20]
6960	3-Hexanone Ethyl propyl ketone	$C_6H_{12}O$ 100.16	589-38-8 -55.5	123.5	4-01-00-03301 0.8118[20]	H_2O 2; EtOH 5; eth 5; ace 3 1.4004[20]

No.	Name Synonym	Mol. Form. Mol. Wt.	CAS RN mp/°C	Merck No. bp/°C	Beil. Ref. den/g cm^{-3}	Solubility n_D
6961	2-Hexanone, 6-bromo-	$C_8H_{11}BrO$ 179.06	10226-29-6	216; 136[90]	3-01-00-02829 1.3494[0]	eth 4; EtOH 4
6962	1-Hexanone, 1-(2,4-dihydroxyphenyl)-	$C_{12}H_{16}O_3$ 208.26	3144-54-5 57	343	4-08-00-01888	H_2O 1; EtOH 3; eth 3; ace 3
6963	2-Hexanone, 3,3-dimethyl- 3,3-Dimethyl-2-hexanone	$C_8H_{16}O$ 128.21	26118-38-7	·149	4-01-00-03350 0.838[0]	ace 4; eth 4; EtOH 4 1.4098[20]
6964	2-Hexanone, 3,4-dimethyl- 3,4-Dimethyl-2-hexanone	$C_8H_{16}O$ 128.21	19550-10-8	158	3-01-00-02882 0.8295[22]	EtOH 3; eth 3; ace 3 1.4193[20]
6965	3-Hexanone, 2,2-dimethyl- tert-Butyl propyl ketone	$C_8H_{16}O$ 128.21	5405-79-8	147	4-01-00-03347 0.8105[25]	EtOH 3; eth 3; ace 3 1.4119[20]
6966	3-Hexanone, 2,5-dimethyl-	$C_8H_{16}O$ 128.21	1888-57-9	147.5	4-01-00-03349 0.8269[0]	ace 4; eth 4; EtOH 4 1.4049[20]
6967	3-Hexanone, 4,4-dimethyl-	$C_8H_{16}O$ 128.21	19550-14-2	148	4-01-00-03350 0.8285[20]	bz 4; EtOH 4; chl 4 1.4208[25]
6968	3-Hexanone, 6-(dimethylamino)-4,4-diphenyl- Normethadone	$C_{20}H_{25}NO$ 295.42	467-85-6	6628 164-7[3]	4-14-00-00297	
6969	2-Hexanone, 5-hydroxy- 5-Hydroxy-2-hexanone	$C_6H_{12}O_2$ 116.16	56745-61-0	203[270]; 80[10]	4-01-00-04019 0.9626[25]	H_2O 4; eth 4; EtOH 4 1.4312[25]
6970	2-Hexanone, 6-hydroxy-	$C_6H_{12}O_2$ 116.16	21856-89-3	227	4-01-00-04020 0.9886[15]	eth 4; EtOH 4 1.4494
6971	3-Hexanone, 4-hydroxy-	$C_6H_{12}O_2$ 116.16	4984-85-4	133[227]; 73[20]	4-01-00-04021 0.956[21]	ace 4; EtOH 4 1.4340[21]
6972	3-Hexanone, 5-hydroxy- 5-Hydroxy-3-hexanone	$C_6H_{12}O_2$ 116.16	33683-44-2	75-8[12]	4-01-00-04021 0.950[20]	1.4280[25]
6973	3-Hexanone, 2-hydroxy-2-methyl- 2-Hexanol-3-one, 2-methyl	$C_7H_{14}O_2$ 130.19	18905-91-4	60-2[15]	4-01-00-04035 0.899[18]	1.419[18]
6974	3-Hexanone, 5-hydroxy-2-methyl- 2-Hexanol-4-one, 5-methyl	$C_7H_{14}O_2$ 130.19	59357-07-2	72-3[9]	4-01-00-04035 0.929[20]	1.4278[20]
6975	3-Hexanone, 4-hydroxy-2,2,5,5-tetramethyl- Pivaloin	$C_{10}H_{20}O_2$ 172.27	815-66-7 81	80[10]	4-01-00-04060	eth 4
6976	2-Hexanone, 3-methyl- 3-Methyl-2-hexanone	$C_7H_{14}O$ 114.19	2550-21-2	143.5	3-01-00-02863 0.828[25]	ace 4; bz 4; eth 4; EtOH 4 1.4035[20]
6977	2-Hexanone, 4-methyl- Methyl 2-methylbutyl ketone	$C_7H_{14}O$ 114.19	105-42-0	144.5	4-01-00-03329 0.8130[20]	H_2O 2; EtOH 4; eth 4; ace 4 1.4081[24]
6978	2-Hexanone, 5-methyl- Methyl isopentyl ketone	$C_7H_{14}O$ 114.19	110-12-3	144	4-01-00-03329 0.888[20]	H_2O 2; EtOH 5; eth 5; ace 4 1.4062[20]
6979	3-Hexanone, 2-methyl-	$C_7H_{14}O$ 114.19	7379-12-6	135	4-01-00-03328 0.8091[20]	EtOH 3; eth 3; ace 4; chl 3 1.4042[20]
6980	3-Hexanone, 4-methyl-	$C_7H_{14}O$ 114.19	17042-16-9	134.5	4-01-00-03329 0.8162[20]	ace 4; bz 4; eth 4; EtOH 4 1.4069[20]
6981	3-Hexanone, 5-methyl-	$C_7H_{14}O$ 114.19	623-56-3	135	4-01-00-03328 0.8090[20]	H_2O 1; EtOH 5; eth 5; ctc 3 1.4047[20]
6982	2-Hexanone, 5-methyl-, oxime	$C_7H_{15}NO$ 129.20	624-44-2	195.5	0-01-00-00701 0.8881[20]	chl 2 1.4448[20]
6983	1-Hexanone, 1-phenyl-	$C_{12}H_{16}O$ 176.26	942-92-7 27	265	4-07-00-00754 0.9576[20]	H_2O 2; EtOH 3; eth 3; ace 3 1.5027[25]
6984	2-Hexanone, 3-phenyl-	$C_{12}H_{16}O$ 176.26	6306-30-5	235.5; 114[13]	3-07-00-01160 0.970[0]	1.5020[20]
6985	3-Hexanone, 2-phenyl- 1-Phenylethyl propyl ketone	$C_{12}H_{16}O$ 176.26	65248-43-3	241	2-07-00-00258 0.941[25]	1.4988[25]
6986	Hexanoyl chloride Caproyl chloride	$C_6H_{11}ClO$ 134.61	142-61-0 -87	1763 153	4-02-00-00928 0.9784[20]	eth 3; ace 3 1.4264[20]
6987	Hexanoyl chloride, 2-ethyl-	$C_8H_{15}ClO$ 162.66	760-67-8	67-8[11]	3-02-00-00805 0.939[25]	1.4335[20]
6988	Hexanoyl chloride, 3-methyl- Caproyl chloride, 3-methyl	$C_7H_{13}ClO$ 148.63	57323-93-0	163; 82[50]	3-02-00-00777 0.967[20]	bz 4 1.4293[25]
6989	Hexanoyl chloride, 4-methyl- Caproyl chloride, 4-methyl	$C_7H_{13}ClO$ 148.63	50599-73-0	167	2-02-00-00299 0.9677[20]	eth 4
6990	Hexasiloxane, tetradecamethyl- Tetradecamethylhexasiloxane	$C_{14}H_{42}O_5Si_6$ 459.00	107-52-8 -59	9131 245.5	4-04-00-04127 0.8910[20]	bz 4 1.3948[20]
6991	Hexatriacontane	$C_{36}H_{74}$ 506.98	630-06-8 76.5	298.4[3]	4-01-00-00599 0.7803[80]	1.4397[80]
6992	1,3,5-Hexatriene, (E)-	C_6H_8 80.13	821-07-8 -12	78.5	4-01-00-01093 0.7369[15]	H_2O 1; EtOH 3; ace 3; chl 3 1.5135[20]
6993	1,3,5-Hexatriene, (Z)-	C_6H_8 80.13	2612-46-6 -12	78	0.7175[20]	H_2O 1; EtOH 3; ace 3; chl 3 1.4577[20]
6994	1,3,5-Hexatriene, 2,5-dimethyl-	C_8H_{12} 108.18	4916-63-6 -9	146	4-01-00-01102 0.7822[20]	H_2O 1; ace 3; MeOH 3; lig 3 1.5122[20]
6995	Hexazinone 1,3,5-Triazine-2,4(1H,3H)-dione, 3-cyclohexyl-6-(dimethylamino)-1-methyl-	$C_{12}H_{20}N_4O_2$ 252.32	51235-04-2 99	4617 dec	1.25	
6996	2-Hexenal, (E)-	$C_6H_{10}O$ 98.14	6728-26-3	146.5; 50[20]	4-01-00-03468 0.8491[20]	1.4480[20]
6997	3-Hexenal, (E)-	$C_6H_{10}O$ 98.14	69112-21-6	42-3[28]	4-01-00-03469 0.8455[22]	H_2O 1; eth 3; ace 3 1.4275[21]
6998	3-Hexenal, (Z)-	$C_6H_{10}O$ 98.14	6789-80-6	121	4-01-00-03469 0.8533[22]	1.4300[21]
6999	1-Hexene	C_6H_{12} 84.16	592-41-6 -139.7	63.4	4-01-00-00828 0.6731[20]	bz 4; eth 4; EtOH 4; peth 4 1.3837[20]

No.	Name Synonym	Mol. Form. Mol. Wt.	CAS RN mp/°C	Merck No. bp/°C	Beil. Ref. den/g cm⁻³	Solubility n_D
7000	2-Hexene, (E)- trans-2-Hexene	C_6H_{12} 84.16	4050-45-7 -133	67.9	4-01-00-00834 0.6732[25]	H_2O 1; EtOH 3; eth 3; bz 3 1.3936[20]
7001	2-Hexene, (Z)- cis-2-Hexene	C_6H_{12} 84.16	7688-21-3 -141.1	68.8	4-01-00-00833 0.6869[20]	H_2O 1; EtOH 3; eth 3; bz 3 1.3979[20]
7002	3-Hexene, (E)-	C_6H_{12} 84.16	13269-52-8 -115.4	67.1	4-01-00-00837 0.6772[20]	H_2O 1; EtOH 3; eth 3; bz 3 1.3943[20]
7003	3-Hexene, (Z)-	C_6H_{12} 84.16	7642-09-3 -137.8	66.4	4-01-00-00837 0.6796[20]	H_2O 1; EtOH 3; eth 3; bz 3 1.3947[20]
7004	1-Hexene, 1-chloro- 1-Chloro-1-hexene	$C_6H_{11}Cl$ 118.61	22922-67-4	121	4-01-00-00831 0.8872[22]	ace 4; bz 4; eth 4; chl 4 1.4300[22]
7005	1-Hexene, 2-chloro- 2-Chloro-1-hexene	$C_6H_{11}Cl$ 118.61	10124-73-9	113	3-01-00-00803 0.8886[25]	ace 4; bz 4; chl 4 1.4278[25]
7006	1-Hexene, 5-chloro- 5-Chloro-1-hexene	$C_6H_{11}Cl$ 118.61	927-54-8	120.7	4-01-00-00832 0.8891[25]	ace 4; bz 4; eth 4; chl 4 1.4305[20]
7007	2-Hexene, 4-chloro- 4-Chloro-2-hexene	$C_6H_{11}Cl$ 118.61	6734-98-1	123	4-01-00-00835 0.8934[20]	ace 4; bz 4; eth 4; chl 4 1.4400[20]
7008	3-Hexene, 1-chloro- 3-Hexen-1-yl chloride	$C_6H_{11}Cl$ 118.61	62706-16-5	61[60]	4-01-00-00838 0.900[24]	ace 4; bz 4; eth 4; chl 4 1.435[24]
7009	3-Hexene, 3-chloro-, (Z)- 3-Hexene, 3-chloro (cis)	$C_6H_{11}Cl$ 118.61	17226-34-5	117	4-01-00-00838 0.9009[20]	ace 4; bz 4; eth 4; chl 4 1.4360[20]
7010	3-Hexene, 1-chloro-4-ethyl-	$C_8H_{15}Cl$ 146.66	82507-04-8	173	2-01-00-00201 0.9102[20]	bz 4; chl 4 1.4524[20]
7011	1-Hexene, 1,2-dichloro-, (E)-	$C_6H_{10}Cl_2$ 153.05	59697-51-7	63-5[22]	3-01-00-00803 1.1167[25]	bz 4; chl 4 1.4576[25]
7012	1-Hexene, 1,2-dichloro-, (Z)-	$C_6H_{10}Cl_2$ 153.05	59697-55-1	88[30]	3-01-00-00803 1.0812[25]	bz 4; chl 4 1.4631[25]
7013	1-Hexene, 2,3-dimethyl- 2,3-Dimethyl-1-hexene	C_8H_{16} 112.22	16746-86-4	110.5	3-01-00-00845 0.7172[25]	1.4113[20]
7014	1-Hexene, 2,4-dimethyl- 2,4-Dimethyl-1-hexene	C_8H_{16} 112.22	16746-87-5	111.2	4-01-00-00887 0.716[25]	1.409[20]
7015	1-Hexene, 2,5-dimethyl- 2,5-Dimethyl-1-hexene	C_8H_{16} 112.22	6975-92-4	111.6	3-01-00-00846 0.7129[25]	1.4105[20]
7016	1-Hexene, 3,3-dimethyl- 3,3-Dimethyl-1-hexene	C_8H_{16} 112.22	3404-77-1	104	4-01-00-00889 0.7099[25]	1.4070[20]
7017	1-Hexene, 3,4-dimethyl- 3,4-Dimethyl-1-hexene	C_8H_{16} 112.22	16745-94-1	112	4-01-00-00889 0.720[25]	1.413[20]
7018	1-Hexene, 3,5-dimethyl-	C_8H_{16} 112.22	7423-69-0	104	3-01-00-00846 0.712[25]	1.404[20]
7019	1-Hexene, 4,4-dimethyl- 4,4-Dimethyl-1-hexene	C_8H_{16} 112.22	1647-08-1	107.2	3-01-00-00847 0.7157[25]	1.4102[20]
7020	1-Hexene, 4,5-dimethyl- 4,5-Dimethyl-1-hexene	C_8H_{16} 112.22	16106-59-5	109	4-01-00-00887 0.724[25]	1.414[20]
7021	1-Hexene, 5,5-dimethyl- 5,5-Dimethyl-1-hexene	C_8H_{16} 112.22	7116-86-1	104	4-01-00-00886 0.705[25]	1.4049[20]
7022	2-Hexene, 2,3-dimethyl- 2,3-Dimethyl-2-hexene	C_8H_{16} 112.22	7145-20-2 -115.1	121.8	4-01-00-00887 0.7366[25]	1.4268[20]
7023	2-Hexene, 2,4-dimethyl- 2,4-Dimethyl-2-hexene	C_8H_{16} 112.22	14255-23-3	110.6	4-01-00-00887 0.7171[25]	1.4118[20]
7024	2-Hexene, 2,5-dimethyl- 2,5-Dimethyl-2-hexene	C_8H_{16} 112.22	3404-78-2	112.2	4-01-00-00888 0.7182[20]	1.4140[20]
7025	2-Hexene, 4,5-dimethyl- 2,3-Dimethyl-4-hexene	C_8H_{16} 112.22	73548-71-7	110	3-01-00-00846 0.721[25]	1.413[20]
7026	2-Hexene, 5,5-dimethyl-, (E)-	C_8H_{16} 112.22	39782-43-9	104.1	4-01-00-00886 0.7023[25]	1.4055[20]
7027	2-Hexene, 5,5-dimethyl-, (Z)- cis-5,5-Dimethyl-2-hexene	C_8H_{16} 112.22	39761-61-0	106.9	4-01-00-00886 0.7125[25]	1.4113[20]
7028	3-Hexene, 2,2-dimethyl-, (E)-	C_8H_{16} 112.22	690-93-7	100.8	4-01-00-00885 0.6995[25]	1.4063[20]
7029	3-Hexene, 2,2-dimethyl-, (Z)-	C_8H_{16} 112.22	690-92-6 -137.4	105.5	4-01-00-00885 0.7086[25]	1.4099[20]
7030	3-Hexene, 2,4-dimethyl-, (E)- 2,4-Dimethyl-trans-3-hexene	C_8H_{16} 112.22	61847-78-7	107.6	4-01-00-00887 0.7101[25]	1.4126[20]
7031	3-Hexene, 2,4-dimethyl-, (Z)- 3-Hexene, 2,4-dimethyl (cis)	C_8H_{16} 112.22	37549-89-6	109	4-01-00-00887 0.7135[25]	1.4140[20]
7032	1-Hexene, 1,1,2,3,3,4,4,5,5,6,6,6-dodecafluoro- Perfluoro-1-hexene	C_6F_{12} 300.05	755-25-9	57.0	4-01-00-00831	chl 4
7033	1-Hexene, 3-ethyl- 3-Propyl-1-pentene	C_8H_{16} 112.22	3404-58-8	110.3	4-01-00-00885 0.711[25]	1.407[20]
7034	3-Hexene, 3-ethyl-	C_8H_{16} 112.22	16789-51-8	116	3-01-00-00845 0.725[25]	1.418[20]
7035	3-Hexene, 1,2,3,4,5,6-hexachloro- 1,2,3,4,5,6-Hexachloro-3-hexene	$C_6H_6Cl_6$ 290.83	1725-74-2 58.5	112[2]	3-01-00-00808	MeOH 4; chl 4
7036	1-Hexene, 2-methyl- 2-Methyl-1-hexene	C_7H_{14} 98.19	6094-02-6 -102.8	92	4-01-00-00863 0.7000[25]	1.4035[20]
7037	1-Hexene, 3-methyl- 3-Methyl-1-hexene	C_7H_{14} 98.19	3404-61-3	83.9	4-01-00-00865 0.6871[25]	1.3965[20]
7038	1-Hexene, 4-methyl- 4-Methyl-1-hexene	C_7H_{14} 98.19	3769-23-1 -141.5	86.7	3-01-00-00830 0.6942[25]	1.4000[20]
7039	1-Hexene, 5-methyl- 5-Methyl-1-hexene	C_7H_{14} 98.19	3524-73-0	85.3	4-01-00-00865 0.6877[25]	1.3967[20]

No.	Name Synonym	Mol. Form. Mol. Wt.	CAS RN mp/°C	Merck No. bp/°C	Beil. Ref. den/g cm^{-3}	Solubility n_D
7040	2-Hexene, 2-methyl- 2-Methyl-2-hexene	C$_7$H$_{14}$ 98.19	2738-19-4 -130.4	95.4	4-01-00-00863 0.7038[25]	1.4106[20]
7041	2-Hexene, 3-methyl-, (Z)-	C$_7$H$_{14}$ 98.19	10574-36-4 -118.5	95.6	4-01-00-00866 0.712[20]	1.4126[20]
7042	2-Hexene, 4-methyl-, (E)-	C$_7$H$_{14}$ 98.19	3683-22-5 -125.7	87.6	4-01-00-00867 0.6925[25]	1.4025[20]
7043	2-Hexene, 4-methyl-, (Z)-	C$_7$H$_{14}$ 98.19	3683-19-0	86.3	4-01-00-00867 0.6952[25]	1.4026[20]
7044	2-Hexene, 5-methyl-, (E)-	C$_7$H$_{14}$ 98.19	7385-82-2 -124.3	88.1	4-01-00-00865 0.6883[25]	1.4006[20]
7045	2-Hexene, 5-methyl-, (Z)-	C$_7$H$_{14}$ 98.19	13151-17-2	89.5	4-01-00-00865 0.697[25]	1.404[20]
7046	3-Hexene, 2-methyl-, (E)-	C$_7$H$_{14}$ 98.19	692-24-0 -141.6	85.9	4-01-00-00864 0.6853[25]	1.4001[20]
7047	3-Hexene, 2-methyl-, (Z)-	C$_7$H$_{14}$ 98.19	15840-60-5	86	4-01-00-00864 0.690[25]	1.401[20]
7048	1-Hexene, 1,1,2-trichloro-	C$_6$H$_9$Cl$_3$ 187.50	53977-99-4	90-3[10]	3-01-00-00803 1.125[25]	eth 4 1.476[25]
7049	2-Hexenoic acid	C$_6$H$_{10}$O$_2$ 114.14	1191-04-4 36.5	216.5	4-02-00-01563 0.965[20]	eth 4 1.4460[40]
7050	3-Hexenoic acid Hydrosorbic acid	C$_6$H$_{10}$O$_2$ 114.14	4219-24-3 12	208	4-02-00-01566 0.9640[23]	1.4935[20]
7051	5-Hexenoic acid 5-Hexanoic acid	C$_6$H$_{10}$O$_2$ 114.14	1577-22-6 -37	203	4-02-00-01562 0.9610[20]	eth 4; EtOH 4 1.4343[20]
7052	4-Hexenoic acid, 6-(1,3-dihydro-4-hydroxy-6-methoxy-7-methyl-3-oxo-5-isobenzofuranyl)-4-methyl-, (E)- Mycophenolic acid	C$_{17}$H$_{20}$O$_6$ 320.34	24280-93-1 141	6238	4-18-00-06513	H$_2$O 1; EtOH 4; eth 4; bz 2
7053	2-Hexenoic acid, ethyl ester Ethyl 2-hexenoate	C$_8$H$_{14}$O$_2$ 142.20	1552-67-6	174.5; 80[22]	3-02-00-01318 0.8986[20]	1.4348[20]
7054	3-Hexenoic acid, ethyl ester Hydrosorbic acid, ethyl ester	C$_8$H$_{14}$O$_2$ 142.20	2396-83-0	166.5	4-02-00-01567 0.8957[20]	1.4255[20]
7055	2-Hexenoic acid, 2-methyl-, (E)-	C$_7$H$_{12}$O$_2$ 128.17	97961-66-5	205; 118[11]	4-02-00-01581 0.9627[20]	eth 3; ace 3 1.4601[20]
7056	2-Hexenoic acid, 4-methyl- 4-Methyl-2-hexenoic acid	C$_7$H$_{12}$O$_2$ 128.17	37549-83-0	125[13]	3-02-00-01336 0.9441[20]	1.4526[20]
7057	2-Hexenoic acid, 5-methyl- 5-Methyl-2-hexenoic acid	C$_7$H$_{12}$O$_2$ 128.17	41653-96-7 16.5	226.5	2-02-00-00411 0.942[20]	1.4425[17]
7058	3-Hexenoic acid, 2-methyl- Hydrosorbic acid, 2-methyl-	C$_7$H$_{12}$O$_2$ 128.17	73513-50-5	122[24]	4-02-00-01582 0.9353[20]	1.4379[20]
7059	3-Hexenoic acid, 3-methyl-	C$_7$H$_{12}$O$_2$ 128.17	35205-71-1	113[10]	3-02-00-01335 0.9549[20]	1.4469[20]
7060	3-Hexenoic acid, 4-methyl- 4-Ethyl-3-pentenoic acid	C$_7$H$_{12}$O$_2$ 128.17	55665-79-7	118[12]	3-02-00-01336 0.9644[16]	1.4512[16]
7061	4-Hexenoic acid, 5-methyl-	C$_7$H$_{12}$O$_2$ 128.17	5636-65-7 -28	217	4-02-00-01581 0.9862[0]	1.4504
7062	3-Hexenoic acid, methyl ester Methyl 3-hexenoate	C$_7$H$_{12}$O$_2$ 128.17	2396-78-3	67-8[34]	4-02-00-01566 0.9132[25]	1.4240[23]
7063	2-Hexenoic acid, 2-methyl-, ethyl ester	C$_9$H$_{16}$O$_2$ 156.22	26311-33-1	72[10]	3-02-00-01333 0.9031[20]	1.4407[20]
7064	3-Hexenoic acid, 2-methyl-, ethyl ester	C$_9$H$_{16}$O$_2$ 156.22	21994-78-5	78[25]	3-02-00-01334 0.8778[20]	1.4237[20]
7065	3-Hexenoic acid, 3-methyl-, ethyl ester	C$_9$H$_{16}$O$_2$ 156.22	21994-77-4	84-5[26]	3-02-00-01335 0.8961[20]	1.4309[20]
7066	1-Hexen-3-ol	C$_6$H$_{12}$O 100.16	4798-44-1	134	4-01-00-02136 0.834[22]	ace 4; eth 4; EtOH 4 1.4297[18]
7067	4-Hexen-2-ol	C$_6$H$_{12}$O 100.16	52387-50-5	137.5	4-01-00-02139 0.8405[18]	1.4392[20]
7068	5-Hexen-2-ol	C$_6$H$_{12}$O 100.16	626-94-8	139	3-01-00-01926 0.8421[16]	H$_2$O 2
7069	2-Hexen-1-ol, (E)-	C$_6$H$_{12}$O 100.16	928-95-0	157	4-01-00-02138 0.8490[16]	1.4340[20]
7070	2-Hexen-1-ol, (Z)-	C$_6$H$_{12}$O 100.16	928-94-9	157	4-01-00-02138 0.8472[20]	H$_2$O 3; EtOH 4; eth 3; ace 3 1.4397[20]
7071	3-Hexen-1-ol, (Z)-	C$_6$H$_{12}$O 100.16	928-96-1	156.5	4-01-00-02141 0.8478[22]	H$_2$O 3; EtOH 4; eth 3 1.4380[20]
7072	4-Hexen-1-ol, (E)-	C$_6$H$_{12}$O 100.16	928-92-7	159	4-01-00-02140 0.8513[20]	1.4402[20]
7073	1-Hexen-3-ol, 3,5-dimethyl-	C$_8$H$_{16}$O 128.21	3329-48-4	146.5	4-01-00-02173 0.8382[20]	1.4342[20]
7074	4-Hexen-3-ol, 2,2-dimethyl- 2-Hexen-4-ol, 5,5-dimethyl	C$_8$H$_{16}$O 128.21	37674-67-2	75-6[50]	4-01-00-02171 0.8308[25]	1.4369[25]
7075	4-Hexen-3-ol, 2,5-dimethyl- 2,5-Dimethyl-4-hexen-3-ol	C$_8$H$_{16}$O 128.21	60703-31-3	162	2-01-00-00491 0.8444[20]	1.4449[20]
7076	4-Hexen-3-ol, 3,5-dimethyl- 2-Hexen-4-ol, 2,4-dimethyl	C$_8$H$_{16}$O 128.21	1569-43-3	48-50[3-6]	4-01-00-02172 0.8600[17]	1.4460[17]
7077	1-Hexen-3-ol, 5-methyl-	C$_7$H$_{14}$O 114.19	4798-46-3	125; 58[15]	3-01-00-01939 0.8306[15]	1.4263[23]
7078	2-Hexen-1-ol, 5-methyl- 5-Methyl-2-hexenol	C$_7$H$_{14}$O 114.19	77053-92-0	169	4-01-00-02158 0.8355[20]	1.4390[20]

No.	Name Synonym	Mol. Form. Mol. Wt.	CAS RN mp/°C	Merck No. bp/°C	Beil. Ref. den/g cm⁻³	Solubility n_D
7079	3-Hexen-2-ol, 2-methyl-	C₇H₁₄O 114.19	18812-62-9	49[11]	4-01-00-02157 0.8536[18]	1.443[18]
7080	3-Hexen-2-ol, 3-methyl-	C₇H₁₄O 114.19	110383-31-8	89[55]	0-01-00-00447 0.8678[9]	1.4487[9]
7081	4-Hexen-3-ol, 2-methyl-	C₇H₁₄O 114.19	4798-60-1	139.5	3-01-00-01939 0.8426[20]	1.438[20]
7082	4-Hexen-3-ol, 3-methyl-	C₇H₁₄O 114.19	60111-14-0	72-3[60]	4-01-00-02160 0.8471[25]	1.4268[16]
7083	5-Hexen-2-ol, 2-methyl- Dimethyl(vinylethyl)carbinol	C₇H₁₄O 114.19	16744-89-1 -22	142.5	4-01-00-02158 0.8397[15]	1.4349[17]
7084	5-Hexen-3-ol, 3-methyl- Methylethylallylcarbinol	C₇H₁₄O 114.19	1569-44-4	140; 61[35]	4-01-00-02160 0.8432[20]	1.4370[20]
7085	5-Hexen-3-ol, 4-methyl- 4-Methyl-5-hexen-3-ol	C₇H₁₄O 114.19	1838-77-3	140.5	3-01-00-01940 0.8452[22]	1.4365[22]
7086	3-Hexen-2-one 1-Butenyl methyl ketone	C₆H₁₀O 98.14	763-93-9	140	4-01-00-03468 0.8655[20]	ace 4; eth 4; EtOH 4 1.4418[20]
7087	4-Hexen-2-one 2-Hexen-5-one	C₆H₁₀O 98.14	25659-22-7	127	3-01-00-02994 0.8520[16]	1.4300[16]
7088	4-Hexen-3-one	C₆H₁₀O 98.14	2497-21-4	138.5	4-01-00-03468 0.8559[20]	EtOH 3; eth 3; ace 4 1.4388[20]
7089	5-Hexen-2-one	C₆H₁₀O 98.14	109-49-9	129.5	4-01-00-03467 0.833[27]	1.4178[27]
7090	3-Hexen-2-one, (E)-	C₆H₁₀O 98.14	4376-23-2	140	4-01-00-03469 0.8665[20]	EtOH 3; eth 3; ace 4 1.4418[20]
7091	3-Hexen-2-one, 3,4-dimethyl-	C₈H₁₄O 126.20	1635-02-5	158; 65[20]	4-01-00-03500 0.8585[20]	1.4418[20]
7092	4-Hexen-2-one, 3,4-dimethyl- 2-Hexen-5-one, 3,4-dimethyl	C₈H₁₄O 126.20	53252-21-4	154	4-01-00-03500 0.8538[19]	1.4377[19]
7093	1-Hexen-3-one, 5-methyl- Isobutyl vinyl ketone	C₇H₁₂O 112.17	2177-32-4	41[22]; 32[10]	4-01-00-03483 0.8400[15]	1.4293[15]
7094	3-Hexen-2-one, 5-methyl- 2-Oxo-5-methylhex-3-ene	C₇H₁₂O 112.17	5166-53-0	77.5[50]	3-01-00-03005 0.8407[22]	1.4395[22]
7095	4-Hexen-2-one, 5-methyl-	C₇H₁₂O 112.17	28332-44-7	76[50]	4-01-00-03482 0.9012[21]	1.4385[20]
7096	4-Hexen-3-one, 2-methyl-	C₇H₁₂O 112.17	53252-19-0	148	4-01-00-03483 0.843[20]	EtOH 3; eth 3; ace 4 1.4345[20]
7097	5-Hexen-2-one, 3-methyl- 1-Methyl-1-allyl acetone	C₇H₁₂O 112.17	2550-22-3	137	4-01-00-03484 0.845[15]	1.4215[25]
7098	5-Hexen-2-one, 4-methyl- 4-Methyl-5-hexen-2-one	C₇H₁₂O 112.17	61675-14-7	137.5	3-01-00-03006 0.8273[25]	1.4193[22]
7099	5-Hexen-2-one, 5-methyl- 5-Hexene-2-one,5-methyl	C₇H₁₂O 112.17	3240-09-3	150	4-01-00-03482 0.8460[20]	ace 4; eth 4; EtOH 4 1.4348[20]
7100	1-Hexen-3-one, 5-methyl-1-phenyl-	C₁₃H₁₆O 188.27	2892-18-4 43	154[25]	4-07-00-01058 0.9509[46]	H₂O 2; EtOH 3; bz 3; chl 3 1.5523[25]
7101	5-Hexen-2-one, 4-phenyl-	C₁₂H₁₄O 174.24	50552-30-2	85-6[1]	3-07-00-01460 0.9848[25]	1.5193[25]
7102	5-Hexen-2-one, 6-phenyl- 6-Phenyl-5-hexen-2-one	C₁₂H₁₄O 174.24	69371-59-1	154[10]; 98[0.3]	3-07-00-01455 0.9967[25]	1.5458[25]
7103	1-Hexen-3-yne Ethylvinylacetylene	C₆H₈ 80.13	13721-54-5	85	4-01-00-01091 0.7492[20]	bz 4; eth 4; peth 4; chl 4 1.4522[20]
7104	1-Hexen-4-yne	C₆H₈ 80.13	5009-11-0	87	4-01-00-01092 0.767[14]	1.446[14]
7105	1-Hexen-5-yne Hex-5-en-1-yne	C₆H₈ 80.13	14548-31-3	70	4-01-00-01092 0.7650[20]	bz 4; eth 4; peth 4; chl 4 1.4318[20]
7106	2-Hexen-4-yne 2-Hexene-4-yne	C₆H₈ 80.13	14092-20-7	88.5	4-01-00-01092 0.7710[20]	1.4918[20]
7107	1-Hexen-3-yne, 5-chloro-5-methyl- Dimethylvinylethynylchloromethane	C₇H₉Cl 128.60	819-44-3	48[28]	4-01-00-01098 0.9375[15]	ace 4; bz 4; eth 4; EtOH 4 1.4778[20]
7108	3-Hexen-1-yne, 3-propyl-	C₉H₁₄ 122.21	688-52-8	136	3-01-00-01049 0.7799[25]	bz 4; peth 4; chl 4 1.4432[25]
7109	Hexitol, 2,5-anhydro-3,4-dideoxy- Tetrahydro-2,5-furandimethanol	C₆H₁₂O₃ 132.16	104-80-3 <-50	9145 265	4-17-00-02052 1.154[20]	H₂O 4; ace 4; bz 4; EtOH 4
7110	D-arabino-Hexose, 2-deoxy- 2-Deoxy-D-glucose	C₆H₁₂O₅ 164.16	154-17-6 146.5	2886	4-01-00-04282	
7111	D-ribo-Hexose, 2,6-dideoxy- Digitoxose	C₆H₁₂O₄ 148.16	527-52-6 112	3147	4-01-00-04191 	H₂O 4; ace 4; py 3; AcOEt 3
7112	ribo-Hexose, 2,6-dideoxy-3-O-methyl- Cymarose	C₇H₁₄O₄ 162.19	579-04-4 101	2769	4-01-00-04193 	H₂O 4; ace 4; EtOH 4
7113	3-Hexynal	C₆H₈O 96.13	89533-67-5	42-3[17]	4-01-00-03544 0.9036[23]	1.4498[22]
7114	1-Hexyne Butylacetylene	C₆H₁₀ 82.15	693-02-7 -131.9	71.3	4-01-00-01006 0.7155[20]	H₂O 1; EtOH 3; eth 3; bz 3 1.3989[20]
7115	2-Hexyne 1-Methyl-2-propylacetylene	C₆H₁₀ 82.15	764-35-2 -89.6	84.5	4-01-00-01009 0.7315[20]	H₂O 1; EtOH 5; eth 5; bz 3 1.4138[20]
7116	3-Hexyne Diethylacetylene	C₆H₁₀ 82.15	928-49-4 -103	81	4-01-00-01009 0.7231[20]	H₂O 1; EtOH 3; eth 3; bz 3 1.4115[20]
7117	3-Hexyne, 2,5-dichloro-2,5-dimethyl- 3-Hexyne, 2,5-dimethyl-2,5-dichloro	C₈H₁₂Cl₂ 179.09	2431-30-3 29	178; 62[14]	4-01-00-01042 1.0118[20]	chl 4
7118	3-Hexyne-2,5-diol	C₆H₁₀O₂ 114.14	3031-66-1	120[11]	3-01-00-02271 1.0180[20]	1.4691[20]

No.	Name Synonym	Mol. Form. Mol. Wt.	CAS RN mp/°C	Merck No. bp/°C	Beil. Ref. den/g cm^{-3}	Solubility n_D
7119	3-Hexyne-2,5-diol, 2,5-dimethyl-	$C_8H_{14}O_2$ 142.20	142-30-3 95	205	4-01-00-02699 0.947[20]	H_2O 3; EtOH 4; eth 4; ace 4
7120	1-Hexyne, 5-methyl- 5-Methyl-1-hexyne	C_7H_{12} 96.17	2203-80-7 -125	92	3-01-00-01000 0.7274[20]	H_2O 1; EtOH 3; eth 3; bz 3 1.4059[20]
7121	2-Hexyne, 5-methyl-	C_7H_{12} 96.17	53566-37-3 -92.9	102.5	4-01-00-01030 0.7378[20]	H_2O 1; eth 3; ace 3; bz 3 1.4176[20]
7122	3-Hexyne, 2-methyl- 1-Ethyl-2-isopropylacetylene	C_7H_{12} 96.17	36566-80-0 -116.7	95.2	4-01-00-01029 0.7263[20]	bz 4; eth 4; chl 4; peth 4 1.4120[20]
7123	2-Hexynoic acid Propiolic acid, propyl-	$C_6H_8O_2$ 112.13	764-33-0 27	126[24]	4-02-00-01699 0.9820[20]	eth 4; EtOH 4; lig 4
7124	1-Hexyn-3-ol	$C_6H_{10}O$ 98.14	105-31-7 -80	142	4-01-00-02234 0.8704[20]	ctc 3 1.4340[25]
7125	2-Hexyn-1-ol	$C_6H_{10}O$ 98.14	764-60-3	100[53]; 80[21]	4-01-00-02235 0.8472[20]	1.4341[20]
7126	3-Hexyn-1-ol 3-Hexynol	$C_6H_{10}O$ 98.14	1002-28-4	163; 65[12]	4-01-00-02236 0.8982[20]	1.4530[20]
7127	5-Hexyn-2-ol	$C_6H_{10}O$ 98.14	23470-12-4	152	4-01-00-02235 0.899[20]	1.4481[20]
7128	1-Hexyn-3-ol, 3-methyl- Ethynylmethylpropylcarbinol	$C_7H_{12}O$ 112.17	4339-05-3	137	4-01-00-02252 0.8620[20]	H_2O 4; eth 4; EtOH 4 1.4338[20]
7129	3-Hexyn-2-ol, 2-methyl- 2-Methyl-3-hexyn-2-ol	$C_7H_{12}O$ 112.17	5075-33-2 -6	146	4-01-00-02250 0.962[25]	eth 4; EtOH 4 1.4392[25]
7130	3-Hexyn-2-one 3-Hexyne-2-one	C_6H_8O 96.13	1679-36-3	77-8[75]	4-01-00-03544 0.8804[20]	1.4400[20]
7131	5-Hexyn-2-one	C_6H_8O 96.13	2550-28-9	149	4-01-00-03543 0.9065[20]	1.4366[20]
7132	L-Histidine Glyoxaline-5-alanine	$C_6H_9N_3O_2$ 155.16	71-00-1 287 dec	4642	4-25-00-04343	H_2O 3; EtOH 2; eth 1; ace 1
7133	L-Histidine, N-β-alanyl- Carnosine	$C_9H_{14}N_4O_3$ 226.24	305-84-0 260	1857	4-25-00-04381	H_2O 4
7134	Homoserine (S)-2-Amino-4-hydroxybutanoic acid	$C_4H_9NO_3$ 119.12	672-15-1 203 dec	4661	4-04-00-03187	H_2O 4; EtOH 2; eth 1; bz 1
7135	L-Homoserine, O-[(aminoiminomethyl)amino]- Canavanine	$C_5H_{12}N_4O_3$ 176.18	543-38-4	1745		H_2O 4
7136	Hydramethylnon Tetrahydro-5,5-dimethyl-2-(1H)- pyrimidinone[3[4-(trifluoromethyl-styryl)- cinnamylidene- hydrazone	$C_{25}H_{24}F_6N_4$ 494.48	67485-29-4 190	4684		
7137	Hydrazine, 1,2-bis(1-methylethyl)-	$C_6H_{16}N_2$ 116.21	3711-34-0	125	4-04-00-03339 0.7894[20]	ace 4; bz 4; eth 4; EtOH 4 1.4173[20]
7138	Hydrazine, 1,2-bis(3-methylphenyl)-	$C_{14}H_{16}N_2$ 212.29	621-26-1 38	224	4-15-00-01220	H_2O 1; EtOH 3; eth 3; bz 3
7139	Hydrazine, 1,2-bis(4-methylphenyl)-	$C_{14}H_{16}N_2$ 212.29	637-47-8 135		4-15-00-01230 0.957[20]	bz 4; eth 4; EtOH 4
7140	Hydrazine, 1,2-bis(2-methylpropyl)-	$C_8H_{20}N_2$ 144.26	3711-37-3	171; 63[18]	2-04-00-00962 0.8002[20]	ace 4; bz 4; eth 4; EtOH 4 1.4276
7141	Hydrazine, (4-bromophenyl)- (p-Bromophenyl)hydrazine	$C_6H_7BrN_2$ 187.04	589-21-9 108	1417	4-15-00-00282	eth 4; EtOH 4; lig 4
7142	Hydrazinecarbothioamide Thiosemicarbazide	CH_5N_3S 91.14	79-19-6 183	9292	4-03-00-00374	H_2O 4; EtOH 4
7143	Hydrazinecarboxamide, monohydrochloride Semicarbazide hydrochloride	CH_6ClN_3O 111.53	563-41-7 175-7 dec	8396	4-03-00-00177	H_2O 4
7144	Hydrazinecarboxamide, 2-[(5-nitro-2- furanyl)methylene]- Nitrofurazone	$C_6H_8N_4O_4$ 198.14	59-87-0 236-40 dec	6521	5-17-09-00335	H_2O 1; EtOH 2; eth 1; DMSO 2
7145	Hydrazinecarboxamide, 2-phenyl- Phenicarbazide	$C_7H_9N_3O$ 151.17	103-03-7 172	7193	4-15-00-00180	H_2O 2; EtOH 3; eth 2; ace 3
7146	Hydrazinecarboxamide, N-phenyl- 4-Phenylsemicarbazide	$C_7H_9N_3O$ 151.17	537-47-3 128	7283	3-12-00-00822	H_2O 2; EtOH 4; eth 1; chl 4
7147	Hydrazinecarboximidamide Aminoguanidine	CH_6N_4 74.09	79-17-4 dec	453	4-03-00-00236	H_2O 4; EtOH 4
7148	Hydrazinecarboxylic acid, ethyl ester	$C_3H_8N_2O_2$ 104.11	4114-31-2 46	198 dec; 93[9]	4-03-00-00174	EtOH 3; eth 3; chl 2
7149	Hydrazinecarboxylic acid, methyl ester	$C_2H_6N_2O_2$ 90.08	6294-89-9 73	108[12]	2-03-00-00078	H_2O 3; EtOH 3; bz 2; peth 1
7150	1,1-Hydrazinedicarboxylic acid, diethyl ester	$C_6H_{12}N_2O_4$ 176.17	5311-96-6 30.5	138[12]	2-03-00-00079	EtOH 4
7151	1,2-Hydrazinedicarboxylic acid, diethyl ester	$C_6H_{12}N_2O_4$ 176.17	4114-28-7 135	250 dec	4-03-00-00175 1.324[8]	eth 4; EtOH 4
7152	Hydrazine, 1,1-diethyl- 1,1-Diethylhydrazine	$C_4H_{12}N_2$ 88.15	616-40-0	99	4-04-00-03335 0.8804[20]	bz 4; eth 4; EtOH 4; chl 4 1.4214[20]
7153	Hydrazine, 1,2-diethyl- 1,2-Diethylhydrazine	$C_4H_{12}N_2$ 88.15	1615-80-1	85.5	4-04-00-03336 0.797[26]	bz 4; eth 4; EtOH 4 1.4204[20]
7154	Hydrazine, 1,1-dimethyl- 1,1-Dimethylhydrazine	$C_2H_8N_2$ 60.10	57-14-7 -58	3236 63.9	4-04-00-03322 0.791[22]	H_2O 4; EtOH 4; eth 4; MeOH 4 1.4075[22]
7155	Hydrazine, 1,2-dimethyl- 1,2-Dimethylhydrazine	$C_2H_8N_2$ 60.10	540-73-8 81	3237	4-04-00-03323 0.8274[20]	H_2O 5; EtOH 5; eth 5 1.4209[20]
7156	Hydrazine, (2,4-dinitrophenyl)- (2,4-Dinitrophenyl)hydrazine	$C_6H_6N_4O_4$ 198.14	119-26-6 194	3277	4-15-00-00380	H_2O 1; EtOH 3; eth 2; bz 2

No.	Name Synonym	Mol. Form. Mol. Wt.	CAS RN mp/°C	Merck No. bp/°C	Beil. Ref. den/g cm^{-3}	Solubility n_D
7157	Hydrazine, 1,1-diphenyl- 1,1-Diphenylhydrazine	$C_{12}H_{12}N_2$ 184.24	530-50-7 50.5	3326 220[40]	4-15-00-00055 1.190[16]	bz 4; eth 4; EtOH 4; chl 4
7158	Hydrazine, 1,2-diphenyl- N,N'-Diphenylhydrazine	$C_{12}H_{12}N_2$ 184.24	122-66-7 131		4-15-00-00056 1.158[16]	EtOH 4; bz 2; DMSO-d_6 2; HOAc 1
7159	Hydrazine, 1-ethyl-1-phenyl- 1-Phenyl-1-ethylhydrazine	$C_8H_{12}N_2$ 136.20	644-21-3 	 237	3-15-00-00074 1.0181[21]	ace 4; bz 4; eth 4; EtOH 4 1.5711[21]
7160	Hydrazine, 1-ethyl-2-phenyl- 1-Ethyl-2-phenylhydrazine	$C_8H_{12}N_2$ 136.20	622-82-2 	241; 110[14]	4-15-00-00054 1.0150[20]	bz 4; eth 4; EtOH 4; chl 4 1.5676[20]
7161	Hydrazine, methyl- Methylhydrazine	CH_6N_2 46.07	60-34-4 -52.4	6001 87.5	4-04-00-03322 	H_2O 3; EtOH 5; eth 3; ctc 3 1.4325[20]
7162	Hydrazine, 1-(3-methylbutyl)-1-phenyl- Hydrazine, 1-isopentyl-1-phenyl	$C_{11}H_{18}N_2$ 178.28	636-10-2 236	 260	0-15-00-00121 0.9588[15]	
7163	Hydrazine, 1-methyl-2-(3-methylphenyl)- Hydrazine, 1-methyl-2-(3-tolyl)	$C_8H_{12}N_2$ 136.20	116836-06-7 60		2-15-00-00229 1.0265[100]	bz 4; EtOH 4
7164	Hydrazine, (3-methylphenyl)-	$C_7H_{10}N_2$ 122.17	536-89-0 	 244 dec	4-15-00-01220 1.057[20]	H_2O 2; EtOH 4; eth 4; bz 4
7165	Hydrazine, 1-methyl-1-phenyl-	$C_7H_{10}N_2$ 122.17	618-40-6 	228; 131[35]	4-15-00-00053 1.0404[20]	H_2O 2; EtOH 4; eth 5; bz 5 1.5691[20]
7166	Hydrazine, 1-methyl-2-phenyl- 1-Methyl-2-phenylhydrazine	$C_7H_{10}N_2$ 122.17	622-36-6 	231; 112[14]	4-15-00-00054 1.0320[20]	bz 4; eth 4; EtOH 4; chl 4 1.5733[20]
7167	Hydrazine, 1-naphthalenyl-	$C_{10}H_{10}N_2$ 158.20	2243-55-2 117	 203[20]	3-15-00-00728 	bz 4; eth 4; EtOH 4; chl 4
7168	Hydrazine, (4-nitrophenyl)- (4-Nitrophenyl)hydrazine	$C_6H_7N_3O_2$ 153.14	100-16-3 158 dec	6545	4-15-00-00317 	H_2O 2; EtOH 3; eth 3; bz 3
7169	Hydrazine, phenyl- Phenylhydrazine	$C_6H_8N_2$ 108.14	100-63-0 19.6	7264 243.5	4-15-00-00050 1.0986[20]	H_2O 3; EtOH 5; eth 5; ace 4 1.6084[10]
7170	Hydrazine, (1-phenylethyl)-	$C_8H_{12}N_2$ 136.20	65-64-5 	69[0.65]	4-15-00-01267 0.9672[25]	 1.5436[25]
7171	Hydrazine, phenyl-, hemihydrate	$C_{12}H_{18}N_4O$ 234.30	6152-31-4 24	120[12]	3-15-00-00067 1.0938[25]	 1.6081[20]
7172	Hydrazine, phenyl-, monohydrochloride	$C_6H_9ClN_2$ 144.60	59-88-1 243-6 dec	7265 sub		H_2O 4; EtOH 4
7173	Hydrazine, triphenyl- Triphenylhydrazine	$C_{18}H_{16}N_2$ 260.34	606-88-2 142 dec		4-15-00-00058 0.869[70]	EtOH 2; bz 2
7174	Hydrazobenzene, 3-methyl- Hydrazine, 1-phenyl-2-(3-tolyl)	$C_{13}H_{14}N_2$ 198.27	621-25-0 61		2-15-00-00229 1.0265[100]	bz 4; EtOH 4; lig 4
7175	Hydrocinnamic acid, β-ethyl-β-methyl- Valeric acid, 3-methyl-3-phenyl	$C_{12}H_{16}O_2$ 192.26	105401-59-0 284	174[14]	4-09-00-01905 1.050[25]	H_2O 3 1.5197[25]
7176	Hydrocyanic acid Hydrogen cyanide	CHN 27.03	74-90-8 -13.4	4722 26	4-02-00-00050 0.6876[20]	H_2O 5; EtOH 5; eth 5 1.2614[20]
7177	Hydroperoxide, cyclohexyl Cyclohexyl hydroperoxide	$C_6H_{12}O_2$ 116.16	766-07-4 -20	42[0.1]	4-06-00-00053 1.019[20]	eth 4; EtOH 4; HOAc 4 1.4645[25]
7178	Hydroperoxide, 1,1-dimethylethyl tert-Butyl hydroperoxide	$C_4H_{10}O_2$ 90.12	75-91-2 6	1569 89 dec; 36[17]	4-01-00-01616 0.8960[20]	H_2O 4; EtOH 3; eth 3; ctc 3 1.4015[20]
7179	Hydroperoxide, ethyl Ethyl hydrogen peroxide	$C_2H_6O_2$ 62.07	3031-74-1 -100	 95	4-01-00-01323 0.9332[20]	H_2O 4; bz 4; eth 4; EtOH 4 1.3800[20]
7180	Hydroperoxide, methyl	CH_4O_2 48.04	3031-73-0 	38-40[65]	4-01-00-01249 1.9967[15]	H_2O 4; bz 4; eth 4; EtOH 4 1.3641[15]
7181	Hydroperoxide, 1-methyl-1-phenylethyl Cumene hydroperoxide	$C_9H_{12}O_2$ 152.19	80-15-9 		4-06-00-03221 	
7182	Hydroprene 2,4-Dodecadienoic acid, 3,7,11-trimethyl-, ethyl ester, (E,E)-	$C_{17}H_{30}O_2$ 266.42	41096-46-2 	174[19]	 0.8955[20]	
7183	Hypochlorous acid, 1,1-dimethylethyl ester tert-Butyl hypochlorite	C_4H_9ClO 108.57	507-40-4 	1570 77.5	4-01-00-01621 0.9583[18]	H_2O 1; eth 4; ace 3; bz 4 1.403[20]
7184	Hypochlorous acid, ethyl ester Ethyl hypochlorite	C_2H_5ClO 80.51	624-85-1 	 37	4-01-00-01324 1.013[-6]	H_2O 1; EtOH 3; eth 5; bz 5
7185	Hypophosphoric acid, tetraethyl ester	$C_8H_{20}O_6P_2$ 274.19	679-37-8 	116-7[2]	4-01-00-01358 1.1283[18]	bz 3; ctc 3 1.4284[20]
7186	Ibogamine, 12-methoxy- Ibogaine	$C_{20}H_{26}N_2O$ 310.44	83-74-9 148	4806	5-23-12-00284 	chl 3
7187	Imazalil 1H-Imidazole, 1-[2-(2,4-dichlorophenyl)-2-(2-propenyloxy)ethyl]-	$C_{14}H_{14}Cl_2N_2O$ 297.18	35554-44-0 50	3537 dec	1.243[23]	
7188	Imazapyr 2-(4-Isopropyl-4-methyl-5-oxo-2-imidazolin-2-yl)nicotinic acid	$C_{13}H_{15}N_3O_3$ 261.28	81334-34-1 171			
7189	Imazaquin (RS)-2-(4-Isopropyl-4-methyl-5-oxo-2-imidazolin-2-yl)quinoline-3-carboxylic acid	$C_{17}H_{17}N_3O_3$ 311.34	81335-37-7 221	4826		
7190	Imazethapyr (RS)-5-Ethyl-2-(4-isopropyl-4-methyl-5-oxo-2-imidazolin-2-yl)nicotinic acid	$C_{15}H_{19}N_3O_3$ 289.33	81335-77-5 173			
7191	Imidazo[4,5-d]imidazole-2,5(1H,3H)-dione, tetrahydro- Acetyleneurea	$C_4H_6N_4O_2$ 142.12	496-46-8 300 dec	87	2-26-00-00260 	H_2O 2; EtOH 1; eth 3; HCl 3
7192	1H-Imidazole 1,3-Diazole	$C_3H_4N_2$ 68.08	288-32-4 90.5	4828 257	5-23-04-00191 1.0303[101]	H_2O 4; EtOH 4; eth 3; ace 3 1.4801[101]
7193	1H-Imidazole, 1,1'-carbonylbis- N,N'-Carbonyldiimidazole	$C_7H_6N_4O$ 162.15	530-62-1 119	1825	5-23-04-00245 	

No.	Name / Synonym	Mol. Form. / Mol. Wt.	CAS RN / mp/°C	Merck No. / bp/°C	Beil. Ref. / den/g cm⁻³	Solubility / n_D
7194	1H-Imidazole-1-carboxylic acid, 2,3-dihydro-3-methyl-2-thioxo-, ethyl ester Carbimazole	$C_7H_{10}N_2O_2S$ 186.23	22232-54-8 123.5	1803	4-24-00-00064	ace 4; chl 4
7195	1H-Imidazole-4,5-dicarboxylic acid	$C_5H_4N_2O_4$ 156.10	570-22-9 290 dec		4-25-00-01051 1.749[25]	H_2O 2; EtOH 1; eth 1; bz 1
7196	1H-Imidazole, 1,2-diethyl- 1,2-Diethylimidazole	$C_7H_{12}N_2$ 124.19	51807-53-5	219	5-23-05-00165 0.9813[25]	H_2O 4; eth 4; EtOH 4
7197	1H-Imidazole, 4,5-dihydro-2-methyl- Lysidine	$C_4H_8N_2$ 84.12	534-26-9 107	5508 196.5	5-23-03-00385	H_2O 4; EtOH 4; eth 1; chl 3
7198	1H-Imidazole, 4,5-dihydro-2-(phenylmethyl)- 2-Imidazolidine, 2-benzyl	$C_{10}H_{12}N_2$ 160.22	59-98-3	9430	5-23-06-00488	
7199	1H-Imidazole, 4,5-dihydro-2,4,5-triphenyl-, cis-	$C_{21}H_{18}N_2$ 298.39	573-33-1 134.5	198 dec	5-23-10-00053	bz 4; eth 4; EtOH 4
7200	1H-Imidazole, 1,2-dimethyl-	$C_5H_8N_2$ 96.13	1739-84-0	206	5-23-05-00039 1.0051[11]	H_2O 4; eth 4; EtOH 4
7201	1H-Imidazole, 1,4-dimethyl- 1,4-Dimethylimidazole	$C_5H_8N_2$ 96.13	6338-45-0 25.8	199	5-23-05-00094 0.996[16]	
7202	1H-Imidazole, 1,2-dimethyl-5-nitro- Dimetridazole	$C_5H_7N_3O_2$ 141.13	551-92-8 138.5	3255	5-23-05-00058	eth 4; EtOH 4
7203	1H-Imidazole-4-ethanamine Histamine	$C_5H_9N_3$ 111.15	51-45-6 83	4640 209[18]	4-25-00-02049	H_2O 3; EtOH 3; eth 2; chl 3
7204	1H-Imidazole-4-ethanaminium, α-carboxy-2,3-dihydro-N,N,N-trimethyl-2-thioxo-, hydroxide, in Ergothioneine	$C_9H_{15}N_3O_2S$ 229.30	497-30-3 290 dec	3611	4-25-00-04487	H_2O 4; EtOH 2; eth 1; ace 2
7205	1H-Imidazole-1-ethanol	$C_5H_8N_2O$ 112.13	1615-14-1 39	203[20]	5-23-04-00309	H_2O 4; EtOH 4
7206	1H-Imidazole-1-ethanol, 2-methyl-5-nitro- Metronidazole	$C_6H_9N_3O_3$ 171.16	443-48-1 160.5	6079	5-23-05-00063	
7207	1H-Imidazole, 1-ethyl-	$C_5H_8N_2$ 96.13	7098-07-9	208	5-23-04-00261 0.999[25]	H_2O 5
7208	1H-Imidazole, 1-methyl-	$C_4H_6N_2$ 82.11	616-47-7 -6	195.5	5-23-04-00256 1.0325[20]	H_2O 4; ace 4; eth 4; EtOH 4 1.4970[20]
7209	1H-Imidazole, 2-methyl-	$C_4H_6N_2$ 82.11	693-98-1 144	267	5-23-05-00035	H_2O 4; EtOH 4
7210	1H-Imidazole, 4-methyl-	$C_4H_6N_2$ 82.11	822-36-6 56	263	5-23-05-00089 1.0416[14]	H_2O 4; EtOH 4 1.5037[14]
7211	1H-Imidazole, 2-nitro- Azomycin	$C_3H_3N_3O_2$ 113.08	527-73-1 287 dec	934	5-23-04-00471	
7212	1H-Imidazole, 1-phenyl-	$C_9H_8N_2$ 144.18	7164-98-9 13	276	5-23-04-00267 1.1397[15]	H_2O 1; eth 4; ace 4; chl 4 1.6025[25]
7213	1H-Imidazole, 2-phenyl-	$C_9H_8N_2$ 144.18	670-96-2 149.3	340	5-23-07-00194	EtOH 4
7214	2H-Imidazole-2-thione, 1,3-dihydro-1-methyl- Methimazole	$C_4H_6N_2S$ 114.17	60-56-0 146	5892 280 dec	5-24-01-00305	H_2O 4; EtOH 3; eth 2; bz 2
7215	1H-Imidazole, 2,4,5-triphenyl-	$C_{21}H_{16}N_2$ 296.37	484-47-9 275	sub	5-23-10-00107	H_2O 1; EtOH 3; eth 3
7216	2,4-Imidazolidinedione Hydantoin	$C_3H_4N_2O_2$ 100.08	461-72-3 220	4678	5-24-05-00188	H_2O 4; EtOH 3; eth 2; alk 3
7217	2,4-Imidazolidinedione, 1,3-dibromo-5,5-dimethyl- Dibromantine	$C_5H_6Br_2N_2O_2$ 285.92	77-48-5 197-9 dec		5-24-05-00374	
7218	2,4-Imidazolidinedione, 5,5-dimethyl-	$C_5H_8N_2O_2$ 128.13	77-71-4 178	sub	5-24-05-00348	H_2O 4; EtOH 4; eth 4; ace 4
7219	2,4-Imidazolidinedione, 5,5-diphenyl- Phenytoin	$C_{15}H_{12}N_2O_2$ 252.27	57-41-0 286	7293	5-24-08-00376	H_2O 1; EtOH 3; eth 2; ace 3
7220	2,4-Imidazolidinedione, 5-(1-ethylpentyl)-3-[(trichloromethyl)thio]- Chlordantoin	$C_{11}H_{17}Cl_3N_2O_2S$ 347.69	5588-20-5	2080		CS_2 3
7221	2,4-Imidazolidinedione, 3-methyl-5-phenyl- 3-Methyl-5-phenylhydantoin	$C_{10}H_{10}N_2O_2$ 190.20	6846-11-3 164.5	6026	5-24-08-00005	chl 3
7222	2,4-Imidazolidinedione, 1-[[(5-nitro-2-furanyl)methylene]amino]- Nitrofurantoin	$C_8H_6N_4O_5$ 238.16	67-20-9 263	6520	5-24-05-00224	
7223	4-Imidazolidinehexanoic acid, 5-methyl-2-oxo-, (4R-cis)- Desthiobiotin	$C_{10}H_{18}N_2O_3$ 214.26	533-48-2 157	2914	4-25-00-01543	H_2O 3
7224	4-Imidazolidinehexanoic acid, 5-methyl-2-oxo-, methyl ester, (4R-cis)- Desthiobiotin, methyl ester	$C_{11}H_{20}N_2O_3$ 228.29	6020-51-5 69.5	194-7[0.03]	4-25-00-01544	EtOH 4
7225	2-Imidazolidinethione Ethylenethiourea	$C_3H_6N_2S$ 102.16	96-45-7 203	3759	5-24-01-00165	H_2O 4; EtOH 3; eth 1; bz 1
7226	Imidazolidinetrione Parabanic acid	$C_3H_2N_2O_3$ 114.06	120-89-8 244 dec	6970 sub 100	5-24-09-00009	H_2O 3; EtOH 4
7227	2-Imidazolidinone Ethyleneurea	$C_3H_6N_2O$ 86.09	120-93-4 131	4830	5-24-01-00022	H_2O 4; EtOH 4; eth 2; chl 2
7228	4-Imidazolidinone, 5,5-diphenyl- Doxenitoin	$C_{15}H_{14}N_2O$ 238.29	3254-93-1 183	3424	5-24-04-00145	
7229	4-Imidazolidinone, 5-(2-methylpropyl)-3-(2-propenyl)-2-thioxo- Albutoin	$C_{10}H_{16}N_2OS$ 212.32	830-89-7 210.5	210	4-24-00-01139	

No.	Name / Synonym	Mol. Form. Mol. Wt.	CAS RN mp/°C	Merck No. bp/°C	Beil. Ref. den/g cm^{-3}	Solubility n_D
7230	4H-Imidazol-4-one, 2-amino-1,5-dihydro-1-methyl- Creatinine	$C_4H_7N_3O$ 113.12	60-27-5 300 dec	2571	4-25-00-03543	H_2O 3; EtOH 2; eth 1; ace 1
7231	Imidodicarbonic acid, diethyl ester	$C_6H_{11}NO_4$ 161.16	19617-44-8 50	226	3-03-00-00050	H_2O 4
7232	Imidodicarbonic diamide Biuret	$C_2H_5N_3O_2$ 103.08	108-19-0 190 dec	1319	4-03-00-00141	H_2O 2; EtOH 4; eth 1
7233	Imidodicarbonimidic diamide Biguanide	$C_2H_7N_5$ 101.11	56-03-1 136	1233 142 dec	4-03-00-00162	H_2O 4; EtOH 3; bz 1; chl 1
7234	Imidodicarbonimidic diamide, N-phenyl- Phenyl biguanide	$C_8H_{11}N_5$ 177.21	102-02-3 143	7247	4-12-00-00771	
7235	1H-Indazole 1H-Benzopyrazole	$C_7H_6N_2$ 118.14	271-44-3 148	4848 269	5-23-06-00156	H_2O 3; EtOH 3; eth 3
7236	1H-Indazole, 3-chloro- 3-Chloroindazole	$C_7H_5ClN_2$ 152.58	29110-74-5 149	sub	5-23-06-00173	H_2O 3; EtOH 4; eth 4; bz 4
7237	1H-Indazole, 1,5-dimethyl- 1,5-Dimethylindazole	$C_9H_{10}N_2$ 146.19	70127-93-4 62.5	260	2-23-00-00157	ace 4; eth 4; EtOH 4; MeOH 4
7238	2H-Indazole, 2,3-dimethyl- 2,3-Dimethylindazole	$C_9H_{10}N_2$ 146.19	50407-18-6 79.5	286	5-23-06-00317	eth 4; EtOH 4
7239	1H-Indazole, 1-methyl- 1-Methylindazole	$C_8H_8N_2$ 132.17	13436-48-1 60.5	231	5-23-06-00159 1.0315[99]	H_2O 1; EtOH 3; ace 3; chl 3
7240	1H-Indazole, 3-methyl- 3-Methylindazole	$C_8H_8N_2$ 132.17	3176-62-3 113	280.5	5-23-06-00312	H_2O 3; EtOH 3; eth 3; chl 3
7241	1H-Indazole, 5-methyl- 5-Methylindazole	$C_8H_8N_2$ 132.17	1776-37-0 117	294	5-23-06-00319	H_2O 3; EtOH 3; eth 3; chl 3
7242	2H-Indazole, 2-methyl- 2-Methyl-2H-indazole	$C_8H_8N_2$ 132.17	4838-00-0 56	261	5-23-06-00165 1.0450[99]	H_2O 1; EtOH 3; eth 3; ace 3
7243	1H-Inden-1-amine, 2,3-dihydro-, (±)- (RS)-1-Aminoindan	$C_9H_{11}N$ 133.19	61949-83-5 221; 96[8]		3-12-00-02798 1.038[15]	H_2O 2; eth 3; ace 3; bz 3 1.5613[20]
7244	1H-Inden-5-amine, 2,3-dihydro-	$C_9H_{11}N$ 133.19	24425-40-9 37.5	248; 131[15]	4-12-00-02929	eth 3; ace 3; bz 3
7245	1H-Indene Indonaphthene	C_9H_8 116.16	95-13-6 -1.8	4851 182	4-05-00-01532 0.9960[25]	H_2O 1; EtOH 5; eth 5; ace 3 1.5768[20]
7246	1H-Indene-1-carboxylic acid 1-Indenecarboxylic acid	$C_{10}H_8O_2$ 160.17	5020-21-3 74	193-5[12]	3-09-00-03068	H_2O 2; EtOH 4; eth 2; bz 2
7247	1H-Indene, 1,2-dibromo-2,3-dihydro- 1,2-Dibromoindan	$C_9H_8Br_2$ 275.97	20357-79-3 32	140[11]	2-05-00-00377 1.7470[25]	
7248	2H-Indene-2,2-dicarboxylic acid, 1,3-dihydro-1,3-dioxo-, diethyl ester Phthaloylmalonic acid, diethyl ester	$C_{15}H_{14}O_6$ 290.27	116836-20-5		1-10-00-00441 1.1896[83]	eth 4; EtOH 4 1.541[84]
7249	1H-Indene, 1,2-dichloro-2,3-dihydro- 1,2-Dichloroindan	$C_9H_8Cl_2$ 187.07	74925-48-7 87-90[2]		3-05-00-01202 1.254[25]	1.5715[23]
7250	1H-Indene, 2,3-dihydro- Indan	C_9H_{10} 118.18	496-11-7 -51.4	4844 177.9	4-05-00-01371 0.9639[20]	H_2O 1; EtOH 5; eth 5; chl 2 1.5378[20]
7251	1H-Indene, 2,3-dihydro-1,1-dimethyl- 1,1-Dimethylindan	$C_{11}H_{14}$ 146.23	4912-92-9	191	4-05-00-01415 0.919[20]	1.5135[25]
7252	1H-Indene, 2,3-dihydro-1,2-dimethyl- 1,2-Dimethylindan	$C_{11}H_{14}$ 146.23	17057-82-8 79-80[10]		4-05-00-01415 0.927[20]	1.5186[20]
7253	1H-Indene, 2,3-dihydro-4,7-dimethyl- 4,7-Dimethylindan	$C_{11}H_{14}$ 146.23	6682-71-9 94-7[10]		4-05-00-01416 0.949[20]	1.5342[20]
7254	1H-Indene, 2,3-dihydro-5,6-dimethyl- 5,6-Dimethylindan	$C_{11}H_{14}$ 146.23	1075-22-5 94[10]		4-05-00-01416 0.9449[20]	1.5360[20]
7255	1H-Indene, 2,3-dihydro-1-methyl- 1-Methylindan	$C_{10}H_{12}$ 132.21	767-58-8 190.6		4-05-00-01397 0.9385[25]	H_2O 1 1.5266[20]
7256	1H-Indene, 2,3-dihydro-2-methyl- 2-Methylindan	$C_{10}H_{12}$ 132.21	824-63-5 195		4-05-00-01397 0.940[25]	H_2O 1 1.5220[20]
7257	1H-Indene, 2,3-dihydro-4-methyl- 4-Methylindan	$C_{10}H_{12}$ 132.21	824-22-6 205.3		4-05-00-01397 0.9577[20]	1.5356[20]
7258	1H-Indene, 2,3-dihydro-5-methyl- 5-Methylindan	$C_{10}H_{12}$ 132.21	874-35-1 202		4-05-00-01398 0.944[25]	1.5336[20]
7259	1H-Indene, 2,3-dihydro-2-phenyl- 2-Phenylindan	$C_{15}H_{14}$ 194.28	22253-11-8 162-3[10]		4-05-00-02198 1.0821[16]	1.5955[15]
7260	1H-Indene, 2,3-dihydro-1,1,4,7-tetramethyl- 1,1,4,7-Tetramethylindan	$C_{13}H_{18}$ 174.29	1078-04-2 114.2[15]		3-05-00-01282 0.934[25]	1.5216[25]
7261	1H-Indene, 2,3-dihydro-1,1,5-trimethyl- 1,1,5-Trimethylindan	$C_{12}H_{16}$ 160.26	40650-41-7 86[10]		4-05-00-01433 0.9119[20]	1.5126[20]
7262	1H-Indene, 2,3-dihydro-1,4,7-trimethyl- 1,4,7-Trimethylindan	$C_{12}H_{16}$ 160.26	54340-87-3 95[10]		4-05-00-01434 0.938[20]	1.5252[20]
7263	1H-Indene, 2,3-dihydro-1,1,3-trimethyl-3-phenyl- Indan, 1-phenyl-1,3,3-trimethyl	$C_{18}H_{20}$ 236.36	3910-35-8 52.5	308.5	4-05-00-02246 1.0009[20]	H_2O 1; EtOH 3; bz 3; MeOH 3 1.5681[20]
7264	1H-Indene-1,3(2H)-dione	$C_9H_6O_2$ 146.15	606-23-5 131-2 dec		4-07-00-02344 1.37[21]	H_2O 2; EtOH 4; eth 3; bz 3
7265	1H-Indene-1,3(2H)-dione, 2-(4-bromophenyl)- Bromindione	$C_{15}H_9BrO_2$ 301.14	1146-98-1 138	1381	4-07-00-02572	
7266	1H-Indene-1,3(2H)-dione, 2-(4-chlorophenyl)- Clorindione	$C_{15}H_9ClO_2$ 256.69	1146-99-2 145.5	2402	4-07-00-02571	bz 4; eth 4; EtOH 4
7267	1H-Indene-1,3(2H)-dione, 2,2-dihydroxy- Ninhydrin	$C_9H_6O_4$ 178.14	485-47-2 241-3 dec	6470	4-07-00-02786	H_2O 4; EtOH 3; eth 2; alk 3

No.	Name / Synonym	Mol. Form. / Mol. Wt.	CAS RN / mp/°C	Merck No. / bp/°C	Beil. Ref. / den/g cm^{-3}	Solubility / n_D
7268	1H-Indene-1,3(2H)-dione, 2-(diphenylacetyl)- / Diphenadione	$C_{23}H_{16}O_3$ / 340.38	82-66-6 / 146.5	3304	4-07-00-02838	ace 4; HOAc 4 / 1.670
7269	1H-Indene-1,3(2H)-dione, 2-(4-methoxyphenyl)- / Anisindione	$C_{16}H_{12}O_3$ / 252.27	117-37-3 / 156.5	698	3-08-00-02931	
7270	1H-Indene-1,3(2H)-dione, 2-phenyl- / Phenindione	$C_{15}H_{10}O_2$ / 222.24	83-12-5 / 150	7196	4-07-00-02570	H_2O 1; EtOH 3; eth 3; ace 3
7271	1H-Indene, 1,3-diphenyl- / 1,3-Diphenylindene	$C_{21}H_{16}$ / 268.36	4467-88-3 / 85	230[15]	3-05-00-02474	ace 4; eth 4
7272	1H-Indene, 2,3-diphenyl- / 2,3-Diphenylindene	$C_{21}H_{16}$ / 268.36	5324-00-5 / 109.3	237[12]	4-05-00-02641	H_2O 1; EtOH 3; eth 3; ace 3
7273	1H-Indene, 1-ethyl-2,3-dihydro- / 1-Ethylindan	$C_{11}H_{14}$ / 146.23	4830-99-3 /	212	4-05-00-01414 / 0.9348[25]	1.5121[25]
7274	1H-Indene, 5-hexyl-2,3-dihydro- / 5-Hexylindan	$C_{15}H_{22}$ / 202.34	54889-55-3 / -34	292.1	4-05-00-01474 / 0.9114[20]	1.5122[20]
7275	1H-Indene, 1-methyl- / 1-Methylindene	$C_{10}H_{10}$ / 130.19	767-59-9 /	199; 82[15]	3-05-00-01371 / 0.970[25]	1.5616[20]
7276	1H-Indene, 2-methyl- / 2-Methylindene	$C_{10}H_{10}$ / 130.19	2177-47-1 / 80	208	4-05-00-01545 / 0.974[25]	H_2O 1; eth 3; ace 3; bz 3 / 1.5652[20]
7277	1H-Indene, 3-methyl- / 3-Methylindene	$C_{10}H_{10}$ / 130.19	767-60-2 /	198	4-05-00-01545 / 0.972[25]	H_2O 1; eth 3; ace 3; bz 3 / 1.5621[20]
7278	1H-Indene, 4-methyl- / 4-Methylindene	$C_{10}H_{10}$ / 130.19	7344-34-5 /	209; 88[13]	2-05-00-00418 / 0.989[25]	1.568[20]
7279	1H-Indene, 6-methyl- / 6-Methylindene	$C_{10}H_{10}$ / 130.19	20232-11-5 /	207; 90[17]	4-05-00-01545 / 0.977[25]	1.566[20]
7280	1H-Indene, 7-methyl- / 7-Methylindene	$C_{10}H_{10}$ / 130.19	7372-92-1 /	209; 90[15]	2-05-00-00418 / 0.989[25]	1.568[20]
7281	1H-Indene, octahydro- / Hexahydroindan	C_9H_{16} / 124.23	496-10-6 / -53	167	4-05-00-00292 / 0.876[25]	1.4702[20]
7282	6H-Indeno[1,2-b]oxirene, 1a,6a-dihydro-	C_9H_8O / 132.16	768-22-9 / 24.5	113[20]; 98[6]	5-17-02-00033 / 1.1255[24]	chl 3
7283	1H-Inden-4-ol, 7-chloro-2,3-dihydro- / Chlorindanol	C_9H_9ClO / 168.62	145-94-8 / 92	2094	4-06-00-03828	
7284	1H-Inden-1-ol, 2,3-dihydro-	$C_9H_{10}O$ / 134.18	6351-10-6 / 54.8	220; 128[12]	4-06-00-03824	bz 4; EtOH 4; chl 4
7285	1H-Inden-5-ol, 2,3-dihydro-	$C_9H_{10}O$ / 134.18	1470-94-6 / 58	253	4-06-00-03829	H_2O 2; EtOH 4; eth 4; sulf 3
7286	1H-Inden-1-one, 2,3-dihydro-	C_9H_8O / 132.16	83-33-0 / 42	243; 129[12]	4-07-00-00999 / 1.0943[40]	H_2O 2; EtOH 4; eth 4; ace 4 / 1.561[25]
7287	2H-Inden-2-one, 1,3-dihydro- / 2-Indanone	C_9H_8O / 132.16	615-13-4 / 59	218 dec	4-07-00-01002 / 1.0712[69]	H_2O 1; EtOH 4; eth 4; ace 4 / 1.538[67]
7288	1H-Inden-1-one, 2,3-dihydro-3,3-dimethyl-	$C_{11}H_{12}O$ / 160.22	26465-81-6 /	130-1[18]	4-07-00-01035 / 1.0320[14]	1.5453[14]
7289	1H-Inden-1-one, 2,3-dihydro-2-methyl-	$C_{10}H_{10}O$ / 146.19	17496-14-9 /	250; 125[18]	4-07-00-01019 / 1.0651[21]	1.553[22]
7290	1H-Inden-1-one, 2,3-dihydro-4-methyl- / 4-Methyl-1-indanone	$C_{10}H_{10}O$ / 146.19	24644-78-8 / 95	144[13]; 120[5]	4-07-00-01020	bz 4; eth 4; EtOH 4
7291	1H-Inden-1-one, 2,3-dihydro-2-phenyl- / 2-Phenyl-1-indanone	$C_{15}H_{12}O$ / 208.26	16619-12-8 / 77.5	344	4-07-00-01673	ace 4; bz 4; EtOH 4
7292	2H-Inden-2-one, octahydro-, cis-	$C_9H_{14}O$ / 138.21	5689-04-3 / 10	225; 113[25]	4-07-00-00174 / 0.9970[20]	EtOH 3; bz 3; lig 3 / 1.4830[20]
7293	2H-Inden-2-one, octahydro-, trans-	$C_9H_{14}O$ / 138.21	16484-17-6 / -12	218	4-07-00-00174 / 0.9807[17]	EtOH 3; bz 3; lig 3 / 1.4769[17]
7294	1H-Indole / 2,3-Benzopyrrole	C_8H_7N / 117.15	120-72-9 / 52.5	4869 / 253.6	5-20-07-00005 / 1.22[25]	H_2O 3; EtOH 4; eth 4; bz 3
7295	1H-Indole-3-acetic acid / Indoleacetic acid	$C_{10}H_9NO_2$ / 175.19	87-51-4 / 168.5	4870	5-22-03-00065	H_2O 1; EtOH 4; eth 3; ace 3
7296	1H-Indole-3-butanoic acid / Indolebutyric acid	$C_{12}H_{13}NO_2$ / 203.24	133-32-4 / 124.5	4871	5-22-03-00140	bz 4; DMSO 3; peth 1
7297	1H-Indole, 2,3-dihydro-	C_8H_9N / 119.17	496-15-1 /	229	5-20-06-00238 / 1.069[20]	H_2O 2; eth 3; ace 3; bz 3 / 1.5923[20]
7298	1H-Indole, 1,3-dimethyl-	$C_{10}H_{11}N$ / 145.20	875-30-9 / 142	258.5	5-20-07-00071	eth 3
7299	1H-Indole, 2,5-dimethyl- / 2,5-Dimethylindole	$C_{10}H_{11}N$ / 145.20	1196-79-8 / 115.3	188[40]	5-20-07-00111	H_2O 1; EtOH 2; eth 3; bz 3
7300	1H-Indole-2,3-dione / Isatin	$C_8H_5NO_2$ / 147.13	91-56-5 / 203 dec	4985	5-21-10-00221	H_2O 3; EtOH 4; eth 2; ace 3
7301	1H-Indole-5,6-dione, 2,3-dihydro-3-hydroxy-1-methyl- / Adrenochrome	$C_9H_9NO_3$ / 179.18	54-06-8 / 125 dec	162	5-21-13-00079	H_2O 4; EtOH 4; eth 1; bz 1
7302	1H-Indole-3-ethanamine / Tryptamine	$C_{10}H_{12}N_2$ / 160.22	61-54-1 / 118	9705 / 137[0.15]	5-22-10-00045	H_2O 1; EtOH 3; eth 1; ace 3
7303	1H-Indole-3-ethanamine, N,N-dimethyl- / N,N-Dimethyltryptamine	$C_{12}H_{16}N_2$ / 188.27	61-50-7 / 46	3251	5-22-10-00048	
7304	1H-Indole-3-ethanamine, 5-methoxy- / 5-Methoxytryptamine	$C_{11}H_{14}N_2O$ / 190.25	608-07-1 / 121.5	5926	5-22-12-00018	
7305	1H-Indole-3-ethanol / Tryptophol	$C_{10}H_{11}NO$ / 161.20	526-55-6 / 59	9708 / 174[2]	5-21-03-00061	ace 4; eth 4; EtOH 4; chl 4
7306	1H-Indole-3-ethanol, 2-(5-ethenyl-1-azabicyclo[2.2.2]oct-2-yl)- / Cinchonamine	$C_{19}H_{24}N_2O$ / 296.41	482-28-0 / 186	2287	4-23-00-02735	H_2O 1; EtOH 4; eth 4; bz 3

No.	Name Synonym	Mol. Form. Mol. Wt.	CAS RN mp/°C	Merck No. bp/°C	Beil. Ref. den/g cm⁻³	Solubility n_D
7307	1H-Indole, 1-ethyl-	$C_{10}H_{11}N$ 145.20	10604-59-8 105	252.5; 122[14]	5-20-07-00014 1.2563[15]	
7308	1H-Indole, 3-ethyl-	$C_{10}H_{11}N$ 145.20	1484-19-1 37	279; 138[6]	5-20-07-00099	bz 4; eth 4; EtOH 4
7309	1H-Indole, 3-ethyl-2-methyl-	$C_{11}H_{13}N$ 159.23	35246-18-5 44.5	292; 156[12]	5-20-07-00136 1.0414[0]	eth 4; EtOH 4
7310	1H-Indole-3-methanamine, N,N-dimethyl- Gramine	$C_{11}H_{14}N_2$ 174.25	87-52-5 138.5	4440	5-22-10-00025	H_2O 1; EtOH 3; eth 3; chl 3
7311	1H-Indole, 7-methoxy-	C_9H_9NO 147.18	3189-22-8	108-10[0.3]	5-21-03-00024 1.126[25]	
7312	1H-Indole, 1-methyl-	C_9H_9N 131.18	603-76-9	237	5-20-07-00012 1.0707[25]	H_2O 1; EtOH 3; eth 3; bz 3
7313	1H-Indole, 2-methyl-	C_9H_9N 131.18	95-20-5 61	272	5-20-07-00059 1.07[20]	H_2O 2; EtOH 4; eth 4; ace 4
7314	1H-Indole, 3-methyl- Skatole	C_9H_9N 131.18	83-34-1 97.5	8503 266	5-20-07-00069	H_2O 3; EtOH 3; eth 3; ace 3
7315	1H-Indole, 4-methyl-	C_9H_9N 131.18	16096-32-5 5	267	5-20-07-00079 1.062[20]	
7316	1H-Indole, 5-methyl-	C_9H_9N 131.18	614-96-0 60	267	5-20-07-00080 1.0202[78]	H_2O 3; EtOH 3; eth 3; bz 3
7317	1H-Indole, 7-methyl- 7-Methylindole	C_9H_9N 131.18	933-67-5 85	266	5-20-07-00081 1.0202[100]	
7318	1H-Indole, 3-methyl-2-phenyl-	$C_{15}H_{13}N$ 207.27	10257-92-8 91.5	280-90[120]	5-20-08-00274	H_2O 1; EtOH 3; bz 3
7319	1H-Indole, octahydro-	$C_8H_{15}N$ 125.21	4375-14-8 143	185.5	5-20-04-00358 0.9472[20] 1.4892[20]	bz 4; eth 4; EtOH 4
7320	1H-Indole, 2-phenyl-	$C_{14}H_{11}N$ 193.25	948-65-2 190.5	250[10]	5-20-08-00232	H_2O 2; eth 3; bz 3; chl 3
7321	1H-Indole, 1-(phenylmethyl)-	$C_{15}H_{13}N$ 207.27	3377-71-7 45.8	172[2]	5-20-07-00016	bz 4; eth 4; EtOH 4; lig 4
7322	1H-Indole, 1,2,3-trimethyl- 1,2,3-Trimethylindole	$C_{11}H_{13}N$ 159.23	1971-46-6 18	283	5-20-07-00102	H_2O 2; EtOH 2; eth 2; bz 2
7323	1H-Indole, 2,3,5-trimethyl- 2,3,5-Trimethylindole	$C_{11}H_{13}N$ 159.23	21296-92-4 121.5	297	5-20-07-00143	EtOH 4; chl 4; lig 4
7324	Indolizine	C_8H_7N 117.15	274-40-8 75	205	5-20-07-00047	H_2O 1; EtOH 3
7325	Indolizine, octahydro-	$C_8H_{15}N$ 125.21	13618-93-4	75[43]	5-20-04-00360 0.9074[10] 1.4748	eth 4; EtOH 4
7326	1H-Indol-5-ol, 3-[2-(dimethylamino)ethyl]- Bufotenine	$C_{12}H_{16}N_2O$ 204.27	487-93-4 146.5	1467 320[0.1]	5-22-12-00026	eth 4; EtOH 4
7327	2H-Indol-2-one, 1,3-dihydro-	C_8H_7NO 133.15	59-48-3 128	227[23]; 195[17]	5-21-08-00007	H_2O 3; EtOH 3; eth 3
7328	2H-Indol-2-one, 1,3-dihydro-1,3-dimethyl-	$C_{10}H_{11}NO$ 161.20	24438-17-3 55	273; 137[11]	5-21-08-00077	eth 4; EtOH 4
7329	2H-Indol-2-one, 1,3-dihydro-3,3-dimethyl-	$C_{10}H_{11}NO$ 161.20	19155-24-9 154.3	300	5-21-08-00113	bz 4; eth 4; EtOH 4
7330	2H-Indol-2-one, 1,3-dihydro-4,7-dimethyl- 4,7-Dimethyl-2-indolinone	$C_{10}H_{11}NO$ 161.20	59022-71-8 159	205-10[15]	4-21-00-03676	ace 4; EtOH 4; chl 4
7331	3H-Indol-3-one, 2-(1,3-dihydro-3-oxo-2H-indol-2-ylidene)-1,2-dihydro- Indigo	$C_{16}H_{10}N_2O_2$ 262.27	482-89-3 390 dec	4855 sub 300	5-24-08-00503	
7332	2H-Indol-2-one, 1-ethyl-1,3-dihydro-3-methyl- 2-Indolinone, 1-ethyl-3-methyl-	$C_{11}H_{13}NO$ 175.23	84258-49-1	283; 105[0.5]	2-21-00-00258 1.557[25]	eth 4; EtOH 4
7333	Indolo[2',3':3,4]pyrido[2,1-b]quinazolin-5(7H)-one, 8,13,13b,14-tetrahydro-14-methyl-, (S)- Evodiamine	$C_{19}H_{17}N_3O$ 303.36	518-17-2 28	3865	4-26-00-00525	
7334	Indolo[2',3':3,4]pyrido[2,1-b]quinazolin-5(7H)-one, 8,13-dihydro- Rutecarpine	$C_{18}H_{13}N_3O$ 287.32	84-26-4 259.5	8271	4-26-00-00528	EtOH 2; ace 2; bz 2
7335	Inosine Hypoxanthine riboside	$C_{10}H_{12}N_4O_5$ 268.23	58-63-9 218 dec	4880	4-26-00-02086	EtOH 4
7336	scyllo-Inositol	$C_6H_{12}O_6$ 180.16	488-59-5 353 dec		4-06-00-07920 1.659[19]	H_2O 2; EtOH 1; eth 1; bz 1
7337	Iodine cyanide Cyanogen iodide	CIN 152.92	506-78-5 146.7	2702 sub 45	4-03-00-00093 2.84[18]	eth 4; EtOH 4
7338	Iprodione 1-Imidazolidinecarboxamide, 3-(3,5-dichlorophenyl)-N-(1-methylethyl)-2,4-dioxo-	$C_{13}H_{13}Cl_2N_3O_3$ 330.17	36734-19-7 136	4964		
7339	Iron carbonyl Iron pentacarbonyl	C_5FeO_5 195.90	13463-40-6 -20	4980 103	1.5[20] 1.453[22]	H_2O 1; EtOH 2; bz 3; ace 3
7340	Isazophos Phosphorothioic acid, O-[2-chloro-l-(l-methylethyl)-lH-imidazol-4-yl] O,O-diethyl ester	$C_9H_{17}ClN_3O_3PS$ 313.74	67329-04-8	4988 170; 100[0.001]	1.22[20]	
7341	5-Isobenzofurancarboxylic acid, 1,3-dihydro-1,3-dioxo- Trimellitic anhydride	$C_9H_4O_5$ 192.13	552-30-7 162	9617 240-3[14]	5-18-08-00562	
7342	1,3-Isobenzofurandione Phthalic anhydride	$C_8H_4O_3$ 148.12	85-44-9 130.8	7346 295	5-17-11-00253 1.527[4]	H_2O 2; EtOH 3; eth 2; ace 3

No.	Name / Synonym	Mol. Form. / Mol. Wt.	CAS RN / mp/°C	Merck No. / bp/°C	Beil. Ref. / den/g cm^{-3}	Solubility / n_D
7343	1,3-Isobenzofurandione, 4,5-dichloro- / Phthalic anhydride, 3,4-dichloro	$C_8H_2Cl_2O_3$ / 217.01	56962-07-3 / 121	329	4-17-00-06142	EtOH 4; chl 4
7344	1,3-Isobenzofurandione, 5,6-dichloro- / 4,5-Dichlorophthalic anhydride	$C_8H_2Cl_2O_3$ / 217.01	942-06-3 / 188	313	5-17-11-00260	eth 4; EtOH 4; tol 4
7345	1,3-Isobenzofurandione, hexahydro- / Hexahydrophthalic anhydride	$C_8H_{10}O_3$ / 154.17	85-42-7 / 32	145[18]	5-17-11-00076	
7346	1,3-Isobenzofurandione, 4,5,6,7-tetrachloro- / Tetrachlorophthalic anhydride	$C_8Cl_4O_3$ / 285.90	117-08-8 / 254.5	sub	5-17-11-00260 / 1.49[275]	eth 2
7347	1,3-Isobenzofurandione, 4,5,6,7-tetrahydro- / 3,4,5,6-Tetrahydrophthalic anhydride	$C_8H_8O_3$ / 152.15	2426-02-0 / 74		5-17-11-00133 / 1.2[105]	EtOH 3; eth 4; ace 3; chl 3
7348	1,3-Isobenzofurandione, 4,5,6,7-tetraiodo-	$C_8I_4O_3$ / 651.70	632-80-4 / 327.5	sub	5-17-11-00266	H_2O 1; EtOH 1; bz 1; HOAc 2
7349	1(3H)-Isobenzofuranone	$C_8H_6O_2$ / 134.13	87-41-2 / 75	290	5-17-10-00007 / 1.1636[99]	H_2O 3; EtOH 4; eth 4; chl 2 / 1.536[99]
7350	1(3H)-Isobenzofuranone, 3,3-bis(3,5-dibromo-4-hydroxyphenyl)- / 3',3'',5',5''-Tetrabromophenolphthalein	$C_{20}H_{10}Br_4O_4$ / 633.91	76-62-0 / 296	9122	5-18-04-00190	H_2O 1; EtOH 2; eth 4; alk 3
7351	1(3H)-Isobenzofuranone, 3,3-bis(4-hydroxy-3,5-diiodophenyl)-	$C_{20}H_{10}I_4O_4$ / 821.91	386-17-4 / 308		4-18-00-01949 / 2.0201[22]	H_2O 1; EtOH 1; eth 1; chl 2
7352	1(3H)-Isobenzofuranone, 3,3-bis[4-hydroxy-2-methyl-5-(1-methylethyl)phenyl]- / Thymolphthalein	$C_{28}H_{30}O_4$ / 430.54	125-20-2 / 253	9336	5-18-04-00194	H_2O 1; EtOH 3; eth 3; ace 3
7353	1(3H)-Isobenzofuranone, 3,3-bis(4-hydroxy-3-methylphenyl)- / o-Cresolphthalein	$C_{22}H_{18}O_4$ / 346.38	596-27-0 / 223	2582	5-18-04-00193	EtOH 4
7354	1(3H)-Isobenzofuranone, 3,3-bis(4-hydroxyphenyl)- / Phenolphthalein	$C_{20}H_{14}O_4$ / 318.33	77-09-8 / 262.5	7208	5-18-04-00188 / 1.277[32]	H_2O 1; EtOH 4; eth 3; ace 4
7355	1(3H)-Isobenzofuranone, 6,7-dimethoxy- / Meconin	$C_{10}H_{10}O_4$ / 194.19	569-31-3 / 102.5	5665	5-18-03-00068	H_2O 2; EtOH 3; eth 3; ace 3
7356	1(3H)-Isobenzofuranone, 6,7-dimethoxy-3-(5,6,7,8-tetrahydro-4-methoxy-6-methyl-1,3-dioxolo[4,5-g]isoquinolin-5 / Noscapine	$C_{22}H_{23}NO_7$ / 413.43	128-62-1 / 176	6638	4-27-00-06838	H_2O 1; EtOH 3; eth 2; ace 4
7357	1(3H)-Isobenzofuranone, 6,7-dimethoxy-3-(5,6,7,8-tetrahydro-6-methyl-1,3-dioxolo[4,5-g]isoquinolin-5-yl)-, [S- / Hydrastine	$C_{21}H_{21}NO_6$ / 383.40	118-08-1 / 132	4688	4-27-00-06833	H_2O 1; ace 3; bz 3
7358	1(3H)-Isobenzofuranone, 3,3-diphenyl-	$C_{20}H_{14}O_2$ / 286.33	596-29-2 / 120	235[15]	5-17-10-00735	chl 2
7359	Isofenphos / Benzoic acid, 2-[[ethoxy[(1-methylethyl)amino]phosphinothioyl]oxy]-, 1-methylethyl ester	$C_{15}H_{24}NO_4PS$ / 345.40	25311-71-1 / <-12	5055 / 120[0.01]	1.134[20]	
7360	Isofulminic acid	CHNO / 43.03	51060-05-0 / stable only at low temp.			
7361	2H-Isoindole-2-acetic acid, 1,3-dihydro-1,3-dioxo-, ethyl ester	$C_{12}H_{11}NO_4$ / 233.22	6974-10-3 / 112.5	300	5-21-10-00431	bz 4; eth 4; EtOH 4; chl 4
7362	1H-Isoindole-1,3(2H)-dione / Phthalimide	$C_8H_5NO_2$ / 147.13	85-41-6 / 238	7347	5-21-10-00270	bz 4
7363	1H-Isoindole-1,3(2H)-dione, 2-(2,6-dioxo-3-piperidinyl)- / Thalidomide	$C_{13}H_{10}N_2O_4$ / 258.23	50-35-1 / 270	9182	5-22-13-00224	py 4; diox 4
7364	1H-Isoindole-1,3(2H)-dione, N-ethyl-	$C_{10}H_9NO_2$ / 175.19	5022-29-7 / 79	285.5	5-21-10-00274	EtOH 3; eth 3
7365	1H-Isoindole-1,3(2H)-dione, 2-methyl-	$C_9H_7NO_2$ / 161.16	550-44-7 / 134	286	5-21-10-00273	H_2O 1; EtOH 2
7366	1H-Isoindole-1,3(2H)-dione, 2-phenyl-	$C_{14}H_9NO_2$ / 223.23	520-03-6 / 210	sub	5-21-10-00292	H_2O 1; EtOH 2; chl 5
7367	1H-Isoindole-1,3(2H)-dione, 2-(phenylmethyl)-	$C_{15}H_{11}NO_2$ / 237.26	2142-01-0 / 116		5-21-10-00299 / 1.343[18]	EtOH 3; DMSO-d$_6$ 2; HOAc 3
7368	1H-Isoindole-1,3(2H)-dione, 3a,4,7,7a-tetrahydro-2-[(trichloromethyl)thio]- / Captan	$C_9H_8Cl_3NO_2S$ / 300.59	133-06-2 / 172.5	1771	5-21-10-00136 / 1.74[25]	chl 4
7369	1H-Isoindol-1-one, 2,3-dihydro-	C_8H_7NO / 133.15	480-91-1 / 151	338; 103[18]	5-21-08-00018	eth 4; EtOH 4; chl 4
7370	L-Isoleucine / 2-Amino-3-methylpentanoic acid	$C_6H_{13}NO_2$ / 131.17	73-32-5 / 284 dec	5064	4-04-00-02775	H_2O 3; EtOH 1
7371	Isopropalin / Benzenamine, 4-(1-methylethyl)-2,6-dinitro-N,N-dipropyl-	$C_{15}H_{23}N_3O_4$ / 309.37	33820-53-0	5089		
7372	5-Isoquinolinamine / 5-Aminoisoquinoline	$C_9H_8N_2$ / 144.18	1125-60-6 / 128	196[18]	5-22-10-00342	chl 2; lig 3
7373	Isoquinoline / Benzopyridine	C_9H_7N / 129.16	119-65-3 / 26.47	5110 / 243.2	5-20-07-00333 / 1.0910[30]	H_2O 1; EtOH 4; eth 5; ace 4 / 1.6148[20]
7374	Isoquinoline, 4-bromo- / 4-Bromoisoquinoline	C_9H_6BrN / 208.06	1532-97-4 / 41.5	282.5	5-20-07-00363	eth 4
7375	5-Isoquinolinecarbonitrile	$C_{10}H_6N_2$ / 154.17	27655-41-0 / 137	sub 100	4-22-00-01208	eth 4; EtOH 4

No.	Name / Synonym	Mol. Form. / Mol. Wt.	CAS RN / mp/°C	Merck No. / bp/°C	Beil. Ref. / den/g cm⁻³	Solubility / n_D
7376	Isoquinoline, 1-chloro- / 1-Chloroisoquinoline	C_9H_6ClN / 163.61	19493-44-8 / 37.5	/ 274.5	5-20-07-00361	H_2O 1; bz 3
7377	Isoquinoline, 1-[(3,4-diethoxyphenyl)methyl]-6,7-diethoxy- / Isoquinoline, 6,7-diethoxy-1(3,4-diethoxybenzyl)	$C_{24}H_{29}NO_4$ / 395.50	486-47-5 / 100	3682	5-21-06-00183	H_2O 1; EtOH 3; eth 2; chl 2
7378	Isoquinoline, 1-[(3,4-dimethoxyphenyl)methyl]-6,7-dimethoxy- / Papaverine	$C_{20}H_{21}NO_4$ / 339.39	58-74-2 / 147.5	6968 / sub 135	5-21-06-00182 / 1.337[20]	H_2O 2; EtOH 4; eth 2; ace 3 / 1.625
7379	Isoquinoline, 1-[(3,4-dimethoxyphenyl)methyl]-1,2,3,4-tetrahydro-6,7-dimethoxy-2-methyl-, (S)- / Laudanosine	$C_{21}H_{27}NO_4$ / 357.45	2688-77-9 / 89	5249	5-21-06-00049	ace 4; eth 4; EtOH 4; chl 4
7380	Isoquinoline, 7-methoxy- / 7-Methoxyisoquinoline	$C_{10}H_9NO$ / 159.19	39989-39-4 / 49	/ 194[50]; 184[34]	5-21-03-00326	H_2O 1; EtOH 3; lig 3
7381	Isoquinoline, 1-methyl- / Isoquinaldine	$C_{10}H_9N$ / 143.19	1721-93-3 / 10	/ 248	5-20-07-00407 / 1.077[20]	H_2O 2; eth 3; ace 3; bz 3 / 1.6095[20]
7382	Isoquinoline, 3-methyl- / 3-Methylisoquinoline	$C_{10}H_9N$ / 143.19	1125-80-0 / 68	/ 249	5-20-07-00410	H_2O 2; eth 3; ace 3; chl 2
7383	Isoquinoline, 4-methyl- / 4-Methylisoquinoline	$C_{10}H_9N$ / 143.19	1196-39-0 / <-75	/ 256	5-20-07-00413	H_2O 2; eth 3; ace 3; bz 3
7384	Isoquinoline, 6-methyl- / 6-Methylisoquinoline	$C_{10}H_9N$ / 143.19	42398-73-2 / 85.5	/ 265.5	5-20-07-00414	ace 4; bz 4; eth 4; EtOH 4
7385	Isoquinoline, 8-methyl- / 8-Methylisoquinoline	$C_{10}H_9N$ / 143.19	62882-00-2 / 205	/ 258	5-20-07-00414	ace 4; bz 4; eth 4
7386	Isoquinoline, 5-nitro- / 5-Nitroisoquinoline	$C_9H_6N_2O_2$ / 174.16	607-32-9 / 110	/ sub	5-20-07-00365	H_2O 2; EtOH 4; eth 4; chl 4
7387	Isoquinoline, 3-(phenylmethyl)- / Isoquinoline, 3-benzyl	$C_{16}H_{13}N$ / 219.29	90210-56-3 / 104	/ 311[23]	5-20-08-00383	bz 4; chl 4
7388	Isoquinoline, 1,2,3,4-tetrahydro-	$C_9H_{11}N$ / 133.19	91-21-4 / <-15	/ 232.5	5-20-06-00320 / 1.0642[24]	H_2O 1; EtOH 3; bz 3; acid 3 / 1.5668[20]
7389	Isoquinoline, 5,6,7,8-tetrahydro- / 5,6,7,8-Tetrahydroisoquinoline	$C_9H_{11}N$ / 133.19	36556-06-6 / 146.5	/ 218; 106[13]	5-20-06-00319 / 1.0504[10]	H_2O 1; ace 3; bz 3; chl 3 / 1.5276[10]
7390	Isoquinoline, 1,2,3,4-tetrahydro-6,7-dimethoxy-1,2-dimethyl-, (±)- / Carnegine	$C_{13}H_{19}NO_2$ / 221.30	490-53-9 /	1854 / 170[1]	5-21-04-00501	H_2O 4; eth 4; EtOH 4
7391	Isoquinoline, 1,2,3,4-tetrahydro-6,7-dimethoxy-1,2-dimethyl- / Isoquinoline, 6,7-dimethoxy-1,2-dimethyl-1,2,3,4-tetrahydro	$C_{13}H_{19}NO_2$ / 221.30	71783-56-7 / 262.5	/ 170[1]; 100[0.01]	5-21-04-00501	H_2O 3; EtOH 3; eth 3; chl 3
7392	7-Isoquinolinol, 1-[(3,4-dimethoxyphenyl)methyl]-1,2,3,4-tetrahydro-6-methoxy-2-methyl-, (S)- / Codamine	$C_{20}H_{25}NO_4$ / 343.42	21040-59-5 / 127	2458	4-21-00-02704	eth 4; EtOH 4; chl 4
7393	8-Isoquinolinol, 1,2,3,4-tetrahydro-6,7-dimethoxy- / Anhalamine	$C_{11}H_{15}NO_3$ / 209.25	643-60-7 / 187.5	682	5-21-05-00513	eth 4; EtOH 4
7394	8-Isoquinolinol, 1,2,3,4-tetrahydro-6,7-dimethoxy-1,2-dimethyl-, (S)- / Pellotine	$C_{13}H_{19}NO_3$ / 237.30	83-14-7 / 111.5	7018	4-21-00-02525	ace 4; eth 4; EtOH 4; peth 4
7395	8-Isoquinolinol, 1,2,3,4-tetrahydro-6,7-dimethoxy-1-methyl-, (S)- / Anhalonidine	$C_{12}H_{17}NO_3$ / 223.27	17627-77-9 / 160.5	683	4-21-00-02524	H_2O 4; EtOH 4
7396	7-Isoquinolinol, 1,2,3,4-tetrahydro-1-[(4-hydroxyphenyl)methyl]-6-methoxy-, (S)- / Coclaurine	$C_{17}H_{19}NO_3$ / 285.34	486-39-5 / 220.5	2455	4-21-00-02605	
7397	6-Isoquinolinol, 1,2,3,4-tetrahydro-7-methoxy-1-methyl-, (S)- / Salsoline	$C_{11}H_{15}NO_2$ / 193.25	89-31-6 / 221.5	8308	5-21-04-00499	H_2O 2; EtOH 1; bz 2
7398	1(2H)-Isoquinolinone, 3,4-dihydro-6,7-dimethoxy- / Corydaldine	$C_{11}H_{13}NO_3$ / 207.23	493-49-2 / 175	2542	5-21-13-00086	H_2O 4; bz 4; eth 4; EtOH 4
7399	Isothiocyanic acid	CHNS / 59.09	3129-90-6			
7400	DL-Isovaline	$C_5H_{11}NO_2$ / 117.15	595-39-1 / 316.5	/ sub 300	4-04-00-02655	H_2O 3; EtOH 3; eth 2; bz 1
7401	L-Isovaline / 2-Amino-2-methylbutyric acid	$C_5H_{11}NO_2$ / 117.15	595-40-4 /	5124	4-04-00-02655	
7402	Isoxazole / 1-Oxa-2-azacyclopentadiene	C_3H_3NO / 69.06	288-14-2 /	/ 95	4-27-00-00959 / 1.078[20]	H_2O 3 / 1.4298[17]
7403	Isoxazole, 3,5-dimethyl- / 3,5-Dimethylisoxazole	C_5H_7NO / 97.12	300-87-8 /	/ 143	4-27-00-00978 / 0.99[25]	/ 1.4421[20]
7404	5-Isoxazolemethanol, 3-methyl-	$C_5H_7NO_2$ / 113.12	14716-89-3 /	/ 130[20]	4-27-00-01745 / 1.132[25]	/ 1.4790[20]
7405	Isoxazole, 5-methyl- / 5-Methylisoxazole	C_4H_5NO / 83.09	5765-44-6 /	/ 122	4-27-00-00966 / 1.023[20]	DMSO 3 / 1.4386[20]
7406	3-Isoxazolidinone, 4-amino-, (R)- / Cycloserine	$C_3H_6N_2O_2$ / 102.09	68-41-7 / 155-6 dec	2758	4-27-00-05549	H_2O 3; MeOH 2
7407	Isoxazolium, 2-ethyl-5-(3-sulfophenyl)-, hydroxide, inner salt / Woodward's Reagent K	$C_{11}H_{11}NO_4S$ / 253.28	4156-16-5 / dec 207	9962		

No.	Name / Synonym	Mol. Form. Mol. Wt.	CAS RN mp/°C	Merck No. bp/°C	Beil. Ref. den/g cm⁻³	Solubility n_D
7408	Kepone 1,3,4-Metheno-2H-cyclobuta[cd]pentalen-2-one, 1,1a,3,3a,4,5,5,5a,5b,6-decachloro-octahydro-	$C_{10}Cl_{10}O$ 490.64	143-50-0 350 dec	2081	1.61[25]	
7409	Ketone, 1-methyl-4-piperidyl phenyl Piperidine, 4-benzoyl-N-methyl	$C_{13}H_{17}NO$ 203.28	92040-00-1 36	190[21]; 133[2]	4-21-00-03689	H_2O 3; EtOH 3; eth 4; ace 3 1.5430[23]
7410	Lactofen (±)-2-Ethoxy-1-ethyl-2-oxoethyl 5-[2-chloro-4-(trifluoromethyl)phenoxy]-2-nitrobenzoate	$C_{19}H_{15}ClF_3NO_7$ 461.78	77501-63-4			
7411	Lanosta-8,24-dien-3-ol, (3β)- Lanosterol	$C_{30}H_{50}O$ 426.73	79-63-0 140.5	5232	4-06-00-04188	eth 4; EtOH 4; chl 4
7412	DL-Leucine	$C_6H_{13}NO_2$ 131.17	328-39-2 293	sub	2-04-00-00870 1.293[18]	H_2O 3; EtOH 2; eth 1
7413	L-Leucine 2-Amino-4-methylpentanoic acid	$C_6H_{13}NO_2$ 131.17	61-90-5 293	5331 sub	4-04-00-02738 1.293[18]	H_2O 2; EtOH 1; eth 1
7414	DL-Leucine, N-glycyl-	$C_8H_{16}N_2O_3$ 188.23	688-14-2 242 dec		4-04-00-02766 1.181[25]	H_2O 4; EtOH 1
7415	Linuron Urea, N'-(3,4-dichlorophenyl)-N-methoxy-N-methyl-	$C_9H_{10}Cl_2N_2O_2$ 249.10	330-55-2 93	5387		
7416	Lup-20(29)-ene-3,28-diol, (3β)- Betulin	$C_{30}H_{50}O_2$ 442.73	473-98-3 250	1212 sub 240	4-06-00-06534	H_2O 1; EtOH 3; eth 3; bz 2
7417	Lup-20(29)-en-3-ol, (3β)- Lupeol	$C_{30}H_{50}O$ 426.73	545-47-1 216	5478	4-06-00-04200 0.9457[218]	H_2O 1; EtOH 4; eth 4; ace 4 1.4910[218]
7418	DL-Lysine 2,6-Diaminohexanoic acid	$C_6H_{14}N_2O_2$ 146.19	70-54-2 224		2-04-00-00858	H_2O 2
7419	L-Lysine 2,6-Diaminohexanoic acid	$C_6H_{14}N_2O_2$ 146.19	56-87-1 224-5 dec	5509	4-04-00-02717	H_2O 4; EtOH 1; eth 1; ace 1
7420	Maneb Manganese, [[1,2-ethanediylbis[carbamodithioato]](2-)]-	$C_4H_6MnN_2S_4$ 265.31	12427-38-2 dec 200	5603		
7421	Manganous dimethyldithiocarbamate Carbamic acid, dimethyldithio-, manganese(2+) salt	$C_8H_{12}MnN_2S_4$ 295.38	15339-36-3			
7422	D-Mannitol Cordycepic acid	$C_6H_{14}O_6$ 182.17	69-65-8 168	5629 295[3.5]	4-01-00-02841 1.489[20]	H_2O 4; EtOH 2; eth 1; py 2 1.3330
7423	D-Mannitol, 1,4:3,6-dianhydro-	$C_6H_{10}O_4$ 146.14	641-74-7 88	274 dec	5-19-03-00201	H_2O 4
7424	D-Mannitol, hexanitrate Mannitol, hexanitrate	$C_6H_8N_6O_{18}$ 452.16	15825-70-4 107	5630 exp	4-01-00-02849 1.8[20]	bz 4; eth 4; EtOH 4
7425	D-Mannose Seminose	$C_6H_{12}O_6$ 180.16	3458-28-4 132 dec	5632	4-01-00-04328 1.539[20]	H_2O 4; EtOH 2; eth 1; bz 1
7426	Matridin-15-one Matrine	$C_{15}H_{24}N_2O$ 248.37	519-02-8 223[6]	5641	5-24-02-00301	H_2O 3; EtOH 4; eth 3; ace 3 1.5286[25]
7427	Mefluidide Acetamide, N-[2,4-dimethyl-5-[[(trifluoromethyl)sulfonyl]amino]phenyl]-	$C_{11}H_{13}F_3N_2O_3S$ 310.30	53780-34-0 184	5684		
7428	m-Menthan-6-ol Cyclohexanol-4-isopropyl-2-methyl	$C_{10}H_{20}O$ 156.27	1490-05-7	119.21[28]	2-06-00-00038 0.9140[20]	eth 4; EtOH 4 1.4666[20]
7429	p-Menth-3-ene, (S)-(-)- Cyclohexene, 1-isopropyl-4-methyl (l)	$C_{10}H_{18}$ 138.25	22564-83-6	167	4-05-00-00301 0.8099[19]	H_2O 1; EtOH 3; eth 3; bz 3 1.4511[20]
7430	Mepiquat chloride Piperidinium, 1,1-dimethyl-, chloride	$C_7H_{16}NCl$ 149.66	24307-26-4 223	5746		
7431	Mercury, (acetato-O)phenyl- Phenylmercuric acetate	$C_8H_8HgO_2$ 336.74	62-38-4 153	7271	4-16-00-01720	chl 3
7432	Mercury, bis(4-methylphenyl)- Bis(p-tolyl)mercury	$C_{14}H_{14}Hg$ 382.86	537-64-4 245.7	3387	4-16-00-01705	
7433	Mercury, dibutyl- Dibutylmercury	$C_8H_{18}Hg$ 314.82	629-35-6	105[10]	4-04-00-04429 1.7779[20]	1.5057[20]
7434	Mercury, diethyl- Diethyl mercury	$C_4H_{10}Hg$ 258.71	627-44-1	159; 57[16]	2.43[20]	eth 3; EtOH 2
7435	Mercury, dimethyl- Dimethyl mercury	C_2H_6Hg 230.66	593-74-8 93	3238	3.17[25]	H_2O 1; EtOH 4; eth 4 1.5452[20]
7436	Mercury, diphenyl- Benzene, mercuriodi-	$C_{12}H_{10}Hg$ 354.80	587-85-9	204[10]	4-16-00-01702 2.318[25]	H_2O 1; EtOH 2; eth 2; bz 3
7437	Mercury, [2-methyl-5-nitrophenolato(2-)-C6,O1]- Nitromersol	$C_7H_5HgNO_3$ 351.71	133-58-4	6531		H_2O 1; ace 2; EtOH 2; alk 3
7438	Merphos Phosphorotrithious acid, S,S,S-tributyl ester	$C_{12}H_{27}PS_3$ 298.52	150-50-5 100	137[0.7]; 176[15]	1.02[20]	
7439	Metalaxyl N-(2,6-Dimethylphenyl)-N-(methoxyacetyl)alanine, methyl ester	$C_{15}H_{21}NO_4$ 279.34	57837-19-1 71	5826		
7440	Metaldehyde Acetaldehyde polymer	$(C_2H_4O)_x$	37273-91-9 246	5827 sub 115	4-19-00-05643	H_2O 1; EtOH 2; eth 2; ace 1
7441	Methamidophos Phosphoramidothioic acid, O,S-dimethyl ester	$C_2H_8NO_2PS$ 141.13	10265-92-6 46	5858	1.31[20]	

No.	Name / Synonym	Mol. Form. / Mol. Wt.	CAS RN / mp/°C	Merck No. / bp/°C	Beil. Ref. / den/g cm^{-3}	Solubility / n_D
7442	Methanamine / Methylamine	CH_5N / 31.06	74-89-5 / -93.4	5938 / -6.3	4-04-00-00118 / *0.656^{25}	H_2O 4; EtOH 3; eth 5; ace 3
7443	Methanamine, 1,1-diethoxy-N,N-dimethyl-	$C_7H_{17}NO_2$ / 147.22	1188-33-6	129	4-04-00-00174 / 0.859^{25}	1.4007^{20}
7444	Methanamine, N,N-dimethyl- / Trimethylamine	C_3H_9N / 59.11	75-50-3 / -117.1	9625 / 2.8	4-04-00-00134 / *0.627^{25}	H_2O 4; EtOH 3; eth 3; bz 3 / 1.3631^0
7445	Methanamine, N,N-dimethyl-, hydrochloride	$C_3H_{10}ClN$ / 95.57	593-81-7 / 277.5		4-04-00-00134	H_2O 4; EtOH 4; chl 4
7446	Methanamine, hydrochloride	CH_6ClN / 67.52	593-51-1 / 227.5	5938 / 227^{15}	4-04-00-00118	H_2O 4; EtOH 4
7447	Methanamine, N-hydroxy- / Hydroxylamine, N-methyl-	CH_5NO / 47.06	593-77-1 / 87.5	/ 62.5^{15}	4-04-00-03299 / 1.0003^{20}	H_2O 4; EtOH 4 / 1.4164^{20}
7448	Methanamine, N-methyl- / Dimethylamine	C_2H_7N / 45.08	124-40-3 / -92.2	3217 / 6.8	4-04-00-00128 / 0.6804^0	H_2O 4; EtOH 3; eth 3 / 1.350^{17}
7449	Methanamine, N-methyl-N-nitro-	$C_2H_6N_2O_2$ / 90.08	4164-28-7 / 58	187	4-04-00-03401 / 1.1090^{72}	H_2O 4; ace 4; eth 4; EtOH 4 / 1.4462^{72}
7450	Methanamine, N-methyl-N-nitroso- / N-Nitrosodimethylamine	$C_2H_6N_2O$ / 74.08	62-75-9	6558 / 154	4-04-00-03384 / 1.0059^{20}	H_2O 3; EtOH 3; eth 3; chl 3 / 1.4358^{20}
7451	Methanamine, N-nitro-	$CH_4N_2O_2$ / 76.06	598-57-2 / 38	$80-5^{10}$	4-04-00-03400 / 1.2433^{49}	H_2O 4; EtOH 4; eth 3; bz 4 / 1.4616^{49}
7452	Methanamine, N-(phenylmethylene)-	C_8H_9N / 119.17	622-29-7	$185; 92^{34}$	4-07-00-00518 / 0.9671^{14}	EtOH 3; eth 3; ace 3; chl 3 / 1.5526^{20}
7453	Methanaminium, N-[4-[bis[4-(dimethylamino)phenyl]methylene]-2,5-cyclohexadien-1-ylidene]-N-methyl, chloride / Crystal violet	$C_{25}H_{30}ClN_3$ / 407.99	548-62-9 / 215 dec	4287	H_2O 4; chl 4	
7454	Methanaminium, 1-carboxy-N,N,N-trimethyl-, hydroxide, inner salt / Betaine	$C_5H_{11}NO_2$ / 117.15	107-43-7 / 293 dec	1201	4-04-00-02369	H_2O 4; EtOH 3; eth 2; chl 2
7455	Methanaminium, N-[4-[[4-(dimethylamino)phenyl]phenylmethylene]-2,5-cyclohexadien-1-ylidene]-N-methyl-, chlorid / Malachite green	$C_{23}H_{25}ClN_2$ / 364.92	569-64-2	5581		H_2O 4; EtOH 4; MeOH 4
7456	Methanaminium, N,N,N-trimethyl-, bromide / Tetramethylammonium bromide	$C_4H_{12}BrN$ / 154.05	64-20-0 / 230 dec		4-04-00-00145 / 1.56^{25}	H_2O 4; EtOH 2; eth 1; bz 1
7457	Methanaminium, N,N,N-trimethyl-, chloride / Tetramethylammonium chloride	$C_4H_{12}ClN$ / 109.60	75-57-0 / 420 dec		4-04-00-00145 / 1.169^{20}	H_2O 3; EtOH 2; eth 1; bz 1
7458	Methanaminium, N,N,N-trimethyl-, iodide / Tetramethylammonium iodide	$C_4H_{12}IN$ / 201.05	75-58-1 / >230 dec	9156	4-04-00-00145 / 1.829^{25}	H_2O 2; EtOH 2; eth 1; ace 2
7459	Methane	CH_4 / 16.04	74-82-8 / -182.4	5863 / -161.5	4-01-00-00003 / 0.4228^{-162}	H_2O 3; EtOH 3; eth 3; ace 2
7460	Methane, azido / Methyl azide	CH_3N_3 / 57.06	624-90-8	20.5; exp 500	0.869^{15}	
7461	Methane, bis(methylthio)- / Bis[methylmercapto]methane	$C_3H_8S_2$ / 108.23	1618-26-4	1267 / 148	4-01-00-03088	
7462	Methane, bromo- / Methyl bromide	CH_3Br / 94.94	74-83-9 / -93.7	5951 / 3.5	4-01-00-00068 / 1.6755^{20}	H_2O 2; EtOH 5; eth 5; chl 5 / 1.4218^{20}
7463	Methane, bromochloro- / Bromochloromethane	CH_2BrCl / 129.38	74-97-5 / -87.9	/ 68.0	4-01-00-00074 / 1.9344^{20}	H_2O 1; EtOH 3; eth 3; ace 3 / 1.4838^{20}
7464	Methane, bromochlorodifluoro- / Bromochlorodifluoromethane	$CBrClF_2$ / 165.36	353-59-3 / -159.5	/ -3.7	4-01-00-00075	
7465	Methane, bromochlorodinitro- / Bromochlorodinitromethane	$CBrClN_2O_4$ / 219.38	33829-48-0 / 9.3	$75-6^{15}$	2-01-00-00044 / 2.0393^{20}	EtOH 4 / 1.4793
7466	Methane, bromochlorofluoro- / Bromochlorofluoromethane	$CHBrClF$ / 147.37	593-98-6 / -115	/ 36	4-01-00-00075 / 1.9771^0	H_2O 1; eth 3; ace 3; chl 3 / 1.4144^{25}
7467	Methane, bromodichloro- / Bromodichloromethane	$CHBrCl_2$ / 163.83	75-27-4 / -57	/ 90	4-01-00-00076 / 1.980^{20}	H_2O 1; EtOH 4; eth 4; ace 4 / 1.4964^{20}
7468	Methane, bromodichlorofluoro- / Bromodichlorofluoromethane	$CBrCl_2F$ / 181.82	353-58-2		4-01-00-00076	
7469	Methane, bromodifluoro- / Bromodifluoromethane	$CHBrF_2$ / 130.92	1511-62-2 / -14.5		4-01-00-00072 / 1.55^{16}	H_2O 3; EtOH 4
7470	Methane, bromofluoro- / Bromofluoromethane	CH_2BrF / 112.93	373-52-4	19	4-01-00-00072	EtOH 3; chl 4
7471	Methane, bromoiodo- / Bromoiodomethane	CH_2BrI / 220.84	557-68-6	139.5	4-01-00-00095 / 2.926^{17}	chl 4 / 1.6410^{20}
7472	Methane, bromomethoxy-	C_2H_5BrO / 124.97	13057-17-5	87	4-01-00-03051 / 1.5976^{20}	1.4562^{20}
7473	Methane, bromonitro- / Bromonitromethane	CH_2BrNO_2 / 139.94	563-70-2	$149; 71^{40}$	4-01-00-00106	EtOH 4 / 1.4880^{20}
7474	Methane, bromotrichloro- / Bromotrichloromethane	$CBrCl_3$ / 198.27	75-62-7 / -5.7	/ 105	4-01-00-00077 / 2.012^{25}	eth 4; EtOH 4 / 1.5065^{20}
7475	Methane, bromotrifluoro- / Bromotrifluoromethane	$CBrF_3$ / 148.91	75-63-8 / -172	/ -57.8	4-01-00-00073 / 1.5800^{20}	chl 4
7476	Methane, bromotrinitro- / Bromotrinitromethane	$CBrN_3O_6$ / 229.93	560-95-2 / 17.5	56^{10}	3-01-00-00116 / 2.0312^{20}	EtOH 4; chl 4 / 1.4808^{20}
7477	Methane, chloro- / Methyl chloride	CH_3Cl / 50.49	74-87-3 / -97.7	5964 / -24.0	4-01-00-00028 / *0.911^{25}	H_2O 3; EtOH 3; eth 5; ace 5 / 1.3389^{20}
7478	Methane, chlorodifluoro- / Chlorodifluoromethane	$CHClF_2$ / 86.47	75-45-6 / -157.4	/ -40.7	4-01-00-00032 / 1.4909^{-69}	H_2O 4; eth 3; ace 3; chl 3

No.	Name Synonym	Mol. Form. Mol. Wt.	CAS RN mp/°C	Merck No. bp/°C	Beil. Ref. den/g cm⁻³	Solubility n_D
7479	Methane, chlorodiiodo-	CHClI₂ 302.28	638-73-3 -4	200 dec; 101[33]	4-01-00-00097 3.1700[0]	ace 4; eth 4; chl 4
7480	Methane, chlorodinitro- Chlorodinitromethane	CHClN₂O₄ 140.48	921-13-1	34-6[13]	3-01-00-00115 1.6125[20]	1.4575[20]
7481	Methane, chlorofluoro- Chlorofluoromethane	CH₂ClF 68.48	593-70-4 -133	-9.1	4-01-00-00032	chl 4
7482	Methane, chloroiodo- Chloroiodomethane	CH₂ClI 176.38	593-71-5	109	4-01-00-00094 2.422[20]	ace 4; bz 4; eth 4; EtOH 4 1.5822[20]
7483	Methane, chloromethoxy- Chloromethyl methyl ether	C₂H₅ClO 80.51	107-30-2 -103.5	2146 59.5	4-01-00-03046 1.063[10]	EtOH 3; eth 3; ace 3; chl 3 1.397[20]
7484	Methane, chloro(methylthio)-	C₂H₅ClS 96.58	2373-51-5	105	4-01-00-03079 1.153[25]	1.4963[20]
7485	Methane, chloronitro- Chloronitromethane	CH₂ClNO₂ 95.49	1794-84-9	122.5	4-01-00-00106 1.466[15]	H₂O 4
7486	Methane, chlorotrifluoro- Chlorotrifluoromethane	CClF₃ 104.46	75-72-9 -181	-81.4	4-01-00-00034	
7487	Methane, chlorotrinitro-	CClN₃O₆ 185.48	1943-16-4 4.5	134 dec; 56[40]	3-01-00-00116 1.6769[20]	eth 4; EtOH 4; chl 4 1.4500[20]
7488	Methanediamine, 1-(2-furanyl)-N,N '-bis(2-furanylmethylene)- Hydrofuramide	C₁₅H₁₂N₂O₃ 268.27	494-47-3 117	4718	5-17-09-00306	H₂O 1; EtOH 4; eth 4
7489	Methanediamine, N,N,N ',N '-tetraethyl-	C₉H₂₂N₂ 158.29	102-53-4	165.8	4-04-00-00338 0.8000[20]	1.4420[25]
7490	Methanediamine, N,N,N ',N '-tetramethyl- Tetramethylmethanediamine	C₅H₁₄N₂ 102.18	51-80-9	83	4-04-00-00153 0.7491[18]	H₂O 3
7491	Methane, diazo- Diazomethane	CH₂N₂ 42.04	334-88-3 -145	2983 -23	4-01-00-03056	eth 4; diox 4
7492	Methane, dibromo- Methylene bromide	CH₂Br₂ 173.83	74-95-3 -52.5	5980 97	4-01-00-00078 2.4969[20]	H₂O 2; EtOH 5; eth 5; ace 5 1.5420[20]
7493	Methane, dibromochloro- Chlorodibromomethane	CHBr₂Cl 208.28	124-48-1 -20	120	4-01-00-00081 2.451[20]	H₂O 1; EtOH 3; eth 3; ace 3 1.5482[20]
7494	Methane, dibromochlorofluoro- Dibromochlorofluoromethane	CBr₂ClF 226.27	353-55-9	80.3	4-01-00-00082 2.3173[22]	1.4570[20]
7495	Methane, dibromodichloro- Dibromodichloromethane	CBr₂Cl₂ 242.72	594-18-3 38	150.2	4-01-00-00082 2.42[25]	H₂O 1; EtOH 3; eth 3; ace 3
7496	Methane, dibromodifluoro- Dibromodifluoromethane	CBr₂F₂ 209.82	75-61-6 -110.1	25	4-01-00-00080	H₂O 3; eth 3; ace 3; bz 3
7497	Methane, dibromodinitro- Dibromodinitromethane	CBr₂N₂O₄ 263.83	2973-00-4 5.5	158 dec; 77[21]	3-01-00-00115 2.4439[20]	EtOH 4 1.5280[25]
7498	Methane, dibromofluoro- Dibromofluoromethane	CHBr₂F 191.83	1868-53-7 -78	64.9	4-01-00-00080 2.421[20]	H₂O 1; EtOH 3; eth 3; ace 3 1.4685[20]
7499	Methane, dichloro- Methylene chloride	CH₂Cl₂ 84.93	75-09-2 -95.1	5982 40	4-01-00-00035 1.3266[20]	H₂O 2; EtOH 5; eth 5; ctc 3 1.4242[20]
7500	Methane, dichlorodifluoro- Dichlorodifluoromethane	CCl₂F₂ 120.91	75-71-8 -158	3053 -29.8	4-01-00-00040	H₂O 3; EtOH 3; eth 3; HOAc 3
7501	Methane, dichlorodinitro- Dichlorodinitromethane	CCl₂N₂O₄ 174.93	1587-41-3	121.5	3-01-00-00115 1.6124[20]	bz 4; eth 4; EtOH 4; chl 4 1.4575[20]
7502	Methane, dichlorofluoro- Dichlorofluoromethane	CHCl₂F 102.92	75-43-4 -135	8.9	4-01-00-00039 1.405[9]	H₂O 3; EtOH 3; eth 3; ctc 3 1.3724[9]
7503	Methane, dichloroiodo- Dichloroiodomethane	CHCl₂I 210.83	594-04-7	132	4-01-00-00095 2.392[20]	ace 4; bz 4; eth 4; EtOH 4 1.5840[20]
7504	Methane, difluoro- Methylene difluoride	CH₂F₂ 52.02	75-10-5 -136	-51.6	4-01-00-00024 *0.960[25]	H₂O 1; EtOH 3
7505	Methane, difluoroiodo- Difluoroiodomethane	CHF₂I 177.92	1493-03-4 -122	21.6	4-01-00-00092 3.238[-19]	
7506	Methane, diiodo- Methylene iodide	CH₂I₂ 267.84	75-11-6 6.1	5985 182	4-01-00 00096 3.3212[20]	H₂O 2; EtOH 3; eth 3; bz 3 1.7425[20]
7507	Methane, dimethoxy- Methylal	C₃H₈O₂ 76.10	109-87-5 -104.8	5936 42	4-01-00-03026 0.8593[20]	ace 4; bz 4; eth 4; EtOH 4 1.3513[20]
7508	Methane, dinitro- Dinitromethane	CH₂N₂O₄ 106.04	625-76-3 <-15	exp 100	4-01-00-00107	H₂O 1; EtOH 3; eth 3
7509	Methanediol, dibenzoate	C₁₅H₁₂O₄ 256.26	5342-31-4 99	225 dec	4-09-00-00448 1.275[22]	ace 4; bz 4; eth 4
7510	Methanediol, 2-furanyl-, diacetate	C₉H₁₀O₅ 198.18	613-75-2 53.3	220	5-17-09-00300	bz 4; eth 4; EtOH 4
7511	Methanediol, (3-nitrophenyl)-, diacetate (ester) Benzaldehyde, 3-nitro, diacetate	C₁₁H₁₁NO₆ 253.21	29949-19-7 72		3-07-00-00900 1.393[25]	ace 4; bz 4; eth 4; EtOH 4
7512	Methanediol, phenyl-, diacetate	C₁₁H₁₂O₄ 208.21	581-55-5 46	220	4-07-00-00516 1.11[20]	bz 4; eth 4; EtOH 4
7513	Methane, diphenyl-m-tolyl- Triphenylmethane, 3-methyl	C₂₀H₁₈ 258.36	603-26-9 62	357	3-05-00-02332 1.07[16]	bz 4; eth 4; chl 4; lig 4
7514	Methanedisulfonic acid Methionic acid	CH₄O₆S₂ 176.17	503-40-2 98	5895	4-01-00-03054	H₂O 1; HNO₃ 3
7515	Methane, fluoro- Methyl fluoride	CH₃F 34.03	593-53-3 -141.8	4103 -78.4	4-01-00-00022 *0.566[25]	H₂O 4; EtOH 4; eth 4; bz 3
7516	Methane, fluorodiiodo-	CHFI₂ 285.83	1493-01-2 -34.5	100	4-01-00-00097 3.1969[22]	eth 4; EtOH 4

No.	Name Synonym	Mol. Form. Mol. Wt.	CAS RN mp/°C	Merck No. bp/°C	Beil. Ref. den/g cm^{-3}	Solubility n_D
7517	Methane, fluoroiodo- Fluoroiodomethane	CH_2FI 159.93	373-53-5	53.4	4-01-00-00092 2.366[20]	ace 4; bz 4; eth 4; chl 4 1.5256[20]
7518	Methane, iodo- Methyl iodide	CH_3I 141.94	74-88-4 -66.4	6002 42.5	4-01-00-00087 2.279[20]	H_2O 2; EtOH 5; eth 5; ace 3 1.5380[20]
7519	Methane, iodomethoxy- (Iodomethyl) methyl ether	C_2H_5IO 171.97	13057-19-7	122	4-01-00-03052 2.030[20]	1.5472[20]
7520	Methane, isocyanato- Methyl isocyanate	C_2H_3NO 57.05	624-83-9 -45	6004 39.5	4-04-00-00247 0.9230[27]	H_2O 4 1.3419[18]
7521	Methane, isothiocyanato- Methyl isothiocyanate	C_2H_3NS 73.12	556-61-6 36	6005 119	4-04-00-00248 1.0691[37]	H_2O 2; EtOH 5; eth 4 1.5258
7522	Methane, nitro- Nitromethane	CH_3NO_2 61.04	75-52-5 -28.5	6532 101.1	4-01-00-00100 1.1371[20]	H_2O 3; EtOH 3; eth 3; ace 3 1.3817[20]
7523	Methane, oxybis- Dimethyl ether	C_2H_6O 46.07	115-10-6 -141.5	5990 -24.8	4-01-00-01245	H_2O 3; EtOH 3; eth 3; ace 3
7524	Methane, oxybis[bromo-	$C_2H_4Br_2O$ 203.86	4497-29-4 -34	154.5	4-01-00-03052 2.2013[20]	
7525	Methane, oxybis[chloro- Bis(chloromethyl) ether	$C_2H_4Cl_2O$ 114.96	542-88-1 -41.5	3058 106	4-01-00-03051 1.323[15]	EtOH 5; eth 5 1.435[21]
7526	Methane, selenobis-	C_2H_6Se 109.03	593-79-3	55	4-01-00-01288 1.4077[15]	eth 4; EtOH 4; chl 4
7527	Methanesulfenyl chloride, trichloro- Perchloromethyl mercaptan	CCl_4S 185.89	594-42-3	147.5	4-03-00-00290 1.6947[20]	eth 3 1.5484[20]
7528	Methanesulfenyl fluoride, trifluoro- Methyl mercaptan, perfluoro	CF_4S 120.07	17742-04-0	147.5	4-01-00-01288 1.6947[20]	eth 4 1.5484[20]
7529	Methane, sulfinylbis- Dimethyl sulfoxide	C_2H_6OS 78.14	67-68-5 18.5	3247 189	4-01-00-01277 1.1014[20]	H_2O 3; EtOH 3; eth 3; ace 3 1.4170[20]
7530	Methanesulfonic acid Methylsulfonic acid	CH_4O_3S 96.11	75-75-2 20	5865 167[10]	4-04-00-00010 1.4812[18]	H_2O 3 1.4317[18]
7531	Methanesulfonic acid, ethyl ester Ethyl methanesulfonate	$C_3H_8O_3S$ 124.16	62-50-0	3782 86[10]	4-04-00-00012	
7532	Methanesulfonic acid, methyl ester Methyl methanesulfonate	$C_2H_6O_3S$ 110.13	66-27-3 20	6010 202.5	4-04-00-00011 1.2943[20]	1.4138[20]
7533	Methanesulfonic acid, trifluoro-, ethyl ester	$C_3H_5F_3O_3S$ 178.13	425-75-2	115	4-03-00-00034 1.3740[0]	eth 3
7534	Methane, sulfonylbis- Dimethyl sulfone	$C_2H_6O_2S$ 94.13	67-71-0 109	3246 238	4-01-00-01279 1.1700[110]	H_2O 3; EtOH 3; bz 3 1.4226
7535	Methanesulfonyl chloride Methanesulfonic acid chloride	CH_3ClO_2S 114.55	124-63-0	5866 162; 55[11]	4-04-00-00027 1.4805[18]	H_2O 1; EtOH 3; eth 3 1.4573[20]
7536	Methanesulfonyl chloride, trichloro-	CCl_4O_2S 217.89	2547-61-7 140.5	170	4-03-00-00036	H_2O 1; EtOH 3; eth 3; CS_2 3
7537	Methanesulfonyl chloride, trifluoro-	$CClF_3O_2S$ 168.52	421-83-0	162; 62[18]	4-03-00-00035	H_2O 1 1.3344[20]
7538	Methanesulfonyl fluoride Methanesulfonic acid fluoride	CH_3FO_2S 98.10	558-25-8	123.5	4-04-00-00027	
7539	Methane, tetrabromo- Carbon tetrabromide	CBr_4 331.63	558-13-4 90.1	189.5	4-01-00-00085 2.9608[100]	H_2O 1; EtOH 3; eth 3; chl 3 1.5942[100]
7540	Methane, tetrachloro- Carbon tetrachloride	CCl_4 153.82	56-23-5 -23	1822 76.8	4-01-00-00056 1.5940[20]	H_2O 1; EtOH 3; eth 5; ace 3 1.4601[20]
7541	Methane, tetrafluoro- Carbon tetrafluoride	CF_4 88.00	75-73-0 -183.5	1823 -128.0	4-01-00-00026 3.034[25]	H_2O 2; bz 3; chl 3
7542	Methane, tetraiodo- Carbon tetraiodide	CI_4 519.63	507-25-5 171	1824 130-40[1-2]	4-01-00-00098 4.23[20]	py 4; chl 4
7543	Methane, tetramethoxy-	$C_5H_{12}O_4$ 136.15	1850-14-2 -5	114	4-03-00-00004 1.023[25]	1.3845[20]
7544	Methane, tetranitro- Tetranitromethane	CN_4O_8 196.03	509-14-8 13.8	9164 126.1	4-01-00-00107 1.6380[20]	H_2O 1; EtOH 3; eth 3 1.4384[20]
7545	Methane, thiobis- Dimethyl sulfide	C_2H_6S 62.14	75-18-3 -98.3	6042 37.3	4-01-00-01275 0.8483[20]	H_2O 1; EtOH 3; eth 3 1.4438[20]
7546	Methanethiol Methyl mercaptan	CH_4S 48.11	74-93-1 -123	5867 5.9	4-01-00-01273 0.8665[20]	H_2O 2; EtOH 4; eth 4; chl 2
7547	Methanethione, diphenyl-	$C_{13}H_{10}S$ 198.29	1450-31-3 53.5	174[14]	4-07-00-01385	EtOH 2; eth 2; bz 4; chl 4
7548	Methane, tribromo- Bromoform	$CHBr_3$ 252.73	75-25-2 8.0	1407 149.1	4-01-00-00082 2.899[15]	H_2O 2; EtOH 5; eth 5; bz 3 1.6005[15]
7549	Methane, tribromochloro- Tribromochloromethane	CBr_3Cl 287.18	594-15-0 55	158.5	4-01-00-00085 2.71[15]	eth 4
7550	Methane, tribromofluoro- Tribromofluoromethane	CBr_3F 270.72	353-54-8 -73.6	108		EtOH 3
7551	Methane, tribromonitro-	CBr_3NO_2 297.73	464-10-8 10	127[18]	4-01-00-00106 2.811[12]	H_2O 1; EtOH 3; eth 3; ace 4 1.5790[20]
7552	Methanetricarboxylic acid, triethyl ester Triethyl methanetricarboxylate	$C_{10}H_{16}O_6$ 232.23	6279-86-3 29	253	4-02-00-02360 1.1084[16]	1.4243[20]
7553	Methanetricarboxylic acid, trimethyl ester Trimethyl methanetricarboxylate	$C_7H_{10}O_6$ 190.15	1186-73-8 46.5	242.7; 128[15]	3-02-00-02023	bz 4; eth 4; EtOH 4; chl 4
7554	Methane, trichloro- Chloroform	$CHCl_3$ 119.38	67-66-3 -63.6	2141 61.1	4-01-00-00042 1.4832[20]	H_2O 2; EtOH 5; eth 5; ace 3 1.4459[20]
7555	Methane, trichlorofluoro- Trichlorofluoromethane	CCl_3F 137.37	75-69-4 -111.1	9553 23.7	4-01-00-00054	
7556	Methane, trichloroiodo- Trichloroiodomethane	CCl_3I 245.27	594-22-9 -7.8	142	4-01-00-00095 2.355[20]	ace 4; bz 4; eth 4; chl 4 1.5854[20]

No.	Name / Synonym	Mol. Form. / Mol. Wt.	CAS RN / mp/°C	Merck No. / bp/°C	Beil. Ref. / den/g cm^{-3}	Solubility / n_D
7557	Methane, trichloronitro- / Chloropicrin	CCl$_3$NO$_2$ / 164.37	76-06-2 / -64	2156 / 112	4-01-00-00106 / 1.6558[20]	H$_2$O 3; EtOH 5; ace 5; bz 5 / 1.4611[20]
7558	Methane, trifluoro- / Fluoroform	CHF$_3$ / 70.01	75-46-7 / -155.1	4102 / -82.1	4-01-00-00024 / *0.673[25]	H$_2$O 3; EtOH 4; ace 3; bz 3
7559	Methane, trifluoroiodo- / Trifluoroiodomethane	CF$_3$I / 195.91	2314-97-8 /	/ -22.5	4-01-00-00092 / 2.3607[-32]	1.3790[-32]
7560	Methane, triiodo- / Iodoform	CHI$_3$ / 393.73	75-47-8 / 119	4926 / 218	4-01-00-00097 / 4.008[25]	H$_2$O 1; EtOH 3; eth 3; ace 3
7561	Methane, trimethoxy-	C$_4$H$_{10}$O$_3$ / 106.12	149-73-5 / 15	/ 104	4-02-00-00022 / 0.9676[20]	EtOH 3; eth 3 / 1.3793[20]
7562	Methane, trinitro- / Trinitromethane	CHN$_3$O$_6$ / 151.04	517-25-9 / 15	9642 / exp	4-01-00-00107 / 1.479[20]	ace 4; EtOH 4 / 1.4451[24]
7563	Methanimidamide / Formamidine	CH$_4$N$_2$ / 44.06	463-52-5 / 81	/ dec	4-02-00-00082	H$_2$O 4; EtOH 4
7564	Methanimidamide, N'-(4-chloro-2-methylphenyl)-N,N-dimethyl- / Chlordimeform	C$_{10}$H$_{13}$ClN$_2$ / 196.68	6164-98-3 / 35	2083 / 156-7[0.4]	/ 1.105[25]	bz 4; eth 4; EtOH 4 / 1.5885[25]
7565	Methanimidamide, N,N'-diphenyl-	C$_{13}$H$_{12}$N$_2$ / 196.25	622-15-1 / 142	/ >250	4-12-00-00372	H$_2$O 2; EtOH 3; eth 4; ace 3
7566	Methanimidamide, N-hydroxy-	CH$_4$N$_2$O / 60.06	624-82-8 / 114.5	/ dec	4-02-00-00084	EtOH 4
7567	Methanimidic acid, N-phenyl, ethyl ester	C$_9$H$_{11}$NO / 149.19	6780-49-0 /	/ 214; 87[10]	4-12-00-00371 / 1.0051[20]	eth 3; bz 3 / 1.5279[20]
7568	1,4-Methanoazulene, decahydro-4,8,8-trimethyl-9-methylene-, [1S-(1α,3aβ,4α,8aβ)]- / Longifolene	C$_{15}$H$_{24}$ / 204.36	475-20-7 /	5446 / 258; 126[15]	4-05-00-01192 / 0.9319[18]	H$_2$O 1; bz 3 / 1.5040[20]
7569	1,4-Methanoazulen-9-ol, decahydro-1,5,5,8a-tetramethyl-, [1R-(1α,3aβ,4α,8aβ,9S*)]	C$_{15}$H$_{26}$O / 222.37	465-24-7 / 112	/ 287 dec	3-06-00-00426 / 1.0441[20]	1.519
7570	1H-3a,7-Methanoazulen-6-ol, octahydro-3,6,8,8-tetramethyl-, / Cedrol	C$_{15}$H$_{26}$O / 222.37	77-53-2 / 86	1919 /	3-06-00-00424 / 0.9479[90]	1.4824[90]
7571	7,14-Methano-2H,6H-dipyrido[1,2-a:1',2'-e][1,5]diazocine, dodecahydro-, / Sparteine	C$_{15}$H$_{26}$N$_2$ / 234.38	90-39-1 / 30.5	8692 / 325; 173[8]	5-23-05-00497 / 1.0196[20]	eth 4; EtOH 4; chl 4 / 1.5312[20]
7572	7,14-Methano-2H,6H-dipyrido[1,2-a:1',2'-e][1,5]diazocin-6-one, dodecahydro-, [7R-(7α,7aβ,14α,14aalp / Aphylline	C$_{15}$H$_{24}$N$_2$O / 248.37	577-37-7 / 54.5	761 / 200 dec	5-24-02-00298	ace 4; bz 4; eth 4; EtOH 4
7573	7,14-Methano-4H,6H-dipyrido[1,2-a:1',2'-e][1,5]diazocin-4-one, 7,7a,8,9,10,11,13,14-octahydro-, / Anagyrine	C$_{15}$H$_{20}$N$_2$O / 244.34	486-89-5 /	660 / 265[12]; 212[4]	5-24-03-00410	H$_2$O 3; EtOH 4; eth 3; bz 3
7574	7,14-Methano-4H,6H-dipyrido[1,2-a:1',2'-e][1,5]diazocin-4-one, dodecahydro-9-hydroxy-, / Hydroxylupanine	C$_{15}$H$_{24}$N$_2$O$_2$ / 264.37	15358-48-2 / 169.5	4760 /	4-25-00-00082	H$_2$O 4; EtOH 4; chl 4
7575	2,5-Methano-2H-furo[3,2-b]pyrrol-6-ol, hexahydro-4-methyl-, (2α,3aβ,5α,6β,6aβ / Scopoline	C$_8$H$_{13}$NO$_2$ / 155.20	487-27-4 / 108.5	8365 / 248	2-27-00-00061 / 1.0891[134]	H$_2$O 3
7576	4,7-Methano-1H-indene, 1,4,5,6,7,8,8-heptachloro-3a,4,7,7a-tetrahydro-, / Heptachlor	C$_{10}$H$_5$Cl$_7$ / 373.32	76-44-8 / 95.5	4576 /	/ 1.57[9]	bz 4; eth 4; EtOH 4; lig 4
7577	4,7-Methano-1H-indene, 3a,4,7,7a-tetrahydro-, (3aα,4α,7α,7aα)-	C$_{10}$H$_{12}$ / 132.21	1755-01-7 / 32	/ 170 dec; 65[14]	4-05-00-01399 / 0.9302[35]	eth 4; EtOH 4 / 1.5050[35]
7578	1,5-Methano-8H-pyrido[1,2-a][1,5]diazocin-8-one, 1,2,3,4,5,6-hexahydro-3-methyl-, (1R)- / Caulophylline	C$_{12}$H$_{18}$N$_2$O / 204.27	486-86-2 / 137	1914 /	5-24-02-00538	H$_2$O 4; ace 4; bz 4; EtOH 4
7579	3,6-Methano-8H-1,5,7-trioxacyclopenta[ij]cycloprop[a]azulene-4,8(3H)-dione, hexahydro-2a-hydroxy / Picrotoxin	C$_{30}$H$_{34}$O$_{13}$ / 602.59	124-87-8 / 203.5	7388 /	4-19-00-05245	py 4; EtOH 4
7580	4,7-Methanoisobenzofuran-1,3-dione, 3a,4,7,7a-tetrahydro-, (3aα,4α,7α,7aα)- / Carbic anhydride	C$_9$H$_8$O$_3$ / 164.16	129-64-6 / 164.5	1801 /	5-17-11-00192 / 1.417[25]	ace 4; bz 4; EtOH 4; chl 4
7581	Methanol / Methyl alcohol	CH$_4$O / 32.04	67-56-1 / -97.6	5868 / 64.6	4-01-00-01227 / 0.7914[20]	H$_2$O 5; EtOH 5; eth 5; ace 5 / 1.3288[20]
7582	Methanol, chloro-, acetate	C$_3$H$_5$ClO$_2$ / 108.52	625-56-9 /	/ 116	4-02-00-00280 / 1.194[20]	eth 4; EtOH 4 / 1.409[20]
7583	Methanol, (1,3,5-triazine-2,4,6-triyltriimino)tris- / Trimethylolmelamine	C$_6$H$_{12}$N$_6$O$_3$ / 216.20	1017-56-7 /	9631 /	4-26-00-01273	
7584	Methanol, (1,3,5-triazine-2,4,6-triyltrinitrilo)hexakis- / Hexamethylolmelamine	C$_9$H$_{18}$N$_6$O$_6$ / 306.28	531-18-0 / 137	4611 /	4-26-00-01274	H$_2$O 4
7585	Methanol, trichloro-, carbonate (2:1)	C$_3$Cl$_6$O$_3$ / 296.75	32315-10-9 / 79	/ 203	4-03-00-00033 / 1.6290[80]	
7586	1,6-Methanonaphthalen-1(2H)-ol, octahydro-4,8a,9,9-tetramethyl-, / Patchouli alcohol	C$_{15}$H$_{26}$O / 222.37	5986-55-0 / 56	7001 /	3-06-00-00426 / 0.9906[65]	H$_2$O 1; EtOH 3; eth 3 / 1.5029[65]

No.	Name Synonym	Mol. Form. Mol. Wt.	CAS RN mp/°C	Merck No. bp/°C	Beil. Ref. den/g cm^{-3}	Solubility n_D
7587	Methanone, 1,3,5-benzenetriyltris[phenyl- 1,3,5-Tribenzoylbenzene	$C_{27}H_{18}O_3$ 390.44	25871-69-6 120	260[1]	4-07-00-02847	chl 3
7588	Methanone, 2-benzofuranylphenyl-	$C_{15}H_{10}O_2$ 222.24	6272-40-8 91	360	5-17-10-00568	H_2O 1
7589	Methanone, bis(3-aminophenyl)-	$C_{13}H_{12}N_2O$ 212.25	611-79-0 173.5	285[11]	4-14-00-00255	eth 4; EtOH 4
7590	Methanone, bis(4-bromophenyl)-	$C_{13}H_8Br_2O$ 340.01	3988-03-2 177	395	4-07-00-01380	bz 4; HOAc 4; chl 4
7591	Methanone, bis(3-chlorophenyl)-	$C_{13}H_8Cl_2O$ 251.11	7094-34-0 124	163[2]	4-07-00-01376	H_2O 1; EtOH 3; eth 3
7592	Methanone, bis(4-chlorophenyl)-	$C_{13}H_8Cl_2O$ 251.11	90-98-2 147.5	353	4-07-00-01376 1.4500[20]	H_2O 1; EtOH 3; eth 4; ace 2
7593	Methanone, bis[4-(dimethylamino)phenyl]- Michler's ketone	$C_{17}H_{20}N_2O$ 268.36	90-94-8 179	6100 360 dec	4-14-00-00255	H_2O 1; EtOH 2; eth 1; bz 4
7594	Methanone, bis(2,4-dimethylphenyl)- 2,2',4,4'-Tetramethylbenzophenone	$C_{17}H_{18}O$ 238.33	3478-88-4	190[10]	4-07-00-01470 1.043[15]	1.5790[25]
7595	Methanone, bis(2-hydroxy-4-methoxyphenyl)- Benzophenone 6	$C_{15}H_{14}O_5$ 274.27	131-54-4 139.5	1109	4-08-00-03505	
7596	Methanone, bis(2-hydroxyphenyl)-	$C_{13}H_{10}O_3$ 214.22	835-11-0 59.5	333	4-08-00-02446	H_2O 1; EtOH 3; eth 3; chl 3
7597	Methanone, bis(4-hydroxyphenyl)-	$C_{13}H_{10}O_3$ 214.22	611-99-4 210		4-08-00-02452 1.133[131]	H_2O 2; EtOH 3; eth 3; ace 3
7598	Methanone, bis(4-iodophenyl)- 4,4'-Diiodobenzophenone	$C_{13}H_8I_2O$ 434.01	5630-56-8 238.5	281[12]	4-07-00-01381 2.2600[20]	
7599	Methanone, bis(4-methylphenyl)-	$C_{15}H_{14}O$ 210.28	611-97-2 96.5	334	4-07-00-01434	ace 4; bz 4; eth 4; EtOH 4
7600	Methanone, (2-bromophenyl)phenyl- 2-Bromobenzophenone	$C_{13}H_9BrO$ 261.12	13047-06-8 42	345	3-07-00-02079	ace 4; EtOH 4; lig 4
7601	Methanone, (3-bromophenyl)phenyl-	$C_{13}H_9BrO$ 261.12	1016-77-9 81	186[6]	4-07-00-01378	EtOH 4
7602	Methanone, (4-bromophenyl)phenyl-	$C_{13}H_9BrO$ 261.12	90-90-4 82.5	350	4-07-00-01378	H_2O 1; EtOH 2; eth 2; bz 2
7603	Methanone, (2-chlorophenyl)(4-chlorophenyl)-	$C_{13}H_8Cl_2O$ 251.11	85-29-0 67	214-5[22]	4-07-00-01376 1.393[14]	EtOH 3; chl 2
7604	Methanone, (4-chlorophenyl)phenyl-	$C_{13}H_9ClO$ 216.67	134-85-0 77.5	332	4-07-00-01375	EtOH 3; eth 3; ace 3; ctc 2
7605	Methanone, cyclobutylphenyl-	$C_{11}H_{12}O$ 160.22	5407-98-7	260	4-07-00-01027 1.0426[25]	1.5472[20]
7606	Methanone, dicyclohexyl-	$C_{13}H_{22}O$ 194.32	119-60-8	159[20]	4-07-00-00262 0.986[0]	eth 3; ace 3; ctc 3 1.4860[20]
7607	Methanone, dicyclopropyl- Dicyclopropyl ketone	$C_7H_{10}O$ 110.16	1121-37-5	161	4-07-00-00139 0.977[25]	1.4670[20]
7608	Methanone, (2,6-dihydroxy-4- methoxyphenyl)phenyl- Cotoin	$C_{14}H_{12}O_4$ 244.25	479-21-0 130.5	2553	4-08-00-03163	ace 4; bz 4; eth 4; EtOH 4
7609	Methanone, (2,4-dihydroxyphenyl)phenyl- Benzoresorcinol	$C_{13}H_{10}O_3$ 214.22	131-56-6 144	1117	4-08-00-02442	H_2O 1; EtOH 3; eth 4; bz 2
7610	Methanone, (3,4-dihydroxyphenyl)(2,4,6- trihydroxyphenyl)- Maclurin	$C_{13}H_{10}O_6$ 262.22	519-34-6 222.5	5519	3-08-00-04226	eth 4; EtOH 4
7611	Methanone, (6,7-dimethoxy-1- isoquinolinyl)(3,4-dimethoxyphenyl)- Papaveraldine	$C_{20}H_{19}NO_5$ 353.37	522-57-6 210.5	6966	5-21-13-00658	H_2O 1; EtOH 2; eth 2; bz 3
7612	Methanone, (2,4-dimethoxyphenyl)phenyl- 2,4-Dimethoxybenzophenone	$C_{15}H_{14}O_3$ 242.27	3555-84-8 87.5	218[10]	4-08-00-02442	H_2O 1; EtOH 4; eth 2; chl 4
7613	Methanone, [4-(dimethylamino)phenyl]phenyl- 4-(Dimethylamino)benzophenone	$C_{15}H_{15}NO$ 225.29	530-44-9 92.5	3222	4-14-00-00248	H_2O 1; EtOH 3; eth 4; chl 3
7614	Methanone, (2,4-dimethylphenyl)phenyl-	$C_{15}H_{14}O$ 210.28	1140-14-3	321.5	4-07-00-01429 1.071[20]	bz 4; eth 4; EtOH 4; chl 4
7615	Methanone, (3,4-dimethylphenyl)phenyl- 3,4-Dimethylbenzophenone	$C_{15}H_{14}O$ 210.28	2571-39-3 47.5	341	4-07-00-01433	bz 4; eth 4; EtOH 4; chl 4
7616	Methanone, diphenyl- Benzophenone	$C_{13}H_{10}O$ 182.22	119-61-9 47.8	1108 305.4	4-07-00-01357 1.6077[19]	H_2O 1; EtOH 4; eth 4; ace 4
7617	Methanone, di-2-thienyl- Di-(2-thienyl) ketone	$C_9H_6OS_2$ 194.28	704-38-1 90	326	5-19-04-00318	ace 4; eth 4
7618	Methanone, 2-furanylphenyl-	$C_{11}H_8O_2$ 172.18	2689-59-0 <-15	285	5-17-10-00289 1.1732[20]	eth 4; EtOH 4 1.6055[20]
7619	Methanone, (2-hydroxy-4-methoxyphenyl)(2- hydroxyphenyl)- Dioxybenzone	$C_{14}H_{12}O_4$ 244.25	131-53-3	3298 170-5[1]	4-08-00-03163	
7620	Methanone, (2-hydroxy-4- methoxyphenyl)phenyl- Oxybenzone	$C_{14}H_{12}O_3$ 228.25	131-57-7 65.5	6907	4-08-00-02442	ctc 3
7621	Methanone, [2-hydroxy-4- (octyloxy)phenyl]phenyl- Octabenzone	$C_{21}H_{26}O_3$ 326.44	1843-05-6 48.5	6662		
7622	Methanone, (2-hydroxyphenyl)phenyl-	$C_{13}H_{10}O_2$ 198.22	117-99-7 40	250[560]	4-08-00-01246	H_2O 1; EtOH 4; eth 4; bz 4
7623	Methanone, (2-methoxyphenyl)phenyl-	$C_{14}H_{12}O_2$ 212.25	2553-04-0 41	210[27]; 195[18]	4-08-00-01246	H_2O 1; EtOH 3; bz 3; HOAc 3

No.	Name Synonym	Mol. Form. Mol. Wt.	CAS RN mp/°C	Merck No. bp/°C	Beil. Ref. den/g cm^{-3}	Solubility n_D
7624	Methanone, (3-methoxyphenyl)phenyl-	$C_{14}H_{12}O_2$ 212.25	6136-67-0 44	343; 201[17]	4-08-00-01262	H_2O 1; EtOH 4; bz 4; HOAc 4
7625	Methanone, (4-methoxyphenyl)phenyl-	$C_{14}H_{12}O_2$ 212.25	611-94-9 61.5	355; 168[12]	4-08-00-01263	H_2O 1; EtOH 4; eth 4; ace 3
7626	Methanone, [2-(methylamino)phenyl]phenyl-	$C_{14}H_{13}NO$ 211.26	1859-76-3 69	282	4-14-00-00244	bz 4; EtOH 4; HOAc 4
7627	Methanone, (4-methyl-2-nitrophenyl)phenyl- 4-Methyl-2-nitrobenzophenone	$C_{14}H_{11}NO_3$ 241.25	100224-75-7 126.5	sub	0-07-00-00442	bz 4; EtOH 4; chl 4
7628	Methanone, (2-methylphenyl)(4-methylphenyl)- 2,4'-Dimethylbenzophenone	$C_{15}H_{14}O$ 210.28	1140-16-5 317		4-07-00-01432 1.074[19]	eth 4; EtOH 4; HOAc 4
7629	Methanone, (2-methylphenyl)-1-naphthalenyl-	$C_{18}H_{14}O$ 246.31	68723-25-1 64	365	3-07-00-02639	eth 4; EtOH 4
7630	Methanone, (4-methylphenyl)(4-nitrophenyl)-	$C_{14}H_{11}NO_3$ 241.25	5350-47-0 124	sub	4-07-00-01406	bz 4; eth 4; EtOH 4; chl 4
7631	Methanone, (2-methylphenyl)phenyl-	$C_{14}H_{12}O$ 196.25	131-58-8 <-18	308; 128[12]	4-07-00-01401 1.1098[20]	H_2O 1; EtOH 4; os 3
7632	Methanone, (3-methylphenyl)phenyl-	$C_{14}H_{12}O$ 196.25	643-65-2	317; 170[9]	4-07-00-01403 1.088[17]	H_2O 1; EtOH 3; eth 3; bz 3
7633	Methanone, (4-methylphenyl)phenyl-	$C_{14}H_{12}O$ 196.25	134-84-9 59.5	228[70]	4-07-00-01403 0.9926[0]	H_2O 1; EtOH 2; eth 3; bz 3
7634	Methanone, 1-naphthalenyl-2-naphthalenyl- 1,2'-Dinaphthyl ketone	$C_{21}H_{14}O$ 282.34	605-79-8 136.5	235[0.6]	4-07-00-01886	chl 4
7635	Methanone, (3-nitrophenyl)phenyl-	$C_{13}H_9NO_3$ 227.22	2243-80-3 95	234[18]	4-07-00-01382	H_2O 1; EtOH 3
7636	Methanone, (4-nitrophenyl)phenyl-	$C_{13}H_9NO_3$ 227.22	1144-74-7 138		4-07-00-01382 1.406[9]	bz 4
7637	Methanone, phenyl-1H-pyrrol-2-yl- Pyrrole, 2-benzoyl	$C_{11}H_9NO$ 171.20	7697-46-3 79	305	5-21-08-00461	bz 4; EtOH 4; peth 4; HOAc 4
7638	Methanone, phenyl-2-pyridinyl-	$C_{12}H_9NO$ 183.21	91-02-1 42	317	5-21-08-00566 1.1556[20]	chl 3
7639	Methanone, phenyl-4-pyridinyl-	$C_{12}H_9NO$ 183.21	14548-46-0 72	315; 170[10]	5-21-08-00576	H_2O 2; EtOH 3; eth 3; bz 3
7640	Methanone, phenyl-4-quinolinyl-	$C_{16}H_{11}NO$ 233.27	54885-00-6 60	222[15]	1-21-00-00320	bz 4; eth 4; EtOH 4
7641	Methanone, phenyl-2-thienyl-	$C_{11}H_8OS$ 188.25	135-00-2 56.5	300	5-17-10-00291 1.1890[54]	H_2O 1; EtOH 3; eth 3 1.6181[54]
7642	Methanone, phenyl-3-thienyl-	$C_{11}H_8OS$ 188.25	6453-99-2 63.5	129[3]	5-17-10-00295	H_2O 1; EtOH 3; eth 3
7643	Methazole 1,2,4-Oxadiazolidine-3,5-dione, 2-(3,4-dichlorophenyl)-4-methyl-	$C_9H_6Cl_2N_2O_3$ 261.06	20354-26-1 123	5876	1.24[25]	
7644	1,3,4-Metheno-1H-cyclobuta[cd]pentalene, 1,1a,2,2,3,3a,4,5,5,5a,5b,6-dodecachlorooctahydro- Mirex	$C_{10}Cl_{12}$ 545.54	2385-85-5 485 dec	6126	3-05-00-01237	bz 4; diox 4
7645	Methidathion Phosphorodithioic acid, S-[(5-methoxy-2-oxo-1,3,4-thiadiazol-3(2H)-yl)methyl], O,O-dimethyl ester	$C_6H_{11}N_2O_4PS_3$ 302.34	950-37-8 39	5891		
7646	Methiocarb Phenol, 3,5-dimethyl-4-(methylthio)-, methylcarbamate	$C_{11}H_{15}NO_2S$ 225.31	2032-65-7 120	5893		
7647	DL-Methionine	$C_5H_{11}NO_2S$ 149.21	59-51-8 281 dec		4-04-00-03190 1.340[25]	H_2O 4; EtOH 2; eth 1; bz 1
7648	L-Methionine 2-Amino-4-methylthiobutanoic acid	$C_5H_{11}NO_2S$ 149.21	63-68-3 281 dec	5896	4-04-00-03189	H_2O 3; EtOH 1; eth 1; ace 1
7649	L-Methionine, N-acetyl- Methionamine	$C_7H_{13}NO_3S$ 191.25	65-82-7 105.5	90	4-04-00-03205	
7650	Methomyl Acetimidic acid, N-[(methylcarbamoyl)oxy]thio-, methyl ester	$C_5H_{10}N_2O_2S$ 162.21	16752-77-5 78	5905	1.2946[24]	
7651	Methoprene 2,4-Dodecadienoic acid, 11-methoxy-3,7,11-trimethyl-, 1-methylethyl ester, (E,E)-	$C_{19}H_{34}O_3$ 310.48	40596-69-8 100[0.05]	5906	0.926[20]	
7652	Methylamine, N-(o-chlorobenzylidene)- Toluene, 2-chloro-α-methylimino	C_8H_8ClN 153.61	17972-08-6 7	78-80[2]	3-07-00-00866	1.5660[20]
7653	Methyl 3,6-dichloro-o-anisate Benzoic acid, 3,6-dichloro-2-methoxy-, methyl ester	$C_9H_8Cl_2O_3$ 235.07	6597-78-0	115[2.4]		
7654	Methylene bispropionate Propanoic acid, methylene ester	$C_7H_{12}O_4$ 160.17	7044-96-4	98[1]; 191[745]	1.053[20]	
7655	Methylisonitrile Methyl isocyanate	C_2H_3N 41.05	593-75-9 -45	59.6 exp	0.756[4]	
7656	Metobromuron Urea, 3-(p-bromophenyl)-1-methoxy-1-methyl-	$C_9H_{11}BrN_2O_2$ 259.10	3060-89-7 95	6061	1.60[20]	
7657	Metolachlor Acetamide, 2-chloro-N-(2-ethyl-6-methylphenyl)-N-(2-methoxy-1-methylethyl)-	$C_{15}H_{22}ClNO_2$ 283.80	51218-45-2 100[0.001]	6067	1.12[20]	

No.	Name / Synonym	Mol. Form. Mol. Wt.	CAS RN mp/°C	Merck No. bp/°C	Beil. Ref. den/g cm^{-3}	Solubility n_D
7658	Metribuzin 1,2,4-Triazin-5(4H)-one, 4-amino-6-(1,1-dimethylethyl)-3-(methylthio)-	$C_8H_{14}N_4OS$ 214.29	21087-64-9 126	6076	1.31[20]	
7659	Mevinphos 2-Butenoic acid, 3-[(dimethoxyphosphinyl)oxy]-, methyl ester,(E)-	$C_7H_{13}O_6P$ 224.15	7786-34-7 21 (E), 6.9 (Z)	6089 101[0.3]		
7660	MGK 264 4,7-Methano-1H-isoindole-1,3(2H)-dione, 2-(2-ethylhexyl)-3a,4,7,7a-tetrahydro-	$C_{17}H_{25}NO_2$ 275.39	113-48-4 <-20	157	1.04	
7661	Molinate 1H-Azepine-1-carbothioic acid, hexahydro-, S-ethyl ester	$C_9H_{17}NOS$ 187.31	2212-67-1	202[10]	1.063[20]	
7662	Molybdenum carbonyl Molybdenum hexacarbonyl	C_6MoO_6 264.00	13939-06-5 dec 150	701		os 3
7663	Monocrotophos (E)-Dimethyl 1-methyl-3-(methylamino)-3-oxo-1-propenyl phosphate	$C_7H_{14}NO_5P$ 223.17	6923-22-4 55	6161 125[0.0005]	1.33[20]	
7664	Morphinan, 6,7-didehydro-4,5-epoxy-3,6-dimethoxy-17-methyl-, (5α)- Dihydrothebaine	$C_{19}H_{23}NO_3$ 313.40	561-25-1 162.5	3164	2-27-00-00110	H_2O 1; EtOH 3; bz 3; AcOEt 3
7665	Morphinan-3,6-diol, 7,8-didehydro-4,5-epoxy-, (5α,6α)- Normorphine	$C_{16}H_{17}NO_3$ 271.32	466-97-7 273	6630		
7666	Morphinan-3,6-diol, 7,8-didehydro-4,5-epoxy-17-methyl- (5α,6α)- Morphine	$C_{17}H_{19}NO_3$ 285.34	57-27-2 255	6186 sub 190	4-27-00-02223	H_2O 1; EtOH 2; eth 1; ace 1
7667	Morphinan-3,6-diol, 7,8-didehydro-4,5-epoxy-17-methyl- (5α,6α)-, diacetate (ester) Diacetylmorphine	$C_{21}H_{23}NO_5$ 369.42	561-27-3 173	2948 272-4[12]	4-27-00-02236 1.56[25]	bz 4; chl 4
7668	Morphinan-3,6-diol, 7,8-didehydro-4,5-epoxy-17-methyl- (5α,6α)-, 17-oxide Morphine, N-oxide	$C_{17}H_{19}NO_4$ 301.34	639-46-3 274.5	6192	4-27-00-02249	H_2O 2; EtOH 2; ace 1; bz 1
7669	Morphinan-6-ol, 7,8-didehydro-4,5-epoxy-3-methoxy-17-methyl-, (5α,6α)- Codeine	$C_{18}H_{21}NO_3$ 299.37	76-57-3 157.5	2459 250[22]; 140[1.5]	2-27-00-00136 1.32[25]	H_2O 3; EtOH 4; eth 3; bz 3
7670	Morphinan-6-ol, 7,8-didehydro-4,5-epoxy-3-methoxy-17-methyl-, (5α,6β)-	$C_{18}H_{21}NO_3$ 299.37	509-64-8 171.5	dec	2-27-00-00175 1.87[4]	1.675
7671	Morphinan-6-ol, 8,14-didehydro-4,5-epoxy-3-methoxy-17-methyl-, (5α,6α)- Neopine	$C_{18}H_{21}NO_3$ 299.37	467-14-1 127.5	6376		H_2O 3; EtOH 3; eth 3; bz 3
7672	Morphinan-8-ol, 6,7-didehydro-4,5-epoxy-3-methoxy-17-methyl-, (5α,8β)- Pseudocodeine	$C_{18}H_{21}NO_3$ 299.37	466-96-6 181.5	7927	2-27-00-00112 1.290[80]	1.574
7673	Morphinan-6-ol, 7,8-didehydro-4,5-epoxy-3-methoxy-17-methyl-, (5α,6α)-, phosphate (1:1) (salt) Codeine, phosphate	$C_{18}H_{24}NO_7P$ 397.36	52-28-8 220-35 dec	2462	4-27-00-02228	EtOH 4; chl 4
7674	Morphinan-6-ol, 4,5-epoxy-3-methoxy-17-methyl-, (5α,6α)- Dihydrocodeine	$C_{18}H_{23}NO_3$ 301.39	125-28-0 112.5	3154 248[15]	2-27-00-00103	
7675	Morphinan-6-one, 7,8-didehydro-4,5-epoxy-14-hydroxy-3-methoxy-17-methyl-, (5α)- Hydroxycodeinone	$C_{18}H_{19}NO_4$ 313.35	508-54-3 275 dec	4752		
7676	Morphinan-6-one, 7,8-didehydro-4-hydroxy-3,7-dimethoxy-17-methyl-, (9α,13α,14α)- Sinomenine	$C_{19}H_{23}NO_4$ 329.40	115-53-7 162	8496	5-21-13-00516	H_2O 2; EtOH 3; eth 2; ace 3
7677	Morphinan-6-one, 7,8-didehydro-4-hydroxy-3-methoxy-17-methyl- Thebainone	$C_{18}H_{21}NO_3$ 299.37	467-98-1 151.5	9203	5-21-13-00268	H_2O 2; EtOH 2; eth 2; ace 3
7678	Morphinan-6-one, 4,5-epoxy-3-methoxy-17-methyl-, (5α)- Hydrocodone	$C_{18}H_{21}NO_3$ 299.37	125-29-1 198	4708	4-27-00-03580	H_2O 1; EtOH 3
7679	Morphinan, 6,7,8,14-tetradehydro-4,5-epoxy-3,6-dimethoxy-17-methyl-, (5α)- Thebaine	$C_{19}H_{21}NO_3$ 311.38	115-37-7 193	9202 sub 91	4-27-00-02271 1.305[20]	H_2O 1; EtOH 4; eth 2; bz 3
7680	4-Morpholinamine	$C_4H_{10}N_2O$ 102.14	4319-49-7	166	4-27-00-00623 1.059[25]	
7681	Morpholine Tetrahydro-1,4-oxazine	C_4H_9NO 87.12	110-91-8 -4.9	6194 128	4-27-00-00015 1.0005[20]	H_2O 5; EtOH 3; eth 3; ace 3 1.4548[20]
7682	Morpholine, 4-acetyl-	$C_6H_{11}NO_2$ 129.16	1696-20-4 14.5	152[50]; 118[12]	4-27-00-00274 1.1145[20]	H_2O 5; EtOH 3; ace 3; ctc 3 1.4827[20]
7683	Morpholine, 4-[3-(4-butoxyphenoxy)propyl]- Pramoxine	$C_{17}H_{27}NO_3$ 293.41	140-65-8	7707 196[6]	4-27-00-00127	
7684	Morpholine, 4-butyl-	$C_8H_{17}NO$ 143.23	1005-67-0 -57.1	213.5	4-27-00-00026 0.9068[20]	H_2O 4; ace 4; bz 4; EtOH 4 1.4451[20]
7685	4-Morpholinecarboxaldehyde	$C_5H_9NO_2$ 115.13	4394-85-8 21	239	4-27-00-00274 1.1520[20]	1.4845[20]

No.	Name Synonym	Mol. Form. Mol. Wt.	CAS RN mp/°C	Merck No. bp/°C	Beil. Ref. den/g cm^{-3}	Solubility n_D
7686	4-Morpholinecarboxamide, N,N'-1,2-ethanediylbis[N-butyl- Dimorpholamine	C$_{20}$H$_{38}$N$_4$O$_4$ 398.55	119-48-2 41.5	3257 229[0.4]	4-27-00-00298	H$_2$O 4
7687	Morpholine, 2,6-dimethyl- 2,6-Dimethylmorpholine	C$_6$H$_{13}$NO 115.18	141-91-3 -88	146.6	4-27-00-00674 0.9329[20]	H$_2$O 5; EtOH 5; ace 3; bz 5 1.4460[20]
7688	Morpholine, 4,4'-dithiobis- 4,4'-Dithiodimorpholine	C$_8$H$_{16}$N$_2$O$_2$S$_2$ 236.36	103-34-4 124.5	3379	4-27-00-00613	chl 3
7689	4-Morpholineethanamine	C$_6$H$_{14}$N$_2$O 130.19	2038-03-1 25.6	205	4-27-00-00370 0.9897[20]	H$_2$O 5; EtOH 5; ace 3; bz 5 1.4715[20]
7690	Morpholine, 4,4'-(1,2-ethanediyl)bis-	C$_{10}$H$_{20}$N$_2$O$_2$ 200.28	1723-94-0 75	285; 160[25]	4-27-00-00375	H$_2$O 4; ace 4; bz 4; EtOH 4
7691	4-Morpholineethanol	C$_6$H$_{13}$NO$_2$ 131.17	622-40-2 -1.6	227	4-27-00-00056 1.0710[20]	H$_2$O 3; EtOH 3; ctc 2 1.4763[20]
7692	4-Morpholineethanol, α-methyl-	C$_7$H$_{15}$NO$_2$ 145.20	2109-66-2 92-4[13]		4-27-00-00119 1.0174[20]	H$_2$O 4; ace 4; bz 4; EtOH 4 1.4638[20]
7693	Morpholine, 4-(2-ethoxyethyl)- Morpholine, N-(2-ethoxyethyl)	C$_8$H$_{17}$NO$_2$ 159.23	622-09-3 -100	206	4-27-00-00057 0.9632[0]	H$_2$O 4; ace 4; bz 4; eth 4
7694	Morpholine, 4-ethyl- N-Ethylmorpholine	C$_6$H$_{13}$NO 115.18	100-74-3 138.5		4-27-00-00023 0.8996[20]	H$_2$O 5; EtOH 5; eth 5; ace 3 1.4400[20]
7695	Morpholine, 4-(2-hydroxybenzoyl)-	C$_{11}$H$_{10}$NO$_3$ 204.21	3202-84-4	8302	4-27-00-00346	DMSO 3
7696	Morpholine, 4-methyl- N-Methylmorpholine	C$_5$H$_{11}$NO 101.15	109-02-4 116		4-27-00-00022 0.9051[20]	H$_2$O 3; EtOH 3; eth 3 1.4332[20]
7697	Morpholine, 3-methyl-2-phenyl- Phenmetrazine	C$_{11}$H$_{15}$NO 177.25	134-49-6 139[12]; 104[1]	7200	4-27-00-01050	
7698	Morpholine, 4-(4-methylphenyl)-	C$_{11}$H$_{15}$NO 177.25	3077-16-5 51	167[30]	2-27-00-00004	eth 4; EtOH 4
7699	Morpholine, 4-nitroso- N-Nitrosomorpholine	C$_4$H$_8$N$_2$O$_2$ 116.12	59-89-2 29	6561 225; 140[25]		H$_2$O 3
7700	Morpholine, 4-(phenylmethyl)-	C$_{11}$H$_{15}$NO 177.25	10316-00-4 194	260.5	4-27-00-00039 1.0387[20]	ace 4; bz 4 1.5302[20]
7701	4-Morpholinepropanamine 4-(3-Aminopropyl)morpholine	C$_7$H$_{16}$N$_2$O 144.22	123-00-2 -15	220; 134[50]	4-27-00-00411 0.9854[20]	H$_2$O 5; EtOH 5; ace 3; bz 5 1.4762[20]
7702	Naled Phosphoric acid, 1,2-dibromo-2,2-dichloroethyl dimethyl ester	C$_4$H$_7$Br$_2$Cl$_2$O$_4$P 380.78	300-76-5 27	6272 110[0.5]	1.96[20]	
7703	Naphthacene 2,3-Benzanthracene	C$_{18}$H$_{12}$ 228.29	92-24-0 357	6288 sub	4-05-00-02545	H$_2$O 2; bz 2; con sulf 3; os 2
7704	2-Naphthacenecarboxamide, 7-chloro-4-(dimethylamino)-1,4,4a,5,5a,6,11,12a-octahydro-3,6,10,12,12a-pentahydroxy-6-methyl-1,11-dioxo, [4S-(4α,4aα,5aα,6β,12aα)]- Chlortetracycline	C$_{22}$H$_{23}$ClN$_2$O$_8$ 478.89	57-62-5 168.5	2193	4-14-00-02629	H$_2$O 1; EtOH 2; eth 1; ace 2
7705	2-Naphthacenecarboxamide, 4-(dimethylamino)-1,4,4a,5,5a,6,11,12a-octahydro-3,5,6,10,12,12a-hexahydroxy-6-methyl-1,11-dioxo-,[4S-(4α,4aα,5aα,6β,12aα)]- Oxytetracycline	C$_{22}$H$_{24}$N$_2$O$_9$ 460.44	79-57-2 184.5	6931	4-14-00-02633 1.634[20]	
7706	2-Naphthacenecarboxamide, 4-(dimethylamino)-1,4,4a,5,5a,6,11,12a-octahydro-3,6,10,12,12a-pentahydroxy-6-methyl-1,11-dioxo-,[4S-(4α,4aα,5aα,6β,12aα)]- Tetracycline	C$_{22}$H$_{24}$N$_2$O$_8$ 444.44	60-54-8 170-5 dec	9130	4-14-00-02625	
7707	1,2-Naphthacenedione 9,10-Naphthacenequinone	C$_{18}$H$_{10}$O$_2$ 258.28	29276-40-2 294	sub	3-07-00-04273	EtOH 4
7708	1-Naphthalenamine 1-Naphthylamine	C$_{10}$H$_9$N 143.19	134-32-7 49.2	6318 300.7	4-12-00-03009 1.0228[20]	chl 3 1.6140[20]
7709	2-Naphthalenamine 2-Naphthylamine	C$_{10}$H$_9$N 143.19	91-59-8 113	6319 306.2	4-12-00-03122 1.6414[98]	H$_2$O 3; EtOH 3; eth 3 1.6493[98]
7710	2-Naphthalenamine, N,N-bis(2-chloroethyl)- Chlornaphazine	C$_{14}$H$_{15}$Cl$_2$N 268.19	494-03-1 55	2107 210[5]	4-12-00-03126	ace 4; bz 4; eth 4; EtOH 4
7711	1-Naphthalenamine, 5-bromo- 1-Amino-5-bromonaphthalene	C$_{10}$H$_8$BrN 222.08	4766-33-0 70	sub	4-12-00-03111	bz 4; eth 4; EtOH 4; lig 4
7712	1-Naphthalenamine, decahydro-	C$_{10}$H$_{19}$N 153.27	7250-95-5 13.5	227.6	3-12-00-00183 0.9508[21]	
7713	1-Naphthalenamine, N,N-diethyl-	C$_{14}$H$_{17}$N 199.30	84-95-7 285		3-12-00-02855 1.013[20]	EtOH 3; eth 3; bz 3; ctc 2 1.5961[20]
7714	1-Naphthalenamine, 5,8-dihydro- 1-Naphthylamine, 5,8-dihydro-	C$_{10}$H$_{11}$N 145.20	32666-56-1 37.5	247[408]	4-12-00-02995	EtOH 4; chl 4
7715	1-Naphthalenamine, N,N-dimethyl- N,N-Dimethyl-1-naphthylamine	C$_{12}$H$_{13}$N 171.24	86-56-6 250; 140[13]	3240	4-12-00-03013 1.0423[20]	H$_2$O 1; EtOH 3; eth 3; ctc 3 1.624[15]
7716	2-Naphthalenamine, N,N-dimethyl-	C$_{12}$H$_{13}$N 171.24	2436-85-3 52.5	305	4-12-00-03125 1.0279[60]	H$_2$O 1; EtOH 3; eth 3 1.6443[53]
7717	1-Naphthalenamine, N,N-diphenyl- 1-Naphthylamine, N,N-diphenyl-	C$_{22}$H$_{17}$N 295.38	61231-45-6 142	335-40[80]	0-12-00-01225	ace 4; bz 4; eth 4; EtOH 4
7718	1-Naphthalenamine, N-ethyl-	C$_{12}$H$_{13}$N 171.24	118-44-5 305; 191[16]		4-12-00-03013 1.060[20]	eth 4; EtOH 4 1.6477[15]

No.	Name Synonym	Mol. Form. Mol. Wt.	CAS RN mp/°C	Merck No. bp/°C	Beil. Ref. den/g cm⁻³	Solubility n_D
7719	2-Naphthalenamine, N-ethyl-	$C_{12}H_{13}N$ 171.24	2437-03-8 236	316.5	4-12-00-03126 1.0545[21]	1.6544[21]
7720	1-Naphthalenamine, 4-methyl-	$C_{11}H_{11}N$ 157.22	4523-45-9 51.5	176[12]	4-12-00-03190	eth 4
7721	1-Naphthalenamine, N-methyl-	$C_{11}H_{11}N$ 157.22	2216-68-4 174	294.5	4-12-00-03013	eth 4; EtOH 4 1.6722[20]
7722	2-Naphthalenamine, N-methyl-	$C_{11}H_{11}N$ 157.22	2216-67-3	317	4-12-00-03125	eth 4; EtOH 4 1.6722[20]
7723	1-Naphthalenamine, N-(2-methylphenyl)- 1-Naphthyl(2-tolyl)amine	$C_{17}H_{15}N$ 233.31	634-41-3 94.5	200[9]	3-12-00-02857	bz 4; eth 4; EtOH 4
7724	1-Naphthalenamine, N-(4-methylphenyl)-	$C_{17}H_{15}N$ 233.31	634-43-5 79	360[528]; 230[10]	4-12-00-03016	bz 4; eth 4
7725	2-Naphthalenamine, N-(2-methylphenyl)- 2-Naphthyl(2-tolyl)amine	$C_{17}H_{15}N$ 233.31	644-15-5 95.5	402.5	1-12-00-00535	ace 4; bz 4; eth 4; EtOH 4
7726	1-Naphthalenamine, N-1-naphthalenyl-	$C_{20}H_{15}N$ 269.35	737-89-3 115	311[15]	4-12-00-03018	H_2O 1; EtOH 3; eth 3; ace 3
7727	2-Naphthalenamine, N-2-naphthalenyl- β,β'-Dinaphthylamine	$C_{20}H_{15}N$ 269.35	532-18-3 172.2	3260 471	4-12-00-03131	H_2O 1; EtOH 2; eth 3; bz 2
7728	2-Naphthalenamine, N-phenyl- N-Phenyl-β-naphthylamine	$C_{16}H_{13}N$ 219.29	135-88-6 108	395.5	4-12-00-03128	H_2O 1; EtOH 3; eth 3; bz 3
7729	2-Naphthalenamine, 1-(phenylazo)- Yellow AB	$C_{16}H_{13}N_3$ 247.30	85-84-7 103	10003		EtOH 4; HOAc 4
7730	2-Naphthalenamine, N-(phenylmethyl)-	$C_{17}H_{15}N$ 233.31	13672-18-9 68	405	4-12-00-03130	bz 4; lig 4
7731	1-Naphthalenamine, 5,6,7,8-tetrahydro-	$C_{10}H_{13}N$ 147.22	2217-41-6 38	279	4-12-00-02934 1.0625[16]	H_2O 2; EtOH 3; eth 3; acid 3 1.5900[20]
7732	Naphthalene	$C_{10}H_8$ 128.17	91-20-3 80.2	6289 217.9	4-05-00-01640 1.0253[20]	H_2O 1; EtOH 3; eth 4; ace 4 1.5898[25]
7733	1-Naphthaleneacetamide Acetic acid, amide, 1-naphthyl	$C_{12}H_{11}NO$ 185.23	86-86-2	sub 180	4-09-00-02426	H_2O 1; eth 3; bz 3; CS_2 3
7734	1-Naphthaleneacetic acid 1-Naphthylacetic acid	$C_{12}H_{10}O_2$ 186.21	86-87-3 135	6290 dec	4-09-00-02424	H_2O 2; EtOH 2; eth 4; ace 4
7735	1-Naphthaleneacetic acid, ethyl ester Ethyl 1-naphthylacetate	$C_{14}H_{14}O_2$ 214.26	2122-70-5 88.5	222[20]; 118[13]	4-09-00-02425	EtOH 3; eth 3
7736	1-Naphthaleneacetonitrile	$C_{12}H_9N$ 167.21	132-75-2 32.5	192[18]; 163[12]	4-09-00-02428	EtOH 3 1.6192[20]
7737	2-Naphthaleneacetonitrile	$C_{12}H_9N$ 167.21	7498-57-9 83	303	3-09-00-03212 1.092[25]	
7738	Naphthalene, 1-azido-	$C_{10}H_7N_3$ 169.19	6921-40-0 12	75[0.001]	1.1713[25]	EtOH 4; eth 5; ace 5 1.6550[25]
7739	Naphthalene, 1-bromo- 1-Naphthyl bromide	$C_{10}H_7Br$ 207.07	90-11-9 -1.8	1413 281	4-05-00-01665 1.4785[20]	H_2O 3; EtOH 5; eth 5; ace 3 1.658[20]
7740	Naphthalene, 2-bromo- 2-Bromonaphthalene	$C_{10}H_7Br$ 207.07	580-13-2 59	1414 281.5	4-05-00-01667 1.605[25]	H_2O 1; EtOH 3; eth 3; bz 3 1.6382[60]
7741	Naphthalene, 1-bromo-4-chloro- 1-Bromo-4-chloronaphthalene	$C_{10}H_6BrCl$ 241.51	53220-82-9 66.5	303	3-05-00-01585	eth 4; HOAc 4
7742	Naphthalene, 1-(bromomethyl)-	$C_{11}H_9Br$ 221.10	3163-27-7 56	183[18]; 167[10]	4-05-00-01693	ace 4; bz 4; eth 4; EtOH 4
7743	Naphthalene, 2-(bromomethyl)-	$C_{11}H_9Br$ 221.10	939-26-4 56	213[100]; 167[14]	4-05-00-01698	EtOH 3; eth 3; chl 3; HOAc 3
7744	Naphthalene, 1-bromo-3-nitro-	$C_{10}H_8BrNO_2$ 252.07	7499-65-2 132.3	sub	4-05-00-01678	eth 4; EtOH 4
7745	Naphthalene, 2-bromo-1-nitro- 2-Bromo-1-nitronaphthalene	$C_{10}H_6BrNO_2$ 252.07	4185-62-0 104.3	>360	4-05-00-01677	bz 4; eth 4; EtOH 4
7746	Naphthalene, 1-butyl-	$C_{14}H_{16}$ 184.28	1634-09-9 -19.8	289.3	4-05-00-01737 0.9738[20]	H_2O 1; EtOH 3; eth 3; ace 3 1.5819[20]
7747	Naphthalene, 2-butyl- 2-Butylnaphthalene	$C_{14}H_{16}$ 186.30	1134-62-9 -5	292	4-05-00-01738 0.9673[20]	ace 4; bz 4; EtOH 4 1.5777[20]
7748	1-Naphthalenecarbonitrile	$C_{11}H_7N$ 153.18	86-53-3 37.5	299	4-09-00-02407 1.1080[25]	H_2O 1; EtOH 4; eth 4; lig 3 1.6298[18]
7749	2-Naphthalenecarbonitrile	$C_{11}H_7N$ 153.18	613-46-7 66	306.5	4-09-00-02419 1.0755[60]	H_2O 2; EtOH 3; eth 3; chl 2
7750	2-Naphthalenecarbonyl chloride	$C_{11}H_7ClO$ 190.63	2243-83-6 51	305	4-09-00-02417	bz 4; eth 4; chl 4
7751	1-Naphthalenecarboxaldehyde	$C_{11}H_8O$ 156.18	66-77-3 33.5	292	4-07-00-01286 1.1503[20]	H_2O 1; EtOH 3; eth 3; ace 3 1.6507[20]
7752	2-Naphthalenecarboxaldehyde	$C_{11}H_8O$ 156.18	66-99-9 62	160[19]	4-07-00-01288 1.0775[99]	H_2O 2; EtOH 4; eth 4; ace 3 1.621[99]
7753	1-Naphthalenecarboxaldehyde, 2-ethoxy-	$C_{13}H_{12}O_2$ 200.24	19523-57-0 115	185-7[25]	3-08-00-01110	EtOH 3; HOAc 3
7754	1-Naphthalenecarboxaldehyde, 2-hydroxy-	$C_{11}H_8O_2$ 172.18	708-06-5 83	192[27]	4-08-00-01160	H_2O 1; EtOH 3; eth 3; aq alk 3
7755	1-Naphthalenecarboxaldehyde, 4-methoxy-	$C_{12}H_{10}O_2$ 186.21	15971-29-6 34.8	212[40]	4-08-00-01170 1.1879[39]	

No.	Name / Synonym	Mol. Form. / Mol. Wt.	CAS RN / mp/°C	Merck No. / bp/°C	Beil. Ref. / den/g cm⁻³	Solubility / n_D
7756	1-Naphthalenecarboxaldehyde, 4-methyl- 1-Naphthaldehyde, 4-methyl-	$C_{12}H_{10}O$ 170.21	33738-48-6 33.5	 175[13]	3-07-00-01972 1.1252[38]	
7757	1-Naphthalenecarboxamide	$C_{11}H_9NO$ 171.20	2243-81-4 205.8	 sub	4-09-00-02405	H_2O 2; EtOH 2; con HCl 3; HOAc 3
7758	1-Naphthalenecarboxylic acid 1-Naphthoic acid	$C_{11}H_8O_2$ 172.18	86-55-5 161	6301 >300	4-09-00-02402 1.398[25]	eth 4; EtOH 4; chl 4 1.46
7759	2-Naphthalenecarboxylic acid 2-Naphthoic acid	$C_{11}H_8O_2$ 172.18	93-09-4 185.5	6302 >300	4-09-00-02414 1.077[100]	H_2O 2; EtOH 3; eth 3; chl 3
7760	1-Naphthalenecarboxylic acid, 5-amino-	$C_{11}H_9NO_2$ 187.20	32018-88-5 212.3	 sub	4-14-00-01812	EtOH 4; HOAc 4
7761	1-Naphthalenecarboxylic acid, 5-bromo-	$C_{11}H_7BrO_2$ 251.08	16726-67-3 261	 sub	3-09-00-03153	H_2O 2; EtOH 2; eth 2; bz 4
7762	1-Naphthalenecarboxylic acid, 5-chloro- 1-Naphthoic acid, 5-chloro-	$C_{11}H_7ClO_2$ 206.63	16650-52-5 245	 sub	3-09-00-03149	H_2O 2; EtOH 4; bz 2; HOAc 2
7763	1-Naphthalenecarboxylic acid, 8-chloro-	$C_{11}H_7ClO_2$ 206.63	4537-00-2 171.5	 sub	3-09-00-03151	H_2O 3; EtOH 3; ace 4; bz 3
7764	1-Naphthalenecarboxylic acid, 3,4-dihydro- 3,4-Dihydro-1-naphthoic acid	$C_{11}H_{10}O_2$ 174.20	3333-23-1 125	 306	4-09-00-02338	H_2O 2; EtOH 3; chl 3; MeOH 3
7765	1-Naphthalenecarboxylic acid, ethyl ester	$C_{13}H_{12}O_2$ 200.24	3007-97-4	 310	4-09-00-02403 1.1264[15]	H_2O 1; EtOH 3 1.5966[15]
7766	1-Naphthalenecarboxylic acid, methyl ester	$C_{12}H_{10}O_2$ 186.21	2459-24-7 59.5	 168[20]; 101[0.04]	4-09-00-02402 1.1290[20]	bz 4; EtOH 4 1.6086[20]
7767	1-Naphthalenecarboxylic acid, 5-nitro-	$C_{11}H_7NO_4$ 217.18	1975-44-6 241.5	 sub	4-09-00-02411	H_2O 1; EtOH 2; eth 2; bz 2
7768	2-Naphthalenecarboxylic acid, 3-amino- 3-Amino-2-naphthoic acid	$C_{11}H_9NO_2$ 187.20	5959-52-4 216.5	464	3-14-00-01341	EtOH 3; eth 3
7769	2-Naphthalenecarboxylic acid, 5-bromo- 5-Bromo-2-naphthoic acid	$C_{11}H_7BrO_2$ 251.08	1013-83-8 270	 sub	3-09-00-03196	bz 4; eth 4; EtOH 4; HOAc 4
7770	2-Naphthalenecarboxylic acid, ethyl ester	$C_{13}H_{12}O_2$ 200.24	3007-91-8 32	 308.5	4-09-00-02415 1.1143[23]	eth 4; EtOH 4; chl 4; HOAc 4 1.5951[23]
7771	2-Naphthalenecarboxylic acid, 1-hydroxy- 1-Hydroxy-2-naphthoic acid	$C_{11}H_8O_3$ 188.18	86-48-6 195	4766	4-10-00-01181	H_2O 2; EtOH 4; eth 4; bz 3
7772	2-Naphthalenecarboxylic acid, 3-hydroxy- 3-Hydroxy-2-naphthoic acid	$C_{11}H_8O_3$ 188.18	92-70-6 222.5	4767	4-10-00-01184	H_2O 2; EtOH 4; eth 4; bz 3
7773	2-Naphthalenecarboxylic acid, 3-hydroxy-, ethyl ester	$C_{13}H_{12}O_3$ 216.24	7163-25-9 85	 291	4-10-00-01186	ace 4; chl 4
7774	2-Naphthalenecarboxylic acid, 3-hydroxy-, hexadecyl ester Hexadecyl 3-hydroxy-2-naphthoate	$C_{27}H_{40}O_3$ 412.61	531-84-0 72.5	4604	3-10-00-01087	bz 4; HOAc 4
7775	2-Naphthalenecarboxylic acid, 3-hydroxy-, methyl ester	$C_{12}H_{10}O_3$ 202.21	883-99-8 75.5	 206	4-10-00-01186	H_2O 1; EtOH 3; os 3
7776	2-Naphthalenecarboxylic acid, methyl ester	$C_{12}H_{10}O_2$ 186.21	2459-25-8 77	 290	4-09-00-02414	bz 4; eth 4; EtOH 4; chl 4
7777	2-Naphthalenecarboxylic acid, 4,4'-methylenebis[3-hydroxy- Pamoic acid	$C_{23}H_{16}O_6$ 388.38	130-85-8 315	6955	4-10-00-02362	
7778	2-Naphthalenecarboxylic acid, 1,2,3,4-tetrahydro- 1,2,3,4-Tetrahydro-2-naphthoic acid	$C_{11}H_{12}O_2$ 176.22	53440-12-3 97.7	 168[15]	3-09-00-02805	bz 4; eth 4; EtOH 4; chl 4
7779	2-Naphthalenecarboxylic acid, 5,6,7,8-tetrahydro-	$C_{11}H_{12}O_2$ 176.22	1131-63-1 155.5	 216[14]	4-09-00-02096	EtOH 4; bz 3; HOAc 3; peth 1
7780	Naphthalene, 1-chloro- 1-Naphthyl chloride	$C_{10}H_7Cl$ 162.62	90-13-1 -2.5	2149 259; 106.5[5]	4-05-00-01658 1.1938[20]	H_2O 1; EtOH 3; eth 3; bz 3 1.6326[20]
7781	Naphthalene, 2-chloro- 2-Chloronaphthalene	$C_{10}H_7Cl$ 162.62	91-58-7 61	2150 256	4-05-00-01660 1.1377[71]	H_2O 1; EtOH 3; eth 3; bz 3 1.6079[13]
7782	Naphthalene, 2-chlorodecahydro- Decalin, 2-chloro	$C_{10}H_{17}Cl$ 172.70	5597-81-9	 112[15]	4-05-00-00313 1.0421[20]	 1.5020[20]
7783	Naphthalene, 1-(chloromethyl)-	$C_{11}H_9Cl$ 176.65	86-52-2 32	 291.5	4-05-00-01692 1.1813[20]	H_2O 1; EtOH 3; ctc 3; peth 3 1.6380[20]
7784	Naphthalene, 1-chloro-2-methyl- 1-Chloro-2-methylnaphthalene	$C_{11}H_9Cl$ 176.65	5859-45-0	 150-2[10]	3-05-00-01632 1.6265[20]	ctc 3
7785	Naphthalene, 2-(chloromethyl)-	$C_{11}H_9Cl$ 176.65	2506-41-4 48.5	 169[20]	4-05-00-01697	H_2O 1; EtOH 3; peth 3
7786	Naphthalene, 1-chloro-5-nitro- 1-Chloro-5-nitronaphthalene	$C_{10}H_6ClNO_2$ 207.62	605-63-0 111	 >360	3-05-00-01597	EtOH 4
7787	Naphthalene, 1-chloro-8-nitro- 1-Chloro-8-nitronaphthalene	$C_{10}H_6ClNO_2$ 207.62	602-37-9 95.8	 175[2]	4-05-00-01676	EtOH 3; bz 3; chl 2; HOAc 3
7788	Naphthalene, 2-chloro-1-nitro- 2-Chloro-1-nitronaphthalene	$C_{10}H_6ClNO_2$ 207.62	4185-63-1 99.5	 >360	4-05-00-01676 1.4854[25]	EtOH 4; eth 4; ace 4; bz 4
7789	Naphthalene, decahydro- Decalin	$C_{10}H_{18}$ 138.25	91-17-8 -43	2839 155.5	3-05-00-00245 0.8965[22]	ace 4; bz 4; eth 4; EtOH 4 1.4810[20]
7790	Naphthalene, decahydro-, cis- cis-Decalin	$C_{10}H_{18}$ 138.25	493-01-6 -42.9	2839 195.8	4-05-00-00310 0.8965[20]	H_2O 1; EtOH 5; eth 4; ace 4 1.4810[20]
7791	Naphthalene, decahydro-, trans- trans-Decalin	$C_{10}H_{18}$ 138.25	493-02-7 -30.3	2839 187.3	4-05-00-00311 0.8699[20]	H_2O 1; EtOH 4; eth 4; ace 4 1.4695[20]

No.	Name Synonym	Mol. Form. Mol. Wt.	CAS RN mp/°C	Merck No. bp/°C	Beil. Ref. den/g cm^{-3}	Solubility n_D
7792	Naphthalene, decahydro-2-methylene-, *trans*- Decalin, 2-methylene (*trans*)	$C_{11}H_{18}$ 150.26	7787-72-6	201; 82[10]	3-05-00-00397 0.8897[20]	1.4841[22]
7793	Naphthalene, 1-decyl- Decane, 1-(1-naphthyl)-	$C_{20}H_{28}$ 268.44	26438-27-7 15	379	4-05-00-01786 0.9322[20]	1.5435[20]
7794	Naphthalene, 2-decyl-	$C_{20}H_{28}$ 268.44	14188-79-5 20	387	4-05-00-01786 0.9253[20]	1.5413[20]
7795	1,2-Naphthalenediamine 1,2-Diaminonaphthalene	$C_{10}H_{10}N_2$ 158.20	938-25-0 98.5	214[13.] 150[0.5]	4-13-00-00337	H_2O 2; EtOH 4; eth 4; chl 4
7796	1,4-Naphthalenediamine 1,4-Diaminonaphthalene	$C_{10}H_{10}N_2$ 158.20	2243-61-0 120		4-13-00-00338	ace 4; bz 4; EtOH 4; chl 4 1.6441[18]
7797	1,5-Naphthalenediamine	$C_{10}H_{10}N_2$ 158.20	2243-62-1 190	sub	4-13-00-00340 1.4[25]	H_2O 3; EtOH 3; eth 3; chl 4
7798	1,8-Naphthalenediamine Naphthalene, 1,8-diamino	$C_{10}H_{10}N_2$ 158.20	479-27-6 66.5	6291 205[12]	4-13-00-00344 1.1265[90]	eth 4; EtOH 4 1.6828[99]
7799	2,3-Naphthalenediamine 2,3-Diaminonaphthalene	$C_{10}H_{10}N_2$ 158.20	771-97-1 199		4-13-00-00346 1.0968[26]	H_2O 2; EtOH 4; eth 3 1.6392[26]
7800	1,4-Naphthalenediamine, 2-methyl- Vitamin K6	$C_{11}H_{12}N_2$ 172.23	83-68-1 113.5	9937	3-13-00-00406	
7801	Naphthalene, 1,3-dibromo- 1,3-Dibromonaphthalene	$C_{10}H_6Br_2$ 285.97	52358-73-3 62.8	290	3-05-00-01586	EtOH 4
7802	Naphthalene, 1,4-dibromo- 1,4-Dibromonaphthalene	$C_{10}H_6Br_2$ 285.97	83-53-4 83	310	4-05-00-01669	H_2O 1; EtOH 3; eth 3; HOAc 2
7803	Naphthalene, 1,5-dibromo- 1,5-Dibromonaphthalene	$C_{10}H_6Br_2$ 285.97	7351-74-8 131.3	326	4-05-00-01669	eth 4; EtOH 4
7804	1,8-Naphthalenedicarbonyl dichloride Naphthaloyl chloride	$C_{12}H_6Cl_2O_2$ 253.08	6423-29-6 85	197[0.2]	3-09-00-04468	bz 4; chl 4
7805	1,8-Naphthalenedicarboxylic acid Naphthalic acid	$C_{12}H_8O_4$ 216.19	518-05-8 260	6299	4-09-00-03534	H_2O 1; EtOH 2; eth 2
7806	1,8-Naphthalenedicarboxylic acid, diethyl ester Naphthalic acid, diethyl ester	$C_{16}H_{16}O_4$ 272.30	58618-39-6 59.5	238[19]; 202[11]	4-09-00-03534 1.1399[70]	
7807	Naphthalene, 1,2-dichloro- 1,2-Dichloronaphthalene	$C_{10}H_6Cl_2$ 197.06	2050-69-3 36	296.5	4-05-00-01661 1.3147[49]	EtOH 3; eth 3 1.5338[49]
7808	Naphthalene, 1,3-dichloro- 1,3-Dichloronaphthalene	$C_{10}H_6Cl_2$ 197.06	2198-75-6 62.3	291	4-05-00-01661	EtOH 3
7809	Naphthalene, 1,4-dichloro- 1,4-Dichloronaphthalene	$C_{10}H_6Cl_2$ 197.06	1825-31-6 67.5	288; 147[12]	4-05-00-01661 1.2997[76]	H_2O 1; EtOH 2; eth 3; ace 4 1.6228[76]
7810	Naphthalene, 1,5-dichloro- 1,5-Dichloronaphthalene	$C_{10}H_6Cl_2$ 197.06	1825-30-5 107	sub	4-05-00-01662 1.4900[20]	H_2O 1; EtOH 2; eth 3
7811	Naphthalene, 1,7-dichloro- 1,7-Dichloronaphthalene	$C_{10}H_6Cl_2$ 197.06	2050-73-9 63.5	285.5	4-05-00-01662 1.2611[100]	EtOH 3; eth 3; bz 3; HOAc 3 1.6092[100]
7812	Naphthalene, 1,8-dichloro- 1,8-Dichloronaphthalene	$C_{10}H_6Cl_2$ 197.06	2050-74-0 89	sub	4-05-00-01662 1.2924[100]	EtOH 3; peth 3 1.6236[100]
7813	Naphthalene, 2,6-dichloro- 2,6-Dichloronaphthalene	$C_{10}H_6Cl_2$ 197.06	2065-70-5 140.5	285	4-05-00-01662	EtOH 2; eth 3; bz 3; chl 3
7814	Naphthalene, 1,2-dihydro-	$C_{10}H_{10}$ 130.19	447-53-0 -8	206.5	4-05-00-01543 0.9974[20]	1.5814[20]
7815	Naphthalene, 1,4-dihydro- Δ 2-Dialin	$C_{10}H_{10}$ 130.19	612-17-9 25	211.5	4-05-00-01544 0.9928[33]	1.5577[20]
7816	Naphthalene, 1,2-dihydro-3-methyl-	$C_{11}H_{12}$ 144.22	2717-44-4	105[13]; 64[0.6]	4-05-00-01552 0.9819[20]	H_2O 1; eth 3; bz 3 1.5751[20]
7817	Naphthalene, 1,2-dihydro-4-methyl-	$C_{11}H_{12}$ 144.22	4373-13-1	112[18]; 84[5]	4-05-00-01551 0.9895[20]	H_2O 1; eth 3; bz 3 1.5758[20]
7818	Naphthalene, 1,2-dimethyl- 1,2-Dimethylnaphthalene	$C_{12}H_{12}$ 156.23	573-98-8 1.6	266.5	4-05-00-01708 1.0179[20]	H_2O 1; eth 3; bz 3 1.6166[20]
7819	Naphthalene, 1,3-dimethyl- 1,3-Dimethylnaphthalene	$C_{12}H_{12}$ 156.23	575-41-7 -6	263	4-05-00-01708 1.0144[20]	H_2O 1; eth 3; bz 3 1.6140[20]
7820	Naphthalene, 1,4-dimethyl- 1,4-Dimethylnaphthalene	$C_{12}H_{12}$ 156.23	571-58-4 7.6	268	4-05-00-01709 1.0166[20]	H_2O 1; EtOH 4; eth 5; ace 5 1.6127[20]
7821	Naphthalene, 1,5-dimethyl- 1,5-Dimethylnaphthalene	$C_{12}H_{12}$ 156.23	571-61-9 82	265	4-05-00-01710	bz 4; eth 4
7822	Naphthalene, 1,6-dimethyl- 1,6-Dimethylnaphthalene	$C_{12}H_{12}$ 156.23	575-43-9 -16.9	264	4-05-00-01711 1.0021[20]	H_2O 1; eth 3; bz 3 1.6166[20]
7823	Naphthalene, 1,7-dimethyl- 1,7-Dimethylnaphthalene	$C_{12}H_{12}$ 156.23	575-37-1 -13.9	263	4-05-00-01711 1.0115[20]	H_2O 1; eth 3; bz 3 1.6083[20]
7824	Naphthalene, 1,8-dimethyl- 1,8-Dimethylnaphthalene	$C_{12}H_{12}$ 156.23	569-41-5 65	270	4-05-00-01712 1.003[20]	H_2O 1; eth 3; bz 3
7825	Naphthalene, 2,3-dimethyl- Guajen	$C_{12}H_{12}$ 156.23	581-40-8 105	268	4-05-00-01713 1.003[20]	bz 4; eth 4 1.5060[20]
7826	Naphthalene, 2,6-dimethyl- 2,6-Dimethylnaphthalene	$C_{12}H_{12}$ 156.23	581-42-0 112	262	4-05-00-01714 1.003[20]	
7827	Naphthalene, 2,7-dimethyl- 2,7-Dimethylnaphthalene	$C_{12}H_{12}$ 156.23	582-16-1 97	265	4-05-00-01715 1.003[20]	
7828	Naphthalene, 1-(1,1-dimethylethyl)-	$C_{14}H_{16}$ 184.28	17085-91-5 -11.3	279	4-05-00-01739 0.9629[20]	1.5726[20]
7829	Naphthalene, 2-(1,1-dimethylethyl)-	$C_{14}H_{16}$ 184.28	2876-35-9 -4	276; 125[4]	4-05-00-01740 0.9674[20]	ace 4; bz 4; EtOH 4 1.5685[20]

No.	Name Synonym	Mol. Form. Mol. Wt.	CAS RN mp/°C	Merck No. bp/°C	Beil. Ref. den/g cm^{-3}	Solubility n_D
7830	Naphthalene, 1,6-dimethyl-4-(1-methylethyl)- Cadalene	$C_{15}H_{18}$ 198.31	483-78-3	1607 294; 149[10]	4-05-00-01758 0.9667[25]	oils 4 1.5785[25]
7831	Naphthalene, 1,3-dinitro- 1,3-Dinitronaphthalene	$C_{10}H_8N_2O_4$ 218.17	606-37-1 148	sub	4-05-00-01680	H_2O 1; EtOH 3; ace 3
7832	Naphthalene, 1,5-dinitro- 1,5-Dinitronaphthalene	$C_{10}H_8N_2O_4$ 218.17	605-71-0 219	sub	4-05-00-01680 1.5860[20]	H_2O 1; EtOH 2; eth 4; ace 2
7833	Naphthalene, 1,8-dinitro- 1,8-Dinitronaphthalene	$C_{10}H_8N_2O_4$ 218.17	602-38-0 173	445 dec	4-05-00-01681	H_2O 1; EtOH 2; ace 3; bz 2
7834	1,3-Naphthalenediol Naphthoresorcinol	$C_{10}H_8O_2$ 160.17	132-86-5 123.5	6316	4-06-00-06543	H_2O 3; EtOH 3; eth 3; ace 2
7835	1,5-Naphthalenediol	$C_{10}H_8O_2$ 160.17	83-56-7 262 dec	sub	4-06-00-06554	H_2O 2; EtOH 2; eth 4; ace 4
7836	1,6-Naphthalenediol	$C_{10}H_8O_2$ 160.17	575-44-0 138	sub	4-06-00-06557	H_2O 2; EtOH 2; eth 3; ace 3
7837	1,7-Naphthalenediol	$C_{10}H_8O_2$ 160.17	575-38-2 180.5	sub	4-06-00-06559	H_2O 2; EtOH 4; eth 4; bz 3
7838	2,6-Naphthalenediol	$C_{10}H_8O_2$ 160.17	581-43-1 220	sub	4-06-00-06566	H_2O 2; EtOH 3; eth 3; ace 3
7839	2,7-Naphthalenediol	$C_{10}H_8O_2$ 160.17	582-17-2 193	sub	4-06-00-06570	H_2O 3; EtOH 3; eth 3; ace 2
7840	1,8-Naphthalenediol, cyclic sulfite 1-Naphthalenesulfonic acid, 8-hydroxy, sulfone	$C_{10}H_6O_3S$ 206.22	21849-97-8 97	>360	2-06-00-00954	bz 4; chl 4
7841	1,4-Naphthalenediol, 2-methyl-, diacetate Menadiol diacetate	$C_{15}H_{14}O_4$ 258.27	573-20-6 113	5710	4-06-00-06581	EtOH 4
7842	1,2-Naphthalenedione 1,2-Naphthoquinone	$C_{10}H_6O_2$ 158.16	524-42-5 146	6314	4-07-00-02417 1.450[25]	H_2O 3; EtOH 3; eth 3; sulf 3
7843	1,4-Naphthalenedione 1,4-Naphthoquinone	$C_{10}H_6O_2$ 158.16	130-15-4 128.5	6315 sub	4-07-00-02422	H_2O 2; EtOH 4; eth 3; bz 3
7844	1,2-Naphthalenedione, 3-bromo- 3-Bromo-1,2-naphthoquinone	$C_{10}H_5BrO_2$ 237.05	7474-83-1 178	sub	4-07-00-02422	bz 4; EtOH 4
7845	1,4-Naphthalenedione, 2-bromo-3-hydroxy-	$C_{10}H_5BrO_3$ 253.05	1203-39-0 202	sub	4-08-00-02364	ace 4
7846	1,4-Naphthalenedione, 2-bromo-3-methyl-	$C_{11}H_7BrO_2$ 251.08	3129-39-3 152.8	sub 100	4-07-00-02431	ace 4; bz 4; eth 4; EtOH 4
7847	1,4-Naphthalenedione, 5-chloro- 5-Chloro-1,4-naphthoquinone	$C_{10}H_5ClO_2$ 192.60	40242-15-7 163	sub	4-07-00-02425	EtOH 4; lig 4
7848	1,4-Naphthalenedione, 2-chloro-3-hydroxy-	$C_{10}H_5ClO_3$ 208.60	1526-73-4 216.5	sub	4-08-00-02363	bz 4; eth 4; EtOH 4
7849	1,2-Naphthalenedione, 3,4-dichloro-	$C_{10}H_4Cl_2O_2$ 227.05	18398-36-2 184	sub	4-07-00-02422	bz 4; chl 4
7850	1,4-Naphthalenedione, 2,3-dichloro- Dichlone	$C_{10}H_4Cl_2O_2$ 227.05	117-80-6 195	3032	4-07-00-02426	H_2O 1; EtOH 2; eth 2; bz 2
7851	1,4-Naphthalenedione, 5,6-dichloro- 1,4-Naphthoquinone, 5,6-dichloro	$C_{10}H_4Cl_2O_2$ 227.05	56961-93-4 181	sub	0-07-00-00730	eth 4
7852	1,4-Naphthalenedione, 2,3-dihydroxy-	$C_{10}H_6O_4$ 190.16	605-37-8 282	sub	4-08-00-02953	H_2O 2; EtOH 2; eth 2; ace 3
7853	1,4-Naphthalenedione, 5,8-dihydroxy-	$C_{10}H_6O_4$ 190.16	475-38-7 232	sub	4-08-00-02946	H_2O 2; EtOH 2; eth 2; HOAc 3
7854	1,4-Naphthalenedione, 5,8-dihydroxy-6-(1-hydroxy-4-methyl-3-pentenyl)-, (S)- Alkannin	$C_{16}H_{18}O_5$ 288.30	23444-65-7 149	243 sub 140	3-08-00-04089	EtOH 4
7855	1,4-Naphthalenedione, 5,8-dihydroxy-6-methoxy-2-methyl-3-(2-oxopropyl)- Javanicin	$C_{15}H_{14}O_6$ 290.27	476-45-9 208 dec	5143	4-08-00-03646	alk 3
7856	1,4-Naphthalenedione, 3,5-dihydroxy-2-methyl-	$C_{11}H_8O_4$ 204.18	478-40-0 181	sub 100	4-08-00-02958	H_2O 3; EtOH 3; eth 3; peth 3
7857	1,4-Naphthalenedione, 2-ethyl-3,5,6,7,8-pentahydroxy- Echinochrome A	$C_{12}H_{10}O_7$ 266.21	517-82-8 220 dec	3470 sub 120	4-08-00-03714	H_2O 2; EtOH 3; eth 4; ace 3
7858	1,4-Naphthalenedione, 6-ethyl-2,3,5,7,8-pentahydroxy-	$C_{12}H_{10}O_7$ 266.21	1471-96-1 220 dec	120[0.0004]	4-08-00-03714	ace 4; eth 4; EtOH 4
7859	1,4-Naphthalenedione, 2-hydroxy- Lawsone	$C_{10}H_6O_3$ 174.16	83-72-7 195-6 dec	5263	4-08-00-02360	EtOH 4; eth 1; bz 1; chl 1
7860	1,4-Naphthalenedione, 5-hydroxy- Juglone	$C_{10}H_6O_3$ 174.16	481-39-0 155	5150 sub	4-08-00-02368	H_2O 1; EtOH 3; eth 3; bz 3
7861	1,4-Naphthalenedione, 2-hydroxy-3-methyl- Phthiocol	$C_{11}H_8O_3$ 188.18	483-55-6 173.5	7352 sub	4-08-00-02375	ace 4; eth 4
7862	1,4-Naphthalenedione, 2-hydroxy-3-(3-methyl-2-butenyl)- Lapachol	$C_{15}H_{14}O_3$ 242.27	84-79-7 139.5	5235	4-08-00-02487	H_2O 1; EtOH 3; eth 3; bz 3
7863	1,4-Naphthalenedione, 5-hydroxy-2-methyl- 1,4-Naphthoquinone, 5-hydroxy-2-methyl	$C_{11}H_8O_3$ 188.18	481-42-5 78.5	7511 sub	4-08-00-02376	ace 4; bz 4; eth 4; EtOH 4
7864	1,4-Naphthalenedione, 2-methyl- Menadione	$C_{11}H_8O_2$ 172.18	58-27-5 107	5714	4-07-00-02430	H_2O 1; EtOH 2; eth 3; bz 3
7865	1,4-Naphthalenedione, 2-methyl-3-(3,7,11,15-tetramethyl-2-hexadecenyl)-, [R-[R*,R*-(E)]]- Vitamin K1	$C_{31}H_{46}O_2$ 450.71	84-80-0 -20	9933 140-5[0.001]	4-07-00-02496 0.964[25]	ace 4; bz 4; eth 4; EtOH 4 1.5250[25]

No.	Name Synonym	Mol. Form. Mol. Wt.	CAS RN mp/°C	Merck No. bp/°C	Beil. Ref. den/g cm^{-3}	Solubility n_D
7866	1,4-Naphthalenedione, 2-(phenylamino)-	C$_{16}$H$_{11}$NO$_2$ 249.27	6628-97-3 193	sub	4-14-00-00399	bz 4; eth 4
7867	1,5-Naphthalenedisulfonic acid Armstrong's acid	C$_{10}$H$_8$O$_6$S$_2$ 288.30	81-04-9 240-5 dec	812	4-11-00-00561 1.493[25]	H$_2$O 4; EtOH 3; eth 1
7868	1,6-Naphthalenedisulfonic acid Eiver-pick acid	C$_{10}$H$_8$O$_6$S$_2$ 288.30	525-37-1 125 dec	6292	4-11-00-00562	H$_2$O 4; EtOH 3; eth 1
7869	2,7-Naphthalenedisulfonic acid Naphthalene-2,7-disulfonic acid	C$_{10}$H$_8$O$_6$S$_2$ 288.30	92-41-1 199	6294	4-11-00-00563	H$_2$O 3; con HCl 2
7870	1,3-Naphthalenedisulfonic acid, 7-amino- Amido-G-Acid	C$_{10}$H$_9$NO$_6$S$_2$ 303.32	86-65-7 274	412	4-14-00-02811	H$_2$O 4; EtOH 4
7871	1,3-Naphthalenedisulfonic acid, 7-hydroxy- 2-Naphthol-6,8-disulfonic acid	C$_{10}$H$_8$O$_7$S$_2$ 304.30	118-32-1	6308	4-11-00-00625	H$_2$O 3
7872	1,6-Naphthalenedisulfonic acid, 4-amino- 1-Naphthylamine-4,7-disulfonic acid	C$_{10}$H$_9$NO$_6$S$_2$ 303.32	85-75-6	6322	2-14-00-00476	H$_2$O 4
7873	1,7-Naphthalenedisulfonic acid, 4-amino- 1-Naphthylamine-4,6-disulfonic acid	C$_{10}$H$_9$NO$_6$S$_2$ 303.32	85-74-5	6321	2-14-00-00477	H$_2$O 4; EtOH 4
7874	2,7-Naphthalenedisulfonic acid, 4-amino-5-hydroxy- 1-Amino-8-naphthol-3,6-disulfonic acid	C$_{10}$H$_9$NO$_7$S$_2$ 319.32	90-20-0	6305	4-14-00-02824	H$_2$O 2; EtOH 2; eth 2
7875	2,7-Naphthalenedisulfonic acid, 4,5-dihydroxy- Chromotropic acid	C$_{10}$H$_8$O$_8$S$_2$ 320.30	148-25-4	2243	4-11-00-00636	H$_2$O 3; EtOH 1; eth 1; alk 3
7876	2,7-Naphthalenedisulfonic acid, 3-hydroxy- 2-Naphthol-3,6-disulfonic acid	C$_{10}$H$_8$O$_7$S$_2$ 304.30	148-75-4 dec	6307	4-11-00-00624	H$_2$O 4; EtOH 4
7877	1,3-Naphthalenedisulfonic acid, 6,6'-[(3,3'-dimethyl[1,1'-biphenyl]-4,4'-diyl)bis(azo)]bis[4-amino-5-hydroxy-, C.I. Direct blue 53	C$_{34}$H$_{24}$N$_6$Na$_4$O$_{14}$S$_4$ 960.82	314-13-6	3863		H$_2$O 3; EtOH 3; acid 3
7878	2,7-Naphthalenedisulfonic acid, 4,5-dihydroxy-3-[(4-nitrophenyl)azo]-, disodium salt	C$_{16}$H$_9$N$_3$Na$_2$O$_{10}$S$_2$ 513.37	548-80-1	2242		
7879	Naphthalene, 1-dodecyl- Dodecane, 1-(1-naphthyl)-	C$_{22}$H$_{32}$ 296.50	26438-28-8 27	415	4-05-00-01790 0.9240[20]	1.5364[20]
7880	Naphthalene, 2-dodecyl- Dodecane, 1-(2-naphthyl)-	C$_{22}$H$_{32}$ 296.50	60899-39-0 26	414	4-05-00-01791 0.9177[20]	1.5343[20]
7881	Naphthalene, 1-ethenyl-	C$_{12}$H$_{10}$ 154.21	826-74-4	124-5[15]	4-05-00-01833 1.036[18]	1.644[20]
7882	Naphthalene, 2-ethenyl- 2-Vinylnaphthalene	C$_{12}$H$_{10}$ 154.21	827-54-3 66	135[18]; 95[2]	4-05-00-01833	H$_2$O 1; EtOH 3; ace 3; bz 3
7883	Naphthalene, 1-ethoxy-	C$_{12}$H$_{12}$O 172.23	5328-01-8 5.5	280.5	4-06-00-04212 1.060[20]	H$_2$O 1; EtOH 4; eth 4 1.5953[25]
7884	Naphthalene, 2-ethoxy- 2-Ethoxynaphthalene	C$_{12}$H$_{12}$O 172.23	93-18-5 37.5	282	4-06-00-04257 1.0640[20]	H$_2$O 1; EtOH 3; tol 3 1.5975[36]
7885	Naphthalene, 1-ethyl- 1-Ethylnaphthalene	C$_{12}$H$_{12}$ 156.23	1127-76-0 -13.9	258.6	4-05-00-01705 1.0082[20]	H$_2$O 1; EtOH 5; eth 5 1.6062[20]
7886	Naphthalene, 2-ethyl- 2-Ethylnaphthalene	C$_{12}$H$_{12}$ 156.23	939-27-5 -7.4	258	4-05-00-01707 0.9922[20]	H$_2$O 1; EtOH 5; eth 5; chl 2 1.5999[20]
7887	Naphthalene, 1-ethyl-5-methyl- 1-Methyl-5-ethylnaphthalene	C$_{13}$H$_{14}$ 170.25	17057-92-0 40	133[10]	3-05-00-01659	1.600[30]
7888	Naphthalene, 1-ethyl-1,2,3,4-tetrahydro- 1-Ethyltetralin	C$_{12}$H$_{16}$ 160.26	13556-58-6	241	4-05-00-01429 0.9528[20]	1.5318[20]
7889	Naphthalene, 2-ethyl-1,2,3,4-tetrahydro- 2-Ethyltetralin	C$_{12}$H$_{16}$ 160.26	32367-54-7 -42.8	235	4-05-00-01429 0.9401[15]	1.5250[15]
7890	Naphthalene, 5-ethyl-1,2,3,4-tetrahydro-	C$_{12}$H$_{16}$ 160.26	42775-75-7 -44.5	245	4-05-00-01428 0.9730[20]	1.540[20]
7891	Naphthalene, 6-ethyl-1,2,3,4-tetrahydro- 6-Ethyltetralin	C$_{12}$H$_{16}$ 160.26	22531-20-0 -70	244	4-05-00-01429 0.9632[17]	1.5414[16]
7892	Naphthalene, 1-ethynyl-	C$_{12}$H$_8$ 152.20	15727-65-8 2	143-4[25]	3-05-00-01935 1.0513[20]	1.6360[20]
7893	Naphthalene, 1-fluoro- 1-Fluoronaphthalene	C$_{10}$H$_7$F 146.16	321-38-0 -9	215; 80[11]	4-05-00-01657 1.1322[20]	H$_2$O 1; EtOH 3; eth 3; bz 3 1.5939[20]
7894	Naphthalene, 2-fluoro- 2-Fluoronaphthalene	C$_{10}$H$_7$F 146.16	323-09-1 61	212; 90[16]	4-05-00-01658	H$_2$O 1; EtOH 3; eth 3; bz 3
7895	Naphthalene, 1-heptyl- 1-Heptylnaphthalene	C$_{17}$H$_{22}$ 226.36	2876-52-0 -8	340	4-05-00-01775 0.9491[20]	1.5582[20]
7896	Naphthalene, 2-heptyl- 2-Heptylnaphthalene	C$_{17}$H$_{22}$ 226.36	2876-45-1 1	341	4-05-00-01775 0.9410[20]	1.5556[20]
7897	Naphthalene, 1,2,3,4,4a,8a-hexahydro- Naphthalene, 1,2,3,4,9,10-hexahydro	C$_{10}$H$_{14}$ 134.22	62690-62-4	197	4-05-00-01082 0.934[23]	bz 4; eth 4 1.5260[16]
7898	Naphthalene, 1,2,4a,5,8,8a-hexahydro-4,7-dimethyl-1-(1-methylethyl)-, [1S-(1α,4aβ,8aα)]-	C$_{15}$H$_{24}$ 204.36	523-47-7	274; 136[11]	3-05-00-01086 0.9230[20]	eth 4; lig 4 1.5059[20]
7899	Naphthalene, 1,2,3,5,6,8a-hexahydro-4,7-dimethyl-1-(1-methylethyl)-, (1S-cis)-	C$_{15}$H$_{24}$ 204.36	483-76-1	125-6[12]	2-05-00-00348 0.9160[15]	1.5089[15]
7900	Naphthalene, 1-hexyl- 1-Hexylnaphthalene	C$_{16}$H$_{20}$ 212.33	2876-53-1 -18	322	4-05-00-01767 0.9566[20]	1.5647[20]
7901	Naphthalene, 2-hexyl-	C$_{16}$H$_{20}$ 212.33	2876-46-2 -3	324	4-05-00-01767 0.9479[20]	1.5620[20]
7902	Naphthalene, 1-iodo-	C$_{10}$H$_7$I 254.07	90-14-2 4.2	302	4-05-00-01670 1.7399[20]	H$_2$O 1; EtOH 5; eth 5; bz 5 1.7026[20]

No.	Name / Synonym	Mol. Form. / Mol. Wt.	CAS RN / mp/°C	Merck No. / bp/°C	Beil. Ref. / den/g cm^{-3}	Solubility / n_D
7903	Naphthalene, 2-iodo-	$C_{10}H_7I$ 254.07	612-55-5 54.5	308	4-05-00-01671 1.6319[99]	H$_2$O 1; EtOH 4; eth 4; HOAc 4 1.6662[99]
7904	Naphthalene, 1-isocyanato- 1-Naphthyl isocyanate	$C_{11}H_7NO$ 169.18	86-84-0	6330 269	4-12-00-03094 1.1774[20]	eth 3; bz 3
7905	Naphthalene, 1-isopropyl-2,4,7-trimethyl- Cadalene, 3-methyl	$C_{16}H_{20}$ 212.33	6995-30-8	129-31[1.5]	3-05-00-01689 0.9721[30]	1.5840[30]
7906	Naphthalene, 1-isothiocyanato- 1-Naphthyl isothiocyanate	$C_{11}H_7NS$ 185.25	551-06-4 58	6331	4-12-00-03094	bz 4; eth 4; EtOH 4; chl 4
7907	2-Naphthalenemethanamine	$C_{11}H_{11}N$ 157.22	2018-90-8 59.5	180[24]; 148[12]	4-12-00-03207	eth 4; EtOH 4
7908	1-Naphthalenemethanol	$C_{11}H_{10}O$ 158.20	4780-79-4 64	304; 163[12]	4-06-00-04332 1.1039[80]	H$_2$O 2; EtOH 4; eth 4
7909	2-Naphthalenemethanol	$C_{11}H_{10}O$ 158.20	1592-38-7 81.3	178[12]	4-06-00-04340	H$_2$O 2; EtOH 3; eth 3
7910	1-Naphthalenemethanol, α,α-diphenyl- Benzhydrol, α-(1-naphthyl)	$C_{23}H_{18}O$ 310.40	630-95-5 136.5	dec	4-06-00-05134	bz 4; eth 4; EtOH 4
7911	1-Naphthalenemethanol, α-methyl-, (±)- (±)-1-(α-Naphthyl)ethanol	$C_{12}H_{12}O$ 172.23	57605-95-5 66	178[15]	4-06-00-04346 1.1190[14]	ace 4; bz 4; EtOH 4; chl 4 1.6188[25]
7912	1-Naphthalenemethanol, α-phenyl-	$C_{17}H_{14}O$ 234.30	642-28-4 87.8	360	4-06-00-04989	H$_2$O 1; EtOH 3; eth 3; bz 3
7913	Naphthalene, 1-methoxy-	$C_{11}H_{10}O$ 158.20	2216-69-5 <-10	269	4-06-00-04211 1.0963[14]	H$_2$O 1; EtOH 3; eth 3; bz 3 1.6940[25]
7914	Naphthalene, 2-methoxy- 2-Methoxynaphthalene	$C_{11}H_{10}O$ 158.20	93-04-9 73.5	5918 274	4-06-00-04257	bz 4; eth 4; chl 4
7915	Naphthalene, 1-methyl- 1-Methylnaphthalene	$C_{11}H_{10}$ 142.20	90-12-0 -30.4	244.7	4-05-00-01687 1.0202[20]	H$_2$O 1; EtOH 4; eth 4; bz 3 1.6170[20]
7916	Naphthalene, 2-methyl- 2-Methylnaphthalene	$C_{11}H_{10}$ 142.20	91-57-6 34.4	241.1	4-05-00-01693 1.0058[20]	H$_2$O 1; EtOH 4; eth 4; bz 3 1.6015[40]
7917	Naphthalene, 1-(3-methylbutoxy)-	$C_{15}H_{18}O$ 214.31	20213-30-3	319; 150[3]	3-06-00-02925 1.0069[14]	1.5705[16]
7918	Naphthalene, 2-(3-methylbutoxy)- Isopentyl β-Naphthyl ether	$C_{15}H_{18}O$ 214.31	635-88-1 26.5	325 dec	4-06-00-04259 1.0155[12]	eth 4; EtOH 4 1.5768[12]
7919	Naphthalene, 1,1'-methylenebis-	$C_{21}H_{16}$ 268.36	607-50-1 109	>360; 270[14]	4-05-00-02646	H$_2$O 2; EtOH 2; eth 3; bz 3
7920	Naphthalene, 1-(1-methylethyl)-	$C_{13}H_{14}$ 170.25	6158-45-8 -16	268	4-05-00-01722 0.9956[20]	1.5952[20]
7921	Naphthalene, 2-(1-methylethyl)-	$C_{13}H_{14}$ 170.25	2027-17-0 14.5	268.2	4-05-00-01723 0.9753[20]	H$_2$O 1; EtOH 4; eth 4; bz 3 1.5848[20]
7922	Naphthalene, 1-methyl-4-(1-methylethyl)-	$C_{14}H_{16}$ 184.28	1680-53-1	145-8[12]	3-05-00-01671 0.9934[14]	1.5907[14]
7923	Naphthalene, 1-methyl-7-(1-methylethyl)-	$C_{14}H_{16}$ 184.28	490-65-3	140[11]	4-05-00-01741 0.9740[20]	H$_2$O 1; eth 3; ace 3; bz 3 1.5833[20]
7924	Naphthalene, 7-methyl-1-(1-methylethyl)- Naphthalene, 1-isopropyl-7-methyl	$C_{14}H_{16}$ 184.28	66577-17-1	282	3-05-00-01672 0.9833[20]	1.5884[20]
7925	Naphthalene, 1-methyl-4-nitro- 4-Methyl-1-nitronaphthalene	$C_{11}H_9NO_2$ 187.20	880-93-3 71.5	182[18]	3-05-00-01624	ace 4; eth 4; EtOH 4
7926	Naphthalene, 2-methyl-1-nitro- 2-Methyl-1-nitronaphthalene	$C_{11}H_9NO_2$ 187.20	881-03-8 81.5	188[20]	4-05-00-01698	H$_2$O 1; EtOH 3; ace 4; os 4
7927	Naphthalene, 1-(1-methylpropyl)-	$C_{14}H_{16}$ 184.28	1680-58-6	273	4-05-00-01738 0.9742[20]	1.5701[20]
7928	Naphthalene, 1-(2-naphthalenyloxy)- 1,2'-Bis-naphthyl ether	$C_{20}H_{14}O$ 270.33	611-49-4 81	264[15]	2-06-00-00600	bz 4; eth 4
7929	Naphthalene, 1-nitro- 1-Nitronaphthalene	$C_{10}H_7NO_2$ 173.17	86-57-7 61	6535 180[14]	4-05-00-01673 1.332[20]	H$_2$O 1; EtOH 4; eth 4; bz 4
7930	Naphthalene, 2-nitro- 2-Nitronaphthalene	$C_{10}H_7NO_2$ 173.17	581-89-5 79	314; 165[15]	4-05-00-01675	H$_2$O 1; EtOH 4; eth 4
7931	Naphthalene, 1-nonyl-	$C_{19}H_{26}$ 254.42	26438-26-8	366	4-05-00-01784 0.9371[20]	1.5477[20]
7932	Naphthalene, 2-nonyl-	$C_{19}H_{26}$ 254.42	61886-67-7 12	372	4-05-00-01784 0.9298[20]	1.5454[20]
7933	Naphthalene, octachloro- Perchloronaphthalene	$C_{10}Cl_8$ 403.73	2234-13-1 197.5	441[7]; 248[0.5]	4-05-00-01665	EtOH 2; bz 4; chl 4; lig 4
7934	Naphthalene, octadecafluorodecahydro- Perfluorodecalin	$C_{10}F_{18}$ 462.08	306-94-5 -10	142	3-05-00-00249	
7935	Naphthalene, octafluoro- Perfluoronaphthalene	$C_{10}F_8$ 272.10	313-72-4 87.5	209	4-05-00-01658	
7936	Naphthalene, 1,2,3,4,4a,5,6,8a-octahydro-7-methyl-4-methylene-1-(1-methylethyl)-, (1α,4aβ,8aα)- γ-Cadinene	$C_{15}H_{24}$ 204.36	39029-41-9	124-8[12]	3-05-00-01088 0.9182[15]	1.3166[20]
7937	Naphthalene, 1-octyl- 1-Octylnaphthalene	$C_{18}H_{24}$ 240.39	2876-51-9 -2	356	4-05-00-01779 0.9427[20]	1.5526[20]
7938	Naphthalene, 2-octyl-	$C_{18}H_{24}$ 240.39	2876-44-0 12	357	4-05-00-01779 0.9350[20]	1.5501[20]
7939	Naphthalene, 1,1'-oxybis- 1-Naphthyl ether	$C_{20}H_{14}O$ 270.33	607-52-3 110	280-5[22]	4-06-00-04215	bz 4; eth 4
7940	Naphthalene, 2,2'-oxybis-	$C_{20}H_{14}O$ 270.33	613-80-9 105	250[19] dec	4-06-00-04260	H$_2$O 1; EtOH 2; eth 4; bz 4

No.	Name Synonym	Mol. Form. Mol. Wt.	CAS RN mp/°C	Merck No. bp/°C	Beil. Ref. den/g cm^{-3}	Solubility n_D
7941	Naphthalene, 1-pentyl-	$C_{15}H_{18}$ 198.31	86-89-5 -22	307	4-05-00-01755 0.9656[20]	1.5725[20]
7942	Naphthalene, 2-pentyl- 2-Pentylnaphthalene	$C_{15}H_{18}$ 198.31	93-22-1 -4	310	4-05-00-01755 0.9561[20]	1.5694[20]
7943	Naphthalene, 1-(pentyloxy)- 1-Naphthyl pentyl ether	$C_{15}H_{18}O$ 214.31	108438-40-0 30	322	4-06-00-04213	bz 4; eth 4; EtOH 4
7944	Naphthalene, 2-(pentyloxy)- 2-Naphthyl pentyl ether	$C_{15}H_{18}O$ 214.31	31059-19-5 24.5	335	3-06-00-02973	bz 4; eth 4; EtOH 4; chl 4 1.5587[30]
7945	Naphthalene, 1-phenoxy-	$C_{16}H_{12}O$ 220.27	3402-76-4 55.5	349.5	4-06-00-04214	H_2O 1; EtOH 2; eth 3; ace 3
7946	Naphthalene, 2-phenoxy-	$C_{16}H_{12}O$ 220.27	19420-29-2 46	335	3-06-00-02975	H_2O 1; eth 3; chl 3; HOAc 3
7947	Naphthalene, 1-phenyl- 1-Phenylnaphthalene	$C_{16}H_{12}$ 204.27	605-02-7 45	334	4-05-00-02411 1.096[20]	H_2O 1; EtOH 4; eth 4; bz 4 1.6664[20]
7948	Naphthalene, 2-phenyl-	$C_{16}H_{12}$ 204.27	612-94-2 103.5	345.5	4-05-00-02412 1.2180[20]	EtOH 3; eth 4; bz 3; chl 3
7949	Naphthalene, 1-(phenylmethyl)-	$C_{17}H_{14}$ 218.30	611-45-0 59.5	350	4-05-00-02420 1.166[17]	H_2O 1; EtOH 2; eth 4; bz 3
7950	Naphthalene, 2-(phenylmethyl)- 2-Benzylnaphthalene	$C_{17}H_{14}$ 218.30	613-59-2 58	350	4-05-00-02420 1.176[12]	bz 4; eth 4; chl 4
7951	Naphthalene, 1-(phenylthio)-	$C_{16}H_{12}S$ 236.34	7570-98-1 41.8	255[43.] 222[11.]	4-06-00-04242 1.167[15]	eth 4; EtOH 4
7952	1-Naphthalenepropanoic acid	$C_{13}H_{12}O_2$ 200.24	3243-42-3 156.5	179[11]	4-09-00-02435	H_2O 3; EtOH 3; bz 3
7953	2-Naphthalenepropanoic acid, 6-hydroxy- Allenolic acid	$C_{13}H_{12}O_3$ 216.24	553-39-9 180.5	247	3-10-00-01113	py 4; EtOH 4; MeOH 4
7954	Naphthalene, 1-(2-propenyl)-	$C_{13}H_{12}$ 168.24	2489-86-3	266	3-05-00-01805 1.0228[20]	EtOH 3; bz 4; chl 3 1.6140[20]
7955	Naphthalene, 1-propoxy-	$C_{13}H_{14}O$ 186.25	20009-26-1	293.5	4-06-00-04213 1.0447[18]	1.5928[18]
7956	Naphthalene, 2-propoxy-	$C_{13}H_{14}O$ 186.25	19718-45-7 41	305	3-06-00-02973	EtOH 4
7957	Naphthalene, 1-propyl-	$C_{13}H_{14}$ 170.25	2765-18-6 -8.6	274.5	4-05-00-01721 0.9897[20]	1.5923[20]
7958	Naphthalene, 2-propyl-	$C_{13}H_{14}$ 170.25	2027-19-2 -3	273.5	4-05-00-01722 0.9770[20]	1.5872[20]
7959	1-Naphthalenesulfonic acid	$C_{10}H_8O_3S$ 208.24	85-47-2 140	6295	4-11-00-00521	H_2O 3; EtOH 3; eth 2
7960	2-Naphthalenesulfonic acid β-Naphthylsulfonic acid	$C_{10}H_8O_3S$ 208.24	120-18-3 91 dec	6296	4-11-00-00527 1.441[25]	H_2O 4; EtOH 4; eth 3; bz 2
7961	1-Naphthalenesulfonic acid, 2-amino- 2-Naphthylamine-1-sulfonic acid	$C_{10}H_9NO_3S$ 223.25	81-16-3	6326	4-14-00-02792	DMSO 3
7962	1-Naphthalenesulfonic acid, 4-amino- 1-Naphthylamine-4-sulfonic acid	$C_{10}H_9NO_3S$ 223.25	84-86-6 dec	6323	4-14-00-02793 1.6703[25]	H_2O 1; EtOH 2; MeOH 3; py 3
7963	1-Naphthalenesulfonic acid, 5-amino- 1-Naphthylamine-5-sulfonic acid	$C_{10}H_9NO_3S$ 223.25	84-89-9	6324	4-14-00-02800	H_2O 3; eth 1
7964	1-Naphthalenesulfonic acid, 7-amino- Badische acid	$C_{10}H_9NO_3S$ 223.25	86-60-2	952	4-14-00-02802	HOAc 4
7965	1-Naphthalenesulfonic acid, 8-amino- 1-Naphthylamine-8-sulfonic acid	$C_{10}H_9NO_3S$ 223.25	82-75-7	6325	4-14-00-02802	gl HOAc 4
7966	1-Naphthalenesulfonic acid, 4-amino-3-hydroxy- 1-Amino-2-naphthol-4-sulfonic acid	$C_{10}H_9NO_4S$ 239.25	116-63-2	466	4-14-00-02825	
7967	2-Naphthalenesulfonic acid, 8-amino- 1,7-Cleve's acid	$C_{10}H_9NO_3S$ 223.25	119-28-8	2349	4-14-00-02805	
7968	1-Naphthalenesulfonic acid, 3,6-bis(1,1-dimethylethyl)-, ethyl ester Ethyl dibunate	$C_{20}H_{28}O_3S$ 348.51	5560-69-0	3745		chl 3
7969	1-Naphthalenesulfonic acid, 3,3'-[[1,1'-biphenyl]-4,4'-diylbis(azo)]bis[4-amino-, disodium salt Congo red	$C_{32}H_{22}N_6Na_2O_6S_2$ 696.68	573-58-0 >360	2493	1-16-00-00342	H_2O 3; EtOH 2; eth 1
7970	1-Naphthalenesulfonic acid, 3,4-dihydro-3,4-dioxo-, sodium salt 1,2-Naphthoquinone-4-sulfonic acid, sodium salt	$C_{10}H_5NaO_5S$ 260.20	521-24-4 287 dec	8597		
7971	1-Naphthalenesulfonic acid, 4-hydroxy- 1-Naphthol-4-sulfonic acid	$C_{10}H_8O_4S$ 224.24	84-87-7 170 dec	6311	4-11-00-00614	H_2O 4; eth 1
7972	1-Naphthalenesulfonic acid, 7-hydroxy- Croceic acid	$C_{10}H_8O_4S$ 224.24	132-57-0	2591	4-11-00-00623	H_2O 3
7973	2-Naphthalenesulfonic acid, 1-hydroxy- 1-Naphthol-2-sulfonic acid	$C_{10}H_8O_4S$ 224.24	567-18-0 >250	6310	4-11-00-00613	H_2O 2; EtOH 3; eth 1; dil HCl 2
7974	2-Naphthalenesulfonic acid, 6-hydroxy- 2-Naphthol-6-sulfonic acid	$C_{10}H_8O_4S$ 224.24	93-01-6 125	6312	4-11-00-00622	H_2O 4; EtOH 4; eth 1; HOAc 3
7975	1-Naphthalenesulfonyl chloride	$C_{10}H_7ClO_2S$ 226.68	85-46-1 68	209[20.] 147[0.9]	3-11-00-00383	bz 4; eth 4; EtOH 4

No.	Name Synonym	Mol. Form. Mol. Wt.	CAS RN mp/°C	Merck No. bp/°C	Beil. Ref. den/g cm^{-3}	Solubility n_D
7976	2-Naphthalenesulfonyl chloride	$C_{10}H_7ClO_2S$ 226.68	93-11-8 81	201[13,] 148[0.5]	4-11-00-00529	H_2O 1; EtOH 3; eth 4; bz 3
7977	1-Naphthalenesulfonyl chloride, 5-(dimethylamino)- Dansyl chloride	$C_{12}H_{12}ClNO_2S$ 269.75	605-65-2 70	2812	4-14-00-02801	
7978	2-Naphthalenesulfonyl chloride, 5,6,7,8-tetrahydro- 6-Tetralinsulfonyl chloride	$C_{10}H_{11}ClO_2S$ 230.71	61551-49-3 58	197[18]	3-11-00-00375	eth 4
7979	Naphthalene, 1,2,3,4-tetrahydro- Tetralin	$C_{10}H_{12}$ 132.21	119-64-2 -35.7	9152 207.6	4-05-00-01388 0.9660[25]	H_2O 1; EtOH 4; eth 4; chl 3 1.5413[20]
7980	Naphthalene, 1,2,3,4-tetrahydro-1,1-dimethyl- 1,1-Dimethyltetralin	$C_{12}H_{16}$ 160.26	1985-59-7	221	4-05-00-01430 0.950[20]	1.5292[20]
7981	Naphthalene, 1,2,3,4-tetrahydro-1,2-dimethyl- 1,2,3,4-Tetrahydro-1,2-dimethylnaphthalene	$C_{12}H_{16}$ 160.26	5195-40-4	235	3-05-00-01262 0.9470[20]	1.5286[20]
7982	Naphthalene, 1,2,3,4-tetrahydro-1,3-dimethyl- Tetralin, 1,3-dimethyl	$C_{12}H_{16}$ 160.26	5195-37-9	234; 78[1]	3-05-00-01263 0.940[20]	1.525[20]
7983	Naphthalene, 1,2,3,4-tetrahydro-1,4-dimethyl- Tetralin, 1,4-dimethyl	$C_{12}H_{16}$ 160.26	4175-54-6	226	4-05-00-01430 0.940[20]	1.525[20]
7984	Naphthalene, 1,2,3,4-tetrahydro-1,5-dimethyl-	$C_{12}H_{16}$ 160.26	21564-91-0	239	3-05-00-01263 0.941[20]	1.526[20]
7985	Naphthalene, 1,2,3,4-tetrahydro-2,2-dimethyl- Tetralin, 2,2-dimethyl	$C_{12}H_{16}$ 160.26	13556-55-3	230	4-05-00-01431 0.935[20]	1.5200[20]
7986	Naphthalene, 1,2,3,4-tetrahydro-2,3-dimethyl- Tetralin, 2,3-dimethyl	$C_{12}H_{16}$ 160.26	21564-92-1 -5		3-05-00-01265 0.940[20]	1.523[20]
7987	Naphthalene, 1,2,3,4-tetrahydro-2,5-dimethyl- Tetralin, 2,5-dimethyl	$C_{12}H_{16}$ 160.26	25419-37-8	236	3-05-00-01263 0.946[20]	1.526[20]
7988	Naphthalene, 1,2,3,4-tetrahydro-2,6-dimethyl-	$C_{12}H_{16}$ 160.26	7524-63-2 20	240; 117[15]	4-05-00-01432 0.941[20]	1.526[20]
7989	Naphthalene, 1,2,3,4-tetrahydro-2,7-dimethyl- Tetralin, 2,7-dimethyl	$C_{12}H_{16}$ 160.26	13065-07-1	237	3-05-00-01265 0.941[20]	1.526[20]
7990	Naphthalene, 1,2,3,4-tetrahydro-2,8-dimethyl- Tetralin, 2,8-dimethyl	$C_{12}H_{16}$ 160.26	25419-36-7	236	3-05-00-01264 0.941[20]	1.526[20]
7991	Naphthalene, 1,2,3,4-tetrahydro-5,6-dimethyl- Tetralin, 5,6-dimethyl	$C_{12}H_{16}$ 160.26	20027-77-4	252	4-05-00-01430 0.975[20]	1.552[20]
7992	Naphthalene, 1,2,3,4-tetrahydro-5,7-dimethyl- Tetralin, 5,7-dimethyl	$C_{12}H_{16}$ 160.26	21693-54-9 -6	253	4-05-00-01430 0.9583[20]	1.5405[20]
7993	Naphthalene, 1,2,3,4-tetrahydro-5,8-dimethyl-	$C_{12}H_{16}$ 160.26	14108-88-4 -4.1	254	4-05-00-01430 0.967[20]	1.547[20]
7994	Naphthalene, 1,2,3,4-tetrahydro-6,7-dimethyl-	$C_{12}H_{16}$ 160.26	1076-61-5 10	252	4-05-00-01431 0.954[20]	1.538[20]
7995	Naphthalene, 1,2,3,4-tetrahydro-1-methyl-	$C_{11}H_{14}$ 146.23	1559-81-5	220.6	4-05-00-01412 0.9583[20]	1.5353[20]
7996	Naphthalene, 1,2,3,4-tetrahydro-2-methyl-	$C_{11}H_{14}$ 146.23	3877-19-8 -43.1	221	4-05-00-01414 0.952[20]	1.531[20]
7997	Naphthalene, 1,2,3,4-tetrahydro-5-methyl-	$C_{11}H_{14}$ 146.23	2809-64-5 -23	234	4-05-00-01412 0.9720[20]	1.5439[20]
7998	Naphthalene, 1,2,3,4-tetrahydro-6-methyl-	$C_{11}H_{14}$ 146.23	1680-51-9 -40	229	4-05-00-01413 0.9537[20]	1.5357[20]
7999	Naphthalene, 1,2,3,4-tetrahydro-1,1,2,6-tetramethyl- Irene	$C_{14}H_{20}$ 188.31	1681-22-7 120-5[10]		4-05-00-01460 0.9332[20]	H_2O 1; ace 3; bz 3 1.5217[20]
8000	Naphthalene, 1,2,3,4-tetrahydro-1,1,6-trimethyl-	$C_{13}H_{18}$ 174.29	475-03-6 240; 90[4]		4-05-00-01445 0.9356[20]	EtOH 3; eth 3; bz 3; chl 3 1.5257[20]
8001	Naphthalene, 1,2,4,7-tetramethyl-	$C_{14}H_{16}$ 184.28	16020-17-0 -3	174-5[12]	4-05-00-01746 1.011[20]	1.6032[20]
8002	Naphthalene, 1,2,5,6-tetramethyl-	$C_{14}H_{16}$ 184.28	2131-43-3 118	152[12]	4-05-00-01746	H_2O 1; EtOH 3; bz 3; peth 3
8003	Naphthalene, 1,3,6,8-tetramethyl- 1,3,6,8-Tetramethylnaphthalene	$C_{14}H_{16}$ 184.28	14558-14-6 84.5	115-6[2]	4-05-00-01746	H_2O 1; EtOH 3
8004	Naphthalene, 1,3,6,8-tetranitro-	$C_{10}H_4N_4O_8$ 308.16	28995-89-3 207	exp	4-05-00-01683 1.6400[0]	bz 4; HOAc 4
8005	Naphthalene, 1,1'-thiobis-	$C_{20}H_{14}S$ 286.40	607-53-4 110	289[15]	4-06-00-04243	bz 4; HOAc 4
8006	Naphthalene, 2,2'-thiobis-	$C_{20}H_{14}S$ 286.40	613-81-0 151	296[15]	4-06-00-04314	EtOH 1; bz 4; CS_2 4; HOAc 4
8007	1-Naphthalenethiol 1-Naphthyl mercaptan	$C_{10}H_8S$ 160.24	529-36-2 285 dec; 161[20]	6297	4-06-00-04241 1.1607[20]	H_2O 2; EtOH 4; eth 4; dil alk 2 1.6802[20]
8008	2-Naphthalenethiol 2-Naphthyl mercaptan	$C_{10}H_8S$ 160.24	91-60-1 81	6298 288	4-06-00-04312 1.550[25]	H_2O 2; EtOH 4; eth 4; lig 4
8009	Naphthalene, 1,2,4-trimethyl- 1,2,4-Trimethylnaphthalene	$C_{13}H_{14}$ 170.25	2717-42-2 55.5	146[12]	4-05-00-01725	EtOH 3; eth 3; bz 3; chl 3
8010	Naphthalene, 1,2,5-trimethyl- 1,2,5-Trimethylnaphthalene	$C_{13}H_{14}$ 170.25	641-91-8 33.5	140[12]	4-05-00-01726 1.010[22]	H_2O 1; EtOH 2; eth 4; bz 4 1.6093[22]
8011	Naphthalene, 1,2,6-trimethyl- 1,2,6-Trimethylnaphthalene	$C_{13}H_{14}$ 170.25	3031-05-8 14	146[10]	4-05-00-01726	H_2O 1; eth 3; bz 3 1.6010[20]
8012	Naphthalene, 1,2,7-trimethyl-	$C_{13}H_{14}$ 170.25	486-34-0	147-8[16]	4-05-00-01727 1.0087[20]	H_2O 1; eth 3; bz 3 1.6097[20]

No.	Name Synonym	Mol. Form. Mol. Wt.	CAS RN mp/°C	Merck No. bp/°C	Beil. Ref. den/g cm^{-3}	Solubility n_D
8013	Naphthalene, 1,3,7-trimethyl- 2,4,6-Trimethylnaphthalene	$C_{13}H_{14}$ 170.25	2131-38-6 12	131-3[9]	4-05-00-01728 0.9801[21]	1.5972[15]
8014	Naphthalene, 1,2,5-trimethyl-8-(1-methylethyl)- Cadalene, 5-methyl	$C_{16}H_{20}$ 212.33	6897-76-3	144-5[1.5]	3-05-00-01689 0.9847[30]	1.5926[30]
8015	Naphthalene, 1,3,8-trimethyl-5-(1-methylethyl)- Cadalene, 8-methyl	$C_{16}H_{20}$ 212.33	6897-88-7	135-6[1.5]	4-05-00-01770 0.9800[30]	1.5876[30]
8016	Naphthalene, 1,3,5-trinitro-	$C_{10}H_5N_3O_6$ 263.17	2243-94-9 124	exp 364	4-05-00-01682	ace 4; EtOH 4; chl 4
8017	1-Naphthalenol 1-Naphthol	$C_{10}H_8O$ 144.17	90-15-3 95	6303 288; 184[40]	4-06-00-04208 1.0989[99]	H_2O 1; EtOH 4; eth 4; ace 3 1.6224[99]
8018	2-Naphthalenol 2-Naphthol	$C_{10}H_8O$ 144.17	135-19-3 123	6304 285	4-06-00-04253 1.28[20]	H_2O 1; EtOH 4; eth 4; bz 3
8019	2-Naphthalenol, 6-amino-	$C_{10}H_9NO$ 159.19	4363-04-6 193	212 dec	3-13-00-01902	H_2O 4; EtOH 4
8020	2-Naphthalenol, 8-amino-	$C_{10}H_9NO$ 159.19	118-46-7 206	sub	3-13-00-01907	H_2O 3; EtOH 4; eth 3; bz 2
8021	2-Naphthalenol, benzoate 2-Naphthyl benzoate	$C_{17}H_{12}O_2$ 248.28	93-44-7 107	6328	4-09-00-00339	H_2O 1; EtOH 3; eth 2; HOAc 2
8022	2-Naphthalenol, 1-bromo-	$C_{10}H_7BrO$ 223.07	573-97-7 84	130	4-06-00-04301	H_2O 1; EtOH 3; eth 3; bz 3
8023	2-Naphthalenol, 6-chloro-	$C_{10}H_7ClO$ 178.62	40604-49-7 116.5	sub	4-06-00-04294	bz 4; eth 4; EtOH 4; chl 4
8024	2-Naphthalenol, decahydro-	$C_{10}H_{18}O$ 154.25	825-51-4	109[14]	1-06-00-00044 0.996[25]	1.4992[20]
8025	1-Naphthalenol, 2,3-dichloro- 2,3-Dichloro-1-naphthol	$C_{10}H_6Cl_2O$ 213.06	71284-96-3 101	sub	0-06-00-00612	bz 4; eth 4; EtOH 4; chl 4
8026	1-Naphthalenol, 2,4-dichloro-	$C_{10}H_6Cl_2O$ 213.06	2050-76-2 107.5	180	4-06-00-04233	bz 4; eth 4; EtOH 4
8027	2-Naphthalenol, 1,4-dimethyl- 2-Naphthol, 1,4-dimethyl-	$C_{12}H_{12}O$ 172.23	4705-94-6 135.5	315.5	4-06-00-04351	eth 4; EtOH 4
8028	2-Naphthalenol, 1-[(2,4-dimethylphenyl)azo]- C.I. Solvent Orange 7	$C_{18}H_{16}N_2O$ 276.34	3118-97-6 166		4-16-00-00234	eth 4; EtOH 4
8029	2-Naphthalenol, 1-[(2,5-dimethylphenyl)azo]- 1-(2,5-Xylylazo)-2-naphthol	$C_{18}H_{16}N_2O$ 276.34	85-82-5 153	9997	4-16-00-00234	
8030	1-Naphthalenol, 4-methyl-	$C_{11}H_{10}O$ 158.20	10240-08-1 85.8	166[13]	4-06-00-04330	H_2O 3; EtOH 3; eth 3; ace 3
8031	1-Naphthalenol, 1-methyl-	$C_{11}H_{10}O$ 158.20	1076-26-2 112	160[10]	4-06-00-04329	H_2O 2; EtOH 4; eth 4; ace 4
8032	1-Naphthalenol, methylcarbamate Carbaryl	$C_{12}H_{11}NO_2$ 201.22	63-25-2 145	1789	4-06-00-04219 1.228[25]	ace 4; DMF 4
8033	2-Naphthalenol, 1-[[2-methyl-4-[(2-methylphenyl)azo]phenyl]azo]- Scarlet red	$C_{24}H_{20}N_4O$ 380.45	85-83-6 186 dec	8349 260 dec	4-16-00-00249	chl 4; peth 4
8034	2-Naphthalenol, 1-nitro- 1-Nitro-2-naphthol	$C_{10}H_7NO_3$ 189.17	550-60-7 104	6536 115[0.05]	4-06-00-04307	H_2O 3; EtOH 4; eth 4; chl 2
8035	2-Naphthalenol, 1-nitroso- 1-Nitroso-2-naphthol	$C_{10}H_7NO_2$ 173.17	131-91-9 109.5	6562	4-07-00-02419	bz 4; eth 4
8036	1-Naphthalenol, 2-(phenylazo)-	$C_{16}H_{12}N_2O$ 248.28	3375-23-3 138	sub	4-16-00-00219	EtOH 4; HOAc 4
8037	2-Naphthalenol, 1-(phenylazo)- Azo, benzene 1'-naphthalene, 2'-hydroxy	$C_{16}H_{12}N_2O$ 248.28	842-07-9		4-16-00-00228	
8038	2-Naphthalenol, 1-[[4-(phenylazo)phenyl]azo]- Sudan III	$C_{22}H_{18}N_4O$ 352.40	85-86-9 195	8858	4-16-00-00248	H_2O 1; EtOH 3; eth 3; ace 3
8039	1-Naphthalenol, 2-(phenylmethyl)-	$C_{17}H_{14}O$ 234.30	36441-32-4 73.5	237[12]	4-06-00-04990	H_2O 1; bz 3; lig 2
8040	2-Naphthalenol, 1-(phenylmethyl)- 2-Naphthol, 1-benzyl-	$C_{17}H_{14}O$ 234.30	36441-31-3 114	247[14]	4-06-00-04988	ace 4; bz 4; eth 4; EtOH 4
8041	1-Naphthalenol, 1,2,3,4-tetrahydro-	$C_{10}H_{12}O$ 148.20	529-33-9	102-4[2]	3-06-00-02457 1.09[25]	1.5638[20]
8042	1-Naphthalenol, 5,6,7,8-tetrahydro-	$C_{10}H_{12}O$ 148.20	529-35-1 70	266; 143[11]	4-06-00-03851 1.0556[75]	
8043	2-Naphthalenol, 1,2,3,4-tetrahydro- 2-Tetralol	$C_{10}H_{12}O$ 148.20	530-91-6 15.5	9153 140[12]	0-06-00-00580	
8044	2-Naphthalenol, 5,6,7,8-tetrahydro-	$C_{10}H_{12}O$ 148.20	1125-78-6 57	275.5	4-06-00-03854 1.0552[65]	
8045	1(2H)-Naphthalenone, 2-acetyl-3,4-dihydro-	$C_{12}H_{12}O_2$ 188.23	17216-08-9 56	135-42[1]	4-07-00-02358	chl 3
8046	1(2H)-Naphthalenone, 3,4-dihydro-	$C_{10}H_{10}O$ 146.19	529-34-0 8	115[6]	4-07-00-01015 1.0988[16]	1.5672[20]
8047	2(1H)-Naphthalenone, 3,4-dihydro-	$C_{10}H_{10}O$ 146.19	530-93-8 18	237	4-07-00-01018 1.1055[27]	H_2O 1; eth 3; bz 3 1.5598[20]
8048	1(2H)-Naphthalenone, 3,4-dihydro-6-methoxy- 6-Methoxy-α-tetralone	$C_{11}H_{12}O_2$ 176.22	1078-19-9 78	5924 171[11]	4-08-00-00904	
8049	1(2H)-Naphthalenone, 3,4-dihydro-2-methyl-	$C_{11}H_{12}O$ 160.22	1590-08-5	127-31[12]	4-07-00-01033 1.057[25]	1.5535[20]
8050	1(2H)-Naphthalenone, 3,4-dihydro-4-methyl-	$C_{11}H_{12}O$ 160.22	19832-98-5	133-4[12]	4-07-00-01031 1.0779[19]	1.5620[19]

No.	Name / Synonym	Mol. Form. / Mol. Wt.	CAS RN / mp/°C	Merck No. / bp/°C	Beil. Ref. / den/g cm^{-3}	Solubility / n_D
8051	1(2H)-Naphthalenone, 7-ethyl-3,4-dihydro- / 1-Tetralone, 7-ethyl	$C_{12}H_{14}O$ / 174.24	22531-06-2	/ 152-3[12]	4-07-00-01049 / 1.0556[17]	bz 4; eth 4; EtOH 4 / 1.5599[17]
8052	2(3H)-Naphthalenone, 4,4a,5,6,7,8-hexahydro-4,4a-dimethyl-6-(1-methylethylidene)-, (4R-cis)-	$C_{15}H_{22}O$ / 218.34	15764-04-2 / 51.5	144[2]	1.0035[20]	ace 4 / 1.5370[20]
8053	1(2H)-Naphthalenone, octahydro-, cis- / 1-Decalone (cis)	$C_{10}H_{16}O$ / 152.24	32166-40-8 / 2	126[20]	4-07-00-00201 / 1.008[20]	1.4936[20]
8054	1(2H)-Naphthalenone, octahydro-, trans-	$C_{10}H_{16}O$ / 152.24	21370-71-8 / 33	122[20]	4-07-00-00201 / 0.986[20]	1.4849[21]
8055	2(1H)-Naphthalenone, octahydro-	$C_{10}H_{16}O$ / 152.24	4832-17-1	96[2.5]	4-07-00-00203 / 0.979[25]	1.4900[20]
8056	Naphtho[2,3-b]-1,4-dithiin-2,3-dicarbonitrile, 5,10-dihydro-5,10-dioxo- / Dithianone	$C_{14}H_4N_2O_2S_2$ / 296.33	3347-22-6 / 220	3375	5-19-08-00189	
8057	Naphtho[1,2-b]furan-2,8(3H,4H)-dione, 3a,5,5a,9b-tetrahydro-3,5a,9-trimethyl-, / α-Santonin	$C_{15}H_{18}O_3$ / 246.31	481-06-1 / 175	8325	5-17-11-00314 / 1.590[25]	H_2O 2; EtOH 2; eth 2; bz 3
8058	Naphtho[1,2-b]furan-2,8(3H,4H)-dione, 3a,5,5a,9b-tetrahydro-4,5-dihydroxy-9-methoxy-3,5a-dimethyl- / Cedrin	$C_{15}H_{18}O_6$ / 294.30	6040-62-6 / 266	1918		EtOH 4; MeOH 4; chl 4
8059	Naphtho[2,3-b]furan-2(3H)-one, 3a,5,6,7,8,8a,9,9a-octahydro-5,8a-dimethyl-3-methylene-, / Alantolactone	$C_{15}H_{20}O_2$ / 232.32	546-43-0 / 76	198 / 275	5-17-10-00097	bz 4; eth 4; EtOH 4; chl 4
8060	Naphtho[1,2-c]furan-1,3-dione, 4,5-dihydro-	$C_{12}H_8O_3$ / 200.19	37845-14-0 / 126.5	227-30[23]	5-17-11-00444	bz 4; MeOH 4
8061	Naphtho[1,2,3,4-def]chrysene / Dibenzo[a,e]pyrene	$C_{24}H_{14}$ / 302.38	192-65-4 / 233.5		4-05-00-02803	EtOH 2; ace 2; bz 2; tol 3
8062	Naphtho[2,3-f]quinoline-7,12-dione, 5,6-dihydroxy- / Alizarine Blue	$C_{17}H_9NO_4$ / 291.26	568-02-5 / 269	239	5-21-13-00590	bz 4; gl HOAc 4
8063	2-Naphthol, 8-chloro-	$C_{10}H_7ClO$ / 178.62	29921-50-4 / 101	307.5	3-06-00-02992	bz 4; eth 4; EtOH 4; chl 4
8064	2-Naphthylamine, 1,4-dimethyl- / Naphthalene, 2-amino-1,4-dimethyl	$C_{12}H_{13}N$ / 171.24	878-93-3 / 75	333	0-12-00-01317	eth 4; EtOH 4
8065	1,5-Naphthyridine / 1,5-Diazanaphthalene	$C_8H_6N_2$ / 130.15	254-79-5 / 75	112[12]	5-23-07-00156 / 1.2100[20]	
8066	1,8-Naphthyridine-3-carboxylic acid, 1-ethyl-1,4-dihydro-7-methyl-4-oxo-	$C_{12}H_{12}N_2O_3$ / 232.24	389-08-2 / 229.5	6273		EtOH 2; eth 2; chl 3
8067	Napropamide / Propanamide, N,N-diethyl-2-(1-naphthalenyloxy)-	$C_{17}H_{21}NO_2$ / 271.36	15299-99-7 / 75	6336		
8068	Nickel carbonyl / Nickel tetracarbonyl	C_4NiO_4 / 170.74	13463-39-3 / -19.3	6411 / 43	1.31[25]	
8069	Nicosulfuron / 2-(4,6-DimethoxypyrimidiN-2-ylcarbamoylsulfamoyl)-N,N-dimethylnicotinamide	$C_{15}H_{18}N_6O_6S$ / 410.41	111991-09-4 / 172			
8070	Nitralin / Benzenamine, 4-(methylsulfonyl)-2,6-dinitro-N,N-dipropyl-	$C_{13}H_{19}N_3O_6S$ / 345.38	4726-14-1 / 150	6486		
8071	Nitrapyrin / Pyridine, 2-chloro-6-(trichloromethyl)-	$C_6H_3Cl_4N$ / 230.91	1929-82-4 / 63	6490 / 136[11]		
8072	Nitric acid, butyl ester / Butyl nitrate	$C_4H_9NO_3$ / 119.12	928-45-0 /	/ 133	4-01-00-01524 / 1.0228[30]	H_2O 1; EtOH 3; eth 3; ctc 2 / 1.4013[23]
8073	Nitric acid, decyl ester / Decyl nitrate	$C_{10}H_{21}NO_3$ / 203.28	2050-78-4	128[11]; 88[1]	4-01-00-01819 / 0.951[0]	EtOH 3; eth 3
8074	Nitric acid, ethyl ester / Ethyl nitrate	$C_2H_5NO_3$ / 91.07	625-58-1 / -94.6	87.2	4-01-00-01327 / 1.1084[20]	H_2O 3; EtOH 5; eth 5 / 1.3852[20]
8075	Nitric acid, methyl ester / Methyl nitrate	CH_3NO_3 / 77.04	598-58-3 / -82.5	6015 / 64.6 exp	4-01-00-01254 / 1.2075[20]	H_2O 2; EtOH 3; eth 3 / 1.3748[20]
8076	Nitric acid, 1-methylethyl ester / Isopropyl nitrate	$C_3H_7NO_3$ / 105.09	1712-64-7 /	100	4-01-00-01475 / 1.034[19]	EtOH 3; eth 3 / 1.3912[16]
8077	Nitric acid, 1-methylpropyl ester	$C_4H_9NO_3$ / 119.12	924-52-7 /	124	4-01-00-01573 / 1.0264[20]	eth 4; EtOH 4 / 1.4015[20]
8078	Nitric acid, 2-methylpropyl ester / Isobutyl nitrate	$C_4H_9NO_3$ / 119.12	543-29-3 /	5031 / 123.4	4-01-00-01595 / 1.0152[20]	1.4028[20]
8079	Nitric acid, octyl ester	$C_8H_{17}NO_3$ / 175.23	629-39-0 /	110-2[20]	4-01-00-01762 / 0.8409[17]	H_2O 2; EtOH 3; eth 3
8080	Nitric acid, propyl ester / Propyl nitrate	$C_3H_7NO_3$ / 105.09	627-13-4 /	7877 / 110	4-01-00-01424 / 1.0538[20]	H_2O 2; EtOH 3; eth 3; ctc 3 / 1.3973[20]
8081	Nitrofen / Benzene, 2,4-dichloro-1-(4-nitrophenoxy)-	$C_{12}H_7Cl_2NO_3$ / 284.10	1836-75-5 / 70	6519		
8082	Nitrofluorfen / Benzene, 2-chloro-1-(4-nitrophenoxy)-4-(trifluoromethyl)-	$C_{13}H_7ClF_3NO_3$ / 317.65	42874-01-1 / 68			
8083	Nitrous acid, butyl ester / Butyl nitrite	$C_4H_9NO_2$ / 103.12	544-16-1 /	1581 / 78	4-01-00-01523 / 0.9114[25]	EtOH 5; eth 5 / 1.3762[20]

No.	Name Synonym	Mol. Form. Mol. Wt.	CAS RN mp/°C	Merck No. bp/°C	Beil. Ref. den/g cm^{-3}	Solubility n_D
8084	Nitrous acid, decyl ester Decyl nitrite	$C_{10}H_{21}NO_2$ 187.28	1653-57-2	105-8[12]	4-01-00-01819	EtOH 3; eth 3 1.4247[20]
8085	Nitrous acid, 1,1-dimethylethyl ester *tert*-Butyl nitrite	$C_4H_9NO_2$ 103.12	540-80-7	1582 63	4-01-00-01622 0.8670[20]	H$_2$O 2; EtOH 3; eth 3; chl 3 1.368[20]
8086	Nitrous acid, 1,2-ethanediyl ester Ethylene glycol, dinitrite	$C_2H_4N_2O_4$ 120.06	629-16-3 <-15	98	4-01-00-02411 1.2156[0]	eth 4; EtOH 4
8087	Nitrous acid, ethyl ester Ethyl nitrite	$C_2H_5NO_2$ 75.07	109-95-5	3786 18	4-01-00-01327 0.899[15]	EtOH 5; eth 5 1.3418[10]
8088	Nitrous acid, heptyl ester Heptyl nitrite	$C_7H_{15}NO_2$ 145.20	629-43-6	155.8	4-01-00-01735 0.8939[0]	eth 4 1.4032[20]
8089	Nitrous acid, hexyl ester Hexyl nitrite	$C_6H_{13}NO_2$ 131.17	638-51-7	129; 52[44]	4-01-00-01699 0.8778[20]	H$_2$O 1; EtOH 3; eth 3 1.3987[20]
8090	Nitrous acid, 3-methylbutyl ester Isoamyl nitrite	$C_5H_{11}NO_2$ 117.15	110-46-3	5005 99.2	4-01-00-01683 0.8828[20]	H$_2$O 2; EtOH 5; eth 5 1.3918[20]
8091	Nitrous acid, methyl ester Methyl nitrite	CH_3NO_2 61.04	624-91-9 -16	-12	4-01-00-01253 0.991[15]	EtOH 3; eth 3
8092	Nitrous acid, 1-methylethyl ester Isopropyl nitrite	$C_3H_7NO_2$ 89.09	541-42-4	5104 40	4-01-00-01474 0.8684[15]	H$_2$O 1; EtOH 3; eth 3
8093	Nitrous acid, 1-methylpropyl ester	$C_4H_9NO_2$ 103.12	924-43-6	68.5	4-01-00-01573 0.8726[20]	eth 4; EtOH 4; chl 4 1.3710[20]
8094	Nitrous acid, 2-methylpropyl ester Isobutyl nitrite	$C_4H_9NO_2$ 103.12	542-56-3	5032 67	4-01-00-01595 0.8699[22]	H$_2$O 2; EtOH 3; eth 3 1.3715[22]
8095	Nitrous acid, octyl ester	$C_8H_{17}NO_2$ 159.23	629-46-9	174.5	4-01-00-01762 0.862[17]	H$_2$O 2; EtOH 4; eth 4 1.4127[20]
8096	Nitrous acid, pentyl ester	$C_5H_{11}NO_2$ 117.15	463-04-7	104.5	4-01-00-01644 0.8817[20]	H$_2$O 2; EtOH 5; eth 5 1.3851[20]
8097	Nitrous acid, phenylmethyl ester Benzyl nitrite	$C_7H_7NO_2$ 137.14	935-05-7	80-3[35]	4-06-00-02568 1.075[25]	1.4989[25]
8098	Nitrous acid, propyl ester Propyl nitrite	$C_3H_7NO_2$ 89.09	543-67-9	7878 48	4-01-00-01424 0.886[20]	H$_2$O 2; EtOH 3; eth 3 1.3604[20]
8099	Nonacosane Celidoniol, deoxy-	$C_{29}H_{60}$ 408.80	630-03-5 63.7	440.8	4-01-00-00591 0.8083[20]	H$_2$O 1; EtOH 4; eth 4; ace 4 1.4529[20]
8100	1-Nonacosanol Nonacosanyl alcohol	$C_{29}H_{60}O$ 424.79	6624-76-6 84.5	200[1.0]	4-01-00-01916	EtOH 3
8101	Nonadecane	$C_{19}H_{40}$ 268.53	629-92-5 32.1	329.9	4-01-00-00560 0.7855[20]	H$_2$O 1; EtOH 2; eth 3; ace 3 1.4409[20]
8102	1,2,3-Nonadecanetricarboxylic acid, 2-hydroxy- Agaricic acid	$C_{22}H_{40}O_7$ 416.56	666-99-9 142 dec	174	4-03-00-01284	H$_2$O 3; EtOH 2; eth 2; bz 1
8103	Nonadecanoic acid	$C_{19}H_{38}O_2$ 298.51	646-30-0 69.4	297[100]; 228[10]	4-02-00-01256 0.8468[70]	H$_2$O 1; EtOH 4; eth 4; bz 4
8104	1-Nonadecanol	$C_{19}H_{40}O$ 284.53	1454-84-8 63.3	166[0.3]	4-01-00-01898	eth 3; ace 3 1.4328[75]
8105	10-Nonadecanone	$C_{19}H_{38}O$ 282.51	504-57-4 65.5	>350; 156[1.1]	4-01-00-03402	H$_2$O 1; EtOH 2; eth 3; ace 3
8106	2-Nonadecanone	$C_{19}H_{38}O$ 282.51	629-66-3 57	266[110]; 165[2]	4-01-00-03400 0.8108[56]	H$_2$O 1; EtOH 2; eth 4; ace 3
8107	1-Nonadecyne	$C_{19}H_{36}$ 264.49	26186-01-6 37.5	327	3-01-00-01030 0.8054[20]	ace 4; bz 4; eth 4 1.4488[20]
8108	2,6-Nonadienal Leaf aldehyde violet	$C_9H_{14}O$ 138.21	26370-28-5	85-7[11]	4-01-00-03560 0.8678[20]	1.4460[20]
8109	2,4-Nonadienal, (E,E)-	$C_9H_{14}O$ 138.21	5910-87-2	97-8[10]	4-01-00-03560 0.862[25]	1.5207[20]
8110	1,8-Nonadiene	C_9H_{16} 124.23	4900-30-5	142.5	4-01-00-01048 0.7511[20]	1.4302[20]
8111	2,7-Nonadiene	C_9H_{16} 124.23	51333-70-1 -72.5	151.5	3-01-00-01013 0.7499[20]	1.4358[20]
8112	2,6-Nonadien-1-ol	$C_9H_{16}O$ 140.23	7786-44-9	96-100[11]	4-01-00-02266 0.8604[25]	1.4598[25]
8113	6,8-Nonadien-2-one, 8-methyl-5-(1-methylethyl)-, [S-(E)]- Solanone	$C_{13}H_{22}O$ 194.32	1937-54-8	8663 60[1]	0.870[20]	1.4755[20]
8114	1,4-Nonadiyne	C_9H_{12} 120.19	6088-94-4	83-4[41]	4-01-00-01125 0.8112[25]	1.4518[25]
8115	1,8-Nonadiyne	C_9H_{12} 120.19	2396-65-8 -27.3	162	4-01-00-01125 0.8158[20]	eth 3; ace 3 1.4490[20]
8116	2,7-Nonadiyne	C_9H_{12} 120.19	31699-35-1 4.3	180	3-01-00-01063 0.8332[20]	1.4674[20]
8117	Nonanal Nonaldehyde	$C_9H_{18}O$ 142.24	124-19-6	191	4-01-00-03352 0.8264[22]	eth 3; chl 3 1.4273[20]
8118	Nonanamide	$C_9H_{19}NO$ 157.26	1120-07-6 99.5	sub	4-02-00-01023 0.8394[110]	H$_2$O 1; EtOH 2; eth 2 1.4248[110]
8119	1-Nonanamine	$C_9H_{21}N$ 143.27	112-20-9 -1	202.2	4-04-00-00777 0.7886[20]	H$_2$O 2; EtOH 3; eth 3; chl 2 1.4336[20]
8120	Nonane	C_9H_{20} 128.26	111-84-2 -53.5	150.8	4-01-00-00447 0.7176[20]	H$_2$O 1; EtOH 4; eth 4; ace 5 1.4054[20]
8121	Nonane, 1-bromo-	$C_9H_{19}Br$ 207.15	693-58-3	88[4]	4-01-00-00451 1.0165[20]	1.4533[20]

No.	Name / Synonym	Mol. Form. / Mol. Wt.	CAS RN / mp/°C	Merck No. / bp/°C	Beil. Ref. / den/g cm^{-3}	Solubility / n_D
8122	Nonane, 5-butyl-	$C_{13}H_{28}$ / 184.37	17312-63-9	217.5	4-01-00-00517 / 0.7635[18]	1.4273[18]
8123	Nonane, 1-chloro-	$C_9H_{19}Cl$ / 162.70	2473-01-0 / -39.4	203.5	4-01-00-00450 / 0.8720[20]	H_2O 1; eth 3; chl 3 / 1.4345[20]
8124	Nonane, 2-chloro- / 2-Chlorononane	$C_9H_{19}Cl$ / 162.70	2216-36-6	190	0-01-00-00166 / 0.8790[20]	chl 4 / 1.4420[20]
8125	Nonane, 5-chloro- / 5-Chlorononane	$C_9H_{19}Cl$ / 162.70	28123-70-8	85-7[14]	2-01-00-00128 / 0.8639[15]	eth 4 / 1.4314[15]
8126	Nonane, 1-chloro-9-fluoro-	$C_9H_{18}ClF$ / 180.69	463-23-0	102[11]	4-01-00-00450 / 0.966[20]	eth 4; EtOH 4 / 1.4301[25]
8127	Nonane, 1,2-dibromo- / 1,2-Dibromononane	$C_9H_{18}Br_2$ / 286.05	73642-91-8	141.5[20]	3-01-00-00506 / 1.3980[20]	1.4942[20]
8128	Nonane, 1,9-dichloro- / 1,9-Dichlorononane	$C_9H_{18}Cl_2$ / 197.15	821-99-8	260; 138[17]	4-01-00-00451 / 1.0173[25]	1.4586[25]
8129	Nonanedinitrile	$C_9H_{14}N_2$ / 150.22	1675-69-0	198-9[25]	4-02-00-02059 / 0.9200[19]	eth 4; EtOH 4; bz 4 / 1.4518[19]
8130	Nonanedioic acid / Azelaic acid	$C_9H_{16}O_4$ / 188.22	123-99-9 / 106.5	921 / 287[100], 225[10]	4-02-00-02055 / 1.225[25]	H_2O 2; EtOH 3; eth 2; bz 2 / 1.4303[111]
8131	Nonanedioic acid, bis(2-ethylbutyl) ester	$C_{21}H_{40}O_4$ / 356.55	105-03-3 / -45	230[5]	4-02-00-02057 / 0.928[25]	ace 4; bz 4; EtOH 4 / 1.443[25]
8132	Nonanedioic acid, bis(2-ethylhexyl) ester	$C_{25}H_{48}O_4$ / 412.65	103-24-2 / -78	237[5]	3-02-00-01787 / 0.915[25]	H_2O 1; EtOH 3; ace 3; bz 3 / 1.446[25]
8133	Nonanedioic acid, diethyl ester	$C_{13}H_{24}O_4$ / 244.33	624-17-9 / -18.5	291.5	4-02-00-02056 / 0.9729[20]	H_2O 1; EtOH 3; eth 3 / 1.4351[20]
8134	Nonanedioic acid, dimethyl ester	$C_{11}H_{20}O_4$ / 216.28	1732-10-1	156[20]	4-02-00-02056 / 1.0082[20]	H_2O 1; EtOH 3; ace 3; bz 3 / 1.4367[20]
8135	Nonanedioic acid, 3-methyl- / Azelaic acid, 3-methyl	$C_{10}H_{18}O_4$ / 202.25	76078-85-8 / 44	168[0.2]	3-02-00-01819	1.4670[19]
8136	1,9-Nonanediol / 1,9-Dihydroxynonane	$C_9H_{20}O_2$ / 160.26	3937-56-2 / 45.8	173[20]; 150[3]	4-01-00-02607	H_2O 2; EtOH 4; eth 4; bz 3
8137	2,4-Nonanedione / Methyl β-oxoheptyl ketone	$C_9H_{16}O_2$ / 156.22	6175-23-1 / -18	105-7[22]	4-01-00-03715 / 0.938[25]	
8138	Nonanedioyl dichloride	$C_9H_{14}Cl_2O_2$ / 225.11	123-98-8	166[18]	3-02-00-01789 / 1.4680[20]	eth 3; bz 4 / 1.4680[20]
8139	Nonane, icosofluoro / Perfluorononane	C_9F_{20} / 488.07	375-96-2	125.3	1.800[25]	
8140	Nonane, 2-methyl- / 2-Methylnonane	$C_{10}H_{22}$ / 142.28	871-83-0 / -74.6	167.1	4-01-00-00473 / 0.7281[20]	H_2O 1; eth 3; bz 3; chl 3 / 1.4099[20]
8141	Nonane, 3-methyl- / 3-Methylnonane	$C_{10}H_{22}$ / 142.28	5911-04-6 / -84.8	167.9	2-01-00-00130 / 0.7354[20]	bz 4; eth 4; chl 4 / 1.4125[20]
8142	Nonane, 4-methyl- / 4-Methylnonane	$C_{10}H_{22}$ / 142.28	17301-94-9 / -99	165.7	4-01-00-00475 / 0.7323[20]	bz 4; eth 4; chl 4 / 1.4123[20]
8143	Nonane, 5-methyl- / 5-Methylnonane	$C_{10}H_{22}$ / 142.28	15869-85-9 / -87.7	165.1	4-01-00-00475 / 0.7326[20]	H_2O 1; eth 3; bz 3; chl 3 / 1.4116[20]
8144	Nonanenitrile	$C_9H_{17}N$ / 139.24	2243-27-8 / -34.2	224.4	4-02-00-01024 / 0.8178[20]	H_2O 1; EtOH 3; eth 3; ctc 2 / 1.4255[20]
8145	1-Nonanethiol / Nonyl mercaptan	$C_9H_{20}S$ / 160.32	1455-21-6 / -20.1	220	4-01-00-01802 / 0.842[25]	1.4548[20]
8146	Nonanoic acid / Pelargonic acid	$C_9H_{18}O_2$ / 158.24	112-05-0 / 12.3	7013 / 254.5	4-02-00-01018 / 0.9052[20]	H_2O 1; EtOH 3; eth 3; chl 3 / 1.4343[19]
8147	Nonanoic acid, butyl ester / Pelargonic acid, butyl ester	$C_{13}H_{26}O_2$ / 214.35	50623-57-9 / -38	122-4[20]	3-02-00-00820 / 0.8520[25]	1.4262[25]
8148	Nonanoic acid, 9-cyano-, methyl ester / Sebacic acid, monomitrile, methyl ester	$C_{11}H_{19}NO_2$ / 197.28	53663-26-6 / 3.5	178[16]	4-02-00-02089 / 0.934[20]	eth 4 / 1.4398[25]
8149	Nonanoic acid, ethenyl ester	$C_{11}H_{20}O_2$ / 184.28	6280-03-1	133.5[50]	4-02-00-01020 / 0.8689[30]	1.4447[30]
8150	Nonanoic acid, ethyl ester / Ethyl nonanoate	$C_{11}H_{22}O_2$ / 186.29	123-29-5 / -36.7	3793 / 227.0	4-02-00-01019 / 0.8657[20]	H_2O 1; EtOH 3; eth 3; ace 3 / 1.4220[20]
8151	Nonanoic acid, 9-fluoro- / Pelargonic acid, 9-fluoro	$C_9H_{17}FO_2$ / 176.23	463-16-1 / 18	89[0.2]	4-02-00-01024	eth 4; EtOH 4 / 1.4289[25]
8152	Nonanoic acid, heptyl ester / Pelargonic acid, heptyl ester	$C_{16}H_{32}O_2$ / 256.43	71605-85-1 / -15.5	210[75]	3-02-00-00821 / 0.8553[25]	1.4350[20]
8153	Nonanoic acid, 3-methyl- / Pelargonic acid, 3-methyl	$C_{10}H_{20}O_2$ / 172.27	35205-79-9	147-8[12]	4-02-00-01057 / 0.9012[20]	1.4342[20]
8154	Nonanoic acid, methyl ester	$C_{10}H_{20}O_2$ / 172.27	1731-84-6	213.5	4-02-00-01019 / 0.8799[15]	H_2O 1; EtOH 3; eth 3; ctc 2 / 1.4214[20]
8155	Nonanoic acid, 3-methyl-, ethyl ester / Pelargonic acid 3-methyl, ethyl ester	$C_{12}H_{24}O_2$ / 200.32	86051-37-8	115[13]	3-02-00-00849 / 0.8653[20]	1.4240[20]
8156	Nonanoic acid, pentyl ester / Pelargonic acid, pentyl ester	$C_{14}H_{28}O_2$ / 228.38	61531-45-1 / -27	130-2[20]	3-02-00-00820 / 0.8506[25]	1.4318[20]
8157	1-Nonanol / Nonyl alcohol	$C_9H_{20}O$ / 144.26	143-08-8 / -5	6598 / 213.3	4-01-00-01798 / 0.8273[20]	H_2O 1; EtOH 3; eth 3; ctc 2 / 1.4333[20]
8158	2-Nonanol, (±)-	$C_9H_{20}O$ / 144.26	74683-66-2 / -35	193.5	4-01-00-01803 / 0.8471[20]	eth 4; EtOH 4 / 1.4353[20]
8159	3-Nonanol, (±)-	$C_9H_{20}O$ / 144.26	74742-08-8 / 22	195; 93[18]	4-01-00-01803 / 0.8250[20]	H_2O 1; EtOH 3; eth 3 / 1.4289[20]
8160	4-Nonanol, (S)-	$C_9H_{20}O$ / 144.26	52708-03-9	192.5; 94[18]	3-01-00-01747 / 0.8282[20]	H_2O 1; EtOH 3; eth 3 / 1.4197[20]
8161	5-Nonanol / Dibutylcarbinol	$C_9H_{20}O$ / 144.26	623-93-8	193; 97[20]	4-01-00-01804 / 0.8356[20]	H_2O 1; EtOH 3 / 1.4289[20]

No.	Name / Synonym	Mol. Form. / Mol. Wt.	CAS RN / mp/°C	Merck No. / bp/°C	Beil. Ref. / den/g cm^{-3}	Solubility / n_D
8162	5-Nonanol, 5-butyl-	$C_{13}H_{28}O$ 200.36	597-93-3 20	232 dec	4-01-00-01863 0.8408[20]	EtOH 4 1.4445[20]
8163	1-Nonanol, 9-chloro- 9-Chlorononan-1-ol	$C_9H_{19}ClO$ 178.70	51308-99-7 28	147[14]	4-01-00-01802 0.8305[12]	eth 4; EtOH 4 1.4575[20]
8164	5-Nonanol, 2,8-dimethyl- 2,8-Dimethylnonan-5-ol	$C_{11}H_{24}O$ 172.31	19780-96-2	105[9]	0-01-00-00428 0.8305[20]	1.4380[12]
8165	1-Nonanol, 9-fluoro-	$C_9H_{19}FO$ 162.25	463-24-1	125-6[15]	4-01-00-01801 0.928[20]	eth 4; EtOH 4 1.4279[25]
8166	1-Nonanol, 3-methyl- 3-Hexylbutanol-1	$C_{10}H_{22}O$ 158.28	22663-64-5	114[14]; 103[9]	4-01-00-01825 0.8342[20]	1.4361[20]
8167	1-Nonanol, 4-methyl-	$C_{10}H_{22}O$ 158.28	1489-47-0	120[17]	4-01-00-01826 0.826[27]	1.4364[20]
8168	1-Nonanol, 8-methyl 8-Methyl-1-nonanol	$C_{10}H_{22}O$ 158.28	55505-26-5	108[10]		
8169	3-Nonanol, 2-methyl-	$C_{10}H_{22}O$ 158.28	26533-33-5	115[28]	4-01-00-01825 0.8290[20]	1.4346[20]
8170	4-Nonanol, 4-methyl- 1-Methyl-1-propylhexanol	$C_{10}H_{22}O$ 158.28	23418-38-4	92-3[15]	4-01-00-01826 0.8245[25]	1.4338[20]
8171	5-Nonanol, 5-methyl- 2-Butyl-2-hexanol	$C_{10}H_{22}O$ 158.28	33933-78-7	84-5[10]	4-01-00-01827 0.8290[20]	1.4341[20]
8172	2-Nonanone Heptyl methyl ketone	$C_9H_{18}O$ 142.24	821-55-6 -7.5	195.3	4-01-00-03353 0.8208[20]	H_2O 1; EtOH 3; eth 3; ace 4 1.4210[20]
8173	3-Nonanone Ethyl hexyl ketone	$C_9H_{18}O$ 142.24	925-78-0 -8	190; 86[20]	4-01-00-03354 0.8241[20]	H_2O 1; EtOH 3; eth 3; ace 4 1.4208[20]
8174	4-Nonanone Propyl amyl ketone	$C_9H_{18}O$ 142.24	4485-09-0	187.5	4-01-00-03354 0.8190[25]	H_2O 1; EtOH 3; eth 3; ace 4 1.4189[20]
8175	5-Nonanone Dibutyl ketone	$C_9H_{18}O$ 142.24	502-56-7 -5.9	188.4	4-01-00-03355 0.8217[20]	H_2O 1; EtOH 3; eth 4; chl 4 1.4195[20]
8176	5-Nonanone, 2-methyl- 2-Methyl-5-nonanone	$C_{10}H_{20}O$ 156.27	22287-02-1	203.5	3-01-00-02900 0.8213[20]	1.4239[20]
8177	Nonanoyl chloride	$C_9H_{17}ClO$ 176.69	764-85-2 -60.5	215.3	4-02-00-01023 0.9463[15]	eth 3; ace 3
8178	Nonasiloxane, eicosamethyl- Eicosamethylnonasiloxane	$C_{20}H_{60}O_8Si_9$ 681.46	2652-13-3 3494	307.5; 198[16]	4-00-00-04130 0.9173[20]	bz 4 1.3980[20]
8179	2-Nonenal 2-Nonen-1-al	$C_9H_{16}O$ 140.23	2463-53-8	56-8[0.1]	4-01-00-03501 0.8418[25]	1.4502[25]
8180	2-Nonenal, (E)-	$C_9H_{16}O$ 140.23	18829-56-6 6595		4-01-00-03502	
8181	6-Nonenamide, N-[(4-hydroxy-3-methoxyphenyl)methyl]-8-methyl-, (E)- Capsaicin	$C_{18}H_{27}NO_3$ 305.42	404-86-4 1767 65	210-20[0.01]	4-13-00-02588	H_2O 1; EtOH 4; eth 3; bz 3
8182	1-Nonene	C_9H_{18} 126.24	124-11-8 -81.3	146.9	4-01-00-00894 0.7253[25]	1.4257[20]
8183	3-Nonene, (E)-	C_9H_{18} 126.24	20063-92-7	147.5	3-01-00-00853 0.732[21]	H_2O 1; eth 3; bz 3; ctc 3 1.4181[21]
8184	4-Nonene, 5-butyl- 5-Butyl-4-nonene	$C_{13}H_{26}$ 182.35	7367-38-6	215.5	4-01-00-00922 0.7745[20]	1.4375[20]
8185	1-Nonene, 2-methyl- 2-Methylnon-1-ene	$C_{10}H_{20}$ 140.27	2980-71-4 -64.2	168.4	4-01-00-00903 0.7412[25]	1.4241[20]
8186	3-Nonene, 2-methyl-	$C_{10}H_{20}$ 140.27	53966-53-3	161.0	4-01-00-00903 0.7340[20]	1.4202[20]
8187	3-Nonenoic acid	$C_9H_{16}O_2$ 156.22	4124-88-3	106[1]	4-02-00-01598 0.9254[20]	1.4454[25]
8188	8-Nonenoic acid	$C_9H_{16}O_2$ 156.22	31642-67-8 5	116-8[1]	4-02-00-01597 0.9146[16]	1.4492[15]
8189	1-Nonen-3-ol 1-Vinylheptanol	$C_9H_{18}O$ 142.24	21964-44-3	193.5	3-01-00-01947 0.824[21]	1.4382[15]
8190	8-Nonen-1-ol	$C_9H_{18}O$ 142.24	13038-21-6	135[20]	3-01-00-01947 0.8394[25]	1.4450[23]
8191	5-Nonen-4-one, 6-methyl-	$C_{10}H_{18}O$ 154.25	7036-98-8	90-2[16]	3-01-00-03022 0.8608[20]	1.4518[20]
8192	6-Nonen-4-one, 6-methyl-	$C_{10}H_{18}O$ 154.25	53252-20-3	94[18]	3-01-00-03022 0.8413[21]	1.4429[21]
8193	1-Nonen-3-yne	C_9H_{14} 122.21	57223-18-4	27-8[4]	4-01-00-01103 0.7602[25]	ace 4; eth 4 1.4487[25]
8194	1-Nonen-4-yne	C_9H_{14} 122.21	31508-12-0	58[22]	4-01-00-01103 0.777[25]	ace 4; eth 4 1.4418[25]
8195	2-Nonen-4-yne	C_9H_{14} 122.21	90644-60-3	70[20]	4-01-00-01048 0.7832[25]	ace 4; eth 4 1.4590[25]
8196	1-Nonyne Heptylacetylene	C_9H_{16} 124.23	3452-09-3 -50	150.8	4-01-00-01047 0.7658[20]	H_2O 1; eth 3; bz 3; ctc 3 1.4217[20]
8197	2-Nonyne	C_9H_{16} 124.23	19447-29-1	161.9	3-01-00-01013 0.7690[20]	H_2O 1; eth 3; lig 3 1.4337[20]
8198	3-Nonyne	C_9H_{16} 124.23	20184-89-8	157.1	3-01-00-01013 0.7616[20]	H_2O 1; eth 3; lig 3 1.4299[20]
8199	4-Nonyne	C_9H_{16} 124.23	20184-91-2	152	4-01-00-01047 0.757[25]	H_2O 1; eth 3; ace 3 1.4296[25]
8200	1-Nonyne, 1-chloro-	$C_9H_{15}Cl$ 158.67	90722-14-8	75-7[15]	3-01-00-01012 0.906[20]	eth 4 1.450[20]
8201	4-Nonyne, 3,3-dimethyl- 3,3-Dimethyl-4-nonyne	$C_{11}H_{20}$ 152.28	29022-31-9	82[40]	3-01-00-01024 0.7667[20]	bz 4; eth 4 1.4317[20]

No.	Name / Synonym	Mol. Form. / Mol. Wt.	CAS RN / mp/°C	Merck No. / bp/°C	Beil. Ref. / den/g cm^{-3}	Solubility / n_D
8202	4-Nonyne, 8-methyl- / 8-Methyl-4-nonyne	C$_{10}$H$_{18}$ / 138.25	70732-43-3	104.5[97]	3-01-00-01018 / 0.7681[20]	bz 4; eth 4 / 1.4311[20]
8203	2-Nonynoic acid	C$_9$H$_{14}$O$_2$ / 154.21	1846-70-4 / -8	154-6[18]	4-02-00-01721 / 0.9525[12]	1.4605[25]
8204	3-Nonynoic acid	C$_9$H$_{14}$O$_2$ / 154.21	56630-33-2 / 17	118[2]	4-02-00-01721	1.4603[25]
8205	5-Nonynoic acid	C$_9$H$_{14}$O$_2$ / 154.21	56630-34-3 / -3	115[1]	4-02-00-01723	1.4558[25]
8206	6-Nonynoic acid	C$_9$H$_{14}$O$_2$ / 154.21	56630-31-0 / 14	130[3]	4-02-00-01723	1.4578[25]
8207	8-Nonynoic acid	C$_9$H$_{14}$O$_2$ / 154.21	30964-01-3 / 19	100[6]	4-02-00-01721	1.4524[25]
8208	1-Nonyn-3-ol	C$_9$H$_{16}$O / 140.23	7383-20-2	85[10]; 63[2]	4-01-00-02265 / 0.8627[20]	1.4440[20]
8209	2-Nonyn-1-ol	C$_9$H$_{16}$O / 140.23	5921-73-3	109[11]	4-01-00-02265 / 0.878[16]	1.4576[16]
8210	24-Norcholan-23-oic acid, (5β)- / Norcholanic acid	C$_{23}$H$_{38}$O$_2$ / 346.55	511-18-2 / 177	6606	3-09-00-02652	
8211	29-Nordammara-1,17(20),24-trien-21-oic acid, 6,16-bis(acetyloxy)-3,7-dioxo-, / Helvolic acid	C$_{33}$H$_{44}$O$_8$ / 568.71	29400-42-8 / 212 dec	4552	4-10-00-04083	H$_2$O 2; EtOH 2; eth 3; ace 3
8212	Norea / Urea, N,N-dimethyl-N'-(octahydro-4,7-methano-1H-indeN-5-yl)-, (3aα,4α, 5α,7α,7aα)-	C$_{13}$H$_{22}$N$_2$O / 222.33	18530-56-8 / 177	6611		
8213	Norflurazon / 3(2H)-Pyridazinone, 4-chloro-5-(methylamino)-2-[3-(trifluoromethyl)phenyl]-	C$_{12}$H$_9$ClF$_3$N$_3$O / 303.67	27314-13-2 / 184	6618		
8214	Norleucine / 2-Aminohexanoic acid	C$_6$H$_{13}$NO$_2$ / 131.17	327-57-1 / 301 dec	6624	4-04-00-02686	H$_2$O 4
8215	DL-Norleucine / Hexanoic acid, 2-amino(DL)	C$_6$H$_{13}$NO$_2$ / 131.17	616-06-8 / 327 dec	6624	3-04-00-01386 / 1.172[25]	H$_2$O 3; EtOH 2; eth 1
8216	19-Norpregna-1,3,5(10)-trien-20-yne-3,17-diol, (17α)- / 17α-Ethinyl-17β-estradiol	C$_{20}$H$_{24}$O$_2$ / 296.41	57-63-6	3689	4-06-00-06877	chl 2
8217	DL-Norvaline	C$_5$H$_{11}$NO$_2$ / 117.15	760-78-1 / 303 sub	6636	4-04-00-02629	H$_2$O 3; EtOH 1; eth 1; chl 1
8218	L-Norvaline / (S)-2-Aminopentanoic acid	C$_5$H$_{11}$NO$_2$ / 117.15	6600-40-4 / 307	6636	4-04-00-02629	H$_2$O 4
8219	Nuarimol / 5-Pyrimidinemethanol, α-(2-chlorophenyl)-α-(4-fluorophenyl)-	C$_{17}$H$_{12}$ClFN$_2$O / 314.75	63284-71-9 / 126			
8220	Octacosane	C$_{28}$H$_{58}$ / 394.77	630-02-4 / 64.5	431.6	4-01-00-00588 / 0.8067[20]	H$_2$O 1; ace 5; bz 3; chl 3 / 1.4330[70]
8221	Octacosanoic acid / Montanic acid	C$_{28}$H$_{56}$O$_2$ / 424.75	506-48-9 / 90.9		4-02-00-01318 / 0.8191[100]	bz 4; chl 4 / 1.4313[100]
8222	1-Octacosanol / Montanyl alcohol	C$_{28}$H$_{58}$O / 410.77	557-61-9 / 83.3	6664 / 200[1]	4-01-00-01915	
8223	9,12-Octadecadienoic acid (Z,Z)- / Linoleic acid	C$_{18}$H$_{32}$O$_2$ / 280.45	60-33-3 / -12	5382 / 229-30[16]	4-02-00-01754 / 0.9022[20]	ace 4; bz 4; eth 4; EtOH 4 / 1.4699[20]
8224	10,12-Octadecadienoic acid, (E,E)-	C$_{18}$H$_{32}$O$_2$ / 280.45	1072-36-2 / 56.5		4-02-00-01752 / 0.8686[70]	1.4689[40]
8225	9,12-Octadecadienoic acid	C$_{18}$H$_{32}$I$_2$ / 502.26	2197-37-7 / -5	229-30[16]	4-02-00-01754 / 0.9022[20]	ace 4; bz 4; eth 4; EtOH 4 / 1.4699[20]
8226	9,12-Octadecadienoic acid (Z,Z)-, ethyl ester / Ethyl linoleate	C$_{20}$H$_{36}$O$_2$ / 308.50	544-35-4	3776 / 272[180]; 212[12]	4-02-00-01757 / 0.8865[20]	eth 4; EtOH 4
8227	9,12-Octadecadienoic acid, ethyl ester	C$_{20}$H$_{36}$O$_2$ / 308.50	7619-08-1	272[180] / 212[12]	2-02-00-00461 / 0.8865[20]	H$_2$O 1; EtOH 3; eth 3
8228	9,12-Octadecadienoic acid (Z,Z)-, methyl ester / Methyl linoleate	C$_{19}$H$_{34}$O$_2$ / 294.48	112-63-0 / -35	6008 / 215[20]	4-02-00-01756 / 0.8886[10]	eth 4; EtOH 4 / 1.4638[20]
8229	9,12-Octadecadienoic acid, methyl ester / Methyl 9,12-octadecadienoate	C$_{19}$H$_{34}$O$_2$ / 294.48	2462-85-3 / -35	215[20] 169[1]	2-02-00-00461 / 0.8886[18]	H$_2$O 1; EtOH 3; eth 3; ctc 2 / 1.4638[20]
8230	7,11-Octadecadiyne	C$_{18}$H$_{30}$ / 246.44	103697-14-9	167-8[7]	3-01-00-01068 / 0.841[19]	1.4698[19]
8231	Octadecanamide	C$_{18}$H$_{37}$NO / 283.50	124-26-5 / 109	250[12]	4-02-00-01240	eth 4; chl 4
8232	Octadecanamide, N-(4-hydroxyphenyl)-	C$_{24}$H$_{41}$NO$_2$ / 375.60	103-99-1 / 133.8	239.5[10]	4-13-00-01113	H$_2$O 1; eth 2; ace 3; bz 2
8233	1-Octadecanamine	C$_{18}$H$_{39}$N / 269.51	124-30-1 / 52.9	346.8	4-04-00-00825 / 0.8618[20]	H$_2$O 1; EtOH 3; eth 3; ace 2 / 1.4522[20]
8234	Octadecane	C$_{18}$H$_{38}$ / 254.50	593-45-3 / 28.2	316.3	4-01-00-00553 / 0.7768[28]	H$_2$O 1; EtOH 2; eth 3; ace 3 / 1.4390[20]
8235	Octadecane, 1-bromo-	C$_{18}$H$_{37}$Br / 333.40	112-89-0 / 28.2	210[10]	4-01-00-00555 / 0.9848[20]	H$_2$O 1; EtOH 3; eth 3; ctc 2 / 1.4631[20]
8236	Octadecane, 1-chloro-	C$_{18}$H$_{37}$Cl / 288.94	3386-33-2 / 28.6	348	4-01-00-00554 / 0.8641[20]	H$_2$O 1; ctc 2 / 1.4531[20]
8237	Octadecane, 1,1-dimethoxy- / Octadecanal, dimethyl acetal	C$_{20}$H$_{42}$O$_2$ / 314.55	14620-55-4	168-70[3]	4-01-00-03397	eth 4; EtOH 4 / 1.4410[25]

No.	Name / Synonym	Mol. Form. / Mol. Wt.	CAS RN / mp/°C	Merck No. / bp/°C	Beil. Ref. / den/g cm^{-3}	Solubility / n_D
8238	Octadecanedioic acid, diethyl ester	$C_{22}H_{42}O_4$ 370.57	1472-90-8 54.5	240[12]	4-02-00-02176	eth 4; EtOH 4
8239	Octadecanedioic acid, 9,10-dihydroxy-, (R*,R*)-(±)- Phloionic acid	$C_{18}H_{34}O_6$ 346.46	23843-52-9 126	7297	4-03-00-01250	
8240	Octadecane, 1-(ethenyloxy)-	$C_{20}H_{40}O$ 296.54	930-02-9 30	182[3]	4-01-00-02057 0.8138[40]	chl 2
8241	Octadecane, 3-ethyl-5-(2-ethylbutyl)-	$C_{26}H_{54}$ 366.71	55282-12-7	229.5[10]	4-01-00-00586 0.8115[20]	1.45239[20]
8242	Octadecane, 1-iodo-	$C_{18}H_{37}I$ 380.40	629-93-6 34	383	4-01-00-00556 1.0994[20]	H_2O 1; EtOH 2; eth 2 1.4810[20]
8243	Octadecanenitrile	$C_{18}H_{35}N$ 265.48	638-65-3 41	362	4-02-00-01242 0.8325[20]	H_2O 1; EtOH 3; eth 4; ace 4 1.4389[45]
8244	1-Octadecanethiol Stearyl mercaptan	$C_{18}H_{38}S$ 286.57	2885-00-9 30	204-10[11]	4-01-00-01894 0.8475[20]	eth 4 1.4645[20]
8245	Octadecane, 9-p-tolyl- Toluene, p-(1-octyldecyl)-	$C_{25}H_{44}$ 344.62	4445-08-3	185[10]	4-05-00-01221 0.8549[20]	1.4811[20]
8246	9,11,13-Octadecanetrienoic acid (Z,Z,Z) Eleostearic acid	$C_{18}H_{30}O_2$ 278.44	3884-88-6 48.5			
8247	Octadecanoic acid Stearic acid	$C_{18}H_{36}O_2$ 284.48	57-11-4 68.8	8761 350 dec; 232[15]	4-02-00-01206 0.9408[20]	H_2O 1; EtOH 2; eth 4; ace 3 1.4299[80]
8248	Octadecanoic acid, aluminum salt	$C_{54}H_{105}AlO_6$ 877.41	637-12-7 118	370	4-02-00-01206	H_2O 1; EtOH 3; peth 3
8249	Octadecanoic acid, anhydride	$C_{36}H_{70}O_3$ 550.95	638-08-4 72		4-02-00-01239 0.8365[82]	H_2O 1; EtOH 1; eth 2; bz 2 1.4362[80]
8250	Octadecanoic acid, 18-bromo- Stearic acid, 18-bromo	$C_{18}H_{35}BrO_2$ 363.38	2536-38-1 75.5	240[4]	2-02-00-00361	bz 4; eth 4; EtOH 4
8251	Octadecanoic acid, butyl ester Butyl stearate	$C_{22}H_{44}O_2$ 340.59	123-95-5 27	1589 343	4-02-00-01219 0.854[25]	H_2O 1; EtOH 3; ace 4 1.4328[50]
8252	Octadecanoic acid, calcium salt	$C_{36}H_{70}CaO_4$ 607.03	1592-23-0 179.5	1710	4-02-00-01206	H_2O 1; EtOH 1; eth 1
8253	Octadecanoic acid, cyclohexyl ester Stearic acid, cyclohexyl ester	$C_{24}H_{46}O_2$ 366.63	104-07-4 44		4-06-00-00038 0.889[15]	eth 4
8254	Octadecanoic acid, 9,10-dihydroxy- 9,10-Dihydroxystearic acid	$C_{18}H_{36}O_4$ 316.48	120-87-6 90	3171	4-03-00-01092	H_2O 1; EtOH 2; eth 2
8255	Octadecanoic acid, 2,3-dihydroxypropyl ester, (±)-	$C_{21}H_{42}O_4$ 358.56	22610-63-5 74		4-02-00-01225 0.9841[20]	H_2O 1; EtOH 2; eth 2; lig 3 1.4400[86]
8256	Octadecanoic acid, 1,2-ethanediyl ester	$C_{38}H_{74}O_4$ 595.00	627-83-8 79	241[20]	4-02-00-01223 0.8581[78]	H_2O 1; EtOH 1; eth 4; ace 4
8257	Octadecanoic acid, ethyl ester	$C_{20}H_{40}O_2$ 312.54	111-61-5 33	199[10]	4-02-00-01218 1.057[20]	H_2O 1; EtOH 3; eth 3; ace 4 1.4349[40]
8258	Octadecanoic acid, hexadecyl ester	$C_{34}H_{68}O_2$ 508.91	1190-63-2 57		4-02-00-01220	ace 4; eth 4; chl 4 1.4410[70]
8259	Octadecanoic acid, 2-[2-[2-(2-hydroxyethoxy)ethoxy]ethoxy]ethyl ester	$C_{26}H_{52}O_6$ 460.70	106-07-0 40	328	1.1285[15]	1.4593[20]
8260	Octadecanoic acid, 2-hydroxyethyl ester	$C_{20}H_{40}O_3$ 328.54	111-60-4 60.5	189-91[3]	4-02-00-01222 0.8780[60]	EtOH 2; eth 3 1.4310[60]
8261	Octadecanoic acid, lead (II) salt Lead stearate	$C_{36}H_{70}O_4Pb$ 774.15	7428-48-0 125		4-02-00-01206 1.4	H_2O 1; Hot EtOH 3; eth 1
8262	Octadecanoic acid, 14-methyl- Stearic acid, 14-methyl	$C_{19}H_{38}O_2$ 298.51	94434-64-7 37.5	182[0.4]	4-02-00-01265 0.9400[20]	
8263	Octadecanoic acid, 17-methyl-	$C_{19}H_{38}O_2$ 298.51	2724-59-6 67.5	180[0.3]	4-02-00-01260 0.8420[70]	
8264	Octadecanoic acid, 9-methyl- Stearic acid, 9-methyl	$C_{19}H_{38}O_2$ 298.51	86073-38-3 40	171[0.1]	4-02-00-01271 0.9980[20]	1.4336[70]
8265	Octadecanoic acid, 3-methylbutyl ester Stearic acid, isopentyl ester	$C_{23}H_{46}O_2$ 354.62	627-88-3 25.5	192[2]	2-02-00-00353 0.855[20]	H_2O 1; EtOH 2; eth 3; ace 3 1.433[50]
8266	Octadecanoic acid, methyl ester	$C_{19}H_{38}O_2$ 298.51	112-61-8 39.1	443; 215[15]	4-02-00-01216 0.8498[40]	eth 4; chl 4 1.4367[40]
8267	Octadecanoic acid, 1-methylethyl ester	$C_{21}H_{42}O_2$ 326.56	112-10-7 28	207[6]	4-02-00-01219 0.8403[38]	ace 4; eth 4; EtOH 4; chl 4
8268	Octadecanoic acid, 2-methylpropyl ester Isobutyl stearate	$C_{22}H_{44}O_2$ 340.59	646-13-9 28.9	5034 223[15]	3-02-00-01017 0.8498[20]	eth 4
8269	Octadecanoic acid, 12-oxo-, ethyl ester Stearic acid, 12-oxo, ethyl ester	$C_{20}H_{38}O_3$ 326.52	88472-61-1 38	199[3]	3-03-00-01294	EtOH 4
8270	Octadecanoic acid, pentyl ester Stearic acid, pentyl ester	$C_{23}H_{46}O_2$ 354.62	6382-13-4 30		4-02-00-01220	eth 4; EtOH 4 1.4342[50]
8271	Octadecanoic acid, phenyl ester Stearic acid, phenyl ester	$C_{24}H_{40}O_2$ 360.58	637-55-8 52	267[15]	4-06-00-00618	H_2O 1; EtOH 3; eth 3
8272	Octadecanoic acid, 1,2,3-propanetriyl ester Tristearin	$C_{57}H_{110}O_6$ 891.50	555-43-1	9669	4-02-00-01233 0.8559[90]	H_2O 1; EtOH 1; ace 3; bz 2 1.4395[80]
8273	Octadecanoic acid, propyl ester	$C_{21}H_{42}O_2$ 326.56	3634-92-2 28.9	186.8[2]	4-02-00-01219 0.8452[38]	ace 4; eth 4; EtOH 4 1.4400[30]
8274	Octadecanoic acid, 9,10,12,13-tetrabromo-, methyl ester Stearic acid, 9,10,12,13-tetrabromo, methyl ester	$C_{19}H_{34}Br_4O_2$ 614.09	62080-86-8 63	215[15]	3-02-00-01049	eth 4; EtOH 4; chl 4 1.4346[45]

No.	Name / Synonym	Mol. Form. / Mol. Wt.	CAS RN / mp/°C	Merck No. / bp/°C	Beil. Ref. / den/g cm^{-3}	Solubility / n_D
8275	Octadecanoic acid, (tetrahydro-2-furanyl)methyl ester / Stearic acid, tetrahydrofurfuryl ester	$C_{23}H_{44}O_3$ / 368.60	6940-09-6 / 22		4-02-00-01248 / 0.914^{25}	eth 4; EtOH 4
8276	Octadecanoic acid, 9,10,18-trihydroxy-, (R*,R*)- / Phloionolic acid	$C_{18}H_{36}O_5$ / 332.48	583-86-8 / 101.5	7298	4-03-00-01119	
8277	Octadecanoic acid, zinc salt	$C_{36}H_{70}O_4Zn$ / 632.34	557-05-1 / 130	10063	4-02-00-01206	H_2O 1; EtOH 1; eth 1
8278	1-Octadecanol / Stearyl alcohol	$C_{18}H_{38}O$ / 270.50	112-92-5 / 59.5	210.5^{15}	4-01-00-01888 / 0.8124^{59}	H_2O 1; EtOH 3; eth 3; ace 2
8279	3-Octadecanol, (±)- / 3-Octadecanol (DL)	$C_{18}H_{38}O$ / 270.50	111897-18-8 / 51	202^{13}	4-01-00-01896 / 0.7858^{80}	1.4290^{80}
8280	3-Octadecanone	$C_{18}H_{36}O$ / 268.48	18261-92-2 / 51	197^{11}	4-01-00-03398	ctc 3
8281	Octadecanoyl chloride	$C_{18}H_{35}ClO$ / 302.93	112-76-5 / 23	215^{15}	4-02-00-01240 / 0.9069^0	EtOH 2 / 1.4523^{24}
8282	9,11,13-Octadecatrienoic acid, (E,E,E)-	$C_{18}H_{30}O_2$ / 278.44	544-73-0 / 71.5	188^1	4-02-00-01787 / 0.8839^{80}	EtOH 4 / 1.5000^{80}
8283	9,12,15-Octadecatrienoic acid, ethyl ester, (Z,Z,Z)-	$C_{20}H_{34}O_2$ / 306.49	1191-41-9	218^{15}	4-02-00-01782 / 0.8919^{20}	eth 4; EtOH 4 / 1.4694^{20}
8284	9,11,13-Octadecatrienoic acid, (E,Z,E)-	$C_{18}H_{30}O_2$ / 278.44	506-23-0 / 49	235^{12} dec; 170^1	4-02-00-01787 / 0.9028^{50}	eth 4; EtOH 4 / 1.5112^{50}
8285	9,12,15-Octadecatrienoic acid, methyl ester, (Z,Z,Z)-	$C_{19}H_{32}O_2$ / 292.46	301-00-8	182^3	4-02-00-01782 / 0.896^{25}	1.4709^{20}
8286	9,11,13-Octadecatrienoic acid, (Z,Z,E)-	$C_{18}H_{30}O_2$ / 278.44	544-72-9 / 62		4-02-00-01787 / 0.9027^{20}	EtOH 4; peth 4 / 1.5114^{50}
8287	9,12,15-Octadecatrienoic acid, (Z,Z,Z)- / Linolenic acid	$C_{18}H_{30}O_2$ / 278.44	463-40-1 / -11	5383 231^{17}, 129$^{0.05}$	4-02-00-01781 / 0.9164^{20}	H_2O 1; EtOH 3; eth 3; bz 2 / 1.4800^{20}
8288	9-Octadecenal / Octadecenyl aldehyde	$C_{18}H_{34}O$ / 266.47	5090-41-5	168-9^3	4-01-00-03533 / 0.8509^{20}	1.4558^{20}
8289	9-Octadecenal, (Z)-	$C_{18}H_{34}O$ / 266.47	2423-10-1	168-9^3	4-01-00-03533 / 0.8509^{20}	1.4558^{20}
8290	9-Octadecenamide, N-phenyl-, (Z)-	$C_{24}H_{39}NO$ / 357.58	5429-85-6 / 41	143.5^{10}	4-12-00-00411	bz 4; eth 4; HOAc 4
8291	1-Octadecene	$C_{18}H_{36}$ / 252.48	112-88-9 / 17.5	179^{15}; 145^8	4-01-00-00930 / 0.7891^{20}	H_2O 1; ace 3; ctc 3 / 1.4448^{20}
8292	9-Octadecene	$C_{18}H_{36}$ / 252.48	5557-31-3 / -30.5	162^9	4-01-00-00932 / 0.7916^{20}	1.4470^{20}
8293	9-Octadecenenitrile, (Z)-	$C_{18}H_{33}N$ / 263.47	112-91-4 / -1	332 dec	4-02-00-01668 / 0.847^{17}	EtOH 4 / 1.4566^{20}
8294	9-Octadecenoic acid (Z)- / Oleic acid	$C_{18}H_{34}O_2$ / 282.47	112-80-1 / 13.4	6788 286^{100}	4-02-00-01641 / 0.8935^{20}	H_2O 1; EtOH 5; eth 5; ace 5 / 1.4582^{20}
8295	11-Octadecenoic acid, (E)- / Vaccenic acid	$C_{18}H_{34}O_2$ / 282.47	693-72-1 / 44	9810	4-02-00-01640	ace 3 / 1.4499^{60}
8296	2-Octadecenoic acid, (E)-	$C_{18}H_{34}O_2$ / 282.47	2825-79-8 / 58.5		4-02-00-01632 / 0.8484^{90}	H_2O 1; EtOH 2; eth 3; bz 3
8297	9-Octadecenoic acid, (E)- / Elaidic acid	$C_{18}H_{34}O_2$ / 282.47	112-79-8 / 45	3497 288^{100}, 234^{15}	4-02-00-01647 / 0.8734^{45}	H_2O 1; EtOH 3; eth 3; bz 3 / 1.4499^{45}
8298	9-Octadecenoic acid (Z)-, butyl ester	$C_{22}H_{42}O_2$ / 338.57	142-77-8 / -26.4	227-8^{15}	4-02-00-01653 / 0.8704^{15}	EtOH 4 / 1.4480^{25}
8299	9-Octadecenoic acid (Z)-, 2,3-dihydroxypropyl ester	$C_{21}H_{40}O_4$ / 356.55	111-03-5 / 35	238-40^3	2-02-00-00439 / 0.9420^{20}	H_2O 1; EtOH 3; eth 3; chl 3 / 1.4626^{20}
8300	6-Octadecenoic acid, 6,7-diiodo- / Diiodotariric acid	$C_{18}H_{32}I_2O_2$ / 534.26	533-86-8 / 50.3	dec	4-02-00-01638	bz 4; eth 4; chl 4
8301	9-Octadecenoic acid (Z)-, ethyl ester	$C_{20}H_{38}O_2$ / 310.52	111-62-6	207^{13}	4-02-00-01651 / 0.8720^{20}	eth 4; EtOH 4 / 1.4515^{20}
8302	9-Octadecenoic acid, ethyl ester, (E)-	$C_{20}H_{38}O_2$ / 310.52	6114-18-7 / 5.8	217-9^{15}	4-02-00-01652 / 0.8664^{25}	eth 4; EtOH 4 / 1.4480^{25}
8303	9-Octadecenoic acid, 12-hydroxy-, [R-(Z)]- / Ricinoleic acid	$C_{18}H_{34}O_3$ / 298.47	141-22-0	8213 227^{10}	4-03-00-01026 / 0.9450^{21}	eth 4; EtOH 4 / 1.4716^{21}
8304	9-Octadecenoic acid, 12-hydroxy-, butyl ester, [R-(Z)]-	$C_{22}H_{42}O_3$ / 354.57	151-13-3	278^{12}	0-03-00-00388 / 0.9058^{22}	eth 4 / 1.4566^{22}
8305	9-Octadecenoic acid, 12-hydroxy-, ethyl ester, [R-(Z)]-	$C_{20}H_{38}O_3$ / 326.52	55066-53-0	258^{13}	4-03-00-01029 / 0.9045^{22}	1.4618^{22}
8306	9-Octadecenoic acid (Z)-, 3-methylbutyl ester / Oleic acid, isopentyl ester	$C_{23}H_{44}O_2$ / 352.60	627-89-4	223-4^{10}	3-02-00-01414 / 0.897^{15}	eth 4; EtOH 4
8307	9-Octadecenoic acid (Z)-, methyl ester / Methyl oleate	$C_{19}H_{36}O_2$ / 296.49	112-62-9 / -19.9	218.5^{20}	4-02-00-01649 / 0.8739^{20}	H_2O 1; EtOH 5; eth 5; chl 3 / 1.4522^{20}
8308	9-Octadecenoic acid, methyl ester, (E)-	$C_{19}H_{36}O_2$ / 296.49	1937-62-8	213-5^{15}	4-02-00-01651 / 0.8730^{20}	eth 4; EtOH 4 / 1.4513^{20}
8309	9-Octadecenoic acid (Z)-, phenylmethyl ester / Oleic acid, benzyl ester	$C_{25}H_{40}O_2$ / 372.59	55130-16-0	237^1	4-06-00-02269 / 0.9302^{25}	eth 4; EtOH 4 / 1.4875^{25}

No.	Name Synonym	Mol. Form. Mol. Wt.	CAS RN mp/°C	Merck No. bp/°C	Beil. Ref. den/g cm^{-3}	Solubility n_D
8310	9-Octadecenoic acid (Z)-, 1,2,3-propanetriyl ester Triolein	$C_{57}H_{104}O_6$ 885.45	122-32-7 -32	9644 235-40[18]	4-02-00-01664 0.8988[40]	H_2O 1; EtOH 2; eth 4; chl 3 1.4621[40]
8311	9-Octadecen-1-ol, (E)-	$C_{18}H_{36}O$ 268.48	506-42-3 36.5	333	4-01-00-02204 0.8338[40]	ace 4; eth 4; EtOH 4 1.4552[40]
8312	9-Octadecen-1-ol, (Z)- Oleyl alcohol	$C_{18}H_{36}O$ 268.48	143-28-2 6.5	6791 205-10[15]	4-01-00-02204 0.8489[20]	H_2O 1; EtOH 3; eth 3; ctc 2 1.4606[20]
8313	1-Octadecyne	$C_{18}H_{34}$ 250.47	629-89-0 22.5	313	4-01-00-01075 0.8025[20]	1.4774[20]
8314	2-Octadecyne	$C_{18}H_{34}$ 250.47	61847-97-0 30	184[15]	4-01-00-01075 0.8016[30]	
8315	9-Octadecyne	$C_{18}H_{34}$ 250.47	35365-59-4 3	163-4[7]	4-01-00-01076 0.8012[20]	1.4488[25]
8316	9-Octadecynoic acid 9-Stearolic acid	$C_{18}H_{32}O_2$ 280.45	506-24-1 48	189[1.8]	4-02-00-01751 	H_2O 1; EtOH 3; eth 4 1.4510[54]
8317	2,6-Octadienal, 3,7-dimethyl-, (E)-	$C_{10}H_{16}O$ 152.24	141-27-5 	229	4-01-00-03569 0.8888[20]	H_2O 1; EtOH 5; eth 5 1.4898[20]
8318	2,6-Octadienal, 3,7-dimethyl-, (Z)-	$C_{10}H_{16}O$ 152.24	106-26-3 	120[20]	4-01-00-03569 0.8869[20]	H_2O 1; EtOH 5; eth 5 1.4869[20]
8319	1,7-Octadiene α,ω-Octadiene	C_8H_{14} 110.20	3710-30-3 	115.5	4-01-00-01038 0.734[20]	1.4245[20]
8320	2,6-Octadiene 1,6-Dimethyldiallyl	C_8H_{14} 110.20	4974-27-0 -76	124.5	4-01-00-01039 0.7406[25]	1.4292[17]
8321	2,6-Octadiene, 1-bromo-3,7-dimethyl-, (E)- trans-Geranyl bromide	$C_{10}H_{17}Br$ 217.15	6138-90-5 	101[12;] 47[0.005]	4-01-00-01059 1.0940[22]	1.5027[20]
8322	2,4-Octadiene, 3,7-dimethyl- 3,7-Dimethylocta-2,4-diene	$C_{10}H_{18}$ 138.25	56523-26-3 	165.5	3-01-00-01020 0.7933[20]	1.456[20]
8323	2,5-Octadiene, 2,6-dimethyl-	$C_{10}H_{18}$ 138.25	116668-48-5 	59.5[12]	3-01-00-01018 0.773[20]	1.4500[20]
8324	2,6-Octadiene, 2,6-dimethyl-, (Z)- cis-2,6-Dimethyl-2,6-octadiene	$C_{10}H_{18}$ 138.25	2492-22-0 	168	4-01-00-01058 0.775[21]	ace 4; eth 4; EtOH 4 1.4498[20]
8325	2,6-Octadiene, 2,7-dimethyl- 2,7-Dimethyl-2,6-octadiene	$C_{10}H_{18}$ 138.25	16736-42-8 -74.4	166	4-01-00-01061 0.7849[20]	1.4481[20]
8326	2,6-Octadiene, 3,6-dimethyl-	$C_{10}H_{18}$ 138.25	116668-49-6 	154	4-01-00-01061 0.7767[20]	1.4445[20]
8327	2,6-Octadiene, 4,5-dimethyl- 4,5-Dimethyl-2,6-octadiene	$C_{10}H_{18}$ 138.25	18476-57-8 -64.8	153.5	3-01-00-01021 0.7611[25]	1.4375[25]
8328	2,4-Octadiene, 7-methyl-	C_9H_{16} 124.23	2216-70-8 	149	2-01-00-00239 0.7521[18]	1.4543[18]
8329	3,5-Octadiene, 4-methyl- 4-Methyl-3,5-octadiene	C_9H_{16} 124.23	36903-95-4 	149.5	2-01-00-00239 0.7640[25]	1.4628[25]
8330	1,6-Octadiene, 7-methyl-3-methylene- β-Myrcene	$C_{10}H_{16}$ 136.24	123-35-3 	6243 167	4-01-00-01108 0.8013[15]	H_2O 1; EtOH 3; eth 3; bz 3 1.4722[20]
8331	2,4-Octadienoic acid, 3,7-dimethyl-	$C_{10}H_{16}O_2$ 168.24	101715-60-0 	151-4[14]	4-02-00-01736 0.959[17]	1.4919
8332	1,6-Octadien-3-ol, 3,7-dimethyl-, (±)- Linalol	$C_{10}H_{18}O$ 154.25	22564-99-4 	198; 86[13]	4-01-00-02278 0.870[15]	1.4627
8333	2,6-Octadien-1-ol, 3,7-dimethyl-, (E)- trans-Geraniol	$C_{10}H_{18}O$ 154.25	106-24-1 <-15	4298 230	4-01-00-02277 0.8894[20]	H_2O 1; EtOH 3; eth 3; ace 3 1.4766[20]
8334	2,6-Octadien-1-ol, 3,7-dimethyl-, (Z)- Nerol	$C_{10}H_{18}O$ 154.25	106-25-2 <-15	6387 225; 125[25]	4-01-00-02276 0.8756[20]	EtOH 4 1.4746[20]
8335	1,6-Octadien-3-ol, 3,7-dimethyl-, acetate, (R)-	$C_{12}H_{20}O_2$ 196.29	16509-46-9 	220; 115[25]	4-02-00-00204 0.8951[20]	eth 4; EtOH 4 1.4544[21]
8336	2,6-Octadien-1-ol, 3,7-dimethyl-, acetate, (Z)-	$C_{12}H_{20}O_2$ 196.29	141-12-8 	134[25;] 93[3]	2-02-00-00153 0.905[15]	1.452[20]
8337	1,6-Octadien-3-ol, 3,7-dimethyl-, formate Linalyl formate	$C_{11}H_{18}O_2$ 182.26	115-99-1 	100-3[10]	4-02-00-00035 0.915[25]	EtOH 4 1.456[20]
8338	2,6-Octadien-1-ol, 3,7-dimethyl-, formate, (E)-	$C_{11}H_{18}O_2$ 182.26	105-86-2 	229 dec; 113[25]	4-02-00-00035 0.9086[25]	H_2O 1; EtOH 4; eth 3; ace 3 1.4659[20]
8339	2,6-Octadien-4-yne, 3,6-diethyl-	$C_{12}H_{18}$ 162.27	100319-48-0 	170	3-01-00-01066 0.8196[20]	ace 4; eth 4 1.4965[20]
8340	2,6-Octadien-4-yne, 3,6-dimethyl- 3,6-Dimethyl-2,6-octadiene-4-yne	$C_{10}H_{14}$ 134.22	3725-07-3 -45	170	4-01-00-01129 0.8071[22]	ace 4; eth 4 1.4998[20]
8341	1,5-Octadien-3-yne, 5-propyl- 5-Propyl-1,5-octadien-3-yne	$C_{11}H_{16}$ 148.25	70732-44-4 	57-8[6]	4-01-00-01129 0.8047[20]	ace 4; eth 4 1.4949[20]
8342	1,7-Octadiyne	C_8H_{10} 106.17	871-84-1 	135.5; 59[35]	4-01-00-01122 0.8169[21]	eth 3 1.4521[18]
8343	2,6-Octadiyne	C_8H_{10} 106.17	764-73-8 27	62[19]	4-01-00-01123 0.828[80]	eth 4 1.4658[30]
8344	3,5-Octadiyne	C_8H_{10} 106.17	16387-70-5 	163.5	4-01-00-01123 0.826[0]	eth 3 1.4968[0]
8345	3,5-Octadiyne, 2,7-dimethyl- Diisopropylbutadiyne	$C_{10}H_{14}$ 134.22	14813-68-4 	74[12]	1-01-00-00128 0.8090[20]	eth 4
8346	Octanal Caprylic aldehyde	$C_8H_{16}O$ 128.21	124-13-0 	1766 171	4-01-00-03337 0.8211[20]	ace 4; bz 4; eth 4; EtOH 4 1.4217[20]
8347	Octanal, 7-hydroxy-3,7-dimethyl-	$C_{10}H_{20}O_2$ 172.27	107-75-5 	103[3]	4-01-00-04058 0.9220[20]	H_2O 2; EtOH 3; ace 3 1.4494[20]
8348	Octanal, oxime	$C_8H_{17}NO$ 143.23	929-55-5 60	112[9]	2-01-00-00758 	ace 4; EtOH 4

No.	Name / Synonym	Mol. Form. / Mol. Wt.	CAS RN / mp/°C	Merck No. / bp/°C	Beil. Ref. / den/g cm^{-3}	Solubility / n_D
8349	Octanamide	$C_8H_{17}NO$ 143.23	629-01-6 108	239	4-02-00-00992 0.8450[110]	H_2O 2; EtOH 4; eth 3; ace 3
8350	1-Octanamine	$C_8H_{19}N$ 129.25	111-86-4 0	179.6	4-04-00-00751 0.7826[20]	H_2O 2; EtOH 4; eth 4; ctc 3 1.4292[20]
8351	2-Octanamine, (±)-	$C_8H_{19}N$ 129.25	44855-57-4 97	164	4-04-00-00762 0.7744[20]	eth 4; EtOH 4 1.4232[25]
8352	2-Octanamine, N-(1-methylheptyl)-	$C_{18}H_{35}N$ 241.46	5412-92-0	283	4-04-00-00764 0.7948[20]	H_2O 1; EtOH 4; eth 4
8353	1-Octanamine, N-methyl-N-octyl-	$C_{17}H_{37}N$ 255.49	4455-26-9 -30.1	158[10]	4-04-00-00753	1.4424[20]
8354	1-Octanamine, N-octyl-	$C_{16}H_{35}N$ 241.46	1120-48-5 35.5	297.5	4-04-00-00753 0.7963[26]	eth 4; EtOH 4 1.4415[26]
8355	Octane	C_8H_{18} 114.23	111-65-9 -56.8	6672 125.6	4-01-00-00412 0.6986[25]	H_2O 1; EtOH 5; eth 3; ace 5 1.3974[20]
8356	Octane, 1-bromo- Octyl bromide	$C_8H_{17}Br$ 193.13	111-83-1 -55	6684 200	4-01-00-00422 1.108[25]	H_2O 1; EtOH 5; eth 5; ctc 2 1.4524[20]
8357	Octane, 2-bromo-, (±)- Octane, 2-bromo (DL)	$C_8H_{17}Br$ 193.13	60251-57-2	188.5	4-01-00-00423 1.0878[25]	H_2O 1; EtOH 5; eth 5 1.4442[25]
8358	Octane, 1-bromo-8-fluoro-	$C_8H_{16}BrF$ 211.12	593-12-4	118-20[22.5]	4-01-00-00424	eth 4; EtOH 4 1.4500[20]
8359	Octane, 1-chloro- Octyl chloride	$C_8H_{17}Cl$ 148.68	111-85-3 -57.8	181.5	4-01-00-00419 0.8738[20]	H_2O 1; EtOH 4; eth 4; ctc 2 1.4305[20]
8360	Octane, 2-chloro-	$C_8H_{17}Cl$ 148.68	628-61-5	172; 75[28]	1-01-00-00060 0.8658[17]	H_2O 1; EtOH 4; eth 3 1.4273[21]
8361	Octane, 1-chloro-8-fluoro-	$C_8H_{16}ClF$ 166.67	593-14-6	87[10]	4-01-00-00420 0.978[20]	ace 4; eth 4; EtOH 4 1.4266[25]
8362	Octane, 3-chloro-3-methyl-	$C_9H_{19}Cl$ 162.70	28320-88-9	73-4[15]	3-01-00-00508 0.8680[25]	ace 4; bz 4; eth 4; chl 4 1.4351[20]
8363	Octane, 4-chloro-4-methyl- 4-Chloro-4-methyloctane	$C_9H_{19}Cl$ 162.70	36903-89-6	71[14.5]	3-01-00-00509 0.8723[20]	eth 4; chl 4 1.4360[20]
8364	Octanedial	$C_8H_{14}O_2$ 142.20	638-54-0	235 dec; 97[3]	4-01-00-03706	H_2O 4; EtOH 4 1.4439[20]
8365	1,8-Octanediamine	$C_8H_{20}N_2$ 144.26	373-44-4 52	225	3-04-00-00612	H_2O 4; eth 4; EtOH 4
8366	Octane, 1,2-dibromo- 1,2-Dibromooctane	$C_8H_{16}Br_2$ 272.02	6269-92-7	241	3-01-00-00468 1.4580[20]	1.4970[20]
8367	Octane, 1,4-dibromo-	$C_8H_{16}Br_2$ 272.02	70690-24-3	125-6[11]	4-01-00-00424 1.466[11]	1.5003[11]
8368	Octane, 1,5-dibromo- 1,5-Dibromooctane	$C_8H_{16}Br_2$ 272.02	17912-17-3	127-8[11]	4-01-00-00424 1.447[16]	1.4968[16]
8369	Octane, 1,8-dibromo- Octamethylene dibromide	$C_8H_{16}Br_2$ 272.02	4549-32-0 15.5	271	4-01-00-00424 1.4594[25]	H_2O 1; eth 3; ctc 3; chl 3 1.4971[25]
8370	Octane, 4,5-dibromo-	$C_8H_{16}Br_2$ 272.02	61539-75-1	84[4.3]	3-01-00-00469 1.4569[20]	1.4981[20]
8371	Octane, 1,8-dichloro- 1,8-Dichlorooctane	$C_8H_{16}Cl_2$ 183.12	2162-99-4	241	4-01-00-00420 1.0248[25]	1.4572[25]
8372	Octane, 2,3-dichloro-	$C_8H_{16}Cl_2$ 183.12	21948-47-0 -55.1	206	4-01-00-00425 0.9712[20]	1.4523[20]
8373	Octane, 2,2-dimethyl- 2,2-Dimethyloctane	$C_{10}H_{22}$ 142.28	15869-87-1	155	4-01-00-00476 0.7208[25]	1.4082[20]
8374	Octane, 2,3-dimethyl- 2,3-Dimethyloctane	$C_{10}H_{22}$ 142.28	7146-60-3	164.3	4-01-00-00476 0.7377[20]	1.4146[20]
8375	Octane, 2,4-dimethyl- 2,4-Dimethyloctane	$C_{10}H_{22}$ 142.28	4032-94-4	156	4-01-00-00477 0.7226[25]	1.4091[20]
8376	Octane, 2,5-dimethyl- 2,5-Dimethyloctane	$C_{10}H_{22}$ 142.28	15869-89-3	158.5	4-01-00-00477 0.7264[25]	1.4112[20]
8377	Octane, 2,6-dimethyl- 2,6-Dimethyloctane	$C_{10}H_{22}$ 142.28	2051-30-1	160.4	2-01-00-00131 0.7313[20]	1.4097[20]
8378	Octane, 2,7-dimethyl-	$C_{10}H_{22}$ 142.28	1072-16-8 -54.9	159.9	4-01-00-00478 0.7202[25]	eth 3; HOAc 3 1.4086[20]
8379	Octane, 3,3-dimethyl- 3,3-Dimethyloctane	$C_{10}H_{22}$ 142.28	4110-44-5	161.2	4-01-00-00479 0.7351[25]	1.4165[20]
8380	Octane, 3,4-dimethyl- 3,4-Dimethyloctane	$C_{10}H_{22}$ 142.28	15869-92-8	163.4	4-01-00-00479 0.7410[25]	1.4182[20]
8381	Octane, 3,5-dimethyl- 3,5-Dimethyloctane	$C_{10}H_{22}$ 142.28	15869-93-9	159.4	4-01-00-00479 0.7329[25]	1.4139[20]
8382	Octane, 3,6-dimethyl- 3,6-Dimethyloctane	$C_{10}H_{22}$ 142.28	15869-94-0	160.8	4-01-00-00479 0.7324[25]	1.4139[20]
8383	Octane, 4,4-dimethyl- 4,4-Dimethyloctane	$C_{10}H_{22}$ 142.28	15869-95-1	157.5	4-01-00-00480 0.7312[25]	1.4144[20]
8384	Octane, 4,5-dimethyl- 4,5-Dimethyloctane	$C_{10}H_{22}$ 142.28	15869-96-2	162.15	4-01-00-00480 0.7432[25]	1.4190[20]
8385	Octanedinitrile Suberonitrile	$C_8H_{12}N_2$ 136.20	629-40-3 -3.5	185[15]	3-02-00-01768 0.954[25]	1.4436[20]
8386	Octanedioic acid Suberic acid	$C_8H_{14}O_4$ 174.20	505-48-6 144	8833 219[20]	4-02-00-02028	H_2O 1; eth 5; bz 5
8387	Octanedioic acid, diethyl ester	$C_{12}H_{22}O_4$ 230.30	2050-23-9 5.9	282.6	4-02-00-02029 0.9811[20]	H_2O 1; EtOH 3; eth 3; ctc 2 1.4328[20]
8388	Octanedioic acid, dimethyl ester	$C_{10}H_{18}O_4$ 202.25	1732-09-8 -3.1	268	4-02-00-02029 1.0217[20]	H_2O 1; EtOH 3; eth 3; ace 3 1.4341[20]

No.	Name / Synonym	Mol. Form. / Mol. Wt.	CAS RN / mp/°C	Merck No. / bp/°C	Beil. Ref. / den/g cm⁻³	Solubility / n_D
8389	1,8-Octanediol	$C_8H_{18}O_2$ 146.23	629-41-4 63	172[20]	4-01-00-02592	H_2O 2; EtOH 4; eth 2; bz 3
8390	2,4-Octanediol	$C_8H_{18}O_2$ 146.23	90162-24-6	117-8[8]	3-01-00-02219 0.918[25]	1.4422[25]
8391	4,5-Octanediol, (±)-	$C_8H_{18}O_2$ 146.23	22520-40-7 30	110[8]	4-01-00-02593	ctc 2 1.4419[25]
8392	1,7-Octanediol, 3,7-dimethyl-	$C_{10}H_{22}O_2$ 174.28	107-74-4	265	3-01-00-02233 0.937[20]	bz 2; tol 2 1.4599[20]
8393	2,4-Octanedione Valerylacetone	$C_8H_{14}O_2$ 142.20	14090-87-0	79-83[20]	4-01-00-03706 0.9233[20]	1.4559[20]
8394	2,7-Octanedione	$C_8H_{14}O_2$ 142.20	1626-09-1 44	110[13]	4-01-00-03707	EtOH 4
8395	3,6-Octanedione	$C_8H_{14}O_2$ 142.20	2955-65-9 35.5	98[14]	4-01-00-03708	EtOH 4
8396	4,5-Octanedione	$C_8H_{14}O_2$ 142.20	5455-24-3	168	4-01-00-03708 0.934[0]	EtOH 4; eth 4; ace 4
8397	4,5-Octanedione, dioxime Dipropylglyoxime	$C_8H_{16}N_2O_2$ 172.23	61050-68-8 186.5	sub	2-01-00-00846	eth 4; EtOH 4
8398	3,5-Octanedione, 6,6,7,7,8,8,8-heptafluoro-2,2-dimethyl-	$C_{10}H_{11}F_7O_2$ 296.18	17587-22-3 38	46-7[5]	1.273[25]	1.3766[20]
8399	2,3-Octanedione, 3-oxime 3-Oximino-2-octanone	$C_8H_{15}NO_2$ 157.21	584-92-9 59	133[13]	0-01-00-00795	eth 4
8400	3,6-Octanedione, 2,2,7,7-tetramethyl- 2,2,7,7-Tetramethyl-3,6-octanedione	$C_{12}H_{22}O_2$ 198.31	27610-88-4 2.5	116[17]; 57[0.5]	4-01-00-03734 0.900[27]	EtOH 4 1.4400[20]
8401	Octane, 1-(ethenyloxy)-	$C_{10}H_{20}O$ 156.27	929-62-4	58[4]	4-01-00-02056 0.8020[20]	1.4268[20]
8402	Octane, 1-ethoxy-	$C_{10}H_{22}O$ 158.28	929-61-3 12.5	186.3	4-01-00-01759 0.7847[20]	EtOH 4 1.4127[20]
8403	Octane, 3-ethyl- 3-Ethyloctane	$C_{10}H_{22}$ 142.28	5881-17-4	166.5	4-01-00-00476 0.7359[25]	1.4156[20]
8404	Octane, 4-ethyl- 4-Ethyloctane	$C_{10}H_{22}$ 142.28	15869-86-0	163.7	4-01-00-00476 0.7343[25]	1.4151[20]
8405	Octane, 1-fluoro- Octyl fluoride	$C_8H_{17}F$ 132.22	463-11-6	142.4	4-01-00-00418 0.8108[20]	1.3935[20]
8406	Octane, 1-iodo-	$C_8H_{17}I$ 240.13	629-27-6 -45.7	225.5	4-01-00-00425 1.3297[20]	EtOH 3; eth 3 1.4889[20]
8407	Octane, 2-iodo-, (±)- (±)-2-Octyl iodide	$C_8H_{17}I$ 240.13	36049-78-2	210; 95[16]	4-01-00-00425 1.3251[20]	H_2O 1; EtOH 3; eth 3; lig 3 1.4896[20]
8408	Octane, 2-methyl- 2-Methyloctane	C_9H_{20} 128.26	3221-61-2 -80.3	143.2	4-01-00-00454 0.7095[25]	H_2O 1; EtOH 3; eth 3; ctc 2 1.4031[20]
8409	Octane, 3-methyl- 3-Methyloctane	C_9H_{20} 128.26	2216-33-3 -107.6	144.2	0.717[25]	1.4040[25]
8410	Octane, 4-methyl- 4-Methyloctane	C_9H_{20} 128.26	2216-34-4 -113.3	142.4	0.716[25]	1.4039[25]
8411	Octanenitrile Caprylnitrile	$C_8H_{15}N$ 125.21	124-12-9 -45.6	205.2	4-02-00-00993 0.8136[20]	eth 4 1.4203[20]
8412	Octane, 1-nitro- 1-Nitrooctane	$C_8H_{17}NO_2$ 159.23	629-37-8 15	208.5	4-01-00-00426 0.9346[20]	1.4322[20]
8413	Octane, 2-nitro-, (+)-	$C_8H_{17}NO_2$ 159.23	116836-12-5	102-5[23]	4-01-00-00426 0.9166[20]	1.4280[20]
8414	Octane, 3-nitro-	$C_8H_7NO_2$ 149.15	4609-92-1	91[25]	4-01-00-00427 0.917[25]	1.4210[25]
8415	Octane, octadecafluoro- Perfluorooctane	C_8F_{18} 438.06	307-34-6	105.9	4-01-00-00418 1.73[20]	1.282[20]
8416	Octane, 1,1'-oxybis-	$C_{16}H_{34}O$ 242.45	629-82-3 -7.6	283	4-01-00-01760 0.8063[20]	H_2O 2; EtOH 3; eth 3; ctc 3 1.4327[20]
8417	Octane, 1,1'-thiobis-	$C_{16}H_{34}S$ 258.51	2690-08-6	180[10]	4-01-00-01768 0.842[25]	1.4610[20]
8418	1-Octanethiol Octyl mercaptan	$C_8H_{18}S$ 146.30	111-88-6 -49.2	199.1	4-01-00-01767 0.8433[20]	EtOH 3; ctc 2 1.4540[20]
8419	2-Octanethiol, (±)- 2-Octanethiol (DL)	$C_8H_{18}S$ 146.30	10435-81-1 -79	186.4	4-01-00-01777 0.8366[20]	bz 4; eth 4; EtOH 4 1.4504[20]
8420	Octane, 2,2,4-trimethyl- Octane, 2,2,4-trimethyl	$C_{11}H_{24}$ 156.31	18932-14-4	168.5	4-01-00-00494 0.7381[20]	1.4155[20]
8421	Octane, 2,4,7-trimethyl 2,4,7-Trimethyloctane	$C_{11}H_{24}$ 156.31	62016-38-0	168.1		
8422	Octanoic acid Caprylic acid	$C_8H_{16}O_2$ 144.21	124-07-2 16.3	1765 239	4-02-00-00982 0.9106[20]	H_2O 2; EtOH 5; chl 5; CH_3CN 5 1.4285[20]
8423	Octanoic acid, 2-amino-, (±)-	$C_8H_{17}NO_2$ 159.23	644-90-6 270	sub	4-04-00-02799	H_2O 2; EtOH 2; eth 2; bz 2
8424	Octanoic acid, anhydride	$C_{16}H_{30}O_3$ 270.41	623-66-5 -1	282.5	3-02-00-00796 0.9065[18]	ace 4; eth 4; EtOH 4 1.4358[18]
8425	Octanoic acid, 2-bromo-	$C_8H_{15}BrO_2$ 223.11	2623-82-7	140[5]; 118[0.5]	4-02-00-01000 1.2785[24]	1.4613[24]
8426	Octanoic acid, 8-bromo-	$C_8H_{15}BrO_2$ 223.11	17696-11-6 38.5	147[2]	4-02-00-01000	bz 4; eth 4; EtOH 4
8427	Octanoic acid, butyl ester	$C_{12}H_{24}O_2$ 200.32	589-75-3 -42.9	240.5	4-02-00-00987 0.8628[20]	ace 4; eth 4; EtOH 4 1.4232[25]

No.	Name Synonym	Mol. Form. Mol. Wt.	CAS RN mp/°C	Merck No. bp/°C	Beil. Ref. den/g cm^{-3}	Solubility n_D
8428	Octanoic acid, ethyl ester Ethyl octanoate	$C_{10}H_{20}O_2$ 172.27	106-32-1 -43.1	3736 208.5	4-02-00-00987 0.866[18]	H$_2$O 1; EtOH 4; eth 4; ctc 2 1.4178[20]
8429	Octanoic acid, 8-fluoro- Caprylic acid, 8-fluoro	$C_8H_{15}FO_2$ 162.20	353-25-3 35	132[4]	4-02-00-00994	ace 4; eth 4; EtOH 4
8430	Octanoic acid, heptyl ester	$C_{15}H_{30}O_2$ 242.40	4265-97-8 -10.6	290.5	3-02-00-00794 0.8596[20]	1.4340[20]
8431	Octanoic acid, hexyl ester	$C_{14}H_{28}O_2$ 228.38	1117-55-1 -30.6	277.4	3-02-00-00794 0.8603[20]	H$_2$O 1; EtOH 3; eth 3; ace 3 1.4323[25]
8432	Octanoic acid, 2-hydroxy-	$C_8H_{16}O_3$ 160.21	617-73-2 70	162[10]	3-03-00-00642	H$_2$O 2; EtOH 4; eth 4; chl 2
8433	Octanoic acid, 3-methyl- Caprylic acid, 3-methyl	$C_9H_{18}O_2$ 158.24	6061-10-5	135[16]	3-02-00-00826 0.899[23]	1.4298[25]
8434	Octanoic acid, methyl ester Methyl caprylate	$C_9H_{18}O_2$ 158.24	111-11-5 -40	192.9	4-02-00-00986 0.8775[20]	H$_2$O 1; EtOH 4; eth 4; ctc 2 1.4170[20]
8435	Octanoic acid, 1-methylethyl ester	$C_{11}H_{22}O_2$ 186.29	5458-59-3 	93.8[10]	4-02-00-00987 0.8555[20]	1.4147[25]
8436	Octanoic acid, 2-methyl-3-oxo-, ethyl ester Caprylic acid, 2-methyl-3-oxo, ethyl ester	$C_{11}H_{20}O_3$ 200.28	10488-94-5	128-9[12]	4-03-00-01629 0.963[0]	EtOH 4
8437	Octanoic acid, octyl ester	$C_{16}H_{32}O_2$ 256.43	2306-88-9 -18.1	306.8	4-02-00-00988 0.8554[20]	ace 4; eth 4; EtOH 4 1.4352[20]
8438	Octanoic acid, pentadecafluoro-, methyl ester	$C_9H_3F_{15}O_2$ 428.10	376-27-2	158	4-02-00-00995 1.684[20]	1.304[27]
8439	Octanoic acid, pentyl ester	$C_{13}H_{26}O_2$ 214.35	638-25-5 -34.8	260.2	3-02-00-00794 0.8613[20]	H$_2$O 1; EtOH 3; eth 3; ace 3 1.4262[25]
8440	Octanoic acid, 1,2,3-propanetriyl ester	$C_{27}H_{50}O_6$ 470.69	538-23-8 10	233	4-02-00-00991 0.9540[20]	H$_2$O 1; EtOH 5; eth 4; bz 4 1.4482[20]
8441	Octanoic acid, propyl ester	$C_{11}H_{22}O_2$ 186.29	624-13-5 -46.2	226.4	4-02-00-00987 0.8659[20]	ace 4; eth 4; EtOH 4 1.4191[25]
8442	Octanoic acid, 3,5,7,8-tetrachloro- 2,2,3,4,4,5,6,6,7,8,8-undecafluoro- Octanoic acid, 3,5,7,8-tetrachloro- 2,2,3,4,4,5,6,6,7,8,8-undecafluoro	$C_8HCl_4F_{11}O_2$ 479.89	2923-68-4 154.0-6.5[10]		4-02-00-00998 1.899[20]	1.3980[20]
8443	1-Octanol Caprylic alcohol	$C_8H_{18}O$ 130.23	111-87-5 -15.5	6674 195.1	4-01-00-01756 0.8262[25]	H$_2$O 1; EtOH 5; eth 5; ctc 3 1.4295[20]
8444	2-Octanol, (±)-	$C_8H_{18}O$ 130.23	4128-31-8 -31.6	6675 180	4-01-00-01770 0.8193[20]	H$_2$O 2; EtOH 3; eth 3; ace 3 1.4203[20]
8445	3-Octanol Ethylamylcarbinol	$C_8H_{18}O$ 130.23	589-98-0 -45	171	3-01-00-01723 0.8258[20]	
8446	4-Octanol, (±)-	$C_8H_{18}O$ 130.23	74778-22-6 -40.7	176.3	4-01-00-01779 0.8186[20]	H$_2$O 2; EtOH 3; eth 3; ctc 2 1.4248[20]
8447	2-Octanol, acetate, (±)- Acetic acid, 2-octyl ester(DL)	$C_{10}H_{20}O_2$ 172.27	74112-36-0 	195	4-02-00-00165 0.8626[14]	eth 4; EtOH 4 1.4146[20]
8448	1-Octanol, 8-chloro-	$C_8H_{17}ClO$ 164.68	23144-52-7 	139[19]	4-01-00-01766 	eth 4; EtOH 4 1.4563[25]
8449	1-Octanol, 3,7-dimethyl-, (±)- Geraniol, tetrahydro (DL)	$C_{10}H_{22}O$ 158.28	59204-02-3 	212.5; 106[13]	4-01-00-01830 0.8308[10]	H$_2$O 1; EtOH 3; eth 3; ace 3 1.4367[20]
8450	2-Octanol, 2,6-dimethyl- Tetrahydromyrcenol	$C_{10}H_{22}O$ 158.28	18479-57-7 	80.5[10]	4-01-00-01829 0.8023[25]	1.4220[25]
8451	3-Octanol, 2,3-dimethyl- Amylisopropylmethylcarbinol	$C_{10}H_{22}O$ 158.28	19781-10-3 	69-70[5]	4-01-00-01828 0.8401[20]	1.4380[20]
8452	3-Octanol, 2,7-dimethyl-	$C_{10}H_{22}O$ 158.28	66719-55-9 	193.5	1-01-00-00214 0.8152[20]	1.4302[20]
8453	3-Octanol, 3,6-dimethyl-	$C_{10}H_{22}O$ 158.28	151-19-9 -67.5	202.2	1-01-00-00214 0.8347[22]	1.4370[20]
8454	3-Octanol, 3,6-dimethyl-(+-)- Linalool, tetrahydro (DL)	$C_{10}H_{22}O$ 158.28	57706-88-4 31.5	196.5	3-01-00-01767 0.8280[20]	EtOH 4 1.4335[20]
8455	4-Octanol, 2,4-dimethyl-	$C_{10}H_{22}O$ 158.28	33933-79-8 	97[27]	4-01-00-01828 0.8238[20]	1.4338[20]
8456	4-Octanol, 2,5-dimethyl-	$C_{10}H_{22}O$ 158.28	66719-53-7 	102-4[34]	1-01-00-00213 0.8125[25]	1.4260[25]
8457	4-Octanol, 2,6-dimethyl-	$C_{10}H_{22}O$ 158.28	66719-54-8 	195	2-01-00-00460 0.8114[20]	1.4270[20]
8458	4-Octanol, 2,7-dimethyl- 2,7-Dimethyloctan-4-ol	$C_{10}H_{22}O$ 158.28	19781-11-4 	96[18]	3-01-00-01769 0.8139[20]	1.4280[20]
8459	3-Octanol, 3-ethyl-	$C_{10}H_{22}O$ 158.28	2051-32-3 	199	3-01-00-01766 0.8361[25]	EtOH 4 1.4390[20]
8460	1-Octanol, 8-fluoro-	$C_8H_{17}FO$ 148.22	408-27-5 	106-7[10]	4-01-00-01766 0.945[20]	eth 4; EtOH 4 1.4248[25]
8461	1-Octanol, 3-methyl- 3-Methyl-1-octanol	$C_9H_{20}O$ 144.26	38514-02-2 	110[25]	4-01-00-01806 0.827[4]	1.4328[25]
8462	1-Octanol, 4-methyl- 4-Methyl-1-octanol	$C_9H_{20}O$ 144.26	38514-03-3 	106[17]	4-01-00-01807 0.820[27]	1.4335[25]
8463	1-Octanol, 7-methyl-	$C_9H_{20}O$ 144.26	2430-22-0 64.5	206	4-01-00-01806 0.8260[25]	1.4316[25]
8464	2-Octanol, 2-methyl- 2-Methyl-2-octanol	$C_9H_{20}O$ 144.26	628-44-4 	178	4-01-00-01805 0.8210[20]	H$_2$O 1; EtOH 3; eth 3 1.4280[20]
8465	2-Octanol, 3-methyl-	$C_9H_{20}O$ 144.26	27644-49-1 	75[15]	3-01-00-01748 0.831[27]	1.437[27]
8466	3-Octanol, 3-methyl- Amylethylmethylcarbinol	$C_9H_{20}O$ 144.26	5340-36-3 	97.5[50]; 36[3]	4-01-00-01807 0.8258[25]	1.4257[25]

No.	Name / Synonym	Mol. Form. / Mol. Wt.	CAS RN mp/°C	Merck No. bp/°C	Beil. Ref. den/g cm^{-3}	Solubility n_D
8467	3-Octanol, 4-methyl-	$C_9H_{20}O$ 144.26	66793-80-4	132-3[20]	3-01-00-01750 0.8437[25]	1.4372[20]
8468	3-Octanol, 6-methyl- 6-Methyl-3-octanol	$C_9H_{20}O$ 144.26	40225-75-0	81-3[15]	3-01-00-01749 0.8320[28]	1.4372[20]
8469	4-Octanol, 4-methyl- 4-Methyl-4-octanol	$C_9H_{20}O$ 144.26	23418-37-3	180; 78[15]	4-01-00-01807 0.8267[20]	1.4327[20]
8470	4-Octanol, 5-methyl-	$C_9H_{20}O$ 144.26	59734-23-5	74-6[9]	1-01-00-00211 0.8156[25]	1.4262[25]
8471	2-Octanone Hexyl methyl ketone	$C_8H_{16}O$ 128.21	111-13-7 -16	4632 172.5	4-01-00-03339 0.820[20]	H_2O 2; EtOH 5; eth 5 1.4151[20]
8472	3-Octanone Ethyl amyl ketone	$C_8H_{16}O$ 128.21	106-68-3	3721 167.5	4-01-00-03341 0.822[25]	H_2O 1; EtOH 5; eth 5 1.4150[20]
8473	4-Octanone Butyl propyl ketone	$C_8H_{16}O$ 128.21	589-63-9	163	4-01-00-03342 0.8146[25]	H_2O 1; EtOH 5; eth 5; ctc 3 1.4173[14]
8474	4-Octanone, 5-hydroxy- Butyroin	$C_8H_{16}O_2$ 144.21	496-77-5 -10	1595 185	4-01-00-04042 0.9107[16]	1.4345[16]
8475	2-Octanone, 3-methyl- 3-Methyl-2-octanone	$C_9H_{18}O$ 142.24	6137-08-2	64-5[18]	4-01-00-03356 0.832[27]	1.424[27]
8476	3-Octanone, 4-methyl- 4-Methyl-3-octanone	$C_9H_{18}O$ 142.24	6137-15-1	62-4[10]	4-01-00-03357 0.820[25]	1.4186[25]
8477	3-Octanone, 7-methyl- 7-Methyl-3-octanone	$C_9H_{18}O$ 142.24	5408-57-1	182.5	3-01-00-02889 0.8353[25]	1.4748
8478	4-Octanone, 3-methyl-	$C_9H_{18}O$ 142.24	20754-04-5	174	4-01-00-03357 0.829[14]	1.4200[14]
8479	4-Octanone, 7-methyl- 7-Methyl-4-octanone	$C_9H_{18}O$ 142.24	20809-46-5	178	4-01-00-03356 0.8239[20]	H_2O 1; EtOH 3; eth 3 1.4210[20]
8480	1-Octanone, 1-phenyl-	$C_{14}H_{20}O$ 204.31	1674-37-9 22.8	285; 164[15]	4-07-00-00798 0.9360[30]	EtOH 3; eth 3
8481	Octanoyl chloride	$C_8H_{15}ClO$ 162.66	111-64-8 -63	195.6	4-02-00-00992 0.9535[15]	eth 3 1.4335[20]
8482	Octasiloxane, octadecamethyl- Octadecamethyloctasiloxane	$C_{18}H_{54}O_7Si_8$ 607.31	556-69-4	153[5.1]	4-04-00-04130 0.913[25]	bz 4; peth 4; lig 4 1.3970[20]
8483	1,3,5,7-Octatetraene Octatetraene	C_8H_{10} 106.17	1482-91-3 50	sub	4-01-00-01124	peth 3; HOAc 3
8484	2,4,6-Octatrienal Octatrienal	$C_8H_{10}O$ 122.17	17609-31-3 56.8	75[3]	4-01-00-03591 0.8891[60]	eth 4; EtOH 4
8485	1,3,6-Octatriene, 3,7-dimethyl- Ocimene	$C_{10}H_{16}$ 136.24	13877-91-3	6661 177 dec; 73[21]	3-01-00-01052 0.800[20]	bz 4; eth 4; EtOH 4; chl 4 1.4862[20]
8486	1,3,7-Octatriene, 3,7-dimethyl-	$C_{10}H_{16}$ 136.24	502-99-8	177 dec	3-01-00-01050 0.8000[20]	H_2O 1; EtOH 3; eth 3; chl 3 1.4862[20]
8487	2,4,6-Octatriene, 2,6-dimethyl-, (E,E)-	$C_{10}H_{16}$ 136.24	3016-19-1 -35.4	188; 91[20]	4-01-00-01106 0.8118[20]	1.5446[20]
8488	2,4,6-Octatriene, 2,6-dimethyl-, (E,Z)-	$C_{10}H_{16}$ 136.24	7216-56-0 -20.6	89[20]	4-01-00-01106 0.8060[20]	1.5446[20]
8489	2,4,6-Octatriene, (E,E,E)-	C_8H_{12} 108.18	15192-80-0 52	147.5	4-01-00-01101 0.7961[23]	EtOH 3; chl 3; lig 3 1.5131[27]
8490	6-Octenal, 3,7-dimethyl- Resorcinol, 4-acetyl-	$C_{10}H_{18}O$ 154.25	106-23-0 47[1]	2331	1-01-00-00387 0.85[20]	H_2O 2; EtOH 3
8491	1-Octene Caprylene	C_8H_{16} 112.22	111-66-0 -101.7	1764 121.2	4-01-00-00874 0.7149[20]	H_2O 1; EtOH 5; eth 3; ace 3 1.4087[20]
8492	2-Octene, (E)-	C_8H_{16} 112.22	13389-42-9 -87.7	125	4-01-00-00879 0.7199[20]	H_2O 1; EtOH 3; eth 3; ace 3 1.4132[20]
8493	2-Octene, (Z)-	C_8H_{16} 112.22	7642-04-8 -100.2	125.6	4-01-00-00878 0.7243[20]	H_2O 1; EtOH 3; eth 3; ace 3 1.4150[20]
8494	3-Octene, (E)-	C_8H_{16} 112.22	14919-01-8 -110	123.3	4-01-00-00880 0.7152[20]	H_2O 1; EtOH 3; eth 3; ace 3 1.4126[20]
8495	3-Octene, (Z)-	C_8H_{16} 112.22	14850-22-7 -126	122.9	4-01-00-00880 0.7159[20]	ace 4; bz 4; eth 4; EtOH 4 1.4135[20]
8496	4-Octene, (E)-	C_8H_{16} 112.22	14850-23-8 -93.8	122.3	4-01-00-00880 0.7141[20]	H_2O 1; EtOH 3; eth 3; ace 3 1.4114[20]
8497	4-Octene, (Z)-	C_8H_{16} 112.22	7642-15-1 -118.7	122.5	4-01-00-00880 0.7212[20]	ace 4; bz 4; eth 4; EtOH 4 1.4148[20]
8498	1-Octene, 2-chloro-	$C_8H_{15}Cl$ 146.66	31283-43-9	169	3-01-00-00840 0.9273[0]	ace 4; bz 4; eth 4
8499	2-Octene, 2-chloro-	$C_8H_{15}Cl$ 146.66	90202-21-4	167.5; 82[50]	4-01-00-00879 0.8914[16]	ace 4; bz 4; eth 4; EtOH 4 1.4424[16]
8500	4-Octene, 4-chloro-, (Z)- 4-Octene, 4-chloro (cis)	$C_8H_{15}Cl$ 146.66	7321-48-4	165.3; 75[50]	4-01-00-00881 0.8912[20]	eth 4 1.4447[20]
8501	2-Octene, 4-chloro-	$C_8H_{15}Cl$ 146.66	116668-50-9	153	3-01-00-00840 0.8924[20]	ace 4; bz 4; eth 4; chl 4 1.4952[20]
8502	1-Octene, 3,7-dimethyl- 2,6-Dimethyl-7-octene	$C_{10}H_{20}$ 140.27	4984-01-4	154	3-01-00-00862 0.7396[20]	1.4212[20]
8503	2-Octene, 2,6-dimethyl- 2,6-Dimethyl-2-octene	$C_{10}H_{20}$ 140.27	4057-42-5	163	3-01-00-00861 0.746[22]	1.425[22]
8504	4-Octene-2,7-dione	$C_8H_{12}O_2$ 140.18	2130-23-6 30.5	72[0.03]	4-01-00-03757	eth 4; EtOH 4
8505	1-Octene, 2-methyl- 2-Methyl-1-octene	C_9H_{18} 126.24	4588-18-5 -77.8	144.8	4-01-00-00896 0.7343[20]	1.4184[20]
8506	2-Octene, 3-methyl-	C_9H_{18} 126.24	113426-22-5	145	3-01-00-00854 0.7409[25]	1.4247[20]

No.	Name Synonym	Mol. Form. Mol. Wt.	CAS RN	mp/°C	Merck No. bp/°C	Beil. Ref. den/g cm⁻³	Solubility n_D
8507	3-Octene, 7-methyl-	C_9H_{18} 126.24	86668-33-9		142	3-01-00-00854 0.7278[20]	1.4168[20]
8508	4-Octene, 2-methyl-	C_9H_{18} 126.24	64501-77-5		139	3-01-00-00854 0.7379[20]	1.4181[20]
8509	1-Octen-4-ol	$C_8H_{12}O$ 124.18	40575-42-6		172; 68[10]	3-01-00-01942 0.8373[20]	1.4383[20]
8510	2-Octen-1-ol	$C_8H_{16}O$ 128.21	22104-78-5		87-9[11]	4-01-00-02164 0.850[20]	1.4470[20]
8511	2-Octen-4-ol	$C_8H_{16}O$ 128.21	4798-61-2		175	3-01-00-01943 0.8393[20]	1.4395
8512	4-Octen-1-ol	$C_8H_{16}O$ 128.21	67700-26-9		95[17]	4-01-00-02166 0.842[20]	1.4462[20]
8513	5-Octen-1-ol	$C_8H_{16}O$ 128.21	90200-83-2		91.5[14]	4-01-00-02166 0.8492[21]	1.4478[21]
8514	6-Octen-1-ol, 3,7-dimethyl-, (±)- (±)-Citronellol	$C_{10}H_{20}O$ 156.27	26489-01-0	2332	224; 99[10]	4-01-00-02188 0.8560[20]	H₂O 2; EtOH 5; eth 5 1.4543[20]
8515	6-Octen-3-ol, 3,7-dimethyl-	$C_{10}H_{20}O$ 156.27	18479-51-1		94[14]	2-01-00-00494 0.8695[15]	1.4569[15]
8516	7-Octen-1-ol, 3,7-dimethyl-, (S)- (-)-α-Citronellol	$C_{10}H_{20}O$ 156.27	6812-78-8	8186	114-5[12]	2-01-00-00497 0.8549[20]	eth 4; EtOH 4 1.4556[20]
8517	1-Octen-4-ol, 4-methyl-	$C_9H_{18}O$ 142.24	62108-06-9		178	2-01-00-00492 0.8440[20]	1.4399
8518	2-Octen-4-ol, 7-methyl-	$C_9H_{18}O$ 142.24	4798-64-5		85[12]	2-01-00-00492 0.8402[14]	1.4449[14]
8519	3-Octen-2-ol, 2-methyl-, (Z)- 3-Octen-2-ol, 2-methyl (cis)	$C_9H_{18}O$ 142.24	18521-07-8		177	4-01-00-02177 0.8378[20]	1.4426[20]
8520	4-Octen-3-ol, 3-methyl-	$C_9H_{18}O$ 142.24	90676-55-4		69-70[10]	4-01-00-02178 0.812[14]	1.4414[14]
8521	5-Octen-4-ol, 5-methyl- 3-Octen-5-ol, 4-methyl	$C_9H_{18}O$ 142.24	36903-92-1		89[16]	2-01-00-00492 0.8468[25]	1.4446[25]
8522	1-Octen-3-one, 5-methyl-	$C_9H_{16}O$ 140.23	116836-13-6		72-3[16]	3-01-00-03018 0.8450[19]	1.4380[19]
8523	5-Octen-4-one, 2-methyl- 2-Methyl-5-octen-4-one	$C_9H_{16}O$ 140.23	17577-93-4		183	4-01-00-03504 0.839[25]	1.4387[25]
8524	5-Octen-4-one, 7-methyl-	$C_9H_{16}O$ 140.23	32064-78-1		68-78[24]	3-01-00-03018 0.9011[12]	1.4748
8525	7-Octen-4-one, 2-methyl- 1-Octen-5-one, 7-methyl	$C_9H_{16}O$ 140.23	54298-97-4		62-3[14]	2-01-00-00801 0.8361[12]	1.4288[12]
8526	1-Octen-3-yne	C_8H_{12} 108.18	17679-92-4		62[60]	4-01-00-01100 0.7830[20]	1.4592[20]
8527	7-Octen-5-yn-4-ol, 4-methyl- 1-Octen-3-yne-5-ol, 5-methyl	$C_9H_{14}O$ 138.21	39118-35-9		80[13]	3-01-00-02034 0.8851[15]	EtOH 4 1.4785[20]
8528	Octhilinone 3(2H)-Isothiazolone, 2-octyl-	$C_{11}H_{19}NOS$ 213.34	26530-20-1	6677	120[0.01]		
8529	1-Octyne Hexylacetylene	C_8H_{14} 110.20	629-05-0	-79.3	126.3	4-01-00-01034 0.7461[20]	H₂O 1; EtOH 3; eth 3 1.4159[20]
8530	2-Octyne Methylpentylacetylene	C_8H_{14} 110.20	2809-67-8	-61.6	137.6	4-01-00-01035 0.7596[20]	H₂O 1; EtOH 3; eth 3 1.4278[20]
8531	3-Octyne	C_8H_{14} 110.20	15232-76-5	-103.9	133.1	4-01-00-01036 0.7529[20]	H₂O 1; EtOH 3; eth 3 1.4250[20]
8532	4-Octyne Dipropylacetylene	C_8H_{14} 110.20	1942-45-6	-101	131.6	4-01-00-01037 0.7509[20]	H₂O 1; EtOH 3; eth 3 1.4248[20]
8533	1-Octyne, 1-chloro- 1-Chloro-1-octyne	$C_8H_{13}Cl$ 144.64	64531-26-6		61-2[17]	3-01-00-01005 0.912[20]	eth 4; EtOH 4 1.445[20]
8534	3-Octyne, 2-chloro-2-methyl- 2-Chloro-2-methyloct-3-yne	$C_9H_{15}Cl$ 158.67	20599-21-7		68[15]	4-01-00-01048 0.8929[20]	H₂O 1; eth 4 1.4480[20]
8535	3-Octyne, 2,2-dimethyl- 2,2-Dimethyl-3-octyne	$C_{10}H_{18}$ 138.25	19482-57-6		79[60]	3-01-00-01018 0.7491[20]	eth 4 1.4270[20]
8536	2-Octyne, 1-methoxy- Methyl (2-octyn-1-yl) ether	$C_9H_{16}O$ 140.23	18495-23-3		77[19]	3-01-00-01996 0.8370[25]	eth 4 1.4380[20]
8537	3-Octyne, 7-methyl- 7-Methyl-3-octyne	C_9H_{16} 124.23	37050-06-9		87[99]	3-01-00-01014 0.7599[20]	eth 4 1.4280[20]
8538	2-Octynoic acid Pentylpropiolic acid	$C_8H_{12}O_2$ 140.18	5663-96-7	4	148[19]; 114[2]	4-02-00-01714 0.9623[13]	1.4595[20]
8539	3-Octynoic acid	$C_8H_{12}O_2$ 140.18	57074-96-1	18	110[2]	4-02-00-01715	1.4577[25]
8540	5-Octynoic acid	$C_8H_{12}O_2$ 140.18	76469-08-4	8	111[2]	4-02-00-01715 0.9787[27]	1.4540[25]
8541	7-Octynoic acid	$C_8H_{12}O_2$ 140.18	10297-09-3	19	123[2]	4-02-00-01714	1.4502[25]
8542	2-Octynoic acid, methyl ester	$C_9H_{14}O_2$ 154.21	111-12-6		107[20]; 94[10]	4-02-00-01715 0.926[20]	1.4464[20]
8543	2-Octyn-1-ol 2-Octynol	$C_8H_{14}O$ 126.20	20739-58-6	-18	98-9[15]	4-01-00-02256 0.8805[20]	eth 4 1.4556[20]
8544	1-Octyn-3-ol, 3-methyl- 1-Octyne-3-ol, 3-methyl	$C_9H_{16}O$ 140.23	23580-51-0		75[10]	4-01-00-02268 0.863[10]	1.443[10]
8545	Olean-12-en-28-oic acid, 3,23-dihydroxy-, (3β,4α)- Hederagenin	$C_{30}H_{48}O_4$ 472.71	465-99-6	4540 333			

No.	Name Synonym	Mol. Form. Mol. Wt.	CAS RN mp/°C	Merck No. bp/°C	Beil. Ref. den/g cm⁻³	Solubility n_D
8546	Olean-12-en-28-oic acid, 3,16-dihydroxy-23-oxo-, (3β,4α,16α)- Quillaic acid	$C_{30}H_{46}O_5$ 486.69	631-01-6 294	8049	3-10-00-04652	ace 4; eth 4; py 4; EtOH 4
8547	Olean-12-en-28-oic acid, 3-hydroxy-, (3β)- Oleanolic acid	$C_{30}H_{48}O_3$ 456.71	508-02-1 310 dec	6787 sub 280	4-10-00-01164	H_2O 1; EtOH 2; eth 2; ace 2
8548	Olean-12-en-3-ol, (3β)- β-Amyrin	$C_{30}H_{50}O$ 426.73	559-70-6 197	654 260⁰⁵	4-06-00-04195	H_2O 1; EtOH 2; eth 3; bz 3
8549	Orcein	$C_{28}H_{24}N_2O_7$ 500.51	1400-62-0	6818	2-06-00-00876	
8550	L-Ornithine (S)-2,5-Diaminopentanoic acid	$C_5H_{12}N_2O_2$ 132.16	70-26-8 140	6826	4-04-00-02644	H_2O 4; EtOH 4
8551	L-Ornithine, N5-(aminocarbonyl)- Citrulline	$C_6H_{13}N_3O_3$ 175.19	372-75-8 235.5	2333	4-04-00-02647	H_2O 3; EtOH 1; MeOH 1
8552	Orthoformic acid, triisopentyl ester Isopentyl alcohol, orthoformate (3:1)	$C_{16}H_{34}O_3$ 274.44	5337-70-2 268 dec		4-02-00-00030 0.8628²⁰	eth 4; EtOH 4 1.4238²⁰
8553	Oryzalin Benzenesulfonamide, 4-(dipropylamino)-3,5-dinitro-	$C_{12}H_{18}N_4O_6S$ 346.36	19044-88-3 141	6840		
8554	7-Oxabicyclo[4.1.0]heptane	$C_6H_{10}O$ 98.14	286-20-4 <-10	131.5	5-17-01-00203 0.9663²⁰	H_2O 1; EtOH 4; eth 4; ace 4 1.4519²⁰
8555	7-Oxabicyclo[4.1.0]heptane, 3-ethenyl- Cyclohexane, 1,2-epoxy-4-vinyl-	$C_8H_{12}O$ 124.18	106-86-5 <-100	169; 70²⁰	4-17-00-00314 0.9581²⁰	1.4700²⁰
8556	7-Oxabicyclo[2.2.1]heptane, 1-methyl-4-(1-methylethyl)-	$C_{10}H_{18}O$ 154.25	470-67-7 1	173.5	5-17-01-00273 0.8997²⁰	H_2O 2; EtOH 5; eth 5; bz 3 1.4562²⁰
8557	7-Oxabicyclo[4.1.0]heptane, 3-oxiranyl- 4-Vinyl-1-cyclohexene dioxide	$C_8H_{12}O_2$ 140.18	106-87-6 <-55	227	5-19-01-00295 1.0966²⁰	H_2O 4 1.4787²⁰
8558	6-Oxabicyclo[3.1.0]hexane	C_5H_8O 84.12	285-67-6	102	5-17-01-00190 0.964²⁵	1.4336²⁰
8559	3-Oxabicyclo[3.2.1]octane-2,4-dione, 1,8,8-trimethyl-, (±)-	$C_{10}H_{14}O_3$ 182.22	595-30-2 221	270	5-17-11-00089 1.194²⁰	bz 4
8560	2-Oxabicyclo[2.2.2]octane, 1,3,3-trimethyl- Eucalyptol	$C_{10}H_{18}O$ 154.25	470-82-6 1.5	3851 176.4	5-17-01-00273 0.9267²⁰	H_2O 1; EtOH 3; eth 3; ctc 2 1.4586²⁰
8561	6-Oxabicyclo[3.2.1]oct-3-ene, 4,7,7-trimethyl-, (±)- (±)-Pinol	$C_{10}H_{16}O$ 152.24	60761-04-8 183.5		4-17-00-00327 0.9515²⁰	eth 4; EtOH 4 1.4695²⁰
8562	Oxacyclohexadecan-2-one Exaltolide	$C_{15}H_{28}O_2$ 240.39	106-02-5	3867	5-17-09-00106	
8563	1,3,4-Oxadiazole 1-Oxa-3,4-diazacyclopentadiene	$C_2H_2N_2O$ 70.05	288-99-3 150	6859		1.4300²⁵
8564	1,2,4-Oxadiazole, 3,5-diphenyl- 3,5-Diphenyl-1,2,4-oxadiazole	$C_{14}H_{10}N_2O$ 222.25	888-71-1 109	290; 210¹⁷	4-27-00-07210	bz 4; eth 4; EtOH 4
8565	Oxadiazon 1,2,4-Oxadiazolidin-3,5-dione, 2-tert-butyl-4-(2,4-dichloro-5-isopropoxyphenyl)-	$C_{15}H_{18}Cl_2N_2O_3$ 345.23	19666-30-9 90	6860		
8566	Oxadixyl N-(2,6-Dimethylphenyl)-2-methoxy-N-(2-oxo-3-oxazolidinyl)acetamide	$C_{14}H_{18}N_2O_4$ 278.31	77732-09-3 104			
8567	16(15H)-Oxaerythrinan-15-one, 1,2,6,7-tetrahydro-14,17-dihydro-3-methoxy-, (3β)- β-Erythroidine	$C_{16}H_{19}NO_3$ 273.33	466-81-9 99.5	3625	4-27-00-03568	H_2O 3; EtOH 4; eth 3; bz 4
8568	Oxamyl Ethanimidothioic acid, 2-(dimethylamino)-N-[[(methylamino)carbonyl]oxy]-2-oxo-, methyl ester	$C_7H_{13}N_3O_3S$ 219.26	23135-22-0 109 dec	6873	0.97²⁵	
8569	1,4-Oxathiane	C_4H_8OS 104.17	15980-15-1 147		5-19-01-00033 1.1174²⁰	H_2O 2
8570	1,2-Oxathiolane, 2,2-dioxide Propane sultone	$C_3H_6O_3S$ 122.14	1120-71-4		5-19-01-00005	chl 3
8571	Oxayohimban-16-carboxylic acid, 16,17-didehydro-19-methyl-, methyl ester, (19α)- Raubasine	$C_{21}H_{24}N_2O_3$ 352.43	483-04-5 258 dec	8129	4-27-00-07927	H_2O 1; MeOH 3
8572	Oxayohimbanium 3,4,5,6,16,17-hexahydro-16-(methoxycarbonyl)-19-methyl-, (19α)- Serpentine	$C_{21}H_{20}N_2O_3$ 348.40	18786-24-8 175	8415		H_2O 1; EtOH 3; eth 3; ace 3
8573	2H-1,3-Oxazine-2,4(3H)-dione, 5,5-diethyldihydro- Diethadione	$C_8H_{13}NO_3$ 171.20	702-54-5 97.5	3096	4-27-00-03263	
8574	2-Oxazolamine, 4,5-dihydro-5-phenyl- Aminorex	$C_9H_{10}N_2O$ 162.19	2207-50-3 137	490		
8575	Oxazole 1,3-Oxazole	C_3H_3NO 69.06	288-42-6 69.5		4-27-00-00960	1.4285¹⁷
8576	Oxazole, 4,5-dihydro-2-methyl-	C_4H_7NO 85.11	1120-64-5 111		4-27-00-00921 1.005²⁵	1.4340²⁰
8577	Oxazole, 2,4-dimethyl- 2,4-Dimethyloxazole	C_5H_7NO 97.12	7208-05-1 108		4-27-00-00980 0.9352¹⁵	H_2O 4; eth 4; EtOH 4 1.4166¹⁵
8578	Oxazole, 2,5-dimethyl- 2,5-Dimethyloxazole	C_5H_7NO 97.12	23012-11-5 117.5		4-27-00-00985 0.9958²¹	H_2O 4 1.4385²¹

No.	Name / Synonym	Mol. Form. / Mol. Wt.	CAS RN / mp/°C	Merck No. / bp/°C	Beil. Ref. / den/g cm^{-3}	Solubility / n_D
8579	Oxazole, 2,4-diphenyl- / 2,4-Diphenyloxazole	$C_{15}H_{11}NO$ / 221.26	838-41-5 / 103	/ 339	4-27-00-01435	EtOH 4; eth 4; bz 4
8580	Oxazole, 2,5-diphenyl-	$C_{15}H_{11}NO$ / 221.26	92-71-7 / 74	/ 360	4-27-00-01435 / 1.0940[100]	H_2O 1; EtOH 4; eth 4; chl 2 / 1.6231[100]
8581	Oxazole, 4,5-diphenyl- / 4,5-Diphenyloxazole	$C_{15}H_{11}NO$ / 221.26	4675-18-7 / 44	/ 193[15.] 132[0.05]	4-27-00-01438	H_2O 2; EtOH 4; eth 4; con acid 3 / 1.6283[100]
8582	Oxazole, 5-methoxy-2-phenyl- / 5-Methoxy-2-phenyloxazole	$C_{10}H_9NO_2$ / 175.19	40527-16-0	/ 140[10]	4-27-00-01958	ctc 3 / 1.584[23]
8583	Oxazole, 4-methyl- / 4-Methyloxazole	C_4H_5NO / 83.09	693-93-6	/ 88	4-27-00-00968 / 1.015[25]	1.4317[20]
8584	2,4-Oxazolidinedione, 5,5-dimethyl- / Dimethadione	$C_5H_7NO_3$ / 129.12	695-53-4 / 76.5	3201	4-27-00-03237	
8585	2,4-Oxazolidinedione, 5,5-dipropyl- / 5,5-Dipropyl-2,4-oxazolidinedione	$C_9H_{15}NO_3$ / 185.22	512-12-9 / 42.5	3352 / 148-50[3]	4-27-00-03269	
8586	2,4-Oxazolidinedione, 5-methyl-3-(2-propenyl)- / Aloxidone	$C_7H_9NO_3$ / 155.15	526-35-2	305 / 138[35]; 86[0.5]	4-27-00-03214	1.4688[25]
8587	2,4-Oxazolidinedione, 3,5,5-trimethyl- / Trimethadione	$C_6H_9NO_3$ / 143.14	127-48-0 / 46	9620 / 78-80[5]	4-27-00-03237	H_2O 3; EtOH 4; eth 4; ace 4
8588	2-Oxazolidinone, 5-[(2-methoxyphenoxy)methyl]- / Mephenoxalone	$C_{11}H_{13}NO_4$ / 223.23	70-07-5 / 144	5739		
8589	2-Oxazolidinone, 3-[[(5-nitro-2-furanyl)methylene]amino]- / Furazolidone	$C_8H_7N_3O_5$ / 225.16	67-45-8 / 255	4210	4-27-00-02530	
8590	Oxepane	$C_6H_{12}O$ / 100.16	592-90-5 /	/ 119	5-17-01-00090 / 0.89[25]	1.4400[20]
8591	2-Oxepanone / Caprolactone	$C_6H_{10}O_2$ / 114.14	502-44-3 / -18	/ 215	5-17-09-00034 / 1.0693[20]	EtOH 3; eth 3; ace 3 / 1.4611[20]
8592	Oxetane / Trimethylene oxide	C_3H_6O / 58.08	503-30-0 / -97	9630 / 47.6	5-17-01-00011 / 0.8930[25]	H_2O 5; EtOH 5; eth 3; ace 4 / 1.3961[20]
8593	Oxetane, 3,3-dimethyl- / 3,3-Dimethyloxetane	$C_5H_{10}O$ / 86.13	6921-35-3	/ 80.6	5-17-01-00085 / 0.834[25]	1.3965[20]
8594	Oxetane, 2-methyl-	C_4H_8O / 72.11	2167-39-7 /	/ 59	5-17-01-00054 / 0.841[25]	1.3885[20]
8595	2-Oxetanone / β-Propiolactone	$C_3H_4O_2$ / 72.06	57-57-8 / -33.4	7832 / 162	5-17-09-00003 / 1.1460[20]	eth 5; chl 3 / 1.4105[20]
8596	3-Oxetanone	$C_3H_4O_2$ / 72.06	6704-31-0			
8597	2-Oxetanone, 3,3-dimethyl-4-(1-methylethylidene)-	$C_8H_{12}O_2$ / 140.18	3173-79-3 / -18	/ 170	5-17-09-00181 / 0.947[25]	1.4380[20]
8598	2-Oxetanone, 4-methylene- / Diketene	$C_4H_4O_2$ / 84.07	674-82-8 / -6.5	/ 126.1	5-17-09-00115 / 1.0877[20]	1.4379[20]
8599	Oxirane / Ethylene oxide	C_2H_4O / 44.05	75-21-8 / -111.7	3758 / 10.6	5-17-01-00003 / 0.8821[10]	H_2O 3; EtOH 3; eth 3; ace 3 / 1.3579[7]
8600	Oxirane, (bromomethyl)-, (±)- / Propane, 3-bromo-1,2-epoxy (DL)	C_3H_5BrO / 136.98	82584-73-4 / -40	/ 137	5-17-01-00023 / 1.615[14]	H_2O 1; EtOH 3; eth 3; bz 3 / 1.4841[20]
8601	Oxirane, 2,2'-[1,4-butanediylbis(oxymethylene)]bis-	$C_{10}H_{18}O_4$ / 202.25	2425-79-8	/ 266; 155[11]	5-17-03-00023 / 1.1[25]	1.4611[20]
8602	Oxirane, (butoxymethyl)- / Butyl glycidyl ether	$C_7H_{14}O_2$ / 130.19	2426-08-6	/ 169; 75[26]	5-17-03-00011 / 0.918[20]	
8603	Oxiranecarboxaldehyde / Glycidaldehyde	$C_3H_4O_2$ / 72.06	765-34-4 / -62	/ 112.5	5-17-09-00006	1.4265[20]
8604	Oxiranecarboxylic acid, 3-methyl-3-phenyl-, ethyl ester	$C_{12}H_{14}O_3$ / 206.24	77-83-8	/ 273.5	5-18-06-00385 / 1.044[20]	1.5182[20]
8605	Oxirane, (chloromethyl)-, (±)- / DL-α-Epichlorohydrin	C_3H_5ClO / 92.52	13403-37-7 / -26	3563 / 117; 62[100]	5-17-01-00020 / 1.1812[20]	H_2O 1; EtOH 5; eth 5; bz 3 / 1.4361[20]
8606	Oxirane, 2-(chloromethyl)-2-methyl-	C_4H_7ClO / 106.55	598-09-4	/ 122	5-17-01-00059 / 1.1011[20]	H_2O 4 / 1.4310[20]
8607	Oxirane, 2,2-dimethyl-	C_4H_8O / 72.11	558-30-5	/ 52	5-17-01-00058 / 0.8112[20]	EtOH 3; eth 3 / 1.3712[22]
8608	Oxirane, 2,3-dimethyl-, cis-	C_4H_8O / 72.11	1758-33-4 / -80	/ 60	5-17-01-00061 / 0.8226[25]	eth 4; ace 4; bz 4; os 4 / 1.3802[20]
8609	Oxirane, 2,3-dimethyl-, trans-	C_4H_8O / 72.11	21490-63-1 / -85	/ 56.5	5-17-01-00061 / 0.8010[25]	eth 4; ace 4; bz 4; os 4 / 1.3736[20]
8610	Oxirane, [(1,1-dimethylethoxy)methyl]-	$C_7H_{14}O_2$ / 130.19	7665-72-7 / -70	/ 152	5-17-03-00011 / 0.898[20]	
8611	Oxirane, [[4-(1,1-dimethylethyl)phenoxy]methyl]-	$C_{13}H_{18}O_2$ / 206.28	3101-60-8	/ 165-70[14]	5-17-03-00017 / 1.038[25]	1.5145[20]
8612	Oxirane, 2-(2,2-dimethylpropyl)-2-methyl-	$C_8H_{16}O$ / 128.21	107-48-2 / -64	/ 140.9	4-17-00-00112 / 0.8272[20]	bz 4; eth 4 / 1.4097[20]
8613	Oxirane, ethenyl-	C_4H_6O / 70.09	930-22-3	/ 68	5-17-01-00180 / 0.9006[25]	EtOH 3; eth 3; bz 3; os 3 / 1.4168[20]
8614	Oxirane, (ethoxymethyl)- / 2,3-Epoxypropyl ethyl ether	$C_5H_{10}O_2$ / 102.13	4016-11-9	/ 128	5-17-03-00010 / 0.9700[20]	H_2O 3; EtOH 3; eth 3; ctc 2 / 1.4320[20]

No.	Name / Synonym	Mol. Form. / Mol. Wt.	CAS RN	mp/°C	Merck No. / bp/°C	Beil. Ref. / den/g cm⁻³	Solubility / n_D
8615	Oxirane, ethyl- / 1,2-Epoxybutane	$C_4H_8O_2$ / 88.11	106-88-7	-150	/ 63.3	5-17-01-00056 / 0.8297^{20}	EtOH 4; eth 5; ace 4; os 4 / 1.3851^{20}
8616	Oxirane, (fluoromethyl)-	C_3H_5FO / 76.07	503-09-3		/ 85.5	5-17-01-00019 / 1.067^{25}	
8617	Oxirane, (iodomethyl)-	C_3H_5IO / 183.98	624-57-7		/ 161; 62^{24}	5-17-01-00024 / 1.982^{24}	eth 4; EtOH 4 / 1.3715^{20}
8618	Oxiranemethanol / Glycidol	$C_3H_6O_2$ / 74.08	556-52-5	4385 / 167 dec; $66^{2.5}$		5-17-03-00009 / 1.1143^{25}	H_2O 5; EtOH 5; eth 3; ace 3 / 1.4287^{20}
8619	Oxiranemethanol, 3-phenyl-	$C_9H_{10}O_2$ / 150.18	21915-53-7	26.5	/ 134^4	4-17-00-01349 / 1.1512^{27}	1.5427^{27}
8620	Oxirane, (methoxymethyl)-	$C_4H_8O_2$ / 88.11	930-37-0		/ 113	5-17-03-00010 / 0.9890^{20}	H_2O 4; ace 4; eth 4; EtOH 4 / 1.4320^{20}
8621	Oxirane, methyl- / Propylene oxide	C_3H_6O / 58.08	75-56-9	-111.9	7869 / 35	5-17-01-00017 / 0.859^0	H_2O 4; eth 4; EtOH 4 / 1.3670^{20}
8622	Oxirane, 2-methyl-2-phenyl-	$C_9H_{10}O$ / 134.18	2085-88-3		/ $83-5^{15}$	5-17-01-00591 / 1.0280^{20}	1.5232^{20}
8623	Oxirane, (phenoxymethyl)- / Phenyl glycidyl ether	$C_9H_{10}O_2$ / 150.18	122-60-1		/ 243	5-17-03-00013 / 1.1109^{21}	1.5307^{21}
8624	Oxirane, phenyl- / Phenylethylene oxide	C_8H_8O / 120.15	96-09-3	-35.6	/ 194.1	5-17-01-00577 / 1.0523^{16}	H_2O 1; EtOH 3; eth 3; chl 3 / 1.5342^{20}
8625	Oxirane, 2,2'-[1,3-phenylenebis(oxymethylene)]bis- / Diglycidyl resorcinol ether	$C_{12}H_{14}O_4$ / 222.24	101-90-6	42.5	/ $143-51^{0.4}$	5-17-03-00024 / 1.2183^{30}	1.5408^{20}
8626	Oxirane, [(2-propenyloxy)methyl]- / Allyl glycidyl ether	$C_6H_{10}O_2$ / 114.14	106-92-3		/ 154	5-17-03-00012 / 0.9698^{20}	1.4332^{20}
8627	Oxirane, tetramethyl-	$C_6H_{12}O$ / 100.16	5076-20-0		/ $90-4^{74.5}$	5-17-01-00108 / 0.8156^{16}	H_2O 3 / 1.3984^{16}
8628	Oxirane, (trichloromethyl)-	$C_3H_3Cl_3O$ / 161.41	3083-23-6		/ 149; 44^{13}	5-17-01-00023 / 1.495^{20}	eth 4; chl 3 / 1.4737^{25}
8629	Oxirane, trifluoro(trifluoromethyl)- / Perfluoropropylene oxide	C_3F_6O / 166.02	428-59-1			5-17-01-00020	
8630	Oxyacanthan, 6',12'-dimethoxy-2,2'-dimethyl-6,7-[methylenebis(oxy)]- / Cepharanthine	$C_{37}H_{38}N_2O_6$ / 606.72	481-49-2	150	1981	4-27-00-09061	
8631	Oxyacanthan-12'-ol, 6,6',7-trimethoxy-2,2'-dimethyl- / Oxyacanthine	$C_{37}H_{40}N_2O_6$ / 608.73	548-40-3	216.5	6906	2-27-00-00894	H_2O 1; EtOH 3; eth 3; bz 3
8632	Oxydemeton-methyl / Phosphorothioic acid, S-[2-(ethylsulfinyl)ethyl] O,O-dimethyl ester	$C_6H_{15}O_4PS_2$ / 246.29	301-12-2	<-20	/ $106^{0.01}$	/ 1.289^{20}	
8633	Oxyfluorfen / Benzene, 2-chloro-1-(3-ethoxy-4-nitrophenoxy)-4-(trifluoromethyl)-	$C_{15}H_{11}ClF_3NO_4$ / 361.70	42874-03-3	84	6916 / 358.2 dec	/ 1.35^{73}	
8634	Paraformaldehyde / Formaldehyde polymer	$(CH_2O)_x$	30525-89-4	164 dec	6974	5-19-09-00103	H_2O 2; EtOH 1; eth 1; alk 3
8635	Pendimethalin / Benzenamine, N-(1-ethylpropyl)-3,4-dimethyl-2,6-dinitro-	$C_{13}H_{19}N_3O_4$ / 281.31	40487-42-1	56	7026 / dec	/ 1.19^{25}	
8636	Pentacene / Benzo[b]naphthacene	$C_{22}H_{14}$ / 278.35	135-48-8	257	7057	4-05-00-02721	H_2O 1; bz 2; $PhNO_2$ 3; os 2
8637	Pentacosane	$C_{25}H_{52}$ / 352.69	629-99-2	54	/ 401.9; 282^{40}	4-01-00-00582 / 0.8012^{20}	bz 3; chl 3 / 1.4491^{20}
8638	Pentacosane, 13-phenyl- / Benzene, (1-dodecyltridecyl)-	$C_{31}H_{56}$ / 428.79	6006-90-2	31.7	/ 235^{10}	4-05-00-01241 / 0.8537^{20}	1.4787^{20}
8639	Pentacosane, 13-undecyl-	$C_{36}H_{74}$ / 506.98	55517-89-0	9.7	/ 307^{10}	4-01-00-00600 / 0.8168^{20}	1.4567^{20}
8640	Pentadecanal / Pentadecaldehyde	$C_{15}H_{30}O$ / 226.40	2765-11-9	24.5	/ 185^{25}	4-01-00-03391	ace 4; eth 4; EtOH 4
8641	1-Pentadecanamine	$C_{15}H_{33}N$ / 227.43	2570-26-5	37.3	/ 307.6	4-04-00-00817 / 0.8104^{20}	eth 4; EtOH 4 / 1.4480^{20}
8642	Pentadecane	$C_{15}H_{32}$ / 212.42	629-62-9	9.9	/ 270.6	4-01-00-00529 / 0.7685^{20}	H_2O 1; EtOH 4; eth 4 / 1.4315^{20}
8643	Pentadecane, 1-bromo-	$C_{15}H_{31}Br$ / 291.32	629-72-1	19	/ 322	4-01-00-00531 / 1.0675^{20}	H_2O 1; ace 3; chl 4 / 1.4611^{20}
8644	Pentadecane, 2,6,10,14-tetramethyl- / Pristane	$C_{19}H_{40}$ / 268.53	1921-70-6		7757 / 296	3-01-00-00570 / 0.783^{20}	bz 4; eth 4; chl 4; peth 4 / 1.4379^{20}
8645	Pentadecanoic acid / Pentadecylic acid	$C_{15}H_{30}O_2$ / 242.40	1002-84-2	52.3	257^{100}; 158^1	4-02-00-01147 / 0.8423^{80}	H_2O 1; EtOH 4; eth 3; ace 4 / 1.4254^{80}
8646	Pentadecanoic acid, methyl ester / Methyl pentadecanoate	$C_{16}H_{32}O_2$ / 256.43	7132-64-1	18.5	/ 153.5	3-02-00-00936 / 0.8618^{25}	EtOH 3; eth 3; os 3 / 1.4390^{25}
8647	2-Pentadecanol / sec-Pentadecyl alcohol	$C_{15}H_{32}O$ / 228.42	1653-34-5	35	/ 299	4-01-00-01871 / 0.8328^{20}	1.4463^{20}
8648	3-Pentadecanol	$C_{15}H_{32}O$ / 228.42	53346-71-7	38.8	/ 388; 163^{12}	4-01-00-01871	1.4227^{80}
8649	1-Pentadecanol, 2-methyl- / 2-Methyl-1-pentadecanol	$C_{16}H_{34}O$ / 242.45	25354-98-7	12.5	/ $185-7^{15}$	3-01-00-01827 / 0.8320^{25}	1.4453^{25}
8650	6-Pentadecanol, 6-methyl-	$C_{16}H_{34}O$ / 242.45	108836-86-8		/ $199-200^{50}$	3-01-00-01828 / 0.8316^{25}	1.4446^{25}

No.	Name Synonym	Mol. Form. Mol. Wt.	CAS RN mp/°C	Merck No. bp/°C	Beil. Ref. den/g cm^{-3}	Solubility n_D
8651	2-Pentadecanone	$C_{15}H_{30}O$ 226.40	2345-28-0 39.5	294	4-01-00-03391 0.8182[39]	
8652	8-Pentadecanone	$C_{15}H_{30}O$ 226.40	818-23-5 43	291	4-01-00-03392 0.8180[39]	EtOH 3; eth 3; bz 3; ctc 3
8653	1-Pentadecene	$C_{15}H_{30}$ 210.40	13360-61-7 -2.8	268.2	4-01-00-00926 0.7764[20]	H_2O 1; ace 3 1.4389[20]
8654	7-Pentadecene	$C_{15}H_{30}$ 210.40	15430-98-5 	114[3.2]	3-01-00-00875 0.7765[20]	1.4420[20]
8655	1-Pentadecen-3-ol	$C_{15}H_{30}O$ 226.40	99814-65-0 27.5	150-2[6]	3-01-00-01961 0.834[32]	1.4481[32]
8656	1-Pentadecyne	$C_{15}H_{28}$ 208.39	765-13-9 10	268	4-01-00-01072 0.7928[20]	ace 4 1.4419[20]
8657	2,4-Pentadienal, 5-phenyl- 5-Phenyl-2,4-pentadienal	$C_{11}H_{10}O$ 158.20	13466-40-5 42.5	160[3]; 133[1.0]	4-07-00-01179	H_2O 1; EtOH 5; eth 4; bz 5
8658	1,2-Pentadiene Ethylallene	C_5H_8 68.12	591-95-7 -137.3	44.9	4-01-00-00993 0.6926[20]	EtOH 5; eth 5; ace 5; bz 5 1.4209[20]
8659	1,4-Pentadiene	C_5H_8 68.12	591-93-5 -148.8	26	4-01-00-00998 0.6608[20]	H_2O 1; EtOH 4; eth 4; ace 4 1.3888[20]
8660	2,3-Pentadiene 1,3-Dimethylallene	C_5H_8 68.12	591-96-8 -125.6	48.2	4-01-00-00999 0.6950[20]	H_2O 1; EtOH 5; eth 5; ace 5 1.4284[20]
8661	1,3-Pentadiene, (E)- trans-Piperylene	C_5H_8 68.12	2004-70-8 -87.4	42	4-01-00-00994 0.6710[25]	1.4301[20]
8662	1,3-Pentadiene, (Z)- cis-Piperylene	C_5H_8 68.12	1574-41-0 -140.8	44.1	4-01-00-00994 0.6910[20]	EtOH 5; eth 5; ace 5; bz 5 1.4363[20]
8663	1,3-Pentadiene, 3-chloro- 3-Chloro-1,3-pentadiene	C_5H_7Cl 102.56	37710-49-9 	100	3-01-00-00962 0.9576[20]	ace 4; bz 4; eth 4 1.4785[20]
8664	1,2-Pentadiene, 1-chloro-3-methyl-	C_6H_9Cl 116.59	32337-74-9 	68-70[100]	3-01-00-00990 0.9562[20]	eth 4
8665	1,3-Pentadiene, 2,4-dimethyl-	C_7H_{12} 96.17	1000-86-8 -114	93.2	4-01-00-01034 0.7343[23]	1.4390[23]
8666	1,2-Pentadiene, 3-methyl- 1-Ethyl-1-methylallene	C_6H_{10} 82.15	7417-48-3 	70	4-01-00-01021 0.710[25]	1.425[20]
8667	1,2-Pentadiene, 4-methyl-	C_6H_{10} 82.15	13643-05-5 	70	2-01-00-00232 0.703[25]	1.424[20]
8668	1,3-Pentadiene, 2-methyl-, (Z)- cis-2-Methyl-1,3-pentadiene	C_6H_{10} 82.15	1501-60-6 -117.6	76	4-01-00-01019 0.714[25]	1.446[20]
8669	1,3-Pentadiene, 3-methyl-	C_6H_{10} 82.15	4549-74-0 	77	4-01-00-01021 0.730[25]	1.452[20]
8670	1,3-Pentadiene, 4-methyl- 1,1-Dimethylbutadiene	C_6H_{10} 82.15	926-56-7 	76.5	4-01-00-01020 0.7181[20]	1.4532[20]
8671	1,4-Pentadiene, 2-methyl- 2-Methyl-1,4-pentadiene	C_6H_{10} 82.15	763-30-4 	56	4-01-00-01020 0.689[25]	1.405[20]
8672	2,3-Pentadiene, 2-methyl-	C_6H_{10} 82.15	3043-33-2 	72	4-01-00-01020 0.706[25]	1.425[20]
8673	2,4-Pentadienenitrile	C_5H_5N 79.10	1615-70-9 -60	136.5	4-02-00-01692 0.8444[21]	1.4880
8674	2,4-Pentadienenitrile, (E)-	C_5H_5N 79.10	2180-68-9 -43	41[13]	4-02-00-01695 0.8576[20]	eth 3; ace 3 1.4986[20]
8675	2,4-Pentadienenitrile, (Z)-	C_5H_5N 79.10	2180-69-0 -64	49[32]; 32[13]	4-02-00-01695 0.8541[26]	eth 3; ace 3 1.4855[20]
8676	2,4-Pentadienoic acid	$C_5H_6O_2$ 98.10	626-99-3 79	112 dec	3-02-00-01451	H_2O 4; EtOH 4; eth 4; bz 4
8677	2,4-Pentadienoic acid, 5-(1,3-benzodioxol-5-yl)- 2,4-Pentadienoic acid, 5(3,4-metylenedioxyphenyl)	$C_{12}H_{10}O_4$ 218.21	5285-18-7 215	7437	2-19-00-00300	H_2O 1; EtOH 3; eth 2; bz 2
8678	2,4-Pentadienoic acid, 5-(1,3-benzodioxol-5-yl)-, (E,E)-	$C_{12}H_{10}O_4$ 218.21	136-72-1 215.8	sub	5-19-07-00382	EtOH 4
8679	2,4-Pentadienoic acid, 5-(1-hydroxy-2,6,6-trimethyl-4-oxo-2-cyclohexen-1-yl)-3-methyl-, [S-(Z,E)]- Abscisic acid	$C_{15}H_{20}O_4$ 264.32	21293-29-8 6 160	sub 120		ace 4; eth 4; chl 4
8680	2,4-Pentadienoic acid, 5-phenyl-, ethyl ester Ethyl cinnamylideneacetate	$C_{13}H_{14}O_2$ 202.25	1552-95-0 26	149-50[4]	2-09-00-00441 1.0467[20]	eth 4; EtOH 4 1.5768[80]
8681	1,4-Pentadien-3-ol	C_5H_8O 84.12	922-65-6 	115.5	4-01-00-02229 0.860[23]	1.4400[17]
8682	1,4-Pentadien-3-one, 1,5-di-2-furanyl-	$C_{13}H_{10}O_3$ 214.22	886-77-1 60.5	181[4]	5-19-04-00459	eth 4; EtOH 4; chl 4
8683	1,4-Pentadien-3-one, 1,5-diphenyl- Dibenzalacetone	$C_{17}H_{14}O$ 234.30	538-58-9 113 dec	2988 dec	4-07-00-01747	H_2O 1; EtOH 2; eth 2; ace 3
8684	1,3-Pentadiyne Methyldiacetylene	C_5H_4 64.09	4911-55-1 -38.5	55	4-01-00-01117 0.7909[20]	H_2O 1; eth 3; bz 3; chl 3 1.4431[21]
8685	1,5-Pentalenedione, hexahydro- Bicyclo[3.3.0]octane-2,6-dione	$C_8H_{10}O_2$ 138.17	62353-69-9 45	86-8[0.2]	3-07-00-03280 1.1290[60]	1.4877[54]
8686	Pentalene, octahydro-, cis-	C_8H_{14} 110.20	1755-05-1 <-80	139	4-05-00-00277 0.8638[25]	H_2O 1; EtOH 3 1.4595[25]
8687	Pentalene, octahydro-, trans-	C_8H_{14} 110.20	5597-89-7 -30	132	4-05-00-00277 0.8624[18]	H_2O 1; EtOH 3 1.4625[18]
8688	2(1H)-Pentalenone, hexahydro- Bicyclo[3.3.0]octane-2-one	$C_8H_{12}O$ 124.18	56180-61-1 	72[13]	4-07-00-00156 1.0097[20]	ace 4; EtOH 4 1.4790[20]

No.	Name Synonym	Mol. Form. Mol. Wt.	CAS RN	Merck No. mp/°C bp/°C	Beil. Ref. den/g cm^{-3}	Solubility n_D
8689	2(1H)-Pentalenone, hexahydro-, cis- cis-Bicyclo[3.3.0]octan-3-one	$C_8H_{12}O$ 124.18	19915-11-8	50$^{2.3}$	4-07-00-00156 1.0660^{25}	1.4766^{25}
8690	Pentanal Valeraldehyde	$C_5H_{10}O$ 86.13	110-62-3 -91.5	9813 103	4-01-00-03268 0.8095^{20}	H_2O 2; EtOH 3; eth 3 1.3944^{20}
8691	Pentanal, 3-hydroxy-2-methyl-	$C_6H_{12}O_2$ 116.16	615-30-5	94-6^{23}	4-01-00-04022 0.986^{25}	H_2O 4; ace 4; eth 4; EtOH 4 1.4502^{20}
8692	Pentanal, 3-hydroxy-2,2,4-trimethyl-	$C_8H_{16}O_2$ 144.21	918-79-6	118-20^{14}	4-01-00-04049 0.9482^{20}	1.4501^{20}
8693	Pentanal, 2-methyl- 2-Methylvaleraldehyde	$C_6H_{12}O$ 100.16	123-15-9	117	4-01-00-03304	H_2O 3; eth 3; ace 3; ctc 2
8694	Pentanal, 3-methyl-	$C_6H_{12}O$ 100.16	15877-57-3	121	3-01-00-02836 0.811^{25}	1.4026^{20}
8695	Pentanal, 2-oxo-	$C_5H_8O_2$ 100.12	7332-93-6	112	4-01-00-03658	H_2O 4; eth 4; EtOH 4 1.4043^{25}
8696	Pentanal, 4-oxo-	$C_5H_8O_2$ 100.12	626-96-0 <-21	187 dec	4-01-00-03659 1.0134^{21}	H_2O 4; ace 4; eth 4; EtOH 4 1.4257^{22}
8697	Pentanal, 2-propyl- Valeraldehyde, 2-propyl-	$C_8H_{16}O$ 128.21	18295-59-5	160	3-01-00-02880 0.8347^{15}	1.4142^{15}
8698	Pentanamide	$C_5H_{11}NO$ 101.15	626-97-1 106	225	4-02-00-00874 0.8735^{110}	H_2O 4; EtOH 4; eth 4; chl 2 1.4183^{110}
8699	Pentanamide, N,N-dimethyl-	$C_7H_{15}NO$ 129.20	6225-06-5 -51	141^{100}	3-04-00-00127 0.896^{25}	H_2O 4; eth 4; EtOH 4 1.4419^{25}
8700	1-Pentanamine Amylamine	$C_5H_{13}N$ 87.16	110-58-7 -55	630 104.3	4-04-00-00674 0.7544^{20}	H_2O 5; EtOH 5; eth 5; ace 4 1.448^{20}
8701	2-Pentanamine	$C_5H_{13}N$ 87.16	625-30-9	92	4-04-00-00689 0.7384^{20}	ace 4; bz 4; eth 4; EtOH 4 1.4027^{20}
8702	3-Pentanamine	$C_5H_{13}N$ 87.16	616-24-0	89	4-04-00-00692 0.7487^{20}	EtOH 3; chl 2 1.4063^{20}
8703	1-Pentanamine, N,N-dimethyl-	$C_7H_{17}N$ 115.22	26153-88-8	123	4-04-00-00675 0.743^{20}	eth 4 1.4083^{20}
8704	2-Pentanamine, 2,4-dimethyl-	$C_7H_{17}N$ 115.22	64379-30-2	122	4-04-00-00749 0.7719^{20}	1.4009^{20}
8705	1-Pentanamine, N,N-dipentyl-	$C_{15}H_{33}N$ 227.43	621-77-2	242.5	4-04-00-00676 0.7907^{20}	H_2O 1; EtOH 3; eth 3; acid 3 1.4366^{20}
8706	1-Pentanamine, 3-methyl- 3-Methylpentylamine	$C_6H_{15}N$ 101.19	42245-37-4	122	3-04-00-00366 0.767^{20}	1.4196^{25}
8707	1-Pentanamine, 4-methyl-	$C_6H_{15}N$ 101.19	5344-20-7 -94.4	125	4-04-00-00727 0.758^{52}	
8708	1-Pentanamine, N-pentyl- Diamylamine	$C_{10}H_{23}N$ 157.30	2050-92-2	202.5	4-04-00-00676 0.7771^{20}	H_2O 2; EtOH 4; eth 5; ace 3 1.4272^{20}
8709	Pentane	C_5H_{12} 72.15	109-66-0 -129.7	7072 36.0	4-01-00-00303 0.6262^{20}	H_2O 2; EtOH 5; eth 5; ace 5 1.3575^{20}
8710	Pentane, 1-bromo- Amyl bromide	$C_5H_{11}Br$ 151.05	110-53-2 -95	637 129.8	4-01-00-00312 1.2182^{20}	H_2O 1; EtOH 3; eth 5; bz 3 1.4447^{20}
8711	Pentane, 2-bromo- 2-Pentyl bromide	$C_5H_{11}Br$ 151.05	107-81-3 -95.5	117.4	4-01-00-00312 1.2075^{20}	bz 4; eth 4; EtOH 4; chl 4 1.4413^{20}
8712	Pentane, 3-bromo- 3-Pentyl bromide	$C_5H_{11}Br$ 151.05	1809-10-5 -126.2	118.6	4-01-00-00313 1.214^{20}	H_2O 1; EtOH 3; eth 3; bz 3 1.4441^{20}
8713	Pentane, 1-bromo-5-fluoro- 1-Bromo-5-fluoropentane	$C_5H_{10}BrF$ 169.04	407-97-6	162	4-01-00-00313 1.3604^{25}	eth 4; EtOH 4 1.4406^{25}
8714	Pentane, 1-bromo-2-methyl- 2-Methylpentyl bromide	$C_6H_{13}Br$ 165.07	25346-33-2	141	4-01-00-00361 1.1624^{20}	eth 4; chl 4 1.4495^{20}
8715	Pentane, 1-bromo-3-methyl- 3-Methylpentyl bromide	$C_6H_{13}Br$ 165.07	51116-73-5	148	4-01-00-00365 1.1829^{20}	H_2O 1; eth 3; chl 4 1.4496^{20}
8716	Pentane, 1-bromo-4-methyl-	$C_6H_{13}Br$ 165.07	626-88-0	145	4-01-00-00361 1.1683^{20}	eth 4; chl 4 1.4490
8717	Pentane, 2-bromo-2-methyl- 1-Bromo-1,1-dimethylbutane	$C_6H_{13}Br$ 165.07	4283-80-1	142.5	4-01-00-00361	eth 4; chl 4 1.442^{23}
8718	Pentane, 3-bromo-3-methyl- 3-Bromo-3-methylpentane	$C_6H_{13}Br$ 165.07	25346-31-0	130; 76^{100}	4-01-00-00365 1.1835^{20}	eth 4; chl 4 1.4525^{20}
8719	Pentane, 1-chloro- Amyl chloride	$C_5H_{11}Cl$ 106.60	543-59-9 -99	642 107.8	4-01-00-00309 0.8820^{20}	H_2O 1; EtOH 5; eth 5; bz 3 1.4127^{20}
8720	Pentane, 2-chloro-, (+)- Pentane, 2-chloro (d)	$C_5H_{11}Cl$ 106.60	29882-57-3 -137	96.9	4-01-00-00309 0.8698^{20}	H_2O 1; EtOH 3; eth 3; bz 3 1.4069^{20}
8721	Pentane, 3-chloro- 1-Ethylpropyl chloride	$C_5H_{11}Cl$ 106.60	616-20-6 -105	97.5	4-01-00-00309 0.8731^{20}	H_2O 1; EtOH 3; eth 3; ace 2 1.4082^{20}
8722	Pentane, 2-chloro-2,3-dimethyl-	$C_7H_{15}Cl$ 134.65	59889-45-1	38-9^{20}	4-01-00-00406 1.4264^{20}	bz 4; eth 4
8723	Pentane, 2-chloro-2,4-dimethyl-	$C_7H_{15}Cl$ 134.65	35951-33-8	129; 33^{20}	4-01-00-00408 0.861^{20}	eth 4 1.4180^{20}
8724	Pentane, 3-chloro-2,3-dimethyl-	$C_7H_{15}Cl$ 134.65	595-38-0	136 dec; 41^{20}	4-01-00-00406 0.882^{22}	eth 4; chl 4 1.4318^{20}
8725	Pentane, 4-chloro-2,2-dimethyl-	$C_7H_{15}Cl$ 134.65	33429-72-0	93^{250}	4-01-00-00404 0.855^{20}	eth 4; chl 4 1.4180^{20}
8726	Pentane, 3-chloro-3-ethyl-	$C_7H_{15}Cl$ 134.65	994-25-2	143.5	4-01-00-00403 0.8856^{20}	eth 4 1.4400^{20}
8727	Pentane, 3-chloro-3-ethyl-2,2-dimethyl- Pentane 3-chloro-2,2-dimethyl-3-ethyl	$C_9H_{19}Cl$ 162.70	86661-53-2	53^6 dec	4-01-00-00462	eth 4; chl 4 1.4528^{25}
8728	Pentane, 1-chloro-5-fluoro-	$C_5H_{10}ClF$ 124.59	407-98-7	143.2	4-01-00-00309 1.0325^{25}	eth 4; EtOH 4 1.4120^{23}

No.	Name Synonym	Mol. Form. Mol. Wt.	CAS RN mp/°C	Merck No. bp/°C	Beil. Ref. den/g cm⁻³	Solubility n_D
8729	Pentane, 2-chloro-2-methyl-	C₆H₁₃Cl 120.62	4325-48-8	110 dec; 36[15]	4-01-00-00360 0.863[20]	eth 4 1.4126[20]
8730	Pentane, 2-chloro-4-methyl- 2-Chloro-4-methylpentane	C₆H₁₃Cl 120.62	25346-32-1	113	3-01-00-00399 0.8610[20]	eth 4 1.4113[20]
8731	Pentane, 3-(chloromethyl)-	C₆H₁₃Cl 120.62	4737-41-1	126; 83[202]	3-01-00-00403 0.8914[20]	bz 4; eth 4; chl 4 1.4222[20]
8732	Pentane, 3-chloro-3-methyl-	C₆H₁₃Cl 120.62	918-84-3	116	4-01-00-00365 0.8900[20]	bz 4; eth 4; chl 4 1.4210[20]
8733	Pentane, 2-chloro-2,4,4-trimethyl-	C₈H₁₇Cl 148.68	6111-88-2 -26	147 dec; 44[16]	4-01-00-00444 0.8746[20]	EtOH 4 1.4308[20]
8734	Pentanedial Glutaraldehyde	C₅H₈O₂ 100.12	111-30-8	4366 188 dec	4-01-00-03659	H₂O 5; EtOH 5; bz 3
8735	1,5-Pentanediamine Cadaverine	C₅H₁₄N₂ 102.18	462-94-2 9	1608 179	4-04-00-01310 0.873[25]	H₂O 3; EtOH 3; eth 2 1.463[20]
8736	1,4-Pentanediamine, N1,N1-diethyl- Novoldiamine	C₉H₂₂N₂ 158.29	140-80-7	6642 201	4-04-00-01308 0.8142[0]	1.4429[20]
8737	1,5-Pentanediamine, N-(6-methoxy-8-quinolinyl)-N'-(1-methylethyl)-	C₁₈H₂₇N₃O 301.43	86-78-2	165-70[0.02]	5-22-12-00126	dil HCl 4 1.5785[25]
8738	1,5-Pentanediaminium, N,N,N,N',N',N'-hexamethyl-, dibromide Ammonium, pentamethylenebis trimethyl, bromide	C₁₁H₂₈Br₂N₂ 348.16	541-20-8 301	7070	4-04-00-01311	H₂O 2
8739	Pentane, 1,3-dibromo- 1,3-Dibromopentane	C₅H₁₀Br₂ 229.94	42474-20-4	193; 64[10]	4-01-00-00314 1.6721[20]	1.5048[207]
8740	Pentane, 1,4-dibromo- 1,4-Dibromopentane	C₅H₁₀Br₂ 229.94	626-87-9 -34.4	146[150]; 99[14]	4-01-00-00314 1.6222[20]	1.5086[20]
8741	Pentane, 1,5-dibromo-	C₅H₁₀Br₂ 229.94	111-24-0 -39.5	222.3	4-01-00-00314 1.7018[20]	H₂O 1; bz 3; ctc 2; chl 3 1.5126[20]
8742	Pentane, 2,3-dibromo- 2,3-Dibromopentane	C₅H₁₀Br₂ 229.94	5398-25-4 -56	178; 91[50]	3-01-00-00347 1.6865[14]	1.5087[20]
8743	Pentane, 2,4-dibromo- 2,4-Dibromopentane	C₅H₁₀Br₂ 229.94	19398-53-9	60[12]	3-01-00-00348 1.6659[20]	1.4987[20]
8744	Pentane, 1,2-dichloro- 1,2-Dichloropentane	C₅H₁₀Cl₂ 141.04	1674-33-5	148.3	3-01-00-00341 1.0872[20]	H₂O 1; EtOH 3; chl 4 1.4485[20]
8745	Pentane, 1,3-dichloro- 1,3-Dichloropentane	C₅H₁₀Cl₂ 141.04	30122-12-4	80.4[60]	3-01-00-00342 1.0834[20]	ace 4; chl 4 1.4485[20]
8746	Pentane, 1,4-dichloro- 1,4-Dichloropentane	C₅H₁₀Cl₂ 141.04	626-92-6	162; 88[60]	4-01-00-00309 1.0840[20]	ace 3; chl 3 1.4503[20]
8747	Pentane, 1,5-dichloro- 1,5-Dichloropentane	C₅H₁₀Cl₂ 141.04	628-76-2 -72.8	179	4-01-00-00310 1.1006[22]	H₂O 1; EtOH 3; eth 3; bz 3 1.4564[20]
8748	Pentane, 2,2-dichloro- 2,2-Dichloropentane	C₅H₁₀Cl₂ 141.04	34887-14-4	128.5	2-01-00-00095 1.040[20]	H₂O 1; eth 3; bz 3; chl 3 1.4342[0]
8749	Pentane, 2,3-dichloro- 2,3-Dichloropentane	C₅H₁₀Cl₂ 141.04	600-11-3 -77.3	139	4-01-00-00310 1.0789[20]	1.4464[20]
8750	Pentane, 2,4-dichloro- 2,4-Dichloropentane	C₅H₁₀Cl₂ 141.04	625-67-2	145	3-01-00-00343 1.063[15]	H₂O 1; eth 4; bz 3; chl 4 1.4471[8]
8751	Pentane, 3,3-dichloro- 3,3-Dichloropentane	C₅H₁₀Cl₂ 141.04	21571-91-5	132; 32[14]	3-01-00-00343 1.053[20]	bz 4; chl 4 1.442[20]
8752	Pentane, 1,2-dichloro-4,4-dimethyl- 1,2-Dichloro-4,4-dimethylpentane	C₇H₁₄Cl₂ 169.09	6065-90-3	175; 58[12]	3-01-00-00445 1.0259[20]	bz 4; chl 4 1.4489[20]
8753	Pentane, 1,5-dichloro-3,3-dimethyl- 1,5-Dichloro-3,3-dimethylpentane	C₇H₁₄Cl₂ 169.09	62496-53-1	135[80]; 58[8]	4-01-00-00409 1.0563[20]	chl 4 1.4652[20]
8754	Pentane, 2,4-dichloro-2,4-dimethyl- 2,4-Dichloro-2,4-dimethylpentane	C₇H₁₄Cl₂ 169.09	33553-93-4 23.5	51-7[8]	4-01-00-00408 1.0292[20]	1.4537[20]
8755	Pentane, 1,1-diethoxy-	C₉H₂₀O₂ 160.26	3658-79-5	59[12]	4-01-00-03269 0.829[22]	1.4029[22]
8756	Pentane, 3,3-diethyl- 3,3-Diethylpentane	C₉H₂₀ 128.26	1067-20-5 -33.1	146.3	4-01-00-00462 0.7536[20]	H₂O 1; eth 3; bz 3; os 3 1.4206[20]
8757	Pentane, 3,3-diethyl-2-methyl- 3,3-Diethyl-2-methylpentane	C₁₀H₂₂ 142.28	52897-16-2	172	4-01-00-00487 0.7755[25]	1.4343[20]
8758	Pentane, 1,5-diiodo- Pentamethylene diiodide	C₅H₁₀I₂ 323.94	628-77-3 9	149[20]; 101[3]	4-01-00-00317 2.1903[15]	H₂O 1; eth 3; chl 3 1.6046[15]
8759	Pentane, 2,4-diiodo-	C₅H₁₀I₂ 323.94	66719-29-7	147[34]	0-01-00-00133 2.195[20]	1.600[20]
8760	Pentane, 2,2-dimethyl- 2,2-Dimethylpentane	C₇H₁₆ 100.20	590-35-2 -123.8	79.2	4-01-00-00403 0.6739[20]	H₂O 1; EtOH 3; eth 3; ace 5 1.3822[20]
8761	Pentane, 2,3-dimethyl- 2,3-Dimethylpentane	C₇H₁₆ 100.20	565-59-3	89.7	3-01-00-00445 0.6951[20]	H₂O 1; EtOH 3; eth 3; ace 5 1.3919[20]
8762	Pentane, 2,4-dimethyl- 2,4-Dimethylpentane	C₇H₁₆ 100.20	108-08-7 -119.9	80.4	4-01-00-00406 0.6727[20]	H₂O 1; EtOH 3; eth 3; ace 5 1.3815[20]
8763	Pentane, 3,3-dimethyl- 3,3-Dimethylpentane	C₇H₁₆ 100.20	562-49-2 -134.9	86.0	4-01-00-00409 0.6936[20]	H₂O 1; EtOH 3; eth 3; ace 5 1.3909[20]
8764	Pentane, 2,2-dimethyl-3-methylene-	C₈H₁₆ 112.22	18231-53-3	110	3-01-00-00848 0.724[25]	1.4159[20]
8765	Pentane, 2,4-dimethyl-3-methylene- 1-Butene, 2-isopropyl-3-methyl-	C₈H₁₆ 112.22	111823-35-9	104	3-01-00-00851 0.718[25]	1.4086[20]

No.	Name / Synonym	Mol. Form. / Mol. Wt.	CAS RN / mp/°C	Merck No. / bp/°C	Beil. Ref. / den/g cm^{-3}	Solubility / n_D
8766	Pentane, 2,4-dimethyl-3-(1-methylethyl)-	$C_{10}H_{22}$ 142.28	13475-79-1 -81.7	157.1	3-01-00-00534 0.7545[25]	1.4246[20]
8767	Pentanedinitrile / Glutaronitrile	$C_5H_6N_2$ 94.12	544-13-8 -29	4368 286	4-02-00-01941 0.9911[15]	EtOH 4; chl 4 1.4295[20]
8768	Pentanedinitrile, 2-bromo-2-(bromomethyl)- / 1,2-Dibromo-2,4-dicyanobutane	$C_6H_6Br_2N_2$ 265.94	35691-65-7 52	3004		H_2O 1; ace 4; bz 4; DMF 4
8769	Pentane, 1,5-dinitro- / 1,5-Dinitropentane	$C_5H_{10}N_2O_4$ 162.15	6848-84-6	134[1.2]	4-01-00-00319	H_2O 1; bz 3 1.461[20]
8770	Pentanedioic acid / Glutaric acid	$C_5H_8O_4$ 132.12	110-94-1 97.8	4367 303 dec	4-02-00-01934 1.429[15]	H_2O 4; EtOH 4; eth 4; bz 1 1.4188[106]
8771	Pentanedioic acid, 2-acetyl-, diethyl ester	$C_{11}H_{18}O_5$ 230.26	1501-06-0	271 dec	4-03-00-01835 1.0712[20]	H_2O 2; EtOH 3; eth 3; ctc 2 1.4420[15]
8772	Pentanedioic acid, diethyl ester	$C_9H_{16}O_4$ 188.22	818-38-2 -24.1	236.5	4-02-00-01937 1.0220[20]	eth 4 1.4241[20]
8773	Pentanedioic acid, 3,3-dimethyl-	$C_7H_{12}O_4$ 160.17	4839-46-7 103.5	126[415]; 89[2]	4-02-00-02023 1.4278[20]	H_2O 4; EtOH 4; eth 4; bz 2
8774	Pentanedioic acid, 3,3-dimethyl-, dimethyl ester	$C_9H_{16}O_4$ 188.22	19184-67-9	103-4[15]	3-02-00-01758 1.0366[20]	ctc 3
8775	Pentanedioic acid, dimethyl ester	$C_7H_{12}O_4$ 160.17	1119-40-0 -42.5	214; 109[21]	4-02-00-01937 1.0876[20]	EtOH 4; eth 4; chl 3 1.4242[20]
8776	Pentanedioic acid, diphenyl ester / Diphenyl glutarate	$C_{17}H_{16}O_4$ 284.31	47172-89-4 54	236.5[15]	4-06-00-00626	ace 4; EtOH 4; lig 4
8777	Pentanedioic acid, 3-ethyl-3-methyl- / 3-Ethyl-3-methylglutaric acid	$C_8H_{14}O_4$ 174.20	5345-01-7 87	261	3-02-00-01781	H_2O 3; EtOH 3; eth 3; bz 3
8778	Pentanedioic acid, 3-hydroxy-, diethyl ester / Diethyl 3-hydroxyglutarate	$C_9H_{16}O_5$ 204.22	32328-03-3	156-7[23]	4-03-00-01149 1.103[25]	1.4368[20]
8779	Pentanedioic acid, 3-methyl- / Glutaric acid, 3-methyl	$C_6H_{10}O_4$ 146.14	626-51-7 87	165-7[0.5]	4-02-00-01992	H_2O 3; EtOH 3; eth 3; bz 2
8780	Pentanedioic acid, monomethyl ester	$C_6H_{10}O_4$ 146.14	1501-27-5	150-1[10]	4-02-00-01937 1.169[25]	1.4381[20]
8781	Pentanedioic acid, 2-oxo- / α-Ketoglutaric acid	$C_5H_6O_5$ 146.10	328-50-7 115.5	5182	4-03-00-01813	H_2O 4; EtOH 4; eth 4; ace 3
8782	Pentanedioic acid, 3-oxo- / Acetonedicarboxylic acid	$C_5H_6O_5$ 146.10	542-05-2 138 dec	60	4-03-00-01816	H_2O 3; EtOH 3; eth 2; bz 1
8783	Pentanedioic acid, 3-oxo-, diethyl ester	$C_9H_{14}O_5$ 202.21	105-50-0	250	4-03-00-01817 1.113[20]	H_2O 2; EtOH 5
8784	Pentanedioic acid, 3-oxo-, dimethyl ester	$C_7H_{10}O_5$ 174.15	1830-54-2	150[25]	4-03-00-01817 1.185[25]	1.4434[20]
8785	1,3-Pentanediol	$C_5H_{12}O_2$ 104.15	3174-67-2	120-2[16]	4-01-00-02538 0.9863[20]	1.4659[20]
8786	1,4-Pentanediol / 4-Hydroxy-1-pentanol	$C_5H_{12}O_2$ 104.15	626-95-9	202; 125[10]	3-01-00-02191 0.9883[20]	H_2O 4; EtOH 4; chl 4 1.4452[23]
8787	1,5-Pentanediol / Pentamethylene glycol	$C_5H_{12}O_2$ 104.15	111-29-5 -18	7073 239	4-01-00-02540 0.9914[20]	H_2O 3; EtOH 3; eth 2; bz 2 1.4494[20]
8788	2,3-Pentanediol	$C_5H_{12}O_2$ 104.15	42027-23-6	187.5; 100[17]	4-01-00-02543 0.9798[19]	H_2O 3; EtOH 3; eth 2 1.4412[25]
8789	2,4-Pentanediol / 2,4-Amylene glycol	$C_5H_{12}O_2$ 104.15	625-69-4	199; 97-8[13]	3-01-00-02195 0.9635[20]	H_2O 4; EtOH 4 1.4349[20]
8790	1,2-Pentanediol, (±)- / 1,2-Pentanediol (DL)	$C_5H_{12}O_2$ 104.15	91049-43-3	211	4-01-00-02538 0.9784[20]	1.4412[19]
8791	1,5-Pentanediol, diacetate / Pentamethylene acetate	$C_9H_{16}O_4$ 188.22	6963-44-6	122-3[3]	4-02-00-00226 1.0296[20]	1.4261[19]
8792	1,3-Pentanediol, 2,2-dimethyl- / 2,2-Dimethyl-1,3-dihydroxypentane	$C_7H_{16}O_2$ 132.20	2157-31-5 62.3	213	0-01-00-00490	EtOH 4
8793	2,4-Pentanediol, 2-methyl- / Hexylene glycol	$C_6H_{14}O_2$ 118.18	107-41-5 -50	4631 197.1	4-01-00-02565 0.923[15]	H_2O 3; EtOH 3; eth 3; ctc 2 1.4276[20]
8794	2,4-Pentanediol, 3-methyl- / 3-Methyl-2,4-pentanediol	$C_6H_{14}O_2$ 118.18	5683-44-3	211.5	4-01-00-02572 0.9640[20]	H_2O 3; EtOH 3 1.4433[20]
8795	1,3-Pentanediol, 2,2,4-trimethyl- / 2,2,4-Trimethyl-1,3-pentanediol	$C_8H_{18}O_2$ 146.23	144-19-4 51.5	235; 81-2[1]	4-01-00-02604 0.936[15]	H_2O 2; EtOH 4; eth 4; bz 3 1.4513[15]
8796	1,4-Pentanediol, 2,2,4-trimethyl-	$C_8H_{18}O_2$ 146.23	80864-10-4 86	210	0-01-00-00493	eth 4; EtOH 4
8797	2,3-Pentanedione / Acetylpropionyl	$C_5H_8O_2$ 100.12	600-14-6	108	4-01-00-03660 0.9565[19]	H_2O 3; EtOH 5; eth 5; ace 5 1.4014[19]
8798	2,4-Pentanedione / Acetylacetone	$C_5H_8O_2$ 100.12	123-54-6 -23	75 138	4-01-00-03662 0.9721[25]	H_2O 4; EtOH 5; eth 5; ace 5 1.4494[20]
8799	2,4-Pentanedione, 3,3-dimethyl- / 3,3-Dimethylacetylacetone	$C_7H_{12}O_2$ 128.17	3142-58-3 19	173	4-01-00-03705 0.9575[20]	eth 4 1.4306[20]
8800	2,3-Pentanedione, dioxime	$C_5H_{10}N_2O_2$ 130.15	4775-86-4 172.5	sub	4-01-00-03661	EtOH 4
8801	2,4-Pentanedione, 3-ethyl-	$C_7H_{12}O_2$ 128.17	1540-34-7	178.5	4-01-00-03703 0.9531[19]	eth 4; EtOH 4; chl 4 1.4408[19]
8802	2,4-Pentanedione, 1,1,1,5,5,5-hexafluoro- / Acetylacetone, hexafluoro	$C_5H_2F_6O_2$ 208.06	1522-22-1	54.1	4-01-00-03681 1.485[20]	1.3333[20]
8803	2,3-Pentanedione, 3-oxime / 3-Oximino-2-pentanone	$C_5H_9NO_2$ 115.13	609-29-0 58.5	183	4-01-00-03661	eth 4; EtOH 4; chl 4
8804	1,4-Pentanedione, 1-phenyl-	$C_{11}H_{12}O_2$ 176.22	583-05-1	162[12]	4-07-00-02160	ace 4 1.5250[30]
8805	2,4-Pentanedione, 3-(2-propenyl)-	$C_8H_{12}O_2$ 140.18	3508-78-9	199; 92[16]	4-01-00-03759 0.974[15]	1.4698[14]

No.	Name Synonym	Mol. Form. Mol. Wt.	CAS RN mp/°C	Merck No. bp/°C	Beil. Ref. den/g cm^{-3}	Solubility n_D
8806	Pentanedioyl dichloride	$C_5H_6Cl_2O_2$ 169.01	2873-74-7	217	4-02-00-01939 1.324[20]	eth 3; chl 2 1.4728[20]
8807	Pentane, dodecafluoro- Perfluoropentane	C_5F_{12} 288.04	678-26-2 -10	29.2	4-01-00-00308	
8808	Pentane, 1-ethoxy- Ethyl pentyl ether	$C_7H_{16}O$ 116.20	17952-11-3	117.6	3-01-00-01602 0.7622[20]	eth 4; EtOH 4 1.3927[20]
8809	Pentane, 3-ethyl- 3-Ethylpentane	C_7H_{16} 100.20	617-78-7 -118.6	93.5	4-01-00-00402 0.6982[20]	H$_2$O 1; EtOH 3; eth 3; ace 5 1.3934[20]
8810	Pentane, 3-ethyl-2,2-dimethyl-	C_9H_{20} 128.26	16747-32-3 -99.3	133.8	4-01-00-00462 0.7438[20]	1.4123[20]
8811	Pentane, 3-ethyl-2,3-dimethyl-	C_9H_{20} 128.26	16747-33-4	144.7	4-01-00-00462 0.7508[25]	1.4221[20]
8812	Pentane, 3-ethyl-2,4-dimethyl-	C_9H_{20} 128.26	1068-87-7 -122.4	136.7	4-01-00-00462 0.7365[20]	1.4131[20]
8813	Pentane, 3-ethyl-2-methyl- 3-Ethyl-2-methylpentane	C_8H_{18} 114.23	609-26-7 -114.9	115.6	4-01-00-00437 0.7193[20]	H$_2$O 1; EtOH 5; eth 3; ace 5 1.4040[20]
8814	Pentane, 3-ethyl-3-methyl- 3-Ethyl-3-methylpentane	C_8H_{18} 114.23	1067-08-9 -90.9	118.2	4-01-00-00438 0.7274[20]	H$_2$O 1; EtOH 5; eth 3; ace 5 1.4078[20]
8815	Pentane, 3-ethyl-2,2,3-trimethyl- 3-Ethyl-2,2,3-trimethylpentane	$C_{10}H_{22}$ 142.28	52897-17-3 -42.5	169.5	4-01-00-00487 0.7780[25]	1.4420[20]
8816	Pentane, 3-ethyl-2,3,4-trimethyl- 3-Ethyl-2,3,4-trimethylpentane	$C_{10}H_{22}$ 142.28	52897-19-5	169.47	4-01-00-00487 0.7735[25]	1.4333[20]
8817	Pentane, 1-fluoro- Pentyl fluoride	$C_5H_{11}F$ 90.14	592-50-7 -120	62.8	4-01-00-00308 0.7907[20]	eth 4; EtOH 4 1.3591[2-]
8818	Pentane, 1-iodo- Amyl iodide	$C_5H_{11}I$ 198.05	628-17-1 -85.6	155	4-01-00-00315 1.5161[20]	chl 3 1.4959[20]
8819	Pentane, 2-iodo-, (±)- Pentane, 2-iodo (DL)	$C_5H_{11}I$ 198.05	52152-72-4	141	4-01-00-00316 1.5096[20]	H$_2$O 1; eth 3; ace 3; bz 3 1.4961[20]
8820	Pentane, 3-iodo- 3-Iodopentane	$C_5H_{11}I$ 198.05	1809-05-8	145.5	3-01-00-00349 1.5176[20]	ace 4; bz 4; eth 4 1.4974[20]
8821	Pentane, 1-isocyano- Pentyl isocyanide	$C_6H_{11}N$ 97.16	18971-59-0 -51.1	155.5	3-04-00-00331 0.806[20]	H$_2$O 1; EtOH 3
8822	Pentane, 1-methoxy- Methyl pentyl ether	$C_6H_{14}O$ 102.18	628-80-8	99	4-01-00-01643 0.759[22]	ace 4; eth 4; EtOH 4 1.3862[22]
8823	Pentane, 2-methyl- Isohexane	C_6H_{14} 86.18	107-83-5 -153.7	60.2	4-01-00-00358 0.650[25]	H$_2$O 1; EtOH 3; eth 3; ace 5 1.3715[20]
8824	Pentane, 3-methyl- 3-Methylpentane	C_6H_{14} 86.18	96-14-0 -162.9	63.2	4-01-00-00363 0.6598[25]	H$_2$O 1; EtOH 3; eth 5; ace 5 1.3765[20]
8825	Pentane, 2-methylene- 2-Ethyl-1-butene	C_6H_{12} 84.16	760-21-4 -131.5	64.7	4-01-00-00850 0.6894[20]	H$_2$O 1; eth 3; ace 3; bz 3 1.3969[20]
8826	Pentane, 1-(methylthio)- Methyl pentyl sulfide	$C_6H_{14}S$ 118.24	1741-83-9 -94	145.1	3-01-00-01608 0.8431[20]	EtOH 3; eth 3; ace 3; bz 3 1.4506[20]
8827	Pentanenitrile Valeronitrile	C_5H_9N 83.13	110-59-8 -96.2	141.3	4-02-00-00875 0.8008[20]	eth 3; ace 3; bz 3; ctc 2 1.3971[20]
8828	Pentanenitrile, 5-bromo-	C_5H_8BrN 162.03	5414-21-1	111[12]	4-02-00-00884 1.386[25]	1.4780[20]
8829	Pentanenitrile, 2-hydroxy-2-propyl-	$C_8H_{15}NO$ 141.21	5699-74-1	119-20[21]	4-03-00-00880 0.9077[18]	EtOH 4 1.4337[18]
8830	Pentanenitrile, 3-imino-2-methyl- Valeronitrile, 3-imino-2-methyl	$C_6H_{10}N_2$ 110.16	95642-49-2 46.3	257.5	0.9525[19]	eth 4; EtOH 4
8831	Pentanenitrile, 4-methyl- Isoamyl cyanide	$C_6H_{11}N$ 97.16	542-54-1 -51	4999 156.5	4-02-00-00946 0.8030[20]	H$_2$O 1; EtOH 3; eth 5; ctc 2 1.4059[20]
8832	Pentane, 1-nitro- 1-Nitropentane	$C_5H_{11}NO_2$ 117.15	628-05-7	172.5	4-01-00-00317 0.9525[20]	EtOH 3; eth 3; bz 3 1.4175[20]
8833	Pentane, 2-nitro- 2-Nitropentane	$C_5H_{11}NO_2$ 117.15	4609-89-6	149; 88[100]	4-01-00-00318 0.938[25]	1.4092[25]
8834	Pentane, 3-nitro- 3-Nitropentane	$C_5H_{11}NO_2$ 117.15	551-88-2	6537 154	4-01-00-00318 0.957[0]	ace 4; eth 4; EtOH 4
8835	Pentane, 1,1'-oxybis- Amyl ether	$C_{10}H_{22}O$ 158.28	693-65-2 -69	646 190	4-01-00-01643 0.7833[20]	H$_2$O 1; EtOH 5; eth 5; chl 3 1.4119[20]
8836	Pentane, 2,2,3,3,4-pentamethyl- 2,2,3,3,4-Pentamethylpentane	$C_{10}H_{22}$ 142.28	16747-44-7 -36.4	166.1	4-01-00-00487 0.7767[25]	1.4361[20]
8837	Pentane, 2,2,3,4,4-pentamethyl- 2,2,3,4,4-Pentamethylpentane	$C_{10}H_{22}$ 142.28	16747-45-8 -38.7	159.3	4-01-00-00487 0.7636[25]	1.4307[20]
8838	Pentane, 1,1,1,5-tetrachloro- 1,1,1,5-Tetrachloropentane	$C_5H_8Cl_4$ 209.93	2467-10-9	112[24]	4-01-00-00311 1.3416[25]	1.4859[25]
8839	Pentane, 2,2,3,3-tetramethyl- 2,2,3,3-Tetramethylpentane	C_9H_{20} 128.26	7154-79-2 -9.8	140.2	4-01-00-00463 0.7530[25]	1.4236[20]
8840	Pentane, 2,2,3,4-tetramethyl- 2,2,3,4-Tetramethylpentane	C_9H_{20} 128.26	1186-53-4 -121.0	133.0	4-01-00-00463 0.7389[20]	1.4147[20]
8841	Pentane, 2,2,4,4-tetramethyl- Di-tert-butylmethane	C_9H_{20} 128.26	1070-87-7 -66.5	122.2	4-01-00-00464 0.7195[20]	H$_2$O 1; EtOH 4; bz 4; os 4 1.4069[20]
8842	Pentane, 2,3,3,4-tetramethyl- 2,3,3,4-Tetramethylpentane	C_9H_{20} 128.26	16747-38-9 -102.1	141.5	4-01-00-00464 0.7547[20]	1.4222[20]
8843	Pentane, 1,1'-thiobis-	$C_{10}H_{22}S$ 174.35	872-10-6 -51.3	86[3.7]	4-01-00-01654 0.8407[20]	H$_2$O 1; eth 3 1.4561[20]
8844	1-Pentanethiol Pentyl mercaptan	$C_5H_{12}S$ 104.22	110-66-7 -75.7	648 126.6	4-01-00-01653 0.850[20]	H$_2$O 1; EtOH 5; eth 5 1.4469[20]
8845	2-Pentanethiol sec-Amylmercaptan	$C_5H_{12}S$ 104.22	2084-19-7 -169	112.9	4-01-00-01662 0.8327[20]	EtOH 3; lig 3 1.4412[20]

No.	Name Synonym	Mol. Form. Mol. Wt.	CAS RN mp/°C	Merck No. bp/°C	Beil. Ref. den/g cm^{-3}	Solubility n_D
8846	3-Pentanethiol 3-Pentyl mercaptan	$C_5H_{12}S$ 104.22	616-31-9 -110.8	105	4-01-00-01665 0.8410[20]	EtOH 3; DMSO-d$_8$ 2 1.4447[20]
8847	Pentane, 2,2,3-trimethyl- 2-tert-Butylbutane	C_8H_{18} 114.23	564-02-3 -112.2	110	4-01-00-00438 0.7161[20]	H$_2$O 1; EtOH 5; eth 5; ace 5 1.4030[20]
8848	Pentane, 2,2,4-trimethyl- Isooctane	C_8H_{18} 114.23	540-84-1 -107.3	5079 99.2	4-01-00-00439 0.6877[25]	H$_2$O 1; EtOH 5; eth 3; ace 5 1.3915[20]
8849	Pentane, 2,3,3-trimethyl- 2,3,3-Trimethylpentane	C_8H_{18} 114.23	560-21-4 -100.9	114.8	4-01-00-00445 0.7262[20]	H$_2$O 1; EtOH 4; eth 5; ace 5 1.4075[20]
8850	Pentane, 2,3,4-trimethyl- 2,3,4-Trimethylpentane	C_8H_{18} 114.23	565-75-3 -109.2	113.5	4-01-00-00446 0.7191[20]	H$_2$O 1; EtOH 4; eth 5; ace 5 1.4042[20]
8851	1,2,5-Pentanetriol 1,4,5-Pentanetriol	$C_5H_{12}O_3$ 120.15	14697-46-2	190-1[13]	4-01-00-02779 1.136[20]	H$_2$O 4; EtOH 4 1.4730[20]
8852	1,3,5-Pentanetriol	$C_5H_{12}O_3$ 120.15	4328-94-3	188[11]; 139[0.1]	4-01-00-02779 1.1291[20]	H$_2$O 4; EtOH 4; eth 3; ace 3 1.4785[20]
8853	Pentane, 1,1,1,2,2,3,4,4,5,5,5-undecafluoro-3- (trifluoromethyl)- Perfluoro-3-methylpentane	C_6F_{14} 338.04	865-71-4 -115	58.4		bz 3
8854	Pentane, 1,1,1,2,2,3,3,4,4,5,5,5-undecafluoro-4- (trifluoromethyl)- Perfluoro(2-methylpentane)	C_6F_{14} 338.04	355-04-4	57.6	4-01-00-00360 1.7326[20]	H$_2$O 1; bz 3 1.2564[22]
8855	Pentanoic acid Valeric acid	$C_5H_{10}O_2$ 102.13	109-52-4 -34	9815 186.1	4-02-00-00868 0.9391[20]	H$_2$O 3; EtOH 3; eth 3; ctc 2 1.4085[20]
8856	Pentanoic acid, 2-acetyl-, ethyl ester	$C_9H_{16}O_3$ 172.22	1540-28-9	224; 90[15]	4-03-00-01602 0.9661[20]	eth 4; EtOH 4 1.4255[20]
8857	Pentanoic acid, 5-amino-	$C_5H_{11}NO_2$ 117.15	660-88-8 157-8 dec	dec	4-04-00-02636	H$_2$O 3; EtOH 2; eth 1; bz 1
8858	Pentanoic acid, anhydride	$C_{10}H_{18}O_3$ 186.25	2082-59-9 -56.1	227	4-02-00-00874 0.924[20]	eth 4; EtOH 4 1.4171[26]
8859	Pentanoic acid, 2-bromo-, (±)-	$C_5H_9BrO_2$ 181.03	2681-92-7	118[12]	4-02-00-00883 1.381[20]	eth 4; EtOH 4
8860	Pentanoic acid, 2-bromo-, ethyl ester	$C_7H_{13}BrO_2$ 209.08	615-83-8	191	3-02-00-00681 1.226[18]	H$_2$O 1; EtOH 3; eth 3 1.4496[20]
8861	Pentanoic acid, 5-bromo-, ethyl ester	$C_7H_{13}BrO_2$ 209.08	14660-52-7	107-8[20]	4-02-00-00884 1.2804[20]	ctc 2 1.4543[20]
8862	Pentanoic acid, butyl ester	$C_9H_{18}O_2$ 158.24	591-68-4 -92.8	185.8	3-02-00-00671 0.8710[15]	H$_2$O 2; EtOH 3; eth 3 1.4128[20]
8863	Pentanoic acid, 2-chloro-, (±)-	$C_5H_9ClO_2$ 136.58	94347-45-2 -15	222	4-02-00-00878 1.141[13]	eth 4; EtOH 4 1.4481[11]
8864	Pentanoic acid, 3-chloro- Valeric acid, 3-chloro	$C_5H_9ClO_2$ 136.58	51637-47-9 33	112[10]	4-02-00-00879 1.1484[20]	eth 4; EtOH 4 1.4462[20]
8865	Pentanoic acid, 4-chloro- 4-Chlorovaleric acid	$C_5H_9ClO_2$ 136.58	32607-54-8	116[10]	4-02-00-00879 1.1514[20]	eth 4; EtOH 4 1.4458[20]
8866	Pentanoic acid, 5-chloro-	$C_5H_9ClO_2$ 136.58	1119-46-6 18	230	4-02-00-00880 1.3416[25]	eth 4; EtOH 4 1.4555[20]
8867	Pentanoic acid, 4-chloro-, ethyl ester Ethyl 4-chlorovalerate	$C_7H_{13}ClO_2$ 164.63	70786-82-2	196	4-02-00-00879 1.0393[20]	eth 4; EtOH 4 1.4310[20]
8868	Pentanoic acid, 5-chloro-, ethyl ester	$C_7H_{13}ClO_2$ 164.63	2323-81-1	205.5	4-02-00-00880 1.0561[20]	H$_2$O 1; EtOH 3; eth 3; chl 2 1.4355[20]
8869	Pentanoic acid, 2-chloroethyl ester	$C_7H_{13}ClO_2$ 164.63	7735-33-3	185	3-02-00-00670 1.040[12]	eth 4 1.4307[11]
8870	Pentanoic acid, 2,3-dibromo- Valeric acid, 2,3-dibromo-	$C_5H_8Br_2O_2$ 259.93	79912-57-5 57	104-6[13]	3-02-00-00683	bz 4; CS$_2$ 4
8871	Pentanoic acid, 2,5-dibromo- Valeric acid, 2,5-dibromo-	$C_5H_8Br_2O_2$ 259.93	1450-81-3	173[15]; 151[5]	3-02-00-00683 1.8629[25]	1.5347[25]
8872	Pentanoic acid, 2,3-dibromo-, ethyl ester Valeric acid, 2,3-dibromo, ethyl ester	$C_7H_{12}Br_2O_2$ 287.98	79912-55-3	117[14]	2-02-00-00269 1.6199[15]	1.4953[15]
8873	Pentanoic acid, 2,5-dibromo-, ethyl ester Valeric acid, 2,5-dibromo-, ethyl ester	$C_7H_{12}Br_2O_2$ 287.98	29823-16-3	133-5[14]	3-02-00-00683 1.6289[25]	1.4947[25]
8874	Pentanoic acid, 2,2-dimethyl-4-oxo-	$C_7H_{12}O_3$ 144.17	470-49-5 76.8	138[15]	3-03-00-01247	H$_2$O 4; bz 4; eth 4; EtOH 4
8875	Pentanoic acid, 2,4-dioxo-	$C_5H_6O_4$ 130.10	5699-58-1 99.5	130[37]	3-03-00-01331	H$_2$O 3; EtOH 3; eth 3; ace 3
8876	Pentanoic acid, 2,4-dioxo-, ethyl ester	$C_7H_{10}O_4$ 158.15	615-79-2 18	214	4-03-00-01777 1.1251[20]	eth 4; EtOH 4 1.4757[17]
8877	Pentanoic acid, 2-ethyl-, (±)- Valeric acid, 2-ethyl (DL)	$C_7H_{14}O_2$ 130.19	116908-83-9	209.2	4-02-00-00975 0.9311[33]	eth 4; EtOH 4
8878	Pentanoic acid, ethyl ester Ethyl valerate	$C_7H_{14}O_2$ 130.19	539-82-2 -91.2	146.1	4-02-00-00872 0.8770[20]	H$_2$O 1; EtOH 5; eth 5; ctc 2 1.4120[20]
8879	Pentanoic acid, 5-fluoro- Valeric acid, 5-fluoro-	$C_5H_9FO_2$ 120.12	407-75-0	83[2]	4-02-00-00876	eth 4; EtOH 4 1.4080[25]
8880	Pentanoic acid, 2-furanylmethyl ester Furfuryl valerate	$C_{10}H_{14}O_3$ 182.22	36701-01-6	228; 82[1]	2-17-00-00115 1.0284[20]	eth 4; EtOH 4
8881	Pentanoic acid, heptyl ester	$C_{12}H_{24}O_2$ 200.32	5451-80-9 -46.4	245.2	4-02-00-00872 0.8623[20]	ace 4; eth 4; EtOH 4 1.4254[15]
8882	Pentanoic acid, hexyl ester	$C_{11}H_{22}O_2$ 186.29	1117-59-5 -63.1	226.3	3-02-00-00671 0.8635[20]	ace 4; eth 4; EtOH 4 1.4228[15]
8883	Pentanoic acid, 2-hydroxy-	$C_5H_{10}O_3$ 118.13	617-31-2 34	sub	4-03-00-00807	H$_2$O 3; EtOH 3; eth 3
8884	Pentanoic acid, 4-hydroxy-, ethyl ester Valeric acid, 4-hydroxy-, ethyl ester	$C_7H_{14}O_3$ 146.19	6149-46-8 85-6[2]		3-03-00-00612 0.9504[25]	eth 4; EtOH 4 1.4265[25]

No.	Name Synonym	Mol. Form. Mol. Wt.	CAS RN mp/°C	Merck No. bp/°C	Beil. Ref. den/g cm⁻³	Solubility n_D
8885	Pentanoic acid, 2-hydroxyethyl ester Valeric acid, 2-hydroxyethyl ester	$C_7H_{14}O_3$ 146.19	16179-36-5	190	3-02-00-00672 0.9839[20]	1.4220[20]
8886	Pentanoic acid, 2-methoxyphenyl ester Guaiacol valerate	$C_{12}H_{16}O_3$ 208.26	531-39-5	4461 265	1.05[25]	bz 4; eth 4; EtOH 4
8887	Pentanoic acid, 2-methyl-, (±)-	$C_6H_{12}O_2$ 116.16	22160-39-0	195.6	4-02-00-00942 0.9230[20]	H_2O 3; EtOH 3; eth 3; ctc 2 1.413[20]
8888	Pentanoic acid, 3-methyl-, (±)- Valeric acid, 3-methyl (DL)	$C_6H_{12}O_2$ 116.16	22160-40-3 -41.6	197.5	4-02-00-00948 0.9262[20]	eth 4; EtOH 4 1.4159[20]
8889	Pentanoic acid, 4-methyl-	$C_6H_{12}O_2$ 116.16	646-07-1 -33	200.5	4-02-00-00944 0.9225[20]	H_2O 2; EtOH 3; eth 3; chl 3 1.4144[20]
8890	Pentanoic acid, 2-methylene- Valeric acid, 2-methylene	$C_6H_{10}O_2$ 114.14	5650-75-9 24.4	214	0-02-00-00437 0.9783[25]	bz 4; eth 4; chl 4
8891	Pentanoic acid, methyl ester Methyl valerate	$C_6H_{12}O_2$ 116.16	624-24-8	127.4	4-02-00-00871 0.8947[20]	H_2O 2; EtOH 5; eth 5; ace 3 1.4003[20]
8892	Pentanoic acid, 4-methyl-, ethyl ester	$C_8H_{16}O_2$ 144.21	25415-67-2	163; 52[10]	4-02-00-00945 0.8705[20]	1.4050[20]
8893	Pentanoic acid, 1-methylethyl ester	$C_8H_{16}O_2$ 144.21	18362-97-5		3-02-00-00671 0.8579[20]	H_2O 1; EtOH 3; eth 3; ace 3 1.4061[20]
8894	Pentanoic acid, 4-methyl-, 2-methylpropyl ester	$C_{10}H_{20}O_2$ 172.27	25415-70-7	172; 76[10]	4-02-00-00945 0.8536[20]	1.4135[20]
8895	Pentanoic acid, 2-methyl-4-oxo-	$C_6H_{10}O_3$ 130.14	6641-83-4	165[40]	4-03-00-01587 1.11[25]	1.4431[20]
8896	Pentanoic acid, 3-methyl-4-oxo-	$C_6H_{10}O_3$ 130.14	6628-79-1 31.5	241.5	3-03-00-01238 1.0932[17]	H_2O 4; EtOH 4 1.4443[17]
8897	Pentanoic acid, 4-methyl-3-oxo-, ethyl ester	$C_8H_{14}O_3$ 158.20	7152-15-0 -9	173	4-03-00-01589 0.98[25]	1.250[20]
8898	Pentanoic acid, 3-methyl-4-oxo-, methyl ester Levulinic acid, 3-methyl-, methyl ester	$C_7H_{12}O_3$ 144.17	25234-83-7	200; 82[10]	1-03-00-00240 1.022[20]	1.4052[73]
8899	Pentanoic acid, 1-methylpropyl ester, (+)- Valeric acid, sec-butyl ester (d)	$C_9H_{18}O_2$ 158.24	116836-32-9	174.5	3-02-00-00671 0.8605[20]	bz 4; eth 4; py 4; EtOH 4 1.4070[20]
8900	Pentanoic acid, 2-methylpropyl ester	$C_9H_{18}O_2$ 158.24	10588-10-0	179	3-02-00-00671 0.8625[25]	H_2O 1; EtOH 5; eth 3; ace 3 1.4046[20]
8901	Pentanoic acid, octyl ester Valeric acid, octyl ester	$C_{13}H_{26}O_2$ 214.35	5451-85-4 -42.5	261.6	3-02-00-00672 0.8615[20]	ace 4; eth 4; EtOH 4 1.4273[15]
8902	Pentanoic acid, 2-oxo-	$C_5H_8O_3$ 116.12	1821-02-9 6.5	179	4-03-00-01558 1.0970[14]	H_2O 2; eth 3; bz 3; chl 3
8903	Pentanoic acid, 4-oxo- Levulinic acid	$C_5H_8O_3$ 116.12	123-76-2 33	5352 245 dec	4-03-00-01560 1.1335[20]	H_2O 4; EtOH 4; eth 4; chl 3 1.4396[20]
8904	Pentanoic acid, 4-oxo-, butyl ester	$C_9H_{16}O_3$ 172.22	2052-15-5	237.5	4-03-00-01563 0.9735[20]	chl 2 1.4290[20]
8905	Pentanoic acid, 2-oxo-, ethyl ester Valeric acid, 2-oxo-, ethyl ester	$C_7H_{12}O_3$ 144.17	50461-74-0	182.5; 70[11]	4-03-00-01558 0.9985[18]	1.4170[18]
8906	Pentanoic acid, 3-oxo-, ethyl ester	$C_7H_{12}O_3$ 144.17	4949-44-4	191	4-03-00-01559 1.0120[20]	bz 4; eth 4; EtOH 4 1.4230[20]
8907	Pentanoic acid, 4-oxo-, ethyl ester Ethyl levulinate	$C_7H_{12}O_3$ 144.17	539-88-8	3775 205.8	4-03-00-01562 1.0111[20]	H_2O 4; EtOH 4 1.4229[20]
8908	Pentanoic acid, 4-oxo-, methyl ester	$C_6H_{10}O_3$ 130.14	624-45-3	196	4-03-00-01562 1.0511[20]	H_2O 2; EtOH 3; eth 5; ace 3 1.4233[20]
8909	Pentanoic acid, 4-oxo-, 1-methylethyl ester Levulinic acid, isopropyl ester	$C_8H_{14}O_3$ 158.20	21884-26-4	209.3	4-03-00-01563 0.9842[20]	ace 4; bz 4; eth 4; EtOH 4 1.4420[20]
8910	Pentanoic acid, 4-oxo-, 2-methylpropyl ester Levulinic acid, isobutyl ester	$C_9H_{16}O_3$ 172.22	3757-32-2	228	4-03-00-01563 0.9669[20]	ace 4; eth 4; EtOH 4 1.4249[20]
8911	Pentanoic acid, 4-oxo-, phenylmethyl ester Levulinic acid, benzyl ester	$C_{12}H_{14}O_3$ 206.24	6939-75-9	132-4[2]	4-06-00-02481 1.0935[20]	tol 4 1.5090[20]
8912	Pentanoic acid, 4-oxo-, propyl ester Levulinic acid, propyl ester	$C_8H_{14}O_3$ 158.20	645-67-0	218	4-03-00-01563 0.9896[20]	H_2O 2; EtOH 3; eth 3; ace 3 1.4258[20]
8913	Pentanoic acid, pentyl ester	$C_{10}H_{20}O_2$ 172.27	2173-56-0 -78.8	203.7	4-02-00-00872 0.8638[20]	H_2O 2; EtOH 5; eth 5 1.4164[20]
8914	Pentanoic acid, propyl ester	$C_8H_{16}O_2$ 144.21	141-06-0 -70.7	167.5	4-02-00-00872 0.8699[20]	H_2O 1; EtOH 3; eth 3; chl 3 1.4065[20]
8915	1-Pentanol Amyl alcohol	$C_5H_{12}O$ 88.15	71-41-0 -78.9	7074 137.9	4-01-00-01640 0.8144[20]	H_2O 1; EtOH 5; eth 5; ace 5 1.4101[20]
8916	2-Pentanol sec-Amyl alcohol	$C_5H_{12}O$ 88.15	6032-29-7 -73	7075 119.3	3-01-00-01609 0.8094[20]	H_2O 4; EtOH 3; eth 3; ctc 3 1.4053[20]
8917	3-Pentanol Diethyl carbinol	$C_5H_{12}O$ 88.15	584-02-1 -69	7076 116.2	4-01-00-01662 0.8203[20]	H_2O 2; EtOH 3; eth 3; ace 3 1.4104[20]
8918	2-Pentanol, acetate, (R)- sec-Amyl acetate (R)	$C_7H_{14}O_2$ 130.19	54638-10-7	142	0-02-00-00132 0.8803[18]	eth 4; EtOH 4 1.4012[20]
8919	3-Pentanol, acetate	$C_7H_{14}O_2$ 130.19	620-11-1	133	4-02-00-00155 0.8712[20]	eth 4; EtOH 4 1.4005[20]
8920	1-Pentanol, 5-amino-	$C_5H_{13}NO$ 103.16	2508-29-4 38.5	221.5	4-04-00-01750 0.9488[17]	H_2O 5; EtOH 5; ace 5 1.4618[17]
8921	3-Pentanol, 2-amino-	$C_5H_{13}NO$ 103.16	116836-16-9	171	1-04-00-00442 0.9289[23]	1.4458
8922	1-Pentanol, 2-amino-4-methyl-, (±)-	$C_6H_{15}NO$ 117.19	16369-17-8	199	2-04-00-00748 0.9173[13]	H_2O 4; EtOH 4
8923	1-Pentanol, 5-chloro-	$C_5H_{11}ClO$ 122.59	5259-98-3	112[12]	4-01-00-01650	eth 4; EtOH 4 1.4518[20]
8924	2-Pentanol, 1-chloro-	$C_5H_{11}ClO$ 122.59	17658-32-1	160; 60[14]	3-01-00-01613 1.035[20]	eth 4; EtOH 4 1.4404[25]

No.	Name / Synonym	Mol. Form. / Mol. Wt.	CAS RN / mp/°C	Merck No. / bp/°C	Beil. Ref. / den/g cm⁻³	Solubility / n_D
8925	3-Pentanol, 1-chloro- / 1-Chloro-3-hydroxypentane	$C_5H_{11}ClO$ / 122.59	32541-33-6 /	173	4-01-00-01665 / 1.0327[25]	eth 4; EtOH 4 / 1.448[25]
8926	1-Pentanol, 2-(diethylamino)-4-methyl-, 4-aminobenzoate (ester), monomethanesulfonate (salt) / Panthesin	$C_{18}H_{32}N_2O_5S$ / 388.53	135-44-4 / 158	5332	4-14-00-01143 /	H_2O 4; EtOH 4
8927	1-Pentanol, 2,4-dimethyl-, (±)- / 1-Pentanol, 2,4-dimethyl (DL)	$C_7H_{16}O$ / 116.20	111768-02-6 /	159	4-01-00-01753 / 0.793[20]	eth 4; EtOH 4 / 1.427[20]
8928	2-Pentanol, 2,4-dimethyl- / Isobutyldimethylcarbinol	$C_7H_{16}O$ / 116.20	625-06-9 / <-20	133.1	4-01-00-01753 / 0.8103[20]	H_2O 1; EtOH 3; eth 3; ctc 3 / 1.4172[20]
8929	3-Pentanol, 2,2-dimethyl- / 2,2-Dimethyl-3-pentanol	$C_7H_{16}O$ / 116.20	3970-62-5 / -5	135	3-01-00-01697 / 0.8253[20]	H_2O 1; EtOH 3; eth 3 / 1.4223[20]
8930	3-Pentanol, 2,3-dimethyl- / 2,3-Dimethyl-3-pentanol	$C_7H_{16}O$ / 116.20	595-41-5 / <-30	139.7	4-01-00-01752 / 0.833[20]	H_2O 1; EtOH 3; eth 3; bz 2 / 1.4287[20]
8931	3-Pentanol, 2,4-dimethyl-	$C_7H_{16}O$ / 116.20	600-36-2 / <-70	138.7	4-01-00-01754 / 0.8288[20]	H_2O 2; EtOH 3; eth 3 / 1.4250[20]
8932	3-Pentanol, 3-ethyl-	$C_7H_{16}O$ / 116.20	597-49-9 / -12.5	142	4-01-00-01750 / 0.8407[22]	H_2O 1; EtOH 3; eth 3 / 1.4294[20]
8933	2-Pentanol, 3-ethyl-2-methyl-	$C_8H_{18}O$ / 130.23	19780-63-3 /	156	4-01-00-01794 / 0.8382[20]	bz 4; eth 4; EtOH 4; lig 4 / 1.4325[20]
8934	3-Pentanol, 3-ethyl-2-methyl- / Diethylisopropylcarbinol	$C_8H_{18}O$ / 130.23	597-05-7 /	160; 56[48]	4-01-00-01794 / 0.8280[20]	H_2O 1; EtOH 3; eth 3 / 1.4372[10]
8935	1-Pentanol, 5-fluoro-	$C_5H_{11}FO$ / 106.14	592-80-3 /	70-1[11]	4-01-00-01647 /	eth 4; EtOH 4 / 1.4057[25]
8936	2-Pentanol, 1,1'-iminobis[2-methyl- / Dipentylamine, 2,2'-dimethyl,2,2'-dihydroxy	$C_{12}H_{27}NO_2$ / 217.35	85733-97-7 /	165-70[15]	0.9264[20]	ace 4; eth 4; EtOH 4 / 1.4584[20]
8937	1-Pentanol, 2-methyl- / 2-Methyl-1-pentanol	$C_6H_{14}O$ / 102.18	105-30-6 /	149	3-01-00-01665 / 0.8263[20]	EtOH 3; eth 3; ace 3; ctc 3 / 1.4182[20]
8938	1-Pentanol, 3-methyl-, (±)- / 1-Pentanol, 3-methyl (DL)	$C_6H_{14}O$ / 102.18	20281-83-8 /	153	4-01-00-01722 / 0.8242[20]	H_2O 1; EtOH 3; eth 3 / 1.4112[23]
8939	1-Pentanol, 4-methyl- / Isohexyl alcohol	$C_6H_{14}O$ / 102.18	626-89-1 /	151.9	4-01-00-01721 / 0.8131[20]	H_2O 1; EtOH 3; eth 3 / 1.4134[20]
8940	2-Pentanol, 2-methyl- / 2-Methyl-2-pentanol	$C_6H_{14}O$ / 102.18	590-36-3 / -103	121.1	4-01-00-01714 / 0.8350[16]	H_2O 2; EtOH 3; eth 3 / 1.4100[20]
8941	2-Pentanol, 3-methyl- / 3-Methyl-2-pentanol	$C_6H_{14}O$ / 102.18	565-60-6 /	134.3	3-01-00-01672 / 0.8307[20]	H_2O 3; EtOH 3; eth 3 / 1.4182[20]
8942	2-Pentanol, 4-methyl- / 4-Methyl-2-pentanol	$C_6H_{14}O$ / 102.18	108-11-2 / -90	131.6	4-01-00-01717 / 0.8075[20]	H_2O 2; EtOH 3; eth 3; ctc 2 / 1.4100[20]
8943	3-Pentanol, 2-methyl- / 2-Methyl-3-pentanol	$C_6H_{14}O$ / 102.18	565-67-3 /	126.5	4-01-00-01716 / 0.8243[20]	H_2O 2; EtOH 5; eth 5 / 1.4175[20]
8944	3-Pentanol, 3-methyl-	$C_6H_{14}O$ / 102.18	77-74-7 / -23.6	122.4	4-01-00-01723 / 0.8286[20]	H_2O 2; EtOH 5; eth 5; ctc 2 / 1.4186[20]
8945	1-Pentanol, 2-methyl-, acetate / Acetic acid, (2-methylpentyl) ester	$C_8H_{16}O_2$ / 144.21	7789-99-3 /	163	4-02-00-00160 / 0.8691[25]	eth 4; EtOH 4
8946	2-Pentanol, 2-methyl-, acetate / 1,1-Dimethylbutyl acetate	$C_8H_{16}O_2$ / 144.21	34859-98-8 /	153; 34[10]	3-02-00-00257 / 0.8798[18]	eth 4; EtOH 4 / 1.4068[10]
8947	2-Pentanol, 4-methyl-, acetate / 1,3-Dimethylbutyl acetate	$C_8H_{16}O_2$ / 144.21	108-84-9 /	147.5	4-02-00-00161 / 0.8805[25]	eth 4; EtOH 4 / 1.3980[20]
8948	3-Pentanol, 2-methyl-, acetate / Acetic acid, (2-methyl-3-pentyl) ester	$C_8H_{16}O_2$ / 144.21	35897-16-6 /	149	4-02-00-00160 / 0.8688[20]	eth 4; EtOH 4
8949	3-Pentanol, 3-methyl-, acetate / Acetic acid, (3-methyl-3-pentyl) ester	$C_8H_{16}O_2$ / 144.21	10250-47-2 /	148	4-02-00-00161 / 0.8818[20]	eth 4; EtOH 4 / 1.4169[18]
8950	1-Pentanol, 2-nitro- / 2-Nitro-1-pentanol	$C_5H_{11}NO_3$ / 133.15	2899-90-3 /	133[28]; 117[10]	3-01-00-01606 / 1.0818[25]	/ 1.4405[25]
8951	2-Pentanol, 1-nitro-	$C_5H_{11}NO_3$ / 133.15	2224-37-5 /	87-8[1]	4-01-00-01661 / 1.0968[20]	/ 1.4493[20]
8952	2-Pentanol, 3-nitro- / 3-Nitro-2-pentanol	$C_5H_{11}NO_3$ / 133.15	5447-99-4 /	60[0.5]	3-01-00-01614 / 1.075[25]	/ 1.4430[20]
8953	1-Pentanol, 2,2,3,3,4,4,5,5-octafluoro- / 2,2,3,3,4,4,5,5-Octafluoro-1-pentanol	$C_5H_4F_8O$ / 232.07	355-80-6 /	6667 140.5	4-01-00-01648 / 1.6647[20]	/ 1.3178[20]
8954	2-Pentanol, 2,4,4-trimethyl- / 2,4,4-Trimethyl-2-pentanol	$C_8H_{18}O$ / 130.23	690-37-9 / -20	147.5	4-01-00-01796 / 0.8225[20]	H_2O 1; EtOH 2; eth 3 / 1.4284[20]
8955	3-Pentanol, 2,2,4-trimethyl- / Isopropyl-tert-butylcarbinol	$C_8H_{18}O$ / 130.23	5162-48-1 / -13	150.5	4-01-00-01795 / 0.8297[20]	/ 1.4288[20]
8956	3-Pentanol, 2,3,4-trimethyl-	$C_8H_{18}O$ / 130.23	3054-92-0 /	156.5	4-01-00-01798 / 0.8492[20]	/ 1.4353[20]
8957	2-Pentanone / Methyl propyl ketone	$C_5H_{10}O$ / 86.13	107-87-9 / -76.9	6032 102.2	4-01-00-03271 / 0.809[20]	H_2O 2; EtOH 5; eth 5; ctc 2 / 1.3895[20]
8958	3-Pentanone / Diethyl ketone	$C_5H_{10}O$ / 86.13	96-22-0 / -39	3111 101.96	4-01-00-03279 / 0.8098[25]	H_2O 4; EtOH 5; eth 5; ctc 5 / 1.3905[25]
8959	2-Pentanone, 4-amino-4-methyl- / Diacetonamine	$C_6H_{13}NO$ / 115.18	625-04-7 /	2942 25[0.14]	3-04-00-00894 /	H_2O 3; EtOH 5; eth 5
8960	1-Pentanone, 1-(4-aminophenyl)-	$C_{11}H_{15}NO$ / 177.25	38237-74-0 / 74.5	160-3[3]	4-14-00-00178 /	H_2O 1; EtOH 3; eth 3
8961	2-Pentanone, 5-chloro-	C_5H_9ClO / 120.58	5891-21-4 /	76[34]	4-01-00-03277 / 1.0523[20]	eth 3; ace 3; ctc 2 / 1.4375[20]
8962	3-Pentanone, 1-chloro-	C_5H_9ClO / 120.58	32830-97-0 /	68[20]	4-01-00-03284 /	eth 4; EtOH 4 / 1.4361[20]
8963	2-Pentanone, 3,5-dichloro- / 3,5-Dichloro-2-pentanone	$C_5H_8Cl_2O$ / 155.02	58371-98-5 /	99[15]; 58[3]	4-01-00-03277 / 1.239[15]	/ 1.4632[20]

No.	Name Synonym	Mol. Form. Mol. Wt.	CAS RN mp/°C	Merck No. bp/°C	Beil. Ref. den/g cm^{-3}	Solubility n_D
8964	2-Pentanone, 5-(diethylamino)-	$C_9H_{19}NO$ 157.26	105-14-6	83-5[15]	4-04-00-01935 0.861[25]	1.4350[20]
8965	2-Pentanone, 4,4-dimethyl-	$C_7H_{14}O$ 114.19	590-50-1 -64	126	4-01-00-03331 0.809[25]	1.4036[20]
8966	3-Pentanone, 2,2-dimethyl- 2,2-Dimethyl-3-pentanone	$C_7H_{14}O$ 114.19	564-04-5 -45	125.6	4-01-00-03331 0.8125[20]	H_2O 2; EtOH 3; eth 3; ace 3 1.4065[20]
8967	3-Pentanone, 2,4-dimethyl- Diisopropyl ketone	$C_7H_{14}O$ 114.19	565-80-0 -69	125.4	4-01-00-03334 0.8108[20]	H_2O 1; EtOH 5; eth 5; bz 3 1.3999[20]
8968	3-Pentanone, 2,4-dimethyl-, oxime Diisopropyl ketone oxime	$C_7H_{15}NO$ 129.20	1113-74-2 34	181; 91.5[21]	4-01-00-03335 0.9022[17]	
8969	3-Pentanone, 1,5-diphenyl- Diphenethyl ketone	$C_{17}H_{18}O$ 238.33	5396-91-8 13.5	352	4-07-00-01462 1.0356[0]	
8970	2-Pentanone, 3-ethyl-4-methyl-	$C_8H_{16}O$ 128.21	71172-57-1	154.5	3-01-00-02883 0.812[20]	H_2O 3; EtOH 5; eth 5; bz 5 1.4105[20]
8971	2-Pentanone, 1-hydroxy- 1-Pentanol-2-one	$C_5H_{10}O_2$ 102.13	64502-89-2	152	4-01-00-04004 0.9860[20]	H_2O 4; eth 4; EtOH 4 1.4234[12]
8972	2-Pentanone, 3-hydroxy-	$C_5H_{10}O_2$ 102.13	3142-66-3	147.5	3-01-00-03220 0.9500[20]	H_2O 4; ace 4; bz 4; eth 4 1.4350[10]
8973	2-Pentanone, 4-hydroxy-	$C_5H_{10}O_2$ 102.13	4161-60-8	177	4-01-00-04005 1.0071[20]	H_2O 3; EtOH 3; eth 3 1.4415[18]
8974	2-Pentanone, 5-hydroxy-	$C_5H_{10}O_2$ 102.13	1071-73-4	209; 117[33]	4-01-00-04006 1.0071[20]	H_2O 5; EtOH 3; eth 3 1.4390[20]
8975	3-Pentanone, 2-hydroxy- 2-Pentanol-3-one	$C_5H_{10}O_2$ 102.13	5704-20-1	152.5	4-01-00-04008 0.9742[20]	eth 4; EtOH 4 1.4128[20]
8976	2-Pentanone, 4-hydroxy-4-methyl- Diacetone alcohol	$C_6H_{12}O_2$ 116.16	123-42-2 -44	2944 167.9	4-01-00-04023 0.9387[20]	H_2O 5; EtOH 5; eth 5; chl 3 1.4213[20]
8977	2-Pentanone, 4-imino-	C_5H_9NO 99.13	870-74-6 43	209	4-01-00-03678 0.9427[0]	H_2O 4; eth 4
8978	2-Pentanone, 4-methoxy-4-methyl- Pentoxone	$C_7H_{14}O_2$ 130.19	107-70-0	160	4-01-00-04024 0.8980[25]	1.418[20]
8979	1-Pentanone, 1-(3-methoxyphenyl)-	$C_{12}H_{16}O_2$ 192.26	20359-55-1	196[30]	4-08-00-00533 1.012[20]	1.5242[31]
8980	2-Pentanone, 3-methyl-, (±)- 2-Pentanone, 3-methyl (DL)	$C_6H_{12}O$ 100.16	55156-16-6	118	4-01-00-03309 0.8130[20]	H_2O 2; EtOH 5; eth 5; chl 3 1.4002[20]
8981	2-Pentanone, 4-methyl- Isobutyl methyl ketone	$C_6H_{12}O$ 100.16	108-10-1 -84	5095 116.5	4-01-00-03305 0.7965[25]	H_2O 2; EtOH 5; eth 5; ace 5 1.3962[20]
8982	3-Pentanone, 2-methyl- Ethyl isopropyl ketone	$C_6H_{12}O$ 100.16	565-69-5	113.5	4-01-00-03304 0.814[18]	H_2O 2; EtOH 4; eth 5; ace 5 1.3975[20]
8983	1-Pentanone, 4-methyl-1-phenyl-	$C_{12}H_{16}O$ 176.26	2050-07-9 -2	255.5	3-07-00-01158 0.9623[15]	ace 4; bz 4; eth 4; EtOH 4 1.533[20]
8984	2-Pentanone, 3-methyl-1-phenyl- Benzyl sec-butyl ketone	$C_{12}H_{16}O$ 176.26	27993-42-6	127-8[17]	4-07-00-00758 0.964[20]	1.5019[20]
8985	2-Pentanone, oxime Methyl propyl ketone oxime	$C_5H_{11}NO$ 101.15	623-40-5	168	4-01-00-03274 0.9095[20]	H_2O 4; eth 4; EtOH 4 1.4450[20]
8986	1-Pentanone, 1-phenyl-	$C_{11}H_{14}O$ 162.23	1009-14-9 -9.4	245	4-07-00-00733 0.986[20]	H_2O 1; EtOH 4; eth 4; ctc 2 1.5158[20]
8987	2-Pentanone, 3-(phenylmethylene)- Methyl α-ethylstyryl ketone	$C_{12}H_{14}O$ 174.24	3437-89-6	136-8[12]	3-07-00-01461 1.0005[22]	1.5650[22]
8988	3-Pentanone, 2,2,4,4-tetramethyl-	$C_9H_{18}O$ 142.24	815-24-7 -25.2	152	4-01-00-03365 0.8240[18]	H_2O 1; EtOH 3; eth 3; ace 3 1.4194[20]
8989	3-Pentanone, 2,2,4-trimethyl- tert-Butyl isopropyl ketone	$C_8H_{16}O$ 128.21	5857-36-3	135.1	4-01-00-03351 0.8065[20]	H_2O 1; eth 3; ace 3 1.4060
8990	Pentanoyl chloride Valeroyl chloride	C_5H_9ClO 120.58	638-29-9 -110	109	4-02-00-00874 1.0155[15]	1.4200[20]
8991	Pentanoyl chloride, 2-chloro-	$C_5H_8Cl_2O$ 155.02	61589-68-2	156	4-02-00-00879 1.1765[20]	eth 4 1.4465[20]
8992	Pentanoyl chloride, 5-chloro-	$C_5H_8Cl_2O$ 155.02	1575-61-7	83[12]	4-02-00-00881 1.210[18]	eth 4 1.4639[20]
8993	Pentanoyl chloride, 2-methyl-, (±)- Valeryl chloride, 2-methyl (DL)	$C_6H_{11}ClO$ 134.61	116908-84-0	141	3-02-00-00741 0.9781[20]	eth 4 1.4330[27]
8994	Pentanoyl chloride, 3-methyl-, (±)- Valeryl chloride, 3-methyl (DL)	$C_6H_{11}ClO$ 134.61	116908-85-1	141	4-02-00-00949 0.9781[20]	eth 4
8995	2,5,8,11,14-Pentaoxapentadecane Tetraethylene glycol dimethyl ether	$C_{10}H_{22}O_5$ 222.28	143-24-8	275.3	4-01-00-02404 1.0114[20]	H_2O 5; EtOH 3; eth 3; ctc 3
8996	Pentaphene 2,3:6,7-Dibenzphenanthrene	$C_{22}H_{14}$ 278.35	222-93-5 257	2992	4-05-00-02721	H_2O 1; EtOH 2; eth 2; bz 3
8997	Pentasiloxane, dodecamethyl- Dodecamethylpentasiloxane	$C_{12}H_{36}O_4Si_5$ 384.84	141-63-9 -80	3401 232; 105[12]	4-04-00-04124 0.8755[20]	ctc 3; CS_2 3 1.3925[20]
8998	1,2,3,5,6-Pentathiepane Lenthionine	$C_2H_4S_5$ 188.38	292-46-6 60.5	5321	5-19-12-00251	
8999	Pentatriacontane	$C_{35}H_{72}$ 492.96	630-07-9 75	490	4-01-00-00598 0.8157[20]	H_2O 1; eth 2; ace 3 1.4568[20]
9000	18-Pentatriacontanone	$C_{35}H_{70}O$ 506.94	504-53-0 89.0	270[0.1]	4-01-00-03413 0.793[95]	H_2O 1; EtOH 2; eth 2; ace 2
9001	4-Pentenal Pent-4-en-1-al	C_5H_8O 84.12	2100-17-6	99	4-01-00-03459 0.852[20]	H_2O 1; eth 3; ace 3 1.4191[20]
9002	2-Pentenal, 2-methyl-	$C_6H_{10}O$ 98.14	623-36-9	136.5	4-01-00-03471 0.8581[20]	H_2O 1; EtOH 3; eth 3; bz 3 1.4488[20]
9003	4-Pentenamide, 2,2-diethyl- Novonal	$C_9H_{17}NO$ 155.24	512-48-1 75.5	6643	3-02-00-01351	eth 4; EtOH 4

No.	Name Synonym	Mol. Form. Mol. Wt.	CAS RN mp/°C	Merck No. bp/°C	Beil. Ref. den/g cm^{-3}	Solubility n_D
9004	1-Pentene α-Amylene	C_5H_{10} 70.13	109-67-1 -165.2	7079 29.9	4-01-00-00808 0.6405[20]	H_2O 1; EtOH 5; eth 5; bz 3 1.3715[20]
9005	2-Pentene, (E)- trans-2-Pentene	C_5H_{10} 70.13	646-04-8 -140.2	7080 36.3	4-01-00-00814 0.6431[25]	H_2O 1; EtOH 5; eth 5; bz 3 1.3793[20]
9006	2-Pentene, (Z)- cis-2-Pentene	C_5H_{10} 70.13	627-20-3 -151.4	7080 36.9	4-01-00-00814 0.6556[20]	H_2O 1; EtOH 5; eth 5; bz 3 1.3830[20]
9007	1-Pentene, 1-bromo- 1-Bromo-1-pentene	C_5H_9Br 149.03	60468-23-7 119	 	3-01-00-00775 1.2606[20]	bz 4; eth 4; chl 4 1.4572[20]
9008	1-Pentene, 2-bromo- 2-Bromo-1-pentene	C_5H_9Br 149.03	31844-95-8 107.5	 	3-01-00-00775 1.228[20]	bz 4; eth 4; lig 4 1.4535[20]
9009	1-Pentene, 3-bromo- 1-Ethylallyl bromide	C_5H_9Br 149.03	53045-71-9 30.5[30]	 	2-01-00-00183 1.2417[25]	ace 4; bz 4; chl 4 1.4626[25]
9010	1-Pentene, 5-bromo-	C_5H_9Br 149.03	1119-51-3 125.5	 	3-01-00-00775 1.2581[20]	 1.4640[20]
9011	2-Pentene, 1-bromo-	C_5H_9Br 149.03	20599-27-3 123.5	 	3-01-00-00783 1.2545[20]	ace 4; bz 4; chl 4 1.4731[20]
9012	2-Pentene, 2-bromo- 2-Bromo-2-pentene	C_5H_9Br 149.03	80204-19-9 111	 	3-01-00-00784 1.275[20]	ace 4; bz 4; chl 4 1.4580[20]
9013	2-Pentene, 3-bromo- 3-Bromo-2-pentene	C_5H_9Br 149.03	21964-23-8 116	 	3-01-00-00784 1.271[20]	ace 4; bz 4; chl 4 1.4628[20]
9014	2-Pentene, 4-bromo-	C_5H_9Br 149.03	1809-26-3 117 dec; 22[9]	 	4-01-00-00816 1.2312[21]	ace 4; bz 4; chl 4 1.4752[21]
9015	2-Pentene, 5-bromo- 3-Pentenyl bromide	C_5H_9Br 149.03	51952-42-2 121.7[621]	 	4-01-00-00816 1.2715[20]	ace 4; bz 4; chl 4 1.4695[20]
9016	1-Pentene, 2-chloro- 2-Chloro-1-pentene	C_5H_9Cl 104.58	42131-85-1 96	 	3-01-00-00774 0.872[5]	eth 4; EtOH 4
9017	1-Pentene, 3-chloro- 1-Ethylallyl chloride	C_5H_9Cl 104.58	24356-00-1 93	 	3-01-00-00774 0.8978[20]	ace 4; eth 4; EtOH 4 1.4254[20]
9018	1-Pentene, 4-chloro- 4-Chloro-1-pentene	C_5H_9Cl 104.58	10524-08-0 97	 	2-01-00-00183 0.9341[5]	H_2O 1; eth 4; chl 4 1.417[15]
9019	1-Pentene, 5-chloro-	C_5H_9Cl 104.58	928-50-7 105	 	4-01-00-00810 0.9125[20]	ace 4; eth 4 1.4297[20]
9020	2-Pentene, 1-chloro- γ-Ethylallyl chloride	C_5H_9Cl 104.58	10071-60-0 108	 	3-01-00-00782 0.908[22]	ace 4; eth 4; EtOH 4; chl 4 1.4352[22]
9021	2-Pentene, 2-chloro- 2-Chloro-2-pentene	C_5H_9Cl 104.58	67747-70-0 96	 	3-01-00-00782 0.9067[20]	ace 4; eth 4 1.4261[20]
9022	2-Pentene, 3-chloro- 3-Chloro-2-pentene	C_5H_9Cl 104.58	34238-52-3 91	 	3-01-00-00782 0.9125[20]	ace 4; eth 4 1.423[24]
9023	2-Pentene, 4-chloro-	C_5H_9Cl 104.58	1458-99-7 103; 47[25]	 	4-01-00-00816 0.8988[20]	ace 4; eth 4; chl 4 1.4332[20]
9024	2-Pentene, 5-chloro-	C_5H_9Cl 104.58	16435-50-0 108	 	4-01-00-00816 0.9043[20]	ace 4; bz 4; chl 4 1.4310[20]
9025	1-Pentene, 3-chloro-2-methyl-	$C_6H_{11}Cl$ 118.61	4104-01-2 122.5	 	3-01-00-00809 	bz 4; eth 4; chl 4 1.4422[20]
9026	2-Pentene, 5-chloro-2-methyl- 5-Chloro-2-methyl-2-pentene	$C_6H_{11}Cl$ 118.61	7712-60-9 133	 	4-01-00-00843 0.9135[20]	ace 4; chl 4
9027	1-Pentene, 1,1,2,3,3,4,4,5,5,5-decafluoro-	C_5F_{10} 250.04	376-87-4 30	 	4-01-00-00810 	chl 4 1.2571[25]
9028	2-Pentene, 2,5-dichloro- 1,4-Dichloro-3-pentene	$C_5H_8Cl_2$ 139.02	20177-02-0 40-1[8]	 	4-01-00-00816 1.1182[15]	chl 4
9029	1-Pentene, 2,3-dimethyl- 2,3-Dimethyl-1-pentene	C_7H_{14} 98.19	3404-72-6 -134.3	 84.3	3-01-00-00832 0.7051[20]	H_2O 1; EtOH 5; eth 5; dil sulf 4 1.4033[20]
9030	1-Pentene, 2,4-dimethyl- 2,4-Dimethyl-1-pentene	C_7H_{14} 98.19	2213-32-3 -124.1	 81.6	4-01-00-00871 0.6943[20]	H_2O 1; EtOH 5; eth 5; bz 3 1.3986[20]
9031	1-Pentene, 3,3-dimethyl- 3,3-Dimethyl-1-pentene	C_7H_{14} 98.19	3404-73-7 -134.3	 77.5	4-01-00-00873 0.6974[20]	H_2O 1; EtOH 5; eth 5; bz 3 1.3984[20]
9032	1-Pentene, 3,4-dimethyl- 3,4-Dimethyl-1-pentene	C_7H_{14} 98.19	7385-78-6 80.8	 	4-01-00-00871 0.6943[25]	 1.3992[20]
9033	1-Pentene, 4,4-dimethyl-	C_7H_{14} 98.19	762-62-9 -136.6	 72.5	4-01-00-00869 0.6827[20]	H_2O 1; EtOH 5; eth 5; bz 3 1.3818[20]
9034	2-Pentene, 2,3-dimethyl- Ethyltrimethylethylene	C_7H_{14} 98.19	10574-37-5 -118.3	 97.5	4-01-00-00870 0.7277[20]	H_2O 1; EtOH 3; eth 3; bz 3 1.4208[20]
9035	2-Pentene, 2,4-dimethyl- 2,4-Dimethyl-2-pentene	C_7H_{14} 98.19	625-65-0 -127.7	 83.4	4-01-00-00872 0.6954[20]	H_2O 1; EtOH 3; eth 3; bz 3 1.4040[20]
9036	2-Pentene, 3,4-dimethyl- 3,4-Dimethyl-2-pentene	C_7H_{14} 98.19	24910-63-2 -124.2	 89	3-01-00-00833 0.7126[20]	H_2O 1; EtOH 3; eth 3; bz 3 1.4070[20]
9037	2-Pentene, 3,4-dimethyl-, (E)-	C_7H_{14} 98.19	4914-92-5 -124.2	 91.5	4-01-00-00871 0.7124[25]	 1.4128[20]
9038	2-Pentene, 3,4-dimethyl-, (Z)-	C_7H_{14} 98.19	4914-91-4 -113.4	 89.3	4-01-00-00871 0.7092[25]	 1.4104[20]
9039	2-Pentene, 4,4-dimethyl-, (E)-	C_7H_{14} 98.19	690-08-4 -115.2	 76.7	4-01-00-00868 0.6889[20]	H_2O 1; EtOH 3; eth 3; bz 3 1.3982[20]
9040	2-Pentene, 4,4-dimethyl-, (Z)-	C_7H_{14} 98.19	762-63-0 -135.4	 80.4	4-01-00-00868 0.6951[25]	 1.4026[20]
9041	2-Pentenedioic acid, diethyl ester Diethyl glutaconate	$C_9H_{14}O_4$ 186.21	2049-67-4 237	 	 1.0496[20]	eth 4; EtOH 4 1.4411[20]
9042	2-Pentenedioic acid, diethyl ester, (E)- trans-2-Pentenedioic acid diethyl ester	$C_9H_{14}O_4$ 186.21	73178-43-5 237; 126[14]	 	4-02-00-02227 1.0496[20]	eth 4; EtOH 4 1.4411[20]

No.	Name / Synonym	Mol. Form. / Mol. Wt.	CAS RN / mp/°C	Merck No. / bp/°C	Beil. Ref. / den/g cm^{-3}	Solubility / n_D
9043	1-Pentene, 2-ethyl- / 2-Ethyl-1-pentene	C_7H_{14} / 98.19	3404-71-5 /	/ 94	4-01-00-00867 / 0.7079^{20}	bz 4; eth 4; EtOH 4 / 1.405^{20}
9044	1-Pentene, 3-ethyl- / 3-Ethyl-1-pentene	C_7H_{14} / 98.19	4038-04-4 / -127.5	/ 84.1	4-01-00-00867 / 0.6917^{25}	1.3982^{20}
9045	2-Pentene, 3-ethyl- / 3-Ethyl-2-pentene	C_7H_{14} / 98.19	816-79-5 /	/ 96	4-01-00-00868 / 0.7204^{20}	H_2O 1; EtOH 3; eth 3; bz 3 / 1.4148^{20}
9046	1-Pentene, 3-ethyl-2-methyl-	C_8H_{16} / 112.22	19780-66-6 / -112.9	/ 109.5	4-01-00-00891 / 0.7262^{20}	1.4140^{20}
9047	1-Pentene, 3-ethyl-3-methyl- / 3-Ethyl-3-methyl-1-pentene	C_8H_{16} / 112.22	6196-60-7 /	/ 112	4-01-00-00891 / 0.7264^{20}	1.418^{20}
9048	1-Pentene, 3-ethyl-4-methyl- / 1-Pentene, 4-methyl-3-ethyl	C_8H_{16} / 112.22	61847-80-1 /	/ 107.5	4-01-00-00890 / 0.7158^{25}	1.4097^{20}
9049	2-Pentene, 3-ethyl-2-methyl- / 3-Ethyl-2-methyl-2-pentene	C_8H_{16} / 112.22	19780-67-7 /	/ 115	4-01-00-00890 / 0.735^{25}	1.4247^{20}
9050	2-Pentene, 3-ethyl-4-methyl-, (E)- / 2-Pentene, 4-methyl-3-ethyl (trans)	C_8H_{16} / 112.22	42067-49-2 /	/ 114.3	4-01-00-00890 / 0.7308^{25}	1.4210^{20}
9051	2-Pentene, 3-ethyl-4-methyl-, (Z)- / 2-Pentene, 4-methyl-3-ethyl (cis)	C_8H_{16} / 112.22	42067-48-1 /	/ 116	4-01-00-00890 / 0.735^{25}	1.424^{20}
9052	1-Pentene, 2-methyl- / 2-Methyl-1-pentene	C_6H_{12} / 84.16	763-29-1 / -135.7	/ 62.1	4-01-00-00841 / 0.6799^{20}	H_2O 1; EtOH 3; bz 3; ctc 2 / 1.3920^{20}
9053	1-Pentene, 3-methyl- / sec-Butylethene	C_6H_{12} / 84.16	760-20-3 / -153	/ 54.2	3-01-00-00812 / 0.6675^{20}	H_2O 1; EtOH 3; bz 3; chl 3 / 1.3841^{20}
9054	1-Pentene, 4-methyl- / Isobutylethene	C_6H_{12} / 84.16	691-37-2 / -153.6	/ 53.9	4-01-00-00846 / 0.6642^{20}	H_2O 1; EtOH 3; bz 3; chl 3 / 1.3828^{20}
9055	2-Pentene, 2-methyl- / 2-Methyl-2-pentene	C_6H_{12} / 84.16	625-27-4 / -135	/ 67.3	4-01-00-00842 / 0.6863^{20}	H_2O 1; EtOH 3; bz 3; ctc 3 / 1.4004^{20}
9056	2-Pentene, 3-methyl-, (E)-	C_6H_{12} / 84.16	616-12-6 / -138.5	/ 70.4	4-01-00-00848 / 0.6930^{25}	H_2O 1; EtOH 3; bz 3; ctc 3 / 1.4045^{20}
9057	2-Pentene, 3-methyl-, (Z)- / (cis)-3-Methyl-2-pentene	C_6H_{12} / 84.16	922-62-3 / -134.8	/ 67.7	4-01-00-00848 / 0.6886^{25}	H_2O 1; EtOH 3; bz 3; chl 3 / 1.4016^{20}
9058	2-Pentene, 4-methyl-, (E)- / 4-Methyl-trans-2-pentene	C_6H_{12} / 84.16	674-76-0 / -140.8	/ 58.6	4-01-00-00844 / 0.6686^{20}	H_2O 1; EtOH 3; bz 3; ctc 2 / 1.3889^{20}
9059	2-Pentene, 4-methyl-, (Z)- / cis-4-Methyl-2-pentene	C_6H_{12} / 84.16	691-38-3 / -134.8	/ 56.3	4-01-00-00844 / 0.6690^{20}	H_2O 1; EtOH 3; bz 3; chl 3 / 1.3800^{20}
9060	1-Pentene, 3-methylene-	C_6H_{10} / 82.15	3404-63-5 /	/ 75	4-01-00-01022 / 0.712^{25}	1.445^{20}
9061	3-Pentenenitrile, (E)- / trans-3-Pentenenitrile	C_5H_7N / 81.12	16545-78-1 /	/ 142.6		
9062	4-Pentenenitrile	C_5H_7N / 81.12	592-51-8 /	/ 140	4-02-00-01543 / 0.8239^{24}	H_2O 1; EtOH 5; eth 5 / 1.4213^{14}
9063	3-Pentenenitrile, 2-hydroxy-, (Z)-	C_5H_7NO / 97.12	116908-78-2 /	/ 139^{70}	4-03-00-01001 / 0.9675^{15}	bz 4; eth 4; EtOH 4; chl 4 / 1.4460^{21}
9064	Pentene, 2,4,4-trimethyl- / Diisobutylene	C_8H_{16} / 112.22	25167-70-8 /	/ 101	4-01-00-00892 /	
9065	1-Pentene, 2,3,3-trimethyl- / 2,3,3-Trimethyl-1-pentene	C_8H_{16} / 112.22	560-23-6 / -69	/ 108.3	4-01-00-00894 / 0.7308^{25}	1.4174^{20}
9066	1-Pentene, 2,3,4-trimethyl- / 2,3,4-Trimethyl-1-pentene	C_8H_{16} / 112.22	565-76-4 /	/ 108	4-01-00-00894 / 0.725^{25}	1.415^{20}
9067	1-Pentene, 2,4,4-trimethyl- / 2,2,4-Trimethyl-4-pentene	C_8H_{16} / 112.22	107-39-1 / -93.5	/ 101.4	4-01-00-00892 / 0.7150^{20}	H_2O 1; eth 3; bz 3; ctc 3 / 1.4086^{20}
9068	1-Pentene, 3,3,4-trimethyl-	C_8H_{16} / 112.22	560-22-5 /	/ 105	3-01-00-00851 / 0.725^{25}	1.4144^{20}
9069	1-Pentene, 3,4,4-trimethyl- / 3,4,4-Trimethyl-1-pentene	C_8H_{16} / 112.22	564-03-4 /	/ 104	4-01-00-00891 / 0.715^{25}	1.412^{20}
9070	2-Pentene, 2,3,4-trimethyl- / 2,3,4-Trimethyl-2-pentene	C_8H_{16} / 112.22	565-77-5 / -113.4	/ 116.5	4-01-00-00894 / 0.7434^{20}	1.4274^{20}
9071	2-Pentene, 2,4,4-trimethyl- / 2,4,4-Trimethyl-2-pentene	C_8H_{16} / 112.22	107-40-4 / -106.3	/ 104.9	4-01-00-00891 / 0.7218^{20}	H_2O 1; eth 3; bz 3; ctc 3 / 1.4160^{20}
9072	2-Pentene, 3,4,4-trimethyl- / 3,4,4-Trimethyl-2-pentene	C_8H_{16} / 112.22	598-96-9 /	/ 112	4-01-00-00891 / 0.735^{25}	EtOH 4 / 1.423^{20}
9073	4-Pentenoic acid / Allylacetic acid	$C_5H_8O_2$ / 100.12	591-80-0 / -22.5	/ 188.5	4-02-00-01542 / 0.9809^{20}	H_2O 2; EtOH 4; eth 4 / 1.4281^{20}
9074	4-Pentenoic acid, 2-acetyl-, ethyl ester	$C_9H_{14}O_3$ / 170.21	610-89-9 /	/ 208	3-03-00-01314 / 0.9898^{20}	EtOH 5; eth 5; bz 5 / 1.4388^{18}
9075	4-Pentenoic acid, 2,2-dimethyl- / 2,2-Dimethylpent-4-enoic acid	$C_7H_{12}O_2$ / 128.17	16386-93-9 /	/ 116^{30}	4-02-00-01585 / 0.933^{20}	1.4338^{20}
9076	2-Pentenoic acid, 2-methyl-, (E)- / 2-Pentenoic acid, 2-methyl (trans)	$C_6H_{10}O_2$ / 114.14	16957-70-3 / 24.4	/ 214; 112^{12}	4-02-00-01568 / 0.9751^{20}	H_2O 2; eth 3; chl 3; CS_2 3 / 1.4513^{20}
9077	2-Pentenoic acid, 3-methyl-, (Z)- / 2-Pentenoic acid, 3-methyl (cis)	$C_6H_{10}O_2$ / 114.14	19866-51-4 / 12	/ 96^5	4-02-00-01571 / 0.9830^{20}	1.4650
9078	2-Pentenoic acid, 4-methyl- / 4,4-Dimethyl-2-butenoic acid	$C_6H_{10}O_2$ / 114.14	10321-71-8 / 35	/ 217	3-02-00-01324 / 0.9529^{21}	ace 4; eth 4; EtOH 4 / 1.4489^{21}
9079	3-Pentenoic acid, 2-methyl-	$C_6H_{10}O_2$ / 114.14	37674-63-8 /	/ 199	3-02-00-01324 / 0.966^{15}	1.4402^{25}
9080	2-Pentenoic acid, 4-methyl-, ethyl ester / Ethyl 4,4-dimethylcrotonate	$C_8H_{14}O_2$ / 142.20	2351-97-5 /	/ 174; 76-7^{25}	3-02-00-01324 / 0.9048^{21}	1.4328^{20}
9081	3-Pentenoic acid, 3-methyl-, ethyl ester / Ethyl 3-methyl-3-pentenoate	$C_8H_{14}O_2$ / 142.20	86623-80-5 /	/ 61^{13}	3-02-00-01328 / 0.9183^{20}	1.4298^{20}
9082	3-Pentenoic acid, 4-methyl-, ethyl ester / Ethyl 4-methyl-3-pentenoate	$C_8H_{14}O_2$ / 142.20	6849-18-9 /	/ 58^{11}	3-02-00-01324 / 0.9134^{17}	1.4329^{17}

No.	Name Synonym	Mol. Form. Mol. Wt.	CAS RN mp/°C	Merck No. bp/°C	Beil. Ref. den/g cm⁻³	Solubility n_D
9083	2-Pentenoic acid, 2-methyl-, methyl ester	$C_7H_{12}O_2$ 128.17	10478-12-3	51[11]	4-02-00-01568 0.920[25]	1.4336
9084	2-Pentenoic acid, 3-methyl-, methyl ester, (E)-	$C_7H_{12}O_2$ 128.17	17447-01-7	79[50]	4-02-00-01571 0.9302[20]	1.4446[20]
9085	2-Pentenoic acid, 3-methyl-, methyl ester, (Z)- cis-3-Methyl-2-pentenoic acid methyl ester	$C_7H_{12}O_2$ 128.17	17447-00-6	74[50]	4-02-00-01571 0.9279[20]	1.4420[20]
9086	3-Pentenoic acid, 3-methyl-, methyl ester Methyl 3-methyl-3-pentenoate	$C_7H_{12}O_2$ 128.17	2258-58-4	74[50]	4-02-00-01573 0.9949[20]	1.4306[20]
9087	1-Penten-3-ol	$C_5H_{10}O$ 86.13	616-25-1	115	1-01-00-00227 0.839[20]	H_2O 2; EtOH 5; eth 5 1.4239[20]
9088	4-Penten-1-ol	$C_5H_{10}O$ 86.13	821-09-0	141	4-01-00-02119 0.8457[20]	H_2O 2; eth 3; ctc 2 1.4309[20]
9089	4-Penten-2-ol	$C_5H_{10}O$ 86.13	625-31-0	116	1-01-00-00227 0.8367[20]	H_2O 4; EtOH 5; eth 5 1.4225[20]
9090	2-Penten-1-ol, (E)-	$C_5H_{10}O$ 86.13	1576-96-1	138	4-01-00-02121 0.8471[20]	EtOH 3; eth 3; ace 3 1.4341[20]
9091	2-Penten-1-ol, (Z)-	$C_5H_{10}O$ 86.13	1576-95-0	138	4-01-00-02121 0.8529[20]	EtOH 3; eth 3; ace 3 1.4354[20]
9092	3-Penten-2-ol, (±)-	$C_5H_{10}O$ 86.13	42569-16-4	121.6; 65[70]	4-01-00-02122 0.8328[25]	ace 4; eth 4; EtOH 4 1.4280[20]
9093	2-Penten-1-ol, 5-(2,3-dimethyltricyclo[2.2.1.02,6]hept-3-yl)-2-methyl-, stereoisomer α-Santalol	$C_{15}H_{24}O$ 220.35	115-71-9 301.5	8321	4-06-00-03521 0.9679[20]	H_2O 1; EtOH 3 1.5023[20]
9094	2-Penten-1-ol, 2-methyl-	$C_6H_{12}O$ 100.16	1610-29-3	167.5	4-01-00-02144 0.8501[24]	1.444[24]
9095	2-Penten-1-ol, 4-methyl-	$C_6H_{12}O$ 100.16	5362-55-0	159	4-01-00-02146 0.8489[16]	1.4403[16]
9096	3-Penten-1-ol, 4-methyl- 4-Methyl-3-pentenol	$C_6H_{12}O$ 100.16	763-89-3	157	4-01-00-02145 0.8600[20]	1.4432[21]
9097	3-Penten-2-ol, 2-methyl- 2-Methyl-3-penten-2-ol	$C_6H_{12}O$ 100.16	63468-05-3	112; 65[72]	4-01-00-02145 0.8347[20]	1.4302[20]
9098	3-Penten-2-ol, 3-methyl- 3-Methyl-3-penten-2-ol	$C_6H_{12}O$ 100.16	2747-53-7	140	4-01-00-02149 0.879[20]	1.4428[17]
9099	3-Penten-2-ol, 4-methyl-	$C_6H_{12}O$ 100.16	4325-82-0	134	4-01-00-02144 0.840[15]	1.9377[15]
9100	4-Penten-2-ol, 2-methyl-	$C_6H_{12}O$ 100.16	624-97-5 -73	119.5	4-01-00-02146 0.8300[20]	H_2O 3; EtOH 3; eth 3; ctc 2 1.4268[20]
9101	4-Penten-2-ol, 3-methyl-	$C_6H_{12}O$ 100.16	1569-59-1	125.5	3-01-00-01933 0.8429[22]	1.4326[22]
9102	4-Penten-2-ol, 4-methyl-	$C_6H_{12}O$ 100.16	2004-67-3	129	4-01-00-02143 0.8436[20]	1.4297[17]
9103	2-Penten-1-ol, 2-methyl-5-(2-methyl-3-methylenebicyclo[2.2.1]hept-2-yl)-, [1S-[1α,2α(Z),4α] β-Santalol	$C_{15}H_{24}O$ 220.35	77-42-9 167-8[10]	8322	4-06-00-03519 0.9750[20]	1.5115[20]
9104	1-Penten-3-one Ethyl vinyl ketone	C_5H_8O 84.12	1629-58-9	103; 44[90]	4-01-00-03457 0.8468[20]	H_2O 1; EtOH 3; eth 3; ace 3 1.4195[20]
9105	3-Penten-2-one, (E)-	C_5H_8O 84.12	3102-33-8	122	4-01-00-03460 0.8624[20]	H_2O 3; eth 3; ace 3; ctc 3 1.4350[20]
9106	1-Penten-3-one, 4,4-dimethyl-1-phenyl-	$C_{13}H_{16}O$ 188.27	538-44-3 43	154[25]	3-07-00-01490 0.9508[46]	1.5523[25]
9107	1-Penten-3-one, 2-methyl-	$C_6H_{10}O$ 98.14	25044-01-3 -69.5	118.5	4-01-00-03470 1.8530[20]	ace 4; EtOH 4 1.4289[20]
9108	3-Penten-2-one, 4-methyl- Mesityl oxide	$C_6H_{10}O$ 98.14	141-79-7 -59	5811 130	4-01-00-03471 0.8653[20]	H_2O 5; EtOH 5; eth 5; ace 3 1.4440[20]
9109	4-Penten-2-one, 4-methyl-	$C_6H_{10}O$ 98.14	3744-02-3 -72.6	124.2	4-01-00-03470 0.8411[20]	
9110	1-Penten-3-one, 1-phenyl-	$C_{11}H_{12}O$ 160.22	3152-68-9 38.5	142[12]	2-07-00-00298 0.8697[20]	H_2O 2; EtOH 4; eth 4; bz 4 1.5684[20]
9111	1-Penten-3-yne Methylvinylacetylene	C_5H_6 66.10	646-05-9	59.5	4-01-00-01087 0.7401[20]	bz 4; eth 4 1.4496[20]
9112	1-Penten-4-yne	C_5H_6 66.10	871-28-3	42.5	4-01-00-01088 0.738[16]	H_2O 1; eth 3; bz 3 1.4125[16]
9113	3-Penten-1-yne, (E)- trans-3-Penten-1-yne	C_5H_6 66.10	2004-69-5	52.2	4-01-00-01085	
9114	3-Penten-1-yne, (Z)- cis-3-Penten-1-yne	C_5H_6 66.10	1574-40-9	44.6		
9115	3-Penten-1-yne, 3-ethyl-	C_7H_{10} 94.16	14272-54-9	96.5	4-01-00-01099 0.7886[25]	bz 4; eth 4 1.4338[25]
9116	1-Penten-3-yne, 2-methyl- Methylisopropenylacetylene	C_6H_8 80.13	926-55-6	81-2[100]	4-01-00-01095	bz 4; eth 4 1.4002[20]
9117	3-Penten-1-yne, 3-methyl- 3-Methyl-3-penten-1-yne	C_6H_8 80.13	1574-33-0	66.5	4-01-00-01096 0.739[20]	eth 3; bz 3 1.4332[20]
9118	4-Penten-2-yn-1-ol 1-Hydroxypent-2-yn-4-ene	C_5H_6O 82.10	2919-05-3	57[10]; 44[2.5]	4-01-00-02303 0.9581[20]	1.4986[20]
9119	Pentitol, 1,2-dideoxy- 1,2,3-Pentanetriol	$C_5H_{12}O_3$ 120.15	5371-48-2	192[63]	4-01-00-02778 1.0849[34]	H_2O 4; EtOH 4; eth 4
9120	Pentonic acid, 5-C-[3,5-bis(1-methylpropyl)-1-cyclopenten-1-yl]-4-deoxy-	$C_{18}H_{32}O_5$ 328.45	491-14-5 196		3-10-00-02045 1.292[14]	EtOH 4

No.	Name / Synonym	Mol. Form. / Mol. Wt.	CAS RN / mp/°C	Merck No. / bp/°C	Beil. Ref. / den/g cm^{-3}	Solubility / n_D
9121	1-Pentyn-3-amine, 3-ethyl-	$C_7H_{13}N$ / 111.19	3234-64-8 /	/ 71-2[90]	4-04-00-01154 / 0.828[25]	1.4409[20]
9122	1-Pentyne / Propylacetylene	C_5H_8 / 68.12	627-19-0 / -90	/ 40.1	4-01-00-00990 / 0.6901[20]	H_2O 1; EtOH 4; eth 5; bz 3 / 1.3852[20]
9123	2-Pentyne	C_5H_8 / 68.12	627-21-4 / -109.3	/ 56.1	4-01-00-00992 / 0.7058[25]	H_2O 1; EtOH 4; eth 5; bz 3 / 1.4039[20]
9124	1-Pentyne, 1-bromo- / 1-Pentynyl bromide	C_5H_7Br / 147.01	14752-60-4 /	/ 44-6[52]	3-01-00-00958 / 1.281[13]	1.4579[13]
9125	1-Pentyne, 3-chloro-3-ethyl- / 3-Chloro-3-ethyl-1-pentyne	$C_7H_{11}Cl$ / 130.62	6080-79-1 /	/ 73-6[100]	4-01-00-01032 / 0.9230[19]	bz 4; eth 4; chl 4 / 1.4437[19]
9126	1-Pentyne, 3-chloro-3-methyl- / 3-Chloro-3-methyl-1-pentyne	C_6H_9Cl / 116.59	14179-94-3 /	/ 102.3	4-01-00-01021 / 0.9163[20]	H_2O 1; eth 3; bz 3; chl 3 / 1.4330[20]
9127	2-Pentyne, 4-chloro-4-methyl-	C_6H_9Cl / 116.59	999-79-1 /	/ 55[70]	4-01-00-01018 /	ace 4; bz 4; chl 4 / 1.4143[20]
9128	1-Pentyne, 3,3-dimethyl- / 3,3-Dimethyl-1-pentyne	C_7H_{12} / 96.17	918-82-1 /	/ 70	4-01-00-01034 / 0.7032[25]	1.3934[20]
9129	1-Pentyne, 4,4-dimethyl- / 4,4-Dimethyl-1-pentyne	C_7H_{12} / 96.17	13361-63-2 / -75.7	/ 76.1	4-01-00-01033 / 0.7142[20]	bz 4; eth 4; chl 4 / 1.3983[20]
9130	2-Pentyne, 4,4-dimethyl-	C_7H_{12} / 96.17	999-78-0 / -82.4	/ 83	4-01-00-01032 / 0.7176[20]	H_2O 1; eth 3; bz 3; ctc 2 / 1.4071[20]
9131	1-Pentyne, 3-ethyl- / 3-Ethyl-1-pentyne	C_7H_{12} / 96.17	21020-26-8 /	/ 84	3-01-00-01002 / 0.7246[25]	bz 4; eth 4; chl 4 / 1.4043[25]
9132	1-Pentyne, 3-ethyl-3-methyl- / 3-Ethyl-3-methyl-1-pentyne	C_8H_{14} / 110.20	919-12-0 /	/ 101.5	4-01-00-01046 / 0.7422[20]	bz 4; eth 4; chl 4 / 1.4110[20]
9133	1-Pentyne, 1-iodo- / 1-Iodo-1-pentyne	C_5H_7I / 194.02	14752-61-5 /	/ 54[23]	4-01-00-00992 / 1.6127[19]	1.5148[19]
9134	1-Pentyne, 3-methyl- / 3-Methyl-1-pentyne	C_6H_{10} / 82.15	922-59-8 /	/ 57.7	4-01-00-01021 / 0.6992[25]	1.3916[20]
9135	1-Pentyne, 4-methyl-	C_6H_{10} / 82.15	7154-75-8 / -104.6	/ 61.2	4-01-00-01019 / 0.7000[25]	H_2O 1; bz 3; chl 3 / 1.3936[20]
9136	2-Pentyne, 4-methyl-	C_6H_{10} / 82.15	21020-27-9 / -110.3	/ 73.1	4-01-00-01018 / 0.7112[25]	bz 4; chl 4 / 1.4057[20]
9137	2-Pentynoic acid / Ethylpropiolic acid	$C_5H_6O_2$ / 98.10	5963-77-9 / 50	/ 122[10]; 81[3]	4-02-00-01693 / 0.978[20]	H_2O 4 / 1.4619[20]
9138	4-Pentynoic acid / Propargylacetic acid	$C_5H_6O_2$ / 98.10	6089-09-4 / 57.7	/ 110[30]; 102[17];	4-02-00-01693 /	eth 4; EtOH 4
9139	2-Pentynoic acid, methyl ester / Methyl 2-pentynoate	$C_6H_8O_2$ / 112.13	24342-04-9 /	/ 72[10]	4-02-00-01694 / 0.937[20]	1.4455[20]
9140	1-Pentyn-3-ol / Ethylethynylcarbinol	C_5H_8O / 84.12	4187-86-4 /	/ 125	4-01-00-02224 / 0.8926[15]	1.4347[15]
9141	2-Pentyn-1-ol	C_5H_8O / 84.12	6261-22-9 /	/ 61-2[15]	4-01-00-02226 / 0.913[17]	1.4518[17]
9142	3-Pentyn-1-ol	C_5H_8O / 84.12	10229-10-4 /	/ 154	4-01-00-02227 / 0.9002[20]	1.4454[20]
9143	4-Pentyn-1-ol	C_5H_8O / 84.12	5390-04-5 /	/ 154	4-01-00-02225 / 0.913[20]	1.4414[20]
9144	1-Pentyn-3-ol, 3,4-dimethyl-	$C_7H_{12}O$ / 112.17	1482-15-1 /	/ 134	4-01-00-02255 / 0.8691[20]	H_2O 4; eth 4; EtOH 4 / 1.4372[20]
9145	1-Pentyn-3-ol, 3-ethyl-	$C_7H_{12}O$ / 112.17	6285-06-9 /	/ 138	4-01-00-02254 / 0.8691[20]	H_2O 2
9146	1-Pentyn-3-ol, 3-methyl- / Meparfynol	$C_6H_{10}O$ / 98.14	77-75-8 / 30.5	5731 / 120.5	2-01-00-00506 / 0.8688[20]	1.4310[20]
9147	2-Pentynoyl chloride, 4,4-dimethyl- / 4,4-Dimethyl-2-pentynoyl chloride	C_7H_9ClO / 144.60	52324-03-5 /	/ 52[20]	2-02-00-00454 / 0.9743[20]	1.4443[20]
9148	Perchloric acid, trichloromethyl ester / Trichloro perchlorate	CCl_4O_4 / 217.82	67632-66-0 / -55	/ exp 40	3-03-00-00037 /	eth 4
9149	Perfluidone / o-Toluidine, 4-(phenylsulfonyl)-N-[(trifluoromethyl)sulfono]-	$C_{14}H_{12}F_3NO_4S_2$ / 379.38	37924-13-3 / 143	7113 /		
9150	Permethrin / Cyclopropanecarboxylic acid, 3-(2,2-dichloroethenyl)-2,2-dimethyl-, (3-phenoxyphenyl), methyl ester	$C_{21}H_{20}Cl_2O_3$ / 391.29	52645-53-1 / 34	7132 / 200[0.01]	1.23[20]	
9151	Peroxide, acetyl benzoyl / Acetozone	$C_9H_8O_4$ / 180.16	644-31-5 / 37	71 / 130[19]	4-09-00-00717 /	eth 4
9152	Peroxide, bis(2,4-dichlorobenzoyl)	$C_{14}H_6Cl_4O_4$ / 380.01	133-14-2 / 106		4-09-00-01001 /	
9153	Peroxide, bis(1,1-dimethylethyl) / DTBP	$C_8H_{18}O_2$ / 146.23	110-05-4 / -40	3446 / 111	4-01-00-01619 / 0.704[20]	H_2O 1; ace 5; ctc 3; os 3 / 1.3890[20]
9154	Peroxide, bis(1,1-dimethylpropyl)	$C_{10}H_{22}O_2$ / 174.28	10508-09-5 /	/ 58.5[14]; 44[10]	4-01-00-01671 / 0.821[20]	1.4095[20]
9155	Peroxide, bis(1-oxododecyl) / Lauroyl peroxide	$C_{24}H_{46}O_4$ / 398.63	105-74-8 / 49		4-02-00-01102 /	H_2O 1; chl 3
9156	Peroxide, diacetyl / Acetyl peroxide	$C_4H_6O_4$ / 118.09	110-22-5 / 30	/ 63[21]	4-02-00-00392 /	eth 4; EtOH 4
9157	Peroxide, dibenzoyl / Benzoyl peroxide	$C_{14}H_{10}O_4$ / 242.23	94-36-0 / 105	1128 / exp	4-09-00-00717 /	H_2O 2; EtOH 3; eth 3; ace 3 / 1.543
9158	Peroxide, diethyl	$C_4H_{10}O_2$ / 90.12	628-37-5 / -70	/ 65	4-01-00-01323 / 0.8240[19]	H_2O 2; EtOH 5; eth 5 / 1.3715[17]

No.	Name / Synonym	Mol. Form. / Mol. Wt.	CAS RN / mp/°C	Merck No. / bp/°C	Beil. Ref. / den/g cm^{-3}	Solubility / n_D
9159	Peroxide, dimethyl	$C_2H_8O_2$ / 62.07	690-02-8 /	/ 14	4-01-00-01249 / 0.8677[0]	EtOH 2; eth 2; tol 3; HOAc 3 / 1.3503[0]
9160	Perthane / Ethane, 1,1-dichloro-2,2-bis(p-ethylphenyl)-	$C_{18}H_{20}Cl_2$ / 307.26	72-56-0 / 56	3050 /		
9161	Perylene / Dibenz[de,kl]anthracene	$C_{20}H_{12}$ / 252.32	198-55-0 / 274	7137 /	4-05-00-02689 / 1.35[25]	H_2O 1; EtOH 2; eth 2; ace 4
9162	9-Phenanthrenamine	$C_{14}H_{11}N$ / 193.25	947-73-9 / 138.3	/ sub	4-12-00-03443 /	eth 2; bz 2; chl 2
9163	Phenanthrene	$C_{14}H_{10}$ / 178.23	85-01-8 / 99.2	7167 / 340	4-05-00-02297 / 0.9800[4]	H_2O 1; EtOH 3; eth 3; ace 3 / 1.5943
9164	Phenanthrene, 9-bromo- / 9-Phenanthryl bromide	$C_{14}H_9Br$ / 257.13	573-17-1 / 64.5	/ >360	4-05-00-02303 / 1.4093[10]	H_2O 1; EtOH 3; eth 3; chl 2
9165	9-Phenanthrenecarboxylic acid / 9-Phenanthroic acid	$C_{15}H_{10}O_2$ / 222.24	837-45-6 / 257.3	/ sub	4-09-00-02674 /	H_2O 1; EtOH 3; eth 3; bz 3
9166	1-Phenanthrenecarboxylic acid, 1,2,3,4,4a,5,6,9,10,10a-decahydro-1,4a-dimethyl-7-(1-methylethyl)-, / Palustric acid	$C_{20}H_{30}O_2$ / 302.46	1945-53-5 / 164.5	6950 /	4-09-00-02174 /	
9167	1-Phenanthrenecarboxylic acid, 1,2,3,4,4a,4b,5,6,10,10a-decahydro-1,4a-dimethyl-7-(1-methylethyl)-, / Abietic acid	$C_{20}H_{30}O_2$ / 302.46	514-10-3 / 173.5	2 / 250[9]	4-09-00-02175 /	ace 4; bz 4; eth 4; EtOH 4
9168	1-Phenanthrenecarboxylic acid, 1,2,3,4,4a,4b,5,6,10,10a-decahydro-1,4a-dimethyl-7-(1-methylethyl)-, methyl ester, [1R-(1α,4aβ,4bα,10aα)]- / Methyl abietate	$C_{21}H_{32}O_2$ / 316.48	127-25-3 /	5930 / 225-6[16]	4-09-00-02176 / 1.049[20]	H_2O 1; EtOH 3; HOAc 3 / 1.5344
9169	1-Phenanthrenecarboxylic acid, 1,2,3,4,4a,4b,5,9,10,10a-decahydro-1,4a-dimethyl-7-(1-methylethyl)-, / Levopimaric acid	$C_{20}H_{30}O_2$ / 302.46	79-54-9 / 150	5348 /	4-09-00-02173 /	
9170	1-Phenanthrenecarboxylic acid, 7-ethenyl-1,2,3,4,4a,4b,5,6,7,9,10,10a-dodecahydro-1,4a,7-trimethyl-, [1R-(1α,4aβ,4bα,7β,10aα)]- / Pimaric acid	$C_{20}H_{30}O_2$ / 302.46	127-27-5 / 218.5	7398 / 282[18]	4-09-00-02179 /	eth 4; py 4; EtOH 4
9171	Phenanthrene, 9,10-dihydro- / 9,10-Dihydrophenanthrene	$C_{14}H_{12}$ / 180.25	776-35-2 / 34.5	/ 168-9[15]	4-05-00-02183 / 1.0757[40]	chl 3 / 1.6415[20]
9172	Phenanthrene, 9,10-dimethyl- / 9,10-Phenanthraquinodimethane	$C_{16}H_{14}$ / 206.29	604-83-1 / 144	/ sub	4-05-00-02335 /	EtOH 2; bz 3; chl 3; HOAc 3
9173	3,4-Phenanthrenediol	$C_{14}H_{10}O_2$ / 210.23	478-71-7 / 143	/ sub 130	3-06-00-05689 /	eth 4; EtOH 4
9174	9,10-Phenanthrenedione / Phenanthrenequinone	$C_{14}H_8O_2$ / 208.22	84-11-7 / 209	7168 /	4-07-00-02565 / 1.405[22]	H_2O 1; EtOH 2; eth 3; bz 2
9175	9,10-Phenanthrenedione, 2-hydroxy-	$C_{14}H_8O_3$ / 224.22	4088-81-7 / 284	/ sub	3-08-00-02929 /	H_2O 4
9176	9,10-Phenanthrenedione, 3-hydroxy- / 9,10-Phenanthraquinone, 3-hydroxy	$C_{14}H_8O_3$ / 224.22	57404-54-3 / 330 dec	/ sub	2-08-00-00398 /	H_2O 4; EtOH 4
9177	9,10-Phenanthrenedione, 1-methyl-7-(1-methylethyl)-	$C_{18}H_{16}O_2$ / 264.32	5398-75-4 / 197.5	/ sub	4-07-00-02595 /	eth 4; EtOH 4
9178	Phenanthrene, dodecahydro- / Dodecahydrophenanthrene	$C_{14}H_{22}$ / 190.33	1322-67-4 /	/ 81-2[1.5]	3-05-00-01074 / 0.9674[20]	ace 4; bz 4 / 1.5102[20]
9179	Phenanthrene, 9-ethyl- / 9-Ethylphenanthrene	$C_{16}H_{14}$ / 206.29	3674-75-7 / 62.5	/ 199-200[1]	4-05-00-02332 / 1.0603[78]	H_2O 1; EtOH 3; bz 3; peth 2 / 1.6582[78]
9180	Phenanthrene, 3-methyl- / 3-Methylphenanthrene	$C_{15}H_{12}$ / 192.26	832-71-3 / 65	/ 350; 145[6]	4-05-00-02314 /	H_2O 1; EtOH 3; ace 3; chl 2
9181	Phenanthrene, 4-methyl- / 4-Methylphenanthrene	$C_{15}H_{12}$ / 192.26	832-64-4 / 53.5	/ 175-80[10]	3-05-00-02154 /	H_2O 1; EtOH 3; ctc 3
9182	Phenanthrene, 9-(1-methylethenyl)- / 9-Isopropenylphenanthrene	$C_{17}H_{14}$ / 218.30	58873-44-2 / 38	/ 163[20]	3-05-00-02240 /	1.6765[22]
9183	Phenanthrene, 1-methyl-7-(1-methylethyl)- / Retene	$C_{18}H_{18}$ / 234.34	483-65-8 / 101	8162 / 390	4-05-00-02368 / 1.035[25]	H_2O 1; EtOH 3; eth 3; bz 3
9184	Phenanthrene, 1,2,3,4,4a,9,10,10a-octahydro- / 1,2,3,4,4a,9,10,10a-Octahydrophenanthrene	$C_{14}H_{18}$ / 186.30	16306-39-1 /	/ 159[15]	4-05-00-01585 / 0.9973[32]	1.5527[19]
9185	Phenanthrene, 1,2,3,4,5,6,7,8-octahydro- / 1,2,3,4,5,6,7,8-Octahydrophenanthrene	$C_{14}H_{18}$ / 186.30	5325-97-3 / 16.7	/ 295	4-05-00-01585 / 1.026[20]	H_2O 1; ace 3; bz 3; CS_2 3 / 1.5569[17]
9186	Phenanthrene, tetradecahydro-	$C_{14}H_{24}$ / 192.34	5743-97-5 /	/ 86-9[2]	4-05-00-00492 / 0.9447[20]	H_2O 1; eth 3; ace 3; bz 3 / 1.5011[20]
9187	Phenanthrene, 1,2,3,4-tetrahydro- / 1,2,3,4-Tetrahydrophenanthrene	$C_{14}H_{14}$ / 182.27	1013-08-7 / 33.5	/ 173[11]	4-05-00-01909 / 1.0601[40]	H_2O 1; EtOH 3; eth 3; ace 3
9188	Phenanthridine	$C_{13}H_9N$ / 179.22	229-87-8 / 107.4	/ 349	5-20-08-00223 /	H_2O 2; EtOH 4; eth 4; ace 3
9189	1,10-Phenanthroline / o-Phenanthroline	$C_{12}H_8N_2$ / 180.21	66-71-7 / 117	7169 / >300	5-23-08-00419 /	H_2O 4; EtOH 3; ace 3; bz 3
9190	1,7-Phenanthroline	$C_{12}H_8N_2$ / 180.21	230-46-6 / 78	/ 360	5-23-08-00417 /	H_2O 3; EtOH 4; eth 1; bz 1
9191	4,7-Phenanthroline	$C_{12}H_8N_2$ / 180.21	230-07-9 / 177	/ sub 100	5-23-08-00430 /	H_2O 3; EtOH 4; eth 2; bz 2
9192	1,10-Phenanthroline, 2,9-dimethyl- / Neocuproine	$C_{14}H_{12}N_2$ / 208.26	484-11-7 / 159.5	6364 /	5-23-08-00527 /	
9193	Phenazine / Dibenzopyrazine	$C_{12}H_8N_2$ / 180.21	92-82-0 / 176.5	7173 /	5-23-08-00389 /	H_2O 2; EtOH 3; eth 2; bz 3

No.	Name Synonym	Mol. Form. Mol. Wt.	CAS RN mp/°C	Merck No. bp/°C	Beil. Ref. den/g cm^{-3}	Solubility n_D
9194	2,3-Phenazinediamine 2,3-Diaminophenazine	$C_{12}H_{10}N_4$ 210.24	655-86-7 264	2960 sub	4-25-00-03028	bz 4; EtOH 4
9195	Phenazine, 2-methyl- Tolazine	$C_{13}H_{10}N_2$ 194.24	1016-94-0 118.5	350	5-23-08-00475	eth 4; EtOH 4; chl 4
9196	Phenazinium, 1-hydroxy-5-methyl-, hydroxide, inner salt Pyocyanine	$C_{13}H_{10}N_2O$ 210.24	85-66-5 133 dec	7965	5-24-04-00100	H_2O 2; EtOH 3; eth 1; ace 3
9197	1-Phenazinol Hemipyocyanine	$C_{12}H_8N_2O$ 196.21	528-71-2 158	4564 sub	5-23-12-00297	H_2O 2; EtOH 2; bz 3; os 4
9198	Phenmedipham Carbamic acid, (3-methylphenyl)-, 3-[(methoxycarbonyl)amino]phenyl ester	$C_{16}H_{16}N_2O_4$ 300.31	13684-63-4 143	7199		
9199	Phenol Hydroxybenzene	C_6H_6O 94.11	108-95-2 40.9	7206 181.8	4-06-00-00531 1.0545[45]	H_2O 3; EtOH 3; eth 4; ace 5 1.5408[41]
9200	Phenol, amino- Hydroxyphenylamine	C_6H_7NO 109.13	27598-85-2 83.5		2-15-00-00004	H_2O 3; EtOH 4; eth 4; bz 4
9201	Phenol, 2-amino- o-Aminophenol	C_6H_7NO 109.13	95-55-6 174	473 sub 153	4-13-00-00805 1.328[25]	H_2O 3; EtOH 4; eth 3; bz 2
9202	Phenol, 3-amino- m-Aminophenol	C_6H_7NO 109.13	591-27-5 123	472 164[11]	4-13-00-00952	H_2O 3; EtOH 4; eth 4; bz 2
9203	Phenol, 4-amino- p-Aminophenol	C_6H_7NO 109.13	123-30-8 187.5	474 110[0.3]	4-13-00-01014	H_2O 2; EtOH 4; bz 1; chl 1
9204	Phenol, 2-amino-4-chloro- 2-Hydroxy-5-chloroaniline	C_6H_6ClNO 143.57	95-85-2 140		4-13-00-00878	DMSO 2
9205	Phenol, 2-amino-4,6-dichloro- 2-Amino-4,6-dichlorophenol	$C_6H_5Cl_2NO$ 178.02	527-62-8 95.5	448 sub 70	4-13-00-00889	
9206	Phenol, 4-amino-2,6-dichloro-	$C_6H_5Cl_2NO$ 178.02	5930-28-9 168	sub	2-13-00-00274	H_2O 1; EtOH 4; eth 4; ace 3
9207	Phenol, 2-amino-4,6-dinitro- Picramic acid	$C_6H_5N_3O_5$ 199.12	96-91-3 169	7379	4-13-00-00909	bz 4; EtOH 4
9208	Phenol, 4-(2-aminoethyl)- Tyramine	$C_8H_{11}NO$ 137.18	51-67-2 164.5	9743 205-7[25]	4-13-00-01788	H_2O 2; EtOH 3; bz 2; xzl 3
9209	Phenol, 2-amino-3-methyl-	C_7H_9NO 123.15	2835-97-4 150	sub	2-13-00-00324	eth 4
9210	Phenol, 2-amino-4-methyl-	C_7H_9NO 123.15	95-84-1 136	sub	4-13-00-01712	H_2O 2; EtOH 3; eth 3; bz 2
9211	Phenol, 3-amino-4-methyl-	C_7H_9NO 123.15	2836-00-2 157.8	sub	4-13-00-01710	eth 4
9212	Phenol, 3-amino-5-methyl- m-Cresol, 5-amino-	C_7H_9NO 123.15	76619-89-1 139	245	4-13-00-01697	EtOH 4
9213	Phenol, 4-amino-2-methyl-	C_7H_9NO 123.15	2835-96-3 176.5	sub	4-13-00-01672	H_2O 2; EtOH 3; eth 3; bz 2
9214	Phenol, 2-amino-3-nitro-	$C_6H_6N_2O_3$ 154.13	603-85-0 216.5	sub	4-13-00-00895	H_2O 4
9215	Phenol, 4-(2-aminopropyl)-, (±)- Hydroxyamphetamine	$C_9H_{13}NO$ 151.21	1518-86-1 125.5	4740	4-13-00-01871	H_2O 3; EtOH 3; bz 3; chl 3
9216	Phenol, 2,2'-azobis-	$C_{12}H_{10}N_2O_2$ 214.22	2050-14-8 173	140[0.001]	2-16-00-00033	H_2O 1; EtOH 2; eth 4; bz 2
9217	Phenol, 2-(2-benzothiazolyl)-	$C_{13}H_9NOS$ 227.29	3411-95-8 131	175-93[3]	4-27-00-02081	EtOH 3
9218	Phenol, 2-(2-benzoxazolyl)-	$C_{13}H_9NO_2$ 211.22	835-64-3 123.5	338	4-27-00-02080	H_2O 2; EtOH 4; eth 3; ace 3
9219	Phenol, 2,4-bis(1,1-dimethylethyl)-	$C_{14}H_{22}O$ 206.33	96-76-4 56.5	263.5	4-06-00-03493	ctc 2; alk 1 1.5080[20]
9220	Phenol, 2,6-bis(1,1-dimethylethyl)-	$C_{14}H_{22}O$ 206.33	128-39-2 39	161[50] 133[20]	4-06-00-03492	EtOH 2; ctc 3; alk 1 1.5001[20]
9221	Phenol, 2,6-bis(1,1-dimethylethyl)-4-ethyl-	$C_{16}H_{26}O$ 234.38	4130-42-1 44	272	4-06-00-03529	alk 1
9222	Phenol, 2,4-bis(1,1-dimethylethyl)-5-methyl- DBMC	$C_{15}H_{24}O$ 220.35	497-39-2 62.1	2829 282	4-06-00-03510 0.912[80]	H_2O 1; EtOH 3; eth 3; ace 3
9223	Phenol, 2,4-bis(1,1-dimethylethyl)-6-methyl-	$C_{15}H_{24}O$ 220.35	616-55-7 51	269	4-06-00-03510 0.891[80]	alk 1
9224	Phenol, 2,6-bis(1,1-dimethylethyl)-4-methyl- 2,6-Di-tert-butyl-4-methylphenol	$C_{15}H_{24}O$ 220.35	128-37-0 71	1548 265	4-06-00-03511 0.8937[75]	H_2O 1; EtOH 3; ace 3; bz 3 1.4859[75]
9225	Phenol, 2,6-bis(1,1-dimethylpropyl)-4-methyl- p-Cresol, 2,6-di-tert-pentyl-	$C_{17}H_{28}O$ 248.41	56103-67-4 	283	4-06-00-03533 0.931[25]	1.4950[20]
9226	Phenol, 2,6-bis(1-methylethyl)- Propofol	$C_{12}H_{18}O$ 178.27	2078-54-8 18	7847 256	4-06-00-03435 1.5140[20]	1.5140[20]
9227	Phenol, 2,6-bis(1-methylpropyl)- 2,6-Di-sec-butylphenol	$C_{14}H_{22}O$ 206.33	5510-99-6 -42	257.5	4-06-00-03490	1.5080[20]
9228	Phenol, 2-bromo-	C_6H_5BrO 173.01	95-56-7 5.6	194.5	4-06-00-01037 1.4924[20]	H_2O 2; EtOH 3; eth 3; chl 2 1.589[20]
9229	Phenol, 3-bromo-	C_6H_5BrO 173.01	591-20-8 33	236.5	4-06-00-01042	H_2O 2; EtOH 4; eth 4; ctc 2
9230	Phenol, 4-bromo-	C_6H_5BrO 173.01	106-41-2 66.4	238	4-06-00-01043 1.840[15]	H_2O 3; EtOH 4; eth 4; chl 3
9231	Phenol, 3-bromo-5-chloro-	C_6H_4BrClO 207.45	56962-04-0 70	256	2-06-00-00187	EtOH 4; HOAc 4
9232	Phenol, 4-bromo-2-chloro- 4-Bromo-2-chlorophenol	C_6H_4BrClO 207.45	3964-56-5 50.5	233.5	3-06-00-00750 1.6170[20]	EtOH 4; eth 4; ace 4; bz 4 1.5859[20]

No.	Name / Synonym	Mol. Form. / Mol. Wt.	CAS RN / mp/°C	Merck No. / bp/°C	Beil. Ref. / den/g cm⁻³	Solubility / n_D
9233	Phenol, 2-bromo-4,6-dichloro-	$C_6H_3BrCl_2O$ 241.90	4524-77-0 68	220[200]	3-06-00-00751	bz 4; eth 4; chl 4
9234	Phenol, 2-bromo-4,6-dinitro-	$C_6H_3BrN_2O_5$ 263.00	2316-50-9 119.3	sub	3-06-00-00871	H_2O 2; EtOH 3; eth 4; bz 4
9235	Phenol, 4-bromo-2,6-dinitro-	$C_6H_3BrN_2O_5$ 263.00	40466-95-3 78	sub	3-06-00-00872	bz 4; eth 4; EtOH 4; chl 4
9236	Phenol, 2-bromo-4-methyl-	C_7H_7BrO 187.04	6627-55-0 56.5	213.5	4-06-00-02143 1.5422[25]	H_2O 2; EtOH 3; bz 3; chl 3 1.5772[20]
9237	Phenol, 2-bromo-5-methyl- 6-Bromo-m-cresol	C_7H_7BrO 187.04	14847-51-9 38.8	209	3-06-00-01320	ace 4; eth 4; EtOH 4
9238	Phenol, 3-bromo-2-methyl-	C_7H_7BrO 187.04	7766-23-6 95	55[4]	0-06-00-00360	ace 4; bz 4; eth 4; EtOH 4
9239	Phenol, 3-bromo-4-methyl-	C_7H_7BrO 187.04	60710-39-6 56	246	4-06-00-02143	ace 4; bz 4; eth 4; EtOH 4
9240	Phenol, 3-bromo-5-methyl-	C_7H_7BrO 187.04	74204-00-5 56.5	161[28]	2-06-00-00357	eth 4; EtOH 4
9241	Phenol, 4-bromo-2-methyl-	C_7H_7BrO 187.04	2362-12-1 64.7	235	4-06-00-02006	ace 4; eth 4; EtOH 4
9242	Phenol, 4-bromo-3-methyl-	C_7H_7BrO 187.04	14472-14-1 63.5	142[23]	4-06-00-02072	eth 4; py 4
9243	Phenol, 2-bromo-3-nitro-	$C_6H_4BrNO_3$ 218.01	101935-40-4 147.5	sub	3-06-00-00844	eth 4; EtOH 4
9244	Phenol, 4-bromo-2-nitro- 4-Bromo-2-nitrophenol	$C_6H_4BrNO_3$ 218.01	7693-52-9 92	sub	4-06-00-01363	bz 4; eth 4; EtOH 4; chl 4
9245	Phenol, 2-butoxy-	$C_{10}H_{14}O_2$ 166.22	39075-90-6	232.5	4-06-00-05566 1.026[25]	1.5113[25]
9246	Phenol, 2-butyl-	$C_{10}H_{14}O$ 150.22	3180-09-4 -20	235	4-06-00-03267 0.975[20]	H_2O 1; EtOH 3; eth 3; alk 3 1.5180[25]
9247	Phenol, 3-butyl-	$C_{10}H_{14}O$ 150.22	4074-43-5	248	4-06-00-03269 0.974[20]	eth 4; EtOH 4
9248	Phenol, 4-butyl-	$C_{10}H_{14}O$ 150.22	1638-22-8 22	248	4-06-00-03269 0.976[22]	H_2O 1; EtOH 3; eth 3; ctc 2 1.5165[25]
9249	Phenol, 4,4'-butylidenebis-	$C_{16}H_{18}O_2$ 242.32	4731-84-4 137	270[12]	4-06-00-06736	bz 4; EtOH 4; tol 4
9250	Phenol, 2-butyl-4-methyl-	$C_{11}H_{16}O$ 164.25	6891-45-8 19	228; 124[15]	4-06-00-03392 0.9680[20]	
9251	Phenol, 2-chloro- o-Chlorophenol	C_6H_5ClO 128.56	95-57-8 9.8	2154 174.9	4-06-00-00782 1.2634[20]	H_2O 2; EtOH 3; eth 3; bz 4 1.5524[20]
9252	Phenol, 3-chloro- m-Chlorophenol	C_6H_5ClO 128.56	108-43-0 32.6	2154 214	4-06-00-00810 1.245[45]	H_2O 2; EtOH 3; eth 3; bz 4 1.5565[40]
9253	Phenol, 4-chloro- p-Chlorophenol	C_6H_5ClO 128.56	106-48-9 42.7	2154 220	4-06-00-00820 1.265[40]	H_2O 1; EtOH 4; eth 4; bz 4 1.5579[40]
9254	Phenol, 2-chloro-4-cyclohexyl- 2-Chloro-4-cyclohexylphenol	$C_{12}H_{15}ClO$ 210.70	3964-61-2 40.5	160[12]; 122[3]	3-06-00-02508 1.1600[15]	ctc 3; CS_2 3
9255	Phenol, 4-chloro-2-cyclopentyl- Dowicide 9	$C_{11}H_{13}ClO$ 196.68	13347-42-7	3420 181-5[18]	3-06-00-02478	
9256	Phenol, 2-chloro-3,4-dimethyl-	C_8H_9ClO 156.61	10283-15-5 27	188	2-06-00-00456 1.5538[20]	peth 4
9257	Phenol, 4-chloro-3,5-dimethyl- Chloroxylenol	C_8H_9ClO 156.61	88-04-0 115	2176 246	4-06-00-03152	H_2O 2; EtOH 3; eth 3; bz 2
9258	Phenol, 4-chloro-2,6-dinitro- 4-Chloro-2,6-dinitrophenol	$C_6H_3ClN_2O_5$ 218.55	88-87-9 81		4-06-00-01386 1.74[22]	eth 4; EtOH 4; chl 4
9259	Phenol, 5-chloro-2,4-dinitro- 3-Chloro-4,6-dinitrophenol	$C_6H_3ClN_2O_5$ 218.55	54715-57-0 92		4-06-00-01385 1.74[22]	eth 4; EtOH 4; chl 4; peth 4
9260	Phenol, 2-chloro-3-methyl-	C_7H_7ClO 142.58	608-26-4 56	230	4-06-00-02062	H_2O 2; HOAc 4; peth 3
9261	Phenol, 2-chloro-4-methyl-	C_7H_7ClO 142.58	6640-27-3	195.5	4-06-00-02135 1.1785[27]	bz 4; eth 4; EtOH 4 1.5200[27]
9262	Phenol, 2-chloro-5-methyl- 6-Chloro-m-cresol	C_7H_7ClO 142.58	615-74-7 45.5	196	3-06-00-01315 1.215[15]	H_2O 4; EtOH 4
9263	Phenol, 2-chloro-6-methyl-	C_7H_7ClO 142.58	87-64-9	189; 80[20]	4-06-00-01984 1.5449[20]	eth 3
9264	Phenol, 3-chloro-2-methyl-	C_7H_7ClO 142.58	3260-87-5 86	225	4-06-00-02000	bz 4; eth 4; EtOH 4
9265	Phenol, 3-chloro-4-methyl-	C_7H_7ClO 142.58	615-62-3 55.5	228	3-06-00-01374	bz 4; eth 4; EtOH 4
9266	Phenol, 4-chloro-2-methyl- 4-Chloro-o-cresol	C_7H_7ClO 142.58	1570-64-5 51	223	4-06-00-01987	H_2O 2; peth 3
9267	Phenol, 4-chloro-3-methyl- 4-Chloro-m-cresol	C_7H_7ClO 142.58	59-50-7 67	2133 235	4-06-00-02064	H_2O 3; EtOH 3; eth 3; chl 2
9268	Phenol, 4-chloro-5-methyl-2-(1-methylethyl)- Chlorothymol	$C_{10}H_{13}ClO$ 184.67	89-68-9 63	2171 258.5	4-06-00-03344	H_2O 4; EtOH 3; eth 3; bz 3
9269	Phenol, 5-chloro-2-nitro- 5-Chloro-2-nitrophenol	$C_6H_4ClNO_3$ 173.56	611-07-4 41	sub	3-06-00-00836	H_2O 2; EtOH 3; eth 3; HOAc 3
9270	Phenol, 4-chloro-2-(phenylmethyl)- Clorophene	$C_{13}H_{11}ClO$ 218.68	120-32-1 48.5	2403 160-2[3.5]	4-06-00-04636 1.185[58]	ctc 3; CS_2 3
9271	Phenol, 4-chloro-2-(2-propenyl)-	C_9H_9ClO 168.62	13997-73-4 48	130-2[15]	4-06-00-03813 1.171[15]	bz 4; EtOH 4
9272	Phenol, 4-cyclohexyl-	$C_{12}H_{16}O$ 176.26	1131-60-8 133	294; 133[4]	4-06-00-03904	H_2O 1; EtOH 4; eth 4; bz 3

No.	Name / Synonym	Mol. Form. / Mol. Wt.	CAS RN mp/°C	Merck No. bp/°C	Beil. Ref. den/g cm^{-3}	Solubility n_D
9273	Phenol, 2-cyclohexyl-4,6-dinitro- / 4,6-Dinitro-o-cyclohexylphenol	$C_{12}H_{14}N_2O_5$ / 266.25	131-89-5	2739	4-06-00-03903	
9274	Phenol, 2,4-diamino- / 2,4-Diaminophenol	$C_6H_8N_2O$ / 124.14	95-86-3 / 78-80 dec	2961	4-13-00-01425	H_2O 4; ace 4; EtOH 4
9275	Phenol, 4,4'-(1,2-diarsenediyl)bis[2-amino-, dihydrochloride / Salvarsan	$C_{12}H_{20}As_2Cl_2O_5$ / 465.04	139-93-5 / 185-95 dec	839	2-16-00-00560	H_2O 4
9276	Phenol, 2,4-dibromo- / 2,4-Dibromophenol	$C_6H_4Br_2O$ / 251.91	615-58-7 / 38	238.5	4-06-00-01061 / 2.0700^{20}	H_2O 2; EtOH 4; eth 4; bz 4
9277	Phenol, 2,6-dibromo- / 2,6-Dibromophenol	$C_6H_4Br_2O$ / 251.91	608-33-3 / 56.5	255; 162[21]	4-06-00-01064	H_2O 3; EtOH 4; eth 4
9278	Phenol, 3,5-dibromo- / 3,5-Dibromophenol	$C_6H_4Br_2O$ / 251.91	626-41-5 / 81	274; 121[3]	4-06-00-01065	H_2O 2; EtOH 3; eth 3; lig 2
9279	Phenol, 2,4-dibromo-6-methyl-	$C_7H_6Br_2O$ / 265.93	609-22-3 / 58	265 dec; 105[4]	3-06-00-01271	chl 3
9280	Phenol, 3,5-dibromo-2-methyl-	$C_7H_6Br_2O$ / 265.93	14122-00-0 / 100.3	285	2-06-00-00334	peth 4
9281	Phenol, 2,4-dibromo-3-methyl-6-(1-methylethyl)- / Thymol, 2,6-dibromo	$C_{10}H_{12}Br_2O$ / 308.01	52262-38-1 / 3.7	160-1[16]	2-06-00-00500 / 1.6568[17]	
9282	Phenol, 2,3-dichloro- / 2,3-Dichlorophenol	$C_6H_4Cl_2O$ / 163.00	576-24-9 / 58		4-06-00-00883	EtOH 3; eth 3; bz 3; lig 3
9283	Phenol, 2,4-dichloro- / 2,4-Dichlorophenol	$C_6H_4Cl_2O$ / 163.00	120-83-2 / 45	3061 / 210	4-06-00-00885	H_2O 2; EtOH 3; eth 3; bz 3
9284	Phenol, 2,5-dichloro- / 2,5-Dichlorophenol	$C_6H_4Cl_2O$ / 163.00	583-78-8 / 59	211	4-06-00-00942	H_2O 2; EtOH 4; eth 4; bz 3
9285	Phenol, 2,6-dichloro- / 2,6-Dichlorophenol	$C_6H_4Cl_2O$ / 163.00	87-65-0 / 68.5	3062 / 220; 82[4]	4-06-00-00949 / 1.653[20]	EtOH 4; eth 4; bz 3; peth 3
9286	Phenol, 3,4-dichloro- / 3,4-Dichlorophenol	$C_6H_4Cl_2O$ / 163.00	95-77-2 / 68	253	4-06-00-00952	H_2O 2; EtOH 4; eth 4; bz 3
9287	Phenol, 3,5-dichloro- / 3,5-Dichlorophenol	$C_6H_4Cl_2O$ / 163.00	591-35-5 / 68	233	4-06-00-00957	H_2O 2; EtOH 4; eth 4; peth 3
9288	Phenol, 2,4-dichloro-, acetate / 2,4-Dichlorophenyl acetate	$C_8H_6Cl_2O_2$ / 205.04	6341-97-5	244.5	3-06-00-00704	EtOH 3; eth 3; chl 3
9289	Phenol, 2,4-dichloro-, benzenesulfonate / Genite	$C_{12}H_8Cl_2O_3S$ / 303.17	97-16-5 / 45.5	4283	4-11-00-00034	ctc 3; CS_2 3
9290	Phenol, 2,4-dichloro-3,5-dimethyl- / Dichloroxylenol	$C_8H_8Cl_2O$ / 191.06	133-53-9 / 83	3066	4-06-00-03156	eth 4
9291	Phenol, 2,6-dichloro-3,5-dimethyl-	$C_8H_8Cl_2O$ / 191.06	1943-54-0 / 87.5	105-10[1]	4-06-00-03157	eth 4; chl 4
9292	Phenol, 2,4-dichloro-3-methyl- / m-Cresol, 2,4-dichloro	$C_7H_6Cl_2O$ / 177.03	17788-00-0 / 58	236; 77[4]	3-06-00-01319	eth 4; chl 4
9293	Phenol, 2,4-dichloro-5-methyl-	$C_7H_6Cl_2O$ / 177.03	1124-07-8 / 73	234	4-06-00-02069	chl 4; peth 3 / 1.572[20]
9294	Phenol, 2,6-dichloro-3-methyl- / 2,6-Dichloro-m-cresol	$C_7H_6Cl_2O$ / 177.03	13481-70-4 / 28	237; 77[4]	3-06-00-01319	eth 4; chl 4
9295	Phenol, 2,6-dichloro-4-methyl-	$C_7H_6Cl_2O$ / 177.03	2432-12-4 / 39	231; 138[28]	4-06-00-02141	eth 4; EtOH 4; HOAc 4
9296	Phenol, 2,6-dichloro-4-nitro-	$C_6H_3ClNO_3$ / 172.55	618-80-4 / 127 exp		4-06-00-01361 / 1.822[25]	eth 4; chl 4
9297	Phenol, 3-(diethylamino)-	$C_{10}H_{15}NO$ / 165.24	91-68-9 / 78	276; 170[15]	4-13-00-00969	H_2O 3; EtOH 3; eth 3; CS_2 3
9298	Phenol, 4,4'-(1,2-diethyl-1,2-ethanediyl)bis-, (R*,S*)- / Hexestrol	$C_{18}H_{22}O_2$ / 270.37	84-16-2 / 186.5	4621	4-06-00-06761	ace 4; eth 4; EtOH 4
9299	Phenol, 4,4'-(1,2-diethyl-1,2-ethanediyl)bis[2-methyl- / Methestrol	$C_{20}H_{26}O_2$ / 298.43	130-73-4 / 145	5888	4-06-00-06786	
9300	Phenol, 4,4'-(1,2-diethyl-1,2-ethenediyl)bis-, (E)- / Diethylstilbestrol	$C_{18}H_{20}O_2$ / 268.36	56-53-1 / 170.5	3118	4-06-00-06856	eth 4; EtOH 4; chl 4
9301	Phenol, 4,4'-(1,2-diethyl-1,2-ethenediyl)bis-, dipropanoate, (E)- / Diethylstilbestrol dipropionate	$C_{24}H_{28}O_4$ / 380.48	130-80-3 / 104	3119	4-06-00-06857	bz 4; eth 4; EtOH 4
9302	Phenol, 4,4'-(1,2-diethylidene-1,2-ethanediyl)bis- / Dienestrol	$C_{18}H_{18}O_2$ / 266.34	84-17-3 / 227.5	3094 / sub 130	4-06-00-06916	ace 4; eth 4; EtOH 4
9303	Phenol, 4,4'-(1,2-diethyl-3-methyl-1,3-propanediyl)bis- / Benzestrol	$C_{20}H_{26}O_2$ / 298.43	85-95-0 / 164	1082	3-06-00-05538	ace 4; eth 4; EtOH 4; HOAc 4
9304	Phenol, 2,6-diiodo-4-nitro- / Disophenol	$C_6H_3I_2NO_3$ / 390.90	305-85-1 / 157	3365	4-06-00-01368	EtOH 4
9305	Phenol, 2,6-dimethoxy-	$C_8H_{10}O_3$ / 154.17	91-10-1 / 56.5	261	4-06-00-07329	eth 4; EtOH 4
9306	Phenol, 3,5-dimethoxy-	$C_8H_{10}O_3$ / 154.17	500-99-2 / 37	199[35]; 170[10]	4-06-00-07362	eth 3; bz 3; lig 2
9307	Phenol, 2,3-dimethyl- / 2,3-Xylenol	$C_8H_{10}O$ / 122.17	526-75-0 / 72.8	9989 / 216.9	4-06-00-03096	H_2O 2; EtOH 3; eth 3 / 1.542[20]

No.	Name Synonym	Mol. Form. Mol. Wt.	CAS RN mp/°C	Merck No. bp/°C	Beil. Ref. den/g cm^{-3}	Solubility n_D
9308	Phenol, 2,4-dimethyl- 2,4-Xylenol	$C_8H_{10}O$ 122.17	105-67-9 24.5	9989 210.9	4-06-00-03126 0.9650[20]	H_2O 2; EtOH 5; eth 5; ctc 3 1.5420[14]
9309	Phenol, 2,5-dimethyl- 2,5-Xylenol	$C_8H_{10}O$ 122.17	95-87-4 74.8	9989 211.1	4-06-00-03164	H_2O 3; EtOH 3; eth 4; chl 2
9310	Phenol, 2,6-dimethyl- 2,6-Xylenol	$C_8H_{10}O$ 122.17	576-26-1 45.7	9989 201.0	4-06-00-03112	H_2O 3; EtOH 3; eth 3; ctc 3
9311	Phenol, 3,4-dimethyl- 3,4-Xylenol	$C_8H_{10}O$ 122.17	95-65-8 60.8	9989 227	4-06-00-03099 0.9830[20]	H_2O 2; EtOH 3; eth 5; ctc 3
9312	Phenol, 3,5-dimethyl- 3,5-Xylenol	$C_8H_{10}O$ 122.17	108-68-9 63.6	9989 221.7	4-06-00-03141 0.9680[20]	H_2O 3; EtOH 3; ctc 3
9313	Phenol, 2,4-dimethyl-, acetate	$C_{10}H_{12}O_2$ 164.20	877-53-2	226	2-06-00-00459 1.0298[15]	eth 4; EtOH 4 1.4990[15]
9314	Phenol, 2,5-dimethyl-, acetate	$C_{10}H_{12}O_2$ 164.20	877-48-5 <-20	237	2-06-00-00467 1.0624[15]	eth 4; EtOH 4
9315	Phenol, 3,4-dimethyl-, acetate	$C_{10}H_{12}O_2$ 164.20	22618-23-1 22	235	4-06-00-03102	bz 4; eth 4; EtOH 4
9316	Phenol, 3,5-dimethyl-, acetate	$C_{10}H_{12}O_2$ 164.20	877-82-7 <-20	237	3-06-00-01757 1.0624[15]	eth 4; EtOH 4
9317	Phenol, 2-(dimethylamino)-	$C_8H_{11}NO$ 137.18	3743-22-4 45	199.5	4-13-00-00815	eth 4; EtOH 4; HOAc 4
9318	Phenol, 3-(dimethylamino)-	$C_8H_{11}NO$ 137.18	99-07-0 86	266.5	4-13-00-00959	H_2O 1; EtOH 3; eth 3; ace 3 1.5895[26]
9319	Phenol, 4-(dimethylamino)-	$C_8H_{11}NO$ 137.18	619-60-3 77	165[30]	4-13-00-01040	H_2O 2; EtOH 3; eth 3
9320	Phenol, 4-[2-(dimethylamino)ethyl]- Hordenine	$C_{10}H_{15}NO$ 165.24	539-15-1 117.5	4666 173[11]	4-13-00-01790	eth 4; EtOH 4; chl 4
9321	Phenol, 2-(1,1-dimethylethyl)-	$C_{10}H_{14}O$ 150.22	88-18-6 -6.8	223	4-06-00-03292 0.9783[20]	EtOH 3; eth 4; ctc 3; alk 3 1.5160[20]
9322	Phenol, 3-(1,1-dimethylethyl)-	$C_{10}H_{14}O$ 150.22	585-34-2 42.3	240	4-06-00-03294	EtOH 3; eth 4; alk 3
9323	Phenol, 4-(1,1-dimethylethyl)- p-tert-Butylphenol	$C_{10}H_{14}O$ 150.22	98-54-4 98	1584 237	4-06-00-03296 0.9080[80]	H_2O 3; EtOH 3; eth 3; chl 3 1.4787[114]
9324	Phenol, 2-(1,1-dimethylethyl)-4,5-dimethyl- 3,4-Xylenol, 6-tert-butyl-	$C_{12}H_{18}O$ 178.27	1445-23-4 46	258.5	3-06-00-02019 0.920[80]	1.5222[20]
9325	Phenol, 2-(1,1-dimethylethyl)-4,6-dimethyl-	$C_{12}H_{18}O$ 178.27	1879-09-0 22.3	249	4-06-00-03442 0.917[80]	alk 1 1.5183[20]
9326	Phenol, 4-(1,1-dimethylethyl)-2,5-dimethyl-	$C_{12}H_{18}O$ 178.27	17696-37-6 71.2	264	3-06-00-02019 0.939[80]	alk 3 1.5311[20]
9327	Phenol, 4-(1,1-dimethylethyl)-2,6-dimethyl-	$C_{12}H_{18}O$ 178.27	879-97-0 82.4	248	3-06-00-02019 0.916[80]	alk 3
9328	Phenol, 2-(1,1-dimethylethyl)-4,6-dinitro- 2-tert-Butyl-4,6-dinitrophenol	$C_{10}H_{12}N_2O_5$ 240.22	1420-07-1		4-06-00-03293	
9329	Phenol, (1,1-dimethylethyl)-4-methoxy- Butylated hydroxyanisole	$C_{11}H_{16}O_2$ 180.25	25013-16-5 51	1547 268		H_2O 1; peth 3; EtOH 3
9330	Phenol, 2-(1,1-dimethylethyl)-4-methyl-	$C_{11}H_{16}O$ 164.25	2409-55-4 51.5	237	4-06-00-03397 0.9247[75]	H_2O 2; ace 3; bz 3; chl 3 1.4969[75]
9331	Phenol, 2-(1,1-dimethylethyl)-5-methyl-	$C_{11}H_{16}O$ 164.25	88-60-8 46.5	127[11]	4-06-00-03400 0.922[80]	H_2O 1; EtOH 3; eth 3; ace 3 1.5250[20]
9332	Phenol, 2-(1,1-dimethylethyl)-6-methyl-	$C_{11}H_{16}O$ 164.25	2219-82-1 31	230	4-06-00-03396 0.9240[80]	1.5195[20]
9333	Phenol, 4-(1,1-dimethylethyl)-2-methyl-	$C_{11}H_{16}O$ 164.25	98-27-1 27.5	237; 132[20]	4-06-00-03398 0.9650[20]	H_2O 1; eth 3; ace 3; bz 3 1.5230[20]
9334	Phenol, 3,6-dimethyl-2-nitro-	$C_8H_9NO_3$ 167.16	71608-10-1 34.5	236 dec; 150[15]	4-06-00-03169	EtOH 4; eth 3; ace 3; bz 3
9335	Phenol, 2,4-dimethyl-, phosphate (3:1)	$C_{24}H_{27}O_4P$ 410.45	3862-12-2	233.5	0-06-00-00488 1.142[38]	H_2O 1; bz 3; chl 3; hx 3 1.5550[20]
9336	Phenol, 2,5-dimethyl-, phosphate (3:1)	$C_{24}H_{27}O_4P$ 410.45	19074-59-0 79.8	260-5[8]	3-06-00-01773 1.197[25]	H_2O 1; EtOH 2; eth 3; bz 3
9337	Phenol, 2,6-dimethyl-, phosphate (3:1)	$C_{24}H_{27}O_4P$ 410.45	121-06-2 137.8	262-4[6]	4-06-00-03121	H_2O 1; EtOH 2; bz 3; hx 2
9338	Phenol, 3,4-dimethyl-, phosphate (3:1)	$C_{24}H_{27}O_4P$ 410.45	3862-11-1 72	260-3[7]	0-06-00-00482	H_2O 1; EtOH 2; bz 3; chl 2
9339	Phenol, 4-(1,1-dimethylpropyl)- p-tert-Pentylphenol	$C_{11}H_{16}O$ 164.25	80-46-6 95	7098 262.5	4-06-00-03383	
9340	Phenol, 4-(1,1-dimethylpropyl)-2-methyl-	$C_{12}H_{18}O$ 178.27	71745-63-6 15	260	3-06-00-02010 0.969[20]	chl 2 1.5240[20]
9341	Phenol, 2,3-dinitro- 2,3-Dinitrophenol	$C_8H_4N_2O_5$ 184.11	66-56-8 144.5		4-06-00-01369 1.681[20]	H_2O 2; EtOH 4; eth 4; bz 3
9342	Phenol, 2,4-dinitro- 2,4-Dinitrophenol	$C_8H_4N_2O_5$ 184.11	51-28-5 115.5	3274 sub	4-06-00-01369 1.683[24]	H_2O 2; EtOH 3; eth 3; ace 3
9343	Phenol, 2,5-dinitro- 2,5-Dinitrophenol	$C_8H_4N_2O_5$ 184.11	329-71-5 108	3275	4-06-00-01383	bz 4; eth 4
9344	Phenol, 2,6-dinitro- 2,6-Dinitrophenol	$C_8H_4N_2O_5$ 184.11	573-56-8 63.5	3276	4-06-00-01383	H_2O 1; EtOH 4; eth 4; ace 4
9345	Phenol, 3,4-dinitro-	$C_8H_4N_2O_5$ 184.11	577-71-9 134		4-06-00-01384 1.672[25]	bz 4; eth 4; EtOH 4
9346	Phenol, 3,5-dinitro- 3,5-Dinitrophenol	$C_8H_4N_2O_5$ 184.11	586-11-8 125.1		4-06-00-01385 1.702[25]	bz 4; eth 4; EtOH 4; chl 4
9347	Phenol, 2,6-dipropyl- 2,6-Dipropylphenol	$C_{12}H_{18}O$ 178.27	6626-32-0 27	256	3-06-00-02013 1.6450[0]	eth 4

No.	Name / Synonym	Mol. Form. / Mol. Wt.	CAS RN / mp/°C	Merck No. / bp/°C	Beil. Ref. / den/g cm⁻³	Solubility / n_D
9348	Phenol, 2-ethenyl-	C_8H_8O 120.15	695-84-1 29.5	101[14]	4-06-00-03771 1.0609[18]	eth 4; EtOH 4 1.5851[20]
9349	Phenol, 3-ethenyl-	C_8H_8O 120.15	620-18-8	114-6[16]	4-06-00-03773 1.0353[21]	1.5804[21]
9350	Phenol, 2-ethoxy-	$C_8H_{10}O_2$ 138.17	94-71-3 29	217	4-06-00-05565 1.0903[25]	H_2O 2; EtOH 5; eth 5; ctc 2
9351	Phenol, 3-ethoxy-	$C_8H_{10}O_2$ 138.17	621-34-1	131[10]	4-06-00-05664 1.0705[4]	H_2O 1; EtOH 3; eth 3; bz 3
9352	Phenol, 4-ethoxy-	$C_8H_{10}O_2$ 138.17	622-62-8 66.5	246.5	4-06-00-05719	H_2O 2; EtOH 4; eth 4; chl 3
9353	Phenol, 2-ethyl- o-Ethylphenol	$C_8H_{10}O$ 122.17	90-00-6 18	7302 204.5	4-06-00-03011 1.0146[25]	ace 4; bz 4; eth 4; EtOH 4 1.5367[20]
9354	Phenol, 3-ethyl- m-Ethylphenol	$C_8H_{10}O$ 122.17	620-17-7 -4	218.4	4-06-00-03016 1.0283[20]	H_2O 2; EtOH 4; eth 4; chl 2
9355	Phenol, 4-ethyl- p-Ethylphenol	$C_8H_{10}O$ 122.17	123-07-9 45.0	217.9	4-06-00-03020	H_2O 2; EtOH 4; eth 4; ace 3 1.5239[25]
9356	Phenol, 3-(ethylamino)-	$C_8H_{11}NO$ 137.18	621-31-8 62	176[12]	0-13-00-00408	bz 4; eth 4; EtOH 4
9357	Phenol, 4-ethyl-2-methoxy-	$C_9H_{12}O_2$ 152.19	2785-89-9 -7	236.5	4-06-00-05927 1.0931[18]	
9358	Phenol, 4-[1-ethyl-2-(4-methoxyphenyl)-1-butenyl]- Mestilbol	$C_{19}H_{22}O_2$ 282.38	7773-60-6 117.5	5818 185-95[0.3]	4-06-00-06856	ace 4; eth 4; EtOH 4
9359	Phenol, 4-[1-ethyl-2-(4-methoxyphenyl)-1-butenyl]-, (E)-	$C_{19}H_{22}O_2$ 282.38	18839-90-2 117	185-95[0.3]	4-06-00-06856	EtOH 4
9360	Phenol, 2-fluoro-	C_6H_5FO 112.10	367-12-4 16.1	151.5	4-06-00-00770 1.120[25]	H_2O 3 1.5144[20]
9361	Phenol, 3-fluoro-	C_6H_5FO 112.10	372-20-3 13.7	178	4-06-00-00772 1.238[25]	1.5140[20]
9362	Phenol, 4-fluoro-	C_6H_5FO 112.10	371-41-5 48	185.5	4-06-00-00773 1.1889[56]	H_2O 2; ace 3; peth 3
9363	Phenol, 4,4'-(3H-2,1-benzoxathiol-3-ylidene)bis-, S,S-dioxide Phenolsulfonphthalein	$C_{19}H_{14}O_5S$ 354.38	143-74-8 >300	7213	5-19-03-00457	H_2O 2; EtOH 2; eth 1; ace 2
9364	Phenol, 4,4'-(3H-2,1-benzoxathiol-3-ylidene)bis[2-bromo-6-methyl-, S,S-dioxide Bromcresol purple	$C_{21}H_{18}Br_2O_5S$ 540.23	115-40-2 241.5	1376	5-19-03-00460	
9365	Phenol, 4,4'-(3H-2,1-benzoxathiol-3-ylidene)bis[2-bromo-3-methyl-6-(1-methylethyl)-, S,S-dioxide Bromthymol blue	$C_{27}H_{28}Br_2O_5S$ 624.39	76-59-5 201	1435	5-19-03-00461	eth 4; EtOH 4
9366	Phenol, 4,4'-(3H-2,1-benzoxathiol-3-ylidene)bis[2,6-dibromo-, S,S-dioxide Bromphenol blue	$C_{19}H_{10}Br_4O_5S$ 669.97	115-39-9 279 dec	1434	5-19-03-00458	H_2O 2; EtOH 3; bz 3; HOAc 3
9367	Phenol, 4,4'-(3H-2,1-benzoxathiol-3-ylidene)bis[2,6-dibromo-3-methyl-, S,S-dioxide Bromcresol green	$C_{21}H_{14}Br_4O_5S$ 698.02	76-60-8 218.5	1375	5-19-03-00460	H_2O 2; EtOH 4; eth 4; bz 3
9368	Phenol, 4,4'-(3H-2,1-benzoxathiol-3-ylidene)bis[2-methyl-, S,S-dioxide Cresol red	$C_{21}H_{18}O_5S$ 382.44	1733-12-6 >300	2583	5-19-03-00460	H_2O 4; EtOH 4
9369	Phenol, 4,4'-(3H-2,1-benzoxathiol-3-ylidene)bis[5-methyl-2-(1-methylethyl)-, S,S-dioxide Thymol blue	$C_{27}H_{30}O_5S$ 466.60	76-61-9 221-4 dec	9334	5-19-03-00462	H_2O 2; EtOH 3; ace 2; bz 2
9370	Phenol, 4-(3-hydroxy-1-propenyl)-2-methoxy- Coniferyl alcohol	$C_{10}H_{12}O_3$ 180.20	458-35-5 74	2499 163-5[3]	2-06-00-01093	H_2O 1; EtOH 3; eth 4; alk 3
9371	Phenol, 2-iodo- o-Iodophenol	C_6H_5IO 220.01	533-58-4 43	4928 186[160]; 91[2]	4-06-00-01070 1.8757[80]	H_2O 3; EtOH 4; eth 4; CS_2 4
9372	Phenol, 3-iodo-	C_6H_5IO 220.01	626-02-8 118	186[100]	4-06-00-01073	H_2O 2; EtOH 3; eth 3
9373	Phenol, 4-iodo- p-Iodophenol	C_6H_5IO 220.01	540-38-5 93.5	4929 139[5] dec	4-06-00-01074 1.8573[112]	H_2O 2; EtOH 4; eth 4
9374	Phenol, 2-iodo-4-methyl- p-Cresol, 2-iodo-	C_7H_7IO 234.04	16188-57-1 35	117[12]	4-06-00-02147 1.684[25]	1.5331[25]
9375	Phenol, 2-iodo-6-methyl-	C_7H_7IO 234.04	24885-45-8 15.5	70-5[2]	4-06-00-02008 1.661[25]	1.610[25]
9376	Phenol, 2-mercapto-	C_6H_6OS 126.18	1121-24-0 5.5	217; 89[8]	4-06-00-05633 1.2371[0]	bz 4; eth 4; EtOH 4
9377	Phenol, 3-mercapto-	C_6H_6OS 126.18	40248-84-8 16.5	168[35]	2-06-00-00827	bz 4; EtOH 4
9378	Phenol, 4-mercapto-	C_6H_6OS 126.18	637-89-8 29.5	167[45]; 135[11]	4-06-00-05790 1.1285[25]	H_2O 3; EtOH 3; alk 3; con sulf 3 1.5101[25]
9379	Phenol, 2-methoxy- Guaiacol	$C_7H_8O_2$ 124.14	90-05-1 32	4457 205	4-06-00-05563 1.1287[21]	H_2O 2; EtOH 3; eth 3; ctc 3 1.5429[20]
9380	Phenol, 3-methoxy-	$C_7H_8O_2$ 124.14	150-19-6 <-17	114[5]	4-06-00-05662 1.131[25]	H_2O 2; EtOH 5; eth 5; chl 2 1.5510[20]

No.	Name Synonym	Mol. Form. Mol. Wt.	CAS RN mp/°C	Merck No. bp/°C	Beil. Ref. den/g cm⁻³	Solubility n_D
9381	Phenol, 4-methoxy- 4-Methoxyphenol	$C_7H_8O_2$ 124.14	150-76-5 57	243	4-06-00-05717	H_2O 3; EtOH 4; eth 4; bz 3
9382	Phenol, 2-methoxy-, acetate	$C_9H_{10}O_3$ 166.18	613-70-7 31.5	123-4[13]	4-06-00-05581 1.1285[25]	H_2O 1; EtOH 3; eth 3 1.5101[25]
9383	Phenol, 2-methoxy-, benzoate Guaiacol benzoate	$C_{14}H_{12}O_3$ 228.25	531-37-3 57.5	4458	4-09-00-00371	eth 4; chl 4
9384	Phenol, 2-methoxy-, carbonate (2:1) Guaiacol carbonate	$C_{15}H_{14}O_5$ 274.27	553-17-3 89	4459	3-06-00-04233	H_2O 1; EtOH 2; eth 3; chl 4
9385	Phenol, 2-methoxy-4-methyl- Creosol	$C_8H_{10}O_2$ 138.17	93-51-6 5.5	2573 221	4-06-00-05878 1.098[20]	eth 4; EtOH 4 1.5353[25]
9386	Phenol, 2-methoxy-3-nitro-	$C_7H_7NO_4$ 169.14	20734-71-8 70	sub	3-06-00-04263	H_2O 2; EtOH 4; chl 3; peth 3
9387	Phenol, 2-methoxy-, phosphate (3:1) Guaiacol phosphate	$C_{21}H_{21}O_7P$ 416.37	563-03-1 91	4460 275-80[3]	3-06-00-04246	ace 4; tol 4; chl 4
9388	Phenol, 2-methoxy-4-(1-propenyl)- Isoeugenol	$C_{10}H_{12}O_2$ 164.20	97-54-1	5054 266	3-06-00-04993 1.080[25]	eth 4; EtOH 4 1.5739[19]
9389	Phenol, 2-methoxy-4-(2-propenyl)- Eugenol	$C_{10}H_{12}O_2$ 164.20	97-53-0 -7.5	3855 253.2	4-06-00-06337 1.0652[20]	H_2O 1; EtOH 5; eth 5; chl 3 1.5405[20]
9390	Phenol, 2-methoxy-5-(2-propenyl)-	$C_{10}H_{12}O_2$ 164.20	501-19-9 8.5	253.5	3-06-00-05024 1.0613[25]	eth 4; EtOH 4 1.5413[20]
9391	Phenol, 2-methoxy-6-(1-propenyl)-	$C_{10}H_{12}O_2$ 164.20	1076-55-7 81	267.5	3-06-00-04991	H_2O 2; EtOH 3; eth 3; ace 3
9392	Phenol, 2-methoxy-6-(2-propenyl)-	$C_{10}H_{12}O_2$ 164.20	579-60-2	250.5	4-06-00-06334 1.062[20]	ace 3; bz 3 1.5545[20]
9393	Phenol, 2-methoxy-4-(1-propenyl)-, acetate, (E)-	$C_{12}H_{14}O_3$ 206.24	5912-87-8 80.5	282.5	4-06-00-06330 1.0251[100]	eth 4 1.5052[100]
9394	Phenol, 2-methoxy-4-(1-propenyl)-, acetate, (Z)- Isoeugenol, acetate (cis)	$C_{12}H_{14}O_3$ 206.24	97412-23-2 160-2[13]		3-06-00-05007 1.0947[19]	H_2O 1; eth 3 1.5418[20]
9395	Phenol, 2-methoxy-4-(2-propenyl)-, acetate	$C_{12}H_{14}O_3$ 206.24	93-28-7 30.5	281; 127-8[6]	3-06-00-05029 1.0806[20]	H_2O 1; EtOH 3; ctc 2 1.5205[20]
9396	Phenol, 2-methoxy-4-(1-propenyl)-, (E)-	$C_{10}H_{12}O_2$ 164.20	5932-68-3 33.5	141-2[13]	4-06-00-06324 1.0852[20]	H_2O 2; EtOH 3; eth 3; chl 3 1.5784[20]
9397	Phenol, 2-methoxy-4-(1-propenyl)-, (Z)-	$C_{10}H_{12}O_2$ 164.20	5912-86-7	133[13]; 80[0.5]	4-06-00-06324 1.0837[20]	H_2O 2; EtOH 3; eth 3 1.5726[20]
9398	Phenol, 2-methoxy-5-[(1,2,3,4-tetrahydro-6,7-dimethoxy-2-methyl-1-isoquinolinyl)methyl]-, (±)- Laudanine	$C_{20}H_{25}NO_4$ 343.42	85-64-3 167	5248	5-21-06-00048 1.26[20]	H_2O 2; EtOH 2; eth 2; bz 3
9399	Phenol, 2-methoxy-5-[(1,2,3,4-tetrahydro-6,7-dimethoxy-2-methyl-1-isoquinolinyl)methyl]-, (R)- Laudanidine	$C_{20}H_{25}NO_4$ 343.42	301-21-3 184.5	5247	5-21-06-00048	H_2O 4; bz 4
9400	Phenol, 2-methyl- o-Cresol	C_7H_8O 108.14	95-48-7 29.8	2580 191.0	4-06-00-01940 1.135[25]	H_2O 3; EtOH 4; eth 4; ace 5 1.5361[20]
9401	Phenol, 3-methyl- m-Cresol	C_7H_8O 108.14	108-39-4 11.8	2579 202.2	4-06-00-02035 1.0341[20]	H_2O 3; EtOH 5; eth 5; ace 5 1.5438[20]
9402	Phenol, 4-methyl- p-Cresol	C_7H_8O 108.14	106-44-5 35.5	2581 201.9	4-06-00-02093 1.0185[40]	H_2O 3; EtOH 5; eth 5; ace 5 1.5312[20]
9403	Phenol, 4-[2-(methylamino)propyl]- Pholedrine	$C_{10}H_{15}NO$ 165.24	370-14-9 161	7304	3-13-00-01710	eth 4; EtOH 4
9404	Phenol, 4-(3-methylbutyl)-	$C_{11}H_{16}O$ 164.25	1805-61-4 29	255; 128[7]	4-06-00-03378 0.9562[23]	1.505[27]
9405	Phenol, 2-methyl-4,6-dinitro- Dinitrocresol	$C_7H_6N_2O_5$ 198.14	534-52-1 86.5	3272	4-06-00-02014	H_2O 2; EtOH 3; eth 3; ace 3
9406	Phenol, 3,3'-methylenebis- Diphenylmethane, 3,3'-dihydroxy	$C_{13}H_{12}O_2$ 200.24	10193-50-7 102.5	235[3]	3-06-00-05411	eth 4; EtOH 4; HOAc 4
9407	Phenol, 4,4'-methylenebis-	$C_{13}H_{12}O_2$ 200.24	620-92-8 162.5	sub	4-06-00-06664	EtOH 3; eth 3; chl 3; DMSO-d_6 2
9408	Phenol, 2,2'-methylenebis[4-chloro- Dichlorophene	$C_{13}H_{10}Cl_2O_2$ 269.13	97-23-4 177.5	3059	4-06-00-06658	H_2O 1; EtOH 3; ace 3
9409	Phenol, 2,2'-methylenebis[3,4,6-trichloro- Hexachlorophene	$C_{13}H_6Cl_6O_2$ 406.91	70-30-4 166.5	4602	4-06-00-06659	H_2O 1; EtOH 3; eth 3; ace 3
9410	Phenol, 2-(1-methylethyl)-	$C_9H_{12}O$ 136.19	88-69-7 15.5	213.5	4-06-00-03210 1.012[20]	H_2O 2; EtOH 3; eth 3; bz 3 1.5315[20]
9411	Phenol, 3-(1-methylethyl)-	$C_9H_{12}O$ 136.19	618-45-1 26	228	4-06-00-03214	eth 4 1.5261[20]
9412	Phenol, 4-(1-methylethyl)-	$C_9H_{12}O$ 136.19	99-89-8 62.3	230; 110[10]	4-06-00-03215 0.990[20]	H_2O 2; EtOH 3; chl 3 1.5228[20]
9413	Phenol, 4,4'-(1-methylethylidene)bis- Bisphenol A	$C_{15}H_{16}O_2$ 228.29	80-05-7 153	1311 220[4]; 222[3]	4-06-00-06717	H_2O 1; EtOH 4; eth 4; bz 4
9414	Phenol, 2-methyl-4-(1-methylethyl)-	$C_{10}H_{14}O$ 150.22	1740-97-2 8.6	230; 83[3]	4-06-00-03330 0.9793[25]	H_2O 3; EtOH 3; bz 3; chl 3 1.5253[20]
9415	Phenol, 2-methyl-5-(1-methylethyl)- Carvacrol	$C_{10}H_{14}O$ 150.22	499-75-2 1	1881 237.7	4-06-00-03331 0.9772[20]	H_2O 2; EtOH 3; eth 3; ace 4 1.5230[20]
9416	Phenol, 2-methyl-6-(1-methylethyl)-	$C_{10}H_{14}O$ 150.22	3228-04-4 -14.5	225.5	3-06-00-01882 0.9782[25]	EtOH 3; bz 3; chl 3; os 3 1.5239[15]
9417	Phenol, 4-methyl-2-(1-methylethyl)-	$C_{10}H_{14}O$ 150.22	4427-56-9 36.5	228.5	4-06-00-03329 0.9910[20]	H_2O 2; EtOH 3; bz 3; chl 3 1.5275[20]

No.	Name Synonym	Mol. Form. Mol. Wt.	CAS RN mp/°C	Merck No. bp/°C	Beil. Ref. den/g cm^{-3}	Solubility n_D
9418	Phenol, 5-methyl-2-(1-methylethyl)- Thymol	$C_{10}H_{14}O$ 150.22	89-83-8 51.5	9333 232.5	4-06-00-03334 0.970[25]	H_2O 1; EtOH 4; eth 4; chl 4 1.5227[20]
9419	Phenol, 2-methyl-5-(1-methylethyl)-, acetate Carvacrol, acetate	$C_{12}H_{18}O_2$ 192.26	6380-28-5	2-06-00-00494 246.5 0.9896[25]	eth 4; EtOH 4 1.4913[28]
9420	Phenol, 5-methyl-2-(1-methylethyl)-, acetate Thymol, acetate	$C_{12}H_{18}O_2$ 192.26	528-79-0	4-06-00-03337 245 1.009[9]	bz 4; eth 4; EtOH 4; chl 4
9421	Phenol, 5-methyl-2-(1-methylpropyl)- 6-sec-Butyl-m-cresol	$C_{11}H_{16}O$ 164.25	29472-95-5	4-06-00-03394 89-90[2] 0.9612[25]	1.5188[20]
9422	Phenol, 2-methyl-4-nitro-	$C_7H_7NO_3$ 153.14	99-53-6 96	188[9]; 132[0.2]	4-06-00-02011	H_2O 2; EtOH 4; eth 4; bz 4
9423	Phenol, 2-methyl-6-nitro-	$C_7H_7NO_3$ 153.14	13073-29-5 70	255 dec; 102[9]	4-06-00-02009	H_2O 1; EtOH 4; eth 4; peth 3
9424	Phenol, 4-methyl-2-nitro-	$C_7H_7NO_3$ 153.14	119-33-5 36.5	125[22]	4-06-00-02149 1.2399[20]	ace 4; bz 4; eth 4; EtOH 4 1.5744[40]
9425	Phenol, 5-methyl-2-pentyl- 6-n-Amyl-m-cresol	$C_{12}H_{18}O$ 178.27	1300-94-3 24	643 137-9[15]	3-06-00-02005	ace 4; eth 4; EtOH 4
9426	Phenol, 4-methyl-2-(phenylazo)-	$C_{13}H_{12}N_2O$ 212.25	952-47-6 108.5	sub	4-16-00-00197	EtOH 3; eth 3; bz 3; chl 3
9427	Phenol, 2-methyl-6-(2-propenyl)-	$C_{10}H_{12}O$ 148.20	3354-58-3	232	4-06-00-03844 0.992[25]	1.5381[20]
9428	Phenol, 4-methyl-2-(1-propenyl)-	$C_{10}H_{12}O$ 148.20	53889-94-4	126-8[14]	4-06-00-03844 1.019[20]	1.5727[20]
9429	Phenol, 4-methyl-2-(2-propenyl)-	$C_{10}H_{12}O$ 148.20	6628-06-4	237	4-06-00-03845 0.96[25]	1.5385[20]
9430	Phenol, 2-(1-methylpropyl)- o-sec-Butylphenol	$C_{10}H_{14}O$ 150.22	89-72-5 16	228; 116[21]	3-06-00-01852 0.9804[25]	1.5200[25]
9431	Phenol, 4-(1-methylpropyl)-	$C_{10}H_{14}O$ 150.22	99-71-8 61.5	241	4-06-00-03279 0.986[20]	H_2O 1; EtOH 3; eth 4; alk 3 1.5182[21]
9432	Phenol, 4-(2-methylpropyl)-	$C_{10}H_{14}O$ 150.22	4167-74-2 51.5	237	4-06-00-03288 0.9778[20]	ace 4; eth 4; EtOH 4 1.5319[25]
9433	Phenol, 4,4'-(1-methylpropylidene)bis- Bisphenol B	$C_{16}H_{18}O_2$ 242.32	77-40-7 120.5	1312	4-06-00-06738	ace 4; MeOH 4
9434	Phenol, 3-methyl-2,4,6-trinitro- 3-Methylpicric acid	$C_7H_5N_3O_7$ 243.13	602-99-3 109.5	exp 150	4-06-00-02079	H_2O 2; EtOH 3; eth 3; ace 3
9435	Phenol, 2-nitro- o-Nitrophenol	$C_6H_5NO_3$ 139.11	88-75-5 44.8	6541 216	4-06-00-01246 1.2942[40]	H_2O 2; EtOH 4; eth 4; ace 4 1.5723[50]
9436	Phenol, 3-nitro- m-Nitrophenol	$C_6H_5NO_3$ 139.11	554-84-7 96.8	6540 194[70]	4-06-00-01269 1.2797[100]	H_2O 2; EtOH 4; eth 4; ace 4
9437	Phenol, 4-nitro- p-Nitrophenol	$C_6H_5NO_3$ 139.11	100-02-7 113.8	6542	4-06-00-01279 1.479[20]	H_2O 2; EtOH 4; eth 4; ace 4
9438	Phenol, 4-nitroso- 4-Nitrosophenol	$C_6H_5NO_2$ 123.11	104-91-6 144 dec	6563	4-07-00-02073	H_2O 2; EtOH 3; eth 3; ace 3
9439	Phenol, pentabromo- Pentabromophenol	C_6HBr_5O 488.59	608-71-9 229.5	sub	3-06-00-00766	H_2O 1; EtOH 3; eth 2; bz 3
9440	Phenol, pentachloro- Pentachlorophenol	C_6HCl_5O 266.34	87-86-5 174	7059 310 dec	4-06-00-01025 1.978[22]	H_2O 2; EtOH 4; eth 4; bz 3
9441	Phenol, 3-pentadecyl-	$C_{21}H_{36}O$ 304.52	501-24-6 53.5	230[8]; 197[1.5]	4-06-00-03556	ace 4; bz 4; EtOH 4
9442	Phenol, pentafluoro- Pentafluorophenol	C_6HF_5O 184.07	771-61-9 32.8	145.6	4-06-00-00782	1.4263[20]
9443	Phenol, pentamethyl- Pentamethylphenol	$C_{11}H_{16}O$ 164.25	2819-86-5 128	267	4-06-00-03412	H_2O 1; EtOH 3
9444	Phenol, 4-pentyl-	$C_{11}H_{16}O$ 164.25	14938-35-3 23	250.5	4-06-00-03370 0.960[20]	eth 4; EtOH 4 1.5272[25]
9445	Phenol, 2-(phenylamino)-	$C_{12}H_{11}NO$ 185.23	644-71-3 69.5	185[20]	4-13-00-00822	H_2O 2; EtOH 3; eth 3; bz 2
9446	Phenol, 3-(phenylamino)-	$C_{12}H_{11}NO$ 185.23	101-18-8 81.5	340	3-13-00-00944	H_2O 2; EtOH 4; eth 4; ace 4
9447	Phenol, 4-(phenylamino)-	$C_{12}H_{11}NO$ 185.23	122-37-2 73	330	4-13-00-01052	H_2O 2; EtOH 4; eth 4; bz 4
9448	Phenol, 4-(phenylazo)-	$C_{12}H_{10}N_2O$ 198.22	1689-82-3 155	225[20] dec	2-16-00-00038	H_2O 1; EtOH 4; eth 4; bz 3
9449	Phenol, 4-(phenylazo)-, acetate (ester)	$C_{14}H_{12}N_2O_2$ 240.26	13102-31-3 89	360 dec	1-16-00-00236	EtOH 4; lig 4
9450	Phenol, 2-[(phenylimino)methyl]-	$C_{13}H_{11}NO$ 197.24	779-84-0 49.5		4-12-00-00339 1.087[25]	H_2O 1; EtOH 3
9451	Phenol, 2-(phenylmethoxy)-	$C_{13}H_{12}O_2$ 200.24	6272-38-4	173-4[13]	4-06-00-05572 1.154[22]	eth 4; EtOH 4 1.5906[18]
9452	Phenol, 3-(phenylmethoxy)-	$C_{13}H_{12}O_2$ 200.24	3769-41-3 69.2	200[5]	4-06-00-05669	H_2O 4
9453	Phenol, 4-(phenylmethoxy)- Monobenzone	$C_{13}H_{12}O_2$ 200.24	103-16-2 122	6159	4-06-00-05728	H_2O 2; EtOH 4; eth 4; ace 3
9454	Phenol, 2-(phenylmethyl)- o-Benzylphenol	$C_{13}H_{12}O$ 184.24	28994-41-4 21	1158 312	4-06-00-04628	ace 4; bz 4; EtOH 4 1.5994[20]
9455	Phenol, 4-(phenylmethyl)- p-Benzylphenol	$C_{13}H_{12}O$ 184.24	101-53-1 84	1159 322	4-06-00-04640	H_2O 3; EtOH 3; eth 3; bz 3
9456	Phenol, 2-(1-propenyl)-	$C_9H_{10}O$ 134.18	6380-21-8 -6	220	3-06-00-02393 1.0246[15]	eth 4 1.5181[20]
9457	Phenol, 2-(2-propenyl)-	$C_9H_{10}O$ 134.18	1745-81-9 -6	220	4-06-00-03807 1.0246[15]	eth 4 1.5181[20]

No.	Name Synonym	Mol. Form. Mol. Wt.	CAS RN mp/°C	Merck No. bp/°C	Beil. Ref. den/g cm⁻³	Solubility n_D
9458	Phenol, 4-(2-propenyl)- Chavicol	$C_9H_{10}O$ 134.18	501-92-8 15.8	2039 238	4-06-00-03817 1.020^{15}	eth 4; EtOH 4; chl 4 1.5441^{18}
9459	Phenol, 2-(1-propenyl)-, (E)-	$C_9H_{10}O$ 134.18	23619-59-2 37.5	230.5	4-06-00-03793 1.044^{16}	eth 4; lig 4 1.5823^{21}
9460	Phenol, 2-propoxy-	$C_9H_{12}O_2$ 152.19	6280-96-2 227		4-06-00-05566 1.0523^{25}	EtOH 4 1.5176^{25}
9461	Phenol, 2-propyl-	$C_9H_{12}O$ 136.19	644-35-9 7	220	4-06-00-03176 1.015^{20}	eth 4; EtOH 4
9462	Phenol, 3-propyl-	$C_9H_{12}O$ 136.19	621-27-2 26	228	4-06-00-03180 0.987^{20}	H_2O 2; EtOH 3; eth 3 1.5223^{20}
9463	Phenol, 4-propyl-	$C_9H_{12}O$ 136.19	645-56-7 22	232.6	4-06-00-03181 1.009^{20}	H_2O 2; EtOH 3; ctc 2 1.5379^{25}
9464	Phenol, 4,4'-propylidenebis-	$C_{15}H_{18}O_2$ 228.29	1576-13-2 132	275^{20}	3-06-00-05457	eth 4; EtOH 4
9465	Phenol, 4,4'-(2-pyridinylmethylene)bis-, diacetate (ester) Bisacodyl	$C_{22}H_{19}NO_4$ 361.40	603-50-9 133.5	1253	5-21-05-00415	
9466	Phenol, 4,4'-sulfonylbis-	$C_{12}H_{10}O_4S$ 250.28	80-09-1 240.5		4-06-00-05809 1.3663^{15}	H_2O 1; EtOH 3; eth 3; bz 2
9467	Phenol, 2,3,4,6-tetrabromo- 2,3,4,6-Tetrabromophenol	$C_6H_2Br_4O$ 409.70	14400-94-3 113.5	sub	3-06-00-00766	bz 4; EtOH 4
9468	Phenol, 2,3,4,5-tetrabromo-6-methyl- 3,4,5,6-Tetrabromo-o-cresol	$C_7H_4Br_4O$ 423.72	576-55-6 208	9120 dec	3-06-00-01272	H_2O 1; EtOH 3; eth 3; bz 3
9469	Phenol, 2,3,4,5-tetrachloro- 2,3,4,5-Tetrachlorophenol	$C_6H_2Cl_4O$ 231.89	4901-51-3 116.5	sub	4-06-00-01020	EtOH 4
9470	Phenol, 2,3,4,6-tetrachloro- 2,3,4,6-Tetrachlorophenate	$C_6H_2Cl_4O$ 231.89	58-90-2 70	150^{15}	4-06-00-01021	H_2O 1; EtOH 3; bz 3; chl 3
9471	Phenol, 2,3,5,6-tetrachloro- 2,3,5,6-Tetrachlorophenol	$C_6H_2Cl_4O$ 231.89	935-95-5 115		4-06-00-01025	H_2O 2; bz 4; lig 3
9472	Phenol, 2,3,5,6-tetrachloro-4-methoxy- Drosophilin A	$C_7H_4Cl_4O_2$ 261.92	484-67-3 116	3441	3-06-00-04436	
9473	Phenol, 2,3,4,5-tetramethyl- Prehnitenol	$C_{10}H_{14}O$ 150.22	488-70-0 85.3	266	4-06-00-03359	H_2O 2; EtOH 4; eth 4; lig 2
9474	Phenol, 2,3,4,6-tetramethyl-	$C_{10}H_{14}O$ 150.22	3238-38-8 80.5	240	4-06-00-03360	EtOH 3
9475	Phenol, 2,3,5,6-tetramethyl-	$C_{10}H_{14}O$ 150.22	527-35-5 118.5	247	4-06-00-03363	chl 3; peth 3; HOAc 3
9476	Phenol, 2,3,4,6-tetranitro- 2,3,4,6-Tetranitrophenol	$C_6H_2N_4O_9$ 274.10	641-16-7 140 dec	exp	2-06-00-00284	H_2O 4; chl 3
9477	Phenol, 2,2'-thiobis[4-chloro- Fenticlor	$C_{12}H_8Cl_2O_2S$ 287.17	97-24-5 174	3947	4-06-00-05645	H_2O 1; EtOH 3; eth 3; gl HOAc 3
9478	Phenol, 2,2'-thiobis[4,6-dichloro- Bithionol	$C_{12}H_6Cl_4O_2S$ 356.06	97-18-7 188	1316	4-06-00-05646 1.73^{25}	ace 4
9479	Phenol, 2,4,6-tribromo- 2,4,6-Tribromophenol	$C_6H_3Br_3O$ 330.80	118-79-6 95.5	9526 286	4-06-00-01067 2.55^{20}	H_2O 2; EtOH 4; eth 3; bz 3
9480	Phenol, 2,4,6-tribromo-3-methyl- 2,4,6-Tribromo-m-cresol	$C_7H_5Br_3O$ 344.83	4619-74-3 84	9524	3-06-00-01324	EtOH 3; eth 3; bz 3; chl 2
9481	Phenol, 2,3,4-trichloro- 2,3,4-Trichlorophenol	$C_6H_3Cl_3O$ 197.45	15950-66-0 83.5	sub	3-06-00-00716	EtOH 3; eth 3; bz 3; alk 3
9482	Phenol, 2,3,5-trichloro- 2,3,5-Trichlorophenol	$C_6H_3Cl_3O$ 197.45	933-78-8 62	$248-9^{250}$	3-06-00-00716	eth 4; EtOH 4
9483	Phenol, 2,3,6-trichloro- 2,3,6-Trichlorophenol	$C_6H_3Cl_3O$ 197.45	933-75-5 58		4-06-00-00962	H_2O 2; EtOH 4; eth 4; bz 4
9484	Phenol, 2,4,5-trichloro- 2,4,5-Trichlorophenol	$C_6H_3Cl_3O$ 197.45	95-95-4 69	9555 247	4-06-00-00962	H_2O 2; EtOH 4; eth 4; bz 4
9485	Phenol, 2,4,6-trichloro- 2,4,6-Trichlorophenol	$C_6H_3Cl_3O$ 197.45	88-06-2 69	9556 246	4-06-00-01005 1.4901^{75}	H_2O 2; EtOH 3; eth 3; HOAc 3
9486	Phenol, 3,4,5-trichloro- 3,4,5-Trichlorophenol	$C_6H_3Cl_3O$ 197.45	609-19-8 101	275	3-06-00-00729	H_2O 2; eth 3; lig 2
9487	Phenol, 2,3,4-trichloro-6-methyl- 4,5,6-Trichloro-o-cresol	$C_7H_5Cl_3O$ 211.47	551-78-0 77	9548	3-06-00-01268	
9488	Phenol, 2,3,6-trichloro-4-methyl- 2,3,6-Trichloro-p-cresol	$C_7H_5Cl_3O$ 211.47	551-77-9 66.5	9546	0-06-00-00404	EtOH 4
9489	Phenol, 2,4,6-trichloro-3-methyl- 2,4,6-Trichloro-m-cresol	$C_7H_5Cl_3O$ 211.47	551-76-8 46	9547 265	4-06-00-02070	EtOH 4; MeOH 4; chl 4
9490	Phenol, 3,4,6-trichloro-2-nitro- 3,4,6-Trichloro-2-nitrophenol	$C_6H_2Cl_3NO_3$ 242.44	82-62-2 92.5	9554	3-06-00-00842	
9491	Phenol, 3-(trifluoromethyl)-	$C_7H_5F_3O$ 162.11	98-17-9 -1.8	178	4-06-00-02060 1.3418^{25}	
9492	Phenol, 2,4,6-triiodo- 2,4,6-Triiodophenol	$C_6H_3I_3O$ 471.80	609-23-4 159.8	sub	4-06-00-01085	H_2O 1; EtOH 2; eth 3; ace 3
9493	Phenol, 2,3,4-trimethyl-	$C_9H_{12}O$ 136.19	526-85-2 81	236	4-06-00-03245	bz 4; eth 4; EtOH 4
9494	Phenol, 2,4,5-trimethyl-	$C_9H_{12}O$ 136.19	496-78-6 72	232	4-06-00-03247	H_2O 1; EtOH 4; eth 4
9495	Phenol, 2,4,6-trimethyl-	$C_9H_{12}O$ 136.19	527-60-6 73	220	4-06-00-03253	eth 4; EtOH 4
9496	Phenol, 2,4,5-trimethyl-, acetate 2,4,5-Trimethylphenyl acetate	$C_{11}H_{14}O_2$ 178.23	69305-42-6 34	245.5	2-06-00-00482	eth 4; EtOH 4

No.	Name / Synonym	Mol. Form. / Mol. Wt.	CAS RN / mp/°C	Merck No. / bp/°C	Beil. Ref. / den/g cm⁻³	Solubility / n_D
9497	Phenol, 2,4,6-trinitro- / Picric acid	$C_6H_3N_3O_7$ / 229.11	88-89-1 / 122.5	7380 / exp 300	4-06-00-01388	H_2O 2; EtOH 3; eth 3; ace 4 / 1.763
9498	Phenol, 2,4,6-tris(1,1-dimethylethyl)-	$C_{18}H_{30}O$ / 262.44	732-26-3 / 131	278	4-06-00-03539 / 0.864[27]	H_2O 1; EtOH 3; ace 3; ctc 3
9499	10H-Phenothiazine / Thiodiphenylamine	$C_{12}H_9NS$ / 199.28	92-84-2 / 187.5	7220 / 371	4-27-00-01214	ace 4; bz 4; eth 4; EtOH 4
9500	10H-Phenothiazine-2-acetic acid, 10-methyl- / Metiazinic acid	$C_{15}H_{13}NO_2S$ / 271.34	13993-65-2 / 144	6057		chl 3
9501	10H-Phenothiazine, 2-chloro-10-[3-(4-methyl-1-piperazinyl)propyl]- / Phenothiazine, 2-chloro-10(3(4-methyl-1-piperazinyl)propyl)	$C_{20}H_{24}ClN_3S$ / 373.95	58-38-8 / 228	7768	4-27-00-01307	
9502	10H-Phenothiazine, 10-[(dimethylamino)acetyl]- / Ahistan	$C_{16}H_{16}N_2OS$ / 284.38	518-61-6 / 144.5	179	4-27-00-01272	
9503	10H-Phenothiazine-10-ethanamine, N,N-diethyl- / Diethazine	$C_{18}H_{22}N_2S$ / 298.45	60-91-3 /	3098 / 167[0.5]	4-27-00-01240	H_2O 1; dil HCl 3
9504	10H-Phenothiazine-10-ethanamine, N,N,α-trimethyl- / Promethazine	$C_{17}H_{20}N_2S$ / 284.43	60-87-7 / 60	7797 / 190-3[0.5]	4-27-00-01253	H_2O 1; dil HCl 4
9505	10H-Phenothiazine-10-propanamine, 2-chloro-N,N-dimethyl- / Chlorpromazine	$C_{17}H_{19}ClN_2S$ / 318.87	50-53-3 /	2186 / 200-5[0.8]	4-27-00-01300	H_2O 1; EtOH 4; eth 4; bz 4
9506	Phenothiazin-5-ium, 3,7-bis(dimethylamino)-, chloride / Methylene blue	$C_{16}H_{18}ClN_3S$ / 319.86	61-73-4 /	5979	4-27-00-05152	H_2O 3; EtOH 3; eth 1; chl 3
9507	10H-Phenoxazine / Phenoxazine	$C_{12}H_9NO$ / 183.21	135-67-1 / 156	7222 / dec	4-27-00-01209	bz 4; eth 4; EtOH 4
9508	3H-Phenoxazin-3-one, 7-hydroxy-, 10-oxide / Resazurin	$C_{12}H_7NO_4$ / 229.19	550-82-3 /	8144 / sub	4-27-00-03594	H_2O 1; EtOH 2; eth 1; alk 3
9509	DL-Phenylalanine	$C_9H_{11}NO_2$ / 165.19	150-30-1 / 284-8 dec	7242 / sub	4-14-00-01553	H_2O 4; EtOH 1; eth 1; bz 1
9510	L-Phenylalanine / (S)-α-Aminobenzenepropanoic acid	$C_9H_{11}NO_2$ / 165.19	63-91-2 / 283 dec	7242	4-14-00-01552	H_2O 4; EtOH 1; eth 1; bz 1
9511	L-Phenylalanine, N-L-α-aspartyl-, 1-methyl ester / Aspartame	$C_{14}H_{18}N_2O_5$ / 294.31	22839-47-0 / 246.5	861		
9512	L-Phenylalanine, ethyl ester	$C_{11}H_{15}NO_2$ / 193.25	3081-24-1 /	/ 148[13]	4-14-00-01556 / 1.065[15]	H_2O 2
9513	Phorate / Phosphorodithioic acid, O,O-diethyl S-[(ethylthio)methyl] ester	$C_7H_{17}O_2PS_3$ / 260.38	298-02-2 / <-15	7305 / 119[0.8]	1.16[25]	
9514	Phosalone / Phosphorodithioic acid, S-[(6-chloro-2-oxo-3(2H)-benzoxazolyl)methyl]O,O-diethyl ester	$C_{12}H_{15}ClNO_4PS_2$ / 367.81	2310-17-0 / 46	7308		
9515	Phosmet / Phosphorodithioic acid, S-[(1,3-dihydro-1,3-dioxo-2H-isoindol-2-yl)methyl] O,O-dimethyl ester	$C_{11}H_{12}NO_4PS_2$ / 317.33	732-11-6 / 72	7311 / dec		
9516	Phosphine, dicyclohexyl- / Dicyclohexylphosphine	$C_{12}H_{23}P$ / 198.29	829-84-5 /	/ 281; 129[8]	4-16-00-00947 / 0.904[25]	1.5163[20]
9517	Phosphine oxide, triethyl-	$C_6H_{15}OP$ / 134.16	597-50-2 / 48	/ 243	4-04-00-03463	H_2O 4; eth 4; EtOH 4
9518	Phosphine oxide, triphenyl-	$C_{18}H_{15}OP$ / 278.29	791-28-6 / 156.5	/ >360	4-16-00-01011 / 1.2124[23]	H_2O 2; EtOH 4; eth 2; bz 4
9519	Phosphine, phenyl- / Monophenylphosphine	C_6H_7P / 110.10	638-21-1 /	/ 160.5	4-16-00-00948 / 1.001[15]	1.5796[20]
9520	Phosphine, tributyl- / Tributylphosphine	$C_{12}H_{27}P$ / 202.32	998-40-3 /	/ 150[50]	4-04-00-03436 / 0.812[25]	1.4619[20]
9521	Phosphine, triethyl- / Triethyl phosphine	$C_6H_{15}P$ / 118.16	554-70-1 / -88	9590 / 129	4-04-00-03431 / 0.8006[19]	H_2O 1; EtOH 5; eth 5 / 1.458[15]
9522	Phosphine, trimethyl- / Trimethylphosphine	C_3H_9P / 76.08	594-09-2 / -85	/ 37.5	4-04-00-03429	H_2O 1; eth 3
9523	Phosphine, triphenyl- / Triphenylphosphine	$C_{18}H_{15}P$ / 262.29	603-35-0 / 80	9657 / 188[1]	4-16-00-00951 / 1.0749[80]	H_2O 1; EtOH 3; eth 4; bz 3 / 1.6358[80]
9524	Phosphine, tris(chloromethyl)- / Tris(chloromethyl)phosphine	$C_3H_6Cl_3P$ / 179.41	13482-62-7 /	/ 100[7]	3-01-00-02608 / 1.414[20]	ace 4; bz 4; eth 4; EtOH 4
9525	Phosphinic acid, diethyl- / Diethylphosphinic acid	$C_4H_{11}O_2P$ / 122.10	813-76-3 / 18.5	/ 320	4-04-00-03474	H_2O 4; EtOH 4; eth 4
9526	Phosphinic acid, diethyl-, anhydride	$C_8H_{20}O_3P_2$ / 226.19	7495-97-8 /	/ 188[14]	4-04-00-03477 / 1.1053[20]	1.4647[20]
9527	Phosphinic acid, diethyl-, ethyl ester / Ethyl diethylphosphinate	$C_6H_{15}O_2P$ / 150.16	4775-09-1 /	/ 95[14]	4-04-00-03475 / 0.9908[20]	eth 4; EtOH 4 / 1.4337[20]
9528	Phosphinic acid, dimethyl- / Dimethylphosphinic acid	$C_2H_7O_2P$ / 94.05	3283-12-3 / 92	/ 377	4-04-00-03470	H_2O 4; EtOH 4; eth 4; bz 3
9529	Phosphinic chloride, diethyl-	$C_4H_{10}ClOP$ / 140.55	1112-37-4 /	/ 104[15]	4-04-00-03476 / 1.1394[20]	1.4647[20]
9530	Phosphonic acid, acetyl-, diethyl ester / O,O-Diethyl acetylphosphonate	$C_6H_{13}O_4P$ / 180.14	919-19-7 /	/ 114-5[20]	4-02-00-00440 / 1.1005[20]	1.4200[26]

No.	Name / Synonym	Mol. Form. / Mol. Wt.	CAS RN / mp/°C	Merck No. / bp/°C	Beil. Ref. / den/g cm⁻³	Solubility / n_D
9531	Phosphonic acid, bis(1,1-dimethylethyl) ester	$C_8H_{19}O_3P$ 194.21	13086-84-5	70-2[10]	4-01-00-01623 0.975[25]	1.4168[25]
9532	Phosphonic acid, dibutyl ester	$C_8H_{19}O_3P$ 194.21	1809-19-4	131[19]; 77[1.5]	4-01-00-01525 0.9485[20]	1.4220[20]
9533	Phosphonic acid, (1,1-dimethylethyl)-	$C_4H_{11}O_3P$ 138.10	4923-84-6 192	dec	4-04-00-03557	H_2O 4; EtOH 4
9534	Phosphonic acid, diphenyl ester	$C_{12}H_{11}O_3P$ 234.19	4712-55-4 12	218-9[26]	4-06-00-00703 1.223[25]	1.5575[20]
9535	Phosphonic acid, ethenyl-, diethyl ester	$C_6H_{13}O_3P$ 164.14	682-30-4	110[2]	4-04-00-03569 1.068[25]	1.4290[20]
9536	Phosphonic acid, ethyl-, diethyl ester Diethyl ethylphosphonate	$C_6H_{15}O_3P$ 166.16	78-38-6	198; 90[16]	4-04-00-03524 1.0259[20]	H_2O 2; EtOH 3; eth 3 1.4163[20]
9537	Phosphonic acid, ethyl-, dimethyl ester	$C_4H_{11}O_3P$ 138.10	6163-75-3	82[18]	4-04-00-03523 1.1029[30]	H_2O 4; bz 4; EtOH 4 1.4128[30]
9538	Phosphonic acid, methyl-	CH_5O_3P 96.02	993-13-5 108.5	dec	4-04-00-03498	H_2O 4; EtOH 4; eth 4; bz 1
9539	Phosphonic acid, methyl-, diethyl ester	$C_5H_{13}O_3P$ 152.13	683-08-9	194	4-04-00-03500 1.0406[30]	H_2O 3; EtOH 3; eth 3; bz 1 1.4101[30]
9540	Phosphonic acid, methyl-, dimethyl ester	$C_3H_9O_3P$ 124.08	756-79-6	181; 79.5[20]	4-04-00-03499 1.1507[20]	H_2O 3; EtOH 3; eth 3 1.4099[30]
9541	Phosphonic acid, methyl-, diphenyl ester Diphenyl methylphosphonate	$C_{13}H_{13}O_3P$ 248.22	7526-26-3 35	205[13]	4-06-00-00703 1.2051[20]	H_2O 1
9542	Phosphonic acid, (1-methylethyl)- 2-Propanephosphonic acid	$C_3H_9O_3P$ 124.08	4721-37-3 74.5	dec	4-04-00-03548	H_2O 4; eth 4; EtOH 4
9543	Phosphonic acid, methyl-, monoethyl ester	$C_3H_9O_3P$ 124.08	1832-53-7	106-7[0.1]	4-04-00-03499 1.1800[20]	1.4258[20]
9544	Phosphonic acid, (2-methylpropyl)-	$C_4H_{11}O_3P$ 138.10	4721-34-0 119	dec	4-04-00-03556	H_2O 4; eth 4; EtOH 4; xyl 4
9545	Phosphonic acid, (phenylmethyl)-, diethyl ester	$C_{11}H_{17}O_3P$ 228.23	1080-32-6	110[2]	4-16-00-01093	ctc 3 1.4930[20]
9546	Phosphonic acid, 1,3-propanediylbis-, tetraethyl ester	$C_{11}H_{26}O_6P_2$ 316.27	22401-25-8	185-6[3]	4-04-00-03582 1.1227[20]	1.4500[20]
9547	Phosphonic acid, propyl-	$C_3H_9O_3P$ 124.08	4672-38-2 73.8	dec	4-04-00-03542	H_2O 4; eth 4; EtOH 4
9548	Phosphonic dichloride, phenyl-	$C_6H_5Cl_2OP$ 194.98	824-72-6 1	258	4-16-00-01074 1.197[25]	DMSO-d_6 2 1.5581[25]
9549	Phosphonothioic acid, methyl-, O,O-bis(1-methylethyl) ester Phosphonic acid, thiono, methyl, diisopropyl ester	$C_7H_{17}O_2PS$ 196.25	66295-45-2	42[1]	4-04-00-03516 0.9885[25]	1.4512[25]
9550	Phosphonothioic acid, methyl-, O,O-dimethyl ester	$C_3H_9O_2PS$ 140.14	681-06-1	43[4]	4-04-00-03514 1.1375[25]	1.4738[25]
9551	Phosphonothioic acid, phenyl-, O-ethyl O-(4-nitrophenyl) ester O-Ethyl O-p-nitrophenyl benzenethiophosphonate	$C_{14}H_{14}NO_4PS$ 323.31	2104-64-5 36	3576	4-16-00-01089 1.27[25]	bz 4; eth 4; EtOH 4 1.5978[30]
9552	Phosphonothioic dichloride, phenyl- Dichlorophenylphosphine sulfide	$C_6H_5Cl_2PS$ 211.05	3497-00-5	205[130]	4-16-00-01090 1.376[13]	
9553	Phosphonous dichloride, (2,5-dimethylphenyl)- Phosphonous dichloride, 2,5-xylyl-	$C_8H_9Cl_2P$ 207.04	57150-66-0 -30	253.5	3-16-00-00848 1.25[18]	
9554	Phosphonous dichloride, (4-ethylphenyl)-	$C_8H_9Cl_2P$ 207.04	5274-50-0	251	4-16-00-00974 1.237[20]	1.584[20]
9555	Phosphonous dichloride, [4-(1-methylethyl)phenyl]- Benzene, 1-chlorophosphino-4-isopropyl	$C_9H_{11}Cl_2P$ 221.07	82906-78-3	269	3-16-00-00848 1.190[12]	ace 4 1.5677[25]
9556	Phosphonous dichloride, phenyl- Dichlorophenylphosphine	$C_6H_5Cl_2P$ 178.98	644-97-3 -51	225; 142[57]	4-16-00-00972 1.356[20]	bz 4 1.6030[20]
9557	Phosphoramidothioic dichloride, dimethyl- Phosphoric acid, amide,thiono,dichloride,N,N-dimethyl	$C_2H_6Cl_2NPS$ 178.02	1498-65-3	77.5[10]	4-04-00-00293 1.3731[20]	1.5390[20]
9558	Phosphoric acid, diethyl ester	$C_4H_{11}O_4P$ 154.10	598-02-7	203 dec	4-01-00-01339 1.186[25]	eth 4 1.4170[20]
9559	Phosphoric acid, dimethyl ester	$C_2H_7O_4P$ 126.05	813-78-5	174 dec	4-01-00-01259 1.335[25]	H_2O 4; ace 4; EtOH 4 1.408[25]
9560	Phosphoric acid, 2-ethylhexyl diphenyl ester	$C_{20}H_{27}O_4P$ 362.41	1241-94-7	232[5]	4-06-00-00718 1.090[25]	1.510[25]
9561	Phosphoric acid, monoethyl ester	$C_2H_7O_4P$ 126.05	1623-14-9	dec	4-01-00-01338 1.430[25]	H_2O 4; ace 4; eth 4; EtOH 4 1.427
9562	Phosphoric acid tributyl ester Tributyl phosphate	$C_{12}H_{27}O_4P$ 266.32	126-73-8	9531 289	4-01-00-01531 0.9727[25]	H_2O 3; EtOH 5; eth 3; bz 3 1.4224[25]
9563	Phosphoric acid, triethyl ester Ethyl phosphate	$C_6H_{15}O_4P$ 182.16	78-40-0 -56.4	9589 215.5	4-01-00-01339 1.0695[20]	H_2O 3; EtOH 4; eth 3; bz 3 1.4053[20]
9564	Phosphoric acid, trimethyl ester	$C_3H_9O_4P$ 140.08	512-56-1 -46	197.2	4-01-00-01259 1.2144[20]	H_2O 4; EtOH 2; eth 3 1.3967[20]
9565	Phosphoric acid, tripentyl ester	$C_{15}H_{33}O_4P$ 308.40	2528-38-3	225[50.] 167[5]	4-01-00-01645 0.9608[20]	H_2O 1; EtOH 3; eth 3; tol 3 1.4319[20]
9566	Phosphoric acid, triphenyl ester Triphenyl phosphate	$C_{18}H_{15}O_4P$ 326.29	115-86-6 50.5	9656 245[11]	4-06-00-00720 1.2055[50]	H_2O 1; EtOH 3; eth 4; bz 4

No.	Name / Synonym	Mol. Form. / Mol. Wt.	CAS RN / mp/°C	Merck No. / bp/°C	Beil. Ref. / den/g cm⁻³	Solubility / n_D
9567	Phosphoric acid, tripropyl ester / Tripropyl phosphate	$C_9H_{21}O_4P$ / 224.24	513-08-6	/ 252	4-01-00-01428 / 1.0121^{20}	H_2O 2; EtOH 3; eth 3; chl 2 / 1.4165^{20}
9568	Phosphoric acid, tris(1-methylethyl) ester	$C_9H_{21}O_4P$ / 224.24	513-02-0	/ 219	4-01-00-01478 / 0.9867^{20}	EtOH 4 / 1.4057^{20}
9569	Phosphoric acid, tris(2-methylphenyl) ester / Tri-o-tolyl phosphate	$C_{21}H_{21}O_4P$ / 368.37	78-30-8 / 11	9676 / 410	4-06-00-01979 / 1.1955^{20}	H_2O 1; EtOH 4; eth 4; ctc 4 / 1.5575^{20}
9570	Phosphoric acid, tris(3-methylphenyl) ester	$C_{21}H_{21}O_4P$ / 368.37	563-04-2 / 25.5	/ 260^{15}	4-06-00-02057 / 1.150^{25}	H_2O 2; EtOH 2; eth 3; ctc 4 / 1.5575^{20}
9571	Phosphoric acid, tris(4-methylphenyl) ester	$C_{21}H_{21}O_4P$ / 368.37	78-32-0 / 77.5	/ 224^{35}	4-06-00-02130 / 1.247^{25}	EtOH 3; eth 3; bz 3; chl 3
9572	Phosphoric acid, tris(2-methylpropyl) ester / Triisobutyl phosphate	$C_{12}H_{27}O_4P$ / 266.32	126-71-6	/ 264	4-01-00-01598 / 0.9681^{20}	H_2O 4; bz 4; eth 4; EtOH 4 / 1.4193^{20}
9573	Phosphoric triamide, hexamethyl- / Tris(dimethylamino)phosphine oxide	$C_6H_{18}N_3OP$ / 179.20	680-31-9 / 7.5	4568 / 232.5	4-04-00-00284 / 1.03^{20}	EtOH 3; eth 3 / 1.4579^{20}
9574	Phosphorochloridic acid, diethyl ester / Diethyl chlorophosphate	$C_4H_{10}ClO_3P$ / 172.55	814-49-3	/ 93.5	4-01-00-01345 / 1.205^{19}	/ 1.4170^{20}
9575	Phosphorochloridic acid, diphenyl ester	$C_{12}H_{10}ClO_3P$ / 268.64	2524-64-3	/ $314-6^{272}$	4-06-00-00737 / 1.296^{25}	tfa 3 / 1.5500^{20}
9576	Phosphorochloridothioic acid, O,O-diethyl ester / Diethyl chlorothiophosphate	$C_4H_{10}ClO_2PS$ / 188.61	2524-04-1	/ 45^3	4-01-00-01352	ctc 3
9577	Phosphorodichloridic acid, ethyl ester / Ethyl dichlorophosphate	$C_2H_5Cl_2O_2P$ / 162.94	1498-51-7	/ $60-5^{10}$	4-01-00-01345	/ 1.4338^{20}
9578	Phosphorodichloridothioic acid, O-(2,4,5-trichlorophenyl) ester / Phenol, 2,4,5-trichloro-, O-phosphorodichloridothioate	$C_6H_2Cl_5OPS$ / 330.38	16805-78-0	/ 110^1	4-06-00-00996 / 1.6653^{20}	/ 1.6084^{20}
9579	Phosphorodithioic acid, O,O,S-triethyl ester	$C_6H_{15}O_2PS_2$ / 214.29	2524-09-6	/ 128^{20}	4-01-00-01402 / 1.1336^0	/ 1.5043
9580	Phosphorofluoridic acid, bis(1-methylethyl) ester / Isoflurophate	$C_6H_{14}FO_3P$ / 184.15	55-91-4	5060 / 62^9	4-01-00-01480 / 1.055^{25}	H_2O 2; eth 3; oils 4; lig 2 / 1.3830^{25}
9581	Phosphorothioic acid, O-(2,4-dichlorophenyl) O,O-diethyl ester / Dichlofenthion	$C_{10}H_{13}Cl_2O_3PS$ / 315.16	97-17-6	3030	4-06-00-00939	ctc 3; CS_2 3
9582	Phosphorothioic acid, O,O-diethyl O-[6-methyl-2-(1-methylethyl)-4-pyrimidinyl] ester / Diazinon	$C_{12}H_{21}N_2O_3PS$ / 304.35	333-41-5	2978 / $85-90^{0.05}$	5-23-11-00187 / 1.1088^{20}	/ 1.4922^{20}
9583	Phosphorothioic acid, O,O-diethyl O-(4-nitrophenyl) ester / Parathion	$C_{10}H_{14}NO_5PS$ / 291.26	56-38-2 / 6.1	6983 / 375	4-06-00-01337 / 1.2681^{20}	H_2O 1; EtOH 5; eth 3; ace 3 / 1.5370^{25}
9584	Phosphorothioic acid, O,O-dimethyl O-(2,4,5-trichlorophenyl) ester / Ronnel	$C_8H_8Cl_3O_3PS$ / 321.55	299-84-3 / 41	8239 / $151-4^{0.4}$	4-06-00-00994 / 1.44^{32}	/ 1.5335^{35}
9585	Phosphorothioic acid, O,O,O-triethyl ester	$C_6H_{15}O_3PS$ / 198.22	126-68-1	/ 100^{16}	4-01-00-01348 / 1.082^{25}	/ 1.4480^{20}
9586	Phosphorous acid, bis(phenylmethyl) ester / Dibenzyl phosphite	$C_{14}H_{15}O_3P$ / 262.25	17176-77-1 / -5	2996 / $162^{0.1}$	4-06-00-02571	/ 1.5521^{18}
9587	Phosphorous acid, diethyl ester	$C_4H_{11}O_3P$ / 138.10	123-22-8	/ 87^{20}	/ 1.0720^{20}	EtOH 3; eth 3 / 1.4101^{20}
9588	Phosphorous acid, dimethyl ester	$C_2H_7O_3P$ / 110.05	868-85-9	/ 170.5	/ 1.2002^{20}	EtOH 3; ctc 2; py 3 / 1.4036^{20}
9589	Phosphorous acid, 2-ethylhexyl diphenyl ester	$C_{20}H_{27}O_3P$ / 346.41	15647-08-2	/ $148-56^{0.15}$	4-06-00-00695 / 1.054^{20}	/ 1.5207^{27}
9590	Phosphorous acid, tributyl ester	$C_{12}H_{27}O_3P$ / 250.32	102-85-2	/ 122^{12}	4-01-00-01527 / 0.9259^{20}	EtOH 3; eth 4; ctc 2 / 1.4321^{19}
9591	Phosphorous acid, triethyl ester / Triethyl phosphite	$C_6H_{15}O_3P$ / 166.16	122-52-1	/ 157.9	4-01-00-01333 / 0.9629^{20}	H_2O 1; EtOH 4; eth 4 / 1.4127^{20}
9592	Phosphorous acid, trimethyl ester / Trimethyl phosphite	$C_3H_9O_3P$ / 124.08	121-45-9	/ 111.5	4-01-00-01256 / 1.0518^{20}	EtOH 4; eth 4; ctc 2 / 1.4095^{20}
9593	Phosphorous acid, triphenyl ester	$C_{18}H_{15}O_3P$ / 310.29	101-02-0 / 25	/ 360	4-06-00-00695 / 1.1842^{20}	H_2O 1; EtOH 4; os 4 / 1.5900^{20}
9594	Phosphorous acid, tripropyl ester / Propyl phosphite	$C_9H_{21}O_3P$ / 208.24	923-99-9	/ 206.5	4-01-00-01426 / 0.9417^{20}	eth 4; EtOH 4 / 1.4282^{20}
9595	Phosphorous acid, tris(1-methylethyl) ester	$C_9H_{21}O_3P$ / 208.24	116-17-6	/ $60-1^{10}$	4-01-00-01476 / 0.9685^{18}	EtOH 3; eth 3; chl 3 / 1.4085^{25}
9596	Phosphorous acid, tris(2-methylphenyl) ester	$C_{21}H_{21}O_3P$ / 352.37	2622-08-4	/ 238^{11} 198^2	4-06-00-01977 / 1.1423^{20}	eth 3; chl 2 / 1.5740^{28}
9597	Phosphorous acid, tris(4-methylphenyl) ester	$C_{21}H_{21}O_3P$ / 352.37	620-42-8	/ $250-5^{10}$	4-06-00-02128 / 1.1280^{25}	eth 4 / 1.5703^{28}
9598	p-Phthalaldehydic acid, 3-hydroxy- / 3-Hydroxybenzoic acid, 4-formyl	$C_8H_6O_4$ / 166.13	619-12-5 / 234	/ sub	4-10-00-03623	eth 4; EtOH 4
9599	Phthalazine / 2,3-Benzodiazine	$C_8H_6N_2$ / 130.15	253-52-1 / 90.5	7344 / 316	5-23-07-00149	H_2O 3; EtOH 3; eth 2; bz 3
9600	1,4-Phthalazinedione, 5-amino-2,3-dihydro- / Luminol	$C_8H_7N_3O_2$ / 177.16	521-31-3 / 330.5	5470	4-25-00-04192	H_2O 1; EtOH 2; eth 2; alk 4
9601	Picene / Benzo[a]chrysene	$C_{22}H_{14}$ / 278.35	213-46-7 / 368	7368 / 519	4-05-00-02724	H_2O 1; EtOH 2; bz 2; chl 2
9602	Piperazine / Diethylenediamine	$C_4H_{10}N_2$ / 86.14	110-85-0 / 106	7431 / 146	5-23-01-00030	H_2O 4; EtOH 3; eth 1; chl 3 / 1.446^{113}

No.	Name / Synonym	Mol. Form. / Mol. Wt.	CAS RN / mp/°C	Merck No. / bp/°C	Beil. Ref. / den/g cm^{-3}	Solubility / n_D
9603	1-Piperazinecarboxamide, N,N-diethyl-4-methyl- / Diethylcarbamazine	$C_{10}H_{21}N_3O$ / 199.30	90-89-1 / 48	3106 / 109-11[3]	4-23-00-00225	
9604	1-Piperazinecarboxylic acid, ethyl ester	$C_7H_{14}N_2O_2$ / 158.20	120-43-4 /	/ 237	5-23-01-00055	H_2O 4; eth 4; EtOH 4 / 1.4760[25]
9605	Piperazine, 1-[(4-chlorophenyl)phenylmethyl]-4-[(3-methylphenyl)methyl]- / Meclizine	$C_{25}H_{27}ClN_2$ / 390.96	569-65-3 /	5657 / 230	5-23-01-00235	CS_2 3
9606	1,4-Piperazinedicarboxylic acid, diethyl ester / N,N'-Piperazinedicarboxylic acid, diethyl ester	$C_{10}H_{18}N_2O_4$ / 230.26	5470-28-0 / 47	/ 315	4-23-00-00254	eth 4; EtOH 4
9607	Piperazine, dihydrochloride / Diethylenediamine dihydrochloride	$C_4H_{12}Cl_2N_2$ / 159.06	142-64-3 /		5-23-01-00030	H_2O 2; EtOH 1
9608	Piperazine, 1,4-dimethyl-	$C_6H_{14}N_2$ / 114.19	106-58-1 /	/ 131	5-23-01-00128 / 0.8600[20]	H_2O 4; EtOH 4; eth 4 / 1.4474[20]
9609	Piperazine, 2,5-dimethyl-, cis- / cis-2,5-Dimethylpiperazine	$C_6H_{14}N_2$ / 114.19	6284-84-0 / 114	/ 162	5-23-03-00304	H_2O 4; EtOH 4; eth 2; bz 2 / 1.4720[20]
9610	Piperazine, 2,5-dimethyl-, trans- / trans-2,5-Dimethylpiperazine	$C_6H_{14}N_2$ / 114.19	2815-34-1 / 118.5	/ 162	5-23-03-00304	H_2O 4; EtOH 4; eth 2; bz 2
9611	Piperazine, 2,6-dimethyl-, cis- / cis-2,6-Dimethylpiperazine	$C_6H_{14}N_2$ / 114.19	21655-48-1 / 114	/ 160	5-23-03-00326	H_2O 3; EtOH 3; eth 1; bz 2
9612	2,5-Piperazinedione / Glycine anhydride	$C_4H_6N_2O_2$ / 114.10	106-57-0 / 312 dec	7434 / sub 260	5-24-05-00294	H_2O 2; EtOH 2; HCl 3
9613	Piperazine, 1,4-diphenyl-	$C_{16}H_{18}N_2$ / 238.33	613-39-8 / 165	/ 300 dec; 232[12]	5-23-01-00173	H_2O 1; EtOH 2
9614	1,4-Piperazinedipropanamine / 1,4-Bis(3-aminopropyl)piperazine	$C_{10}H_{24}N_4$ / 200.33	7209-38-3 / 15	/ 150-2[2]	5-23-01-00356 / 0.973[25]	1.5015[20]
9615	1-Piperazineethanamine / 1-(2-Aminoethyl)piperazine	$C_6H_{15}N_3$ / 129.21	140-31-8 /	/ 220	5-23-01-00257 / 0.985[25]	1.4983[20]
9616	1-Piperazineethanol	$C_6H_{14}N_2O$ / 130.19	103-76-4 /	/ 246	5-23-01-00406 / 1.061[25]	1.5065[20]
9617	1-Piperazineethanol, 4-[3-(2-chloro-10H-phenothiazin-10-yl)propyl]- / Perphenazine	$C_{21}H_{26}ClN_3OS$ / 403.98	58-39-9 / 97	7135 /	4-27-00-01307	
9618	Piperazine, 1-methyl-	$C_5H_{12}N_2$ / 100.16	109-01-3 /	/ 138	5-23-01-00082	H_2O 4; eth 4; EtOH 4 / 1.4378[20]
9619	Piperazine, 2-methyl- / 2-Methylpiperazine	$C_5H_{12}N_2$ / 100.16	109-07-9 / 62	/ 153	5-23-03-00267	H_2O 4; EtOH 3; eth 3; bz 3
9620	Piperazine, 1-phenyl-	$C_{10}H_{14}N_2$ / 162.23	92-54-6 /	/ 286.5; 161[15]	5-23-01-00160 / 1.0621[20]	H_2O 1; EtOH 5; eth 5; chl 3 / 1.5875[20]
9621	Piperazine, 1-(phenylmethyl)-	$C_{11}H_{16}N_2$ / 176.26	2759-28-6 /	/ 145-7[12]	5-23-01-00193	H_2O 3; EtOH 3; eth 3; chl 2 / 1.5430[28]
9622	Piperazine, 1,2,4-trimethyl- / 1,2,4-Trimethylpiperazine	$C_7H_{16}N_2$ / 128.22	120-85-4 /	/ 149.5	5-23-03-00271	ctc 3 / 1.4433[20]
9623	Piperazinium, 4-(2-cyclohexyl-2-hydroxy-2-phenylethyl)-1,1-dimethyl-, methyl sulfate (salt) / Hexocyclium methyl sulfate	$C_{21}H_{36}N_2O_5S$ / 428.59	115-63-9 /	4627 /		
9624	1-Piperidinamine	$C_5H_{12}N_2$ / 100.16	2213-43-6 /	/ 147	5-20-03-00516 / 0.928[25]	1.4750[20]
9625	1-Piperidinamine, 2,6-dimethyl-	$C_7H_{16}N_2$ / 128.22	39135-39-2 /	/ 65-80[30]	5-20-04-00187 / 0.865[25]	1.4650[20]
9626	3-Piperidinamine, 1-ethyl-	$C_7H_{16}N_2$ / 128.22	6789-94-2 /	/ 155; 82[40]	5-22-08-00026 / 0.923[25]	1.4715[20]
9627	4-Piperidinamine, 1-methyl-N-phenyl-N-(phenylmethyl)- / Bamipine	$C_{19}H_{24}N_2$ / 280.41	4945-47-5 / 115	966 /	5-22-08-00036	
9628	4-Piperidinamine, 2,2,6,6-tetramethyl-	$C_9H_{20}N_2$ / 156.27	36768-62-4 / 17	/ 188.5	5-22-08-00162 / 0.912[25]	1.4706[20]
9629	Piperidine / Azacyclohexane	$C_5H_{11}N$ / 85.15	110-89-4 / -11.03	7438 / 106.2	5-20-02-00003 / 0.8606[20]	H_2O 5; EtOH 5; eth 3; ace 3 / 1.4530[20]
9630	Piperidine, 1-acetyl-	$C_7H_{13}NO$ / 127.19	618-42-8 / -13.4	/ 226.5	5-20-02-00440 / 1.011[9]	H_2O 4; EtOH 4 / 1.4790[25]
9631	Piperidine, 1-acetyl-3-methyl-	$C_8H_{15}NO$ / 141.21	4593-16-2 / -13.6	/ 239	5-20-04-00107 / 0.9684[25]	H_2O 4 / 1.4731[25]
9632	Piperidine, 1-[5-(1,3-benzodioxol-5-yl)-1-oxo-2,4-pentadienyl]-, (E,E)- / Piperine	$C_{17}H_{19}NO_3$ / 285.34	94-62-2 / 131.5	7442 /	5-20-03-00469	H_2O 1; EtOH 3; eth 2; bz 3
9633	Piperidine, 1-[5-(1,3-benzodioxol-5-yl)-1-oxo-2,4-pentadienyl]-, (Z,Z)- / Chavicine	$C_{17}H_{19}NO_3$ / 285.34	495-91-0 /	2038 /	5-20-03-00469	eth 4; EtOH 4; peth 4
9634	Piperidine, 1-benzoyl-	$C_{12}H_{15}NO$ / 189.26	776-75-0 / 49	/ 320.5	5-20-02-00450	H_2O 1; EtOH 3; eth 3; ctc 2
9635	Piperidine, 1-butyl-	$C_9H_{19}N$ / 141.26	4945-48-6 /	/ 176	5-20-02-00029 / 0.8245[20]	1.4467[20]
9636	Piperidine, 3-butyl- / 3-Butylpiperidine	$C_9H_{19}N$ / 141.26	13603-21-9 /	/ 196.5	5-20-04-00230 / 0.862[18]	eth 4; EtOH 4; chl 4
9637	Piperidine, 4-butyl- / 4-n-Butylpiperidine	$C_9H_{19}N$ / 141.26	24152-39-4 /	/ 194	4-20-00-01635 / 0.879[23]	1.4472[23]
9638	1-Piperidinecarboxaldehyde	$C_6H_{11}NO$ / 113.16	2591-86-8 / -30.8	/ 222.5	5-20-02-00432 / 1.0158[25]	H_2O 5; EtOH 5; eth 5; bz 5 / 1.4805[25]

No.	Name Synonym	Mol. Form. Mol. Wt.	CAS RN mp/°C	Merck No. bp/°C	Beil. Ref. den/g cm^{-3}	Solubility n_D
9639	2-Piperidinecarboxamide, N-(2,6-dimethylphenyl)-1-methyl- Mepivacaine	$C_{15}H_{22}N_2O$ 246.35	96-88-8 96-88-8	5748	5-22-01-00223	CS_2 3
9640	2-Piperidinecarboxylic acid Pipecolic acid	$C_6H_{11}NO_2$ 129.16	535-75-1 264	7425	5-22-01-00220	H_2O 3; EtOH 2
9641	3-Piperidinecarboxylic acid Nipecotic acid	$C_6H_{11}NO_2$ 129.16	498-95-3 261 dec	6478	5-22-01-00235	H_2O 4
9642	4-Piperidinecarboxylic acid Isonipecotic acid	$C_6H_{11}NO_2$ 129.16	498-94-2 336	5075	5-22-01-00244	
9643	4-Piperidinecarboxylic acid, ethyl ester	$C_8H_{15}NO_2$ 157.21	1126-09-6 100-1[10]		4-22-00-00128	H_2O 4; bz 4; eth 4; EtOH 4 1.4591[20]
9644	4-Piperidinecarboxylic acid, methyl ester Methyl isonipecotate	$C_7H_{13}NO_2$ 143.19	2971-79-1 107-10[22]		4-22-00-00128	H_2O 4; bz 4; eth 4; EtOH 4 1.4635[25]
9645	4-Piperidinecarboxylic acid, 1-methyl-, methyl ester Methyl N-methylisonipecotate	$C_8H_{15}NO_2$ 157.21	1690-75-1 98[20]; 84[10]		5-22-01-00245	H_2O 3; EtOH 3; eth 3; bz 3 1.4539[24]
9646	2,6-Piperidinediethanol, 1-methyl-α,α'-diphenyl-, [2α(R*),6α(S*)]- Lobelanidine	$C_{22}H_{29}NO_2$ 339.48	552-72-7 150	5429	4-21-00-02386	H_2O 1; EtOH 3; eth 2; ace 4
9647	Piperidine, 2,5-diethyl-	$C_9H_{19}N$ 141.26	116836-21-6 190		0-20-00-00128 0.8722[25]	eth 4; chl 4
9648	Piperidine, 1,2-dimethyl-, (±)- Piperidine, N,2-dimethyl (DL)	$C_7H_{15}N$ 113.20	2512-81-4 127.5		5-20-04-00072 0.824[15]	H_2O 4; eth 4; EtOH 4 1.4395[20]
9649	Piperidine, 2,5-dimethyl- 2,5-Dimethylpiperidine	$C_7H_{15}N$ 113.20	34893-50-0 137		5-20-04-00179 0.8317[20]	EtOH 4 1.4452[25]
9650	Piperidine, 2,6-dimethyl-	$C_7H_{15}N$ 113.20	504-03-0 127		5-20-04-00180 0.8158[25]	H_2O 5; EtOH 5; eth 5; ctc 2 1.4377[20]
9651	Piperidine, 3,3-dimethyl- 3,3-Dimethylpiperidine	$C_7H_{15}N$ 113.20	1193-12-0 137		5-20-04-00189	EtOH 4 1.4452[25]
9652	Piperidine, 3,5-dimethyl- 3,5-Lupetidine	$C_7H_{15}N$ 113.20	35794-11-7 144		5-20-04-00191 0.853[25]	1.4454[20]
9653	Piperidine, 1-(1,1-dimethylethyl)- Piperidine, N-tert-butyl	$C_9H_{19}N$ 141.26	14446-69-6 166		5-20-02-00031 0.8465[20]	eth 4; EtOH 4 1.4532[20]
9654	2,4-Piperidinedione, 3,3-diethyl- Piperidione	$C_9H_{15}NO_2$ 169.22	77-03-2 104	7439	5-21-09-00609	H_2O 4; EtOH 4; chl 4; MeOH 4
9655	2,6-Piperidinedione, 4-[2-(3,5-dimethyl-2-oxocyclohexyl)-2-hydroxyethyl]-, Cycloheximide	$C_{15}H_{23}NO_4$ 281.35	66-81-9 119	2734	5-21-13-00434	EtOH 4
9656	2,6-Piperidinedione, 4-ethyl-4-methyl- Bemegride	$C_8H_{13}NO_2$ 155.20	64-65-3 126.5	1036 sub 100	5-21-09-00601	chl 3
9657	Piperidine, 1-[2-[3-(2,2-diphenylethyl)-1,2,4-oxadiazol-5-yl]ethyl]-, monohydrochloride Prenoxdiazine hydrochloride	$C_{23}H_{28}ClN_3O$ 397.95	982-43-4 186.5	7743		
9658	Piperidine, 1-(3,3-diphenylpropyl)- Fenpiprane	$C_{20}H_{25}N$ 279.43	3540-95-2 41.5	3934 210-20[8]	5-20-02-00091	
9659	Piperidine, 1,1'-dithiobis-	$C_{10}H_{20}N_2S_2$ 232.41	10220-20-9 64.2	160.1[11]	5-20-03-00487	acid 1
9660	Piperidine, 1-dodecyl-	$C_{17}H_{35}N$ 253.47	5917-47-5 161[5]; 115[0.6]		4-20-00-00321 0.8378[20]	1.4588[20]
9661	Piperidine, 1,1'-(1,2-ethanediyl)bis-	$C_{12}H_{24}N_2$ 196.34	1932-04-3 -0.5	265	5-20-03-00148 0.9160[25]	1.4853[25]
9662	1-Piperidineethanol	$C_7H_{15}NO$ 129.20	3040-44-6 17.9	202; 90[12]	5-20-02-00101 0.9703[25]	H_2O 5; EtOH 4 1.4749[20]
9663	2-Piperidineethanol 2-(2-Hydroxyethyl)piperidine	$C_7H_{15}NO$ 129.20	1484-84-0 69	202; 145[36]	5-21-01-00121 1.01[27]	H_2O 4
9664	4-Piperidineethanol 4-(2-Hydroxyethyl)piperidine	$C_7H_{15}NO$ 129.20	622-26-4 132.5	227.5	4-21-00-00093 1.0059[15]	H_2O 4; eth 4; EtOH 4 1.4907[20]
9665	1-Piperidineethanol, 2-(diphenylmethyl)- Diphemethoxidine	$C_{20}H_{25}NO$ 295.42	13862-07-2 106.5	3303 180-1[0.1]	5-20-08-00173	
9666	Piperidine, 3-ethenyl- Piperidine, 3-vinyl-	$C_7H_{13}N$ 111.19	57502-49-5 153.5		4-20-00-01946 0.9274[25]	1.4731[25]
9667	Piperidine, 1-ethyl- N-Ethylpiperidine	$C_7H_{15}N$ 113.20	766-09-6 130.8		5-20-02-00025 0.8237[20]	1.4480[20]
9668	Piperidine, 2-ethyl-, (±)- Piperidine, 2-ethyl (DL)	$C_7H_{15}N$ 113.20	78738-37-1 142.5		5-20-04-00169 0.8649[0]	1.4494[21]
9669	Piperidine, 3-ethyl-, (±)- Piperidine, 3-ethyl (DL)	$C_7H_{15}N$ 113.20	59433-08-8 152.6		5-20-04-00173 0.8565[23]	ace 4 1.4531[20]
9670	Piperidine, 4-ethyl- 4-Ethylpiperidine	$C_7H_{15}N$ 113.20	3230-23-7 154		5-20-04-00174 0.8759[25]	1.4503[25]
9671	Piperidine, 1-(1-ethyl-1-methylpropyl)- Piperidine, N-(3-methyl-3-pentyl)	$C_{11}H_{23}N$ 169.31	14045-26-2 214		5-20-02-00033 0.8614[20]	1.4637[20]
9672	Piperidine, 1-hexyl- 1-Hexylpiperidine	$C_{11}H_{23}N$ 169.31	7335-01-5 219.2		5-20-02-00032 0.8292[20]	1.4522[20]
9673	4-Piperidinemethanamine 4-(Aminomethyl)piperidine	$C_6H_{14}N_2$ 114.19	7144-05-0 25	200; 31[10]	5-22-08-00128	1.4900[20]
9674	2-Piperidinemethanol	$C_6H_{13}NO$ 115.18	3433-37-2 69	104[10]; 80[1]	5-21-01-00098	chl 2
9675	3-Piperidinemethanol	$C_6H_{13}NO$ 115.18	4606-65-9 61	106-7[3.5]	5-21-01-00107 1.0263[20]	chl 2 1.4964[20]

No.	Name / Synonym	Mol. Form. / Mol. Wt.	CAS RN / mp/°C	Merck No. / bp/°C	Beil. Ref. / den/g cm⁻³	Solubility / n_D
9676	2-Piperidinemethanol, α,α-diphenyl- / Pipradrol	$C_{18}H_{21}NO$ / 267.37	467-60-7	7455	5-21-04-00070	
9677	2-Piperidinemethanol, α-ethyl-, (R*,S*)-(±)- / Conhydrine (DL) (lower m.p.)	$C_8H_{17}NO$ / 143.23	3238-62-8 / 121	2495 / 226	5-21-01-00138	H_2O 2; bz 4; eth 4; EtOH 4
9678	3-Piperidinemethanol, 1-methyl-	$C_7H_{15}NO$ / 129.20	7583-53-1		5-21-01-00108	
9679	Piperidine, 1-methyl- / N-Methylpiperidine	$C_6H_{13}N$ / 99.18	626-67-5	142.5; 114²⁰	5-20-02-00021 / 1.013²⁵	1.4772²⁰
9680	Piperidine, 2-methyl-, (±)-	$C_6H_{13}N$ / 99.18	3000-79-1 / -4.9	107	5-20-04-00071 / 0.8159²⁰	H_2O 4; EtOH 5; eth 5; ctc 3 / 1.4355²⁰
9681	Piperidine, 3-methyl-, (±)- / Piperidine, 3-methyl (DL)	$C_6H_{13}N$ / 99.18	53152-98-0 / -24	118	5-20-04-00099 / 0.8436²⁴	H_2O 4; EtOH 3; eth 3; chl 2 / 1.4459²⁰
9682	Piperidine, 4-methyl-	$C_6H_{13}N$ / 99.18	626-58-4	125.5	5-20-04-00116 / 0.8446²⁶	H_2O 4; chl 2 / 1.4470²⁰
9683	Piperidine, 1,1'-methylenebis-	$C_{11}H_{22}N_2$ / 182.31	880-09-1	130	5-20-04-00116 / 0.8674²⁵	H_2O 4; chl 2 / 1.4458²⁰
9684	Piperidine, 1-(1-methylethyl)-	$C_8H_{17}N$ / 127.23	766-79-0	122-3¹⁵	5-20-02-00293 / 0.915²⁵	1.4820²⁰
9685	Piperidine, 1-(2-methylpropyl)-	$C_9H_{19}N$ / 141.26	10315-89-6	149.5	5-20-02-00028 / 0.8389²⁰	H_2O 4 / 1.4491²⁰
9686	Piperidine, 1-methyl-2-propyl-, (S)- / Methylconiine	$C_9H_{19}N$ / 141.26	35305-13-6	160.5 / 5968	5-20-02-00030 / 0.8161²⁵	H_2O 4 / 1.4382²⁵
9687	Piperidine, 1-nitro-	$C_5H_{10}N_2O_2$ / 130.15	7119-94-0 / -5.5	174	5-20-04-00206 / 0.8326²²	ace 4; EtOH 4 / 1.4538¹²
9688	Piperidine, 1-nitroso- / N-Nitrosopiperidine	$C_5H_{10}N_2O$ / 114.15	100-75-4	245; 121²⁰	5-20-03-00531 / 1.1519²⁶	1.4954²⁶
9689	Piperidine, 1-nonyl-	$C_{14}H_{29}N$ / 211.39	30538-80-8	219; 109²⁰	5-20-03-00530 / 1.0631¹⁸	H_2O 3; HCl 3 / 1.4933¹⁸
9690	Piperidine, 1-octyl-	$C_{13}H_{27}N$ / 197.36	7335-02-6	135-7¹¹	4-20-00-00319 / 0.8313²⁵	1.4538²⁵
9691	Piperidine, 1,2,2,6,6-pentamethyl- / Pempidine	$C_{10}H_{21}N$ / 155.28	79-55-0	137¹³; 89¹ / 7022	5-20-02-00034 / 0.8324²⁰	1.4544²⁰
9692	Piperidine, 1-pentyl-	$C_{10}H_{21}N$ / 155.28	10324-58-0	147	5-20-04-00238 / 0.8580⁰	1.4550²¹
9693	Piperidine, 1-phenyl-	$C_{11}H_{15}N$ / 161.25	4096-20-2 / 4.7	198.2	5-20-02-00031 / 0.8282²⁰	1.4498²⁰
9694	Piperidine, 3-phenyl-	$C_{11}H_{15}N$ / 161.25	3973-62-4 / 14.5	258	5-20-02-00057 / 0.9944²⁵	EtOH 4; eth 4; bz 4; chl 4 / 1.5598²⁵
9695	Piperidine, 4-phenyl- / 4-Phenylpiperidine	$C_{11}H_{15}N$ / 161.25	771-99-3 / 60.5	257; 140¹⁹ / 257	5-20-06-00405 / 1.0010²⁵	1.5473²⁵
9696	Piperidine, 1-(1-phenylcyclohexyl)- / Phencyclidine	$C_{17}H_{25}N$ / 243.39	77-10-1 / 46.5	7179 / 135-7¹·⁰	5-20-06-00408 / 0.9996¹⁶	chl 3
9697	Piperidine, 1-(1-phenylethyl)-	$C_{13}H_{19}N$ / 189.30	7529-63-7 / -13	272	5-20-02-00078	
9698	Piperidine, 1-(phenylmethyl)-	$C_{12}H_{17}N$ / 175.27	2905-56-8	245	5-20-02-00064 / 0.945²⁵	1.5218²⁵
9699	Piperidine, 2-(phenylmethyl)- / 2-Benzylpiperidine	$C_{12}H_{17}N$ / 175.27	32838-55-4 / 32	267.5	5-20-02-00062 / 0.9625¹⁶	1.5227²⁰
9700	Piperidine, 4-(phenylmethyl)- / 4-Benzylpiperidine	$C_{12}H_{17}N$ / 175.27	31252-42-3 / 16.8	270; 150¹⁷	5-20-06-00438 / 0.9660²⁵	eth 4; EtOH 4
9701	Piperidine, 4,4'-(1,3-propanediyl)bis-	$C_{13}H_{26}N_2$ / 210.36	16898-52-5 / 67.1	329	5-20-06-00439 / 0.9970²⁰	H_2O 1; EtOH 3; eth 3 / 1.5337²⁵
9702	Piperidine, 4,4'-(1,3-propanediyl)bis[1-methyl- / Propane, 1,3-bis(1-methyl-4-piperidyl)	$C_{15}H_{30}N_2$ / 238.42	64168-11-2 / 13.7	215⁵⁰	5-23-04-00101	H_2O 4
9703	1-Piperidinepropanenitrile	$C_8H_{14}N_2$ / 138.21	3088-41-3 / -6.8	145⁵⁰	4-23-00-00536 / 0.8962²⁵	1.4804²⁵
9704	1-Piperidinepropanoic acid, ethyl ester / Piperidine, N-carboxyethyl, ethyl ester	$C_{10}H_{19}NO_2$ / 185.27	19653-33-9	139⁵⁰	5-20-03-00047 / 0.9403²⁵	1.4676²⁵
9705	1-Piperidinepropanol, α-cyclohexyl-α-phenyl-, hydrochloride / Trihexyphenidyl hydrochloride	$C_{20}H_{32}ClNO$ / 337.93	52-49-3 / 258.5	9607	5-20-03-00042	H_2O 4 / 1.4525²⁵
9706	Piperidine, 2-(1-propenyl)-, (±)-	$C_8H_{15}N$ / 125.21	5913-85-9 / 6	168	5-20-04-00346 / 0.8716¹⁵	eth 4; EtOH 4
9707	Piperidine, 1-propyl-	$C_8H_{17}N$ / 127.23	5470-02-0	151.5	5-20-02-00027 / 0.8231²⁰	H_2O 4; eth 4; EtOH 4 / 1.4446²⁰
9708	Piperidine, 2-propyl-, (S)- / Coniine	$C_8H_{17}N$ / 127.23	458-88-8 / -2	2500 / 166.5	5-20-04-00205 / 0.8440²⁰	H_2O 2; EtOH 5; eth 4; bz 3 / 1.4512²²
9709	Piperidine, 4-propyl- / 4-Propylpiperidine	$C_8H_{17}N$ / 127.23	22398-09-0	173	5-20-04-00208 / 0.864²²	EtOH 3; chl 2 / 1.4465²³
9710	Piperidine, 2,2,4,6-tetramethyl- / 2,2,4,6-Tetramethylpiperidine	$C_9H_{19}N$ / 141.26	6292-82-6	156	4-20-00-01638 / 0.816²⁵	1.4374²⁵
9711	Piperidine, 2,2,6,6-tetramethyl- / Norpempidine	$C_9H_{19}N$ / 141.26	768-66-1 / 28	156	5-20-04-00237 / 0.8367¹⁶	eth 4 / 1.4455²⁰
9712	Piperidine, 2,2,4-trimethyl-	$C_8H_{17}N$ / 127.23	101257-71-0	148	4-20-00-01624 / 0.832¹⁵	eth 4; EtOH 4 / 1.4458²⁰
9713	Piperidine, 2,3,6-trimethyl- / 2,3,6-Trimethylpiperidine	$C_8H_{17}N$ / 127.23	50402-72-7	36⁵	4-20-00-01625 / 0.830²²	1.4434²⁰
9714	Piperidine, 2,4,6-trimethyl- / 2,4,6-Trimethylpiperidine	$C_8H_{17}N$ / 127.23	21974-48-1 / <20	145	5-20-04-00220 / 0.8315¹⁹	H_2O 2 / 1.4412²⁰
9715	3-Piperidinol, 1-ethyl- / 1-Ethyl-3-piperidinol	$C_7H_{15}NO$ / 129.20	13444-24-1	3799 / 93-5¹⁵	5-21-01-00039	1.4777¹⁴

No.	Name / Synonym	Mol. Form. / Mol. Wt.	CAS RN / mp/°C	Merck No. / bp/°C	Beil. Ref. / den/g cm^{-3}	Solubility / n_D
9716	3-Piperidinol, 1-methyl-	$C_6H_{13}NO$ 115.18	3554-74-3	76-8[11]	5-21-01-00029 0.999[25]	1.4735[20]
9717	4-Piperidinol, 1-methyl-	$C_6H_{13}NO$ 115.18	106-52-5 29	200	5-21-01-00052	1.4775[20]
9718	3-Piperidinol, 6-propyl-, (3S-trans)- Pseudoconhydrine	$C_8H_{17}NO$ 143.23	140-55-6 106	7928 236	5-21-01-00138	H_2O 4; eth 4; EtOH 4
9719	2-Piperidinone	C_5H_9NO 99.13	675-20-7 39.5	256	5-21-06-00396	H_2O 4; EtOH 4; eth 4; dil acid 3
9720	4-Piperidinone, 1-acetyl-	$C_7H_{11}NO_2$ 141.17	32161-06-1	218; 124[0.2]	5-21-06-00426 1.146[25]	1.5026[20]
9721	2-Piperidinone, 1-methyl-	$C_6H_{11}NO$ 113.16	931-20-4	105[12]; 94[9]	5-21-06-00398 1.0263[25]	1.4820[20]
9722	2-Piperidinone, 3-methyl-	$C_6H_{11}NO$ 113.16	3768-43-2 55.3	249.5	5-21-06-00479	bz 4; eth 4; EtOH 4
9723	3-Piperidinone, 1-methyl-	$C_6H_{11}NO$ 113.16	5519-50-6	63-4[13]	5-21-06-00417 0.9655[25]	1.4559[25]
9724	4-Piperidinone, 1-methyl-	$C_6H_{11}NO$ 113.16	1445-73-4	56-8[11]	5-21-06-00419 0.9696[25]	1.4580[25]
9725	4-Piperidinone, 1,2,3,6-tetramethyl- 1,2,5,6-Tetramethyl-4-piperidinone	$C_9H_{17}NO$ 155.24	53381-90-1	86-8[7.5]	4-21-00-03272 0.9499[20]	1.4680[20]
9726	4-Piperidinone, 2,2,6,6-tetramethyl-	$C_9H_{17}NO$ 155.24	826-36-8 36	205	5-21-06-00538	H_2O 3; EtOH 3; eth 3; chl 2
9727	Piperonyl butoxide 1,3-Benzodioxole, 5-[[2-(2-butoxyethoxy)ethoxy]methyl]-6-propyl-	$C_{19}H_{30}O_5$ 338.44	51-03-6	7446 180[1]	1.05[25]	
9728	Pirimicarb Carbamic acid, dimethyl-, 2-(dimethylamino)-5,6-dimethyl-4-pyrimidinyl ester	$C_{11}H_{18}N_4O_2$ 238.29	23103-98-2 90.5	7468		
9729	Pirimiphos-ethyl Phosphorothioic acid, O-[2-(diethylamino)-6-methyl-4-pyrimidinyl] O,O-diethyl ester	$C_{13}H_{24}N_3O_3PS$ 333.39	23505-41-1	7469 dec >130	1.14[20]	
9730	Pirimiphos-methyl Phosphorothioic acid, O-[2-(diethylamino)-6-methyl-4-pyrimidinyl] O,O-dimethyl ester	$C_{11}H_{20}N_3O_3PS$ 305.34	29232-93-7 15	dec	1.17[20]	
9731	Plumbane, tetraethyl Tetraethyl lead	$C_8H_{20}Pb$ 323.45	78-00-2	9136 200 dec	1.653[20]	H_2O 1; bz 3; EtOH 2 1.5198[20]
9732	Plumbane, tetramethyl Tetramethyl lead	$C_4H_{12}Pb$ 267.34	75-74-1 -27.5	110	1.995[20]	
9733	Plumbane, tetraphenyl-	$C_{24}H_{20}Pb$ 515.62	595-89-1 228.3	126[13]	4-16-00-01615 1.5298[20]	chl 3
9734	21H,23H-Porphine	$C_{20}H_{14}N_4$ 310.36	101-60-0 360	7574 sub 300	4-26-00-01900 1.336[25]	H_2O 1; EtOH 2; eth 1; ace 1
9735	21H,23H-Porphine-2,18-dipropanoic acid, 7,12-bis(1-hydroxyethyl)-3,8,13,17-tetramethyl- Hematoporphyrin	$C_{34}H_{38}N_4O_6$ 598.70	14459-29-1 172.5	4555	4-26-00-03157	H_2O 1; EtOH 3; eth 2; chl 2
9736	Pregna-4,6-diene-21-carboxylic acid, 17-hydroxy-3-oxo-, γ-lactone, (17α)- Canrenone	$C_{22}H_{28}O_3$ 340.46	976-71-6 150	1753	5-17-11-00476	
9737	Pregna-1,4-diene-3,20-dione, 9-fluoro-11,17,21-trihydroxy-16-methyl-, (11β,16α)- Dexamethasone	$C_{22}H_{29}FO_5$ 392.47	50-02-2 262	2922		
9738	Pregna-1,4-diene-3,20-dione, 11,17,21-trihydroxy-, (11β)- Prednisolone	$C_{21}H_{28}O_5$ 360.45	50-24-8 235	7719	4-08-00-03467	
9739	Pregna-1,4-diene-3,11,20-trione, 17,21-dihydroxy- Prednisone	$C_{21}H_{28}O_5$ 358.43	53-03-2 233-5 dec	7727	4-08-00-03531	
9740	Pregnane, (5α)- Allopregnane	$C_{21}H_{36}$ 288.52	641-85-0 84.5	257	4-05-00-01215	
9741	Pregnane, (5β)- 17β-Ethyletiocholane	$C_{21}H_{36}$ 288.52	481-26-5 83.5	7731	4-05-00-01215 1.032[15]	H_2O 1; chl 3; MeOH 3
9742	Pregnane-3,20-diol, (3α,5β,20S)- Pregnanediol	$C_{21}H_{36}O_2$ 320.52	80-92-2 243.5	7732	4-06-00-06111 1.15[25]	EtOH 2; eth 2; ace 3; os 2
9743	Pregnane-3,20-dione, (5β)- 5β-Pregnane-3,20-dione	$C_{21}H_{32}O_2$ 316.48	128-23-4 123	7733	4-07-00-02198	H_2O 1; EtOH 4; eth 3; ace 3
9744	Pregnane-11,20-dione, 3,17,21-trihydroxy-, (3α,5β)- Tetrahydrocortisone	$C_{21}H_{32}O_5$ 364.48	53-05-4 190	9143	4-08-00-03395	
9745	Pregnan-20-one, 3-hydroxy-, (3α,5β)- Pregnan-3α-ol-20-one	$C_{21}H_{34}O_2$ 318.50	128-20-1 149.5	7734		EtOH 4
9746	Pregn-4-en-18-al, 11,21-dihydroxy-3,20-dioxo-, (11β)- Aldosterone	$C_{21}H_{28}O_5$ 360.45	52-39-1 166.5	218	4-08-00-03491	
9747	Pregn-4-ene-3,20-dione, 21-(acetyloxy)-11,17-dihydroxy-, (11β)- Hydrocortisone acetate	$C_{23}H_{23}O_6$ 395.43	50-03-3 223 dec	4711	4-08-00-03424 1.289[20]	
9748	Pregn-4-ene-3,20-dione, 11,21-dihydroxy-, (11β)- Corticosterone	$C_{21}H_{30}O_4$ 346.47	50-22-6 181	2532 190[0.01]	4-08-00-02907	H_2O 1; EtOH 3; eth 3; ace 3

No.	Name Synonym	Mol. Form. Mol. Wt.	CAS RN mp/°C	Merck No. bp/°C	Beil. Ref. den/g cm⁻³	Solubility n_D
9749	Pregn-4-ene-3,20-dione, 17,21-dihydroxy-11-Deoxy-17-hydroxycorticosterone	$C_{21}H_{30}O_4$ 346.47	152-58-9 215	2912	4-08-00-02913	ace 4; EtOH 4; chl 4
9750	Pregn-4-ene-3,20-dione, 17-hydroxy-17α-Hydroxyprogesterone	$C_{21}H_{30}O_3$ 330.47	68-96-2	4773	4-08-00-02189	chl 2
9751	Pregn-4-ene-3,20-dione, 21-hydroxy-Deoxycorticosterone	$C_{21}H_{30}O_3$ 330.47	64-85-7 141.5	2882	4-08-00-02195	H_2O 2; EtOH 4; eth 2; ace 4
9752	Pregn-4-ene-3,20-dione, 17-hydroxy-6-methyl-, (6α)-Medroxyprogesterone	$C_{22}H_{32}O_3$ 344.49	520-85-4 214.5	5677	4-08-00-02211	chl 4
9753	Pregn-4-ene-3,20-dione, 11,17,21-trihydroxy-, (11β)-Hydrocortisone	$C_{21}H_{30}O_5$ 362.47	50-23-7 220	4710	4-08-00-03422	H_2O 2; EtOH 3; diox 3; HOAc 3
9754	Pregn-4-ene-3,11,20-trione, 17,21-dihydroxy-Cortisone	$C_{21}H_{28}O_5$ 360.45	53-06-5 222	2533	4-08-00-03480	H_2O 2; EtOH 3; eth 2; ace 3
9755	Pregn-4-ene-3,11,20-trione, 21-hydroxy-11-Dehydrocorticosterone	$C_{21}H_{28}O_4$ 344.45	72-23-1 183.5	2859	4-08-00-02998	H_2O 1; EtOH 3; ace 3; bz 3
9756	Pregn-5-en-20-one, 3-hydroxy-, (3β)-Pregnenolone	$C_{21}H_{32}O_2$ 316.48	145-13-1 192	7739	4-08-00-01019	
9757	Pregn-4-en-3-one, 11,17,20,21-tetrahydroxy-, (11β,20R)-4-Pregnene-11β,17α,20β,21-tetrol-3-one	$C_{21}H_{32}O_5$ 364.48	116-58-5 125 dec	7736	4-08-00-03384	ace 4; EtOH 4
9758	Pregn-4-en-20-yn-3-one, 17-hydroxy-, (17α)-Ethisterone	$C_{21}H_{28}O_2$ 312.45	434-03-7 272	3696	4-08-00-01225	
9759	Procyazine Propanenitrile, 2-[[4-chloro-6-(cyclopropylamino)-1,3,5-triazin-2-yl]amino]-2-methyl-	$C_{10}H_{13}ClN_6$ 252.71	32889-48-8 170			
9760	Procymidone N-(3,5-Dichlorophenyl)-1,2-dimethylcyclopropane-1,2-dicarboximide	$C_{13}H_{11}Cl_2NO_2$ 284.14	32809-16-8 166		1.452[25]	
9761	Prodiamine 1,3-Benzenediamine, 2,4-dinitro-N3,N3-dipropyl-6-(trifluoromethyl)-	$C_{13}H_{17}F_3N_4O_4$ 350.30	29091-21-2 124	7773	1.47[25]	
9762	Profenofos Phosphorothioic acid, O-(4-bromo-2-chlorophenyl) O-ethyl S-propyl ester	$C_{11}H_{15}BrClO_3PS$ 373.63	41198-08-7	110[0.001]	1.455[20]	
9763	Profluralin Benzenamine, N-(cyclopropylmethyl)-2,6-dinitro-N-propyl-4-(trifluoromethyl)-	$C_{14}H_{16}F_3N_3O_4$ 347.29	26399-36-0 34	7781		
9764	L-Proline 2-Pyrrolidinecarboxylic acid	$C_5H_9NO_2$ 115.13	147-85-3 221 dec	7790	5-22-01-00031	H_2O 4; EtOH 2; eth 1; ace 2
9765	DL-Proline	$C_5H_9NO_2$ 115.13	609-36-9 205 dec		5-22-01-00033	H_2O 4; EtOH 4
9766	L-Proline, 1-acetyl-4-hydroxy-, trans-Oxaceprol	$C_7H_{11}NO_4$ 173.17	33996-33-7 132	6857	5-22-05-00013	H_2O 4; MeOH 4
9767	L-Proline, 4-hydroxy-, trans-4-Hydroxy-L-proline	$C_5H_9NO_3$ 131.13	51-35-4 274	4775	5-22-05-00007	H_2O 4; EtOH 2
9768	L-Proline, 5-oxo-L-Pyroglutamic acid	$C_5H_7NO_3$ 129.12	98-79-3 162	8012	5-22-06-00007	DMSO 3
9769	Promecarb Phenol, 3-methyl-5-(1-methylethyl)-, methylcarbamate	$C_{12}H_{17}NO_2$ 207.27	2631-37-0 87	7794 117[0.01]		
9770	Prometryn 1,3,5-Triazine-2,4-diamine, N,N'-bis(1-methylethyl)-6-(methylthio)-	$C_{10}H_{19}N_5S$ 241.36	7287-19-6 119	7800	1.157[20]	
9771	Propachlor Acetamide, 2-chloro-N-(1-methylethyl)-N-phenyl-	$C_{11}H_{14}ClNO$ 211.69	1918-16-7 77	7805 110[0.03]	1.242[25]	
9772	1,2-Propadiene Allene	C_3H_4 40.06	463-49-0 -136.2	-34.4	4-01-00-00966 *0.584[25]	bz 4; peth 4 1.4168
9773	1,2-Propadiene-1,3-dione Carbon suboxide	C_3O_2 68.03	504-64-3 -107	1821 6.8	4-01-00-03764 1.114[0]	eth 3; bz 3; CS_2 3 1.4538[0]
9774	Propanal Propionaldehyde	C_3H_6O 58.08	123-38-6 -80	7835 48	4-01-00-03165 0.8657[25]	H_2O 3; EtOH 5; eth 5 1.3636[20]
9775	Propanal, 2-bromo-	C_3H_5BrO 136.98	19967-57-8	109.5	4-01-00-03177 1.592[20]	eth 4 1.4813[20]
9776	Propanal, 2-bromo-2-methyl-	C_4H_7BrO 151.00	13206-46-7	113	4-01-00-03267 1.383[25]	1.4518[25]
9777	Propanal, 2-chloro-	C_3H_5ClO 92.52	683-50-1	86	3-01-00-02691 1.182[15]	bz 4; eth 4 1.431[17]
9778	Propanal, 3-chloro-	C_3H_5ClO 92.52	19434-65-2 <20	130; 45[10]	4-01-00-03174 1.268[15]	eth 4; EtOH 4 1.475[25]
9779	Propanal, 2-chloro-2-methyl-	C_4H_7ClO 106.55	917-93-1	90	4-01-00-03267 1.053[15]	eth 4; EtOH 4 1.4160[16]
9780	Propanal, 2,3-dibromo-	$C_3H_4Br_2O$ 215.87	5221-17-0	73-5[10]	4-01-00-03178 2.198[15]	eth 4 1.5082[20]
9781	Propanal, 2,3-dichloro-	$C_3H_4Cl_2O$ 126.97	10140-89-3	73[50]; 48[14]	4-01-00-03174 1.400[20]	1.4762[20]
9782	Propanal, 2,3-dihydroxy-, (±)-	$C_3H_6O_3$ 90.08	56-82-6 145	140-50[0.8]	3-01-00-03282 1.453[18]	H_2O 3; EtOH 2; eth 2; bz 1

No.	Name / Synonym	Mol. Form. / Mol. Wt.	CAS RN / mp/°C	Merck No. / bp/°C	Beil. Ref. / den/g cm^{-3}	Solubility / n_D
9783	Propanal, 2,2-dimethyl- / Pivaldehyde	$C_5H_{10}O$ / 86.13	630-19-3 / 6	/ 77.5	4-01-00-03295 / 0.7923[17]	EtOH 3; eth 3 / 1.3791[20]
9784	Propanal, 2-(hydroxyimino)-, oxime	$C_3H_6N_2O_2$ / 102.09	1804-15-5 / 157	/ sub	4-01-00-03633	H_2O 2; EtOH 3; eth 3; bz 3
9785	Propanal, 2-methyl- / Isobutanal	C_4H_8O / 72.11	78-84-2 / -65.9	5038 / 64.5	4-01-00-03262 / 0.7891[20]	H_2O 3; eth 3; ace 3; ctc 2 / 1.3730[20]
9786	Propanal, 3-(methylthio)- / 3-(Methylthio)propionaldehyde	C_4H_8OS / 104.17	3268-49-3	/ 62[11]	4-01-00-03974	
9787	Propanal, oxime	C_3H_7NO / 73.09	627-39-4 / 40	/ 131.5	4-01-00-03170 / 0.9258[20]	1.4287[20]
9788	Propanal, 2-oxo- / Pyruvaldehyde	$C_3H_4O_2$ / 72.06	78-98-8 / 72	8030	4-01-00-03631 / 1.0455[20]	EtOH 3; eth 3; bz 3 / 1.4002[18]
9789	Propanal, 2-oxo-, 1-oxime / Isonitrosoacetone	$C_3H_5NO_2$ / 87.08	306-44-5 / 69	5076 / sub	3-01-00-03092 / 1.0744[67]	H_2O 3; eth 3; bz 2; ctc 2
9790	Propanal, 2,2,3-trichloro-	$C_3H_3Cl_3O$ / 161.41	7789-90-4	/ 63-5[45]	3-01-00-02693 / 1.470[25]	eth 4 / 1.473[25]
9791	Propanamide / Propionamide	C_3H_7NO / 73.09	79-05-0 / 81.3	7836 / 213	4-02-00-00725 / 0.9262[110]	H_2O 4; EtOH 4; eth 4; chl 4 / 1.4180[110]
9792	Propanamide, 2-bromo-2-methyl- / Isobutyramide, 2-bromo	C_4H_8BrNO / 166.02	7462-74-0 / 148	/ 145[17]	4-02-00-00863	EtOH 4; chl 4
9793	Propanamide, N,N-diethyl-	$C_7H_{15}NO$ / 129.20	1114-51-8	/ 191	4-04-00-00353 / 0.8972[20]	EtOH 4 / 1.4425[20]
9794	Propanamide, N,N-diethyl-2,2-dimethyl- / Pivalamide, N,N-diethyl	$C_9H_{19}NO$ / 157.26	24331-72-4	/ 206	3-04-00-00211 / 0.891[15]	eth 4; EtOH 4
9795	Propanamide, N-(2,3-dihydro-1,5-dimethyl-3-oxo-2-phenyl-1H-pyrazol-4-yl)-2-(dimethylamino)- / Aminopropylon	$C_{16}H_{22}N_4O_2$ / 302.38	3690-04-8 / 181	484	4-25-00-03595	H_2O 4
9796	Propanamide, N-(4-ethoxyphenyl)-2-hydroxy- / p-Lactophenetide	$C_{11}H_{15}NO_3$ / 209.25	539-08-2 / 118	5220	3-13-00-01135	H_2O 3; EtOH 4; eth 2; bz 2
9797	Propanamide, 2-hydroxy-, (±)-	$C_3H_7NO_2$ / 89.09	65144-02-7 / 75.5		4-03-00-00674 / 1.1381[80]	H_2O 4; EtOH 4
9798	Propanamide, 2-methyl-	C_4H_9NO / 87.12	563-83-7 / 128	/ 218	4-02-00-00852	chl 3
9799	Propanamide, N-methyl-	C_4H_9NO / 87.12	1187-58-2 / -30.9	/ 148	4-04-00-00183 / 0.9305[25]	1.4345[25]
9800	Propanamide, N-phenyl-	$C_9H_{11}NO$ / 149.19	620-71-3 / 105.5	/ 222.2	4-12-00-00385 / 1.175[25]	H_2O 2; EtOH 4; eth 4
9801	Propanamide, N-phenyl-N-[1-(2-phenylethyl)-4-piperidinyl]- / Fentanyl	$C_{22}H_{28}N_2O$ / 336.48	437-38-7 / 87.5	3944	5-22-08-00049	
9802	Propanamide, N-propyl-	$C_6H_{13}NO$ / 115.18	3217-86-5 / 154	/ 215; 108[9]	4-04-00-00477 / 0.8995[25]	H_2O 2; eth 2
9803	1-Propanamine / Propylamine	C_3H_9N / 59.11	107-10-8 / -83	7855 / 47.2	4-04-00-00464 / 0.7173[20]	H_2O 3; EtOH 4; eth 4; ace 4 / 1.3870[20]
9804	2-Propanamine / Isopropylamine	C_3H_9N / 59.11	75-31-0 / -95.1	5097 / 31.7	4-04-00-00504 / 0.6891[20]	H_2O 5; EtOH 5; eth 5; ace 4 / 1.3742[20]
9805	1-Propanamine, N,N-dichloro- / N,N-Dichloropropylamine	$C_3H_7Cl_2N$ / 128.00	69947-01-9	/ 117	0-04-00-00145 / 1.1454[23]	1.4525[23]
9806	1-Propanamine, 2,2-dimethyl-	$C_5H_{13}N$ / 87.16	5813-64-9	/ 82	4-04-00-00707 / 0.7455[20]	eth 4 / 1.4023[20]
9807	1-Propanamine, N,N-dimethyl- / Dimethylpropylamine	$C_5H_{13}N$ / 87.16	926-63-6	/ 66	4-04-00-00467 / 0.7152[20]	bz 4; eth 4; EtOH 4 / 1.3860[20]
9808	1-Propanamine, N,N-dipropyl- / Tripropylamine	$C_9H_{21}N$ / 143.27	102-69-2 / -93.5	/ 156	4-04-00-00470 / 0.7558[20]	eth 4; EtOH 4 / 1.4181[20]
9809	1-Propanamine, N-ethyl-	$C_5H_{13}N$ / 87.16	20193-20-8	/ 81	4-04-00-00468 / 0.7204[17]	H_2O 2; ace 4; EtOH 4 / 1.3858[25]
9810	2-Propanamine, N-ethyl- / Ethylisopropylamine	$C_5H_{13}N$ / 87.16	19961-27-4	/ 69.6	4-04-00-00508	1.3872[25]
9811	2-Propanamine, N-ethyl-N-(1-methylethyl)-	$C_8H_{19}N$ / 129.25	7087-68-5	/ 126.5	4-04-00-00511 / 0.742[25]	ctc 3 / 1.4138[20]
9812	1-Propanamine, N-ethyl-N-propyl-	$C_8H_{19}N$ / 129.25	20634-92-8	/ 138	2-04-00-00622 / 0.807[24]	ace 4; bz 4; eth 4
9813	1-Propanamine, 1,1,2,2,3,3,3-heptafluoro-N,N-bis(heptafluoropropyl)-	$C_9F_{21}N$ / 521.07	338-83-0 /	/ 130	4-02-00-00742 / 1.822[4]	1.279[25]
9814	1-Propanamine, N-hydroxy-N-propyl- / N,N-Dipropylhydroxylamine	$C_6H_{15}NO$ / 117.19	7446-43-7 / 29	/ 154.5	4-04-00-03308	bz 4; eth 4; EtOH 4; chl 4
9815	2-Propanamine, N,N-methanetetraylbis-	$C_7H_{14}N_2$ / 126.20	693-13-0	/ 147	4-04-00-00531 / 0.806[25]	1.4320[20]
9816	1-Propanamine, 3-methoxy-	$C_4H_{11}NO$ / 89.14	5332-73-0	/ 117.5	4-04-00-01623 / 0.8727[20]	H_2O 3; ace 3; bz 3; ctc 3 / 1.4391[20]
9817	1-Propanamine, 2-methyl- / Isobutylamine	$C_4H_{11}N$ / 73.14	78-81-9 / -86.7	5016 / 67.7	4-04-00-00625 / 0.724[25]	1.3988[19]
9818	2-Propanamine, 2-methyl- / tert-Butylamine	$C_4H_{11}N$ / 73.14	75-64-9 / -66.9	1545 / 44.0	4-04-00-00657 / 0.6958[20]	H_2O 5; EtOH 5; eth 5; chl 3 / 1.3784[20]
9819	2-Propanamine, N-methyl- / Isopropylmethylamine	$C_4H_{11}N$ / 73.14	4747-21-1	/ 50.4		
9820	1-Propanamine, 2-methyl-N,N-bis(2-methylpropyl)- / Triisobutylamine	$C_{12}H_{27}N$ / 185.35	1116-40-1 / -21.8	/ 191.5	4-04-00-00631 / 0.7684[20]	eth 4; EtOH 4 / 1.4252[17]

No.	Name / Synonym	Mol. Form. / Mol. Wt.	CAS RN / mp/°C	Merck No. / bp/°C	Beil. Ref. / den/g cm⁻³	Solubility / n_D
9821	2-Propanamine, N-(1-methylethyl)- Diisopropylamine	$C_6H_{15}N$ 101.19	108-18-9 -61	3181 83.9	4-04-00-00510 0.7153[20]	ace 4; bz 4; eth 4; EtOH 4 1.3924[20]
9822	2-Propanamine, N-(1-methylethyl)-N-nitroso-	$C_6H_{14}N_2O$ 130.19	601-77-4 48	194.5	4-04-00-03388 0.9422[20]	H_2O 2; EtOH 3; eth 3; bz 3
9823	1-Propanamine, 2-methyl-N-(2-methylpropyl)- Diisobutylamine	$C_8H_{19}N$ 129.25	110-96-3 -73.5	139.6	4-04-00-00630 1.4090[20]	H_2O 2; EtOH 3; eth 3; ace 3
9824	1-Propanamine, N-nitro-	$C_3H_8N_2O_2$ 104.11	627-07-6 -21	128-9[40]	4-04-00-03405 1.1046[15]	eth 4; EtOH 4 1.4610[20]
9825	1-Propanamine, N-nitroso-N-propyl- Dipropylnitrosamine	$C_6H_{14}N_2O$ 130.19	621-64-7	206; 113[40]	4-04-00-03388 0.9163[20]	H_2O 2; EtOH 5; eth 5 1.4437[20]
9826	1-Propanamine, N-propyl- Dipropylamine	$C_6H_{15}N$ 101.19	142-84-7 -63	3350 109.3	4-04-00-00469 0.7400[20]	H_2O 3; EtOH 3; eth 5; ace 4 1.4050[20]
9827	2-Propanamine, N-propyl Isopropylpropylamine	$C_6H_{15}N$ 101.19	21968-17-2	96.9		
9828	1-Propanamine, 3-(triethoxysilyl)-	$C_9H_{23}NO_3Si$ 221.37	919-30-2	122-3[31]	4-04-00-04273 0.9506[20]	1.4225[20]
9829	1-Propanamine, N,N,2-trimethyl-	$C_6H_{15}N$ 101.19	7239-24-9	80.5	4-04-00-00627 0.7097[20]	H_2O 4 1.3907[20]
9830	1-Propanaminium, 2-(acetyloxy)-N,N,N-trimethyl-, chloride Methacholine chloride	$C_8H_{18}ClNO_2$ 195.69	62-51-1 172	5847		H_2O 4; EtOH 4; chl 4
9831	1-Propanaminium, 2-hydroxy-N,N,N-trimethyl-, chloride (2-Hydroxypropyl)trimethyl ammonium chloride	$C_6H_{16}ClNO$ 153.65	2382-43-6 165	dec	3-04-00-00754	H_2O 4; EtOH 4
9832	2-Propanaminium, N-methyl-N-(1-methylethyl)-N-[2-[(9H-xanthen-9-ylcarbonyl)oxy]ethyl]-, bromide Propantheline bromide	$C_{23}H_{30}BrNO_3$ 448.40	50-34-0 159.5	7816	5-18-06-00590	
9833	1-Propanaminium, N,N,N-tripropyl-, iodide Tetrapropylammonium iodide	$C_{12}H_{28}IN$ 313.27	631-40-3 280 dec		4-04-00-00471 1.3138[25]	H_2O 4; EtOH 3; eth 2; chl 4
9834	Propane	C_3H_8 44.10	74-98-6 -187.6	7809 -42.1	4-01-00-00176 *0.493[25]	H_2O 3; EtOH 3; eth 4; ace 2
9835	Propane, 2,2-bis(ethylsulfonyl)- Sulfonmethane	$C_7H_{16}O_4S_2$ 228.33	115-24-2 125.8	8939 300 dec	3-01-00-02754	bz 4; EtOH 4; chl 4
9836	Propane, 1-bromo- Propyl bromide	C_3H_7Br 122.99	106-94-5 -110	7857 71.1	4-01-00-00205 1.3537[20]	H_2O 2; EtOH 3; eth 3; ace 3 1.4343[20]
9837	Propane, 2-bromo- Isopropyl bromide	C_3H_7Br 122.99	75-26-3 -89	5098 59.5	4-01-00-00208 1.3140[20]	H_2O 2; EtOH 5; eth 5; ace 3 1.4251[20]
9838	Propane, 1-bromo-2-chloro- 1-Bromo-2-chloropropane	C_3H_6BrCl 157.44	3017-96-7	118	4-01-00-00212 1.531[20]	ace 4; bz 4; eth 4; EtOH 4 1.4745[20]
9839	Propane, 1-bromo-3-chloro- 1-Bromo-3-chloropropane	C_3H_6BrCl 157.44	109-70-6 -58.9	143.3	4-01-00-00212 1.5969[20]	H_2O 1; EtOH 4; eth 4; chl 4 1.4864[20]
9840	Propane, 2-bromo-1-chloro- 2-Bromo-1-chloropropane	C_3H_6BrCl 157.44	3017-95-6	118	4-01-00-00212 1.537[20]	H_2O 1; EtOH 4; eth 4; ace 3 1.4795[20]
9841	Propane, 2-bromo-2-chloro- 2-Bromo-2-chloropropane	C_3H_6BrCl 157.44	2310-98-7	95	4-01-00-00213 1.474[22]	ace 4; bz 4; eth 4; EtOH 4 1.4575[20]
9842	Propane, 1-bromo-3-chloro-2-methyl-	C_4H_8BrCl 171.46	6974-77-2	155	4-01-00-00298 1.4839[20]	1.4796[25]
9843	Propane, 1-bromo-2,2-dimethyl-	$C_5H_{11}Br$ 151.05	630-17-1	106	4-01-00-00337 1.1997[20]	H_2O 1; EtOH 3; eth 3; ace 3 1.4370[20]
9844	Propane, 1-bromo-3-fluoro-	C_3H_6BrF 140.98	352-91-0	101.4	4-01-00-00210 1.542[25]	bz 4; eth 4; EtOH 4; chl 4 1.4290[25]
9845	Propane, 1-bromo-2-methyl- Isobutyl bromide	C_4H_9Br 137.02	78-77-3 -119	5019 91.1	4-01-00-00294 1.27[15]	EtOH 4; eth 4; ace 4; bz 4 1.4348[20]
9846	Propane, 2-bromo-2-methyl- tert-Butyl bromide	C_4H_9Br 137.02	507-19-7 -16.2	1555 73.3	4-01-00-00295 1.4278[20]	H_2O 1; ctc 2 1.4278[20]
9847	Propane, 2-bromo-2-nitro-	$C_3H_6BrNO_2$ 167.99	5447-97-2	153; 73-5[50]	4-01-00-00233 1.6562[25]	eth 4; EtOH 4
9848	Propane, 1-chloro- Propyl chloride	C_3H_7Cl 78.54	540-54-5 -122.8	7859 46.5	4-01-00-00189 0.8899[20]	H_2O 2; EtOH 5; eth 5; bz 3 1.3879[20]
9849	Propane, 2-chloro- Isopropyl chloride	C_3H_7Cl 78.54	75-29-6 -117.2	5099 35.7	4-01-00-00191 0.8617[20]	H_2O 2; EtOH 5; eth 5; bz 3 1.3777[20]
9850	Propane, 3-chloro-1,1-diethoxy-	$C_7H_{15}ClO_2$ 166.65	35573-93-4	84[25]	3-01-00-02592 0.9951[19]	ace 4; bz 4 1.4268[20]
9851	Propane, 1-chloro-2,2-difluoro- 2,2-Difluoropropyl chloride	$C_3H_5ClF_2$ 114.52	420-99-5 -56.2	55	4-01-00-00193 1.2001[20]	eth 3; bz 3; chl 3 1.3520[20]
9852	Propane, 3-chloro-1,1-dimethoxy-	$C_5H_{11}ClO_2$ 138.59	35502-06-8	86[100]; 51[19]	3-01-00-02692 1.064[15]	1.4163[20]
9853	Propane, 1-chloro-2,2-dimethyl-	$C_5H_{11}Cl$ 106.60	753-89-9 -20	84.3	4-01-00-00336 0.8660[20]	bz 4; eth 4; EtOH 4; chl 4 1.4044[20]
9854	Propane, 1-(1-chloroethoxy)- 1-Chloroethyl propyl ether	$C_5H_{11}ClO$ 122.59	692-35-3	114	3-01-00-02655 0.9322[20]	eth 4 1.4013[20]
9855	Propane, 2-[(2-chloroethyl)thio]-2-methyl-	$C_6H_{13}ClS$ 152.69	4303-44-0 -49	81-2[30]	3-01-00-01592 1.000[25]	
9856	Propane, 1-chloro-3-fluoro- 1-Chloro-3-fluoropropane	C_3H_6ClF 96.53	462-38-4	80	4-01-00-00193 1.3871[25]	bz 4; eth 4; EtOH 4; chl 4
9857	Propane, 1-chloro-3-iodo-	C_3H_6ClI 204.44	6940-76-7	171	4-01-00-00226 1.904[20]	H_2O 1; eth 3; bz 3; ctc 2 1.5472[20]
9858	Propane, 1-(chloromethoxy)-	C_4H_9ClO 108.57	3587-57-3	109	3-01-00-02589 0.9884[20]	eth 4; EtOH 4 1.4125[20]

No.	Name / Synonym	Mol. Form. / Mol. Wt.	CAS RN / mp/°C	Merck No. / bp/°C	Bell. Ref. / den/g cm^{-3}	Solubility / n_D
9859	Propane, 1-chloro-2-methoxy-	C$_4$H$_9$ClO / 108.57	5390-72-7 /	/ 103.5	3-01-00-01470 / 1.009[20]	1.4137[20]
9860	Propane, 1-chloro-3-methoxy-	C$_4$H$_9$ClO / 108.57	36215-07-3 /	/ 111	4-01-00-01441 / 1.0013[20]	1.4131[20]
9861	Propane, 2-chloro-1-methoxy-	C$_4$H$_9$ClO / 108.57	5390-71-6 /	/ 99	3-01-00-01424 / 0.9946[20]	1.4075[20]
9862	Propane, 1-chloro-2-methyl- / Isobutyl chloride	C$_4$H$_9$Cl / 92.57	513-36-0 / -130.3	5022 / 68.5	4-01-00-00287 / 0.8773[20]	eth 3; ace 3; ctc 2; chl 3 / 1.3984[20]
9863	Propane, 2-chloro-2-methyl- / tert-Butyl chloride	C$_4$H$_9$Cl / 92.57	507-20-0 / -26	1562 / 50.9	4-01-00-00288 / 0.8420[20]	H$_2$O 2; EtOH 5; eth 5; bz 3 / 1.3857[20]
9864	Propane, 1-chloro-3-(methylthio)- / 3-Chloropropyl methyl sulfide	C$_4$H$_9$ClS / 124.63	13012-59-4 /	/ 71-2[29]	4-01-00-01458 / 1.0844[20]	1.4833[20]
9865	Propane, 2-chloro-1-(methylthio)- / 2-Chloropropyl methyl sulfide	C$_4$H$_9$ClS / 124.63	19987-13-4 /	/ 67[37]	4-01-00-01457 / 1.076[20]	1.4905[20]
9866	Propane, 1-chloro-1-nitro- / 1-Chloro-1-nitropropane	C$_3$H$_6$ClNO$_2$ / 123.54	600-25-9 /	/ 142	4-01-00-00232 / 1.207[20]	H$_2$O 2; EtOH 3; eth 3; chl 2 / 1.4251[20]
9867	Propane, 1-chloro-2-nitro-	C$_3$H$_6$ClNO$_2$ / 123.54	2425-66-3 /	/ 172.5	4-01-00-00233 / 1.245[22]	eth 4; EtOH 4; chl 4 / 1.4432[25]
9868	Propane, 1-chloro-3-nitro- / 1-Chloro-3-nitropropane	C$_3$H$_6$ClNO$_2$ / 123.54	16694-52-3 /	/ 197 dec; 115[40]	3-01-00-00259 / 1.267[20]	eth 4; EtOH 4
9869	Propane, 2-chloro-1-nitro-	C$_3$H$_6$ClNO$_2$ / 123.54	503-76-4 /	/ 172	4-01-00-00232 / 1.2361[15]	eth 4; EtOH 4 / 1.4447[20]
9870	Propane, 2-chloro-2-nitro- / 2-Chloro-2-nitropropane	C$_3$H$_6$ClNO$_2$ / 123.54	594-71-8 /	/ 134 dec; 57[50]	4-01-00-00233 / 1.230[19]	H$_2$O 2; EtOH 3; eth 3; ctc 3 / 1.4378[19]
9871	Propane, 3-chloro-1,1,1-trifluoro-	C$_3$H$_4$ClF$_3$ / 132.51	460-35-5 / -106.5	/ 45.1	4-01-00-00193 / 1.3253[20]	1.3350[20]
9872	1,2-Propanediamine / Propylenediamine	C$_3$H$_{10}$N$_2$ / 74.13	78-90-0 /	7865 / 119.5	2-04-00-00698 / 0.878[15]	H$_2$O 4; eth 1 / 1.4460[20]
9873	1,3-Propanediamine / 1,3-Diaminopropane	C$_3$H$_{10}$N$_2$ / 74.13	109-76-2 /	/ 139.8	4-04-00-01258 / 0.884[25]	H$_2$O 3; EtOH 5; eth 5 / 1.4600[20]
9874	1,3-Propanediamine, N-(3-aminopropyl)- / Bis(3-aminopropyl)amine	C$_6$H$_{17}$N$_3$ / 131.22	56-18-8 / -14	/ 151[50]	4-04-00-01278 / 0.938[25]	chl 3 / 1.4810[20]
9875	1,3-Propanediamine, N-(3-aminopropyl)-N-methyl-	C$_7$H$_{19}$N$_3$ / 145.25	105-83-9 /	/ 232.5; 112[6]	4-04-00-01279 / 0.9023[20]	1.4705[25]
9876	1,3-Propanediamine, N,N-diethyl- / 3-(Diethylamino)propylamine	C$_7$H$_{18}$N$_2$ / 130.23	104-78-9 /	/ 168.5	4-04-00-01260 / 0.822[20]	1.443[20]
9877	1,3-Propanediamine, N,N-diethyl-N'-(6-methoxy-8-quinolinyl)- / Plasmocid	C$_{17}$H$_{25}$N$_3$O / 287.41	551-01-9 /	7495 / 182[1.0]	5-22-12-00126 / 1.0569[24]	1.5855[24]
9878	1,2-Propanediamine, 2-methyl-	C$_4$H$_{12}$N$_2$ / 88.15	811-93-8 /	/ 123	4-04-00-01306 / 0.841[25]	ctc 3 / 1.4410[20]
9879	1,3-Propanediamine, N,N,N',N'-tetramethyl-	C$_7$H$_{18}$N$_2$ / 130.23	110-95-2 /	/ 144	4-04-00-01259 / 0.7837[18]	H$_2$O 5; EtOH 5; eth 5
9880	Propane, 1,1-dibromo- / 1,1-Dibromopropane	C$_3$H$_6$Br$_2$ / 201.89	598-17-4 /	/ 133.5	4-01-00-00215 / 1.982[20]	eth 4; EtOH 4; chl 4 / 1.5100[20]
9881	Propane, 1,2-dibromo- / Propylene dibromide	C$_3$H$_6$Br$_2$ / 201.89	78-75-1 / -55.2	7866 / 141.9	4-01-00-00215 / 1.9324[20]	EtOH 3; eth 3; ctc 2; chl 3 / 1.5201[20]
9882	Propane, 1,3-dibromo- / 1,3-Dibromopropane	C$_3$H$_6$Br$_2$ / 201.89	109-64-8 / -34.2	9628 / 167.3	4-01-00-00216 / 1.9822[20]	EtOH 3; eth 3; ctc 2; chl 3 / 1.5232[20]
9883	Propane, 2,2-dibromo- / 2,2-Dibromopropane	C$_3$H$_6$Br$_2$ / 201.89	594-16-1 /	/ 115	4-01-00-00217 / 1.7825[20]	eth 4; EtOH 4; chl 4
9884	Propane, 1,3-dibromo-2,2-bis(bromomethyl)-	C$_5$H$_8$Br$_4$ / 387.73	3229-00-3 / 163	/ 305.5	4-01-00-00337 / 2.596[15]	EtOH 3; eth 2; bz 3; chl 2
9885	Propane, 1,2-dibromo-3-chloro- / Dibromochloropropane	C$_3$H$_5$Br$_2$Cl / 236.33	96-12-8 /	3003 / 196	3-01-00-00250 / 2.093[14]	1.553[14]
9886	Propane, 1,1-dibromo-2,2-dimethyl- / Neopentane, 1,1-dibromo	C$_5$H$_{10}$Br$_2$ / 229.94	2443-91-6 /	/ 64-5[43]	0-01-00-00141 / 1.7880[0]	1.5085
9887	Propane, 1,3-dibromo-2,2-dimethyl-	C$_5$H$_{10}$Br$_2$ / 229.94	5434-27-5 /	/ 187 dec; 80[26]	4-01-00-00337 / 1.7049[0]	1.5090
9888	Propane, 1,2-dibromo-3-methoxy-	C$_4$H$_8$Br$_2$O / 231.91	5836-66-8 /	/ 185	4-01-00-01447 / 1.832[12]	eth 4 / 1.5123[20]
9889	Propane, 1,2-dibromo-2-methyl-	C$_4$H$_8$Br$_2$ / 215.92	594-34-3 / 10.5	/ 150	4-01-00-00298 / 1.7827[20]	EtOH 3; eth 3; chl 3 / 1.5119[20]
9890	Propane, 1,3-dibromo-2-methyl-	C$_4$H$_8$Br$_2$ / 215.92	28148-04-1 /	/ 177.5; 62[13]	4-01-00-00299 / 1.8204[20]	1.5068[20]
9891	Propane, 1,1-dichloro- / Propylidene chloride	C$_3$H$_6$Cl$_2$ / 112.99	78-99-9 /	7874 / 88.1	4-01-00-00195 / 1.1321[20]	EtOH 3; eth 3; bz 3; chl 3 / 1.4289[20]
9892	Propane, 1,2-dichloro-, (±)- / Propane, 1,2-dichloro (DL)	C$_3$H$_6$Cl$_2$ / 112.99	26198-63-0 / -100.4	7867 / 96.4	4-01-00-00195 / 1.1560[20]	H$_2$O 2; EtOH 3; eth 3; bz 3 / 1.4394[20]
9893	Propane, 1,3-dichloro- / Trimethylene dichloride	C$_3$H$_6$Cl$_2$ / 112.99	142-28-9 / -99.5	/ 120.9	4-01-00-00196 / 1.1876[20]	H$_2$O 2; EtOH 4; eth 4; bz 3 / 1.4487[20]
9894	Propane, 2,2-dichloro-	C$_3$H$_6$Cl$_2$ / 112.99	594-20-7 / -33.8	/ 69.3	4-01-00-00196 / 1.1136[20]	H$_2$O 1; EtOH 3; eth 5; bz 3 / 1.4148[20]
9895	Propane, 1,3-dichloro-2,2-bis(chloromethyl)-	C$_5$H$_8$Cl$_4$ / 209.93	3228-99-7 / 97	/ 110[12]	4-01-00-00336 /	H$_2$O 1; eth 3; ctc 3; chl 3

No.	Name / Synonym	Mol. Form. / Mol. Wt.	CAS RN / mp/°C	Merck No. / bp/°C	Beil. Ref. / den/g cm^{-3}	Solubility / n_D
9896	Propane, 1,3-dichloro-2-(ethoxymethoxy)- Formaldehyde, (2,2'-dichloroisopropyl) ethyl acetal	$C_6H_{12}Cl_2O_2$ 187.07	89583-61-9	96-8[16]	3-01-00-02574 1.181[17]	eth 4 1.4491[17]
9897	Propane, 1,2-dichloro-2-fluoro-	$C_3H_5Cl_2F$ 130.98	420-97-3 -91.7	88.6	4-01-00-00197 1.2624[20]	H_2O 2; ace 3; bz 3 1.4099[20]
9898	Propane, 1,2-dichloro-1,1,2,3,3,3-hexafluoro- 1,2-Dichlorohexafluoropropane	$C_3Cl_2F_6$ 220.93	661-97-2	34.1		
9899	Propane, 1,1-dichloro-2-methyl- Isobutane, 1,1-dichloro	$C_4H_8Cl_2$ 127.01	598-76-5	105.5	4-01-00-00292 1.011[12]	bz 4; eth 4; EtOH 4; chl 4 1.4330[25]
9900	Propane, 1,2-dichloro-2-methyl- Isobutane, 1,2-dichloro	$C_4H_8Cl_2$ 127.01	594-37-6	106.5	4-01-00-00292 1.093[20]	H_2O 1; EtOH 5; eth 5; ace 5 1.4370[20]
9901	Propane, 1,3-dichloro-2-methyl-	$C_4H_8Cl_2$ 127.01	616-19-3	136.4	3-01-00-00319 1.131[20]	H_2O 1; EtOH 3; eth 4; bz 4 1.4488[25]
9902	Propane, 1,1-dichloro-1-nitro- 1,1-Dichlor-1-nitropropane	$C_3H_5Cl_2NO_2$ 157.98	595-44-8	145	3-01-00-00259 1.312[20]	ctc 3
9903	Propane, 1,1-diethoxy-	$C_7H_{16}O_2$ 132.20	4744-08-5	123	4-01-00-03168 0.8232[20]	H_2O 3; EtOH 4; eth 4; ace 3 1.3924[19]
9904	Propane, 2,2-diethoxy-	$C_7H_{16}O_2$ 132.20	126-84-1	114	4-01-00-03200 0.8200[21]	EtOH 3; eth 4; ace 3; bz 3 1.3891[20]
9905	Propane, 1,3-difluoro- 1,3-Difluoropropane	$C_3H_6F_2$ 80.08	462-39-5	41.6	3-01-00-00218 1.0057[25]	bz 4 1.3190[26]
9906	Propane, 2,2-difluoro- Dimethyldifluoromethane	$C_3H_6F_2$ 80.08	420-45-1 -104.8	-0.4	4-01-00-00187 0.9205[20]	
9907	Propane, 1,3-diiodo- Trimethylene diiodide	$C_3H_6I_2$ 295.89	627-31-6 -20	227 dec; 110[19]	4-01-00-00228 2.5754[20]	H_2O 1; eth 3; ctc 3; chl 3 1.6423[20]
9908	Propane, 2,2-diiodo- 2,2-Diiodopropane	$C_3H_6I_2$ 295.89	630-13-7	173; 156[500]	3-01-00-00255 2.5754[20]	eth 4; chl 4 1.651[20]
9909	Propane, 2,2-dimethoxy-	$C_5H_{12}O_2$ 104.15	77-76-9 -47	83	4-01-00-03199 0.847[25]	1.3780[20]
9910	Propane, 2,2-dimethyl- Neopentane	C_5H_{12} 72.15	463-82-1 -16.6	6372 9.4	4-01-00-00333 *0.5852[25]	H_2O 1; EtOH 3; eth 3; ctc 3 1.3476[6]
9911	Propanedinitrile Malononitrile	$C_3H_2N_2$ 66.06	109-77-3 32	5592 218.5	4-02-00-01892 1.1910[20]	H_2O 3; EtOH 4; eth 4; ace 3 1.4146[34]
9912	Propanedinitrile, dichloro-	$C_3Cl_2N_2$ 134.95	13063-43-9	97	3-02-00-01638 1.312[25]	1.4312[20]
9913	Propanedinitrile, dimethyl-	$C_5H_6N_2$ 94.12	7321-55-3 31.5	169.5; 62[22]	3-02-00-01704	H_2O 3; EtOH 2
9914	Propanedinitrile, (ethoxymethylene)-	$C_6H_6N_2O$ 122.13	123-06-8 66	160[12]	4-03-00-01194	EtOH 3; eth 3; chl 2
9915	Propanedinitrile, oxo- Carbonyl cyanide	C_3N_2O 80.05	1115-12-4 -36	65.5	4-03-00-01806 1.124[20]	eth 3; ace 3; ctc 3; chl 3 1.3919[20]
9916	Propanedinitrile, phenyl- Phenylmalononitrile	$C_9H_6N_2$ 142.16	3041-40-5 70.5	152[21]	4-09-00-03341	EtOH 4
9917	Propanedinitrile, (phenylmethyl)-	$C_{10}H_8N_2$ 156.19	1867-37-4 91	174[23]	4-09-00-03359	bz 4; eth 4; EtOH 4
9918	Propane, 1,1-dinitro- 1,1-Dinitropropane	$C_3H_6N_2O_4$ 134.09	601-76-3 -42	184	4-01-00-00234 1.2610[25]	alk 3 1.4339[20]
9919	Propane, 1,3-dinitro- 1,3-Dinitropropane	$C_3H_6N_2O_4$ 134.09	6125-21-9 -21.4	103[1]	4-01-00-00234 1.353[28]	H_2O 1; eth 3 1.4654[20]
9920	Propane, 2,2-dinitro- 2,2-Dinitropropane	$C_3H_6N_2O_4$ 134.09	595-49-3 53	185.5	4-01-00-00234 1.30[25]	H_2O 2
9921	Propanedioic acid Malonic acid	$C_3H_4O_4$ 104.06	141-82-2 135 dec	5591 sub	4-02-00-01874 1.619[10]	H_2O 4; EtOH 3; eth 3; bz 1
9922	Propanedioic acid, (acetylamino)-, diethyl ester	$C_9H_{15}NO_5$ 217.22	1068-90-2 96.3	185[20]	4-04-00-02993	H_2O 2; EtOH 3; eth 2; tfa 3
9923	Propanedioic acid, acetyl-, diethyl ester	$C_9H_{14}O_5$ 202.21	570-08-1	232	4-03-00-01819 1.0834[26]	H_2O 1; ace 3 1.4435[25]
9924	Propanedioic acid, amino-, diethyl ester Diethyl aminomalonate	$C_7H_{13}NO_4$ 175.18	6829-40-9	122-3[16]	4-04-00-02991 1.100[16]	H_2O 4; EtOH 4; eth 4; ace 3 1.4353[16]
9925	Propanedioic acid, bis(1,1-dimethylethyl) ester Di-tert-butyl malonate	$C_{11}H_{20}O_4$ 216.28	541-16-2 -6	3021 113[31]; 66[2]	4-02-00-01884 1.4184[20]	ace 3; chl 3 1.4184[29]
9926	Propanedioic acid, bis(phenylmethyl)-, diethyl ester Diethyl dibenzylmalonate	$C_{21}H_{24}O_4$ 340.42	597-55-7 14	234-5[23]	3-09-00-04552 1.093[20]	eth 4; EtOH 4
9927	Propanedioic acid, bis(phenylmethyl) ester	$C_{17}H_{16}O_4$ 284.31	15014-25-2	188[0.2]	4-06-00-02270 1.137[25]	1.5447[20]
9928	Propanedioic acid, bromo-, diethyl ester	$C_7H_{11}BrO_4$ 239.07	685-87-0 -54	254 dec	4-02-00-01904 1.4022[25]	H_2O 1; EtOH 5; eth 5; ace 3 1.4521[20]
9929	Propanedioic acid, butyl- n-Butylmalonic acid	$C_7H_{12}O_4$ 160.17	534-59-8 104.5	1574	4-02-00-02011	H_2O 4; EtOH 3; eth 3
9930	Propanedioic acid, butyl-, diethyl ester	$C_{11}H_{20}O_4$ 216.28	133-08-4	237.5	4-02-00-02011	EtOH 4; eth 4 1.4250[20]
9931	Propanedioic acid, chloro-, diethyl ester	$C_7H_{11}ClO_4$ 194.61	14064-10-9	222	4-02-00-01903 1.2040[20]	H_2O 1; EtOH 5; eth 5; chl 5 1.4327[20]
9932	Propanedioic acid, (3-chloropropyl)-, diethyl ester	$C_{10}H_{17}ClO_4$ 236.70	18719-43-2	147-9[10]	4-02-00-01992	H_2O 1; EtOH 3; eth 3; ctc 2 1.4429[20]

No.	Name / Synonym	Mol. Form. / Mol. Wt.	CAS RN / mp/°C	Merck No. / bp/°C	Bell. Ref. / den/g cm^{-3}	Solubility / n_D
9933	Propanedioic acid, (2-cyanoethyl)-, diethyl ester	$C_{10}H_{15}NO_4$ 213.23	17216-62-5	115-6[2]	4-02-00-02365 1.078[25]	1.4368[20]
9934	Propanedioic acid, cyclohexyl-, diethyl ester	$C_{13}H_{22}O_4$ 242.32	2163-44-2	163-5[20]	3-09-00-03834 1.0281[19]	ace 4; bz 4; eth 4; EtOH 4 1.4478[25]
9935	Propanedioic acid, 2-cyclopenten-1-yl-, diethyl ester Malonic acid, 2-cyclopentenyl, diethyl ester	$C_{12}H_{18}O_4$ 226.27	53608-93-8	141[10]	3-09-00-03951 1.0507[20]	H_2O 1; eth 3; ace 3 1.4536[20]
9936	Propanedioic acid, cyclopentylidene-, diethyl ester Malonic acid, cyclopentylidene, diethyl ester	$C_{12}H_{18}O_4$ 226.27	41589-42-8	140[10]	3-09-00-03952 1.0616[20]	ace 4; eth 4 1.4724[20]
9937	Propanedioic acid, dibutyl-, diethyl ester Diethyl dibutylmalonate	$C_{15}H_{28}O_4$ 272.38	596-75-8	153-4[14]	4-02-00-02119 0.9457[20]	H_2O 1; EtOH 3; eth 3; ctc 3 1.4341[20]
9938	Propanedioic acid, dibutyl ester	$C_{11}H_{20}O_4$ 216.28	1190-39-2 -83	251.5	4-02-00-01884 0.9824[20]	H_2O 1; EtOH 3; eth 3; ace 3 1.4262[20]
9939	Propanedioic acid, diethyl- Diethylmalonic acid	$C_7H_{12}O_4$ 160.17	510-20-3 127 dec	3114	4-02-00-02026	H_2O 4; EtOH 4; eth 4; bz 2
9940	Propanedioic acid, diethyl-, diethyl ester Diethyl 2,2-diethylmalonate	$C_{11}H_{20}O_4$ 216.28	77-25-8	3746 230	4-02-00-02026 0.9643[30]	H_2O 1; EtOH 5; eth 5; ctc 3 1.4240[20]
9941	Propanedioic acid, diethyl ester Diethyl malonate	$C_7H_{12}O_4$ 160.17	105-53-3 -50	3779 200	4-02-00-01881 1.0551[20]	H_2O 2; EtOH 5; eth 5; ace 4 1.4139[20]
9942	Propanedioic acid, dimethyl-, diethyl ester Diethyl dimethylmalonate	$C_9H_{16}O_4$ 188.22	1619-62-1 -30.4	197	4-02-00-01955 0.9964[20]	H_2O 1; EtOH 5; eth 5; ctc 2 1.4129[20]
9943	Propanedioic acid, dimethyl ester Methyl malonate	$C_5H_8O_4$ 132.12	108-59-8 -61.9	6009 181.4	4-02-00-01880 1.528[20]	H_2O 2; EtOH 5; eth 5; ace 4 1.4135[20]
9944	Propanedioic acid, di-2-propenyl-, diethyl ester	$C_{13}H_{20}O_4$ 240.30	3195-24-2	243.5	3-02-00-02003 0.9943[20]	EtOH 3; eth 3; ace 3; ctc 2 1.4445[22]
9945	Propanedioic acid, dipropyl ester Dipropyl malonate	$C_9H_{16}O_4$ 188.22	1117-19-7 -77.1	229	3-02-00-01619 1.0097[20]	ace 4; bz 4; eth 4; EtOH 4 1.4206[20]
9946	Propanedioic acid, di-2-propynyl-, diethyl ester Diethyl dipropargylmalonate	$C_{13}H_{16}O_4$ 236.27	2689-88-5 45.5	129[11]	4-02-00-02340	chl 3
9947	Propanedioic acid, (ethoxymethylene)-, diethyl ester	$C_{10}H_{18}O_5$ 216.23	87-13-8	280 dec; 165[19]	4-03-00-01192	H_2O 1; EtOH 3; eth 3; chl 2 1.4600[20]
9948	Propanedioic acid, ethyl-	$C_5H_8O_4$ 132.12	601-75-2 114	180[0.05]	4-02-00-01952	H_2O 4; EtOH 3; eth 3; ace 1
9949	Propanedioic acid, ethyl-, diethyl ester	$C_9H_{16}O_4$ 188.22	133-13-1	208; 98[12]	4-02-00-01953 1.0047[20]	H_2O 2; EtOH 4; eth 4; ace 3 1.4166[20]
9950	Propanedioic acid, ethylidene-, diethyl ester	$C_9H_{14}O_4$ 186.21	1462-12-0	115-8[17]	4-02-00-02234 1.0194[17]	eth 4; EtOH 4 1.4308[17]
9951	Propanedioic acid, ethyl(1-methylethyl)-, diethyl ester Malonic acid, ethylisopropyl-, diethyl ester	$C_{12}H_{22}O_4$ 230.30	2049-66-3	234; 109[11]	4-02-00-02052	ace 4; eth 4; EtOH 4 1.4280[25]
9952	Propanedioic acid, ethylphenyl-, diethyl ester	$C_{15}H_{20}O_4$ 264.32	76-67-5	170[19]	4-09-00-03385 1.071[20]	H_2O 1; EtOH 3; eth 3; chl 2 1.4896[25]
9953	Propanedioic acid, (formylamino)-, diethyl ester	$C_8H_{13}NO_5$ 203.19	6326-44-9 48.5	173[11]	4-04-00-02992	EtOH 4
9954	Propanedioic acid, hexadecyl-, diethyl ester	$C_{23}H_{44}O_4$ 384.60	41433-81-2	238[14]	4-02-00-02183 0.9118[20]	EtOH 4 1.4433[20]
9955	Propanedioic acid, hexyl-, diethyl ester	$C_{13}H_{24}O_4$ 244.33	5398-10-7	269; 143[15]	3-02-00-01791 0.9577[21]	ace 4; bz 4; eth 4; EtOH 4 1.4278[21]
9956	Propanedioic acid, hydroxy- Tartronic acid	$C_3H_4O_5$ 120.06	80-69-3 157	9042 sub	4-03-00-01120	H_2O 3; EtOH 3; eth 2
9957	Propanedioic acid, hydroxy-, diethyl ester	$C_7H_{12}O_5$ 176.17	13937-08-1 -2.5	221	3-03-00-00905 1.1520[15]	ace 4; bz 4; eth 4; EtOH 4
9958	Propanedioic acid, (hydroxyimino)-, diethyl ester	$C_7H_{11}NO_5$ 189.17	6829-41-0	172[12]	4-03-00-01805 1.1821[18]	ace 4; bz 4; eth 4; EtOH 4 1.4544[18]
9959	Propanedioic acid, (hydroxymethyl)- β-Isomalic acid	$C_4H_6O_5$ 134.09	4360-96-7 142 dec	170 dec	0-03-00-00441	H_2O 4; EtOH 4; eth 4
9960	Propanedioic acid, methyl-	$C_4H_6O_4$ 118.09	516-05-2 135 dec		4-02-00-01932 1.455[20]	H_2O 4; EtOH 4; eth 4; bz 2
9961	Propanedioic acid, (3-methylbutyl)-, diethyl ester	$C_{12}H_{22}O_4$ 230.30	5398-08-3	241	3-02-00-01777 0.9580[25]	ace 4; eth 4; EtOH 4 1.4255[25]
9962	Propanedioic acid, methyl-, diethyl ester	$C_8H_{14}O_4$ 174.20	609-08-5	201	4-02-00-01932 1.0225[20]	H_2O 2; EtOH 4; eth 4; ace 3 1.4126[20]
9963	Propanedioic acid, methyl-, dimethyl ester	$C_6H_{10}O_4$ 146.14	609-02-9	174	3-02-00-01681 1.0977[20]	ace 4; eth 4; EtOH 4; chl 4 1.4128[20]
9964	Propanedioic acid, (1-methylethyl)-, diethyl ester	$C_{10}H_{18}O_4$ 202.25	759-36-4	215	4-02-00-02001 0.9961[20]	H_2O 2; EtOH 4; eth 4; ctc 2 1.4188[21]
9965	Propanedioic acid, (1-methylethylidene)-	$C_8H_8O_4$ 144.13	4441-90-1 170.5	dec	3-02-00-01948	EtOH 4
9966	Propanedioic acid, (1-methylethylidene)-, diethyl ester Diethyl isopropylidenemalonate	$C_{10}H_{16}O_4$ 200.23	6802-75-1	176.5; 116[14]	4-02-00-02244 1.0282[18]	ace 4; EtOH 4 1.4486[17]

No.	Name Synonym	Mol. Form. Mol. Wt.	CAS RN mp/°C	Merck No. bp/°C	Bell. Ref. den/g cm^{-3}	Solubility n_D
9967	Propanedioic acid, (1-methylpropyl)-, diethyl ester	C$_{11}$H$_{20}$O$_4$ 216.28	83-27-2		4-02-00-02019 0.988[15]	H$_2$O 1; EtOH 4; eth 4; ctc 2 1.4248[20]
9968	Propanedioic acid, (2-methylpropyl)-	C$_7$H$_{12}$O$_4$ 160.17	4361-06-2 112 dec	dec	4-02-00-02023	eth 4; EtOH 4
9969	Propanedioic acid, (2-methylpropyl)-, diethyl ester Diethyl isobutylmalonate	C$_{11}$H$_{20}$O$_4$ 216.28	10203-58-4		3-02-00-01756 0.9804[20]	H$_2$O 1; EtOH 4; eth 4; chl 3 1.4236[20]
9970	Propanedioic acid, 1-naphthalenyl-, diethyl ester	C$_{17}$H$_{18}$O$_4$ 286.33	6341-60-2 61.3	176[1.0]	4-09-00-03536	EtOH 3; bz 3; chl 2
9971	Propanedioic acid, octyl-	C$_{11}$H$_{20}$O$_4$ 216.28	760-55-4 116		4-02-00-02114 1.173[17]	ace 4; EtOH 4
9972	Propanedioic acid, (1-oxobutyl)-, diethyl ester	C$_{11}$H$_{18}$O$_5$ 230.26	21633-79-4	249.5	4-03-00-01833 1.056[20]	1.4451[20]
9973	Propanedioic acid, oxo-, diethyl ester	C$_7$H$_{10}$O$_5$ 174.15	609-09-6 -30	210; 105[19]	3-03-00-01356 1.1419[16]	H$_2$O 4; EtOH 3; eth 3; chl 3 1.4310[22]
9974	Propanedioic acid, pentyl-, diethyl ester	C$_{12}$H$_{22}$O$_4$ 230.30	6065-59-4	134-6[14]	4-02-00-02034 0.9652[20]	H$_2$O 1; EtOH 4; eth 4; ctc 2 1.4253[20]
9975	Propanedioic acid, phenyl-, diethyl ester	C$_{13}$H$_{16}$O$_4$ 236.27	83-13-6 16.5	205 dec; 168[12]	4-09-00-03339 1.0950[20]	ace 4; EtOH 4 1.4977[20]
9976	Propanedioic acid, phenyl-, dimethyl ester	C$_{11}$H$_{12}$O$_4$ 208.21	37434-59-6 51.6	147[13]	4-09-00-03339	eth 4; EtOH 4
9977	Propanedioic acid, (phenylmethyl)-, diethyl ester	C$_{14}$H$_{18}$O$_4$ 250.29	607-81-8	300	4-09-00-03357 1.076[15]	H$_2$O 1; chl 2 1.4872[20]
9978	Propanedioic acid, (phenylmethylene)-, diethyl ester	C$_{14}$H$_{16}$O$_4$ 248.28	5292-53-5 32	216[30]; 196[14]	4-09-00-03461 1.1045[20]	H$_2$O 1; EtOH 3; eth 3; ace 3 1.5389[20]
9979	Propanedioic acid, 2-propenyl-	C$_6$H$_8$O$_4$ 144.13	2583-25-7 105	180 dec	3-02-00-01945	H$_2$O 4; eth 4; EtOH 4
9980	Propanedioic acid, 2-propenyl-, diethyl ester	C$_{10}$H$_{16}$O$_4$ 200.23	2049-80-1	222.5; 93[6]	4-02-00-02240 1.0098[20]	H$_2$O 1; EtOH 4; eth 4; ctc 3 1.4305[20]
9981	Propanedioic acid, propyl-, diethyl ester	C$_{10}$H$_{18}$O$_4$ 202.25	2163-48-6	221; 114[22]	4-02-00-01991 0.9873[20]	H$_2$O 2; EtOH 4; eth 4 1.4197[20]
9982	1,2-Propanediol 1,2-Propylene glycol	C$_3$H$_8$O$_2$ 76.10	57-55-6 -60	7868 187.6	3-01-00-02142 1.0361[20]	H$_2$O 5; EtOH 5; eth 3; bz 3 1.4324[20]
9983	1,3-Propanediol Trimethylene glycol	C$_3$H$_8$O$_2$ 76.10	504-63-2 -26.7	9629 214.4	4-01-00-02493 1.0538[20]	H$_2$O 5; EtOH 5; eth 4; bz 2 1.4398[20]
9984	1,2-Propanediol, 3-amino-, (±)- 1,2-Propanediol, 3-amino(DL)	C$_3$H$_9$NO$_2$ 91.11	13552-31-3	265 dec; 145[9]	3-04-00-00839 1.1752[20]	H$_2$O 3; EtOH 3; eth 1; bz 1 1.4910[25]
9985	1,3-Propanediol, 2-amino-2-ethyl- 2-Amino-2-ethyl-1,3-propanediol	C$_5$H$_{13}$NO$_2$ 119.16	115-70-8 37.5	451 152-3[10]	3-04-00-00850 1.099[20]	H$_2$O 5 1.490[20]
9986	1,3-Propanediol, 2-amino-2-(hydroxymethyl)- Tromethamine	C$_4$H$_{11}$NO$_3$ 121.14	77-86-1 171.5	9684 219-20[10]	4-04-00-01903	H$_2$O 4; MeOH 3
9987	1,3-Propanediol, 2-amino-2-methyl- 2-Amino-2-methyl-1,3-propanediol	C$_4$H$_{11}$NO$_2$ 105.14	115-69-5 110	460 151-2[10]	4-04-00-01881	H$_2$O 4; EtOH 3
9988	1,3-Propanediol, 2,2-bis[(acetyloxy)methyl]-, diacetate Pentaerythrityl tetraacetate	C$_{13}$H$_{20}$O$_8$ 304.30	597-71-7 83.5	7065	4-02-00-00264 1.273[18]	H$_2$O 3; EtOH 4; eth 4
9989	1,3-Propanediol, 2,2-bis(hydroxymethyl)- Pentaerythritol	C$_5$H$_{12}$O$_4$ 136.15	115-77-5 260	7062 sub	4-01-00-02812	H$_2$O 3; eth 1; bz 1 1.548
9990	1,3-Propanediol, 2,2-bis[(nitrooxy)methyl]-, dinitrate (ester) Pentaerythrityl tetranitrate	C$_5$H$_8$N$_4$O$_{12}$ 316.14	78-11-5 140.5	7066	4-01-00-02816 1.773[20]	H$_2$O 2; EtOH 2; eth 2; ace 4
9991	1,3-Propanediol, 2-bromo-2-nitro- Bronopol	C$_3$H$_8$BrNO$_4$ 199.99	52-51-7 131.5	1437	4-01-00-02501	
9992	1,3-Propanediol, 2-butyl-2-ethyl-	C$_9$H$_{20}$O$_2$ 160.26	115-84-4 43.8	262	4-01-00-02611 0.927[50]	H$_2$O 2; EtOH 3; ace 2 1.4587[25]
9993	1,3-Propanediol, 2-butyl-2-methyl-	C$_8$H$_{18}$O$_2$ 146.23	3121-83-3 43.8	262	4-01-00-02599 0.927[50]	EtOH 4 1.4587[25]
9994	1,2-Propanediol, 3-chloro- α-Chlorohydrin	C$_3$H$_7$ClO$_2$ 110.54	96-24-2	2145 213 dec; 116[11]	4-01-00-02484 1.325[18]	H$_2$O 4; EtOH 3; eth 3 1.4809[20]
9995	1,3-Propanediol, 2-chloro- Glycerol β-chlorohydrin	C$_3$H$_7$ClO$_2$ 110.54	497-04-1	146[18]	4-01-00-02499 1.3219[20]	H$_2$O 4; ace 4; EtOH 4 1.4831[20]
9996	1,2-Propanediol, 3-chloro-, diacetate	C$_7$H$_{11}$ClO$_4$ 194.61	869-50-1	245	3-02-00-00313 1.199[25]	eth 4; EtOH 4 1.4407[20]
9997	1,2-Propanediol, 3-chloro-, dinitrate Clonitrate	C$_3$H$_5$ClN$_2$O$_6$ 200.54	2612-33-1	2390 192.5	4-01-00-02487 1.5112[9]	ace 4; EtOH 4; chl 4
9998	1,2-Propanediol, 3-chloro-2-methyl- Isobutylene glycol, 3-chloro	C$_4$H$_9$ClO$_2$ 124.57	597-33-1	114-7[20]	4-01-00-02536 1.2362[20]	H$_2$O 4; eth 4; EtOH 4 1.4748[20]
9999	1,2-Propanediol, 3-(2-chlorophenoxy)-	C$_9$H$_{11}$ClO$_3$ 202.64	5112-21-0 71.5	250[19]	4-06-00-00790	H$_2$O 3; eth 4; bz 3
10000	1,2-Propanediol, 3-(4-chlorophenoxy)- Chlorphenesin	C$_9$H$_{11}$ClO$_3$ 202.64	104-29-0 78	2178 214-5[19]	4-06-00-00831	H$_2$O 1; EtOH 4; eth 4; bz 3

No.	Name Synonym	Mol. Form. Mol. Wt.	CAS RN mp/°C	Merck No. bp/°C	Bell. Ref. den/g cm^{-3}	Solubility n_D
10001	1,2-Propanediol, 3-(4-chlorophenoxy)-, 1-carbamate Chlorphenesin carbamate	$C_{10}H_{12}ClNO_4$ 245.66	886-74-8 90	2179		ace 4; EtOH 4; diox 4
10002	1,2-Propanediol, diacetate	$C_7H_{12}O_4$ 160.17	623-84-7 	190.5	4-02-00-00220 1.059[20]	H_2O 4; EtOH 3; eth 3 1.4173[20]
10003	1,3-Propanediol, diacetate	$C_7H_{12}O_4$ 160.17	628-66-0 	209.5	4-02-00-00221 1.070[14]	H_2O 4; EtOH 3 1.4192
10004	1,3-Propanediol, 2,2-diethyl-	$C_7H_{16}O_2$ 132.20	115-76-4 61.5	240.5	4-01-00-02589 1.050[20]	H_2O 4; EtOH 4; eth 4; chl 3 1.4574[25]
10005	1,3-Propanediol, 2,2-dimethyl- Neopentyl glycol	$C_5H_{12}O_2$ 104.15	126-30-7 130	6374 208	4-01-00-02551	H_2O 3; EtOH 4; eth 4; bz 3
10006	1,3-Propanediol, 2-ethyl-2-(hydroxymethyl)- Trimethylolpropane	$C_6H_{14}O_3$ 134.18	77-99-6 58	160[5]	4-01-00-02786	H_2O 4; EtOH 4
10007	1,3-Propanediol, 2-ethyl-2-methyl-	$C_6H_{14}O_2$ 118.18	77-84-9 40.3	225.8	4-01-00-02574	chl 3
10008	1,3-Propanediol, 2-ethyl-2-nitro- 2-Nitro-2-ethyl-1,3-propanediol	$C_5H_{11}NO_4$ 149.15	597-09-1 57.5	dec	4-01-00-02550	H_2O 4; eth 4; EtOH 4
10009	1,2-Propanediol, 3-(hexadecyloxy)-	$C_{19}H_{40}O_3$ 316.52	53584-29-5 64	205[1.0]	3-01-00-02322	ace 4; chl 4; peth 4
10010	1,2-Propanediol, 3-(hexadecyloxy)-, (S)- Chimyl alcohol	$C_{19}H_{40}O_3$ 316.52	506-03-6 64	2049 120[0.005]	3-01-00-02322	ace 4; peth 4; chl 4
10011	1,3-Propanediol, 2-(hydroxymethyl)-2-methyl-	$C_5H_{12}O_3$ 120.15	77-85-0 204	135-7[15]	4-01-00-02780	H_2O 5; EtOH 5; eth 1; bz 1
10012	1,3-Propanediol, 2-(hydroxymethyl)-2-nitro- Tris(hydroxymethyl)nitromethane	$C_4H_9NO_5$ 151.12	126-11-4 165	9667 dec	4-01-00-02777	H_2O 4; eth 4; EtOH 4
10013	1,2-Propanediol, 3-mercapto- Thioglycerol	$C_3H_8O_2S$ 108.16	96-27-5 	9264 100-1[1]	3-01-00-02339 1.2455[20]	H_2O 2; EtOH 5; eth 2; ace 4 1.5268[20]
10014	1,2-Propanediol, 3-methoxy-	$C_4H_{10}O_3$ 106.12	623-39-2 	220	4-01-00-02755 1.114[20]	H_2O 4; EtOH 4; eth 3; ace 4 1.442[25]
10015	1,3-Propanediol, 2-methoxy- Glycerol, 2-methyl ether	$C_4H_{10}O_3$ 106.12	761-06-8 	232	3-01-00-02317 1.124[25]	H_2O 4; ace 4; eth 4; EtOH 4 1.4505[12]
10016	1,2-Propanediol, 3-(2-methoxyphenoxy)- Guaifenesin	$C_{10}H_{14}O_4$ 198.22	93-14-1 78.5	4465 215[19.] 127[0.2]	4-06-00-05576	H_2O 3; EtOH 4; bz 3; chl 3
10017	1,2-Propanediol, 2-methyl-	$C_4H_{10}O_2$ 90.12	558-43-0 	176	4-01-00-02533 1.0024[20]	H_2O 4; eth 4; EtOH 4 1.4350[20]
10018	1,3-Propanediol, 2-methyl-2-nitro-	$C_4H_9NO_4$ 135.12	77-49-6 150.1	dec	4-01-00-02537	H_2O 4; EtOH 4; DMSO-d$_6$ 2
10019	1,2-Propanediol, 3-(2-methylphenoxy)- Mephenesin	$C_{10}H_{14}O_3$ 182.22	59-47-2 70 dec	5737	4-06-00-01952	H_2O 2; EtOH 3; eth 2
10020	1,3-Propanediol, 2-methyl-2-propyl-	$C_7H_{16}O_2$ 132.20	78-26-2 62.5	234; 121[10]	4-01-00-02585	H_2O 3; chl 2; os 3; hx 3
10021	1,3-Propanediol, 2-methyl-2-propyl-, dicarbamate Meprobamate	$C_9H_{18}N_2O_4$ 218.25	57-53-4 105	5751	4-03-00-00073	bz 4; eth 4; EtOH 4
10022	1,2-Propanediol, 1-nitrate	$C_3H_7NO_4$ 121.09	20266-65-3 	148-9[4]	4-01-00-02476 1.2417[20]	1.4368[20]
10023	1,2-Propanediol, 3-(octadecyloxy)- Batyl alcohol	$C_{21}H_{44}O_3$ 344.58	544-62-7 70.5	1021 217[2]	4-01-00-02758	eth 4
10024	Propanediol, oxybis-	$C_6H_{14}O_5$ 166.17	59113-36-9 	267[15]; 225[3]	3-01-00-02327 1.2751[20]	H_2O 4; EtOH 4 1.4890[20]
10025	1,2-Propanediol, 3,3'-oxybis-	$C_6H_{14}O_5$ 166.17	627-82-7 	267[15]; 225[3]	3-01-00-02327 1.2751[20]	H_2O 4; EtOH 4 1.4890[20]
10026	1,2-Propanediol, 3-phenoxy- Phenylglyceryl ether	$C_9H_{12}O_3$ 168.19	538-43-2 67.5	7261 200[22]	4-06-00-00589 1.225[20]	H_2O 4; bz 4; eth 4; EtOH 4
10027	1,3-Propanedione, 1,3-diphenyl- Dibenzoylmethane	$C_{15}H_{12}O_2$ 224.26	120-46-7 70.5	2991	4-07-00-02512	EtOH 3; eth 3; chl 3; dil NaOH 3
10028	1,2-Propanedione, 1-phenyl-	$C_9H_8O_2$ 148.16	579-07-7 <20	222; 102[12]	4-07-00-02143 1.1006[20]	H_2O 3; EtOH 3; eth 3 1.537[10]
10029	Propanedioyl dichloride	$C_3H_2Cl_2O_2$ 140.95	1663-67-8 	58[26]	4-02-00-01887 1.4509[20]	eth 3; AcOEt 3 1.4639[20]
10030	1,2-Propanedithiol	$C_3H_8S_2$ 108.23	814-67-5 	152	4-01-00-02492 1.08[20]	chl 3 1.532[20]
10031	1,3-Propanedithiol Trimethylene dimercaptan	$C_3H_8S_2$ 108.23	109-80-8 -79	7811 172.9	4-01-00-02503 1.0772[20]	H_2O 2; EtOH 5; eth 5; bz 5 1.5392[20]
10032	Propane, 1,1'-[1,2-ethanediylbis(oxy)]bis[2-methyl- Ethylene glycol diisobutyl ether	$C_{10}H_{22}O_2$ 174.28	54922-86-0 	171.3	4-01-00-02382 0.8200[20]	1.4021[20]
10033	Propane, 1-(ethenyloxy)-	$C_5H_{10}O$ 86.13	764-47-6 	65	4-01-00-02052 0.7674[20]	1.3908[20]
10034	Propane, 2-(ethenyloxy)-	$C_5H_{10}O$ 86.13	926-65-8 -140	55.5	4-01-00-02052 0.7534[20]	ace 4; bz 4; eth 4; EtOH 4 1.3840[20]
10035	Propane, 1-(ethenyloxy)-2-methyl- Isobutyl vinyl ether	$C_6H_{12}O$ 100.16	109-53-5 -112	83	4-01-00-02054 0.7645[20]	H_2O 2; EtOH 4; eth 5; ace 4 1.3966[20]
10036	Propane, 2-(ethenyloxy)-2-methyl-	$C_6H_{12}O$ 100.16	926-02-3 -112	75	4-01-00-02055 0.7691[20]	1.3922[20]
10037	Propane, 1-ethoxy- Ethyl propyl ether	$C_5H_{12}O$ 88.15	628-32-0 -127.5	63.2	4-01-00-01421 0.7386[20]	eth 4; EtOH 4; HOAc 4 1.3695[20]
10038	Propane, 2-ethoxy- Ethyl isopropyl ether	$C_5H_{12}O$ 88.15	625-54-7 	54.1	4-01-00-01471 0.720[25]	H_2O 3; EtOH 5; eth 5; ace 3 1.3698[25]

No.	Name Synonym	Mol. Form. Mol. Wt.	CAS RN mp/°C	Merck No. bp/°C	Beil. Ref. den/g cm⁻³	Solubility n_D
10039	Propane, 1-ethoxy-2-methyl-	$C_6H_{14}O$ 102.18	627-02-1	81	4-01-00-01593 0.751[20]	H_2O 1; EtOH 5; eth; ace 3 1.3739[25]
10040	Propane, 2-ethoxy-2-methyl- Ethyl tert-butyl ether	$C_6H_{14}O$ 102.18	637-92-3 -94	3732 73.1	4-01-00-01615 0.7519[25]	H_2O 1; EtOH 4; eth 4 1.3794[20]
10041	Propane, 1,1'-[ethylidenebis(oxy)]bis[2-methyl-	$C_{10}H_{22}O_2$ 174.28	5669-09-0	171.3	4-01-00-03107 0.8200[20]	1.4021[20]
10042	Propane, 1-(ethylthio)- Ethyl propyl sulfide	$C_5H_{12}S$ 104.22	4110-50-3 -117	118.6	4-01-00-01451 0.8370[20]	EtOH 3 1.4462[20]
10043	Propane, 2-(ethylthio)- Ethyl isopropyl sulfide	$C_5H_{12}S$ 104.22	5145-99-3 -122.2	107.5	4-01-00-01501 0.8246[20]	
10044	Propane, 1-(ethylthio)-2-methyl- Ethyl isobutyl sulfide	$C_6H_{14}S$ 118.24	1613-45-2	134.2	4-01-00-01606 0.8306[20]	chl 3 1.4450[20]
10045	Propane, 1-(ethynyloxy)-	C_5H_8O 84.12	42842-08-0	75	4-01-00-02213 0.8080[20]	eth 4; EtOH 4 1.3935[20]
10046	Propane, 1-fluoro- Propyl fluoride	C_3H_7F 62.09	460-13-9 -159	2.5	4-01-00-00187 0.7956[20]	H_2O 2; EtOH 4; eth 4 1.3115[20]
10047	Propane, 2-fluoro- Isopropyl fluoride	C_3H_7F 62.09	420-26-8		4-01-00-00187	
10048	Propane, 1,1,1,2,2,3,3-heptachloro- 1,1,1,2,2,3,3-Heptachloropropane	C_3HCl_7 285.21	594-89-8 29.4	247.5	4-01-00-00205 1.8048[34]	ctc 2; chl 3
10049	Propane, 1,1,1,2,3,3,3-heptachloro- 1,1,1,2,3,3,3-Heptachloropropane	C_3HCl_7 285.21	3849-33-0 11	249	4-01-00-00205 1.7921[34]	chl 4 1.5427[21]
10050	Propane, 1,1,1,2,3,3,3-heptafluoro-2-(trifluoromethyl)- Perfluoroisobutane	C_4F_{10} 238.03	354-92-7	0		
10051	Propane, 1,1,1,2,2,3-hexachloro- 1,1,1,2,2,3-Hexachloropropane	$C_3H_2Cl_6$ 250.77	24425-97-6 26	216	4-01-00-00204 1.7187[25]	1.5282[25]
10052	Propane, 1,1,1,3,3,3-hexachloro- 1,1,1,3,3,3-Hexachloropropane	$C_3H_2Cl_6$ 250.77	3607-78-1 -27	206	4-01-00-00205 1.6800[20]	1.5179[20]
10053	Propane, 1,1,2,2,3,3-hexachloro- 1,1,2,2,3,3-Hexachloropropane	$C_3H_2Cl_6$ 250.77	15600-01-8	218.5	4-01-00-00205 1.7137[34]	1.5262[18]
10054	Propane, 1-iodo- Propyl iodide	C_3H_7I 169.99	107-08-4 -101.3	7875 102.6	4-01-00-00222 1.7489[20]	H_2O 2; EtOH 5; eth 5; ctc 2 1.5058[20]
10055	Propane, 2-iodo- Isopropyl iodide	C_3H_7I 169.99	75-30-9 -90	5102 89.5	4-01-00-00223 1.7042[20]	H_2O 2; EtOH 5; eth 5; bz 5 1.5028[20]
10056	Propane, 1-iodo-2,2-dimethyl-	$C_5H_{11}I$ 198.05	15501-33-4	128 dec	4-01-00-00338 1.4940[20]	H_2O 1; EtOH 3; eth 3 1.4890[20]
10057	Propane, 1-iodo-2-methyl- Isobutyl iodide	C_4H_9I 184.02	513-38-2	5027 121.1	4-01-00-00299 1.6035[20]	
10058	Propane, 2-iodo-2-methyl- tert-Butyl iodide	C_4H_9I 184.02	558-17-8 -38.2	100.1	4-01-00-00300 1.571[25]	EtOH 5; eth 5 1.4918[20]
10059	Propane, 1-isocyanato- Propyl isocyanate	C_4H_7NO 85.11	110-78-1	83.5	4-04-00-00491 0.908[25]	1.3970[20]
10060	Propane, 2-isocyanato- Isopropyl isocyanate	C_4H_7NO 85.11	1795-48-8	74.5	4-04-00-00530 0.866[25]	1.3825[20]
10061	Propane, 1-isocyanato-2-methyl- Isobutyl isocyanate	C_5H_9NO 99.13	1873-29-6	106	4-04-00-00653	
10062	Propane, 2-isocyanato-2-methyl- tert-Butyl isocyanate	C_5H_9NO 99.13	1609-86-5 85.5		4-04-00-00669 0.8670[7]	1.4061[20]
10063	Propane, 2-isocyano- Isopropyl isocyanide	C_4H_7N 69.11	598-45-8	87	0-04-00-00154 0.7596[25]	H_2O 1; EtOH 5; eth 5
10064	Propane, 1-isothiocyanato-	C_4H_7NS 101.17	628-30-8	153	4-04-00-00491 0.9781[16]	H_2O 2; EtOH 5; eth 5 1.5085[16]
10065	Propane, 1-isothiocyanato-2-methyl-	C_5H_9NS 115.20	591-82-2	160	4-04-00-00653 0.9631[14]	1.5005[14]
10066	Propane, 2-isothiocyanato-2-methyl- tert-Butyl isothiocyanate	C_5H_9NS 115.20	590-42-1 10.5	140	4-04-00-00669 0.9187[10]	
10067	Propane, 1-isothiocyanato-3-(methylsulfonyl)- Cheirolin	$C_5H_9NO_2S_2$ 179.26	505-34-0 47.5	2040 200[3]	4-04-00-01664	EtOH 4; chl 4
10068	Propane, 1,1',1'',1'''-[methanetetrayltetrakis(oxy)]tetrakis-	$C_{13}H_{28}O_4$ 248.36	597-72-8	224.2	4-03-00-00007 0.897[20]	EtOH 4; eth 4 1.4100[20]
10069	Propane, 1-methoxy- Methyl propyl ether	$C_4H_{10}O$ 74.12	557-17-5	6031 39.1	4-01-00-01421 0.7356[13]	H_2O 3; EtOH 5; eth 5; ace 3 1.3579[25]
10070	Propane, 2-methoxy- Isopropyl methyl ether	$C_4H_{10}O$ 74.12	598-53-8	30.7	4-01-00-01471 0.7237[15]	H_2O 2; EtOH 5; eth 5 1.3576[20]
10071	Propane, 1-methoxy-2-methyl- Isobutyl methyl ether	$C_5H_{12}O$ 88.15	625-44-5	58.6	4-01-00-01593 0.7311[20]	eth 4; EtOH 4
10072	Propane, 2-methoxy-2-methyl- tert-Butyl methyl ether	$C_5H_{12}O$ 88.15	1634-04-4 -108.6	5954 55.2	4-01-00-01615 0.7405[20]	H_2O 3; EtOH 4; eth 4 1.3690[20]
10073	Propane, 2-methyl- Isobutane	C_4H_{10} 58.12	75-28-5 -138.3	-11.7	4-01-00-00282 *0.5510[25]	H_2O 2; EtOH 3; eth 3; chl 3 1.3518-[25]
10074	Propane, 1,1'-[methylenebis(oxy)]bis-	$C_7H_{16}O_2$ 132.20	505-84-0 -97.3	140.5	4-01-00-03029 0.8345[20]	ace 4; bz 4; eth 4; EtOH 4 1.3939[19]
10075	Propane, 1,1'-[methylenebis(oxy)]bis[2-methyl-	$C_9H_{20}O_2$ 160.26	2568-91-4	165.5	4-01-00-03030 0.8268[20]	1.4108[17]
10076	Propane, 1-(1-methylethoxy)-	$C_6H_{14}O$ 102.18	627-08-7	83	4-01-00-01471 0.7370[20]	H_2O 2; EtOH 4; eth 3; ace 3 1.376[21]
10077	Propane, 1-[(1-methylethyl)thio]- Isopropyl propyl sulfide	$C_6H_{14}S$ 118.24	5008-73-1	132.1	4-01-00-01501 0.8269[20]	

No.	Name Synonym	Mol. Form. Mol. Wt.	CAS RN mp/°C	Merck No. bp/°C	Beil. Ref. den/g cm^{-3}	Solubility n_D
10078	Propane, 1,1',1''-[methylidynetris(oxy)]tris-	C$_{10}$H$_{22}$O$_3$ 190.28	621-76-1	191; 93[30]	4-02-00-00027 0.8805[20]	eth 4; EtOH 4 1.4072[20]
10079	Propane, 2,2',2''-[methylidynetris(oxy)]tris-	C$_{10}$H$_{22}$O$_3$ 190.28	4447-60-3	167	4-02-00-00028 0.8621[20]	eth 4; EtOH 4 1.4000[20]
10080	Propane, 1,1',1''-[methylidynetris(oxy)]tris[2-methyl- Orthoformic acid, triisobutyl ester	C$_{13}$H$_{28}$O$_3$ 232.36	16754-49-7	223	4-02-00-00029 0.8582[20]	eth 4; EtOH 4 1.4120[20]
10081	Propane, 2-methyl-2-(1-methylethoxy)-	C$_7$H$_{16}$O 116.20	17348-59-3 -88	87.6	4-01-00-01616 0.7365[25]	chl 3
10082	Propane, 2-methyl-1-(methylthio)-	C$_5$H$_{12}$S 104.22	5008-69-5	112.5	4-01-00-01606 0.8335[20]	EtOH 3; eth 3; ace 3 1.4433[20]
10083	Propane, 2-methyl-2-(methylthio)- tert-Butyl methyl sulfide	C$_5$H$_{12}$S 104.22	6163-64-0	99	4-01-00-01636	
10084	Propane, 2-methyl-1-nitro-	C$_4$H$_9$NO$_2$ 103.12	625-74-1	140.5	4-01-00-00301 0.9597[25]	H$_2$O 2; EtOH 5; eth 5 1.4066[25]
10085	Propane, 2-methyl-2-nitro-	C$_4$H$_9$NO$_2$ 103.12	594-70-7 26.23	127.16	4-01-00-00301 0.9501[28]	EtOH 5; eth 5; ace 5; bz 5 1.4015[20]
10086	Propane, 2-methyl-1-propoxy-	C$_7$H$_{16}$O 116.20	15268-49-2	107	4-01-00-01594 0.7549[20]	eth 4; EtOH 4 1.3852[25]
10087	Propane, 2-methyl-2-propoxy-	C$_7$H$_{16}$O 116.20	29072-93-3	100	3-01-00-01578 0.7472[25]	1.3830[25]
10088	Propane, 1-(methylthio)- Methyl propyl sulfide	C$_4$H$_{10}$S 90.19	3877-15-4 -113	95.6	4-01-00-01450 0.8424[20]	H$_2$O 3; EtOH 3; eth 3; ace 3 1.4442[20]
10089	Propane, 2-(methylthio)- Isopropyl methyl sulfide	C$_4$H$_{10}$S 90.19	1551-21-9 -101.5	84.8	4-01-00-01500 0.8291[20]	EtOH 3; eth 3; ace 3 1.4932[20]
10090	Propanenitrile Ethyl cyanide	C$_3$H$_5$N 55.08	107-12-0 -92.8	7839 97.1	4-02-00-00728 0.7818[20]	H$_2$O 4; EtOH 3; eth 3; ace 3 1.3655[20]
10091	Propanenitrile, 2-(acetyloxy)-	C$_5$H$_7$NO$_2$ 113.12	15657-96-2 -75	172.5	4-03-00-00676 1.0278[20]	H$_2$O 4 1.4027[20]
10092	Propanenitrile, 3-amino- 3-Aminopropionitrile	C$_3$H$_6$N$_2$ 70.09	151-18-8	481 185; 88[20]	4-04-00-02530 0.9584[20]	1.4396[20]
10093	Propanenitrile, 2,2'-azobis[2-methyl- 2,2'-Azobis(isobutyronitrile)	C$_8$H$_{12}$N$_4$ 164.21	78-67-1	931	4-04-00-03377	H$_2$O 1; EtOH 2; eth 2
10094	Propanenitrile, 2-bromo-	C$_3$H$_4$BrN 133.98	19481-82-4	59[24]	4-02-00-00764 1.5505[20]	ace 4; eth 4 1.4585[20]
10095	Propanenitrile, 3-bromo-	C$_3$H$_4$BrN 133.98	2417-90-5	92[25]; 69[7]	4-02-00-00766 1.6152[20]	EtOH 4; eth 4; ctc 2 1.4800[20]
10096	Propanenitrile, 2-bromo-2-methyl-	C$_4$H$_6$BrN 148.00	41658-69-9	139.5	4-02-00-00863 1.4796[15]	ace 4; eth 4; EtOH 4 1.4379[15]
10097	Propanenitrile, 3-chloro- β-Chloropropionitrile	C$_3$H$_4$ClN 89.52	542-76-7 -51	2160 175.5	4-02-00-00751 1.1573[20]	ctc 2 1.4360[20]
10098	Propanenitrile, 3-chloro-2-hydroxy-2-methyl- Lactonitrile, 3-chloro-2-methyl-	C$_4$H$_6$ClNO 119.55	33401-05-7	110[27]	3-03-00-00599 1.2027[15]	H$_2$O 4; EtOH 4; ace 4 1.4356[11]
10099	Propanenitrile, 2,3-dichloro-	C$_3$H$_3$Cl$_2$N 123.97	2601-89-0	58-9[8]	4-02-00-00757 1.3500[20]	1.4640[20]
10100	Propanenitrile, 2,2-dimethyl- tert-Butyl cyanide	C$_5$H$_9$N 83.13	630-18-2 15	106.1	4-02-00-00913 0.7586[25]	1.3774[20]
10101	Propanenitrile, 3-(dimethylamino)- 3-(Dimethylamino)propionitrile	C$_5$H$_{10}$N$_2$ 98.15	1738-25-6	173	4-04-00-02533 0.8705[20]	
10102	Propanenitrile, 2-ethoxy- 2-Ethoxypropionitrile	C$_5$H$_9$NO 99.13	14631-45-9	136	4-03-00-00675 0.8743[20]	eth 4; EtOH 4 1.3890[22]
10103	Propanenitrile, 3-ethoxy-	C$_5$H$_9$NO 99.13	2141-62-0	171	4-03-00-00708 0.9285[15]	eth 4; EtOH 4 1.4068[20]
10104	Propanenitrile, 2-(β-D-glucopyranosyloxy)-2-methyl- Linamarin	C$_{10}$H$_{17}$NO$_6$ 247.25	554-35-8 145	5375	5-17-07-00397	ace 4
10105	Propanenitrile, 2-hydroxy- Acetaldehyde, cyanohydrin	C$_3$H$_5$NO 71.08	78-97-7 -40	183	3-03-00-00451 0.9877[20]	H$_2$O 5; EtOH 5; eth 3; chl 3 1.4058[18]
10106	Propanenitrile, 3-hydroxy- Hydracrylonitrile	C$_3$H$_5$NO 71.08	109-78-4 -46	3751 221	4-03-00-00708 1.0404[25]	H$_2$O 5; EtOH 5; eth 2; chl 3 1.4248[20]
10107	Propanenitrile, 3-[2-(2-hydroxyethoxy)ethoxy]-	C$_7$H$_{13}$NO$_3$ 159.19	10143-54-1	186[9]	4-03-00-00713 1.089[32]	1.4452[22]
10108	Propanenitrile, 2-hydroxy-2-methyl- Acetone cyanohydrin	C$_4$H$_7$NO 85.11	75-86-5 -19	59 82[23]	4-03-00-00785 0.932[19]	H$_2$O 4; EtOH 4; eth 4; ace 3 1.3992[20]
10109	Propanenitrile, 2-methoxy- 2-Methoxypropionitrile	C$_4$H$_7$NO 85.11	33695-59-9	119	4-03-00-00675 0.8928[20]	EtOH 4 1.3818[20]
10110	Propanenitrile, 3-methoxy-	C$_4$H$_7$NO 85.11	110-67-8	163	4-03-00-00708 0.9379[20]	EtOH 3; eth 3; chl 3 1.4043[20]
10111	Propanenitrile, 2-methyl- Isobutyronitrile	C$_4$H$_7$N 69.11	78-82-0 -71.5	103.9	4-02-00-00853 0.7704[20]	H$_2$O 2; EtOH 4; eth 4; ace 3 1.3720[20]
10112	Propanenitrile, 3-(methylamino)-	C$_4$H$_8$N$_2$ 84.12	693-05-0	101-4[49]	3-04-00-01264 0.8992[20]	H$_2$O 3; ace 3; bz 3; chl 3 1.4320[20]
10113	Propanenitrile, 2-oxo-	C$_3$H$_3$NO 69.06	631-57-2	92.3	4-03-00-01515 0.9745[20]	eth 3; ace 3; CH$_3$CN 3 1.3764[20]
10114	Propanenitrile, 3,3'-oxybis-	C$_6$H$_8$N$_2$O 124.14	1656-48-0	110-12[0.5]	4-03-00-00720 1.043[25]	1.4405[20]
10115	Propanenitrile, 3,3,3-trichloro-2-hydroxy- Chlorocyanohydrin	C$_3$H$_2$Cl$_3$NO 174.41	513-96-2 61	2134 217 dec	4-03-00-00680	H$_2$O 4; eth 4; EtOH 4
10116	Propane, 1-nitro- 1-Nitropropane	C$_3$H$_7$NO$_2$ 89.09	108-03-2 -108	6548 131.1	4-01-00-00229 0.9961[25]	H$_2$O 2; EtOH 5; eth 5; chl 3 1.4018[20]

No.	Name / Synonym	Mol. Form. / Mol. Wt.	CAS RN mp/°C	Merck No. bp/°C	Beil. Ref. den/g cm⁻³	Solubility n_D
10117	Propane, 2-nitro- 2-Nitropropane	$C_3H_7NO_2$ 89.09	79-46-9 -91.3	6549 120.2	4-01-00-00230 0.9821^{25}	H_2O 2; chl 3 1.3944^{20}
10118	Propane, octafluoro- Perfluoropropane	C_3F_8 188.02	76-19-7 -147.6	-36.6	4-01-00-00189	
10119	Propane, 1,1'-oxybis- Dipropyl ether	$C_6H_{14}O$ 102.18	111-43-3 -126.1	7870 90.0	4-01-00-01422 0.7466^{20}	eth 4; EtOH 4 1.3809^{20}
10120	Propane, 2,2'-oxybis- Diisopropyl ether	$C_6H_{14}O$ 102.18	108-20-3 -86.8	5100 68.5	4-01-00-01471 0.7241^{20}	H_2O 2; EtOH 5; eth 5; ace 3 1.3679^{20}
10121	Propane, 1,1'-oxybis[2-chloro-	$C_6H_{12}Cl_2O$ 171.07	54460-96-7	188	2-01-00-00370 1.109^{20}	EtOH 3; eth 3 1.4467^{20}
10122	Propane, 1,1'-oxybis[3-chloro-	$C_6H_{12}Cl_2O$ 171.07	629-36-7	216; 90.5^{11}	4-01-00-01442 1.138^{20}	EtOH 3; eth 3 1.4158^{20}
10123	Propane, 2,2'-oxybis[1-chloro- 2,2'-Dichlorodiisopropyl ether	$C_6H_{12}Cl_2O$ 171.07	108-60-1	187	3-01-00-00470 1.103^{20}	H_2O 1; EtOH 5; eth 5; ace 5 1.4505^{20}
10124	Propane, 2,2'-oxybis[2-methyl- Di-tert-Butyl ether	$C_8H_{18}O$ 130.23	6163-66-2	3020 107.2	4-01-00-01616 0.7658^{20}	1.3949^{20}
10125	Propane, 1,1,1,2,3-pentachloro- 1,1,1,2,3-Pentachloropropane	$C_3H_3Cl_5$ 216.32	21700-31-2	179.5	4-01-00-00203	eth 4; EtOH 4 1.5130^{20}
10126	Propane, 1,1,2,3,3-pentachloro- 1,1,2,3,3-Pentachloropropane	$C_3H_3Cl_5$ 216.32	15104-61-7	199; 99^{20}	4-01-00-00204 1.6086^{39}	H_2O I; EtOH 4; eth 3; chl 3 1.5131^{17}
10127	Propane, 1,1,1,2,3-pentachloro-2-methyl- Isobutane, 1,1,1,2,3-pentachloro	$C_4H_5Cl_5$ 230.35	4749-31-9 73.5	215	4-01-00-00294 1.5686^{25}	chl 4 1.5165^{20}
10128	Propane, 1,1,1,2,2-pentafluoro 1,1,1,2,2-Pentafluoropropane	$C_3H_3F_5$ 134.05	1814-88-6	-17.4		
10129	Propane, 1,1'-sulfinylbis-	$C_6H_{14}OS$ 134.24	4253-91-2 22.5	80^2	4-01-00-01453 0.9654^{20}	eth 4; EtOH 4 1.4663^{20}
10130	2-Propanesulfonic acid Isopropanesulfonic acid	$C_3H_8O_3S$ 124.16	14159-48-9 -37	$159^{1.4}$	4-04-00-00042 1.187^{25}	H_2O 4 1.4332^{20}
10131	Propane, 1,1'-sulfonylbis-	$C_6H_{14}O_2S$ 150.24	598-03-8 29.5		4-01-00-01453 1.0278^{50}	H_2O 2; EtOH 3; eth 3 1.4456^{30}
10132	Propane, 1,1'-sulfonylbis[2-methyl-	$C_8H_{18}O_2S$ 178.30	10495-45-1 17	265	4-01-00-01607 1.0056^{18}	
10133	1-Propanesulfonyl chloride	$C_3H_7ClO_2S$ 142.61	10147-36-1	180 dec; 77^{12}	4-04-00-00039 1.2826^{15}	1.452^{20}
10134	1-Propanesulfonyl chloride, 2-methyl-	$C_4H_9ClO_2S$ 156.63	35432-36-1	190	4-04-00-00049 1.2014^{20}	1.4520^{25}
10135	Propane, 1,1,1,2-tetrabromo-	$C_3H_4Br_4$ 359.68	62127-49-5	122^{15}	3-01-00-00251 2.679^{20}	1.6187^{20}
10136	Propane, 1,1,2,2-tetrabromo- 1,1,2,2-Tetrabromopropane	$C_3H_4Br_4$ 359.68	34570-59-7	$105-7^9$	4-01-00-00222 2.687^{20}	1.6166^{20}
10137	Propane, 1,1,3,3-tetrabromo-	$C_3H_4Br_4$ 359.68	51525-97-4	$154-6^{19}$	2-01-00-00077 2.702^{21}	1.6225^{21}
10138	Propane, 1,2,2,3-tetrabromo- 1,2,2,3-Tetrabromopropane	$C_3H_4Br_4$ 359.68	54268-02-9 10.5	169^{80}; 115^9	3-01-00-00251 2.703^{20}	1.6200^{20}
10139	1,1,2,3-Propanetetracarboxylic acid, tetraethyl ester Tetraethyl 1,1,2,3-propanetetracarboxylate	$C_{15}H_{24}O_8$ 332.35	635-03-0	$203-4^{18}$	3-02-00-02077 1.1184^{20}	EtOH 4 1.4395^{20}
10140	1,1,3,3-Propanetetracarboxylic acid, tetraethyl ester 1,1,3,3-Tetracarbethoxypropane	$C_{15}H_{24}O_8$ 332.35	2121-66-6 -30	305 dec; 195^8	4-02-00-02417 1.116^{20}	EtOH 3 1.4398^{20}
10141	Propane, 1,1,1,2-tetrachloro- 1,1,1,2-Tetrachloropropane	$C_3H_4Cl_4$ 181.88	812-03-3 -64	152.5	4-01-00-00201 1.473^{20}	H_2O 1; EtOH 4; eth 3; chl 3 1.4887^{20}
10142	Propane, 1,1,1,3-tetrachloro- 1,1,1,3-Tetrachloropropane	$C_3H_4Cl_4$ 181.88	1070-78-6	157	4-01-00-00201 1.4509^{20}	H_2O 1; EtOH 4; eth 4; bz 4 1.4825^{20}
10143	Propane, 1,1,2,2-tetrachloro- 1,1,2,2-Tetrachloropropane	$C_3H_4Cl_4$ 181.88	13116-60-4	153	4-01-00-00201 1.47^{12}	H_2O 1; EtOH 5; eth 3; chl 3 1.4850^{25}
10144	Propane, 1,1,2,3-tetrachloro- 1,1,2,3-Tetrachloropropane	$C_3H_4Cl_4$ 181.88	18495-30-2	179.5	4-01-00-00202 1.513^{17}	H_2O 1; EtOH 3; eth 4; chl 3 1.5037^{17}
10145	Propane, 1,2,2,3-tetrachloro- 1,2,2,3-Tetrachloropropane	$C_3H_4Cl_4$ 181.88	13116-53-5	165	4-01-00-00202 1.500^{18}	H_2O 1; EtOH 4; eth 4; chl 3 1.4940^{18}
10146	Propane, 1,1,2,3-tetrachloro-2-(chloromethyl)- 1,1,2,3-Tetrachloro-2-(chloromethyl)propane	$C_4H_5Cl_5$ 230.35	66997-45-3	227; 95^9	3-01-00-00321 1.5686^{25}	bz 4; EtOH 4; chl 4 1.5165^{25}
10147	Propane, 1,1,1,2-tetrachloro-2-methyl- Isobutane, 1,1,1,2-tetrachloro	$C_4H_6Cl_4$ 195.90	7086-07-9 178.5	192^{117}	4-01-00-00293	eth 4; EtOH 4; chl 4
10148	Propane, 1,1,2,3-tetrachloro-2-methyl-	$C_4H_6Cl_4$ 195.90	18963-01-4 -46	190.5	4-01-00-00294 1.4393^{25}	bz 4; eth 4; chl 4 1.4963^{20}
10149	Propane, 1,2,2,3-tetrachloro-1,1,3,3-tetrafluoro- Propane, 1,2,2,3-tetrachloro-tetrafluoro	$C_3Cl_4F_4$ 253.84	677-68-9 -42.9	112	4-01-00-00203 1.7199^{20}	1.3958^{20}
10150	Propane, 1,1,3,3-tetramethoxy-	$C_7H_{16}O_4$ 164.20	102-52-3	183; 66^{12}	4-01-00-03635 0.997^{25}	1.4081^{20}
10151	Propane, 1,1'-thiobis- Dipropyl sulfide	$C_6H_{14}S$ 118.24	111-47-7 -102.5	7881 142.9	4-01-00-01452 0.814^{17}	H_2O 1; EtOH 3; eth 3 1.4487^{20}
10152	Propane, 2,2'-thiobis- Diisopropyl sulfide	$C_6H_{14}S$ 118.24	625-80-9 -78.1	120.1	4-01-00-01502 0.8142^{20}	H_2O 1; EtOH 3; eth 3 1.4438^{20}
10153	Propane, 1,1'-thiobis[3-chloro- Bis(3-chloropropyl) sulfide	$C_6H_{12}Cl_2S$ 187.13	55882-21-8	162^{43}	3-01-00-01438 1.1774^{25}	eth 4; EtOH 4; tol 4 1.5075^{20}
10154	Propane, 1,1'-thiobis[2-methyl- Diisobutyl sulfide	$C_8H_{18}S$ 146.30	592-65-4 -105.5	5035 171	4-01-00-01607 0.8363^{10}	

No.	Name Synonym	Mol. Form. Mol. Wt.	CAS RN mp/°C	Merck No. bp/°C	Beil. Ref. den/g cm^{-3}	Solubility n_D
10155	Propane, 2,2'-thiobis[2-methyl- Di-*tert*-butyl sulfide	$C_8H_{18}S$ 146.30	107-47-1 -9.0	149.1	4-01-00-01637 0.815[25]	1.4506[20]
10156	1-Propanethiol Propyl mercaptan	C_3H_8S 76.16	107-03-9 -113.3	67.8	4-01-00-01449 0.8411[20]	H_2O 2; EtOH 3; eth 3; ace 3 1.4380[20]
10157	2-Propanethiol Isopropyl mercaptan	C_3H_8S 76.16	75-33-2 -130.5	52.6	4-01-00-01498 0.8143[20]	H_2O 2; EtOH 5; eth 5; ace 4 1.4255[20]
10158	1-Propanethiol, 2-methyl- Isobutyl mercaptan	$C_4H_{10}S$ 90.19	513-44-0 <-70	5030 88.5	4-01-00-01605 0.8357[20]	H_2O 2; EtOH 4; eth 4; ace 4 1.4387[20]
10159	2-Propanethiol, 2-methyl- *tert*-Butyl mercaptan	$C_4H_{10}S$ 90.19	75-66-1 -0.5	1577 64.3	4-01-00-01634 0.7943[25]	H_2O 1; ctc 3; hp 3 1.4232[20]
10160	Propane, 1,1,1-tribromo-	$C_3H_5Br_3$ 280.78	62127-61-1	77-83[25]	4-01-00-00221 2.35[25]	1.5606[25]
10161	Propane, 1,1,2-tribromo-	$C_3H_5Br_3$ 280.78	14602-62-1	200.5	3-01-00-00251 2.3547[20]	H_2O 1; EtOH 3; eth 4; chl 3 1.5790[20]
10162	Propane, 1,2,2-tribromo-	$C_3H_5Br_3$ 280.78	14476-30-3	190.5	2-01-00-00077 2.2984[20]	eth 4; EtOH 4; chl 4 1.5670[20]
10163	Propane, 1,2,3-tribromo- 1,2,3-Tribromopropane	$C_3H_5Br_3$ 280.78	96-11-7 16.9	9527 222.1	4-01-00-00221 2.4208[20]	H_2O 1; EtOH 4; eth 4; ctc 2 1.5862[20]
10164	Propane, 1,2,3-tribromo-2-methyl-	$C_4H_7Br_3$ 294.81	631-28-7	178[235]; 108[18]	4-01-00-00299 2.2105[14]	H_2O 1; eth 3; chl 3 1.5701[21]
10165	1,2,3-Propanetricarboxylic acid Tricarballylic acid	$C_8H_8O_6$ 176.13	99-14-9 166	9534	4-02-00-02366	H_2O 4; EtOH 4; eth 2
10166	1,2,3-Propanetricarboxylic acid, 2-(acetyloxy)-, triethyl ester	$C_{14}H_{22}O_8$ 318.32	77-89-4	214[40]	3-03-00-01105 1.135[25]	1.4380
10167	1,2,3-Propanetricarboxylic acid, 2-hydroxy- Citric acid	$C_8H_8O_7$ 192.13	77-92-9 153	2328 dec	4-03-00-01272 1.665[20]	H_2O 4; EtOH 4; eth 3; bz 1
10168	1,2,3-Propanetricarboxylic acid, 2-hydroxy-, tributyl ester Butyl citrate	$C_{18}H_{32}O_7$ 360.45	77-94-1 -20	1564 233[22]	4-03-00-01276 1.043[20]	 1.4460[20]
10169	1,2,3-Propanetricarboxylic acid, 2-hydroxy-, triethyl ester	$C_{12}H_{20}O_7$ 276.29	77-93-0	 294	4-03-00-01276 1.1369[20]	H_2O 1; EtOH 3; eth 3; ctc 2 1.4455[20]
10170	1,2,3-Propanetricarboxylic acid, 2-hydroxy-, trimethyl ester Trimethyl citrate	$C_9H_{14}O_7$ 234.21	1587-20-8 79.3	 285; 176[16]	4-03-00-01276 	eth 4; EtOH 4
10171	Propane, 1,1,1-trichloro- 1,1,1-Trichloropropane	$C_3H_5Cl_3$ 147.43	7789-89-1	108	4-01-00-00198 1.287[23]	H_2O 1; EtOH 3; eth 4; chl 4
10172	Propane, 1,1,2-trichloro- 1,1,2-Trichloropropane	$C_3H_5Cl_3$ 147.43	598-77-6	132.0; 117[500]	4-01-00-00199 1.372[15]	H_2O 1; EtOH 3; eth 4; ctc 2
10173	Propane, 1,1,3-trichloro- 1,1,3-Trichloropropane	$C_3H_5Cl_3$ 147.43	20395-25-9 -59	145.5	4-01-00-00199 1.3557[20]	eth 4; EtOH 4; chl 4 1.4718[20]
10174	Propane, 1,2,2-trichloro- 1,2,2-Trichloropropane	$C_3H_5Cl_3$ 147.43	3175-23-3	124	4-01-00-00199 1.318[25]	H_2O 1; EtOH 3; eth 3; chl 4 1.4609[20]
10175	Propane, 1,2,3-trichloro- 1,2,3-Trichloropropane	$C_3H_5Cl_3$ 147.43	96-18-4 -14.7	157	4-01-00-00199 1.3889[20]	H_2O 2; EtOH 3; eth 3; ctc 2 1.4852[20]
10176	Propane, 1,2,3-trichloro-2-(chloromethyl)-	$C_4H_6Cl_4$ 195.90	18963-00-3	211; 87[9]	3-01-00-00321 1.5036[25]	chl 4 1.508[20]
10177	Propane, 1,1,2-trichloro-2-methyl- Isobutane, 1,1,2-trichloro	$C_4H_7Cl_3$ 161.46	29559-52-2 -6	163	4-01-00-00293 1.2588[20]	eth 4; chl 4 1.4666[20]
10178	Propane, 1,2,3-trichloro-2-methyl-	$C_4H_7Cl_3$ 161.46	1871-58-5	162.5	4-01-00-00293 1.3012[25]	chl 4 1.4765[20]
10179	Propane, 1,1,1-trichloro-2,2,3,3,3-pentafluoro-	$C_3Cl_3F_5$ 237.38	4259-43-2 -80	74	4-01-00-00200 1.643[25]	
10180	Propane, 1,2,2-trichloro-1,1,3,3,3-pentafluoro-	$C_3Cl_3F_5$ 237.38	1599-41-3 -4.3	72.0	3-01-00-00233 1.6681[20]	1.3519[20]
10181	Propane, 1,2,3-trichloro-1,1,2,3,3-pentafluoro-	$C_3Cl_3F_5$ 237.38	76-17-5 -72	73.7	4-01-00-00200 1.6631[20]	1.3512[20]
10182	Propane, 1,1,1-triethoxy-	$C_9H_{20}O_3$ 176.26	115-80-0	171	4-02-00-00707 	eth 4; EtOH 4 1.4000[25]
10183	Propane, 1,1,3-triethoxy-	$C_9H_{20}O_3$ 176.26	7789-92-6	185 dec; 78[14]	4-01-00-03971 0.898[19]	H_2O 2; EtOH 3; ctc 3 1.4067[20]
10184	Propane, 1,1,1-trifluoro-3-nitro- Propane, 1-nitro-3,3,3-trifluoro	$C_3H_4F_3NO_2$ 143.07	461-35-8	135	4-01-00-00232 1.4178[20]	eth 4 1.3549[20]
10185	Propane, 1,2,3-trimethoxy- Glycerol trimethyl ether	$C_6H_{14}O_3$ 134.18	20637-49-4	148	4-01-00-02755 0.9460[15]	H_2O 5; eth 3; ace 4; bz 3 1.4055[15]
10186	1,2,3-Propanetriol Glycerol	$C_3H_8O_3$ 92.09	56-81-5 18.2	4379 290	4-01-00-02751 1.2613[20]	H_2O 5; EtOH 5; eth 2; bz 1 1.4746[20]
10187	1,2,3-Propanetriol, 1-acetate, (±)- Glycerol, 1-monoacetate(*DL*)	$C_5H_{10}O_4$ 134.13	93713-40-7	158[165] 130[3]	4-02-00-00251 1.2060[20]	H_2O 3; EtOH 3; eth 2; bz 1 1.4157[20]
10188	1,2,3-Propanetriol, 1-(4-aminobenzoate) Glyceryl *p*-aminobenzoate	$C_{10}H_{13}NO_4$ 211.22	136-44-7	4382		H_2O 1; EtOH 3
10189	1,2,3-Propanetriol, diacetate Diacetin	$C_7H_{12}O_5$ 176.17	25395-31-7	2941 259	4-02-00-00252 1.184[16]	H_2O 4; eth 4; EtOH 4 1.44[20]
10190	1,2,3-Propanetriol, 1,3-diacetate 1,3-Diacetin	$C_7H_{12}O_5$ 176.17	105-70-4	2941 173[40]	4-02-00-00252 1.179[15]	H_2O 4; EtOH 4; eth 2; CS_2 1 1.4395[20]

No.	Name / Synonym	Mol. Form. / Mol. Wt.	CAS RN / mp/°C	Merck No. / bp/°C	Beil. Ref. / den/g cm⁻³	Solubility / n_D
10191	1,2,3-Propanetriol, 1,3-dinitrate	$C_3H_6N_2O_7$ 182.09	623-87-0 26	148^{15}; $116^{0.6}$	2-01-00-00591 1.523^{20}	H_2O 4; eth 4; EtOH 4 1.4715^{20}
10192	1,2,3-Propanetriol, 1-nitrate	$C_3H_7NO_5$ 137.09	624-43-1 61	157.5	4-01-00-02761 1.4164^{20}	H_2O 4; EtOH 4; eth 2 1.4698^{20}
10193	1,2,3-Propanetriol, 2-nitrate	$C_3H_7NO_5$ 137.09	620-12-2 54	157.5	2-01-00-00591 1.40^{22}	H_2O 4; eth 4; EtOH 4
10194	1,2,3-Propanetriol, 1-phenyl- Glycerol, 1-phenyl	$C_9H_{12}O_3$ 168.19	63157-81-3 100.5	185	4-06-00-07407 1.2213^{13}	1.5606^{15}
10195	1,2,3-Propanetriol, triacetate Triacetin	$C_9H_{14}O_6$ 218.21	102-76-1 -78	9504 259	4-02-00-00253 1.1583^{20}	H_2O 2; EtOH 5; eth 5; ace 4 1.4301^{20}
10196	1,2,3-Propanetriol, tribenzoate	$C_{24}H_{20}O_6$ 404.42	614-33-5 76		3-09-00-00666 1.228^{12}	H_2O 1; EtOH 3; eth 4; ace 4
10197	1,2,3-Propanetriol, trinitrate Nitroglycerin	$C_3H_5N_3O_9$ 227.09	55-63-0 13.5	6528 $93^{0.31}$; exp 218	4-01-00-02762 1.5931^{20}	H_2O 2; EtOH 3; eth 5; ace 4 1.4786^{12}
10198	1,2,3-Propanetriol, tripropanoate	$C_{12}H_{20}O_6$ 260.29	139-45-7	175-6^{20}	4-02-00-00717 1.099^{20}	H_2O 1; EtOH 3; eth 4; chl 3 1.4318^{19}
10199	Propanetrione, diphenyl-	$C_{15}H_{10}O_3$ 238.24	643-75-4 70	289^{175}; 179^4	4-07-00-02822	eth 4
10200	Propanil Propanamide, N-(3,4-dichlorophenyl)-	$C_9H_9Cl_2NO$ 218.08	709-98-8 92	7814	1.25^{25}	
10201	Propanoic acid Propionic acid	$C_3H_6O_2$ 74.08	79-09-4 -20.7	7837 141.1	4-02-00-00695 0.9930^{20}	H_2O 5; EtOH 5; eth 3; chl 2 1.3809^{20}
10202	Propanoic acid, 2-(acetyloxy)-, (±)-	$C_5H_8O_4$ 132.12	3853-80-3	167-70^{78}	4-03-00-00638 1.1758^{20}	bz 4; EtOH 4 1.4240^{20}
10203	Propanoic acid, anhydride Propionic anhydride	$C_6H_{10}O_3$ 130.14	123-62-6 -45	7838 170; 67.5^{18}	4-02-00-00722 1.0110^{20}	eth 5; ctc 2 1.4038^{20}
10204	Propanoic acid, 2-bromo-, (±)-	$C_3H_5BrO_2$ 152.98	10327-08-9 25.7	203.5	4-02-00-00695 1.700^{20}	H_2O 4; EtOH 4; eth 4; chl 2 1.4753^{20}
10205	Propanoic acid, 3-bromo- β-Bromopropionic acid	$C_3H_5BrO_2$ 152.98	590-92-1 62.5	1421 140-2^{45}	4-02-00-00764 1.48^{25}	H_2O 3; EtOH 3; eth 3; bz 3
10206	Propanoic acid, 3-bromo-, butyl ester	$C_7H_{13}BrO_2$ 209.08	6973-79-1	130^{26}	4-02-00-00765 1.4549^{20}	ace 4; eth 4; EtOH 4 1.3051^{20}
10207	Propanoic acid, 2-bromo-, ethyl ester Ethyl α-bromopropionate	$C_5H_9BrO_2$ 181.03	535-11-5	3731 160 dec; 71^{26}	4-02-00-00762 1.4135^{20}	H_2O 1; EtOH 5; eth 5; chl 3 1.4490^{20}
10208	Propanoic acid, 3-bromo-, ethyl ester	$C_5H_9BrO_2$ 181.03	539-74-2	179; 65^{15}	4-02-00-00765 1.4123^{18}	EtOH 3; eth 3; ace 3; ctc 2 1.4516^{20}
10209	Propanoic acid, 2-bromo-2-methyl- α-Bromoisobutyric acid	$C_4H_7BrO_2$ 167.00	2052-01-9 48.5	1408 199; 115^{24}	4-02-00-00862 1.4969^{60}	
10210	Propanoic acid, 3-bromo-, 3-methylbutyl ester Isopentyl 3-bromopropionate	$C_8H_{15}BrO_2$ 223.11	100983-12-8	110-1^{11}	2-02-00-00231 1.2217^{15}	eth 4; EtOH 4 1.4556^9
10211	Propanoic acid, 2-bromo-, methyl ester, (±)-	$C_4H_7BrO_2$ 167.00	57885-43-5	144	4-02-00-00762 1.4966^{25}	eth 4; EtOH 4 1.4451^{22}
10212	Propanoic acid, 3-bromo-, methyl ester	$C_4H_7BrO_2$ 167.00	3395-91-3	105.5^{60}	4-02-00-00765 1.4897^{15}	EtOH 3; eth 3; ace 3; os 4 1.4542^{20}
10213	Propanoic acid, 2-bromo-2-methyl-, ethyl ester	$C_6H_{11}BrO_2$ 195.06	600-00-0	163	4-02-00-00862 1.3263^{20}	H_2O 1; EtOH 3; eth 5 1.4446^{20}
10214	Propanoic acid, butyl ester Butyl propionate	$C_7H_{14}O_2$ 130.19	590-01-2 -89	1587 146.8	4-02-00-00708 0.8754^{20}	H_2O 2; EtOH 5; eth 5; ctc 2 1.4014^{20}
10215	Propanoic acid, 2-chloro-	$C_3H_5ClO_2$ 108.52	598-78-7	185	4-02-00-00745 1.2585^{20}	H_2O 5; EtOH 5; eth 5; ace 3 1.4380^{20}
10216	Propanoic acid, 3-chloro- β-Chloropropionic acid	$C_3H_5ClO_2$ 108.52	107-94-8 41	2159 204 dec	4-02-00-00748	H_2O 3; EtOH 3; eth 5; chl 3
10217	Propanoic acid, 2-chloro-, butyl ester	$C_7H_{13}ClO_2$ 164.63	54819-86-2	184	4-02-00-00746 1.0253^{20}	eth 4 1.4263^{20}
10218	Propanoic acid, 3-chloro-, butyl ester	$C_7H_{13}ClO_2$ 164.63	27387-79-7	104^{22}	4-02-00-00749 1.0370^{20}	H_2O 4; eth 4 1.4321^{20}
10219	Propanoic acid, 3-chloro-2,2-dimethyl-	$C_5H_9ClO_2$ 136.58	13511-38-1 41.5	110^{10}	4-02-00-00914	ctc 4
10220	Propanoic acid, 2-chloro-, ethyl ester Ethyl α-chloropropionate	$C_5H_9ClO_2$ 136.58	535-13-7	3743 147	4-02-00-00746 1.0793^{20}	H_2O 1; EtOH 5; eth 5; ctc 2 1.4178^{20}
10221	Propanoic acid, 3-chloro-, ethyl ester	$C_5H_9ClO_2$ 136.58	623-71-2	162	4-02-00-00749 1.1086^{20}	H_2O 2; EtOH 5; eth 5; os 3 1.4254^{20}
10222	Propanoic acid, 2-chloro-, methyl ester Methyl 2-chloropropionate	$C_4H_7ClO_2$ 122.55	17639-93-9	132.5	4-02-00-00746 1.0750^{25}	
10223	Propanoic acid, 2-chloro-, 1-methylethyl ester Isopropyl 2-chloropropionate	$C_6H_{11}ClO_2$ 150.60	40058-87-5	151.5	4-02-00-00746 1.0315^{20}	H_2O 1; EtOH 3; eth 3 1.4149^{20}
10224	Propanoic acid, 2-chloro-, 2-methylpropyl ester, (R)- Propionic acid, 2-chloro, isobutyl ester (d)	$C_7H_{13}ClO_2$ 164.63	114489-96-2	176	0-02-00-00248 1.0312^{20}	1.4247^{20}
10225	Propanoic acid, 3-chloro-, methyl ester	$C_4H_7ClO_2$ 122.55	6001-87-2	156	4-02-00-00748 1.1861^{15}	EtOH 3; ctc 3 1.4263^{20}
10226	Propanoic acid, 3-chloro-, 3-methylbutyl ester	$C_8H_{15}ClO_2$ 178.66	62108-70-7	208; 87^{12}	4-02-00-00749 1.0171^{20}	eth 4; EtOH 4 1.4343^{20}
10227	Propanoic acid, 3-chloro-, 2-methylpropyl ester	$C_7H_{13}ClO_2$ 164.63	62108-68-3	191.3	4-02-00-00749 1.0323^{20}	eth 4; EtOH 4 1.4295^{20}

No.	Name / Synonym	Mol. Form. / Mol. Wt.	CAS RN / mp/°C	Merck No. / bp/°C	Beil. Ref. / den/g cm^{-3}	Solubility / n_D
10228	Propanoic acid, 2-chloro-2-methyl-	$C_4H_7ClO_2$ 122.55	594-58-1 31	118[50]; 93[20]	4-02-00-00858 1.1790[20]	EtOH 4 1.450[20]
10229	Propanoic acid, 2-chloro-2-methyl-, methyl ester	$C_5H_9ClO_2$ 136.58	22421-97-2	135	4-02-00-00859 1.0883[15]	eth 4 1.4122[21]
10230	Propanoic acid, 3-chloro-2-methyl-	$C_4H_7ClO_2$ 122.55	16674-04-7	128-33[50]	4-02-00-00860 1.0153[20]	eth 4; EtOH 4 1.4310[20]
10231	Propanoic acid, 2-(4-chlorophenoxy)-2-methyl-, ethyl ester Clofibrate	$C_{12}H_{15}ClO_3$ 242.70	637-07-0	2374 148-50[20]	4-06-00-00851	
10232	Propanoic acid, 3-chloro-, propyl ester	$C_6H_{11}ClO_2$ 150.60	62108-66-1	180	4-02-00-00749 1.0656[20]	eth 4; EtOH 4 1.4290[20]
10233	Propanoic acid, 2-cyano-	$C_4H_5NO_2$ 99.09	632-07-5 35	143[11]	3-02-00-01683 1.1400[20]	
10234	Propanoic acid, 3-cyano-, methyl ester	$C_5H_7NO_2$ 113.12	4107-62-4	211	3-02-00-01672	chl 3 1.4250[20]
10235	Propanoic acid, cyclohexyl ester	$C_9H_{16}O_2$ 156.22	6222-35-1	193; 93[35]	4-06-00-00037 0.9359[20]	H_2O 1; EtOH 3; eth 3; ace 3 1.4403[20]
10236	Propanoic acid, 2-diazo-, ethyl ester Ethyl 2-diazopropionate	$C_5H_8N_2O_2$ 128.13	6111-99-5	65-8[41]	2-03-00-00403 1.086[12]	1.4472[18]
10237	Propanoic acid, 2-diazo-, methyl ester Methyl 2-diazopropionate	$C_4H_6N_2O_2$ 114.10	34757-14-7	43-5[11]	2-03-00-00403 1.1011[13]	1.4487[20]
10238	Propanoic acid, 2,3-dibromo-	$C_3H_4Br_2O_2$ 231.87	600-05-5 66.5	160[20]; 138[12]	4-02-00-00767	bz 4; eth 4; EtOH 4
10239	Propanoic acid, 2,3-dibromo-, butyl ester Butyl 2,3-dibromopropionate	$C_7H_{12}Br_2O_2$ 287.98	21179-48-6	237.5	4-02-00-00768 1.6107[20]	1.4890[20]
10240	Propanoic acid, 2,3-dibromo-, ethyl ester	$C_5H_8Br_2O_2$ 259.93	3674-13-3	214.5	4-02-00-00767 1.7966[20]	EtOH 3; eth 3 1.5007[20]
10241	Propanoic acid, 2,3-dibromo-, methyl ester	$C_4H_6Br_2O_2$ 245.90	1729-67-5	206	4-02-00-00767 1.9333[20]	EtOH 3 1.5127[20]
10242	Propanoic acid, 2,3-dibromo-2-methyl-, methyl ester	$C_5H_8Br_2O_2$ 259.93	3673-79-8	203	4-02-00-00864 1.805[20]	1.505[20]
10243	Propanoic acid, 2,3-dibromo-, propyl ester Propyl 2,3-dibromopropionate	$C_6H_{10}Br_2O_2$ 273.95	79762-76-8	222.5	3-02-00-00572 1.6799[20]	1.4950[20]
10244	Propanoic acid, 2,2-dichloro- Dalapon	$C_3H_4Cl_2O_2$ 142.97	75-99-0	2806 187.5; 92[14]	4-02-00-00753 1.389[12]	H_2O 4; EtOH 4; eth 3; ctc 3
10245	Propanoic acid, 2,3-dichloro-	$C_3H_4Cl_2O_2$ 142.97	565-64-0 50	210; 130[26]	4-02-00-00756	H_2O 3; EtOH 3; eth 3; peth 2 1.4650[20]
10246	Propanoic acid, 2,3-dichloro-, ethyl ester	$C_5H_8Cl_2O_2$ 171.02	6628-21-3	183.5	4-02-00-00756 1.2401[20]	eth 4; EtOH 4 1.4482[20]
10247	Propanoic acid, 3,3-dichloro-2-hydroxy-2-methyl-	$C_4H_6Cl_2O_3$ 173.00	31340-44-0 82.5	dec	2-03-00-00224	H_2O 3; EtOH 3; eth 3
10248	Propanoic acid, 2,3-dichloro-, methyl ester	$C_4H_6Cl_2O_2$ 157.00	3674-09-7	92[50]	4-02-00-00756 1.3282[20]	ace 4; eth 4; EtOH 4
10249	Propanoic acid, 2,3-dichloro-, 1-methylethyl ester	$C_6H_{10}Cl_2O_2$ 185.05	54774-99-1	61-2[5]	4-02-00-00756 1.2010[25]	1.4470
10250	Propanoic acid, 2-(2,4-dichlorophenoxy)- Dichlorprop	$C_9H_8Cl_2O_3$ 235.07	120-36-5 117.5	3068	3-06-00-00707	H_2O 2; EtOH 3; eth 3; lig 2
10251	Propanoic acid, 2,2-dichloro-3,3,3-trifluoro-, methyl ester Propionic acid, 2,2-dichloro-3,3,3-trifluoro-, methyl ester	$C_4H_3Cl_2F_3O_2$ 210.97	378-68-7	116-7[625]	4-02-00-00758 1.5092[20]	1.3806[20]
10252	Propanoic acid, 2,3-dihydroxy- Glyceric acid	$C_3H_6O_4$ 106.08	473-81-4	4378 dec	4-03-00-01050	H_2O 4; ace 4; EtOH 4
10253	Propanoic acid, 2,3-dihydroxy-, ethyl ester, (±)- Glyceric acid, ethyl ester (DL)	$C_5H_{10}O_4$ 134.13	89300-69-6	235; 120[14]	4-03-00-01052 1.1897[15]	H_2O 4; eth 4; EtOH 4
10254	Propanoic acid, 2,3-dihydroxy-, methyl ester, (±)- Methyl DL-glycerate	$C_4H_8O_4$ 120.11	15909-76-9	241.5; 119[14]	4-03-00-01052 1.2802[15]	H_2O 4; EtOH 4 1.4502[20]
10255	Propanoic acid, 2,3-dihydroxy-2-methyl-, ethyl ester Glyceric acid, 2-methyl, ethyl ester	$C_6H_{12}O_4$ 148.16	67535-07-3	95[10]	3-03-00-00860 1.114[25]	1.4370[25]
10256	Propanoic acid, 2,3-dihydroxy-2-methyl-, methyl ester Glyceric acid, 2-methyl-, methyl ester	$C_5H_{10}O_4$ 134.13	19860-56-1	89[10]	4-03-00-01060 1.185[25]	1.4438[25]
10257	Propanoic acid, 2,3-dihydroxy-, propyl ester	$C_6H_{12}O_4$ 148.16	116435-95-1	132-4[3]	3-03-00-00854 1.1537[20]	1.4503[20]
10258	Propanoic acid, 2,2-dimethyl- Pivalic acid	$C_5H_{10}O_2$ 102.13	75-98-9 35	7482 164	4-02-00-00908 0.905[50]	H_2O 2; EtOH 4; eth 4 1.3931[30]
10259	Propanoic acid, 2,2-dimethyl-, ethyl ester Ethyl 2,2-dimethylpropionate	$C_7H_{14}O_2$ 130.19	3938-95-2 -89.5	118.4	4-02-00-00910 0.856[20]	EtOH 3; eth 3 1.3906[20]
10260	Propanoic acid, 2,2-dimethyl-, methyl ester Methyl 2,2-dimethylpropionate	$C_6H_{12}O_2$ 116.16	598-98-1	101.1	4-02-00-00909 0.891[0]	eth 4; EtOH 4 1.3905[20]
10261	Propanoic acid, ethenyl ester Vinyl propionate	$C_5H_8O_2$ 100.12	105-38-4	91.2	4-02-00-00711	

No.	Name / Synonym	Mol. Form. / Mol. Wt.	CAS RN mp/°C	Merck No. bp/°C	Beil. Ref. den/g cm⁻³	Solubility n_D
10262	Propanoic acid, 3-ethoxy-, ethyl ester	$C_7H_{14}O_3$ 146.19	763-69-9	$48-9^5$	4-03-00-00697 0.92^{25}	
10263	Propanoic acid, ethyl ester Ethyl propionate	$C_5H_{10}O_2$ 102.13	105-37-3 -73.9	3801 99.1	4-02-00-00705 0.8917^{20}	H_2O 2; EtOH 5; eth 5; ace 3 1.3839^{20}
10264	Propanoic acid, heptyl ester	$C_{10}H_{20}O_2$ 172.27	2216-81-1 -50.9	210	4-02-00-00710 0.8679^{20}	H_2O 1; EtOH 3; eth 3; ace 3 1.4201^{15}
10265	Propanoic acid, hexyl ester	$C_9H_{18}O_2$ 158.24	2445-76-3 -57.5	190	4-02-00-00709 0.8698^{20}	H_2O 1; EtOH 3; eth 3; ace 3 1.4162^{15}
10266	Propanoic acid, 2-hydroxy-, (±)- Lactic acid (DL)	$C_3H_6O_3$ 90.08	598-82-3 18	5215 122^{15}	2-03-00-00192 1.2060^{21}	H_2O 4; eth 2; EtOH 4 1.4392^{20}
10267	Propanoic acid, 3-hydroxy- Hydracrylic acid	$C_3H_6O_3$ 90.08	503-66-2	4681 dec	4-03-00-00689	H_2O 4; EtOH 3; eth 5 1.4489^{20}
10268	Propanoic acid, 2-hydroxy-, butyl ester, (R)- Lactic acid, butyl ester (d)	$C_7H_{14}O_3$ 146.19	34451-18-8	77^{10}	3-03-00-00450 0.9744^{27}	eth 4; EtOH 4
10269	Propanoic acid, 2-hydroxy-, ethyl ester, (±)- Ethyl lactate	$C_5H_{10}O_3$ 118.13	2676-33-7 -26	3773 154.5	4-03-00-00640 1.0328^{20}	H_2O 4; eth 4; EtOH 4 1.4124^{20}
10270	Propanoic acid, 3-hydroxy-, ethyl ester	$C_5H_{10}O_3$ 118.13	623-72-3	187.5	2-03-00-00213 1.059^{20}	1.4271^{23}
10271	Propanoic acid, 2-hydroxy-, hexadecyl ester Cetyl lactate	$C_{19}H_{38}O_3$ 314.51	35274-05-6 41	2022 $219^{10}; 170^1$	4-03-00-00660	1.4410^{40}
10272	Propanoic acid, 3-hydroxy-2-(hydroxymethyl)-2-methyl- Dimethylolpropionic acid	$C_5H_{10}O_4$ 134.13	4767-03-7 190	3241	4-03-00-01066	
10273	Propanoic acid, 2-hydroxy-, methyl ester, (±)- Lactic acid, methyl ester (DL)	$C_4H_8O_3$ 104.11	2155-30-8	144.8	4-03-00-00640 1.0928^{20}	H_2O 4; eth 4; EtOH 4 1.4141^{20}
10274	Propanoic acid, 2-hydroxy-, 3-methylbutyl ester	$C_8H_{16}O_3$ 160.21	19329-89-6	202.4	3-03-00-00483 0.9589^{25}	eth 4; EtOH 4 1.4240^{25}
10275	Propanoic acid, 2-hydroxy-, 1-methylethyl ester Isopropyl lactate	$C_6H_{12}O_3$ 132.16	617-51-6	167	3-03-00-00479 0.9980^{20}	H_2O 4; bz 4; eth 4; EtOH 4 1.4082^{25}
10276	Propanoic acid, 3-hydroxy-, methyl ester	$C_4H_8O_3$ 104.11	6149-41-3	179	3-03-00-00525 1.105^{16}	H_2O 5; EtOH 5; eth 5 1.43^{23}
10277	Propanoic acid, 3-hydroxy-, 1-methylethyl ester Hydracrylic acid, isopropyl ester	$C_6H_{12}O_3$ 132.16	84098-45-3	$128.5^{82}; 95^{12}$	1-03-00-00113 1.058^{25}	1.4303^{23}
10278	Propanoic acid, 2-hydroxy-2-methyl-	$C_4H_8O_3$ 104.11	594-61-6 82.5	212	4-03-00-00782	H_2O 4; EtOH 4; eth 4; bz 2
10279	Propanoic acid, 2-hydroxy-2-methyl-, ethyl ester	$C_6H_{12}O_3$ 132.16	80-55-7	150	4-03-00-00783 0.987^{20}	H_2O 5; EtOH 5; ctc 3 1.4080^{20}
10280	Propanoic acid, 2-hydroxy-2-methyl-, methyl ester	$C_5H_{10}O_3$ 118.13	2110-78-3	137	4-03-00-00783	H_2O 4; EtOH 4 1.4056^{20}
10281	Propanoic acid, 2-hydroxy-, 2-propenyl ester	$C_6H_{10}O_3$ 130.14	5349-55-3	$56-60^8$	3-03-00-00486 1.0452^{20}	py 3 1.4369^{20}
10282	Propanoic acid, 2-iodo-, (±)-	$C_3H_5IO_2$ 199.98	18791-49-6 46	$93-6^{0.2}$	3-02-00-00573 2.073^{18}	H_2O 2; EtOH 4; eth 4
10283	Propanoic acid, 3-iodo-, ethyl ester	$C_5H_9IO_2$ 228.03	6414-69-3	202	4-02-00-00770 1.6226^{30}	1.4961^{25}
10284	Propanoic acid, 3-iodo-, methyl ester	$C_4H_7IO_2$ 214.00	5029-66-3	188	3-02-00-00574 1.8407^7	EtOH 4
10285	Propanoic acid, 2-mercapto- Thiolactic acid	$C_3H_8O_2S$ 106.15	79-42-5 12	9269 $106-7^{15}$	4-03-00-00682 1.1938^{20}	H_2O 4; eth 4; EtOH 4 1.4810^{20}
10286	Propanoic acid, 3-mercapto-	$C_3H_8O_2S$ 106.15	107-96-0 18	$111^{15}; 86^3$	4-03-00-00726 1.218^{21}	H_2O 3; EtOH 3; eth 3; ctc 3 1.494^{20}
10287	Propanoic acid, 3-mercapto-, methyl ester	$C_4H_8O_2S$ 120.17	2935-90-2	$54-5^{14}$	3-03-00-00553 1.085^{25}	1.4640^{20}
10288	Propanoic acid, 2-mercapto-2-methyl-	$C_4H_8O_2S$ 120.17	4695-31-2 47	101^{15}	4-03-00-00797	H_2O 4
10289	Propanoic acid, 3-methoxy-, methyl ester	$C_5H_{10}O_3$ 118.13	3852-09-3	142.8	4-03-00-00693 1.0139^{15}	1.4030^{20}
10290	Propanoic acid, 2-methyl- Isobutyric acid	$C_4H_8O_2$ 88.11	79-31-2 -46	5039 154.4	4-02-00-00843 0.9681^{20}	H_2O 4; EtOH 5; eth 5; ctc 2 1.3930^{20}
10291	Propanoic acid, 2-methyl-, anhydride Isobutyric anhydride	$C_8H_{14}O_3$ 158.20	97-72-3 -53.5	$183; 89^{32}$	4-02-00-00851 0.9535^{20}	eth 5; chl 3 1.4061^{19}
10292	Propanoic acid, 2-methyl-, 2-chloroethyl ester 2-Chloroethyl isobutyrate	$C_6H_{11}ClO_2$ 150.60	33662-96-3	170	4-02-00-00847 1.062^{25}	eth 4; EtOH 4 1.4109^{16}
10293	Propanoic acid, 2-methyl-, cyclohexyl ester Isobutyric acid, cyclohexyl ester	$C_{10}H_{18}O_2$ 170.25	1129-47-1	204	3-06-00-00024 0.9489^0	eth 4; EtOH 4
10294	Propanoic acid, 2-methyl-, 1,1-dimethylethyl ester	$C_8H_{16}O_2$ 144.21	16889-72-8	126.7	3-02-00-00648	ace 4; eth 4; EtOH 4 1.3921^{20}
10295	Propanoic acid, 2-methyl-, 3,7-dimethyl-2,6-octadienyl ester, (E)-	$C_{14}H_{24}O_2$ 224.34	2345-26-8	$135-7^{13}$	2-02-00-00281 0.8997^{15}	1.4576^{20}
10296	Propanoic acid, methyl ester Methyl propionate	$C_4H_8O_2$ 88.11	554-12-1 -87.5	6030 79.8	4-02-00-00704 0.9150^{20}	H_2O 2; EtOH 5; eth 5; ace 3 1.3775^{20}
10297	Propanoic acid, 2-methyl-, ethenyl ester	$C_6H_{10}O_2$ 114.14	2424-98-8	104.5	4-02-00-00848 0.8932^{20}	1.4061^{20}

No.	Name Synonym	Mol. Form. Mol. Wt.	CAS RN mp/°C	Merck No. bp/°C	Beil. Ref. den/g cm^{-3}	Solubility n_D
10298	Propanoic acid, 2-methyl-, ethyl ester Ethyl isobutanoate	$C_6H_{12}O_2$ 116.16	97-62-1 -88.2	3770 110.1	4-02-00-00846 0.868[20]	H_2O 2; EtOH 5; eth 5; ace 3 1.3869[18]
10299	Propanoic acid, 1-methylethyl ester Isopropyl propionate	$C_6H_{12}O_2$ 116.16	637-78-5	109.5	4-02-00-00708 0.8660[20]	H_2O 2; EtOH 5; eth 5 1.3872[20]
10300	Propanoic acid, 2-methyl-, 2-furanylmethyl ester	$C_9H_{12}O_3$ 168.19	6270-55-9 85-6[15]		2-17-00-00115 1.0313[20]	eth 4; EtOH 4
10301	Propanoic acid, 2-methyl-, methyl ester Methyl isobutyrate	$C_5H_{10}O_2$ 102.13	547-63-7 -84.7	6003 92.5	4-02-00-00846 0.8906[20]	H_2O 2; EtOH 5; eth 5; ace 3 1.3840[20]
10302	Propanoic acid, 2-methyl-, 3-methylbutyl ester	$C_9H_{18}O_2$ 158.24	2050-01-3	168.5	3-02-00-00649 0.8627[20]	H_2O 2; EtOH 3; eth 3; ace 3
10303	Propanoic acid, 2-methyl-, 1-methylethyl ester Isopropyl isobutyrate	$C_7H_{14}O_2$ 130.19	617-50-5	120.7	4-02-00-00847 0.8471[21]	H_2O 1; EtOH 3; eth 3; ace 3
10304	Propanoic acid, 2-methyl-, 2-methylpropyl ester Isobutyl isobutanoate	$C_8H_{16}O_2$ 144.21	97-85-8 -80.7	5028 148.6	4-02-00-00847 0.8542[20]	H_2O 2; EtOH 3; eth 5; ace 3 1.3999[20]
10305	Propanoic acid, 2-methyl-, 2-nitropentyl ester Isobutyric acid, (2-nitropentyl) ester	$C_9H_{17}NO_4$ 203.24	116530-63-3 249.5; 121[10]		3-02-00-00648 1.0310[20]	ace 4; eth 4 1.4315[20]
10306	Propanoic acid, 2-methyl-, 2-phenylethyl ester	$C_{12}H_{16}O_2$ 192.26	103-48-0 122-4[15]		4-06-00-03073 0.9950[15]	1.4871[20]
10307	Propanoic acid, 2-methyl-, phenylmethyl ester	$C_{11}H_{14}O_2$ 178.23	103-28-6 114-5[20]		4-06-00-02266 1.0075[15]	1.4883[20]
10308	Propanoic acid, 2-methyl-, propyl ester Propyl isobutyrate	$C_7H_{14}O_2$ 130.19	644-49-5	135.4	4-02-00-00847 0.8843[0]	H_2O 2; EtOH 4; ace 3 1.3955[20]
10309	Propanoic acid, 1-methylpropyl ester	$C_7H_{14}O_2$ 130.19	591-34-4	133	4-02-00-00709 0.8657[20]	EtOH 3; eth 3 1.3952[20]
10310	Propanoic acid, 2-methylpropyl ester Isobutyl propionate	$C_7H_{14}O_2$ 130.19	540-42-1 -71.4	5033 137	4-02-00-00709 0.888[0]	H_2O 2; EtOH 4; eth 4; ace 3 1.3973[20]
10311	Propanoic acid, 3-(methylthio)-, methyl ester	$C_5H_{10}O_2S$ 134.20	13532-18-8 74-5[13]		4-03-00-00737 1.077[25]	1.4650[20]
10312	Propanoic acid, 2-nitro-, ethyl ester	$C_5H_9NO_4$ 147.13	2531-80-8	190.5	4-02-00-00771 1.4210[20]	bz 4; eth 4; EtOH 4
10313	Propanoic acid, octyl ester	$C_{11}H_{22}O_2$ 186.29	142-60-9 -42.6	228	4-02-00-00710 0.8663[20]	H_2O 1; EtOH 3; eth 3; bz 3 1.4221[15]
10314	Propanoic acid, 2-oxo- Pyruvic acid	$C_3H_4O_3$ 88.06	127-17-3 13.8	8032 165 dec; 54[10]	4-03-00-01505 1.2272[20]	H_2O 5; EtOH 5; eth 5; ace 3 1.4280[20]
10315	Propanoic acid, 2-oxo-, ethyl ester	$C_5H_8O_3$ 116.12	617-35-6 -50	155	4-03-00-01513 1.059[15]	H_2O 2; EtOH 5; eth 5; ace 3 1.4052[20]
10316	Propanoic acid, 2-oxo-, methyl ester	$C_4H_6O_3$ 102.09	600-22-6	135.5	4-01-00-04146 1.154[0]	H_2O 2; EtOH 5; eth 5; ace 3 1.4046[25]
10317	Propanoic acid, pentafluoro-, methyl ester Pentafluoropropionic acid, methyl ester	$C_4H_3F_5O_2$ 178.06	378-75-6 59.5		4-02-00-00740 1.390[25]	1.2869[25]
10318	Propanoic acid, pentyl ester	$C_8H_{16}O_2$ 144.21	624-54-4 -73.1	168.6	4-02-00-00709 0.8761[25]	H_2O 1; EtOH 5; eth 5; bz 3 1.4096[15]
10319	Propanoic acid, 2-phenoxy-, (±)-	$C_9H_{10}O_3$ 166.18	1912-21-6 115.5	265.5	4-06-00-00642 1.1865[20]	eth 4; EtOH 4 1.5184[20]
10320	Propanoic acid, 3-phenoxy-	$C_9H_{10}O_3$ 166.18	7170-38-9 97.5	240; 188[26]	4-06-00-00643	H_2O 3; chl 2; lig 3
10321	Propanoic acid, 2-phenoxy-, ethyl ester	$C_{11}H_{14}O_3$ 194.23	42412-84-0	243.5	4-06-00-00643 1.360[17]	H_2O 1; EtOH 3; eth 3 1.360[17]
10322	Propanoic acid, 3-phenoxy-, ethyl ester	$C_{11}H_{14}O_3$ 194.23	22409-91-2 24	142[11]	4-06-00-00644 1.0745[20]	1.5007[18]
10323	Propanoic acid, phenyl ester	$C_9H_{10}O_2$ 150.18	637-27-4 20	211	4-06-00-00615 1.0436[25]	H_2O 1; EtOH 4; eth 4; bz 3 1.4980[20]
10324	Propanoic acid, 2-phenylethyl ester 2-Phenethyl propionate	$C_{11}H_{14}O_2$ 178.23	122-70-3		3-06-00-01709	
10325	Propanoic acid, 2-propenyl ester	$C_6H_{10}O_2$ 114.14	2408-20-0	124	4-02-00-00711 0.9140[20]	EtOH 3; eth 3; ace 3 1.4105[20]
10326	Propanoic acid, propyl ester Propyl propionate	$C_6H_{12}O_2$ 116.16	106-36-5 -75.9	7880 122.5	4-02-00-00707 0.8809[20]	H_2O 2; EtOH 5; eth 5; ace 3 1.3935[20]
10327	Propanoic acid, sodium salt	$C_3H_5NaO_2$ 96.06	137-40-6	8623	4-02-00-00695	H_2O 2
10328	Propanoic acid, 2,3,3,3-tetrafluoro-, ethyl ester Ethyl 2,3,3,3-tetrafluoropropionate	$C_5H_6F_4O_2$ 174.10	399-92-8 108.5		4-02-00-00737 1.289[20]	1.3260[20]
10329	Propanoic acid, 3,3'-thiobis- 3,3'-Thiodipropionic acid	$C_6H_{10}O_4S$ 178.21	111-17-1 129	9261	4-03-00-00735	H_2O 4; EtOH 4
10330	Propanoic acid, 3,3'-thiobis-, diethyl ester	$C_{10}H_{18}O_4S$ 234.32	673-79-0 121[2]		4-03-00-00741 1.095[25]	1.4655[20]
10331	Propanoic acid, 3,3'-thiobis-, dimethyl ester	$C_8H_{14}O_4S$ 206.26	4131-74-2 148[18]		4-03-00-00738 1.198[25]	1.4740[20]
10332	Propanoic acid, 2,2,3-trichloro-	$C_3H_3Cl_3O_2$ 177.41	3278-46-4 65.5	140[40]	4-02-00-00759	H_2O 3; EtOH 3; bz 3; CS_2 3
10333	Propanoic acid, 3,3,3-trichloro-2-hydroxy- Lactic acid, 3,3,3-trichloro-	$C_3H_3Cl_3O_3$ 193.41	599-01-9 125	140[45]	3-03-00-00501	H_2O 4; eth 4; EtOH 4; chl 4
10334	1-Propanol Propyl alcohol	C_3H_8O 60.10	71-23-8 -126.1	7854 97.2	4-01-00-01413 0.7997[25]	H_2O 5; EtOH 5; eth 5; ace 3 1.3850[20]
10335	2-Propanol Isopropyl alcohol	C_3H_8O 60.10	67-63-0 -89.5	5096 82.3	4-01-00-01461 0.7809[25]	H_2O 5; EtOH 5; eth 5; ace 3 1.3776[20]

No.	Name / Synonym	Mol. Form. / Mol. Wt.	CAS RN / mp/°C	Merck No. / bp/°C	Beil. Ref. / den/g cm⁻³	Solubility / n_D
10336	1-Propanol, 2-amino-, (±)-	C_3H_9NO / 75.11	6168-72-5	/ 174.5	3-04-00-00737	H_2O 4; EtOH 4; eth 4; chl 2 / 1.4502[20]
10337	1-Propanol, 3-amino- / Propanolamine	C_3H_9NO / 75.11	156-87-6 / 11	/ 187.5	4-04-00-01623 / 0.9824[26]	H_2O 3; EtOH 3; eth 3 / 1.4617[20]
10338	2-Propanol, 1-amino-, (±)- / DL-1-Amino-2-propanol	C_3H_9NO / 75.11	1674-56-2 / 1.7	/ 159.4	4-04-00-01665 / 0.9611[20]	H_2O 5; EtOH 5; eth 5; ace 5 / 1.4479[20]
10339	2-Propanol, 1-amino-3-(diethylamino)- / 3-(Diethylamino)-2-hydroxypropylamine	$C_7H_{18}N_2O$ / 146.23	6322-01-6	/ 223	3-04-00-00767 / 0.931[20]	EtOH 3; eth 3; chl 2 / 1.465[20]
10340	2-Propanol, 1-[(2-aminoethyl)amino]- / N-(2-Hydroxypropyl)ethylenediamine	$C_5H_{14}N_2O$ / 118.18	123-84-2	450 / 94[3]	4-04-00-01684 / 0.9837[25]	1.4738[20]
10341	1-Propanol, 2-amino-2-methyl- / 2-Aminoisobutanol	$C_4H_{11}NO$ / 89.14	124-68-5 / 25.5	461 / 165.5	4-04-00-01740 / 0.934[20]	H_2O 5; ctc 3 / 1.449[20]
10342	1-Propanol, 2-[(2-amino-2-methylpropyl)amino]-2-methyl- / 1-Propanol, 2-[2-(2-amino-2-methylpropyl)amino]-2-methyl-	$C_8H_{20}N_2O$ / 160.26	72622-74-3 / 46	/ 231	3-04-00-00788 / 0.9360[20]	chl 2 / 1.4635[25]
10343	2-Propanol, 1,3-bis(dimethylamino)-	$C_7H_{18}N_2O$ / 146.23	5966-51-8	/ 181.5	4-04-00-01695 / 0.8788[20]	H_2O 4 / 1.4418[20]
10344	1-Propanol, 3-bromo-	C_3H_7BrO / 138.99	627-18-9	105[185]; 62[5]	4-01-00-01446 / 1.5374[20]	H_2O 3; EtOH 5; eth 5 / 1.4834[15]
10345	2-Propanol, 1-bromo-	C_3H_7BrO / 138.99	19686-73-8	/ 146.5	3-01-00-01474 / 1.5585[30]	H_2O 3; EtOH 4; eth 4 / 1.4801[20]
10346	2-Propanol, 1-butoxy-	$C_7H_{16}O_2$ / 132.20	5131-66-8	171.5; 71[20]	4-01-00-02471 / 0.882[20]	EtOH 3; eth 3; bz 3; ctc 3 / 1.4168[20]
10347	1-Propanol, 2-chloro- / Propylene chlorohydrin	C_3H_7ClO / 94.54	78-89-7	7863 / 133.5	4-01-00-01440 / 1.103[20]	H_2O 4; eth 4; EtOH 4 / 1.4390[20]
10348	1-Propanol, 3-chloro- / 3-Chloro-1-propanol	C_3H_7ClO / 94.54	627-30-5	/ 165	4-01-00-01441 / 1.1309[20]	H_2O 4; EtOH 3; eth 3; ctc 2 / 1.4459[20]
10349	2-Propanol, 1-chloro- / sec-Propylene chlorohydrin	C_3H_7ClO / 94.54	127-00-4	7864 / 127	4-01-00-01490 / 1.113[20]	H_2O 5; EtOH 5; eth 5; ctc 2 / 1.4392[20]
10350	1-Propanol, 3-chloro-, acetate	$C_5H_9ClO_2$ / 136.58	628-09-1	164; 62[10]	4-02-00-00140 / 1.250[19]	H_2O 1; EtOH 3; eth 3; ace 3 / 1.431[20]
10351	2-Propanol, 1-chloro-, acetate	$C_5H_9ClO_2$ / 136.58	623-60-9	/ 149.5	4-02-00-00142 / 1.0788[20]	eth 4; EtOH 4 / 1.4223[20]
10352	1-Propanol, 2-chloro-2-methyl- / Isobutyl alcohol, 2-chloro	C_4H_9ClO / 108.57	558-38-3	132 dec; 60[50]	4-01-00-01603 / 1.0472[20]	1.4388[20]
10353	1-Propanol, 3-chloro-2-methyl- / Isobutyl alcohol, 3-chloro	C_4H_9ClO / 108.57	10317-10-9	76-8[11]	3-01-00-01564 / 1.083[25]	eth 4; EtOH 4 / 1.4460[25]
10354	2-Propanol, 1-chloro-2-methyl- / Isobutylene chlorohydrin	C_4H_9ClO / 108.57	558-42-9 / -20	/ 128.5	4-01-00-01628 / 1.0628[20]	H_2O 4; EtOH 4 / 1.4380[24]
10355	2-Propanol, 1-chloro-3-(1-methylethoxy)- / 2-Propanol, 1-chloro-3-isopropoxy-	$C_6H_{13}ClO_2$ / 152.62	4288-84-0	87[20]	4-01-00-02485 / 1.0530[25]	EtOH 3; eth 3 / 1.4370[25]
10356	2-Propanol, 1-chloro-3-propoxy-	$C_6H_{13}ClO_2$ / 152.62	6943-58-4	92-5[15]	3-01-00-02153 / 1.0526[25]	ace 4; eth 4 / 1.4378[25]
10357	2-Propanol, 1,3-dibromo-	$C_3H_8Br_2O$ / 217.89	96-21-9	219 dec; 105[16]	4-01-00-01496 / 2.1201[25]	ace 4; eth 4; EtOH 4 / 1.5495[25]
10358	1-Propanol, 2,3-dibromo-, phosphate (3:1) / Tris(2,3-dibromopropyl) phosphate	$C_9H_{15}Br_6O_4P$ / 697.61	126-72-7	9665		chl 3
10359	1-Propanol, 2,3-dichloro-	$C_3H_6Cl_2O$ / 128.99	616-23-9	/ 184	4-01-00-01442 / 1.3607[20]	H_2O 2; EtOH 5; eth 5; ace 5 / 1.4819[20]
10360	1-Propanol, 3,3-dichloro-	$C_3H_6Cl_2O$ / 128.99	83682-72-8	82-3[20]	3-01-00-01427 / 1.316[25]	1.467[20]
10361	2-Propanol, 1,1-dichloro-	$C_3H_6Cl_2O$ / 128.99	53894-19-2	/ 147	2-01-00-00383 / 1.3334[22]	ace 4; eth 4; EtOH 4
10362	2-Propanol, 1,3-dichloro- / 1,3-Dichloro-2-propanol	$C_3H_6Cl_2O$ / 128.99	96-23-1	3063 / 176	4-01-00-01491 / 1.3506[17]	H_2O 4; EtOH 4; eth 5; ace 3 / 1.4837[20]
10363	2-Propanol, 1,3-dichloro-, acetate	$C_5H_8Cl_2O_2$ / 171.02	3674-10-0	/ 205	3-02-00-00233 / 1.281[20]	eth 4; EtOH 4; chl 4 / 1.4542[20]
10364	2-Propanol, 1,1-dichloro-2-methyl- / 2,2-Dichloro-1,1-dimethylethanol	$C_4H_8Cl_2O$ / 143.01	4773-53-9 / 8	/ 152.5	4-01-00-01629 / 1.236[19]	bz 4; EtOH 4 / 1.4598[19]
10365	2-Propanol, 1,3-dichloro-2-methyl- / β-Methylglycerol α,γ-dichlorohydrin	$C_4H_8Cl_2O$ / 143.01	597-32-0	/ 174.5	4-01-00-01629 / 1.2745[20]	H_2O 3; EtOH 4; ace 3 / 1.4744[21]
10366	1-Propanol, 2-(diethylamino)- / 2-Diethylamino-1-propanol	$C_7H_{17}NO$ / 131.22	611-12-1	168; 78[12]	4-04-00-01618 / 0.8649[27]	EtOH 3; eth 3; ace 3; bz 3 / 1.4332[20]
10367	1-Propanol, 3-(diethylamino)-	$C_7H_{17}NO$ / 131.22	622-93-5	/ 189.5	4-04-00-01633 / 0.860[20]	EtOH 3; eth 3; ace 3; bz 3 / 1.4439[20]
10368	2-Propanol, 1-(diethylamino)-, (±)- / 2-Propanol, 1-diethylamino (DL)	$C_7H_{17}NO$ / 131.22	78738-36-0 / 13.5	159; 63[22]	3-04-00-00759 / 0.8511[20]	H_2O 4; ace 4; EtOH 4 / 1.4255[20]
10369	2-Propanol, 1,3-difluoro- / 1,3-Difluoro-2-propanol	$C_3H_6F_2O$ / 96.08	453-13-4	54-5[34]	4-01-00-01489 / 1.24[25]	1.3725[20]
10370	1-Propanol, 2,3-dimercapto- / Dimercaprol	$C_3H_8OS_2$ / 124.23	59-52-9	3196 / 82-4[0.8]	4-01-00-02770 / 1.2463[20]	EtOH 3; eth 3; chl 2; oils 3 / 1.5749[20]
10371	2-Propanol, 1,3-dimercapto-	$C_3H_8OS_2$ / 124.23	584-04-3	94[12]	4-01-00-02773 / 1.2386[20]	1.5700[20]
10372	2-Propanol, 1,3-dimethoxy- / Glycerol 1,3-dimethyl ether	$C_5H_{12}O_3$ / 120.15	623-69-8	/ 169	3-01-00-02318 / 1.0085[20]	H_2O 5; EtOH 4; eth 3 / 1.4192[20]
10373	1-Propanol, 2,2-dimethyl- / Neopentyl alcohol	$C_5H_{12}O$ / 88.15	75-84-3 / 52.5	6373 / 113.5	4-01-00-01690 / 0.812[20]	H_2O 2; EtOH 4; eth 4; ctc 3

No.	Name Synonym	Mol. Form. Mol. Wt.	CAS RN mp/°C	Merck No. bp/°C	Beil. Ref. den/g cm⁻³	Solubility n_D
10374	1-Propanol, 2-(dimethylamino)- 2-Dimethylamino-1-propanol	$C_5H_{13}NO$ 103.16	15521-18-3	150.3	4-04-00-01617 0.8820^{26}	H_2O 3
10375	1-Propanol, 3-(dimethylamino)-	$C_5H_{13}NO$ 103.16	3179-63-3	163.5	4-04-00-01628 0.872^{25}	ctc 3 1.4360^{20}
10376	2-Propanol, 1-(dimethylamino)-	$C_5H_{13}NO$ 103.16	108-16-7	124.5	3-04-00-00756 0.837^{25}	ctc 3 1.4193^{20}
10377	2-Propanol, 1,3-diphenoxy-	$C_{15}H_{16}O_3$ 244.29	622-04-8 81.5	224^{17}; 175^2	4-06-00-00590 1.179^{24}	bz 4; eth 4; EtOH 4; chl 4
10378	2-Propanol, 1,3-diphenoxy-, acetate Glycerol, 1,3-diphenyl ether-2-acetate	$C_{17}H_{18}O_4$ 286.33	71159-31-4 33	190^{160}	3-06-00-00583	bz 4; eth 4; EtOH 4; chl 4
10379	2-Propanol, 1,1',1'',1'''-(1,2-ethanediyldinitrilo)tetrakis- ENTPROL	$C_{14}H_{32}N_2O_4$ 292.42	102-60-3 3551		4-04-00-01685 1.030^{25}	chl 2 1.478^{25}
10380	1-Propanol, 2-ethoxy- Propylene glycol β-monoethyl ether	$C_5H_{12}O_2$ 104.15	19089-47-5	140.5	3-01-00-02147 0.9044^{20}	H_2O 3; EtOH 3; eth 3 1.4122^{20}
10381	2-Propanol, 1-ethoxy- 1-Ethoxy-2-propanol	$C_5H_{12}O_2$ 104.15	1569-02-4	131	4-01-00-02471 0.9028^{20}	H_2O 3; EtOH 3; eth 3 1.4075^{20}
10382	1-Propanol, 3-fluoro- 3-Hydroxypropyl fluoride	C_3H_7FO 78.09	462-43-1	127.8	4-01-00-01437 1.0390^{25}	H_2O 4; eth 4; EtOH 4 1.3771^{25}
10383	2-Propanol, 1-[(2-hydroxyethyl)amino]-	$C_5H_{13}NO_2$ 119.16	6579-55-1	$120-3^1$	4-04-00-01678 1.0455^{20}	H_2O 3; EtOH 3; ace 3 1.4695^{20}
10384	2-Propanol, 1,1'-iminobis- Diisopropanolamine	$C_6H_{15}NO_2$ 133.19	110-97-4 44.5	250; 151^{23}	3-04-00-00761 0.9892^{20}	H_2O 3; EtOH 3; eth 2
10385	1-Propanol, 3-iodo-	C_3H_7IO 185.99	627-32-7	115^{38}; 88^4	4-01-00-01448 1.9976^{20}	1.5585^{20}
10386	1-Propanol, 2-methoxy- 2-Methoxy-1-propanol	$C_4H_{10}O_2$ 90.12	1589-47-5	130	4-01-00-02471 0.938^{20}	1.4070^{20}
10387	2-Propanol, 1-methoxy- 1-Methoxy-2-propanol	$C_4H_{10}O_2$ 90.12	107-98-2	119	3-01-00-02146 0.9620^{20}	1.4034^{20}
10388	1-Propanol, 2-methyl- Isobutyl alcohol	$C_4H_{10}O$ 74.12	78-83-1 -108	5015 107.8	4-01-00-01588 0.8018^{20}	H_2O 2; EtOH 3; eth 3; ace 3 1.3955^{20}
10389	2-Propanol, 2-methyl- tert-Butyl alcohol	$C_4H_{10}O$ 74.12	75-65-0 25.4	1542 82.4	4-01-00-01609 0.7887^{20}	H_2O 5; EtOH 5; eth 5; chl 3 1.3878^{20}
10390	2-Propanol, 1-(1-methylethoxy)- 1-Isopropoxy-2-propanol	$C_6H_{14}O_2$ 118.18	3944-36-3	137.5	4-01-00-02471 0.879^{20}	1.4070^{20}
10391	1-Propanol, 2-methyl-2-nitro-	$C_4H_9NO_3$ 119.12	76-39-1 89.5	94^{10}	4-01-00-01604	H_2O 2; EtOH 4; eth 4; chl 3
10392	1-Propanol, 3-[methyl(phenylmethyl)amino]-	$C_{11}H_{17}NO$ 179.26	5814-42-6	$132-5^4$	4-12-00-02208 1.001^{25}	ctc 2; CS_2 3 1.5230^{20}
10393	2-Propanol, 1,1',1''-nitrilotris-	$C_9H_{21}NO_3$ 191.27	122-20-3 45	$170-80^{10}$	4-04-00-01680 1.02^{0}	H_2O 3; EtOH 3; chl 2
10394	1-Propanol, 2-nitro-	$C_3H_7NO_3$ 105.09	2902-96-7	100^{12}	4-01-00-01448 1.1841^{25}	H_2O 3; EtOH 3; eth 3; chl 2 1.4379^{20}
10395	1-Propanol, 3-nitro-	$C_3H_7NO_3$ 105.09	25182-84-7	$138-40^{32}$	3-01-00-01429 1.173^{13}	H_2O 4; eth 4; EtOH 4
10396	2-Propanol, 1-nitro- 1-Nitro-2-propanol	$C_3H_7NO_3$ 105.09	3156-73-8	112^{13}; 68^1	4-01-00-01496 1.1906^{20}	H_2O 3; EtOH 3; eth 3; bz 3 1.4383^{20}
10397	Propanol, oxybis- Dipropylene glycol	$C_6H_{14}O_3$ 134.18	25265-71-8	230.5	4-01-00-02473 1.0206^{20}	H_2O 5; EtOH 3
10398	1-Propanol, 2-phenoxy-	$C_9H_{12}O_2$ 152.19	4169-04-4	244	4-06-00-00582 0.9801^{25}	EtOH 3; eth 3 1.4760^{25}
10399	2-Propanol, 1-phenoxy-	$C_9H_{12}O_2$ 152.19	770-35-4	134.5^{20}	4-06-00-00582 1.0622^{20}	1.5232^{20}
10400	1-Propanol, 3-(phenylmethoxy)-	$C_{10}H_{14}O_2$ 166.22	4799-68-2	172^{43}	4-06-00-02243 1.0474^{20}	EtOH 3; eth 3
10401	2-Propanol, 1-propoxy- 1,2-Propylene glycol 1-propyl ether	$C_6H_{14}O_2$ 118.18	1569-01-3	150	3-01-00-02147 0.8886^{20}	1.4130^{20}
10402	2-Propanol, 1,1,3,3-tetrachloro- 1,1,3,3-Tetrachloro-2-propanol	$C_3H_4Cl_4O$ 197.87	18992-39-7	$80-90^{14}$	3-01-00-01474 1.612^{20}	eth 3 1.5133^{20}
10403	2-Propanol, 1,1,1,3-tetrachloro-	$C_3H_4Cl_4O$ 197.87	88947-16-4	$95-6^{17}$	3-01-00-01474 1.610^{20}	eth 4 1.5145^{20}
10404	1-Propanol, 2,2,3,3-tetrafluoro- 2,2,3,3-Tetrafluoro-1-propanol	$C_3H_4F_4O$ 132.06	76-37-9 -15	9139 109.5	4-01-00-01438 1.4853^{20}	EtOH 3; ace 3; chl 3 1.3197^{20}
10405	2-Propanol, 1,1,1-tribromo-2-methyl- Tribromo-tert-butyl alcohol	$C_4H_7Br_3O$ 310.81	76-08-4 169	9523 sub	3-01-00-01588	H_2O 2; EtOH 3; eth 3; chl 2
10406	2-Propanol, 1,1,1-trichloro- 1,1,1-Trichloro-2-propanol	$C_3H_5Cl_3O$ 163.43	76-00-6 50.5	9557 163; $53-5^{12}$	3-01-00-01474	ace 4; bz 4; eth 4; EtOH 4
10407	2-Propanol, 1,1,1-trichloro-2-methyl- Chlorobutanol	$C_4H_7Cl_3O$ 177.46	57-15-8 97	2129 167	4-01-00-01629	H_2O 1; EtOH 3; eth 3; ace 3
10408	2-Propanol, 1,1,1-trichloro-3-nitro- 2-Propanol, 1-nitro-3,3,3-trichloro	$C_3H_4Cl_3NO_3$ 208.43	760-40-7 45.5	$105^{3.5}$	4-01-00-01497	eth 4; EtOH 4
10409	2-Propanol, 1,1,1-trifluoro-, (±)- 2-Propanol, 1,1,1-trifluoro(DL)	$C_3H_5F_3O$ 114.07	17556-48-8 -52	78	4-01-00-01489 1.2632^{25}	EtOH 4; eth 4; ace 3; bz 3 1.3130^{25}
10410	1-Propanol, 3-(trimethylsilyl)-	$C_6H_{16}OSi$ 132.28	2917-47-7	141; 82^{24}	4-04-00-03953 0.822^{25}	1.4298^{20}
10411	2-Propanone Acetone	C_3H_6O 58.08	67-64-1 -94.8	58 56.05	4-01-00-03180 0.7845^{25}	H_2O 5; EtOH 5; eth 5; ace 5 1.3588^{20}
10412	2-Propanone, 1-(acetyloxy)-	$C_5H_8O_3$ 116.12	592-20-1	171; 63^{11}	4-02-00-00297 1.0757^{20}	H_2O 4; eth 4; EtOH 4 1.4141^{20}
10413	1-Propanone, 1-(4-aminophenyl)- p-Aminopropiophenone	$C_9H_{11}NO$ 149.19	70-69-9 140	482	4-14-00-00139	DMSO 3

No.	Name / Synonym	Mol. Form. / Mol. Wt.	CAS RN / mp/°C	Merck No. / bp/°C	Beil. Ref. / den/g cm^{-3}	Solubility / n_D
10414	1-Propanone, 2-amino-1-phenyl-	$C_9H_{11}NO$ 149.19	5265-18-9 46.5	$93^{0.8}$	3-14-00-00147	ace 4; eth 4; EtOH 4
10415	2-Propanone, 1,3-bis(1-methyl-2-pyrrolidinyl)-, (R*,S*)- Cuscohygrine	$C_{13}H_{24}N_2O$ 224.35	454-14-8	2685 $169^{23}; 122^2$	5-24-01-00458 0.9733^{20}	H_2O 4; bz 4; eth 4; EtOH 4 1.4832^{20}
10416	2-Propanone, 1-bromo- Bromoacetone	C_3H_5BrO 136.98	598-31-2 -36.5	1389 138; 31.5^8	4-01-00-03223 1.634^{23}	H_2O 2; EtOH 3; eth 3; ace 3 1.4697^{15}
10417	2-Propanone, 1-bromo-3-chloro- 1-Bromo-3-chloroacetone	C_3H_4BrClO 171.42	53535-68-5 34.5	179	4-01-00-03224	eth 4; EtOH 4
10418	1-Propanone, 1-(4-bromophenyl)-	C_9H_9BrO 213.07	10342-83-3 48	$169^{15}; 137^2$	4-07-00-00684	H_2O 1; EtOH 3; eth 3; ace 3
10419	1-Propanone, 2-bromo-1-phenyl-	C_9H_9BrO 213.07	2114-00-3	247.5	4-07-00-00684 1.4298^{20}	H_2O 1; EtOH 3; eth 3; ace 3 1.5720^{20}
10420	2-Propanone, 1-chloro- Chloroacetone	C_3H_5ClO 92.52	78-95-5 -44.5	2113 119	4-01-00-03215 1.15^{20}	H_2O 3; EtOH 3; eth 3; chl 3
10421	2-Propanone, 1-chloro-1,1,3,3,3-pentafluoro- Chloropentafluoroacetone	C_3ClF_5O 182.48	79-53-8 -133	8		
10422	1-Propanone, 1-(4-chlorophenyl)-	C_9H_9ClO 168.62	6285-05-8 37.3	$135^{31}; 114^2$	4-07-00-00683	H_2O 1; EtOH 3; chl 2; CS_2 3
10423	1-Propanone, 1-cyclohexyl- Cyclohexyl ethyl ketone	$C_9H_{16}O$ 140.23	1123-86-0	196	4-07-00-00064 0.9105^{20}	1.4530^{20}
10424	2-Propanone, 1,3-dibromo-	$C_3H_4Br_2O$ 215.87	816-39-7 26	97^{22}	4-01-00-03225 2.1670^{18}	eth 4; CS_2 4
10425	2-Propanone, 1,1-dichloro- 1,1-Dichloroacetone	$C_3H_4Cl_2O$ 126.97	513-88-2	3038 120	4-01-00-03218 1.304^{18}	H_2O 2; EtOH 3; eth 5
10426	2-Propanone, 1,3-dichloro- 1,3-Dichloroacetone	$C_3H_4Cl_2O$ 126.97	534-07-6 45	3039 173.4	4-01-00-03219 1.3826^{46}	H_2O 3; EtOH 3; eth 3 1.4716^{40}
10427	2-Propanone, 1-(diethylamino)-	$C_7H_{15}NO$ 129.20	1620-14-0	157 dec	4-04-00-01925 0.8620^{20}	H_2O 5; EtOH 5; eth 5; ctc 3 1.4249^{20}
10428	1-Propanone, 2-(diethylamino)-1-phenyl- Diethylpropion	$C_{13}H_{19}NO$ 205.30	90-84-6	3116	4-14-00-00144	
10429	2-Propanone, 1,3-dihydroxy- Dihydroxyacetone	$C_3H_6O_3$ 90.08	96-26-4 90	3166	4-01-00-04119	H_2O 3; EtOH 3; eth 3; ace 3
10430	1-Propanone, 1-(2,4-dihydroxyphenyl)-	$C_9H_{10}O_3$ 166.18	5792-36-9 99	176^6	4-08-00-01837	bz 4; eth 4; EtOH 4
10431	1-Propanone, 2,2-dimethyl-1-phenyl-	$C_{11}H_{14}O$ 162.23	938-16-9	220	4-07-00-00740 0.963^{26}	ace 3 1.5086^{19}
10432	2-Propanone, 1,1-diphenyl-	$C_{15}H_{14}O$ 210.28	781-35-1 46	$307; 174^{10}$	4-07-00-01426	EtOH 3; eth 3; bz 3; chl 3 1.5361^{16}
10433	2-Propanone, 1,3-diphenyl-	$C_{15}H_{14}O$ 210.28	102-04-5 35	331	4-07-00-01420 1.195^0	H_2O 1; EtOH 3; eth 3; peth 3
10434	1-Propanone, 3-(3-ethenyl-4-piperidinyl)-1-(6-methoxy-4-quinolinyl)-, (3R-cis)- Viquidil	$C_{20}H_{24}N_2O_2$ 324.42	84-55-9 60	9908	4-25-00-00181	eth 4; EtOH 4; chl 4
10435	1-Propanone, 3-(3-ethenyl-4-piperidinyl)-1-(4-quinolinyl)-, (3R-cis)- Cinchotoxine	$C_{19}H_{22}N_2O$ 294.40	69-24-9 59	2291	4-24-00-00651	H_2O 1; EtOH 4; eth 4; ace 4
10436	1-Propanone, 3-ethoxy-1-phenyl-	$C_{11}H_{14}O_2$ 178.23	34008-71-4 11.5	135^{18}	4-08-00-00453 1.0386^{20}	1.5190^{20}
10437	2-Propanone, 1-fluoro-	C_3H_5FO 76.07	430-51-3	77	4-01-00-03213 1.0288^{20}	1.3700^{20}
10438	1-Propanone, 1-(2-furanyl)-	$C_7H_8O_2$ 124.14	3194-15-8 28	88^{14}	5-17-09-00415 1.0626^{28}	eth 3; ctc 2 1.4922^{25}
10439	2-Propanone, 1-(2-furanyl)-	$C_7H_8O_2$ 124.14	6975-60-6 29	179.5	5-17-09-00418 1.104^{20}	1.5035^{20}
10440	2-Propanone, 1,1,1,3,3,3-hexachloro- Hexachloroacetone	C_3Cl_6O 264.75	116-16-5 -2	203	4-01-00-03223 1.7434^{12}	H_2O 2; ace 3 1.5112^{20}
10441	2-Propanone, 1,1,1,3,3,3-hexafluoro- Perfluoroacetone	C_3F_6O 166.02	684-16-2 -125	-27.4	4-01-00-03215	
10442	2-Propanone, 1-(1H-indol-3-yl)- 3-Indolylacetone	$C_{11}H_{11}NO$ 173.21	1201-26-9 116	4873	5-21-08-00327	
10443	2-Propanone, 1-hydroxy-	$C_3H_6O_2$ 74.08	116-09-6 -17	145.5	4-01-00-03977 1.0805^{20}	H_2O 4; EtOH 4; eth 4 1.4295^{20}
10444	1-Propanone, 1-(2-hydroxy-3-methylphenyl)- 2-Methyl-6-propionylphenol	$C_{10}H_{12}O_2$ 164.20	3338-15-6 22.5	124^{13}	4-08-00-00517 1.0900^{20}	ctc 3
10445	1-Propanone, 1-(2-hydroxy-4-methylphenyl)- 2'-Hydroxy-4'-methylpropiophenone	$C_{10}H_{12}O_2$ 164.20	2886-52-4 44	115^{10}	4-08-00-00519	chl 3
10446	1-Propanone, 1-(2-hydroxy-5-methylphenyl)-	$C_{10}H_{12}O_2$ 164.20	938-45-4 2	$129-30^{16.5}$	4-08-00-00517 1.0841^{14}	chl 3 1.549^{13}
10447	1-Propanone, 1-(2-hydroxy-5-methylphenyl)-2-methyl-	$C_{11}H_{14}O_2$ 178.23	64207-03-0	$251; 125^{11}$	4-08-00-00545 1.046^{16}	1.538^{16}
10448	1-Propanone, 1-(4-hydroxy-3-methylphenyl)-2-methyl- o-Cresol, 4-isobutyryl	$C_{11}H_{14}O_2$ 178.23	73206-57-2 122	182^{12}	2-08-00-00130	eth 4; EtOH 4
10449	1-Propanone, 1-(2-hydroxyphenyl)-	$C_9H_{10}O_2$ 150.18	610-99-1	$150^{80}; 115^{15}$	4-08-00-00436	H_2O 2; EtOH 3; eth 3; ctc 3 1.5501^{20}
10450	1-Propanone, 1-(4-hydroxyphenyl)- Paroxypropione	$C_9H_{10}O_2$ 150.18	70-70-2 149	6992	4-08-00-00441	H_2O 2; EtOH 3; ace 2

No.	Name Synonym	Mol. Form. Mol. Wt.	CAS RN mp/°C	Merck No. bp/°C	Beil. Ref. den/g cm⁻³	Solubility n_D
10451	1-Propanone, 2-hydroxy-1-phenyl-	$C_9H_{10}O_2$ 150.18	5650-40-8	251	4-08-00-00450 1.1085^{18}	1.536^{23}
10452	1-Propanone, 3-hydroxy-1-phenyl-	$C_9H_{10}O_2$ 150.18	5650-41-9	$98^{0.2}$	4-08-00-00453 1.1170^{20}	1.5450^{20}
10453	1-Propanone, 3-(4-hydroxyphenyl)-1-(2,4,6-trihydroxyphenyl)- Phloretin	$C_{15}H_{14}O_5$ 274.27	60-82-2 263-4 dec	7299	4-08-00-03518	H_2O 2; EtOH 5; eth 1; ace 3
10454	2-Propanone, 1-iodo-	C_3H_5IO 183.98	3019-04-3	62^{12}	4-01-00-03226 2.17^{15}	EtOH 3
10455	2-Propanone, 1-methoxy-	$C_4H_8O_2$ 88.11	5878-19-3	116	4-01-00-03977 0.957^{25}	1.3970^{20}
10456	1-Propanone, 1-(3-methoxyphenyl)- 3-Methoxypropiophenone	$C_{10}H_{12}O_2$ 164.20	37951-49-8	259	4-08-00-00441 1.0812^{25}	eth 4; EtOH 4 1.5230^{25}
10457	1-Propanone, 2-methoxy-1-phenyl- Propiophenone, 2-methoxy-	$C_{10}H_{12}O_2$ 164.20	6493-83-0	$79-80^1$	4-08-00-00451 1.0704^{20}	1.5225^{20}
10458	1-Propanone, 3-methoxy-1-phenyl-	$C_{10}H_{12}O_2$ 164.20	55563-72-9	$125-6^{16}$	4-08-00-00453 1.0602^{20}	1.5250^{20}
10459	2-Propanone, 1-(methoxyphenyl)- 2-Propanone, (methoxyphenyl)-	$C_{10}H_{12}O_2$ 164.20	31116-84-4 <-15	$267-9\ 142^{14}$	2-08-00-00104 1.0670^{18}	H_2O 2; EtOH 3; eth 3 1.5253^{20}
10460	2-Propanone, 1-(3-methoxyphenyl)-	$C_{10}H_{12}O_2$ 164.20	3027-13-2	259	4-08-00-00459 1.0812^6	EtOH 3; eth 3 1.5230^{25}
10461	2-Propanone, 1-(4-methoxyphenyl)-	$C_{10}H_{12}O_2$ 164.20	122-84-9 <-15	268	4-08-00-00460 1.0694^{17}	eth 4; EtOH 4 1.5253^{20}
10462	1-Propanone, 2-methyl-1,2-di-3-pyridinyl- Metyrapone	$C_{14}H_{14}N_2O$ 226.28	54-36-4 50.5	6083	5-24-04-00048	
10463	2-Propanone, (1-methylethylidene)hydrazone	$C_6H_{12}N_2$ 112.17	627-70-3 -12.5	133	4-01-00-03207 0.8390^{20}	H_2O 5; EtOH 5; eth 5; ace 3 1.4535^{20}
10464	1-Propanone, 2-methyl-1-(1-naphthalenyl)- Isopropyl 1-naphthyl ketone	$C_{14}H_{14}O$ 198.26	61838-78-6	307	3-07-00-01985 1.0761^{25}	bz 4; eth 4; EtOH 4 1.5948^{20}
10465	1-Propanone, 2-methyl-1-(2-naphthalenyl)- 2-Isobutyronaphthone	$C_{14}H_{14}O$ 198.26	107574-57-2	313	3-07-00-01985 1.0617^{25}	bz 4; eth 4; EtOH 4
10466	1-Propanone, 1-(2-methylphenyl)-	$C_{10}H_{12}O$ 148.20	2040-14-4 -27.6	219.5	4-07-00-00720 1.0119^0	
10467	1-Propanone, 1-(3-methylphenyl)-	$C_{10}H_{12}O$ 148.20	51772-30-6	$235; 132^{31}$	4-07-00-00721 1.0059^0	H_2O 1; EtOH 3; eth 3; ace 3
10468	1-Propanone, 1-(4-methylphenyl)-	$C_{10}H_{12}O$ 148.20	5337-93-9 7.2	236	4-07-00-00722 0.9926^{20}	H_2O 1; EtOH 3; eth 3; ace 3 1.5278^{20}
10469	1-Propanone, 2-methyl-1-phenyl-	$C_{10}H_{12}O$ 148.20	611-70-1 -1.3	220	4-07-00-00716 0.9863^{11}	eth 4; EtOH 4 1.5172^{20}
10470	2-Propanone, 1-(1-methyl-2-piperidinyl)-, (±)-	$C_9H_{17}NO$ 155.24	18747-42-7	$96-8^{13}$	5-21-06-00523 0.9478^{20}	H_2O 4; lig 4 1.4674^{20}
10471	2-Propanone, 1-(1-methyl-2-pyrrolidinyl)-, (R)- Hygrine	$C_8H_{15}NO$ 141.21	496-49-1	4788 76.5^{11}	5-21-06-00511	EtOH 4; chl 4 1.4555^{20}
10472	1-Propanone, 2-methyl-1-(2-thienyl)- Isopropyl 2-thienyl ketone	$C_8H_{10}OS$ 154.23	36448-60-9	229	5-17-09-00439 1.0894^{20}	1.5405^{20}
10473	1-Propanone, 1-(1-naphthalenyl)-	$C_{13}H_{12}O$ 184.24	2876-63-3	306	4-07-00-01298 1.0971^{20}	1.6108^{20}
10474	1-Propanone, 1-(2-naphthalenyl)-	$C_{13}H_{12}O$ 184.24	6315-96-4 60	313	4-07-00-01299	EtOH 4; chl 4
10475	2-Propanone, 1-nitro-	$C_3H_5NO_3$ 103.08	10230-68-9 50.3	103^{24}	4-01-00-03226	bz 4; eth 4; EtOH 4
10476	2-Propanone, oxime Acetoxime	C_3H_7NO 73.09	127-06-0 61	68 $136; 61^{20}$	4-01-00-03202 0.9113^{62}	H_2O 3; EtOH 3; eth 3; chl 3 1.4156^{20}
10477	2-Propanone, 1,1,1,3,3-pentabromo- Pentabromoacetone	C_3HBr_5O 452.56	79-49-2 79.5	7056 sub	4-01-00-03226	H_2O 1; EtOH 4; eth 4; ace 4
10478	2-Propanone, 1,1,1,3,3-pentachloro-	C_3HCl_5O 230.30	1768-31-6 2.1	$192^{253}; 98^{40}$	4-01-00-03222 1.69^{15}	
10479	2-Propanone, 1-phenoxy-	$C_9H_{10}O_2$ 150.18	621-87-4	229.5	4-06-00-00604 1.0903^{20}	eth 3; ace 3 1.5228^{20}
10480	1-Propanone, 1-phenyl- Propiophenone	$C_9H_{10}O$ 134.18	93-55-0 18.6	7842 217.5	4-07-00-00680 1.0096^{20}	H_2O 1; EtOH 3; eth 3; chl 3 1.5269^{20}
10481	2-Propanone, 1-phenyl- Phenylacetone	$C_9H_{10}O$ 134.18	103-79-7 -15	7240 216.5	4-07-00-00687 1.0157^{20}	H_2O 1; EtOH 4; eth 4; bz 5 1.5168^{20}
10482	2-Propanone, phenylhydrazone Acetone, phenylhydrazone	$C_9H_{12}N_2$ 148.21	103-02-6 42	163^{50}	4-15-00-00060	EtOH 3; eth 3; dil acid 3
10483	1-Propanone, 1-phenyl-, oxime	$C_9H_{11}NO$ 149.19	2157-50-8 54.8	$245; 165^{38}$	4-07-00-00682 1.0639^{18}	H_2O 1; EtOH 3; eth 3; chl 2
10484	1-Propanone, 1-(2-piperidinyl)- Piperidine, 2(1-propionyl)	$C_8H_{15}NO$ 141.21	97073-23-9	94^{18}	5-21-06-00522 0.9380^{20}	1.4621^{20}
10485	2-Propanone, 1-(2-piperidinyl)-, (R)- (-)-Pelletierine	$C_8H_{15}NO$ 141.21	2858-66-4	7015 $195; 125^{100}$	5-21-06-00522 0.988^{20}	eth 4; EtOH 4; chl 4 1.4683^{20}
10486	2-Propanone, 1,1,1,3-tetrachloro-	$C_3H_2Cl_4O$ 195.86	16995-35-0 65	183	4-01-00-03222 1.624^{25}	H_2O 2; eth 4; ace 4 1.497^{18}
10487	2-Propanone, 1,1,3,3-tetrachloro-	$C_3H_2Cl_4O$ 195.86	632-21-3	183	4-01-00-03222	H_2O 4; EtOH 4; eth 4; ace 4 1.4944^{20}
10488	2-Propanone, 1,1,1-trichloro-	$C_3H_3Cl_3O$ 161.41	918-00-3	$149; 28^{10}$	4-01-00-03221 1.435^{20}	H_2O 1; EtOH 4; eth 4 1.4635^{17}
10489	1-Propanone, 1-(2,4,6-trihydroxyphenyl)- Flopropione	$C_9H_{10}O_4$ 182.18	2295-58-1 175.5	4039	3-08-00-03413	eth 4; EtOH 4

No.	Name / Synonym	Mol. Form. / Mol. Wt.	CAS RN / mp/°C	Merck No. / bp/°C	Beil. Ref. / den/g cm⁻³	Solubility / n_D
10490	Propanoyl bromide / Propionic acid bromide	C_3H_5BrO / 136.98	598-22-1	/ 102	4-02-00-00724 / 1.5210^{16}	eth 4 / 1.4578^{16}
10491	Propanoyl bromide, 2-bromo-2-methyl-	$C_4H_6Br_2O$ / 229.90	20769-85-1	/ 163	3-02-00-00661 / 1.4067^{14}	ace 4; CS_2 4
10492	Propanoyl bromide, 2-methyl-	C_4H_7BrO / 151.00	2736-37-0	/ 117	4-02-00-00852 / 1.4067^{15}	/ 1.4552^{25}
10493	Propanoyl chloride / Propionyl chloride	C_3H_5ClO / 92.52	79-03-8 / -94	7840 / 80	4-02-00-00724 / 1.0646^{20}	eth 3 / 1.4032^{20}
10494	Propanoyl chloride, 2-(acetyloxy)-, (±)- / Lactyl chloride, acetate (DL)	$C_5H_7ClO_3$ / 150.56	55057-45-9	/ 150 dec; 56^{11}	4-03-00-00674 / 1.1920^{17}	/ 1.4241^{17}
10495	Propanoyl chloride, 2-bromo-	C_3H_4BrClO / 171.42	7148-74-5	/ 132	4-02-00-00764 / 1.697^{11}	eth 3; ctc 2; chl 3 / 1.4780^{20}
10496	Propanoyl chloride, 2-chloro-, (S)- / Propanoic acid, chloride, 2-chloro(d)	$C_3H_4Cl_2O$ / 126.97	70110-24-6	/ 110	4-02-00-00747 / 1.2394^7	/ 1.4400^{20}
10497	Propanoyl chloride, 3-chloro-	$C_3H_4Cl_2O$ / 126.97	625-36-5	/ 144	4-02-00-00750 / 1.3307^{13}	H_2O 2; EtOH 4; eth 4; chl 4 / 1.4549^{20}
10498	Propanoyl chloride, 2-chloro-2-methyl-	$C_4H_6Cl_2O$ / 141.00	13222-26-9	/ 126.5	4-02-00-00859 /	eth 4 / 1.4369^{20}
10499	Propanoyl chloride, 3-chloro-2-methyl- / Isobutyryl chloride, 3-chloro	$C_4H_6Cl_2O$ / 141.00	7623-10-1 / -90	/ 92	4-02-00-00860 / 1.0174^{20}	eth 4 / 1.4079^{20}
10500	Propanoyl chloride, 2,2-dichloro-	$C_3H_3Cl_3O$ / 161.41	26073-26-7	/ 118; 70^{89}	4-02-00-00755 / 1.4062^{20}	/ 1.4524^{20}
10501	Propanoyl chloride, 2,3-dichloro-	$C_3H_3Cl_3O$ / 161.41	7623-13-4	/ $52-4^{16}$	4-02-00-00757 / 1.4757^{20}	/ 1.4764^{20}
10502	Propanoyl chloride, 3,3-dichloro- / 3,3-Dichloropropionyl chloride	$C_3H_3Cl_3O$ / 161.41	17880-86-3	/ $43-4^{10}$	3-02-00-00562 / 1.4557^{20}	eth 4; Diox 4 / 1.4738^{20}
10503	Propanoyl chloride, 2,2-dimethyl- / Pivalic acid chloride	C_5H_9ClO / 120.58	3282-30-2	/ 107	4-02-00-00912 / 1.003^{20}	eth 4 / 1.4139^{20}
10504	Propanoyl chloride, 2-methyl- / Isobutyric acid chloride	C_4H_7ClO / 106.55	79-30-1 / -90	/ 92	4-02-00-00852 /	eth 3 / 1.4079^{20}
10505	Propanoyl chloride, 2-phenoxy-	$C_9H_9ClO_2$ / 184.62	122-35-0	/ 147; 116^{10}	4-06-00-00643 / 1.1865^{20}	eth 3 / 1.5178^{20}
10506	Propanoyl fluoride / Propionyl fluoride	C_3H_5FO / 76.07	430-71-7	/ 44	4-02-00-00724 / 0.972^{15}	/ 1.329^{13}
10507	Propargite / Sulfurous acid, 2-[4-(1,1-dimethylethyl)phenoxy]cyclohexyl 2-propynyl ester	$C_{19}H_{26}O_4S$ / 350.48	2312-35-8	7818 /	/ 1.10^{25}	
10508	Propazine / 1,3,5-Triazine-2,4-diamine, 6-chloro-N,N'-bis(1-methylethyl)-	$C_9H_{16}ClN_5$ / 229.71	139-40-2 / 213	7822 /	/ 1.162^{20}	
10509	2-Propenal / Acrolein	C_3H_4O / 56.06	107-02-8 / -87.7	122 / 52.6	4-01-00-03435 / 0.840^{20}	H_2O 4; EtOH 3; eth 3; ace 3 / 1.4017^{20}
10510	2-Propenal, 3-bromo-3-phenyl-	C_9H_7BrO / 211.06	14804-59-2	/ $144-6^{12}$	3-07-00-01383 / 1.492^{20}	eth 4
10511	2-Propenal, 2-chloro-	C_3H_3ClO / 90.51	683-51-2	/ 40^{30}	4-01-00-03440 / 1.199^{20}	eth 4; EtOH 4 / 1.463^{20}
10512	2-Propenal, 3-(2-furanyl)-	$C_7H_6O_2$ / 122.12	623-30-3 / 54	/ 135^{14}	5-17-09-00570 /	H_2O 1; EtOH 5; eth 3; chl 2
10513	2-Propenal, 3-(2-furanyl)-, (E)-	$C_7H_6O_2$ / 122.12	39511-08-5 / 50	/ 143^{37}; 100^{12}	5-17-09-00570 /	bz 4; eth 4; EtOH 4 / 1.5286^{20}
10514	2-Propenal, 3-(4-hydroxy-3-methoxyphenyl)-	$C_{10}H_{10}O_3$ / 178.19	458-36-6 / 84	/	4-08-00-01979 / 1.1562^{102}	bz 4; eth 4; EtOH 4
10515	2-Propenal, 2-methyl- / Methacrolein	C_4H_6O / 70.09	78-85-3	/ 68.4	4-01-00-03455 / 0.840^{25}	H_2O 5; EtOH 5; eth 5 / 1.4144^{20}
10516	2-Propenal, 2-methyl-3-phenyl-	$C_{10}H_{10}O$ / 146.19	101-39-3	/ 150^{100}	2-07-00-00291 / 1.0407^{17}	/ 1.6057^{17}
10517	2-Propenal, 3-(4-methylphenyl)-	$C_{10}H_{10}O$ / 146.19	1504-75-2 / 41.5	/ 154^{25}	4-07-00-01013 / 0.9670^{23}	EtOH 4
10518	2-Propenal, 3-phenyl-, (E)- / trans-Cinnamaldehyde	C_9H_8O / 132.16	14371-10-9 / -7.5	/ 253 dec; 127^{16}	4-07-00-00984 / 1.0497^{20}	H_2O 2; EtOH 3; eth 3; chl 3 / 1.6195^{20}
10519	2-Propenamide / Acrylamide	C_3H_5NO / 71.08	79-06-1 / 84.5	123 / 192.6	4-02-00-01471 /	H_2O 4; EtOH 3; eth 3; ace 3
10520	2-Propenamide, N-(3,4-dichlorophenyl)-2-methyl- / Dicryl	$C_{10}H_9Cl_2NO$ / 230.09	2164-09-2 / 128	3078 /		ace 4; EtOH 4
10521	2-Propenamide, N-(1,1-dimethyl-3-oxobutyl)- / Diacetone acrylamide	$C_9H_{15}NO_2$ / 169.22	2873-97-4	2943 /		chl 3
10522	2-Propen-1-amine / Allylamine	C_3H_7N / 57.10	107-11-9 / -88.2	285 / 53.3	4-04-00-01057 / 0.758^{20}	H_2O 5; EtOH 5; eth 5; chl 3 / 1.4205^{20}
10523	2-Propen-1-amine, N,N-di-2-propenyl- / Triallylamine	$C_9H_{15}N$ / 137.22	102-70-5 / 94	/ 155.5	4-04-00-01061 / 0.809^{20}	EtOH 3; eth 3; ace 3; bz 3 / 1.4502^{20}
10524	2-Propen-1-amine, N-methyl-	C_4H_9N / 71.12	627-37-2	/ 65	4-04-00-01058 /	H_2O 4; ace 4; eth 4; EtOH 4 / 1.4065^{20}
10525	2-Propen-1-amine, N-(1-methylethyl)-	$C_6H_{13}N$ / 99.18	35000-22-7	/ 96.5	4-04-00-01059 / 0.7400^{30}	/ 1.4140^{25}
10526	2-Propen-1-amine, N-2-propenyl- / Diallylamine	$C_6H_{11}N$ / 97.16	124-02-7	2951 / 111	4-04-00-01060 /	EtOH 3; eth 3 / 1.4387^{20}

No.	Name / Synonym	Mol. Form. / Mol. Wt.	CAS RN / mp/°C	Merck No. / bp/°C	Beil. Ref. / den/g cm⁻³	Solubility / n_D
10527	2-Propen-1-amine, N-propyl-	$C_6H_{13}N$ 99.18	5666-21-7	112	0-04-00-00207 0.7708[18]	H_2O 4 1.4140[25]
10528	1-Propene Propylene	C_3H_6 42.08	115-07-1 -185.2	7862 -47.6	4-01-00-00725 *0.505[25]	H_2O 4; EtOH 4; HOAc 4 1.3567[-70]
10529	1-Propene, 1-bromo-, (Z)-	C_3H_5Br 120.98	590-13-6 -113	57.8	4-01-00-00754 1.4291[20]	H_2O 1; eth 3; ace 3; chl 3 1.4560[20]
10530	1-Propene, 2-bromo-	C_3H_5Br 120.98	557-93-7 -126	48.4	4-01-00-00754 1.3965[16]	H_2O 1; eth 3; ace 3; chl 3 1.44671[6]
10531	1-Propene, 3-bromo- Allyl bromide	C_3H_5Br 120.98	106-95-6 -119	286 70.1	4-01-00-00754 1.398[20]	H_2O 1; EtOH 5; eth 5; ctc 3 1.4697[20]
10532	1-Propene, 3-bromo-3,3-difluoro- Propylene, 3-bromo-3,3-difluoro	$C_3H_3BrF_2$ 156.96	420-90-6	42	4-01-00-00756 1.543[25]	1.3773[25]
10533	1-Propene, 1-chloro-, (E)- trans-1-Chloropropene	C_3H_5Cl 76.53	16136-85-9 -99	37.4	4-01-00-00737 0.9349[20]	H_2O 1; eth 3; ace 3; bz 3 1.4054[20]
10534	1-Propene, 1-chloro-, (Z)- cis-1-Chloropropene	C_3H_5Cl 76.53	16136-84-8 -134.8	32.8	4-01-00-00737 0.9347[20]	H_2O 1; eth 3; ace 3; bz 3 1.4055[20]
10535	1-Propene, 2-chloro- Isopropenyl chloride	C_3H_5Cl 76.53	557-98-2 -137.4	22.6	4-01-00-00737 0.9017[20]	H_2O 1; eth 3; ace 3; bz 3 1.3973[20]
10536	1-Propene, 3-chloro- Allyl chloride	C_3H_5Cl 76.53	107-05-1 -134.5	287 45.1	4-01-00-00738 0.9376[20]	H_2O 1; EtOH 5; eth 5; ace 5 1.4157[20]
10537	1-Propene, 3-chloro-2-(chloromethyl)-	$C_4H_6Cl_2$ 125.00	1871-57-4 -14	138	4-01-00-00805 1.1782[20]	EtOH 4; chl 4 1.4753
10538	1-Propene, 2-chloro-3-isothiocyanato-	C_4H_4ClNS 133.60	14214-31-4	182	0-04-00-00219 1.27[12]	EtOH 2
10539	1-Propene, 1-chloro-2-methyl- 1-Chloro-2-methylpropene	C_4H_7Cl 90.55	513-37-1	2147 68	4-01-00-00803 0.9186[20]	chl 3 1.4221[20]
10540	1-Propene, 2-chloro-2-methyl- 3-Chloro-2-methylpropene	C_4H_7Cl 90.55	563-47-3	2148 71.5	4-01-00-00803 0.9165[20]	EtOH 5; eth 5; ace 3; chl 4 1.4291[20]
10541	1-Propene, 2-chloro-1,1,3,3,3-pentafluoro-	C_3ClF_5 166.48	2804-50-4 -130	6.8	4-01-00-00741 1.5920[20]	
10542	1-Propene, 1,1-dibromo-	$C_3H_4Br_2$ 199.87	13195-80-7	125	4-01-00-00759 1.9767[20]	H_2O 2; bz 3; ctc 3; chl 3 1.5260[20]
10543	1-Propene, 1,3-dibromo-, (E)- Propylene, 1,3-dibromo (trans)	$C_3H_4Br_2$ 199.87	32121-07-6	66[25]	4-01-00-00760 1.9791[20]	1.5495[20]
10544	1-Propene, 1,3-dibromo-, (Z)- Propylene, 1,3-dibromo (cis)	$C_3H_4Br_2$ 199.87	32121-06-5	60[25]	4-01-00-00760 2.0599[20]	1.5550[20]
10545	1-Propene, 2,3-dibromo- 2,3-Dibromopropene	$C_3H_4Br_2$ 199.87	513-31-5	3009 141; 37.7[11]	4-01-00-00760 2.0345[25]	H_2O 1; eth 3; ace 3; chl 3 1.5416[25]
10546	1-Propene, 1,3-dibromo-2-methyl- Isobutylene, 1,3-dibromo	$C_4H_6Br_2$ 213.90	35911-17-2	102[100]	3-01-00-00769 1.8939[4]	1.5396[15]
10547	1-Propene, 1,1-dichloro-	$C_3H_4Cl_2$ 110.97	563-58-6	76.5	4-01-00-00742 1.1864[25]	H_2O 1; eth 3; ace 3; chl 3 1.4430[25]
10548	1-Propene, 1,2-dichloro-, (E)-	$C_3H_4Cl_2$ 110.97	7069-38-7	77	4-01-00-00742 1.1818[20]	H_2O 1; EtOH 4; ctc 4; MeOH 4 1.4471[20]
10549	1-Propene, 1,2-dichloro-, (Z)-	$C_3H_4Cl_2$ 110.97	6923-20-2	93	4-01-00-00742	H_2O 1; ace 3; bz 3; chl 3 1.4549[20]
10550	1-Propene, 1,3-dichloro-, (E)- trans-1,3-Dichloropropylene	$C_3H_4Cl_2$ 110.97	10061-02-6	3064 112	4-01-00-00744 1.217[20]	H_2O 1; eth 3; bz 3; chl 3 1.4730[20]
10551	1-Propene, 1,3-dichloro-, (Z)- cis-1,3-Dichloropropylene	$C_3H_4Cl_2$ 110.97	10061-01-5	3064 104.3	4-01-00-00743 1.224[20]	H_2O 1; eth 3; bz 3; chl 3 1.4682[20]
10552	1-Propene, 2,3-dichloro- 2,3-Dichloropropene	$C_3H_4Cl_2$ 110.97	78-88-6 10	94	4-01-00-00744 1.211[20]	H_2O 1; EtOH 5; eth 3; bz 3 1.4603[20]
10553	1-Propene, 3,3-dichloro-	$C_3H_4Cl_2$ 110.97	563-57-5	84.4	4-01-00-00745 1.175[20]	bz 4; eth 4; EtOH 4 1.4510[20]
10554	1-Propene, 1,1-dichloro-2-fluoro- Propylene, 1,1-dichloro-2-fluoro	$C_3H_3Cl_2F$ 128.96	430-95-5	78	4-01-00-00745 1.3026[25]	1.4196[25]
10555	1-Propene, 1,1-dichloro-2-methyl-	$C_4H_6Cl_2$ 125.00	6065-93-6 -43	108.5	4-01-00-00804 1.1447[20]	bz 4; eth 4; chl 4 1.4580[20]
10556	1-Propene, 3,3-dichloro-2-methyl-	$C_4H_6Cl_2$ 125.00	22227-75-4	108	4-01-00-00805 1.136[24]	bz 4; eth 4; chl 4 1.4523[24]
10557	1-Propene, 1,2-dichloro-1,3,3,3-tetrafluoro-	$C_3Cl_2F_4$ 182.93	431-53-8 -137	47.3	4-01-00-00747 1.5468[25]	1.3511[20]
10558	Propene, 1,2-dichloro-3,3,3-trifluoro- Propylene, 1,2-dichloro-3,3,3-trifluoro	$C_3HCl_2F_3$ 164.94	431-27-6 -109.2	53.7	4-01-00-00746 1.4653[20]	1.3670[20]
10559	1-Propene, 1,1-diethoxy-	$C_7H_{14}O_2$ 130.19	21504-43-8	133.5	4-01-00-03434 0.8628[25]	1.4083[25]
10560	1-Propene, 3,3-diethoxy- Acrolein, diethyl acetal	$C_7H_{14}O_2$ 130.19	3054-95-3	123.5	4-01-00-03437 0.8543[15]	H_2O 2; EtOH 5; eth 5 1.4000[20]
10561	1-Propene, 1,1-difluoro 1,1-Difluoro-1-propene	$C_3H_4F_2$ 78.06	430-63-7		4-01-00-03437	
10562	1-Propene, 3,3-dimethoxy-	$C_5H_{10}O_2$ 102.13	6044-68-4	88	4-01-00-03437 0.862[25]	1.3954[20]
10563	2-Propene-1,1-diol, diacetate	$C_7H_{10}O_4$ 158.15	869-29-4 -37.6	180	4-02-00-00291 1.0760[20]	ace 4; bz 4; eth 4; EtOH 4 1.4193[20]
10564	2-Propene-1,1-diol, 2-methyl-, diacetate Methacrolein diacetate	$C_8H_{12}O_4$ 172.18	10476-95-6	191	4-02-00-00292	1.4241[20]
10565	1-Propene, 3-(ethenyloxy)-	C_5H_8O 84.12	3917-15-5	66	4-01-00-02085 0.8050[20]	H_2O 1; eth 3; ace 3; chl 3 1.4062[20]

No.	Name Synonym	Mol. Form. Mol. Wt.	CAS RN mp/°C	Merck No. bp/°C	Beil. Ref. den/g cm^{-3}	Solubility n_D
10566	1-Propene, 3-ethoxy- Allyl ethyl ether	$C_5H_{10}O$ 86.13	557-31-3	291 67.6	4-01-00-02083 0.7651[20]	H_2O 1; EtOH 5; eth 5; ace 3 1.3881[20]
10567	1-Propene, 1-fluoro- (Z) cis-1-Fluoropropene	C_3H_5F 60.07	19184-10-2			
10568	1-Propene, 1-fluoro- (E) trans-1-Fluoropropene	C_3H_5F 60.07	20327-65-5			
10569	1-Propene, 2-fluoro- 2-Fluoropropene	C_3H_5F 60.07	1184-60-7	-24	4-01-00-00733	
10570	1-Propene, 3-fluoro- 3-Fluoropropene	C_3H_5F 60.07	818-92-8	-3	4-01-00-00733	H_2O 2; EtOH 4; eth 4; chl 3
10571	1-Propene, 1,1,2,3,3,3-hexachloro- Hexachloropropylene	C_3Cl_6 248.75	1888-71-7 -72.9	209.5	4-01-00-00753 1.7632[20]	H_2O 1; ctc 3; chl 3 1.5091[20]
10572	1-Propene, 1,1,2,3,3,3-hexafluoro- Perfluoropropene	C_3F_6 150.02	116-15-4 -156.5	-29.6	4-01-00-00735	1.583[-40]
10573	1-Propene, 3-iodo- Allyl iodide	C_3H_5I 167.98	556-56-9 -99.3	292 103	4-01-00-00761 1.848[12]	H_2O 1; EtOH 3; eth 3; chl 3 1.5540[21]
10574	1-Propene, 3-isocyano- Allyl isocyanide	C_4H_5N 67.09	2835-21-4	98	0-04-00-00208 0.794[17]	eth 4; EtOH 4
10575	1-Propene, 3-isothiocyanato- Allyl isothiocyanate	C_4H_5NS 99.16	57-06-7 -80	293 152	4-04-00-01081 1.0126[20]	bz 4; eth 4; EtOH 4 1.5306[20]
10576	1-Propene, 3-methoxy-	C_4H_8O 72.11	627-40-7	44	4-01-00-02083 0.77[11]	H_2O 1; EtOH 5; eth 5; ace 3 1.3778[20]
10577	1-Propene, 3-methoxy-2-methyl-	$C_5H_{10}O$ 86.13	22418-49-1	68	3-01-00-01904 0.7698[20]	1.3964[20]
10578	1-Propene, 2-methyl- Isobutene	C_4H_8 56.11	115-11-7 -140.4	5024 -6.9	4-01-00-00796 *0.589[25]	H_2O 1; EtOH 4; eth 4; bz 3 1.3926[-25]
10579	1-Propene, 3-(1-methylethoxy)-	$C_6H_{12}O$ 100.16	6140-80-3	83.5	4-01-00-02083 0.7764[20]	ace 4; eth 4; EtOH 4 1.3946[20]
10580	1-Propene, 2-methyl-, tetramer Propene, 2-methyl(tetramer)	$C_{16}H_{32}$ 224.43	15220-85-6 -98	244; 109[15]	3-01-00-00763 0.7944[20]	1.4482[20]
10581	1-Propene, 3-(methylthio)-	C_4H_8S 88.17	10152-76-8	92	4-01-00-02095 0.8767[20]	1.4714[20]
10582	1-Propene, 2-methyl-, trimer Isobutylene (trimer)	$C_{12}H_{24}$ 168.32	7758-94-7 -76	180	3-01-00-00763 0.7590[20]	1.4314[20]
10583	2-Propenenitrile Acrylonitrile	C_3H_3N 53.06	107-13-1 -83.5	125 77.3	4-02-00-01473 0.8060[20]	ace 4; bz 4; eth 4; EtOH 4 1.3911[20]
10584	2-Propenenitrile, 2-chloro-	C_3H_2ClN 87.51	920-37-6 -65	88.5	4-02-00-01484 1.096[25]	1.4290[20]
10585	2-Propenenitrile, 3-(2-furanyl)- 2-Furanacrylonitrile	C_7H_5NO 119.12	7187-01-1 38	4208 96	4-18-00-04152	tol 4 1.5824[25]
10586	2-Propenenitrile, 2-methyl- Methacrylonitrile	C_4H_5N 67.09	126-98-7 -35.8	5850 90.3	4-02-00-01539 0.8001[20]	H_2O 1; EtOH 5; eth 5; ace 5 1.4003[20]
10587	2-Propenenitrile, 3-phenyl-, (E)-	C_9H_7N 129.16	1885-38-7 22	263.8	4-09-00-02027 1.0304[20]	H_2O 1; EtOH 3; ace 3; ctc 3 1.6013[20]
10588	2-Propenenitrile, 3-phenyl-, (Z)-	C_9H_7N 129.16	24840-05-9 -4.4	249; 139[30]	4-09-00-02027 1.0289[20]	H_2O 1; EtOH 3; bz 4 1.5843[20]
10589	1-Propene, 1-nitro-	$C_3H_5NO_2$ 87.08	3156-70-5	60[34]; 37[10]	3-01-00-00715 1.0661[20]	eth 3; ace 3; chl 3 1.4527[20]
10590	1-Propene, 2-nitro-	$C_3H_5NO_2$ 87.08	4749-28-4	57[100]; 32[30]	4-01-00-00764 1.0643[20]	eth 3; ace 3; chl 3 1.4358[20]
10591	1-Propene, 3-nitro-	$C_3H_5NO_2$ 87.08	625-46-7	127.5	4-01-00-00764 1.051[22]	eth 4; EtOH 4
10592	1-Propene, 3,3'-oxybis- Allyl ether	$C_6H_{10}O$ 98.14	557-40-4 -6	290 94	4-01-00-02086 0.8260[20]	H_2O 1; EtOH 5; eth 5; ace 4 1.4163[20]
10593	1-Propene, 1,1,2,3,3-pentachloro-	C_3HCl_5 214.30	1600-37-9	185	4-01-00-00753 1.6317[34]	eth 4 1.5313[20]
10594	1-Propene, 3-propoxy-	$C_6H_{12}O$ 100.16	1471-03-0	91	4-01-00-02083 0.7764[20]	ace 4; eth 4; EtOH 4 1.3919[20]
10595	2-Propene-1-sulfinothioic acid, S-2-propenyl ester Allicin	$C_6H_{10}OS_2$ 162.28	539-86-6 dec	249	4-04-00-00007 1.112[20]	H_2O 4 1.561[20]
10596	1-Propene, 3,3'-sulfinylbis-	$C_6H_{10}OS$ 130.21	14180-63-3 23.5	114[12]; 89[2.6]	4-01-00-02097 1.0261[20]	EtOH 3; eth 3; ace 3 1.5115[20]
10597	1-Propene, 3,3'-sulfonylbis-	$C_6H_{10}O_2S$ 146.21	16841-48-8	109[3]	4-01-00-02098 1.1213[20]	1.4893[20]
10598	1-Propene, 1,2,3,3-tetrachloro-	$C_3H_2Cl_4$ 179.86	20589-85-9	162	3-01-00-00708 1.537[19]	bz 4; chl 4 1.5121[19]
10599	1-Propene, 3,3'-thiobis- Diallyl sulfide	$C_6H_{10}S$ 114.21	592-88-1 -85	295 138.6	4-01-00-02097 0.8887[27]	eth 4; EtOH 4 1.4870[25]
10600	2-Propene-1-thiol	C_3H_6S 74.15	870-23-5	65	4-01-00-02095 0.925[23]	H_2O 1; EtOH 5; eth 5; chl 3 1.4832[10]
10601	1-Propene-1,2,3-tricarboxylic acid, triethyl ester, (E)- Aconitic acid, triethyl ester (trans)	$C_{12}H_{18}O_6$ 258.27	68077-28-1	275 dec; 159[9]	4-02-00-02405 1.1064[20]	eth 4; EtOH 4
10602	1-Propene, 1,1,2-trichloro-	$C_3H_3Cl_3$ 145.41	21400-25-9	118	4-01-00-00747 1.382[20]	H_2O 1; EtOH 3; eth 3; bz 3 1.4827[20]
10603	1-Propene, 1,2,3-trichloro-	$C_3H_3Cl_3$ 145.41	96-19-5	142	4-01-00-00748 1.412[20]	H_2O 1; EtOH 4; eth 4; bz 3 1.5030[20]
10604	1-Propene, 3,3,3-trichloro-	$C_3H_3Cl_3$ 145.41	2233-00-3 -30	114.5	4-01-00-00749 1.367[20]	H_2O 1; EtOH 3; eth 3; bz 3 1.4827[20]

No.	Name Synonym	Mol. Form. Mol. Wt.	CAS RN mp/°C	Merck No. bp/°C	Beil. Ref. den/g cm^{-3}	Solubility n_D
10605	1-Propene, 1,1,3-trichloro-2-methyl-	$C_4H_5Cl_3$ 159.44	31702-33-7	156	4-01-00-00805 1.346[20]	ace 4; bz 4; chl 4 1.4990[20]
10606	1-Propene, 3,3,3-trichloro-2-methyl-	$C_4H_5Cl_3$ 159.44	4749-27-3	133	4-01-00-00805 1.293[20]	H_2O 1; ace 3; bz 3; chl 4 1.4770[20]
10607	1-Propene, 1,1,2-trichloro-3,3,3-trifluoro-	$C_3Cl_3F_3$ 199.39	431-52-7 -114.6	88.1	4-01-00-00751 1.617[20]	
10608	1-Propene, 3,3,3-trifluoro- 3,3,3-Trifluoropropene	$C_3H_3F_3$ 96.05	677-21-4	-17	4-01-00-00734 0	
10609	2-Propenoic acid Acrylic acid	$C_3H_4O_2$ 72.06	79-10-7 12.3	124 141	4-02-00-01455 1.0511[20]	H_2O 5; EtOH 5; eth 5; ace 3 1.4224[20]
10610	2-Propenoic acid, 2-bromo-	$C_3H_3BrO_2$ 150.96	10443-65-9 71.5	dec	4-02-00-01486	H_2O 4; EtOH 4
10611	2-Propenoic acid, 2-bromoethyl ester	$C_5H_7BrO_2$ 179.01	4823-47-6	156.5; 53[5]	4-02-00-01462 1.4581[25]	1.4660[25]
10612	2-Propenoic acid, 3-bromo-2-methyl- Methacrylic acid, 3-bromo	$C_4H_5BrO_2$ 164.99	89123-63-7 64	229	0-02-00-00424	H_2O 4
10613	2-Propenoic acid, 2-bromo-3-phenyl-, ethyl ester	$C_{11}H_{11}BrO_2$ 255.11	57296-00-1	295	4-09-00-02042 1.3885[25]	1.5845[25]
10614	2-Propenoic acid, 3-bromo-3-phenyl-, ethyl ester, (Z)- Cinnamic acid, β-bromo-, ethyl ester	$C_{11}H_{11}BrO_2$ 255.11	1504-70-7	167[12]	3-09-00-02733 1.4044[10]	1.5841[10]
10615	2-Propenoic acid, 2-bromo-3-phenyl-, methyl ester Methyl α-bromocinnamate	$C_{10}H_9BrO_2$ 241.08	24127-62-6 23	120-1[0.6]	4-09-00-02042 1.4975[20]	
10616	2-Propenoic acid, butyl ester Butyl acrylate	$C_7H_{12}O_2$ 128.17	141-32-2 -64.6	1539 145	4-02-00-01463 0.8998[20]	H_2O 1; EtOH 3; eth 3; ace 3 1.4185[20]
10617	2-Propenoic acid, 3-chloro-, (E)-	$C_3H_3ClO_2$ 106.51	2345-61-1 86	94[18]	4-02-00-01481	EtOH 3; eth 3; chl 2
10618	2-Propenoic acid, 3-chloro-, (Z)-	$C_3H_3ClO_2$ 106.51	1609-93-4 63.5	106[17]	4-02-00-01481	EtOH 3; eth 3
10619	2-Propenoic acid, chloro-, ethyl ester Acrylic acid, chloro-, ethyl ester	$C_5H_7ClO_2$ 134.56	30600-19-2	51-3[18]	3-02-00-01245 1.1404[20]	EtOH 4; eth 4 1.4384[20]
10620	2-Propenoic acid, 2-chloroethyl ester	$C_5H_7ClO_2$ 134.56	2206-89-5	51-3[18]	3-02-00-01225 1.1604[20]	eth 4; EtOH 4 1.4384[20]
10621	2-Propenoic acid, 2-chloro-, methyl ester Methyl 2-chloroacrylate	$C_4H_5ClO_2$ 120.54	80-63-7	52[51]	4-02-00-01482 1.189[20]	eth 4 1.4420[20]
10622	2-Propenoic acid, 2-chloro-3-phenyl-, methyl ester, (E)- Cinnamic acid, α-chloro-, methyl ester, (E)-	$C_{10}H_9ClO_2$ 196.63	14737-94-1 33	108-9[0.5]	4-09-00-02035	ace 4
10623	2-Propenoic acid, 3-chloro-3-phenyl-, methyl ester Cinnamic acid, β-chloro-, methyl ester	$C_{10}H_9ClO_2$ 196.63	87541-87-5 29	113-4[0.5]	1.2248[21]	ace 4 1.5781[21]
10624	2-Propenoic acid, 2-cyano-, methyl ester Mecrylate	$C_5H_5NO_2$ 111.10	137-05-3	5667 47-8[2]	4-02-00-02226 1.1012[20]	1.4430
10625	2-Propenoic acid, 2-cyano-3-phenyl-, ethyl ester	$C_{12}H_{11}NO_2$ 201.22	2025-40-3 51	188[15]	2-09-00-00640 1.1076[25]	ace 4; chl 4 1.5033
10626	2-Propenoic acid, cyclohexyl ester	$C_9H_{14}O_2$ 154.21	3066-71-5	183; 88[20]	4-06-00-00038 1.0275[20]	H_2O 1; EtOH 5; eth 5; chl 3 1.4673[20]
10627	2-Propenoic acid, 2,3-dibromo-, (Z)-	$C_3H_2Br_2O_2$ 229.86	24557-10-6 85.5	148[6]	3-02-00-01249	H_2O 2; EtOH 3; eth 3; bz 2
10628	2-Propenoic acid, 3,3-dibromo- Acrylic acid, 3,3-dibromo-	$C_3H_2Br_2O_2$ 229.86	1578-21-8 86.3	245	4-02-00-01487	eth 4; EtOH 4
10629	2-Propenoic acid, 2,3-dibromo-3-phenyl-, (Z)-	$C_9H_6Br_2O_2$ 305.95	708-82-7 100	124[0.5]	4-09-00-02043	H_2O 1; EtOH 3; eth 3; chl 3
10630	2-Propenoic acid, 3,3-dichloro- Acrylic acid, 3,3-dichloro-	$C_3H_2Cl_2O_2$ 140.95	1561-20-2 77.5	sub	4-02-00-01484	eth 4; chl 4
10631	2-Propenoic acid, 2-(diethylamino)ethyl ester	$C_9H_{17}NO_2$ 171.24	2426-54-2 <-60	81[10]	4-04-00-01477 0.937[20]	1.4376[25]
10632	2-Propenoic acid, 3-(3,4-dihydroxyphenyl)- Caffeic acid	$C_9H_8O_4$ 180.16	331-39-5 225 dec	1634	2-10-00-00294	EtOH 4
10633	2-Propenoic acid, 2-(dimethylamino)ethyl ester	$C_7H_{13}NO_2$ 143.19	2439-35-2 <-60	95[50]	4-04-00-01431 0.938[20]	
10634	2-Propenoic acid, 2-ethoxyethyl ester	$C_7H_{12}O_3$ 144.17	106-74-1 -47	174	4-02-00-01469 0.983[20]	1.4274[20]
10635	2-Propenoic acid, ethyl ester Ethyl acrylate	$C_5H_8O_2$ 100.12	140-88-5 -71.2	3715 99.4	4-02-00-01460 0.9234[20]	H_2O 2; EtOH 5; eth 5; chl 3 1.4068[20]
10636	2-Propenoic acid, 2-ethylhexyl ester 2-Ethylhexyl acrylate	$C_{11}H_{20}O_2$ 184.28	103-11-7 -90	123-7[60]	4-02-00-01467 0.880[25]	1.4332[25]
10637	2-Propenoic acid, 2-fluoro-3-phenyl- Cinnamic acid, α-fluoro-	$C_9H_7FO_2$ 166.15	350-90-3 157.6	290	1-09-00-00237	EtOH 3; eth 3
10638	2-Propenoic acid, 3-(2-furanyl)- 2-Furanacrylic acid	$C_7H_6O_3$ 138.12	539-47-9 141	4207 286	5-18-06-00306	eth 4; EtOH 4
10639	2-Propenoic acid, 3-(2-furanyl)-, ethyl ester	$C_9H_{10}O_3$ 166.18	623-20-1 24.5	232	5-18-06-00307 1.0891[25]	H_2O 1; EtOH 5; eth 5; chl 2 1.5286[20]
10640	2-Propenoic acid, 3-(2-furanyl)-, methyl ester	$C_8H_8O_3$ 152.15	623-18-7 36	227; 117[23]	4-18-00-04144 1.4447[20]	bz 4; eth 4; EtOH 4

No.	Name Synonym	Mol. Form. Mol. Wt.	CAS RN mp/°C	Merck No. bp/°C	Beil. Ref. den/g cm^{-3}	Solubility n_D
10641	2-Propenoic acid, 3-(2-furanyl)-, pentyl ester Acrylic acid, 3-(2-furyl), pentyl ester	$C_{12}H_{16}O_3$ 208.26	2438-19-9	116-8[2]	4-18-00-04146 1.0322[20]	eth 4; EtOH 4 1.5289[14]
10642	2-Propenoic acid, 3-(2-furanyl)-, propyl ester Acrylic acid, 3-(2-furyl), propyl ester	$C_{10}H_{12}O_3$ 180.20	623-22-3	91-4[3]	4-18-00-04146 1.0744[20]	bz 4; eth 4; EtOH 4 1.5392[24]
10643	2-Propenoic acid, hexyl ester	$C_9H_{16}O_2$ 156.22	2499-95-8 -45	40[1]	4-02-00-01466 0.878[20]	
10644	2-Propenoic acid, 3-(4-hydroxyphenyl)- p-Coumaric acid	$C_9H_8O_3$ 164.16	7400-08-0 211.5	2561	4-10-00-01005	eth 4; EtOH 4
10645	2-Propenoic acid, 2-isocyano-3-phenyl-, ethyl ester Malonomononitrile, benzylidene,monoethyl ester	$C_{12}H_{11}NO_2$ 201.22	52744-85-1 51	188[15]; 118[0.4]	1.0762[25]	H_2O 1; EtOH 2; eth 3; bz 4 1.5033
10646	2-Propenoic acid, isodecyl ester	$C_{13}H_{24}O_2$ 212.33	1330-61-6 -100	158[50]	0.885[20]	1.4416[20]
10647	2-Propenoic acid, 3-(2-methoxyphenyl)-, methyl ester	$C_{11}H_{12}O_3$ 192.21	15854-58-7	304; 161-3[3]	3-10-00-00835 1.1366[17]	1.5854[16]
10648	2-Propenoic acid, 2-methyl- Methacrylic acid	$C_4H_8O_2$ 86.09	79-41-4 16	5849 162.5	4-02-00-01518 1.0153[20]	H_2O 3; EtOH 5; eth 5; chl 3 1.4314[20]
10649	2-Propenoic acid, 2-methyl-, anhydride Methacrylic acid anhydride	$C_8H_{10}O_3$ 154.17	760-93-0	89[5]	4-02-00-01537	EtOH 5; eth 5 1.4540[20]
10650	2-Propenoic acid, 2-methyl-, butyl ester Butyl methacrylate	$C_8H_{14}O_2$ 142.20	97-88-1	160	4-02-00-01525 0.8936[20]	eth 4; EtOH 4 1.4240[20]
10651	2-Propenoic acid, methylbutyl ester Propenoic acid, 2-methyl,butyl ester	$C_8H_{14}O_2$ 142.20	90638-48-5	160; 71[37]	3-02-00-01286 0.8936[20]	H_2O 1; EtOH 5; eth 5; ctc 2 1.4240[20]
10652	2-Propenoic acid, 2-methylbutyl ester Acrylic acid, (2-methylbutyl) ester	$C_8H_{14}O_2$ 142.20	44914-03-6	160; 45[10]	3-02-00-01228 0.8936[20]	eth 4; EtOH 4 1.4240[20]
10653	2-Propenoic acid, 2-methyl-, cyclohexyl ester	$C_{10}H_{18}O_2$ 168.24	101-43-9	210	4-06-00-00039 0.9626[20]	1.4578[20]
10654	2-Propenoic acid, 2-methyl-, 2-(dimethylamino)ethyl ester β-Dimethylaminoethyl methacrylate	$C_8H_{15}NO_2$ 157.21	2867-47-2	62-5[6]	4-04-00-01432	
10655	2-Propenoic acid, methyl ester Methyl acrylate	$C_4H_8O_2$ 86.09	96-33-3 <-75	5935 80.7	4-02-00-01457 0.9535[20]	H_2O 2; EtOH 3; eth 3; ace 3 1.4040[20]
10656	2-Propenoic acid, 2-methyl-, 1,2-ethanediyl ester	$C_{10}H_{14}O_4$ 198.22	97-90-5 -40	260	4-02-00-01532 1.053[20]	bz 4; EtOH 4; lig 4 1.4532[25]
10657	2-Propenoic acid, 2-methyl-, 1,2-ethanediylbis(oxy-2,1-ethanediyl) ester	$C_{14}H_{22}O_6$ 286.33	109-16-0		4-02-00-01531 1.072[25]	ace 4; eth 4; EtOH 4; peth 4 1.4595[25]
10658	2-Propenoic acid, 2-methyl-, ethyl ester Ethyl methacrylate	$C_6H_{10}O_2$ 114.14	97-63-2	117	4-02-00-01523 0.9135[20]	H_2O 2; EtOH 5; eth 5; chl 2 1.4147[20]
10659	2-Propenoic acid, 2-methyl-, 2-ethylhexyl ester 2-Ethylhexyl methacrylate	$C_{12}H_{22}O_2$ 198.31	688-84-6	110[14]	4-02-00-01528 0.880[25]	1.436[25]
10660	2-Propenoic acid, 2-methyl-, hexyl ester	$C_{10}H_{18}O_2$ 170.25	142-09-6	162; 86[17]	4-02-00-01527 0.880[25]	ace 4; bz 4; eth 4; EtOH 4 1.429[25]
10661	2-Propenoic acid, 2-methyl-, 2-hydroxyethyl ester	$C_8H_{10}O_3$ 130.14	868-77-9	67[3.5]	4-02-00-01530 1.034[25]	1.4515[20]
10662	2-Propenoic acid, 2-methyl-, 2-hydroxypropyl ester	$C_7H_{12}O_3$ 144.17	923-26-2	90[9]	4-02-00-01532 1.066[25]	1.4458[20]
10663	2-Propenoic acid, 2-methyl-, methyl ester Methyl methacrylate	$C_5H_8O_2$ 100.12	80-62-6 -48	5849 100.5	4-02-00-01519 0.9440[20]	H_2O 2; EtOH 5; eth 5; ace 5 1.4142[20]
10664	2-Propenoic acid, 2-methyl-, 1-methylethyl ester	$C_7H_{12}O_2$ 128.17	4655-34-9	125	4-02-00-01525 0.8847[20]	ace 4; bz 4; eth 4; EtOH 4 1.4122[20]
10665	2-Propenoic acid, 2-methyl-, 1-methylpropyl ester	$C_8H_{14}O_2$ 142.20	2998-18-7	147; 35[4]	4-02-00-01526 0.886[20]	1.4179[20]
10666	2-Propenoic acid, 2-methyl-, 2-methylpropyl ester Isobutyl methacrylate	$C_8H_{14}O_2$ 142.20	97-86-9	155	4-02-00-01526 0.8858[20]	H_2O 1; EtOH 5; eth 5 1.4199[20]
10667	2-Propenoic acid, 2-methyl-, octadecyl ester	$C_{22}H_{42}O_2$ 338.57	32360-05-7	195[6]	4-02-00-01529 0.880[25]	1.429[25]
10668	2-Propenoic acid, 2-methyl-, oxiranylmethyl ester	$C_7H_{10}O_3$ 142.15	106-91-2	75[10]	5-17-03-00035 1.07[25]	bz 4; eth 4; EtOH 4 1.448[25]
10669	2-Propenoic acid, 2-methyl-, oxydi-2,1-ethanediyl ester	$C_{12}H_{18}O_5$ 242.27	2358-84-1	>200	4-02-00-01531 1.056[20]	1.4571[25]
10670	2-Propenoic acid, 2-methyl-3-phenyl-, (Z)-	$C_{10}H_{10}O_2$ 162.19	15250-29-0 94	288	4-09-00-02063	H_2O 2; EtOH 4; eth 4; bz 4
10671	2-Propenoic acid, 2-methyl-, 2-propenyl ester	$C_7H_{10}O_2$ 126.16	96-05-9	55[30]; 60[43]	4-02-00-01529 0.927[20]	1.4360[20]
10672	2-Propenoic acid, 2-methyl-, propyl ester Propyl methacrylate	$C_7H_{12}O_2$ 128.17	2210-28-8	141	4-02-00-01524 0.9022[20]	H_2O 1; EtOH 5; eth 5 1.4190[20]
10673	2-Propenoic acid, 2-methylpropyl ester Isobutyl acrylate	$C_7H_{12}O_2$ 128.17	106-63-8 -61	132	4-02-00-01465 0.8896[20]	H_2O 2; EtOH 3; eth 3; MeOH 3 1.4150[20]

No.	Name / Synonym	Mol. Form. / Mol. Wt.	CAS RN / mp/°C	Merck No. / bp/°C	Beil. Ref. / den/g cm⁻³	Solubility / n_D
10674	2-Propenoic acid, 2-methyl-, (tetrahydro-2-furanyl)methyl ester	$C_9H_{14}O_3$ / 170.21	2455-24-5	265; 81[4]	4-17-00-01105 / 1.040[25]	1.4554[25]
10675	2-Propenoic acid, 3-(1-naphthalenyl)-, (E)- / Acrylic acid, 3-(1-naphtyl) (trans)	$C_{13}H_{10}O_2$ / 198.22	2006-14-6 / 211.5	sub	4-09-00-02486	H_2O 2; EtOH 3; eth 3; chl 3
10676	2-Propenoic acid, 3-(2-nitrophenyl)-, (E)-	$C_9H_7NO_4$ / 193.16	1013-96-3 / 244.3	sub	4-09-00-02043	H_2O 1; EtOH 3; bz 1
10677	2-Propenoic acid, 3-(2-nitrophenyl)-, methyl ester, (E)- / Cinnamic acid, 2-nitro, methyl ester (trans)	$C_{10}H_9NO_4$ / 207.19	39228-29-0 / 73	187[15]	4-09-00-02044	EtOH 4
10678	2-Propenoic acid, 3-(3-nitrophenyl)-, methyl ester, (E)- / Cinnamic acid, 3-nitro, methyl ester (trans)	$C_{10}H_9NO_4$ / 207.19	659-04-1 / 125.3	dec	3-09-00-02742	bz 4; eth 4; chl 4
10679	2-Propenoic acid, 3-(4-nitrophenyl)-, methyl ester, (E)-	$C_{10}H_9NO_4$ / 207.19	637-57-0 / 162	283.5	4-09-00-02046	EtOH 4
10680	2-Propenoic acid, oxiranylmethyl ester	$C_6H_8O_3$ / 128.13	106-90-1	53[10]	4-17-00-01005 / 1.1[20]	bz 4 / 1.4490[20]
10681	2-Propenoic acid, 2-phenoxyethyl ester	$C_{11}H_{12}O_3$ / 192.21	48145-04-6	110[2]	3-06-00-00572 / 1.090[25]	ace 4; eth 4; chl 4
10682	2-Propenoic acid, 3-phenyl-, (E)- / trans-Cinnamic acid	$C_9H_8O_2$ / 148.16	140-10-3 / 133	2300 / 300	4-09-00-02002 / 1.2475[4]	H_2O 1; EtOH 4; eth 3; ace 3
10683	2-Propenoic acid, 3-phenyl-, (Z)-	$C_9H_8O_2$ / 148.16	102-94-3 / 42		4-09-00-02001	EtOH 4; HOAc 4; lig 4
10684	2-Propenoic acid, 3-phenyl-, anhydride / Cinnamic anhydride	$C_{18}H_{14}O_3$ / 278.31	538-56-7 / 136	2300		bz 4
10685	2-Propenoic acid, 3-phenyl-, ethyl ester, (E)-	$C_{11}H_{12}O_2$ / 176.22	4192-77-2 / 10	2300 / 271.5	4-09-00-02006 / 1.0491[20]	H_2O 1; EtOH 4; eth 4; ace 4 / 1.5598[20]
10686	2-Propenoic acid, 3-phenyl-, methyl ester, (E)-	$C_{10}H_{10}O_2$ / 162.19	1754-62-7 / 36.5	2300 / 261.9	4-09-00-02005 / 1.042[36]	H_2O 1; EtOH 4; eth 4; ace 3 / 1.5766[22]
10687	2-Propenoic acid, 3-phenyl-, 1-methylethyl ester, (E)- / Cinnamic acid, isopropyl ester (trans)	$C_{12}H_{14}O_2$ / 190.24	60512-85-8	269	4-09-00-02008 / 1.0320[20]	H_2O 1; EtOH 3; eth 3; ace 3 / 1.5455[20]
10688	2-Propenoic acid, 3-phenyl-, phenylmethyl ester	$C_{10}H_{10}O_2$ / 162.19	2495-35-4	228	3-06-00-01481 / 1.0573[20]	H_2O 1; EtOH 3; eth 3; ace 3 / 1.5143[20]
10689	2-Propenoic acid, 3-phenyl-, phenyl ester, (E)- / Cinnamic acid, phenyl ester (trans)	$C_{15}H_{12}O_2$ / 224.26	25695-77-6 / 74.3	205[15]	4-09-00-02011 / 1.0650[100]	bz 4; eth 4
10690	2-Propenoic acid, 3-phenyl-, phenylmethyl ester, (E)- / Cinnamic acid, benzyl ester (trans)	$C_{16}H_{14}O_2$ / 238.29	78277-23-3 / 39	350 dec; 244[5]	4-09-00-02012 / 1.109[15]	H_2O 1; EtOH 3; eth 3; bz 2
10691	2-Propenoic acid, 3-phenyl-, 3-phenyl-2-propenyl ester / Cinnamyl cinnamate	$C_{18}H_{16}O_2$ / 264.32	122-69-0 / 44	2307	4-09-00-02013 / 1.1565[4]	H_2O 1; EtOH 3; eth 4; chl 3
10692	2-Propenoic acid, 3-phenyl-, 2-propenyl ester / Cinnamic acid, allyl ester	$C_{12}H_{12}O_2$ / 188.23	1866-31-5	268 dec; 163[17]	4-09-00-02010 / 1.048[23]	H_2O 1; EtOH 4; eth 5; ctc 2 / 1.530[20]
10693	2-Propenoic acid, 3-phenyl-, propyl ester, (E)- / Cinnamic acid, propyl ester (trans)	$C_{12}H_{14}O_2$ / 190.24	74513-58-9	285	4-09-00-02007 / 1.0433[0]	H_2O 1
10694	2-Propenoic acid, 2-propenyl ester	$C_6H_8O_2$ / 112.13	999-55-3	121	4-02-00-01468 / 0.9441[20]	H_2O 2; EtOH 3; eth 3; acid 3 / 1.4320[20]
10695	2-Propenoic acid, (tetrahydro-2-furanyl)methyl ester	$C_8H_{12}O_3$ / 156.18	2399-48-6 / <-60	96[6]	4-17-00-01104 / 1.061[20]	
10696	2-Propenoic acid, 2,3,3-trichloro-	$C_3HCl_3O_2$ / 175.40	2257-35-4 / 76	222	4-02-00-01486	H_2O 3; EtOH 3; eth 3; ctc 3 / 1.5271[18]
10697	2-Propen-1-ol / Allyl alcohol	C_3H_6O / 58.08	107-18-6 / -129	284 / 97.0	4-01-00-02079 / 0.8540[20]	H_2O 5; EtOH 5; eth 5; chl 3 / 1.4135[20]
10698	1-Propen-2-ol, acetate / Isopropenyl acetate	$C_5H_8O_2$ / 100.12	108-22-5 / -92.9	5091 / 94	4-02-00-00179 / 0.9090[20]	H_2O 2; EtOH 3; eth 4; ace 3 / 1.4033[20]
10699	2-Propen-1-ol, 2-bromo-	C_3H_5BrO / 136.98	598-19-6	154; 62[11]	4-01-00-02094 / 1.621[18]	eth 4; chl 4 / 1.500[18]
10700	2-Propen-1-ol, 2-chloro-	C_3H_5ClO / 92.52	5976-47-6	135	4-01-00-02091 / 1.1618[20]	EtOH 3; ctc 2 / 1.4588[20]
10701	2-Propen-1-ol, 3-chloro- / Allyl alcohol, 3-chloro	C_3H_5ClO / 92.52	29560-84-7	147	4-01-00-02090 / 1.1769[20]	1.4738[20]
10702	2-Propen-1-ol, 2-methyl- / Methallyl alcohol	C_4H_8O / 72.11	513-42-8	114.5	4-01-00-02114 / 0.8515[20]	H_2O 4; EtOH 5; eth 5 / 1.4255[20]
10703	2-Propen-1-ol, 2-methyl-, carbonate (2:1) / Bis-(2-methallyl)carbonate	$C_9H_{14}O_3$ / 170.21	64057-79-0 / 201.3	66[3]	4-03-00-00011 / 0.943[25]	1.4371[20]
10704	2-Propen-1-ol, 3-phenyl-, (E)-	$C_9H_{10}O$ / 134.18	4407-36-7 / 34	257.5	4-06-00-03799 / 1.0440[20]	H_2O 2; EtOH 4; eth 4; chl 2 / 1.5819[20]
10705	2-Propen-1-ol, 3-phenyl-, (Z)-	$C_9H_{10}O$ / 134.18	4510-34-3 / 34	257.5	3-06-00-02401 / 1.0440[20]	eth 4; EtOH 4 / 1.5819[20]
10706	2-Propen-1-ol, 3-phenyl-, acetate, (E)-	$C_{11}H_{12}O_2$ / 176.22	21040-45-9	145[15]; 114[1]	4-06-00-03801 / 1.0567[20]	H_2O 1; EtOH 3; eth 3; ace 3 / 1.5425[20]
10707	2-Propen-1-ol, 3-phenyl-, formate	$C_{10}H_{10}O_2$ / 162.19	104-65-4 / 0	252	0-06-00-00571 / 1.086[25]	
10708	2-Propen-1-one, 1,3-diphenyl- / Chalcone	$C_{15}H_{12}O$ / 208.26	94-41-7 / 59	2028 / 346 dec	4-07-00-01658 / 1.0712[62]	bz 4; EtOH 4; chl 4

No.	Name Synonym	Mol. Form. Mol. Wt.	CAS RN mp/°C	Merck No. bp/°C	Beil. Ref. den/g cm^{-3}	Solubility n_D
10709	2-Propen-1-one, 1,3-diphenyl-, (E)-	$C_{15}H_{12}O$ 208.26	614-47-1 59	346 dec	4-07-00-01658 1.0712[62]	H_2O 1; EtOH 2; eth 3; bz 3
10710	2-Propen-1-one, 3-(2-furanyl)-1-phenyl-	$C_{13}H_{10}O_2$ 198.22	717-21-5 47	317	5-17-10-00371 1.1140[20]	EtOH 3; eth 3
10711	2-Propen-1-one, 3-hydroxy-1,3-diphenyl-	$C_{15}H_{12}O_2$ 224.26	1704-15-0 79.8	220[18.] 140[1.0]	4-07-00-02512	eth 4; EtOH 4; chl 4
10712	2-Propen-1-one, 3-(3-methoxyphenyl)-1-phenyl-	$C_{16}H_{14}O_2$ 238.29	5470-91-7 65	247[12]	1-08-00-00579	ace 4; bz 4; eth 4; EtOH 4
10713	2-Propen-1-one, 3-(4-methoxyphenyl)-1-phenyl-	$C_{16}H_{14}O_2$ 238.29	959-33-1 79	187[19]	2-08-00-00218	H_2O 1; EtOH 4; eth 3; ctc 3
10714	2-Propenoyl chloride Acrylic acid chloride	C_3H_3ClO 90.51	814-68-6	75.5	4-02-00-01471 1.1136[20]	chl 4 1.4343[20]
10715	2-Propenoyl chloride, 2-methyl- Methacrylic acid chloride	C_4H_5ClO 104.54	920-46-7 -60	96	4-02-00-01537 1.0871[20]	eth 3; ace 3; chl 3 1.4435[20]
10716	2-Propenoyl chloride, 3-phenyl- Cinnamoyl chloride	C_9H_7ClO 166.61	102-92-1 36	2303 257	2-09-00-00395 1.1617[45]	lig 4 1.614[42]
10717	2-Propenoyl chloride, 3-phenyl-, (E)-	C_9H_7ClO 166.61	17082-09-6 37.5	257.5	2-09-00-00390 1.1617[45]	H_2O 1; EtOH 3; ctc 3; lig 3 1.614[42]
10718	2-Propenoyl chloride, 2,3,3-trichloro-	C_3Cl_4O 193.84	815-58-7	158	4-02-00-01486	bz 4 1.5271[18]
10719	Propetamphos 2-Butenoic acid, 3-[[(ethylamino)methoxyphosphinothioyl]oxy]-, 1-methylethyl ester, (E)-	$C_{10}H_{20}NO_4PS$ 281.31	31218-83-4	7827 88[0.005]	1.1294[20]	
10720	Propiconazole 1-[[2-(2,4-Dichlorophenyl)-4-propyl-1,3-dioxolan-2-yl]methyl]-1H-1,2,4-triazole	$C_{15}H_{17}Cl_2N_3O_2$ 342.22	60207-90-1	180[0.1]	1.27[20]	
10721	Propionic acid, 2,2-dichloro-3,3,3-trifluoro-, propyl ester Propanoic acid, 2,2-dichloro-3,3,3-trifluoro, propyl ester	$C_6H_7Cl_2F_3O_2$ 239.02	357-49-3	144.5[625]	4-02-00-00758 1.3531[20]	1.3888[20]
10722	Propoxur Phenol, 2-(1-methylethoxy)-, methylcarbamate	$C_{11}H_{15}NO_3$ 209.25	114-26-1 87	7849 dec	1.12[20]	
10723	2-Propynal Propargyl aldehyde	C_3H_2O 54.05	624-67-9	60	4-01-00-03537 0.9152[20]	H_2O 5; EtOH 3; eth 3; ace 3 1.4033[25]
10724	2-Propynal, 3-phenyl-	C_9H_6O 130.15	2579-22-8	127[28.] 104[11.]	4-07-00-01173 1.0680[12]	1.6079[12]
10725	2-Propyn-1-amine	C_3H_5N 55.08	2450-71-7	83	4-04-00-01135 0.803[25]	
10726	2-Propyn-1-amine, N-methyl-	C_4H_7N 69.11	35161-71-8	83	0-04-00-00228 0.819[25]	1.4480[20]
10727	1-Propyne Methylacetylene	C_3H_4 40.06	74-99-7 -102.7	-23.2	4-01-00-00958 *0.607[25]	H_2O 2; EtOH 4; bz 3; chl 3 1.3863[-40]
10728	1-Propyne, 3-bromo- Propargyl bromide	C_3H_3Br 118.96	106-96-7	89	4-01-00-00964 1.579[19]	EtOH 3; eth 3; bz 3; ctc 3 1.4922[20]
10729	1-Propyne, 3-chloro- Propargyl chloride	C_3H_3Cl 74.51	624-65-7	7820 58	4-01-00-00963 1.030[25]	H_2O 1; EtOH 5; eth 5; bz 3 1.4349[20]
10730	1-Propyne, 1,3-dibromo- Propyne, 1,3-dibromo-	$C_3H_2Br_2$ 197.86	627-16-7	73-4[30]	4-01-00-00965 2.1894[20]	eth 4; chl 4 1.5690[20]
10731	1-Propyne, 3,3-diethoxy-	$C_7H_{12}O_2$ 128.17	10160-87-9	139	4-01-00-03538 0.8942[22]	ace 4; eth 4; EtOH 4; chl 4 1.4140[20]
10732	1-Propyne, 1-ethoxy-	C_5H_8O 84.12	14273-06-4	84	3-01-00-01969 0.8276[20]	eth 4; EtOH 4 1.4039[20]
10733	1-Propyne, 3-ethoxy-	C_5H_8O 84.12	628-33-1	82	4-01-00-02215 0.8326[20]	eth 4; EtOH 4 1.4039[20]
10734	1-Propyne, 1-iodo-	C_3H_3I 165.96	624-66-8	110	4-01-00-00965 2.08[22]	eth 4; EtOH 4
10735	1-Propyne, 3-iodo-	C_3H_3I 165.96	659-86-9	115	4-01-00-00965 2.0177[25]	eth 3
10736	1-Propyne, 3-methoxy-	C_4H_6O 70.09	627-41-8	63	4-01-00-02215 0.83[12]	eth 4; EtOH 4 1.5035[20]
10737	2-Propynenitrile Cyanoacetylene	C_3HN 51.05	1070-71-9 5	42.5	4-02-00-01689 0.8167[17]	H_2O 2; EtOH 3 1.3868[25]
10738	2-Propynoic acid Propiolic acid	$C_3H_2O_2$ 70.05	471-25-0 9	7833 144 dec; 72[50]	4-02-00-01687 1.1380[20]	H_2O 4; eth 4; EtOH 4; chl 4 1.4306[20]
10739	2-Propynoic acid, ethyl ester	$C_5H_8O_2$ 98.10	623-47-2	120	4-02-00-01688 0.9560[25]	H_2O 1; EtOH 4; eth 4; chl 4 1.4105[20]
10740	2-Propynoic acid, 3-(2-nitrophenyl)- o-Nitrophenylpropiolic acid	$C_9H_5NO_4$ 191.14	530-85-8 157 dec	6546 exp	4-09-00-02330	eth 4; EtOH 4
10741	2-Propynoic acid, 3-phenyl-, ethyl ester	$C_{11}H_{10}O_2$ 174.20	2216-94-6	265; 128[1.6]	4-09-00-02328 1.055[25]	eth 3 1.5520[20]
10742	2-Propynoic acid, 3-phenyl-, methyl ester	$C_{10}H_8O_2$ 160.17	4891-38-7 26	158[48.] 132[16.]	4-09-00-02328 1.0830[25]	1.5618[25]

No.	Name / Synonym	Mol. Form. / Mol. Wt.	CAS RN / mp/°C	Merck No. / bp/°C	Beil. Ref. / den/g cm⁻³	Solubility / n_D
10743	2-Propyn-1-ol / Propargyl alcohol	C_3H_4O / 56.06	107-19-7 / -51.8	7819 / 113.6	4-01-00-02214 / 0.9478[20]	H_2O 3; EtOH 5; eth 5; chl 3 / 1.4322[20]
10744	2-Propyn-1-ol, acetate / Propargyl acetate	$C_5H_6O_2$ / 98.10	627-09-8 /	/ 121.5	4-02-00-00197 / 0.9982[20]	H_2O 2; EtOH 3; eth 3 / 1.4187[20]
10745	2-Propyn-1-ol, 3-phenyl-	C_9H_8O / 132.16	1504-58-1 /	/ 137-8[15]	4-06-00-04065 / 1.07[18]	eth 3; ace 3; bz 3 / 1.5873[28]
10746	Propyzamide / Benzamide, 3,5-dichloro-N-(1,1-dimethyl-2-propynyl)-	$C_{12}H_{11}Cl_2NO$ / 256.13	23950-58-5 / 155	7886 /		
10747	Pteridine / Pyrazino[2,3-d]pyrimidine	$C_6H_4N_4$ / 132.12	91-18-9 / 139.5	7947 / sub 125	4-26-00-01770 /	H_2O 4; EtOH 3; eth 2; bz 2
10748	2,4(1H,3H)-Pteridinedione / Lumazine	$C_6H_4N_4O_2$ / 164.12	487-21-8 / 348.5	5467 /	4-26-00-02489 /	HOAc 4
10749	4,6-Pteridinedione, 2-amino-1,5-dihydro- / Xanthopterin	$C_6H_5N_5O_2$ / 179.14	119-44-8 / >410 dec	9973 / 98-100[18]	4-26-00-04000 / 1.559[25]	H_2O 1; EtOH 2; eth 2; acid 4
10750	2,4,7-Pteridinetriamine, 6-phenyl- / Triamterene	$C_{12}H_{11}N_7$ / 253.27	396-01-0 /	9515 /	4-26-00-03837 /	
10751	1H-Purin-6-amine / Adenine	$C_5H_5N_5$ / 135.13	73-24-5 / 360 dec	141 / sub 220	4-26-00-03561 /	H_2O 3; EtOH 2; eth 1; chl 1
10752	1H-Purine / 7H-Imidazo[4,5-d]pyrimidine	$C_5H_4N_4$ / 120.11	120-73-0 / 216.5	7959 /	4-26-00-01736 /	H_2O 4; EtOH 4; eth 2; ace 3
10753	1H-Purine, 6-chloro- / 6-Chloropurine	$C_5H_3ClN_4$ / 154.56	87-42-3 / 175-7 dec	2161 /	4-26-00-01742 /	
10754	1H-Purine-2,6-diamine / 2,6-Diaminopurine	$C_5H_6N_6$ / 150.14	1904-98-9 / 302	2963 /	4-26-00-03785 /	
10755	1H-Purine-2,6-dione, 3,7-dihydro- / Xanthine	$C_5H_4N_4O_2$ / 152.11	69-89-6 / dec	9968 / sub	4-26-00-02327 /	
10756	1H-Purine-2,6-dione, 3,7-dihydro-1,3-dimethyl- / Theophylline	$C_7H_8N_4O_2$ / 180.17	58-55-9 / 273	9212 /	4-26-00-02331 /	H_2O 4; EtOH 2; eth 2; chl 2
10757	1H-Purine-2,6-dione, 3,7-dihydro-3,7-dimethyl- / Theobromine	$C_7H_8N_4O_2$ / 180.17	83-67-0 / 357	9209 / sub 290	4-26-00-02336 /	H_2O 2; EtOH 2; eth 1; bz 1
10758	1H-Purine-2,6-dione, 3,7-dihydro-7-(2-hydroxyethyl)-1,3-dimethyl- / Etofylline	$C_9H_{12}N_4O_3$ / 224.22	519-37-9 / 158	3835 /	4-26-00-02357 /	H_2O 4; EtOH 3; eth 2; bz 2
10759	1H-Purine-2,6-dione, 3,7-dihydro-8-methoxy-1,3,7-trimethyl-	$C_9H_{12}N_4O_3$ / 224.22	569-34-6 / 176		4-26-00-02727 / 1.399[25]	H_2O 3; EtOH 4; bz 3
10760	1H-Purine-2,6-dione, 3,7-dihydro-1,3,7-trimethyl- / Caffeine	$C_8H_{10}N_4O_2$ / 194.19	58-08-2 / 238	1635 / sub 90	4-26-00-02338 / 1.23[19]	H_2O 2; EtOH 2; eth 1; ace 2
10761	1H-Purine-2,6-dione, 7-(2,3-dihydroxypropyl)-3,7-dihydro-1,3-dimethyl- / Dyphylline	$C_{10}H_{14}N_4O_4$ / 254.25	479-18-5 / 161.5	3459 /	4-26-00-02370 /	
10762	6H-Purine-6-thione, 2-amino-1,7-dihydro- / Thioguanine	$C_5H_5N_5S$ / 167.19	154-42-7 / >360	9266 /	4-26-00-03926 /	
10763	6H-Purine-6-thione, 1,7-dihydro- / 6-Mercaptopurine	$C_5H_4N_4S$ / 152.18	50-44-2 / 313 dec	5762 /		H_2O 1; alk 3
10764	1H-Purine-2,6,8(3H)-trione, 7,9-dihydro- / Uric acid	$C_5H_4N_4O_3$ / 168.11	69-93-2 / dec	9790 / dec	4-26-00-02619 / 1.89[25]	H_2O 1; EtOH 1; eth 1; alk 4
10765	1H-Purine-2,6,8(3H)-trione, 4,9-dihydro-3-methyl- / 3-Methyluric acid	$C_6H_8N_4O_3$ / 182.14	605-99-2 / >350		4-26-00-02621 / 1.6104[25]	1.6334[25]
10766	1H-Purine-2,6,8(3H)-trione, 7,9-dihydro-7-methyl- / 7-Methyluric acid	$C_6H_8N_4O_3$ / 182.14	612-37-3 / 370-80 dec		4-26-00-02622 / 1.706[25]	H_2O 2; NaOH 3
10767	2H-Purin-2-one, 6-amino-1,3-dihydro- / Isoguanine	$C_5H_5N_5O$ / 151.13	3373-53-3 / >360			
10768	6H-Purin-6-one, 2-amino-1,7-dihydro- / Guanine	$C_5H_5N_5O$ / 151.13	73-40-5 / 360 dec	4477 / sub	4-26-00-03890 /	H_2O 1; EtOH 2; eth 2; alk 3
10769	6H-Purin-6-one, 2-amino-1,7-dihydro-7-methyl / 7-Methylguanine	$C_6H_7N_5O$ / 165.15	578-76-7 / 370			
10770	6H-Purin-6-one, 1,7-dihydro- / Hypoxanthine	$C_5H_4N_4O$ / 136.11	68-94-0 / 150 dec	4805 /	4-26-00-02081 /	H_2O 2; alk 3; dil acid 3
10771	Pyran	C_5H_6O / 82.10	33941-07-0 /	/ 80	0-17-00-00036 /	EtOH 3; eth 3; bz 3 / 1.4559[20]
10772	2H-Pyran-5-carboxylic acid, 3,4-dihydro-3-(hydroxymethyl)-2-methyl-4-[(9-methyl-9H-pyrido[3,4-b]indol-1-yl)methyl]-, methyl ester, [2S-(2α,3β,4β)]- / Alstonidine	$C_{22}H_{24}N_2O_4$ / 380.44	25394-75-6 / 189	313 /	4-27-00-07962 /	ace 4; EtOH 4
10773	2H-Pyran-5-carboxylic acid, 2-oxo- / Coumalic acid	$C_6H_4O_4$ / 140.10	500-05-0 / 205-10 dec	2558 / 218[120]	5-18-08-00120 /	H_2O 2; EtOH 3; eth 2; ace 2
10774	2H-Pyran-4-carboxylic acid, tetrahydro- / Pyran-4-carboxylic acid, tetrahydro-	$C_6H_{10}O_3$ / 130.14	5337-03-1 / 88	/ 144[11]	5-18-06-00023 /	H_2O 3; peth 2
10775	4H-Pyran-2,6-dicarboxylic acid, 3-hydroxy-4-oxo- / Meconic acid	$C_7H_4O_7$ / 200.10	497-59-6 / 120 dec	5664 / 120 dec	5-18-09-00319 /	H_2O 2; EtOH 3; eth 2; ace 2
10776	4H-Pyran-2,6-dicarboxylic acid, 4-oxo- / Chelidonic acid	$C_7H_4O_6$ / 184.11	99-32-1 / 262	2042 /	5-18-08-00646 /	H_2O 3; EtOH 2

PHYSICAL CONSTANTS OF ORGANIC COMPOUNDS (continued)

No.	Name / Synonym	Mol. Form. Mol. Wt.	CAS RN mp/°C	Merck No. bp/°C	Beil. Ref. den/g cm⁻³	Solubility n_D
10777	2H-Pyran, 3,4-dihydro- / 3,4-Dihydro-2H-pyran	C_5H_8O 84.12	110-87-2	86	5-17-01-00181 0.921^{19}	H_2O 3; EtOH 3; chl 2 1.4402^{19}
10778	2H-Pyran, 3,6-dihydro-4-methyl-	$C_6H_{10}O$ 98.14	16302-35-5	117.5	5-17-01-00194 0.912^{25}	1.4495^{20}
10779	2H-Pyran-2,4(3H)-dione, 3-acetyl-6-methyl- / Dehydroacetic acid	$C_8H_8O_4$ 168.15	520-45-6 109	2855 270	4-17-00-06699	H_2O 4; EtOH 2; eth 4; chl 2
10780	2H-Pyran-2,6(3H)-dione, dihydro-	$C_5H_6O_3$ 114.10	108-55-4 56.3	158^{15}	5-17-11-00009 1.4110^{20}	
10781	2H-Pyran-2,6(3H)-dione, dihydro-3-phenyl-	$C_{11}H_{10}O_3$ 190.20	2959-96-8 95	224^{13}	5-17-11-00279	EtOH 4
10782	2H-Pyran-2,6(3H)-dione, dihydro-4-phenyl-	$C_{11}H_{10}O_3$ 190.20	4160-80-9 105.6	217^{15}	5-17-11-00279	bz 4; eth 4; chl 4
10783	2H-Pyran, 2-ethoxy-3,4-dihydro-	$C_7H_{12}O_2$ 128.17	103-75-3	42^{16}	5-17-03-00196 0.957^{25}	1.4394^{20}
10784	2H-Pyran-2-methanol, tetrahydro-	$C_6H_{12}O_2$ 116.16	100-72-1	185	5-17-03-00136 1.027^{25}	1.458^{20}
10785	4H,5H-Pyrano[4,3-b]pyran-4,5-dione, 2,3-dihydro-3-hydroxy-2-methyl-7-(1-propenyl)-, [2S-[2α,3β,7(E)]]- / Radicinin	$C_{12}H_{12}O_5$ 236.22	10088-95-6 221.5	8117	4-19-00-02746	chl 2
10786	Pyrano[3,2-c][2]benzopyran-6(2H)-one, 3,4,4a,10b-tetrahydro-3,4,8,10-tetrahydroxy-2-(hydroxymethyl)-9-methoxy-[2R-(2α,3β,4α,4aα,10bβ)]- / Bergenin	$C_{14}H_{16}O_9$ 328.28	477-90-7 238	1174	5-19-07-00132	H_2O 4; EtOH 4
10787	2H-Pyran-4-ol, tetrahydro-	$C_5H_{10}O_2$ 102.13	2081-44-9	87^{15}	5-17-03-00111 1.071^{25}	1.4600^{20}
10788	2H-Pyran-2-one	$C_5H_4O_2$ 96.09	504-31-4 8.5	207.5	5-17-09-00288 1.200^{20}	H_2O 5; ace 4 1.5270^{25}
10789	4H-Pyran-4-one	$C_5H_4O_2$ 96.09	108-97-4 32.5	212.5	5-17-09-00290 1.190^{25}	H_2O 4; EtOH 3; eth 4; bz 3 1.5238
10790	2H-Pyran-2-one, 6-[2-(1,3-benzodioxol-5-yl)ethenyl]-5,6-dihydro-4-methoxy-, [R-(E)]- / Methysticin	$C_{15}H_{14}O_5$ 274.27	495-85-2 137	6056	5-19-10-00456	
10791	2H-Pyran-2-one, 5,6-dihydro-6-methyl-, (S)- / Parasorbic acid	$C_6H_8O_2$ 112.13	10048-32-5	6981 100^{15}	5-17-09-00130 1.079^{18}	H_2O 4; eth 4; EtOH 4 1.4736^{20}
10792	2H-Pyran-2-one, 4,6-dimethyl-	$C_7H_8O_2$ 124.14	675-09-2 51.5	245	5-17-09-00409	H_2O 4; eth 4; EtOH 4
10793	4H-Pyran-4-one, 2,6-dimethyl-	$C_7H_8O_2$ 124.14	1004-36-0 132	251; 140^{25}	5-17-09-00410 0.9953^{137}	H_2O 3; EtOH 3; eth 3; ace 3
10794	2H-Pyran-2-one, 3-hydroxy-	$C_5H_4O_3$ 112.08	496-64-0 93.8	$112^{20};$ 102^{15}	5-18-01-00095	H_2O 3; EtOH 3; eth 3; bz 3
10795	4H-Pyran-4-one, 5-hydroxy-2-(hydroxymethyl)- / Kojic acid	$C_6H_6O_4$ 142.11	501-30-4 153.5	5197	5-18-02-00516	H_2O 2; EtOH 3; eth 3; ace 3
10796	4H-Pyran-4-one, 3-hydroxy-2-methyl- / Maltol	$C_6H_6O_3$ 126.11	118-71-8 161.5	5594 sub 93	5-18-01-00114	H_2O 2; eth 2; bz 2; chl 4
10797	4H-Pyran-4-one, 2-methoxy-3,5-dimethyl-6-[tetrahydro-4-[2-methyl-3-(4-nitrophenyl)-2-propenylidene]-2-furanyl] / Aureothin	$C_{22}H_{23}NO_6$ 397.43	2825-00-5 158	897	5-19-06-00099	ace 4; EtOH 4; chl 4
10798	2H-Pyran-2-one, tetrahydro-	$C_5H_8O_2$ 100.12	542-28-9 -12.5	219	5-17-09-00017 1.0794^{20}	H_2O 3; EtOH 5; eth 5; ctc 2 1.4503^{20}
10799	4H-Pyran-4-one, tetrahydro-	$C_5H_8O_2$ 100.12	29943-42-8	166.5	5-17-09-00021 1.084^{25}	1.4520^{20}
10800	2H-Pyran, tetrahydro- / Oxane	$C_5H_{10}O$ 86.13	142-68-7 -45	9149 88	5-17-01-00064 0.8814^{20}	EtOH 3; eth 3; bz 3; ctc 3 1.4200^{20}
10801	Pyrazine / 1,4-Diazine	$C_4H_4N_2$ 80.09	290-37-9 55	7971 115	5-23-05-00351 1.0311^{61}	H_2O 3; EtOH 3; eth 3; ace 3 1.4953^{61}
10802	Pyrazinecarboxamide / Pyrazinamide	$C_5H_5N_3O$ 123.11	98-96-4 192	7970 sub	4-25-00-00772	H_2O 3; EtOH 3
10803	Pyrazinecarboxylic acid / Pyrazinoic acid	$C_5H_4N_2O_2$ 124.10	98-97-5 225 dec	7973 sub	4-25-00-00771	
10804	2,3-Pyrazinedicarboxylic acid / 2,3-Dicarboxypyrazine	$C_6H_4N_2O_4$ 168.11	89-01-0 193 dec	7972	4-25-00-01064	H_2O 4; EtOH 2; eth 2; ace 3
10805	Pyrazine, 2,3-dimethyl- / 2,3-Dimethylpyrazine	$C_6H_8N_2$ 108.14	5910-89-4	156	5-23-05-00402 1.0281^0	H_2O 3; EtOH 3; eth 3
10806	Pyrazine, 2,5-dimethyl- / 2,5-Dimethylpyrazine	$C_6H_8N_2$ 108.14	123-32-0 15	155	5-23-05-00403 0.9887^{20}	H_2O 5; EtOH 5; eth 5; ace 3 1.4980^{20}
10807	Pyrazine, 2,6-dimethyl- / 2,6-Dimethylpyrazine	$C_6H_8N_2$ 108.14	108-50-9 47.5	155.6	5-23-05-00406 0.9647^{50}	H_2O 3; EtOH 3; eth 3; ctc 2
10808	Pyrazineethanol	$C_6H_8N_2O$ 124.14	6705-31-3	129^{10}	5-23-11-00173 1.163^{20}	1.5378^{20}
10809	Pyrazine, 3-ethyl-2,5-dimethyl-	$C_8H_{12}N_2$ 136.20	13360-65-1	180.5	5-23-05-00440 0.9657^{24}	H_2O 2; EtOH 2; eth 2 1.5014^{24}
10810	Pyrazine, methyl- / 2-Methylpyrazine	$C_5H_6N_2$ 94.12	109-08-0 -29	137	5-23-05-00386 1.03^{20}	H_2O 5; EtOH 5; eth 5; ace 3 1.5042^{20}
10811	2(1H)-Pyrazinone, 1-hydroxy-6-(1-methylpropyl)-3-(2-methylpropyl)-, (+)- / Aspergillic acid	$C_{12}H_{20}N_2O_2$ 224.30	490-02-8 98	863	4-24-00-00235	bz 4; eth 4; EtOH 4

No.	Name Synonym	Mol. Form. Mol. Wt.	CAS RN mp/°C	Merck No. bp/°C	Beil. Ref. den/g cm^{-3}	Solubility n_D
10812	1H-Pyrazol-4-amine 4-Aminopyrazole	$C_3H_5N_3$ 83.09	28466-26-4 81	sub	4-25-00-02029	H_2O 4; EtOH 4; chl 4
10813	1H-Pyrazol-4-amine, 3-methyl-1-phenyl- Pyrazole, 4-amino-3-methyl-1-phenyl-	$C_{10}H_{11}N_3$ 173.22	103095-51-8 88.8	314.5	4-25-00-02033	bz 4; eth 4; EtOH 4; chl 4
10814	1H-Pyrazol-5-amine, 3-methyl-1-phenyl-	$C_{10}H_{11}N_3$ 173.22	1131-18-6 116	333	4-25-00-02034	H_2O 3; EtOH 3; bz 2; chl 3
10815	1H-Pyrazole 1,2-Diazole	$C_3H_4N_2$ 68.08	288-13-1 68	7974 187	5-23-04-00122	H_2O 3; EtOH 3; eth 3; bz 3 1.4203
10816	1H-Pyrazole, 4-bromo-1,3-dimethyl- 4-Bromo-1,3-dimethylpyrazole	$C_5H_7BrN_2$ 175.03	5775-82-6	76-7[10]	5-23-05-00022 1.4975[15]	ace 4; eth 4; EtOH 4 1.5214[15]
10817	1H-Pyrazole, 4-bromo-3-methyl- 4-Bromo-3-methylpyrazole	$C_4H_5BrN_2$ 161.00	13808-64-5 76.5		5-23-05-00022 1.5638[100]	ace 4; eth 4; EtOH 4 1.5182[100]
10818	1H-Pyrazole, 3-chloro-1,5-dimethyl- 3-Chloro-1,5-dimethylpyrazole	$C_5H_7ClN_2$ 130.58	51500-32-4 47.5	211	2-23-00-00049 1.0823[100]	ace 4; bz 4; eth 4; EtOH 4 1.4648[100]
10819	1H-Pyrazole, 4-chloro-3,5-dimethyl-	$C_5H_7ClN_2$ 130.58	15953-73-8 117.5	221	5-23-05-00151	ace 4; bz 4; eth 4; EtOH 4
10820	1H-Pyrazole, 5-chloro-1,3-dimethyl- 5-Chloro-1,3-dimethylpyrazole	$C_5H_7ClN_2$ 130.58	54454-10-3	157.5	5-23-05-00019 1.1367[18]	H_2O 4 1.4877[18]
10821	1H-Pyrazole, 3-chloro-5-methyl-	$C_4H_5ClN_2$ 116.55	15953-45-4 118.5	258; 138[15]	5-23-05-00019	H_2O 4
10822	1H-Pyrazole, 4,5-dihydro- 2-Pyrazoline	$C_3H_6N_2$ 70.09	109-98-8	7975 144	5-23-03-00361 1.020[17]	H_2O 4; eth 4; EtOH 4 1.4796[17]
10823	1H-Pyrazole, 4,5-dihydro-1-phenyl- 1-Phenyl-2-pyrazoline	$C_9H_{10}N_2$ 146.19	936-53-8 52	273; 151[17]	5-23-03-00363 1.0689[58]	H_2O 1; EtOH 3; eth 3; bz 3 1.6015[58]
10824	1H-Pyrazole, 4,5-dihydro-3-phenyl- 2-Pyrazoline, 3-phenyl	$C_9H_{10}N_2$ 146.19	936-48-1 45.8	164[17]	5-23-06-00396 1.0892[54]	
10825	1H-Pyrazole, 1,3-dimethyl-	$C_5H_8N_2$ 96.13	694-48-4	137	5-23-05-00012 0.9561[17]	H_2O 4 1.4734[15]
10826	1H-Pyrazole, 1,5-dimethyl-	$C_5H_8N_2$ 96.13	694-31-5	149	5-23-05-00006 0.9813[17]	H_2O 4; EtOH 4; eth 4 1.4782[16]
10827	1H-Pyrazole, 3,4-dimethyl-	$C_5H_8N_2$ 96.13	2820-37-3 58	110[11]	5-23-05-00108 0.9325[99]	EtOH 3; eth 3; ace 3; bz 3
10828	1H-Pyrazole, 3,5-dimethyl-	$C_5H_8N_2$ 96.13	67-51-6 107.5	218	5-23-05-00110 0.8839[16]	H_2O 3; EtOH 4; eth 4; ace 3
10829	1H-Pyrazole, 3,5-dimethyl-1-phenyl-	$C_{11}H_{12}N_2$ 172.23	1131-16-4	273; 145[12.5]	5-23-05-00127 1.0566[20]	eth 4; EtOH 4; chl 4 1.5738[19]
10830	1H-Pyrazole, 1,3-diphenyl- 1,3-Diphenylpyrazole	$C_{15}H_{12}N_2$ 220.27	4492-01-7 85.8	341[270]	5-23-07-00178 1.0794[101]	
10831	1H-Pyrazole, 1,5-diphenyl- 1,5-Diphenylpyrazole	$C_{15}H_{12}N_2$ 220.27	6831-89-6 55.5	338; 167[12]	5-23-07-00176 1.0696[100]	bz 4; eth 4; EtOH 4
10832	1H-Pyrazole, 1-ethyl-	$C_5H_8N_2$ 96.13	2817-71-2	139	5-23-04-00135 0.9537[20]	ace 4; bz 4; eth 4; EtOH 4 1.4700[20]
10833	1H-Pyrazole, 1-ethyl-3-methyl- 1-Ethyl-3-methylpyrazole	$C_6H_{10}N_2$ 110.16	30433-57-9 115	152	5-23-05-00013 0.936[20]	
10834	1H-Pyrazole, 1-methyl- 1-Methylpyrazole	$C_4H_6N_2$ 82.11	930-36-9	127	5-23-04-00134 0.9929[13]	1.4787[13]
10835	1H-Pyrazole, 3-methyl-	$C_4H_6N_2$ 82.11	1453-58-3 36.5	204; 108[25]	5-23-05-00004 1.0203[16]	H_2O 5; EtOH 5; eth 5 1.4915[20]
10836	1H-Pyrazole, 4-methyl- 4-Methylpyrazole	$C_4H_6N_2$ 82.11	7554-65-6	10146 206; 95[13]	5-23-05-00031 1.015[20]	
10837	1H-Pyrazole, 1-methyl-3-phenyl-	$C_{10}H_{10}N_2$ 158.20	3463-26-1 56	280.5	5-23-07-00178 1.0232[99]	
10838	1H-Pyrazole, 3-methyl-1-phenyl-	$C_{10}H_{10}N_2$ 158.20	1128-54-7 37	255	5-23-05-00013 1.076[20]	
10839	1H-Pyrazole, 1,3,4-trimethyl- 1,3,4-Trimethylpyrazole	$C_6H_{10}N_2$ 110.16	15802-99-0	160	5-23-05-00109 0.9587[17]	1.4866[17]
10840	1H-Pyrazole, 1,3,5-trimethyl- 1,3,5-Trimethylpyrazole	$C_6H_{10}N_2$ 110.16	1072-91-9	170	5-23-05-00125 0.9130[57]	1.4589[57]
10841	1H-Pyrazole, 1,4,5-trimethyl- 1,4,5-Trimethylpyrazole	$C_6H_{10}N_2$ 110.16	15802-97-8	176.5	5-23-05-00108 0.9685[17]	1.4849[17]
10842	1H-Pyrazole, 3,4,5-trimethyl-	$C_6H_{10}N_2$ 110.16	5519-42-6 138.5	233	5-23-05-00198	bz 4; eth 4; EtOH 4
10843	3,5-Pyrazolidinedione, 1,4-diphenyl- Phenopyrazone	$C_{15}H_{12}N_2O_2$ 252.27	3426-01-5 233.5	7217	5-24-08-00003	
10844	3-Pyrazolidinone, 1-phenyl- 1-Phenyl-3-pyrazolidinone	$C_9H_{10}N_2O$ 162.19	92-43-3 126	7281	5-24-01-00011	eth 1; lig 1
10845	3H-Pyrazol-3-one, 4-amino-1,2-dihydro-1,5-dimethyl-2-phenyl- Ampyrone	$C_{11}H_{13}N_3O$ 203.24	83-07-8 109	625	4-25-00-03554	H_2O 3; EtOH 3; eth 2; bz 3
10846	3H-Pyrazol-3-one, 4-bromo-1,2-dihydro-1,5-dimethyl-2-phenyl- 4-Bromoantipyrine	$C_{11}H_{11}BrN_2O$ 267.13	5426-65-3 122	300[9]	5-24-01-00343 1.5900[20]	H_2O 3; EtOH 4; eth 3; chl 4
10847	3H-Pyrazol-3-one, 1,2-dihydro-1,5-dimethyl-2-phenyl- Antipyrine	$C_{11}H_{12}N_2O$ 188.23	60-80-0 114	748 319	5-24-01-00330	H_2O 4; EtOH 4
10848	3H-Pyrazol-3-one, 2,4-dihydro-5-methyl-4-nitro-2-(4-nitrophenyl)- Picrolonic acid	$C_{10}H_8N_4O_5$ 264.20	550-74-3 116	7383 dec		H_2O 2; EtOH 3; eth 3; MeOH 3

No.	Name / Synonym	Mol. Form. / Mol. Wt.	CAS RN / mp/°C	Merck No. / bp/°C	Beil. Ref. / den/g cm^{-3}	Solubility / n_D
10849	3H-Pyrazol-3-one, 1,2-dihydro-5-methyl-2-phenyl-	$C_{10}H_{10}N_2O$ 174.20	19735-89-8 128	287[105]; 191[17]	5-24-01-00317 1.2600[20]	H_2O 3; EtOH 3; bz 2; peth 1 1.637
10850	3H-Pyrazol-3-one, 1,2-dihydro-2-(1-methyl-4-piperidinyl)-5-phenyl-4-(phenylmethyl)- Benzpiperylon	$C_{22}H_{25}N_3O$ 347.46	53-89-4 181-3 dec	1131	5-24-04-00398	
10851	3H-Pyrazol-3-one, 4-(dimethylamino)-1,2-dihydro-1,5-dimethyl-2-phenyl- Aminopyrine	$C_{13}H_{17}N_3O$ 231.30	58-15-1 134.5	488	4-24-00-00096	H_2O 4; bz 4; EtOH 4
10852	4H-Pyrazolo[3,4-d]pyrimidin-4-one, 1,5-dihydro- Allopurinol	$C_5H_4N_4O$ 136.11	315-30-0 350	278	4-26-00-02076	
10853	Pyrene Benzo[def]phenanthrene	$C_{16}H_{10}$ 202.26	129-00-0 151.2	7977 404	4-05-00-02467 1.271[23]	H_2O 1; EtOH 3; eth 3; bz 3
10854	1-Pyrenecarboxylic acid 3-Pyrenecarboxylic acid	$C_{17}H_{10}O_2$ 246.27	19694-02-1 274	sub	3-09-00-03575	eth 4; chl 4
10855	Pyridazine 1,2-Diazabenzene	$C_4H_4N_2$ 80.09	289-80-5 -8	7982 208	5-23-05-00321 1.1035[23]	H_2O 5; EtOH 5; eth 4; ace 4 1.5218[20]
10856	Pyridazine, 3-methyl- 3-Methyl-1,2-diazine	$C_5H_6N_2$ 94.12	1632-76-4 184	214	5-23-05-00371 1.0450[26]	1.5145[20]
10857	2-Pyridinamine 2-Aminopyridine	$C_5H_6N_2$ 94.12	504-29-0 57.5	486 105[20]	5-22-08-00280	EtOH 3; eth 3; ace 3; bz 3
10858	3-Pyridinamine β-Aminopyridine	$C_5H_6N_2$ 94.12	462-08-8 64.5	487 252	5-22-09-00003	H_2O 3; EtOH 3; eth 3; lig 2
10859	4-Pyridinamine 4-Aminopyridine	$C_5H_6N_2$ 94.12	504-24-5 158.5	273	5-22-09-00106	H_2O 3; EtOH 4; eth 3; bz 3
10860	3-Pyridinamine, 6-butoxy-	$C_9H_{14}N_2O$ 166.22	539-23-1	148-50[12]	4-22-00-05574 1.037[25]	EtOH 3; eth 3; chl 2; dil acid 3 1.5373[20]
10861	2-Pyridinamine, 5-chloro-	$C_5H_5ClN_2$ 128.56	1072-98-6 137	127[11]	5-22-08-00451	H_2O 3; EtOH 3; peth 1; lig 1
10862	3-Pyridinamine, 2,6-dichloro- Pyridine, 3-amino-2,6-dichloro	$C_5H_4Cl_2N_2$ 163.01	62476-56-6 119	110[0.3]	4-22-00-04093	eth 4; EtOH 4
10863	2-Pyridinamine, N,N-dimethyl-	$C_7H_{10}N_2$ 122.17	5683-33-0 182	196	5-22-08-00287 1.014[14]	EtOH 3; eth 3; bz 3 1.5663[20]
10864	4-Pyridinamine, 2,6-dimethyl-	$C_7H_{10}N_2$ 122.17	3512-80-9 192.5	246	5-22-09-00417	H_2O 3; EtOH 3; eth 2; ace 3
10865	3-Pyridinamine, 6-methoxy-	$C_8H_8N_2O$ 124.14	6628-77-9 30	125[10]; 87[1]	5-22-11-00408	1.5745[20]
10866	2-Pyridinamine, 3-methyl-	$C_6H_8N_2$ 108.14	1603-40-3 33.5	222; 95[8]	5-22-09-00272	H_2O 4; EtOH 3; eth 3; ace 3
10867	2-Pyridinamine, 4-methyl- 2-Amino-4-picoline	$C_6H_8N_2$ 108.14	695-34-1 100	478 115-7[11]	5-22-09-00325	H_2O 4; EtOH 3; eth 3; ace 3
10868	2-Pyridinamine, 6-methyl-	$C_6H_8N_2$ 108.14	1824-81-3 41	208.5	5-22-09-00210	H_2O 4; EtOH 3; eth 3; ace 3
10869	2-Pyridinamine, N-methyl-	$C_6H_8N_2$ 108.14	4597-87-9 15	200.5	5-22-08-00286 1.048[29]	H_2O 3; EtOH 4; eth 4; bz 3
10870	3-Pyridinamine, 4-methyl-	$C_6H_8N_2$ 108.14	3430-27-1 106	255	5-22-09-00344	H_2O 2; EtOH 2; eth 2; ace 2
10871	2-Pyridinamine, N-2-pyridinyl-	$C_{10}H_9N_3$ 171.20	1202-34-2 90.5	307.5	5-22-08-00415	H_2O 2; EtOH 4; eth 4; ace 4
10872	Pyridine Azine	C_5H_5N 79.10	110-86-1 -41.6	7983 115.2	5-20-05-00160 0.9819[20]	H_2O 5; EtOH 5; eth 5; ace 5 1.5095[20]
10873	1(4H)-Pyridineacetic acid, 3,5-diiodo-4-oxo-, propyl ester Propyliodone	$C_{10}H_{11}I_2NO_3$ 447.01	587-61-1 186	7876	5-21-07-00164	
10874	Pyridine, 2,6-bis(1,1-dimethylethyl)- 2,6-Di-tert-butylpyridine	$C_{13}H_{21}N$ 191.32	585-48-8	3022 120-1[20]	5-20-06-00181	
10875	Pyridine, 2-bromo-	C_5H_4BrN 158.00	109-04-6	193; 75[13]	5-20-05-00422 1.657[15]	H_2O 2; EtOH 3; eth 3; ctc 3 1.5734[20]
10876	Pyridine, 3-bromo- 3-Bromopyridine	C_5H_4BrN 158.00	626-55-1 -27.3	173; 69[18]	5-20-05-00426 1.645[0]	H_2O 3; EtOH 4; eth 4 1.5694[20]
10877	Pyridine, 4-bromo- 4-Bromopyridine	C_5H_4BrN 158.00	1120-87-2 0.5	28-30[0.4]	5-20-05-00431 1.6450[0]	ace 3; bz 3 1.5694[20]
10878	Pyridine, 2-butyl- 2-Butylpyridine	$C_9H_{13}N$ 135.21	5058-19-5 97	192	5-20-06-00118 0.9135[15]	
10879	Pyridine, 4-butyl-	$C_9H_{13}N$ 135.21	5335-75-1	209; 98[20]	5-20-06-00120 0.9157[20]	1.4946[20]
10880	Pyridine, 4-(1-butylpentyl)-	$C_{14}H_{23}N$ 205.34	2961-47-9	181[50]	5-20-06-00189 0.8878[25]	1.4846[25]
10881	2-Pyridinecarbonitrile	$C_6H_4N_2$ 104.11	100-70-9 29	224.5	5-22-02-00019 1.0810[25]	H_2O 3; EtOH 4; eth 4; bz 4 1.5242[25]
10882	3-Pyridinecarbonitrile	$C_6H_4N_2$ 104.11	100-54-9 51	206.9; 170[300]	5-22-02-00115 1.1590[25]	H_2O 4; EtOH 4; eth 4; bz 4
10883	3-Pyridinecarbonitrile, 1,2-dihydro-4-methoxy-1-methyl-2-oxo- Ricinine	$C_8H_8N_2O_2$ 164.16	524-40-3 201.5	8212 sub 170	5-22-07-00318	H_2O 3; EtOH 2; bz 2; chl 3

No.	Name Synonym	Mol. Form. Mol. Wt.	CAS RN mp/°C	Merck No. bp/°C	Beil. Ref. den/g cm⁻³	Solubility n_D
10884	3-Pyridinecarbonitrile, 1,2-dihydro-1-methyl-2-oxo-	$C_7H_8N_2O$ 134.14	767-88-4 140	243[28]	5-22-06-00129	EtOH 4
10885	4-Pyridinecarbothioamide, 2-ethyl- Ethionamide	$C_8H_{10}N_2S$ 166.25	536-33-4 163	3692	5-22-02-00360	
10886	4-Pyridinecarbothioamide, 2-propyl- Protionamide	$C_9H_{12}N_2S$ 180.27	14222-60-7 136.7	7905	5-22-02-00376	
10887	2-Pyridinecarboxaldehyde	C_6H_5NO 107.11	1121-60-4	180; 62[13]	5-21-07-00293 1.1255[18]	H_2O 3; EtOH 3; eth 3; ctc 2 1.5389[18]
10888	3-Pyridinecarboxaldehyde	C_6H_5NO 107.11	500-22-1	89.5[14]	5-21-07-00334 1.1394[25]	H_2O 3; EtOH 3; eth 2; ace 3
10889	4-Pyridinecarboxaldehyde	C_6H_5NO 107.11	872-85-5	77-8[12]	5-21-07-00351	H_2O 3; eth 3; ctc 3 1.5423[20]
10890	3-Pyridinecarboxamide	$C_6H_6N_2O$	98-92-0	6398	5-22-02-00080	H_2O 4; EtOH 4; chl 2; glycerol 4
	Niacinamide	122.13	130	157[0.0005]	1.400[25]	1.466
10891	3-Pyridinecarboxamide, N,N-diethyl- Nikethamide	$C_{10}H_{14}N_2O$ 178.23	59-26-7 25	6459 280 dec; 175[25]	5-22-02-00084 1.060[25]	DMSO-d_6 2 1.525[20]
10892	4-Pyridinecarboxamide, N,N-diethyl- Isonicotinic acid diethylamide	$C_{10}H_{14}N_2O$ 178.23	530-40-5	5074 119-20[1]	5-22-02-00196	H_2O 4; ace 4; eth 4; EtOH 4 1.525[20]
10893	2-Pyridinecarboxylic acid Picolinic acid	$C_6H_5NO_2$ 123.11	98-98-6 136.5	7375 sub	5-22-02-00003	H_2O 2; EtOH 3; eth 1; bz 2
10894	3-Pyridinecarboxylic acid Nicotinic acid	$C_6H_5NO_2$ 123.11	59-67-6 236.6	6435 sub	5-22-02-00057 1.473[25]	H_2O 2; EtOH 2; eth 2
10895	4-Pyridinecarboxylic acid Isonicotinic acid	$C_6H_5NO_2$ 123.11	55-22-1 315	5073 sub 260	5-22-02-00188	H_2O 2; EtOH 2; eth 2; bz 2
10896	2-Pyridinecarboxylic acid, 4-amino-3,5,6-trichloro- Picloram	$C_6H_3Cl_3N_2O_2$ 241.46	1918-02-1 218.5	7370	5-22-13-00585	
10897	3-Pyridinecarboxylic acid, 6-amino- 6-Aminonicotinic acid	$C_6H_6N_2O_2$ 138.13	3167-49-5 312	468	4-22-00-06726	
10898	2-Pyridinecarboxylic acid, 5-butyl- Fusaric acid	$C_{10}H_{13}NO_2$ 179.22	536-69-6 97	4227	5-22-02-00384	
10899	3-Pyridinecarboxylic acid, 1,6-dihydro-6-oxo-	$C_6H_5NO_3$ 139.11	5006-66-6 310 dec	sub	5-22-06-00119	H_2O 2; EtOH 1; eth 1; bz 1
10900	4-Pyridinecarboxylic acid, 1,2-dihydro-6-hydroxy-2-oxo- Citrazinic acid	$C_6H_5NO_4$ 155.11	99-11-6 >300 dec	2327	5-22-07-00024	H_2O 3; alk 3; HCl 2
10901	2-Pyridinecarboxylic acid, ethyl ester	$C_8H_9NO_2$ 151.17	2524-52-9 1	243	5-22-02-00004 1.1194[20]	H_2O 4; eth 4; EtOH 4 1.5104[20]
10902	3-Pyridinecarboxylic acid, ethyl ester	$C_8H_9NO_2$ 151.17	614-18-6 8.5	224	5-22-02-00060 1.1070[20]	H_2O 4; EtOH 4; eth 4; bz 4 1.5024[20]
10903	4-Pyridinecarboxylic acid, ethyl ester	$C_8H_9NO_2$ 151.17	1570-45-2 23	219.5	5-22-02-00190 1.0091[15]	H_2O 2; EtOH 3; eth 4; bz 3 1.5017[20]
10904	4-Pyridinecarboxylic acid, [1-(2-furanyl)ethylidene]hydrazide Furonazide	$C_{12}H_{11}N_3O_2$ 229.24	3460-67-1 202.3	4220	5-22-02-00264	
10905	4-Pyridinecarboxylic acid, hydrazide Isoniazid	$C_6H_7N_3O$ 137.14	54-85-3 171.4	5071	5-22-02-00219	H_2O 4; EtOH 4
10906	2-Pyridinecarboxylic acid, methyl ester Methyl nicotinate	$C_7H_7NO_2$ 137.14	93-60-7 42.5	6014 204	5-22-02-00059	H_2O 3; EtOH 3; bz 3
10907	3-Pyridinecarboxylic acid, 1-methylethyl ester	$C_9H_{11}NO_2$ 165.19	553-60-6	126[30]; 92.5[5]	5-22-02-00060 1.0624[20]	1.4926[20]
10908	4-Pyridinecarboxylic acid, methyl ester	$C_7H_7NO_2$ 137.14	2459-09-8 16.1	208	5-22-02-00189 1.1599[20]	H_2O 2; EtOH 3; eth 3; bz 3 1.5135[20]
10909	3-Pyridinecarboxylic acid, 1-oxide Oxiniacic acid	$C_6H_5NO_3$ 139.11	2398-81-4 254-5 dec	6895	5-22-02-00140	H_2O 4; MeOH 4
10910	4-Pyridinecarboxylic acid, 2-[3-oxo-3-[(phenylmethyl)amino]propyl]hydrazide Nialamide	$C_{16}H_{18}N_4O_2$ 298.34	51-12-7 151.6	6399	5-22-02-00251	
10911	3-Pyridinecarboxylic acid, phenylmethyl ester Nicotinic acid benzyl ester	$C_{13}H_{11}NO_2$ 213.24	94-44-0	6436 170[3]	5-22-02-00062	
10912	3-Pyridinecarboxylic acid, 1,2,5,6-tetrahydro- Guvacine	$C_6H_9NO_2$ 127.14	498-96-4 295 dec	4495	5-22-01-00322	H_2O 4
10913	3-Pyridinecarboxylic acid, 1,2,5,6-tetrahydro-1-methyl- Arecaidine	$C_7H_{11}NO_2$ 141.17	499-04-7 232 dec	801	5-22-01-00322	H_2O 4; EtOH 1; eth 1; bz 1
10914	3-Pyridinecarboxylic acid, 1,2,5,6-tetrahydro-1-methyl-, methyl ester Arecoline	$C_8H_{13}NO_2$ 155.20	63-75-2	802 209	5-22-01-00322 1.0485[20]	H_2O 5; EtOH 5; eth 5; chl 3 1.486-[20]
10915	Pyridine, 2-chloro- 2-Chloropyridine	C_5H_4ClN 113.55	109-09-1	170	5-20-05-00402 1.205[15]	H_2O 2; EtOH 3; eth 3 1.5320[20]
10916	Pyridine, 3-chloro- 3-Chloropyridine	C_5H_4ClN 113.55	626-60-8	148; 86[100]	5-20-05-00406	H_2O 2 1.5304[20]
10917	Pyridine, 4-chloro- 4-Chloropyridine	C_5H_4ClN 113.55	626-61-9 -43.5	147.5	5-20-05-00410 1.2000[25]	H_2O 3; EtOH 5
10918	Pyridine, 2-chloro-6-methoxy-	C_6H_6ClNO 143.57	17228-64-7	185.5	5-21-02-00018 1.207[25]	1.5263[20]
10919	Pyridine, 2-chloro-4-methyl-	C_6H_6ClN 127.57	3678-62-4 115	192	5-20-05-00579 1.1459[20]	1.5293[8]

No.	Name / Synonym	Mol. Form. / Mol. Wt.	CAS RN / mp/°C	Merck No. / bp/°C	Beil. Ref. / den/g cm^{-3}	Solubility / n_D
10920	Pyridine, 2-chloro-6-methyl-	C_6H_6ClN 127.57	18368-63-3	183.5; 64[10]	5-20-05-00498 1.167[25]	1.5270[20]
10921	Pyridine, 2-[(4-chlorophenyl)methyl]-	$C_{12}H_{10}ClN$ 203.67	4350-41-8	181-3[20]	5-20-07-00558 1.39[25]	1.5868[20]
10922	Pyridine, 4-(3-cyclohexen-1-yl)- 4-(Cyclohex-3-enyl)pyridine	$C_{11}H_{13}N$ 159.23	70644-46-1 22.1	226	4-20-00-03239 1.0222[25]	1.5466[25]
10923	2,3-Pyridinediamine	$C_5H_7N_3$ 109.13	452-58-4 120.8	148-50[5]	5-22-11-00241	H_2O 3; EtOH 3; bz 3
10924	2,5-Pyridinediamine 2,5-Diaminopyridine	$C_5H_7N_3$ 109.13	4318-76-7 110.3	182[12]	5-22-11-00252	H_2O 4; EtOH 4
10925	2,6-Pyridinediamine	$C_5H_7N_3$ 109.13	141-86-6 121.5	285; 148[5]	5-22-11-00255	H_2O 2; ace 2
10926	Pyridine, 3,5-dibromo- 3,5-Dibromopyridine	$C_5H_3Br_2N$ 236.89	625-92-3 112	222	5-20-05-00436	H_2O 2; EtOH 3; eth 3
10927	2,3-Pyridinedicarboxylic acid Quinolinic acid	$C_7H_5NO_4$ 167.12	89-00-9 228.5	8102	5-22-04-00115	H_2O 2; EtOH 1; eth 1; bz 1
10928	2,4-Pyridinedicarboxylic acid	$C_7H_5NO_4$ 167.12	499-80-9 249		5-22-04-00119 0.942[25]	H_2O 3; EtOH 3; eth 1; bz 1
10929	2,5-Pyridinedicarboxylic acid Isocinchomeronic acid	$C_7H_5NO_4$ 167.12	100-26-5 254	5043	5-22-04-00124	H_2O 3; EtOH 2; eth 1; bz 1
10930	3,4-Pyridinedicarboxylic acid Cinchomeronic acid	$C_7H_5NO_4$ 167.12	490-11-9 256	2285 sub	5-22-04-00135	H_2O 2; EtOH 2; eth 1; bz 2
10931	3,5-Pyridinedicarboxylic acid	$C_7H_5NO_4$ 167.12	499-81-0 324	sub	5-22-04-00139	H_2O 1; eth 2; HCl 3; HOAc 2
10932	3,5-Pyridinedicarboxylic acid, 2,6-dimethyl-, diethyl ester	$C_{13}H_{17}NO_4$ 251.28	1149-24-2 71	301; 208[40]	5-22-04-00162	H_2O 1; EtOH 3; eth 3; bz 3
10933	Pyridine, 2,4-diethyl- 2,4-Diethylpyridine	$C_9H_{13}N$ 135.21	626-21-1 112	187.5	4-20-00-02825 0.9338[25]	
10934	3,4-Pyridinedimethanol, 5-hydroxy-6-methyl-, hydrochloride Pyridoxine hydrochloride	$C_8H_{12}ClNO_3$ 205.64	58-56-0 207	7995 sub	5-21-05-00492	H_2O 4
10935	Pyridine, 2,6-dimethoxy- 2,6-Dimethoxypyridine	$C_7H_9NO_2$ 139.15	6231-18-1	179	5-21-04-00416 1.053[25]	1.5029[20]
10936	Pyridine, 2,3-dimethyl- 2,3-Lutidine	C_7H_9N 107.16	583-61-9	161.2	5-20-06-00015 0.9319[25]	H_2O 3; EtOH 3; eth 3 1.5057[20]
10937	Pyridine, 2,4-dimethyl- 2,4-Lutidine	C_7H_9N 107.16	108-47-4 -64	158.5	5-20-06-00019 0.9309[20]	H_2O 4; EtOH 4; eth 4; ace 3 1.5010[20]
10938	Pyridine, 2,5-dimethyl- 2,5-Lutidine	C_7H_9N 107.16	589-93-5 -16	157.0	5-20-06-00027 0.9297[20]	H_2O 2; EtOH 4; eth 5; ace 3 1.5006[20]
10939	Pyridine, 2,6-dimethyl- 2,6-Lutidine	C_7H_9N 107.16	108-48-5 -6.1	5485 144.1	5-20-06-00032 0.9226[20]	H_2O 5; EtOH 2; eth 3; ace 3 1.4953[20]
10940	Pyridine, 3,4-dimethyl- 3,4-Lutidine	C_7H_9N 107.16	583-58-4 -11	179.1	5-20-06-00054 0.9281[20]	H_2O 2; EtOH 3; eth 3; ace 3 1.5096[20]
10941	Pyridine, 3,5-dimethyl- 3,5-Lutidine	C_7H_9N 107.16	591-22-0 -6.6	172	5-20-06-00060 0.9419[20]	H_2O 3; EtOH 3; eth 3; ace 3 1.5061[20]
10942	Pyridine, 2-(1,1-dimethylethyl)- 2-tert-Butylpyridine	$C_9H_{13}N$ 135.21	5944-41-2 -33	170	5-20-06-00121	1.4891[20]
10943	Pyridine, 3-(1,1-dimethylethyl)- 3-tert-Butylpyridine	$C_9H_{13}N$ 135.21	38031-78-6 -44	194.8	5-20-06-00122	1.4965[20]
10944	Pyridine, 4-(1,1-dimethylethyl)-	$C_9H_{13}N$ 135.21	3978-81-2 -41	196.5	5-20-06-00123 0.915[25]	ctc 3; CS_2 3 1.4958[20]
10945	Pyridine, 2,6-dimethyl-, 1-oxide	C_7H_9NO 123.15	1073-23-0	115-9[18]	5-20-06-00040 1.073[25]	1.5706[20]
10946	2,4(1H,3H)-Pyridinedione, 3,3-diethyl- Pyrithyldione	$C_9H_{13}NO_2$ 167.21	77-04-3 90.7	8005	5-21-10-00086	
10947	Pyridine, 2,6-diphenyl- 2,6-Diphenylpyridine	$C_{17}H_{13}N$ 231.30	3558-69-8 82	397	5-20-08-00459	eth 4; EtOH 4
10948	2-Pyridineethanamine	$C_7H_{10}N_2$ 122.17	2706-56-1	131.5[50]	5-22-09-00368 1.0220[25]	1.5335[25]
10949	4-Pyridineethanamine	$C_7H_{10}N_2$ 122.17	13258-63-4	121[10]	4-22-00-04201 1.0302[25]	H_2O 4 1.5381[25]
10950	2-Pyridineethanol	C_7H_9NO 123.15	103-74-2 -7.8	190[200]; 170[100]	5-21-02-00206 1.091[25]	H_2O 4; EtOH 4; eth 2; chl 4 1.5366[20]
10951	2-Pyridineethanol, α-methyl- 2-(2-Hydroxypropyl)pyridine	$C_8H_{11}NO$ 137.18	5307-19-7 33	124[20]; 110[10]	5-21-02-00234	H_2O 4; EtOH 4; chl 4
10952	Pyridine, 2-ethenyl-	C_7H_7N 105.14	100-69-6	159.5	5-20-06-00211 0.9983[20]	H_2O 2; EtOH 4; eth 4; ace 4 1.5495[20]
10953	Pyridine, 3-ethenyl-	C_7H_7N 105.14	1121-55-7	162	5-20-06-00212	H_2O 2; EtOH 3; eth 3 1.5530[20]
10954	Pyridine, 4-ethenyl-	C_7H_7N 105.14	100-43-6	121[150]; 65[15]	5-20-06-00213 0.9800[20]	H_2O 3; EtOH 3; eth 2; chl 3 1.5449[20]
10955	Pyridine, 5-ethenyl-2-methyl- 5-Vinyl-2-picoline	C_8H_9N 119.17	140-76-1		5-20-06-00218	
10956	Pyridine, 2-ethyl- 2-Ethylpyridine	C_7H_9N 107.16	100-71-0 -63.1	148.6	5-20-06-00003 0.950[25]	H_2O 3; EtOH 5; eth 4; ace 4 1.4964[20]
10957	Pyridine, 3-ethyl- 3-Ethylpyridine	C_7H_9N 107.16	536-78-7 -76.9	3802 165	5-20-06-00007 0.9539[25]	H_2O 3; EtOH 3; eth 3; ace 4 1.5021[20]

No.	Name / Synonym	Mol. Form. Mol. Wt.	CAS RN mp/°C	Merck No. bp/°C	Beil. Ref. den/g cm⁻³	Solubility n_D
10958	Pyridine, 4-ethyl- / 4-Ethylpyridine	C_7H_9N 107.16	536-75-4 -90.5	3803 168.3	5-20-06-00010 0.9417[20]	H_2O 3; EtOH 3; eth 3; ace 4 1.5009[20]
10959	Pyridine, 2-ethyl-3,5-dimethyl-	$C_9H_{13}N$ 135.21	1123-96-2	198; 86[15]	5-20-06-00129 0.9338[25]	H_2O 3; EtOH 4; eth 4; ace 4
10960	Pyridine, 3-ethyl-2,6-dimethyl-	$C_9H_{13}N$ 135.21	23580-52-1	83-5[23]	5-20-06-00129 0.9120[18]	H_2O 2
10961	Pyridine, 4-ethyl-2,6-dimethyl-	$C_9H_{13}N$ 135.21	36917-36-9 97	187.5	5-20-06-00130 0.9089[25]	1.4964[25]
10962	Pyridine, 2-ethyl-4-methyl-	$C_8H_{11}N$ 121.18	2150-18-7	175	5-20-06-00086 0.9238[20]	ace 4; eth 4; EtOH 4
10963	Pyridine, 2-ethyl-6-methyl-	$C_8H_{11}N$ 121.18	1122-69-6	160.5	5-20-06-00089 0.9207[25]	H_2O 2; EtOH 3; eth 3; ace 4 1.4920[25]
10964	Pyridine, 3-ethyl-4-methyl- / 3-Ethyl-4-picoline	$C_8H_{11}N$ 121.18	529-21-5	3796 198	5-20-06-00090 0.9286[17]	H_2O 2; EtOH 3; eth 3; ace 4
10965	Pyridine, 4-ethyl-2-methyl- / 4-Ethyl-2-picoline	$C_8H_{11}N$ 121.18	536-88-9	3797 179	5-20-06-00085 0.9130[25]	ace 4; bz 4; eth 4; EtOH 4
10966	Pyridine, 5-ethyl-2-methyl- / 5-Ethyl-2-picoline	$C_8H_{11}N$ 121.18	104-90-5	3798 178.3	5-20-06-00086 0.9202[20]	H_2O 2; EtOH 3; eth 3; ace 4 1.4971[20]
10967	Pyridine, 2-(1-ethylpropyl)-	$C_{10}H_{15}N$ 149.24	7399-50-0	195.4	5-20-06-00145 0.8981[20]	1.4850[25]
10968	Pyridine, 4-(1-ethylpropyl)- / 4-(3-Pentyl)pyridine	$C_{10}H_{15}N$ 149.24	35182-51-5 125.5	217; 80[12]	5-20-06-00145 0.9085[25]	1.40905[25]
10969	Pyridine, hydrochloride / Pyridinium chloride	C_5H_6ClN 115.56	628-13-7 146	222	5-20-05-00160	H_2O 4; EtOH 4; chl 4
10970	Pyridine, 2-iodo- / 2-Iodopyridine	C_5H_4IN 205.00	5029-67-4	93[13]	5-20-05-00438 1.9735[20]	EtOH 3; eth 3; ace 3; bz 3 1.6366[20]
10971	2-Pyridinemethanamine	$C_6H_8N_2$ 108.14	3731-51-9	203; 91[17]	5-22-09-00232 1.0525[25]	H_2O 4 1.5431[25]
10972	3-Pyridinemethanamine	$C_6H_8N_2$ 108.14	3731-52-0 -21.1	226	5-22-09-00301 1.064[20]	H_2O 4; eth 4; EtOH 4 1.5520[20]
10973	4-Pyridinemethanamine	$C_6H_8N_2$ 108.14	3731-53-1 -7.6	230; 103[11]	5-22-09-00345 1.072[20]	H_2O 4 1.5495[25]
10974	2-Pyridinemethanamine, N-(2-pyridinylmethyl)-	$C_{12}H_{13}N_3$ 199.26	1539-42-0	200[10]	5-22-09-00267 1.1074[25]	1.5757[25]
10975	2-Pyridinemethanol	C_6H_7NO 109.13	586-98-1	112[16]; 102.5[8]	5-21-02-00150 1.1317[20]	H_2O 5; EtOH 4; eth 4; ace 4 1.5444[20]
10976	3-Pyridinemethanol / Nicotinyl alcohol	C_6H_7NO 109.13	100-55-0 -6.5	6438 266	5-21-02-00172 1.131[20]	H_2O 4; eth 4 1.5455[20]
10977	4-Pyridinemethanol / 4-Picolyl alcohol	C_6H_7NO 109.13	586-95-8 53	140-2[12]	5-21-02-00191	chl 3
10978	3-Pyridinemethanol, 4-(aminomethyl)-5-hydroxy-6-methyl-, dihydrochloride / Pyridoxamine dihydrochloride	$C_8H_{14}Cl_2N_2O_2$ 241.12	524-36-7 226-7 dec	7993	5-22-12-00324	H_2O 4; EtOH 2
10979	Pyridine, 2-(2-methoxyethyl)- / Metyridine	$C_8H_{11}NO$ 137.18	114-91-0	6084 203; 96[17]	5-21-02-00206 0.988[20]	H_2O 4; EtOH 4 1.4975[20]
10980	Pyridine, 2-methyl- / 2-Picoline	C_6H_7N 93.13	109-06-8 -66.7	7372 129.3	5-20-05-00464 0.9443[20]	H_2O 4; EtOH 5; eth 5; ace 4 1.4957[20]
10981	Pyridine, 3-methyl- / 3-Picoline	C_6H_7N 93.13	108-99-6 -18.1	7373 144.1	5-20-05-00506 0.9566[20]	H_2O 5; EtOH 5; eth 5; ace 4 1.5040[20]
10982	Pyridine, 4-methyl- / 4-Picoline	C_6H_7N 93.13	108-89-4 3.66	7374 145.3	5-20-05-00543 0.9548[20]	H_2O 5; EtOH 5; eth 5; ace 3 1.5037[20]
10983	Pyridine, 2-[2-(1-methylethoxy)ethyl]- / Pyridine, 2-(2-isopropoxyethyl)-	$C_{10}H_{15}NO$ 165.24	70715-19-4	133[50]	0.9502[25]	H_2O 4 1.4820[25]
10984	Pyridine, 2-(1-methylethyl)-	$C_8H_{11}N$ 121.18	644-98-4 -141	159.8	5-20-06-00080 0.9342[25]	H_2O 2; EtOH 5; eth 5; ace 4 1.4915[20]
10985	Pyridine, 3-(1-methylethyl)- / 3-Isopropylpyridine	$C_8H_{11}N$ 121.18	6304-18-3 -45.8	177.5	5-20-06-00081 0.9217[16]	
10986	Pyridine, 4-(1-methylethyl)-	$C_8H_{11}N$ 121.18	696-30-0 -54.9	178	5-20-06-00082 0.9382[25]	H_2O 2; EtOH 5; eth 5; ace 4 1.4962[20]
10987	Pyridine, 3-(1-methyl-1H-pyrrol-2-yl)-	$C_{10}H_{10}N_2$ 158.20	487-19-4	281; 150[16]	5-23-07-00227 1.241[20]	EtOH 4 1.6057[20]
10988	Pyridine, 3-methyl-, 1-oxide	C_6H_7NO 109.13	1003-73-2 39	148[15]	5-20-05-00517	chl 3
10989	Pyridine, 2-(4-methylphenyl)-	$C_{12}H_{11}N$ 169.23	4467-06-5	170-80[20]	5-20-07-00569 0.999[25]	1.6125[20]
10990	Pyridine, 2-methyl-5-phenyl-	$C_{12}H_{11}N$ 169.23	3256-88-0	189[50]	5-20-07-00573 1.0590[25]	1.6055[25]
10991	Pyridine, 3-methyl-2-phenyl- / 3-Picoline, 2-phenyl-	$C_{12}H_{11}N$ 169.23	10273-90-2	148-9[16]	5-20-07-00571 1.065[25]	1.6026[20]
10992	Pyridine, 3-(1-methyl-2-pyrrolidinyl)-, (S)- / Nicotine	$C_{10}H_{14}N_2$ 162.23	54-11-5 -79	6434 247; 125[18]	5-23-06-00064 1.0097[20]	H_2O 5; EtOH 4; eth 4; chl 4 1.5282[20]
10993	Pyridine, 4-nitro-, 1-oxide / 4-Nitropyridine N-oxide	$C_5H_4N_2O_3$ 140.10	1124-33-0 160.5		5-20-05-00448	
10994	Pyridine, 1-oxide / Pyridine N-oxide	C_5H_5NO 95.10	694-59-7 65.5	7984 146-7[13]	5-20-05-00217	
10995	Pyridine, pentachloro-	C_5Cl_5N 251.33	2176-62-7 125.5	280	5-20-05-00422	bz 4; EtOH 4; lig 4
10996	Pyridine, 2-(3-pentenyl)-	$C_{10}H_{13}N$ 147.22	2057-43-4	216; 93[12]	5-20-06-00355 0.9234[25]	1.5076[25]

No.	Name / Synonym	Mol. Form. / Mol. Wt.	CAS RN / mp/°C	Merck No. / bp/°C	Beil. Ref. / den/g cm^{-3} / n_D	Solubility
10997	Pyridine, 2-phenyl- / 2-Phenylpyridine	$C_{11}H_9N$ / 155.20	1008-89-5	/ 271	5-20-07-00545 / 1.0833[25] / 1.6210[20]	H_2O 2; EtOH 5; eth 5
10998	Pyridine, 3-phenyl- / 3-Phenylpyridine	$C_{11}H_9N$ / 155.20	1008-88-4 / 164	/ 272	5-20-07-00548 / 1.6123[25]	H_2O 2; EtOH 3; eth 3
10999	Pyridine, 4-phenyl- / 4-Phenylpyridine	$C_{11}H_9N$ / 155.20	939-23-1 / 77.5	/ 281	5-20-07-00549	H_2O 3; EtOH 3; eth 3
11000	Pyridine, 2-(2-phenylethenyl)-, (E)-	$C_{13}H_{11}N$ / 181.24	538-49-8 / 91.5	325; 194[14]	5-20-08-00057 / 1.0147[100]	H_2O 1; EtOH 3; eth 3; bz 3
11001	Pyridine, 4-(2-phenylethenyl)-, (E)-	$C_{13}H_{11}N$ / 181.24	5097-93-8 / 131	208[33]	5-20-08-00063	H_2O 1; EtOH 3; eth 3; chl 3
11002	Pyridine, 2-(2-phenylethenyl)-, (Z)-	$C_{13}H_{11}N$ / 181.24	1519-59-1 / -50	141[10]	4-20-00-03874	EtOH 3; eth 3; bz 3
11003	Pyridine, 2-(2-phenylethyl)-	$C_{13}H_{13}N$ / 183.25	2116-62-3 / -3	/ 289	5-20-07-00588 / 1.0465[0]	
11004	Pyridine, 2-(phenylmethyl)- / 2-Benzylpyridine	$C_{12}H_{11}N$ / 169.23	101-82-6 / 12.5	277; 149[16]	5-20-07-00556 / 1.067[0] / 1.5785[20]	H_2O 1; EtOH 3; eth 3; chl 3
11005	Pyridine, 3-(phenylmethyl)- / 3-Benzylpyridine	$C_{12}H_{11}N$ / 169.23	620-95-1 / 34	/ 287	5-20-07-00560 / 1.061[25]	eth 4; EtOH 4
11006	Pyridine, 4-(phenylmethyl)-	$C_{12}H_{11}N$ / 169.23	2116-65-6 / 12.4	288; 180[31]	5-20-07-00561 / 1.061[20] / 1.5818[20]	H_2O 1; EtOH 3; eth 4; ctc 3
11007	Pyridine, 4-(3-phenylpropyl)-	$C_{14}H_{15}N$ / 197.28	2057-49-0	322; 150[5]	5-20-07-00607 / 1.024[25] / 1.5616[25]	bz 4; eth 4; py 4; EtOH 4
11008	Pyridine, 3-(2-piperidinyl)-, (S)- / Anabasine	$C_{10}H_{14}N_2$ / 162.23	494-52-0 / 9	655 / 276	5-23-06-00096 / 1.0436[20]	H_2O 5; EtOH 3; eth 3; bz 3
11009	Pyridine, 4-(4-piperidinyl)-	$C_{10}H_{14}N_2$ / 162.23	581-45-3 / 79	/ 292	5-23-06-00115	H_2O 2; EtOH 4; eth 3; bz 3
11010	2-Pyridinepropanamine, N,N-dimethyl-γ-phenyl- / Pheniramine	$C_{16}H_{20}N_2$ / 240.35	86-21-5	7198 / 181[13]; 135[0.5]	5-22-10-00487 / 1.0081[25] / 1.5519[25]	bz 4; eth 4; EtOH 4; chl 4
11011	1(4H)-Pyridinepropanoic acid, α-amino-3-hydroxy-4-oxo-, (S)- / Mimosine	$C_8H_{10}N_2O_4$ / 198.18	500-44-7 / 228 dec	6118	5-21-12-00107	H_2O 2; EtOH 1; eth 1; ace 1
11012	2-Pyridinepropanol / 2-(3-Hydroxypropyl)pyridine	$C_8H_{11}NO$ / 137.18	2859-68-9 / 34	/ 260.2; 116[4]	5-21-02-00236 / 1.060[25] / 1.5298[20]	H_2O 4
11013	3-Pyridinepropanol	$C_8H_{11}NO$ / 137.18	2859-67-8	/ 284; 130[3]	5-21-02-00237 / 1.063[25] / 1.5313[20]	H_2O 4
11014	Pyridine, 2-(2-propenyl)-	C_8H_9N / 119.17	2835-33-8	/ 190; 63[15]	5-20-06-00216 / 0.959[20]	eth 4; EtOH 4
11015	Pyridine, 2-propyl-	$C_8H_{11}N$ / 121.18	622-39-9 / 2	/ 167	5-20-06-00076 / 0.9119[20] / 1.4925[20]	H_2O 2; EtOH 5; eth 5; ace 4
11016	Pyridine, 4-propyl-	$C_8H_{11}N$ / 121.18	1122-81-2	/ 185	5-20-06-00078 / 0.938[15] / 1.4966[20]	eth 4; EtOH 4
11017	Pyridine, 3-(2-pyrrolidinyl)-, (S)- / Nornicotine	$C_9H_{12}N_2$ / 148.21	494-97-3	6631 / 270	4-23-00-00998 / 1.073[19] / 1.5378[18]	H_2O 4; ace 4; eth 4; EtOH 4
11018	3-Pyridinesulfonic acid / 3-Pyridylsulfonic acid	$C_5H_5NO_3S$ / 159.17	636-73-7 / 357 dec		5-22-07-00552 / 1.713[25]	H_2O 4; EtOH 2; eth 1
11019	Pyridine, 2,3,5,6-tetrachloro- / 2,3,5,6-Tetrachloropyridine	C_5HCl_4N / 216.88	2402-79-1 / 90.5	250.5	5-20-05-00421	eth 4; EtOH 4; peth 4
11020	Pyridine, 1,2,3,6-tetrahydro- / Δ 3-Piperidine	C_5H_9N / 83.13	694-05-3 / -48	/ 108	5-20-04-00282 / 0.911[25] / 1.4800[20]	chl 3
11021	Pyridine, 2,3,4,5-tetrahydro-6-propyl- / γ-Coniceine	$C_8H_{15}N$ / 125.21	1604-01-9 / 174	2497	5-20-04-00345 / 0.8753[15]	
11022	Pyridine, 2,3,4,6-tetramethyl- / 2,3,4,6-Tetramethylpyridine	$C_9H_{13}N$ / 135.21	20820-82-0	/ 203	5-20-06-00132 / 0.9322[25] / 1.4661[16]	eth 4; EtOH 4
11023	Pyridine, 2-(trichloromethyl)- / 2-(Trichloromethyl)pyridine	$C_6H_4Cl_3N$ / 196.46	4377-37-1 / -10	125-6[25]	5-20-05-00499 / 1.4526[25] / 1.5087[25]	
11024	Pyridine, 2,3,4-trimethyl- / 2,3,4-Collidine	$C_8H_{11}N$ / 121.18	2233-29-6	/ 192.5	5-20-06-00091 / 0.9127[15] / 1.5596[25]	H_2O 3; EtOH 3; eth 3
11025	Pyridine, 2,3,5-trimethyl- / 2,3,5-Collidine	$C_8H_{11}N$ / 121.18	695-98-7	/ 184	5-20-06-00092 / 0.9352[19] / 1.5150[20]	H_2O 2; EtOH 3; eth 3; ace 3
11026	Pyridine, 2,3,6-trimethyl- / 2,3,6-Collidine	$C_8H_{11}N$ / 121.18	1462-84-6	/ 171.6	5-20-06-00092 / 0.9220[25] / 1.5057[25]	H_2O 3; EtOH 3; eth 3; ace 3
11027	Pyridine, 2,4,5-trimethyl- / 2,4,5-Collidine	$C_8H_{11}N$ / 121.18	1122-39-0	/ 188	5-20-06-00093 / 0.9330[25] / 1.5053[20]	ace 4; bz 4; eth 4; EtOH 4
11028	Pyridine, 2,4,6-trimethyl- / 2,4,6-Collidine	$C_8H_{11}N$ / 121.18	108-75-8 / -46	9632 / 170.6	5-20-06-00093 / 0.9166[22] / 1.4959[25]	H_2O 3; EtOH 3; eth 3; ace 3
11029	Pyridinium, 3-carboxy-1-methyl-, hydroxide, inner salt / Trigonelline	$C_7H_7NO_2$ / 137.14	535-83-1	9606	5-22-02-00143	H_2O 4
11030	Pyridinium, 1-hexadecyl-, chloride / Hexadecyl pyridinium chloride	$C_{21}H_{38}ClN$ / 339.99	123-03-5 / 80	2024	5-20-05-00233	H_2O 4; chl 4
11031	2-Pyridinol	C_5H_5NO / 95.10	72762-00-6 / 107.8		5-21-07-00106 / 1.3910[20]	H_2O 4; bz 4; EtOH 4
11032	4-Pyridinol	C_5H_5NO / 95.10	626-64-2 / 149.8	>350; 257[10]	5-21-07-00152	H_2O 3; EtOH 3; eth 1; bz 1
11033	2(1H)-Pyridinone	C_5H_5NO / 95.10	142-08-5 / 107.8	/ 280	5-21-07-00106 / 1.3910[20]	H_2O 3; EtOH 2; bz 3
11034	2(1H)-Pyridinone, hydrazone	$C_5H_7N_3$ / 109.13	4930-98-7 / 46.6	185[140]; 90[1]	5-22-14-00486	chl 3

No.	Name / Synonym	Mol. Form. / Mol. Wt.	CAS RN / mp/°C	Merck No. / bp/°C	Beil. Ref. / den/g cm^{-3}	Solubility / n_D
11035	2(1H)-Pyridinone, 1-methyl-	C_6H_7NO 109.13	694-85-9 31	250	5-21-07-00109 1.1120[20]	H_2O 5; peth 2; lig 2
11036	2(1H)-Pyridinone, 3-methyl-	C_6H_7NO 109.13	1003-56-1 141.5	289	5-21-07-00199	H_2O 4; EtOH 4; chl 4
11037	2(1H)-Pyridinone, 4-methyl-	C_6H_7NO 109.13	13466-41-6 130	308	5-21-07-00201	H_2O 3; EtOH 3; eth 2; bz 2
11038	3H-Pyrido[3,4-b]indole, 4,9-dihydro-7-methoxy-1-methyl- Harmaline	$C_{13}H_{14}N_2O_{13}$ 406.26	304-21-2 250 dec	4528	5-23-12-00148	H_2O 2; EtOH 2; eth 2; chl 3
11039	9H-Pyrido[3,4-b]indole, 7-methoxy-1-methyl- Harmine	$C_{13}H_{12}N_2O$ 212.25	442-51-3 273 sub	4531	5-23-12-00237	H_2O 2; EtOH 2; eth 2; chl 2
11040	9H-Pyrido[3,4-b]indole, 1-methyl- Harman	$C_{12}H_{10}N_2$ 182.22	486-84-0 236.5	4530	5-23-08-00261	
11041	9H-Pyrido[3,4-b]indole, 1-[(2-methylphenyl)methyl]-	$C_{19}H_{16}N_2$ 272.35	525-15-5 217.8	150[0.01]	5-23-09-00536	EtOH 3; eth 3; bz 3; chl 3
11042	2-Pyrimidinamine	$C_4H_5N_3$ 95.10	109-12-6 127.5	sub	4-25-00-02071	H_2O 3; chl 2
11043	4-Pyrimidinamine, 2-chloro-N,N,6-trimethyl- Crimidine	$C_7H_{10}ClN_3$ 171.63	535-89-7 87	2590 140-7[4]	4-25-00-02184	EtOH 4
11044	2-Pyrimidinamine, 4,5-dimethyl- 2-Amino-4,5-dimethylpyrimidine	$C_6H_9N_3$ 123.16	1193-74-4 215.8	sub	4-25-00-02203	ace 4; bz 4; EtOH 4; chl 4
11045	4-Pyrimidinamine, 2,6-dimethyl- Kyanmethin	$C_6H_9N_3$ 123.16	461-98-3 183	5205 sub	4-25-00-02192	H_2O 2; EtOH 2; bz 2; chl 2
11046	2-Pyrimidinamine, 4-methyl-	$C_5H_7N_3$ 109.13	108-52-1 160.3	sub	4-25-00-02152	H_2O 3; EtOH 3; chl 2
11047	2-Pyrimidinamine, 5-methyl-	$C_5H_7N_3$ 109.13	50840-23-8 193.5	sub	5-25-10-00191	EtOH 4
11048	4-Pyrimidinamine, 6-methyl-	$C_5H_7N_3$ 109.13	3435-28-7 197	sub	4-25-00-02182	H_2O 3; EtOH 3
11049	Pyrimidine 1,3-Diazine	$C_4H_4N_2$ 80.09	289-95-2 22	7998 123.8	5-23-05-00334	H_2O 5; EtOH 3 1.4998[20]
11050	4-Pyrimidinecarboxylic acid, 1,2,3,6-tetrahydro-2,6-dioxo- Orotic acid	$C_5H_4N_2O_4$ 156.10	65-86-1 345.5	6828	4-25-00-01759	H_2O 2; os 1
11051	2,4-Pyrimidinediamine, 5-(4-chlorophenyl)-6-ethyl- Pyrimethamine	$C_{12}H_{13}ClN_4$ 248.71	58-14-0 233.5	7997	4-25-00-03014	
11052	Pyrimidine, 2,4-dichloro-5-methyl- 2,4-Dichloro-5-methylpyrimidine	$C_5H_4Cl_2N_2$ 163.01	1780-31-0 26	235	5-23-05-00385	H_2O 2; EtOH 4; eth 4; bz 4
11053	Pyrimidine, 2,4-dichloro-6-methyl- 2,6-Dichloro-4-methylpyrimidine	$C_5H_4Cl_2N_2$ 163.01	5424-21-5 46.5	219	5-23-05-00381	bz 4; eth 4; EtOH 4; chl 4
11054	Pyrimidine, 2,4-dichloro-5-nitro- 2,4-Dichloro-5-nitropyrimidine	$C_4HCl_2N_3O_2$ 193.98	49845-33-2 29.3	154[58] 135[17]	5-23-05-00350	EtOH 2; eth 2
11055	Pyrimidine, 2,4-dimethyl- 2,4-Dimethylpyrimidine	$C_6H_8N_2$ 108.14	14331-54-5 -4	150.5	5-23-05-00397 1.168[14]	1.4880[25]
11056	Pyrimidine, 2,5-dimethyl- 2,5-Dimethylpyrimidine	$C_6H_8N_2$ 108.14	22868-76-4 19	59[13]	5-23-05-00399	H_2O 4
11057	Pyrimidine, 4,5-dimethyl- 4,5-Dimethylpyrimidine	$C_6H_8N_2$ 108.14	694-81-5 3	177	5-23-05-00399	H_2O 4
11058	Pyrimidine, 4,6-dimethyl- 4,6-Dimethylpyrimidine	$C_6H_8N_2$ 108.14	1558-17-4 25	159	5-23-05-00400	H_2O 4 1.4880[20]
11059	2,4(1H,3H)-Pyrimidinedione Uracil	$C_4H_4N_2O_2$ 112.09	66-22-8 338	9761	5-24-06-00043	H_2O 2; EtOH 4; eth 4
11060	2,4(1H,3H)-Pyrimidinedione, 6-amino-3-ethyl-1-(2-propenyl)- Aminometradine	$C_9H_{13}N_3O_2$ 195.22	642-44-4 143	463	4-25-00-04113	
11061	2,4(1H,3H)-Pyrimidinedione, 5-[bis(2-chloroethyl)amino]- Uracil mustard	$C_8H_{11}Cl_2N_3O_2$ 252.10	66-75-1 206 dec	9762	4-25-00-04127	H_2O 2
11062	2,4(1H,3H)-Pyrimidinedione, 5-bromo- 5-Bromouracil	$C_4H_3BrN_2O_2$ 190.98	51-20-7 310	1430	5-24-06-00322	
11063	4,6(1H,5H)-Pyrimidinedione, dihydro-5-(2-methylpropyl)-5-(2-propenyl)-2-thioxo-, monosodium salt Barbituric acid, 5-allyl-5-isobutyl-2-thio, sodium salt	$C_{11}H_{15}N_2NaO_2S$ 262.31	510-90-7	1518		H_2O 4; EtOH 2; eth 1; bz 1
11064	4,6(1H,5H)-Pyrimidinedione, 5-ethyldihydro-5-(2-methyl-2-propenyl)-2-thioxo- Methallatal	$C_{10}H_{14}N_2O_2S$ 226.30	115-56-0 160.5	5855	5-24-09-00224	
11065	4,6(1H,5H)-Pyrimidinedione, 5-ethyldihydro-5-(1-methylpropyl)-2-thioxo- Barbituric acid, 5-sec-butyl-5-ethyl-2-thio-	$C_{10}H_{16}N_2O_2S$ 228.32	2095-57-0 169	9250	5-24-09-00164	
11066	4,6(1H,5H)-Pyrimidinedione, 5-ethyldihydro-5-phenyl- Primidone	$C_{12}H_{14}N_2O_2$ 218.26	125-33-7 281.5	7753	5-24-08-00102	
11067	2,4(1H,3H)-Pyrimidinedione, 5-fluoro- 5-Fluorouracil	$C_4H_3FN_2O_2$ 130.08	51-21-8 283	4109 sub 190	5-24-06-00271	
11068	2,4(1H,3H)-Pyrimidinedione, 5-methyl- Thymine	$C_5H_6N_2O_2$ 126.11	65-71-4 316	9332	5-24-07-00055	H_2O 2; EtOH 2; eth 2

No.	Name / Synonym	Mol. Form. / Mol. Wt.	CAS RN / mp/°C	Merck No. / bp/°C	Beil. Ref. / den/g cm^{-3}	Solubility / n_D
11069	2,4(1H,3H)-Pyrimidinedione, 6-methyl- / 6-Methyluracil	$C_5H_6N_2O_2$ / 126.11	626-48-2 / 270-80 dec	6051	5-24-07-00006	H_2O 3; EtOH 3; eth 2; tfa 2
11070	Pyrimidine, 2-methyl- / 2-Methyl-1,3-diazine	$C_5H_6N_2$ / 94.12	5053-43-0 / -4	138	5-23-05-00376	H_2O 5
11071	Pyrimidine, 4-methyl- / 4-Methyl-1,3-diazine	$C_5H_6N_2$ / 94.12	3438-46-8 / 32	142	5-23-05-00379	H_2O 5
11072	Pyrimidine, 5-methyl- / 5-Methyl-1,3-diazine	$C_5H_6N_2$ / 94.12	2036-41-1 / 30.5	153	5-23-05-00383	1.030^{16} / 1.500^{20} / H_2O 4
11073	2,4,5,6(1H,3H)-Pyrimidinetetrone / Alloxan	$C_4H_2N_2O_4$ / 142.07	50-71-5 / 256 dec	281 sub	5-24-09-00382	H_2O 4; EtOH 3; ace 3; bz 3
11074	2,4,5,6(1H,3H)-Pyrimidinetetrone, 5-oxime / Violuric acid	$C_4H_3N_3O_4$ / 157.09	87-39-8 / 203-4 dec	9904	5-24-09-00386	H_2O 2; EtOH 3
11075	Pyrimidine, 2,4,6-trichloro- / 2,4,6-Trichloropyrimidine	$C_4HCl_3N_2$ / 183.42	3764-01-0 / 22.5	212.5	5-23-05-00347	1.570^{20}
11076	2,4,6(1H,3H,5H)-Pyrimidinetrione / Barbituric acid	$C_4H_4N_2O_3$ / 128.09	67-52-7 / 248	973 260 dec	5-24-09-00078	H_2O 3; EtOH 2; eth 3
11077	2,4,6(1H,3H,5H)-Pyrimidinetrione, 5-amino- / Uramil	$C_4H_5N_3O_3$ / 143.10	118-78-5 / >400	9763	4-25-00-04228	H_2O 3; eth 1; bz 1; chl 3
11078	2,4,6(1H,3H,5H)-Pyrimidinetrione, 5-(2-bromo-2-propenyl)-5-(1-methylethyl)- / Propallylonal	$C_{10}H_{13}BrN_2O_3$ / 289.13	545-93-7 / 181	7807	4-24-00-01994	H_2O 2; EtOH 4; eth 2; ace 4
11079	2,4,6(1H,3H,5H)-Pyrimidinetrione, 5-(2-bromo-2-propenyl)-1-methyl-5-(1-methylethyl)- / Narcobarbital	$C_{11}H_{15}BrN_2O_3$ / 303.16	125-55-3 / 115	6342	4-24-00-01995	H_2O 2; EtOH 3; py 3
11080	2,4,6(1H,3H,5H)-Pyrimidinetrione, 5-(2-bromo-2-propenyl)-5-(1-methylpropyl)- / Butallylonal	$C_{11}H_{15}BrN_2O_3$ / 303.16	1142-70-7 / 131.5	1503	5-24-09-00231	eth 4; EtOH 4
11081	2,4,6(1H,3H,5H)-Pyrimidinetrione, 5-butyl-1-cyclohexyl- / Bucolome	$C_{14}H_{22}N_2O_3$ / 266.34	841-73-6 / 84	1452 185-7$^{0.8}$	5-24-09-00132	
11082	2,4,6(1H,3H,5H)-Pyrimidinetrione, 5-butyl-5-ethyl- / Butethal	$C_{10}H_{16}N_2O_3$ / 212.25	77-28-1 / 128.5	1515	5-24-09-00161	
11083	2,4,6(1H,3H,5H)-Pyrimidinetrione, 5-(1-cyclohexen-1-yl)-1,5-dimethyl- / Hexobarbital	$C_{12}H_{16}N_2O_3$ / 236.27	56-29-1 / 146.5	4625	5-24-09-00253	
11084	2,4,6(1H,3H,5H)-Pyrimidinetrione, 5-(1-cyclohexen-1-yl)-5-ethyl- / Cyclobarbital	$C_{12}H_{16}N_2O_3$ / 236.27	52-31-3 / 173	2717	5-24-09-00258	H_2O 1; EtOH 4; eth 3; dil alk 3
11085	2,4,6(1H,3H,5H)-Pyrimidinetrione, 5-(2-cyclopenten-1-yl)-5-(2-propenyl)- / Cyclopentobarbital	$C_{12}H_{14}N_2O_3$ / 234.25	76-68-6 / 139.5	2751	5-24-09-00269	H_2O 2; EtOH 4
11086	2,4,6(1H,3H,5H)-Pyrimidinetrione, 5,5-dibromo-	$C_4H_2Br_2N_2O_3$ / 285.88	511-67-1 / 235 dec		4-24-00-01881	eth 4; EtOH 4
11087	2,4,6(1H,3H,5H)-Pyrimidinetrione, 5,5-diethyl- / Barbital	$C_8H_{12}N_2O_3$ / 184.19	57-44-3 / 190	972	5-24-09-00137 / 1.220^{25}	H_2O 2; EtOH 3; eth 3; ace 3
11088	2,4,6(1H,3H,5H)-Pyrimidinetrione, 5,5-diethyl-1-methyl- / Metharbital	$C_9H_{14}N_2O_3$ / 198.22	50-11-3 / 150.5	5873	5-24-09-00144	H_2O 3; chl 2
11089	2,4,6(1H,3H,5H)-Pyrimidinetrione, 5,5-diethyl-1-phenyl- / Phenetharbital	$C_{14}H_{16}N_2O_3$ / 260.29	357-67-5 / 178	7183	4-24-00-01908	EtOH 4
11090	2,4,6(1H,3H,5H)-Pyrimidinetrione, 5-(2,2-dimethylpropyl)-5-(2-propenyl)- / Nealbarbital	$C_{12}H_{18}N_2O_3$ / 238.29	561-83-1 / 156	6350	5-24-09-00238	ace 4; eth 4; EtOH 4
11091	2,4,6(1H,3H,5H)-Pyrimidinetrione, 5,5-di-2-propenyl- / Allobarbital	$C_{10}H_{12}N_2O_3$ / 208.22	52-43-7 / 172	252	5-24-09-00245	H_2O 2; EtOH 3; eth 3; bz 3
11092	2,4,6(1H,3H,5H)-Pyrimidinetrione, 5-ethyl-5-(3-methylbutyl)- / Amobarbital	$C_{11}H_{18}N_2O_3$ / 226.28	57-43-2 / 157	601	5-24-09-00173	bz 4; EtOH 4; chl 4
11093	2,4,6(1H,3H,5H)-Pyrimidinetrione, 5-ethyl-5-(1-methylethyl)- / Probarbital	$C_9H_{14}N_2O_3$ / 198.22	76-76-6 / 203	7759	5-24-09-00159	H_2O 2; EtOH 1; eth 3
11094	2,4,6(1H,3H,5H)-Pyrimidinetrione, 5-ethyl-1-methyl-5-phenyl- / Mephobarbital	$C_{13}H_{14}N_2O_3$ / 246.27	115-38-8 / 176	5742	5-24-09-00294	H_2O 2; EtOH 4; eth 2; chl 3
11095	2,4,6(1H,3H,5H)-Pyrimidinetrione, 5-ethyl-5-phenyl- / Phenobarbital	$C_{12}H_{12}N_2O_3$ / 232.24	50-06-6 / 174	7201	5-24-09-00286	H_2O 1; EtOH 3; eth 3; bz 1
11096	2,4,6(1H,3H,5H)-Pyrimidinetrione, 5-ethyl-5-(1-piperidinyl)- / 5-Ethyl-5-(1-piperidyl)barbituric acid	$C_{11}H_{17}N_3O_3$ / 239.27	509-87-5 / 215	3800	4-25-00-04239	ace 4; eth 4; EtOH 4
11097	2,4,6(1H,3H,5H)-Pyrimidinetrione, 5-[(hexahydro-2,4,6-trioxo-5-pyrimidinyl)imino]-, monoammonium salt / Murexide	$C_8H_{10}N_8O_7$ / 302.20	3051-09-0	6215	4-25-00-04236	H_2O 2; EtOH 1; eth 1; alk 3
11098	2,4,6(1H,3H,5H)-Pyrimidinetrione, 5-(1-methylethyl)-5-(2-propenyl)- / Aprobarbital	$C_{10}H_{14}N_2O_3$ / 210.23	77-02-1 / 141	782	5-24-09-00224	ace 4; eth 4; EtOH 4; chl 4

No.	Name Synonym	Mol. Form. Mol. Wt.	CAS RN mp/°C	Merck No. bp/°C	Beil. Ref. den/g cm^{-3}	Solubility n_D
11099	2,4,6(1H,3H,5H)-Pyrimidinetrione, 1-methyl-5-(1-methylethyl)-5-(2-propenyl)- Enallylpropymal	C$_{11}$H$_{16}$N$_2$O$_3$ 224.26	1861-21-8 56.5	3523 176-8[12]	5-24-09-00225	bz 4; eth 4; EtOH 4; chl 4
11100	2,4,6(1H,3H,5H)-Pyrimidinetrione, 5-methyl-5-phenyl- Phenylmethylbarbituric acid	C$_{11}$H$_{10}$N$_2$O$_3$ 218.21	76-94-8 220	7275	5-24-09-00283	H$_2$O 1; EtOH 3; eth 3; alk 3
11101	2,4,6(1H,3H,5H)-Pyrimidinetrione, 5-(2-methylpropyl)-5-(2-propenyl)- Butalbital	C$_{11}$H$_{16}$N$_2$O$_3$ 224.26	77-26-9 138.5	1502	5-24-09-00231	H$_2$O 2; EtOH 3; eth 3; ace 3
11102	2,4,6(1H,3H,5H)-Pyrimidinetrione, 5-nitro- 5-Nitrobarbituric acid	C$_4$H$_3$N$_3$O$_5$ 173.09	480-68-2 180.5	6507	5-24-09-00105	H$_2$O 3; EtOH 3; eth 1
11103	2,4,6(1H,3H,5H)-Pyrimidinetrione, 5-phenyl-5-(2-propenyl)- Phenallymal	C$_{13}$H$_{12}$N$_2$O$_3$ 244.25	115-43-5 156.5	7162	5-24-09-00334	H$_2$O 2; EtOH 4; eth 4; bz 2
11104	2(1H)-Pyrimidinone, 4-amino- Cytosine	C$_4$H$_5$N$_3$O 111.10	71-30-7 320-5 dec	2801	4-25-00-03654	H$_2$O 3; EtOH 2; eth 1; chl 2
11105	2(1H)-Pyrimidinone, 4-amino-5-(hydroxymethyl)- 5-Hydroxymethylcytosine	C$_5$H$_7$N$_3$O$_2$ 141.13	1123-95-1 >300 dec			
11106	2(1H)-Pyrimidinone, 4-amino-5-methyl- 5-Methylcytosine	C$_5$H$_7$N$_3$O 125.13	554-01-8 270 dec	5970	4-25-00-03727	H$_2$O 3; EtOH 2; eth 1; acid 3
11107	4(1H)-Pyrimidinone, 2,6-diamino-5-(β-D-glucopyranosyloxy)- Vicine	C$_{10}$H$_{16}$N$_4$O$_7$ 304.26	152-93-2 239-42 dec	9880	4-25-00-04285	H$_2$O 2; EtOH 2; acid 4; alk 4
11108	4(1H)-Pyrimidinone, 2,3-dihydro-6-methyl-2-thioxo- Methylthiouracil	C$_5$H$_6$N$_2$OS 142.18	56-04-2 330 dec	6048 sub	5-24-07-00049	H$_2$O 1; EtOH 2; eth 2; bz 2
11109	4(1H)-Pyrimidinone, 2,3-dihydro-6-propyl-2-thioxo- Propylthiouracil	C$_7$H$_{10}$N$_2$OS 170.24	51-52-5 219	7882	5-24-07-00257	H$_2$O 2; EtOH 2; eth 1; bz 1
11110	4(1H)-Pyrimidinone, 2,3-dihydro-2-thioxo- 2-Thiouracil	C$_4$H$_4$N$_2$OS 128.15	141-90-2 >340 dec	9298	5-24-06-00361	H$_2$O 2; EtOH 2; anh HF 3
11111	5H-Pyrimido[4,5-d]thiazolo[3,2-a]pyrimidine-8-ethanol, 2,7-dimethyl- Thiochrome	C$_{12}$H$_{14}$N$_4$OS 262.34	92-35-3 228.8	sub	4-27-00-09599	H$_2$O 3; EtOH 2; eth 2; ace 2
11112	1H-Pyrrole Azole	C$_4$H$_5$N 67.09	109-97-7 -23.42	8025 129.79	5-20-05-00003 0.9698[20]	H$_2$O 2; EtOH 3; eth 3; ace 3 1.5085[20]
11113	1H-Pyrrole, 1-butyl-	C$_8$H$_{13}$N 123.20	589-33-3	170.5	5-20-05-00010 0.8747[20]	1.4727[20]
11114	1H-Pyrrole-2-carboxaldehyde	C$_5$H$_5$NO 95.10	1003-29-8 46.5	218	5-21-07-00169	chl 2; os 4; lig 2 1.5939[16]
11115	1H-Pyrrole-3-carboxylic acid, 2,4-dimethyl-, ethyl ester	C$_9$H$_{13}$NO$_2$ 167.21	2199-51-1 78.5	291	5-22-01-00427	eth 4; EtOH 4
11116	1H-Pyrrole-3-carboxylic acid, 2,5-dimethyl-, ethyl ester	C$_9$H$_{13}$NO$_2$ 167.21	2199-52-2 117.5	291; 130[15]	5-22-01-00432	EtOH 4
11117	1H-Pyrrole, 3,4-diethyl-2-methyl- 3,4-Diethyl-2-methylpyrrole	C$_9$H$_{15}$N 137.22	34874-30-1 202.5		4-20-00-02171 0.9100[17]	1.4988[17]
11118	1H-Pyrrole, 2,5-dihydro- 3-Pyrroline	C$_4$H$_7$N 69.11	109-96-6 90.5	8028	5-20-04-00274 0.9097[20]	H$_2$O 4; ace 4; eth 4; EtOH 4 1.4664[20]
11119	1H-Pyrrole, 2,4-dimethyl- 2,4-Dimethylpyrrole	C$_6$H$_9$N 95.14	625-82-1 168		5-20-05-00063 0.9236[20]	H$_2$O 2; EtOH 4; eth 4; bz 4 1.5048[20]
11120	1H-Pyrrole, 2,5-dimethyl-	C$_6$H$_9$N 95.14	625-84-3 171; 51[8]		5-20-05-00064 0.9353[20]	H$_2$O 1; EtOH 4; eth 4 1.5036[20]
11121	1H-Pyrrole-2,5-dione	C$_4$H$_3$NO$_2$ 97.07	541-59-3 94 sub		5-21-10-00003 1.2493[106]	H$_2$O 3; EtOH 3; eth 3
11122	1H-Pyrrole-2,5-dione, 1-ethyl- N-Ethylmaleimide	C$_6$H$_7$NO$_2$ 125.13	128-53-0 45.5	3778	5-21-10-00006	H$_2$O 2; EtOH 4; eth 4; chl 3
11123	1H-Pyrrole-2,5-dione, 1-phenyl- N-Phenylmaleimide	C$_{10}$H$_7$NO$_2$ 173.17	941-69-5 90.5	7270 162[12]	5-21-10-00009	bz 4; eth 4; EtOH 4
11124	1H-Pyrrole, 1-ethyl-	C$_6$H$_9$N 95.14	617-92-5 129.5		5-20-05-00010 0.9009[20]	EtOH 4 1.4841[20]
11125	1H-Pyrrole, 2-ethyl-	C$_6$H$_9$N 95.14	1551-06-0 164		5-20-05-00061 0.9042[20]	H$_2$O 1; EtOH 3; chl 3 1.4942[20]
11126	1H-Pyrrole, 3-ethyl-2,4-dimethyl-	C$_8$H$_{13}$N 123.20	517-22-6 0	199; 96[16]	5-20-05-00099 0.913[20]	H$_2$O 2; EtOH 3; eth 3; bz 3 1.4961[20]
11127	1H-Pyrrole, 4-ethyl-2,3-dimethyl-	C$_8$H$_{13}$N 123.20	491-18-9 16.5	200; 113[16]	4-20-00-02152 0.915[20]	H$_2$O 4
11128	1H-Pyrrole, 3-ethyl-4-methyl-	C$_7$H$_{11}$N 109.17	488-92-6 3	70[11]	5-20-05-00083 0.9059[20]	eth 4; EtOH 4 1.4913[20]
11129	Pyrrole, 2-isopropyl-	C$_7$H$_{11}$N 109.17	7696-51-7	172	5-20-05-00081 0.908[25]	eth 4; EtOH 4 1.491[25]
11130	1H-Pyrrole, 1-methyl-	C$_5$H$_7$N 81.12	96-54-8 -56.32	112.81	5-20-05-00008 0.9145[15]	H$_2$O 1; EtOH 5; eth 5 1.4875[20]
11131	1H-Pyrrole, 2-methyl-	C$_5$H$_7$N 81.12	636-41-9 -35.6	147.6	5-20-05-00053 0.9446[15]	H$_2$O 1; EtOH 5; eth 5 1.5035[16]
11132	1H-Pyrrole, 3-methyl- 3-Methylpyrrole	C$_5$H$_7$N 81.12	616-43-3 -48.4	142.9; 45[11]	5-20-05-00056	EtOH 5; eth 5 1.4970[20]
11133	1H-Pyrrole, 2,2'-methylenebis-	C$_9$H$_{10}$N$_2$ 146.19	21211-65-4 73	164[12]	5-23-06-00468	bz 4; eth 4; EtOH 4

No.	Name / Synonym	Mol. Form. / Mol. Wt.	CAS RN / mp/°C	Merck No. / bp/°C	Beil. Ref. / den/g cm^{-3}	Solubility / n_D
11134	1H-Pyrrole, 1-phenyl-	$C_{10}H_9N$ 143.19	635-90-5 62	234	5-20-05-00012	H_2O 1; EtOH 3; eth 3; ace 3
11135	1H-Pyrrole, 2-phenyl- 2-Phenylpyrrole	$C_{10}H_9N$ 143.19	3042-22-6 129	272	5-20-07-00369	H_2O 1; EtOH 4; eth 4; bz 4
11136	1H-Pyrrole, 1-(phenylmethyl)-	$C_{11}H_{11}N$ 157.22	2051-97-0 15	247	5-20-05-00013 1.0183[20]	H_2O 1; EtOH 4; eth 4 1.5655[24]
11137	1H-Pyrrole-1-propanenitrile	$C_7H_8N_2$ 120.15	43036-06-2	132-3[10]	5-20-05-00023 1.048[25]	1.5103[20]
11138	1H-Pyrrole, 1-propyl-	$C_7H_{11}N$ 109.17	5145-64-2	145.5	5-20-05-00010 0.8833[20]	eth 4; EtOH 4
11139	1H-Pyrrole, 3-propyl- 3-Propylpyrrole	$C_7H_{11}N$ 109.17	1551-09-3	90[30]; 49[2]	5-20-05-00081	eth 3; ace 3; chl 2 1.4900[25]
11140	1H-Pyrrole, 2,3,4,5-tetraiodo- Iodopyrrole	C_4HI_4N 570.68	87-58-1 150 dec	4934	5-20-05-00045	ace 4; eth 4; chl 4
11141	1H-Pyrrole, 2,3,4,5-tetramethyl-	$C_8H_{13}N$ 123.20	1003-90-3 111	130[7]	5-20-05-00099	EtOH 5; eth 5; ace 5; bz 4
11142	1H-Pyrrole, 1,2,5-trimethyl- 1,2,5-Trimethylpyrrole	$C_7H_{11}N$ 109.17	930-87-0	171	5-20-05-00065 0.807[25]	1.4969[20]
11143	Pyrrolidine Azacyclopentane	C_4H_9N 71.12	123-75-1 -57.8	8026 86.5	5-20-01-00162 0.8586[20]	H_2O 5; EtOH 3; eth 3; bz 2 1.4431[20]
11144	3-Pyrrolidineacetic acid, 2-carboxy-4-(1-methylethenyl)-, [2S-(2α,3β,4β)]- Kainic acid	$C_{10}H_{15}NO_4$ 213.23	487-79-6	5157	4-22-00-01523	
11145	1-Pyrrolidineacetic acid, 2,5-dioxo-, ethyl ester Aminoacetic acid, N-succinyl, ethyl ester	$C_8H_{11}NO_4$ 185.18	14181-05-6 67	198[32]	5-21-09-00474	H_2O 4; EtOH 4
11146	Pyrrolidine, 1-butyl-	$C_8H_{17}N$ 127.23	767-10-2	154.5	5-20-01-00170 0.816[25]	EtOH 4 1.4373[25]
11147	Pyrrolidine, 2-butyl- 2-Butylpyrrolidine	$C_8H_{17}N$ 127.23	3446-98-8	174; 67[18]	5-20-04-00222 0.827[20]	H_2O 4; EtOH 4 1.4490[20]
11148	1,2-Pyrrolidinedicarboxylic acid, 1-(phenylmethyl) ester, (S)-	$C_{13}H_{15}NO_4$ 249.27	1148-11-4 77		5-22-01-00084	chl 2 1.5310[20]
11149	Pyrrolidine, 2,2-dimethyl- 2,2-Dimethylpyrrolidine	$C_6H_{13}N$ 99.18	35018-15-6	106	4-20-00-01529 0.8211[25]	1.4304[25]
11150	Pyrrolidine, 2,4-dimethyl-	$C_6H_{13}N$ 99.18	13603-04-8	116	4-20-00-01537 0.8297[20]	H_2O 4; eth 4; EtOH 4 1.4325[20]
11151	Pyrrolidine, 2,5-dimethyl-, cis- cis-2,5-Dimethylpyrrolidine	$C_6H_{13}N$ 99.18	39713-71-8	106.5	5-20-04-00146 0.8205[20]	H_2O 4; eth 4; EtOH 4 1.4299[20]
11152	2,5-Pyrrolidinedione Succinimide	$C_4H_5NO_2$ 99.09	123-56-8 126.5	8842 287 dec	5-21-09-00438 1.418[25]	H_2O 3; EtOH 2; eth 2; ace 2
11153	2,5-Pyrrolidinedione, 1-bromo- N-Bromosuccinimide	$C_4H_4BrNO_2$ 177.99	128-08-5 174	1428	5-21-09-00543 2.098[25]	H_2O 2; eth 2; ace 4
11154	2,5-Pyrrolidinedione, 1-chloro- N-Chlorosuccinimide	$C_4H_4ClNO_2$ 133.53	128-09-6 150	2165	5-21-09-00543 1.65[25]	H_2O 2; EtOH 2; ace 3; bz 2
11155	2,5-Pyrrolidinedione, 1-(chloromethyl)-	$C_5H_6ClNO_2$ 147.56	54553-14-9 58	158[12]	5-21-09-00460	ace 4; bz 4
11156	2,5-Pyrrolidinedione, 1,3-dimethyl-3-phenyl- Methsuximide	$C_{12}H_{13}NO_2$ 203.24	77-41-8 52.5	5928 121-2[0.1]	5-21-11-00209	
11157	2,5-Pyrrolidinedione, 1-ethyl-	$C_6H_9NO_2$ 127.14	2314-78-5 26	236	5-21-09-00442	eth 4; EtOH 4
11158	2,5-Pyrrolidinedione, 3-ethyl-3-methyl- Ethosuximide	$C_7H_{11}NO_2$ 141.17	77-67-8 64.5	3704	5-21-09-00595	H_2O 4
11159	2,5-Pyrrolidinedione, 1-iodo- N-Iodosuccinimide	$C_4H_4INO_2$ 224.99	516-12-1 200.5	4936	5-21-09-00544 2.245[25]	H_2O 4; EtOH 3; eth 2; ace 3
11160	2,5-Pyrrolidinedione, 1-methyl-	$C_5H_7NO_2$ 113.12	1121-07-9 71	234	5-21-09-00441	H_2O 3; EtOH 3; eth 4
11161	2,5-Pyrrolidinedione, 1-methyl-3-phenyl- Phensuximide	$C_{11}H_{11}NO_2$ 189.21	86-34-0 72	7231	5-21-11-00188	EtOH 4; MeOH 4
11162	2,5-Pyrrolidinedione, 1-phenyl- Succinanil	$C_{10}H_9NO_2$ 175.19	83-25-0 156	8838 400	5-21-09-00446 1.356[25]	H_2O 1; EtOH 3; eth 3
11163	2,5-Pyrrolidinedione, 1-(phenylmethyl)- N-Benzylsuccinimide	$C_{11}H_{11}NO_2$ 189.21	2142-06-5 103	393	5-21-09-00449	bz 4; eth 4; EtOH 4; chl 4
11164	1-Pyrrolidineethanamine	$C_6H_{14}N_2$ 114.19	7154-73-6	66-70[23]	5-20-01-00405 0.901[25]	
11165	1-Pyrrolidineethanol	$C_6H_{13}NO$ 115.18	2955-88-6	79-81[13]	4-20-00-00092 0.985[25]	1.4687[20]
11166	Pyrrolidine, 1-ethyl-	$C_6H_{13}N$ 99.18	7335-06-0	104	5-20-01-00168 0.8156[20]	1.4713[20]
11167	2-Pyrrolidinemethanamine, 1-ethyl-	$C_7H_{16}N_2$ 128.22	26116-12-1	58-60[16]	5-22-08-00096 0.887[25]	1.4113[15]
11168	Pyrrolidine, 1-methyl- N-Methylpyrrolidine	$C_5H_{11}N$ 85.15	120-94-5	81	5-20-01-00166 0.8188[20]	H_2O 4; eth 4 1.4665[20] 1.4247[20]
11169	Pyrrolidine, 1-[2-(3-methylbutoxy)-2-phenylethyl]- Amixetrine	$C_{17}H_{27}NO$ 261.41	24622-72-8	507 121[2]	5-20-01-00521	1.4978[22]
11170	Pyrrolidine, 1-nitroso- N-Nitrosopyrrolidine	$C_4H_8N_2O$ 100.12	930-55-2	6564 214	5-20-01-00182 1.085[25]	1.4880[25]
11171	Pyrrolidine, 1-phenyl-	$C_{10}H_{13}N$ 147.22	4096-21-3	119-20[12]	1.0260[25]	eth 3 1.5813[20]

No.	Name / Synonym	Mol. Form. / Mol. Wt.	CAS RN / mp/°C	Merck No. / bp/°C	Beil. Ref. / den/g cm^{-3}	Solubility / n_D
11172	Pyrrolidine, 1-(phenylmethyl)-	$C_{11}H_{15}N$ 161.25	29897-82-3	$45^{0.7}$	5-20-01-00184 0.965^{25}	1.5270^{20}
11173	Pyrrolidine, 1-propyl-	$C_7H_{15}N$ 113.20	7335-07-1 105	132.5	5-20-01-00169 0.8171^{20}	1.4389^{20}
11174	Pyrrolidine, 3-propyl-	$C_7H_{15}N$ 113.20	116632-47-4	159	4-20-00-01598 0.8450^{20}	1.4469^{20}
11175	Pyrrolidine, 2,2,4-trimethyl- 2,2,4-Trimethylpyrrolidine	$C_7H_{15}N$ 113.20	35018-28-1	119	5-20-04-00198 0.8063^{25}	1.4259^{25}
11176	Pyrrolidine, 2,2,5-trimethyl-	$C_7H_{15}N$ 113.20	6496-48-6	112	4-20-00-01603 0.7980^{25}	1.4223^{25}
11177	Pyrrolidinium, 2-carboxy-1,1-dimethyl-, hydroxide, inner salt, (S)- Stachydrine	$C_7H_{13}NO_2$ 143.19	471-87-4 235	8729		H_2O 4; EtOH 4; EtOH 4
11178	Pyrrolidinium, 2-carboxy-4-hydroxy-1,1-dimethyl-, hydroxide, inner salt, (2S-trans)- Betonicine	$C_7H_{13}NO_3$ 159.19	515-25-3 252 dec	1209		EtOH 4
11179	Pyrrolidinium, 3-[(cyclopentylhydroxyphenylacetyl)oxy]-1,1-dimethyl-, bromide Glycopyrrolate	$C_{19}H_{28}BrNO_3$ 398.34	596-51-0 192.5	4397	5-21-01-00015	
11180	2-Pyrrolidinone 2-Pyrrolidone	C_4H_7NO 85.11	616-45-5 25	8027 251; 133^{12}	5-21-06-00317 1.120^{20}	H_2O 4; EtOH 4; eth 4; bz 4 1.4806^{30}
11181	2-Pyrrolidinone, 1,5-dimethyl-	$C_6H_{11}NO$ 113.16	5075-92-3	217; 87.5^{10}	5-21-06-00431	H_2O 4; eth 4 1.4650^{20}
11182	2-Pyrrolidinone, 3,3-dimethyl- 3,3-Dimethyl-2-pyrrolidone	$C_6H_{11}NO$ 113.16	4831-43-0 66	237	0-21-00-00242	H_2O 4; ace 4; eth 4; EtOH 4
11183	2-Pyrrolidinone, 1-ethenyl-	C_6H_9NO 111.14	88-12-0 13.5	193^{400}; 93^{11}	5-21-06-00330 1.04^{20}	
11184	2-Pyrrolidinone, 1-(2-hydroxyethyl)-	$C_6H_{11}NO_2$ 129.16	3445-11-2 20	295	5-21-06-00341 1.1435^{20}	
11185	2-Pyrrolidinone, 1-methyl- N-Methyl-2-pyrrolidone	C_5H_9NO 99.13	872-50-4 -24	202	5-21-06-00321 1.0230^{25}	H_2O 4; eth 3; ace 3; chl 3 1.4684^{20}
11186	2-Pyrrolidinone, 5-methyl-	C_5H_9NO 99.13	108-27-0 43	248	5-21-06-00431 1.0458^{20}	
11187	3-Pyrrolidinone, 1-methyl-	C_5H_9NO 99.13	68165-06-0	$46-7^{18}$	5-21-06-00389 0.9646^{25}	1.4431^{25}
11188	2-Pyrrolidinone, 1-[4-(1-pyrrolidinyl)-2-butynyl]- Oxotremorine	$C_{12}H_{18}N_2O$ 206.29	70-22-4 	6904 $124^{0.1}$	5-21-06-00370 0.991^{25}	1.5160^{20}
11189	2-Pyrroline, 2-methyl-	C_5H_9N 83.13	100791-95-5	96	0-20-00-00135 0.8995^{22}	H_2O 4; eth 4; EtOH 4
11190	Pyrrolo[1,2-a]pyrimidine, 2,3,4,6,7,8-hexahydro-	$C_7H_{12}N_2$ 124.19	3001-72-7	$95-8^7$	5-23-05-00239 1.005^{25}	1.5190^{20}
11191	Pyrrolo[2,3-b]indol-5-ol, 1,2,3,3a,8,8a-hexahydro-1,3a,8-trimethyl-, methylcarbamate (ester), (3aS-cis)- Physostigmine	$C_{15}H_{21}N_3O_2$ 275.35	57-47-6 105.5	7357	5-23-11-00401	H_2O 2; EtOH 3; eth 3; bz 3
11192	Pyrrolo[2,1-b]quinazolin-3-ol, 1,2,3,9-tetrahydro-, (R)- Vasicine	$C_{11}H_{12}N_2O$ 188.23	6159-55-3 211.5	9842	5-23-12-00040	H_2O 2; EtOH 3; eth 2; ace 3
11193	1,1':2',1'':2'',1'''-Quaterphenyl	$C_{24}H_{18}$ 306.41	641-96-3 118.5	420; $193^{1.0}$	4-05-00-02729	H_2O 1; EtOH 2; eth 3; ace 3
11194	1,1':4',1'':4'',1'''-Quaterphenyl	$C_{24}H_{18}$ 306.41	135-70-6 320	428^{18}	4-05-00-02730	H_2O 1; EtOH 1; eth 1; bz 3
11195	Quinazoline 1,3-Benzodiazine	$C_8H_6N_2$ 130.15	253-82-7 48	8062 241	5-23-07-00129	H_2O 4; EtOH 3; eth 3; ace 3
11196	Quinazoline, 3,4-dihydro-3-phenyl- Orexine	$C_{14}H_{12}N_2$ 208.26	612-97-5 96.5	dec	4-23-00-01108 1.290^4	bz 4; eth 4; EtOH 4; chl 4
11197	6-Quinazolinesulfonamide, 7-chloro-2-ethyl-1,2,3,4-tetrahydro-4-oxo- Quinethazone	$C_{10}H_{12}ClN_3O_3S$ 289.74	73-49-4	8067		tfa 3
11198	4(1H)-Quinazolinone, 2,3-dihydro-1-(4-morpholinylacetyl)-3-phenyl- Moquizone	$C_{20}H_{21}N_3O_3$ 351.41	19395-58-5 136	6174		chl 3
11199	4(3H)-Quinazolinone, 2-methyl-3-(2-methylphenyl)- Methaqualone	$C_{16}H_{14}N_2O$ 250.30	72-44-6 120	5872	5-24-03-00132	eth 4; EtOH 4; chl 4
11200	4(1H)-Quinazolinone, 1-methyl-2-(phenylmethyl)- Glycosine	$C_{16}H_{14}N_2O$ 250.30	6873-15-0 161.5	4398	5-24-04-00363	
11201	Quinclorac 3,7-Dichloroquinoline-8-carboxylic acid	$C_{10}H_5Cl_2NO_2$ 242.06	84087-01-4 274		1.75	
11202	2-Quinolinamine 2-Aminoquinoline	$C_9H_8N_2$ 144.18	580-22-3 131.5	sub	5-22-10-00220	H_2O 4; EtOH 3; eth 3; ace 3
11203	4-Quinolinamine	$C_9H_8N_2$ 144.18	578-68-7 154.8	180^{12}	5-22-10-00241	H_2O 3; EtOH 4; eth 4; bz 3
11204	5-Quinolinamine	$C_9H_8N_2$ 144.18	611-34-7 110	310; 184^{10}	5-22-10-00297	H_2O 2; EtOH 4; eth 4; bz 3
11205	6-Quinolinamine 6-Aminoquinoline	$C_9H_8N_2$ 144.18	580-15-4 114	187^{12}	5-22-10-00303	H_2O 2; EtOH 3; eth 2; NH_3 3

No.	Name Synonym	Mol. Form. Mol. Wt.	CAS RN mp/°C	Merck No. bp/°C	Beil. Ref. den/g cm⁻³	Solubility n_D
11206	8-Quinolinamine	$C_9H_8N_2$ 144.18	578-66-5 70	157^{19}	5-22-10-00316	H_2O 4; EtOH 4
11207	4-Quinolinamine, 3-bromo- 4-Amino-3-bromoquinoline	$C_9H_7BrN_2$ 223.07	36825-36-2 203	sub	4-22-00-04662	eth 4; EtOH 4
11208	2-Quinolinamine, 4-methyl- Lepidine, 2-amino-	$C_{10}H_{10}N_2$ 158.20	27063-27-0 133	320	5-22-10-00375	eth 4; EtOH 4; chl 4
11209	3-Quinolinamine, 2-methyl-	$C_{10}H_{10}N_2$ 158.20	21352-22-7 160	278; 198^{16}	5-22-10-00347	bz 4; eth 4; EtOH 4
11210	4-Quinolinamine, 2-methyl-	$C_{10}H_{10}N_2$ 158.20	6628-04-2 169	333	5-22-10-00347	H_2O 2; EtOH 4; eth 4; ace 4
11211	7-Quinolinamine, 2-methyl- 7-Amino-2-methylquinoline	$C_{10}H_{10}N_2$ 158.20	64334-96-9 148	304	5-22-10-00364	ace 4; bz 4; eth 4; EtOH 4
11212	8-Quinolinamine, 6-methyl- 8-Amino-6-methylquinoline	$C_{10}H_{10}N_2$ 158.20	68420-93-9 73	sub	4-22-00-04822	ace 4; bz 4; eth 4; EtOH 4
11213	6-Quinolinamine, N,N,2-trimethyl-	$C_{12}H_{14}N_2$ 186.26	92-99-9 101	319	2-22-00-00361	ctc 3; CS_2 3
11214	Quinoline 1-Azanaphthalene	C_9H_7N 129.16	91-22-5 -14.78	8097 237.1	5-20-07-00276 1.0977^{15}	H_2O 1; EtOH 5; eth 5; ace 5 1.6268^{20}
11215	Quinoline, 6-amino-3-bromo-	$C_9H_7BrN_2$ 223.07	7101-96-4 106	sub	5-22-10-00312	EtOH 4
11216	Quinoline, 2-[2-[(1,3-benzodioxol-5-yl)ethyl]-4-methoxy- Cusparine	$C_{19}H_{17}NO_3$ 307.35	529-92-0 92(form a); 116(form b)	2687	2-27-00-00545	ace 4; bz 4; eth 4; EtOH 4
11217	Quinoline, 3-bromo- 3-Bromoquinoline	C_9H_6BrN 208.06	5332-24-1 13.3	275	5-20-07-00318	chl 3; HOAc 4 1.6641^{20}
11218	Quinoline, 5-bromo- 5-Bromoquinoline	C_9H_6BrN 208.06	4964-71-0 52	280; 106^{12}	5-20-07-00319	acid 3
11219	Quinoline, 6-bromo- 6-Bromoquinoline	C_9H_6BrN 208.06	5332-25-2 24	281	5-20-07-00319	EtOH 3; eth 3; acid 3
11220	Quinoline, 7-bromo- 7-Bromoquinoline	C_9H_6BrN 208.06	4965-36-0 34	290	4-20-00-03390	EtOH 3; eth 3; acid 4
11221	Quinoline, 6-bromo-5-nitro- 6-Bromo-5-nitroquinoline	$C_9H_5BrN_2O_2$ 253.05	98203-04-4 130	sub	0-20-00-00376	eth 4; EtOH 4
11222	Quinoline, 2-butyl- 2-Butylquinoline	$C_{13}H_{15}N$ 185.27	7661-39-4	153^{14}	5-20-07-00482 1.003^{20}	1.5699^{20}
11223	2-Quinolinecarbonitrile 2-Cyanoquinoline	$C_{10}H_6N_2$ 154.17	1436-43-7 95	160^{23}	5-22-03-00190	H_2O 3; EtOH 4; eth 4; bz 4
11224	4-Quinolinecarbonitrile	$C_{10}H_6N_2$ 154.17	2973-27-5 103.5	242.5	5-22-03-00206	ace 4; bz 4; eth 4; EtOH 4
11225	4-Quinolinecarboxaldehyde Cinchoninaldehyde	$C_{10}H_7NO$ 157.17	4363-93-3 51	122^4	5-21-08-00443	eth 4; tol 4
11226	4-Quinolinecarboxamide, 2-butoxy-N-[2-(diethylamino)ethyl]-, monohydrochloride Dibucaine hydrochloride	$C_{20}H_{30}ClN_3O_2$ 379.93	61-12-1 94 dec	3016	5-22-05-00278	chl 3
11227	2-Quinolinecarboxylic acid Quinaldic acid	$C_{10}H_7NO_2$ 173.17	93-10-7 156	8054	5-22-03-00183	H_2O 3; bz 4
11228	7-Quinolinecarboxylic acid	$C_{10}H_7NO_2$ 173.17	1078-30-4 249.5	sub	5-22-03-00217	EtOH 4
11229	8-Quinolinecarboxylic acid 8-Carboxyquinoline	$C_{10}H_7NO_2$ 173.17	86-59-9 187	8099 sub	5-22-03-00217	EtOH 4
11230	2-Quinolinecarboxylic acid, 4,8-dihydroxy- Xanthurenic acid	$C_{10}H_7NO_4$ 205.17	59-00-7 289	9977	4-22-00-02513	H_2O 1; EtOH 3; eth 2; bz 2
11231	1(2H)-Quinolinecarboxylic acid, 2-ethoxy-, ethyl ester EEDQ	$C_{14}H_{17}NO_3$ 247.29	16357-59-8 56.5	3488 $125-8^{0.1}$	5-21-03-00028	chl 3
11232	2-Quinolinecarboxylic acid, 4-hydroxy- Kynurenic acid	$C_{10}H_7NO_3$ 189.17	492-27-3 282.5	5206	5-22-06-00280	H_2O 2; EtOH 3; eth 1; os 2
11233	4-Quinolinecarboxylic acid, 3-hydroxy-2-phenyl- Oxycinchophen	$C_{16}H_{11}NO_3$ 265.27	485-89-2 206-7 dec	6910	4-22-00-02383	bz 4; EtOH 4; HOAc 4
11234	4-Quinolinecarboxylic acid, 6-methoxy- Quininic acid	$C_{11}H_9NO_3$ 203.20	86-68-0 285 dec	8092 sub	4-22-00-02297	H_2O 2; EtOH 3; eth 2; bz 2
11235	4-Quinolinecarboxylic acid, 2-phenyl- Cinchophen	$C_{16}H_{11}NO_2$ 249.27	132-60-5 214.5	2290	5-22-03-00484	H_2O 1; EtOH 3; eth 3; ace 2
11236	4-Quinolinecarboxylic acid, 2-phenyl-, 2-propenyl ester	$C_{19}H_{15}NO_2$ 289.33	524-34-5 31.5	262^{15}	4-22-00-01361	ace 4; eth 4; EtOH 4
11237	Quinoline, 2-chloro- 2-Chloroquinoline	C_9H_6ClN 163.61	612-62-4 38	266; 153^{22}	5-20-07-00312 1.2464^{25}	H_2O 1; EtOH 4; eth 4; bz 3 1.634^{25}
11238	Quinoline, 4-chloro- 4-Chloroquinoline	C_9H_6ClN 163.61	611-35-8 34.5	262; 130^{15}	5-20-07-00314 1.251^{25}	H_2O 2; EtOH 4; eth 4; dil HCl 3
11239	Quinoline, 5-chloro- 5-Chloroquinoline	C_9H_6ClN 163.61	635-27-8 45	256	5-20-07-00314	EtOH 3
11240	Quinoline, 6-chloro- 6-Chloroquinoline	C_9H_6ClN 163.61	612-57-7 43.8	263	5-20-07-00315	1.6110^{56}
11241	Quinoline, 7-chloro- 7-Chloroquinoline	C_9H_6ClN 163.61	612-61-3 31.5	267.5	5-20-07-00315 1.2158^{58}	H_2O 2; EtOH 4; eth 4; ace 4 1.6108^{58}
11242	Quinoline, 8-chloro- 8-Chloroquinoline	C_9H_6ClN 163.61	611-33-6 -20	288.5	5-20-07-00315 1.2834^{14}	H_2O 3; EtOH 4; eth 4; ace 4 1.6408^{14}

No.	Name / Synonym	Mol. Form. / Mol. Wt.	CAS RN / mp/°C	Merck No. / bp/°C	Beil. Ref. / den/g cm^{-3}	Solubility / n_D
11243	Quinoline, 2-chloro-4-methyl-	$C_{10}H_8ClN$ 177.63	634-47-9 59	296	5-20-07-00396	H_2O 1; EtOH 4; eth 4; chl 4
11244	Quinoline, 2-chloro-8-methyl- 2-Chloro-8-methylquinoline	$C_{10}H_8ClN$ 177.63	4225-85-8 61	287	5-20-07-00407	bz 4; eth 4; EtOH 4; chl 4
11245	Quinoline, 4-chloro-2-methyl-	$C_{10}H_8ClN$ 177.63	4295-06-1 42.5	269.5	5-20-07-00383 0.881[25]	H_2O 2; EtOH 3; eth 3 1.6224[20]
11246	Quinoline, decahydro-, cis- cis-Decahydroquinoline	$C_9H_{17}N$ 139.24	10343-99-4 -40	206; 90[20]	5-20-04-00405 0.9426[20]	H_2O 2; EtOH 3; eth 3 1.4926[20]
11247	Quinoline, decahydro-, trans-(±)- Quinoline, decahydro (trans, DL)	$C_9H_{17}N$ 139.24	105728-23-2 48	204	5-20-04-00405 0.9610[22]	ace 4; bz 4; eth 4; EtOH 4 1.4692[58]
11248	Quinoline, 2,4-dichloro- 2,4-Dichloroquinoline	$C_9H_5Cl_2N$ 198.05	703-61-7 67.5	281	4-20-00-03382	bz 4; eth 4; EtOH 4; chl 4
11249	Quinoline, 2,7-dichloro- 2,7-Dichloroquinoline	$C_9H_5Cl_2N$ 198.05	613-77-4 120	100[2]	4-20-00-03383	eth 4; EtOH 4
11250	Quinoline, 4,5-dichloro-	$C_9H_5Cl_2N$ 198.05	21617-18-5 118	134[5.5]	4-20-00-03383	eth 4; EtOH 4
11251	Quinoline, 4,7-dichloro- 4,7-Dichloroquinoline	$C_9H_5Cl_2N$ 198.05	86-98-6 93	148[10]	5-20-07-00316	chl 2
11252	Quinoline, 5,8-dichloro-	$C_9H_5Cl_2N$ 198.05	703-32-2 97.5	sub	5-20-07-00317	eth 4; EtOH 4
11253	Quinoline, 2-[2-(3,4-dimethoxyphenyl)ethyl]-4-methoxy- Galipine	$C_{20}H_{21}NO_3$ 323.39	525-68-8 115.5	4247	4-21-00-02656	ace 4; bz 4; eth 4; EtOH 4
11254	Quinoline, 2,3-dimethyl- 2,3-Dimethylquinoline	$C_{11}H_{11}N$ 157.22	1721-89-7 68.5	263	5-20-07-00431 1.1013[25]	eth 4; EtOH 4; lig 4
11255	Quinoline, 2,4-dimethyl- 4-Methylquinaldine	$C_{11}H_{11}N$ 157.22	1198-37-4	265	5-20-07-00431 1.0611[15]	H_2O 2; EtOH 4; eth 4; chl 2 1.6075[20]
11256	Quinoline, 2,5-dimethyl-	$C_{11}H_{11}N$ 157.22	26190-82-9 61	264.5	4-20-00-03524	bz 4; eth 4; EtOH 4
11257	Quinoline, 2,6-dimethyl-	$C_{11}H_{11}N$ 157.22	877-43-0 60	266.5	5-20-07-00433	H_2O 2; EtOH 2; eth 2; bz 4
11258	Quinoline, 2,7-dimethyl- m-Toluquinaldine	$C_{11}H_{11}N$ 157.22	93-37-8 61	264.5	5-20-07-00435	H_2O 2; EtOH 3; eth 3; chl 3
11259	Quinoline, 2,8-dimethyl-	$C_{11}H_{11}N$ 157.22	1463-17-8 27	255.3	5-20-07-00436 1.0394[20]	H_2O 2; EtOH 4; eth 4; ctc 3 1.6022[20]
11260	Quinoline, 3,4-dimethyl- 3,4-Dimethylquinoline	$C_{11}H_{11}N$ 157.22	2436-92-2 73.5	291; 114[4]	5-20-07-00437	eth 4; EtOH 4
11261	Quinoline, 3,8-dimethyl- 3,8-Dimethylquinoline	$C_{11}H_{11}N$ 157.22	20668-29-5	270	5-20-07-00437 1.049[20]	1.6063[20]
11262	Quinoline, 4,6-dimethyl- 6-Methyllepidine	$C_{11}H_{11}N$ 157.22	826-77-7 22	273.5; 140[12]	5-20-07-00437 1.0577[20]	eth 4; EtOH 4 1.6100[20]
11263	Quinoline, 5,8-dimethyl- 5,8-Dimethylquinoline	$C_{11}H_{11}N$ 157.22	2623-50-9 4.5	266	5-20-07-00438 1.070[21]	eth 4; EtOH 4
11264	Quinoline, 6,8-dimethyl- 6,8-Dimethylquinoline	$C_{11}H_{11}N$ 157.22	2436-93-3	268.5	5-20-07-00439 1.0665[4]	eth 4; EtOH 4
11265	Quinoline, 6-ethoxy-1,2-dihydro-2,2,4-trimethyl- Ethoxyquin	$C_{14}H_{19}NO$ 217.31	91-53-2	3710 123-5[2]	5-21-03-00095 1.026[25]	1.569[25]
11266	Quinoline, 2-ethyl- 2-Ethylquinoline	$C_{11}H_{11}N$ 157.22	1613-34-9	245.5; 134[16]	5-20-07-00428 1.050[17]	eth 4; EtOH 4; chl 4 1.5979[23]
11267	Quinoline, 3-ethyl- 3-Ethylquinoline	$C_{11}H_{11}N$ 157.22	1873-54-7	266	5-20-07-00428 1.0508[20]	1.6160[20]
11268	Quinoline, 6-ethyl- 6-Ethylquinoline	$C_{11}H_{11}N$ 157.22	19655-60-8	135-6[14]	5-20-07-00429 1.043[25]	1.6009[25]
11269	Quinoline, 2-ethyl-3-methyl- 2-Ethyl-3-methylquinoline	$C_{12}H_{13}N$ 171.24	27356-52-1 57	269; 140[12]	5-20-07-00456	bz 4; eth 4; EtOH 4
11270	Quinoline, 2-ethyl-4-methyl- Lepidine, 2-ethyl-	$C_{12}H_{13}N$ 171.24	33357-44-7	272.5	5-20-07-00456 1.025[20]	1.5941[20]
11271	Quinoline, 3-ethyl-6-methyl-	$C_{12}H_{13}N$ 171.24	105688-74-2	284.5	4-20-00-03555 1.0280[20]	1.5955[20]
11272	Quinoline, 4-iodo- 4-Iodoquinoline	C_9H_6IN 255.06	16560-43-3 98.5	sub	5-20-07-00322	eth 4; EtOH 4
11273	Quinoline, 6-iodo-	C_9H_6IN 255.06	13327-31-6 91	120[1.0]	5-20-07-00322	eth 4; EtOH 4
11274	Quinoline, 4-methoxy- 4-Methoxyquinoline	$C_{10}H_9NO$ 159.19	607-31-8 41	245	5-21-03-00229	H_2O 1; EtOH 3; eth 3; bz 3
11275	Quinoline, 6-methoxy- 6-Methoxyquinoline	$C_{10}H_9NO$ 159.19	5263-87-6 26.5	306; 153[12]	5-21-03-00244 1.152[20]	EtOH 3; eth 3; chl 3; dil HCl 3
11276	Quinoline, 8-methoxy- 8-Methoxyquinoline	$C_{10}H_9NO$ 159.19	938-33-0 49.5	283; 164[14]	5-21-03-00254 1.034[29]	EtOH 3; eth 3; bz 3; peth 3
11277	Quinoline, 2-methyl- Quinaldine	$C_{10}H_9N$ 143.19	91-63-4 -1.5	8055 246.5	5-20-07-00375 1.06[25]	H_2O 2; EtOH 3; eth 3; ace 3 1.6116[20]
11278	Quinoline, 3-methyl- 3-Methylquinoline	$C_{10}H_9N$ 143.19	612-58-8 16.5	259.8	5-20-07-00388 1.0673[20]	ace 4; eth 4; EtOH 4 1.6171[20]
11279	Quinoline, 4-methyl- Lepidine	$C_{10}H_9N$ 143.19	491-35-0 9.5	5324 262	5-20-07-00389 1.083[20]	H_2O 2; EtOH 3; eth 3; ace 3 1.6200[20]
11280	Quinoline, 5-methyl- 5-Methylquinoline	$C_{10}H_9N$ 143.19	7661-55-4 19	262.7	5-20-07-00399 1.0832[20]	H_2O 2; EtOH 5; eth 5; ace 3 1.6219[20]
11281	Quinoline, 6-methyl- 6-Methylquinoline	$C_{10}H_9N$ 143.19	91-62-3 -22	258.6	5-20-07-00400 1.0654[20]	H_2O 2; EtOH 3; eth 3; ace 3 1.6157[20]

No.	Name / Synonym	Mol. Form. Mol. Wt.	CAS RN mp/°C	Merck No. bp/°C	Beil. Ref. den/g cm⁻³	Solubility n_D
11282	Quinoline, 7-methyl- / m-Toluquinoline	$C_{10}H_9N$ 143.19	612-60-2 39	257.6	5-20-07-00402 1.0609[20]	H₂O 2; EtOH 3; eth 3; ace 3 1.6150[20]
11283	Quinoline, 8-methyl- / 8-Methylquinoline	$C_{10}H_9N$ 143.19	611-32-5 -80	247.5	5-20-07-00405 1.0719[20]	H₂O 2; EtOH 5; eth 5; ace 3 1.6164[20]
11284	Quinoline, 5-nitro- / 5-Nitroquinoline	$C_9H_6N_2O_2$ 174.16	607-34-1 74	sub	5-20-07-00325	H₂O 2; EtOH 3; bz 3; chl 2
11285	Quinoline, 6-nitro- / 6-Nitroquinoline	$C_9H_6N_2O_2$ 174.16	613-50-3 153.5	170[0.2]	5-20-07-00326	H₂O 3; EtOH 3; eth 2; bz 4
11286	Quinoline, 7-nitro- / 7-Nitroquinoline	$C_9H_6N_2O_2$ 174.16	613-51-4 132.5	sub	5-20-07-00327	H₂O 3; EtOH 3; eth 3; chl 4
11287	Quinoline, 4-nitro-, 1-oxide / 4-Nitroquinoline oxide	$C_9H_6N_2O_3$ 190.16	56-57-5 154		5-20-07-00324	
11288	Quinoline, 2-phenyl- / 2-Phenylquinoline	$C_{15}H_{11}N$ 205.26	612-96-4 86	363; 194[6]	5-20-08-00355	H₂O 2; EtOH 4; eth 4; ace 4
11289	Quinoline, 3-phenyl- / 3-Phenylquinoline	$C_{15}H_{11}N$ 205.26	1666-96-2 52	205[12]	5-20-08-00359	H₂O 2; EtOH 3; eth 3; ace 4
11290	Quinoline, 4-phenyl- / 4-Phenylquinoline	$C_{15}H_{11}N$ 205.26	605-03-8 61.5	260[77]; 156[0.3]	5-20-08-00360	H₂O 1; EtOH 4; eth 4; ace 4
11291	Quinoline, 6-phenyl- / 6-Phenylquinoline	$C_{15}H_{11}N$ 205.26	612-95-3 110	260[77]	5-20-08-00361 1.1945[20]	H₂O 2; EtOH 4; eth 2; ace 4
11292	Quinoline, 2-propyl- / 2-Propylquinoline	$C_{12}H_{13}N$ 171.24	1613-32-7	130-1[10]	5-20-07-00453 1.038[17]	1.5886[23]
11293	5-Quinolinesulfonic acid, 8-ethoxy- / Actinoquinol	$C_{11}H_{11}NO_4S$ 253.28	15301-40-3 286 dec	134	5-22-07-00581	alk 3
11294	5-Quinolinesulfonic acid, 8-hydroxy- / 8-Hydroxy-5-quinolinesulfonic acid	$C_9H_7NO_4S$ 225.22	84-88-8 322.5	4780	5-22-07-00581	H₂O 2
11295	5-Quinolinesulfonic acid, 8-hydroxy-7-iodo- / Ferron	$C_9H_6INO_4S$ 351.12	547-91-1 260 dec	4757	5-22-07-00582	H₂O 2; EtOH 2; eth 1; bz 1
11296	Quinoline, 1,2,3,4-tetrahydro-	$C_9H_{11}N$ 133.19	635-46-1 20	251	5-20-06-00293 1.0588[20]	H₂O 3; EtOH 5; eth 5; chl 3 1.606[19]
11297	Quinoline, 5,6,7,8-tetrahydro- / 2,3-Cyclohexenopyridine	$C_9H_{11}N$ 133.19	10500-57-9	222	5-20-06-00292 1.0304[13]	H₂O 2; EtOH 3; eth 3; ace 3 1.5435[20]
11298	Quinoline, 1,2,3,4-tetrahydro-6-methoxy-	$C_{10}H_{13}NO$ 163.22	120-15-0 42.5	284; 128[1]	5-21-02-00314	chl 3 1.5718[20]
11299	Quinoline, 1,2,3,4-tetrahydro-1-methyl-	$C_{10}H_{13}N$ 147.22	491-34-9	246	5-20-06-00294 1.022[20]	EtOH 4 1.5802[23]
11300	Quinoline, 2,3,6-trimethyl-	$C_{12}H_{13}N$ 171.24	2437-73-2 86.5	285	4-20-00-03557	eth 4; EtOH 4
11301	Quinoline, 2,3,8-trimethyl-	$C_{12}H_{13}N$ 171.24	4945-28-2	283	4-20-00-03557 1.0069[20]	H₂O 1; EtOH 3; eth 3; bz 2
11302	Quinoline, 2,4,6-trimethyl- / 4,6-Dimethylquinaldine	$C_{12}H_{13}N$ 171.24	2243-89-2 68	280	5-20-07-00458	ace 4; bz 4; eth 4; EtOH 4
11303	Quinoline, 2,4,7-trimethyl- / 2,4,7-Trimethylquinoline	$C_{12}H_{13}N$ 171.24	71633-43-7 63.5	280.5	5-20-07-00458 1.0337[20]	H₂O 2; EtOH 4; eth 4; ace 4 1.5973[24]
11304	Quinoline, 2,5,7-trimethyl-	$C_{12}H_{13}N$ 171.24	102871-67-0	287; 107[1.2]	4-20-00-03559	H₂O 1; EtOH 4; eth 4 1.5980[20]
11305	5-Quinolinol / 5-Hydroxyquinoline	C_9H_7NO 145.16	578-67-6 226 dec	sub	5-21-03-00240	H₂O 3; EtOH 2; eth 2; bz 3
11306	6-Quinolinol	C_9H_7NO 145.16	580-16-5 195	360	5-21-03-00244	H₂O 1; EtOH 2; eth 2; bz 1
11307	7-Quinolinol / 7-Hydroxyquinoline	C_9H_7NO 145.16	580-20-1 239	sub	5-21-03-00249	EtOH 4
11308	8-Quinolinol / 8-Hydroxyquinoline	C_9H_7NO 145.16	148-24-3 75.5	4778 267	5-21-03-00252 1.034[20]	H₂O 1; EtOH 4; eth 1; ace 3
11309	8-Quinolinol, benzoate (ester) / Benzoxiquine	$C_{16}H_{11}NO_2$ 249.27	86-75-9	1122	5-21-03-00259	chl 2
11310	8-Quinolinol, 5-bromo-	C_9H_6BrNO 224.06	1198-14-7 124	sub	5-21-03-00288	H₂O 4; EtOH 4; bz 4; chl 4
11311	8-Quinolinol, 5-chloro- / Cloxyquin	C_9H_6ClNO 179.61	130-16-5 130	2416	5-21-03-00279	
11312	8-Quinolinol, 5-chloro-7-iodo- / Iodochlorhydroxyquin	C_9H_5ClINO 305.50	130-26-7 178.5	4924	5-21-03-00294	EtOH 2; HOAc 3
11313	8-Quinolinol, 5,7-dibromo- / Broxyquinoline	$C_9H_5Br_2NO$ 302.95	521-74-4 196	1441 sub	5-21-03-00290	H₂O 1; EtOH 3; eth 2; ace 3
11314	8-Quinolinol, 5,7-dichloro- / Chloroxine	$C_9H_5Cl_2NO$ 214.05	773-76-2 179.5	2175	5-21-03-00286	EtOH 2; ace 2; bz 3; chl 2
11315	8-Quinolinol, 5,7-diiodo- / Iodoquinol	$C_9H_5I_2NO$ 396.95	83-73-8 210	4935	5-21-03-00296	H₂O 2; EtOH 4; eth 2; bz 2
11316	4-Quinolinol, 2,8-dimethyl-	$C_{11}H_{11}NO$ 173.21	15644-80-1 262.3	sub	5-21-08-00321	EtOH 4
11317	6-Quinolinol, 2,4-dimethyl- / 2,4-Dimethyl-6-hydroxyquinoline	$C_{11}H_{11}NO$ 173.21	64165-34-0 214	360 dec	2-21-00-00068	ace 4; EtOH 4
11318	8-Quinolinol, 2,4-dimethyl- / Quinoline, 2,4-dimethyl-8-hydroxy	$C_{11}H_{11}NO$ 173.21	115310-98-0 65	281	4-21-00-01292	ace 4; bz 4; eth 4; EtOH 4
11319	6-Quinolinol, 2-methyl-	$C_{10}H_9NO$ 159.19	613-21-8 213	304.5	5-21-03-00339 1.1665[25]	eth 4; EtOH 4
11320	7-Quinolinol, 6-methyl- / Quinoline, 7-hydroxy-6-methyl	$C_{10}H_9NO$ 159.19	84583-53-9 244	sub	0-21-00-00111	bz 4; eth 4; EtOH 4
11321	8-Quinolinol, 2-methyl-	$C_{10}H_9NO$ 159.19	826-81-3 73.8	267	5-21-03-00341	H₂O 1; EtOH 3; eth 3; bz 3

No.	Name / Synonym	Mol. Form. / Mol. Wt.	CAS RN / mp/°C	Merck No. / bp/°C	Beil. Ref. / den/g cm⁻³	Solubility / n_D
11322	8-Quinolinol, 6-methyl-	$C_{10}H_9NO$ 159.19	20984-33-2 95.5	sub	4-21-00-01275	EtOH 4
11323	8-Quinolinol, 5-nitro- Nitroxoline	$C_9H_6N_2O_3$ 190.16	4008-48-4 180	6577	5-21-03-00297	
11324	8-Quinolinol, sulfate (2:1) (salt) 8-Hydroxyquinoline sulfate	$C_9H_8NO_3S$ 210.23	134-31-6 177.5	4779	5-21-03-00252	H_2O 4; EtOH 3; eth 1
11325	2(1H)-Quinolinone Carbostyril	C_9H_7NO 145.16	59-31-4 199.5	1831 sub	5-21-08-00217	H_2O 2; EtOH 4; eth 4
11326	2(1H)-Quinolinone, 4-bromo-	C_9H_6BrNO 224.06	938-39-6 266.5	sub	5-21-08-00229	EtOH 4
11327	2(1H)-Quinolinone, 3,4-dihydro- Hydrocarbostyril	C_9H_9NO 147.18	553-03-7 163.5	4702 201[45]	5-21-08-00057	eth 4; EtOH 4
11328	2(1H)-Quinolinone, 1-methyl-	$C_{10}H_9NO$ 159.19	606-43-9 74	325	5-21-08-00218	H_2O 2; EtOH 3; eth 3; ace 3
11329	2(1H)-Quinolinone, 6-methyl-	$C_{10}H_9NO$ 159.19	4053-34-3 237	240[12]	5-21-08-00289	ace 4; bz 4; eth 4; EtOH 4
11330	4(1H)-Quinolinone, 1-methyl- Echinopsine	$C_{10}H_9NO$ 159.19	83-54-5	3472	5-21-08-00210	H_2O 3; EtOH 4; eth 2; bz 4
11331	2H-Quinolizine-1-methanol, octahydro-, (1R-trans)- Lupinine	$C_{10}H_{19}NO$ 169.27	486-70-4 70	5479 270	5-21-01-00338	H_2O 3; EtOH 3; eth 3; bz 3
11332	6-Quinoxalinamine 6-Aminoquinoxaline	$C_8H_7N_3$ 145.16	6298-37-9 159	sub	4-25-00-02611	H_2O 4; EtOH 4; eth 3; bz 2
11333	Quinoxaline 1,4-Benzodiazine	$C_8H_6N_2$ 130.15	91-19-0 28	8107 229.5	5-23-07-00135 1.1334[48]	H_2O 3; EtOH 5; eth 5; ace 5 1.6231[48]
11334	Quinoxaline, 6-chloro- 6-Chloroquinoxaline	$C_8H_5ClN_2$ 164.59	5448-43-1 64	117[10]	5-23-07-00144	H_2O 2
11335	Quinoxaline, 2,6-dimethyl-	$C_{10}H_{10}N_2$ 158.20	60814-29-1 54	268	0-23-00-00192	H_2O 4; ace 4; bz 4; EtOH 4
11336	Quinoxaline, 6-methoxy- 6-Methoxyquinoxaline	$C_9H_8N_2O$ 160.18	6639-82-3 58.9	128[7]	5-23-11-00461	ace 4; eth 4
11337	Quinoxaline, 2-methyl- 2-Methylquinoxaline	$C_9H_8N_2$ 144.18	7251-61-8 180.5	244	5-23-07-00219	H_2O 5; EtOH 4; eth 5; ace 5
11338	Quinoxaline, 6-methyl- 6-Methylquinoxaline	$C_9H_8N_2$ 144.18	6344-72-5 218.5	249; 141.5[29]	5-23-07-00223 1.1164[20]	H_2O 5; EtOH 5; eth 5; ace 5 1.6211[18]
11339	Quinoxaline, 1,2,3,4-tetrahydro- 1,2,3,4-Tetrahydroquinoxaline	$C_8H_{10}N_2$ 134.18	3476-89-9 99	289	5-23-06-00024	H_2O 3; EtOH 4; eth 4; bz 4
11340	(-)-cis-Resmethrin Cyclopropanecarboxylic acid, 2,2-dimethyl-3-(2-methyl-1-propenyl)-, [5-(phenylmethyl)-3-furanyl] methyl ester	$C_{22}H_{26}O_3$ 338.45	10453-86-8 75			
11341	Retinol Vitamin A	$C_{20}H_{30}O$ 286.46	68-26-8 63.5	9918 137-8[10-6]	4-06-00-04133	H_2O 1; EtOH 3; eth 3; ace 3
11342	Rheadan, 8-methoxy-16-methyl-2,3:10,11-bis[methylenebis(oxy)]-, (8β)- Rheadine	$C_{21}H_{21}NO_6$ 383.40	2718-25-4 257	8174 sub	4-27-00-06897	
11343	Ribitol Adonitol	$C_5H_{12}O_5$ 152.15	488-81-3 104	157	4-01-00-02832	H_2O 3; EtOH 3; eth 1; lig 1
11344	Riboflavin Riboflavine	$C_{17}H_{20}N_4O_6$ 376.37	83-88-5 280 dec	8201	4-26-00-02542	H_2O 1; EtOH 2; eth 1; ace 1
11345	D-Ribose	$C_5H_{10}O_5$ 150.13	50-69-1 88	8205	4-01-00-04211	H_2O 3; EtOH 2
11346	Sarpagan-17-al Vellosimine	$C_{19}H_{20}N_2O$ 292.38	6874-98-2 305.5	9846 sub 180		
11347	Sarpagan-10,17-diol Sarpagine	$C_{19}H_{22}N_2O_2$ 310.40	482-68-8 320	8336	5-23-13-00401	H_2O 1; EtOH 3
11348	9,10-Secocholesta-5,7,10(19)-trien-3-ol, (3β,5Z,7E)- Vitamin D3	$C_{27}H_{44}O$ 384.65	67-97-0 84.5	9929	4-06-00-04149	H_2O 1; os 3
11349	9,10-Secoergosta-5(10),6,8,22-tetraen-3-ol, (3β,6E,22E)- Tachysterol	$C_{28}H_{44}O$ 396.66	115-61-7	9002	4-06-00-04404	H_2O 1; EtOH 3; eth 3; ace 3
11350	9,10-Secoergosta-5,7,10(19),22-tetraen-3-ol, (3β,5Z,7E,22E)- Vitamin D2	$C_{28}H_{44}O$ 396.66	50-14-6 116.5	9928 sub	4-06-00-04404	H_2O 1; EtOH 3; eth 3; ace 3
11351	16,19-Secostrychnidine-10,16-dione, 4-hydroxy-19-methyl- Vomicine	$C_{22}H_{24}N_2O_4$ 380.44	125-15-5 282	9946	4-27-00-07873	EtOH 2; eth 2; ace 2; chl 4
11352	9,10-Secostrychnidin-10-oic acid, 2,3-dihydro-4-nitro-2,3-dioxo- Cacotheline	$C_{21}H_{21}N_3O_7$ 427.41	561-20-6 >300	1604	4-27-00-08014	H_2O 2
11353	Selenanthrene Selanthrene	$C_{12}H_8Se_2$ 310.12	262-30-6 181	223[11]	5-19-02-00057	EtOH 2; eth 2; CS_2 3
11354	Selenourea Carbamimidoselenoic acid	CH_4N_2Se 123.02	630-10-4 200 dec		4-03-00-00435	H_2O 4
11355	D-Serine	$C_3H_7NO_3$ 105.09	312-84-5 229 dec	dec	4-04-00-03118	H_2O 4; EtOH 1; eth 1; bz 1
11356	DL-Serine	$C_3H_7NO_3$ 105.09	302-84-1 246 dec		4-04-00-03119 1.603[22]	H_2O 3; EtOH 1; eth 1; bz 1
11357	L-Serine (S)-2-Amino-3-hydroxypropanoic acid	$C_3H_7NO_3$ 105.09	56-45-1 228 dec	8411 sub 150	4-04-00-03118 1.6[22]	H_2O 4; EtOH 1; eth 1; bz 1

No.	Name / Synonym	Mol. Form. / Mol. Wt.	CAS RN / mp/°C	Merck No. / bp/°C	Beil. Ref. / den/g cm⁻³	Solubility / n_D
11358	Sesone / Ethanol, 2-(2,4-dichlorophenoxy)-, hydrogen sulfate, sodium salt	$C_8H_8Cl_2NaO_5S$ 310.11	136-78-7 245 dec	3372		
11359	Sethoxydim / (±)-2-[1-(Ethoxyimino)butyl]-5-[2-(ethylthio)propyl]-3-hydroxy-2-cyclohexen-1-one	$C_{17}H_{29}NO_3S$ 327.49	74051-80-2	8424 $>90^{0.00003}$	1.043^{25}	
11360	Silacyclohexane, 1,1-diphenyl- / 1,1-Diphenylsilacyclohexane	$C_{17}H_{20}Si$ 252.43	18002-79-4		4-27-00-09833	
11361	Silanamine, pentamethyl-	$C_5H_{15}NSi$ 117.27	2083-91-2	$193-8^5$	1.0319^{25}	1.5779^{20}
11362	Silanamine, 1,1,1-trimethyl-*N*-phenyl-*N*-(trimethylsilyl)-	$C_{12}H_{23}NSi_2$ 237.49	4147-89-1 16.5	86	4-04-00-04010 0.7400^{20}	1.4379^{24}
11363	Silanamine, 1,1,1-trimethyl-*N*-(trimethylsilyl)- / Hexamethyldisilazane	$C_6H_{19}NSi_2$ 161.39	999-97-3	$98-9^{11}$ 10158	0.8951^{23} 4-04-00-04014	1.4855^{20}
11364	Silane, [1,1'-biphenyl]-4-yltrichloro-	$C_{12}H_9Cl_3Si$ 287.65	18030-61-0	125 $209-11^{30}$	0.7741^{25} 4-16-00-01572 1.3190^{20}	1.4090^{20} ctc 3; CS_2 3
11365	Silane, (bromomethyl)chlorodimethyl- / (Bromomethyl)dimethylchlorosilane	$C_3H_8BrClSi$ 187.54	16532-02-8	131	4-04-00-04024 1.375^{25}	
11366	Silane, (bromomethyl)trimethyl- / Bromomethyltrimethylsilane	$C_4H_{11}BrSi$ 167.12	18243-41-9	116.5	4-04-00-03878 1.170^{25}	1.4630^{25}
11367	Silane, (4-bromophenoxy)trimethyl-	$C_9H_{13}BrOSi$ 245.19	17878-44-3	126^{25}	4-06-00-01056 1.2619^{20}	1.4460^{20}
11368	Silane, bromotriethyl-	$C_6H_{15}BrSi$ 195.17	1112-48-7	$66-7^{24}$	4-04-00-04051 1.1403^{20}	1.5145^{20}
11369	Silane, butylchlorodimethyl-	$C_6H_{15}ClSi$ 150.72	1000-50-6	139	4-04-00-04068 0.8751^{20}	1.4561^{20}
11370	Silane, butyltrichloro- / Trichlorobutylsilane	$C_4H_9Cl_3Si$ 191.56	7521-80-4	148.5	4-04-00-04246 1.1606^{20}	1.5145^{20} eth 3; bz 3; tol 3; AcOEt 3
11371	Silane, butyltrimethyl- / Trimethylbutylsilane	$C_7H_{18}Si$ 130.31	1000-49-3	115	4-04-00-03907 0.7353^0	1.4363^{20} H_2O 1
11372	Silane, chloro(chloromethyl)dimethyl-	$C_3H_8Cl_2Si$ 143.09	1719-57-9	115.5	4-04-00-04024 1.0865^{20}	1.4360^{20}
11373	Silane, chloro(dichloromethyl)dimethyl- / (Dichloromethyl)dimethylchlorosilane	$C_3H_7Cl_3Si$ 177.53	18171-59-0 -48	149	4-04-00-04031 1.2369^{20}	1.461^{20}
11374	Silane, chlorodimethylphenyl-	$C_8H_{11}ClSi$ 170.71	768-33-2	$80-4^{16}$	4-16-00-01475 1.032^{20}	1.5082^{20}
11375	Silane, chlorodimethyl-2-propenyl-	$C_5H_{11}ClSi$ 134.68	4028-23-3	111	4-04-00-04080 0.8964^{20}	1.4195^{20}
11376	Silane, chloroethenyldimethyl-	C_4H_9ClSi 120.65	1719-58-0	83.5	4-04-00-04080 0.8744^{20}	1.4141^{20}
11377	Silane, chloroethyldimethyl-	$C_4H_{11}ClSi$ 122.67	6917-76-6	89.5	4-04-00-04035 0.8675^{20}	1.4105^{20}
11378	Silane, chloromethyl / Chloromethylsilane	CH_5ClSi 80.59	993-00-0 -135	$7; -45^{63}$		
11379	Silane, (chloromethyl)dimethylphenyl-	$C_9H_{13}ClSi$ 184.74	1833-51-8	225	4-16-00-01376 1.024^{20}	ctc 3; CS_2 3
11380	Silane, chloromethyldiphenyl-	$C_{13}H_{13}ClSi$ 232.78	144-79-6	295	4-16-00-01479 1.1277^{20}	1.5742^{20}
11381	Silane, chloromethylphenyl-	C_7H_9ClSi 156.69	1631-82-9	113^{100}	4-16-00-01473 1.0540^{20}	1.5171^{20}
11382	Silane, (chloromethyl)trimethyl-	$C_4H_{11}ClSi$ 122.67	2344-80-1	98.5	4-04-00-03877 0.879^{25}	1.4175^{20}
11383	Silane, (4-chlorophenoxy)trimethyl-	$C_9H_{13}ClOSi$ 200.74	17005-59-3	$214; 101^{14}$	4-06-00-00878 1.0320^{20}	1.4930^{20}
11384	Silane, chlorophenyl- / Phenylchlorosilane	C_8H_7ClSi 142.66	4206-75-1	162.5	4-16-00-01472 1.0683^{20}	1.5340^{20}
11385	Silane, (3-chlorophenyl)trimethyl-	$C_9H_{13}ClSi$ 184.74	4405-42-9	206.5	4-16-00-01386 1.0071^{20}	1.5108^{20}
11386	Silane, (3-chloropropyl)trimethyl-	$C_6H_{15}ClSi$ 150.72	2344-83-4	151	4-04-00-03905 0.8789^{20}	1.4319^{20}
11387	Silane, chlorotriethoxy-	$C_6H_{15}ClO_3Si$ 198.72	4667-99-6 -51	156	4-01-00-01363 1.030^{20}	EtOH 4 1.3999^{20}
11388	Silane, chlorotriethyl-	$C_6H_{15}ClSi$ 150.72	994-30-9	144.5	4-04-00-04051 0.8967^{20}	1.4314^{20}
11389	Silane, chlorotrimethyl- / Trimethylchlorosilane	C_3H_9ClSi 108.64	75-77-4 -40	60	4-04-00-04007 0.856^{25}	1.3870^{20}
11390	Silane, dichloro(chloromethyl)methyl- / Dichloro(chloromethyl)methylsilane	$C_2H_5Cl_3Si$ 163.51	1558-33-4	121.5	4-04-00-04135 1.2858^{20}	1.4500^{20}
11391	Silane, dichloro(dichloromethyl)methyl- / Dichloromethyl methyl dichlorosilane	$C_2H_4Cl_4Si$ 197.95	1558-31-2	149	4-04-00-04144 1.4116^{20}	1.4700^{20}
11392	Silane, dichlorodiethoxy-	$C_4H_{10}Cl_2O_2Si$ 189.11	4667-38-3 -130	135.9	4-01-00-01364 1.1290^{20}	EtOH 4
11393	Silane, dichlorodiethyl- / Dichlorodiethylsilane	$C_4H_{10}Cl_2Si$ 157.11	1719-53-5 -96.5	129 dec	4-04-00-04158 1.0504^{20}	1.4309^{20}
11394	Silane, dichlorodimethyl- / Dichlorodimethylsilane	$C_2H_6Cl_2Si$ 129.06	75-78-5 -16	70.3	4-04-00-04111 1.064^{25}	H_2O 6; EtOH 6 1.4038^{20}
11395	Silane, dichlorodiphenyl- / Dichlorodiphenylsilane	$C_{12}H_{10}Cl_2Si$ 253.20	80-10-4	305	4-16-00-01526 1.204^{25}	EtOH 3; eth 3; ace 3; bz 3 1.5800^{20}

No.	Name Synonym	Mol. Form. Mol. Wt.	CAS RN mp/°C	Merck No. bp/°C	Beil. Ref. den/g cm⁻³	Solubility n_D
11396	Silane, dichloroethenylmethyl-	$C_3H_6Cl_2Si$ 141.07	124-70-9	92.5	4-04-00-04184 1.0868^{20}	H_2O 6 1.4270^{20}
11397	Silane, dichloroethylmethyl- Methylethyldichlorosilane	$C_3H_8Cl_2Si$ 143.09	4525-44-4	101	4-04-00-04152 1.0630^{20}	1.4197^{20}
11398	Silane, dichloromethyl- Dichloromethylsilane	CH_4Cl_2Si 115.03	75-54-7 -93	41	4-04-00-04096 1.1052^{5}	
11399	Silane, dichloromethyl(1-methylethyl)-	$C_4H_{10}Cl_2Si$ 157.11	18236-89-0	121.5	4-04-00-04171 1.0385^{20}	1.4270^{20}
11400	Silane, dichloromethyl(4-methylphenyl)-	$C_8H_{10}Cl_2Si$ 205.16	18236-57-2 57		1.1619^{20}	1.5170^{20}
11401	Silane, dichloromethylphenyl- Dichloromethylphenylsilane	$C_7H_8Cl_2Si$ 191.13	149-74-6	206.5	4-16-00-01517 1.1866^{20}	1.5180^{20}
11402	Silane, dichloromethyl-2-propenyl-	$C_4H_8Cl_2Si$ 155.10	1873-92-3	119.5	4-04-00-04188 1.0758^{20}	1.4419^{20}
11403	Silane, dichlorophenyl- Phenyldichlorosilane	$C_6H_6Cl_2Si$ 177.10	1631-84-1	181	4-16-00-01514 1.221^{25}	H_2O 6
11404	Silane, diethenyldiphenyl-	$C_{16}H_{16}Si$ 236.39	17937-68-7	$130-1^{0.05}$	4-16-00-01368 1.0092^{25}	1.5350^{25}
11405	Silane, diethoxydimethyl- Dimethyldiethoxysilane	$C_6H_{16}O_2Si$ 148.28	78-62-6 -87	114	4-04-00-04101 0.865^{25}	ctc 3 1.3811^{20}
11406	Silane, diethoxydiphenyl- Diphenyldiethoxysilane	$C_{16}H_{20}O_2Si$ 272.42	2553-19-7	167^{15}	4-16-00-01524 1.0329^{20}	1.5269^{20}
11407	Silane, diethoxymethylphenyl-	$C_{11}H_{18}O_2Si$ 210.35	775-56-4	218	4-16-00-01515 0.9627^{20}	1.4690^{20}
11408	Silane, diethoxymethyl-2-propenyl- Silane, allyldiethoxymethyl-	$C_8H_{18}O_2Si$ 174.32	18388-45-9	155	4-04-00-04187 0.8572^{25}	1.4104^{20}
11409	Silane, diethyl- 3-Silapentane	$C_4H_{12}Si$ 88.22	542-91-6 -134.3	3117 57	4-04-00-03894 0.6843^{20}	H_2O 1 1.3921^{20}
11410	Silane, diethyldifluoro- Diethyldifluorosilane	$C_4H_{10}F_2Si$ 124.21	358-06-5 -78.7	61	4-04-00-04157 0.9348^{20}	1.3385^{20}
11411	Silane, difluorodiphenyl-	$C_{12}H_{10}F_2Si$ 220.29	312-40-3	157^{50}	4-16-00-01525 1.145^{17}	1.5221^{25}
11412	Silane, dimethoxydimethyl-	$C_4H_{12}O_2Si$ 120.22	1112-39-6	82	4-04-00-04100 0.8646^{20}	H_2O 6 1.3708^{20}
11413	Silane, dimethoxydiphenyl-	$C_{14}H_{16}O_2Si$ 244.37	6843-66-9	161^{15}	4-16-00-01524 1.0771^{20}	1.5447^{20}
11414	Silane, dimethyl- 2-Silapropane	C_2H_8Si 60.17	1111-74-6 -150	-20	4-04-00-03874 0.68^{-80}	
11415	Silane, dimethyldiphenoxy- Diphenoxydimethylsilane	$C_{14}H_{16}O_2Si$ 244.37	3440-02-6 -23	$130-2^5$	4-06-00-00764 1.0599^{25}	1.5330^{20}
11416	Silane, dimethyldiphenyl- Diphenyl(dimethyl)silane	$C_{14}H_{16}Si$ 212.37	778-24-5	277; 173^{45}	4-16-00-01366 0.9867^{20}	1.5644^{20}
11417	Silane, dimethyldi-2-propenyl-	$C_8H_{16}Si$ 140.30	1113-12-8	68^{50}	4-04-00-03929 0.7679^{20}	1.4420^{20}
11418	Silane, dimethylphenyl-	$C_8H_{12}Si$ 136.27	766-77-8	156.5	4-16-00-01360 0.8891^{20}	H_2O 1 1.4995^{20}
11419	Silanediol, dimethyl-, diacetate	$C_6H_{12}O_4Si$ 176.24	2182-66-3 -12.5	165	4-04-00-04108 1.0540^{20}	1.4030^{20}
11420	Silane, diphenyl- Diphenylsilane	$C_{12}H_{12}Si$ 184.31	775-12-2	$95-7^{13}$	4-16-00-01366 0.9969^{20}	ctc 3; CS_2 3 1.5800^{20}
11421	Silane, 1,2-ethenediylbis[trimethyl-, (E)-	$C_8H_{20}Si_2$ 172.42	18178-59-1 -18.5	145.5	4-04-00-03948 0.7589^{20}	H_2O 1 1.4310^{20}
11422	Silane, ethenyldiethoxymethyl-	$C_7H_{16}O_2Si$ 160.29	5507-44-8	133	4-04-00-04183 0.8620^{20}	1.4001^{20}
11423	Silane, ethenylethoxydimethyl-	$C_6H_{14}OSi$ 130.26	5356-83-2	99	4-04-00-04079 0.790^{20}	1.3983^{20}
11424	Silane, ethenyltriethoxy-	$C_8H_{18}O_3Si$ 190.31	78-08-0	148^{535}; 62^{20}	4-04-00-04256 0.9027^{20}	chl 3 1.3960^{25}
11425	Silane, ethenyltrimethyl-	$C_5H_{12}Si$ 100.24	754-05-2	55	4-04-00-03922 0.6910^{20}	H_2O 1 1.3914^{20}
11426	Silane, ethenyltris(1-methylethoxy)-	$C_{11}H_{24}O_3Si$ 232.40	18023-33-1	179.5; 77^{20}	4-04-00-04257 0.862^{25}	ctc 3 1.3981^{20}
11427	Silane, ethenyltris(2-propenyloxy)- Tris(allyloxy)vinylsilane	$C_{11}H_{18}O_3Si$ 226.35	17988-31-7	$90.3^{7.5}$	4-04-00-04257 0.9435^{25}	1.4380^{20}
11428	Silane, ethoxytriethyl-	$C_8H_{20}OSi$ 160.33	597-67-1	154.5	4-04-00-04042 0.8160^{20}	eth 4; EtOH 4 1.4140^{20}
11429	Silane, ethoxytrifluoro- Ethoxytrifluorosilicon	$C_2H_5F_3OSi$ 130.14	460-55-9 -122	-7	3-01-00-01336 1.3200^{-63}	
11430	Silane, ethoxytrimethyl-	$C_5H_{14}OSi$ 118.25	1825-62-3	76	4-04-00-03994 0.7573^{20}	H_2O 1; EtOH 3; eth 3; ace 3 1.3741^{20}
11431	Silane, ethoxytriphenyl-	$C_{20}H_{20}OSi$ 304.46	1516-80-9 65	344	4-16-00-01481	chl 3
11432	Silane, ethyltrifluoro- Trifluoroethylsilane	$C_2H_5F_3Si$ 114.14	353-89-9 -105	-4.4	4-04-00-04226 1.227^{-76}	
11433	Silane, ethyltrimethoxy- Ethyltrimethoxysilane	$C_5H_{14}O_3Si$ 150.25	5314-55-6	124.3	4-04-00-04223 0.9488^{20}	EtOH 4 1.3838^{20}
11434	Silane, 1,2-ethynediylbis[trimethyl-	$C_8H_{18}Si_2$ 170.40	14630-40-1 26	134	4-04-00-03950 0.770^{20}	1.413^{20}
11435	Silane, ethynyl- Ethynylsilane	C_2H_4Si 56.14	1066-27-9	-22.5		

No.	Name / Synonym	Mol. Form. / Mol. Wt.	CAS RN / mp/°C	Merck No. / bp/°C	Beil. Ref. / den/g cm^{-3}	Solubility / n_D
11436	Silane, methoxy / Methyl silyl ether	CH_8OSi / 62.14	2171-96-2			
11437	Silane, methyl- / Methylsilane	CH_8Si / 46.14	992-94-9 / -156.5	-57.5	4-04-00-03873	
11438	Silane, methyldiphenyl-	$C_{13}H_{14}Si$ / 198.34	776-76-1	93.5[1]	4-16-00-01366 / 0.9973[20]	ctc 3 / 1.5694[20]
11439	Silane, methylenebis-	CH_8Si_2 / 76.25	1759-88-2	15	4-01-00-03072 / 0.6979[4]	1.4115[4]
11440	Silane, methylenebis[trichloro-	$CH_2Cl_6Si_2$ / 282.91	4142-85-2	183; 64[10]	4-01-00-03077 / 1.5567[20]	eth 4; EtOH 4 / 1.4740[20]
11441	Silane, (2-methylphenoxy)triphenyl-	$C_{25}H_{22}OSi$ / 366.53	18858-65-6	192	4-16-00-01482 / 0.9287[20]	1.4830[20]
11442	Silane, methylphenyl-	$C_7H_{10}Si$ / 122.24	766-08-5	140	4-16-00-01360 / 0.8895[20]	1.5058[20]
11443	Silane, methyltriphenoxy- / Methyltriphenoxysilane	$C_{19}H_{18}O_3Si$ / 322.44	3439-97-2	269[100]; 210[12]	4-06-00-00765 / 1.135[20]	1.5599[20]
11444	Silane, methyltriphenyl- / Triphenylmethylsilane	$C_{19}H_{18}Si$ / 274.44	791-29-7 / 69.5	198[9]	4-16-00-01370 / 1.0880[20]	ctc 3; CS_2 3
11445	Silane, methyltri-p-tolyl- / Silane, methyl tri-(4-tolyl)	$C_{22}H_{24}Si$ / 316.52	18752-92-6 / 92	198[4]	4-16-00-01399	chl 2
11446	Silane, phenyl-	C_6H_8Si / 108.22	694-53-1	119	4-16-00-01360 / 0.8681[20]	H_2O 1 / 1.5125[20]
11447	Silane, [1,3-phenylenebis(oxy)]bis[trimethyl-	$C_{12}H_{22}O_2Si_2$ / 254.48	4520-29-0	240	4-06-00-05683 / 0.950[20]	1.4748[20]
11448	Silane, phenyltripropyl- / Phenyltripropylsilane	$C_{15}H_{26}Si$ / 234.46	78938-11-1	146-7[18]	4-16-00-01362 / 0.8799[20]	1.4950[20]
11449	Silane, tetraethenyl- / Tetravinylsilane	$C_8H_{12}Si$ / 136.27	1112-55-6	130.2	4-04-00-03925 / 0.7999[20]	1.4625[20]
11450	Silane, tetraethyl- / Tetraethylsilane	$C_8H_{20}Si$ / 144.33	631-36-7	154.7	4-04-00-03895 / 0.7658[20]	H_2O 1 / 1.4268[20]
11451	Silane, tetramethyl- / Tetramethylsilane	$C_4H_{12}Si$ / 88.22	75-76-3 / -99.0	26.6	4-04-00-03875 / 0.648[19]	H_2O 1; EtOH 4; eth 4; sulf 1 / 1.3587[20]
11452	Silane, tetraphenyl- / Tetraphenylsilane	$C_{24}H_{20}Si$ / 336.51	1048-08-4 / 236.5	228[3]	4-16-00-01372 / 1.078[20]	ctc 3; CS_2 3
11453	Silane, tribromomethyl-	CH_3Br_3Si / 282.83	4095-09-4 / -28.4	132	4-04-00-04213 / 2.2130[25]	1.5152[25]
11454	Silane, tributyl- / Tributylsilane	$C_{12}H_{28}Si$ / 200.44	998-41-4	221	4-04-00-03908 / 0.7794[20]	1.4380[20]
11455	Silane, tributylphenyl- / Silane, phenyl-tributyl	$C_{18}H_{32}Si$ / 276.54	18510-29-7	140-2[4]	4-16-00-01362 / 0.8753[20]	1.4915[20]
11456	Silane, trichloro(chloromethyl)- / (Chloromethyl)trichlorosilane	CH_2Cl_4Si / 183.92	1558-25-4	118	4-01-00-03074 / 1.4650[20]	1.4555[20]
11457	Silane, trichloro(4-chlorophenyl)-	$C_6H_4Cl_4Si$ / 245.99	825-94-5	115-7[20]	4-16-00-01563 / 1.4316[20]	1.5418[20]
11458	Silane, trichloro(3-chloropropyl)-	$C_3H_6Cl_4Si$ / 211.98	2550-06-3	181.5	4-04-00-04243 / 1.3590[20]	1.4668[20]
11459	Silane, trichloro(dichloromethyl)- / (Dichloromethyl)trichlorosilane	$CHCl_5Si$ / 218.37	1558-24-3	145	4-02-00-00088 / 1.5518[20]	1.4714[20]
11460	Silane, trichlorododecyl- / Dodecyltrichlorosilane	$C_{12}H_{25}Cl_3Si$ / 303.77	4484-72-4	155[10]	4-04-00-04255	1.4581[20]
11461	Silane, trichloroethenyl- / Vinyltrichlorosilane	$C_2H_3Cl_3Si$ / 161.49	75-94-5 / -95	91.5	4-04-00-04258 / 1.2426[20]	chl 4 / 1.4295[20]
11462	Silane, trichloroethoxy-	$C_2H_5Cl_3OSi$ / 179.50	1825-82-7 / -135	101.9	4-01-00-01364 / 1.2274[20]	EtOH 4 / 1.4045[20]
11463	Silane, trichloroethyl- / Ethyltrichlorosilane	$C_2H_5Cl_3Si$ / 163.51	115-21-9 / -105.6	100.5	4-04-00-04227 / 1.2373[20]	ctc 3 / 1.4256[20]
11464	Silane, trichlorohexyl- / Hexyltrichlorosilane	$C_6H_{13}Cl_3Si$ / 219.61	928-65-4	190	4-04-00-04251 / 1.1100[20]	H_2O 6
11465	Silane, trichloromethyl- / Methyltrichlorosilane	CH_3Cl_3Si / 149.48	75-79-6 / -90	65.6	4-04-00-04212 / 1.273[20]	H_2O 6; EtOH 6 / 1.4106[20]
11466	Silane, trichloro(1-methylethyl)-	$C_3H_7Cl_3Si$ / 177.53	4170-46-1 / -87.7	121	4-04-00-04244 / 1.1934[20]	H_2O 6 / 1.4319[20]
11467	Silane, trichloro(2-methylphenyl)-	$C_7H_7Cl_3Si$ / 225.58	13835-81-9 / -25.5	226.7	4-16-00-01565 / 1.306[25]	1.5336[25]
11468	Silane, trichloro(3-methylphenyl)-	$C_7H_7Cl_3Si$ / 225.58	13688-75-0 / -26	226.5	4-16-00-01565 / 1.306[25]	1.5336[25]
11469	Silane, trichloro(2-methylpropyl)-	$C_4H_9Cl_3Si$ / 191.56	18169-57-8	143.3	4-04-00-04248 / 1.154[20]	H_2O 6
11470	Silane, trichlorooctadecyl- / Octadecyltrichlorosilane	$C_{18}H_{37}Cl_3Si$ / 387.94	112-04-9	223[10]	4-04-00-04256 / 0.984[25]	1.4602[20]
11471	Silane, trichlorooctyl- / Octyltrichlorosilane	$C_8H_{17}Cl_3Si$ / 247.67	5283-66-9	232	4-04-00-04253	H_2O 6; EtOH 6; ctc 3 / 1.4480[20]
11472	Silane, trichloropentyl- / Amyltrichlorosilane	$C_5H_{11}Cl_3Si$ / 205.59	107-72-2	172; 60.5[15]	4-04-00-04249 / 1.1330[20]	1.4503[20]
11473	Silane, trichlorophenyl- / Phenyltrichlorosilane	$C_6H_5Cl_3Si$ / 211.55	98-13-5	201	4-16-00-01560 / 1.321[20]	ctc 3; chl 3; CS_2 3 / 1.5230[20]
11474	Silane, trichloro(2-phenylethyl)-	$C_8H_9Cl_3Si$ / 239.60	940-41-0	242; 98-9[5]	4-16-00-01568 / 1.2397[20]	1.5185[20]
11475	Silane, trichloro-2-propenyl- / Allyltrichlorosilane	$C_3H_5Cl_3Si$ / 175.52	107-37-9 / 35	117.5	4-04-00-04261 / 1.2011[20]	1.4460[20]

No.	Name / Synonym	Mol. Form. / Mol. Wt.	CAS RN / mp/°C	Merck No. / bp/°C	Beil. Ref. / den/g cm^{-3}	Solubility / n_D
11476	Silane, trichloropropyl- / Propyltrichlorosilane	$C_3H_7Cl_3Si$ / 177.53	141-57-1 /	/ 123.5	4-04-00-04239 / 1.195[20]	/ 1.4310[20]
11477	Silane, triethoxy- / Triethoxysilane	$C_6H_{16}O_3Si$ / 164.28	998-30-1 /	/ 133.5	4-01-00-01359 / 0.8745[20]	/
11478	Silane, triethoxyethyl-	$C_8H_{20}O_3Si$ / 192.33	78-07-9 /	/ 158.5	4-04-00-04223 / 0.8963[20]	H$_2$O 1; EtOH 5; eth 5; chl 3 / 1.3955[20]
11479	Silane, triethoxymethyl-	$C_7H_{18}O_3Si$ / 178.30	2031-67-6 /	/ 142	4-04-00-04204 / 0.8948[25]	/ 1.3832[20]
11480	Silane, triethoxypentyl-	$C_{11}H_{26}O_3Si$ / 234.41	2761-24-2 /	/ 95[13]	4-04-00-04248 / 0.8862[20]	/ 1.4059[20]
11481	Silane, triethoxyphenyl-	$C_{12}H_{20}O_3Si$ / 240.37	780-69-8 /	/ 112-3[10]	4-16-00-01556 / 0.996[25]	/ 1.4604[20]
11482	Silane, triethoxy-2-propenyl-	$C_9H_{20}O_3Si$ / 204.34	2550-04-1 /	/ 100[50]	4-04-00-04261 / 0.9030[20]	/ 1.4072[20]
11483	Silane, triethyl- / Triethylsilane	$C_6H_{16}Si$ / 116.28	617-86-7 /	/ 109	4-04-00-03895 / 0.7302[20]	H$_2$O 1; sulf 1 / 1.447[20]
11484	Silane, triethylfluoro-	$C_6H_{15}FSi$ / 134.27	358-43-0 /	/ 110	4-04-00-04051 / 0.8354[25]	peth 4 / 1.3900[25]
11485	Silane, triethylphenyl-	$C_{12}H_{20}Si$ / 192.38	2987-77-1 /	/ 236	4-16-00-01362 / 0.8915[20]	/ 1.4999[20]
11486	Silane, trifluorophenyl-	$C_6H_5F_3Si$ / 162.19	368-47-8 / -18	/ 101.5	4-16-00-01559 / 1.2169[20]	bz 4; EtOH 4 / 1.4110[20]
11487	Silane, trimethoxymethyl-	$C_4H_{12}O_3Si$ / 136.22	1185-55-3 /	/ 102.5	4-04-00-04203 / 0.9548[20]	chl 3 / 1.3696[20]
11488	Silane, trimethoxyphenyl-	$C_9H_{14}O_3Si$ / 198.29	2996-92-1 /	/ 130[45]	4-16-00-01556 / 1.064[20]	ctc 3; CS$_2$ 3 / 1.4734[20]
11489	Silane, trimethyl- / Trimethylsilane	$C_3H_{10}Si$ / 74.20	993-07-7 / -135.9	/ 6.7	/	/
11490	Silane, trimethyl(4-methylphenyl)-	$C_{10}H_{16}Si$ / 164.32	3728-43-6 /	/ 192; 73[10]	4-16-00-01396 / 0.860[25]	/ 1.4900[20]
11491	Silane, trimethyl(2-methylpropyl)-	$C_7H_{18}Si$ / 130.31	1118-09-8 /	/ 108.5	1-04-00-00580 / 0.7330[0]	/
11492	Silane, trimethylphenoxy-	$C_9H_{14}OSi$ / 166.30	1529-17-5 / -55	/ 119	4-06-00-00762 / 0.8681[20]	/ 1.5125[20]
11493	Silane, trimethylphenyl-	$C_9H_{14}Si$ / 150.30	768-32-1 /	/ 169.5	4-16-00-01361 / 0.8722[20]	ctc 3; CS$_2$ 3 / 1.4907[20]
11494	Silane, trimethyl(phenylmethyl)-	$C_{10}H_{16}Si$ / 164.32	770-09-2 /	/ 190.5	4-16-00-01401 / 0.8933[20]	/ 1.4941[20]
11495	Silane, trimethyl-2-propenyl-	$C_6H_{14}Si$ / 114.26	762-72-1 /	/ 86	4-04-00-03927 / 0.7158[25]	H$_2$O 1 / 1.4074[20]
11496	Silane, trimethylpropyl- / Trimethylpropylsilane	$C_6H_{16}Si$ / 116.28	3510-70-1 /	/ 89	4-04-00-03901 / 0.7196[0]	/ 1.3929[20]
11497	Silane, trimethyl[4-[(trimethylsilyl)oxy]phenyl]- / Silane, (oxy-p-phenylene)bis[trimethyl-	$C_{12}H_{22}OSi_2$ / 238.48	18036-81-2 /	/ 132[25]	4-16-00-01441 / 0.900[25]	/ 1.4794[25]
11498	Silanetriol, ethenyl-, triacetate	$C_8H_{12}O_6Si$ / 232.27	4130-08-9 /	/ 115[10]	4-04-00-04258 / 1.167[20]	/ 1.4226[20]
11499	Silanetriol, methyl-, triacetate	$C_7H_{12}O_6Si$ / 220.25	4253-34-3 / 40.5	/ 110-12[17]	4-04-00-04208 / 1.1750[20]	/ 1.4083[20]
11500	Silane, tripropyl- / Tripropylsilane	$C_9H_{22}Si$ / 158.36	998-29-8 /	/ 172	4-04-00-03902 / 0.7723[0]	H$_2$O 1 / 1.4280[20]
11501	Silanol, ethyldimethyl- / Silanol, dimethyl ethyl	$C_4H_{12}OSi$ / 104.22	5906-73-0 /	/ 120; 58[50]	4-04-00-04034 / 0.8332[20]	/ 1.4070[20]
11502	Silanol, methyldiphenyl-	$C_{13}H_{14}OSi$ / 214.34	778-25-6 / 167	/ 184[24]; 148[3]	4-16-00-01478 / 1.0840[25]	ctc 3; CS$_2$ 3 /
11503	Silanol, triethyl-	$C_6H_{16}OSi$ / 132.28	597-52-4 /	/ 154	4-04-00-04040 / 0.8647[20]	H$_2$O 1; EtOH 5; eth 5 / 1.4329[20]
11504	Silanol, triphenyl-	$C_{18}H_{16}OSi$ / 276.41	791-31-1 / 154.8	/	4-16-00-01480 / 1.1777[20]	ctc 3; CS$_2$ 3 /
11505	Silicic acid (H4SiO4), tetrabutyl ester	$C_{16}H_{36}O_4Si$ / 320.54	4766-57-8 /	/ 120[3]	4-01-00-01541 / 0.8990[20]	/ 1.4128[20]
11506	Silicic acid (H4SiO4), tetraethyl ester / Ethyl silicate	$C_8H_{20}O_4Si$ / 208.33	78-10-4 / -82.5	3805 / 168.8	4-01-00-01360 / 0.9320[20]	H$_2$O 6 / 1.3928[20]
11507	Silicic acid (H4SiO4), tetrakis(2-ethylbutyl) ester	$C_{24}H_{52}O_4Si$ / 432.76	78-13-7 /	/	4-01-00-01725 / 0.8920[20]	H$_2$O 1; EtOH 2; eth 3; bz 3 / 1.4307[20]
11508	Silicic acid (H4SiO4), tetramethyl ester / Methyl silicate	$C_4H_{12}O_4Si$ / 152.22	681-84-5 / -2	/ 121	4-01-00-01266 / 1.0232[20]	EtOH 4 / 1.3683[20]
11509	Silicic acid (H4SiO4), tetraphenyl ester	$C_{24}H_{20}O_4Si$ / 400.51	1174-72-7 / 49	/ 417; 236[1]	4-06-00-00766 / 1.1412[80]	/
11510	Silicic acid (H4SiO4), tetrapropyl ester	$C_{12}H_{28}O_4Si$ / 264.44	682-01-9 /	/ 226	4-01-00-01435 / 0.9158[20]	ctc 3; CS$_2$ 3 / 1.4012[20]
11511	Silicic acid (H4SiO4), triethyl phenyl ester	$C_{12}H_{20}O_4Si$ / 256.37	18023-36-4 /	/ 120-2[10]	4-06-00-00765 / 1.0283[20]	/ 1.4525[20]
11512	Silvex / Propanoic acid, 2-(2,4,5-trichlorophenoxy)-	$C_9H_7Cl_3O_3$ / 269.51	93-72-1 / 181.6	8483 /	/	/
11513	Simazine / 1,3,5-Triazine-2,4-diamine, 6-chloro-N,N'-diethyl-	$C_7H_{12}ClN_5$ / 201.66	122-34-9 / 226	8485 /	/ 1.302[20]	/
11514	Sodium dimethyldithiocarbamate / Carbamodithioic acid, dimethyl-, sodium salt	$C_3H_7NNaS_2$ / 144.22	128-04-1 /	/	/	/

No.	Name / Synonym	Mol. Form. / Mol. Wt.	CAS RN / mp/°C	Merck No. / bp/°C	Beil. Ref. / den/g cm⁻³	Solubility / n_D
11515	Solanid-5-ene-3,12-diol, (3β,12α)- / Rubijervine	$C_{27}H_{43}NO_2$ / 413.64	79-58-3 / 242	8264	5-21-05-00208	bz 4; EtOH 4; chl 4
11516	Solanid-5-ene-3,18-diol, (3β)- / Isorubijervine	$C_{27}H_{43}NO_2$ / 413.64	468-45-1 / 242.5	5111	5-21-05-00208	bz 4; chl 4
11517	Sonar / 2-Naphthalenol, 1,1'-methylenebis-, monosodium salt	$C_{21}H_{16}NaO_2$ / 323.35	38827-66-6 / 132			
11518	DL-Sorbose	$C_6H_{12}O_6$ / 180.16	65732-90-3 / 162.5	8681	4-01-00-04410 / 1.638[17]	H₂O 4; MeOH 4
11519	L-Sorbose / L-Sorbinose	$C_6H_{12}O_6$ / 180.16	87-79-6 / 165	8681	4-01-00-04412 / 1.612[17]	H₂O 3; EtOH 2; eth 2; MeOH 2
11520	Spiro[benzofuran-2(3H),1'-[2]cyclohexene]-3,4'-dione, 7-chloro-2',4,6-trimethoxy- / Griseofulvin, (+)-	$C_{17}H_{17}ClO_6$ / 352.77	126-07-8 / 220	4453	5-18-05-00150	H₂O 1; EtOH 2; eth 2; ace 2
11521	Spiro[4.5]dec-6-en-8-one, 6,10-dimethyl-2-(1-methylethylidene)-, (5R-cis)-	$C_{15}H_{22}O$ / 218.34	18444-79-6 / 44.5	141-2[2]	4-07-00-00822 / 1.0001[20]	ace 3 / 1.5309[20]
11522	Spiro[isobenzofuran-1(3H),9'-[9H]xanthen]-3-one, 3',6'-dihydroxy- / Fluorescein	$C_{20}H_{12}O_5$ / 332.31	2321-07-5 / 315 dec	4085	5-19-06-00456	H₂O 2; EtOH 2; eth 2; ace 4
11523	Spiro[isobenzofuran-1(3H),9'-[9H]xanthen]-3-one, 3',6'-dihydroxy-4',5'-diiodo- / Fluorescein, 4',5'-diiodo-	$C_{20}H_{10}I_2O_5$ / 584.11	38577-97-8	3174	5-19-06-00466	
11524	Spiro[isobenzofuran-1(3H),9'-[9H]xanthen]-3-one, 3',6'-dihydroxy-2',4',5',7'-tetraiodo-, disodium salt / Erythrosin	$C_{20}H_8I_4O_5$ / 835.90	16423-68-0	3640	4-19-00-02923	eth 4; EtOH 4
11525	Spiro[isobenzofuran-1(3H),9'-[9H]xanthen]-3-one, 2',4',5',7'-tetrabromo-3',6'-dihydroxy-, disodium salt / Eosin	$C_{20}H_8Br_4Na_2O_5$ / 691.86	17372-87-1 / 295.5	3555		EtOH 4
11526	Spiro[isobenzofuran-1(3H),9'-[9H]xanthen]-3-one, 3',4',5',6'-tetrahydroxy- / Gallein	$C_{20}H_{12}O_5$ / 332.31	2103-64-2 / >300	4250	5-19-07-00083	ace 4; EtOH 4
11527	Spiropentane / Spiro[2.2]pentane	C_5H_8 / 68.12	157-40-4 / -134.6	39	4-05-00-00218 / 0.7266[20]	1.4120[20]
11528	Spirosolan-3-ol, (3β,5α,22β,25S)- / Tomatidine	$C_{27}H_{45}NO_2$ / 415.66	77-59-8 / 210.5	9470		EtOH 3; eth 3
11529	Spirosol-5-en-3-ol, (3β,22α,25R)- / Solasodine	$C_{27}H_{43}NO_2$ / 413.64	126-17-0 / 202	8665	4-27-00-02000	EtOH 3; eth 2; ace 3; bz 3
11530	Spirostan-2,3-diol, (2α,3β,5α,25R)- / Gitogenin	$C_{27}H_{44}O_4$ / 432.64	511-96-6 / 271.5	4325	5-19-03-00338	H₂O 1; EtOH 3; eth 2; chl 3
11531	Spirostan-3-ol, (3β,5β,25S)- / Sarsasapogenin	$C_{27}H_{44}O_3$ / 416.64	126-19-2 / 200.5	8338	5-19-02-00672	EtOH 3; ace 3; bz 3; chl 3
11532	Spirostan-3-ol, (3β,5α,25R)- / Tigogenin	$C_{27}H_{44}O_3$ / 416.64	77-60-1 / 205.5	9367		EtOH 3; eth 3; ace 3; ctc 3
11533	Spirostan-3-ol, (3β,5β,25R)- / Smilagenin	$C_{27}H_{44}O_3$ / 416.64	126-18-1 / 185	8508	5-19-02-00673	ace 4; bz 4; EtOH 4
11534	Spirostan-12-one, 3-hydroxy-, (3β,5α,25R)- / Hecogenin	$C_{27}H_{42}O_4$ / 430.63	467-55-0 / 266.5	4537	5-19-05-00592	ace 4; eth 4; EtOH 4
11535	Spirostan-2,3,15-triol, (2α,3β,5α,15β,25R)- / Digitogenin	$C_{27}H_{44}O_5$ / 448.64	511-34-2 / 281.5	3143	5-19-03-00603	chl 4
11536	Spirost-5-en-3-ol, (3β,25R)- / Diosgenin	$C_{27}H_{42}O_3$ / 414.63	512-04-9 / 205.5	3289	5-19-03-00030	EtOH 4
11537	Spiro[5.5]undecane / Spir[5.]undecane	$C_{11}H_{20}$ / 152.28	180-43-8	207	4-05-00-00328 / 0.8783[20]	1.4731
11538	Stannane, (acetyloxy)tributyl- / Acetoxytributylstannane	$C_{14}H_{30}O_2Sn$ / 349.10	56-36-0 / 84.7			
11539	Stannane, (acetyloxy)triphenyl- / Triphenyltin acetate	$C_{20}H_{18}O_2Sn$ / 409.07	900-95-8 / 121.5			
11540	Stannane, chlorotrimethyl-	C_3H_9ClSn / 199.27	1066-45-1 / 38.5	148		H₂O 3; chl 3; os 3
11541	Stannane, chlorotriphenyl- / Triphenyltin chloride	$C_{18}H_{15}ClSn$ / 385.48	639-58-7 / 103.5			chl 3
11542	Stannane, chlorotripropyl-	$C_9H_{21}ClSn$ / 283.43	2279-76-7 / -23.5	123[13]	1.2678[28]	ctc 3; os 3 / 1.49102[28]
11543	Stannane, tetraethenyl-	$C_8H_{12}Sn$ / 226.89	1112-56-7 /	55-7[17]	4-04-00-04316 / 1.2851[25]	ctc 3 / 1.4993[25]
11544	Stannane, tetramethyl- / Tetramethylstannane	$C_4H_{12}Sn$ / 178.85	594-27-4 / -54.8	78	4-04-00-04307 / 1.314[25]	H₂O 1; ctc 3; CS₂ 3; os 3 / 1.4386
11545	Stannane, tetraphenyl-	$C_{24}H_{20}Sn$ / 427.13	595-90-4 / 228	420	4-16-00-01592	chl 2
11546	Stannane, tetrapropyl-	$C_{12}H_{28}Sn$ / 291.06	2176-98-9 / -109.1	228	4-04-00-04310 / 1.1065[20]	
11547	Stibine, dichlorophenyl-	$C_6H_5Cl_2Sb$ / 269.77	5035-52-9 / 69.5	110-5[10]	4-16-00-01205	ace 4; EtOH 4; HOAc 4 / 1.4745[20]
11548	Stibine oxide, dihydroxyphenyl- / Benzenestibonic acid	$C_6H_7O_3Sb$ / 248.88	535-46-6 / 139	1077	3-16-00-01178	

No.	Name / Synonym	Mol. Form. / Mol. Wt.	CAS RN / mp/°C	Merck No. / bp/°C	Beil. Ref. / den/g cm^{-3}	Solubility / n_D
11549	Stibine, triethyl- / Triethylstibine	$C_6H_{15}Sb$ / 208.94	617-85-6 / -98	/ 161.4	4-04-00-03685 / 1.3224[15]	H_2O 1; EtOH 3; eth 3
11550	Stibine, trimethyl- / Trimethylstibine	C_3H_9Sb / 166.86	594-10-5 / -62	/ 80.6	4-04-00-03684 / 1.523[15]	H_2O 1; EtOH 3; eth 3; CS_2 3 / 1.42[15]
11551	Stibine, triphenyl-	$C_{18}H_{15}Sb$ / 353.07	603-36-1 / 53.5	/ >360	4-16-00-01198 / 1.4343[25]	H_2O 1; EtOH 3; eth 4; ace 4 / 1.6948[42]
11552	Stibine, tris(3-methylphenyl)-	$C_{21}H_{21}Sb$ / 395.15	35569-54-1 / 72		4-16-00-01199 / 1.3957[16]	bz 4; eth 4; EtOH 4; chl 4
11553	Stibine, tris(4-methylphenyl)-	$C_{21}H_{21}Sb$ / 395.15	5395-43-7 / 127.5	/ 218[3]	4-16-00-01199 / 1.3595[16]	bz 4; eth 4; chl 4
11554	Stigmasta-5,7-dien-3-ol, (3β)- / 7-Dehydrositosterol	$C_{29}H_{48}O$ / 412.70	521-04-0 / 144.5	2863	3-06-00-02855	bz 4; eth 4; EtOH 4
11555	Stigmasta-5,22-dien-3-ol, (3β,22E)- / Stigmasterol	$C_{29}H_{48}O$ / 412.70	83-48-7 / 170	8771	4-06-00-04170	bz 4; eth 4; EtOH 4
11556	Stigmasta-7,24(28)-dien-3-ol, 4-methyl-, (3β,4α,5α,24Z)- / α1-Sitosterol	$C_{30}H_{50}O$ / 426.73	474-40-8 / 166	8499	3-06-00-02876	EtOH 4; chl 4
11557	Stigmastan-3-ol, (3β,5α)- / Stigmastanol	$C_{29}H_{52}O$ / 416.73	83-45-4 / 144	8770	3-06-00-02172	
11558	Stigmast-5-en-3-ol, (3β)- / β-Sitosterol	$C_{29}H_{50}O$ / 414.72	83-46-5 / 140	8500	4-06-00-04037	EtOH 3; eth 3; HOAc 3
11559	Stigmast-5-en-3-ol, (3β,24S)- / γ-Sitosterol	$C_{29}H_{50}O$ / 414.72	83-47-6	8501		
11560	Stilbene, α-ethyl- / 1-Butene, 1,2-diphenyl	$C_{16}H_{16}$ / 208.30	22692-70-2 / 57	/ 296.5	4-05-00-02206 / 1.0124[18]	1.593[18]
11561	D-Streptamine, O-2-deoxy-2-(methylamino)-α-L-glucopyranosyl-(12)-O-5-deoxy-3-C-formyl-α-L-lyxofuranosyl-(1-4)-N,N'55-bis(aminoiminomethyl)-, sulfate (2:3)(salt) / Streptomycin, sulfate	$C_{42}H_{84}N_{14}O_{36}S_3$ / 1457.40	3810-74-0		5-18-11-00082	
11562	Streptomycin / N-Methyl-L-glucosamidinostreptosidostreptidine	$C_{21}H_{39}N_7O_{12}$ / 581.58	57-92-1	8786		
11563	Strychnidin-10-one / Strychnine	$C_{21}H_{22}N_2O_2$ / 334.42	57-24-9 / 287	8822 / 270[5]	4-27-00-07537 / 1.36[20]	H_2O 2; EtOH 2; eth 1; ace 2
11564	Strychnidin-10-one, 2,3-dimethoxy- / Brucine	$C_{23}H_{26}N_2O_4$ / 394.47	357-57-3 / 178	1443	4-27-00-07875	H_2O 2; EtOH 4; eth 2; bz 2
11565	Strychnidin-10-one, mononitrate / Strychnine nitrate	$C_{21}H_{23}N_3O_5$ / 397.43	66-32-0 / 295	8823	2-27-00-00723 / 1.627[25]	H_2O 4; MeOH 4
11566	Strychnidin-10-one, sulfate (2:1) / Strychnine, sulfate	$C_{42}H_{46}N_4O_8S$ / 766.92	60-41-3 / 200 dec	8826		H_2O 3; EtOH 3; eth 1; chl 2
11567	Succinimide, N-(ethoxymethyl)-	$C_7H_{11}NO_3$ / 157.17	98431-97-1 / 31.5	/ 262	2-21-00-00304	H_2O 4; EtOH 4
11568	Sulfamic acid, cyclohexyl- / Cyclamic acid	$C_6H_{13}NO_3S$ / 179.24	100-88-9 / 169.5	2707	4-12-00-00102	alk 4
11569	Sulfamide, tetramethyl-	$C_4H_{12}N_2O_2S$ / 152.22	3768-63-6 / 73	/ 225	4-04-00-00270	EtOH 4
11570	Sulfide, m-tolyl p-tolyl / 3,4'-Ditolyl sulfide	$C_{14}H_{14}S$ / 214.33	107770-92-3 / 28	/ 179[11]	4-06-00-02173	eth 4; EtOH 4
11571	Sulfoxylic acid, diethyl ester	$C_4H_{10}O_2S$ / 122.19	10297-38-8	/ 118	3-01-00-01314 / 0.9940[20]	H_2O 3; EtOH 4; eth 4; ace 4 / 1.4234[20]
11572	Sulfur cyanide	C_2N_2S / 84.10	627-52-1 / 63.5	/ sub 30	4-03-00-00339	eth 4; EtOH 4
11573	Sulfuric acid, diethyl ester / Diethyl sulfate	$C_4H_{10}O_4S$ / 154.19	64-67-5 / -24	3120 / 208	4-01-00-01326 / 1.172[25]	H_2O 1; EtOH 5; eth 5 / 1.3989[20]
11574	Sulfuric acid, dimethyl ester / Dimethyl sulfate	$C_2H_6O_4S$ / 126.13	77-78-1 / -27	3244 / 188 dec; 76[15]	4-01-00-01251 / 1.3322[20]	H_2O 3; EtOH 5; eth 3; bz 3 / 1.3874[20]
11575	Sulfuric acid, dipropyl ester	$C_6H_{14}O_4S$ / 182.24	598-05-0	/ 121[30]	4-01-00-01423 / 1.1064[20]	peth 4 / 1.4135[20]
11576	Sulfuric acid, monoethyl ester / Ethyl sulfate	$C_2H_6O_4S$ / 126.13	540-82-9	3808 / 280 dec	4-01-00-01324 / 1.3657[20]	H_2O 4 / 1.4105[20]
11577	Sulfuric acid, monomethyl ester / Methyl sulfate	CH_4O_4S / 112.11	75-93-4 / <-30	6041 / 135 dec	4-01-00-01250	H_2O 4; eth 4; EtOH 4
11578	Sulfurous acid, bis(2-methylpropyl) ester	$C_8H_{18}O_3S$ / 194.30	18748-27-1	/ 210; 92-4[13]	4-01-00-01595 / 0.9862[20]	1.4268[20]
11579	Sulfurous acid, 2-chloroethyl 2-[4-(1,1-dimethylethyl)phenoxy]-1-methylethyl ester / Aramite	$C_{15}H_{23}ClO_4S$ / 334.86	140-57-8 / -37.3	794 / 195[2]	4-06-00-03301 / 1.143[20]	ace 4; bz 4; eth 4; EtOH 4 / 1.5100[20]
11580	Sulfurous acid, dibutyl ester / Butyl sulfite	$C_8H_{18}O_3S$ / 194.30	626-85-7	/ 230	4-01-00-01522 / 0.9957[20]	EtOH 3; eth 3 / 1.4310[20]
11581	Sulfurous acid, diethyl ester / Ethyl sulfite	$C_4H_{10}O_3S$ / 138.19	623-81-4	/ 158; 51[13]	4-01-00-01324 / 1.1[20]	EtOH 3; eth 3 / 1.4310[20]
11582	Sulfurous acid, dimethyl ester	$C_2H_6O_3S$ / 110.13	616-42-2	/ 126	4-01-00-01250 / 1.2129[20]	H_2O 3; EtOH 3; eth 3 / 1.4083[20]
11583	Sulfurous diamide, tetramethyl-	$C_4H_{12}N_2OS$ / 136.22	3768-60-3 / 31	/ 209; 74[13]	4-04-00-00269	eth 4
11584	Sulfur, pentafluoro(trifluoromethyl)-, (OC-6-21)-	CF_8S / 196.06	373-80-8 / -87	/ -20	4-03-00-00035	H_2O 1; EtOH 1; CS_2 3

No.	Name Synonym	Mol. Form. Mol. Wt.	CAS RN mp/°C	Merck No. bp/°C	Beil. Ref. den/g cm^{-3}	Solubility n_D
11585	Sulfuryl chloride isocyanate	CClNO$_3$S 141.53	1189-71-5 -44	107	4-03-00-00086 1.626[25]	1.4467[20]
11586	Sutan Carbamothioic acid, bis(2-methylpropyl)-, S-ethyl ester	C$_{11}$H$_{23}$NOS 217.38	2008-41-5	1546 138[21]	0.9402[25]	
11587	D-Tagatose Tagatose, D-	C$_6$H$_{12}$O$_6$ 180.16	87-81-0 134.5	9005	4-01-00-04414	H$_2$O 4
11588	Tartar emetic Antimonate(2-), bis[μ-[2,3-dihydroxybutanedioato(4-)-O1,O2:O3,O4]]di-, dipotassium, trihydrate, stereoisomer	C$_8$H$_{10}$K$_2$O$_{15}$Sb$_2$ 667.87	28300-74-5	732	2.6	
11589	Tebuconazole (RS)-1-p-Chlorophenyl-4,4-dimethyl-3-(1H-1,2,4-triazol-1-ylmethyl)pentan-3-ol	C$_{16}$H$_{23}$ClN$_3$O 308.83	107534-96-3 102.4			
11590	Tebuthiuron Urea, N-[5-(1,1-dimethylethyl)-1,3,4-thiadiazol-2-yl]-N,N'-dimethyl-	C$_9$H$_{16}$N$_4$OS 228.32	34014-18-1 163 dec	9053		
11591	Terbacil 2,4(1H,3H)-Pyrimidinedione, 5-chloro-3-(1,1-dimethylethyl)-6-methyl-	C$_9$H$_{13}$ClN$_2$O$_2$ 216.67	5902-51-2 176	9085 sub 175	1.34[25]	
11592	Terbufos Phosphorodithioic acid, S-[[(1,1-dimethylethyl)thio]methyl] O,O-diethyl ester	C$_9$H$_{21}$O$_2$PS$_3$ 288.44	13071-79-9 -29.2	9088 69[0.01]	1.105[24]	
11593	Terbuthylazine 1,3,5-Triazine-2,4-diamine, 6-chloro-N-(1,1-dimethylethyl)-N'-ethyl-	C$_9$H$_{16}$ClN$_5$ 229.71	5915-41-3 178		1.188[20]	
11594	Terbutryn 1,3,5-Triazine-2,4-diamine, N-(1,1-dimethylethyl)-N'-ethyl-6-(methylthio)-	C$_{10}$H$_{19}$N$_5$S 241.36	886-50-0 104	157[0.06]	1.115[20]	
11595	1,1':2',1"-Terphenyl o-Terphenyl	C$_{18}$H$_{14}$ 230.31	84-15-1 56.2	332	4-05-00-02478	H$_2$O 1; ace 3; bz 3; chl 3
11596	1,1':3',1"-Terphenyl m-Terphenyl	C$_{18}$H$_{14}$ 230.31	92-06-8 87	363	4-05-00-02480	H$_2$O 1; EtOH 3; eth 3; bz 3
11597	1,1':4',1"-Terphenyl p-Terphenyl	C$_{18}$H$_{14}$ 230.31	92-94-4 210.1	376	4-05-00-02483	H$_2$O 1; EtOH 2; eth 3; bz 3
11598	1,1':3',1"-Terphenyl, 5'-phenyl-	C$_{24}$H$_{18}$ 306.41	612-71-5 176	462	4-05-00-02732 1.199[30]	H$_2$O 1; EtOH 3; eth 3; bz 4
11599	Terrazole 1,2,4-Thiadiazole, 5-ethoxy-3-(trichloromethyl)-	C$_5$H$_5$Cl$_3$N$_2$OS 247.53	2593-15-9 19.9	95[1]	1.503[25]	
11600	2,2':5',2"-Terthiophene α-Terthienyl	C$_{12}$H$_8$S$_3$ 248.39	1081-34-1	9106	5-19-09-00226	
11601	1,3,5,7-Tetraazatricyclo[3.3.1.1.3,7]decane Methenamine	C$_6$H$_{12}$N$_4$ 140.19	100-97-0 >250	5879 sub	4-26-00-01680 1.331[-5]	H$_2$O 4; EtOH 3; eth 2; ace 3
11602	2,6,10,14,18,22-Tetracosahexaene, 2,6,10,15,19,23-hexamethyl-, (all-E)- Squalene	C$_{30}$H$_{50}$ 410.73	111-02-4 <-20	8727 280[17]; 241[4]	3-01-00-01071 0.8584[20]	H$_2$O 1; EtOH 2; eth 3; ace 3 1.4990[20]
11603	Tetracosane	C$_{24}$H$_{50}$ 338.66	646-31-1 54	391.3	4-01-00-00578 0.7991[20]	H$_2$O 1; EtOH 2; eth 4 1.4283[70]
11604	Tetracosane, 11-decyl-	C$_{34}$H$_{70}$ 478.93	55429-84-0 10.8	301[10]	4-01-00-00598 0.8161[20]	1.4556[20]
11605	Tetracosane, 3-ethyl-	C$_{26}$H$_{54}$ 366.71	55282-17-2 30.1	255.5[10]	4-01-00-00584 0.7949[40]	1.4436[40]
11606	Tetracosane, 2,6,10,15,19,23-hexamethyl- Squalane	C$_{30}$H$_{62}$ 422.82	111-01-3 -38	8726 350	4-01-00-00593 0.8115[15]	H$_2$O 1; EtOH 2; eth 3; ace 2 1.4530[15]
11607	Tetracosanoic acid Lignoceric acid	C$_{24}$H$_{48}$O$_2$ 368.64	557-59-5 87.5	5364 272[10]	4-02-00-01301 0.8207[100]	bz 4; eth 4 1.4287[100]
11608	15-Tetracosenoic acid	C$_{24}$H$_{46}$O$_2$ 366.63	14490-79-0 61		4-02-00-01681	EtOH 3; chl 3; ace 3
11609	5,9-Tetradecadien-7-yne, 6,9-dimethyl-	C$_{16}$H$_{26}$ 218.38	101427-54-7	95-8[0.3]	4-01-00-01133 0.8241[20]	bz 4; eth 4 1.4866[20]
11610	6,8-Tetradecadiyne	C$_{14}$H$_{22}$ 190.33	16387-72-7 2	118-9[4]	3-01-00-01067 0.8699[16]	eth 4
11611	Tetradecanamide	C$_{14}$H$_{29}$NO 227.39	638-58-4 104	217[12]	4-02-00-01138	EtOH 4
11612	1-Tetradecanamine	C$_{14}$H$_{31}$N 213.41	2016-42-4 83.1	291.2	4-04-00-00812 0.8079[20]	H$_2$O 1; EtOH 4; eth 4; ace 3 1.4463[20]
11613	Tetradecane	C$_{14}$H$_{30}$ 198.39	629-59-4 5.8	253.5	4-01-00-00520 0.7628[20]	H$_2$O 1; EtOH 4; eth 4; ctc 3 1.4290[20]
11614	Tetradecane, 1-bromo-	C$_{14}$H$_{29}$Br 277.29	112-71-0 5.6	307	4-01-00-00523 1.0170[20]	ace 4; bz 4; EtOH 4 1.4603[20]
11615	Tetradecane, 1-chloro-	C$_{14}$H$_{29}$Cl 232.84	2425-54-9 4.9	292	3-01-00-00550 0.8665[20]	H$_2$O 1; EtOH 3; ace 4; bz 4 1.4473[20]
11616	Tetradecane, 1,14-dibromo- Tetradecamethylene dibromide	C$_{14}$H$_{28}$Br$_2$ 356.18	37688-96-3 50.4	190[8]	4-01-00-00523	eth 4; EtOH 4; chl 4
11617	Tetradecane, 1,1-dimethoxy- Myristaldehyde, dimethyl acetal	C$_{16}$H$_{34}$O$_2$ 258.44	14620-53-2	134-6[4]	4-01-00-03389	eth 4; EtOH 4 1.4342[25]
11618	1,14-Tetradecanediol	C$_{14}$H$_{30}$O$_2$ 230.39	19812-64-7 85.8	200[9]	4-01-00-02631	eth 4; EtOH 4

No.	Name / Synonym	Mol. Form. / Mol. Wt.	CAS RN / mp/°C	Merck No. / bp/°C	Beil. Ref. / den/g cm^{-3}	Solubility / n_D
11619	Tetradecanenitrile / Myristonitrile	$C_{14}H_{27}N$ / 209.38	629-63-0 / 19	226[100]; 119[1]	4-02-00-01139 / 0.8281[19]	H_2O 1; EtOH 5; eth; ace 5 / 1.4392[23]
11620	1-Tetradecanesulfonic acid / Tetradecylsulfonic acid	$C_{14}H_{30}O_3S$ / 278.46	7314-37-6 / 65		4-04-00-00066 / 0.9996[25]	H_2O 4
11621	1-Tetradecanethiol	$C_{14}H_{30}S$ / 230.46	2079-95-0	176-80[22]	4-01-00-01867 / 0.8469[20]	H_2O 1; EtOH 3; eth 3; ctc 3 / 1.4597[20]
11622	Tetradecanoic acid / Myristic acid	$C_{14}H_{28}O_2$ / 228.38	544-63-8 / 53.9	6246 / 250[100]	4-02-00-01126 / 0.8622[54]	H_2O 1; EtOH 3; eth 2; ace 3 / 1.4723[70]
11623	Tetradecanoic acid, anhydride	$C_{28}H_{54}O_3$ / 438.73	626-29-9 / 53.4		4-02-00-01138 / 0.8502[70]	eth 4; EtOH 4 / 1.4335[70]
11624	Tetradecanoic acid, ethyl ester	$C_{16}H_{32}O_2$ / 256.43	124-06-1 / 12.3	295	4-02-00-01131 / 0.8573[25]	H_2O 1; EtOH 3; eth 2; ctc 3 / 1.4362[20]
11625	Tetradecanoic acid, methyl ester	$C_{15}H_{30}O_2$ / 242.40	124-10-7 / 19	295; 157[7]	4-02-00-01131 / 0.8671[20]	H_2O 1; EtOH 5; eth 5; ace 5 / 1.425[45]
11626	Tetradecanoic acid, 1-methylethyl ester / Isopropyl myristate	$C_{17}H_{34}O_2$ / 270.46	110-27-0 /	5103 / 193[20]; 140[2]	4-02-00-01132 / 0.8532[20]	H_2O 1; EtOH 3; eth 3; ace 4 / 1.4325[25]
11627	Tetradecanoic acid, phenylmethyl ester	$C_{21}H_{34}O_2$ / 318.50	31161-71-4 / 20.5	229.3[11]	2-06-00-00417 / 0.9293[25]	bz 4; eth 4; EtOH 4; chl 4
11628	Tetradecanoic acid, 1,2,3-propanetriyl ester / Trimyristin	$C_{45}H_{86}O_6$ / 723.17	555-45-3 / 56.5	9638 / 311	4-02-00-01135 / 0.8848[60]	H_2O 1; EtOH 2; eth 3; ace 3 / 1.4428[60]
11629	Tetradecanoic acid, propyl ester	$C_{17}H_{34}O_2$ / 270.46	14303-70-9	147[2]	4-02-00-01132 / 0.8592[20]	ace 4; bz 4; eth 4; EtOH 4 / 1.4356[25]
11630	1-Tetradecanol / Tetradecyl alcohol	$C_{14}H_{30}O$ / 214.39	112-72-1 / 39.5	6248 / 289	4-01-00-01864 / 0.8236[38]	H_2O 1; EtOH 4; eth 4; ace 4
11631	2-Tetradecanol	$C_{14}H_{30}O$ / 214.39	4706-81-4 / 34	284	4-01-00-01867 / 0.8315[20]	1.4444[20]
11632	3-Tetradecanol	$C_{14}H_{30}O$ / 214.39	1653-32-3 / 31.5	173[25]; 146[10]	4-01-00-01868 / 0.8098[53]	bz 4; eth 4; EtOH 4 / 1.4340[45]
11633	2-Tetradecanone / Dodecylmethylketone	$C_{14}H_{28}O$ / 212.38	2345-27-9 / 33.5	205[100]; 134[13]	4-01-00-03389	H_2O 1; EtOH 3; ace 3; os 3
11634	3-Tetradecanone	$C_{14}H_{28}O$ / 212.38	629-23-2 / 34	152[16]	4-01-00-03389	H_2O 1; EtOH 3; ace 3; os 3
11635	Tetradecanoyl chloride / Myristoyl chloride	$C_{14}H_{27}ClO$ / 246.82	112-64-1 / -1	171[16]	4-02-00-01138 / 0.9078[25]	eth 3
11636	1-Tetradecene	$C_{14}H_{28}$ / 196.38	1120-36-1 / -12	233	4-01-00-00924 / 0.7745[25]	H_2O 1; EtOH 4; eth 4; bz 3 / 1.4351[20]
11637	4-Tetradecenoic acid / Tuduic acid	$C_{14}H_{26}O_2$ / 226.36	544-65-0 / 18.5	185-8[13]	4-02-00-01626 / 0.9024[20]	bz 4; peth 4 / 1.4559[20]
11638	5-Tetradecenoic acid / Physoteric acid	$C_{14}H_{26}O_2$ / 226.36	544-66-1 / 20	190-5[15]	3-02-00-01373 / 0.9046[20]	1.4552[20]
11639	9-Tetradecenoic acid	$C_{14}H_{26}O_2$ / 226.36	13147-06-3 / -4	144[0.6]	4-02-00-01626 / 0.9018[20]	1.4519[20]
11640	2-Tetradecyne	$C_{14}H_{26}$ / 194.36	638-60-8 / 6.5	252.5	0-01-00-00262 / 0.8000[20]	eth 4; EtOH 4
11641	7-Tetradecyne	$C_{14}H_{26}$ / 194.36	35216-11-6	144[30]	3-01-00-01027 / 0.7991[20]	eth 4; EtOH 4 / 1.4330[25]
11642	2,5,8,11-Tetraoxadodecane / Triglyme	$C_8H_{18}O_4$ / 178.23	112-49-2 / -45	9604 / 216	4-01-00-02401 / 0.986[20]	H_2O 4; bz 4 / 1.4224[20]
11643	2,4,8,10-Tetraoxaspiro[5.5]undecane	$C_7H_{12}O_4$ / 160.17	126-54-5 / 48.3	147[53]; 68[1]	5-19-11-00342	H_2O 4; ace 4; eth 4; EtOH 4
11644	Tetraphenylene / Tetrabenzocyclooctatetraene	$C_{24}H_{16}$ / 304.39	212-74-8 / 233	sub 200	4-05-00-02773	EtOH 3; eth 2; AcOEt 3; $PhNO_2$ 3
11645	Tetraphosphoric acid, hexaethyl ester / Ethyl tetraphosphate	$C_{12}H_{30}O_{13}P_4$ / 506.26	757-58-4 / -40	150 dec	1.2917[27]	ace 4; bz 4; EtOH 4 / 1.4273[27]
11646	Tetrasiloxane, decamethyl- / Decamethyltetrasiloxane	$C_{10}H_{30}O_3Si_4$ / 310.69	141-62-8 / -76	2843 / 194	4-04-00-04119 / 0.8536[25]	H_2O 1; EtOH 2; bz 3; peth 3 / 1.3895[20]
11647	Tetrasiloxane, 1,1,1,3,5,7,7,7-octamethyl- / 1,1,1,3,5,7,7,7-Octamethyltetrasiloxane	$C_8H_{26}O_3Si_4$ / 282.63	16066-09-4	170	4-04-00-04098 / 0.8559[20]	1.3854[20]
11648	Tetrasul / p-Chlorophenyl 2,4,5-trichlorophenyl sulfide	$C_{12}H_6Cl_4S$ / 324.06	2227-13-6			
11649	Tetrasulfide, bis(1,1-dimethylethyl) / Di-tert-Butyl tetrasulfide	$C_8H_{18}S_4$ / 242.49	5943-35-1 / 2.3	70[0.2]	4-01-00-01638 / 1.0690[20]	1.5660[20]
11650	Tetratriacontane	$C_{34}H_{70}$ / 478.93	14167-59-0 / 72.6	285.4[3]	4-01-00-00597 / 0.7728[90]	1.4296[90]
11651	1,2,4,5-Tetrazine / sym-Tetrazine	$C_2H_2N_4$ / 82.06	290-96-0 / 99	sub	4-26-00-01710	H_2O 3; EtOH 3; eth; sulf 3
11652	1H-Tetrazole	CH_2N_4 / 70.05	288-94-8 / 157	sub	4-26-00-01652 / 1.4060[20]	H_2O 2
11653	2H-Tetrazolium, 2,3,5-triphenyl-, chloride / Triphenyltetrazolium chloride	$C_{19}H_{15}ClN_4$ / 334.81	298-96-4 / 243 dec	9658	4-26-00-01774	H_2O 3; EtOH 3; eth 1; ace 3
11654	5H-Tetrazolo[1,5-a]azepine, 6,7,8,9-tetrahydro- / Pentylenetetrazole	$C_6H_{10}N_4$ / 138.17	54-95-5 / 59.5	7097 / 194[12]	4-26-00-01712	H_2O 4; EtOH 4; eth 3; ace 4
11655	4-Thia-1-azabicyclo[3.2.0]heptane-2-carboxylic acid, 3,3-dimethyl-7-oxo-6-[(phenylacetyl)amino]- / Penicillin G, sodium salt	$C_{16}H_{17}N_2NaO_4S$ / 356.38	69-57-8	1157	4-27-00-05861	

No.	Name / Synonym	Mol. Form. / Mol. Wt.	CAS RN / mp/°C	Merck No. / bp/°C	Beil. Ref. / den/g cm^{-3}	Solubility / n_D
11656	Thiabendazole / 1H-Benzimidazole, 2-(4-thiazolyl)-	$C_{10}H_7N_3S$ / 201.25	148-79-8	9217 / sub 305		
11657	1,2,4-Thiadiazole / 1-Thia-2,4-diazacyclopentadiene	$C_2H_2N_2S$ / 86.12	288-92-6 / -33	/ 121	4-27-00-07093 / 1.3298[20]	
11658	1,2,5-Thiadiazole / Piazthiole	$C_2H_2N_2S$ / 86.12	288-39-1 / -50.1	/ 94	/ 1.268[25]	1.5150[25]
11659	1,3,4-Thiadiazole, 2,5-dimethyl- / 2,5-Dimethyl-1,3,4-thiadiazole	$C_4H_6N_2S$ / 114.17	27464-82-0 / 65	/ 202.5	4-27-00-07097	H_2O 2; EtOH 2; eth 2
11660	Thianthrene	$C_{12}H_8S_2$ / 216.33	92-85-3 / 159.3	/ 365	5-19-02-00049 / 1.4420[20]	H_2O 1; EtOH 2; eth 3; bz 3
11661	Thianthrene, 2,7-dimethyl- / Mesulphen	$C_{14}H_{12}S_2$ / 244.38	135-58-0 / 123	5821 / 184[3]	5-19-02-00072	ace 4; eth 4; peth 4; chl 4
11662	2-Thiazolamine / 2-Aminothiazole	$C_3H_4N_2S$ / 100.14	96-50-4 / 93	494 / 140[11]	4-27-00-04574	H_2O 2; EtOH 2; eth 2; chl 2
11663	2-Thiazolamine, 5-[(4-aminophenyl)sulfonyl]- / Thiazolsulfone	$C_9H_9N_3O_2S_2$ / 255.32	473-30-3 / 219-21 dec	9237	4-27-00-05401	ace 4; eth 4; EtOH 4; diox 4
11664	2-Thiazolamine, 4,5-dihydro-	$C_3H_6N_2S$ / 102.16	1779-81-3 / 85.3	/ dec	4-27-00-04555	H_2O 4; EtOH 4; bz 4; chl 4
11665	2-Thiazolamine, 4-methyl- / 2-Amino-4-methylthiazole	$C_4H_6N_2S$ / 114.17	1603-91-4 / 45.5	462 / 125[20]; 70[0.4]	4-27-00-04687	H_2O 4; EtOH 4; eth 4
11666	2-Thiazolamine, 5-nitro- / 2-Amino-5-nitrothiazole	$C_3H_3N_3O_2S$ / 145.14	121-66-4 / 202 dec	469	4-27-00-04675	
11667	Thiazole	C_3H_3NS / 85.13	288-47-1 /	9235 / 118	4-27-00-00960 / 1.1998[17]	H_2O 2; EtOH 3; eth 3; ace 3 / 1.5969[20]
11668	Thiazole, 2-bromo- / 2-Bromo-1,3-thiazole	C_3H_2BrNS / 164.03	3034-53-5 /	/ 171	4-27-00-00962 / 1.82[25]	1.5927[20]
11669	5-Thiazolecarboxylic acid, 4-methyl- / 4-Methyl-5-thiazolecarboxylic acid	$C_5H_5NO_2S$ / 143.17	20485-41-0 / 280 dec	/ sub 250	4-27-00-04008	H_2O 2; EtOH 3; eth 3; bz 2
11670	Thiazole, 5-(2-chloroethyl)-4-methyl- / Clomethiazole	C_6H_8ClNS / 161.65	533-45-9 /	2382 / 92[7]	4-27-00-00990 / 1.233[25]	
11671	2,4-Thiazolediamine, 5-phenyl- / Amiphenazole	$C_9H_9N_3S$ / 191.26	490-55-1 / 163-4 dec	498	4-27-00-05139	
11672	Thiazole, 4,5-dihydro-2-methyl-	C_4H_7NS / 101.17	2346-00-1 / -101	/ 145	4-27-00-00921 / 1.067[25]	1.5200[20]
11673	Thiazole, 4,5-dihydro-2,4,4-trimethyl-	$C_6H_{11}NS$ / 129.23	4145-94-2 /	/ 147	4-27-00-00931 / 0.969[25]	1.4825[25]
11674	Thiazole, 2,4-dimethyl- / 2,4-Dimethylthiazole	C_5H_7NS / 113.18	541-58-2 /	3249 / 146; 71[50]	4-27-00-00981 / 1.0562[15]	H_2O 2; EtOH 3; eth 3; chl 3 / 1.5091[20]
11675	Thiazole, 4,5-dimethyl- / 4,5-Dimethylthiazole	C_5H_7NS / 113.18	3581-91-7 / 83.5	/ 158	4-27-00-00986 / 1.0699[20]	eth 4; EtOH 4
11676	Thiazole, 2,4-diphenyl- / 2,4-Diphenylthiazole	$C_{15}H_{11}NS$ / 237.33	1826-14-8 / 92.5	/ >360	4-27-00-01435 / 1.1554[98]	eth 4; EtOH 4
11677	5-Thiazoleethanol, 4-methyl- / 4-Methyl-5-thiazoleethanol	C_6H_9NOS / 143.21	137-00-8 /	6046 / 135[7]	4-27-00-01754 / 1.196[24]	H_2O 4; EtOH 3; eth 3; bz 3
11678	Thiazole, 2-methyl- / 2-Methylthiazole	C_4H_5NS / 99.16	3581-87-1 /	/ 128.5	4-27-00-00967	H_2O 5; EtOH 3; ace 3 / 1.510
11679	Thiazole, 4-methyl- / 4-Methylthiazole	C_4H_5NS / 99.16	693-95-8 /	/ 133.3	4-27-00-00969 / 1.112[25]	H_2O 3; EtOH 3; eth 3
11680	2(3H)-Thiazolethione, 4-methyl-	$C_4H_5NS_2$ / 131.22	5685-06-3 / 89.3	/ 188[3]	4-27-00-02584	EtOH 4
11681	Thiazolidine	C_3H_7NS / 89.16	504-78-9 /	/ 164.5	4-27-00-00009 / 1.131[25]	H_2O 5; EtOH 3; eth 4; ace 4 / 1.551[20]
11682	4-Thiazolidinecarboxylic acid / Timonacic	$C_4H_7NO_2S$ / 133.17	444-27-9 / 196.5	9375		H_2O 4
11683	2,4-Thiazolidinedione	$C_3H_3NO_2S$ / 117.13	2295-31-0 / 128	/ 179[19]	4-27-00-03144	eth 4
11684	4-Thiazolidinone, 5-[[4-(dimethylamino)phenyl]methylene]-2-thioxo- / p-(Dimethylamino)benzal-5-rhodanine	$C_{12}H_{12}N_2OS_2$ / 264.37	536-17-4 / 270 dec	3220	4-27-00-05697	H_2O 1; EtOH 2; eth 4; ace 3
11685	4-Thiazolidinone, 2-thioxo- / Rhodanine	$C_3H_3NOS_2$ / 133.19	141-84-4 / 170	8184	4-27-00-03188 / 0.868[25]	H_2O 2; EtOH 4; eth 4
11686	Thiazolium, 3-[(4-amino-2-methyl-5-pyrimidinyl)methyl]-5-(2-hydroxyethyl)-4-methyl- chloride, monohydrochlorid / Thiamine hydrochloride	$C_{12}H_{18}Cl_2N_4OS$ / 337.27	67-03-8 / 248 dec	9222	4-27-00-01764	H_2O 4; EtOH 2; eth 1; ace 2
11687	Thidiazuron / Urea, N-phenyl-N'-1,2,3-thiadiazol-5-yl-	$C_9H_8N_4OS$ / 220.25	51707-55-2 / 211 dec			
11688	Thieno[2,3-b]thiophene	$C_6H_4S_2$ / 140.23	250-84-0 / 6.5	/ 225	5-19-01-00431	eth 4
11689	1H-Thieno[3,4-d]imidazole-4-pentanoic acid, hexahydro-2-oxo-, [3aS-(3aα,4β,6aα)]- / Biotin	$C_{10}H_{16}N_2O_3S$ / 244.31	58-85-5 / 232 dec	1244	4-27-00-07979	H_2O 3; EtOH 2; chl 2
11690	Thiepane	$C_6H_{12}S$ / 116.23	4753-80-4 / 0.5	/ 173.5	5-17-01-00090 / 0.9883[20]	H_2O 1; eth 3; ace 3; chl 3 / 1.5044[18]
11691	Thietane / Trimethylene sulfide	C_3H_6S / 74.15	287-27-4 / -73.2	/ 95	5-17-01-00014 / 1.0200[20]	H_2O 1; EtOH 4; ace 3; bz 4 / 1.5102[20]
11692	Thietane, 1,1-dioxide / Trimethylene sulfone	$C_3H_6O_2S$ / 106.15	5687-92-3 / 75.5	/ 91.2[14]	5-17-01-00015	H_2O 3; EtOH 2; eth 2; peth 2 / 1.5156[20]
11693	Thietane, 2-methyl-	C_4H_8S / 88.17	17837-41-1 /	/ 107	5-17-01-00055 / 0.9571[20]	H_2O 1; EtOH 3; eth 3; ace 3 / 1.4852[20]

No.	Name / Synonym	Mol. Form. / Mol. Wt.	CAS RN / mp/°C	Merck No. / bp/°C	Beil. Ref. / den/g cm^{-3}	Solubility / n_D
11694	Thietane, 1-oxide / 1,3-Propanesulfone	$C_3H_8O_3S$ / 122.14	13153-11-2 / 31.5	180[30]	5-17-01-00015 / 1.392[25]	
11695	Thiirane	C_2H_4S / 60.12	420-12-2 /	57 dec	5-17-01-00009 / 1.0368[0]	EtOH 2; eth 2; ace 3; chl 3 / 1.4935[20]
11696	Thiirane, methyl-	C_3H_6S / 74.15	1072-43-1 / -91	72.5	5-17-01-00025 / 0.941[20]	chl 3 / 1.472[20]
11697	Thiobencarb / Carbamothioic acid, diethyl-, S-[(4-chlorophenyl)methyl] ester	$C_{12}H_{16}CINOS$ / 257.78	28249-77-6 / 3.3	127[0.008]	/ 1.16[20]	
11698	Thioboric acid, trimethyl ester	$C_3H_9BS_3$ / 152.11	997-49-9 / 5	218.2	4-01-00-01287 / 1.126[20]	ace 4; eth 4 / 1.5788[20]
11699	Thiocyanic acid, butyl ester / Butyl thiocyanate	C_5H_9NS / 115.20	628-83-1 /	186	4-03-00-00329 / 0.9563[15]	H_2O 1; EtOH 3; eth 3 / 1.4360[20]
11700	Thiocyanic acid, chloromethyl ester	C_2H_2CINS / 107.56	3268-79-9 /	185	2-03-00-00124 / 1.37[15]	os 5
11701	Thiocyanic acid, 1,1-dimethylethyl ester / tert-Butyl thiocyanate	C_5H_9NS / 115.20	37985-18-5 / 10.5	140 dec; 40[10]	4-03-00-00330 / 0.9187[10]	
11702	Thiocyanic acid, 1,2-ethanediyl ester	$C_4H_4N_2S_2$ / 144.22	629-17-4 / 90	dec	4-03-00-00333 / 1.4200[0]	H_2O 2; EtOH 3; eth 3; ace 4
11703	Thiocyanic acid, ethyl ester / Ethyl thiocyanate	C_3H_5NS / 87.15	542-90-5 / -85.5	146	4-03-00-00328 / 1.007[23]	H_2O 1; EtOH 5; eth 5; chl 3 / 1.4684[15]
11704	Thiocyanic acid, heptyl ester / Heptyl thiocyanate	$C_8H_{15}NS$ / 157.28	5416-94-4 /	235	4-03-00-00331 / 0.92[20]	eth 4; EtOH 4
11705	Thiocyanic acid, methyl ester / Methyl thiocyanate	C_2H_3NS / 73.12	556-64-9 / -5	6047 / 132.9	4-03-00-00327 / 1.0678[25]	H_2O 2; EtOH 5; eth 5; ctc 3 / 1.4669[25]
11706	Thiocyanic acid, 1-methylethyl ester / Isopropyl thiocyanate	C_4H_7NS / 101.17	625-59-2 /	153	4-03-00-00329 / 0.9784[20]	H_2O 1; EtOH 5; eth 5
11707	Thiocyanic acid, 2-methylpropyl ester / Isobutyl thiocyanate	C_5H_9NS / 115.20	591-84-4 / -59	5036 / 175.4	4-03-00-00330	eth 4; EtOH 4
11708	Thiocyanic acid, octyl ester	$C_9H_{17}NS$ / 171.31	19942-78-0 / 105	141-2[19]	4-03-00-00331 / 0.9149[25]	H_2O 1; EtOH 4; eth 4; bz 3 / 1.4649[20]
11709	Thiocyanic acid, phenyl ester / Phenyl thiocyanate	C_7H_5NS / 135.19	5285-87-0 /	232.5	4-06-00-01536 / 1.153[18]	H_2O 1; EtOH 3; eth 3
11710	Thiocyanic acid, phenylmethyl ester	C_8H_7NS / 149.22	3012-37-1 / 43	232	4-06-00-02680	H_2O 1; EtOH 3; eth 3; chl 3
11711	Thiocyanic acid, potassium salt / KSCN	CKNS / 97.18	333-20-0 /	7687	4-03-00-00299	
11712	Thiocyanic acid, 2-propenyl ester	C_4H_5NS / 99.16	764-49-8 /	161	4-03-00-00332 / 1.056[15]	eth 4; EtOH 4
11713	Thiodicarb / Dimethyl N,N'-[thiobis[[(methylimino)carbonyl]oxyl]]bis[ethanimidothioate]	$C_{10}H_{18}N_4O_4S_3$ / 354.48	59669-26-0 / 173	9258	/ 1.4[20]	
11714	Thiodicarbonic diamide, tetraethyl- / Sulfiram	$C_{10}H_{20}N_2S_3$ / 264.48	95-05-6 /	8928 / 225-40[3]	4-04-00-00397 / 1.12[20]	chl 3
11715	Thiodicarbonic diamide, tetramethyl-	$C_6H_{12}N_2S_3$ / 208.37	97-74-5 /	109.5	4-04-00-00238 / 1.37[25]	H_2O 1; EtOH 3; eth 2; ace 3
11716	Thiodiphosphoric acid, tetraethyl ester / Sulfotep	$C_8H_{20}O_5P_2S_2$ / 322.32	3689-24-5 /	8945 / 136-9[2]	4-01-00-01351 / 1.196[25]	H_2O 1; EtOH 3 / 1.4753[25]
11717	Thiofanox / 2-Butanone, 3,3-dimethyl-1-(methylthio)-, O-[(methylamino)carbonyl]oxime	$C_9H_{18}N_2O_2S$ / 218.32	39196-18-4 / 57			
11718	Thioimidodicarbonic diamide / 2,4-Dithiobiuret	$C_2H_5N_3S_2$ / 135.21	541-53-7 / 181 dec	3378	4-03-00-00356	ace 4
11719	Thiomorpholine / Thiamorpholine	C_4H_9NS / 103.19	123-90-0 /	9229 / 175; 110[100]	4-27-00-00636 / 1.0882[20]	H_2O 4; ace 4; eth 4; EtOH 4 / 1.5386[20]
11720	Thionazin / Phosphorothioic acid, O,O-diethyl O-pyrazinyl ester	$C_8H_{13}N_2O_3PS$ / 248.24	297-97-2 / -1.7	9275 / 80		
11721	Thioperoxydicarbonic diamide ([[(H2N)C(S)]2S2), tetraethyl- / Disulfiram	$C_{10}H_{20}N_2S_4$ / 296.55	97-77-8 / 71.5	3370 / 117[17]	4-04-00-00398	H_2O 1; EtOH 3; eth 2; chl 4
11722	Thioperoxydicarbonic diamide ([[(H2N)C(S)]2S2), tetramethyl- / Thiram	$C_6H_{12}N_2S_4$ / 240.44	137-26-8 / 155.6	9304 / 129[20]	4-04-00-00242	chl 4
11723	Thiophanate-methyl / Carbamic acid, [1,2-phenylenebis(iminocarbonothioyl)]bis-, dimethyl ester	$C_{12}H_{14}N_4O_4S_2$ / 342.40	23564-05-8 / 172 dec			
11724	Thiophene / Thiofuran	C_4H_4S / 84.14	110-02-1 / -39.4	9283 / 84.0	5-17-01-00297 / 1.0649[20]	EtOH 5; eth 5; ace 5; bz 5 / 1.5289[20]
11725	2-Thiopheneacetonitrile	C_6H_5NS / 123.18	20893-30-5 /	115-20[22]	5-18-06-00208 / 1.157[25]	/ 1.5425[20]
11726	Thiophene, 2-bromo- / 2-Thienyl bromide	C_4H_3BrS / 163.04	1003-09-4 /	150	5-17-01-00305 / 1.684[20]	H_2O 1; eth 4; ace 4; ctc 3 / 1.5868[20]
11727	Thiophene, 3-bromo-	C_4H_3BrS / 163.04	872-31-1 /	159.5	5-17-01-00306 / 1.735[20]	H_2O 1; ace 3; bz 3; chl 2 / 1.5919[20]
11728	Thiophene, 2-bromo-5-chloro- / 2-Bromo-5-chlorothiophene	C_4H_2BrCIS / 197.48	2873-18-9 / -20	70[18]	5-17-01-00307 / 1.803[25]	eth 4 / 1.5925[25]

No.	Name Synonym	Mol. Form. Mol. Wt.	CAS RN	mp/°C	Merck No. bp/°C	Beil. Ref. den/g cm^{-3}	Solubility n_D
11729	Thiophene, 2-bromo-3-methyl- 2-Bromo-3-methylthiophene	C_5H_5BrS 177.06	14282-76-9		176; 27[1.8]	5-17-01-00333 1.5844[18]	bz 4; eth 4 1.5714[20]
11730	Thiophene, 2-bromo-5-methyl-	C_5H_5BrS 177.06	765-58-2		178; 29[1.8]	5-17-01-00327 1.5529[20]	eth 3; bz 3 1.5673[20]
11731	Thiophene, 2-butyl- 2-Butylthiophene	$C_8H_{12}S$ 140.25	1455-20-5		181.5	5-17-01-00369 0.9537[20]	1.5090[20]
11732	2-Thiophenecarbonitrile 2-Cyanothiophene	C_5H_3NS 109.15	1003-31-2		192	5-18-06-00171 1.172[25]	chl 3 1.5629[20]
11733	2-Thiophenecarboxaldehyde	C_5H_4OS 112.15	98-03-3		197; 85[16]	5-17-09-00349 1.2127[21]	H_2O 1; EtOH 4; eth 3; chl 2 1.5920[20]
11734	3-Thiophenecarboxaldehyde 3-Formylthiophene	C_5H_4OS 112.15	498-62-4		86.7[20]	5-17-09-00373	H_2O 1; EtOH 4; eth 3 1.5855[20]
11735	2-Thiophenecarboxaldehyde, 5-chloro-	C_5H_3ClOS 147.60	7283-96-7		77.5[5]	5-17-09-00360	chl 2 1.6036[25]
11736	2-Thiophenecarboxaldehyde, 5-methyl-	C_6H_6OS 126.18	13679-70-4		114[25]	5-17-09-00406	chl 3 1.5825[20]
11737	2-Thiophenecarboxylic acid 2-Carboxythiophene	$C_5H_4O_2S$ 128.15	527-72-0	129.5	9284 260 dec	5-18-06-00158	H_2O 4; EtOH 4; eth 4; chl 3
11738	3-Thiophenecarboxylic acid 3-Thenoic acid	$C_5H_4O_2S$ 128.15	88-13-1	138	9206	5-18-06-00199	H_2O 3
11739	2-Thiophenecarboxylic acid, ethyl ester	$C_7H_8O_2S$ 156.21	2810-04-0		218	5-18-06-00159 1.1623[16]	EtOH 3; ace 3; ctc 2; os 3 1.5248[20]
11740	Thiophene, 2-chloro- 2-Thienyl chloride	C_4H_3ClS 118.59	96-43-5	-71.9	128.3	5-17-01-00303 1.2863[20]	H_2O 1; EtOH 5; eth 5; chl 2 1.5487[20]
11741	Thiophene, 2-chloro-5-methyl- 2-Chloro-5-methylthiophene	C_5H_5ClS 132.61	17249-82-0		155; 55[19]	5-17-01-00326 1.2147[25]	ace 4; bz 4; eth 4; EtOH 4 1.5372[20]
11742	Thiophene, 2,3-dibromo- 2,3-Dibromothiophene	$C_4H_2Br_2S$ 241.93	3140-93-0	-17.5	218.5; 89[13]	5-17-01-00308	
11743	Thiophene, 2,4-dibromo- 2,4-Dibromothiophene	$C_4H_2Br_2S$ 241.93	3140-92-9	-26.3	210	5-17-01-00308 2.1488[20]	1.6304[22]
11744	Thiophene, 2,5-dibromo- 2,5-Dibromothiophene	$C_4H_2Br_2S$ 241.93	3141-27-3	-6	210.3	5-17-01-00308 2.142[23]	H_2O 4; EtOH 4; ctc 3 1.6288[20]
11745	2,5-Thiophenedicarboxylic acid 2,5-Dicarboxythiophene	$C_6H_4O_4S$ 172.16	4282-31-9	359	sub 150	5-18-07-00049	H_2O 2; EtOH 3; eth 3
11746	Thiophene, 2,3-dichloro- 2,3-Dichlorothiophene	$C_4H_2Cl_2S$ 153.03	17249-79-5	-26.2	173.5	5-17-01-00304 1.4605[20]	H_2O 1; EtOH 5; eth 3; ace 3 1.5651[20]
11747	Thiophene, 2,4-dichloro- 2,4-Dichlorothiophene	$C_4H_2Cl_2S$ 153.03	17249-75-1	-34	167.6	5-17-01-00304 1.4553[20]	H_2O 1; EtOH 4; eth 3; ace 3 1.5660[20]
11748	Thiophene, 2,5-dichloro- 5-Chloro-2-thienyl chloride	$C_4H_2Cl_2S$ 153.03	3172-52-9	-40.5	162	5-17-01-00304 1.4422[20]	H_2O 1; EtOH 5; eth 5; ctc 3 1.5626[20]
11749	Thiophene, 3,4-dichloro- 3,4-Dichlorothiophene	$C_4H_2Cl_2S$ 153.03	17249-76-2	1	185	5-17-01-00305 1.4867[20]	H_2O 1; EtOH 4; eth 4; ace 3 1.5821[19]
11750	Thiophene, 2,3-dihydro- 2,3-Dihydrothiophene	C_4H_6S 86.16	1120-59-8		112.1	5-17-01-00308	
11751	Thiophene, 2,5-dihydro- 2,5-Dihydrothiophene	C_4H_6S 86.16	1708-32-3		122.4		
11752	Thiophene, 2,5-dihydro-, 1,1-dioxide 3-Sulfolene	$C_4H_6O_2S$ 118.16	77-79-2	64.5	8935	5-17-01-00177	chl 3
11753	Thiophene, 2,5-diiodo- 2,5-Diiodothiophene	$C_4H_2I_2S$ 335.93	625-88-7	41.5	139[15]	5-17-01-00313	H_2O 1; EtOH 4
11754	Thiophene, 2,3-dimethyl- 2,3-Dimethylthiophene	C_6H_8S 112.20	632-16-6	-49	141.6	5-17-01-00342 1.0021[20]	H_2O 1; EtOH 4; eth 4; bz 3 1.5192[20]
11755	Thiophene, 2,4-dimethyl- 2,4-Dimethylthiophene	C_6H_8S 112.20	638-00-6		140.7	5-17-01-00344 0.9938[20]	H_2O 1; EtOH 3; eth 3; bz 3 1.5104[20]
11756	Thiophene, 2,5-dimethyl- 2,5-Dimethylthiophene	C_6H_8S 112.20	638-02-8	-62.6	136.5	5-17-01-00345 0.9850[20]	H_2O 1; EtOH 3; eth 3; bz 3 1.5129[20]
11757	Thiophene, 3,4-dimethyl- 3,4-Dimethylthiophene	C_6H_8S 112.20	632-15-5		145	5-17-01-00347 0.993[25]	H_2O 1; EtOH 3; eth 4 1.5206[20]
11758	Thiophene, 2,5-dinitro- 2,5-Dinitrothiophene	$C_4H_2N_2O_4S$ 174.14	59434-05-8	81	290	5-17-01-00315	eth 4; EtOH 4
11759	Thiophene, 2-ethyl- 2-Ethylthiophene	C_6H_8S 112.20	872-55-9		134	5-17-01-00339 0.9930[20]	H_2O 1; EtOH 4; eth 4 1.5122[20]
11760	Thiophene, 3-ethyl- 3-Ethylthiophene	C_6H_8S 112.20	1795-01-3	-89.1	136	5-17-01-00341 0.9980[20]	H_2O 1; EtOH 3; eth 4 1.5146[20]
11761	Thiophene, 2-ethyl-3-methyl- 2-Ethyl-3-methylthiophene	$C_7H_{10}S$ 126.22	31805-48-8		161	5-17-01-00357 0.9815[20]	1.5105[20]
11762	Thiophene, 2-ethyl-5-methyl- 2-Ethyl-5-methylthiophene	$C_7H_{10}S$ 126.22	40323-88-4	-68.4	160	5-17-01-00357 0.9663[20]	1.5073[20]
11763	Thiophene, 3-ethyl-4-methyl-	$C_7H_{10}S$ 126.22	66577-03-5		165.5	4-17-00-00301 0.9976[20]	1.5186[20]
11764	Thiophene, 4-ethyl-2-methyl- 5-Methyl-3-ethylthiophene	$C_7H_{10}S$ 126.22	13678-54-1	-60	163	4-17-00-00299 0.9742[20]	1.5098[20]
11765	Thiophene, 2-ethyltetrahydro-	$C_6H_{12}S$ 116.23	1551-32-2		157	5-17-01-00096 0.944[20]	ace 4; bz 4; eth 4; EtOH 4 1.490[20]
11766	Thiophene, 3-ethyltetrahydro- 3-Ethyltetrahydrothiophene	$C_6H_{12}S$ 116.23	62184-67-2		165	5-17-01-00096 0.950[20]	1.491[20]
11767	Thiophene, 2-iodo-	C_4H_3IS 210.04	3437-95-4	-40	181	5-17-01-00309 2.0595[25]	EtOH 4; eth 4; chl 2 1.6465[25]
11768	2-Thiophenemethanol	C_5H_6OS 114.17	636-72-6		207; 86[10]	5-17-03-00358 1.2053[16]	EtOH 3; ace 3 1.5280[20]

No.	Name / Synonym	Mol. Form. / Mol. Wt.	CAS RN / mp/°C	Merck No. / bp/°C	Beil. Ref. / den/g cm^{-3}	Solubility / n_D
11769	Thiophene, 2-methyl- / 2-Methylthiophene	C$_5$H$_6$S / 98.17	554-14-3 / -63.4	112.6	5-17-01-00324 / 1.0193[20]	H$_2$O 1; EtOH 5; eth 5; ace 5 / 1.5203[20]
11770	Thiophene, 3-methyl- / 3-Thiotolene	C$_5$H$_6$S / 98.17	616-44-4 / -69	115.5	5-17-01-00331 / 1.0218[20]	H$_2$O 1; EtOH 5; eth 5; ace 5 / 1.5204[20]
11771	Thiophene, 2-(3-methylbutyl)- / 2-Isopentylthiophene	C$_9$H$_{14}$S / 154.28	26963-33-7 /	74-5[11]	4-17-00-00316 / 0.9481[12]	1.5014[12]
11772	Thiophene, 2-(1-methylethyl)- / 2-Isopropylthiophene	C$_7$H$_{10}$S / 126.22	4095-22-1 /	153	5-17-01-00356 / 0.9678[20]	1.5038[20]
11773	Thiophene, 3-(1-methylethyl)- / 3-Isopropylthiophene	C$_7$H$_{10}$S / 126.22	29488-27-5 /	157	5-17-01-00356 / 0.9733[20]	1.5052[20]
11774	Thiophene, 2-methyl-5-phenyl- / 2-Methyl-5-phenylthiophene	C$_{11}$H$_{10}$S / 174.27	5069-26-1 / 51	271	5-17-02-00153	EtOH 4; eth 4; lig 4
11775	Thiophene, 2-nitro- / 2-Nitrothiophene	C$_4$H$_3$NO$_2$S / 129.14	609-40-5 / 46.5	224.5	5-17-01-00313 / 1.3644[43]	H$_2$O 1; EtOH 4; alk 3; peth 2
11776	Thiophene, 3-nitro- / 3-Nitrothiophene	C$_4$H$_3$NO$_2$S / 129.14	822-84-4 / 78.5	225	5-17-01-00313	H$_2$O 1; EtOH 4; ctc 2; peth 3
11777	Thiophene-2-ol	C$_4$H$_4$OS / 100.14	17236-58-7 / 9	218	1.255[20]	H$_2$O 3; ace 3; chl 3 / 1.5644[20]
11778	Thiophene, 2-propyl- / 2-Propylthiophene	C$_7$H$_{10}$S / 126.22	1551-27-5 /	158	5-17-01-00355 / 0.9683[20]	H$_2$O 1; EtOH 3; eth 3; bz 3 / 1.5076[20]
11779	Thiophene, 3-propyl- / 3-Propylthiophene	C$_7$H$_{10}$S / 126.22	1518-75-8 /	163.2	5-17-01-00355 / 0.9738[20]	H$_2$O 1; EtOH 3; eth 3; bz 3 / 1.5057[20]
11780	2-Thiophenesulfonyl chloride	C$_4$H$_3$ClO$_2$S$_2$ / 182.65	16629-19-9 / 28	100[6]	5-18-09-00426	eth 3
11781	Thiophene, tetrabromo- / 2,3,4,5-Tetrabromothiophene	C$_4$Br$_4$S / 399.73	3958-03-0 / 117.5	326	5-17-01-00309	H$_2$O 1; EtOH 3; eth 4
11782	Thiophene, tetrachloro-	C$_4$Cl$_4$S / 221.92	6012-97-1 / 30.5	233.4	5-17-01-00305 / 1.7036[30]	H$_2$O 1; EtOH 4; eth 5 / 1.5915[30]
11783	Thiophene, tetrahydro- / Thiacyclopentane	C$_4$H$_8$S / 88.17	110-01-0 / -96.1	121.0	5-17-01-00036 / 0.9987[20]	H$_2$O 1; EtOH 5; eth 5; ace 5 / 1.4871[18]
11784	Thiophene, tetrahydro-2,4-dimethyl-, 1,1-dioxide / 2,4-Dimethylsulfolane	C$_6$H$_{12}$O$_2$S / 148.23	1003-78-7 / -3	3245 / 281	5-17-01-00097 / 1.1362[20]	lig 4 / 1.4732[20]
11785	Thiophene, tetrahydro-2,5-dimethyl-, cis-	C$_6$H$_{12}$S / 116.23	5161-13-7 / -89	142.3	5-17-01-00098 / 0.9222[20]	ace 4; bz 4; eth 4; EtOH 4 / 1.4799[20]
11786	Thiophene, tetrahydro-, 1,1-dioxide / Sulfolane	C$_4$H$_8$O$_2$S / 120.17	126-33-0 / 27.6	8934 / 287.3	5-17-01-00039 / 1.2723[18]	chl 3 / 1.4833[18]
11787	Thiophene, tetrahydro-2-methyl-	C$_5$H$_{10}$S / 102.20	1795-09-1 / -100.7	134	5-17-01-00080 / 0.9564[18]	H$_2$O 1; EtOH 4; eth 4; ace 4 / 1.4886[15]
11788	Thiophene, tetrahydro-3-methyl-	C$_5$H$_{10}$S / 102.20	4740-00-5 /	139	5-17-01-00082 / 0.9596[18]	H$_2$O 1; EtOH 4; eth 4; ace 4 / 1.4886[18]
11789	Thiophene, tetrahydro-3-methyl-, 1,1-dioxide / 3-Methylsulfolane	C$_5$H$_{10}$O$_2$S / 134.20	872-93-5 / 1	276	5-17-01-00082 / 1.188[25]	1.4772[20]
11790	Thiophene, 2,3,5-tribromo- / 2,3,5-Tribromothiophene	C$_4$HBr$_3$S / 320.83	3141-24-0 / 29	260	5-17-01-00309	chl 3
11791	Thiophene, 2,3,5-trichloro- / 2,3,5-Trichlorothiophene	C$_4$HCl$_3$S / 187.48	17249-77-3 / -16.1	198.7	5-17-01-00305 / 1.5856[20]	H$_2$O 1; EtOH 5; eth 5 / 1.5791[20]
11792	Thiophene, 2,3,4-trimethyl- / 2,3,4-Trimethylthiophene	C$_7$H$_{10}$S / 126.22	1795-04-6 / -26.2	162; 75[30]	5-17-01-00359 / 0.995[20]	1.5208[20]
11793	Thiophene, 2,3,5-trimethyl- / 2,3,5-Trimethylthiophene	C$_7$H$_{10}$S / 126.22	1795-05-7 /	164.5	5-17-01-00359 / 0.9753[20]	H$_2$O 1; EtOH 4; eth 4; ace 4 / 1.5112[20]
11794	2(3H)-Thiophenone, dihydro-	C$_4$H$_6$OS / 102.16	1003-10-7 /	39-40[1]	5-17-09-00012 / 1.18[25]	1.5230[20]
11795	2H-Thiopyran, tetrahydro- / Thiacyclohexane	C$_5$H$_{10}$S / 102.20	1613-51-0 / 19	141.8	5-17-01-00068 / 0.9861[20]	H$_2$O 1; EtOH 3; eth 3; ace 3 / 1.5067[20]
11796	2H-Thiopyran, tetrahydro-2-methyl-	C$_6$H$_{12}$S / 116.23	5161-16-0 / -58	153	5-17-01-00093 / 0.9428[20]	1.4905[20]
11797	2H-Thiopyran, tetrahydro-3-methyl-	C$_6$H$_{12}$S / 116.23	5258-50-4 / -60	158	5-17-01-00094 / 0.9473[20]	1.4922[20]
11798	2H-Thiopyran, tetrahydro-4-methyl-	C$_6$H$_{12}$S / 116.23	5161-17-1 / -28	158.6	5-17-01-00094 / 0.9471[20]	1.4923[20]
11799	Thiourea / Thiocarbamide	CH$_4$N$_2$S / 76.12	62-56-6 / 182	9299	4-03-00-00342 / 1.405[25]	H$_2$O 3; EtOH 3; eth 1
11800	Thiourea, N,N'-bis[4-(3-methylbutoxy)phenyl]- / Tiocarlide	C$_{23}$H$_{32}$N$_2$O$_2$S / 400.59	910-86-1 / 146	9382	4-13-00-01187	
11801	Thiourea, (2-chlorophenyl)- / 1-(2-Chlorophenyl)thiourea	C$_7$H$_7$ClN$_2$S / 186.66	5344-82-1 / 146		3-12-00-01296	bz 4; EtOH 4
11802	Thiourea, N-(3,5-dichlorophenyl)-N'-(4-fluorophenyl)- / Loflucarban	C$_{13}$H$_9$Cl$_2$FN$_2$S / 315.20	790-69-2 / 163.5	5440	4-12-00-01275	
11803	Thiourea, N,N'-diethyl-	C$_5$H$_{12}$N$_2$S / 132.23	105-55-5 / 78	dec	4-04-00-00375	H$_2$O 3; EtOH 3; eth 4; ctc 2
11804	Thiourea, N,N'-dimethyl- / N,N'-Dimethylthiourea	C$_3$H$_8$N$_2$S / 104.18	534-13-4 / 62	3250	4-04-00-00217	H$_2$O 4; EtOH 4; eth 2; ace 4
11805	Thiourea, N,N'-diphenyl- / sym-Diphenylthiourea	C$_{13}$H$_{12}$N$_2$S / 228.32	102-08-9 / 154.5	3337	4-12-00-00810 / 1.32[25]	H$_2$O 2; EtOH 4; eth 4; chl 4
11806	Thiourea, (2-methylphenyl)- / o-Tolylthiourea	C$_8$H$_{10}$N$_2$S / 166.25	614-78-8 / 162		4-12-00-01766	H$_2$O 4; EtOH 4; eth 2

No.	Name Synonym	Mol. Form. Mol. Wt.	CAS RN mp/°C	Merck No. bp/°C	Beil. Ref. den/g cm^{-3}	Solubility n_D
11807	Thiourea, 1-naphthalenyl- ANTU	$C_{11}H_{10}N_2S$ 202.28	86-88-4 198	755	4-12-00-03086	H_2O 1; EtOH 2; eth 2; ace 2
11808	Thiourea, phenyl- Phenylthiourea	$C_7H_8N_2S$ 152.22	103-85-5 154	7286	4-12-00-00804	H_2O 2; EtOH 3; NaOH 3
11809	Thiourea, 2-propenyl- Thiosinamine	$C_4H_8N_2S$ 116.19	109-57-9 78	9293	4-04-00-01072 1.217[20]	H_2O 3; EtOH 3; eth 2; bz 1 1.5936[78]
11810	Thiourea, tetramethyl-	$C_5H_{12}N_2S$ 132.23	2782-91-4 79.3	245	4-04-00-00232	H_2O 3; EtOH 3; eth 2; chl 3
11811	9H-Thioxanthene Dibenzothiapyran	$C_{13}H_{10}S$ 198.29	261-31-4 128.5	9300 341	5-17-02-00254	chl 3
11812	9H-Thioxanthen-9-one Thioxanthone	$C_{13}H_8OS$ 212.27	492-22-8 209	9301 373	5-17-10-00437	H_2O 1; EtOH 2; bz 3; chl 3
11813	L-Threonine [R-(R*,S*)]-2-Amino-3-hydroxybutanoic acid	$C_4H_9NO_3$ 119.12	72-19-5 256 dec	9316	4-04-00-03171	H_2O 3; EtOH 1; eth 1; chl 1
11814	Thujane, (1S,4R,5S)-(+)- Sabinane (d)	$C_{10}H_{18}$ 138.25	7712-66-5 157		4-05-00-00318 0.8139[20]	1.4376[20]
11815	4(10)-Thujene, (+)- 4(10)-Thujene (d)	$C_{10}H_{16}$ 136.24	2514-91-2 164		4-05-00-00451 0.842[20]	ace 4; bz 4; eth 4; EtOH 4 1.4678[20]
11816	Thymidine Thymine 2-desoxyriboside	$C_{10}H_{14}N_2O_5$ 242.23	50-89-5 186.5	9331	5-24-07-00109	H_2O 3; EtOH 3; ace 3; chl 2
11817	Toluene, 3-chloro-2-iodo-	C_7H_6ClI 252.48	5100-98-1 27.5	123[14]	3-05-00-00726	1.608[23]
11818	o-Toluenesulfonic acid, thio-, S-phenyl ester Benzenesulfonic acid, thiolo,2-methyl,phenyl ester	$C_{13}H_{12}O_2S_2$ 264.37	96097-69-7 59.5	241.6	4-11-00-00237	H_2O 2; EtOH 4; eth 3; chl 2 1.53412[20]
11819	Triacontane	$C_{30}H_{62}$ 422.82	638-68-6 65.8	449.7	4-01-00-00592 0.8097[20]	H_2O 1; EtOH 2; eth 3; bz 4 1.4352[70]
11820	Triacontanoic acid	$C_{30}H_{60}O_2$ 452.81	506-50-3 93.6		4-02-00-01321	bz 4; CS_2 4; chl 4 1.4323[100]
11821	1-Triacontanol Myricyl alcohol	$C_{30}H_{62}O$ 438.82	593-50-0 88	9506	4-01-00-01918 0.777[95]	bz 4; eth 4; EtOH 4
11822	Triadimenol Mercury, chloro(2-methoxyethyl)-	C_3H_7ClHgO 295.13	123-88-6			
11823	Triallate Carbamothioic acid, bis(1-methylethyl)-, S-(2,3,3-trichloro-2-propenyl) ester	$C_{10}H_{16}Cl_3NOS$ 304.67	2303-17-5 29	9510 117[0.0003]	1.273[25]	
11824	Triasulfuron 2-(2-Chloroethoxy)-N-[[(4-methoxy-6-methyl-1,3,5-triaziN-2-yl)amino]carbonyl]benzene sulfonamide	$C_{14}H_{16}ClN_5O_5S$ 401.83	82097-50-5 186			
11825	3,5,7-Triaza-1-azoniatricyclo[3.3.1.13,7]decane, 1-(2-propenyl)-, iodide Hexamethylenetetramine, allyl iodide	$C_9H_{17}IN_4$ 308.17	36895-62-2	5880		
11826	1-Triazene, 1,3-diphenyl- Diazoaminobenzene	$C_{12}H_{11}N_3$ 197.24	136-35-6 98	2981	4-16-00-00904	H_2O 1; EtOH 4; eth 4; bz 4
11827	1,3,5-Triazine	$C_3H_3N_3$ 81.08	290-87-9 86	114	4-26-00-00063 1.38[25]	EtOH 3; eth 3
11828	1,3,5-Triazine-2,4-diamine, 6-chloro-N,N,N',N'-tetraethyl-	$C_{11}H_{20}ClN_5$ 257.77	580-48-3 27	154-6[9]	4-26-00-01208 1.0956[20]	bz 4; chl 4; EtOH 4; lig 4 1.5320[20]
11829	1,3,5-Triazine-2,4-diamine, 6-phenyl- Benzoguanamine	$C_9H_9N_5$ 187.20	91-76-9 226.5	1099	4-26-00-01244	EtOH 3; eth 3; tfa 2
11830	1,3,5-Triazine-2,4-diamine, N-phenyl- Amanozine	$C_9H_9N_5$ 187.20	537-17-7 235.5	379	4-26-00-01195	
11831	1,2,4-Triazine-3,5(2H,4H)-dione, 6-methyl- 6-Azathymine	$C_4H_5N_3O_2$ 127.10	932-53-6 211	919	4-26-00-00556	H_2O 3; EtOH 3; ace 3
11832	1,2,4-Triazine-3,5(2H,4H)-dione, 2-β-D-ribofuranosyl- 6-Azauridine	$C_8H_{11}N_3O_6$ 245.19	54-25-1 158	920	4-26-00-00554	H_2O 3
11833	1,3,5-Triazine, hexahydro-1,3,5-trinitro- Cyclonite	$C_3H_6N_6O_6$ 222.12	121-82-4 205.5	2742	4-26-00-00022 1.82[20]	H_2O 1; EtOH 1; eth 2; ace 3
11834	1,3,5-Triazine, hexahydro-1,3,5-triphenyl-	$C_{21}H_{21}N_3$ 315.42	91-78-1 144	185; 60[29]	4-26-00-00007	H_2O 1; EtOH 3; ace 3
11835	1,3,5-Triazine-2,4,6-triamine Melamine	$C_3H_6N_6$ 126.12	108-78-1 345 dec	5691 sub	4-26-00-01253 1.573[16]	H_2O 2; EtOH 2; eth 1 1.872[20]
11836	1,3,5-Triazine-2,4,6-tricarbonitrile	C_6N_6 156.11	7615-57-8 119	262; 119[1.0]	1-26-00-00091	bz 4
11837	1,3,5-Triazine, 2,4,6-trichloro- Cyanuric acid trichloride	$C_3Cl_3N_3$ 184.41	108-77-0 154	192	4-26-00-00066	EtOH 4
11838	1,3,5-Triazine-2,4,6(1H,3H,5H)-trione, 1,3-dichloro- Dichlorocyanuric acid	$C_3HCl_2N_3O_3$ 197.96	2782-57-2			
11839	1,3,5-Triazine-2,4,6(1H,3H,5H)-trione, dihydrate Cyanuric acid, dihydrate	$C_3H_7N_3O_5$ 165.11	6202-04-6 >360 dec	dec	4-26-00-00632 2.50[20]	H_2O 3; EtOH 2; eth 1; ace 1 2.500[20]
11840	1,3,5-Triazine-2,4,6(1H,3H,5H)-trione, 1,3,5-trichloro- Symclosene	$C_3Cl_3N_3O_3$ 232.41	87-90-1 246.7 dec	8993	4-26-00-00642	

No.	Name / Synonym	Mol. Form. / Mol. Wt.	CAS RN / mp/°C	Merck No. / bp/°C	Beil. Ref. / den/g cm^{-3}	Solubility / n_D
11841	1,3,5-Triazine-2,4,6(1H,3H,5H)-trione, 1,3,5-trimethyl-	$C_6H_9N_3O_3$ 171.16	827-16-7 176.5	274	4-26-00-00635	H_2O 1; EtOH 3
11842	1,3,5-Triazine-2,4,6(1H,3H,5H)-trione, 1,3,5-tri-2-propenyl-	$C_{12}H_{15}N_3O_3$ 249.27	1025-15-6 20.5	149[4]; 105[0.5]	4-26-00-00637 1.1590[20]	
11843	1,3,5-Triazine, 2,4,6-tri-2-pyridinyl- 2,4,6-Tripyridyl-s-triazine	$C_{18}H_{12}N_6$ 312.33	3682-35-7 210	9664	4-26-00-04192	
11844	1,3,5-Triazine, 2,4,6-tris(1-aziridinyl)- Triethylenemelamine	$C_9H_{12}N_6$ 204.23	51-18-3	9586	4-26-00-01268	
11845	1H-1,2,4-Triazol-3-amine Amitrole	$C_2H_4N_4$ 84.08	61-82-5 159	506	4-26-00-01073	H_2O 4; EtOH 1; eth 1; ace 1
11846	1H-1,2,3-Triazole	$C_2H_3N_3$ 69.07	288-36-8 23	204	4-26-00-00029 1.1861[25]	H_2O 3; eth 3; ace 3; os 3 1.4854[25]
11847	1H-1,2,4-Triazole, 1-butyl 1-Butyl-s-triazole	$C_6H_{11}N_3$ 125.17	6086-22-2			
11848	4H-1,2,4-Triazole, 4-cyclohexyl-3-ethyl- Hexazole	$C_{10}H_{17}N_3$ 179.27	4671-03-8 89.5	4618 227[10]	4-26-00-00049	H_2O 4; bz 4; chl 4
11849	1H-1,2,3-Triazole, 1,5-dimethyl- 1,5-Dimethyl-1,2,3-triazole	$C_4H_7N_3$ 97.12	15922-53-9 -4	255	1-26-00-00005	H_2O 4
11850	1,2,4-Triazolidine-3,5-dione Urazole	$C_2H_3N_3O_2$ 101.06	3232-84-6 249-50 dec	9780	4-26-00-00538	
11851	7H-1,2,3-Triazolo[4,5-d]pyrimidin-7-one, 5-amino-1,4-dihydro- 8-Azaguanine	$C_4H_4N_6O$ 152.12	134-58-7 300	912	4-26-00-04171	
11852	S,S,S-Tributyl phosphorotrithioate S,S,S-Tributyl trithiophosphate	$C_{12}H_{27}OPS_3$ 314.52	78-48-8 <-25	150[0.3]	1.057[20]	
11853	Trichlorfon Phosphonic acid, (2,2,2-trichloro-1-hydroxyethyl)-, dimethyl ester	$C_4H_8Cl_3O_4P$ 257.44	52-68-6 77	9536 100[0.1]	1.73[20]	
11854	Triclopyr Acetic acid, [(3,5,6-trichloro-2-pyridinyl)oxy]-	$C_7H_4Cl_3NO_3$ 256.47	55335-06-3 149	9572 dec 290		
11855	Tricosane	$C_{23}H_{48}$ 324.63	638-67-5 47.6	380	4-01-00-00576 0.7785[48]	H_2O 1; EtOH 2; eth 3; ctc 3 1.4468[20]
11856	Tricosane, 2-methyl- 2-Methyltricosane	$C_{24}H_{50}$ 338.66	1928-30-9 37.6	205[3]	3-01-00-00576 0.7539[90]	1.4201[90]
11857	12-Tricosanone Diundecyl ketone	$C_{23}H_{46}O$ 338.62	540-09-0 70.2		4-01-00-03408 0.8086[69]	bz 4; eth 4; chl 4 1.4283[80]
11858	Tricyclazole 1,2,4-Triazolo[3,4-b]benzothiazole, 5-methyl-	$C_9H_7N_3S$ 189.24	41814-78-2 187			
11859	Tricyclo[3.3.1.13,7]decan-1-amine Amantadine	$C_{10}H_{17}N$ 151.25	768-94-5 180	380		H_2O 2
11860	Tricyclo[3.3.1.13,7]decane Adamantane	$C_{10}H_{16}$ 136.24	281-23-2 268	140 sub	4-05-00-00469 1.07[25]	bz 3; ctc 3 1.568
11861	Tricyclo[4.4.0.02,7]dec-3-ene, 1,3-dimethyl-8-(1-methylethyl)-, stereoisomer Copaene	$C_{15}H_{24}$ 204.36	3856-25-5 248.5	2510	4-05-00-01189 0.8996[20]	H_2O 1; eth 3; ace 3; HOAc 3 1.4894[20]
11862	Tricyclo[2.2.1.02,6]heptane-1-carboxylic acid, 7,7-dimethyl- Tricyclenic acid	$C_{10}H_{14}O_2$ 166.22	512-60-7 151.3	263	4-09-00-00246	bz 4; EtOH 4; peth 4
11863	Tricyclo[2.2.1.02,6]heptane-1-carboxylic acid, 7,7-dimethyl-, ethyl ester Tricyclenic acid, ethyl ester	$C_{12}H_{18}O_2$ 194.27	56694-98-5 100-1[10]		2-09-00-00064 1.0143[20]	1.4230[20]
11864	Tricyclo[2.2.1.02,6]heptane-1-carboxylic acid, 7,7-dimethyl-, methyl ester Tricyclenic acid, methyl ester	$C_{11}H_{16}O_2$ 180.25	56694-97-4 45.5	99[14]	2-09-00-00064 1.0255[42]	1.4695[42]
11865	Tricyclo[2.2.1.02,6]heptane, 3,3-dimethyl-	C_9H_{14} 122.21	473-02-9 42.5	138	4-05-00-00430 0.8710[40]	bz 4; eth 4; EtOH 4 1.4514[40]
11866	Tricyclo[2.2.1.02,6]heptane, 1,3,3-trimethyl- Cyclofenchene	$C_{10}H_{16}$ 136.24	488-97-1	145	3-05-00-00392 0.859[20]	1.4503[22]
11867	Tricyclo[2.2.1.02,6]heptane, 1,7,7-trimethyl-	$C_{10}H_{16}$ 136.24	508-32-7 67.5	152.5	4-05-00-00468 0.8668[80]	1.4296[80]
11868	1,12-Tridecadiyne	$C_{13}H_{20}$ 176.30	38628-39-6 -3	115.5[12]	2-01-00-00248 0.8262[21]	1.454[20]
11869	Tridecanal	$C_{13}H_{26}O$ 198.35	10486-19-8 14	156[13]	4-01-00-03386 0.8356[18]	H_2O 1; EtOH 3; os 4 1.4384[18]
11870	1-Tridecanamine Tridecane, 1-amino	$C_{13}H_{29}N$ 199.38	2869-34-3 27.4	275.8	4-04-00-00810 0.8049[20]	H_2O 2; EtOH 3; eth 3 1.4443[20]
11871	Tridecane	$C_{13}H_{28}$ 184.37	629-50-5 -5.3	235.4	4-01-00-00512 0.7564[20]	H_2O 1; EtOH 4; eth 4; ctc 3 1.4256[20]
11872	Tridecane, 1-bromo-	$C_{13}H_{27}Br$ 263.26	765-09-3 6.2	296	4-01-00-00514 1.0177[20]	H_2O 1; chl 4 1.4593[20]
11873	Tridecane, 1-bromo-12-methyl- Tridecane, 2-methyl-13-bromo-	$C_{14}H_{29}Br$ 277.29	111772-90-8	120-2[3]	3-01-00-00551 1.0241[20]	1.4598[20]
11874	Tridecane, 1,13-dibromo- Tridecamethylene dibromide	$C_{13}H_{26}Br_2$ 342.16	31772-05-1 9	188-92[12]	3-01-00-00548 1.276[15]	eth 4; chl 4 1.4880[27]
11875	1,12-Tridecanediol	$C_{13}H_{28}O_2$ 216.36	99706-68-0 60.5	188[8]	2-01-00-00563	EtOH 4
11876	1,13-Tridecanediol	$C_{13}H_{28}O_2$ 216.36	13362-52-2 76.5	195[10]	4-01-00-02630	EtOH 4

No.	Name / Synonym	Mol. Form. / Mol. Wt.	CAS RN / mp/°C	Merck No. / bp/°C	Beil. Ref. / den/g cm⁻³	Solubility / n_D
11877	Tridecane, 2,2,4,10,12,12-hexamethyl-7-(3,5,5-trimethylhexyl)-	$C_{28}H_{58}$ 394.77	3035-75-4	166[0.5]	4-01-00-00590 0.8137[20]	1.4558[20]
11878	Tridecane, 7-hexyl-	$C_{19}H_{40}$ 268.53	7225-66-3 -28.3	170.5[10]	4-01-00-00562 0.7877[20]	1.4409[20]
11879	Tridecane, 7-methyl- 7-Methyltridecane	$C_{14}H_{30}$ 198.39	26730-14-3 -37.2	115.5[10]	4-01-00-00525 0.7634[20]	1.4291[20]
11880	Tridecanenitrile	$C_{13}H_{25}N$ 195.35	629-60-7 9.7	293	3-02-00-00906 0.8257[20]	EtOH 4; eth 4 1.4378[20]
11881	Tridecanoic acid	$C_{13}H_{26}O_2$ 214.35	638-53-9 41.5	236[100]; 140[1]	4-02-00-01117 0.8458[80]	H₂O 1; EtOH 4; eth 4; ace 3 1.4286[60]
11882	Tridecanoic acid, methyl ester Methyl tridecanoate	$C_{14}H_{28}O_2$ 228.38	1731-88-0 6.5	92[1]	4-02-00-01118	EtOH 5; ctc 3 1.4405[20]
11883	1-Tridecanol Tridecyl alcohol	$C_{13}H_{28}O$ 200.36	112-70-9 32.5	152[14]	4-01-00-01860 0.8223[31]	H₂O 1; EtOH 3; eth 3
11884	2-Tridecanol 2-Hydroxytridecane	$C_{13}H_{28}O$ 200.36	1653-31-2 23	161[30]; 145[12]	4-01-00-01861	bz 4; eth 4; EtOH 4 1.4188[70]
11885	3-Tridecanol	$C_{13}H_{28}O$ 200.36	10289-68-6 32	139[12]	4-01-00-01861 0.8139[46]	
11886	1-Tridecanol, 12-methyl- 12-Methyl-1-tridecanol	$C_{14}H_{30}O$ 214.39	21987-21-3 10.5	145-50[6]	4-01-00-01869 0.8414[20]	1.4437[20]
11887	3-Tridecanol, 2-methyl-	$C_{14}H_{30}O$ 214.39	98930-89-3 20	274	1-01-00-00219 0.8390[20]	1.4460[20]
11888	4-Tridecanol, 2-methyl Acetic acid, trifluoro-,(1,2-dimethyl-1-ethylbutyl) ester	$C_{14}H_{30}O$ 214.39	36691-77-7	147-8[14]	4-01-00-01869	chl 2 1.4404[25]
11889	2-Tridecanone Methyl undecyl ketone	$C_{13}H_{26}O$ 198.35	593-08-8 30.5	263	4-01-00-03386 0.8217[30]	H₂O 1; EtOH 4; eth 4; ace 4 1.4318[20]
11890	3-Tridecanone	$C_{13}H_{26}O$ 198.35	1534-26-5 31	140[17]	4-01-00-03387	H₂O 1; ace 3
11891	7-Tridecanone Dihexyl ketone	$C_{13}H_{26}O$ 198.35	462-18-0 33	261	4-01-00-03315 0.825[30]	EtOH 3; eth 4; chl 3; lig 3
11892	3-Tridecanone, 2-methyl- Decyl isopropyl ketone	$C_{14}H_{28}O$ 212.38	40239-35-8 47	267; 142[17]	4-01-00-03390 0.8314[20]	
11893	1-Tridecene	$C_{13}H_{26}$ 182.35	2437-56-1 -13	232.8	4-01-00-00921 0.7658[20]	H₂O 1; EtOH 4; eth 4; bz 3 1.4340[20]
11894	2-Tridecenoic acid	$C_{13}H_{24}O_2$ 212.33	6969-16-0	167-71[2]	4-02-00-01622 0.8995[30]	1.4612[20]
11895	12-Tridecenoic acid (10-Undecenyl)acetic acid	$C_{13}H_{24}O_2$ 212.33	6006-06-0 38.5	192[20]	2-02-00-00422	bz 4; eth 4; EtOH 4; peth 4
11896	12-Tridecenoic acid, methyl ester	$C_{14}H_{26}O_2$ 226.36	29780-00-5	143[8]	4-02-00-01622 0.8803[20]	1.4438[20]
11897	1-Tridecyne	$C_{13}H_{24}$ 180.33	26186-02-7	94.5[25]	4-01-00-01069 0.7729[20]	bz 4; eth 4 1.4309[20]
11898	Tridemorph Morpholine, 2,6-dimethyl-4-tridecyl-	$C_{19}H_{39}NO$ 297.52	24602-86-6	9576 141[1.3]	0.86	
11899	Tridiphane (±)-2-(3,5-Dichlorophenyl)-2-(2,2,2-trichloroethyl)oxirane	$C_{10}H_7Cl_5O$ 320.43	58138-08-2 42.8			
11900	Triflumizole (E)-1-[1-[[4-Chloro-2-(trifluoromethyl)phenyl]imino]-2-propoxyethyl]-1H-imidazole	$C_{15}H_{15}ClF_3N_3O$ 345.75	68694-11-1 63.5			
11901	Trifluralin Benzenamine, 2,6-dinitro-N,N-dipropyl-4-(trifluoromethyl)-	$C_{13}H_{16}F_3N_3O_4$ 335.28	1582-09-8 49	9598 140[4.2]		
11902	Triforine Formamide, N,N'-[1,4-piperazinediylbis(2,2,2-trichloroethylidene)]bis-	$C_{10}H_{14}Cl_6N_4O_2$ 434.96	26644-46-2 155 dec	9602		
11903	2,5,7-Trioxabicyclo[2.2.1]heptane, 1,3,3,4,6,6-hexamethyl- Bicylo[2.2.1]heptane, 1,3,3,4,6,6-hexamethyl-2,5,7-trioxa	$C_{10}H_{18}O_3$ 186.25	7045-89-8	166	5-19-09-00139 0.9702[17]	H₂O 1
11904	2H,4aH-1,4,5-Trioxadicyclopent[a,hi]indene-7-carboxylic acid, 3-ethylidene-3,3a,7a,9b-tetrahydro-2-oxo-, methyl ester, [3aS-(3E,3aα,4aβ,7aβ,9aR*,9bβ)]- Plumericin	$C_{15}H_{14}O_6$ 290.27	77-16-7	7512	5-19-11-00097	chl 3
11905	1,3,5-Trioxane Formaldehyde, trimer	$C_3H_6O_3$ 90.08	110-88-3 60.2	9646 114.5	5-19-09-00103 1.17[65]	H₂O 4; EtOH 3; eth 3; bz 3
11906	1,3,5-Trioxane, 2,4,6-triethenyl- Metacrolein	$C_9H_{12}O_3$ 168.19	588-01-2 50	170	5-19-09-00165	eth 4; EtOH 4
11907	1,3,5-Trioxane-2,4,6-triimine Cyamelide	$C_3H_3N_3O_3$ 129.08	462-02-2 dec	2689 dec	3-03-00-00030 1.127[15]	eth 4; EtOH 4
11908	1,3,5-Trioxane, 2,4,6-trimethyl- Paraldehyde	$C_6H_{12}O_3$ 132.16	123-63-7 12.6	6975 124.3	5-19-09-00112 0.9943[20]	H₂O 2; EtOH 5; eth 5; chl 5 1.4049[20]

No.	Name / Synonym	Mol. Form. / Mol. Wt.	CAS RN / mp/°C	Merck No. / bp/°C	Beil. Ref. / den/g cm^{-3}	Solubility / n_D
11909	2,6,7-Trioxa-1-phosphabicyclo[2.2.2]octane, 4-ethyl- / Trimethylolpropane phosphite	$C_6H_{11}O_3P$ / 162.13	824-11-3 / 53.7			chl 3
11910	Triphenylene / 9,10-Benzophenanthrene	$C_{18}H_{12}$ / 228.29	217-59-4 / 199	9654 / 425	4-05-00-02556	H_2O 1; EtOH 3; bz 4; chl 4
11911	Triphenyltin hydroxide / Stannane, hydroxytriphenyl-	$C_{18}H_{16}OSn$ / 367.03	76-87-9 / 119	9659	1.54^{20}	
11912	Tripropylamine, eicosafluoro-2-(trifluoromethyl)- / Dipropylisobutylamine, perfluoro	$C_{10}F_{23}N$ / 571.08	307-11-9 / 147		4-02-00-00858 / 1.84^{25}	eth 4; EtOH 4 / 1.283^{25}
11913	Trisiloxane, 1,5-dichloro-1,1,3,3,5,5-hexamethyl-	$C_6H_{18}Cl_2O_2Si_3$ / 277.37	3582-71-6 / -53	/ 184	4-04-00-04123 / 1.018^{20}	H_2O 6
11914	Trisiloxane, octamethyl- / Octamethyltrisiloxane	$C_8H_{24}O_2Si_3$ / 236.53	107-51-7 / -80	6669 / 153; 50-2^{17}	4-04-00-04115 / 0.8200^{20}	EtOH 2; bz 3; peth 3 / 1.3840^{20}
11915	Trisiloxane, 1,1,3,5,5-pentamethyl-1,3,5-triphenyl- / 1,1,3,5,5-Pentamethyl-1,3,5-triphenyltrisiloxane	$C_{23}H_{30}O_2Si_3$ / 422.75	80-14-8 / 169$^{0.7}$		3-16-00-01210 / 1.0227^{20}	bz 4; lig 4 / 1.5280^{20}
11916	Trisulfide, diethyl	$C_4H_{10}S_3$ / 154.32	3600-24-6 / 96-7^{26}		4-01-00-01398 / 1.114^{20}	1.5689^{13}
11917	Trisulfide, di-2-propenyl	$C_6H_{10}S_3$ / 178.34	2050-87-5 / 112-22^{16}		0-01-00-00441 / 1.0845^{15}	eth 4
11918	Tritetracontane	$C_{43}H_{88}$ / 605.17	7098-21-7 / 85.5	332^3	3-01-00-00593 / 0.7812^{90}	1.4340^{90}
11919	1,3,5-Trithiane	$C_3H_6S_3$ / 138.28	291-21-4 / 220	sub	5-19-09-00105 / 1.6374^{24}	H_2O 2; EtOH 2; eth 2; bz 3
11920	1,3,5-Trithiane, 2,4,6-trimethyl- / Thioacetaldehyde trimer	$C_6H_{12}S_3$ / 180.36	2765-04-0	9245	4-19-00-04718	
11921	1,3,5-Trithiane, 2,4,6-trimethyl-, (2α,4α,6β)-	$C_6H_{12}S_3$ / 180.36	23769-39-3 / 101	/ 246.5	2-19-00-00399	H_2O 1; EtOH 3; eth 3; ace 3
11922	L-Tryptophan / l-α-Aminoindole-3-propionic acid	$C_{11}H_{12}N_2O_2$ / 204.23	73-22-3 / 289	9707	5-22-14-00014	H_2O 2; EtOH 3; eth 1; chl 1
11923	L-Tryptophan, N-methyl- / L-Abrine	$C_{12}H_{14}N_2O_2$ / 218.26	526-31-8 / 295 dec	5	5-22-14-00040	H_2O 2; MeOH 2; eth 1; alk 3
11924	Tubocuraran-7',12'-diol, 6,6'-dimethoxy-2,2'-dimethyl-, (1'α)- / Bebeerine	$C_{36}H_{38}N_2O_6$ / 594.71	477-60-1 / 221	1024	2-27-00-00896	EtOH 3; eth 3; ace 4; chl 4
11925	Tubocuraran-7',12'-diol, 6,6'-dimethoxy-2,2'-dimethyl-, (1β)- / Curine	$C_{36}H_{38}N_2O_6$ / 594.71	436-05-5 / 221	2682	4-27-00-08725	ace 4; bz 4; py 4
11926	L-Tyrosine / L-Phenylalanine, 4-hydroxy-	$C_9H_{11}NO_3$ / 181.19	60-18-4 / 343 dec	9747 / sub	4-14-00-02264	H_2O 2; EtOH 1; eth 1; HOAc 2
11927	L-Tyrosine, 3,5-dibromo- / Tyrosine, 3,5-dibromo(L)	$C_9H_9Br_2NO_3$ / 338.98	300-38-9 / 245	3014	4-14-00-02365	H_2O 2; EtOH 2; eth 1; alk 3
11928	Tyrosine, 3,5-diiodo- / 3,5-Diiodotyrosine	$C_9H_9I_2NO_3$ / 432.98	66-02-4 / 213 dec	3177	2-14-00-00384	H_2O 2; EtOH 1; eth 1; bz 1
11929	L-Tyrosine, 3-hydroxy- / Levodopa	$C_9H_{11}NO_4$ / 197.19	59-92-7 / 277 dec	5344	4-14-00-02492	H_2O 3; EtOH 1; eth 1; ace 1
11930	L-Tyrosine, O-(4-hydroxy-3,5-diiodophenyl)-3,5-diiodo- / Thyroxine	$C_{15}H_{11}I_4NO_4$ / 776.87	51-48-9 / 235.5	9348	4-14-00-02373	H_2O 2; EtOH 1; bz 1
11931	L-Tyrosine, 3-hydroxy-α-methyl- / Methyldopa	$C_{10}H_{13}NO_4$ / 211.22	555-30-6 / 300 dec	5974		
11932	L-Tyrosine, 3-iodo- / 3-Iodotyrosine	$C_9H_{10}INO_3$ / 307.09	70-78-0 / 204-6 dec	4937	4-14-00-02369	
11933	L-Tyrosine, N-methyl- / Surinamine	$C_{10}H_{13}NO_3$ / 195.22	537-49-5 / 293	8988	3-14-00-01513	
11934	1,10-Undecadiyne	$C_{11}H_{16}$ / 148.25	4117-15-1 / -17	83^{12}	2-01-00-00248 / 0.8182^{21}	ace 4; bz 4 / 1.453^{21}
11935	Undecanal	$C_{11}H_{22}O$ / 170.30	112-44-7 / -4	117^{18}	4-01-00-03374 / 0.8251^{23}	H_2O 1; EtOH 3; eth 3 / 1.4520^{20}
11936	Undecanal, 2-methyl-	$C_{12}H_{24}O$ / 184.32	110-41-8 /	114^{10}	4-01-00-03383 / 0.830^{15}	H_2O 2; EtOH 3; eth 3 / 1.4321^{20}
11937	1-Undecanamine	$C_{11}H_{25}N$ / 171.33	7307-55-3 / 17	242	4-04-00-00792 / 0.7979^{20}	H_2O 3; EtOH 3; eth 1; ctc 2 / 1.4398^{20}
11938	Undecane / Hendecane	$C_{11}H_{24}$ / 156.31	1120-21-4 / -25.6	195.9	4-01-00-00487 / 0.7402^{20}	H_2O 1; EtOH 5; eth 5 / 1.4398^{20}
11939	Undecane, 1-bromo-11-fluoro- / 1-Bromo-11-fluoroundecane	$C_{11}H_{22}BrF$ / 253.20	463-33-2 /	95$^{0.6}$	4-01-00-00490	eth 4; EtOH 4 / 1.4518^{25}
11940	Undecane, 1,11-dibromo- / 1,11-Dibromoundecane	$C_{11}H_{22}Br_2$ / 314.10	16696-65-4 / 166	170-5^{15}	3-01-00-00536 / 1.332^{15}	
11941	Undecanenitrile / Decyl cyanide	$C_{11}H_{21}N$ / 167.29	2244-07-7 /	253	3-02-00-00860 / 0.8254^{30}	H_2O 1; EtOH 3; eth 3; ctc 3 / 1.4293^{30}
11942	1-Undecanethiol / Undecyl mercaptan	$C_{11}H_{24}S$ / 188.38	5332-52-5 / -3	257.4	4-01-00-01838 / 0.8448^{20}	1.4585^{20}
11943	Undecanoic acid	$C_{11}H_{22}O_2$ / 186.29	112-37-8 / 28.6	280	4-02-00-01068 / 0.8907^{20}	H_2O 1; EtOH 4; eth 3; ace 4 / 1.4294^{45}

No.	Name / Synonym	Mol. Form. / Mol. Wt.	CAS RN / mp/°C	Merck No. / bp/°C	Beil. Ref. / den/g cm⁻³	Solubility / n_D
11944	Undecanoic acid, 11-bromo- / 11-Bromoundecanoic acid	$C_{11}H_{21}BrO_2$ / 265.19	2834-05-1 / 57	188[18]	4-02-00-01073	ace 4; bz 4; eth 4; EtOH 4
11945	Undecanoic acid, 2-bromo- / 2-Bromoundecanoic acid	$C_{11}H_{21}BrO_2$ / 265.19	2623-84-9 / 10	180[14]	4-02-00-01073 / 1.1586[24]	
11946	Undecanoic acid, ethyl ester / Ethyl undecylate	$C_{13}H_{26}O_2$ / 214.35	627-90-7 / -15	131[14]	3-02-00-00858 / 0.8633[20]	H_2O 1; EtOH 3; eth 3; ace 3 / 1.4285[20]
11947	Undecanoic acid, 11-fluoro-	$C_{11}H_{21}FO_2$ / 204.28	463-17-2 / 36	113[0.25]	4-02-00-01071	eth 4; EtOH 4; lig 4
11948	Undecanoic acid, 11-hydroxy- / 11-Hydroxyundecanoic acid	$C_{11}H_{22}O_3$ / 202.29	3669-80-5 / 65.5	148-50[4]	4-03-00-00903	eth 4; EtOH 4
11949	Undecanoic acid, 11-hydroxy-, methyl ester / Methyl 11-hydroxyhendecanoate	$C_{12}H_{24}O_3$ / 216.32	24724-07-0 / 27.5	168[8]	2-03-00-00243 / 0.9542[20]	bz 4; eth 4; EtOH 4
11950	1-Undecanol	$C_{11}H_{24}O$ / 172.31	112-42-5 / 19	243	4-01-00-01835 / 0.8298[20]	H_2O 1; EtOH 3; eth 4 / 1.4392[20]
11951	2-Undecanol, (±)-	$C_{11}H_{24}O$ / 172.31	113666-64-1 / 12	228	4-01-00-01838 / 0.8268[19]	H_2O 1; EtOH 3; eth 3 / 1.4369[20]
11952	3-Undecanol, (R)-	$C_{11}H_{24}O$ / 172.31	107494-37-1 / 17	229	2-01-00-00462 / 0.8295[20]	bz 4; eth 4; EtOH 4 / 1.4367[20]
11953	5-Undecanol / 1-Butylheptyl alcohol	$C_{11}H_{24}O$ / 172.31	37493-70-2 / -3.5	229	4-01-00-01839 / 0.8292[20]	eth 4; EtOH 4 / 1.4354[24]
11954	6-Undecanol	$C_{11}H_{24}O$ / 172.31	23708-56-7 / 25	228	4-01-00-01839 / 0.8334[20]	H_2O 1; EtOH 3; ace 3 / 1.4374[20]
11955	1-Undecanol, 2-methyl- / 2-Methyl-1-undecanol	$C_{12}H_{26}O$ / 186.34	10522-26-6	129-31[12]	4-01-00-01855 / 0.8300[15]	eth 4; EtOH 4 / 1.4382[20]
11956	3-Undecanol, 2-methyl- / 3-Hydroxy-2-methylundecane	$C_{12}H_{26}O$ / 186.34	60671-36-5	135[26]	1-01-00-00217 / 0.8327[20]	1.4405[20]
11957	5-Undecanol, 2-methyl-	$C_{12}H_{26}O$ / 186.34	33978-71-1	130-3[20]	4-01-00-01855 / 0.8235[20]	eth 4; EtOH 4 / 1.4346[20]
11958	2-Undecanone / Methyl nonyl ketone	$C_{11}H_{22}O$ / 170.30	112-12-9 / 15	231.5	4-01-00-03374 / 0.8250[20]	H_2O 1; EtOH 3; eth 3; ace 3 / 1.4291[20]
11959	3-Undecanone / Ethyl octyl ketone	$C_{11}H_{22}O$ / 170.30	2216-87-7 / 12	227	4-01-00-03375 / 0.8272[20]	H_2O 1; EtOH 3; eth 3; ctc 2 / 1.4296[20]
11960	4-Undecanone / Heptyl propyl ketone	$C_{11}H_{22}O$ / 170.30	14476-37-0 / 4.5	106[13]	4-01-00-03375 / 0.8274[25]	eth 4; EtOH 4 / 1.4248[24]
11961	5-Undecanone / Butyl hexyl ketone	$C_{11}H_{22}O$ / 170.30	33083-83-9 / 2	227	4-01-00-03375 / 0.8278[19]	H_2O 1; EtOH 3; eth 3; ctc 3 / 1.4275[18]
11962	6-Undecanone / Diamyl ketone	$C_{11}H_{22}O$ / 170.30	927-49-1 / 14.5	228	4-01-00-03376 / 0.8308[20]	H_2O 1; EtOH 4; eth 4 / 1.4270[20]
11963	4-Undecanone, 7-ethyl-2-methyl-	$C_{14}H_{28}O$ / 212.38	6976-00-7	252.5	3-01-00-02920 / 0.8362[20]	H_2O 1; EtOH 4; eth 4; ace 4 / 1.4370[20]
11964	Undecasiloxane, tetracosamethyl- / Tetracosamethylhendecasiloxane	$C_{24}H_{72}O_{10}Si_{11}$ / 829.77	107-53-9	9128 / 322.8; 202[47]	3-04-00-01881 / 0.9247[25]	bz 4 / 1.3994[20]
11965	3,5,9-Undecatrien-2-one, 6,10-dimethyl- / Pseudoionone	$C_{13}H_{20}O$ / 192.30	141-10-6	7931 / 143-5[12]	3-01-00-03067 / 0.8984[20]	EtOH 3; eth 3; chl 3; MeOH 3 / 1.5335[20]
11966	1-Undecene	$C_{11}H_{22}$ / 154.30	821-95-4 / -49.2	192.7	4-01-00-00910 / 0.7503[20]	H_2O 1; eth 3; chl 3; lig 3 / 1.4261[20]
11967	3-Undecene	$C_{11}H_{22}$ / 154.30	60669-40-1 / -62	193.5	0.7516[20]	1.4290[20]
11968	2-Undecene, (E)-	$C_{11}H_{22}$ / 154.30	693-61-8 / -48.3	192.5	4-01-00-00911 / 0.7528[20]	1.4292[20]
11969	2-Undecene, (Z)-	$C_{11}H_{22}$ / 154.30	821-96-5 / -66.5	196.1	0.7576[20]	
11970	4-Undecene, (E)-	$C_{11}H_{22}$ / 154.30	693-62-9 / -63.7	193	0.7508[20]	1.4285[20]
11971	4-Undecene, (Z)-	$C_{11}H_{22}$ / 154.30	821-98-7 / -97	192.6	0.7541[20]	1.4302[20]
11972	5-Undecene, (E)-	$C_{11}H_{22}$ / 154.30	764-97-6 / -61.1	192	4-01-00-00911 / 0.7497[20]	eth 4; chl 4; lig 4 / 1.4285[20]
11973	5-Undecene, (Z)-	$C_{11}H_{22}$ / 154.30	764-96-5 / -106.5	192.3	4-01-00-00911 / 0.7537[20]	1.4302[20]
11974	10-Undecenoic acid / Undecylenic acid	$C_{11}H_{20}O_2$ / 184.28	112-38-9 / 24.5	9760 / 275	4-02-00-01612 / 0.9072[24]	H_2O 1; EtOH 3; eth 3; ctc 2 / 1.4486[24]
11975	9-Undecenoic acid, (E)- / trans-9-Undecenoic acid	$C_{11}H_{20}O_2$ / 184.28	37973-84-5 / 19	273	4-02-00-01616 / 0.9118[25]	eth 4 / 1.4519[20]
11976	9-Undecenoic acid, (Z)-	$C_{11}H_{20}O_2$ / 184.28	116836-30-7 / 2.5	156[4]; 119[0.8]	4-02-00-01616 / 0.9150[20]	1.4530[20]
11977	10-Undecenoic acid, ethyl ester	$C_{13}H_{24}O_2$ / 212.33	692-86-4 / -38	264.5	4-02-00-01614 / 0.8827[15]	H_2O 1; EtOH 3; eth 3; ctc 2 / 1.4449[25]
11978	10-Undecenoic acid, methyl ester	$C_{12}H_{22}O_2$ / 198.31	111-81-9 / -27.5	248	4-02-00-01613 / 0.889[15]	H_2O 1; EtOH 3; eth 3; ctc 2 / 1.4393[20]
11979	10-Undecen-1-ol	$C_{11}H_{22}O$ / 170.30	112-43-6 / -2	250	4-01-00-02194 / 0.8495[15]	H_2O 1; EtOH 3; eth 3; ctc 2 / 1.4500[20]
11980	2-Undecen-4-ol	$C_{11}H_{22}O$ / 170.30	22381-86-8	114-5[13]	4-01-00-02194 / 0.843[22]	1.4485[22]
11981	9-Undecen-1-ol	$C_{11}H_{22}O$ / 170.30	112-46-9 / 6	248.5	3-01-00-01957 / 0.8507[15]	1.4535[19]
11982	10-Undecenoyl chloride	$C_{11}H_{19}ClO$ / 202.72	38460-95-6	120-2[10]	4-02-00-01615 / 0.944[20]	1.454[20]
11983	1-Undecen-3-yne	$C_{11}H_{18}$ / 150.26	74744-28-8	74[9]	3-01-00-01054 / 0.7962[20]	1.4606[20]

No.	Name / Synonym	Mol. Form. / Mol. Wt.	CAS RN / mp/°C	Merck No. / bp/°C	Beil. Ref. / den/g cm^{-3}	Solubility / n_D
11984	1-Undecyne	$C_{11}H_{20}$ 152.28	2243-98-3 -25	196	4-01-00-01064 0.7728[20]	ace 4; bz 4; eth 4; EtOH 4 1.4306[20]
11985	2-Undecyne	$C_{11}H_{20}$ 152.28	60212-29-5 -30.1	204.2	0-01-00-00261 0.7827[20]	1.4391[20]
11986	4-Undecyne	$C_{11}H_{20}$ 152.28	60212-31-9 -74.7	198.5	3-01-00-01022 0.7752[20]	1.4369
11987	5-Undecyne	$C_{11}H_{20}$ 152.28	2294-72-6 -74.1	198	4-01-00-01064 0.7753[20]	1.4369[20]
11988	6-Undecynoic acid	$C_{11}H_{18}O_2$ 182.26	55182-83-7	124-5[0.17]	4-02-00-01739 0.9537[25]	1.4566[25]
11989	9-Undecynoic acid	$C_{11}H_{18}O_2$ 182.26	22202-65-9 61	170[15.] 132[0.1]	3-02-00-01471	H$_2$O 4; eth 4; EtOH 4
11990	9-Undecynoic acid, methyl ester Methyl 9-undecynoate	$C_{12}H_{20}O_2$ 196.29	18937-76-3	93-4[1]	3-02-00-01472 0.9245[17]	1.4732[19]
11991	10-Undecynoic acid, methyl ester Methyl 10-undecynoate	$C_{12}H_{20}O_2$ 196.29	2777-66-4	121[15]	4-02-00-01738 0.9177[20]	1.4421[21]
11992	Urea Carbamide	CH_4N_2O 60.06	57-13-6 132.7	9781 dec	4-03-00-00094 1.3230[20]	H$_2$O 4; EtOH 4; eth 1; bz 1 1.484
11993	Urea, (aminoiminomethyl)-	$C_2H_6N_4O$ 102.10	141-83-3 105	160 dec	4-03-00-00155	H$_2$O 3; EtOH 2; eth 1; bz 1
11994	Urea, N,N'-bis(2-chloroethyl)-N-nitroso- Carmustine	$C_5H_9Cl_2N_3O_2$ 214.05	154-93-8 31	1852		H$_2$O 4; EtOH 4
11995	Urea, N,N'-bis(hydroxymethyl)- Oxymethurea	$C_3H_8N_2O_3$ 120.11	140-95-4 126	6921 149[25]	4-03-00-00107	H$_2$O 3; EtOH 3; eth 1; MeOH 3
11996	Urea, N,N-bis(2-methylpropyl)- 1,1-Diisobutylurea	$C_9H_{20}N_2O$ 172.27	77464-06-3 73	180[25]	0-04-00-00170	eth 4; EtOH 4
11997	Urea, N,N'-bis(4-nitrophenyl)- 4,4'-Dinitrocarbanilide	$C_{13}H_{10}N_4O_5$ 302.25	587-90-6 312 dec	3271	4-12-00-01646	
11998	Urea, (4-bromophenyl)-	$C_7H_7BrN_2O$ 215.05	1967-25-5 226	260	4-12-00-01511	H$_2$O 2; EtOH 4; eth 4; bz 4
11999	Urea, N'-(4-chlorophenyl)-N,N-dimethyl- Monuron	$C_9H_{11}ClN_2O$ 198.65	150-68-5 170.5	6169	4-12-00-01191	H$_2$O 1; EtOH 2; ace 2
12000	Urea, N'-(3,4-dichlorophenyl)-N,N-dimethyl- Diuron	$C_9H_{10}Cl_2N_2O$ 233.10	330-54-1 158	3388	4-12-00-01263	
12001	Urea, N,N'-diethyl-	$C_5H_{12}N_2O$ 116.16	623-76-7 112.5	263	4-04-00-00370 1.0415[25]	H$_2$O 4; EtOH 4; eth 4 1.4616[40]
12002	Urea, N,N-diethyl-	$C_5H_{12}N_2O$ 116.16	634-95-7 75	94-6[0.02]	4-04-00-00380	H$_2$O 4; EtOH 4; eth 3; bz 4
12003	Urea, N,N'-diethyl-N,N'-diphenyl- N,N'-Diethylcarbanilide	$C_{17}H_{20}N_2O$ 268.36	85-98-3 79	3107	4-12-00-00844	H$_2$O 1; EtOH 4; chl 3
12004	Urea, N,N'-dimethyl-	$C_3H_8N_2O$ 88.11	96-31-1 108	269	4-04-00-00207 1.142[25]	H$_2$O 4; EtOH 4; eth 1; chl 2
12005	Urea, N,N-dimethyl-	$C_3H_8N_2O$ 88.11	598-94-7 183.5		4-04-00-00224 1.2555[25]	H$_2$O 3; EtOH 4; eth 1; tfa 2
12006	Urea, N,N'-dimethyl-N,N'-diphenyl-	$C_{15}H_{16}N_2O$ 240.30	611-92-7 122	350	4-12-00-00839	H$_2$O 4; EtOH 4; eth 2; ace 4
12007	Urea, N,N-dimethyl-N'-[3-(trifluoromethyl)phenyl]- Fluometuron	$C_{10}H_{11}F_3N_2O$ 232.21	2164-17-2 164	4080		ace 4; EtOH 4
12008	Urea, N,N'-di-1-naphthalenyl-	$C_{21}H_{16}N_2O$ 312.37	607-56-7 296	sub	4-12-00-03078	py 4
12009	Urea, (2,5-dioxo-4-imidazolidinyl)- Allantoin	$C_4H_6N_4O_3$ 158.12	97-59-6 239	246	4-25-00-04071	H$_2$O 2; EtOH 3; eth 1; MeOH 1
12010	Urea, N,N'-diphenyl- Carbanilide	$C_{13}H_{12}N_2O$ 212.25	102-07-8 236	1787 262	4-12-00-00741 1.239[25]	H$_2$O 2; EtOH 2; eth 3; ace 2
12011	Urea, N,N-diphenyl- 1,1-Diphenylurea	$C_{13}H_{12}N_2O$ 212.25	603-54-3 189	dec	4-12-00-00852 1.276[25]	H$_2$O 2; EtOH 3; eth 1; chl 3
12012	Urea, N,N'-dipropyl-	$C_7H_{16}N_2O$ 144.22	623-95-0 105	255	4-04-00-00482	eth 4; EtOH 4
12013	Urea, (4-ethoxyphenyl)- Dulcin	$C_9H_{12}N_2O_2$ 180.21	150-69-6 173.5	3448 dec	4-13-00-01154	H$_2$O 2; EtOH 3; AcOEt 4
12014	Urea, ethyl-	$C_3H_8N_2O$ 88.11	625-52-5 92.5	dec	4-04-00-00369 1.2130[18]	H$_2$O 4; EtOH 4; eth 3; bz 4
12015	Urea, N-ethyl-N-nitroso- N-Ethyl-N-nitrosourea	$C_3H_7N_3O_2$ 117.11	759-73-9 98-100 dec	3788	3-04-00-00233	chl 3
12016	Urea, hydroxy- Hydroxyurea	$CH_4N_2O_2$ 76.06	127-07-1 141	4785 dec	4-03-00-00170	H$_2$O 4
12017	Urea, methyl-	$C_2H_6N_2O$ 74.08	598-50-5 103	dec	4-04-00-00205 1.2040[0]	H$_2$O 4; EtOH 4; eth 1; bz 1
12018	Urea, N-methyl-N,N'-diphenyl-	$C_{14}H_{14}N_2O$ 226.28	612-01-1 106	203	4-12-00-00839	H$_2$O 2; EtOH 2; eth 4; bz 3
12019	Urea, N-methyl-N-nitroso- Methylnitrosourea	$C_2H_5N_3O_2$ 103.08	684-93-5 123-4 dec		4-04-00-03385	H$_2$O 2; EtOH 2; eth 2
12020	Urea, monohydrochloride Urea hydrochloride	CH_5ClN_2O 96.52	506-89-8 145 dec	9782	4-03-00-00094	H$_2$O 3
12021	Urea, mononitrate Urea nitrate	$CH_5N_3O_4$ 123.07	124-47-0 152 dec	9784	4-03-00-00094 1.690[20]	EtOH 4

No.	Name / Synonym	Mol. Form. / Mol. Wt.	CAS RN / mp/°C	Merck No. / bp/°C	Beil. Ref. / den/g cm^{-3}	Solubility / n_D
12022	Urea, 1-naphthalenyl- / 1-Naphthylurea	$C_{11}H_{10}N_2O$ / 186.21	6950-84-1 / 219.5		4-12-00-03076	H_2O 1; EtOH 4; eth 3
12023	Urea, 2-naphthalenyl- / 2-Naphthylurea	$C_{11}H_{10}N_2O$ / 186.21	13114-62-0 / 219		4-12-00-03149	EtOH 4
12024	Urea, nitro- / Nitrourea	$CH_3N_3O_3$ / 105.05	556-89-8 / 158-9 dec	6573	4-03-00-00248	ace 4; EtOH 4
12025	Urea, (4-nitrophenyl)- / 4-Nitrophenylurea	$C_7H_7N_3O_3$ / 181.15	556-10-5 / 238	6547	4-12-00-01645	H_2O 4; EtOH 4
12026	Urea, phenyl- / Phenylurea	$C_7H_8N_2O$ / 136.15	64-10-8 / 147	7290 / 238	4-12-00-00734 / 1.302^{25}	H_2O 2; EtOH 3; eth 2; AcOEt 3
12027	Urea, (phenylmethyl)- / Benzylurea	$C_8H_{10}N_2O$ / 150.18	538-32-9 / 148	1165 / 200 dec	4-12-00-02253	ace 4; EtOH 4
12028	Urea, 2-propenyl- / Allylurea	$C_4H_8N_2O$ / 100.12	557-11-9 / 85	296	4-04-00-01070	H_2O 5; EtOH 5; eth 2; chl 2
12029	Urea, tetraethyl- / Urea, 1,1,3,3-tetraethyl-	$C_9H_{20}N_2O$ / 172.27	1187-03-7 /	209	4-04-00-00380 / 0.919^{20}	H_2O 1; alk 1; acid 1 / 1.4474^{20}
12030	Urea, tetramethyl- / Tetramethylurea	$C_5H_{12}N_2O$ / 116.16	632-22-4 / -1.2	9160 / 176.5	4-04-00-00225 / 0.9687^{20}	EtOH 2; eth 2; ctc 2 / 1.4496^{23}
12031	Urea, tetraphenyl- / Tetraphenylurea	$C_{25}H_{20}N_2O$ / 364.45	632-89-3 / 183		4-12-00-00853 / 1.222^{25}	H_2O 1; EtOH 2
12032	Urea, trimethyl- / Trimethylurea	$C_4H_{10}N_2O$ / 102.14	632-14-4 / 75.5	232.5	4-04-00-00224 / 1.1900^{20}	H_2O 3; EtOH 3; eth 2; bz 2
12033	Uridine / 1-β-D-Ribofuranosyluracil	$C_9H_{12}N_2O_6$ / 244.20	58-96-8 / 165	9792	5-24-06-00132	H_2O 3; EtOH 3; py 3
12034	Uridine, 2'-deoxy-5-fluoro- / Floxuridine	$C_9H_{11}FN_2O_5$ / 246.20	50-91-9 /	4045	5-24-06-00291	
12035	5'-Uridylic acid / Uridine 5'-phosphoric acid	$C_9H_{13}N_2O_9P$ / 324.18	58-97-9 / 202 dec	9796	5-24-06-00173	H_2O 4; MeOH 3
12036	Urs-12-ene-27,28-dioic acid, 3-hydroxy-, (3β)- / Quinovic acid	$C_{30}H_{46}O_5$ / 486.69	465-74-7 / 298 dec	8104	4-10-00-02197	
12037	Urs-12-en-28-oic acid, 3-hydroxy-, (3β)- / Ursolic acid	$C_{30}H_{48}O_3$ / 456.71	77-52-1 / 284	9802	4-10-00-01160	ace 4; eth 4; chl 4
12038	Urs-12-en-3-ol, (3β)- / α-Amyrin	$C_{30}H_{50}O$ / 426.73	638-95-9 / 186	653 / 243$^{0.5}$	4-06-00-04191	EtOH 3; eth 3; bz 3; chl 3
12039	Valeric acid, 3-chloro-, ethyl ester	$C_7H_{13}ClO_2$ / 164.63	6513-13-9 /	189	3-02-00-00678 / 1.0330^{20}	eth 4; EtOH 4 / 1.4278^{20}
12040	DL-Valine	$C_5H_{11}NO_2$ / 117.15	516-06-3 / 298 dec	9818 / sub	3-04-00-01370 / 1.316^{25}	H_2O 3; EtOH 2; eth 2; bz 2
12041	L-Valine / 2-Aminoisovaleric acid	$C_5H_{11}NO_2$ / 117.15	72-18-4 / 315	9818 / sub	4-04-00-02659 / 1.23^{25}	H_2O 4
12042	D-Valine, 3-[(2-amino-2-carboxyethyl)dithio]-, (R)- / Penicillamine cysteine disulfide	$C_8H_{16}N_2O_4S_2$ / 268.36	18840-45-4 / 195	7030	4-04-00-03231	
12043	D-Valine, 3-mercapto- / Penicillamine	$C_5H_{11}NO_2S$ / 149.21	52-67-5 / 198.5	7029	4-04-00-03228	
12044	Veratraman-3,23-diol, 14,15,16,17-tetradehydro-, (3β,23β)- / Veratramine	$C_{27}H_{39}NO_2$ / 409.61	60-70-8 / 206	9853	5-21-05-00297	EtOH 3; bz 3; chl 3; dil acid 3
12045	Veratraman-11-one, 17,23-epoxy-3-(β-D-glucopyranosyloxy)-, (3β,23β)- / Pseudojervine	$C_{33}H_{49}NO_8$ / 587.75	36069-05-3 / 304-5 dec	7932	4-27-00-03592	H_2O 1; EtOH 3; eth 1; bz 1
12046	Veratraman-11-one, 17,23-epoxy-3-hydroxy-, (3β,23β)- / Jervine	$C_{27}H_{39}NO_3$ / 425.61	469-59-0 / 243 dec	5145	4-27-00-03590	H_2O 1; EtOH 3; eth 2; ace 3
12047	Vernolate / Carbamothioic acid, dipropyl-, S-propyl ester	$C_{10}H_{21}NOS$ / 203.35	1929-77-7 /	9866 / 150^{30}	/ 0.952^{20}	
12048	Vinclozolin / 2,4-Oxazolidinedione, 3-(3,5-dichlorophenyl)-5-ethenyl-5-methyl-	$C_{12}H_9Cl_2NO_3$ / 286.11	50471-44-8 / 108	9890 / 131$^{0.05}$	/ 1.51	
12049	9H-Xanthene / 10H-9-Oxaanthracene	$C_{13}H_{10}O$ / 182.22	92-83-1 / 100.5	311	5-17-02-00252	H_2O 1; EtOH 2; eth 3; bz 3
12050	9H-Xanthen-9-one / Xanthone	$C_{13}H_8O_2$ / 196.21	90-47-1 / 174	9971 / 351; 146^3	5-17-10-00430	H_2O 1; EtOH 3; eth 3; bz 3
12051	9H-Xanthen-9-one, 1,3-dihydroxy-	$C_{13}H_8O_4$ / 228.20	3875-68-1 / 259	/ sub	5-18-04-00004	EtOH 4
12052	9H-Xanthen-9-one, 2,7-dihydroxy- / Xanthone, 2,7-dihydroxy	$C_{13}H_8O_4$ / 228.20	64632-72-0 / 330	/ sub	1-18-00-00357	eth 4; EtOH 4
12053	9H-Xanthen-9-one, 3,6-dihydroxy-	$C_{13}H_8O_4$ / 228.20	1214-24-0 / 350 dec	/ sub	5-18-04-00010	EtOH 4; HOAc 4
12054	9H-Xanthen-9-one, 1,7-dihydroxy-3-methoxy- / Gentisin	$C_{14}H_{10}O_5$ / 258.23	437-50-3 / 266.5	4291	5-18-04-00497	H_2O 1; EtOH 4; ace 1; py 3
12055	peri-Xanthenoxanthene / Dinaphthalene dioxide	$C_{20}H_{10}O_2$ / 282.30	191-28-6 / 243.5	400$^{20-25}$	5-19-02-00242	bz 3; chl 3
12056	Xanthylium, 9-(2-carboxyphenyl)-3,6-bis(diethylamino)-, chloride / Rhodamine, B	$C_{28}H_{32}ClN_2O_3$ / 480.03	81-88-9 / 165	8182	5-19-08-00669	H_2O 3; EtOH 3; eth 3; bz 3
12057	o-Xylene, 4-propyl-	$C_{11}H_{16}$ / 148.25	3982-66-9 / -20	208.9	3-05-00-01005 / 0.8715^{20}	ace 4; bz 4; eth 4; EtOH 4 / 1.5000^{20}

No.	Name Synonym	Mol. Form. Mol. Wt.	CAS RN mp/°C	Merck No. bp/°C	Bell. Ref. den/g cm⁻³	Solubility n_D
12058	Xylitol Xylite	C₅H₁₂O₅ 152.15	87-99-0 93.5	9992 216	4-01-00-02832	H₂O 4; py 4; EtOH 4
12059	*D*-Xylose	C₅H₁₀O₅ 150.13	58-86-6 90.5	9995	4-01-00-04223 1.525²⁰	H₂O 4; EtOH 3; eth 2
12060	Yohimban-16-carboxylic acid, 11,17-dimethoxy-18-[[1-oxo-3-(3,4,5-trimethoxyphenyl)-2-propenyl]oxy]-, methyl ester, (3β,16β,17α,18β,20α)- Rescinnamine	C₃₅H₄₂N₂O₉ 634.73	24815-24-5 238.5	8146	4-25-00-01323	H₂O 1; EtOH 2; ace 3; chl 3
12061	Yohimban-16-carboxylic acid, 11,17-dimethoxy-18-[(3,4,5-trimethoxybenzoyl)oxy]-, methyl ester, Reserpine	C₃₃H₄₀N₂O₉ 608.69	50-55-5 264.5	8149	4-25-00-01319	H₂O 2; EtOH 3; eth 2; ace 2
12062	Yohimban-16-carboxylic acid, 17-hydroxy-, methyl ester, (16α,17α)- Yohimbine	C₂₁H₂₆N₂O₃ 354.45	146-48-5 241	10011 sub 160	4-25-00-01237	H₂O 2; EtOH 3; eth 3; bz 2
12063	Yohimban-16-carboxylic acid, 18-hydroxy-11,17-dimethoxy-, (3β,16β,17α,18β,20 Reserpic acid	C₂₂H₂₈N₂O₅ 400.47	83-60-3 242	8147	4-25-00-01305	
12064	Yohimban-16-carboxylic acid, 17-hydroxy-18-[(3,4,5-trimethoxybenzoyl)oxy]-, methyl ester, (3β,16β Raunescine	C₃₁H₃₈N₂O₈ 564.64	117-73-7 165	8130	4-25-00-01281	H₂O 1; EtOH 3; chl 3; HOAc 3
12065	Yohimban-16-carboxylic acid, 17-methoxy-18-[(3,4,5-trimethoxybenzoyl)oxy]-, methyl ester, (3β,16β Deserpidine	C₃₂H₃₈N₂O₈ 578.66	131-01-1 230.5	2901	4-25-00-01282	H₂O 1; EtOH 3; chl 3
12066	Zinc, dimethyl Dimethyl zinc	C₂H₆Zn 95.46	544-97-8 -40	3252 46	1.386¹⁰	eth 3; peth 5
12067	Zinc, [[1,2-ethanediylbis[carbamodithioato]] (2-)]- Zineb	C₄H₆N₂S₄Zn 275.76	12122-67-7 157 dec	10071		
12068	Ziram Zinc, bis(dimethylcarbamodithioato-*S,S'*)-, (T-4)-	C₆H₁₂N₂S₄Zn 305.83	137-30-4			

STRUCTURAL FORMULAS OF ORGANIC COMPOUNDS

In numeric order as they occur in the Table of Physical Constants of Organic Compounds

47

48

49

50

51

52

53

54

55

56

57

58

59

60

61

62

63

64

65

66

67

68

69

70

71

72

73

74

75

76

77

78

79

80

81

82

83

84

85

86

87

88

89

90

91

92

93

95

96

97

98

99

100

101

102

103

104

105

107

108

109

111

112

113

114

115

116

117

118

119

120

121

123

124

125

126

127

128

129

130

131

132

141

150

161

169

133

142

151

162

170

134

143

152

163

171

135

144

153

164

172

136

145

154

165

173

137

146

157

166

174

138

147

158

167

175

139

148

159

168

176

140

149

160

177

178

179

180

181

182

183

184

185

186

187

188

189

190

191

192

193

194

195

196

197

198

199

200

201

202

203

204

205

206

207

208

209

211

212

213

214

215

216

217

218

219

220

221

222

223

224

225

STRUCTURAL FORMULAS OF ORGANIC COMPOUNDS (continued)

In numeric order as they occur in the Table of Physical Constants of Organic Compounds

226

227

228

229

230

231

232

233

235

236

238

239

240

241

242

243

244

245

246

247

248

249

250

252

253

254

255

256

257

258

259

260

261

262

263

264

265

266

267

268

269

270

271

272

273

274

277

278

279

280

281

282

283

284

285

286

287

288

289

290

291

292

293

294

295

296

297

298

299

300

301

302

303

304

306

312

313

314

315

316

317

318

319

320

321

322

323

324

325

326

327

329

331

332

333

334

335

336

337

338

339

340

341

342

343

344

345

346

347

348

349

350

351

352

353

354

355

356

357

358

359

360

362

372

382

390

398

364

373

383

391

399

365

374

384

392

400

366

375

385

393

401

367

376

394

402

368

377

386

395

403

369

379

387

396

404

370

380

388

397

405

371

381

389

406

407

408

409

410

411

412

413

414

415

416

417

418

419

420

421

422

423

424

425

426

427

428

429

430

431

432

433

434

435

436

437

438

439

440

441

442

443

444

445

446

447

448

449

450

451

452

453

454

455

456

457

458

459

460

461

462

463

464

465

466

467

468

470

471

472

473

474

475

476

477

478

479

480

481

482

483

484

485

486

487

488

489

490

491

492

493

494

495

496

497

498

499

500

501

502

503

504

505

506

507

509

510

511

512

513

514

515

516

517

518

519

520

521

522

523

524

525

526

527

528

529

530

531

532

533

535

536

537

538

539

540

541

542

543

544

545

546

547

548

549

550

551

552

553

554

555

556

557

558

559

560

561

562

563

564

565

566

567

568

569

570

571

572

573

574

575

576

577

578

579

580

581

582

583

584

585

586

587

588

589

590

591

592

593

594

595

596

597

598

599

600

601

603

604

605

606

607

608

609

610

611

612

613

614

615

616

617

618

619

620

621

622

623

624

625

626

627

628

629

630

631

632

633

635

636

637

638

639

640

641

642

643

644

645

646

647

648

649

650

651

652

653

654

655

656

657

658

659

660

661

662

663

666

667

668

669

670

671

672

673

674

675

676

677

678

679

680

681

682

683

684

685

686

687

688

689

690

692

693

694

695

696

697

698

699

700

701

702

703

704

705

706

707

708

709

710

711

712

713

714

715

716

718

719

720

721

722

723

724

725

726

727

728

729

730

731

732

733

734

735

736

737

738

740

741

742

743

744

745

746

747

748

749

750

751

752

753

754

755

756

757

758

759

760

761

762

763

764

765

766

767

768

769

770

771

772

773

775

776

777

778

779

780

781

782

783

784

785

786

787

788

789

790

791

793

794

795

796

797

798

799

800

801

802

803

804

805

806

807

808

809

810

811

812

813

814

815

816

817

818

819

820

822

823

824

825

826

828

829

830

831

832

833

834

835

836

837

838

839

840

841

842

843

844

845

846

847

848

849

850

851

852

853

854

855

856

857

858

859

860

861

862

863

864

867

868

869

870

871

872

873

874

875

877

878

879

880

881

882

883

884

886

887

888

890

891

892

893

894

895

896

897

898

899

900

901

902

903

904

905

906

907

908

909

910

911

912

913

914

915

916

917

918

919

920

921

923

924

925

926

927

928

930

931

932

933

934

935

937

938

939

940

941

942

944

945

946

947

948

949

951

952

953

954

956

957

959

960

961

962

963

964

965

966

967

968

969

970

971

972

973

974

975

976

977

978

985

993

1001

1010

979

986

994

1002

1011

980

987

995

1003

1012

981

988

996

1004

1013

982

989

997

1005

1014

983

990

998

1006

1015

991

999

1007

1016

984

992

1000

1008

1017

1025

1032

1040

1047

1018

1026

1033

1041

1048

1019

1027

1034

1042

1049

1020

1028

1035

1043

1050

1021

1029

1036

1044

1051

1022

1030

1037

1045

1052

1023

1031

1038

1046

1053

1024

1039

1054

1055

1056

1057

1058

1059

1060

1061

1062

1063

1064

1065

1066

1067

1068

1069

1070

1071

1072

1073

1074

1075

1076

1077

1078

1079

1080

1081

1082

1083

1084

1085

1086

1087

1088

1089

1090

1091

1092

1093

1094

1095

1096

1097

1098

1099

1100

1101

1102

1103

1104

1105

1106

1108

1109

1110

1111

1112

1113

1114

1115

1116

1117

1118

1119

1120

1121

1122

1123

1124

1125

1126

1127

1128

1129

1130

1131

1132

1133

1134

1143

1153

1161

1171

1135

1144

1154

1162

1172

1136

1145

1155

1163

1173

1137

1146

1156

1164

1175

1138

1147

1157

1166

1176

1139

1148

1158

1167

1177

1140

1149

1159

1168

1178

1141

1150

1160

1169

1179

1142

1151

1170

1180

1181

1191

1200

1208

1216

1182

1192

1201

1209

1217

1183

1193

1202

1210

1218

1184

1194

1203

1211

1219

1185

1195

1204

1212

1220

1186

1196

1205

1213

1221

1187

1197

1206

1214

1222

1189

1198

1207

1215

1223

1190

1199

1224

1225

1226

1227

1228

1229

1231

1232

1234

1235

1236

1237

1238

1239

1240

1241

1242

1243

1244

1245

1246

1247

1248

1249

1250

1251

1252

1253

1254

1255

1256

1257

1258

1259

1260

1261

1262

1263

1264

1265

1266

1267

1275

1281

1288

1297

1268

1276

1289

1298

1282

1270

1277

1290

1299

1283

1271

1278

1291

1300

1284

1272

1279

1292

1301

1285

1273

1293

1302

1286

1280

1294

1274

1287

1295

1303

1304

1305

1306

1307

1308

1309

1310

1311

1312

1313

1314

1315

1316

1317

1318

1319

1320

1321

1322

1323

1324

1325

1327

1328

1329

1330

1331

1332

1333

1334

1335

1336

1337

1338

1339

1340

1341

1342

1343

1351

1359

1368

1377

1344

1352

1361

1369

1378

1345

1353

1362

1370

1379

1346

1354

1363

1371

1380

1347

1355

1364

1372

1381

1348

1356

1365

1373

1382

1349

1357

1366

1374

1383

1350

1358

1367

1375

1384

1376

1385

1386

1387

1388

1389

1390

1391

1392

1393

1394

1395

1396

1397

1398

1399

1400

1401

1402

1403

1404

1405

1406

1407

1408

1409

1410

1411

1412

1413

1414

1415

1416

1417

1418

1419

1420

1421

1422

1423

1424

1425

1426

1427

1428

1429

1430

1431

1439

1449

1459

1468

1432

1440

1450

1460

1469

1433

1441

1451

1461

1470

1434

1442

1452

1462

1471

1435

1443

1453

1463

1472

1436

1444

1454

1464

1473

1437

1445

1455

1465

1474

1438

1447

1456

1466

1475

1448

1457

1467

1476

1458

1477

1478

1488

1495

1503

1512

1479

1489

1496

1504

1513

1480

1490

1497

1505

1514

1481

1491

1498

1506

1515

1482

1492

1499

1507

1516

1483

1493

1500

1508

1517

1484

1494

1501

1509

1518

1486

1502

1510

1519

1487

1511

1520

1521

1522

1523

1524

1525

1526

1527

1528

1529

1530

1531

1532

1533

1534

1535

1536

1537

1538

1539

1540

1541

1542

1543

1544

1545

1546

1547

1548

1549

1550

1551

1552

1553

1554

1555

1556

1557

1558

1560

1561

1562

1563

1564

1565

1566

1567

1568

1569

1570

1571

1572

1573

1574

1575

1576

1577

1578

1579

1580

1581

1583

1584

1585

1586

1587

1588

1589

1590

1591

1592

1593

1594

1595

1596

1597

1604

1613

1621

1631

1598

1606

1614

1622

1632

1599

1607

1615

1623

1633

1600

1608

1616

1624

1634

1601

1609

1617

1626

1635

1602

1610

1618

1627

1636

1603

1612

1619

1628

1637

1620

1629

1638

1630

1639

1640

1641

1642

1643

1644

1645

1646

1647

1648

1649

1650

1651

1652

1653

1654

1655

1656

1657

1658

1659

1660

1661

1662

1663

1664

1665

1666

1667

1668

1669

1670

1671

1672

1673

1674

1675

1676

1677

1678

1679

1680

1681

1682

1683

1684

1693

1702

1710

1719

1685

1694

1703

1711

1720

1686

1695

1704

1712

1721

1687

1696

1705

1713

1722

1689

1697

1706

1714

1723

1690

1698

1707

1715

1724

1691

1699

1708

1716

1725

1700

1709

1717

1692

1701

1718

1726

1727

1728

1729

1730

1731

1732

1733

1734

1735

1736

1737

1738

1739

1740

1741

1742

1743

1744

1745

1746

1747

1748

1749

1750

1751

1752

1753

1754

1755

1756

1757

1758

1759

1760

1761

1762

1763

1764

1765

1766

1767

1768

1769

1770

1771

1772

1773

1774

1775

1776

1777

1778

1779

1780

1781

1782

1783

1784

1785

1786

1787

1788

1789

1790

1791

1792

1793

1794

1795

1796

1797

1798

1799

1800

1801

1802

1803

1804

1805

1806

1807

1808

1809

1810

1818

1827

1835

1843

1811

1819

1828

1836

1844

1812

1820

1829

1837

1845

1813

1821

1830

1838

1846

1814

1822

1831

1839

1847

1815

1823

1832

1840

1848

1816

1826

1833

1841

1849

1817

1834

1842

1850

1857

1864

1872

1880

1851

1858

1865

1873

1881

1852

1859

1866

1874

1882

1853

1860

1867

1875

1883

1854

1861

1868

1876

1884

1855

1862

1869

1877

1885

1856

1863

1870

1878

1886

1871

1879

1887

1895

1903

1911

1919

1888

1896

1904

1912

1920

1889

1897

1905

1913

1921

1890

1898

1906

1914

1923

1891

1899

1907

1915

1924

1892

1900

1908

1916

1925

1893

1901

1909

1917

1926

1894

1902

1910

1918

1968

1969

1970

1971

1972

1973

1974

1975

1976

1977

1978

1979

1980

1981

1982

1983

1984

1985

1986

1987

1988

1989

1990

1991

1992

1993

1995

1996

1997

1998

1999

2000

2001

2002

2003

2004

2005

2006

2007

2009

2010

2011

2012

2013

2014

2015

2016

2017

2018

2019

2021

2022

2023

2024

2025

2026

2027

2028

2029

2030

2031

2033

2034

2035

2036

2037

2038

2039

2040

2041

2042

2044

2045

2047

2048

2049

2050

2051

2052

2053

2054

2055

2056

2057

2058

2059

2060

2061

2062

2063

2064

2065

2066

2067

2068

2069

2070

2071

2072

2073

2074

2075

2076

2077

2078

2079

2080

2081

2083

2084

2085

2086

2087

2088

2089

2090

2091

2092

2093

2094

2095

2097

2098

2099

2100

2101

2102

2103

2104

2105

2106

2107

2108

2110

2111

2113

2114

2116

2117

2118

2119

2121

2122

2123

2124

2125

2126

2127

2129

2130

2131

2132

2133

2134

2137

2138

2139

2140

2141

2142

2143

2144

2145

2146

2147

2148

2149

2150

2151

2152

2153

2154

2155

2156

2157

2158

2159

2160

2161

2162

2163

2164

2165

2166

2167

2168

2169

2170

2171

2172

2173

2174

2175

2176

2177

2178

2179

2181

2182

2183

2184

2185

2186

2187

2188

2189

2230

2231

2232

2233

2234

2235

2236

2237

2238

2239

2240

2242

2243

2244

2245

2246

2247

2248

2249

2250

2251

2252

2253

2254

2255

2256

2257

2258

2259

2260

2261

2262

2264

2265

2266

2267

2268

2269

2270

2271

2272

2276

2277

2278

2279

2280

2281

2283

2284

2285

2286

2287

2289

2290

2291

2292

2293

2294

2295

2296

2297

2299

2300

2301

2302

2303

2304

2305

2306

2308

2309

2310

2311

2312

2313

2314

2316

2324

2331

2339

2347

2317

2325

2332

2340

2348

2318

2326

2333

2341

2352

2319

2327

2334

2342

2354

2335

2343

2355

2321

2328

2344

2336

2356

2322

2329

2345

2357

2330

2337

2346

2358

2323

2359

2360

2367

2375

2383

2391

2361

2368

2376

2384

2392

2362

2369

2377

2385

2393

2363

2370

2378

2386

2394

2364

2371

2379

2387

2395

2365

2372

2380

2388

2396

2366

2373

2381

2389

2397

2374

2382

2390

2398

2399

2400

2401

2402

2403

2404

2405

2406

2407

2408

2409

2410

2411

2412

2413

2414

2415

2416

2417

2418

2419

2420

2421

2422

2423

2424

2425

2426

2427

2428

2429

2430

2431

2432

2434

2435

2436

2437

2438

2439

2440

2441

2442

2443

2444

2445

2446

2447

2448

2449

2451

2452

2453

2454

2455

2456

2457

2458

2459

2460

2461

2462

2463

2464

2465

2466

2468

2469

2470

2471

2472

2474

2475

2476

2477

2478

2479

2480

2481

2482

2483

2484

2485

2486

2487

2488

2489

2490

2491

2492

2493

2494

2495

2496

2497

2498

2499

2500

2501

2502

2503

2504

2505

2506

2507

2508

2509

2510

2511

2512

2513

2514

2515

2516

2517

2518

2519

2520

2521

2522

2523

2524

2525

2526

2527

2528

2529

2530

2532

2533

2534

2535

2536

2537

2538

2539

2540

2541

2542

2543

2544

2545

2546

2547

2548

2549

2550

2551

2552

2553

2554

2555

2556

2557

2558

2559

2560

2561

2562

2563

2564

2565

2566

2567

2568

2569

2570

2571

2572

2573

2574

2575

2576

2577

2578

2579

2580

2581

2582

2583

2584

2585

2586

2587

2588

2589

2590

2591

2592

2593

2594

2595

2596

2597

2598

2599

2600

2601

2602

2603

2604

2605

2606

2607

2608

2609

2610

2611

2612

2613

2614

2615

2616

2617

2618

2619

2620

2621

2622

2624

2625

2626

2627

2628

2629

2630

2631

2632

2633

2634

2635

2636

2637

2638

2639

2640

2641

2642

2643

2644

2645

2646

2647

2648

2649

2650

2651

2652

2653

2654

2655

2656

2657

2658

2659

2660

2661

2662

2663

2664

2665

2666

2667

2668

2669

2670

2671

2672

2673

2674

2675

2676

2677

2678

2686

2693

2700

2707

2679

2687

2694

2701

2708

2680

2688

2695

2702

2709

2681

2689

2696

2703

2710

2682

2690

2697

2704

2711

2683

2691

2698

2705

2712

2684

2692

2699

2706

2713

2685

2714

2715

2716

2717

2718

2719

2720

2721

2722

2723

2724

2726

2727

2728

2729

2730

2731

2732

2733

2734

2735

2736

2737

2738

2739

2740

2741

2742

2744

2745

2746

2747

2748

2749

2750

2751

2752

2753

2754

2755

2756

2757

2758

2759

2760

2761

2762

2763

2764

2765

2766

2767

2768

2769

2770

2771

2772

2773

2774

2775

2776

2777

2778

2780

2781

2782

2783

2784

2785

2787

2788

2789

2790

2791

2792

2793

2795

2797

2798

2799

2800

2801

2802

2803

2804

2805

2806

2807

2808

2809

2810

2811

2812

2813

2814

2815

2816

2817

2818

2819

2820

2822

2825

2826

2827

2828

2829

2830

2831

2832

2833

2834

2835

2836

2837

2838

2847

2856

2865

2873

2839

2849

2866

2874

2857

2840

2850

2858

2867

2875

2841

2851

2859

2868

2876

2842

2852

2860

2869

2877

2843

2853

2861

2870

2878

2844

2854

2862

2871

2879

2845

2855

2863

2872

2880

2864

2881

2882

2883

2884

2885

2886

2887

2888

2889

2890

2891

2892

2893

2894

2895

2896

2897

2898

2899

2900

2901

2902

2903

2904

2905

2906

2907

2908

2909

2910

2911

2912

2913

2914

2915

2916

2918

2919

2928

2937

2951

2960

2920

2929

2939

2952

2961

2921

2930

2941

2953

2962

2922

2932

2944

2954

2963

2923

2933

2945

2955

2964

2925

2934

2946

2957

2966

2926

2935

2947

2958

2967

2927

2936

2948

2959

2968

2950

2969

2970

2971

2972

2973

2974

2975

2976

2977

2978

2980

2981

2982

2983

2984

2985

2986

2987

2988

2989

2990

2991

2992

2993

2994

2995

2996

2997

2998

2999

3000

3001

3002

3003

3004

3005

3007

3008

3009

3010

3011

3013

3014

3015

3016

3017

3018

3019

3020

3021

3022

3023

3024

3025

3026

3027

3028

3029

3030

3031

3032

3033

3034

3035

3037

3038

3039

3040

3041

3042

3043

3044

3045

3046

3050

3051

3052

3053

3054

3055

3056

3057

3058

3059

3060

3061

3062

3063

3064

3065

3066

3070

3071

3072

3073

3074

3075

3076

3077

3078

3079

3080

3081

3082

3083

3084

3085

3086

3087

3089

3090

3091

3092

3093

3094

3095

3096

3098

3099

3100

3101

3102

3103

3104

3105

3108

3109

3111

3112

3113

3114

3115

3116

3117

3118

3119

3120

3121

3122

3123

3124

3125

3126

3127

3128

3129

3130

3131

3132

3134

3135

3136

3137

3138

3139

3140

3141

3143

3144

3145

3146

3147

3148

3150

3151

3152

3153

3154

3155

3156

3157

3158

3159

3160

3161

3162

3163

3164

3166

3167

3168

3169

3170

3171

3172

3173

3174

3175

3176

3177

3178

3179

3180

3181

3182

3183

3184

3185

3186

3188

3189

3190

3191

3192

3194

3195

3196

3197

3198

3199

3200

3201

3202

3203

3205

3206

3207

3208

3209

3210

3211

3212

3213

3214

3215

3216

3217

3218

3220

3221

3222

3223

3224

3225

3226

3227

3228

3229

3230

STRUCTURAL FORMULAS OF ORGANIC COMPOUNDS (continued)

In numeric order as they occur in the Table of Physical Constants of Organic Compounds

3231

3232

3233

3234

3235

3236

3237

3238

3239

3240

3241

3242

3243

3244

3245

3246

3247

3248

3249

3250

3251

3252

3253

3254

3255

3256

3257

3258

3259

3262

3263

3264

3265

3266

3267

3268

3269

3270

3271

3272

3273

3274

3275

3276

3284

3292

3301

3311

3277

3285

3293

3302

3312

3278

3286

3294

3303

3313

3279

3287

3295

3304

3316

3280

3288

3296

3305

3317

3281

3289

3297

3307

3318

3298

3282

3290

3299

3308

3319

3283

3291

3300

3309

3320

3310

3321

3322

3323

3324

3325

3326

3327

3328

3329

3330

3331

3332

3333

3334

3336

3339

3340

3341

3342

3343

3344

3345

3346

3347

3348

3349

3350

3351

3352

3353

3354

3355

3356

3357

3359

3360

3361

3362

3363

3364

3373

3383

3395

3404

3374

3384

3396

3405

3365

3385

3406

3375

3397

3366

3376

3407

3387

3398

3367

3377

3388

3399

3408

3378

3368

3390

3400

3369

3379

3409

3391

3401

3370

3381

3392

3410

3402

3371

3382

3403

3411

3394

3372

STRUCTURAL FORMULAS OF ORGANIC COMPOUNDS (continued)

In numeric order as they occur in the Table of Physical Constants of Organic Compounds

3412

3414

3415

3416

3417

3418

3419

3420

3421

3422

3423

3426

3427

3428

3429

3430

3431

3433

3434

3435

3436

3437

3438

3439

3440

3441

3442

3443

3444

3445

3446

3447

3448

3449

3450

3451

3452

3453

3454

3455

3456

3457

3458

3459

3470

3479

3488

3498

3460

3471

3480

3489

3499

3461

3472

3481

3490

3500

3462

3473

3482

3492

3502

3464

3474

3483

3493

3503

3465

3475

3484

3494

3504

3467

3476

3485

3495

3505

3468

3477

3486

3497

3506

3469

3478

3487

3507

3508

3509

3510

3511

3512

3513

3514

3515

3516

3517

3518

3519

3520

3521

3522

3523

3524

3525

3526

3527

3528

3529

3530

3531

3532

3533

3534

3535

3537

3538

3539

3540

3541

3542

3543

3544

3545

3546

3547

3548

3549

3551

3552

3553

3563

3573

3582

3594

3555

3565

3574

3584

3595

3556

3566

3575

3585

3596

3557

3567

3576

3586

3597

3558

3568

3577

3587

3598

3560

3569

3578

3588

3599

3570

3590

3561

3571

3579

3601

3562

3572

3580

3592

3604

3593

3581

STRUCTURAL FORMULAS OF ORGANIC COMPOUNDS (continued)

In numeric order as they occur in the Table of Physical Constants of Organic Compounds

3605

3616

3624

3632

3641

3606

3617

3633

3642

3608

3618

3625

3634

3643

3626

3609

3619

3627

3635

3644

3636

3612

3620

3628

3637

3645

3613

3621

3629

3638

3646

3614

3622

3630

3639

3647

3615

3623

3631

3640

3648

3649

3650

3651

3653

3654

3655

3656

3657

3658

3659

3660

3661

3662

3663

3664

3665

3666

3667

3668

3669

3670

3671

3673

3675

3676

3677

3678

3679

3680

3681

3682

3683

3684

3685

3688

3690

3691

3692

3693

3695

3696

3697

3698

3699

3700

3702

3704

3705

3706

3707

3708

3709

3710

3711

3712

3713

3714

3715

3716

3717

3718

3720

3721

3722

3723

3724

3725

3726

3727

3728

3729

3730

3731

3732

3733

3734

3735

3736

3737

3738

3747

3755

3763

3773

3739

3748

3756

3764

3774

3740

3749

3757

3765

3775

3741

3750

3758

3766

3776

3742

3751

3759

3767

3777

3743

3752

3760

3769

3778

3744

3753

3761

3770

3779

3745

3754

3762

3771

3781

3746

3772

3782

3783

3792

3800

3808

3817

3784

3793

3809

3819

3785

3794

3801

3810

3820

3786

3795

3802

3812

3821

3787

3796

3803

3813

3822

3788

3797

3804

3814

3823

3789

3798

3805

3815

3824

3790

3799

3806

3816

3825

3791

3807

3826

3827

3828

3829

3830

3831

3832

3833

3834

3835

3836

3837

3838

3839

3840

3841

3842

3843

3844

3845

3846

3847

3848

3849

3850

3851

3852

3856

3857

3858

3859

3860

3861

3862

3864

3865

3866

3868

3869

3870

3871

3873

3874

3875

3887

3895

3905

3915

3877

3888

3896

3907

3916

3878

3889

3897

3908

3917

3879

3890

3898

3909

3918

3891

3899

3910

3919

3883

3892

3900

3911

3920

3884

3893

3901

3912

3921

3885

3894

3902

3913

3922

3886

3904

3914

3923

3933

3942

3954

3964

3925

3943

3955

3966

3934

3926

3944

3956

3969

3935

3927

3947

3957

3970

3936

3929

3948

3959

3971

3937

3960

3930

3949

3972

3938

3961

3939

3950

3931

3973

3940

3962

3953

3932

3963

3941

3976

3978

3979

3980

3982

3989

3990

3991

3992

3993

3994

3995

3997

3998

3999

4000

4002

4003

4005

4006

4007

4008

4009

4012

4013

4014

4015

4016

4017

4018

4019

4020

4021

4022

4023

4024

4025

4028

4029

4030

4031

4032

4033

4034

4044

4066

4077

4084

4035

4046

4067

4078

4085

4036

4079

4086

4037

4047

4068

4087

4080

4038

4048

4069

4081

4088

4071

4039

4049

4082

4040

4050

4089

4075

4083

4090

4065

4043

4076

4091

4092

4093

4094

4095

4096

4097

4098

4099

4100

4101

4102

4103

4106

4107

4108

4110

4111

4113

4116

4118

4119

4120

4121

4122

4123

4124

4125

4126

4127

4128

4130

4133

4134

4137

4138

4139

4140

4141

4142

4143

4144

4146

4147

4150

4151

4152

4153

4154

4155

4156

4157

4159

4161

4162

4163

4164

4165

4166

4167

4168

4169

4170

4171

4172

4173

4174

4175

4176

4177

4178

4179

4180

4181

4182

4183

4184

4185

4186

4187

4188

4189

4190

4191

4192

4193

4194

4195

4196

4197

4198

4199

4200

4201

4202

4203

4204

4205

4206

4207

4208

4209

4210

4211

4212

4213

4214

4215

4216

4217

4218

4219

4220

4221

4222

4223

4224

4225

3-428

4226

4227

4228

4229

4230

4231

4233

4234

4235

4236

4237

4238

4239

4240

4241

4242

4243

4244

4245

4246

4247

4248

4249

4250

4251

4252

4253

4254

4255

4256

4257

4258

4259

4260

4261

4262

4263

4264

4265

4266

4267

4268

4269

4270

4271

4273

4274

4275

4276

4277

4278

4279

4280

4281

4282

4283

4284

4285

4286

4287

4288

4289

4290

4291

4292

4293

4294

4295

4296

4297

4298

4299

4300

4301

4302

4303

4304

4305

4306

4307

4308

4309

4310

4311

4312

4313

4314

4315

4316

4317

4318

4319

4320

4322

4323

4324

4325

4326

4327

4328

4329

4330

4331

4332

4333

4334

4335

4336

4337

4339

4340

4341

4342

4343

4344

4345

4346

4347

4348

4349

4350

4351

4352

4353

4354

4356

4357

4358

4359

4360

4361

4371

4380

4389

4398

4362

4372

4381

4390

4399

4373

4382

4391

4400

4364

4374

4383

4392

4401

4365

4384

4393

4402

4375

4385

4394

4403

4366

4376

4386

4404

4367

4395

4387

4396

4405

4368

4377

4388

4397

4406

4369

4378

4370

4379

4407

4408

4409

4410

4411

4413

4414

4415

4416

4417

4418

4419

4420

4421

4422

4423

4424

4425

4426

4427

4428

4429

4430

4431

4432

4433

4434

4435

4436

4437

4438

4439

4440

4441

4442

4443

4444

4445

4446

4447

4448

4449

4450

4451

4452

4453

4454

4455

4467

4477

4488

4498

4456

4468

4478

4489

4499

4457

4469

4479

4490

4500

4458

4470

4480

4491

4501

4459

4471

4481

4492

4502

4460

4472

4482

4493

4503

4461

4473

4483

4494

4504

4462

4474

4484

4495

4463

4475

4485

4496

4505

4464

4476

4486

4497

4465

4487

4466

4507

4508

4509

4510

4511

4513

4514

4515

4516

4517

4518

4519

4521

4522

4523

4525

4526

4527

4528

4529

4530

4531

4532

4533

4534

4535

4536

4537

4538

4539

4540

4541

4542

4543

4544

4545

4546

4547

4548

4549

4550

4551

4552

4553

4554

4563

4573

4582

4592

4593

4555

4564

4574

4583

4594

4556

4565

4575

4585

4595

4576

4586

4557

4566

4577

4587

4596

4558

4568

4578

4588

4597

4559

4569

4579

4589

4598

4570

4580

4590

4599

4560

4571

4562

4591

4600

4572

4581

4601

4602

4603

4604

4605

4606

4607

4608

4609

4610

4611

4612

4613

4614

4615

4616

4617

4618

4619

4620

4621

4622

4623

4624

4625

4626

4627

4628

4629

4630

4631

4632

4633

4634

4635

4636

4637

4638

4639

4640

4642

4643

4644

4645

4646

4647

4648

4649

4650

4651

4652

4653

4654

4655

4656

4657

4658

4659

4660

4661

4662

4663

4664

4665

4666

4667

4668

4669

4670

4671

4672

4673

4674

4675

4676

4677

4678

4679

4680

4681

4682

4683

4684

4685

4686

4687

4688

4689

4690

4691

4692

4694

4695

4696

4697

4698

4699

4700

4701

4702

4703

4704

4705

4706

4707

4708

4709

4710

4711

4712

4713

4714

4715

4716

4718

4719

4720

4721

4722

4723

4724

4725

4726

4727

4728

4729

4731

4732

4733

4734

4735

4736

4737

4738

4739

4740

4741

4750

4760

4771

4780

4742

4751

4761

4772

4781

4743

4752

4762

4773

4782

4744

4753

4763

4774

4783

4745

4754

4764

4775

4784

4746

4755

4765

4776

4785

4747

4756

4766

4777

4786

4748

4757

4767

4778

4789

4749

4758

4768

4779

4790

4759

4770

4791

4792

4793

4794

4796

4797

4798

4799

4800

4801

4802

4803

4804

4805

4806

4807

4808

4809

4810

4811

4812

4813

4814

4815

4816

4817

4818

4819

4820

4821

4822

4823

4824

4825

4826

4827

4828

4829

4830

4831

4832

4833

4834

4835

4836

4837

4838

4839

4840

4841

4842

4843

4844

4845

4846

4847

4848

4849

4850

4851

4852

4853

4854

4855

4856

4857

4858

4859

4860

4861

4862

4863

4864

4865

4866

4867

4868

4869

4870

4871

4872

4873

4874

4875

4876

4877

4878

4879

4880

4881

4882

4883

4884

4885

4886

4887

4888

4889

4890

4891

4892

4893

4894

4895

4896

4897

4898

4899

4900

4901

4902

4903

4904

4905

4906

4907

4908

4909

4910

4911

4912

4913

4914

4915

4916

4917

4918

4919

4920

4921

4922

4923

4924

4925

4926

4927

4928

4929

4930

4931

4932

4933

4934

4935

4936

4937

4938

4939

4940

4941

4942

4943

4944

4945

4946

4947

4948

4949

4950

4951

4952

4953

4954

4955

4956

4957

4958

4959

4960

4961

4962

4963

4964

4966

4967

4968

4969

4970

4971

4972

4973

4974

4975

4976

4977

4978

4979

4980

4981

4982

4983

4984

4985

4986

4987

4988

4989

4990

4991

4992

4993

4994

4995

4996

4997

4998

4999

5000

5001

5002

5003

5004

5005

5006

5008

5009

5010

5011

5012

5013

5014

5015

5016

5017

5018

5020

5021

5022

5023

5024

5026

5027

5028

5029

5030

5032

5033

5034

5035

5036

5037

5038

5039

5040

5041

5042

5043

5044

5045

5046

5047

5048

5049

5050

5051

5052

5053

5054

5055

5056

5057

5058

5059

5060

5061

5062

5063

5064

5065

5066

5067

5068

5069

5070

5071

5072

5073

5074

5075

5076

5077

5078

5080

5081

5082

5083

5084

5085

5086

5087

5088

5089

5090

5091

5092

5093

5094

5095

5097

5098

5099

5100

5101

5102

5103

5104

5105

5106

5109

5110

5111

5112

5113

5114

5115

5116

5118

5120

5121

5122

5123

5124

5125

5126

5127

5128

5129

5130

5131

5132

5133

5134

5135

5136

5137

5140

5141

5142

5143

5144

5145

5146

5147

5148

5149

5150

5151

5152

5153

5154

5155

5156

5157

5158

5159

5160

5161

5162

5163

5164

5165

5166

5167

5168

5169

5170

5177

5184

5194

5203

5204

5171

5178

5186

5195

5205

5179

5187

5196

5206

5172

5188

5197

5207

5180

5189

5173

5200

5208

5181

5190

5201

5209

5174

5192

5182

5202

5210

5175

5193

5176

5183

5211

5212

5213

5214

5215

5217

5219

5220

5221

5222

5224

5225

5226

5227

5228

5229

5230

5232

5233

5234

5235

5236

5237

5238

5239

5240

5241

5242

5243

5244

5245

5246

5247

5248

5249

5250

5251

5252

5253

5254

5257

5258

5265

5275

5285

5296

5259

5266

5276

5286

5297

5260

5267

5277

5287

5298

5261

5268

5278

5288

5299

5262

5269

5279

5289

5300

5263

5270

5280

5290

5301

5264

5271

5281

5291

5302

5272

5282

5292

5303

5274

5283

5293

5304

5284

5294

5305

5295

5307

5308

5309

5310

5311

5312

5314

5315

5316

5317

5318

5319

5320

5321

5322

5323

5324

5325

5326

5327

5328

5329

5330

5331

5332

5333

5334

5335

5337

5338

5339

5340

5341

5342

5343

5344

5345

5346

5347

5348

5349

5350

5351

5352

5354

5355

5356

5357

5358

5359

5361

5362

5363

5364

5365

5366

5367

5368

5369

5370

5371

5372

5373

5374

5376

5379

5383

5384

5385

5386

5387

5391

5393

5394

5396

5397

5398

5399

5400

5403

5404

5405

5406

5409

5410

5411

5420

5430

5448

5457

5412

5421

5431

5449

5458

5413

5422

5432

5450

5459

5414

5436

5451

5460

5415

5423

5441

5452

5461

5416

5424

5442

5453

5462

5417

5426

5454

5463

5427

5445

5455

5464

5419

5429

5456

5465

5466

5474

5483

5490

5500

5475

5484

5501

5467

5491

5485

5492

5502

5468

5476

5493

5503

5469

5477

5486

5505

5470

5478

5487

5495

5506

5479

5471

5488

5496

5507

5480

5472

5481

5489

5497

5473

5482

5498

5508

5499

5509

5510

5511

5512

5513

5514

5515

5516

5517

5518

5519

5520

5521

5522

5523

5524

5525

5526

5527

5528

5529

5530

5531

5532

5533

5534

5535

5536

5537

5538

5539

5540

5541

5542

5543

5544

5545

5546

5547

5548

5549

5550

5551

5552

5553

5554

5555

5556

5557

5558

5559

5560

5562

5563

5564

5565

5567

5569

5570

5571

5572

5573

5574

5576

5577

5578

5579

5580

5582

5583

5584

5585

5586

5588

5589

5590

5591

5592

5593

5595

5596

5597

5598

5599

5600

5601

5602

5603

5604

5605

5606

5607

5608

5609

5610

5611

5612

5613

5614

5615

5616

5617

5618

5619

5620

5621

5622

5626

5627

5628

5629

5630

5631

5632

5633

5634

5635

5636

5637

5638

5639

5640

5641

5642

5643

5653

5660

5667

5678

5644

5654

5661

5668

5679

5645

5655

5662

5669

5680

5646

5672

5681

5656

5663

5673

5682

5647

5657

5664

5674

5683

5648

5650

5658

5665

5675

5684

5651

5659

5666

5676

5652

5677

5685

STRUCTURAL FORMULAS OF ORGANIC COMPOUNDS (continued)

In numeric order as they occur in the Table of Physical Constants of Organic Compounds

5686

5696

5704

5712

5723

5687

5697

5705

5713

5724

5688

5698

5706

5714

5726

5689

5699

5707

5715

5727

5690

5700

5708

5716

5728

5691

5701

5709

5717

5729

5692

5702

5710

5718

5730

5693

5703

5711

5719

5731

5694

5720

5695

5721

5722

5732

5733

5734

5735

5736

5737

5738

5739

5740

5741

5742

5743

5744

5745

5746

5747

5748

5749

5750

5751

5752

5753

5754

5755

5756

5757

5758

5759

5760

5761

5762

5763

5764

5765

5766

5767

5768

5769

5770

5771

5772

5773

5774

5775

5776

5777

5779

5780

5781

5782

5783

5784

5786

5787

5788

5789

5790

5791

5792

5793

5794

5795

5796

5797

5798

5800

5801

5802

5803

5805

5806

5807

5808

5809

5810

5811

5812

5813

5814

5815

5816

5817

5818

5819

5820

5821

5822

5823

5824

5826

5827

5828

5829

5830

5831

5832

5833

5834

5835

5836

5837

5838

5839

5840

5841

5842

5843

5844

5845

5846

5847

5848

5849

5850

5851

5852

5853

5855

5856

5857

5858

5859

5860

5861

5862

5863

5864

5865

5866

5867

5868

5869

5870

5871

5880

5888

5895

5902

5872

5881

5889

5896

5903

5873

5882

5890

5897

5904

5874

5883

5891

5898

5905

5875

5884

5892

5899

5906

5876

5885

5893

5900

5907

5877

5886

5894

5901

5908

5879

5887

5909

5910

5911

5912

5913

5914

5915

5916

5917

5918

5919

5920

5921

5922

5923

5924

5925

5926

5927

5929

5930

5931

5932

5933

5934

5935

5936

5937

5938

5939

5940

5941

5942

5943

5944

5945

5946

5947

5948

5950

5960

5970

5981

5989

5951

5961

5971

5982

5990

5952

5973

5991

5962

5953

5974

5984

5992

5963

5954

5975

5985

5964

5977

5993

5955

5986

5965

5978

5994

5987

5956

5979

5967

5995

5968

5980

5957

5988

5996

5997

5999

6000

6001

6002

6004

6005

6006

6007

6008

6009

6010

6011

6012

6013

6014

6015

6016

6017

6018

6019

6020

6021

6024

6025

6026

6027

6028

6029

6031

6032

6033

6034

6035

6036

6037

6039

6040

6041

6042

6043

6044

6045

6046

6047

6049

6050

6051

6052

6053

6055

6056

6057

6058

6059

6060

6061

6062

6063

6064

6066

6067

6068

6069

6070

6071

6072

6073

6074

6075

6076

6077

6081

6082

6083

6084

6085

6086

6087

6088

6089

6090

6091

6092

6099

6109

6117

6126

6093

6101

6110

6118

6127

6094

6102

6111

6119

6128

6095

6103

6112

6120

6129

6096

6104

6113

6121

6130

6097

6105

6114

6122

6131

6098

6106

6115

6123

6132

6107

6116

6124

6133

6108

6125

6134

6135

6137

6138

6139

6140

6141

6142

6143

6145

6146

6147

6148

6149

6150

6151

6152

6153

6154

6155

6156

6157

6158

6159

6160

6163

6164

6165

6166

6167

6168

6169

6170

6171

6172

6174

6175

6176

6180

6181

6183

6184

6185

6186

6187

6188

6190

6191

6192

6194

6206

6221

6240

6260

6197

6208

6241

6262

6199

6212

6222

6243

6263

6200

6214

6227

6249

6265

6201

6216

6228

6250

6266

6202

6217

6229

6251

6267

6203

6218

6252

6268

6204

6238

6254

6272

6220

6239

6255

6275

6287

6296

6307

6320

6308

6321

6288

6297

6309

6322

6276

6310

6323

6289

6298

6311

6277

6312

6324

6291

6299

6313

6325

6278

6300

6326

6279

6292

6301

6314

6327

6315

6280

6294

6302

6316

6328

6295

6304

6317

6329

6282

6305

6318

6284

6306

6319

6330

6331

6341

6352

6363

6372

6332

6342

6353

6364

6373

6333

6343

6354

6365

6374

6334

6344

6355

6366

6375

6335

6345

6356

6367

6376

6336

6346

6357

6368

6377

6337

6347

6358

6369

6378

6338

6348

6359

6370

6379

6339

6349

6360

6371

6380

6340

6350

6361

6381

6351

6362

6383

6384

6385

6386

6387

6388

6389

6390

6391

6392

6393

6394

6395

6396

6397

6398

6399

6400

6401

6402

6403

6404

6405

6406

6407

6408

6409

6410

6411

6412

6413

6414

6415

6416

6417

6418

6419

6420

6421

6422

6423

6424

6425

6426

6427

6428

6429

6430

6431

6432

6433

6434

6435

6436

6437

6438

6439

6440

6441

6442

6443

6444

6445

6446

6447

6448

6449

6450

6451

6452

6453

6454

6455

6456

6457

6459

6461

6462

6463

6464

6465

6466

6467

6468

6469

6470

6471

6472

6473

6474

6475

6476

6477

6478

6479

6480

6481

6482

6483

6484

6485

6486

6487

6488

6489

6490

6491

6492

6493

6494

6495

6496

6497

6498

6499

6500

6501

6502

6503

6504

6505

6506

6508

6509

6510

6511

6512

6513

6514

6515

6517

6518

6519

6520

6521

6522

6523

6524

6525

6526

6527

6528

6529

6530

6531

6532

6533

6534

6535

6536

6537

6538

6539

6540

6541

6542

6543

6544

6545

6546

6547

6548

6549

6550

6551

6552

6553

6554

6555

6556

6557

6558

6559

6560

6561

6562

6563

6564

6565

6566

6567

6568

6570

6572

6573

6574

6575

6578

6579

6580

6581

6582

6583

6584

6585

6586

6587

6588

6589

6590

6591

6592

6593

6594

6595

6596

6597

6598

6599

6600

6601

6602

6603

6604

6605

6606

6607

6608

6609

6610

6611

6612

6613

6614

6615

6616

6617

6618

6619

6620

6621

6622

6623

6625

STRUCTURAL FORMULAS OF ORGANIC COMPOUNDS (continued)

In numeric order as they occur in the Table of Physical Constants of Organic Compounds

6626

6627

6628

6629

6630

6631

6632

6633

6634

6635

6636

6637

6638

6639

6640

6641

6642

6643

6644

6647

6649

6650

6651

6652

6653

6654

6655

6656

6657

6658

6659

6660

6661

6662

6663

6664

6665

6666

6667

6668

6669

6670

6671

6672

6673

6674

6675

6676

6677

6678

6679

6680

6681

6682

6683

6684

6685

6686

6687

6688

6689

6690

6691

6692

6693

6694

6695

6696

6697

6698

6699

6700

6701

6702

6703

6705

6706

6707

6708

6709

6710

6711

6712

6713

6714

6715

6716

6717

6718

6719

6720

6721

6722

6723

6724

6725

6726

6727

6737

6746

6757

6767

6728

6738

6747

6758

6768

6729

6739

6748

6759

6769

6730

6740

6749

6760

6770

6731

6750

6771

6741

6732

6751

6761

6772

6733

6742

6752

6762

6773

6743

6734

6753

6763

6744

6735

6754

6764

6774

6745

6755

6756

6766

6775

6736

6765

6776

6777

6778

6779

6781

6782

6783

6784

6785

6786

6787

6788

6789

6790

6791

6792

6793

6794

6795

6796

6797

6798

6799

6800

6801

6802

6803

6804

6805

6806

6807

6808

6809

6810

6811

6812

6813

6814

6815

6816

6817

6818

6819

6820

6821

6822

STRUCTURAL FORMULAS OF ORGANIC COMPOUNDS (continued)

In numeric order as they occur in the Table of Physical Constants of Organic Compounds

6823

6824

6825

6826

6827

6828

6829

6830

6831

6832

6833

6834

6835

6836

6837

6838

6839

6840

6841

6842

6843

6844

6845

6846

6847

6848

6849

6850

6851

6852

6853

6854

6855

6856

6857

6858

6859

6860

6861

6862

6863

6864

6865

6866

6868

6869

6870

6872

6873

6874

6883

6892

6904

6914

6875

6884

6893

6905

6915

6876

6885

6894

6906

6916

6877

6886

6895

6907

6917

6878

6887

6896

6908

6918

6879

6888

6897

6909

6919

6880

6889

6898

6910

6921

6881

6890

6899

6911

6922

6882

6891

6900

6912

6923

6901

6913

6924

6932

6941

6949

6957

6925

6933

6942

6950

6958

6926

6934

6943

6951

6959

6927

6935

6944

6952

6960

6928

6936

6945

6953

6961

6929

6937

6946

6954

6962

6930

6938

6947

6955

6963

6939

6948

6956

6964

6931

6940

6965

6973

6981

6989

6998

6966

6974

6982

6990

6999

6967

6975

6983

6991

7000

6968

6976

6984

6992

7001

6969

6977

6985

6993

7002

6970

6978

6986

6994

7003

6971

6979

6987

6995

7004

6972

6980

6988

6996

7005

6997

7007

In numeric order as they occur in the Table of Physical Constants of Organic Compounds

7008

7009

7010

7011

7012

7013

7014

7015

7016

7017

7018

7019

7020

7021

7022

7023

7024

7025

7026

7027

7028

7029

7030

7031

7032

7033

7034

7035

7036

7037

7038

7039

7040

7041

7042

7043

7044

7045

7046

7047

7048

7049

7050

7051

7053

7054

7055

7056

7057

7058

STRUCTURAL FORMULAS OF ORGANIC COMPOUNDS (continued)

In numeric order as they occur in the Table of Physical Constants of Organic Compounds

7060

7061

7062

7063

7064

7065

7066

7067

7068

7069

7070

7071

7072

7073

7074

7075

7076

7077

7078

7079

7080

7081

7082

7083

7084

7085

7086

7087

7088

7089

7090

7091

7092

7093

7094

7095

7096

7097

7098

7099

7100

7101

7102

7103

7104

7105

7106

7107

7108

7109

7113

7114

7115

7116

7117

7118

7119

7120

7121

7122

7123

7124

7125

7126

7127

7128

7129

7130

7131

7132

7136

7137

7138

7140

7142

7144

7145

7146

7147

7148

7149

7150

7151

7152

7153

7154

7155

7156

7158

7159

7160

7161

7164

7165

7167

7168

7170

7172

7175

7176

7177

7178

7179

7180

7181

7182

7183

7184

7185

7186

7187

7188

7189

7190

7192

7193

7194

7195

7196

7197

7198

7199

7200

7201

7202

7203

7205

7206

7207

7208

7209

7210

7211

7212

7213

7214

7215

7216

7217

7218

7219

7222

7224

7225

7226

7227

7228

7230

7231

7232

7233

7235

7236

7237

7238

7239

7240

7241

7242

7243

7244

7245

7246

7247

7249

7250

7251

7252

7253

7254

7255

7257

7258

7259

7260

7272

7285

7294

7302

7262

7273

7286

7295

7303

7263

7274

7287

7296

7304

7264

7276

7289

7297

7305

7267

7277

7290

7298

7307

7268

7278

7291

7299

7308

7269

7279

7292

7300

7271

7281

7282

7293

7284

7309

7310

7319

7327

7337

7346

7312

7320

7328

7338

7347

7313

7321

7329

7339

7350

7314

7322

7330

7340

7351

7315

7323

7331

7342

7354

7316

7324

7335

7343

7355

7317

7325

7336

7344

7356

7326

7345

7357

7367

7375

7386

7403

7359

7368

7376

7387

7404

7361

7369

7378

7388

7405

7362

7370

7380

7389

7406

7363

7371

7381

7390

7408

7364

7372

7383

7398

7409

7365

7374

7384

7399

7410

7385

7402

7411

STRUCTURAL FORMULAS OF ORGANIC COMPOUNDS (continued)

In numeric order as they occur in the Table of Physical Constants of Organic Compounds

7412

7413

7415

7416

7417

7418

7419

7420

7421

7422

7423

7425

7426

7427

7430

7431

7432

7433

7434

7435

7436

7438

7439

7441

7442

7443

7444

7446

7447

7448

7449

7450

7451

7452

7454

7455

7459

7460

7461

3-495

7462

7469

7477

7485

7494

7463

7470

7478

7486

7495

7464

7471

7479

7487

7496

7465

7472

7480

7489

7497

7466

7473

7481

7490

7498

7467

7474

7482

7491

7499

7468

7475

7483

7492

7500

7476

7484

7493

7501

7509

7519

7528

7535

7502

7510

7520

7529

7536

7512

7521

7503

7513

7530

7537

7522

7504

7515

7531

7538

7523

7505

7516

7524

7532

7539

7506

7517

7525

7533

7540

7507

7518

7526

7534

7508

7527

7541

7542

7549

7556

7564

7573

7543

7550

7557

7565

7574

7544

7551

7558

7567

7575

7545

7552

7559

7568

7576

7546

7553

7560

7569

7577

7547

7554

7561

7570

7578

7548

7555

7562

7571

7572

7579

7580

7581

7593

7604

7618

7631

7582

7594

7605

7619

7632

7633

7585

7595

7606

7621

7586

7597

7607

7622

7635

7587

7598

7609

7623

7636

7588

7599

7612

7624

7639

7590

7600

7614

7625

7640

7591

7601

7615

7626

7641

7592

7602

7616

7629

7642

7617

7643

7644

7645

7646

7647

7648

7649

7650

7651

7653

7654

7655

7656

7657

7658

7659

7660

7661

7662

7663

7665

7666

7667

7669

7670

7671

7672

7674

7675

7678

7679

7680

7681

7682

7683

STRUCTURAL FORMULAS OF ORGANIC COMPOUNDS (continued)

In numeric order as they occur in the Table of Physical Constants of Organic Compounds

7684

7685

7686

7687

7689

7690

7691

7692

7693

7694

7696

7697

7698

7699

7700

7701

7702

7703

7704

7705

7706

7708

7709

7710

7712

7713

7714

7715

7716

7717

7718

7720

7721

7722

7723

7724

7725

7727

7728

7730

7731

7732

7733

3-501

STRUCTURAL FORMULAS OF ORGANIC COMPOUNDS (continued)

In numeric order as they occur in the Table of Physical Constants of Organic Compounds

7734

7735

7736

7737

7738

7739

7740

7742

7743

7746

7747

7749

7750

7751

7752

7753

7754

7756

7758

7759

7764

7765

7766

7770

7771

7772

7773

7775

7776

7778

7779

7780

7781

7782

7783

7784

7785

7789

7790

7791

7792

7793

7794

7795

7796

7797

7798

In numeric order as they occur in the Table of Physical Constants of Organic Compounds

7799

7801

7802

7803

7804

7806

7807

7808

7809

7810

7811

7812

7813

7814

7815

7816

7817

7818

7819

7820

7821

7822

7823

7824

7825

7826

7827

7828

7829

7830

7831

7832

7833

7834

7835

7836

7838

7839

7842

7843

7846

7848

7850

7852

7853

7859

STRUCTURAL FORMULAS OF ORGANIC COMPOUNDS (continued)

In numeric order as they occur in the Table of Physical Constants of Organic Compounds

7934

7935

7936

7937

7938

7939

7941

7942

7943

7944

7945

7946

7948

7949

7950

7951

7952

7954

7955

7956

7957

7958

7959

7966

7975

7976

7977

7978

7979

7980

7981

7982

7983

7984

7985

7986

7987

7988

7989

7990

7991

7992

7993

7994

7995

7996

7997

7998

7999

8000

8001

8002

8003

8005

8006

8007

8008

8009

8010

8011

8012

8013

8015

8017

8018

8020

8022

8024

8026

8027

8028

8030

8031

8032

8033

8034

8035

8037

8038

8039

8041

8043

8044

8045

8046

8047

8048

8050

8052

8054

8055

8056

8057

8059

8061

8062

8063

8064

8071

8080

8090

8103

8065

8072

8081

8091

8104

8066

8073

8082

8092

8105

8074

8083

8106

8067

8075

8093

8107

8084

8094

8108

8068

8076

8085

8095

8109

8077

8086

8096

8110

8069

8078

8087

8097

8111

8070

8079

8088

8098

8112

8089

8099

8113

8101

8114

8115

8126

8136

8146

8157

8116

8127

8137

8147

8158

8117

8128

8138

8148

8159

8118

8129

8139

8149

8160

8119

8130

8140

8150

8161

8120

8131

8141

8151

8162

8121

8132

8142

8152

8163

8122

8133

8143

8154

8164

8123

8134

8144

8155

8165

8124

8135

8145

8156

8166

8125

8167

8168

8169

8170

8171

8172

8173

8174

8175

8176

8177

8178

8179

8180

8181

8182

8183

8184

8185

8186

8187

8188

8189

8190

8191

8192

8193

8194

8195

8196

8197

8198

8199

8200

8201

8202

8203

8204

8205

8206

8207

8208

8209

8212

8213

8214

8215

8216

8218

8219

8220

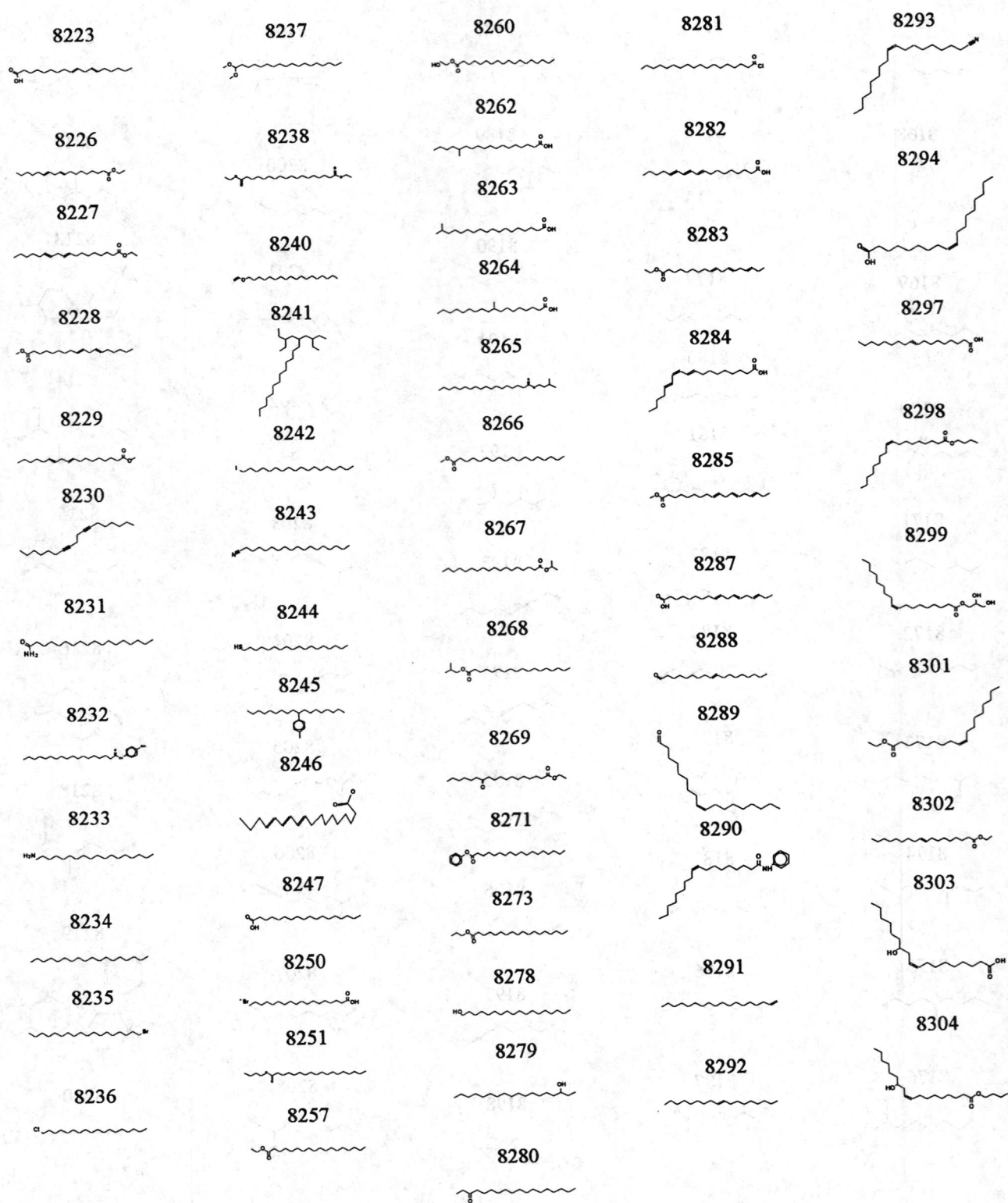

8305

8306

8307

8308

8309

8310

8311

8312

8313

8314

8315

8316

8317

8318

8319

8320

8321

8322

8324

8325

8327

8328

8329

8330

8331

8332

8333

8334

8335

8336

8337

8338

8339

8340

8341

8342

8343

8344

8345

8346

8347

8348

8349

8350

8351

8352

8353

8354

8355

8356

8357

8358

8359

8360

STRUCTURAL FORMULAS OF ORGANIC COMPOUNDS (continued)

In numeric order as they occur in the Table of Physical Constants of Organic Compounds

8361

8363

8364

8365

8366

8367

8368

8369

8370

8371

8372

8373

8374

8375

8376

8377

8378

8379

8380

8381

8382

8383

8384

8385

8386

8387

8388

8389

8390

8391

8392

8393

8394

8395

8396

8398

8399

8400

8401

8402

8403

8404

8405

8406

8407

8408

8409

8410

8411

8412

8415

8416

8417

8418

8419

8420

8421

8422

8424

8425

8426

8427

8428

8429

8430

8431

8432

8433

8434

8435

8436

8437

8438

8439

8440

8441

8442

8443

8444

8445

8446

8447

8449

8450

8451

8452

8453

8454

8455

8456

8457

8458

8459

8460

8461

8462

8464

8465

8466

8468

8469

8470

8471

8472

8473

8474

8475

8476

8477

8478

8479

8480

8481

8482

8484

8485

8486

8487

8488

8489

8490

8491

8492

8493

8494

8495

8496

8497

8498

8499

8500

8502

8503

8504

8505

8507

8508

8509

8510

8511

8512

8513

8514

8515

8516

8517

8518

8519

8520

8521

8523

8524

8525

8526

8527

8528

8529

8530

8531

8532

8533

8534

8535

8536

8537

8538

8539

8540

8541

8542

8543

8544

8545

8547

8548

8552

8553

8554

8555

8556

8557

8558

8559

8560

8562

8563

8564

8565

8566

8568

8569

8570

8571

8574

8575

8576

8577

8578

8579

8580

8581

8582

8583

8584

8585

8586

8587

8588

8598

8609

8619

8629

8599

8610

8620

8589

8600

8611

8621

8632

8590

8612

8622

8633

8591

8601

8613

8623

8635

8592

8602

8614

8624

8593

8603

8615

8625

8636

8594

8604

8616

8626

8637

8595

8605

8617

8627

8638

8596

8607

8618

8628

8639

8597

8608

8640

8641

8642

8643

8644

8645

8646

8647

8648

8649

8650

8651

8652

8653

8654

8655

8656

8657

8658

8659

8660

8661

8662

8663

8664

8665

8666

8667

8668

8669

8670

8671

8672

8673

8674

8675

8676

8679

8680

8681

8682

8683

8684

8685

8686

8687

8688

8689

8690

8691

8692

8693

8694

8695

8696

8697

STRUCTURAL FORMULAS OF ORGANIC COMPOUNDS (continued)

In numeric order as they occur in the Table of Physical Constants of Organic Compounds

8698	8707	8716	8726	8734

8699   8708

8700   8709   8717   8727   8735

8701   8710   8718   8728   8736

8702   8711   8719   8729   8737

8703   8712   8720   8730   8739

8704   8713   8721   8731   8740

8705   8714   8722   8732   8741

8706   8715   8723   8733   8742

8743

8744

8745

8746

8747

8748

8749

8750

8751

8752

8753

8754

8755

8756

8757

8758

8760

8761

8762

8763

8764

8765

8766

8767

8768

8769

8770

8771

8772

8773

8774

8775

8776

8777

8778

8780

8781

8783

8784

8785

8786

8787

8788

8789

8790

8791

8792

8793

8794

In numeric order as they occur in the Table of Physical Constants of Organic Compounds

8795

8805

8817

8829

8838

8818

8806

8839

8796

8819

8830

8840

8807

8841

8797

8820

8831

8808

8842

8798

8821

8832

8843

8809

8844

8799

8822

8810

8823

8833

8845

8811

8824

8834

8846

8801

8812

8825

8802

8813

8826

8835

8847

8814

8827

8836

8803

8815

8837

8848

8828

8804

8816

8849

STRUCTURAL FORMULAS OF ORGANIC COMPOUNDS (continued)

In numeric order as they occur in the Table of Physical Constants of Organic Compounds

8850

8851

8852

8853

8854

8855

8856

8857

8858

8859

8860

8861

8862

8865

8866

8867

8868

8869

8870

8871

8872

8873

8874

8875

8876

8878

8879

8881

8882

8884

8885

8886

8887

8888

8889

8891

8892

8893

8894

8895

8896

8897

8898

8900

8903

8904

8905

8915

8927

8935

8944

8906

8916

8928

8936

8945

8907

8917

8929

8937

8946

8908

8918

8930

8938

8947

8909

8919

8931

8939

8920

8950

8910

8922

8932

8940

8952

8911

8923

8941

8912

8924

8933

8942

8953

8913

8925

8934

8943

8954

8914

8955

8964

8972

8980

8989

8965

8990

8956

8973

8981

8957

8966

8974

8982

8991

8967

8975

8992

8958

8983

8959

8968

8976

8984

8995

8985

8996

8961

8977

8969

8997

8986

8998

8962

8978

8970

8987

8999

8963

8979

9000

8971

8988

9001

9002

9004

9005

9006

9007

9008

9009

9010

9011

9012

9013

9014

9015

9016

9017

9018

9019

9020

9021

9022

9023

9024

9025

9026

9027

9028

9029

9030

9031

9032

9033

9034

9035

9036

9037

9038

9039

9040

9041

9042

9043

9044

9045

9046

9047

9048

9049

9050

9051

9052

9053

9054

9055

9056

9057

9058

9059

9060

9061

9062

9065

9066

9067

9068

9069

9070

9071

9072

9073

9074

9075

9076

9077

9078

9079

9080

9081

9082

9083

9084

9085

9086

9087

9088

9089

9090

9091

9092

9093

9094

9095

9096

9097

9098

9099

9100

9101

9102

9103

9104

9105

9106

9107

9108

9117

9126

9135

9143

9109

9118

9136

9144

9110

9119

9128

9137

9145

9111

9121

9129

9138

9146

9112

9122

9130

9139

9147

9113

9123

9131

9140

9149

9114

9124

9132

9141

9150

9115

9125

9133

9142

9151

9116

9134

9153

9167

9182

9192

9206

9154

9168

9183

9193

9208

9156

9170

9184

9195

9210

9158

9171

9185

9198

9211

9159

9172

9186

9199

9212

9160

9173

9187

9201

9213

9161

9174

9188

9202

9216

9162

9179

9189

9204

9217

9164

9180

9190

9205

9181

9191

9218

9219

9220

9221

9223

9224

9225

9226

9227

9228

9230

9231

9232

9233

9249

9250

9251

9253

9254

9255

9256

9257

9260

9262

9263

9264

9265

9266

9267

9270

9271

9272

9273

9276

9277

9279

9280

9282

9283

9284

9285

9287

9295

9307

9308

9317

9327

9288

9296

9309

9319

9328

9289

9297

9310

9320

9330

9290

9298

9312

9321

9331

9291

9300

9313

9322

9333

9292

9305

9314

9323

9334

9294

9306

9315

9324

9335

9316

9326

9336

9344

9355

9364

9377

9337

9345

9356

9365

9378

9338

9348

9357

9370

9379

9340

9349

9358

9371

9381

9341

9350

9359

9372

9382

9342

9351

9360

9374

9385

9343

9352

9362

9375

9387

9353

9376

9388

STRUCTURAL FORMULAS OF ORGANIC COMPOUNDS (continued)

In numeric order as they occur in the Table of Physical Constants of Organic Compounds

9389

9391

9392

9394

9395

9396

9397

9401

9402

9403

9404

9405

9407

9409

9410

9411

9412

9413

9414

9416

9417

9418

9419

9420

9421

9423

9424

9425

9426

9427

9428

9430

9431

9432

9434

9442

9452

9462

9473

9435

9443

9454

9463

9474

9436

9445

9455

9465

9475

9438

9446

9456

9466

9479

9439

9447

9457

9469

9482

9448

9458

9471

9483

9440

9451

9459

9472

9484

9441

9461

9485

9497

9510

9520

9530

9486

9499

9512

9521

9532

9489

9501

9513

9522

9535

9492

9503

9514

9523

9536

9493

9504

9515

9524

9537

9494

9505

9517

9525

9539

9495

9507

9518

9527

9540

9496

9519

9528

9541

9529

9543

STRUCTURAL FORMULAS OF ORGANIC COMPOUNDS (continued)

In numeric order as they occur in the Table of Physical Constants of Organic Compounds

9545

9546

9548

9549

9550

9551

9553

9554

9555

9556

9557

9558

9559

9560

9562

9563

9564

9565

9566

9567

9568

9569

9570

9571

9572

9573

9574

9575

9576

9577

9578

9579

9580

9581

9582

9583

9584

9585

9586

9587

9588

9589

9590

9591

9592

9593

9594

9595

9596

9597

9599

9600

9601

9602

9603

9604

9605

9606

9608

9609

9610

9611

9612

9613

9614

9615

9616

9617

9618

9619

9620

9621

9622

9624

9625

9626

9627

9628

9629

9630

9631

9632

9634

9644

9656

9666

9675

9635

9645

9658

9667

9676

9636

9648

9659

9668

9678

9637

9649

9660

9669

9679

9638

9650

9661

9670

9680

9639

9651

9662

9671

9681

9642

9652

9663

9673

9682

9643

9655

9664

9674

9683

3-536

9684

9692

9701

9702

9703

9710

9718

9685

9693

9704

9711

9719

9686

9695

9687

9696

9688

9697

9689

9698

9690

9699

9691

9700

9705

9706

9707

9708

9709

9712

9713

9714

9715

9716

9717

9720

9721

9722

9723

9724

9726

9727

9728

9729

9730

9731

9732

9733

9734

9736

9738

9739

9740

9741

9743

9745

9746

9748

9749

9750

9751

9752

9753

9754

9756

9758

9759

9760

9761

9762

9763

9764

9765

9766

9767

9768

9769

9770

9771

9772

9773

9774

9775

9784

9798

9808

9817

9776

9785

9799

9809

9818

9777

9786

9800

9810

9819

9778

9787

9801

9811

9820

9779

9788

9803

9812

9821

9780

9790

9804

9813

9822

9781

9791

9805

9814

9823

9782

9792

9806

9815

9824

9783

9793

9807

9816

9825

STRUCTURAL FORMULAS OF ORGANIC COMPOUNDS (continued)

In numeric order as they occur in the Table of Physical Constants of Organic Compounds

9826

9827

9828

9829

9832

9834

9835

9836

9837

9838

9839

9840

9841

9842

9843

9844

9845

9846

9847

9848

9849

9850

9851

9852

9853

9854

9855

9856

9857

9858

9859

9860

9861

9862

9863

9864

9865

9866

9867

9868

9869

9870

9871

9872

9873

9874

9875

9876

9885

9894

9903

9913

9877

9886

9895

9904

9914

9878

9887

9896

9905

9915

9879

9888

9897

9906

9916

9880

9889

9898

9907

9917

9881

9890

9899

9908

9918

9882

9891

9900

9909

9919

9883

9892

9901

9910

9920

9884

9893

9902

9911

9912

STRUCTURAL FORMULAS OF ORGANIC COMPOUNDS (continued)

In numeric order as they occur in the Table of Physical Constants of Organic Compounds

9921

9931

9939

9948

9958

9922

9932

9940

9949

9959

9923

9933

9941

9950

9960

9924

9934

9942

9951

9961

9925

9935

9943

9952

9962

9926

9936

9944

9953

9963

9927

9937

9945

9954

9964

9928

9946

9955

9966

9930

9938

9947

9957

9967

9978

9989

9999

10007

9969

9980

9990

10000

10008

9970

9981

9991

10001

10009

9972

9982

9992

10002

10010

9973

9983

9994

10003

10011

9974

9985

9994

10004

10012

9975

9986

9995

10005

10013

9976

9987

9997

10006

10014

9977

9998

10016

10017

10018

10019

10020

10021

10022

10023

10025

10026

10027

10028

10029

10030

10031

10032

10033

10034

10035

10036

10037

10038

10039

10040

10041

10042

10043

10044

10045

10046

10047

10048

10049

10050

10051

10052

10053

10054

10055

10056

10057

10058

10059

10060

10061

10062

10063

10064

10065

10075

10084

10093

10102

10066

10076

10085

10094

10103

10067

10077

10086

10095

10105

10068

10078

10087

10096

10106

10069

10079

10088

10097

10107

10070

10080

10089

10098

10108

10071

10081

10090

10099

10109

10072

10082

10091

10100

10110

10073

10083

10092

10101

10111

10074

10112

10121

10130

10139

10147

10113

10122

10131

10140

10148

10114

10123

10132

10141

10149

10115

10124

10133

10142

10150

10116

10125

10134

10143

10151

10117

10126

10135

10144

10152

10118

10127

10136

10145

10153

10119

10128

10146

10154

10120

10129

10138

10155

10156

10157

10158

10159

10160

10161

10162

10163

10164

10166

10167

10168

10169

10170

10171

10172

10173

10174

10175

10176

10177

10178

10179

10180

10181

10182

10183

10184

10185

10186

10187

10188

10189

10190

10191

10192

10193

10194

10195

10196

10197

10198

10199

10200

10201

10202

10203

10204

10205

10207

10208

10209

10210

10211

10212

10213

10214

10215

10216

10217

10218

10219

10220

10221

10222

10223

10225

10226

10227

10228

10229

10230

10231

10232

10233

10234

10235

10236

10237

10238

10239

10240

10241

10242

10243

10244

10245

10246

10248

10249

10250

10251

10253

10254

10255

10256

10258

10259

10260

10261

10262

10263

10264

10265

10268

10269

10270

10271

10273

10274

10275

10276

10277

10278

10279

10280

10281

10282

10283

10285

10286

10287

10288

10289

10290

10291

STRUCTURAL FORMULAS OF ORGANIC COMPOUNDS (continued)

In numeric order as they occur in the Table of Physical Constants of Organic Compounds

10292

10293

10294

10295

10296

10297

10298

10299

10300

10301

10302

10303

10304

10306

10307

10308

10309

10310

10311

10312

10313

10314

10315

10316

10317

10318

10319

10320

10321

10322

10323

10324

10325

10326

10328

10329

10330

10331

10332

10333

10334

10335

10336

STRUCTURAL FORMULAS OF ORGANIC COMPOUNDS (continued)

In numeric order as they occur in the Table of Physical Constants of Organic Compounds

10337

10346

10355

10364

10373

10338

10347

10356

10365

10374

10339

10348

10357

10366

10375

10340

10349

10358

10367

10376

10341

10350

10359

10368

10378

10342

10351

10360

10369

10380

10343

10352

10361

10370

10381

10344

10353

10362

10371

10382

10345

10354

10363

10372

10383

3-551

10384

10385

10386

10387

10388

10389

10390

10391

10393

10394

10395

10396

10398

10399

10401

10402

10404

10405

10406

10407

10408

10409

10410

10411

10412

10413

10414

10415

10416

10417

10418

10419

10420

10421

10422

10423

10424

10425

10426

10427

10428

10429

10440

10448

10460

10469

10430

10441

10449

10461

10470

10431

10442

10450

10462

10471

10432

10443

10451

10463

10472

10433

10444

10452

10464

10473

10436

10445

10454

10465

10474

10437

10446

10455

10466

10475

10438

10447

10457

10467

10476

10439

10458

10468

10478

10486

10493

10501

10509

10479

10487

10494

10502

10510

10480

10488

10495

10503

10511

10481

10489

10496

10504

10512

10482

10490

10497

10505

10514

10483

10491

10498

10506

10515

10484

10492

10500

10507

10516

10485

10508

10517

10518

STRUCTURAL FORMULAS OF ORGANIC COMPOUNDS (continued)

In numeric order as they occur in the Table of Physical Constants of Organic Compounds

10519

10520

10521

10522

10523

10524

10525

10526

10527

10528

10529

10530

10531

10532

10533

10534

10535

10536

10537

10538

10539

10540

10541

10542

10543

10544

10545

10546

10547

10548

10549

10550

10551

10552

10553

10554

10555

10556

10557

10558

10559

10560

10561

10562

10563

10564

10565

10566

10567

10568

10569

10570

10571

10572

10573

10574

10575

10576

10577

10578

10579

10580

10581

10582

10583

10584

10585

10586

10587

10588

10589

10590

10591

10592

10593

10594

10596

10597

10598

10599

10600

10601

10602

10603

10604

10605

10606

10607

10618

10628

10638

10649

10608

10620

10629

10639

10650

10609

10621

10631

10640

10652

10611

10622

10632

10643

10653

10613

10624

10633

10644

10654

10614

10625

10634

10645

10655

10616

10626

10635

10647

10656

10617

10627

10636

10648

10657

10637

10658

10659

10668

10680

10692

10702

10660

10670

10681

10693

10703

10661

10671

10683

10694

10704

10662

10672

10685

10695

10706

10663

10673

10686

10696

10707

10664

10674

10687

10697

10708

10665

10677

10688

10698

10709

10666

10679

10689

10699

10710

10691

10700

10711

10721

10730

10739

10750

10713

10722

10731

10743

10751

10714

10723

10732

10744

10752

10715

10724

10733

10745

10753

10716

10725

10734

10746

10755

10717

10726

10735

10747

10756

10718

10727

10736

10748

10757

10719

10728

10737

10749

10720

10729

10738

10758

10768

10779

10789

10799

10759

10769

10780

10790

10800

10781

10791

10801

10760

10770

10782

10792

10802

10761

10773

10783

10793

10803

10774

10784

10794

10805

10763

10775

10787

10795

10806

10764

10777

10788

10796

10807

10767

10778

10798

10809

10810

10811

10813

10814

10815

10816

10818

10819

10820

10821

10822

10823

10824

10825

10826

10827

10828

10829

10830

10831

10832

10833

10834

10835

10836

10837

10838

10839

10840

10841

10842

10844

10845

10846

10847

10849

10851

10852

10853

10855

10856

10858

10859

10860

10861

10862

10863

10864

10865

10866

10868

10869

10870

10871

10872

10873

10875

10876

10877

10878

10879

10880

10882

10883

10884

10886

10887

10888

10890

10891

10892

10893

10894

10895

10896

10897

10898

10899

10901

10903

10905

10906

10907

10908

10909

10911

10914

10915

10916

10917

10918

10920

10921

10922

10923

10924

10925

10926

10927

10929

10932

10933

10935

10936

10937

10938

10939

10940

10941

10942

10943

10944

10945

10946

10947

10948

10949

10959

10968

10977

10988

10951

10960

10969

10980

10989

10952

10961

10970

10981

10990

10953

10962

10972

10982

10991

10954

10963

10973

10983

10992

10965

10984

10955

10974

10994

10985

10956

10966

10975

10995

10987

10958

10967

10976

10996

10997

10998

10999

11001

11002

11003

11004

11005

11006

11007

11008

11009

11010

11012

11013

11014

11016

11017

11018

11019

11020

11021

11023

11024

11025

11026

11027

11028

11029

11031

11032

11033

11034

11035

11036

11038

11039

11040

11041

11042

11055

11067

11080

11090

11043

11056

11068

11081

11091

11045

11057

11069

11082

11092

11049

11058

11071

11083

11093

11050

11059

11072

11084

11094

11051

11062

11075

11085

11095

11052

11064

11076

11087

11098

11054

11066

11078

11088

11099

11100

11101

11103

11104

11105

11106

11109

11110

11112

11113

11114

11115

11116

11117

11118

11119

11120

11121

11122

11124

11125

11126

11127

11128

11129

11130

11131

11132

11133

11134

11135

11136

11137

11138

11139

11141

11142

11143

11146

11147

11148

11149

11150

11151

11152

11153

11154

11155

11156

11157

11158

11160

11161

11163

11164

11165

11166

11167

11168

11169

11170

11171

11172

11173

11175

11176

11180

11181

11182

11183

11184

11185

11186

11187

11188

11189

11190

11191

11193

11194

11195

11208

11219

11231

11242

11199

•11209

11220

11232

11243

11201

11210

11222

11236

11244

11202

11211

11223

11237

11245

11203

11213

11224

11238

11246

11204

11214

11225

11239

11247

11206

11217

11227

11240

11248

11218

11230

11241

11250

11251

11262

11270

11280

11288

11254

11263

11271

11281

11289

11255

11264

11274

11282

11290

11257

11265

11275

11283

11291

11258

11266

11276

11284

11292

11259

11267

11277

11285

11296

11260

11268

11278

11286

11297

11269

11279

11287

11298

11261

11299

11307

11318

11330

11340

11300

11308

11319

11331

11341

11301

11311

11320

11333

11342

11302

11312

11321

11334

11343

11303

11313

11325

11335

11344

11304

11314

11327

11337

11345

11305

11315

11328

11338

11348

11306

11317

11329

11339

11350

11351

11361

11369

11377

11385

11353

11362

11370

11378

11386

11354

11363

11371

11380

11387

11356

11364

11372

11381

11388

11357

11365

11373

11382

11389

11358

11366

11374

11383

11390

11359

11367

11375

11384

11391

11360

11368

11376

11392

11393

11394

11395

11396

11397

11398

11399

11400

11401

11402

11403

11404

11405

11406

11407

11408

11409

11410

11411

11412

11413

11414

11415

11417

11418

11419

11420

11421

11422

11423

11424

11425

11426

11427

11428

11429

11430

11431

11432

11433

11434

11444

11452

11460

11461

11468

11435

11445

11453

11462

11469

11436

11446

11454

11463

11470

11437

11447

11455

11464

11471

11438

11456

11465

11472

11448

11439

11457

11466

11474

11440

11449

11458

11467

11475

11443

11450

11459

11476

11451

11477

STRUCTURAL FORMULAS OF ORGANIC COMPOUNDS (continued)

In numeric order as they occur in the Table of Physical Constants of Organic Compounds

11478

11479

11480

11481

11482

11483

11484

11485

11486

11487

11488

11489

11490

11491

11492

11493

11494

11495

11496

11497

11498

11499

11500

11501

11502

11503

11504

11505

11506

11507

11508

11509

11510

11511

11512

11513

11514

11517

11519

11520

11521

11527

11541

11551

11564

11575

11528

11542

11555

11567

11576

11529

11543

11556

11568

11577

11531

11544

11558

11569

11578

11532

11545

11559

11570

11579

11536

11546

11560

11571

11580

11537

11549

11562

11573

11581

11539

11550

11563

11574

11582

11540

11583

11593

11603

11617

11631

11604

11618

11632

11584

11594

11605

11619

11633

11585

11606

11621

11634

11595

11607

11622

11635

11586

11596

11609

11624

11636

11610

11625

11637

11588

11598

11611

11626

11638

11599

11612

11627

11640

11590

11613

11628

11641

11600

11614

11629

11591

11615

11642

11601

11616

11630

11643

11592

11602

11644

11656

11666

11675

11683

11646

11657

11667

11676

11684

11647

11658

11668

11677

11685

11648

11659

11670

11678

11687

11649

11660

11671

11679

11688

11650

11661

11672

11680

11689

11651

11662

11673

11681

11690

11652

11664

11682

11691

11654

11665

11674

11692

11693

11703

11713

11723

11731

11694

11704

11714

11724

11695

11705

11715

11725

11732

11696

11706

11716

11726

11733

11697

11707

11717

11727

11734

11698

11708

11719

11728

11735

11699

11709

11720

11729

11736

11700

11710

11721

11730

11737

11701

11712

11722

11738

11739

11748

11756

11764

11773

11740

11749

11757

11765

11774

11741

11750

11758

11767

11775

11742

11751

11759

11768

11776

11743

11752

11760

11769

11777

11744

11753

11761

11770

11778

11746

11754

11762

11771

11779

11747

11755

11763

11772

11780

11781

11789

11797

11807

11819

11782

11790

11798

11808

11822

11783

11791

11799

11809

11823

11784

11792

11800

11810

11824

11785

11793

11801

11812

11826

11786

11794

11803

11813

11827

11787

11795

11804

11816

11828

11788

11796

11805

11817

11829

11806

11831

11842

11854

11865

11878

11833

11845

11855

11866

11879

11834

11846

11856

11867

11880

11835

11847

11857

11868

11881

11836

11848

11858

11869

11882

11837

11849

11859

11870

11883

11840

11852

11860

11871

11884

11841

11853

11861

11872

11885

11862

11873

11886

11863

11874

11887

11875

11888

11877

11889

11890

11891

11892

11893

11894

11895

11897

11898

11899

11900

11901

11902

11903

11904

11905

11906

11908

11909

11910

11911

11912

11913

11914

11915

11916

11917

11919

11920

11921

11922

11926

11929

11931

11934

11935

11936

11937

11938

11939

11940

11941

11942

11943

11944

11945

11946

11947

11948

11949

11950

11962

11974

11987

11999

11951

11963

11975

11988

12000

11953

11964

11977

11990

12001

11954

11965

11978

11991

12002

11955

11966

11979

11992

12004

11956

11967

11980

11993

12005

11957

11968

11981

11994

11958

11969

11982

11995

12006

11959

11970

11983

12007

11960

11971

11984

11996

11961

11972

11985

11973

11986

12009

12018

12032

12043

12058

12010

12019

12033

12044

12059

12012

12026

12034

12047

12061

12013

12027

12037

12048

12062

12014

12028

12038

12049

12065

12015

12029

12039

12050

12066

12016

12030

12040

12055

12067

12017

12031

12041

12056

12068

12057

β-Aminopyridine	: 10858
4-Aminopyridine	: 10859
Aminopyrine	: 10851
2-Aminoquinoline	: 11202
4-Aminoquinoline	: 11203
5-Aminoquinoline	: 11204
6-Aminoquinoline	: 11205
8-Aminoquinoline	: 11206
6-Aminoquinoxaline	: 11332
Aminorex	: 8574
4-Amino-1-β-D-ribofuranosyl-2(1H)-pyrimidinone	: 5021
p-Aminosalicylic acid	: 2498
5-Aminosalicylic acid	: 2499
p-Aminosalicylic acid hydrazide	: 2500
α-Aminosuccinamic acid	: 468
L-Aminosuccinic acid	: 471
2-Amino-1,3-thiazole	: 11662
2-Aminothiazole	: 11662
Aminoxafen	: 8574
Aminozide	: 3334
Amiphenazole	: 11671
Amitrole	: 11845
Amixetrine	: 11169
Amizepin	: 5144
Ammonium, acetate	: 106
Ammonium, (3-carbamoyl-3,3-diphenylpropyl) diisopropyl methyl,iodide	: 2120
Ammonium, diethyl(2-hydroxyethyl)methyl-, bromide, xanthene-9-carboxylate	: 5445
Ammonium, dimethyl ethyl 3-hydroxyphenyl,chloride	: 865
Ammonium formate	: 6003
Ammonium, pentamethylenebis trimethyl, bromide	: 8738
Ammonium, phenyl trimethyl,iodide	: 866
Ammonium, tetraethyl,bromide	: 5447
Ammonium, tetramethyl,bromide	: 7456
Ammonium, tetramethyl,chloride	: 7457
Ammonium, tetramethyl,iodide	: 7458
Ammonium, tetrapropyl,iodide	: 9833
Amobarbital	: 11092
Amolanone	: 2463
Amphecloral	: 1583
Amphetamine	: 1578
Ampyrone	: 10845
Amudane	: 11520
Amygdalin	: 954
Amyl acetate	: 227
sec-Amyl acetate (R)	: 8918
Amylacetylene	: 6609
tert-Amyl alcohol	: 3628
Amyl alcohol	: 8915
sec-Amyl alcohol	: 8916
iso-Amylamine	: 3223
Amylamine	: 8700
Amylbenzene	: 2107
d-Amyl bromide	: 3238
tert-Amyl bromide	: 3240
Amyl bromide	: 8710
Amyl butyrate	: 3579
Amyl caproate	: 6907
tert-Amyl carbamate	: 3634
Amyl chloride	: 8719
6-n-Amyl-m-cresol	: 9425
Amylene	: 3801
α-Amylene	: 9004
Amylene dichloride	: 3286
2,4-Amylene glycol	: 8789
Amyl ether	: 8835
Amyl ethyl ketone	: 8472
Amylethylmethylcarbinol	: 8466
Amyl formate	: 6015
Amyl heptanoate	: 6470
Amyl hexanoate	: 6907
Amyl hexyl ketone	: 5326
Amyl iodide	: 8818
Amylisopropylmethylcarbinol	: 8451
tert-Amyl isovalerate	: 3546
Amyl ketone	: 11962
Amyl mercaptan	: 8844
sec-Amylmercaptan	: 8845
Amyl nitrite	: 8090
	: 8096
Amylol	: 8915

4-Amylresorcinol	: 1544
Amyl salicylate	: 2649
Amyl stearate	: 8270
Amyl sulfide	: 8843
Amyltrichlorosilane	: 11472
Amyl valerate	: 8913
β-Amyrin	: 12038
α-Amyrin	: 11008
Anabasine	: 11008
Anagyrine	: 7573
Analeptin	: 1887
Anastil	: 9379
Anatabine	: 3105
Anchovyxanthin, all-trans-	: 4057
Androfluorone	: 355
4-Androstene-3,17-dione	: 352
Androsterone	: 348
Anemonin	: 5185
Anesthesin	: 2497
Anethole	: 1942
Angelic acid	: 3845
Anhalamine	: 7393
Anhalonidine	: 7395
Anhydroecgonine	: 491
Anhydron	: 2851
Aniline	: 633
Aniline, 2-bromo	: 638
Aniline, 3-bromo	: 639
Aniline, 4-bromo	: 640
Aniline, 2-bromo-6-nitro-	: 653
Aniline, 4-butyl	: 655
Aniline, N-butyl-	: 656
Aniline, 2-tert-butyl	: 714
Aniline, 4-tert-butyl	: 715
Aniline, N-tert-butyl-	: 716
Aniline-4-sec-butyl	: 813
Aniline, N-sec-butyl-	: 816
Aniline-3-carboxylic acid	: 2485
Aniline-4-carboxylic acid	: 2486
Aniline, 2-chloro	: 658
Aniline, 3-chloro	: 659
Aniline, 4-chloro	: 660
Aniline, 2-chloro, hydrochloride	: 665
Aniline, 2,4-dibromo-	: 681
Aniline, 2,6-dibromo-	: 682
Aniline, 3,4-dibromo-	: 683
Aniline, 2,3-dichloro-	: 685
Aniline, 2,4-dichloro-	: 686
Aniline, 2,5-dichloro-	: 687
Aniline, 2,6-dichloro-	: 688
Aniline, 3,4-dichloro-	: 689
Aniline, 3,5-dichloro-	: 690
Aniline, 2,6-dichloro-4-ethoxy	: 691
Aniline, 3,4-diethoxy-	: 693
Aniline, N,N-diethyl-4-nitroso-	: 699
Aniline, 2,4-difluoro-	: 700
Aniline, 2,4-diiodo-	: 701
Aniline, 2,6-diisopropyl-	: 636
Aniline, 2,6-dimethoxy-	: 703
Aniline, N,N-dimethyl-p-[(α,α,α-trifluoro-m-tolyl)azo]-	: 723
Aniline, 2,3-dinitro-	: 724
Aniline, 2,6-dinitro-	: 726
Aniline, 3,5-dinitro-	: 727
Aniline, hydrochloride	: 756
Aniline, 2-methylthio	: 819
Aniline nitrate	: 821
Aniline, 5-nitro-2-propoxy	: 827
p-Anilinesulfonamide	: 2187
2-Anilinesulfonic acid	: 2215
3-Anilinesulfonic acid	: 2216
Aniline-4-sulfonic acid	: 2217
Aniline, 2,3,4,5-tetramethyl	: 847
Aniline, 2,3,4-trichloro-	: 852
p-Anilinobenzenesulfonic acid	: 2234
o-Anilinobenzoic acid	: 2715
Anilotic acid	: 2647
p-Anisaldehyde	: 585
Anise alcohol	: 1891
p-Anisic acid	: 2666
m-Anisic acid, 2-(methylamino)-, methyl ester	: 2672
o-Anisidine	: 766
m-Anisidine	: 767
p-Anisidine	: 768
m-Anisidine, 4-nitro-	: 776
Anisindione	: 7269

Anisole	: 1915
Anisole, 3,5-dinitro-	: 1924
Anisole, 2,5-dinitro-	: 1926
Anisole, 2,3-dinitro	: 1922
Anisole, 4-nitro	: 1937
Anisole, 2,3,4,6-tetrabromo	: 2256
Anisole, 3,4,5-tribromo	: 2310
Anisole, 2,4,5-tribromo	: 2311
Anisole, 2,3,5-tribromo	: 2312
o-(p-Anisoyl)benzoic acid	: 2667
p-Anisoyl chloride	: 2901
p-Anisyl alcohol	: 1891
o-Anisyl benzoate	: 9383
1-(p-Anisyl)ethanol	: 1895
[6]Annulene	: 867
[8]Annulene	: 4742
Anthorine	: 475
Anthracene, 9,10-dibenzyl-	: 367
Anthracene, 9,3-dimethyl	: 378
Anthragallol	: 415
Anthralin	: 430
Anthranil	: 2390
o-Anthranilic acid	: 2484
Anthranilic acid, N-methyl-, ethyl ester	: 2680
Anthranilic acid, 3-nitro-	: 2506
Anthranilic acid, N-(α,α,α-trifluoro-m-tolyl)-	: 2729
Anthranol	: 432
Anthraquinone	: 379
Anthraquinone, 2-amino-1-hydroxy	: 383
Anthraquinone, 2,7-dibromo-	: 390
9,10-Anthraquinone, 1,7-dihydroxy	: 394
Anthrarobin	: 429
Anthrarufin	: 393
1-Anthroic acid	: 368
2-Anthroic acid	: 369
9-Anthroic acid	: 370
Anthrone	: 433
Anthropodeoxycholic acid	: 4097
Anticanitic vitamin	: 2486
Antifebrin	: 86
Antimonate(2-), bis[μ-[2,3-dihydroxybutanedioato(4-)-O1,O2:O3,O4]]di-, dipotassium, trihydrate, stereoisomer	: 11588
Antipyrine	: 10847
Antipyrine, 4-bromo-	: 10846
ANTU	: 11807
Apacil	: 2498
Apacizin	: 2500
Aphylline	: 7572
Apigenin	: 2806
Apigenin 4'-methyl ether	: 2809
Apiole	: 2423
Apoatropine	: 928
Apocodeine	: 5160
Apocynin	: 5824
Apomorphine	: 5155
6aα-Aporphine-2,9-diol, 1,10-dimethoxy-	: 5156
6aα-Aporphine, 1,2,9,10-tetramethoxy-	: 5157
6aα-Aporphin-11-ol, 10-methoxy-1,2-(methylenedioxy)-	: 2470
6aα-Aporphin-11-ol, 1,2,10-trimethoxy-	: 5161
Aprobarbital	: 11098
Arabinose (D)(β)	: 436
Arachic alcohol	: 5358
Arachidic acid	: 5355
Arachidonic acid	: 5361
Aramite	: 11579
Arborine	: 11200
Arbutin	: 6224
Arecaidine	: 10913
Arecoline	: 10914
Argobyl	: 10489
Armstrong's acid	: 7867
Arsanilic acid	: 456
Arsine, difluoromethyl-	: 463
Arsine, oxybis[dimethyl-	: 452
Ascaridole	: 5184
Ascensil	: 10867
Asculetine	: 2804
Asmacoril	: 2840
Aspartame	: 9511

2,3,1',8'-Binaphthylene	: 2739
Bineopentyl	: 6849
Bineophyl	: 2281
Biochanin B	: 2830
Biotin	: 11689
Biphenyl	: 3008
p-Biphenylacetic acid	: 3009
p-Biphenylamine	: 3011
4-Biphenylamine, 4'-methyl-	: 3015
Biphenyl, 2-amino-6-methyl	: 3013
Biphenyl, 4-amino-4'-methyl	: 3015
2,4'-Biphenyldiamine	: 3029
2,2'-Biphenyldicarboxylic acid	: 3038
Biphenyl, 2,2'-dimethoxy-	: 3045
Biphenyl, 3,3'-dimethoxy-	: 3046
Biphenyl, 2,2'-dimethoxy-5,5'-dimethyl	: 3048
Biphenyl, 2,5'-dimethoxy-2',5-dimethyl	: 3049
Biphenyl, 2,3-dimethyl-	: 3052
Biphenyl, 2,4-dimethyl-	: 3054
Biphenyl, 3,4-dimethyl-	: 3059
Biphenyl, 3,5-dimethyl-	: 3060
4,4'-Biphenyldisulfonic acid	: 3070
o-Biphenylenemethane	: 5967
2,2'-Biphenylene oxide	: 5163
Biphenyl, 3-ethoxy	: 3071
2-Biphenylol	: 3088
4-Biphenylol	: 3090
2-Biphenylol, 3-chloro-	: 3092
2,3'-Bipiperidyl	: 3098
Bipropargyl	: 6716
Bipropionyl	: 6806
3,3'-Bipyridyl	: 3102
3,4'-Bipyridyl	: 3103
4,4'-Bipyridyl	: 3104
2,3'-Bipyridyl	: 3100
2,3'-Biquinolyl	: 3107
5,5'-Biresorcinol	: 3097
Birodan	: 2136
Bis(4-acetamidophenyl) sulfone	: 91
N,N '-Bis(acetoacetyl)-3,3'-dimethylbenzidine	: 3204
Bisacodyl	: 9465
Bis(β-amino-β-carboxyethyl) disulfide	: 5020
Bis(p-aminophenyl) ether	: 832
Bis(4-aminophenyl) sulfide	: 850
Bis(3-aminopropyl)amine	: 9874
1,4-Bis(3-aminopropyl)piperazine	: 9614
N,N '-Bis(3-aminopropyl)-1,4-tetramethylenediamine	: 3259
1,3-Bis(bromomethyl)tetramethyl-disiloxane	: 5229
1,2-Bis(chlorocarbonyl)ethane	: 3366
Bis(2-chloroethyl) carbonate	: 5697
Bis(2-chloroethyl) ether	: 5611
Bis(2-chloroethyl) sulfide	: 5645
Bis(chloromethyl) ether	: 7525
Bis(p-chlorophenoxy)methane	: 1965
1,1-Bis(4-chlorophenyl)ethane	: 1678
1,1-Bis(4-chlorophenyl)ethanol	: 1850
3,6-Bis(2-chlorophenyl)-1,2,4,5-tetrazine	: 4139
1,1-Bis(4-chlorophenyl)-2,2,2-trichloroethanol	: 1851
1,1-Bis(p-chlorophenyl)-2,2,2-trichloroethanol	: 1851
Bis(3-chloropropyl) sulfide	: 10153
Bis(cyclohexyl)carbodiimide	: 4291
Bis(cyclopentadienyl)iron	: 5960
1,3-Bis-(dichloromethyl)tetramethyl disiloxane	: 5231
Bis(2,2-diethoxyethyl) ether	: 5612
Bis-O,O-diethylphosphoric anhydride	: 5221
Bis(diethylthiocarbamoyl) sulfide	: 11714
Bis(dimethylamino)phosphonous anhydride	: 5220
2,3-Bis(ethoxycarbonyl)furan	: 6082
Bis-(2-ethoxyethyl) amine	: 5425
Bis(ethoxymethyl)ether	: 5614
Bis(2-ethylhexyl) adipate	: 6781
Bis(2-ethylhexyl) decanedioate	: 5043
Bis(2-ethylhexyl) phthalate	: 1334
Bis(2-ethylhexyl) sebacate	: 5043
2,2-Bis(ethylsulfonyl)propane	: 9835

Bis(2-furyl)glyoxime	: 5583
Bis-(3-hydroxy-2-butyl)amine	: 3624
Bishydroxycoumarin	: 2839
Bis(hydroxymethyl)acetylene	: 3911
Bis-(2-methallyl)carbonate	: 10703
Bis(methoxydimethylsilyl) oxide	: 5234
Bis(2-methoxyethyl) carbonate	: 5736
Bis(2-methylbutyl) sulfide	: 3435
Bis[methylmercapto]methane	: 7461
Bis(methylthio)methane	: 7461
1,2'-Bis-naphthyl ether	: 7928
Bis(m-nitrophenyl) disulfide	: 5247
Bisphenol A	: 9413
Bisphenol B	: 9433
1,2-Bis(phenylamino)ethane	: 5508
Bis-(1-phenylethyl)ether (DL)	: 2081
Bis(1,2,2,2-tetrachloroethyl) ether	: 5615
1,5-Bis(2-tetrahydrofuryl)-3-pentanol	: 6166
Bis(p-tolyl)mercury	: 7432
Bis(tributyltin) oxide	: 5241
Bithionol	: 9478
Bitoscanate	: 1432
Δ 1,1'-Biurea	: 5124
Biuret	: 7232
Bivinyl	: 3146
Bixin	: 5108
Boldenone	: 343
Boldine	: 5156
Boldine dimethyl ether	: 5157
Bolusanthin	: 2801
Borane, triisopentyl-	: 3122
Borane-trimethylamine (1:1)	: 3132
Borine, dimethoxy-	: 3117
Borine, triisopentyl-	: 3122
Bornane, 2-chloro-	: 2919
10-Bornanesulfonic acid, 2-oxo-, (1S,4R)-(+)-	: 2929
3-Bornanol, acetate, (1S,3R,4S)-(-)-	: 2947
(+)-2-Bornanone	: 2961
2-Bornanone, 3-acetyl-17O-	: 2950
Borneol, acetate	: 2948
Bornyl chloride	: 2920
Bornyl formate (d)	: 2949
Bornyl isovalerate (d)	: 3565
d-Bornyl isovalerate	: 3565
Boron fluoride-diethyl ether compound	: 3137
Boron trifluoride compd. with methyl ether	: 3138
Boron trifluoride etherate	: 3137
Boron trimethoxide	: 3126
Brassidic acid	: 5282
Brilliant green	: 5444
Bristurin	: 2858
Bromacetocarbamide	: 3195
ω-Bromacetophenone	: 5782
Bromal	: 20
Bromcarbamide	: 3196
Bromcholitin	: 5157
Bromcresol green	: 9367
Bromcresol purple	: 9364
Bromdiethylacetylurea	: 3195
Bromindione	: 7265
Bromine cyanide	: 4165
Bromisovalum	: 3196
5-Bromoacenaphthene	: 4
N-Bromoacetamide	: 36
p-Bromoacetanilide	: 37
Bromoacetic acid	: 111
Bromoacetic acid, propyl ester	: 120
Bromoacetone	: 10416
p-Bromoacetophenone	: 5781
Bromoacetyl bromide	: 292
Bromoacetylene	: 5937
5-Bromoacetylsalicylic acid	: 2481
o-Bromoaniline	: 638
m-Bromoaniline	: 639
p-Bromoaniline	: 640
5-Bromoanthranilic acid	: 2487
4-Bromoantipyrine	: 10846
Bromoaspirin	: 2481
Bromobenzene	: 998
p-Bromobenzenesulfonyl chloride	: 2244
p-Bromobenzoic acid	: 2520
2-Bromobenzophenone	: 7600
p-Bromobenzyl bromide	: 1000
p-Bromobenzyl bromide	: 1008

α-Bromobenzyl cyanide	: 947
4-Bromo-1,2-butadiene	: 3147
1-Bromo-1,3-butadiene	: 3148
2-Bromo-1-butene	: 3709
3-Bromo-1-butene	: 3710
(±)-2-Bromobutyric acid	: 3468
DL-α-Bromobutyric acid	: 3468
(RS)-α-Bromobutyric acid	: 3468
Bromochloroacetic acid	: 112
Bromochloroacetic acid, ethyl ester	: 113
1-Bromo-3-chloroacetone	: 10417
2-Bromo-1-chlorobutane	: 3234
Bromochlorodifluoromethane	: 7464
Bromochlorodinitromethane	: 7465
1-Bromo-2-chloroethane	: 5455
Bromochlorofluoromethane	: 7466
Bromochloromethane	: 7463
1-Bromo-4-chloronaphthalene	: 7741
4-Bromo-2-chlorophenol	: 9232
1-Bromo-2-chloropropane	: 9838
1-Bromo-3-chloropropane	: 9839
2-Bromo-1-chloropropane	: 9840
2-Bromo-2-chloropropane	: 9841
2-Bromo-5-chlorothiophene	: 11728
4-Bromo-3-chloro-3,4,4-trifluoro-1-butene	: 3715
1-Bromo-2-chloro-1,1,2-trifluoroethane	: 5457
6-Bromo-m-cresol	: 9237
Bromocycloheptane	: 4222
2-Bromo-1-cyclohexanol	: 4503
Bromocyclopentane	: 4777
2-Bromo-1-decene	: 5086
1-Bromo-2-decene	: 5087
2-Bromodibenzofuran	: 5164
3-Bromodibenzofuran	: 5165
Bromodichlorofluoromethane	: 7468
Bromodichloromethane	: 7467
2-Bromo-2,2-diethylacetamide	: 3198
1-Bromo-1,2-difluoroethylene	: 5883
Bromodifluoromethane	: 7469
2-Bromo-2,3-dimethylbutane	: 3236
1-Bromo-1,1-dimethylbutane	: 8717
4-Bromo-1,3-dimethylpyrazole	: 10816
1-Bromo-2,3-dinitrobenzene	: 1024
Bromodiphenylmethane	: 1058
2-Bromoethanesulfonyl chloride	: 5628
2-Bromoethanol	: 5685
2-Bromoethyl ethyl ether	: 5461
2-Bromo-2-ethylisovaleramide	: 3199
2-Bromofluorene	: 5968
Bromofluoroacetic acid	: 116
1-Bromo-4-fluorobutane	: 3237
1-Bromo-10-fluorodecane	: 5035
1-Bromo-12-fluorododecane	: 5294
1-Bromo-6-fluorohexane	: 6735
Bromofluoromethane	: 7470
1-Bromo-5-fluoropentane	: 8713
1-Bromo-11-fluoroundecane	: 11939
Bromoform	: 7548
2-Bromofumaric acid	: 3770
3-Bromofuran	: 6034
1-Bromoheptadecane	: 6307
2-Bromoheptane	: 6357
4-Bromoheptane	: 6358
2-Bromoheptanoic acid	: 6457
1-Bromo-1-heptyne	: 6612
1-Bromo-2-heptyne	: 6613
2-Bromohexane	: 6733
3-Bromohexane	: 6734
1-Bromo-3-iodobenzene	: 1047
m-Bromoiodobenzene	: 1047
1-Bromo-2-iodoethane	: 5463
Bromoiodomethane	: 7471
α-Bromoisobutyric acid	: 10209
4-Bromoisoquinoline	: 7374
β-Bromoisovaleric acid	: 3472
2-Bromoisovalerylurea	: 3196
Bromomethane	: 7462
Bromomethoxymethane	: 7472
2-Bromo-p-methylacetophenone	: 5779
2-Bromo-5-methylaniline	: 648
(+)-1-Bromo-2-methylbutane	: 3238
(S)-1-Bromo-2-methylbutane	: 3238
2-Bromo-3-methyl-1-butene	: 3716
(Bromomethyl)chlorodimethylsilane	: 11365
(Bromomethyl)dimethylchlorosilane	: 11365

Carbamic acid, [1-[(butylamino)carbonyl]-1H-benzimidazol-2-yl]-, methyl ester	: 515
Carbamic acid, sec-butyl-, ethyl ester	: 3960
Carbamic acid, butyl-, 2-(hydroxymethyl)-2-methylpentyl ester, carbamate	: 3933
Carbamic acid, N-sec-butyl, methyl ester	: 3962
Carbamic acid, (3-chlorophenyl)-, 4-chloro-2-butynyl ester	: 510
Carbamic acid, N,N-diethyl	: 3941
Carbamic acid, dimethyl-, 2-(dimethylamino)-5,6-dimethyl-4-pyrimidinyl ester	: 9728
Carbamic acid, dimethyldithio-, manganese(2+) salt	: 7421
Carbamic acid, N-ethyl, butyl ester	: 3947
Carbamic acid, N-ethyl-N-nitro, butyl ester	: 3951
Carbamic acid, N-ethyl-N-nitro, ethyl ester	: 3950
Carbamic acid, isobutyl ester	: 3959
Carbamic acid, isopentyl-, ethyl ester	: 3952
Carbamic acid, isopropyl ester	: 3955
Carbamic acid, N-isopropyl-N-nitro, ethyl ester	: 3957
Carbamic acid, (3-methylphenyl)-, 3-[(methoxycarbonyl)amino]phenyl ester	: 9198
Carbamic acid, [3-[[(phenylamino)-carbonyl]oxy]phenyl]-, ethyl ester	: 5104
Carbamic acid, [1,2-phenylenebis-(iminocarbonothioyl)]bis-, dimethyl ester	: 11723
Carbamic acid, 3-phenylpropyl ester	: 2149
Carbamic acid, 1-phenyl-2-propynyl ester	: 1877
Carbamic acid, N-propyl-N-nitro, ethyl ester	: 3964
Carbamide	: 11992
Carbamimidoselenoic acid	: 11354
Carbamodithioic acid, diethyl-, 2-chloro-2-propenyl ester	: 4065
Carbamodithioic acid, dimethyl-, sodium salt	: 11514
Carbamonitrile	: 4155
Carbamothioic acid, bis(2-methylpropyl)-, S-ethyl ester	: 11586
Carbamothioic acid, bis(1-methylethyl)-, S-(2,3,3-trichloro-2-propenyl) ester	: 11823
Carbamothioic acid, bis(1-methylethyl)-, S-(2,3-dichloro-2-propenyl) ester	: 5106
Carbamothioic acid, cyclohexyl-ethyl-, S-ethyl ester	: 4168
Carbamothioic acid, diethyl-, S-[(4-chlorophenyl)methyl] ester	: 11697
Carbamothioic acid, diethyl-, S-ethyl ester	: 5930
Carbamothioic acid, dipropyl-, S-ethyl ester	: 5372
Carbamothioic acid, dipropyl-, S-propyl ester	: 12047
Carbamoyl chloride, N,N-diethyl	: 3972
Carbamoylcholine chloride	: 5440
5-Carbamoyl-5H-dibenz[b,f]azepine	: 5144
Carbamyl chloride	: 3971
Carbanilaldehyde	: 5999
Carbanilide	: 12010
Carbarsone	: 455
Carbaryl	: 8032
Carbazide	: 4009
(N-Carbazolyl)acetic acid	: 3982
Carbazone, 1,5-diphenyl	: 5122
Carbic anhydride	: 7580
Carbimazole	: 7194
Carbitol	: 5720
Carbitol acetate	: 5721
Carbobenzoxy chloride	: 4027
Carbocysteine	: 5017
Carbodiimide, dicyclohexyl-	: 4291
Carbohydrazide	: 4009
Carbomethene	: 5927

o-Carbomethoxyaniline	: 2503
2-Carbomethoxyphenol	: 2634
Carbon bisulfide	: 3993
Carbon dibromide dichloride	: 7495
Carbon diselenide	: 4036
Carbonic acid, bis(2-chloroethyl) ester	: 5697
Carbonic acid, bis(2-ethoxyethyl) ester	: 5719
Carbonic acid, bis(2-methoxyethyl) ester	: 5736
Carbonic acid, bis(2-methoxyphenyl) ester	: 9384
Carbonic acid, bis(2-methylallyl) ester	: 10703
Carbonic acid, dithio-, cyclic S,S-(6-methyl-2,3-quinoxalinediyl) ester	: 4072
Carbonic acid, monochloride,monoethyl ester	: 4018
Carbonic anhydride	: 3992
Carbon oxide	: 4011
Carbon oxysulfide	: 4035
Carbon suboxide	: 9773
Carbon tetrabromide	: 7539
Carbon tetrachloride	: 7540
Carbon tetrafluoride	: 7541
Carbon tetraiodide	: 7542
Carbonyl bromide	: 4006
Carbonyl chloride	: 4007
Carbonyl chloride fluoride	: 4005
Carbonyl cyanide	: 9915
N,N'-Carbonyldiimidazole	: 7193
Carbonyl fluoride	: 4008
Carbonyl sulfide	: 4035
Carbostyril	: 11325
3-Carboxyaniline	: 2485
4-Carboxyaniline	: 2486
9-Carboxyanthracene	: 370
4-Carboxybenzenesulfonamide	: 2508
3-Carboxycoumarin	: 2775
4-Carboxy-1,3-dimethylbenzene	: 2580
2-Carboxyethylamine	: 323
2-Carboxyfuran	: 6050
4-Carboxymethylbiphenyl	: 3009
S-(Carboxymethyl)-L-cysteine	: 5017
3-[(Carboxymethyl)thio]-L-alanine	: 5017
m-Carboxynitrobenzene	: 2704
p-Carboxynitrobenzene	: 2705
4-Carboxyphenol	: 2613
4-Carboxyphenylmethane-sulfonanilide	: 2721
4-Carboxyphthalic acid	: 2319
3-Carboxypropylamine	: 3464
8-Carboxyquinoline	: 11229
2-Carboxythiophene	: 11737
Carbromal	: 3195
Carbromide	: 3198
Carbutamide	: 2192
Carbyloxime	: 6024
Cardilate	: 3432
Cardine	: 3540
Carditoxin	: 4043
Cardrase	: 2865
Δ4-Carene (l)	: 4050
Carfimate	: 1877
Carinamide	: 2721
Carminic acid	: 373
Carmustine	: 11994
Carnegine	: 7390
Carnosine	: 7133
γ-Carotene	: 4052
α-Carotene	: 4054
β-Carotene, all-trans-	: 4051
α-Carotene, 6',7'-didehydro-5,6-epoxy-4',5,5',6,7,8-hexahydro-3,3'	: 4055
Carvacrol	: 9415
Carvacrol, acetate	: 9419
Carvacryl acetate	: 9419
Carvenol	: 4698
Carvenone (DL)	: 4717
(+)-Carvone	: 4715
(-)-β-Caryophyllene	: 2996
β-Caryophyllene	: 2996
trans-Caryophyllene	: 2996
Caryophyllene	: 2996
Carzenide	: 2508
Casimiroin	: 5218

Catapyrin	: 11060
Catechol, 5-iodo	: 1523
Caulophylline	: 7578
CDT	: 4212
Cedrin	: 8058
Cedrol	: 7570
Celidoniol, deoxy-	: 8099
Cellocidin	: 3902
Cellosolve	: 5717
Cephaeline	: 5367
Cepharanthine	: 8630
Cerotic acid	: 6624
Cetane	: 6632
1-Cetene	: 6667
Cetyl alcohol	: 6659
Cetylamine	: 6631
Cetyl bromide	: 6633
Cetyl chloride	: 6634
α-Cetylcitric acid	: 8102
Cetyl iodide	: 6640
Cetyl lactate	: 10271
Cetyl mercaptan	: 6643
Cetyl palmitate	: 6651
Cetyl stearate	: 8258
Cevadic acid	: 3844
Cevadine	: 4069
CFC 11	: 7555
CFC 12	: 7500
CFC 13	: 7486
CFC 13B1	: 7475
CFC 21	: 7502
CFC 22	: 7478
CFC 113	: 5663
CFC 114	: 5538
CFC 115	: 5484
CFC 123	: 5540
CFC 123a	: 5539
C16 Guerbet fatty acid	: 5065
Chalcone	: 10708
Champacol	: 507
Chaulmoogric acid	: 4936
Chavicine	: 9633
Chavicol	: 9458
Cheirolin	: 10067
Chelerythrine	: 2433
Chelidonic acid	: 10776
Chenic acid	: 4097
Chenodeoxycholic acid	: 4097
Chenodiol	: 4097
Chimyl alcohol	: 10010
Chitosamine	: 6233
Chloral	: 21
Chloral alcoholate	: 5769
Chloral ammonia	: 5680
Chloral, diethyl acetal	: 5657
Chloral formamide	: 6000
Chloral hydrate	: 5582
Chloralide	: 5212
Chloramben	: 2492
Chlorambucil	: 1101
Chloraminophenamide	: 1558
Chloramphenicol	: 52
Chloramphenicol palmitate	: 6648
Chloranil	: 4268
Chloranilic acid	: 4258
Chlorocyanohydrin	: 10115
Chlordantoin	: 7220
Chlordimeform	: 7564
Chlorfenac	: 944
Chlorfenethol	: 1850
Chlorfenson	: 2221
Chlorindanol	: 7283
Chlorine cyanide	: 4166
Chlornaphazine	: 7710
5-Chloroacenaphthene	: 5
2-Chloroacetaldehyde	: 12
Chloroacetaldehyde, trimer	: 13
Chloroacetamide	: 41
Chloroacetamide, N-3-chloroallyl	: 43
Chloroacetic acid (β)	: 123
Chloroacetic acid, methyl ester	: 138
Chloroacetic acid, phenyl ester	: 147
Chloroacetic acid, propyl ester	: 149
Chloroacetic anhydride	: 124
Chloroacetone	: 10420
m-Chloroacetophenone	: 5788
p-Chloroacetophenone	: 5789

Coumarin-3-carboxylic acid	: 2775
Coumarin, 3-chloro-7-hydroxy-4-methyl, O-ester with O,O-diethylphosphorothioate	: 4147
Coumarin, 3-ethyl-	: 2820
Coumarin, 7-hydroxy-	: 2826
Coumarin, 7-hydroxy-6-methoxy-	: 2829
cis-o-Coumarinic acid lactone	: 2792
Coumarin, 3-methyl-	: 2833
Coumarin, 5-methyl-	: 2834
Coumarin, 7-methyl-	: 2836
Coumarin, 8-methyl-	: 2837
Coumarone	: 2438
Coumarone-2-carboxylic acid	: 2439
Coumestrol	: 2466
Creatine	: 6254
Creatinine	: 7230
Creosol	: 9385
o-Cresol	: 9400
m-Cresol	: 9401
p-Cresol	: 9402
m-Cresol, 5-amino-	: 9212
m-Cresol, 6-bromo-	: 9237
m-Cresol, 6-sec-butyl-	: 9421
m-Cresol, 2,4-dichloro	: 9292
m-Cresol, 2,6-dichloro-	: 9294
p-Cresol, 2,6-di-tert-pentyl-	: 9225
p-Cresol, 2-iodo-	: 9374
o-Cresol, 4-isobutyryl-	: 10448
o-Cresol, 3-isopropyl, methyl ether	: 363
m-Cresol, 6-pentyl-	: 9425
o-Cresol, α-phenyl-	: 9454
p-Cresol, α-phenyl-	: 9455
o-Cresolphthalein	: 7353
Cresol red	: 9368
o-Cresol, 4,5,6-trichloro-	: 9487
p-Cresol, 2,3,6-trichloro-	: 9488
m-Cresol, 2,4,6-trichloro-	: 9489
o-Cresotic acid	: 2629
m-Cresotic acid	: 2630
p-Cresotic acid	: 2631
o-Cresyl acetate	: 211
m-Cresyl acetate	: 212
Crimidine	: 11043
Crithminic acid	: 2678
Croceic acid	: 7972
Crocetin (trans)	: 5107
Croconic acid	: 4939
Crodimyl	: 2838
Cromaril	: 2840
trans-Crotonaldehyde	: 3695
Crotonaldehyde, 2-chloro-	: 3697
Crotonaldehyde-2-chloro, dimethylacetal	: 3730
Crotonaldehyde, diethyl acetal	: 3758
Crotonchloral	: 3191
trans-Crotonic acid	: 3814
cis-Crotonic acid	: 3815
Crotonic acid, allyl ester	: 3856
Crotonic acid anhydride	: 3818
Crotonic acid, 4-chloro-	: 3822
Crotonic acid, 2-chloro-, ethyl ester	: 3823
Crotonic acid, 3-chloro-, ethyl ester	: 3826
Crotonic acid, 4-chloro-, ethyl ester	: 3829
Crotonic acid, 4-chloro-, methyl ester	: 3833
Crotonic acid, 2,3-dimethyl-, ethyl ester	: 3836
Crotonic acid, 2-ethyl-, stereoisomer	: 3837
trans-Crotyl alcohol	: 3859
cis-Crotyl alcohol	: 3860
Cryptocavine	: 2435
Cryptopine	: 2435
all-trans-β-Cryptoxanthin	: 4061
Cryptoxanthin	: 4061
Cryptoxanthol, all-trans-	: 4061
Crystal violet	: 7453
Cubebin	: 6136
Cumene	: 1975
Cumene hydroperoxide	: 7181
3-p-Cumenyl-2-methylpropionaldehyde	: 2117
Cumic acid	: 2689
Cumic alcohol	: 1902
Cumidine	: 789
Cuminaldehyde	: 590

Cuminic acid	: 2689
Cuminol	: 1902
Cumotocopherol	: 2789
α-Cumyl alcohol	: 1857
Cupferron	: 759
Cupreine	: 4119
Cupreol	: 11558
ar-Curcumene	: 1468
Curcumin	: 6334
Curine	: 11925
Curling factor	: 11520
Cuscohygrine	: 10415
Cusparine	: 11216
Cyacetacide	: 153
Cyamelide	: 11907
Cyanidelonon 1522	: 2817
Cyanoacetamide	: 49
Cyanoacetic acid	: 150
Cyanoacetic acid, diethyl, ethyl ester	: 3495
2-Cyanoacetophenone	: 2124
Cyanoacetylene	: 10737
N-Cyanoamine	: 4155
Cyanobenzene	: 2740
2-Cyanocaproic acid, ethyl ester	: 6881
Cyanocyclobutane	: 4173
1-Cyanocyclohexene	: 4630
Cyanocyclopentane	: 4779
1-Cyanocyclopentene	: 4907
4-Cyano-2,6-dibromophenol	: 2751
2-Cyanoethylamine	: 10092
(2-Cyanoethyl)benzene	: 2123
Cyanoethylene	: 10583
Cyano(4-fluoro-3-phenoxyphenyl)methyl 3-(2,2-dichloroethenyl)-2,2-dimethylcyclopropanecarboxylate	
1-Cyano-2-(2-furyl)ethene	: 10585
Cyanogen	: 5550
Cyanogenamide	: 4155
Cyanogen iodide	: 7337
Cyanogen nitride	: 4155
Cyanoiodomethane	: 283
Cyanomethane	: 271
Cyanomethyl methyl sulfide	: 286
Cyanomethyl phenyl ketone	: 2124
Cyano(3-phenoxyphenyl)methyl-3-(2,2-dibromoethenyl)-2,2-dimethylcyclopropanecarboxylate	: 5102
Cyano(3-phenoxyphenyl)methyl 2,2,3,3-tetramethylcyclo-propanecarboxylate	: 5953
1-Cyano-2-phenylethane	: 2123
1-Cyanopropane	: 3403
2-Cyano-2-propanol	: 10108
3-Cyano-1-propene	: 3803
2-Cyanoquinoline	: 11223
2-Cyanothiophene	: 11732
2-Cyanotoluene	: 2763
m-Cyanotoluene	: 2764
p-Cyanotoluene	: 2765
1-Cyanoundecane	: 5303
Cyanuric acid, dihydrate	: 11839
Cyanuric acid trichloride	: 11837
Cyclamal	: 2117
Cyclamic acid	: 11568
Cyclobarbital	: 11084
Cyclobutaneacetic acid, 3-carboxy-2,2-dimethyl-, diethyl ester	: 4171
Cyclobutane-1-carboxylic acid	: 4174
Cyclobutane-1,1-dicarboxylic acid, diethyl ester	: 4180
Cyclobutane-1,2-dicarboxylic acid, dimethyl ester (trans)	: 4181
Cyclobutane-1,3-dicarboxylic acid (cis), 2,2-dimethyl	: 4183
Cyclobutane-1,3-dicarboxylic acid, 2,2-dimethyl, dimethyl ester	: 4184
Cyclobutane, 1,2-divinyl-, trans-	: 4187
Cyclobutylamine	: 4169
Cyclobutylcarboxylic acid	: 4174
Cyclobutyl hydroxide	: 4196
Cyclobutyl methyl ketone	: 5792
Cyclobutyrol	: 4309
Cyclocapron	: 4333
β-Cyclocitral	: 4636

1,5,9-Cyclododecatriene, (cis, trans, trans)	: 4213
cis,cis,cis-Cyclododeca-1,5,9-triene	: 4214
Cyclododecanone	: 4211
Cyclofenchene	: 11866
Cycloform	: 2505
β-Cyclogeraniol	: 4666
α-Cyclogeraniol	: 4667
Cyclogeraniolane	: 4486
α-Cyclogeraniolene	: 4691
Cyclogeranyl alcohol	: 4666
Cycloglycylglycine	: 9612
9-Cycloheptadecene-1-one	: 4217
Cycloheptyl bromide	: 4222
Cycloheptylcarboxylic acid	: 4223
1,3-Cyclohexadiene, 3,5-dimethyl	: 4256
1,3-Cyclohexadiene, 5-methyl (DL)	: 4272
Cyclohexanacetic acid, 2-methyl	: 4311
Cyclohexanacetic acid, 3-methyl	: 4312
Cyclohexane, 1,3-bis(aminomethyl)	: 4362
Cyclohexane, 1,3-bis-(hydroxymethyl)	: 4363
Cyclohexane, sec-butyl-	: 4448
Cyclohexane, decafluoro-1,2-bis(trifluoromethyl)-	: 4342
Cyclohexane-1,2-dicarboxylic acid, diethyl ester (trans, DL)	: 4350
Cyclohexane-1,3-dicarboxylic acid, diethyl ester (cis)	: 4351
Cyclohexane-1,3-dicarboxylic acid, dimethyl ester (cis)	: 4352
Cyclohexane-1,3-dicarboxylic acid, dimethyl ester (cis)	: 4353
Cyclohexane-1,4-dicarboxylic acid, diethyl ester (trans)	: 4356
Cyclohexane-1,4-dicarboxylic acid, dimethyl ester (cis)	: 4357
Cyclohexane-1,4-dicarboxylic acid, diethyl ester (cis)	: 4355
Cyclohexane, 2,4-diisopropenyl-1-methyl-1-vinyl-, stereoisomer	: 4389
Cyclohexane, 1,2-dimethyl, perfluoro	: 4342
Cyclohexane, 1,2-epoxy-4-vinyl	: 8555
Cyclohexane, 1,2,3,4,5,6-hexachloro-, (α, DL)	: 4408
Cyclohexane hydroperoxide	: 7177
Cyclohexane, 1-hydroxymethyl-2-methyl (cis)	: 4422
Cyclohexane, 1-hydroxymethyl-2-methyl (trans)	: 4423
Cyclohexane, 1-isopropenyl-3-methyl	: 4435
Cyclohexane, isopropyl-	: 4431
Cyclohexane, 1-methylamino-4-methyl	: 4723
Cyclohexane, 1-methyl-2-ethyl (trans)	: 4396
Cyclohexane, 2-methyl-1-ethylidene	: 4393
Cyclohexane, 2-nitrophenyl	: 1251
1-L-1,3,4/2,5-Cyclohexanepentol	: 4074
Cyclohexanol, acetate	: 155
Cyclohexanol, 2-allyl-, trans-	: 4556
Cyclohexanol, 2-amino (trans, DL)	: 4501
Cyclohexanol, 1,2-dimethyl-, stereoisomer	: 4508
	: 4509
Cyclohexanol, 2-ethyl (trans, DL)	: 4523
Cyclohexanol, 2-isopropyl-, cis-	: 4539
Cyclohexanol, 4-isopropyl-, cis-	: 4540
Cyclohexanol, 4-isopropyl-, trans-	: 4541
Cyclohexanol-4-isopropyl-2-methyl	: 7428
Cyclohexanol, 2-methyl (cis, DL)	: 4531
Cyclohexanol, 2-phenyl (cis, DL)	: 4553
Cyclohexanol, 2-phenyl (trans, DL)	: 4554
Cyclohexanone-4-carboxylic acid, 2-methyl	: 4336
Cyclohexanone, 2,4-dimethyl (trans, d)	: 4583
Cyclohexanone, 2-isopropyl-	: 4599
Cyclohexanone, 4-isopropyl-	: 4600
Cyclohexanone, 2-isopropyl-5-methyl (l)	: 4606
1-Cyclohexene-1-carboxaldehyde, 4-isopropenyl-	: 4633
1-Cyclohexene-1-carboxaldehyde, 4-isopropyl-	: 4635

Di-*tert*-butyl malonate	: 9925
Dibutylmercury	: 7433
Di-*tert*-butylmethane	: 8841
2,6-Di-*tert*-butyl-4-methylphenol	: 9224
Dibutylnitrosamine	: 3212
Dibutyl oxide	: 3415
2,6-Di-*sec*-butylphenol	: 9227
Dibutyl phthalate	: 1339
2,6-Di-*tert*-butylpyridine	: 10874
Dibutyl sebacate	: 5044
Di-*tert*-butyl succinate	: 3296
Di-*sec*-butyl succinate	: 3297
Dibutyl succinate	: 3301
Di-*tert*-butyl sulfide	: 10155
Dibutyl sulfide	: 3433
Dibutyl sulfite	: 11580
Di-*tert*-Butyl tetrasulfide	: 11649
Dicamba	: 2567
1,3-Dicarboxyacetone	: 8782
2,2'-Dicarboxybiphenyl	: 3038
2,2'-Dicarboxydiphenylamine	: 2656
2,3-Dicarboxypyrazine	: 10804
2,5-Dicarboxythiophene	: 11745
Dicentrine	: 2468
Dichlobenil	: 2752
Dichlofenthion	: 9581
Dichlone	: 7850
Dichloramine-T	: 2210
Dichloroacetic acid	: 162
Dichloroacetic acid butyl ester	: 163
1,1-Dichloroacetone	: 10425
1,3-Dichloroacetone	: 10426
Dichloroacetonitrile	: 278
Dichloroacetyl chloride	: 297
Dichloroacetylene	: 5940
β,β-Dichloroacrylic acid	: 10630
2,4-Dichloro-6-aminophenol	: 9205
2,3-Dichloroaniline	: 685
2,4-Dichloroaniline	: 686
2,5-Dichloroaniline	: 687
2,6-Dichloroaniline	: 688
3,4-Dichloroaniline	: 689
3,5-Dichloroaniline	: 690
3,6-Dichloro-*o*-anisic acid	: 2567
2,5-Dichlorobenzaldehyde	: 535
3,4-Dichlorobenzaldehyde	: 536
3,5-Dichlorobenzaldehyde	: 537
o-Dichlorobenzene	: 1354
m-Dichlorobenzene	: 1355
p-Dichlorobenzene	: 1356
3,4-Dichlorobenzenecarboxaldehyde	: 536
o-Dichlorobenzidine	: 3032
2,4-Dichlorobenzoic acid	: 2561
2,5-Dichlorobenzoic acid	: 2562
2,6-Dichlorobenzoic acid	: 2563
3,5-Dichlorobenzoic acid	: 2564
2,6-Dichlorobenzonitrile	: 2752
2,6-Dichloro-*p*-benzoquinone-4-chloroimide	: 4280
2,4-Dichlorobenzoyl chloride	: 2893
3,4-Dichlorobenzoyl chloride	: 2894
2,4-Dichlorobenzyl alcohol	: 1853
2,2-Dichloro-1,1-bis(4-chlorophenyl)ethylene	: 1374
1,1-Dichloro-2,2-bis(*p*-chlorophenyl)ethane	: 1378
Dichlorobromomethane	: 7467
1,1-Dichloro-1,3-butadiene	: 3156
1,2-Dichloro-1,3-butadiene	: 3157
1,1-Dichlorobutane	: 3276
3,4-Dichlorobutane	: 3277
1,2-Dichlorobutane	: 3277
1,3-Dichlorobutane	: 3278
1,4-Dichlorobutane	: 3279
2,2-Dichlorobutane	: 3280
1,4-Dichloro-2,3-butanediol	: 3349
1,3-Dichloro-2-butanone	: 3654
1,3-Dichloro-1-butene	: 3743
2,3-Dichloro-1-butene	: 3744
3,4-Dichloro-1-butene	: 3745
1,2-Dichloro-3-butene	: 3745
1,1-Dichloro-2-butene	: 3746
1,2-Dichloro-2-butene	: 3747
1,4-Dichloro-*trans*-2-butene	: 3750
3,4-Dichlorobutyric acid	: 3926
Dichloro(chloromethyl)methylsilane	: 11390
2,6-Dichloro-*m*-cresol	: 9294

4,4-Dichlorocrotonic acid	: 4152
2,6-Dichlorocyanobenzene	: 2752
Dichlorocyanuric acid	: 11838
1,1-Dichlorocyclohexane	: 4358
cis-1,2-Dichlorocyclohexane	: 4359
cis-1,4-Dichlorocyclohexane	: 4361
1,10-Dichlorodecane	: 5040
2,2'-Dichloro-4,4'-diaminobiphenyl	: 3032
Dichloro(dichloromethyl)methyl-silane	: 11391
2,3-Dichloro-5,6-dicyano-1,4-benzoquinone	: 4251
2,3-Dichloro-5,6-dicyanobenzoquinone	: 4251
Dichlorodicyano-*p*-benzoquinone	: 4251
Dichlorodicyanoquinone	: 4251
Dichlorodiethylsilane	: 11393
1,2-Dichloro-1,1-difluoroethane	: 5531
1,2-Dichloro-1,2-difluoroethane	: 5532
1,1-Dichloro-2,2-difluoroethylene	: 5897
1,2-Dichloro-1,2-difluoroethylene	: 5898
Dichlorodifluoromethane	: 7500
2,5-Dichloro-3,6-dihydroxy-*p*-benzoquinone	: 4258
2,2'-Dichlorodiisopropyl ether	: 10123
2,2-Dichloro-1,1-dimethylethanol	: 10364
2,5-Dichloro-2,5-dimethylhexane	: 6765
1,2-Dichloro-4,4-dimethylpentane	: 8752
1,5-Dichloro-3,3-dimethylpentane	: 8753
2,4-Dichloro-2,4-dimethylpentane	: 8754
Dichlorodimethylsilane	: 11394
Dichlorodinitromethane	: 7501
Dichlorodiphenylsilane	: 11395
2,2-Dichloroethanol	: 5704
2,5-Dichloroethylbenzene	: 1377
1,1-Dichloroethylene	: 5894
trans-1,2-Dichloroethylene	: 5895
cis-1,2-Dichloroethylene	: 5896
Dichloroethylmethylsilane	: 11397
2,7-Dichlorofluorene	: 5972
Dichlorofluoroacetonitrile	: 279
1,1-Dichloro-1-fluoroethane	: 5534
1,1-Dichloro-2-fluoroethylene	: 5900
Dichlorofluoromethane	: 7502
1,1-Dichloroheptane	: 6377
1,2-Dichloroheptane	: 6378
1,7-Dichloroheptane	: 6379
2,2-Dichloroheptane	: 6380
2,3-Dichlorohexafluoro-2-butene	: 3754
1,2-Dichlorohexafluoropropane	: 9898
1,2-Dichlorohexane	: 6761
1,6-Dichlorohexane	: 6762
2,2-Dichlorohexane	: 6763
2,3-Dichlorohexane	: 6764
(±)-2,5-Dichlorohexane	: 6766
Di(5-Chloro-2-hydroxyphenyl)methane	: 9408
1,2-Dichloroindan	: 7249
Dichloroiodomethane	: 7503
2,4-Dichloromesitylene	: 1402
Dichloromethane	: 7499
1,1-Dichloro-3-methylbutane	: 3282
1,2-Dichloro-2-methylbutane	: 3283
2,4-Dichloro-2-methylbutane	: 3284
2,3-Dichloro-2-methylbutane	: 3286
1,2-Dichloro-3-methylbutane	: 3285
3,3-Dichloro-2-methyl-1-butene	: 3755
1,3-Dichloro-2-methyl-2-butene	: 3756
1,4-Dichloro-2-methyl-2-butene	: 3757
Dichloromethyl cyanide	: 278
(Dichloromethyl)dimethylchloro-silane	: 11373
Dichloromethyl methyl dichlorosilane	: 11391
Dichloromethylphenylsilane	: 11401
1,1-Dichloro-2-methylpropane	: 9899
1,2-Dichloro-2-methylpropane	: 9900
1,3-Dichloro-2-methylpropane	: 9901
2,4-Dichloro-5-methylpyrimidine	: 11052
2,6-Dichloro-4-methylpyrimidine	: 11053
Dichloromethylsilane	: 11398
(Dichloromethyl)trichlorosilane	: 11459
1,2-Dichloronaphthalene	: 7807
1,3-Dichloronaphthalene	: 7808
1,4-Dichloronaphthalene	: 7809
1,5-Dichloronaphthalene	: 7810
1,7-Dichloronaphthalene	: 7811

1,8-Dichloronaphthalene	: 7812
2,6-Dichloronaphthalene	: 7813
2,3-Dichloro-1-naphthol	: 8025
2,6-Dichloro-4-nitroaniline	: 692
2,3-Dichloronitrobenzene	: 1391
2,4-Dichloronitrobenzene	: 1396
2,5-Dichloronitrobenzene	: 1395
2,6-Dichloronitrobenzene	: 1393
3,4-Dichloronitrobenzene	: 1392
3,5-Dichloronitrobenzene	: 1394
1,1-Dichloro-1-nitroethane	: 5536
1,1-Dichloro-1-nitropropane	: 9902
2,4-Dichloro-5-nitropyrimidine	: 11054
1,9-Dichlorononane	: 8128
2,4-Dichloro-6-(*o*-chloroanilino)-*s*-triazine	: 360
1,8-Dichlorooctane	: 8371
1,2-Dichloropentane	: 8744
1,3-Dichloropentane	: 8745
1,4-Dichloropentane	: 8746
1,5-Dichloropentane	: 8747
2,2-Dichloropentane	: 8748
2,3-Dichloropentane	: 8749
2,4-Dichloropentane	: 8750
3,3-Dichloropentane	: 8751
3,5-Dichloro-2-pentanone	: 8963
2,5-Dichloro-2-pentene	: 9028
1,4-Dichloro-3-pentene	: 9028
Dichlorophene	: 9408
2,3-Dichlorophenol	: 9282
2,4-Dichlorophenol	: 9283
2,5-Dichlorophenol	: 9284
2,6-Dichlorophenol	: 9285
3,4-Dichlorophenol	: 9286
3,5-Dichlorophenol	: 9287
2,4-Dichlorophenyl acetate	: 9288
Dichlorophenylarsine	: 462
N-(3,5-Dichlorophenyl)-1,2-dimethylcyclopropane-1,2-dicarboximide	: 9760
Dichlorophenylphosphine	: 9556
Dichlorophenylphosphine sulfide	: 9552
1-[[2-(2,4-Dichlorophenyl)-4-propyl-1,3-dioxolan-2-yl]methyl]-1H-1,2,4-triazole	: 10720
Dichlorophenylsilane	: 11403
(±)-2-(3,5-Dichlorophenyl)-2-(2,2,2-trichloroethyl)oxirane	: 11899
4,5-Dichlorophthalic anhydride	: 7344
1,1-Dichloropropane	: 9891
1,3-Dichloropropane	: 9893
1,3-Dichloro-2-propanol	: 10362
2,3-Dichloropropene	: 10552
3,3-Dichloropropionyl chloride	: 10502
N,N-Dichloropropylamine	: 9805
1,1-Dichloropropylene	: 10547
trans-1,2-Dichloropropylene	: 10548
cis-1,2-Dichloropropylene	: 10549
trans-1,3-Dichloropropylene	: 10550
cis-1,3-Dichloropropylene	: 10551
2,3-Dichloropropylene	: 10552
3,3-Dichloropropylene	: 10553
3,6-Dichloro-2-pyridinecarboxylic acid	: 4142
2,4-Dichloroquinoline	: 11248
2,7-Dichloroquinoline	: 11249
4,7-Dichloroquinoline	: 11251
3,7-Dichloroquinoline-8-carboxylic acid	: 11201
2,6-Dichloroquinone-4-chloroimide	: 4280
2,3-Dichlorostyrene	: 1367
2,6-Dichlorostyrene	: 1369
3,5-Dichlorostyrene	: 1370
2,5-Dichlorostyrene	: 1371
2,4-Dichlorostyrene	: 1373
p-Dichlorosulfamoylbenzoic Acid	: 2565
p-(*N,N*-Dichlorosulfamyl)benzoic acid	: 2565
2,5-Dichloroterephthalic acid	: 1349
1,2-Dichlorotetrafluorobenzene	: 1398
1,3-Dichlorotetrafluorobenzene	: 1399
1,4-Dichlorotetrafluorobenzene	: 1400
1,1-Dichlorotetrafluoroethane	: 5537
1,2-Dichloro-1,1,2,2-tetrafluoroethane	: 5538
2,3-Dichlorothiophene	: 11746
2,4-Dichlorothiophene	: 11747

2-Ethyl-2-methylcyclohexanone	: 4595
1-Ethyl-4-methylcyclohexene	: 4663
1-Ethyl-1-methylcyclopentane	: 4814
cis-1-Ethyl-2-methylcyclopentane	: 4815
cis-1-Ethyl-3-methylcyclopentane	: 4817
trans-1-Ethyl-3-methylcyclopentane	: 4818
2-Ethyl-5-methylcyclopentanone	: 4883
cis-1-Ethyl-2-methylcyclopropane	: 4976
Ethyl methyl ether	: 5602
N-Ethyl-N'-(1-methylethyl)-6-(methylthio)1,3,5-triazine-2,4-diamine	: 341
2-Ethyl-5-methylfuran	: 6106
3-Ethyl-3-methylglutaric acid	: 8777
3-Ethyl-4-methylheptane	: 6411
3-Ethyl-5-methylheptane	: 6412
4-Ethyl-2-methylheptane	: 6413
4-Ethyl-3-methylheptane	: 6414
4-Ethyl-4-methylheptane	: 6415
5-Ethyl-5-methyl-3-heptyne	: 6619
3-Ethyl-2-methylhexane	: 6825
3-Ethyl-3-methylhexane	: 6826
4-Ethyl-3-methylhexane	: 6827
4-Ethyl-4-methylhexane	: 6827
4-Ethyl-2-methylhexane	: 6828
Ethyl 3-methylhexanoate	: 6899
Ethyl 4-methylhexanoate	: 6900
2-Ethyl-3-methylhexanoic acid	: 6886
3-Ethyl-3-methylhexanoic acid	: 6887
Ethyl 4-methyl-5-oxohexanoate	: 6902
3-Ethyl-2-methylpentane	: 8813
3-Ethyl-3-methylpentane	: 8814
3-Ethyl-3-methyl-1-pentene	: 9047
3-Ethyl-3-methyl-2-pentene	: 9049
Ethyl 4-methyl-2-pentenoate	: 9080
Ethyl 3-methyl-3-pentenoate	: 9081
Ethyl 4-methyl-3-pentenoate	: 9082
3-Ethyl-3-methyl-1-pentyne	: 9132
1-Ethyl-1-methylpropylbenzene	: 1691
1-Ethyl-3-methylpyrazole	: 10833
2-Ethyl-3-methylquinoline	: 11269
2-Ethyl-4-methylquinoline	: 11270
Ethyl methyl sulfide	: 5606
2-Ethyl-3-methylthiophene	: 11761
2-Ethyl-5-methylthiophene	: 11762
N-Ethylmorpholine	: 7694
1-Ethylnaphthalene	: 7885
2-Ethylnaphthalene	: 7886
Ethyl 1-naphthylacetate	: 7735
Ethyl nitrate	: 8074
Ethyl nitrite	: 8087
Ethyl nitroacetate	: 218
O-Ethyl O-p-nitrophenyl benzenethiophosphonate	: 9551
2-Ethyl-2-nitro-1,3-propanediol	: 10008
N-Ethyl-N-nitrosoethanamine	: 5432
N-Ethyl-N-nitrosourea	: 12015
Ethyl nonanoate	: 8150
N-Ethylnorephedrine	: 1871
3-Ethyloctane	: 8403
4-Ethyloctane	: 8404
Ethyl octanoate	: 8428
Ethyl octyl ketone	: 11959
Ethyl oenanthate	: 6462
Ethyl orthoformate	: 5604
Ethyl 3-oxo-3-phenylpropanoate	: 2141
Ethylparaben	: 2621
3-Ethylpentane	: 8809
2-Ethyl-2-pentanol	: 6946
2-Ethyl-1-pentene	: 9043
3-Ethyl-1-pentene	: 9044
3-Ethyl-2-pentene	: 9045
4-Ethyl-3-pentenoic acid	: 7060
Ethyl pentyl ether	: 8808
3-Ethyl-1-pentyne	: 9131
9-Ethylphenanthrene	: 9179
o-Ethylphenol	: 9353
m-Ethylphenol	: 9354
p-Ethylphenol	: 9355
Ethyl phenoxyacetate	: 229
Ethyl 4-phenoxybutyrate	: 3581
Ethyl 2-(4-phenoxyphenoxy)ethylcarbamate	: 5952
Ethyl phenylacetate	: 899
Ethyl N-phenylcarbamate	: 3965
1-Ethyl-1-phenylhydrazine	: 7159
1-Ethyl-2-phenylhydrazine	: 7160

Ethyl 5-phenyl-2,4-pentadienoate	: 8680
Ethyl N-phenylurethane	: 3965
Ethyl phosphate	: 9563
3-Ethyl-4-picoline	: 10964
4-Ethyl-2-picoline	: 10965
5-Ethyl-2-picoline	: 10966
N-Ethylpiperidine	: 9667
4-Ethylpiperidine	: 9670
1-Ethyl-3-piperidinol	: 9715
5-Ethyl-5-(1-piperidyl)barbituric acid	: 11096
Ethyl 2-propenyl ether	: 10566
Ethylpropiolic acid	: 9137
Ethyl propionate	: 10263
1-Ethyl-2-propylacetylene	: 6611
1-Ethyl-4-propylbenzene	: 1703
1-Ethylpropyl chloride	: 8721
Ethyl propyl ether	: 10037
Ethyl propyl ketone	: 6960
Ethyl propyl sulfide	: 10042
2-Ethylpyridine	: 10956
3-Ethylpyridine	: 10957
4-Ethylpyridine	: 10958
2-Ethylquinoline	: 11266
3-Ethylquinoline	: 11267
6-Ethylquinoline	: 11268
Ethyl salicylate	: 2619
Ethyl silicate	: 11506
Ethyl succinate	: 3303
Ethyl sulfate	: 11576
Ethyl sulfite	: 11581
Ethyl sulfocyanate	: 11703
Ethylsulfonal	: 3230
Ethylsulfonic acid	: 5622
2-(Ethylsulfonyl)ethanol	: 5725
Ethylsulfonylethyl alcohol	: 5725
Ethyl 2,3,3,3-tetrafluoropropionate	: 10328
2-Ethyltetrahydrofuran	: 6107
3-Ethyltetrahydrothiophene	: 11766
1-Ethyltetralin	: 7888
2-Ethyltetralin	: 7889
6-Ethyltetralin	: 7891
Ethyl tetraphosphate	: 11645
Ethyl thiocyanate	: 11703
2-(Ethylthio)ethanol	: 5726
2-Ethylthiophene	: 11759
3-Ethylthiophene	: 11760
3-Ethylthiophenol	: 2297
o-Ethyltoluene	: 1685
m-Ethyltoluene	: 1686
p-Ethyltoluene	: 1687
Ethyl p-toluenesulfonate	: 2231
N-Ethyl-4-toluidine	: 748
Ethyl m-tolyl sulfide	: 1706
Ethyl tosylate	: 2231
Ethyl tribromoacetate	: 242
Ethyl trichloroacetate	: 248
Ethyl(trichloromethyl)carbinol	: 3643
Ethyltrichlorosilane	: 11463
Ethyltrifluorosilane	: 11432
Ethyltrimethoxysilane	: 11433
1-Ethyl-2,3,5-trimethylbenzene	: 1709
1-Ethyl-2,4,5-trimethylbenzene	: 1710
5-Ethyl-1,2,3-trimethylbenzene	: 1713
Ethyltrimethylethylene	: 9034
3-Ethyl-2,2,3-trimethylpentane	: 8815
3-Ethyl-2,3,4-trimethylpentane	: 8816
Ethyl undecanoate	: 11946
Ethyl undecylate	: 11946
Ethyl urethane	: 3948
O-Ethylurethane	: 3948
Ethyl valerate	: 8878
Ethylvanillin	: 555
Ethylvinylacetylene	: 7103
Ethyl vinyl ether	: 5905
Ethyl vinyl ketone	: 9104
3-Ethyl-o-xylene	: 1670
Ethynylbenzene	: 1715
α-Ethynylbenzyl alcohol	: 1876
α-Ethynylbenzyl carbamate	: 1877
1-Ethynyl-1-cyclopentanol	: 4866
Ethynyldimethylcarbinol	: 3922
Ethynylestradiol	: 8216
Ethynyl methyl ketone	: 3923
Ethynylmethylpropylcarbinol	: 7128
Ethynylsilane	: 11435
Etiocholanic acid	: 346
Etofylline	: 10758

Eucalyptol	: 8560
Eucarvone	: 4220
Eucazulen	: 505
4αH-Eudesma-5,11(13)-dien-12-oic acid, 8β-hydroxy-, γ-lactone	: 8059
Eugenol	: 9389
Euonymit	: 6194
Euparin	: 5825
Euprax	: 7229
Evasprin	: 1913
Evodiamine	: 7333
Exaltol	: 4760
Exaltolide	: 8562
Fagarine	: 6179
(E,E)-α-Farnesene	: 5329
α-Farnesene	: 5329
β-Farnesene	: 5331
Felbinac	: 3009
Fenamic acid	: 2715
Fenamisal	: 2502
Fenchane(d)	: 2931
α-Fenchene (DL)	: 2923
l-α-Fenchyl alcohol	: 2941
Fenibut	: 2128
Fenicol	: 1870
Fenpiprane	: 9658
Fentanyl	: 9801
Fenticlor	: 9477
Fenyramidol	: 1913
Ferrocene, 1,1'-bis(3-phenyl-1,1,3,3-tetramethyldisiloxanyl)	: 5961
Ferron	: 11295
Fisetin	: 2813
Flacitran	: 2814
Flavan	: 2782
Flavanone, 4'-methoxy-3',5,7-trihydroxy	: 2801
Flavanone, 3',5,7-hydroxy-4'-methoxy-, 7-(6-O-α-L-rhamnosyl-D-glucoside)	: 2794
Flavone	: 2840
Flexin	: 2876
Flopropione	: 10489
Floxuridine	: 12034
Flufenamic acid	: 2729
Fluimucetin	: 5015
Fluometuron	: 12007
Fluonilid	: 11802
Fluorakil 100	: 63
3',6'-Fluorandiol	: 11522
Fluorene	: 5967
Fluorene, 2-bromo-	: 5968
9H-Fluorene-9-carboxylic acid, 2-chloro-9-hydroxy-	: 4079
Fluorene, 9-(2-chlorobenzylidene)	: 5970
Fluorene, 9-methyl-	: 5974
N-2-Fluorenylacetamide	: 64
Fluorescein	: 11522
Fluorescein, 4',5'-diiodo	: 11523
Fluorescin	: 2572
Fluorine cyanide	: 4167
Fluoroacetic acid amide	: 63
Fluoroacetic acid, methyl ester	: 192
Fluoroacetylene	: 5943
o-Fluoroaniline	: 752
m-Fluoroaniline	: 753
p-Fluoroaniline	: 754
Fluorobenzene	: 1717
4-Fluorobenzeneacetic acid	: 901
p-Fluorobenzoic acid	: 2605
Fluorobromomethane	: 7470
2-Fluoro-1,3-butadiene	: 3162
1-Fluorobutane	: 3378
2-Fluorobutane	: 3379
4-Fluoro-1-butanol	: 3621
4-Fluorobutyric acid	: 3927
Fluorochlorobromomethane	: 7466
α-Fluorocinnamic acid	: 10637
Fluorocyclohexane	: 4403
1-Fluorodecane	: 5052
10-Fluorodecanoic acid	: 5064
10-Fluoro-1-decanol	: 5074
Fluorodichloroacetonitrile	: 279
1-Fluoro-1,1-dichloroethane	: 5534
Fluorodichloromethane	: 7502
Fluorodiiodomethane	: 7516
Fluoroethanoic acid	: 189

Fluoroform	: 7558
Fluoroformaldehyde	: 6019
1-Fluorohexadecane	: 6639
1-Fluorohexane	: 6829
2-Fluorohexane	: 6830
9α-Fluoro-11β-hydroxy-17-methyltestosterone	: 355
Fluoroiodomethane	: 7517
Fluoromethane	: 7515
1-Fluoronaphthalene	: 7893
α-Fluoronaphthalene	: 7893
2-Fluoronaphthalene	: 7894
β-Fluoronaphthalene	: 7894
o-Fluoronitrobenzene	: 1733
m-Fluoronitrobenzene	: 1734
p-Fluoronitrobenzene	: 1735
3-Fluoro-4-nitrotoluene	: 1730
8-Fluorooctanoic acid	: 8429
1-Fluoropentane	: 8817
5-Fluoropentanoic acid	: 8879
(p-Fluorophenyl)acetic acid	: 901
p-Fluorophenylamine	: 754
Fluoroplex	: 11067
Fluoroprene	: 3162
2-Fluoropropane	: 10047
3-Fluoro-1-propanol	: 10382
2-Fluoropropene	: 10569
3-Fluoropropene	: 10570
cis-1-Fluoropropene	: 10567
trans-1-Fluoropropene	: 10568
4-(Fluorosulfonyl)aniline	: 2252
o-Fluorotoluene	: 1726
m-Fluorotoluene	: 1727
p-Fluorotoluene	: 1728
Fluorotrichloromethane	: 7555
5-Fluorouracil	: 11067
5-Fluorouracil 2'-deoxyriboside	: 12034
Fluoxymesterone	: 355
Fluracil	: 11067
Folic acid	: 6243
Foligan	: 10852
Folinic acid	: 6244
Follestrol	: 5406
Follicyclin	: 5399
Formaldehyde, (2,2'-dichloroisopropyl) ethyl acetal	: 9896
Formaldehyde dimethyl mercaptal	: 7461
Formaldehyde, trimer	: 11905
Formalin	: 5987
Formamide, 1,1'-azobis-	: 5124
Formamide, N,N '-[1,4-piperazinediylbis(2,2,2-trichloroethylidene)]bis-	: 11902
Formamidine	: 7563
Formanilide	: 5999
Formic acid, isopropyl ester	: 6011
Formononetin	: 2830
N-Formylaniline	: 5999
Formylcyclooctane	: 4738
Formylformic acid	: 223
2-Formylfuran	: 6041
Formylpteroic acid	: 2494
4-Formylquinoline	: 11225
N5-Formyl-5,6,7,8-tetrahydrofolic acid	: 6244
5-Formyl-5,6,7,8-tetrahydrofolic acid	: 6244
3-Formylthiophene	: 11734
1-Formyl-2,6,6-trimethyl-1-cyclohexene	: 4636
Fraxetin	: 2808
Fraxin	: 2823
Freon 11	: 7555
Freon 12	: 7500
Freon 12B1	: 7464
Freon 12B2	: 7496
Freon 13	: 7486
Freon 13B1	: 7475
Freon 13I1	: 7559
Freon 14	: 7541
Freon 21	: 7502
Freon 22	: 7478
Freon 23	: 7558
Freon 31	: 7481
Freon 32	: 7504
Freon 41	: 7515
Freon 111	: 5618
Freon 112	: 5637

Freon 112a	: 5618
Freon 113	: 5663
Freon 114	: 5538
Freon 114a	: 5537
Freon 114B2	: 5525
Freon 115	: 5484
Freon 116	: 5596
Freon 121	: 5638
Freon 122	: 5659
Freon 123	: 5540
Freon 132	: 5532
Freon 133a	: 5489
Freon 152	: 5545
Freon 215	: 10179
Freon 218	: 10118
Freon C 318	: 4194
Freon 1110	: 5917
Freon 1111	: 5923
Frigen 11	: 7555
Frigen 11A	: 7555
Frigen 12	: 7500
Frigen 13	: 7486
Frigen 22	: 7478
Frigen 113	: 5663
Frigen 113A	: 5663
Frigen 114	: 5538
Frigen 114A	: 5537
Frigen 122	: 5659
Fuchsin, acid	: 2218
Fucoxanthin	: 4055
Fulvene	: 4767
Fulvene, 6-vinyl(trans)	: 4769
Fumaric acid	: 3764
Fumaric acid, bromo-	: 3770
Fumaric acid, diallyl ester	: 3781
Fumaric acid dichloride	: 3795
Fumaric acid, diisobutyl ester	: 3767
Fumaric acid, diisopropyl ester	: 3766
Fumaric acid, dipropyl ester	: 3783
Fumarine	: 3109
Fumaronitrile, chloro-	: 3763
Fumaryl chloride	: 3795
Fumigacin	: 8211
Fumigatin	: 4265
Furacrylic acid	: 10638
2-Furalacetic acid	: 10638
Furaldehyde	: 6041
2-Furanacrylic acid	: 10638
2-Furanacrylonitrile	: 10585
Furan, 2-tert-butyl-	: 6090
2-Furancarbinol	: 6118
Furan, 2(2-cyanovinyl)	: 10585
2-Furanmethanimine, 5-nitro-N-(2-oxo-3-oxazolidinyl)-	: 8589
2-Furanpropionic acid, tetrahydro-	: 6164
Furan, tetrahydro-2-ethyl	: 6107
2-Furanylmethanol	: 6118
Furazolidone	: 8589
Furfural	: 6041
Furfuran	: 6025
Furfuryl alcohol	: 6118
Furfuryl alcohol, tetrahydro-, stearate	: 8275
Furfurylamine	: 6113
Furfurylideneacetic acid	: 10638
Furfuryl phenyl ether	: 6158
Furfuryl valerate	: 8880
Furilazone	: 10904
α-Furildioxime	: 5583
2-Furoic acid	: 6050
β-Furoic acid, 5-bromo, ethyl ester	: 6190
2-Furoic acid, 5-chloro-, ethyl ester	: 6053
α-Furoic acid, hexyl ester	: 6058
3-Furoic acid, 5-methyl-	: 6065
α-Furoic acid, octyl ester	: 6067
2-Furoic acid, pentyl ester	: 6068
Furole	: 6041
Furonazide	: 10904
2-Furylacetic acid, butyl ester	: 6027
2-Furylacetonitrile	: 6031
3-(2-Furyl)acrylonitrile	: 10585
2-Furylcarbinol	: 6118
2-Furyl iodide	: 6110
2-Furyl isopropyl ether	: 6112
2-Furyl methyl ketone isonicotinoylhydrazone	: 10904
2-Furyl octyl ether	: 6135

2-Furyl phenyl ether	: 6157
2-Furyl phenyl sulfide	: 6160
Fusaric acid	: 10898
Fussol	: 63
Gainex	: 2384
Galegine	: 6283
Galipine	: 11253
Gallacetophenone	: 5873
Gallamine triethiodide	: 5441
Gallein	: 11526
Gallic acid	: 2731
Gallic acid, 3-gallate	: 2575
Gallodesoxycholic acid	: 4097
Gallomide blue	: 8062
m-Galloylgallic acid	: 2575
Gamarex	: 3464
Gardol	: 6269
Gastrodigenin	: 1884
Geissoschizoline	: 4154
Gelseminic acid	: 2829
Genetron 132b-B2	: 5522
Genetron 11	: 7555
Genetron 23	: 7558
Genetron 218	: 10118
Genistein	: 2807
Genistein, 4'-β-D-glucopyranoside	: 2824
Genisteol	: 2807
Genite	: 9289
Gentianic acid	: 12054
Gentisic acid	: 2570
Gentisin	: 12054
Gentisyl alcohol	: 1520
trans-Geraniol	: 8333
Geraniol, tetrahydro (DL)	: 8449
Geranyl alcohol	: 8333
trans-Geranyl bromide	: 8321
Germine	: 4068
Geronic acid	: 6460
Gibberellic acid	: 6204
Gibberellin A3	: 6204
Gibberellin X	: 6204
Gibbs' reagent	: 4280
Gingerone	: 3663
Gitogenin	: 11530
Gitoxigenin	: 4048
Gitoxin	: 4042
d-Glaucine	: 5157
Glucamine	: 6207
Gluconic acid	: 6209
D-Gluconic acid-1,5-lactone	: 6212
Gluconic acid, sodium salt	: 6213
δ-D-Gluconolactone	: 6212
1,5-Gluconolactone	: 6212
Glucopyranoside, genistein-4', β-D-	: 2824
D-Glucosaccharic acid	: 6205
D-Glucosamine	: 6233
L-Glucose, methylamino	: 6215
α-D-Glucose, pentamethyl ether	: 6228
Glucurone	: 6239
D-Glucuronolactone	: 6239
Glutaconic acid, diethyl ester	: 9041
α-Glutamic acid	: 6241
Glutamic acid 5-amide	: 6249
Glutamic acid, N-[p-[[(2-amino-5-formyl-5,6,7,8-tetrahydro-4-hydroxy-6-pteridinyl)methyl]amino]benzoyl]-, L-	: 6244
Glutamic acid, 5-[2-(α-hydroxy-p-tolyl)hydrazide], L-	: 6247
Glutamic acid, monosodium salt	: 6248
γ-L-Glutamyl-L-cysteinylglycine	: 6264
Glutaral	: 8734
Glutaraldehyde	: 8734
Glutaric acid	: 8770
Glutaric acid, 3(3-benzoxy-4-methoxyphenyl)	: 6892
Glutaric acid dialdehyde	: 8734
Glutaric acid dinitrile	: 8767
Glutaric acid, diphenyl ester	: 8776
Glutaric acid, 3-ethyl-3-methyl-	: 8777
Glutaric acid, 3-hydroxy-, diethyl ester	: 8778
Glutaric acid, 3-methyl	: 8779
Glutaronitrile	: 8767
Glutatione	: 6264
Glycamine	: 6207
Glyceric acid	: 10252

Glyceric acid, ethyl ester (DL)	: 10253
Glyceric acid, methyl ester, DL-	: 10254
Glyceric acid, 2-methyl, ethyl ester	: 10255
Glyceric acid, 2-methyl-, methyl ester	: 10256
Glycerin	: 10186
Glycerin tributyrate	: 3586
Glycerol	: 10186
Glycerol acetonide	: 5207
Glycerol β-chlorohydrin	: 9995
Glycerol diacetate	: 10189
Glycerol 1,3-dimethyl ether	: 10372
Glycerol, 1,3-diphenyl ether-2-acetate	: 10378
Glycerol guaiacolate	: 10016
Glycerol, 1-laurate (DL)	: 5310
Glycerol, 1-(2-methoxyphenyl)ether	: 10016
Glycerol, 2-methyl ether	: 10015
Glycerol, 1-monoacetate(DL)	: 10187
Glycerol, 1-monododecanoate(DL)	: 5310
Glycerol, 1-phenyl	: 10194
Glycerol tributanoate	: 3586
Glycerol trimethyl ether	: 10185
Glyceryl p-aminobenzoate	: 10188
Glycidaldehyde	: 8603
Glycidol	: 8618
Glycidyl allyl ether	: 8626
Glycine anhydride	: 9612
Glycine, N,N-bis(phosphonomethyl)-	: 6277
Glycine, N-(chloroacetyl)-N-(2,6-diethylphenyl)-, ethyl ester	: 5173
Glycine cholate	: 6273
Glycine, N-(phosphonomethyl)-	: 6276
Glycinol	: 5675
Glycinonitrile, N,N-dimethyl-	: 281
Glycocholic acid	: 6273
Glycocyamine	: 6253
Glycolaldehyde	: 15
Glycol diacetate	: 5571
Glycolic acid	: 196
Glycolic aldehyde	: 15
Glycol salicylate	: 2622
Glyconitrile	: 282
Glycopyrrolate	: 11179
Glycopyrronium bromide	: 11179
Glycosine	: 11200
Glycosthene	: 6250
Glycurone	: 6239
Glycyl alcohol	: 10186
α-Glycylglycine	: 6265
Glycyrrhetinic acid glycoside	: 6230
Glycyrrhizic acid	: 6230
Glycyrrhizin	: 6230
Glyoxal	: 5491
Glyoxaldiureine	: 7191
Glyoxaline-5-alanine	: 7132
Glyoxylaldehyde	: 5491
Glyoxylic acid	: 223
Glyoxylic diureide	: 12009
Gramine	: 7310
Grasex	: 21
Griseofulvin, (+)-	: 11520
Guaiacol	: 9379
Guaiacol benzoate	: 9383
Guaiacol carbonate	: 9384
Guaiacol glycerol ether	: 10016
Guaiacol phosphate	: 9387
Guaiacol valerate	: 505
Guaiazulene	: 507
Guai-1(5)-en-11-ol	: 507
Guaifenesin	: 10016
Guaiol	: 507
Guajen	: 7825
Guanidine, dodecyl-, monoacetate	: 5347
Guanidinoacetic acid	: 6253
Guanine	: 10768
Guanyl glycine	: 6253
Guanyl hydrazine	: 7147
Gulose, D-	: 6290
Guvacine	: 10912
Gynecormone	: 5403
Halazone	: 2565
Halon 2302	: 5526
Haloperidol	: 3651
Halotestin	: 355
Halothane	: 5458
Harmaline	: 11038

Harman	: 11040
Harmine	: 11039
Hecogenin	: 11534
Hederagenin	: 8545
Helenin	: 8059
Helicin	: 562
Heliotropic acid	: 2418
Heliotropin	: 2417
Helvolic acid	: 8211
Hematein	: 628
Hematin	: 5959
Hematoidin	: 3000
Hematoporphyrin	: 9735
Hematoxylin	: 629
Hemimellitic acid	: 2318
Hemin	: 5958
Hemipyocyanine	: 9197
Hendecane	: 11938
Heneicosylbenzene	: 1741
Hepin	: 6873
Heptachlor	: 7576
1,1,1,2,2,3,3-Heptachloropropane	: 10048
1,1,1,2,3,3,3-Heptachloropropane	: 10049
Heptacosylbenzene	: 1742
Heptadecane, 1-cyclohexyl-	: 4404
Heptadecane, 9-phenethyl-	: 2070
Heptadecane, 1-phenyl-	: 1743
Heptadecyl alcohol	: 6317
Heptadecylbenzene	: 1743
Heptadecylcyclohexane	: 4404
1,6-Heptadiene-4-ol, 4-methyl	: 6338
1,6-Heptadiene-3-yne	: 6341
Heptaldehyde	: 6344
Heptamethylcyclotetrasiloxane	: 4992
Heptamethylene chlorohydrin	: 6477
Heptamethylene dibromide	: 6374
Heptane, 1-cyclopentyl-	: 4820
Heptane, 1,1-diphenyl-	: 1746
Heptane, 4-isopropyl-	: 6427
Heptane, 2-methyl-3-ethyl	: 6409
4-Heptanol, 3-methyl(DL)	: 6504
4-Heptanol-2-one, 3-methyl	: 6518
3-Heptanol-5-one, 3-methyl	: 6519
4-Heptanol, 4-phenyl	: 1867
1-Heptanol, 1-phenyl-	: 1881
3-Heptanol, 7-phenyl	: 2102
cis-2-Heptene	: 6535
trans-3-Heptene	: 6536
cis-3-Heptene	: 6537
1-Heptene, 1,1-diphenyl-	: 1744
2-Heptene, 3-methyl (trans)	: 6549
2-Heptene, 3-methyl (cis)	: 6550
2-Heptene, 4-methyl (trans)	: 6551
2-Heptene, 4-methyl (cis)	: 6552
1-Heptene-4-ol	: 6572
1-Heptene-4-ol, 4-methyl	: 6573
cis-2-Hepten-1-ol	: 6588
1-Hepten-5-one	: 6586
3-Hepten-2-one (cis)	: 6588
2-Hepten-6-one, 4-ethyl-2-methyl	: 6593
6-Hepten-4-yne-3-ol, 3-ethyl	: 6608
Heptyl acetate	: 193
Heptylacetylene	: 8196
Heptyl alcohol	: 6472
Heptylamine	: 6349
Heptyl bromide	: 6356
2-Heptyl bromide	: 6357
4-Heptyl bromide	: 6358
Heptyl chloride	: 6360
Heptylcyclopentane	: 4820
Heptyl formate	: 6008
Heptyl heptanoate	: 6464
Heptylidene acetone	: 5093
Heptyl mercaptan	: 6436
Heptyl methyl ketone	: 8172
1-Heptylnaphthalene	: 7895
2-Heptylnaphthalene	: 7896
Heptyl nitrite	: 8088
Heptyl propyl ketone	: 11960
Heptyl thiocyanate	: 11704
2-Heptynyl bromide	: 6613
4-Heptynyl chloride	: 6616
Herbicide 634	: 4900
Hesperetin	: 2800
Hesperetin 7-rutinoside	: 2794
Hesperidin	: 2794
Hexabromoethane	: 5594

Hexachlorane	: 4409
Hexachloroacetone	: 10440
Hexachlorobenzene	: 1749
Hexachloro-1,3-butadiene	: 3163
γ-Hexachlorocyclohexane	: 4409
Hexachloroethane	: 5595
1,2,3,4,5,6-Hexachloro-3-hexene	: 7035
Hexachlorophene	: 9409
1,1,1,2,2,3-Hexachloropropane	: 10051
1,1,1,3,3,3-Hexachloropropane	: 10052
1,1,2,2,3,3-Hexachloropropane	: 10053
Hexachloropropylene	: 10571
Hexacosane, 1-phenyl-	: 1750
Hexacose	: 6707
Hexacosylcyclohexane	: 4410
Hexacosylcyclopentane	: 4821
Hexadecamethylcyclooctasiloxane	: 4741
Hexadecamethylene dibromide	: 6635
Hexadecamethylheptasiloxane	: 6528
Hexadecanal dimethyl acetal	: 6636
Hexadecane, 1-cyclohexyl	: 4411
Hexadecane, 1-cyclopentyl-	: 4822
Hexadecylcyclopentane	: 4822
Hexadecyl 3-hydroxy-2-naphthoate	: 7774
Hexadecyl mercaptan	: 6643
Hexadecyl pyridinium chloride	: 11030
1,5-Hexadiene, 2-methyl-3-ethyl	: 6694
1,3-Hexadiene, 1-phenyl	: 1753
1,5-Hexadiene, 1-phenyl-	: 1754
1,3-Hexadiene, 6-phenyl	: 1755
1,5-Hexadiene, 3-phenyl	: 1625
1,5-Hexadiene-3-yne	: 6713
1,3-Hexadien-5-ol	: 6708
1,5-Hexadienylbenzene	: 1754
Hexaethylbenzene	: 1756
1,1,1,3,3,3-Hexaethyldisiloxane	: 5235
Hexafluorenium bromide	: 6753
Hexafluoro-2,4-pentanedione	: 8802
Hexafluoropropylene oxide	: 8629
1,2,3,4,5,6-Hexahydroanthracene	: 422
Hexahydroaplotaxene	: 6322
Hexahydroazepine	: 492
Hexahydrobenzene	: 4305
Hexahydrobenzoic acid	: 4332
Hexahydrobenzyl alcohol	: 4419
Hexahydroequilenin	: 5402
Hexahydroindan	: 7281
Hexahydronicotinic acid	: 9641
Hexahydrophthalic anhydride	: 7345
Hexahydropyrazine	: 9602
3,3',4',5,5',7-Hexahydroxyflavylium chloride	: 2849
3,3',4',5,5',7-Hexahydroxy-2-phenylbenzopyrylium chloride	: 2849
Hexamethylbenzene	: 1758
Hexamethylcyclohexasiloxane	: 4625
1,1,3,3,5,5-Hexamethylcyclotrisilazane	: 5001
Hexamethylcyclotrisiloxane	: 5002
Hexamethyldisilane	: 5228
Hexamethyldisilazane	: 11363
Hexamethyldisiloxane	: 5236
Hexamethylenediamine	: 6749
Hexamethylene diiodide	: 6769
Hexamethylene diisocyanate	: 6770
Hexamethylene glycol	: 6796
Hexamethylenetetramine, allyl iodide	: 11825
Hexamethylenimine	: 492
Hexamethylolmelamine	: 7584
Hexane, 1-cyclohexyl-	: 4414
Hexane, 1-cyclopentyl-	: 4823
Hexane, 2,5-dibromo(DL)	: 6759
3,4-Dibromohexane	: 6760
Hexane, 2,5-dimethyl-2,5-diphenyl	: 2281
Hexane, 2,2-dimethyl-3-ethyl	: 6816
Hexane, 2,2-dimethyl-4-ethyl	: 6821
Hexane, 2,3-dimethyl-3-ethyl	: 6817
Hexane, 2,3-dimethyl-4-ethyl	: 6822
Hexane, 2,3-dimethyl-5-ethyl	: 6818
Hexane, 2,4-dimethyl-3-ethyl	: 6823
Hexane, 2,4-dimethyl-4-ethyl	: 6823
Hexane, 2,5-dimethyl-3-ethyl	: 6819
Hexane, 3,3-dimethyl-4-ethyl	: 6824
Hexane, 3,4-dimethyl-3-ethyl	: 6820
Hexane, 2-iodo (l)	: 6832
Hexane, 3-isopropyl-2-methyl-	: 6839
Hexane, 2-methyl-3-isopropyl	: 6839

4-Hydroxy-1-pentanol	: 8786
1-Hydroxy-2-pentanone	: 8971
2-Hydroxy-3-pentanone	: 8975
3-Hydroxy-1-pentyne	: 9140
1-Hydroxypent-2-yn-4-ene	: 9118
3-Hydroxy-9,10-phenanthraquinone	: 9176
2-(β-Hydroxyphenethyl-amino)pyridine	: 1913
Hydroxyphenylamine	: 9200
o-Hydroxyphenylamine	: 9201
m-Hydroxyphenylamine	: 9202
N-(4-Hydroxyphenyl)glycine	: 6267
1-Hydroxy-3-phenylpropane	: 2148
1-(m-Hydroxyphenyl)-1-propanol	: 1872
DL-2-Hydroxy-3-phenylpropionic acid	: 2133
4-Hydroxyphenylpropionic acid	: 2134
3-Hydroxy-3-phenylprop-1-yne	: 1876
4-Hydroxyphenylsulfonic acid	: 2226
17α-Hydroxyprogesterone	: 9750
4-Hydroxy-L-proline	: 9767
3-Hydroxy-1-propene	: 10697
2-Hydroxypropionitrile	: 10105
(3-Hydroxypropyl)benzene	: 2148
1-Hydroxy-1-propylcyclohexane	: 4557
3-Hydroxypropylene oxide	: 8618
N-(2-Hydroxypropyl)ethylenediamine	: 10340
3-Hydroxypropyl fluoride	: 10382
2-(2-Hydroxypropyl)pyridine	: 10951
2-(3-Hydroxypropyl)pyridine	: 11012
(2-Hydroxypropyl)trimethyl ammonium chloride	: 9831
4-Hydroxypyrazolo[3,4-d]pyrimidine	: 10852
5-Hydroxyquinoline	: 11305
7-Hydroxyquinoline	: 11307
8-Hydroxyquinoline	: 11308
8-Hydroxyquinoline sulfate	: 11324
8-Hydroxy-5-quinolinesulfonic acid	: 11294
p-Hydroxysalicylic acid	: 2569
5-Hydroxysalicylic acid	: 2570
2α-Hydroxytigogenin	: 11530
2-Hydroxy-m-toluic acid	: 2629
2-Hydroxy-p-toluic acid	: 2630
6-Hydroxy-m-toluic acid	: 2631
2-Hydroxytridecane	: 11884
11-Hydroxyundecanoic acid	: 11948
4-Hydroxyundecanoic acid lactone	: 6152
Hydroxyurea	: 12016
Hydroxyzine	: 5700
Hygrine	: 10471
Hymecromone	: 2831
Hyodeoxycholic acid	: 4096
Hyoscyamine	: 917
Hypothiazide	: 2855
Hypoxanthine	: 10770
Hypoxanthine riboside	: 7335
Ibogaine	: 7186
Ibrotamide	: 3199
Imecromone	: 2831
1H-Imidazole, 1-[2-(2,4-dichlorophenyl)-2-(2-propenyloxy)ethyl]-	: 7187
1H-Imidazole, 2-heptadecyl-4,5-dihydro-, monoacetate	: 6275
2-Imidazolidine, 2-benzyl	: 7198
1-Imidazolidinecarboxamide, 3-(3,5-dichlorophenyl)-N-(1-methylethyl)-2,4-dioxo-	: 7338
7H-Imidazo[4,5-d]pyrimidine	: 10752
2-Iminobenzothiazoline	: 2859
Iminodiacetic acid	: 6260
2,2'-Iminodibenzoic acid	: 2656
Iminourea	: 6278
Imperatorin	: 6189
Indaconitine	: 307
Indan	: 7250
Indan, 1-amino (DL)	: 7243
Indan, 1,2-dibromo-	: 7247
Indan, 1,2-dichloro	: 7249
Indan, 1,1-diemthyl	: 7251
Indan, 1,2-dimethyl-	: 7252
Indan, 4,7-dimethyl-	: 7253
Indan, 5,6-dimethyl-	: 7254
Indan, 1-ethyl-	: 7273
Indan, 5-hexyl-	: 7274
Indan, 1-methyl-	: 7255
Indan, 2-methyl-	: 7256

Indan, 4-methyl-	: 7257
Indan, 5-methyl-	: 7258
2-Indanone	: 7287
1-Indanone, 4-methyl-	: 7290
1-Indanone, 2-phenyl-	: 7291
Indan, 2-phenyl-	: 7259
Indan, 1-phenyl-1,3,3-trimethyl	: 7263
Indan, 1,1,4,7-tetramethyl-	: 7260
Indanthrene	: 434
Indan, 1,1,5-trimethyl-	: 7261
Indan, 1,4,7-trimethyl	: 7262
1-Indenecarboxylic acid	: 7246
Indene, 1,3-diphenyl-	: 7271
Indene, 2,3-diphenyl-	: 7272
Indene, 4-methyl-	: 7278
Indene, 6-methyl-	: 7279
Indene, 7-methyl-	: 7280
Indigo	: 7331
Indigo Yellow	: 2843
Indoleacetic acid	: 7295
Indolebutyric acid	: 7296
Indole, 7-methyl-	: 7317
Indole, 1,2,3-trimethyl-	: 7322
Indole, 2,3,5-trimethyl-	: 7323
2-Indolinone, 1-ethyl-3-methyl-	: 7332
1H-Indolo[3,2,1-de]pyrido[3,2,1-ij][1,5]naphthyridine, eburnamenine-14-carboxylic acid deriv.	: 5349
Indolo[4,3-fg]quinoline, ergoline-8-carboxamide deriv.	: 5373
3-Indolylacetone	: 10442
Indonaphthene	: 7245
Indoxazene	: 2389
Inosine, 2-amino-	: 6289
Iodipamide	: 2594
5-Iodoacenaphthene	: 7
Iodoacetic acid	: 200
Iodoacetonitrile	: 283
Iodoalphionic acid	: 2135
p-Iodoaniline	: 761
o-Iodoanisole	: 1775
Iodobenzene	: 1766
p-Iodobenzenesulfonyl chloride	: 2247
o-Iodobenzoic acid	: 2657
4-Iodo-1,2-butadiene	: 3166
4-Iodocatechol	: 1523
Iodochlorhydroxyquin	: 11312
Iodocyanide (ICN)	: 7337
Iodocyclohexane	: 4415
Iodocyclopentane	: 4824
2-Iodo-p-cymene	: 1781
1-Iododecane	: 5053
Iododifluoromethane	: 7505
1-Iodo-3,5-dimethylbenzene	: 1771
4-Iododiphenyl	: 3079
1-Iodododecane	: 5302
Iodoethane	: 5597
2-Iodoethanol	: 5731
Iodoethene	: 5907
Iodoethylene	: 5907
Iodoform	: 7560
2-Iodofuran	: 6110
3-Iodofuran	: 6111
1-Iodoheptane	: 6419
2-Iodoheptane	: 6420
1-Iodo-1-heptyne	: 6620
1-Iodohexadecane	: 6640
1-Iodohexane	: 6831
Iodomethane	: 7518
2-Iodo-4-methylaniline	: 762
4-Iodo-2-methylbutane	: 3384
(Iodomethyl) methyl ether	: 7519
1-Iodonaphthalene	: 7902
1-Iodooctadecane	: 8242
1-Iodooctane	: 8406
1-Iodopentane	: 8818
3-Iodopentane	: 8820
1-Iodo-1-pentyne	: 9133
o-Iodophenol	: 9371
p-Iodophenol	: 9373
1-Iodopropane	: 10054
2-Iodopropane	: 10055
3-Iodopropene	: 10573
3-Iodopropylene	: 10573
1-Iodopropyne	: 10734
3-Iodopropyne	: 10735
2-Iodopyridine	: 10970

Iodopyrrole	: 11140
Iodoquinol	: 11315
4-Iodoquinoline	: 11272
Iodostarin	: 8300
N-Iodosuccinimide	: 11159
Iodotrifluoroethylene	: 5925
Iodotrifluoromethane	: 7559
3-Iodotyrosine	: 11932
3-Iodoveratrole	: 1767
4-Iodo-o-xylene	: 1774
β-Ionol	: 3873
α-Ionol	: 3874
Irene	: 7999
Iron dicyclopentadienyl	: 5960
β-Irone	: 3883
α-Irone	: 3884
Iron pentacarbonyl	: 7339
Iron, tris(dimethylcarbamodithioato-S,S')-, (OC-6-11)-	: 5957
Isatin	: 7300
Isoamylaldehyde	: 3186
Isoamylamine	: 3220
Isoamyl benzoate	: 3632
Isoamyl bromide	: 3239
Isoamyl chloride	: 3252
Isoamyl cyanide	: 8831
Isoamyl ether	: 3418
Isoamyl iodide	: 3384
Isoamyl isovalerate	: 3551
Isoamyl 3-methylbutyrate	: 3551
Isoamyl nitrate	: 3637
Isoamyl nitrite	: 8090
Isoamyl salicylate	: 2633
Isoamyl sulfide	: 3436
Isobutamben	: 2505
Isobutanal	: 9785
Isobutane	: 10073
Isobutane, 1,1-dichloro	: 9899
Isobutane, 1,2-dichloro	: 9900
Isobutane, 1,1,1,2,3-pentachloro	: 10127
Isobutane, 1,1,1,2-tetrachloro	: 10147
Isobutane, 1,1,2-trichloro	: 10177
Isobutene	: 10578
2-Isobutoxyethanol	: 5746
1-Isobutoxy-2-nitrobenzene	: 2034
Isobutryic anhydride	: 10291
Isobutyl acetate	: 215
Isobutyl acrylate	: 10673
Isobutyl alcohol	: 10388
Isobutyl alcohol, 2-chloro	: 10352
Isobutyl alcohol, 3-chloro	: 10353
Isobutylamine	: 9817
Isobutyl p-aminobenzoate	: 2505
Isobutylbenzene	: 2039
Isobutyl bromide	: 9845
Isobutyl butanoate	: 3563
Isobutyl carbamate	: 3959
Isobutyl chloride	: 9862
Isobutyl chlorocarbonate	: 4024
Isobutyl chloroformate	: 4024
Isobutyl cyanide	: 3410
Isobutylcyclopentane	: 4838
Isobutyldimethylcarbinol	: 8928
Isobutylene chlorohydrin	: 10354
Isobutylene, 1,3-dibromo	: 10546
Isobutylene glycol, 3-chloro	: 9998
Isobutylene (trimer)	: 10582
Isobutylethene	: 9054
Isobutyl formate	: 6013
Isobutylhydrochlorothiazide	: 2856
Isobutyl iodide	: 10057
Isobutyl isobutanoate	: 10304
Isobutyl isocyanate	: 10061
Isobutyl isovalerate	: 3555
Isobutyl ketone	: 6514
Isobutyl mercaptan	: 10158
Isobutyl methacrylate	: 10666
Isobutyl methyl ether	: 10071
Isobutyl methyl ketone	: 8981
Isobutyl nitrate	: 8078
Isobutyl nitrite	: 8094
Isobutyloxycarbonyl chloride	: 4024
Isobutyl propionate	: 10310
Isobutyl propyl ketone	: 6524
4-Isobutylresorcinol	: 1539
Isobutyl salicylate	: 2643
Isobutyl stearate	: 8268

Isobutyl thiocyanate	: 11707
Isobutyl 2-tolyl ketone	: 3669
Isobutyl 4-tolyl ketone	: 3670
Isobutyl urethane	: 3961
Isobutyl vinyl ether	: 10035
Isobutyl vinyl ketone	: 7093
Isobutyramide, 2-bromo	: 9792
Isobutyric acid	: 10290
Isobutyric acid chloride	: 10504
Isobutyric acid, 2-chloroethyl ester	: 10292
Isobutyric acid, cyclohexyl ester	: 10293
Isobutyric acid, (2-nitropentyl) ester	: 10305
Isobutyric acid, 3-phenyl	: 2137
1-Isobutyronaphthone	: 10464
2-Isobutyronaphthone	: 10465
Isobutyronitrile	: 10111
Isobutyronitrile, 2-hydroxy-3-chloro	: 10098
Isobutyryl chloride, 3-chloro	: 10499
Isocaine	: 2505
Isocamphoric acid, (DL)	: 3931
Isochroman	: 2776
Isochrysene	: 11910
Isocinchomeronic acid	: 10929
Isocitrate lactone	: 6079
Isocorybulbine	: 5150
L-(+)-Isocorydine	: 5161
Isocorydine	: 5161
Isocoumarin	: 2791
Isodurene	: 2279
Isoestradiol	: 5401
8-Isoestrone	: 5407
Isoeugenol	: 9388
Isoeugenol, acetate (cis)	: 9394
Isoflavone	: 2841
Isoflavone, 4',7-dihydroxy-	: 2827
Isoflavone, 4',5-dihydroxy-7-methoxy-	: 2828
Isoflurophate	: 9580
Isogeronic acid	: 6461
Isoguanine	: 10767
Isohexane	: 8823
Isohexyl alcohol	: 8939
1H-Isoindole-1,3(2H)-dione, 2-[(trichloromethyl)thio]-	: 5984
Isolysergic acid	: 5375
β-Isomalic acid	: 9959
Isomeptene	: 6532
Isoniazid	: 10905
Isonicotinic acid	: 10895
Isonicotinic acid diethylamide	: 10892
Isonicotinic acid, (α-methylfurfurylidene)hydrazide	: 10904
Isonipecotic acid	: 9642
Isonipecotic acid, methyl ester	: 9644
Isonipecotic acid, 1-methyl-, methyl ester	: 9645
Isonitrosoacetone	: 9789
Isonitrosoacetophenone	: 876
β-Isonitrosopropane	: 10476
Isooctane	: 8848
Isopentadiene	: 3171
Isopentanal	: 3186
Isopentane	: 3397
Isopentanoic acid	: 3539
Isopentanoyl chloride	: 3693
Isopentyl acetate	: 3631
Isopentyl alcohol	: 3627
Isopentyl alcohol, orthoformate (3:1)	: 8552
Isopentyl alcohol, stearate	: 8265
Isopentylamine	: 3220
Isopentyl bromide	: 3239
Isopentyl 3-bromopropionate	: 10210
Isopentyl butanoate	: 3544
Isopentyl chloride	: 3252
N-Isopentylethanolamine	: 5739
Isopentyl ether	: 3418
Isopentyl formate	: 3636
Isopentyl iodide	: 3384
Isopentyl isovalerate	: 3551
Isopentyl mercaptan	: 3440
Isopentyl 3-methylbutanoate	: 3551
Isopentyl β-Naphthyl ether	: 7918
Isopentyl oleate	: 8306
Isopentyl propanoate	: 3639
Isopentyl salicylate	: 2633
2-Isopentylthiophene	: 11771
Isopentyl vinyl ether	: 3371

Isophorol	: 4702
Isophorone	: 4721
Isophthalamide, N,N,N',N'-tetraethyl	: 1326
Isophthalic acid	: 1328
Isophthalic acid, diethyl ester	: 1346
Isophytol	: 6671
Isopilosine	: 6144
Isoprene	: 3171
Isopropanesulfonic acid	: 10130
Isopropenyl acetate	: 10698
Isopropenylacetylene	: 3892
Isopropenyl carbinol	: 10702
Isopropenyl chloride	: 10535
Isopropenyl chloroformate	: 4022
4-Isopropenyl-1-cyclohexenecarboxaldehyde	: 4633
4-Isopropenyl-1-methylcyclohexene	: 4677
Isopropenyl methyl ketone	: 3879
9-Isopropenylphenanthrene	: 9182
2-Isopropoxyethanol	: 5740
1-Isopropoxy-2-propanol	: 10390
N-Isopropylacetanilide	: 74
Isopropyl acetate	: 209
Isopropylacetic acid	: 3539
Isopropyl acetoacetate	: 3574
Isopropyl alcohol	: 10335
Isopropylamine	: 9804
2-Isopropylaniline	: 788
N-Isopropylaniline	: 790
Isopropylbenzene hydroperoxide	: 7181
p-Isopropylbenzoic acid	: 2689
α-Isopropylbenzyl alcohol	: 1901
4-Isopropylbenzyl alcohol	: 1902
Isopropyl borate	: 3128
Isopropyl bromide	: 9837
α-Isopropyl-α-bromobutyramide	: 3199
Isopropyl-tert-butylcarbinol	: 8955
Isopropyl butyrate	: 3549
Isopropyl carbamate	: 3955
Isopropyl carbanilate	: 3966
Isopropyl chloride	: 9849
Isopropyl chloroacetate	: 139
Isopropyl 3-chlorocarbanilate	: 3939
Isopropyl chloroformate	: 4023
Isopropyl N-(3-chlorophenyl)carbamate	: 3939
Isopropyl 2-chloropropionate	: 10223
Isopropyl trans-cinnamate	: 10687
Isopropylcyclohexane	: 4431
cis-2-Isopropylcyclohexanol	: 4539
cis-4-Isopropylcyclohexanol	: 4540
trans-4-Isopropylcyclohexanol	: 4541
2-Isopropylcyclohexanone	: 4599
4-Isopropylcyclohexanone	: 4600
Isopropylcyclopentane	: 4828
2-Isopropylcyclopentanone	: 4887
7-Isopropyl-1,4-dimethylazulene	: 505
N-Isopropylethanolamine	: 5741
Isopropyl fluoride	: 10047
Isopropyl formate	: 6011
Isopropyl heptafluorobutyrate	: 3526
4-Isopropylheptane	: 6427
Isopropylidene glycerol	: 5207
Isopropyl iodide	: 10055
Isopropyl isobutyrate	: 10303
Isopropyl isocyanate	: 10060
Isopropyl isocyanide	: 10063
Isopropyl isonitrile	: 10063
Isopropyl lactate	: 10275
Isopropyl mercaptan	: 10157
Isopropylmethylamine	: 9819
4-Isopropyl-1-methylcyclopentene	: 4927
Isopropyl methyl ether	: 10070
2-Isopropyl-5-methyl-1-hexanol	: 6948
2-(4-Isopropyl-4-methyl-5-oxo-2-imidazolin-2-yl)nicotinic acid	: 7188
(RS)-2-(4-Isopropyl-4-methyl-5-oxo-2-imidazolin-2-yl)quinoline-3-carboxylic acid	: 7189
Isopropyl methyl sulfide	: 10089
Isopropyl myristate	: 11626
Isopropyl 1-naphthyl ketone	: 10464
Isopropyl nitrate	: 8076
Isopropyl nitrite	: 8092
2-Isopropyl-1-pentene	: 6836
Isopropyl phenylcarbamate	: 3966

Isopropylphenylcarbinol	: 1901
3-(p-Isopropylphenyl)-2-methylpropionaldehyde	: 2117
Isopropyl phenyl urethane	: 3966
Isopropyl propionate	: 10299
Isopropylpropylamine	: 9827
Isopropyl propyl sulfide	: 10077
3-Isopropylpyridine	: 10985
Isopropyl salicylate	: 2635
Isopropyl 2-thienyl ketone	: 10472
Isopropyl thiocyanate	: 11706
2-Isopropylthiophene	: 11772
3-Isopropylthiophene	: 11773
Isoquinaldine	: 7381
Isoquinoline, 5-amino-	: 7372
Isoquinoline, 3-benzyl	: 7387
Isoquinoline, 3-butyl-1-[2-(dimethylamino)ethoxy]-	: 5411
Isoquinoline, 6,7-diethoxy-1(3,4-diethoxybenzyl)	: 7377
Isoquinoline, 6,7-dimethoxy-1,2-dimethyl-1,2,3,4-tetrahydro	: 7391
Isoquinoline, 8-methoxy-2-methyl-6,7-methylenedioxy-1,2,3,4-tetrahydro	: 5214
Isorubijervine	: 11516
(+)-Isothebaine	: 5158
Isothebaine	: 5158
3(2H)-Isothiazolone, 2-octyl-	: 8528
Isothiocyanic acid, acetyl	: 303
Isothiocyanic acid, tert-butyl ester	: 10066
Isovaleraldehyde	: 3186
Isovaleramide	: 3206
Isovaleric acid	: 3539
Isovaleric acid chloride	: 3693
Isovaleric aldehyde	: 3186
Isovaleric amide	: 3206
Isovalerylamine	: 3220
Isovaleryl chloride	: 3693
Isovaleryl diethylamide	: 3201
Isoxylaldehyde	: 548
Itaconic acid	: 3328
Jasmone	: 4949
Javanicin	: 7855
Jervine	: 12046
Jilkon	: 6200
Juglone	: 7860
Julolidine	: 2737
Julolidine, 1,6-dioxo	: 2736
Kaempferol	: 2843
Kainic acid	: 11144
Kelthane	: 1851
Ketamine	: 4579
Ketene	: 5927
Ketene, diethyl-	: 3877
β-Ketobutyranilide	: 3207
α-Ketoglutaric acid	: 8781
3-Ketoglutaric acid	: 8782
Ketohydroxyestrin	: 5406
Ketone, 2-benzofuranyl methyl	: 5775
Ketone, 2-benzofuranyl phenyl	: 7588
Ketone, benzyl phenyl	: 5812
Ketone, 5-bromo-2-thienyl methyl	: 5783
Ketone, α-chlorobenzyl phenyl	: 5785
Ketone, 5-chloro-2-thienyl methyl	: 5791
Ketone, cyano phenyl	: 968
Ketone, cyclobutyl methyl	: 5792
Ketone, cyclobutyl phenyl	: 7605
Ketone, 1-cyclohexen-1-yl methyl	: 5793
Ketone, cyclohexyl methyl	: 5794
Ketone, cyclopentyl methyl	: 5795
Ketone, cyclopropyl methyl	: 5796
Ketone, dicyclohexyl	: 7606
Ketone, dicyclopropyl	: 7607
Ketone, 6,7-dimethoxy-1-isoquinolyl 3,4-dimethoxyphenyl	: 7611
Ketone, 10-[3-(dimethylamino)-propyl]phenothiazin-2-yl methyl	: 5807
Ketone, 2-furyl methyl	: 5817
Ketone, 2-furyl phenyl	: 7618
Ketone, 1-hydroxycyclohexyl methyl	: 5821
Ketone, 6-hydroxy-2-isopropenyl-5-benzofuranyl methyl	: 5825
Ketone, 3-hydroxy-2-thienyl methyl	: 5835
Ketone, indol-3-yl methyl	: 5819
Ketone, methyl 1-methylcyclohexyl	: 5843
Ketone, methyl 3-methylcyclohexyl	: 5844

Ketone, methyl 4-methylcyclohexyl	: 5845
Ketone, methyl 1-methylcyclopentyl	: 5846
Ketone, methyl 2-methylcyclopentyl	: 5847
Ketone, methyl 6-methyl-3-pyridyl	: 5855
Ketone, methyl 1-methylpyrrol-2-yl	: 5849
Ketone, methyl 5-methyl-2-thienyl	: 5856
Ketone, methyl 2-pyridyl	: 5867
Ketone, methyl 3-pyridyl	: 5868
Ketone, methyl 4-pyridyl	: 5869
Ketone, methyl pyrrol-2-yl	: 5820
Ketone, methyl styryl	: 3881
Ketone, methyl 2-thienyl	: 5870
Ketone, 1-naphthyl o-tolyl	: 7629
Ketone, phenyl 2-pyridyl	: 7638
Ketone, phenyl 4-pyridyl	: 7639
Ketone, phenyl 2-thienyl	: 7641
Ketone, phenyl 3-thienyl	: 7642
2-Ketosuccinic acid	: 3339
Khellin	: 6185
Kojic acid	: 10795
KSCN	: 11711
Kyanmethin	: 11045
Kynurenic acid	: 11232
Kynurenine	: 1102
Lactic acid (DL)	: 10266
Lactic acid, butyl ester (d)	: 10268
Lactic acid, isopropyl ester	: 10275
Lactic acid, methyl ester (DL)	: 10273
Lactic acid, 3,3,3-trichloro-	: 10333
Lactobionic acid	: 6210
Lactonic acid	: 6210
Lactonitrile	: 10105
Lactonitrile, 3-chloro-2-methyl-	: 10098
Lactonitrile, 3,3,3-trichloro-	: 10115
p-Lactophenetide	: 9796
Lactyl chloride, acetate (DL)	: 10494
Lamoryl	: 11520
Lanex	: 12007
Lanosterol	: 7411
Lanoxin	: 4041
L-Lanthionine	: 5016
Lapachol	: 7862
Lappaconitine	: 309
Laricic acid	: 8102
Laudanidine	: 9399
Laudanine	: 9398
Laudanosine	: 7379
Lauraldehyde	: 5288
Lauraldehyde, dimethyl acetal	: 5297
Laureline	: 2469
Lauric acid	: 5305
Lauric acid, phenyl ester	: 5317
Lauric acid, propyl ester	: 5320
Laurolenic acid	: 4912
Laurone	: 11857
Lauronitrile	: 5303
Lauroyl peroxide	: 9155
Lauryl alcohol	: 5321
Lauryl bromide	: 5293
Lawsone	: 7859
Leaf aldehyde violet	: 8108
Ledol	: 4989
Lemonol	: 8333
Lenacil	: 4900
Lenthionine	: 8998
Lepidine	: 11279
Lepidine, 2-amino-	: 11208
Lepidine, 2-ethyl-	: 11270
Lergitin	: 5506
Leucethane	: 3948
Leucoline	: 7373
Levodopa	: 11929
Levopimaric acid	: 9169
Levulinic acid	: 8903
Levulinic acid, benzyl ester	: 8911
Levulinic acid, isopropyl ester	: 8909
Levulinic acid, 3-methyl-, methyl ester	: 8898
Levulinic acid, propyl ester	: 8912
β-Levulose	: 6023
Lichenic acid	: 3764
Lidocaine	: 54
Lignoceric acid	: 11607
Limettin	: 2818
α-Limonene	: 4677
Limonene	: 4677
Linalool formate	: 8337

Linalool, tetrahydro (DL)	: 8454
Linalyl formate	: 8337
Linamarin	: 10104
Linarigenin	: 2809
β-Lindane	: 4406
δ-Lindane	: 4407
Lindane	: 4409
Linderic acid	: 5338
Linoleic acid	: 8223
Linolenic acid	: 8287
Lithocholic acid	: 4099
Lobelanidine	: 9646
Lobelanine	: 5854
Lobeline	: 5834
Loflucarban	: 11802
Longifolene	: 7568
Luciculine	: 5007
Lumazine	: 10748
Lumichrome	: 2472
Luminol	: 9600
Lumisterol	: 5382
Lunacrine	: 6182
Lupeol	: 7417
3,5-Lupetidine	: 9652
Lupinine	: 11331
Lupulon	: 4281
trans-Lutein	: 4058
Luteolin	: 2814
2,3-Lutidine	: 10936
2,4-Lutidine	: 10937
2,5-Lutidine	: 10938
2,6-Lutidine	: 10939
3,4-Lutidine	: 10940
3,5-Lutidine	: 10941
Lycomarasmine	: 472
trans-Lycopene	: 4053
Lycoremine	: 6200
Lycorine	: 6201
Lycoxanthin	: 4060
Lysergamide	: 5374
Lysergic acid	: 5376
D-Lysergic acid D-propanolamide	: 5373
Lysidine	: 7197
Lysilan	: 5140
Maclurin	: 7610
Magenta I	: 634
Magneson	: 1542
Malachite green	: 7455
Malathion	: 3316
Maleamic acid	: 3817
Maleanilic acid	: 3853
Maleic acid	: 3765
Maleic acid, chloro-, diethyl ester	: 3771
Maleic acid, diallyl ester	: 3782
Maleic acid, dihydroxy-	: 3777
Maleic anhydride	: 6091
Maleimide, N-phenyl-	: 11123
Malidone	: 8586
Malilum	: 11091
Malonic acid	: 9921
Malonic acid, amino-, diethyl ester	: 9924
Malonic acid, 2-cyclopentenyl, diethyl ester	: 9935
Malonic acid, cyclopentylidene, diethyl ester	: 9936
Malonic acid, dibenzyl-, diethyl ester	: 9926
Malonic acid, dibutyl-, diethyl ester	: 9937
Malonic acid, diethyl-, diethyl ester	: 9940
Malonic acid, dimethyl-, diethyl ester	: 9942
Malonic acid, dipropyl ester	: 9945
Malonic acid, di-2-propynyl-, diethyl ester	: 9946
Malonic acid, ethylisopropyl-, diethyl ester	: 9951
Malonic acid, (hydroxymethyl)-	: 9959
Malonic acid, isobutyl-, diethyl ester	: 9969
Malonic acid, isopropylidene-, diethyl ester	: 9966
Malonomononitrile, benzylidene,monoethyl ester	: 10645
Malononitrile	: 9911
Malononitrile, phenyl-	: 9916
Maltol	: 10796
Malvidin chloride	: 2847
Mandelic acid isoamyl ester	: 912
Mandelonitrile glucoside	: 955

Manganese, [[1,2-ethanediylbis-[carbamodithioato]](2-)]-	: 7420
Mannite	: 7422
Mannitol, hexanitrate	: 7424
Margaric acid	: 6314
Margaric aldehyde	: 6304
Matrine	: 7426
MCPA	: 141
Mechlorethamine	: 5413
Meclizine	: 9605
Meconic acid	: 10775
Meconin	: 7355
Mecrylate	: 10624
Medroxyprogesterone	: 9752
Melamine	: 11835
Melampyrin	: 6194
Melatonin	: 69
Meldrum's acid	: 5190
Melezitose	: 6221
Mellitene	: 1758
Mellitic acid	: 1748
Menadiol diacetate	: 7841
Menadione	: 7864
p-Mentha-1,8-dien-7-al	: 4633
p-Mentha-1(7),2-diene	: 4673
p-Menthane, 4-bromo	: 4321
trans-(-)-p-Menthan-3-one	: 4606
m-Menth-8-ene	: 4435
p-Menth-2-ene, 1,4-epidioxy-	: 5184
p-Menth-3-en-2-ol	: 4698
(±)-Menthol	: 4548
Menthol(DL)	: 4548
Menthone(d)	: 4605
l-Menthone	: 4606
(-)-Menthone	: 4606
Meparfynol	: 9146
Mephenesin	: 10019
Mephenoxalone	: 8588
Mephobarbital	: 11094
Mepivacaine	: 9639
Meprobamate	: 10021
β-Mercaptoalanine	: 5014
Mercaptobenzene	: 2289
Mercaptobenzimidazole	: 2386
2-Mercaptobenzoic acid	: 2663
2-Mercaptobenzothiazole	: 2866
1-Mercaptobutane	: 3437
Mercaptocyclopentane	: 4852
2-Mercaptoethanol	: 5732
3-Mercaptohexane	: 6858
Mercaptomethane	: 7546
6-Mercaptopurine	: 10763
Mercapturic acid	: 5015
Mercury, chloro(2-methoxyethyl)-	: 11822
Mesaconic acid	: 3785
Mesaconic acid, diethyl ester	: 3787
Mesalamine	: 2499
Mescaline	: 1586
Mesitylamine	: 859
Mesitylene	: 2359
Mesitylene, α-bromo-	: 1057
Mesitylene, 2,4-dibromo-	: 1313
Mesitylene, 2,4-dichloro-	: 1402
Mesitylenic acid	: 2583
Mesityl oxide	: 9108
Mesoxalylcarbamide	: 11073
Mesoxalylurea	: 11073
Mestanolone	: 350
Mestilbol	: 9358
Mesulphen	: 11661
Metacrolein	: 11906
Metaldehyde II (tetramer)	: 19
Metanilic acid	: 2216
Methacetin	: 71
Methacholine chloride	: 9830
Methacrolein	: 10515
Methacrolein diacetate	: 10564
Methacrylic acid	: 10648
Methacrylic acid anhydride	: 10649
Methacrylic acid, 3-bromo	: 10612
Methacrylic acid chloride	: 10715
Methacrylic acid, 2-ethylhexyl ester	: 10659
Methadone hydrochloride	: 6516
Methallatal	: 11064
Methallyl alcohol	: 10702
α-Methallyl bromide	: 3710
Methaminodiazepoxide	: 2409

trans-2-Methylcyclohexylamine	: 4293
3-Methylcyclohexyl bromide	: 4320
Methyl cyclohexylideneacetate	: 156
Methyl cyclohexyl ketone	: 5794
2-Methylcyclopentadecanone	: 4762
Methylcyclopentane	: 4826
3-Methylcyclopentaneacetic acid	: 4776
1-Methylcyclopentanecarboxylic acid	: 4783
cis-2-Methylcyclopentanol	: 4868
trans-2-Methylcyclopentanol	: 4869
3-Methylcyclopentanol	: 4870
(±)-3-Methylcyclopentanone	: 4886
3-Methylcyclopentene	: 4929
3-Methyl-2-cyclopentenone	: 4945
Methylcyclopropane	: 4981
Methyl cyclopropanecarboxylate	: 4963
2-Methylcyclopropanecarboxylic acid	: 4964
Methyl cyclopropylcarboxylate	: 4963
Methylcysteine hydrochloride	: 5019
5-Methylcytosine	: 11106
2-Methyldecane	: 5054
3-Methyldecane	: 5055
4-Methyldecane	: 5056
Methyl decyl ketone	: 5325
Methyldiacetylene	: 8684
1-Methyldiallyl	: 6328
Methyldiazepinone	: 2410
3-Methyl-1,2-diazine	: 10856
2-Methyl-1,3-diazine	: 11070
4-Methyl-1,3-diazine	: 11071
5-Methyl-1,3-diazine	: 11072
Methyl 2-diazopropionate	: 10237
Methyl 3,5-dibromosalicylate	: 2559
Methyl dichloroacetate	: 166
5-Methyl-2,4-dichloropyrimidine	: 11052
N-Methyldiethanolamine	: 5742
3-Methyldigitoxose	: 7112
Methyl dihydrocinnamate	: 2139
α-Methyldihydrocinnamic acid	: 2137
Methyl 2,4-dimethylbenzoate	: 2589
Methyl α,α-dimethylhydrocinnamate	: 2129
Methyl N-(2,6-dimethylphenyl)-N-(phenylacetyl)-DL-alaninate	: 513
Methyl 2,2-dimethylpropionate	: 10260
Methyldiphenylamine	: 807
Methyl 11-dodecenoate	: 5341
Methyldopa	: 11931
Methyleneaminoacetonitrile	: 285
2,2'-Methylenebiphenyl	: 5967
4,4'-Methylenebis[2-chloroaniline]	: 785
3,3'-Methylenebis[6-hydroxybenzoic acid]	: 2686
3,3'-Methylenebis[4-hydroxycoumarin]	: 2839
Methylene blue	: 9506
Methylene bromide	: 7492
α-Methylene butyrolactone	: 6147
Methylene chloride	: 7499
exo-Methylenecyclobutane	: 4193
Methylenecyclohexane	: 4428
Methylenecyclopentane	: 4827
Methylene difluoride	: 7504
1,2-Methylenedioxy-9,10-dimethoxy-6aα-aporphine	: 2468
1,2-(Methylenedioxy)-4-propylbenzene	: 2431
5,5'-Methylenedisalicylic acid	: 2686
Methylenedisulfonic acid	: 7514
Methylene iodide	: 7506
3-Methylene-2-norbornanone	: 2958
Methylene oxide	: 5987
Methylenesuccinic acid	: 3328
(1R,2S)-N-Methylephedrine	: 1863
1-Methylethanesulfonic acid	: 10130
N-Methyl-2-ethanolamine	: 5738
Methyl ethoxyacetate	: 182
Methyl(2-ethoxyethyl)amine	: 5426
Methylethylallene	: 6679
Methylethylallylcarbinol	: 7084
cis-1-Methyl-3-ethylcyclohexane	: 4397
cis-1-Methyl-4-ethylcyclohexane	: 4399
trans-1-Methyl-4-ethylcyclohexane	: 4400
1-Methyl-1-ethylcyclopentane	: 4814
cis-1-Methyl-2-ethylcyclopentane	: 4815
1-Methyl-cis-3-ethylcyclopentane	: 4817

1-Methyl-trans-3-ethylcyclopentane	: 4818
cis-1-Methyl-2-ethylcyclopropane	: 4976
trans-1-Methyl-2-ethylcyclopropane	: 4977
Methylethyldichlorosilane	: 11397
2-Methyl-5-ethylfuran	: 6106
3-Methyl-3-ethylhexane	: 6826
Methyl ethyl ketone	: 3645
1-Methyl-5-ethylnaphthalene	: 7887
2-Methyl-3-ethylpentane	: 8813
3-Methyl-3-ethylpentane	: 8814
Methyl α-ethylstyryl ketone	: 8987
5-Methyl-3-ethylthiophene	: 11764
Methyl ethynyl ketone	: 3923
9-Methylfluorene	: 5974
Methyl fluoride	: 7515
Methyl fluoroacetate	: 192
Methylfluoroform	: 5665
N-Methylformamide	: 5995
Methyl formate	: 6010
2-Methylfuran	: 6130
3-Methylfuran	: 6131
5-Methyl-3-furancarboxylic acid	: 6065
N-Methylglucamine	: 6208
N-Methyl-L-glucosamidinostreptosidostreptidine	: 11562
N-Methyl-α-L-glucosamine	: 6215
α-Methylglucoside	: 6227
Methyl DL-glycerate	: 10254
1-Methylglycerol	: 3458
β-Methylglycerol α,γ-dichlorohydrin	: 10365
7-Methylguanine	: 10769
2-Methylheptadecane	: 6310
9-Methylheptadecane	: 6311
3-Methyl-1,5-heptadiene	: 6335
3-Methylheptan-2,4-diol	: 6401
2-Methylheptane	: 6422
4-Methylheptane	: 6424
Methyl heptanoate	: 6466
3-Methyl-1-heptanol	: 6492
4-Methyl-1-heptanol	: 6493
6-Methyl-1-heptanol	: 6494
2-Methyl-2-heptanol	: 6495
3-Methyl-2-heptanol	: 6496
6-Methyl-2-heptanol	: 6497
2-Methyl-3-heptanol	: 6500
4-Methyl-3-heptanol	: 6501
6-Methyl-4-heptanol	: 6503
2-Methyl-4-heptanol	: 6503
4-Methyl-4-heptanol	: 6505
3-Methyl-2-heptanone	: 6520
2-Methyl-3-heptanone	: 6522
2-Methyl-4-heptanone	: 6524
3-Methyl-4-heptanone	: 6525
2-Methyl-1-heptene	: 6543
3-Methyl-1-heptene	: 6544
4-Methyl-1-heptene	: 6545
5-Methyl-1-heptene	: 6546
6-Methyl-1-heptene	: 6547
2-Methyl-2-heptene	: 6548
3-Methyl-trans-2-heptene	: 6549
4-Methyl-trans-2-heptene	: 6551
2-Methyl-5-heptene	: 6553
2-Methyl-3-heptene	: 6554
3-Methyl-3-heptene	: 6555
4-Methyl-3-heptene	: 6556
2-Methyl-4-heptene	: 6558
6-Methyl-2-hepten-4-ol	: 6575
5-Methyl-2-hepten-4-one	: 6596
6-Methyl-4-hepten-2-one	: 6599
2-Methyl-3-hepten-6-one	: 6599
5-Methylhept-4-en-3-one	: 6600
3-Methyl-5-hepten-2-one	: 6601
5-Methylhept-5-en-3-one	: 6605
Methyl hexacosanoate	: 6625
Methyl 2,4-hexadienoate	: 6702
2-Methylhexa-4,5-dione	: 6809
3-Methylhexanal	: 6720
2-Methylhexane	: 6834
(+)-3-Methylhexane	: 6835
Methylhexaneamine	: 6730
3-Methyl-2,5-hexanedione	: 6810
(R)-2-Methyl-1-hexanol	: 6938
5-Methyl-1-hexanol	: 6941
3-Methyl-2-hexanol	: 6943
5-Methyl-2-hexanol	: 6944
(±)-2-Methyl-3-hexanol	: 6945
3-Methyl-3-hexanol	: 6946

(±)-5-Methyl-3-hexanol	: 6947
3-Methyl-2-hexanone	: 6976
4-Methyl-2-hexanone	: 6977
2-Methyl-1-hexene	: 7036
3-Methyl-1-hexene	: 7037
4-Methyl-1-hexene	: 7038
5-Methyl-1-hexene	: 7039
2-Methyl-2-hexene	: 7040
Methyl 3-hexenoate	: 7062
4-Methyl-2-hexenoic acid	: 7056
5-Methyl-2-hexenoic acid	: 7057
5-Methyl-2-hexenol	: 7078
4-Methyl-5-hexen-3-ol	: 7085
4-Methyl-5-hexen-2-one	: 7098
5-Methyl-1-hexyne	: 7120
2-Methyl-3-hexyne	: 7122
2-Methyl-3-hexyn-2-ol	: 7129
2-Methyl-2H-indazole	: 7242
Methylhydrazine	: 7161
Methyl hydrocinnamate	: 2139
α-Methylhydrocinnamic acid	: 2137
Methyl hydrogen sulfate	: 11577
Methyl p-hydroxybenzoate	: 2637
Methyl 11-hydroxyhendecanoate	: 11949
Methyl 11-hydroxyundecanoate	: 11949
1-Methylindan	: 7255
2-Methylindan	: 7256
4-Methylindan	: 7257
5-Methylindan	: 7258
4-Methyl-1-indanone	: 7290
1-Methylindazole	: 7239
3-Methylindazole	: 7240
5-Methylindazole	: 7241
1-Methylindene	: 7275
2-Methylindene	: 7276
3-Methylindene	: 7277
4-Methylindene	: 7278
6-Methylindene	: 7279
7-Methylindene	: 7280
7-Methylindole	: 7317
Methyl iodide	: 7518
6-Methyl-α-ionone	: 3884
Methylisobutylcarbinol acetate	: 8947
Methyl isobutyrate	: 10301
Methyl isocrotonate	: 3848
Methyl isocyanate	: 7520
	: 7655
Methyl isonipecotate	: 9644
Methyl isopentyl ketone	: 6978
Methylisopropenylacetylene	: 9116
Methyl isopropenyl carbinol	: 3871
α-Methyl-p-isopropylhydrocinnamaldehyde	: 2117
Methyl isopropyl ketone	: 3668
6-Methyl-3-isopropylsalicylic acid	: 2642
1-Methylisoquinoline	: 7381
3-Methylisoquinoline	: 7382
4-Methylisoquinoline	: 7383
6-Methylisoquinoline	: 7384
8-Methylisoquinoline	: 7385
Methyl isothiocyanate	: 7521
Methyl isovalerate	: 3552
5-Methylisoxazole	: 7405
Methyl ketone	: 10411
2-Methyllactonitrile	: 10108
Methyl laurate	: 5315
6-Methyllepidine	: 11262
Methyl linoleate	: 8228
Methyl malonate	: 9943
Methyl mercaptan	: 7546
Methyl mercaptan, perfluoro	: 7528
Methyl methacrylate	: 10663
Methyl methanesulfonate	: 7532
3-Methyl-6-methoxy-2,5-dihydroxy-1,4-benzoquinone	: 4260
4-Methyl-4-methoxy-2-pentanone	: 8978
Methyl (2-methylbutyl) ether	: 3394
Methyl 2-methylbutyl ketone	: 6977
Methyl N-methylisonipecotinate	: 9645
Methyl 4-methyl-5-oxohexanoate	: 6903
Methyl 3-methyl-4-oxopentanoate	: 8898
Methyl 3-methyl-3-pentenoate	: 9086
N-Methylmorpholine	: 7696
1-Methylnaphthalene	: 7915
α-Methylnaphthalene	: 7915
2-Methylnaphthalene	: 7916
β-Methylnaphthalene	: 7916

p-Nitroso-N,N-dimethylaniline	: 720
p-Nitrosodiphenylamine	: 829
N-Nitroso-N-diphenylamine	: 830
N-Nitrosomorpholine	: 7699
1-Nitroso-2-naphthol	: 8035
4-Nitrosophenol	: 9438
N-Nitrosopiperidine	: 9688
N-Nitrosopyrrolidine	: 11170
5-Nitrothiazolylamine-2	: 11666
2-Nitrothiophene	: 11775
3-Nitrothiophene	: 11776
o-Nitrotoluene	: 2000
m-Nitrotoluene	: 2001
p-Nitrotoluene	: 2002
5-Nitro-o-toluidine	: 801
2-Nitro-4-(trifluoromethyl)-1-chlorobenzene	: 1217
Nitrourea	: 12024
Nitrous acid, benzyl ester	: 8097
Nitrous acid, isopropyl ester	: 8092
Nitroxoline	: 11323
4-Nitro-o-xylene	: 1475
Nonacosanyl alcohol	: 8100
6,8-Nonadien-2-one, 5-isopropyl-8-methyl-, (E)-(S)-(+)-	: 8113
Nonaldehyde	: 8117
2-Nonen-1-al	: 8179
Nonyl acetate	: 220
Nonyl alcohol	: 8157
Nonyl mercaptan	: 8145
Nonyl methyl ketone	: 11958
19-Norandrostenolone phenylpropionate	: 5408
Norboral	: 2192
2-Norbornanecarboxaldehyde	: 2916
Norbornane, 7,7-dimethyl-2-methylene-, (±)-	: 2923
2-Norbornanone, 3-methylene-	: 2958
2-Norbornene, 2,3-dimethyl-	: 2968
Norcarane	: 2915
Norcholanic acid	: 8210
Nordazepam	: 2411
Nordihydroguaiaretic acid	: 1515
Norephedrine, N-ethyl	: 1871
Norepinephrine	: 1500
Norhyoscyamine	: 916
Normethadone	: 6968
Normorphine	: 7665
Normotiroide	: 11665
Normuscol	: 4760
Nornicotine	: 11017
Norpempidine	: 9711
cis-Norpinic acid	: 4183
Norpinic acid, dimethyl ester	: 4184
Norsulfazole	: 2206
Nortestosterone phenylpropionate	: 5408
Norvalamine	: 3209
Noscapine	: 7356
Novoldiamine	: 8736
Novonal	: 9003
Novoscabin	: 2720
Ocimene	: 8485
Octaacetyl sucrose	: 6229
Octabenzone	: 7621
Octachlorocyclopentene	: 4932
Octacosyl alcohol	: 8222
Octacosylbenzene	: 2065
Octacosylcyclohexane	: 4458
Octadecamethylcyclononasiloxane	: 4726
Octadecamethyloctasiloxane	: 8482
Octadecanal, dimethyl acetal	: 8237
Octadecane, 1-cyclohexyl-	: 4459
Octadecane, 1-cyclopentyl-	: 4842
1,18-Octadecanedicarboxylic acid	: 5352
Octadecane, 1-phenyl-	: 2066
3-Octadecanol (DL)	: 8279
Octadecenyl aldehyde	: 8288
Octadecylbenzene	: 2066
Octadecyl mercaptan	: 8244
Octadecyltrichlorosilane	: 11470
α,ω-Octadiene	: 8319
2,2,3,3,4,4,5,5-Octafluoro-1-pentanol	: 8953
1,2,3,4,4a,9,10,10-Octahydrophenanthrene	: 9184
1,2,3,4,5,6,7,8-Octahydrophenanthrene	: 9185

Octalene	: 5177
Octamethylcyclotetrasiloxane	: 4993
(all-E)-2,6,10,14,19,23,27,31-Octamethyl-2,6,8,10,12,14,16,18,20,22,24,26,30-dotriacontatridecaene	: 4053
Octamethylene dibromide	: 8369
Octamethylpyrophosphoramide	: 5220
1,1,1,3,5,7,7,7-Octamethyltetrasiloxane	: 11647
Octamethyltrisiloxane	: 11914
Octamylamine	: 6354
Octane, 2-bromo (DL)	: 8357
Octane, 1-cyclohexyl-	: 4460
Octane, 1-cyclopentyl-	: 4843
Octane, 2-phenyl-	: 1981
Octane, 1-phenyl-	: 2067
2-Octanethiol (DL)	: 8419
Octane, 2,2,4-trimethyl	: 8420
Octanoic acid, 3,5,7,8-tetrachloro-2,2,3,4,4,5,6,6,7,8,8-undecafluoro	: 8442
Octatetraene	: 8483
Octatrienal	: 8484
4-Octene, 4-chloro (cis)	: 8500
3-Octen-2-ol, 2-methyl (cis)	: 8519
3-Octen-5-ol, 4-methyl	: 8521
1-Octen-5-one, 7-methyl	: 8525
1-Octen-3-yne-5-ol, 5-methyl	: 8527
Octodecylcyclohexane	: 4459
Octylacetylene	: 5097
Octyl alcohol	: 8443
Octyl bromide	: 8356
Octyl chloride	: 8359
9-Octyldocosane	: 5275
Octyl enanthate	: 6468
Octyl fluoride	: 8405
Octyl heptanoate	: 6468
(±)-2-Octyl iodide	: 8407
Octyl mercaptan	: 8418
Octyl methyl ketone	: 5075
1-Octylnaphthalene	: 7937
Octyl phthalate	: 1344
Octyltrichlorosilane	: 11471
1-Octyne-3-ol, 3-methyl	: 8544
2-Octynol	: 8543
Oenanthylidene acetone	: 5093
Oestra-1,3,5(10)-triene-3,17β-diol	: 5399
Oleanolic acid	: 8547
Oleic acid	: 8294
Oleic acid, benzyl ester	: 8309
Oleic acid, isopentyl ester	: 8306
Oleyl alcohol	: 8312
Opianic acid	: 2608
Orcinol	: 1531
Orcinolcarboxylic acid	: 2573
Oretic	: 2855
Orexine	: 11196
Orotic acid	: 11050
Oroxylin A	: 2810
o-Orsellinic acid	: 2573
Orthanilic acid	: 2215
Orthesin	: 2497
Orthocaine	: 2501
Orthoform	: 2501
Orthoformic acid, triisobutyl ester	: 10080
Ouabain	: 4040
Ovahormon benzoate	: 5403
Ovex	: 2221
10H-9-Oxaanthracene	: 12049
1-Oxa-2-azacyclopentadiene	: 7402
1-Oxa-2-aza-1H-indene	: 2389
1-Oxa-3-azaindene	: 2877
7-Oxabicyclo[4.1.0]heptane, β,β-carotene-3,3'-diol deriv.	: 4056
7-Oxabicyclo[2.2.1]heptane-2,3-dicarboxylic acid, disodium salt	: 5370
Oxaceprol	: 9766
Oxacyclohexane	: 10800
Oxacyclopentadiene	: 6025
Oxacyclotetradecane-2,10-dione, 4-[[(2,6-dideoxy-3-C-methyl-3-O-methyl-α-L-ribo-hexopyranosyl)oxy]-14-eth	: 5396
1-Oxa-3,4-diazacyclopentadiene	: 8563
1,2,4-Oxadiazolidin-3,5-dione, 2-tert-butyl-4-(2,4-dichloro-5-isopropoxyphenyl)	: 8565

1,2,4-Oxadiazolidine-3,5-dione, 2-(3,4-dichlorophenyl)-4-methyl-	: 7643
Oxalacetic acid	: 3339
Oxalacetic ester	: 3340
Oxalic acid	: 5553
Oxalic acid, diallyl ester	: 5564
Oxalic acid dichloride	: 5585
Oxalic acid, dicyclohexyl ester	: 5559
Oxalic acid dihydrate	: 5561
Oxalic acid, diisobutyl ester	: 5557
Oxalic acid, disodium salt	: 5566
Oxalic acid, lead salt	: 5568
Oxalyl chloride	: 5585
Oxamic acid	: 104
Oxamide	: 5493
Oxane	: 10800
1,4-OxathiiN-3-carboxamide, 5,6-dihydro-2-methyl-N-phenyl-	: 4039
1,3-Oxazole	: 8575
2,4-Oxazolidinedione, 3-allyl-5-methyl-	: 8586
2,4-Oxazolidinedione, 3-(3,5-dichlorophenyl)-5-ethenyl-5-methyl-	: 12048
2-Oxazoline, 2-amino-5-phenyl-	: 8574
Oxetane, gluconic acid deriv.	: 6211
3-Oximino-2-octanone	: 8399
3-Oximino-2-pentanone	: 8803
1-Oxindene	: 2438
Oxindole, 4,7-dimethyl	: 7330
Oxindole, 1-methyl-3-ethyl	: 7332
Oxiniacic acid	: 10909
2-Oxo-3-acetyltetrahydrofuran	: 6139
(+)-2-Oxo-10-bornanesulfonic acid	: 2929
2-Oxobutanedioic acid diethyl ester	: 3340
18-Oxocorticosterone	: 9746
Oxoethanoic acid	: 223
β-Oxoglutaric acid	: 8782
2-Oxo-2H-1-benzopyran-3-carboxylic acid	: 2775
Oxole	: 6025
2-Oxo-5-methylhex-3-ene	: 7094
Oxophenylarsine	: 447
2-Oxosuccinic acid	: 3339
Oxotremorine	: 11188
Oxyacanthine	: 8631
Oxybenzone	: 7620
Oxycholin	: 4101
Oxycinchophen	: 11233
Oxydiethylene acetate	: 5754
Oxydimorphine	: 3002
Oxymethurea	: 11995
Oxyphenonium bromide	: 5443
Oxytetracycline	: 7705
Palmitaldehyde, dimethyl acetal	: 6636
Palmitic acid	: 6644
Palmitic acid, benzyl ester	: 6655
Palmitoleic acid	: 6670
Palustric acid	: 9166
Pamacyl	: 2498
Pamoic acid	: 7777
Panthesin	: 8926
Pantocaine	: 2531
Pantocid	: 2565
Pantolactone	: 6143
Pantothenic acid	: 329
D(+)-Pantothenyl alcohol	: 3202
d-Pantothenyl alcohol	: 3202
Papaveraldine	: 7611
Papaverine	: 7378
Parabanic acid	: 7226
Paracetamin	: 29
Paraldehyde	: 11908
Paraldol	: 5191
Parasorbic acid	: 10791
Parasympatol	: 1887
Parathion	: 9583
Paraxenol	: 3090
Pargyline	: 1823
Paroxypropione	: 10450
Parsley camphor	: 2423
Patchouli alcohol	: 7586
Patulin	: 6183
Pebulate	: 3976
Pelargonic acid	: 8146
Pelargonic acid, butyl ester	: 8147
Pelargonic acid, 9-fluoro	: 8151
Pelargonic acid, heptyl ester	: 8152

Pelargonic acid, 3-methyl	: 8153
Pelargonic acid 3-methyl, ethyl ester	: 8155
Pelargonic acid, pentyl ester	: 8156
Pelargonidin chloride	: 2848
(-)-Pelletierine	: 10485
Pellotine	: 7394
Pempidine	: 9691
Penicillamine	: 12043
Penicillamine cysteine disulfide	: 12042
Penicillic acid	: 6701
Penicillin G, sodium salt	: 11655
Pentabromoacetone	: 10477
1,2,3,4,5-Pentabromobenzene	: 2082
Pentabromobenzene	: 2082
1,1,1,2,2-Pentabromoethane	: 5616
Pentabromophenol	: 9439
1,2,3,4,5-Pentachlorobenzene	: 2084
Pentachlorobenzene	: 2084
Pentachlorobenzoic acid	: 2712
1,2,2,3,4-Pentachlorobutane	: 3420
1,1,1,4,4-Pentachloro-2-butene	: 3805
Pentachloroethane	: 5617
Pentachlorophenol	: 9440
1,1,1,2,3-Pentachloropropane	: 10125
1,1,2,3,3-Pentachloropropane	: 10126
Pentacosylbenzene	: 2088
Pentacosylcyclohexane	: 4462
1H-Pentadecafluoroheptane	: 6432
Pentadecaldehyde	: 8640
Pentadecane, 1-cyclopentyl-	: 4844
Pentadecanoic acid, 15-hydroxy,ω-lactone	: 8562
sec-Pentadecyl alcohol	: 8647
1-Pentadecylamine	: 8641
3-Pentadecylcatechol	: 1543
Pentadecylic acid	: 8645
Pentadecyl methyl ketone	: 6320
1,3-Pentadiene-1-carboxylic acid	: 6698
2,4-Pentadienoic acid, 5(3,4-metylenedioxyphenyl)	: 8677
Pentaerythritol	: 9989
Pentaerythrityl tetraacetate	: 9988
Pentaerythrityl tetranitrate	: 9990
Pentaethylbenzene	: 2090
1,2,3,4,5-Pentafluorobenzene	: 2091
Pentafluorobenzene	: 2091
Pentafluorophenol	: 9442
1,1,1,2,2-Pentafluoropropane	: 10128
Pentafluoropropionic acid, methyl ester	: 10317
2,3,4,5,6-Pentafluorotoluene	: 2094
3,5,7,2',4'-Pentahydroxyflavone	: 2816
3,5,7,3',4'-Pentahydroxyflavone	: 2817
Pentaiodobenzene	: 2096
Pentalin	: 5617
1,2,3,4,5-Pentamethylbenzene	: 2097
Pentamethylbenzene	: 2097
Pentamethylbenzoic acid	: 2713
2,4,6,8,10-Pentamethylcyclopentasiloxane	: 4902
Pentamethylene	: 4774
Pentamethylene acetate	: 8791
Pentamethylene chloride	: 8747
Pentamethylene diiodide	: 8758
Pentamethylene glycol	: 8787
Pentamethylene oxide	: 10800
2,2,4,6,6-Pentamethylheptane	: 6433
2,2,3,3,4-Pentamethylpentane	: 8836
2,2,3,4,4-Pentamethylpentane	: 8837
Pentamethylphenol	: 9443
1,1,3,5,5-Pentamethyl-1,3,5-triphenyltrisiloxane	: 11915
Pentane, 1-amino-3-methyl	: 8706
3-Pentanecarboxylic acid	: 3514
Pentane, 2-chloro (d)	: 8720
Pentane 3-chloro-2,2-dimethyl-3-ethyl	: 8727
Pentane, 1,5-dicyclohexyl-	: 4464
1,2-Pentanediol (DL)	: 8790
1,5-Pentanedione	: 8734
2,4-Pentanedione, bis,beryllium(II)complex	: 2912
Pentane, 1,3-diphenyl-	: 1697
Pentane, 1,5-diphenyl-	: 2098
Pentane, 1,1-diphenyl-	: 2108
Pentane, 2-iodo (DL)	: 8819
Pentane, 3-isopropyl-2,4-dimethyl-	: 8766

Pentane, 2-methyl-3,3-diethyl	: 8757
Pentane, 3-methyl-3-phenyl	: 1691
Pentane, 3-methyl-1-phenyl	: 2010
Pentane, 2-methyl-2-phenyl	: 1455
Pentane, perfluoro	: 8807
Pentane, perfluoro,2-trifluoro methyl	: 8854
Pentane, 3-phenyl	: 1701
Pentane, 1-phenyl	: 2107
Pentane, 2,2,3-trimethyl-3-ethyl	: 8815
Pentane, 2,3,4-trimethyl-3-ethyl	: 8816
1,4,5-Pentanetriol	: 8851
1,2,3-Pentanetriol	: 9119
Pentanoic acid, 4-[(17β)-androstan-17-yl]-	: 4094
Pentanoic acid, 5-chloro-4-hydroxy,γ-lactone	: 6141
3-Pentanol, 1,5-bis(tetrahydro-2-furyl)-	: 6166
1-Pentanol, 2,4-dimethyl (DL)	: 8927
1-Pentanol, 3-methyl (DL)	: 8938
1-Pentanol-2-one	: 8971
1-Pentanol-4-one	: 8974
2-Pentanol-3-one	: 8975
3-Pentanol-2-one	: 8972
1-Pentanol, 1-phenyl-	: 1846
1-Pentanol, 5-phenyl-	: 2101
2-Pentanol, 2-phenyl-	: 1904
2-Pentanol, 5-phenyl-	: 1106
3-Pentanol, 1-phenyl (d)	: 2152
3-Pentanol, 3-phenyl-	: 1855
2-Pentanone, 3-benzylidene-	: 8987
2-Pentanone, 3-methyl (DL)	: 8980
Pentasa	: 2499
Pentatriacontylcyclopentane	: 4845
Pent-4-en-1-al	: 9001
trans-2-Pentene	: 9005
cis-2-Pentene	: 9006
trans-2-Pentenedioic acid diethyl ester	: 9042
1-Pentene, 2-ethyl-4-methyl-	: 6837
1-Pentene, 2-isopropyl-	: 6836
2-Pentene, 3-methyl (cis)	: 9057
1-Pentene, 3-methyl-3-ethyl	: 9047
1-Pentene, 4-methyl-3-ethyl	: 9048
2-Pentene, 4-methyl-3-ethyl	: 9049
2-Pentene, 4-methyl-3-ethyl (trans)	: 9050
2-Pentene, 4-methyl-3-ethyl (cis)	: 9051
trans-3-Pentenenitrile	: 9061
1-Pentene-3-one	: 9104
2-Pentene, 3-phenyl-	: 1699
2-Pentene, 2-phenyl-	: 1950
1-Pentene, 5-phenyl-	: 2106
1-Pentene, 2-propyl-	: 6426
2-Pentenoic acid, 2-methyl (trans)	: 9076
2-Pentenoic acid, 3-methyl (cis)	: 9077
3-Pentenyl bromide	: 9015
trans-3-Penten-1-yne	: 9113
cis-3-Penten-1-yne	: 9114
Pentetic acid	: 6257
Pentoxone	: 8978
Pentyl acetate	: 227
sec-Pentyl acetate (R)	: 8918
sec-Pentyl alcohol	: 8916
	: 8917
tert-Pentyl bromide	: 3240
2-Pentyl bromide	: 8711
3-Pentyl bromide	: 8712
Pentyl butanoate	: 3579
Pentyl butyrate	: 3579
Pentylcyclohexane	: 4485
Pentylenetetrazole	: 11654
Pentyl fluoride	: 8817
Pentyl formate	: 6015
Pentyl heptanoate	: 6470
Pentyl isocyanide	: 8821
Pentyl ketone	: 11962
Pentyl mercaptan	: 8844
3-Pentyl mercaptan	: 8846
2-Pentylnaphthalene	: 7942
Pentyl nitrite	: 8096
Pentyl nonanoate	: 8156
p-tert-Pentylphenol	: 9339
Pentyl phenyl ether	: 2109
Pentylpropiolic acid	: 8538
4-(3-Pentyl)pyridine	: 10968
1-Pentynyl bromide	: 9124
Perbenzoic acid	: 1126

Perchloroacetone	: 10440
Perchlorobenzoic acid	: 2712
Perchlorocyclopentadiene	: 4766
Perchlorocyclopentene	: 4932
Perchloromethyl mercaptan	: 7527
Perchloronaphthalene	: 7933
Perchloropropylene	: 10571
Pereirine	: 4154
Perfluoroacetone	: 10441
Perfluorobenzene	: 1757
Perfluorobutane	: 3255
Perfluoro-2-butene	: 3804
Perfluoro-2-butyltetrahydrofuran	: 6109
Perfluorocyclobutane	: 4194
Perfluorocyclohexane	: 4381
Perfluorocyclohexene	: 4643
Perfluorodecalin	: 7934
Perfluorodecane	: 5050
Perfluoro-2,3-dimethylbutane	: 3414
Perfluoroethane	: 5596
Perfluoroheptane	: 6418
Perfluoro-1-heptene	: 6560
Perfluorohexane	: 6843
Perfluoro-1-hexene	: 7032
Perfluoroisobutane	: 10050
Perfluoro(methylcyclohexane)	: 4499
Perfluoro(2-methylpentane)	: 8854
Perfluoro-3-methylpentane	: 8853
Perfluoronaphthalene	: 7935
Perfluorononane	: 8139
Perfluorooctane	: 8415
Perfluoropentane	: 8807
Perfluoropropane	: 10118
Perfluoropropene	: 10572
Perfluoropropylene oxide	: 8629
Perfluorotoluene	: 2095
Perillaldehyde	: 4633
Perillyl aldehyde	: 4633
Permethylbenzoic acid	: 2713
Permethyldisilane	: 5228
1,4-Peroxido-p-menthene-2	: 5184
Peroxyacetic acid	: 5619
Perphenazine	: 9617
Peucedanin	: 6188
Phellandral	: 4635
β-Phellandrene	: 4673
Phenacemide	: 879
Phenacetin	: 60
Phenacyl acetate	: 5771
Phenacylamine	: 5774
Phenacyl bromide	: 5782
Phenacyl cyanide	: 2124
Phenallymal	: 11103
Phenamizole	: 11671
9,10-Phenanthraquinodimethane	: 9172
9,10-Phenanthraquinone, 3-hydroxy	: 9176
Phenanthrene, 9-isopropenyl-	: 9182
Phenanthrenequinone	: 9174
9-Phenanthroic acid	: 9165
o-Phenanthroline	: 9189
9-Phenanthryl bromide	: 9164
Phenazine, 2,3-diamino-	: 9194
Phenbenzamine	: 5506
Phencarbamide	: 3977
Phencyclidine	: 9696
[v-Phenenyltris(oxyethylene)]-tris[triethylammonium iodide]	: 5441
Phenetharbital	: 11089
Phenethyl alcohol	: 1592
Phenethyl alcohol, 2-ethyl	: 1603
Phenethyl alcohol, 4-ethyl	: 1604
Phenethyl alcohol, α,α,β-trimethyl-	: 1616
Phenethylamine, N-benzyl-N,α-dimethyl-, (+)-	: 1575
Phenethylamine, 4-bromo	: 1570
Phenethylamine, 2-methoxy	: 1577
2-Phenethyl propionate	: 10324
o-Phenetidine	: 736
p-Phenetidine	: 737
o-Phenetidine, 5-nitro	: 739
Phenetole	: 1157
Phenetole, m-chloro-	: 1650
Phenetole, 2,6-dinitro-	: 1665
Phenetole, 2-propenyl	: 1667
Phenetole, 2-propyl	: 7145
Phenicarbazide	: 2979
Phenicin	

Propane, 1,1-dicyclohexyl-	: 4472
Propane, 1,3-dicyclohexyl-2-ethyl	: 4402
Propane, 1,3-dicyclohexyl-2-methyl-	: 4447
1,2-Propanediol, 3-amino(DL)	: 9984
1,3-Propanediol, 2-methyl-2-propyl,N-butyl carbamate,carbamate	: 3933
Propane, 1,2-diphenoxy	: 1970
Propane, 1,3-dipheny-1-ethyl	: 1697
Propane, 1,3-diphenyl-	: 2121
Propane, 1,1-diphenyl-	: 2171
Propanenitrile, 2-[[4-chloro-6-(cyclopropylamino)-1,3,5-triazin-2-yl]amino]-2-methyl	: 9759
Propanenitrile, 2-[[4-chloro-6-(ethylamino)-1,3,5-triazin-2-yl]amino]-2-methyl	: 4161
Propane, 1-nitro-3,3,3-trifluoro	: 10184
2-Propanephosphonic acid	: 9542
1,3-Propanesulfone	: 11694
Propane sultone	: 8570
Propane, 1,2,2,3-tetrachloro-tetrafluoro	: 10149
Propanoic acid, 3-amino-	: 323
Propanoic acid, 2-amino-3-mercapto-, (R)-	: 5014
Propanoic acid, chloride, 2-chloro(d)	: 10496
Propanoic acid, 2-(3-chlorophenoxy)-	: 4141
Propanoic acid, 2-[4-(2,4-dichlorophenoxy)phenoxy]-, methyl ester	: 5171
Propanoic acid, 2,2-dichloro-3,3,3-trifluoro, propyl ester	: 10721
Propanoic acid, 3,3'-dithiobis[2-amino-, [R-(R*,R*)]-	: 5020
Propanoic acid, methylene ester	: 7654
Propanoic acid, 2-(2,4,5-trichlorophenoxy)-	: 11512
Propanolamine, N-pantoyl-	: 3202
1-Propanol, 2-[2-(2-amino-2-methylpropyl)amino]-2-methyl-	: 10342
2-Propanol, 1-chloro-3-isopropoxy-	: 10355
2-Propanol, 1-diethylamino (DL)	: 10368
2-Propanol, 1,1',1'',1'''-(ethylenedinitrilo)tetra-	: 10379
2-Propanol, 1-isopropoxy-	: 10390
2-Propanol, 1-nitro-3,3,3-trichloro	: 10408
2-Propanol, 1,1,1-trifluoro(DL)	: 10409
2-Propanone, acetyl-	: 8798
2-Propanone, (methoxyphenyl)-	: 10459
Propantheline bromide	: 9832
Propargylacetic acid	: 9138
Propargyl alcohol	: 10743
Propargyl aldehyde	: 10723
Propargyl bromide	: 10728
Propargyl chloride	: 10729
Propargylic acid, (2-nitrophenyl)	: 10740
2-Propene-1,2-dicarboxylic acid	: 3328
Propene-3,3-diol, 2-methyl,diacetate	: 10564
Propene, 2-methyl(tetramer)	: 10580
2-Propene-1-ol, 1-phenyl	: 1868
Propenoic acid, 2-chloro,ethyl ester	: 10619
Propenoic acid, 2-methyl,butyl ester	: 10651
Propenoic acid, 3(1-naphthyl)(trans)	: 10675
2-Propenyl alcohol	: 10697
2-Propenyl bromide	: 10531
2-Propenyl chloride	: 10536
2-Propenyl isothiocyanate	: 10575
N-2-Propenylurea	: 12028
Propesin	: 2507
Propham	: 3966
Propiocine	: 5396
β-Propiolactone	: 8595
Propiolic acid	: 10738
Propiolic acid, propyl-	: 7123
Propionaldehyde	: 9774
Propionaldehyde, 3-(2-tolyl)	: 2115
Propionamide	: 9791
Propionamide, N-antipyrinyl-2-(dimethylamino)-	: 9795
Propionic acid	: 10201
Propionic acid bromide	: 10490
Propionic acid, 3-bromo, isopentyl ester	: 10210
Propionic acid, 2-chloro, isobutyl ester (d)	: 10224

Propionic acid, 2-diazo-, ethyl ester	: 10236
Propionic acid, 2-diazo-, methyl ester	: 10237
Propionic acid, 2,3-dibromo-, butyl ester	: 10239
Propionic acid, 2,3-dibromo-, propyl ester	: 10243
Propionic acid, 2,2-dichloro-3,3,3-trifluoro-, methyl ester	: 10251
Propionic acid, 3-(diethylamino), ethyl ester	: 328
Propionic acid, 2,3-dihydroxy-2-methyl-, methyl ester	: 10256
Propionic acid, 2,3-diphenyl (DL)	: 2143
Propionic acid, 3,3'-iminodi-N-methyl, diethyl ester	: 330
Propionic acid, 3,3'-(methylimino)-di-, diethyl ester	: 330
Propionic acid, 2,3,3,3-tetrafluoro-, ethyl ester	: 10328
Propionic acid, 3-(2-tetrahydrofuryl)	: 6164
Propionic acid, 3,3,3-trichloro-2-(2,2,2-trichloro-1-hydroxyethoxy)-, γ-lactone	: 5212
Propionic anhydride	: 10203
Propionitrile, 2-ethoxy-	: 10102
Propionitrile, 2-methoxy-	: 10109
Propionylacetone	: 6804
Propionyl bromide	: 10490
Propionyl chloride	: 10493
Propionyl chloride, 3,3-dichloro	: 10502
Propionyl fluoride	: 10506
Propiophenone	: 10480
Propiophenone, 2'-hydroxy-3'-methyl-	: 10444
Propiophenone, 2'-hydroxy-4'-methyl-	: 10445
Propiophenone, 3-methoxy	: 10456
Propiophenone, 2-methoxy-	: 10457
Propofol	: 9226
2-Propoxyethanol	: 5764
1-Propoxy-2-propanol	: 10401
N-Propylacetanilide	: 88
Propyl acetate	: 236
Propylacetylene	: 9122
Propyl alcohol	: 10334
Propylallene	: 6676
Propylamine	: 9803
Propylamine, N,N-dichloro	: 9805
Propyl p-aminobenzoate	: 2507
Propyl amyl ketone	: 8174
Propylbenzene	: 2169
Propyl benzenesulfonate	: 2235
Propyl benzoate	: 2724
Propyl bromide	: 9836
Propyl bromoacetate	: 120
Propyl butanoate	: 3588
Propyl-tert-butylcarbinol	: 6925
Propyl carbamate	: 3969
Propyl carbonate	: 4002
Propyl chloride	: 9848
Propyl chloroacetate	: 149
1-Propyl-3-(p-chlorobenzenesulfonyl)urea	: 2208
Propyl chlorocarbonate	: 4028
trans-Propyl cinnamate	: 10693
Propyl cyanide	: 3403
Propylcyclohexane	: 4471
Propyl cyclohexanecarboxylate	: 4337
trans-4-Propylcyclohexanol	: 4560
cis-2-Propylcyclohexanol	: 4558
trans-2-Propylcyclohexanol	: 4559
4-Propylcyclohexanone	: 4613
2-Propylcyclohexanone	: 4612
trans-2-Propylcyclopentanol	: 4874
Propylcyclopentane	: 4849
cis-2-Propylcyclopentanol	: 4873
2-Propylcyclopentanone	: 4894
3-Propylcyclopentanone	: 4895
1-Propyl-1-cyclopentene	: 4935
Propylcyclopropane	: 4984
Propyl 2,3-dibromopropionate	: 10243
Propyldimethylamine	: 9807
Propylene	: 10528
Propylene, 2-bromo-3-cyclohexyl	: 4322
Propylene, 3-bromo-3,3-difluoro	: 10532
Propylene chlorohydrin	: 10347

sec-Propylene chlorohydrin	: 10349
Propylenediamine	: 9872
Propylene dibromide	: 9881
Propylene, 1,3-dibromo (trans)	: 10543
Propylene, 1,3-dibromo (cis)	: 10544
Propylenedicarboxylic acid	: 3328
Propylene, 1,1-dichloro-2-fluoro	: 10554
Propylene, 1,2-dichloro-3,3,3-trifluoro	: 10558
1,2-Propylene glycol	: 9982
1,3-Propylene glycol	: 9983
Propylene glycol β-monoethyl ether	: 10380
1,2-Propylene glycol 1-propyl ether	: 10401
Propylene oxide	: 8821
1,2-Propylenimine	: 501
N-Propylethanolamine	: 5765
Propylethylene	: 9004
Propyl fluoride	: 10046
Propyl formate	: 6018
Propylformic acid	: 3460
2-Propylfuran	: 6167
Propyl gallate	: 2732
4-Propylheptane	: 6434
Propyl heptanoate	: 6471
4-Propyl-3-heptene	: 6559
1-Propylheptyl alcohol	: 5072
Propyl hexahydrobenzoate	: 4337
Propylhexedrine	: 4385
Propyl hexyl ketone	: 5077
Propyl p-hydroxybenzoate	: 2653
Propylidene chloride	: 9891
Propyl iodide	: 10054
Propyliodone	: 10873
Propyl isobutyrate	: 10308
Propyl isocyanate	: 10059
Propyl isopropyl sulfide	: 10077
Propyl isovalerate	: 3562
Propyl laurate	: 5320
Propyl levulinate	: 8912
Propyl mercaptan	: 10156
Propyl methacrylate	: 10672
Propyl nitrate	: 8080
Propyl nitrite	: 8098
5-Propyl-1,5-octadien-3-yne	: 8341
Propylparaben	: 2653
2-Propylpentanal	: 8697
2-Propyl-1-pentene	: 6426
3-Propyl-1-pentene	: 7033
Propyl phosphite	: 9594
4-Propylpiperidine	: 9709
Propyl propionate	: 10326
3-Propylpyrrole	: 11139
2-Propylquinoline	: 11292
Propyl sulfide, 3,3'-dichloro	: 10153
2-Propylthiophene	: 11778
3-Propylthiophene	: 11779
Propylthiouracil	: 11109
N-Propyl-p-toluidine	: 815
Propyltrichlorosilane	: 11476
Propyl trifluoroacetate	: 269
Propyl 3,4,5-trihydroxybenzoate	: 2732
Propyne, 1,3-dibromo-	: 10730
Protionamide	: 10886
Protoanemonin	: 6155
Protocatechualdehyde	: 542
Protocatechuic acid	: 2571
Protocatechuic acid methylene ether	: 2418
Protopine	: 3109
Protoverine	: 4070
Provitamin D3	: 4103
Prunetin	: 2828
Prunetol	: 2807
Pseudoaconitine	: 308
Pseudocinchonine	: 4122
Pseudocodeine	: 7672
Pseudoconhydrine	: 9718
Pseudocumene	: 2358
Pseudoionone	: 11965
Pseudojervine	: 12045
Pseudomorphine	: 3002
Pseudopelletierine	: 479
Pseudotropine	: 486
Pteroylglutamic acid	: 6243
Puleganic acid	: 4784
Pulegene	: 4931
Pulegone	: 4608

Purodigin	: 4043
Purostrophan	: 4040
Purpurin	: 416
Purpurogallin	: 2408
Putrescine	: 3258
Pyocyanine	: 9196
Pyran-4-carboxylic acid, tetrahydro-	: 10774
Pyrazinamide	: 10802
Pyrazinoic acid	: 10803
Pyrazino[2,3-d]pyrimidine	: 10747
N1-2-Pyrazinylsulfanilamide	: 2202
Pyrazole, 4-amino-3-methyl-1-phenyl-	: 10813
1H-Pyrazole, 3-chloro-5-methyl-4-nitro-	: 4083
2-Pyrazoline	: 10822
2-Pyrazoline, 1-phenyl-	: 10823
2-Pyrazoline, 3-phenyl	: 10824
1H-Pyrazolium, 1,2-dimethyl-3,5-diphenyl-, methyl sulfate	: 5175
3-Pyrenecarboxylic acid	: 10854
Pyrethrin 1	: 4960
Pyrethrin 2	: 4962
Pyribenzamine	: 5505
3(2H)-Pyridazinone, 5-amino-4-chloro-2-phenyl-	: 4080
3(2H)-Pyridazinone, 4-chloro-5-(methylamino)-2-[3-(trifluoromethyl)phenyl]-	: 8213
Pyridine, 3-amino-2,6-dichloro	: 10862
Pyridine, 2-benzyl-	: 11004
Pyridine, 3-benzyl-	: 11005
Pyridine, 2-tert-butyl-	: 10942
Pyridine, 3-tert-butyl-	: 10943
Pyridine-3-carbinol	: 10976
4-Pyridinecarbinol	: 10977
Pyridine, 2-chloro-6-(trichloromethyl)-	: 8071
Pyridine, 2,6-di-tert-butyl-	: 10874
Pyridine, 2-[[2-(dimethylamino)ethyl]-2-thenylamino]-	: 5507
Pyridine, 2(2-dimethylaminoethyl-4-methoxybenzylamino)	: 5511
Pyridine, 2-(2-hydroxypropyl)	: 10951
Pyridine, 2-(2-isopropoxyethyl)	: 10983
Pyridine, 3-isopropyl-	: 10985
Pyridine N-oxide	: 10994
Pyridine, 4-(3-pentyl)	: 10968
Pyridinium chloride	: 10969
4(1H)-Pyridinone, 1-methyl-3-phenyl-5-[3-(trifluoromethyl)phenyl]-	: 5981
1,5-Pyridopyridine	: 8065
Pyridoxamine dihydrochloride	: 10978
Pyridoxine hydrochloride	: 10934
α-Pyridylamine	: 10857
3-Pyridylamine	: 10858
2-(2-Pyridyl)pyridine	: 3099
4-(4-Pyridyl)pyridine	: 3104
5-[p-(2-Pyridylsulfamyl)phenylazo]salicylic acid	: 2654
N1-2-Pyridylsulfanilamide	: 2203
3-Pyridylsulfonic acid	: 11018
Pyrilamine	: 5511
Pyrimethamine	: 11051
Pyrimidine, 2-amino-4,5-dimethyl-	: 11044
2,4(1H,3H)-Pyrimidinedione, 5-bromo-6-methyl-3-(1-methylpropyl)-	: 3139
2,4(1H,3H)-Pyrimidinedione, 5-chloro-3-(1,1-dimethylethyl)-6-methyl-	: 11591
5-Pyrimidinemethanol, α-(2-chlorophenyl)-α-(4-chlorophenyl)-	: 5948
5-Pyrimidinemethanol, α-(2-chlorophenyl)-α-(4-fluorophenyl)-	: 8219
4(1H)-Pyrimidinone, 5-butyl-2-(ethylamino)-6-methyl-	: 5932
N1-2-Pyrimidinylsulfanilamide	: 2204
Pyrithyldione	: 10946
Pyrocalciferol	: 5381
Pyrocarbonic acid diethyl ester	: 5169
Pyrocatechol	: 1497
Pyrochol	: 4095
Pyrocoll	: 5223
Pyrogallic acid	: 2366
Pyrogallol	: 2366

L-Pyroglutamic acid	: 9768
Pyrolan	: 3945
Pyromellitic acid	: 2257
Pyromucic acid	: 6050
Pyropentylene	: 4764
Pyrrole, 2-benzoyl	: 7637
Pyrrole-3-propionic acid, 2-[[3-(2-carboxyethyl)-4-methyl-5-[[(3-methyl-5-oxo-4-vinyl-3-pyrrolin-2-ylidene)meth	: 3001
2-Pyrrolidinecarboxylic acid	: 9764
2-Pyrrolidone	: 11180
2-Pyrrolidone, 3,3-dimethyl	: 11182
3-Pyrroline	: 11118
Pyruvaldehyde	: 9788
Pyruvic acid	: 10314
Pyrylene	: 9111
Quercetin	: 2817
Quercetin 7-methyl ether	: 2815
Quercetin-3-L-rhamnoside	: 2796
D-Quercitol	: 4074
Quercitrin	: 2796
Quillaic acid	: 8546
Quinaldic acid	: 11227
Quinaldine	: 11277
Quinaldinic acid	: 11227
Quinaldonitrile	: 11223
Quinalizarin	: 413
Quinamine	: 6178
4-Quinazolinone, 2-benzyl-1-methyl	: 11200
Quinethazone	: 11197
Quinic acid	: 4338
Quinidine	: 4126
Quinine	: 4127
Quinine carbonate	: 4128
Quinine, formate	: 4129
Quinine hydrochloride	: 4130
Quinine salicylate	: 4131
Quinine sulfate	: 4132
Quininic acid	: 11234
Quininone	: 4134
Quinizarin	: 392
Quinoline, 2-amino-	: 11202
Quinoline, 6-amino-	: 11205
Quinoline, 4-amino-3-bromo	: 11207
Quinoline, 2-amino-4-methyl	: 11208
Quinoline, 7-amino-2-methyl	: 11211
Quinoline, 8-amino-6-methyl-	: 11212
Quinoline, decahydro (trans, DL)	: 11247
Quinoline, 2,4-dimethyl-6-hydroxy	: 11317
Quinoline, 2,4-dimethyl-8-hydroxy	: 11318
Quinoline, 5-hydroxy	: 11305
Quinoline, 7-hydroxy	: 11307
Quinoline, 7-hydroxy-6-methyl	: 11320
Quinoline, 4-methoxy-2-[3,4-(methylenedioxy)phenethyl]-	: 11216
Quinolinic acid	: 10927
p-Quinone	: 4257
Quinone	: 4257
Quinone tetrachloride	: 4268
Quinovic acid	: 12036
Quinovose	: 6235
Quinoxaline, 6-amino-	: 11332
N1-(2-Quinoxalinyl)sulfanilamide	: 2205
Quintozene	: 2087
Quinuclidine	: 480
3-Quinuclidinol	: 485
Racemic tartaric acid	: 3311
Radicinin	: 10785
Raffinose	: 6219
Randox	: 44
Raubasine	: 8571
Raunescine	: 12064
Reductic acid	: 4942
Reductinic acid	: 4942
Refrigerant 10	: 7540
Refrigerant 11	: 7555
Refrigerant 12	: 7500
Refrigerant 13	: 7486
Refrigerant 13B1	: 7475
Refrigerant 14	: 7541
Refrigerant 20	: 7554
Refrigerant 21	: 7502
Refrigerant 22	: 7478
Refrigerant 31	: 7481
Refrigerant 32	: 7504
Refrigerant 50	: 7459

Refrigerant 113	: 5663
Refrigerant 114	: 5538
Refrigerant 121a	: 5639
Refrigerant 122a	: 5660
Refrigerant 122b	: 5658
Refrigerant 124	: 5486
Refrigerant 133	: 5488
Refrigerant 133b	: 5487
Refrigerant 134	: 5641
Refrigerant 142	: 5471
Refrigerant 142b	: 5470
Refrigerant 245	: 10128
Refrigerant C317	: 4177
Renazide	: 2851
Renese	: 2857
Resacetophenone	: 5803
Resazurin	: 9508
Rescinnamine	: 12060
Reserpic acid	: 12063
Reserpine	: 12061
Resorcinol	: 1498
Resorcinol, 4-acetyl	: 8490
Resorcinol diglycidyl ether	: 8625
Resorcinol dipropyl ether	: 1556
Resorcinol, 5-iodo	: 1524
Resorcinol, 4-isobutyl-	: 1539
Resorcinolphthalein	: 11522
β-Resorcylaldehyde	: 541
β-Resorcylic acid	: 2569
Retene	: 9183
Reychler's acid	: 2929
Rhamnetin	: 2815
Rhamnol	: 11558
Rheadine	: 11342
Rhein	: 371
Rhizopterin	: 2494
Rhodamine, B	: 12056
Rhodanine	: 11685
Rhodanine, 5-[p-(dimethylamino)benzylidene]-	: 11684
(-)-Rhodinol	: 8516
Rhodizonic acid	: 4686
Riboflavine	: 11344
Riboflavin lumichrome	: 2472
β-D-Ribofuranoside, adenine-9	: 319
β-D-Ribofuranoside, guanine-9	: 6289
1-β-D-Ribofuranosyluracil	: 12033
Ricinine	: 10883
Ricinoleic acid	: 8303
Risocaine	: 2507
Ronnel	: 9584
Roquessine	: 4143
p-Rosolic acid	: 4278
Rotenone	: 2783
Rubeanic acid	: 5588
Rubijervine	: 11515
Rubixanthin	: 4062
Rufigallol	: 406
Rutecarpine	: 7334
Rutinose	: 6236
Sabadine	: 4067
Sabinane (d)	: 11814
Sabinene (DL)	: 2981
D-Saccharic acid	: 6205
Saccharin	: 2388
Safranal	: 4250
Safrole	: 2429
Salicin	: 6222
Salicyl alcohol	: 1882
Salicylaldehyde	: 564
Salicylaldoxime	: 579
Salicylamide	: 616
Salicylamide O-acetic acid	: 103
Salicylanilide	: 617
Salicylhydroxamic acid	: 613
Salicylic acid	: 2612
p-Salicylic acid	: 2613
Salicylic acid, 4-benzamido-	: 2514
Salicylic acid, 3-chloro, ethyl ester	: 2541
Salicylic acid, 3,5-dibromo-, methyl ester	: 2559
Salicylic ether	: 2619
Salicylsulfonic acid	: 2655
Saligenol	: 1882
Salirepol	: 1520
Salsalate	: 2616
Salsoline	: 7397

2,2,4,6-Tetramethylpiperidine	: 9710
2,2,6,6-Tetramethylpiperidine	: 9711
1,2,5,6-Tetramethyl-4-piperidinone	: 9725
Tetramethylputrescine	: 3263
2,3,4,6-Tetramethylpyridine	: 11022
Tetramethylsilane	: 11451
Tetramethyl silicate	: 11508
Tetramethylstannane	: 11544
Tetramethylsuccinonitrile	: 3292
2,2,4,4-Tetramethyltetrahydrofuran	: 6174
N,N,N',N'-Tetramethyltetra-methylenediamine	: 3263
Tetramethylurea	: 12030
Tetranitrin	: 3432
Tetranitromethane	: 7544
Tetranitropentaerythritol	: 9990
2,3,4,6-Tetranitrophenol	: 9476
2,4,5,6-	: 11073
Tetraoxohexahydropyrimidine	
1,1,1,2-Tetraphenylethane	: 1618
1,1,2,2-Tetraphenyl-1,2-ethanediol	: 5580
1,1,2,2-Tetraphenylethylene glycol	: 5580
2,3,4,5-Tetraphenylfuran	: 6175
Tetraphenylfuran	: 6175
Tetraphenylmethane	: 1831
Tetraphenylsilane	: 11452
Tetraphenylurea	: 12031
Tetraphthene	: 9
Tetrapropylammonium iodide	: 9833
Tetratriacontylcyclohexane	: 4479
Tetravinylsilane	: 11449
sym-Tetrazine	: 11651
Tetrolaldehyde	: 3895
Tetrole	: 6025
Tetroquinone	: 4269
Thalidomide	: 7363
Thebaine	: 7679
Thebainone	: 7677
3-Thenoic acid	: 11738
Thenylpyramine	: 5507
Theobromine	: 10757
Theophylline	: 10756
THF glycol	: 7109
1-Thia-2-azaindene	: 2387
Thiabutazide	: 2856
Thiacetazone	: 34
Thiachromine-8-ethanol, 2,7-dimethyl-	: 11111
Thiacyclohexane	: 11795
Thiacyclopentane	: 11783
1-Thia-2,4-diazacyclopentadiene	: 11657
1,2,4-Thiadiazole, 5-ethoxy-3-(trichloromethyl)-	: 11599
1,3,4-Thiadiazole-2-sulfonic acid,amide,5-acetamido	: 32
1-Thiaindene	: 2399
Thiamine hydrochloride	: 11686
Thiamorpholine	: 11719
Thianaphthene	: 2399
Thiazolsulfone	: 11663
2-Thiazolylamine	: 11662
4'-(2-Thiazolylsulfamoyl)succinanilic acid	: 3578
2-Thienonitrile	: 11732
2-Thienyl bromide	: 11726
2-Thienyl chloride	: 11740
Thioacetaldehyde trimer	: 11920
Thioacetamide	: 5643
Thioacetic acid	: 5646
Thiobenzyl alcohol	: 1832
Thiobutyl alcohol	: 3437
Thiocarbamide	: 11799
Thiochrome	: 11111
Thiocyanatobenzene	: 11709
Thiocyanic acid, tert-butyl ester	: 11701
Thiocyanic acid, isopropyl ester	: 11706
1-Thiocyanobutane	: 11699
Thiocyclohexanone	: 4481
2,2'-Thiodiethanol	: 5766
Thiodiglycolic acid	: 240
Thiodiphenylamine	: 9499
3,3'-Thiodipropionic acid	: 10329
Thiofuran	: 11724
Thioglycerol	: 10013
Thioglycolic acid	: 202
Thioguanine	: 10762

Thiolactic acid	: 10285
Thiomedon	: 7849
Thionalide	: 68
Thionaphthene-2-carboxylic acid	: 2400
Thiophene, 2-isobutyryl	: 10472
Thiophene, 2-isopentyl-	: 11771
Thiophene, 2-isopropyl-	: 11772
Thiophene, 3-isopropyl-	: 11773
Thiophenol	: 2289
Thiophenol, 3-ethyl	: 2297
Thiophosgene	: 4032
o-Thiosalicylic acid	: 2663
Thioserine	: 5014
Thiosemicarbazide	: 7142
Thiosinamine	: 11809
Thiotax	: 2866
3-Thiotolene	: 11770
2-Thiouracil	: 11110
Thioxanthone	: 11812
Thiram	: 11722
THP	: 10800
Threose (D-)	: 3193
4(10)-Thujene, (±)-	: 2981
4(10)-Thujene (d)	: 11815
4(10)-Thujene-3-ol	: 2983
Thujic acid	: 4241
Thujopsene	: 4954
Thymine	: 11068
Thymine 2-desoxyriboside	: 11816
Thymol	: 9418
Thymol, acetate	: 9420
Thymol blue	: 9369
2-Thymolcarboxylic acid	: 2642
Thymol, 2,6-dibromo	: 9281
Thymolphthalein	: 7352
o-Thymotic acid	: 2642
Thymyl iodide	: 1782
Thyroxine	: 11930
Tiglic acid	: 3844
Tigogenin	: 11532
Timonacic	: 11682
Tiocarlide	: 11800
δ-Tocopherol	: 2785
α-Tocopherol	: 2787
β-Tocopherol	: 2789
γ-Tocopherol	: 2790
α-Tocopheryl acetate	: 2788
Tokoferol acetate	: 2788
Tolazine	: 9195
Tolbutamide	: 2207
3,3'-Tolidine	: 3034
o-Tolidine	: 3034
o-Tolualdehyde	: 587
p-Tolualdehyde	: 589
α-Tolualdehyde	: 868
m-Tolualdehyde, 2-hydroxy	: 575
o-Toluamide	: 619
Toluene	: 1947
Toluene, 3-bromo-2-nitro-	: 1068
Toluene, 3-bromo-5-nitro-	: 1069
Toluene, 3-bromo-4-nitro-	: 1073
Toluene, 4-bromo-2-nitro-	: 1074
Toluene, 2-bromo-6-nitro-	: 1070
o-Toluenecarbonitrile	: 2763
Toluene, 4-chloro-3-iodo-	: 1173
Toluene, 5-chloro-2-iodo	: 1174
Toluene, 2-chloro-α-methylimino	: 7652
Toluene, 3-chloro-5-nitro-	: 1207
Toluene-3,4-diamine	: 1276
Toluene-2,6-diamine	: 1277
Toluene-2,4-diamine	: 1278
Toluene, 2,3-dibromo	: 1302
Toluene, 3,4-dibromo	: 1303
Toluene, 2,6-dibromo	: 1304
Toluene, 2,5-dibromo-	: 1305
Toluene, 2,3-dichloro-	: 1383
Toluene, 2,5-dichloro-	: 1387
Toluene, 2,4-dichloro-α-hydroxy	: 1853
Toluene, 2,5-diethoxy	: 1410
Toluene, α,4-dihydroxy	: 1884
Toluene 2,4-diisocyanate	: 1431
Toluene-2,6-diisocyanate	: 1430
Toluene, 2,4-diisopropyl-	: 1948
Toluene-α,α-diol, m-nitro-, diacetate (ester)	: 7511
Toluene-3,4-dithiol	: 1564

Toluene, 3-fluoro-4-nitro-	: 1730
Toluene, 4-fluoro-3-nitro	: 1729
Toluene, 3-(1-hydroxyethyl)	: 1858
Toluene, α-hydroxy-4-methoxy-α-methyl	: 1895
Toluene, α-hydroxy-4-nitro	: 1908
m-Toluenenitrile	: 2764
p-Toluenenitrile	: 2765
Toluene, p-(1-octyldecyl)-	: 8245
Toluene, α,α,2,3,4-pentachloro-	: 2326
Toluene, α,α,2,4,5-pentachloro-	: 2327
Toluene, 4-propoxy	: 2032
o-Toluenesulfinic acid	: 2181
p-Toluenesulfinic acid	: 2182
p-Toluenesulfonamide, N,N-dichloro-	: 2210
p-Toluenesulfonic acid	: 2229
p-Toluenesulfonic acid chloride	: 2250
p-Toluenesulfonic acid, ethyl ester	: 2231
o-Toluenesulfonyl chloride	: 2249
p-Toluenesulfonyl chloride	: 2250
Toluene, 2,3,4,6-tetrachloro-	: 2262
Toluene, 2,3,5,6-tetrachloro-	: 2263
α-Toluenethiol, α-methyl-, (S)-(-)-	: 1834
Toluene, 2,4,6-tribromo-	: 2314
Toluene, 2,3,4-trichloro-	: 2334
Toluene, 3,4,5-trichloro-	: 2335
Toluene, 2,4,5-trichloro-	: 2336
Toluene, 2,3,5-trichloro	: 2337
Toluene, 2,4,6-triiodo-	: 2352
Toluene, 2,3,4-trinitro-	: 2046
Toluene, 2,4,5-trinitro-	: 2047
o-Toluic acid	: 2676
m-Toluic acid	: 2677
p-Toluic acid	: 2678
o-Toluidine	: 778
m-Toluidine	: 779
p-Toluidine	: 780
m-Toluidine, N-benzyl	: 1824
p-Toluidine, N-benzyl	: 1825
m-Toluidine, 6-bromo-	: 648
o-Toluidine hydrochloride	: 791
p-Toluidine, 2-iodo-	: 762
o-Toluidine, 5-iodo	: 763
o-Toluidine, 4-(phenylsulfonyl)-N-[(trifluoromethyl)sulfono]-	: 9149
p-Toluidine, N-propyl-	: 815
o-Tolunitrile	: 2763
m-Tolunitrile	: 2764
p-Tolunitrile	: 2765
m-Toluquinaldine	: 11258
m-Toluquinoline	: 11282
2-(p-Toluyl)benzoic acid	: 2684
o-Tolyl alcohol	: 1897
p-Tolyl alcohol	: 1899
2-o-Tolylaminoethanol	: 5744
2-p-Tolylaminoethanol	: 5745
p-Tolylcarbinol	: 1899
1-(4-Tolyl)ethanol	: 1859
m-Tolylmethanol	: 1898
p-Tolylsulfonylmethylnitrosamide	: 2212
o-Tolylthiourea	: 11806
Tomatidine	: 11528
Tomatine	: 6196
Topazone	: 8589
Torularhodin	: 4059
Tosic acid	: 2229
p-Tosyl chloride	: 2250
Tranexamic acid	: 4333
trans-1,3-Dimethylcyclohexanol	: 4511
Traumatic acid	: 5336
Trehalose	: 6220
Triacetamide	: 50
Triacetin	: 10195
N,N,N-Triacetylammonia	: 50
Triacontyl alcohol	: 11821
Triacontylcyclohexane	: 4482
Triacontylcyclopentane	: 4853
Triallylamine	: 10523
1,2,3-Triaminobenzene	: 2305
Triamterene	: 10750
1,2,3-Triaza-1H-indene	: 2871
1,2,4-Triazanaphthalene	: 2870
1,3,5-Triazine-2,4-diamine, N,N'-bis(1-methylethyl)-6-(methylthio)-	: 9770

1,3,5-Triazine-2,4-diamine, 6-chloro-N,N'-bis(1-methylethyl)-	: 10508
1,3,5-Triazine-2,4-diamine, 6-chloro-N-cyclopropyl-N'-(1-methylethyl)-	: 5013
1,3,5-Triazine-2,4-diamine, 6-chloro-N,N'-diethyl-	: 11513
1,3,5-Triazine-2,4-diamine, 6-chloro-N-(1,1-dimethylethyl)-N'-ethyl-	: 11593
1,3,5-Triazine-2,4-diamine, N-(1,1-dimethylethyl)-N'-ethyl-6-(methylthio)-	: 11594
1,3,5-Triazine-2,4-diamine, N-ethyl-6-methoxy-N'-(1-methylpropyl)-	: 3143
1,3,5-Triazine-2,4-diamine, 6-(ethylthio)-N,N'-bis(1-methylethyl)-	: 5222
1,3,5-Triazine-2,4(1H,3H)-dione, 3-cyclohexyl-6-(dimethylamino)-1-methyl-	: 6995
1,2,4-Triazin-5(4H)-one, 4-amino-6-(1,1-dimethylethyl)-3-(methylthio)-	: 7658
1,2,4-Triazolo[3,4-b]benzothiazole, 5-methyl-	: 11858
Tribenzobicyclo[2.2.2]octatriene	: 2375
1,3,5-Tribenzoylbenzene	: 7587
Tribromoacetamide	: 94
Tribromoacetic acid	: 241
Tribromoacetic acid, ethyl ester	: 242
2,4,6-Tribromoaniline	: 851
2,4,5-Tribromoanisole	: 2311
1,2,3-Tribromobenzene	: 2307
1,2,4-Tribromobenzene	: 2308
1,3,5-Tribromobenzene	: 2309
1,2,4-Tribromobutane	: 3445
1,2,2-Tribromobutane	: 3443
1,2,3-Tribromobutane	: 3444
1,3,3-Tribromobutane	: 3446
2,2,3-Tribromobutane	: 3447
Tribromo-tert-butyl alcohol	: 10405
Tribromochloromethane	: 7549
2,4,6-Tribromo-m-cresol	: 9480
1,1,2-Tribromocyclobutane	: 3442
1,1,2-Tribromoethane	: 5653
Tribromoethanol	: 5767
Tribromofluoromethane	: 7550
Tribromomethane	: 7548
3,4,5-Tribromonitrobenzene	: 2315
Tribromonitromethane	: 7551
2,4,6-Tribromophenol	: 9479
1,2,3-Tribromopropane	: 10163
2,3,5-Tribromothiophene	: 11790
2,4,6-Tribromotoluene	: 2314
Tribromsalan	: 609
Tributylamine	: 3213
Tributyl phosphate	: 9562
Tributylphosphine	: 9520
Tributylsilane	: 11454
Tributyltin acetate	: 11538
S,S,S-Tributyl trithiophosphate	: 11852
Tributyrin	: 3586
Tributyryl glyceride	: 3586
Tricarballylic acid	: 10165
Tricarbethoxymethane	: 7552
1,2,3-Tricarboxybenzene	: 2318
1,2,4-Tricarboxybenzene	: 2319
2,2,2-Trichloroacetaldehyde	: 21
Trichloroacetaldehyde hydrate	: 5582
Trichloroacetic acid	: 243
Trichloroacetic acid, ethyl ester	: 248
Trichloroacetic acid, (2-methoxyethyl) ester	: 250
Trichloroacetic acid, methyl ester	: 253
Trichloroacetic anhydride	: 244
Trichloroacetonitrile	: 288
Trichloroacetyl bromide	: 293
Trichloroacetyl chloride	: 300
2,3,4-Trichloroaniline	: 852
2,4,6-Trichloroanisole	: 2332
1,2,3-Trichlorobenzene	: 2321
1,2,4-Trichlorobenzene	: 2322
1,3,5-Trichlorobenzene	: 2323
2,3,6-Trichlorobenzoic acid	: 2727
2,4,5-Trichlorobenzoic acid	: 2728
2,2,2-Trichloro-1,1-bis(4-chlorophenyl)ethanol	: 1851
Trichlorobromomethane	: 7474
1,2,3-Trichloro-1,3-butadiene	: 3174
1,2,3-Trichlorobutane	: 3449

1,2,4-Trichlorobutane	: 3450
2,2,3-Trichlorobutane	: 3451
1,1,3-Trichlorobutane	: 3448
1,1,1-Trichloro-2-butanol	: 3643
2,3,4-Trichloro-1-butene	: 3809
1,2,4-Trichloro-2-butene	: 3810
Trichlorobutylsilane	: 11370
2,2,3-Trichlorobutyraldehyde	: 3191
α,α,β-Trichlorobutyraldehyde	: 3191
2,2,3-Trichlorobutyric acid	: 3590
Trichloro(chloromethyl)silane	: 11456
4,5,6-Trichloro-o-cresol	: 9487
2,3,6-Trichloro-p-cresol	: 9488
2,4,6-Trichloro-m-cresol	: 9489
Trichloro(dichloromethyl)silane	: 11459
1,1,1-Trichloro-2,2-difluoroethane	: 5658
1,2,2-Trichloro-1,2-difluoroethane	: 5660
1,1,1-Trichloro-2,2-dihydroxyethane	: 5582
Trichlorododecylsilane	: 11460
2,2,2-Trichloroethanal	: 21
1,1,1-Trichloroethane	: 5655
1,1,2-Trichloroethane	: 5656
2,2,2-Trichloro-1,1-ethanediol	: 5582
2,2,2-Trichloroethanol	: 5768
Trichloroethylene	: 5922
Trichlorofluoromethane	: 7555
Trichlorohexylsilane	: 11464
Trichloroiodomethane	: 7556
Trichloromethane	: 7554
Trichloromethiazide	: 2854
(Trichloromethyl)benzene	: 2333
1,1,2-Trichloro-2-methylpropane	: 10177
2-(Trichloromethyl)pyridine	: 11023
3,4,6-Trichloro-2-nitrophenol	: 9490
Trichloro perchlorate	: 9148
2,3,4-Trichlorophenol	: 9481
2,3,5-Trichlorophenol	: 9482
2,3,6-Trichlorophenol	: 9483
2,4,5-Trichlorophenol	: 9484
2,4,6-Trichlorophenol	: 9485
3,4,5-Trichlorophenol	: 9486
1,1,1-Trichloropropane	: 10171
1,1,2-Trichloropropane	: 10172
1,1,3-Trichloropropane	: 10173
1,2,2-Trichloropropane	: 10174
1,2,3-Trichloropropane	: 10175
1,1,1-Trichloro-2-propanol	: 10406
2,4,6-Trichloropyrimidine	: 11075
3',4',5-Trichlorosalicylanilide	: 608
2,3,5-Trichlorothiophene	: 11791
2,3,4-Trichlorotoluene	: 2334
2,4,5-Trichlorotoluene	: 2336
2,3,5-Trichlorotoluene	: 2337
2,2',2''-Trichlorotriethylamine	: 5412
1,3,5-Trichloro-2,4,6-trifluorobenzene	: 2340
1,1,1-Trichlorotrifluoroethane	: 5662
1,1,2-Trichlorotrifluoroethane	: 5663
Trichloro-1,3,5-trihydroxybenzene	: 2371
Trichlorovinylsilane	: 11461
Trichocid	: 85
Tricosylbenzene	: 2341
Tricosylcyclohexane	: 4483
Tricromyl	: 2838
Tricyclenic acid	: 11862
Tricyclenic acid, ethyl ester	: 11863
Tricyclenic acid, methyl ester	: 11864
Tridecamethylene dibromide	: 11874
Tridecane, 1-amino-	: 11870
Tridecane, 1-cyclohexyl-	: 4484
Tridecane, 1-cyclopentyl-	: 4854
Tridecane, 2-methyl-13-bromo-	: 11873
Tridecane, 7-phenyl-	: 1764
Tridecyl alcohol	: 11883
Tridecylcyclohexane	: 4484
Trientine	: 5499
Triethanolamine	: 5751
1,3,5-Triethoxybenzene	: 2343
Triethoxysilane	: 11477
Triethylamine	: 5417
Triethylantimony	: 11549
Triethyl arsenate	: 439
Triethyl arsenite	: 440
Triethylarsine	: 448
1,2,4-Triethylbenzene	: 2344
1,3,5-Triethylbenzene	: 2345
Triethylborane	: 3118

Triethylenediamine	: 5111
Triethylene glycol	: 5713
Triethylene glycol, (3-aminopropyl) ether	: 5678
Triethylene glycol, diacetate	: 5714
Triethylenemelamine	: 11844
Triethylenetetramine	: 5499
Triethyl methanetricarboxylate	: 7552
Triethyl phosphate	: 9563
Triethyl phosphine	: 9521
Triethyl phosphite	: 9591
Triethylphosphorus	: 9521
Triethylsilane	: 11483
Triethylstibine	: 11549
2,4,6-Triethyl-2,4,6-trimethylcyclotrisiloxane	: 5003
2,4,6-Triethyl-2,4,6-triphenylcyclotrisiloxane	: 5004
Trifluoroacetic acid	: 263
Trifluoroacetic acid, propyl ester	: 269
Trifluoroacetonitrile	: 289
1,3,5-Trifluorobenzene	: 2347
sym-Trifluorobenzene	: 2347
Trifluoroboron etherate	: 3137
Trifluorobromomethane	: 7475
Trifluorochloromethane	: 7486
1,1,2-Trifluoroethane	: 5666
2,2,2-Trifluoroethanol	: 5770
Trifluoroethylene	: 5924
Trifluoroethylsilane	: 11432
Trifluoroiodomethane	: 7559
Trifluoromethane	: 7558
o-(Trifluoromethyl)aniline	: 855
m-(Trifluoromethyl)aniline	: 856
(Trifluoromethyl)benzene	: 2348
3'-Trifluoromethyldiphenylamine-2-carboxylic acid	: 2729
Trifluoromethylhydrothiazide	: 2858
3-(Trifluoromethyl)-1-nitrobenzene	: 2061
N-(m-Trifluoromethylphenyl)-2-aminobenzoic acid	: 2729
N-(3-Trifluoromethylphenyl)-N',N'-dimethylurea	: 12007
(±)-2-[4-[[5-(Trifluoromethyl)-2-pyridinyl]oxy]phenoxy]propanoic acid, butyl ester	: 5962
(R)-2-[4-(5-Trifluoromethyl-2-pyridyloxy)phenoxy]propionic acid, butyl ester	: 5963
3,3,3-Trifluoropropene	: 10608
Triglycol diacetate	: 5714
Triglyme	: 11642
Trigonelline	: 11029
Trihexyphenidyl hydrochloride	: 9705
2',3',4'-Trihydroxyacetophenone	: 5873
1,3,5-Trihydroxybenzene	: 2368
3,4,5-Trihydroxybenzoic acid	: 2731
1,2,3-Trihydroxybutane	: 3458
Trihydroxyestrin	: 5404
4',5,7-Trihydroxyflavone	: 2806
1,2,6-Trihydroxyhexane	: 6868
4',5,7-Trihydroxyisoflavone	: 2807
3',5,7-Trihydroxy-4'-methoxyflavanone	: 2800
1,2,3-Trihydroxypropane	: 10186
2,4,6-Trihydroxypropiophenone	: 10489
1,2,3-Triiodobenzene	: 2349
1,2,4-Triiodobenzene	: 2350
1,3,5-Triiodobenzene	: 2351
Triiodomethane	: 7560
2,4,6-Triiodophenol	: 9492
3,3',5-Triiodothyropropionic acid	: 2136
2,4,6-Triiodotoluene	: 2352
Triisobutylamine	: 9820
Triisobutyl phosphate	: 9572
1,2,4-Triisopropylbenzene	: 2372
3,7,12-Triketocholanic acid	: 4101
Trimellitic acid	: 2319
Trimellitic anhydride	: 7341
Trimethadione	: 8587
3,4,5-Trimethoxybenzyl alcohol	: 1914
Trimethoxyborane	: 3126
Trimethoxyphosphine	: 9592
1,1,1-Trimethylacetone	: 3657
Trimethyl aluminum	: 339
Trimethylamine	: 7444
Trimethylamineborane	: 3132

Trimethylantimony	: 11550
Trimethylarsine	: 449
1,2,3-Trimethylbenzene	: 2357
Trimethylborane	: 3119
Trimethyl borate	: 3126
2,2,3-Trimethylbutane	: 3457
Trimethylbutylsilane	: 11371
Trimethylchlorosilane	: 11389
Trimethyl citrate	: 10170
1,1,2-Trimethylcycloheptane	: 4227
1,1,4-Trimethylcycloheptane	: 4228
3,3,7-Trimethylcycloheptanol	: 4233
2,6,6-Trimethylcycloheptanone	: 4238
1,4,4-Trimethyl-1-cycloheptene	: 4247
1,1,2-Trimethylcyclohexane	: 4485
1,1,3-Trimethylcyclohexane	: 4486
1,1,4-Trimethylcyclohexane	: 4487
1,2,3-Trimethylcyclohexane	: 4488
trans-1,3,5-Trimethylcyclohexane	: 4497
1,1,3-Trimethyl-2-cyclohexanol	: 4566
2,3,6-Trimethylcyclohexanol	: 4568
cis-3,3,5-Trimethylcyclohexanol	: 4569
trans-3,5,5-Trimethylcyclohexanol	: 4570
trans-3,3,5-Trimethylcyclohexanol	: 4570
1,2,2-Trimethylcyclohexanol	: 4562
1,2,6-Trimethylcyclohexanol	: 4563
1,3,5-Trimethylcyclohexanol	: 4564
2,2,5-Trimethylcyclohexanol	: 4565
2,2,3-Trimethylcyclohexanone	: 4614
2,2,4-Trimethylcyclohexanone	: 4615
2,3,6-Trimethylcyclohexanone	: 4618
2,4,4-Trimethylcyclohexanone	: 4619
2,4,5-Trimethylcyclohexanone	: 4620
3,3,5-Trimethylcyclohexanone	: 4622
3,4,4-Trimethylcyclohexanone	: 4623
2,2,5-Trimethylcyclohexanone	: 4616
2,4,6-Trimethylcyclohexanone	: 4621
1,4,4-Trimethylcyclohexene	: 4689
1,5,5-Trimethylcyclohexene	: 4691
1,2,3-Trimethylcyclohexene	: 4687
1,3,5-Trimethylcyclohexene	: 4688
1,4,5-Trimethylcyclohexene	: 4690
1,6,6-Trimethylcyclohexene	: 4692
2,4,4-Trimethyl-2-cyclohexen-1-ol	: 4701
3,6,6-Trimethyl-2-cyclohexenone	: 4722
4-(2,6,6-Trimethyl-1-cyclohexen-1-yl)-3-buten-2-ol	: 3873
1,1,2-Trimethylcyclopentane	: 4855
1,1,3-Trimethylcyclopentane	: 4856
1,2,2-Trimethylcyclopentanol	: 4875
1,2,4-Trimethylcyclopentanol	: 4876
1,2,5-Trimethylcyclopentanol	: 4877
2,3,5-Trimethylcyclopentanol	: 4898
2,2,4-Trimethylcyclopentanone	: 4896
2,2,5-Trimethylcyclopentanone	: 4897
2,4,4-Trimethylcyclopentanone	: 4899
1,5,5-Trimethylcyclopentene	: 4938
1,1,2-Trimethylcyclopropane	: 4985
trans-3,7,11-Trimethyl-1,3,6,10-dodecatetraene	: 5329
trans-2,6,10-Trimethyl-2,6,9,11-dodecatetraene	: 5329
Trimethylene	: 4955
Trimethylene chlorohydrin	: 10348
Trimethylene dichloride	: 9893
Trimethylene diiodide	: 9907
Trimethylene dimercaptan	: 10031
Trimethylene glycol	: 9983
Trimethylene oxide	: 8592
Trimethylene sulfide	: 11691
Trimethylene sulfone	: 11692
2,2,3-Trimethylheptane	: 6437
2,2,4-Trimethylheptane	: 6438
2,2,5-Trimethylheptane	: 6439
2,2,6-Trimethylheptane	: 6440
2,3,3-Trimethylheptane	: 6441
2,3,4-Trimethylheptane	: 6442
2,3,5-Trimethylheptane	: 6443
2,3,6-Trimethylheptane	: 6444
2,4,4-Trimethylheptane	: 6445
2,4,5-Trimethylheptane	: 6446
2,4,6-Trimethylheptane	: 6447
2,5,5-Trimethylheptane	: 6448
3,3,4-Trimethylheptane	: 6449
3,3,5-Trimethylheptane	: 6450
3,4,4-Trimethylheptane	: 6451
3,4,5-Trimethylheptane	: 6452

2,2,3-Trimethylhexane	: 6859
2,2,4-Trimethylhexane	: 6860
2,2,5-Trimethylhexane	: 6861
2,3,3-Trimethylhexane	: 6862
2,3,4-Trimethylhexane	: 6863
2,3,5-Trimethylhexane	: 6864
2,4,4-Trimethylhexane	: 6865
3,3,4-Trimethylhexane	: 6866
3,5,5-Trimethylhexan-3-ol	: 6958
1,1,5-Trimethylindan	: 7261
1,4,7-Trimethylindan	: 7262
1,2,3-Trimethylindole	: 7322
2,3,5-Trimethylindole	: 7323
Trimethyl methanetricarboxylate	: 7553
1,2,4-Trimethylnaphthalene	: 8009
1,2,5-Trimethylnaphthalene	: 8010
1,2,6-Trimethylnaphthalene	: 8011
2,4,6-Trimethylnaphthalene	: 8013
2,4,7-Trimethyloctane	: 8421
Trimethylolmelamine	: 7583
Trimethylolpropane	: 10006
Trimethylolpropane phosphite	: 11909
2,2,3-Trimethylpentane	: 8847
2,3,3-Trimethylpentane	: 8849
2,3,4-Trimethylpentane	: 8850
2,2,4-Trimethyl-1,3-pentanediol	: 8795
2,4,4-Trimethyl-2-pentanol	: 8954
2,2,4-Trimethyl-3-pentanol	: 8955
2,4,4-Trimethyl-1-pentene	: 9064
2,3,3-Trimethyl-1-pentene	: 9065
2,3,4-Trimethyl-1-pentene	: 9066
2,2,4-Trimethyl-4-pentene	: 9067
3,4,4-Trimethyl-1-pentene	: 9069
2,3,4-Trimethyl-2-pentene	: 9070
2,4,4-Trimethyl-2-pentene	: 9071
3,4,4-Trimethyl-2-pentene	: 9072
2,4,5-Trimethylphenyl acetate	: 9496
Trimethylphosphine	: 9522
Trimethyl phosphite	: 9592
1,2,4-Trimethylpiperazine	: 9622
2,3,6-Trimethylpiperidine	: 9713
2,4,6-Trimethylpiperidine	: 9714
1,1,2-Trimethylpropyl bromide	: 3236
Trimethylpropylsilane	: 11496
1,3,4-Trimethylpyrazole	: 10839
1,3,5-Trimethylpyrazole	: 10840
1,4,5-Trimethylpyrazole	: 10841
2,3,4-Trimethylpyridine	: 11024
2,3,5-Trimethylpyridine	: 11025
2,3,6-Trimethylpyridine	: 11026
2,4,5-Trimethylpyridine	: 11027
1,2,5-Trimethylpyrrole	: 11142
2,2,4-Trimethylpyrrolidine	: 11175
2,4,6-Trimethylquinoline	: 11302
2,4,7-Trimethylquinoline	: 11303
Trimethylsilane	: 11489
(Trimethylsilyl)methyl bromide	: 11366
Trimethylstibine	: 11550
2,3,4-Trimethylthiophene	: 11792
2,3,5-Trimethylthiophene	: 11793
5,7,8-Trimethyltocol	: 2787
1,1,1-Trimethyl-3,3,3-triphenyldisiloxane	: 5239
Trimethylurea	: 12032
Trimyristin	: 11628
Trinitroacetonitrile	: 290
2,3,5-Trinitroanisole	: 364
sym-Trinitrobenzene	: 2365
2,4,6-Trinitrobenzoic acid	: 2735
Trinitro-m-cresol	: 9434
2,4,6-Trinitro-3,5-dimethyl-tert-butylbenzene	: 1461
2,4,7-Trinitrofluorenone	: 5979
Trinitromethane	: 7562
2,3,4-Trinitrotoluene	: 2046
2,4,5-Trinitrotoluene	: 2047
2,4,6-Trinitrotoluene	: 2048
28,29,30-Trinorlanostane	: 4105
Triolein	: 8310
Trional	: 3230
Triopron	: 2136
s-Trioxane, 2,4,6-trivinyl-	: 11906
3,6,9-Trioxaundecane	: 5612
3,7,12-Trioxocholanic acid	: 4101
Tripalmitin	: 6656
Tripelennamine	: 5505
Triphenylamine	: 728

Triphenylarsane	: 450
Triphenylarsine	: 450
Triphenylbismuth	: 3111
Triphenylcarbinol	: 1866
Triphenylhydrazine	: 7173
Triphenylmethane	: 1982
Triphenylmethane, 3-methyl	: 7513
Triphenylmethanol	: 1866
Triphenylmethylsilane	: 11444
Triphenyl phosphate	: 9566
Triphenylphosphine	: 9523
Triphenyltetrazolium chloride	: 11653
Triphenyltin acetate	: 11539
Triphenyltin chloride	: 11541
Tripropylamine	: 9808
Tripropyl phosphate	: 9567
Tripropyl phosphite	: 9594
Tripropylsilane	: 11500
Triptycene	: 2375
2,4,6-Tripyridyl-s-triazine	: 11843
Tris(allyloxy)vinylsilane	: 11427
Tris-BP	: 10358
Tris-(2-chloroethyl) amine	: 5412
Tris(chloromethyl)phosphine	: 9524
Tris(2,3-dibromopropyl) phosphate	: 10358
Tris(dimethylamino)phosphine oxide	: 9573
Tris(hydroxymethyl)nitromethane	: 10012
Tristearin	: 8272
Trithioacetaldehyde	: 11920
Trithiocarbonic acid	: 4034
Tri-o-tolyl phosphate	: 9569
Trityl alcohol	: 1866
Tromasedan	: 2385
Tromethamine	: 9986
Tropacocaine	: 488
Tropane	: 484
Tropic acid	: 915
Tropilidene	: 4240
Tropine	: 489
Tryptamine	: 7302
Tryptophol	: 7305
Tsuduranine	: 5159
Tsuzuic acid	: 11637
Tuaminoheptane	: 6350
Tubatoxin	: 2783
Tuduic acid	: 11637
D-Turanose	: 6022
Tybamate	: 3933
Tylenol	: 67
Tyramine	: 9208
Tyrosine, 3,5-dibromo(L)	: 11927
Ugurol	: 4333
Umbelliferone	: 2826
Umbelliferone glucoside	: 2821
Undecafluorocyclohexane	: 4498
trans-9-Undecenoic acid	: 11975
(10-Undecenyl)acetic acid	: 11895
Undecylenic acid	: 11974
Undecyl ketone	: 11857
Undecyl mercaptan	: 11942
Uneicosane, 11-phenyl	: 6297
Unidigin	: 4043
Untriacontane	: 6299
Uracil	: 11059
Uracil mustard	: 11061
Uramil	: 11077
Urazole	: 11850
Urea, N'-(4-bromo-3-chlorophenyl)-N-methoxy-N-methyl-	: 4076
Urea, 1-(2-bromo-2-ethylbutanoyl)	: 3195
Urea, 1-(2-bromo-3-methylbutanoyl)	: 3196
Urea, 3-(p-bromophenyl)-1-methoxy-1-methyl-	: 7656
Urea, N'-[4-(4-chlorophenoxy)phenyl]-N,N-dimethyl-	: 4088
Urea, N,N'-diamino-	: 4009
Urea, N'-(3,4-dichlorophenyl)-N-methoxy-N-methyl-	: 7415
Urea, 1,1-diisobutyl	: 11996
Urea, N-[5-(1,1-dimethylethyl)-1,3,4-thiadiazol-2-yl]-N,N'-dimethyl-	: 11590
Urea, N,N-dimethyl-N'-(octahydro-4,7-methano-1H-indeN-5-yl)-, (3aα,4α,5α,7α,7aα)-	: 8212
Urea hydrochloride	: 12020
Urea nitrate	: 12021

$CBrClF_2$
7464 - Methane, bromochlorodifluoro-
$CBrClN_2O_4$
7465 - Methane, bromochlorodinitro-
$CBrCl_2F$
7468 - Methane, bromodichlorofluoro-
$CBrCl_3$
7474 - Methane, bromotrichloro-
$CBrF_3$
7475 - Methane, bromotrifluoro-
$CBrN$
4165 - Cyanogen bromide
$CBrN_3O_6$
7476 - Methane, bromotrinitro-
CBr_2ClF
7494 - Methane, dibromochlorofluoro-
CBr_2Cl_2
7495 - Methane, dibromodichloro-
CBr_2F_2
7496 - Methane, dibromodifluoro-
$CBr_2N_2O_4$
7497 - Methane, dibromodinitro-
CBr_2O
4006 - Carbonic dibromide
CBr_3Cl
7549 - Methane, tribromochloro-
CBr_3F
7550 - Methane, tribromofluoro-
CBr_3NO_2
7551 - Methane, tribromonitro-
CBr_4
7539 - Methane, tetrabromo-
$CClFO$
4005 - Carbonic chloride fluoride
$CClF_3$
7486 - Methane, chlorotrifluoro-
$CClF_3O_2S$
7537 - Methanesulfonyl chloride, trifluoro-
$CClN$
4166 - Cyanogen chloride
$CClNO_3S$
11585 - Sulfuryl chloride isocyanate
$CClN_3O_6$
7487 - Methane, chlorotrinitro-
CCl_2F_2
7500 - Methane, dichlorodifluoro-
$CCl_2N_2O_4$
7501 - Methane, dichlorodinitro-
CCl_2O
4007 - Carbonic dichloride
CCl_2S
4032 - Carbonothioic dichloride
CCl_3F
7555 - Methane, trichlorofluoro-
CCl_3I
7556 - Methane, trichloroiodo-
CCl_3NO_2
7557 - Methane, trichloronitro-
CCl_4
7540 - Methane, tetrachloro-
CCl_4O_2S
7536 - Methanesulfonyl chloride, trichloro-
CCl_4O_4
9148 - Perchloric acid, trichloromethyl ester
CCl_4S
7527 - Methanesulfenyl chloride, trichloro-
CFN
4167 - Cyanogen fluoride
CF_2O
4008 - Carbonic difluoride
CF_3I
7559 - Methane, trifluoroiodo-
CF_4
7541 - Methane, tetrafluoro-
CF_4S
7528 - Methanesulfenyl fluoride, trifluoro-

CF_8S
11584 - Sulfur, pentafluoro(trifluoromethyl)-, (OC-6-21)-
$CHBrClF$
7466 - Methane, bromochlorofluoro-
$CHBrCl_2$
7467 - Methane, bromodichloro-
$CHBrF_2$
7469 - Methane, bromodifluoro-
$CHBr_2Cl$
7493 - Methane, dibromochloro-
$CHBr_2F$
7498 - Methane, dibromofluoro-
$CHBr_3$
7548 - Methane, tribromo-
$CHClF_2$
7478 - Methane, chlorodifluoro-
$CHClI_2$
7479 - Methane, chlorodiiodo-
$CHClN_2O_4$
7480 - Methane, chlorodinitro-
$CHCl_2F$
7502 - Methane, dichlorofluoro-
$CHCl_2I$
7503 - Methane, dichloroiodo-
$CHCl_3$
7554 - Methane, trichloro-
$CHCl_5Si$
11459 - Silane, trichloro(dichloromethyl)-
$CHFI_2$
7516 - Methane, fluorodiiodo-
$CHFO$
6019 - Formyl fluoride
CHF_2I
7505 - Methane, difluoroiodo-
CHF_3
7558 - Methane, trifluoro-
CHI_3
7560 - Methane, triiodo-
CHN
7176 - Hydrocyanic acid
$CHNO$
4162 - Cyanic acid
6024 - Fulminic acid
7360 - Isofulminic acid
$CHNS$
7399 - Isothiocyanic acid
CHN_3O_6
7562 - Methane, trinitro-
$CHNaO_3$
4004 - Carbonic acid monosodium salt
CH_2BrCl
7463 - Methane, bromochloro-
CH_2BrF
7470 - Methane, bromofluoro-
CH_2BrI
7471 - Methane, bromoiodo-
CH_2BrNO_2
7473 - Methane, bromonitro-
CH_2Br_2
7492 - Methane, dibromo-
CH_2ClF
7481 - Methane, chlorofluoro-
CH_2ClI
7482 - Methane, chloroiodo-
CH_2ClNO
3971 - Carbamic chloride
CH_2ClNO_2
7485 - Methane, chloronitro-
CH_2Cl_2
7499 - Methane, dichloro-
CH_2Cl_4Si
11456 - Silane, trichloro(chloromethyl)-
$CH_2Cl_6Si_2$
11440 - Silane, methylenebis[trichloro-
$CH_2Cu_2O_5$
511 - Basic copper carbonate

CH_2FI
7517 - Methane, fluoroiodo-
CH_2F_2
7504 - Methane, difluoro-
CH_2I_2
7506 - Methane, diiodo-
CH_2N_2
4155 - Cyanamide
7491 - Methane, diazo-
$CH_2N_2O_4$
7508 - Methane, dinitro-
CH_2N_4
11652 - 1H-Tetrazole
CH_2O
5987 - Formaldehyde
$(CH_2O)_x$
8634 - Paraformaldehyde
CH_2O_2
6002 - Formic acid
CH_2S_3
4034 - Carbonotrithioic acid
CH_3AsF_2
463 - Arsonous difluoride, methyl-
CH_3Br
7462 - Methane, bromo-
CH_3Br_3Si
11453 - Silane, tribromomethyl-
CH_3Cl
7477 - Methane, chloro-
CH_3ClO_2S
7535 - Methanesulfonyl chloride
CH_3ClO_3S
4086 - Chlorosulfuric acid, methyl ester
CH_3Cl_3Si
11465 - Silane, trichloromethyl-
CH_3F
7515 - Methane, fluoro-
CH_3FO_2S
7538 - Methanesulfonyl fluoride
CH_3I
7518 - Methane, iodo-
CH_3NO
5988 - Formaldehyde, oxime
5989 - Formamide
CH_3NO_2
7522 - Methane, nitro-
8091 - Nitrous acid, methyl ester
CH_3NO_3
8075 - Nitric acid, methyl ester
CH_3N_3
7460 - Methane, azido-
$CH_3N_3O_3$
12024 - Urea, nitro-
CH_4
7459 - Methane
CH_4Cl_2Si
11398 - Silane, dichloromethyl-
CH_4N_2
7563 - Methanimidamide
CH_4N_2O
7566 - Methanimidamide, N-hydroxy-
11992 - Urea
$CH_4N_2O_2$
7451 - Methanamine, N-nitro-
12016 - Urea, hydroxy-
CH_4N_2S
11799 - Thiourea
CH_4N_2Se
11354 - Selenourea
$CH_4N_4O_2$
6287 - Guanidine, nitro-
CH_4O
7581 - Methanol
CH_4O_2
7180 - Hydroperoxide, methyl
CH_4O_3S
7530 - Methanesulfonic acid

CH_4O_4S
11577 - Sulfuric acid, monomethyl ester
$CH_4O_6S_2$
7514 - Methanedisulfonic acid
CH_4S
7546 - Methanethiol
CH_5As
446 - Arsine, methyl-
CH_5AsO_3
458 - Arsonic acid, methyl-
CH_5ClN_2O
12020 - Urea, monohydrochloride
CH_5ClSi
11378 - Silane, chloromethyl
CH_5N
7442 - Methanamine
CH_5NO
7447 - Methanamine, N-hydroxy-
CH_5NO_2
6003 - Formic acid, ammonium salt
CH_5N_3
6278 - Guanidine
$CH_5N_3O_4$
12021 - Urea, mononitrate
CH_5N_3S
7142 - Hydrazinecarbothioamide
CH_5O_3P
9538 - Phosphonic acid, methyl-
CH_6ClN
7446 - Methanamine, hydrochloride
CH_6ClN_3
6285 - Guanidine, monohydrochloride
CH_6ClN_3O
7143 - Hydrazinecarboxamide, monohydrochloride
CH_6N_2
7161 - Hydrazine, methyl-
CH_6N_4
7147 - Hydrazinecarboximidamide
CH_6N_4O
4009 - Carbonic dihydrazide
$CH_6N_4O_3$
6286 - Guanidine, mononitrate
CH_6N_4S
4033 - Carbonothioic dihydrazide
CH_6OSi
11436 - Silane, methoxy
CH_6Si
11437 - Silane, methyl-
CH_8B_2S
5168 - Diborane(6), mu-(methylthio)-
CH_8Si_2
11439 - Silane, methylenebis-
CIN
7337 - Iodine cyanide
CI_4
7542 - Methane, tetraiodo-
$CKNS$
11711 - Thiocyanic acid, potassium salt
$CNNaO$
4164 - Cyanic acid, sodium salt
CN_4O_8
7544 - Methane, tetranitro-
CO
4011 - Carbon monoxide
COS
4035 - Carbon oxide sulfide
CO_2
3992 - Carbon dioxide
CS_2
3993 - Carbon disulfide
CSe_2
4036 - Carbon selenide
C_2BrCl_3O
293 - Acetyl bromide, trichloro-
C_2Br_2
5939 - Ethyne, dibromo-

$C_2Br_2ClF_3$
5519 - Ethane, 1,2-dibromo-1-chloro-1,2,2-trifluoro-
$C_2Br_2Cl_4$
5524 - Ethane, 1,2-dibromo-1,1,2,2-tetrachloro-
$C_2Br_2F_4$
5525 - Ethane, 1,2-dibromo-1,1,2,2-tetrafluoro-
C_2Br_4
5914 - Ethene, tetrabromo-
C_2Br_6
5594 - Ethane, hexabromo-
C_2CaN_2
3930 - Calcium cyanide
C_2ClF_3
5889 - Ethene, chlorotrifluoro-
C_2ClF_5
5484 - Ethane, chloropentafluoro-
C_2Cl_2
5940 - Ethyne, dichloro-
C_2Cl_2FN
279 - Acetonitrile, dichlorofluoro-
$C_2Cl_2F_2$
5897 - Ethene, 1,1-dichloro-2,2-difluoro-
5898 - Ethene, 1,2-dichloro-1,2-difluoro-
$C_2Cl_2F_4$
5537 - Ethane, 1,1-dichloro-1,2,2,2-tetrafluoro-
5538 - Ethane, 1,2-dichloro-1,1,2,2-tetrafluoro-
$C_2Cl_2O_2$
5585 - Ethanedioyl dichloride
C_2Cl_3F
5923 - Ethene, trichlorofluoro-
$C_2Cl_3F_3$
5662 - Ethane, 1,1,1-trichloro-2,2,2-trifluoro-
5663 - Ethane, 1,1,2-trichloro-1,2,2-trifluoro-
C_2Cl_3N
288 - Acetonitrile, trichloro-
C_2Cl_4
5917 - Ethene, tetrachloro-
$C_2Cl_4F_2$
5636 - Ethane, 1,1,1,2-tetrachloro-2,2-difluoro-
5637 - Ethane, 1,1,2,2-tetrachloro-1,2-difluoro-
C_2Cl_4O
300 - Acetyl chloride, trichloro-
$C_2Cl_4O_2$
4029 - Carbonochloridic acid, trichloromethyl ester
C_2Cl_5F
5618 - Ethane, pentachlorofluoro-
C_2Cl_6
5595 - Ethane, hexachloro-
C_2F_3I
5925 - Ethene, trifluoroiodo-
C_2F_3N
289 - Acetonitrile, trifluoro-
C_2F_4
5918 - Ethene, tetrafluoro-
$C_2F_4N_2O_4$
5642 - Ethane, 1,1,2,2-tetrafluoro-1,2-dinitro-
C_2F_6
5596 - Ethane, hexafluoro-
$C_2F_6S_2$
5249 - Disulfide, bis(trifluoromethyl)
C_2HBr
5937 - Ethyne, bromo-
$C_2HBrClF_3$
5457 - Ethane, 1-bromo-2-chloro-1,1,2-trifluoro-
5458 - Ethane, 2-bromo-2-chloro-1,1,1-trifluoro-
C_2HBrCl_2
5882 - Ethene, 2-bromo-1,1-dichloro-
5935 - Ethylene, 1-bromo-1,2-dichloro-, trans-
C_2HBrF_2
5883 - Ethene, 1-bromo-1,2-difluoro-
$C_2HBr_2F_3$
5526 - Ethane, 1,2-dibromo-1,1,2-trifluoro-
C_2HBr_2N
277 - Acetonitrile, dibromo-
C_2HBr_3
5921 - Ethene, tribromo-

C_2HBr_3O
20 - Acetaldehyde, tribromo-
$C_2HBr_3O_2$
241 - Acetic acid, tribromo-
C_2HBr_5
5616 - Ethane, pentabromo-
C_2HCl
5938 - Ethyne, chloro-
C_2HClF_2
5885 - Ethene, 2-chloro-1,1-difluoro-
$C_2HClF_2O_2$
127 - Acetic acid, chlorodifluoro-
C_2HClF_3I
5490 - Ethane, 1-chloro-1,1,2-trifluoro-2-iodo-
C_2HClF_4
5485 - Ethane, 1-chloro-1,1,2,2-tetrafluoro-
5486 - Ethane, 1-chloro-1,2,2,2-tetrafluoro-
C_2HCl_2F
5900 - Ethene, 1,1-dichloro-2-fluoro-
C_2HCl_2FO
296 - Acetyl chloride, chlorofluoro-
$C_2HCl_2F_3$
5539 - Ethane, 1,2-dichloro-1,1,2-trifluoro-
5540 - Ethane, 2,2-dichloro-1,1,1-trifluoro-
C_2HCl_2N
278 - Acetonitrile, dichloro-
C_2HCl_3
5922 - Ethene, trichloro-
$C_2HCl_3F_2$
5658 - Ethane, 1,1,1-trichloro-2,2-difluoro
5659 - Ethane, 1,2,2-trichloro-1,1-difluoro-
5660 - Ethane, 1,2,2-trichloro-1,2-difluoro-
C_2HCl_3O
21 - Acetaldehyde, trichloro-
297 - Acetyl chloride, dichloro-
$C_2HCl_3O_2$
243 - Acetic acid, trichloro-
4016 - Carbonochloridic acid, dichloromethyl ester
C_2HCl_4F
5638 - Ethane, 1,1,2,2-tetrachloro-1-fluoro-
5639 - Ethane, 1,1,1,2-tetrachloro-2-fluoro-
C_2HCl_5
5617 - Ethane, pentachloro-
C_2HF
5943 - Ethyne, fluoro-
C_2HF_3
5924 - Ethene, trifluoro-
$C_2HF_3O_2$
263 - Acetic acid, trifluoro-
C_2H_2
5936 - Ethyne
C_2H_2BrCl
5881 - Ethene, 1-bromo-2-chloro-
$C_2H_2BrClO_2$
112 - Acetic acid, bromochloro-
$C_2H_2BrFO_2$
116 - Acetic acid, bromofluoro-
$C_2H_2BrF_3$
5466 - Ethane, 2-bromo-1,1,1-trifluoro-
$C_2H_2Br_2$
5890 - Ethene, 1,1-dibromo-
5891 - Ethene, 1,2-dibromo-, (E)-
5892 - Ethene, 1,2-dibromo-, (Z)-
$C_2H_2Br_2Cl_2$
5520 - Ethane, 1,2-dibromo-1,1-dichloro-
5521 - Ethane, 1,2-dibromo-1,2-dichloro-
$C_2H_2Br_2F_2$
5522 - Ethane, 1,2-dibromo-1,1-difluoro-
$C_2H_2Br_2O$
292 - Acetyl bromide, bromo-
$C_2H_2Br_2O_2$
159 - Acetic acid, dibromo-
$C_2H_2Br_3NO$
94 - Acetamide, 2,2,2-tribromo-
$C_2H_2Br_4$
5631 - Ethane, 1,1,1,2-tetrabromo-

$C_2H_4O_5S$
238 - Acetic acid, sulfo-
C_2H_4S
11695 - Thiirane
$C_2H_4S_2$
5589 - Ethane(dithioic) acid
$C_2H_4S_5$
8998 - 1,2,3,5,6-Pentathiepane
C_2H_4Si
11435 - Silane, ethynyl
$C_2H_5AsF_2$
445 - Arsine, ethyldifluoro-
C_2H_5Br
5454 - Ethane, bromo-
C_2H_5BrO
5685 - Ethanol, 2-bromo-
7472 - Methane, bromomethoxy-
C_2H_5Cl
5467 - Ethane, chloro-
C_2H_5ClO
5693 - Ethanol, 2-chloro-
7184 - Hypochlorous acid, ethyl ester
7483 - Methane, chloromethoxy-
$C_2H_5ClO_2S$
4087 - Chlorosulfurous acid, ethyl ester
5627 - Ethanesulfonyl chloride
$C_2H_5ClO_3S$
4085 - Chlorosulfuric acid, ethyl ester
C_2H_5ClS
5651 - Ethanethiol, 2-chloro-
7484 - Methane, chloro(methylthio)-
$C_2H_5Cl_2O_2P$
9577 - Phosphorodichloridic acid, ethyl ester
$C_2H_5Cl_3OSi$
11462 - Silane, trichloroethoxy-
$C_2H_5Cl_3Si$
11390 - Silane, dichloro(chloromethyl)methyl-
11463 - Silane, trichloroethyl-
C_2H_5F
5593 - Ethane, fluoro-
C_2H_5FO
5727 - Ethanol, 2-fluoro-
$C_2H_5F_3OSi$
11429 - Silane, ethoxytrifluoro-
$C_2H_5F_3Si$
11432 - Silane, ethyltrifluoro-
C_2H_5I
5597 - Ethane, iodo-
C_2H_5IO
5731 - Ethanol, 2-iodo-
7519 - Methane, iodomethoxy-
C_2H_5N
498 - Aziridine
C_2H_5NO
17 - Acetaldehyde, oxime
22 - Acetamide
5995 - Formamide, N-methyl-
$C_2H_5NO_2$
3953 - Carbamic acid, methyl ester
5607 - Ethane, nitro-
6250 - Glycine
8087 - Nitrous acid, ethyl ester
$C_2H_5NO_3$
5752 - Ethanol, 2-nitro-
8074 - Nitric acid, ethyl ester
C_2H_5NS
5643 - Ethanethioamide
$C_2H_5N_3O$
5681 - Ethanol, 2-azido-
$C_2H_5N_3O_2$
105 - Acetic acid, aminooxo-, hydrazide
7232 - Imidodicarbonic diamide
12019 - Urea, N-methyl-N-nitroso-
$C_2H_5N_3S_2$
11718 - Thioimidodicarbonic diamide
$C_2H_5N_5O_3$
6284 - Guanidine, N-methyl-N´-nitro-N-nitroso-

C_2H_6
5448 - Ethane
C_2H_6AsCl
453 - Arsinous chloride, dimethyl-
$C_2H_6BF_3O$
3138 - Boron, trifluoro[oxybis[methane]]-, (T-4)-
C_2H_6Cd
3929 - Cadmium, dimethyl-
$C_2H_6ClO_3P$
5929 - Ethephon
$C_2H_6Cl_2NPS$
9557 - Phosphoramidothioic dichloride, dimethyl-
$C_2H_6Cl_2Si$
11394 - Silane, dichlorodimethyl-
C_2H_6Hg
7435 - Mercury, dimethyl-
$C_2H_6N_2$
5125 - Diazene, dimethyl-
5126 - Diazene, dimethyl-, (E)-
$C_2H_6N_2O$
195 - Acetic acid, hydrazide
7450 - Methanamine, N-methyl-N-nitroso-
12017 - Urea, methyl-
$C_2H_6N_2O_2$
7149 - Hydrazinecarboxylic acid, methyl ester
7449 - Methanamine, N-methyl-N-nitro-
$C_2H_6N_4O$
11993 - Urea, (aminoiminomethyl)-
$C_2H_6N_4O_2$
5562 - Ethanedioic acid, dihydrazide
C_2H_6O
5672 - Ethanol
7523 - Methane, oxybis-
C_2H_6OS
5732 - Ethanol, 2-mercapto-
7529 - Methane, sulfinylbis-
$C_2H_6O_2$
5569 - 1,2-Ethanediol
7179 - Hydroperoxide, ethyl
9159 - Peroxide, dimethyl
$C_2H_6O_2S$
7534 - Methane, sulfonylbis-
$C_2H_6O_3S$
5622 - Ethanesulfonic acid
7532 - Methanesulfonic acid, methyl ester
11582 - Sulfurous acid, dimethyl ester
$C_2H_6O_4S$
11574 - Sulfuric acid, dimethyl ester
11576 - Sulfuric acid, monoethyl ester
$C_2H_6O_6$
5561 - Ethanedioic acid, dihydrate
$C_2H_6O_6S_2$
5587 - 1,2-Ethanedisulfonic acid
C_2H_6S
5648 - Ethanethiol
7545 - Methane, thiobis-
$C_2H_6S_2$
5254 - Disulfide, dimethyl
5590 - 1,2-Ethanedithiol
C_2H_6Se
7526 - Methane, selenobis-
C_2H_6Zn
12066 - Zinc, dimethyl
C_2H_7As
442 - Arsine, dimethyl-
444 - Arsine, ethyl-
$C_2H_7AsO_2$
451 - Arsinic acid, dimethyl-
$C_2H_7AsO_3$
457 - Arsonic acid, ethyl-
$C_2H_7BO_2$
3117 - Borane, dimethoxy-
3133 - Boronic acid, ethyl-
$C_2H_7ClN_2$
5671 - Ethanimidamide, monohydrochloride
C_2H_7N
5410 - Ethanamine

7448 - Methanamine, N-methyl-
C_2H_7NO
5434 - Ethanamine, N-hydroxy-
5674 - Ethanol, 1-amino-
5675 - Ethanol, 2-amino-
$C_2H_7NO_2$
106 - Acetic acid, ammonium salt
$C_2H_7NO_3S$
5623 - Ethanesulfonic acid, 2-amino-
C_2H_7NS
5649 - Ethanethiol, 2-amino-
$C_2H_7N_3O$
6266 - Glycine, hydrazide
$C_2H_7N_5$
7233 - Imidodicarbonimidic diamide
$C_2H_7O_2P$
9528 - Phosphinic acid, dimethyl-
$C_2H_7O_3P$
9588 - Phosphorous acid, dimethyl ester
$C_2H_7O_4P$
9559 - Phosphoric acid, dimethyl ester
9561 - Phosphoric acid, monoethyl ester
C_2H_8ClN
5433 - Ethanamine, hydrochloride
$C_2H_8NO_2PS$
7441 - Methamidophos
$C_2H_8N_2$
5496 - 1,2-Ethanediamine
7154 - Hydrazine, 1,1-dimethyl-
7155 - Hydrazine, 1,2-dimethyl-
$C_2H_8N_2O$
5729 - Ethanol, 2-hydrazino-
C_2H_8Si
11414 - Silane, dimethyl-
$C_2H_{10}N_2O$
5513 - 1,2-Ethanediamine, monohydrate
C_2I_2
5941 - Ethyne, diiodo-
C_2I_4
5919 - Ethene, tetraiodo-
C_2N_2
5550 - Ethanedinitrile
C_2N_2S
11572 - Sulfur cyanide
$C_2N_4O_6$
290 - Acetonitrile, trinitro-
$C_2Na_2O_4$
5566 - Ethanedioic acid, disodium salt
C_2O_4Pb
5568 - Ethanedioic acid, lead salt
C_3ClF_5
10541 - 1-Propene, 2-chloro-1,1,3,3,3-pentafluoro-
C_3ClF_5O
10421 - 2-Propanone, 1-chloro-1,1,3,3,3-pentafluoro-
$C_3Cl_2F_4$
10557 - 1-Propene, 1,2-dichloro-1,3,3,3-tetrafluoro-
$C_3Cl_2F_6$
9898 - Propane, 1,2-dichloro-1,1,2,3,3,3-hexafluoro-
$C_3Cl_2N_2$
9912 - Propanedinitrile, dichloro-
$C_3Cl_3F_3$
10607 - 1-Propene, 1,1,2-trichloro-3,3,3-trifluoro-
$C_3Cl_3F_5$
10179 - Propane, 1,1,1-trichloro-2,2,3,3,3-pentafluoro-
10180 - Propane, 1,2,2-trichloro-1,1,3,3,3-pentafluoro-
10181 - Propane, 1,2,3-trichloro-1,1,2,3,3-pentafluoro-
$C_3Cl_3N_3$
11837 - 1,3,5-Triazine, 2,4,6-trichloro-
$C_3Cl_3N_3O_3$
11840 - 1,3,5-Triazine-2,4,6(1H,3H,5H)-trione, 1,3,5-trichloro-
$C_3Cl_4F_4$
10149 - Propane, 1,2,2,3-tetrachloro-1,1,3,3-tetrafluoro-

C_3Cl_4O
  10718 - 2-Propenoyl chloride, 2,3,3-trichloro-
C_3Cl_6
  10571 - 1-Propene, 1,1,2,3,3,3-hexachloro-
C_3Cl_6O
  10440 - 2-Propanone, 1,1,1,3,3,3-hexachloro-
$C_3Cl_6O_2$
  262 - Acetic acid, trichloro-, trichloromethyl ester
$C_3Cl_6O_3$
  7585 - Methanol, trichloro-, carbonate (2:1)
C_3F_6
  10572 - 1-Propene, 1,1,2,3,3,3-hexafluoro-
C_3F_6O
  8629 - Oxirane, trifluoro(trifluoromethyl)-
  10441 - 2-Propanone, 1,1,1,3,3,3-hexafluoro-
C_3F_8
  10118 - Propane, octafluoro-
C_3HBr_5O
  10477 - 2-Propanone, 1,1,1,3,3-pentabromo-
$C_3HCl_2F_3$
  10558 - Propene, 1,2-dichloro-3,3,3-trifluoro-
$C_3HCl_2N_3O_3$
  11838 - 1,3,5-Triazine-2,4,6(1H,3H,5H)-trione, 1,3-
    dichloro-
$C_3HCl_3O_2$
  10696 - 2-Propenoic acid, 2,3,3-trichloro-
C_3HCl_5
  10593 - 1-Propene, 1,1,2,3,3-pentachloro-
C_3HCl_5O
  10478 - 2-Propanone, 1,1,1,3,3-pentachloro-
C_3HCl_7
  10048 - Propane, 1,1,1,2,2,3,3-heptachloro-
  10049 - Propane, 1,1,1,2,3,3,3-heptachloro-
C_3HN
  10737 - 2-Propynenitrile
C_3H_2BrNS
  11668 - Thiazole, 2-bromo-
$C_3H_2Br_2$
  10730 - 1-Propyne, 1,3-dibromo-
$C_3H_2Br_2O_2$
  10627 - 2-Propenoic acid, 2,3-dibromo-, (Z)-
  10628 - 2-Propenoic acid, 3,3-dibromo-
C_3H_2ClN
  10584 - 2-Propenenitrile, 2-chloro-
$C_3H_2Cl_2O_2$
  10029 - Propanedioyl dichloride
  10630 - 2-Propenoic acid, 3,3-dichloro-
$C_3H_2Cl_3NO$
  10115 - Propanenitrile, 3,3,3-trichloro-2-hydroxy-
$C_3H_2Cl_4$
  10598 - 1-Propene, 1,2,3,3-tetrachloro-
$C_3H_2Cl_4O$
  10486 - 2-Propanone, 1,1,1,3-tetrachloro-
  10487 - 2-Propanone, 1,1,3,3-tetrachloro-
$C_3H_2Cl_6$
  10051 - Propane, 1,1,1,2,2,3-hexachloro-
  10052 - Propane, 1,1,1,3,3,3-hexachloro-
  10053 - Propane, 1,1,2,2,3,3-hexachloro-
$C_3H_2N_2$
  9911 - Propanedinitrile
$C_3H_2N_2O_3$
  7226 - Imidazolidinetrione
C_3H_2O
  10723 - 2-Propynal
$C_3H_2O_2$
  10738 - 2-Propynoic acid
C_3H_3Br
  10728 - 1-Propyne, 3-bromo-
$C_3H_3BrF_2$
  10532 - 1-Propene, 3-bromo-3,3-difluoro-
$C_3H_3BrO_2$
  10610 - 2-Propenoic acid, 2-bromo-
C_3H_3Cl
  10729 - 1-Propyne, 3-chloro-
C_3H_3ClO
  10511 - 2-Propenal, 2-chloro-
  10714 - 2-Propenoyl chloride

$C_3H_3ClO_2$
  10617 - 2-Propenoic acid, 3-chloro-, (E)-
  10618 - 2-Propenoic acid, 3-chloro-, (Z)-
$C_3H_3ClO_3$
  145 - Acetic acid, chlorooxo-, methyl ester
$C_3H_3Cl_2F$
  10554 - 1-Propene, 1,1-dichloro-2-fluoro-
$C_3H_3Cl_2N$
  10099 - Propanenitrile, 2,3-dichloro-
$C_3H_3Cl_3$
  10602 - 1-Propene, 1,1,2-trichloro-
  10603 - 1-Propene, 1,2,3-trichloro-
  10604 - 1-Propene, 3,3,3-trichloro-
$C_3H_3Cl_3O$
  8628 - Oxirane, (trichloromethyl)-
  9790 - Propanal, 2,2,3-trichloro-
  10488 - 2-Propanone, 1,1,1-trichloro-
  10500 - Propanoyl chloride, 2,2-dichloro-
  10501 - Propanoyl chloride, 2,3-dichloro-
  10502 - Propanoyl chloride, 3,3-dichloro-
$C_3H_3Cl_3O_2$
  253 - Acetic acid, trichloro-, methyl ester
  10332 - Propanoic acid, 2,2,3-trichloro-
$C_3H_3Cl_3O_3$
  10333 - Propanoic acid, 3,3,3-trichloro-2-hydroxy-
$C_3H_3Cl_5$
  10125 - Propane, 1,1,1,2,3-pentachloro-
  10126 - Propane, 1,1,2,3,3-pentachloro-
$C_3H_3F_3$
  10608 - 1-Propene, 3,3,3-trifluoro-
$C_3H_3F_5$
  10128 - Propane, 1,1,1,2,2-pentafluoro
C_3H_3I
  10734 - 1-Propyne, 1-iodo-
  10735 - 1-Propyne, 3-iodo-
C_3H_3N
  10583 - 2-Propenenitrile
C_3H_3NO
  7402 - Isoxazole
  8575 - Oxazole
  10113 - Propanenitrile, 2-oxo-
C_3H_3NOS
  303 - Acetyl isothiocyanate
$C_3H_3NOS_2$
  11685 - 4-Thiazolidinone, 2-thioxo-
$C_3H_3NO_2$
  150 - Acetic acid, cyano-
$C_3H_3NO_2S$
  11683 - 2,4-Thiazolidinedione
C_3H_3NS
  11667 - Thiazole
$C_3H_3N_3$
  11827 - 1,3,5-Triazine
$C_3H_3N_3O_2$
  7211 - 1H-Imidazole, 2-nitro-
$C_3H_3N_3O_2S$
  11666 - 2-Thiazolamine, 5-nitro-
$C_3H_3N_3O_3$
  11907 - 1,3,5-Trioxane-2,4,6-triimine
C_3H_4
  4990 - Cyclopropene
  9772 - 1,2-Propadiene
  10727 - 1-Propyne
C_3H_4BrClO
  10417 - 2-Propanone, 1-bromo-3-chloro-
  10495 - Propanoyl chloride, 2-bromo-
C_3H_4BrN
  10094 - Propanenitrile, 2-bromo-
  10095 - Propanenitrile, 3-bromo-
$C_3H_4Br_2$
  10542 - 1-Propene, 1,1-dibromo-
  10543 - 1-Propene, 1,3-dibromo-, (E)-
  10544 - 1-Propene, 1,3-dibromo-, (Z)-
  10545 - 1-Propene, 2,3-dibromo-
$C_3H_4Br_2O$
  9780 - Propanal, 2,3-dibromo-
  10424 - 2-Propanone, 1,3-dibromo-

$C_3H_4Br_2O_2$
  10238 - Propanoic acid, 2,3-dibromo-
$C_3H_4Br_4$
  10135 - Propane, 1,1,1,2-tetrabromo-
  10136 - Propane, 1,1,2,2-tetrabromo-
  10137 - Propane, 1,1,3,3-tetrabromo-
  10138 - Propane, 1,2,2,3-tetrabromo-
$C_3H_4ClFO_2$
  134 - Acetic acid, chlorofluoro-, methyl ester
$C_3H_4ClF_3$
  9871 - Propane, 3-chloro-1,1,1-trifluoro-
C_3H_4ClN
  10097 - Propanenitrile, 3-chloro-
$C_3H_4Cl_2$
  10547 - 1-Propene, 1,1-dichloro-
  10548 - 1-Propene, 1,2-dichloro-, (E)-
  10549 - 1-Propene, 1,2-dichloro-, (Z)-
  10550 - 1-Propene, 1,3-dichloro-, (E)-
  10551 - 1-Propene, 1,3-dichloro-, (Z)-
  10552 - 1-Propene, 2,3-dichloro-
  10553 - 1-Propene, 3,3-dichloro-
$C_3H_4Cl_2O$
  9781 - Propanal, 2,3-dichloro-
  10425 - 2-Propanone, 1,1-dichloro-
  10426 - 2-Propanone, 1,3-dichloro-
  10496 - Propanoyl chloride, 2-chloro-, (S)-
  10497 - Propanoyl chloride, 3-chloro-
$C_3H_4Cl_2O_2$
  166 - Acetic acid, dichloro-, methyl ester
  4013 - Carbonochloridic acid, 2-chloroethyl ester
  10244 - Propanoic acid, 2,2-dichloro-
  10245 - Propanoic acid, 2,3-dichloro-
$C_3H_4Cl_3NO_2$
  6000 - Formamide, N-(2,2,2-trichloro-
    1-hydroxyethyl)-
$C_3H_4Cl_3NO_3$
  10408 - 2-Propanol, 1,1,1-trichloro-3-nitro-
$C_3H_4Cl_4$
  10141 - Propane, 1,1,1,2-tetrachloro-
  10142 - Propane, 1,1,1,3-tetrachloro-
  10143 - Propane, 1,1,2,2-tetrachloro-
  10144 - Propane, 1,1,2,3-tetrachloro-
  10145 - Propane, 1,2,2,3-tetrachloro-
$C_3H_4Cl_4O$
  10402 - 2-Propanol, 1,1,3,3-tetrachloro-
  10403 - 2-Propanol, 1,1,1,3-tetrachloro-
$C_3H_4F_2$
  10561 - 1-Propene, 1,1-difluoro
$C_3H_4F_3NO_2$
  10184 - Propane, 1,1,1-trifluoro-3-nitro-
$C_3H_4F_4O$
  10404 - 1-Propanol, 2,2,3,3-tetrafluoro-
$C_3H_4N_2$
  285 - Acetonitrile, (methyleneamino)-
  7192 - 1H-Imidazole
  10815 - 1H-Pyrazole
$C_3H_4N_2O$
  49 - Acetamide, 2-cyano-
$C_3H_4N_2O_2$
  7216 - 2,4-Imidazolidinedione
$C_3H_4N_2S$
  11662 - 2-Thiazolamine
C_3H_4O
  4988 - Cyclopropanone
  5944 - Ethyne, methoxy-
  10509 - 2-Propenal
  10743 - 2-Propyn-1-ol
$C_3H_4O_2$
  6006 - Formic acid, ethenyl ester
  8595 - 2-Oxetanone
  8596 - 3-Oxetanone
  8603 - Oxiranecarboxaldehyde
  9788 - Propanal, 2-oxo-
  10609 - 2-Propenoic acid
$C_3H_4O_3$
  5211 - 1,3-Dioxolan-2-one
  10314 - Propanoic acid, 2-oxo-

$C_3H_4O_4$
9921 - Propanedioic acid
$C_3H_4O_5$
9956 - Propanedioic acid, hydroxy-
C_3H_5Br
10529 - 1-Propene, 1-bromo-, (Z)-
10530 - 1-Propene, 2-bromo-
10531 - 1-Propene, 3-bromo-
C_3H_5BrO
8600 - Oxirane, (bromomethyl)-, (±)-
9775 - Propanal, 2-bromo-
10416 - 2-Propanone, 1-bromo-
10490 - Propanoyl bromide
10699 - 2-Propen-1-ol, 2-bromo-
$C_3H_5BrO_2$
117 - Acetic acid, bromo-, methyl ester
10204 - Propanoic acid, 2-bromo-, (±)-
10205 - Propanoic acid, 3-bromo-
$C_3H_5Br_2Cl$
9885 - Propane, 1,2-dibromo-3-chloro-
$C_3H_5Br_3$
10160 - Propane, 1,1,1-tribromo-
10161 - Propane, 1,1,2-tribromo-
10162 - Propane, 1,2,2-tribromo-
10163 - Propane, 1,2,3-tribromo-
C_3H_5Cl
10533 - 1-Propene, 1-chloro-, (E)-
10534 - 1-Propene, 1-chloro-, (Z)-
10535 - 1-Propene, 2-chloro-
10536 - 1-Propene, 3-chloro-
$C_3H_5ClF_2$
9851 - Propane, 1-chloro-2,2-difluoro-
$C_3H_5ClN_2O_6$
9997 - 1,2-Propanediol, 3-chloro-, dinitrate
C_3H_5ClO
8605 - Oxirane, (chloromethyl)-, (±)-
9777 - Propanal, 2-chloro-
9778 - Propanal, 3-chloro-
10420 - 2-Propanone, 1-chloro-
10493 - Propanoyl chloride
10700 - 2-Propen-1-ol, 2-chloro-
10701 - 2-Propen-1-ol, 3-chloro-
$C_3H_5ClO_2$
138 - Acetic acid, chloro-, methyl ester
299 - Acetyl chloride, methoxy-
4018 - Carbonochloridic acid, ethyl ester
7582 - Methanol, chloro-, acetate
10215 - Propanoic acid, 2-chloro-
10216 - Propanoic acid, 3-chloro-
$C_3H_5Cl_2F$
9897 - Propane, 1,2-dichloro-2-fluoro-
$C_3H_5Cl_2NO_2$
3940 - Carbamic acid, dichloro-, ethyl ester
9902 - Propane, 1,1-dichloro-1-nitro-
$C_3H_5Cl_3$
10171 - Propane, 1,1,1-trichloro-
10172 - Propane, 1,1,2-trichloro-
10173 - Propane, 1,1,3-trichloro-
10174 - Propane, 1,2,2-trichloro-
10175 - Propane, 1,2,3-trichloro-
$C_3H_5Cl_3O$
10406 - 2-Propanol, 1,1,1-trichloro-
$C_3H_5Cl_3Si$
11475 - Silane, trichloro-2-propenyl-
C_3H_5F
10567 - 1-Propene, 1-fluoro- (Z)-
10568 - 1-Propene, 1-fluoro- (E)-
10569 - 1-Propene, 2-fluoro-
10570 - 1-Propene, 3-fluoro-
C_3H_5FO
8616 - Oxirane, (fluoromethyl)-
10437 - 2-Propanone, 1-fluoro-
10506 - Propanoyl fluoride
$C_3H_5FO_2$
192 - Acetic acid, fluoro-, methyl ester
$C_3H_5F_3O$
10409 - 2-Propanol, 1,1,1-trifluoro-, (±)-

$C_3H_5F_3O_3S$
7533 - Methanesulfonic acid, trifluoro-, ethyl ester
C_3H_5I
10573 - 1-Propene, 3-iodo-
C_3H_5IO
8617 - Oxirane, (iodomethyl)-
10454 - 2-Propanone, 1-iodo-
$C_3H_5IO_2$
10282 - Propanoic acid, 2-iodo-, (±)-
C_3H_5N
5599 - Ethane, isocyano-
10090 - Propanenitrile
10725 - 2-Propyn-1-amine
C_3H_5NO
284 - Acetonitrile, methoxy-
496 - 2-Azetidinone
4163 - Cyanic acid, ethyl ester
5598 - Ethane, isocyanato-
10105 - Propanenitrile, 2-hydroxy-
10106 - Propanenitrile, 3-hydroxy-
10519 - 2-Propenamide
$C_3H_5NO_2$
9789 - Propanal, 2-oxo-, 1-oxime
10589 - 1-Propene, 1-nitro-
10590 - 1-Propene, 2-nitro-
10591 - 1-Propene, 3-nitro-
$C_3H_5NO_3$
10475 - 2-Propanone, 1-nitro-
C_3H_5NS
286 - Acetonitrile, (methylthio)-
5600 - Ethane, isothiocyanato-
11703 - Thiocyanic acid, ethyl ester
$C_3H_5N_3$
10812 - 1H-Pyrazol-4-amine
$C_3H_5N_3O$
153 - Acetic acid, cyano-, hydrazide
$C_3H_5N_3O_9$
10197 - 1,2,3-Propanetriol, trinitrate
$C_3H_5NaO_2$
10327 - Propanoic acid, sodium salt
C_3H_6
4955 - Cyclopropane
10528 - 1-Propene
C_3H_6BrCl
9838 - Propane, 1-bromo-2-chloro-
9839 - Propane, 1-bromo-3-chloro-
9840 - Propane, 2-bromo-1-chloro-
9841 - Propane, 2-bromo-2-chloro-
C_3H_6BrF
9844 - Propane, 1-bromo-3-fluoro-
$C_3H_6BrNO_2$
9847 - Propane, 2-bromo-2-nitro-
$C_3H_6BrNO_4$
9991 - 1,3-Propanediol, 2-bromo-2-nitro-
$C_3H_6Br_2$
9880 - Propane, 1,1-dibromo-
9881 - Propane, 1,2-dibromo-
9882 - Propane, 1,3-dibromo-
9883 - Propane, 2,2-dibromo-
$C_3H_6Br_2O$
10357 - 2-Propanol, 1,3-dibromo-
C_3H_6ClF
9856 - Propane, 1-chloro-3-fluoro-
C_3H_6ClI
9857 - Propane, 1-chloro-3-iodo-
C_3H_6ClNO
3973 - Carbamic chloride, dimethyl-
$C_3H_6ClNO_2$
9866 - Propane, 1-chloro-1-nitro-
9867 - Propane, 1-chloro-2-nitro-
9868 - Propane, 1-chloro-3-nitro-
9869 - Propane, 2-chloro-1-nitro-
9870 - Propane, 2-chloro-2-nitro-
C_3H_6ClNS
3980 - Carbamothioic chloride, dimethyl-
$C_3H_6Cl_2$
9891 - Propane, 1,1-dichloro-

9892 - Propane, 1,2-dichloro-, (±)-
9893 - Propane, 1,3-dichloro-
9894 - Propane, 2,2-dichloro-
$C_3H_6Cl_2O$
10359 - 1-Propanol, 2,3-dichloro-
10360 - 1-Propanol, 3,3-dichloro-
10361 - 2-Propanol, 1,1-dichloro-
10362 - 2-Propanol, 1,3-dichloro-
$C_3H_6Cl_2Si$
11396 - Silane, dichloroethenylmethyl-
$C_3H_6Cl_3P$
9524 - Phosphine, tris(chloromethyl)-
$C_3H_6Cl_4Si$
11458 - Silane, trichloro(3-chloropropyl)-
$C_3H_6F_2$
9905 - Propane, 1,3-difluoro-
9906 - Propane, 2,2-difluoro-
$C_3H_6F_2O$
10369 - 2-Propanol, 1,3-difluoro-
$C_3H_6I_2$
9907 - Propane, 1,3-diiodo-
9908 - Propane, 2,2-diiodo-
$C_3H_6N_2$
4157 - Cyanamide, dimethyl-
10092 - Propanenitrile, 3-amino-
10822 - 1H-Pyrazole, 4,5-dihydro-
$C_3H_6N_2O$
7227 - 2-Imidazolidinone
$C_3H_6N_2OS$
33 - Acetamide, N-(aminothioxomethyl)-
$C_3H_6N_2O_2$
27 - Acetamide, N-(aminocarbonyl)-
7406 - 3-Isoxazolidinone, 4-amino-, (R)-
9784 - Propanal, 2-(hydroxyimino)-, oxime
$C_3H_6N_2O_4$
3963 - Carbamic acid, nitro-, ethyl ester
9918 - Propane, 1,1-dinitro-
9919 - Propane, 1,3-dinitro-
9920 - Propane, 2,2-dinitro-
$C_3H_6N_2O_7$
10191 - 1,2,3-Propanetriol, 1,3-dinitrate
$C_3H_6N_2S$
7225 - 2-Imidazolidinethione
11664 - 2-Thiazolamine, 4,5-dihydro-
$C_3H_6N_6$
11835 - 1,3,5-Triazine-2,4,6-triamine
$C_3H_6N_6O_6$
11833 - 1,3,5-Triazine, hexahydro-1,3,5-trinitro-
C_3H_6O
5908 - Ethene, methoxy-
8592 - Oxetane
8621 - Oxirane, methyl-
9774 - Propanal
10411 - 2-Propanone
10697 - 2-Propen-1-ol
$C_3H_6O_2$
16 - Acetaldehyde, methoxy-
208 - Acetic acid, methyl ester
5204 - 1,3-Dioxolane
6007 - Formic acid, ethyl ester
8618 - Oxiranemethanol
10201 - Propanoic acid
10443 - 2-Propanone, 1-hydroxy-
$C_3H_6O_2S$
204 - Acetic acid, mercapto-, methyl ester
216 - Acetic acid, (methylthio)-
5909 - Ethene, (methylsulfonyl)-
10285 - Propanoic acid, 2-mercapto-
10286 - Propanoic acid, 3-mercapto-
11692 - Thietane, 1,1-dioxide
$C_3H_6O_3$
198 - Acetic acid, hydroxy-, methyl ester
205 - Acetic acid, methoxy-
4000 - Carbonic acid, dimethyl ester
9782 - Propanal, 2,3-dihydroxy-, (±)-
$C_3H_6O_2$
10266 - Propanoic acid, 2-hydroxy-, (±)-

$C_3H_6O_3$
  10267 - Propanoic acid, 3-hydroxy-
  10429 - 2-Propanone, 1,3-dihydroxy-
  11905 - 1,3,5-Trioxane
$C_3H_6O_3S$
  8570 - 1,2-Oxathiolane, 2,2-dioxide
  11694 - Thietane, 1-oxide
$C_3H_6O_4$
  10252 - Propanoic acid, 2,3-dihydroxy-
C_3H_6S
  5910 - Ethene, (methylthio)-
  10600 - 2-Propene-1-thiol
  11691 - Thietane
  11696 - Thiirane, methyl-
$C_3H_6S_2$
  5266 - 1,3-Dithiolane
$C_3H_6S_3$
  11919 - 1,3,5-Trithiane
C_3H_7Br
  9836 - Propane, 1-bromo-
  9837 - Propane, 2-bromo-
C_3H_7BrO
  5464 - Ethane, (bromomethoxy)-
  5465 - Ethane, 1-bromo-2-methoxy-
  10344 - 1-Propanol, 3-bromo-
  10345 - 2-Propanol, 1-bromo-
C_3H_7Cl
  9848 - Propane, 1-chloro-
  9849 - Propane, 2-chloro-
C_3H_7ClHgO
  11822 - Triadimenol
C_3H_7ClO
  5478 - Ethane, (chloromethoxy)-
  5479 - Ethane, 1-chloro-1-methoxy-
  5480 - Ethane, 1-chloro-2-methoxy-
  10347 - 1-Propanol, 2-chloro-
  10348 - 1-Propanol, 3-chloro-
  10349 - 2-Propanol, 1-chloro-
$C_3H_7ClO_2$
  9994 - 1,2-Propanediol, 3-chloro-
  9995 - 1,3-Propanediol, 2-chloro-
$C_3H_7ClO_2S$
  10133 - 1-Propanesulfonyl chloride
$C_3H_7ClO_3S$
  5699 - Ethanol, 2-chloro-, methanesulfonate
C_3H_7ClS
  5481 - Ethane, 1-chloro-2-(methylthio)-
$C_3H_7Cl_2N$
  9805 - 1-Propanamine, N,N-dichloro-
$C_3H_7Cl_3Si$
  11373 - Silane, chloro(dichloromethyl)dimethyl-
  11466 - Silane, trichloro(1-methylethyl)-
  11476 - Silane, trichloropropyl-
C_3H_7F
  10046 - Propane, 1-fluoro-
  10047 - Propane, 2-fluoro-
C_3H_7FO
  10382 - 1-Propanol, 3-fluoro-
C_3H_7I
  10054 - Propane, 1-iodo-
  10055 - Propane, 2-iodo-
C_3H_7IO
  10385 - 1-Propanol, 3-iodo-
C_3H_7N
  494 - Azetidine
  500 - Aziridine, 1-methyl-
  501 - Aziridine, 2-methyl-
  4953 - Cyclopropanamine
  10522 - 2-Propen-1-amine
$C_3H_7NNaS_2$
  11514 - Sodium dimethyldithiocarbamate
C_3H_7NO
  72 - Acetamide, N-methyl-
  5992 - Formamide, N,N-dimethyl-
  5994 - Formamide, N-ethyl-
  9787 - Propanal, oxime
  9791 - Propanamide

  10476 - 2-Propanone, oxime
C_3H_7NOS
  3978 - Carbamothioic acid, O-ethyl ester
  3979 - Carbamothioic acid, S-ethyl ester
$C_3H_7NO_2$
  323 - β-Alanine
  324 - DL-Alanine
  325 - L-Alanine
  3948 - Carbamic acid, ethyl ester
  6268 - Glycine, N-methyl-
  8092 - Nitrous acid, 1-methylethyl ester
  8098 - Nitrous acid, propyl ester
  9797 - Propanamide, 2-hydroxy-, (±)-
  10116 - Propane, 1-nitro-
  10117 - Propane, 2-nitro-
$C_3H_7NO_2S$
  5014 - L-Cysteine
$C_3H_7NO_3$
  8076 - Nitric acid, 1-methylethyl ester
  8080 - Nitric acid, propyl ester
  10394 - 1-Propanol, 2-nitro-
  10395 - 1-Propanol, 3-nitro-
  10396 - 2-Propanol, 1-nitro-
  11355 - D-Serine
  11356 - DL-Serine
  11357 - L-Serine
$C_3H_7NO_4$
  10022 - 1,2-Propanediol, 1-nitrate
$C_3H_7NO_5$
  10192 - 1,2,3-Propanetriol, 1-nitrate
  10193 - 1,2,3-Propanetriol, 2-nitrate
C_3H_7NS
  11681 - Thiazolidine
$C_3H_7N_3O_2$
  6253 - Glycine, N-(aminoiminomethyl)-
  12015 - Urea, N-ethyl-N-nitroso-
$C_3H_7N_3O_5$
  11839 - 1,3,5-Triazine-2,4,6(1H,3H,5H)-trione, dihydrate
C_3H_8
  9834 - Propane
$C_3H_8BrClSi$
  11365 - Silane, (bromomethyl)chlorodimethyl-
$C_3H_8Cl_2Si$
  11372 - Silane, chloro(chloromethyl)dimethyl-
  11397 - Silane, dichloroethylmethyl-
$C_3H_8NO_5P$
  6276 - Glyphosate
$C_3H_8N_2O$
  12004 - Urea, N,N'-dimethyl-
  12005 - Urea, N,N-dimethyl-
  12014 - Urea, ethyl-
$C_3H_8N_2O_2$
  326 - Alanine, 3-amino-
  7148 - Hydrazinecarboxylic acid, ethyl ester
  9824 - 1-Propanamine, N-nitro-
$C_3H_8N_2O_3$
  11995 - Urea, N,N'-bis(hydroxymethyl)-
$C_3H_8N_2S$
  11804 - Thiourea, N,N'-dimethyl-
C_3H_8O
  5602 - Ethane, methoxy-
  10334 - 1-Propanol
  10335 - 2-Propanol
C_3H_8OS
  5749 - Ethanol, 2-(methylthio)-
$C_3H_8OS_2$
  10370 - 1-Propanol, 2,3-dimercapto-
  10371 - 2-Propanol, 1,3-dimercapto-
$C_3H_8O_2$
  5733 - Ethanol, 2-methoxy-
  7507 - Methane, dimethoxy-
  9982 - 1,2-Propanediol
  9983 - 1,3-Propanediol
$C_3H_8O_2S$
  10013 - 1,2-Propanediol, 3-mercapto-

$C_3H_8O_3$
  10186 - 1,2,3-Propanetriol
$C_3H_8O_3S$
  7531 - Methanesulfonic acid, ethyl ester
  10130 - 2-Propanesulfonic acid
C_3H_8S
  5606 - Ethane, (methylthio)-
  10156 - 1-Propanethiol
  10157 - 2-Propanethiol
$C_3H_8S_2$
  7461 - Methane, bis(methylthio)-
  10030 - 1,2-Propanedithiol
  10031 - 1,3-Propanedithiol
C_3H_9Al
  339 - Aluminum, trimethyl
C_3H_9As
  449 - Arsine, trimethyl-
$C_3H_9AsO_3$
  461 - Arsonic acid, propyl-
C_3H_9B
  3119 - Borane, trimethyl-
$C_3H_9BO_2$
  3136 - Boronic acid, propyl-
$C_3H_9BO_3$
  3126 - Boric acid, trimethyl ester
C_3H_9BS
  3129 - Borinic acid, dimethylthio-, methyl ester
$C_3H_9BS_3$
  11698 - Thioboric acid, trimethyl ester
C_3H_9ClSi
  11389 - Silane, chlorotrimethyl-
C_3H_9ClSn
  11540 - Stannane, chlorotrimethyl-
C_3H_9N
  7444 - Methanamine, N,N-dimethyl-
  9803 - 1-Propanamine
  9804 - 2-Propanamine
C_3H_9NO
  5738 - Ethanol, 2-(methylamino)-
  10336 - 1-Propanol, 2-amino-, (±)-
  10337 - 1-Propanol, 3-amino-
  10338 - 2-Propanol, 1-amino-, (±)-
$C_3H_9NO_2$
  9984 - 1,2-Propanediol, 3-amino-, (±)-
$C_3H_9NO_3S$
  5624 - Ethanesulfonic acid, 2-(methylamino)-
$C_3H_9O_2PS$
  9550 - Phosphonothioic acid, methyl-, O,O-dimethyl ester
$C_3H_9O_3P$
  9540 - Phosphonic acid, methyl-, dimethyl ester
  9542 - Phosphonic acid, (1-methylethyl)-
  9543 - Phosphonic acid, methyl-, monoethyl ester
  9547 - Phosphonic acid, propyl-
  9592 - Phosphorous acid, trimethyl ester
$C_3H_9O_4P$
  9564 - Phosphoric acid, trimethyl ester
C_3H_9P
  9522 - Phosphine, trimethyl-
C_3H_9Sb
  11550 - Stibine, trimethyl-
$C_3H_{10}ClN$
  5435 - Ethanamine, N-methyl-, hydrochloride
  7445 - Methanamine, N,N-dimethyl-, hydrochloride
$C_3H_{10}N_2$
  5512 - 1,2-Ethanediamine, N-methyl-
  9872 - 1,2-Propanediamine
  9873 - 1,3-Propanediamine
$C_3H_{10}Si$
  11489 - Silane, trimethyl-
$C_3H_{12}BN$
  3120 - Borane, trimethyl-, monoammoniate
  3132 - Boron, (N,N-dimethylmethanamine)trihydro-, (T-4)-
C_3N_2O
  9915 - Propanedinitrile, oxo-

C_3O_2
9773 - 1,2-Propadiene-1,3-dione
$C_4Br_2F_8$
3275 - Butane, 1,4-dibromo-1,1,2,2,3,3,4,4-octafluoro-
C_4Br_4S
11781 - Thiophene, tetrabromo-
C_4ClF_7
4177 - Cyclobutane, 1-chloro-1,2,2,3,3,4,4-heptafluoro-
C_4ClF_7O
3692 - Butanoyl chloride, heptafluoro-
$C_4Cl_2F_4$
4199 - Cyclobutene, 1,2-dichloro-3,3,4,4-tetrafluoro-
$C_4Cl_2F_6$
3754 - 2-Butene, 2,3-dichloro-1,1,1,4,4,4-hexafluoro-
$C_4Cl_2F_8$
3287 - Butane, 2,3-dichloro-1,1,1,2,3,4,4,4-octafluoro-
$C_4Cl_3F_7$
3452 - Butane, 2,2,3-trichloro-1,1,1,3,4,4,4-heptafluoro-
$C_4Cl_4F_4$
3808 - 1-Butene, 1,3,4,4-tetrachloro-1,2,3,4-tetrafluoro-
C_4Cl_4S
11782 - Thiophene, tetrachloro-
C_4Cl_6
3163 - 1,3-Butadiene, 1,1,2,3,4,4-hexachloro-
$C_4Cl_6O_3$
244 - Acetic acid, trichloro-, anhydride
$C_4F_4O_3$
6099 - 2,5-Furandione, 3,3,4,4-tetrafluorodihydro-
C_4F_6
3164 - 1,3-Butadiene, 1,1,2,3,4,4-hexafluoro-
3914 - 2-Butyne, 1,1,1,4,4,4-hexafluoro-
4200 - Cyclobutene, hexafluoro-
$C_4F_6O_3$
264 - Acetic acid, trifluoro-, anhydride
C_4F_8
3804 - 2-Butene, 1,1,1,2,3,4,4,4-octafluoro-
4194 - Cyclobutane, octafluoro-
C_4F_{10}
3255 - Butane, decafluoro-
10050 - Propane, 1,1,1,2,3,3,3-heptafluoro-2-(trifluoromethyl)-
C_4HBr_3S
11790 - Thiophene, 2,3,5-tribromo-
C_4HClN_2
3763 - 2-Butenedinitrile, 2-chloro-, (Z)-
C_4HClO_3
6092 - 2,5-Furandione, 3-chloro-
$C_4HCl_2N_3O_2$
11054 - Pyrimidine, 2,4-dichloro-5-nitro-
$C_4HCl_3N_2$
11075 - Pyrimidine, 2,4,6-trichloro-
$C_4HCl_3O_2$
3796 - 2-Butenedioyl dichloride, 2-chloro-, (Z)-
C_4HCl_3S
11791 - Thiophene, 2,3,5-trichloro-
$C_4HCl_7O_2$
5192 - 1,4-Dioxane, 2,2,3,3,5,5,6-heptachloro-
$C_4HF_7O_2$
3521 - Butanoic acid, heptafluoro-
C_4HI_4N
11140 - 1H-Pyrrole, 2,3,4,5-tetraiodo-
C_4H_2
3177 - 1,3-Butadiyne
C_4H_2BrClS
11728 - Thiophene, 2-bromo-5-chloro-
$C_4H_2Br_2N_2O_3$
11086 - 2,4,6(1H,3H,5H)-Pyrimidinetrione, 5,5-dibromo-
$C_4H_2Br_2O$
6077 - Furan, 2,3-dibromo-
6078 - Furan, 2,5-dibromo-

$C_4H_2Br_2S$
11742 - Thiophene, 2,3-dibromo-
11743 - Thiophene, 2,4-dibromo-
11744 - Thiophene, 2,5-dibromo-
$C_4H_2Cl_2O_2$
3795 - 2-Butenedioyl dichloride, (E)-
$C_4H_2Cl_2S$
11746 - Thiophene, 2,3-dichloro-
11747 - Thiophene, 2,4-dichloro-
11748 - Thiophene, 2,5-dichloro-
11749 - Thiophene, 3,4-dichloro-
$C_4H_2Cl_4$
3173 - 1,3-Butadiene, 1,2,3,4-tetrachloro-
$C_4H_2Cl_6$
3798 - 2-Butene, 1,1,2,3,4,4-hexachloro-
$C_4H_2Cl_8O$
5615 - Ethane, 1,1'-oxybis[1,2,2,2-tetrachloro-
$C_4H_2I_2S$
11753 - Thiophene, 2,5-diiodo-
$C_4H_2N_2$
3762 - 2-Butenedinitrile, (E)-
$C_4H_2N_2O_4$
11073 - 2,4,5,6(1H,3H)-Pyrimidinetetrone
$C_4H_2N_2O_4S$
11758 - Thiophene, 2,5-dinitro-
$C_4H_2O_3$
6091 - 2,5-Furandione
$C_4H_3BrClF_3$
3715 - 1-Butene, 4-bromo-3-chloro-3,4,4-trifluoro-
$C_4H_3BrN_2O_2$
11062 - 2,4(1H,3H)-Pyrimidinedione, 5-bromo-
C_4H_3BrO
6033 - Furan, 2-bromo-
6034 - Furan, 3-bromo-
$C_4H_3BrO_4$
3769 - 2-Butenedioic acid, 2-bromo-, (E)-
3770 - 2-Butenedioic acid, 2-bromo-, (Z)-
C_4H_3BrS
11726 - Thiophene, 2-bromo-
11727 - Thiophene, 3-bromo-
C_4H_3Cl
3890 - 1-Buten-3-yne, 4-chloro-
C_4H_3ClO
6073 - Furan, 2-chloro-
6074 - Furan, 3-chloro-
$C_4H_3ClO_2S_2$
11780 - 2-Thiophenesulfonyl chloride
C_4H_3ClS
11740 - Thiophene, 2-chloro-
$C_4H_3Cl_2F_3O_2$
10251 - Propanoic acid, 2,2-dichloro-3,3,3-trifluoro-, methyl ester
$C_4H_3Cl_3$
3174 - 1,3-Butadiene, 1,2,3-trichloro-
$C_4H_3Cl_5$
3805 - 2-Butene, 1,1,1,4,4-pentachloro-
$C_4H_3Cl_7$
3380 - Butane, 1,1,2,2,3,4,4-heptachloro-
$C_4H_3FN_2O_2$
11067 - 2,4(1H,3H)-Pyrimidinedione, 5-fluoro-
$C_4H_3F_5O_2$
10317 - Propanoic acid, pentafluoro-, methyl ester
$C_4H_3F_7O$
3622 - 1-Butanol, 2,2,3,3,4,4,4-heptafluoro-
C_4H_3IO
6110 - Furan, 2-iodo-
6111 - Furan, 3-iodo-
C_4H_3IS
11767 - Thiophene, 2-iodo-
$C_4H_3NO_2$
11121 - 1H-Pyrrole-2,5-dione
$C_4H_3NO_2S$
11775 - Thiophene, 2-nitro-
11776 - Thiophene, 3-nitro-
$C_4H_3NO_3$
6134 - Furan, 2-nitro-

$C_4H_3N_3O_4$
11074 - 2,4,5,6(1H,3H)-Pyrimidinetetrone, 5-oxime
$C_4H_3N_3O_5$
11102 - 2,4,6(1H,3H,5H)-Pyrimidinetrione, 5-nitro-
C_4H_4
3889 - 1-Buten-3-yne
$C_4H_4BrNO_2$
11153 - 2,5-Pyrrolidinedione, 1-bromo-
$C_4H_4Br_2$
3904 - 2-Butyne, 1,4-dibromo-
$C_4H_4Br_4$
4195 - Cyclobutane, 1,2,3,4-tetrabromo-
$C_4H_4ClNO_2$
11154 - 2,5-Pyrrolidinedione, 1-chloro-
C_4H_4ClNS
10538 - 1-Propene, 2-chloro-3-isothiocyanato-
$C_4H_4ClN_3O_2$
4083 - 5-Chloro-3-methyl-4-nitro-1H-pyrazole
$C_4H_4Cl_2$
3156 - 1,3-Butadiene, 1,1-dichloro-
3157 - 1,3-Butadiene, 1,2-dichloro-
3158 - 1,3-Butadiene, 2,3-dichloro-
3905 - 2-Butyne, 1,4-dichloro-
$C_4H_4Cl_2O_2$
3366 - Butanedioyl dichloride
3687 - Butanoyl chloride, 4-chloro-3-oxo-
4152 - Crotonic acid, 4,4-dichloro-
$C_4H_4Cl_2O_3$
124 - Acetic acid, chloro-, anhydride
$C_4H_4Cl_2O_4$
3302 - Butanedioic acid, 2,3-dichloro-, (±)-
$C_4H_4Cl_4$
3806 - 1-Butene, 1,3,4,4-tetrachloro-
3807 - 1-Butene, 2,3,3,4-tetrachloro-
$C_4H_4Cl_4O_2$
246 - Acetic acid, trichloro-, 2-chloroethyl ester
$C_4H_4INO_2$
11159 - 2,5-Pyrrolidinedione, 1-iodo-
$C_4H_4I_2$
3906 - 2-Butyne, 1,4-diiodo-
$C_4H_4N_2$
3291 - Butanedinitrile
10801 - Pyrazine
10855 - Pyridazine
11049 - Pyrimidine
$C_4H_4N_2OS$
11110 - 4(1H)-Pyrimidinone, 2,3-dihydro-2-thioxo-
$C_4H_4N_2O_2$
3902 - 2-Butynediamide
11059 - 2,4(1H,3H)-Pyrimidinedione
$C_4H_4N_2O_3$
11076 - 2,4,6(1H,3H,5H)-Pyrimidinetrione
$C_4H_4N_2S_2$
11702 - Thiocyanic acid, 1,2-ethanediyl ester
$C_4H_4N_6O$
11851 - 7H-1,2,3-Triazolo[4,5-d]pyrimidin-7-one, 5-amino-1,4-dihydro-
$C_4H_4Na_2O_6$
3314 - Butanedioic acid, 2,3-dihydroxy- [R-(R*,R*)]-, disodium salt
C_4H_4O
3895 - 2-Butynal
3923 - 3-Butyn-2-one
6025 - Furan
C_4H_4OS
11777 - Thiophene-2-ol
$C_4H_4O_2$
3917 - 2-Butynoic acid
5202 - 1,4-Dioxin
8598 - 2-Oxetanone, 4-methylene-
$C_4H_4O_3$
6093 - 2,5-Furandione, dihydro-
$C_4H_4O_4$
3764 - 2-Butenedioic acid (E)-
3765 - 2-Butenedioic acid (Z)-
$C_4H_4O_5$
3339 - Butanedioic acid, oxo-

C₄H₄O₆

$C_4H_4O_6$
3777 - 2-Butenedioic acid, 2,3-dihydroxy-, (Z)-

C_4H_4S
11724 - Thiophene

C_4H_5Br
3147 - 1,2-Butadiene, 4-bromo-
3148 - 1,3-Butadiene, 1-bromo-
3149 - 1,3-Butadiene, 2-bromo-

$C_4H_5BrN_2$
10817 - 1H-Pyrazole, 4-bromo-3-methyl-

C_4H_5BrO
3696 - 2-Butenal, 2-bromo-

$C_4H_5BrO_2$
10612 - 2-Propenoic acid, 3-bromo-2-methyl-

$C_4H_5BrO_4$
3299 - Butanedioic acid, bromo-, (±)-
3300 - Butanedioic acid, bromo-

$C_4H_5Br_2FO_2$
161 - Acetic acid, dibromofluoro-, ethyl ester

$C_4H_5Br_3O_2$
242 - Acetic acid, tribromo-, ethyl ester

C_4H_5Cl
3150 - 1,2-Butadiene, 4-chloro-
3151 - 1,3-Butadiene, 1-chloro-
3152 - 1,3-Butadiene, 2-chloro-
3899 - 1-Butyne, 3-chloro-
3900 - 2-Butyne, 1-chloro-

$C_4H_5ClN_2$
10821 - 1H-Pyrazole, 3-chloro-5-methyl-

C_4H_5ClO
3697 - 2-Butenal, 2-chloro-
3887 - 2-Butenoyl chloride
10715 - 2-Propenoyl chloride, 2-methyl-

$C_4H_5ClO_2$
3820 - 2-Butenoic acid, 3-chloro-, (E)-
3821 - 2-Butenoic acid, 3-chloro-, (Z)-
3822 - 2-Butenoic acid, 4-chloro-
4022 - Carbonochloridic acid, 1-methylethenyl ester
4150 - Crotonic acid, 2-chloro-
4151 - Crotonic acid, 4-chloro-, (E)-
10621 - 2-Propenoic acid, 2-chloro-, methyl ester

$C_4H_5ClO_3$
144 - Acetic acid, chlorooxo-, ethyl ester

$C_4H_5Cl_3$
3809 - 1-Butene, 2,3,4-trichloro-
3810 - 2-Butene, 1,2,4-trichloro-
10605 - 1-Propene, 1,1,3-trichloro-2-methyl-
10606 - 1-Propene, 3,3,3-trichloro-2-methyl-

$C_4H_5Cl_3O$
3191 - Butanal, 2,2,3-trichloro-

$C_4H_5Cl_3O_2$
248 - Acetic acid, trichloro-, ethyl ester
3590 - Butanoic acid, 2,2,3-trichloro-

$C_4H_5Cl_3O_3$
249 - Acetic acid, trichloro-, 2-hydroxyethyl ester

$C_4H_5Cl_5$
3419 - Butane, 1,1,2,3,4-pentachloro-
3420 - Butane, 1,2,2,3,4-pentachloro-
10127 - Propane, 1,1,1,2,3-pentachloro-2-methyl-
10146 - Propane, 1,1,2,3-tetrachloro-2-(chloromethyl)-

C_4H_5F
3162 - 1,3-Butadiene, 2-fluoro-

$C_4H_5F_3O_2$
267 - Acetic acid, trifluoro-, ethyl ester

C_4H_5I
3165 - 1,3-Butadiene, 2-iodo-
3166 - 1,2-Butadiene, 4-iodo-

C_4H_5N
3802 - 2-Butenenitrile, (E)-
3803 - 3-Butenenitrile
4956 - Cyclopropanecarbonitrile
10574 - 1-Propene, 3-isocyano-
10586 - 2-Propenenitrile, 2-methyl-
11112 - 1H-Pyrrole

C_4H_5NO
7405 - Isoxazole, 5-methyl-

8583 - Oxazole, 4-methyl-

$C_4H_5NO_2$
154 - Acetic acid, cyano-, methyl ester
4030 - Carbonocyanidic acid, ethyl ester
10233 - Propanoic acid, 2-cyano-
11152 - 2,5-Pyrrolidinedione

$C_4H_5NO_3$
3817 - 2-Butenoic acid, 4-amino-4-oxo-, (Z)-

C_4H_5NS
10575 - 1-Propene, 3-isothiocyanato-
11678 - Thiazole, 2-methyl-
11679 - Thiazole, 4-methyl-
11712 - Thiocyanic acid, 2-propenyl ester

$C_4H_5N_2S$
11680 - 2(3H)-Thiazolethione, 4-methyl-

$C_4H_5N_3$
11042 - 2-Pyrimidinamine

$C_4H_5N_3O$
11104 - 2(1H)-Pyrimidinone, 4-amino-

$C_4H_5N_3O_2$
11831 - 1,2,4-Triazine-3,5(2H,4H)-dione, 6-methyl-

$C_4H_5N_3O_3$
11077 - 2,4,6(1H,3H,5H)-Pyrimidinetrione, 5-amino-

C_4H_6
3145 - 1,2-Butadiene
3146 - 1,3-Butadiene
3897 - 1-Butyne
3898 - 2-Butyne
4198 - Cyclobutene

$C_4H_6BaO_4$
110 - Acetic acid, barium salt

C_4H_6BrClO
3682 - Butanoyl chloride, 2-bromo-

$C_4H_6BrClO_2$
113 - Acetic acid, bromochloro-, ethyl ester

C_4H_6BrN
3404 - Butanenitrile, 4-bromo-
10096 - Propanenitrile, 2-bromo-2-methyl-

$C_4H_6Br_2$
3738 - 1-Butene, 1,1-dibromo-
3739 - 1-Butene, 1,2-dibromo-
3740 - 1-Butene, 3,4-dibromo-
3741 - 2-Butene, 1,3-dibromo-
3742 - 2-Butene, 1,4-dibromo-, (E)-
4178 - Cyclobutane, 1,1-dibromo-
4179 - Cyclobutane, 1,2-dibromo-
10546 - 1-Propene, 1,3-dibromo-2-methyl-

$C_4H_6Br_2O$
3653 - 2-Butanone, 3,4-dibromo-
5893 - Ethene, 1,1-dibromo-2-ethoxy-
10491 - Propanoyl bromide, 2-bromo-2-methyl-

$C_4H_6Br_2O_2$
160 - Acetic acid, dibromo-, ethyl ester
3497 - Butanoic acid, 2,3-dibromo-
10237 - Propanoic acid, 2,3-dibromo-, methyl ester

$C_4H_6Br_4$
3424 - Butane, 1,1,4,4-tetrabromo-
3425 - Butane, 1,2,2,3-tetrabromo-
3426 - Butane, 1,2,3,4-tetrabromo-, DL-

$C_4H_6CaO_4$
122 - Acetic acid, calcium salt

$C_4H_6ClFO_2$
133 - Acetic acid, chlorofluoro-, ethyl ester

C_4H_6ClN
3405 - Butanenitrile, 4-chloro-

C_4H_6ClNO
10098 - Propanenitrile, 3-chloro-2-hydroxy-2-methyl-

$C_4H_6Cl_2$
3743 - 1-Butene, 1,3-dichloro-
3744 - 1-Butene, 2,3-dichloro-
3745 - 1-Butene, 3,4-dichloro-
3746 - 2-Butene, 1,1-dichloro-
3747 - 2-Butene, 1,2-dichloro-
3748 - 2-Butene, 1,3-dichloro-, (E)-
3749 - 2-Butene, 1,3-dichloro-, (Z)-
3750 - 2-Butene, 1,4-dichloro-, (E)-
3751 - 2-Butene, 1,4-dichloro-, (Z)-

3752 - 2-Butene, 2,3-dichloro-, (E)-
3753 - 2-Butene, 2,3-dichloro-, (Z)-
10537 - 1-Propene, 3-chloro-2-(chloromethyl)-
10555 - 1-Propene, 1,1-dichloro-2-methyl-
10556 - 1-Propene, 3,3-dichloro-2-methyl-

$C_4H_6Cl_2O$
3182 - Butanal, 2,3-dichloro-
3654 - 2-Butanone, 1,3-dichloro-
3655 - 2-Butanone, 3,4-dichloro-
3683 - Butanoyl chloride, 2-chloro-
3684 - Butanoyl chloride, 3-chloro-
3685 - Butanoyl chloride, 4-chloro-
5899 - Ethene, 1,2-dichloro-1-ethoxy-
10498 - Propanoyl chloride, 2-chloro-2-methyl-
10499 - Propanoyl chloride, 3-chloro-2-methyl-

$C_4H_6Cl_2O_2$
126 - Acetic acid, chloro-, 2-chloroethyl ester
164 - Acetic acid, dichloro-, ethyl ester
3500 - Butanoic acid, 2,2-dichloro-
3926 - Butyric acid, 3,4-dichloro-
4015 - Carbonochloridic acid, 3-chloropropyl ester
5188 - 1,4-Dioxane, 2,3-dichloro-
10248 - Propanoic acid, 2,3-dichloro-, methyl ester

$C_4H_6Cl_2O_3$
165 - Acetic acid, dichloro-, 2-hydroxyethyl ester
10247 - Propanoic acid, 3,3-dichloro-2-hydroxy-2-methyl-

$C_4H_6Cl_3NO$
97 - Acetamide, 2,2,2-trichloro-N,N-dimethyl-

$C_4H_6Cl_4$
3427 - Butane, 1,1,1,2-tetrachloro-
3428 - Butane, 1,2,2,3-tetrachloro-
3429 - Butane, 1,2,3,3-tetrachloro-
10147 - Propane, 1,1,1,2-tetrachloro-2-methyl-
10148 - Propane, 1,1,2,3-tetrachloro-2-methyl-
10176 - Propane, 1,2,3-trichloro-2-(chloromethyl)-

$C_4H_6F_2O_2$
176 - Acetic acid, difluoro-, ethyl ester

$C_4H_6F_3NO$
99 - Acetamide, 2,2,2-trifluoro-N,N-dimethyl-

$C_4H_6MnN_2S_4$
7420 - Maneb

$C_4H_6N_2$
7208 - 1H-Imidazole, 1-methyl-
7209 - 1H-Imidazole, 2-methyl-
7210 - 1H-Imidazole, 4-methyl-
10834 - 1H-Pyrazole, 1-methyl-
10835 - 1H-Pyrazole, 3-methyl-
10836 - 1H-Pyrazole, 4-methyl-

$C_4H_6N_2O$
6176 - Furazan, dimethyl-

$C_4H_6N_2O_2$
158 - Acetic acid, diazo-, ethyl ester
9612 - 2,5-Piperazinedione
10237 - Propanoic acid, 2-diazo-, methyl ester

$C_4H_6N_2S$
7214 - 2H-Imidazole-2-thione, 1,3-dihydro-1-methyl-
11659 - 1,3,4-Thiadiazole, 2,5-dimethyl-
11665 - 2-Thiazolamine, 4-methyl-

$C_4H_6N_2S_4Zn$
12067 - Zinc, [[1,2-ethanediylbis [carbamodithioato]](2-)]-

$C_4H_6N_4O_2$
7191 - Imidazo[4,5-d]imidazole-2,5(1H,3H)-dione, tetrahydro-

$C_4H_6N_4O_3$
12009 - Urea, (2,5-dioxo-4-imidazolidinyl)-

$C_4H_6N_4O_3S_2$
32 - Acetamide, N-[5-(aminosulfonyl)-1,3,4-thiadiazol-2-yl]-

$C_4H_6N_4O_{12}$
3432 - 1,2,3,4-Butanetetrol, tetranitrate, (R*,S*)-

C_4H_6O
3175 - 2,3-Butadien-1-ol
3695 - 2-Butenal, (E)-
3875 - 3-Buten-2-one
3919 - 2-Butyn-1-ol

5609 - Ethane, 1,1'-oxybis[2-bromo-
9888 - Propane, 1,2-dibromo-3-methoxy-

C₄H₈ClF

3250 - Butane, 1-chloro-4-fluoro-

C₄H₈Cl₂

3276 - Butane, 1,1-dichloro-
3277 - Butane, 1,2-dichloro-
3278 - Butane, 1,3-dichloro-
3279 - Butane, 1,4-dichloro-
3280 - Butane, 2,2-dichloro-
3281 - Butane, 2,3-dichloro-, (R*,R*)-(±)-
9899 - Propane, 1,1-dichloro-2-methyl-
9900 - Propane, 1,2-dichloro-2-methyl-
9901 - Propane, 1,3-dichloro-2-methyl-

C₄H₈Cl₂O

5468 - Ethane, 1-chloro-1-(2-chloroethoxy)-
5533 - Ethane, 1,2-dichloro-1-ethoxy-
5610 - Ethane, 1,1'-oxybis[1-chloro-
5611 - Ethane, 1,1'-oxybis[2-chloro-
10364 - 2-Propanol, 1,1-dichloro-2-methyl-
10365 - 2-Propanol, 1,3-dichloro-2-methyl-

C₄H₈Cl₂O₂

3349 - 2,3-Butanediol, 1,4-dichloro-

C₄H₈Cl₂S

5645 - Ethane, 1,1'-thiobis[2-chloro-

C₄H₈Cl₂Si

11402 - Silane, dichloromethyl-2-propenyl-

C₄H₈Cl₃O₄P

11853 - Trichlorfon

C₄H₈I₂

3288 - Butane, 1,4-diiodo-

C₄H₈N₂

281 - Acetonitrile, (dimethylamino)-
3903 - 2-Butyne-1,4-diamine
7197 - 1H-Imidazole, 4,5-dihydro-2-methyl-
10112 - Propanenitrile, 3-(methylamino)-

C₄H₈N₂O

11170 - Pyrrolidine, 1-nitroso-
12028 - Urea, 2-propenyl-

C₄H₈N₂O₂

73 - Acetamide, N-[(methylamino)carbonyl]-
3257 - Butanediamide
3360 - 2,3-Butanedione, dioxime
7699 - Morpholine, 4-nitroso-

C₄H₈N₂O₃

468 - L-Asparagine
3932 - Carbamic acid, (aminocarbonyl)-, ethyl ester
6265 - Glycine, N-glycyl-

C₄H₈N₂O₄

3293 - Butane, 1,4-dinitro-

C₄H₈N₂S

11809 - Thiourea, 2-propenyl-

C₄H₈O

3178 - Butanal
3645 - 2-Butanone
3858 - 3-Buten-2-ol, (±)-
3859 - 2-Buten-1-ol, (E)-
3860 - 2-Buten-1-ol, (Z)-
3861 - 3-Buten-1-ol
4196 - Cyclobutanol
4980 - Cyclopropane, methoxy-
5905 - Ethene, ethoxy-
6168 - Furan, tetrahydro-
8594 - Oxetane, 2-methyl-
8607 - Oxirane, 2,2-dimethyl-
8608 - Oxirane, 2,3-dimethyl-, cis-
8609 - Oxirane, 2,3-dimethyl-, trans-
9785 - Propanal, 2-methyl-
10576 - 1-Propene, 3-methoxy-
10702 - 2-Propen-1-ol, 2-methyl-

C₄H₈OS

5647 - Ethanethioic acid, S-ethyl ester
8569 - 1,4-Oxathiane
9786 - Propanal, 3-(methylthio)-

C₄H₈O₂

14 - Acetaldehyde, ethoxy-
186 - Acetic acid ethyl ester

3184 - Butanal, 3-hydroxy-
3460 - Butanoic acid
3660 - 2-Butanone, 1-hydroxy-
3661 - 2-Butanone, 3-hydroxy-
3662 - 2-Butanone, 4-hydroxy-
3792 - 2-Butene-1,4-diol, (E)-
3793 - 2-Butene-1,4-diol, (Z)-
3794 - 2-Butene-1,2-diol
5186 - 1,3-Dioxane
5187 - 1,4-Dioxane
5210 - 1,3-Dioxolane, 2-methyl-
5716 - Ethanol, 2-(ethenyloxy)-
6011 - Formic acid, 1-methylethyl ester
6018 - Formic acid, propyl ester
6137 - 3-Furanol, tetrahydro-
8615 - Oxirane, ethyl-
8620 - Oxirane, (methoxymethyl)-
10290 - Propanoic acid, 2-methyl-
10296 - Propanoic acid, methyl ester
10455 - 2-Propanone, 1-methoxy-

C₄H₈O₂S

188 - Acetic acid, (ethylthio)-
203 - Acetic acid, mercapto-, ethyl ester
3537 - Butanoic acid, 3-mercapto-
10287 - Propanoic acid, 3-mercapto-, methyl ester
10288 - Propanoic acid, 2-mercapto-2-methyl-
11786 - Thiophene, tetrahydro-, 1,1-dioxide

C₄H₈O₃

180 - Acetic acid, ethoxy-
197 - Acetic acid, hydroxy-, ethyl ester
207 - Acetic acid, methoxy-, methyl ester
3530 - Butanoic acid, 2-hydroxy-, (±)-
3531 - Butanoic acid, 3-hydroxy-, (±)-
4003 - Carbonic acid, ethyl methyl ester
5577 - 1,2-Ethanediol, monoacetate
10273 - Propanoic acid, 2-hydroxy-, methyl ester, (±)-
10276 - Propanoic acid, 3-hydroxy-, methyl ester
10278 - Propanoic acid, 2-hydroxy-2-methyl-

C₄H₈O₃S

5198 - 1,3,2-Dioxathiane, 4-methyl-, 2-oxide

C₄H₈O₄

3192 - Butanal, 2,3,4-trihydroxy-, [R-(R*,R*)]-
3193 - Butanal, 2,3,4-trihydroxy-, [S-(R*,S*)]-
3679 - 2-Butanone, 1,3,4-trihydroxy-, (S)-
10254 - Propanoic acid, 2,3-dihydroxy-, methyl ester, (±)-

C₄H₈S

10581 - 1-Propene, 3-(methylthio)-
11693 - Thietane, 2-methyl-
11783 - Thiophene, tetrahydro-

C₄H₈S₂

5261 - 1,2-Dithiane
5262 - 1,3-Dithiane
5263 - 1,4-Dithiane

C₄H₉Br

3231 - Butane, 1-bromo-
3232 - Butane, 2-bromo-, (±)-
9845 - Propane, 1-bromo-2-methyl-
9846 - Propane, 2-bromo-2-methyl-

C₄H₉BrO

3600 - 1-Butanol, 3-bromo-
3601 - 2-Butanol, 3-bromo-, (R*,S*)-(±)-
5461 - Ethane, 1-bromo-2-ethoxy-

C₄H₉BrO₂

5460 - Ethane, 2-bromo-1,1-dimethoxy-

C₄H₉Cl

3241 - Butane, 1-chloro-
3242 - Butane, 2-chloro-, (±)-
9862 - Propane, 1-chloro-2-methyl-
9863 - Propane, 2-chloro-2-methyl-

C₄H₉ClO

1605 - Benzeneethanol, α-ethyl-α-methyl-, (±)-
3602 - 1-Butanol, 2-chloro-
3604 - 1-Butanol, 4-chloro-
3605 - 2-Butanol, 1-chloro-
3606 - 2-Butanol, 3-chloro-
3607 - 2-Butanol, 4-chloro-

5473 - Ethane, 1-chloro-1-ethoxy-
5474 - Ethane, 1-chloro-2-ethoxy-
7183 - Hypochlorous acid, 1,1-dimethylethyl ester
9858 - Propane, 1-(chloromethoxy)-
9859 - Propane, 1-chloro-2-methoxy-
9860 - Propane, 1-chloro-3-methoxy-
9861 - Propane, 2-chloro-1-methoxy-
10352 - 1-Propanol, 2-chloro-2-methyl-
10353 - 1-Propanol, 3-chloro-2-methyl-
10354 - 2-Propanol, 1-chloro-2-methyl-

C₄H₉ClO₂

5472 - Ethane, 2-chloro-1,1-dimethoxy-
5698 - Ethanol, 2-(2-chloroethoxy)-
9998 - 1,2-Propanediol, 3-chloro-2-methyl-

C₄H₉ClO₂S

10134 - 1-Propanesulfonyl chloride, 2-methyl-

C₄H₉ClS

9864 - Propane, 1-chloro-3-(methylthio)-
9865 - Propane, 2-chloro-1-(methylthio)-

C₄H₉ClSi

11376 - Silane, chloroethenyldimethyl-

C₄H₉Cl₃Si

11370 - Silane, butyltrichloro-
11469 - Silane, trichloro(2-methylpropyl)-

C₄H₉F

3378 - Butane, 1-fluoro-
3379 - Butane, 2-fluoro-

C₄H₉FO

3621 - 1-Butanol, 4-fluoro-

C₄H₉I

3381 - Butane, 1-iodo-
3382 - Butane, 2-iodo-, (±)-
10057 - Propane, 1-iodo-2-methyl-
10058 - Propane, 2-iodo-2-methyl-

C₄H₉N

4169 - Cyclobutanamine
10524 - 2-Propen-1-amine, N-methyl-
11143 - Pyrrolidine

C₄H₉NO

55 - Acetamide, N,N-dimethyl-
61 - Acetamide, N-ethyl-
499 - 1-Aziridineethanol
3188 - Butanal, oxime
3194 - Butanamide
3675 - 2-Butanone, oxime
7681 - Morpholine
9798 - Propanamide, 2-methyl-
9799 - Propanamide, N-methyl-

C₄H₉NO₂

65 - Acetamide, N-(2-hydroxyethyl)-
331 - Alanine, 2-methyl-
3412 - Butane, 1-nitro-
3413 - Butane, 2-nitro-, (±)-
3463 - Butanoic acid, 2-amino-, (±)-
3464 - Butanoic acid, 4-amino-
3954 - Carbamic acid, methyl-, ethyl ester
3955 - Carbamic acid, 1-methylethyl ester
3969 - Carbamic acid, propyl ester
6261 - Glycine, N,N-dimethyl-
6263 - Glycine, ethyl ester
8083 - Nitrous acid, butyl ester
8085 - Nitrous acid, 1,1-dimethylethyl ester
8093 - Nitrous acid, 1-methylpropyl ester
8094 - Nitrous acid, 2-methylpropyl ester
10084 - Propane, 2-methyl-1-nitro-
10085 - Propane, 2-methyl-2-nitro-

C₄H₉NO₃

3465 - Butanoic acid, 3-amino-4-hydroxy-
3640 - 1-Butanol, 2-nitro-
3641 - 1-Butanol, 4-nitro-
3642 - 2-Butanol, 3-nitro-
7134 - Homoserine
8072 - Nitric acid, butyl ester
8077 - Nitric acid, 1-methylpropyl ester
8078 - Nitric acid, 2-methylpropyl ester
10391 - 1-Propanol, 2-methyl-2-nitro-
11813 - L-Threonine

C$_4$H$_9$NO$_4$
  10018 - 1,3-Propanediol, 2-methyl-2-nitro-
C$_4$H$_9$NO$_5$
  10012 - 1,3-Propanediol, 2-(hydroxymethyl)-2-nitro-
C$_4$H$_9$NS
  11719 - Thiomorpholine
C$_4$H$_9$NS$_2$
  5264 - 4H-1,3,5-Dithiazine, dihydro-5-methyl-
C$_4$H$_9$N$_3$O$_2$
  6254 - Glycine, N-(aminoiminomethyl)-N-methyl-
C$_4$H$_{10}$
  3229 - Butane
  10073 - Propane, 2-methyl-
C$_4$H$_{10}$BF$_3$O
  3137 - Boron, trifluoro[1,1'-oxybis[ethane]]-, (T-4)-
C$_4$H$_{10}$ClNO$_2$S
  5019 - L-Cysteine, methyl ester, hydrochloride
C$_4$H$_{10}$ClOP
  9529 - Phosphinic chloride, diethyl-
C$_4$H$_{10}$ClO$_2$PS
  9576 - Phosphorochloridothioic acid, O,O-diethyl
        ester
C$_4$H$_{10}$ClO$_3$P
  9574 - Phosphorochloridic acid, diethyl ester
C$_4$H$_{10}$Cl$_2$O$_2$Si
  11392 - Silane, dichlorodiethoxy-
C$_4$H$_{10}$Cl$_2$Si
  11393 - Silane, dichlorodiethyl-
  11399 - Silane, dichloromethyl(1-methylethyl)-
C$_4$H$_{10}$F$_2$Si
  11410 - Silane, diethyldifluoro-
C$_4$H$_{10}$Hg
  7434 - Mercury, diethyl-
C$_4$H$_{10}$NO$_3$PS
  10 - Acephate
C$_4$H$_{10}$N$_2$
  9602 - Piperazine
C$_4$H$_{10}$N$_2$O
  28 - Acetamide, N-(2-aminoethyl)-
  3529 - Butanoic acid, hydrazide
  5432 - Ethanamine, N-ethyl-N-nitroso-
  7680 - 4-Morpholinamine
  12032 - Urea, trimethyl-
C$_4$H$_{10}$N$_2$O$_2$
  5431 - Ethanamine, N-ethyl-N-nitro-
C$_4$H$_{10}$N$_2$O$_4$
  469 - DL-Asparagine, monohydrate
C$_4$H$_{10}$O
  3593 - 1-Butanol
  3594 - 2-Butanol
  5608 - Ethane, 1,1'-oxybis-
  10069 - Propane, 1-methoxy-
  10070 - Propane, 2-methoxy-
  10388 - 1-Propanol, 2-methyl-
  10389 - 2-Propanol, 2-methyl-
C$_4$H$_{10}$OS
  5621 - Ethane, 1,1'-sulfinylbis-
  5726 - Ethanol, 2-(ethylthio)-
C$_4$H$_{10}$O$_2$
  3344 - 1,2-Butanediol, (±)-
  3345 - 1,3-Butanediol
  3346 - 1,4-Butanediol
  3347 - 2,3-Butanediol, (R*,R*)-(±)-
  5548 - Ethane, 1,1-dimethoxy-
  5549 - Ethane, 1,2-dimethoxy-
  5717 - Ethanol, 2-ethoxy-
  7178 - Hydroperoxide, 1,1-dimethylethyl-
  9158 - Peroxide, diethyl
  10017 - 1,2-Propanediol, 2-methyl-
  10386 - 1-Propanol, 2-methoxy-
  10387 - 2-Propanol, 1-methoxy-
C$_4$H$_{10}$O$_2$S
  5626 - Ethane, 1,1'-sulfonylbis-
  5766 - Ethanol, 2,2'-thiobis-
  11571 - Sulfoxylic acid, diethyl ester
C$_4$H$_{10}$O$_3$
  3458 - 1,2,3-Butanetriol

  3459 - 1,2,4-Butanetriol
  5753 - Ethanol, 2,2'-oxybis-
  7561 - Methane, trimethoxy-
  10014 - 1,2-Propanediol, 3-methoxy-
  10015 - 1,3-Propanediol, 2-methoxy-
C$_4$H$_{10}$O$_3$S
  5725 - Ethanol, 2-(ethylsulfonyl)-
  11581 - Sulfurous acid, diethyl ester
C$_4$H$_{10}$O$_4$
  3431 - 1,2,3,4-Butanetetrol, (R*,S*)-
C$_4$H$_{10}$O$_4$S
  11573 - Sulfuric acid, diethyl ester
C$_4$H$_{10}$S
  3437 - 1-Butanethiol
  3438 - 2-Butanethiol, (±)-
  5644 - Ethane, 1,1'-thiobis-
  10088 - Propane, 1-(methylthio)-
  10089 - Propane, 2-(methylthio)-
  10158 - 1-Propanethiol, 2-methyl-
  10159 - 2-Propanethiol, 2-methyl-
C$_4$H$_{10}$S$_2$
  3368 - 1,4-Butanedithiol
  5253 - Disulfide, diethyl
  5452 - Ethane, 1,2-bis(methylthio)-
C$_4$H$_{10}$S$_3$
  5652 - Ethanethiol, 2,2'-thiobis-
  11916 - Trisulfide, diethyl
C$_4$H$_{10}$Se
  5620 - Ethane, 1,1'-selenobis-
C$_4$H$_{10}$Te
  5630 - Ethane, 1,1'-tellurobis-
C$_4$H$_{11}$As
  441 - Arsine, diethyl-
C$_4$H$_{11}$BrSi
  11366 - Silane, (bromomethyl)trimethyl-
C$_4$H$_{11}$ClSi
  11377 - Silane, chloroethyldimethyl-
  11382 - Silane, (chloromethyl)trimethyl-
C$_4$H$_{11}$N
  3209 - 1-Butanamine
  3210 - 2-Butanamine, (±)-
  5421 - Ethanamine, N,N-dimethyl-
  5427 - Ethanamine, N-ethyl-
  9817 - 1-Propanamine, 2-methyl-
  9818 - 2-Propanamine, 2-methyl-
  9819 - 2-Propanamine, N-methyl-
C$_4$H$_{11}$NO
  3595 - 1-Butanol, 2-amino-, (±)-
  3596 - 1-Butanol, 3-amino-
  3597 - 1-Butanol, 4-amino-
  3598 - 2-Butanol, 3-amino-
  5423 - Ethanamine, 2-ethoxy-
  5429 - Ethanamine, N-ethyl-N-hydroxy-
  5709 - Ethanol, 2-(dimethylamino)-
  5722 - Ethanol, 2-(ethylamino)-
  9816 - 1-Propanamine, 3-methoxy-
  10341 - 1-Propanol, 2-amino-2-methyl-
C$_4$H$_{11}$NO$_2$
  5419 - Ethanamine, 2,2-dimethoxy-
  5730 - Ethanol, 2,2'-iminobis-
  9987 - 1,3-Propanediol, 2-amino-2-methyl-
C$_4$H$_{11}$NO$_3$
  9986 - 1,3-Propanediol, 2-amino-2-(hydroxymethyl)-
C$_4$H$_{11}$NO$_8$P$_2$
  6277 - Glyphosine
C$_4$H$_{11}$O$_2$P
  9525 - Phosphinic acid, diethyl-
C$_4$H$_{11}$O$_3$P
  9533 - Phosphonic acid, (1,1-dimethylethyl)-
  9537 - Phosphonic acid, ethyl-, dimethyl ester
  9544 - Phosphonic acid, (2-methylpropyl)-
  9587 - Phosphorous acid, diethyl ester
C$_4$H$_{11}$O$_4$P
  9558 - Phosphoric acid, diethyl ester
C$_4$H$_{12}$As$_2$
  5110 - Diarsine, tetramethyl-

C$_4$H$_{12}$As$_2$O
  452 - Arsinous acid, dimethyl-, anhydride
C$_4$H$_{12}$BFN$_2$
  3114 - Boranediamine, 1-fluoro-N,N',N'-
        tetramethyl-
C$_4$H$_{12}$BN
  3113 - Boranamine, N,N,1,1-tetramethyl-
C$_4$H$_{12}$BrN
  7456 - Methanaminium, N,N,N-trimethyl-, bromide
C$_4$H$_{12}$ClN
  5428 - Ethanamine, N-ethyl-, hydrochloride
  7457 - Methanaminium, N,N,N-trimethyl-, chloride
C$_4$H$_{12}$Cl$_2$N$_2$
  9607 - Piperazine, dihydrochloride
C$_4$H$_{12}$Cl$_2$OSi$_2$
  5232 - Disiloxane, 1,3-dichloro-1,1,3,3-tetramethyl-
C$_4$H$_{12}$Cl$_2$Si$_2$
  5226 - Disilane, 1,2-dichloro-1,1,2,2-tetramethyl-
C$_4$H$_{12}$F$_2$Si$_2$
  5227 - Disilane, 1,2-difluoro-1,1,2,2-tetramethyl-
C$_4$H$_{12}$IN
  7458 - Methanaminium, N,N,N-trimethyl-, iodide
C$_4$H$_{12}$N$_2$
  3258 - 1,4-Butanediamine
  3261 - 2,3-Butanediamine, (R*,R*)-(±)-
  5503 - 1,2-Ethanediamine, N,N'-dimethyl-
  5509 - 1,2-Ethanediamine, N-ethyl-
  7152 - Hydrazine, 1,1-diethyl-
  7153 - Hydrazine, 1,2-diethyl-
  9878 - 1,2-Propanediamine, 2-methyl-
C$_4$H$_{12}$N$_2$O
  5676 - Ethanol, 2-[(2-aminoethyl)amino]-
C$_4$H$_{12}$N$_2$OS
  11583 - Sulfurous diamide, tetramethyl-
C$_4$H$_{12}$N$_2$O$_2$S
  11569 - Sulfamide, tetramethyl-
C$_4$H$_{12}$OSi
  11501 - Silanol, ethyldimethyl-
C$_4$H$_{12}$O$_2$Si
  11412 - Silane, dimethoxydimethyl-
C$_4$H$_{12}$O$_3$Si
  11487 - Silane, trimethoxymethyl-
C$_4$H$_{12}$O$_4$
  19 - Acetaldehyde, tetramer
C$_4$H$_{12}$O$_4$Si
  11508 - Silicic acid (H$_4$SiO$_4$), tetramethyl ester
C$_4$H$_{12}$Pb
  9732 - Plumbane, tetramethyl
C$_4$H$_{12}$Si
  11409 - Silane, diethyl-
  11451 - Silane, tetramethyl-
C$_4$H$_{12}$Sn
  11544 - Stannane, tetramethyl-
C$_4$H$_{13}$N$_3$
  5497 - 1,2-Ethanediamine, N-(2-aminoethyl)-
C$_4$H$_{14}$Cl$_2$N$_2$
  3260 - 1,4-Butanediamine, dihydrochloride
C$_4$H$_{14}$OSi$_2$
  5237 - Disiloxane, 1,1,3,3-tetramethyl-
C$_4$H$_{16}$O$_4$Si$_4$
  4999 - Cyclotetrasiloxane, 2,4,6,8-tetramethyl-
C$_4$N$_2$
  3908 - 2-Butynedinitrile
C$_4$NiO$_4$
  8068 - Nickel carbonyl
C$_5$ClF$_2$F$_6$
  4917 - Cyclopentene, 1,2-dichloro-3,3,4,4,5,5-
        hexafluoro-
C$_5$Cl$_5$N
  10995 - Pyridine, pentachloro-
C$_5$Cl$_6$
  4766 - 1,3-Cyclopentadiene, 1,2,3,4,5,5-hexachloro-
C$_5$Cl$_8$
  4932 - Cyclopentene, octachloro-
C$_5$F$_8$
  3172 - 1,3-Butadiene, 1,1,2,4,4-pentafluoro-3-
        (trifluoromethyl)-

C5F10
9027 - 1-Pentene, 1,1,2,3,3,4,4,5,5,5-decafluoro-
C5F12
8807 - Pentane, dodecafluoro-
C5FeO5
7339 - Iron carbonyl
C5HCl4N
11019 - Pyridine, 2,3,5,6-tetrachloro-
C5H2Cl6O3
5212 - 1,3-Dioxolan-4-one, 2,5-bis(trichloromethyl)-
C5H2F6O2
8802 - 2,4-Pentanedione, 1,1,1,5,5,5-hexafluoro-
C5H2O5
4939 - 4-Cyclopentene-1,2,3-trione, 4,5-dihydroxy-
C5H3BrO2
6043 - 2-Furancarboxaldehyde, 5-bromo-
C5H3Br2N
10926 - Pyridine, 3,5-dibromo-
C5H3ClN4
10753 - 1H-Purine, 6-chloro-
C5H3ClO2
6040 - 2-Furancarbonyl chloride
C5H3F7O2
3525 - Butanoic acid, heptafluoro-, methyl ester
C5H3NO
6039 - 2-Furancarbonitrile
C5H3NO4
6046 - 2-Furancarboxaldehyde, 5-nitro-
C5H3NO5
6066 - 2-Furancarboxylic acid, 5-nitro-
C5H3NS
11732 - 2-Thiophenecarbonitrile
C5H4
8684 - 1,3-Pentadiyne
C5H4BrN
10875 - Pyridine, 2-bromo-
10876 - Pyridine, 3-bromo-
10877 - Pyridine, 4-bromo-
C5H4ClN
10915 - Pyridine, 2-chloro-
10916 - Pyridine, 3-chloro-
10917 - Pyridine, 4-chloro-
C5H4ClOS
11735 - 2-Thiophenecarboxaldehyde, 5-chloro-
C5H4Cl2N2
10862 - 3-Pyridinamine, 2,6-dichloro-
11052 - Pyrimidine, 2,4-dichloro-5-methyl-
11053 - Pyrimidine, 2,4-dichloro-6-methyl-
C5H4Cl2O2
3367 - Butanedioyl dichloride, methylene-
C5H4F8O
8953 - 1-Pentanol, 2,2,3,3,4,4,5,5-octafluoro-
C5H4IN
10970 - Pyridine, 2-iodo-
C5H4N2O2
10803 - Pyrazinecarboxylic acid
C5H4N2O3
10993 - Pyridine, 4-nitro-, 1-oxide
C5H4N2O4
7195 - 1H-Imidazole-4,5-dicarboxylic acid
11050 - 4-Pyrimidinecarboxylic acid, 1,2,3,6-tetrahydro-2,6-dioxo-
C5H4N4
10752 - 1H-Purine
C5H4N4O
10770 - 6H-Purin-6-one, 1,7-dihydro-
10852 - 4H-Pyrazolo[3,4-d]pyrimidin-4-one, 1,5-dihydro-
C5H4N4O2
10755 - 1H-Purine-2,6-dione, 3,7-dihydro-
C5H4N4O3
10764 - 1H-Purine-2,6,8(3H)-trione, 7,9-dihydro-
C5H4N4S
10763 - 6H-Purine-6-thione, 1,7-dihydro-
C5H4OS
11733 - 2-Thiophenecarboxaldehyde
11734 - 3-Thiophenecarboxaldehyde

C5H4O2
6041 - 2-Furancarboxaldehyde
6042 - 3-Furancarboxaldehyde
6155 - 2(5H)-Furanone, 5-methylene-
10788 - 2H-Pyran-2-one
10789 - 4H-Pyran-4-one
C5H4O2S
11737 - 2-Thiophenecarboxylic acid
11738 - 3-Thiophenecarboxylic acid
C5H4O3
6050 - 2-Furancarboxylic acid
6051 - 3-Furancarboxylic acid
6095 - 2,5-Furandione, dihydro-3-methylene-
6098 - 2,5-Furandione, 3-methyl-
10794 - 2H-Pyran-2-one, 3-hydroxy-
C5H4O4
6054 - 3-Furancarboxylic acid, 4,5-dihydro-5-oxo-
C5H5BrO
6035 - Furan, 2-(bromomethyl)-
C5H5BrS
11729 - Thiophene, 2-bromo-3-methyl-
11730 - Thiophene, 2-bromo-5-methyl-
C5H5ClN2
10861 - 2-Pyridinamine, 5-chloro-
C5H5ClO
6075 - Furan, 2-(chloromethyl)-
C5H5ClS
11741 - Thiophene, 2-chloro-5-methyl-
C5H5Cl3N2OS
11599 - Terrazole
C5H5N
8673 - 2,4-Pentadienenitrile
8674 - 2,4-Pentadienenitrile, (E)-
8675 - 2,4-Pentadienenitrile, (Z)-
10872 - Pyridine
C5H5NO
10994 - Pyridine, 1-oxide
11031 - 2-Pyridinol
11032 - 4-Pyridinol
11033 - 2(1H)-Pyridinone
11114 - 1H-Pyrrole-2-carboxaldehyde
C5H5NO2
6047 - 2-Furancarboxaldehyde, oxime, (E)-
6048 - 2-Furancarboxaldehyde, oxime, (Z)-
10624 - 2-Propenoic acid, 2-cyano-, methyl ester
C5H5NO2S
11669 - 5-Thiazolecarboxylic acid, 4-methyl-
C5H5NO3S
11018 - 3-Pyridinesulfonic acid
C5H5N3O
10802 - Pyrazinecarboxamide
C5H5N3O3S
85 - Acetamide, N-(5-nitro-2-thiazolyl)-
C5H5N5
10751 - 1H-Purin-6-amine
C5H5N5O
10767 - 2H-Purin-2-one, 6-amino-1,3-dihydro-
10768 - 6H-Purin-6-one, 2-amino-1,7-dihydro-
C5H5N5S
10762 - 6H-Purine-6-thione, 2-amino-1,7-dihydro-
C5H6
3892 - 1-Buten-3-yne, 2-methyl-
4764 - 1,3-Cyclopentadiene
9111 - 1-Penten-3-yne
9112 - 1-Penten-4-yne
9113 - 3-Penten-1-yne, (E)-
9114 - 3-Penten-1-yne, (Z)-
C5H6Br2N2O2
7217 - 2,4-Imidazolidinedione, 1,3-dibromo-5,5-dimethyl-
C5H6ClN
10969 - Pyridine, hydrochloride
C5H6ClNO2
11155 - 2,5-Pyrrolidinedione, 1-(chloromethyl)-
C5H6Cl2O2
8806 - Pentanedioyl dichloride

C5H6F4O2
10328 - Propanoic acid, 2,3,3,3-tetrafluoro-, ethyl ester
C5H6N2
8767 - Pentanedinitrile
9913 - Propanedinitrile, dimethyl-
10810 - Pyrazine, methyl-
10856 - Pyridazine, 3-methyl-
10857 - 2-Pyridinamine
10858 - 3-Pyridinamine
10859 - 4-Pyridinamine
11070 - Pyrimidine, 2-methyl-
11071 - Pyrimidine, 4-methyl-
11072 - Pyrimidine, 5-methyl-
C5H6N2OS
11108 - 4(1H)-Pyrimidinone, 2,3-dihydro-6-methyl-2-thioxo-
C5H6N2O2
11068 - 2,4(1H,3H)-Pyrimidinedione, 5-methyl-
11069 - 2,4(1H,3H)-Pyrimidinedione, 6-methyl-
C5H6N6
10754 - 1H-Purine-2,6-diamine
C5H6O
3891 - 1-Buten-3-yne, 1-methoxy-
4941 - 2-Cyclopenten-1-one
6130 - Furan, 2-methyl-
6131 - Furan, 3-methyl-
9118 - 4-Penten-2-yn-1-ol
10771 - Pyran
C5H6OS
6117 - 2-Furanmethanethiol
11768 - 2-Thiophenemethanol
C5H6O2
6118 - 2-Furanmethanol
6127 - Furan, 2-methoxy-
6147 - 2(3H)-Furanone, dihydro-3-methylene-
6148 - 2(3H)-Furanone, dihydro-5-methylene-
6153 - 2(3H)-Furanone, 5-methyl-
6154 - 2(5H)-Furanone, 5-methyl-
8676 - 2,4-Pentadienoic acid
9137 - 2-Pentynoic acid
9138 - 4-Pentynoic acid
10739 - 2-Propynoic acid, ethyl ester
10744 - 2-Propyn-1-ol, acetate
C5H6O3
4942 - 2-Cyclopenten-1-one, 2,3-dihydroxy-
6094 - 2,5-Furandione, dihydro-3-methyl-
10780 - 2H-Pyran-2,6(3H)-dione, dihydro-
C5H6O4
3328 - Butanedioic acid, methylene-
3785 - 2-Butenedioic acid, 2-methyl-, (E)-
3786 - 2-Butenedioic acid, 2-methyl-, (Z)-
4970 - 1,2-Cyclopropanedicarboxylic acid, trans-(±)-
8875 - Pentanoic acid, 2,4-dioxo-
C5H6O5
8781 - Pentanedioic acid, 2-oxo-
8782 - Pentanedioic acid, 3-oxo-
C5H6S
11769 - Thiophene, 2-methyl-
11770 - Thiophene, 3-methyl-
C5H7Br
9124 - 1-Pentyne, 1-bromo-
C5H7BrN2
10816 - 1H-Pyrazole, 4-bromo-1,3-dimethyl-
C5H7BrO2
10611 - 2-Propenoic acid, 2-bromoethyl ester
C5H7Cl
3153 - 1,3-Butadiene, 1-chloro-2-methyl-
3154 - 1,3-Butadiene, 1-chloro-3-methyl-
3155 - 1,3-Butadiene, 2-chloro-3-methyl-
3901 - 1-Butyne, 3-chloro-3-methyl-
4914 - Cyclopentene, 3-chloro-
8663 - 1,3-Pentadiene, 3-chloro-
C5H7ClN2
10818 - 1H-Pyrazole, 3-chloro-1,5-dimethyl-
10819 - 1H-Pyrazole, 4-chloro-3,5-dimethyl-
10820 - 1H-Pyrazole, 5-chloro-1,3-dimethyl-

C₅H₇ClO
3888 - 2-Butenoyl chloride, 3-methyl-
4880 - Cyclopentanone, 2-chloro-

C₅H₇ClO₂
3830 - 2-Butenoic acid, 2-chloro-, methyl ester, (E)-
3831 - 2-Butenoic acid, 3-chloro-, methyl ester, (E)-
3832 - 2-Butenoic acid, 3-chloro-, methyl ester, (Z)-
3833 - 2-Butenoic acid, 4-chloro-, methyl ester
6141 - 2(3H)-Furanone, 5-(chloromethyl)dihydro-
10619 - 2-Propenoic acid, chloro-, ethyl ester
10620 - 2-Propenoic acid, 2-chloroethyl ester

C₅H₇ClO₃
10494 - Propanoyl chloride, 2-(acetyloxy)-, (±)-

C₅H₇Cl₂NO
43 - Acetamide, 2-chloro-N-3-chloroallyl-

C₅H₇Cl₃O₂
254 - Acetic acid, trichloro-, 1-methylethyl ester
261 - Acetic acid, trichloro-, propyl ester

C₅H₇Cl₃O₃
250 - Acetic acid, trichloro-, 2-methoxyethyl ester

C₅H₇F₃O₂
269 - Acetic acid, trifluoro-, propyl ester

C₅H₇I
9133 - 1-Pentyne, 1-iodo-

C₅H₇N
4173 - Cyclobutanecarbonitrile
9061 - 3-Pentenenitrile, (E)
9062 - 4-Pentenenitrile
11130 - 1H-Pyrrole, 1-methyl-
11131 - 1H-Pyrrole, 2-methyl-
11132 - 1H-Pyrrole, 3-methyl-

C₅H₇NO
6113 - 2-Furanmethanamine
7403 - Isoxazole, 3,5-dimethyl-
8577 - Oxazole, 2,4-dimethyl-
8578 - Oxazole, 2,5-dimethyl-
9063 - 3-Pentenenitrile, 2-hydroxy-, (Z)-

C₅H₇NO₂
152 - Acetic acid, cyano-, ethyl ester
7404 - 5-Isoxazolemethanol, 3-methyl-
10091 - Propanenitrile, 2-(acetyloxy)-
10234 - Propanoic acid, 3-cyano-, methyl ester
11160 - 2,5-Pyrrolidinedione, 1-methyl-

C₅H₇NO₃
8584 - 2,4-Oxazolidinedione, 5,5-dimethyl-
9768 - L-Proline, 5-oxo-

C₅H₇NS
11674 - Thiazole, 2,4-dimethyl-
11675 - Thiazole, 4,5-dimethyl-

C₅H₇N₃
10923 - 2,3-Pyridinediamine
10924 - 2,5-Pyridinediamine
10925 - 2,6-Pyridinediamine
11034 - 2(1H)-Pyridinone, hydrazone
11046 - 2-Pyrimidinamine, 4-methyl-
11047 - 2-Pyrimidinamine, 5-methyl-
11048 - 4-Pyrimidinamine, 6-methyl-

C₅H₇N₃O
11106 - 2(1H)-Pyrimidinone, 4-amino-5-methyl-

C₅H₇N₃O₂
7202 - 1H-Imidazole, 1,2-dimethyl-5-nitro-
11105 - 2(1H)-Pyrimidinone, 4-amino-5-
    (hydroxymethyl)-

C₅H₈
3170 - 1,2-Butadiene, 3-methyl-
3171 - 1,3-Butadiene, 2-methyl-
3916 - 1-Butyne, 3-methyl-
4193 - Cyclobutane, methylene-
4201 - Cyclobutene, 1-methyl-
4903 - Cyclopentene
8658 - 1,2-Pentadiene
8659 - 1,4-Pentadiene
8660 - 2,3-Pentadiene
8661 - 1,3-Pentadiene, (E)-
8662 - 1,3-Pentadiene, (Z)-
9122 - 1-Pentyne
9123 - 2-Pentyne

11527 - Spiropentane

C₅H₈BrN
8828 - Pentanenitrile, 5-bromo-

C₅H₈Br₂O₂
8870 - Pentanoic acid, 2,3-dibromo-
8871 - Pentanoic acid, 2,5-dibromo-
10240 - Propanoic acid, 2,3-dibromo-, ethyl ester
10242 - Propanoic acid, 2,3-dibromo-2-methyl-,
    methyl ester

C₅H₈Br₄
9884 - Propane, 1,3-dibromo-2,2-bis(bromomethyl)-

C₅H₈ClFO₂
135 - Acetic acid, chlorofluoro-, propyl ester

C₅H₈Cl₂
3729 - 1-Butene, 3-chloro-2-(chloromethyl)-
3755 - 1-Butene, 3,3-dichloro-2-methyl-
3756 - 2-Butene, 1,3-dichloro-2-methyl-
3757 - 2-Butene, 1,4-dichloro-2-methyl-
9028 - 2-Pentene, 2,5-dichloro-

C₅H₈Cl₂O
3686 - Butanoyl chloride, 2-chloro-3-methyl-
8963 - 2-Pentanone, 3,5-dichloro-
8991 - Pentanoyl chloride, 2-chloro-
8992 - Pentanoyl chloride, 5-chloro-

C₅H₈Cl₂O₂
167 - Acetic acid, dichloro-, 1-methylethyl ester
3502 - Butanoic acid, 3,4-dichloro-, methyl ester
10246 - Propanoic acid, 2,3-dichloro-, ethyl ester
10363 - 2-Propanol, 1,3-dichloro-, acetate

C₅H₈Cl₂O₃
5697 - Ethanol, 2-chloro-, carbonate (2:1)

C₅H₈Cl₄
8838 - Pentane, 1,1,1,5-tetrachloro-
9895 - Propane, 1,3-dichloro-2,2-bis(chloromethyl)-

C₅H₈NNaO₄
6248 - L-Glutamic acid, monosodium salt

C₅H₈N₂
7200 - 1H-Imidazole, 1,2-dimethyl-
7201 - 1H-Imidazole, 1,4-dimethyl-
7207 - 1H-Imidazole, 1-ethyl-
10825 - 1H-Pyrazole, 1,3-dimethyl-
10826 - 1H-Pyrazole, 1,5-dimethyl-
10827 - 1H-Pyrazole, 3,4-dimethyl-
10828 - 1H-Pyrazole, 3,5-dimethyl-
10832 - 1H-Pyrazole, 1-ethyl-

C₅H₈N₂O
7205 - 1H-Imidazole-1-ethanol

C₅H₈N₂O₂
7218 - 2,4-Imidazolidinedione, 5,5-dimethyl-
10236 - Propanoic acid, 2-diazo-, ethyl ester

C₅H₈N₂O₃
40 - Acetamide, N,N'-carbonylbis-

C₅H₈N₄O₁₂
9990 - 1,3-Propanediol, 2,2-bis[(nitrooxy)methyl]-,
    dinitrate (ester)

C₅H₈O
3167 - 1,2-Butadiene, 4-methoxy-
3168 - 1,3-Butadiene, 1-methoxy-
3169 - 1,3-Butadiene, 2-methoxy-
3698 - 2-Butenal, 2-methyl-
3699 - 2-Butenal, 2-methyl-, (E)-
3700 - 2-Butenal, 3-methyl-
3879 - 3-Buten-2-one, 3-methyl-
3915 - 3-Butyne, 1-methoxy-
3922 - 3-Butyn-2-ol, 2-methyl-
4878 - Cyclopentanone
5796 - Ethanone, 1-cyclopropyl-
8558 - 6-Oxabicyclo[3.1.0]hexane
8681 - 1,4-Pentadien-3-ol
9001 - 4-Pentenal
9104 - 1-Penten-3-one
9105 - 3-Penten-2-one, (E)-
9140 - 1-Pentyn-3-ol
9141 - 2-Pentyn-1-ol
9142 - 3-Pentyn-1-ol
9143 - 4-Pentyn-1-ol
10045 - Propane, 1-(ethynyloxy)-

10565 - 1-Propene, 3-(ethenyloxy)-
10732 - 1-Propyne, 1-ethoxy-
10733 - 1-Propyne, 3-ethoxy-
10777 - 2H-Pyran, 3,4-dihydro-

C₅H₈O₂
235 - Acetic acid, 2-propenyl ester
3844 - 2-Butenoic acid, 2-methyl-, (E)-
3845 - 2-Butenoic acid, 2-methyl-, (Z)-
3846 - 2-Butenoic acid, 3-methyl-
3847 - 2-Butenoic acid, methyl ester, (E)-
3848 - 2-Butenoic acid, methyl ester, (Z)-
4174 - Cyclobutanecarboxylic acid
4963 - Cyclopropanecarboxylic acid, methyl ester
4964 - Cyclopropanecarboxylic acid, 2-methyl-
6049 - 2-Furancarboxaldehyde, tetrahydro-
6145 - 2(3H)-Furanone, dihydro-3-methyl-
6146 - 2(3H)-Furanone, dihydro-5-methyl-, (±)-
8695 - Pentanal, 2-oxo-
8696 - Pentanal, 4-oxo-
8734 - Pentanedial
8797 - 2,3-Pentanedione
8798 - 2,4-Pentanedione
9073 - 4-Pentenoic acid
10261 - Propanoic acid, ethenyl ester
10635 - 2-Propenoic acid, ethyl ester
10663 - 2-Propenoic acid, 2-methyl-, methyl ester
10698 - 1-Propen-2-ol, acetate
10798 - 2H-Pyran-2-one, tetrahydro-
10799 - 4H-Pyran-4-one, tetrahydro-

C₅H₈O₃
3556 - Butanoic acid, 3-methyl-2-oxo-
3573 - Butanoic acid, 3-oxo-, methyl ester
6072 - 2-Furancarboxylic acid, tetrahydro-
8902 - Pentanoic acid, 2-oxo-
8903 - Pentanoic acid, 4-oxo-
10315 - Propanoic acid, 2-oxo-, ethyl ester
10412 - 2-Propanone, 1-(acetyloxy)-

C₅H₈O₄
3325 - Butanedioic acid, methyl-, (±)-
3336 - Butanedioic acid, monomethyl ester
5567 - Ethanedioic acid, ethyl methyl ester
8770 - Pentanedioic acid
9943 - Propanedioic acid, dimethyl ester
9948 - Propanedioic acid, ethyl-
10202 - Propanoic acid, 2-(acetyloxy)-, (±)-

C₅H₉Br
3716 - 1-Butene, 2-bromo-3-methyl-
3717 - 1-Butene, 1-bromo-3-methyl-
3718 - 2-Butene, 2-bromo-3-methyl-
4777 - Cyclopentane, bromo-
9007 - 1-Pentene, 1-bromo-
9008 - 1-Pentene, 2-bromo-
9009 - 1-Pentene, 3-bromo-
9010 - 1-Pentene, 5-bromo-
9011 - 2-Pentene, 1-bromo-
9012 - 2-Pentene, 2-bromo-
9013 - 2-Pentene, 3-bromo-
9014 - 2-Pentene, 4-bromo-
9015 - 2-Pentene, 5-bromo-

C₅H₉BrO
6036 - Furan, 2-(bromomethyl)tetrahydro-

C₅H₉BrO₂
120 - Acetic acid, bromo-, propyl ester
3471 - Butanoic acid, 2-bromo-3-methyl-, (±)-
3472 - Butanoic acid, 3-bromo-3-methyl-
3473 - Butanoic acid, 2-bromo-, methyl ester
3474 - Butanoic acid, 4-bromo-, methyl ester
8859 - Pentanoic acid, 2-bromo-, (±)-
10207 - Propanoic acid, 2-bromo-, ethyl ester
10208 - Propanoic acid, 3-bromo-, ethyl ester

C₅H₉Cl
3732 - 1-Butene, 1-chloro-2-methyl-
3733 - 1-Butene, 1-chloro-3-methyl-
3734 - 1-Butene, 3-chloro-2-methyl-
3735 - 1-Butene, 3-chloro-3-methyl-
3736 - 2-Butene, 1-chloro-3-methyl-
3737 - 2-Butene, 2-chloro-3-methyl-

4789 - Cyclopentane, chloro-
9016 - 1-Pentene, 2-chloro-
9017 - 1-Pentene, 3-chloro-
9018 - 1-Pentene, 4-chloro-
9019 - 1-Pentene, 5-chloro-
9020 - 2-Pentene, 1-chloro-
9021 - 2-Pentene, 2-chloro-
9022 - 2-Pentene, 3-chloro-
9023 - 2-Pentene, 4-chloro-
9024 - 2-Pentene, 5-chloro-

C_5H_9ClO
3603 - 1-Butanol, 3-chloro-
3649 - 2-Butanone, 3-chloro-3-methyl-
3693 - Butanoyl chloride, 3-methyl-
3694 - Butanoyl chloride, 2-methyl-, (±)-
8961 - 2-Pentanone, 5-chloro-
8962 - 3-Pentanone, 1-chloro-
8990 - Pentanoyl chloride
10503 - Propanoyl chloride, 2,2-dimethyl-

$C_5H_9ClO_2$
139 - Acetic acid, chloro-, 1-methylethyl ester
149 - Acetic acid, chloro-, propyl ester
3486 - Butanoic acid, 2-chloro-2-methyl-
3487 - Butanoic acid, 2-chloro-3-methyl-
3488 - Butanoic acid, 2-chloro-, methyl ester
3489 - Butanoic acid, 4-chloro-, methyl ester
4012 - Carbonochloridic acid, butyl ester
4024 - Carbonochloridic acid, 2-methylpropyl ester
8863 - Pentanoic acid, 2-chloro-, (±)-
8864 - Pentanoic acid, 3-chloro-
8865 - Pentanoic acid, 4-chloro-
8866 - Pentanoic acid, 5-chloro-
10219 - Propanoic acid, 3-chloro-2,2-dimethyl-
10220 - Propanoic acid, 2-chloro-, ethyl ester
10221 - Propanoic acid, 3-chloro-, ethyl ester
10229 - Propanoic acid, 2-chloro-2-methyl-, methyl ester
10350 - 1-Propanol, 3-chloro-, acetate
10351 - 2-Propanol, 1-chloro-, acetate

$C_5H_9ClO_3$
137 - Acetic acid, chloro-, 2-methoxyethyl ester
4017 - Carbonochloridic acid, 2-ethoxyethyl ester

$C_5H_9Cl_2N_3O_2$
11994 - Urea, N,N'-bis(2-chloroethyl)-N-nitroso-

$C_5H_9Cl_3$
3453 - Butane, 1,2,3-trichloro-2-methyl-
3454 - Butane, 2,2,3-trichloro-3-methyl-

$C_5H_9FO_2$
8879 - Pentanoic acid, 5-fluoro-

C_5H_9I
4824 - Cyclopentane, iodo-

$C_5H_9IO_2$
10283 - Propanoic acid, 3-iodo-, ethyl ester

C_5H_9N
3387 - Butane, 1-isocyano-
3409 - Butanenitrile, 2-methyl-
3410 - Butanenitrile, 3-methyl-
3896 - 3-Butyn-2-amine, 2-methyl-
8827 - Pentanenitrile
10100 - Propanenitrile, 2,2-dimethyl-
11020 - Pyridine, 1,2,3,6-tetrahydro-
11189 - 2-Pyrroline, 2-methyl-

C_5H_9NO
4892 - Cyclopentanone, oxime
8977 - 2-Pentanone, 4-imino-
9719 - 2-Piperidinone
10061 - Propane, 1-isocyanato-2-methyl-
10062 - Propane, 2-isocyanato-2-methyl-
10102 - Propanenitrile, 2-ethoxy-
10103 - Propanenitrile, 3-ethoxy-
11185 - 2-Pyrrolidinone, 1-methyl-
11186 - 2-Pyrrolidinone, 5-methyl-
11187 - 3-Pyrrolidinone, 1-methyl-

$C_5H_9NO_2$
25 - Acetamide, N-acetyl-N-methyl-
4839 - Cyclopentane, nitro-
7685 - 4-Morpholinecarboxaldehyde

8803 - 2,3-Pentanedione, 3-oxime
9764 - L-Proline
9765 - DL-Proline

$C_5H_9NO_2S_2$
10067 - Propane, 1-isothiocyanato-3-(methylsulfonyl)-

$C_5H_9NO_3$
9767 - L-Proline, 4-hydroxy-, trans-

$C_5H_9NO_3S$
5015 - L-Cysteine, N-acetyl-

$C_5H_9NO_4$
6240 - DL-Glutamic acid
6241 - L-Glutamic acid
10312 - Propanoic acid, 2-nitro-, ethyl ester

$C_5H_9NO_4S$
5017 - L-Cysteine, S-(carboxymethyl)-

$C_5H_9NO_5$
6246 - DL-Glutamic acid, 3-hydroxy-

C_5H_9NS
3388 - Butane, 1-isothiocyanato-
3389 - Butane, 2-isothiocyanato-, (±)-
10065 - Propane, 1-isothiocyanato-2-methyl-
10066 - Propane, 2-isothiocyanato-2-methyl-
11699 - Thiocyanic acid, butyl ester
11701 - Thiocyanic acid, 1,1-dimethylethyl ester
11707 - Thiocyanic acid, 2-methylpropyl ester

$C_5H_9N_3$
7203 - 1H-Imidazole-4-ethanamine

C_5H_{10}
3799 - 1-Butene, 2-methyl-
3800 - 1-Butene, 3-methyl-
3801 - 2-Butene, 2-methyl-
4192 - Cyclobutane, methyl-
4774 - Cyclopentane
4971 - Cyclopropane, 1,1-dimethyl-
4972 - Cyclopropane, 1,2-dimethyl-, cis-
4973 - Cyclopropane, 1,2-dimethyl-, trans-
4974 - Cyclopropane, ethyl-
9004 - 1-Pentene
9005 - 2-Pentene, (E)-
9006 - 2-Pentene, (Z)-

$C_5H_{10}BrF$
8713 - Pentane, 1-bromo-5-fluoro-

$C_5H_{10}Br_2$
3272 - Butane, 1,2-dibromo-2-methyl-
3273 - Butane, 1,4-dibromo-2-methyl-, (R)-
3274 - Butane, 2,3-dibromo-2-methyl-
8739 - Pentane, 1,3-dibromo-
8740 - Pentane, 1,4-dibromo-
8741 - Pentane, 1,5-dibromo-
8742 - Pentane, 2,3-dibromo-
8743 - Pentane, 2,4-dibromo-
9886 - Propane, 1,1-dibromo-2,2-dimethyl-
9887 - Propane, 1,3-dibromo-2,2-dimethyl-

$C_5H_{10}ClF$
8728 - Pentane, 1-chloro-5-fluoro-

$C_5H_{10}ClNO$
3972 - Carbamic chloride, diethyl-

$C_5H_{10}Cl_2$
3282 - Butane, 1,1-dichloro-3-methyl-
3283 - Butane, 1,2-dichloro-2-methyl-
3284 - Butane, 1,3-dichloro-3-methyl-
3285 - Butane, 1,4-dichloro-2-methyl-
3286 - Butane, 2,3-dichloro-2-methyl-
8744 - Pentane, 1,2-dichloro-
8745 - Pentane, 1,3-dichloro-
8746 - Pentane, 1,4-dichloro-
8747 - Pentane, 1,5-dichloro-
8748 - Pentane, 2,2-dichloro-
8749 - Pentane, 2,3-dichloro-
8750 - Pentane, 2,4-dichloro-
8751 - Pentane, 3,3-dichloro-

$C_5H_{10}I_2$
8758 - Pentane, 1,5-diiodo-
8759 - Pentane, 2,4-diiodo-

$C_5H_{10}N_2$
4156 - Cyanamide, diethyl-

10101 - Propanenitrile, 3-(dimethylamino)-

$C_5H_{10}N_2O$
9688 - Piperidine, 1-nitroso-

$C_5H_{10}N_2O_2$
8800 - 2,3-Pentanedione, dioxime
9687 - Piperidine, 1-nitro-

$C_5H_{10}N_2O_2S$
7650 - Methomyl

$C_5H_{10}N_2O_3$
6249 - L-Glutamine

$C_5H_{10}N_2O_4$
3950 - Carbamic acid, ethylnitro-, ethyl ester
8769 - Pentane, 1,5-dinitro-

$C_5H_{10}O$
3185 - Butanal, 2-methyl-, (±)-
3186 - Butanal, 3-methyl-
3668 - 2-Butanone, 3-methyl-
3869 - 3-Buten-1-ol, 3-methyl-
3870 - 3-Buten-2-ol, 2-methyl-
3871 - 3-Buten-2-ol, 3-methyl-
4864 - Cyclopentanol
4979 - Cyclopropanemethanol, α-methyl-
6171 - Furan, tetrahydro-2-methyl-
6172 - Furan, tetrahydro-3-methyl-
8593 - Oxetane, 3,3-dimethyl-
8690 - Pentanal
8957 - 2-Pentanone
8958 - 3-Pentanone
9087 - 1-Penten-3-ol
9088 - 4-Penten-1-ol
9089 - 4-Penten-2-ol
9090 - 2-Penten-1-ol, (E)-
9091 - 2-Penten-1-ol, (Z)-
9092 - 3-Penten-2-ol, (±)-
9783 - Propanal, 2,2-dimethyl-
10033 - Propane, 1-(ethenyloxy)-
10034 - Propane, 2-(ethenyloxy)-
10566 - 1-Propene, 3-ethoxy-
10577 - 1-Propene, 3-methoxy-2-methyl-
10800 - 2H-Pyran, tetrahydro-

$C_5H_{10}OS_2$
4031 - Carbonodithioic acid, S,S-diethyl ester

$C_5H_{10}O_2$
209 - Acetic acid, 1-methylethyl ester
236 - Acetic acid, propyl ester
3538 - Butanoic acid, 2-methyl-, (±)-
3539 - Butanoic acid, 3-methyl-
3547 - Butanoic acid, methyl ester
3664 - 2-Butanone, 4-hydroxy-3-methyl-
4806 - 1,3-Cyclopentanediol, cis-
4807 - 1,3-Cyclopentanediol, trans-
5193 - 1,3-Dioxane, 4-methyl-
5763 - Ethanol, 2-(2-propenyloxy)-
6004 - Formic acid, butyl ester
6012 - Formic acid, 1-methylpropyl ester
6013 - Formic acid, 2-methylpropyl ester
6124 - 2-Furanmethanol, tetrahydro-
8614 - Oxirane, (ethoxymethyl)-
8855 - Pentanoic acid
8971 - 2-Pentanone, 1-hydroxy-
8972 - 2-Pentanone, 3-hydroxy-
8973 - 2-Pentanone, 4-hydroxy-
8974 - 2-Pentanone, 5-hydroxy-
8975 - 3-Pentanone, 2-hydroxy-
10258 - Propanoic acid, 2,2-dimethyl-
10263 - Propanoic acid, ethyl ester
10301 - Propanoic acid, 2-methyl-, methyl ester
10562 - 1-Propene, 3,3-dimethoxy-
10787 - 2H-Pyran-4-ol, tetrahydro-

$C_5H_{10}O_2S$
10311 - Propanoic acid, 3-(methylthio)-, methyl ester
11789 - Thiophene, tetrahydro-3-methyl-, 1,1-dioxide

$C_5H_{10}O_3$
182 - Acetic acid, ethoxy-, methyl ester
199 - Acetic acid, hydroxy-, propyl ester
206 - Acetic acid, methoxy-, ethyl ester
3535 - Butanoic acid, 3-hydroxy-3-methyl-

3999 - Carbonic acid, diethyl ester
5197 - 1,3-Dioxan-5-ol, 2-methyl-
5208 - 1,3-Dioxolane-4-methanol, 2-methyl-
5734 - Ethanol, methoxy-, acetate
5735 - Ethanol, 2-methoxy-, acetate
8883 - Pentanoic acid, 2-hydroxy-
10269 - Propanoic acid, 2-hydroxy-, ethyl ester, (±)-
10270 - Propanoic acid, 3-hydroxy-, ethyl ester
10280 - Propanoic acid, 2-hydroxy-2-methyl-, methyl ester
10289 - Propanoic acid, 3-methoxy-, methyl ester

$C_5H_{10}O_4$
10187 - 1,2,3-Propanetriol, 1-acetate, (±)-
10253 - Propanoic acid, 2,3-dihydroxy-, ethyl ester, (±)-
10256 - Propanoic acid, 2,3-dihydroxy-2-methyl-, methyl ester
10272 - Propanoic acid, 3-hydroxy-2-(hydroxymethyl)-2-methyl-

$C_5H_{10}O_5$
436 - β-D-Arabinopyranose
11345 - D-Ribose
12059 - D-Xylose

$C_5H_{10}S$
4852 - Cyclopentanethiol
11787 - Thiophene, tetrahydro-2-methyl-
11788 - Thiophene, tetrahydro-3-methyl-
11795 - 2H-Thiopyran, tetrahydro-

$C_5H_{11}Br$
3238 - Butane, 1-bromo-2-methyl-, (S)-
3239 - Butane, 1-bromo-3-methyl-
3240 - Butane, 2-bromo-2-methyl-
8710 - Pentane, 1-bromo-
8711 - Pentane, 2-bromo-
8712 - Pentane, 3-bromo-
9843 - Propane, 1-bromo-2,2-dimethyl-

$C_5H_{11}Cl$
3251 - Butane, 1-chloro-2-methyl-, (±)-
3252 - Butane, 1-chloro-3-methyl-
3253 - Butane, 2-chloro-2-methyl-
8719 - Pentane, 1-chloro-
8720 - Pentane, 2-chloro-, (+)-
8721 - Pentane, 3-chloro-
9853 - Propane, 1-chloro-2,2-dimethyl-

$C_5H_{11}ClO$
3610 - 2-Butanol, 1-chloro-2-methyl-
3611 - 2-Butanol, 3-chloro-2-methyl-
8923 - 1-Pentanol, 5-chloro-
8924 - 2-Pentanol, 1-chloro-
8925 - 3-Pentanol, 1-chloro-
9854 - Propane, 1-(1-chloroethoxy)-

$C_5H_{11}ClO_2$
9852 - Propane, 3-chloro-1,1-dimethoxy-

$C_5H_{11}ClSi$
11375 - Silane, chlorodimethyl-2-propenyl-

$C_5H_{11}Cl_2N$
5413 - Ethanamine, 2-chloro-N-(2-chloroethyl)-N-methyl-

$C_5H_{11}Cl_3Si$
11472 - Silane, trichloropentyl-

$C_5H_{11}F$
8817 - Pentane, 1-fluoro-

$C_5H_{11}FO$
8935 - 1-Pentanol, 5-fluoro-

$C_5H_{11}I$
3383 - Butane, 1-iodo-2-methyl-, (S)-
3384 - Butane, 1-iodo-3-methyl-
3385 - Butane, 2-iodo-2-methyl-
8818 - Pentane, 1-iodo-
8819 - Pentane, 2-iodo-, (±)-
8820 - Pentane, 3-iodo-
10056 - Propane, 1-iodo-2,2-dimethyl-

$C_5H_{11}N$
4771 - Cyclopentanamine
9629 - Piperidine
11168 - Pyrrolidine, 1-methyl-

$C_5H_{11}NO$
3187 - Butanal, 3-methyl-, oxime
3206 - Butanamide, 3-methyl-
5991 - Formamide, N,N-diethyl-
6116 - 2-Furanmethanamine, tetrahydro-
7696 - Morpholine, 4-methyl-
8698 - Pentanamide
8985 - 2-Pentanone, oxime

$C_5H_{11}NO_2$
3400 - Butane, 3-methyl-1-nitro-
3935 - Carbamic acid, butyl ester
3941 - Carbamic acid, diethyl ester
3949 - Carbamic acid, ethyl-, ethyl ester
3959 - Carbamic acid, 2-methylpropyl ester
7400 - DL-Isovaline
7401 - L-Isovaline
7454 - Methanaminium, 1-carboxy-N,N,N-trimethyl-, hydroxide, inner salt
8090 - Nitrous acid, 3-methylbutyl ester
8096 - Nitrous acid, pentyl ester
8217 - DL-Norvaline
8218 - L-Norvaline
8832 - Pentane, 1-nitro-
8833 - Pentane, 2-nitro-
8834 - Pentane, 3-nitro-
8857 - Pentanoic acid, 5-amino-
12040 - DL-Valine
12041 - L-Valine

$C_5H_{11}NO_2S$
7647 - DL-Methionine
7648 - L-Methionine
12043 - D-Valine, 3-mercapto-

$C_5H_{11}NO_3$
3637 - 1-Butanol, 3-methyl-, nitrate
3638 - 1-Butanol, 3-methyl-2-nitro-
8950 - 1-Pentanol, 2-nitro-
8951 - 2-Pentanol, 1-nitro-
8952 - 2-Pentanol, 3-nitro-

$C_5H_{11}NO_4$
10008 - 1,3-Propanediol, 2-ethyl-2-nitro-

C_5H_{12}
3397 - Butane, 2-methyl-
8709 - Pentane
9910 - Propane, 2,2-dimethyl-

$C_5H_{12}NO_3PS_2$
5009 - Cygon

$C_5H_{12}N_2$
9618 - Piperazine, 1-methyl-
9619 - Piperazine, 2-methyl-
9624 - 1-Piperidinamine

$C_5H_{12}N_2O$
12001 - Urea, N,N'-diethyl-
12002 - Urea, N,N-diethyl-
12030 - Urea, tetramethyl-

$C_5H_{12}N_2O_2$
8550 - L-Ornithine

$C_5H_{12}N_2S$
11803 - Thiourea, N,N'-diethyl-
11810 - Thiourea, tetramethyl-

$C_5H_{12}N_4O_3$
7135 - L-Homoserine, O-[(aminoiminomethyl)amino]-

$C_5H_{12}O$
3391 - Butane, 1-methoxy-
3392 - Butane, 2-methoxy-, (S)-
3393 - Butane, 2-methoxy-, (±)-
3626 - 1-Butanol, 2-methyl-
3627 - 1-Butanol, 3-methyl-
3628 - 2-Butanol, 2-methyl-
3629 - 2-Butanol, 3-methyl-, (±)-
8915 - 1-Pentanol
8916 - 2-Pentanol
8917 - 3-Pentanol
10037 - Propane, 1-ethoxy-
10038 - Propane, 2-ethoxy-
10071 - Propane, 1-methoxy-2-methyl-
10072 - Propane, 2-methoxy-2-methyl-
10373 - 1-Propanol, 2,2-dimethyl-

$C_5H_{12}O_2$
3352 - 1,2-Butanediol, 3-methyl-
3353 - 1,3-Butanediol, 2-methyl-
3354 - 1,3-Butanediol, 3-methyl-
3355 - 2,3-Butanediol, 2-methyl-
3625 - 1-Butanol, 3-methoxy-
5591 - Ethane, 1-ethoxy-2-methoxy-
5603 - Ethane, 1,1'-[methylenebis(oxy)]bis-
5740 - Ethanol, 2-(1-methylethoxy)-
5764 - Ethanol, 2-propoxy-
8785 - 1,3-Pentanediol
8786 - 1,4-Pentanediol
8787 - 1,5-Pentanediol
8788 - 2,3-Pentanediol
8789 - 2,4-Pentanediol
8790 - 1,2-Pentanediol, (±)-
9909 - Propane, 2,2-dimethoxy-
10005 - 1,3-Propanediol, 2,2-dimethyl-
10380 - 1-Propanol, 2-ethoxy-
10381 - 2-Propanol, 1-ethoxy-

$C_5H_{12}O_3$
5669 - Ethane, 1,1,1-trimethoxy-
5737 - Ethanol, 2-(2-methoxyethoxy)-
8851 - 1,2,5-Pentanetriol
8852 - 1,3,5-Pentanetriol
9119 - Pentitol, 1,2-dideoxy-
10011 - 1,3-Propanediol, 2-(hydroxymethyl)-2-methyl-
10372 - 2-Propanol, 1,3-dimethoxy-

$C_5H_{12}O_4$
7543 - Methane, tetramethoxy-
9989 - 1,3-Propanediol, 2,2-bis(hydroxymethyl)-

$C_5H_{12}O_5$
11343 - Ribitol
12058 - Xylitol

$C_5H_{12}S$
3402 - Butane, 1-(methylthio)-
3439 - 1-Butanethiol, 2-methyl-, (S)-
3440 - 1-Butanethiol, 3-methyl-
3441 - 2-Butanethiol, 2-methyl-
8844 - 1-Pentanethiol
8845 - 2-Pentanethiol
8846 - 3-Pentanethiol
10042 - Propane, 1-(ethylthio)-
10043 - Propane, 2-(ethylthio)-
10082 - Propane, 2-methyl-1-(methylthio)-
10083 - Propane, 2-methyl-2-(methylthio)-

$C_5H_{12}Si$
11425 - Silane, ethenyltrimethyl-

$C_5H_{13}Cl_2N$
5442 - Ethanaminium, 2-chloro-N,N,N-trimethyl-, chloride

$C_5H_{13}N$
3220 - 1-Butanamine, 3-methyl-
3221 - 1-Butanamine, N-methyl-
3222 - 2-Butanamine, 2-methyl-
3223 - 2-Butanamine, 3-methyl-
5430 - Ethanamine, N-ethyl-N-methyl-
8700 - 1-Pentanamine
8701 - 2-Pentanamine
8702 - 3-Pentanamine
9806 - 1-Propanamine, 2,2-dimethyl-
9807 - 1-Propanamine, N,N-dimethyl-
9809 - 1-Propanamine, N-ethyl-
9810 - 2-Propanamine, N-ethyl-

$C_5H_{13}NO$
3599 - 1-Butanol, 2-amino-3-methyl-, (±)-
5426 - Ethanamine, 2-ethoxy-N-methyl-
5741 - Ethanol, 2-[(1-methylethyl)amino]-
5765 - Ethanol, 2-(propylamino)-
5878 - Ethenaminium, N,N,N-trimethyl-, hydroxide
8920 - 1-Pentanol, 5-amino-
8921 - 3-Pentanol, 2-amino-
10374 - 1-Propanol, 2-(dimethylamino)-
10375 - 1-Propanol, 3-(dimethylamino)-
10376 - 2-Propanol, 1-(dimethylamino)-

$C_5H_{13}NO_2$
5420 - Ethanamine, 2,2-dimethoxy-*N*-methyl-
5742 - Ethanol, 2,2'-(methylimino)bis-
9985 - 1,3-Propanediol, 2-amino-2-ethyl-
10383 - 2-Propanol, 1-[(2-hydroxyethyl)amino]-

$C_5H_{13}O_3P$
9539 - Phosphonic acid, methyl-, diethyl ester

$C_5H_{14}CINO$
5446 - Ethanaminium, 2-hydroxy-*N,N,N*-trimethyl-, chloride

$C_5H_{14}N_2$
7490 - Methanediamine, *N,N,N',N'*-tetramethyl-
8735 - 1,5-Pentanediamine

$C_5H_{14}N_2O$
10340 - 2-Propanol, 1-[(2-aminoethyl)amino]-

$C_5H_{14}OSi$
11430 - Silane, ethoxytrimethyl-

$C_5H_{14}O_3Si$
11433 - Silane, ethyltrimethoxy-

$C_5H_{15}NSi$
11361 - Silanamine, pentamethyl-

$C_5H_{20}O_5Si_5$
4902 - Cyclopentasiloxane, 2,4,6,8,10-pentamethyl-

C_6BrF_5
1078 - Benzene, bromopentafluoro-

$C_6Br_2F_4$
1312 - Benzene, 1,2-dibromo-3,4,5,6-tetrafluoro-

C_6CIF_5
1220 - Benzene, chloropentafluoro-

$C_6Cl_2F_4$
1398 - Benzene, 1,2-dichloro-3,4,5,6-tetrafluoro
1399 - Benzene, 1,3-dichloro-2,4,5,6-tetrafluoro-
1400 - Benzene, 1,4-dichloro-2,3,5,6-tetrafluoro-

$C_6Cl_2F_8$
4647 - Cyclohexene, 1,2-dichloro-3,3,4,4,5,5,6,6-octafluoro-

$C_6Cl_3F_3$
2340 - Benzene, 1,3,5-trichloro-2,4,6-trifluoro-

$C_6Cl_4O_2$
4268 - 2,5-Cyclohexadiene-1,4-dione, 2,3,5,6-tetrachloro-

$C_6Cl_5NO_2$
2087 - Benzene, pentachloronitro-

C_6Cl_6
1749 - Benzene, hexachloro-

C_6Cl_{10}
6689 - 1,5-Hexadiene, 1,1,2,3,3,4,4,5,6,6-decachloro-

C_6F_5I
2092 - Benzene, pentafluoroiodo-

C_6F_6
1757 - Benzene, hexafluoro-

C_6F_8
4276 - 1,3-Cyclohexadiene, 1,2,3,4,5,5,6,6-octafluoro-
4277 - 1,4-Cyclohexadiene, 1,2,3,3,4,5,6,6-octafluoro-

C_6F_{10}
4643 - Cyclohexene, decafluoro-

C_6F_{12}
4381 - Cyclohexane, dodecafluoro-
7032 - 1-Hexene, 1,1,2,3,3,4,4,5,5,6,6,6-dodecafluoro-

C_6F_{14}
3414 - Butane, 1,1,1,2,3,4,4,4-octafluoro-2,3-bis(trifluoromethyl)-
6843 - Hexane, tetradecafluoro-
8853 - Pentane, 1,1,1,2,2,3,4,4,5,5,5-undecafluoro-3-(trifluoromethyl)-
8854 - Pentane, 1,1,1,2,2,3,3,4,5,5,5-undecafluoro-4-(trifluoromethyl)-

$C_6F_{15}N$
5436 - Ethanamine, 1,1,2,2,2-pentafluoro-*N,N*-bis(pentafluoroethyl)-

C_6HBr_5
2082 - Benzene, pentabromo-

C_6HBr_5O
9439 - Phenol, pentabromo-

$C_6HCl_3F_8O_2$
6910 - Hexanoic acid, 3,5,6-trichloro-2,2,3,4,4,5,6,6-octafluoro-

$C_6HCl_4NO_2$
2264 - Benzene, 1,2,4,5-tetrachloro-3-nitro-

C_6HCl_5
2084 - Benzene, pentachloro-

C_6HCl_5O
9440 - Phenol, pentachloro-

C_6HF_5
2091 - Benzene, pentafluoro-

C_6HF_5O
9442 - Phenol, pentafluoro-

C_6HF_5S
2304 - Benzenethiol, pentafluoro-

C_6HF_{11}
4498 - Cyclohexane, undecafluoro-

C_6HI_5
2096 - Benzene, pentaiodo-

$C_6H_2Br_2CINO$
4279 - 2,5-Cyclohexadien-1-one, 2,6-dibromo-4-(chloroimino)-

$C_6H_2Br_3NO_2$
2315 - Benzene, 1,2,3-tribromo-5-nitro-
2316 - Benzene, 1,3,5-tribromo-2-nitro-

$C_6H_2Br_4$
2253 - Benzene, 1,2,3,5-tetrabromo-
2254 - Benzene, 1,2,4,5-tetrabromo-

$C_6H_2Br_4O$
9467 - Phenol, 2,3,4,6-tetrabromo-

$C_6H_2Br_4O_2$
1551 - 1,4-Benzenediol, 2,3,5,6-tetrabromo-

$C_6H_2CIN_3O_6$
1244 - Benzene, 2-chloro-1,3,5-trinitro-

$C_6H_2Cl_2O_4$
4258 - 2,5-Cyclohexadiene-1,4-dione, 2,5-dichloro-3,6-dihydroxy-

$C_6H_2Cl_3NO$
4280 - 2,5-Cyclohexadien-1-one, 2,6-dichloro-4-(chloroimino)-

$C_6H_2Cl_3NO_2$
2338 - Benzene, 1,2,3-trichloro-5-nitro-
2339 - Benzene, 1,2,4-trichloro-5-nitro-

$C_6H_2Cl_3NO_3$
9490 - Phenol, 3,4,6-trichloro-2-nitro-

$C_6H_2Cl_4$
2259 - Benzene, 1,2,3,4-tetrachloro-
2260 - Benzene, 1,2,4,5-tetrachloro-

$C_6H_2Cl_4O$
9469 - Phenol, 2,3,4,5-tetrachloro-
9470 - Phenol, 2,3,4,6-tetrachloro-
9471 - Phenol, 2,3,5,6-tetrachloro-

$C_6H_2Cl_5OPS$
9578 - Phosphorodichloridothioic acid, *O*-(2,4,5-trichlorophenyl) ester

$C_6H_2F_4$
2270 - Benzene, 1,2,3,4-tetrafluoro-
2271 - Benzene, 1,2,3,5-tetrafluoro-
2272 - Benzene, 1,2,4,5-tetrafluoro-

$C_6H_2I_3NO_2$
2353 - Benzene, 1,2,3-triiodo-5-nitro-

$C_6H_2I_4$
2273 - Benzene, 1,2,3,4-tetraiodo-
2274 - Benzene, 1,2,3,5-tetraiodo-
2275 - Benzene, 1,2,4,5-tetraiodo-

$C_6H_2N_2O_8$
4259 - 2,5-Cyclohexadiene-1,4-dione, 2,5-dihydroxy-3,6-dinitro-

$C_6H_2N_4O_5$
2874 - 1,2,3-Benzoxadiazole, 5,7-dinitro-

$C_6H_2N_4O_9$
9476 - Phenol, 2,3,4,6-tetranitro-

$C_6H_2O_6$
4686 - 5-Cyclohexene-1,2,3,4-tetrone, 5,6-dihydroxy-

$C_6H_3BrCl_2$
1010 - Benzene, 1-bromo-2,3-dichloro-
1011 - Benzene, 1-bromo-2,4-dichloro-
1012 - Benzene, 1-bromo-3,5-dichloro-
1013 - Benzene, 2-bromo-1,3-dichloro-
1014 - Benzene, 2-bromo-1,4-dichloro-
1015 - Benzene, 4-bromo-1,2-dichloro-

$C_6H_3BrCl_2O$
9233 - Phenol, 2-bromo-4,6-dichloro-

$C_6H_3BrN_2O_4$
1024 - Benzene, 1-bromo-2,3-dinitro-

$C_6H_3BrN_2O_5$
9234 - Phenol, 2-bromo-4,6-dinitro-
9235 - Phenol, 4-bromo-2,6-dinitro-

$C_6H_3Br_2Cl$
1292 - Benzene, 1,2-dibromo-3-chloro-
1293 - Benzene, 1,2-dibromo-4-chloro-
1294 - Benzene, 1,3-dibromo-2-chloro-
1295 - Benzene, 1,4-dibromo-2-chloro-
1296 - Benzene, 2,4-dibromo-1-chloro-

$C_6H_3Br_2F$
1300 - Benzene, 2,4-dibromo-1-fluoro-

$C_6H_3Br_2NO_2$
1307 - Benzene, 1,2-dibromo-4-nitro-
1308 - Benzene, 1,3-dibromo-2-nitro-
1309 - Benzene, 1,3-dibromo-5-nitro-
1310 - Benzene, 1,4-dibromo-2-nitro-
1311 - Benzene, 2,4-dibromo-1-nitro-

$C_6H_3Br_3$
2307 - Benzene, 1,2,3-tribromo-
2308 - Benzene, 1,2,4-tribromo-
2309 - Benzene, 1,3,5-tribromo-

$C_6H_3Br_3O$
9479 - Phenol, 2,4,6-tribromo-

$C_6H_3CINO_3$
9296 - Phenol, 2,6-dichloro-4-nitro-

$C_6H_3CIN_2O_4$
1144 - Benzene, 1-chloro-2,4-dinitro-
1145 - Benzene, 2-chloro-1,3-dinitro-

$C_6H_3CIN_2O_4S$
2178 - Benzenesulfenyl chloride, 2,4-dinitro-

$C_6H_3CIN_2O_5$
9258 - Phenol, 4-chloro-2,6-dinitro-
9259 - Phenol, 5-chloro-2,4-dinitro-

$C_6H_3Cl_2I$
1380 - Benzene, 1,4-dichloro-2-iodo-

$C_6H_3Cl_2NO_2$
1391 - Benzene, 1,2-dichloro-3-nitro-
1392 - Benzene, 1,2-dichloro-4-nitro-
1393 - Benzene, 1,3-dichloro-2-nitro-
1394 - Benzene, 1,3-dichloro-5-nitro-
1395 - Benzene, 1,4-dichloro-2-nitro-
1396 - Benzene, 2,4-dichloro-1-nitro-
4142 - Clopyralid

$C_6H_3Cl_3$
2321 - Benzene, 1,2,3-trichloro-
2322 - Benzene, 1,2,4-trichloro-
2323 - Benzene, 1,3,5-trichloro-

$C_6H_3Cl_3N_2O_2$
10896 - 2-Pyridinecarboxylic acid, 4-amino-3,5,6-trichloro-

$C_6H_3Cl_3O$
9481 - Phenol, 2,3,4-trichloro-
9482 - Phenol, 2,3,5-trichloro-
9483 - Phenol, 2,3,6-trichloro-
9484 - Phenol, 2,4,5-trichloro-
9485 - Phenol, 2,4,6-trichloro-
9486 - Phenol, 3,4,5-trichloro-

$C_6H_3Cl_3O_3$
2371 - 1,3,5-Benzenetriol, 2,4,6-trichloro-

$C_6H_3Cl_4N$
8071 - Nitrapyrin

$C_6H_3FN_2O_4$
1718 - Benzene, 1-fluoro-2,4-dinitro-

$C_6H_3F_2NO_2$
1426 - Benzene, 2,4-difluoro-1-nitro-

$C_6H_3F_3$
2346 - Benzene, 1,2,4-trifluoro-
2347 - Benzene, 1,3,5-trifluoro-
$C_6H_3F_7O_2$
3523 - Butanoic acid, heptafluoro-, ethenyl ester
$C_6H_3I_2NO_3$
9304 - Phenol, 2,6-diiodo-4-nitro-
$C_6H_3I_3$
2349 - Benzene, 1,2,3-triiodo-
2350 - Benzene, 1,2,4-triiodo-
2351 - Benzene, 1,3,5-triiodo-
$C_6H_3I_3O$
9492 - Phenol, 2,4,6-triiodo-
$C_6H_3N_3O_6$
2365 - Benzene, 1,3,5-trinitro-
$C_6H_3N_3O_7$
9497 - Phenol, 2,4,6-trinitro-
$C_6H_3N_3O_8$
1554 - 1,3-Benzenediol, 2,4,6-trinitro-
C_6H_4BrCl
1004 - Benzene, 1-bromo-2-chloro-
1005 - Benzene, 1-bromo-3-chloro-
1006 - Benzene, 1-bromo-4-chloro-
C_6H_4BrClO
9231 - Phenol, 3-bromo-5-chloro-
9232 - Phenol, 4-bromo-2-chloro-
$C_6H_4BrClO_2S$
2244 - Benzenesulfonyl chloride, 4-bromo-
$C_6H_4BrCl_2N$
641 - Benzenamine, 2-bromo-4,6-dichloro-
C_6H_4BrF
1043 - Benzene, 1-bromo-3-fluoro-
1044 - Benzene, 1-bromo-4-fluoro-
C_6H_4BrI
1046 - Benzene, 1-bromo-2-iodo-
1047 - Benzene, 1-bromo-3-iodo-
1048 - Benzene, 1-bromo-4-iodo-
$C_6H_4BrNO_2$
1075 - Benzene, 1-bromo-2-nitro-
1076 - Benzene, 1-bromo-3-nitro-
1077 - Benzene, 1-bromo-4-nitro-
$C_6H_4BrNO_3$
9243 - Phenol, 2-bromo-3-nitro-
9244 - Phenol, 4-bromo-2-nitro-
$C_6H_4BrN_3$
977 - Benzene, 1-azido-4-bromo-
$C_6H_4BrN_3O_4$
646 - Benzenamine, 2-bromo-4,6-dinitro-
$C_6H_4Br_2$
1288 - Benzene, 1,2-dibromo-
1289 - Benzene, 1,3-dibromo-
1290 - Benzene, 1,4-dibromo-
$C_6H_4Br_2O$
9276 - Phenol, 2,4-dibromo-
9277 - Phenol, 2,6-dibromo-
9278 - Phenol, 3,5-dibromo-
$C_6H_4Br_3N$
851 - Benzenamine, 2,4,6-tribromo-
C_6H_4ClF
1166 - Benzene, 1-chloro-2-fluoro-
1167 - Benzene, 1-chloro-3-fluoro-
1168 - Benzene, 1-chloro-4-fluoro-
$C_6H_4ClFO_2S$
2246 - Benzenesulfonyl chloride, 4-fluoro-
C_6H_4ClI
1171 - Benzene, 1-chloro-2-iodo-
1172 - Benzene, 1-chloro-4-iodo-
$C_6H_4ClIO_2S$
2247 - Benzenesulfonyl chloride, 4-iodo-
$C_6H_4ClNO_2$
1214 - Benzene, 1-chloro-2-nitro-
1215 - Benzene, 1-chloro-3-nitro-
1216 - Benzene, 1-chloro-4-nitro-
$C_6H_4ClNO_2S$
2179 - Benzenesulfenyl chloride, 4-nitro-
$C_6H_4ClNO_3$
9269 - Phenol, 5-chloro-2-nitro-

$C_6H_4ClN_3$
978 - Benzene, 1-azido-4-chloro-
$C_6H_4ClO_2P$
2413 - 1,3,2-Benzodioxaphosphole, 2-chloro-
$C_6H_4Cl_2$
1354 - Benzene, 1,2-dichloro-
1355 - Benzene, 1,3-dichloro-
1356 - Benzene, 1,4-dichloro-
$C_6H_4Cl_2N_2O_2$
692 - Benzenamine, 2,6-dichloro-4-nitro-
$C_6H_4Cl_2O$
9282 - Phenol, 2,3-dichloro-
9283 - Phenol, 2,4-dichloro-
9284 - Phenol, 2,5-dichloro-
9285 - Phenol, 2,6-dichloro-
9286 - Phenol, 3,4-dichloro-
9287 - Phenol, 3,5-dichloro-
$C_6H_4Cl_2O_2$
1510 - 1,3-Benzenediol, 4,6-dichloro-
$C_6H_4Cl_2O_2S$
2245 - Benzenesulfonyl chloride, 4-chloro-
$C_6H_4Cl_3N$
852 - Benzenamine, 2,3,4-trichloro-
853 - Benzenamine, 2,4,5-trichloro-
854 - Benzenamine, 2,4,6-trichloro-
11023 - Pyridine, 2-(trichloromethyl)-
$C_6H_4Cl_4Si$
11457 - Silane, trichloro(4-chlorophenyl)-
C_6H_4FI
1719 - Benzene, 1-fluoro-2-iodo-
1720 - Benzene, 1-fluoro-4-iodo-
$C_6H_4FNO_2$
1733 - Benzene, 1-fluoro-2-nitro-
1734 - Benzene, 1-fluoro-3-nitro-
1735 - Benzene, 1-fluoro-4-nitro-
$C_6H_4F_2$
1422 - Benzene, 1,2-difluoro-
1423 - Benzene, 1,3-difluoro-
1424 - Benzene, 1,4-difluoro-
$C_6H_4INO_2$
1783 - Benzene, 1-iodo-2-nitro-
1784 - Benzene, 1-iodo-3-nitro-
1785 - Benzene, 1-iodo-4-nitro-
$C_6H_4IN_3$
979 - Benzene, 1-azido-2-iodo-
$C_6H_4I_2$
1427 - Benzene, 1,2-diiodo-
1428 - Benzene, 1,3-diiodo-
1429 - Benzene, 1,4-diiodo-
$C_6H_4N_2$
10881 - 2-Pyridinecarbonitrile
10882 - 3-Pyridinecarbonitrile
$C_6H_4N_2O_4$
1491 - Benzene, 1,2-dinitro-
1492 - Benzene, 1,3-dinitro-
1493 - Benzene, 1,4-dinitro-
10804 - 2,3-Pyrazinedicarboxylic acid
$C_6H_4N_2O_5$
9341 - Phenol, 2,3-dinitro-
9342 - Phenol, 2,4-dinitro-
9343 - Phenol, 2,5-dinitro-
9344 - Phenol, 2,6-dinitro-
9345 - Phenol, 3,4-dinitro-
9346 - Phenol, 3,5-dinitro-
$C_6H_4N_2O_6$
1517 - 1,3-Benzenediol, 2,4-dinitro-
$C_6H_4N_4$
10747 - Pteridine
$C_6H_4N_4O_2$
2873 - 1H-Benzotriazole, 5-nitro-
10748 - 2,4(1H,3H)-Pteridinedione
$C_6H_4N_4O_6$
863 - Benzenamine, 2,4,6-trinitro-
C_6H_4O
6108 - Furan, 2-ethynyl-
$C_6H_4O_2$
4257 - 2,5-Cyclohexadiene-1,4-dione

$C_6H_4O_4$
10773 - 2H-Pyran-5-carboxylic acid, 2-oxo-
$C_6H_4O_4S$
11745 - 2,5-Thiophenedicarboxylic acid
$C_6H_4O_5$
6079 - 2,3-Furandicarboxylic acid
6080 - 2,4-Furandicarboxylic acid
6081 - 2,5-Furandicarboxylic acid
$C_6H_4O_6$
4269 - 2,5-Cyclohexadiene-1,4-dione, 2,3,5,6-
tetrahydroxy-
$C_6H_4S_2$
11688 - Thieno[2,3-b]thiophene
$C_6H_5AsCl_2$
462 - Arsonous dichloride, phenyl-
$C_6H_5AsI_2$
464 - Arsonous diiodide, phenyl-
C_6H_5AsO
447 - Arsine, oxophenyl-
$C_6H_5BF_2$
3116 - Borane, difluorophenyl-
$C_6H_5BO_2$
2412 - 1,3,2-Benzodioxaborole
C_6H_5Br
998 - Benzene, bromo-
$C_6H_5BrN_2O_2$
653 - Benzenamine, 2-bromo-6-nitro-
654 - Benzenamine, 4-bromo-2-nitro-
C_6H_5BrO
9228 - Phenol, 2-bromo-
9229 - Phenol, 3-bromo-
9230 - Phenol, 4-bromo-
C_6H_5BrOS
5783 - Ethanone, 1-(5-bromo-2-thienyl)-
$C_6H_5BrO_2$
1502 - 1,3-Benzenediol, 4-bromo-
1503 - 1,4-Benzenediol, 2-bromo-
C_6H_5BrS
2292 - Benzenethiol, 4-bromo-
$C_6H_5Br_2N$
681 - Benzenamine, 2,4-dibromo-
682 - Benzenamine, 2,6-dibromo-
683 - Benzenamine, 3,4-dibromo-
C_6H_5Cl
1131 - Benzene, chloro-
C_6H_5ClFN
664 - Benzenamine, 2-chloro-6-fluoro-
$C_6H_5ClN_2O_2$
677 - Benzenamine, 5-chloro-2-nitro-
C_6H_5ClO
9251 - Phenol, 2-chloro-
9252 - Phenol, 3-chloro-
9253 - Phenol, 4-chloro-
C_6H_5ClOS
2185 - Benzenesulfinyl chloride
5791 - Ethanone, 1-(5-chloro-2-thienyl)-
$C_6H_5ClO_2$
1504 - 1,2-Benzenediol, 3-chloro-
1505 - 1,2-Benzenediol, 4-chloro-
1506 - 1,3-Benzenediol, 5-chloro-
1507 - 1,4-Benzenediol, 2-chloro-
$C_6H_5ClO_2S$
2242 - Benzenesulfonyl chloride
$C_6H_5ClO_3S$
2220 - Benzenesulfonic acid, 4-chloro-
C_6H_5ClS
2293 - Benzenethiol, 2-chloro-
2294 - Benzenethiol, 3-chloro-
2295 - Benzenethiol, 4-chloro-
$C_6H_5ClS_2$
1563 - 1,3-Benzenedithiol, 4-chloro-
$C_6H_5Cl_2N$
685 - Benzenamine, 2,3-dichloro-
686 - Benzenamine, 2,4-dichloro-
687 - Benzenamine, 2,5-dichloro-
688 - Benzenamine, 2,6-dichloro-
689 - Benzenamine, 3,4-dichloro-

690 - Benzenamine, 3,5-dichloro-

$C_6H_5Cl_2NO$
9205 - Phenol, 2-amino-4,6-dichloro-
9206 - Phenol, 4-amino-2,6-dichloro-

$C_6H_5Cl_2OP$
9548 - Phosphonic dichloride, phenyl-

$C_6H_5Cl_2P$
9556 - Phosphonous dichloride, phenyl-

$C_6H_5Cl_2PS$
9552 - Phosphonothioic dichloride, phenyl-

$C_6H_5Cl_2Sb$
11547 - Stibine, dichlorophenyl-

$C_6H_5Cl_3Si$
11473 - Silane, trichlorophenyl-

C_6H_5F
1717 - Benzene, fluoro-

C_6H_5FO
9360 - Phenol, 2-fluoro-
9361 - Phenol, 3-fluoro-
9362 - Phenol, 4-fluoro-

$C_6H_5FO_2S$
2251 - Benzenesulfonyl fluoride

$C_6H_5F_2N$
700 - Benzenamine, 2,4-difluoro-

$C_6H_5F_3Si$
11486 - Silane, trifluorophenyl-

$C_6H_5F_7O_2$
3524 - Butanoic acid, heptafluoro-, ethyl ester

C_6H_5I
1766 - Benzene, iodo-

C_6H_5IO
9371 - Phenol, 2-iodo-
9372 - Phenol, 3-iodo-
9373 - Phenol, 4-iodo-

$C_6H_5IO_2$
1523 - 1,2-Benzenediol, 4-iodo-
1524 - 1,3-Benzenediol, 5-iodo-

$C_6H_5I_2N$
701 - Benzenamine, 2,4-diiodo-

C_6H_5NO
2059 - Benzene, nitroso-
6031 - 2-Furanacetonitrile
10887 - 2-Pyridinecarboxaldehyde
10888 - 3-Pyridinecarboxaldehyde
10889 - 4-Pyridinecarboxaldehyde

C_6H_5NOS
844 - Benzenamine, N-sulfinyl-

$C_6H_5NO_2$
2049 - Benzene, nitro-
9438 - Phenol, 4-nitroso-
10893 - 2-Pyridinecarboxylic acid
10894 - 3-Pyridinecarboxylic acid
10895 - 4-Pyridinecarboxylic acid

$C_6H_5NO_3$
9435 - Phenol, 2-nitro-
9436 - Phenol, 3-nitro-
9437 - Phenol, 4-nitro-
10899 - 3-Pyridinecarboxylic acid, 1,6-dihydro-6-oxo-
10909 - 3-Pyridinecarboxylic acid, 1-oxide

$C_6H_5NO_4$
1541 - 1,3-Benzenediol, 4-nitro-
10900 - 4-Pyridinecarboxylic acid, 1,2-dihydro-6-hydroxy-2-oxo-

C_6H_5NS
11725 - 2-Thiopheneacetonitrile

$C_6H_5N_3$
976 - Benzene, azido-
2871 - 1H-Benzotriazole

$C_6H_5N_3O_4$
724 - Benzenamine, 2,3-dinitro-
725 - Benzenamine, 2,4-dinitro-
726 - Benzenamine, 2,6-dinitro-
727 - Benzenamine, 3,5-dinitro-

$C_6H_5N_3O_5$
9207 - Phenol, 2-amino-4,6-dinitro-

$C_6H_5N_5O_2$
10749 - 4,6-Pteridinedione, 2-amino-1,5-dihydro-

C_6H_6
867 - Benzene
4767 - 1,3-Cyclopentadiene, 5-methylene-
6712 - 1,3-Hexadien-5-yne
6713 - 1,5-Hexadien-3-yne
6715 - 1,4-Hexadiyne
6716 - 1,5-Hexadiyne
6717 - 2,4-Hexadiyne

$C_6H_6AsNO_5$
459 - Arsonic acid, (4-nitrophenyl)-

C_6H_6BrN
638 - Benzenamine, 2-bromo-
639 - Benzenamine, 3-bromo-
640 - Benzenamine, 4-bromo-

$C_6H_6Br_2N_2$
8768 - Pentanedinitrile, 2-bromo-2-(bromomethyl)-

C_6H_6ClN
658 - Benzenamine, 2-chloro-
659 - Benzenamine, 3-chloro-
660 - Benzenamine, 4-chloro-
10919 - Pyridine, 2-chloro-4-methyl-
10920 - Pyridine, 2-chloro-6-methyl-

C_6H_6ClNO
9204 - Phenol, 2-amino-4-chloro-
10918 - Pyridine, 2-chloro-6-methoxy-

$C_6H_6ClN_2O_4$
1559 - 1,3-Benzenedisulfonamide, 4,5-dichloro-

$C_6H_6Cl_2Si$
11403 - Silane, dichlorophenyl-

$C_6H_6Cl_4$
6697 - 2,4-Hexadiene, 1,3,4,6-tetrachloro-

$C_6H_6Cl_6$
4406 - Cyclohexane, 1,2,3,4,5,6-hexachloro-, (1α,2β,3α,4β,5α,6β)-
4407 - Cyclohexane, 1,2,3,4,5,6-hexachloro-, (1α,2α,3α,4β,5α,6β)-
4408 - Cyclohexane, 1,2,3,4,5,6-hexachloro-, (1α,2α,3β,4α,5β,6β)-(±)-
4409 - Cyclohexane, 1,2,3,4,5,6-hexachloro-, (1α,2α,3β,4α,5α,6β)-
7035 - 3-Hexene, 1,2,3,4,5,6-hexachloro-

C_6H_6FN
752 - Benzenamine, 2-fluoro-
753 - Benzenamine, 3-fluoro-
754 - Benzenamine, 4-fluoro-

$C_6H_6FNO_2S$
2252 - Benzenesulfonyl fluoride, 4-amino-

C_6H_6IN
760 - Benzenamine, 3-iodo-
761 - Benzenamine, 4-iodo-

$C_6H_6N_2O$
828 - Benzenamine, 4-nitroso-
9914 - Propanedinitrile, (ethoxymethylene)-
10890 - 3-Pyridinecarboxamide

$C_6H_6N_2O_2$
822 - Benzenamine, 2-nitro-
823 - Benzenamine, 3-nitro-
824 - Benzenamine, 4-nitro-
825 - Benzenamine, N-nitro-
10897 - 3-Pyridinecarboxylic acid, 6-amino-

$C_6H_6N_2O_3$
9214 - Phenol, 2-amino-3-nitro-

$C_6H_6N_4O_3$
10765 - 1H-Purine-2,6,8(3H)-trione, 4,9-dihydro-3-methyl-
10766 - 1H-Purine-2,6,8(3H)-trione, 7,9-dihydro-7-methyl-

$C_6H_6N_4O_4$
7144 - Hydrazinecarboxamide, 2-[(5-nitro-2-furanyl)methylene]-
7156 - Hydrazine, (2,4-dinitrophenyl)-

C_6H_6O
6102 - Furan, 2-ethenyl-
9199 - Phenol

C_6H_6OS
5870 - Ethanone, 1-(2-thienyl)-
9376 - Phenol, 2-mercapto-

9377 - Phenol, 3-mercapto-
9378 - Phenol, 4-mercapto-
11736 - 2-Thiophenecarboxaldehyde, 5-methyl-

$C_6H_6O_2$
1497 - 1,2-Benzenediol
1498 - 1,3-Benzenediol
1499 - 1,4-Benzenediol
5817 - Ethanone, 1-(2-furanyl)-
6045 - 2-Furancarboxaldehyde, 5-methyl-

$C_6H_6O_2S$
2180 - Benzenesulfinic acid
5835 - Ethanone, 1-(3-hydroxy-2-thienyl)-

$C_6H_6O_2Se$
2174 - Benzeneseleninic acid

$C_6H_6O_3$
2366 - 1,2,3-Benzenetriol
2367 - 1,2,4-Benzenetriol
2368 - 1,3,5-Benzenetriol
6026 - 2-Furanacetic acid
6044 - 2-Furancarboxaldehyde, 5-(hydroxymethyl)-
6059 - 2-Furancarboxylic acid, methyl ester
6060 - 2-Furancarboxylic acid, 5-methyl-
6063 - 3-Furancarboxylic acid, methyl ester
6065 - 3-Furancarboxylic acid, 5-methyl-
6097 - 2,5-Furandione, 3,4-dimethyl-
10796 - 4H-Pyran-4-one, 3-hydroxy-2-methyl-

$C_6H_6O_3S$
2214 - Benzenesulfonic acid

$C_6H_6O_4$
3910 - 2-Butynedioic acid, dimethyl ester
10795 - 4H-Pyran-4-one, 5-hydroxy-2-(hydroxymethyl)-

$C_6H_6O_4S$
2225 - Benzenesulfonic acid, 2-hydroxy-
2226 - Benzenesulfonic acid, 4-hydroxy-

$C_6H_6O_7S_2$
1560 - 1,3-Benzenedisulfonic acid, 4-hydroxy-

C_6H_6S
2289 - Benzenethiol

$C_6H_6S_2$
1561 - 1,2-Benzenedithiol
1562 - 1,3-Benzenedithiol

C_6H_6Se
2177 - Benzeneselenol

$C_6H_7AsO_3$
460 - Arsonic acid, phenyl-

$C_6H_7BO_2$
3135 - Boronic acid, phenyl-

$C_6H_7BrN_2$
7141 - Hydrazine, (4-bromophenyl)-

$C_6H_7ClN_2$
1264 - 1,2-Benzenediamine, 4-chloro-

C_6H_7ClO
6711 - 2,4-Hexadienoyl chloride, (E,E)-

$C_6H_7ClO_4$
3773 - 2-Butenedioic acid, 2-chloro-, dimethyl ester, (E)-

C_6H_7ClSi
11384 - Silane, chlorophenyl-

$C_6H_7Cl_2F_3O_2$
10721 - Propionic acid, 2,2-dichloro-3,3,3-trifluoro-, propyl ester

$C_6H_7Cl_2N$
665 - Benzenamine, 2-chloro-, hydrochloride

$C_6H_7F_3O_3$
3592 - Butanoic acid, 4,4,4-trifluoro-3-oxo-, ethyl ester

$C_6H_7KO_2$
6704 - 2,4-Hexadienoic acid, potassium salt, (E,E)-

C_6H_7N
633 - Benzenamine
4907 - 1-Cyclopentenecarbonitrile
10980 - Pyridine, 2-methyl-
10981 - Pyridine, 3-methyl-
10982 - Pyridine, 4-methyl-

C_6H_7NO
757 - Benzenamine, N-hydroxy-

5820 - Ethanone, 1-(1H-pyrrol-2-yl)-

C_6H_7NO
9200 - Phenol, amino-
9201 - Phenol, 2-amino-
9202 - Phenol, 3-amino-
9203 - Phenol, 4-amino-
10975 - 2-Pyridinemethanol
10976 - 3-Pyridinemethanol
10977 - 4-Pyridinemethanol
10988 - Pyridine, 3-methyl-, 1-oxide
11035 - 2(1H)-Pyridinone, 1-methyl-
11036 - 2(1H)-Pyridinone, 3-methyl-
11037 - 2(1H)-Pyridinone, 4-methyl-

$C_6H_7NO_2$
11122 - 1H-Pyrrole-2,5-dione, 1-ethyl-

$C_6H_7NO_3S$
2215 - Benzenesulfonic acid, 2-amino-
2216 - Benzenesulfonic acid, 3-amino-
2217 - Benzenesulfonic acid, 4-amino-

$C_6H_7NO_4$
6129 - Furan, 2-(methoxymethyl)-5-nitro-

C_6H_7NS
2290 - Benzenethiol, 2-amino-
2291 - Benzenethiol, 4-amino-

$C_6H_7N_3O$
10905 - 4-Pyridinecarboxylic acid, hydrazide

$C_6H_7N_3O_2$
1281 - 1,2-Benzenediamine, 4-nitro-
7168 - Hydrazine, (4-nitrophenyl)-

$C_6H_7N_5O$
10769 - 6H-Purin-6-one, 2-amino-1,7-dihydro-7-methyl-

$C_6H_7O_3Sb$
11548 - Stibine oxide, dihydroxyphenyl-

C_6H_7P
9519 - Phosphine, phenyl-

C_6H_8
4172 - Cyclobutane, 1,2-bis(methylene)-
4248 - 1,3-Cyclohexadiene
4249 - 1,4-Cyclohexadiene
6992 - 1,3,5-Hexatriene, (E)-
6993 - 1,3,5-Hexatriene, (Z)-
7103 - 1-Hexen-3-yne
7104 - 1-Hexen-4-yne
7105 - 1-Hexen-5-yne
7106 - 2-Hexen-4-yne
9116 - 1-Penten-3-yne, 2-methyl-
9117 - 3-Penten-1-yne, 3-methyl-

$C_6H_8AsNO_3$
456 - Arsonic acid, (4-aminophenyl)-

C_6H_8ClN
756 - Benzenamine, hydrochloride

C_6H_8ClNS
11670 - Thiazole, 5-(2-chloroethyl)-4-methyl-

$C_6H_8ClN_3O_4S_2$
1558 - 1,3-Benzenedisulfonamide, 4-amino-6-chloro-

$C_6H_8Cl_2$
6690 - 2,4-Hexadiene, 1,3-dichloro-

$C_6H_8Cl_2O_3$
3503 - Butanoic acid, 2,2-dichloro-3-oxo-, ethyl ester

$C_6H_8Cl_2O_4$
129 - Acetic acid, chloro-, 1,2-ethanediyl ester
5554 - Ethanedioic acid, bis(2-chloroethyl) ester

$C_6H_8N_2$
1260 - 1,2-Benzenediamine
1261 - 1,3-Benzenediamine
1262 - 1,4-Benzenediamine
6777 - Hexanedinitrile
7169 - Hydrazine, phenyl-
10805 - Pyrazine, 2,3-dimethyl-
10806 - Pyrazine, 2,5-dimethyl-
10807 - Pyrazine, 2,6-dimethyl-
10866 - 2-Pyridinamine, 3-methyl-
10867 - 2-Pyridinamine, 4-methyl-
10868 - 2-Pyridinamine, 6-methyl-
10869 - 2-Pyridinamine, N-methyl-
10870 - 3-Pyridinamine, 4-methyl-
10971 - 2-Pyridinemethanamine

10972 - 3-Pyridinemethanamine
10973 - 4-Pyridinemethanamine
11055 - Pyrimidine, 2,4-dimethyl-
11056 - Pyrimidine, 2,5-dimethyl-
11057 - Pyrimidine, 4,5-dimethyl-
11058 - Pyrimidine, 4,6-dimethyl-

$C_6H_8N_2O$
9274 - Phenol, 2,4-diamino-
10114 - Propanenitrile, 3,3'-oxybis-
10808 - Pyrazineethanol
10865 - 3-Pyridinamine, 6-methoxy-

$C_6H_8N_2O_2S$
2187 - Benzenesulfonamide, 4-amino-

$C_6H_8N_2O_3$
821 - Benzenamine, nitrate

$C_6H_8N_2O_3S$
2224 - Benzenesulfonic acid, 4-hydrazino-

$C_6H_8N_6O_{18}$
7424 - D-Mannitol, hexanitrate

C_6H_8O
4703 - 2-Cyclohexen-1-one
4704 - 3-Cyclohexen-1-one
4908 - 1-Cyclopentene-1-carboxaldehyde
4944 - 2-Cyclopenten-1-one, 2-methyl-
4945 - 2-Cyclopenten-1-one, 3-methyl-
4946 - 2-Cyclopenten-1-one, 5-methyl-
6088 - Furan, 2,4-dimethyl-
6089 - Furan, 2,5-dimethyl-
6105 - Furan, 2-ethyl-
6675 - 2,4-Hexadienal, (E,E)-
7113 - 3-Hexynal
7130 - 3-Hexyn-2-one
7131 - 5-Hexyn-2-one

$C_6H_8O_2$
3176 - 1,3-Butadien-1-ol, acetate
3918 - 2-Butynoic acid, ethyl ester
4375 - 1,2-Cyclohexanedione
4376 - 1,3-Cyclohexanedione
4377 - 1,4-Cyclohexanedione
4909 - 1-Cyclopentene-1-carboxylic acid
6103 - Furan, 2-ethoxy-
6121 - 2-Furanmethanol, α-methyl-
6122 - 2-Furanmethanol, 5-methyl-
6128 - Furan, 2-(methoxymethyl)-
6698 - 2,4-Hexadienoic acid
7123 - 2-Hexynoic acid
9139 - 2-Pentynoic acid, methyl ester
10694 - 2-Propenoic acid, 2-propenyl ester
10791 - 2H-Pyran-2-one, 5,6-dihydro-6-methyl-, (S)-

$C_6H_8O_3$
6139 - 2(3H)-Furanone, 3-acetyldihydro-
10680 - 2-Propenoic acid, oxiranylmethyl ester

$C_6H_8O_4$
3778 - 2-Butenedioic acid (E)-, dimethyl ester
3779 - 2-Butenedioic acid (Z)-, dimethyl ester
3791 - 2-Butenedioic acid (E)-, monoethyl ester
4182 - 1,3-Cyclobutanedicarboxylic acid, cis-
4185 - 1,3-Cyclobutanedicarboxylic acid, trans-
5190 - 1,3-Dioxane-4,6-dione, 2,2-dimethyl-
9965 - Propanedioic acid, (1-methylethylidene)-
9979 - Propanedioic acid, 2-propenyl-

$C_6H_8O_5$
467 - L-Ascorbic acid, 6-deoxy-

$C_6H_8O_6$
465 - DL-Ascorbic acid
466 - L-Ascorbic acid
6239 - D-Glucuronic acid, γ-lactone
10165 - 1,2,3-Propanetricarboxylic acid

$C_6H_8O_7$
10167 - 1,2,3-Propanetricarboxylic acid, 2-hydroxy-

C_6H_8S
11754 - Thiophene, 2,3-dimethyl-
11755 - Thiophene, 2,4-dimethyl-
11756 - Thiophene, 2,5-dimethyl-
11757 - Thiophene, 3,4-dimethyl-
11759 - Thiophene, 2-ethyl-
11760 - Thiophene, 3-ethyl-

C_6H_8Si
11446 - Silane, phenyl-

C_6H_9Br
4627 - Cyclohexene, 1-bromo-
4628 - Cyclohexene, 3-bromo-

C_6H_9BrO
4574 - Cyclohexanone, 2-bromo-

$C_6H_9BrO_2$
3819 - 2-Butenoic acid, 4-bromo-, ethyl ester, (E)-

$C_6H_9BrO_3$
3477 - Butanoic acid, 2-bromo-3-oxo-, ethyl ester
3478 - Butanoic acid, 4-bromo-3-oxo-, ethyl ester

C_6H_9Cl
4642 - Cyclohexene, 1-chloro-
6688 - 1,3-Hexadiene, 3-chloro-
8664 - 1,2-Pentadiene, 1-chloro-3-methyl-
9126 - 1-Pentyne, 3-chloro-3-methyl-
9127 - 2-Pentyne, 4-chloro-4-methyl-

$C_6H_9ClN_2$
7172 - Hydrazine, phenyl-, monohydrochloride

C_6H_9ClO
4577 - Cyclohexanone, 2-chloro-
4578 - Cyclohexanone, 4-chloro-

$C_6H_9ClO_2$
3823 - 2-Butenoic acid, 2-chloro-, ethyl ester
3824 - 2-Butenoic acid, 2-chloro-, ethyl ester, (E)-
3825 - 2-Butenoic acid, 2-chloro-, ethyl ester, (Z)-
3826 - 2-Butenoic acid, 3-chloro-, ethyl ester
3827 - 2-Butenoic acid, 3-chloro-, ethyl ester, (E)-
3828 - 2-Butenoic acid, 3-chloro-, ethyl ester, (Z)-
3829 - 2-Butenoic acid, 4-chloro-, ethyl ester

$C_6H_9ClO_3$
3494 - Butanoic acid, 4-chloro-3-oxo-, ethyl ester
3570 - Butanoic acid, 3-oxo-, 2-chloroethyl ester

$C_6H_9Cl_3$
7048 - 1-Hexene, 1,1,2-trichloro-

$C_6H_9Cl_3O_2$
245 - Acetic acid, trichloro-, butyl ester
247 - Acetic acid, trichloro-, 1,1-dimethylethyl ester
251 - Acetic acid, trichloro-, 1-methylpropyl ester
252 - Acetic acid, trichloro-, 2-methylpropyl ester
3591 - Butanoic acid, 2,2,3-trichloro-, ethyl ester

$C_6H_9Cl_3O_3$
13 - Acetaldehyde, chloro-, trimer

$C_6H_9F_3O_2$
265 - Acetic acid, trifluoro-, butyl ester
266 - Acetic acid, trifluoro-, 1,1-dimethylethyl ester

C_6H_9N
4779 - Cyclopentanecarbonitrile
11119 - 1H-Pyrrole, 2,4-dimethyl-
11120 - 1H-Pyrrole, 2,5-dimethyl-
11124 - 1H-Pyrrole, 1-ethyl-
11125 - 1H-Pyrrole, 2-ethyl-

C_6H_9NO
6115 - 2-Furanmethanamine, N-methyl-
11183 - 2-Pyrrolidinone, 1-ethenyl-

C_6H_9NOS
11677 - 5-Thiazoleethanol, 4-methyl-

$C_6H_9NO_2$
10912 - 3-Pyridinecarboxylic acid, 1,2,5,6-tetrahydro-
11157 - 2,5-Pyrrolidinedione, 1-ethyl-

$C_6H_9NO_3$
50 - Acetamide, N,N-diacetyl-
8587 - 2,4-Oxazolidinedione, 3,5,5-trimethyl-

$C_6H_9NO_6$
6256 - Glycine, N,N-bis(carboxymethyl)-

$C_6H_9N_3$
2305 - 1,2,3-Benzenetriamine
2306 - 1,2,4-Benzenetriamine
11044 - 2-Pyrimidinamine, 4,5-dimethyl-
11045 - 4-Pyrimidinamine, 2,6-dimethyl-

$C_6H_9N_3O_2$
759 - Benzenamine, N-hydroxy-N-nitroso-, ammonium salt
7132 - L-Histidine

$C_6H_9N_3O_3$
7206 - 1H-Imidazole-1-ethanol, 2-methyl-5-nitro-

11841 - 1,3,5-Triazine-2,4,6(1H,3H,5H)-trione, 1,3,5-
trimethyl-

C_6H_{10}

3159 - 1,3-Butadiene, 2,3-dimethyl-
3907 - 1-Butyne, 3,3-dimethyl-
4626 - Cyclohexene
4827 - Cyclopentane, methylene-
4928 - Cyclopentene, 1-methyl-
4929 - Cyclopentene, 3-methyl-
4930 - Cyclopentene, 4-methyl-
4982 - Cyclopropane, (1-methylethenyl)-
6676 - 1,2-Hexadiene
6677 - 1,4-Hexadiene
6678 - 1,5-Hexadiene
6679 - 2,3-Hexadiene
6680 - 2,4-Hexadiene
6681 - 1,3-Hexadiene, (E)-
6682 - 1,3-Hexadiene, (Z)-
6683 - 1,4-Hexadiene, (E)-
6684 - 1,4-Hexadiene, (Z)-
6685 - 2,4-Hexadiene, (E,E)-
6686 - 2,4-Hexadiene, (E,Z)-
6687 - 2,4-Hexadiene, (Z,Z)-
7114 - 1-Hexyne
7115 - 2-Hexyne
7116 - 3-Hexyne
8666 - 1,2-Pentadiene, 3-methyl-
8667 - 1,2-Pentadiene, 4-methyl-
8668 - 1,3-Pentadiene, 2-methyl-, (Z)-
8669 - 1,3-Pentadiene, 3-methyl-
8670 - 1,3-Pentadiene, 4-methyl-
8671 - 1,4-Pentadiene, 2-methyl-
8672 - 2,3-Pentadiene, 2-methyl-
9060 - 1-Pentene, 3-methylene-
9134 - 1-Pentyne, 3-methyl-
9135 - 1-Pentyne, 4-methyl-
9136 - 2-Pentyne, 4-methyl-

$C_6H_{10}Br_2$

4345 - Cyclohexane, 1,1-dibromo-
4346 - Cyclohexane, 1,2-dibromo-, cis-
4347 - Cyclohexane, 1,2-dibromo-, trans-(±)-
4348 - Cyclohexane, 1,3-dibromo-, trans-
4349 - Cyclohexane, 1,4-dibromo-, cis-

$C_6H_{10}Br_2O_2$

3498 - Butanoic acid, 2,3-dibromo-, ethyl ester
3499 - Butanoic acid, 2,4-dibromo-, ethyl ester
10243 - Propanoic acid, 2,3-dibromo-, propyl ester

$C_6H_{10}ClFO_2$

132 - Acetic acid, chlorofluoro-, butyl ester

$C_6H_{10}Cl_2$

4358 - Cyclohexane, 1,1-dichloro-
4359 - Cyclohexane, 1,2-dichloro-, cis-
4360 - Cyclohexane, 1,2-dichloro-, trans-(±)-
4361 - Cyclohexane, 1,4-dichloro-, cis-
7011 - 1-Hexene, 1,2-dichloro-, (E)-
7012 - 1-Hexene, 1,2-dichloro-, (Z)-

$C_6H_{10}Cl_2O_2$

163 - Acetic acid, dichloro-, butyl ester
10249 - Propanoic acid, 2,3-dichloro-, 1-methylethyl
ester

$C_6H_{10}Cl_3NO$

96 - Acetamide, 2,2,2-trichloro-N,N-diethyl-

$C_6H_{10}N_2$

8830 - Pentanenitrile, 3-imino-2-methyl-
10833 - 1H-Pyrazole, 1-ethyl-3-methyl-
10839 - 1H-Pyrazole, 1,3,4-trimethyl-
10840 - 1H-Pyrazole, 1,3,5-trimethyl-
10841 - 1H-Pyrazole, 1,4,5-trimethyl-
10842 - 1H-Pyrazole, 3,4,5-trimethyl-

$C_6H_{10}N_2O_2$

4379 - 1,2-Cyclohexanedione, dioxime

$C_6H_{10}N_4$

11654 - 5H-Tetrazolo[1,5-a]azepine, 6,7,8,9-
tetrahydro-

$C_6H_{10}N_4O_3S_2$

3197 - Butanamide, N-[5-(aminosulfonyl)-1,3,4-
thiadiazol-2-yl]-

$C_6H_{10}O$

3160 - 1,3-Butadiene, 1-ethoxy-
3161 - 1,3-Butadiene, 2-ethoxy-
3377 - Butane, 1-(ethynyloxy)-
3877 - 1-Buten-1-one, 2-ethyl-
4571 - Cyclohexanone
4694 - 2-Cyclohexen-1-ol
4695 - 3-Cyclohexen-1-ol
4780 - Cyclopentanecarboxaldehyde
4885 - Cyclopentanone, 2-methyl-
4886 - Cyclopentanone, 3-methyl-, (±)-
5792 - Ethanone, 1-cyclobutyl-
6705 - 1,4-Hexadien-3-ol
6706 - 1,5-Hexadien-3-ol
6707 - 2,4-Hexadien-1-ol
6708 - 3,5-Hexadien-2-ol
6709 - 2,4-Hexadien-1-ol, (E,E)-
6996 - 2-Hexenal, (E)-
6997 - 3-Hexenal, (E)-
6998 - 3-Hexenal, (Z)-
7086 - 3-Hexen-2-one
7087 - 4-Hexen-2-one
7088 - 4-Hexen-3-one
7089 - 5-Hexen-2-one
7090 - 3-Hexen-2-one, (E)-
7124 - 1-Hexyn-3-ol
7125 - 2-Hexyn-1-ol
7126 - 3-Hexyn-1-ol
7127 - 5-Hexyn-2-ol
8554 - 7-Oxabicyclo[4.1.0]heptane
9002 - 2-Pentenal, 2-methyl-
9107 - 1-Penten-3-one, 2-methyl-
9108 - 3-Penten-2-one, 4-methyl-
9109 - 4-Penten-2-one, 4-methyl-
9146 - 1-Pentyn-3-ol, 3-methyl-
10592 - 1-Propene, 3,3'-oxybis-
10778 - 2H-Pyran, 3,6-dihydro-4-methyl-

$C_6H_{10}OS$

10596 - 1-Propene, 3,3'-sulfinylbis-

$C_6H_{10}OS_2$

10595 - 2-Propene-1-sulfinothioic acid, S-2-propenyl
ester

$C_6H_{10}O_2$

3513 - Butanoic acid, ethenyl ester
3835 - 3-Butenoic acid, 2,2-dimethyl-
3837 - 2-Butenoic acid, 2-ethyl-, (E)-
3838 - 2-Butenoic acid, ethyl ester
3839 - 2-Butenoic acid, ethyl ester, (E)-
3840 - 2-Butenoic acid, ethyl ester, (Z)-
3841 - 3-Butenoic acid, ethyl ester
3851 - 2-Butenoic acid, 2-methyl-, methyl ester, (E)-
3852 - 2-Butenoic acid, 3-methyl-, methyl ester
3862 - 2-Buten-1-ol, acetate
4176 - Cyclobutanecarboxylic acid, methyl ester
4781 - Cyclopentanecarboxylic acid
4961 - Cyclopropanecarboxylic acid, ethyl ester
6150 - 2(3H)-Furanone, 5-ethyldihydro-
6693 - 1,5-Hexadiene-3,4-diol, (R*,R*)-(±)-
6748 - Hexanedial
6804 - 2,4-Hexanedione
6805 - 2,5-Hexanedione
6806 - 3,4-Hexanedione
7049 - 2-Hexenoic acid
7050 - 3-Hexenoic acid
7051 - 5-Hexenoic acid
7118 - 3-Hexyne-2,5-diol
8591 - 2-Oxepanone
8626 - Oxirane, [(2-propenyloxy)methyl]-
8890 - Pentanoic acid, 2-methylene-
9076 - 2-Pentenoic acid, 2-methyl-, (E)-
9077 - 2-Pentenoic acid, 3-methyl-, (Z)-
9078 - 2-Pentenoic acid, 4-methyl-
9079 - 3-Pentenoic acid, 3-methyl-
10297 - Propanoic acid, 2-methyl-, ethenyl ester
10325 - Propanoic acid, 2-propenyl ester
10658 - 2-Propenoic acid, 2-methyl-, ethyl ester

$C_6H_{10}O_2S$

3589 - Butanoic acid, 3-thioxo-, ethyl ester
10597 - 1-Propene, 3,3'-sulfonylbis-

$C_6H_{10}O_2S_2$

5453 - Ethanebis(thioic) acid, S,S-diethyl ester

$C_6H_{10}O_3$

3510 - Butanoic acid, 3,3-dimethyl-2-oxo-
3558 - Butanoic acid, 2-methyl-3-oxo-, methyl ester
3571 - Butanoic acid, 3-oxo-, ethyl ester
3572 - Butanoic acid, 4-oxo-, ethyl ester
3843 - 3-Butenoic acid, 2-hydroxy-, ethyl ester
5206 - 1,3-Dioxolane-4-carboxaldehyde, 2,2-
dimethyl-
6087 - Furan, 2,5-dihydro-2,5-dimethoxy-
6143 - 2(3H)-Furanone, dihydro-3-hydroxy-4,4-
dimethyl-, (R)-
6808 - 2,5-Hexanedione, 3-hydroxy-
6905 - Hexanoic acid, 5-oxo-
8895 - Pentanoic acid, 2-methyl-4-oxo-
8896 - Pentanoic acid, 3-methyl-4-oxo-
8908 - Pentanoic acid, 4-oxo-, methyl ester
10203 - Propanoic acid, anhydride
10281 - Propanoic acid, 2-hydroxy-, 2-propenyl ester
10661 - 2-Propenoic acid, 2-methyl-, 2-hydroxyethyl
ester
10774 - 2H-Pyran-4-carboxylic acid, tetrahydro-

$C_6H_{10}O_4$

102 - Acetic acid, (acetyloxy)-, ethyl ester
3317 - Butanedioic acid, 2,3-dimethyl-
3318 - Butanedioic acid, dimethyl ester
3322 - Butanedioic acid, ethyl-, (R)-
3335 - Butanedioic acid, monoethyl ester
5560 - Ethanedioic acid, diethyl ester
5570 - 1,1-Ethanediol, diacetate
5571 - 1,2-Ethanediol, diacetate
6779 - Hexanedioic acid
7423 - D-Mannitol, 1,4:3,6-dianhydro-
8779 - Pentanedioic acid, 3-methyl-
8780 - Pentanedioic acid, monomethyl ester
9963 - Propanedioic acid, methyl-, dimethyl ester

$C_6H_{10}O_4S$

10329 - Propanoic acid, 3,3'-thiobis-

$C_6H_{10}O_4S_2$

5179 - Dimethipin

$C_6H_{10}O_5$

5169 - Dicarbonic acid, diethyl ester

$(C_6H_{10}O_5)_x$

6274 - Glycogen

$C_6H_{10}O_6$

3308 - Butanedioic acid, 2,3-dihydroxy-, dimethyl
ester, (R*,R*)-(±)-
3315 - Butanedioic acid, 2,3-dihydroxy- [R-(R*,R*)]-,
monoethyl ester
6211 - Gluconic acid, β-lactone
6212 - D-Gluconic acid, δ-lactone

$C_6H_{10}O_7$

6199 - D-Galacturonic acid
6238 - D-Glucuronic acid

$C_6H_{10}O_8$

6193 - Galactaric acid
6205 - D-Glucaric acid

$C_6H_{10}S$

4481 - Cyclohexanethione
10599 - 1-Propene, 3,3'-thiobis-

$C_6H_{10}S_3$

11917 - Trisulfide, di-2-propenyl

$C_6H_{11}Br$

4315 - Cyclohexane, bromo-

$C_6H_{11}BrN_2O_2$

3196 - Butanamide, N-(aminocarbonyl)-2-bromo-3-
methyl-

$C_6H_{11}BrO$

4503 - Cyclohexanol, 2-bromo-
6961 - 2-Hexanone, 6-bromo-

$C_6H_{11}BrO_2$

114 - Acetic acid, bromo-, 1,1-dimethylethyl ester
118 - Acetic acid, bromo-, 2-methylpropyl ester

3469 - Butanoic acid, 2-bromo-, ethyl ester
3470 - Butanoic acid, 4-bromo-, ethyl ester
3476 - Butanoic acid, 2-bromo-3-methyl-, methyl ester
6875 - Hexanoic acid, 2-bromo-, (±)-
6876 - Hexanoic acid, 6-bromo-
10213 - Propanoic acid, 2-bromo-2-methyl-, ethyl ester

$C_6H_{11}Cl$
3731 - 2-Butene, 1-chloro-2,3-dimethyl-
4339 - Cyclohexane, chloro-
7004 - 1-Hexene, 1-chloro-
7005 - 1-Hexene, 2-chloro-
7006 - 1-Hexene, 5-chloro-
7007 - 2-Hexene, 4-chloro-
7008 - 3-Hexene, 1-chloro-
7009 - 3-Hexene, 3-chloro-, (Z)-
9025 - 1-Pentene, 3-chloro-2-methyl-
9026 - 2-Pentene, 5-chloro-2-methyl-

$C_6H_{11}ClO$
3688 - Butanoyl chloride, 2,2-dimethyl-
3689 - Butanoyl chloride, 2,3-dimethyl-
3690 - Butanoyl chloride, 3,3-dimethyl-
3691 - Butanoyl chloride, 2-ethyl-
4505 - Cyclohexanol, 2-chloro-, trans-
4506 - Cyclohexanol, 2-chloro-, cis-(±)-
4507 - Cyclohexanol, 4-chloro-, trans-
6986 - Hexanoyl chloride
8993 - Pentanoyl chloride, 2-methyl-, (±)-
8994 - Pentanoyl chloride, 3-methyl-, (±)-

$C_6H_{11}ClO_2$
125 - Acetic acid, chloro-, butyl ester
128 - Acetic acid, chloro-, 1,1-dimethylethyl ester
142 - Acetic acid, chloro-, 1-methylpropyl ester
143 - Acetic acid, chloro-, 2-methylpropyl ester
3483 - Butanoic acid, 2-chloro-, ethyl ester
3484 - Butanoic acid, 3-chloro-, ethyl ester
3485 - Butanoic acid, 4-chloro-, ethyl ester
3730 - 2-Butene, 1-chloro-1,1-dimethoxy-
4021 - Carbonochloridic acid, 3-methylbutyl ester
4025 - Carbonochloridic acid, pentyl ester
10223 - Propanoic acid, 2-chloro-, 1-methylethyl ester
10232 - Propanoic acid, 3-chloro-, propyl ester
10292 - Propanoic acid, 2-methyl-, 2-chloroethyl ester

$C_6H_{11}Cl_3O_2$
5657 - Ethane, 1,1,1-trichloro-2,2-diethoxy-

$C_6H_{11}F$
4403 - Cyclohexane, fluoro-

$C_6H_{11}FO_2$
190 - Acetic acid, fluoro-, 1,1-dimethylethyl ester
6888 - Hexanoic acid, 6-fluoro-

$C_6H_{11}I$
4415 - Cyclohexane, iodo-

$C_6H_{11}IO_2$
3928 - Butyric acid, 2-iodo-, ethyl ester

$C_6H_{11}N$
478 - 7-Azabicyclo[4.1.0]heptane
3407 - Butanenitrile, 2-ethyl-
6840 - Hexanenitrile
8821 - Pentane, 1-isocyano-
8831 - Pentanenitrile, 4-methyl-
10526 - 2-Propen-1-amine, N-2-propenyl-

$C_6H_{11}NO$
493 - 2H-Azepin-2-one, hexahydro-
3386 - Butane, 1-isocyanato-3-methyl-
4610 - Cyclohexanone, oxime
9638 - 1-Piperidinecarboxaldehyde
9721 - 2-Piperidinone, 1-methyl-
9722 - 2-Piperidinone, 3-methyl-
9723 - 3-Piperidinone, 1-methyl-
9724 - 4-Piperidinone, 1-methyl-
11181 - 2-Pyrrolidinone, 1,5-dimethyl-
11182 - 2-Pyrrolidinone, 3,3-dimethyl-

$C_6H_{11}NO_2$
24 - Acetamide, N-acetyl-N-ethyl-

3816 - 2-Butenoic acid, 3-amino-, ethyl ester, (E)-
4454 - Cyclohexane, nitro-
4782 - Cyclopentanecarboxylic acid, 1-amino-
4830 - Cyclopentane, 1-methyl-1-nitro-
4831 - Cyclopentane, 1-methyl-2-nitro-
4840 - Cyclopentane, (nitromethyl)-
7682 - Morpholine, 4-acetyl-
9640 - 2-Piperidinecarboxylic acid
9641 - 3-Piperidinecarboxylic acid
9642 - 4-Piperidinecarboxylic acid
11184 - 2-Pyrrolidinone, 1-(2-hydroxyethyl)-

$C_6H_{11}NO_3S$
333 - L-Alanine, 3-(2-propenylsulfinyl)-, (S)-

$C_6H_{11}NO_4$
7231 - Imidodicarbonic acid, diethyl ester

$C_6H_{11}NS$
3390 - Butane, 1-isothiocyanato-3-methyl-
11673 - Thiazole, 4,5-dihydro-2,4,4-trimethyl-

$C_6H_{11}N_2O_4PS_3$
7645 - Methidathion

$C_6H_{11}N_3$
11847 - 1H-1,2,4-Triazole, 1-butyl

$C_6H_{11}NaO_7$
6213 - D-Gluconic acid, monosodium salt

$C_6H_{11}O_3P$
11909 - 2,6,7-Trioxa-1-phosphabicyclo[2.2.2]octane, 4-ethyl-

C_6H_{12}
3759 - 1-Butene, 2,3-dimethyl-
3760 - 1-Butene, 3,3-dimethyl-
3761 - 2-Butene, 2,3-dimethyl-
4188 - Cyclobutane, 1,2-dimethyl-, trans-
4189 - Cyclobutane, 1,3-dimethyl-, cis-
4190 - Cyclobutane, 1,3-dimethyl-, trans-
4191 - Cyclobutane, ethyl-
4305 - Cyclohexane
4826 - Cyclopentane, methyl-
4975 - Cyclopropane, 1-ethyl-1-methyl-
4976 - Cyclopropane, 1-ethyl-2-methyl-, cis-
4977 - Cyclopropane, 1-ethyl-2-methyl-, trans-
4983 - Cyclopropane, (1-methylethyl)-
4984 - Cyclopropane, propyl-
4985 - Cyclopropane, 1,1,2-trimethyl-
4986 - Cyclopropane, 1,2,3-trimethyl-, (1α,2α,3α)-
4987 - Cyclopropane, 1,2,3-trimethyl-, (1α,2α,3β)-
6999 - 1-Hexene
7000 - 2-Hexene, (E)-
7001 - 2-Hexene, (Z)-
7002 - 3-Hexene, (E)-
7003 - 3-Hexene, (Z)-
8825 - Pentane, 3-methylene-
9052 - 1-Pentene, 2-methyl-
9053 - 1-Pentene, 3-methyl-
9054 - 1-Pentene, 4-methyl-
9055 - 2-Pentene, 2-methyl-
9056 - 2-Pentene, 3-methyl-, (E)-
9057 - 2-Pentene, 3-methyl-, (Z)-
9058 - 2-Pentene, 4-methyl-, (E)-
9059 - 2-Pentene, 4-methyl-, (Z)-

$C_6H_{12}BrF$
6735 - Hexane, 1-bromo-6-fluoro-

$C_6H_{12}BrNO$
3198 - Butanamide, 2-bromo-2-ethyl-

$C_6H_{12}Br_2$
3269 - Butane, 1,2-dibromo-2,3-dimethyl-
3270 - Butane, 1,4-dibromo-2,3-dimethyl-
3271 - Butane, 2,3-dibromo-2,3-dimethyl-
6754 - Hexane, 1,2-dibromo-
6755 - Hexane, 1,4-dibromo-
6756 - Hexane, 1,5-dibromo-
6757 - Hexane, 1,6-dibromo-
6758 - Hexane, 2,3-dibromo-
6759 - Hexane, 2,5-dibromo-, (R*,R*)-(±)-
6760 - Hexane, 3,4-dibromo-

$C_6H_{12}ClF$
6742 - Hexane, 1-chloro-6-fluoro-

$C_6H_{12}ClNO$
47 - Acetamide, 2-chloro-N-(1-methylpropyl)-

$C_6H_{12}ClNO_2$
46 - Acetamide, 2-chloro-N-(3-methoxypropyl)-

$C_6H_{12}Cl_2$
6761 - Hexane, 1,2-dichloro-
6762 - Hexane, 1,6-dichloro-
6763 - Hexane, 2,2-dichloro-
6764 - Hexane, 2,3-dichloro-
6766 - Hexane, 2,5-dichloro-, (R*,R*)-(±)-

$C_6H_{12}Cl_2O$
10121 - Propane, 1,1'-oxybis[2-chloro-
10122 - Propane, 1,1'-oxybis[3-chloro-
10123 - Propane, 2,2'-oxybis[1-chloro-

$C_6H_{12}Cl_2O_2$
5449 - Ethane, 1,2-bis(2-chloroethoxy)-
5530 - Ethane, 1,1-dichloro-2,2-diethoxy-
5592 - Ethane, 1,1'-[ethylidenebis(oxy)]bis[2-chloro-
9896 - Propane, 1,3-dichloro-2-(ethoxymethoxy)-

$C_6H_{12}Cl_2S$
10153 - Propane, 1,1'-thiobis[3-chloro-

$C_6H_{12}Cl_3N$
5412 - Ethanamine, 2-chloro-N,N-bis(2-chloroethyl)-

$C_6H_{12}Cl_3O_3P$
5702 - Ethanol, 2-chloro-, phosphite (3:1)

$C_6H_{12}Cl_3O_4P$
5701 - Ethanol, 2-chloro-, phosphate (3:1)

$C_6H_{12}I_2$
6769 - Hexane, 1,6-diiodo-

$C_6H_{12}MnN_2S_4$
7421 - Manganous dimethyldithiocarbamate

$C_6H_{12}N_2$
273 - Acetonitrile, (butylamino)-
280 - Acetonitrile, (diethylamino)-
5111 - 1,4-Diazabicyclo[2.2.2]octane
10463 - 2-Propanone, (1-methylethylidene)hydrazone

$C_6H_{12}N_2O_2$
5494 - Ethanediamide, N,N'-diethyl-

$C_6H_{12}N_2O_3$
3334 - Butanedioic acid, mono (2,2-dimethylhydrazide)

$C_6H_{12}N_2O_4$
3957 - Carbamic acid, (1-methylethyl)nitro-, ethyl ester
3964 - Carbamic acid, nitropropyl-, ethyl ester
6778 - Hexane, 1,6-dinitro-
7150 - 1,1-Hydrazinedicarboxylic acid, diethyl ester
7151 - 1,2-Hydrazinedicarboxylic acid, diethyl ester

$C_6H_{12}N_2O_4S$
5016 - L-Cysteine, S-(2-amino-2-carboxyethyl)-, (R)-

$C_6H_{12}N_2O_4S_2$
5020 - L-Cystine

$C_6H_{12}N_2S_3$
11715 - Thiodicarbonic diamide ([(H2N)C(S)]2S), tetramethyl-

$C_6H_{12}N_2S_4$
11722 - Thioperoxydicarbonic diamide ([(H2N)C(S)]2S2), tetramethyl-

$C_6H_{12}N_2S_4Zn$
12068 - Ziram

$C_6H_{12}N_4$
11601 - 1,3,5,7-Tetraazatricyclo[3.3.1.13,7]decane

$C_6H_{12}N_6O_3$
7583 - Methanol, (1,3,5-triazine-2,4,6-triyltriimino)tris-

$C_6H_{12}O$
3183 - Butanal, 2-ethyl-
3369 - Butane, 1-(ethenyloxy)-
3370 - Butane, 2-(ethenyloxy)-
3657 - 2-Butanone, 3,3-dimethyl-
4500 - Cyclohexanol
4825 - Cyclopentanemethanol
4867 - Cyclopentanol, 1-methyl-
4868 - Cyclopentanol, 2-methyl-, cis-
4869 - Cyclopentanol, 2-methyl-, trans-
4870 - Cyclopentanol, 3-methyl-
6107 - Furan, 2-ethyltetrahydro-

6718 - Hexanal
6959 - 2-Hexanone
6960 - 3-Hexanone
7066 - 1-Hexen-3-ol
7067 - 4-Hexen-2-ol
7068 - 5-Hexen-2-ol
7069 - 2-Hexen-1-ol, (E)-
7070 - 2-Hexen-1-ol, (Z)-
7071 - 3-Hexen-1-ol, (Z)-
7072 - 4-Hexen-1-ol, (E)-
8590 - Oxepane
8627 - Oxirane, tetramethyl-
8693 - Pentanal, 2-methyl-
8694 - Pentanal, 3-methyl-
8980 - 2-Pentanone, 3-methyl-, (±)-
8981 - 2-Pentanone, 4-methyl-
8982 - 3-Pentanone, 2-methyl-
9094 - 2-Penten-1-ol, 2-methyl-
9095 - 2-Penten-1-ol, 4-methyl-
9096 - 3-Penten-1-ol, 4-methyl-
9097 - 3-Penten-2-ol, 2-methyl-
9098 - 3-Penten-2-ol, 3-methyl-
9099 - 3-Penten-2-ol, 4-methyl-
9100 - 4-Penten-2-ol, 2-methyl-
9101 - 4-Penten-2-ol, 3-methyl-
9102 - 4-Penten-2-ol, 4-methyl-
10035 - Propane, 1-(ethenyloxy)-2-methyl-
10036 - Propane, 2-(ethenyloxy)-2-methyl-
10579 - 1-Propene, 3-(1-methylethoxy)-
10594 - 1-Propene, 3-propoxy-

$C_6H_{12}O_2$
121 - Acetic acid, butyl ester
177 - Acetic acid, 1,1-dimethylethyl ester
214 - Acetic acid, 1-methylpropyl ester
215 - Acetic acid, 2-methylpropyl ester
3506 - Butanoic acid, 2,2-dimethyl-
3507 - Butanoic acid, 2,3-dimethyl-
3508 - Butanoic acid, 3,3-dimethyl-
3514 - Butanoic acid, 2-ethyl-
3515 - Butanoic acid, ethyl ester
3552 - Butanoic acid, 3-methyl-, methyl ester
3636 - 1-Butanol, 3-methyl-, formate
4372 - 1,2-Cyclohexanediol, cis-
4373 - 1,2-Cyclohexanediol, trans-(±)-
4374 - 1,4-Cyclohexanediol, trans-
5189 - 1,3-Dioxane, 2,4-dimethyl-
5901 - Ethene, 1,1-diethoxy-
6015 - Formic acid, pentyl ester
6170 - Furan, tetrahydro-2-(methoxymethyl)-
6870 - Hexanoic acid
6969 - 2-Hexanone, 5-hydroxy-
6970 - 2-Hexanone, 6-hydroxy-
6971 - 3-Hexanone, 4-hydroxy-
6972 - 3-Hexanone, 5-hydroxy-
7177 - Hydroperoxide, cyclohexyl
8691 - Pentanal, 3-hydroxy-2-methyl-
8887 - Pentanoic acid, 2-methyl-, (±)-
8888 - Pentanoic acid, 3-methyl-, (±)-
8889 - Pentanoic acid, 4-methyl-
8891 - Pentanoic acid, methyl ester
8976 - 2-Pentanone, 4-hydroxy-4-methyl-
10260 - Propanoic acid, 2,2-dimethyl-, methyl ester
10298 - Propanoic acid, 2-methyl-, ethyl ester
10299 - Propanoic acid, 1-methylethyl ester
10326 - Propanoic acid, propyl ester
10784 - 2H-Pyran-2-methanol, tetrahydro-

$C_6H_{12}O_2S$
11784 - Thiophene, tetrahydro-2,4-dimethyl-, 1,1-dioxide

$C_6H_{12}O_3$
181 - Acetic acid, ethoxy-, ethyl ester
3532 - Butanoic acid, 2-hydroxy-, ethyl ester, (±)-
3533 - Butanoic acid, 3-hydroxy-, ethyl ester, (±)-
3534 - Butanoic acid, 3-hydroxy-, ethyl ester
5207 - 1,3-Dioxolane-4-methanol, 2,2-dimethyl-
5718 - Ethanol, 2-ethoxy-, acetate
6169 - Furan, tetrahydro-2,5-dimethoxy-

6891 - Hexanoic acid, 2-hydroxy-, (±)-
7109 - Hexitol, 2,5-anhydro-3,4-dideoxy-
10275 - Propanoic acid, 2-hydroxy-, 1-methylethyl ester
10277 - Propanoic acid, 3-hydroxy-, 1-methylethyl ester
10279 - Propanoic acid, 2-hydroxy-2-methyl-, ethyl ester
11908 - 1,3,5-Trioxane, 2,4,6-trimethyl-

$C_6H_{12}O_4$
5673 - Ethanol, 2-[2-(acetyloxy)ethoxy]-
7111 - D-ribo-Hexose, 2,6-dideoxy-
10255 - Propanoic acid, 2,3-dihydroxy-2-methyl-, ethyl ester
10257 - Propanoic acid, 2,3-dihydroxy-, propyl ester

$C_6H_{12}O_4Si$
11419 - Silanediol, dimethyl-, diacetate

$C_6H_{12}O_5$
4074 - D-chiro-Inositol, 2-deoxy-
6235 - D-Glucose, 6-deoxy-
7110 - D-arabino-Hexose, 2-deoxy-

$C_6H_{12}O_5S$
184 - Acetic acid, (ethoxysulfonyl)-, ethyl ester

$C_6H_{12}O_6$
337 - D-Allose
338 - D-Altrose
4073 - DL-chiro-Inositol
6023 - β-D-Fructose
6197 - D-Galactose
6232 - β-D-Glucose
6290 - D-Gulose
7336 - scyllo-Inositol
7425 - D-Mannose
11518 - DL-Sorbose
11519 - L-Sorbose
11587 - D-Tagatose

$C_6H_{12}O_7$
6209 - D-Gluconic acid

$C_6H_{12}S$
4480 - Cyclohexanethiol
11690 - Thiepane
11765 - Thiophene, 2-ethyltetrahydro-
11766 - Thiophene, 3-ethyltetrahydro-
11785 - Thiophene, tetrahydro-2,5-dimethyl-, cis-
11796 - 2H-Thiopyran, tetrahydro-2-methyl-
11797 - 2H-Thiopyran, tetrahydro-3-methyl-
11798 - 2H-Thiopyran, tetrahydro-4-methyl-

$C_6H_{12}S_3$
11920 - 1,3,5-Trithiane, 2,4,6-trimethyl-
11921 - 1,3,5-Trithiane, 2,4,6-trimethyl-, (2α,4α,6β)-

$C_6H_{13}Br$
3235 - Butane, 1-bromo-3,3-dimethyl-
3236 - Butane, 2-bromo-2,3-dimethyl-
6732 - Hexane, 1-bromo-
6733 - Hexane, 2-bromo-
6734 - Hexane, 3-bromo-
8714 - Pentane, 1-bromo-2-methyl-
8715 - Pentane, 1-bromo-3-methyl-
8716 - Pentane, 1-bromo-4-methyl-
8717 - Pentane, 2-bromo-2-methyl-
8718 - Pentane, 3-bromo-3-methyl-

$C_6H_{13}BrO_2$
5459 - Ethane, 2-bromo-1,1-diethoxy-

$C_6H_{13}Cl$
3245 - Butane, 1-chloro-2,3-dimethyl-
3246 - Butane, 1-chloro-3,3-dimethyl-
3247 - Butane, 2-chloro-2,3-dimethyl-
3248 - Butane, 3-chloro-2,2-dimethyl-
6737 - Hexane, 1-chloro-
6738 - Hexane, 2-chloro-
6739 - Hexane, 3-chloro-
8729 - Pentane, 2-chloro-2-methyl-
8730 - Pentane, 2-chloro-4-methyl-
8731 - Pentane, 3-(chloromethyl)-
8732 - Pentane, 3-chloro-3-methyl-

$C_6H_{13}ClO$
3249 - Butane, 1-(2-chloroethoxy)-

3608 - 2-Butanol, 1-chloro-2,3-dimethyl-
3609 - 2-Butanol, 1-chloro-3,3-dimethyl-
6916 - 1-Hexanol, 6-chloro-
6917 - 2-Hexanol, 1-chloro-
6918 - 3-Hexanol, 1-chloro-
6919 - 3-Hexanol, 5-chloro-

$C_6H_{13}ClO_2$
3244 - Butane, 3-chloro-1,1-dimethoxy-
5469 - Ethane, 2-chloro-1,1-diethoxy-
10355 - 2-Propanol, 1-chloro-3-(1-methylethoxy)-
10356 - 2-Propanol, 1-chloro-3-propoxy-

$C_6H_{13}ClS$
9855 - Propane, 2-[[(2-chloroethyl)thio]-2-methyl-

$C_6H_{13}Cl_3Si$
11464 - Silane, trichlorohexyl-

$C_6H_{13}F$
6829 - Hexane, 1-fluoro-
6830 - Hexane, 2-fluoro-

$C_6H_{13}FO$
6936 - 1-Hexanol, 6-fluoro-

$C_6H_{13}I$
6831 - Hexane, 1-iodo-
6832 - Hexane, 2-iodo-, (R)-

$C_6H_{13}IO_2$
5543 - Ethane, 1,1-diethoxy-2-iodo-

$C_6H_{13}N$
492 - 1H-Azepine, hexahydro-
4283 - Cyclohexanamine
4772 - Cyclopentanamine, 1-methyl-
4773 - Cyclopentanamine, 2-methyl-
9679 - Piperidine, 1-methyl-
9680 - Piperidine, 2-methyl-, (±)-
9681 - Piperidine, 3-methyl-, (±)-
9682 - Piperidine, 4-methyl-
10525 - 2-Propen-1-amine, N-(1-methylethyl)-
10527 - 2-Propen-1-amine, N-propyl-
11149 - Pyrrolidine, 2,2-dimethyl-
11150 - Pyrrolidine, 2,4-dimethyl-
11151 - Pyrrolidine, 2,5-dimethyl-, cis-
11166 - Pyrrolidine, 1-ethyl-

$C_6H_{13}NO$
38 - Acetamide, N-butyl-
53 - Acetamide, N,N-diethyl-
3203 - Butanamide, N,N-dimethyl-
4501 - Cyclohexanol, 2-amino-, trans-(±)-
6721 - Hexanamide
7687 - Morpholine, 2,6-dimethyl-
7694 - Morpholine, 4-ethyl-
8959 - 2-Pentanone, 4-amino-4-methyl-
9674 - 2-Piperidinemethanol
9675 - 3-Piperidinemethanol
9716 - 3-Piperidinol, 1-methyl-
9717 - 4-Piperidinol, 1-methyl-
9802 - Propanamide, N-propyl-
11165 - 1-Pyrrolidineethanol

$C_6H_{13}NO_2$
332 - β-Alanine, N-methyl-, ethyl ester
3633 - 1-Butanol, 3-methyl-, carbamate
3634 - 2-Butanol, 2-methyl-, carbamate
3937 - Carbamic acid, butyl-, methyl ester
3943 - Carbamic acid, (1,1-dimethylethyl)-, methyl ester
3956 - Carbamic acid, (1-methylethyl)-, ethyl ester
3962 - Carbamic acid, (1-methylpropyl)-, methyl ester
3970 - Carbamic acid, propyl-, ethyl ester
6120 - 2-Furanmethanol, 5-(aminomethyl)tetrahydro-
6841 - Hexane, 1-nitro-
6873 - Hexanoic acid, 6-amino-
7370 - L-Isoleucine
7412 - DL-Leucine
7413 - L-Leucine
7691 - 4-Morpholineethanol
8089 - Nitrous acid, hexyl ester
8214 - Norleucine
8215 - DL-Norleucine

$C_6H_{13}NO_3S$
11568 - Sulfamic acid, cyclohexyl-

$C_6H_{13}NO_4$
6259 - Glycine, N,N-bis(2-hydroxyethyl)-
$C_6H_{13}NO_5$
6233 - D-Glucose, 2-amino-2-deoxy-
$C_6H_{13}NS_2$
5265 - 4H-1,3,5-Dithiazine, dihydro-2,4,6-trimethyl-, (2α,4α,6α)-
$C_6H_{13}N_3$
6283 - Guanidine, (3-methyl-2-butenyl)-
$C_6H_{13}N_3O_3$
8551 - L-Ornithine, N5-(aminocarbonyl)-
$C_6H_{13}O_3P$
9535 - Phosphonic acid, ethenyl-, diethyl ester
$C_6H_{13}O_4P$
9530 - Phosphonic acid, acetyl-, diethyl ester
C_6H_{14}
3289 - Butane, 2,2-dimethyl-
3290 - Butane, 2,3-dimethyl-
6731 - Hexane
8823 - Pentane, 2-methyl-
8824 - Pentane, 3-methyl-
$C_6H_{14}Cl_4OSi_2$
5231 - Disiloxane, 1,3-bis(dichloromethyl)-1,1,3,3-tetramethyl-
$C_6H_{14}FO_3P$
9580 - Phosphorofluoridic acid, bis(1-methylethyl) ester
$C_6H_{14}N_2$
5132 - Diazene, dipropyl
5141 - 1H-1,4-Diazepine, hexahydro-1-methyl-
9608 - Piperazine, 1,4-dimethyl-
9609 - Piperazine, 2,5-dimethyl-, cis-
9610 - Piperazine, 2,5-dimethyl-, trans-
9611 - Piperazine, 2,6-dimethyl-, cis-
9673 - 4-Piperidinemethanamine
11164 - 1-Pyrrolidineethanamine
$C_6H_{14}N_2O$
7689 - 4-Morpholineethanamine
9616 - 1-Piperazineethanol
9822 - 2-Propanamine, N-(1-methylethyl)-N-nitroso-
9825 - 1-Propanamine, N-nitroso-N-propyl-
$C_6H_{14}N_2O_2$
7418 - DL-Lysine
7419 - L-Lysine
$C_6H_{14}N_4O_2$
437 - DL-Arginine
438 - L-Arginine
$C_6H_{14}O$
3372 - Butane, 1-ethoxy-
3373 - Butane, 2-ethoxy-
3394 - Butane, 1-methoxy-2-methyl-
3395 - Butane, 1-methoxy-3-methyl-
3396 - Butane, 2-methoxy-2-methyl-
3615 - 1-Butanol, 2,2-dimethyl-
3616 - 1-Butanol, 2,3-dimethyl-, (±)-
3617 - 1-Butanol, 3,3-dimethyl-
3618 - 2-Butanol, 2,3-dimethyl-
3619 - 2-Butanol, 3,3-dimethyl-, (±)-
3620 - 1-Butanol, 2-ethyl-
6911 - 1-Hexanol
6912 - 2-Hexanol, (±)-
6913 - 2-Hexanol, (R)-
6914 - 3-Hexanol, (±)-
6915 - 3-Hexanol, (R)-
8822 - Pentane, 1-methoxy-
8937 - 1-Pentanol, 2-methyl-
8938 - 1-Pentanol, 3-methyl-, (±)-
8939 - 1-Pentanol, 4-methyl-
8940 - 2-Pentanol, 2-methyl-
8941 - 2-Pentanol, 3-methyl-
8942 - 2-Pentanol, 4-methyl-
8943 - 3-Pentanol, 2-methyl-
8944 - 3-Pentanol, 2-methyl-
10039 - Propane, 1-ethoxy-2-methyl-
10040 - Propane, 2-ethoxy-2-methyl-
10076 - Propane, 1-(1-methylethoxy)-
10119 - Propane, 1,1'-oxybis-

10120 - Propane, 2,2'-oxybis-
$C_6H_{14}OS$
10129 - Propane, 1,1'-sulfinylbis-
$C_6H_{14}OSi$
11423 - Silane, ethenylethoxydimethyl-
$C_6H_{14}O_2$
3350 - 2,3-Butanediol, 2,3-dimethyl-
5541 - Ethane, 1,1-diethoxy-
5542 - Ethane, 1,2-diethoxy-
5687 - Ethanol, 2-butoxy-
5746 - Ethanol, 2-(2-methylpropoxy)-
6793 - 1,3-Hexanediol
6794 - 1,4-Hexanediol
6795 - 1,5-Hexanediol
6796 - 1,6-Hexanediol
6797 - 2,3-Hexanediol
6798 - 2,4-Hexanediol
6799 - 2,5-Hexanediol
8793 - 2,4-Pentanediol, 2-methyl-
8794 - 2,4-Pentanediol, 3-methyl-
10007 - 1,3-Propanediol, 2-ethyl-2-methyl-
10390 - 2-Propanol, 1-(1-methylethoxy)-
10401 - 2-Propanol, 1-propoxy-
$C_6H_{14}O_2S$
10131 - Propane, 1,1'-sulfonylbis-
$C_6H_{14}O_3$
5613 - Ethane, 1,1'-oxybis[2-methoxy-
5614 - Ethane, 1,1'-[oxybis(methyleneoxy)]bis-
5705 - Ethanol, 2,2-diethoxy-
5720 - Ethanol, 2-(2-ethoxyethoxy)-
6867 - 1,2,5-Hexanetriol
6868 - 1,2,6-Hexanetriol
6869 - 2,3,4-Hexanetriol
10006 - 1,3-Propanediol, 2-ethyl-2-(hydroxymethyl)-
10185 - Propane, 1,2,3-trimethoxy-
10397 - Propanol, oxybis-
$C_6H_{14}O_4$
5713 - Ethanol, 2,2'-[1,2-ethanediylbis(oxy)]bis-
$C_6H_{14}O_4S$
11575 - Sulfuric acid, dipropyl ester
$C_6H_{14}O_5$
10024 - Propanediol, oxybis-
10025 - 1,2-Propanediol, 3,3'-oxybis-
$C_6H_{14}O_6$
6194 - Galactitol
6206 - D-Glucitol
7422 - D-Mannitol
$C_6H_{14}S$
3376 - Butane, 1-(ethylthio)-
6856 - 1-Hexanethiol
6857 - 2-Hexanethiol
6858 - 3-Hexanethiol
8826 - Pentane, 1-(methylthio)-
10044 - Propane, 1-(ethylthio)-2-methyl-
10077 - Propane, 1-[(1-methylethyl)thio]-
10151 - Propane, 1,1'-thiobis-
10152 - Propane, 2,2'-thiobis-
$C_6H_{14}S_2$
5244 - Disulfide, bis(1-methylethyl)
5259 - Disulfide, dipropyl
5450 - Ethane, 1,1-bis(ethylthio)-
5451 - Ethane, 1,2-bis(ethylthio)-
6812 - 1,6-Hexanedithiol
$C_6H_{14}Si$
11495 - Silane, trimethyl-2-propenyl-
$C_6H_{15}As$
448 - Arsine, triethyl-
$C_6H_{15}AsO_3$
440 - Arsenous acid, triethyl ester
$C_6H_{15}AsO_4$
439 - Arsenic acid, triethyl ester
$C_6H_{15}B$
3118 - Borane, triethyl
$C_6H_{15}BO_3$
3125 - Boric acid, triethyl ester
$C_6H_{15}BrSi$
11368 - Silane, bromotriethyl-

$C_6H_{15}ClN_2O_2$
5440 - Ethanaminium, 2-[(aminocarbonyl)oxy]-N,N,N-trimethyl-, chloride
$C_6H_{15}ClO_3Si$
11387 - Silane, chlorotriethoxy-
$C_6H_{15}ClSi$
11369 - Silane, butylchlorodimethyl-
11386 - Silane, (3-chloropropyl)trimethyl-
11388 - Silane, chlorotriethyl-
$C_6H_{15}FSi$
11484 - Silane, triethylfluoro-
$C_6H_{15}N$
3215 - 1-Butanamine, N,N-dimethyl-
3216 - 2-Butanamine, 2,3-dimethyl-
3217 - 2-Butanamine, 3,3-dimethyl-
3218 - 1-Butanamine, N-ethyl-
3219 - 2-Butanamine, N-ethyl-, (±)-
5417 - Ethanamine, N,N-diethyl-
6723 - 1-Hexanamine
6724 - 2-Hexanamine, (±)-
8706 - 1-Pentanamine, 3-methyl-
8707 - 1-Pentanamine, 4-methyl-
9821 - 2-Propanamine, N-(1-methylethyl)-
9826 - 1-Propanamine, N-propyl-
9827 - 2-Propanamine, N-propyl
9829 - 1-Propanamine, N,N,2-trimethyl-
$C_6H_{15}NO$
5424 - Ethanamine, 2-ethoxy-N,N-dimethyl-
5691 - Ethanol, 2-(butylamino)-
5706 - Ethanol, 2-(diethylamino)-
5710 - Ethanol, 2-[(1,1-dimethylethyl)amino]-
5747 - Ethanol, 2-[(2-methylpropyl)amino]-
8922 - 1-Pentanol, 2-amino-4-methyl-, (±)-
9814 - 1-Propanamine, N-hydroxy-N-propyl-
$C_6H_{15}NO_2$
3623 - 2-Butanol, 1-[(2-hydroxyethyl)amino]-
5414 - Ethanamine, 2,2-diethoxy-
5723 - Ethanol, 2,2'-(ethylimino)bis-
10384 - 2-Propanol, 1,1'-iminobis-
$C_6H_{15}NO_3$
5751 - Ethanol, 2,2',2''-nitrilotris-
$C_6H_{15}NO_5$
6207 - D-Glucitol, 1-amino-1-deoxy-
$C_6H_{15}NS$
5650 - Ethanethiol, 2-(butylamino)-
$C_6H_{15}N_2$
5510 - 1,2-Ethanediamine, N'-ethyl-N,N-dimethyl-
$C_6H_{15}N_3$
9615 - 1-Piperazineethanamine
$C_6H_{15}OP$
9517 - Phosphine oxide, triethyl-
$C_6H_{15}O_2P$
9527 - Phosphinic acid, diethyl-, ethyl ester
$C_6H_{15}O_2PS_2$
9579 - Phosphorodithioic acid, O,O,S-triethyl ester
$C_6H_{15}O_3P$
9536 - Phosphonic acid, ethyl-, diethyl ester
9591 - Phosphorous acid, triethyl ester
$C_6H_{15}O_3PS$
9585 - Phosphorothioic acid, O,O,O-triethyl ester
$C_6H_{15}O_3PS_2$
5103 - Demeton S methyl
$C_6H_{15}O_4P$
9563 - Phosphoric acid, triethyl ester
$C_6H_{15}O_4PS_2$
8632 - Oxydemeton-methyl
$C_6H_{15}P$
9521 - Phosphine, triethyl-
$C_6H_{15}Sb$
11549 - Stibine, triethyl-
$C_6H_{16}Br_2OSi_2$
5229 - Disiloxane, 1,3-bis(bromomethyl)-1,1,3,3-tetramethyl-
$C_6H_{16}ClN$
5418 - Ethanamine, N,N-diethyl-, hydrochloride

$C_6H_{16}ClNO$
9831 - 1-Propanaminium, 2-hydroxy-N,N,N-trimethyl-, chloride

$C_6H_{16}Cl_2OSi_2$
5230 - Disiloxane, 1,3-bis(chloromethyl)-1,1,3,3-tetramethyl-

$C_6H_{16}N_2$
5501 - 1,2-Ethanediamine, N,N'-diethyl-
5502 - 1,2-Ethanediamine, N,N-diethyl-
5516 - 1,2-Ethanediamine, N,N,N',N'-tetramethyl-
6749 - 1,6-Hexanediamine
7137 - Hydrazine, 1,2-bis(1-methylethyl)-

$C_6H_{16}N_2O$
5677 - Ethanol, 2-[(2-amino-2-methylpropyl)amino]-

$C_6H_{16}N_2O_2$
5715 - Ethanol, 2,2'-(1,2-ethanediyldiimino)bis-

$C_6H_{16}OSi$
10410 - 1-Propanol, 3-(trimethylsilyl)-
11503 - Silanol, triethyl-

$C_6H_{16}O_2Si$
11405 - Silane, diethoxydimethyl-

$C_6H_{16}O_3Si$
11477 - Silane, triethoxy-

$C_6H_{16}Si$
11483 - Silane, triethyl-
11496 - Silane, trimethylpropyl-

$C_6H_{17}N_3$
9874 - 1,3-Propanediamine, N-(3-aminopropyl)-

$C_6H_{18}AlO_9P_3$
6020 - Fosetyl-Al

$C_6H_{18}Cl_2O_2Si_3$
11913 - Trisiloxane, 1,5-dichloro-1,1,3,3,5,5-hexamethyl-

$C_6H_{18}N_3OP$
9573 - Phosphoric triamide, hexamethyl-

$C_6H_{18}N_4$
5499 - 1,2-Ethanediamine, N,N'-bis(2-aminoethyl)-

$C_6H_{18}OSi_2$
5236 - Disiloxane, hexamethyl-

$C_6H_{18}O_3Si_2$
5234 - Disiloxane, 1,3-dimethoxy-1,1,3,3-tetramethyl-

$C_6H_{18}O_3Si_3$
5002 - Cyclotrisiloxane, hexamethyl-

$C_6H_{18}Si_2$
5228 - Disilane, hexamethyl-

$C_6H_{19}NSi_2$
11363 - Silanamine, 1,1,1-trimethyl-N-(trimethylsilyl)-

$C_6H_{21}N_3Si_3$
5001 - Cyclotrisilazane, 2,2,4,4,6,6-hexamethyl-

$C_6H_{24}O_6Si_6$
4625 - Cyclohexasiloxane, 2,4,6,8,10,12-hexamethyl-

$C_6H_{26}O_8$
3351 - 2,3-Butanediol, 2,3-dimethyl-, hexahydrate

C_6MoO_6
7662 - Molybdenum carbonyl

C_6N_4
5915 - Ethenetetracarbonitrile

C_6N_6
11836 - 1,3,5-Triazine-2,4,6-tricarbonitrile

C_7F_5N
2767 - Benzonitrile, pentafluoro-

C_7F_8
2095 - Benzene, pentafluoro(trifluoromethyl)-

C_7F_{14}
4499 - Cyclohexane, undecafluoro(trifluoromethyl)-
6560 - 1-Heptene, 1,1,2,3,3,4,4,5,5,6,6,7,7,7-tetradecafluoro-

C_7F_{16}
6418 - Heptane, hexadecafluoro-

$C_7HCl_5O_2$
2712 - Benzoic acid, pentachloro-

C_7HCl_7
2085 - Benzene, pentachloro(dichloromethyl)-

C_7HF_5O
596 - Benzaldehyde, pentafluoro-

C_7HF_{15}
6432 - Heptane, 1,1,1,2,2,3,3,4,4,5,5,6,6,7,7-pentadecafluoro-

$C_7H_3Br_2NO$
2751 - Benzonitrile, 3,5-dibromo-4-hydroxy-

$C_7H_3Br_5$
2083 - Benzene, pentabromomethyl-

$C_7H_3ClF_3NO_2$
1217 - Benzene, 1-chloro-2-nitro-4-(trifluoromethyl)-
1218 - Benzene, 1-chloro-4-nitro-2-(trifluoromethyl)-
1219 - Benzene, 4-chloro-1-nitro-2-(trifluoromethyl)-

$C_7H_3Cl_2NO_5$
2896 - Benzoyl chloride, 3,5-dinitro-

$C_7H_3Cl_2N$
2752 - Benzonitrile, 2,6-dichloro-

$C_7H_3Cl_3O$
2893 - Benzoyl chloride, 2,4-dichloro-
2894 - Benzoyl chloride, 3,4-dichloro-

$C_7H_3Cl_3O_2$
2727 - Benzoic acid, 2,3,6-trichloro-
2728 - Benzoic acid, 2,4,5-trichloro-

$C_7H_3Cl_5$
1401 - Benzene, 1,2-dichloro-4-(trichloromethyl)-
2086 - Benzene, pentachloromethyl-
2326 - Benzene, 1,2,3-trichloro-4-(dichloromethyl)-
2327 - Benzene, 1,2,4-trichloro-5-(dichloromethyl)-

$C_7H_3F_5$
2094 - Benzene, pentafluoromethyl-

$C_7H_3F_5O$
2093 - Benzene, pentafluoromethoxy-

$C_7H_3N_3O_8$
2735 - Benzoic acid, 2,4,6-trinitro-

C_7H_4BrClO
2887 - Benzoyl chloride, 2-bromo-
2888 - Benzoyl chloride, 4-bromo-

$C_7H_4BrF_3$
1089 - Benzene, 1-bromo-2-(trifluoromethyl)-
1090 - Benzene, 1-bromo-3-(trifluoromethyl)-
1091 - Benzene, 1-bromo-4-(trifluoromethyl)-

C_7H_4BrN
2744 - Benzonitrile, 2-bromo-
2745 - Benzonitrile, 3-bromo-
2746 - Benzonitrile, 4-bromo-

C_7H_4BrNO
1049 - Benzene, 1-bromo-4-isocyanato-

$C_7H_4BrNO_4$
2527 - Benzoic acid, 2-bromo-4-nitro-
2528 - Benzoic acid, 2-bromo-5-nitro-
2529 - Benzoic acid, 4-bromo-3-nitro-
2530 - Benzoic acid, 5-bromo-2-nitro-

$C_7H_4Br_2O_2$
534 - Benzaldehyde, 3,5-dibromo-2-hydroxy-
2555 - Benzoic acid, 2,4-dibromo-
2556 - Benzoic acid, 2,5-dibromo-
2557 - Benzoic acid, 2,6-dibromo-

$C_7H_4Br_2O_3$
2558 - Benzoic acid, 3,5-dibromo-2-hydroxy-

$C_7H_4Br_2O_5$
2560 - Benzoic acid, 2,6-dibromo-3,4,5-trihydroxy-

$C_7H_4Br_4O$
2256 - Benzene, 1,2,3,5-tetrabromo-4-methoxy-
9468 - Phenol, 2,3,4,5-tetrabromo-6-methyl-

C_7H_4ClFO
2897 - Benzoyl chloride, 2-fluoro-
2898 - Benzoyl chloride, 3-fluoro-
2899 - Benzoyl chloride, 4-fluoro-

$C_7H_4ClF_3$
1240 - Benzene, 1-chloro-2-(trifluoromethyl)-
1241 - Benzene, 1-chloro-3-(trifluoromethyl)-
1242 - Benzene, 1-chloro-4-(trifluoromethyl)-

C_7H_4ClN
2747 - Benzonitrile, 2-chloro-
2748 - Benzonitrile, 3-chloro-
2749 - Benzonitrile, 4-chloro-

C_7H_4ClNO
1176 - Benzene, 1-chloro-2-isocyanato-
2878 - Benzoxazole, 2-chloro-

$C_7H_4ClNO_2$
2883 - 2(3H)-Benzoxazolone, 5-chloro-

$C_7H_4ClNO_3$
2905 - Benzoyl chloride, 3-nitro-
2906 - Benzoyl chloride, 4-nitro-

$C_7H_4ClNO_4$
2544 - Benzoic acid, 2-chloro-3-nitro-
2545 - Benzoic acid, 2-chloro-5-nitro-
2546 - Benzoic acid, 3-chloro-2-nitro-
2547 - Benzoic acid, 4-chloro-3-nitro-

C_7H_4ClNS
1177 - Benzene, 1-chloro-4-isothiocyanato-
2862 - Benzothiazole, 2-chloro-

$C_7H_4Cl_2O$
535 - Benzaldehyde, 2,5-dichloro-
536 - Benzaldehyde, 3,4-dichloro-
537 - Benzaldehyde, 3,5-dichloro-
2890 - Benzoyl chloride, 2-chloro-
2891 - Benzoyl chloride, 3-chloro-
2892 - Benzoyl chloride, 4-chloro-

$C_7H_4Cl_2O_2$
2561 - Benzoic acid, 2,4-dichloro-
2562 - Benzoic acid, 2,5-dichloro-
2563 - Benzoic acid, 2,6-dichloro-
2564 - Benzoic acid, 3,5-dichloro-

$C_7H_4Cl_2O_3$
2566 - Benzoic acid, 3,5-dichloro-2-hydroxy-

$C_7H_4Cl_2F$
1736 - Benzene, 1-fluoro-2-(trichloromethyl)-

$C_7H_4Cl_3NO_3$
11854 - Triclopyr

$C_7H_4Cl_4$
1237 - Benzene, 1-chloro-2-(trichloromethyl)-
1238 - Benzene, 1-chloro-3-(trichloromethyl)-
1239 - Benzene, 1-chloro-4-(trichloromethyl)-
1359 - Benzene, 1,2-dichloro-4-(dichloromethyl)-
2262 - Benzene, 1,2,3,5-tetrachloro-4-methyl-
2263 - Benzene, 1,2,4,5-tetrachloro-3-methyl-
2324 - Benzene, 1,2,4-trichloro-5-(chloromethyl)-

$C_7H_4Cl_4O_2$
9472 - Phenol, 2,3,5,6-tetrachloro-4-methoxy-

C_7H_4FN
2756 - Benzonitrile, 4-fluoro-

C_7H_4FNS
1721 - Benzene, 1-fluoro-3-isothiocyanato-

$C_7H_4F_3NO_2$
2060 - Benzene, 1-nitro-2-(trifluoromethyl)-
2061 - Benzene, 1-nitro-3-(trifluoromethyl)-

$C_7H_4F_4$
1737 - Benzene, 1-fluoro-2-(trifluoromethyl)-
1738 - Benzene, 1-fluoro-3-(trifluoromethyl)-
1739 - Benzene, 1-fluoro-4-(trifluoromethyl)-

$C_7H_4I_2O_3$
2617 - Benzoic acid, 2-hydroxy-3,5-diiodo-
2618 - Benzoic acid, 4-hydroxy-3,5-diiodo-

$C_7H_4N_2O_3$
1788 - Benzene, 1-isocyanato-2-nitro-
1789 - Benzene, 1-isocyanato-3-nitro-
1790 - Benzene, 1-isocyanato-4-nitro-

$C_7H_4N_2O_5$
551 - Benzaldehyde, 2,4-dinitro-

$C_7H_4N_2O_6$
2590 - Benzoic acid, 3,4-dinitro-
2591 - Benzoic acid, 3,5-dinitro-

$C_7H_4O_4S$
2875 - 3H-2,1-Benzoxathiol-3-one, 1,1-dioxide

$C_7H_4O_6$
10776 - 4H-Pyran-2,6-dicarboxylic acid, 4-oxo-

$C_7H_4O_7$
10775 - 4H-Pyran-2,6-dicarboxylic acid, 3-hydroxy-4-oxo-

C_7H_5BrO
526 - Benzaldehyde, 2-bromo-
527 - Benzaldehyde, 3-bromo-
528 - Benzaldehyde, 4-bromo-
2885 - Benzoyl bromide

C7H5BrO2
2518 - Benzoic acid, 2-bromo-
2519 - Benzoic acid, 3-bromo-
2520 - Benzoic acid, 4-bromo-

C7H5BrO3
2524 - Benzoic acid, 5-bromo-2-hydroxy-

C7H5Br3
1291 - Benzene, 1,3-dibromo-5-(bromomethyl)-
2314 - Benzene, 1,3,5-tribromo-2-methyl-

C7H5Br3O
2310 - Benzene, 1,2,3-tribromo-5-methoxy-
2311 - Benzene, 1,2,4-tribromo-5-methoxy-
2312 - Benzene, 1,2,5-tribromo-3-methoxy-
2313 - Benzene, 1,3,5-tribromo-2-methoxy-
9480 - Phenol, 2,4,6-tribromo-3-methyl-

C7H5ClN2
7236 - 1H-Indazole, 3-chloro-

C7H5ClN2O
2876 - 2-Benzoxazolamine, 5-chloro-

C7H5ClO
529 - Benzaldehyde, 2-chloro-
530 - Benzaldehyde, 3-chloro-
531 - Benzaldehyde, 4-chloro-
2886 - Benzoyl chloride

C7H5ClO2
532 - Benzaldehyde, 3-chloro-4-hydroxy-
533 - Benzaldehyde, 5-chloro-2-hydroxy-
2534 - Benzoic acid, 2-chloro-
2535 - Benzoic acid, 3-chloro-
2536 - Benzoic acid, 4-chloro-
2900 - Benzoyl chloride, 2-hydroxy-
4026 - Carbonochloridic acid, phenyl ester

C7H5ClO3
1127 - Benzenecarboperoxoic acid, 3-chloro-
2540 - Benzoic acid, 3-chloro-4-hydroxy-

C7H5Cl2F
1379 - Benzene, (dichlorofluoromethyl)-

C7H5Cl2FN2O3
5982 - Fluroxypyr

C7H5Cl2NO2
1135 - Benzene, 1-chloro-4-(chloromethyl)-2-nitro-
2491 - Benzoic acid, 2-amino-3,6-dichloro-
2492 - Benzoic acid, 3-amino-2,5-dichloro-

C7H5Cl2NO4S
2565 - Benzoic acid, 4-[(dichloroamino)sulfonyl]-

C7H5Cl3
1357 - Benzene, 1,2-dichloro-4-(chloromethyl)-
1358 - Benzene, 1,3-dichloro-5-(chloromethyl)-
2333 - Benzene, (trichloromethyl)-
2334 - Benzene, 1,2,3-trichloro-4-methyl-
2335 - Benzene, 1,2,3-trichloro-5-methyl-
2336 - Benzene, 1,2,4-trichloro-5-methyl-
2337 - Benzene, 1,2,5-trichloro-3-methyl-

C7H5Cl3O
2330 - Benzene, 1,2,4-trichloro-3-methoxy-
2331 - Benzene, 1,2,4-trichloro-5-methoxy-
2332 - Benzene, 1,3,5-trichloro-2-methoxy-
9487 - Phenol, 2,3,4-trichloro-6-methyl-
9488 - Phenol, 2,3,6-trichloro-4-methyl-
9489 - Phenol, 2,4,6-trichloro-3-methyl-

C7H5FO
559 - Benzaldehyde, 2-fluoro-
560 - Benzaldehyde, 3-fluoro-
561 - Benzaldehyde, 4-fluoro-
2909 - Benzoyl fluoride

C7H5FO2
2603 - Benzoic acid, 2-fluoro-
2604 - Benzoic acid, 3-fluoro-
2605 - Benzoic acid, 4-fluoro-

C7H5F3
2348 - Benzene, (trifluoromethyl)-

C7H5F3O
9491 - Phenol, 3-(trifluoromethyl)-

C7H5HgNO3
7437 - Mercury, [2-methyl-5-nitrophenolato(2-)-C6,O1]-

C7H5IO
580 - Benzaldehyde, 2-iodo-
581 - Benzaldehyde, 3-iodo-
582 - Benzaldehyde, 4-iodo-
2910 - Benzoyl iodide

C7H5IO2
2657 - Benzoic acid, 2-iodo-
2658 - Benzoic acid, 3-iodo-
2659 - Benzoic acid, 4-iodo-

C7H5IO3
2623 - Benzoic acid, 4-hydroxy-3-iodo-
2624 - Benzoic acid, 5-hydroxy-2-iodo-

C7H5I3
2352 - Benzene, 1,3,5-triiodo-2-methyl-

C7H5N
2740 - Benzonitrile

C7H5NO
1786 - Benzene, isocyanato-
2389 - 1,2-Benzisoxazole
2390 - 2,1-Benzisoxazole
2759 - Benzonitrile, 2-hydroxy-
2877 - Benzoxazole
10585 - 2-Propenenitrile, 3-(2-furanyl)-

C7H5NOS
2868 - 2(3H)-Benzothiazolone

C7H5NO2
2881 - 2-Benzoxazolol
2882 - 2(3H)-Benzoxazolone

C7H5NO3
591 - Benzaldehyde, 2-nitro-
592 - Benzaldehyde, 3-nitro-
593 - Benzaldehyde, 4-nitro-

C7H5NO3S
2388 - 1,2-Benzisothiazol-3(2H)-one, 1,1-dioxide

C7H5NO4
2703 - Benzoic acid, 2-nitro-
2704 - Benzoic acid, 3-nitro-
2705 - Benzoic acid, 4-nitro-
10927 - 2,3-Pyridinedicarboxylic acid
10928 - 2,4-Pyridinedicarboxylic acid
10929 - 2,5-Pyridinedicarboxylic acid
10930 - 3,4-Pyridinedicarboxylic acid
10931 - 3,5-Pyridinedicarboxylic acid

C7H5NO5
2646 - Benzoic acid, 2-hydroxy-3-nitro-
2647 - Benzoic acid, 2-hydroxy-5-nitro-

C7H5NS
1792 - Benzene, isothiocyanato-
2387 - 1,2-Benzisothiazole
2861 - Benzothiazole
11709 - Thiocyanic acid, phenyl ester

C7H5NS2
2866 - 2(3H)-Benzothiazolethione

C7H5N3
2870 - 1,2,4-Benzotriazine

C7H5N3O
2884 - Benzoyl azide

C7H5N3O5
614 - Benzamide, 3,5-dinitro-

C7H5N3O6
2046 - Benzene, 1-methyl-2,3,4-trinitro-
2047 - Benzene, 1-methyl-2,4,5-trinitro-
2048 - Benzene, 2-methyl-1,3,5-trinitro-

C7H5N3O7
364 - Anisole, 2,3,5-trinitro-
1946 - Benzene, 2-methoxy-1,3,5-trinitro-
9434 - Phenol, 3-methyl-2,4,6-trinitro-

C7H5N5O8
818 - Benzenamine, N-methyl-N,2,4,6-tetranitro-

C7H5NaO2
2725 - Benzoic acid, sodium salt

C7H6BrCl
1007 - Benzene, 1-bromo-3-(chloromethyl)-
1008 - Benzene, 1-bromo-4-(chloromethyl)-

C7H6BrNO
604 - Benzamide, 2-bromo-
605 - Benzamide, 3-bromo-

C7H6BrNO2
1066 - Benzene, 1-bromo-2-methyl-4-nitro-
1067 - Benzene, 1-(bromomethyl)-3-nitro-
1068 - Benzene, 1-bromo-3-methyl-2-nitro-
1069 - Benzene, 1-bromo-3-methyl-5-nitro-
1070 - Benzene, 1-bromo-4-methyl-2-nitro-
1071 - Benzene, 2-bromo-1-methyl-3-nitro-
1072 - Benzene, 2-bromo-1-methyl-4-nitro-
1073 - Benzene, 2-bromo-4-methyl-1-nitro-
1074 - Benzene, 4-bromo-1-methyl-2-nitro-
2487 - Benzoic acid, 2-amino-5-bromo-

C7H6BrNO3
606 - Benzamide, 5-bromo-N,2-dihydroxy-

C7H6Br2
999 - Benzene, 1-bromo-2-(bromomethyl)-
1000 - Benzene, 1-bromo-4-(bromomethyl)-
1301 - Benzene, (dibromomethyl)-
1302 - Benzene, 1,2-dibromo-3-methyl-
1303 - Benzene, 1,2-dibromo-4-methyl-
1304 - Benzene, 1,3-dibromo-2-methyl-
1305 - Benzene, 1,4-dibromo-2-methyl-
1306 - Benzene, 2,4-dibromo-1-methyl-

C7H6Br2O
9279 - Phenol, 2,4-dibromo-6-methyl-
9280 - Phenol, 3,5-dibromo-2-methyl-

C7H6ClF
1169 - Benzene, 1-chloro-3-fluoro-2-methyl-
1170 - Benzene, 2-chloro-4-fluoro-1-methyl-
1194 - Benzene, 1-(chloromethyl)-2-fluoro-
1195 - Benzene, 1-(chloromethyl)-4-fluoro-

C7H6ClI
1173 - Benzene, 1-chloro-1-iodo-4-methyl-
1174 - Benzene, 4-chloro-1-iodo-2-methyl-
1175 - Benzene, 4-chloro-2-iodo-1-methyl-
11817 - Toluene, 3-chloro-2-iodo-

C7H6ClNO2
1203 - Benzene, 1-(chloromethyl)-2-nitro-
1204 - Benzene, 1-chloro-2-methyl-3-nitro-
1205 - Benzene, 1-chloro-2-methyl-4-nitro-
1206 - Benzene, 1-(chloromethyl)-3-nitro-
1207 - Benzene, 1-chloro-3-methyl-5-nitro-
1208 - Benzene, 1-(chloromethyl)-4-nitro-
1209 - Benzene, 1-chloro-4-methyl-2-nitro-
1210 - Benzene, 2-chloro-1-methyl-4-nitro-
1211 - Benzene, 2-chloro-4-methyl-1-nitro-
1212 - Benzene, 4-chloro-1-methyl-2-nitro-
1213 - Benzene, 4-chloro-2-methyl-1-nitro-
2490 - Benzoic acid, 5-amino-2-chloro-

C7H6ClN3O4S2
2852 - 2H-1,2,4-Benzothiadiazine-7-sulfonamide, 6-chloro-, 1,1-dioxide

C7H6Cl2
1132 - Benzene, 1-chloro-2-(chloromethyl)-
1133 - Benzene, 1-chloro-3-(chloromethyl)-
1134 - Benzene, 1-chloro-4-(chloromethyl)-
1382 - Benzene, (dichloromethyl)-
1383 - Benzene, 1,2-dichloro-3-methyl-
1384 - Benzene, 1,2-dichloro-4-methyl-
1385 - Benzene, 1,3-dichloro-2-methyl-
1386 - Benzene, 1,3-dichloro-5-methyl-
1387 - Benzene, 1,4-dichloro-2-methyl-
1388 - Benzene, 2,4-dichloro-1-methyl-

C7H6Cl2O
1381 - Benzene, 2,4-dichloro-1-methoxy-
1853 - Benzenemethanol, 2,4-dichloro-
9292 - Phenol, 2,4-dichloro-3-methyl-
9293 - Phenol, 2,4-dichloro-5-methyl-
9294 - Phenol, 2,6-dichloro-3-methyl-
9295 - Phenol, 2,6-dichloro-4-methyl-

C7H6Cl2S
1136 - Benzene, 1-chloro-4-[(chloromethyl)thio]-

C7H6Cl4O2
4770 - 1,3-Cyclopentadiene, 1,2,3,4-tetrachloro-5,5-dimethoxy-

C7H6FNO2
1729 - Benzene, 1-fluoro-4-methyl-2-nitro-
1730 - Benzene, 2-fluoro-4-methyl-1-nitro-

1731 - Benzene, 4-fluoro-1-methyl-2-nitro-
1732 - Benzene, 4-fluoro-2-methyl-1-nitro-

C₇H₆F₂

1425 - Benzene, (difluoromethyl)-

C₇H₆F₃N

855 - Benzenamine, 2-(trifluoromethyl)-
856 - Benzenamine, 3-(trifluoromethyl)-
857 - Benzenamine, 4-(trifluoromethyl)-

C₇H₆N₂

2376 - 1H-Benzimidazole
2741 - Benzonitrile, 2-amino-
2742 - Benzonitrile, 3-amino-
7235 - 1H-Indazole

C₇H₆N₂O

10884 - 3-Pyridinecarbonitrile, 1,2-dihydro-1-methyl-
2-oxo-

C₇H₆N₂O₃

624 - Benzamide, 2-nitro-
625 - Benzamide, 3-nitro-

C₇H₆N₂O₄

1956 - Benzene, 1-methyl-2,4-dinitro-
1957 - Benzene, 1-methyl-3,5-dinitro-
1958 - Benzene, 2-methyl-1,3-dinitro-
1959 - Benzene, 2-methyl-1,4-dinitro-
1960 - Benzene, 4-methyl-1,2-dinitro-
2506 - Benzoic acid, 2-amino-3-nitro-

C₇H₆N₂O₅

1922 - Benzene, 1-methoxy-2,3-dinitro-
1923 - Benzene, 1-methoxy-2,4-dinitro-
1924 - Benzene, 1-methoxy-3,5-dinitro-
1925 - Benzene, 2-methoxy-1,3-dinitro-
1926 - Benzene, 2-methoxy-1,4-dinitro-
1927 - Benzene, 4-methoxy-1,2-dinitro-
9405 - Phenol, 2-methyl-4,6-dinitro-

C₇H₆N₂S

2386 - 2H-Benzimidazole-2-thione, 1,3-dihydro-
2859 - 2-Benzothiazolamine
2860 - 6-Benzothiazolamine

C₇H₆N₄O

7193 - 1H-Imidazole, 1,1′-carbonylbis-

C₇H₆O

525 - Benzaldehyde
4242 - 2,4,6-Cycloheptatrien-1-one

C₇H₆OS

1130 - Benzenecarbothioic acid

C₇H₆OS₂

1125 - Benzenecarbodithioic acid, 2-hydroxy-

C₇H₆O₂

564 - Benzaldehyde, 2-hydroxy-
565 - Benzaldehyde, 3-hydroxy-
566 - Benzaldehyde, 4-hydroxy-
2415 - 1,3-Benzodioxole
2475 - Benzoic acid
4243 - 2,4,6-Cycloheptatrien-1-one, 2-hydroxy-
4266 - 2,5-Cyclohexadiene-1,4-dione, 2-methyl-
10512 - 2-Propenal, 3-(2-furanyl)-
10513 - 2-Propenal, 3-(2-furanyl)-, (E)-

C₇H₆O₂S

2663 - Benzoic acid, 2-mercapto-

C₇H₆O₃

540 - Benzaldehyde, 2,3-dihydroxy-
541 - Benzaldehyde, 2,4-dihydroxy-
542 - Benzaldehyde, 3,4-dihydroxy-
1126 - Benzenecarboperoxoic acid
2612 - Benzoic acid, 2-hydroxy-
2613 - Benzoic acid, 4-hydroxy-
10638 - 2-Propenoic acid, 3-(2-furanyl)-

C₇H₆O₄

2568 - Benzoic acid, 2,3-dihydroxy-
2569 - Benzoic acid, 2,4-dihydroxy-
2570 - Benzoic acid, 2,5-dihydroxy-
2571 - Benzoic acid, 3,4-dihydroxy-
6183 - 4H-Furo[3,2-c]pyran-2(6H)-one, 4-hydroxy-

C₇H₆O₅

2730 - Benzoic acid, 2,3,4-trihydroxy-
2731 - Benzoic acid, 3,4,5-trihydroxy-

C₇H₆O₆S

2655 - Benzoic acid, 2-hydroxy-5-sulfo-

C₇H₇BF₂

3115 - Borane, difluoro(4-methylphenyl)-

C₇H₇Br

1053 - Benzene, (bromomethyl)-
1054 - Benzene, 1-bromo-2-methyl-
1055 - Benzene, 1-bromo-3-methyl-
1056 - Benzene, 1-bromo-4-methyl-

C₇H₇BrN₂O

11998 - Urea, (4-bromophenyl)-

C₇H₇BrO

1050 - Benzene, 1-bromo-2-methoxy-
1051 - Benzene, 1-bromo-3-methoxy-
1052 - Benzene, 1-bromo-4-methoxy-
9236 - Phenol, 2-bromo-4-methyl-
9237 - Phenol, 2-bromo-5-methyl-
9238 - Phenol, 3-bromo-4-methyl-
9239 - Phenol, 3-bromo-4-methyl-
9240 - Phenol, 3-bromo-5-methyl-
9241 - Phenol, 4-bromo-2-methyl-
9242 - Phenol, 4-bromo-3-methyl-

C₇H₇BrO₂

1843 - Benzenemethanol, 5-bromo-2-hydroxy-

C₇H₇BrO₃

6190 - 3-Furoic acid, 5-bromo-, ethyl ester

C₇H₇Cl

1182 - Benzene, (chloromethyl)-
1183 - Benzene, 1-chloro-2-methyl-
1184 - Benzene, 1-chloro-3-methyl-
1185 - Benzene, 1-chloro-4-methyl-

C₇H₇ClN₂S

11801 - Thiourea, (2-chlorophenyl)-

C₇H₇ClO

1178 - Benzene, 1-chloro-2-methoxy-
1179 - Benzene, 1-chloro-3-methoxy-
1180 - Benzene, 1-chloro-4-methoxy-
1848 - Benzenemethanol, 2-chloro-
1849 - Benzenemethanol, 4-chloro-
9260 - Phenol, 2-chloro-3-methyl-
9261 - Phenol, 2-chloro-4-methyl-
9262 - Phenol, 2-chloro-5-methyl-
9263 - Phenol, 2-chloro-6-methyl-
9264 - Phenol, 3-chloro-2-methyl-
9265 - Phenol, 3-chloro-4-methyl-
9266 - Phenol, 4-chloro-2-methyl-
9267 - Phenol, 4-chloro-3-methyl-

C₇H₇ClOS

2186 - Benzenesulfinyl chloride, 4-methyl-

C₇H₇ClO₂S

2249 - Benzenesulfonyl chloride, 2-methyl-
2250 - Benzenesulfonyl chloride, 4-methyl-

C₇H₇ClO₃

6053 - 2-Furancarboxylic acid, 5-chloro-, ethyl ester

C₇H₇ClO₃S

2248 - Benzenesulfonyl chloride, 2-methoxy-

C₇H₇ClS

1833 - Benzenemethanethiol, 4-chloro-

C₇H₇Cl₂N

1803 - Benzenemethanamine, 2,4-dichloro-

C₇H₇Cl₂NO₂S

2210 - Benzenesulfonamide, N,N-dichloro-4-methyl-

C₇H₇Cl₃NO₃PS

4090 - Chlorpyrifos-methyl

C₇H₇Cl₃Si

11467 - Silane, trichloro(2-methylphenyl)-
11468 - Silane, trichloro(3-methylphenyl)-

C₇H₇F

1725 - Benzene, (fluoromethyl)-
1726 - Benzene, 1-fluoro-2-methyl-
1727 - Benzene, 1-fluoro-3-methyl-
1728 - Benzene, 1-fluoro-4-methyl-

C₇H₇FO

1722 - Benzene, 1-fluoro-2-methoxy-
1723 - Benzene, 1-fluoro-3-methoxy-
1724 - Benzene, 1-fluoro-4-methoxy-
1879 - Benzenemethanol, 3-fluoro-

1880 - Benzenemethanol, 4-fluoro-

C₇H₇F₇O₂

3526 - Butanoic acid, heptafluoro-, 1-methylethyl
ester

C₇H₇I

1777 - Benzene, (iodomethyl)-
1778 - Benzene, 1-iodo-2-methyl-
1779 - Benzene, 1-iodo-3-methyl-
1780 - Benzene, 1-iodo-4-methyl-

C₇H₇IO

1775 - Benzene, 1-iodo-2-methoxy-
1776 - Benzene, 1-iodo-4-methoxy-
1888 - Benzenemethanol, 2-iodo-
9374 - Phenol, 2-iodo-4-methyl-
9375 - Phenol, 2-iodo-6-methyl-

C₇H₇N

10952 - Pyridine, 2-ethenyl-
10953 - Pyridine, 3-ethenyl-
10954 - Pyridine, 4-ethenyl-

C₇H₇NO

594 - Benzaldehyde, oxime, (E)-
595 - Benzaldehyde, oxime, (Z)-
600 - Benzamide
5867 - Ethanone, 1-(2-pyridinyl)-
5868 - Ethanone, 1-(3-pyridinyl)-
5869 - Ethanone, 1-(4-pyridinyl)-
5999 - Formamide, N-phenyl-

C₇H₇NO₂

579 - Benzaldehyde, 2-hydroxy-, oxime
616 - Benzamide, 2-hydroxy-
2000 - Benzene, 1-methyl-2-nitro-
2001 - Benzene, 1-methyl-3-nitro-
2002 - Benzene, 1-methyl-4-nitro-
2052 - Benzene, (nitromethyl)-
2484 - Benzoic acid, 2-amino-
2485 - Benzoic acid, 3-amino-
2486 - Benzoic acid, 4-amino-
8097 - Nitrous acid, phenylmethyl ester
10906 - 3-Pyridinecarboxylic acid, methyl ester
10908 - 4-Pyridinecarboxylic acid, methyl ester
11029 - Pyridinium, 3-carboxy-1-methyl-, hydroxide,
inner salt

C₇H₇NO₂S

2043 - Benzene, 1-(methylthio)-2-nitro-
2044 - Benzene, 1-(methylthio)-4-nitro-

C₇H₇NO₃

613 - Benzamide, N,2-dihydroxy-
1906 - Benzenemethanol, 2-nitro-
1907 - Benzenemethanol, 3-nitro-
1908 - Benzenemethanol, 4-nitro-
1935 - Benzene, 1-methoxy-2-nitro-
1936 - Benzene, 1-methoxy-3-nitro-
1937 - Benzene, 1-methoxy-4-nitro-
2498 - Benzoic acid, 4-amino-2-hydroxy-
2499 - Benzoic acid, 5-amino-2-hydroxy-
9422 - Phenol, 2-methyl-4-nitro-
9423 - Phenol, 2-methyl-6-nitro-
9424 - Phenol, 4-methyl-2-nitro-

C₇H₇NO₄

9386 - Phenol, 2-methoxy-3-nitro-

C₇H₇NO₄S

2508 - Benzoic acid, 4-(aminosulfonyl)-

C₇H₇N₃

980 - Benzene, (azidomethyl)-
981 - Benzene, 1-azido-2-methyl-
982 - Benzene, 1-azido-4-methyl-
2872 - 1H-Benzotriazole, 1-methyl-

C₇H₇N₃O₃

12025 - Urea, (4-nitrophenyl)-

C₇H₈

1947 - Benzene, methyl-
2913 - Bicyclo[2.2.1]hepta-2,5-diene
4240 - 1,3,5-Cycloheptatriene
4765 - 1,3-Cyclopentadiene, 5-ethylidene-
6341 - 1,6-Heptadien-3-yne
6342 - 1,5-Heptadiyne
6343 - 1,6-Heptadiyne

C₇H₈BrN
647 - Benzenamine, 2-bromo-4-methyl-
648 - Benzenamine, 2-bromo-5-methyl-
649 - Benzenamine, 3-bromo-4-methyl-
650 - Benzenamine, 4-bromo-2-methyl-
651 - Benzenamine, 4-bromo-3-methyl-
652 - Benzenamine, 5-bromo-2-methyl-
1799 - Benzenemethanamine, 4-bromo-

C₇H₈ClN
668 - Benzenamine, 2-chloro-4-methyl-
669 - Benzenamine, 2-chloro-5-methyl-
670 - Benzenamine, 2-chloro-N-methyl-
671 - Benzenamine, 3-chloro-2-methyl-
672 - Benzenamine, 3-chloro-4-methyl-
673 - Benzenamine, 4-chloro-2-methyl-
674 - Benzenamine, 4-chloro-3-methyl-
675 - Benzenamine, 4-chloro-N-methyl-
676 - Benzenamine, 5-chloro-2-methyl-

C₇H₈ClNO
666 - Benzenamine, 2-chloro-5-methoxy-
667 - Benzenamine, 4-chloro-2-methoxy-

C₇H₈ClN₃O₄S₂
2855 - 2H-1,2,4-Benzothiadiazine-7-sulfonamide, 6-chloro-3,4-dihydro-, 1,1-dioxide

C₇H₈Cl₂Si
11401 - Silane, dichloromethylphenyl-

C₇H₈FN
755 - Benzenamine, 4-fluoro-2-methyl-

C₇H₈IN
762 - Benzenamine, 2-iodo-4-methyl-
763 - Benzenamine, 5-iodo-2-methyl-

C₇H₈N₂
563 - Benzaldehyde, hydrazone
11137 - 1H-Pyrrole-1-propanenitrile

C₇H₈N₂O
89 - Acetamide, N-3-pyridinyl-
601 - Benzamide, 2-amino-
806 - Benzenamine, N-methyl-N-nitroso-
2611 - Benzoic acid, hydrazide
12026 - Urea, phenyl-

C₇H₈N₂O₂
799 - Benzenamine, 2-methyl-3-nitro-
800 - Benzenamine, 2-methyl-4-nitro-
801 - Benzenamine, 2-methyl-5-nitro-
802 - Benzenamine, 2-methyl-6-nitro-
803 - Benzenamine, 4-methyl-2-nitro-
804 - Benzenamine, N-methyl-2-nitro-
805 - Benzenamine, N-methyl-4-nitro-
2552 - Benzoic acid, 2,3-diamino-
2553 - Benzoic acid, 2,4-diamino-
2554 - Benzoic acid, 3,5-diamino-

C₇H₈N₂O₃
775 - Benzenamine, 2-methoxy-5-nitro-
776 - Benzenamine, 3-methoxy-4-nitro-
777 - Benzenamine, 5-methoxy-2-nitro-

C₇H₈N₂S
11808 - Thiourea, phenyl-

C₇H₈N₄O₂
10756 - 1H-Purine-2,6-dione, 3,7-dihydro-1,3-dimethyl-
10757 - 1H-Purine-2,6-dione, 3,7-dihydro-3,7-dimethyl-

C₇H₈O
1836 - Benzenemethanol
1915 - Benzene, methoxy-
9400 - Phenol, 2-methyl-
9401 - Phenol, 3-methyl-
9402 - Phenol, 4-methyl-

C₇H₈OS
2040 - Benzene, (methylsulfinyl)-
2299 - Benzenethiol, 3-methoxy-
2300 - Benzenethiol, 4-methoxy-
5856 - Ethanone, 1-(5-methyl-2-thienyl)-

C₇H₈O₂
1527 - 1,2-Benzenediol, 3-methyl-
1528 - 1,2-Benzenediol, 4-methyl-
1529 - 1,3-Benzenediol, 2-methyl-
1530 - 1,3-Benzenediol, 4-methyl-
1531 - 1,3-Benzenediol, 5-methyl-
1532 - 1,4-Benzenediol, 2-methyl-
1882 - Benzenemethanol, 2-hydroxy-
1883 - Benzenemethanol, 3-hydroxy-
1884 - Benzenemethanol, 4-hydroxy-
9379 - Phenol, 2-methoxy-
9380 - Phenol, 3-methoxy-
9381 - Phenol, 4-methoxy-
10438 - 1-Propanone, 1-(2-furanyl)-
10439 - 2-Propanone, 1-(2-furanyl)-
10792 - 2H-Pyran-2-one, 4,6-dimethyl-
10793 - 4H-Pyran-4-one, 2,6-dimethyl-

C₇H₈O₂S
2181 - Benzenesulfinic acid, 2-methyl-
2182 - Benzenesulfinic acid, 4-methyl-
11739 - 2-Thiophenecarboxylic acid, ethyl ester

C₇H₈O₃
1520 - 1,4-Benzenediol, 2-(hydroxymethyl)-
1525 - 1,2-Benzenediol, 3-methoxy-
1526 - 1,3-Benzenediol, 5-methoxy-
2369 - 1,2,3-Benzenetriol, methyl-
6029 - 2-Furanacetic acid, methyl ester
6056 - 2-Furancarboxylic acid, ethyl ester
6119 - 2-Furanmethanol, acetate
6161 - 2-Furanpropanoic acid

C₇H₈O₃S
2227 - Benzenesulfonic acid, methyl ester
2228 - Benzenesulfonic acid, 2-methyl-
2229 - Benzenesulfonic acid, 4-methyl-

C₇H₈S
1832 - Benzenemethanethiol
2041 - Benzene, (methylthio)-
2301 - Benzenethiol, 2-methyl-
2302 - Benzenethiol, 3-methyl-
2303 - Benzenethiol, 4-methyl-

C₇H₈S₂
1564 - 1,2-Benzenedithiol, 4-methyl-

C₇H₉AsN₂O₄
455 - Arsonic acid,
        [4-[(aminocarbonyl)amino]phenyl]-

C₇H₉BO₂
3134 - Boronic acid, (2-methylphenyl)-

C₇H₉Cl
7107 - 1-Hexen-3-yne, 5-chloro-5-methyl-

C₇H₉ClO
9147 - 2-Pentynoyl chloride, 4,4-dimethyl-

C₇H₉ClSi
11381 - Silane, chloromethylphenyl-

C₇H₉N
778 - Benzenamine, 2-methyl-
779 - Benzenamine, 3-methyl-
780 - Benzenamine, 4-methyl-
781 - Benzenamine, N-methyl-
1795 - Benzenemethanamine
4629 - 3-Cyclohexene-1-carbonitrile
4630 - 1-Cyclohexenecarbonitrile
4905 - 1-Cyclopentene-1-acetonitrile
10936 - Pyridine, 2,3-dimethyl-
10937 - Pyridine, 2,4-dimethyl-
10938 - Pyridine, 2,5-dimethyl-
10939 - Pyridine, 2,6-dimethyl-
10940 - Pyridine, 3,4-dimethyl-
10941 - Pyridine, 3,5-dimethyl-
10956 - Pyridine, 2-ethyl-
10957 - Pyridine, 3-ethyl-
10958 - Pyridine, 4-ethyl-

C₇H₉NO
758 - Benzenamine, N-hydroxy-4-methyl-
766 - Benzenamine, 2-methoxy-
767 - Benzenamine, 3-methoxy-
768 - Benzenamine, 4-methoxy-
1837 - Benzenemethanol, 2-amino-
5849 - Ethanone, 1-(1-methyl-1H-pyrrol-2-yl)-
9209 - Phenol, 2-amino-3-methyl-
9210 - Phenol, 2-amino-4-methyl-
9211 - Phenol, 3-amino-4-methyl-
9212 - Phenol, 3-amino-5-methyl-
9213 - Phenol, 4-amino-2-methyl-
10945 - Pyridine, 2,6-dimethyl-, 1-oxide
10950 - 2-Pyridineethanol

C₇H₉NO₂
2509 - Benzoic acid, ammonium salt
10935 - Pyridine, 2,6-dimethoxy-

C₇H₉NO₃
8586 - 2,4-Oxazolidinedione, 5-methyl-3-(2-propenyl)-

C₇H₉NS
819 - Benzenamine, 2-(methylthio)-
820 - Benzenamine, 4-(methylthio)-

C₇H₉N₃O
7145 - Hydrazinecarboxamide, 2-phenyl-
7146 - Hydrazinecarboxamide, N-phenyl-

C₇H₉N₃O₂
2500 - Benzoic acid, 4-amino-2-hydroxy-, hydrazide

C₇H₉N₃O₂S₂
2189 - Benzenesulfonamide, 4-amino-N-(aminothioxomethyl)-

C₇H₉N₃O₃S
2190 - Benzenesulfonamide, 4-amino-N-(aminocarbonyl)-

C₇H₁₀
4219 - 1,3-Cycloheptadiene
4271 - 1,3-Cyclohexadiene, 2-methyl-
4272 - 1,3-Cyclohexadiene, 5-methyl-, (±)-
6529 - 1,3,5-Heptatriene
6606 - 1-Hepten-3-yne
9115 - 3-Penten-1-yne, 3-ethyl-

C₇H₁₀ClN
791 - Benzenamine, 2-methyl-, hydrochloride
792 - Benzenamine, N-methyl-, hydrochloride

C₇H₁₀ClN₃
11043 - 4-Pyrimidinamine, 2-chloro-N,N,6-trimethyl-

C₇H₁₀N₂
1275 - 1,2-Benzenediamine, 3-methyl-
1276 - 1,2-Benzenediamine, 4-methyl-
1277 - 1,3-Benzenediamine, 2-methyl-
1278 - 1,3-Benzenediamine, 4-methyl-
1279 - 1,4-Benzenediamine, 2-methyl-
1280 - 1,4-Benzenediamine, N-methyl-
1796 - Benzenemethanamine, 2-amino-
4159 - Cyanamide, di-2-propenyl-
6392 - Heptanedinitrile
7164 - Hydrazine, (3-methylphenyl)-
7165 - Hydrazine, 1-methyl-1-phenyl-
7166 - Hydrazine, 1-methyl-2-phenyl-
10863 - 2-Pyridinamine, N,N-dimethyl-
10864 - 4-Pyridinamine, 2,6-dimethyl-
10948 - 2-Pyridineethanamine
10949 - 4-Pyridineethanamine

C₇H₁₀N₂O
1273 - 1,2-Benzenediamine, 4-methoxy-
1274 - 1,3-Benzenediamine, 4-methoxy-

C₇H₁₀N₂OS
11109 - 4(1H)-Pyrimidinone, 2,3-dihydro-6-propyl-2-thioxo-

C₇H₁₀N₂O₂S
7194 - 1H-Imidazole-1-carboxylic acid, 2,3-dihydro-3-methyl-2-thioxo-, ethyl ester

C₇H₁₀N₄O₂S
2191 - Benzenesulfonamide, 4-amino-N-(aminoiminomethyl)-

C₇H₁₀O
4631 - 1-Cyclohexene-1-carboxaldehyde
4632 - 3-Cyclohexene-1-carboxaldehyde
4710 - 2-Cyclohexen-1-one, 2-methyl-
4711 - 2-Cyclohexen-1-one, 3-methyl-
4712 - 2-Cyclohexen-1-one, 5-methyl-
4713 - 3-Cyclohexen-1-one, 3-methyl-
4714 - 3-Cyclohexen-1-one, 4-methyl-
4866 - Cyclopentanol, 1-ethynyl-
4943 - 2-Cyclopenten-1-one, 2,4-dimethyl-
6106 - Furan, 2-ethyl-5-methyl-
6167 - Furan, 2-propyl-

6325 - 2,4-Heptadienal, (E,E)-
6339 - 3,5-Heptadien-2-one
7607 - Methanone, dicyclopropyl-

$C_7H_{10}O_2$
3856 - 2-Butenoic acid, 2-propenyl ester
3857 - 2-Butenoic acid, 2-propenyl ester, (E)-
4225 - 1,2-Cycloheptanedione
4637 - 1-Cyclohexene-1-carboxylic acid
4638 - 3-Cyclohexene-1-carboxylic acid
4879 - Cyclopentanone, 2-acetyl-
4904 - 3-Cyclopentene-1-acetic acid
6104 - Furan, 2-(ethoxymethyl)-
6112 - Furan, 2-isopropoxy-
6702 - 2,4-Hexadienoic acid, methyl ester
6703 - 2,4-Hexadienoic acid, methyl ester, (E,E)-
10671 - 2-Propenoic acid, 2-methyl-, 2-propenyl ester

$C_7H_{10}O_3$
3577 - Butanoic acid, 3-oxo-, 2-propenyl ester
6454 - 2,4,6-Heptanetrione
10668 - 2-Propenoic acid, 2-methyl-, oxiranylmethyl ester

$C_7H_{10}O_4$
3330 - Butanedioic acid, methylene-, dimethyl ester
3331 - Butanedioic acid, (1-methylethylidene)-
3789 - 2-Butenedioic acid, 2-methyl-, dimethyl ester, (E)-
3790 - 2-Butenedioic acid, 2-methyl-, dimethyl ester, (Z)-
4793 - 1,3-Cyclopentanedicarboxylic acid, cis-
4969 - 1,2-Cyclopropanedicarboxylic acid, dimethyl ester
8876 - Pentanoic acid, 2,4-dioxo-, ethyl ester
10563 - 2-Propene-1,1-diol, diacetate

$C_7H_{10}O_5$
8784 - Pentanedioic acid, 3-oxo-, dimethyl ester
9973 - Propanedioic acid, oxo-, diethyl ester

$C_7H_{10}O_6$
7553 - Methanetricarboxylic acid, trimethyl ester

$C_7H_{10}S$
11761 - Thiophene, 2-ethyl-3-methyl-
11762 - Thiophene, 2-ethyl-5-methyl-
11763 - Thiophene, 3-ethyl-4-methyl-
11764 - Thiophene, 3-ethyl-2-methyl-
11772 - Thiophene, 2-(1-methylethyl)-
11773 - Thiophene, 3-(1-methylethyl)-
11778 - Thiophene, 2-propyl-
11779 - Thiophene, 3-propyl-
11792 - Thiophene, 2,3,4-trimethyl-
11793 - Thiophene, 2,3,5-trimethyl-

$C_7H_{10}Si$
11442 - Silane, methylphenyl-

$C_7H_{11}Br$
6612 - 1-Heptyne, 1-bromo-
6613 - 2-Heptyne, 1-bromo-

$C_7H_{11}BrO_4$
9928 - Propanedioic acid, bromo-, diethyl ester

$C_7H_{11}Cl$
2918 - Bicyclo[2.2.1]heptane, 2-chloro-
6614 - 1-Heptyne, 1-chloro-
6615 - 2-Heptyne, 7-chloro-
6616 - 3-Heptyne, 7-chloro-
9125 - 1-Pentyne, 3-chloro-3-ethyl-

$C_7H_{11}ClO$
4330 - Cyclohexanecarbonyl chloride

$C_7H_{11}ClO_3$
3492 - Butanoic acid, 2-chloro-2-methyl-3-oxo-, ethyl ester

$C_7H_{11}ClO_4$
9931 - Propanedioic acid, chloro-, diethyl ester
9996 - 1,2-Propanediol, 3-chloro-, diacetate

$C_7H_{11}Cl_3O_2$
255 - Acetic acid, trichloro-, 3-methylbutyl ester

$C_7H_{11}I$
6620 - 1-Heptyne, 1-iodo-

$C_7H_{11}N$
4328 - Cyclohexanecarbonitrile
11128 - 1H-Pyrrole, 3-ethyl-4-methyl-

11129 - Pyrrole, 2-isopropyl-
11138 - 1H-Pyrrole, 1-propyl-
11139 - 1H-Pyrrole, 3-propyl-
11142 - 1H-Pyrrole, 1,2,5-trimethyl-

$C_7H_{11}NO$
4329 - Cyclohexanecarbonitrile, 1-hydroxy-
4416 - Cyclohexane, isocyanato-

$C_7H_{11}NO_2$
151 - Acetic acid, cyano-, butyl ester
9720 - 4-Piperidinone, 1-acetyl-
10913 - 3-Pyridinecarboxylic acid, 1,2,5,6-tetrahydro-1-methyl-
11158 - 2,5-Pyrrolidinedione, 3-ethyl-3-methyl-

$C_7H_{11}NO_3$
11567 - Succinimide, N-(ethoxymethyl)-

$C_7H_{11}NO_4$
9766 - L-Proline, 1-acetyl-4-hydroxy-, trans-

$C_7H_{11}NO_5$
9958 - Propanedioic acid, (hydroxyimino)-, diethyl ester

$C_7H_{11}NS$
4417 - Cyclohexane, isothiocyanato-

C_7H_{12}
2914 - Bicyclo[2.2.1]heptane
2915 - Bicyclo[4.1.0]heptane
2980 - Bicyclo[3,1,0]hexane, 1-methyl-
4244 - Cycloheptene
4428 - Cyclohexane, methylene-
4670 - Cyclohexene, 1-methyl-
4671 - Cyclohexene, 3-methyl-, (±)-
4672 - Cyclohexene, 4-methyl-
4813 - Cyclopentane, ethylidene-
4918 - Cyclopentene, 1,2-dimethyl-
4919 - Cyclopentene, 1,3-dimethyl-
4920 - Cyclopentene, 1,4-dimethyl-
4921 - Cyclopentene, 1,5-dimethyl-
4922 - Cyclopentene, 4,4-dimethyl-
4924 - Cyclopentene, 1-ethyl-
4925 - Cyclopentene, 3-ethyl-
4926 - Cyclopentene, 4-ethyl-
6326 - 1,2-Heptadiene
6327 - 1,4-Heptadiene
6328 - 1,5-Heptadiene
6329 - 2,4-Heptadiene
6609 - 1-Heptyne
6610 - 2-Heptyne
6611 - 3-Heptyne
6695 - 1,5-Hexadiene, 2-methyl-
6696 - 2,4-Hexadiene, 2-methyl-
7120 - 1-Hexyne, 5-methyl-
7121 - 2-Hexyne, 5-methyl-
7122 - 3-Hexyne, 2-methyl-
8665 - 1,3-Pentadiene, 2,4-dimethyl-
9128 - 1-Pentyne, 3,3-dimethyl-
9129 - 1-Pentyne, 4,4-dimethyl-
9130 - 2-Pentyne, 4,4-dimethyl-
9131 - 1-Pentyne, 3-ethyl-

$C_7H_{12}Br_2O_2$
8872 - Pentanoic acid, 2,3-dibromo-, ethyl ester
8873 - Pentanoic acid, 2,5-dibromo-, ethyl ester
10239 - Propanoic acid, 2,3-dibromo-, butyl ester

$C_7H_{12}ClN_5$
11513 - Simazine

$C_7H_{12}N_2$
7196 - 1H-Imidazole, 1,2-diethyl-
11190 - Pyrrolo[1,2-a]pyrimidine, 2,3,4,6,7,8-hexahydro-

$C_7H_{12}O$
4234 - Cycloheptanone
4331 - Cyclohexanecarboxaldehyde
4537 - Cyclohexanol, 2-methylene-
4596 - Cyclohexanone, 2-methyl-, (±)-
4597 - Cyclohexanone, 3-methyl-, (±)-
4598 - Cyclohexanone, 4-methyl-
4882 - Cyclopentanone, 2,5-dimethyl-
5795 - Ethanone, 1-cyclopentyl-
6336 - 1,5-Heptadien-4-ol

6337 - 1,6-Heptadien-4-ol
6530 - 2-Heptenal
6531 - 3-Heptenal
6583 - 1-Hepten-3-one
6584 - 4-Hepten-2-one
6585 - 6-Hepten-2-one
6586 - 6-Hepten-3-one
6587 - 3-Hepten-2-one, (E)-
6588 - 3-Hepten-2-one, (Z)-
6589 - 5-Hepten-2-one, (E)-
7093 - 1-Hexen-3-one, 5-methyl-
7094 - 3-Hexen-2-one, 5-methyl-
7095 - 4-Hexen-3-one, 5-methyl-
7096 - 4-Hexen-3-one, 2-methyl-
7097 - 5-Hexen-2-one, 3-methyl-
7098 - 5-Hexen-2-one, 4-methyl-
7099 - 5-Hexen-2-one, 5-methyl-
7128 - 1-Hexyn-3-ol, 3-methyl-
7129 - 3-Hexyn-2-ol, 2-methyl-
9144 - 1-Pentyn-3-ol, 3,4-dimethyl-
9145 - 1-Pentyn-3-ol, 3-ethyl-

$C_7H_{12}O_2$
3587 - Butanoic acid, 2-propenyl ester
3849 - 2-Butenoic acid, 2-methyl-, ethyl ester, (E)-
3850 - 2-Butenoic acid, 3-methyl-, ethyl ester
4175 - Cyclobutanecarboxylic acid, ethyl ester
4332 - Cyclohexanecarboxylic acid
4775 - Cyclopentaneacetic acid
4783 - Cyclopentanecarboxylic acid, 1-methyl-
4865 - Cyclopentanol, acetate
6005 - Formic acid, cyclohexyl ester
6402 - 2,3-Heptanedione
6403 - 2,6-Heptanedione
6561 - 2-Heptenoic acid
6562 - 4-Heptenoic acid
6563 - 5-Heptenoic acid
6564 - 6-Heptenoic acid
6809 - 2,3-Hexanedione, 5-methyl-
6810 - 2,5-Hexanedione, 3-methyl-
7055 - 2-Hexenoic acid, 2-methyl-, (E)-
7056 - 2-Hexenoic acid, 4-methyl-
7057 - 2-Hexenoic acid, 5-methyl-
7058 - 3-Hexenoic acid, 2-methyl-
7059 - 3-Hexenoic acid, 3-methyl-
7060 - 3-Hexenoic acid, 4-methyl-
7061 - 4-Hexenoic acid, 5-methyl-
7062 - 3-Hexenoic acid, methyl ester
8799 - 2,4-Pentanedione, 3,3-dimethyl-
8801 - 2,4-Pentanedione, 3-ethyl-
9075 - 4-Pentenoic acid, 2,2-dimethyl-
9083 - 2-Pentenoic acid, 2-methyl-, methyl ester
9084 - 2-Pentenoic acid, 3-methyl-, methyl ester, (E)-
9085 - 2-Pentenoic acid, 3-methyl-, methyl ester, (Z)-
9086 - 3-Pentenoic acid, 3-methyl-, methyl ester
10616 - 2-Propenoic acid, butyl ester
10664 - 2-Propenoic acid, 2-methyl-, 1-methylethyl ester
10672 - 2-Propenoic acid, 2-methyl-, propyl ester
10673 - 2-Propenoic acid, 2-methylpropyl ester
10731 - 1-Propyne, 3,3-diethoxy-
10783 - 2H-Pyran, 2-ethoxy-3,4-dihydro-

$C_7H_{12}O_3$
3519 - Butanoic acid, 2-ethyl-3-oxo-, methyl ester
3557 - Butanoic acid, 2-methyl-3-oxo-, ethyl ester
3574 - Butanoic acid, 3-oxo-, 1-methylethyl ester
6125 - 2-Furanmethanol, tetrahydro-, acetate
6164 - 2-Furanpropanoic acid, tetrahydro-
6469 - Heptanoic acid, 6-oxo-
6901 - Hexanoic acid, 4-methyl-5-oxo-
8874 - Pentanoic acid, 2,2-dimethyl-4-oxo-
8898 - Pentanoic acid, 3-methyl-4-oxo-, methyl ester
8905 - Pentanoic acid, 2-oxo-, ethyl ester
8906 - Pentanoic acid, 3-oxo-, ethyl ester
8907 - Pentanoic acid, 4-oxo-, ethyl ester
10634 - 2-Propenoic acid, 2-ethoxyethyl ester
10662 - 2-Propenoic acid, 2-methyl-, 2-hydroxypropyl ester

C₇H₁₂O₄

3323 - Butanedioic acid, ethyl methyl ester
3327 - Butanedioic acid, methyl-, dimethyl ester
3338 - Butanedioic acid, monopropyl ester
6393 - Heptanedioic acid
6788 - Hexanedioic acid, 2-methyl-
6789 - Hexanedioic acid, 3-methyl-, (±)-
6792 - Hexanedioic acid, monomethyl ester
7654 - Methylene bispropionate
8773 - Pentanedioic acid, 3,3-dimethyl-
8775 - Pentanedioic acid, dimethyl ester
9929 - Propanedioic acid, butyl-
9939 - Propanedioic acid, diethyl-
9941 - Propanedioic acid, diethyl ester
9968 - Propanedioic acid, (2-methylpropyl)-
10002 - 1,2-Propanediol, diacetate
10003 - 1,3-Propanediol, diacetate
11643 - 2,4,8,10-Tetraoxaspiro[5.5]undecane

C₇H₁₂O₅

9957 - Propanedioic acid, hydroxy-, diethyl ester
10189 - 1,2,3-Propanetriol, diacetate
10190 - 1,2,3-Propanetriol, 1,3-diacetate

C₇H₁₂O₆

4338 - Cyclohexanecarboxylic acid, 1,3,4,5-tetrahydroxy-, [1R-(1α,3α,4α,5β)]-

C₇H₁₂O₆Si

11499 - Silanetriol, methyl-, triacetate

C₇H₁₃Br

4222 - Cycloheptane, bromo-
4318 - Cyclohexane, (bromomethyl)-
4319 - Cyclohexane, 1-bromo-1-methyl-
4320 - Cyclohexane, 1-bromo-3-methyl-

C₇H₁₃BrN₂O₂

3195 - Butanamide, N-(aminocarbonyl)-2-bromo-2-ethyl-

C₇H₁₃BrO

4317 - Cyclohexane, 1-bromo-2-methoxy-

C₇H₁₃BrO₂

3475 - Butanoic acid, 2-bromo-3-methyl-, ethyl ester
6457 - Heptanoic acid, 2-bromo-
6458 - Heptanoic acid, 7-bromo-
8860 - Pentanoic acid, 2-bromo-, ethyl ester
8861 - Pentanoic acid, 5-bromo-, ethyl ester
10206 - Propanoic acid, 3-bromo-, butyl ester

C₇H₁₃Cl

6538 - 1-Heptene, 1-chloro-
6539 - 1-Heptene, 7-chloro-
6540 - 3-Heptene, 4-chloro-

C₇H₁₃ClO

6513 - 2-Heptanone, 1-chloro-
6527 - Heptanoyl chloride
6988 - Hexanoyl chloride, 3-methyl-
6989 - Hexanoyl chloride, 4-methyl-

C₇H₁₃ClO₂

3490 - Butanoic acid, 2-chloro-3-methyl-, ethyl ester
3491 - Butanoic acid, 2-chloro-2-methyl-, ethyl ester
8867 - Pentanoic acid, 4-chloro-, ethyl ester
8868 - Pentanoic acid, 5-chloro-, ethyl ester
8869 - Pentanoic acid, 2-chloroethyl ester
10217 - Propanoic acid, 2-chloro-, butyl ester
10218 - Propanoic acid, 3-chloro-, butyl ester
10224 - Propanoic acid, 2-chloro-, 2-methylpropyl ester, (R)-
10227 - Propanoic acid, 3-chloro-, 2-methylpropyl ester
12039 - Valeric acid, 3-chloro-, ethyl ester

C₇H₁₃FO₂

6463 - Heptanoic acid, 7-fluoro-

C₇H₁₃N

480 - 1-Azabicyclo[2.2.2]octane
6428 - Heptanenitrile
9121 - 1-Pentyn-3-amine, 3-ethyl-
9666 - Piperidine, 3-ethenyl-

C₇H₁₃NO

485 - 1-Azabicyclo[2.2.2]octan-3-ol
5990 - Formamide, N-cyclohexyl-
9630 - Piperidine, 1-acetyl-

C₇H₁₃NO₂

4443 - Cyclohexane, 1-methyl-1-nitro-
4444 - Cyclohexane, 1-methyl-2-nitro-
4445 - Cyclohexane, 1-methyl-4-nitro-
4455 - Cyclohexane, (nitromethyl)-
9644 - 4-Piperidinecarboxylic acid, methyl ester
10633 - 2-Propenoic acid, 2-(dimethylamino)ethyl ester
11177 - Pyrrolidinium, 2-carboxy-1,1-dimethyl-, hydroxide, inner salt, (S)-

C₇H₁₃NO₃

10107 - Propanenitrile, 3-[2-(2-hydroxyethoxy)ethoxy]-
11178 - Pyrrolidinium, 2-carboxy-4-hydroxy-1,1-dimethyl-, hydroxide, inner salt, (2S-trans)-

C₇H₁₃NO₃S

7649 - L-Methionine, N-acetyl-

C₇H₁₃NO₄

6245 - L-Glutamic acid, 5-ethyl ester
9924 - Propanedioic acid, amino-, diethyl ester

C₇H₁₃N₃O₃S

8568 - Oxamyl

C₇H₁₃O₆P

7659 - Mevinphos

C₇H₁₄

3797 - 1-Butene, 2-ethyl-3-methyl-
3812 - 1-Butene, 2,3,3-trimethyl-
4221 - Cycloheptane
4426 - Cyclohexane, methyl-
4800 - Cyclopentane, 1,1-dimethyl-
4801 - Cyclopentane, 1,2-dimethyl-, cis-
4802 - Cyclopentane, 1,2-dimethyl-, trans-
4803 - Cyclopentane, 1,3-dimethyl-, cis-
4804 - Cyclopentane, 1,3-dimethyl-, trans-
4812 - Cyclopentane, ethyl-
6533 - 1-Heptene
6534 - 2-Heptene, (E)-
6535 - 2-Heptene, (Z)-
6536 - 3-Heptene, (E)-
6537 - 3-Heptene, (Z)-
7036 - 1-Hexene, 2-methyl-
7037 - 1-Hexene, 3-methyl-
7038 - 1-Hexene, 4-methyl-
7039 - 1-Hexene, 5-methyl-
7040 - 2-Hexene, 2-methyl-
7041 - 2-Hexene, 3-methyl-, (Z)-
7042 - 2-Hexene, 4-methyl-, (E)-
7043 - 2-Hexene, 4-methyl-, (Z)-
7044 - 2-Hexene, 5-methyl-, (E)-
7045 - 2-Hexene, 5-methyl-, (Z)-
7046 - 3-Hexene, 2-methyl-, (E)-
7047 - 3-Hexene, 2-methyl-, (Z)-
9029 - 1-Pentene, 2,3-dimethyl-
9030 - 1-Pentene, 2,4-dimethyl-
9031 - 1-Pentene, 3,3-dimethyl-
9032 - 1-Pentene, 3,4-dimethyl-
9033 - 1-Pentene, 4,4-dimethyl-
9034 - 2-Pentene, 2,3-dimethyl-
9035 - 2-Pentene, 2,4-dimethyl-
9036 - 2-Pentene, 3,4-dimethyl-
9037 - 2-Pentene, 3,4-dimethyl-, (E)-
9038 - 2-Pentene, 3,4-dimethyl-, (Z)-
9039 - 2-Pentene, 4,4-dimethyl-, (E)-
9040 - 2-Pentene, 4,4-dimethyl-, (Z)-
9043 - 1-Pentene, 2-ethyl-
9044 - 1-Pentene, 3-ethyl-
9045 - 2-Pentene, 3-ethyl-

C₇H₁₄BrF

6359 - Heptane, 1-bromo-7-fluoro-

C₇H₁₄BrNO

3199 - Butanamide, 2-bromo-2-ethyl-3-methyl-

C₇H₁₄Br₂

6372 - Heptane, 1,2-dibromo-
6373 - Heptane, 1,5-dibromo-
6374 - Heptane, 1,7-dibromo-
6375 - Heptane, 2,3-dibromo-
6376 - Heptane, 3,4-dibromo-

C₇H₁₄ClF

6365 - Heptane, 1-chloro-7-fluoro-

C₇H₁₄Cl₂

6377 - Heptane, 1,1-dichloro-
6378 - Heptane, 1,2-dichloro-
6379 - Heptane, 1,7-dichloro-
6380 - Heptane, 2,2-dichloro-
6381 - Heptane, 4,4-dichloro-
8752 - Pentane, 1,2-dichloro-4,4-dimethyl-
8753 - Pentane, 1,5-dichloro-3,3-dimethyl-
8754 - Pentane, 2,4-dichloro-2,4-dimethyl-

C₇H₁₄NO₅P

7663 - Monocrotophos

C₇H₁₄N₂

9815 - 2-Propanamine, N,N′-methanetetraylbis-

C₇H₁₄N₂O₂

9604 - 1-Piperazinecarboxylic acid, ethyl ester

C₇H₁₄N₂O₂S

334 - Aldicarb

C₇H₁₄N₂O₄

3944 - Carbamic acid, (1,1-dimethylethyl)nitro-, ethyl ester
3951 - Carbamic acid, ethylnitro-, methyl ester

C₇H₁₄N₂O₄S

335 - Aldoxycarb

C₇H₁₄N₂O₄S₂

5018 - L-Cysteine, S,S′-methylenebis-

C₇H₁₄O

3371 - Butane, 1-(ethenyloxy)-3-methyl-
4229 - Cycloheptanol
4419 - Cyclohexanemethanol
4425 - Cyclohexane, methoxy-
4530 - Cyclohexanol, 1-methyl-
4531 - Cyclohexanol, 2-methyl-, cis-(±)-
4532 - Cyclohexanol, 2-methyl-, trans-(±)-
4533 - Cyclohexanol, 3-methyl-, cis-(±)-
4534 - Cyclohexanol, 3-methyl-, trans-(±)-
4535 - Cyclohexanol, 4-methyl-, cis-
4536 - Cyclohexanol, 4-methyl-, trans-
4811 - Cyclopentaneethanol
6173 - Furan, tetrahydro-2-propyl-
6344 - Heptanal
6510 - 2-Heptanone
6511 - 3-Heptanone
6512 - 4-Heptanone
6566 - 1-Hepten-4-ol
6567 - 2-Hepten-1-ol, (E)-
6568 - 2-Hepten-1-ol, (Z)-
6569 - 2-Hepten-4-ol, (±)-
6570 - 3-Hepten-2-ol
6571 - 6-Hepten-2-ol
6572 - 6-Hepten-3-ol
6720 - Hexanal, 3-methyl-
6976 - 2-Hexanone, 3-methyl-
6977 - 2-Hexanone, 4-methyl-
6978 - 2-Hexanone, 5-methyl-
6979 - 3-Hexanone, 2-methyl-
6980 - 3-Hexanone, 4-methyl-
6981 - 3-Hexanone, 5-methyl-
7077 - 1-Hexen-3-ol, 5-methyl-
7078 - 2-Hexen-1-ol, 5-methyl-
7079 - 3-Hexen-2-ol, 2-methyl-
7080 - 3-Hexen-2-ol, 3-methyl-
7081 - 4-Hexen-3-ol, 2-methyl-
7082 - 4-Hexen-3-ol, 3-methyl-
7083 - 5-Hexen-2-ol, 2-methyl-
7084 - 5-Hexen-2-ol, 3-methyl-
7085 - 5-Hexen-3-ol, 4-methyl-
8965 - 2-Pentanone, 4,4-dimethyl-
8966 - 3-Pentanone, 2,2-dimethyl-
8967 - 3-Pentanone, 2,4-dimethyl-

C₇H₁₄O₂

227 - Acetic acid, pentyl ester
3516 - Butanoic acid, 2-ethyl-2-methyl-
3517 - Butanoic acid, 2-ethyl-, methyl ester
3548 - Butanoic acid, 3-methyl-, ethyl ester
3549 - Butanoic acid, 1-methylethyl ester

3550 - Butanoic acid, 2-methyl-, ethyl ester, (S)-
3588 - Butanoic acid, propyl ester
3630 - 1-Butanol, 2-methyl-, acetate
3631 - 1-Butanol, 3-methyl-, acetate
6009 - Formic acid, hexyl ester
6455 - Heptanoic acid
6517 - 4-Heptanone, 2-hydroxy-
6893 - Hexanoic acid, 2-methyl-, (±)-
6894 - Hexanoic acid, 3-methyl-
6895 - Hexanoic acid, 4-methyl-, (±)-
6896 - Hexanoic acid, 5-methyl-
6898 - Hexanoic acid, methyl ester
6973 - 3-Hexanone, 2-hydroxy-2-methyl-
6974 - 3-Hexanone, 5-hydroxy-2-methyl-
8602 - Oxirane, (butoxymethyl)-
8610 - Oxirane, [(1,1-dimethylethoxy)methyl]-
8877 - Pentanoic acid, 2-ethyl-, (±)-
8878 - Pentanoic acid, ethyl ester
8918 - 2-Pentanol, acetate, (R)-
8919 - 3-Pentanol, acetate
8978 - 2-Pentanone, 4-methoxy-4-methyl-
10214 - Propanoic acid, butyl ester
10259 - Propanoic acid, 2,2-dimethyl-, ethyl ester
10303 - Propanoic acid, 2-methyl-, 1-methylethyl ester
10308 - Propanoic acid, 2-methyl-, propyl ester
10309 - Propanoic acid, 1-methylpropyl ester
10310 - Propanoic acid, 2-methylpropyl ester
10559 - 1-Propene, 1,1-diethoxy-
10560 - 1-Propene, 3,3-diethoxy-

$C_7H_{14}O_3$
3995 - Carbonic acid, bis(1-methylethyl) ester
4002 - Carbonic acid, dipropyl ester
8884 - Pentanoic acid, 4-hydroxy-, ethyl ester
8885 - Pentanoic acid, 2-hydroxyethyl ester
10262 - Propanoic acid, 2-hydroxy-, ethyl ester
10268 - Propanoic acid, 2-hydroxy-, butyl ester, (R)-

$C_7H_{14}O_4$
3505 - Butanoic acid, 2,3-dihydroxypropyl ester
7112 - ribo-Hexose, 2,6-dideoxy-3-O-methyl-

$C_7H_{14}O_5$
5736 - Ethanol, 2-methoxy-, carbonate (2:1)
6198 - Galactose, 6-deoxy-3-O-methyl-

$C_7H_{14}O_6$
6227 - α-D-Glucopyranoside, methyl

$C_7H_{15}Br$
6356 - Heptane, 1-bromo-
6357 - Heptane, 2-bromo-
6358 - Heptane, 4-bromo-

$C_7H_{15}BrO$
6736 - Hexane, 1-bromo-6-methoxy-

$C_7H_{15}Cl$
3254 - Butane, 2-chloro-2,3,3-trimethyl-
6360 - Heptane, 1-chloro-
6361 - Heptane, 2-chloro-
6362 - Heptane, 3-chloro-
6363 - Heptane, 4-chloro-
6743 - Hexane, 1-chloro-3-methyl-
6744 - Hexane, 2-chloro-2-methyl-
6745 - Hexane, 2-chloro-5-methyl-
6746 - Hexane, 3-chloro-3-methyl-
8722 - Pentane, 2-chloro-2,3-dimethyl-
8723 - Pentane, 2-chloro-2,4-dimethyl-
8724 - Pentane, 3-chloro-2,3-dimethyl-
8725 - Pentane, 4-chloro-2,2-dimethyl-
8726 - Pentane, 3-chloro-3-ethyl-

$C_7H_{15}ClO$
6477 - 1-Heptanol, 7-chloro-
6478 - 2-Heptanol, 1-chloro-

$C_7H_{15}ClO_2$
9850 - Propane, 3-chloro-1,1-diethoxy-

$C_7H_{15}F$
6417 - Heptane, 1-fluoro-

$C_7H_{15}FO$
6491 - 1-Heptanol, 7-fluoro-

$C_7H_{15}I$
6419 - Heptane, 1-iodo-

6420 - Heptane, 2-iodo-
$C_7H_{15}N$
502 - Azocine, octahydro-
4292 - Cyclohexanamine, 2-methyl-, cis-
4293 - Cyclohexanamine, 2-methyl-, trans-
4294 - Cyclohexanamine, 3-methyl-
4295 - Cyclohexanamine, 3-methyl-, trans-
4296 - Cyclohexanamine, 4-methyl-, cis-
4297 - Cyclohexanamine, 4-methyl-, trans-
4298 - Cyclohexanamine, N-methyl-
4418 - Cyclohexanemethanamine
9648 - Piperidine, 1,2-dimethyl-, (±)-
9649 - Piperidine, 2,5-dimethyl-
9650 - Piperidine, 2,6-dimethyl-
9651 - Piperidine, 3,3-dimethyl-
9652 - Piperidine, 3,5-dimethyl-
9667 - Piperidine, 1-ethyl-
9668 - Piperidine, 2-ethyl-, (±)-
9669 - Piperidine, 3-ethyl-, (±)-
9670 - Piperidine, 4-ethyl-
11173 - Pyrrolidine, 1-propyl-
11174 - Pyrrolidine, 3-propyl-
11175 - Pyrrolidine, 2,2,4-trimethyl-
11176 - Pyrrolidine, 2,2,5-trimethyl-

$C_7H_{15}NO$
6346 - Heptanal, oxime
6348 - Heptanamide
6982 - 2-Hexanone, 5-methyl-, oxime
8699 - Pentanamide, N,N-dimethyl-
8968 - 3-Pentanone, 2,4-dimethyl-, oxime
9662 - 1-Piperidineethanol
9663 - 2-Piperidineethanol
9664 - 4-Piperidineethanol
9678 - 3-Piperidinemethanol, 1-methyl-
9715 - 3-Piperidinol, 1-ethyl-
9793 - Propanamide, N,N-diethyl-
10427 - 2-Propanone, 1-(diethylamino)-

$C_7H_{15}NOS$
5930 - Ethiolate

$C_7H_{15}NO_2$
3541 - Butanoic acid, 3-(methylamino)-, ethyl ester
3936 - Carbamic acid, butyl-, ethyl ester
3942 - Carbamic acid, (1,1-dimethylethyl)-, ethyl ester
3947 - Carbamic acid, ethyl-, butyl ester
3960 - Carbamic acid, (1-methylpropyl)-, ethyl ester
3961 - Carbamic acid, (2-methylpropyl)-, ethyl ester
6429 - Heptane, 1-nitro-
7692 - 4-Morpholineethanol, α-methyl-
8088 - Nitrous acid, heptyl ester

$C_7H_{15}NO_5$
6215 - α-L-Glucopyranose, 2-(methylamino)-2-deoxy-

C_7H_{16}
3457 - Butane, 2,2,3-trimethyl-
6355 - Heptane
6834 - Hexane, 2-methyl-
6835 - Hexane, 3-methyl-, (S)-
6941 - 1-Hexanol, 5-methyl-
8760 - Pentane, 2,2-dimethyl-
8761 - Pentane, 2,3-dimethyl-
8762 - Pentane, 2,4-dimethyl-
8763 - Pentane, 3,3-dimethyl-
8809 - Pentane, 3-ethyl-

$C_7H_{16}BrNO_2$
5438 - Ethanaminium, 2-(acetyloxy)-N,N,N-trimethyl-, bromide

$C_7H_{16}NCl$
7430 - Mepiquat chloride

$C_7H_{16}ClNO_2$
5439 - Ethanaminium, 2-(acetyloxy)-N,N,N-trimethyl-, chloride

$C_7H_{16}N_2$
9622 - Piperazine, 1,2,4-trimethyl-
9625 - 1-Piperidinamine, 2,6-dimethyl-
9626 - 3-Piperidinamine, 1-ethyl-
11167 - 2-Pyrrolidinemethanamine, 1-ethyl-

$C_7H_{16}N_2O$
7701 - 4-Morpholinepropanamine
12012 - Urea, N,N'-dipropyl-

$C_7H_{16}O$
3374 - Butane, 1-ethoxy-3-methyl-
3375 - Butane, 2-ethoxy-2-methyl-
3399 - Butane, 1-(1-methylethoxy)-
3421 - Butane, 1-propoxy-
3644 - 2-Butanol, 2,3,3-trimethyl-
6472 - 1-Heptanol
6473 - 2-Heptanol, (±)-
6474 - 3-Heptanol, (S)-
6475 - 4-Heptanol
6833 - Hexane, 1-methoxy-
6937 - 1-Hexanol, 2-methyl-, (±)-
6938 - 1-Hexanol, 2-methyl-, (R)-
6939 - 1-Hexanol, 3-methyl-, (±)-
6940 - 1-Hexanol, 4-methyl-, (±)-
6942 - 2-Hexanol, 2-methyl-
6943 - 2-Hexanol, 3-methyl-
6944 - 2-Hexanol, 5-methyl-
6945 - 3-Hexanol, 2-methyl-, (±)-
6946 - 3-Hexanol, 3-methyl-
6947 - 3-Hexanol, 5-methyl-, (±)-
8808 - Pentane, 1-ethoxy-
8927 - 1-Pentanol, 2,4-dimethyl-, (±)-
8928 - 2-Pentanol, 2,4-dimethyl-
8929 - 3-Pentanol, 2,2-dimethyl-
8930 - 3-Pentanol, 2,3-dimethyl-
8931 - 3-Pentanol, 2,4-dimethyl-
8932 - 3-Pentanol, 3-ethyl-
10004 - Propane, 2-methyl-2-(1-methylethoxy)-
10086 - Propane, 2-methyl-1-propoxy-
10087 - Propane, 2-methyl-2-propoxy-

$C_7H_{16}O_2$
5757 - Ethanol, 2-(pentyloxy)-
6397 - 1,4-Heptanediol
6398 - 1,5-Heptanediol
6399 - 1,7-Heptanediol
6400 - 2,4-Hexanediol, 2-methyl-
6803 - 2,4-Hexanediol, 2-methyl-
8792 - 1,3-Pentanediol, 2,2-dimethyl-
9903 - Propane, 1,1-diethoxy-
9904 - Propane, 2,2-diethoxy-
10004 - 1,3-Propanediol, 2,2-diethyl-
10020 - 1,3-Propanediol, 2-methyl-2-propyl-
10074 - Propane, 1,1'-[methylenebis(oxy)]bis-
10346 - 2-Propanol, 1-butoxy-

$C_7H_{16}O_2Si$
11422 - Silane, ethenyldiethoxymethyl-

$C_7H_{16}O_3$
3455 - Butane, 1,1,3-trimethoxy-
3456 - Butane, 1,3,3-trimethoxy-
5604 - Ethane, 1,1',1''-[methylidynetris(oxy)]tris-
6453 - 1,4,7-Heptanetriol

$C_7H_{16}O_4$
10150 - Propane, 1,1,3,3-tetramethoxy-

$C_7H_{16}O_4S_2$
9835 - Propane, 2,2-bis(ethylsulfonyl)-

$C_7H_{16}S$
6436 - 1-Heptanethiol

$C_7H_{16}S_3$
5605 - Ethane, 1,1',1''-[methylidynetris(thio)]tris-

$C_7H_{17}N$
6349 - 1-Heptanamine
6350 - 2-Heptanamine
6351 - 4-Heptanamine
6728 - 1-Hexanamine, 3-methyl-
6729 - 1-Hexanamine, 4-methyl-
6730 - 2-Hexanamine, 4-methyl-
8703 - Pentanamine, N,N-dimethyl-
8704 - 2-Pentanamine, 2,4-dimethyl-

$C_7H_{17}NO$
5739 - Ethanol, 2-[(3-methylbutyl)amino]-
10366 - 1-Propanol, 2-(diethylamino)-
10367 - 1-Propanol, 3-(diethylamino)-

$C_8H_8O_4$
895 - Benzeneacetic acid, 2,5-dihydroxy-
2573 - Benzoic acid, 2,4-dihydroxy-6-methyl-
2625 - Benzoic acid, 4-hydroxy-3-methoxy-
4262 - 2,5-Cyclohexadiene-1,4-dione, 2,6-dimethoxy-
4265 - 2,5-Cyclohexadiene-1,4-dione, 3-hydroxy-2-methoxy-5-methyl-
5873 - Ethanone, 1-(2,3,4-trihydroxyphenyl)-
10779 - 2H-Pyran-2,4(3H)-dione, 3-acetyl-6-methyl-

$C_8H_8O_5$
4260 - 2,5-Cyclohexadiene-1,4-dione, 2,5-dihydroxy-3-methoxy-6-methyl-
6083 - 2,5-Furandicarboxylic acid, dimethyl ester

C_8H_9Br
1017 - Benzene, 1-bromo-2,3-dimethyl-
1018 - Benzene, 1-bromo-2,4-dimethyl-
1019 - Benzene, 1-bromo-3,5-dimethyl-
1020 - Benzene, 2-bromo-1,3-dimethyl-
1021 - Benzene, 2-bromo-1,4-dimethyl-
1022 - Benzene, 4-bromo-1,2-dimethyl-
1034 - Benzene, (1-bromoethyl)-, (±)-
1035 - Benzene, 1-bromo-2-ethyl-
1036 - Benzene, 1-bromo-3-ethyl-
1037 - Benzene, 1-bromo-4-ethyl-
1038 - Benzene, (2-bromoethyl)-
1039 - Benzene, (1-bromoethyl)-, (R)-
1062 - Benzene, 1-(bromomethyl)-2-methyl-
1063 - Benzene, 1-(bromomethyl)-3-methyl-
1064 - Benzene, 1-(bromomethyl)-4-methyl-

C_8H_9BrO
1032 - Benzene, 1-bromo-4-ethoxy-
1033 - Benzene, (2-bromoethoxy)-
1595 - Benzeneethanol, 4-bromo-
1844 - Benzenemethanol, α-(bromomethyl)-
1845 - Benzenemethanol, 2-bromo-α-methyl-

$C_8H_9BrO_2$
1016 - Benzene, 4-bromo-1,2-dimethoxy-

C_8H_9Cl
1137 - Benzene, 1-chloro-2,3-dimethyl-
1138 - Benzene, 1-chloro-2,4-dimethyl-
1139 - Benzene, 2-chloro-1,3-dimethyl-
1140 - Benzene, 2-chloro-1,4-dimethyl-
1141 - Benzene, 4-chloro-1,2-dimethyl-
1160 - Benzene, (1-chloroethyl)-, (±)-
1161 - Benzene, 1-chloro-2-ethyl-
1162 - Benzene, 1-chloro-3-ethyl-
1163 - Benzene, 1-chloro-4-ethyl-
1164 - Benzene, (2-chloroethyl)-
1198 - Benzene, 1-(chloromethyl)-2-methyl-
1199 - Benzene, 1-(chloromethyl)-3-methyl-
1200 - Benzene, 1-(chloromethyl)-4-methyl-

C_8H_9ClO
1156 - Benzene, 1-chloro-2-ethoxy-
1157 - Benzene, 1-chloro-3-ethoxy-
1158 - Benzene, 1-chloro-4-ethoxy-
1159 - Benzene, (2-chloroethoxy)-
1181 - Benzene, [(chloromethoxy)methyl]-
1197 - Benzene, 1-(chloromethyl)-4-methoxy-
1596 - Benzeneethanol, 3-chloro-
1597 - Benzeneethanol, 4-chloro-
1852 - Benzenemethanol, α-(chloromethyl)-
9256 - Phenol, 2-chloro-3,4-dimethyl-
9257 - Phenol, 4-chloro-3,5-dimethyl-

C_8H_9ClS
1165 - Benzene, [(2-chloroethyl)thio]-

$C_8H_9Cl_2NO$
691 - Benzenamine, 2,6-dichloro-4-ethoxy-

$C_8H_9Cl_2P$
9553 - Phosphonous dichloride, (2,5-dimethylphenyl)-
9554 - Phosphonous dichloride, (4-ethylphenyl)-

$C_8H_9Cl_3Si$
11474 - Silane, trichloro(2-phenylethyl)-

C_8H_9FO
1654 - Benzene, 1-ethoxy-2-fluoro-
1655 - Benzene, 1-ethoxy-3-fluoro-
1656 - Benzene, 1-ethoxy-4-fluoro-

C_8H_9I
1680 - Benzene, 1-ethyl-2-iodo-
1681 - Benzene, 1-ethyl-4-iodo-
1769 - Benzene, 1-iodo-2,3-dimethyl-
1770 - Benzene, 1-iodo-2,4-dimethyl-
1771 - Benzene, 1-iodo-3,5-dimethyl-
1772 - Benzene, 2-iodo-1,3-dimethyl-
1773 - Benzene, 2-iodo-1,4-dimethyl-
1774 - Benzene, 4-iodo-1,2-dimethyl-

C_8H_9IO
1657 - Benzene, 1-ethoxy-4-iodo-

$C_8H_9IO_2$
1767 - Benzene, 1-iodo-2,3-dimethoxy-
1768 - Benzene, 4-iodo-1,2-dimethoxy-

C_8H_9N
733 - Benzenamine, 2-ethenyl-
734 - Benzenamine, 3-ethenyl-
735 - Benzenamine, 4-ethenyl-
2966 - Bicyclo[2.2.1]hept-5-ene-2-carbonitrile
7297 - 1H-Indole, 2,3-dihydro-
7452 - Methanamine, N-(phenylmethylene)-
10955 - Pyridine, 5-ethenyl-2-methyl-
11014 - Pyridine, 2-(2-propenyl)-

C_8H_9NO
86 - Acetamide, N-phenyl-
619 - Benzamide, 2-methyl-
620 - Benzamide, N-methyl-
878 - Benzeneacetamide
5772 - Ethanone, 1-(3-aminophenyl)-
5773 - Ethanone, 1-(4-aminophenyl)-
5774 - Ethanone, 2-amino-1-phenyl-
5855 - Ethanone, 1-(6-ethyl-3-pyridinyl)-
5866 - Ethanone, 1-phenyl-, oxime
5996 - Formamide, N-(2-methylphenyl)-
5997 - Formamide, N-(3-methylphenyl)-
5998 - Formamide, N-methyl-N-phenyl-

$C_8H_9NO_2$
67 - Acetamide, N-(4-hydroxyphenyl)-
618 - Benzamide, 4-methoxy-
885 - Benzeneacetic acid, α-amino-
886 - Benzeneacetic acid, 4-amino-
1474 - Benzene, 1,2-dimethyl-3-nitro-
1475 - Benzene, 1,2-dimethyl-4-nitro-
1476 - Benzene, 1,3-dimethyl-2-nitro-
1477 - Benzene, 1,3-dimethyl-5-nitro-
1478 - Benzene, 1,4-dimethyl-2-nitro-
1479 - Benzene, 2,4-dimethyl-1-nitro-
1692 - Benzene, 1-ethyl-2-nitro-
1693 - Benzene, 1-ethyl-3-nitro-
1694 - Benzene, 1-ethyl-4-nitro-
2004 - Benzene, 1-methyl-4-(nitromethyl)-
2051 - Benzene, (2-nitroethyl)-
2426 - 1,3-Benzodioxole-5-methanamine
2503 - Benzoic acid, 2-amino-, methyl ester
2504 - Benzoic acid, 3-amino-, methyl ester
2679 - Benzoic acid, 2-(methylamino)-
6270 - Glycine, N-phenyl-
10901 - 2-Pyridinecarboxylic acid, ethyl ester
10902 - 3-Pyridinecarboxylic acid, ethyl ester
10903 - 4-Pyridinecarboxylic acid, ethyl ester

$C_8H_9NO_3$
1612 - Benzeneethanol, 2-nitro-
1662 - Benzene, 1-ethoxy-2-nitro-
1663 - Benzene, 1-ethoxy-3-nitro-
1664 - Benzene, 1-ethoxy-4-nitro-
1934 - Benzene, 4-methoxy-1-methyl-2-nitro-
2501 - Benzoic acid, 3-amino-4-hydroxy-, methyl ester
6267 - Glycine, N-(4-hydroxyphenyl)-
9334 - Phenol, 3,6-dimethyl-2-nitro-

$C_8H_9NO_4$
1445 - Benzene, 1,2-dimethoxy-4-nitro-
1446 - Benzene, 1,3-dimethoxy-5-nitro-
1447 - Benzene, 1,4-dimethoxy-2-nitro-
1448 - Benzene, 2,4-dimethoxy-1-nitro-

C_8H_{10}
1452 - Benzene, 1,2-dimethyl-

1453 - Benzene, 1,3-dimethyl-
1454 - Benzene, 1,4-dimethyl-
1668 - Benzene, ethyl-
4749 - 1,3,5-Cyclooctatriene
4750 - 1,3,6-Cyclooctatriene
4768 - 1,3-Cyclopentadiene, 5-(1-methylethylidene)-
6714 - 1,5-Hexadien-3-yne, 2,5-dimethyl-
8342 - 1,7-Octadiyne
8343 - 2,6-Octadiyne
8344 - 3,5-Octadiyne
8483 - 1,3,5,7-Octatetraene

$C_8H_{10}BrN$
643 - Benzenamine, 2-bromo-N,N-dimethyl-
644 - Benzenamine, 3-bromo-N,N-dimethyl-
645 - Benzenamine, 4-bromo-N,N-dimethyl-
1570 - Benzeneethanamine, 4-bromo-

$C_8H_{10}ClN$
662 - Benzenamine, 2-chloro-N,N-dimethyl-
663 - Benzenamine, 4-chloro-N,N-dimethyl-
1802 - Benzenemethanamine, 3-chloro-N-methyl-

$C_8H_{10}Cl_2Si$
11400 - Silane, dichloromethyl(4-methylphenyl)-

$C_8H_{10}K_2O_{15}Sb_2$
11588 - Tartar emetic

$C_8H_{10}N_2$
18 - Acetaldehyde, phenylhydrazone
5134 - Diazene, ethylphenyl-
11339 - Quinoxaline, 1,2,3,4-tetrahydro-

$C_8H_{10}N_2O$
29 - Acetamide, N-(4-aminophenyl)-
720 - Benzenamine, N,N-dimethyl-4-nitroso-
750 - Benzenamine, N-ethyl-N-nitroso-
12027 - Urea, (phenylmethyl)-

$C_8H_{10}N_2O_2$
718 - Benzenamine, N,N-dimethyl-2-nitro-
719 - Benzenamine, N,N-dimethyl-3-nitro-

$C_8H_{10}N_2O_3$
739 - Benzenamine, 2-ethoxy-5-nitro-

$C_8H_{10}N_2O_3S$
30 - Acetamide, N-[(4-aminophenyl)sulfonyl]-
31 - Acetamide, N-[4-(aminosulfonyl)phenyl]-
2212 - Benzenesulfonamide, N,4-dimethyl-N-nitroso-

$C_8H_{10}N_2O_4$
11011 - 1(4H)-Pyridinepropanoic acid, α-amino-3-hydroxy-4-oxo-, (S)-

$C_8H_{10}N_2O_4S$
474 - Asulam

$C_8H_{10}N_2S$
10885 - 4-Pyridinecarbothioamide, 2-ethyl-
11806 - Thiourea, (2-methylphenyl)-

$C_8H_{10}N_4O_2$
10760 - 1H-Purine-2,6-dione, 3,7-dihydro-1,3,7-trimethyl-

$C_8H_{10}N_6O_7$
11097 - 2,4,6(1H,3H,5H)-Pyrimidinetrione, 5-[(hexahydro-2,4,6-trioxo-5-pyrimidinyl)imino]-, monoammonium salt

$C_8H_{10}Na_2O_5$
5370 - Endothall, disodium salt

$C_8H_{10}O$
1592 - Benzeneethanol
1649 - Benzene, ethoxy-
1896 - Benzenemethanol, α-methyl-
1897 - Benzenemethanol, 2-methyl-
1898 - Benzenemethanol, 3-methyl-
1899 - Benzenemethanol, 4-methyl-
1930 - Benzene, (methoxymethyl)-
1931 - Benzene, 1-methoxy-2-methyl-
1932 - Benzene, 1-methoxy-3-methyl-
1933 - Benzene, 1-methoxy-4-methyl-
2916 - Bicyclo[2.2.1]heptane-2-carboxaldehyde
2958 - Bicyclo[2.2.1]heptan-2-one, 3-methylene-
2967 - Bicyclo[2.2.1]hept-5-ene-2-carboxaldehyde
8484 - 2,4,6-Octatrienal
9307 - Phenol, 2,3-dimethyl-
9308 - Phenol, 2,4-dimethyl-
9309 - Phenol, 2,5-dimethyl-

9310 - Phenol, 2,6-dimethyl-
9311 - Phenol, 3,4-dimethyl-
9312 - Phenol, 3,5-dimethyl-
9353 - Phenol, 2-ethyl-
9354 - Phenol, 3-ethyl-
9355 - Phenol, 4-ethyl-

$C_8H_{10}OS$
2296 - Benzenethiol, 4-ethoxy-
10472 - 1-Propanone, 2-methyl-1-(2-thienyl)-

$C_8H_{10}O_2$
1434 - 1,3-Benzenedimethanol
1435 - 1,4-Benzenedimethanol
1436 - Benzene, 1,2-dimethoxy-
1437 - Benzene, 1,3-dimethoxy-
1438 - Benzene, 1,4-dimethoxy-
1511 - 1,2-Benzenediol, 4,5-dimethyl-
1512 - 1,3-Benzenediol, 2,5-dimethyl-
1513 - 1,3-Benzenediol, 4,5-dimethyl-
1514 - 1,3-Benzenediol, 4,6-dimethyl-
1518 - 1,3-Benzenediol, 4-ethyl-
1606 - Benzeneethanol, 3-hydroxy-
1889 - Benzenemethanol, 2-methoxy-
1890 - Benzenemethanol, 3-methoxy-
1891 - Benzenemethanol, 4-methoxy-
3659 - 2-Butanone, 4-(2-furanyl)-
5579 - 1,2-Ethanediol, 1-phenyl-
5758 - Ethanol, 2-phenoxy-
8685 - 1,5-Pentalenedione, hexahydro-
9350 - Phenol, 2-ethoxy-
9351 - Phenol, 3-ethoxy-
9352 - Phenol, 4-ethoxy-
9385 - Phenol, 2-methoxy-4-methyl-

$C_8H_{10}O_2S$
1704 - Benzene, (ethylsulfonyl)-

$C_8H_{10}O_3$
1885 - Benzenemethanol, 2-hydroxy-3-methoxy-
1886 - Benzenemethanol, 4-hydroxy-3-methoxy-
3818 - 2-Butenoic acid, anhydride
6028 - 2-Furanacetic acid, ethyl ester
6061 - 2-Furancarboxylic acid, 1-methylethyl ester
6064 - 3-Furancarboxylic acid, methyl-, ethyl ester
6071 - 2-Furancarboxylic acid, propyl ester
6123 - 2-Furanmethanol, propanoate
6162 - 2-Furanpropanoic acid, methyl ester
7345 - 1,3-Isobenzofurandione, hexahydro-
9305 - Phenol, 2,6-dimethoxy-
9306 - Phenol, 3,5-dimethoxy-
10649 - 2-Propenoic acid, 2-methyl-, anhydride

$C_8H_{10}O_3S$
2223 - Benzenesulfonic acid, ethyl ester
2232 - Benzenesulfonic acid, 4-methyl-, methyl ester

$C_8H_{10}O_4$
3909 - 2-Butenedioic acid, diethyl ester
3913 - 2-Butyne-1,4-diol, diacetate
5564 - Ethanedioic acid, di-2-propenyl ester
6701 - 2,5-Hexadienoic acid, 3-methoxy-5-methyl-4-oxo-

$C_8H_{10}S$
1705 - Benzene, (ethylthio)-
1834 - Benzenemethanethiol, α-methyl-, (S)-
2042 - Benzene, [(methylthio)methyl]-
2297 - Benzenethiol, 3-ethyl-
2298 - Benzenethiol, 4-ethyl-

$C_8H_{11}ClO_4$
3771 - 2-Butenedioic acid, 2-chloro-, diethyl ester, (E)-
3772 - 2-Butenedioic acid, 2-chloro-, diethyl ester, (Z)-

$C_8H_{11}ClSi$
11374 - Silane, chlorodimethylphenyl-

$C_8H_{11}Cl_2N_3O_2$
11061 - 2,4(1H,3H)-Pyrimidinedione, 5-[bis(2-chloroethyl)amino]-

$C_8H_{11}Cl_3O_7$
6231 - β-D-Glucopyranosiduronic acid, 2,2,2-trichloroethyl

$C_8H_{11}N$
276 - Acetonitrile, cyclohexylidene-
704 - Benzenamine, 2,3-dimethyl-
705 - Benzenamine, 2,4-dimethyl-
706 - Benzenamine, 2,5-dimethyl-
707 - Benzenamine, 2,6-dimethyl-
708 - Benzenamine, 3,4-dimethyl-
709 - Benzenamine, 3,5-dimethyl-
710 - Benzenamine, N,2-dimethyl-
711 - Benzenamine, N,3-dimethyl-
712 - Benzenamine, N,4-dimethyl-
713 - Benzenamine, N,N-dimethyl-
740 - Benzenamine, 2-ethyl-
741 - Benzenamine, 3-ethyl-
742 - Benzenamine, 4-ethyl-
743 - Benzenamine, N-ethyl-
1569 - Benzeneethanamine
1814 - Benzenemethanamine, α-methyl-, (±)-
1815 - Benzenemethanamine, 2-methyl-
1816 - Benzenemethanamine, 3-methyl-
1817 - Benzenemethanamine, 4-methyl-
1818 - Benzenemethanamine, N-methyl-
10962 - Pyridine, 2-ethyl-4-methyl-
10963 - Pyridine, 2-ethyl-6-methyl-
10964 - Pyridine, 3-ethyl-4-methyl-
10965 - Pyridine, 4-ethyl-2-methyl-
10966 - Pyridine, 5-ethyl-2-methyl-
10984 - Pyridine, 2-(1-methylethyl)-
10985 - Pyridine, 3-(1-methylethyl)-
10986 - Pyridine, 4-(1-methylethyl)-
11015 - Pyridine, 2-propyl-
11016 - Pyridine, 4-propyl-
11024 - Pyridine, 2,3,4-trimethyl-
11025 - Pyridine, 2,3,5-trimethyl-
11026 - Pyridine, 2,3,6-trimethyl-
11027 - Pyridine, 2,4,5-trimethyl-
11028 - Pyridine, 2,4,6-trimethyl-

$C_8H_{11}NO$
736 - Benzenamine, 2-ethoxy-
737 - Benzenamine, 4-ethoxy-
770 - Benzenamine, 2-methoxy-5-methyl-
771 - Benzenamine, 3-methoxy-4-methyl-
772 - Benzenamine, 4-methoxy-2-methyl-
773 - Benzenamine, 5-methoxy-2-methyl-
774 - Benzenamine, 2-methoxy-6-methyl-
1593 - Benzeneethanol, β-amino-
1594 - Benzeneethanol, 2-amino-
1812 - Benzenemethanamine, 2-methoxy-
1813 - Benzenemethanamine, 4-methoxy-
1839 - Benzenemethanol, α-(aminomethyl)-
5759 - Ethanol, 2-(phenylamino)-
9208 - Phenol, 4-(2-aminoethyl)-
9317 - Phenol, 2-(dimethylamino)-
9318 - Phenol, 3-(dimethylamino)-
9319 - Phenol, 4-(dimethylamino)-
9356 - Phenol, 3-(ethylamino)-
10951 - 2-Pyridineethanol, α-methyl-
10979 - Pyridine, 2-(2-methoxyethyl)-
11012 - 2-Pyridinepropanol
11013 - 3-Pyridinepropanol

$C_8H_{11}NO_2$
702 - Benzenamine, 2,5-dimethoxy-
703 - Benzenamine, 2,6-dimethoxy-

$C_8H_{11}NO_2S$
2211 - Benzenesulfonamide, N,4-dimethyl-

$C_8H_{11}NO_3$
1500 - 1,2-Benzenediol, 4-(2-amino-1-hydroxyethyl)-, (R)-

$C_8H_{11}NO_4$
11145 - 1-Pyrrolidineacetic acid, 2,5-dioxo-, ethyl ester

$C_8H_{11}N_3O_6$
11832 - 1,2,4-Triazine-3,5(2H,4H)-dione, 2-β-D-ribofuranosyl-

$C_8H_{11}N_5$
7234 - Imidodicarbonimidic diamide, N-phenyl-

$C_8H_{11}O_4$
5191 - 1,3-Dioxane-2-ethanol, 4-hydroxy-α,6-dimethyl-

C_8H_{12}
2993 - Bicyclo[4.2.0]oct-2-ene
4186 - Cyclobutane, 1,2-diethenyl-, cis-
4187 - Cyclobutane, 1,2-diethenyl-, trans-
4252 - 1,3-Cyclohexadiene, 1,3-dimethyl-
4253 - 1,3-Cyclohexadiene, 1,4-dimethyl-
4254 - 1,3-Cyclohexadiene, 1,5-dimethyl-
4255 - 1,3-Cyclohexadiene, 2,5-dimethyl-
4256 - 1,3-Cyclohexadiene, 2,6-dimethyl-
4314 - Cyclohexane, 1,2-bis(methylene)-
4657 - Cyclohexene, 1-ethenyl-
4658 - Cyclohexene, 4-ethenyl-
4732 - 1,4-Cyclooctadiene
4733 - 1,5-Cyclooctadiene
4734 - 1,5-Cyclooctadiene, (Z,Z)-
4755 - Cyclooctyne
6994 - 1,3,5-Hexatriene, 2,5-dimethyl-
8489 - 2,4,6-Octatriene, (E,E,E)-
8526 - 1-Octen-3-yne

$C_8H_{12}ClN$
717 - Benzenamine, N,N-dimethyl-, hydrochloride

$C_8H_{12}ClNO$
44 - Acetamide, 2-chloro-N,N-di-2-propenyl-

$C_8H_{12}ClNO_3$
10934 - 3,4-Pyridinedimethanol, 5-hydroxy-6-methyl-, hydrochloride

$C_8H_{12}Cl_2$
7117 - 3-Hexyne, 2,5-dichloro-2,5-dimethyl-

$C_8H_{12}N_2$
1267 - 1,2-Benzenediamine, N,N-dimethyl-
1268 - 1,3-Benzenediamine, N,N-dimethyl-
1269 - 1,2-Benzenediamine, N,N'-dimethyl-
1270 - 1,4-Benzenediamine, N,N-dimethyl-
1433 - 1,3-Benzenedimethanamine
3292 - Butanedinitrile, tetramethyl-
7159 - Hydrazine, 1-ethyl-1-phenyl-
7160 - Hydrazine, 1-ethyl-2-phenyl-
7163 - Hydrazine, 1-methyl-2-(3-methylphenyl)-
7170 - Hydrazine, (1-phenylethyl)-
8385 - Octanedinitrile
10809 - Pyrazine, 3-ethyl-2,5-dimethyl-

$C_8H_{12}N_2O$
1272 - 1,2-Benzenediamine, 4-ethoxy-

$C_8H_{12}N_2O_2$
6770 - Hexane, 1,6-diisocyanato-

$C_8H_{12}N_2O_3$
11087 - 2,4,6(1H,3H,5H)-Pyrimidinetrione, 5,5-diethyl-

$C_8H_{12}N_4$
10093 - Propanenitrile, 2,2'-azobis[2-methyl-

$C_8H_{12}O$
4529 - Cyclohexanol, 1-ethynyl-
4594 - Cyclohexanone, 2-ethylidene-
4705 - 2-Cyclohexen-1-one, 2,3-dimethyl-
4706 - 2-Cyclohexen-1-one, 2,5-dimethyl-
4707 - 2-Cyclohexen-1-one, 3,5-dimethyl-
4708 - 2-Cyclohexen-1-one, 3,6-dimethyl-
4754 - 3-Cycloocten-1-one
4947 - 2-Cyclopenten-1-one, 3-(1-methylethyl)-
5793 - Ethanone, 1-(1-cyclohexen-1-yl)-
6090 - Furan, 2-(1,1-dimethylethyl)-
8509 - 1-Octen-4-ol
8555 - 7-Oxabicyclo[4.1.0]heptane, 3-ethenyl-
8688 - 2(1H)-Pentalenone, hexahydro-
8689 - 2(1H)-Pentalenone, hexahydro-, cis-

$C_8H_{12}O_2$
4378 - 1,3-Cyclohexanedione, 5,5-dimethyl-
4572 - Cyclohexanone, 2-acetyl-
4640 - 3-Cyclohexene-1-carboxylic acid, methyl ester
6699 - 2,4-Hexadienoic acid, ethyl ester
6700 - 2,4-Hexadienoic acid, ethyl ester, (E,E)-
8504 - 4-Octene-2,7-dione
8538 - 2-Octynoic acid
8539 - 3-Octynoic acid

8540 - 5-Octynoic acid
8541 - 7-Octynoic acid
8557 - 7-Oxabicyclo[4.1.0]heptane, 3-oxiranyl-
8597 - 2-Oxetanone, 3,3-dimethyl-4-
(1-methylethylidene)-
8805 - 2,4-Pentanedione, 3-(2-propenyl)-

$C_8H_{12}O_3$
4336 - Cyclohexanecarboxylic acid, 3-methyl-4-oxo-
4785 - Cyclopentanecarboxylic acid, 2-oxo-, ethyl
ester
10695 - 2-Propenoic acid, (tetrahydro-2-
furanyl)methyl ester

$C_8H_{12}O_4$
3462 - Butanoic acid, 2-acetyl-3-oxo-, ethyl ester
3775 - 2-Butenedioic acid (E)-, diethyl ester
3776 - 2-Butenedioic acid (Z)-, diethyl ester
4181 - 1,2-Cyclobutanedicarboxylic acid, dimethyl
ester, trans-
4183 - 1,3-Cyclobutanedicarboxylic acid, 2,2-
dimethyl-, cis-
4354 - 1,4-Cyclohexanedicarboxylic acid, trans-
6030 - 3-Furanacetic acid, tetrahydro-2,2-dimethyl-5-
oxo-
10564 - 2-Propene-1,1-diol, 2-methyl-, diacetate

$C_8H_{12}O_5$
3340 - Butanedioic acid, oxo-, diethyl ester

$C_8H_{12}O_6Si$
11498 - Silanetriol, ethenyl-, triacetate

$C_8H_{12}S$
11731 - Thiophene, 2-butyl-

$C_8H_{12}Si$
11418 - Silane, dimethylphenyl-
11449 - Silane, tetraethenyl-

$C_8H_{12}Sn$
11543 - Stannane, tetraethenyl-

$C_8H_{13}Cl$
8533 - 1-Octyne, 1-chloro-

$C_8H_{13}F_3O_2$
268 - Acetic acid, trifluoro-, 2-methylpentyl ester

$C_8H_{13}N$
4290 - Cyclohexanamine, 1-ethynyl-
4313 - Cyclohexaneacetonitrile
11113 - 1H-Pyrrole, 1-butyl-
11126 - 1H-Pyrrole, 3-ethyl-2,4-dimethyl-
11127 - 1H-Pyrrole, 4-ethyl-2,3-dimethyl-
11141 - 1H-Pyrrole, 2,3,4,5-tetramethyl-

$C_8H_{13}NO$
490 - 8-Azabicyclo[3.2.1]octan-3-one, 8-methyl-

$C_8H_{13}NO_2$
7575 - 2,5-Methano-2H-furo[3,2-b]pyrrol-6-ol,
hexahydro-4-methyl-, (2α,3aβ,5α,6β,6aβ
9656 - 2,6-Piperidinedione, 4-ethyl-4-methyl-
10914 - 3-Pyridinecarboxylic acid, 1,2,5,6-tetrahydro-
1-methyl-, methyl ester

$C_8H_{13}NO_3$
8573 - 2H-1,3-Oxazine-2,4(3H)-dione, 5,5-
diethyldihydro-

$C_8H_{13}NO_5$
9953 - Propanedioic acid, (formylamino)-, diethyl
ester

$C_8H_{13}N_2O_3PS$
11720 - Thionazin

C_8H_{14}
2991 - Bicyclo[4.2.0]octane
4245 - Cycloheptene, 1-methyl-
4246 - Cycloheptene, 3-methyl-
4388 - Cyclohexane, ethenyl-
4391 - Cyclohexane, ethylidene-
4434 - Cyclohexane, 1-methyl-4-methylene-
4648 - Cyclohexene, 1,2-dimethyl-
4649 - Cyclohexene, 1,3-dimethyl-
4650 - Cyclohexene, 1,4-dimethyl-
4651 - Cyclohexene, 1,5-dimethyl-
4652 - Cyclohexene, 1,6-dimethyl-
4653 - Cyclohexene, 3,3-dimethyl-
4654 - Cyclohexene, 3,5-dimethyl-
4655 - Cyclohexene, 3,6-dimethyl-

4656 - Cyclohexene, 4,4-dimethyl-
4660 - Cyclohexene, 1-ethyl-
4661 - Cyclohexene, 3-ethyl-
4662 - Cyclohexene, 4-ethyl-
4751 - Cyclooctene, (E)-
4752 - Cyclooctene, (Z)-
4848 - Cyclopentane, 2-propenyl-
4935 - Cyclopentene, 1-propyl-
4937 - Cyclopentene, 1,2,3-trimethyl-
4938 - Cyclopentene, 1,5,5-trimethyl-
6335 - 1,5-Heptadiene, 3-methyl-
6691 - 1,5-Hexadiene, 2,5-dimethyl-
6692 - 2,4-Hexadiene, 2,5-dimethyl-
8319 - 1,7-Octadiene
8320 - 2,6-Octadiene
8529 - 1-Octyne
8530 - 2-Octyne
8531 - 3-Octyne
8532 - 4-Octyne
8686 - Pentalene, octahydro-, cis-
8687 - Pentalene, octahydro-, trans-
9132 - 1-Pentyne, 3-ethyl-3-methyl-

$C_8H_{14}ClNS_2$
4065 - CDEC

$C_8H_{14}ClN_5$
476 - Atrazine

$C_8H_{14}Cl_2N_2O_2$
10978 - 3-Pyridinemethanol, 4-(aminomethyl)-5-
hydroxy-6-methyl-, dihydrochloride

$C_8H_{14}Cl_2O_2$
3501 - Butanoic acid, 3,4-dichloro-, butyl ester

$C_8H_{14}N_2$
275 - Acetonitrile, (cyclohexylamino)-
9703 - 1-Piperidinepropanenitrile

$C_8H_{14}N_4OS$
7658 - Metribuzin

$C_8H_{14}O$
4235 - Cycloheptanone, 2-methyl-
4580 - Cyclohexanone, 2,2-dimethyl-
4581 - Cyclohexanone, 2,3-dimethyl-, cis-
4582 - Cyclohexanone, 2,3-dimethyl-, trans-
4583 - Cyclohexanone, 2,4-dimethyl-, (2S-trans)-
4584 - Cyclohexanone, 2,4-dimethyl-, cis-
4585 - Cyclohexanone, 2,5-dimethyl-, trans-(±)-
4586 - Cyclohexanone, 2,6-dimethyl-
4587 - Cyclohexanone, 3,3-dimethyl-
4588 - Cyclohexanone, 3,4-dimethyl-
4589 - Cyclohexanone, 3,5-dimethyl-
4590 - Cyclohexanone, 4,4-dimethyl-
4593 - Cyclohexanone, 4-ethyl-
4740 - Cyclooctanone
4753 - 2-Cycloocten-1-ol
4883 - Cyclopentanone, 2-ethyl-5-methyl-
4887 - Cyclopentanone, 2-(1-methylethyl)-
4894 - Cyclopentanone, 2-propyl-
4895 - Cyclopentanone, 3-propyl-
4896 - Cyclopentanone, 2,2,4-trimethyl-
4897 - Cyclopentanone, 2,2,5-trimethyl-
4898 - Cyclopentanone, 2,3,5-trimethyl-
4899 - Cyclopentanone, 2,4,4-trimethyl-
5794 - Ethanone, 1-cyclohexyl-
5846 - Ethanone, 1-(1-methylcyclopentyl)-
5847 - Ethanone, 1-(2-methylcyclopentyl)-
6338 - 1,6-Heptadien-4-ol, 4-methyl-
6595 - 1-Hepten-5-one, 5-methyl-
6596 - 2-Hepten-4-one, 5-methyl-
6597 - 2-Hepten-4-one, 6-methyl-
6598 - 3-Hepten-2-one, 6-methyl-
6599 - 4-Hepten-2-one, 6-methyl-
6600 - 4-Hepten-3-one, 6-methyl-
6601 - 5-Hepten-2-one, 3-methyl-
6602 - 5-Hepten-2-one, 5-methyl-
6603 - 5-Hepten-2-one, 6-methyl-
6604 - 5-Hepten-3-one, 2-methyl-
6605 - 5-Hepten-3-one, 5-methyl-
7091 - 3-Hexen-2-one, 3,4-dimethyl-
7092 - 4-Hexen-2-one, 3,4-dimethyl-

8543 - 2-Octyn-1-ol

$C_8H_{14}O_2$
155 - Acetic acid, cyclohexyl ester
3836 - 2-Butenoic acid, 2,3-dimethyl-, ethyl ester
4223 - Cycloheptanecarboxylic acid
4306 - Cyclohexaneacetic acid
4335 - Cyclohexanecarboxylic acid, methyl ester
4776 - Cyclopentaneacetic acid, 3-methyl-
4847 - Cyclopentanepropanoic acid
5821 - Ethanone, 1-(1-hydroxycyclohexyl)-
6140 - 2(3H)-Furanone, 5-butyldihydro-
6565 - 2-Heptenoic acid, 6-methyl-
6807 - 3,4-Hexanedione, 2,5-dimethyl-
6892 - Hexanoic acid, 3-hydroxy-2,2-dimethyl-, β-
lactone
7053 - 2-Hexenoic acid, ethyl ester
7054 - 3-Hexenoic acid, ethyl ester
7119 - 3-Hexyne-2,5-diol, 2,5-dimethyl-
8364 - Octanedial
8393 - 2,4-Octanedione
8394 - 2,7-Octanedione
8395 - 3,6-Octanedione
8396 - 4,5-Octanedione
9080 - 2-Pentenoic acid, 4-methyl-, ethyl ester
9081 - 3-Pentenoic acid, 3-methyl-, ethyl ester
9082 - 3-Pentenoic acid, 4-methyl-, ethyl ester
10650 - 2-Propenoic acid, 2-methyl-, butyl ester
10651 - 2-Propenoic acid, methylbutyl ester
10652 - 2-Propenoic acid, 2-methylbutyl ester
10665 - 2-Propenoic acid, 2-methyl-, 1-methylpropyl
ester
10666 - 2-Propenoic acid, 2-methyl-, 2-methylpropyl
ester

$C_8H_{14}O_3$
3467 - Butanoic acid, anhydride
3511 - Butanoic acid, 2,2-dimethyl-3-oxo-, ethyl ester
3518 - Butanoic acid, 2-ethyl-3-oxo-, ethyl ester
3569 - Butanoic acid, 3-oxo-, butyl ester
6126 - 2-Furanmethanol, tetrahydro-, propanoate
6903 - Hexanoic acid, 4-methyl-5-oxo-, methyl ester
6906 - Hexanoic acid, 5-oxo-, ethyl ester
8897 - Pentanoic acid, 4-methyl-3-oxo-, ethyl ester
8909 - Pentanoic acid, 4-oxo-, 1-methylethyl ester
8912 - Pentanoic acid, 4-oxo-, propyl ester
10291 - Propanoic acid, 2-methyl-, anhydride

$C_8H_{14}O_4$
3303 - Butanedioic acid, diethyl ester
3333 - Butanedioic acid, monobutyl ester
3343 - Butanedioic acid, tetramethyl-
3348 - 1,4-Butanediol, diacetate
5556 - Ethanedioic acid, bis(1-methylethyl) ester
5565 - Ethanedioic acid, dipropyl ester
5576 - 1,2-Ethanediol, dipropanoate
6786 - Hexanedioic acid, dimethyl ester
6791 - Hexanedioic acid, monoethyl ester
8386 - Octanedioic acid
8777 - Pentanedioic acid, 3-ethyl-3-methyl-
9962 - Propanedioic acid, methyl-, diethyl ester

$C_8H_{14}O_4S$
10331 - Propanoic acid, 3,3'-thiobis-, dimethyl ester

$C_8H_{14}O_5$
5754 - Ethanol, 2,2'-oxybis-, diacetate

$C_8H_{14}O_6$
3307 - Butanedioic acid, 2,3-dihydroxy-, diethyl ester,
(R*,R*)-(±)-

$C_8H_{15}Br$
4316 - Cyclohexane, (2-bromoethyl)-

$C_8H_{15}BrO_2$
4149 - Crotonaldehyde, 2-bromo-, diethyl acetal
6877 - Hexanoic acid, 2-bromo-, ethyl, (±)-
6878 - Hexanoic acid, 5-bromo-, ethyl ester
6879 - Hexanoic acid, 6-bromo-, ethyl ester
8425 - Octanoic acid, 2-bromo-
8426 - Octanoic acid, 8-bromo-
10210 - Propanoic acid, 3-bromo-, 3-methylbutyl
ester

$C_8H_{15}Cl$
6541 - 2-Heptene, 6-chloro-2-methyl-
7010 - 3-Hexene, 1-chloro-4-ethyl-
8498 - 1-Octene, 2-chloro-
8499 - 2-Octene, 2-chloro-
8500 - 4-Octene, 4-chloro-, (Z)-
8501 - 2-Octene, 4-chloro-

$C_8H_{15}ClO$
6987 - Hexanoyl chloride, 2-ethyl-
8481 - Octanoyl chloride

$C_8H_{15}ClO_2$
10226 - Propanoic acid, 3-chloro-, 3-methylbutyl ester

$C_8H_{15}FO_2$
8429 - Octanoic acid, 8-fluoro-

$C_8H_{15}N$
484 - 8-Azabicyclo[3.2.1]octane, 8-methyl-
7319 - 1H-Indole, octahydro-
7325 - Indolizine, octahydro-
8411 - Octanenitrile
9706 - Piperidine, 2-(1-propenyl)-, (±)-
11021 - Pyridine, 2,3,4,5-tetrahydro-6-propyl-

$C_8H_{15}NO$
486 - 8-Azabicyclo[3.2.1]octan-3-ol, 8-methyl-, exo-
489 - 8-Azabicyclo[3.2.1]octan-3-ol, 8-methyl-, endo-
3408 - Butanenitrile, 2-hydroxy-3-methyl-2-(1-methylethyl)-
8829 - Pentanenitrile, 2-hydroxy-2-propyl-
9631 - Piperidine, 1-acetyl-3-methyl-
10471 - 2-Propanone, 1-(1-methyl-2-pyrrolidinyl)-, (R)-
10484 - 1-Propanone, 1-(2-piperidinyl)-
10485 - 2-Propanone, 1-(2-piperidinyl)-, (R)-

$C_8H_{15}NO_2$
3406 - Butanenitrile, 4,4-diethoxy-
4333 - Cyclohexanecarboxylic acid, 4-(aminomethyl)-, trans-
8399 - 2,3-Octanedione, 3-oxime
9643 - 4-Piperidinecarboxylic acid, ethyl ester
9645 - 4-Piperidinecarboxylic acid, 1-methyl-, methyl ester
10654 - 2-Propenoic acid, 2-methyl-, 2-(dimethylamino)ethyl ester

$C_8H_{15}NO_3$
6871 - Hexanoic acid, 6-(acetylamino)-

$C_8H_{15}NS$
11704 - Thiocyanic acid, heptyl ester

C_8H_{16}
4226 - Cycloheptane, methyl-
4364 - Cyclohexane, 1,1-dimethyl-
4365 - Cyclohexane, 1,2-dimethyl-, cis-
4366 - Cyclohexane, 1,2-dimethyl-, trans-
4367 - Cyclohexane, 1,3-dimethyl-, cis-
4368 - Cyclohexane, 1,4-dimethyl-, cis-
4370 - Cyclohexane, 1,3-dimethyl-, trans-
4371 - Cyclohexane, 1,4-dimethyl-, trans-
4390 - Cyclohexane, ethyl-
4736 - Cyclooctane
4814 - Cyclopentane, 1-ethyl-1-methyl-
4815 - Cyclopentane, 1-ethyl-2-methyl-, cis-
4816 - Cyclopentane, 1-ethyl-2-methyl-, trans-
4817 - Cyclopentane, 1-ethyl-3-methyl-, cis-
4818 - Cyclopentane, 1-ethyl-3-methyl-, trans-
4828 - Cyclopentane, (1-methylethyl)-
4849 - Cyclopentane, propyl-
4855 - Cyclopentane, 1,1,2-trimethyl-
4856 - Cyclopentane, 1,1,3-trimethyl-
4857 - Cyclopentane, 1,2,3-trimethyl-, (1α,2β,3α)-
4858 - Cyclopentane, 1,2,3-trimethyl-, (1α,2α,3α)-
4859 - Cyclopentane, 1,2,3-trimethyl-, (1α,2α,3β)-
4860 - Cyclopentane, 1,2,4-trimethyl-, (1α,2β,4α)-
4861 - Cyclopentane, 1,2,4-trimethyl-, (1α,2α,4α)-
4862 - Cyclopentane, 1,2,4-trimethyl-, (1α,2α,4β)-
6425 - Heptane, 3-methylene-
6426 - Heptane, 4-methylene-
6543 - 1-Heptene, 2-methyl-
6544 - 1-Heptene, 3-methyl-

6545 - 1-Heptene, 4-methyl-
6546 - 1-Heptene, 5-methyl-
6547 - 1-Heptene, 6-methyl-
6548 - 2-Heptene, 2-methyl-
6549 - 2-Heptene, 3-methyl-, (E)-
6550 - 2-Heptene, 3-methyl-, (Z)-
6551 - 2-Heptene, 4-methyl-, (E)-
6552 - 2-Heptene, 4-methyl-, (Z)-
6553 - 2-Heptene, 6-methyl-
6554 - 3-Heptene, 2-methyl-
6555 - 3-Heptene, 3-methyl-
6556 - 3-Heptene, 4-methyl-
6557 - 3-Heptene, 5-methyl-
6558 - 3-Heptene, 6-methyl-
6836 - Hexane, 2-methyl-3-methylene-
6837 - Hexane, 2-methyl-4-methylene-
6838 - Hexane, 3-methyl-4-methylene-
7013 - 1-Hexene, 2,3-dimethyl-
7014 - 1-Hexene, 2,4-dimethyl-
7015 - 1-Hexene, 2,5-dimethyl-
7016 - 1-Hexene, 3,3-dimethyl-
7017 - 1-Hexene, 3,4-dimethyl-
7018 - 1-Hexene, 3,5-dimethyl-
7019 - 1-Hexene, 4,4-dimethyl-
7020 - 1-Hexene, 4,5-dimethyl-
7021 - 1-Hexene, 5,5-dimethyl-
7022 - 2-Hexene, 2,3-dimethyl-
7023 - 2-Hexene, 2,4-dimethyl-
7024 - 2-Hexene, 2,5-dimethyl-
7025 - 2-Hexene, 4,4-dimethyl-
7026 - 2-Hexene, 5,5-dimethyl-, (E)-
7027 - 2-Hexene, 5,5-dimethyl-, (Z)-
7028 - 3-Hexene, 2,2-dimethyl-, (E)-
7029 - 3-Hexene, 2,2-dimethyl-, (Z)-
7030 - 3-Hexene, 2,4-dimethyl-, (E)-
7031 - 3-Hexene, 2,4-dimethyl-, (Z)-
7033 - 1-Hexene, 3-ethyl-
7034 - 3-Hexene, 3-ethyl-
8491 - 1-Octene
8492 - 2-Octene, (E)-
8493 - 2-Octene, (Z)-
8494 - 3-Octene, (E)-
8495 - 3-Octene, (Z)-
8496 - 4-Octene, (E)-
8497 - 4-Octene, (Z)-
8764 - Pentane, 2,2-dimethyl-3-methylene-
8765 - Pentane, 2,4-dimethyl-3-methylene-
9046 - 1-Pentene, 3-ethyl-2-methyl-
9047 - 1-Pentene, 3-ethyl-3-methyl-
9048 - 1-Pentene, 3-ethyl-4-methyl-
9049 - 2-Pentene, 3-ethyl-2-methyl-
9050 - 2-Pentene, 3-ethyl-4-methyl-, (E)-
9051 - 2-Pentene, 3-ethyl-4-methyl-, (Z)-
9064 - Pentene, 2,4,4-trimethyl-
9065 - 1-Pentene, 2,3,3-trimethyl-
9066 - 1-Pentene, 2,3,4-trimethyl-
9067 - 1-Pentene, 2,4,4-trimethyl-
9068 - 1-Pentene, 3,3,4-trimethyl-
9069 - 1-Pentene, 3,4,4-trimethyl-
9070 - 2-Pentene, 2,3,4-trimethyl-
9071 - 2-Pentene, 2,4,4-trimethyl-
9072 - 2-Pentene, 3,4,4-trimethyl-

$C_8H_{16}BrF$
8358 - Octane, 1-bromo-8-fluoro-

$C_8H_{16}Br_2$
8366 - Octane, 1,2-dibromo-
8367 - Octane, 1,4-dibromo-
8368 - Octane, 1,5-dibromo-
8369 - Octane, 1,8-dibromo-
8370 - Octane, 4,5-dibromo-

$C_8H_{16}ClF$
8361 - Octane, 1-chloro-8-fluoro-

$C_8H_{16}ClNO$
42 - Acetamide, 2-chloro-N,N-bis(1-methylethyl)-
45 - Acetamide, 2-chloro-N-hexyl-

$C_8H_{16}Cl_2$
6765 - Hexane, 2,5-dichloro-2,5-dimethyl-

8371 - Octane, 1,8-dichloro-
8372 - Octane, 2,3-dichloro-

$C_8H_{16}NO_5P$
5172 - Dicrotophos

$C_8H_{16}N_2$
3673 - 2-Butanone, (1-methylpropylidene)hydrazone

$C_8H_{16}N_2O_2$
8397 - 4,5-Octanedione, dioxime

$C_8H_{16}N_2O_2S_2$
7688 - Morpholine, 4,4'-dithiobis-

$C_8H_{16}N_2O_3$
7414 - DL-Leucine, N-glycyl-

$C_8H_{16}N_2O_4S_2$
12042 - D-Valine, 3-[(2-amino-2-carboxyethyl)dithio]-, (R)-

$C_8H_{16}O$
4230 - Cycloheptanol, 1-methyl-
4231 - Cycloheptanol, 2-methyl-
4232 - Cycloheptanol, 4-methyl-
4387 - Cyclohexaneethanol
4420 - Cyclohexanemethanol, α-methyl-
4421 - Cyclohexanemethanol, α-methyl-, (S)-
4422 - Cyclohexanemethanol, 2-methyl-, cis-
4423 - Cyclohexanemethanol, 2-methyl-, trans-
4424 - Cyclohexanemethanol, 4-methyl-
4508 - Cyclohexanol, 1,2-dimethyl-, cis-
4509 - Cyclohexanol, 1,2-dimethyl-, trans-
4510 - Cyclohexanol, 1,3-dimethyl-, cis-
4511 - Cyclohexanol, 1,3-dimethyl-, trans-
4512 - Cyclohexanol, 1,4-dimethyl-
4513 - Cyclohexanol, 2,2-dimethyl-
4514 - Cyclohexanol, 2,4-dimethyl-
4515 - Cyclohexanol, 2,6-dimethyl-
4516 - Cyclohexanol, 3,3-dimethyl-
4517 - Cyclohexanol, 3,4-dimethyl-
4518 - Cyclohexanol, 3,5-dimethyl-
4519 - Cyclohexanol, 4,4-dimethyl-
4522 - Cyclohexanol, 1-ethyl-
4523 - Cyclohexanol, 2-ethyl-, trans-(±)-
4524 - Cyclohexanol, 2-ethyl-, cis-(±)-
4525 - Cyclohexanol, 4-ethyl-
4739 - Cyclooctanol
4872 - Cyclopentanol, 1-propyl-
4873 - Cyclopentanol, 2-propyl-, cis-
4874 - Cyclopentanol, 2-propyl-, trans-
4875 - Cyclopentanol, 1,2,2-trimethyl-
4876 - Cyclopentanol, 1,2,4-trimethyl-
4877 - Cyclopentanol, 1,2,5-trimethyl-
6085 - Furan, 2,2-diethyltetrahydro-
6174 - Furan, tetrahydro-2,2,4,4-tetramethyl-
6345 - Heptanal, 5-methyl-
6520 - 2-Heptanone, 3-methyl-
6521 - 2-Heptanone, 6-methyl-
6522 - 3-Heptanone, 2-methyl-
6523 - 3-Heptanone, 6-methyl-
6524 - 4-Heptanone, 2-methyl-
6525 - 4-Heptanone, 3-methyl-
6573 - 1-Hepten-4-ol, 4-methyl-
6574 - 2-Hepten-4-ol, 5-methyl-
6575 - 2-Hepten-4-ol, 6-methyl-
6576 - 3-Hepten-2-ol, 2-methyl-
6577 - 4-Hepten-2-ol, 3-methyl-
6578 - 4-Hepten-3-ol, 3-methyl-
6579 - 4-Hepten-3-ol, 4-methyl-
6580 - 5-Hepten-2-ol, 6-methyl-
6581 - 6-Hepten-1-ol, 3-methyl-
6582 - 6-Hepten-2-ol, 2-methyl-
6719 - Hexanal, 2-ethyl-
6813 - Hexane, 1-(ethenyloxy)-
6963 - 2-Hexanone, 3,3-dimethyl-
6964 - 2-Hexanone, 3,4-dimethyl-
6965 - 3-Hexanone, 2,2-dimethyl-
6966 - 3-Hexanone, 2,5-dimethyl-
6967 - 3-Hexanone, 4,4-dimethyl-
7073 - 1-Hexen-3-ol, 3,5-dimethyl-
7074 - 4-Hexen-3-ol, 2,2-dimethyl-
7075 - 4-Hexen-3-ol, 2,5-dimethyl-

7076 - 4-Hexen-3-ol, 3,5-dimethyl-
8346 - Octanal
8471 - 2-Octanone
8472 - 3-Octanone
8473 - 4-Octanone
8510 - 2-Octen-1-ol
8511 - 2-Octen-4-ol
8512 - 4-Octen-1-ol
8513 - 5-Octen-1-ol
8612 - Oxirane, 2-(2,2-dimethylpropyl)-2-methyl-
8697 - Pentanal, 2-propyl-
8970 - 2-Pentanone, 3-ethyl-4-methyl-
8989 - 3-Pentanone, 2,2,4-trimethyl-

$C_8H_{16}O_2$
185 - Acetic acid, 2-ethylbutyl ester
194 - Acetic acid, hexyl ester
3479 - Butanoic acid, butyl ester
3509 - Butanoic acid, 1,1-dimethylethyl ester
3553 - Butanoic acid, 3-methyl-, 1-methylethyl ester
3562 - Butanoic acid, 3-methyl-, propyl ester
3563 - Butanoic acid, 2-methylpropyl ester
3564 - Butanoic acid, 1-methylpropyl ester, (S)-
3639 - 1-Butanol, 3-methyl-, propanoate
3758 - 2-Butene, 1,1-diethoxy-
4363 - 1,3-Cyclohexanedimethanol
6008 - Formic acid, heptyl ester
6466 - Heptanoic acid, methyl ester
6518 - 2-Heptanone, 4-hydroxy-3-methyl-
6519 - 3-Heptanone, 5-hydroxy-5-methyl-
6882 - Hexanoic acid, 2-ethyl-
6883 - Hexanoic acid, 3-ethyl-
6884 - Hexanoic acid, ethyl ester
8422 - Octanoic acid
8474 - 4-Octanone, 5-hydroxy-
8692 - Pentanal, 3-hydroxy-2,2,4-trimethyl-
8892 - Pentanoic acid, 4-methyl-, ethyl ester
8893 - Pentanoic acid, 1-methylethyl ester
8914 - Pentanoic acid, propyl ester
8945 - 1-Pentanol, 2-methyl-, acetate
8946 - 2-Pentanol, 2-methyl-, acetate
8947 - 2-Pentanol, 4-methyl-, acetate
8948 - 3-Pentanol, 2-methyl-, acetate
8949 - 3-Pentanol, 2-methyl-, acetate
10294 - Propanoic acid, 2-methyl-, 1,1-dimethylethyl
       ester
10304 - Propanoic acid, 2-methyl-, 2-methylpropyl
       ester
10318 - Propanoic acid, pentyl ester

$C_8H_{16}O_3$
3536 - Butanoic acid, 2-hydroxy-, 2-methylpropyl
       ester
8432 - Octanoic acid, 2-hydroxy-
10274 - Propanoic acid, 2-hydroxy-, 3-methylbutyl
       ester

$C_8H_{16}O_4$
174 - Acetic acid, diethoxy-, ethyl ester
5721 - Ethanol, 2-(2-ethoxyethoxy)-, acetate

$C_8H_{16}Si$
11417 - Silane, dimethyldi-2-propenyl-

$C_8H_{17}Br$
8356 - Octane, 1-bromo-
8357 - Octane, 2-bromo-, (±)-

$C_8H_{17}Cl$
6366 - Heptane, 2-chloro-2-methyl-
6367 - Heptane, 2-chloro-6-methyl-
6368 - Heptane, 3-(chloromethyl)-
6369 - Heptane, 3-chloro-3-methyl-
6370 - Heptane, 4-chloro-4-methyl-
6740 - Hexane, 2-chloro-2,5-dimethyl-
6741 - Hexane, 2-chloro-2,3-dimethyl-
8359 - Octane, 1-chloro-
8360 - Octane, 2-chloro-
8733 - Pentane, 2-chloro-2,4,4-trimethyl-

$C_8H_{17}ClO$
8448 - 1-Octanol, 8-chloro-

$C_8H_{17}ClO_2$
3243 - Butane, 3-chloro-1,1-diethoxy-

$C_8H_{17}Cl_3Si$
11471 - Silane, trichlorooctyl-

$C_8H_{17}F$
8405 - Octane, 1-fluoro-

$C_8H_{17}FO$
8460 - 1-Octanol, 8-fluoro-

$C_8H_{17}I$
8406 - Octane, 1-iodo-
8407 - Octane, 2-iodo-, (±)-

$C_8H_{17}N$
503 - 1H-Azonine, octahydro-
4286 - Cyclohexanamine, N,N-dimethyl-
4287 - Cyclohexanamine, 2-ethyl-
4288 - Cyclohexanamine, N-ethyl-
4723 - Cyclohexylamine, N,4-dimethyl-
4735 - Cyclooctanamine
9684 - Piperidine, 1-(1-methylethyl)-
9707 - Piperidine, 1-propyl-
9708 - Piperidine, 2-propyl-, (S)-
9709 - Piperidine, 4-propyl-
9712 - Piperidine, 2,2,4-trimethyl-
9713 - Piperidine, 2,3,6-trimethyl-
9714 - Piperidine, 2,4,6-trimethyl-
11146 - Pyrrolidine, 1-butyl-
11147 - Pyrrolidine, 2-butyl-

$C_8H_{17}NO$
58 - Acetamide, N,N-dipropyl-
3200 - Butanamide, N,N-diethyl-
3656 - 2-Butanone, 4-(diethylamino)-
4520 - Cyclohexanol, 3-(dimethylamino)-
7684 - Morpholine, 4-butyl-
8348 - Octanal, oxime
8349 - Octanamide
9677 - 2-Piperidinemethanol, α-ethyl-, (R*,S*)-(±)-
9718 - 3-Piperidinol, 6-propyl-, (3S-trans)-

$C_8H_{17}NO_2$
3952 - Carbamic acid, (3-methylbutyl)-, ethyl ester
7693 - Morpholine, 4-(2-ethoxyethyl)-
8095 - Nitrous acid, octyl ester
8412 - Octane, 1-nitro-
8413 - Octane, 2-nitro-, (+)-
8423 - Octanoic acid, 2-amino-, (±)-

$C_8H_{17}NO_3$
8079 - Nitric acid, octyl ester

C_8H_{18}
3430 - Butane, 2,2,3,3-tetramethyl-
6422 - Heptane, 2-methyl-
6423 - Heptane, 3-methyl-, (S)-
6424 - Heptane, 4-methyl-
6771 - Hexane, 2,2-dimethyl-
6772 - Hexane, 2,3-dimethyl-
6773 - Hexane, 2,4-dimethyl-
6774 - Hexane, 2,5-dimethyl-
6775 - Hexane, 3,3-dimethyl-
6776 - Hexane, 3,4-dimethyl-
6815 - Hexane, 3-ethyl-
8355 - Octane
8813 - Pentane, 3-ethyl-2-methyl-
8814 - Pentane, 3-ethyl-3-methyl-
8847 - Pentane, 2,2,3-trimethyl-
8848 - Pentane, 2,2,4-trimethyl-
8849 - Pentane, 2,3,3-trimethyl-
8850 - Pentane, 2,3,4-trimethyl-

$C_8H_{18}ClNO_2$
9830 - 1-Propanaminium, 2-(acetyloxy)-N,N,N-
       trimethyl-, chloride

$C_8H_{18}Hg$
7433 - Mercury, dibutyl-

$C_8H_{18}N_2$
4362 - 1,3-Cyclohexanedimethanamine
5123 - Diazene, dibutyl

$C_8H_{18}N_2O$
3212 - 1-Butanamine, N-butyl-N-nitroso-

$C_8H_{18}O$
3401 - Butane, 1-(2-methylpropoxy)-
3415 - Butane, 1,1'-oxybis-
3416 - Butane, 2,2'-oxybis-, (±)-

6421 - Heptane, 1-methoxy-
6492 - 1-Heptanol, 3-methyl-
6493 - 1-Heptanol, 4-methyl-
6494 - 1-Heptanol, 6-methyl-
6495 - 2-Heptanol, 2-methyl-
6496 - 2-Heptanol, 3-methyl-
6497 - 2-Heptanol, 5-methyl-
6498 - 2-Heptanol, 6-methyl-
6499 - 3-Heptanol, 2-methyl-, (±)-
6500 - 3-Heptanol, 3-methyl-
6501 - 3-Heptanol, 4-methyl-
6502 - 3-Heptanol, 5-methyl-
6503 - 4-Heptanol, 2-methyl-
6504 - 4-Heptanol, 3-methyl-, (R*,S*)-(±)-
6505 - 4-Heptanol, 4-methyl-
6814 - Hexane, 1-ethoxy-
6921 - 1-Hexanol, 2,5-dimethyl-
6922 - 2-Hexanol, 2,3-dimethyl-
6923 - 2-Hexanol, 2,5-dimethyl-
6924 - 2-Hexanol, 3,4-dimethyl-
6925 - 3-Hexanol, 2,2-dimethyl-
6926 - 3-Hexanol, 2,3-dimethyl-
6927 - 3-Hexanol, 2,5-dimethyl-
6928 - 3-Hexanol, 3,5-dimethyl-, (±)-
6929 - 3-Hexanol, 4,4-dimethyl-
6930 - 1-Hexanol, 2-ethyl-
6931 - 3-Hexanol, 3-ethyl-
8443 - 1-Octanol
8444 - 2-Octanol, (±)-
8445 - 3-Octanol
8446 - 4-Octanol, (±)-
8933 - 2-Pentanol, 3-ethyl-2-methyl-
8934 - 3-Pentanol, 3-ethyl-2-methyl-
8954 - 2-Pentanol, 2,4,4-trimethyl-
8955 - 3-Pentanol, 2,2,4-trimethyl-
8956 - 3-Pentanol, 2,3,4-trimethyl-
10124 - Propane, 2,2'-oxybis[2-methyl-

$C_8H_{18}OS$
3422 - Butane, 1,1'-sulfinylbis-

$C_8H_{18}OSi_2$
5233 - Disiloxane, 1,3-diethenyl-1,1,3,3-tetramethyl-

$C_8H_{18}O_2$
5586 - Ethane, 1,1-dipropoxy-
5728 - Ethanol, 2-(hexyloxy)-
6401 - 2,4-Heptanediol, 3-methyl-
6801 - 2,5-Hexanediol, 2,5-dimethyl-
6802 - 1,3-Hexanediol, 2-ethyl-
8389 - 1,8-Octanediol
8390 - 2,4-Octanediol
8391 - 4,5-Octanediol, (±)-
8795 - 1,3-Pentanediol, 2,2,4-trimethyl-
8796 - 1,4-Pentanediol, 2,2,4-trimethyl-
9153 - Peroxide, bis(1,1-dimethylethyl)
9993 - 1,3-Propanediol, 2-butyl-2-methyl-

$C_8H_{18}O_2S$
3423 - Butane, 1,1'-sulfonylbis-
10132 - Propane, 1,1'-sulfonylbis[2-methyl-

$C_8H_{18}O_2Si$
11408 - Silane, diethoxymethyl-2-propenyl-

$C_8H_{18}O_3$
5612 - Ethane, 1,1'-oxybis[2-ethoxy-
5664 - Ethane, 1,1,1-triethoxy-
5688 - Ethanol, 2-(2-butoxyethoxy)-
6920 - 1-Hexanol, 3,5-dimethoxy-

$C_8H_{18}O_3S$
11578 - Sulfurous acid, bis(2-methylpropyl) ester
11580 - Sulfurous acid, dibutyl ester

$C_8H_{18}O_3Si$
11424 - Silane, ethenyltriethoxy-

$C_8H_{18}O_4$
11642 - 2,5,8,11-Tetraoxadodecane

$C_8H_{18}O_4S_2$
3230 - Butane, 2,2-bis(ethylsulfonyl)-

$C_8H_{18}O_5$
5756 - Ethanol, 2,2'-[oxybis(2,1-ethanediyloxy)]bis-

$C_8H_{18}S$
3433 - Butane, 1,1'-thiobis-

3434 - Butane, 2,2'-thiobis-
8418 - 1-Octanethiol
8419 - 2-Octanethiol, (±)-
10154 - Propane, 1,1'-thiobis[2-methyl-
10155 - Propane, 2,2'-thiobis[2-methyl-

$C_8H_{18}S_2$
5242 - Disulfide, bis(1,1-dimethylethyl)
5252 - Disulfide, dibutyl

$C_8H_{18}S_4$
11649 - Tetrasulfide, bis(1,1-dimethylethyl)

$C_8H_{18}Si_2$
11434 - Silane, 1,2-ethynediylbis[trimethyl-

$C_8H_{19}N$
3211 - 1-Butanamine, N-butyl-
3227 - 2-Butanamine, N-(1-methylpropyl)-
6353 - 2-Heptanamine, 6-methyl-, (±)-
6726 - 1-Hexanamine, 2-ethyl-
8350 - 1-Octanamine
8351 - 2-Octanamine, (±)-
9811 - 2-Propanamine, N-ethyl-N-(1-methylethyl)-
9812 - 1-Propanamine, N-ethyl-N-propyl-
9823 - 1-Propanamine, 2-methyl-N-(2-methylpropyl)-

$C_8H_{19}NO$
5683 - Ethanol, 2-[bis(1-methylethyl)amino]-
5712 - Ethanol, 2-(dipropylamino)-

$C_8H_{19}NO_2$
3214 - 1-Butanamine, 4,4-diethoxy-
3624 - 2-Butanol, 3,3'-iminobis-
5416 - Ethanamine, 2,2-diethoxy-N,N-dimethyl-
5425 - Ethanamine, 2-ethoxy-N-(2-ethoxyethyl)-
5692 - Ethanol, 2,2'-(butylimino)bis-
5707 - Ethanol, 2-[2-(diethylamino)ethoxy]-

$C_8H_{19}O_2PS_2$
5934 - Ethoprop

$C_8H_{19}O_2PS_3$
5260 - Disulfoton

$C_8H_{19}O_3P$
9531 - Phosphonic acid, bis(1,1-dimethylethyl) ester
9532 - Phosphonic acid, dibutyl ester

$C_8H_{20}As_2$
5109 - Diarsine, tetraethyl-

$C_8H_{20}BrN$
5447 - Ethanaminium, N,N,N-triethyl-, bromide

$C_8H_{20}N_2$
3263 - 1,4-Butanediamine, N,N,N',N'-tetramethyl-
6751 - 2,5-Hexanediamine, 2,5-dimethyl-
7140 - Hydrazine, 1,2-bis(2-methylpropyl)-
8365 - 1,8-Octanediamine

$C_8H_{20}N_2O$
10342 - 1-Propanol, 2-[(2-amino-
2-methylpropyl)amino]-2-methyl-

$C_8H_{20}OSi$
11428 - Silane, ethoxytriethyl-

$C_8H_{20}O_3P_2$
9526 - Phosphinic acid, diethyl-, anhydride

$C_8H_{20}O_3Si$
11478 - Silane, triethoxyethyl-

$C_8H_{20}O_4Si$
11506 - Silicic acid (H4SiO4), tetraethyl ester

$C_8H_{20}O_5P_2S_2$
11716 - Thiodiphosphoric acid ([(HO)2P(S)]2O),
tetraethyl ester

$C_8H_{20}O_6P_2$
7185 - Hypophosphoric acid, tetraethyl ester

$C_8H_{20}O_7P_2$
5221 - Diphosphoric acid, tetraethyl ester

$C_8H_{20}Pb$
9731 - Plumbane, tetraethyl

$C_8H_{20}Si$
11450 - Silane, tetraethyl-

$C_8H_{20}Si_2$
11421 - Silane, 1,2-ethenediylbis[trimethyl-, (E)-

$C_8H_{23}N_5$
5498 - 1,2-Ethanediamine, N-(2-aminoethyl)-N'-[2-
[(2-aminoethyl)amino]ethyl]-

$C_8H_{24}N_4O_3P_2$
5220 - Diphosphoramide, octamethyl-

$C_8H_{24}O_2Si_3$
11914 - Trisiloxane, octamethyl-

$C_8H_{24}O_4Si_4$
4993 - Cyclotetrasiloxane, octamethyl-

$C_8H_{26}O_3Si_4$
11647 - Tetrasiloxane, 1,1,1,3,5,7,7,7-octamethyl-

$C_8I_4O_3$
7348 - 1,3-Isobenzofurandione, 4,5,6,7-tetraiodo-

C_9F_{20}
8139 - Nonane, icosofluoro

$C_9F_{21}N$
9813 - 1-Propanamine, 1,1,2,2,3,3,3-heptafluoro-
N,N-bis(heptafluoropropyl)-

$C_9H_3Cl_3O_3$
2317 - 1,3,5-Benzenetricarbonyl trichloride

$C_9H_3F_{15}O_2$
8438 - Octanoic acid, pentadecafluoro-, methyl ester

$C_9H_4Cl_3NO_2S$
5984 - Folpet

$C_9H_4O_5$
7341 - 5-Isobenzofurancarboxylic acid, 1,3-dihydro-
1,3-dioxo-

$C_9H_5BrN_2O_2$
11221 - Quinoline, 6-bromo-5-nitro-

$C_9H_5Br_2NO$
11313 - 8-Quinolinol, 5,7-dibromo-

C_9H_5ClINO
11312 - 8-Quinolinol, 5-chloro-7-iodo-

$C_9H_5ClO_6$
2320 - 1,3,5-Benzenetricarboxylic acid, 2-chloro-

$C_9H_5Cl_2N$
11248 - Quinoline, 2,4-dichloro-
11249 - Quinoline, 2,7-dichloro-
11250 - Quinoline, 4,5-dichloro-
11251 - Quinoline, 4,7-dichloro-
11252 - Quinoline, 5,8-dichloro-

$C_9H_5Cl_2NO$
11314 - 8-Quinolinol, 5,7-dichloro-

$C_9H_5Cl_3N_4$
360 - Anilazine

$C_9H_5I_2NO$
11315 - 8-Quinolinol, 5,7-diiodo-

$C_9H_5NO_4$
10740 - 2-Propynoic acid, 3-(2-nitrophenyl)-

C_9H_6BrN
7374 - Isoquinoline, 4-bromo-
11217 - Quinoline, 3-bromo-
11218 - Quinoline, 5-bromo-
11219 - Quinoline, 6-bromo-
11220 - Quinoline, 7-bromo-

C_9H_6BrNO
11310 - 8-Quinolinol, 5-bromo-
11326 - 2(1H)-Quinolinone, 4-bromo-

$C_9H_6Br_2O_2$
10629 - 2-Propenoic acid, 2,3-dibromo-3-phenyl-,
(Z)-

C_9H_6ClN
7376 - Isoquinoline, 1-chloro-
11237 - Quinoline, 2-chloro-
11238 - Quinoline, 4-chloro-
11239 - Quinoline, 5-chloro-
11240 - Quinoline, 6-chloro-
11241 - Quinoline, 7-chloro-
11242 - Quinoline, 8-chloro-

C_9H_6ClNO
11311 - 8-Quinolinol, 5-chloro-

$C_9H_6Cl_2N_2O_3$
7643 - Methazole

$C_9H_6Cl_6O_3S$
5369 - Endosulfan

C_9H_6IN
11272 - Quinoline, 4-iodo-
11273 - Quinoline, 6-iodo-

$C_9H_6INO_4S$
11295 - 5-Quinolinesulfonic acid, 8-hydroxy-7-iodo-

$C_9H_6N_2$
9916 - Propanedinitrile, phenyl-

$C_9H_6N_2O_2$
1430 - Benzene, 1,3-diisocyanato-2-methyl-
1431 - Benzene, 2,4-diisocyanato-1-methyl-
7386 - Isoquinoline, 5-nitro-
11284 - Quinoline, 5-nitro-
11285 - Quinoline, 6-nitro-
11286 - Quinoline, 7-nitro-

$C_9H_6N_2O_3$
11287 - Quinoline, 4-nitro-, 1-oxide
11323 - 8-Quinolinol, 5-nitro-

C_9H_6O
10724 - 2-Propynal, 3-phenyl-

$C_9H_6OS_2$
7617 - Methanone, di-2-thienyl-

$C_9H_6O_2$
2791 - 1H-2-Benzopyran-1-one
2792 - 2H-1-Benzopyran-2-one
2793 - 4H-1-Benzopyran-4-one
7264 - 1H-Indene-1,3(2H)-dione

$C_9H_6O_2S$
2400 - Benzo[b]thiophene-2-carboxylic acid

$C_9H_6O_3$
2439 - 2-Benzofurancarboxylic acid
2775 - 2H-1-Benzopyran-3-carboxylic acid, 2-oxo-
2825 - 2H-1-Benzopyran-2-one, 6-hydroxy-
2826 - 2H-1-Benzopyran-2-one, 7-hydroxy-

$C_9H_6O_4$
2804 - 2H-1-Benzopyran-2-one, 6,7-dihydroxy-
2805 - 2H-1-Benzopyran-2-one, 7,8-dihydroxy-
7267 - 1H-Indene-1,3(2H)-dione, 2,2-dihydroxy-

$C_9H_6O_6$
2318 - 1,2,3-Benzenetricarboxylic acid
2319 - 1,2,4-Benzenetricarboxylic acid

$C_9H_7BrN_2$
11207 - 4-Quinolinamine, 3-bromo-
11215 - Quinoline, 6-amino-3-bromo-

C_9H_7BrO
10510 - 2-Propenal, 3-bromo-3-phenyl-

$C_9H_7BrO_4$
2481 - Benzoic acid, 2-(acetyloxy)-5-bromo-

C_9H_7ClO
10716 - 2-Propenoyl chloride, 3-phenyl-
10717 - 2-Propenoyl chloride, 3-phenyl-, (E)-

$C_9H_7Cl_3O_3$
11512 - Silvex

$C_9H_7FO_2$
10637 - 2-Propenoic acid, 2-fluoro-3-phenyl-

C_9H_7N
7373 - Isoquinoline
10587 - 2-Propenenitrile, 3-phenyl-, (E)-
10588 - 2-Propenenitrile, 3-phenyl-, (Z)-
11214 - Quinoline

C_9H_7NO
2124 - Benzenepropanenitrile, β-oxo-
11305 - 5-Quinolinol
11306 - 6-Quinolinol
11307 - 7-Quinolinol
11308 - 8-Quinolinol
11325 - 2(1H)-Quinolinone

$C_9H_7NO_2$
7365 - 1H-Isoindole-1,3(2H)-dione, 2-methyl-

$C_9H_7NO_4$
10676 - 2-Propenoic acid, 3-(2-nitrophenyl)-, (E)-

$C_9H_7NO_4S$
11294 - 5-Quinolinesulfonic acid, 8-hydroxy-

$C_9H_7N_3S$
11858 - Tricyclazole

C_9H_8
2173 - Benzene, 1-propynyl-
7245 - 1H-Indene

$C_9H_8Br_2$
7247 - 1H-Indene, 1,2-dibromo-2,3-dihydro-

$C_9H_8Cl_2$
1397 - Benzene, (3,3-dichloro-2-propenyl)-
7249 - 1H-Indene, 1,2-dichloro-2,3-dihydro-

5852 - Ethanone, 1-(3-methylphenyl)-
5853 - Ethanone, 1-(4-methylphenyl)-
7284 - 1H-Inden-1-ol, 2,3-dihydro-
7285 - 1H-Inden-5-ol, 2,3-dihydro-
8622 - Oxirane, 2-methyl-2-phenyl-
9456 - Phenol, 2-(1-propenyl)-
9457 - Phenol, 2-(2-propenyl)-
9458 - Phenol, 4-(2-propenyl)-
9459 - Phenol, 2-(1-propenyl)-, (E)-
10480 - 1-Propanone, 1-phenyl-
10481 - 2-Propanone, 1-phenyl-
10704 - 2-Propen-1-ol, 3-phenyl-, (E)-
10705 - 2-Propen-1-ol, 3-phenyl-, (Z)-

$C_9H_{10}O_2$

211 - Acetic acid, 2-methylphenyl ester
212 - Acetic acid, 3-methylphenyl ester
213 - Acetic acid, 4-methylphenyl ester
233 - Acetic acid, phenylmethyl ester
552 - Benzaldehyde, 2-ethoxy-
553 - Benzaldehyde, 3-ethoxy-
554 - Benzaldehyde, 4-ethoxy-
586 - Benzaldehyde, 4-methoxy-3-methyl-
870 - Benzeneacetaldehyde, 2-methoxy-
871 - Benzeneacetaldehyde, 4-methoxy-
924 - Benzeneacetic acid, α-methyl-, (±)-
925 - Benzeneacetic acid, 3-methyl-
926 - Benzeneacetic acid, 4-methyl-
929 - Benzeneacetic acid, methyl ester
1547 - 1,2-Benzenediol, 3-(2-propenyl)-
2126 - Benzenepropanoic acid
2580 - Benzoic acid, 2,4-dimethyl-
2581 - Benzoic acid, 2,5-dimethyl-
2582 - Benzoic acid, 2,6-dimethyl-
2583 - Benzoic acid, 3,5-dimethyl-
2600 - Benzoic acid, 2-ethyl-
2601 - Benzoic acid, 3-ethyl-
2602 - Benzoic acid, ethyl ester
2696 - Benzoic acid, 2-methyl-, methyl ester
2697 - Benzoic acid, 3-methyl-, methyl ester
2698 - Benzoic acid, 4-methyl-, methyl ester
5826 - Ethanone, 1-(2-hydroxy-4-methylphenyl)-
5827 - Ethanone, 1-(2-hydroxy-5-methylphenyl)-
5828 - Ethanone, 1-(4-hydroxy-2-methylphenyl)-
5840 - Ethanone, 1-(3-methoxyphenyl)-
5841 - Ethanone, 1-(4-methoxyphenyl)-
5842 - Ethanone, 2-methoxy-1-phenyl-
8619 - Oxiranemethanol, 3-phenyl-
8623 - Oxirane, (phenoxymethyl)-
10323 - Propanoic acid, phenyl ester
10449 - 1-Propanone, 1-(2-hydroxyphenyl)-
10450 - 1-Propanone, 1-(4-hydroxyphenyl)-
10451 - 1-Propanone, 2-hydroxy-1-phenyl-
10452 - 1-Propanone, 3-hydroxy-1-phenyl-
10479 - 2-Propanone, 1-phenoxy-

$C_9H_{10}O_3$

230 - Acetic acid, phenoxy-, methyl ester
543 - Benzaldehyde, 2,4-dimethoxy-
544 - Benzaldehyde, 2,5-dimethoxy-
545 - Benzaldehyde, 3,4-dimethoxy-
546 - Benzaldehyde, 3,5-dimethoxy-
555 - Benzaldehyde, 3-ethoxy-4-hydroxy-
556 - Benzaldehyde, 5-ethoxy-2-hydroxy-
910 - Benzeneacetic acid, α-hydroxy-, methyl ester, (±)-
914 - Benzeneacetic acid, α-hydroxy-α-methyl-
915 - Benzeneacetic acid, α-(hydroxymethyl)-
921 - Benzeneacetic acid, 2-methoxy-
922 - Benzeneacetic acid, 4-methoxy-
2133 - Benzenepropanoic acid, α-hydroxy-, (±)-
2134 - Benzenepropanoic acid, 4-hydroxy-
2596 - Benzoic acid, 2-ethoxy-
2597 - Benzoic acid, 3-ethoxy-
2619 - Benzoic acid, 2-hydroxy-, ethyl ester
2620 - Benzoic acid, 3-hydroxy-, ethyl ester
2621 - Benzoic acid, 4-hydroxy-, ethyl ester
2638 - Benzoic acid, 2-hydroxy-3-methyl-, methyl ester

2639 - Benzoic acid, 2-hydroxy-4-methyl-, methyl ester
2640 - Benzoic acid, 2-hydroxy-5-methyl-, methyl ester
2673 - Benzoic acid, 2-methoxy-, methyl ester
2674 - Benzoic acid, 3-methoxy-, methyl ester
2675 - Benzoic acid, 4-methoxy-, methyl ester
5578 - 1,2-Ethanediol, monobenzoate
5823 - Ethanone, 1-(2-hydroxy-4-methoxyphenyl)-
5824 - Ethanone, 1-(4-hydroxy-3-methoxyphenyl)-
9382 - Phenol, 2-methoxy-, acetate
10319 - Propanoic acid, 2-phenoxy-, (±)-
10320 - Propanoic acid, 3-phenoxy-
10430 - 1-Propanone, 1-(2,4-dihydroxyphenyl)-
10639 - 2-Propenoic acid, 3-(2-furanyl)-, ethyl ester

$C_9H_{10}O_4$

567 - Benzaldehyde, 2-hydroxy-4,6-dimethoxy-
568 - Benzaldehyde, 3-hydroxy-4,5-dimethoxy-
569 - Benzaldehyde, 4-hydroxy-3,5-dimethoxy-
913 - Benzeneacetic acid, 4-hydroxy-3-methoxy-
2576 - Benzoic acid, 3,4-dimethoxy-
2577 - Benzoic acid, 3,5-dimethoxy-
2622 - Benzoic acid, 2-hydroxy-, 2-hydroxyethyl ester
2627 - Benzoic acid, 4-hydroxy-3-methoxy-, methyl ester
6163 - 2-Furanpropanoic acid, β-oxo-, ethyl ester
10489 - 1-Propanone, 1-(2,4,6-trihydroxyphenyl)-

$C_9H_{10}O_5$

896 - Benzeneacetic acid, α,4-dihydroxy-3-methoxy-
7510 - Methanediol, 2-furanyl-, diacetate

$C_9H_{11}Br$

1057 - Benzene, 1-(bromomethyl)-3,5-dimethyl-
1059 - Benzene, 1-bromo-4-(1-methylethyl)-
1060 - Benzene, (2-bromo-1-methylethyl)-
1086 - Benzene, (1-bromopropyl)-, (±)-
1087 - Benzene, (2-bromopropyl)-
1088 - Benzene, (3-bromopropyl)-
1092 - Benzene, 1-bromo-2,3,5-trimethyl-
1093 - Benzene, 1-bromo-2,4,5-trimethyl-
1094 - Benzene, 2-bromo-1,3,5-trimethyl-

$C_9H_{11}BrN_2O_2$

7656 - Metobromuron

$C_9H_{11}BrO$

1085 - Benzene, (3-bromopropoxy)-

$C_9H_{11}Cl$

1186 - Benzene, 1-(chloromethyl)-2,4-dimethyl-
1189 - Benzene, (1-chloro-1-methylethyl)-
1190 - Benzene, 1-chloro-2-(1-methylethyl)-
1191 - Benzene, 1-(chloromethyl)-4-ethyl-
1192 - Benzene, 1-chloro-4-(1-methylethyl)-
1193 - Benzene, (2-chloro-1-methylethyl)-
1234 - Benzene, (2-chloropropyl)-, (R)-
1235 - Benzene, (3-chloropropyl)-
1243 - Benzene, 2-chloro-1,3,5-trimethyl-

$C_9H_{11}ClN_2O$

11999 - Urea, N'-(4-chlorophenyl)-N,N-dimethyl-

$C_9H_{11}ClO$

1233 - Benzene, (3-chloropropoxy)-

$C_9H_{11}ClO_3$

9999 - 1,2-Propanediol, 3-(2-chlorophenoxy)-
10000 - 1,2-Propanediol, 3-(4-chlorophenoxy)-

$C_9H_{11}ClS$

1236 - Benzene, [(3-chloropropyl)thio]-

$C_9H_{11}Cl_2P$

9555 - Phosphonous dichloride, [4-(1-methylethyl)phenyl]-

$C_9H_{11}Cl_3NO_3PS$

4089 - Chlorpyrifos

$C_9H_{11}F$

1740 - Benzene, 2-fluoro-1,3,5-trimethyl-

$C_9H_{11}FN_2O_5$

12034 - Uridine, 2'-deoxy-5-fluoro-

$C_9H_{11}N$

787 - Benzenamine, 2-(1-methylethenyl)-
841 - Benzenamine, N-2-propenyl-
5437 - Ethanamine, N-(phenylmethylene)-
7243 - 1H-Inden-1-amine, 2,3-dihydro-, (±)-

7244 - 1H-Inden-5-amine, 2,3-dihydro-
7388 - Isoquinoline, 1,2,3,4-tetrahydro-
7389 - Isoquinoline, 5,6,7,8-tetrahydro-
11296 - Quinoline, 1,2,3,4-tetrahydro-
11297 - Quinoline, 5,6,7,8-tetrahydro-

$C_9H_{11}NO$

76 - Acetamide, N-(2-methylphenyl)-
77 - Acetamide, N-(3-methylphenyl)-
78 - Acetamide, N-(4-methylphenyl)-
79 - Acetamide, N-methyl-N-phenyl-
87 - Acetamide, N-(phenylmethyl)-
550 - Benzaldehyde, 4-(dimethylamino)-
7567 - Methanimidic acid, N-phenyl-, ethyl ester
9800 - Propanamide, N-phenyl-
10413 - 1-Propanone, 1-(4-aminophenyl)-
10414 - 1-Propanone, 2-amino-1-phenyl-
10483 - 1-Propanone, 1-phenyl-, oxime

$C_9H_{11}NO_2$

70 - Acetamide, N-(2-methoxyphenyl)-
71 - Acetamide, N-(4-methoxyphenyl)-
615 - Benzamide, 2-ethoxy-
1977 - Benzene, 1-(1-methylethyl)-2-nitro-
1978 - Benzene, 1-(1-methylethyl)-4-nitro-
2003 - Benzene, (1-methyl-1-nitroethyl)-
2362 - Benzene, 1,2,4-trimethyl-5-nitro-
2363 - Benzene, 1,3,5-trimethyl-2-nitro-
2424 - 1,3-Benzodioxole-5-ethanamine
2495 - Benzoic acid, 2-amino-, ethyl ester
2496 - Benzoic acid, 3-amino-, ethyl ester
2497 - Benzoic acid, 4-amino-, ethyl ester
2584 - Benzoic acid, 2-(dimethylamino)-
2585 - Benzoic acid, 3-(dimethylamino)-
2681 - Benzoic acid, 2-(methylamino)-, methyl ester
3958 - Carbamic acid, methylphenyl-, methyl ester
3965 - Carbamic acid, phenyl-, ethyl ester
9509 - DL-Phenylalanine
9510 - L-Phenylalanine
10907 - 3-Pyridinecarboxylic acid, 1-methylethyl ester

$C_9H_{11}NO_3$

5804 - Ethanone, 1-(3,4-dihydroxyphenyl)-2-(methylamino)-
11926 - L-Tyrosine

$C_9H_{11}NO_4$

1443 - Benzene, 1-(dimethoxymethyl)-3-nitro-
1444 - Benzene, 1-(dimethoxymethyl)-4-nitro-
11929 - L-Tyrosine, 3-hydroxy-

C_9H_{12}

1685 - Benzene, 1-ethyl-2-methyl-
1686 - Benzene, 1-ethyl-3-methyl-
1687 - Benzene, 1-ethyl-4-methyl-
1975 - Benzene, (1-methylethyl)-
2169 - Benzene, propyl-
2357 - Benzene, 1,2,3-trimethyl-
2358 - Benzene, 1,2,4-trimethyl-
2359 - Benzene, 1,3,5-trimethyl-
8114 - 1,4-Nonadiyne
8115 - 1,8-Nonadiyne
8116 - 2,7-Nonadiyne

$C_9H_{12}NO_5PS$

5951 - Fenitrothion

$C_9H_{12}N_2$

10482 - 2-Propanone, phenylhydrazone
11017 - Pyridine, 3-(2-pyrrolidinyl)-, (S)-

$C_9H_{12}N_2O_2$

12013 - Urea, (4-ethoxyphenyl)-

$C_9H_{12}N_2O_3$

827 - Benzenamine, 5-nitro-2-propoxy-

$C_9H_{12}N_2O_6$

12033 - Uridine

$C_9H_{12}N_2S$

10886 - 4-Pyridinecarbothioamide, 2-propyl-

$C_9H_{12}N_4O_3$

10758 - 1H-Purine-2,6-dione, 3,7-dihydro-7-(2-hydroxyethyl)-1,3-dimethyl-
10759 - 1H-Purine-2,6-dione, 3,7-dihydro-8-methoxy-1,3,7-trimethyl-

$C_9H_{12}N_6$
11844 - 1,3,5-Triazine, 2,4,6-tris(1-aziridinyl)-

$C_9H_{12}O$
1607 - Benzeneethanol, α-methyl-
1608 - Benzeneethanol, β-methyl-
1609 - Benzeneethanol, 2-methyl-
1610 - Benzeneethanol, 4-methyl-
1658 - Benzene, (ethoxymethyl)-
1659 - Benzene, 1-ethoxy-2-methyl-
1660 - Benzene, 1-ethoxy-3-methyl-
1661 - Benzene, 1-ethoxy-4-methyl-
1682 - Benzene, 1-ethyl-2-methoxy-
1683 - Benzene, 1-ethyl-3-methoxy-
1684 - Benzene, 1-ethyl-4-methoxy-
1857 - Benzenemethanol, α,α-dimethyl-
1858 - Benzenemethanol, α,3-dimethyl-
1859 - Benzenemethanol, α,4-dimethyl-
1860 - Benzenemethanol, 2,4-dimethyl-
1861 - Benzenemethanol, 3,5-dimethyl-
1870 - Benzenemethanol, α-ethyl-
1916 - Benzene, 1-methoxy-2,3-dimethyl-
1917 - Benzene, 1-methoxy-2,4-dimethyl-
1918 - Benzene, 1-methoxy-3,5-dimethyl-
1919 - Benzene, 2-methoxy-1,3-dimethyl-
1920 - Benzene, 2-methoxy-1,4-dimethyl-
1921 - Benzene, 4-methoxy-1,2-dimethyl-
1974 - Benzene, (1-methylethoxy)-
2148 - Benzenepropanol
2168 - Benzene, propoxy-
9410 - Phenol, 2-(1-methylethyl)-
9411 - Phenol, 3-(1-methylethyl)-
9412 - Phenol, 4-(1-methylethyl)-
9461 - Phenol, 2-propyl-
9462 - Phenol, 3-propyl-
9463 - Phenol, 4-propyl-
9493 - Phenol, 2,3,4-trimethyl-
9494 - Phenol, 2,4,5-trimethyl-
9495 - Phenol, 2,4,6-trimethyl-

$C_9H_{12}O_2$
1439 - Benzene, 1,2-dimethoxy-3-methyl-
1440 - Benzene, 1,2-dimethoxy-4-methyl-
1441 - Benzene, 1,3-dimethoxy-5-methyl-
1535 - 1,2-Benzenediol, 4-(1-methylethyl)-
1548 - 1,2-Benzenediol, 4-propyl-
1549 - 1,3-Benzenediol, 4-propyl-
1550 - 1,3-Benzenediol, 5-propyl-
1553 - 1,3-Benzenediol, 2,4,6-trimethyl-
1872 - Benzenemethanol, α-ethyl-3-hydroxy-
1893 - Benzenemethanol, 2-methoxy-α-methyl-
1894 - Benzenemethanol, 3-methoxy-α-methyl-
1895 - Benzenemethanol, 4-methoxy-α-methyl-
5761 - Ethanol, 2-(phenylmethoxy)-
7181 - Hydroperoxide, 1-methyl-1-phenylethyl
9357 - Phenol, 4-ethyl-2-methoxy-
9460 - Phenol, 2-propoxy-
10398 - 1-Propanol, 2-phenoxy-
10399 - 2-Propanol, 1-phenoxy-

$C_9H_{12}O_3$
1856 - Benzenemethanol, 3,4-dimethoxy-
2354 - Benzene, 1,2,3-trimethoxy-
2355 - Benzene, 1,3,5-trimethoxy-
3520 - Butanoic acid, 2-furanylmethyl ester
6052 - 2-Furancarboxylic acid, butyl ester
6055 - 3-Furancarboxylic acid, 2,5-dimethyl-, ethyl ester
6062 - 2-Furancarboxylic acid, 2-methylpropyl ester
10026 - 1,2-Propanediol, 3-phenoxy-
10194 - 1,2,3-Propanetriol, 1-phenyl-
10300 - Propanoic acid, 2-methyl-, 2-furanylmethyl ester
11906 - 1,3,5-Trioxane, 2,4,6-triethenyl-

$C_9H_{12}O_3S$
2231 - Benzenesulfonic acid, 4-methyl-, ethyl ester
2235 - Benzenesulfonic acid, propyl ester

$C_9H_{12}S$
1706 - Benzene, 1-(ethylthio)-3-methyl-
1707 - Benzene, 1-(ethylthio)-4-methyl-

1980 - Benzene, [(1-methylethyl)thio]-
2125 - Benzenepropanethiol
2172 - Benzene, (propylthio)-

$C_9H_{13}BrN_2O_2$
3139 - Bromacil

$C_9H_{13}BrOSi$
11367 - Silane, (4-bromophenoxy)trimethyl-

$C_9H_{13}ClN_2O_2$
11591 - Terbacil

$C_9H_{13}ClN_6$
4161 - Cyanazine

$C_9H_{13}ClOSi$
11383 - Silane, (4-chlorophenoxy)trimethyl-

$C_9H_{13}ClSi$
11379 - Silane, (chloromethyl)dimethylphenyl-
11385 - Silane, (3-chlorophenyl)trimethyl-

$C_9H_{13}N$
745 - Benzenamine, 2-ethyl-6-methyl-
746 - Benzenamine, N-ethyl-2-methyl-
747 - Benzenamine, N-ethyl-3-methyl-
748 - Benzenamine, N-ethyl-4-methyl-
749 - Benzenamine, N-ethyl-N-methyl-
788 - Benzenamine, 2-(1-methylethyl)-
789 - Benzenamine, 4-(1-methylethyl)-
790 - Benzenamine, N-(1-methylethyl)-
842 - Benzenamine, 2-propyl-
843 - Benzenamine, N-propyl-
858 - Benzenamine, 2,4,5-trimethyl-
859 - Benzenamine, 2,4,6-trimethyl-
860 - Benzenamine, N,N,2-trimethyl-
861 - Benzenamine, N,N,3-trimethyl-
862 - Benzenamine, N,N,4-trimethyl-
1578 - Benzeneethanamine, α-methyl-, (±)-
1579 - Benzeneethanamine, β-methyl-
1580 - Benzeneethanamine, 4-methyl-
1581 - Benzeneethanamine, N-methyl-
1806 - Benzenemethanamine, α,α-dimethyl-
1807 - Benzenemethanamine, N,N-dimethyl-
1809 - Benzenemethanamine, α-methyl-
1810 - Benzenemethanamine, N-ethyl-
10878 - Pyridine, 2-butyl-
10879 - Pyridine, 4-butyl-
10933 - Pyridine, 2,4-diethyl-
10942 - Pyridine, 2-(1,1-dimethylethyl)-
10943 - Pyridine, 3-(1,1-dimethylethyl)-
10944 - Pyridine, 4-(1,1-dimethylethyl)-
10959 - Pyridine, 2-ethyl-3,5-dimethyl-
10960 - Pyridine, 3-ethyl-2,6-dimethyl-
10961 - Pyridine, 4-ethyl-2,6-dimethyl-
11022 - Pyridine, 2,3,4,6-tetramethyl-

$C_9H_{13}NO$
769 - Benzenamine, 2-methoxy-N,N-dimethyl-
1577 - Benzeneethanamine, 2-methoxy-
1808 - Benzenemethanamine, 2-ethoxy-
1862 - Benzenemethanol, 4-(dimethylamino)-
4467 - Cyclohexanepropanenitrile, 2-oxo-
5743 - Ethanol, 2-(methylphenylamino)-
5744 - Ethanol, 2-[(2-methylphenyl)amino]-
5745 - Ethanol, 2-[(4-methylphenyl)amino]-
5762 - Ethanol, 2-[(phenylmethyl)amino]-
9215 - Phenol, 4-(2-aminopropyl)-, (±)-

$C_9H_{13}NO_2$
491 - 8-Azabicyclo[3.2.1]oct-2-ene-2-carboxylic acid, 8-methyl-, (1R)-
1533 - 1,2-Benzenediol, 4-[2-(methylamino)ethyl]-
1805 - Benzenemethanamine, 3,4-dimethoxy-
1887 - Benzenemethanol, 4-hydroxy-α-[(methylamino)methyl]-
10946 - 2,4(1H,3H)-Pyridinedione, 3,3-diethyl-
11115 - 1H-Pyrrole-3-carboxylic acid, 2,4-dimethyl-, ethyl ester
11116 - 1H-Pyrrole-3-carboxylic acid, 2,5-dimethyl-, ethyl ester

$C_9H_{13}NO_3$
1521 - 1,2-Benzenediol, 4-[1-hydroxy-2-(methylamino)ethyl]-, (R)-

$C_9H_{13}N_2O_9P$
12035 - 5'-Uridylic acid

$C_9H_{13}N_3O_2$
11060 - 2,4(1H,3H)-Pyrimidinedione, 6-amino-3-ethyl-1-(2-propenyl)-

$C_9H_{13}N_3O_5$
5021 - Cytidine

C_9H_{14}
2968 - Bicyclo[2.2.1]hept-2-ene, 2,3-dimethyl-
4473 - Cyclohexane, 2-propynyl-
4729 - Cyclononyne
6607 - 1-Hepten-4-yne, 6,6-dimethyl-
7108 - 3-Hexen-1-yne, 3-propyl-
8193 - 1-Nonen-3-yne
8194 - 1-Nonen-4-yne
8195 - 2-Nonen-4-yne
11865 - Tricyclo[2.2.1.0^{2,6}]heptane, 3,3-dimethyl-

$C_9H_{14}ClNO$
1838 - Benzenemethanol, α-(1-aminoethyl)-, hydrochloride, (R*,S*)-(±)-

$C_9H_{14}ClN_5$
5013 - Cyprazine

$C_9H_{14}Cl_2O_2$
8138 - Nonanedioyl dichloride

$C_9H_{14}IN$
866 - Benzenaminium, N,N,N-trimethyl-, iodide

$C_9H_{14}N_2$
8129 - Nonanedinitrile

$C_9H_{14}N_2O$
10860 - 3-Pyridinamine, 6-butoxy-

$C_9H_{14}N_2O_3$
11088 - 2,4,6(1H,3H,5H)-Pyrimidinetrione, 5,5-diethyl-1-methyl-
11093 - 2,4,6(1H,3H,5H)-Pyrimidinetrione, 5-ethyl-5-(1-methylethyl)-

$C_9H_{14}N_3O_8P$
5022 - 3'-Cytidylic acid

$C_9H_{14}N_4O_3$
7133 - L-Histidine, N-β-alanyl-

$C_9H_{14}O$
2954 - Bicyclo[2.2.1]heptan-2-one, 3,3-dimethyl-, (1S)-
2955 - Bicyclo[3.1.1]heptan-2-one, 6,6-dimethyl-, (1R)-
4719 - 2-Cyclohexen-1-one, 3,4,4-trimethyl-
4720 - 2-Cyclohexen-1-one, 3,4,6-trimethyl-
4721 - 2-Cyclohexen-1-one, 3,5,5-trimethyl-
4722 - 2-Cyclohexen-1-one, 3,6,6-trimethyl-
4948 - 2-Cyclopenten-1-one, 5-methyl-2-(1-methylethyl)-
6340 - 2,5-Heptadien-4-one, 2,6-dimethyl-
6608 - 6-Hepten-4-yn-3-ol, 3-ethyl-
7292 - 2H-Inden-2-one, octahydro-, cis-
7293 - 2H-Inden-2-one, octahydro-, trans-
8108 - 2,6-Nonadienal
8109 - 2,4-Nonadienal, (E,E)-
8527 - 7-Octen-5-yn-4-ol, 4-methyl-

$C_9H_{14}OSi$
11492 - Silane, trimethylphenoxy-

$C_9H_{14}O_2$
156 - Acetic acid, cyclohexylidene-, methyl ester
4639 - 3-Cyclohexene-1-carboxylic acid, ethyl ester
4910 - 1-Cyclopentene-1-carboxylic acid, 2,3,3-trimethyl-
4912 - 2-Cyclopentene-1-carboxylic acid, 1,2,3-trimethyl-
4913 - 3-Cyclopentene-1-carboxylic acid, 2,2,3-trimethyl-
6037 - Furan, 2-(butoxymethyl)-
8203 - 2-Nonynoic acid
8204 - 3-Nonynoic acid
8205 - 5-Nonynoic acid
8206 - 6-Nonynoic acid
8207 - 8-Nonynoic acid
8542 - 2-Octynoic acid, methyl ester
10626 - 2-Propenoic acid, cyclohexyl ester

$C_9H_{14}O_3$
6084 - Furan, 2-(diethoxymethyl)-
9074 - 4-Pentenoic acid, 2-acetyl, ethyl ester
10674 - 2-Propenoic acid, 2-methyl-, (tetrahydro-2-furanyl)methyl ester
10703 - 2-Propen-1-ol, 2-methyl-, carbonate (2:1)

$C_9H_{14}O_3Si$
11488 - Silane, trimethoxyphenyl-

$C_9H_{14}O_4$
3329 - Butanedioic acid, methylene-, diethyl ester
3787 - 2-Butenedioic acid, 2-methyl-, diethyl ester, (E)-
3788 - 2-Butenedioic acid, 2-methyl-, diethyl ester, (Z)-
4792 - 1,2-Cyclopentanedicarboxylic acid, dimethyl ester, trans-(±)-
4967 - 1,1-Cyclopropanedicarboxylic acid, diethyl ester
4968 - 1,2-Cyclopropanedicarboxylic acid, diethyl ester, cis-
9041 - 2-Pentenedioic acid, diethyl ester
9042 - 2-Pentenedioic acid, diethyl ester, (E)-
9950 - Propanedioic acid, ethylidene-, diethyl ester

$C_9H_{14}O_5$
3324 - Butanedioic acid, formyl-, diethyl ester
8783 - Pentanedioic acid, 3-oxo-, diethyl ester
9923 - Propanedioic acid, acetyl-, diethyl ester

$C_9H_{14}O_6$
10195 - 1,2,3-Propanetriol, triacetate

$C_9H_{14}O_7$
10170 - 1,2,3-Propanetricarboxylic acid, 2-hydroxy-, trimethyl ester

$C_9H_{14}S$
11771 - Thiophene, 2-(3-methylbutyl)-

$C_9H_{14}Si$
11493 - Silane, trimethylphenyl-

$C_9H_{15}AlO_9$
340 - Aluminum, tris(2-hydroxypropanoato-O1,O2)-

$C_9H_{15}Br$
4322 - Cyclohexane, (2-bromo-2-propenyl)-

$C_9H_{15}Br_6O_4P$
10358 - 1-Propanol, 2,3-dibromo-, phosphate (3:1)

$C_9H_{15}Cl$
8200 - 1-Nonyne, 1-chloro-
8534 - 3-Octyne, 2-chloro-2-methyl-

$C_9H_{15}N$
10523 - 2-Propen-1-amine, N,N-di-2-propenyl-
11117 - 1H-Pyrrole, 3,4-diethyl-2-methyl-

$C_9H_{15}NO$
479 - 9-Azabicyclo[3.3.1]nonan-3-one, 9-methyl-

$C_9H_{15}NO_2$
3495 - Butanoic acid, 2-cyano-2-ethyl, ethyl ester
6881 - Hexanoic acid, 2-cyano-, ethyl ester
9654 - 2,4-Piperidinedione, 3,3-diethyl-
10521 - 2-Propenamide, N-(1,1-dimethyl-3-oxobutyl)-

$C_9H_{15}NO_3$
483 - 8-Azabicyclo[3.2.1]octane-2-carboxylic acid, 3-hydroxy-8-methyl-, [1R-(exo,exo)]-
8585 - 2,4-Oxazolidinedione, 5,5-dipropyl-

$C_9H_{15}NO_5$
9922 - Propanedioic acid, (acetylamino)-, diethyl ester

$C_9H_{15}N_3O_2S$
7204 - 1H-Imidazole-4-ethanaminium, α-carboxy-2,3-dihydro-N,N,N-trimethyl-2-thioxo-, hydroxide, in

$C_9H_{15}N_3O_7$
472 - L-Aspartic acid, N-[2-[(2-amino-2-oxoethyl)amino]-2-carboxyethyl]-

C_9H_{16}
2990 - Bicyclo[3.3.1]nonane
2992 - Bicyclo[2.2.2]octane, 2-methyl-
4393 - Cyclohexane, 1-ethylidene-2-methyl-
4470 - Cyclohexane, 2-propenyl-
4663 - Cyclohexene, 1-ethyl-4-methyl-
4687 - Cyclohexene, 1,2,3-trimethyl-
4688 - Cyclohexene, 1,3,5-trimethyl-
4689 - Cyclohexene, 1,4,4-trimethyl-
4690 - Cyclohexene, 1,4,5-trimethyl-
4691 - Cyclohexene, 1,5,5-trimethyl-
4692 - Cyclohexene, 1,6,6-trimethyl-
4693 - Cyclohexene, 1,5,6-trimethyl-
4727 - Cyclononene, (E)-
4728 - Cyclononene, (Z)-
4906 - Cyclopentene, 1-butyl-
4927 - Cyclopentene, 4-isopropyl-1-methyl-
4931 - Cyclopentene, 3-methyl-1-(1-methylethyl)-
6330 - 1,3-Heptadiene, 2,6-dimethyl-
6331 - 1,5-Heptadiene, 2,6-dimethyl-
6332 - 2,4-Heptadiene, 2,6-dimethyl-
6333 - 2,4-Heptadiene, 3,5-dimethyl-
6617 - 3-Heptyne, 2,6-dimethyl-
6618 - 3-Heptyne, 5,5-dimethyl-
6694 - 1,5-Hexadiene, 3-ethyl-2-methyl-
7281 - 1H-Indene, octahydro-
8110 - 1,8-Nonadiene
8111 - 2,7-Nonadiene
8196 - 1-Nonyne
8197 - 2-Nonyne
8198 - 3-Nonyne
8199 - 4-Nonyne
8328 - 2,4-Octadiene, 7-methyl-
8329 - 3,5-Octadiene, 4-methyl-
8537 - 3-Octyne, 7-methyl-

$C_9H_{16}ClN_5$
10508 - Propazine
11593 - Terbuthylazine

$C_9H_{16}N_4OS$
11590 - Tebuthiuron

$C_9H_{16}O$
2938 - Bicyclo[2.2.1]heptan-2-ol, 1,7-dimethyl-, (exo,syn)-
4555 - Cyclohexanol, 1-(2-propenyl)-
4556 - Cyclohexanol, 2-(2-propenyl)-, trans-
4595 - Cyclohexanone, 2-ethyl-2-methyl-
4599 - Cyclohexanone, 2-(1-methylethyl)-
4600 - Cyclohexanone, 4-(1-methylethyl)-
4612 - Cyclohexanone, 2-propyl-
4613 - Cyclohexanone, 4-propyl-
4614 - Cyclohexanone, 2,2,3-trimethyl-
4615 - Cyclohexanone, 2,2,4-trimethyl-
4616 - Cyclohexanone, 2,2,5-trimethyl-
4617 - Cyclohexanone, 2,2,6-trimethyl-
4618 - Cyclohexanone, 2,3,6-trimethyl-
4619 - Cyclohexanone, 2,4,4-trimethyl-
4620 - Cyclohexanone, 2,4,5-trimethyl-
4621 - Cyclohexanone, 2,4,6-trimethyl-
4622 - Cyclohexanone, 3,3,5-trimethyl-
4623 - Cyclohexanone, 3,4,4-trimethyl-
4700 - 2-Cyclohexen-1-ol, 1,3,5-trimethyl-
4701 - 2-Cyclohexen-1-ol, 2,4,4-trimethyl-
4702 - 2-Cyclohexen-1-ol, 3,5,5-trimethyl-
4725 - Cyclononanone
4738 - Cyclooctanecarboxaldehyde
4884 - Cyclopentanone, 4-isopropyl-2-methyl-
4889 - Cyclopentanone, 2-methyl-2-(1-methylethyl)-
4890 - Cyclopentanone, 2-methyl-5-(1-methylethyl)-
4891 - Cyclopentanone, 4-methyl-2-(1-methylethyl)-
5843 - Ethanone, 1-(1-methylcyclohexyl)-
5844 - Ethanone, 1-(3-methylcyclohexyl)-
5845 - Ethanone, 1-(4-methylcyclohexyl)-
8112 - 2,6-Nonadien-1-ol
8179 - 2-Nonenal
8180 - 2-Nonenal, (E)-
8208 - 1-Nonyn-3-ol
8209 - 2-Nonyn-1-ol
8522 - 1-Octen-3-one, 5-methyl-
8523 - 5-Octen-4-one, 4-methyl-
8524 - 5-Octen-4-one, 7-methyl-
8525 - 7-Octen-4-one, 2-methyl-
8536 - 2-Octyne, 1-methoxy-
8544 - 1-Octyn-3-ol, 3-methyl-
10423 - 1-Propanone, 1-cyclohexyl-

$C_9H_{16}O_2$
4310 - Cyclohexaneacetic acid, methyl ester
4311 - Cyclohexaneacetic acid, 2-methyl-
4312 - Cyclohexaneacetic acid, 3-methyl-
4334 - Cyclohexanecarboxylic acid, ethyl ester
4468 - Cyclohexanepropanoic acid
6149 - 2(3H)-Furanone, dihydro-5-pentyl-
7063 - 2-Hexenoic acid, 2-methyl-, ethyl ester
7064 - 3-Hexenoic acid, 2-methyl-, ethyl ester
7065 - 3-Hexenoic acid, 2-methyl-, ethyl ester
8137 - 2,4-Nonanedione
8187 - 3-Nonenoic acid
8188 - 8-Nonenoic acid
10235 - Propanoic acid, cyclohexyl ester
10643 - 2-Propenoic acid, hexyl ester

$C_9H_{16}O_3$
3461 - Butanoic acid, 2-acetyl-3-methyl, ethyl ester
6165 - 2-Furanpropanoic acid, tetrahydro-, ethyl ester
6460 - Heptanoic acid, 2,2-dimethyl-6-oxo-
6461 - Heptanoic acid, 4,4-dimethyl-6-oxo-
6902 - Hexanoic acid, 4-methyl-5-oxo-, ethyl ester
8856 - Pentanoic acid, 2-acetyl-, ethyl ester
8904 - Pentanoic acid, 4-oxo-, butyl ester
8910 - Pentanoic acid, 4-oxo-, 2-methylpropyl ester

$C_9H_{16}O_4$
3326 - Butanedioic acid, methyl-, diethyl ester
3337 - Butanedioic acid, monopentyl ester
6395 - Heptanedioic acid, dimethyl ester
6396 - Heptanedioic acid, monoethyl ester
8130 - Nonanedioic acid
8772 - Pentanedioic acid, diethyl ester
8774 - Pentanedioic acid, 3,3-dimethyl-, dimethyl ester
8791 - 1,5-Pentanediol, diacetate
9942 - Propanedioic acid, dimethyl-, diethyl ester
9945 - Propanedioic acid, dipropyl ester
9949 - Propanedioic acid, ethyl-, diethyl ester

$C_9H_{16}O_5$
8778 - Pentanedioic acid, 3-hydroxy-, diethyl ester

$C_9H_{17}ClN_3O_3PS$
7340 - Isazophos

$C_9H_{17}ClO$
8177 - Nonanoyl chloride

$C_9H_{17}FO_2$
8151 - Nonanoic acid, 9-fluoro-

$C_9H_{17}IN_4$
11825 - 3,5,7-Triaza-1-azoniatricyclo[3.3.1.13,7]decane, 1-(2-propenyl)-, iodide

$C_9H_{17}N$
4304 - Cyclohexanamine, N-2-propenyl-
8144 - Nonanenitrile
11246 - Quinoline, decahydro-, cis-
11247 - Quinoline, decahydro-, trans-(±)-

$C_9H_{17}NO$
4591 - Cyclohexanone, 2-[(dimethylamino)methyl]-
9003 - 4-Pentenamide, 2,2-diethyl-
9725 - 4-Piperidinone, 1,2,3,6-tetramethyl-
9726 - 4-Piperidinone, 2,2,6,6-tetramethyl-
10470 - 2-Propanone, 1-(1-methyl-2-piperidinyl)-, (±)-

$C_9H_{17}NOS$
7661 - Molinate

$C_9H_{17}NO_2$
10631 - 2-Propenoic acid, 2-(diethylamino)ethyl ester

$C_9H_{17}NO_4$
10305 - Propanoic acid, 2-methyl-, 2-nitropentyl ester

$C_9H_{17}NO_5$
329 - β-Alanine, N-(2,4-dihydroxy-3,3-dimethyl-1-oxobutyl)-, (R)-

$C_9H_{17}NS$
11708 - Thiocyanic acid, octyl ester

$C_9H_{17}N_5S$
341 - Ametryn

C_9H_{18}
4394 - Cyclohexane, 1-ethyl-1-methyl-
4395 - Cyclohexane, 1-ethyl-2-methyl-, cis-
4396 - Cyclohexane, 1-ethyl-2-methyl-, trans-

4397 - Cyclohexane, 1-ethyl-3-methyl-, *cis*-
4398 - Cyclohexane, 1-ethyl-3-methyl-, *trans*-
4399 - Cyclohexane, 1-ethyl-4-methyl-, *cis*-
4400 - Cyclohexane, 1-ethyl-4-methyl-, *trans*-
4431 - Cyclohexane, (1-methylethyl)-
4471 - Cyclohexane, propyl-
4485 - Cyclohexane, 1,1,2-trimethyl-
4486 - Cyclohexane, 1,1,3-trimethyl-
4487 - Cyclohexane, 1,1,4-trimethyl-
4488 - Cyclohexane, 1,2,3-trimethyl-
4489 - Cyclohexane, 1,2,3-trimethyl-, (1α,2β,3α)-
4490 - Cyclohexane, 1,2,3-trimethyl-, (1α,2α,3α)-
4491 - Cyclohexane, 1,2,3-trimethyl-, (1α,2α,3β)-
4492 - Cyclohexane, 1,2,4-trimethyl-, (1α,2α,4α)-
4493 - Cyclohexane, 1,2,4-trimethyl-, (1α,2β,4α)-
4494 - Cyclohexane, 1,2,4-trimethyl-, (1α,2α,4β)-
4495 - Cyclohexane, 1,2,4-trimethyl-, (1α,2β,4β)-
4496 - Cyclohexane, 1,3,5-trimethyl-, (1α,3α,5α)-
4497 - Cyclohexane, 1,3,5-trimethyl-, (1α,3α,5β)-
4724 - Cyclononane
4778 - Cyclopentane, butyl-
4796 - Cyclopentane, 1,1-diethyl-
4797 - Cyclopentane, 1,2-diethyl-, *cis*-
4798 - Cyclopentane, 1,2-diethyl-, *trans*-
4799 - Cyclopentane, 1,3-diethyl-
4805 - Cyclopentane, (1,1-dimethylethyl)-
4829 - Cyclopentane, 1-methyl-2-(1-methylethyl)-
4832 - Cyclopentane, (1-methylpropyl)-
4833 - Cyclopentane, 1-methyl-1-propyl-
4834 - Cyclopentane, 1-methyl-2-propyl-, *cis*-
4835 - Cyclopentane, 1-methyl-2-propyl-, *trans*-
4836 - Cyclopentane, 1-methyl-3-propyl-, *cis*-
4837 - Cyclopentane, 1-methyl-3-propyl-, *trans*-
4838 - Cyclopentane, (2-methylpropyl)-
4851 - Cyclopentane, 1,1,3,3-tetramethyl-
6542 - 2-Heptene, 2,3-dimethyl-
8182 - 1-Nonene
8183 - 3-Nonene, (*E*)-
8505 - 1-Octene, 2-methyl-
8506 - 2-Octene, 3-methyl-
8507 - 3-Octene, 7-methyl-
8508 - 4-Octene, 2-methyl-

C$_9$H$_{18}$Br$_2$
8127 - Nonane, 1,2-dibromo-

C$_9$H$_{18}$ClF
8126 - Nonane, 1-chloro-9-fluoro-

C$_9$H$_{18}$Cl$_2$
6382 - Heptane, 2,6-dichloro-2,6-dimethyl-
8128 - Nonane, 1,9-dichloro-

C$_9$H$_{18}$FeN$_3$S$_6$
5957 - Ferbam

C$_9$H$_{18}$NO$_3$PS$_2$
6021 - Fosthiazate

C$_9$H$_{18}$N$_2$O$_2$S
11717 - Thiofanox

C$_9$H$_{18}$N$_2$O$_4$
3938 - Carbamic acid, *N*-butyl-*N*-nitro,butyl ester
10021 - 1,3-Propanediol, 2-methyl-2-propyl-, dicarbamate

C$_9$H$_{18}$N$_6$O$_6$
7584 - Methanol, (1,3,5-triazine-2,4,6-triyltrinitrilo)hexakis-

C$_9$H$_{18}$O
4526 - Cyclohexanol, 1-ethyl-2-methyl-
4527 - Cyclohexanol, 1-ethyl-3-methyl-
4538 - Cyclohexanol, 1-(1-methylethyl)-
4539 - Cyclohexanol, 2-(1-methylethyl)-, *cis*-
4540 - Cyclohexanol, 4-(1-methylethyl)-, *cis*-
4541 - Cyclohexanol, 4-(1-methylethyl)-, *trans*-
4557 - Cyclohexanol, 1-propyl-
4558 - Cyclohexanol, 2-propyl-, *cis*-
4559 - Cyclohexanol, 2-propyl-, *trans*-
4560 - Cyclohexanol, 4-propyl-, *trans*-
4561 - Cyclohexanol, 4-propyl-, *cis*-
4562 - Cyclohexanol, 1,2,2-trimethyl-
4563 - Cyclohexanol, 1,2,6-trimethyl-
4564 - Cyclohexanol, 1,3,5-trimethyl-

4565 - Cyclohexanol, 2,2,5-trimethyl-
4566 - Cyclohexanol, 2,2,6-trimethyl-
4567 - Cyclohexanol, 2,3,3-trimethyl-
4568 - Cyclohexanol, 2,3,6-trimethyl-
4569 - Cyclohexanol, 3,3,5-trimethyl-, *cis*-
4570 - Cyclohexanol, 3,3,5-trimethyl-, *trans*-
6514 - 4-Heptanone, 2,6-dimethyl-
6515 - 4-Heptanone, 3,5-dimethyl-
8117 - Nonanal
8172 - 2-Nonanone
8173 - 3-Nonanone
8174 - 4-Nonanone
8175 - 5-Nonanone
8189 - 1-Nonen-3-ol
8190 - 8-Nonen-1-ol
8475 - 2-Octanone, 3-methyl-
8476 - 3-Octanone, 4-methyl-
8477 - 3-Octanone, 7-methyl-
8478 - 4-Octanone, 3-methyl-
8479 - 4-Octanone, 5-methyl-
8517 - 1-Octen-4-ol, 4-methyl-
8518 - 2-Octen-4-ol, 7-methyl-
8519 - 3-Octen-2-ol, 2-methyl-, (*Z*)-
8520 - 4-Octen-3-ol, 4-methyl-
8521 - 5-Octen-4-ol, 5-methyl-
8988 - 3-Pentanone, 2,2,4,4-tetramethyl-

C$_9$H$_{18}$O$_2$
193 - Acetic acid, heptyl ester
3543 - Butanoic acid, 3-methyl-, butyl ester
3544 - Butanoic acid, 3-methylbutyl ester
3545 - Butanoic acid, 2-methyl-, butyl ester
3555 - Butanoic acid, 3-methyl-, 2-methylpropyl ester
3579 - Butanoic acid, pentyl ester
6014 - Formic acid, octyl ester
6462 - Heptanoic acid, ethyl ester
6476 - 4-Heptanol, acetate
6886 - Hexanoic acid, 2-ethyl-3-methyl-
6887 - Hexanoic acid, 3-ethyl-3-methyl-
6899 - Hexanoic acid, 2-methyl-, ethyl ester
6900 - Hexanoic acid, 4-methyl-, ethyl ester
6909 - Hexanoic acid, propyl ester
8146 - Nonanoic acid
8433 - Octanoic acid, 3-methyl-
8434 - Octanoic acid, methyl ester
8862 - Pentanoic acid, butyl ester
8899 - Pentanoic acid, 1-methylpropyl ester, (+)-
8900 - Pentanoic acid, 2-methylpropyl ester
10265 - Propanoic acid, hexyl ester
10302 - Propanoic acid, 2-methyl-, 3-methylbutyl ester

C$_9$H$_{18}$O$_3$
3994 - Carbonic acid, bis(1,1-dimethylethyl) ester
3997 - Carbonic acid, bis(2-methylpropyl) ester
3998 - Carbonic acid, dibutyl ester

C$_9$H$_{18}$O$_5$
5719 - Ethanol, 2-ethoxy-, carbonate (2:1)

C$_9$H$_{19}$Br
8121 - Nonane, 1-bromo-

C$_9$H$_{19}$Cl
6364 - Heptane, 3-chloro-3-ethyl-
6747 - Hexane, 3-chloro-2,2,3-trimethyl-
8123 - Nonane, 1-chloro-
8124 - Nonane, 2-chloro-
8125 - Nonane, 5-chloro-
8362 - Octane, 3-chloro-3-methyl-
8363 - Octane, 4-chloro-4-methyl-
8727 - Pentane, 3-chloro-3-ethyl-2,2-dimethyl-

C$_9$H$_{19}$ClO
8163 - 1-Nonanol, 9-chloro-

C$_9$H$_{19}$FO
8165 - 1-Nonanol, 9-fluoro-

C$_9$H$_{19}$N
4303 - Cyclohexanamine, *N*-(1-methylethyl)-
4810 - Cyclopentaneethanamine, *N*,α-dimethyl-
6532 - 5-Hepten-2-amine, *N*,6-dimethyl-
9635 - Piperidine, 1-butyl-
9636 - Piperidine, 3-butyl-

9637 - Piperidine, 4-butyl-
9647 - Piperidine, 2,5-diethyl-
9653 - Piperidine, 1-(1,1-dimethylethyl)-
9685 - Piperidine, 1-(2-methylpropyl)-
9686 - Piperidine, 1-methyl-2-propyl-, (*S*)-
9710 - Piperidine, 2,2,4,6-tetramethyl-
9711 - Piperidine, 2,2,6,6-tetramethyl-

C$_9$H$_{19}$NO
3201 - Butanamide, *N*,*N*-diethyl-3-methyl-
4144 - Conhydrine, *N*-methyl-
8118 - Nonanamide
8964 - 2-Pentanone, 5-(diethylamino)-
9794 - Propanamide, *N*,*N*-diethyl-2,2-dimethyl-

C$_9$H$_{19}$NOS
5372 - EPTC

C$_9$H$_{19}$NO$_2$
328 - β-Alanine, *N*,*N*-diethyl-, ethyl ester
3934 - Carbamic acid, butyl-, butyl ester

C$_9$H$_{19}$NO$_4$
3202 - Butanamide, 2,4-dihydroxy-*N*-(3-hydroxypropyl)-3,3-dimethyl-, (*R*)-

C$_9$H$_{20}$
6383 - Heptane, 2,2-dimethyl-
6384 - Heptane, 2,3-dimethyl-
6385 - Heptane, 2,4-dimethyl-
6386 - Heptane, 2,5-dimethyl-
6387 - Heptane, 2,6-dimethyl-
6388 - Heptane, 3,3-dimethyl-
6389 - Heptane, 3,4-dimethyl-
6390 - Heptane, 3,5-dimethyl-
6391 - Heptane, 4,4-dimethyl-
6407 - Heptane, 3-ethyl-
6408 - Heptane, 4-ethyl-
6825 - Hexane, 3-ethyl-2-methyl-
6826 - Hexane, 3-ethyl-3-methyl-
6827 - Hexane, 3-ethyl-4-methyl-
6828 - Hexane, 4-ethyl-2-methyl-
6859 - Hexane, 2,2,3-trimethyl-
6860 - Hexane, 2,2,4-trimethyl-
6861 - Hexane, 2,2,5-trimethyl-
6862 - Hexane, 2,3,3-trimethyl-
6863 - Hexane, 2,3,4-trimethyl-
6864 - Hexane, 2,3,5-trimethyl-
6865 - Hexane, 2,4,4-trimethyl-
6866 - Hexane, 3,3,4-trimethyl-
8120 - Nonane
8408 - Octane, 2-methyl-
8409 - Octane, 3-methyl-
8410 - Octane, 4-methyl-
8756 - Pentane, 3,3-diethyl-
8810 - Pentane, 3-ethyl-2,2-dimethyl-
8811 - Pentane, 3-ethyl-2,3-dimethyl-
8812 - Pentane, 3-ethyl-2,4-dimethyl-
8839 - Pentane, 2,2,3,3-tetramethyl-
8840 - Pentane, 2,2,3,4-tetramethyl-
8841 - Pentane, 2,2,4,4-tetramethyl-
8842 - Pentane, 2,3,3,4-tetramethyl-

C$_9$H$_{20}$N$_2$
9628 - 4-Piperidinamine, 2,2,6,6-tetramethyl-

C$_9$H$_{20}$N$_2$O
11996 - Urea, *N*,*N*-bis(2-methylpropyl)-
12029 - Urea, tetraethyl-

C$_9$H$_{20}$O
6406 - Heptane, 1-ethoxy-
6479 - 2-Heptanol, 2,6-dimethyl-
6480 - 3-Heptanol, 2,3-dimethyl-
6481 - 3-Heptanol, 2,6-dimethyl-
6482 - 3-Heptanol, 3,5-dimethyl-
6483 - 3-Heptanol, 3,6-dimethyl-
6484 - 4-Heptanol, 2,4-dimethyl-
6485 - 4-Heptanol, 2,6-dimethyl-
6486 - 4-Heptanol, 3,5-dimethyl-
6487 - 4-Heptanol, 4-ethyl-
6932 - 1-Hexanol, 2-ethyl-3-methyl-
6933 - 3-Hexanol, 2-ethyl-2-methyl-
6934 - 3-Hexanol, 3-ethyl-2-methyl-
6935 - 3-Hexanol, 3-ethyl-5-methyl-

6951 - 1-Hexanol, 3,5,5-trimethyl-
6952 - 2-Hexanol, 2,3,4-trimethyl-
6953 - 3-Hexanol, 2,2,3-trimethyl-
6954 - 3-Hexanol, 2,3,5-trimethyl-
6955 - 3-Hexanol, 2,4,4-trimethyl-
6956 - 3-Hexanol, 2,5,5-trimethyl-
6957 - 3-Hexanol, 3,4,4-trimethyl-
6958 - 3-Hexanol, 3,5,5-trimethyl-
8157 - 1-Nonanol
8158 - 2-Nonanol, (±)-
8159 - 3-Nonanol, (±)-
8160 - 4-Nonanol, (S)-
8161 - 5-Nonanol
8461 - 1-Octanol, 3-methyl-
8462 - 1-Octanol, 4-methyl-
8463 - 1-Octanol, 7-methyl-
8464 - 2-Octanol, 2-methyl-
8465 - 2-Octanol, 3-methyl-
8466 - 3-Octanol, 3-methyl-
8467 - 3-Octanol, 4-methyl-
8468 - 3-Octanol, 6-methyl-
8469 - 4-Octanol, 4-methyl-
8470 - 4-Octanol, 5-methyl-

$C_9H_{20}O_2$
3398 - Butane, 1,1'-[methylenebis(oxy)]bis-
8136 - 1,9-Nonanediol
8755 - Pentane, 1,1-diethoxy-
9992 - 1,3-Propanediol, 2-butyl-2-ethyl-
10075 - Propane, 1,1'-[methylenebis(oxy)]bis[2-methyl-

$C_9H_{20}O_3$
10182 - Propane, 1,1,1-triethoxy-
10183 - Propane, 1,1,3-triethoxy-

$C_9H_{20}O_3Si$
11482 - Silane, triethoxy-2-propenyl-

$C_9H_{20}O_4$
5601 - Ethane, 1,1',1''1'''-[methanetetrayl tetrakis(oxy)]tetrakis-

$C_9H_{20}S$
8145 - 1-Nonanethiol

$C_9H_{21}B$
3121 - Borane, tripropyl-

$C_9H_{21}BO_3$
3127 - Boric acid, tripropyl ester
3128 - Boric acid, tris(1-methylethyl) ester

$C_9H_{21}ClSn$
11542 - Stannane, chlorotripropyl-

$C_9H_{21}N$
8119 - 1-Nonanamine
9808 - 1-Propanamine, N,N-dipropyl-

$C_9H_{21}NO_3$
10393 - 2-Propanol, 1,1',1''-nitrilotris-

$C_9H_{21}NO_4$
5678 - Ethanol, 2-[2-[2-(3-aminopropoxy) ethoxy]ethoxy]-

$C_9H_{21}O_2PS_3$
11592 - Terbufos

$C_9H_{21}O_3P$
9594 - Phosphorous acid, tripropyl ester
9595 - Phosphorous acid, tris(1-methylethyl) ester

$C_9H_{21}O_4P$
9567 - Phosphoric acid, tripropyl ester
9568 - Phosphoric acid, tris(1-methylethyl) ester

$C_9H_{22}N_2$
7489 - Methanediamine, N,N,N',N'-tetraethyl-
8736 - 1,4-Pentanediamine, N1,N1-diethyl-

$C_9H_{22}O_4P_2S_4$
5931 - Ethion

$C_9H_{22}Si$
11500 - Silane, tripropyl-

$C_9H_{23}NO_3Si$
9828 - 1-Propanamine, 3-(triethoxysilyl)-

$C_9H_{24}O_3Si_3$
5003 - Cyclotrisiloxane, 2,4,6-triethyl-2,4,6-trimethyl-

$C_9H_{24}O_4Si_4$
4991 - Cyclotetrasiloxane, ethenylheptamethyl-

$C_{10}Cl_8$
7933 - Naphthalene, octachloro-

$C_{10}Cl_{10}O$
7408 - Kepone

$C_{10}Cl_{12}$
7644 - 1,3,4-Metheno-1H-cyclobuta[cd]pentalene, 1,1a,2,2,3,3a,4,5,5,5a,5b, 6-dodecachlorooctahydro-

$C_{10}F_8$
7935 - Naphthalene, octafluoro-

$C_{10}F_{18}$
7934 - Naphthalene, octadecafluorodecahydro-

$C_{10}F_{22}$
5050 - Decane, docosafluoro-

$C_{10}F_{23}N$
11912 - Tripropylamine, eicosafluoro-2-(trifluoromethyl)-

$C_{10}H_4Cl_2O_2$
7849 - 1,2-Naphthalenedione, 3,4-dichloro-
7850 - 1,4-Naphthalenedione, 2,3-dichloro-
7851 - 1,4-Naphthalenedione, 5,6-dichloro-

$C_{10}H_4N_4O_8$
8004 - Naphthalene, 1,3,6,8-tetranitro-

$C_{10}H_5BrO_2$
7844 - 1,2-Naphthalenedione, 3-bromo-

$C_{10}H_5BrO_3$
7845 - 1,4-Naphthalenedione, 2-bromo-3-hydroxy-

$C_{10}H_5ClO_2$
7847 - 1,4-Naphthalenedione, 5-chloro-

$C_{10}H_5ClO_3$
7848 - 1,4-Naphthalenedione, 2-chloro-3-hydroxy-

$C_{10}H_5Cl_2NO_2$
11201 - Quinclorac

$C_{10}H_5Cl_7$
7576 - 4,7-Methano-1H-indene, 1,4,5,6,7,8,8-heptachloro-3a,4,7,7a-tetrahydro-

$C_{10}H_5Cl_7O$
6301 - Heptachlor epoxide

$C_{10}H_5N_3O_6$
8016 - Naphthalene, 1,3,5-trinitro-

$C_{10}H_5NaO_5S$
7970 - 1-Naphthalenesulfonic acid, 3,4-dihydro-3,4-dioxo-, sodium salt

$C_{10}H_6BrCl$
7741 - Naphthalene, 1-bromo-4-chloro-

$C_{10}H_6BrNO_2$
7744 - Naphthalene, 1-bromo-3-nitro-
7745 - Naphthalene, 2-bromo-1-nitro-

$C_{10}H_6Br_2$
7801 - Naphthalene, 1,3-dibromo-
7802 - Naphthalene, 1,4-dibromo-
7803 - Naphthalene, 1,5-dibromo-

$C_{10}H_6ClNO_2$
7786 - Naphthalene, 1-chloro-5-nitro-
7787 - Naphthalene, 1-chloro-8-nitro-
7788 - Naphthalene, 2-chloro-1-nitro-

$C_{10}H_6Cl_2$
7807 - Naphthalene, 1,2-dichloro-
7808 - Naphthalene, 1,3-dichloro-
7809 - Naphthalene, 1,4-dichloro-
7810 - Naphthalene, 1,5-dichloro-
7811 - Naphthalene, 1,7-dichloro-
7812 - Naphthalene, 1,8-dichloro-
7813 - Naphthalene, 2,6-dichloro-

$C_{10}H_6Cl_2O$
8025 - 1-Naphthalenol, 2,3-dichloro-
8026 - 1-Naphthalenol, 2,4-dichloro-

$C_{10}H_6Cl_4O_4$
1353 - 1,2-Benzenedicarboxylic acid, 3,4,5,6-tetrachloro-, monoethyl ester
5180 - Dimethyl tetrachloroterephthalate

$C_{10}H_6Cl_8$
4077 - Chlordane

$C_{10}H_6N_2$
7375 - 5-Isoquinolinecarbonitrile
11223 - 2-Quinolinecarbonitrile
11224 - 4-Quinolinecarbonitrile

$C_{10}H_6N_2OS_2$
4072 - Chinomethionat

$C_{10}H_6N_2O_2$
5223 - 5H,10H-Dipyrrolo[1,2-a:1',2'-d]pyrazine-5,10-dione

$C_{10}H_6N_2O_4$
7831 - Naphthalene, 1,3-dinitro-
7832 - Naphthalene, 1,5-dinitro-
7833 - Naphthalene, 1,8-dinitro-

$C_{10}H_6O_2$
7842 - 1,2-Naphthalenedione
7843 - 1,4-Naphthalenedione

$C_{10}H_6O_3$
7859 - 1,4-Naphthalenedione, 2-hydroxy-
7860 - 1,4-Naphthalenedione, 5-hydroxy-

$C_{10}H_6O_3S$
7840 - 1,8-Naphthalenediol, cyclic sulfite

$C_{10}H_6O_4$
7852 - 1,4-Naphthalenedione, 2,3-dihydroxy-
7853 - 1,4-Naphthalenedione, 5,8-dihydroxy-

$C_{10}H_6O_8$
2257 - 1,2,4,5-Benzenetetracarboxylic acid

$C_{10}H_7Br$
7739 - Naphthalene, 1-bromo-
7740 - Naphthalene, 2-bromo-

$C_{10}H_7BrO$
8022 - 2-Naphthalenol, 1-bromo-

$C_{10}H_7Cl$
7780 - Naphthalene, 1-chloro-
7781 - Naphthalene, 2-chloro-

$C_{10}H_7ClO$
8023 - 2-Naphthalenol, 6-chloro-
8063 - 2-Naphthol, 8-chloro-

$C_{10}H_7ClO_2S$
7975 - 1-Naphthalenesulfonyl chloride
7976 - 2-Naphthalenesulfonyl chloride

$C_{10}H_7Cl_5O$
11899 - Tridiphane

$C_{10}H_7F$
7893 - Naphthalene, 1-fluoro-
7894 - Naphthalene, 2-fluoro-

$C_{10}H_7I$
7902 - Naphthalene, 1-iodo-
7903 - Naphthalene, 2-iodo-

$C_{10}H_7NO$
11225 - 4-Quinolinecarboxaldehyde

$C_{10}H_7NO_2$
7929 - Naphthalene, 1-nitro-
7930 - Naphthalene, 2-nitro-
8035 - 1-Naphthalenol, 1-nitroso-
11123 - 1H-Pyrrole-2,5-dione, 1-phenyl-
11227 - 2-Quinolinecarboxylic acid
11228 - 7-Quinolinecarboxylic acid
11229 - 8-Quinolinecarboxylic acid

$C_{10}H_7NO_3$
8034 - 2-Naphthalenol, 1-nitro-
11232 - 2-Quinolinecarboxylic acid, 4-hydroxy-

$C_{10}H_7NO_4$
11230 - 2-Quinolinecarboxylic acid, 4,8-dihydroxy-

$C_{10}H_7N_3$
7738 - Naphthalene, 1-azido-

$C_{10}H_7N_3S$
11656 - Thiabendazole

$C_{10}H_8$
504 - Azulene
7732 - Naphthalene

$C_{10}H_8BrN$
7711 - 1-Naphthalenamine, 5-bromo-

$C_{10}H_8ClN$
11243 - Quinoline, 2-chloro-4-methyl-
11244 - Quinoline, 2-chloro-8-methyl-
11245 - Quinoline, 4-chloro-2-methyl-

$C_{10}H_8ClN_3O$
4080 - Chloridazon

$C_{10}H_8N_2$
3099 - 2,2'-Bipyridine
3100 - 2,3'-Bipyridine

3101 - 2,4'-Bipyridine
3102 - 3,3'-Bipyridine
3103 - 3,4'-Bipyridine
3104 - 4,4'-Bipyridine
9917 - Propanedinitrile, (phenylmethyl)-

$C_{10}H_8N_2O_4$
5583 - Ethanedione, di-2-furanyl, dioxime

$C_{10}H_8N_4O_5$
10848 - 3H-Pyrazol-3-one, 2,4-dihydro-5-methyl-4-nitro-2-(4-nitrophenyl)-

$C_{10}H_8O$
3924 - 3-Butyn-2-one, 4-phenyl-
6159 - Furan, 2-phenyl-
8017 - 1-Naphthalenol
8018 - 2-Naphthalenol

$C_{10}H_8OS$
6160 - Furan, 2-(phenylthio)-

$C_{10}H_8O_2$
2833 - 2H-1-Benzopyran-2-one, 3-methyl-
2834 - 2H-1-Benzopyran-2-one, 5-methyl-
2835 - 2H-1-Benzopyran-2-one, 6-methyl-
2836 - 2H-1-Benzopyran-2-one, 7-methyl-
2837 - 2H-1-Benzopyran-2-one, 8-methyl-
2838 - 4H-1-Benzopyran-4-one, 3-methyl-
5775 - Ethanone, 1-(2-benzofuranyl)-
6157 - Furan, 2-phenoxy-
7246 - 1H-Indene-1-carboxylic acid
7834 - 1,3-Naphthalenediol
7835 - 1,5-Naphthalenediol
7836 - 1,6-Naphthalenediol
7837 - 1,7-Naphthalenediol
7838 - 2,6-Naphthalenediol
7839 - 2,7-Naphthalenediol
10742 - 2-Propynoic acid, 3-phenyl-, methyl ester

$C_{10}H_8O_3$
2831 - 2H-1-Benzopyran-2-one, 7-hydroxy-4-methyl-
6096 - 2,5-Furandione, dihydro-3-phenyl-, (±)-

$C_{10}H_8O_3S$
7959 - 1-Naphthalenesulfonic acid
7960 - 2-Naphthalenesulfonic acid

$C_{10}H_8O_4$
2829 - 2H-1-Benzopyran-2-one, 7-hydroxy-6-methoxy-
5185 - 1,7-Dioxadispiro[4.0.4.2]dodeca-3,9-diene-2,8-dione, trans-
6057 - 2-Furancarboxylic acid, 2-furanylmethyl ester

$C_{10}H_8O_4S$
7971 - 1-Naphthalenesulfonic acid, 4-hydroxy-
7972 - 1-Naphthalenesulfonic acid, 7-hydroxy-
7973 - 2-Naphthalenesulfonic acid, 1-hydroxy-
7974 - 2-Naphthalenesulfonic acid, 6-hydroxy-

$C_{10}H_8O_5$
2808 - 2H-1-Benzopyran-2-one, 7,8-dihydroxy-6-methoxy-

$C_{10}H_8O_6S_2$
7867 - 1,5-Naphthalenedisulfonic acid
7868 - 1,6-Naphthalenedisulfonic acid
7869 - 2,7-Naphthalenedisulfonic acid

$C_{10}H_8O_7S_2$
7871 - 1,3-Naphthalenedisulfonic acid, 7-hydroxy-
7876 - 2,7-Naphthalenedisulfonic acid, 3-hydroxy-

$C_{10}H_8O_8S_2$
7875 - 2,7-Naphthalenedisulfonic acid, 4,5-dihydroxy-

$C_{10}H_8S$
8007 - 1-Naphthalenethiol
8008 - 2-Naphthalenethiol

$C_{10}H_9BrO_2$
10615 - 2-Propenoic acid, 2-bromo-3-phenyl-, methyl ester

$C_{10}H_9ClN_4O_2S$
2193 - Benzenesulfonamide, 4-amino-N-(6-chloro-3-pyridazinyl)-

$C_{10}H_9ClO$
4957 - Cyclopropanecarbonyl chloride, 2-phenyl-, trans-

$C_{10}H_9ClO_2$
10622 - 2-Propenoic acid, 2-chloro-3-phenyl-, methyl ester, (E)-
10623 - 2-Propenoic acid, 3-chloro-3-phenyl-, methyl ester

$C_{10}H_9ClO_3$
972 - Benzeneacetyl chloride, α-(acetyloxy)-

$C_{10}H_9Cl_2NO$
10520 - 2-Propenamide, N-(3,4-dichlorophenyl)-2-methyl-

$C_{10}H_9Cl_4O_4P$
6202 - Gardona

$C_{10}H_9N$
7381 - Isoquinoline, 1-methyl-
7382 - Isoquinoline, 3-methyl-
7383 - Isoquinoline, 4-methyl-
7384 - Isoquinoline, 6-methyl-
7385 - Isoquinoline, 8-methyl-
7708 - 1-Naphthalenamine
7709 - 2-Naphthalenamine
11134 - 1H-Pyrrole, 1-phenyl-
11135 - 1H-Pyrrole, 2-phenyl-
11277 - Quinoline, 2-methyl-
11278 - Quinoline, 3-methyl-
11279 - Quinoline, 4-methyl-
11280 - Quinoline, 5-methyl-
11281 - Quinoline, 6-methyl-
11282 - Quinoline, 7-methyl-
11283 - Quinoline, 8-methyl-

$C_{10}H_9NO$
5819 - Ethanone, 1-(1H-indol-3-yl)-
7380 - Isoquinoline, 7-methoxy-
8019 - 2-Naphthalenol, 6-amino-
8020 - 2-Naphthalenol, 8-amino-
11274 - Quinoline, 4-methoxy-
11275 - Quinoline, 6-methoxy-
11276 - Quinoline, 8-methoxy-
11319 - 6-Quinolinol, 2-methyl-
11320 - 7-Quinolinol, 6-methyl-
11321 - 8-Quinolinol, 2-methyl-
11322 - 8-Quinolinol, 6-methyl-
11328 - 2(1H)-Quinolinone, 1-methyl-
11329 - 2(1H)-Quinolinone, 6-methyl-
11330 - 4(1H)-Quinolinone, 1-methyl-

$C_{10}H_9NO_2$
1877 - Benzenemethanol, α-ethynyl-, carbamate
7295 - 1H-Indole-3-acetic acid
7364 - 1H-Isoindole-1,3(2H)-dione, N-ethyl-
8582 - Oxazole, 5-methoxy-2-phenyl-
11162 - 2,5-Pyrrolidinedione, 1-phenyl-

$C_{10}H_9NO_3$
3853 - 2-Butenoic acid, 4-oxo-4-(phenylamino)-, (Z)-

$C_{10}H_9NO_3S$
7961 - 1-Naphthalenesulfonic acid, 2-amino-
7962 - 1-Naphthalenesulfonic acid, 4-amino-
7963 - 1-Naphthalenesulfonic acid, 5-amino-
7964 - 1-Naphthalenesulfonic acid, 7-amino-
7965 - 1-Naphthalenesulfonic acid, 8-amino-
7967 - 2-Naphthalenesulfonic acid, 8-amino-

$C_{10}H_9NO_4$
10677 - 2-Propenoic acid, 3-(2-nitrophenyl)-, methyl ester, (E)-
10678 - 2-Propenoic acid, 3-(3-nitrophenyl)-, methyl ester, (E)-
10679 - 2-Propenoic acid, 3-(4-nitrophenyl)-, methyl ester, (E)-

$C_{10}H_9NO_4S$
7966 - 1-Naphthalenesulfonic acid, 4-amino-3-hydroxy-

$C_{10}H_9NO_6S_2$
7870 - 1,3-Naphthalenedisulfonic acid, 7-amino-
7872 - 1,6-Naphthalenedisulfonic acid, 4-amino-
7873 - 1,7-Naphthalenedisulfonic acid, 4-amino-

$C_{10}H_9NO_7S_2$
7874 - 2,7-Naphthalenedisulfonic acid, 4-amino-5-hydroxy-

$C_{10}H_9N_3$
10871 - 2-Pyridinamine, N-2-pyridinyl-

$C_{10}H_{10}$
1097 - Benzene, 1,3-butadienyl-, (E)-
1098 - Benzene, 1,3-butadienyl-, (Z)-
1124 - Benzene, 3-butynyl-
1403 - Benzene, 1,2-diethenyl-
1404 - Benzene, 1,3-diethenyl-
1405 - Benzene, 1,4-diethenyl-
1967 - Benzene, (1-methylene-2-propenyl)-
7275 - 1H-Indene, 1-methyl-
7276 - 1H-Indene, 2-methyl-
7277 - 1H-Indene, 3-methyl-
7278 - 1H-Indene, 4-methyl-
7279 - 1H-Indene, 6-methyl-
7280 - 1H-Indene, 7-methyl-
7814 - Naphthalene, 1,2-dihydro-
7815 - Naphthalene, 1,4-dihydro-

$C_{10}H_{10}Cl_2O_3$
3925 - Butyrac 118

$C_{10}H_{10}Fe$
5960 - Ferrocene

$C_{10}H_{10}N_2$
7167 - Hydrazine, 1-naphthalenyl-
7795 - 1,2-Naphthalenediamine
7796 - 1,4-Naphthalenediamine
7797 - 1,5-Naphthalenediamine
7798 - 1,8-Naphthalenediamine
7799 - 2,3-Naphthalenediamine
10837 - 1H-Pyrazole, 1-methyl-3-phenyl-
10838 - 1H-Pyrazole, 3-methyl-1-phenyl-
10987 - Pyridine, 3-(1-methyl-1H-pyrrol-2-yl)-
11208 - 2-Quinolinamine, 4-methyl-
11209 - 3-Quinolinamine, 2-methyl-
11210 - 4-Quinolinamine, 2-methyl-
11211 - 7-Quinolinamine, 2-methyl-
11212 - 8-Quinolinamine, 6-methyl-
11335 - Quinoxaline, 2,6-dimethyl-

$C_{10}H_{10}N_2O$
10849 - 3H-Pyrazol-3-one, 1,2-dihydro-5-methyl-2-phenyl-

$C_{10}H_{10}N_2O_2$
7221 - 2,4-Imidazolidinedione, 3-methyl-5-phenyl-

$C_{10}H_{10}N_4O_2S$
2202 - Benzenesulfonamide, 4-amino-N-pyrazinyl-
2204 - Benzenesulfonamide, 4-amino-N-2-pyrimidinyl-

$C_{10}H_{10}O$
1878 - Benzenemethanol, α-ethynyl-α-methyl-
2445 - Benzofuran, 2,5-dimethyl-
2446 - Benzofuran, 2,6-dimethyl-
2447 - Benzofuran, 2,7-dimethyl-
2448 - Benzofuran, 3,5-dimethyl-
2449 - Benzofuran, 3,6-dimethyl-
2450 - Benzofuran, 4,6-dimethyl-
2451 - Benzofuran, 4,7-dimethyl-
2452 - Benzofuran, 5,6-dimethyl-
2453 - Benzofuran, 5,7-dimethyl-
2454 - Benzofuran, 6,7-dimethyl-
3701 - 2-Butenal, 3-phenyl-
3880 - 2-Buten-1-one, 1-phenyl-
3881 - 3-Buten-2-one, 4-phenyl-
3882 - 3-Buten-2-one, 4-phenyl-, (E)-
7289 - 1H-Inden-1-one, 2,3-dihydro-2-methyl-
7290 - 1H-Inden-1-one, 2,3-dihydro-4-methyl-
8046 - 1(2H)-Naphthalenone, 3,4-dihydro-
8047 - 2(1H)-Naphthalenone, 3,4-dihydro-
10516 - 2-Propenal, 2-methyl-3-phenyl-
10517 - 2-Propenal, 3-(4-methylphenyl)-

$C_{10}H_{10}O_2$
2118 - Benzenepropanal, α-methyl-β-oxo-
2429 - 1,3-Benzodioxole, 5-(2-propenyl)-
2430 - 1,3-Benzodioxole, 5-(1-propenyl)-, (E)-
2722 - Benzoic acid, 2-propenyl ester
2803 - 4H-1-Benzopyran-4-one, 2,3-dihydro-6-methyl-
3362 - 1,3-Butanedione, 1-phenyl-

3854 - 2-Butenoic acid, 3-phenyl-, (Z)-
3855 - 3-Butenoic acid, 4-phenyl-
5865 - Ethanone, 1,1'-(1,3-phenylene)bis-
10670 - 2-Propenoic acid, 2-methyl-3-phenyl-, (Z)-
10686 - 2-Propenoic acid, 3-phenyl-, methyl ester, (E)-
10688 - 2-Propenoic acid, phenylmethyl ester
10707 - 2-Propen-1-ol, 3-phenyl-, formate

$C_{10}H_{10}O_2S_2$
6101 - Furan, 2,2'-[dithiobis(methylene)]bis-

$C_{10}H_{10}O_3$
935 - Benzeneacetic acid, α-oxo, ethyl ester
2142 - Benzenepropanoic acid, β-oxo-, methyl ester
2479 - Benzoic acid, 4-acetyl-, methyl ester
5771 - Ethanone, 2-(acetyloxy)-1-phenyl-
6156 - Furan, 2,2'-[oxybis(methylene)]bis-
10514 - 2-Propenal, 3-(4-hydroxy-3-methoxyphenyl)-

$C_{10}H_{10}O_4$
1343 - 1,2-Benzenedicarboxylic acid, dimethyl ester
1347 - 1,3-Benzenedicarboxylic acid, dimethyl ester
1348 - 1,3-Benzenedicarboxylic acid, 4,6-dimethyl-
1351 - 1,4-Benzenedicarboxylic acid, dimethyl ester
1508 - 1,2-Benzenediol, diacetate
1509 - 1,4-Benzenediol, diacetate
2416 - 1,3-Benzodioxole-5-acetic acid, methyl ester
2419 - 1,3-Benzodioxole-5-carboxylic acid, ethyl ester
2482 - Benzoic acid, 2-(acetyloxy)-, methyl ester
3341 - Butanedioic acid, phenyl-, (±)-
7355 - 1(3H)-Isobenzofuranone, 6,7-dimethoxy-

$C_{10}H_{10}O_5$
2608 - Benzoic acid, 6-formyl-2,3-dimethoxy-

$C_{10}H_{11}Br$
1001 - Benzene, (3-bromo-3-butenyl)-
1002 - Benzene, (4-bromo-2-butenyl)-
3719 - 2-Butene, 2-bromo-3-phenyl-

$C_{10}H_{11}Cl$
1152 - Benzene, 1-(1-chloroethenyl)-2,4-dimethyl-
1188 - Benzene, 1-(2-chloro-1-methylethenyl)-4-methyl-

$C_{10}H_{11}ClO$
974 - Benzeneacetyl chloride, α-ethyl-
1231 - Benzene, 1-(3-chloro-2-propenyl)-4-methoxy-
3650 - 1-Butanone, 4-chloro-1-phenyl-

$C_{10}H_{11}ClO_2$
891 - Benzeneacetic acid, α-chloro-, ethyl ester, (S)-

$C_{10}H_{11}ClO_2S$
7978 - 2-Naphthalenesulfonyl chloride, 5,6,7,8-tetrahydro-

$C_{10}H_{11}F_3N_2O$
12007 - Urea, N,N-dimethyl-N'-[3-(trifluoromethyl)phenyl]-

$C_{10}H_{11}F_7O_2$
8398 - 3,5-Octanedione, 6,6,7,7,8,8,8-heptafluoro-2,2-dimethyl-

$C_{10}H_{11}I_2NO_3$
10873 - 1(4H)-Pyridineacetic acid, 3,5-diiodo-4-oxo-, propyl ester

$C_{10}H_{11}N$
951 - Benzeneacetonitrile, α-ethyl-
7298 - 1H-Indole, 1,3-dimethyl-
7299 - 1H-Indole, 2,5-dimethyl-
7307 - 1H-Indole, 1-ethyl-
7308 - 1H-Indole, 3-ethyl-
7714 - 1-Naphthalenamine, 5,8-dihydro-

$C_{10}H_{11}NO$
3411 - Butanenitrile, 4-phenoxy-
7305 - 1H-Indole-3-ethanol
7328 - 2H-Indol-2-one, 1,3-dihydro-1,3-dimethyl-
7329 - 2H-Indol-2-one, 1,3-dihydro-3,3-dimethyl-
7330 - 2H-Indol-2-one, 1,3-dihydro-4,7-dimethyl-

$C_{10}H_{11}NO_2$
26 - Acetamide, N-acetyl-N-phenyl-
3207 - Butanamide, 3-oxo-N-phenyl-
6114 - 2-Furanmethanamine, N-(2-furanylmethyl)-

$C_{10}H_{11}NO_3$
224 - Acetic acid, oxo(phenylamino)-, ethyl ester

327 - DL-Alanine, N-benzoyl-
3575 - Butanoic acid, 4-oxo-4-(phenylamino)-

$C_{10}H_{11}N_3$
10813 - 1H-Pyrazol-4-amine, 3-methyl-1-phenyl-
10814 - 1H-Pyrazol-5-amine, 3-methyl-1-phenyl-

$C_{10}H_{11}N_3O_2S_2$
2200 - Benzenesulfonamide, 4-amino-N-(4-methyl-2-thiazolyl)-

$C_{10}H_{11}N_3O_3S$
2198 - Benzenesulfonamide, 4-amino-N-(5-methyl-3-isoxazolyl)-

$C_{10}H_{12}$
1109 - Benzene, 1-butenyl-, (E)-
1110 - Benzene, 1-butenyl-, (Z)-
1111 - Benzene, 2-butenyl-
1112 - Benzene, 3-butenyl-
1245 - Benzene, cyclobutyl-
1626 - Benzene, 1-ethenyl-2,4-dimethyl-
1627 - Benzene, 1-ethenyl-3,5-dimethyl-
1628 - Benzene, 2-ethenyl-1,4-dimethyl-
1629 - Benzene, 4-ethenyl-1,2-dimethyl-
1630 - Benzene, 1-ethenyl-2-ethyl-
1631 - Benzene, 1-ethenyl-3-ethyl-
1632 - Benzene, 1-ethenyl-4-ethyl-
1955 - Benzene, (1-methylcyclopropyl)-
1968 - Benzene, (1-methylenepropyl)-
1984 - Benzene, 1-methyl-2-(1-methylethenyl)-
1985 - Benzene, 1-methyl-3-(1-methylethenyl)-
1986 - Benzene, 1-methyl-4-(1-methylethenyl)-
2023 - Benzene, 1-methyl-2-(2-propenyl)-
2024 - Benzene, (2-methyl-1-propenyl)-
2025 - Benzene, (1-methyl-1-propenyl)-, (E)-
2029 - Benzene, (1-methyl-1-propenyl)-, (Z)-
4746 - 1,3,5,7-Cyclooctatetraene, 1-ethyl-
7255 - 1H-Indene, 2,3-dihydro-1-methyl-
7256 - 1H-Indene, 2,3-dihydro-2-methyl-
7257 - 1H-Indene, 2,3-dihydro-4-methyl-
7258 - 1H-Indene, 2,3-dihydro-5-methyl-
7577 - 4,7-Methano-1H-indene, 3a,4,7,7a-tetrahydro-, (3aα,4α,7α,7aα)-
7979 - Naphthalene, 1,2,3,4-tetrahydro-

$C_{10}H_{12}Br_2O$
9281 - Phenol, 2,4-dibromo-3-methyl-6-(1-methylethyl)-

$C_{10}H_{12}ClNO_2$
3939 - Carbamic acid, (3-chlorophenyl)-, 1-methylethyl ester

$C_{10}H_{12}ClNO_4$
10001 - 1,2-Propanediol, 3-(4-chlorophenoxy)-, 1-carbamate

$C_{10}H_{12}ClN_3O_3S$
11197 - 6-Quinazolinesulfonamide, 7-chloro-2-ethyl-1,2,3,4-tetrahydro-4-oxo-

$C_{10}H_{12}N_2$
2381 - 1H-Benzimidazole, 1-ethyl-2-methyl-
3105 - 2,3'-Bipyridine, 1,2,3,6-tetrahydro-, (S)-
7198 - 1H-Imidazole, 4,5-dihydro-2-(phenylmethyl)-
7302 - 1H-Indole-3-ethanamine

$C_{10}H_{12}N_2O_3$
1102 - Benzenebutanoic acid, α,2-diamino-γ-oxo-
11091 - 2,4,6(1H,3H,5H)-Pyrimidinetrione, 5,5-di-2-propenyl-

$C_{10}H_{12}N_2O_3S$
518 - Bentazon

$C_{10}H_{12}N_2O_4$
2282 - Benzene, 1,2,4,5-tetramethyl-3,6-dinitro-

$C_{10}H_{12}N_2O_5$
5183 - Dinoseb
9328 - Phenol, 2-(1,1-dimethylethyl)-4,6-dinitro-

$C_{10}H_{12}N_3O_3PS_2$
497 - Azinphos-methyl

$C_{10}H_{12}N_4OS$
34 - Acetamide, N-[4-[[(aminothioxomethyl)hydrazono]methyl]phenyl]-

$C_{10}H_{12}N_4O_5$
7335 - Inosine

$C_{10}H_{12}O$
590 - Benzaldehyde, 4-(1-methylethyl)-
599 - Benzaldehyde, 2,4,6-trimethyl-
1651 - Benzene, (2-ethoxyethenyl)-
1869 - Benzenemethanol, α-ethenyl-α-methyl-
1911 - Benzenemethanol, α-2-propenyl-
1941 - Benzene, 1-methoxy-2-(1-propenyl)-
1942 - Benzene, 1-methoxy-4-(1-propenyl)-
1943 - Benzene, 1-methoxy-4-(2-propenyl)-
1944 - Benzene, 1-methoxy-4-(1-propenyl)-, (E)-
2026 - Benzene, 1-methyl-2-(2-propenyloxy)-
2027 - Benzene, 1-methyl-3-(2-propenyloxy)-
2028 - Benzene, 1-methyl-4-(2-propenyloxy)-
2115 - Benzenepropanal, 2-methyl-
2116 - Benzenepropanal, 4-methyl-
2157 - Benzenepropanol, γ-methylene-
2778 - 2H-1-Benzopyran, 3,4-dihydro-2-methyl-
2779 - 2H-1-Benzopyran, 3,4-dihydro-4-methyl-
2780 - 2H-1-Benzopyran, 3,4-dihydro-6-methyl-
2781 - 2H-1-Benzopyran, 3,4-dihydro-7-methyl-
3676 - 1-Butanone, 1-phenyl-
3677 - 2-Butanone, 1-phenyl-
3678 - 2-Butanone, 4-phenyl-
3872 - 2-Buten-1-ol, 3-phenyl-
5809 - Ethanone, 1-(2,4-dimethylphenyl)-
5810 - Ethanone, 1-(2,5-dimethylphenyl)-
5811 - Ethanone, 1-(3,4-dimethylphenyl)-
8041 - 1-Naphthalenol, 1,2,3,4-tetrahydro-
8042 - 1-Naphthalenol, 5,6,7,8-tetrahydro-
8043 - 2-Naphthalenol, 1,2,3,4-tetrahydro-
8044 - 2-Naphthalenol, 5,6,7,8-tetrahydro-
9427 - Phenol, 2-methyl-6-(2-propenyl)-
9428 - Phenol, 4-methyl-2-(1-propenyl)-
9429 - Phenol, 4-methyl-2-(2-propenyl)-
10466 - 1-Propanone, 1-(2-methylphenyl)-
10467 - 1-Propanone, 1-(3-methylphenyl)-
10468 - 1-Propanone, 1-(4-methylphenyl)-
10469 - 1-Propanone, 2-methyl-1-phenyl-

$C_{10}H_{12}O_2$
232 - Acetic acid, 2-phenylethyl ester
897 - Benzeneacetic acid, α-ethyl-
899 - Benzeneacetic acid, ethyl ester
1100 - Benzenebutanoic acid
2137 - Benzenepropanoic acid, α-methyl-
2138 - Benzenepropanoic acid, β-methyl-, (±)-
2139 - Benzenepropanoic acid, methyl ester
2431 - 1,3-Benzodioxole, 5-propyl-
2589 - Benzoic acid, 2,4-dimethyl-, methyl ester
2688 - Benzoic acid, 2-(1-methylethyl)-
2689 - Benzoic acid, 4-(1-methylethyl)-
2690 - Benzoic acid, 2-methyl-, ethyl ester
2691 - Benzoic acid, 3-methyl-, ethyl ester
2692 - Benzoic acid, 4-methyl-, ethyl ester
2693 - Benzoic acid, 1-methylethyl ester
2723 - Benzoic acid, o-propyl-
2724 - Benzoic acid, propyl ester
3584 - Butanoic acid, phenyl ester
3665 - 1-Butanone, 1-(2-hydroxyphenyl)-
3666 - 1-Butanone, 2-hydroxy-1-phenyl-
3667 - 1-Butanone, 4-hydroxy-1-phenyl-
4241 - 1,3,6-Cycloheptatriene-1-carboxylic acid, 5,5-dimethyl-
4267 - 2,5-Cyclohexadiene-1,4-dione, 2-methyl-5-(1-methylethyl)-
4270 - 2,5-Cyclohexadiene-1,4-dione, 2,3,5,6-tetramethyl-
5195 - 1,3-Dioxane, 2-phenyl-
5196 - 1,3-Dioxane, 2-phenyl-
5814 - Ethanone, 2-ethoxy-1-phenyl-
9313 - Phenol, 2,4-dimethyl-, acetate
9314 - Phenol, 2,5-dimethyl-, acetate
9315 - Phenol, 3,4-dimethyl-, acetate
9316 - Phenol, 3,5-dimethyl-, acetate
9388 - Phenol, 2-methoxy-4-(1-propenyl)-
9389 - Phenol, 2-methoxy-4-(2-propenyl)-
9390 - Phenol, 2-methoxy-5-(2-propenyl)-
9391 - Phenol, 2-methoxy-6-(1-propenyl)-

9392 - Phenol, 2-methoxy-6-(2-propenyl)-
9396 - Phenol, 2-methoxy-4-(1-propenyl)-, (E)-
9397 - Phenol, 2-methoxy-4-(1-propenyl)-, (Z)-
10444 - 1-Propanone, 1-(2-hydroxy-3-methylphenyl)-
10445 - 1-Propanone, 1-(2-hydroxy-4-methylphenyl)-
10446 - 1-Propanone, 1-(2-hydroxy-5-methylphenyl)-
10456 - 1-Propanone, 1-(3-methoxyphenyl)-
10457 - 1-Propanone, 2-methoxy-1-phenyl-
10458 - 1-Propanone, 3-methoxy-1-phenyl-
10459 - 2-Propanone, 1-(2-methoxyphenyl)-
10460 - 2-Propanone, 1-(3-methoxyphenyl)-
10461 - 2-Propanone, 1-(4-methoxyphenyl)-

$C_{10}H_{12}O_3$
229 - Acetic acid, phenoxy-, ethyl ester
557 - Benzaldehyde, 4-ethoxy-3-methoxy-
869 - Benzeneacetaldehyde, 3,4-dimethoxy-
907 - Benzeneacetic acid, α-hydroxy-, ethyl ester, (±)-
909 - Benzeneacetic acid, α-hydroxy-, ethyl ester
1892 - Benzenemethanol, 4-methoxy-, acetate
2635 - Benzoic acid, 2-hydroxy-, 1-methylethyl ester
2652 - Benzoic acid, 2-hydroxy-, propyl ester
2653 - Benzoic acid, 4-hydroxy-, propyl ester
2669 - Benzoic acid, 2-methoxy-, ethyl ester
2670 - Benzoic acid, 3-methoxy-, ethyl ester
2671 - Benzoic acid, 4-methoxy-, ethyl ester
3580 - Butanoic acid, 2-phenoxy-
5209 - 1,3-Dioxolane-4-methanol, 2-phenyl-
5805 - Ethanone, 1-(3,4-dimethoxyphenyl)-
9370 - Phenol, 4-(3-hydroxy-1-propenyl)-2-methoxy-
10642 - 2-Propenoic acid, 3-(2-furanyl)-, propyl ester

$C_{10}H_{12}O_4$
2574 - Benzoic acid, 2,4-dihydroxy-6-methyl-, ethyl ester
2579 - Benzoic acid, 3,4-dimethoxy-, methyl ester
2626 - Benzoic acid, 4-hydroxy-3-methoxy-, ethyl ester
3781 - 2-Butenedioic acid (E)-, di-2-propenyl ester
3782 - 2-Butenedioic acid (Z)-, di-2-propenyl ester
5371 - 4,7-Epoxyisobenzofuran-1,3-dione, hexahydro-3a,7a-dimethyl-, (3aα,4β,7β,7aα)-

$C_{10}H_{12}O_5$
2732 - Benzoic acid, 3,4,5-trihydroxy-, propyl ester
2733 - Benzoic acid, 2,4,5-trimethoxy-
2734 - Benzoic acid, 3,4,5-trimethoxy-
6082 - 3,4-Furandicarboxylic acid, diethyl ester

$C_{10}H_{13}Br$
1023 - Benzene, 1-bromo-4-(1,1-dimethylethyl)-
1065 - Benzene, 2-bromo-1-methyl-4-(1-methylethyl)-

$C_{10}H_{13}BrN_2O_3$
11078 - 2,4,6(1H,3H,5H)-Pyrimidinetrione, 5-(2-bromo-2-propenyl)-5-(1-methylethyl)-

$C_{10}H_{13}BrO$
1003 - Benzene, (4-bromobutoxy)-

$C_{10}H_{13}Cl$
1142 - Benzene, 1-chloro-4-(1,1-dimethylethyl)-
1143 - Benzene, (2-chloro-1,1-dimethylethyl)-
1201 - Benzene, 2-chloro-1-methyl-4-(1-methylethyl)-
1202 - Benzene, 2-chloro-4-methyl-1-(1-methylethyl)-

$C_{10}H_{13}ClN_2$
7564 - Methanimidamide, N'-(4-chloro-2-methylphenyl)-N,N-dimethyl-

$C_{10}H_{13}ClN_2O_3S$
2208 - Benzenesulfonamide, 4-chloro-N-[(propylamino)carbonyl]-

$C_{10}H_{13}ClN_6$
9759 - Procyazine

$C_{10}H_{13}ClO$
1598 - Benzeneethanol, 4-chloro-α,α-dimethyl-
9268 - Phenol, 4-chloro-5-methyl-2-(1-methylethyl)-

$C_{10}H_{13}Cl_2O_3PS$
9581 - Phosphorothioic acid, O-(2,4-dichlorophenyl) O,O-diethyl ester

$C_{10}H_{13}I$
1781 - Benzene, 2-iodo-1-methyl-4-(1-methylethyl)-
1782 - Benzene, 2-iodo-4-methyl-1-(1-methylethyl)-

$C_{10}H_{13}N$
7731 - 1-Naphthalenamine, 5,6,7,8-tetrahydro-
10996 - Pyridine, 2-(3-pentenyl)-
11171 - Pyrrolidine, 1-phenyl-
11299 - Quinoline, 1,2,3,4-tetrahydro-1-methyl-

$C_{10}H_{13}NO$
56 - Acetamide, N-(2,4-dimethylphenyl)-
62 - Acetamide, N-ethyl-N-phenyl-
75 - Acetamide, N-methyl-N-(2-methylphenyl)-
627 - Benzamide, N,N,4-trimethyl-
881 - Benzeneacetamide, α-ethyl-
3208 - Butanamide, N-phenyl-
5806 - Ethanone, 1-[3-(dimethylamino)phenyl]-
11298 - Quinoline, 1,2,3,4-tetrahydro-6-methoxy-

$C_{10}H_{13}NO_2$
59 - Acetamide, N-(2-ethoxyphenyl)-
60 - Acetamide, N-(4-ethoxyphenyl)-
1467 - Benzene, 1-(1,1-dimethylethyl)-3-nitro-
1991 - Benzene, 1-methyl-4-(1-methylethyl)-2-nitro-
2128 - Benzenepropanol, β-(aminomethyl)-
2149 - Benzenepropanol, carbamate
2507 - Benzoic acid, 4-amino-, propyl ester
2680 - Benzoic acid, 2-(methylamino)-, ethyl ester
3205 - Butanamide, N-(4-hydroxyphenyl)-
3966 - Carbamic acid, phenyl-, 1-methylethyl ester
3967 - Carbamic acid, (phenylmethyl)-, ethyl ester
6271 - Glycine, N-phenyl-, ethyl ester
10898 - 2-Pyridinecarboxylic acid, 5-butyl-

$C_{10}H_{13}NO_3$
2034 - Benzene, 1-(2-methylpropoxy)-2-nitro-
2672 - Benzoic acid, 3-methoxy-2-(methylamino)-, methyl ester
11933 - L-Tyrosine, N-methyl-

$C_{10}H_{13}NO_4$
10188 - 1,2,3-Propanetriol, 1-(4-aminobenzoate)
11931 - L-Tyrosine, 3-hydroxy-α-methyl-

$C_{10}H_{13}N_3O_4$
3894 - N-sec-Butyl-2,6-dinitrobenzene

$C_{10}H_{13}N_5O_4$
319 - Adenosine

$C_{10}H_{13}N_5O_5$
6289 - Guanosine

$C_{10}H_{14}$
1118 - Benzene, butyl-
1413 - Benzene, 1,2-diethyl-
1414 - Benzene, 1,3-diethyl-
1415 - Benzene, 1,4-diethyl-
1459 - Benzene, (1,1-dimethylethyl)-
1670 - Benzene, 1-ethyl-2,3-dimethyl-
1671 - Benzene, 1-ethyl-2,4-dimethyl-
1672 - Benzene, 1-ethyl-3,5-dimethyl-
1673 - Benzene, 2-ethyl-1,3-dimethyl-
1674 - Benzene, 2-ethyl-1,4-dimethyl-
1675 - Benzene, 4-ethyl-1,2-dimethyl-
1988 - Benzene, 1-methyl-2-(1-methylethyl)-
1989 - Benzene, 1-methyl-3-(1-methylethyl)-
1990 - Benzene, 1-methyl-4-(1-methylethyl)-
2035 - Benzene, 1-methylpropyl
2036 - Benzene, 1-methyl-2-propyl-
2037 - Benzene, 1-methyl-3-propyl-
2038 - Benzene, 1-methyl-4-propyl-
2039 - Benzene, (2-methylpropyl)-
2278 - Benzene, 1,2,3,4-tetramethyl-
2279 - Benzene, 1,2,3,5-tetramethyl-
2280 - Benzene, 1,2,4,5-tetramethyl-
7897 - Naphthalene, 1,2,3,4,4a,8a-hexahydro-
8340 - 2,6-Octadien-4-yne, 3,6-dimethyl-
8345 - 3,5-Octadiyne, 2,7-dimethyl-

$C_{10}H_{14}BeO_4$
2912 - Beryllium, bis(2,4-pentanedionato-O,O')-, (T-4)-

$C_{10}H_{14}BrN$
642 - Benzenamine, 4-bromo-N,N-diethyl-

$C_{10}H_{14}Br_2O$
2953 - Bicyclo[2.2.1]heptan-2-one, 3,3-dibromo-1,7,7-trimethyl-, (1R)-

$C_{10}H_{14}ClN$
661 - Benzenamine, 4-chloro-N,N-diethyl-

$C_{10}H_{14}Cl_6N_4O_2$
11902 - Triforine

$C_{10}H_{14}NO_5PS$
9583 - Phosphorothioic acid, O,O-diethyl O-(4-nitrophenyl) ester

$C_{10}H_{14}N_2$
3189 - Butanal, phenylhydrazone
9620 - Piperazine, 1-phenyl-
10992 - Pyridine, 3-(1-methyl-2-pyrrolidinyl)-, (S)-
11008 - Pyridine, 3-(2-piperidinyl)-, (S)-
11009 - Pyridine, 4-(4-piperidinyl)-

$C_{10}H_{14}N_2O$
699 - Benzenamine, N,N-diethyl-4-nitroso-
10891 - 3-Pyridinecarboxamide, N,N-diethyl-
10892 - 4-Pyridinecarboxamide, N,N-diethyl-

$C_{10}H_{14}N_2O_2$
698 - Benzenamine, N,N-diethyl-4-nitro-

$C_{10}H_{14}N_2O_2S$
11064 - 4,6(1H,5H)-Pyrimidinedione, 5-ethyldihydro-5-(2-methyl-2-propenyl)-2-thioxo-

$C_{10}H_{14}N_2O_3$
11098 - 2,4,6(1H,3H,5H)-Pyrimidinetrione, 5-(1-methylethyl)-5-(2-propenyl)-

$C_{10}H_{14}N_2O_5$
11816 - Thymidine

$C_{10}H_{14}N_4O_4$
10761 - 1H-Purine-2,6-dione, 7-(2,3-dihydroxypropyl)-3,7-dihydro-1,3-dimethyl-

$C_{10}H_{14}N_5O_7P$
320 - 5'-Adenylic acid

$C_{10}H_{14}O$
1114 - Benzene, butoxy-
1458 - Benzene, (1,1-dimethylethoxy)-
1600 - Benzeneethanol, α,α-dimethyl-
1602 - Benzeneethanol, β-ethyl-
1603 - Benzeneethanol, 2-ethyl-
1604 - Benzeneethanol, 4-ethyl-
1653 - Benzene, 1-ethoxy-4-ethyl-
1874 - Benzenemethanol, α-ethyl-α-methyl-, (R)-
1901 - Benzenemethanol, α-(1-methylethyl)-
1902 - Benzenemethanol, 4-(1-methylethyl)-
1912 - Benzenemethanol, α-propyl-, (R)-
1945 - Benzene, 1-methoxy-4-propyl-
1987 - Benzene, 1-methyl-3-(1-methylethoxy)-
2030 - Benzene, (1-methylpropoxy)-
2031 - Benzene, (2-methylpropoxy)-
2032 - Benzene, 1-methyl-4-propoxy-
2155 - Benzenepropanol, α-methyl-
2156 - Benzenepropanol, γ-methyl-
2465 - Benzofuran, 4,5,6,7-tetrahydro-3,6-dimethyl-
2956 - Bicyclo[3.1.1]heptan-3-one, 6,6-dimethyl-2-methylene-, (1S)-
2977 - Bicyclo[3.1.1]hept-2-en-6-one, 2,7,7-trimethyl-
2978 - Bicyclo[3.1.1]hept-3-en-2-one, 4,6,6-trimethyl-, (1R)-
2987 - Bicyclo[3.1.0]hex-3-en-2-one, 4-methyl-1-(1-methylethyl)-
4220 - 2,4-Cycloheptadien-1-one, 2,6,6-trimethyl-
4250 - 1,3-Cyclohexadiene-1-carboxaldehyde, 2,6,6-trimethyl-
4633 - 1-Cyclohexene-1-carboxaldehyde, 4-(1-methylethenyl)-
4634 - 1-Cyclohexene-1-carboxaldehyde, 4-(1-methylethenyl)-, (R)-
4715 - 2-Cyclohexen-1-one, 2-methyl-5-(1-methylethenyl)-, (S)-
4718 - 2-Cyclohexen-1-one, 3-methyl-6-(1-methylethylidene)-
4881 - Cyclopentanone, 2-cyclopentylidene-
6133 - Furan, 3-(4-methyl-3-pentenyl)-
9246 - Phenol, 2-butyl-
9247 - Phenol, 3-butyl-
9248 - Phenol, 4-butyl-
9321 - Phenol, 2-(1,1-dimethylethyl)-

9322 - Phenol, 3-(1,1-dimethylethyl)-
9323 - Phenol, 4-(1,1-dimethylethyl)-
9414 - Phenol, 2-methyl-4-(1-methylethyl)-
9415 - Phenol, 2-methyl-5-(1-methylethyl)-
9416 - Phenol, 2-methyl-6-(1-methylethyl)-
9417 - Phenol, 4-methyl-2-(1-methylethyl)-
9418 - Phenol, 5-methyl-2-(1-methylethyl)-
9430 - Phenol, 2-(1-methylpropyl)-
9431 - Phenol, 4-(1-methylpropyl)-
9432 - Phenol, 4-(2-methylpropyl)-
9473 - Phenol, 2,3,4,5-tetramethyl-
9474 - Phenol, 2,3,4,6-tetramethyl-
9475 - Phenol, 2,3,5,6-tetramethyl-

$C_{10}H_{14}O_2$
1406 - Benzene, 1,2-diethoxy-
1407 - Benzene, 1,3-diethoxy-
1408 - Benzene, 1,4-diethoxy-
1516 - 1,2-Benzenediol, 4-(1,1-dimethylethyl)-
1536 - 1,2-Benzenediol, 3-methyl-6-(1-methylethyl)-
1538 - 1,4-Benzenediol, 2-methyl-5-(1-methylethyl)-
1539 - 1,3-Benzenediol, 4-(2-methylpropyl)-
1552 - 1,4-Benzenediol, 2,3,5,6-tetramethyl-
1652 - Benzene, (2-ethoxyethoxy)-
1873 - Benzenemethanol, α-ethyl-4-methoxy-
2154 - Benzenepropanol, 4-methoxy-
2926 - Bicyclo[2.2.1]heptane-2,3-dione, 1,7,7-trimethyl-, (1R)-
3356 - 1,2-Butanediol, 2-phenyl-
3357 - 1,3-Butanediol, 1-phenyl-
3358 - 1,3-Butanediol, 3-phenyl-
4758 - Cyclopenta[c]pyran-1(4aH)-one, 5,6,7,7a-tetrahydro-4,7-dimethyl-
6076 - Furan, 2-(cyclohexyloxy)-
9245 - Phenol, 2-butoxy-
10400 - 1-Propanol, 3-(phenylmethoxy)-
11862 - Tricyclo[2.2.1.0²,⁶]heptane-1-carboxylic acid, 7,7-dimethyl-

$C_{10}H_{14}O_3$
1599 - Benzeneethanol, 2,5-dimethoxy-
2153 - Benzenepropanol, 4-hydroxy-3-methoxy-
6027 - 2-Furanacetic acid, butyl ester
6068 - 2-Furancarboxylic acid, pentyl ester
8559 - 3-Oxabicyclo[3.2.1]octane-2,4-dione, 1,8,8-trimethyl-, (±)-
8880 - Pentanoic acid, 2-furanylmethyl ester
10019 - 1,2-Propanediol, 3-(2-methylphenoxy)-

$C_{10}H_{14}O_3S$
2219 - Benzenesulfonic acid, sec-butyl ester
2233 - Benzenesulfonic acid, 4-methyl-, propyl ester

$C_{10}H_{14}O_4$
1914 - Benzenemethanol, 3,4,5-trimethoxy-
4646 - 1-Cyclohexene-1,4-dicarboxylic acid, dimethyl ester
10016 - 1,2-Propanediol, 3-(2-methoxyphenoxy)-
10656 - 2-Propenoic acid, 2-methyl-, 1,2-ethanediyl ester

$C_{10}H_{15}BrO$
2951 - Bicyclo[2.2.1]heptan-2-one, 1-(bromomethyl)-7,7-dimethyl-, (1S)-
2952 - Bicyclo[2.2.1]heptan-2-one, 3-bromo-1,7,7-trimethyl-, (1R-endo)-

$C_{10}H_{15}N$
655 - Benzenamine, 4-butyl-
656 - Benzenamine, N-butyl-
694 - Benzenamine, 2,6-diethyl-
695 - Benzenamine, N,N-diethyl-
714 - Benzenamine, 2-(1,1-dimethylethyl)-
715 - Benzenamine, 4-(1,1-dimethylethyl)-
716 - Benzenamine, N-(1,1-dimethylethyl)-
793 - Benzenamine, 2-methyl-5-(1-methylethyl)-
794 - Benzenamine, 4-methyl-N-(1-methylethyl)-
812 - Benzenamine, 2-(1-methylpropyl)-
813 - Benzenamine, 3-(1-methylpropyl)-
814 - Benzenamine, 4-(2-methylpropyl)-
815 - Benzenamine, 4-methyl-N-propyl-
816 - Benzenamine, N-(1-methylpropyl)-
817 - Benzenamine, N-(2-methylpropyl)-

847 - Benzenamine, 2,3,4,5-tetramethyl-
848 - Benzenamine, 2,3,4,6-tetramethyl-
849 - Benzenamine, N,N,2,6-tetramethyl-
1572 - Benzeneethanamine, α,α-dimethyl-
1573 - Benzeneethanamine, N,α-dimethyl-, (S)-
1574 - Benzeneethanamine, N,β-dimethyl-
1819 - Benzenemethanamine, 4-(1-methylethyl)-
1820 - Benzenemethanamine, N-(1-methylethyl)-
1830 - Benzenemethanamine, α-propyl-
2119 - Benzenepropanamine, α-methyl-
10967 - Pyridine, 2-(1-ethylpropyl)-
10968 - Pyridine, 4-(1-ethylpropyl)-

$C_{10}H_{15}NO$
1840 - Benzenemethanol, α-(1-aminopropyl)-
1900 - Benzenemethanol, α-[1-(methylamino)ethyl]-, (R*,S*)-(±)-
9297 - Phenol, 3-(diethylamino)-
9320 - Phenol, 4-[2-(dimethylamino)ethyl]-
9403 - Phenol, 4-[2-(methylamino)propyl]-
10983 - Pyridine, 2-[2-(1-methylethoxy)ethyl]-

$C_{10}H_{15}NO_2$
693 - Benzenamine, 3,4-diethoxy-
1571 - Benzeneethanamine, 3,4-dimethoxy-
5760 - Ethanol, 2,2'-(phenylimino)bis-

$C_{10}H_{15}NO_4$
9933 - Propanedioic acid, (2-cyanoethyl)-, diethyl ester
11144 - 3-Pyrrolidineacetic acid, 2-carboxy-4-(1-methylethenyl)-, [2S-(2α,3β,4β)]-

$C_{10}H_{15}N_3$
6281 - Guanidine, N,N'-dimethyl-N''-(phenylmethyl)-

$C_{10}H_{15}OPS_2$
5986 - Fonofos

$C_{10}H_{15}O_3PS_2$
5955 - Fenthion

$C_{10}H_{16}$
2921 - Bicyclo[2.2.1]heptane, 2,2-dimethyl-3-methylene-, (±)-
2922 - Bicyclo[2.2.1]heptane, 2,2-dimethyl-5-methylene-
2923 - Bicyclo[2.2.1]heptane, 7,7-dimethyl-2-methylene-, (±)-
2924 - Bicyclo[3.1.1]heptane, 6,6-dimethyl-2-methylene-, (1R)-
2925 - Bicyclo[3.1.1]heptane, 6,6-dimethyl-2-methylene-
2970 - Bicyclo[2.2.1]hept-2-ene, 1,7,7-trimethyl-
2971 - Bicyclo[3.1.1]hept-2-ene, 2,6,6-trimethyl-, (1S)-
2972 - Bicyclo[3.1.1]hept-2-ene, 2,6,6-trimethyl-, (±)-
2973 - Bicyclo[3.1.1]hept-2-ene, 2,6,6-trimethyl
2974 - Bicyclo[4.1.0]hept-3-ene, 3,7,7-trimethyl-, (1S)-
2981 - Bicyclo[3.1.0]hexane, 4-methylene-1-(1-methylethyl)-, (±)-
2986 - Bicyclo[3.1.0]hex-2-ene, 2-methyl-5-(1-methylethyl)-
4050 - 4-Carene, (1S,3R,6R)-(-)-
4209 - Cyclodecyne
4273 - 1,3-Cyclohexadiene, 1-methyl-4-(1-methylethyl)-
4274 - 1,3-Cyclohexadiene, 2-methyl-5-(1-methylethyl)-, (±)-
4275 - 1,4-Cyclohexadiene, 1-methyl-4-(1-methylethyl)-
4673 - Cyclohexene, 3-methylene-6-(1-methylethyl)-
4674 - Cyclohexene, 3-methylene-6-(1-methylethyl)-, (+)-
4675 - Cyclohexene, 4-methylene-1-(1-methylethyl)-
4676 - Cyclohexene, 1-methyl-3-(1-methylethenyl)-
4677 - Cyclohexene, 1-methyl-4-(1-methylethenyl)-
4678 - Cyclohexene, 1-methyl-5-(1-methylethenyl)-
4679 - Cyclohexene, 5-methyl-1-(1-methylethenyl)-, (±)-
4684 - Cyclohexene, 1-methyl-4-(1-methylethylidene)-
5094 - 1-Decen-3-yne

5095 - 1-Decen-4-yne
5096 - 2-Decen-4-yne
8330 - 1,6-Octadiene, 7-methyl-3-methylene-
8485 - 1,3,6-Octatriene, 3,7-dimethyl-
8486 - 1,3,7-Octatriene, 3,7-dimethyl-
8487 - 2,4,6-Octatriene, 2,6-dimethyl-, (E,E)-
8488 - 2,4,6-Octatriene, 2,6-dimethyl-, (E,Z)-
11815 - 4(10)-Thujene, (+)-
11860 - Tricyclo[3.3.1.1³,⁷]decane
11866 - Tricyclo[2.2.1.0²,⁶]heptane, 1,3,3-trimethyl-
11867 - Tricyclo[2.2.1.0²,⁶]heptane, 1,7,7-trimethyl-

$C_{10}H_{16}ClNO$
865 - Benzenaminium, N-ethyl-3-hydroxy-N,N-dimethyl-, chloride

$C_{10}H_{16}Cl_2O_2$
5049 - Decanedioyl dichloride

$C_{10}H_{16}Cl_3NOS$
11823 - Triallate

$C_{10}H_{16}NO_5PS_2$
5946 - Famphur

$C_{10}H_{16}N_2$
1265 - 1,3-Benzenediamine, N,N-diethyl-
1266 - 1,4-Benzenediamine, N,N-diethyl-
1286 - 1,2-Benzenediamine, N,N,N',N'-tetramethyl-
1287 - 1,4-Benzenediamine, N,N,N',N'-tetramethyl-
5041 - Decanedinitrile

$C_{10}H_{16}N_2OS$
7229 - 4-Imidazolidinone, 5-(2-methylpropyl)-3-(2-propenyl)-2-thioxo-

$C_{10}H_{16}N_2O_2S$
11065 - 4,6(1H,5H)-Pyrimidinedione, 5-ethyldihydro-5-(1-methylpropyl)-2-thioxo-

$C_{10}H_{16}N_2O_3$
11082 - 2,4,6(1H,3H,5H)-Pyrimidinetrione, 5-butyl-5-ethyl-

$C_{10}H_{16}N_2O_3S$
11689 - 1H-Thieno[3,4-d]imidazole-4-pentanoic acid, hexahydro-2-oxo-, [3aS-(3aα,4β,6aα)]-

$C_{10}H_{16}N_2O_8$
6262 - Glycine, N,N'-1,2-ethanediylbis[N-(carboxymethyl)-

$C_{10}H_{16}N_4O_7$
11107 - 4(1H)-Pyrimidinone, 2,6-diamino-5-(β-D-glucopyranosyloxy)-

$C_{10}H_{16}O$
2960 - Bicyclo[2.2.1]heptan-2-one, 1,3,3-trimethyl-, (±)-
2961 - Bicyclo[2.2.1]heptan-2-one, 1,7,7-trimethyl-, (1R)-
2963 - Bicyclo[2.2.1]heptan-2-one, 4,7,7-trimethyl-, (1S)-
2964 - Bicyclo[3.1.1]heptan-2-one, 4,6,6-trimethyl-, [1R-(1α,4α,5α)]-
2965 - Bicyclo[3.1.1]heptan-3-one, 2,6,6-trimethyl-, [1S-(1α,2β,5α)]-
2975 - Bicyclo[3.1.1]hept-3-en-2-ol, 4,6,6-trimethyl-, (1α,2β,5α)-
2976 - Bicyclo[3.1.1]hept-3-en-2-ol, 4,6,6-trimethyl-, (1α,2α,5α)-
2983 - Bicyclo[3.1.0]hexan-3-ol, 4-methylene-1-(1-methylethyl)-
2984 - Bicyclo[3.1.0]hexan-3-ol, 4-methylene-1-(1-methylethyl)-, [1S-(1α,3β,5α)]-
2995 - [1,1'-Bicyclopentyl]-2-one
4576 - Cyclohexanone, 2-butylidene-
4601 - Cyclohexanone, 2-methyl-5-(1-methylethenyl)-, (2S-trans)-
4602 - Cyclohexanone, 5-methyl-2-(1-methylethenyl)-, trans-
4603 - Cyclohexanone, 5-methyl-2-(1-methylethenyl)-, trans-
4607 - Cyclohexanone, 5-methyl-2-(1-methylethylidene)-
4608 - Cyclohexanone, 5-methyl-2-(1-methylethylidene)-, (R)-
4635 - 1-Cyclohexene-1-carboxaldehyde, 4-(1-methylethyl)-

4636 - 1-Cyclohexene-1-carboxaldehyde, 2,6,6-
 trimethyl-
4664 - 1-Cyclohexene-1-methanol, 4-(1-
 methylethenyl)-
4696 - 2-Cyclohexen-1-ol, 2-methyl-5-(1-
 methylethyl)-
4716 - 2-Cyclohexen-1-one, 3-methyl-6-(1-
 methylethyl)-, (±)-
4717 - 2-Cyclohexen-1-one, 6-methyl-3-(1-
 methylethyl)-, (±)-
8053 - 1(2H)-Naphthalenone, octahydro-, cis-
8054 - 1(2H)-Naphthalenone, octahydro-, trans-
8055 - 2(1H)-Naphthalenone, octahydro-
8317 - 2,6-Octadienal, 3,7-dimethyl-, (E)-
8318 - 2,6-Octadienal, 3,7-dimethyl-, (Z)-
8561 - 6-Oxabicyclo[3.2.1]oct-3-ene, 4,7,7-trimethyl-,
 (±)-

$C_{10}H_{16}O_2$
2957 - Bicyclo[2.2.1]heptan-2-one, 3-hydroxy-1,7,7-
 trimethyl-
3834 - 2-Butenoic acid, 3-cyclohexyl-
4203 - 1,2-Cyclodecanedione
4307 - Δ1,α-Cyclohexaneacetic acid, 3,3-dimethyl-
4709 - 2-Cyclohexen-1-one, 2-hydroxy-3-methyl-6-(1-
 methylethyl)-
4911 - 1-Cyclopentene-1-carboxylic acid, 2,3,3-
 trimethyl-, methyl ester
4959 - Cyclopropanecarboxylic acid, 2,2-dimethyl-3-
 (2-methyl-1-propenyl)-, (1R-cis)-
5025 - 2,4-Decadienoic acid, (E,E)-
5184 - 2,3-Dioxabicyclo[2.2.2]oct-5-ene, 1-methyl-4-
 (1-methylethyl)-
8331 - 2,4-Octadienoic acid, 3,7-dimethyl-
10653 - 2-Propenoic acid, 2-methyl-, cyclohexyl ester

$C_{10}H_{16}O_4$
3766 - 2-Butenedioic acid (E)-, bis(1-methylethyl)
 ester
3783 - 2-Butenedioic acid (E)-, dipropyl ester
3784 - 2-Butenedioic acid (Z)-, dipropyl ester
3931 - Camphoric acid, trans-(±)-
4180 - 1,1-Cyclobutanedicarboxylic acid, diethyl ester
4184 - 1,3-Cyclobutanedicarboxylic acid, 2,2-
 dimethyl-, dimethyl ester
4352 - 1,3-Cyclohexanedicarboxylic acid, dimethyl
 ester, cis-
4353 - 1,3-Cyclohexanedicarboxylic acid, dimethyl
 ester, trans-
4357 - 1,4-Cyclohexanedicarboxylic acid, dimethyl
 ester, cis-
4794 - 1,3-Cyclopentanedicarboxylic acid, 1,2,2-
 trimethyl-, (1R-cis)-
9966 - Propanedioic acid, (1-methylethylidene)-,
 diethyl ester
9980 - Propanedioic acid, 2-propenyl-, diethyl ester

$C_{10}H_{16}O_4S$
2929 - Bicyclo[2.2.1]heptane-1-methanesulfonic acid,
 7,7-dimethyl-2-oxo-, (1S)-

$C_{10}H_{16}O_5$
3295 - Butanedioic acid, acetyl-, diethyl ester
9947 - Propanedioic acid, (ethoxymethylene)-, diethyl
 ester

$C_{10}H_{16}O_6$
7552 - Methanetricarboxylic acid, triethyl ester

$C_{10}H_{16}Si$
11490 - Silane, trimethyl(4-methylphenyl)-
11494 - Silane, trimethyl(phenylmethyl)-

$C_{10}H_{17}Br$
8321 - 2,6-Octadiene, 1-bromo-3,7-dimethyl-, (E)-

$C_{10}H_{17}Cl$
2919 - Bicyclo[2.2.1]heptane, 2-chloro-1,7,7-
 trimethyl-
2920 - Bicyclo[2.2.1]heptane, 2-chloro-1,7,7-
 trimethyl-, endo-
7782 - Naphthalene, 2-chlorodecahydro-

$C_{10}H_{17}ClO_4$
9932 - Propanedioic acid, (3-chloropropyl)-, diethyl
 ester

$C_{10}H_{17}Cl_2NOS$
5106 - Diallate

$C_{10}H_{17}N$
11859 - Tricyclo[3.3.1.13,7]decan-1-amine

$C_{10}H_{17}NO$
2962 - Bicyclo[2.2.1]heptan-2-one, 1,7,7-trimethyl-,
 oxime, (1S)-

$C_{10}H_{17}NO_6$
6258 - Glycine, N,N-bis(ethoxycarbonyl)-, ethyl ester
10104 - Propanenitrile, 2-(β-D-glucopyranosyloxy)-2-
 methyl-

$C_{10}H_{17}N_2O_4PS$
5945 - Etrimfos

$C_{10}H_{17}N_3$
11848 - 4H-1,2,4-Triazole, 4-cyclohexyl-3-ethyl-

$C_{10}H_{17}N_3O_6S$
6264 - Glycine, N-(N-L-γ-glutamyl-L-cysteinyl)-

$C_{10}H_{18}$
2931 - Bicyclo[2.2.1]heptane, 1,3,3-trimethyl-, (+)-
2932 - Bicyclo[2.2.1]heptane, 1,7,7-trimethyl-
2933 - Bicyclo[2.2.1]heptane, 2,2,3-trimethyl-
2934 - Bicyclo[3.1.1]heptane, 2,6,6-trimethyl-,
 (1α,2α,5α)-(±)-
2935 - Bicyclo[3.1.1]heptane, 2,6,6-trimethyl-, [1R-
 (1α,2β,5α)]-
2936 - Bicyclo[4.1.0]heptane, 3,7,7-trimethyl-, [1S-
 (1α,3β,6α)]-
2937 - Bicyclo[4.1.0]heptane, 3,7,7-trimethyl-, [1S-
 (1α,3α,6α)]-
2982 - Bicyclo[3.1.0]hexane, 4-methyl-1-(1-
 methylethyl)-
4207 - Cyclodecene, (E)-
4208 - Cyclodecene, (Z)-
4247 - Cycloheptene, 1,4,4-trimethyl-
4435 - Cyclohexane, 1-methyl-3-(1-methylethenyl)-
4441 - Cyclohexane, 1-methyl-3-
 (1-methylethylidene)-
4442 - Cyclohexane, 1-methyl-4-
 (1-methylethylidene)-
4680 - Cyclohexene, 1-methyl-4-(1-methylethyl)-
4681 - Cyclohexene, 1-methyl-4-(1-methylethyl)-,
 (R)-
4682 - Cyclohexene, 3-methyl-6-(1-methylethyl)-,
 (3R-trans)-
4683 - Cyclohexene, 4-methyl-1-(1-methylethyl)-,
 (R)-
4923 - Cyclopentene, 1,5-dimethyl-4-(1-methylethyl)-
4934 - Cyclopentene, 1-pentyl-
5023 - 1,3-Decadiene
5024 - 1,9-Decadiene
5097 - 1-Decyne
5098 - 3-Decyne
5099 - 4-Decyne
5100 - 5-Decyne
6619 - 3-Heptyne, 5-ethyl-5-methyl-
7429 - p-Menth-3-ene, (S)-(-)-
7789 - Naphthalene, decahydro-
7790 - Naphthalene, decahydro-, cis-
7791 - Naphthalene, decahydro-, trans-
8202 - 4-Nonyne, 8-methyl-
8322 - 2,4-Octadiene, 3,7-dimethyl-
8323 - 2,5-Octadiene, 2,6-dimethyl-
8324 - 2,6-Octadiene, 2,6-dimethyl-, (Z)-
8325 - 2,6-Octadiene, 2,7-dimethyl-
8326 - 2,6-Octadiene, 3,6-dimethyl-
8327 - 2,6-Octadiene, 4,5-dimethyl-
8535 - 3-Octyne, 2,2-dimethyl-
11814 - Thujane, (1S,4R,5S)-(+)-

$C_{10}H_{18}N_2O_3$
7223 - 4-Imidazolidinehexanoic acid, 5-methyl-2-
 oxo-, (4R-cis)-

$C_{10}H_{18}N_2O_4$
9606 - 1,4-Piperazinedicarboxylic acid, diethyl ester

$C_{10}H_{18}N_4O_4S_3$
11713 - Thiodicarb

$C_{10}H_{18}O$
2930 - Bicyclo[3.1.1]heptane-2-methanol, 6,6-
 dimethyl-, [1S-(1α,2β,5α)]-
2940 - Bicyclo[2.2.1]heptan-2-ol, 1,3,3-trimethyl-,
 endo-(±)-
2941 - Bicyclo[2.2.1]heptan-2-ol, 1,3,3-trimethyl-,
 (1S-endo)-
2942 - Bicyclo[2.2.1]heptan-2-ol, 1,7,7-trimethyl-,
 endo-(±)-
2943 - Bicyclo[2.2.1]heptan-2-ol, 1,7,7-trimethyl-,
 exo-(±)-
2944 - Bicyclo[3.1.1]heptan-2-ol, 4,6,6-trimethyl-,
 [1R-(1α,2α,4α,5α)]-
2985 - Bicyclo[3.1.0]hexan-3-ol, 4-methyl-1-(1-
 methylethyl)-
2994 - [1,1'-Bicyclopentyl]-2-ol
3652 - 1-Butanone, 1-cyclohexyl-
4205 - Cyclodecanone
4237 - Cycloheptanone, 2,2,6-trimethyl-
4238 - Cycloheptanone, 2,6,6-trimethyl-
4542 - Cyclohexanol, 1-methyl-4-(1-methylethyl)-
4543 - Cyclohexanol, 2-methyl-5-(1-methylethenyl)-,
 [1S-(1α,2β,5α)]-
4544 - Cyclohexanol, 5-methyl-2-(1-methylethyl)-,
 [1R-(1α,2β,5α)]-
4575 - Cyclohexanone, 2-butyl-
4604 - Cyclohexanone, 2-methyl-5-(1-methylethyl)-,
 (2S-trans)-
4605 - Cyclohexanone, 5-methyl-2-(1-methylethyl)-,
 (2R-cis)-
4606 - Cyclohexanone, 5-methyl-2-(1-methylethyl)-,
 (2S-trans)-
4609 - Cyclohexanone, 4-(2-methylpropyl)-
4665 - 1-Cyclohexene-1-methanol, 2,4,4-trimethyl-
4666 - 1-Cyclohexene-1-methanol, 2,6,6-trimethyl-
4667 - 2-Cyclohexene-1-methanol, 2,6,6-trimethyl-
4668 - 3-Cyclohexene-1-methanol, α,α,4-trimethyl-,
 (±)-
4697 - 2-Cyclohexen-1-ol, 3-methyl-6-(1-
 methylethyl)-
4698 - 2-Cyclohexen-1-ol, 6-methyl-3-(1-
 methylethyl)-
4699 - 3-Cyclohexen-1-ol, 4-methyl-1-(1-
 methylethyl)-
4978 - Cyclopropanemethanol, 2,2-dimethyl-3-(2-
 methyl-1-propenyl)-
5080 - 2-Decenal
5081 - 3-Decenal
5093 - 3-Decen-2-one
6590 - 3-Hepten-2-one, 3-ethyl-4-methyl-
6591 - 4-Hepten-2-one, 3-ethyl-4-methyl-
6592 - 4-Hepten-3-one, 5-ethyl-4-methyl-
6593 - 5-Hepten-2-one, 4-ethyl-6-methyl-
6594 - 5-Hepten-3-one, 5-ethyl-4-methyl-
8024 - 2-Naphthalenol, decahydro-
8191 - 5-Nonen-4-one, 6-methyl-
8192 - 6-Nonen-4-one, 6-methyl-
8332 - 1,6-Octadien-3-ol, 3,7-dimethyl-, (±)-
8333 - 2,6-Octadien-1-ol, 3,7-dimethyl-, (E)-
8334 - 2,6-Octadien-1-ol, 3,7-dimethyl-, (Z)-
8490 - 6-Octenal, 3,7-dimethyl-
8556 - 7-Oxabicyclo[2.2.1]heptane, 1-methyl-4-(1-
 methylethyl)-
8560 - 2-Oxabicyclo[2.2.2]octane, 1,3,3-trimethyl-

$C_{10}H_{18}O_2$
3496 - Butanoic acid, cyclohexyl ester
4206 - Cyclodecanone, 2-hydroxy-
4224 - Cycloheptanecarboxylic acid, ethyl ester
4308 - Cyclohexaneacetic acid, ethyl ester
4337 - Cyclohexanecarboxylic acid, propyl ester
4784 - Cyclopentanecarboxylic acid, 2-methyl-5-(1-
 methylethyl)-
4786 - Cyclopentanecarboxylic acid, 1,2,2,3-
 tetramethyl-
5088 - 2-Decenoic acid
5089 - 3-Decenoic acid

5090 - 4-Decenoic acid
5091 - 9-Decenoic acid
6811 - 3,4-Hexanedione, 2,2,5,5-tetramethyl-
10293 - Propanoic acid, 2-methyl-, cyclohexyl ester
10660 - 2-Propenoic acid, 2-methyl-, hexyl ester

$C_{10}H_{18}O_3$

3504 - Butanoic acid, 2,2-diethyl-3-oxo-, ethyl ester
3542 - Butanoic acid, 3-methyl-, anhydride
4309 - Cyclohexaneacetic acid, α-ethyl-1-hydroxy-
6872 - Hexanoic acid, 2-acetyl-, ethyl ester
8858 - Pentanoic acid, anhydride
11903 - 2,5,7-Trioxabicyclo[2.2.1]heptane,
1,3,3,4,6,6-hexamethyl-

$C_{10}H_{18}O_4$

3321 - Butanedioic acid, dipropyl ester
3512 - Butanoic acid, 1,2-ethanediyl ester
5042 - Decanedioic acid
5557 - Ethanedioic acid, bis(2-methylpropyl) ester
5558 - Ethanedioic acid, dibutyl ester
6785 - Hexanedioic acid, diethyl ester
8135 - Nonanedioic acid, 3-methyl-
8388 - Octanedioic acid, dimethyl ester
8601 - Oxirane, 2,2'-[1,4-butanediylbis
(oxymethylene)]bis-
9964 - Propanedioic acid, (1-methylethyl)-, diethyl
ester
9981 - Propanedioic acid, propyl-, diethyl ester

$C_{10}H_{18}O_4S$

10330 - Propanoic acid, 3,3'-thiobis-, diethyl ester

$C_{10}H_{18}O_6$

3305 - Butanedioic acid, 2,3-dihydroxy-, bis(1-
methylethyl) ester, (R*,R*)-(±)-
3310 - Butanedioic acid, 2,3-dihydroxy- [R-(R*,R*)]-,
dipropyl ester
5714 - Ethanol, 2,2'-[1,2-ethanediylbis(oxy)]bis-,
diacetate

$C_{10}H_{19}Br$

4321 - Cyclohexane, 1-bromo-4-methyl-1-(1-
methylethyl)-
5086 - 1-Decene, 2-bromo-
5087 - 2-Decene, 1-bromo-

$C_{10}H_{19}BrO_2$

5061 - Decanoic acid, 2-bromo-

$C_{10}H_{19}ClO$

5078 - Decanoyl chloride

$C_{10}H_{19}FO_2$

5064 - Decanoic acid, 10-fluoro-

$C_{10}H_{19}N$

2927 - Bicyclo[2.2.1]heptane-1-methanamine, 7,7-
dimethyl-
2928 - Bicyclo[3.1.1]heptane-2-methanamine, 6,6-
dimethyl-, [1S-(1α,2β,5α)]-
5057 - Decanenitrile
7712 - 1-Naphthalenamine, decahydro-

$C_{10}H_{19}NO$

11331 - 2H-Quinolizine-1-methanol, octahydro-, (1R-
trans)-

$C_{10}H_{19}NO_2$

9704 - 1-Piperidinepropanoic acid, ethyl ester

$C_{10}H_{19}N_5O$

3143 - sec-Bumeton

$C_{10}H_{19}N_5S$

9770 - Prometryn
11594 - Terbutryn

$C_{10}H_{19}O_6PS_2$

3316 - Butanedioic acid,
[(dimethoxyphosphinothioyl)thio]-,
diethyl ester

$C_{10}H_{20}$

4202 - Cyclodecane
4227 - Cycloheptane, 1,1,2-trimethyl-
4228 - Cycloheptane, 1,1,4-trimethyl-
4325 - Cyclohexane, butyl-
4369 - Cyclohexane, (1,1-dimethylethyl)-
4436 - Cyclohexane, 1-methyl-2-(1-methylethyl)-
4437 - Cyclohexane, 1-methyl-3-(1-methylethyl)-
4438 - Cyclohexane, 1-methyl-4-(1-methylethyl)-

4439 - Cyclohexane, 1-methyl-4-(1-methylethyl)-, cis-
4440 - Cyclohexane, 1-methyl-4-(1-methylethyl)-,
trans-
4448 - Cyclohexane, (1-methylpropyl)-
4449 - Cyclohexane, 1-methyl-1-propyl-
4450 - Cyclohexane, 1-methyl-2-propyl-
4451 - Cyclohexane, 1-methyl-3-propyl-
4452 - Cyclohexane, 1-methyl-4-propyl-
4453 - Cyclohexane, (2-methylpropyl)-
4477 - Cyclohexane, 1,1,3,4-tetramethyl-
4478 - Cyclohexane, 1,1,3,5-tetramethyl-, cis-
4846 - Cyclopentane, pentyl-
5082 - 1-Decene
5083 - 4-Decene
5084 - 5-Decene, (E)-
5085 - 5-Decene, (Z)-
6559 - 3-Heptene, 4-propyl-
8185 - 1-Nonene, 2-methyl-
8186 - 3-Nonene, 2-methyl-
8502 - 1-Octene, 3,7-dimethyl-
8503 - 2-Octene, 2,6-dimethyl-

$C_{10}H_{20}BrF$

5035 - Decane, 1-bromo-10-fluoro-

$C_{10}H_{20}Br_2$

5038 - Decane, 1,10-dibromo-
5039 - Decane, 5,6-dibromo-

$C_{10}H_{20}ClF$

5037 - Decane, 1-chloro-10-fluoro-

$C_{10}H_{20}Cl_2$

5040 - Decane, 1,10-dichloro-

$C_{10}H_{20}NO_4PS$

10719 - Propetamphos

$C_{10}H_{20}N_2$

3098 - 2,3'-Bipiperidine

$C_{10}H_{20}N_2O_2$

7690 - Morpholine, 4,4'-(1,2-ethanediyl)bis-

$C_{10}H_{20}N_2S_2$

9659 - Piperidine, 1,1'-dithiobis-

$C_{10}H_{20}N_2S_3$

11714 - Thiodicarbonic diamide ([(H2N)C(S)]2S),
tetraethyl-

$C_{10}H_{20}N_2S_4$

11721 - Thioperoxydicarbonic
diamide ([(H2N)C(S)]2S2), tetraethyl-

$C_{10}H_{20}O$

4204 - Cyclodecanol
4233 - Cycloheptanol, 2,6,6-trimethyl-
4324 - Cyclohexanebutanol
4469 - Cyclohexanepropanol, α-methyl-
4504 - Cyclohexanol, 2-butyl-, trans-
4521 - Cyclohexanol, 2-(1,1-dimethylethyl)-
4546 - Cyclohexanol, 1-methyl-4-(1-methylethyl)-
4547 - Cyclohexanol, 5-methyl-2-(1-methylethyl)-,
[1R-(1α,2α,5β)]-
4548 - Cyclohexanol, 5-methyl-2-(1-methylethyl)-,
(1α,2β,5α)-(±)-
4550 - Cyclohexanol, 2-methyl-1-propyl-
4551 - Cyclohexanol, 3-methyl-1-propyl-
5030 - Decanal
5075 - 2-Decanone
5076 - 3-Decanone
5077 - 4-Decanone
5092 - 9-Decen-1-ol
6405 - Heptane, 3-[(ethenyloxy)methyl]-
7428 - m-Menthan-6-ol
8176 - 5-Nonanone, 2-methyl-
8401 - Octane, 1-(ethenyloxy)-
8514 - 6-Octen-1-ol, 3,7-dimethyl-, (±)-
8515 - 6-Octen-3-ol, 3,7-dimethyl-
8516 - 7-Octen-1-ol, 3,7-dimethyl-, (S)-

$C_{10}H_{20}O_2$

187 - Acetic acid, 2-ethylhexyl ester
222 - Acetic acid, octyl ester
3528 - Butanoic acid, hexyl ester
3546 - Butanoic acid, 3-methyl-, 1,1-dimethylpropyl
ester
3551 - Butanoic acid, 3-methyl-, 3-methylbutyl ester

3559 - Butanoic acid, 1-methylpentyl ester, (S)-
5059 - Decanoic acid
6471 - Heptanoic acid, propyl ester
6506 - 2-Heptanol, 6-methyl-, acetate
6507 - 3-Heptanol, 6-methyl-, acetate
6880 - Hexanoic acid, butyl ester
6885 - Hexanoic acid, 2-ethyl-, ethyl ester
6975 - 3-Hexanone, 4-hydroxy-2,2,5,5-tetramethyl-
8153 - Nonanoic acid, 3-methyl-
8154 - Nonanoic acid, methyl ester
8347 - Octanal, 7-hydroxy-3,7-dimethyl-
8428 - Octanoic acid, ethyl ester
8447 - 2-Octanol, acetate, (±)-
8894 - Pentanoic acid, 4-methyl-, 2-methylpropyl
ester
8913 - Pentanoic acid, pentyl ester
10264 - Propanoic acid, heptyl ester

$C_{10}H_{20}O_3$

3811 - 2-Butene, 1,1,3-triethoxy-

$C_{10}H_{20}O_4$

5689 - Ethanol, 2-(2-butoxyethoxy)-, acetate

$C_{10}H_{21}Br$

5034 - Decane, 1-bromo-

$C_{10}H_{21}Cl$

5036 - Decane, 1-chloro-

$C_{10}H_{21}ClO$

5073 - 1-Decanol, 10-chloro-

$C_{10}H_{21}F$

5052 - Decane, 1-fluoro-

$C_{10}H_{21}FO$

5074 - 1-Decanol, 10-fluoro-

$C_{10}H_{21}I$

5053 - Decane, 1-iodo-

$C_{10}H_{21}N$

4285 - Cyclohexanamine, N,N-diethyl-
4385 - Cyclohexaneethanamine, N,α-dimethyl-
9691 - Piperidine, 1,2,2,6,6-pentamethyl-
9692 - Piperidine, 1-pentyl-

$C_{10}H_{21}NO$

4502 - Cyclohexanol, 3-(aminomethyl)-3,5,5-
trimethyl-
5031 - Decanamide

$C_{10}H_{21}NOS$

3976 - Carbamothioic acid, butylethyl-, S-propyl ester
12047 - Vernolate

$C_{10}H_{21}NO_2$

8084 - Nitrous acid, decyl ester

$C_{10}H_{21}NO_3$

8073 - Nitric acid, decyl ester

$C_{10}H_{21}N_3O$

9603 - 1-Piperazinecarboxamide, N,N-diethyl-4-
methyl-

$C_{10}H_{22}$

5033 - Decane
6409 - Heptane, 3-ethyl-2-methyl-
6410 - Heptane, 3-ethyl-3-methyl-
6411 - Heptane, 3-ethyl-4-methyl-
6412 - Heptane, 3-ethyl-5-methyl-
6413 - Heptane, 4-ethyl-2-methyl-
6414 - Heptane, 4-ethyl-3-methyl-
6415 - Heptane, 4-ethyl-4-methyl-
6416 - Heptane, 5-ethyl-2-methyl-
6427 - Heptane, 4-(1-methylethyl)-
6434 - Heptane, 4-propyl-
6437 - Heptane, 2,2,3-trimethyl-
6438 - Heptane, 2,2,4-trimethyl-
6439 - Heptane, 2,2,5-trimethyl-
6440 - Heptane, 2,2,6-trimethyl-
6441 - Heptane, 2,3,3-trimethyl-
6442 - Heptane, 2,3,4-trimethyl-
6443 - Heptane, 2,3,5-trimethyl-
6444 - Heptane, 2,3,6-trimethyl-
6445 - Heptane, 2,4,4-trimethyl-
6446 - Heptane, 2,4,5-trimethyl-
6447 - Heptane, 2,4,6-trimethyl-
6448 - Heptane, 2,5,5-trimethyl-
6449 - Heptane, 3,3,4-trimethyl-

6450 - Heptane, 3,3,5-trimethyl-
6451 - Heptane, 3,4,4-trimethyl-
6452 - Heptane, 3,4,5-trimethyl-
6767 - Hexane, 3,3-diethyl-
6768 - Hexane, 3,4-diethyl-
6816 - Hexane, 3-ethyl-2,2-dimethyl-
6817 - Hexane, 3-ethyl-2,3-dimethyl-
6818 - Hexane, 3-ethyl-2,4-dimethyl-
6819 - Hexane, 3-ethyl-2,5-dimethyl-
6820 - Hexane, 3-ethyl-3,4-dimethyl-
6821 - Hexane, 4-ethyl-2,2-dimethyl-
6822 - Hexane, 4-ethyl-2,3-dimethyl-
6823 - Hexane, 4-ethyl-2,4-dimethyl-
6824 - Hexane, 4-ethyl-3,3-dimethyl-
6839 - Hexane, 2-methyl-3-(1-methylethyl)-
6844 - Hexane, 2,2,3,3-tetramethyl-
6845 - Hexane, 2,2,3,4-tetramethyl-
6846 - Hexane, 2,2,3,5-tetramethyl-
6847 - Hexane, 2,2,4,4-tetramethyl-
6848 - Hexane, 2,2,4,5-tetramethyl-
6849 - Hexane, 2,2,5,5-tetramethyl-
6850 - Hexane, 2,3,3,4-tetramethyl-
6851 - Hexane, 2,3,3,5-tetramethyl-
6852 - Hexane, 2,3,4,4-tetramethyl-
6853 - Hexane, 2,3,4,5-tetramethyl-
6854 - Hexane, 3,3,4,4-tetramethyl-
8140 - Nonane, 2-methyl-
8141 - Nonane, 3-methyl-
8142 - Nonane, 4-methyl-
8143 - Nonane, 5-methyl-
8373 - Octane, 2,2-dimethyl-
8374 - Octane, 2,3-dimethyl-
8375 - Octane, 2,4-dimethyl-
8376 - Octane, 2,5-dimethyl-
8377 - Octane, 2,6-dimethyl-
8378 - Octane, 2,7-dimethyl-
8379 - Octane, 3,3-dimethyl-
8380 - Octane, 3,4-dimethyl-
8381 - Octane, 3,5-dimethyl-
8382 - Octane, 3,6-dimethyl-
8383 - Octane, 4,4-dimethyl-
8384 - Octane, 4,5-dimethyl-
8403 - Octane, 3-ethyl-
8404 - Octane, 4-ethyl-
8757 - Pentane, 3,3-diethyl-2-methyl-
8766 - Pentane, 2,4-dimethyl-3-(1-methylethyl)-
8815 - Pentane, 3-ethyl-2,2,3-trimethyl-
8816 - Pentane, 3-ethyl-2,3,4-trimethyl-
8836 - Pentane, 2,2,3,3,4-pentamethyl-
8837 - Pentane, 2,2,3,4,4-pentamethyl-

$C_{10}H_{22}O$
3418 - Butane, 1,1'-oxybis[3-methyl-
5070 - 1-Decanol
5071 - 2-Decanol, (±)-
5072 - 4-Decanol
6488 - 3-Heptanol, 3-ethyl-2-methyl-
6489 - 3-Heptanol, 3-ethyl-6-methyl-
6490 - 3-Heptanol, 5-ethyl-4-methyl-
6508 - 4-Heptanol, 4-propyl-
6509 - 4-Heptanol, 2,4,6-trimethyl-
6948 - 1-Hexanol, 5-methyl-2-(1-methylethyl)-
6950 - 3-Hexanol, 2,2,5,5-tetramethyl-
8166 - 1-Nonanol, 4-methyl-
8167 - 1-Nonanol, 4-methyl-
8168 - 1-Nonanol, 8-methyl
8169 - 3-Nonanol, 4-methyl-
8170 - 4-Nonanol, 4-methyl-
8171 - 5-Nonanol, 5-methyl-
8402 - Octane, 1-ethoxy-
8449 - 1-Octanol, 3,7-dimethyl-, (±)-
8450 - 2-Octanol, 2,6-dimethyl-
8451 - 3-Octanol, 2,3-dimethyl-
8452 - 3-Octanol, 2,6-dimethyl-
8453 - 3-Octanol, 3,6-dimethyl-
8454 - 3-Octanol, 3,6-dimethyl-(+-)-
8455 - 4-Octanol, 2,4-dimethyl-
8456 - 4-Octanol, 2,5-dimethyl-

8457 - 4-Octanol, 2,6-dimethyl-
8458 - 4-Octanol, 2,7-dimethyl-
8459 - 3-Octanol, 3-ethyl-
8835 - Pentane, 1,1'-oxybis-

$C_{10}H_{22}O_2$
5047 - 1,10-Decanediol
5048 - 2,9-Decanediol
5527 - Ethane, 1,2-dibutoxy
6800 - 3,4-Hexanediol, 3,4-diethyl-
8392 - 1,7-Octanediol, 3,7-dimethyl-
9154 - Peroxide, bis(1,1-dimethylpropyl)
10032 - Propane, 1,1'-[1,2-ethanediylbis(oxy)]bis[2-
    methyl-
10041 - Propane, 1,1'-[ethylidenebis(oxy)]bis[2-
    methyl-

$C_{10}H_{22}O_3$
10078 - Propane, 1,1',1''-[methylidynetris(oxy)]tris-
10079 - Propane, 2,2',2''-[methylidynetris(oxy)]tris-

$C_{10}H_{22}O_4$
5690 - Ethanol, 2-[2-(2-butoxyethoxy)ethoxy]-

$C_{10}H_{22}O_5$
8995 - 2,5,8,11,14-Pentaoxapentadecane

$C_{10}H_{22}S$
3435 - Butane, 1,1'-thiobis[2-methyl-
3436 - Butane, 1,1'-thiobis[3-methyl-
5058 - 1-Decanethiol
8843 - Pentane, 1,1'-thiobis-

$C_{10}H_{22}S_2$
5243 - Disulfide, bis(3-methylbutyl)
5257 - Disulfide, dipentyl

$C_{10}H_{23}N$
3226 - 1-Butanamine, 3-methyl-N-(3-methylbutyl)-
5032 - 1-Decanamine
8708 - 1-Pentanamine, N-pentyl-

$C_{10}H_{23}NO$
5684 - Ethanol, 2-[bis(2-methylpropyl)amino]-
5703 - Ethanol, 2-(dibutylamino)-

$C_{10}H_{23}NO_2$
5415 - Ethanamine, 2,2-diethoxy-N,N-diethyl-

$C_{10}H_{24}N_2$
5515 - 1,2-Ethanediamine, N,N,N',N'-tetraethyl-
6752 - 1,6-Hexanediamine, N,N,N',N'-tetramethyl-

$C_{10}H_{24}N_4$
9614 - 1,4-Piperazinedipropanamine

$C_{10}H_{26}N_4$
3259 - 1,4-Butanediamine, N,N'-bis(3-aminopropyl)-

$C_{10}H_{30}O_3Si_4$
11646 - Tetrasiloxane, decamethyl-

$C_{10}H_{30}O_5Si_5$
4901 - Cyclopentasiloxane, decamethyl-

$C_{11}H_7BrO_2$
7761 - 1-Naphthalenecarboxylic acid, 5-bromo-
7769 - 2-Naphthalenecarboxylic acid, 5-bromo-
7846 - 1,4-Naphthalenedione, 2-bromo-3-methyl-

$C_{11}H_7ClO$
7750 - 2-Naphthalenecarbonyl chloride

$C_{11}H_7ClO_2$
7762 - 1-Naphthalenecarboxylic acid, 5-chloro-
7763 - 1-Naphthalenecarboxylic acid, 8-chloro-

$C_{11}H_7N$
7748 - 1-Naphthalenecarbonitrile
7749 - 2-Naphthalenecarbonitrile

$C_{11}H_7NO$
7904 - Naphthalene, 1-isocyanato-

$C_{11}H_7NO_4$
7767 - 1-Naphthalenecarboxylic acid, 5-nitro-

$C_{11}H_7NS$
7906 - Naphthalene, 1-isothiocyanato-

$C_{11}H_8O$
7751 - 1-Naphthalenecarboxaldehyde
7752 - 2-Naphthalenecarboxaldehyde

$C_{11}H_8OS$
7641 - Methanone, phenyl-2-thienyl-
7642 - Methanone, phenyl-3-thienyl-

$C_{11}H_8O_2$
7618 - Methanone, 2-furanylphenyl-
7754 - 1-Naphthalenecarboxaldehyde, 2-hydroxy-

7758 - 1-Naphthalenecarboxylic acid
7759 - 2-Naphthalenecarboxylic acid
7864 - 1,4-Naphthalenedione, 2-methyl-

$C_{11}H_8O_3$
7771 - 2-Naphthalenecarboxylic acid, 1-hydroxy-
7772 - 2-Naphthalenecarboxylic acid, 3-hydroxy-
7861 - 1,4-Naphthalenedione, 2-hydroxy-3-methyl-
7863 - 1,4-Naphthalenedione, 5-hydroxy-2-methyl-

$C_{11}H_8O_4$
7856 - 1,4-Naphthalenedione, 3,5-dihydroxy-2-
    methyl-

$C_{11}H_8O_5$
2408 - 5H-Benzocyclohepten-5-one, 2,3,4,6-
    tetrahydroxy-

$C_{11}H_9Br$
7742 - Naphthalene, 1-(bromomethyl)-
7743 - Naphthalene, 2-(bromomethyl)-

$C_{11}H_9Cl$
7783 - Naphthalene, 1-(chloromethyl)-
7784 - Naphthalene, 1-chloro-2-methyl-
7785 - Naphthalene, 2-(chloromethyl)-

$C_{11}H_9Cl_2NO_2$
510 - Barban

$C_{11}H_9N$
10997 - Pyridine, 2-phenyl-
10998 - Pyridine, 3-phenyl-
10999 - Pyridine, 4-phenyl-

$C_{11}H_9NO$
7637 - Methanone, phenyl-1H-pyrrol-2-yl-
7757 - 1-Naphthalenecarboxamide

$C_{11}H_9NO_2$
7760 - 1-Naphthalenecarboxylic acid, 5-amino-
7768 - 2-Naphthalenecarboxylic acid, 3-amino-
7925 - Naphthalene, 1-methyl-4-nitro-
7926 - Naphthalene, 2-methyl-1-nitro-

$C_{11}H_9NO_3$
11234 - 4-Quinolinecarboxylic acid, 6-methoxy-

$C_{11}H_{10}$
7915 - Naphthalene, 1-methyl-
7916 - Naphthalene, 2-methyl-

$C_{11}H_{10}NO_3$
7695 - Morpholine, 4-(2-hydroxybenzoyl)-

$C_{11}H_{10}N_2O$
90 - Acetamide, N-4-quinolinyl-
12022 - Urea, 1-naphthalenyl-
12023 - Urea, 2-naphthalenyl-

$C_{11}H_{10}N_2O_3$
11100 - 2,4,6(1H,3H,5H)-Pyrimidinetrione, 5-methyl-
    5-phenyl-

$C_{11}H_{10}N_2S$
11807 - Thiourea, 1-naphthalenyl-

$C_{11}H_{10}O$
4951 - 2-Cyclopenten-1-one, 3-phenyl-
7908 - 1-Naphthalenemethanol
7909 - 2-Naphthalenemethanol
7913 - Naphthalene, 1-methoxy-
7914 - Naphthalene, 2-methoxy-
8030 - 1-Naphthalenol, 4-methyl-
8031 - 2-Naphthalenol, 1-methyl-
8657 - 2,4-Pentadienal, 5-phenyl-

$C_{11}H_{10}O_2$
2820 - 2H-1-Benzopyran-2-one, 3-ethyl-
6158 - Furan, 2-(phenoxymethyl)-
7764 - 1-Naphthalenecarboxylic acid, 3,4-dihydro-
10741 - 2-Propynoic acid, 3-phenyl-, ethyl ester

$C_{11}H_{10}O_3$
2440 - 2-Benzofurancarboxylic acid, ethyl ester
10781 - 2H-Pyran-2,6(3H)-dione, dihydro-3-phenyl-
10782 - 2H-Pyran-2,6(3H)-dione, dihydro-4-phenyl-

$C_{11}H_{10}O_4$
2818 - 2H-1-Benzopyran-2-one, 5,7-dimethoxy-

$C_{11}H_{10}S$
11774 - Thiophene, 2-methyl-5-phenyl-

$C_{11}H_{11}BrN_2O$
10846 - 3H-Pyrazol-3-one, 4-bromo-1,2-dihydro-1,5-
    dimethyl-2-phenyl-

C₁₁H₁₁BrO₂

10613 - 2-Propenoic acid, 2-bromo-3-phenyl-, ethyl ester

10614 - 2-Propenoic acid, 3-bromo-3-phenyl-, ethyl ester, (Z)-

C₁₁H₁₁Cl₃O₃

259 - Acetic acid, (2,4,5-trichlorophenoxy)-, 1-methylethyl ester

C₁₁H₁₁N

7720 - 1-Naphthalenamine, 4-methyl-
7721 - 1-Naphthalenamine, N-methyl-
7722 - 2-Naphthalenamine, N-methyl-
7907 - 2-Naphthalenemethanamine
11136 - 1H-Pyrrole, 1-(phenylmethyl)-
11254 - Quinoline, 2,3-dimethyl-
11255 - Quinoline, 2,4-dimethyl-
11256 - Quinoline, 2,5-dimethyl-
11257 - Quinoline, 2,6-dimethyl-
11258 - Quinoline, 2,7-dimethyl-
11259 - Quinoline, 2,8-dimethyl-
11260 - Quinoline, 3,4-dimethyl-
11261 - Quinoline, 3,8-dimethyl-
11262 - Quinoline, 4,6-dimethyl-
11263 - Quinoline, 5,8-dimethyl-
11264 - Quinoline, 6,8-dimethyl-
11266 - Quinoline, 2-ethyl-
11267 - Quinoline, 3-ethyl-
11268 - Quinoline, 6-ethyl-

C₁₁H₁₁NO

10442 - 2-Propanone, 1-(1H-indol-3-yl)-
11316 - 4-Quinolinol, 2,8-dimethyl-
11317 - 6-Quinolinol, 2,4-dimethyl-
11318 - 8-Quinolinol, 2,4-dimethyl-

C₁₁H₁₁NO₂

893 - Benzeneacetic acid, α-cyano-, ethyl ester
11161 - 2,5-Pyrrolidinedione, 1-methyl-3-phenyl-
11163 - 2,5-Pyrrolidinedione, 1-(phenylmethyl)-

C₁₁H₁₁NO₄S

7407 - Isoxazolium, 2-ethyl-5-(3-sulfophenyl)-, hydroxide, inner salt
11293 - 5-Quinolinesulfonic acid, 8-ethoxy-

C₁₁H₁₁NO₆

7511 - Methanediol, (3-nitrophenyl)-, diacetate (ester)

C₁₁H₁₁N₃O₂S

2203 - Benzenesulfonamide, 4-amino-N-2-pyridinyl-

C₁₁H₁₂

7816 - Naphthalene, 1,2-dihydro-3-methyl-
7817 - Naphthalene, 1,2-dihydro-4-methyl-

C₁₁H₁₂Cl₂N₂O₅

52 - Acetamide, 2,2-dichloro-N-[2-hydroxy-1-(hydroxymethyl)-2-(4-nitrophenyl)ethyl]-, [R-(R*,R*)]-

C₁₁H₁₂Cl₂O₃

173 - Acetic acid, (2,4-dichlorophenoxy)-, 1-methylethyl ester

C₁₁H₁₂Cl₃N

1583 - Benzeneethanamine, α-methyl-N-(2,2,2-trichloroethylidene)-

C₁₁H₁₂NO₄PS₂

9515 - Phosmet

C₁₁H₁₂N₂

7800 - 1,4-Naphthalenediamine, 2-methyl-
10829 - 1H-Pyrazole, 3,5-dimethyl-1-phenyl-

C₁₁H₁₂N₂O

10847 - 3H-Pyrazol-3-one, 1,2-dihydro-1,5-dimethyl-2-phenyl-
11192 - Pyrrolo[2,1-b]quinazolin-3-ol, 1,2,3,9-tetrahydro-, (R)-

C₁₁H₁₂N₂O₂

11922 - L-Tryptophan

C₁₁H₁₂N₂O₆

2592 - Benzoic acid, 3,5-dinitro-, butyl ester

C₁₁H₁₂N₄O₂S

2199 - Benzenesulfonamide, 4-amino-N-(4-methyl-2-pyrimidinyl)-

C₁₁H₁₂N₄O₃S

2197 - Benzenesulfonamide, 4-amino-N-(6-methoxy-3-pyridazinyl)-

C₁₁H₁₂O

2407 - 5H-Benzocyclohepten-5-one, 6,7,8,9-tetrahydro-
3190 - Butanal, 2-(phenylmethylene)-
7288 - 1H-Inden-1-one, 2,3-dihydro-3,3-dimethyl-
7605 - Methanone, cyclobutylphenyl-
8049 - 1(2H)-Naphthalenone, 3,4-dihydro-2-methyl-
8050 - 1(2H)-Naphthalenone, 3,4-dihydro-4-methyl-
9110 - 1-Penten-3-one, 1-phenyl-

C₁₁H₁₂O₂

7778 - 2-Naphthalenecarboxylic acid, 1,2,3,4-tetrahydro-
7779 - 2-Naphthalenecarboxylic acid, 5,6,7,8-tetrahydro-
8048 - 1(2H)-Naphthalenone, 3,4-dihydro-6-methoxy-
8804 - 1,4-Pentanedione, 1-phenyl-
10685 - 2-Propenoic acid, 3-phenyl-, ethyl ester, (E)-
10706 - 2-Propen-1-ol, 3-phenyl-, acetate, (E)-

C₁₁H₁₂O₃

2141 - Benzenepropanoic acid, β-oxo-, ethyl ester
2428 - 1,3-Benzodioxole, 4-methoxy-6-(2-propenyl)-
6138 - 2-Furanol, tetrahydro-, benzoate
10647 - 2-Propenoic acid, 3-(2-methoxyphenyl)-, methyl ester
10681 - 2-Propenoic acid, 2-phenoxyethyl ester

C₁₁H₁₂O₄

7512 - Methanediol, phenyl-, diacetate
9976 - Propanedioic acid, phenyl-, dimethyl ester

C₁₁H₁₃ClF₃N₃O₄S₃

2857 - 2H-1,2,4-Benzothiadiazine-7-sulfonamide, 6-chloro-3,4-dihydro-2-methyl-3-[[(2,2,2-trifluoroethyl)thio]methyl]-

C₁₁H₁₃ClO

2889 - Benzoyl chloride, 4-butyl-
2895 - Benzoyl chloride, 4-(1,1-dimethylethyl)-
9255 - Phenol, 4-chloro-2-cyclopentyl-

C₁₁H₁₃ClO₃

3493 - Butanoic acid, 4-(4-chloro-2-methylphenoxy)-

C₁₁H₁₃F₃N₂O₃S

7427 - Mefluidide

C₁₁H₁₃F₃N₄O₄

5181 - Dinitramine

C₁₁H₁₃N

963 - Benzeneacetonitrile, α-(1-methylethyl)-
970 - Benzeneacetonitrile, α-propyl-
1823 - Benzenemethanamine, N-methyl-N-2-propynyl-
7309 - 1H-Indole, 3-ethyl-2-methyl-
7322 - 1H-Indole, 1,2,3-trimethyl-
7323 - 1H-Indole, 2,3,5-trimethyl-
10922 - Pyridine, 4-(3-cyclohexen-1-yl)-

C₁₁H₁₃NO

3703 - 3-Butenamide, 3-methyl-4-phenyl-
7332 - 2H-Indol-2-one, 1-ethyl-1,3-dihydro-3-methyl-

C₁₁H₁₃NO₂

5215 - 1,3-Dioxolo[4,5-g]isoquinoline, 5,6,7,8-tetrahydro-6-methyl-

C₁₁H₁₃NO₃

5217 - 1,3-Dioxolo[4,5-g]isoquinolin-5-ol, 5,6,7,8-tetrahydro-6-methyl-
7398 - 1(2H)-Isoquinolinone, 3,4-dihydro-6,7-dimethoxy-

C₁₁H₁₃NO₄

514 - Bendiocarb
2706 - Benzoic acid, 4-nitro-, butyl ester
8588 - 2-Oxazolidinone, 5-[(2-methoxyphenoxy)methyl]-

C₁₁H₁₃N₃O

10845 - 3H-Pyrazol-3-one, 4-amino-1,2-dihydro-1,5-dimethyl-2-phenyl-

C₁₁H₁₃N₃O₃S

2195 - Benzenesulfonamide, 4-amino-N-(3,4-dimethyl-5-isoxazolyl)-

C₁₁H₁₄

1254 - Benzene, cyclopentyl-
1643 - Benzene, 1-ethenyl-4-(1-methylethyl)-
1647 - Benzene, 2-ethenyl-1,3,5-trimethyl-
1699 - Benzene, (1-ethyl-1-propenyl)-
1700 - Benzene, (1-ethyl-2-propenyl)-
1950 - Benzene, (1-methyl-1-butenyl)-
2104 - Benzene, 1-pentenyl-
2105 - Benzene, 3-pentenyl-
2106 - Benzene, 4-pentenyl-
4748 - 1,3,5,7-Cyclooctatetraene, propyl-
7251 - 1H-Indene, 2,3-dihydro-1,1-dimethyl-
7252 - 1H-Indene, 2,3-dihydro-1,2-dimethyl-
7253 - 1H-Indene, 2,3-dihydro-4,7-dimethyl-
7254 - 1H-Indene, 2,3-dihydro-5,6-dimethyl-
7273 - 1H-Indene, 1-ethyl-2,3-dihydro-
7995 - Naphthalene, 1,2,3,4-tetrahydro-1-methyl-
7996 - Naphthalene, 1,2,3,4-tetrahydro-2-methyl-
7997 - Naphthalene, 1,2,3,4-tetrahydro-5-methyl-
7998 - Naphthalene, 1,2,3,4-tetrahydro-6-methyl-

C₁₁H₁₄ClNO

9771 - Propachlor

C₁₁H₁₄ClNO₂

607 - Benzamide, N-butyl-4-chloro-2-hydroxy-

C₁₁H₁₄N₂

7310 - 1H-Indole-3-methanamine, N,N-dimethyl-

C₁₁H₁₄N₂O

7304 - 1H-Indole-3-ethanamine, 5-methoxy-

C₁₁H₁₄N₂O₄

84 - Acetamide, N-(5-nitro-2-propoxyphenyl)-

C₁₁H₁₄O

598 - Benzaldehyde, 2,3,5,6-tetramethyl-
1665 - Benzene, 1-ethenoxy-2-(1-propenyl)-
1666 - Benzene, (3-ethoxy-1-propenyl)-
3671 - 1-Butanone, 3-methyl-1-phenyl-
3672 - 1-Butanone, 1-(4-methylphenyl)-
4871 - Cyclopentanol, 1-phenyl-
5848 - Ethanone, 1-[4-(1-methylethyl)phenyl]-
5875 - Ethanone, 1-(2,4,5-trimethylphenyl)-
5876 - Ethanone, 1-(2,4,6-trimethylphenyl)-
5877 - Ethanone, 1-(3,4,5-trimethylphenyl)-
8986 - 1-Pentanone, 1-phenyl-
10431 - 1-Propanone, 2,2-dimethyl-1-phenyl-

C₁₁H₁₄O₂

900 - Benzeneacetic acid, α-ethyl-, methyl ester
930 - Benzeneacetic acid, α-(1-methylethyl)-
943 - Benzeneacetic acid, α-propyl-, (±)-
1104 - Benzenebutanoic acid, γ-methyl-
1449 - Benzene, 1,2-dimethoxy-4-(1-propenyl)-
1450 - Benzene, 1,2-dimethoxy-4-(2-propenyl)-
1451 - Benzene, 1,2-dimethoxy-4-(1-propenyl)-, (Z)-
2099 - Benzenepentanoic acid
2130 - Benzenepropanoic acid, β,β-dimethyl-
2131 - Benzenepropanoic acid, ethyl ester
2532 - Benzoic acid, butyl ester
2588 - Benzoic acid, 4-(1,1-dimethylethyl)-
2694 - Benzoic acid, 4-(1-methylethyl)-, methyl ester
2701 - Benzoic acid, 2-methylpropyl ester
3585 - Butanoic acid, phenylmethyl ester
5194 - 1,3-Dioxane, 4-methyl-4-phenyl-
9496 - Phenol, 2,4,5-trimethyl-, acetate
10307 - Propanoic acid, 2-methyl-, phenylmethyl ester
10324 - Propanoic acid, 2-phenylethyl ester
10436 - 1-Propanone, 3-ethoxy-1-phenyl-
10447 - 1-Propanone, 1-(2-hydroxy-5-methylphenyl)-2-methyl-
10448 - 1-Propanone, 1-(4-hydroxy-3-methylphenyl)-2-methyl-

C₁₁H₁₄O₃

538 - Benzaldehyde, 3,4-diethoxy-
923 - Benzeneacetic acid, 4-methoxy-, ethyl ester
1128 - Benzenecarboperoxoic acid, 1,1-dimethylethyl ester
2598 - Benzoic acid, 2-ethoxy-, ethyl ester
2599 - Benzoic acid, 4-ethoxy-, ethyl ester
2614 - Benzoic acid, 2-hydroxy-, butyl ester

2615 - Benzoic acid, 4-hydroxy-, butyl ester
2641 - Benzoic acid, 2-hydroxy-3-methyl-6-(1-methylethyl)-
2642 - Benzoic acid, 2-hydroxy-6-methyl-3-(1-methylethyl)-
2643 - Benzoic acid, 2-hydroxy-, 2-methylpropyl ester
3663 - 2-Butanone, 4-(4-hydroxy-3-methoxyphenyl)-
10321 - Propanoic acid, 2-phenoxy-, ethyl ester
10322 - Propanoic acid, 3-phenoxy-, ethyl ester

$C_{11}H_{14}O_4$
2578 - Benzoic acid, 3,4-dimethoxy-, ethyl ester
5874 - Ethanone, 1-(2,3,4-trimethoxyphenyl)-

$C_{11}H_{15}BrClO_3PS$
9762 - Profenofos

$C_{11}H_{15}BrN_2O_3$
11079 - 2,4,6(1H,3H,5H)-Pyrimidinetrione, 5-(2-bromo-2-propenyl)-1-methyl-5-(1-methylethyl)-
11080 - 2,4,6(1H,3H,5H)-Pyrimidinetrione, 5-(2-bromo-2-propenyl)-5-(1-methylpropyl)-

$C_{11}H_{15}Cl_2O_3PS_2$
4092 - Chlorthiophos

$C_{11}H_{15}N$
9693 - Piperidine, 1-phenyl-
9694 - Piperidine, 3-phenyl-
9695 - Piperidine, 4-phenyl-
11172 - Pyrrolidine, 1-(phenylmethyl)-

$C_{11}H_{15}NO$
74 - Acetamide, N-(1-methylethyl)-N-phenyl-
88 - Acetamide, N-phenyl-N-propyl-
539 - Benzaldehyde, 4-(diethylamino)-
882 - Benzeneacetamide, α-(1-methylethyl)-
7697 - Morpholine, 3-methyl-2-phenyl-
7698 - Morpholine, 4-(4-methylphenyl)-
7700 - Morpholine, 4-(phenylmethyl)-
8960 - 1-Pentanone, 1-(4-aminophenyl)-

$C_{11}H_{15}NO_2$
2488 - Benzoic acid, 4-amino-, butyl ester
2505 - Benzoic acid, 4-amino-, 2-methylpropyl ester
3968 - Carbamic acid, phenyl-, 2-methylpropyl ester
6272 - Glycine, N-(phenylmethyl)-, ethyl ester
7397 - 6-Isoquinolinol, 1,2,3,4-tetrahydro-7-methoxy-1-methyl-, (S)-
9512 - L-Phenylalanine, ethyl ester

$C_{11}H_{15}NO_2S$
7646 - Methiocarb

$C_{11}H_{15}NO_3$
7393 - 8-Isoquinolinol, 1,2,3,4-tetrahydro-6,7-dimethoxy-
9796 - Propanamide, N-(4-ethoxyphenyl)-2-hydroxy-
10722 - Propoxur

$C_{11}H_{15}N_2NaO_2S$
11063 - 4,6(1H,5H)-Pyrimidinedione, dihydro-5-(2-methylpropyl)-5-(2-propenyl)-2-thioxo-, monosodium salt

$C_{11}H_{16}$
1121 - Benzene, 1-butyl-2-methyl-
1122 - Benzene, 1-butyl-3-methyl-
1123 - Benzene, 1-butyl-4-methyl-
1416 - Benzene, 1,2-diethyl-3-methyl-
1417 - Benzene, 1,2-diethyl-4-methyl-
1418 - Benzene, 1,3-diethyl-2-methyl-
1419 - Benzene, 1,3-diethyl-5-methyl-
1420 - Benzene, 1,4-diethyl-2-methyl-
1421 - Benzene, 2,4-diethyl-1-methyl-
1464 - Benzene, 1-(1,1-dimethylethyl)-2-methyl-
1465 - Benzene, 1-(1,1-dimethylethyl)-3-methyl-
1466 - Benzene, 1-(1,1-dimethylethyl)-4-methyl-
1469 - Benzene, 1,2-dimethyl-3-(1-methylethyl)-
1470 - Benzene, 1,2-dimethyl-4-(1-methylethyl)-
1471 - Benzene, 1,3-dimethyl-4-(1-methylethyl)-
1472 - Benzene, 2,4-dimethyl-1-(1-methylethyl)-
1480 - Benzene, (1,1-dimethylpropyl)-
1481 - Benzene, (1,2-dimethylpropyl)-
1482 - Benzene, 1,3-dimethyl-2-propyl-
1483 - Benzene, 1,3-dimethyl-5-propyl-
1484 - Benzene, (2,2-dimethylpropyl)-
1485 - Benzene, 1,2-dimethyl-3-propyl-

1486 - Benzene, 1,4-dimethyl-2-propyl-
1487 - Benzene, 2,4-dimethyl-1-propyl-
1688 - Benzene, 1-ethyl-2-(1-methylethyl)-
1689 - Benzene, 1-ethyl-3-(1-methylethyl)-
1690 - Benzene, 1-ethyl-4-(1-methylethyl)-
1701 - Benzene, (1-ethylpropyl)-
1702 - Benzene, 1-ethyl-2-propyl-
1703 - Benzene, 1-ethyl-4-propyl-
1708 - Benzene, 1-ethyl-2,3,4-trimethyl-
1709 - Benzene, 1-ethyl-2,3,5-trimethyl-
1710 - Benzene, 1-ethyl-2,4,5-trimethyl-
1711 - Benzene, 2-ethyl-1,3,4-trimethyl-
1712 - Benzene, 2-ethyl-1,3,5-trimethyl-
1713 - Benzene, 5-ethyl-1,2,3-trimethyl-
1952 - Benzene, (1-methylbutyl)-
1953 - Benzene, (2-methylbutyl)-
1954 - Benzene, (3-methylbutyl)-
1994 - Benzene, 1-methyl-2-(1-methylpropyl)-
1995 - Benzene, 1-methyl-2-(2-methylpropyl)-
1996 - Benzene, 1-methyl-3-(1-methylpropyl)-
1997 - Benzene, 1-methyl-3-(2-methylpropyl)-
1998 - Benzene, 1-methyl-4-(1-methylpropyl)-
1999 - Benzene, 1-methyl-4-(2-methylpropyl)-
2097 - Benzene, pentamethyl-
2107 - Benzene, pentyl-
8341 - 1,5-Octadien-3-yne, 5-propyl-
11934 - 1,10-Undecadiyne
12057 - o-Xylene, 4-propyl-

$C_{11}H_{16}ClN_3O_2$
6001 - Formetanate hydrochloride

$C_{11}H_{16}ClN_3O_4S_2$
2856 - 2H-1,2,4-Benzothiadiazine-7-sulfonamide, 6-chloro-3,4-dihydro-3-(2-methylpropyl)-, 1,1-dioxide

$C_{11}H_{16}ClO_2PS_3$
4037 - Carbophenothion

$C_{11}H_{16}N_2$
9621 - Piperazine, 1-(phenylmethyl)-

$C_{11}H_{16}N_2O_2$
6151 - 2(3H)-Furanone, 3-ethyldihydro-4-[(1-methyl-1H-imidazol-5-yl)methyl]-, (3S-cis)-

$C_{11}H_{16}N_2O_3$
11099 - 2,4,6(1H,3H,5H)-Pyrimidinetrione, 1-methyl-5-(1-methylethyl)-5-(2-propenyl)-
11101 - 2,4,6(1H,3H,5H)-Pyrimidinetrione, 5-(2-methylpropyl)-5-(2-propenyl)-

$C_{11}H_{16}O$
363 - Anisole, 3-isopropyl-2-methyl-
1106 - Benzenebutanol, α-methyl-
1107 - Benzenebutanol, β-methyl-, (±)-
1115 - Benzene, (butoxymethyl)-
1116 - Benzene, 1-butoxy-3-methyl-
1117 - Benzene, 1-butoxy-4-methyl-
1462 - Benzene, 1-(1,1-dimethylethyl)-4-methoxy-
1611 - Benzeneethanol, β-(1-methylethyl)-
1616 - Benzeneethanol, α,α,β-trimethyl-
1617 - Benzeneethanol, α,β,β-trimethyl-
1667 - Benzene, 1-ethoxy-2-propyl-
1846 - Benzenemethanol, α-butyl-
1855 - Benzenemethanol, α,α-diethyl-
1864 - Benzenemethanol, α-(1,1-dimethylethyl)-
1865 - Benzenemethanol, 4-(1,1-dimethylethyl)-
1904 - Benzenemethanol, α-methyl-α-propyl-
1905 - Benzenemethanol, α-(2-methylpropyl)-
2033 - Benzene, [(2-methylpropoxy)methyl]-
2101 - Benzenepentanol
2109 - Benzene, (pentyloxy)-
2150 - Benzenepropanol, α,α-dimethyl-
2151 - Benzenepropanol, β,β-dimethyl-
2152 - Benzenepropanol, α-ethyl-, (S)-
4949 - 2-Cyclopenten-1-one, 3-methyl-2-(2-pentenyl)-, (Z)-
9250 - Phenol, 2-butyl-4-methyl-
9330 - Phenol, 2-(1,1-dimethylethyl)-4-methyl-
9331 - Phenol, 2-(1,1-dimethylethyl)-5-methyl-
9332 - Phenol, 2-(1,1-dimethylethyl)-6-methyl-
9333 - Phenol, 4-(1,1-dimethylethyl)-2-methyl-

9339 - Phenol, 4-(1,1-dimethylpropyl)-
9404 - Phenol, 4-(3-methylbutyl)-
9421 - Phenol, 5-methyl-2-(1-methylpropyl)-
9443 - Phenol, pentamethyl-
9444 - Phenol, 4-pentyl-

$C_{11}H_{16}O_2$
1105 - Benzenebutanol, 4-methoxy-
1410 - Benzene, 1,4-diethoxy-2-methyl-
1442 - Benzene, 1,4-dimethoxy-2-(1-methylethyl)-
1534 - 1,3-Benzenediol, 4-(3-methylbutyl)-
1544 - 1,3-Benzenediol, 4-pentyl-
9329 - Phenol, (1,1-dimethylethyl)-4-methoxy-
11864 - Tricyclo[2.2.1.0²,⁶]heptane-1-carboxylic acid, 7,7-dimethyl-, methyl ester

$C_{11}H_{16}O_3$
1409 - Benzene, (diethoxymethoxy)-
2917 - Bicyclo[2.2.1]heptane-2-carboxylic acid, 4,7,7-trimethyl-3-oxo-
4641 - 2-Cyclohexene-1-carboxylic acid, 2,6-dimethyl-4-oxo-, ethyl ester
6058 - 2-Furancarboxylic acid, hexyl ester

$C_{11}H_{16}O_3S$
2230 - Benzenesulfonic acid, 4-methyl-, butyl ester

$C_{11}H_{16}O_4$
4915 - 1-Cyclopentene-1,2-dicarboxylic acid, diethyl ester
4916 - 1-Cyclopentene-1,3-dicarboxylic acid, diethyl ester

$C_{11}H_{17}Cl_3N_2O_2S$
7220 - 2,4-Imidazolidinedione, 5-(1-ethylpentyl)-3-[(trichloromethyl)thio]-

$C_{11}H_{17}N$
362 - Aniline, p-isopentyl-
696 - Benzenamine, N,N-diethyl-2-methyl-
697 - Benzenamine, N,N-diethyl-4-methyl-
783 - Benzenamine, N-(3-methylbutyl)-
833 - Benzenamine, 2,3,4,5,6-pentamethyl-
1576 - Benzeneethanamine, N-ethyl-α-methyl-

$C_{11}H_{17}NO$
1863 - Benzenemethanol, α-[1-(dimethylamino)ethyl]-, [R-(R*,S*)]-
1871 - Benzenemethanol, α-[1-(ethylamino)ethyl]-, (R*,S*)-
5724 - Ethanol, 2-[ethyl(3-methylphenyl)amino]-
10392 - 1-Propanol, 3-[methyl(phenylmethyl)amino]-

$C_{11}H_{17}NO_3$
1522 - 1,2-Benzenediol, 4-[1-hydroxy-2-[(1-methylethyl)amino]ethyl]-
1586 - Benzeneethanamine, 3,4,5-trimethoxy-

$C_{11}H_{17}N_3O_3$
11096 - 2,4,6(1H,3H,5H)-Pyrimidinetrione, 5-ethyl-5-(1-piperidinyl)-

$C_{11}H_{17}N_3O_3S$
2192 - Benzenesulfonamide, 4-amino-N-[(butylamino)carbonyl]-

$C_{11}H_{17}O_3P$
9545 - Phosphonic acid, (phenylmethyl)-, diethyl ester

$C_{11}H_{17}O_4PS_2$
5954 - Fensulfothion

$C_{11}H_{18}$
7792 - Naphthalene, decahydro-2-methylene-, trans-
11983 - 1-Undecen-3-yne

$C_{11}H_{18}N_2$
5504 - 1,2-Ethanediamine, N,N-dimethyl-N'-(phenylmethyl)-
7162 - Hydrazine, 1-(3-methylbutyl)-1-phenyl-

$C_{11}H_{18}N_2O_3$
11092 - 2,4,6(1H,3H,5H)-Pyrimidinetrione, 5-ethyl-5-(3-methylbutyl)-

$C_{11}H_{18}N_4O_2$
9728 - Pirimicarb

$C_{11}H_{18}O$
2959 - Bicyclo[2.2.1]heptan-2-one, 1,4,7,7-tetramethyl-
2969 - Bicyclo[3.1.1]hept-2-ene-2-ethanol, 6,6-dimethyl-

4950 - 2-Cyclopenten-1-one, 3-methyl-2-pentyl-

$C_{11}H_{18}O_2$

2949 - Bicyclo[2.2.1]heptan-2-ol, 1,7,7-trimethyl-, formate, (1R-endo)-

3131 - 3-Bornanecarboxylic acid

4965 - Cyclopropanecarboxylic acid, 2,2-dimethyl-3-(2-methyl-1-propenyl)-, methyl ester, trans-(±)-

5026 - 2,4-Decadienoic acid, methyl ester, (E,E)-

5027 - 2,4-Decadienoic acid, methyl ester, (E,Z)-

5028 - 2,4-Decadienoic acid, methyl ester, (Z,E)-

5029 - 2,4-Decadienoic acid, methyl ester, (Z,Z)-

8337 - 1,6-Octadien-3-ol, 3,7-dimethyl-, formate

8338 - 2,6-Octadien-1-ol, 3,7-dimethyl-, formate, (E)-

11988 - 6-Undecynoic acid

11989 - 9-Undecynoic acid

$C_{11}H_{18}O_2Si$

11407 - Silane, diethoxymethylphenyl-

$C_{11}H_{18}O_3Si$

11427 - Silane, ethenyltris(2-propenyloxy)-

$C_{11}H_{18}O_4$

4791 - 1,1-Cyclopentanedicarboxylic acid, diethyl ester

$C_{11}H_{18}O_5$

8771 - Pentanedioic acid, 2-acetyl-, diethyl ester

9972 - Propanedioic acid, (1-oxobutyl)-, diethyl ester

$C_{11}H_{18}O_6$

5654 - 1,1,2-Ethanetricarboxylic acid, triethyl ester

$C_{11}H_{19}ClO$

11982 - 10-Undecenoyl chloride

$C_{11}H_{19}NOS$

8528 - Octhilinone

$C_{11}H_{19}NO_2$

8148 - Nonanoic acid, 9-cyano-, methyl ester

$C_{11}H_{19}N_3O$

5932 - Ethirimol

$C_{11}H_{20}$

4340 - Cyclohexane, cyclopentyl-

8201 - 4-Nonyne, 3,3-dimethyl-

11537 - Spiro[5.5]undecane

11984 - 1-Undecyne

11985 - 2-Undecyne

11986 - 4-Undecyne

11987 - 5-Undecyne

$C_{11}H_{20}ClN_5$

11828 - 1,3,5-Triazine-2,4-diamine, 6-chloro-N,N,N′,N′-tetraethyl-

$C_{11}H_{20}N_2O_3$

7224 - 4-Imidazolidinehexanoic acid, 5-methyl-2-oxo-, methyl ester, (4R-cis)-

$C_{11}H_{20}N_3O_3PS$

9730 - Pirimiphos-methyl

$C_{11}H_{20}O$

4592 - Cyclohexanone, 4-(1,1-dimethylpropyl)-

4888 - Cyclopentanone, 3-methyl-2-(3-methylbutyl)-

$C_{11}H_{20}O_2$

6152 - 2(3H)-Furanone, 5-heptyldihydro-

6404 - 3,5-Heptanedione, 2,2,6,6-tetramethyl-

8149 - Nonanoic acid, ethenyl ester

10636 - 2-Propenoic acid, 2-ethylhexyl ester

11974 - 10-Undecenoic acid

11975 - 9-Undecenoic acid, (E)-

11976 - 9-Undecenoic acid, (Z)-

$C_{11}H_{20}O_3$

8436 - Octanoic acid, 2-methyl-3-oxo-, ethyl ester

$C_{11}H_{20}O_4$

6394 - Heptanedioic acid, diethyl ester

6790 - Hexanedioic acid, 3-methyl-, diethyl ester

8134 - Nonanedioic acid, dimethyl ester

9925 - Propanedioic acid, bis(1,1-dimethylethyl) ester

9930 - Propanedioic acid, butyl-, diethyl ester

9938 - Propanedioic acid, dibutyl ester

9940 - Propanedioic acid, diethyl-, diethyl ester

9967 - Propanedioic acid, (1-methylpropyl)-, diethyl ester

9969 - Propanedioic acid, (2-methylpropyl)-, diethyl ester

9971 - Propanedioic acid, octyl-

$C_{11}H_{20}O_{10}$

6234 - D-Glucose, 6-O-α-L-arabinopyranosyl-

6237 - D-Glucose, 6-O-β-D-xylopyranosyl-

$C_{11}H_{21}BrO_2$

11944 - Undecanoic acid, 11-bromo-

11945 - Undecanoic acid, 2-bromo-

$C_{11}H_{21}FO_2$

11947 - Undecanoic acid, 11-fluoro-

$C_{11}H_{21}N$

11941 - Undecanenitrile

$C_{11}H_{21}NOS$

4168 - Cycloate

$C_{11}H_{21}NO_4$

330 - β-Alanine, N-(3-ethoxy-3-oxopropyl)-N-methyl-, ethyl ester

$C_{11}H_{21}N_5S$

5222 - Dipropetryn

$C_{11}H_{22}$

4427 - Cyclohexane, (3-methylbutyl)-

4465 - Cyclohexane, pentyl-

4823 - Cyclopentane, hexyl-

11966 - 1-Undecene

11967 - 3-Undecene

11968 - 2-Undecene, (E)-

11969 - 2-Undecene, (Z)-

11970 - 4-Undecene, (E)-

11971 - 4-Undecene, (Z)-

11972 - 5-Undecene, (E)-

11973 - 5-Undecene, (Z)-

$C_{11}H_{22}BrF$

11939 - Undecane, 1-bromo-11-fluoro-

$C_{11}H_{22}Br_2$

11940 - Undecane, 1,11-dibromo-

$C_{11}H_{22}N_2$

9683 - Piperidine, 1,1′-methylenebis-

$C_{11}H_{22}O$

11935 - Undecanal

11958 - 2-Undecanone

11959 - 3-Undecanone

11960 - 4-Undecanone

11961 - 5-Undecanone

11962 - 6-Undecanone

11979 - 10-Undecen-1-ol

11980 - 2-Undecen-4-ol

11981 - 9-Undecen-1-ol

$C_{11}H_{22}O_2$

220 - Acetic acid, nonyl ester

3527 - Butanoic acid, heptyl ester

5066 - Decanoic acid, methyl ester

6459 - Heptanoic acid, butyl ester

6467 - Heptanoic acid, 2-methylpropyl ester

6897 - Hexanoic acid, 3-methylbutyl ester

6907 - Hexanoic acid, pentyl ester

8150 - Nonanoic acid, ethyl ester

8435 - Octanoic acid, 1-methylethyl ester

8441 - Octanoic acid, propyl ester

8882 - Pentanoic acid, hexyl ester

10313 - Propanoic acid, octyl ester

11943 - Undecanoic acid

$C_{11}H_{22}O_3$

3635 - 1-Butanol, 3-methyl-, carbonate (2:1)

11948 - Undecanoic acid, 11-hydroxy-

$C_{11}H_{22}O_6$

6228 - α-D-Glucopyranoside, methyl 2,3,4,6-tetra-O-methyl-

$C_{11}H_{23}N$

9671 - Piperidine, 1-(1-ethyl-1-methylpropyl)-

9672 - Piperidine, 1-hexyl-

$C_{11}H_{23}NOS$

11586 - Sutan

$C_{11}H_{24}$

5054 - Decane, 2-methyl-

5055 - Decane, 3-methyl-

5056 - Decane, 4-methyl-

8420 - Octane, 2,2,4-trimethyl-

8421 - Octane, 2,4,7-trimethyl

11938 - Undecane

$C_{11}H_{24}O$

8164 - 5-Nonanol, 2,8-dimethyl-

11950 - 1-Undecanol

11951 - 2-Undecanol, (±)-

11952 - 3-Undecanol, (R)-

11953 - 5-Undecanol

11954 - 6-Undecanol

$C_{11}H_{24}O_3Si$

11426 - Silane, ethenyltris(1-methylethoxy)-

$C_{11}H_{24}S$

11942 - 1-Undecanethiol

$C_{11}H_{25}N$

11937 - 1-Undecanamine

$C_{11}H_{26}O_3Si$

11480 - Silane, triethoxypentyl-

$C_{11}H_{26}O_6P_2$

9546 - Phosphonic acid, 1,3-propanediylbis-, tetraethyl ester

$C_{11}H_{28}Br_2N_2$

8738 - 1,5-Pentanediaminium, N,N,N,N′,N′,N′-hexamethyl-, dibromide

$C_{12}F_{10}$

3027 - 1,1′-Biphenyl, 2,2′,3,3′,4,4′,5,5′,6,6′-decafluoro-

$C_{12}F_{27}N$

3228 - 1-Butanamine, 1,1,2,2,3,3,4,4,4-nonafluoro-N,N-bis(nonafluorobutyl)-

$C_{12}H_4N_6O_{12}S$

2288 - Benzene, 1,1′-thiobis[2,4,6-trinitro-

$C_{12}H_5N_7O_{12}$

864 - Benzenamine, 2,4,6-trinitro-N-(2,4,6-trinitrophenyl)-

$C_{12}H_6Cl_2O_2$

7804 - 1,8-Naphthalenedicarbonyl dichloride

$C_{12}H_6Cl_4O_2S$

2325 - Benzene, 1,2,4-trichloro-5-[(4-chlorophenyl)sulfonyl]-

9478 - Phenol, 2,2′-thiobis[4,6-dichloro-

$C_{12}H_6Cl_4S$

11648 - Tetrasul

$C_{12}H_6O_2$

8 - 1,2-Acenaphthylenedione

$C_{12}H_6O_{12}$

1748 - Benzenehexacarboxylic acid

$C_{12}H_7BrO$

5164 - Dibenzofuran, 2-bromo-

5165 - Dibenzofuran, 3-bromo-

$C_{12}H_7Cl_2NO_3$

8081 - Nitrofen

$C_{12}H_7NO_3$

5166 - Dibenzofuran, 3-nitro-

$C_{12}H_7NO_4$

9508 - 3H-Phenoxazin-3-one, 7-hydroxy-, 10-oxide

$C_{12}H_8$

3 - Acenaphthylene

7892 - Naphthalene, 1-ethynyl-

$C_{12}H_8Br_2$

3037 - 1,1′-Biphenyl, 4,4′-dibromo-

$C_{12}H_8Br_2O$

2072 - Benzene, 1,1′-oxybis[4-bromo-

$C_{12}H_8Br_2O_2S$

2238 - Benzene, 1,1′-sulfonylbis[4-bromo-

$C_{12}H_8Br_2S$

2284 - Benzene, 1,1′-thiobis[4-bromo-

$C_{12}H_8Cl_2$

3040 - 1,1′-Biphenyl, 3,3′-dichloro-

3041 - 1,1′-Biphenyl, 4,4′-dichloro-

$C_{12}H_8Cl_2O$

2073 - Benzene, 1,1′-oxybis[4-chloro-

$C_{12}H_8Cl_2O_2S$

2239 - Benzene, 1,1′-sulfonylbis[4-chloro-

9477 - Phenol, 2,2′-thiobis[4-chloro-

$C_{12}H_8Cl_2O_3S$

2221 - Benzenesulfonic acid, 4-chloro-, 4-chlorophenyl ester

9289 - Phenol, 2,4-dichloro-, benzenesulfonate

$C_{12}H_{11}N_3$
837 - Benzenamine, 4-(phenylazo)-
11826 - 1-Triazene, 1,3-diphenyl-

$C_{12}H_{11}N_3O_2$
10904 - 4-Pyridinecarboxylic acid, [1-(2-furanyl)ethylidene]hydrazide

$C_{12}H_{11}N_7$
10750 - 2,4,7-Pteridinetriamine, 6-phenyl-

$C_{12}H_{11}O_3P$
9534 - Phosphonic acid, diphenyl ester

$C_{12}H_{12}$
7818 - Naphthalene, 1,2-dimethyl-
7819 - Naphthalene, 1,3-dimethyl-
7820 - Naphthalene, 1,4-dimethyl-
7821 - Naphthalene, 1,5-dimethyl-
7822 - Naphthalene, 1,6-dimethyl-
7823 - Naphthalene, 1,7-dimethyl-
7824 - Naphthalene, 1,8-dimethyl-
7825 - Naphthalene, 2,3-dimethyl-
7826 - Naphthalene, 2,6-dimethyl-
7827 - Naphthalene, 2,7-dimethyl-
7885 - Naphthalene, 1-ethyl-
7886 - Naphthalene, 2-ethyl-

$C_{12}H_{12}N_2 2Br$
5224 - Diquat dibromide

$C_{12}H_{12}ClNO_2S$
7977 - 1-Naphthalenesulfonyl chloride, 5-(dimethylamino)-

$C_{12}H_{12}ClN_5O_4S$
4091 - Chlorsulfuron

$C_{12}H_{12}N_2$
1282 - 1,2-Benzenediamine, N-phenyl-
1283 - 1,4-Benzenediamine, N-phenyl-
3028 - [1,1'-Biphenyl]-2,2'-diamine
3029 - [1,1'-Biphenyl]-2,4'-diamine
3030 - [1,1'-Biphenyl]-3,3'-diamine
3031 - [1,1'-Biphenyl]-4,4'-diamine
7157 - Hydrazine, 1,1-diphenyl-
7158 - Hydrazine, 1,2-diphenyl-

$C_{12}H_{12}N_2O$
832 - Benzenamine, 4,4'-oxybis-

$C_{12}H_{12}N_2OS$
845 - Benzenamine, 4,4'-sulfinylbis-

$C_{12}H_{12}N_2OS_2$
11684 - 4-Thiazolidinone, 5-[[4-(dimethylamino)phenyl]methylene]-2-thioxo-

$C_{12}H_{12}N_2O_2S$
846 - Benzenamine, 4,4'-sulfonylbis-

$C_{12}H_{12}N_2O_3$
8066 - 1,8-Naphthyridine-3-carboxylic acid, 1-ethyl-1,4-dihydro-7-methyl-4-oxo-
11095 - 2,4,6(1H,3H,5H)-Pyrimidinetrione, 5-ethyl-5-phenyl-

$C_{12}H_{12}N_2S$
850 - Benzenamine, 4,4'-thiobis-

$C_{12}H_{12}N_4$
635 - Benzenamine, 4,4'-azobis-

$C_{12}H_{12}O$
6710 - 3,5-Hexadien-2-one, 6-phenyl-
7883 - Naphthalene, 1-ethoxy-
7884 - Naphthalene, 2-ethoxy-
7911 - 1-Naphthalenemethanol, α-methyl-, (±)-
8027 - 2-Naphthalenol, 1,4-dimethyl-

$C_{12}H_{12}O_2$
8045 - 1(2H)-Naphthalenone, 2-acetyl-3,4-dihydro-
10692 - 2-Propenoic acid, 3-phenyl-, 2-propenyl ester

$C_{12}H_{12}O_5$
2845 - 2H-1-Benzopyran-2-one, 6,7,8-trimethoxy-
10785 - 4H,5H-Pyrano[4,3-b]pyran-4,5-dione, 2,3-dihydro-3-hydroxy-2-methyl-7-(1-propenyl)-, [2S-[2α,3β,7(E)]]-

$C_{12}H_{12}O_6$
2370 - 1,2,4-Benzenetriol, triacetate

$C_{12}H_{12}Si$
11420 - Silane, diphenyl-

$C_{12}H_{13}ClF_3N_3O_4$
5964 - Fluchloralin

$C_{12}H_{13}ClN_4$
1284 - 1,3-Benzenediamine, 4-(phenylazo)-, monohydrochloride
11051 - 2,4-Pyrimidinediamine, 5-(4-chlorophenyl)-6-ethyl-

$C_{12}H_{13}Cl_3O_3$
258 - Acetic acid, (2,4,5-trichlorophenoxy)-, butyl ester
260 - Acetic acid, (2,4,5-trichlorophenoxy)-, 1-methylpropyl ester

$C_{12}H_{13}N$
3989 - 1H-Carbazole, 2,3,4,9-tetrahydro-
7715 - 1-Naphthalenamine, N,N-dimethyl-
7716 - 2-Naphthalenamine, N,N-dimethyl-
7718 - 1-Naphthalenamine, N-ethyl-
7719 - 2-Naphthalenamine, N-ethyl-
8064 - 2-Naphthylamine, 1,4-dimethyl-
11269 - Quinoline, 2-ethyl-3-methyl-
11270 - Quinoline, 2-ethyl-4-methyl-
11271 - Quinoline, 3-ethyl-6-methyl-
11292 - Quinoline, 2-propyl-
11300 - Quinoline, 2,3,6-trimethyl-
11301 - Quinoline, 2,3,8-trimethyl-
11302 - Quinoline, 2,4,6-trimethyl-
11303 - Quinoline, 2,4,7-trimethyl-
11304 - Quinoline, 2,5,7-trimethyl-

$C_{12}H_{13}NO$
5750 - Ethanol, 2-(2-naphthalenylamino)-

$C_{12}H_{13}NO_2$
7296 - 1H-Indole-3-butanoic acid
11156 - 2,5-Pyrrolidinedione, 1,3-dimethyl-3-phenyl-

$C_{12}H_{13}NO_2S$
4039 - Carboxin

$C_{12}H_{13}N_3$
1263 - 1,4-Benzenediamine, N-(4-aminophenyl)-
10974 - 2-Pyridinemethanamine, N-(2-pyridinylmethyl)-

$C_{12}H_{13}N_3O_4S_2$
2188 - Benzenesulfonamide, 4-amino-N-[4-(aminosulfonyl)phenyl]-

$C_{12}H_{14}$
9 - Acenaphthylene, 1,2,2a,3,4,5-hexahydro-
1246 - Benzene, 1-cyclohexen-1-yl-
1247 - Benzene, 2-cyclohexen-1-yl-
1625 - Benzene, (1-ethenyl-3-butenyl)-
1753 - Benzene, 1,3-hexadienyl-
1754 - Benzene, 1,5-hexadienyl-
1755 - Benzene, 3,5-hexadienyl-

$C_{12}H_{14}ClNO_2$
4140 - Clomazone

$C_{12}H_{14}ClN_5O_2S$
2209 - Benzenesulfonamide, 4-[(2,4-diaminophenyl)azo]-

$C_{12}H_{14}Cl_2O_3$
171 - Acetic acid, (2,4-dichlorophenoxy)-, butyl ester

$C_{12}H_{14}Cl_3O_4P$
4078 - Chlorfenvinphos

$C_{12}H_{14}N_2$
5514 - 1,2-Ethanediamine, N-1-naphthalenyl-
11213 - 6-Quinolinamine, N,N,2-trimethyl-

$C_{12}H_{14}N_2O_2$
11066 - 4,6(1H,5H)-Pyrimidinedione, 5-ethyldihydro-5-phenyl-
11923 - L-Tryptophan, N-methyl-

$C_{12}H_{14}N_2O_3$
11085 - 2,4,6(1H,3H,5H)-Pyrimidinetrione, 5-(2-cyclopenten-1-yl)-5-(2-propenyl)-

$C_{12}H_{14}N_2O_5$
6242 - L-Glutamic acid, N-(4-aminobenzoyl)-
9273 - Phenol, 2-cyclohexyl-4,6-dinitro-

$C_{12}H_{14}N_4OS$
11111 - 5H-Pyrimido[4,5-d]thiazolo[3,2-a]pyrimidine-8-ethanol, 2,7-dimethyl-

$C_{12}H_{14}N_4O_2S$
2196 - Benzenesulfonamide, 4-amino-N-(4,6-dimethyl-2-pyrimidinyl)-

$C_{12}H_{14}N_4O_4S$
2194 - Benzenesulfonamide, 4-amino-N-(2,6-dimethoxy-4-pyrimidinyl)-

$C_{12}H_{14}N_4O_4S_2$
11723 - Thiophanate-methyl

$C_{12}H_{14}O$
2460 - Benzofuran, 4-methyl-7-(1-methylethyl)-
4893 - Cyclopentanone, 2-(phenylmethyl)-
7101 - 5-Hexen-2-one, 4-phenyl-
7102 - 5-Hexen-2-one, 6-phenyl-
8051 - 1(2H)-Naphthalenone, 7-ethyl-3,4-dihydro-
8987 - 2-Pentanone, 3-(phenylmethylene)-

$C_{12}H_{14}O_2$
4135 - Cinnamic acid, β-propyl-
6038 - Furan, 2,2'-sec-butylidenedi-
10687 - 2-Propenoic acid, 3-phenyl-, 1-methylethyl ester, (E)-
10693 - 2-Propenoic acid, 3-phenyl-, propyl ester, (E)-

$C_{12}H_{14}O_3$
884 - Benzeneacetic acid, α-acetyl-, ethyl ester
2100 - Benzenepentanoic acid, δ-oxo-, methyl ester
8604 - Oxiranecarboxylic acid, 3-methyl-3-phenyl-, ethyl ester
8911 - Pentanoic acid, 4-oxo-, phenylmethyl ester
9393 - Phenol, 2-methoxy-4-(1-propenyl)-, acetate, (E)-
9394 - Phenol, 2-methoxy-4-(1-propenyl)-, acetate, (Z)-
9395 - Phenol, 2-methoxy-4-(2-propenyl)-, acetate

$C_{12}H_{14}O_4$
1259 - 1,4-Benzenediacetic acid, dimethyl ester
1342 - 1,2-Benzenedicarboxylic acid, diethyl ester
1346 - 1,3-Benzenedicarboxylic acid, diethyl ester
1350 - 1,4-Benzenedicarboxylic acid, diethyl ester
2422 - 1,3-Benzodioxole, 4,7-dimethoxy-5-(1-propenyl)-, (E)-
2423 - 1,3-Benzodioxole, 4,7-dimethoxy-5-(2-propenyl)-
8625 - Oxirane, 2,2'-[1,3-phenylenebis(oxymethylene)]bis-

$C_{12}H_{15}Br$
1009 - Benzene, 1-bromo-4-cyclohexyl-

$C_{12}H_{15}ClNO_4PS_2$
9514 - Phosalone

$C_{12}H_{15}ClO$
2907 - Benzoyl chloride, 4-pentyl-
9254 - Phenol, 2-chloro-4-cyclohexyl-

$C_{12}H_{15}ClO_2$
2908 - Benzoyl chloride, 4-(pentyloxy)-

$C_{12}H_{15}ClO_3$
10231 - Propanoic acid, 2-(4-chlorophenoxy)-2-methyl-, ethyl ester

$C_{12}H_{15}N$
964 - Benzeneacetonitrile, α-(2-methylpropyl)-
1800 - Benzenemethanamine, N-3-butynyl-N-methyl-
2737 - 1H,5H-Benzo[ij]quinolizine, 2,3,6,7-tetrahydro-

$C_{12}H_{15}NO$
9634 - Piperidine, 1-benzoyl-

$C_{12}H_{15}NO_2$
1251 - Benzene, 1-cyclohexyl-2-nitro-
1252 - Benzene, 1-cyclohexyl-4-nitro-

$C_{12}H_{15}NO_3$
3991 - Carbofuran
5214 - 1,3-Dioxolo[4,5-g]isoquinoline, 5,6,7,8-tetrahydro-4-methoxy-6-methyl-

$C_{12}H_{15}NO_4$
5216 - 1,3-Dioxolo[4,5-g]isoquinolin-5-ol, 5,6,7,8-tetrahydro-4-methoxy-6-methyl-

$C_{12}H_{15}N_3O_3$
11842 - 1,3,5-Triazine-2,4,6(1H,3H,5H)-trione, 1,3,5-tri-2-propenyl-

$C_{12}H_{15}N_3O_6$
1461 - Benzene, 1-(1,1-dimethylethyl)-3,5-dimethyl-2,4,6-trinitro-

$C_{12}H_{16}$
1248 - Benzene, cyclohexyl-
1760 - Benzene, 2-hexenyl-
1761 - Benzene, 3-hexenyl-
1762 - Benzene, 5-hexenyl-
1973 - Benzene, 1-(1-methylethenyl)-4-(1-methylethyl)-
2008 - Benzene, (1-methyl-1-pentenyl)-
4744 - 1,3,5,7-Cyclooctatetraene, 1-butyl-
7261 - 1H-Indene, 2,3-dihydro-1,1,5-trimethyl-
7262 - 1H-Indene, 2,3-dihydro-1,4,7-trimethyl-
7888 - Naphthalene, 1-ethyl-1,2,3,4-tetrahydro-
7889 - Naphthalene, 2-ethyl-1,2,3,4-tetrahydro-
7890 - Naphthalene, 5-ethyl-1,2,3,4-tetrahydro-
7891 - Naphthalene, 6-ethyl-1,2,3,4-tetrahydro-
7980 - Naphthalene, 1,2,3,4-tetrahydro-1,1-dimethyl-
7981 - Naphthalene, 1,2,3,4-tetrahydro-1,2-dimethyl-
7982 - Naphthalene, 1,2,3,4-tetrahydro-1,3-dimethyl-
7983 - Naphthalene, 1,2,3,4-tetrahydro-1,4-dimethyl-
7984 - Naphthalene, 1,2,3,4-tetrahydro-1,5-dimethyl-
7985 - Naphthalene, 1,2,3,4-tetrahydro-2,2-dimethyl-
7986 - Naphthalene, 1,2,3,4-tetrahydro-2,3-dimethyl-
7987 - Naphthalene, 1,2,3,4-tetrahydro-2,5-dimethyl-
7988 - Naphthalene, 1,2,3,4-tetrahydro-2,6-dimethyl-
7989 - Naphthalene, 1,2,3,4-tetrahydro-2,7-dimethyl-
7990 - Naphthalene, 1,2,3,4-tetrahydro-2,8-dimethyl-
7991 - Naphthalene, 1,2,3,4-tetrahydro-5,6-dimethyl-
7992 - Naphthalene, 1,2,3,4-tetrahydro-5,7-dimethyl-
7993 - Naphthalene, 1,2,3,4-tetrahydro-5,8-dimethyl-
7994 - Naphthalene, 1,2,3,4-tetrahydro-6,7-dimethyl-

$C_{12}H_{16}ClNOS$
11697 - Thiobencarb

$C_{12}H_{16}N_2$
7303 - 1H-Indole-3-ethanamine, N,N-dimethyl-

$C_{12}H_{16}N_2O$
7326 - 1H-Indol-5-ol, 3-[2-(dimethylamino)ethyl]-
7578 - 1,5-Methano-8H-pyrido[1,2-a][1,5]diazocin-8-one, 1,2,3,4,5,6-hexahydro-3-methyl-, (1R)-

$C_{12}H_{16}N_2O_3$
11083 - 2,4,6(1H,3H,5H)-Pyrimidinetrione, 5-(1-cyclohexen-1-yl)-1,5-dimethyl-
11084 - 2,4,6(1H,3H,5H)-Pyrimidinetrione, 5-(1-cyclohexen-1-yl)-5-ethyl-

$C_{12}H_{16}N_2O_5$
1463 - Benzene, 1-(1,1-dimethylethyl)-2-methoxy-4-methyl-3,5-dinitro-

$C_{12}H_{16}O$
1253 - Benzene, (cyclohexyloxy)-
3669 - 1-Butanone, 3-methyl-1-(2-methylphenyl)-
3670 - 1-Butanone, 3-methyl-1-(4-methylphenyl)-
4552 - Cyclohexanol, 1-phenyl-
4553 - Cyclohexanol, 2-phenyl-, cis-(±)-
4554 - Cyclohexanol, 2-phenyl-, trans-(±)-
5808 - Ethanone, 1-[4-(1,1-dimethylethyl)phenyl]-
5850 - Ethanone, 1-[2-methyl-5-(1-methylethyl)phenyl]-
6983 - 1-Hexanone, 1-phenyl-
6984 - 2-Hexanone, 3-phenyl-
6985 - 3-Hexanone, 2-phenyl-
8983 - 1-Pentanone, 4-methyl-1-phenyl-
8984 - 2-Pentanone, 3-methyl-1-phenyl-
9272 - Phenol, 4-cyclohexyl-

$C_{12}H_{16}O_2$
890 - Benzeneacetic acid, α-butyl-
932 - Benzeneacetic acid, 2-methylpropyl ester
1759 - Benzenehexanoic acid
2129 - Benzenepropanoic acid, α,α-dimethyl-, methyl ester
2140 - Benzenepropanoic acid, 1-methylethyl ester
2146 - Benzenepropanoic acid, β-propyl-
2685 - Benzoic acid, 3-methyl-, butyl ester
2713 - Benzoic acid, pentamethyl-
3560 - Butanoic acid, 3-methyl-, phenylmethyl ester
3632 - 1-Butanol, 3-methyl-, benzoate

7175 - Hydrocinnamic acid, β-ethyl-β-methyl-
8979 - 1-Pentanone, 1-(3-methoxyphenyl)-
9419 - Phenol, 2-methyl-5-(1-methylethyl)-, acetate
9420 - Phenol, 5-methyl-2-(1-methylethyl)-, acetate
10306 - Propanoic acid, 2-methyl-, 2-phenylethyl ester

$C_{12}H_{16}O_3$
2356 - Benzene, 1,2,4-trimethoxy-5-(1-propenyl)-
2633 - Benzoic acid, 2-hydroxy-, 3-methylbutyl ester
2649 - Benzoic acid, 2-hydroxy-, pentyl ester
2668 - Benzoic acid, 4-methoxy-, butyl ester
3581 - Butanoic acid, 4-phenoxy-, ethyl ester
3582 - Butanoic acid, 2-phenoxyethyl ester
6962 - 1-Hexanone, 1-(2,4-dihydroxyphenyl)-
8886 - Pentanoic acid, 2-methoxyphenyl ester
10641 - 2-Propenoic acid, 3-(2-furanyl)-, pentyl ester

$C_{12}H_{16}O_4$
2628 - Benzoic acid, 4-hydroxy-3-methoxy-, 2-methylpropyl ester

$C_{12}H_{16}O_7$
6224 - β-D-Glucopyranoside, 4-hydroxyphenyl

$C_{12}H_{17}BrO$
1045 - Benzene, 1-bromo-4-(hexyloxy)-

$C_{12}H_{17}N$
680 - Benzenamine, N-cyclohexyl-
9698 - Piperidine, 1-(phenylmethyl)-
9699 - Piperidine, 2-(phenylmethyl)-
9700 - Piperidine, 4-(phenylmethyl)-

$C_{12}H_{17}NO$
39 - Acetamide, N-butyl-N-phenyl-
612 - Benzamide, N,N-diethyl-3-methyl-
6722 - Hexanamide, N-phenyl-

$C_{12}H_{17}NO_2$
9769 - Promecarb

$C_{12}H_{17}NO_3$
611 - Benzamide, N,N-diethyl-4-hydroxy-3-methoxy-
880 - Benzeneacetamide, 4-butoxy-N-hydroxy-
7395 - 8-Isoquinolinol, 1,2,3,4-tetrahydro-6,7-dimethoxy-1-methyl-, (S)-

$C_{12}H_{17}N_3O_4$
6247 - L-Glutamic acid, 5-[2-[4-(hydroxymethyl)phenyl]hydrazide]

$C_{12}H_{18}$
992 - Benzene, 1,2-bis(1-methylethyl)-
993 - Benzene, 1,3-bis(1-methylethyl)-
994 - Benzene, 1,4-bis(1-methylethyl)-
1455 - Benzene, (1,1-dimethylbutyl)-
1460 - Benzene, 1-(1,1-dimethylethyl)-3,5-dimethyl-
1473 - Benzene, 2,4-dimethyl-1-(1-methylpropyl)-
1557 - Benzene, 1,4-dipropyl-
1669 - Benzene, (1-ethylbutyl)-
1691 - Benzene, (1-ethyl-1-methylpropyl)-
1758 - Benzene, hexamethyl-
1763 - Benzene, hexyl-
2009 - Benzene, (1-methylpentyl)-
2010 - Benzene, (3-methylpentyl)-
2011 - Benzene, (4-methylpentyl)-
2344 - Benzene, 1,2,4-triethyl-
2345 - Benzene, 1,3,5-triethyl-
2361 - Benzene, 1,2,4-trimethyl-5-(1-methylethyl)-
4212 - 1,5,9-Cyclododecatriene
4213 - 1,5,9-Cyclododecatriene, (Z,E,E)-
4214 - 1,5,9-Cyclododecatriene, (Z,Z,Z)-
5286 - 3,9-Dodecadiyne
5287 - 5,7-Dodecadiyne
8339 - 2,6-Octadien-4-yne, 3,6-diethyl-

$C_{12}H_{18}Cl_2N_4OS$
11686 - Thiazolium, 3-[(4-amino-2-methyl-5-pyrimidinyl)methyl]-5-(2-hydroxyethyl)-4-methyl- chloride, monohydrochlorid

$C_{12}H_{18}N_2O$
11188 - 2-Pyrrolidinone, 1-[4-(1-pyrrolidinyl)-2-butynyl]-

$C_{12}H_{18}N_2O_3$
11090 - 2,4,6(1H,3H,5H)-Pyrimidinetrione, 5-(2,2-dimethylpropyl)-5-(2-propenyl)-

$C_{12}H_{18}N_2O_3S$
2207 - Benzenesulfonamide, N-[(butylamino)carbonyl]-4-methyl-

$C_{12}H_{18}N_4O$
7171 - Hydrazine, phenyl, hemihydrate

$C_{12}H_{18}N_4O_6S$
8553 - Oryzalin

$C_{12}H_{18}O$
1488 - Benzene, 1-(1,1-dimethylpropyl)-4-methoxy-
1765 - Benzene, (hexyloxy)-
1847 - Benzenemethanol, α-butyl-α-methyl-
1875 - Benzenemethanol, α-ethyl-α-propyl-
1909 - Benzenemethanol, α-pentyl-
1951 - Benzene, [(3-methylbutoxy)methyl]-
2103 - Benzenepentanol, α-methyl-
2159 - Benzenepropanol, γ-propyl-
6949 - 3-Hexanol, 1-phenyl-
9226 - Phenol, 2,6-bis(1-methylethyl)-
9324 - Phenol, 2-(1,1-dimethylethyl)-4,5-dimethyl-
9325 - Phenol, 2-(1,1-dimethylethyl)-4,6-dimethyl-
9326 - Phenol, 4-(1,1-dimethylethyl)-2,5-dimethyl-
9327 - Phenol, 4-(1,1-dimethylethyl)-2,6-dimethyl-
9340 - Phenol, 4-(1,1-dimethylpropyl)-2-methyl-
9347 - Phenol, 2,6-dipropyl-
9425 - Phenol, 5-methyl-2-pentyl-

$C_{12}H_{18}O_2$
1519 - 1,3-Benzenediol, 4-hexyl-
1555 - Benzene, 1,2-dipropoxy-
1556 - Benzene, 1,3-dipropoxy-
2950 - Bicyclo[2.2.1]heptan-2-one, 3-(acetyl-17O)-1,7,7-trimethyl-
11863 - Tricyclo[2.2.1.0²,⁶]heptane-1-carboxylic acid, 7,7-dimethyl-, ethyl ester

$C_{12}H_{18}O_3$
2343 - Benzene, 1,3,5-triethoxy-

$C_{12}H_{18}O_4$
4644 - 1-Cyclohexene-1,2-dicarboxylic acid, diethyl ester
4645 - 1-Cyclohexene-1,3-dicarboxylic acid, diethyl ester
9935 - Propanedioic acid, 2-cyclopenten-1-yl-, diethyl ester
9936 - Propanedioic acid, cyclopentylidene-, diethyl ester

$C_{12}H_{18}O_5$
10669 - 2-Propenoic acid, 2-methyl-, oxydi-2,1-ethanediyl ester

$C_{12}H_{18}O_6$
10601 - 1-Propene-1,2,3-tricarboxylic acid, triethyl ester, (E)-

$C_{12}H_{19}ClNO_3P$
4153 - Crufomate

$C_{12}H_{19}N$
636 - Benzenamine, 2,6-bis(1-methylethyl)-
730 - Benzenamine, N,N-dipropyl-

$C_{12}H_{19}NO$
738 - Benzenamine, 3-ethoxy-N,N-diethyl-

$C_{12}H_{20}$
5342 - 1-Dodecen-3-yne

$C_{12}H_{20}As_2Cl_2O_5$
9275 - Phenol, 4,4'-(1,2-diarsenediyl)bis[2-amino-, dihydrochloride

$C_{12}H_{20}N_2O_2$
10811 - 2(1H)-Pyrazinone, 1-hydroxy-6-(1-methylpropyl)-3-(2-methylpropyl)-, (+)-

$C_{12}H_{20}N_4O_2$
6995 - Hexazinone

$C_{12}H_{20}O$
2989 - [1,1'-Bicyclohexyl]-2-one
6032 - Furan, 2,5-bis(1,1-dimethylethyl)-

$C_{12}H_{20}O_2$
2945 - Bicyclo[2.2.1]heptan-2-ol, 1,7,7-trimethyl-, acetate, endo-(±)-
2946 - Bicyclo[2.2.1]heptan-2-ol, 1,7,7-trimethyl-, acetate, exo-(±)-
2947 - Bicyclo[2.2.1]heptan-2-ol, 4,7,7-trimethyl-, acetate, (1R-endo)-

4545 - Cyclohexanol, 5-methyl-2-(1-methylethenyl)-, acetate, [1R-(1α,2β,5α)]-
4669 - 3-Cyclohexene-1-methanol, α,α,4-trimethyl-, acetate
6135 - Furan, 2-(octyloxy)-
8335 - 1,6-Octadien-3-ol, 3,7-dimethyl-, acetate, (R)-
8336 - 2,6-Octadien-1-ol, 3,7-dimethyl-, acetate, (Z)-
11990 - 9-Undecynoic acid, methyl ester
11991 - 10-Undecynoic acid, methyl ester

$C_{12}H_{20}O_3Si$
11481 - Silane, triethoxyphenyl-

$C_{12}H_{20}O_4$
3767 - 2-Butenedioic acid (E)-, bis(2-methylpropyl) ester
3774 - 2-Butenedioic acid (E)-, dibutyl ester
4350 - 1,2-Cyclohexanedicarboxylic acid, diethyl ester, trans-(±)-
4351 - 1,3-Cyclohexanedicarboxylic acid, diethyl ester, cis-
4355 - 1,4-Cyclohexanedicarboxylic acid, diethyl ester, cis-
4356 - 1,4-Cyclohexanedicarboxylic acid, diethyl ester, trans-
4795 - 1,3-Cyclopentanedicarboxylic acid, 1,2,2-trimethyl-, dimethyl ester, (1R-cis)-
5336 - 2-Dodecenedioic acid, (E)-

$C_{12}H_{20}O_4Si$
11511 - Silicic acid (H_4SiO_4), triethyl phenyl ester

$C_{12}H_{20}O_6$
10198 - 1,2,3-Propanetriol, tripropanoate

$C_{12}H_{20}O_7$
10169 - 1,2,3-Propanetricarboxylic acid, 2-hydroxy-, triethyl ester

$C_{12}H_{20}Si$
11485 - Silane, triethylphenyl-

$C_{12}H_{21}NO$
4611 - Cyclohexanone, 2-(1-piperidinylmethyl)-

$C_{12}H_{21}N_2O_3PS$
9582 - Phosphorothioic acid, O,O-diethyl O-[6-methyl-2-(1-methylethyl)-4-pyrimidinyl] ester

$C_{12}H_{22}$
2988 - 1,1'-Bicyclohexyl
4215 - Cyclododecene, (E)-
4216 - Cyclododecene, (Z)-
5101 - 4-Decyne, 3,3-dimethyl-
5285 - 1,11-Dodecadiene
5343 - 1-Dodecyne
5344 - 2-Dodecyne
5345 - 3-Dodecyne
5346 - 6-Dodecyne

$C_{12}H_{22}O$
4211 - Cyclododecanone
4461 - Cyclohexane, 1,1'-oxybis-

$C_{12}H_{22}OSi_2$
11497 - Silane, trimethyl[4-[(trimethylsilyl)oxy]phenyl]-

$C_{12}H_{22}O_2$
2948 - Bicyclo[2.2.1]heptan-2-ol, 1,7,7-trimethyl-, acetate, endo-
4412 - Cyclohexanehexanoic acid
4549 - Cyclohexanol, 5-methyl-2-(1-methylethyl)-, acetate, [1R-(1α,2β,5α)]-
4787 - Cyclopentanecarboxylic acid, 3-isopropyl-1-methyl-
4788 - Cyclopentanecarboxylic acid, 2-methyl-5-(1-methylethyl)-, ethyl ester
5337 - 2-Dodecenoic acid
5338 - 4-Dodecenoic acid
5339 - 5-Dodecenoic acid
5340 - 11-Dodecenoic acid
8400 - 3,6-Octanedione, 2,2,7,7-tetramethyl-
10659 - 2-Propenoic acid, 2-methyl-, 2-ethylhexyl ester
11978 - 10-Undecenoic acid, methyl ester

$C_{12}H_{22}O_2Si_2$
11447 - Silane, [1,3-phenylenebis(oxy)]bis[trimethyl-

$C_{12}H_{22}O_3$
210 - Acetic acid, [[5-methyl-2-(1-methylethyl)cyclohexyl]oxy]-, [1R-(1α,2β,5α)]-
6874 - Hexanoic acid, anhydride

$C_{12}H_{22}O_4$
3296 - Butanedioic acid, bis(1,1-dimethylethyl) ester
3297 - Butanedioic acid, bis(1-methylpropyl) ester
3301 - Butanedioic acid, dibutyl ester
3332 - Butanedioic acid, (1-methylpropyl)-, diethyl ester
5046 - Decanedioic acid, dimethyl ester
5299 - Dodecanedioic acid
5555 - Ethanedioic acid, bis(3-methylbutyl) ester
6782 - Hexanedioic acid, bis(1-methylethyl) ester
6787 - Hexanedioic acid, dipropyl ester
8387 - Octanedioic acid, diethyl ester
9951 - Propanedioic acid, ethyl(1-methylethyl)-, diethyl ester
9961 - Propanedioic acid, (3-methylbutyl)-, diethyl ester
9974 - Propanedioic acid, pentyl-, diethyl ester

$C_{12}H_{22}O_6$
3313 - Butanedioic acid, 2,3-dihydroxy- [R-(R^*,R^*)]-, dibutyl ester

$C_{12}H_{22}O_{10}$
6236 - D-Glucose, 6-O-(6-deoxy-α-L-mannopyranosyl)-

$C_{12}H_{22}O_{11}$
6022 - D-Fructopyranose, 3-O-α-D-glucopyranosyl-
6214 - β-D-Glucopyranose, 4-O-β-D-galactopyranosyl-
6218 - α-D-Glucopyranoside, β-D-fructofuranosyl
6220 - α-D-Glucopyranoside, α-D-glucopyranosyl

$C_{12}H_{22}O_{12}$
6210 - D-Gluconic acid, 4-O-β-D-galactopyranosyl-

$C_{12}H_{23}BrO_2$
5307 - Dodecanoic acid, 2-bromo-

$C_{12}H_{23}ClO$
5328 - Dodecanoyl chloride

$C_{12}H_{23}N$
4284 - Cyclohexanamine, N-cyclohexyl-
5303 - Dodecanenitrile

$C_{12}H_{23}NSi_2$
11362 - Silanamine, 1,1,1-trimethyl-N-phenyl-N-(trimethylsilyl)-

$C_{12}H_{23}P$
9516 - Phosphine, dicyclohexyl-

$C_{12}H_{24}$
4210 - Cyclododecane
4414 - Cyclohexane, hexyl-
4446 - Cyclohexane, 1-methyl-2-pentyl-
4737 - Cyclooctane, butyl-
4820 - Cyclopentane, heptyl-
5335 - 1-Dodecene
10582 - 1-Propene, 2-methyl-, trimer

$C_{12}H_{24}BrF$
5294 - Dodecane, 1-bromo-12-fluoro-

$C_{12}H_{24}Br_2$
5296 - Dodecane, 1,12-dibromo-

$C_{12}H_{24}N_2$
9661 - Piperidine, 1,1'-(1,2-ethanediyl)bis-

$C_{12}H_{24}O$
5051 - Decane, 1-(ethenyloxy)-
5288 - Dodecanal
5325 - 2-Dodecanone
5326 - 6-Dodecanone
11936 - Undecanal, 2-methyl-

$C_{12}H_{24}O_2$
157 - Acetic acid, decyl ester
3566 - Butanoic acid, octyl ester
5063 - Decanoic acid, ethyl ester
5305 - Dodecanoic acid
6470 - Heptanoic acid, pentyl ester
6890 - Hexanoic acid, hexyl ester
8155 - Nonanoic acid, 3-methyl-, ethyl ester
8427 - Octanoic acid, butyl ester

8881 - Pentanoic acid, heptyl ester

$C_{12}H_{24}O_3$
11949 - Undecanoic acid, 11-hydroxy-, methyl ester

$C_{12}H_{24}O_4Si_4$
4996 - Cyclotetrasiloxane, 2,4,6,8-tetraethenyl-2,4,6,8-tetramethyl-

$C_{12}H_{25}Br$
5293 - Dodecane, 1-bromo-

$C_{12}H_{25}Cl$
5295 - Dodecane, 1-chloro-

$C_{12}H_{25}Cl_3Si$
11460 - Silane, trichlorododecyl-

$C_{12}H_{25}I$
5302 - Dodecane, 1-iodo-

$C_{12}H_{25}NO$
5289 - Dodecanamide

$C_{12}H_{26}$
5292 - Dodecane
6433 - Heptane, 2,2,4,6,6-pentamethyl-

$C_{12}H_{26}O$
5321 - 1-Dodecanol
5322 - 2-Dodecanol
5323 - 3-Dodecanol
5324 - 6-Dodecanol
6842 - Hexane, 1,1'-oxybis-
11955 - 1-Undecanol, 2-methyl-
11956 - 3-Undecanol, 2-methyl-
11957 - 5-Undecanol, 2-methyl-

$C_{12}H_{26}O_2$
5301 - 1,12-Dodecanediol

$C_{12}H_{26}O_3$
3417 - Butane, 1,1'-[oxybis(2,1-ethanediyloxy)]bis-

$C_{12}H_{26}O_6P_2S_4$
5201 - Dioxathion

$C_{12}H_{26}S$
5304 - 1-Dodecanethiol
6855 - Hexane, 1,1'-thiobis-

$C_{12}H_{27}B$
3123 - Borane, tris(2-methylpropyl)-

$C_{12}H_{27}BO_3$
3124 - Boric acid, tributyl ester

$C_{12}H_{27}N$
3213 - 1-Butanamine, N,N-dibutyl-
5291 - 1-Decanamine
6727 - 1-Hexanamine, N-hexyl-
9820 - 1-Propanamine, 2-methyl-N,N-bis(2-methylpropyl)-

$C_{12}H_{27}NO$
5682 - Ethanol, 2-[bis(3-methylbutyl)amino]-

$C_{12}H_{27}NO_2$
8936 - 2-Pentanol, 1,1'-iminobis[2-methyl-

$C_{12}H_{27}OPS_3$
11852 - S,S,S-Tributyl phosphorotrithioate

$C_{12}H_{27}O_3P$
9590 - Phosphorous acid, tributyl ester

$C_{12}H_{27}O_4P$
9562 - Phosphoric acid tributyl ester
9572 - Phosphoric acid, tris(2-methylpropyl) ester

$C_{12}H_{27}P$
9520 - Phosphine, tributyl-

$C_{12}H_{27}PS_3$
7438 - Merphos

$C_{12}H_{28}IN$
9833 - 1-Propanaminium, N,N,N-tripropyl-, iodide

$C_{12}H_{28}N_2$
3262 - 1,4-Butanediamine, N,N,N',N'-tetraethyl-

$C_{12}H_{28}O_4Si$
11510 - Silicic acid (H_4SiO_4), tetrapropyl ester

$C_{12}H_{28}Si$
11454 - Silane, tributyl-

$C_{12}H_{28}Sn$
11546 - Stannane, tetrapropyl-

$C_{12}H_{30}OSi_2$
5235 - Disiloxane, hexaethyl-

$C_{12}H_{30}O_{13}P_4$
11645 - Tetraphosphoric acid, hexaethyl ester

C₁₂H₃₂O₄Si₄
 4997 - Cyclotetrasiloxane, 2,4,6,8-tetraethyl-2,4,6,8-tetramethyl-

C₁₂H₃₆O₄Si₅
 8997 - Pentasiloxane, dodecamethyl-

C₁₂H₃₆O₆Si₆
 4624 - Cyclohexasiloxane, dodecamethyl-

C₁₃H₅N₃O₇
 5979 - 9H-Fluoren-9-one, 2,4,7-trinitro-

C₁₃H₆Cl₆O₂
 9409 - Phenol, 2,2'-methylenebis[3,4,6-trichloro-

C₁₃H₇ClF₃NO₃
 8082 - Nitrofluorfen

C₁₃H₇ClO
 5976 - 9H-Fluoren-9-one, 2-chloro-

C₁₃H₇Cl₄NO₂
 610 - Benzamide, 3,5-dichloro-N-(3,4-dichlorophenyl)-2-hydroxy-

C₁₃H₇F₃N₂O₅
 5980 - Fluorodifen

C₁₃H₇NO₃
 5977 - 9H-Fluoren-9-one, 2-nitro-

C₁₃H₈Br₂O
 7590 - Methanone, bis(4-bromophenyl)-

C₁₃H₈Br₃NO₂
 609 - Benzamide, 3,5-dibromo-N-(4-bromophenyl)-2-hydroxy-

C₁₃H₈ClN
 315 - Acridine, 9-chloro-

C₁₃H₈Cl₂
 5972 - 9H-Fluorene, 2,7-dichloro-

C₁₃H₈Cl₂O
 7591 - Methanone, bis(3-chlorophenyl)-
 7592 - Methanone, bis(4-chlorophenyl)-
 7603 - Methanone, (2-chlorophenyl)(4-chlorophenyl)-

C₁₃H₈Cl₃NO₂
 608 - Benzamide, 5-chloro-N-(3,4-dichlorophenyl)-2-hydroxy-

C₁₃H₈I₂O
 7598 - Methanone, bis(4-iodophenyl)-

C₁₃H₈O
 5975 - 9H-Fluoren-9-one

C₁₃H₈OS
 11812 - 9H-Thioxanthen-9-one

C₁₃H₈O₂
 12050 - 9H-Xanthen-9-one

C₁₃H₈O₄
 12051 - 9H-Xanthen-9-one, 1,3-dihydroxy-
 12052 - 9H-Xanthen-9-one, 2,7-dihydroxy-
 12053 - 9H-Xanthen-9-one, 3,6-dihydroxy-

C₁₃H₉Br
 5968 - 9H-Fluorene, 2-bromo-

C₁₃H₉BrO
 7600 - Methanone, (2-bromophenyl)phenyl-
 7601 - Methanone, (3-bromophenyl)phenyl-
 7602 - Methanone, (4-bromophenyl)phenyl-

C₁₃H₉ClO
 7604 - Methanone, (4-chlorophenyl)phenyl-

C₁₃H₉Cl₂FN₂S
 11802 - Thiourea, N-(3,5-dichlorophenyl)-N'-(4-fluorophenyl)-

C₁₃H₉N
 314 - Acridine
 2437 - Benzo[f]quinoline
 2473 - Benzo[g]quinoline
 2474 - Benzo[h]quinoline
 3020 - [1,1'-Biphenyl]-2-carbonitrile
 9188 - Phenanthridine

C₁₃H₉NOS
 9217 - Phenol, 2-(2-benzothiazolyl)-

C₁₃H₉NO₂
 9218 - Phenol, 2-(2-benzoxazolyl)-

C₁₃H₉NO₃
 7635 - Methanone, (3-nitrophenyl)phenyl-
 7636 - Methanone, (4-nitrophenyl)phenyl-

C₁₃H₉NS
 2864 - Benzothiazole, 2-phenyl-

C₁₃H₉N₃O₅
 2648 - Benzoic acid, 2-hydroxy-5-[(4-nitrophenyl)azo]-

C₁₃H₁₀
 5967 - 9H-Fluorene

C₁₃H₁₀Br₂
 1962 - Benzene, 1,1'-methylenebis[2-bromo-

C₁₃H₁₀Cl₂
 1389 - Benzene, 1,1'-(dichloromethylene)bis-
 1963 - Benzene, 1,1'-methylenebis[4-chloro-

C₁₃H₁₀Cl₂O₂
 1965 - Benzene, 1,1'-[methylenebis(oxy)]bis[4-chloro-
 9408 - Phenol, 2,2'-methylenebis[4-chloro-

C₁₃H₁₀Cl₂S
 4075 - Chlorbenside

C₁₃H₁₀N₂
 312 - 4-Acridinamine
 313 - 9-Acridinamine
 764 - Benzenamine, N,N'-methanetetraylbis-
 2383 - 1H-Benzimidazole, 1-phenyl-
 2384 - 1H-Benzimidazole, 2-phenyl-
 4158 - Cyanamide, diphenyl-
 9195 - Phenazine, 2-methyl-

C₁₃H₁₀N₂O
 9196 - Phenazinium, 1-hydroxy-5-methyl-, hydroxide, inner salt

C₁₃H₁₀N₂O₄
 7363 - 1H-Isoindole-1,3(2H)-dione, 2-(2,6-dioxo-3-piperidinyl)-

C₁₃H₁₀N₄O₅
 11997 - Urea, N,N'-bis(4-nitrophenyl)-

C₁₃H₁₀O
 7616 - Methanone, diphenyl-
 12049 - 9H-Xanthene

C₁₃H₁₀O₂
 597 - Benzaldehyde, 3-phenoxy-
 2716 - Benzoic acid, phenyl ester
 3021 - [1,1'-Biphenyl]-2-carboxylic acid
 3022 - [1,1'-Biphenyl]-4-carboxylic acid
 7622 - Methanone, (2-hydroxyphenyl)phenyl-
 10675 - 2-Propenoic acid, 3-(1-naphthalenyl)-, (E)-
 10710 - 2-Propen-1-one, 3-(2-furanyl)-1-phenyl-

C₁₃H₁₀O₃
 2650 - Benzoic acid, 2-hydroxy-, phenyl ester
 2714 - Benzoic acid, 2-phenoxy-
 4001 - Carbonic acid, diphenyl ester
 7596 - Methanone, bis(2-hydroxyphenyl)-
 7597 - Methanone, bis(4-hydroxyphenyl)-
 7609 - Methanone, (2,4-dihydroxyphenyl)phenyl-
 8682 - 1,4-Pentadien-3-one, 1,5-di-2-furanyl-

C₁₃H₁₀O₄
 6186 - 5H-Furo[3,2-g][1]benzopyran-5-one, 4-methoxy-7-methyl-

C₁₃H₁₀O₆
 7610 - Methanone, (3,4-dihydroxyphenyl)(2,4,6-trihydroxyphenyl)-

C₁₃H₁₀S
 7547 - Methanethione, diphenyl-
 11811 - 9H-Thioxanthene

C₁₃H₁₁Br
 1058 - Benzene, 1,1'-(bromomethylene)bis-

C₁₃H₁₁Cl
 1187 - Benzene, 1,1'-(chloromethylene)bis-
 1224 - Benzene, 1-chloro-4-(phenylmethyl)-
 3026 - Biphenyl, 2-chloro-2'-methyl-

C₁₃H₁₁ClO
 9270 - Phenol, 4-chloro-2-(phenylmethyl)-

C₁₃H₁₁Cl₂NO₂
 9760 - Procymidone

C₁₃H₁₁N
 840 - Benzenamine, N-(phenylmethylene)-
 1835 - Benzenemethanimine, α-phenyl-
 3986 - 9H-Carbazole, 3-methyl-
 3987 - 9H-Carbazole, 9-methyl-
 11000 - Pyridine, 2-(2-phenylethenyl)-, (E)-
 11001 - Pyridine, 4-(2-phenylethenyl)-, (E)-

 11002 - Pyridine, 2-(2-phenylethenyl)-, (Z)-

C₁₃H₁₁NO
 626 - Benzamide, N-phenyl-
 5993 - Formamide, N,N-diphenyl-
 9450 - Phenol, 2-[(phenylimino)methyl]-

C₁₃H₁₁NO₂
 617 - Benzamide, 2-hydroxy-N-phenyl-
 2715 - Benzoic acid, 2-(phenylamino)-
 10911 - 3-Pyridinecarboxylic acid, phenylmethyl ester

C₁₃H₁₁NO₃
 2005 - Benzene, 1-methyl-2-(2-nitrophenoxy)-
 2006 - Benzene, 1-methyl-4-(4-nitrophenoxy)-
 2502 - Benzoic acid, 4-amino-2-hydroxy-, phenyl ester
 6179 - Furo[2,3-b]quinoline, 4,8-dimethoxy-

C₁₃H₁₁NS
 1129 - Benzenecarbothioamide, N-phenyl-

C₁₃H₁₁N₃
 316 - 3,6-Acridinediamine

C₁₃H₁₂
 632 - 1H-Benz[e]indene, 2,3-dihydro-
 1961 - Benzene, 1,1'-methylenebis-
 3082 - 1,1'-Biphenyl, 2-methyl-
 3083 - 1,1'-Biphenyl, 3-methyl-
 3084 - 1,1'-Biphenyl, 4-methyl-
 7954 - Naphthalene, 1-(2-propenyl)-

C₁₃H₁₂Cl₂N₂
 785 - Benzenamine, 4,4'-methylenebis[2-chloro-

C₁₃H₁₂Cl₂O₄
 168 - Acetic acid, [2,3-dichloro-4-(2-methylene-1-oxobutyl)phenoxy]-

C₁₃H₁₂N₂
 5138 - Diazene, (3-methylphenyl)phenyl-
 5139 - Diazene, (4-methylphenyl)phenyl-
 5971 - 9H-Fluorene-2,7-diamine
 7565 - Methanimidamide, N,N'-diphenyl-

C₁₃H₁₂N₂O
 2717 - Benzoic acid, 2-phenylhydrazide
 5135 - Diazene, (2-methoxyphenyl)phenyl-
 5136 - Diazene, (3-methoxyphenyl)phenyl-
 5137 - Diazene, (4-methoxyphenyl)phenyl-
 7589 - Methanone, bis(3-aminophenyl)-
 9426 - Phenol, 4-methyl-2-(phenylazo)-
 11039 - 9H-Pyrido[3,4-b]indole, 7-methoxy-1-methyl-
 12010 - Urea, N,N'-diphenyl-
 12011 - Urea, N,N-diphenyl-

C₁₃H₁₂N₂O₃
 11103 - 2,4,6(1H,3H,5H)-Pyrimidinetrione, 5-phenyl-5-(2-propenyl)-

C₁₃H₁₂N₂O₃S
 603 - Benzamide, N-[(4-aminophenyl)sulfonyl]-

C₁₃H₁₂N₂S
 11805 - Thiourea, N,N'-diphenyl-

C₁₃H₁₂N₄O
 5122 - Diazenecarboxylic acid, phenyl-, 2-phenylhydrazide

C₁₃H₁₂N₄S
 5121 - Diazenecarbothioic acid, phenyl-, 2-phenylhydrazide

C₁₃H₁₂O
 1910 - Benzenemethanol, α-phenyl-
 2012 - Benzene, 1-methyl-2-phenoxy-
 2013 - Benzene, 1-methyl-3-phenoxy-
 3080 - 1,1'-Biphenyl, 2-methoxy-
 3081 - 1,1'-Biphenyl, 4-methoxy-
 5978 - 9H-Fluoren-9-one, 1,2,3,4-tetrahydro-
 9454 - Phenol, 2-(phenylmethyl)-
 9455 - Phenol, 4-(phenylmethyl)-
 10473 - 1-Propanone, 1-(1-naphthalenyl)-
 10474 - 1-Propanone, 1-(2-naphthalenyl)-

C₁₃H₁₂O₂
 1546 - 1,4-Benzenediol, 2-(phenylmethyl)-
 1938 - Benzene, 1-methoxy-2-phenoxy-
 7753 - 1-Naphthalenecarboxaldehyde, 2-ethoxy-
 7765 - 1-Naphthalenecarboxylic acid, ethyl ester
 7770 - 2-Naphthalenecarboxylic acid, ethyl ester
 7952 - 1-Naphthalenepropanoic acid

8113 - 6,8-Nonadien-2-one, 8-methyl-5-(1-methylethyl)-, [S-(E)]-

C₁₃H₂₂O₄
4171 - Cyclobutaneacetic acid, 3-(ethoxycarbonyl)-2,2-dimethyl-, ethyl ester
9934 - Propanedioic acid, cyclohexyl-, diethyl ester

C₁₃H₂₄
4429 - Cyclohexane, 1,1'-methylenebis-
11897 - 1-Tridecyne

C₁₃H₂₄N₂O
10415 - 2-Propanone, 1,3-bis(1-methyl-2-pyrrolidinyl)-, (R*,S*)-

C₁₃H₂₄N₃O₃PS
9729 - Pirimiphos-ethyl

C₁₃H₂₄O₂
5341 - 11-Dodecenoic acid, methyl ester
10646 - 2-Propenoic acid, isodecyl ester
11894 - 2-Tridecenoic acid
11895 - 12-Tridecenoic acid
11977 - 10-Undecenoic acid, ethyl ester

C₁₃H₂₄O₃
6166 - 2-Furanpropanol, tetrahydro-α-[2-(tetrahydro-2-furanyl)ethyl]-

C₁₃H₂₄O₄
8133 - Nonanedioic acid, diethyl ester
9955 - Propanedioic acid, hexyl-, diethyl ester

C₁₃H₂₅BrO₂
5309 - Dodecanoic acid, 2-bromo-, methyl ester

C₁₃H₂₅N
11880 - Tridecanenitrile

C₁₃H₂₆
4327 - Cyclohexane, 2-butyl-1,1,3-trimethyl-
4405 - Cyclohexane, heptyl-
4843 - Cyclopentane, octyl-
8184 - 4-Nonene, 5-butyl-
11893 - 1-Tridecene

C₁₃H₂₆Br₂
11874 - Tridecane, 1,13-dibromo-

C₁₃H₂₆N₂
4299 - Cyclohexanamine, 4,4'-methylenebis-
4300 - Cyclohexanamine, 4,4'-methylenebis-, [cis(cis)]-
4301 - Cyclohexanamine, 4,4'-methylenebis-, [trans(cis)]-
4302 - Cyclohexanamine, 4,4'-methylenebis-, [trans(trans)]-
9701 - Piperidine, 4,4'-(1,3-propanediyl)bis-

C₁₃H₂₆N₂O₄
3933 - Carbamic acid, butyl-,2-[[(aminocarbonyl)oxy]methyl]-2-methylpentyl ester

C₁₃H₂₆O
11869 - Tridecanal
11889 - 2-Tridecanone
11890 - 3-Tridecanone
11891 - 7-Tridecanone

C₁₃H₂₆O₂
5067 - Decanoic acid, 1-methylethyl ester
5069 - Decanoic acid, propyl ester
5314 - Dodecanoic acid, 2-methyl-
5315 - Dodecanoic acid, methyl ester
6465 - Heptanoic acid, hexyl ester
6889 - Hexanoic acid, heptyl ester
8147 - Nonanoic acid, butyl ester
8439 - Octanoic acid, pentyl ester
8901 - Pentanoic acid, octyl ester
11881 - Tridecanoic acid
11946 - Undecanoic acid, ethyl ester

C₁₃H₂₇Br
11872 - Tridecane, 1-bromo-

C₁₃H₂₇N
9690 - Piperidine, 1-octyl-

C₁₃H₂₈
8122 - Nonane, 5-butyl-
11871 - Tridecane

C₁₃H₂₈O
8162 - 5-Nonanol, 5-butyl-
11883 - 1-Tridecanol

11884 - 2-Tridecanol
11885 - 3-Tridecanol

C₁₃H₂₈O₂
11875 - 1,12-Tridecanediol
11876 - 1,13-Tridecanediol

C₁₃H₂₈O₃
10080 - Propane, 1,1',1''-[methylidynetris(oxy)]tris[2-methyl-

C₁₃H₂₈O₄
10068 - Propane, 1,1',1'',1'''-[methanetetrayltetrakis(oxy)]tetrakis-

C₁₃H₂₉N
6354 - 2-Heptanamine, 6-methyl-N-(3-methylbutyl)-
11870 - 1-Tridecanamine

C₁₄H₄N₂O₂S₂
8056 - Naphtho[2,3-b]-1,4-dithiin-2,3-dicarbonitrile, 5,10-dihydro-5,10-dioxo-

C₁₄H₄N₄O₁₂
401 - 9,10-Anthracenedione, 1,8-dihydroxy-2,4,5,7-tetranitro-

C₁₄H₄O₆
2784 - [2]Benzopyrano[6,5,4-def][2]benzopyran-1,3,6,8-tetrone

C₁₄H₆Br₂O₂
390 - 9,10-Anthracenedione, 2,7-dibromo-

C₁₄H₆Cl₄O₄
9152 - Peroxide, bis(2,4-dichlorobenzoyl)

C₁₄H₆N₂O₆
405 - 9,10-Anthracenedione, 1,5-dinitro-

C₁₄H₇BrO₂
385 - 9,10-Anthracenedione, 1-bromo-
386 - 9,10-Anthracenedione, 2-bromo-

C₁₄H₇ClF₃NO₅
304 - Acifluorfen

C₁₄H₇ClO₂
387 - 9,10-Anthracenedione, 1-chloro-
388 - 9,10-Anthracenedione, 2-chloro-

C₁₄H₇NO₄
410 - 9,10-Anthracenedione, 1-nitro-
411 - 9,10-Anthracenedione, 2-nitro-

C₁₄H₇NO₆
399 - 9,10-Anthracenedione, 1,2-dihydroxy-3-nitro-
400 - 9,10-Anthracenedione, 1,2-dihydroxy-4-nitro-

C₁₄H₈Br₂
375 - Anthracene, 9,10-dibromo-

C₁₄H₈Cl₂N₄
4139 - Clofentezine

C₁₄H₈Cl₄
1374 - Benzene, 1,1'-(dichloroethenylidene)bis[4-chloro-

C₁₄H₈O₂
379 - 9,10-Anthracenedione
9174 - 9,10-Phenanthrenedione

C₁₄H₈O₃
407 - 9,10-Anthracenedione, 1-hydroxy-
408 - 9,10-Anthracenedione, 2-hydroxy-
5145 - Dibenz[c,e]oxepin-5,7-dione
5969 - 9H-Fluorene-2-carboxylic acid, 9-oxo-
9175 - 9,10-Phenanthrenedione, 2-hydroxy-
9176 - 9,10-Phenanthrenedione, 3-hydroxy-

C₁₄H₈O₄
391 - 9,10-Anthracenedione, 1,2-dihydroxy-
392 - 9,10-Anthracenedione, 1,4-dihydroxy-
393 - 9,10-Anthracenedione, 1,5-dihydroxy-
394 - 9,10-Anthracenedione, 1,7-dihydroxy-
395 - 9,10-Anthracenedione, 1,8-dihydroxy-
396 - 9,10-Anthracenedione, 2,7-dihydroxy-

C₁₄H₈O₅
415 - 9,10-Anthracenedione, 1,2,3-trihydroxy-
416 - 9,10-Anthracenedione, 1,2,4-trihydroxy-
417 - 9,10-Anthracenedione, 1,2,6-trihydroxy-
418 - 9,10-Anthracenedione, 1,2,7-trihydroxy-

C₁₄H₈O₆
412 - 9,10-Anthracenedione, 1,2,5,6-tetrahydroxy-
413 - 9,10-Anthracenedione, 1,2,5,8-tetrahydroxy-
414 - 9,10-Anthracenedione, 1,3,5,7-tetrahydroxy-

C₁₄H₈O₈
406 - 9,10-Anthracenedione, 1,2,3,5,6,7-hexahydroxy-

C₁₄H₉Br
9164 - Phenanthrene, 9-bromo-

C₁₄H₉Cl
374 - Anthracene, 1-chloro-

C₁₄H₉ClF₂N₂O₂
5176 - Diflubenzuron

C₁₄H₉ClO₃
4079 - Chlorflurecol

C₁₄H₉Cl₂NO₅
2997 - Bifenox

C₁₄H₉Cl₅
2328 - Benzene, 1,1'-(2,2,2-trichloroethylidene)bis[4-chloro-

C₁₄H₉Cl₅O
1851 - Benzenemethanol, 4-chloro-α-(4-chlorophenyl)-α-(trichloromethyl)-

C₁₄H₉NO₂
380 - 9,10-Anthracenedione, 1-amino-
381 - 9,10-Anthracenedione, 2-amino-
426 - Anthracene, 9-nitro-
7366 - 1H-Isoindole-1,3(2H)-dione, 2-phenyl-

C₁₄H₉NO₃
383 - 9,10-Anthracenedione, 2-amino-1-hydroxy-

C₁₄H₉NO₄
382 - 9,10-Anthracenedione, 3-amino-1,2-dihydroxy-

C₁₄H₁₀
366 - Anthracene
1714 - Benzene, 1,1'-(1,2-ethynediyl)bis-
9163 - Phenanthrene

C₁₄H₁₀Cl₂
1365 - Benzene, 1,1'-(1,2-dichloro-1,2-ethenediyl)bis-, (E)-
1366 - Benzene, 1,1'-(1,2-dichloro-1,2-ethenediyl)bis-, (Z)-

C₁₄H₁₀Cl₄
1378 - Benzene, 1,1'-(2,2-dichloroethylidene)bis[4-chloro-

C₁₄H₁₀F₃NO₂
2729 - Benzoic acid, 2-[[3-(trifluoromethyl)phenyl]amino]-

C₁₄H₁₀N₂O
8564 - 1,2,4-Oxadiazole, 3,5-diphenyl-

C₁₄H₁₀N₂O₂
389 - 9,10-Anthracenedione, 1,5-diamino-

C₁₄H₁₀N₂O₄
1494 - Benzene, 1,1'-(1,2-dinitro-1,2-ethenediyl)bis-, (Z)-
1496 - Benzene, 2,4-dinitro-1-(2-phenylethenyl)-, (E)-

C₁₄H₁₀O
431 - 1-Anthracenol
432 - 9-Anthracenol
433 - 9(10H)-Anthracenone
5928 - Ethenone, diphenyl-

C₁₄H₁₀O₂
5584 - Ethanedione, diphenyl-
9173 - 3,4-Phenanthrenediol

C₁₄H₁₀O₂S₂
5251 - Disulfide, dibenzoyl

C₁₄H₁₀O₃
429 - 1,2,10-Anthracenetriol
430 - 1,8,9-Anthracenetriol
2510 - Benzoic acid, anhydride
2511 - Benzoic acid, 3-benzoyl-
2512 - Benzoic acid, 4-benzoyl-

C₁₄H₁₀O₄
3038 - [1,1'-Biphenyl]-2,2'-dicarboxylic acid
9157 - Peroxide, dibenzoyl

C₁₄H₁₀O₅
2616 - Benzoic acid, 2-hydroxy-, 2-carboxyphenyl ester
12054 - 9H-Xanthen-9-one, 1,7-dihydroxy-3-methoxy-

$C_{14}H_{10}O_6$

2979 - [Bi-1,4-cyclohexadien-1-yl]-3,3',6,6'-tetrone, 2,2'-dihydroxy-4,4'-dimethyl-

$C_{14}H_{10}O_9$

2575 - Benzoic acid, 3,4-dihydroxy-5-[(3,4,5-trihydroxybenzoyl)oxy]-

$C_{14}H_{11}Br$

1030 - Benzene, 1,1'-(bromoethenylidene)bis-
1080 - Benzene, 1-bromo-2-(2-phenylethenyl)-, (E)-
1081 - Benzene, 1-bromo-2-(2-phenylethenyl)-, (Z)-

$C_{14}H_{11}Cl$

1146 - Benzene, 1,1'-(1-chloro-1,2-ethenediyl)bis-, (E)-
1147 - Benzene, 1,1'-(1-chloro-1,2-ethenediyl)bis-, (Z)-
1222 - Benzene, 1-chloro-2-(2-phenylethenyl)-, (E)-
1223 - Benzene, 1-chloro-3-(2-phenylethenyl)-, (E)-

$C_{14}H_{11}ClO$

975 - Benzeneacetyl chloride, α-phenyl-
5785 - Ethanone, 2-chloro-1,2-diphenyl-

$C_{14}H_{11}N$

365 - 2-Anthracenamine
969 - Benzeneacetonitrile, α-phenyl-
7320 - 1H-Indole, 2-phenyl-
9162 - 9-Phenanthrenamine

$C_{14}H_{11}NO$

3983 - 9H-Carbazole, 9-acetyl-

$C_{14}H_{11}NO_2$

2055 - Benzene, 1-nitro-2-(2-phenylethenyl)-, (E)-
2056 - Benzene, 1-nitro-2-(2-phenylethenyl)-, (Z)-
3982 - 9H-Carbazole-9-acetic acid

$C_{14}H_{11}NO_3$

2513 - Benzoic acid, 3-(benzoylamino)-
7627 - Methanone, (4-methyl-2-nitrophenyl)phenyl-
7630 - Methanone, (4-methylphenyl)(4-nitrophenyl)-

$C_{14}H_{11}NO_4$

2514 - Benzoic acid, 4-(benzoylamino)-2-hydroxy-
2656 - Benzoic acid, 2,2'-iminobis-

$C_{14}H_{12}$

376 - Anthracene, 9,10-dihydro-
1619 - Benzene, 1,1'-(1,2-ethenediyl)bis-, (E)-
1620 - Benzene, 1,1'-(1,2-ethenediyl)bis-, (Z)-
1636 - Benzene, 1,1'-ethenylidenebis-
5974 - 9H-Fluorene, 9-methyl-
9171 - Phenanthrene, 9,10-dihydro-

$C_{14}H_{12}Cl_2$

1678 - Benzene, 1,1'-ethylidenebis[4-chloro-

$C_{14}H_{12}Cl_2O$

1850 - Benzenemethanol, 4-chloro-α-(4-chlorophenyl)-α-methyl-

$C_{14}H_{12}F_3NO_4S_2$

9149 - Perfluidone

$C_{14}H_{12}N_2$

2385 - 1H-Benzimidazole, 2-(phenylmethyl)-
9192 - 1,10-Phenanthroline, 2,9-dimethyl-
11196 - Quinazoline, 3,4-dihydro-3-phenyl-

$C_{14}H_{12}N_2O_2$

5495 - Ethanediamide, N,N'-diphenyl-
9449 - Phenol, 4-(phenylazo)-, acetate (ester)

$C_{14}H_{12}N_2S$

782 - Benzenamine, 4-(6-methyl-2-benzothiazolyl)-

$C_{14}H_{12}N_4O_2S$

2205 - Benzenesulfonamide, 4-amino-N-2-quinoxalinyl-

$C_{14}H_{12}O$

877 - Benzeneacetaldehyde, α-phenyl-
5776 - Ethanone, 1-[1,1'-biphenyl]-4-yl-
5812 - Ethanone, 1,2-diphenyl-
7631 - Methanone, (2-methylphenyl)phenyl-
7632 - Methanone, (3-methylphenyl)phenyl-
7633 - Methanone, (4-methylphenyl)phenyl-

$C_{14}H_{12}O_2$

936 - Benzeneacetic acid, α-phenyl-
1545 - 1,3-Benzenediol, 5-(2-phenylethenyl)-, (E)-
2699 - Benzoic acid, 2-methylphenyl ester
2700 - Benzoic acid, 4-methylphenyl ester
2718 - Benzoic acid, 2-(phenylmethyl)-

2719 - Benzoic acid, 4-(phenylmethyl)-
2720 - Benzoic acid, phenylmethyl ester
3009 - [1,1'-Biphenyl]-4-acetic acid
3091 - [1,1'-Biphenyl]-4-ol, acetate
5822 - Ethanone, 2-hydroxy-1,2-diphenyl-, (±)-
7623 - Methanone, (2-methoxyphenyl)phenyl-
7624 - Methanone, (3-methoxyphenyl)phenyl-
7625 - Methanone, (4-methoxyphenyl)phenyl-

$C_{14}H_{12}O_3$

918 - Benzeneacetic acid, α-hydroxy-α-phenyl-
2397 - 2H,8H-Benzo[1,2-b:5,4-b']dipyran-2-one, 8,8-dimethyl-
2651 - Benzoic acid, 2-hydroxy-, phenylmethyl ester
7620 - Methanone, (2-hydroxy-4-methoxyphenyl)phenyl-
9383 - Phenol, 2-methoxy-, benzoate

$C_{14}H_{12}O_4$

7608 - Methanone, (2,6-dihydroxy-4-methoxyphenyl)phenyl-
7619 - Methanone, (2-hydroxy-4-methoxyphenyl)(2-hydroxyphenyl)-

$C_{14}H_{12}O_5$

6185 - 5H-Furo[3,2-g][1]benzopyran-5-one, 4,9-dimethoxy-7-methyl-

$C_{14}H_{12}S_2$

11661 - Thianthrene, 2,7-dimethyl-

$C_{14}H_{13}N$

809 - Benzenamine, 2-methyl-N-(phenylmethylene)-
810 - Benzenamine, 3-methyl-N-(phenylmethylene)-
811 - Benzenamine, 4-methyl-N-(phenylmethylene)-
3985 - 9H-Carbazole, 9-ethyl-

$C_{14}H_{13}NO$

35 - Acetamide, N-[1,1'-biphenyl]-2-yl-
57 - Acetamide, N,N-diphenyl-
621 - Benzamide, N-(2-methylphenyl)-
622 - Benzamide, N-(3-methylphenyl)-
623 - Benzamide, N-(4-methylphenyl)-
7626 - Methanone, [2-(methylamino)phenyl]phenyl-

$C_{14}H_{13}NO_4$

6181 - Furo[2,3-b]quinoline, 4,7,8-trimethoxy-

$C_{14}H_{13}NO_4S$

2721 - Benzoic acid, 4-[[(phenylmethyl)sulfonyl]amino]-

$C_{14}H_{14}$

1587 - Benzene, 1,1'-(1,2-ethanediyl)bis-
1677 - Benzene, 1,1'-ethylidenebis-
2014 - Benzene, 1-methyl-2-(phenylmethyl)-
2015 - Benzene, 1-methyl-3-(phenylmethyl)-
2016 - Benzene, 1-methyl-4-(phenylmethyl)-
3050 - 1,1'-Biphenyl, 2,2'-dimethyl-
3051 - 1,1'-Biphenyl, 2,3'-dimethyl-
3052 - 1,1'-Biphenyl, 2,3-dimethyl-
3053 - 1,1'-Biphenyl, 2,4'-dimethyl-
3054 - 1,1'-Biphenyl, 2,4-dimethyl-
3055 - 1,1'-Biphenyl, 2,5-dimethyl-
3056 - 1,1'-Biphenyl, 2,6-dimethyl-
3057 - 1,1'-Biphenyl, 3,3'-dimethyl-
3058 - 1,1'-Biphenyl, 3,4'-dimethyl-
3059 - 1,1'-Biphenyl, 3,4-dimethyl-
3060 - 1,1'-Biphenyl, 3,5-dimethyl-
3061 - 1,1'-Biphenyl, 4,4'-dimethyl-
3072 - 1,1'-Biphenyl, 2-ethyl-
3073 - 1,1'-Biphenyl, 3-ethyl-
9187 - Phenanthrene, 1,2,3,4-tetrahydro-

$C_{14}H_{14}Cl_2N_2O$

7187 - Imazalil

$C_{14}H_{14}Hg$

7432 - Mercury, bis(4-methylphenyl)-

$C_{14}H_{14}NO_4PS$

9551 - Phosphonothioic acid, phenyl-, O-ethyl O-(4-nitrophenyl) ester

$C_{14}H_{14}N_2$

732 - Benzenamine, 4,4'-(1,2-ethenediyl)bis-, (E)-
5114 - Diazene, bis(2-methylphenyl)-
5115 - Diazene, bis(3-methylphenyl)-, (E)-

$C_{14}H_{14}N_2O$

5116 - Diazene, bis(2-methylphenyl)-, 1-oxide, (Z)-

5117 - Diazene, bis(2-methylphenyl)-, 1-oxide, (E)-
5118 - Diazene, bis(3-methylphenyl)-, 1-oxide, (Z)-
5119 - Diazene, bis(3-methylphenyl)-, 1-oxide, (E)-
5133 - Diazene, (4-ethoxyphenyl)phenyl-
10462 - 1-Propanone, 2-methyl-1,2-di-3-pyridinyl-
12018 - Urea, N-methyl-N,N'-diphenyl-

$C_{14}H_{14}N_3NaO_3S$

2222 - Benzenesulfonic acid, 4-[[4-(dimethylamino)phenyl]azo]-, sodium salt

$C_{14}H_{14}O$

1613 - Benzeneethanol, α-phenyl-
1614 - Benzeneethanol, α-phenyl-, (S)-
1615 - Benzeneethanol, β-phenyl-
2076 - Benzene, 1,1'-oxybis[2-methyl-
2077 - Benzene, 1,1'-oxybis[3-methyl-
2078 - Benzene, 1,1'-oxybis[4-methyl-
2079 - Benzene, 1,1'-[oxybis(methylene)]bis-
3071 - 1,1'-Biphenyl, 3-ethoxy-
3674 - 1-Butanone, 1-(1-naphthalenyl)-
10464 - 1-Propanone, 2-methyl-1-(1-naphthalenyl)-
10465 - 1-Propanone, 2-methyl-1-(2-naphthalenyl)-

$C_{14}H_{14}OS$

2184 - Benzene, 1,1'-[sulfinylbis(methylene)]bis-

$C_{14}H_{14}O_2$

1588 - Benzene, 1,1'-[1,2-ethanediylbis(oxy)]bis-
3045 - 1,1'-Biphenyl, 2,2'-dimethoxy-
3046 - 1,1'-Biphenyl, 3,3'-dimethoxy-
3047 - 1,1'-Biphenyl, 4,4'-dimethoxy-
3067 - [1,1'-Biphenyl]-2,2'-diol, 3,3'-dimethyl-
3068 - [1,1'-Biphenyl]-2,2'-diol, 5,5'-dimethyl-
5575 - 1,2-Ethanediol, 1,2-diphenyl-, (R*,R*)-(±)-
7735 - 1-Naphthaleneacetic acid, ethyl ester

$C_{14}H_{14}O_2S$

2240 - Benzene, 1,1'-sulfonylbis[4-methyl-
2241 - Benzene, 1,1'-[sulfonylbis(methylene)]bis-

$C_{14}H_{14}O_3$

1929 - Benzene, 1-methoxy-2-(3-methoxyphenoxy)-
2075 - Benzene, 1,1'-oxybis[2-methoxy-
2786 - 2H-1-Benzopyran-7-ol, 3,4-dihydro-3-(4-hydroxyphenyl)-, (S)-

$C_{14}H_{14}O_8$

2258 - 1,2,4,5-Benzenetetracarboxylic acid, tetramethyl ester

$C_{14}H_{14}S$

1993 - Benzene, 1-methyl-2-[(4-methylphenyl)thio]-
2285 - Benzene, 1,1'-thiobis[2-methyl-
2286 - Benzene, 1,1'-thiobis[4-methyl-
2287 - Benzene, 1,1'-[thiobis(methylene)]bis-
11570 - Sulfide, m-tolyl p-tolyl

$C_{14}H_{14}S_2$

5245 - Disulfide, bis(3-methylphenyl)
5246 - Disulfide, bis(4-methylphenyl)
5248 - Disulfide, bis(phenylmethyl)

$C_{14}H_{14}Se$

2176 - Benzene, 1,1'-selenobis[4-methyl-

$C_{14}H_{15}Cl_2N$

7710 - 2-Naphthalenamine, N,N-bis(2-chloroethyl)-

$C_{14}H_{15}N$

751 - Benzenamine, N-ethyl-N-phenyl-
795 - Benzenamine, 2-methyl-N-(2-methylphenyl)-
796 - Benzenamine, 3-methyl-N-(3-methylphenyl)-
797 - Benzenamine, 4-methyl-N-(4-methylphenyl)-
838 - Benzenamine, 2-(2-phenylethyl)-
1584 - Benzeneethanamine, α-phenyl-
1821 - Benzenemethanamine, N-(2-methylphenyl)-
1824 - Benzenemethanamine, N-(3-methylphenyl)-
1825 - Benzenemethanamine, N-(4-methylphenyl)-
1828 - Benzenemethanamine, N-(phenylmethyl)-
11007 - Pyridine, 4-(3-phenylpropyl)-

$C_{14}H_{15}N_3$

721 - Benzenamine, N,N-dimethyl-4-(phenylazo)-
798 - Benzenamine, 2-methyl-4-[(2-methylphenyl)azo]-

$C_{14}H_{15}O_3P$

9586 - Phosphorous acid, bis(phenylmethyl) ester

$C_{14}H_{16}$

422 - Anthracene, 1,2,3,4,5,6-hexahydro-

11632 - 3-Tetradecanol
11886 - 1-Tridecanol, 12-methyl-
11887 - 3-Tridecanol, 2-methyl-
11888 - 4-Tridecanol, 2-methyl

C₁₄H₃₀O₂
5297 - Dodecane, 1,1-dimethoxy-
5298 - Dodecane, 1,12-dimethoxy-
11618 - 1,14-Tetradecanediol

C₁₄H₃₀O₂Sn
11538 - Stannane, (acetyloxy)tributyl-

C₁₄H₃₀O₃S
11620 - 1-Tetradecanesulfonic acid

C₁₄H₃₀S
6435 - Heptane, 1,1'-thiobis-
11621 - 1-Tetradecanethiol

C₁₄H₃₁N
6352 - 1-Heptanamine, N-heptyl-
11612 - 1-Tetradecanamine

C₁₄H₃₂N₂
6750 - 1,6-Hexanediamine, N,N'-dibutyl-

C₁₄H₃₂N₂O₄
10379 - 2-Propanol, 1,1',1'',1'''-
(1,2-ethanediyldinitrilo)tetrakis-

C₁₄H₄₂O₅Si₆
6990 - Hexasiloxane, tetradecamethyl-

C₁₄H₄₂O₇Si₇
4239 - Cycloheptasiloxane, tetradecamethyl-

C₁₅H₈O₄
372 - 2-Anthracenecarboxylic acid, 9,10-dihydro-
9,10-dioxo-

C₁₅H₈O₅
2466 - 6H-Benzofuro[3,2-c][1]benzopyran-6-one, 3,9-
dihydroxy-

C₁₅H₈O₆
371 - 2-Anthracenecarboxylic acid, 9,10-dihydro-4,5-
dihydroxy-9,10-dioxo-

C₁₅H₉BrO₂
7265 - 1H-Indene-1,3(2H)-dione, 2-(4-bromophenyl)-

C₁₅H₉ClO₂
7266 - 1H-Indene-1,3(2H)-dione, 2-(4-chlorophenyl)-

C₁₅H₁₀ClF₃N₂O₆S
5985 - Fomesafen

C₁₅H₁₀O₂
368 - 1-Anthracenecarboxylic acid
369 - 2-Anthracenecarboxylic acid
370 - 9-Anthracenecarboxylic acid
409 - 9,10-Anthracenedione, 2-methyl-
2840 - 4H-1-Benzopyran-4-one, 2-phenyl-
2841 - 4H-1-Benzopyran-4-one, 3-phenyl-
7270 - 1H-Indene-1,3(2H)-dione, 2-phenyl-
7588 - Methanone, 2-benzofuranylphenyl-
9165 - 9-Phenanthrenecarboxylic acid

C₁₅H₁₀O₃
10199 - Propanetrione, diphenyl-

C₁₅H₁₀O₄
398 - 9,10-Anthracenedione, 1,8-dihydroxy-3-methyl-
2811 - 4H-1-Benzopyran-4-one, 5,7-dihydroxy-2-
phenyl-
2827 - 4H-1-Benzopyran-4-one, 7-hydroxy-3-(4-
hydroxyphenyl)-

C₁₅H₁₀O₅
397 - 9,10-Anthracenedione, 1,8-dihydroxy-3-
(hydroxymethyl)-
419 - 9,10-Anthracenedione, 1,3,8-trihydroxy-6-
methyl-
1332 - 1,2-Benzenedicarboxylic acid, 3-benzoyl-
2533 - Benzoic acid, 4,4'-carbonylbis-
2806 - 4H-1-Benzopyran-4-one, 5,7-dihydroxy-2-(4-
hydroxyphenyl)-
2807 - 4H-1-Benzopyran-4-one, 5,7-dihydroxy-3-(4-
hydroxyphenyl)-
2844 - 4H-1-Benzopyran-4-one, 5,6,7-trihydroxy-2-
phenyl-

C₁₅H₁₀O₆
2813 - 4H-1-Benzopyran-4-one, 2-
(3,4-dihydroxyphenyl)-3,7-dihydroxy-

2814 - 4H-1-Benzopyran-4-one,
2-(3,4-dihydroxyphenyl)-5,7-dihydroxy-
2842 - 4H-1-Benzopyran-4-one, 3,5,7-trihydroxy-2-
(2-hydroxyphenyl)-
2843 - 4H-1-Benzopyran-4-one, 3,5,7-trihydroxy-2-
(4-hydroxyphenyl)-

C₁₅H₁₀O₇
2816 - 4H-1-Benzopyran-4-one,
2-(2,4-dihydroxyphenyl)-3,5,7-trihydroxy-
2817 - 4H-1-Benzopyran-4-one,
2-(3,4-dihydroxyphenyl)-3,5,7-trihydroxy-

C₁₅H₁₁ClF₃NO₄
8633 - Oxyfluorfen

C₁₅H₁₁ClN₂O
2411 - 2H-1,4-Benzodiazepin-2-one, 7-chloro-1,3-
dihydro-5-phenyl-

C₁₅H₁₁ClO₅
2848 - 1-Benzopyrylium, 3,5,7-trihydroxy-2-(4-
hydroxyphenyl)-, chloride

C₁₅H₁₁ClO₇
2849 - 1-Benzopyrylium, 3,5,7-trihydroxy-2-(3,4,5-
trihydroxyphenyl)-, chloride

C₁₅H₁₁I₃O₄
2136 - Benzenepropanoic acid, 4-(4-hydroxy-3-
iodophenoxy)-3,5-diiodo-

C₁₅H₁₁I₄NO₄
11930 - L-Tyrosine, O-(4-hydroxy-3,5-diiodophenyl)-
3,5-diiodo-

C₁₅H₁₁N
2768 - Benzonitrile, 4-(2-phenylethenyl)-, (Z)-
11288 - Quinoline, 2-phenyl-
11289 - Quinoline, 3-phenyl-
11290 - Quinoline, 4-phenyl-
11291 - Quinoline, 6-phenyl-

C₁₅H₁₁NO
8579 - Oxazole, 2,4-diphenyl-
8580 - Oxazole, 2,5-diphenyl-
8581 - Oxazole, 4,5-diphenyl-

C₁₅H₁₁NO₂
384 - 9,10-Anthracenedione, 1-amino-2-methyl-
7367 - 1H-Isoindole-1,3(2H)-dione, 2-(phenylmethyl)-

C₁₅H₁₁NS
11676 - Thiazole, 2,4-diphenyl-

C₁₅H₁₂
423 - Anthracene, 1-methyl-
424 - Anthracene, 2-methyl-
425 - Anthracene, 9-methyl-
9180 - Phenanthrene, 3-methyl-
9181 - Phenanthrene, 4-methyl-

C₁₅H₁₂I₂O₃
2135 - Benzenepropanoic acid, 4-hydroxy-3,5-diiodo-
α-phenyl-

C₁₅H₁₂N₂
10830 - 1H-Pyrazole, 1,3-diphenyl-
10831 - 1H-Pyrazole, 1,5-diphenyl-

C₁₅H₁₂N₂O
5144 - 5H-Dibenz[b,f]azepine-5-carboxamide

C₁₅H₁₂N₂O₂
7219 - 2,4-Imidazolidinedione, 5,5-diphenyl-
10843 - 3,5-Pyrazolidinedione, 1,4-diphenyl-

C₁₅H₁₂N₂O₃
7488 - Methanediamine, 1-(2-furanyl)-N,N'-bis(2-
furanylmethylene)-

C₁₅H₁₂N₆O₄
2494 - Benzoic acid, 4-[[(2-amino-1,4-dihydro-4-oxo-
6-pteridinyl)methyl]formylamino]-

C₁₅H₁₂O
5146 - 10H-Dibenzo[a,d]cyclohepten-10-one, 5,11-
dihydro-
5147 - 5H-Dibenzo[a,d]cyclohepten-5-one, 10,11-
dihydro-
7291 - 1H-Inden-1-one, 2,3-dihydro-2-phenyl-
10708 - 2-Propen-1-one, 1,3-diphenyl-
10709 - 2-Propen-1-one, 1,3-diphenyl-, (E)-

C₁₅H₁₂O₂
941 - Benzeneacetic acid, α-(phenylmethylene)-
10027 - 1,3-Propanedione, 1,3-diphenyl-

10689 - 2-Propenoic acid, 3-phenyl-, phenyl ester,
(E)-
10711 - 2-Propen-1-one, 3-hydroxy-1,3-diphenyl-

C₁₅H₁₂O₃
2516 - Benzoic acid, 2-benzoyl-, methyl ester
2684 - Benzoic acid, 2-(4-methylbenzoyl)-

C₁₅H₁₂O₄
2483 - Benzoic acid, 2-(acetyloxy)-, phenyl ester
2667 - Benzoic acid, 2-(4-methoxybenzoyl)-
7509 - Methanediol, dibenzoate

C₁₅H₁₂O₅
2802 - 4H-1-Benzopyran-4-one, 2,3-dihydro-5,7-
dihydroxy-2-(4-hydroxyphenyl)-, (S)-

C₁₅H₁₂O₆
2686 - Benzoic acid, 3,3'-methylenebis[6-hydroxy-
2812 - 4H-1-Benzopyran-4-one,
2-(3,4-dihydroxyphenyl)-2,3-dihydro-5,7-
dihydroxy-, (S)-

C₁₅H₁₃N
7318 - 1H-Indole, 3-methyl-2-phenyl-
7321 - 1H-Indole, 1-(phenylmethyl)-

C₁₅H₁₃NO
64 - Acetamide, N-9H-fluoren-2-yl-

C₁₅H₁₃NO₂S
9500 - 10H-Phenothiazine-2-acetic acid, 10-methyl-

C₁₅H₁₄
1966 - Benzene, 1,1'-(methylene-
1,2-ethanediyl)bis-
1971 - Benzene, 1,1'-(1-methyl-1,2-ethenediyl)bis-
2161 - Benzene, 1,1'-(1-propene-1,3-diyl)bis-
2166 - Benzene, 1,1'-(1-propenylidene)bis-
7259 - 1H-Indene, 2,3-dihydro-2-phenyl-

C₁₅H₁₄F₃N₃
723 - Benzenamine, N,N-dimethyl-4-[[3-
(trifluoromethyl)phenyl]azo]-

C₁₅H₁₄N₂
765 - Benzenamine, N,N'-methanetetraylbis[4-
methyl-

C₁₅H₁₄N₂O
7228 - 4-Imidazolidinone, 5,5-diphenyl-

C₁₅H₁₄N₄O₂S
2201 - Benzenesulfonamide, 4-amino-N-(1-phenyl-
1H-pyrazol-5-yl)-

C₁₅H₁₄O
1939 - Benzene, 1-methoxy-4-(2-phenylethenyl)-,
(E)-
2782 - 2H-1-Benzopyran, 3,4-dihydro-2-phenyl-
7599 - Methanone, bis(4-methylphenyl)-
7614 - Methanone, (2,4-dimethylphenyl)phenyl-
7615 - Methanone, (3,4-dimethylphenyl)phenyl-
7628 - Methanone, (2-methylphenyl)
(4-methylphenyl)-
10432 - 2-Propanone, 1,1-diphenyl-
10433 - 2-Propanone, 1,3-diphenyl-

C₁₅H₁₄O₂
931 - Benzeneacetic acid, α-methyl-α-phenyl-
2143 - Benzenepropanoic acid, α-phenyl-, (±)-
5839 - Ethanone, 2-methoxy-1,2-diphenyl-

C₁₅H₁₄O₃
920 - Benzeneacetic acid, α-hydroxy-α-phenyl-,
methyl ester
3996 - Carbonic acid, bis(2-methylphenyl) ester
7612 - Methanone, (2,4-dimethoxyphenyl)phenyl-
7862 - 1,4-Naphthalenedione, 2-hydroxy-3-(3-methyl-
2-butenyl)-

C₁₅H₁₄O₄
2398 - 2H,8H-Benzo[1,2-b:5,4-b']dipyran-2-one, 5-
methoxy-8,8-dimethyl-
6188 - 7H-Furo[3,2-g][1]benzopyran-7-one, 3-
methoxy-2-(1-methylethyl)-
7841 - 1,4-Naphthalenediol, 2-methyl-, diacetate

C₁₅H₁₄O₅
7595 - Methanone, bis(2-hydroxy-4-methoxyphenyl)-
9384 - Phenol, 2-methoxy-, carbonate (2:1)
10453 - 1-Propanone, 3-(4-hydroxyphenyl)-1-(2,4,6-
trihydroxyphenyl)-

10790 - 2H-Pyran-2-one, 6-[2-(1,3-benzodioxol-5-yl)ethenyl]-5,6-dihydro-4-methoxy-, [*R-(E)*]-

C15H14O6

2846 - 2H-1-Benzopyran-3,5,7-triol, 2-(3,4-dihydroxyphenyl)-3,4-dihydro-, (2*S-cis*)-

7248 - 2H-Indene-2,2-dicarboxylic acid, 1,3-dihydro-1,3-dioxo-, diethyl ester

7855 - 1,4-Naphthalenedione, 5,8-dihydroxy-6-methoxy-2-methyl-3-(2-oxopropyl)-

11904 - 2H,4aH-1,4,5-Trioxadicyclopent[a,hi]indene-7-carboxylic acid, 3-ethylidene-3,3a,7a,9b-tetrahydro-2-oxo-, methyl ester, [3aS-(3E,3aα,4aβ,7aβ,9aR*,9bβ)]-

C15H15ClF3N3O

11900 - Triflumizole

C15H15ClN2O2

4088 - Chloroxuron

C15H15ClN4O6S

4081 - Chlorimuron

C15H15NO

2771 - Benzophenone, 3-(dimethylamino)-

7613 - Methanone, [4-(dimethylamino)phenyl]phenyl-

C15H15NO2

3946 - Carbamic acid, diphenyl-, ethyl ester

C15H15N3O

317 - 3,9-Acridinediamine, 7-ethoxy-

C15H15N3O2

2587 - Benzoic acid, 2-[[4-(dimethylamino)phenyl]azo]-

C15H16

1696 - Benzene, 1-ethyl-4-(phenylmethyl)-

1964 - Benzene, 1,1′-methylenebis[4-methyl-

1969 - Benzene, 1,1′-(1-methyl-1,2-ethanediyl)bis-

1976 - Benzene, 1,1′-(1-methylethylidene)bis-

2121 - Benzene, 1,1′-(1,3-propanediyl)bis-

2171 - Benzene, 1,1′-propylidenebis-

C15H16F5NO2S2

5267 - Dithiopyr

C15H16N2O

12006 - Urea, N,N′-dimethyl-N,N′-diphenyl-

C15H16O

1940 - Benzene, 1-methoxy-4-(1-phenylethyl)-

2158 - Benzenepropanol, β-phenyl-

C15H16O2

1970 - Benzene, 1,1′-[(1-methyl-1,2-ethanediyl)bis(oxy)]bis-

2122 - Benzene, 1,1′-[1,3-propanediylbis(oxy)]bis-

9413 - Phenol, 4,4′-(1-methylethylidene)bis-

9464 - Phenol, 4,4′-propylidenebis-

C15H16O3

10377 - 2-Propanol, 1,3-diphenoxy-

C15H16O8

2821 - 2H-1-Benzopyran-2-one, 7-(β-D-glucopyranosyloxy)-

C15H16O9

2822 - 2H-1-Benzopyran-2-one, 6-(β-D-glucopyranosyloxy)-7-hydroxy-

C15H17Cl2N3O2

10720 - Propiconazole

C15H17N

1811 - Benzenemethanamine, N-ethyl-N-phenyl-

C15H17N3

6279 - Guanidine, N,N′-bis(2-methylphenyl)-

C15H18

505 - Azulene, 1,4-dimethyl-7-(1-methylethyl)-

506 - Azulene, 4,8-dimethyl-2-(1-methylethyl)-

7830 - Naphthalene, 1,6-dimethyl-4-(1-methylethyl)-

7941 - Naphthalene, 1-pentyl-

7942 - Naphthalene, 2-pentyl-

C15H18Cl2N2O3

8565 - Oxadiazon

C15H18N2O6

3003 - Binapacryl

C15H18N6O6S

8069 - Nicosulfuron

C15H18O

7917 - Naphthalene, 1-(3-methylbutoxy)-

7918 - Naphthalene, 2-(3-methylbutoxy)-

7943 - Naphthalene, 1-(pentyloxy)-

7944 - Naphthalene, 2-(pentyloxy)-

C15H18O2

2939 - Bicyclo[2.2.1]heptan-2-ol, 1-phenyl-, acetate

C15H18O3

8057 - Naphtho[1,2-b]furan-2,8(3H,4H)-dione, 3a,5,5a,9b-tetrahydro-3,5a,9-trimethyl-

C15H18O6

8058 - Naphtho[1,2-b]furan-2,8(3H,4H)-dione, 3a,5,5a,9b-tetrahydro-4,5-dihydroxy-9-methoxy-3,5a-dimethyl-

C15H19NO2

487 - 8-Azabicyclo[3.2.1]octan-3-ol, 8-methyl-, benzoate (ester), endo-

488 - 8-Azabicyclo[3.2.1]octan-3-ol, 8-methyl-, benzoate (ester), exo-

C15H19N3O3

7190 - Imazethapyr

C15H20N2O

7573 - 7,14-Methano-4H,6H-dipyrido[1,2-a1′,2′-e][1,5]diazocin-4-one, 7,7a,8,9,10,11,13,14-octahydro-,

C15H20O2

6293 - Helenine

8059 - Naphtho[2,3-b]furan-2(3H)-one, 3a,5,6,7,8,8a,9,9a-octahydro-5,8a-dimethyl-3-methylene-,

C15H20O4

8679 - 2,4-Pentadienoic acid, 5-(1-hydroxy-2,6,6-trimethyl-4-oxo-2-cyclohexen-1-yl)-3-methyl-, [S-(Z,E)]-

9952 - Propanedioic acid, ethylphenyl-, diethyl ester

C15H21NO4

7439 - Metalaxyl

C15H21N3O2

11191 - Pyrrolo[2,3-b]indol-5-ol, 1,2,3,3a,8,8a-hexahydro-1,3a,8-trimethyl-, methylcarbamate (ester), (3aS-cis)-

C15H22

1468 - Benzene, 1-(1,5-dimethyl-4-hexenyl)-4-methyl-

7274 - 1H-Indene, 5-hexyl-2,3-dihydro-

C15H22ClNO2

7657 - Metolachlor

C15H22N2O

9639 - 2-Piperidinecarboxamide, N-(2,6-dimethylphenyl)-1-methyl-

C15H22O

8052 - 2(3H)-Naphthalenone, 4,4a,5,6,7,8-hexahydro-4,4a-dimethyl-6-(1-methylethylidene)-, (4R-cis)-

11521 - Spiro[4.5]dec-6-en-8-one, 6,10-dimethyl-2-(1-methylethylidene)-, (5R-cis)-

C15H23ClO4S

11579 - Sulfurous acid, 2-chloroethyl 2-[4-(1,1-dimethylethyl)phenoxy]-1-methylethyl ester

C15H23NO4

9655 - 2,6-Piperidinedione, 4-[2-(3,5-dimethyl-2-oxocyclohexyl)-2-hydroxyethyl]-,

C15H23N3O4

7371 - Isopropalin

C15H24

2064 - Benzene, nonyl-

2372 - Benzene, 1,2,4-tris(1-methylethyl)-

2373 - Benzene, 1,3,5-tris(1-methylethyl)-

2996 - Bicyclo[7.2.0]undec-4-ene, 4,11,11-trimethyl-8-methylene-, [1R-(1R*,4E,9S*)]-

4066 - Cedrene

4389 - Cyclohexane, 1-ethenyl-1-methyl-2,4-bis(1-methylethenyl)-, (1α,2β,4β)-

4659 - Cyclohexene, 6-ethenyl-6-methyl-1-(1-methylethyl)-3-(1-methylethylidene)-, (S)-

4685 - Cyclohexene, 1-methyl-4-(5-methyl-1-methylene-4-hexenyl)-, (S)-

4954 - Cyclopropa[d]naphthalene, 1,1a,4,4a,5,6,7,8-octahydro-2,4a,8,8-tetramethyl-, [1aS-(1aα,4aβ,8aR*)]-

5006 - 1,4,8-Cycloundecatriene, 2,6,6,9-tetramethyl-, (E,E,E)-

5329 - 1,3,6,10-Dodecatetraene, 3,7,11-trimethyl-, (E,E)-

5331 - 1,6,10-Dodecatriene, 7,11-dimethyl-3-methylene-, (E)-

7568 - 1,4-Methanoazulene, decahydro-4,8,8-trimethyl-9-methylene-, [1S-(1α,3aβ,4α,8aβ)]-

7898 - Naphthalene, 1,2,4a,5,8,8a-hexahydro-4,7-dimethyl-1-(1-methylethyl)-, [1S-(1α,4aβ,8aα)]-

7899 - Naphthalene, 1,2,3,5,6,8a-hexahydro-4,7-dimethyl-1-(1-methylethyl)-, (1S-cis)-

7936 - Naphthalene, 1,2,3,4,4a,5,6,8a-octahydro-7-methyl-4-methylene-1-(1-methylethyl)-, (1α,4aβ,8aα)-

11861 - Tricyclo[4.4.0.02,7]dec-3-ene, 1,3-dimethyl-8-(1-methylethyl)-, stereoisomer

C15H24NO4PS

7359 - Isofenphos

C15H24N2O

7426 - Matridin-15-one

7572 - 7,14-Methano-2H,6H-dipyrido[1,2-a1′,2′-e][1,5]diazocin-6-one, dodecahydro-, [7R-(7α,7aβ,14α,14aalp]

C15H24N2O2

7574 - 7,14-Methano-4H,6H-dipyrido[1,2-a1′,2′-e][1,5]diazocin-4-one, dodecahydro-9-hydroxy-,

C15H24O

5330 - 2,6,10-Dodecatrienal, 3,7,11-trimethyl-

9093 - 2-Penten-1-ol, 5-(2,3-dimethyltricyclo[2.2.1.02,6]hept-3-yl)-2-methyl-, stereoisomer

9103 - 2-Penten-1-ol, 2-methyl-5-(2-methyl-3-methylenebicyclo[2.2.1]hept-2-yl)-, [1S-[1α,2α(Z),4α]

9222 - Phenol, 2,4-bis(1,1-dimethylethyl)-5-methyl-

9223 - Phenol, 2,4-bis(1,1-dimethylethyl)-6-methyl-

9224 - Phenol, 2,6-bis(1,1-dimethylethyl)-4-methyl-

C15H24O8

10139 - 1,1,2,3-Propanetetracarboxylic acid, tetraethyl ester

10140 - 1,1,3,3-Propanetetracarboxylic acid, tetraethyl ester

C15H25ClN2O2

2531 - Benzoic acid, 4-(butylamino)-, 2-(dimethylamino)ethyl ester, monohydrochloride

C15H26N2

7571 - 7,14-Methano-2H,6H-dipyrido[1,2-a1′,2′-e][1,5]diazocine, dodecahydro-,

C15H26O

507 - 5-Azulenemethanol, 1,2,3,4,5,6,7,8-octahydro-α,α,3,8-tetramethyl-

4989 - 1H-Cycloprop[e]azulen-4-ol, decahydro-1,1,4,7tetramethyl-, (1aR,4R,4aS,7R,7aS,7bS)-

5332 - 1,6,10-Dodecatrien-3-ol, 3,7,11-trimethyl-, [S-(Z)]-

5333 - 2,6,10-Dodecatrien-1-ol, 3,7,11-trimethyl-, (E,E)-

5334 - 2,6,10-Dodecatrien-1-ol, 3,7,11-trimethyl-, (Z,E)-

7569 - 1,4-Methanoazulen-9-ol, decahydro-1,5,5,8a-tetramethyl-, [1R-(1α,3aβ,4α,8aβ,9S*)]

7570 - 1H-3a,7-Methanoazulen-6-ol, octahydro-3,6,8,8-tetramethyl-,

7586 - 1,6-Methanonaphthalen-1(2H)-ol, octahydro-4,8a,9,9-tetramethyl-,

C15H26O2

3565 - Butanoic acid, 3-methyl-, 1,7,7-trimethylbicyclo[2.2.1]hept-2-yl ester, (1R-endo)-

C15H26O6

3586 - Butanoic acid, 1,2,3-propanetriyl ester

$C_{15}H_{26}Si$
11448 - Silane, phenyltripropyl-

$C_{15}H_{28}$
4430 - Cyclohexane, 1,1'-(1-methyl-1,2-ethanediyl)bis-
4432 - Cyclohexane, 1,1'-(1-methylethylidene)bis-
4466 - Cyclohexane, 1,1'-(1,3-propanediyl)bis-
4472 - Cyclohexane, 1,1'-propylidenebis-
8656 - 1-Pentadecyne

$C_{15}H_{28}NNaO_3$
6269 - Glycine, N-methyl-N-(1-oxododecyl)-, sodium salt

$C_{15}H_{28}O$
4761 - Cyclopentadecanone

$C_{15}H_{28}O_2$
3554 - Butanoic acid, 3-methyl-, 5-methyl-2-(1-methylethyl)cyclohexyl ester
8562 - Oxacyclohexadecan-2-one

$C_{15}H_{28}O_4$
9937 - Propanedioic acid, dibutyl-, diethyl ester

$C_{15}H_{30}$
4401 - Cyclohexane, 1-ethyl-1-methyl-2,4-bis(1-methylethyl)-, [1R-(1α,2β,4β)]-
4457 - Cyclohexane, nonyl-
4759 - Cyclopentadecane
4790 - Cyclopentane, decyl-
8653 - 1-Pentadecene
8654 - 7-Pentadecene

$C_{15}H_{30}N_2$
9702 - Piperidine, 4,4'-(1,3-propanediyl)bis[1-methyl-

$C_{15}H_{30}O$
4760 - Cyclopentadecanol
8640 - Pentadecanal
8651 - 2-Pentadecanone
8652 - 8-Pentadecanone
8655 - 1-Pentadecen-3-ol

$C_{15}H_{30}O_2$
5316 - Dodecanoic acid, 1-methylethyl ester
5320 - Dodecanoic acid, propyl ester
6468 - Heptanoic acid, octyl ester
8430 - Octanoic acid, heptyl ester
8645 - Pentadecanoic acid
11625 - Tetradecanoic acid, methyl ester

$C_{15}H_{30}O_4$
5310 - Dodecanoic acid, 2,3-dihydroxypropyl ester, (±)-

$C_{15}H_{31}Br$
8643 - Pentadecane, 1-bromo-

$C_{15}H_{32}$
8642 - Pentadecane

$C_{15}H_{32}O$
8647 - 2-Pentadecanol
8648 - 3-Pentadecanol

$C_{15}H_{33}B$
3122 - Borane, tris(3-methylbutyl)-

$C_{15}H_{33}N$
3224 - 1-Butanamine, 2-methyl-N,N-bis(2-methylbutyl)-
3225 - 1-Butanamine, 3-methyl-N,N-bis(3-methylbutyl)-
8641 - 1-Pentadecanamine
8705 - 1-Pentanamine, N,N-dipentyl-

$C_{13}H_{29}N_3•C_2H_4O_2$
5347 - Dodine

$C_{15}H_{33}O_4P$
9565 - Phosphoric acid, tripentyl ester

$C_{16}H_8O_2S_2$
2404 - Benzo[b]thiophen-3(2H)-one, 2-(3-oxobenzo[b]thien-2(3H)-ylidene)-

$C_{16}H_9N_3Na_2O_{10}S_2$
7878 - 2,7-Naphthalenedisulfonic acid, 4,5-dihydroxy-3-[(4-nitrophenyl)azo]-, disodium salt

$C_{16}H_{10}$
5966 - Fluoranthene
10853 - Pyrene

$C_{16}H_{10}Cl_4O_5$
5153 - 11H-Dibenzo[b,e][1,4]dioxepin-11-one, 2,4,7,9-tetrachloro-3-hydroxy-8-methoxy-1,6-dimethyl-

$C_{16}H_{10}N_2$
2394 - Benzo[a]phenazine

$C_{16}H_{10}N_2O_2$
7331 - 3H-Indol-3-one, 2-(1,3-dihydro-3-oxo-2H-indol-2-ylidene)-1,2-dihydro-

$C_{16}H_{11}N$
2405 - 7H-Benzo[c]carbazole

$C_{16}H_{11}NO$
7640 - Methanone, phenyl-4-quinolinyl-

$C_{16}H_{11}NO_2$
7866 - 1,4-Naphthalenedione, 2-(phenylamino)-
11235 - 4-Quinolinecarboxylic acid, 2-phenyl-
11309 - 8-Quinolinol, benzoate (ester)

$C_{16}H_{11}NO_3$
11233 - 4-Quinolinecarboxylic acid, 3-hydroxy-2-phenyl-

$C_{16}H_{12}$
7947 - Naphthalene, 1-phenyl-
7948 - Naphthalene, 2-phenyl-

$C_{16}H_{12}N_2O$
8036 - 1-Naphthalenol, 2-(phenylazo)-
8037 - 2-Naphthalenol, 1-(phenylazo)-

$C_{16}H_{12}O$
6100 - Furan, 2,5-diphenyl-
7945 - Naphthalene, 1-phenoxy-
7946 - Naphthalene, 2-phenoxy-

$C_{16}H_{12}O_2$
402 - 9,10-Anthracenedione, 1,4-dimethyl-
403 - 9,10-Anthracenedione, 2,3-dimethyl-
404 - 9,10-Anthracenedione, 2,6-dimethyl-

$C_{16}H_{12}O_3$
7269 - 1H-Indene-1,3(2H)-dione, 2-(4-methoxyphenyl)-

$C_{16}H_{12}O_4$
2830 - 4H-1-Benzopyran-4-one, 7-hydroxy-3-(4-methoxyphenyl)-
3780 - 2-Butenedioic acid (Z)-, diphenyl ester

$C_{16}H_{12}O_5$
2809 - 4H-1-Benzopyran-4-one, 5,7-dihydroxy-2-(4-methoxyphenyl)-
2810 - 4H-1-Benzopyran-4-one, 5,7-dihydroxy-6-methoxy-2-phenyl-
2828 - 4H-1-Benzopyran-4-one, 5-hydroxy-3-(4-hydroxyphenyl)-7-methoxy-

$C_{16}H_{12}O_6$
628 - Benz[b]indeno[1,2-d]pyran-9(6H)-one, 6a,7-dihydro-3,4,6a,10-tetrahydroxy-

$C_{16}H_{12}O_7$
2815 - 4H-1-Benzopyran-4-one, 2-(3,4-dihydroxyphenyl)-3,5-dihydroxy-7-methoxy-

$C_{16}H_{12}S$
7951 - Naphthalene, 1-(phenylthio)-

$C_{16}H_{13}ClN_2O$
2410 - 2H-1,4-Benzodiazepin-2-one, 7-chloro-1,3-dihydro-1-methyl-5-phenyl-

$C_{16}H_{13}N$
7387 - Isoquinoline, 3-(phenylmethyl)-
7728 - 2-Naphthalenamine, N-phenyl-

$C_{16}H_{13}N_3$
7729 - 2-Naphthalenamine, 1-(phenylazo)-

$C_{16}H_{14}$
377 - Anthracene, 1,3-dimethyl-
378 - Anthracene, 2,10-dimethyl-
420 - Anthracene, 9-ethyl-
1095 - Benzene, 1,1'-(1,3-butadiene-1,4-diyl)bis-, (E,E)-
1096 - Benzene, 1,1'-(1,3-butadiene-1,4-diyl)bis-, (Z,Z)-
9172 - Phenanthrene, 9,10-dimethyl-
9179 - Phenanthrene, 9-ethyl-

$C_{16}H_{14}ClN_3O$
2409 - 3H-1,4-Benzodiazepin-2-amine, 7-chloro-N-methyl-5-phenyl-, 4-oxide

$C_{16}H_{14}Cl_2O_3$
4082 - Chlorobenzilate

$C_{16}H_{14}Cl_2O_4$
5171 - Diclofop-methyl

$C_{16}H_{14}N_2O$
11199 - 4(3H)-Quinazolinone, 2-methyl-3-(2-methylphenyl)-
11200 - 4(1H)-Quinazolinone, 1-methyl-2-(phenylmethyl)-

$C_{16}H_{14}O$
3876 - 2-Buten-1-one, 1,3-diphenyl-

$C_{16}H_{14}O_2$
10690 - 2-Propenoic acid, 3-phenyl-, phenylmethyl ester, (E)-
10712 - 2-Propen-1-one, 3-(3-methoxyphenyl)-1-phenyl-
10713 - 2-Propen-1-one, 3-(4-methoxyphenyl)-1-phenyl-

$C_{16}H_{14}O_3$
887 - Benzeneacetic acid, anhydride
2515 - Benzoic acid, 2-benzoyl-, ethyl ester
2682 - Benzoic acid, 2-methyl-, anhydride
2683 - Benzoic acid, 3-methyl-, anhydride

$C_{16}H_{14}O_4$
3039 - [1,1'-Biphenyl]-2,2'-dicarboxylic acid, dimethyl ester
3320 - Butanedioic acid, diphenyl ester
5572 - 1,2-Ethanediol, dibenzoate
6189 - 7H-Furo[3,2-g][1]benzopyran-7-one, 9-[(3-methyl-2-butenyl)oxy]-

$C_{16}H_{14}O_6$
629 - Benz[b]indeno[1,2-d]pyran-3,4,6a,9,10(6H)-pentol, 7,11b-dihydro-, cis-(+)-
2800 - 4H-1-Benzopyran-4-one, 2,3-dihydro-5,7-dihydroxy-2-(3-hydroxy-4-methoxyphenyl)-, (S)-
2801 - 4H-1-Benzopyran-4-one, 2,3-dihydro-5,7-dihydroxy-3-(3-hydroxy-4-methoxyphenyl)-

$C_{16}H_{15}ClO_2$
892 - Benzeneacetic acid, α-chloro-α-phenyl-, ethyl ester

$C_{16}H_{15}Cl_3O_2$
2329 - Benzene, 1,1'-(2,2,2-trichloroethylidene)bis[4-methoxy-

$C_{16}H_{16}$
421 - Anthracene, 9-ethyl-9,10-dihydro-
1113 - Benzene, 1,1'-(1-butenylidene)bis-
1456 - Benzene, 1,1'-(1,2-dimethyl-1,2-ethenediyl)bis-, (E)-
1457 - Benzene, 1,1'-(1,2-dimethyl-1,2-ethenediyl)bis-, (Z)-
1622 - Benzene, 1,1'-(1,2-ethenediyl)bis[2-methyl-
2022 - Benzene, 1,1'-(3-methyl-1-propene-1,3-diyl)bis-
11560 - Stilbene, α-ethyl-

$C_{16}H_{16}N_2OS$
9502 - 10H-Phenothiazine, 10-[(dimethylamino)acetyl]-

$C_{16}H_{16}N_2O_2$
5375 - Ergoline-8-carboxylic acid, 9,10-didehydro-6-methyl-, (8α)-
5376 - Ergoline-8-carboxylic acid, 9,10-didehydro-6-methyl-, (8β)-

$C_{16}H_{16}N_2O_4$
5104 - Desmedipham
9198 - Phenmedipham

$C_{16}H_{16}N_2O_4S$
91 - Acetamide, N,N'-(sulfonyldi-4,1-phenylene)bis-

$C_{16}H_{16}O$
3658 - 2-Butanone, 3,3-diphenyl-

$C_{16}H_{16}O_2$
939 - Benzeneacetic acid, α-phenyl-, ethyl ester
940 - Benzeneacetic acid, 2-phenylethyl ester
1621 - Benzene, 1,1'-(1,2-ethenediyl)bis[4-methoxy-

2144 - Benzenepropanoic acid, α-(phenylmethyl)-
2145 - Benzenepropanoic acid, phenylmethyl ester
3019 - [1,1'-Biphenyl]-4-butanoic acid
5813 - Ethanone, 2-ethoxy-1,2-diphenyl-

$C_{16}H_{16}O_3$
919 - Benzeneacetic acid, α-hydroxy-α-phenyl-, ethyl ester
3583 - Butanoic acid, 2-phenoxy-, phenyl ester

$C_{16}H_{16}O_4$
7806 - 1,8-Naphthalenedicarboxylic acid, diethyl ester

$C_{16}H_{16}O_5$
7854 - 1,4-Naphthalenedione, 5,8-dihydroxy-6-(1-hydroxy-4-methyl-3-pentenyl)-, (S)-

$C_{16}H_{16}Si$
11404 - Silane, diethenyldiphenyl-

$C_{16}H_{17}N$
722 - Benzenamine, N,N-dimethyl-4-(2-phenylethenyl)-, (E)-
3984 - 9H-Carbazole, 9-butyl-

$C_{16}H_{17}NO$
5219 - Diphenamid

$C_{16}H_{17}NO_3$
7665 - Morphinan-3,6-diol, 7,8-didehydro-4,5-epoxy-, (5α,6α)-

$C_{16}H_{17}NO_4$
6201 - Galanthan-1,2-diol, 3,12-didehydro-9,10-[methylenebis(oxy)]-, (1α,2β)-

$C_{16}H_{17}N_2NaO_4S$
11655 - 4-Thia-1-azabicyclo[3.2.0]heptane-2-carboxylic acid, 3,3-dimethyl-7-oxo-6-[(phenylacetyl)amino]-

$C_{16}H_{17}N_3O$
5374 - Ergoline-8-carboxamide, 9,10-didehydro-6-methyl-, (8β)-

$C_{16}H_{18}$
1099 - Benzene, 1,1'-(1,4-butanediyl)bis-
1120 - Benzene, 1,1'-butylidenebis-
1589 - Benzene, 1,1'-(1,2-ethanediyl)bis[3-methyl-
1590 - Benzene, 1,1'-(1,2-ethanediyl)bis[4-methyl-
1676 - Benzene, 1-ethyl-1,2-ethanediyl)bis-
1979 - Benzene, 1-(1-methylethyl)-4-(phenylmethyl)-
2020 - Benzene, 1,1'-(1-methyl-1,3-propanediyl)bis-, (±)-
2021 - Benzene, 1,1'-(2-methyl-1,3-propanediyl)bis-
2111 - Benzene, 1-(phenylmethyl)-4-propyl-
3095 - 1,1'-Biphenyl, 2,2',4,4'-tetramethyl-
3096 - 1,1'-Biphenyl, 2,2',5,5'-tetramethyl-

$C_{16}H_{18}ClN_3S$
9506 - Phenothiazin-5-ium, 3,7-bis(dimethylamino)-, chloride

$C_{16}H_{18}N_2$
9613 - Piperazine, 1,4-diphenyl-

$C_{16}H_{18}N_2O_2$
5112 - Diazene, bis(2-ethoxyphenyl)-
5113 - Diazene, bis(4-ethoxyphenyl)-

$C_{16}H_{18}N_2O_3$
6144 - 2(3H)-Furanone, dihydro-3-(hydroxyphenylmethyl)-4-[(1-methyl-1H-imidazol-5-yl)methyl]-,

$C_{16}H_{18}N_4O_2$
10910 - 4-Pyridinecarboxylic acid, 2-[3-oxo-3-[(phenylmethyl)amino]propyl]hydrazide

$C_{16}H_{18}N_4O_7S$
516 - Bensulfuron-methyl

$C_{16}H_{18}O$
1108 - Benzenebutanol, β-phenyl-, (R)-
2074 - Benzene, 1,1'-oxybis[2-ethyl-
2080 - Benzene, 1,1'-(oxydi-2,1-ethanediyl)bis-
2081 - Benzene, 1,1'-(oxydiethylidene)bis-, (R*,R*)-(±)-

$C_{16}H_{18}O_2$
3048 - 1,1'-Biphenyl, 2,2'-dimethoxy-5,5'-dimethyl-
3049 - 1,1'-Biphenyl, 2,5'-dimethoxy-2'-5-dimethyl-
3069 - [1,1'-Biphenyl]-4,4'-diol, 3,3',5,5'-tetramethyl-
9249 - Phenol, 4,4'-butylidenebis-
9433 - Phenol, 4,4'-(1-methylpropylidene)bis-

$C_{16}H_{18}O_{10}$
2823 - 2H-1-Benzopyran-2-one, 8-(β-D-glucopyranosyloxy)-7-hydroxy-6-methoxy-

$C_{16}H_{19}Cl_2N$
1801 - Benzenemethanamine, N-(2-chloroethyl)-N-(phenylmethyl)-, hydrochloride

$C_{16}H_{19}N$
744 - Benzenamine, 2-ethyl-N-(2-ethylphenyl)-
1585 - Benzeneethanamine, N-(2-phenylethyl)-
1822 - Benzenemethanamine, α-methyl-N-(1-phenylethyl)-

$C_{16}H_{19}NO_3$
6182 - Furo[2,3-b]quinolin-4(2H)-one, 3,9-dihydro-8-methoxy-9-methyl-2-(1-methylethyl)-, (R)-
8567 - 16(15H)-Oxaerythrinan-15-one, 1,2,6,7-tetrahydro-14,17-dihydro-3-methoxy-, (3β)-

$C_{16}H_{19}NO_4$
481 - 8-Azabicyclo[3.2.1]octane-2-carboxylic acid, 3-(benzoyloxy)-8-methyl-, [1R-(exo,exo)]-

$C_{16}H_{20}$
7900 - Naphthalene, 1-hexyl-
7901 - Naphthalene, 2-hexyl-
7905 - Naphthalene, 1-isopropyl-2,4,7-trimethyl-
8014 - Naphthalene, 1,2,5-trimethyl-8-(1-methylethyl)-
8015 - Naphthalene, 1,3,8-trimethyl-5-(1-methylethyl)-

$C_{16}H_{20}N_2$
3036 - [1,1'-Biphenyl]-4,4'-diamine, tetramethyl-
5500 - 1,2-Ethanediamine, N,N'-bis(phenylmethyl)-
11010 - 2-Pyridinepropanamine, N,N-dimethyl-γ-phenyl-

$C_{16}H_{20}N_4$
361 - Aniline, 3,3'-azobis[N,N-dimethyl-

$C_{16}H_{20}O_2Si$
11406 - Silane, diethoxydiphenyl-

$C_{16}H_{20}O_6P_2S_3$
1 - Abate

$C_{16}H_{21}NO_3$
911 - Benzeneacetic acid, α-hydroxy-, 8-methyl-8-azabicyclo[3.2.1]oct-3-yl ester, endo-(±)-
916 - Benzeneacetic acid, α-(hydroxymethyl)-, 8-azabicyclo[3.2.1]oct-3-yl ester, [3(S)-endo]-

$C_{16}H_{21}N_3$
5505 - 1,2-Ethanediamine, N,N-dimethyl-N'-(phenylmethyl)-N'-2-pyridinyl-

$C_{16}H_{22}ClNO_3$
5173 - Diethatyl-ethyl

$C_{16}H_{22}N_4O_2$
9795 - Propanamide, N-(2,3-dihydro-1,5-dimethyl-3-oxo-2-phenyl-1H-pyrazol-4-yl)-2-(dimethylamino)-

$C_{16}H_{22}OSi_2$
5238 - Disiloxane, 1,1,3,3-tetramethyl-1,3-diphenyl-

$C_{16}H_{22}O_4$
1336 - 1,2-Benzenedicarboxylic acid, bis(2-methylpropyl) ester
1338 - 1,4-Benzenedicarboxylic acid, bis(2-methylpropyl) ester
1339 - 1,2-Benzenedicarboxylic acid, dibutyl ester

$C_{16}H_{22}O_6$
1333 - 1,2-Benzenedicarboxylic acid, bis(2-ethoxyethyl) ester

$C_{16}H_{22}O_8$
6226 - β-D-Glucopyranoside, 4-(3-hydroxy-1-propenyl)-2-methoxyphenyl

$C_{16}H_{22}O_{11}$
6216 - α-D-Glucopyranose, pentaacetate
6217 - β-D-Glucopyranose, pentaacetate

$C_{16}H_{23}ClN_3O$
11589 - Tebuconazole

$C_{16}H_{24}N_2O_2$
1325 - 1,2-Benzenedicarboxamide, N,N,N',N'-tetraethyl-
1326 - 1,3-Benzenedicarboxamide, N,N,N',N'-tetraethyl-

$C_{16}H_{25}NO_2$
898 - Benzeneacetic acid, α-ethyl-, 2-(diethylamino)ethyl ester

$C_{16}H_{26}$
1258 - Benzene, decyl-
2090 - Benzene, pentaethyl-
6627 - 6,10-Hexadecadiyne
11609 - 5,9-Tetradecadien-7-yne, 6,9-dimethyl-

$C_{16}H_{26}O$
9221 - Phenol, 2,6-bis(1,1-dimethylethyl)-4-ethyl-

$C_{16}H_{26}O_2$
1501 - 1,4-Benzenediol, 2,5-bis(1,1-dimethylpropyl)-
6666 - 6,10,14-Hexadecatrienoic acid

$C_{16}H_{28}O_2$
4940 - 2-Cyclopentene-1-undecanoic acid, (R)-

$C_{16}H_{30}$
4323 - Cyclohexane, 1,1'-(1,4-butanediyl)bis-
4326 - Cyclohexane, 1,1'-butylidenebis-
4447 - Cyclohexane, 1,1'-(2-methyl-1,3-propanediyl)bis-
6673 - 1-Hexadecyne
6674 - 2-Hexadecyne

$C_{16}H_{30}O$
4762 - Cyclopentadecanone, 2-methyl-
4763 - Cyclopentadecanone, 3-methyl-

$C_{16}H_{30}O_2$
6668 - 7-Hexadecenoic acid
6669 - 9-Hexadecenoic acid
6670 - 9-Hexadecenoic acid, (Z)-

$C_{16}H_{30}O_3$
8424 - Octanoic acid, anhydride

$C_{16}H_{31}ClO$
6665 - Hexadecanoyl chloride

$C_{16}H_{31}N$
6641 - Hexadecanenitrile

$C_{16}H_{32}$
4344 - Cyclohexane, decyl-
4863 - Cyclopentane, undecyl-
6667 - 1-Hexadecene
10580 - 1-Propene, 2-methyl-, tetramer

$C_{16}H_{32}Br_2$
6635 - Hexadecane, 1,16-dibromo-

$C_{16}H_{32}O$
6628 - Hexadecanal
6663 - 3-Hexadecanone

$C_{16}H_{32}O_2$
5065 - Decanoic acid, 2-hexyl-
6644 - Hexadecanoic acid
8152 - Nonanoic acid, heptyl ester
8437 - Octanoic acid, octyl ester
8646 - Pentadecanoic acid, methyl ester
11624 - Tetradecanoic acid, ethyl ester

$C_{16}H_{32}O_4$
5313 - Dodecanoic acid, 2-(2-hydroxyethoxy)ethyl ester

$C_{16}H_{33}Br$
6633 - Hexadecane, 1-bromo-

$C_{16}H_{33}Cl$
6634 - Hexadecane, 1-chloro-

$C_{16}H_{33}F$
6639 - Hexadecane, 1-fluoro-

$C_{16}H_{33}I$
6640 - Hexadecane, 1-iodo-

$C_{16}H_{33}NO$
5290 - Dodecanamide, N,N-diethyl-
6629 - Hexadecanamide

$C_{16}H_{34}$
6632 - Hexadecane

$C_{16}H_{34}O$
6431 - Heptane, 3,3'-[oxybis(methylene)]bis-
6659 - 1-Hexadecanol
6660 - 2-Hexadecanol
6661 - 3-Hexadecanol
8416 - Octane, 1,1'-oxybis-
8649 - 1-Pentadecanol, 2-methyl-
8650 - 6-Pentadecanol, 6-methyl-

$C_{16}H_{34}O_2$
11617 - Tetradecane, 1,1-dimethoxy-

$C_{16}H_{34}O_3$
8552 - Orthoformic acid, triisopentyl ester

$C_{16}H_{34}S$
6643 - 1-Hexadecanethiol
8417 - Octane, 1,1'-thiobis-

$C_{16}H_{35}N$
6631 - 1-Hexadecanamine
8352 - 2-Octanamine, N-(1-methylheptyl)-
8354 - 1-Octanamine, N-octyl-

$C_{16}H_{36}O_4Si$
11505 - Silicic acid (H_4SiO_4), tetrabutyl ester

$C_{16}H_{48}O_6Si_7$
6528 - Heptasiloxane, hexadecamethyl-

$C_{16}H_{48}O_8Si_8$
4741 - Cyclooctasiloxane, hexadecamethyl-

$C_{17}H_9NO_4$
8062 - Naphtho[2,3-f]quinoline-7,12-dione, 5,6-dihydroxy-

$C_{17}H_{10}O$
6292 - 7H-Benz[de]anthracen-7-one

$C_{17}H_{10}O_2$
10854 - 1-Pyrenecarboxylic acid

$C_{17}H_{11}N$
630 - Benz[c]acridine

$C_{17}H_{12}$
2393 - 11H-Benzo[a]fluorene

$C_{17}H_{12}ClFN_2O$
8219 - Nuarimol

$C_{17}H_{12}Cl_2N_2O$
5948 - Fenarimol

$C_{17}H_{12}O_2$
8021 - 2-Naphthalenol, benzoate

$C_{17}H_{12}O_3$
2644 - Benzoic acid, 2-hydroxy-, 1-naphthalenyl ester
2645 - Benzoic acid, 2-hydroxy-, 2-naphthalenyl ester

$C_{17}H_{12}O_7$
6184 - 1H,12H-Furo[3',2':4,5]furo[2,3-h]pyrano[3,4-c][1]benzopyran-1,12-dione, 3,4,7a,10a-tetrahydro-5-meth

$C_{17}H_{13}N$
10947 - Pyridine, 2,6-diphenyl-

$C_{17}H_{13}N_3O_5S_2$
2726 - Benzoic acid, 2-[[[4-[(2-thiazolylamino)sulfonyl]phenyl]amino]carbonyl]-

$C_{17}H_{14}$
4756 - 15H-Cyclopenta[a]phenanthrene, 16,17-dihydro-
7949 - Naphthalene, 1-(phenylmethyl)-
7950 - Naphthalene, 2-(phenylmethyl)-
9182 - Phenanthrene, 9-(1-methylethenyl)-

$C_{17}H_{14}O$
7912 - 1-Naphthalenemethanol, α-phenyl-
8039 - 1-Naphthalenol, 2-(phenylmethyl)-
8040 - 2-Naphthalenol, 1-(phenylmethyl)-
8683 - 1,4-Pentadien-3-one, 1,5-diphenyl-

$C_{17}H_{14}O_6$
4757 - Cyclopenta[c]furo[3',2':4,5]furo[2,3-h][1]benzopyran-1,11-dione, 2,3,3a,8,9,9a-hexahydro-4-methoxy-, (6aR-

$C_{17}H_{15}ClF_3NO_4$
6291 - Haloxyfop-(2-ethoxyethyl)

$C_{17}H_{15}ClO_7$
2847 - 1-Benzopyrylium, 3,5,7-trihydroxy-2-(4-hydroxy-3,5-dimethylphenyl)-, chloride

$C_{17}H_{15}N$
7723 - 1-Naphthalenamine, N-(2-methylphenyl)-
7724 - 1-Naphthalenamine, N-(4-methylphenyl)-
7725 - 2-Naphthalenamine, N-(2-methylphenyl)-
7730 - 2-Naphthalenamine, N-(phenylmethyl)-

$C_{17}H_{16}Br_2O_3$
3140 - Bromopropylate

$C_{17}H_{16}O_2$
942 - Benzeneacetic acid, α-(phenylmethylene)-, ethyl ester

$C_{17}H_{16}O_3$
8776 - Pentanedioic acid, diphenyl ester
9927 - Propanedioic acid, bis(phenylmethyl) ester

$C_{17}H_{17}ClO_6$
11520 - Spiro[benzofuran-2(3H),1'-[2]cyclohexene]-3,4'-dione, 7-chloro-2',4,6-trimethoxy-

$C_{17}H_{17}NO_2$
5155 - 4H-Dibenzo[de,g]quinoline-10,11-diol, 5,6,6a,7-tetrahydro-6-methyl-, (R)-

$C_{17}H_{17}N_3O_3$
7189 - Imazaquin

$C_{17}H_{18}O$
7594 - Methanone, bis(2,4-dimethylphenyl)-
8969 - 3-Pentanone, 1,5-diphenyl-

$C_{17}H_{18}O_4$
1103 - Benzenebutanoic acid, 4-hydroxy-γ-(4-hydroxyphenyl)-γ-methyl-
9970 - Propanedioic acid, 1-naphthalenyl-, diethyl ester
10378 - 2-Propanol, 1,3-diphenoxy-, acetate

$C_{17}H_{19}ClN_2S$
9505 - 10H-Phenothiazine-10-propanamine, 2-chloro-N,N-dimethyl-

$C_{17}H_{19}NO_3$
7396 - 7-Isoquinolinol, 1,2,3,4-tetrahydro-1-[(4-hydroxyphenyl)methyl]-6-methoxy-, (S)-
7666 - Morphinan-3,6-diol, 7,8-didehydro-4,5-epoxy-17-methyl- (5α,6α)-
9632 - Piperidine, 1-[5-(1,3-benzodioxol-5-yl)-1-oxo-2,4-pentadienyl]-, (E,E)-
9633 - Piperidine, 1-[5-(1,3-benzodioxol-5-yl)-1-oxo-2,4-pentadienyl]-, (Z,Z)-

$C_{17}H_{19}NO_4$
5952 - Fenoxycarb
7668 - Morphinan-3,6-diol, 7,8-didehydro-4,5-epoxy-17-methyl- (5α,6α)-, 17-oxide

$C_{17}H_{20}$
1697 - Benzene, 1,1'-(1-ethyl-1,3-propanediyl)bis-
1698 - Benzene, 1,1'-(2-ethyl-1,3-propanediyl)bis-
2098 - Benzene, 1,1'-(1,5-pentanediyl)bis-
2108 - Benzene, 1,1'-pentylidenebis-

$C_{17}H_{20}N_2O$
7593 - Methanone, bis[4-(dimethylamino)phenyl]-
12003 - Urea, N,N'-diethyl-N,N'-diphenyl-

$C_{17}H_{20}N_2S$
9504 - 10H-Phenothiazine-10-ethanamine, $N,N,$α-trimethyl-

$C_{17}H_{20}N_4O_6$
11344 - Riboflavin

$C_{17}H_{20}O_6$
7052 - 4-Hexenoic acid, 6-(1,3-dihydro-4-hydroxy-6-methoxy-7-methyl-3-oxo-5-isobenzofuranyl)-4-methyl-, (E)-

$C_{17}H_{20}Si$
11360 - Silacyclohexane, 1,1-diphenyl-

$C_{17}H_{21}N$
1575 - Benzeneethanamine, $N,$α-dimethyl-N-(phenylmethyl)-, (+)-
1804 - Benzenemethanamine, N,N-diethyl-α-phenyl-

$C_{17}H_{21}NO$
5422 - Ethanamine, 2-(diphenylmethoxy)-N,N-dimethyl-

$C_{17}H_{21}NO_2$
928 - Benzeneacetic acid, α-methylene-, 8-methyl-8-azabicyclo[3.2.1]oct-3-yl ester, endo-
8067 - Napropamide

$C_{17}H_{21}NO_3$
6200 - Galanthamine

$C_{17}H_{21}NO_4$
482 - 8-Azabicyclo[3.2.1]octane-2-carboxylic acid, 3-(benzoyloxy)-8-methyl-, methyl ester, [1R-(exo,exo)]-

908 - Benzeneacetic acid, α-(hydroxymethyl)-, 9-methyl-3-oxa-9-azatricyclo[3.3.1.02,4]non-7-yl ester, [(7S)-(1α,2β,4β,5α,7β)]-

$C_{17}H_{21}N_3$
657 - Benzenamine, 4,4'-carbonimidoylbis[N,N-dimethyl-

$C_{17}H_{22}$
7895 - Naphthalene, 1-heptyl-
7896 - Naphthalene, 2-heptyl-

$C_{17}H_{22}N_2$
786 - Benzenamine, 4,4'-methylenebis[N,N-dimethyl-
5506 - 1,2-Ethanediamine, N,N-dimethyl-N'-phenyl-N'-(phenylmethyl)-

$C_{17}H_{23}NO_3$
902 - Benzeneacetic acid, α-(hydroxymethyl)- 8-methyl-8-azabicyclo[3.2.1]oct-3-yl ester endo-(±)-
917 - Benzeneacetic acid, α-(hydroxymethyl)-, 8-methyl-8-azabicyclo[3.2.1]oct-3-yl ester, [3(S)-endo]-

$C_{17}H_{23}N_3O$
5511 - 1,2-Ethanediamine, N-[(4-methoxyphenyl)methyl]-N',N'-dimethyl-N-2-pyridinyl-

$C_{17}H_{24}N_2O$
5411 - Ethanamine, 2-[(3-butyl-1-isoquinolinyl)oxy]-N,N-dimethyl-

$C_{17}H_{24}O_9$
6225 - β-D-Glucopyranoside, 4-(3-hydroxy-1-propenyl)-2,6-dimethoxyphenyl

$C_{17}H_{25}N$
9696 - Piperidine, 1-(1-phenylcyclohexyl)-

$C_{17}H_{25}NO_2$
7660 - MGK 264

$C_{17}H_{25}N_3O$
9877 - 1,3-Propanediamine, N,N-diethyl-N'-(6-methoxy-8-quinolinyl)-

$C_{17}H_{26}ClNO_2$
3144 - Butachlor

$C_{17}H_{26}ClNO_3S$
4138 - Clethodim

$C_{17}H_{26}O_4$
4261 - 2,5-Cyclohexadiene-1,4-dione, 2,5-dihydroxy-3-undecyl-

$C_{17}H_{27}NO$
11169 - Pyrrolidine, 1-[2-(3-methylbutoxy)-2-phenylethyl]-

$C_{17}H_{27}NO_2$
7683 - Morpholine, 4-[3-(4-butoxyphenoxy)propyl]-

$C_{17}H_{28}$
2374 - Benzene, undecyl-

$C_{17}H_{28}O$
9225 - Phenol, 2,6-bis(1,1-dimethylpropyl)-4-methyl-

$C_{17}H_{29}NO_3S$
11359 - Sethoxydim

$C_{17}H_{30}O$
4217 - 9-Cycloheptadecen-1-one
4218 - 9-Cycloheptadecen-1-one, (Z)-

$C_{17}H_{30}O_2$
7182 - Hydroprene

$C_{17}H_{32}$
4402 - Cyclohexane, 1,1'-(2-ethyl-1,3-propanediyl)bis-
4464 - Cyclohexane, 1,1'-(1,5-pentanediyl)bis-

$C_{17}H_{32}O_2$
6324 - 9-Heptadecenoic acid

$C_{17}H_{33}N$
6312 - Heptadecanenitrile

$C_{17}H_{34}$
4809 - Cyclopentane, dodecyl-
6322 - 1-Heptadecene

$C_{17}H_{34}Br_2$
6308 - Heptadecane, 1,17-dibromo-

$C_{17}H_{34}O$
6304 - Heptadecanal
6320 - 2-Heptadecanone
6321 - 9-Heptadecanone

C19H24N2
9627 - 4-Piperidinamine, 1-methyl-N-phenyl-N-(phenylmethyl)-

C19H24N2O
4121 - Cinchonan-9-ol, 10,11-dihydro-, (8α,9R)-
4122 - Cinchonan-9-ol, 10,11-dihydro-, (9S)-
7306 - 1H-Indole-3-ethanol, 2-(5-ethenyl-1-azabicyclo[2.2.2]oct-2-yl)-

C19H24N2OS
3977 - Carbamothioic acid, diphenyl, S-[2-(diethylamino)ethyl] ester

C19H24N2O2
6177 - 3aH-Furo[2,3-b]indol-3a-ol, 8a-(5-ethenyl-1-azabicyclo[2.2.2]oct-2-yl)-2,3,8,8a-tetrahydro-
6178 - 3aH-Furo[2,3-b]indol-3a-ol, 8a-(5-ethenyl-1-azabicyclo[2.2.2]oct-2-yl)-2,3,8,8a-tetrahydro-

C19H24O3
353 - Androst-4-ene-3,11,17-trione

C19H26
7931 - Naphthalene, 1-nonyl-
7932 - Naphthalene, 2-nonyl-

C19H26N2O
4154 - Curan-17-ol, (16α)-

C19H26O2
343 - Androsta-1,4-dien-3-one, 17-hydroxy-, (17β)-
352 - Androst-4-ene-3,17-dione

C19H26O3
336 - Allethrin

C19H26O4S
10507 - Propargite

C19H27ClO2
354 - Androst-4-en-3-one, 4-chloro-17-hydroxy-, (17β)-

C19H27NO3
2396 - 2H-Benzo[a]quinolizin-2-one, 1,3,4,6,7,11b-hexahydro-9,10-dimethoxy-3-(2-methylpropyl)-

C19H28BrNO3
11179 - Pyrrolidinium, 3-[(cyclopentylhydroxyphenylacetyl)oxy]-1,1-dimethyl-, bromide

C19H28O2
356 - Androst-4-en-3-one, 17-hydroxy-, (17β)-

C19H30O2
347 - Androstan-17-one, 3-hydroxy-, (3β,5α)-
348 - Androstan-17-one, 3-hydroxy-, (3α,5α)-
349 - Androstan-3-one, 17-hydroxy-, (5α,17β)-
5318 - Dodecanoic acid, phenylmethyl ester

C19H30O5
9727 - Piperonyl butoxide

C19H32
345 - Androstane

C20H27O4P
9560 - Phosphoric acid, 2-ethylhexyl diphenyl ester

C20H28
7793 - Naphthalene, 1-decyl-
7794 - Naphthalene, 2-decyl-

C20H28O2
344 - Androsta-1,4-dien-3-one, 17-hydroxy-17-methyl-, (17β)-

C20H28O3S
7968 - 1-Naphthalenesulfonic acid, 3,6-bis(1,1-dimethylethyl)-, ethyl ester

C20H29FO3
355 - Androst-4-en-3-one, 9-fluoro-11,17-dihydroxy-17-methyl-, (11β,17β)-

C20H30ClN3O2
11226 - 4-Quinolinecarboxamide, 2-butoxy-N-[2-(diethylamino)ethyl]-, monohydrochloride

C20H30O
11341 - Retinol

C20H30O2
358 - Androst-4-en-3-one, 17-hydroxy-17-methyl-, (17β)-
9166 - 1-Phenanthrenecarboxylic acid, 1,2,3,4,4a,5,6,9,10,10a-decahydro-1,4a-dimethyl-7-(1-methylethyl)-

9167 - 1-Phenanthrenecarboxylic acid, 1,2,3,4,4a,4b,5,6,10,10a-decahydro-1,4a-dimethyl-7-(1-methylethyl)-,
9169 - 1-Phenanthrenecarboxylic acid, 1,2,3,4,4a,4b,5,9,10,10a-decahydro-1,4a-dimethyl-7-(1-methylethyl)-,
9170 - 1-Phenanthrenecarboxylic acid, 7-ethenyl-1,2,3,4,4a,4b,5,6,7,9,10,10a-dodecahydro-1,4a,7-trimethyl-, [1R-(1α,4aβ,4bα,7β,10aα)]-

C20H32ClNO
9705 - 1-Piperidinepropanol, α-cyclohexyl-α-phenyl-, hydrochloride

C20H32N2O3S
4038 - Carbosulfan

C20H32O2
346 - Androstane-17-carboxylic acid, (5β,17β)-
350 - Androstan-3-one, 17-hydroxy-17-methyl-, (5α,17β)-
5360 - 5,8,11,14-Eicosatetraenoic acid
5361 - 5,8,11,14-Eicosatetraenoic acid, (all-Z)-

C20H34
2266 - Benzene, tetradecyl-

C20H34O2
8283 - 9,12,15-Octadecatrienoic acid, ethyl ester, (Z,Z,Z)-

C20H36O2
8226 - 9,12-Octadecadienoic acid (Z,Z)-, ethyl ester
8227 - 9,12-Octadecadienoic acid, ethyl ester

C20H38
5366 - 1-Eicosyne

C20H38N4O4
7686 - 4-Morpholinecarboxamide, N,N'-1,2-ethanediylbis[N-butyl-

C20H38O2
4952 - Cycloprate
5363 - 11-Eicosenoic acid
5364 - 9-Eicosenoic acid
5365 - 9-Eicosenoic acid, (Z)-
8301 - 9-Octadecenoic acid (Z)-, ethyl ester
8302 - 9-Octadecenoic acid, ethyl ester, (E)-

C20H38O3
5060 - Decanoic acid, anhydride
8269 - 9-Octadecanoic acid, 12-oxo-, ethyl ester
8305 - 9-Octadecenoic acid, 12-hydroxy-, ethyl ester, [R-(Z)]-

C20H38O4
5352 - Eicosanedioic acid

C20H40
4475 - Cyclohexane, tetradecyl-
4844 - Cyclopentane, pentadecyl-
5362 - 1-Eicosene

C20H40O
6671 - 1-Hexadecen-3-ol, 3,7,11,15-tetramethyl-
6672 - 2-Hexadecen-1-ol, 3,7,11,15-tetramethyl-, [R-[R*,R*-(E)]]-
8240 - Octadecane, 1-(ethenyloxy)-

C20H40O2
221 - Acetic acid, octadecyl ester
5062 - Decanoic acid, decyl ester
5355 - Eicosanoic acid
6647 - Hexadecanoic acid, butyl ester
6658 - Hexadecanoic acid, 3,7,11,15-tetramethyl-
8257 - Octadecanoic acid, ethyl ester

C20H40O3
8260 - Octadecanoic acid, 2-hydroxyethyl ester

C20H42
5350 - Eicosane

C20H42O
5358 - 1-Eicosanol
5359 - 2-Eicosanol

C20H42O2
8237 - Octadecane, 1,1-dimethoxy-

C20H48O4Si4
4995 - Cyclotetrasiloxane, 2,4,6,8-tetrabutyl-2,4,6,8-tetramethyl-

C20H60O8Si9
8178 - Nonasiloxane, eicosamethyl-

C21H12O
5152 - 13H-Dibenzo[a,h]fluoren-13-one

C21H13N
5143 - Dibenz[a,j]acridine

C21H14
5148 - 13H-Dibenzo[a,g]fluorene

C21H14Br4O5S
9367 - Phenol, 4,4'-(3H-2,1-benzoxathiol-3-ylidene)bis[2,6-dibromo-3-methyl-, S,S-dioxide

C21H14O
7634 - Methanone, 1-naphthalenyl-2-naphthalenyl-

C21H16
2392 - Benz[j]aceanthrylene, 1,2-dihydro-3-methyl-
7271 - 1H-Indene, 1,3-diphenyl-
7272 - 1H-Indene, 2,3-diphenyl-
7919 - Naphthalene, 1,1'-methylenebis-

C21H16Br2O5S
9364 - Phenol, 4,4'-(3H-2,1-benzoxathiol-3-ylidene)bis[2-bromo-6-methyl-, S,S-dioxide

C21H16N2
7215 - 1H-Imidazole, 2,4,5-triphenyl-

C21H16N2O
12008 - Urea, N,N'-di-1-naphthalenyl-

C21H16NaO2
11517 - Sonar

C21H18N2
7199 - 1H-Imidazole, 4,5-dihydro-2,4,5-triphenyl-, cis-

C21H18O5S
9368 - Phenol, 4,4'-(3H-2,1-benzoxathiol-3-ylidene)bis[2-methyl-, S,S-dioxide

C21H19NO5
2433 - [1,3]Benzodioxolo[5,6-c]phenanthridinium, 1,2-dimethoxy-12-methyl-

C21H20
2110 - Benzene, 1,1'-(phenylmethylene)bis[4-methyl-

C21H20Cl2O3
9150 - Permethrin

C21H20N2O
8572 - Oxayohimbanium, 3,4,5,6,16,17-hexadehydro-16-(methoxycarbonyl)-19-methyl-, (19α)-

C21H20O6
6334 - 1,6-Heptadiene-3,5-dione, 1,7-bis(4-hydroxy-3-methoxyphenyl)-, (E,E)-

C21H20O10
2824 - 4H-1-Benzopyran-4-one, 3-[4-(β-D-glucopyranosyloxy)phenyl]-5,7-dihydroxy-

C21H20O11
2796 - 4H-1-Benzopyran-4-one, 3-[(6-deoxy-α-L-mannopyranosyl)oxy]-2-(3,4-dihydroxyphenyl)-5,7-dihydroxy-

C21H21N
1798 - Benzenemethanamine, N,N-bis(phenylmethyl)-

C21H21NO5
3108 - Bis[1,3]benzodioxolo[4,5-c:5',6'-g]azecin-13(5H)-one, 4,6,7,14-tetrahydro-5,14-dimethyl-, (R)-

C21H21NO6
7357 - 1(3H)-Isobenzofuranone, 6,7-dimethoxy-3-(5,6,7,8-tetrahydro-6-methyl-1,3-dioxolo[4,5-g]isoquinolin-5-yl)-, [S-
11342 - Rheadan, 8-methoxy-16-methyl-2,3 10,11-bis[methylenebis(oxy)]-, (8β)-

C21H21N3
11834 - 1,3,5-Triazine, hexahydro-1,3,5-triphenyl-

C21H21N3O7
11352 - 9,10-Secostrychnidin-10-oic acid, 2,3-dihydro-4-nitro-2,3-dioxo-

C21H21O3P
9596 - Phosphorous acid, tris(2-methylphenyl) ester
9597 - Phosphorous acid, tris(4-methylphenyl) ester

C21H21O4P
9569 - Phosphoric acid, tris(2-methylphenyl) ester
9570 - Phosphoric acid, tris(3-methylphenyl) ester
9571 - Phosphoric acid, tris(4-methylphenyl) ester

$C_{21}H_{21}O_7P$
9387 - Phenol, 2-methoxy-, phosphate (3:1)

$C_{21}H_{21}Sb$
11552 - Stibine, tris(3-methylphenyl)-
11553 - Stibine, tris(4-methylphenyl)-

$C_{21}H_{22}N_2O_2$
11563 - Strychnidin-10-one

$C_{21}H_{23}ClFNO_2$
3651 - 1-Butanone, 4-[4-(4-chlorophenyl)-4-hydroxy-1-piperidinyl]-1-(4-fluorophenyl)-

$C_{21}H_{23}NO_5$
2435 - Benzo[e]-1,3-dioxolo[4,5-l][2]benzazecin-12(5H)-one, 4,6,7,13-tetrahydro-9,10-dimethoxy-5-methyl-
7667 - Morphinan-3,6-diol, 7,8-didehydro-4,5-epoxy-17-methyl- (5α,6α)-, diacetate (ester)

$C_{21}H_{23}NO_6$
92 - Acetamide, N-(5,6,7,9-tetrahydro-10-hydroxy-1,2,3-trimethoxy-9-oxobenzo[a]heptalen-7-yl)-, (S)-

$C_{21}H_{23}N_3O_5$
11565 - Strychnidin-10-one, mononitrate

$C_{21}H_{24}N_2O_3$
8571 - Oxayohimban-16-carboxylic acid, 16,17-didehydro-19-methyl-, methyl ester, (19α)-

$C_{21}H_{24}OSi_2$
5239 - Disiloxane, 1,1,1-trimethyl-3,3,3-triphenyl-

$C_{21}H_{24}O_3Si_3$
5005 - Cyclotrisiloxane, 2,4,6-trimethyl-2,4,6-triphenyl-

$C_{21}H_{24}O_4$
9926 - Propanedioic acid, bis(phenylmethyl)-, diethyl ester

$C_{21}H_{24}O_7$
3540 - Butanoic acid, 2-methyl-, 10-(acetyloxy)-9,10-dihydro-8,8-dimethyl-2-oxo-2H,8H-benzo[1,2-b 3,4-b′]dipyran-9-yl

$C_{21}H_{25}NO_4$
5150 - 6H-Dibenzo[a,g]quinolizin-2-ol, 5,8,13,13a-tetrahydro-3,9,10-trimethoxy-13-methyl-, (13S-trans)-
5151 - 6H-Dibenzo[a,g]quinolizin-3-ol, 5,8,13,13a-tetrahydro-2,9,10-trimethoxy-13-methyl-, (13S-trans)-
5157 - 4H-Dibenzo[de,g]quinoline, 5,6,6a,7-tetrahydro-1,2,9,10-tetramethoxy-6-methyl-, (S)-

$C_{21}H_{26}BrNO_3$
5445 - Ethanaminium, N,N-diethyl-N-methyl-2-[(9H-xanthen-9-ylcarbonyl)oxy]-, bromide

$C_{21}H_{26}ClN_3OS$
9617 - 1-Piperazineethanol, 4-[3-(2-chloro-10H-phenothiazin-10-yl)propyl]-

$C_{21}H_{26}N_2O_3$
5349 - Eburnamenine-14-carboxylic acid, 14,15-dihydro-14-hydroxy-, methyl ester, (3α,14β,16α)-
12062 - Yohimban-16-carboxylic acid, 17-hydroxy-, methyl ester, (16α,17α)-

$C_{21}H_{26}N_2O_4$
4129 - Cinchonan-9-ol, 6′-methoxy-, (8α,9R)-, monoformate (salt)

$C_{21}H_{26}O_2$
5154 - 6H-Dibenzo[b,d]pyran-1-ol, 6,6,9-trimethyl-3-pentyl-

$C_{21}H_{26}O_3$
7621 - Methanone, [2-hydroxy-4-(octyloxy) phenyl]phenyl-

$C_{21}H_{26}O_5$
9739 - Pregna-1,4-diene-3,11,20-trione, 17,21-dihydroxy-

$C_{21}H_{27}ClN_2O_2$
5700 - Ethanol, 2-[2-[4-[(4-chlorophenyl) phenylmethyl]-1-piperazinyl]ethoxy]-

$C_{21}H_{27}NO_4$
7379 - Isoquinoline, 1-[(3,4-dimethoxyphenyl)methyl]-1,2,3,4-tetrahydro-6,7-dimethoxy-2-methyl-, (S)-

$C_{21}H_{28}ClNO$
6516 - 3-Heptanone, 6-(dimethylamino)-4,4-diphenyl-, hydrochloride

$C_{21}H_{28}O_2$
9758 - Pregn-4-en-20-yn-3-one, 17-hydroxy-, (17α)-

$C_{21}H_{28}O_3$
4960 - Cyclopropanecarboxylic acid, 2,2-dimethyl-3-(2-methyl-1-propenyl)-, 2-methyl-4-oxo-3-(2,4-pentadienyl)-2-cyclopenten-1-yl ester, [1R-[1α[S*(Z)],3β]]-

$C_{21}H_{28}O_4$
9755 - Pregn-4-ene-3,11,20-trione, 21-hydroxy-

$C_{21}H_{28}O_5$
9738 - Pregna-1,4-diene-3,20-dione, 11,17,21-trihydroxy-, (11β)-
9746 - Pregn-4-en-18-al, 11,21-dihydroxy-3,20-dioxo-, (11β)-
9754 - Pregn-4-ene-3,11,20-trione, 17,21-dihydroxy-

$C_{21}H_{30}O_2$
1537 - 1,3-Benzenediol, 2-[3-methyl-6-(1-methylethenyl)-2-cyclohexen-1-yl]-5-pentyl-, (1R-trans)-

$C_{21}H_{30}O_3$
9750 - Pregn-4-ene-3,20-dione, 17-hydroxy-
9751 - Pregn-4-ene-3,20-dione, 21-hydroxy-

$C_{21}H_{30}O_4$
9748 - Pregn-4-ene-3,20-dione, 11,21-dihydroxy-, (11β)-
9749 - Pregn-4-ene-3,20-dione, 17,21-dihydroxy-

$C_{21}H_{30}O_5$
4282 - 2,4-Cyclohexadien-1-one, 3,5,6-trihydroxy-4,6-bis(3-methyl-2-butenyl)-2-(3-methyl-1-oxobutyl)-, (R)-
9753 - Pregn-4-ene-3,20-dione, 11,17,21-trihydroxy-, (11β)-

$C_{21}H_{32}O_2$
9168 - 1-Phenanthrenecarboxylic acid, 1,2,3,4,4a,4b,5,6,10,10a-decahydro-1,4a-dimethyl-7-(1-methylethyl)-, methyl ester, [1R-(1α,4aβ,4bα,10aα)]-
9743 - Pregnane-3,20-dione, (5β)-
9756 - Pregn-5-en-20-one, 3-hydroxy-, (3β)-

$C_{21}H_{32}O_3$
9744 - Pregnane-11,20-dione, 3,17,21-trihydroxy-, (3α,5β)-
9757 - Pregn-4-en-3-one, 11,17,20,21-tetrahydroxy-, (11β,20R)-

$C_{21}H_{34}BrNO_2$
5443 - Ethanaminium, 2-[(cyclohexylhydroxyphenylacetyl) oxy]-N,N-diethyl-N-methyl-, bromide

$C_{21}H_{34}O_2$
9745 - Pregnan-20-one, 3-hydroxy-, (3α,5β)-
11627 - Tetradecanoic acid, phenylmethyl ester

$C_{21}H_{36}$
2089 - Benzene, pentadecyl-
9740 - Pregnane, (5α)-
9741 - Pregnane, (5β)-

$C_{21}H_{36}N_2O_5S$
9623 - Piperazinium, 4-(2-cyclohexyl-2-hydroxy-2-phenylethyl)-1,1-dimethyl-, methyl sulfate (salt)

$C_{21}H_{36}O$
9441 - Phenol, 3-pentadecyl-

$C_{21}H_{36}O_2$
1543 - 1,2-Benzenediol, 3-pentadecyl-
9742 - Pregnane-3,20-diol, (3α,5β,20S)-

$C_{21}H_{38}ClN$
11030 - Pyridinium, 1-hexadecyl-, chloride

$C_{21}H_{38}O_6$
6908 - Hexanoic acid, 1,2,3-propanetriyl ester

$C_{21}H_{39}N_7O_{12}$
11562 - Streptomycin

$C_{21}H_{40}O_4$
8131 - Nonanedioic acid, bis(2-ethylbutyl) ester
8299 - 9-Octadecenoic acid (Z)-, 2,3-dihydroxypropyl ester

$C_{21}H_{41}O_4$
5357 - Eicosanoic acid, methyl ester

$C_{21}H_{42}$
4463 - Cyclohexane, pentadecyl-
4822 - Cyclopentane, hexadecyl-

$C_{21}H_{42}O_2$
8267 - Octadecanoic acid, 1-methylethyl ester
8273 - Octadecanoic acid, propyl ester

$C_{21}H_{42}O_4$
8255 - Octadecanoic acid, 2,3-dihydroxypropyl ester, (±)-

$C_{21}H_{44}$
6294 - Heneicosane

$C_{21}H_{44}O_3$
10023 - 1,2-Propanediol, 3-(octadecyloxy)-

$C_{22}H_{12}$
2471 - Benzo[ghi]perylene

$C_{22}H_{14}$
5142 - Dibenz[a,h]anthracene
8636 - Pentacene
8996 - Pentaphene
9601 - Picene

$C_{22}H_{16}N_4O$
8038 - 2-Naphthalenol, 1-[[4-(phenylazo)phenyl]azo]-

$C_{22}H_{17}N$
7717 - 1-Naphthalenamine, N,N-diphenyl-

$C_{22}H_{18}Cl_2FNO_3$
5008 - Cyfluthrin

$C_{22}H_{18}O_4$
1337 - 1,2-Benzenedicarboxylic acid, bis(phenylmethyl) ester
7353 - 1(3H)-Isobenzofuranone, 3,3-bis(4-hydroxy-3-methylphenyl)-

$C_{22}H_{19}Br_2NO_3$
5102 - Deltamethrin

$C_{22}H_{19}Cl_2NO_3$
5012 - Cypermethrin

$C_{22}H_{19}NO_4$
9465 - Phenol, 4,4′-(2-pyridinylmethylene)bis-, diacetate (ester)

$C_{22}H_{20}O_{13}$
373 - 2-Anthracenecarboxylic acid, 7-β-D-glucopyranosyl-9,10-dihydro-3,5,6,8-tetrahydroxy-1-methyl-9,10-dioxo-

$C_{22}H_{21}NO_2$
1797 - Benzenemethanamine, N-[bis (4-methoxyphenyl)methylene]-

$C_{22}H_{22}O_8$
6191 - Furo[3′,4′:6,7]naphtho[2,3-d]-1,3-dioxol-6(5aH)-one, 5,8,8a,9-tetrahydro-9-hydroxy-5-(3,4,5-trimethoxyphenyl)-
6192 - Furo[3′,4′:6,7]naphtho[2,3-d]-1,3-dioxol-6(5aH)-one, 5,8,8a,9-tetrahydro-9-hydroxy-5-(3,4,5-trimetho

$C_{22}H_{23}ClN_2O_8$
7704 - 2-Naphthacenecarboxamide, 7-chloro-4-(dimethylamino)-1,4,4a,5,5a,6,11,12a-octahydro-3,6,10,12,12a-pentahydroxy-6-methyl-1,11-dioxo-, [4S (4α,4aα,5aα,6β,12aα)]-

$C_{22}H_{23}NO_3$
5953 - Fenpropathrin

$C_{22}H_{23}NO_6$
10797 - 4H-Pyran-4-one, 2-methoxy-3,5-dimethyl-6-[tetrahydro-4-[2-methyl-3-(4-nitrophenyl)-2-propenylidene]-2-furanyl]

$C_{22}H_{23}NO_7$
7356 - 1(3H)-Isobenzofuranone, 6,7-dimethoxy-3-(5,6,7,8-tetrahydro-4-methoxy-6-methyl-1,3-dioxolo[4,5-g]isoquinolin-5

$C_{22}H_{24}N_2O_4$
3204 - Butanamide, N,N′-(3,3′-dimethyl[1,1′-biphenyl]-4,4′-diyl)bis[3-oxo-

10772 - 2H-Pyran-5-carboxylic acid, 3,4-dihydro-3-(hydroxymethyl)-2-methyl-4-[(9-methyl-9H-pyrido[3,4-b]indol-1-yl)methyl]-, methyl ester, [2S-(2α,3β,4β)]-

11351 - 16,19-Secostrychnidine-10,16-dione, 4-hydroxy-19-methyl-

C22H24N2O8
7706 - 2-Naphthacenecarboxamide, 4-(dimethylamino)-1,4,4a,5,5a,6,11,12a-octahydro-3,6,10,12,12a-pentahydroxy-6-methyl-1,11-dioxo-, [4S-(4α,4aα,5aα,6β,12aα)]-

C22H24N2O9
7705 - 2-Naphthacenecarboxamide, 4-(dimethylamino)-1,4,4a,5,5a,6,11,12a-octahydro-3,5,6,10,12,12a-hexahydroxy-6-methyl-1,11-dioxo-,[4S-(4α,4aα,5aα,6β,12aα)]-

C22H24Si
11445 - Silane, methyltri-p-tolyl-

C22H25NO2
5854 - Ethanone, 2,2′-(1-methyl-2,6-piperidinediyl)bis[1-phenyl-, cis-

C22H25NO6
93 - Acetamide, N-(5,6,7,9-tetrahydro-1,2,3,10-tetramethoxy-9-oxobenzo[a]heptalen-7-yl)-, (S)-

C22H25N3O
10850 - 3H-Pyrazol-3-one, 1,2-dihydro-2-(1-methyl-4-piperidinyl)-5-phenyl-4-(phenylmethyl)-

C22H26N2O3
4146 - Corynan-16-carboxylic acid, 16,17,18,19-tetradehydro-17-methoxy-, methyl ester, (16E)-

C22H26O3
11340 - (-)-cis-Resmethrin

C22H27NO2
5834 - Ethanone, 2-[6-(2-hydroxy-2-phenylethyl)-1-methyl-2-piperidinyl]-1-phenyl-, [2R-[2α,6α(S*)]]-

C22H27NO4
5149 - 6H-Dibenzo[a,g]quinolizine, 5,8,13,13a-tetrahydro-2,3,9,10-tetramethoxy-13-methyl-, (13S-trans)-

C22H28N2O
9801 - Propanamide, N-phenyl-N-[1-(2-phenylethyl)-4-piperidinyl]-

C22H28N2O5
12063 - Yohimban-16-carboxylic acid, 18-hydroxy-11,17-dimethoxy-, (3β,16β,17α,18β,20

C22H28O3
9736 - Pregna-4,6-diene-21-carboxylic acid, 17-hydroxy-3-oxo-, γ-lactone, (17α)-

C22H28O5
4962 - Cyclopropanecarboxylic acid, 3-(3-methoxy-2-methyl-3-oxo-1-propenyl)-2,2-dimethyl-, 2-methyl-4-oxo-3-(2,4-pentadienyl)-2-cyclopenten-1-yl ester [1R-[1α[S*(Z)],3β(E)]]-

C22H29FO5
9737 - Pregna-1,4-diene-3,20-dione, 9-fluoro-11,17,21-trihydroxy-16-methyl-, (11β,16α)-

C22H29NO2
9646 - 2,6-Piperidinediethanol, 1-methyl-α,α′-diphenyl-, [2α(R*),6α(S*)]-

C22H30N2O2
473 - Aspidospermidine, 1-acetyl-17-methoxy-

C22H30N2O2
4730 - 2,4(1H)-Cyclo-3,4-secoakuammilanium, 3,17-dihydroxy-16-(methoxycarbonyl)-4-methyl-, (3β,16R)-

C22H32
7879 - Naphthalene, 1-dodecyl-
7880 - Naphthalene, 2-dodecyl-

C22H32O3
359 - Androst-4-en-3-one, 17-(1-oxopropoxy)-, (17β)-
9752 - Pregn-4-ene-3,20-dione, 17-hydroxy-6-methyl-, (6α)-

C22H33NO2
475 - Atisine

C22H35NO3
5007 - 7,20-Cycloveatchane-1,12,15-triol, 21-ethyl-4-methyl-16-methylene-, (1α,12α,15β)-

C22H36O
6664 - 1-Hexadecanone, 1-phenyl-

C22H37NO
6630 - Hexadecanamide, N-phenyl-

C22H38
1751 - Benzene, hexadecyl-

C22H38O
1752 - Benzene, (hexadecyloxy)-

C22H40O7
8102 - 1,2,3-Nonadecanetricarboxylic acid, 2-hydroxy-

C22H42O2
5281 - 13-Docosenoic acid
5282 - 13-Docosenoic acid, (E)-
5283 - 13-Docosenoic acid, (Z)-
8298 - 9-Octadecenoic acid (Z)-, butyl ester
10667 - 2-Propenoic acid, 2-methyl-, octadecyl ester

C22H42O3
8304 - 9-Octadecenoic acid, 12-hydroxy-, butyl ester, [R-(Z)]-

C22H42O4
6781 - Hexanedioic acid, bis(2-ethylhexyl) ester
8238 - Octadecanedioic acid, diethyl ester

C22H44
4411 - Cyclohexane, hexadecyl-
5280 - 1-Docosene

C20H40N2•C2H4O2
6275 - Glyodin

C22H44O
5284 - 13-Docosen-1-ol, (Z)-

C22H44O2
5276 - Docosanoic acid
5356 - Eicosanoic acid, ethyl ester
8251 - Octadecanoic acid, butyl ester
8268 - Octadecanoic acid, 2-methylpropyl ester

C22H46
5268 - Docosane

C22H46O
5279 - 1-Docosanol

C22H66O9Si10
5079 - Decasiloxane, docosamethyl-

C23H16O3
7268 - 1H-Indene-1,3(2H)-dione, 2-(diphenylacetyl)-

C23H16O6
7777 - 2-Naphthalenecarboxylic acid, 4,4′-methylenebis[3-hydroxy-

C23H18O
7910 - 1-Naphthalenemethanol, α,α-diphenyl-

C23H19ClF3NO3
5010 - Cyhalothrin

C23H21ClO3
1154 - Benzene, 1,1′,1″-(1-chloro-1-ethenyl-2-ylidene)tris[4-methoxy-

C23H22ClF3O2
2998 - Bifenthrin

C23H22O6
2783 - [1]Benzopyrano[3,4-b]furo[2,3-h][1]benzopyran-6(6aH)-one, 1,2,12,12a-tetrahydro-8,9-dimethoxy-2-(1-methylethen

C23H22O7
3110 - 3H-Bis[1]benzopyrano[3,4-b:6′,5′-e]pyran-7(7aH)-one, 13,13a-dihydro-7a-hydroxy-9,10-dimethoxy-3,3-dimethyl

C23H23O6
9747 - Pregn-4-ene-3,20-dione, 21-(acetyloxy)-11,17-dihydroxy-, (11β)-

C23H25ClN2
7455 - Methanaminium, N-[4-[[4-(dimethylamino)phenyl]phenylmethylene]-2,5-cyclohexadien-1-ylidene]-N-methyl-, chlorid

C23H26N2O4
11564 - Strychnidin-10-one, 2,3-dimethoxy-

C23H27NO8
2586 - Benzoic acid, 6-[[6-[2-(dimethylamino)ethyl]-4-methoxy-1,3-benzodioxol-5-yl]acetyl]-2,3-dimethoxy-

C23H28ClN3O
9657 - Piperidine, 1-[2-[3-(2,2-diphenylethyl)-1,2,4-oxadiazol-5-yl]ethyl]-, monohydrochloride

C23H30BrNO2
9832 - 2-Propanaminium, N-methyl-N-(1-methylethyl)-N-[2-[(9H-xanthen-9-ylcarbonyl)oxy]ethyl]-, bromide

C23H30O2Si3
11915 - Trisiloxane, 1,1,3,5,5-pentamethyl-1,3,5-triphenyl-

C23H32N2O2S
11800 - Thiourea, N,N′-bis[4-(3-methylbutoxy)phenyl]-

C23H32O6
4049 - Card-20(22)-enolide, 3,5,14-trihydroxy-19-oxo-, (3β,5β)-

C23H33IN2O
2120 - Benzenepropanaminium, γ-(aminocarbonyl)-N-methyl-N,N-bis(1-methylethyl)-γ-phenyl-, iod

C23H34O4
4044 - Card-20(22)-enolide, 3,14-dihydroxy-, (3β,5β)-

C23H34O5
4046 - Card-20(22)-enolide, 3,11,14-trihydroxy-, (3β,5β,11α)-
4047 - Card-20(22)-enolide, 3,12,14-trihydroxy-, (3β,5β,12β)-
4048 - Card-20(22)-enolide, 3,14,16-trihydroxy-, (3β,5β,16β)-

C23H36O3
351 - Androstan-3-one, 2-methyl-17-(1-oxopropoxy)-, (2α,5α,17β)-

C23H38O2
6655 - Hexadecanoic acid, phenylmethyl ester
8210 - 24-Norcholan-23-oic acid, (5β)-

C23H40
1743 - Benzene, heptadecyl-

C23H44O2
8306 - 9-Octadecenoic acid (Z)-, 3-methylbutyl ester

C23H44O3
8275 - Octadecanoic acid, (tetrahydro-2-furanyl)methyl ester

C23H44O4
9954 - Propanedioic acid, hexadecyl-, diethyl ester

C23H46
4404 - Cyclohexane, heptadecyl-
4842 - Cyclopentane, octadecyl-

C23H46O
11857 - 12-Tricosanone

C23H46O2
5278 - Docosanoic acid, methyl ester
8265 - Octadecanoic acid, 3-methylbutyl ester
8270 - Octadecanoic acid, pentyl ester

C23H48
6309 - Heptadecane, 9-hexyl-
11855 - Tricosane

C24H12
4145 - Coronene

C24H14
2850 - Benzo[rst]pentaphene
8061 - Naphtho[1,2,3,4-def]chrysene

C24H16
11644 - Tetraphenylene

C24H18
11193 - 1,1′:2′,1″:2″,1‴-Quaterphenyl
11194 - 1,1′:4′,1″:4″,1‴-Quaterphenyl
11598 - 1,1′:3′,1″-Terphenyl, 5′-phenyl-

C24H20N2
3035 - [1,1′-Biphenyl]-4,4′-diamine, N,N′-diphenyl-

C24H20N4O
8033 - 2-Naphthalenol, 1-[[2-methyl-4-[(2-methylphenyl)azo]phenyl]azo]-

C24H20O4Si
11509 - Silicic acid (H4SiO4), tetraphenyl ester
C24H20O6
10196 - 1,2,3-Propanetriol, tribenzoate
C24H20Pb
9733 - Plumbane, tetraphenyl-
C24H20Si
11452 - Silane, tetraphenyl-
C24H20Sn
11545 - Stannane, tetraphenyl-
C24H27O4P
9335 - Phenol, 2,4-dimethyl-, phosphate (3:1)
9336 - Phenol, 2,5-dimethyl-, phosphate (3:1)
9337 - Phenol, 2,6-dimethyl-, phosphate (3:1)
9338 - Phenol, 3,4-dimethyl-, phosphate (3:1)
C24H28O4
9301 - Phenol, 4,4'-(1,2-diethyl-1,2-ethenediyl)bis-, dipropanoate, (E)-
C24H29NO4
7377 - Isoquinoline, 1-[(3,4-diethoxyphenyl)methyl]-6,7-diethoxy-
C24H30O3Si3
5004 - Cyclotrisiloxane, 2,4,6-triethyl-2,4,6-triphenyl-
C24H34O5
4101 - Cholan-24-oic acid, 3,7,12-trioxo-, (5β)-
C24H38O4
1334 - 1,2-Benzenedicarboxylic acid, bis(2-ethylhexyl) ester
1344 - 1,2-Benzenedicarboxylic acid, dioctyl ester
C24H39NO
8290 - 9-Octadecenamide, N-phenyl-, (Z)-
C24H40N2
4143 - Con-5-enin-3-amine, N,N-dimethyl-, (3β)-
C24H40N8O4
5711 - Ethanol, 2,2',2'',2'''-[(4,8-di-1-piperidinylpyrimido[5,4-d]pyrimidine-
C24H40O2
1257 - Benzenedecanoic acid, ι-octyl-
2063 - Benzenenonanoic acid, θ-nonyl-
4094 - Cholan-24-oic acid
8271 - Octadecanoic acid, phenyl ester
C24H40O3
4099 - Cholan-24-oic acid, 3-hydroxy-, (3α,5β)-
C24H40O4
4095 - Cholan-24-oic acid, 3,12-dihydroxy-, (3α,5β,12α)-
4096 - Cholan-24-oic acid, 3,6-dihydroxy-, (3α,5β,6α)-
4097 - Cholan-24-oic acid, 3,7-dihydroxy-, (3α,5β,7α)-
4098 - Cholan-24-oic acid, 3,7-dihydroxy-, (3α,5β,7β)-
C24H40O5
4100 - Cholan-24-oic acid, 3,7,12-trihydroxy-, (3α,5β,7α,12α)-
C24H41NO2
8232 - Octadecanamide, N-(4-hydroxyphenyl)-
C24H42
2066 - Benzene, octadecyl-
4093 - Cholane
C24H46O2
8253 - Octadecanoic acid, cyclohexyl ester
11608 - 15-Tetracosenoic acid
C24H46O3
5306 - Dodecanoic acid, anhydride
C24H46O4
5353 - Eicosanedioic acid, diethyl ester
9155 - Peroxide, bis(1-oxododecyl)
C24H48
4459 - Cyclohexane, octadecyl-
C24H48O2
5277 - Docosanoic acid, ethyl ester
11607 - Tetracosanoic acid
C24H50
11603 - Tetracosane
11856 - Tricosane, 2-methyl-

C24H52O4Si
11507 - Silicic acid (H4SiO4), tetrakis(2-ethylbutyl) ester
C24H54OSn2
5241 - Distannoxane, hexabutyl-
C24H72O10Si11
11964 - Undecasiloxane, tetracosamethyl-
C25H20
1831 - Benzene, 1,1',1'',1'''-methanetetrayltetrakis-
C25H20N2O
12031 - Urea, tetraphenyl-
C25H22ClNO3
5956 - Fenvalerate
C25H22OSi
11441 - Silane, (2-methylphenoxy)triphenyl-
C25H24F6N4
7136 - Hydramethylnon
C25H27ClN2
9605 - Piperazine, 1-[(4-chlorophenyl)phenylmethyl]-4-[(3-methylphenyl)methyl]-
C25H28O3
5403 - Estra-1,3,5(10)-triene-3,17-diol (17β)-, 3-benzoate
C25H30ClN3
7453 - Methanaminium, N-[4-[bis[4-(dimethylamino)phenyl]methylene]-2,5-cyclohexadien-1-ylidene]-N-methyl, chloride
C25H30O4
5108 - 6,6'-Diapo-ψ,ψ-carotenedioic acid, monomethyl ester, 9-cis-
C25H40O2
8309 - 9-Octadecenoic acid (Z)-, phenylmethyl ester
C25H41NO9
305 - Aconitane-3,8,13,14,15-pentol, 20-ethyl-1,6,16-trimethoxy-4-(methoxymethyl)-,
C25H44
2062 - Benzene, nonadecyl-
2070 - Benzene, (3-octylundecyl)-
8245 - Octadecane, 9-p-tolyl-
C25H48O4
8132 - Nonanedioic acid, bis(2-ethylhexyl) ester
C25H50
4456 - Cyclohexane, nonadecyl-
5351 - Eicosane, 1-cyclopentyl-
6323 - 8-Heptadecene, 9-octyl-
C25H52
6313 - Heptadecane, 9-octyl-
8637 - Pentacosane
C26H16
6621 - Hexacene
C26H20
1623 - Benzene, 1,1',1'',1'''-(1,2-ethenediylidene)tetrakis-
C26H22
1591 - Benzene, 1,1',1'',1'''-(1,2-ethanediylidene)tetrakis-
1618 - Benzene, 1,1',1'',1'''-(1-ethanyl-2-ylidyne)tetrakis-
C26H22ClF3N2O3
5983 - Fluvalinate
C26H22O2
5580 - 1,2-Ethanediol, 1,1,2,2-tetraphenyl-
C26H23F2NO4
5965 - Flucythrinate
C26H36O6
3141 - Bufa-20,22-dienolide, 16-(acetyloxy)-3,14-dihydroxy-, (3β,5β,16β)-
C26H38
1567 - Benzene, 1,1'-dodecylidenebis[4-methyl-
2267 - Benzene, 1,1'-tetradecylidenebis-
C26H38O4
4281 - 2,4-Cyclohexadien-1-one, 3,5-dihydroxy-2,6,6-tris(3-methyl-2-butenyl)-4-(3-methyl-1-oxobutyl)-
C26H43NO6
6273 - Glycine, N-[(3α,5β,7α,12α)-3,7,12-trihydroxy-24-oxocholan-24-yl]-

C26H45NO7S
5625 - Ethanesulfonic acid, 2-[[(3α,5β,7α,12α)-3,7,12-trihydroxy-24-oxocholan-24-yl]amino]-
C26H46
1119 - Benzene, (1-butylhexadecyl)-
1568 - Benzene, eicosyl-
1695 - Benzene, (1-ethyloctadecyl)-
2007 - Benzene, (1-methylnonadecyl)-
2068 - Benzene, (1-octyldodecyl)-
2170 - Benzene, (1-propylheptadecyl)-
C26H50
4476 - Cyclohexane, 1,1'-tetradecylidenebis-
C26H50O4
5043 - Decanedioic acid, bis(2-ethylhexyl) ester
5311 - Dodecanoic acid, 1,2-ethanediyl ester
C26H52
4384 - Cyclohexane, eicosyl-
4819 - Cyclopentane, heneicosyl-
C26H52O2
6624 - Hexacosanoic acid
C26H54O5
8259 - Octadecanoic acid, 2-[2-[2-(2-hydroxyethoxy)ethoxy]ethoxy]ethyl ester
C26H54
5269 - Docosane, 5-butyl-
5270 - Docosane, 7-butyl-
5271 - Docosane, 9-butyl-
5272 - Docosane, 11-butyl-
6296 - Heneicosane, 11-(2,2-dimethylpropyl)-
6622 - Hexacosane
6637 - Hexadecane, 6,11-dipentyl-
8241 - Octadecane, 3-ethyl-5-(2-ethylbutyl)-
11605 - Tetracosane, 3-ethyl-
C26H54O
6626 - 1-Hexacosanol
C27H18O3
7587 - Methanone, 1,3,5-benzenetriyltris[phenyl-
C27H20O12
2773 - 1H-2-Benzopyran-3-carboxylic acid, 6-[2-(4,9-dihydro-8-hydroxy-5,7-dimethoxy-4,9-dioxonaphtho[2,3-b]furan-2-yl
C27H28Br2O5S
9365 - Phenol, 4,4'-(3H-2,1-benzoxathiol-3-ylidene)bis[2-bromo-3-methyl-6-(1-methylethyl)-, S,S-dioxide
C27H30O5S
9369 - Phenol, 4,4'-(3H-2,1-benzoxathiol-3-ylidene)bis[5-methyl-2-(1-methylethyl)-, S,S-dioxide
C27H32N2O6
4131 - Cinchonan-9-ol, 6'-methoxy-, (8α,9R)-, mono(2-hydroxybenzoate) (salt)
C27H32O14
2795 - 4H-1-Benzopyran-4-one, 7-[[2-O-(6-deoxy-α-L-mannopyranosyl)-β-D-glucopyranosyl]oxy]-2,3-dihydro-5-hy
C27H34N2O4S
5444 - Ethanaminium, N-[4-[[4-(diethylamino)phenyl]phenylmethylene]-2,5-cyclohexadien-1-ylidene]-N-ethyl-, sulfate (1
C27H34O3
5408 - Estr-4-en-3-one, 17-(1-oxo-3-phenylpropoxy)-, (17β)-
C27H39NO2
12044 - Veratraman-3,23-diol, 14,15,16,17-tetradehydro-, (3β,23β)-
C27H39NO3
12046 - Veratraman-11-one, 17,23-epoxy-3-hydroxy-, (3β,23β)-
C27H40O3
7774 - 2-Naphthalenecarboxylic acid, 3-hydroxy-, hexadecyl ester

C$_{27}$H$_{42}$Cl$_2$N$_2$O$_6$
6648 - Hexadecanoic acid, 2-[(dichloroacetyl)amino]-3-hydroxy-3-(4-nitrophenyl)propyl ester, [R-(R*,R*)]-

C$_{27}$H$_{42}$O$_3$
11536 - Spirost-5-en-3-ol, (3β,25R)-

C$_{27}$H$_{42}$O$_4$
11534 - Spirostan-12-one, 3-hydroxy-, (3β,5α,25R)-

C$_{27}$H$_{43}$NO$_2$
11515 - Solanid-5-ene-3,12-diol, (3β,12α)-
11516 - Solanid-5-ene-3,18-diol, (3β)
11529 - Spirosol-5-en-3-ol, (3β,22α,25R)-

C$_{27}$H$_{43}$NO$_8$
4068 - Cevane-3,4,7,14,15,16,20-heptol, 4,9-epoxy-, (3β,4α,7α,15α,16β)-

C$_{27}$H$_{43}$NO$_9$
4070 - Cevane-3,4,6,7,14,15,16,20-octol, 4,9-epoxy-, (3β,4α,6α,7α,15α,16β

C$_{27}$H$_{44}$
4102 - Cholesta-3,5-diene

C$_{27}$H$_{44}$O
4103 - Cholesta-5,7-dien-3-ol, (3β)
4104 - Cholesta-8,24-dien-3-ol, (3β,5α)-
11348 - 9,10-Secocholesta-5,7,10(19)-trien-3-ol, (3β,5Z,7E)-

C$_{27}$H$_{44}$O$_3$
11531 - Spirostan-3-ol, (3β,5β,25S)-
11532 - Spirostan-3-ol, (3β,5α,25R)-
11533 - Spirostan-3-ol, (3β,5β,25R)-

C$_{27}$H$_{44}$O$_4$
11530 - Spirostan-2,3-diol, (2α,3β,5α,25R)-

C$_{27}$H$_{44}$O$_5$
11535 - Spirostan-2,3,15-triol, (2α,3β,5α,15β,25R)-

C$_{27}$H$_{45}$NO$_2$
11528 - Spirosolan-3-ol, (3β,5α,22β,25S)-

C$_{27}$H$_{46}$
6298 - 10-Heneicosene, 11-phenyl-

C$_{27}$H$_{46}$O
4110 - Cholest-5-en-3-ol (3β)
4111 - Cholest-4-en-3-ol, (3β)-
4112 - Cholest-5-en-3-ol, (3α)-

C$_{27}$H$_{46}$O$_2$
2785 - 2H-1-Benzopyran-6-ol, 3,4-dihydro-2,8-dimethyl-2-(4,8,12-trimethyltridecyl)-, [2R-[2R*(4R*,8R*)]]-

C$_{27}$H$_{48}$
1741 - Benzene, heneicosyl-
4105 - Cholestane, (5α)-
4106 - Cholestane, (5β)-
6297 - Heneicosane, 11-phenyl-

C$_{27}$H$_{48}$O
4107 - Cholestanol
4108 - Cholestan-3-ol, (3α,5α)-
4109 - Cholestan-3-ol, (3β,5β)-

C$_{27}$H$_{50}$O$_6$
8440 - Octanoic acid, 1,2,3-propanetriyl ester

C$_{27}$H$_{54}$
4808 - Cyclopentane, docosyl-

C$_{27}$H$_{54}$O
6303 - 14-Heptacosanone

C$_{27}$H$_{54}$O$_2$
6625 - Hexacosanoic acid, methyl ester

C$_{27}$H$_{56}$
6302 - Heptacosane

C$_{28}$H$_{14}$N$_2$O$_4$
434 - 5,9,14,18-Anthrazinetetrone, 6,15-dihydro-

C$_{28}$H$_{20}$O
6175 - Furan, tetraphenyl-

C$_{28}$H$_{22}$
367 - Anthracene, 9,10-bis(phenylmethyl)-

C$_{28}$H$_{22}$O$_3$
937 - Benzeneacetic acid, α-phenyl-, anhydride

C$_{28}$H$_{24}$N$_2$O$_7$
8549 - Orcein

C$_{28}$H$_{30}$O$_4$
7352 - 1(3H)-Isobenzofuranone, 3,3-bis[4-hydroxy-2-methyl-5-(1-methylethyl)phenyl]-

C$_{28}$H$_{32}$ClN$_2$O$_3$
12056 - Xanthylium, 9-(2-carboxyphenyl)-3,6-bis(diethylamino)-, chloride

C$_{28}$H$_{32}$O$_4$Si$_4$
5000 - Cyclotetrasiloxane, 2,4,6,8-tetramethyl-2,4,6,8-tetraphenyl-

C$_{28}$H$_{34}$O$_{15}$
2794 - 4H-1-Benzopyran-4-one, 7-[[6-O-(6-deoxy-α-L-mannopyranosyl)-β-D-glucopyranosyl]oxy]-2,3-dih

C$_{28}$H$_{38}$N$_2$O$_4$
5367 - Emetan-6'-ol, 7',10,11-trimethoxy-

C$_{28}$H$_{38}$O$_{19}$
6229 - α-D-Glucopyranoside, 1,3,4,6-tetra-O-acetyl-β-D-fructofuranosyl, tetraacetate

C$_{28}$H$_{42}$O
5380 - Ergosta-5,7,9(11),22-tetraen-3-ol, (3β,22E)-

C$_{28}$H$_{44}$O
5381 - Ergosta-5,7,22-trien-3-ol, (3β,10α,22E)-
5382 - Ergosta-5,7,22-trien-3-ol, (3β,9β,10α,22E)-
5383 - Ergosta-5,7,22-trien-3-ol, (3β,22E)-
11349 - 9,10-Secoergosta-5(10),6,8,22-tetraen-3-ol, (3β,6E,22E)-
11350 - 9,10-Secoergosta-5,7,10(19),22-tetraen-3-ol, (3β,5Z,7E,22E)-

C$_{28}$H$_{48}$O
357 - Androst-4-en-3-one, 17-hydroxy-7,17-dimethyl-, (7β,17β)-
5384 - Ergost-7-en-3-ol, (3β,5α)-
5385 - Ergost-8(14)-en-3-ol, (3β,5α)-
5386 - Ergost-5-en-3-ol, (3β,24R)-

C$_{28}$H$_{48}$O$_2$
2789 - 2H-1-Benzopyran-6-ol, 3,4-dihydro-2,5,8-trimethyl-2-(4,8,12-trimethyltridecyl)-
2790 - 2H-1-Benzopyran-6-ol, 3,4-dihydro-2,7,8-trimethyl-2-(4,8,12-trimethyltridecyl)-

C$_{28}$H$_{50}$
1565 - Benzene, docosyl-
5377 - Ergostane, (5α)-
5378 - Ergostane, (5β)-

C$_{28}$H$_{50}$N$_2$O$_4$
4064 - Carpaine

C$_{28}$H$_{50}$O
5379 - Ergostan-3-ol, (3β,5α)-

C$_{28}$H$_{54}$O$_3$
11623 - Tetradecanoic acid, anhydride

C$_{28}$H$_{56}$
4380 - Cyclohexane, docosyl-

C$_{28}$H$_{56}$O$_2$
8221 - Octacosanoic acid

C$_{28}$H$_{58}$
5274 - Docosane, 7-hexyl-
5354 - Eicosane, 9-octyl-
8220 - Octacosane
11877 - Tridecane, 2,2,4,10,12,12-hexamethyl-7-(3,5,5-trimethylhexyl)-

C$_{28}$H$_{58}$O
8222 - 1-Octacosanol

C$_{29}$H$_{40}$N$_2$O$_4$
5368 - Emetan, 6',7',10,11-tetramethoxy-

C$_{29}$H$_{44}$O$_{12}$
4040 - Card-20(22)-enolide, 3-[(6-deoxy-α-L-mannopyranosyl)oxy]-1,5,11,14,19-pentahydroxy-, (1β,3β

C$_{29}$H$_{47}$NO$_8$
4067 - Cevane-3,4,12,14,16,17,20-heptol, 3-acetate, (3β,4α,16β)-

C$_{29}$H$_{48}$O
11554 - Stigmasta-5,7-dien-3-ol, (3β)-
11555 - Stigmasta-5,22-dien-3-ol, (3β,22E)-

C$_{29}$H$_{50}$O
11558 - Stigmast-5-en-3-ol, (3β)-
11559 - Stigmast-5-en-3-ol, (3β,24S)-

C$_{29}$H$_{50}$O$_2$
2787 - 2H-1-Benzopyran-6-ol, 3,4-dihydro-2,5,7,8-tetramethyl-2-(4,8,12-trimethyltridecyl)-, [2R-[2R*(4R*,8R*)]]-

C$_{29}$H$_{52}$
2341 - Benzene, tricosyl-

C$_{29}$H$_{52}$O
11557 - Stigmastan-3-ol, (3β,5α)-

C$_{29}$H$_{58}$
4483 - Cyclohexane, tricosyl-

C$_{29}$H$_{60}$
8099 - Nonacosane

C$_{29}$H$_{60}$O
8100 - 1-Nonacosanol

C$_{30}$H$_{34}$O$_{13}$
7579 - 3,6-Methano-8H-1,5,7-trioxacyclopenta[ij]cycloprop[a]azulene-4,8(3H)-dione, hexahydro-2a-hydroxy

C$_{30}$H$_{37}$N$_5$O$_5$
5393 - Ergotaman-3',6',18-trione, 12'-hydroxy-2'-methyl-5'-(2-methylpropyl)-,

C$_{30}$H$_{40}$O$_2$
435 - 8'-Apo-β,ψ-carotenal, 3-hydroxy-, (3R)-

C$_{30}$H$_{42}$FeO$_2$Si$_4$
5961 - Ferrocene, 1,1'-bis(1,1,3,3-tetramethyl-3-phenyldisiloxanyl)-

C$_{30}$H$_{46}$O$_5$
8546 - Olean-12-en-28-oic acid, 3,16-dihydroxy-23-oxo-, (3β,4α,16α)-
12036 - Urs-12-ene-27,28-dioic acid, 3-hydroxy-, (3β)-

C$_{30}$H$_{48}$O$_3$
8547 - Olean-12-en-28-oic acid, 3-hydroxy-, (3β)-
12037 - Urs-12-en-28-oic acid, 3-hydroxy-, (3β)-

C$_{30}$H$_{48}$O$_4$
8545 - Olean-12-en-28-oic acid, 3,23-dihydroxy-, (3β,4α)-

C$_{30}$H$_{50}$
11602 - 2,6,10,14,18,22-Tetracosahexaene, 2,6,10,15,19,23-hexamethyl-, (all-E)-

C$_{30}$H$_{50}$O
7411 - Lanosta-8,24-dien-3-ol, (3β)-
7417 - Lup-20(29)-en-3-ol, (3β)-
8548 - Olean-12-en-3-ol, (3β)-
11556 - Stigmasta-7,24(28)-dien-3-ol, 4-methyl-, (3β,4α,5α,24Z)-
12038 - Urs-12-en-3-ol, (3β)-

C$_{30}$H$_{50}$O$_2$
7416 - Lup-20(29)-ene-3,28-diol, (3β)-

C$_{30}$H$_{54}$
2265 - Benzene, tetracosyl-

C$_{30}$H$_{60}$
4474 - Cyclohexane, tetracosyl-

C$_{30}$H$_{60}$I$_3$N$_3$O$_3$
5441 - Ethanaminium, 2,2',2''-[1,2,3-benzenetriyltris(oxy)]tris[N,N,N-triethyl-, tr

C$_{30}$H$_{60}$O$_2$
11820 - Triacontanoic acid

C$_{30}$H$_{62}$
5275 - Docosane, 9-octyl-
11606 - Tetracosane, 2,6,10,15,19,23-hexamethyl-
11819 - Triacontane

C$_{30}$H$_{62}$O
11821 - 1-Triacontanol

C$_{31}$H$_{36}$N$_2$O$_8$
12064 - Yohimban-16-carboxylic acid, 17-hydroxy-18-[(3,4,5-trimethoxybenzoyl)oxy]-, methyl ester, (3β,16β

C$_{31}$H$_{39}$N$_5$O$_5$
5387 - Ergotaman-3',6',18-trione, 12'-hydroxy-2',5'-bis(1-methylethyl)-,
5388 - Ergotaman-3',6',18-trione, 12'-hydroxy-2',5'-bis(1-methylethyl)-,

C$_{31}$H$_{46}$O$_2$
7865 - 1,4-Naphthalenedione, 2-methyl-3-(3,7,11,15-tetramethyl-2-hexadecenyl)-, [R-[R*,R*-(E)]]-

C$_{31}$H$_{52}$O$_3$
2788 - 2H-1-Benzopyran-6-ol, 3,4-dihydro-2,5,7,8-tetramethyl-2-(4,8,12-trimethyltridecyl)-, acetate, [2R-[2R*(4R*,8R*

$C_{31}H_{56}$
2088 - Benzene, pentacosyl-
8638 - Pentacosane, 13-phenyl-

$C_{31}H_{62}$
4462 - Cyclohexane, pentacosyl-
4821 - Cyclopentane, hexacosyl-

$C_{31}H_{62}O$
6300 - 16-Hentriacontanone

$C_{31}H_{64}$
6295 - Heneicosane, 11-decyl-
6299 - Hentriacontane

$C_{32}H_{22}N_6Na_2O_6S_2$
7969 - 1-Naphthalenesulfonic acid, 3,3'-[[1,1'-biphenyl]-4,4'-diylbis(azo)]bis[4-amino-, disodium salt

$C_{32}H_{38}N_2O_8$
12065 - Yohimban-16-carboxylic acid, 17-methoxy-18-[(3,4,5-trimethoxybenzoyl)oxy]-, methyl ester, (3β,16β

$C_{32}H_{40}O_4Si_4$
4998 - Cyclotetrasiloxane, 2,4,6,8-tetraethyl-2,4,6,8-tetraphenyl-

$C_{32}H_{41}N_5O_5$
5389 - Ergotaman-3',6',18-trione, 12'-hydroxy-2'-(1-methylethyl)-5'-(2-methylpropyl)-

$C_{32}H_{44}N_2O_8$
309 - Aconitane-4,8,9-triol, 20-ethyl-1,14,16-trimethoxy-, 4-[2-(acetylamino)benzoate], (1α,14α,16β)-

$C_{32}H_{44}O_{12}$
3142 - Bufa-4,20,22-trienolide, 6-(acetyloxy)-3-(β-D-glucopyranosyloxy)-8,14-dihydroxy-, (3β,6β)-

$C_{32}H_{49}NO_9$
4069 - Cevane-3,4,12,14,16,17,20-heptol, 4,9-epoxy-, 3-(2-methyl-2-butenoate), [3β(Z),4α,16β]-

$C_{32}H_{58}$
1750 - Benzene, hexacosyl-

$C_{32}H_{62}O_3$
6646 - Hexadecanoic acid, anhydride

$C_{32}H_{64}$
4410 - Cyclohexane, hexacosyl-

$C_{32}H_{64}O_2$
6651 - Hexadecanoic acid, hexadecyl ester

$C_{32}H_{66}$
5273 - Docosane, decyl-
5348 - Dotriacontane

$C_{32}H_{66}O$
6642 - Hexadecane, 1,1'-oxybis-

$C_{33}H_{34}N_4O_6$
3001 - 21H-Biline-8,12-dipropanoic acid, 3,18-diethenyl-1,19,22,24-tetrahydro-2,7,13,17-tetramethyl-1,19-dioxo-

$C_{33}H_{35}N_5O_5$
5395 - Ergotaman-3',6',18-trione, 12'-hydroxy-2'-methyl-5'-(phenylmethyl)-,

$C_{33}H_{36}N_4O_6$
3000 - 21H-Biline-8,12-dipropanoic acid, 2,17-diethenyl-1,10,19,22,23,24-hexahydro-3,7,13,18-tetramethyl-1,19-dioxo-

$C_{33}H_{40}N_2O_9$
12061 - Yohimban-16-carboxylic acid, 11,17-dimethoxy-18-[(3,4,5-trimethoxybenzoyl)oxy]-, methyl ester,

$C_{33}H_{41}N_5O_5$
5390 - Ergotaman-3',6',18-trione, 12'-hydroxy-2'-(1-methylethyl)-5'-(2-methylpropyl)-, (5'α,8α)-

$C_{33}H_{44}O_8$
8211 - 29-Nordammara-1,17(20),24-trien-21-oic acid, 6,16-bis(acetyloxy)-3,7-dioxo-,

$C_{33}H_{45}NO_9$
311 - Aconitane-8,13,14-triol, 1,6,16-trimethoxy-4-(methoxymethyl)-20-methyl-, 8-acetate 14-benzoate,

$C_{33}H_{49}NO_8$
12045 - Veratraman-11-one, 17,23-epoxy-3-(β-D-glucopyranosyloxy)-, (3β,23β)-

$C_{33}H_{55}N_5O_5$
5394 - Ergotaman-3',6',18-trione, 12'-hydroxy-2'-methyl-5'-(phenylmethyl)-, (5'α)-

$C_{33}H_{56}$
1742 - Benzene, heptacosyl-

$C_{34}H_{24}N_6Na_4O_{14}S_4$
7877 - 1,3-Naphthalenedisulfonic acid, 6,6'-[(3,3'-dimethyl[1,1'-biphenyl]-4,4'-diyl)bis(azo)]bis[4-amino-5-hydroxy-,

$C_{34}H_{32}ClFeN_4O_4$
5958 - Ferrate(2-), chloro[7,12-diethenyl-3,8,13,17-tetramethyl-21H,23H-porphine-2,18-dipropanoato(4-)-N21,N22,N23,

$C_{34}H_{33}FeN_4O_5$
5959 - Ferrate(2-), [7,12-diethenyl-3,8,13,17-tetramethyl-21H,23H-porphine-2,18-dipropanoato(4-)-N21,N22,N23,N24

$C_{34}H_{36}N_2O_6$
3002 - [2,2'-Bimorphinan]-3,3',6,6'-tetrol, 7,7',8,8'-tetradehydro-4,5:4',5'-diepoxy-17,17'dimethyl-

$C_{34}H_{38}N_4O_6$
9735 - 21H,23H-Porphine-2,18-dipropanoic acid, 7,12-bis(1-hydroxyethyl)-3,8,13,17-tetramethyl-

$C_{34}H_{47}NO_{10}$
307 - Aconitane-3,8,13,14-tetrol, 20-ethyl-1,6,16-trimethoxy-4-(methoxymethyl)-, 8-acetate 14-benzoate-, (1α,3alp

$C_{34}H_{47}NO_{11}$
306 - Aconitane-3,8,13,14,15-pentol, 20-ethyl-1,6,16-trimethoxy-4-(methoxymethyl)-, 8-acetate 14-benzoate,

$C_{34}H_{62}$
2065 - Benzene, octacosyl-

$C_{34}H_{66}O_4$
6649 - Hexadecanoic acid, 1,2-ethanediyl ester

$C_{34}H_{68}$
4458 - Cyclohexane, octacosyl-

$C_{34}H_{68}O_2$
8258 - Octadecanoic acid, hexadecyl ester

$C_{34}H_{70}$
11604 - Tetracosane, 11-decyl-
11650 - Tetratriacontane

$C_{35}H_{39}N_5O_5$
5391 - Ergotaman-3',6',18-trione, 12'-hydroxy-2'-(1-methylethyl)-5'-(phenylmethyl)-
5392 - Ergotaman-3',6',18-trione, 12'-hydroxy-2'-(1-methylethyl)-5'-(phenylmethyl)-

$C_{35}H_{42}N_2O_9$
12060 - Yohimban-16-carboxylic acid, 11,17-dimethoxy-18-[[1-oxo-3-(3,4,5-trimethoxyphenyl)-2-propenyl]oxy]-, methyl ester, (3β,16β,17α,18β,20α)-

$C_{35}H_{54}O_{14}$
4045 - Card-20(22)-enolide, 3-[(6-O-β-D-glucopyranosyl-β-D-glucopyranosyl)oxy]-14-hydroxy-,

$C_{35}H_{70}$
4853 - Cyclopentane, triacontyl-

$C_{35}H_{70}O$
9000 - 18-Pentatriacontanone

$C_{35}H_{72}$
8999 - Pentatriacontane

$C_{36}H_{38}N_2O_6$
11924 - Tubocuraran-7',12'-diol, 6,6'-dimethoxy-2,2'-dimethyl-, (1'α)-
11925 - Tubocuraran-7',12'-diol, 6,6'-dimethoxy-2,2'-dimethyl-, (1β)-

$C_{36}H_{42}Br_2N_2$
6753 - 1,6-Hexanediaminium, N,N'-di-9H-fluoren-9-yl-N,N,N',N'-tetramethyl-dibromide

$C_{36}H_{51}NO_{11}$
310 - Aconitane-8,13,14-triol, 20-ethyl-1,6,16-trimethoxy-4-(methoxymethyl)-, 8-acetate 14-(3,4-dimethoxybenzoate),

$C_{36}H_{51}NO_{12}$
308 - Aconitane-3,8,13,14-tetrol, 20-ethyl-1,6,16-trimethoxy-4-(methoxymethyl)-, 8-acetate 14-(3,4-dimethoxybenzoate),(1α,3α,6α,14α,16β)-

$C_{36}H_{70}CaO_4$
8252 - Octadecanoic acid, calcium salt

$C_{36}H_{70}O_3$
8249 - Octadecanoic acid, anhydride

$C_{36}H_{70}O_4Pb$
8261 - Octadecanoic acid, lead salt

$C_{36}H_{70}O_4Zn$
8277 - Octadecanoic acid, zinc salt

$C_{36}H_{72}$
4482 - Cyclohexane, triacontyl-

$C_{36}H_{74}$
6991 - Hexatriacontane
8639 - Pentacosane, 13-undecyl-

$C_{37}H_{38}N_2O_6$
8630 - Oxyacanthan, 6',12'-dimethoxy-2,2'-dimethyl-6,7-[methylenebis(oxy)]-

$C_{37}H_{40}N_2O_6$
2911 - Berbaman-12-ol, 6,6',7-trimethoxy-2,2'-dimethyl-
8631 - Oxyacanthan-12'-ol, 6,6',7-trimethoxy-2,2'-dimethyl-

$C_{37}H_{67}NO_{13}$
5396 - Erythromycin

$C_{38}H_{74}O_4$
8256 - Octadecanoic acid, 1,2-ethanediyl ester

$C_{38}H_{76}$
4383 - Cyclohexane, dotriacontyl-

$C_{38}H_{78}$
6623 - Hexacosane, 13-dodecyl-

$C_{39}H_{74}O_6$
5319 - Dodecanoic acid, 1,2,3-propanetriyl ester

$C_{40}H_{36}$
4054 - β,ε-Carotene, (6'R)-

$C_{40}H_{50}N_4O_8S$
4132 - Cinchonan-9-ol, 6'-methoxy-, (8α,9R)-, sulfate (2:1) (salt)

$C_{40}H_{52}O_2$
4059 - β,ψ-Caroten-16'-oic acid, 3',4'-didehydro-

$C_{40}H_{56}$
4051 - β,β-Carotene
4052 - β,ψ-Carotene
4053 - ψ,ψ-Carotene

$C_{40}H_{56}O$
4060 - ψ,ψ-Caroten-16-ol
4061 - β,β-Caroten-3-ol, (3R)-
4062 - β,ψ-Caroten-3-ol, (3R)-

$C_{40}H_{56}O_2$
4057 - β,β-Carotene-3,3'-diol, (3R,3'R)-
4058 - β,ε-Carotene-3,3'-diol, (3R,3'R,6'R)-

$C_{40}H_{56}O_3$
4063 - β,κ-Caroten-6'-one, 3,3'-dihydroxy-, (3R,3'S,5'R)-

$C_{40}H_{56}O_4$
4056 - β,β-Carotene-3,3'-diol, 5,6:5',6'-diepoxy-5,5',6,6'-tetrahydro-

$C_{40}H_{60}O_6$
4055 - β,β-Carotene, 3'-(acetyloxy)-6',7'-didehydro-5,6-epoxy-5,5',6,6',7,8-hexahydro-3,5'-dihydroxy-8-oxo-,

$C_{40}H_{80}$
4479 - Cyclohexane, tetratriacontyl-
4845 - Cyclopentane, pentatriacontyl-

$C_{41}H_{46}N_4O_5$
4128 - Cinchonan-9-ol, 6'-methoxy-, carbonate (2:1) (esteR), (8α,9R)-

$C_{41}H_{64}O_{13}$
4043 - Card-20(22)-enolide, 3-[(O-2,6-dideoxy-β-D-ribo-hexopyranosyl-(1-4)-O-2,6-dideoxy-β-D-ribo-hexopyranosyl-(1-4)-2,6-dideoxy-β-D, 14-dihydroxy

$C_{41}H_{64}O_{14}$
4041 - Card-20(22)-enolide, 3-[(O-2,6-dideoxy-β-D-ribo-hexopyranosyl-(1-4)-O-2,6-dideoxy-β-D-ribo-hexopyranosyl-(1-4)-2,6-dideoxy-β-D, 12,14-dihydroxy

4042 - Card-20(22)-enolide, 3-[(O-2,6-dideoxy-β-D-ribo-hexopyranosyl-(1-4)-O-2,6-dideoxy-β-D-ribo-hexopyranosyl-(1-4)-2,6-dideoxy-β-D, 14,16-dihydroxy

$C_{42}H_{46}N_4O_8S$
11566 - Strychnidin-10-one, sulfate (2:1)

$C_{42}H_{62}O_{16}$
6230 - α-D-Glucopyranosiduronic acid, (3β,20β)-20-carboxy-11-oxo-30-norolean-12-en-3-yl 2-O-

$C_{42}H_{84}$
4413 - Cyclohexane, hexatriacontyl-

$C_{42}H_{84}N_{14}O_{36}S_3$
11561 - D-Streptamine, O-2-deoxy-2-(methylamino)-α-L-glucopyranosyl-(12)-O-5-deoxy-3-C-formyl-α-L-lyxofuranosyl-(1-4)-N,N,N′-55-bis(aminoiminomethyl)-, sulfate (2:3)(salt)

$C_{43}H_{88}$
11918 - Tritetracontane

$C_{45}H_{86}O_6$
11628 - Tetradecanoic acid, 1,2,3-propanetriyl ester

$C_{48}H_{40}O_4Si_4$
4994 - Cyclotetrasiloxane, octaphenyl-

$C_{48}H_{72}O_{14}$
477 - Avermectin B1a

$C_{48}H_{93}AlO_6$
6645 - Hexadecanoic acid, aluminum salt

$C_{50}H_{83}NO_{21}$
6196 - β-D-Galactopyranoside, (3β,5α,22β,25S)-spirosolan-3-yl O-β-

$C_{51}H_{98}O_6$
6656 - Hexadecanoic acid, 1,2,3-propanetriyl ester

$C_{54}H_{105}AlO_6$
8248 - Octadecanoic acid, aluminum salt

$C_{56}H_{92}O_{29}$
6195 - β-D-Galactopyranoside, (2α,3β,5α,15β,25R)-2,15-dihydroxyspirostan-3-yl

$C_{57}H_{104}O_6$
8310 - 9-Octadecenoic acid (Z)-, 1,2,3-propanetriyl ester

$C_{57}H_{110}O_6$
8272 - Octadecanoic acid, 1,2,3-propanetriyl ester

$C_{60}H_{78}OSn_2$
5950 - Fenbutatin oxide

$C_{66}H_{103}N_{17}O_{16}S$
508 - Bacitracin

CAS NUMBER INDEX OF ORGANIC COMPOUNDS

CAS	Page	CAS	Page	CAS	Page	CAS	Page	CAS	Page
50-00-0	5987	54-95-5	11654	58-72-0	1648	63-74-1	2187	71-30-7	11104
50-01-1	6285	55-10-7	896	58-73-1	5422	63-75-2	10914	71-36-3	3593
50-02-2	9737	55-18-5	5432	58-74-2	7378	63-91-2	9510	71-41-0	8915
50-03-3	9747	55-21-0	600	58-85-5	11689	63-98-9	879	71-43-2	867
50-06-6	11095	55-22-1	10895	58-86-6	12059	64-04-0	1569	71-44-3	3259
50-10-2	5443	55-38-9	5955	58-89-9	4409	64-10-8	12026	71-55-6	5655
50-11-3	11088	55-43-6	1801	58-90-2	9470	64-17-5	5672	71-63-6	4043
50-14-6	11350	55-63-0	10197	58-93-5	2855	64-18-6	6002	71-81-8	2120
50-22-6	9748	55-73-2	6281	58-94-6	2852	64-19-7	100	71-91-0	5447
50-23-7	9753	55-91-4	9580	58-95-7	2788	64-20-0	7456	72-14-0	2206
50-24-8	9738	56-03-1	7233	58-96-8	12033	64-65-3	9656	72-18-4	12041
50-27-1	5404	56-04-2	11108	58-97-9	12035	64-67-5	11573	72-19-5	11813
50-28-2	5399	56-12-2	3464	59-00-7	11230	64-69-7	200	72-23-1	9755
50-29-3	2328	56-18-8	9874	59-02-9	2787	64-77-7	2207	72-43-5	2329
50-30-6	2563	56-23-5	7540	59-23-4	6197	64-85-7	9751	72-44-6	11199
50-31-7	2727	56-25-7	5371	59-26-7	10891	64-86-8	93	72-48-0	391
50-32-8	2395	56-29-1	11083	59-30-3	6243	65-29-2	5441	72-54-8	1378
50-34-0	9832	56-33-7	5238	59-31-4	11325	65-45-2	616	72-55-9	1374
50-35-1	7363	56-35-9	5241	59-40-5	2205	65-46-3	5021	72-56-0	9160
50-36-2	482	56-36-0	11538	59-46-1	2493	65-49-6	2498	72-63-9	344
50-42-0	938	56-38-2	9583	59-47-2	10019	65-64-5	7170	73-22-3	11922
50-44-2	10763	56-40-6	6250	59-48-3	7327	65-71-4	11068	73-24-5	10751
50-50-0	5403	56-41-7	325	59-49-4	2882	65-82-7	7649	73-31-4	69
50-53-3	9505	56-45-1	11357	59-50-7	9267	65-85-0	2475	73-32-5	7370
50-55-5	12061	56-49-5	2392	59-51-8	7647	65-86-1	11050	73-40-5	10768
50-69-1	11345	56-53-1	9300	59-52-9	10370	66-02-4	11928	73-49-4	11197
50-70-4	6206	56-54-2	4126	59-66-5	32	66-22-8	11059	74-11-3	2536
50-71-5	11073	56-55-3	519	59-67-6	10894	66-23-9	5438	74-31-7	1271
50-78-2	2480	56-57-5	11287	59-87-0	7144	66-25-1	6718	74-39-5	1542
50-79-3	2562	56-72-4	4147	59-88-1	7172	66-27-3	7532	74-79-3	438
50-81-7	466	56-75-7	52	59-89-2	7699	66-28-4	4049	74-82-8	7459
50-82-8	2728	56-81-5	10186	59-92-7	11929	66-32-0	11565	74-83-9	7462
50-84-0	2561	56-82-6	9782	59-98-3	7198	66-56-8	9341	74-84-0	5448
50-85-1	2630	56-84-8	471	60-00-4	6262	66-71-7	9189	74-85-1	5879
50-89-5	11816	56-85-9	6249	60-01-5	3586	66-75-1	11061	74-86-2	5936
50-91-9	12034	56-86-0	6241	60-09-3	837	66-76-2	2839	74-87-3	7477
51-03-6	9727	56-87-1	7419	60-10-6	5121	66-77-3	7751	74-88-4	7518
51-12-7	10910	56-89-3	5020	60-11-7	721	66-81-9	9655	74-89-5	7442
51-17-2	2376	57-00-1	6254	60-12-8	1592	66-99-9	7752	74-90-8	7176
51-18-3	11844	57-06-7	10575	60-18-4	11926	67-03-8	11686	74-93-1	7546
51-20-7	11062	57-08-9	6871	60-23-1	5649	67-20-9	7222	74-95-3	7492
51-21-8	11067	57-10-3	6644	60-24-2	5732	67-43-6	6257	74-96-4	5454
51-26-3	2136	57-11-4	8247	60-27-5	7230	67-45-8	8589	74-97-5	7463
51-28-5	9342	57-13-6	11992	60-29-7	5608	67-47-0	6044	74-98-6	9834
51-34-3	908	57-14-7	7154	60-31-1	5439	67-48-1	5446	74-99-7	10727
51-35-4	9767	57-15-8	10407	60-32-2	6873	67-51-6	10828	75-00-3	5467
51-36-5	2564	57-24-9	11563	60-33-3	8223	67-52-7	11076	75-01-4	5884
51-41-2	1500	57-27-2	7666	60-34-4	7161	67-56-1	7581	75-02-5	5906
51-43-4	1521	57-41-0	7219	60-35-5	22	67-63-0	10335	75-03-6	5597
51-45-6	7203	57-43-2	11092	60-41-3	11566	67-64-1	10411	75-04-7	5410
51-48-9	11930	57-44-3	11087	60-51-5	5009	67-66-3	7554	75-05-8	271
51-52-5	11109	57-47-6	11191	60-54-8	7706	67-68-5	7529	75-07-0	11
51-55-8	902	57-50-1	6218	60-56-0	7214	67-71-0	7534	75-08-1	5648
51-63-8	1582	57-53-4	10021	60-57-1	5178	67-72-1	5595	75-09-2	7499
51-66-1	71	57-55-6	9982	60-70-8	12044	67-97-0	11348	75-10-5	7504
51-67-2	9208	57-57-8	8595	60-80-0	10847	68-11-0	202	75-11-6	7506
51-75-2	5413	57-62-5	7704	60-82-2	10453	68-12-2	5992	75-12-7	5989
51-79-6	3948	57-63-6	8216	60-87-7	9504	68-26-8	11341	75-15-0	3993
51-80-9	7490	57-66-9	2595	60-91-3	9503	68-35-9	2204	75-17-2	5988
51-83-2	5440	57-67-0	2191	61-00-7	5807	68-36-0	996	75-18-3	7545
52-28-8	7673	57-68-1	2196	61-12-1	11226	68-41-7	7406	75-19-4	4955
52-31-3	11084	57-71-6	3361	61-19-8	320	68-88-2	5700	75-21-8	8599
52-39-1	9746	57-74-9	4077	61-50-7	7303	68-94-0	10770	75-22-9	3132
52-43-7	11091	57-85-2	359	61-54-1	7302	68-96-2	9750	75-24-1	339
52-49-3	9705	57-87-4	5383	61-73-4	9506	69-24-9	10435	75-25-2	7548
52-51-7	9991	57-88-5	4110	61-78-9	6252	69-57-8	11655	75-26-3	9837
52-52-8	4782	57-91-0	5400	61-80-3	2876	69-65-8	7422	75-27-4	7467
52-67-5	12043	57-92-1	11562	61-82-5	11845	69-72-7	2612	75-28-5	10073
52-68-6	11853	57-97-6	520	61-90-5	7413	69-89-6	10755	75-29-6	9849
52-85-7	5946	58-00-4	5155	62-23-7	2705	69-91-0	885	75-30-9	10055
52-86-8	3651	58-05-9	6244	62-38-4	7431	69-93-2	10764	75-31-0	9804
52-90-4	5014	58-08-2	10760	62-44-2	60	70-07-5	8588	75-33-2	10157
53-03-2	9739	58-14-0	11051	62-50-0	7531	70-11-1	5782	75-34-3	5528
53-05-4	9744	58-15-1	10851	62-51-1	9830	70-18-8	6264	75-35-4	5894
53-06-5	9754	58-18-4	358	62-53-3	633	70-22-4	11188	75-36-5	294
53-16-7	5406	58-22-0	356	62-54-4	122	70-25-7	6284	75-37-6	5544
53-41-8	348	58-25-3	2409	62-55-5	5643	70-26-8	8550	75-38-7	5902
53-46-3	5445	58-27-5	7864	62-56-6	11799	70-29-1	5621	75-39-8	5674
53-70-3	5142	58-32-2	5711	62-57-7	331	70-30-4	9409	75-43-4	7502
53-89-4	10850	58-38-8	9501	62-59-9	4069	70-34-8	1718	75-44-5	4007
53-96-3	64	58-39-9	9617	62-73-7	5170	70-47-3	468	75-45-6	7478
54-04-6	1586	58-46-8	2396	62-75-9	7450	70-54-2	7418	75-46-7	7558
54-06-8	7301	58-54-8	168	62-76-0	5566	70-69-9	10413	75-47-8	7560
54-11-5	10992	58-55-9	10756	62-90-0	5408	70-70-2	10450	75-50-3	7444
54-25-1	11832	58-56-0	10934	63-05-8	352	70-78-0	11932	75-52-5	7522
54-36-4	10462	58-61-7	319	63-25-2	8032	71-00-1	7132	75-54-7	11398
54-85-3	10905	58-63-9	7335	63-68-3	7648	71-23-8	10334	75-55-8	501

88-06-2	: 9485	90-80-2	: 6212	93-15-2	: 1450	95-50-1	: 1354	97-30-3	: 6227
88-09-5	: 3514	90-81-3	: 1900	93-16-3	: 1449	95-51-2	: 658	97-39-2	: 6279
88-10-8	: 3972	90-84-6	: 10428	93-18-5	: 7884	95-52-3	: 1726	97-53-0	: 9389
88-12-0	: 11183	90-86-8	: 1903	93-22-1	: 7942	95-53-4	: 778	97-54-1	: 9388
88-13-1	: 11738	90-89-1	: 9603	93-25-4	: 921	95-54-5	: 1260	97-56-3	: 798
88-14-2	: 6050	90-90-4	: 7602	93-26-5	: 70	95-55-6	: 9201	97-59-6	: 12009
88-15-3	: 5870	90-94-8	: 7593	93-28-7	: 9395	95-56-7	: 9228	97-62-1	: 10298
88-16-4	: 1240	90-98-2	: 7592	93-35-6	: 2826	95-57-8	: 9251	97-63-2	: 10658
88-17-5	: 855	90-99-3	: 1187	93-37-8	: 11258	95-59-0	: 5188	97-65-4	: 3328
88-18-6	: 9321	91-00-9	: 1826	93-39-0	: 2821	95-63-6	: 2358	97-72-3	: 10291
88-20-0	: 2228	91-01-0	: 1910	93-42-5	: 68	95-64-7	: 708	97-74-5	: 11715
88-21-1	: 2215	91-02-1	: 7638	93-44-7	: 8021	95-65-8	: 9311	97-77-8	: 11721
88-60-8	: 9331	91-08-7	: 1430	93-50-5	: 667	95-66-9	: 1138	97-85-8	: 10304
88-65-3	: 2518	91-10-1	: 9305	93-51-8	: 9385	95-68-1	: 705	97-86-9	: 10666
88-67-5	: 2657	91-13-4	: 983	93-52-7	: 1299	95-69-2	: 673	97-88-1	: 10650
88-68-6	: 601	91-14-5	: 1403	93-53-8	: 872	95-70-5	: 1279	97-90-5	: 10656
88-69-7	: 9410	91-16-7	: 1436	93-54-9	: 1870	95-71-6	: 1532	97-94-9	: 3118
88-72-2	: 2000	91-17-8	: 7789	93-55-0	: 10480	95-72-7	: 1140	97-95-0	: 3620
88-73-3	: 1214	91-18-9	: 10747	93-56-1	: 5579	95-73-8	: 1388	97-96-1	: 3183
88-74-4	: 822	91-19-0	: 11333	93-58-3	: 2687	95-74-9	: 672	97-97-2	: 5472
88-75-5	: 9435	91-20-3	: 7732	93-59-4	: 1126	95-75-0	: 1384	97-99-4	: 6124
88-85-7	: 5183	91-21-4	: 7388	93-60-7	: 10906	95-76-1	: 689	98-00-0	: 6118
88-87-9	: 9258	91-22-5	: 11214	93-61-8	: 5998	95-77-2	: 9286	98-01-1	: 6041
88-88-0	: 1244	91-23-6	: 1935	93-71-0	: 44	95-78-3	: 706	98-02-2	: 6117
88-89-1	: 9497	91-40-7	: 2715	93-72-1	: 11512	95-79-4	: 676	98-03-3	: 11733
88-95-9	: 1318	91-49-6	: 39	93-76-5	: 256	95-80-7	: 1278	98-04-4	: 866
88-96-0	: 1324	91-53-2	: 11265	93-78-7	: 259	95-81-8	: 669	98-05-5	: 460
88-99-3	: 1327	91-56-5	: 7300	93-79-8	: 258	95-82-9	: 687	98-06-6	: 1459
89-00-9	: 10927	91-57-6	: 7916	93-88-9	: 1574	95-83-0	: 1264	98-07-7	: 2333
89-01-0	: 10804	91-58-7	: 7781	93-89-0	: 2602	95-84-1	: 9210	98-08-8	: 2348
89-05-4	: 2257	91-59-8	: 7709	93-90-3	: 5743	95-85-2	: 9204	98-09-9	: 2242
89-31-6	: 7397	91-60-1	: 8008	93-91-4	: 3362	95-86-3	: 9274	98-11-3	: 2214
89-37-2	: 3975	91-62-3	: 11281	93-97-0	: 2510	95-87-4	: 9309	98-13-5	: 11473
89-39-4	: 1447	91-63-4	: 11277	93-98-1	: 626	95-92-1	: 5560	98-15-7	: 1241
89-54-3	: 2490	91-64-5	: 2792	93-99-2	: 2716	95-93-2	: 2280	98-16-8	: 856
89-55-4	: 2524	91-65-6	: 4285	94-02-0	: 2141	95-94-3	: 2260	98-17-9	: 9491
89-56-5	: 2631	91-66-7	: 695	94-07-5	: 1887	95-95-4	: 9484	98-19-1	: 1460
89-57-6	: 2499	91-68-9	: 9297	94-08-6	: 2692	96-00-4	: 4206	98-27-1	: 9333
89-58-7	: 1478	91-73-6	: 1829	94-09-7	: 2497	96-01-5	: 4203	98-29-3	: 1516
89-59-8	: 1212	91-76-9	: 11829	94-11-1	: 173	96-02-6	: 6092	98-46-4	: 2061
89-60-1	: 1209	91-78-1	: 11834	94-12-2	: 2507	96-04-8	: 6402	98-50-0	: 456
89-61-2	: 1395	91-80-5	: 5507	94-13-3	: 2653	96-05-9	: 10671	98-51-1	: 1466
89-62-3	: 803	91-81-6	: 5505	94-14-4	: 2505	96-09-3	: 8624	98-54-4	: 9323
89-68-8	: 9268	91-84-9	: 5511	94-20-2	: 2208	96-11-7	: 10163	98-56-6	: 1242
89-69-0	: 2339	91-88-3	: 5724	94-25-7	: 2488	96-12-8	: 9885	98-58-8	: 2244
89-71-4	: 2696	91-96-3	: 3204	94-26-8	: 2615	96-14-0	: 8824	98-59-9	: 2250
89-72-5	: 9430	92-06-8	: 11596	94-30-4	: 2671	96-18-4	: 10175	98-60-2	: 2245
89-73-6	: 613	92-13-7	: 6151	94-33-7	: 5578	96-19-5	: 10603	98-61-3	: 2247
89-74-7	: 5809	92-24-0	: 7703	94-36-0	: 9157	96-21-9	: 10357	98-62-4	: 2252
89-75-8	: 2893	92-35-3	: 11111	94-41-7	: 10708	96-22-0	: 8958	98-66-8	: 2220
89-79-2	: 4544	92-36-4	: 782	94-44-0	: 10911	96-23-1	: 10362	98-67-9	: 2226
89-82-7	: 4608	92-41-1	: 7869	94-46-2	: 3632	96-24-2	: 9994	98-71-5	: 2224
89-83-8	: 9418	92-43-3	: 10844	94-49-5	: 5572	96-26-4	: 10429	98-72-6	: 459
89-84-9	: 5803	92-48-8	: 2835	94-53-1	: 2418	96-27-5	: 10013	98-73-7	: 2588
89-86-1	: 2569	92-51-3	: 2988	94-58-6	: 2431	96-29-7	: 3675	98-79-3	: 9768
89-87-2	: 1479	92-52-4	: 3008	94-59-7	: 2429	96-31-1	: 12004	98-80-6	: 3135
89-92-9	: 1062	92-54-6	: 9620	94-62-2	: 9632	96-32-2	: 117	98-81-7	: 1025
89-93-0	: 1815	92-59-1	: 1811	94-65-5	: 4612	96-33-3	: 10655	98-82-8	: 1975
89-95-2	: 1897	92-61-5	: 2829	94-67-7	: 579	96-34-4	: 138	98-83-9	: 1972
89-96-3	: 1161	92-62-6	: 316	94-68-8	: 746	96-35-5	: 198	98-85-1	: 1896
89-98-5	: 529	92-66-0	: 3018	94-69-9	: 5996	96-37-7	: 4826	98-86-2	: 5864
90-00-6	: 9353	92-67-1	: 3011	94-70-2	: 736	96-40-2	: 4914	98-87-3	: 1382
90-01-7	: 1882	92-69-3	: 3090	94-71-3	: 9350	96-41-3	: 4864	98-88-4	: 2886
90-02-8	: 564	92-70-6	: 7772	94-74-6	: 141	96-43-5	: 11740	98-89-5	: 4332
90-04-0	: 766	92-71-7	: 8580	94-75-7	: 169	96-45-7	: 7225	98-91-9	: 1130
90-05-5	: 9379	92-82-0	: 9193	94-80-4	: 171	96-47-9	: 6171	98-92-0	: 10890
90-11-9	: 7739	92-83-1	: 12049	94-81-5	: 3493	96-48-0	: 6142	98-94-2	: 4286
90-12-0	: 7915	92-84-2	: 9499	94-82-6	: 3925	96-49-1	: 5211	98-95-3	: 2049
90-13-1	: 7780	92-85-3	: 11660	94-96-2	: 6802	96-50-4	: 11662	98-96-4	: 10802
90-14-2	: 7902	92-86-4	: 3037	95-00-1	: 1803	96-54-8	: 11130	98-97-5	: 10803
90-15-3	: 8017	92-87-5	: 3031	95-01-2	: 541	96-76-4	: 9219	98-98-6	: 10893
90-19-7	: 2815	92-91-1	: 5776	95-05-6	: 11714	96-77-5	: 1560	99-02-5	: 5788
90-20-0	: 7874	92-92-2	: 3022	95-06-7	: 4065	96-80-0	: 5683	99-03-6	: 5772
90-26-6	: 881	92-93-3	: 3087	95-11-4	: 2966	96-82-2	: 6210	99-04-7	: 2677
90-27-7	: 897	92-94-4	: 11597	95-13-6	: 7245	96-88-8	: 9639	99-05-8	: 2485
90-33-5	: 2831	92-99-9	: 11213	95-14-7	: 2871	96-91-3	: 9207	99-07-0	: 9318
90-39-1	: 7571	93-01-6	: 7974	95-15-8	: 2399	96-97-9	: 2647	99-08-1	: 2001
90-41-5	: 3010	93-02-7	: 544	95-16-9	: 2861	96-99-1	: 2547	99-09-2	: 823
90-42-6	: 2989	93-03-8	: 1856	95-20-5	: 7313	97-00-7	: 1144	99-11-6	: 10900
90-43-7	: 3088	93-04-9	: 7914	95-21-6	: 2880	97-02-9	: 725	99-12-7	: 1477
90-44-8	: 433	93-05-0	: 1266	95-25-0	: 2883	97-05-2	: 2655	99-14-9	: 10165
90-45-9	: 313	93-07-2	: 2576	95-43-2	: 3193	97-16-5	: 9289	99-17-2	: 6223
90-47-1	: 12050	93-08-3	: 5858	95-45-4	: 3360	97-17-6	: 9581	99-20-7	: 6220
90-59-5	: 534	93-09-4	: 7759	95-46-5	: 1054	97-18-7	: 9478	99-30-9	: 692
90-65-3	: 6701	93-10-7	: 11227	95-47-6	: 1452	97-23-4	: 9408	99-31-0	: 1331
90-69-7	: 5834	93-11-8	: 7976	95-48-7	: 9400	97-24-5	: 9477	99-32-1	: 10776
90-74-4	: 6236	93-14-1	: 10016	95-49-8	: 1183	97-25-6	: 6231	99-33-2	: 2896

CAS	No.	CAS	No.	CAS	No.	CAS	No.	CAS	No.	CAS	No.
99-34-3	: 2591	100-71-0	: 10956	103-19-5	: 5246	105-07-7	: 2758	106-96-7	: 10728
99-35-4	: 2365	100-72-1	: 10784	103-23-1	: 6781	105-13-5	: 1891	106-97-8	: 3229
99-36-5	: 2697	100-74-3	: 7694	103-24-2	: 8132	105-14-6	: 8964	106-98-9	: 3704
99-45-6	: 5804	100-75-4	: 9688	103-25-3	: 2139	105-30-6	: 8937	106-99-0	: 3146
99-48-9	: 4696	100-76-5	: 480	103-28-6	: 10307	105-31-7	: 7124	107-00-6	: 3897
99-50-3	: 2571	100-79-8	: 5207	103-29-7	: 1587	105-34-0	: 154	107-02-8	: 10509
99-51-4	: 1475	100-80-1	: 1641	103-30-0	: 1619	105-36-2	: 115	107-03-9	: 10156
99-52-5	: 800	100-81-2	: 1816	103-32-2	: 1827	105-37-3	: 10263	107-04-0	: 5456
99-53-6	: 9422	100-83-4	: 565	103-34-4	: 7688	105-38-4	: 10261	107-05-1	: 10536
99-54-7	: 1392	100-84-5	: 1932	103-37-7	: 3585	105-39-5	: 130	107-06-2	: 5529
99-55-8	: 801	100-86-7	: 1600	103-38-8	: 3560	105-40-8	: 3954	107-07-3	: 5693
99-56-9	: 1281	100-88-9	: 11568	103-43-5	: 3298	105-41-9	: 6730	107-08-4	: 10054
99-59-2	: 775	100-97-0	: 11601	103-44-6	: 6405	105-42-0	: 6977	107-10-8	: 9803
99-61-6	: 592	101-01-9	: 6288	103-45-7	: 232	105-45-3	: 3573	107-11-9	: 10522
99-62-7	: 993	101-02-0	: 9593	103-48-0	: 10306	105-46-4	: 214	107-12-0	: 10090
99-63-8	: 1319	101-05-3	: 360	103-49-1	: 1828	105-48-6	: 139	107-13-1	: 10583
99-65-0	: 1492	101-10-0	: 4141	103-50-4	: 2079	105-50-0	: 8783	107-14-2	: 274
99-71-4	: 9431	101-14-4	: 785	103-55-9	: 5504	105-53-3	: 9941	107-15-3	: 5496
99-73-0	: 5777	101-17-7	: 678	103-63-9	: 1038	105-54-4	: 3515	107-16-4	: 282
99-75-2	: 2698	101-18-8	: 9446	103-65-1	: 2169	105-55-5	: 11803	107-18-6	: 10697
99-76-3	: 2637	101-21-3	: 3939	103-67-3	: 1818	105-56-6	: 152	107-19-7	: 10743
99-77-4	: 2709	101-27-9	: 510	103-69-5	: 743	105-57-7	: 5541	107-20-0	: 12
99-82-1	: 4438	101-31-5	: 917	103-70-8	: 5999	105-58-8	: 3999	107-21-1	: 5569
99-84-3	: 4675	101-38-2	: 4280	103-71-9	: 1786	105-59-9	: 5742	107-22-2	: 5491
99-85-4	: 4275	101-39-3	: 10516	103-72-0	: 1792	105-60-2	: 493	107-25-5	: 5908
99-86-5	: 4273	101-40-6	: 4385	103-73-1	: 1649	105-66-8	: 3588	107-29-9	: 17
99-87-6	: 1990	101-41-7	: 929	103-74-2	: 10950	105-67-9	: 9308	107-30-2	: 7483
99-88-7	: 789	101-43-9	: 10653	103-75-3	: 10783	105-68-0	: 3639	107-31-3	: 6010
99-89-8	: 9412	101-53-1	: 9455	103-76-4	: 9616	105-70-4	: 10190	107-34-6	: 461
99-90-1	: 5781	101-54-2	: 1283	103-79-7	: 10481	105-72-6	: 3512	107-35-7	: 5623
99-91-2	: 5789	101-55-3	: 1079	103-80-0	: 971	105-74-8	: 9155	107-37-9	: 11475
99-92-3	: 5773	101-57-5	: 2234	103-81-1	: 878	105-75-9	: 3774	107-39-1	: 9067
99-93-4	: 5832	101-60-0	: 9734	103-82-2	: 883	105-83-9	: 9875	107-40-4	: 9071
99-94-5	: 2678	101-61-1	: 786	103-83-3	: 1807	105-86-2	: 8338	107-41-5	: 8793
99-96-7	: 2613	101-76-8	: 1963	103-84-4	: 86	105-99-7	: 6784	107-43-7	: 7454
99-97-8	: 862	101-77-9	: 784	103-85-5	: 11808	106-02-5	: 8562	107-46-0	: 5236
99-98-9	: 1270	101-80-4	: 832	103-88-8	: 37	106-07-0	: 8259	107-47-1	: 10155
99-99-0	: 2002	101-81-5	: 1961	103-89-9	: 78	106-19-4	: 6787	107-48-2	: 8612
100-00-5	: 1216	101-82-6	: 11004	103-90-2	: 67	106-23-0	: 8490	107-49-3	: 5221
100-01-6	: 824	101-83-7	: 4284	103-95-7	: 2117	106-24-1	: 8333	107-50-6	: 4239
100-02-7	: 9437	101-84-8	: 2071	103-99-1	: 8232	106-25-2	: 8334	107-51-7	: 11914
100-06-1	: 5841	101-90-6	: 8625	104-01-8	: 922	106-26-3	: 8318	107-52-8	: 6990
100-07-2	: 2901	101-91-7	: 3205	104-03-0	: 933	106-27-4	: 3544	107-53-9	: 11964
100-09-4	: 2666	101-97-3	: 899	104-04-1	: 83	106-28-5	: 5333	107-59-5	: 128
100-10-7	: 550	101-99-5	: 3965	104-06-3	: 34	106-30-9	: 6462	107-68-6	: 5624
100-12-9	: 1694	102-01-2	: 3207	104-07-4	: 8253	106-31-0	: 3467	107-70-0	: 8978
100-13-0	: 1645	102-02-3	: 7234	104-09-6	: 874	106-32-1	: 8428	107-72-2	: 11472
100-14-1	: 1208	102-04-5	: 10433	104-13-2	: 655	106-33-2	: 5312	107-74-4	: 8392
100-15-2	: 805	102-06-7	: 6282	104-15-4	: 2229	106-35-4	: 6511	107-75-5	: 8347
100-16-3	: 7168	102-07-8	: 12010	104-21-2	: 1892	106-36-5	: 10326	107-80-2	: 3266
100-17-4	: 1937	102-08-9	: 11805	104-22-3	: 2213	106-37-6	: 1290	107-81-3	: 8711
100-18-5	: 994	102-09-0	: 4001	104-29-0	: 10000	106-38-7	: 1056	107-82-4	: 3239
100-19-6	: 5861	102-13-6	: 932	104-36-9	: 1314	106-39-8	: 1006	107-83-5	: 8823
100-20-5	: 1320	102-14-7	: 3575	104-45-0	: 1945	106-40-1	: 640	107-84-6	: 3252
100-21-0	: 1329	102-20-5	: 940	104-46-1	: 1942	106-41-2	: 9230	107-85-7	: 3220
100-22-1	: 1287	102-25-0	: 2345	104-47-2	: 958	106-42-3	: 1454	107-86-8	: 3700
100-25-4	: 1493	102-27-2	: 747	104-50-7	: 6140	106-43-4	: 1185	107-87-9	: 8957
100-26-5	: 10929	102-45-4	: 4810	104-51-8	: 1118	106-44-5	: 9402	107-88-0	: 3345
100-28-7	: 1790	102-47-6	: 1357	104-52-9	: 1235	106-45-6	: 2303	107-89-1	: 3184
100-29-8	: 1664	102-50-1	: 772	104-53-0	: 2114	106-46-7	: 1356	107-91-5	: 49
100-36-7	: 5502	102-51-2	: 1273	104-57-4	: 6016	106-47-8	: 660	107-92-6	: 3460
100-37-8	: 5706	102-52-3	: 10150	104-61-0	: 6149	106-48-9	: 9253	107-93-7	: 3814
100-39-0	: 1053	102-53-4	: 7489	104-63-2	: 5762	106-49-0	: 780	107-94-8	: 10216
100-40-3	: 4658	102-54-5	: 5960	104-65-4	: 10707	106-50-3	: 1262	107-95-9	: 323
100-41-4	: 1668	102-56-7	: 702	104-66-5	: 1588	106-51-4	: 4257	107-96-0	: 10286
100-42-5	: 1624	102-60-3	: 10379	104-67-6	: 6152	106-52-5	: 9717	107-97-1	: 6268
100-43-6	: 10954	102-69-2	: 9808	104-72-3	: 1258	106-53-6	: 2292	107-98-2	: 10387
100-44-7	: 1182	102-70-5	: 10523	104-75-6	: 6726	106-54-7	: 2295	108-01-0	: 5709
100-45-8	: 4629	102-71-6	: 5751	104-76-7	: 6930	106-57-0	: 9612	108-03-2	: 10116
100-46-9	: 1795	102-76-1	: 10195	104-78-9	: 9876	106-58-1	: 9608	108-05-4	: 179
100-47-0	: 2740	102-79-4	: 5692	104-80-3	: 7109	106-63-8	: 10673	108-08-7	: 8762
100-49-2	: 4419	102-81-8	: 5703	104-81-4	: 1064	106-65-0	: 3318	108-10-1	: 8981
100-50-5	: 4632	102-82-9	: 3213	104-82-5	: 1200	106-68-3	: 8472	108-11-2	: 8942
100-51-6	: 1836	102-85-2	: 9590	104-83-6	: 1134	106-69-4	: 6868	108-12-3	: 3693
100-52-7	: 525	102-86-3	: 6725	104-84-7	: 1817	106-70-7	: 6898	108-16-7	: 10376
100-53-8	: 1832	102-92-1	: 10716	104-85-8	: 2765	106-73-0	: 6466	108-18-9	: 9821
100-54-9	: 10882	102-94-3	: 10683	104-87-0	: 589	106-74-1	: 10634	108-19-0	: 7232
100-55-0	: 10976	102-97-6	: 1820	104-88-1	: 531	106-79-6	: 5046	108-20-3	: 10120
100-60-7	: 4298	103-01-5	: 6270	104-90-5	: 10966	106-86-5	: 8555	108-21-4	: 209
100-61-8	: 781	103-02-6	: 10482	104-91-6	: 9438	106-87-6	: 8557	108-22-5	: 10698
100-63-0	: 7169	103-03-7	: 7145	104-92-7	: 1052	106-88-7	: 8615	108-23-6	: 4023
100-64-1	: 4610	103-05-9	: 2150	104-93-8	: 1933	106-90-1	: 10680	108-24-7	: 107
100-65-2	: 757	103-09-3	: 187	104-94-9	: 768	106-91-2	: 10668	108-27-0	: 11186
100-66-3	: 1915	103-11-7	: 10636	104-96-1	: 820	106-92-3	: 8626	108-28-1	: 6155
100-68-5	: 2041	103-12-8	: 2209	105-03-3	: 8131	106-93-4	: 5518	108-30-5	: 6093
100-69-6	: 10952	103-16-2	: 9453	105-05-5	: 1415	106-94-5	: 9836	108-31-6	: 6091
100-70-9	: 10881	103-17-3	: 4075	105-06-6	: 1405	106-95-6	: 10531	108-32-7	: 5213

108-36-1	: 1289	109-90-0	: 5598	111-17-1	: 10329	112-51-6	: 5257	117-37-3	: 7269
108-37-2	: 1005	109-92-2	: 5905	111-18-2	: 6752	112-52-7	: 5295	117-39-5	: 2817
108-38-3	: 1453	109-93-3	: 5912	111-19-3	: 5049	112-53-8	: 5321	117-73-7	: 12064
108-39-4	: 9401	109-94-4	: 6007	111-20-6	: 5042	112-54-9	: 5288	117-78-2	: 372
108-40-7	: 2302	109-95-5	: 8087	111-21-7	: 5714	112-55-0	: 5304	117-79-3	: 381
108-41-8	: 1184	109-96-6	: 11118	111-24-0	: 8741	112-57-2	: 5498	117-80-6	: 7850
108-42-9	: 659	109-97-7	: 11112	111-25-1	: 6732	112-58-3	: 6842	117-81-7	: 1334
108-43-0	: 9252	109-98-8	: 10822	111-26-2	: 6723	112-60-7	: 5756	117-84-0	: 1344
108-44-1	: 779	109-99-9	: 6168	111-27-3	: 6911	112-61-8	: 8266	117-93-1	: 66
108-45-2	: 1261	110-00-9	: 6025	111-28-4	: 6707	112-62-9	: 8307	117-99-7	: 7622
108-46-3	: 1498	110-01-0	: 11783	111-29-5	: 8787	112-63-0	: 8228	118-00-3	: 6289
108-47-4	: 10937	110-02-1	: 11724	111-30-8	: 8734	112-64-1	: 11635	118-08-1	: 7357
108-48-5	: 10939	110-03-2	: 6801	111-31-9	: 6856	112-66-3	: 178	118-10-5	: 4120
108-50-9	: 10807	110-04-3	: 5587	111-34-2	: 3369	112-67-4	: 6665	118-32-1	: 7871
108-52-1	: 11046	110-05-4	: 9153	111-40-0	: 5497	112-70-9	: 11883	118-34-3	: 6225
108-55-4	: 10780	110-06-5	: 5242	111-41-1	: 5676	112-71-0	: 11614	118-41-2	: 2734
108-56-5	: 3340	110-12-3	: 6978	111-42-2	: 5730	112-72-1	: 11630	118-44-5	: 7718
108-57-6	: 1404	110-13-4	: 6805	111-43-3	: 10119	112-73-2	: 3417	118-46-7	: 8020
108-59-8	: 9943	110-14-5	: 3257	111-44-4	: 5611	112-76-5	: 8281	118-55-8	: 2650
108-60-1	: 10123	110-15-6	: 3294	111-45-5	: 5763	112-79-8	: 8297	118-58-1	: 2651
108-64-5	: 3548	110-16-7	: 3765	111-46-6	: 5753	112-80-1	: 8294	118-61-6	: 2619
108-67-8	: 2359	110-17-8	: 3764	111-47-7	: 10151	112-82-3	: 6633	118-69-4	: 1385
108-68-9	: 9312	110-18-9	: 5516	111-48-8	: 5766	112-85-6	: 5276	118-71-8	: 10796
108-69-0	: 709	110-19-0	: 215	111-49-9	: 492	112-86-7	: 5283	118-74-1	: 1749
108-70-3	: 2323	110-22-5	: 9156	111-51-3	: 3263	112-88-9	: 8291	118-75-2	: 4268
108-73-6	: 2368	110-27-0	: 11626	111-55-7	: 5571	112-89-0	: 8235	118-76-3	: 4686
108-75-8	: 11028	110-38-3	: 5063	111-56-8	: 5307	112-91-4	: 8293	118-78-5	: 11077
108-77-0	: 11837	110-39-4	: 3566	111-60-4	: 8260	112-92-5	: 8278	118-79-6	: 9479
108-78-1	: 11835	110-40-7	: 5045	111-61-5	: 8257	112-95-8	: 5350	118-83-2	: 1219
108-82-7	: 6485	110-41-8	: 11936	111-62-6	: 8301	113-00-8	: 6278	118-90-1	: 2676
108-83-8	: 6514	110-42-9	: 5066	111-64-8	: 8481	113-15-5	: 5394	118-91-2	: 2534
108-84-9	: 8947	110-43-0	: 6510	111-65-9	: 8355	113-48-4	: 7660	118-92-3	: 2484
108-85-0	: 4315	110-44-1	: 6698	111-66-0	: 8491	114-07-8	: 5396	118-93-4	: 5830
108-86-1	: 998	110-45-2	: 3636	111-68-2	: 6349	114-25-0	: 3001	118-96-7	: 2048
108-87-2	: 4426	110-46-3	: 8090	111-69-3	: 6777	114-26-1	: 10722	119-13-1	: 2785
108-88-3	: 1947	110-49-6	: 5735	111-70-6	: 6472	114-91-0	: 10979	119-20-0	: 2494
108-89-4	: 10982	110-52-1	: 3267	111-71-7	: 6344	115-07-1	: 10528	119-26-6	: 7156
108-90-7	: 1131	110-53-2	: 8710	111-74-0	: 5501	115-10-6	: 7523	119-27-7	: 1923
108-91-8	: 4283	110-54-3	: 6731	111-75-1	: 5691	115-11-7	: 10578	119-28-8	: 7967
108-93-0	: 4500	110-56-5	: 3279	111-76-2	: 5687	115-17-3	: 20	119-33-5	: 9424
108-94-1	: 4571	110-57-6	: 3750	111-77-3	: 5737	115-18-4	: 3870	119-36-8	: 2634
108-95-2	: 9199	110-58-7	: 8700	111-78-4	: 4733	115-19-5	: 3922	119-41-5	: 225
108-97-4	: 10789	110-59-8	: 8827	111-81-9	: 11978	115-20-8	: 5768	119-44-8	: 10749
108-98-5	: 2289	110-60-1	: 3258	111-82-0	: 5315	115-21-9	: 11463	119-48-2	: 7686
108-99-6	: 10981	110-61-2	: 3291	111-83-1	: 8356	115-24-2	: 9835	119-59-5	: 845
109-01-3	: 9618	110-62-3	: 8690	111-84-2	: 8120	115-25-3	: 4194	119-60-8	: 7606
109-02-4	: 7696	110-63-4	: 3346	111-85-3	: 8359	115-29-7	: 5369	119-61-9	: 7616
109-04-6	: 10875	110-65-6	: 3911	111-86-4	: 8350	115-32-2	: 1851	119-64-2	: 7979
109-06-8	: 10980	110-66-7	: 8844	111-87-5	: 8443	115-37-7	: 7679	119-65-3	: 7373
109-07-9	: 9619	110-67-8	: 10110	111-88-6	: 8418	115-38-8	: 11094	119-67-5	: 2607
109-08-0	: 10810	110-68-9	: 3221	111-90-0	: 5720	115-39-9	: 9366	119-68-6	: 2679
109-09-1	: 10915	110-69-0	: 3188	111-92-2	: 3211	115-40-2	: 9364	119-84-6	: 2798
109-12-6	: 11042	110-70-3	: 5503	111-96-6	: 5613	115-43-5	: 11103	119-90-4	: 3033
109-16-0	: 10657	110-71-4	: 5549	112-04-9	: 11470	115-53-7	: 7676	119-93-7	: 3034
109-19-3	: 3543	110-72-5	: 5509	112-05-0	: 8146	115-56-0	: 11064	120-07-0	: 5760
109-21-7	: 3479	110-73-6	: 5722	112-06-1	: 193	115-61-7	: 11349	120-12-7	: 366
109-43-3	: 5044	110-74-7	: 6018	112-10-7	: 8287	115-63-9	: 9623	120-14-9	: 545
109-49-9	: 7089	110-75-8	: 5887	112-12-9	: 11958	115-69-5	: 9987	120-15-0	: 11298
109-52-4	: 8855	110-76-9	: 5423	112-13-0	: 5078	115-70-8	: 9985	120-18-3	: 7960
109-53-5	: 10035	110-77-0	: 5726	112-14-1	: 222	115-71-9	: 9093	120-20-7	: 1571
109-56-8	: 5741	110-78-1	: 10059	112-15-2	: 5721	115-76-4	: 10004	120-21-8	: 539
109-57-9	: 11809	110-80-5	: 5717	112-16-3	: 5328	115-77-5	: 9989	120-22-9	: 699
109-59-1	: 5740	110-81-6	: 5253	112-17-4	: 157	115-80-0	: 10182	120-23-0	: 217
109-60-4	: 236	110-82-7	: 4305	112-20-9	: 8119	115-84-4	: 9992	120-25-2	: 557
109-61-5	: 4028	110-83-8	: 4626	112-23-2	: 6008	115-86-6	: 9566	120-29-6	: 489
109-63-7	: 3137	110-85-0	: 9602	112-24-3	: 5499	115-90-2	: 5954	120-32-1	: 9270
109-64-8	: 9882	110-86-1	: 10872	112-25-4	: 5728	115-96-8	: 5701	120-33-2	: 2691
109-65-9	: 3231	110-87-2	: 10777	112-26-5	: 5449	115-99-1	: 8337	120-36-5	: 10250
109-66-0	: 8709	110-88-3	: 11905	112-27-6	: 5713	116-06-3	: 334	120-43-4	: 9604
109-67-1	: 9004	110-89-4	: 9629	112-29-8	: 5034	116-09-6	: 10443	120-46-7	: 10027
109-69-3	: 3241	110-91-8	: 7681	112-30-1	: 5070	116-14-3	: 5918	120-47-8	: 2621
109-70-6	: 9839	110-93-0	: 6603	112-31-2	: 5030	116-15-4	: 10572	120-48-9	: 2706
109-73-9	: 3209	110-94-1	: 8770	112-32-3	: 6014	116-16-5	: 10440	120-50-3	: 2701
109-74-0	: 3403	110-95-2	: 9879	112-34-5	: 5688	116-17-6	: 9595	120-51-4	: 2720
109-75-1	: 3803	110-96-3	: 9823	112-36-7	: 5612	116-26-7	: 4250	120-55-8	: 5755
109-76-2	: 9873	110-97-4	: 10384	112-37-8	: 11943	116-29-0	: 2325	120-57-0	: 2417
109-77-3	: 9911	110-99-6	: 226	112-38-9	: 11974	116-38-1	: 865	120-61-6	: 1351
109-78-4	: 10106	111-01-3	: 11606	112-39-0	: 6653	116-43-8	: 3578	120-66-1	: 76
109-79-5	: 3437	111-02-4	: 11602	112-40-3	: 5292	116-44-9	: 2202	120-71-8	: 770
109-80-8	: 10031	111-03-5	: 8299	112-41-4	: 5335	116-54-1	: 166	120-72-9	: 7294
109-81-9	: 5512	111-06-8	: 6647	112-42-5	: 11950	116-58-5	: 9757	120-73-0	: 10752
109-82-0	: 285	111-11-5	: 8434	112-43-6	: 11979	116-63-2	: 7966	120-75-2	: 2863
109-83-1	: 5738	111-12-6	: 8542	112-44-7	: 11935	117-08-8	: 7346	120-80-9	: 1497
109-84-2	: 5729	111-13-7	: 8471	112-46-9	: 11981	117-10-2	: 395	120-82-1	: 2322
109-86-4	: 5733	111-14-8	: 6455	112-47-0	: 5047	117-12-4	: 393	120-83-2	: 9283
109-87-5	: 7507	111-15-9	: 5718	112-48-1	: 5527	117-18-0	: 2264	120-85-4	: 9622
109-89-7	: 5427	111-16-0	: 6393	112-49-2	: 11642	117-34-0	: 936	120-87-6	: 8254

120-89-8	: 7226	123-11-5	: 585	126-68-1	: 9585	133-59-5	: 2249	140-80-7	: 8736
120-92-3	: 4878	123-15-9	: 8693	126-71-6	: 9572	133-67-5	: 2854	140-82-9	: 5707
120-93-4	: 7227	123-19-3	: 6512	126-72-7	: 10358	133-90-4	: 2492	140-87-4	: 153
120-94-5	: 11168	123-20-6	: 3513	126-73-8	: 9562	133-91-5	: 2617	140-88-5	: 10635
120-97-8	: 1559	123-22-8	: 9587	126-81-8	: 4378	134-04-3	: 2848	140-95-4	: 11995
121-06-2	: 9337	123-25-1	: 3303	126-84-1	: 9904	134-11-2	: 2596	141-03-7	: 3301
121-14-2	: 1956	123-29-5	: 8150	126-98-7	: 10586	134-20-3	: 2503	141-04-8	: 6783
121-17-5	: 1217	123-30-8	: 9203	126-99-8	: 3152	134-31-6	: 11324	141-05-9	: 3776
121-21-1	: 4960	123-31-9	: 1499	127-00-4	: 10349	134-32-7	: 7708	141-06-0	: 8914
121-29-9	: 4962	123-32-0	: 10806	127-06-0	: 10476	134-49-6	: 7697	141-10-6	: 11965
121-30-2	: 1558	123-35-3	: 8330	127-07-1	: 12016	134-55-4	: 2483	141-12-8	: 8336
121-32-4	: 555	123-38-6	: 9774	127-08-2	: 234	134-58-7	: 11851	141-20-8	: 8303
121-33-5	: 574	123-39-7	: 5995	127-09-3	: 237	134-62-3	: 612	141-27-5	: 8317
121-34-6	: 2625	123-42-2	: 8976	127-17-3	: 10314	134-81-6	: 5584	141-28-6	: 6785
121-43-7	: 3126	123-43-3	: 238	127-18-4	: 5917	134-84-9	: 7633	141-32-2	: 10616
121-44-8	: 5417	123-45-5	: 239	127-19-5	: 55	134-85-0	: 7604	141-43-5	: 5675
121-45-9	: 9592	123-51-3	: 3627	127-25-3	: 9168	134-96-3	: 569	141-46-8	: 15
121-46-0	: 2913	123-54-6	: 8798	127-27-5	: 9170	135-00-2	: 7641	141-57-1	: 11476
121-47-1	: 2216	123-56-8	: 11152	127-29-7	: 308	135-01-3	: 1413	141-62-8	: 11646
121-57-3	: 2217	123-62-6	: 10203	127-40-2	: 4058	135-02-4	: 583	141-63-9	: 8997
121-59-5	: 455	123-63-7	: 11908	127-48-0	: 8587	135-09-1	: 2858	141-66-2	: 5172
121-60-8	: 2243	123-66-0	: 6884	127-63-9	: 2237	135-19-3	: 8018	141-75-3	: 3681
121-61-9	: 31	123-72-8	: 3178	127-66-2	: 1878	135-20-6	: 759	141-78-6	: 186
121-66-4	: 11666	123-73-9	: 3695	127-69-5	: 2195	135-44-4	: 8926	141-79-7	: 9108
121-69-7	: 713	123-75-1	: 11143	127-71-9	: 603	135-48-8	: 8636	141-82-2	: 9921
121-71-1	: 5831	123-76-2	: 8903	127-77-9	: 2199	135-58-0	: 11661	141-83-3	: 11993
121-72-2	: 861	123-77-3	: 5124	127-91-3	: 2925	135-67-1	: 9507	141-84-4	: 11685
121-73-3	: 1215	123-80-8	: 5576	128-04-1	: 11514	135-68-2	: 3012	141-86-6	: 10925
121-75-5	: 3316	123-82-0	: 6350	128-08-5	: 11153	135-70-6	: 11194	141-90-2	: 11110
121-79-9	: 2732	123-83-1	: 5510	128-09-6	: 11154	135-88-6	: 7728	141-91-3	: 7687
121-81-3	: 614	123-84-2	: 10340	128-13-2	: 4098	135-97-7	: 486	141-93-5	: 1414
121-82-4	: 11833	123-86-4	: 121	128-20-1	: 9745	135-98-8	: 2035	141-97-9	: 3571
121-86-8	: 1210	123-88-6	: 11822	128-23-4	: 9743	136-35-5	: 11826	142-04-1	: 756
121-89-1	: 5860	123-90-0	: 11719	128-27-8	: 5381	136-44-7	: 10188	142-08-5	: 11033
121-90-4	: 2905	123-91-1	: 5187	128-33-6	: 4104	136-47-0	: 2531	142-09-6	: 10660
121-91-5	: 1328	123-92-2	: 3631	128-37-0	: 9224	136-60-7	: 2532	142-26-7	: 65
121-92-6	: 2704	123-93-3	: 240	128-39-2	: 9220	136-72-1	: 8678	142-28-9	: 9893
121-98-2	: 2675	123-95-5	: 8251	128-50-7	: 2969	136-77-6	: 1519	142-29-0	: 4903
122-00-9	: 5853	123-98-8	: 8138	128-53-0	: 11122	136-78-7	: 11358	142-30-3	: 7119
122-01-0	: 2892	123-99-9	: 8130	128-62-1	: 7356	136-79-8	: 739	142-47-2	: 6248
122-03-2	: 590	124-02-7	: 10526	128-68-7	: 2979	136-80-1	: 5744	142-50-7	: 5332
122-04-3	: 2906	124-04-9	: 6779	129-00-0	: 10853	136-95-8	: 2859	142-60-9	: 10313
122-07-6	: 5420	124-06-1	: 11624	129-43-1	: 407	137-00-8	: 11677	142-61-0	: 6986
122-09-8	: 1572	124-07-2	: 8422	129-44-2	: 389	137-04-2	: 665	142-62-1	: 6870
122-11-2	: 2194	124-09-4	: 6749	129-64-6	: 7580	137-05-3	: 10624	142-64-3	: 9607
122-14-5	: 5951	124-10-7	: 11625	129-66-8	: 2735	137-06-4	: 2301	142-68-7	: 10800
122-20-3	: 10393	124-11-8	: 8182	129-79-3	: 5979	137-07-5	: 2290	142-73-4	: 6260
122-25-8	: 2686	124-12-9	: 8411	130-15-4	: 7843	137-16-6	: 6269	142-77-8	: 8298
122-28-1	: 82	124-13-0	: 8346	130-16-5	: 11311	137-17-7	: 858	142-82-5	: 6355
122-32-7	: 8310	124-17-4	: 5689	130-26-7	: 11312	137-19-9	: 1510	142-83-6	: 6675
122-34-9	: 11513	124-18-5	: 5033	130-73-4	: 9299	137-26-8	: 11722	142-84-7	: 9826
122-35-0	: 10505	124-19-6	: 8117	130-80-3	: 9301	137-30-4	: 12068	142-91-6	: 6654
122-37-2	: 9447	124-21-0	: 5425	130-85-8	: 7777	137-32-6	: 3626	142-92-7	: 194
122-39-4	: 836	124-22-1	: 5291	130-86-9	: 3109	137-40-6	: 10327	142-96-1	: 3415
122-40-7	: 6347	124-26-5	: 8231	130-89-2	: 4130	137-43-9	: 4777	143-07-7	: 5305
122-42-9	: 3966	124-30-1	: 8233	130-90-5	: 4129	137-58-6	: 54	143-08-8	: 8157
122-46-3	: 212	124-38-9	: 3992	130-95-0	: 4127	137-41-0	: 2508	143-10-2	: 5058
122-48-5	: 3663	124-40-3	: 7448	131-01-1	: 12065	138-52-3	: 6222	143-13-5	: 220
122-51-0	: 5604	124-42-5	: 5671	131-09-9	: 388	138-53-4	: 955	143-15-7	: 5293
122-52-1	: 9591	124-47-0	: 12021	131-11-3	: 1343	138-86-3	: 4677	143-16-8	: 6727
122-57-6	: 3881	124-48-1	: 7493	131-28-2	: 2586	138-87-4	: 4542	143-22-6	: 5690
122-59-8	: 228	124-58-3	: 458	131-53-3	: 7619	138-89-6	: 720	143-24-8	: 8995
122-60-1	: 8623	124-63-0	: 7535	131-54-4	: 7595	139-13-9	: 6256	143-27-1	: 6631
122-62-3	: 5043	124-68-5	: 10341	131-56-8	: 7609	139-40-2	: 10508	143-28-2	: 8312
122-66-7	: 7158	124-70-9	: 11396	131-57-7	: 7620	139-45-7	: 10198	143-50-0	: 7408
122-69-0	: 10691	124-73-2	: 5525	131-58-8	: 7631	139-65-1	: 850	143-62-4	: 4044
122-70-3	: 10324	124-80-1	: 4067	131-73-7	: 864	139-66-2	: 2283	143-74-8	: 9363
122-73-6	: 1951	124-83-4	: 4794	131-89-5	: 9273	139-85-5	: 542	144-19-4	: 8795
122-78-1	: 868	124-87-8	: 7579	131-91-9	: 8035	139-87-7	: 5723	144-49-0	: 189
122-79-2	: 231	125-15-5	: 11351	132-57-0	: 7972	139-93-5	: 9275	144-55-8	: 4004
122-80-5	: 29	125-20-2	: 7352	132-60-5	: 11235	140-08-9	: 5702	144-62-7	: 5553
122-84-9	: 10461	125-24-6	: 3002	132-64-9	: 5163	140-10-3	: 10682	144-68-3	: 4057
122-87-2	: 6267	125-28-0	: 7674	132-65-0	: 5167	140-11-4	: 233	144-79-6	: 11380
122-88-3	: 146	125-29-1	: 7678	132-66-1	: 2702	140-18-1	: 148	144-80-9	: 30
122-95-2	: 1408	125-33-7	: 11066	132-75-2	: 7736	140-22-7	: 4010	144-83-2	: 2203
122-97-4	: 2148	125-55-3	: 11079	132-86-5	: 7834	140-25-0	: 5318	145-13-1	: 9756
122-98-5	: 5759	126-00-1	: 1103	133-06-2	: 7368	140-28-3	: 5500	145-73-8	: 5370
122-99-6	: 5758	126-07-8	: 11520	133-07-3	: 5984	140-29-4	: 945	145-94-8	: 7283
123-00-2	: 7701	126-11-4	: 10012	133-08-4	: 9930	140-31-8	: 9615	146-06-5	: 4128
123-01-3	: 1566	126-14-7	: 6229	133-11-9	: 2502	140-39-6	: 213	146-48-5	: 12062
123-02-4	: 2342	126-17-0	: 11529	133-13-1	: 9949	140-40-9	: 85	147-71-7	: 3312
123-03-5	: 11030	126-18-1	: 11533	133-14-2	: 9152	140-55-6	: 9718	147-73-9	: 3304
123-04-6	: 6368	126-19-2	: 11531	133-26-6	: 6188	140-57-8	: 11579	147-82-0	: 851
123-05-7	: 6719	126-29-4	: 4056	133-32-4	: 7296	140-65-8	: 7683	147-85-3	: 9764
123-06-8	: 9914	126-30-7	: 10005	133-37-9	: 3311	140-67-0	: 1943	147-93-3	: 2663
123-07-9	: 9355	126-33-0	: 11786	133-53-9	: 9290	140-76-1	: 10955	148-03-8	: 2789
123-08-0	: 566	126-54-5	: 11643	133-58-4	: 7437	140-77-2	: 4847		

CAS	No.	CAS	No.	CAS	No.	CAS	No.	CAS	No.
503-66-2	10267	512-04-9	11536	521-04-0	11554	529-86-2	432	537-29-1	916
503-74-2	3539	512-12-9	8585	521-11-9	350	529-92-0	11216	537-45-1	4279
503-76-4	9869	512-13-0	2941	521-12-0	351	530-14-3	5818	537-46-2	1573
503-80-0	452	512-16-3	4309	521-18-6	349	530-22-3	2464	537-47-3	7146
503-93-5	4220	512-48-1	9003	521-24-4	7970	530-40-5	10892	537-49-5	11933
504-02-9	4376	512-56-1	9564	521-31-3	9600	530-43-8	6648	537-64-4	7432
504-03-0	9650	512-60-7	11862	521-35-7	5154	530-44-9	7613	537-65-5	1263
504-15-4	1531	512-69-6	6219	521-74-4	11313	530-48-3	1636	537-91-7	5247
504-20-1	6340	512-85-6	5184	521-85-7	3108	530-50-7	7157	537-92-8	77
504-24-5	10859	513-02-0	9568	522-12-3	2796	530-55-2	4262	538-08-9	4159
504-29-0	10857	513-08-6	9567	522-27-0	5583	530-62-1	7193	538-23-8	8440
504-31-4	10788	513-12-2	5725	522-57-6	7611	530-78-9	2729	538-24-9	5319
504-53-0	9000	513-23-5	2985	522-66-7	4123	530-85-8	10740	538-32-9	12027
504-57-4	8105	513-31-5	10545	522-75-8	2404	530-91-6	8043	538-39-6	1590
504-61-0	3859	513-35-9	3801	523-27-3	375	530-93-8	8047	538-41-0	635
504-63-2	9983	513-36-0	9862	523-31-9	1337	531-02-2	3097	538-43-2	10026
504-64-3	9773	513-37-1	10539	523-47-7	7898	531-18-0	7584	538-44-3	9106
504-78-9	11681	513-38-2	10057	523-80-8	2423	531-29-3	6226	538-49-8	11000
505-20-4	5261	513-42-8	10702	524-15-2	6179	531-37-3	9383	538-51-2	840
505-22-6	5186	513-44-0	10158	524-30-1	2823	531-39-5	8886	538-56-7	10684
505-23-7	5262	513-81-5	3159	524-34-5	11236	531-75-9	2822	538-58-9	8683
505-29-3	5263	513-86-0	3661	524-36-7	10978	531-81-7	2775	538-62-5	5122
505-32-8	6671	513-88-2	10425	524-40-3	10883	531-84-0	7774	538-64-7	3768
505-34-0	10067	513-92-8	5919	524-42-5	7842	531-91-9	3035	538-68-1	2107
505-48-6	8386	513-96-2	10115	524-63-0	4119	531-95-3	2786	538-74-9	2287
505-60-2	5645	514-10-3	9167	524-80-1	3982	532-18-3	7727	538-75-0	4291
505-84-0	10074	514-12-5	2953	524-96-9	1797	532-24-1	490	538-81-8	1095
505-92-0	5338	514-92-1	4059	525-06-4	5928	532-27-4	5790	538-86-3	1930
506-03-6	10010	515-12-8	4401	525-15-5	11041	532-32-1	2725	538-93-2	2039
506-12-7	6314	515-25-3	11178	525-37-1	7868	532-48-9	5825	539-03-7	48
506-23-0	8284	515-30-0	914	525-64-4	5971	532-54-7	876	539-08-2	9796
506-24-1	8316	515-40-2	1143	525-68-8	11253	532-82-1	1284	539-15-1	9320
506-30-9	5355	515-46-8	2223	525-82-6	2840	532-96-7	2717	539-23-1	10860
506-31-0	5364	515-49-1	2189	526-08-9	2201	533-18-6	211	539-30-0	1658
506-32-1	5361	515-59-3	2200	526-31-8	11923	533-24-4	1544	539-47-9	10638
506-33-2	5282	515-82-2	6000	526-35-2	8586	533-30-2	2860	539-52-6	6133
506-42-3	8311	515-83-3	5769	526-55-6	7305	533-32-4	3201	539-74-2	10208
506-46-7	6624	515-84-4	248	526-73-8	2357	533-45-9	11670	539-80-0	4242
506-48-9	8221	515-88-8	2944	526-75-0	9307	533-48-2	7223	539-82-2	8878
506-50-3	11820	515-90-2	2964	526-84-1	3777	533-50-6	3679	539-86-6	10595
506-52-5	6626	515-94-6	326	526-85-2	9493	533-58-4	9371	539-88-8	8907
506-68-3	4165	515-96-8	105	526-86-3	4263	533-68-6	3469	539-89-9	3961
506-77-4	4166	516-05-2	9960	526-95-4	6209	533-70-0	701	539-90-2	3563
506-78-5	7337	516-06-3	12040	526-99-8	6193	533-73-3	2367	539-92-4	3997
506-80-9	4036	516-12-1	11159	527-07-1	6213	533-75-5	4243	540-07-8	6907
506-82-1	3929	516-78-9	5384	527-17-3	4270	533-86-8	8300	540-08-9	6321
506-85-4	6024	516-85-8	5380	527-18-4	1552	533-98-2	3265	540-09-0	11857
506-89-8	12020	516-95-0	4108	527-35-5	9475	534-00-9	3238	540-10-3	6651
506-93-4	6286	517-04-4	5401	527-52-6	7111	534-07-6	10426	540-18-1	3579
506-96-7	291	517-06-6	5407	527-53-7	2279	534-13-4	11804	540-36-3	1424
507-02-8	302	517-07-7	5402	527-55-9	1513	534-15-6	5548	540-37-4	761
507-09-5	5646	517-09-9	5397	527-60-6	9495	534-22-5	6130	540-38-5	9373
507-19-7	9846	517-10-2	4111	527-61-7	4264	534-26-9	7197	540-42-1	10310
507-20-0	9863	517-22-6	11126	527-62-8	9205	534-52-1	9405	540-47-6	4980
507-25-5	7542	517-23-7	6139	527-69-5	6040	534-59-8	9929	540-51-2	5685
507-32-4	457	517-25-9	7562	527-72-0	11737	534-84-9	4611	540-54-5	9848
507-36-8	3240	517-28-2	629	527-73-1	7211	534-85-0	1282	540-61-4	272
507-40-4	7183	517-60-2	1748	527-84-4	1988	535-11-5	10207	540-63-6	5590
507-42-6	5581	517-66-8	2468	527-85-5	619	535-13-7	10220	540-67-0	5602
507-45-9	3286	517-82-8	7857	527-89-9	1125	535-15-9	164	540-69-2	6003
507-47-1	5680	517-92-0	401	527-93-5	5153	535-46-6	11548	540-73-8	7155
507-60-8	3142	517-97-5	5159	528-21-2	5873	535-75-1	9640	540-80-7	8085
508-02-1	8547	518-05-8	7805	528-23-4	4784	535-77-3	1989	540-82-9	11576
508-32-7	11867	518-17-2	7333	528-29-0	1491	535-80-8	2535	540-84-1	8848
508-44-1	5185	518-28-5	6192	528-44-9	2319	535-83-1	11029	540-88-5	177
508-54-3	7675	518-44-3	2572	528-45-0	2590	535-87-5	2554	540-97-6	4624
508-65-6	4068	518-61-6	9502	528-48-3	2813	535-89-7	11043	541-01-5	6528
509-12-6	2938	518-69-4	5149	528-53-0	2849	536-08-3	2575	541-02-6	4901
509-14-8	7544	518-75-2	2774	528-71-2	9197	536-17-4	11684	541-05-9	5002
509-15-9	6203	518-77-4	5151	528-75-6	551	536-25-4	2501	541-16-2	9925
509-20-6	305	518-82-1	419	528-76-7	2178	536-33-4	10885	541-20-8	8738
509-64-8	7670	519-02-8	7426	528-79-0	9420	536-38-9	5778	541-28-6	3384
509-87-5	11096	519-05-1	2608	528-81-4	487	536-50-5	1859	541-31-1	3440
510-15-6	4082	519-09-5	481	529-00-0	4602	536-57-2	2182	541-33-3	3276
510-20-3	9939	519-34-6	7610	529-08-8	506	536-59-4	4664	541-35-5	3194
510-90-7	11063	519-37-9	10758	529-16-8	2968	536-60-7	1902	541-41-3	4018
511-07-9	5392	519-44-8	1517	529-17-9	484	536-66-3	2689	541-42-4	8092
511-08-0	5391	519-72-2	1804	529-19-1	2763	536-69-6	10898	541-46-8	3206
511-09-1	5389	519-73-3	1982	529-20-4	587	536-74-3	1715	541-47-9	3846
511-10-4	5390	519-87-9	57	529-21-5	10964	536-75-4	10958	541-50-4	3568
511-18-2	8210	520-03-6	7366	529-28-2	1775	536-78-7	10957	541-53-7	11718
511-20-6	5377	520-18-3	2843	529-33-9	8041	536-88-9	10965	541-58-2	11674
511-21-7	5378	520-26-3	2794	529-34-0	8046	536-89-0	7164	541-59-3	11121
511-34-2	11535	520-33-2	2800	529-35-1	8042	536-90-3	767	541-73-1	1355
511-67-1	11086	520-36-5	2806	529-36-2	8007	536-95-8	2721	541-88-8	124
511-70-6	3198	520-45-6	10779	529-64-6	915	537-17-7	11830	541-91-3	4763
511-96-6	11530	520-85-4	9752	529-65-7	62	537-26-8	488	542-05-2	8782

542-08-5	: 3574	551-09-7	: 5514	557-89-1	: 453	573-33-1	: 7199	583-53-9	: 1288
542-10-9	: 5570	551-62-2	: 2270	557-91-5	: 5517	573-56-8	: 9344	583-55-1	: 1046
542-15-4	: 821	551-76-8	: 9489	557-93-7	: 10530	573-58-0	: 7969	583-58-4	: 10940
542-18-7	: 4339	551-77-9	: 9488	557-98-2	: 10535	573-97-7	: 8022	583-61-9	: 10936
542-28-9	: 10798	551-78-0	: 9487	557-99-3	: 301	573-98-8	: 7818	583-68-6	: 647
542-37-0	: 3546	551-88-2	: 8834	558-13-4	: 7539	574-09-4	: 5813	583-69-7	: 1503
542-46-1	: 4218	551-92-8	: 7202	558-17-8	: 10058	574-12-9	: 2841	583-70-0	: 1018
542-50-7	: 6303	552-16-9	: 2703	558-25-8	: 7538	574-39-0	: 3983	583-71-1	: 1022
542-52-9	: 3998	552-30-7	: 7341	558-30-5	: 8807	574-84-5	: 2808	583-75-5	: 650
542-54-1	: 8831	552-32-9	: 81	558-37-2	: 3760	575-37-1	: 7823	583-78-8	: 9284
542-55-2	: 6013	552-41-0	: 5823	558-38-3	: 10352	575-38-2	: 7837	583-86-8	: 8276
542-56-3	: 8094	552-45-4	: 1198	558-42-9	: 10354	575-41-7	: 7819	584-02-1	: 8917
542-58-5	: 5695	552-58-9	: 2812	558-43-0	: 10017	575-43-9	: 7822	584-04-3	: 10371
542-59-6	: 5577	552-59-0	: 2828	559-70-6	: 8548	575-44-0	: 7836	584-12-3	: 6033
542-69-8	: 3381	552-70-5	: 479	560-08-7	: 3931	575-74-6	: 607	584-27-0	: 3331
542-76-7	: 10097	552-72-7	: 9646	560-21-4	: 8849	575-22-4	: 1020	584-70-3	: 621
542-81-4	: 5481	552-79-4	: 1863	560-22-5	: 9068	578-23-8	: 1017	584-79-2	: 336
542-85-8	: 5600	552-82-9	: 807	560-23-6	: 9065	578-24-9	: 9282	584-84-9	: 1431
542-88-1	: 7525	552-89-6	: 591	560-95-2	: 7476	578-26-1	: 9310	584-87-2	: 2609
542-90-5	: 11703	552-94-3	: 2616	561-07-9	: 311	576-55-6	: 9468	584-90-7	: 5114
542-91-6	: 11409	553-03-7	: 11327	561-20-6	: 11352	576-83-0	: 1094	584-92-9	: 8399
542-92-7	: 4764	553-17-3	: 9384	561-25-1	: 7664	577-16-2	: 5851	584-94-1	: 6772
543-20-4	: 3366	553-19-5	: 2397	561-27-3	: 7667	577-19-5	: 1075	584-98-5	: 3299
543-21-5	: 3902	553-20-8	: 84	561-83-1	: 11090	577-27-5	: 4989	584-99-6	: 3769
543-24-8	: 6251	553-26-4	: 3104	561-94-4	: 5393	577-33-3	: 429	585-32-0	: 1806
543-27-1	: 4024	553-39-9	: 7953	562-49-2	: 8763	577-37-7	: 7572	585-34-2	: 9322
543-28-2	: 3959	553-60-6	: 10907	562-74-3	: 4699	577-55-9	: 992	585-48-8	: 10874
543-29-3	: 8078	553-69-5	: 1913	563-03-1	: 9387	577-56-0	: 2476	585-68-2	: 6054
543-38-4	: 7135	553-79-7	: 827	563-04-2	: 9570	577-59-3	: 5863	585-74-0	: 5852
543-59-9	: 8719	553-82-2	: 1381	563-12-2	: 5931	577-71-9	: 9345	585-76-2	: 2519
543-67-9	: 8098	553-86-6	: 2461	563-16-6	: 6775	577-91-3	: 2135	585-79-5	: 1076
543-75-9	: 5203	553-90-2	: 5563	563-41-7	: 7143	578-07-4	: 312	585-81-9	: 2632
543-80-6	: 110	553-94-6	: 1021	563-54-1	: 3800	578-54-4	: 740	586-11-8	: 9346
543-83-9	: 6283	553-97-9	: 4266	563-46-2	: 3799	578-57-4	: 1050	586-27-6	: 4698
543-86-2	: 3633	554-00-7	: 686	563-47-3	: 10540	578-58-5	: 1931	586-37-8	: 5840
543-87-3	: 3637	554-01-8	: 11106	563-52-0	: 3723	578-66-5	: 11206	586-38-9	: 2665
544-00-3	: 3226	554-12-1	: 10296	563-57-5	: 10553	578-67-6	: 11305	586-39-0	: 1644
544-01-4	: 3418	554-14-3	: 11769	563-58-6	: 10547	578-68-7	: 11203	586-42-5	: 2477
544-02-5	: 3436	554-21-2	: 5212	563-70-2	: 7473	578-76-7	: 10769	586-61-8	: 1059
544-10-5	: 6737	554-35-8	: 10104	563-78-0	: 3759	579-04-4	: 7112	586-62-9	: 4684
544-13-8	: 8787	554-68-7	: 5418	563-79-1	: 3761	579-07-7	: 10028	586-75-4	: 2888
544-16-1	: 8083	554-70-1	: 9521	563-80-4	: 3668	579-10-2	: 79	586-76-5	: 2520
544-25-2	: 4240	554-84-7	: 9436	563-83-7	: 9798	579-18-0	: 2511	586-77-6	: 645
544-35-4	: 8226	555-03-3	: 1936	563-84-8	: 3606	579-21-5	: 5854	586-78-7	: 1077
544-40-1	: 3433	555-10-2	: 4673	564-00-1	: 3007	579-44-2	: 5822	586-84-5	: 6129
544-62-7	: 10023	555-16-8	: 593	564-02-3	: 8847	579-60-2	: 9392	586-89-0	: 2478
544-63-8	: 11622	555-21-5	: 967	564-03-4	: 9069	579-66-8	: 694	586-95-8	: 10977
544-65-0	: 11637	555-30-6	: 11931	564-04-5	: 8966	579-75-9	: 2684	586-96-9	: 2059
544-66-1	: 11638	555-35-1	: 6645	564-36-3	: 5387	579-92-0	: 2656	586-98-1	: 10975
544-72-9	: 8286	555-43-1	: 8272	564-37-4	: 5388	579-94-2	: 183	587-02-0	: 741
544-73-0	: 8282	555-44-2	: 6656	565-00-4	: 2921	580-02-9	: 2482	587-03-1	: 1898
544-76-3	: 6632	555-45-3	: 11628	565-53-7	: 161	580-13-2	: 7740	587-04-2	: 530
544-77-4	: 6640	555-57-7	: 1823	565-59-3	: 8761	580-15-4	: 11205	587-54-2	: 2513
544-85-4	: 5348	555-59-9	: 3853	565-60-6	: 8941	580-16-5	: 11306	587-61-1	: 10873
544-97-8	: 12066	555-77-1	: 5412	565-63-9	: 3845	580-20-1	: 11307	587-85-9	: 7436
545-06-2	: 288	555-89-5	: 1965	565-64-0	: 10245	580-22-3	: 11202	587-90-6	: 11997
545-26-6	: 4048	556-10-5	: 12025	565-67-3	: 8943	580-48-3	: 11828	588-01-2	: 11906
545-47-1	: 7417	556-22-9	: 6275	565-69-5	: 8982	580-51-8	: 3089	588-46-5	: 87
545-93-7	: 11078	556-24-1	: 3552	565-75-3	: 8850	581-08-8	: 59	588-47-6	: 817
546-06-5	: 4143	556-27-4	: 333	565-76-4	: 9066	581-40-8	: 7825	588-52-3	: 5113
546-33-8	: 5004	556-50-3	: 6265	565-77-5	: 9070	581-42-0	: 7826	588-63-6	: 1085
546-43-0	: 8059	556-52-5	: 8618	565-80-0	: 8967	581-43-1	: 7838	588-67-0	: 1115
546-45-2	: 5005	556-56-9	: 10573	567-18-0	: 7973	581-45-3	: 11009	588-72-7	: 1029
546-56-5	: 4994	556-61-6	: 7521	568-02-5	: 8062	581-46-4	: 3102	588-73-8	: 1031
546-97-4	: 5926	556-64-9	: 11705	568-21-8	: 5158	581-47-5	: 3101	588-96-5	: 1032
547-25-1	: 6022	556-67-2	: 4993	568-93-4	: 399	581-49-7	: 3105	589-08-2	: 1581
547-44-4	: 2190	556-68-3	: 4741	568-99-0	: 383	581-50-0	: 3100	589-09-3	: 841
547-52-4	: 2188	556-69-4	: 8482	569-08-4	: 394	581-55-5	: 7512	589-10-6	: 1033
547-58-0	: 2222	556-70-7	: 5079	569-31-3	: 7355	581-89-5	: 7930	589-14-0	: 6120
547-63-7	: 10301	556-71-8	: 4726	569-34-6	: 10759	582-08-1	: 5128	589-15-1	: 1000
547-65-9	: 6147	556-82-1	: 3869	569-41-5	: 7824	582-16-1	: 7827	589-16-2	: 742
547-81-9	: 5405	556-88-7	: 6287	569-51-2	: 2318	582-17-3	: 7839	589-17-3	: 1008
547-91-1	: 11295	556-89-8	: 12024	569-57-3	: 1154	582-22-9	: 1579	589-18-4	: 1899
548-40-3	: 8631	556-96-7	: 1019	569-64-2	: 7455	582-24-1	: 5833	589-21-9	: 7141
548-51-6	: 2642	557-00-6	: 3562	569-65-3	: 9605	582-33-2	: 2496	589-29-7	: 1435
548-62-9	: 7453	557-05-1	: 8277	569-77-7	: 2408	582-60-5	: 2380	589-33-3	: 11113
548-80-1	: 7878	557-11-9	: 12028	570-08-1	: 9923	582-61-0	: 2884	589-38-8	: 6960
548-98-1	: 4093	557-17-5	: 10069	570-22-9	: 7195	582-62-7	: 3671	589-40-2	: 6012
550-10-7	: 5214	557-24-4	: 3817	570-24-1	: 802	582-77-4	: 622	589-43-5	: 6773
550-24-3	: 4261	557-25-5	: 3505	571-58-4	: 7820	582-78-5	: 623	589-44-6	: 3465
550-44-7	: 7365	557-30-2	: 5492	571-61-9	: 7821	583-03-9	: 1846	589-53-7	: 6424
550-60-7	: 8034	557-31-3	: 10566	572-59-8	: 4125	583-04-0	: 2722	589-55-9	: 6475
550-74-3	: 10848	557-40-4	: 10592	572-83-8	: 386	583-05-1	: 8804	589-59-3	: 3555
550-82-3	: 9508	557-59-5	: 11607	572-93-0	: 396	583-33-5	: 6052	589-63-9	: 8473
550-97-0	: 2644	557-61-9	: 8222	573-17-1	: 9164	583-39-1	: 2386	589-75-3	: 8427
551-01-9	: 9877	557-66-4	: 5433	573-20-6	: 7841	583-48-2	: 6776	589-87-7	: 1048
551-06-4	: 7906	557-68-6	: 7471	573-26-2	: 75	583-50-6	: 3192	589-92-4	: 4598

589-93-5	: 10938	593-59-9	: 444	597-90-0	: 6487	603-39-4	: 2110	608-93-5	: 2084
589-98-0	: 8445	593-60-2	: 5880	597-93-3	: 8162	603-45-2	: 4278	608-96-8	: 2096
590-01-2	: 10214	593-61-3	: 5937	597-96-6	: 6946	603-50-9	: 9465	608-98-0	: 1553
590-02-3	: 125	593-63-5	: 5938	598-01-6	: 6505	603-52-1	: 3946	609-02-9	: 9963
590-11-4	: 5892	593-66-8	: 5907	598-02-7	: 9558	603-54-3	: 12011	609-08-5	: 9962
590-12-5	: 5891	593-70-4	: 7481	598-03-8	: 10131	603-69-0	: 3462	609-09-6	: 9973
590-13-6	: 10529	593-71-5	: 7482	598-04-9	: 3423	603-71-4	: 2363	609-11-0	: 3498
590-16-9	: 5463	593-74-8	: 7435	598-05-0	: 11575	603-76-9	: 7312	609-12-1	: 3475
590-18-1	: 3706	593-75-9	: 7655	598-09-4	: 8606	603-81-6	: 2552	609-13-2	: 3477
590-19-2	: 3145	593-77-1	: 7447	598-16-3	: 5921	603-83-8	: 799	609-14-3	: 3557
590-26-1	: 5904	593-79-3	: 7526	598-17-4	: 9880	603-85-0	: 9214	609-19-8	: 9486
590-35-2	: 8760	593-81-7	: 7445	598-19-6	: 10699	604-53-5	: 3004	609-22-3	: 9279
590-36-3	: 8940	593-88-4	: 449	598-21-0	: 292	604-61-5	: 2515	609-23-4	: 9492
590-42-1	: 10066	593-90-8	: 3119	598-22-1	: 10490	604-68-2	: 6216	609-25-6	: 2369
590-50-1	: 8965	593-92-0	: 5890	598-23-2	: 3916	604-69-3	: 6217	609-26-7	: 8813
590-60-3	: 3634	593-95-3	: 4006	598-25-4	: 3170	604-83-1	: 9172	609-29-0	: 8803
590-66-9	: 4364	593-96-4	: 5455	598-31-2	: 10416	604-88-6	: 1756	609-31-4	: 3640
590-67-0	: 4530	593-98-6	: 7466	598-38-9	: 5704	605-01-6	: 2090	609-36-9	: 9765
590-73-8	: 6771	594-02-5	: 5546	598-45-8	: 10063	605-02-7	: 7947	609-39-2	: 6134
590-86-3	: 3186	594-03-6	: 5589	598-50-5	: 12017	605-03-8	: 11290	609-40-5	: 11775
590-90-9	: 3662	594-04-7	: 7503	598-53-8	: 10070	605-27-6	: 411	609-46-1	: 2225
590-92-1	: 10205	594-08-1	: 4034	598-55-0	: 3953	605-32-3	: 408	609-65-4	: 2890
590-93-2	: 3917	594-09-2	: 9522	598-56-1	: 5421	605-37-8	: 7852	609-72-3	: 860
591-07-1	: 27	594-10-5	: 11550	598-57-2	: 7451	605-39-0	: 3050	609-73-4	: 1783
591-08-2	: 33	594-11-6	: 4981	598-58-3	: 8075	605-42-5	: 390	610-02-6	: 2730
591-09-3	: 108	594-15-0	: 7549	598-61-8	: 4192	605-50-5	: 1335	610-15-1	: 624
591-11-7	: 6154	594-16-1	: 9883	598-74-3	: 3223	605-54-9	: 1333	610-16-2	: 2584
591-12-8	: 6153	594-18-3	: 7495	598-76-5	: 9899	605-63-0	: 7786	610-17-3	: 718
591-17-3	: 1055	594-20-7	: 9894	598-77-6	: 10172	605-65-2	: 7977	610-25-3	: 2047
591-18-4	: 1047	594-22-9	: 7556	598-78-7	: 10215	605-71-0	: 7832	610-34-4	: 2707
591-19-5	: 639	594-27-4	: 11544	598-82-3	: 10266	605-79-8	: 7634	610-39-9	: 1960
591-20-8	: 9229	594-34-3	: 9889	598-88-9	: 5898	605-81-2	: 6228	610-46-8	: 377
591-22-0	: 10941	594-36-5	: 3253	598-92-5	: 5482	605-82-3	: 421	610-48-0	: 423
591-27-5	: 9202	594-37-6	: 9900	598-94-7	: 12005	605-83-4	: 420	610-50-4	: 431
591-31-1	: 584	594-38-7	: 3385	598-96-9	: 9072	605-85-6	: 5859	610-66-2	: 965
591-34-4	: 10309	594-39-8	: 3222	598-98-1	: 10260	605-99-2	: 10765	610-67-3	: 1662
591-35-5	: 9287	594-42-3	: 7527	598-99-2	: 253	606-17-7	: 2594	610-69-5	: 219
591-47-9	: 4672	594-44-5	: 5627	599-01-9	: 10333	606-18-8	: 2506	610-71-9	: 2556
591-49-1	: 4670	594-45-6	: 5622	599-04-2	: 6143	606-20-2	: 1958	610-72-0	: 2581
591-50-4	: 1766	594-47-8	: 94	599-66-6	: 2240	606-21-3	: 1145	610-89-9	: 9074
591-60-6	: 3569	594-51-4	: 3274	599-70-2	: 1704	606-22-4	: 726	610-91-3	: 2362
591-62-8	: 3936	594-52-5	: 3236	599-79-1	: 2654	606-23-5	: 7264	610-97-9	: 2660
591-68-4	: 8862	594-56-9	: 3812	599-91-7	: 2233	606-27-9	: 2710	610-99-1	: 10449
591-76-4	: 6834	594-57-0	: 3247	599-97-3	: 5657	606-28-0	: 2516	611-00-7	: 2555
591-78-6	: 6959	594-58-1	: 10228	599-99-5	: 242	606-35-9	: 1946	611-01-8	: 2580
591-80-0	: 9073	594-60-5	: 3618	600-00-0	: 10213	606-37-1	: 7831	611-03-0	: 2553
591-82-2	: 10065	594-61-6	: 10278	600-05-5	: 10238	606-43-9	: 11328	611-06-3	: 1396
591-84-4	: 11707	594-65-0	: 95	600-06-6	: 3245	606-45-1	: 2673	611-07-4	: 9269
591-87-7	: 235	594-70-7	: 10085	600-07-7	: 3538	606-46-2	: 696	611-10-9	: 4785
591-93-5	: 8659	594-71-8	: 9870	600-11-3	: 8749	606-81-5	: 3063	611-12-1	: 10366
591-95-7	: 8658	594-72-9	: 5536	600-13-5	: 4150	606-88-2	: 7173	611-13-2	: 6059
591-96-8	: 8660	594-73-0	: 5594	600-14-6	: 8797	606-97-3	: 3093	611-14-3	: 1685
592-01-8	: 3930	594-81-0	: 3271	600-15-7	: 3530	606-99-5	: 751	611-15-4	: 1640
592-13-2	: 6774	594-82-1	: 3430	600-17-9	: 3842	607-00-1	: 5993	611-19-8	: 1132
592-20-1	: 10412	594-83-2	: 3644	600-18-0	: 3567	607-31-8	: 11274	611-20-1	: 2759
592-22-3	: 5250	594-89-8	: 10048	600-22-6	: 10316	607-32-9	: 7386	611-21-2	: 710
592-27-8	: 6422	595-30-2	: 8559	600-25-9	: 9866	607-34-1	: 11284	611-32-5	: 11283
592-34-7	: 4012	595-37-9	: 3506	600-30-6	: 3497	607-42-1	: 368	611-33-6	: 11242
592-35-8	: 3935	595-38-0	: 8724	600-36-2	: 8931	607-50-1	: 7919	611-34-7	: 11204
592-41-6	: 6999	595-39-1	: 7400	600-40-8	: 5551	607-52-3	: 7939	611-35-8	: 11238
592-42-7	: 6678	595-40-4	: 7401	601-75-2	: 9948	607-53-4	: 8005	611-43-8	: 3051
592-44-9	: 6676	595-41-5	: 8930	601-76-3	: 9918	607-56-7	: 12008	611-45-0	: 7949
592-45-0	: 6677	595-44-8	: 9902	601-77-4	: 9822	607-81-8	: 9977	611-49-4	: 7928
592-46-1	: 6680	595-49-3	: 9920	601-84-3	: 2557	607-85-2	: 2635	611-61-0	: 3053
592-49-4	: 6679	595-89-1	: 9733	601-88-7	: 1393	607-86-3	: 2682	611-62-1	: 3065
592-50-7	: 8817	595-90-4	: 11545	602-03-9	: 724	607-90-9	: 2652	611-69-8	: 1901
592-51-8	: 9062	596-27-0	: 7353	602-21-1	: 1353	607-91-0	: 2428	611-70-1	: 10469
592-55-2	: 5461	596-29-2	: 7358	602-27-7	: 1489	607-97-6	: 3518	611-73-4	: 934
592-57-4	: 4248	596-43-0	: 1061	602-29-9	: 2046	607-99-8	: 2313	611-79-0	: 7589
592-65-4	: 10154	596-51-0	: 11179	602-37-9	: 7787	608-07-1	: 7304	611-92-7	: 12006
592-76-7	: 6533	596-75-8	: 9937	602-38-0	: 7833	608-21-9	: 2307	611-94-9	: 7625
592-80-3	: 8935	597-04-6	: 3511	602-55-1	: 428	608-23-1	: 1137	611-95-0	: 2512
592-82-5	: 3388	597-05-7	: 8934	602-56-2	: 318	608-25-3	: 1529	611-97-2	: 7599
592-84-7	: 6004	597-09-1	: 10008	602-60-8	: 426	608-26-4	: 9280	611-99-4	: 7597
592-88-1	: 10599	597-12-6	: 6221	602-64-2	: 415	808-27-5	: 685	612-00-0	: 1677
592-90-5	: 8590	597-32-0	: 10365	602-65-3	: 418	608-28-6	: 1772	612-01-1	: 12018
593-03-3	: 6661	597-33-1	: 9998	602-82-4	: 1332	608-29-7	: 2349	612-08-8	: 5109
593-08-8	: 11889	597-35-3	: 5626	602-92-6	: 2560	608-30-0	: 682	612-12-4	: 986
593-12-4	: 8358	597-49-9	: 8932	602-96-0	: 2364	608-31-1	: 688	612-13-5	: 2750
593-14-6	: 8361	597-50-2	: 9517	602-99-3	: 9434	608-32-2	: 2305	612-15-7	: 1637
593-45-3	: 8234	597-52-4	: 11503	603-26-9	: 7513	608-33-3	: 9277	612-16-8	: 1889
593-49-7	: 6302	597-55-7	: 9926	603-32-7	: 450	608-50-4	: 2360	612-17-9	: 7815
593-50-0	: 11821	597-67-1	: 11428	603-33-8	: 3111	608-66-2	: 6194	612-19-1	: 2600
593-51-1	: 7446	597-71-7	: 9988	603-34-9	: 728	608-69-5	: 3308	612-22-6	: 1692
593-52-2	: 446	597-72-8	: 10068	603-35-0	: 9523	608-71-9	: 9439	612-23-7	: 1203
593-53-3	: 7515	597-76-2	: 6931	603-36-1	: 11551	608-89-9	: 3315	612-25-9	: 1906
593-57-7	: 442	597-77-3	: 6935	603-38-3	: 729	608-90-2	: 2082	612-28-2	: 804

612-35-1	: 2718	615-57-6	: 681	618-85-9	: 1957	622-04-8	: 10377	623-96-1	: 4002
612-37-3	: 10766	615-58-7	: 9276	618-87-1	: 727	622-08-2	: 5761	623-97-2	: 5697
612-55-5	: 7903	615-59-8	: 1305	618-89-3	: 2525	622-09-3	: 7693	624-03-3	: 6649
612-57-7	: 11240	615-60-1	: 1141	618-91-7	: 2661	622-15-1	: 7565	624-04-4	: 5311
612-58-8	: 11278	615-62-3	: 9265	618-95-1	: 2711	622-16-2	: 764	624-08-8	: 6319
612-60-2	: 11282	615-65-6	: 668	618-98-4	: 2708	622-24-2	: 1164	624-09-9	: 6464
612-61-3	: 11241	615-67-8	: 1507	619-02-3	: 4601	622-26-4	: 9664	624-13-5	: 8441
612-62-4	: 11237	615-68-9	: 2350	619-12-5	: 9598	622-29-7	: 7452	624-16-8	: 5077
612-64-6	: 750	615-71-4	: 2306	619-15-8	: 1959	622-31-1	: 594	624-17-9	: 8133
612-71-5	: 11598	615-74-7	: 9262	619-20-5	: 2601	622-32-2	: 595	624-20-4	: 6754
612-75-9	: 3057	615-79-2	: 8876	619-23-8	: 1206	622-36-6	: 7166	624-24-8	: 8891
612-76-0	: 3066	615-81-6	: 5556	619-25-0	: 1907	622-37-7	: 976	624-29-3	: 4368
612-78-2	: 3005	615-83-8	: 8860	619-31-8	: 719	622-38-8	: 1705	624-31-7	: 1780
612-81-7	: 3107	615-84-9	: 5494	619-33-0	: 5530	622-39-9	: 11015	624-38-4	: 1429
612-94-2	: 7948	615-85-0	: 5453	619-39-6	: 5068	622-40-2	: 7691	624-41-9	: 3630
612-95-3	: 11291	615-87-2	: 1298	619-41-0	: 5779	622-42-4	: 2052	624-42-0	: 6523
612-96-4	: 11288	615-89-4	: 1514	619-42-1	: 2526	622-45-7	: 155	624-43-1	: 10192
612-97-5	: 11196	615-98-5	: 5565	619-44-3	: 2662	622-47-9	: 926	624-44-2	: 6982
613-03-6	: 2370	615-99-6	: 5564	619-52-3	: 4683	622-57-1	: 748	624-45-3	: 8908
613-08-1	: 369	616-02-4	: 6098	619-58-9	: 2659	622-60-6	: 1661	624-48-6	: 3779
613-12-7	: 424	616-06-8	: 8215	619-60-3	: 9319	622-61-7	: 1158	624-49-7	: 3778
613-13-8	: 365	616-12-6	: 9056	619-73-8	: 1908	622-62-8	: 9352	624-54-4	: 10318
613-21-8	: 11319	616-19-3	: 9901	619-82-9	: 4354	622-63-9	: 1707	624-57-7	: 8617
613-29-6	: 684	616-20-6	: 8721	619-84-1	: 2585	622-78-6	: 1794	624-60-2	: 5435
613-31-0	: 376	616-21-7	: 3277	619-99-8	: 6815	622-79-7	: 980	624-61-3	: 5939
613-33-2	: 3061	616-23-9	: 10359	620-02-0	: 6045	622-80-0	: 843	624-64-6	: 3705
613-37-6	: 3081	616-24-0	: 8702	620-03-1	: 6047	622-82-2	: 7160	624-65-7	: 10729
613-39-8	: 9613	616-25-1	: 9087	620-05-3	: 1777	622-85-5	: 2168	624-66-8	: 10734
613-42-3	: 3094	616-27-3	: 3646	620-11-1	: 8919	622-86-6	: 1159	624-67-9	: 10723
613-43-4	: 5112	616-28-4	: 3912	620-12-2	: 10193	622-93-5	: 10367	624-70-4	: 5477
613-45-6	: 543	616-31-9	: 8846	620-13-3	: 1063	622-96-8	: 1687	624-72-6	: 5545
613-46-7	: 7749	616-38-6	: 4000	620-14-4	: 1686	622-97-9	: 1642	624-73-7	: 5547
613-48-9	: 697	616-39-7	: 5430	620-16-6	: 1162	622-98-0	: 1163	624-74-8	: 5941
613-50-3	: 11285	616-40-0	: 7152	620-17-7	: 9354	623-00-7	: 2746	624-75-9	: 283
613-51-4	: 11286	616-42-2	: 11582	620-18-8	: 9349	623-03-0	: 2749	624-76-0	: 5731
613-59-2	: 7950	616-43-3	: 11132	620-19-9	: 1199	623-05-2	: 1884	624-79-3	: 5599
613-69-4	: 552	616-44-4	: 11770	620-20-2	: 1133	623-08-5	: 712	624-81-7	: 5434
613-70-7	: 9382	616-45-5	: 11180	620-22-4	: 2764	623-09-6	: 1280	624-82-8	: 7566
613-75-2	: 7510	616-47-7	: 7208	620-23-5	: 588	623-10-9	: 758	624-83-9	: 7520
613-77-4	: 11249	616-55-7	: 9223	620-24-6	: 1883	623-12-1	: 1180	624-85-1	: 7184
613-78-5	: 2645	616-91-1	: 5015	620-32-6	: 2241	623-15-4	: 3878	624-89-5	: 5606
613-80-9	: 7940	617-00-5	: 795	620-40-6	: 1798	623-17-6	: 6119	624-90-8	: 7460
613-81-0	: 8006	617-02-7	: 2699	620-42-8	: 9597	623-18-7	: 10640	624-91-9	: 8091
613-84-3	: 577	617-05-0	: 2626	620-43-9	: 3224	623-19-8	: 6123	624-92-0	: 5254
613-88-7	: 956	617-09-4	: 3996	620-47-3	: 2015	623-20-1	: 10639	624-95-3	: 3617
613-89-8	: 5774	617-30-1	: 6797	620-63-3	: 3561	623-21-2	: 3520	624-96-4	: 3284
613-90-1	: 968	617-31-2	: 8883	620-71-3	: 9800	623-22-3	: 10642	624-97-5	: 9100
613-91-2	: 5866	617-33-4	: 160	620-72-4	: 119	623-24-5	: 985	625-01-4	: 4085
613-93-4	: 620	617-35-6	: 10315	620-73-5	: 147	623-25-6	: 988	625-04-7	: 8959
613-94-5	: 2611	617-45-8	: 470	620-80-4	: 3576	623-26-7	: 1316	625-06-9	: 8928
613-97-8	: 749	617-50-5	: 10303	620-81-5	: 5495	623-27-8	: 1323	625-08-1	: 3535
614-00-6	: 806	617-51-6	: 10275	620-82-6	: 5559	623-30-3	: 10512	625-23-0	: 6942
614-16-4	: 2124	617-52-7	: 3330	620-83-7	: 2016	623-34-7	: 3285	625-25-2	: 6495
614-18-6	: 10902	617-53-8	: 3789	620-85-9	: 1696	623-36-9	: 9002	625-27-4	: 9055
614-21-1	: 5862	617-54-9	: 3790	620-86-0	: 2719	623-39-2	: 10014	625-28-5	: 3410
614-27-7	: 2142	617-60-7	: 5309	620-88-2	: 2054	623-40-5	: 8985	625-30-9	: 8701
614-29-9	: 1613	617-65-2	: 6240	620-92-8	: 9407	623-42-7	: 3547	625-31-0	: 9089
614-33-5	: 10196	617-73-2	: 8432	620-93-9	: 797	623-43-8	: 3847	625-36-5	: 10497
614-34-6	: 2700	617-75-4	: 448	620-94-0	: 2286	623-46-1	: 5533	625-37-6	: 3702
614-39-1	: 602	617-78-7	: 8809	620-95-1	: 11005	623-47-2	: 10739	625-38-7	: 3813
614-45-9	: 1128	617-80-1	: 3407	620-99-5	: 5670	623-48-3	: 201	625-44-5	: 10071
614-47-1	: 10709	617-83-4	: 4156	621-08-9	: 2184	623-49-4	: 4030	625-45-6	: 205
614-54-0	: 1881	617-84-5	: 5991	621-14-7	: 3320	623-50-7	: 197	625-46-7	: 10591
614-68-6	: 1787	617-85-6	: 11549	621-15-8	: 6722	623-51-8	: 203	625-47-8	: 5483
614-71-1	: 1659	617-86-7	: 11483	621-23-8	: 2355	623-53-0	: 4003	625-48-9	: 5752
614-72-2	: 1156	617-88-9	: 6075	621-25-0	: 7174	623-56-3	: 6981	625-50-3	: 61
614-75-5	: 904	617-89-0	: 6113	621-26-1	: 7138	623-59-6	: 73	625-52-5	: 12014
614-78-8	: 11806	617-90-3	: 6039	621-27-2	: 9462	623-60-9	: 10351	625-54-7	: 10038
614-96-0	: 7316	617-92-5	: 11124	621-31-8	: 9356	623-65-4	: 6646	625-55-8	: 6011
614-99-3	: 6056	617-94-7	: 1857	621-32-9	: 1660	623-66-5	: 8424	625-56-9	: 7582
615-05-4	: 1274	618-31-5	: 1301	621-34-1	: 9351	623-68-7	: 3818	625-57-0	: 3978
615-09-8	: 6163	618-32-6	: 2885	621-36-3	: 925	623-69-8	: 10372	625-58-1	: 8074
615-10-1	: 6071	618-34-8	: 1148	621-37-4	: 905	623-70-1	: 3839	625-59-2	: 11706
615-11-2	: 6057	618-36-0	: 1814	621-50-1	: 966	623-71-2	: 10221	625-60-5	: 5647
615-13-4	: 7287	618-38-2	: 2910	621-51-2	: 2597	623-72-3	: 10270	625-65-0	: 9035
615-18-9	: 2878	618-40-6	: 7165	621-52-3	: 1663	623-73-4	: 158	625-66-1	: 3282
615-20-3	: 2862	618-41-7	: 2180	621-59-0	: 572	623-76-7	: 12001	625-67-2	: 8750
615-30-5	: 8691	618-42-8	: 9630	621-62-5	: 5489	623-78-9	: 3949	625-68-3	: 3481
615-36-1	: 638	618-45-1	: 9411	621-63-6	: 5705	623-80-3	: 4031	625-69-4	: 8789
615-37-2	: 1778	618-46-2	: 2891	621-64-7	: 9825	623-81-4	: 11581	625-71-8	: 3531
615-38-3	: 4531	618-51-9	: 2658	621-70-5	: 6908	623-84-7	: 10002	625-74-1	: 10084
615-39-4	: 4532	618-62-2	: 1394	621-72-7	: 2385	623-85-8	: 3970	625-76-3	: 7508
615-41-8	: 1171	618-65-5	: 562	621-76-1	: 10078	623-86-9	: 102	625-77-4	: 23
615-42-9	: 1427	618-68-8	: 2144	621-77-2	: 8705	623-87-0	: 10191	625-80-9	: 10152
615-52-1	: 5567	618-71-3	: 2593	621-87-4	: 10479	623-91-6	: 3775	625-82-1	: 11119
615-54-3	: 2308	618-76-8	: 2618	621-95-4	: 731	623-93-8	: 8161	625-84-3	: 11120
615-55-4	: 683	618-80-4	: 9296	622-00-4	: 3306	623-95-0	: 12012	625-86-5	: 6089

681-84-5	: 11508	699-08-1	: 1657	764-96-5	: 11973	779-84-0	: 9450
682-01-9	: 11510	699-09-2	: 2296	764-97-6	: 11972	780-69-8	: 11481
682-30-4	: 9535	699-17-2	: 3659	765-03-7	: 5343	781-35-1	: 10432
683-08-9	: 9539	699-30-9	: 6099	765-05-9	: 5051	782-05-8	: 1457
683-50-1	: 9777	700-12-9	: 2097	765-09-3	: 11872	782-06-9	: 1456
683-51-2	: 10511	700-75-4	: 769	765-13-9	: 8656	784-50-9	: 5969
683-68-1	: 5521	700-88-9	: 1254	765-27-5	: 5366	786-19-6	: 4037
683-72-7	: 51	701-57-5	: 2044	765-30-0	: 4953	790-69-2	: 11802
684-16-2	: 10441	701-97-3	: 4468	765-34-4	: 8603	791-28-6	: 9518
684-84-4	: 3353	702-28-3	: 4969	765-42-4	: 4979	791-29-7	: 11444
684-93-5	: 12019	702-54-5	: 8573	765-43-5	: 5796	791-31-1	: 11504
685-63-2	: 3164	703-32-2	: 11252	765-47-9	: 4918	793-23-7	: 995
685-73-4	: 6199	703-61-7	: 11248	765-58-2	: 11730	799-53-1	: 5239
685-87-0	: 9928	703-80-0	: 5819	765-85-5	: 4176	804-63-7	: 4132
685-91-6	: 53	704-01-8	: 1286	765-87-7	: 4375	811-93-8	: 9878
688-14-2	: 7414	704-38-1	: 7617	766-00-7	: 4811	811-97-2	: 5640
688-52-8	: 7108	704-79-0	: 3854	766-05-2	: 4328	812-01-1	: 4086
688-71-1	: 3127	706-78-5	: 4932	766-07-4	: 7177	812-03-3	: 10141
688-74-4	: 3124	706-79-6	: 4917	766-08-5	: 11442	813-76-3	: 9525
688-84-6	: 10659	708-06-5	: 7754	766-09-6	: 9667	813-78-5	: 9559
689-89-4	: 6703	708-76-9	: 567	766-20-1	: 5189	814-49-3	: 9574
689-97-4	: 3889	708-82-7	: 10629	766-39-2	: 6097	814-67-5	: 10030
690-02-8	: 9159	709-09-1	: 1445	766-46-1	: 1040	814-68-6	: 10714
690-08-4	: 9039	709-98-8	: 10200	766-51-8	: 1178	814-78-8	: 3879
690-37-9	: 8954	710-43-0	: 4968	766-53-0	: 2992	815-17-8	: 3510
690-92-6	: 7029	711-79-5	: 5829	766-77-8	: 11418	815-24-7	: 8988
690-93-7	: 7028	712-48-1	: 454	766-79-0	: 9684	815-58-7	: 10718
691-37-2	: 9054	713-36-0	: 2014	766-81-4	: 1041	815-66-7	: 6975
691-38-3	: 9059	716-79-0	: 2384	766-84-7	: 2748	816-11-5	: 3517
691-83-8	: 3788	717-21-5	: 10710	766-90-5	: 2164	816-39-7	: 10424
692-24-0	: 7046	717-74-8	: 2373	766-92-7	: 2042	816-79-5	: 9045
692-35-3	: 9854	719-79-9	: 1120	766-93-8	: 5990	817-91-4	: 6493
692-42-2	: 441	723-46-6	: 2198	766-94-9	: 1646	817-95-8	: 181
692-45-5	: 6006	723-62-6	: 370	766-96-1	: 1042	818-23-5	: 8652
692-50-2	: 3914	726-42-1	: 765	767-03-3	: 4904	818-38-2	: 8772
692-86-4	: 11977	726-44-3	: 2122	767-10-2	: 11146	818-92-8	: 10570
693-02-7	: 7114	732-11-6	: 9515	767-12-4	: 4516	819-07-8	: 5668
693-05-0	: 10112	732-26-3	: 9498	767-54-4	: 4570	819-44-3	: 7107
693-13-0	: 9815	737-89-3	: 7726	767-58-8	: 7255	819-93-2	: 3502
693-23-2	: 5299	741-58-2	: 517	767-59-9	: 7275	821-06-7	: 3742
693-39-0	: 6429	742-20-1	: 2853	767-60-2	: 7277	821-07-8	: 6992
693-54-9	: 5075	747-90-0	: 4102	767-88-4	: 10884	821-08-9	: 6713
693-58-3	: 8121	750-90-3	: 4131	767-99-7	: 2029	821-09-0	: 9088
693-61-8	: 11968	753-89-9	: 9853	768-00-3	: 2025	821-10-3	: 3905
693-62-9	: 11970	754-05-2	: 11425	768-22-9	: 7282	821-11-4	: 3792
693-65-2	: 8835	755-25-9	: 7032	768-32-1	: 11493	821-25-0	: 6377
693-72-1	: 8295	756-79-6	: 9540	768-33-2	: 11374	821-55-6	: 8172
693-89-0	: 4928	757-58-4	: 11645	768-39-8	: 3115	821-67-0	: 5132
693-93-6	: 8583	759-05-7	: 3556	768-49-0	: 2024	821-76-1	: 6379
693-95-8	: 11679	759-36-4	: 9964	768-52-5	: 790	821-95-4	: 11966
693-98-1	: 7209	759-73-9	: 12015	768-56-9	: 1112	821-96-5	: 11969
694-05-3	: 11020	759-94-4	: 5372	768-66-1	: 9711	821-98-7	: 11971
694-28-0	: 4880	760-20-3	: 9053	768-94-5	: 11859	821-99-8	: 8128
694-31-5	: 10826	760-21-4	: 8825	769-06-2	: 849	822-06-0	: 6770
694-35-9	: 4925	760-23-6	: 3745	769-25-5	: 1647	822-23-1	: 221
694-48-4	: 10825	760-40-7	: 10408	769-68-6	: 951	822-28-6	: 6638
694-53-1	: 11446	760-55-4	: 9971	769-92-6	: 715	822-35-5	: 4198
694-59-7	: 10994	760-67-8	: 6987	770-09-2	: 11494	822-36-6	: 7210
694-80-4	: 1004	760-78-1	: 8217	770-35-4	: 10399	822-50-4	: 4802
694-81-5	: 11057	760-79-2	: 3203	771-56-2	: 2094	822-66-2	: 4695
694-85-9	: 11035	760-93-0	: 10649	771-61-9	: 9442	822-67-3	: 4694
695-06-7	: 6150	761-06-8	: 10015	771-62-0	: 2304	822-84-4	: 11776
695-12-5	: 4388	762-21-0	: 3909	771-97-1	: 7799	822-85-5	: 4574
695-28-3	: 4653	762-42-5	: 3910	771-98-2	: 1246	822-87-7	: 4577
695-34-1	: 10867	762-49-2	: 5462	771-99-3	: 9695	823-17-6	: 4654
695-53-4	: 8584	762-50-5	: 5476	772-00-9	: 5196	823-40-5	: 1277
695-84-1	: 9348	762-62-9	: 9033	772-01-0	: 5195	823-76-7	: 5794
695-98-7	: 11025	762-63-0	: 9040	772-17-8	: 2138	824-11-3	: 11909
696-28-6	: 462	762-72-1	: 11495	773-76-2	: 11314	824-22-6	: 7257
696-29-7	: 4431	763-29-1	: 9052	773-82-0	: 2767	824-42-0	: 575
696-30-0	: 10986	763-30-4	: 8671	774-40-3	: 909	824-47-5	: 1193
696-41-3	: 581	763-69-9	: 10262	775-12-2	: 11420	824-55-5	: 1186
696-44-6	: 711	763-89-3	: 9096	775-51-9	: 4277	824-63-5	: 7256
696-59-3	: 6169	763-93-9	: 7086	775-56-4	: 11407	824-72-6	: 9548
696-62-8	: 1776	764-01-2	: 3919	776-35-2	: 9171	824-94-2	: 1197
696-63-9	: 2300	764-13-6	: 6692	776-74-9	: 1058	825-25-2	: 4881
696-71-9	: 4739	764-33-0	: 7123	776-75-0	: 9634	825-51-4	: 8024
697-11-0	: 4200	764-35-2	: 7115	776-76-1	: 11438	825-94-5	: 11457
697-86-9	: 641	764-42-1	: 3762	777-22-0	: 1981	826-18-6	: 2104
698-00-0	: 643	764-47-6	: 10033	777-37-7	: 1218	826-36-8	: 9726
698-01-1	: 662	764-48-7	: 5716	778-22-3	: 1976	826-73-3	: 2407
698-27-1	: 576	764-49-8	: 11712	778-24-5	: 11416	826-74-4	: 7881
698-63-5	: 6046	764-56-7	: 6342	778-25-6	: 11502	826-77-7	: 11262
698-69-1	: 663	764-60-3	: 7125	778-28-9	: 2230	826-81-3	: 11321
698-87-3	: 1607	764-73-8	: 8343	778-66-5	: 2166	827-08-7	: 1312
698-88-4	: 1372	764-85-2	: 8177	779-02-2	: 425	827-15-6	: 2092
699-02-5	: 1610	764-93-2	: 5097	779-51-1	: 1971	827-16-7	: 11841

827-52-1	: 1248
827-54-3	: 7882
828-01-3	: 2133
829-83-4	: 443
829-84-5	: 9516
830-13-7	: 4211
830-89-7	: 7229
831-81-2	: 1224
831-91-4	: 2113
832-64-4	: 9181
832-71-3	: 9180
834-12-8	: 341
835-11-0	: 7596
835-64-3	: 9218
836-30-6	: 826
837-45-6	: 9165
838-41-5	: 8579
838-45-9	: 1697
838-95-9	: 722
841-73-6	: 11081
842-07-9	: 8037
846-48-0	: 343
848-53-3	: 5140
865-71-4	: 8853
868-18-8	: 3314
868-77-9	: 10661
868-85-9	: 9588
869-29-4	: 10563
869-50-1	: 9996
870-23-5	: 10600
870-63-3	: 3717
870-74-6	: 8977
871-28-3	: 9112
871-83-0	: 8140
871-84-1	: 8342
872-05-9	: 5082
872-10-6	: 8843
872-31-1	: 11727
872-50-4	: 11185
872-53-7	: 4780
872-55-9	: 11759
872-85-5	: 10889
872-93-5	: 11789
873-32-5	: 2747
873-49-4	: 1256
873-66-5	: 2163
873-76-7	: 1849
873-94-9	: 4622
874-23-7	: 4572
874-35-1	: 7258
874-41-9	: 1671
874-60-2	: 2904
874-63-5	: 1918
874-79-3	: 2172
874-90-8	: 2762
875-30-9	: 7298
875-40-1	: 2262
875-51-4	: 654
875-59-2	: 5828
876-05-1	: 4793
877-09-8	: 2261
877-11-2	: 2086
877-43-0	: 11257
877-44-1	: 2344
877-48-5	: 9314
877-53-2	: 9313
877-65-6	: 1865
877-82-7	: 9316
878-93-3	: 8064
879-97-0	: 9327
880-09-1	: 9683
880-93-3	: 7925
881-03-8	: 7926
881-67-4	: 1444
882-25-7	: 4148
882-33-7	: 5258
883-93-2	: 2864
883-99-8	: 7775
886-50-0	: 11594
886-58-8	: 1979
886-74-8	: 10001
886-77-1	: 8682
888-71-1	: 8564
897-78-9	: 4573
900-95-8	: 11539
910-86-1	: 11800
917-61-3	: 4164

CAS	No.	CAS	No.	CAS	No.	CAS	No.	CAS	No.
917-92-0	3907	932-16-1	5849	982-43-4	9657	1070-87-7	8841	1119-49-9	38
917-93-1	9779	932-32-1	670	992-94-9	11437	1071-26-7	6383	1119-51-3	9010
918-00-3	10488	932-39-8	4797	993-00-0	11378	1071-73-4	8974	1119-60-4	6564
918-07-0	3254	932-40-1	4798	993-07-7	11489	1071-81-4	6849	1119-65-9	6610
918-79-6	8692	932-43-4	4834	993-13-5	9538	1071-83-6	6276	1120-07-6	8118
918-82-1	9128	932-44-5	4835	994-05-8	3396	1071-94-9	6589	1120-16-7	5289
918-84-3	8732	932-53-6	11831	994-25-2	8726	1071-98-3	3908	1120-21-4	11938
919-12-0	9132	932-56-9	4235	994-30-9	11388	1072-05-5	6387	1120-28-1	5357
919-19-7	9530	932-66-1	5793	994-49-0	5235	1072-16-8	8378	1120-29-2	5287
919-30-2	9828	932-77-4	1007	996-98-5	5562	1072-21-5	6748	1120-36-1	11636
919-86-8	5103	932-96-7	675	997-49-9	11698	1072-36-2	8224	1120-48-5	8354
919-94-8	3375	933-05-1	4865	998-29-8	11500	1072-39-5	5281	1120-56-5	4193
920-37-6	10584	933-21-1	4728	998-30-1	11477	1072-43-1	11696	1120-59-8	11750
920-46-7	10715	933-36-8	4616	998-40-3	9520	1072-44-2	500	1120-62-3	4929
921-08-4	3487	933-48-2	4569	998-41-4	11454	1072-52-2	499	1120-64-5	8576
921-13-1	7480	933-67-5	7317	998-93-6	6358	1072-53-3	5199	1120-71-4	8570
921-47-1	6863	933-75-5	9483	998-95-8	6363	1072-83-9	5820	1120-72-5	4885
922-28-1	6389	933-78-8	9482	999-21-3	3782	1072-91-9	10840	1120-73-6	4944
922-55-4	5016	933-88-0	2902	999-52-0	6362	1072-98-6	10861	1120-87-2	10877
922-59-8	9134	933-98-2	1670	999-55-3	10694	1073-06-9	1043	1120-97-4	5193
922-62-3	9057	934-00-9	1525	999-78-0	9130	1073-07-0	4732	1121-07-9	11160
922-65-6	8681	934-34-9	2868	999-79-1	9127	1073-23-0	10945	1121-18-2	4710
923-06-8	3300	934-53-2	1189	999-81-5	5442	1073-67-2	1151	1121-24-0	9376
923-26-2	10662	934-74-7	1672	999-97-3	11363	1074-11-9	1376	1121-37-5	7607
923-99-9	9594	934-80-5	1675	1000-49-3	11371	1074-17-5	2036	1121-55-7	10953
924-16-3	3212	935-05-7	8097	1000-50-6	11369	1074-43-7	2037	1121-60-4	10887
924-41-4	6706	935-07-9	18	1000-86-8	8665	1074-55-1	2038	1121-92-2	502
924-43-6	8093	935-08-0	5134	1001-53-2	28	1074-58-4	275	1121-95-5	4875
924-50-5	3852	935-12-6	6164	1001-89-4	6361	1074-92-6	1464	1122-20-9	4705
924-52-7	8077	935-13-7	6161	1002-22-5	7126	1075-22-5	7254	1122-25-4	4594
925-15-5	3321	935-31-9	4208	1002-69-3	5036	1075-38-3	1465	1122-39-0	11027
925-78-0	8173	935-95-5	9471	1002-84-2	8645	1075-74-7	2106	1122-42-5	1773
926-02-3	10036	936-48-1	10824	1003-03-8	4771	1076-26-2	8031	1122-54-9	5869
926-26-1	3296	936-53-8	10823	1003-09-4	11726	1076-55-7	9391	1122-60-7	4454
926-55-6	9116	936-58-3	1911	1003-10-7	11794	1076-61-5	7994	1122-62-9	5867
926-56-7	8670	936-72-1	2026	1003-29-8	11114	1076-66-0	1625	1122-69-6	10963
926-63-6	9807	937-14-4	1127	1003-30-1	6107	1077-16-3	1763	1122-81-2	11016
926-64-7	281	937-20-2	5784	1003-31-2	11732	1078-04-2	7260	1122-82-3	4417
926-65-8	10034	937-32-6	2179	1003-56-1	11036	1078-19-9	8048	1122-83-4	844
926-82-9	6390	937-33-7	716	1003-64-1	4391	1078-21-3	2128	1122-91-4	528
927-49-1	11962	938-16-9	10431	1003-73-2	10988	1078-30-4	11228	1123-00-8	4775
927-54-8	7006	938-25-0	7795	1003-78-7	11784	1078-71-3	1745	1123-09-7	4707
927-62-8	3215	938-33-0	11276	1003-90-3	11141	1079-71-6	427	1123-27-9	5821
927-68-4	5686	938-39-6	11326	1004-36-0	10793	1080-16-6	5130	1123-34-8	4555
927-73-1	3724	938-45-4	10446	1004-66-6	1919	1080-32-6	9545	1123-84-8	1371
927-74-2	3920	938-73-8	615	1004-77-9	4599	1081-34-1	11600	1123-85-9	1608
927-80-0	5942	939-23-1	10999	1005-64-7	1109	1081-75-0	2121	1123-86-0	10423
928-40-5	6795	939-26-4	7743	1005-67-0	7684	1081-77-2	2064	1123-95-1	11105
928-45-0	8072	939-27-5	7886	1006-00-0	1874	1083-56-3	1099	1123-96-2	10959
928-49-4	7116	939-48-0	2693	1006-31-1	2263	1086-80-2	2472	1124-05-6	1362
928-50-7	9019	939-52-6	3650	1007-26-7	1484	1088-11-5	2411	1124-07-8	9293
928-51-8	3604	939-55-9	5696	1007-32-5	3677	1093-58-9	354	1124-20-5	1985
928-56-3	5886	939-83-3	2766	1008-88-4	10998	1095-90-5	6516	1124-27-2	4442
928-65-4	11464	939-87-7	4957	1008-89-5	10997	1111-74-6	11414	1124-33-0	10993
928-68-7	6521	940-41-0	11474	1009-14-9	8986	1111-97-3	3901	1125-60-6	7372
928-80-3	5076	940-49-8	2033	1009-67-2	2137	1112-37-4	9529	1125-74-2	1547
928-92-7	7072	940-54-5	3189	1009-81-0	1754	1112-39-6	11412	1125-78-6	8044
928-94-9	7070	941-69-5	11123	1009-93-4	5001	1112-48-7	11368	1125-80-0	7382
928-95-0	7069	941-98-0	5857	1010-48-6	2130	1112-55-6	11449	1126-09-6	9643
928-96-1	7071	942-01-8	3989	1012-72-2	990	1112-56-7	11543	1126-18-7	4575
929-55-5	8348	942-06-3	7344	1012-84-6	2712	1113-12-8	11417	1126-46-1	2543
929-61-3	8402	942-92-7	6983	1013-08-7	9187	1113-30-0	3113	1126-75-6	2031
929-62-4	8401	943-14-6	2528	1013-83-8	7769	1113-68-4	25	1126-78-9	656
929-77-1	5278	943-15-7	1991	1013-88-3	1835	1113-74-2	8968	1126-79-0	1114
930-02-9	8240	943-27-1	5808	1013-96-3	10676	1114-09-6	3302	1127-76-0	7885
930-18-7	4972	944-22-9	5986	1016-77-9	7601	1114-51-8	9793	1128-08-1	4950
930-21-2	496	945-51-7	2183	1016-94-0	9195	1114-71-2	3976	1128-54-7	10838
930-22-3	8613	947-73-9	9162	1017-56-7	7583	1114-76-7	3200	1128-76-3	2538
930-27-8	6131	947-84-2	3021	1024-57-3	6301	1115-11-3	3698	1129-47-1	10293
930-28-9	4789	947-91-1	877	1025-15-6	11842	1115-12-4	9915	1129-50-6	3208
930-30-3	4941	948-32-3	2372	1048-08-4	11452	1115-30-6	3295	1129-89-1	4216
930-36-9	10834	948-65-2	7320	1056-77-5	6175	1116-24-1	58	1131-16-4	10829
930-37-0	8620	948-97-0	1966	1066-27-9	11435	1116-39-8	3123	1131-18-6	10814
930-55-2	11170	948-98-1	1146	1066-45-1	11540	1116-40-1	9820	1131-60-8	9272
930-66-5	4642	948-99-2	1147	1067-08-9	8814	1116-61-6	3121	1131-62-0	5805
930-68-7	4703	949-87-1	5139	1067-20-5	8756	1117-19-7	9945	1131-63-1	7779
930-87-0	11142	950-37-8	7645	1068-19-5	6391	1117-55-1	8431	1132-21-4	2577
930-89-2	4815	951-86-0	1365	1068-57-1	195	1117-59-5	8882	1132-39-4	2175
930-90-5	4816	952-47-6	9426	1068-87-7	8812	1118-09-8	11491	1132-66-7	1765
931-10-2	4293	952-97-6	2058	1068-90-2	9922	1118-68-9	6261	1133-80-8	5968
931-20-4	9721	955-83-9	6100	1069-53-0	6864	1118-71-4	6404	1134-23-2	4168
931-56-6	4425	957-51-7	5219	1070-14-0	6705	1118-84-9	3577	1134-62-9	7747
931-77-1	4319	959-33-1	10713	1070-32-2	6492	1119-06-8	6465	1135-12-2	839
931-87-3	4752	961-11-5	6202	1070-34-4	3335	1119-19-3	3895	1140-14-3	7614
931-89-5	4751	961-71-7	5506	1070-71-9	10737	1119-33-1	6245	1140-16-5	7628
931-97-5	4329	964-68-1	2533	1070-78-6	10142	1119-40-0	8775	1142-70-7	11080
932-01-4	4519	976-71-6	9736	1070-83-3	3508	1119-46-6	8866	1142-97-8	2236

CAS	Page	CAS	Page	CAS	Page	CAS	Page	CAS	Page
2935-90-2	: 10287	3087-62-5	: 6211	3280-08-8	: 1616	3463-92-1	: 4064	3728-43-6	: 11490
2941-19-7	: 1830	3088-41-3	: 9703	3280-51-1	: 6708	3468-99-3	: 939	3730-60-7	: 6923
2941-20-0	: 1809	3096-47-7	: 5976	3282-30-2	: 10503	3476-89-9	: 11339	3731-51-9	: 10971
2941-55-1	: 5930	3096-52-4	: 5977	3283-12-3	: 9528	3478-88-4	: 7594	3731-52-0	: 10972
2943-70-6	: 5231	3101-60-8	: 8611	3288-04-8	: 1793	3497-00-5	: 9552	3731-53-1	: 10973
2944-49-2	: 1916	3102-33-8	: 9105	3289-28-9	: 4334	3508-00-7	: 6307	3735-90-8	: 3977
2947-60-6	: 961	3112-88-7	: 2112	3290-06-0	: 1358	3508-78-9	: 8805	3741-00-2	: 4846
2947-61-7	: 962	3113-98-2	: 4421	3296-05-7	: 978	3508-94-9	: 930	3741-38-6	: 5200
2955-65-9	: 8395	3118-97-6	: 8028	3299-38-5	: 3656	3510-70-1	: 11496	3742-38-9	: 4926
2955-88-6	: 11165	3121-83-3	: 9993	3319-15-1	: 1895	3512-80-9	: 10864	3742-42-5	: 4662
2959-96-8	: 10781	3128-06-1	: 6905	3320-83-0	: 1176	3521-91-3	: 6566	3743-22-4	: 9317
2961-47-9	: 10880	3128-07-2	: 6469	3320-86-3	: 1788	3522-94-9	: 6861	3744-02-3	: 9109
2969-81-5	: 3470	3129-39-3	: 7846	3320-87-4	: 1789	3524-62-7	: 5839	3757-32-2	: 8910
2971-79-1	: 9644	3129-90-6	: 7399	3321-50-4	: 4386	3524-73-0	: 7039	3761-94-2	: 4230
2973-00-4	: 7497	3132-99-8	: 527	3329-48-4	: 7073	3524-75-2	: 4848	3763-39-1	: 4991
2973-27-5	: 11224	3140-92-9	: 11743	3329-56-4	: 3377	3528-17-4	: 2869	3763-55-1	: 4062
2977-31-3	: 1617	3140-93-0	: 11742	3333-23-1	: 7764	3535-67-9	: 1925	3764-01-0	: 11075
2978-58-7	: 3896	3141-12-6	: 440	3333-44-6	: 6138	3536-29-6	: 2158	3768-43-2	: 9722
2979-19-3	: 4587	3141-24-0	: 11790	3333-52-6	: 3292	3538-65-6	: 3529	3768-60-3	: 11583
2980-71-4	: 8185	3141-27-3	: 11744	3337-71-1	: 474	3540-95-2	: 9658	3768-63-6	: 11569
2983-26-8	: 5523	3142-58-3	: 8799	3338-15-6	: 10444	3544-25-0	: 946	3769-23-1	: 7038
2983-37-1	: 6885	3142-66-3	: 8972	3344-70-5	: 5296	3547-04-4	: 1678	3769-41-3	: 9452
2984-40-9	: 3806	3142-84-5	: 6608	3347-22-6	: 8056	3552-01-0	: 4538	3773-93-1	: 5208
2987-53-3	: 819	3144-16-9	: 2929	3350-30-9	: 4725	3554-74-3	: 9716	3774-03-6	: 5197
2987-77-1	: 11485	3144-54-5	: 6962	3350-78-5	: 3888	3555-84-8	: 7612	3779-29-1	: 4180
2996-92-1	: 11488	3147-55-5	: 2558	3351-28-8	: 4117	3558-69-8	: 10947	3780-58-3	: 6894
2998-18-7	: 10665	3152-68-9	: 9110	3351-86-8	: 4055	3567-38-2	: 1877	3788-32-7	: 4838
3000-79-1	: 9680	3153-36-4	: 3485	3352-87-2	: 5290	3570-55-6	: 5652	3790-71-4	: 5334
3001-72-7	: 11190	3153-37-5	: 3489	3354-58-3	: 9427	3570-58-9	: 5699	3810-26-2	: 4951
3002-24-2	: 6804	3158-70-5	: 10589	3355-17-7	: 3900	3574-40-1	: 3157	3810-74-0	: 11561
3007-91-8	: 7770	3156-73-8	: 10396	3358-28-9	: 6174	3574-43-4	: 5682	3835-64-1	: 1864
3007-97-4	: 7765	3156-74-9	: 3641	3368-16-9	: 941	3581-87-1	: 11678	3837-38-5	: 404
3008-39-7	: 4225	3163-07-3	: 1541	3373-53-3	: 10767	3581-91-7	: 11675	3839-50-7	: 4765
3010-02-4	: 280	3163-27-7	: 7742	3375-23-3	: 8036	3582-71-6	: 11913	3840-31-1	: 1914
3010-04-6	: 273	3167-49-5	: 10897	3377-71-7	: 7321	3586-12-7	: 835	3848-12-2	: 126
3012-37-1	: 11710	3172-52-9	: 11748	3377-86-4	: 6733	3586-14-9	: 2013	3849-33-0	: 10049
3016-19-1	: 8487	3173-53-3	: 4416	3377-87-5	: 6734	3587-57-3	: 9858	3850-30-4	: 3217
3017-67-2	: 5086	3173-79-3	: 8597	3383-96-8	: 1	3587-60-8	: 1181	3852-09-3	: 10289
3017-68-3	: 3714	3174-67-2	: 8785	3384-04-1	: 1233	3588-30-5	: 3169	3853-80-3	: 10202
3017-70-7	: 3718	3174-83-2	: 2157	3386-33-2	: 8236	3593-00-8	: 6563	3856-25-5	: 11861
3017-71-8	: 3713	3175-23-3	: 10174	3393-64-4	: 3664	3594-37-4	: 3364	3857-25-8	: 6122
3017-95-6	: 9840	3176-62-3	: 7240	3393-78-0	: 2284	3598-14-9	: 287	3862-11-1	: 9338
3017-96-7	: 9838	3178-22-1	: 4369	3395-79-7	: 1443	3600-24-6	: 11916	3862-12-2	: 9335
3018-09-5	: 5881	3178-23-2	: 4429	3395-91-3	: 10212	3607-78-1	: 10052	3864-18-4	: 3052
3018-12-0	: 278	3178-24-3	: 4466	3399-21-1	: 4357	3609-53-8	: 2479	3867-18-3	: 733
3019-04-3	: 10454	3178-29-8	: 6434	3400-45-1	: 4781	3610-02-4	: 2403	3875-51-2	: 4828
3019-20-3	: 1980	3179-63-3	: 10375	3402-74-2	: 2006	3613-42-1	: 367	3875-52-3	: 4805
3019-25-8	: 5792	3180-09-4	: 9246	3402-76-4	: 7945	3615-21-2	: 2377	3875-68-1	: 12051
3022-41-1	: 4209	3187-94-8	: 6073	3404-57-7	: 6558	3616-56-6	: 5416	3877-15-4	: 10088
3024-72-4	: 2894	3188-13-4	: 5478	3404-58-8	: 7033	3616-57-7	: 5415	3877-19-8	: 7996
3027-13-2	: 10460	3189-22-8	: 7311	3404-61-3	: 7037	3634-92-2	: 8273	3878-55-5	: 3336
3031-05-8	: 8011	3194-15-8	: 10438	3404-63-5	: 9060	3637-01-2	: 5811	3884-88-6	: 8246
3031-66-1	: 7118	3195-24-2	: 9944	3404-67-9	: 6838	3637-61-4	: 4825	3910-35-8	: 7263
3031-73-0	: 7180	3196-15-4	: 3473	3404-71-5	: 9043	3638-35-5	: 4983	3913-71-1	: 5080
3031-74-1	: 7179	3199-61-9	: 2440	3404-72-6	: 9029	3638-64-0	: 5911	3913-85-7	: 5088
3032-32-4	: 2491	3202-84-4	: 7695	3404-73-7	: 9031	3643-64-9	: 4688	3916-64-1	: 6339
3034-53-5	: 11668	3208-16-0	: 6105	3404-77-1	: 7016	3657-07-6	: 245	3917-15-5	: 10565
3035-75-4	: 11877	3208-22-8	: 6173	3404-78-2	: 7024	3658-79-5	: 8755	3920-53-4	: 6453
3036-66-6	: 3168	3209-22-1	: 1391	3404-80-6	: 6837	3669-80-5	: 11948	3937-45-9	: 4422
3040-44-6	: 9662	3211-48-1	: 4183	3411-95-8	: 9217	3673-79-8	: 10242	3937-46-0	: 4423
3041-40-5	: 9916	3212-75-7	: 4753	3416-24-8	: 6233	3674-09-7	: 10248	3937-56-2	: 8136
3042-22-6	: 11135	3213-79-4	: 1269	3419-71-4	: 4689	3674-10-0	: 10383	3938-95-2	: 10259
3042-50-0	: 1486	3217-86-5	: 9802	3424-93-9	: 618	3674-13-3	: 10240	3938-96-3	: 206
3043-33-2	: 8672	3218-02-8	: 4418	3426-01-5	: 10843	3674-75-7	: 9179	3943-74-6	: 2627
3047-38-9	: 4907	3221-61-2	: 8408	3430-27-1	: 10870	3675-68-1	: 3442	3943-77-9	: 2578
3051-09-0	: 11097	3228-04-4	: 9416	3433-37-2	: 9674	3675-69-2	: 3443	3944-36-3	: 10390
3054-92-0	: 8956	3228-99-7	: 9895	3433-80-5	: 999	3678-62-4	: 10919	3955-26-8	: 2324
3054-95-3	: 10560	3229-00-3	: 9884	3435-28-7	: 11048	3680-02-2	: 5909	3958-03-0	: 11781
3055-14-9	: 1949	3230-23-7	: 9670	3437-89-6	: 8987	3681-78-5	: 5320	3958-38-1	: 4727
3058-41-7	: 2043	3232-84-6	: 11850	3437-95-4	: 11767	3682-35-7	: 11843	3958-57-4	: 1067
3060-89-7	: 7856	3234-84-8	: 9121	3438-46-8	: 11071	3683-19-0	: 7043	3959-07-7	: 1799
3062-81-5	: 3122	3238-38-8	: 9474	3439-97-2	: 11443	3683-22-5	: 7042	3962-77-4	: 1926
3066-71-5	: 10626	3238-40-2	: 6081	3440-02-6	: 11415	3689-24-5	: 11716	3963-78-8	: 382
3068-00-6	: 3459	3238-62-8	: 9677	3445-11-2	: 11184	3690-04-8	: 9795	3964-56-5	: 9232
3073-66-3	: 4486	3238-75-3	: 5712	3446-98-8	: 11147	3697-27-6	: 4115	3964-58-7	: 2540
3073-92-5	: 3421	3240-09-3	: 7099	3452-07-1	: 5362	3699-01-2	: 2019	3964-61-2	: 9254
3074-61-1	: 4935	3243-42-3	: 7952	3452-09-3	: 8196	3710-30-3	: 8319	3968-85-2	: 1953
3074-64-4	: 6542	3244-88-0	: 2218	3452-97-9	: 6951	3710-43-8	: 6088	3970-35-2	: 2544
3074-71-3	: 6384	3252-43-5	: 277	3454-07-7	: 1632	3710-84-7	: 5429	3970-62-5	: 8929
3074-75-7	: 6828	3254-93-1	: 7228	3457-58-7	: 2003	3711-34-0	: 7137	3971-28-6	: 4363
3074-76-8	: 6826	3256-88-0	: 10990	3458-28-4	: 7425	3711-37-3	: 7140	3972-56-3	: 1142
3074-77-9	: 6827	3260-87-5	: 9264	3460-18-2	: 1310	3721-95-7	: 4174	3972-65-4	: 1023
3075-84-1	: 3096	3261-62-9	: 1580	3460-20-6	: 2315	3722-73-4	: 2781	3973-62-4	: 9694
3077-16-5	: 7698	3268-49-3	: 9786	3460-24-0	: 1295	3722-74-5	: 2780	3974-99-0	: 254
3081-24-1	: 9512	3268-79-9	: 11700	3460-67-1	: 10904	3725-05-1	: 6714	3976-34-9	: 3056
3083-23-6	: 8628	3277-26-7	: 5237	3463-26-1	: 10837	3725-07-3	: 8340	3976-36-1	: 3095
3085-53-8	: 5997	3278-46-4	: 10332	3463-40-9	: 2316	3725-30-2	: 4750	3978-81-2	: 10944

CAS	No.	CAS	No.	CAS	No.	CAS	No.	CAS	No.	CAS	No.
3982-64-7	1483	4200-95-7	6305	4388-88-9	6811	4562-36-1	4042	4806-58-0	4986		
3982-66-9	12057	4205-23-6	6290	4389-53-1	2641	4573-05-1	4252	4806-59-1	4987		
3982-87-0	1712	4206-75-1	11384	4390-75-4	3408	4573-09-5	4897	4806-61-5	4191		
3988-03-2	7590	4208-49-5	6070	4392-24-9	1084	4588-18-5	8505	4810-09-7	6544		
3991-61-5	2012	4208-64-4	6121	4392-30-7	1245	4593-16-2	9631	4815-57-0	1557		
4001-73-4	604	4209-24-9	5787	4393-06-0	1868	4594-78-9	4467	4823-47-6	10611		
4008-48-4	11323	4209-90-9	6925	4394-11-0	3103	4597-87-9	10869	4829-04-3	5266		
4016-11-9	8614	4214-28-2	1770	4394-85-8	7685	4599-94-4	5152	4830-99-3	7273		
4018-65-9	1504	4215-86-5	2011	4395-73-7	1819	4604-28-8	1155	4831-43-0	11182		
4021-75-4	3612	4218-48-8	1690	4395-79-3	1201	4606-07-9	4961	4832-17-1	8055		
4028-23-3	11375	4219-24-3	7050	4395-80-6	1202	4606-65-9	9675	4835-11-4	6750		
4032-86-4	6388	4219-49-2	6652	4395-81-7	1782	4609-89-6	8833	4838-00-0	7242		
4032-92-2	6445	4224-70-8	6876	4396-98-9	1847	4609-92-1	8414	4839-46-7	8773		
4032-93-3	6444	4225-85-8	11244	4397-05-1	1842	4619-74-3	9480	4850-28-6	4862		
4032-94-4	8375	4228-00-6	5317	4398-65-6	6744	4620-70-6	5710	4850-32-2	4832		
4038-04-4	9044	4229-91-8	6167	4403-69-4	1796	4625-24-5	2980	4860-03-1	6634		
4041-09-2	4882	4253-34-3	11499	4405-42-9	11385	4628-21-1	3726	4877-77-4	1081		
4043-71-4	2430	4253-89-8	5244	4407-36-7	10704	4630-20-0	3986	4877-93-4	3045		
4044-65-9	1432	4253-91-2	10129	4412-16-2	5337	4630-82-4	4335	4884-24-6	2995		
4048-30-0	6578	4254-02-8	4779	4413-16-5	1249	4635-59-0	3685	4884-25-7	2994		
4048-31-1	6574	4255-62-3	4590	4422-95-1	2317	4636-16-2	5838	4887-30-3	6904		
4048-32-2	6581	4258-93-9	4449	4426-11-3	4173	4645-15-2	4461	4891-38-7	10742		
4049-81-4	6695	4259-00-1	4855	4426-51-1	5198	4654-39-1	1595	4894-61-5	3725		
4050-45-7	7000	4259-43-2	10179	4427-56-9	9417	4655-34-9	10664	4894-62-6	6335		
4053-34-3	11329	4264-29-3	2055	4428-13-1	1601	4662-96-8	1589	4897-84-1	3474		
4054-38-0	4219	4265-25-2	2456	4433-10-7	3054	4663-22-3	4982	4900-30-5	8110		
4057-42-5	8503	4265-97-8	8430	4433-11-8	3059	4667-38-3	11392	4901-51-3	9469		
4065-80-9	4537	4268-36-4	3933	4433-63-0	3133	4667-99-6	11387	4904-61-4	4212		
4068-78-4	2542	4271-30-1	6242	4435-14-7	4313	4669-01-6	4844	4911-55-1	8684		
4074-24-2	3322	4279-22-5	3280	4435-18-1	276	4670-56-8	2639	4911-65-3	1236		
4074-43-5	9247	4279-70-3	1993	4435-50-1	3458	4671-03-8	11848	4912-92-9	7251		
4079-52-1	5842	4279-76-9	1716	4436-59-3	4238	4672-38-2	9547	4914-91-4	9038		
4088-60-2	3860	4280-28-8	1927	4436-96-8	1867	4675-18-7	8581	4914-92-5	9037		
4088-81-7	9175	4282-24-0	6079	4437-18-7	6035	4676-51-1	3326	4915-21-3	6028		
4091-39-8	3647	4282-28-4	6080	4437-20-1	6101	4685-47-6	1921	4915-22-4	6029		
4095-09-4	11453	4282-31-9	11745	4437-22-3	6156	4694-12-6	4899	4915-23-5	6027		
4095-22-1	11772	4282-32-0	6083	4437-23-4	6158	4695-31-2	10288	4916-63-6	6994		
4096-20-2	9693	4282-40-0	6419	4437-50-7	6810	4701-36-4	1699	4920-84-7	1448		
4096-21-3	11171	4283-80-1	8717	4437-51-8	6806	4702-34-5	2797	4920-99-4	1689		
4096-34-8	4704	4285-42-1	3974	4439-20-7	5715	4705-34-4	1621	4923-77-7	4395		
4100-80-5	6094	4286-47-9	6778	4439-24-1	5746	4705-94-6	8027	4923-78-8	4396		
4101-68-2	5038	4286-49-1	3293	4441-57-0	4324	4706-81-4	11631	4923-84-6	9533		
4104-01-2	9025	4288-84-0	10355	4441-90-1	9965	4706-89-2	1472	4926-76-5	4398		
4107-62-4	10234	4291-79-6	4450	4442-79-9	4387	4712-55-4	9534	4926-78-7	4399		
4110-44-5	8379	4291-80-9	4451	4443-55-4	4384	4721-34-0	9544	4926-90-3	4394		
4110-50-3	10042	4291-81-0	4452	4444-12-6	6666	4721-37-3	9542	4930-98-7	11034		
4110-77-4	1153	4291-98-9	4934	4445-06-1	4459	4726-14-1	8070	4944-94-9	632		
4113-12-6	6642	4292-19-7	5302	4445-07-2	2066	4727-17-7	4760	4945-28-2	11301		
4114-28-7	7151	4292-75-5	4414	4445-08-3	8245	4730-22-7	6498	4945-47-5	9627		
4114-31-2	7148	4292-92-6	4465	4447-60-3	10079	4731-34-4	2076	4945-48-6	9635		
4117-15-1	11934	4295-06-1	11245	4455-26-9	8353	4731-84-4	9249	4946-13-8	2298		
4124-31-6	244	4301-39-7	6053	4457-00-5	4823	4734-90-1	4754	4949-44-4	8906		
4124-88-3	8187	4303-44-0	9855	4463-33-6	1439	4736-48-5	4214	4957-14-6	1964		
4126-78-7	4226	4313-03-5	6325	4467-06-5	10989	4737-41-1	8731	4964-71-0	11218		
4127-45-1	4985	4318-37-0	5141	4467-88-3	7271	4740-00-5	11788	4965-36-0	11220		
4128-31-8	8444	4318-76-7	10924	4468-42-2	1669	4744-08-5	9903	4972-29-6	2185		
4130-08-9	11498	4319-49-7	7680	4471-05-0	1909	4747-05-1	3161	4974-21-4	246		
4130-42-1	9221	4325-48-8	8729	4481-08-7	6198	4747-07-3	6833	4974-27-0	8320		
4131-74-2	10331	4325-49-9	6366	4481-30-5	1481	4747-15-3	1928	4975-21-7	6176		
4132-71-2	1442	4325-82-0	9099	4482-03-5	3076	4747-21-1	9819	4984-01-4	8502		
4132-77-8	1470	4325-97-7	5651	4484-72-4	11460	4749-27-3	10606	4984-85-4	6971		
4142-85-2	11440	4328-94-3	8852	4484-80-4	251	4749-28-4	10590	4985-70-0	374		
4143-41-3	5126	4339-05-3	7128	4485-09-0	8174	4749-31-9	10127	4985-85-7	5679		
4145-94-2	11673	4340-76-5	5359	4485-13-6	6559	4753-75-7	6115	4996-48-9	6068		
4147-51-7	5222	4341-76-8	3918	4485-16-9	6556	4753-80-4	11690	5006-66-6	10899		
4147-89-1	11362	4342-61-4	5226	4489-23-0	1082	4755-56-6	4247	5008-52-6	5007		
4156-16-5	7407	4346-18-3	3584	4491-19-4	307	4755-77-5	144	5008-69-5	10082		
4160-52-5	3672	4350-41-8	10921	4492-01-7	10830	4766-33-0	7711	5008-73-1	10077		
4160-80-9	10782	4351-54-6	6005	4493-42-9	5027	4766-57-8	11505	5009-11-0	7104		
4161-60-8	8973	4354-56-7	4412	4497-29-4	7524	4767-03-7	10272	5009-27-8	4988		
4164-28-7	7449	4358-59-2	3848	4510-34-3	10705	4771-47-5	2546	5020-21-3	7246		
4166-46-5	6926	4358-75-2	3216	4516-69-2	4856	4771-80-6	4638	5022-29-7	7364		
4167-74-2	9432	4358-87-6	910	4518-10-9	2504	4773-53-9	10364	5026-76-6	6547		
4167-77-5	4791	4358-88-7	907	4520-29-0	11447	4775-09-1	9527	5029-66-3	10284		
4169-04-4	10398	4360-12-7	321	4523-45-9	7720	4775-86-4	8800	5029-87-4	10970		
4170-24-5	3480	4360-96-7	9959	4524-77-0	9233	4780-79-4	7908	5035-52-9	11547		
4170-46-1	11466	4361-06-2	9968	4525-36-4	6165	4784-77-4	3712	5053-43-0	11070		
4171-83-9	2057	4362-01-0	4135	4525-44-4	11397	4789-40-6	6032	5058-19-5	10878		
4173-44-8	6710	4363-04-6	8019	4534-74-1	4525	4795-29-3	6116	5069-26-1	11774		
4175-54-6	7983	4363-93-3	11225	4535-86-4	5684	4795-86-2	2935	5075-33-2	7129		
4179-19-5	1441	4373-13-1	7817	4537-00-2	7763	4798-44-1	7066	5075-92-3	11181		
4180-23-8	1944	4375-14-8	7319	4537-05-7	2285	4798-46-3	7077	5076-20-0	8627		
4185-62-0	7745	4376-23-2	7090	4542-61-4	3117	4798-60-1	7081	5077-67-8	3660		
4185-63-1	7788	4377-37-1	11023	4549-31-9	6374	4798-61-2	8511	5090-41-5	8288		
4187-86-4	9140	4383-05-5	1885	4549-32-0	8369	4798-62-3	6575	5097-93-8	11001		
4187-87-5	1876	4383-18-0	1904	4549-74-0	8669	4798-64-5	8518	5100-98-1	11817		
4192-77-2	10685	4388-87-8	6807	4553-07-5	893	4799-68-2	10400	5103-42-4	2999		

CAS	Ref	CAS	Ref	CAS	Ref	CAS	Ref	CAS	Ref
5112-21-0	: 9999	5362-55-0	: 9095	5535-49-9	: 1165	5842-00-2	: 5650	6094-02-6	: 7036
5113-93-9	: 4682	5363-64-4	: 6813	5556-07-0	: 5835	5846-43-5	: 4559	6102-15-4	: 4641
5131-66-8	: 10346	5367-28-2	: 1213	5557-31-3	: 8292	5856-77-9	: 3688	6108-61-8	: 6687
5132-75-2	: 6468	5368-81-0	: 2674	5558-29-2	: 963	5856-82-6	: 3680	6109-22-4	: 422
5137-45-1	: 5591	5370-25-2	: 5783	5558-31-6	: 964	5857-36-3	: 8989	6111-61-1	: 3643
5145-64-2	: 11138	5371-48-2	: 9119	5558-66-7	: 931	5857-86-3	: 4558	6111-78-0	: 521
5145-99-3	: 10043	5379-20-4	: 1627	5558-78-1	: 970	5859-05-0	: 7784	6111-88-2	: 8733
5150-80-1	: 6341	5380-40-5	: 6069	5560-69-0	: 7968	5867-45-8	: 89	6111-99-5	: 10236
5150-93-6	: 3333	5390-04-5	: 9143	5563-78-0	: 4547	5870-61-1	: 5882	6114-18-7	: 8302
5153-67-3	: 2050	5390-71-6	: 9861	5581-35-1	: 1583	5870-93-9	: 3527	6117-80-2	: 3793
5159-41-1	: 1888	5390-72-7	: 9859	5581-75-9	: 1759	5876-87-9	: 5285	6118-14-5	: 3858
5160-99-6	: 1997	5392-86-9	: 1375	5582-82-1	: 6500	5877-55-4	: 809	6119-92-2	: 5182
5161-04-6	: 1999	5395-43-7	: 11553	5586-15-2	: 5256	5877-58-7	: 810	6120-13-4	: 2919
5161-13-7	: 11785	5395-75-5	: 5451	5587-63-3	: 4910	5878-19-3	: 10455	6125-21-9	: 9919
5161-16-0	: 11796	5396-24-7	: 149	5587-79-1	: 4948	5881-17-4	: 8403	6125-24-2	: 2051
5161-17-1	: 11798	5396-38-3	: 1462	5588-20-5	: 7220	5882-44-0	: 717	6127-92-0	: 3827
5162-44-7	: 3711	5396-58-7	: 3355	5589-96-8	: 112	5891-21-4	: 8961	6127-93-1	: 3828
5162-48-1	: 8955	5396-91-8	: 8969	5597-27-3	: 2958	5897-76-7	: 1840	6130-75-2	: 2331
5166-35-8	: 3734	5398-08-3	: 9961	5597-81-9	: 7782	5902-51-2	: 11591	6131-24-4	: 6835
5166-53-0	: 7094	5398-10-7	: 9955	5597-89-7	: 8687	5906-73-0	: 11501	6131-25-5	: 6423
5171-84-6	: 6854	5398-25-4	: 8742	5598-13-0	: 4090	5908-87-2	: 5277	6134-66-3	: 3503
5182-44-5	: 1596	5398-75-4	: 9177	5609-09-6	: 6587	5910-87-2	: 8109	6135-29-1	: 262
5183-77-7	: 4347	5399-02-0	: 6312	5614-32-4	: 3160	5910-89-4	: 10805	6136-67-0	: 7624
5183-79-9	: 4360	5405-13-0	: 1821	5617-41-4	: 4405	5911-04-6	: 8141	6137-08-2	: 8475
5194-50-3	: 6686	5405-15-2	: 1825	5617-42-5	: 4820	5911-08-0	: 4966	6137-15-1	: 8476
5194-51-4	: 6685	5405-17-4	: 1824	5617-92-5	: 4978	5912-58-3	: 5694	6138-90-5	: 8321
5195-37-9	: 7982	5405-41-4	: 3534	5630-56-8	: 7598	5912-86-7	: 9397	6139-84-0	: 3181
5195-40-4	: 7981	5405-79-8	: 6965	5634-22-0	: 1565	5912-87-8	: 9393	6140-17-6	: 2045
5208-49-1	: 4050	5406-12-2	: 2116	5634-30-0	: 4809	5913-85-9	: 9706	6140-65-4	: 4908
5209-18-7	: 2161	5406-18-8	: 2154	5636-65-7	: 7061	5915-41-3	: 11593	6140-80-3	: 10579
5209-33-6	: 5	5407-98-7	: 7605	5637-96-7	: 2070	5917-47-5	: 9660	6142-95-6	: 1412
5216-25-1	: 1239	5408-57-1	: 8477	5638-26-6	: 6336	5921-54-0	: 3673	6149-41-3	: 10276
5216-32-0	: 1366	5408-86-6	: 3268	5648-29-3	: 5614	5921-73-3	: 8209	6149-46-8	: 8884
5217-05-0	: 4783	5410-97-9	: 5166	5650-40-8	: 10451	5921-84-6	: 6476	6152-31-4	: 7171
5221-17-0	: 9780	5411-50-7	: 1307	5650-41-9	: 10452	5930-28-9	: 9206	6153-16-8	: 4674
5223-59-6	: 1676	5411-56-3	: 1845	5650-75-9	: 8890	5932-68-3	: 9396	6153-56-6	: 5561
5234-68-4	: 4039	5412-92-0	: 8352	5661-71-2	: 503	5943-35-1	: 11649	6158-45-8	: 7920
5251-93-4	: 524	5413-05-8	: 884	5663-96-7	: 8538	5944-41-2	: 10942	6159-55-3	: 11192
5258-50-4	: 11797	5414-19-7	: 5609	5666-21-7	: 10527	5951-67-7	: 4659	6161-50-8	: 3046
5259-65-4	: 4714	5414-21-1	: 8828	5668-93-9	: 3073	5959-52-4	: 7768	6162-79-4	: 3951
5259-98-3	: 8923	5416-94-4	: 11704	5669-09-0	: 10041	5963-77-9	: 9137	6163-64-0	: 10083
5263-87-6	: 11275	5419-55-6	: 3128	5675-22-9	: 6327	5965-66-2	: 6214	6163-66-2	: 10124
5265-18-9	: 10414	5421-04-5	: 912	5675-51-4	: 5301	5966-51-8	: 10343	6163-75-3	: 9537
5271-38-5	: 5749	5424-21-5	: 11053	5676-58-4	: 2879	5976-47-6	: 10700	6164-98-3	: 7564
5271-39-6	: 1679	5426-65-3	: 10846	5683-33-0	: 10863	5984-58-7	: 6353	6165-44-2	: 4323
5272-02-6	: 6369	5429-85-6	: 8290	5683-44-3	: 8794	5985-23-9	: 6246	6166-86-5	: 4902
5274-50-0	: 9554	5432-61-1	: 4289	5685-06-3	: 11680	5986-55-0	: 7586	6166-87-6	: 4625
5281-18-5	: 563	5432-85-9	: 4600	5687-92-3	: 11692	6001-87-2	: 10225	6168-72-5	: 10336
5283-66-9	: 11471	5434-27-5	: 9887	5689-04-3	: 7292	6004-60-0	: 5795	6174-95-4	: 5916
5285-18-7	: 8677	5441-51-0	: 4593	5697-85-8	: 838	6006-06-0	: 11895	6175-23-1	: 8137
5285-87-0	: 11709	5441-52-1	: 4518	5699-58-1	: 8875	6006-33-3	: 4484	6175-49-1	: 5325
5292-21-7	: 4306	5445-24-9	: 4557	5699-74-1	: 8829	6006-34-4	: 4854	6195-92-2	: 4886
5292-43-3	: 114	5447-97-2	: 9847	5703-21-9	: 869	6006-90-2	: 8638	6196-58-3	: 5757
5292-53-5	: 9978	5447-99-4	: 8952	5703-26-4	: 871	6006-95-7	: 4463	6196-60-7	: 9047
5307-19-7	: 10951	5448-43-1	: 11334	5703-52-6	: 2146	6012-97-1	: 11782	6202-04-6	: 11839
5311-96-6	: 7150	5451-80-9	: 8881	5704-20-1	: 8975	6020-51-5	: 7224	6213-90-7	: 3821
5314-37-4	: 3070	5451-85-4	: 8901	5715-23-1	: 4517	6031-02-3	: 2009	6214-25-1	: 3832
5314-55-6	: 11433	5452-37-9	: 4735	5728-52-9	: 3009	6032-29-7	: 8916	6214-28-4	: 3820
5324-00-5	: 7272	5452-75-5	: 4308	5736-03-8	: 5206	6035-49-0	: 2845	6222-35-1	: 10235
5325-97-3	: 9185	5453-44-1	: 3857	5743-97-5	: 9186	6040-62-6	: 8058	6223-78-5	: 6765
5326-81-8	: 45	5453-80-5	: 2967	5744-03-6	: 5973	6044-68-4	: 10562	6225-06-5	: 8699
5327-44-6	: 1924	5454-28-4	: 6459	5750-00-5	: 3248	6050-13-1	: 5145	6231-18-1	: 10935
5328-01-8	: 7883	5454-79-5	: 4533	5756-43-4	: 6814	6051-43-0	: 2489	6236-88-0	: 4400
5332-06-9	: 3404	5455-24-3	: 8396	5763-61-1	: 1805	6051-52-1	: 1869	6258-66-8	: 1833
5332-24-1	: 11217	5458-48-0	: 1252	5765-44-6	: 7405	6057-80-9	: 3019	6261-22-9	: 9141
5332-25-2	: 11219	5458-59-3	: 8435	5773-56-8	: 1614	6061-06-9	: 3156	6265-26-5	: 6166
5332-52-5	: 11942	5459-58-5	: 151	5775-82-6	: 10816	6061-10-5	: 8433	6267-24-9	: 5605
5332-73-0	: 9816	5459-93-8	: 4288	5779-94-2	: 548	6064-27-3	: 5326	6268-37-7	: 1228
5335-75-1	: 10879	5460-32-2	: 1768	5779-95-3	: 549	6065-59-4	: 9974	6269-92-7	: 8366
5337-03-1	: 10774	5461-51-8	: 74	5781-53-3	: 145	6065-82-3	: 174	6270-16-2	: 3642
5337-70-2	: 8552	5465-09-8	: 4588	5787-31-5	: 3232	6065-90-3	: 8752	6270-34-4	: 6061
5337-72-4	: 4515	5465-13-4	: 2282	5792-36-9	: 10430	6065-93-6	: 10555	6270-55-9	: 10300
5337-93-9	: 10468	5468-97-3	: 1598	5794-88-7	: 2487	6068-69-5	: 816	6270-56-0	: 6104
5339-85-5	: 1594	5469-19-2	: 1093	5798-75-4	: 2523	6069-71-2	: 2954	6272-38-4	: 9451
5340-36-3	: 8466	5470-02-0	: 9707	5798-79-8	: 947	6069-97-2	: 2937	6272-40-8	: 7588
5340-41-0	: 6953	5470-28-0	: 9606	5798-88-9	: 3472	6069-98-3	: 4439	6274-16-4	: 3950
5342-31-4	: 7509	5470-47-3	: 882	5798-94-7	: 606	6078-26-8	: 310	6279-86-3	: 7552
5344-20-7	: 8707	5470-91-7	: 10712	5802-82-4	: 6625	6080-79-1	: 9125	6280-03-1	: 8149
5344-55-8	: 3590	5472-38-8	: 3324	5805-76-5	: 2381	6086-22-2	: 11847	6280-96-2	: 9460
5344-82-1	: 11801	5500-21-0	: 4956	5807-64-7	: 3039	6088-94-4	: 8114	6280-98-4	: 1555
5344-90-1	: 1837	5502-88-5	: 4680	5807-76-1	: 1096	6089-09-4	: 9138	6284-40-8	: 6208
5345-01-7	: 8777	5503-12-8	: 4634	5809-07-4	: 6103	6089-12-9	: 3600	6284-84-0	: 9609
5349-18-8	: 2032	5507-44-8	: 11422	5813-64-9	: 9806	6089-13-0	: 4716	6285-05-8	: 10422
5349-55-3	: 10281	5510-99-6	: 9227	5814-42-6	: 10392	6091-52-7	: 5326	6285-06-9	: 9145
5349-60-0	: 1873	5515-83-3	: 328	5814-85-7	: 1969	6091-58-3	: 3351	6287-38-3	: 536
5350-47-0	: 7630	5519-42-6	: 10842	5836-66-8	: 9888	6091-64-1	: 2521	6290-24-0	: 2598
5356-83-2	: 11423	5519-50-6	: 9723	5837-78-5	: 3849	6093-68-1	: 2825	6290-49-9	: 207

CAS	No.	CAS	No.	CAS	No.	CAS	No.	CAS	No.
6292-82-6	9710	6630-33-7	526	6943-58-4	10356	7225-64-1	6313	7567-22-8	4743
6294-31-1	6855	6639-30-1	2336	6946-29-8	2500	7225-66-3	11878	7568-92-5	1593
6294-89-9	7149	6639-82-3	11336	6946-35-6	2668	7235-40-7	4051	7568-93-6	1839
6298-37-9	11332	6640-27-3	9261	6946-88-9	3338	7236-47-7	3703	7570-26-5	5552
6302-94-9	5264	6640-77-3	2685	6950-43-2	2530	7239-24-9	9829	7570-98-1	7951
6304-18-3	10985	6641-83-4	8895	6950-84-1	12022	7242-17-3	3780	7572-29-4	5940
6306-30-5	6984	6669-13-2	1458	6951-08-2	2419	7250-95-5	7712	7583-53-1	9678
6308-98-1	1585	6682-71-9	7253	6952-59-6	2745	7251-61-8	11337	7611-43-0	472
6309-50-8	5308	6688-11-5	4738	6959-06-4	3624	7252-83-7	5460	7611-86-1	3721
6310-09-4	5791	6693-29-4	4302	6963-44-6	8791	7283-69-4	3767	7611-87-2	3720
6310-21-0	714	6693-30-7	4301	6964-04-1	6401	7283-70-7	3766	7614-93-9	2022
6311-60-0	1309	6693-31-8	4300	6967-43-7	3623	7283-96-7	11735	7615-57-8	11836
6314-28-9	2400	6697-12-7	6664	6969-16-0	11894	7287-19-6	9770	7616-22-0	2790
6315-03-3	2695	6703-78-2	6298	6971-81-3	1890	7287-81-2	1858	7619-08-1	8227
6315-60-2	330	6703-80-6	6297	6973-79-1	10206	7291-33-0	97	7623-01-0	4997
6315-96-4	10474	6703-82-8	4819	6974-10-3	7361	7297-25-8	3432	7623-10-1	10499
6319-21-7	5135	6704-31-0	8596	6974-77-2	9842	7300-03-0	6555	7623-11-2	3683
6319-40-0	2529	6705-31-3	10808	6975-60-6	10439	7307-55-3	11937	7623-13-4	10501
6320-03-2	2293	6709-22-4	4913	6975-92-4	7015	7311-34-4	546	7642-04-8	8493
6320-40-7	2314	6709-39-3	6331	6975-98-0	5054	7314-06-9	732	7642-09-3	7003
6322-01-6	10339	6728-26-3	6996	6975-99-1	5346	7314-37-6	11620	7642-10-6	6537
6322-49-2	3648	6734-98-1	7007	6976-00-7	11963	7315-32-4	2425	7642-15-1	8497
6326-44-9	9953	6738-23-4	1917	6976-72-3	6889	7318-67-4	6684	7658-08-4	8235
6338-45-0	7201	6739-34-0	4337	6982-25-8	3347	7319-00-8	6683	7661-39-4	11222
6341-60-2	9970	6740-88-1	4579	6983-79-5	5108	7321-48-4	8500	7661-55-4	11280
6341-97-5	9288	6742-54-7	2374	6986-48-7	5610	7321-55-3	9913	7665-72-7	8610
6344-28-1	5786	6748-95-4	436	6995-30-8	7905	7328-33-8	5026	7667-55-2	4491
6344-72-5	11338	6753-98-6	5006	6995-79-5	4374	7332-93-6	8695	7667-58-5	4494
6346-09-4	3214	6765-39-5	6322	6996-92-5	2174	7335-01-5	9672	7667-59-6	4493
6351-10-6	7284	6776-19-8	3840	6998-82-9	4352	7335-02-6	9690	7667-60-9	4495
6361-23-5	535	6780-13-8	5513	7005-72-3	1221	7335-06-0	11166	7681-84-7	6049
6372-01-6	3831	6780-49-0	7567	7012-16-0	5972	7335-07-1	11173	7683-59-2	1522
6374-70-5	5146	6781-42-6	5865	7013-11-8	3744	7335-25-3	2537	7688-21-3	7001
6378-11-6	4087	6781-98-2	1139	7021-04-7	889	7335-26-4	2669	7693-52-9	9244
6378-65-0	6890	6784-18-5	4890	7035-03-2	957	7335-27-5	2539	7696-51-7	11129
6380-21-8	9456	6785-23-5	4863	7036-98-8	8191	7344-34-5	7278	7697-46-3	7637
6380-24-1	1451	6789-80-6	6998	7040-43-9	6090	7351-74-8	7803	7700-17-6	4137
6380-28-5	9419	6789-88-4	2610	7044-96-4	7654	7357-93-9	3797	7712-60-9	9026
6380-34-3	464	6789-94-2	9626	7045-89-8	11903	7359-72-0	2334	7712-66-5	11814
6382-13-4	8270	6790-27-8	5345	7058-01-7	4448	7367-38-6	8184	7731-28-4	4535
6402-36-4	5336	6790-47-2	4307	7065-46-5	3690	7369-50-8	1693	7731-29-5	4536
6414-69-3	10283	6795-75-1	5935	7069-38-7	10548	7371-67-7	4181	7735-33-3	8869
6423-02-5	6758	6802-75-1	9966	7081-78-9	5473	7372-85-2	3055	7740-33-2	3589
6423-29-6	7804	6812-38-0	4411	7086-07-9	10147	7372-92-1	7280	7745-91-7	649
6436-90-4	6272	6812-39-1	4822	7087-68-5	9811	7379-12-6	6979	7745-93-9	1072
6443-91-0	5944	6812-78-8	8516	7094-26-0	4485	7383-20-2	8208	7756-94-7	10582
6443-92-1	6535	6814-58-0	4083	7094-27-1	4487	7383-90-6	3058	7766-23-6	9238
6453-99-2	7642	6818-07-1	6901	7094-34-0	7591	7385-78-6	9032	7771-44-0	5360
6482-24-2	5465	6829-40-9	9924	7098-07-9	7207	7385-82-2	7044	7773-60-6	9358
6482-34-4	3995	6829-41-0	9958	7098-21-7	11918	7391-39-1	6881	7778-87-2	6471
6493-77-2	4640	6831-89-6	10831	7101-96-4	11215	7398-82-5	989	7779-80-8	6467
6493-83-0	10457	6832-98-0	6630	7116-86-1	7021	7399-49-7	1984	7781-98-8	2620
6496-48-6	11176	6836-38-0	5324	7119-92-8	5431	7399-50-0	10967	7782-21-0	943
6498-47-1	3868	6843-66-9	11413	7119-94-0	9687	7400-08-0	10644	7785-26-4	2971
6513-13-9	12039	6846-11-3	7221	7132-64-1	8646	7403-66-9	42	7786-34-7	7659
6526-72-3	1977	6848-84-6	8769	7133-68-8	3358	7415-31-8	3748	7786-44-9	8112
6531-35-7	403	6849-18-9	9082	7137-56-6	1251	7417-19-8	1599	7787-72-6	7792
6538-02-9	5379	6850-36-8	4287	7144-05-0	9673	7417-48-3	8666	7789-89-1	10171
6553-48-6	4187	6850-57-3	1812	7145-20-2	7022	7423-69-0	7018	7789-90-4	9790
6556-12-3	6238	6852-54-6	5437	7146-60-3	8374	7425-45-8	3483	7789-92-6	10183
6568-58-7	4731	6861-64-9	7	7148-74-5	10495	7425-47-0	3928	7789-99-3	8945
6573-52-0	4729	6871-44-9	4730	7149-70-4	1066	7425-48-1	3484	9005-79-2	6274
6579-55-1	10383	6873-15-0	11200	7149-75-9	674	7428-48-0	8261	10008-73-8	6148
6592-85-4	5217	6874-84-2	11346	7152-15-0	8897	7433-56-9	5084	10021-92-8	4353
6597-78-0	7653	6876-23-9	4366	7152-88-7	2628	7433-78-5	5085	10022-60-3	6780
6600-40-4	8218	6886-16-4	6921	7153-14-2	4576	7439-33-0	4185	10024-74-5	1822
6607-66-5	3456	6890-03-5	4520	7154-66-7	2887	7443-55-2	4534	10031-82-0	554
6609-56-9	2760	6891-45-8	9250	7154-73-6	11164	7446-43-7	9814	10031-87-5	185
6609-57-0	2754	6894-69-5	4912	7154-75-8	9135	7459-46-3	5654	10042-29-2	268
6617-04-5	4311	6897-76-3	8014	7154-79-2	8839	7462-74-0	9792	10048-32-5	10791
6622-76-0	3851	6897-88-7	8015	7154-80-5	6450	7466-38-8	5133	10061-01-5	10551
6624-76-6	8100	6912-05-6	4923	7163-25-9	7773	7474-10-4	4720	10061-02-6	10550
6626-15-9	1502	6917-76-6	11377	7164-98-9	7212	7474-83-1	7844	10071-60-0	9020
6626-32-0	9347	6921-35-3	8593	7168-55-0	3048	7488-99-5	4054	10075-38-4	3749
6627-55-0	9236	6921-40-0	7738	7169-34-8	2462	7493-82-5	6470	10088-95-6	10785
6627-72-1	2942	6921-64-8	5826	7170-38-9	10320	7495-97-8	9526	10124-73-9	7005
6627-74-3	4667	6923-20-2	10549	7187-01-1	10585	7498-57-9	7737	10130-87-7	2248
6628-00-8	4304	6923-22-4	7663	7193-78-4	5844	7499-65-2	7744	10130-91-3	4566
6628-04-2	11210	6931-71-1	6800	7200-25-1	437	7521-80-4	11370	10136-52-4	6324
6628-06-4	9429	6933-10-4	651	7205-90-5	1136	7523-44-6	3865	10138-10-0	3572
6628-18-8	5452	6938-94-9	6782	7208-05-1	8577	7524-63-2	7988	10138-89-3	3455
6628-21-3	10246	6939-75-9	8911	7208-92-6	5554	7525-62-4	1631	10140-89-3	9781
6628-77-9	10865	6940-09-6	8275	7209-38-3	9614	7526-26-3	9541	10143-54-1	10107
6628-79-1	8896	6940-76-7	9857	7214-50-8	4712	7529-63-7	9697	10143-60-9	6431
6628-80-4	4505	6940-78-9	3233	7216-56-0	8488	7539-25-5	364	10147-36-1	10133
6628-97-3	7866	6941-69-1	129	7220-26-0	6818	7554-65-6	10836	10151-95-8	80
6629-96-5	5192	6942-99-0	1313	7220-81-7	4757	7564-63-8	1630	10152-76-8	10581

CAS	No.	CAS	No.	CAS	No.	CAS	No.	CAS	No.
10160-87-9	: 10731	10574-36-4	: 7041	13402-36-3	: 4748	14077-58-8	: 298	15068-90-3	: 2950
10160-99-3	: 6627	10574-37-5	: 9034	13402-37-4	: 4744	14090-87-0	: 8393	15104-61-7	: 10126
10166-08-2	: 873	10575-87-8	: 6378	13403-37-7	: 8605	14092-20-7	: 7106	15111-56-5	: 4639
10193-50-7	: 9406	10577-44-3	: 1941	13417-43-1	: 3735	14108-88-4	: 7993	15116-17-3	: 1534
10203-08-4	: 537	10588-10-0	: 8900	13423-15-9	: 6172	14116-69-9	: 6234	15121-84-3	: 1612
10203-28-8	: 5322	10602-34-3	: 3758	13436-48-1	: 7239	14122-00-0	: 9280	15150-39-7	: 140
10203-30-2	: 5323	10604-59-8	: 7307	13444-24-1	: 9715	14159-48-9	: 10130	15164-44-0	: 582
10203-58-4	: 9969	10605-21-7	: 3990	13463-39-3	: 8068	14167-59-0	: 11650	15192-80-0	: 8489
10210-64-7	: 2912	10606-73-2	: 891	13463-40-6	: 7339	14179-94-3	: 9126	15220-85-6	: 10580
10212-74-5	: 3960	11024-24-1	: 6195	13466-40-5	: 8657	14180-63-3	: 10596	15232-76-5	: 8531
10220-20-9	: 9659	11028-42-5	: 4066	13466-41-6	: 11037	14181-05-6	: 11145	15232-96-9	: 1247
10222-95-4	: 2361	12069-69-1	: 511	13471-42-6	: 6915	14188-79-5	: 7794	15240-91-2	: 1800
10226-29-6	: 6961	12122-67-7	: 12067	13475-76-8	: 5272	14214-31-4	: 10538	15250-29-0	: 10670
10229-10-4	: 9142	12321-04-9	: 5961	13475-77-9	: 5354	14222-60-7	: 10886	15268-49-2	: 10086
10230-68-9	: 10475	12427-38-2	: 7420	13475-78-0	: 6416	14248-22-7	: 2129	15299-99-7	: 8067
10233-13-3	: 5316	13012-59-4	: 9864	13475-79-1	: 8766	14252-42-7	: 5450	15301-40-3	: 11293
10236-47-2	: 2795	13014-24-9	: 1401	13475-81-5	: 6844	14255-23-3	: 7023	15320-72-6	: 3833
10240-08-1	: 8030	13019-20-0	: 6522	13475-82-6	: 6433	14272-54-9	: 9115	15329-10-9	: 4708
10250-47-2	: 8949	13019-22-2	: 5092	13481-70-4	: 9294	14273-06-4	: 10732	15333-22-9	: 3829
10257-92-8	: 7318	13023-00-2	: 3500	13482-62-7	: 9524	14282-76-9	: 11729	15339-36-3	: 7421
10259-22-0	: 2670	13024-80-1	: 1750	13491-79-7	: 4521	14287-61-7	: 3507	15356-70-4	: 4548
10265-92-6	: 7441	13027-43-5	: 1650	13505-34-5	: 6403	14296-80-1	: 4172	15358-48-2	: 7574
10273-90-2	: 10991	13030-26-7	: 2778	13511-38-1	: 10219	14303-70-9	: 11629	15361-99-6	: 2277
10276-09-2	: 3835	13038-21-6	: 8190	13513-82-1	: 1893	14304-30-4	: 5087	15409-60-6	: 4591
10283-15-5	: 9256	13045-99-3	: 3798	13529-27-6	: 6084	14315-14-1	: 2402	15411-43-5	: 734
10289-68-6	: 11885	13047-06-8	: 7600	13532-18-8	: 10311	14321-27-8	: 1810	15430-98-5	: 8854
10292-98-5	: 2963	13054-87-0	: 3595	13545-04-5	: 3317	14331-54-5	: 11055	15457-05-3	: 5980
10293-06-8	: 2952	13057-17-5	: 7472	13552-31-3	: 9984	14352-61-5	: 4310	15466-93-0	: 4511
10297-09-3	: 8541	13057-19-7	: 7519	13556-55-3	: 7985	14371-10-9	: 10518	15466-94-1	: 4510
10297-38-8	: 11571	13063-43-9	: 9912	13556-58-6	: 7888	14379-80-7	: 6266	15469-77-9	: 5089
10299-06-6	: 6867	13065-07-1	: 7989	13602-13-6	: 3747	14400-94-3	: 9467	15489-90-4	: 5959
10302-94-0	: 3772	13071-79-9	: 11592	13603-04-8	: 11150	14401-73-1	: 5797	15501-33-4	: 10056
10307-61-6	: 3550	13073-29-5	: 9423	13603-21-9	: 9636	14411-75-7	: 1471	15519-73-0	: 3068
10309-06-5	: 2959	13073-35-3	: 991	13618-93-4	: 7325	14436-32-9	: 5091	15521-18-3	: 10374
10311-74-7	: 1622	13086-84-5	: 9531	13632-93-4	: 1416	14444-77-0	: 1409	15543-70-1	: 4965
10311-84-9	: 5105	13099-50-8	: 1227	13632-94-5	: 1420	14446-69-6	: 9653	15570-12-4	: 2299
10312-83-1	: 16	13102-31-3	: 9449	13632-95-6	: 1418	14452-30-3	: 5836	15600-01-8	: 10053
10315-89-6	: 9685	13105-52-7	: 3934	13643-05-5	: 8667	14459-29-1	: 9735	15606-95-8	: 439
10316-00-4	: 7700	13114-62-0	: 12023	13665-04-8	: 5798	14472-14-1	: 9242	15644-80-1	: 11316
10317-10-9	: 10353	13116-53-5	: 10145	13672-18-9	: 7730	14476-30-3	: 10162	15647-08-2	: 9589
10321-71-8	: 9078	13116-60-4	: 10143	13678-54-1	: 11764	14476-37-0	: 11960	15647-11-7	: 4502
10323-40-7	: 3471	13121-70-5	: 5011	13679-46-4	: 6128	14484-64-1	: 5957	15657-96-2	: 10091
10324-58-0	: 9692	13129-23-2	: 6063	13679-70-4	: 11736	14490-79-0	: 11608	15679-02-4	: 4188
10327-08-9	: 10204	13138-51-7	: 3429	13679-74-8	: 5856	14548-31-3	: 7105	15706-73-7	: 3545
10334-26-6	: 2926	13147-06-3	: 11639	13684-56-5	: 5104	14548-46-0	: 7639	15721-05-8	: 4992
10339-67-0	: 6602	13151-04-7	: 6546	13684-63-4	: 9198	14558-14-6	: 8003	15726-15-5	: 6525
10340-91-7	: 1791	13151-05-8	: 6545	13688-75-0	: 11468	14575-93-0	: 2965	15727-65-8	: 7892
10342-83-3	: 10418	13151-17-2	: 7045	13698-16-3	: 3940	14595-35-8	: 3783	15764-04-2	: 8052
10343-99-4	: 11246	13151-34-3	: 5055	13698-87-8	: 1326	14596-92-0	: 6682	15764-16-6	: 547
10373-81-6	: 2957	13153-11-2	: 11694	13706-86-0	: 6809	14602-62-1	: 10161	15797-21-4	: 4795
10394-35-1	: 2378	13163-73-0	: 4604	13721-54-5	: 7103	14620-52-1	: 5297	15802-97-8	: 10841
10403-60-8	: 3451	13165-77-0	: 2181	13732-80-4	: 1417	14620-53-2	: 11617	15802-99-0	: 10839
10410-35-2	: 2448	13172-91-3	: 6557	13794-14-4	: 3580	14620-55-4	: 8237	15825-70-4	: 7424
10420-90-3	: 6712	13176-46-0	: 3478	13799-90-1	: 1349	14630-40-1	: 11434	15826-80-9	: 2981
10420-91-4	: 6715	13194-48-4	: 5934	13808-64-5	: 10817	14631-45-9	: 10102	15840-60-5	: 7047
10424-29-0	: 3341	13195-80-7	: 10542	13811-01-3	: 4274	14660-52-7	: 8861	15843-48-8	: 5568
10428-64-5	: 3272	13206-46-7	: 9776	13828-34-7	: 4441	14676-29-0	: 6409	15854-58-7	: 10647
10435-81-1	: 8419	13222-26-9	: 10498	13831-30-6	: 101	14685-29-1	: 4995	15869-80-4	: 6407
10436-75-6	: 794	13249-58-6	: 1030	13835-81-9	: 11467	14686-13-6	: 6534	15869-85-9	: 8143
10439-23-3	: 2186	13250-46-9	: 303	13855-77-1	: 3964	14686-14-7	: 6536	15869-86-0	: 8404
10443-65-9	: 10610	13254-34-7	: 6479	13862-07-2	: 9665	14689-97-5	: 5592	15869-87-1	: 8373
10453-86-8	: 11340	13258-63-4	: 10949	13865-48-0	: 2018	14697-46-2	: 8851	15869-89-3	: 8376
10463-48-6	: 3740	13269-52-8	: 7002	13877-91-3	: 8485	14716-89-3	: 7404	15869-92-8	: 8380
10473-14-0	: 3871	13279-87-3	: 3448	13898-58-3	: 2514	14719-83-6	: 2548	15869-93-9	: 8381
10476-95-6	: 10564	13290-74-9	: 1205	13898-73-2	: 4678	14720-74-2	: 6438	15869-94-0	: 8382
10478-02-1	: 2592	13313-91-2	: 261	13905-48-1	: 4320	14721-66-5	: 6658	15869-95-1	: 8383
10478-12-3	: 9083	13325-10-5	: 3597	13909-73-4	: 5874	14737-94-1	: 10622	15869-96-2	: 8384
10486-19-8	: 11869	13327-31-6	: 11273	13937-08-1	: 9957	14752-60-4	: 9124	15870-10-7	: 6543
10487-71-5	: 3887	13329-40-3	: 5837	13939-06-5	: 7662	14752-61-5	: 9133	15874-03-0	: 6637
10487-96-4	: 4871	13347-42-7	: 9255	13956-29-1	: 1537	14752-75-1	: 1743	15877-57-3	: 8694
10488-25-2	: 4539	13351-61-6	: 2151	13963-35-4	: 2017	14804-32-1	: 1682	15890-36-5	: 4541
10488-94-5	: 8436	13351-73-0	: 2872	13979-28-7	: 3836	14804-59-2	: 10510	15890-40-1	: 4859
10489-97-1	: 4345	13356-08-6	: 5950	13984-57-1	: 6906	14813-68-4	: 8345	15901-49-2	: 5003
10495-45-1	: 10132	13360-57-1	: 4076	13993-65-2	: 9500	14845-35-3	: 4706	15909-76-9	: 10254
10498-35-8	: 4359	13360-61-7	: 8653	13997-70-1	: 1232	14845-55-7	: 4887	15918-08-8	: 8426
10500-57-9	: 11297	13360-63-9	: 3218	13997-73-4	: 9271	14847-51-9	: 9237	15922-53-9	: 11849
10503-96-5	: 3249	13360-65-1	: 10809	14007-64-8	: 898	14850-22-7	: 8495	15932-80-6	: 4607
10508-09-5	: 9154	13361-35-8	: 143	14010-23-2	: 6315	14850-23-8	: 8496	15950-66-0	: 9481
10519-06-9	: 1117	13361-36-9	: 137	14021-92-2	: 5048	14852-31-4	: 6660	15953-45-4	: 10821
10519-33-2	: 5093	13361-63-2	: 9129	14045-26-2	: 9671	14869-39-7	: 5814	15953-73-8	: 10819
10521-91-2	: 2101	13362-52-2	: 11876	14062-18-1	: 923	14919-01-8	: 8494	15971-29-6	: 7755
10522-26-6	: 11955	13367-29-8	: 2993	14062-78-3	: 627	14924-53-9	: 4175	15972-60-8	: 322
10524-08-0	: 9018	13388-93-7	: 5846	14064-10-9	: 9931	14938-35-3	: 9444	15980-15-1	: 8569
10528-67-3	: 4469	13389-42-9	: 8492	14064-43-8	: 1223	14963-40-7	: 4946	16003-14-8	: 6160
10544-63-5	: 3838	13398-94-2	: 1606	14064-68-7	: 2768	14979-39-6	: 6501	16009-13-5	: 5958
10568-38-4	: 1683	13402-32-9	: 1308	14072-86-7	: 4656	14996-78-2	: 4236	16020-17-0	: 8001
10574-17-1	: 2030	13402-35-2	: 4746	14073-97-3	: 4606	15014-25-2	: 9927	16021-20-8	: 1702

CAS	No.	CAS	No.	CAS	No.	CAS	No.	CAS	No.
16066-09-4	: 11647	16789-46-1	: 6825	17745-45-8	: 3136	18913-31-0	: 3175	19780-66-6	: 9046
16096-32-5	: 7315	16789-51-8	: 7034	17773-65-8	: 3737	18917-91-4	: 340	19780-87-7	: 9049
16106-59-5	: 7020	16790-49-1	: 3197	17788-00-0	: 9292	18932-14-4	: 8420	19780-90-6	: 6798
16109-68-5	: 5120	16805-78-0	: 9578	17789-14-9	: 5205	18936-17-9	: 3409	19780-96-2	: 8164
16133-49-6	: 777	16813-18-6	: 6318	17804-35-2	: 515	18937-76-3	: 11990	19781-10-3	: 8451
16136-84-8	: 10534	16837-43-7	: 2118	17837-41-1	: 11693	18963-00-3	: 10176	19781-11-4	: 8458
16136-85-9	: 10533	16841-48-8	: 10597	17849-38-6	: 1848	18963-01-4	: 10148	19781-68-1	: 4976
16147-07-2	: 1446	16874-33-2	: 6072	17851-27-3	: 1710	18970-44-0	: 1688	19781-69-2	: 4977
16177-46-1	: 4186	16883-48-0	: 4860	17878-44-3	: 11367	18971-59-0	: 8821	19781-73-8	: 4404
16179-36-5	: 8885	16889-72-8	: 10294	17880-86-3	: 10502	18979-60-7	: 1549	19781-77-2	: 6572
16183-46-3	: 3666	16898-52-5	: 9701	17912-17-3	: 8368	18979-62-9	: 1539	19812-64-7	: 11618
16188-57-1	: 9374	16932-49-3	: 2753	17937-68-7	: 11404	18992-39-7	: 10402	19814-71-2	: 2077
16197-90-3	: 3822	16939-57-4	: 1097	17952-11-3	: 8808	18992-80-8	: 5806	19819-98-8	: 1609
16245-79-7	: 831	16957-70-3	: 9076	17972-08-6	: 7652	18999-28-5	: 6561	19821-84-2	: 6612
16246-07-4	: 3947	16980-61-3	: 4512	17988-31-7	: 11427	19037-72-0	: 4922	19832-98-5	: 8050
16292-88-9	: 776	16995-35-0	: 10486	18002-79-4	: 11360	19044-88-3	: 8553	19860-56-1	: 10256
16302-35-5	: 10778	17005-59-3	: 11383	18023-33-1	: 11426	19074-59-0	: 9336	19866-51-4	: 9077
16306-39-1	: 9184	17015-11-1	: 6914	18023-36-4	: 11511	19089-47-5	: 10380	19879-11-9	: 4508
16308-92-2	: 1860	17021-26-0	: 357	18030-61-0	: 11364	19113-78-1	: 6928	19879-12-0	: 4509
16315-07-4	: 1922	17042-16-9	: 6980	18036-81-2	: 11497	19145-60-9	: 487	19889-37-3	: 3516
16326-97-9	: 4806	17057-82-8	: 7252	18169-57-8	: 11469	19145-96-1	: 4356	19890-00-7	: 2956
16326-98-0	: 4807	17057-88-4	: 3060	18171-59-0	: 11373	19155-24-9	: 7329	19891-74-8	: 4060
16357-59-8	: 11231	17057-92-0	: 7887	18178-59-1	: 11421	19163-05-4	: 3129	19902-08-0	: 2924
16369-05-4	: 3599	17059-44-8	: 1485	18181-80-1	: 3140	19177-04-9	: 1987	19915-11-8	: 8689
16369-17-8	: 8922	17059-45-9	: 1482	18187-24-1	: 5234	19184-10-2	: 10567	19942-78-0	: 11708
16369-21-4	: 5765	17059-52-8	: 2459	18231-53-3	: 8764	19184-67-9	: 8774	19947-22-9	: 1700
16386-93-9	: 9075	17071-47-5	: 3401	18236-57-2	: 11400	19221-28-4	: 3165	19961-27-4	: 9810
16387-70-5	: 8344	17082-09-6	: 10717	18236-89-0	: 11399	19230-27-4	: 1294	19967-57-8	: 9775
16387-72-7	: 11610	17082-12-1	: 5129	18240-50-1	: 6114	19246-38-9	: 4346	19987-13-4	: 9865
16409-46-4	: 3554	17085-91-5	: 7828	18243-41-9	: 11366	19246-39-0	: 3601	20009-26-1	: 7955
16419-60-6	: 3134	17091-40-6	: 5747	18261-92-2	: 8280	19269-28-4	: 6720	20024-90-2	: 1703
16420-13-6	: 3980	17094-21-2	: 3558	18262-85-6	: 1709	19317-11-4	: 5330	20027-77-4	: 7991
16423-68-0	: 11524	17096-37-6	: 3796	18281-05-5	: 5356	19329-89-6	: 10274	20030-32-4	: 4690
16426-64-5	: 2527	17102-64-6	: 6709	18282-40-1	: 1680	19354-27-9	: 6170	20063-92-7	: 8183
16432-53-4	: 6794	17113-33-6	: 6159	18282-59-2	: 1015	19374-46-0	: 4857	20089-07-0	: 3439
16433-43-5	: 1104	17176-77-1	: 9586	18295-59-5	: 8697	19393-45-4	: 3773	20098-48-0	: 2338
16435-50-0	: 9024	17199-29-0	: 903	18309-32-5	: 2978	19393-92-1	: 1013	20177-02-0	: 9028
16452-01-0	: 771	17206-52-9	: 4595	18338-40-4	: 3449	19395-58-5	: 11198	20184-89-8	: 8198
16484-17-6	: 7293	17216-08-9	: 8045	18362-36-2	: 578	19396-83-9	: 2916	20184-91-2	: 8199
16491-15-9	: 4921	17216-62-5	: 9933	18362-97-5	: 8893	19398-47-1	: 3614	20185-55-1	: 2694
16503-53-0	: 2147	17226-28-7	: 3416	18368-63-3	: 10920	19398-53-9	: 8743	20193-20-8	: 9809
16509-46-9	: 8335	17226-34-5	: 7009	18381-45-8	: 3406	19398-61-9	: 1387	20213-30-3	: 7917
16518-62-0	: 644	17228-64-7	: 10918	18388-45-9	: 11408	19398-75-5	: 4799	20231-81-6	: 4045
16520-62-0	: 1124	17236-58-7	: 11777	18397-07-4	: 4154	19398-77-7	: 6768	20232-11-5	: 7279
16532-02-8	: 11365	17249-75-1	: 11747	18398-36-2	: 7849	19402-87-0	: 3988	20237-34-7	: 6681
16538-93-5	: 4737	17249-76-2	: 11749	18423-69-3	: 6136	19404-07-0	: 5191	20244-71-7	: 5888
16545-78-1	: 9061	17249-77-3	: 11791	18441-43-5	: 2458	19420-29-2	: 7946	20247-89-6	: 2008
16560-43-3	: 11272	17249-79-5	: 11746	18444-79-6	: 11521	19434-65-2	: 9778	20266-65-3	: 10022
16580-23-7	: 4436	17249-82-0	: 11741	18476-57-8	: 8327	19438-10-9	: 2636	20278-84-6	: 6446
16580-24-8	: 4437	17283-45-3	: 2946	18479-51-1	: 8515	19447-29-1	: 8197	20278-85-7	: 6443
16582-38-0	: 1207	17297-04-0	: 1108	18479-57-7	: 8450	19472-74-3	: 948	20278-87-9	: 6449
16587-71-6	: 4592	17299-41-1	: 4719	18492-37-0	: 2960	19481-82-4	: 10094	20278-88-0	: 6451
16605-36-0	: 4435	17301-94-9	: 8142	18495-23-3	: 8536	19482-57-6	: 8535	20278-89-1	: 6452
16619-12-8	: 7291	17302-01-1	: 6410	18495-26-6	: 6613	19489-10-2	: 4397	20279-53-2	: 6062
16629-19-9	: 11780	17302-02-2	: 6767	18495-30-2	: 10144	19493-31-3	: 3026	20281-83-8	: 8938
16631-63-3	: 4833	17302-04-4	: 6415	18510-29-7	: 11455	19493-44-8	: 7376	20281-85-0	: 3616
16650-52-5	: 7762	17312-63-9	: 8122	18521-07-8	: 8519	19523-57-0	: 7753	20281-86-1	: 6912
16661-99-7	: 4349	17348-59-3	: 10081	18523-48-3	: 109	19549-71-4	: 6480	20281-91-8	: 3619
16672-87-0	: 5929	17356-19-3	: 4866	18530-30-8	: 2917	19549-73-6	: 6481	20291-91-2	: 6816
16674-04-7	: 10230	17372-87-1	: 11525	18530-56-8	: 8212	19549-74-7	: 6482	20291-95-6	: 6439
16694-52-3	: 9868	17406-45-0	: 6196	18589-29-2	: 6420	19549-77-0	: 6484	20327-65-5	: 10568
16696-65-4	: 11940	17432-37-0	: 598	18598-63-5	: 5019	19549-79-2	: 6486	20333-41-9	: 5245
16714-77-5	: 3654	17447-00-6	: 9085	18649-64-4	: 6108	19549-84-9	: 6515	20354-26-1	: 7643
16726-67-3	: 7761	17447-01-7	: 9084	18668-68-3	: 3147	19549-97-4	: 6617	20357-79-3	: 7247
16736-42-8	: 8325	17478-66-9	: 5138	18685-70-6	: 4073	19550-03-9	: 6922	20359-55-1	: 8979
16737-44-3	: 1983	17484-36-5	: 1934	18699-48-4	: 1338	19550-05-1	: 6924	20395-25-9	: 10173
16744-89-1	: 7083	17496-14-9	: 7289	18719-43-2	: 9932	19550-07-3	: 6927	20442-79-9	: 3078
16745-94-1	: 7017	17534-15-5	: 1561	18720-65-5	: 6502	19550-09-5	: 6929	20474-93-5	: 3856
16746-02-4	: 6836	17556-48-8	: 10409	18729-48-1	: 4870	19550-10-8	: 6964	20485-41-0	: 11669
16746-86-4	: 7013	17577-93-4	: 8523	18747-42-7	: 10470	19550-14-2	: 6967	20589-85-9	: 10598
16746-87-5	: 7014	17587-22-3	: 8398	18748-27-1	: 11578	19550-40-4	: 4655	20599-21-7	: 8534
16747-25-4	: 6859	17609-31-3	: 8484	18752-92-6	: 11445	19550-48-2	: 4920	20599-27-3	: 9011
16747-26-5	: 6860	17618-76-7	: 6554	18758-34-4	: 4998	19564-40-0	: 2115	20634-92-8	: 9812
16747-28-7	: 6862	17627-77-9	: 7395	18773-77-8	: 4160	19594-02-6	: 1652	20637-49-4	: 10185
16747-30-1	: 6865	17639-93-9	: 10222	18786-24-8	: 8572	19613-76-4	: 1024	20668-29-5	: 11261
16747-31-2	: 6866	17640-26-5	: 182	18787-64-9	: 6663	19617-44-8	: 7231	20699-48-3	: 3261
16747-32-3	: 8810	17648-05-4	: 2281	18791-49-6	: 10282	19653-33-9	: 9704	20731-93-5	: 1875
16747-33-4	: 8811	17655-74-2	: 1651	18794-84-8	: 5331	19655-60-8	: 11268	20734-71-8	: 9386
16747-38-9	: 8842	17658-32-1	: 8924	18812-82-9	: 7079	19666-30-9	: 8565	20739-58-6	: 8543
16747-42-5	: 6848	17672-88-7	: 2422	18829-56-6	: 8180	19686-73-8	: 10345	20748-86-1	: 6400
16747-44-7	: 8836	17679-92-4	: 8526	18839-90-2	: 9359	19689-18-0	: 5083	20754-04-5	: 8478
16747-45-8	: 8837	17696-11-6	: 8426	18840-45-4	: 12042	19694-02-1	: 10854	20769-85-1	: 10491
16747-50-5	: 4814	17696-37-6	: 9326	18854-52-3	: 5586	19700-97-1	: 6693	20809-46-5	: 8479
16749-11-4	: 4361	17696-64-9	: 142	18858-65-6	: 11441	19718-45-7	: 7956	20820-82-0	: 11022
16751-59-0	: 6351	17704-30-2	: 165	18869-72-2	: 6311	19735-89-8	: 10849	20830-75-5	: 4041
16752-77-5	: 7650	17715-00-3	: 4473	18904-54-6	: 4146	19752-55-7	: 1012	20850-43-5	: 2421
16754-49-7	: 10080	17742-04-0	: 7528	18905-91-4	: 6973	19780-63-3	: 8933	20893-30-5	: 11725

CAS Number	Page	CAS Number	Page	CAS Number	Page	CAS Number	Page	CAS Number	Page
20972-43-4	5131	22401-25-8	9546	24342-04-9	9139	25871-69-6	7587	29091-05-2	5181
20984-33-2	11322	22409-91-2	10322	24346-56-3	2132	26073-26-7	10500	29091-21-2	9761
21020-24-6	3899	22418-49-1	10577	24356-00-1	9017	26106-95-6	3602	29106-32-9	4936
21020-26-8	9131	22421-97-2	10229	24393-70-2	2943	26116-12-1	11167	29110-74-5	7236
21020-27-9	9136	22445-41-6	1771	24395-10-6	6571	26118-38-7	6963	29113-63-1	6055
21040-45-9	10706	22483-09-6	5419	24398-88-7	2522	26120-52-5	4253	29136-19-4	2062
21040-59-5	7392	22520-40-7	8391	24410-50-2	2449	26153-88-8	8703	29172-20-1	6111
21087-29-6	1230	22524-32-9	113	24410-51-3	2446	26171-83-5	3344	29204-02-2	5365
21087-64-9	7658	22531-06-2	8051	24410-52-4	2452	26186-01-6	8107	29232-93-7	9730
21102-13-6	6952	22531-20-0	7891	24425-40-9	7244	26186-02-7	11897	29276-40-2	7707
21129-27-1	4546	22539-65-7	1469	24425-97-6	10051	26190-82-9	11256	29289-13-2	762
21179-48-6	10239	22545-12-6	1603	24438-17-3	7328	26198-63-0	9892	29342-44-7	6740
21211-65-4	11133	22545-13-7	1604	24446-63-7	942	26225-79-6	5933	29342-53-8	2918
21232-53-1	361	22564-83-6	7429	24544-04-5	636	26259-45-0	3143	29394-58-9	3383
21266-88-6	6375	22564-99-4	8332	24545-81-1	2987	26260-02-6	580	29400-42-8	8211
21266-90-0	6376	22587-21-1	4543	24549-06-2	745	26303-90-2	5090	29472-95-5	9421
21293-29-8	8679	22610-63-5	8255	24557-10-6	10627	26311-17-1	5424	29488-27-5	11773
21296-92-4	7323	22618-23-1	9315	24579-70-2	3941	26311-33-1	7063	29538-77-0	4507
21299-26-3	4578	22621-74-5	2947	24580-48-1	4550	26330-51-8	3476	29554-49-2	4745
21326-62-5	3611	22646-79-3	4646	24580-52-7	4551	26340-58-9	4151	29555-02-0	4964
21352-22-7	11209	22663-64-5	8166	24602-86-6	11898	26370-28-5	8108	29559-27-1	2004
21370-71-8	8054	22672-74-8	5150	24612-75-7	4477	26399-36-0	9763	29559-52-2	10177
21391-98-0	4635	22692-70-2	11560	24618-31-3	4317	26438-26-6	7931	29560-84-7	10701
21400-25-9	10602	22717-57-3	2640	24622-72-8	11169	26438-27-7	7793	29604-75-9	1296
21436-44-2	2683	22726-00-7	605	24634-61-5	6704	26438-28-8	7879	29606-79-9	4603
21472-86-6	2335	22734-04-9	4905	24644-78-8	7290	26464-32-4	3488	29624-17-7	4348
21490-63-1	8609	22767-95-9	2140	24677-78-9	540	26465-81-6	7288	29653-38-1	3926
21504-43-8	10559	22767-96-0	2145	24716-09-4	890	26472-41-3	4282	29682-41-5	1380
21531-91-9	6793	22768-19-0	6550	24724-07-0	11949	26473-49-4	3537	29780-00-5	11896
21535-97-7	2457	22768-20-3	6549	24734-68-7	2125	26489-01-0	8514	29823-16-3	8873
21564-91-0	7984	22781-23-3	514	24764-97-4	3179	26513-20-2	1265	29843-58-1	3757
21564-92-1	7986	22839-47-0	9511	24796-87-0	4503	26530-20-1	8528	29865-90-5	568
21570-35-4	6503	22868-76-4	11056	24815-24-5	12060	26531-85-1	6237	29882-57-3	8720
21571-91-5	8751	22900-08-9	4540	24840-05-9	10588	26533-33-5	8169	29883-15-6	954
21617-18-5	11250	22922-67-4	7004	24844-28-8	4556	26549-24-6	6913	29897-82-3	11172
21633-79-4	9972	22924-15-8	553	24885-45-8	9375	26549-25-7	6474	29916-45-8	3269
21655-48-1	9611	22927-13-5	558	24887-75-0	345	26608-06-0	5165	29921-50-4	8063
21693-54-9	7992	22936-86-3	5013	24910-63-2	9036	26644-46-2	11902	29943-42-8	10799
21700-31-2	10125	23010-00-6	3733	24932-49-8	1297	26730-14-3	11879	29949-19-7	7511
21702-79-4	2559	23010-04-0	3283	24948-66-1	5095	26754-48-3	6030	29972-79-0	5341
21725-46-2	4161	23012-11-5	8578	24964-64-5	2757	26771-11-9	4959	30090-17-6	814
21777-84-4	1992	23048-13-7	4943	24965-84-2	4596	26963-33-7	11771	30122-12-4	8745
21849-97-8	7840	23063-36-7	1390	24973-49-7	3020	27063-27-0	11208	30168-50-4	4830
21856-89-3	6970	23074-36-4	3709	25006-60-4	167	27129-86-8	1057	30273-11-1	813
21884-26-4	8909	23079-65-4	1116	25013-16-5	9329	27129-87-9	1861	30361-33-2	5025
21889-88-3	6585	23097-98-5	6618	25044-01-3	9107	27267-69-2	2773	30389-18-5	4290
21890-35-7	3148	23103-98-2	9728	25057-89-0	518	27314-13-2	8213	30433-57-9	10833
21915-53-7	8619	23132-52-7	1467	25109-28-8	1009	27356-52-1	11269	30515-28-7	6458
21923-23-9	4092	23135-22-0	8568	25109-57-3	3653	27387-79-7	10218	30525-89-4	8634
21948-47-0	8372	23144-52-7	8448	25117-74-2	2755	27464-82-0	11659	30538-80-8	9689
21964-23-8	9013	23184-66-9	3144	25118-28-9	6755	27532-76-9	378	30544-34-4	6077
21964-44-3	8189	23204-80-0	1229	25144-04-1	4869	27598-85-2	9200	30560-19-1	10
21968-17-2	9827	23287-26-5	2638	25144-05-2	4868	27610-88-4	8400	30600-19-2	10619
21974-48-1	9714	23308-82-9	1894	25148-87-2	3756	27644-49-1	8465	30614-77-8	6082
21984-93-0	6065	23378-11-2	3732	25167-70-8	9064	27655-41-0	7375	30673-60-0	5069
21987-21-3	11886	23399-70-4	1174	25172-42-3	4873	27701-01-5	3943	30685-95-1	3886
21994-77-4	7065	23418-37-3	8469	25172-43-4	4874	27779-01-7	4158	30689-41-9	4916
21994-78-5	7064	23418-38-4	8170	25182-84-7	10395	27831-13-6	1629	30752-19-3	1045
22029-76-1	3873	23422-53-9	6001	25186-47-4	1386	27876-94-4	5107	30926-60-4	1188
22037-28-1	6034	23431-44-3	2028	25201-40-5	6338	27993-42-6	8984	30964-01-3	8207
22037-73-6	3710	23438-77-9	4722	25229-42-9	3834	28056-54-4	4896	31053-99-3	1092
22038-57-9	3830	23444-65-7	7854	25234-83-7	8898	28123-70-8	8125	31059-19-5	7944
22056-82-2	14	23470-12-4	7127	25245-33-4	1767	28148-04-1	9890	31116-84-4	10459
22077-55-0	2176	23505-41-1	9729	25265-71-8	10397	28249-77-6	11697	31161-71-4	11627
22104-78-5	8510	23511-70-8	3582	25309-64-2	1681	28261-97-4	5273	31202-12-7	363
22118-12-3	3682	23564-05-8	11723	25311-71-1	7359	28300-74-5	11588	31218-83-4	10719
22128-62-7	4014	23578-35-0	6751	25312-34-9	3874	28320-88-9	8362	31252-42-3	9700
22128-63-8	4016	23580-51-0	8544	25312-65-6	4094	28320-89-0	6364	31283-43-9	8498
22139-77-1	1545	23580-52-1	10960	25346-31-0	8718	28332-44-7	7095	31340-44-0	10247
22144-60-1	1912	23617-71-2	2589	25346-32-1	8730	28466-26-4	10812	31366-00-4	1713
22160-12-9	6893	23619-59-2	9459	25346-33-2	8714	28467-92-7	3190	31367-46-1	6496
22160-39-0	8887	23676-09-7	2599	25354-97-6	5065	28469-92-3	1369	31508-08-4	6607
22160-40-3	8888	23708-56-7	11954	25354-98-7	8649	28664-03-1	4171	31508-12-0	8194
22160-41-4	6895	23733-68-8	4717	25394-76-6	10772	28685-60-1	3958	31543-75-6	1306
22202-65-9	11989	23769-39-3	11921	25395-22-6	103	28715-26-6	2451	31599-60-7	1769
22224-92-6	5947	23843-52-9	8239	25395-31-7	10189	28788-62-7	2889	31599-61-8	1774
22227-75-4	10556	23947-60-6	5932	25414-22-6	6127	28823-41-8	6696	31642-67-8	8188
22232-54-8	7194	23950-58-5	10746	25415-67-2	8892	28832-55-5	3180	31656-92-5	981
22253-11-8	7259	24127-62-6	10615	25415-70-7	8894	28905-12-6	6232	31699-35-1	8116
22287-02-1	8176	24152-39-4	9637	25419-36-7	7990	28921-35-9	6064	31702-33-7	10605
22319-28-4	6592	24247-53-8	3696	25419-38-7	7987	28994-41-4	9454	31720-69-1	4562
22331-38-0	5351	24264-06-2	3877	25542-62-5	6879	28995-89-3	8004	31766-07-1	2771
22349-03-7	4456	24280-93-1	7052	25611-78-3	1584	29003-73-4	163	31772-05-1	11874
22364-68-7	960	24306-18-1	6323	25641-98-9	1340	29022-31-9	8201	31805-48-8	11761
22374-89-6	2119	24307-26-4	7430	25659-22-7	7087	29030-84-0	6689	31844-95-8	9008
22381-86-8	11980	24331-72-4	9794	25695-77-6	10689	29040-46-8	2445	31844-96-9	3716
22398-09-0	9709	24332-22-7	6584	25790-55-0	3150	29072-93-3	10087	31849-78-2	3708

CAS	No.	CAS	No.	CAS	No.	CAS	No.	CAS	No.
52914-66-6	4762	55130-16-0	8309	57968-71-5	3307	61828-19-1	4479	66235-62-9	3608
52918-63-5	5102	55156-16-6	8980	58138-08-2	11899	61828-21-5	4413	66295-45-2	9549
53022-14-3	3565	55182-83-7	11988	58138-81-1	3669	61828-25-9	1742	66395-23-1	4585
53045-71-9	9009	55268-62-7	1567	58167-01-4	3305	61828-26-0	2065	66398-52-5	3466
53068-88-5	6902	55268-63-8	2267	58190-94-6	3491	61838-78-6	10464	66577-03-5	11763
53078-85-6	648	55282-10-5	6296	58210-03-0	4568	61847-78-7	7030	66577-17-1	7924
53109-18-5	4933	55282-12-7	8241	58371-98-5	8963	61847-80-1	9048	66610-39-7	3392
53152-98-0	9681	55282-14-9	5271	58474-80-9	5081	61847-97-0	8314	66719-29-7	8759
53172-84-2	1950	55282-15-0	5270	58588-28-6	6919	61886-67-7	7932	66719-37-7	6488
53175-28-3	3697	55282-16-1	5269	58593-78-5	1563	61892-85-1	6808	66719-40-2	6489
53178-20-4	3242	55282-17-2	11605	58616-95-8	4970	61949-83-5	7243	66719-53-7	8456
53188-23-1	6023	55283-68-6	5409	58618-39-6	7806	61971-91-3	3670	66719-54-8	8457
53220-82-9	7741	55285-14-8	4038	58625-77-7	3655	62016-13-1	6839	66719-55-9	8452
53252-08-7	6591	55289-35-5	1070	58766-17-9	6745	62016-38-0	8421	66731-94-0	6490
53252-19-0	7096	55290-64-7	5179	58873-44-2	9182	62016-48-2	3394	66793-72-4	6956
53252-20-3	8192	55319-83-0	5275	59020-74-5	2447	62016-53-9	4842	66793-74-6	6957
53252-21-4	7092	55320-06-4	6295	59022-71-8	7330	62016-55-1	4808	66793-80-4	8467
53346-71-7	8648	55334-08-2	4476	59113-36-9	10024	62016-58-4	4821	66793-92-8	6955
53381-90-1	9725	55335-06-3	11854	59204-02-3	8449	62059-56-7	4351	66794-02-3	6933
53440-12-3	7778	55373-86-9	5274	59255-95-7	653	62080-86-8	8274	66794-03-4	6934
53491-32-0	2441	55429-84-0	11604	59357-07-2	6974	62088-36-2	4663	66794-04-5	6932
53535-68-5	10417	55449-46-2	1234	59368-15-9	4443	62108-06-9	8517	66810-87-5	6958
53566-37-3	7121	55454-22-3	6568	59376-54-4	184	62108-66-1	10232	66898-39-3	2820
53584-29-5	10009	55505-26-5	8168	59433-08-8	9669	62108-68-3	10227	66997-45-3	10146
53588-92-4	5464	55517-73-2	6623	59434-05-8	11758	62108-70-7	10226	67191-37-1	1665
53608-85-8	4322	55517-89-0	8639	59654-13-6	6832	62127-46-2	3446	67329-04-8	7340
53608-93-8	9935	55563-72-9	10458	59669-26-0	11713	62127-47-3	3447	67451-76-7	4312
53660-21-2	6478	55665-79-7	7060	59697-51-7	7011	62127-49-5	10135	67485-29-4	7136
53663-23-3	2353	55696-02-1	3944	59697-55-1	7012	62127-61-1	10160	67535-07-3	10255
53663-26-6	8148	55702-49-3	3894	59734-23-5	8470	62154-77-2	2297	67590-77-6	1760
53776-69-5	2081	55751-54-7	812	59756-60-4	5981	62155-41-3	2111	67632-66-0	9148
53778-43-1	4975	55775-41-2	3182	59777-92-3	4231	62168-25-6	3264	67700-22-5	2159
53780-34-0	7427	55789-02-1	1872	59889-45-1	8722	62184-67-2	11766	67700-26-9	8512
53859-78-2	4213	55877-01-5	6790	59905-55-4	2983	62184-82-1	4919	67747-70-0	9021
53878-96-9	3903	55882-21-8	10153	59956-48-8	118	62184-83-2	4938	67885-08-9	3166
53889-94-4	9428	55944-70-2	6477	60012-83-7	2102	62261-05-6	3957	67952-57-2	6506
53894-19-2	10361	56072-27-6	6518	60096-09-5	6398	62353-69-9	8685	68057-83-0	3532
53966-53-3	8186	56103-67-4	9225	60111-14-0	7082	62476-56-6	10862	68077-28-1	10601
53977-99-4	7048	56106-37-7	1556	60168-88-9	5948	62496-53-1	8753	68107-05-1	6724
54105-76-9	4427	56180-61-1	8688	60207-90-1	10720	62521-69-1	3453	68165-06-0	11187
54106-40-0	2005	56216-12-7	3823	60212-29-5	11985	62624-30-0	465	68266-67-1	1360
54131-44-1	2927	56245-02-4	2034	60212-31-9	11986	62690-62-4	7897	68359-37-5	5008
54244-90-5	6590	56456-42-9	1496	60251-57-2	8357	62692-40-4	1001	68384-70-3	3541
54268-02-9	10138	56523-26-3	8322	60291-32-9	4408	62706-16-5	7008	68420-93-9	11212
54298-97-4	8525	56630-31-0	8206	60468-23-7	9007	62882-00-2	7385	68694-11-1	11900
54305-87-2	6764	56630-33-2	8204	60512-85-8	10687	63019-98-7	3014	68723-25-1	7629
54340-87-3	7262	56630-34-3	8205	60669-40-1	11967	63157-81-3	10194	68750-16-3	5677
54383-22-1	4373	56688-75-6	4671	60671-36-5	11956	63284-71-9	8219	69112-21-6	6997
54410-69-4	2010	56694-97-4	11864	60698-31-9	6157	63370-90-1	4776	69190-65-4	3811
54411-01-7	4446	56694-98-5	11863	60703-31-3	7075	63468-05-3	9097	69248-35-7	3332
54429-56-0	5628	56739-95-8	1781	60710-39-6	9239	63790-14-7	1086	69295-21-2	3729
54454-10-3	10820	56745-61-0	6969	60761-04-8	8561	63927-44-6	6877	69305-42-6	9496
54460-46-7	4952	56800-09-0	3746	60814-29-1	11335	64057-79-0	10703	69321-60-4	1304
54460-96-7	10121	56908-88-4	1291	60836-07-9	6509	64161-50-8	2951	69371-59-1	7102
54462-68-9	6759	56911-77-4	2074	60899-39-0	7880	64165-34-0	11317	69377-81-7	5982
54462-70-3	3270	56920-82-2	6037	60956-23-2	1303	64168-11-2	9702	69498-28-8	3273
54467-95-7	979	56961-23-0	2371	60956-24-3	1293	64207-03-0	10447	69542-91-2	4514
54484-63-8	1377	56961-24-1	2320	60956-26-5	1074	64257-84-7	5953	69564-68-7	2881
54527-68-3	3570	56961-32-1	2541	61050-68-8	8397	64287-26-9	1285	69668-88-8	6588
54553-14-9	11155	56961-77-4	1010	61228-10-2	6619	64334-96-9	11211	69704-44-5	3262
54573-13-6	6620	56961-82-1	2326	61231-45-6	7717	64339-43-1	1524	69745-49-9	4692
54630-50-1	6497	56961-84-3	1359	61531-45-1	8156	64346-45-8	3807	69770-98-5	4891
54638-10-7	8918	56961-86-5	2337	61539-75-1	8370	64379-30-2	8704	69806-50-4	5962
54644-23-4	1231	56961-93-4	7851	61551-49-3	7978	64501-77-5	8508	69813-63-4	1970
54676-39-0	4327	56962-04-0	9231	61563-25-5	1302	64502-89-2	8971	69833-18-7	469
54715-57-0	9259	56962-07-3	7343	61589-68-2	8991	64531-26-6	8533	69947-01-9	9805
54737-45-0	1080	57074-96-1	8539	61592-80-0	1962	64632-72-0	12052	70110-24-0	10496
54774-99-1	10249	57129-90-9	6146	61675-14-7	7098	64653-59-4	744	70116-68-6	3629
54819-86-2	10217	57150-66-0	9553	61764-94-1	6370	64902-72-3	4091	70124-77-5	5965
54829-48-0	6110	57179-35-8	573	61792-07-2	260	64930-16-1	3741	70127-93-4	7237
54833-31-7	4464	57223-18-4	8193	61827-85-8	1487	64965-91-9	2453	70202-92-5	4915
54833-34-0	4402	57296-00-1	10613	61827-86-9	1708	65144-02-7	9797	70396-13-3	6615
54837-90-0	815	57323-93-0	6988	61827-87-0	1711	65248-43-3	6985	70401-56-8	2779
54852-73-2	3071	57386-83-1	1402	61827-89-2	5286	65423-25-8	5340	70644-46-1	10922
54862-92-9	6517	57392-55-9	255	61827-93-8	4853	65438-35-9	13	70690-24-3	8367
54885-00-6	7640	57403-35-7	1135	61827-98-3	4845	65469-88-7	3357	70715-19-4	10983
54889-55-3	7274	57404-54-3	9176	61828-04-4	2341	65530-93-0	6728	70732-43-3	8202
54890-00-5	4326	57456-98-1	3185	61828-05-5	2265	65732-90-3	11518	70732-44-4	8341
54922-86-0	10032	57502-49-5	9666	61828-06-6	2088	65786-09-6	6380	70732-45-5	5101
54934-90-6	4432	57526-28-0	3694	61828-07-7	4380	65786-11-0	6539	70786-82-2	8867
54934-91-7	4472	57576-09-7	4545	61828-08-8	4483	65927-60-8	6954	71159-31-4	10378
55030-56-9	3739	57605-95-9	7911	61828-09-9	4474	66050-98-4	6938	71172-57-1	8970
55038-01-2	2455	57706-88-4	8454	61828-10-2	4462	66056-06-2	19	71173-15-4	2939
55057-45-9	10494	57772-57-3	2624	61828-11-3	4410	66225-16-9	6552	71200-79-8	3763
55066-53-0	8305	57837-19-1	7439	61828-13-5	4458	66225-17-0	6551	71284-96-3	8025
55073-86-4	6950	57885-43-5	10211	61828-15-7	4482	66225-35-2	6803	71297-97-7	5118
55124-79-3	6309	57933-83-2	4022	61828-17-9	4383			71605-85-1	8152

DIAMAGNETIC SUSCEPTIBILITY OF SELECTED ORGANIC COMPOUNDS

When a material is placed in a magnetic field H, a magnetization M is induced in the material which is related to H by $M = \kappa H$, where κ is called the volume susceptibility. Since H and M have the same dimensions, κ is dimensionless. A more useful parameter is the molar susceptibility χ_m, defined by

$$\chi_m = \kappa V_m = \kappa\, M/\rho$$

where V_m is the molar volume of the substance, M the molar mass, and ρ the mass density. When the cgs system is used, the customary unit for χ_m is $cm^3\ mol^{-1}$; the corresponding SI unit is $m^3\ mol^{-1}$. Substances with no unpaired electrons are called diamagnetic; they have negative values of χ_m.

This table gives values of the diamagnetic susceptibility for about 400 common organic compounds. All values refer to room temperature and atmospheric pressure and to the physical form that is stable under these conditions. Substances are arranged by molecular formula in Hill order. A more extensive table may be found in Reference 1.

In keeping with customary practice, the molar susceptibility is given here in units appropriate to the cgs system. These values should be multiplied by 4π to obtain values for use in SI equations (where the magnetic field strength H has units of $A\ m^{-1}$).

REFERENCES

1. *Landolt-Börnstein, Numerical Data and Functional Relationships in Science and Technology, New Series*, II/16, *Diamagnetic Susceptibility*, Gupta, R. R., Ed., Springer-Verlag, Heidelberg, 1986.
2. Barter, C., Meisenheimer, R. G., and Stevenson, D. P., *J. Phys. Chem.* 64, 1312, 1960.
3. Broersma, S., *J. Chem. Phys.* 17, 873, 1949.

Molecular Formula	Compound	$-\chi_m/10^{-6}\ cm^3\ mol^{-1}$	Molecular Formula	Compound	$-\chi_m/10^{-6}\ cm^3\ mol^{-1}$
$CBrCl_3$	Bromotrichloromethane	73.2	C_2HCl_3O	Trichloroacetaldehyde	73.0
CBr_4	Tetrabromomethane	93.7	C_2HCl_3O	Dichloroacetyl chloride	69.0
$CClF_3$	Chlorotrifluoromethane	45.3	$C_2HCl_3O_2$	Trichloroacetic acid	73.0
$CClN$	Cyanogen chloride	32.4	C_2HCl_5	Pentachloroethane	99.1
CCl_2F_2	Dichlorodifluoromethane	52.2	$C_2HF_3O_2$	Trifluoroacetic acid	43.3
CCl_2O	Carbonyl chloride	47.9	C_2H_2	Acetylene	20.8
CCl_3F	Trichlorofluoromethane	58.7	$C_2H_2Br_4$	1,1,2,2-Tetrabromoethane	123.4
CCl_3NO_2	Trichloronitromethane	75.3	$C_2H_2Cl_2$	1,1-Dichloroethylene	49.2
CCl_4	Tetrachloromethane	66.8	$C_2H_2Cl_2$	cis-1,2-Dichloroethylene	51.0
$CHBrCl_2$	Bromodichloromethane	66.3	$C_2H_2Cl_2$	trans-1,2-Dichloroethylene	48.9
$CHBr_3$	Tribromomethane	82.6	$C_2H_2Cl_4$	1,1,2,2-Tetrachloroethane	89.8
$CHCl_3$	Trichloromethane	58.9	C_2H_3Cl	Chloroethylene	35.9
CHI_3	Triiodomethane	117.1	C_2H_3ClO	Acetyl chloride	39.3
CH_2BrCl	Bromochloromethane	55.1	C_2H_3N	Acetonitrile	27.8
CH_2Br_2	Dibromomethane	65.1	C_2H_4	Ethylene	18.8
CH_2Cl_2	Dichloromethane	46.6	$C_2H_4Br_2$	1,2-Dibromoethane	78.9
CH_2I_2	Diiodomethane	93.1	$C_2H_4Cl_2$	1,1-Dichloroethane	57.4
CH_2N_2	Cyanamide	24.8	$C_2H_4Cl_2$	1,2-Dichloroethane	59.6
CH_2O	Formaldehyde	18.6	C_2H_4O	Acetaldehyde	22.2
CH_2O_2	Formic acid	19.9	C_2H_4O	Ethylene oxide	30.5
CH_3Br	Bromomethane	42.8	$C_2H_4O_2$	Acetic acid	31.8
CH_3Cl	Chloromethane	32.0	$C_2H_4O_2$	Methyl formate	31.1
CH_3F	Fluoromethane	17.8	C_2H_5Br	Bromoethane	78.8
CH_3I	Iodomethane	57.2	C_2H_5Cl	Chloroethane	69.9
CH_3NO	Formamide	23.0	C_2H_5I	Iodoethane	69.1
CH_3NO_2	Nitromethane	21.0	C_2H_5NO	Acetamide	33.9
CH_4	Methane	17.4	$C_2H_5NO_2$	Nitroethane	35.4
CH_4N_2O	Urea	33.5	$C_2H_5NO_2$	Glycine	39.6
CH_4O	Methanol	21.4	C_2H_6	Ethane	26.8
CH_5N	Methylamine	27.0	C_2H_6O	Ethanol	33.7
CI_4	Tetraiodomethane	136	C_2H_6O	Dimethyl ether	26.3
CN_4O_8	Tetranitromethane	43.0	$C_2H_6O_2$	Ethylene glycol	38.9
C_2ClF_3	Chlorotrifluoroethylene	49.1	C_2H_6S	Ethanethiol	47.0
C_2Cl_4	Tetrachloroethylene	81.6	C_2H_6S	Dimethyl sulfide	44.9
C_2Cl_6	Hexachloroethane	112.8	$C_2H_8N_2$	1,2-Ethanediamine	46.5
C_2HCl_3	Trichloroethylene	65.8	C_2N_2	Cyanogen	21.6

Molecular Formula	Compound	$-\chi_m/10^{-6} \text{ cm}^3 \text{ mol}^{-1}$	Molecular Formula	Compound	$-\chi_m/10^{-6} \text{ cm}^3 \text{ mol}^{-1}$
C_3H_4	Allene	25.3	$C_4H_8O_2$	1,4-Dioxane	52.2
$C_3H_4O_2$	Vinyl formate	34.7	C_4H_9Br	1-Bromobutane	77.1
C_3H_5Br	3-Bromopropene	58.6	C_4H_9Br	1-Bromo-2-methylpropane	79.9
C_3H_5Cl	2-Chloropropene	47.8	C_4H_9Cl	1-Chlorobutane	67.1
C_3H_5Cl	3-Chloropropene	47.8	C_4H_9Cl	2-Chlorobutane	67.4
C_3H_5N	Propanenitrile	38.6	C_4H_9I	1-Iodobutane	93.6
C_3H_6	Propene	30.7	C_4H_9N	Pyrrolidine	54.8
C_3H_6	Cyclopropane	39.2	C_4H_9NO	Morpholine	55.0
C_3H_6O	Allyl alcohol	36.7	C_4H_{10}	Butane	50.3
C_3H_6O	Propanal	34.2	C_4H_{10}	Isobutane	50.5
C_3H_6O	Acetone	33.8	$C_4H_{10}O$	1-Butanol	56.4
C_3H_6O	Methyloxirane	42.5	$C_4H_{10}O$	2-Butanol	57.6
$C_3H_6O_2$	Propanoic acid	43.2	$C_4H_{10}O$	2-Methyl-1-propanol	57.6
$C_3H_6O_2$	Ethyl formate	42.4	$C_4H_{10}O$	2-Methyl-2-propanol	56.6
C_3H_7Br	1-Bromopropane	65.6	$C_4H_{10}O$	Diethyl ether	55.5
C_3H_7Br	2-Bromopropane	65.1	$C_4H_{10}O_2$	1,3-Butanediol	61.8
C_3H_7Cl	1-Chloropropane	56.0	$C_4H_{10}O_2$	1,4-Butanediol	61.8
C_3H_7I	1-Iodopropane	84.3	$C_4H_{10}S$	1-Butanethiol	70.2
C_3H_7N	Allylamine	40.1	$C_4H_{11}N$	Butylamine	58.9
$C_3H_7NO_2$	1-Nitropropane	45.0	$C_4H_{11}N$	Isobutylamine	59.8
$C_3H_7NO_2$	2-Nitropropane	45.4	$C_4H_{11}N$	Diethylamine	56.8
$C_3H_7NO_2$	Ethyl carbamate	57.0	$C_5H_4O_2$	Furfural	47.2
C_3H_8	Propane	38.6	C_5H_5N	Pyridine	48.7
C_3H_8O	1-Propanol	44.8	$C_5H_6O_2$	Furfuryl alcohol	61.0
C_3H_8O	2-Propanol	45.7	$C_5H_7NO_2$	Ethyl cyanoacetate	67.3
$C_3H_8O_2$	1,3-Propylene glycol	50.2	C_5H_8	2-Methyl-1,3-butadiene	46.0
$C_3H_8O_2$	Dimethoxymethane	47.3	C_5H_8O	Cyclopentanone	51.6
$C_3H_8O_3$	Glycerol	57.1	$C_5H_8O_2$	Methyl methacrylate	57.3
$C_4H_2O_3$	Maleic anhydride	35.8	$C_5H_8O_2$	2,4-Pentanedione	54.9
$C_4H_4N_2$	Pyrazine	37.8	C_5H_{10}	1-Pentene	54.6
$C_4H_4N_2$	Pyrimidine	43.1	C_5H_{10}	2-Methyl-2-butene	54.7
C_4H_4O	Furan	43.1	C_5H_{10}	Cyclopentane	56.2
$C_4H_4O_3$	Succinic anhydride	47.5	$C_5H_{10}O$	Cyclopentanol	64.0
$C_4H_4O_4$	Maleic acid	49.6	$C_5H_{10}O$	Pentanal	57.5
$C_4H_4O_4$	Fumaric acid	49.1	$C_5H_{10}O$	2-Pentanone	57.5
C_4H_4S	Thiophene	57.3	$C_5H_{10}O$	3-Pentanone	57.7
C_4H_5N	Pyrrole	48.6	$C_5H_{10}O_2$	Pentanoic acid	66.5
C_4H_6	1,2-Butadiene	35.6	$C_5H_{10}O_2$	3-Methylbutanoic acid	67.7
C_4H_6	1,3-Butadiene	32.1	$C_5H_{10}O_2$	Butyl formate	65.8
$C_4H_6O_2$	Vinyl acetate	46.4	$C_5H_{10}O_2$	Isobutyl formate	66.8
$C_4H_6O_3$	Acetic anhydride	52.8	$C_5H_{10}O_2$	Propyl acetate	65.9
$C_4H_6O_4$	Succinic acid	58.0	$C_5H_{10}O_2$	Isopropyl acetate	67.0
$C_4H_6O_4$	Dimethyl oxalate	55.7	$C_5H_{10}O_2$	Ethyl propanoate	66.3
C_4H_7N	Butanenitrile	50.4	$C_5H_{10}O_2$	Tetrahydrofurfuryl alcohol	69.4
C_4H_8	1-Butene	41.0	$C_5H_{10}O_3$	Diethyl carbonate	75.4
C_4H_8	cis-2-Butene	42.6	$C_5H_{11}N$	Piperidine	64.2
C_4H_8	trans-2-Butene	43.3	C_5H_{12}	Pentane	61.5
C_4H_8	Isobutene	40.8	C_5H_{12}	Isopentane	63.0
C_4H_8	Cyclobutane	40.0	C_5H_{12}	Neopentane	63.0
C_4H_8O	Ethyl vinyl ether	47.9	$C_5H_{12}O$	1-Pentanol	67.0
C_4H_8O	1,2-Epoxybutane	54.8	$C_5H_{12}O$	2-Pentanol	69.1
C_4H_8O	Butanal	45.9	$C_5H_{12}O_2$	1,5-Pentanediol	73.5
C_4H_8O	2-Butanone	45.6	$C_5H_{13}N$	Pentylamine	69.3
$C_4H_8O_2$	Butanoic acid	55.2	C_6Cl_6	Hexachlorobenzene	147.0
$C_4H_8O_2$	2-Methylpropanoic acid	56.1	$C_6H_4ClNO_2$	1-Chloro-2-nitrobenzene	75.5
$C_4H_8O_2$	Propyl formate	55.0	$C_6H_4ClNO_2$	1-Chloro-3-nitrobenzene	77.2
$C_4H_8O_2$	Ethyl acetate	54.1	$C_6H_4ClNO_2$	1-Chloro-4-nitrobenzene	74.7
$C_4H_8O_2$	Methyl propanoate	54.5	$C_6H_4Cl_2$	o-Dichlorobenzene	84.4

Molecular Formula	Compound	$-\chi_m/10^{-6}$ cm^3 mol^{-1}	Molecular Formula	Compound	$-\chi_m/10^{-6}$ cm^3 mol^{-1}
$C_6H_4Cl_2$	m-Dichlorobenzene	84.1	$C_6H_{14}O$	Dipropyl ether	79.4
$C_6H_4Cl_2$	p-Dichlorobenzene	81.7	$C_6H_{14}O_2$	1,6-Hexanediol	84.3
$C_6H_4O_2$	p-Benzoquinone	36	$C_6H_{14}O_2$	1,1-Diethoxyethane	81.4
C_6H_5Br	Bromobenzene	78.4	$C_6H_{14}O_6$	D-Glucitol	107.8
C_6H_5Cl	Chlorobenzene	69.5	$C_6H_{15}N$	Triethylamine	83.3
C_6H_5ClO	o-Chlorophenol	77.3	C_7H_5N	Benzonitrile	65.2
C_6H_5ClO	m-Chlorophenol	77.6	C_7H_6O	Benzaldehyde	60.7
C_6H_5ClO	p-Chlorophenol	77.7	$C_7H_6O_2$	Salicylaldehyde	66.8
C_6H_5F	Fluorobenzene	58.4	$C_7H_6O_3$	Salicylic acid	75
C_6H_5I	Iodobenzene	92.0	C_7H_7Br	p-Bromotoluene	88.7
$C_6H_5NO_2$	Nitrobenzene	61.9	C_7H_7Cl	o-Chlorotoluene	82.4
$C_6H_5NO_3$	o-Nitrophenol	68.9	C_7H_7Cl	m-Chlorotoluene	79.7
$C_6H_5NO_3$	m-Nitrophenol	65.9	C_7H_7Cl	p-Chlorotoluene	80.3
$C_6H_5NO_3$	p-Nitrophenol	66.9	C_7H_7Cl	(Chloromethyl)benzene	81.6
C_6H_6	Benzene	54.8	C_7H_7NO	Benzamide	72.0
C_6H_6ClN	o-Chloroaniline	79.5	$C_7H_7NO_2$	o-Nitrotoluene	72.2
C_6H_6ClN	m-Chloroaniline	76.6	$C_7H_7NO_2$	m-Nitrotoluene	72.7
C_6H_6ClN	p-Chloroaniline	76.7	$C_7H_7NO_2$	p-Nitrotoluene	73.3
$C_6H_6N_2O_2$	o-Nitroaniline	67.4	C_7H_8	Toluene	65.6
$C_6H_6N_2O_2$	m-Nitroaniline	69.7	C_7H_8O	o-Cresol	73.3
$C_6H_6N_2O_2$	p-Nitroaniline	68.0	C_7H_8O	m-Cresol	72.2
C_6H_6O	Phenol	60.6	C_7H_8O	p-Cresol	72.4
$C_6H_6O_2$	p-Hydroquinone	64.7	C_7H_8O	Benzyl alcohol	71.8
$C_6H_6O_2$	Pyrocatechol	68.2	C_7H_8O	Anisole	72.2
$C_6H_6O_2$	Resorcinol	67.2	C_7H_9N	o-Methylaniline	74.9
C_6H_7N	Aniline	62.4	C_7H_9N	m-Methylaniline	74.6
C_6H_7N	4-Methylpyridine	59.8	C_7H_9N	p-Methylaniline	72.5
C_6H_8	1,4-Cyclohexadiene	48.7	C_7H_9N	N-Methylaniline	74.1
$C_6H_8N_2$	o-Phenylenediamine	72.5	C_7H_9N	2,4-Dimethylpyridine	71.3
$C_6H_8N_2$	m-Phenylenediamine	70.4	C_7H_9N	2,6-Dimethylpyridine	72.5
$C_6H_8N_2$	p-Phenylenediamine	70.7	C_7H_9NO	o-Methoxyaniline [o-Anisidine]	79.1
C_6H_{10}	1,5-Hexadiene	55.1	$C_7H_{12}O_4$	Diethyl malonate	92.6
C_6H_{10}	1-Hexyne	64.5	C_7H_{14}	1-Heptene	77.8
C_6H_{10}	Cyclohexene	58.0	C_7H_{14}	Cycloheptane	73.9
$C_6H_{10}O$	Cyclohexanone	62.0	C_7H_{14}	Methylcyclohexane	78.9
$C_6H_{10}O_3$	Ethyl acetoacetate	71.7	$C_7H_{14}O$	1-Heptanal	81.0
$C_6H_{10}O_4$	Diethyl oxalate	81.7	$C_7H_{14}O$	2-Heptanone	80.5
C_6H_{12}	1-Hexene	66.4	$C_7H_{14}O$	3-Heptanone	80.7
C_6H_{12}	2,3-Dimethyl-2-butene	65.9	$C_7H_{14}O$	4-Heptanone	80.5
C_6H_{12}	Cyclohexane	68	$C_7H_{14}O$	2,4-Dimethyl-3-pentanone	81.1
C_6H_{12}	Methylcyclopentane	70.2	$C_7H_{14}O_2$	Heptanoic acid	89.0
$C_6H_{12}O$	Hexanal	69.4	$C_7H_{14}O_2$	Pentyl acetate	88.9
$C_6H_{12}O$	2-Hexanone	69.2	$C_7H_{14}O_2$	Isopentyl acetate	89.4
$C_6H_{12}O$	3-Hexanone	69.0	$C_7H_{14}O_2$	Butyl propanoate	89.1
$C_6H_{12}O$	4-Methyl-2-pentanone	69.7	$C_7H_{14}O_2$	Ethyl 3-methylbutanoate	91.1
$C_6H_{12}O$	Cyclohexanol	73.4	C_7H_{16}	Heptane	85.2
$C_6H_{12}O_2$	Hexanoic acid	78.1	C_7H_{16}	3-Ethylpentane	86.2
$C_6H_{12}O_2$	Isopentyl formate	78.4	C_7H_{16}	2,2-Dimethylpentane	87.0
$C_6H_{12}O_2$	Isobutyl acetate	78.7	C_7H_{16}	2,3-Dimethylpentane	87.5
$C_6H_{12}O_2$	Propyl propanoate	77.7	C_7H_{16}	2,4-Dimethylpentane	87.5
$C_6H_{12}O_3$	Paraldehyde	86.1	C_7H_{16}	3,3-Dimethylpentane	89.5
C_6H_{14}	Hexane	74.1	$C_7H_{16}O$	1-Heptanol	91.7
C_6H_{14}	2-Methylpentane	75.3	$C_7H_{16}O$	4-Heptanol	92.1
C_6H_{14}	3-Methylpentane	75.5	$C_8H_4O_3$	Phthalic anhydride	66.7
C_6H_{14}	2,2-Dimethylbutane	76.2	$C_8H_6O_4$	Phthalic acid	83.6
C_6H_{14}	2,3-Dimethylbutane	76.2	$C_8H_6O_4$	Isophthalic acid	84.6
$C_6H_{14}O$	1-Hexanol	79.5	$C_8H_6O_4$	Terephthalic acid	83.5
$C_6H_{14}O$	4-Methyl-2-pentanol	80.4	C_8H_7N	Benzeneacetonitrile	76.9

Molecular Formula	Compound	$-\chi_m/10^{-6}$ cm^3 mol^{-1}	Molecular Formula	Compound	$-\chi_m/10^{-6}$ cm^3 mol^{-1}
C_8H_7N	Indole	85.0	$C_{10}H_{10}O_2$	Safrole	97.5
C_8H_8	Styrene	68.2	$C_{10}H_{10}O_4$	Dimethyl terephthalate	101.6
C_8H_8O	Acetophenone	72.5	$C_{10}H_{14}$	Butylbenzene	100.7
$C_8H_8O_2$	o-Toluic acid	84.3	$C_{10}H_{14}$	tert-Butylbenzene	101.8
$C_8H_8O_2$	m-Toluic acid	83.0	$C_{10}H_{14}$	Isobutylbenzene	101.7
$C_8H_8O_2$	p-Toluic acid	82.4	$C_{10}H_{14}$	p-Cymene	102.8
$C_8H_8O_2$	Benzeneacetic acid	82.4	$C_{10}H_{14}$	1,2,4,5-Tetramethylbenzene	101.2
$C_8H_8O_2$	Methyl benzoate	81.6	$C_{10}H_{14}O$	p-tert-Butylphenol	108.0
$C_8H_8O_3$	Methyl salicylate	86.6	$C_{10}H_{15}N$	N,N-Diethylaniline	107.9
C_8H_{10}	Ethylbenzene	77.3	$C_{10}H_{16}$	d-Limonene	98.0
C_8H_{10}	o-Xylene	77.7	$C_{10}H_{16}$	α-Pinene	100.7
C_8H_{10}	m-Xylene	76.4	$C_{10}H_{16}$	β-Pinene	101.9
C_8H_{10}	p-Xylene	77.0	$C_{10}H_{16}O$	Camphor, (+)	103.0
$C_8H_{10}O$	Phenetole	84.5	$C_{10}H_{18}$	cis-Decahydronaphthalene	107.0
$C_8H_{11}N$	N-Ethylaniline	85.6	$C_{10}H_{18}$	trans-Decahydronaphthalene	107.6
$C_8H_{11}N$	N,N-Dimethylaniline	85.1	$C_{10}H_{22}$	Decane	119.5
$C_8H_{11}N$	2,4,6-Trimethylpyridine	83.1	$C_{11}H_{10}$	1-Methylnaphthalene	102.9
$C_8H_{14}O_4$	Ethyl succinate	105.0	$C_{11}H_{10}$	2-Methylnaphthalene	102.7
C_8H_{16}	1-Octene	88.8	$C_{11}H_{24}$	Undecane	131.8
C_8H_{16}	Cyclooctane	85.3	$C_{12}H_8$	Acenaphthylene	111.6
$C_8H_{16}O_2$	Octanoic acid	99.5	$C_{12}H_9N$	Carbazole	119.9
$C_8H_{16}O_2$	Hexyl acetate	100.9	$C_{12}H_{10}$	Acenaphthene	109.9
$C_8H_{17}Cl$	1-Chlorooctane	114.9	$C_{12}H_{10}$	Biphenyl	103.3
C_8H_{18}	Octane	96.6	$C_{12}H_{10}N_2$	Azobenzene	106.8
C_8H_{18}	4-Methylheptane	97.3	$C_{12}H_{11}N$	Diphenylamine	108.4
C_8H_{18}	3-Ethylhexane	97.8	$C_{12}H_{14}O_4$	Diethyl phthalate	127.5
C_8H_{18}	3,4-Dimethylhexane	99.1	$C_{12}H_{18}$	Hexamethylbenzene	122.5
C_8H_{18}	2,2,4-Trimethylpentane	99.1	$C_{12}H_{24}O_2$	Dodecanoic acid	113.0
C_8H_{18}	2,3,4-Trimethylpentane	99.8	$C_{13}H_9N$	Acridine	118.8
$C_8H_{18}O$	1-Octanol	101.6	$C_{13}H_{10}O$	Benzophenone	109.6
$C_8H_{19}N$	Dibutylamine	103.7	$C_{13}H_{12}$	Diphenylmethane	116.0
C_9H_7N	Quinoline	86.1	$C_{13}H_{28}$	Tridecane	153.7
C_9H_7N	Isoquinoline	83.9	$C_{14}H_8O_2$	9,10-Anthracenedione	113.0
C_9H_8	Indene	83	$C_{14}H_{10}$	Anthracene	129.8
C_9H_{10}	Isopropenylbenzene	80.0	$C_{14}H_{10}$	Phenanthrene	127.6
$C_9H_{10}O_2$	Ethyl benzoate	93.8	$C_{14}H_{10}$	Diphenylacetylene	116
$C_9H_{10}O_2$	Benzyl acetate	93.2	$C_{14}H_{10}O_2$	Benzil	106.8
C_9H_{12}	Propylbenzene	89.1	$C_{14}H_{12}O_2$	Benzyl benzoate	132.2
C_9H_{12}	Isopropylbenzene [Cumene]	89.5	$C_{14}H_{14}$	1,2-Diphenylethane	127.8
C_9H_{12}	1,3,5-Trimethylbenzene [Mesitylene]	92.3	$C_{14}H_{28}O_2$	Tetradecanoic acid [Myristic acid]	176.0
C_9H_{18}	1-Nonene	100.1	$C_{14}H_{30}$	Tetradecane	166.2
$C_9H_{18}O$	2,6-Dimethyl-4-heptanone	104.3	$C_{16}H_{10}$	Pyrene	147
C_9H_{20}	Nonane	108.1	$C_{16}H_{32}O_2$	Hexadecanoic acid [Palmitic acid]	198.6
$C_{10}H_7Br$	1-Bromonaphthalene	123.6	$C_{16}H_{34}$	Hexadecane	187.6
$C_{10}H_7Cl$	1-Chloronaphthalene	107.6	$C_{16}H_{34}O$	1-Hexadecanol	183.5
$C_{10}H_8$	Naphthalene	91.6	$C_{18}H_{12}$	Chrysene	148.0
$C_{10}H_8$	Azulene	123.7	$C_{18}H_{14}$	o-Terphenyl	150.4
$C_{10}H_8O$	1-Naphthol	96.2	$C_{18}H_{14}$	m-Terphenyl	155.5
$C_{10}H_8O$	2-Naphthol	96.8	$C_{18}H_{14}$	p-Terphenyl	156.0
$C_{10}H_9N$	1-Naphthalenamine	92.5	$C_{18}H_{34}O_2$	cis-9-Octadecenoic acid [Oleic acid]	208.5
$C_{10}H_9N$	2-Naphthalenamine	98.0	$C_{18}H_{36}O_2$	Octadecanoic acid [Steric acid]	220.8
			$C_{20}H_{12}$	Perylene	167.5

Section 4
Properties of the Elements and Inorganic Compounds

THE ELEMENTS
C. R. Hammond

One of the most striking facts about the elements is their unequal distribution and occurrence in nature. Present knowledge of the chemical composition of the universe, obtained from the study of the spectra of stars and nebulae, indicates that hydrogen is by far the most abundant element and may account for more than 90% of the atoms or about 75% of the mass of the universe. Helium atoms make up most of the remainder. All of the other elements together contribute only slightly to the total mass.

The chemical composition of the universe is undergoing continuous change. Hydrogen is being converted into helium, and helium is being changed into heavier elements. As time goes on, the ratio of heavier elements increases relative to hydrogen. Presumably, the process is not reversible.

Burbidge, Burbidge, Fowler, and Hoyle, and more recently, Peebles, Penzias, and others have studied the synthesis of elements in stars. To explain all of the features of the nuclear abundance curve — obtained by studies of the composition of the earth, meteorites, stars, etc. — it is necessary to postulate that the elements were originally formed by at least eight different processes: (1) hydrogen burning, (2) helium burning, (3) χ process, (4) e process, (5) s process, (6) r process, (7) p process, and (8) the X process. The X process is thought to account for the existence of light nuclei such as D, Li, Be, and B. Common metals such as Fe, Cr, Ni, Cu, Ti, Zn, etc. were likely produced early in the history of our galaxy. It is also probable that most of the heavy elements on earth and elsewhere in the universe were originally formed in supernovae, or in the hot interior of stars.

Studies of the solar spectrum have led to the identification of 67 elements in the sun's atmosphere; however, all elements cannot be identified with the same degree of certainty. Other elements may be present in the sun, although they have not yet been detected spectroscopically. The element helium was discovered on the sun before it was found on earth. Some elements such as scandium are relatively more plentiful in the sun and stars than here on earth.

Minerals in lunar rocks brought back from the moon on the Apollo missions consist predominantly of *plagioclase* {(Ca,Na)(Al,Si)O_4O_8} and *pyroxene* {(Ca,Mg,Fe)$_2$Si$_2$O$_6$} — two minerals common in terrestrial volcanic rock. No new elements have been found on the moon that cannot be accounted for on earth; however, three minerals, *armalcolite* {(Fe,Mg)Ti$_2$O$_5$}, *pyroxferroite* {CaFe$_6$(SiO$_3$)$_7$}, and *tranquillityite* {Fe$_8$(Zr,Y)Ti$_3$Si$_3$O$_2$}, are new. The oldest known terrestrial rocks are about 4 billion years old. One rock, known as the "Genesis Rock," brought back from the Apollo 15 Mission, is about 4.15 billion years old. This is only about one-half billion years younger than the supposed age of the moon and solar system. Lunar rocks appear to be relatively enriched in refractory elements such as chromium, titanium, zirconium, and the rare earths, and impoverished in volatile elements such as the alkali metals, in chlorine, and in noble metals such as nickel, platinum, and gold.

Even older than the "Genesis Rock" are *carbonaceous chondrites,* a type of meteorite that has fallen to earth and has been studied. These are some of the most primitive objects of the solar system yet found. The grains making up these objects probably condensed directly out the gaseous nebula from which the sun and planets were born. Most of the condensation of the grains probably was completed within 50,000 years of the time the disk of the nebula was first formed — about 4.6 billion years ago. It is now thought that this type of meteorite may contain a small percentage of presolar dust grains. The relative abundances of the elements of these meteorites are about the same as the abundances found in the solar chromosphere.

The X-ray fluorescent spectrometer sent with the Viking I spacecraft to Mars shows that the Martian soil contains about 12 to 16% iron, 14 to 15% silicon, 3 to 8% calcium, 2 to 7% aluminum, and one-half to 2% titanium. The gas chromatograph — mass spectrometer on Viking II found no trace of organic compounds that should be present if life ever existed there.

F. W. Clarke and others have carefully studied the composition of rocks making up the crust of the earth. Oxygen accounts for about 47% of the crust, by weight, while silicon comprises about 28% and aluminum about 8%. These elements, plus iron, calcium, sodium, potassium, and magnesium, account for about 99% of the composition of the crust.

Many elements such as tin, copper, zinc, lead, mercury, silver, platinum, antimony, arsenic, and gold, which are so essential to our needs and civilization, are among some of the rarest elements in the earth's crust. These are made available to us only by the processes of concentration in ore bodies. Some of the so-called *rare-earth* elements have been found to be much more plentiful than originally thought and are about as abundant as uranium, mercury, lead, or bismuth. The least abundant rare-earth or *lanthanide* element, thulium, is now believed to be more plentiful on earth than silver, cadmium, gold, or iodine, for example. Rubidium, the 16th most abundant element, is more plentiful than chlorine while its compounds are little known in chemistry and commerce.

It is now thought that at least 24 elements are essential to living matter. The four most abundant in the human body are hydrogen, oxygen, carbon, and nitrogen. The seven next most common, in order of abundance, are calcium, phosphorus, chlorine, potassium, sulfur, sodium, and magnesium. Iron, copper, zinc, silicon, iodine, cobalt, manganese, molybdenum, fluorine, tin, chromium, selenium, and vanadium are needed and play a role in living matter. Boron is also thought essential for some plants, and it is possible that aluminum, nickel, and germanium may turn out to be necessary.

Ninety-one elements occur naturally on earth. Minute traces of plutonium-244 have been discovered in rocks mined in Southern California. This discovery supports the theory that heavy elements were produced during creation of the solar system. While technetium and promethium have not yet been found naturally on earth, they have been found to be present in stars. Technetium has been identified in the spectra of certain "late" type stars, and promethium lines have been identified in the spectra of a faintly visible star HR465 in Andromeda. Promethium must have been made very recently near the star's surface for no known isotope of this element has a half-life longer than 17.7 years.

It has been suggested that californium is present in certain stellar explosions known as supernovae; however, this has not been proved. At present no elements are found elsewhere in the universe that cannot be accounted for here on earth.

All atomic mass numbers from 1 to 238 are found naturally on earth except for masses 5 and 8. About 285 relatively stable and 67 naturally radioactive isotopes occur on earth totaling 352. In addition, the neutron, technetium, promethium, and the transuranic elements (lying beyond uranium) up to Element 112 have been produced artificially. Scientists at G.S.I., Darmstadt, Germany, in February, 1996, reported they had found evidence of an isotope of Element 112. Laboratory processes have now extended the radioactive element mass numbers beyond 238 to 277. Each element from atomic numbers 1 to 108 is known to have at least one radioactive isotope. As of January 1996, 3018 isotopes and isomers were thought to be known and recognized. Many stable and radioactive isotopes are now produced and distributed by the Oak Ridge National Laboratory, Oak Ridge, Tenn., U.S.A., to customers licensed by the U.S. Department of Energy.

The nucleus of an atom is characterized by the number of protons it contains, denoted by Z, and by the number of neutrons, N. Isotopes of an element have the same value of Z, but different values of N. The *mass number A*, is the sum of Z and N. For example, Uranium-238 has a mass number of 238, and contains 92 protons and 146 neutrons.

THE ELEMENTS (continued)

There is evidence that the definition of chemical elements must be broadened to include the electron. Several compounds known as *electrides,* have recently been made of alkaline metal elements and electrons. A relatively stable combination of a positron and electron, known as *positronium,* has also been studied.

In addition to the proton, neutron, and electron, there are considerably more than 100 other fundamental particles which have been discovered or hypothesized. The majority of these fall into one of two classes, *leptons* or *hadrons.* The leptons comprise just four known particles, the *electron,* the *muon* (μ meson), and two kinds of *neutrinos.* The muon is essentially similar to the electron and has a charge of –1, but it is 200 times heavier. The neutrino is either of two stable particles of small (probably zero) rest mass, carrying no charge. Also there are four *antileptons,* identical to the corresponding leptons in some respects, such as mass, but they have properties exactly opposite those of the leptons. The *positron,* for example, is an antilepton, with a charge of +1. Leptons cannot be broken into smaller units and are considered to be elementary. On the other hand, hadrons are complex and thought to have internal structure. Protons and neutrons, which make up atomic nuclei, are hadrons.

Elementary particle physics is not yet clearly understood, but groupings and arrangements of these particles have been made resembling the periodic table of chemical elements. This has led to the speculation that hadrons are composed of simpler components called *quarks.* Quarks are presumed to be elementary particles. Quarks come in a variety of "flavors." Two kinds of quarks — up and down — combine to form the protons and neutrons of ordinary matter. Three other quarks, known as {strange", "charm", and "bottom", are found in particle accelerators, and one of these may exist in the center of very dense, massive stars. Theorists have thought that a sixth "top" quark, much more massive than the others, should exist to complete the quark family of sub-atomic particles. In 1994, scientists using Fermilab's Tevatron accelerator may have "glimpsed" to top quark. There is presently no evidence that quarks exist in isolation. Many physicists now hold that all the matter and energy in the universe is controlled by four fundamental natural forces: the electromagnetic force, gravity, a weak nuclear force, and a strong nuclear force. Each of these natural forces is passed back and forth among the basic particles of matter by unique force-carrying particles. The electromagnetic force is carried by the *photon,* the weak nuclear force by the intermediate vector *boson,* and gravity by the *gravitron.* There is now evidence of the existence of a particle, known as the *gluon,* that binds quarks together by carrying the strong nuclear force. *Hadrons* are thought to be particles composed of quarks and gluons.

The available evidence leads to the conclusion that elements 89 (actinium) through 103 (lawrencium) are chemically similar to the rare-earth or lanthanide elements (elements 57 to 71, inclusive). These elements therefore have been named *actinides* after the first member of this series. Those elements beyond uranium that have been produced artificially have the following names and symbols: neptunium, 93 (Np); plutonium, 94 (Pu); americium, 95 (Am); curium, 96 (Cm); berkelium, 97 (Bk); californium, 98 (Cf); einsteinium, 99 (Es); fermium, 100 (Fm); mendelevium, 101 (Md); nobelium, 102 (No); and lawrencium, 103 (Lr). It is now claimed that Elements 104 through 112 have been produced and identified. In August 1997, the International Union of Pure and Applied Chemistry (IUPAC) gave final approval to the following names for elements 104 to 109: element 104 — rutherfordium (Rf); element 105 — dubnium (Db); element 106 — seaborgium (Sg); element 107 — bohrium (Bh); element 108 — hassium (Hs); and element 109 — meitnerium (Mt). Elements beyond this may be designated by temporary names: unonnilium (Uun) for 110, unononunium (Uuu) for 111, ununbium (Uub) for 112, etc.

Element 104 is expected to have chemical properties similar to those of hafnium and would not be a member of the actinide series. Element 105 probably would have chemical properties similar to those of tantalum, Element 106 similar to tungsten, Element 107 similar to rhenium, and Element 109 similar to iridium, etc.

There is still thought to be a possibility of producing elements beyond Element 112 either by bombardment of heavy isotopic targets with heavy ions, or by the irradiation of uranium or transuranic elements with the instantaneous high flux of neutrons produced by underground nuclear explosions. The limit will be set by the yields of the nuclear reactions and by the half-lives of radioactive decay. It has been suggested that Elements 102 and 103 have abnormally short lives only because they are in a pocket of instability, and that this region of instability might begin to "heal" around Element 105. If so, it may be possible to produce heavier isotopes with longer half-lives. It has also been suggested that Element 114, with a mass number of 298, and Element 126, with a mass number of 310, may be sufficiently stable to make discovery and identification possible. Searches have been made by workers in a number of laboratories for Element 110 and its neighboring elements in naturally occurring platinum. Recent studies of the xenon component (^{131}Xe to ^{136}Xe) of certain carbonaceous chondritic meteorites suggest that Elements 113, 114, or 115 may have been its progenitor.

There are many claims in the literature of the existence of various allotropic modifications of the elements, some of which are based on doubtful or incomplete evidence. Also, the physical properties of an element may change drastically by the presence of small amounts of impurities. With new methods of purification, which are now able to produce elements with 99.9999% purity, it has been necessary to restudy the properties of the elements. For example, the melting point of thorium changes by several hundred degrees by the presence of a small percentage of ThO_2 as an impurity. Ordinary commercial tungsten is brittle and can be worked only with difficulty. Pure tungsten, however, can be cut with a hacksaw, forged, spun, drawn, or extruded. In general, the value of a physical property given here applies to the pure element, when it is known.

Many of the chemical elements and their compounds are toxic and should be handled with due respect and care. In recent years there has been a greatly increased knowledge and awareness of the health hazards associated with chemicals, radioactive materials, and other agents. Anyone working with the elements and certain of their compounds should become thoroughly familiar with the proper safeguards to be taken. Information on specific hazards and recommended exposure limits may be found in Section 16. Reference should be made to publications such as the following:

1. *Code of Federal Regulations,* Title 29, Labor, chapter XVII, section 1910.93 of subpart G, redesignated as 1910.1000 at 40 FR (Federal Register) 23072. May 28, 1975; amended at 41 FR 11505, March 19, 1976; 41 FR 35184, August 20, 1976; FR 46784, October 22, 1976; 42 FR 3304, January 18, 1977 (corrections and additional amendments and corrections as issued, U.S. Government Printing Office, Supt. of Documents, Washington, D.C.

2. *Code of Federal Regulations,* Title 10, Energy, Chapter 1, Nuclear Regulatory Commission, section 20.103 — 108; 20.201 — 207; 20.301 — 305; 20.401 — 409; 20.501 — 502; 20.601; appendices, corrections, and amendments.

3. *Occupational Safety and Health Reporter* (latest edition with amendments and corrections), Bureau of National Affairs, Washington, D.C.

4. *Atomic Energy Law Reporter,* Commerce Clearing House, Chicago, IL.

5. *Nuclear Regulation Reporter,* Commerce Clearing House, Chicago, IL.

6. *TLVs® Threshold Limit Values for Chemical Substances and Physical Agents* is issued annually be the American Conference of Governmental Industrial Hygienists, Cincinnati, Ohio.

7. *The Sigma-Aldrich Library of Regulatory and Safety Data*, Vol. 3 by Robert E. Lenga and Kristine L. Volonpal, published by the Sigma Chemical Co. and The Aldrich Chemical Co., Inc., 1993, (pp. 3211—3254).

Actinium — (Gr. *aktis, aktinos,* beam or ray), Ac; at. wt. (227); at. no. 89; m.p. 1051°C, b.p. 3200 ± 300°C (est.); sp. gr. 10.07 (calc.). Discovered by Andre Debierne in 1899 and independently by F. Giesel in 1902. Occurs naturally in association with uranium minerals. Thirty two isotopes and isomers are now recognized. All are radioactive. Actinium-227, a decay product of uranium-235, is an alpha and beta emitter with a 21.77-year half-life. Its principal decay products are thorium-227 (18.72-day half-life), radium-223 (11.4-day half-life), and a number of short-lived products including radon, bismuth, polonium, and lead isotopes. In equilibrium with its decay products, it is a powerful source of alpha rays. Actinium metal has been prepared by the reduction of actinium fluoride with lithium vapor at about 1100 to 1300°C. The chemical behavior of actinium is similar to that of the rare earths, particularly lanthanum. Purified actinium comes into equilibrium with its decay products at the end of 185 days, and then decays according to its 21.77-year half-life. It is about 150 times as active as radium, making it of value in the production of neutrons.

Aluminum — (L. *alumen, alum*), Al; at. wt. 26.981539(5); at. no. 13; f.p. 660.323°C; b.p. 2519°C; sp. gr. 2.6989 (20°C); valence 3. The ancient Greeks and Romans used *alum* in medicine as an astringent, and as a mordant in dyeing. In 1761 de Morveau proposed the name *alumine* for the base in alum, and Lavoisier, in 1787, thought this to be the oxide of a still undiscovered metal. Wohler is generally credited with having isolated the metal in 1827, although an impure form was prepared by Oersted two years earlier. In 1807, Davy proposed the name *alumium* for the metal, undiscovered at that time, and later agreed to change it to *aluminum*. Shortly thereafter, the name *aluminium* was adopted to conform with the "ium" ending of most elements, and this spelling is now in use elsewhere in the world. *Aluminium* was also the accepted spelling in the U.S. until 1925, at which time the American Chemical Society officially decided to use the name *aluminum* thereafter in their publications. The method of obtaining aluminum metal by the electrolysis of alumina dissolved in *cryolite* was discovered in 1886 by Hall in the U.S. and at about the same time by Heroult in France. Cryolite, a natural ore found in Greenland, is no longer widely used in commercial production, but has been replaced by an artificial mixture of sodium, aluminum, and calcium fluorides. *Bauxite,* an impure hydrated oxide ore, is found in large deposits in Jamaica, Australia, Surinam, Guyana, Arkansas, and elsewhere. The Bayer process is most commonly used today to refine bauxite so it can be accommodated in the Hall-Heroult refining process, used to make most aluminum. Aluminum can now be produced from clay, but the process is not economically feasible at present. Aluminum is the most abundant metal to be found in the earth's crust (8.1%), but is never found free in nature. In addition to the minerals mentioned above, it is found in feldspars, granite, and in many other common minerals. Seventeen isotopes and isomers are known. Natural aluminum is made of one isotope, ^{27}Al. Pure aluminum, a silvery-white metal, possesses many desirable characteristics. It is light, nontoxic, has a pleasing appearance, can easily be formed, machined, or cast, has a high thermal conductivity, and has excellent corrosion resistance. It is nonmagnetic and nonsparking, stands second among metals in the scale of malleability, and sixth in ductility. It is extensively used for kitchen utensils, outside building decoration, and in thousands of industrial applications where a strong, light, easily constructed material is needed. Although its electrical conductivity is only about 60% that of copper, it is used in electrical transmission lines because of its light weight. Pure aluminum is soft and lacks strength, but it can be alloyed with small amounts of copper, magnesium, silicon, manganese, and other elements to impart a variety of useful properties. These alloys are of vital importance in the construction of modern aircraft and rockets. Aluminum, evaporated in a vacuum, forms a highly reflective coating for both visible light and radiant heat. These coatings soon form a thin layer of the protective oxide and do not deteriorate as do silver coatings. They have found application in coatings for telescope mirrors, in making decorative paper, packages, toys, and in many other uses. The compounds of greatest importance are aluminum oxide, the sulfate, and the soluble sulfate with potassium (alum). The oxide, alumina, occurs naturally as ruby, sapphire, corundum, and emery, and is used in glassmaking and refractories. Synthetic ruby and sapphire have found application in the construction of lasers for producing coherent light. In 1852, the price of aluminum was about $1200/kg, and just before Hall's discovery in 1886, about $25/kg. The price rapidly dropped to 60¢ and has been as low as 33¢/kg. The price in December 1995 was about $1.70/kg.

Americium — (the Americas), Am; at. wt. 243; at. no. 95; m.p. 1176°C; b.p. 2011°C; sp. gr. 13.67 (20°C); valence 2, 3, 4, 5, or 6. Americium was the fourth transuranium element to be discovered; the isotope ^{241}Am was identified by Seaborg, James, Morgan, and Ghiorso late in 1944 at the wartime Metallurgical Laboratory of the University of Chicago as the result of successive neutron capture reactions by plutonium isotopes in a nuclear reactor:

$$^{239}\text{Pu}(n,\gamma) \rightarrow\, ^{240}\text{Pu}(n,\gamma) \rightarrow\, ^{241}\text{Pu} \overset{\beta}{\longrightarrow}\, ^{241}\text{Am}$$

Since the isotope ^{241}Am can be prepared in relatively pure form by extraction as a decay product over a period of years from strongly neutron-bombarded plutonium, ^{241}Pu, this isotope is used for much of the chemical investigation of this element. Better suited is the isotope ^{243}Am due to its longer half-life (7.37×10^3 years as compared to 432.2 years for ^{241}Am). A mixture of the isotopes ^{241}Am, ^{242}Am, and ^{243}Am can be prepared by intense neutron irradiation of ^{241}Am according to the reactions ^{241}Am (n, γ) → ^{242}Am (n, γ) → ^{243}Am. Nearly isotopically pure ^{243}Am can be prepared by a sequence of neutron bombardments and chemical separations as follows: neutron bombardment of ^{241}Am yields ^{242}Pu by the reactions ^{241}Am (n, γ) → ^{242}Am → ^{242}Pu, after chemical separation the ^{242}Pu can be transformed to ^{243}Am via the reactions ^{242}Pu (n, γ) → ^{243}Pu → ^{243}Am, and the ^{243}Am can be chemically separated. Fairly pure ^{242}Pu can be prepared more simply by very intense neutron irradiation of ^{239}Pu as the result of successive neutron-capture reactions. Sixteen radioactive isotopes and isomers are now recognized. Americium metal has been prepared by reducing the trifluoride with barium vapor at 1000 to 1200°C or the dioxide by lanthanum metal. The luster of freshly prepared americium metal is white and more silvery than plutonium or neptunium prepared in the same manner. It appears to be more malleable than uranium or neptunium and tarnishes slowly in dry air at room temperature. Americium is thought to exist in two forms: an alpha form which has a double hexagonal close-packed structure and a loose-packed cubic beta form. Americium must be handled with great care to avoid personal contamination. As little as 0.03 μCi of ^{241}Am is the maximum permissible total body burden. The alpha activity from ^{241}Am is about three times that of radium. When gram quantities of ^{241}Am are handled, the intense gamma activity makes exposure a serious problem. Americium dioxide, AmO_2, is the most important oxide. AmF_3, AmF_4, $AmCl_3$, $AmBr_3$, AmI_3, and other compounds have been prepared. The isotope ^{241}Am has been used as a portable source for gamma radiography. It has also been used as a radioactive glass thickness gage for the flat glass industry, and as a source of ionization for smoke detectors. Americium-243 is available from the Oak Ridge National Laboratory at a cost of $160/mg plus packing charges.

Antimony — (Gr. *anti* plus *monos* — a metal not found alone), Sb; at. wt. 121.760(1); at. no. 51; m.p. 630.63°C; b.p. 1587°C; sp. gr. 6.691 (20°C); valence 0, –3, +3, or +5. Antimony was recognized in compounds by the ancients and was known as a metal at the beginning of the 17th century and possibly much earlier. It is not abundant, but is found in over 100 mineral species. It is sometimes found native, but more frequently as the sulfide, *stibnite* (Sb_2S_3); it is also found as antimonides of the heavy metals, and as oxides. It is extracted from the sulfide by roasting to the oxide, which is reduced by salt and scrap iron; from its oxides it is also prepared by reduction with carbon. Two allotropic forms of antimony exist: the normal stable, metallic form, and the amorphous gray form. The so-called explosive antimony is an ill-defined material always containing an appreciable amount of halogen; therefore, it no longer warrants consideration as a separate allotrope. The yellow form, obtained by oxidation of *stibine,* SbH_3, is probably impure, and is not a distinct form. Natural antimony is made of two stable isotopes, ^{121}Sb and ^{123}Sb. Forty four other radioactive isotopes and isomers are now recognized. Metallic antimony is an extremely brittle metal of a flaky, crystalline texture. It is bluish white and has a metallic luster. It is not acted on by air at room temperature, but burns brilliantly when heated with the formation of white fumes of Sb_2O_3. It is a poor conductor of heat and electricity, and has a hardness of 3 to 3.5. Antimony, available commercially with a purity of 99.999 + %, is finding use in semiconductor technology for making infrared detectors, diodes, and Hall-effect devices. Commercial-grade antimony is widely used in alloys with percentages ranging from 1 to 20. It greatly increases the hardness and mechanical strength of lead. Batteries, antifriction alloys, type metal, small arms and tracer bullets, cable sheathing, and minor products use about half the metal produced. Compounds taking up the other half are oxides, sulfides, sodium antimonate, and antimony trichloride. These are used in manufacturing flame-proofing compounds, paints, ceramic enamels, glass, and pottery. Tartar emetic (hydrated potassium antimonyl tartate) has been used in medicine. Antimony and many of its compounds are toxic. Antimony costs about $80/kg (99.5%), and about 75¢/g (99.999%).

Argon — (Gr. *argos,* inactive), Ar; at. wt. 39.948(1); at. no. 18; triple pt. –189.3442; b.p. –185.89°C; density 1.7837 g/l. Its presence in air was suspected by Cavendish in 1785, discovered by Lord Rayleigh and Sir William Ramsay in 1894. The gas is prepared by fractionation of liquid air, the atmosphere containing 0.94% argon. The atmosphere of Mars contains 1.6% of ^{40}Ar and 5 p.p.m. of ^{36}Ar. Argon is two and one half times as soluble in water as nitrogen, having about the same solubility as oxygen. It is recognized by the characteristic lines in the red end of the spectrum. It is used in electric light bulbs and in fluorescent tubes at a pressure of about 400 Pa, and in filling photo tubes, glow tubes, etc. Argon is also used as an inert gas shield for arc welding and cutting, as a blanket for the production of titanium and other reactive elements, and as a protective atmosphere for growing silicon and germanium crystals. Argon is colorless and odorless, both as a gas and liquid. It is available in high-purity form. Commercial argon is available at a cost of about 3¢ per cubic foot. Argon is considered to be a very inert gas and is not known to form true chemical compounds, as do krypton, xenon, and radon. However, it does form a hydrate having a dissociation pressure of 105 atm at 0°C. Ion molecules such as $(ArKr)^+$, $(ArXe)^+$, $(NeAr)^+$ have been observed spectroscopically. Argon also forms a clathrate with β-hydroquinone. This clathrate is stable and can be stored for a considerable time, but a true chemical bond does not exist. Van der Waals' forces act to hold the argon. Naturally occurring argon is a mixture of three isotopes. Seventeen other radioactive isotopes are now known to exist. Commercial argon is priced at about $70/300 cu. ft. or 8.5 cu. meters.

Arsenic — (L. *arsenicum,* Gr. *arsenikon,* yellow orpiment, identified with *arsenikos,* male, from the belief that metals were different sexes; Arabic, *Az-zernikh,* the orpiment from Persian *zerni-zar,* gold), As; at. wt. 74.92159(2); at. no. 33; valence –3, 0, +3 or +5. Elemental arsenic occurs in two solid modifications: yellow, and gray or metallic, with specific gravities of 1.97, and 5.73, respectively. Gray arsenic, the ordinary stable form, has a m.p. of 817°C (28 atm) and sublimes at 614°C. Several other allotropic forms of arsenic are reported in the literature. It is believed that Albertus Magnus obtained the element in 1250 A.D. In 1649 Schroeder published two methods of preparing the element. It is found native, in the sulfides *realgar* and *orpiment,* as arsenides and sulfarsenides of heavy metals, as the oxide, and as arsenates. *Mispickel,* arsenopyrite, (FeSAs) is the most common mineral, from which on heating the arsenic sublimes leaving ferrous sulfide. The element is a steel gray, very brittle, crystalline, semimetallic solid; it tarnishes in air, and when heated is rapidly oxidized to arsenous oxide (As_2O_3) with the odor of garlic. Arsenic and its compounds are poisonous. These values, however, are being studied, and may be lowered. Arsenic is also used in bronzing, pyrotechny, and for hardening and improving the sphericity of shot. The most important compounds are white arsenic (As_2O_3), the sulfide, Paris green $3Cu(AsO_2)_2 \cdot Cu(C_2H_3O_2)_2$, calcium arsenate, and lead arsenate; the last three have been used as agricultural insecticides and poisons. Marsh's test makes use of the formation and ready decomposition of arsine (AsH_3). Arsenic is available in high-purity form. It is finding increasing uses as a doping agent in solid-state devices such as transistors. Gallium arsenide is used as a laser material to convert electricity directly into coherent light. Natural arsenic is made of one isotope ^{75}As. Twenty-five other radioactive isotopes and isomers are known. Arsenic (99%) costs about $175/kg. Purified arsenic (99.9995%) costs about $2/gm.

Astatine — (Gr. *astatos,* unstable), At; at. wt. (210); at. no. 85; m.p. 302°C; b.p. 337°C (est.); valence probably 1, 3, 5, or 7. Synthesized in 1940 by D. R. Corson, K. R. MacKenzie, and E. Segre at the University of California by bombarding bismuth with alpha particles. The longest-lived isotopes, ^{210}At, have a half-life of only 8.1 hours. Thirty-four isotopes and isomers are now known. Minute quantities of ^{215}At, ^{218}At, and ^{219}At exist in equilibrium in nature with naturally occurring uranium and thorium isotopes, and traces of ^{217}At are equilibrium with ^{233}U and ^{239}Np resulting from interaction of thorium and uranium with naturally produced neutrons. The total amount of astatine present in the earth's crust, however, is less than 1 oz. Astatine can be produced by bombarding bismuth with energetic alpha particles to obtain the relatively long-lived $^{209-211}$At, which can be distilled from the target by heating it in air. Only about 0.05 μg of astatine has been prepared to date. The "time of flight" mass spectrometer has been used to confirm that this highly radioactive halogen behaves chemically very much like other halogens, particularly iodine. The interhalogen compounds AtI, AtBr, and AtCl are known to form, but it is not yet known if astatine forms diatomic astatine molecules. HAt and CH_3At (methyl astatide) have been detected. Astatine is said to be more metallic that iodine, and, like iodine, it probably accumulates in the thyroid gland. Workers at the Brookhaven National Laboratory have recently used reactive scattering in crossed molecular beams to identify and measure elementary reactions involving astatine.

Barium — (Gr. *barys,* heavy), Ba; at. wt. 137.327(7), at. no. 56; m.p. 727°C; b.p. 1897°C; sp. gr. 3.5 (20°C); valence 2. Baryta was distinguished from lime by Scheele in 1774; the element was discovered by Sir Humphrey Davy in 1808. It is found only in combination with other elements, chiefly in *barite* or *heavy spar* (sulfate) and *witherite* (carbonate) and is prepared by electrolysis of the chloride. Barium is a metallic element, soft, and when pure is silvery white like lead; it belongs to the alkaline earth group, resembling calcium chemically. The metal oxidizes very easily and should be kept under petroleum or other suitable oxygen-free liquids to exclude air. It is decomposed by water or alcohol. The metal is used as a "getter" in vacuum tubes. The most important compounds are the peroxide (BaO_2), chloride, sulfate, carbonate, nitrate, and chlorate. Lithopone, a pigment containing

THE ELEMENTS (continued)

barium sulfate and zinc sulfide, has good covering power, and does not darken in the presence of sulfides. The sulfate, as permanent white or *blanc fixe,* is also used in paint, in X-ray diagnostic work, and in glassmaking. *Barite* is extensively used as a weighting agent in oilwell drilling fluids, and also in making rubber. The carbonate has been used as a rat poison, while the nitrate and chlorate give colors in pyrotechny. The impure sulfide phosphoresces after exposure to the light. The compounds and the metal are not expensive. Barium metal (99.7 + % pure) costs about 40¢/gm. All barium compounds that are water or acid soluble are poisonous. Naturally occurring barium is a mixture of seven stable isotopes. Thirty nine other radioactive isotopes and isomers are known to exist.

Berkelium — (*Berkeley,* home of the University of California), Bk; at. wt. (247); at. no. 97; m.p. 1050°C; valence 3 or 4; sp. gr. 14 (est.). Berkelium, the eighth member of the actinide transition series, was discovered in December 1949 by Thompson, Ghiorso, and Seaborg, and was the fifth transuranium element synthesized. It was produced by cyclotron bombardment of milligram amounts of ^{241}Am with helium ions at Berkeley, California. The first isotope produced had a mass number of 243 and decayed with a half-life of 4.5 hours. Eleven isotopes are now known and have been synthesized. The existence of ^{249}Bk, with a half-life of 320 days, makes it feasible to isolate berkelium in weighable amounts so that its properties can be investigated with macroscopic quantities. One of the first visible amounts of a pure berkelium compound, berkelium chloride, was produced in 1962. It weighed 3 billionth of a gram. Berkelium probably has not yet been prepared in elemental form, but it is expected to be a silvery metal, easily soluble in dilute mineral acids, and readily oxidized by air or oxygen at elevated temperatures to form the oxide. X-ray diffraction methods have been used to identify the following compounds: BkO_2, BkO_3, BkF_3, $BkCl$, and $BkOCl$. As with other actinide elements, berkelium tends to accumulate in the skeletal system. The maximum permissible body burden of ^{249}Bk in the human skeleton is about 0.0004 µg. Because of its rarity, berkelium presently has no commercial or technological use. Berkelium-249 is available from O.R.N.L. at a cost of $160/µg plus packing charges.

Beryllium — (Gr. *beryllos, beryl;* also called Glucinium or Glucinum, Gr. *glykys,* sweet), Be; at. wt. 9.012182(3); at. no. 4; m.p. 1287°C; b.p. 2471°C; sp. gr. 1.848 (20°C); valence 2. Discovered as the oxide by Vauquelin in beryl and in emeralds in 1798. The metal was isolated in 1828 by Wohler and by Bussy independently by the action of potassium on beryllium chloride. Beryllium is found in some 30 mineral species, the most important of which are *bertrandite, beryl, chrysoberyl,* and *phenacite. Aquamarine* and *emerald* are precious forms of *beryl. Beryl* (3BeO · Al$_2$O$_3$ · 6SiO$_2$) and *bertrandite* (4BeO · 2SiO$_2$ · H$_2$O) are the most important commercial sources of the element and its compounds. Most of the metal is now prepared by reducing beryllium fluoride with magnesium metal. Beryllium metal did not become readily available to industry until 1957. The metal, steel gray in color, has many desirable properties. It is one of the lightest of all metals, and has one of the highest melting points of the light metals. Its modulus of elasticity is about one third greater than that of steel. It resists attack by concentrated nitric acid, has excellent thermal conductivity, and is nonmagnetic. It has a high permeability to X-rays, and when bombarded by alpha particles, as from radium or polonium, neutrons are produced in the ratio of about 30 neutrons/million alpha particles. At ordinary temperatures beryllium resists oxidation in air, although its ability to scratch glass is probably due to the formation of a thin layer of the oxide. Beryllium is used as an alloying agent in producing beryllium copper which is extensively used for springs, electrical contacts, spot-welding electrodes, and nonsparking tools. It has found application as a structural material for high-speed aircraft, missiles, spacecraft, and communication satellites. It is being used in the windshield frame, brake discs, support beams, and other structural components of the space shuttle. Because beryllium is relatively transparent to X-rays, ultra-thin Be-foil is finding use in X-ray lithography for reproduction of microminiature integrated circuits. Natural beryllium is made of ^{9}Be and is stable. Eight other radioactive isotopes are known.

Beryllium is used in nuclear reactors as a reflector or moderator for it has a low thermal neutron absorption cross section. It is used in gyroscopes, computer parts, and instruments where lightness, stiffness, and dimensional stability are required. The oxide has a very high melting point and is also used in nuclear work and ceramic applications. Beryllium and its salts are toxic and should be handled with the greatest of care. Beryllium and its compounds should not be tasted to verify the sweetish nature of beryllium (as did early experimenters). The metal, its alloys, and its salts can be handled safely if certain work codes are observed, but no attempt should be made to work with beryllium before becoming familiar with proper safeguards. Beryllium metal is available at a cost of about $2.50/g (99.5% pure).

Bismuth — (Ger. *Weisse Masse,* white mass; later *Wisuth* and *Bisemutum*), Bi; at. wt. 208.98037(3); at. no. 83; m.p. 271.4°C; b.p. 1564 ± 5°C; sp. gr. 9.747 (20°C); valence 3 or 5. In early times bismuth was confused with tin and lead. Claude Geoffroy the Younger showed it to be distinct from lead in 1753. It is a white crystalline, brittle metal with a pinkish tinge. It occurs native. The most important ores are *bismuthinite* or bismuth glance (Bi$_2$S$_3$) and *bismite* (Bi$_2$O$_3$). Peru, Japan, Mexico, Bolivia, and Canada are major bismuth producers. Much of the bismuth produced in the U.S. is obtained as a by-product in refining lead, copper, tin, silver, and gold ores. Bismuth is the most diamagnetic of all metals, and the thermal conductivity is lower than any metal, except mercury. It has a high electrical resistance, and has the highest Hall effect of any metal (i.e., greatest increase in electrical resistance when placed in a magnetic field). "Bismanol" is a permanent magnet of high coercive force, made of MnBi, by the U.S. Naval Surface Weapons Center. Bismuth expands 3.32% on solidification. This property makes bismuth alloys particularly suited to the making of sharp castings of objects subject to damage by high temperatures. With other metals such as tin, cadmium, etc., bismuth forms low-melting alloys which are extensively used for safety devices in fire detection and extinguishing systems. Bismuth is used in producing malleable irons and is finding use as a catalyst for making acrylic fibers. When bismuth is heated in air it burns with a blue flame, forming yellow fumes of the oxide. The metal is also used as a thermocouple material, and has found application as a carrier for U^{235} or U^{233} fuel in atomic reactors. Its soluble salts are characterized by forming insoluble basic salts on the addition of water, a property sometimes used in detection work. Bismuth oxychloride is used extensively in cosmetics. Bismuth subnitrate and subcarbonate are used in medicine. Natural bismuth contains only one isotope ^{209}Bi. Forty one isotopes and isomers of bismuth are known. Bismuth metal costs about $90/kg (99.999%).

Bohrium — See Element 107.

Boron — (Ar. *Buraq,* Pers. *Burah*), B; at. wt. 10.81(5); at. no. 5; m.p. 2075°C; b.p. 4000°C; sp. gr. of crystals 2.34, of amorphous variety 2.37; valence 3. Boron compounds have been known for thousands of years, but the element was not discovered until 1808 by Sir Humphry Davy and by Gay-Lussac and Thenard. The element is not found free in nature, but occurs as orthoboric acid usually in certain volcanic spring waters and as borates in *borax* and *colemanite. Ulexite,* another boron mineral, is interesting as it is nature's own version of "fiber optics." Important sources of boron are the ores *rasorite (kernite)* and *tincal (borax ore).* Both of these ores are found in the Mojave Desert. *Tincal* is the most important source of boron from the Mojave. Extensive *borax* deposits are also found in Turkey. Boron exists naturally as 19.9% ^{10}B isotope and 80.1% ^{11}B isotope. Eleven isotopes of boron are known. High-purity crystalline boron may be prepared by the vapor phase reduction of boron trichloride or tribromide with hydrogen on electrically heated filaments. The impure, or amorphous, boron, a brownish-black powder, can be obtained by heating the trioxide with magnesium powder. Boron of 99.9999% purity has been produced and is available commercially. Elemental boron has an energy band gap of 1.50 to 1.56 eV, which is higher than that of either silicon or germanium. It has interesting optical characteristics, transmitting portions of the infrared, and is a poor conductor of electricity at room temperature, but a good conductor at high temperature. Amorphous boron is used in pyrotechnic flares to provide a distinctive green color, and in rockets as an igniter. By far the most commercially important boron compound in terms of dollar sales is $Na_2B_4O_7 \cdot 5H_2O$. This pentahydrate is used in very large quantities in the manufacture of insulation fiberglass and sodium perborate bleach. Boric acid is also an important boron compound with major markets in textile fiberglass and in cellulose insulation as a flame retardant. Next in order of importance is borax ($Na_2B_4O_7 \cdot 10H_2O$) which is used principally in laundry products. Use of borax as a mild antiseptic is minor in terms of dollars and tons. Boron compounds are also extensively used in the manufacture of borosilicate glasses. Other boron compounds show promise in treating arthritis. The isotope boron-10 is used as a control for nuclear reactors, as a shield for nuclear radiation, and in instruments used for detecting neutrons. Boron nitride has remarkable properties and can be used to make a material as hard as diamond. The nitride also behaves like an electrical insulator but conducts heat like a metal. It also has lubricating properties similar to graphite. The hydrides are easily oxidized with considerable energy liberation, and have been studied for use as rocket fuels. Demand is increasing for boron filaments, a high-strength, lightweight material chiefly employed for advanced aerospace structures. Boron is similar to carbon in that it has a capacity to form stable covalently bonded molecular networks. Carboranes, metalloboranes, phosphacarboranes, and other families comprise thousands of compounds. Crystalline boron (99%) costs about $8/g. Amorphous boron costs about $4/g. Elemental boron and the borates are not considered to be toxic, and they do not require special care in handling. However, some of the more exotic boron hydrogen compounds are definitely toxic and do require care.

Bromine — (Gr. *bromos,* stench), Br; at. wt. 79.904(1); at. no. 35; m.p. –7.2°C; b.p. 58.78°C; density of gas 7.59 g/l, liquid 3.12 (20°C); valence 1, 3, 5, or 7. Discovered by Balard in 1826, but not prepared in quantity until 1860. A member of the halogen group of elements, it is obtained from natural brines from wells in Michigan and Arkansas. Little bromine is extracted today from seawater, which contains only about 85 ppm. Bromine is the only liquid nonmetallic element. It is a heavy, mobile, reddish-brown liquid, volatilizing readily at room temperature to a red vapor with a strong disagreeable odor, resembling chlorine, and having a very irritating effect on the eyes and throat; it is readily soluble in water or carbon disulfide, forming a red solution, is less active than chlorine but more so than iodine; it unites readily with many elements and has a bleaching action; when spilled on the skin it produces painful sores. It presents a serious health hazard, and maximum safety precautions should be taken when handling it. Much of the bromine output in the U.S. was used in the production of ethylene dibromide, a lead scavenger used in making gasoline antiknock compounds. Lead in gasoline, however, has been drastically reduced, due to environmental considerations. This will greatly affect future production of bromine. Bromine is also used in making fumigants, flameproofing agents, water purification compounds, dyes, medicinals, sanitizers, inorganic bromides for photography, etc. Organic bromides are also important. Natural bromine is made of two isotopes, ^{79}Br and ^{81}Br. Thirty three isotopes and isomers are known. Bromine costs about $100/kg.

Cadmium — (L. *cadmia;* Gr. *kadmeia* — ancient name for calamine, zinc carbonate), Cd; at. wt. 112.411(8); at. no. 48; m.p. 321.07°C; b.p. 767°C; sp. gr. 8.65 (20°C); valence 2. Discovered by Stromeyer in 1817 from an impurity in zinc carbonate. Cadmium most often occurs in small quantities associated with zinc ores, such as *sphalerite* (ZnS). *Greenockite* (CdS) is the only mineral of any consequence bearing cadmium. Almost all cadmium is obtained as a by-product in the treatment of zinc, copper, and lead ores. It is a soft, bluish-white metal which is easily cut with a knife. It is similar in many respects to zinc. It is a component of some of the lowest melting alloys; it is used in bearing alloys with low coefficients of friction and great resistance to fatigue; it is used extensively in electroplating, which accounts for about 60% of its use. It is also used in many types of solder, for standard E.M.F. cells, for Ni-Cd batteries, and as a barrier to control atomic fission. Cadmium compounds are used in black and white television phosphors and in blue and green phosphors for color TV tubes. It forms a number of salts, of which the sulfate is most common; the sulfide is used as a yellow pigment. Cadmium and solutions of its compounds are toxic. Failure to appreciate the toxic properties of cadmium may cause workers to be unwittingly exposed to dangerous fumes. Some silver solders, for example, contain cadmium and should be handled with care. Serious toxicity problems have been found from long-term exposure and work with cadmium plating baths. In 1927 the International Conference on Weights and Measures redefined the meter in terms of the wavelength of the red cadmium spectral line (i.e. 1 m = 1,553,164.13 wavelengths). This definition has been changed (see under Krypton). The current price of cadmium is about $100/kg (99.5%). It is available in high purity form for about $300/kg. Natural cadmium is made of eight isotopes. Thirty four other isotopes and isomers are now known and recognized.

Calcium — (L. *calx,* lime), Ca; at. wt. 40.078(4); at. no. 20; m.p. 842±2°C; b.p. 1484°C; sp. gr. 1.55 (20°C); valence 2. Though lime was prepared by the Romans in the first century under the name calx, the metal was not discovered until 1808. After learning that Berzelius and Pontin prepared calcium amalgam by electrolyzing lime in mercury, Davy was able to isolate the impure metal. Calcium is a metallic element, fifth in abundance in the earth's crust, of which it forms more than 3%. It is an essential constituent of leaves, bones, teeth, and shells. Never found in nature uncombined, it occurs abundantly as *limestone* ($CaCO_3$), *gypsum* ($CaSO_4 \cdot 2H_2O$), and *fluorite* (CaF_2); *apatite* is the fluorophosphate or chlorophosphate of calcium. The metal has a silvery color, is rather hard, and is prepared by electrolysis of the fused chloride to which calcium fluoride is added to lower the melting point. Chemically it is one of the alkaline earth elements; it readily forms a white coating of oxide in air, reacts with water, burns with a yellow-red flame, forming largely the oxide. The metal is used as a reducing agent in preparing other metals such as thorium, uranium, zirconium, etc., and is used as a deoxidizer, desulfurizer, and inclusion modifier for various ferrous and nonferrous alloys. It is also used as an alloying agent for aluminum, beryllium, copper, lead, and magnesium alloys, and serves as a "getter" for residual gases in vacuum tubes, etc. Its natural and prepared compounds are widely used. Quicklime (CaO), made by heating limestone and changed into slaked lime by the careful addition of water, is the great cheap base of chemical industry with countless uses. Mixed with sand it hardens as mortar and plaster by taking up carbon dioxide from the air. Calcium from limestone is an important element in Portland cement. The solubility of the carbonate in water containing carbon dioxide causes the formation of caves with stalactites and stalagmites and is responsible for hardness in water. Other important compounds are the carbide (CaC_2), chloride ($CaCl_2$), cyanamide ($CaCN_2$), hypochlorite ($Ca(OCl)_2$), nitrate ($Ca(NO_3)_2$), and sulfide (CaS). Natural calcium contains six isotopes. Thirteen other radioactive isotopes are known. Metallic calcium (99.5%) costs about $200/kg.

Californium — (State and University of California), Cf; at. wt. (251); m.p. (est) 900°C; at. no. 98. Californium, the sixth transuranium element to be discovered, was produced by Thompson, Street, Ghiorso, and Seaborg in 1950 by bombarding microgram quantities of ^{242}Cm with 35 MeV helium ions in the Berkeley 60-inch cyclotron. Californium (III) is the only ion stable in aqueous solutions, all attempts to reduce or oxidize californium (III) having failed. The isotope ^{249}Cf results from the beta decay of ^{249}Bk while the heavier isotopes are produced by intense neutron irradiation by the reactions:

$$^{249}\text{Bk}(n, \gamma) \rightarrow {}^{250}\text{Bk} \xrightarrow{\beta} {}^{250}\text{Cf} \text{ and } {}^{249}\text{Cf}(n, \gamma) \rightarrow {}^{250}\text{Cf}$$

followed by

$$^{250}\text{Cf}(n, \gamma) \rightarrow {}^{251}\text{Cf}(n, \gamma) \rightarrow {}^{252}\text{Cf}$$

The existence of the isotopes ^{249}Cf, ^{250}Cf, ^{251}Cf, and ^{252}Cf makes it feasible to isolate californium in weighable amounts so that its properties can be investigated with macroscopic quantities. Californium-252 is a very strong neutron emitter. One microgram releases 170 million neutrons per minute, which presents biological hazards. Proper safeguards should be used in handling californium. Eighteen isotopes of californium are now recognized. ^{249}Cf and ^{252}Cf have half-lives of 351 years and 900 years, respectively. In 1960 a few tenths of a microgram of californium trichloride, $CfCl_3$, californium oxychloride, $CfOCl$, and californium oxide, Cf_2O_3, were first prepared. Reduction of californium to its metallic state has not yet been accomplished. Because californium is a very efficient source of neutrons, many new uses are expected for it. It has already found use in neutron moisture gages and in well-logging (the determination of water and oil-bearing layers). It is also being used as a portable neutron source for discovery of metals such as gold or silver by on-the-spot activation analysis. ^{252}Cf is now being offered for sale by the Oak Ridge National Laboratory (O.R.N.L.) at a cost of \$50/µg and ^{249}Cf at a cost of \$160/µg plus packing charges. It has been suggested that californium may be produced in certain stellar explosions, called *supernovae,* for the radioactive decay of ^{254}Cf (55-day half-life) agrees with the characteristics of the light curves of such explosions observed through telescopes. This suggestion, however, is questioned.

Carbon — (L. *carbo,* charcoal), C; at. wt. 12.011(1); at. no. 6; m.p. ~3550°C, graphite sublimes at 3825°C; triple point; (graphite-liquid-gas), 4492°C at a pressure of 10.3 MPa and (graphite-diamond-liquid), 3830—3930° at a pressure of 12—13 GPa; sp. gr. amorphous 1.8 to 2.1, graphite 1.9 to 2.3, diamond 3.15 to 3.53 (depending on variety); gem diamond 3.513 (25°C); valence 2, 3, or 4. Carbon, an element of prehistoric discovery, is very widely distributed in nature. It is found in abundance in the sun, stars, comets, and atmospheres of most planets. Carbon in the form of microscopic diamonds is found in some meteorites. Natural diamonds are found in *kimberlite* of ancient volcanic "pipes," such as found in South Africa, Arkansas, and elsewhere. Diamonds are now also being recovered from the ocean floor off the Cape of Good Hope. About 30% of all industrial diamonds used in the U.S. are now made synthetically. The energy of the sun and stars can be attributed at least in part to the well-known carbon-nitrogen cycle. Carbon is found free in nature in three allotropic forms: amorphous, graphite, and diamond. A fourth form, known as "white" carbon, is now thought to exist. Graphite is one of the softest known materials while diamond is one of the hardest. Graphite exists in two forms: alpha and beta. These have identical physical properties, except for their crystal structure. Naturally occurring graphites are reported to contain as much as 30% of the rhombohedral (beta) form, whereas synthetic materials contain only the alpha form. The hexagonal alpha type can be converted to the beta by mechanical treatment, and the beta form reverts to the alpha on heating it above 1000°C. In 1969 a new allotropic form of carbon was produced during the sublimation of pyrolytic graphite at low pressures. Under free-vaporization conditions above ~2550 K, "white" carbon forms as small transparent crystals on the edges of the basal planes of graphite. The interplanar spacings of "white" carbon are identical to those of carbon form noted in the graphitic gneiss from the Ries (meteoritic) Crater of Germany. "White" carbon is a transparent birefringent material. Little information is presently available about this allotrope. Of recent interest is the discovery of all-carbon molecules, known as "buckyballs" or fullerenes, which have a number of unusual properties. These interesting molecules, consisting of 60 or 70 carbon atoms linked together, seem capable of withstanding great pressure and trapping foreign atoms inside their network of carbon. They are said to be capable of magnetism and superconductivity and have potential as a nonlinear optical material. Buckyball films are reported to remain superconductive at temperatures as high as 45 K. In combination, carbon is found as carbon dioxide in the atmosphere of the earth and dissolved in all natural waters. It is a component of great rock masses in the form of carbonates of calcium (limestone), magnesium, and iron. Coal, petroleum, and natural gas are chiefly hydrocarbons. Carbon is unique among the elements in the vast number and variety of compounds it can form. With hydrogen, oxygen, nitrogen, and other elements, it forms a very large number of compounds, carbon atom often being linked to carbon atom. There are close to ten million known carbon compounds, many thousands of which are vital to organic and life processes. Without carbon, the basis for life would be impossible. While it has been thought that silicon might take the place of carbon in forming a host of similar compounds, it is now not possible to form stable compounds with very long chains of silicon atoms. The atmosphere of Mars contains 96.2% CO_2. Some of the most important compounds of carbon are carbon dioxide (CO_2), carbon monoxide (CO), carbon disulfide (CS_2), chloroform ($CHCl_3$), carbon tetrachloride (CCl_4), methane (CH_4), ethylene (C_2H_4), acetylene (C_2H_2), benzene (C_6H_6), ethyl alcohol (C_2H_5OH), acetic acid (CH_3COOH), and their derivatives. Carbon has thirteen isotopes. Natural carbon consists of 98.89% ^{12}C and 1.11% ^{13}C. In 1961 the International Union of Pure and Applied Chemistry adopted the isotope carbon-12 as the basis for atomic weights. Carbon-14, an isotope with a half-life of 5715 years, has been widely used to date such materials as wood, archeological specimens, etc.

Cerium — (named for the asteroid *Ceres,* which was discovered in 1801 only 2 years before the element), Ce; at. wt. 140.115(4); at. no. 58; m.p. 798°C; b.p. 3424°C; sp. gr. 6.770 (25°C); valence 3 or 4. Discovered in 1803 by Klaproth and by Berzelius and Hisinger; metal prepared by Hillebrand and Norton in 1875. Cerium is the most abundant of the metals of the so-called rare earths. It is found in a number of minerals including *allanite* (also known as *orthite*), *monazite, bastnasite, cerite,* and *samarskite.* Monazite and bastnasite are presently the two most important sources of cerium. Large deposits of monazite found on the beaches of Travancore, India, in river sands in Brazil, and deposits of *allanite* in the western United States, and *bastnasite* in Southern California will supply cerium, thorium, and the other rare-earth metals for many years to come. Metallic cerium is prepared by metallothermic reduction techniques, such as by reducing cerous fluoride with calcium, or by electrolysis of molten cerous chloride or other cerous halides. The metallothermic technique is used to produce high-purity cerium. Cerium is especially interesting because of its variable electronic structure. The energy of the inner 4f level is nearly the same as that of the outer or valence electrons, and only small amounts of energy are required to change the relative occupancy of these electronic levels. This gives rise to dual valency states. For example, a volume change of about 10% occurs when cerium is subjected to high pressures or low temperatures. It appears that the valence changes from about 3 to 4 when it is cooled or compressed. The low temperature behavior of cerium is complex. Four allotropic modifications are thought to exist: cerium at room temperature and at atmospheric pressure is known as γ cerium. Upon cooling to −16°C, γ cerium changes to β cerium. The remaining γ cerium starts to change to α cerium when cooled to −172°C, and the transformation is complete at −269°C. α Cerium has a density of 8.16; δ cerium exists above 726°C. At atmospheric pressure, liquid cerium is more dense than its solid form at the melting point. Cerium is an iron-gray lustrous metal. It is malleable, and oxidizes very readily at room temperature, especially in moist air. Except for europium, cerium is the most reactive of the "rare-earth" metals. It slowly decomposes in cold water, and rapidly in hot water. Alkali solutions and dilute and concentrated acids attack the metal rapidly. The pure metal is likely to ignite if scratched with a knife. Ceric salts are orange red or yellowish; cerous salts are usually white. Cerium is a component of misch metal, which is extensively used in the manufacture of pyrophoric alloys for cigarette lighters, etc. Natural cerium is stable and contains four isotopes. Thirty five other radioactive isotopes and isomers are known. While cerium is not radioactive, the impure commercial grade may contain traces of thorium, which is radioactive. The oxide is an important constituent of incandescent gas mantles and it is emerging as a hydrocarbon catalyst in "self-cleaning" ovens. In this application it can be incorporated into oven walls to prevent the collection of cooking residues. As ceric sulfate it finds extensive use as a volumetric oxidizing agent in quantitative analysis. Cerium compounds are used in the manufacture of glass, both as a component and as a decolorizer. The oxide is finding increased use as a glass polishing agent instead of rouge, for it is much faster than rouge in polishing glass surfaces. Cerium, with other rare earths, is used in carbon-arc lighting, especially in the motion picture industry. It is also finding use as an important catalyst in petroleum refining and in metallurgical and nuclear applications. In small lots, 99.9% cerium costs about $4/g (99.9%).

Cesium — (L. *caesius*, sky blue), Cs; at. wt. 132.90543(5); at. no. 55; m.p. $28.44 \pm 0.01°C$; b.p. 671°C; sp. gr. 1.873 (20°C); valence 1. Cesium was discovered spectroscopically by Bunsen and Kirchhoff in 1860 in mineral water from Durkheim. Cesium, an alkali metal, occurs in *lepidolite*, *pollucite* (a hydrated silicate of aluminum and cesium), and in other sources. One of the world's richest sources of cesium is located at Bernic Lake, Manitoba. The deposits are estimated to contain 300,000 tons of pollucite, averaging 20% cesium. It can be isolated by electrolysis of the fused cyanide and by a number of other methods. Very pure, gas-free cesium can be prepared by thermal decomposition of cesium azide. The metal is characterized by a spectrum containing two bright lines in the blue along with several others in the red, yellow, and green. It is silvery white, soft, and ductile. It is the most electropositive and most alkaline element. Cesium, gallium, and mercury are the only three metals that are liquid at room temperature. Cesium reacts explosively with cold water, and reacts with ice at temperatures above $-116°C$. Cesium hydroxide, the strongest base known, attacks glass. Because of its great affinity for oxygen the metal is used as a "getter" in electron tubes. It is also used in photoelectric cells, as well as a catalyst in the hydrogenation of certain organic compounds. The metal has recently found application in ion propulsion systems. Cesium is used in atomic clocks, which are accurate to 5 s in 300 years. Its chief compounds are the chloride and the nitrate. Cesium has 52 isotopes and isomers with masses ranging from 112 to 148. The present price of cesium is about $40/g (99.98%).

Chlorine — (Gr. *chloros*, greenish yellow), Cl; at. wt. 35.4527(9); at. no. 17; m.p. $-101.5°C$; b.p. $-34.04°C$; density 3.214 g/l; sp. gr. 1.56 ($-33.6°C$); valence 1, 3, 5, or 7. Discovered in 1774 by Scheele, who thought it contained oxygen; named in 1810 by Davy, who insisted it was an element. In nature it is found in the combined state only, chiefly with sodium as common salt (NaCl), *carnallite* ($KMgCl_3 \cdot 6H_2O$), and *sylvite* (KCl). It is a member of the halogen (salt-forming) group of elements and is obtained from chlorides by the action of oxidizing agents and more often by electrolysis; it is a greenish-yellow gas, combining directly with nearly all elements. At 10°C one volume of water dissolves 3.10 volumes of chlorine, at 30°C only 1.77 volumes. Chlorine is widely used in making many everyday products. It is used for producing safe drinking water the world over. Even the smallest water supplies are now usually chlorinated. It is also extensively used in the production of paper products, dyestuffs, textiles, petroleum products, medicines, antiseptics, insecticides, foodstuffs, solvents, paints, plastics, and many other consumer products. Most of the chlorine produced is used in the manufacture of chlorinated compounds for sanitation, pulp bleaching, disinfectants, and textile processing. Further use is in the manufacture of chlorates, chloroform, carbon tetrachloride, and in the extraction of bromine. Organic chemistry demands much from chlorine, both as an oxidizing agent and in substitution, since it often brings desired properties in an organic compound when substituted for hydrogen, as in one form of synthetic rubber. Chlorine is a respiratory irritant. The gas irritates the mucous membranes and the liquid burns the skin. As little as 3.5 ppm can be detected as an odor, and 1000 ppm is likely to be fatal after a few deep breaths. It was used as a war gas in 1915. Natural chlorine contains two isotopes. Sixteen other isotopes and isomers are known.

Chromium — (Gr. *chroma*, color), Cr; at. wt. 51.9961(6); at. no. 24; m.p. 1907°C; b.p. 2671°C; sp. gr. 7.18 to 7.20 (20°C); valence chiefly 2, 3, or 6. Discovered in 1797 by Vauquelin, who prepared the metal the next year, chromium is a steel-gray, lustrous, hard metal that takes a high polish. The principal ore is *chromite* ($FeCr_2O_4$), which is found in Zimbabwe, Russia, Transvaal, Turkey, Iran, Albania, Finland, Democratic Republic of Madagascar, and the Philippines. The metal is usually produced by reducing the oxide with aluminum. Chromium is used to harden steel, to manufacture stainless steel, and to form many useful alloys. Much is used in plating to produce a hard, beautiful surface and to prevent corrosion. Chromium is used to give glass an emerald green color. It finds wide use as a catalyst. All compounds of chromium are colored; the most important are the chromates of sodium and potassium (K_2CrO_4) and the dichromates ($K_2Cr_2O_7$) and the potassium and ammonium chrome alums, as $KCr(SO_4)_2 \cdot 12H_2O$. The dichromates are used as oxidizing agents in quantitative analysis, also in tanning leather. Other compounds are of industrial value; lead chromate is chrome yellow, a valued pigment. Chromium compounds are used in the textile industry as mordants, and by the aircraft and other industries for anodizing aluminum. The refractory industry has found chromite useful for forming bricks and shapes, as it has a high melting point, moderate thermal expansion, and stability of crystalline structure. Many chromium compounds are toxic and should be handled with proper safeguards. Natural chromium contains four isotopes. Sixteen other isotopes are known. Chromium metal (99.99%) costs about $200/kg. Commercial grade chromium (99%) costs about $75/kg.

Cobalt — (*Kobald*, from the German, goblin or evil spirit, *cobalos*, Greek, mine), Co; at. wt. 58.93320(1); at. no. 27; m.p. 1495°C; b.p. 2927°C; sp. gr. 8.9 (20°C); valence 2 or 3. Discovered by Brandt about 1735. Cobalt occurs in the mineral *cobaltite, smaltite,* and *erythrite*, and is often associated with nickel, silver, lead, copper, and iron ores, from which it is most frequently obtained as a by-product. It is also present in meteorites. Important ore deposits are found in Zaire, Morocco, and Canada. The U.S. Geological Survey has announced that the bottom of the north central Pacific Ocean may have cobalt-rich deposits at relatively shallow depths in waters close to the Hawaiian Islands and other U.S. Pacific territories. Cobalt is a brittle, hard metal, closely resembling iron and nickel in appearance. It has a magnetic permeability of about two thirds that of iron. Cobalt tends to exist as a mixture of two allotropes over a wide temperature range; the β-form predominates below 400°C, and the α above that temperature. The transformation is sluggish and accounts in part for the wide variation in reported data on physical properties of cobalt. It is alloyed with iron, nickel and other metals to make Alnico, an alloy of unusual magnetic strength with many important uses. Stellite alloys, containing cobalt, chromium, and tungsten, are used for high-speed, heavy-duty, high temperature cutting tools, and for dies. Cobalt is also used in other magnet steels and stainless steels, and in alloys used in jet turbines and gas turbine generators. The metal is used in electroplating because of its appearance, hardness, and resistance to oxidation. The salts have been used for centuries for the production of brilliant and permanent blue colors in porcelain, glass, pottery, tiles, and enamels. It is the principal ingredient in Sevre's and Thenard's blue. A solution of the chloride ($CoCl_2 \cdot 6H_2O$) is used as sympathetic ink. The cobalt ammines are of interest; the oxide and the nitrate are important. Cobalt carefully used in the form of the chloride, sulfate, acetate, or nitrate has been found effective in correcting a certain mineral deficiency disease in animals. Soils should contain 0.13 to 0.30 ppm of cobalt for proper animal nutrition. Cobalt is found in Vitamin B-12, which is essential for human nutrition. Cobalt-60, an artificial isotope, is an important gamma ray source, and is extensively used as a tracer and a radiotherapeutic agent. Single compact sources of Cobalt-60 vary from about $1 to $10/curie, depending on quantity and specific activity. Twenty six isotopes and isomers of cobalt are known.

Columbium — See Niobium.

Copper — (L. *cuprum,* from the island of Cyprus), Cu; at. wt. 63.546(3); at. no. 29; f.p. 1084.62 °C; b.p. 2562°C; sp. gr. 8.96 (20°C); valence 1 or 2. The discovery of copper dates from prehistoric times. It is said to have been mined for more than 5000 years. It is one of man's most important metals. Copper is reddish colored, takes on a bright metallic luster, and is malleable, ductile, and a good conductor of heat and electricity (second only to silver in electrical conductivity). The electrical industry is one of the greatest users of copper. Copper occasionally occurs native, and is found in many minerals such as *cuprite, malachite, azurite, chalcopyrite,* and *bornite.* Large copper ore deposits are found in the U.S., Chile, Zambia, Zaire, Peru, and Canada. The most important copper ores are the sulfides, oxides, and carbonates. From these, copper is obtained by smelting, leaching, and by electrolysis. Its alloys, brass and bronze, long used, are still very important; all American coins are now copper alloys; monel and gun metals also contain copper. The most important compounds are the oxide and the sulfate, blue vitriol; the latter has wide use as an agricultural poison and as an algicide in water purification. Copper compounds such as Fehling's solution are widely used in analytical chemistry in tests for sugar. High-purity copper (99.999 + %) is available commercially. Natural copper contains two isotopes. Twenty five other radioactive isotopes and isomers are known.

Curium — (Pierre and Marie Curie), Cm; at. wt. (247); at. no. 96; m.p. 1345±40°C; sp. gr. 13.51 (calc.); valence 3 and 4. Although curium follows americium in the periodic system, it was actually known before americium and was the third transuranium element to be discovered. It was identified by Seaborg, James, and Ghiorso in 1944 at the wartime Metallurgical Laboratory in Chicago as a result of helium-ion bombardment of ^{239}Pu in the Berkeley, California, 60-inch cyclotron. Visible amounts (30 µg) of ^{242}Cm, in the form of the hydroxide, were first isolated by Werner and Perlman of the University of California in 1947. In 1950, Crane, Wallmann, and Cunningham found that the magnetic susceptibility of microgram samples of CmF$_3$ was of the same magnitude as that of GdF$_3$. This provided direct experimental evidence for assigning an electronic configuration to Cm^{+3}. In 1951, the same workers prepared curium in its elemental form for the first time. Sixteen isotopes of curium are now known. The most stable, ^{247}Cm, with a half-life of 16 million years, is so short compared to the earth's age that any primordial curium must have disappeared long ago from the natural scene. Minute amounts of curium probably exist in natural deposits of uranium, as a result of a sequence of neutron captures and β decays sustained by the very low flux of neutrons naturally present in uranium ores. The presence of natural curium, however, has never been detected. ^{242}Cm and ^{244}Cm are available in multigram quantities. ^{248}Cm has been produced only in milligram amounts. Curium is similar in some regards to gadolinium, its rare-earth homolog, but it has a more complex crystal structure. Curium is silver in color, is chemically reactive, and is more electropositive than aluminum. CmO$_2$, Cm$_2$O$_3$, CmF$_3$, CmF$_4$, CmCl$_3$, CmBr$_3$, and CmI$_3$ have been prepared. Most compounds of trivalent curium are faintly yellow in color. ^{242}Cm generates about three watts of thermal energy per gram. This compares to one-half watt per gram of ^{238}Pu. This suggests use for curium as a power source. ^{244}Cm is now offered for sale at $160/mg plus packing charges. ^{248}Cm is available at a cost of $160/µg, plus packing charges, from the O.R.N.L. Curium absorbed into the body accumulates in the bones, and is therefore very toxic as its radiation destroys the red-cell forming mechanism. The maximum permissible total body burden of ^{244}Cm (soluble) in a human being is 0.3 µCi (microcurie).

Deuterium, an isotope of hydrogen — see Hydrogen.

Dubnium — see Element 105.

Dysprosium — (Gr. *dysprositos,* hard to get at), Dy; at. wt. 162.50(3); at. no. 66; m.p. 1412°C; b.p. 2567°C; sp. gr. 8.551 (25°C); valence 3. Dysprosium was discovered in 1886 by Lecoq de Boisbaudran, but not isolated. Neither the oxide nor the metal was available in relatively pure form until the development of ion-exchange separation and metallographic reduction techniques by Spedding and associates about 1950. Dysprosium occurs along with other so-called rare-earth or lanthanide elements in a variety of minerals such as *xenotime, fergusonite, gadolinite, euxenite, polycrase,* and *blomstrandine.* The most important sources, however, are from *monazite* and *bastnasite.* Dysprosium can be prepared by reduction of the trifluoride with calcium. The element has a metallic, bright silver luster. It is relatively stable in air at room temperature, and is readily attacked and dissolved, with the evolution of hydrogen, by dilute and concentrated mineral acids. The metal is soft enough to be cut with a knife and can be machined without sparking if overheating is avoided. Small amounts of impurities can greatly affect its physical properties. While dysprosium has not yet found many applications, its thermal neutron absorption cross-section and high melting point suggest metallurgical uses in nuclear control applications and for alloying with special stainless steels. A dysprosium oxide-nickel cermet has found use in cooling nuclear reactor rods. This cermet absorbs neutrons readily without swelling or contracting under prolonged neutron bombardment. In combination with vanadium and other rare earths, dysprosium has been used in making laser materials. Dysprosium-cadmium chalcogenides, as sources of infrared radiation, have been used for studying chemical reactions. The cost of dysprosium metal has dropped in recent years since the development of ion-exchange and solvent extraction techniques, and the discovery of large ore bodies. Thirty two isotopes and isomers are now known. The metal costs about $4/g (99.9% purity).

Einsteinium — (Albert Einstein), Es; at. wt. (252); m.p. 860°C (est.); at. no. 99. Einsteinium, the seventh transuranic element of the actinide series to be discovered, was identified by Ghiorso and co-workers at Berkeley in December 1952 in debris from the first large thermonuclear explosion, which took place in the Pacific in November 1952. The isotope produced was the 20-day ^{253}Es isotope. In 1961, a sufficient amount of einsteinium was produced to permit separation of a macroscopic amount of ^{253}Es. This sample weighed about 0.01 µg. A special magnetic-type balance was used in making this determination. ^{253}Es so produced was used to produce mendelevium (Element 101). About 3 µg of einsteinium has been produced at Oak Ridge National Laboratories by irradiating for several years kilogram quantities of ^{239}Pu in a reactor to produce ^{242}Pu. This was then fabricated into pellets of plutonium oxide and aluminum powder, and loaded into target rods for an initial 1-year irradiation at the Savannah River Plant, followed by irradiation in a HFIR (High Flux Isotopic Reactor). After 4 months in the HFIR the targets were removed for chemical separation of the einsteinium from californium. Seventeen isotopes and isomers of einsteinium are now recognized. ^{254}Es has the longest half-life (276 days). Tracer studies using ^{253}Es show that einsteinium has chemical properties typical of a heavy trivalent, actinide element.

Element 104 — (at. wt. 261). In 1964, workers of the Joint Nuclear Research Institute at Dubna (U.S.S.R.) bombarded plutonium with accelerated 113 to 115 MeV neon ions. By measuring fission tracks in a special glass with a microscope, they detected an isotope that decays by spontaneous fission. They suggested that this isotope, which had a half-life of 0.3 ± 0.1 s might be ^{260}104, produced by the following reaction:

$$^{242}_{94}\text{Pu} + ^{22}_{10}\text{Ne} \rightarrow ^{260}104 + 4n$$

Element 104, the first *transactinide* element, is expected to have chemical properties similar to those of hafnium. It would, for example, form a relatively volatile compound with chlorine (a tetrachloride). The Soviet scientists have performed experiments aimed at chemical identification, and have attempted to show that the 0.3-s activity is more volatile than that of the relatively nonvolatile actinide trichlorides. This experiment does not fulfill the test of chemically separating the new element from all others, but it provides important evidence for evaluation. New data, reportedly issued by Soviet scientists, have reduced the half-life of the isotope they worked with from 0.3 to 0.15 s. The Dubna scientists suggest the name *kurchatovium* and symbol *Ku* for Element 104, in honor of Igor Vasilevich Kurchatov (1903—1960), late Head of Soviet Nuclear Research. The Dubna Group also has proposed the name *dubnium* for Element 104. In 1969, Ghiorso, Nurmia, Harris, K. A. Y. Eskola, and P. L. Eskola of the University of California at Berkeley reported they had positively identified two, and possibly three, isotopes of Element 104. The group also indicated that after repeated attempts so far they had been unable to produce isotope ^{260}104 reported by the Dubna groups in 1964. The discoveries at Berkeley were made by bombarding a target of ^{249}Cf with ^{12}C nuclei of 71 MeV, and ^{13}C nuclei of 69 MeV. The combination of ^{12}C with ^{249}Cf followed by instant emission of four neutrons produced Element ^{257}104. This isotope has a half-life of 4 to 5 s, decaying by emitting an alpha particle into ^{253}No, with a half-life of 105 s. The same reaction, except with the emission of three neutrons, was thought to have produced ^{258}104 with a half-life of about 1/100 s. Element ^{259}104 is formed by the merging of a ^{13}C nuclei with ^{249}Cf, followed by emission of three neutrons. This isotope has a half-life of 3 to 4 s, and decays by emitting an alpha particle into ^{255}No, which has a half-life of 185 s. Thousands of atoms of ^{257}104 and ^{259}104 have been detected. The Berkeley group believe their identification of ^{258}104 was correct. As of January 1995 it was thought that eleven isotopes of Element 104 have been identified. The Berkeley group proposed for the new element the name *rutherfordium* (symbol Rf), in honor of Ernest Rutherford, New Zealand physicist. This name was formally adapted by IUPAC in 1997.

Element 105 — In 1967 G. N. Flerov reported that a Soviet team working at the Joint Institute for Nuclear Research at Dubna may have produced a few atoms of ^{260}105 and ^{261}105 by bombarding ^{243}Am with ^{22}Ne. Their evidence was based on time-coincidence measurements of alpha energies. More recently, it was reported that early in 1970 Dubna scientists synthesized Element 105 and that by the end of April 1970 "had investigated all the types of decay of the new element and had determined its chemical properties." The Soviet group proposed the name *joliotium* for Element 105. In late April 1970, it was announced that Ghiorso, Nurmia, Harris, K. A. Y. Eskola, and P. L. Eskola, working at the University of California at Berkeley, had positively identified Element 105. The discovery was made by bombarding a target of ^{249}Cf with a beam of 84 MeV nitrogen nuclei in the Heavy Ion Linear Accelerator (HILAC). When a ^{15}N nuclear is absorbed by a ^{249}Cf nucleus, four neutrons are emitted and a new atom of ^{260}105 with a half-life of 1.6 s is formed. While the first atoms of Element 105 are said to have been detected conclusively on March 5, 1970, there is evidence that Element 105 had been formed in Berkeley experiments a year earlier by the method described. Ghiorso and his associates have attempted to confirm Soviet findings by more sophisticated methods without success.

In October 1971, it was announced that two new isotopes of Element 105 were synthesized with the heavy ion linear accelerator by A. Ghiorso and co-workers at Berkeley. Element ^{261}105 was produced both by bombarding ^{250}Cf with ^{15}N and by bombarding ^{249}Bk with ^{16}O. The isotope emits 8.93-MeV α particles and decays to ^{257}Lr with a half-life of about 1.8 s. Element ^{262}105 was produced by bombarding ^{249}Bk with ^{18}O. It emits 8.45 MeV α particles and decays to ^{258}Lr with a half-life of about 40 s. Eight isotopes of Element 105 are now recognized. In 1997, the International Union of Pure and Applied Chemistry adopted the name *dubnium* for Element 105.

Element 106 — The discovery of Element 106 took place in 1974 almost simultaneously at the Lawrence-Berkeley Laboratory and at the Joint Institute for Nuclear Research at Dubna (near Moscow). The Berkeley Group, under direction of Ghiorso, used the Super-Heavy Ion Linear Accelerator (Super HILAC) as a source of heavy ^{18}O ions to bombard a 259-µg target of ^{249}Cf. This resulted in the production and positive identification of ^{263}106, which decayed with a half-life of 0.9 ± 0.2 s by the emission of alpha particles as follows:

$$^{263}106 \xrightarrow{\alpha} {^{259}104} \xrightarrow{\alpha} {^{255}\text{No}} \xrightarrow{\alpha}.$$

The Dubna Team, directed by Flerov and Organessian, produced heavy ions of ^{54}Cr with their 310-cm heavy-ion cyclotron to bombard ^{207}Pb and ^{208}Pb and found a product that decayed with a half-life of 7 ms. They assigned ^{259}106 to this isotope. It is now thought six isotopes of Element 106 have been identified. Two of the isotopes are believed to have half-lives of about 30 s. In 1997, IUPAC adopted the name *seaborgium* for Element 106.

Element 107 — This element, expected to have chemical properties similar to rhenium, was synthesized and unambiguously identified in 1981, using the Universal Linear Accelerator (UNILAC) at the Gesellschaft für Schwerionenforschung (G.S.I.) in Darmstadt, Germany. The discovery team was led by Armbruster and Münzenberg. The reaction producing the element was proposed and applied earlier by a Dubna Group led by Oganessian in 1976. A target of ^{209}Bi was bombarded by a beam of ^{54}Cr ions. In 1983 experiments at Dubna using the 157 -inch cyclotron, produced ^{262}107 by the reaction ^{209}Bi + ^{54}Cr. The alpha decay of ^{246}Cf, the sixth member in the decay chain of ^{262}107, served to establish a 1-neutron reaction channel. The IUPAC has recently adopted the name *bohrium* with symbol Bh for Element 107. Three isotopes of 107 are now recognized.

Element 108 — This element was first synthesized and identified in 1984 by the same G.S.I. Darmstadt group who first identified Elements 107 and 109. Presumably this element will have chemical properties similar to osmium. Isotope ^{265}108 was produced using a beam of ^{58}Fe projectiles, produced by the Universal Linear Accelerator (UNILAC) to bombard a ^{208}Pb target. Discovery of Elements 107 and 109 was made using detection of isotopes with odd proton and neutron numbers. Elements having even atomic numbers are thought to be less stable against spontaneous fusion than odd elements. The production of ^{265}108 in the same reaction as was used at G.S.I. was confirmed at Dubna with detection of the 7th member of the decay chain ^{253}Es. Isotopes of Element 108 were believed to decay by spontaneous fission, explaining why 109 was produced before 108. Isotope ^{265}108 and ^{266}108 are thought to decay to ^{261}106, which in turn decays to ^{257}104 and ^{253}No. The IUPAC adopted the name *hassium*, after the German state of Hesse, in 1997.

Element 109 — On August 29, 1992, Element 109 was made and identified by physicists at the Heavy Ion Research Laboratory (G.S.I.), Darmstadt, Germany, by bombarding a target of ^{209}Bi with accelerated nuclei of ^{58}Fe. The production of Element 109 has been extremely small. It took a week of target bombardment (10^{11} nuclear encounters) to produce a single atom of 109. Oganessian and his team at Dubna in 1994 repeated the Darmstadt experiment using a 10-fold irradiation dose. One fission event from seven alpha decays of 109 was observed, thus indirectly confirming the existence of isotope ^{266}109. The IUPAC has adopted the name *meitnerium* (Mt) for this element, honoring L. Meitner, in 1997.

Element 110 — In 1987 Oganessian, et al., at Dubna, claimed discovery of this element. Their experiments indicated the spontaneous fissioning nuclide ^{272}110 with a half-life of 10 ms. More recently a group led by Armbruster at G.S.I. in Darmstadt, Germany, reported evidence of ^{269}110, which was produced by bombarding lead for many days with more than 10^{18} nickel atoms. A detector searched each collision for Element 110's distinct decay sequence. On November 9, 1994, evidence of 110 was detected. Berkeley scientists, in 1991, performed similar experiments and reported evidence of 110, but this was not confirmed. Workers at Dubna have experiments underway to produce ^{273}110 by bombarding plutonium with sulfur atoms. Other experiments at G.S.I. are now searching for heavier isotopes. Several years ago the IUPAC suggested the use of the temporary name *ununnilium* for Element 110 when it was found.

Element 111 — On December 20, 1994, scientists at GSI Darmstadt, Germany announced they had detected three atoms of a new element with 111 protons and 161 neutrons. This element was made by bombarding ^{83}Bi with ^{28}Ni. Signals of Element 111 appeared for less than 0.002 sec, then decayed into lighter elements including Element ^{268}109 and Element ^{264}107. These isotopes had not previously been observed. A name for Element 111 has not been suggested although IUPAC has suggested a temporary name of Unununium.

Element 112 — In late February 1996, Siguard Hofmann and his collaborators at GSI Darmstadt announced their discovery of Element 112, having 112 protons and 165 neutrons, with an atomic mass of 277. This element was made by bombarding a lead target with high-energy zinc ions. A single nucleus of Element 112 was detected, which decayed after less than 0.001 sec by emitting an α particle, consisting of two protons and two neutrons. This created Element 110_{273}, which in turn decayed by emitting an α particle to form a new isotope of Element 108 and so on. Evidence indicates that nuclei with 162 neutrons are held together more strongly than nuclei with a smaller or larger number of neutrons. This suggests a narrow "peninsula" of relatively stable isotopes around Element 114. GSI scientists are experimenting to bombard targets with ions heavier than zinc to produce Elements 113 and 114. A name has not yet been suggested for Element 112, although the IUPAC suggested the temproary name of ununbium when the element was discovered.

Erbium — (*Ytterby*, a town in Sweden), Er; at. wt. 167.26(3); at. no. 68; m.p. 1529°C; b.p. 2868°C; sp. gr. 9.066 (25°C); valence 3, Erbium, one of the so-called rare-earth elements of the lanthanide series, is found in the minerals mentioned under dysprosium above. In 1842 Mosander separated "yttria," found in the mineral *gadolinite*, into three fractions which he called *yttria*, *erbia*, and *terbia*. The names *erbia* and *terbia* became confused in this early period. After 1860, Mosander's *terbia* was known as *erbia*, and after 1877, the earlier known *erbia* became *terbia*. The *erbia* of this period was later shown to consist of five oxides, now known as *erbia*, *scandia*, *holmia*, *thulia* and *ytterbia*. By 1905 Urbain and James independently succeeded in isolating fairly pure Er_2O_3. Klemm and Bommer first produced reasonably pure erbium metal in 1934 by reducing the anhydrous chloride with potassium vapor. The pure metal is soft and malleable and has a bright, silvery, metallic luster. As with other rare-earth metals, its properties depend to a certain extent on the impurities present. The metal is fairly stable in air and does not oxidize as rapidly as some of the other rare-earth metals. Naturally occurring erbium is a mixture of six isotopes, all of which are stable. Twenty four radioactive isotopes of erbium are also recognized. Recent production techniques, using ion-exchange reactions, have resulted in much lower prices of the rare-earth metals and their compounds in recent years. The cost of 99.9% erbium metal is about $4/gm. Erbium is finding nuclear and metallurgical uses. Added to vanadium, for example, erbium lowers the hardness and improves workability. Most of the rare-earth oxides have sharp absorption bands in the visible, ultraviolet, and near infrared. This property, associated with the electronic structure, gives beautiful pastel colors to many of the rare-earth salts. Erbium oxide gives a pink color and has been used as a colorant in glasses and porcelain enamel glazes.

Europium — (Europe), Eu; at. wt. 151.965(9); at. no. 63; m.p. 822°C; b.p. 1529°C; sp. gr. 5.244 (25°C); valence 2 or 3. In 1890 Boisbaudran obtained basic fractions from samarium-gadolinium concentrates which had spark spectral lines not accounted for by samarium or gadolinium. These lines subsequently have been shown to belong to europium. The discovery of europium is generally credited to Demarcay, who separated the rare earth in reasonably pure form in 1901. The pure metal was not isolated until recent years. Europium is now prepared by mixing Eu_2O_3 with a 10%-excess of lanthanum metal and heating the mixture in a tantalum crucible under high vacuum. The element is collected as a silvery-white metallic deposit on the walls of the crucible. As with other rare-earth metals, except for lanthanum, europium ignites in air at about 150 to 180°C. Europium is about as hard as lead and is quite ductile. It is the most reactive of the rare-earth metals, quickly oxidizing in air. It resembles calcium in its reaction with water. *Bastnasite* and *monazite* are the principal ores containing europium. Europium has been identified spectroscopically in the sun and certain stars. Europium isotopes are good neutron absorbers and are being studied for use in nuclear control applications. Europium oxide is now widely used as a phosphor activator and europium-activated yttrium vanadate is in commercial use as the red phosphor in color TV tubes. Europium-doped plastic has been used as a laser material. With the development of ion-exchange techniques and special processes, the cost of the metal has been greatly reduced in recent years. Natural europium contains two stable isotopes. Thirty five other radioactive isotopes and isomers are known. Europium is one of the rarest and most costly of the rare-earth metals. It is priced at about $50/g (99.9% pure).

Fermium — (Enrico Fermi), Fm; at. wt. (257); at. no. 100; m.p. 1527°C. Fermium, the eighth transuranium element of the actinide series to be discovered, was identified by Ghiorso and co-workers in 1952 in the debris from a thermonuclear explosion in the Pacific in work involving the University of California Radiation Laboratory, the Argonne National Laboratory, and the Los Alamos Scientific Laboratory. The isotope produced was the 20-hour ^{255}Fm. During 1953 and early 1954, while discovery of elements 99 and 100 was withheld from publication for security reasons, a group from the Nobel Institute of Physics in Stockholm bombarded ^{238}U with ^{16}O ions, and isolated a 30-min α-emitter, which they ascribed to ^{250}100, without claiming discovery of the element. This isotope has since been identified positively, and the 30-min half-life confirmed. The chemical properties of fermium have been studied solely with tracer amounts, and in normal aqueous media only the (III) oxidation state appears to exist. The isotope ^{254}Fm and heavier isotopes can be produced by intense neutron irradiation of lower elements such as plutonium by a process of successive neutron capture interspersed with beta decays until these mass numbers and atomic numbers are reached. Twenty isotopes and isomers of fermium are known to exist. ^{257}Fm, with a half-life of about 100.5 days, is the longest lived. ^{250}Fm, with a half-life of 30 min, has been shown to be a product of decay of Element ^{254}102. It was by chemical identification of ^{250}Fm that production of Element 102 (nobelium) was confirmed.

Fluorine — (L. and F. *fluere*, flow, or flux), F; at. wt. 18.9984032(9); at. no. 9; m.p. –219.62°C (1 atm); b.p. –188.12°C (1 atm); density 1.696 g/L (0°C, 1 atm); liq. den. at b.p. 1.50 g/cm³; valence 1. In 1529, Georigius Agricola described the use of fluorspar as a flux, and as early as 1670 Schwandhard found that glass was etched when exposed to fluorspar treated with acid. Scheele and many later investigators, including Davy, Gay-Lussac, Lavoisier, and Thenard, experimented with hydrofluoric acid, some experiments ending in tragedy. The element was finally isolated in 1886 by Moisson after nearly 74 years of continuous effort. Fluorine occurs chiefly in *fluorspar* (CaF_2) and *cryolite* (Na_3AlF_6), but is rather widely distributed in other minerals. It is a member of the halogen family of elements, and is obtained by electrolyzing a solution of potassium hydrogen fluoride in anhydrous hydrogen fluoride in a vessel of metal or transparent fluorspar. Modern commercial production methods are essentially variations on the procedures first used by Moisson. Fluorine is the most electronegative and reactive of all elements. It is a pale yellow, corrosive gas, which reacts with practically all organic and inorganic substances. Finely divided metals, glass, ceramics, carbon, and even water burn in fluorine with a bright flame. Until World War II, there was no commercial production of elemental fluorine. The atom bomb project and nuclear energy applications, however, made it necessary to produce large quantities. Safe handling techniques have now been developed and it is possible at present to transport liquid fluorine by the ton. Fluorine and its compounds are used in producing uranium (from the hexafluoride) and more than 100 commercial fluorochemicals, including many well-known high-temperature plastics. Hydrofluoric acid is extensively used for etching the glass of light bulbs, etc. Fluorochloro hydrocarbons are extensively used in air conditioning and refrigeration. It has been suggested that fluorine can be substituted for hydrogen wherever it occurs in organic compounds, which could lead to an astronomical number of new fluorine compounds. The presence of fluorine as a soluble fluoride in drinking water to the extent of 2 ppm may cause mottled enamel in teeth, when used by children acquiring permanent teeth; in smaller amounts, however, fluorides are said to be beneficial and used in water supplies to prevent dental cavities. Elemental fluorine has been studied as a rocket propellant as it has an exceptionally high specific impulse value. Compounds of fluorine with rare gases have now been confirmed. Fluorides of xenon, radon, and krypton are among those known. Elemental fluorine and the fluoride ion are highly toxic. The free element has a characteristic pungent odor, detectable in concentrations as low as 20 ppb, which is below the safe working level. The recommended maximum allowable concentration for a daily 8-hour time-weighted exposure is 1 ppm. Fluorine is known to have thirteen isotopes.

Francium — (France), Fr; at. no. 87; at. wt. (223); m.p. 27°C; b.p. 677°C; valence 1. Discovered in 1939 by Mlle. Marguerite Perey of the Curie Institute, Paris. Francium, the heaviest known member of the alkali metal series, occurs as a result of an alpha disintegration of actinium. It can also be made artificially by bombarding thorium with protons. While it occurs naturally in uranium minerals, there is probably less than an ounce of francium at any time in the total crust of the earth. It has the highest equivalent weight of any element, and is the most unstable of the first 101 elements of the periodic system. Thirty-five isotopes and isomers of francium are recognized. The longest lived ^{223}Fr(Ac, K), a daughter of ^{227}Ac, has a half-life of 21.8 min. This is the only isotope of francium occurring in nature. Because all known isotopes of francium are highly unstable, knowledge of the chemical properties of this element comes from radiochemical techniques. No weighable quantity of the element has been prepared or isolated. The chemical properties of francium most closely resemble cesium.

Gadolinium — (*gadolinite*, a mineral named for Gadolin, a Finnish chemist), Gd; at. wt. 157.25(3); at. no. 64; m.p. 1313°C; b.p. 3273°C; sp. gr. 7.901 (25°C); valence 3. Gadolinia, the oxide of gadolinium, was separated by Marignac in 1880 and Lecoq de Boisbaudran independently isolated the element from Mosander's "yttria" in 1886. The element was named for the mineral *gadolinite* from which this rare earth was originally obtained. Gadolinium is found in several other minerals, including *monazite* and *bastnasite,* which are of commercial importance. The element has been isolated only in recent years. With the development of ion-exchange and solvent extraction techniques, the availability and price of gadolinium and the other rare-earth metals have greatly improved. Thirty isotopes and isomers of gadolinium are now recognized; seven are stable and occur naturally. The metal can be prepared by the reduction of the anhydrous fluoride with metallic calcium. As with other related rare-earth metals, it is silvery white, has a metallic luster, and is malleable and ductile. At room temperature, gadolinium crystallizes in the hexagonal, close-packed α form. Upon heating to 1235°C, α gadolinium transforms into the β form, which has a body-centered cubic structure. The metal is relatively stable in dry air, but in moist air it tarnishes with the formation of a loosely adhering oxide film which spalls off and exposes more surface to oxidation. The metal reacts slowly with water and is soluble in dilute acid. Gadolinium has the highest thermal neutron capture cross-section of any known element (49,000 barns). Natural gadolinium is a mixture of seven isotopes. Two of these, ^{155}Gd and ^{157}Gd, have excellent capture characteristics, but they are present naturally in low concentrations. As a result, gadolinium has a very fast burnout rate and has limited use as a nuclear control rod material. It has been used in making gadolinium yttrium garnets, which have microwave applications. Compounds of gadolinium are used in making phosphors for color TV tubes. The metal has unusual superconductive properties. As little as 1% gadolinium has been found to improve the workability and resistance of iron, chromium, and related alloys to high temperatures and oxidation. Gadolinium ethyl sulfate has extremely low noise characteristics and may find use in duplicating the performance of amplifiers, such as the maser. The metal is ferromagnetic. Gadolinium is unique for its high magnetic moment and for its special Curie temperature (above which ferromagnetism vanishes) lying just at room temperature. This suggests uses as a magnetic component that senses hot and cold. The price of the metal is $2/g (99.9% purity).

Gallium — (L. *Gallia*, France; also from Latin, *gallus,* a translation of Lecoq, a cock); Ga; at. wt. 69.723(1); at. no. 31; m.p. 29.7646°C; b.p. 2204°C; sp. gr. 5.904 (29.6°C) solid; sp. gr. 6.095 (29.6°C) liquid; valence 2 or 3. Predicted and described by Mendeleev as ekaaluminum, and discovered spectroscopically by Lecoq de Boisbaudran in 1875, who in the same year obtained the free metal by electrolysis of a solution of the hydroxide in KOH. Gallium is often found as a trace element in *diaspore, sphalerite, germanite, bauxite,* and *coal.* Some flue dusts from burning coal have been shown to contain as much as 1.5% gallium. It is the only metal, except for mercury, cesium, and rubidium, which can be liquid near room temperatures; this makes possible its use in high-temperature thermometers. It has one of the longest liquid ranges of any metal and has a low vapor pressure even at high temperatures. There is a strong tendency for gallium to supercool below its freezing point. Therefore, seeding may be necessary to initiate solidification. Ultra-pure gallium has a beautiful, silvery appearance, and the solid metal exhibits a conchoidal fracture similar to glass. The metal expands 3.1% on solidifying; therefore, it should not be stored in glass or metal containers, as they may break as the metal solidifies. Gallium wets glass or porcelain, and forms a brilliant mirror when it is painted on glass. It is widely used in doping semiconductors and producing solid-state devices such as transistors. High-purity gallium is attacked only slowly by mineral acids. Magnesium gallate containing divalent impurities such as Mn^{+2} is finding use in commercial ultraviolet activated powder phosphors. Gallium arsenide is capable of converting electricity directly into coherent light. Gallium readily alloys with most metals, and has been used as a component in low-melting alloys. Its toxicity appears to be of a low order, but should be handled with care until more data are forthcoming. Natural gallium contains two stable isotopes. Twenty three other isotopes, one of which is an isomer, are known. The metal can be supplied in ultrapure form (99.99999+%). The cost is about $4/g.

Germanium — (L. *Germania,* Germany), Ge; at. no. 32; m.p. 938.25°C; b.p. 2833°C; sp. gr. 5.323 (25°C); valence 2 and 4. Predicted by Mendeleev in 1871 as ekasilicon, and discovered by Winkler in 1886. The metal is found in *argyrodite,* a sulfide of germanium and silver; in *germanite,* which contains 8% of the element; in zinc ores; in coal; and in other minerals. The element is frequently obtained commercially from flue dusts of smelters processing zinc ores, and has been recovered from the by-products of combustion of certain coals. Its presence in coal insures a large reserve of the element in the years to come. Germanium can be separated from other metals by fractional distillation of its volatile tetrachloride. The tetrachloride may then be hydrolyzed to give GeO_2; the dioxide can be reduced with hydrogen to give the metal. Recently developed zone-refining techniques permit the production of germanium of ultra-high purity. The element is a gray-white metalloid, and in its pure state is crystalline and brittle, retaining its luster in air at room temperature. It is a very important semiconductor material. Zone-refining techniques have led to production of crystalline germanium for semiconductor use with an impurity of only one part in 10^{10}. Doped with arsenic, gallium, or other elements, it is used as a transistor element in thousands of electronic applications. Its application as a semiconductor element now provides the largest use for germanium. Germanium is also finding many other applications including use as an alloying agent, as a phosphor in fluorescent lamps, and as a catalyst. Germanium and germanium oxide are transparent to the infrared and are used in infrared spectroscopes and other optical equipment, including extremely sensitive infrared detectors. Germanium oxide's high index of refraction and dispersion has made it useful as a component of glasses used in wide-angle camera lenses and microscope objectives. The field of organogermanium chemistry is becoming increasingly important. Certain germanium compounds have a low mammalian toxicity, but a marked activity against certain bacteria, which makes them of interest as chemotherapeutic agents. The cost of germanium is about $3/g (99.999% purity). Twenty nine isotopes and isomers are known, five of which occur naturally.

Gold — (Sanskrit *Jval;* Anglo-Saxon *gold*), Au (L. *aurum,* gold); at. wt. 196.96654(3); at. no. 79; m.p. 1064.18°C; b.p. 2856°C; sp. gr. ~19.3 (20°C); valence 1 or 3. Known and highly valued from earliest times, gold is found in nature as the free metal and in tellurides; it is very widely distributed and is almost always associated with quartz or pyrite. It occurs in veins and alluvial deposits, and is often separated from rocks and other minerals by sluicing and panning operations. About two thirds of the world's gold output comes from South Africa, and about two thirds of the total U.S. production comes from South Dakota and Nevada. The metal is recovered from its ores by cyaniding, amalgamating, and smelting processes. Refining is also frequently done by electrolysis. Gold occurs in sea water to the extent of 0.1 to 2 mg/ton, depending on the location where the sample is taken. As yet, no method has been found for recovering gold from sea water profitably. It is estimated that all the gold in the world, so far refined, could be placed in a single cube 60 ft on a side. Of all the elements, gold in its pure state is undoubtedly the most beautiful. It is metallic, having a yellow color when in a mass, but when finely divided it may be black, ruby, or purple. The Purple of Cassius is a delicate test for auric gold. It is the most malleable and ductile metal; 1 oz. of gold can be beaten out to 300 ft^2. It is a soft metal and is usually alloyed to give it more strength. It is a good conductor of heat and electricity, and is unaffected by air and most reagents. It is used in coinage and is a standard for monetary systems in many countries. It is also extensively used for jewelry, decoration, dental work, and for plating. It is used for coating certain space satellites, as it is a good reflector of infrared and is inert. Gold, like other precious metals, is measured in troy weight; when alloyed with other metals, the term *carat* is used to express the amount of gold present, 24 carats being pure gold. For many years the value of gold was set by the U.S. at $20.67/troy ounce; in 1934 this value was fixed by law at $35.00/troy ounce, 9/10th fine. On March 17, 1968, because of a gold crisis, a two-tiered pricing system was established whereby gold was still used to settle international accounts at the old $35.00/troy ounce price while the price of gold on the private market would be allowed to fluctuate. Since this time, the price of gold on the free market has fluctuated widely. The price of gold on the free market reached a price of $620/troy oz. in January 1980. The most common gold compounds are auric chloride ($AuCl_3$) and chlorauric acid ($HAuCl_4$), the latter being used in photography for toning the silver image. Gold has forty six recognized isotopes and isomers; ^{198}Au, with a half-life of 2.7 days, is used for treating cancer and other diseases. Disodium aurothiomalate is administered intramuscularly as a treatment for arthritis. A mixture of one part nitric acid with three of hydrochloric acid is called *aqua regia* (because it dissolved gold, the King of Metals). Gold is available commercially with a purity of 99.999+%. For many years the temperature assigned to the freezing point of gold has been 1063.0°C; this has served as a calibration point for the International Temperature Scales (ITS-27 and ITS-48) and the International Practical Temperature Scale (IPTS-48). In 1968, a new International Practical Temperature Scale (IPTS-68) was adopted, which demanded that the freezing point of gold be changed to 1064.43°C. In 1990 a new International Temperature Scale (ITS-90) was adopted bringing the t.p. (triple point) of H_2O (t_{90} (°C)) to 0.01°C and the freezing point of gold to 1064.18°C. The specific gravity of gold has been found to vary considerably depending on temperature, how the metal is precipitated, and cold-worked. As of January 1996, gold was priced at about $390/troy oz. ($12.50/g).

Hafnium — (*Hafnia,* Latin name for Copenhagen), Hf; at. wt. 178.49(2); at. no. 72; m.p. 2233 ± 20°C; b.p. 4603°C; sp. gr. 13.31 (20°C); valence 4. Hafnium was thought to be present in various minerals and concentrations many years prior to its discovery, in 1923, credited to D. Coster and G. von Hevesey. On the basis of the Bohr theory, the new element was expected to be associated with zirconium. It was finally identified in *zircon* from Norway, by means of X-ray spectroscopic analysis. It was named in honor of the city in which the discovery was made. Most zirconium minerals contain 1 to 5% hafnium. It was originally separated from zirconium by repeated recrystallization of the double ammonium or potassium fluorides by von Hevesey and Jantzen. Metallic hafnium was first prepared by van Arkel and deBoer by passing the vapor of the tetraiodide over a heated tungsten filament. Almost all hafnium metal now produced is made by reducing the tetrachloride with magnesium or with sodium (Kroll Process). Hafnium is a ductile metal with a brilliant silver luster. Its properties are considerably influenced by the impurities of zirconium present. Of all the elements, zirconium and hafnium are two of the most difficult to separate. Their chemistry is almost identical, however, the density of zirconium is about half that of hafnium. Very pure hafnium has been produced, with zirconium being the major impurity. Natural hafnium contains six isotopes, one of which is slightly radioactive. Hafnium has a total of 40 recognized isotopes and isomers. Because hafnium has a good absorption cross section for thermal neutrons (almost 600 times that of zirconium), has excellent mechanical properties, and is extremely corrosion resistant, it is used for reactor control rods. Such rods are used in nuclear submarines. Hafnium has been successfully alloyed with iron, titanium, niobium, tantalum, and other metals. Hafnium carbide is the most refractory binary composition known, and the nitride is the most refractory of all known metal nitrides (m.p. 3310°C). Hafnium is used in gas-filled and incandescent lamps, and is an efficient "getter" for scavenging oxygen and nitrogen. Finely divided hafnium is pyrophoric and can ignite spontaneously in air. Care should be taken when machining the metal or when handling hot sponge hafnium. At 700°C hafnium rapidly absorbs hydrogen to form the composition $HfH_{1.86}$. Hafnium is resistant to concentrated alkalis, but at elevated temperatures reacts with oxygen, nitrogen, carbon, boron, sulfur, and silicon. Halogens react directly to form tetrahalides. The price of the metal is in the broad range of $1/g to $3/g, depending on purity and quantity. The yearly demand for hafnium in the U.S. is now in excess of 50,000 kg.

Hahnium — A name previously used for Element 105, now named *dubnium.*

Hassium — see Element 108.

Helium — (Gr. *helios,* the sun), He; at. wt. 4.002602(2); at. no. 2; m.p. below — 272.2°C (26 atm); b.p. — 268.93°C; density 0.1785 g/l (0°C,

1 atm); liquid density 7.62 lb/ft^3 at. b.p.; valence usually 0. Evidence of the existence of helium was first obtained by Janssen during the solar eclipse of 1868 when he detected a new line in the solar spectrum; Lockyer and Frankland suggested the name *helium* for the new element; in 1895, Ramsay discovered helium in the uranium mineral *clevite*, and it was independently discovered in cleveite by the Swedish chemists Cleve and Langlet about the same time. Rutherford and Royds in 1907 demonstrated that α particles are helium nuclei. Except for hydrogen, helium is the most abundant element found throughout the universe. Helium is extracted from natural gas; all natural gas contains at least trace quantities of helium. It has been detected spectroscopically in great abundance, especially in the hotter stars, and it is an important component in both the proton-proton reaction and the carbon cycle, which account for the energy of the sun and stars. The fusion of hydrogen into helium provides the energy of the hydrogen bomb. The helium content of the atmosphere is about 1 part in 200,000. While it is present in various radioactive minerals as a decay product, the bulk of the Free World's supply is obtained from wells in Texas, Oklahoma, and Kansas. The only known helium extraction plants, outside the United States, in 1996 were in Eastern Europe (Poland), the Russia, China, and India. The cost of helium fell from $2500/ft^3 in 1915 to 1.5¢/ft^3 in 1940. Commercial helium in 1996 was priced at about $70/300 cu. ft. (8.5 cu. meters). Helium has the lowest melting point of any element and has found wide use in cryogenic research, as its boiling point is close to absolute zero. Its use in the study of superconductivity is vital. Using liquid helium, Kurti and co-workers, and others, have succeeded in obtaining temperatures of a few microkelvins by the adiabatic demagnetization of copper nuclei, starting from about 0.01 K. Seven isotopes of helium are known. Liquid helium (He^4) exists in two forms: He^{4}I and He^{4}II, with a sharp transition point at 2.174 K (3.83 cm Hg). He^{4}I (above this temperature) is a normal liquid, but He^{4}II (below it) is unlike any other known substance. It expands on cooling; its conductivity for heat is enormous; and neither its heat conduction nor viscosity obeys normal rules. It has other peculiar properties. Helium is the only liquid that cannot be solidified by lowering the temperature. It remains liquid down to absolute zero at ordinary pressures, but it can readily be solidified by increasing the pressure. Solid ^{3}He and ^{4}He are unusual in that both can readily be changed in volume by more than 30% by application of pressure. The specific heat of helium gas is unusually high. The density of helium vapor at the normal boiling point is also very high, with the vapor expanding greatly when heated to room temperature. Containers filled with helium gas at 5 to 10 K should be treated as though they contained liquid helium due to the large increase in pressure resulting from warming the gas to room temperature. While helium normally has a 0 valence, it seems to have a weak tendency to combine with certain other elements. Means of preparing helium diflouride have been studied, and species such as HeNe and the molecular ions He$^+$ and He^{++} have been investigated. Helium is widely used as an inert gas shield for arc welding; as a protective gas in growing silicon and germanium crystals, and in titanium and zirconium production; as a cooling medium for nuclear reactors, and as a gas for supersonic wind tunnels. A mixture of helium and oxygen is used as an artificial atmosphere for divers and others working under pressure. Different ratios of He/O$_2$ are used for different depths at which the diver is operating. Helium is extensively used for filling balloons as it is a much safer gas than hydrogen. One of the recent largest uses for helium has been for pressuring liquid fuel rockets. A Saturn booster such as used on the Apollo lunar missions required about 13 million ft^3 of helium for a firing, plus more for checkouts. Liquid helium's use in magnetic resonance imaging (MRI) continues to increase as the medical profession accepts and develops new uses for the equipment. This equipment is providing accurate diagnoses of problems where exploratory surgery has previously been required to determine problems. Another medical application that is being developed uses MRI to determine by blood analysis whether a patient has any form of cancer. Lifting gas applications are increasing. Various companies in addition to Goodyear, are now using "blimps" for advertising. The Navy and the Air Force are investigating the use of airships to provide early warning systems to detect low-flying cruise missiles. The Drug Enforcement Agency has used radar-equipped blimps to detect drug smugglers along the southern border of the U.S. In addition, NASA is currently using helium-filled balloons to sample the atmosphere in Antarctica to determine what is depleting the ozone layer that protects Earth from harmful U.V. radiation. Research on and development of materials which become superconductive at temperatures well above the boiling point of helium could have a major impact on the demand for helium. Less costly refrigerants having boiling points considerably higher could replace the present need to cool such superconductive materials to the boiling point of helium.

Holmium — (L. *Holmia* , for Stockholm), Ho; at. wt. 164.93032(3); at. no 67; m.p. 1474°C; b.p. 2700°C; sp. gr. 8.795 (25°C); valence + 3. The spectral absorption bands of holmium were noticed in 1878 by the Swiss chemists Delafontaine and Soret, who announced the existence of an "Element X". Cleve, of Sweden, later independently discovered the element while working on erbia earth. The element is named after Cleve's native city. Pure holmia, the yellow oxide, was prepared by Homberg in 1911. Holmium occurs in *gadolinite, monazite,* and in other rare-earth minerals. It is commercially obtained from monazite, occurring in that mineral to the extent of about 0.05%. It has been isolated by the reduction of its anhydrous chloride or fluoride with calcium metal. Pure holmium has a metallic to bright silver luster. It is relatively soft and malleable, and is stable in dry air at room temperature, but rapidly oxidizes in moist air and at elevated temperatures. The metal has unusual magnetic properties. Few uses have yet been found for the element. The element, as with other rare earths, seems to have a low acute toxic rating. Natural holmium is not radioactive. Holmium has 46 other isotopes known, all of which are radioactive. The price of 99.9% holmium metal is about $15/g.

Hydrogen — (Gr. *hydro*, water, and *genes*, forming), H; at. wt. 1.00794(7); at. no. 1; m.p. –259.34°C; b.p. –252.87°C; density 0.08988 g/l; density (liquid) 70.8 g/l (–253°C); density (solid) 70.6 g/l (–262°C); valence 1. Hydrogen was prepared many years before it was recognized as a distinct substance by Cavendish in 1766. It was named by Lavoisier. Hydrogen is the most abundant of all elements in the universe, and it is thought that the heavier elements were, and still are, being built from hydrogen and helium. It has been estimated that hydrogen makes up more than 90% of all the atoms or three quarters of the mass of the universe. It is found in the sun and most stars, and plays an important part in the proton-proton reaction and carbon-nitrogen cycle, which accounts for the energy of the sun and stars. It is thought that hydrogen is a major component of the planet Jupiter and that at some depth in the planet's interior the pressure is so great that solid molecular hydrogen is converted into solid metallic hydrogen. In 1973, it was reported that a group of Russian experimenters may have produced metallic hydrogen at a pressure of 2.8 Mbar. At the transition the density changed from 1.08 to 1.3 g/cm^3. Earlier, in 1972, a Livermore (California) group also reported on a similar experiment in which they observed a pressure-volume point centered at 2 Mbar. It has been predicted that metallic hydrogen may be metastable; others have predicted it would be a superconductor at room temperature. On earth, hydrogen occurs chiefly in combination with oxygen in water, but it is also present in organic matter such as living plants, petroleum, coal, etc. It is present as the free element in the atmosphere, but only to the extent of less than 1 ppm by volume. It is the lightest of all gases, and combines with other elements, sometimes explosively, to form compounds. Great quantities of hydrogen are required commercially for the fixation of nitrogen from the air in the Haber ammonia process and for the hydrogenation of fats and oils. It is also used in large quantities in methanol production, in hydrodealkylation, hydrocracking, and hydrodesulfurization. It is also used as a rocket fuel, for welding, for production of hydrochloric acid, for the reduction of metallic ores, and for filling balloons. The lifting power of 1 ft^3 of hydrogen gas is about 0.076 lb at 0°C, 760 mm pressure. Production of hydrogen in the U.S. alone now amounts to about 3 billion cubic feet per year. It is prepared by the action of steam on heated carbon, by decomposition of certain hydrocarbons with heat, by the electrolysis of water, or by the displacement from acids by certain metals. It is also produced by the action of sodium or potassium hydroxide on aluminum. Liquid hydrogen is important in cryogenics and in the study of superconductivity, as its melting point is only a 20 degrees above absolute zero. The ordinary isotope of hydrogen, H, is known as *protium*. In 1932, Urey announced the discovery of a stable isotope, deuterium (^{2}H or D) with an atomic weight of 2. Deuterium is present in natural hydrogen to the extent of 0.015%. Two years later an unstable isotope, tritium (H), with an atomic weight of 3 was discovered. Tritium has a half-life of about 12.5 years. Tritium atoms are also present in hydrogen but in much smaller proportion. Tritium is readily produced in nuclear reactors and is used in the production of the hydrogen bomb. It is also used as a radioactive agent in making luminous paints, and as a tracer. Deuterium gas is readily available, without permit, at about $1/l. Heavy water, deuterium oxide (D_2O), which is used as a moderator to slow down neutrons, is available without permit at a cost of 6c to $1/g, depending on quantity and purity. Quite apart from isotopes, it has been shown that hydrogen gas under ordinary conditions is a mixture of two kinds of molecules, known as *ortho-* and *para*-hydrogen, which differ from one another by the spins of their electrons and nuclei. Normal hydrogen at room temperature contains 25% of the *para* form and 75% of the *ortho* form. The *ortho* form cannot be prepared in the pure state. Since the two forms differ in energy, the physical properties also differ. The melting and boiling points of *para*hydrogen are about 0.1°C lower than those of normal hydrogen. Consideration is being given to an entire economy based on solar- and nuclear-generated hydrogen. Located in remote regions, power plants would electrolyze sea water; the hydrogen produced would travel to distant cities by pipelines. Pollution-free hydrogen could replace natural gas, gasoline, etc., and could serve as a reducing agent in metallurgy, chemical processing, refining, etc. It could also be used to convert trash into methane and ethylene. Public acceptance, high capital investment, and the high present cost of hydrogen with respect to present fuels are but a few of the problems facing establishment of such an economy.

Indium — (from the brilliant indigo line in its spectrum), In; at. wt. 114.818(3); at. no. 49; f.p. 156.5985°C; b.p. 2072°C; sp. gr. 7.31 (20°C); valence 1, 2, or 3. Discovered by Reich and Richter, who later isolated the metal. Indium is most frequently associated with zinc materials, and it is from these that most commercial indium is now obtained; however, it is also found in iron, lead, and copper ores. Until 1924, a gram or so constituted the world's supply of this element in isolated form. It is probably about as abundant as silver. About 4 million troy ounces of indium are now produced annually in the Free World. Canada is presently producing more than 1,000,000 troy ounces annually. The present cost of indium is about $2 to $10/g, depending on quantity and purity. It is available in ultrapure form. Indium is a very soft, silvery-white metal with a brilliant luster. The pure metal gives a high-pitched "cry" when bent. It wets glass, as does gallium. It has found application in making low-melting alloys; an alloy of 24% indium-76% gallium is liquid at room temperature. It is used in making bearing alloys, germanium transistors, rectifiers, thermistors, and photoconductors. It can be plated onto metal and evaporated onto glass, forming a mirror as good as that made with silver but with more resistance to atmospheric corrosion. There is evidence that indium has a low order of toxicity; however, care should be taken until further information is available. Sixty seven isotopes and isomers are now recognized (more than any other element). Natural indium contains two isotopes. One is stable. The other, ^{115}In, comprising 95.71% of natural indium is slightly radioactive with a very long half-life.

Iodine — (Gr. *iodes*, violet), I; at. wt. 126.90447(3); at. no. 53; m.p. 113.7°C; b.p. 184.4°C; density of the gas 11.27 g/l; sp. gr. solid 4.93 (20°C); valence 1, 3, 5, or 7. Discovered by Courtois in 1811. Iodine, a halogen, occurs sparingly in the form of iodides in sea water from which it is assimilated by seaweeds, in Chilean saltpeter and nitrate-bearing earth, known as *caliche* in brines from old sea deposits, and in brackish waters from oil and salt wells. Ultrapure iodine can be obtained from the reaction of potassium iodide with copper sulfate. Several other methods of isolating the element are known. Iodine is a bluish-black, lustrous solid, volatilizing at ordinary temperatures into a blue-violet gas with an irritating odor; it forms compounds with many elements, but is less active than the other halogens, which displace it from iodides. Iodine exhibits some metallic-like properties. It dissolves readily in chloroform, carbon tetrachloride, or carbon disulfide to form beautiful purple solutions. It is only slightly soluble in water. Iodine compounds are important in organic chemistry and very useful in medicine. Forty two isotopes and isomers are recognized. Only one stable isotope, ^{127}I is found in nature. The artificial radioisotope ^{131}I, with a half-life of 8 days, has been used in treating the thyroid gland. The most common compounds are the iodides of sodium and potassium (KI) and the iodates (KIO$_3$). Lack of iodine is the cause of goiter. Iodides, and thyroxin which contains iodine, are used internally in medicine, and a solution of KI and iodine in alcohol is used for external wounds. Potassium iodide finds use in photography. The deep blue color with starch solution is characteristic of the free element. Care should be taken in handling and using iodine, as contact with the skin can cause lesions; iodine vapor is intensely irritating to the eyes and mucous membranes. Elemental iodine costs about 25 to 75¢/g depending on purity and quantity.

Iridium — (L. *iris*, rainbow), Ir; at. wt. 192.217(3); at. no. 77; m.p. 2446°C; b.p. 4428°C; sp. gr. 22.42 (17°C); valence 3 or 4. Discovered in 1803 by Tennant in the residue left when crude platinum is dissolved by aqua regia. The name iridium is appropriate, for its salts are highly colored. Iridium, a metal of the platinum family, is white, similar to platinum, but with a slight yellowish cast. It is very hard and brittle, making it very hard to machine, form, or work. It is the most corrosion-resistant metal known, and was used in making the standard meter bar of Paris, which is a 90% platinum-10% iridium alloy. This meter bar was replaced in 1960 as a fundamental unit of length (see under Krypton). Iridium is not attacked by any of the acids nor by aqua regia, but is attacked by molten salts, such as NaCl and NaCN. Iridium occurs uncombined in nature with platinum and other metals of this family in alluvial deposits. It is recovered as a by-product from the nickel mining industry. The presence of iridium has recently been used in examining the Cretaceous-Tertiary (K-T) boundary. Meteorites contain small amounts of iridium. Because iridium is found widely distributed at the K-T boundary, it has been suggested that a large meteorite or asteroid collided with the earth, killing the dinosaurs, and creating a large dust cloud and crater. Searches for such a crater point to one in the Yucatan, known as Chicxulub. Iridium has found use in making crucibles and apparatus for use at high temperatures. It is also used for electrical contacts. Its principal use is as a hardening agent for platinum. With osmium, it forms an alloy which is used for tipping pens and compass bearings. The specific gravity of iridium is only very slightly lower than that of osmium, which has been generally credited as being the heaviest known element. Calculations of the densities of iridium and osmium from the space lattices gives values of 22.65 and 22.61 g/cm^3, respectively. These values may be more reliable than actual physical measurements. Natural iridium contains two isotopes. Forty two other isotopes, all radioactive, are now recognized. At present, therefore, we know that either iridium or osmium is the densest known element, but the data do not yet allow selection between the two. Iridium costs about $50/g.

Iron — (Anglo-Saxon, *iron*), Fe (L. *ferrum*); at. wt. 55.845(2); at. no. 26; m.p. 1538°C; b.p. 2861°C; sp. gr. 7.874 (20°C); valence 2, 3, 4, or 6. The use of iron is prehistoric. Genesis mentions that Tubal-Cain, seven generations from Adam, was "an instructor of every artificer in brass and iron." A remarkable iron pillar, dating to about A.D. 400, remains standing today in Delhi, India. This solid shaft of wrought iron is about 7$^1/_4$ m high by 40 cm in diameter. Corrosion to the pillar has been minimal although it has been exposed to the weather since its erection. Iron is a relatively abundant element in the universe. It is found in the sun and many types of stars in considerable quantity. Its nuclei are very stable. Iron is found native as a principal component of a class of meteorites known as "siderites, and is a minor constituent of the other two classes. The core of the earth, 2150 miles in radius, is thought to be largely composed of iron with about 10% occluded hydrogen. The metal is the fourth most abundant element, by weight, making up the crust of the earth. The most common ore is *hematite* (Fe_2O_3), which is frequently seen as *black sands* along beaches and banks of streams. *Taconite* is becoming increasingly important as a commercial ore. Common iron is a mixture of four isotopes. Ten other isotopes are known to exist. Iron is a vital constituent of plant and animal life, and appears in hemoglobin. The pure metal is not often encountered in commerce, but is usually alloyed with carbon or other metals. The pure metal is very reactive chemically, and rapidly corrodes, especially in moist air or at elevated temperatures. It has four allotropic forms, or ferrites, known as α, β, γ, and δ, with transition points at 700, 928, and 1530°C. The α form is magnetic, but when transformed into the β form, the magnetism disappears although the lattice remains unchanged. The relations of these forms are peculiar. Pig iron is an alloy containing about 3% carbon with varying amounts of S, Si, Mn, and P. It is hard, brittle, fairly fusible, and is used to produce other alloys, including steel. Wrought iron contains only a few tenths of a percent of carbon, is tough, malleable, less fusible, and has usually a "fibrous" structure. Carbon steel is an alloy of iron with carbon, with small amounts of Mn, S, P, and Si. Alloy steels are carbon steels with other additives such as nickel, chromium, vanadium, etc. Iron is the cheapest and most abundant, useful, and important of all metals. Natural iron contains four isotopes and isomers. Twenty one other isotopes and isomers, all radioactive, are now recognized.

Krypton — (Gr. *kryptos*, hidden), Kr; at. wt. 83.80(1); at. no. 36; m.p. –157.36°C; b.p. –153.22 ± 0.10°C; density 3.733 g/l (0°C); valence usually 0. Discovered in 1898 by Ramsay and Travers in the residue left after liquid air had nearly boiled away. Krypton is present in the air to the extent of about 1 ppm. The atmosphere of Mars has been found to contain 0.3 ppm of krypton. It is one of the "noble" gases. It is characterized by its brilliant green and orange spectral lines. Naturally occurring krypton contains six stable isotopes. Twenty four other unstable isotopes and isomers are now recognized. The spectral lines of krypton are easily produced and some are very sharp. In 1960 it was internationally agreed that the fundamental unit of length, the meter, should be defined in terms of the orange-red spectral line of ^{86}Kr. This replaced the standard meter of Paris, which was defined in terms of a bar made of a platinum-iridium alloy. In October 1983 the meter, which originally was defined as being one ten millionth of a quadrant of the earth's polar circumference, was again redefined by the International Bureau of Weights and Measures as being the length of path traveled by light in a vacuum during a time interval of 1/299,792,458 of a second. Solid krypton is a white crystalline substance with a face-centered cubic structure which is common to all the "rare gases". While krypton is generally thought of as a rare gas that normally does not combine with other elements to form compounds, it now appears that the existence of some krypton compounds is established. Krypton difluoride has been prepared in gram quantities and can be made by several methods. A higher fluoride of krypton and a salt of an oxyacid of krypton also have been reported. Molecule-ions of $ArKr^+$ and KrH^+ have been identified and investigated, and evidence is provided for the formation of KrXe or $KrXe^+$. Krypton clathrates have been prepared with hydroquinone and phenol. ^{85}Kr has found recent application in chemical analysis. By imbedding the isotope in various solids, *kryptonates* are formed. The activity of these kryptonates is sensitive to chemical reactions at the surface. Estimates of the concentration of reactants are therefore made possible. Krypton is used in certain photographic flash lamps for high-speed photography. Uses thus far have been limited because of its high cost. Krypton gas presently costs about $690/100 l.

Lanthanum — (Gr. *lanthanein*, to lie hidden), La; at. wt. 138.9055(2); at. no. 57; m.p. 918°C; b.p. 3464°C; sp. gr. 6.145 (25°C); valence 3. Mosander in 1839 extracted a new earth *lanthana*, from impure cerium nitrate, and recognized the new element. Lanthanum is found in rare-earth minerals such as *cerite, monazite, allanite,* and *bastnasite*. Monazite and bastnasite are principal ores in which lanthanum occurs in percentages up to 25 and 38%, respectively. Misch metal, used in making lighter flints, contains about 25% lanthanum. Lanthanum was isolated in relatively pure form in 1923. Iron-exchange and solvent extraction techniques have led to much easier isolation of the so-called "rare-earth" elements. The availability of lanthanum and other rare earths has improved greatly in recent years. The metal can be produced by reducing the anhydrous fluoride with calcium. Lanthanum is silvery white, malleable, ductile, and soft enough to be cut with a knife. It is one of the most reactive of the rare-earth metals. It oxidizes rapidly when exposed to air. Cold water attacks lanthanum slowly, and hot water attacks it much more rapidly. The metal reacts directly with elemental carbon, nitrogen, boron, selenium, silicon, phosphorus, sulfur, and with halogens. At 310°C, lanthanum changes from a hexagonal to a face-centered cubic structure, and at 865°C it again transforms into a body-centered cubic structure. Natural lanthanum is mixture of two isotopes, one of which is stable and one of which is radioactive with a very long half-life. Twenty nine other radioactive isotopes are recognized. Rare-earth compounds containing lanthanum are extensively used in carbon lighting applications, especially by the motion picture industry for studio lighting and projection. This application consumes about 25% of the rare-earth compounds produced. La_2O_3 improves the alkali resistance of glass, and is used in making special optical glasses. Small amounts of lanthanum, as an additive, can be used to produce nodular cast iron. There is current interest in hydrogen sponge alloys containing lanthanum. These alloys take up to 400 times their own volume of hydrogen gas, and the process is reversible. Heat energy is released every time they do so; therefore these alloys have possibilities in energy conservation systems. Lanthanum and its compounds have a low to moderate acute toxicity rating; therefore, care should be taken in handling them. The metal costs about $2/g (99.9%).

Lawrencium — (Ernest O. Lawrence, inventor of the cyclotron), Lr; at. no. 103; at. mass no. (262); valence + 3(?). This member of the 5f transition elements (actinide series) was discovered in March 1961 by A. Ghiorso, T. Sikkeland, A. E. Larsh, and R. M. Latimer. A 3-μg californium target, consisting of a mixture of isotopes of mass number 249, 250, 251, and 252, was bombarded with either ^{10}B or ^{11}B. The electrically charged transmutation nuclei recoiled with an atmosphere of helium and were collected on a thin copper conveyor tape which was then moved to place collected atoms in front of a series of solid-state detectors. The isotope of element 103 produced in this way decayed by emitting an 8.6-MeV alpha particle with a half-life of 8 s. In 1967, Flerov and associates of the Dubna Laboratory reported their inability to detect an alpha emitter with a half-life of 8 s which was assigned by the Berkeley group to $^{257}103$. This assignment has been changed to ^{258}Lr or ^{259}Lr. In 1965, the Dubna workers found a longer-lived lawrencium isotope, ^{256}Lr, with a half-life of 35 s. In 1968, Ghiorso and associates at Berkeley were able to use a few atoms of this isotope to study the oxidation behavior of lawrencium. Using solvent extraction techniques and working very rapidly, they extracted lawrencium ions from a buffered aqueous solution into an organic solvent, completing each extraction in about 30 s. It was found that lawrencium behaves differently from dipositive nobelium and more like the tripositive elements earlier in the actinide series. Ten isotopes of lawrencium are now recognized.

Lead — (Anglo-Saxon *lead*), Pb (L. *plumbum*); at. wt. 207.2(1); at. no. 82; m.p. 327.46°C; b.p. 1749°C; sp. gr. 11.35 (20°C); valence 2 or 4. Long known, mentioned in Exodus. The alchemists believed lead to be the oldest metal and associated it with the planet Saturn. Native lead occurs in nature, but it is rare. Lead is obtained chiefly from *galena* (PbS) by a roasting process. *Anglesite* ($PbSO_4$), *cerussite* ($PbCO_3$), and *minim* (Pb_3O_4) are other common lead minerals. Lead is a bluish-white metal of bright luster, is very soft, highly malleable, ductile, and a poor conductor of electricity. It is very resistant to corrosion; lead pipes bearing the insignia of Roman emperors, used as drains from the baths, are still in service. It is used in containers for corrosive liquids (such as sulfuric acid) and may be toughened by the addition of a small percentage of antimony or other metals. Natural lead is a mixture of four stable isotopes: ^{204}Pb (1.4%), ^{206}Pb (24.1%), ^{207}Pb (22.1%), and ^{208}Pb (52.4%). Lead isotopes are the end products of each of the three series of naturally occurring radioactive elements: ^{206}Pb for the uranium series, ^{207}Pb for the actinium series, and ^{208}Pb for the thorium series. Forty other isotopes of lead, all of which are radioactive, are recognized. Its alloys include solder, type metal, and various antifriction metals. Great quantities of lead, both as the metal and as the dioxide, are used in storage batteries. Much metal also goes into cable covering, plumbing, ammunition, and in the manufacture of lead tetraethyl. The metal is very effective as a sound absorber, is used as a radiation shield around X-ray equipment and nuclear reactors, and is used to absorb vibration. White lead, the basic carbonate, sublimed white lead ($PbSO_4$) chrome yellow ($PbCrO_4$), red lead (Pb_3O_4), and other lead compounds are used extensively in paints, although in recent years the use of lead in paints has been drastically curtailed to eliminate or reduce health hazards. Lead oxide is used in producing fine "crystal glass" and "flint glass" of a high index of refraction for achromatic lenses. The nitrate and the acetate are soluble salts. Lead salts such as lead arsenate have been used as insecticides, but their use in recent years has been practically eliminated in favor of less harmful organic compounds. Care must be used in handling lead as it is a cumulative poison. Environmental concern with lead poisoning has resulted in a national program to eliminate the lead in gasoline. Lead is priced at about $1/kg (99.9%).

Lithium — (Gr. *lithos*, stone), Li; at. wt. 6.941(2); at. no. 3; m.p. 180.5°C; b.p. 1342°C; sp. gr. 0.534 (20°C); valence 1. Discovered by Arfvedson in 1817. Lithium is the lightest of all metals, with a density only about half that of water. It does not occur free in nature; combined it is found in small amounts in nearly all igneous rocks and in the waters of many mineral springs. *Lepidolite, spodumene, petalite,* and *amblygonite* are the more important minerals containing it. Lithium is presently being recovered from brines of Searles Lake, in California, and from Nevada, Chile, and Argentina. Large deposits of spodumene are found in North Carolina. The metal is produced electrolytically from the fused chloride. Lithium is silvery in appearance, much like Na and K, other members of the alkali metal series. It reacts with water, but not as vigorously as sodium. Lithium imparts a beautiful crimson color to a flame, but when the metal burns strongly the flame is a dazzling white. Since World War II, the production of lithium metal and its compounds has increased greatly. Because the metal has the highest specific heat of any solid element, it has found use in heat transfer applications; however, it is corrosive and requires special handling. The metal has been used as an alloying agent, is of interest in synthesis of organic compounds, and has nuclear applications. It ranks as a leading contender as a battery anode material as it has a high electrochemical potential. Lithium is used in special glasses and ceramics. The glass for the 200-inch telescope at Mt. Palomar contains lithium as a minor ingredient. Lithium chloride is one of the most hygroscopic materials known, and it, as well as lithium bromide, is used in air conditioning and industrial drying systems. Lithium stearate is used as an all-purpose and high-temperature lubricant. Other lithium compounds are used in dry cells and storage batteries. Seven isotopes of lithium are recognized. Natural lithium contains two isotopes. The metal is priced at about $500/kg (99.9%).

Lutetium — (Lutetia, ancient name for Paris, sometimes called *cassiopeium* by the Germans), Lu; at. wt. 174.967(1); at. no. 71; m.p. 1663°C; b.p. 3402°C; sp. gr. 9.841 (25°C); valence 3. In 1907, Urbain described a process by which Marignac's ytterbium (1879) could be separated into the two elements, ytterbium (neoytterbium)and lutetium. These elements were identical with "aldebaranium" and "cassiopeium," independently discovered by von Welsbach about the same time. Charles James of the University of New Hampshire also independently prepared the very pure oxide, *lutecia*, at this time. The spelling of the element was changed from *lutecium* to *lutetium* in 1949. Lutetium occurs in very small amounts in nearly all minerals containing yttrium, and is present in *monazite* to the extent of about 0.003%, which is a commercial source. The pure metal has been isolated only in recent years and is one of the most difficult to prepare. It can be prepared by the reduction of anhydrous $LuCl_3$ or LuF_3 by an alkali or alkaline earth metal. The metal is silvery white and relatively stable in air. While new techniques, including ion-exchange reactions, have been developed to separate the various rare-earth elements, lutetium is still the most costly of all rare earths. It is priced at about $100/g. ^{176}Lu occurs naturally (97.41%) with ^{175}Lu (2.59%), which is radioactive with a very long half-life. It is radioactive with a half-life of about 4×10^{10} years. Lutetium has 49 isotopes and isomers that are now recognized. Stable lutetium nuclides, which emit pure beta radiation after thermal neutron activation, can be used as catalysts in cracking, alkylation, hydrogenation, and polymerization. Virtually no other commercial uses have been found yet for lutetium. While lutetium, like other rare-earth metals, is thought to have a low toxicity rating, it should be handled with care until more information is available.

Magnesium — (*Magnesia,* district in Thessaly) Mg; at. wt. 24.3050(6); at. no. 12; m.p. 650°C; b.p. 1090°C; sp. gr. 1.738 (20°C); valence 2. Compounds of magnesium have long been known. Black recognized magnesium as an element in 1755. It was isolated by Davy in 1808, and prepared in coherent form by Bussy in 1831. Magnesium is the eighth most abundant element in the earth's crust. It does not occur uncombined, but is found in large deposits in the form of *magnesite, dolomite,* and other minerals. The metal is now principally obtained in the U.S. by electrolysis of fused magnesium chloride derived from brines, wells, and sea water. Magnesium is a light, silvery-white, and fairly tough metal. It tarnishes slightly in air, and finely divided magnesium readily ignites upon heating in air and burns with a dazzling white flame. It is used in flashlight photography, flares, and pyrotechnics, including incendiary bombs. It is one third lighter than aluminium, and in alloys is essential for airplane and missile contruction. The metal improves the mechanical, fabrication, and welding characteristics of aluminum when used as an alloying agent. Magnesium is used in producing nodular graphite in cast iron,and is used as an additive to conventional propellants. It is also used as a reducing agent in the production of pure uranium and other metals from their salts. The hydroxide (*milk of magnesia*), chloride, sulfate (*Epsom salts*), and citrate are used in medicine. Dead-burned magnesite is employed for refractory purposes such as brick and liners in furnaces and converters. Organic magnesium compounds (Grignard's reagents) are important. Magnesium is an important element in both plant and animal life. Chlorophylls are magnesium-centered porphyrins. The adult daily requirement of magnesium is about 300 mg/day, but this is affected by various factors. Great care should be taken in handling magnesium metal, especially in the finely divided state, as serious fires can occur. Water should not be used on burning magnesium or on magnesium fires. Natural magnesium contains three isotopes. Twelve other isotopes are recognized. Magnesium metal costs about $170/kg (99.99%).

Manganese — (L. *magnes,* magnet, from magnetic properties of pyrolusite; It. *manganese,* corrupt form of *magnesia*), Mn; at. wt. 54.93805(1); at. no. 25; m.p. 1246±3°C; b.p. 2061°C; sp. gr. 7.21 to 7.44, depending on allotropic form; valence 1, 2, 3, 4, 6, or 7. Recognized by Scheele, Bergman, and others as an element and isolated by Gahn in 1774 by reduction of the dioxide with carbon. Manganese minerals are widely distributed; oxides, silicates, and carbonates are the most common. The discovery of large quantities of manganese nodules on the floor of the oceans holds promise as a source of manganese. These nodules contain about 24% manganese together with many other elements in lesser abundance. Most manganese today is obtained from ores found in the Ukraine, Brazil, Australia, Republic of So. Africa, Gabon, China, and India. *Pyrolusite* (MnO_2) and *rhodochrosite* ($MnCO_3$) are among the most common manganese minerals. The metal is obtained by reduction of the oxide with sodium, magnesium, aluminum, or by electrolysis. It is gray-white, resembling iron, but is harder and very brittle. The metal is reactive chemically, and decomposes cold water slowly.

Manganese is used to form many important alloys. In steel, manganese improves the rolling and forging qualities, strength, toughness, stiffness, wear resistance, hardness, and hardenability. With aluminum and antimony, especially with small amounts of copper, it forms highly ferromagnetic alloys. Manganese metal is ferromagnetic only after special treatment. The pure metal exists in four allotropic forms. The alpha form is stable at ordinary temperature; gamma manganese, which changes to alpha at ordinary temperatures, is said to be flexible, soft, easily cut, and capable of being bent. The dioxide (pyrolusite) is used as a depolarizer in dry cells, and is used to "decolorize" glass that is colored green by impurities of iron. Manganese by itself colors glass an amethyst color, and is responsible for the color of true amethyst. The dioxide is also used in the preparation of oxygen and chlorine, and in drying black paints. The permanganate is a powerful oxidizing agent and is used in quantitative analysis and in medicine. Manganese is widely distributed throughout the animal kingdom. It is an important trace element and may be essential for utilization of vitamin B$_1$. Twenty three isotopes and isomers are known. Manganese metal (99.95%) is priced at about $400/kg. Metal of 99.6% purity is priced at about $60/kg.

Meitnerium — See Element 109.

Mendelevium — (Dmitri Mendeleev), Md; at. wt. (258); at. no. 101; m.p. 827°C; valence +2, +3. Mendelevium, the ninth transuranium element of the actinide series to be discovered, was first identified by Ghiorso, Harvey, Choppin, Thompson, and Seaborg early in 1955 as a result of the bombardment of the isotope ^{253}Es with helium ions in the Berkeley 60-inch cyclotron. The isotope produced was ^{256}Md, which has a half-life of 78 min. This first identification was notable in that ^{256}Md was synthesized on a one-atom-at-a-time basis. Sixteen isotopes and isomers are now recognized. ^{258}Md has a half-life of 51.5 days. This isotope has been produced by the bombardment of an isotope of einsteinium with ions of helium. It now appears possible that eventually enough ^{258}Md can be made so that some of its physical properties can be determined. ^{256}Md has been used to elucidate some of the chemical properties of mendelevium in aqueous solution. Experiments seem to show that the element possesses a moderately stable dipositive (II) oxidation state in addition to the tripositive (III) oxidation state, which is characteristic of actinide elements.

Mercury — (Planet *Mercury*), Hg (*hydrargyrum*, liquid silver); at. wt. 200.59(2); at. no. 80; t.p. –38.8344°C; b.p. 356.73°C; sp. gr. 13.546 (20°C); valence 1 or 2. Known to ancient Chinese and Hindus; found in Egyptian tombs of 1500 B.C. Mercury is the only common metal liquid at ordinary temperatures. It only rarely occurs free in nature. The chief ore is *cinnabar* (HgS). Spain and Italy produce about 50% of the world's supply of the metal. The commercial unit for handling mercury is the "flask," which weighs 76 lb (34.46 kg) and is priced at about $250. The metal is obtained by heating cinnabar in a current of air and by condensing the vapor. It is a heavy, silvery-white metal; a rather poor conductor of heat, as compared with other metals, and a fair conductor of electricity. It easily forms alloys with many metals, such as gold, silver, and tin, which are called *amalgams*. Its ease in amalgamating with gold is made use of in the recovery of gold from its ores. The metal is widely used in laboratory work for making thermometers, barometers, diffusion pumps, and many other instruments. It is used in making mercury-vapor lamps and advertising signs, etc. and is used in mercury switches and other electrical apparatus. Other uses are in making pesticides, mercury cells for caustic soda and chlorine production, dental preparations, antifouling paint, batteries, and catalysts. The most important salts are mercuric chloride HgCl$_2$ (corrosive sublimate — a violent poison), mercurous chloride Hg$_2$Cl$_2$(calomel, occasionally still used in medicine), mercury fulminate (Hg(ONC)$_2$), a detonator widely used in explosives, and mercuric sulfide (HgS, vermillion, a high-grade paint pigment). Organic mercury compounds are important. It has been found that an electrical discharge causes mercury vapor to combine with neon, argon, krypton, and xenon. These products, held together with van der Waals' forces, correspond to HgNe, HgAr, HgKr, and HgXe. Mercury is a virulent poison and is readily absorbed through the respiratory tract, the gastrointestinal tract, or through unbroken skin. It acts as a cumulative poison and dangerous levels are readily attained in air. Air saturated with mercury vapor at 20°C contains a concentration that exceeds the toxic limit many times. The danger increases at higher temperatures. *It is therefore important that mercury be handled with care.* Containers of mercury should be securely covered and spillage should be avoided. If it is necessary to heat mercury or mercury compounds, it should be done in a well-ventilated hood. Methyl mercury is a dangerous pollutant and is now widely found in water and streams. The triple point of mercury, –38.8344°C, is a fixed point on the International Temperature Scale (ITS-90). Native mercury contains seven isotopes. Thirty five other isotopes and isomers are known.

Molybdenum — (Gr. *molybdos*, lead), Mo; at. wt. 95.94(1); at. no. 42; m.p. 2623°C; b.p. 4639°C; sp. gr. 10.22 (20°C); valence 2, 3, 4?, 5?, or 6. Before Scheele recognized molybdenite as a distinct ore of a new element in 1778, it was confused with graphite and lead ore. The metal was prepared in an impure form in 1782 by Hjelm. Molybdenum does not occur native, but is obtained principally from *molybdenite* (MoS$_2$). *Wulfenite* (PbMoO$_4$) and *Powellite* (Ca(MoW)O$_4$) are also minor commercial ores. Molybdenum is also recovered as a by-product of copper and tungsten mining operations. The metal is prepared from the powder made by the hydrogen reduction of purified molybdic trioxide or ammonium molybdate. The metal is silvery white, very hard, but is softer and more ductile than tungsten. It has a high elastic modulus, and only tungsten and tantalum, of the more readily available metals, have higher melting points. It is a valuable alloying agent, as it contributes to the hardenability and toughness of quenched and tempered steels. It also improves the strength of steel at high temperatures. It is used in certain nickel-based alloys, such as the "Hastelloys®" which are heat-resistant and corrosion-resistant to chemical solutions. Molybdenum oxidizes at elevated temperatures. The metal has found recent application as electrodes for electrically heated glass furnaces and forehearths. The metal is also used in nuclear energy applications and for missile and aircraft parts. Molybdenum is valuable as a catalyst in the refining of petroleum. It has found application as a filament material in electronic and electrical applications. Molybdenum is an essential trace element in plant nutrition. Some lands are barren for lack of this element in the soil. Molybdenum sulfide is useful as a lubricant, especially at high temperatures where oils would decompose. Almost all ultra-high strength steels with minimum yield points up to 300,000 psi(lb/in.^{2}) contain molybdenum in amounts from 0.25 to 8%. Natural molybdenum contains seven isotopes. Twenty three other isotopes and isomers are known, all of which are radioactive. Molybdenum metal costs about $1/g (99.999% purity). Commercial molybdenum metal (99.9%) costs about $200/kg.

Neodymium — (Gr. *neos*, new, and *didymos*, twin), Nd; at. wt. 144.24(3); at. no. 60; m.p. 1021°C; b.p. 3074°C; sp. gr. 7.008 (25°C); valence 3. In 1841, Mosander, extracted from *cerite* a new rose-colored oxide, which he believed contained a new element. He named the element *didymium*, as it was *an inseparable twin brother of lanthanum*. In 1885 von Welsbach separated didymium into two new elemental components, *neodymia* and *praseodymia*, by repeated fractionation of ammonium didymium nitrate. While the free metal is in *misch metal*, long known and used as a pyrophoric alloy for light flints, the element was not isolated in relatively pure form until 1925. Neodymium is present in misch metal to the extent of about 18%. It is present in the minerals *monazite* and *bastnasite*, which are principal sources of rare-earth metals. The element may be obtained by separating neodymium salts from other rare earths by ion-exchange or solvent extraction techniques, and by reducing anhydrous halides such as NdF$_3$ with calcium metal. Other separation techniques are possible. The metal has a bright silvery metallic luster. Neodymium is one of the more reactive rare-earth metals and quickly tarnishes in air, forming an oxide that spalls off and exposes metal to oxidation. The metal, therefore, should be kept under light mineral oil or sealed in a plastic material. Neodymium exists in two allotropic forms, with a transformation from a double hexagonal to a body-centered cubic structure taking place at 863°C. Natural neodymium is a mixture of seven isotopes, one of which has a very long half-life. Twenty seven other radioactive isotopes and isomers are recognized. Didymium, of which neodymium is a component, is used for coloring glass to make welder's goggles. By itself, neodymium colors glass delicate shades ranging from pure violet through wine-red and warm gray. Light transmitted through such glass shows unusually sharp absorption bands. The glass has been used in astronomical work to produce sharp bands by which spectral lines may be calibrated. Glass containing neodymium can be used as a laser material to produce coherent light. Neodymium salts are also used as a colorant for enamels. The element is also being used with iron and boron to produce extremely strong magnets having energy densities as high as 27 to 35 million gauss oersteds. These are the most compact magnets commercially available. The price of the metal is about $2/g. Neodymium has a low-to-moderate acute toxic rating. As with other rare earths, neodymium should be handled with care.

Neon — (Gr. *neos*, new), Ne; at. wt. 20.1797(6); at. no. 10; t.p. –248.5938°C; b.p. –246.08°C (1 atm); density of gas 0.89990 g/l (1 atm, 0°C); density of liquid at b.p. 1.207 g/cm^3; valence 0. Discovered by Ramsay and Travers in 1898. Neon is a rare gaseous element present in the atmosphere to the extent of 1 part in 65,000 of air. It is obtained by liquefaction of air and separated from the other gases by fractional distillation. Natural neon is a mixture of three isotopes. Six other unstable isotopes are known. It is very inert element; however, it is said to form a compound with fluorine. It is still questionable if true compounds of neon exist, but evidence is mounting in favor of their existence. The following ions are known from optical and mass spectrometric studies: Ne$^+$, (NeAr)$^+$, (NeH)$^+$, and (HeNe$^+$). Neon also forms an unstable hydrate. In a vacuum discharge tube, neon glows reddish orange. Of all the rare gases, the discharge of neon is the most intense at ordinary voltages and currents. Neon is used in making the common neon advertising signs, which accounts for its largest use. It is also used to make high-voltage indicators, lightning arrestors, wave meter tubes, and TV tubes. Neon and helium are used in making gas lasers. Liquid neon is now commercially available and is finding important application as an economical cryogenic refrigerant. It has over 40 times more refrigerating capacity per unit volume than liquid helium and more than three times that of liquid hydrogen. It is compact, inert, and is less expensive than helium when it meets refrigeration requirements. Neon costs about $800/80 cu. ft. (2265 l).

Neptunium — (Planet *Neptune*), Np; at. wt. (237); at. no. 93; m.p. 644°C; b.p. 3902°C (est.); sp. gr. 20.25 (20°C); valence 3, 4, 5, and 6. Neptunium was the first synthetic transuranium element of the actinide series discovered; the isotope ^{239}Np was produced by McMillan and Abelson in 1940 at Berkeley, California, as the result of bombarding uranium with cyclotron-produced neutrons. The isotope ^{237}Np (half-life of 2.14 × 10^6 years) is currently obtained in gram quantities as a by-product from nuclear reactors in the production of plutonium. Twenty isotopes and isomers of neptunium are now recognized. Trace quantities of the element are actually found in nature due to transmutation reactions in uranium ores produced by the neutrons which are present. Neptunium is prepared by the reduction of NpF$_3$ with barium or lithium vapor at about 1200°C. Neptunium metal has a silvery appearance, is chemically reactive, and exists in at least three structural modifications: α-neptunium, orthorhombic, density 20.25 g/cm^3, β-neptunium (above 280°C), tetragonal, density (313°C) 19.36 g/cm^3; γ-neptunium (above 577°C), cubic, density (600°C) 18.0 g/cm^3. Neptunium has four ionic oxidation states in solution: Np^{+3} (pale purple), analogous to the rare earth ion Pm^{+3}, Np^{+4} (yellow green); NpO$^+$ (green blue); and NpO^{++} (pale pink). These latter oxygenated species are in contrast to the rare earths which exhibit only simple ions of the (II), (III), and (IV) oxidation states in aqueous solution. The element forms tri- and tetrahalides such as NpF$_3$, NpF$_4$, NpCl$_4$, NpBr$_3$, NpI$_3$, and oxides of various compositions such as are found in the uranium-oxygen system, including Np$_3$O$_8$ and NpO$_2$. Fifteen isotopes of neptunium are now recognized. The O.R.N.L. has ^{237}Np available for sale to its licensees and for export. This isotope can be used as a component in neutron detection instruments. It is offered at a price of $660/g plus packing costs.

Nickel — (Ger. *Nickel*, Satan or Old Nick's and from *kupfernickel*, Old Nick's copper), Ni; at. wt. 58.6934(2); at. no. 28; m.p. 1455°C; b.p. 2913°C; sp. gr. 8.902 (25°C); valence 0, 1, 2, 3. Discovered by Cronstedt in 1751 in kupfernickel (*niccolite*). Nickel is found as a constituent in most meteorites and often serves as one of the criteria for distinguishing a meteorite from other minerals. Iron meteorites, or *siderites*, may contain iron alloyed with from 5 to nearly 20% nickel. Nickel is obtained commercially from *pentlandite* and *pyrrhotite* of the Sudbury region of Ontario, a district that produces much of the world's nickel. It is now thought that the Sudbury deposit is the result of an ancient meteorite impact. Other deposits of nickel are found in Russia, New Caledonia, Australia, Cuba, Indonesia, and elsewhere. Nickel is silvery white and takes on a high polish. It is hard, malleable, ductile, somewhat ferromagnetic, and a fair conductor of heat and electricity. It belongs to the iron-cobalt group of metals and is chiefly valuable for the alloys it forms. It is extensively used for making stainless steel and other corrosion-resistant alloys such as Invar®, Monel®, Inconel®, and the Hastelloys®. Tubing made of a copper-nickel alloy is extensively used in making desalination plants for converting sea water into fresh water. Nickel is also now used extensively in coinage and in making nickel steel for armor plate and burglar-proof vaults, and is a component in Nichrome®, Permalloy®, and constantan. Nickel added to glass gives a green color. Nickel plating is often used to provide a protective coating for other metals, and finely divided nickel is a catalyst for hydrogenating vegetable oils. It is also used in ceramics, in the manufacture of Alnico magnets, and in the Edison® storage battery. The sulfate and the oxides are important compounds. Natural nickel is a mixture of five stable isotopes; nineteen other unstable isotopes are known. Nickel sulfide fume and dust is recognized as having carcinogenic potential. Nickel metal (99.9%) is priced at about $100/kg or less in larger quantities.

Niobium — (*Niobe*, daughter of Tantalus), Nb; or Columbium (*Columbia*, name for America); at. wt. 92.90638(2); at. no. 41; m.p. 2477 ± 10°C; b.p. 4744°C, sp. gr. 8.57 (20°C); valence 2, 3, 4?, 5. Discovered in 1801 by Hatchett in an ore sent to England more that a century before by John Winthrop the Younger, first governor of Connecticut. The metal was first prepared in 1864 by Blomstrand, who reduced the chloride by heating it in a hydrogen atmosphere. The name *niobium* was adopted by the International Union of Pure and Applied Chemistry in 1950 after 100 years of controversy. Many leading chemical societies and government organizations refer to it by this name. Most metallurgists, leading metal societies, and all but one of the leading U.S. commercial producers, however, still refer to the metal as "columbium". The element is found in *niobite*(or *columbite*), *niobite-tantalite*, *pyrochlore*, and *euxenite*. Large deposits of niobium have been found associated with *carbonatites* (carbon-silicate rocks), as a constituent of *pyrochlore*. Extensive ore reserves are found in Canada, Brazil, Nigeria, Zaire, and in Russia. The metal can be isolated from tantalum, and prepared in several ways. It is a shiny, white, soft, and ductile metal, and takes on a bluish cast when exposed to air at room temperatures for a long time. The metal starts to oxidize in air at 200°C, and when processed at even moderate temperatures must be placed in a protective atmosphere. It is used in arc-welding rods for stabilized grades of stainless steel. Thousands of pounds of niobium have been used in advance air frame systems such as were used in the Gemini space program. The element has superconductive properties; superconductive magnets have been made with Nb-Zr wire, which retains its superconductivity in strong magnetic fields. This type of application offers hope of direct large-scale generation of electric power. Natural niobium is composed of only one isotope, ^{93}Nb. Forty one other isotopes and isomers of niobium are now recognized. Niobium metal (99.9% pure) is priced at about $400/kg.

Nitrogen — (L. *nitrum*, Gr. *nitron*, native soda; genes, *forming*, N; at. wt. 14.00674(7); at. no. 7; m.p. −210.00°C; b.p. −195.8°C; density 1.2506 g/l; sp. gr. liquid 0.808 (−195.8°C), solid 1.026 (−252°C); valence 3 or 5. Discovered by Daniel Rutherford in 1772, but Scheele, Cavendish, Priestley, and others about the same time studied "burnt or dephlogisticated air," as air without oxygen was then called. Nitrogen makes up 78% of the air, by volume. The atmosphere of Mars, by comparison, is 2.6% nitrogen. The estimated amount of this element in our atmosphere is more than 4000 trillion tons. From this inexhaustible source it can be obtained by liquefaction and fractional distillation. Nitrogen molecules give the orange-red, blue-green, blue-violet, and deep violet shades to the aurora. The element is so inert that Lavoisier named it *azote*, meaning without life, yet its compounds are so active as to be most important in foods, poisons, fertilizers, and explosives. Nitrogen can be also easily prepared by heating a water solution of ammonium nitrite. Nitrogen, as a gas, is colorless, odorless, and a generally inert element. As a liquid it is also colorless and odorless, and is similar in appearance to water. Two allotropic forms of solid nitrogen exist, with the transition from the α to the β form taking place at −237°C. When nitrogen is heated, it combines directly with magnesium, lithium, or calcium; when mixed with oxygen and subjected to electric sparks, it forms first nitric oxide (NO) and then the dioxide (NO_2); when heated under pressure with a catalyst with hydrogen, ammonia is formed (Haber process). The ammonia thus formed is of the utmost importance as it is used in fertilizers, and it can be oxidized to nitric acid (Ostwald process). The ammonia industry is the largest consumer of nitrogen. Large amounts of gas are also used by the electronics industry, which uses the gas as a blanketing medium during production of such components as transistors, diodes, etc. Large quantities of nitrogen are used in annealing stainless steel and other steel mill products. The drug industry also uses large quantities. Nitrogen is used as a refrigerant both for the immersion freezing of food products and for transportation of foods. Liquid nitrogen is also used in missile work as a purge for components, insulators for space chambers, etc., and by the oil industry to build up great pressures in wells to force crude oil upward. Sodium and potassium nitrates are formed by the decomposition of organic matter with compounds of the metals present. In certain dry areas of the world these saltpeters are found in quantity. Ammonia, nitric acid, the nitrates, the five oxides (N_2O, NO, N_2O_3, NO_2, and N_2O_5), TNT, the cyanides, etc. are but a few of the important compounds. Nitrogen gas prices vary from 2¢ to $2.75 per 100 ft^3 (2.83 cu. meters), depending on purity, etc. Production of elemental nitrogen in the U.S. is more than 9 million short tons per year. Natural nitrogen contains two isotopes, ^{14}N and ^{15}N. Ten other isotopes are known.

Nobelium — (Alfred Nobel, discoverer of dynamite), No; at. wt. (259); at. no. 102; valence +2, +3. Nobelium was unambiguously discovered and identified in April 1958 at Berkeley by A. Ghiorso, T. Sikkeland, J. R. Walton, and G. T. Seaborg, who used a new double-recoil technique. A heavy-ion linear accelerator (HILAC) was used to bombard a thin target of curium (95% ^{244}Cm and 4.5% ^{246}Cm) with ^{12}C ions to produce 102^{254} according to the ^{246}Cm (^{12}C, $4n$) reaction. Earlier in 1957 workers of the U.S., Britain, and Sweden announced the discovery of an isotope of Element 102 with a 10-min half-life at 8.5 MeV, as a result of bombarding ^{244}Cm with ^{13}C nuclei. On the basis of this experiment the name *nobelium* was assigned and accepted by the Commission on Atomic Weights of the International Union of Pure and Applied Chemistry. The acceptance of the name was premature, for both Russian and American efforts now completely rule out the possibility of any isotope of Element 102 having a half-life of 10 min in the vicinity of 8.5 MeV. Early work in 1957 on the search for this element, in Russia at the Kurchatov Institute, was marred by the assignment of 8.9 ± 0.4 MeV alpha radiation with a half-life of 2 to 40 sec, which was too indefinite to support claim to discovery. Confirmatory experiments at Berkeley in 1966 have shown the existence of ^{254}102 with a 55-s half-life, ^{252}102 with a 2.3-s half-life, and ^{257}102 with a 25-s half-life. Twelve isotopes are now recognized, one of which — ^{255}102 has a half-life of 3.1 min. In view of the discover's traditional right to name an element, the Berkeley group, in 1967, suggested that the hastily given name *nobelium,* along with the symbol No, be retained.

Osmium — (Gr. *osme*, a smell), Os; at. wt. 190.23(3); at. no. 76; m.p. 3033 ± 30°C; b.p. 5012 ± 100°C; sp. gr. 22.57; valence 0 to +8, more usually +3, +4, +6, and +8. Discovered in 1803 by Tennant in the residue left when crude platinum is dissolved by *aqua regia*. Osmium occurs in *iridosmine* and in platinum-bearing river sands of the Urals, North America, and South America. It is also found in the nickel-bearing ores of Sudbury, Ontario, region along with other platinum metals. While the quantity of platinum metals in these ores is very small, the large tonnages of nickel ores processed make commercial recovery possible. The metal is lustrous, bluish white, extremely hard, and brittle even at high temperatures. It has the highest melting point and the lowest vapor pressure of the platinum group. The metal is very difficult to fabricate, but the powder can be sintered in a hydrogen atmosphere at a temperature of 2000°C. The solid metal is not affected by air at room temperature, but the powdered or spongy metal slowly gives off osmium tetroxide, which is a powerful oxidizing agent and has a strong smell. The tetroxide is highly toxic, and boils at 130°C (760 mm). Concentrations in air as low as 10^{-7} g/m^3 can cause lung congestion, skin damage, or eye damage. The tetroxide has been used to detect fingerprints and to stain fatty tissue for microscope slides. The metal is almost entirely used to produce very hard alloys, with other metals of the platinum group, for fountain pen tips, instrument pivots, phonograph needles, and electrical contacts. The price of 99.9% pure osmium powder — the form usually supplied commercially — is about $100/g, depending on quantity and supplier. Natural osmium contains seven isotopes, one of which, ^{186}Os, is radioactive with a very long half-life. Thirty four other isotopes and isomers are known, all of which are radioactive. The measured densities of iridium and osmium seem to indicate that osmium is slightly more dense than iridium, so osmium has generally been credited with being the heaviest known element. Calculations of the density from the space lattice, which may be more reliable for these elements than actual measurements, however, give a density of 22.65 for iridium compared to 226.61 for osmium. At present, therefore, we know either iridium or osmium is the heaviest element, but the data do not allow selection between the two.

Oxygen — (Gr. *oxys*, sharp, acid, and *genes*, forming; acid former), O; at. wt. 15.9994(3); at. no. 8; t.p. −218.7916°C; b.p. −182.95°C; valence 2. For many centuries, workers occasionally realized air was composed of more than one component. The behavior of oxygen and nitrogen as components of air led to the advancement of the phlogiston theory of combustion, which captured the minds of chemists for a century. Oxygen was prepared by several workers, including Bayen and Borch, but they did not know how to collect it, did not study its properties, and did not recognize it as an elementary substance. Priestley is generally credited with its discovery, although Scheele also discovered it independently. Oxygen is the third most abundant element found in the sun, and it plays a part in the carbon-nitrogen cycle, one process thought to give the sun and stars their energy. Oxygen under excited conditions is responsible for the bright red and yellow-green colors of the aurora. Oxygen, as a gaseous element, forms 21% of the atmosphere by volume from which it can be obtained by liquefaction and fractional distillation. The atmosphere of Mars contains about 0.15% oxygen. The element and its compounds make up 49.2%, by weight, of the earth's crust. About two thirds of the human body and nine tenths of water is oxygen. In the laboratory it can be prepared by the electrolysis of water or by heating potassium chlorate with manganese dioxide as a catalyst. The gas is colorless, odorless, and tasteless. The liquid and solid forms are a pale blue color and are strongly paramagnetic. Ozone (O_3), a highly active compound, is formed by the action of an electrical discharge or ultraviolet light on oxygen. Ozone's presence in the atmosphere (amounting to the equivalent of a layer 3 mm thick at ordinary pressures and temperatures) is of vital importance in preventing harmful ultraviolet rays of the sun from reaching the earth's surface. There has been recent concern that pollutants in the atmosphere may have a detrimental effect on this ozone layer. Ozone is toxic and exposure should not exceed 0.2 mg/m^3 (8-hour time-weighted average — 40-hour work week). Undiluted ozone has a bluish color. Liquid ozone is bluish black, and solid ozone is violet-black. Oxygen is very reactive and capable of combining with most elements. It is a component of hundreds of thousands of organic compounds. It is essential for respiration of all plants and animals and for practically all combustion. In hospitals it is frequently used to aid respiration of patients. Its atomic weight was used as a standard of comparison for each of the other elements until 1961 when the International Union of Pure and Applied Chemistry adopted carbon 12 as the new basis. Oxygen has thirteen recognized isotopes. Natural oxygen is a mixture of three isotopes. Oxygen 18 occurs naturally, is stable, and is available commercially. Water (H_2O with 1.5% ^{18}O) is also available. Commercial oxygen consumption in the U.S. is estimated to be 20 million short tons per year and the demand is expected to increase substantially in the next few years. Oxygen enrichment of steel blast furnaces accounts for the greatest use of the gas. Large quantities are also used in making synthesis gas for ammonia and methanol, ethylene oxide, and for oxy-acetylene welding. Air separation plants produce about 99% of the gas, electrolysis plants about 1%. The gas costs 5¢/ft^3 ($1.75/cu. meters) in small quantities.

Palladium — (named after the asteroid *Pallas*, discovered about the same time; Gr. *Pallas*, goddess of wisdom), Pd. at. wt. 106.42(1) at. no. 46; m.p. 1554.9°C; b.p. 2963°C; sp. gr. 1202 (20°C); valence 2, 3, or 4. Discovered in 1803 by Wollaston. Palladium is found along with platinum and other metals of the platinum group in placer deposits of Russia, South and North America, Ethiopia, and Australia. Natural palladium contains six stable isotopes. Twenty five other isotopes are recognized, all of which are radioactive. It is also found associated with the nickel-copper deposits of South Africa and Ontario. Its separation from the platinum metals depends upon the type of ore in which it is found. It is a steel-white metal, does not tarnish in air, and is the least dense and lowest melting of the platinum group of metals. When annealed, it is soft and ductile; cold working greatly increases its strength and hardness. Palladium is attacked by nitric and sulfuric acid. At room temperatures the metal has the unusual property of absorbing up to 900 times its own volume of hydrogen, possibly forming Pd_2H. It is not yet clear if this a true compound. Hydrogen readily diffuses through heated palladium and this provides a means of purifying the gas. Finely divided palladium is a good catalyst and is used for hydrogenation and dehydrogenation reactions. It is alloyed and used in jewelry trades. White gold is an alloy of gold decolorized by the addition of palladium. Like gold, palladium can be beaten into leaf as thin as 1/250,000 in. The metal is used in dentistry, watchmaking, and in making surgical instruments and electrical contacts. The metal sells for about $140/troy oz. ($4.50/g).

Phosphorus — (Gr. *phosphoros*, light bearing; ancient name for the planet Venus when appearing before sunrise), P; at. wt. 30.973762(4); at. no. 15; m.p. (white) 44.15°C; b.p. 280.5°C; sp. gr. (white) 1.82 (red) 2.20, (black) 2.25 to 2.69; valence 3 or 5. Discovered in 1669 by Brand, who prepared it from urine. Phosphorus exists in four or more allotropic forms: white (or yellow), red, and black (or violet). White phosphorus has two modifications: α and β with a transition temperature at –3.8°C. Never found free in nature, it is widely distributed in combination with minerals. Seventeen isotopes of phosphorus are recognized. *Phosphate* rock, which contains the mineral *apatite*, an impure tri-calcium phosphate, is an important source of element. Large deposits are found in the U.S.S.R., in Morocco, and in Florida, Tennessee, Utah, Idaho, and elsewhere. Phosphorus in an essential ingredient of all cell protoplasm, nervous tissue, and bones. Ordinary phosphorus is a waxy white solid; when pure it is colorless and transparent. It is insoluble in water, but soluble in carbon disulfide. It takes fire spontaneously in air, burning to the pentoxide. It is very poisonous, 50 mg constituting an approximate fatal dose. Exposure to white phosphorus should not exceed 0.1 mg/m^3 (8-hour time-weighted average — 40-hour work week). White phosphorus should be kept under water, as it is dangerously reactive in air, and it should be handled with forceps, as contact with the skin may cause severe burns. When exposed to sunlight or when heated in its own vapor to 250°C, it is converted to the red variety, which does not phosphoresce in air as does the white variety. This form does not ignite spontaneously and it is not as dangerous as white phosphorus. It should, however, be handled with care as it does convert to the white form at some temperatures and it emits highly toxic fumes of the oxides of phosphorus when heated. The red modification is fairly stable, sublimes with a vapor pressure of 1 atm at 417°C, and is used in the manufacture of safety matches, pyrotechnics, pesticides, incendiary shells, smoke bombs, tracer bullets, etc. White phosphorus may be made by several methods. By one process, tri-calcium phosphate, the essential ingredient of phosphate rock, is heated in the presence of carbon and silica in an electric furnace or fuel-fired furnace. Elementary phosphorus is liberated as vapor and may be collected under water. If desired, the phosphorus vapor and carbon monoxide produced by the reaction can be oxidized at once in the presence of moisture to produce phosphoric acid, an important compound in making super-phosphate fertilizers. In recent years, concentrated phosphoric acids, which may contain as much as 70 to 75% P_2O_5 content, have become of great importance to agriculture and farm production. World-wide demand for fertilizers has caused record phosphate production. Phosphates are used in the production of special glasses, such as those used for sodium lamps. Bone-ash, calcium phosphate, is also used to produce fine chinaware and to produce mono-calcium phosphate used in baking powder. Phosphorus is also important in the production of steels, phosphor bronze, and many other products. Trisodium phosphate is important as a cleaning agent, as a water softener, and for preventing boiler scale and corrosion of pipes and boiler tubes. Organic compounds of phosphorus are important. Amorphous (red) phosphorus costs about $60/kg (99.5%).

Platinum — (Sp. *platina*, silver), Pt; at. wt. 195.08(3); at. no. 78; m.p. 1768.4°C; b.p. 3825 ± 100°C; sp. gr. 21.45 (20°C); valence 1?, 2, 3, or 4. Discovered in South America by Ulloa in 1735 and by Wood in 1741. The metal was used by pre-Columbian Indians. Platinum occurs native, accompanied by small quantities of iridium, osmium, palladium, ruthenium, and rhodium, all belonging to the same group of metals. These are found in the alluvial deposits of the Ural mountains, of Columbia, and of certain western American states. *Sperrylite* ($PtAs_2$), occurring with the nickel-bearing deposits of Sudbury, Ontario, is the source of a considerable amount of metal. The large production of nickel offsets there being only one part of the platinum metals in two million parts of ore. Platinum is a beautiful silvery-white metal, when pure, and is malleable and ductile. It has a coefficient of expansion almost equal to that of soda-lime-silica glass, and is therefore used to make sealed electrodes in glass systems. The metal does not oxidize in air at any temperature, but is corroded by halogens, cyanides, sulfur, and caustic alkalis. It is insoluble in hydrochloric and nitric acid, but dissolves when they are mixed as *aqua regia*, forming chloroplatinic acid (H_2PtCl_6), an important compound. Natural platinum contains six isotopes, one of which, ^{190}Pt, is radioactive with a long half-life. Thirty five other radioactive isotopes and isomers are recognized. The metal is extensively used in jewelry, wire, and vessels for laboratory use, and in many valuable instruments including thermocouple elements. It is also used for electrical contacts, corrosion-resistant apparatus, and in dentistry. Platinum-cobalt alloys have magnetic properties. One such alloy made of 76.7% Pt and 23.3% Co, by weight, is an extremely powerful magnet that offers a B-H (max) almost twice that of Alnico V. Platinum resistance wires are used for constructing high-temperature electric furnaces. The metal is used for coating missile nose cones, jet engine fuel nozzles, etc., which must perform reliably for long periods of time at high temperatures. The metal, like palladium, absorbs large volumes, of hydrogen, retaining it at ordinary temperatures but giving it up at red heat. In the finely divided state platinum is an excellent catalyst, having long been used in the contact process for producing sulfuric acid. It is also used as a catalyst in cracking petroleum products. There is also much current interest in the use of platinum as a catalyst in fuel cells and in its use as antipollution devices for automobiles. Platinum anodes are extensively used in cathodic protection systems for large ships and ocean-going vessels, pipelines, steel piers, etc. Pure platinum wire will glow red hot when placed in the vapor of methyl alcohol. It acts here as a catalyst, converting the alcohol to formaldehyde. This phenomenon has been used commercially to produce cigarette lighters and hand warmers. Hydrogen and oxygen explode in the presence of platinum. The price of platinum has varied widely; more than a century ago it was used to adulterate gold. It was nearly eight times as valuable as gold in 1920. The price in January 1996 was about $400/troy oz. ($13/g), about the same price as gold.

Plutonium — (Planet *pluto*), Pu; at. wt. (244); at. no. 94; sp. gr. (α modification) 19.84 (25°C); m.p. 640°C; b.p. 3228°C; valence 3, 4, 5, or 6. Plutonium was the second transuranium element of the actinide series to be discovered. The isotope ^{238}Pu was produced in 1940 by Seaborg, McMillan, Kennedy, and Wahl by deuteron bombardment of uranium in the 60-inch cyclotron at Berkeley, California. Plutonium also exists in trace quantities in naturally occurring uranium ores. It is formed in much the same manner as neptunium, by irradiation of natural uranium with the neutrons which are present. By far of greatest importance is the isotope Pu^{239}, with a half-life of 24,100 years, produced in extensive quantities in nuclear reactors from natural uranium:

$$^{238}U(n, \gamma) \longrightarrow ^{239}U \xrightarrow{\beta} ^{239}Np \xrightarrow{\beta} ^{239}Pu$$

Eighteen isotopes of plutonium are now known. Plutonium has assumed the position of dominant importance among the transuranium elements because of its successful use as an explosive ingredient in nuclear weapons and the place which it holds as a key material in the development of industrial use of nuclear power. One kilogram is equivalent to about 22 million kilowatt hours of heat energy. The complete detonation of a kilogram of plutonium produces an explosion equal to about 20,000 tons of chemical explosive. Its importance depends on the nuclear property of being readily fissionable with neutrons and its availability in quantity. The world's nuclear-power reactors are now producing about 20,000 kg of plutonium/yr. By 1982 it was estimated that about 300,000 kg had accumulated. The various nuclear applications of plutonium are well known. ^{238}Pu has been used in the Apollo lunar missions to power seismic and other equipment on the lunar surface. As with neptunium and uranium, plutonium metal can be prepared by reduction of the trifluoride with alkaline-earth metals. The metal has a silvery appearance and takes on a yellow tarnish when slightly oxidized. It is chemically reactive. A relatively large piece of plutonium is warm to the touch because of the energy given off in alpha decay. Larger pieces will produce enough heat to boil water. The metal readily dissolves in concentrated hydrochloric acid, hydroiodic acid, or perchloric acid with formation of the Pu^{+3} ion. The metal exhibits six allotropic modifications having various crystalline structures. The densities of these vary from 16.00 to 19.86 g/cm^3. Plutonium also exhibits four ionic valence states in aqueous solutions: Pu^{+3}(blue lavender), Pu^{+4} (yellow brown), PuO$^+$ (pink?), and PuO^{+2} (pink orange). The ion PuO$^+$ is unstable in aqueous solutions, disproportionating into Pu^{+4} and PuO^{+2}. The Pu^{+4} thus formed, however, oxidizes the PuO$^+$ into PuO^{+2}, itself being reduced to Pu^{+3}, giving finally Pu^{+3} and PuO^{+2}. Plutonium forms binary compounds with oxygen: PuO, PuO$_2$, and intermediate oxides of variable composition; with the halides: PuF$_3$, PuF$_4$, PuCl$_3$, PuBr$_3$, PuI$_3$; with carbon, nitrogen, and silicon: PuC, PuN, PuSi$_2$. Oxyhalides are also well known: PuOCl, PuOBr, PuOI. Because of the high rate of emission of alpha particles and the element being specifically absorbed by bone marrow, plutonium, as well as all of the other transuranium elements except neptunium, are radiological poisons and must be handled with very special equipment and precautions. Plutonium is a very dangerous radiological hazard. Precautions must also be taken to prevent the unintentional formation of a critical mass. Plutonium in liquid solution is more likely to become critical than solid plutonium. The shape of the mass must also be considered where criticality is concerned. Plutonium-238 is available to authorized users from the O.R.N.L. at a cost of about $7.50/mg (97%) plus packing costs of $1250 per package.

Polonium — (Poland, native country of Mme. Curie), Po; at. wt. (209); at. no. 84; m.p. 254°C; b.p. 962°C; sp. gr. (alpha modification) 9.32; valence –2, 0, +2, +3(?), +4, and +6. Polonium was the first element discovered by Mme. Curie in 1898, while seeking the cause of radioactivity of pitchblende from Joachimsthal, Bohemia. The electroscope showed it separating with bismuth. Polonium is also called Radium F. Polonium is a very rare natural element. Uranium ores contain only about 100 μg of the element per ton. Its abundance is only about 0.2% of that of radium. In 1934, it was found that when natural bismuth (^{209}Bi) was bombarded by neutrons, ^{210}Bi, the parent of polonium, was obtained. Milligram amounts of polonium may now be prepared this way, by using the high neutron fluxes of nuclear reactors. Polonium-210 is a low-melting, fairly volatile metal, 50% of which is vaporized in air in 45 hours at 55°C. It is an alpha emitter with a half-life of 138.39 days. A milligram emits as many alpha particles as 5 g of radium. The energy released by its decay is so large (140 W/g) that a capsule containing about half a gram reaches a temperature above 500°C. The capsule also presents a contact gamma-ray dose rate of 0.012 Gy/h. A few curies (1 curie = 3.7×10^{10} Bq) of polonium exhibit a blue glow, caused by excitation of the surrounding gas. Because almost all alpha radiation is stopped within the solid source and its container, giving up its energy, polonium has attracted attention for uses as a lightweight heat source for thermoelectric power in space satellites. Thirty six isotopes and isomers of polonium are known, with atomic masses ranging from 192 to 218. All are radioactive. Polonium-210 is the most readily available. Isotopes of mass 209 (half-life 102 years) and mass 208 (half-life 2.9 years) can be prepared by alpha, proton, or deuteron bombardment of lead or bismuth in a cyclotron, but these are expensive to produce. Metallic polonium has been prepared from polonium hydroxide and some other polonium compounds in the presence of concentrated aqueous or anhydrous liquid ammonia. Two allotropic modifications are known to exist. Polonium is readily dissolved in dilute acids, but is only slightly soluble in alkalis. Polonium salts of organic acids char rapidly; halide amines are reduced to the metal. Polonium can be mixed or alloyed with beryllium to provide a source of neutrons. It has been used in devices for eliminating static charges in textile mills, etc.; however, beta sources are more commonly used and are less dangerous. It is also used on brushes for removing dust from photographic films. The polonium for these is carefully sealed and controlled, minimizing hazards to the user. Polonium-210 is very dangerous to handle in even milligram or microgram amounts, and special equipment and strict control is necessary. Damage arises from the complete absorption of the energy of the alpha particle into tissue. The maximum permissible body burden for ingested polonium is only 0.03 μCi, which represents a particle weighing only 6.8×10^{-12} g. Weight for weight it is about 2.5×10^{11} times as toxic as hydrocyanic acid. The maximum allowable concentration for soluble polonium compounds in air is about 2×10^{11} μCi/cm^3. Polonium-209 is available commercially on special order from the Oak Ridge National Laboratory at a cost of $3195/μCi.

Potassium — (English, *potash* — pot ashes; L. *kalium*, Arab. *qali*, alkali), K; at. wt. 39.0983(1); at. no. 19; m.p. 63.28°C; b.p. 759°C; sp. gr. 0.862 (20°C); valence 1. Discovered in 1807 by Davy, who obtained it from caustic potash (KOH); this was the first metal isolated by electrolysis. The metal is the seventh most abundant and makes up about 2.4% by weight of the earth's crust. Most potassium minerals are insoluble and the metal is obtained from them only with great difficulty. Certain minerals, however, such as *sylvite, carnallite, langbeinite,* and *polyhalite* are found in ancient lake and sea beds and form rather extensive deposits from which potassium and its salts can readily be obtained. Potash is mined in Germany, New Mexico, California, Utah, and elsewhere. Large deposits of potash, found at a depth of some 1000 m in Saskatchewan, promise to be important in coming years. Potassium is also found in the ocean, but is present only in relatively small amounts, compared to sodium. The greatest demand for potash has been in its use for fertilizers. Potassium is an essential constituent for plant growth and it is found in most soils. Potassium is never found free in nature, but is obtained by electrolysis of the hydroxide, much in the same manner as prepared by Davy. Thermal methods also are commonly used to produce potassium (such as by reduction of potassium compounds with CaC_2, C, Si, or Na). It is one of the most reactive and electropositive of metals. Except for lithium, it is the lightest known metal. It is soft, easily cut with a knife, and is silvery in appearance immediately after a fresh surface is exposed. It rapidly oxidizes in air and should be preserved in a mineral oil. As with other metals of the alkali group, it decomposes in water with the evolution of hydrogen. It catches fire spontaneously on water. Potassium and its salts impart a violet color to flames. Twenty one isotopes, one of which is an isomer, of potassium are known. Ordinary potassium is composed of three isotopes, one of which is ^{40}K (0.0117%), a radioactive isotope with a half-life of 1.28×10^9 years. The radioactivity presents no appreciable hazard. An alloy of sodium and potassium (NaK) is used as a heat-transfer medium. Many potassium salts are of utmost importance, including the hydroxide, nitrate, carbonate, chloride, chlorate, bromide, iodide, cyanide, sulfate, chromate, and dichromate. Metallic potassium is available commercially for about $650/kg (98% purity) or $10/g (99.95% purity).

Praseodymium — (Gr. *prasios,* green, and *didymos,* twin), Pr; at. wt. 140.90765(3); at. no. 59; m.p. 931°C; b.p. 3520°C; sp. gr. 6.773; valence 3. In 1841 Mosander extracted the rare earth *didymia* from *lanthana;* in 1879, Lecoq de Boisbaudran isolated a new earth, *samaria,* from didymia obtained from the mineral *samarskite.* Six years later, in 1885, von Welsbach separated didymia into two others, *praseodymia* and *neodymia,* which gave salts of different colors. As with other rare earths, compounds of these elements in solution have distinctive sharp spectral absorption bands or lines, some of which are only a few Angstroms wide. The element occurs along with other rare-earth elements in a variety of minerals. *Monazite* and *bastnasite* are the two principal commercial sources of the rare-earth metals. Ion-exchange and solvent extraction techniques have led to much easier isolation of the rare earths and the cost has dropped greatly in the past few years. Thirty six isotopes and isomers are now recognized. Praseodymium can be prepared by several methods, such as by calcium reduction of the anhydrous chloride of fluoride. Misch metal, used in making cigarette lighters, contains about 5% praseodymium metal. Praseodymium is soft, silvery, malleable, and ductile. It was prepared in relatively pure form in 1931. It is somewhat more resistant to corrosion in air than europium, lanthanum, cerium, or neodymium, but it does develop a green oxide coating that spalls off when exposed to air. As with other rare-earth metals it should be kept under a light mineral oil or sealed in plastic. The rare-earth oxides, including Pr_2O_3, are among the most refractory substances known. Along with other rare earths, it is widely used as a core material for carbon arcs used by the motion picture industry for studio lighting and projection. Salts of praseodymium are used to color glasses and enamels; when mixed with certain other materials, praseodymium produces an intense and unusually clean yellow color in glass. Didymium glass, of which praseodymium is a component, is a colorant for welder's goggles. The metal (99.9% pure) is priced at about $2/g.

Promethium — (*Prometheus,* who, according to mythology, stole fire from heaven), Pm; at. no. 61; at. wt. (145); m.p. 1042°C; b.p. 3000°C (est.); sp. gr. 7.264 (25°C); valence 3. In 1902 Branner predicted the existence of an element between neodymium and samarium, and this was confirmed by Moseley in 1914. Unsuccessful searches were made for this predicted element over two decades, and various investigators proposed the names "illinium", "florentium", and "cyclonium" for this element. In 1941, workers at Ohio State University irradiated neodymium and praseodymium with neutrons, deuterons, and alpha particles, resp., and produced several new radioactivities, which most likely were those of element 61. Wu and Segre, and Bethe, in 1942, confirmed the formation; however, chemical proof of the production of element 61 was lacking because of the difficulty in separating the rare earths from each other at that time. In 1945, Marinsky, Glendenin, and Coryell made the first chemical identification by use of ion-exchange chromatography. Their work was done by fission of uranium and by neutron bombardment of neodymium. These investigators named the newly discovered element. Searches for the element on earth have been fruitless, and it now appears that promethium is completely missing from the earth's crust. Promethium, however, has been reported to be in the spectrum of the star HR^{465} in Andromeda. This element is being formed recently near the star's surface, for no known isotope of promethium has a half-life longer than 17.7 years. Thirty five isotopes and isomers of promethium, with atomic masses from 130 to 158 are now known. Promethium-145, with a half-life of 17.7 years, is the most useful. Promethium-145 has a specific activity of 940 Ci/g. It is a soft beta emitter; although no gamma rays are emitted, X-radiation can be generated when beta particles impinge on elements of a high atomic number, and great care must be taken in handling it. Promethium salts luminesce in the dark with a pale blue or greenish glow, due to their high radioactivity. Ion-exchange methods led to the preparation of about 10 g of promethium from atomic reactor fuel processing wastes in early 1963. Little is yet generally known about the properties of metallic promethium. Two allotropic modifications exist. The element has applications as a beta source for thickness gages, and it can be absorbed by a phosphor to produce light. Light produced in this manner can be used for signs or signals that require dependable operation; it can be used as a nuclear-powered battery by capturing light in photocells which convert it into electric current. Such a battery, using ^{147}Pm, would have a useful life of about 5 years. It is being used for fluorescent lighting starter sand coatings for self-luminous watch dials. Promethium shows promise as a portable X-ray source, and it may become useful as a heat source to provide auxiliary power for space probes and satellites. More than 30 promethium compounds have been prepared. Most are colored. Promethium-147 is available upon special order from the Idaho National Engineering Laboratory, Idaho Falls, ID, or from the Westinghouse Hanford Co., Richland, WA.

Protactinium — (Gr. *protos,* first), Pa; at. wt. 231.03588(2); at. no. 91; m.p. 1572°C; sp. gr. 15.37 (calc.); valence 4 or 5. The first isotope of element 91 to be discovered was ^{234}Pa, also known as UX$_2$, a short-lived member of the naturally occurring ^{238}U decay series. It was identified by K. Fajans and O. H. Gohring in 1913 and they named the new element *brevium.* When the longer-lived isotope ^{231}Pa was identified by Hahn and Meitner in 1918, the name protactinium was adopted as being more consistent with the characteristics of the most abundant isotope. Soddy, Cranson, and Fleck were also active in this work. The name *protoactinium* was shortened to *protactinium* in 1949. In 1927, Grosse prepared 2 mg of a white powder, which was shown to be Pa_2O_5. Later, in 1934, from 0.1 g of pure Pa_2O_5 he isolated the element by two methods, one of which was by converting the oxide to an iodide and "cracking" it in a high vacuum by an electrically heated filament by the reaction

$$2PaI_5 \rightarrow 2Pa + 5I_2$$

Protactinium has a bright metallic luster which it retains for some time in air. The element occurs in *pitchblende* to the extent of about 1 part ^{231}Pa to

10 million of ore. Ores from Zaire have about 3 ppm. Protactinium has twenty four isotopes and isomers, the most common of which is ^{231}Pr with a half-life of 32,500 years. A number of protactinium compounds are known, some of which are colored. The element is superconductive below 1.4 K. The element is a dangerous toxic material and requires precautions similar to those used when handling plutonium. In 1959 and 1961, it was announced that the Great Britain Atomic Energy Authority extracted by a 12-stage process 125 g of 99.9% protactinium, the world's only stock of the metal for many years to come. The extraction was made from 60 tons of waste material at a cost of about $500,000. Protactinium is one of the rarest and most expensive naturally occurring elements.

Radium — (L. *radius*, ray), Ra; at. wt. (226); at. no. 88; m.p. 700°C; b.p. 1140°C; sp. gr. 5; valence 2. Radium was discovered in 1898 by M. and Mme. Curie in the *pitchblende* or *uraninite* of North Bohemia, where it occurs. There is about 1 g of radium in 7 tons of pitchblende. The element was isolated in 1911 by Mme. Curie and Debierne by the electrolysis of a solution of pure radium chloride, employing a mercury cathode; on distillation in an atmosphere of hydrogen this amalgam yielded the pure metal. Originally, radium was obtained from the rich pitchblende ore found at Joachimsthal, Bohemia. The *carnotite* sands of Colorado furnish some radium, but richer ores are found in the Republic of Zaire and the Great Bear Lake region of Canada. Radium is present in all uranium minerals, and could be extracted, if desired, from the extensive wastes of uranium processing. Large uranium deposits are located in Ontario, New Mexico, Utah, Australia, and elsewhere. Radium is obtained commercially as the bromide or chloride; it is doubtful if any appreciable stock of the isolated element now exists. The pure metal is brilliant white when freshly prepared, but blackens on exposure to air, probably due to formation of the nitride. It exhibits luminescence, as do its salts; it decomposes in water and is somewhat more volatile than barium. It is a member of the alkaline-earth group of metals. Radium imparts a carmine red color to a flame. Radium emits alpha, beta, and gamma rays and when mixed with beryllium produce neutrons. One gram of ^{226}Ra undergoes 3.7×10^{10} disintegrations per s. The *curie (Ci)* is defined as that amount of radioactivity which has the same disintegration rate as 1 g of ^{226}Ra. Twenty nine isotopes and one isomer are now known; radium 226, the common isotope, has a half-life of 1599 years. One gram of radium produces about 0.0001 ml (stp) of emanation, or radon gas, per day. This is pumped from the radium and sealed in minute tubes, which are used in the treatment of cancer and other diseases. One gram of radium yields about 4186 kJ per year. Radium is used in producing self-luminous paints, neutron sources, and in medicine for the treatment of disease. Some of the more recently discovered radioisotopes, such as ^{60}Co, are now being used in place of radium. Some of these sources are much more powerful, and others are safer to use. Radium loses about 1% of its activity in 25 years, being transformed into elements of lower atomic weight. Lead is a final product of disintegration. Stored radium should be ventilated to prevent build-up of radon. Inhalation, injection, or body exposure to radium can cause cancer and other body disorders. The maximum permissible burden in the total body for ^{226}Ra is 7400 becquerel.

Radon — (from *radium*; called *niton* at first, L. *nitens*, shining), Rn; at. wt. (222); at. no. 86; m.p. –71°C; b.p. –61.7°C; density of gas 9.73 g/l; sp. gr. liquid 4.4 at –62°C, solid 4; valence usually 0. The element was discovered in 1900 by Dorn, who called it *radium emanation*. In 1908 Ramsay and Gray, who named it *niton*, isolated the element and determined its density, finding it to be the heaviest known gas. It is essentially inert and occupies the last place in the zero group of gases in the Periodic Table. Since 1923, it has been called radon. Thirty nine isotopes and isomers are known. Radon-222, coming from radium, has a half-life of 3.823 days and is an alpha emitter; Radon-220, emanating naturally from thorium and called *thoron*, has a half-life of 55.6 s and is also an alpha emitter. Radon-219 emanates from actinium and is called *actinon*. It has a half-life of 3.96 s and is also an alpha emitter. It is estimated that every square mile of soil to a depth of 6 inch contains about 1 g of radium, which releases radon in tiny amounts to the atmosphere. Radon is present in some spring waters, such as those at Hot Springs, Arkansas. On the average, one part of radon is present to 1×10^{21} part of air. At ordinary temperatures radon is a colorless gas; when cooled below the freezing point, radon exhibits a brilliant phosphorescence which becomes yellow as the temperature is lowered and orange-red at the temperature of liquid air. It has been reported that fluorine reacts with radon, forming radon fluoride. Radon clathrates have also been reported. Radon is still produced for therapeutic use by a few hospitals by pumping it from a radium source and sealing it in minute tubes, called seeds or needles, for application to patients. This practice has now been largely discontinued as hospitals can order the seeds directly from suppliers, who make up the seeds with the desired activity for the day of use. Radon is available at a cost of about $4/m Ci. Care must be taken in handling radon, as with other radioactive materials. The main hazard is from inhalation of the element and its solid daughters, which are collected on dust in the air. Good ventilation should be provided where radium, thorium, or actinium is stored to prevent build-up of this element. Radon build-up is a health consideration in uranium mines. Recently radon build-up in homes has been a concern. Many deaths from lung cancer are caused by radon exposure. In the U.S. it is recommended that remedial action be taken if the air in homes exceeds 4 pCi/l.

Rhenium — (L. *Rhenus*, Rhine), Re; at. wt. 186.207(1); at. no. 75; m.p. 3186°C; b.p. 5596°C; sp. gr. 21.02 (20°C); valence –1, +1, 2, 3, 4, 5, 6, 7. Discovery of rhenium is generally attributed to Noddack, Tacke, and Berg, who announced in 1925 they had detected the element in platinum ores and *columbite*. They also found the element in *gadolinite* and *molybdenite*. By working up 660 kg of molybdenite they were able in 1928 to extract 1 g of rhenium. The price in 1928 was $10,000/g. Rhenium does not occur free in nature or as a compound in a distinct mineral species. It is, however, widely spread throughout the earth's crust to the extent of about 0.001 ppm. Commercial rhenium in the U.S. today is obtained from molybdenite roaster-flue dusts obtained from copper-sulfide ores mined in the vicinity of Miami, Arizona, and elsewhere in Arizona and Utah. Some molybdenites contain from 0.002 to 0.2% rhenium. More than 150,000 troy ounces of rhenium are now being produced yearly in the United States. The total estimated Free World reserve of rhenium metal is 3500 tons. Natural rhenium is a mixture of two isotopes, one of which has a very long half-life. Thirty nine other unstable isotopes are recognized. Rhenium metal is prepared by reducing ammonium perrhenate with hydrogen at elevated temperatures. The element is silvery white with a metallic luster; its density is exceeded only by that of platinum, iridium, and osmium, and its melting point is exceeded only by that of tungsten and carbon. It has other useful properties. The usual commercial form of the element is a powder, but it can be consolidated by pressing and resistance-sintering in a vacuum or hydrogen atmosphere. This produces a compact shape in excess of 90% of the density of the metal. Annealed rhenium is very ductile, and can be bent, coiled, or rolled. Rhenium is used as an additive to tungsten and molybdenum-based alloys to impart useful properties. It is widely used for filaments for mass spectrographs and ion gages. Rhenium-molybdenum alloys are superconductive at 10 K. Rhenium is also used as an electrical contact material as it has good wear resistance and withstands arc corrosion. Thermocouples made of Re-W are used for measuring temperatures up to 2200°C, and rhenium wire has been used in photoflash lamps for photography. Rhenium catalysts are exceptionally resistant to poisoning from nitrogen, sulfur, and phosphorus, and are used for hydrogenation of fine chemicals, hydrocracking, reforming, and disproportionation of olefins. Rhenium costs about $12/g (99.99% pure). Little is known of its toxicity; therefore, it should be handled with care until more data are available.

Rhodium — (Gr. *rhodon*, rose), Rh; at. wt. 102.90550(3); at. no. 45; m.p. 1964 ± 3°C; b.p. 3695 ± 100°C; sp. gr. 12.41 (20°C); valence 2, 3, 4, 5, and 6. Wollaston discovered rhodium in 1803-4 in crude platinum ore he presumably obtained from South America. Rhodium occurs native with other platinum metals in river sands of the Urals and in North and South America. It is also found with other platinum metals in the copper-nickel sulfide ores of the Sudbury, Ontario region. Although the quantity occurring here is very small, the large tonnages of nickel processed make the recovery commercially feasible. The annual world production of rhodium is only 7 or 8 tons. The metal is silvery white and at red heat slowly changes in air to the sesquioxide. At higher temperatures it converts back to the element. Rhodium has a higher melting point and lower density than platinum. Its major use is as an alloying agent to harden platinum and palladium. Such alloys are used for furnace windings, thermocouple elements, bushings for glass fiber production, electrodes for aircraft spark plugs, and laboratory crucibles. It is useful as an electrical contact material as it has a low electrical resistance, a low and stable contact resistance, and is highly resistant to corrosion. Plated rhodium, produced by electroplating or evaporation, is exceptionally hard and is used for optical instruments. It has a high reflectance and is hard and durable. Rhodium is also used for jewelry, for decoration, and as a catalyst. Forty four isotopes and isomers are now known. Soluble salts should not exceed 0.01 mg/m³. Rhodium metal (powder) costs about $300/g (99.9% pure).

Rubidium — (L. *rubidus*, deepest red), Rb; at. wt. 85.4678(3); at. no. 37; m.p. 39.31°C; b.p. 688°C; sp. gr. (solid) 1.532 (20°C), (liquid) 1.475 (39°C); valence 1, 2, 3, 4. Discovered in 1861 by Bunsen and Kirchoff in the mineral *lepidolite* by use of the spectroscope. The element is much more abundant than was thought several years ago. It is now considered to be the 16th most abundant element in the earth's crust. Rubidium occurs in *pollucite, carnallite, leucite,* and *zinnwaldite,* which contains traces up to 1%, in the form of the oxide. It is found in lepidolite to the extent of about 1.5%, and is recovered commercially from this source. Potassium minerals, such as those found at Searles Lake, California, and potassium chloride recovered from brines in Michigan also contain the element and are commercial sources. It is also found along with cesium in the extensive deposits of *pollucite* at Bernic Lake, Manitoba. Rubidium can be liquid at room temperature. It is a soft, silvery-white metallic element of the alkali group and is the second most electropositive and alkaline element. It ignites spontaneously in air and reacts violently in water, setting fire to the liberated hydrogen. As with other alkali metals, it forms amalgams with mercury and it alloys with gold, cesium, sodium, and potassium. It colors a flame yellowish violet. Rubidium metal can be prepared by reducing rubidium chloride with calcium, and by a number of other methods. It must be kept under a dry mineral oil or in a vacuum or inert atmosphere. Thirty five isotopes and isomers of rubidium are known. Naturally occurring rubidium is made of two isotopes, ^{85}Rb and ^{87}Rb. Rubidium-87 is present to the extent of 27.83% in natural rubidium and is a beta emitter with a half-life of 4.9×10^{10} years. Ordinary rubidium is sufficiently radioactive to expose a photographic film in about 30 to 60 days. Rubidium forms four oxides: Rb_2O, Rb_2O_2, Rb_2O_3, Rb_2O_4. Because rubidium can be easily ionized, it has been considered for use in "ion engines" for space vehicles; however, cesium is somewhat more efficient for this purpose. It is also proposed for use as a working fluid for vapor turbines and for use in a thermoelectric generator using the magnetohydro-dynamic principle where rubidium ions are formed by heat at high temperature and passed through a magnetic field. These conduct electricity and act like an armature of a generator thereby generating an electric current. Rubidium is used as a getter in vacuum tubes and as a photocell component. It has been used in making special glasses. $RbAg_4I_5$ is important, as it has the highest room conductivity of any known ionic crystal. At 20°C its conductivity is about the same as dilute sulfuric acid. This suggests use in thin film batteries and other applications. The present cost in small quantities is about $20/g (99.8% pure).

Ruthenium — (L. *Ruthenia*, Russia), Ru; at. wt. 101.07(2); at. no. 44, m.p. 2334°C; b.p. 4150°C; sp. gr. 12.41 (20°C); valence 0, 1, 2, 3, 4, 5, 6, 7, 8. Berzelius and Osann in 1827 examined the residues left after dissolving crude platinum from the Ural mountains in *aqua regia*. While Berzelius found no unusual metals, Osann thought he found three new metals, one of which he named ruthenium. In 1844 Klaus, generally recognized as the discoverer, showed that Osann's ruthenium oxide was very impure and that it contained a new metal. Klaus obtained 6 g of ruthenium from the portion of crude platinum that is insoluble in *aqua regia*. A member of the platinum group, ruthenium occurs native with other members of the group in ores found in the Ural mountains and in North and South America. It is also found along with other platinum metals in small but commercial quantities in *pentlandite* of the Sudbury, Ontario, nickel-mining region, and in *pyroxinite* deposits of South Africa. Natural ruthenium contains seven isotopes. Twenty one other isotopes and isomers are known, all of which are radioactive. The metal is isolated commercially by a complex chemical process, the final stage of which is the hydrogen reduction of ammonium ruthenium chloride, which yields a powder. The powder is consolidated by powder metallurgy techniques or by argon-arc welding. Ruthenium is a hard, white metal and has four crystal modifications. It does not tarnish at room temperatures, but oxidizes in air at about 800°C. The metal is not attacked by hot or cold acids or *aqua regia*, but when potassium chlorate is added to the solution, it oxidizes explosively. It is attacked by halogens, hydroxides, etc. Ruthenium can be plated by electrodeposition or by thermal decomposition methods. The metal is one of the most effective hardeners for platinum and palladium, and is alloyed with these metals to make electrical contacts for severe wear resistance. A ruthenium-molybdenum alloy is said to be superconductive at 10.6 K. The corrosion resistance of titanium is improved a hundredfold by addition of 0.1% ruthenium. It is a versatile catalyst. Hydrogen sulfide can be split catalytically by light using an aqueous suspension of CdS particles loaded with ruthenium dioxide. It is thought this may have application to removal of H_2S from oil refining and other industrial processes. Compounds in at least eight oxidation states have been found, but of these, the +2. +3. and +4 states are the most common. Ruthenium tetroxide, like osmium tetroxide, is highly toxic. In addition, it may explode. Ruthenium compounds show a marked resemblance to those of osmium. The metal is priced at about $30/g (99.95% pure).

Rutherfordium — See Element 104.

Samarium — (*Samarskite* a mineral), Sm; at. wt. 150.36(3); at. no. 62; m.p. 1074°C; b.p. 1794°C; sp. gr (α) 7.520 (25°C); valence 2 or 3. Discovered spectroscopically by its sharp absorption lines in 1879 by Lecoq de Boisbaudran in the mineral *samarskite*, named in honor of a Russian mine official, Col. Samarski. Samarium is found along with other members of the rare-earth elements in many minerals, including *monazite* and *bastnasite*, which are commercial sources. It occurs in monazite to the extent of 2.8%. While *misch metal* containing about 1% of samarium metal, has long been used, samarium has not been isolated in relatively pure form until recent years. Ion-exchange and solvent extraction techniques have recently simplified separation of the rare earths from one another; more recently, electrochemical deposition, using an electrolytic solution of lithium citrate and a mercury electrode, is said to be a simple, fast, and highly specific way to separate the rare earths. Samarium metal can be produced by reducing the oxide with barium or lanthanum. Samarium has a bright silver luster and is reasonably stable in air. Three crystal modifications of the metal exist, with transformations at 734 and 922°C. The metal ignites in air at about 150°C. Thirty two isotopes and isomers of samarium are now recognized. Natural samarium is a mixture of seven isotopes, three of which are unstable but have long half-lives. Samarium, along with other rare earths, is used for carbon-arc lighting for the motion picture industry. The sulfide has excellent high-temperature stability and good thermoelectric efficiencies up to 1100°C. $SmCo_5$ has been used in making a new permanent magnet material with the highest resistance to demagnetization of any known material. It is said to have an intrinsic coercive force as high as 2200 kA/m. Samarium oxide has been used in optical glass to absorb the infrared.

Samarium is used to dope calcium fluoride crystals for use in optical masers or lasers. Compounds of the metal act as sensitizers for phosphors excited in the infrared; the oxide exhibits catalytic properties in the dehydration and dehydrogenation of ethyl alcohol. It is used in infrared absorbing glass and as a neutron absorber in nuclear reactors. The metal is priced at about $2/g (99.9%). Little is known of the toxicity of samarium; therefore, it should be handled carefully.

Scandium — (L. *Scandia*, Scandinavia), Sc; at. wt. 44.955910(9); at. no. 21; m.p. 1541°C; b.p. 2836°C; sp. gr. 2.989 (25°C); valence 3. On the basis of the Periodic System, Mendeleev predicted the existence of *ekaboron*, which would have an atomic weight between 40 of calcium and 48 of titanium. The element was discovered by Nilson in 1878 in the minerals *euxenite* and *gadolinite*, which had not yet been found anywhere except in Scandinavia. By processing 10 kg of euxenite and other residues of rare-earth minerals, Nilson was able to prepare about 2 g of scandium oxide of high purity. Cleve later pointed out that Nilson's scandium was identical with Mendeleev's ekaboron. Scandium is apparently a much more abundant element in the sun and certain stars than here on earth. It is about the 23rd most abundant element in the sun, compared to the 50th most abundant on earth. It is widely distributed on earth, occurring in very minute quantities in over 800 mineral species. The blue color of beryl (aquamarine variety) is said to be due to scandium. It occurs as a principal component in the rare mineral *thortveitite*, found in Scandinavia and Malagasy. It is also found in the residues remaining after the extraction of tungsten from Zinnwald *wolframite*, and in *wiikite* and *bazzite*. Most scandium is presently being recovered from *thortveitite* or is extracted as a by-product from uranium mill tailings. Metallic scandium was first prepared in 1937 by Fischer, Brunger, and Grieneisen, who electrolyzed a eutectic melt of potassium, lithium, and scandium chlorides at 700 to 800°C. Tungsten wire and a pool of molten zinc served as the electrodes in a graphite crucible. Pure scandium is now produced by reducing scandium fluoride with calcium metal. The production of the first pound of 99% pure scandium metal was announced in 1960. Scandium is a silver-white metal which develops a slightly yellowish or pinkish cast upon exposure to air. It is relatively soft, and resembles yttrium and the rare-earth metals more than it resembles aluminum or titanium. It is a very light metal and has a much higher melting point than aluminum, making it of interest to designers of spacecraft. Scandium is not attacked by a 1:1 mixture of conc. HNO_3 and 48% HF. Scandium reacts rapidly with many acids. Nineteen isotopes and isomers of scandium are recognized. The metal is expensive, costing about $120/g with a purity of about 99.9%. Scandium oxide costs about $40/g. About 20 kg of scandium (as Sc_2O_3) are now being used yearly in the U.S. to produce high-intensity lights, and the radioactive isotope ^{46}Sc is used as a tracing agent in refinery crackers for crude oil, etc. Scandium iodide added to mercury vapor lamps produces a highly efficient light source resembling sunlight, which is important for indoor or night-time color TV. Little is yet known about the toxicity of scandium; therefore, it should be handled with care.

Seaborgium — See Element 106.

Selenium — (Gr. *Selene*, moon), Se; at. wt. 78.96(3); at. no. 34; m.p. (gray) 221°C; b.p. (gray) 685°C; sp. gr. (gray) 4.79, (vitreous) 4.28; valence −2, +4, or +6. Discovered by Berzelius in 1817, who found it associated with tellurium, named for the earth. Selenium is found in a few rare minerals, such as *crooksite* and *clausthalite*. In years past it has been obtained from flue dusts remaining from processing copper sulfide ores, but the anode muds from electrolytic copper refineries now provide the source of most of the world's selenium. Selenium is recovered by roasting the muds with soda or sulfuric acid, or by smelting them with soda and niter. Selenium exists in several allotropic forms. Three are generally recognized, but as many as six have been claimed. Selenium can be prepared with either an amorphous or crystalline structure. The color of amorphous selenium is either red, in powder form, or black, in vitreous form. Crystalline monoclinic selenium is a deep red; crystalline hexagonal selenium, the most stable variety, is a metallic gray. Natural selenium contains six stable isotopes. Twenty four other isotopes and isomers have been characterized. The element is a member of the sulfur family and resembles sulfur both in its various forms and in its compounds. Selenium exhibits both photovoltaic action, where light is converted directly into electricity, and photoconductive action, where the electrical resistance decreases with increased illumination. These properties make selenium useful in the production of photocells and exposure meters for photographic use, as well as solar cells. Selenium is also able to convert a.c. electricity to d.c., and is extensively used in rectifiers. Below its melting point selenium is a p-type semiconductor and is finding many uses in electronic and solid-state applications. It is used in Xerography for reproducing and copying documents, letters, etc. It is used by the glass industry to decolorize glass and to make ruby-colored glasses and enamels. It is also used as a photographic toner, and as an additive to stainless steel. Elemental selenium has been said to be practically nontoxic and is considered to be an essential trace element; however, hydrogen selenide and other selenium compounds are extremely toxic, and resemble arsenic in their physiological reactions. Hydrogen selenide in a concentration of 1.5 ppm is intolerable to man. Selenium occurs in some soils in amounts sufficient to produce serious effects on animals feeding on plants, such as locoweed, grown in such soils. Selenium is priced at about $150/kg. It is also available in high-purity form at a cost of about $250/kg (99.999%).

Silicon — (L. *silex, silicis*, flint), Si; at. wt. 28.0855(3); at. no. 14; m.p. 1414°C; b.p. 3265°C; sp. gr. 2.33 (25°C); valence 4. Davy in 1800 thought silica to be a compound and not an element; later in 1811, Gay Lussac and Thenard probably prepared impure amorphous silicon by heating potassium with silicon tetrafluoride. Berzelius, generally credited with the discovery, in 1824 succeeded in preparing amorphous silicon by the same general method as used earlier, but he purified the product by removing the fluosilicates by repeated washings. Deville in 1854 first prepared crystalline silicon, the second allotropic form of the element. Silicon is present in the sun and stars and is a principal component of a class of meteorites known as "aerolites". It is also a component of *tektites*, a natural glass of uncertain origin. Natural silicon contains three isotopes. Fourteen other radioactive isotopes are recognized. Silicon makes up 25.7% of the earth's crust, by weight, and is the second most abundant element, being exceeded only by oxygen. Silicon is not found free in nature, but occurs chiefly as the oxide and as silicates. *Sand, quartz, rock crystal, amethyst, agate, flint, jasper,* and *opal* are some of the forms in which the oxide appears. *Granite, hornblende, asbestos, feldspar, clay mica,* etc. are but a few of the numerous silicate minerals. Silicon is prepared commercially by heating silica and carbon in an electric furnace, using carbon electrodes. Several other methods can be used for preparing the element. Amorphous silicon can be prepared as a brown powder, which can be easily melted or vaporized. Crystalline silicon has a metallic luster and grayish color. The Czochralski process is commonly used to produce single crystals of silicon used for solid-state or semiconductor devices. Hyperpure silicon can be prepared by the thermal decomposition of ultra-pure trichlorosilane in a hydrogen atmosphere, and by a vacuum float zone process. This product can be doped with boron, gallium, phosphorus, or arsenic to produce silicon for use in transistors, solar cells, rectifiers, and other solid-state devices which are used extensively in the electronics and space-age industries. Hydrogenated amorphous silicon has shown promise in producing economical cells for converting solar energy into electricity. Silicon is a relatively inert element, but it is attacked by halogens and dilute alkali. Most acids except hydrofluoric, do not affect it. Silicones are important products of silicon. They may be prepared by hydrolyzing a silicon organic chloride, such as dimethyl silicon chloride. Hydrolysis and condensation of various substituted chlorosilanes can be used to produce a very great number of polymeric products, or silicones, ranging from liquids to hard, glasslike solids with many useful properties. Elemental silicon transmits more than 95% of all wavelengths of infrared, from 1.3 to 6.7 µm. Silicon is one of man's most useful elements. In the form of sand

and clay it is used to make concrete and brick; it is a useful refractory material for high-temperature work, and in the form of silicates it is used in making enamels, pottery, etc. Silica, as sand, is a principal ingredient of glass, one of the most inexpensive of materials with excellent mechanical, optical, thermal, and electrical properties. Glass can be made in a very great variety of shapes, and is used as containers, window glass, insulators, and thousands of other uses. Silicon tetrachloride can be used to iridize glass. Silicon is important in plant and animal life. Diatoms in both fresh and salt water extract silica from the water to build up their cell walls. Silica is present in ashes of plants and in the human skeleton. Silicon is an important ingredient in steel; silicon carbide is one of the most important abrasives and has been used in lasers to produce coherent light of 4560 Å. Regular grade silicon (99.5%) costs about $140/kg. Silicon 99.96% pure costs about $250/kg; hyperpure silicon may cost as much as $400/kg. Miners, stonecutters, and other engaged in work where siliceous dust is breathed in large quantities often develop a serious lung disease known as *silicosis*.

Silver — (Anglo-Saxon, *Seolfor siolfur*), Ag (L. argentum), at. wt. 107.8682(2); at. no. 47; f.p. 961.78°C; b.p. 2162°C; sp. gr. 10.50 (20°C); valence 1, 2. Silver has been known since ancient times. It is mentioned in Genesis. Slag dumps in Asia Minor and on islands in the Aegean Sea indicate that man learned to separate silver from lead as early as 3000 B.C. Silver occurs native and in ores such as *argentite* (Ag_2S) and *horn silver* (AgCl); lead, lead-zinc, copper, gold, and copper-nickel ores are principal sources. Mexico, Canada, Peru, and the U.S. are the principal silver producers in the western hemisphere. Silver is also recovered during electrolytic refining of copper. Commercial fine silver contains at least 99.9% silver. Purities of 99.999+% are available commercially. Pure silver has a brilliant white metallic luster. It is a little harder than gold and is very ductile and malleable, being exceeded only by gold and perhaps palladium. Pure silver has the highest electrical and thermal conductivity of all metals, and possesses the lowest contact resistance. It is stable in pure air and water, but tarnishes when exposed to ozone, hydrogen sulfide, or air containing sulfur. The alloys of silver are important. Sterling silver is used for jewelry, silverware, etc. where appearance is paramount. This alloy contains 92.5% silver, the remainder being copper or some other metal. Silver is of utmost importance in photography, about 30% of the U.S. industrial consumption going into this application. It is used for dental alloys. Silver is used in making solder and brazing alloys, electrical contacts, and high capacity silver-zinc and silver-cadmium batteries. Silver paints are used for making printed circuits. It is used in mirror production and may be deposited on glass or metals by chemical deposition, electrodeposition, or by evaporation. When freshly deposited, it is the best reflector of visible light known, but is rapidly tarnishes and loses much of its reflectance. It is a poor reflector of ultraviolet. Silver fulminate ($Ag_2C_2N_2O_2$), a powerful explosive, is sometimes formed during the silvering process. Silver iodide is used in seeding clouds to produce rain. Silver chloride has interesting optical properties as it can be made transparent; it also is a cement for glass. Silver nitrate, or *lunar caustic*, the most important silver compound, is used extensively in photography. While silver itself is not considered to be toxic, most of its salts are poisonous. Natural silver contains two stable isotopes. Forty nine other radioactive isotopes and isomers are known. Silver compounds can be absorbed in the circulatory system and reduced silver deposited in the various tissues of the body. A condition, known as *argyria*, results, with a greyish pigmentation of the skin and mucous membranes. Silver has germicidal effects and kills many lower organisms effectively without harm to higher animals. Silver for centuries has been used traditionally for coinage by many countries of the world. In recent times, however, consumption of silver has at times greatly exceeded the output. In 1939, the price of silver was fixed by the U.S. Treasury at 71¢/troy oz., and at 90.5¢/troy oz. in 1946. In November 1961 the U.S. Treasury suspended sales of nonmonetized silver, and the price stabilized for a time at about $1.29, the melt-down value of silver U.S. coins. The Coinage Act of 1965 authorized a change in the metallic composition of the three U.S. subsidiary denominations to clad or composite type coins. This was the first change in U.S. coinage since the monetary system was established in 1792. Clad dimes and quarters are made of an outer layer of 75% Cu and 25% Ni bonded to a central core of pure Cu. The composition of the one- and five-cent pieces remains unchanged. One-cent coins are 95% Cu and 5% Zn. Five-cent coins are 75% Cu and 25% Ni. Old silver dollars are 90% Ag and 10% Cu. Earlier subsidiary coins of 90% Ag and 10% Cu officially were to circulate alongside the clad coins; however, in practice they have largely disappeared (Gresham's Law), as the value of the silver is now greater than their exchange value. Silver coins of other countries have largely been replaced with coins made of other metals. On June 24, 1968, the U.S. Government ceased to redeem U.S. Silver Certificates with silver. Since that time, the price of silver has fluctuated widely. As of January 1996, the price of silver was about $5.30/troy oz. (17¢/g); however the price has fluctuated considerably due to market instability.

Sodium — (English, *soda*; Medieval Latin, *sodanum*, headache remedy), Na (L. *natrium*); at. wt. 22.989768(6); at. no. 11; m.p. 97.72 ± 0.03°C; b.p. 883°C; sp. gr. 0.971 (20°C); valence 1. Long recognized in compounds, sodium was first isolated by Davy in 1807 by electrolysis of caustic soda. Sodium is present in fair abundance in the sun and stars. The D lines of sodium are among the most prominent in the solar spectrum. Sodium is the sixth most abundant element on earth, comprising about 2.6% of the earth's crust; it is the most abundant of the alkali group of metals of which it is a member. The most common compound is sodium chloride, but it occurs in many other minerals, such as *soda niter, cryolite, amphibole, zeolite, sodalite*, etc. It is a very reactive element and is never found free in nature. It is now obtained commercially by the electrolysis of absolutely dry fused sodium chloride. This method is much cheaper than that of electrolyzing sodium hydroxide, as was used several years ago. Sodium is a soft, bright, silvery metal which floats on water, decomposing it with the evolution of hydrogen and the formation of the hydroxide. It may or may not ignite spontaneously on water, depending on the amount of oxide and metal exposed to the water. It normally does not ignite in air at temperatures below 115°C. Sodium should be handled with respect, as it can be dangerous when improperly handled. Metallic sodium is vital in the manufacture of sodamide and esters, and in the preparation of organic compounds. The metal may be used to improve the structure of certain alloys, to descale metal, to purify molten metals, and as a heat transfer agent. An alloy of sodium with potassium, NaK, is also an important heat transfer agent. Sodium compounds are important to the paper, glass, soap, textile, petroleum, chemical, and metal industries. Soap is generally a sodium salt of certain fatty acids. The importance of common salt to animal nutrition has been recognized since prehistoric times. Among the many compounds that are of the greatest industrial importance are common salt (NaCl), soda ash (Na_2CO_3), baking soda ($NaHCO_3$), caustic soda (NaOH), Chile saltpeter ($NaNO_3$), di- and tri-sodium phosphates, sodium thiosulfate (hypo, $Na_2S_2O_3 \cdot 5H_2O$), and borax ($Na_2B_4O_7 \cdot 10H_2O$). Seventeen isotopes of sodium are recognized. Metallic sodium is priced at about $250/kg (99.95%). On a volume basis, it is the cheapest of all metals. Sodium metal should be handled with great care. It should be kept in an inert atmosphere and contact with water and other substances with which sodium reacts should be avoided.

Strontium — (*Strontian*, town in Scotland), Sr; at. wt. 87.62(1); at. no. 38; m.p. 777°C; b.p. 1382°C; sp. gr. 2.54; valence 2. Isolated by Davy by electrolysis in 1808; however, Adair Crawford in 1790 recognized a new mineral (strontianite) as differing from other barium minerals (baryta). Strontium is found chiefly as *celestite* ($SrSO_4$) and *strontianite* ($SrCO_3$). The metal can be prepared by electrolysis of the fused chloride mixed with potassium chloride, or is made by reducing strontium oxide with aluminum in a vacuum at a temperature at which strontium distills off. Three allotropic forms of the metal exist, with transition points at 235 and 540°C. Strontium is softer than calcium and decomposes water more vigorously. It does not absorb nitrogen below 380°C. It should be kept under mineral oil to prevent oxidation. Freshly cut strontium has a silvery appearance, but rapidly turns a yellowish color with the formation of the oxide. The finely divided metal ignites spontaneously in air. Volatile strontium salts impart a beautiful crimson color to flames, and these salts are used in pyrotechnics and in the production of flares. Natural strontium is a mixture of four stable isotopes. Twenty six other unstable isotopes and isomers are known to exist. Of greatest importance is ^{90}Sr with a half-life of 29 years. It is a product of nuclear fallout and presents a health problem. This isotope is one of the best long-lived high-energy beta emitters known, and is used in SNAP (Systems for Nuclear Auxiliary Power) devices. These devices hold promise for use in space vehicles, remote weather stations, navigational buoys, etc., where a lightweight, long-lived, nuclear-electric power source is needed. The major use for strontium at present is in producing glass for color television picture tubes. It has also found use in producing ferrite magnets and in refining zinc. Strontium titanate is an interesting optical material as it has an extremely high refractive index and an optical dispersion greater than that of diamond. It has been used as a gemstone, but it is very soft. It does not occur naturally. Strontium metal (99% pure) costs about \$1/g while the metal, 99.95% pure, sells for about \$10/g.

Sulfur — (Sanskrit, *sulvere*; L. *sulphurium*), S; at. wt. 32.066(6); at. no. 16; m.p. 115.21°C; b.p. 444.60°C; sp. gr. (rhombic) 2.07, (monoclinic) 1.957 (20°C); valence 2, 4, or 6. Known to the ancients; referred to in Genesis as *brimstone*. Sulfur is found in meteorites. A dark area near the crater Aristarchus on the moon has been studied by R. W. Wood with ultraviolet light. This study suggests strongly that it is a sulfur deposit. Sulfur occurs native in the vicinity of volcanoes and hot springs. It is widely distributed in nature as *iron pyrites, galena, sphalerite, cinnabar, stibnite, gypsum, Epsom salts, celestite, barite,*etc. Sulfur is commercially recovered from wells sunk into the salt domes along the Gulf Coast of the U.S. It is obtained from these wells by the Frasch process, which forces heated water into the wells to melt the sulfur, which is then brought to the surface. Sulfur also occurs in natural gas and petroleum crudes and must be removed from these products. Formerly this was done chemically, which wasted the sulfur. New processes now permit recovery, and these sources promise to be very important. Large amounts of sulfur are being recovered from Alberta gas fields. Sulfur is a pale yellow, odorless, brittle solid, which is insoluble in water but soluble in carbon disulfide. In every state, whether gas, liquid or solid, elemental sulfur occurs in more than one allotropic form or modification; these present a confusing multitude of forms whose relations are not yet fully understood. Amorphous or "plastic" sulfur is obtained by fast cooling of the crystalline form. X-ray studies indicate that amorphous sulfur may have a helical structure with eight atoms per spiral. Crystalline sulfur seems to be made of rings, each containing eight sulfur atoms, which fit together to give a normal X-ray pattern. Seventeen isotopes of sulfur are now recognized. Four occur in natural sulfur, none of which is radioactive. A finely divided form of sulfur, known as *flowers of sulfur*, is obtained by sublimation. Sulfur readily forms sulfides with many elements. Sulfur is a component of black gunpowder, and is used in the vulcanization of natural rubber and a fungicide. It is also used extensively is making phosphatic fertilizers. A tremendous tonnage is used to produce sulfuric acid, the most important manufactured chemical. It is used in making sulfite paper and other papers, as a fumigant, and in the bleaching of dried fruits. The element is a good electrical insulator. Organic compounds containing sulfur are very important. Calcium sulfate, ammonium sulfate, carbon disulfide, sulfur dioxide, and hydrogen sulfide are but a few of the many other important compounds of sulfur. Sulfur is essential to life. It is a minor constituent of fats, body fluids, and skeletal minerals. Carbon disulfide, hydrogen sulfide, and sulfur dioxide should be handled carefully. Hydrogen sulfide in small concentrations can be metabolized, but in higher concentrations it quickly can cause death by respiratory paralysis. It is insidious in that it quickly deadens the sense of smell. Sulfur dioxide is a dangerous component in atmospheric air pollution. In 1975, University of Pennsylvania scientists reported synthesis of polymeric sulfur nitride, which has the properties of a metal, although it contains no metal atoms. The material has unusual optical and electrical properties. High-purity sulfur is commercially available in purities of 99.999+%, at a cost of about \$50/100 g.

Tantalum — (Gr. *Tantalos*, mythological character, father of *Niobe*), Ta; at. wt. 180.9479(1); at. no. 73; m.p. 3017°C; b.p. 5458 ± 100°C; sp. gr. 16.654; valence 2?, 3, 4?, or 5. Discovered in 1802 by Ekeberg, but many chemists thought niobium and tantalum were identical elements until Rose, in 1844, and Marignac, in1866, showed that niobic and tantalic acids were two different acids. The early investigators only isolated the impure metal. The first relatively pure ductile tantalum was produced by von Bolton in 1903. Tantalum occurs principally in the mineral *columbite-tantalite* (Fe, Mn)(Nb, Ta)$_2$O$_6$. Tantalum ores are found in Australia, Brazil, Mozambique, Thailand, Portugal, Nigeria, Zaire, and Canada. Separation of tantalum from niobium requires several complicated steps. Several methods are used to commercially produce the element, including electrolysis of molten potassium fluorotantalate, reduction of potassium fluorotantalate with sodium, or reacting tantalum carbide with tantalum oxide. Thirty four isotopes and isomers of tantalum are known to exist. Natural tantalum contains two isotopes, one of which is radioactive with a very long half-life. Tantalum is a gray, heavy, and very hard metal. When pure, it is ductile and can be drawn into fine wire, which is used as a filament for evaporating metals such as aluminum. Tantalum is almost completely immune to chemical attack at temperatures below 150°C, and is attacked only by hydrofluoric acid, acidic solutions containing the fluoride ion, and free sulfur trioxide. Alkalis attack it only slowly. At high temperatures, tantalum becomes much more reactive. The element has a melting point exceeded only by tungsten and rhenium. Tantalum is used to make a variety of alloys with desirable properties such as high melting point, high strength, good ductility, etc. Scientists at Los Alamos have produced a tantalum carbide graphite composite material, which is said to be one of the hardest materials ever made. The compound has a melting point of 3738°C. Tantalum has good "gettering" ability at high temperatures, and tantalum oxide films are stable and have good rectifying and dielectric properties. Tantalum is used to make electrolytic capacitors and vacuum furnace parts, which account for about 60% of its use. The metal is also widely used to fabricate chemical process equipment, nuclear reactors, and aircraft and missile parts. Tantalum is completely immune to body liquids and is a nonirritating metal. It has, therefore, found wide use in making surgical appliances. Tantalum oxide is used to make special glass with high index of refraction for camera lenses. The metal has many other uses. The price of (99.9%) tantalum is about \$900/kg. The metal of 99.995% purity sells for about \$2/g.

Technetium — (Gr. *technetos*, artificial), Tc; at. wt. (98); at. no. 43; m.p. 2157°C; b.p. 4265°C; sp. gr. 11.50 (calc.); valence 0, +2, +4, +5, +6, and +7. Element 43 was predicted on the basis of the periodic table, and was erroneously reported as having been discovered in 1925, at which time it was named *masurium*. The element was actually discovered by Perrier and Segre in Italy in 1937. It was found in a sample of molybdenum, which was bombarded by deuterons in the Berkeley cyclotron, and which E. Lawrence sent to these investigators. Technetium was the first element to be produced artificially. Since its discovery, searches for the element in terrestrial materials have been made without success. If it does exist, the concentration must be very small. Technetium has been found in the spectrum of S-, M-, and N-type stars, and its presence in stellar matter is leading to new theories of the production of heavy elements in the stars. Thirty one isotopes and isomers of technetium, with atomic masses ranging from 86 to 113, are known. ^{97}Tc has a half-life of 2.6×10^6 years. ^{98}Tc has a half-life of 4.2×10^6 years. The isomeric isotope ^{95m}Tc, with a half-life of 61 days, is useful for tracer work, as it produces energetic gamma rays. Technetium metal has been produced in kilogram quantities. The metal was first prepared by passing hydrogen gas at 1100°C over Tc$_2$S$_7$. It is now conveniently prepared by the reduction of ammonium pertechnetate with hydrogen. Technetium is a silvery-gray metal that tarnishes slowly in moist air. Until 1960, technetium was available only in small amounts and the price was

THE ELEMENTS (continued)

as high as \$2800/g. It is now commercially available to holders of O.R.N.L. permits at a price of \$74/g plus packing charges. The chemistry of technetium is said to be similar to that of rhenium. Technetium dissolves in nitric acid, aqua regia, and conc. sulfuric acid, but is not soluble in hydrochloric acid of any strength. The element is a remarkable corrosion inhibitor for steel. It is reported that mild carbon steels may be effectively protected by as little as 55 ppm of $KTcO_4$ in aerated distilled water at temperatures up to 250°C. This corrosion protection is limited to closed systems, since technetium is radioactive and must be confined. ^{99}Tc has a specific activity of 6.2×10^8 Bq/g. Activity of this level must not be allowed to spread. ^{99}Tc is a contamination hazard and should be handled in a glove box. The metal is an excellent superconductor at 11 K and below.

Tellurium — (L. *tellus*, earth), Te; at. wt. 127.60(3); at. no. 52; m.p. 449.51 ± 0.3°C; b.p. 988°C; sp. gr. 6.24 (20°C); valence 2, 4, or 6. Discovered by Muller von Reichenstein in 1782; named by Klaproth, who isolated it in 1798. Tellurium is occasionally found native, but is more often found as the telluride of gold (*calaverite*), and combined with other metals. It is recovered commercially from the anode muds produced during the electrolytic refining of blister copper. The U.S., Canada, Peru, and Japan are the largest Free World producers of the element. Crystalline tellurium has a silvery-white appearance, and when pure exhibits a metallic luster. It is brittle and easily pulverized. Amorphous tellurium is formed by precipitating tellurium from a solution of telluric or tellurous acid. Whether this form is truly amorphous, or made of minute crystals, is open to question. Tellurium is a p-type semiconductor, and shows greater conductivity in certain directions, depending on alignment of the atoms. Its conductivity increases slightly with exposure to light. It can be doped with silver, copper, gold, tin, or other elements. In air, tellurium burns with a greenish-blue flame, forming the dioxide. Molten tellurium corrodes iron, copper, and stainless steel. Tellurium and its compounds are probably toxic and should be handled with care. Workmen exposed to as little as 0.01 mg/m³ of air, or less, develop "tellurium breath," which has a garlic-like odor. Forty two isotopes and isomers of tellurium are known, with atomic masses ranging from 106 to 138. Natural tellurium consists of eight isotopes, two of which are radioactive with very long half-lives. Tellurium improves the machinability of copper and stainless steel, and its addition to lead decreases the corrosive action of sulfuric acid on lead and improves its strength and hardness. Tellurium is used as a basic ingredient in blasting caps, and is added to cast iron for chill control. Tellurium is used in ceramics. Bismuth telluride has been used in thermoelectric devices. Tellurium costs about 20¢/g, with a purity of about 99.5%. The metal with a purity of 99.9999% costs about \$2/g.

Terbium — (*Ytterby*, village in Sweden), Tb; at. wt. 158.92534(3); at. no. 65; m.p. 1356°C; b.p. 3230°C; sp. gr. 8.230; valence 3, 4. Discovered by Mosander in 1843. Terbium is a member of the lanthanide or "rare earth" group of elements. It is found in *cerite, gadolinite,* and other minerals along with other rare earths. It is recovered commercially from *monazite* in which it is present to the extent of 0.03%, from *xenotime*, and from *euxenite*, a complex oxide containing 1% of more of terbia. Terbium has been isolated only in recent years with the development of ion-exchange techniques for separating the rare-earth elements. As with other rare earths, it can be produced by reducing the anhydrous chloride or fluoride with calcium metal in a tantalum crucible. Calcium and tantalum impurities can be removed by vacuum remelting. Other methods of isolation are possible. Terbium is reasonably stable in air. It is a silver-gray metal, and is malleable, ductile, and soft enough to be cut with a knife. Two crystal modifications exist, with a transformation temperature of 1289°C. Forty one isotopes and isomers are recognized. The oxide is a chocolate or dark maroon color. Sodium terbium borate is used as a laser material and emits coherent light at 0.546 μm. Terbium is used to dope calcium fluoride, calcium tungstate, and strontium molybdate, used in solid-state devices. The oxide has potential application as an activator for green phosphors used in color TV tubes. It can be used with ZrO_2 as a crystal stabilizer of fuel cells which operate at elevated temperature. Few other uses have been found. The element is priced at about \$30/g (99.9%). Little is known of the toxicity of terbium. It should be handled with care as with other lanthanide elements.

Thallium — (Gr. *thallos*, a green shoot or twig), Tl; at. wt. 204.3833(2); at. no. 81; m.p. 304°C; b.p. 1473 ± 10°C; sp. gr. 11.85 (20°C); valence 1, or 3. Thallium was discovered spectroscopically in 1861 by Crookes. The element was named after the beautiful green spectral line, which identified the element. The metal was isolated both by Crookes and Lamy in 1862 about the same time. Thallium occurs in *crooksite, lorandite,* and *hutchinsonite*. It is also present in *pyrites* and is recovered from the roasting of this ore in connection with the production of sulfuric acid. It is also obtained from the smelting of lead and zinc ores. Extraction is somewhat complex and depends on the source of the thallium. Manganese nodules, found on the ocean floor, contain thallium. When freshly exposed to air, thallium exhibits a metallic luster, but soon develops a bluish-gray tinge, resembling lead in appearance. A heavy oxide builds up on thallium if left in air, and in the presence of water the hydroxide is formed. The metal is very soft and malleable. It can be cut with a knife. forty seven isotopic forms of thallium, with atomic masses ranging from 179 to 210 are recognized. Natural thallium is a mixture of two isotopes. The element and its compounds are toxic and should be handled carefully. Contact of the metal with skin is dangerous, and when melting the metal adequate ventilation should be provided. Thallium is suspected of carcinogenic potential for man. Thallium sulfate has been widely employed as a rodenticide and ant killer. It is odorless and tasteless, giving no warning of its presence. Its use, however, has been prohibited in the U.S. since 1975 as a household insecticide and rodenticide. The electrical conductivity of thallium sulfide changes with exposure to infrared light, and this compound is used in photocells. Thallium bromide-iodide crystals have been used as infrared optical materials. Thallium has been used, with sulfur or selenium and arsenic, to produce low melting glasses which become fluid between 125 and 150°C. These glasses have properties at room temperatures similar to ordinary glasses and are said to be durable and insoluble in water. Thallium oxide has been used to produce glasses with a high index of refraction. Thallium has been used in treating ringworm and other skin infections; however, its use has been limited because of the narrow margin between toxicity and therapeutic benefits. A mercury-thallium alloy, which forms a eutectic at 8.5% thallium, is reported to freeze at −60°C, some 20° below the freezing point of mercury. Thallium metal (99.999%) costs about \$1/g.

Thorium — (*Thor*, Scandinavian god of war), Th; at. wt. 232.0381(1); at. no. 90; m.p. 1750°C; b.p. 4788°C; sp. gr. 11.72; valence +2(?), +3(?), +4. Discovered by Berzelius in 1828. Thorium occurs in *thorite* ($ThSiO_4$) and in *thorianite* ($ThO_2 + UO_2$). Large deposits of thorium minerals have been reported in New England and elsewhere, but these have not yet been exploited. Thorium is now thought to be about three times as abundant as uranium and about as abundant as lead or molybdenum. The metal is a source of nuclear power. There is probably more energy available for use from thorium in the minerals of the earth's crust than from both uranium and fossil fuels. Any sizable demand for thorium as a nuclear fuel is still several years in the future. Work has been done in developing thorium cycle converter-reactor systems. Several prototypes, including the HTGR (high-temperature gas-cooled reactor) and MSRE (molten salt converter reactor experiment), have operated. While the HTGR reactors are efficient, they are not expected to become important commercially for many years because of certain operating difficulties. Thorium is recovered commercially from the mineral *monazite*, which contains from 3 to 9% ThO_2 along with rare-earth minerals. Much of the internal heat the earth produces has been attributed to thorium and uranium. Several methods are available for producing thorium metal: it can be obtained by reducing thorium oxide with calcium, by electrolysis of anhydrous thorium chloride in a fused mixture of sodium and potassium chlorides, by calcium reduction of thorium tetrachloride mixed with anhydrous zinc chloride, and by reduction of thorium tetrachloride with an alkali metal. Thorium was originally assigned a position in Group IV of the periodic table. Because of its atomic weight, valence, etc., it is now considered to be the second member of the *actinide* series of elements. When pure, thorium is a silvery-white metal which is air-stable and retains its luster for several months. When contaminated with the oxide, thorium slowly tarnishes in air, becoming gray and finally black. The physical properties of thorium are greatly influenced by the degree of contamination with the oxide. The purest specimens often contain several tenths of a percent of the oxide. High-purity thorium has been made. Pure thorium is soft, very ductile, and can be cold-rolled, swaged, and drawn. Thorium is dimorphic, changing at 1400°C from a cubic to a body-centered cubic structure. Thorium oxide has a melting point of 3300°C, which is the highest of all oxides. Only a few elements, such as tungsten, and a few compounds, such as tantalum carbide, have higher melting points. Thorium is slowly attacked by water, but does not dissolve readily in most common acids, except hydrochloric. Powdered thorium metal is often pyrophoric and should be carefully handled. When heated in air, thorium turnings ignite and burn brilliantly with a white light. The principal use of thorium has been in the preparation of the Welsbach mantle, used for portable gas lights. These mantles, consisting of thorium oxide with about 1% cerium oxide and other ingredients, glow with a dazzling light when heated in a gas flame. Thorium is an important alloying element in magnesium, imparting high strength and creep resistance at elevated temperatures. Because thorium has a low work-function and high electron emission, it is used to coat tungsten wire used in electronic equipment. The oxide is also used to control the grain size of tungsten used for electric lamps; it is also used for high-temperature laboratory crucibles. Glasses containing thorium oxide have a high refractive index and low dispersion. Consequently, they find application in high quality lenses for cameras and scientific instruments. Thorium oxide has also found use as a catalyst in the conversion of ammonia to nitric acid, in petroleum cracking, and in producing sulfuric acid. Twenty seven isotopes of thorium are known with atomic masses ranging from 212 to 237. All are unstable. ^{232}Th occurs naturally and has a half-life of 1.4×10^{10} years. It is an alpha emitter. ^{232}Th goes through six alpha and four beta decay steps before becoming the stable isotope ^{208}Pb. ^{232}Th is sufficiently radioactive to expose a photographic plate in a few hours. Thorium disintegrates with the production of "thoron" (^{220}Rn), which is an alpha emitter and presents a radiation hazard. Good ventilation of areas where thorium is stored or handled is therefore essential. Thorium metal (99.8%) costs about $15/g.

Thulium — (*Thule*, the earliest name for Scandinavia), Tm; at. wt. 168.93421(3); at. no. 69; m.p. 1545°C; b.p. 1950°C; sp. gr. 9.321 (25°C); valence 3. Discovered in 1879 by Cleve. Thulium occurs in small quantities along with other rare earths in a number of minerals. It is obtained commercially from *monazite*, which contains about 0.007% of the element. Thulium is the least abundant of the rare earth elements, but with new sources recently discovered, it is now considered to be about as rare as silver, gold, or cadmium. Ion-exchange and solvent extraction techniques have recently permitted much easier separation of the rare earths, with much lower costs. Only a few years ago, thulium metal was not obtainable at any cost; in 1996 the oxide cost $20/g. Thulium metal powder now costs $60/g (99.9%). Thulium can be isolated by reduction of the oxide with lanthanum metal or by calcium reduction of the anhydrous fluoride. The pure metal has a bright, silvery luster. It is reasonably stable in air, but the metal should be protected from moisture in a closed container. The element is silver-gray, soft, malleable, and ductile, and can be cut with a knife. Thirty eight isotopes and isomers are known, with atomic masses ranging from 146 to 176. Natural thulium, which is 100% ^{169}Tm, is stable. Because of the relatively high price of the metal, thulium has not yet found many practical applications. ^{169}Tm bombarded in a nuclear reactor can be used as a radiation source in portable X-ray equipment. ^{171}Tm is potentially useful as an energy source. Natural thulium also has possible use in *ferrites* (ceramic magnetic materials) used in microwave equipment. As with other lanthanides, thulium has a low-to-moderate acute toxic rating. It should be handled with care.

Tin — (anglo-Saxon, *tin*), Sn (L. *stannum*); at. wt. 118.710(7); at. no. 50; f.p. 231.928°C; b.p. 2602°C; sp. gr. (gray) 5.75, (white) 7.31; valence 2, 4. Known to the ancients. Tin is found chiefly in *cassiterite* (SnO_2). Most of the world's supply comes from Malaysia, Bolivia, China, Indonesia, Russia, Zaire, Thailand, and Nigeria. The U.S. produces almost none, although occurrences have been found in Alaska and Colorado. Tin is obtained by reducing the ore with coal in a reverberatory furnace. Ordinary tin is composed of ten stable isotopes; thirty five unstable isotopes and isomers are also known. Ordinary tin is a silver-white metal, is malleable, somewhat ductile, and has a highly crystalline structure. Due to the breaking of these crystals, a "tin cry" is heard when a bar is bent. The element has two allotropic forms at normal pressure. On warming, gray, or α tin, with a cubic structure, changes at 13.2°C into white, or β tin, the ordinary form of the metal. White tin has a tetragonal structure. When tin is cooled below 13.2°C, it changes slowly from white to gray. This change is affected by impurities such as aluminum and zinc, and can be prevented by small additions of antimony or bismuth. This change from the α to β form is called the tin pest. There are few if any uses for gray tin. Tin takes a high polish and is used to coat other metals to prevent corrosion or other chemical action. Such tin plate over steel is used in the so-called tin can for preserving food. Alloys of tin are very important. Soft solder, type metal, fusible metal, pewter, bronze, bell metal, Babbitt metal, White metal, die casting alloy, and phosphor bronze are some of the important alloys using tin. Tin resists distilled sea and soft tap water, but is attacked by strong acids, alkalis, and acid salts. Oxygen in solution accelerates the attack. When heated in air, tin forms SnO_2, which is feebly acid, forming stannate salts with basic oxides. The most important salt is the chloride ($SnCl_2 \cdot H_2O$), which is used as a reducing agent and as a mordant in calico printing. Tin salts sprayed onto glass are used to produce electrically conductive coatings. These have been used for panel lighting and for frost-free windshields. Most window glass is now made by floating molten glass on molten tin (float glass) to produce a flat surface (Pilkington process). Of recent interest is a crystalline tin-niobium alloy that is superconductive at very low temperatures. This promises to be important in the construction of superconductive magnets that generate enormous field strengths but use practically no power. Such magnets, made of tin-niobium wire, weigh but a few pounds and produce magnetic fields that, when started with a small battery, are comparable to that of a 100 ton electromagnet operated continuously with a large power supply. The small amount of tin found in canned foods is quite harmless. The agreed limit of tin content in U.S. foods is 300 mg/kg. The trialkyl and triaryl tin compounds are used as biocides and must be handled carefully. Over the past 25 years the price of commercial tin has varied from 50¢/lb ($1.10/kg) to its present price of about $4.30/lb ($9.50/kg) as of January 1996. Tin with a purity of 99.9999% is available at a price of about $4/g.

Titanium — (L. *Titans*, the first sons of the Earth, myth.), Ti; at. wt. 47.867(1); at. no. 22; m.p. 1668 ± 10°C; b.p. 3287°C; sp. gr. 4.54; valence 2, 3, or 4. Discovered by Gregor in 1791; named by Klaproth in 1795. Impure titanium was prepared by Nilson and Pettersson in 1887; however, the pure metal (99.9%) was not made until 1910 by Hunter by heating $TiCl_4$ with sodium in a steel bomb. Titanium is present in meteorites and in the sun. Rocks obtained during the Apollo 17 lunar mission showed presence of 12.1% TiO_2. Analyses of rocks obtained during earlier Apollo missions show lower percentages. Titanium oxide bands are prominent in the spectra of M-type stars. The element is the ninth most abundant in the crust of the earth. Titanium is almost always present in igneous rocks and in the sediments derived from them. It occurs in the minerals *rutile, ilmenite,* and *sphene,* and is present in titanates and in many iron ores. Deposits of ilmenite and rutile are found in Florida, California, Tennessee, New York, and elsewhere. Titanium is present in the ash of coal, in plants, and in the human body. The metal was a laboratory curiosity until Kroll, in 1946, showed that titanium could be produced commercially by reducing titanium tetrachloride with magnesium. This method is largely used for producing the metal today. The

THE ELEMENTS (continued)

metal can be purified by decomposing the iodide. Titanium, when pure, is a lustrous, white metal. It has a low density, good strength, is easily fabricated, and has excellent corrosion resistance. It is ductile only when it is free of oxygen. The metal burns in air and is the only element that burns in nitrogen. Titanium is resistant to dilute sulfuric and hydrochloric acid, most organic acids, moist chlorine gas, and chloride solutions. Natural titanium consists of five isotopes with atomic masses from 46 to 50. All are stable. Fifteen other unstable isotopes are known. The metal is dimorphic. The hexagonal α form changes to the cubic β form very slowly at about 880°C. The metal combines with oxygen at red heat, and with chlorine at 550°C. Titanium is important as an alloying agent with aluminum, molybdenum, manganese, iron, and other metals. Alloys of titanium are principally used for aircraft and missiles where lightweight strength and ability to withstand extremes of temperature are important. Titanium is as strong as steel, but 45% lighter. It is 60% heavier than aluminum, but twice as strong. Titanium has potential use in desalination plants for converting sea water into fresh water. The metal has excellent resistance to sea water and is used for propeller shafts, rigging, and other parts of ships exposed to salt water. A titanium anode coated with platinum has been used to provide cathodic protection from corrosion by salt water. Titanium metal is considered to be physiologically inert; however, titanium powder may be a carcinogenic hazard. When pure, titanium dioxide is relatively clear and has an extremely high index of refraction with an optical dispersion higher than diamond. It is produced artificially for use as a gemstone, but it is relatively soft. Star sapphires and rubies exhibit their asterism as a result of the presence of TiO_2. Titanium dioxide is extensively used for both house paint and artist's paint, as it is permanent and has good covering power. Titanium oxide pigment accounts for the largest use of the element. Titanium paint is an excellent reflector of infrared, and is extensively used in solar observatories where heat causes poor seeing conditions. Titanium tetrachloride is used to iridize glass. This compound fumes strongly in air and has been used to produce smoke screens. The price of titanium metal (99.8%) is about $550/kg.

Tungsten — (Swedish, *tung sten*, heavy stone); also known as *wolfram* (from *wolframite*, said to be named from *wolf rahm* or *spumi lupi*, because the ore interfered with the smelting of tin and was supposed to devour the tin), W; at. wt. 183.84(1); at. no. 74; m.p. 3422 ± 20°C; b.p. 5555°C; sp. gr. 19.3 (20°C); valence 2, 3, 4, 5, or 6. In 1779 Peter Woulfe examined the mineral now known as *wolframite* and concluded it must contain a new substance. Scheele, in 1781, found that a new acid could be made from *tung sten* (a name first applied about 1758 to a mineral now known as *scheelite*). Scheele and Berman suggested the possibility of obtaining a new metal by reducing this acid. The de Elhuyar brothers found an acid in *wolframite* in 1783 that was identical to the acid of *tungsten*(tungstic acid) of Scheele, and in that year they succeeded in obtaining the element by reduction of this acid with charcoal. Tungsten occurs in *wolframite*, (Fe, Mn)WO₄; *scheelite*, CaWO₄; *huebnerite*, MnWO₄; and *ferberite*, FeWO₄. Important deposits of tungsten occur in California, Colorado, South Korea, Bolivia, Russia, and Portugal. China is reported to have about 75% of the world's tungsten resources. Natural tungsten contains five stable isotopes. Thirty two other unstable isotopes and isomers are recognized. The metal is obtained commercially by reducing tungsten oxide with hydrogen or carbon. Pure tungsten is a steel-gray to tin-white metal. Very pure tungsten can be cut with a hacksaw, and can be forged, spun, drawn, and extruded. The impure metal is brittle and can be worked only with difficulty. Tungsten has the highest melting point of all metals, and at temperatures over 1650°C has the highest tensile strength. The metal oxidizes in air and must be protected at elevated temperatures. It has excellent corrosion resistance and is attacked only slightly by most mineral acids. The thermal expansion is about the same as borosilicate glass, which makes the metal useful for glass-to-metal seals. Tungsten and its alloys are used extensively for filaments for electric lamps, electron and television tubes, and for metal evaporation work; for electrical contact points for automobile distributors; X-ray targets; windings and heating elements for electrical furnaces; and for numerous spacecraft and high-temperature applications. High-speed tool steels, Hastelloy®, Stellite®, and many other alloys contain tungsten. Tungsten carbide is of great importance to the metal-working, mining, and petroleum industries. Calcium and magnesium tungstates are widely used in fluorescent lighting; other salts of tungsten are used in the chemical and tanning industries. Tungsten disulfide is a dry, high-temperature lubricant, stable to 500°C. Tungsten bronzes and other tungsten compounds are used in paints. Tungsten powder (99.95%) costs about $325/kg.

Uranium — (Planet *Uranus*), U; at. wt. 238.0289(1); at. no. 92; m.p. 1135°C; b.p. 4131°C; sp. gr. ~18.95; valence 2, 3, 4, 5, or 6. Yellow-colored glass, containing more than 1% uranium oxide and dating back to 79 A.D., has been found near Naples, Italy. Klaproth recognized an unknown element in *pitchblende* and attempted to isolate the metal in 1789. The metal apparently was first isolated in 1841 by Peligot, who reduced the anhydrous chloride with potassium. Uranium is not as rare as it was once thought. It is now considered to be more plentiful than mercury, antimony, silver, or cadmium, and is about as abundant as molybdenum or arsenic. It occurs in numerous minerals such as *pitchblende, uraninite, carnotite, autunite, uranophane, davidite*, and *tobernite*. It is also found in *phosphate rock, lignite, monazite sands*, and can be recovered commercially from these sources. Large deposits of uranium ore occur in Utah, Colorado, New Mexico, Canada, and elsewhere. The U.S.D.O.E. purchases uranium in the form of acceptable U_3O_8 concentrates. This incentive program has greatly increased the known uranium reserves. Uranium can be made by reducing uranium halides with alkali or alkaline earth metals or by reducing uranium oxides by calcium, aluminum, or carbon at high temperatures. The metal can also be produced by electrolysis of KUF_5 or UF_4, dissolved in a molten mixture of $CaCl_2$ and NaCl. High-purity uranium can be prepared by the thermal decomposition of uranium halides on a hot filament. Uranium exhibits three crystallographic modifications as follows:

$$\alpha \xrightarrow{688°C} \beta \xrightarrow{776°C} \gamma$$

Uranium is a heavy, silvery-white metal which is pyrophoric when finely divided. It is a little softer than steel, and is attacked by cold water in a finely divided state. It is malleable, ductile, and slightly paramagnetic. In air, the metal becomes coated with a layer of oxide. Acids dissolve the metal, but it is unaffected by alkalis. Uranium has twenty three isotopes, one of which is an isomer and all of which are radioactive. Naturally occurring uranium contains 99.2745% by weight ^{238}U, 0.720% ^{235}U, and 0.0055% ^{234}U. Studies show that the percentage weight of ^{235}U in natural uranium varies by as much as 0.1%, depending on the source. The U.S.D.O.E. has adopted the value of 0.711 as being their "official" percentage of ^{235}U in natural uranium. Natural uranium is sufficiently radioactive to expose a photographic plate in an hour or so. Much of the internal heat of the earth is thought to be attributable to the presence of uranium and thorium. ^{238}U with a half-life of 4.46×10^9 years, has been used to estimate the age of igneous rocks. The origin of uranium, the highest member of the naturally occurring elements — except perhaps for traces of neptunium or plutonium — is not clearly understood, although it may be presumed that uranium is a decay product of elements of higher atomic weight, which may have once been present on earth or elsewhere in the universe. These original elements may have been formed as a result of a primordial "creation," known as "the big bang," in a supernova, or in some other stellar processes. Uranium is of great importance as a nuclear fuel. ^{238}U can be converted into fissionable plutonium by the following reactions:

$$^{238}U(n,\gamma) \rightarrow {}^{239}U \xrightarrow{\beta} {}^{239}Np \xrightarrow{\beta} {}^{239}Pu$$

This nuclear conversion can be brought about in "breeder" reactors where it is possible to produce more new fissionable material than the fissionable material used in maintaining the chain reaction. ^{235}U is of even greater importance, for it is the key to the utilization of uranium. ^{235}U, while occurring in natural uranium to the extent of only 0.71%, is so fissionable with slow neutrons that a self-sustaining fission chain reaction can be made to occur in a reactor constructed from natural uranium and a suitable moderator, such as heavy water or graphite, alone. ^{235}U can be concentrated by gaseous diffusion and other physical processes, if desired, and used directly as a nuclear fuel, instead of natural uranium, or used as an explosive. Natural uranium, slightly enriched with ^{235}U by a small percentage, is used to fuel nuclear power reactors for the generation of electricity. Natural thorium can be irradiated with neutrons as follows to produce the important isotope ^{233}U.

$$^{232}Th(n,\gamma) \rightarrow {}^{233}Th \xrightarrow{\beta} {}^{233}Pa \xrightarrow{\beta} {}^{233}U$$

While thorium itself is not fissionable, ^{233}U is, and in this way may be used as a nuclear fuel. One pound of completely fissioned uranium has the fuel value of over 1500 tons of coal. The uses of nuclear fuels to generate electrical power, to make isotopes for peaceful purposes, and to make explosives are well known. The estimated world-wide production of the 430 nuclear power reactors in operation in January 1994 amounted to about 338,000 megawatts. Uranium in the U.S.A. is controlled by the U.S. Nuclear Regulatory Commission. New uses are being found for "depleted" uranium, i.e., uranium with the percentage of ^{235}U lowered to about 0.2%. It has found use in inertial guidance devices, gyro compasses, counterweights for aircraft control surfaces, as ballast for missile reentry vehicles, and as a shielding material. Uranium metal is used for X-ray targets for production of high-energy X-rays; the nitrate has been used as photographic toner, and the acetate is used in analytical chemistry. Crystals of uranium nitrate are triboluminescent. Uranium salts have also been used for producing yellow "vaseline" glass and glazes. Uranium and its compounds are highly toxic, both from a chemical and radiological standpoint. Finely divided uranium metal, being pyrophoric, presents a fire hazard. The maximum permissible total body burden of natural uranium (based on radiotoxicity) is 0.2 μCi for soluble compounds. Recently, the natural presence of uranium in many soils has become of concern to homeowners because of the generation of radon and its daughters (see under Radon). Uranium metal is available commercially at a cost of about $200/kg (99.7%) in air-tight glass under argon.

Unnilquadium, etc. — See under Elements 104 to 112.

Vanadium — (Scandinavian goddess, *Vanadis*), V; at. wt. 50.9415(1); at. no. 23; m.p. 1910 ± 10°C; b.p. 3407°C; sp. gr. 6.11 (18.7°C); valence 2, 3, 4, or 5. Vanadium was first discovered by del Rio in 1801. Unfortunately, a French chemist incorrectly declared del Rio's new element was only impure chromium; del Rio thought himself to be mistaken and accepted the French chemist's statement. The element was rediscovered in 1830 by Sefstrom, who named the element in honor of the Scandinavian goddess *Vanadis* because of its beautiful multicolored compounds. It was isolated in nearly pure form by Roscoe, in 1867, who reduced the chloride with hydrogen. Vanadium of 99.3 to 99.8% purity was not produced until 1927. Vanadium is found in about 65 different minerals among which are *carnotite, roscoelite, vanadinite,* and *patronite* important sources of the metal. Vanadium is also found in phosphate rock and certain iron ores, and is present in some crude oils in the form of organic complexes. It is also found in small percentages in meteorites. Commercial production from petroleum ash holds promise as an important source of the element. High-purity ductile vanadium can be obtained by reduction of vanadium trichloride with magnesium or with magnesium-sodium mixtures. Much of the vanadium metal being produced is now made by calcium reduction of V_2O_5 in a pressure vessel, an adaption of a process developed by McKechnie and Seybolt. Natural vanadium is a mixture of two isotopes, ^{50}V (0.25%) and ^{51}V (99.75%). ^{50}V is slightly radioactive, having a long half-life. Seventeen other unstable isotopes are recognized. Pure vanadium is a bright white metal, and is soft and ductile. It has good corrosion resistance to alkalis, sulfuric and hydrochloric acid, and salt water, but the metal oxidizes readily above 660°C. The metal has good structural strength and a low fission neutron cross section, making it useful in nuclear applications. Vanadium is used in producing rust resistant, spring, and highspeed tool steels. It is an important carbide stabilizer in making steels. About 80% of the vanadium now produced is used as ferrovanadium or as a steel additive. Vanadium foil is used as a bonding agent in cladding titanium to steel. Vanadium pentoxide is used in ceramics and as a catalyst. It is also used in producing a superconductive magnet with a field of 175,000 gauss. Vanadium and its compounds are toxic and should be handled with care. Ductile vanadium is commercially available. Commercial vanadium metal, of about 95% purity, costs about $50kg. Vanadium metal (99.7%) costs about $1.50/g or $700/kg.

Wolfram — see Tungsten.

Xenon — (Gr. *xenon*, stranger), Xe; at. wt. 131.29(2); at. no. 54; m.p. –111.75°C; b.p. –108.0°C; density (gas) 5.887 ± 0.009 g/l, sp. gr (liquid) 3.52 (–109°C); valence usually 0. Discovered by Ramsay and Travers in 1898 in the residue left after evaporating liquid air components. Xenon is a member of the so-called noble or "inert" gases. It is present in the atmosphere to the extent of about one part in twenty million. Xenon is present in the Martian atmosphere to the extent of 0.08 ppm. The element is found in the gases evolved from certain mineral springs, and is commercially obtained by extraction from liquid air. Natural xenon is composed of nine stable isotopes. In addition to these, thirty five unstable isotopes and isomers have been characterized. Before 1962, it had generally been assumed that xenon and other noble gases were unable to form compounds. Evidence has been mounting in the past few years that xenon, as well as other members of the zero valence elements, do form compounds. Among the "compounds" of xenon now reported are xenon hydrate, sodium perxenate, xenon deuterate, difluoride, tetrafluoride, hexafluoride, and $XePtF_6$ and $XeRhF_6$. Xenon trioxide, which is highly explosive, has been prepared. More than 80 xenon compounds have been made with xenon chemically bonded to fluorine and oxygen. Some xenon compounds are colored. Metallic xenon has been produced, using several hundred kilobars of pressure. Xenon in a vacuum tube produces a beautiful blue glow when excited by an electrical discharge. The gas is used in making electron tubes, stroboscopic lamps, bactericidal lamps, and lamps used to excite ruby lasers for generating coherent light. Xenon is used in the atomic energy field in bubble chambers, probes, and other applications where its high molecular weight is of value. The perxenates are used in analytical chemistry as oxidizing agents. ^{133}Xe and ^{135}Xe are produced by neutron irradiation in air cooled nuclear reactors. ^{133}Xe has useful applications as a radioisotope. The element is available in sealed glass containers for about $20/l of gas at standard pressure. Xenon is not toxic, but its compounds are highly toxic because of their strong oxidizing characteristics.

Ytterbium — (Ytterby, village in Sweden), Yb; at. wt. 173.04(3); at. no. 70; m.p. 819°C; b.p. 1196°C; sp. gr (α) 6.903 (β) 6.966; valence 2, 3. Marignac in 1878 discovered a new component, which he called *ytterbia*, in the earth then known as *erbia*. In 1907, Urbain separated ytterbia into two components, which he called *neoytterbia* and *lutecia*. The elements in these earths are now known as *ytterbium* and *lutetium*, respectively. These elements are identical with *aldebaranium* and *cassiopeium*, discovered independently and at about the same time by von Welsbach. Ytterbium occurs along with other rare earths in a number of rare minerals. It is commercially recovered principally from *monazite sand*, which contains about 0.03%. Ion-exchange and solvent extraction techniques developed in recent years have greatly simplified the separation of the rare earths from one another. The element was first prepared by Klemm and Bonner in 1937 by reducing ytterbium trichloride with potassium. Their metal was mixed, however, with KCl. Daane, Dennison, and Spedding prepared a much purer form in 1953 from which the chemical and physical properties of the element could be determined. Ytterbium has a bright silvery luster, is soft, malleable, and quite ductile. While the element is fairly stable, it should be kept in closed containers to protect it from air and moisture. Ytterbium is readily attacked and dissolved by dilute and concentrated mineral acids and reacts slowly with water. Ytterbium has three allotropic forms with transformation points at −13° and 795°C. The beta form is a room-temperature, face-centered, cubic modification, while the high-temperature gamma form is a body-centered cubic form. Another body-centered cubic phase has recently been found to be stable at high pressures at room temperatures. The beta form ordinarily has metallic-type conductivity, but becomes a semiconductor when the pressure is increased above 16,000 atm. The electrical resistance increases tenfold as the pressure is increased to 39,000 atm and drops to about 80% of its standard temperature-pressure resistivity at a pressure of 40,000 atm. Natural ytterbium is a mixture of seven stable isotopes. Twenty six other unstable isotopes and isomers are known. Ytterbium metal has possible use in improving the grain refinement, strength, and other mechanical properties of stainless steel. One isotope is reported to have been used as a radiation source as a substitute for a portable X-ray machine where electricity is unavailable. Few other uses have been found. Ytterbium metal is commercially available with a purity of about 99.9% for about $5/g. Ytterbium has a low acute toxic rating, but may present a carcinogenic hazard.

Yttrium — (*Ytterby*, village in Sweden near Vauxholm), Y; at. wt. 88.90585(2); at. no. 39; m.p. 1522°C; b.p. 3345°C; sp. gr. 4.469 (25°C); valence 3. *Yttria*, which is an earth containing yttrium, was discovered by Gadolin in 1794. *Ytterby* is the site of a quarry which yielded many unusually minerals containing rare earths and other elements. This small town, near Stockholm, bears the honor of giving names to *erbium, terbium,* and *ytterbium* as well as *yttrium*. In 1843 Mosander showed that yttria could be resolved into the oxides (or earths) of three elements. The name yttria was reserved for the most basic one; the others were named *erbia* and *terbia*. Yttrium occurs in nearly all of the rare-earth minerals. Analysis of lunar rock samples obtained during the Apollo missions show a relatively high yttrium content. It is recovered commercially from *monazite sand*, which contains about 3%, and from *bastnasite*, which contains about 0.2%. Wohler obtained the impure element in 1828 by reduction of the anhydrous chloride with potassium. The metal is now produced commercially by reduction of the fluoride with calcium metal. It can also be prepared by other techniques. Yttrium has a silver-metallic luster and is relatively stable in air. Turnings of the metal, however, ignite in air if their temperature exceeds 400°C, and finely divided yttrium is very unstable in air. Yttrium oxide is one of the most important compounds of yttrium and accounts for the largest use. It is widely used in making YVO_4 europium, and Y_2O_3 europium phosphors to give the red color in color television tubes. Many hundreds of thousands of pounds are now used in this application. Yttrium oxide also is used to produce yttrium-iron-garnets, which are very effective microwave filters. Yttrium iron, aluminum, and gadolinium garnets, with formulas such as $Y_3Fe_5O_{12}$ and $Y_3Al_5O_{12}$, have interesting magnetic properties. Yttrium iron garnet is also exceptionally efficient as both a transmitter and transducer of acoustic energy. Yttrium aluminum garnet, with a hardness of 8.5, is also finding use as a gemstone (simulated diamond). Small amounts of yttrium (0.1 to 0.2%) can be used to reduce the grain size in chromium, molybdenum, zirconium, and titanium, and to increase strength of aluminum and magnesium alloys. Alloys with other useful properties can be obtained by using yttrium as an additive. The metal can be used as a deoxidizer for vanadium and other nonferrous metals. The metal has a low cross section for nuclear capture. ^{90}Y, one of the isotopes of yttrium, exists in equilibrium with its parent ^{90}Sr, a product of atomic explosions. Yttrium has been considered for use as a nodulizer for producing nodular cast iron, in which the graphite forms compact nodules instead of the usual flakes. Such iron has increased ductility. Yttrium is also finding application in laser systems and as a catalyst for ethylene polymerization.It has also potential use in ceramic and glass formulas, as the oxide has a high melting point and imparts shock resistance and low expansion characteristics to glass. Natural yttrium contains but one isotope, ^{89}Y. Thirty seven other unstable isotopes and isomers have been characterized. Yttrium metal of 99.9% purity is commercially available at a cost of about $4/g.

Zinc — (Ger. *Zink*, of obscure origin), Zn; at. wt. 65.39(2); at. no. 30; f.p. 419.527°C; b.p. 907°C; sp. gr. 7.133 (25°C); valence 2. Centuries before zinc was recognized as a distinct element, zinc ores were used for making brass. Tubal-Cain, seven generations from Adam, is mentioned as being an "instructor in every artificer in brass and iron." An alloy containing 87% zinc has been found in prehistoric ruins in Transylvania. Metallic zinc was produced in the 13th century A.D. in India by reducing calamine with organic substances such as wool. The metal was rediscovered in Europe by Marggraf in 1746, who showed that it could be obtained by reducing *calamine* with charcoal. The principal ores of zinc are *sphalerite* or *blende* (sulfide), *smithsonite* (carbonate), *calamine* (silicate), and *franklinite* (zinc, manganese, iron oxide). Zinc can be obtained by roasting its ores to form the oxide and by reduction of the oxide with coal or carbon, with subsequent distillation of the metal. Other methods of extraction are possible. Naturally occurring zinc contains five stable isotopes. Twenty three other unstable isotopes and isomers are recognized. Zinc is a bluish-white, lustrous metal. It is brittle at ordinary temperatures but malleable at 100 to 150°C. It is a fair conductor of electricity, and burns in air at high red heat with evolution of white clouds of the oxide. The metal is employed to form numerous alloys with other metals. Brass, nickel silver, typewriter metal, commercial bronze, spring brass, German silver, soft solder, and aluminum solder are some of the more important alloys. Large quantities of zinc are used to produce die castings, used extensively by the automotive, electrical, and hardware industries. An alloy called *Prestal®*, consisting of 78% zinc and 22% aluminum is reported to be almost as strong as steel but as easy to mold as plastic. It is said to be so plastic that it can be molded into form by relatively inexpensive die casts made of ceramics and cement. It exhibits superplasticity. Zinc is also extensively used to galvanize other metals such as iron to prevent corrosion. Neither zinc nor zirconium is ferromagnetic; but $ZrZn_2$ exhibits ferromagnetism at temperatures below 35 K. Zinc oxide is a unique and very useful material to modern civilization. It is widely used in the manufacture of paints, rubber products, cosmetics, pharmaceuticals, floor coverings, plastics, printing inks, soap, storage batteries, textiles, electrical equipment, and other products. It has unusual electrical, thermal, optical, and solid-state properties that have not yet been fully investigated. Lithopone, a mixture of zinc sulfide and barium sulfate, is an important pigment. Zinc sulfide is used in making luminous dials, X-ray and TV screens, and fluorescent lights. The chloride and chromate are also important compounds. Zinc is an essential element in the growth of human beings and animals. Tests show that zinc-deficient animals require 50% more food to gain the same weight as an animal supplied with sufficient zinc. Zinc is not considered to be toxic, but when freshly formed ZnO is inhaled a disorder known as the *oxide shakes* or *zinc chills* sometimes occurs. It is recommended that where zinc oxide is encountered good ventilation be provided. The commercial price of zinc was roughly 50¢/lb ($1.10 kg) in January 1996. Zinc metal with a purity of 99.9999% is priced at about 50¢/g.

Zirconium — (Syriac, *zargun*, color of gold), Zr; at. wt. 91.224(2); at. no. 40; m.p. 1855 ± 2°C; b.p. 4409°C; sp. gr. 6.506 (20°C); valence +2, +3, and +4. The name *zircon* may have originated from the Syriac word *zargono,* which describes the color of certain gemstones now known as *zircon, jargon, hyacinth, jacinth,* or *ligure.*This mineral, or its variations, is mentioned in biblical writings. These minerals were not known to contain this element until Klaproth, in 1789, analyzed a jargon from Ceylon and found a new earth, which Werner named zircon (*silex circonius*), and Klaproth called *Zirkonerde (zirconia)*. The impure metal was first isolated by Berzelius in 1824 by heating a mixture of potassium and potassium zirconium fluoride in a small iron tube. Pure zirconium was first prepared in 1914. Very pure zirconium was first produced in 1925 by van Arkel and de Boer by an iodide decomposition process they developed. Zirconium is found in abundance in S-type stars, and has been identified in the sun and meteorites. Analyses of lunar rock samples obtained during the various Apollo missions to the moon show a surprisingly high zirconium oxide content, compared with terrestial rocks. Naturally occurring zirconium contains five isotopes. Twenty six other radioactive isotopes and isomers are known to exist. *Zircon*, $ZrSiO_4$, the principal ore, is found in deposits in Florida, South Carolina, Australia, and Brazil. *Baddeleyite*, found in Brazil, is an important zirconium mineral. It is principally pure ZrO_2in crystalline form having a hafnium content of about 1%. Zirconium also occurs in some 30 other recognized mineral species. Zirconium is produced commercially by reduction of the chloride with magnesium (the Kroll Process), and by other methods. It is a grayish-white lustrous metal. When finely divided, the metal may ignite spontaneously in air, especially at elevated temperatures. The solid metal is much more difficult to ignite. The inherent toxicity of zirconium compounds is low. Hafnium is invariably found in zirconium ores, and the separation is difficult. Commercial-grade zirconium contains from 1 to 3% hafnium. Zirconium has a low absorption cross section for neutrons, and is therefore used for nuclear energy applications, such as for cladding fuel elements. Commercial nuclear power generation now takes more than 90% of zirconium metal production. Reactors of the size now being made may use as much as a half-million lineal feet of zirconium alloy tubing. Reactor-grade zirconium is essentially free of hafnium. *Zircaloy*® is an important alloy developed specifically for nuclear applications. Zirconium is exceptionally resistant to corrosion by many common acids and alkalis, by sea water, and by other agents. It is used extensively by the chemical industry where corrosive agents are employed. Zirconium is used as a getter in vacuum tubes, as an alloying agent in steel, in surgical appliances, photoflash bulbs, explosive primers, rayon spinnerets, lamp filaments, etc. It is used in poison ivy lotions in the form of the carbonate as it combines with *urushiol*. With niobium, zirconium is superconductive at low temperatures and is used to make superconductive magnets, which offer hope of direct large-scale generation of electric power. Alloyed with zinc, zirconium becomes magnetic at temperatures below 35 K. Zirconium oxide (zircon) has a high index of refraction and is used as a gem material. The impure oxide, zirconia, is used for laboratory crucibles that will withstand heat shock, for linings of metallurgical furnaces, and by the glass and ceramic industries as a refractory material. Its use as a refractory material accounts for a large share of all zirconium consumed. Zirconium of about 99.8% purity is available at a cost of about $170/kg.

PHYSICAL CONSTANTS OF INORGANIC COMPOUNDS

The compounds in this table were selected on the basis of their laboratory and industrial importance, as well as their value in illustrating trends in the variation of physical properties with position in the periodic table. An effort has been made to include the most frequently encountered inorganic substances; organometallics are not covered, with the exception of metal salts of organic acids. Many, if not most, of the compounds that are solids at ambient temperature can exist in more than one crystalline modification. The information given here applies to the most stable or common crystalline form. In cases where two or more forms are of practical importance, separate entries will be found in the table.

Compounds are arranged primarily in alphabetical order by the most commonly used name. However, adjustments are made in many instances in order to bring closely related compounds together. For example, hydrides of elements such as boron, silicon, and germanium are grouped together immediately following the entry for the parent element, since they would otherwise be scattered throughout the table. Likewise, the oxoacids of an element are given in one group when a strict alphabetical order would separate them (e.g., sulfuric acid and fluorosulfuric acid). Whenever deviations from alphabetical order occur, the names are included in the Synonym Index to provide an easy way to locate these substances.

The data fields in the table are described below:

- **No.:** An identification number used in the indexes.
- **Name:** Common name for the substance. The valence state of a metallic element is indicated by a Roman numeral , e.g., copper in the +2 state is written as copper(II) rather than cupric (but names beginning with cupric, ferrous, etc. are included in the synonym index).
- **Formula:** The simplest descriptive formula is given, but this does not necessarily specify the actual structure of the compound. For example, aluminum chloride is designated as $AlCl_3$, even though a more accurate representation of the structure in the solid phase (and, under some conditions, in the gas phase) is Al_2Cl_6. A few exceptions are made, such as the use of Hg_2^{+2} for the mercury(I) ion.
- **CAS RN:** Chemical Abstracts Service Registry Number. An asterisk * following the CAS RN for a hydrate indicates that the number refers to the anhydrous compound. In most cases the generic CAS RN for the compound is given rather than the number for a specific crystalline form or mineral.
- **Mol. Wt.:** Molecular weight (relative molar mass) as calculated with the 1993 IUPAC Recommended Atomic Weights. The number of decimal places corresponds to the number of places in the atomic weight of the least accurately known element (e.g., one place for lead compounds, two places for compounds of selenium, germanium, etc.); a maximum of three places is given. For compounds of radioactive elements for which IUPAC makes no recommendation, the mass number of the isotope with the longest half-life is used, and the result is rounded to the nearest integer.
- **Physical form:** The crystal system is given ,when available, for compounds that are solid at room temperature, together with color and other descriptive features. Abbreviations are listed below.
- **mp:** Normal melting point in °C. A "tp" following the number indicates a triple point, where solid, liquid, and gas are in equilibrium at a pressure greater than one atmosphere (i.e., the normal melting point does not exist). When available, the triple point pressure is listed.
- **bp:** Normal boiling point in °C (referred to 101.325 kPa or 760 mmHg pressure). An "sp" following the number indicates a sublimation point, where the pressure of the vapor in equilibrium with the solid reaches 101.325 kPa. See Reference 8, p. 23, for further discussion of sublimation points and triple points. Note: The abbreviation "subl" indicates that there is a perceptible sublimation pressure above the solid, but does not imply any quantitative information on the pressure-temperature behavior.
- **den:** Density. Values for solids and liquids are always in grams per cubic centimeter and can be assumed to refer to temperatures near room temperature unless otherwise stated. Values for gases are the calculated ideal gas densities in grams per liter at 25°C and 101.325 kPa; the unit is always specified with a gas value.
- **Solubility:** Qualitative information on the solubility in various solvents is given. The abbreviations are listed below.
- **Other Data:** The codes refer to other tables in this handbook which include an entry for the compound in question. These codes are:

  a. Data on enthalpy of formation, entropy, and heat capacity in the table "Standard Thermodynamic Properties of Chemical Substances" (Section 5).
  b   Critical temperature, pressure, and volume in the table "Critical Constants, Boiling Points, and Melting Points of Selected Compounds" (Section 6).
  c. Table of "Enthalpy of Fusion" (Section 6).
  d. Table of "Enthalpy of Vaporization" (Section 6).
  e. Vapor pressure data in one or more of the vapor pressure tables in Section 6.
  f. Optical constants and hardness in the table "Physical and Optical Properties of Minerals" (Section 4).

A Synonym Index and a CAS Registry Number Index follow the table.

LIST OF ABBREVIATIONS

Ac	acetyl	gl	glass, glassy	rhom	rhombohedral
ace	acetone	grn	green	s	soluble in
acid	acid solutions	hc	hydrocarbon solvents	silv	silvery
alk	alkaline solutions	hex	hexagonal	sl	slightly soluble in
amorp	amorphous	hp	heptane	soln	solution
anh	anhydrous	hx	hexane	sp	sublimation point
aq	aqueous	hyd	hydrate	stab	stable
blk	black	hyg	hygroscopic	subl	sublimes
brn	brown	i	insoluble in	temp	temperature
bz	benzene	liq	liquid	tetr	tetragonal
chl	chloroform	MeOH	methanol	thf	tetrahydrofuran
col	colorless	mono	monoclinic	tol	toluene
conc	concentrated	octahed	octahedral	tp	triple point
cry	crystals, crystalline	oran	orange	trans	transition, transformation
cub	cubic	orth	orthorhombic	tricl	triclinic
cyhex	cyclohexane	os	organic solvents	trig	trigonal
dec	decomposes	peth	petroleum ether	unstab	unstable
dil	dilute	pow	powder	viol	violet
diox	dioxane	prec	precipitate	visc	viscous
eth	ethyl ether	pur	purple	vs	very soluble in
EtOH	ethanol	py	pyridine	wh	white
exp	explodes, explosive	reac	reacts with	xyl	xylene
flam	flammable	refrac	refractory	yel	yellow

REFERENCES

1. Phillips, S.L., and Perry, D.L., *Handbook of Inorganic Compounds*, CRC Press, Boca Raton, FL, 1995.
2. Trotman-Dickenson, A.F., Executive Editor, *Comprehensive Inorganic Chemistry*, Vol. 1-5, Pergamon Press, Oxford, 1973.
3. Greenwood, N.N., and Earnshaw, A., *Chemistry of the Elements*, Pergamon Press, Oxford, 1984.
4. Budavari, S., Editor, *The Merck Index, Eleventh Edition*, Merck & Co., Rahway, NJ, 1989.
5. *GMELIN Handbook of Inorganic and Organometallic Chemistry*, Springer-Verlag, Heidelberg.
6. Chase, M.W., Davies, C.A., Downey, J.R., Frurip, D.J., McDonald, R.A., and Syverud, A.N.; *JANAF Thermochemical Tables, Third Edition*, *J. Phys. Chem. Ref. Data*, Vol. 14, Suppl. 1, 1985.
7. Donnay, J.D.H., and Ondik, H.M., *Crystal Data Determinative Tables, Third Edition, Volumes 2 and 4, Inorganic Compounds*, Joint Committee on Powder Diffraction Standards, Swarthmore, PA, 1973.
8. Lide, D.R., and Kehiaian, H.V., *CRC Handbook of Thermophysical and Thermochemical Data*, CRC Press, Boca Raton, FL, 1994.
9. *Kirk-Othmer Concise Encyclopedia of Chemical Technology*, Wiley-Interscience, New York, 1985.
10. *Dictionary of Inorganic Compounds*, Chapman & Hall, New York, 1992.
11. Massalski, T.B., Editor, *Binary Alloy Phase Diagrams*, American Society for Metals, Metals Park, Ohio, 1986.
12. *Landolt-Börnstein, Numerical Data and Functional Relationships in Science and Technology, Sixth Edition, II/4, Caloric Quantities of State*, Springer-Verlag, Heidelberg, 1961.
13. Deer, W.A., Howie, R.A., and Zussman, J., *An Introduction to the Rock-Forming Minerals, 2nd Edition*, Longman Scientific & Technical, Harlow, Essex, 1992.
14. Carmichael, R.S., *Practical Handbook of Physical Properties of Rocks and Minerals*, CRC Press, Boca Raton, FL, 1989.
15. Dinsdale, A.T., "SGTE Data for Pure Elements", *CALPHAD*, 15, 317-425, 1991.
16. Madelung, O., *Semiconductors: Group IV Elements and III-IV Compounds*, Springer-Verlag, Heidelberg, 1991.
17. Daubert, T.E., Danner, R.P., Sibul, H.M., and Stebbins, C.C., *Physical and Thermodynamic Properties of Pure Compounds: Data Compilation*, extant 1994 (core with 4 supplements), Taylor & Francis, Bristol, PA.

No.	Name / Formula	CAS RN / Mol. Wt.	Physical Form	mp/°C / den/g cm^{-3}	bp/°C / Other Data	Solubility
1	Actinium / Ac	7440-34-8 / 227	silv metal; cub	1051 / 10	3198 / a	
2	Actinium bromide / AcBr$_3$	33689-81-5 / 467	wh hex cry	5.85		s H$_2$O
3	Actinium chloride / AcCl$_3$	22986-54-5 / 333	wh hex cry	4.81		
4	Actinium fluoride / AcF$_3$	33689-80-4 / 284	wh hex cry	7.88		i H$_2$O
5	Actinium iodide / AcI$_3$	33689-82-6 / 608	wh cry	9.19		s H$_2$O
6	Actinium oxide / Ac$_2$O$_3$	12002-61-8 / 502	wh hex cry	1977		i H$_2$O
7	Aluminum / Al	7429-90-5 / 26.982	silv-wh metal; cub cry	660.32 / 2.70	2519 / a,c,d,e	i H$_2$O; s acid, alk
8	Aluminum ammonium sulfate	7784-25-0 / 237.148	wh powder			sl H$_2$O; i EtOH
9	Aluminum ammonium sulfate dodecahydrate / AlNH$_4$(SO$_4$)$_2$·12H$_2$O	7784-26-1 / 453.331	col cry or powder	94.5 / 1.65	dec >280	s H$_2$O; i EtOH
10	Aluminum antimonide / AlSb	25152-52-7 / 148.742	cub cry	1065 / 4.26		
11	Aluminum arsenide / AlAs	22831-42-1 / 101.903	oran cub cry; hyg	1740 / 3.76		
12	Aluminum borate / 2Al$_2$O$_3$·B$_2$O$_3$	11121-16-7 / 273.543	needles	≈1050		i H$_2$O
13	Aluminum boride / AlB$_2$	12041-50-8 / 48.604	powder	dec >920 / 3.19		s dil HCl
14	Aluminum borohydride / Al(BH$_4$)$_3$	16962-07-5 / 71.510	flam liq	-64.5	44.5 / a,d	reac H$_2$O
15	Aluminum bromate nonahydrate / Al(BrO$_3$)$_3$·9H$_2$O	11126-81-1* / 572.826	wh hyg cry	62	dec >100	s H$_2$O
16	Aluminum bromide / AlBr$_3$	7727-15-3 / 266.694	wh-yel monocl cry; hyg	97.5 / 3.2	255 / a,b,c,d,e	reac H$_2$O; s bz, tol
17	Aluminum bromide hexahydrate / AlBr$_3$·6H$_2$O	7784-11-4 / 374.785	col-yel hyg cry	93 / 2.54		s H$_2$O, EtOH, CS$_2$
18	Aluminum carbide / Al$_4$C$_3$	1299-86-1 / 143.959	yel hex cry	2100 / 2.36	dec >2200	reac H$_2$O
19	Aluminum chlorate nonahydrate / Al(ClO$_3$)$_3$·9H$_2$O	15477-33-5 / 439.472	hyg cry			vs H$_2$O; s EtOH
20	Aluminum chloride / AlCl$_3$	7446-70-0 / 133.340	wh hex cry or powder; hyg	192.6 / 2.48	a,b,c,e	reac H$_2$O; s bz, ctc, chl
21	Aluminum chloride hexahydrate / AlCl$_3$·6H$_2$O	7784-13-6 / 241.431	col hyg cry	dec 100 / 2.398		vs H$_2$O; s EtOH, eth
22	Aluminum diacetate / Al(OH)(C$_2$H$_3$O$_2$)$_2$	142-03-0 / 162.079	wh amorp powder			i H$_2$O
23	Aluminum ethoxide / Al(C$_2$H$_5$O)$_3$	555-75-9 / 162.165	liq, condenses to wh solid	140		reac H$_2$O; sl xyl
24	Aluminum fluoride / AlF$_3$	7784-18-1 / 83.977	wh hex cry	≈2250 tp (220 MPa) / 3.10	1276 sp / a,c,e	sl H$_2$O
25	Aluminum fluoride monohydrate / AlF$_3$·H$_2$O	32287-65-3 / 101.992	orth cry			sl H$_2$O
26	Aluminum fluoride trihydrate / AlF$_3$·3H$_2$O	15098-87-0 / 138.023	wh hyg cry	2.17		sl H$_2$O
27	Aluminum hexafluorosilicate nonahydrate / Al$_2$(SiF$_6$)$_3$·9H$_2$O	17099-70-6 / 642.329	hex prisms	dec >500 / 1.914		s H$_2$O
28	Aluminum hydride / AlH$_3$	7784-21-6 / 30.006	col hex cry	dec >150	a	reac H$_2$O
29	Aluminum hydroxide / Al(OH)$_3$	21645-51-2 / 78.004	wh amorp powder	2.42	f	i H$_2$O; s alk, acid
30	Aluminum hydroxychloride / Al$_2$(OH)$_5$Cl·2H$_2$O	1327-41-9 / 210.483	gl solid			s H$_2$O
31	Aluminum hypophosphite / Al(H$_2$PO$_2$)$_3$	7784-22-7 / 221.948	cry powder	dec 220		i H$_2$O; s alk, acid
32	Aluminum iodide / AlI$_3$	7784-23-8 / 407.695	wh leaflets	188.32 / 3.98	382 / a,b,c,d,e	reac H$_2$O
33	Aluminum iodide hexahydrate / AlI$_3$·6H$_2$O	10090-53-6 / 515.786	yel hyg cry powder			vs H$_2$O; s EtOH, eth
34	Aluminum lactate / Al(C$_3$H$_5$O$_3$)$_3$	18917-91-4 / 294.195	powder			vs H$_2$O
35	Aluminum nitrate nonahydrate / Al(NO$_3$)$_3$·9H$_2$O	7784-27-2 / 375.134	wh hyg cry; monocl	73 / 1.72	dec 135	vs H$_2$O, EtOH; i pyr
36	Aluminum nitride / AlN	24304-00-5 / 40.989	blue-wh hex cry	3000 / 3.255	a	reac H$_2$O
37	Aluminum oleate / Al(C$_{18}$H$_{33}$O$_2$)$_3$	688-37-9 / 871.358	yel solid			i H$_2$O; s EtOH, bz
38	Aluminum phosphate / AlPO$_4$	7784-30-7 / 121.953	wh rhomb plates	>1460 / 2.56	a	i H$_2$O; sl acid

No.	Name / Formula	CAS RN / Mol. Wt.	Physical Form	mp/°C den/g cm^{-3}	bp/°C Other Data	Solubility
39	Aluminum metaphosphate $Al(PO_3)_3$	32823-06-6 263.898	col powder; tetr	≈1525 2.78		i H_2O
40	Aluminum oxide Al_2O_3	1344-28-1 101.961	wh powder; hex	2054 3.97	≈3000 a,c,e,f	i H_2O, os; sl alk
41	Aluminum oxyhydroxide $AlO(OH)$	14457-84-2 59.989	ortho cry	3.44	f	i H_2O; s acid, alk
42	Aluminum palmitate $Al(C_{15}H_{31}COO)_3$	555-35-1 793.244	wh-yel powder			i H_2O, EtOH; s peth
43	Aluminum perchlorate nonahydrate $Al(ClO_4)_3 \cdot 9H_2O$	14452-39-2 487.470	wh hyg cry	82 dec 2.0		
44	Aluminum phosphide AlP	20859-73-8 57.956	grn or yel cub cry	2550 2.40	a	reac H_2O
45	Aluminum selenide Al_2Se_3	1302-82-5 290.84	yel-brown powder	960 3.437		reac H_2O
46	Aluminum silicate Al_2SiO_5	12183-80-1 162.046	gray-grn cry	3.145	f	
47	Aluminum silicate dihydrate $Al_2O_3 \cdot 2SiO_2 \cdot 2H_2O$	1332-58-7 258.161	wh-yel powder; tricl	2.59		i H_2O, acid, alk
48	Aluminum stearate $Al(C_{18}H_{35}O_2)_3$	637-12-7 877.406	wh powder	115 1.070		i H_2O, EtOH, eth; s alk
49	Aluminum sulfate $Al_2(SO_4)_3$	10043-01-3 342.154	wh cry	dec 1040		s H_2O; i EtOH
50	Aluminum sulfate octadecahydrate $Al_2(SO_4)_3 \cdot 18H_2O$	7784-31-8 666.429	col monocl cry	dec 86 1.69	f	s H_2O
51	Aluminum sulfide Al_2S_3	1302-81-4 150.161	yel-gray powder	1100 2.02	a	
52	Aluminum telluride Al_2Te_3	12043-29-7 436.76	gray-blk hex cry	≈895 4.5		
53	Aluminum thiocyanate $Al(CNS)_3$	538-17-0 201.233	yel powder			s H_2O; i EtOH, eth
54	Americium Am	7440-35-9 243	silv metal; hex or cub	1176 12	2011 a,c	s acid
55	Americium(III) oxide Am_2O_3	12254-64-7 534	tan hex cry	11.77		s acid
56	Americium(III) bromide $AmBr_3$	14933-38-1 483	wh orth cry	6.85		s H_2O
57	Americium(III) chloride $AmCl_3$	13464-46-5 349	pink hex cry	500 5.87		
58	Americium(III) fluoride AmF_3	13708-80-0 300	pink hex cry	1393 9.53		
59	Americium(III) iodide AmI_3	13813-47-3 624	yel ortho cry	≈950 6.9		
60	Americium(IV) fluoride AmF_4	15947-41-8 319	tan monocl cry	7.23		
61	Americium(IV) oxide AmO_2	12005-67-3 275	blk cub cry	dec >1000 11.68		s acid
62	Ammonia NH_3	7664-41-7 17.031	col gas	-77.74 0.747 g/L	-33.33 a,b,c,d,e	vs H_2O; s EtOH, eth
63	Ammonium acetate $NH_4C_2H_3O_2$	631-61-8 77.084	wh hyg cry	114 1.073		vs H_2O
64	Ammonium azide NH_4N_3	12164-94-2 60.059	ortho cry; flam	160 1.346	exp a,e	s H_2O
65	Ammonium benzoate $NH_4C_7H_5O_2$	1863-63-4 139.154	wh cry or powder	198 1.26		s H_2O; sl EtOH
66	Ammonium bimalate $NH_4OOCCH_2CH(OH)COOH$	5972-71-4 151.119	ortho cry	160 1.15		s H_2O; sl EtOH
67	Ammonium borate tetrahydrate $(NH_4)_2B_4O_7 \cdot 4H_2O$	12228-87-4 263.377	tetr cry			s H_2O; i EtOH
68	Ammonium bromide NH_4Br	12124-97-9 97.943	wh hyg tetr cry	542 dec 2.429	396 sp a,e	s H_2O, EtOH, ace; sl eth
69	Ammonium caprylate $NH_4C_8H_{15}O_2$	5972-76-9 161.245	hyg monocl cry	≈75		reac H_2O; s EtOH; i chl, bz
70	Ammonium carbamate NH_2COONH_4	1111-78-0 78.071	cry powder			vs H_2O; s EtOH
71	Ammonium carbonate $(NH_4)_2CO_3$	506-87-6 96.086	col cry powder	dec 58		
72	Ammonium cerium(III) sulfate tetrahydrate $NH_4Ce(SO_4)_2 \cdot 4H_2O$	21995-38-0* 422.342	monocl cry			s H_2O
73	Ammonium cerium(IV) nitrate $(NH_4)_2Ce(NO_3)_6$	16774-21-3 548.222	red-oran cry			vs H_2O
74	Ammonium chlorate NH_4ClO_3	10192-29-7 101.490	wh cry	102 exp 1.80		s H_2O
75	Ammonium chloride NH_4Cl	12125-02-9 53.492	col cub cry	520 tp (dec) 1.519	338 sp a,b,e	s H_2O
76	Ammonium chromate $(NH_4)_2CrO_4$	7788-98-9 152.071	yel cry	dec 185 1.90		s H_2O; sl ace, MeOH; i EtOH

No.	Name / Formula	CAS RN / Mol. Wt.	Physical Form	mp/°C / den/g cm⁻³	bp/°C / Other Data	Solubility
77	Ammonium chromic sulfate dodecahydrate $NH_4Cr(SO_4)_2 \cdot 12H_2O$	10022-47-6 478.345	blue-viol cry	94 dec 1.72		s H_2O; sl EtOH
78	Ammonium cobalt(II) phosphate $CoNH_4PO_4$	14590-13-7 171.943	red-viol powder; hyg			i H_2O; s acid
79	Ammonium cobalt(II) sulfate hexahydrate $(NH_4)_2Co(SO_4)_2 \cdot 6H_2O$	13586-38-4 395.229	red monocl prisms	1.90		s H_2O; i EtOH
80	Ammonium copper(II) chloride dihydrate $2NH_4Cl \cdot CuCl_2 \cdot 2H_2O$	10060-13-6 277.464	blue-grn tetr cry	≈115 dec 1.99		s H_2O, EtOH
81	Ammonium cyanide NH_4CN	12211-52-8 44.056	col tetr cry	dec 1.10	a,e	vs H_2O
82	Ammonium dichromate $(NH_4)_2Cr_2O_7$	7789-09-5 252.065	oran-red monocl cry; hyg	dec 180 2.155		vs H_2O
83	Ammonium dihydrogen arsenate $NH_4H_2AsO_4$	13462-93-6 158.975	tetr cry	300 dec 2.311		vs H_2O
84	Ammonium dihydrogen phosphate $NH_4H_2PO_4$	7722-76-1 115.026	wh tetr cry	190 1.80		vs H_2O; sl EtOH; i ace
85	Ammonium dithiocarbamate NH_4NH_2CSS	513-74-6 110.204	yel ortho cry	99 dec 1.45		s H_2O
86	Ammonium ferric chromate $NH_4Fe(CrO_4)_2$	7789-08-4 305.871	red powder			i H_2O
87	Ammonium ferric oxalate trihydrate $(NH_4)_3Fe(C_2O_4)_3 \cdot 3H_2O$	13268-42-3 428.065	grn monocl cry; hyg	≈160 dec 1.780		vs H_2O; i EtOH
88	Ammonium ferric sulfate dodecahydrate $NH_4Fe(SO_4)_2 \cdot 12H_2O$	10138-04-2 482.194	col to viol cry	≈37 1.71		vs H_2O; i EtOH
89	Ammonium ferricyanide trihydrate $(NH_4)_3Fe(CN)_6 \cdot 3H_2O$	14221-48-8* 320.113	red cry			s H_2O; i EtOH
90	Ammonium ferrocyanide trihydrate $(NH_4)_4Fe(CN)_6 \cdot 3H_2O$	14481-29-9* 338.151	yel cry	dec		s H_2O; i EtOH
91	Ammonium ferrous sulfate hexahydrate $(NH_4)_2Fe(SO_4)_2 \cdot 6H_2O$	10045-89-3 392.141	blue-grn monocl cry	dec ≈100 1.86		s H_2O; i EtOH
92	Ammonium fluoride NH_4F	12125-01-8 37.037	wh hex cry; hyg	dec 1.015		vs H_2O; sl EtOH
93	Ammonium fluoroborate NH_4BF_4	13826-83-0 104.844	wh powder; orth	487 dec 1.871	a	sl H_2O
94	Ammonium fluorosulfonate NH_4SO_3F	13446-08-7 117.101	col needles	245		s H_2O, EtOH, MeOH
95	Ammonium formate NH_4CHO_2	540-69-2 63.057	hyg cry	116 1.27		vs H_2O; s EtOH
96	Ammonium hexachloroiridate(IV) $(NH_4)_2IrCl_6$	16940-92-4 441.010	blk cry powder	dec 2.856		sl H_2O
97	Ammonium hexachloroosmiate(IV) $(NH_4)_2OsCl_6$	12125-08-5 439.02	red cry or powder	2.93	subl	s H_2O, EtOH
98	Ammonium hexachloropalladate(IV) $(NH_4)_2PdCl_6$	19168-23-1 355.21	red-brn hyg cry	dec 2.418		
99	Ammonium hexachloroplatinate(IV) $(NH_4)_2PtCl_6$	16919-58-7 443.87	red-oran cub cry	dec 380 3.065		sl H_2O; i EtOH
100	Ammonium hexafluoroaluminate $(NH_4)_3AlF_6$	7784-19-2 195.087	cub cry			s H_2O
101	Ammonium hexafluorogallate $(NH_4)_3GaF_6$	14639-94-2 237.828	col cub cry	1.78 dec >200 2.10		
102	Ammonium hexafluorogermanate $(NH_4)_2GeF_6$	16962-47-3 222.68	wh cry	380 2.564	subl	s H_2O; i EtOH
103	Ammonium hexafluorophosphate NH_4PF_6	16941-11-0 163.003	wh cub cry	dec 58 2.180		vs H_2O; s ace, EtOH, MeOH
104	Ammonium hexafluorosilicate $(NH_4)_2SiF_6$	16919-19-0 178.153	wh cub or trig cry	dec 2.011	a	s H_2O; i EtOH, ace
105	Ammonium hydrogen arsenate $(NH_4)_2HAsO_4$	7784-44-3 176.004	wh powder	1.99		s H_2O
106	Ammonium hydrogen borate trihydrate $NH_4HB_4O_7 \cdot 3H_2O$	10135-84-9 228.332	col cry	≈2.5		s H_2O
107	Ammonium hydrogen carbonate NH_4HCO_3	1066-33-7 79.056	col or wh prisms	107 dec 1.586	a	s H_2O; i EtOH, bz
108	Ammonium hydrogen citrate $(NH_4)_2HC_6H_5O_7$	3012-65-5 226.186	col cry	1.48		vs H_2O; sl EtOH
109	Ammonium hydrogen fluoride NH_4HF_2	1341-49-7 57.044	wh orth cry	125 1.50	240 dec	s H_2O
110	Ammonium hydrogen oxalate monohydrate $NH_4HC_2O_4 \cdot H_2O$	5972-72-5* 125.082	col rhomb cry	dec 1.56		sl H_2O, EtOH
111	Ammonium hydrogen phosphate $(NH_4)_2HPO_4$	7783-28-0 132.055	wh cry	155 dec 1.619		s H_2O; i EtOH, ace
112	Ammonium hydrogen sulfate NH_4HSO_4	7803-63-6 115.111	wh hyg cry	147 1.78	a	vs H_2O; i EtOH, ace, py
113	Ammonium hydrogen sulfide NH_4HS	12124-99-1 51.113	wh tetr or orth cry	dec 1.17	e	vs H_2O; sl ace; i bz, eth
114	Ammonium hydrogen sulfite NH_4HSO_3	10192-30-0 99.111	col cry	dec 2.03	a	vs H_2O

No.	Name Formula	CAS RN Mol. Wt.	Physical Form	mp/°C den/g cm⁻³	bp/°C Other Data	Solubility
115	Ammonium hydrogen tartrate $NH_4OOCCH(OH)CH(OH)COOH$	3095-65-6 167.118	wh cry	1.68		sl H_2O; s alk; i EtOH
116	Ammonium hydroxide NH_4OH	1336-21-6 35.046	exists only in soln		a	
117	Ammonium hypophosphite $NH_4H_2PO_2$	7803-65-8 83.028	wh hyg cry	dec		vs H_2O; sl EtOH; i ace
118	Ammonium iodate NH_4IO_3	13446-09-8 192.941	wh powder	150 3.3		vs H_2O; sl EtOH, MeOH
119	Ammonium iodide NH_4I	12027-06-4 144.943	wh tetr cry; hyg	551 dec 2.514	405 sp a,c,e	s H_2O, EtOH; sl MeOH; i ace, eth
120	Ammonium lactate $NH_4C_3H_5O_3$	52003-58-4 107.109	col cry	92		s H_2O; i EtOH
121	Ammonium metatungstate hexahydrate $(NH_4)_6W_7O_{24} \cdot 6H_2O$	12028-48-7 1887.19	wh cry			sl H_2O
122	Ammonium metavanadate NH_4VO_3	7803-55-6 116.979	wh-yel cry	dec 2.326		s H_2O; i EtOH
123	Ammonium molybdate(IV) tetrahydrate $(NH_4)_6Mo_7O_{24} \cdot 4H_2O$	12054-85-2 1235.86	col or grn-yel cry	dec 2.498		s H_2O
124	Ammonium nickel chloride hexahydrate $NH_4NiCl_3 \cdot 6H_2O$	16122-03-5* 291.181	grn hyg cry	1.65		sl H_2O; i EtOH
125	Ammonium nickel sulfate hexahydrate $(NH_4)_2Ni(SO_4)_2 \cdot 6H_2O$	7785-20-8 394.989	blue-grn cry	dec 1.923		vs H_2O; sl MeOH
126	Ammonium nitrate NH_4NO_3	6484-52-2 80.043	wh hyg cry; orth	dec 210; exp 1.72	a,c	s H_2O, EtOH
127	Ammonium nitroferricyanide $(NH_4)_2Fe(CN)_5NO$	14402-70-1 252.017	red-brn cry			s H_2O; sl ace
128	Ammonium oleate $NH_4C_{18}H_{33}O_2$	544-60-5 299.498	yel-brn paste	21		sl H_2O, EtOH
129	Ammonium oxalate monohydrate $(NH_4)_2C_2O_4 \cdot H_2O$	6009-70-7 142.111	wh orth cry	dec 1.50		s H_2O; sl bz, xyl; i ace, EtOH, ctc
130	Ammonium palmitate $NH_4C_{15}H_{31}COO$	593-26-0 273.460	yel-wh powder	22		vs H_2O
131	Ammonium pentachlorozincate $(NH_4)_3ZnCl_5$	14639-98-6 296.77	hyg orth cry	1.81		vs H_2O; s MeOH; sl EtOH, ace; i eth
132	Ammonium perchlorate NH_4ClO_4	7790-98-9 117.490	wh orth cry	dec, exp 1.95	a	sl H_2O
133	Ammonium permanganate NH_4MnO_4	13446-10-1 136.975	purp rhomb cry	dec 70 2.22		vs H_2O
134	Ammonium peroxydisulfate $(NH_4)_2S_2O_8$	7727-54-0 228.204	monocl cry or wh powder	dec 1.982		sl H_2O
135	Ammonium perrhenate NH_4ReO_4	13598-65-7 268.244	col powder	3.97		s H_2O
136	Ammonium phosphite, dibasic, monohydrate $(NH_4)_2HPO_3 \cdot H_2O$	51503-61-8 134.071	hyg cry			sl H_2O
137	Ammonium phosphomolybdate monohydrate $(NH_4)_3PO_4 \cdot 12MoO_3 \cdot H_2O$	54723-94-3 1894.36	yel cry or powder	dec		sl H_2O
138	Ammonium phosphotungstate dihydrate $(NH_4)_3PO_4 \cdot 12WO_3 \cdot 2H_2O$	1311-90-6 2967.18	cry powder			sl H_2O
139	Ammonium picrate $NH_4C_6H_2N_3O_7$	131-74-8 246.137	yel orth cry	exp 1.72		vs H_2O; s EtOH
140	Ammonium salicylate $NH_4C_7H_5O_3$	528-94-9 155.153	wh cry powder			vs H_2O; i EtOH, ace
141	Ammonium selenate $(NH_4)_2SeO_4$	7783-21-3 179.04	wh monocl cry	dec 2.194		s H_2O
142	Ammonium selenite $(NH_4)_2SeO_3$	7783-19-9 163.04	wh or red hyg cry	dec		sl H_2O, bz; s EtOH, MeOH; i ace
143	Ammonium stearate $NH_4C_{18}H_{35}O_2$	1002-89-7 301.514	yel-wh powder	22 0.89		vs H_2O; sl EtOH
144	Ammonium sulfamate $NH_4NH_2SO_3$	7773-06-0 114.125	wh hyg cry	131	dec 160	vs H_2O; i EtOH, ace
145	Ammonium sulfate $(NH_4)_2SO_4$	7783-20-2 132.141	wh or brn orth cry	dec 280 1.77	a,f	s H_2O, EtOH, alk
146	Ammonium sulfide $(NH_4)_2S$	12135-76-1 68.143	yel-oran cry	dec ≈0		s H_2O; i EtOH, ace
147	Ammonium sulfite monohydrate $(NH_4)_2SO_3 \cdot H_2O$	7783-11-1 134.156	col cry	dec 1.41		s H_2O
148	Ammonium tartrate $(NH_4)_2C_4H_4O_6$	3164-29-2 184.148	wh cry	dec 1.601		
149	Ammonium tellurate $(NH_4)_2TeO_4$	13453-06-0 227.68	wh powder	dec 3.024		s H_2O, eth
150	Ammonium tetrachloroaluminate NH_4AlCl_4	7784-14-7 186.832	wh hyg solid	304		s H_2O; i EtOH
151	Ammonium tetrachloroplatinate(II) $(NH_4)_2PtCl_4$	13820-41-2 372.97	red cry	dec 2.936		vs H_2O
152	Ammonium tetrachlorozincate $(NH_4)_2ZnCl_4$	14639-97-5 243.28	wh orth plates; hyg	150 dec 1.879		

No.	Name / Formula	CAS RN / Mol. Wt.	Physical Form	mp/°C / den/g cm⁻³	bp/°C / Other Data	Solubility
153	Ammonium tetrathiotungstate $(NH_4)_2WS_4$	13862-78-7 / 348.18	oran cry	dec / 2.71		s H_2O
154	Ammonium thiocyanate NH_4SCN	1762-95-4 / 76.122	col hyg cry	≈149 / 1.30	dec	vs H_2O, EtOH; s ace; i chl
155	Ammonium thiosulfate $(NH_4)_2S_2O_3$	7783-18-8 / 148.207	wh cry	dec 150 / 1.678		vs H_2O; i EtOH, eth
156	Ammonium titanium oxalate monohydrate $(NH_4)_2TiO(C_2O_4)_2 \cdot H_2O$	10580-03-7 / 293.997	hyg cry			vs H_2O
157	Ammonium uranate(VI) $(NH_4)_2U_2O_7$	7783-22-4 / 624.131	red-yel amorp powder			i H_2O, akl; s acid
158	Ammonium uranium fluoride $UO_2(NH_4)_3F_5$	18433-40-4 / 419.135	grn-yel monocl cry			s H_2O; i EtOH
159	Ammonium valerate $NH_4C_4H_9COO$	42739-38-8 / 119.164	hyg cry	108		vs H_2O, EtOH; s eth
160	Ammonium zirconyl carbonate dihydrate $(NH_4)_3ZrOH(CO_3)_3 \cdot 2H_2O$	12616-24-9* / 362.405	prisms; unstable			s H_2O
161	Antimony Sb	7440-36-0 / 121.760	silv metal; hex	630.63 / 6.68	1587 / a,c,e,f	i dil acid
162	Stibine SbH_3	7803-52-3 / 124.784	col gas; flam	-88 / 5.475 g/L	-17 / a,d	sl H_2O; s EtOH
163	Antimony arsenide $SbAs$	12322-34-8 / 196.682	hex cry	≈680 / 6.0		
164	Antimony(III) bromide $SbBr_3$	7789-61-9 / 361.472	yel orth cry; hyg	96.6 / 4.35	280 / a,b,d,e	reac H_2O; s ace, bz, chl
165	Antimony(III) chloride $SbCl_3$	10025-91-9 / 228.118	col orth cry; hyg	73.4 / 3.14	220.3 / a,b,c,d,e	vs H_2O; s acid, EtOH, bz, ace
166	Antimony(III) fluoride SbF_3	7783-56-4 / 178.755	wh orth cry; hyg	292 / 4.38	≈345 / a	vs H_2O
167	Antimony(III) iodide SbI_3	7790-44-5 / 502.473	red rhomb cry	168 / 4.92	401 / a,b,d,e	reac H_2O; s EtOH, ace; i ctc
168	Antimony(III) oxide (senarmontite) Sb_2O_3	1309-64-4 / 291.518	col cub cry	570 trans / 5.58	1425 / e,f	sl H_2O; i os
169	Antimony(III) oxide (valentinite) Sb_2O_3	1309-64-4 / 291.518	wh orth cry	655 / 5.7	1425 / e,f	sl H_2O; i os
170	Antimony(III) oxychloride $SbOCl$	7791-08-4 / 173.212	wh momo cry	170 dec		reac H_2O; i EtOH, eth
171	Antimony(III) selenide Sb_2Se_3	1315-05-5 / 480.40	grn orth cry	611 / 5.81		sl H_2O
172	Antimony(III) sulfate $Sb_2(SO_4)_3$	7446-32-4 / 531.711	wh cry powder; hyg	dec / 3.62		sl H_2O
173	Antimony(III) sulfide Sb_2S_3	1345-04-6 / 339.718	gray-blk orth cry	550 / 4.562	f	i H_2O; s conc HCl
174	Antimony(III) telluride Sb_2Te_3	1327-50-0 / 626.32	gray cry	620 / 6.5		
175	Antimony(III,V) oxide Sb_2O_4	1332-81-6 / 307.518	yel orth cry	/ 6.64	f	
176	Antimony(V) chloride $SbCl_5$	7647-18-9 / 299.024	col or yel liq	4 / 2.34	140 dec / e	reac H_2O; s chl, ctc
177	Antimony(V) fluoride SbF_5	7783-70-2 / 216.752	hyg visc liq	8.3 / 3.10	141	reac H_2O
178	Antimony(V) dichlorotrifluoride $SbCl_2F_3$	7791-16-4 / 249.660	visc liq			reac H_2O
179	Antimony(V) oxide Sb_2O_5	1314-60-9 / 323.517	yel powder; cub	dec / 3.78	a	sl H_2O
180	Antimony(V) sulfide Sb_2S_5	1315-04-4 / 403.850	oran-yel powder	75 dec / 4.120		i H_2O; s acid, alk
181	Argon Ar	7440-37-1 / 39.948	col gas	-189.35 / 1.753 g/L	-185.85 / a,b,c,d,e	sl H_2O
182	Arsenic As	7440-38-2 / 74.922	gray metal; rhomb	817 tp / 5.75	614 sp / a,b,c,e,f	i H_2O
183	Arsine AsH_3	7784-42-1 / 77.946	col gas	-116 / 3.420 g/L	-62.5 / a,b,d,e	sl H_2O
184	Diarsine As_2H_4	15942-63-9 / 153.875	unstable liq		≈100	
185	Arsenic acid H_3AsO_4	7778-39-4 / 141.944	exists only in soln			
186	Arsenic acid hemihydrate $H_3AsO_4 \cdot 0.5H_2O$	7778-39-4* / 150.951	wh hyg cry	35.5 / ≈2	a	vs H_2O, EtOH
187	Arsenious acid $H_3AsO_3 \cdot$	13464-58-9 / 125.944	exists only in soln			
188	Arsenic diiodide As_2I_4	13770-56-4 / 657.461	red cry	137		reac H_2O; s os
189	Arsenic hemiselenide As_2Se	1303-35-1 / 228.80	blk cry			i H_2O, os; dec acid, alk
190	Arsenic sulfide As_4S_4	12279-90-2 / 427.950	red monocl cry	320 / 3.5	565 / f	i H_2O; sl bz; s alk

No.	Name Formula	CAS RN Mol. Wt.	Physical Form	mp/°C den/g cm^{-3}	bp/°C Other Data	Solubility
191	Arsenic(III) bromide $AsBr_3$	7784-33-0 314.634	yel orth cry; hyg	31.1 3.40	221 a,c,d,e	reac H_2O; s hc, ctc; vs eth, bz
192	Arsenic(III) chloride $AsCl_3$	7784-34-1 181.280	col liq	-16 2.150	130 a,b,c,d,e	reac H_2O; vs chl, ctc, eth
193	Arsenic(III) fluoride AsF_3	7784-35-2 131.917	col liq	-5.9 2.7	57.8 a,c,d,e	reac H_2O; s EtOH, eth, bz
194	Arsenic(III) iodide AsI_3	7784-45-4 455.635	red hex cry	140.9 4.73	424 a,d	sl H_2O, EtOH, eth; s bz. tol
195	Arsenic(III) oxide (arsenolite) As_2O_3	1327-53-3 197.841	wh cub cry	274 3.86	460 e,f	sl H_2O
196	Arsenic(III) oxide (claudetite) As_2O_3	1327-53-3 197.841	wh monocl cry	313 3.74	460 e,f	sl H_2O; s dil acid, alk; i EtOH
197	Arsenic(III) selenide As_2Se_3	1303-36-2 386.72	brn-blk solid	260 4.75		i H_2O; s alk
198	Arsenic(III) sulfide As_2S_3	1303-33-9 246.041	yel-oran monocl cry	310 3.46	707 a,f	i H_2O; s alk
199	Arsenic(III) telluride As_2Te_3	12044-54-1 532.64	blk monocl cry	621 6.50		
200	Arsenic(V) chloride $AsCl_5$	22441-45-8 252.186	stable at low temp	≈-50 dec		
201	Arsenic(V) fluoride AsF_5	7784-36-3 169.914	col gas	-79.8 7.456 g/L	-52.8 d,e	reac H_2O; s EtOH, bz, eth
202	Arsenic(V) oxide As_2O_5	1303-28-2 229.840	wh amorp powder	315 4.32	a	vs H_2O, EtOH
203	Arsenic(V) selenide As_2Se_5	1303-37-3 544.64	blk solid	dec		i H_2O, EtOH, eth; s alk
204	Arsenic(V) sulfide As_2S_5	1303-34-0 310.173	brn-yel amorp solid	dec		i H_2O; s alk
205	Astatine At	7440-68-8 210	cry	302	a	s HNO_3, os
206	Barium Ba	7440-39-3 137.327	silv-yel metal; cub	727 3.62	1897 a,c,d,e	reac H_2O; sl EtOH
207	Barium acetate $Ba(C_2H_3O_2)_2$	543-80-6 255.417	wh powder	 2.47		vs H_2O
208	Barium acetate monohydrate $Ba(C_2H_3O_2)_2 \cdot H_2O$	5908-64-5 273.432	wh cry	dec 110 2.19		vs H_2O; sl EtOH
209	Barium aluminate $BaAl_2O_4$	12004-04-5 255.288	hex cry	1827		
210	Barium azide $Ba(N_3)_2$	18810-58-7 221.367	monocl cry; exp	dec ≈120 2.936		s H_2O; sl EtOH; i eth
211	Barium bromate monohydrate $Ba(BrO_3)_2 \cdot H_2O$	10326-26-8 411.147	wh monocl cry	dec 260 3.99		sl H_2O; i EtOH
212	Barium bromide $BaBr_2$	10553-31-8 297.135	wh orth cry	857 4.781	1835 a,c	vs H_2O
213	Barium bromide dihydrate $BaBr_2 \cdot 2H_2O$	7791-28-8 333.166	wh cry	dec 75 3.7		vs H_2O; s MeOH; i EtOH, ace, diox
214	Barium carbide BaC_2	50813-65-5 161.349	gray tetr cry	dec 3.74		reac H_2O
215	Barium carbonate $BaCO_3$	513-77-9 197.336	wh orth cry	1555 4.2865	a,f	i H_2O; s acid
216	Barium chlorate monohydrate $Ba(ClO_3)_2 \cdot H_2O$	10294-38-9 322.244	wh monocl cry	414 3.179		s H_2O, acid; sl EtOH, ace
217	Barium chloride $BaCl_2$	10361-37-2 208.232	wh orth cry; hyg	962 3.9	1560 a,c	s H_2O
218	Barium chloride dihydrate $BaCl_2 \cdot 2H_2O$	10326-27-9 244.263	wh monocl cry	dec ≈120 3.097		s H_2O; i EtOH
219	Barium chromate(V) $Ba_3(CrO_4)_2$	12345-14-1 643.968	grn-blk hex cry	 5.25		s H_2O
220	Barium chromate(VI) $BaCrO_4$	10294-40-3 253.321	yel orth cry	dec 4.50		i H_2O; reac acid
221	Barium citrate monohydrate $Ba_3(C_6H_5O_7)_2 \cdot H_2O$	512-25-4* 808.199	gray-wh cry			s H_2O, acid
222	Barium cyanide $Ba(CN)_2$	542-62-1 189.362	wh cry powder			vs H_2O; s EtOH
223	Barium dichromate dihydrate $BaCr_2O_7 \cdot 2H_2O$	10031-16-0 389.346	brn-red needles	dec		reac H_2O
224	Barium dithionate dihydrate $Ba(SO_3)_2 \cdot 2H_2O$	13845-17-5 333.486	cry	 4.54		s H_2O; sl EtOH
225	Barium ferrocyanide hexahydrate $Ba_2Fe(CN)_6 \cdot 6H_2O$	13821-06-2* 594.696	yel monocl cry	dec 80		i H_2O, EtOH
226	Barium fluoride BaF_2	7787-32-8 175.324	wh cub cry	1368 4.893	2260 a,c	sl H_2O
227	Barium formate $Ba(CHO_2)_2$	541-43-5 227.363	cry	 3.21		s H_2O; i EtOH
228	Barium hexaboride BaB_6	12046-08-1 202.193	blk cub cry	2070 4.36		i H_2O; s acid; i EtOH
229	Barium hexafluorosilicate $BaSiF_6$	17125-80-3 279.403	wh orth needles	dec 300 4.29		i H_2O, EtOH; sl acid

No.	Name / Formula	CAS RN / Mol. Wt.	Physical Form	mp/°C / den/g cm⁻³	bp/°C / Other Data	Solubility
230	Barium hydride / BaH_2	13477-09-3 / 139.343	gray orth cry	1200 / 4.16		reac H_2O
231	Barium Hydrogen phosphate / $BaHPO_4$	10048-98-3 / 233.306	wh cry powder	dec 400 / 4.16	a	i H_2O; s dil acid
232	Barium hydrosulfide / $Ba(HS)_2$	25417-81-6 / 203.475	yel hyg cry			s H_2O
233	Barium hydrosulfide tetrahydrate / $Ba(HS)_2 \cdot 4H_2O$	12230-74-9 / 275.536	yel rhomb cry	dec 50		s H_2O
234	Barium hydroxide / $Ba(OH)_2$	17194-00-2 / 171.342	wh powder	408		sl H_2O
235	Barium hydroxide monohydrate / $Ba(OH)_2 \cdot H_2O$	22326-55-2 / 189.357	wh powder		a,c	sl H_2O; s acid
236	Barium hydroxide octahydrate / $Ba(OH)_2 \cdot 8H_2O$	12230-71-6 / 315.464	wh monocl cry	3.743 / dec 78		sl H_2O
237	Barium hypophosphite monohydrate / $Ba(H_2PO_2)_2 \cdot H_2O$	14871-79-5* / 285.320	monocl plates	2.18		s H_2O; i EtOH
238	Barium iodate / $Ba(IO_3)_2$	10567-69-8 / 487.132	wh cry powder	2.90 / dec 476		sl H_2O
239	Barium iodate monohydrate / $Ba(IO_3)_2 \cdot H_2O$	7787-34-0 / 505.148	cry	5.23 / dec 130		sl H_2O; s acid; i EtOH
240	Barium iodide / BaI_2	13718-50-8 / 391.136	wh orth cry	5.00 / 711		vs H_2O
241	Barium iodide dihydrate / $BaI_2 \cdot 2H_2O$	7787-33-9 / 427.167	col cry	5.15 / 740 dec	a,c	vs H_2O; s EtOH, ace
242	Barium manganate(VI) / $BaMnO_4$	7787-35-1 / 256.263	grn-gray hyg cry	5.0		reac H_2O
243	Barium metaborate monohydrate / $Ba(BO_2)_2 \cdot H_2O$	26124-86-7 / 240.962	wh powder	4.85 / >900		sl H_2O
244	Barium molybdate / $BaMoO_4$	7787-37-3 / 297.27	wh powder	3.3 / 1450		i H_2O
245	Barium niobate / $Ba(NbO_3)_2$	12009-14-2 / 419.136	yel orth cry	4.975 / 1455		i H_2O
246	Barium nitrate / $Ba(NO_3)_2$	10022-31-8 / 261.336	wh cub cry	5.44 / 590		s H_2O; sl EtOH, ace
247	Barium nitride / Ba_3N_2	12047-79-9 / 439.994	yel-brn cry	3.24	a	reac H_2O
248	Barium nitrite / $Ba(NO_2)_2$	13465-94-6 / 229.338	col hex cry	4.78 / 267		vs H_2O; i EtOH
249	Barium nitrite monohydrate / $Ba(NO_2)_2 \cdot H_2O$	7787-38-4 / 247.353	yel-wh hex cry	3.234 / 217 dec	a	i H_2O
250	Barium oxalate / BaC_2O_4	516-02-9 / 225.347	wh powder	3.18 / 400 dec		i H_2O; s acid
251	Barium oxalate monohydrate / $BaC_2O_4 \cdot H_2O$	13463-22-4 / 243.362	wh cry powder	2.658		sl H_2O; s dil acid, EtOH; i ace
252	Barium oxide / BaO	1304-28-5 / 153.326	wh-yel powder; cub and hex	2.66 / 1973	a,c	vs H_2O, EtOH
253	Barium perchlorate / $Ba(ClO_4)_2$	13465-95-7 / 336.227	col hex cry	5.72(cub) / 505		s H_2O, MeOH; sl EtOH, ace; i eth
254	Barium perchlorate trihydrate / $Ba(ClO_4)_2 \cdot 3H_2O$	10294-39-0 / 390.273	col cry	3.20		sl H_2O; reac EtOH
255	Barium permanganate / $Ba(MnO_4)_2$	7787-36-2 / 375.198	brn-viol cry	2.74 / dec 200		i H_2O; reac dil acid
256	Barium peroxide / BaO_2	1304-29-6 / 169.326	gray-wh tetr cry	3.77 / 450 dec		i H_2O; sl acid
257	Barium metaphosphate / $Ba(PO_3)_2$	13466-20-1 / 295.271	wh powder	4.96 / 1560		vs H_2O
258	Barium potassium chromate / $BaK_2(CrO_4)_2$	27133-66-0 / 447.511	yel hex cry			sl H_2O; s acid
259	Barium pyrophosphate / $Ba_2P_2O_7$	13466-21-2 / 448.597	wh powder	3.63		i H_2O
260	Barium selenate / $BaSeO_4$	7787-41-9 / 280.29	wh rhomb cry	3.9 / dec		reac H_2O
261	Barium selenide / $BaSe$	1304-39-8 / 216.29	cub cry powder	4.75 / 1780		i H_2O
262	Barium selenite / $BaSeO_3$	13718-59-7 / 264.29	solid	5.02		i H_2O; s acid
263	Barium disilicate / $BaSi_2O_5$	12650-28-1 / 273.495	wh orth cry	3.70 / 1420		reac H_2O
264	Barium metasilicate / $BaSiO_3$	13255-26-0 / 213.411	col rhomb powder	4.40 / 1605		i H_2O
265	Barium silicide / $BaSi_2$	1304-40-1 / 193.498	gray lumps	5.40 / 1180		sl H_2O
266	Barium sodium niobate / $Ba_2Na(NbO_3)_5$	12323-03-4 / 1002.167	wh orth cry	7.24 / 1437		sl H_2O; s acid
267	Barium stannate / $BaSnO_3$	12009-18-6 / 304.035	cub cry			
268	Barium stannate trihydrate / $BaSnO_3 \cdot 3H_2O$	12009-18-6* / 358.081	wh cry powder			

No.	Name Formula	CAS RN Mol. Wt.	Physical Form	mp/°C den/g cm^{-3}	bp/°C Other Data	Solubility
269	Barium stearate $Ba(C_{18}H_{35}O_2)_2$	6865-35-6 704.277	wh powder	160 1.145		i H_2O, alc
270	Barium sulfate $BaSO_4$	7727-43-7 233.391	wh orth cry	1580 4.49	a,c,f	i H_2O, EtOH
271	Barium sulfide BaS	21109-95-5 169.393	col cub cry or gray powder	2230 4.3	a	sl H_2O
272	Barium sulfite $BaSO_3$	7787-39-5 217.391	wh monocl cry	dec 4.44		i H_2O, EtOH
273	Barium tartrate $BaC_4H_4O_6$	5908-81-6 285.399	wh cry	 2.98		s H_2O; i EtOH
274	Barium tetracyanoplatinate(II) tetrahydrate $BaPt(CN)_4·4H_2O$	13755-32-3 508.54	yel powder or cry	 2.076		sl H_2O; i EtOH
275	Barium tetraiodomercurate(II) $BaHgI_4$	10048-99-4 845.54	yel-red hyg cry			vs H_2O, EtOH
276	Barium thiocyanate $Ba(SCN)_2$	2092-17-3 253.494	hyg cry			vs H_2O; s ace, MeOH, EtOH
277	Barium thiocyanate dihydrate $Ba(SCN)_2·2H_2O$	2092-17-3* 289.525	hyg wh cry			s H_2O, EtOH
278	Barium thiocyanate trihydrate $Ba(SCN)_2·3H_2O$	68016-36-4 307.540	wh needles; hyg	 2.286		s H_2O, EtOH
279	Barium thiosulfate BaS_2O_3	35112-53-9 249.457	wh cry powder	dec 220		sl H_2O; i EtOH
280	Barium thiosulfate monohydrate $BaS_2O_3·H_2O$	7787-40-8 267.473	wh cry powder	dec 3.5		sl H_2O; i EtOH
281	Barium titanate $BaTiO_3$	12047-27-7 233.192	wh tetr cry	1625 6.02		i H_2O
282	Barium tungstate $BaWO_4$	7787-42-0 385.17	wh tetr cry	 5.04		i H_2O
283	Barium uranium oxide BaU_2O_7	10380-31-1 725.381	oran-yel powder			i H_2O; s acid
284	Barium orthovanadate $Ba_3(VO_4)_2$	39416-30-3 641.859	hex cry	707 5.14		
285	Barium zirconate $BaZrO_3$	12009-21-1 276.549	gray-wh cub cry	2500 5.52		i H_2O, alk; sl acid
286	Berkelium (α form) Bk	7440-40-6 247	hex	1050 14.78	a	
287	Berkelium (β form) Bk	7440-40-6 247	cub cry	986 13.25	a	
288	Beryllium Be	7440-41-7 9.012	hex	1287 1.85	2471 a,c,e	s acid, alk
289	Beryllium acetate $Be(C_2H_3O_2)_2$	543-81-7 127.102	wh cry	dec 60		i H_2O, EtOH
290	Beryllium acetylacetonate $Be(C_5H_7O_2)_2$	10210-64-7 207.231	monocl cry powder	108 1.168	270	i H_2O, vs EtOH, eth
291	Beryllium aluminate $BeAl_2O_4$	12004-06-7 126.973	orth cry	 3.65	f	
292	Beryllium aluminum metasilicate $Be_3Al_2(SiO_3)_6$	1302-52-9 537.502	col or grn-yel cry; hex	 2.64	f	
293	Beryllium basic acetate $Be_4O(C_2H_3O_2)_6$	1332-52-1 406.316	wh cry	285 1.25	330	i H_2O; s eth, os
294	Beryllium boride BeB_2	12228-40-9 30.634	refrac solid	>1970		
295	Beryllium borohydride $Be(BH_4)_2$	17440-85-6 36.682	solid	dec 125; flam	subl e	reac H_2O
296	Beryllium bromide $BeBr_2$	7787-46-4 168.820	orth cry; hyg	508 3.465	520 a,c,e	vs H_2O; s EtOH, pyr
297	Beryllium carbide Be_2C	506-66-1 30.035	red cub cry	dec >2100 1.90		reac H_2O
298	Beryllium carbonate tetrahydrate $BeCO_3·4H_2O$	60883-64-9 93.085	unstable solid			
299	Beryllium chloride $BeCl_2$	7787-47-5 79.917	wh-yel orth cry; hyg	415 1.90	482 a,c,d,e	vs H_2O; s EtOH, eth, py; i bz, tol
300	Beryllium fluoride BeF_2	7787-49-7 47.009	tetr cry or gl; hyg	552 2.1	1169 a,c	vs H_2O; sl EtOH
301	Beryllium formate $Be(CHO_2)_2$	1111-71-3 99.048	powder	dec >250		reac H_2O; i os
302	Beryllium hydride BeH_2	7787-52-2 11.028	wh amorp solid	dec 250 0.65		reac H_2O; i eth, tol
303	Beryllium hydroxide $Be(OH)_2$	13327-32-7 43.027	amorp powder or cry	 1.92	a	sl H_2O, alk; s acid
304	Beryllium iodide BeI_2	7787-53-3 262.821	hyg needles	470 4.32	487 a,c,d,e	reac H_2O; s EtOH
305	Beryllium nitrate trihydrate $Be(NO_3)_2·3H_2O$	13597-99-4 187.068	yel-wh hyg cry	≈30	dec	vs H_2O; s EtOH
306	Beryllium nitride Be_3N_2	1304-54-7 55.050	gray refrac cry; cub	2200 2.71		reac acid, alk
307	Beryllium oxide BeO	1304-56-9 25.011	wh amorp powder	2578 3.01	a,c	i H_2O; sl acid, alk

No.	Name / Formula	CAS RN / Mol. Wt.	Physical Form	mp/°C / den/g cm^{-3}	bp/°C / Other Data	Solubility
308	Beryllium perchlorate tetrahydrate $Be(ClO_4)_2 \cdot 4H_2O$	7787-48-6 / 279.974	hyg cry			vs H_2O
309	Beryllium selenate tetrahydrate $BeSeO_4 \cdot 4H_2O$	10039-31-3 / 224.03	orth cry	dec 100 / 2.03		vs H_2O
310	Beryllium sulfate $BeSO_4$	13510-49-1 / 105.076	col tetr cry; hyg	1127 / 2.5		s H_2O
311	Beryllium sulfate tetrahydrate $BeSO_4 \cdot 4H_2O$	7787-56-6 / 177.137	col tetr cry	dec ≈100 / 1.71	a	vs H_2O; i EtOH
312	Beryllium sulfide BeS	13598-22-6 / 41.078	col cub cry	dec / 2.36		reac hot H_2O
313	Bismuth Bi	7440-69-9 / 208.980	gray-wh soft metal	271.40 / 9.79	1564 / a	s acid
314	Bismuth basic carbonate $(BiO)_2CO_3$	5892-10-4 / 509.969	wh powder	6.86	a,c,d,e,f	i H_2O; s acid
315	Bismuth bromide $BiBr_3$	7787-58-8 / 448.692	yel cub cry	218 / 5.72	453 / b,d,e	reac H_2O; s dil acid, ace; i EtOH
316	Bismuth chloride $BiCl_3$	7787-60-2 / 315.338	yel-wh cub cry; hyg	230 / 4.75	447 / a,b,c,d,e	reac H_2O; s acid, EtOH, ace
317	Bismuth citrate $BiC_6H_5O_7$	813-93-4 / 398.082	wh powder	3.458		i H_2O; sl EtOH
318	Bismuth fluoride BiF_3	7787-61-3 / 265.975	wh-gray cub cry	725 / 8.3	900	i H_2O
319	Bismuth pentafluoride BiF_5	7787-62-4 / 303.972	wh tetr needles; hyg	154 / 5.55	230	reac H_2O
320	Bismuth hydride BiH_3	18288-22-7 / 212.004	col gas; unstable	-67 / 9.303 g/L	≈17	
321	Bismuth hydroxide $Bi(OH)_3$	10361-43-0 / 260.002	wh-yel amorp powder	4.962		i H_2O; s acid
322	Bismuth iodide BiI_3	7787-64-6 / 589.693	blk hex cry	408.6 / 5.778	542 / a	i H_2O; s EtOH
323	Bismuth molybdate $Bi_2(MoO_4)_3$	51898-99-8 / 897.77	monocl cry	5.95	a	
324	Bismuth nitrate pentahydrate $Bi(NO_3)_3 \cdot 5H_2O$	10035-06-0 / 485.071	col tricl cry; hyg	≈75 dec / 2.83		reac H_2O; s ace; i EtOH
325	Bismuth oleate $Bi(C_{18}H_{33}O_2)_3$	52951-38-9 / 1053.356	soft yel-brn solid			i H_2O; s eth; sl bz
326	Bismuth orthovanadate $BiVO_4$	14059-33-7 / 323.920	oran-yel orth cry	trans 500 / 6.25		
327	Bismuth oxalate $Bi_2(C_2O_4)_3$	6591-55-5 / 682.020	wh powder			i H_2O, EtOH; s dil acid
328	Bismuth oxide Bi_2O_3	1304-76-3 / 465.959	yel monocl cry or powder	817 / 8.9	1890 / a	i H_2O, s acid
329	Bismuth oxybromide $BiOBr$	7787-57-7 / 304.883	col tetr cry	8.08		i H_2O, EtOH; s acid
330	Bismuth oxychloride $BiOCl$	7787-59-9 / 260.432	wh tetr cry	7.72	a	i H_2O
331	Bismuth oxyiodide $BiOI$	7787-63-5 / 351.883	red tetr cry	>300 dec / 7.92		i H_2O, EtOH, chl; s HCl
332	Bismuth oxynitrate $BiONO_3$	10361-46-3 / 286.985	wh powder	260 dec / 4.93		i H_2O, EtOH; s acid
333	Bismuth phosphate $BiPO_4$	10049-01-1 / 303.951	monocl cry	6.32		sl H_2O, dil acid; i EtOH
334	Bismuth selenide Bi_2Se_3	12068-69-8 / 654.84	blk hex cry	710 dec / 7.5		i H_2O
335	Bismuth stannate pentahydrate $Bi_2(SnO_3)_3 \cdot 5H_2O$	12777-45-6 / 1008.162	wh cry			i H_2O
336	Bismuth sulfate $Bi_2(SO_4)_3$	7787-68-0 / 706.152	wh needles or powder	405 dec / 5.08	a	reac H_2O, EtOH
337	Bismuth sulfide Bi_2S_3	1345-07-9 / 514.159	blk-brn orth cry	850 / 6.78	a,f	i H_2O; s acid
338	Bismuth telluride Bi_2Te_3	1304-82-1 / 800.76	gray hex plates	580 / 7.74	f	i H_2O; s EtOH
339	Bismuth tetroxide Bi_2O_4	12048-50-9 / 481.959	red-oran powder	305 / 5.6		reac H_2O
340	Bismuth titanate $Bi_4(TiO_4)_3$	12048-51-0 / 1171.515	wh orth cry	7.85		
341	Boron B	7440-42-8 / 10.811	blk rhomb cry	2075 / 2.34	4000 / a,c,d	i H_2O
342	Diborane(6) B_2H_6	19287-45-7 / 27.670	col gas; flam	-165.5 / 1.214 g/L	-92.4 / a,b,d,e	reac H_2O
343	Tetraborane(10) B_4H_{10}	18283-93-7 / 53.323	col gas	-121 / 2.340 g/L	18 / a,d,e	reac H_2O
344	Pentaborane(9) B_5H_9	19624-22-7 / 63.126	flam liq	-46.6 / 0.60	60 / a,e	reac hot H_2O
345	Pentaborane(11) B_5H_{11}	18433-84-6 / 65.142	col liq; unstable	-123	63 / a,d,e	reac H_2O
346	Hexaborane(10) B_6H_{10}	23777-80-2 / 74.945	col liq	-62.3 / 0.67	108 dec / a	reac hot H_2O

No.	Name Formula	CAS RN Mol. Wt.	Physical Form	mp/°C den/g cm^{-3}	bp/°C Other Data	Solubility
347	Hexaborane(12) B_6H_{12}	12008-19-4 76.961	col liq; unstable	-82	≈80	reac H_2O
348	Nonaborane(15) B_9H_{15}	19465-30-6 112.418	col liq	2.6		sl H_2O; s EtOH, bz, CS_2, ctc
349	Decaborane(14) $B_{10}H_{14}$	17702-41-9 122.221	wh orth cry	99.6 0.94	≈213 e	reac H_2O
350	Borane carbonyl BH_3CO	13205-44-2 41.845	col gas	-137	-64 e	reac H_2O
351	Borazine $B_3N_3H_6$	6569-51-3 80.501	col liq	-58 0.824	53 a	sl H_2O, EtOH
352	Boric acid (orthoboric acid) $B(OH)_3$	10043-35-3 61.833	col tricl cry	170.9 1.5	a	s H_2O
353	Metaboric acid (α form) HBO_2	13460-50-9 43.818	col orth cry; hyg	176 1.784	a	s H_2O
354	Metaboric acid (β form) HBO_2	13460-50-9 43.818	col monocl cry; hyg	201 2.045	a	s H_2O
355	Metaboric acid (γ form) HBO_2	13460-50-9 43.818	col cub cry	236 2.487	a	vs H_2O, EtOH
356	Fluoroboric acid HBF_4	16872-11-0 87.813	col liq	≈1.8	130 dec	
357	Boron arsenide BAs	12005-69-5 85.733	cub cry	dec 920 5.22		reac H_2O, EtOH
358	Boron bromide BBr_3	10294-33-4 250.523	col liq; hyg	-45 2.6	91 a,b,d,e	i H_2O, acid
359	Boron carbide B_4C	12069-32-8 55.255	hard blk cry	2350 2.50	>3500	reac H_2O, EtOH
360	Boron chloride BCl_3	10294-34-5 117.169	col liq or gas	-107 5.141 g/L	12.65 a,b,c,d,e	reac H_2O
361	Boron chloride B_2Cl_4	13701-67-2 163.433	col liq; flam	-92.6	65 a	s H_2O
362	Boron fluoride BF_3	7637-07-2 67.806	col gas	-126.8 2.975 g/L	-101 a,b,c,d,e	reac H_2O
363	Boron fluoride B_2F_4	13965-73-6 97.616	col gas; flam	-56 4.283 g/L	-34 a,d	
364	Boron iodide BI_3	13517-10-7 391.524	needles	43 3.96	210 a,b,d	i H_2O, acid
365	Boron nitride BN	10043-11-5 24.818	wh powder; hex or cub cry	2967 2.18	a	sl H_2O; s EtOH
366	Boron oxide B_2O_3	1303-86-2 69.620	col gl or hex cry; hyg	450 2.55	a,c	reac H_2O, acid
367	Boron phosphide BP	20205-91-8 41.785	red cub cry or powder	dec 1125		
368	Boron sulfide B_2S_3	12007-33-9 117.820	yel amorp solid	softens ≈320 ≈1.7	a	sl H_2O
369	Bromine Br_2	7726-95-6 159.808	red liq	-7.2 3.1028	58.8 a,b,c,d,e	s H_2O
370	Bromic acid $HBrO_3$	7789-31-3 128.910	stable only in aq soln			
371	Bromine oxide Br_2O	21308-80-5 175.807	brn solid	-17.5 dec		
372	Bromine dioxide BrO_2	21255-83-4 111.903	unstable yel cry	dec ≈0	a	
373	Bromine azide BrN_3	13973-87-0 121.924	red cry; exp	≈45	exp	
374	Bromine chloride BrCl	13863-41-7 115.357	unstable red-brn gas	≈-66 5.062 g/L	≈5 dec a	reac H_2O; s eth, CS_2
375	Bromine fluoride BrF	13863-59-7 98.902	unstable red-brn gas	≈-33 4.340 g/L	≈20 dec a,d	
376	Bromine trifluoride BrF_3	7787-71-5 136.899	col hyg liq	8.77 2.803	125.8 a,d	reac H_2O
377	Bromine pentafluoride BrF_5	7789-30-2 174.896	col liq	-60.5 2.460	40.76 a,c,d,e	reac H_2O (exp)
378	Cadmium Cd	7440-43-9 112.411	silv-wh metal	321.07 8.69	767 a,c,d,e	i H_2O; reac acid
379	Cadmium acetate $Cd(C_2H_3O_2)_2$	543-90-8 230.501	col cry	255 2.34		s H_2O, EtOH
380	Cadmium acetate dihydrate $Cd(C_2H_3O_2)_2 \cdot 2H_2O$	5743-04-4 266.530	wh cry	dec 130 2.01		vs H_2O; s EtOH
381	Cadmium antimonide CdSb	12014-29-8 234.171	orth cry	456 6.92		
382	Cadmium arsenide Cd_3As_2	12006-15-4 487.076	gray tetr cry	721 6.25		
383	Cadmium azide $Cd(N_3)_2$	14215-29-3 196.451	yel-wh orth cry; exp	exp 3.24		
384	Cadmium bromide $CdBr_2$	7789-42-6 272.219	wh hex powder or flakes; hyg	568 5.19	844 a,c,d	s H_2O; sl ace, eth
385	Cadmium bromide tetrahydrate $CdBr_2 \cdot 4H_2O$	13464-92-1 344.281	wh-yel cry			s H_2O, ace, EtOH

No.	Name Formula	CAS RN Mol. Wt.	Physical Form	mp/°C den/g cm^{-3}	bp/°C Other Data	Solubility
386	Cadmium carbonate $CdCO_3$	513-78-0 172.420	wh hex cry	dec 500 4.258		i H_2O; s acid
387	Cadmium chlorate dihydrate $Cd(ClO_3)_2 \cdot 2H_2O$	22750-54-5* 315.343	col hyg cry	80 dec 2.28	a	vs H_2O
388	Cadmium chloride $CdCl_2$	10108-64-2 183.316	rhomb cry; hyg	564 4.08	960 a,c,d,e	vs H_2O; s ace; sl EtOH; i eth
389	Cadmium chloride hemipentahydrate $CdCl_2 \cdot 2.5H_2O$	7790-78-5 228.354	wh rhomb leaflets	3.327		vs H_2O; s ace
390	Cadmium chloride monohydrate $CdCl_2 \cdot H_2O$	34330-64-8 201.331	wh cry			vs H_2O
391	Cadmium chromate $CdCrO_4$	14312-00-6 228.405	yel orth cry			i H_2O
392	Cadmium cyanide $Cd(CN)_2$	542-83-6 164.446	wh cub cry	4.5		sl H_2O
393	Cadmium fluoride CdF_2	7790-79-6 150.408	cub cry	1110 6.33	1748 a,c,d,e	sl H_2O; s acid; i EtOH
394	Cadmium hydroxide $Cd(OH)_2$	21041-95-2 146.426	wh trig or hex cry	dec 130 4.79	a	i H_2O, s dil acid
395	Cadmium iodate $Cd(IO_3)_2$	7790-81-0 462.216	wh powder	6.48		i H_2O; s HNO_3
396	Cadmium iodide CdI_2	7790-80-9 366.220	hex flakes	387 5.64	742 a,c,d,e	s H_2O, EtOH, eth, ace
397	Cadmium metasilicate $CdSiO_3$	13477-19-5 188.495	grn monocl cry	1252 5.10		
398	Cadmium molybdate $CdMoO_4$	13972-68-4 272.35	col tetr cry	≈900 dec 5.4		i H_2O; s acid
399	Cadmium niobate $Cd_2Nb_2O_7$	12187-14-3 522.631	cub cry	≈1410 6.28		i H_2O
400	Cadmium nitrate $Cd(NO_3)_2$	10325-94-7 236.420	wh cub cry; hyg	350 3.6		vs H_2O; s EtOH
401	Cadmium nitrate tetrahydrate $Cd(NO_3)_2 \cdot 4H_2O$	10022-68-1 308.482	col orth cry; hyg	59.5 2.45		vs H_2O; s EtOH, ace
402	Cadmium oxalate CdC_2O_4	814-88-0 200.431	wh solid	3.32		i H_2O
403	Cadmium oxalate trihydrate $CdC_2O_4 \cdot 3H_2O$	20712-42-9 254.477	wh amorp powder	340 dec		i H_2O, EtOH; s dil acid
404	Cadmium oxide CdO	1306-19-0 128.410	brn cub cry	8.15	1559 sp a,e	i H_2O; s dil acid
405	Cadmium perchlorate hexahydrate $Cd(ClO_4)_2 \cdot 6H_2O$	10326-28-0 419.403	wh hex cry	2.37		vs H_2O
406	Cadmium phosphide Cd_3P_2	12014-28-7 399.181	gr tetr needles	700 5.96		s dil HCl
407	Cadmium selenate dihydrate $CdSeO_4 \cdot 2H_2O$	10060-09-0 291.40	orth cry	dec 100 3.62		vs H_2O
408	Cadmium selenide $CdSe$	1306-24-7 191.37	wh cub cry	1240 5.81		i H_2O
409	Cadmium sulfate $CdSO_4$	10124-36-4 208.475	col orth cry	1000 4.69	a	vs H_2O; i EtOH
410	Cadmium sulfate monohydrate $CdSO_4 \cdot H_2O$	7790-84-3 226.490	monocl cry	105 3.79		
411	Cadmium sulfate octahydrate $CdSO_4 \cdot 8H_2O$	15244-35-6 352.597	col monocl cry	dec 40 3.08		vs H_2O
412	Cadmium sulfide CdS	1306-23-6 144.477	yel-oran cub cry	1750 4.83	a,f	i H_2O; s acid
413	Cadmium telluride $CdTe$	1306-25-8 240.01	brn-blk cub cry	1042 6.2	a	i H_2O, dil acid
414	Cadmium titanate $CdTiO_3$	12014-14-1 208.276	orth cry	6.5		
415	Cadmium tungstate $CdWO_4$	7790-85-4 360.25	wh monocl cry	8.0		i H_2O, acid; s NH_4OH
416	Calcium Ca	7440-70-2 40.078	silv-wh metal	842 1.54	1484 a,c,e	reac H_2O; i bz
417	Calcium acetate $Ca(C_2H_3O_2)_2$	62-54-4 158.168	wh hyg cry	dec 160 1.50		s H_2O; sl EtOH
418	Calcium acetate monohydrate $Ca(C_2H_3O_2)_2 \cdot H_2O$	5743-26-0 176.183	wh needles or powder	dec ≈150		s H_2O; sl EtOH
419	Calcium aluminate $CaAl_2O_4$	12042-68-1 158.039	wh monocl cry	1605 2.98		reac H_2O
420	Calcium aluminate (β form) $Ca_3Al_2O_6$	12042-78-3 270.193	wh cub cry; refr	1535 3.04		i H_2O
421	Calcium arsenate $Ca_3(AsO_4)_2$	7778-44-1 398.072	wh powder	dec 3.6		sl H_2O; s dil acid
422	Calcium boride CaB_6	12007-99-7 104.944	refrac solid	2235 2.49		
423	Calcium bromide $CaBr_2$	7789-41-5 199.886	rhomb cry; hyg	742 3.38	1815 a,c	vs H_2O; s EtOH, ace
424	Calcium bromide hexahydrate $CaBr_2 \cdot 6H_2O$	13477-28-6 307.977	wh hyg powder	38 dec 2.29		vs H_2O

No.	Name Formula	CAS RN Mol. Wt.	Physical Form	mp/°C den/g cm⁻³	bp/°C Other Data	Solubility
425	Calcium carbide CaC_2	75-20-7 64.100	gray-blk orth cry	2300 2.22	a	reac H_2O
426	Calcium carbonate (aragonite) $CaCO_3$	471-34-1 100.087	wh orth cry or powder	825 dec 2.83	a,c,f	i H_2O; s dil acid
427	Calcium carbonate (calcite) $CaCO_3$	471-34-1 100.087	wh hex cry or powder	1330 2.71	a,c,f	i H_2O; s dil acid
428	Calcium chlorate $Ca(ClO_3)_2$	10137-74-3 206.979	wh cry	340		vs H_2O
429	Calcium chlorate dihydrate $Ca(ClO_3)_2 \cdot 2H_2O$	10035-05-9 243.010	wh monocl cry; hyg	100 dec 2.711		vs H_2O; s EtOH
430	Calcium chloride $CaCl_2$	10043-52-4 110.983	wh cub cry or powder; hyg	775 2.15	1935.5 a,c	vs H_2O, EtOH
431	Calcium chloride dihydrate $CaCl_2 \cdot 2H_2O$	10035-04-8 147.014	hyg flakes or powder	175 dec 1.85		vs H_2O, EtOH
432	Calcium chloride hexahydrate $CaCl_2 \cdot 6H_2O$	7774-34-7 219.074	wh hex cry; hyg	30 dec 1.71		vs H_2O
433	Calcium chloride monohydrate $CaCl_2 \cdot H_2O$	13477-29-7 128.998	wh hyg cry	260 dec 2.24		s H_2O, EtOH
434	Calcium chromate dihydrate $CaCrO_4 \cdot 2H_2O$	13765-19-0 192.102	yel orth cry	 2.50		s H_2O
435	Calcium cyanamide $CaCN_2$	156-62-7 80.102	col hex cry	≈1340 2.29	subl	reac H_2O
436	Calcium cyanide $Ca(CN)_2$	592-01-8 92.113	wh rhomb cry; hyg		a	s H_2O, EtOH
437	Calcium dichromate trihydrate $CaCr_2O_7 \cdot 3H_2O$	14307-33-6* 310.112	red-oran cry	dec 100 2.37		vs H_2O; reac EtOH; i eth, ctc
438	Calcium dihydrogen phosphate monohydrate $Ca(H_2PO_4)_2 \cdot H_2O$	10031-30-8 252.068	col tricl plates	dec ≈100 2.22		sl H_2O; s dil acid
439	Calcium fluoride CaF_2	7789-75-5 78.075	wh cub cry or powder	1418 3.18	2533.4 a,c,f	i H_2O; sl acid
440	Calcium formate $Ca(CHO_2)_2$	544-17-2 130.114	orth cry	300 dec 2.02		s H_2O; i EtOH
441	Calcium hexafluorosilicate dihydrate $CaSiF_6 \cdot 2H_2O$	16925-39-6 218.185	col tetr cry	 2.25		i H_2O, ace; reac hot H_2O
442	Calcium hydride CaH_2	7789-78-8 42.094	gray orth cry or powder	1000 1.7	a	reac H_2O, EtOH
443	Calcium hydrogen phosphate $CaHPO_4$	7757-93-9 136.057	wh tricl cry	dec 2.92	f	i H_2O, EtOH
444	Calcium hydrogen phosphate dihydrate $CaHPO_4 \cdot 2H_2O$	7789-77-7 172.088	monocl cry	dec ≈100 2.31		i H_2O, EtOH; s dil acid
445	Calcium hydroxide $Ca(OH)_2$	1305-62-0 74.093	soft hex cry	≈2.2	a	sl H_2O; s acid
446	Calcium hypophosphite $Ca(H_2PO_2)_2$	7789-79-9 170.055	wh monocl cry	dec 300		s H_2O; i EtOH
447	Calcium iodate $Ca(IO_3)_2$	7789-80-2 389.883	wh monocl cry	 4.52		sl H_2O; s HNO_3; i EtOH
448	Calcium iodide CaI_2	10102-68-8 293.887	hyg hex cry	783 3.96	a,c	vs H_2O; s MeOH, EtOH, ace; i eth
449	Calcium iodide hexahydrate $CaI_2 \cdot 6H_2O$	71626-98-7 401.978	wh hex needles or powder	42 dec 2.55		vs H_2O, EtOH
450	Calcium molybdate $CaMoO_4$	7789-82-4 200.02	wh tetr cry	965 dec 4.35	f	i H_2O, EtOH; s conc acid
451	Calcium nitrate $Ca(NO_3)_2$	10124-37-5 164.087	wh cub cry; hyg	561 2.5	a	vs H_2O; s EtOH, MeOH, ace
452	Calcium nitrate tetrahydrate $Ca(NO_3)_2 \cdot 4H_2O$	13477-34-4 236.149	wh cry	dec ≈40 1.82		vs H_2O; s EtOH, ace
453	Calcium nitride Ca_3N_2	12013-82-0 148.247	red-brn cub cry	1195 2.67		s H_2O, acid; i EtOH
454	Calcium nitrite $Ca(NO_2)_2$	13780-06-8 132.089	wh-yel hex cry; hyg	 2.23		vs H_2O; sl EtOH
455	Calcium oxalate CaC_2O_4	563-72-4 128.098	wh cry powder	 2.2	a	i H_2O
456	Calcium oxalate monohydrate $CaC_2O_4 \cdot H_2O$	5794-28-5 146.113	cub cry	dec 200 2.2	f	i H_2O; s dil acid
457	Calcium oxide CaO	1305-78-8 56.077	gray-wh cub cry	2899 3.34	a,c	reac H_2O; s acid
458	Calcium perchlorate $Ca(ClO_4)_2$	13477-36-6 238.978	wh cry	dec 270 2.65		vs H_2O; s EtOH
459	Calcium permanganate $Ca(MnO_4)_2$	10118-76-0 277.949	purp hyg cry	 2.4		vs H_2O; reac EtOH
460	Calcium peroxide CaO_2	1305-79-9 72.077	wh-yel tetr cry; hyg	dec ≈200 2.9		sl H_2O; s acid
461	Calcium phosphate $Ca_3(PO_4)_2$	7758-87-4 310.177	wh amorp powder	1670 3.14	a	i H_2O, EtOH; s dil acid
462	Calcium dihydrogen phosphate monohydrate $Ca(H_2PO_4)_2 \cdot H_2O$	7758-23-8* 252.068	col hyg cry	dec ≈100 2.22		s H_2O, acid

No.	Name / Formula	CAS RN / Mol. Wt.	Physical Form	mp/°C; den/g cm⁻³	bp/°C; Other Data	Solubility
463	Calcium phosphide Ca_3P_2	1305-99-3; 182.182	red-brn hyg cry	≈1600; 2.51		reac H_2O; i EtOH, eth
464	Calcium pyrophosphate $Ca_2P_2O_7$	7790-76-3; 254.099	wh powder	1353; 3.09		i H_2O; s dil acid
465	Calcium selenate dihydrate $CaSeO_4 \cdot 2H_2O$	7790-74-1; 219.07	wh monocl cry	2.75		s H_2O
466	Calcium selenide $CaSe$	1305-84-6; 119.04	wh-brn cub cry	3.8		reac H_2O
467	Calcium metasilicate $CaSiO_3$	1344-95-2; 116.162	wh monocl cry	2.92		i H_2O
468	Calcium silicide $CaSi_2$	12013-56-8; 96.249	gray hex cry	1040; 2.50	f	i cold H_2O; reac hot H_2O; s acid
469	Calcium silicide $CaSi$	12013-55-7; 68.164	orth cry	1324; 2.39		
470	Calcium sulfate $CaSO_4$	7778-18-9; 136.142	orth cry	1460; 2.96		sl H_2O
471	Calcium sulfate dihydrate $CaSO_4 \cdot 2H_2O$	10101-41-4; 172.172	monocl cry or powder	dec 150; 2.32	a,c,f	s H_2O; i os
472	Calcium sulfate hemihydrate $CaSO_4 \cdot 0.5H_2O$	10034-76-1; 145.149	wh powder		f	s H_2O
473	Calcium sulfide CaS	20548-54-3; 72.144	wh-yel cub cry; hyg	2525; 2.59		sl H_2O; i EtOH
474	Calcium sulfite dihydrate $CaSO_3 \cdot 2H_2O$	10257-55-3; 156.173	wh powder		a,f	sl H_2O, EtOH; s acid
475	Calcium telluride $CaTe$	12013-57-9; 167.68	cub cry	4.87		
476	Calcium tetrahydroaluminate $Ca(AlH_4)_2$	16941-10-9; 102.105	gray powder; flam			reac H_2O; s thf; i eth, bz
477	Calcium thiocyanate tetrahydrate $Ca(SCN)_2 \cdot 4H_2O$	2092-16-2; 228.307	hygr cry	dec 160		vs H_2O; s EtOH, ace
478	Calcium thiosulfate hexahydrate $CaS_2O_3 \cdot 6H_2O$	10124-41-1; 260.300	tricl cry	dec 45; 1.87		s H_2O; i EtOH
479	Calcium titanate $CaTiO_3$	12049-50-2; 135.943	cub cry	1980; 3.98		
480	Calcium tungstate $CaWO_4$	7790-75-2; 287.92	wh tetr cry	1620; 6.06	f	i H_2O; s hot acid
481	Californium Cf	7440-71-3; 251	hex or cub metal	900; 15.1	f	
482	Carbon (diamond) C	7782-40-3; 12.011	col cub cry	4440 (12.4 GPa); 3.513	a; e,f	i H_2O
483	Carbon (graphite) C	7440-44-0; 12.011	soft blk hex cry	4492 tp (10.3 MPa); 2.2	3825 sp; a,c,f	i H_2O
484	Carbon (fullerene-C_{60}) C_{60}	99685-96-8; 720.660	yel needles or plates	>280		s os
485	Carbon (fullerene-C_{70}) C_{70}	115383-22-7; 840.770	red-brn solid	>280		s bz, tol
486	Fullerene fluoride $C_{60}F_{60}$	134929-59-2; 1860.564	col plates	287		vs ace; s thf; i chl
487	Carbon monoxide CO	630-08-0; 28.010	col gas	-205; 1.229 g/L	-191.5; a,b,c,d,e	sl H_2O; s chl, EtOH
488	Carbon dioxide CO_2	124-38-9; 44.010	col gas	-56.57 tp; 1.931 g/L	-78.4 sp; a,b,c,e	s H_2O
489	Carbon diselenide CSe_2	506-80-9; 169.93	yel liq	-45.5; 2.6626	125.5; a,e	i H_2O; vs ctc, tol
490	Carbon disulfide CS_2	75-15-0; 76.143	col or yel liq	-111.5; 1.2555	46; a,b,c,d,e	i H_2O; vs EtOH, bz, os
491	Carbon oxyselenide $COSe$	1603-84-5; 106.97	col gas; unstable	-122; 4.694 g/L	-21.5; e	reac H_2O
492	Carbon oxysulfide COS	463-58-1; 60.076	col gas	-138.8; 2.636 g/L	-50; a,b,e	s H_2O, EtOH
493	Carbon suboxide C_3O_2	504-64-3; 68.032	col gas	-111.3; 2.985 g/L	6.8; e	reac H_2O
494	Carbon sulfide telluride $CSTe$	10340-06-4; 171.68	red-yel liq; unstable			reac H_2O
495	Carbonyl bromide $COBr_2$	593-95-3; 187.818	col liq	2.5	64.5; a	reac H_2O
496	Carbonyl chloride $COCl_2$	75-44-5; 98.915	col gas	-127.9; 4.340 g/L	8; a	sl H_2O; s bz, tol
497	Carbonyl fluoride COF_2	353-50-4; 66.007	col gas	-111.26; 2.896 g/L	-84.57; a,b,c,e	reac H_2O
498	Cyanogen C_2N_2	460-19-5; 52.035	col gas	-27.9; 2.283 g/L	-21.1; a	sl H_2O, eth; s EtOH
499	Cyanogen bromide $BrCN$	506-68-3; 105.922	wh hyg needles	52; 2.005	61.5; a,b,d,e	s H_2O, EtOH, eth
500	Cyanogen chloride $ClCN$	506-77-4; 61.471	col gas	-6.55; 2.697 g/L	13; a	s H_2O, EtOH, eth

No.	Name Formula	CAS RN Mol. Wt.	Physical Form	mp/°C den/g cm^{-3}	bp/°C Other Data	Solubility
501	Cyanogen fluoride FCN	1495-50-7 45.016	col gas	-82 1.975 g/L	-46 a	
502	Cyanogen iodide ICN	506-78-5 152.922	col needles	146.7 1.84	a	s H$_2$O, EtOH, eth
503	Cerium Ce	7440-45-1 140.115	silv metal; cub or hex	798 6.770	3443 a,c	s dil acid
504	Cerium boride CeB$_6$	12008-02-5 204.981	blue refrac solid; hex	2550 4.87		i H$_2$O, HCl
505	Cerium carbide CeC$_2$	12012-32-7 164.137	red hex cry	2250 5.47		reac H$_2$O
506	Cerium nitride CeN	25764-08-3 154.122	refrac cub cry	2557 7.89		
507	Cerium silicide CeSi$_2$	12014-85-6 196.286	tetr cry	1620 5.31		i H$_2$O
508	Cerium(II) hydride CeH$_2$	13569-50-1 142.131	cub cry	5.45		reac H$_2$O
509	Cerium(II) iodide CeI$_2$	19139-47-0 393.924	bronze cry	808		
510	Cerium(II) sulfide CeS	12014-82-3 172.181	yel cub cry	2445 5.9	a	
511	Cerium(III) bromide CeBr$_3$	14457-87-5 379.827	wh hex cry; hyg	733	1457	s H$_2$O
512	Cerium(III) bromide heptahydrate CeBr$_3$·7H$_2$O	14457-87-5* 505.934	col hyg needles	732		s H$_2$O, EtOH
513	Cerium(III) carbide Ce$_2$C$_3$	12115-63-8 316.263	yel-brn cub cry	1505 6.9		i H$_2$O; s dil acid
514	Cerium(III) carbonate hydrate Ce$_2$(CO$_3$)$_3$·5H$_2$O	72520-94-6 550.334	wh powder			s H$_2$O, EtOH
515	Cerium(III) chloride CeCl$_3$	7790-86-5 246.473	wh hex cry	817 3.97	a	vs H$_2$O, EtOH
516	Cerium(III) chloride heptahydrate CeCl$_3$·7H$_2$O	18618-55-8 372.580	yel orth cry; hyg	dec 90		i H$_2$O
517	Cerium(III) fluoride CeF$_3$	7758-88-5 197.110	wh hex cry; hyg	1430 6.157		s H$_2$O
518	Cerium(III) iodide CeI$_3$	7790-87-6 520.828	yel orth cry; hyg	766		vs H$_2$O; s EtOH
519	Cerium(III) iodide nonahydrate CeI$_3$·9H$_2$O	7790-87-6* 682.966	wh-red cry			i H$_2$O; s acid
520	Cerium(III) oxide Ce$_2$O$_3$	1345-13-7 328.228	yel-grn cub cry	2230 6.2	a	s H$_2$O
521	Cerium(III) sulfate octahydrate Ce$_2$(SO$_4$)$_3$·8H$_2$O	13454-94-9 712.543	wh orth cry	dec ≈250 2.87		i H$_2$O
522	Cerium(III) sulfide Ce$_2$S$_3$	12014-93-6 376.428	red cub cry	2450 5.02		i H$_2$O
523	Cerium(IV) fluoride CeF$_4$	10060-10-3 216.109	wh hyg powder	dec ≈600 4.77		i H$_2$O, dil acid
524	Cerium(IV) oxide CeO$_2$	1306-38-3 172.114	wh-yel powder; cub	2400 7.65	a	reac H$_2$O
525	Cerium(IV) sulfate tetrahydrate Ce(SO$_4$)$_2$·4H$_2$O	10294-42-5 404.304	yel-oran orth cry	dec 180 3.91		reac H$_2$O
526	Cesium Cs	7440-46-2 132.905	silv-wh metal	28.44 1.93	671 a,c,e	
527	Cesium amide CsNH$_2$	22205-57-8 148.928	wh tetr cry	3.70	a	
528	Cesium azide CsN$_3$	22750-57-8 174.925	hyg tetr cry; exp	326 ≈3.5		sl H$_2$O
529	Cesium bromate CsBrO$_3$	13454-75-6 260.807	col hex cry	4.11		vs H$_2$O; s EtOH; i ace
530	Cesium bromide CsBr	7787-69-1 212.809	wh cub cry; hyg	636 4.43	≈1300 a,e	s H$_2$O, EtOH, eth
531	Cesium carbonate Cs$_2$CO$_3$	534-17-8 325.820	wh monocl cry; hyg	792 4.24	a	sl H$_2$O
532	Cesium chlorate CsClO$_3$	13763-67-2 216.356	col hex cry	3.57		vs H$_2$O; s EtOH
533	Cesium chloride CsCl	7647-17-8 168.358	wh cub cry; hyg	645 3.988	1297 a,c,e	vs H$_2$O
534	Cesium cyanide CsCN	21159-32-0 158.923	wh cub cry; hyg	350 3.34		vs H$_2$O; s MeOH; i diox, py
535	Cesium fluoride CsF	13400-13-0 151.903	wh cub cry; hyg	703 4.64	a,c,e	reac H$_2$O
536	Cesium hydride CsH	58724-12-2 133.913	wh cub cry; flam	dec ≈170 3.42	a	
537	Cesium hydrogen fluoride CsHF$_2$	12280-52-3 171.910	tetr cry	170 3.86		vs H$_2$O; s EtOH
538	Cesium hydroxide CsOH	21351-79-1 149.912	wh-yel hyg cry	342.3 3.68	a	vs H$_2$O; s EtOH, MeOH, ace
539	Cesium iodide CsI	7789-17-5 259.809	col cub cry; hyg	621 4.51	≈1280 a,e	

No.	Name Formula	CAS RN Mol. Wt.	Physical Form	mp/°C den/g cm^{-3}	bp/°C Other Data	Solubility
540	Cesium metaborate $CsBO_2$	92141-86-1 175.715	cub cry	732 ≈3.7	a	
541	Cesium nitrate $CsNO_3$	7789-18-6 194.910	wh hex or cub cry	414 3.66	a	s H_2O, ace; sl EtOH
542	Cesium oxide Cs_2O	20281-00-9 281.810	yel-oran hex cry	490 4.65	a	vs H_2O
543	Cesium perchlorate $CsClO_4$	13454-84-7 232.356	wh orth cry; hyg	250 3.327	a	sl H_2O
544	Cesium superoxide CsO_2	12018-61-0 164.904	yel tetr cry	432 3.77	a	reac H_2O
545	Cesium sulfate Cs_2SO_4	10294-54-9 361.875	wh orth cry or hex prisms; hyg	1005 4.24	a,c	vs H_2O; i EtOH, ace, py
546	Chlorine Cl_2	7782-50-5 70.905	grn-yel gas	-101.5 3.111 g/L	-34.04 a,b,c,d,e	sl H_2O
547	Hypochlorous acid HOCl	7790-92-3 52.460	grn-yel; stable only in aq soln		a	s H_2O
548	Perchloric acid $HClO_4$	7601-90-3 100.459	col hyg liq	-112 1.77	dec ≈90 a	s H_2O
549	Chlorine oxide Cl_2O	7791-21-1 86.904	yel-brn gas	-120.6 3.813 g/L	2.2 a,d,e	vs H_2O
550	Chlorine dioxide ClO_2	10049-04-4 67.452	oran-grn gas	-59 2.960 g/L	11 a,d,e	sl H_2O
551	Chlorine trioxide Cl_2O_3	17496-59-2 118.903	dark brn solid	exp <25		
552	Chlorine hexoxide Cl_2O_6	12442-63-6 166.901	red liq	3.5	≈200 e	reac H_2O
553	Chlorine heptoxide Cl_2O_7	10294-48-1 182.901	col oily liq; exp	-91.5 1.9	82 e	reac H_2O
554	Chlorine fluoride ClF	7790-89-8 54.451	col gas	-155.6 2.389 g/L	-101.1 a,d,e	reac H_2O
555	Chlorine trifluoride ClF_3	7790-91-2 92.448	gas	-76.34 4.057 g/L	11.75 a,d,e	reac H_2O
556	Chlorine pentafluoride ClF_5	13637-63-3 130.445	col gas	-103 5.724 g/L	-13.1 b	
557	Chlorine trioxide fluoride (perchloryl fluoride) ClO_3F	7616-94-6 102.449	col gas	-147 4.495 g/L	-46.75 a,b,d	
558	Chlorine perchlorate $ClOClO_3$	27218-16-2 134.903	unstable liq	-117 1.81(0°C)	dec ≈25	
559	Chromium Cr	7440-47-3 51.996	blue-wh metal; cub	1907 7.15	2671 a,c,e	reac dil acid
560	Chromium acetylacetonate $Cr(CH_3COCHCOCH_3)_3$	21679-31-2 349.324	red monocl cry	208 1.34	345	i H_2O; s bz
561	Chromium antimonide CrSb	12053-12-2 173.756	hex cry	1110 7.11		
562	Chromium arsenide Cr_2As	12254-85-2 178.914	tetr cry	7.04		
563	Chromium boride CrB	12006-79-0 62.807	refrac orth cry	2100 6.1		
564	Chromium boride CrB_2	12007-16-8 73.618	refrac solid; hex	2200 5.22		
565	Chromium boride Cr_5B_3	12007-38-4 292.414	tetr cry	1900 6.10		
566	Chromium carbide Cr_3C_2	12012-35-0 180.010	gray orth cry	1895 6.68		
567	Chromium carbonyl $Cr(CO)_6$	13007-92-6 220.058	col orth cry	dec 130 1.77	subl e	i H_2O, EtOH; s eth, chl
568	Chromium nitride Cr_2N	12053-27-9 117.999	hex cry	1650 6.8		
569	Chromium nitride CrN	24094-93-7 66.003	gray cub cry	dec 1080 5.9		
570	Chromium phosphide CrP	26342-61-0 82.970	orth cry	5.25		
571	Chromium selenide CrSe	12053-13-3 130.96	hex cry	≈1500 6.1		
572	Chromium silicide Cr_3Si	12018-36-9 184.074	cub cry	1770 6.4		
573	Chromium silicide $CrSi_2$	12018-09-6 108.167	gray hex cry	1490 4.91		
574	Chromium(II) acetate monohydrate $Cr(C_2H_3O_2)_2 \cdot H_2O$	628-52-4* 188.101	red monocl cry	1.79		sl H_2O
575	Chromium(II) bromide $CrBr_2$	10049-25-9 211.804	wh monocl cry; aq soln blue	842 4.236	a	s H_2O, EtOH
576	Chromium(II) chloride $CrCl_2$	10049-05-5 122.901	hyg needles; aq soln blue	814 2.88	1300 a,c,d	s H_2O
577	Chromium(II) chloride tetrahydrate $Cr(H_2O)_4Cl_2 \cdot 4H_2O$	13931-94-7 267.023	blue hyg cry	dec 51		s H_2O

No.	Name Formula	CAS RN Mol. Wt.	Physical Form	mp/°C den/g cm⁻³	bp/°C Other Data	Solubility
578	Chromium(II) fluoride CrF_2	10049-10-2 89.993	blue-grn monocl cry	894 3.79	a	sl H_2O; i EtOH
579	Chromium(II) iodide CrI_2	13478-28-9 305.805	red-brn cry; hyg	868 5.1	a	
580	Chromium(II) oxalate monohydrate $CrC_2O_4 \cdot H_2O$	814-90-4* 158.031	yel-grn powder	2.468		sl H_2O
581	Chromium(II) sulfate pentahydrate $CrSO_4 \cdot 5H_2O$	13825-86-0 238.136	blue cry			s H_2O, dil acid; sl EtOH; i ace
582	Chromium(II,III) oxide Cr_3O_4	12018-34-7 219.986	cub cry	6.1	a	
583	Chromium(III) acetate hexahydrate $Cr(C_2H_3O_2)_3 \cdot 6H_2O$	1066-30-4* 285.226	blue needles			s H_2O
584	Chromium(III) bromide $CrBr_3$	10031-25-1 291.708	dark grn hex cry	1130 4.68		s hot H_2O
585	Chromium(III) bromide hexahydrate $Cr(H_2O)_6Br_3$	10031-25-1* 399.799	viol hyg cry			s H_2O; i EtOH, eth
586	Chromium(III) bromide hexahydrate $CrBr_3(H_2O)_4 \cdot 2H_2O$	18721-05-6 399.799	grn hyg cry			s H_2O, EtOH
587	Chromium(III) chloride $CrCl_3$	10025-73-7 158.354	purp hex plates	1152 2.87	dec 1300 a	sl H_2O
588	Chromium(III) chloride hexahydrate $[CrCl_2(H_2O)_4]Cl \cdot 2H_2O$	10060-12-5 266.445	grn monocl cry; hyg			s H_2O, EtOH; sl ace; i eth
589	Chromium(III) fluoride CrF_3	7788-97-8 108.991	grn needles	1400 3.8	a	i H_2O; EtOH
590	Chromium(III) fluoride trihydrate $CrF_3 \cdot 3H_2O$	16671-27-5 163.037	grn hex cry	2.2		sl H_2O
591	Chromium(III) hydroxide trihydrate $Cr(OH)_3 \cdot 3H_2O$	1308-14-1 157.063	blue-grn powder			i H_2O; s acid
592	Chromium(III) iodide CrI_3	13569-75-0 432.709	dark grn hex cry	dec 500 5.32	a	sl H_2O
593	Chromium(III) nitrate $Cr(NO_3)_3$	13548-38-4 238.011	grn hyg powder	dec >60		vs H_2O
594	Chromium(III) nitrate nonahydrate $Cr(NO_3)_3 \cdot 9H_2O$	7789-02-8 400.148	grn-blk monocl cry	66.3 1.80	dec >100	vs H_2O
595	Chromium(III) oxide Cr_2O_3	1308-38-9 151.990	grn hex cry	2330 5.22	≈3000 a,c	i H_2O, EtOH; sl acid, alk
596	Chromium(III) phosphate $CrPO_4$	7789-04-0 146.967	blue orth cry	>1800 4.6		i H_2O, acid, aqua regia
597	Chromium(III) phosphate hemiheptahydrate $CrPO_4 \cdot 3.5H_2O$	84359-31-9 210.021	blue-grn powder	2.15		i H_2O; s acid
598	Chromium(III) phosphate hexahydrate $CrPO_4 \cdot 6H_2O$	84359-31-9 255.059	viol cry	dec >500 2.121		i H_2O; s acid, alk
599	Chromium(III) potassium sulfate dodecahydrate $CrK(SO_4)_2 \cdot 12H_2O$	7788-99-0 499.405	viol-blk cub cry	89 dec 1.83		s H_2O; i EtOH
600	Chromium(III) sulfate $Cr_2(SO_4)_3$	10101-53-8 392.183	red-brn hex cry	3.1		i H_2O, acid
601	Chromium(III) sulfide Cr_2S_3	12018-22-3 200.190	brn-blk hex cry	3.8		
602	Chromium(III) telluride Cr_2Te_3	12053-39-3 486.79	hex cry	≈1300 7.0		
603	Chromium(IV) chloride $CrCl_4$	15597-88-3 193.807	gas, stable at high temp	8.504 g/L	dec >600	
604	Chromium(IV) fluoride CrF_4	10049-11-3 127.990	grn cry	277		
605	Chromium(IV) oxide CrO_2	12018-01-8 83.995	brn-blk tetr powder	dec ~400 4.89	a	i H_2O; s acid
606	Chromium(V) fluoride CrF_5	14884-42-5 146.988	red orth cry	34	117	
607	Chromium(VI) fluoride CrF_6	13843-28-2 165.986	yel solid; stable at low temp	dec -100		
608	Chromium(VI) oxide CrO_3	1333-82-0 99.994	red orth cry	197 2.7	dec ≈250	vs H_2O
609	Chromyl chloride CrO_2Cl_2	14977-61-8 154.900	red liq	-96.5 1.91	117 a,d,e	reac H_2O; s ctc, chl, bz
610	Cobalt Co	7440-48-4 58.933	gray metal; hex or cub	1495 8.86	2927 a,c	s dil acid
611	Cobalt antimonide CoSb	12052-42-5 180.69	hex cry	1202 8.8		
612	Cobalt arsenic sulfide CoAsS	12254-82-9 165.92	silv-wh solid	≈6.1	f	
613	Cobalt arsenide CoAs	27016-73-5 133.855	orth cry	1180 8.22		
614	Cobalt arsenide $CoAs_2$	12044-42-7 208.776	monocl cry	7.2		
615	Cobalt arsenide $CoAs_3$	12256-04-1 283.698	cub cry	942 6.84		

No.	Name Formula	CAS RN Mol. Wt.	Physical Form	mp/°C den/g cm^{-3}	bp/°C Other Data	Solubility
616	Cobalt boride Co_2B	12045-01-1 128.677	refrac solid	1280 8.1		
617	Cobalt boride CoB	12006-77-8 69.744	refrac solid	1460 7.25		reac H_2O, HNO_3
618	Cobalt carbonyl $Co_2(CO)_8$	10210-68-1 341.949	oran cry	51 dec 1.78		i H_2O; s EtOH, eth, CS_2
619	Cobalt phosphide Co_2P	12134-02-0 148.840	gray needles	1386 6.4		i H_2O; s HNO_3
620	Cobalt silicide $CoSi_2$	12017-12-8 115.10	gray cub cry	1326 4.9		s hot HCl
621	Cobalt disulfide CoS_2	12013-10-4 123.07	cub cry	4.3		
622	Cobalt dodecacarbonyl $Co_4(CO)_{12}$	17786-31-1 571.858	blk cry	dec 60 2.09		
623	Cobalt(II) acetate $Co(C_2H_3O_2)_2$	71-48-7 177.023	pink cry			vs H_2O; s EtOH
624	Cobalt(II) acetate tetrahydrate $Co(C_2H_3O_2)_2 \cdot 4H_2O$	6147-53-1 249.083	red monocl cry	1.705		s H_2O, EtOH, dil acid
625	Cobalt(II) aluminate $CoAl_2O_4$	13820-62-7 176.894	blue cub cry	4.37		i H_2O
626	Cobalt(II) arsenate octahydrate $Co_3(AsO_4)_2 \cdot 8H_2O$	24719-19-5 598.760	red monocl needles	dec 1000 3.0		i H_2O; s dil acid
627	Cobalt(II) bromate hexahydrate $Co(BrO_3)_2 \cdot 6H_2O$	13476-01-2 422.829	viol cry	≈2.5		vs H_2O
628	Cobalt(II) bromide $CoBr_2$	7789-43-7 218.741	grn hex cry; hyg	678 4.91	a	vs H_2O; s MeOH, EtOH, ace
629	Cobalt(II) bromide hexahydrate $CoBr_2 \cdot 6H_2O$	13762-12-4 326.832	red hyg cry	47 dec 2.46	dec 100	s H_2O
630	Cobalt(II) carbonate $CoCO_3$	513-79-1 118.942	pink rhomb cry	4.2		i H_2O, EtOH
631	Cobalt(II) chloride $CoCl_2$	7646-79-9 129.838	blue hyg leaflets	740 3.36	1049 a,c,e	s H_2O, EtOH, eth, ace, py
632	Cobalt(II) chloride dihydrate $CoCl_2 \cdot 2H_2O$	16544-92-6 165.869	viol-blue cry	2.477		
633	Cobalt(II) chloride hexahydrate $CoCl_2 \cdot 6H_2O$	7791-13-1 237.929	pink-red monocl cry	87 dec 1.924		vs H_2O; s EtOH, ace, eth
634	Cobalt(II) chromate $CoCrO_4$	24613-38-5 174.927	yel-brn orth cry	≈4.0		i H_2O; s acid
635	Cobalt(II) chromite $CoCr_2O_4$	13455-25-9 226.923	blue-grn cub cry	5.14		i H_2O, conc acid
636	Cobalt(II) cyanide $Co(CN)_2$	542-84-7 110.968	blue hyg cry	1.872		i H_2O
637	Cobalt(II) cyanide dihydrate $Co(CN)_2 \cdot 2H_2O$	20427-11-6 146.999	pink-brn needles			i H_2O, acid
638	Cobalt(II) ferricyanide $Co_3[Fe(CN)_6]_2$	14049-81-1 600.703	red needles			i H_2O, HCl; s NH_4OH
639	Cobalt(II) fluoride CoF_2	10026-17-2 96.930	red tetr cry	1127 4.46	≈1400 a,c	sl H_2O; s acid
640	Cobalt(II) fluoride tetrahydrate $CoF_2 \cdot 4H_2O$	13817-37-3 168.992	red orth cry	dec 2.22		s H_2O
641	Cobalt(II) formate dihydrate $Co(CHO_2)_2 \cdot 2H_2O$	6424-20-0 184.999	red cry powder	dec 140 2.13		s H_2O; i EtOH
642	Cobalt(II) hexafluorosilicate hexahydrate $CoSiF_6 \cdot 6H_2O$	12021-68-0 309.100	pale red cry	2.087		vs H_2O
643	Cobalt(II) hydroxide $Co(OH)_2$	21041-93-0 92.948	blue-grn cry	dec ≈160 3.60	a	sl H_2O; s acid
644	Cobalt(II) iodate $Co(IO_3)_2$	13455-28-2 408.738	blk-viol needles	dec 200 5.09		sl H_2O
645	Cobalt(II) iodide CoI_2	15238-00-3 312.742	blk hex cry; hyg	520 5.60	a	vs H_2O
646	Cobalt(II) iodide hexahydrate $CoI_2 \cdot 6H_2O$	15238-00-3* 420.833	red hex prisms	dec 130 2.90		s H_2O, EtOH, eth, ace
647	Cobalt(II) titanate $CoTiO_3$	12017-01-5 154.798	grn rhomb cry	5.0		
648	Cobalt(II) molybdate $CoMoO_4$	13762-14-6 218.87	blk monocl cry	1040 4.7		
649	Cobalt(II) nitrate $Co(NO_3)_2$	10141-05-6 182.942	pale red powder	dec 100 2.49	a	vs H_2O
650	Cobalt(II) nitrate hexahydrate $Co(NO_3)_2 \cdot 6H_2O$	10026-22-9 291.034	red monocl cry; hyg	≈55 1.88		vs H_2O; s EtOH
651	Cobalt(II) oxalate CoC_2O_4	814-89-1 146.953	pink powder	250 dec 3.02		i H_2O; s acid, NH_4OH
652	Cobalt(II) oxalate dihydrate $CoC_2O_4 \cdot 2H_2O$	814-89-1* 166.984	pink needles	dec		i H_2O; sl acid; s NH_4OH
653	Cobalt(II) oxide CoO	1307-96-6 74.932	gray cub cry	1830 6.44	a	i H_2O; s acid
654	Cobalt(II) perchlorate $Co(ClO_4)_2$	13455-31-7 257.833	red needles	3.33		vs H_2O; i EtOH, ace

No.	Name Formula	CAS RN Mol. Wt.	Physical Form	mp/°C den/g cm⁻³	bp/°C Other Data	Solubility
655	Cobalt(II) phosphate octahydrate Co₃(PO₄)₂•8H₂O	10294-50-5 510.865	pink amorp powder	2.77		i H₂O; s acid
656	Cobalt(II) potassium sulfate hexahydrate CoK₂(SO₄)₂•6H₂O	10026-20-7 437.349	red monocl cry	dec 75 2.22		vs H₂O
657	Cobalt(II) selenate pentahydrate CoSeO₄•5H₂O	14590-19-3 291.97	red tricl cry	dec 2.51		vs H₂O
658	Cobalt(II) selenide CoSe	1307-99-9 137.89	yel hex cry	1055 7.65		i H₂O, alk; s aqua regia
659	Cobalt(II) selenite dihydrate CoSeO₃•2H₂O	19034-13-0 221.92	blue-red powder			i H₂O
660	Cobalt(II) orthosilicate Co₂SiO₄	12017-08-2 209.950	red-viol orth cry	1345 4.63		i H₂O; s dil HCl
661	Cobalt(II) stannate Co₂SnO₄	12139-93-4 300.574	grn-blue cub cry	6.30		i H₂O; s alk
662	Cobalt(II) sulfate CoSO₄	10124-43-3 154.997	red orth cry	>700 3.71	a	s H₂O
663	Cobalt(II) sulfate heptahydrate CoSO₄•7H₂O	10026-24-1 281.103	pink monocl cry	41 dec 2.03		vs H₂O; sl EtOH, MeOH
664	Cobalt(II) sulfate monohydrate CoSO₄•H₂O	13455-34-0 173.012	red monocl cry	3.08		sl H₂O
665	Cobalt(II) sulfide CoS	1317-42-6 91.00	blk amorp powder	1182 5.45	a	i H₂O; s acid
666	Cobalt(II) telluride CoTe	12017-13-9 186.53	hex cry	≈8.8		
667	Cobalt(II) thiocyanate trihydrate Co(SCN)₂•3H₂O	97126-35-7 229.146	viol rhomb cry			s H₂O, EtOH, eth, ace
668	Cobalt(II) tungstate CoWO₄	12640-47-0 306.77	blue monocl cry	≈7.8		i H₂O; s hot conc acid
669	Cobalt(II,III) oxide Co₃O₄	1308-06-1 240.798	blk cub cry	dec 900 6.11	a	i H₂O; s acid, alk
670	Cobalt(III) acetate Co(C₂H₃O₂)₃	917-69-1 236.066	grn hyg cry	dec 100		s H₂O, EtOH
671	Cobalt(III) ammonium tetranitrodiammine NH₄[Co(NH₃)₂(NO₂)₄]	13600-89-0 295.054	red-brn orth cry	1.97		s H₂O
672	Cobalt(III) fluoride CoF₃	10026-18-3 115.928	brn hex cry	927 3.88		reac H₂O
673	Cobalt(III) hexammine chloride Co(NH₃)₆Cl₃	10534-89-1 267.474	red monocl cry	1.71		s H₂O; i EtOH
674	Cobalt(III) hydroxide Co(OH)₃	1307-86-4 109.955	brn powder	dec ≈4		i H₂O; s acid
675	Cobalt(III) nitrate Co(NO₃)₃	15520-84-0 244.948	grn cub cry; hyg	≈3.0		s H₂O; reac os
676	Cobalt(III) oxide Co₂O₃	1308-04-9 165.864	gray-blk powder	dec 895 5.18		i H₂O; s conc acid
677	Cobalt(III) oxide monohydrate Co₂O₃•H₂O	12016-80-7 183.880	brn-blk hex cry	dec 150		i H₂O; s acid
678	Cobalt(III) potassium nitrite sesquihydrate CoK₃(NO₂)₆•1.5H₂O	13782-01-9* 479.284	yel cub cry	2.6		sl H₂O; reac acid; i EtOH
679	Cobalt(III) sulfide Co₂S₃	1332-71-4 214.064	blk cub cry	4.8	a	reac acid
680	Cobalt(III) titanate Co₂TiO₄	12017-38-8 229.731	grn-blk cub cry	5.1		s conc HCl
681	Copper Cu	7440-50-8 63.546	red metal; cub	1084.62 8.96	2562 a,c,e,f	sl dil acid
682	Copper acetylacetonate Cu(CH₃COCHCOCH₃)₂	13395-16-9 261.765	blue powder	284 dec	subl	sl H₂O; s chl
683	Copper nitride Cu₃N	1308-80-1 204.645	cub cry	300 dec 5.84		
684	Copper phosphide CuP₂	12019-11-3 125.494	monocl cry	≈900 4.20		
685	Copper(I) acetate CuC₂H₃O₂	598-54-9 122.591	col cry	dec	subl	reac H₂O
686	Copper(I) acetylide Cu₂C₂	1117-94-8 151.114	red amorp powder; exp			
687	Copper(I) azide CuN₃	14336-80-2 105.566	tetr cry; exp			
688	Copper(I) bromide CuBr	7787-70-4 143.450	wh cub cry; hyg	497 4.98	1345 a,e	sl H₂O; i ace
689	Copper(I) chloride CuCl	7758-89-6 98.999	wh cub cry	430 4.14	≈1400 a,c,e,f	sl H₂O; i EtOH, ace
690	Copper(I) cyanide CuCN	544-92-3 89.564	wh powder or grn orth cry	474 2.9	dec a	i H₂O, EtOH; s KCN soln
691	Copper(I) fluoride CuF	13478-41-6 82.544	cub cry	7.1		
692	Copper(I) hydride CuH	13517-00-5 64.554	red-brn solid	dec 60		

No.	Name Formula	CAS RN Mol. Wt.	Physical Form	mp/°C den/g cm^{-3}	bp/°C Other Data	Solubility
693	Copper(I) iodide CuI	7681-65-4 190.450	wh cub cry	606 5.67	≈1290 a,f	i H$_2$O, dil acid
694	Copper(I) mercury iodide Cu$_2$HgI$_4$	13876-85-2 835.30	red cry powder	trans ≈60 (brn)		i H$_2$O, EtOH
695	Copper(I) oxide Cu$_2$O	1317-39-1 143.091	red-brn cub cry	1235 6.0	dec 1800 a,f	i H$_2$O
696	Copper(I) selenide Cu$_2$Se	20405-64-5 206.05	blue-blk tetr cry	1113 6.84		i H$_2$O; s acid
697	Copper(I) sulfide Cu$_2$S	22205-45-4 159.158	blue-blk orth cry	≈1100 5.6	a,f	i H$_2$O; sl acid
698	Copper(I) sulfite hemihydrate Cu$_2$SO$_3$·0.5H$_2$O	13982-53-1* 216.164	wh-yel hex cry			sl H$_2$O; s acid, alk; i EtOH, eth
699	Copper(I) telluride Cu$_2$Te	12019-52-2 254.69	blue hex cry	1127 4.6		
700	Copper(I) thiocyanate CuSCN	1111-67-7 121.630	wh-yel amorp powder	1084 2.85		i H$_2$O, dil acid, EtOH, ace; s eth
701	Copper(I,II) sulfite dihydrate Cu$_2$SO$_3$·CuSO$_3$·2H$_2$O	13814-81-8 386.797	red prisms or powder			i H$_2$O, EtOH; s HCl
702	Copper(II) acetate Cu(C$_2$H$_3$O$_2$)$_2$	142-71-2 181.636	blue-grn hyg powder			
703	Copper(II) acetate metaarsenite Cu(C$_2$H$_3$O$_2$)$_2$·3Cu(AsO$_2$)$_2$	12002-03-8 1013.796	grn cry powder			i H$_2$O; reac acid
704	Copper(II) acetate monohydrate Cu(C$_2$H$_3$O$_2$)$_2$·H$_2$O	6046-93-1 199.651	grn monocl cry	115 1.88	dec 240	s H$_2$O, EtOH; sl eth
705	Copper(II) acetylide CuC$_2$	12540-13-5 87.568	brn-blk solid; exp	exp 100		
706	Copper(II) arsenate Cu$_3$(AsO$_4$)$_2$	10103-61-4 468.476	blue-grn cry			i H$_2$O, EtOH; s dil acid
707	Copper(II) arsenite CuHAsO$_3$	10290-12-7 187.474	yel-grn powder			i H$_2$O, EtOH; s acid
708	Copper(II) azide Cu(N$_3$)$_2$	14215-30-6 147.586	brn orth cry; exp	≈2.6		
709	Copper(II) basic acetate Cu(C$_2$H$_3$O$_2$)$_2$·CuO·6H$_2$O	52503-64-7 369.272	blue-grn cry or powder			sl H$_2$O, EtOH; s dil acid, NH$_4$OH
710	Copper(II) borate Cu(BO$_2$)$_2$	39290-85-2 149.166	blue-grn powder	3.859		i H$_2$O; s acid
711	Copper(II) bromide CuBr$_2$	7789-45-9 223.354	blk monocl cry; hyg	498 4.710	900 a	vs H$_2$O; s EtOH, ace; i bz, eth
712	Copper(II) butanoate monohydrate Cu(C$_4$H$_7$O$_2$)$_2$·H$_2$O	540-16-9 255.758	grn monocl plates			s H$_2$O, diox, bz; sl EtOH
713	Copper(II) carbonate hydroxide CuCO$_3$·Cu(OH)$_2$	12069-69-1 221.116	grn monocl cry	200 dec 4.0		i H$_2$O, EtOH; s dil acid
714	Copper(II) chlorate hexahydrate Cu(ClO$_3$)$_2$·6H$_2$O	14721-21-2 338.539	blue-grn hyg cry	65	dec 100	vs H$_2$O, EtOH
715	Copper(II) chloride CuCl$_2$	7447-39-4 134.451	yel-brn monocl cry; hyg	630 dec 3.4	a,c	s H$_2$O, EtOH, ace
716	Copper(II) chloride dihydrate CuCl$_2$·2H$_2$O	10125-13-0 170.482	grn-blue orth cry; hyg	100 dec 2.51		vs H$_2$O, EtOH, MeOH; s ace; i eth
717	Copper(II) chromate CuCrO$_4$	13548-42-0 179.540	red-brn cry			i H$_2$O; s EtOH
718	Copper(II) chromite CuCr$_2$O$_4$	12018-10-9 231.536	gray-blk tetr cry	5.4		i H$_2$O, dil acid
719	Copper(II) citrate hemipentahydrate Cu$_2$C$_6$H$_4$O$_7$·2.5H$_2$O	10402-15-0 360.223	blue-grn cry	dec 100		sl H$_2$O; s dil acid
720	Copper(II) cyanide Cu(CN)$_2$	14763-77-0 115.581	grn powder			i H$_2$O; s acid, alk
721	Copper(II) dichromate dihydrate CuCr$_2$O$_7$·2H$_2$O	13675-47-3 315.565	red-brn tricl cry	2.286		vs H$_2$O
722	Copper(II) ferrocyanide Cu$_2$Fe(CN)$_6$	13601-13-3 339.043	red-br cub cry or powder	2.2		i H$_2$O, acid, os
723	Copper(II) ferrous sulfide CuFeS$_2$	1308-56-1 183.523	yel tetr cry	950 4.2	f	i H$_2$O, HCl; s HNO$_3$
724	Copper(II) fluoride CuF$_2$	7789-19-7 101.543	wh monocl cry	836 4.23	1676 a,c	sl H$_2$O
725	Copper(II) fluoride dihydrate CuF$_2$·2H$_2$O	13454-88-1 137.574	blue monocl cry	dec 130 2.934		sl H$_2$O
726	Copper(II) formate Cu(CHO$_2$)$_2$	544-19-4 153.582	blue cry			s H$_2$O; i os
727	Copper(II) formate tetrahydrate Cu(CHO$_2$)$_2$·4H$_2$O	5893-61-8 225.642	blue monocl cry			s H$_2$O; sl EtOH; i os
728	Copper(II) hexafluorosilicate tetrahydrate CuSiF$_6$·4H$_2$O	12062-24-7 277.684	blue monocl cry	dec 2.56		vs H$_2$O; sl EtOH
729	Copper(II) hydroxide Cu(OH)$_2$	20427-59-2 97.561	blue-grn powder	3.37	a	i H$_2$O; s acid, conc alk
730	Copper(II) molybdate CuMoO$_4$	13767-34-5 223.48	grn cry	820 dec 3.4		i H$_2$O
731	Copper(II) nitrate Cu(NO$_3$)$_2$	3251-23-8 187.555	blue-grn orth cry; hyg	255	subl a	s H$_2$O, diox; reac eth

No.	Name Formula	CAS RN Mol. Wt.	Physical Form	mp/°C den/g cm^{-3}	bp/°C Other Data	Solubility
732	Copper(II) nitrate hexahydrate $Cu(NO_3)_2 \cdot 6H_2O$	13478-38-1 295.647	blue rhomb cry; hyg 	 2.07		vs H_2O; s EtOH
733	Copper(II) nitrate trihydrate $Cu(NO_3)_2 \cdot 3H_2O$	10031-43-3 241.602	blue rhomb cry	114 2.32	170 dec	vs H_2O, EtOH
734	Copper(II) oleate $Cu(C_{18}H_{33}O_2)_2$	1120-44-1 626.464	blue-grn solid			i H_2O; sl EtOH; s eth
735	Copper(II) oxalate CuC_2O_4	814-91-5 151.566	blue-wh powder	dec 310		i H_2O, EtOH, eth; s NH_4OH
736	Copper(II) oxide CuO	1317-38-0 79.545	blk powder or monocl cry	1446 6.31	a,c	i H_2O, EtOH; s dil acid
737	Copper(II) perchlorate $Cu(ClO_4)_2$	13770-18-8 262.446	grn hyg cry	dec 130		s H_2O, eth, diox; i bz, ctc
738	Copper(II) perchlorate hexahydrate $Cu(ClO_4)_2 \cdot 6H_2O$	10294-46-9 370.538	blue monocl cry; hyg	82 2.22	dec 120	vs H_2O, EtOH, HOAc, ace; sl eth
739	Copper(II) phosphate trihydrate $Cu_3(PO_4)_2 \cdot 3H_2O$	10031-48-8 434.627	blue-grn orth cry			i H_2O; s acid, NH_4OH
740	Copper(II) selenate pentahydrate $CuSeO_4 \cdot 5H_2O$	10031-45-5 296.58	blue tricl cry	dec 80 2.56		s H_2O, acid, NH_4OH; sl ace; i EtOH
741	Copper(II) selenide CuSe	1317-41-5 142.51	blue-blk needles or plates	550 dec 5.99	a	reac acid
742	Copper(II) selenite dihydrate $CuSeO_3 \cdot 2H_2O$	15168-20-4 226.54	blue orth cry	 3.31		i H_2O; s acid, NH_4OH
743	Copper(II) stearate $Cu(C_{18}H_{35}O_2)_2$	660-60-6 630.496	blue-grn amorp powder	≈250		i H_2O, EtOH, eth; s py
744	Copper(II) sulfate $CuSO_4$	7758-98-7 159.610	wh-grn amorp powder or rhomb cry	560 dec 3.60	a	s H_2O, i EtOH
745	Copper(II) sulfate pentahydrate $CuSO_4 \cdot 5H_2O$	7758-99-8 249.686	blue tricl cry	dec 110 2.286	f	vs H_2O; s MeOH; sl EtOH
746	Copper(II) sulfate, basic $Cu_3(OH)_4SO_4$	1332-14-5 354.732	grn rhomb cry	 3.88		i H_2O
747	Copper(II) sulfide CuS	1317-40-4 95.612	blk hex cry	trans 507 4.76	a,f	i H_2O, EtOH, dil acid, alk
748	Copper(II) tartrate trihydrate $CuC_4H_4O_6 \cdot 3H_2O$	815-82-7 265.664	blue-grn powder			sl H_2O; s acid, alk
749	Copper(II) telluride CuTe	12019-23-7 191.15	yel orth cry	trans ≈400 7.09		
750	Copper(II) tungstate $CuWO_4$	13587-35-4 311.38	yel-brn powder	 7.5	a	
751	Copper(II) tungstate dihydrate $CuWO_4 \cdot 2H_2O$	13587-35-4* 347.41	grn powder			i H_2O; sl HOAc; reac conc acid
752	Curium Cm	7440-51-9 247	silv metal; hex or cub	1345 13.51	≈3100 a	
753	Dysprosium Dy	7429-91-6 162.50	silv metal; hex	1412 8.55	2567 a,c	s dil acid
754	Dysprosium boride DyB_4	12310-43-9 205.74	tetr cry	2500 6.98		
755	Dysprosium nitride DyN	12019-88-4 176.51	cub cry	 9.93		
756	Dysprosium silicide $DySi_2$	12133-07-2 218.67	orth cry	 5.2		
757	Dysprosium(II) chloride $DyCl_2$	13767-31-2 233.41	blk cry	721 dec		reac H_2O
758	Dysprosium(II) iodide DyI_2	36377-94-3 416.31	purp cry	659		reac H_2O
759	Dysprosium(III) bromide $DyBr_3$	14456-48-5 402.21	wh hyg cry	879		s H_2O
760	Dysprosium(III) chloride $DyCl_3$	10025-74-8 268.86	yel cry	680 3.67	a	s H_2O
761	Dysprosium(III) fluoride DyF_3	13569-80-7 219.50	grn cry	1154		
762	Dysprosium(III) hydride DyH_3	13537-09-2 165.52	hex cry	 7.1		
763	Dysprosium(III) iodide DyI_3	15474-63-2 543.21	grn cry	978		
764	Dysprosium(III) oxide Dy_2O_3	1308-87-8 373.00	wh cub cry	2408 7.81	a	s acid
765	Dysprosium(III) sulfide Dy_2S_3	12133-10-7 421.20	red-brn monocl cry	 6.08		
766	Einsteinium Es	7429-92-7 252	metal; cub	860	a	
767	Erbium Er	7440-52-0 167.26	silv metal; hex	1529 9.07	2868 a,c	i H_2O; s acid
768	Erbium boride ErB_4	12310-44-0 210.50	tetr cry	2450 7.0		
769	Erbium bromide $ErBr_3$	13536-73-7 406.97	viol hyg cry	923		s H_2O
770	Erbium chloride $ErCl_3$	10138-41-7 273.62	viol monocl cry; hyg	776 4.1	a	s H_2O

No.	Name Formula	CAS RN Mol. Wt.	Physical Form	mp/°C den/g cm^{-3}	bp/°C Other Data	Solubility
771	Erbium chloride hexahydrate ErCl$_3$•6H$_2$O	10025-75-9 381.71	pink hyg cry	dec		s H$_2$O; sl EtOH
772	Erbium fluoride ErF$_3$	13760-83-3 224.26	pink orth cry	1147 7.8	a	i H$_2$O
773	Erbium hydride ErH$_3$	13550-53-3 170.28	hex cry	≈7.6		
774	Erbium iodide ErI$_3$	13813-42-8 547.97	viol hex cry; hyg	1014 ≈5.5		s H$_2$O
775	Erbium nitride ErN	12020-21-2 181.27	cub cry	10.6		
776	Erbium oxide Er$_2$O$_3$	12061-16-4 382.52	pink powder	2418 8.64	a	i H$_2$O; s acid
777	Erbium silicide ErSi$_2$	12020-28-9 223.43	orth cry	7.26		
778	Erbium sulfate Er$_2$(SO$_4$)$_3$	13478-49-4 622.71	hyg powder	dec 3.68		s H$_2$O
779	Erbium sulfate octahydrate Er$_2$(SO$_4$)$_3$•8H$_2$O	10031-52-4 766.83	pink monocl cry	dec 3.20		s H$_2$O
780	Erbium sulfide Er$_2$S$_3$	12159-66-9 430.72	red-brn monocl cry	1730 6.07		
781	Erbium telluride Er$_2$Te$_3$	12020-39-2 717.32	orth cry	1213 7.11		
782	Europium Eu	7440-53-1 151.965	soft silv metal; cub	822 5.24	1529 a,c	reac H$_2$O
783	Europium boride EuB$_6$	12008-05-8 216.831	cub cry	≈2600 4.91		
784	Europium nitride EuN	12020-58-5 165.972	cub cry	8.7		
785	Europium silicide EuSi$_2$	12434-24-1 208.136	tetr cry	1500 5.46		
786	Europium(II) bromide EuBr$_2$	13780-48-8 311.773	wh cry	683		s H$_2$O
787	Europium(II) chloride EuCl$_2$	13769-20-5 222.870	wh orth cry	731 4.9		s H$_2$O
788	Europium(II) fluoride EuF$_2$	14077-39-5 189.962	grn-yel cub cry	≈1380 6.5		
789	Europium(II) iodide EuI$_2$	22015-35-6 405.774	grn cry	580		s H$_2$O
790	Europium(II) selenide EuSe	12020-66-5 230.93	brn cub cry	6.45		
791	Europium(II) sulfide EuS	12020-65-4 184.031	cub cry	5.7		
792	Europium(II) telluride EuTe	12020-69-8 279.57	blk cub cry	1526 6.48		
793	Europium(III) bromide EuBr$_3$	13759-88-1 391.677	gray cry	dec		s H$_2$O
794	Europium(III) chloride EuCl$_3$	10025-76-0 258.323	grn-yel needles	623 4.89	a	
795	Europium(III) chloride hexahydrate EuCl$_3$•6H$_2$O	13759-92-7 366.414	wh-yel hyg cry	850 4.89		s H$_2$O
796	Europium(III) fluoride EuF$_3$	13765-25-8 208.960	wh hyg cry	1276		i H$_2$O
797	Europium(III) nitrate hexahydrate Eu(NO$_3$)$_3$•6H$_2$O	10031-53-5 446.071	wh-pink hyg cry	dec 85		s H$_2$O
798	Europium(III) oxide Eu$_2$O$_3$	1308-96-9 351.928	pink powder	2350 7.42	a	i H$_2$O; s acid
799	Europium(III) sulfate Eu$_2$(SO$_4$)$_3$	13537-15-0 592.121	pale pink cry	4.99		sl H$_2$O
800	Europium(III) sulfate octahydrate Eu$_2$(SO$_4$)$_3$•8H$_2$O	10031-52-4 736.243	pink cry	dec 375		sl H$_2$O
801	Fermium Fm	7440-72-4 257	metal	1527	a	
802	Fluorine F$_2$	7782-41-4 37.997	pale yel gas	-219.66 1.667 g/L	-188.12 a,b,c,d,e	reac H$_2$O
803	Fluorine monoxide F$_2$O	7783-41-7 53.996	col gas	-223.8 2.369 g/L	-144.75 a,b,d,e	sl H$_2$O
804	Fluorine dioxide F$_2$O$_2$	7783-44-0 69.996	gas, stable only at low temp	-154 3.071 g/L	-57 a,d	
805	Fluorine nitrate FNO$_3$	7789-26-6 81.003	col gas	-175 3.554 g/L	-46	reac H$_2$O, EtOH, eth; s ace
806	Fluorine perchlorate FOClO$_3$	10049-03-3 118.449	col gas; exp	-167.3 5.197 g/L	-16	reac H$_2$O
807	Francium Fr	7440-73-5 223	short-lived alkali metal	27	a	
808	Gadolinium Gd	7440-54-2 157.25	silv metal; hex	1313 7.90	3273 a,c	s dil acid
809	Gadolinium boride GdB$_6$	12008-06-9 222.12	blk-brn cub cry	2510 5.31		

No.	Name Formula	CAS RN Mol. Wt.	Physical Form	mp/°C den/g cm^{-3}	bp/°C Other Data	Solubility
810	Gadolinium nitride GdN	25764-15-2 171.26	cub cry	9.10		
811	Gadolinium silicide GdSi$_2$	12134-75-7 213.42	orth cry	5.9		
812	Gadolinium(II) iodide GdI$_2$	13814-72-7 411.06	bronze cry	831		
813	Gadolinium(II) selenide GdSe	12024-81-6 236.21	cub cry	2170 8.1		
814	Gadolinium(III) bromide GdBr$_3$	13818-75-2 396.96	wh monocl cry; hyg	770 4.56		
815	Gadolinium(III) chloride GdCl$_3$	10138-52-0 263.61	wh monocl cry; hyg	609 4.52	a	s H$_2$O
816	Gadolinium(III) chloride hexahydrate GdCl$_3$·6H$_2$O	19423-81-5 371.70	col hyg cry	2.424		s H$_2$O
817	Gadolinium(III) fluoride GdF$_3$	13765-26-9 214.25	wh cry	1231	a	
818	Gadolinium(III) iodide GdI$_3$	13572-98-0 537.96	yel cry	925		
819	Gadolinium(III) nitrate hexahydrate Gd(NO$_3$)$_3$·6H$_2$O	19598-90-4 451.36	hyg tricl cry	91 dec 2.33		s H$_2$O, EtOH
820	Gadolinium(III) nitrate pentahydrate Gd(NO$_3$)$_3$·5H$_2$O	52788-53-1 433.34	wh cry	92 dec 2.41		i H$_2$O
821	Gadolinium(III) oxide Gd$_2$O$_3$	12064-62-9 362.50	wh hyg powder	2420 7.07	a	i H$_2$O; s acid
822	Gadolinium(III) sulfate octahydrate Gd$_2$(SO$_4$)$_3$·8H$_2$O	13450-87-8 746.81	col monocl cry	dec 400 4.14		s H$_2$O
823	Gadolinium(III) sulfide Gd$_2$S$_3$	12134-77-9 410.70	yel cub cry	6.1		
824	Gadolinium(III) telluride Gd$_2$Te$_3$	12160-99-5 697.30	orth cry	1255 7.7		
825	Gallium Ga	7440-55-3 69.723	silv liq or gray orth cry	29.76 5.91	2204 a,c,d,e	reac alk
826	Gallium antimonide GaSb	12064-03-8 191.483	cub cry	712 5.6137	a,c	
827	Gallium arsenide GaAs	1303-00-0 144.645	gray cub cry	1238 5.3176	a	
828	Gallium nitride GaN	25617-97-4 83.730	gray hex cry	600 6.1	dec >600 a	
829	Gallium phosphide GaP	12063-98-8 100.697	yel cub cry	1457 4.138	a	
830	Gallium suboxide Ga$_2$O	12024-20-3 155.445	brn powder	>660 4.77	dec >800 a	
831	Gallium(II) chloride GaCl$_2$	24597-12-4 140.628	wh orth cry	172.4 2.74	535	
832	Gallium(II) selenide GaSe	12024-11-2 148.68	hex cry	960 5.03		
833	Gallium(II) sulfide GaS	12024-10-1 101.789	hex cry	965 3.86		
834	Gallium(II) telluride GaTe	12024-14-5 197.32	monocl cry	824 5.44		
835	Gallium(III) acetylacetonate Ga(CH$_3$COCHCOCH$_3$)$_3$	14405-43-7 367.051	wh powder	193 1.42	subl	
836	Gallium(III) bromide GaBr$_3$	13450-88-9 309.435	wh orth cry	121.5 3.69	279 a,b,c,d	
837	Gallium(III) chloride GaCl$_3$	13450-90-3 176.081	col needles or gl solid	77.9 2.47	201 a,b,c,d,e	
838	Gallium(III) fluoride GaF$_3$	7783-51-9 126.718	wh powder or col needles	>1000 4.47	a	i H$_2$O
839	Gallium(III) fluoride trihydrate GaF$_3$·3H$_2$O	22886-66-4 180.764	wh cry	dec >140		sl H$_2$O
840	Gallium(III) hydride GaH$_3$	13572-93-5 72.747	visc liq	-15	dec ≈0	
841	Gallium(III) hydroxide Ga(OH)$_3$	12023-99-3 120.745	unstable prec		a	
842	Gallium(III) iodide GaI$_3$	13450-91-4 450.436	monocl cry	212 4.5	340 a,b,c,d	
843	Gallium(III) nitrate Ga(NO$_3$)$_3$	13494-90-1 255.738	wh cry powder			s H$_2$O, EtOH, eth
844	Gallium(III) oxide Ga$_2$O$_3$	12024-21-4 187.444	wh cry	1807 ≈6.0	a	s hot acid
845	Gallium(III) oxide hydroxide GaOOH	20665-52-5 102.730	orth cry	5.23		
846	Gallium(III) selenide Ga$_2$Se$_3$	12024-24-7 376.33	cub cry	937 4.92		
847	Gallium(III) sulfate Ga$_2$(SO$_4$)$_3$	13494-91-2 427.637	hex cry			
848	Gallium(III) sulfate octadecahydrate Ga$_2$(SO$_4$)$_3$·18H$_2$O	13780-42-2 751.912	octahed cry			s H$_2$O, EtOH

No.	Name Formula	CAS RN Mol. Wt.	Physical Form	mp/°C den/g cm⁻³	bp/°C Other Data	Solubility
849	Gallium(III) sulfide Ga_2S_3	12024-22-5 235.644	monocl cry	1090 3.7		
850	Gallium(III) telluride Ga_2Te_3	12024-27-0 522.25	cub cry	790 5.57		
851	Germanium Ge	7440-56-4 72.61	gray-wh cub cry	938.25 5.3234	2833 a,c,d	i H_2O, dil acid, alk
852	Germanium nitride Ge_3N_4	12065-36-0 273.86	orth cry	dec 900		i H_2O, acid, aqua regia
853	Germane GeH_4	7782-65-2 76.64	col gas; flam	-165 3.363 g/L	-88.1 a,b,d,e	i H_2O
854	Bromogermane GeH_3Br	13569-43-2 155.54	col liq	-32 2.34	52 a	reac H_2O
855	Chlorogermane GeH_3Cl	13637-65-5 111.09	col liq	-52 1.75	28 a	reac H_2O
856	Chlorotrifluorogermane GeF_3Cl	14188-40-0 165.06	gas	-66.2 7.243 g/L	-20.3	
857	Dibromogermane GeH_2Br_2	13769-36-3 234.43	col liq	-15 2.80	89	reac H_2O
858	Dichlorogermane GeH_2Cl_2	15230-48-5 145.53	col liq	-68 1.90	69.5	reac H_2O
859	Dichlorodifluorogermane GeF_2Cl_2	24422-21-7 181.51	col gas	-51.8 7.965 g/L	-2.8	
860	Fluorogermane GeH_3F	13537-30-9 94.63	col gas	 4.152 g/L	 a	reac H_2O
861	Iodogermane GeH_3I	13573-02-9 202.54	liq	-15	≈90 a	reac H_2O
862	Tribromogermane $GeHBr_3$	14779-70-5 313.33	col liq	-25	dec	reac H_2O
863	Trichlorogermane $GeHCl_3$	1184-65-2 179.98	liq	-71 1.93	75.3 e	reac H_2O
864	Trichlorofluorogermane $GeCl_3F$	24422-20-6 197.97	liq	-49.8	37.5	
865	Methylgermane GeH_3CH_3	1449-65-6 90.67	col gas	-158 3.979 g/L	-23	
866	Digermane Ge_2H_6	13818-89-8 151.27	col liq; flam	-109	30.8 a,d,e	
867	Trigermane Ge_3H_8	14691-44-2 225.89	col liq	-105.6 2.20(-105°C)	110.5 a,d,e	i H_2O
868	Germanium(II) bromide $GeBr_2$	24415-00-7 232.42	yel monocl cry	122	dec 150	reac H_2O
869	Germanium(II) chloride $GeCl_2$	10060-11-4 143.51	wh-yel hyg powder	dec		reac H_2O; s eth, bz
870	Germanium(II) fluoride GeF_2	13940-63-1 110.61	wh orth cry; hyg	110 3.64	dec 130	reac H_2O
871	Germanium(II) iodide GeI_2	13573-08-5 326.42	oran-yel hex cry	dec 550 5.4		reac H_2O
872	Germanium(II) oxide GeO	20619-16-3 88.61	blk solid	dec 700	 a	
873	Germanium(II) selenide GeSe	12065-10-0 151.57	gray orth cry or brn powder	667 5.6		
874	Germanium(II) sulfide GeS	12025-32-0 104.68	gray orth cry	615 4.1	 a	
875	Germanium(II) telluride GeTe	12025-39-7 200.21	cub cry	725 6.16		i H_2O; s conc HNO_3
876	Germanium(IV) bromide $GeBr_4$	13450-92-5 392.23	wh cry	26.1 3.132	186.35 a,b,d,e	reac H_2O
877	Germanium(IV) chloride $GeCl_4$	10038-98-9 214.42	col liq	-49.5 1.88	86.55 a,b,d,e	reac H_2O; s bz, eth, EtOH, ctc
878	Germanium(IV) fluoride GeF_4	7783-58-6 148.60	col gas	-15 tp 6.521 g/L	-36.5 sp a	reac H_2O
879	Germanium(IV) iodide GeI_4	13450-95-8 580.23	red-oran cub cry	144 4.4	377 a,b	reac H_2O
880	Germanium(IV) oxide GeO_2	1310-53-8 104.61	wh hex cry	1115 4.25	 a	i H_2O
881	Germanium(IV) selenide $GeSe_2$	12065-11-1 230.53	yel-oran orth ccry	707 dec 4.56		
882	Germanium(IV) sulfide GeS_2	12025-34-2 136.74	blk orth cry	530 3.01		
883	Gold Au	7440-57-5 196.967	soft yel metal	1064.18 19.3	2856 a,c,d,e,f	s aqua regia
884	Bromoauric acid pentahydrate $HAuBr_4 \cdot 5H_2O$	17083-68-0 607.667	red-brn hyg cry	27		s H_2O, EtOH
885	Chloroauric acid tetrahydrate $HAuCl_4 \cdot 4H_2O$	16903-35-8 411.847	yel monocl cry; hyg	≈3.9		vs H_2O, EtOH; s eth
886	Gold(I) bromide AuBr	10294-27-6 276.871	yel-gray tetr cry	dec 165 8.20	 a	i H_2O
887	Gold(I) chloride AuCl	10294-29-8 232.420	yel orth cry	dec 289 7.6	 a	i H_2O

No.	Name Formula	CAS RN Mol. Wt.	Physical Form	mp/°C den/g cm⁻³	bp/°C Other Data	Solubility
888	Gold(I) cyanide $AuCN$	506-65-0 222.985	yel hex cry	dec 7.2		i H_2O, EtOH, eth, dil acid
889	Gold(I) iodide AuI	10294-31-2 323.871	yel-grn powder; tetr	dec 120 8.25	a	i H_2O; s CN soln
890	Gold(I) sulfide Au_2S	1303-60-2 425.999	brn-blk cub cry; unstable	240 dec ≈11		i H_2O, acid; s aqua regia
891	Gold(III) bromide $AuBr_3$	10294-28-7 436.679	red-br monocl cry	dec ≈160	a	s H_2O, EtOH
892	Gold(III) chloride $AuCl_3$	13453-07-1 303.325	red monocl cry	dec >160 4.7	a	
893	Gold(III) cyanide trihydrate $Au(CN)_3 \cdot 3H_2O$	535-37-5* 329.066	wh hyg cry	dec 50		vs H_2O; sl EtOH
894	Gold(III) fluoride AuF_3	14720-21-9 253.962	oran-yel hex cry	>300 6.75	subl a	
895	Gold(III) hydroxide $Au(OH)_3$	1303-52-2 247.989	brn powder	dec ≈100		i H_2O; s acid
896	Gold(III) iodide AuI_3	31032-13-0 577.680	unstable grn powder			
897	Gold(III) oxide Au_2O_3	1303-58-8 441.931	brn powder	dec ≈150		i H_2O; s acid
898	Gold(III) selenate $Au_2(SeO_4)_3$	10294-32-3 822.81	yel cry			i H_2O; s acid
899	Gold(III) selenide Au_2Se_3	1303-62-4 630.81	blk amorp solid	dec 4.65		s aqua regia
900	Gold(III) sulfide Au_2S_3	1303-61-3 490.131	unstable blk powder	dec 200		
901	Hafnium Hf	7440-58-6 178.49	gray metal; hex	2233 13.3	4603 a,c	s HF
902	Hafnium boride HfB_2	12007-23-7 200.11	gray hex cry	3100 10.5		
903	Hafnium bromide $HfBr_4$	13777-22-5 498.11	wh cub cry	424 tp 4.90	323 sp b	
904	Hafnium carbide HfC	12069-85-1 190.50	refrac cub cry	≈3000 12.2		
905	Hafnium chloride $HfCl_4$	13499-05-3 320.30	wh monocl cry	432 tp	317 sp a,b	reac H_2O
906	Hafnium fluoride HfF_4	13709-52-9 254.48	wh monocl cry	>970 7.1	970 sp a	
907	Hafnium hydride HfH_2	12770-26-2 180.51	refrac tetr cry	11.4		
908	Hafnium iodide HfI_4	13777-23-6 686.11	yel-oran cub cry	449 tp 5.6	394 sp b	
909	Hafnium nitride HfN	25817-87-2 192.50	yel-brn cub cry	3305 13.8		
910	Hafnium oxide HfO_2	12055-23-1 210.49	wh cub cry	2774 9.68	a	i H_2O
911	Hafnium oxychloride octahydrate $HfOCl_2 \cdot 8H_2O$	14456-34-9 409.52	wh tetr cry	dec		s H_2O
912	Hafnium phosphide HfP	12325-59-6 209.46	hex cry	9.78		
913	Hafnium selenide $HfSe_2$	12162-21-9 336.41	brn hex cry	7.46		
914	Hafnium orthosilicate $HfSiO_4$	13870-13-8 270.57	tetr cry	7.0		
915	Hafnium silicide $HfSi_2$	12401-56-8 234.66	gray orth cry	≈1700 7.6		
916	Hafnium sulfate $Hf(SO_4)_2$	15823-43-5 370.62	wh cry	dec >500		
917	Hafnium sulfide HfS_2	18855-94-2 242.62	purp-brn hex cry	6.03		
918	Helium He	7440-59-7 4.003	col gas	0.176 g/L	-268.93 a,b,d,e	sl H_2O; i EtOH
919	Holmium Ho	7440-60-0 164.930	silv metal; hex	1474 8.80	2700 a,c	s dil acid
920	Holmium bromide $HoBr_3$	13825-76-8 404.642	yel hyg cry	919	1470	
921	Holmium chloride $HoCl_3$	10138-62-2 271.288	yel monocl cry; hyg	718 3.7	1500 a	s H_2O
922	Holmium fluoride HoF_3	13760-78-6 221.925	pink-yel orth cry; hyg	1143 7.664	>2200 a	s H_2O
923	Holmium iodide HoI_3	13813-41-7 545.643	yel hex cry	994 5.4		
924	Holmium nitride HoN	12029-81-1 178.937	cub cry	10.6		
925	Holmium oxide Ho_2O_3	12055-62-8 377.859	yel cub cry	2415 8.41	a	s acid
926	Holmium silicide $HoSi_2$	12136-24-2 221.10	hex cry	7.1		

No.	Name Formula	CAS RN Mol. Wt.	Physical Form	mp/°C den/g cm^{-3}	bp/°C Other Data	Solubility
927	Holmium sulfide Ho_2S_3	12162-59-3 426.059	yel-oran monocl cry	 5.92		
928	Hydrazine N_2H_4	302-01-2 32.045	col oily liq	1.4 1.0036	113.55 a,b,c,d	vs H_2O, EtOH, MeOH
929	Hydrazine hydrate $N_2H_4 \cdot H_2O$	7803-57-8 50.060	fuming liq	-51.7 1.030	119	vs H_2O, EtOH; i chl, eth
930	Hydrazine hydrobromide $N_2H_4 \cdot HBr$	13775-80-9 112.957	wh monocl cry flakes	84 2.3	dec ≈190	s H_2O, EtOH
931	Hydrazine hydrochloride $N_2H_4 \cdot HCl$	2644-70-4 68.506	wh orth cry	89 1.5	dec 240	s H_2O; i os
932	Hydrazine dihydrochloride $N_2H_4 \cdot 2HCl$	5341-61-7 104.966	wh orth cry	198 dec 1.42		s H_2O; sl EtOH
933	Hydrazine hydroiodide $N_2H_4 \cdot HI$	10039-55-1 159.957	hyg cry	125		s H_2O
934	Hydrazine nitrate $N_2H_4 \cdot HNO_3$	13464-97-6 95.058	monocl cry; exp	70		vs H_2O
935	Hydrazine sulfate $N_2H_4 \cdot H_2SO_4$	10034-93-2 130.125	col orth cry	254 1.378		sl H_2O; i EtOH
936	Hydrazoic acid HN_3	7782-79-8 43.028	col liq; exp	-80	35.7 a,d	s H_2O
937	Hydroxylamine H_2NOH	7803-49-8 33.030	wh orth flakes or needles	33.1 1.21	58 a	vs H_2O, MeOH
938	Hydroxylamine sulfate $(H_2NOH)_2 \cdot H_2SO_4$	10039-54-0 164.139	cry	170		vs H_2O
939	Hydrogen H_2	1333-74-0 2.016	col gas; flam	-259.34 0.088 g/L	-252.87 a,b,c,d,e	sl H_2O
940	Hydrogen bromide HBr	10035-10-6 80.912	col gas	-86.81 3.550 g/L	-66.38 a,b,c,d,e	vs H_2O; s EtOH
941	Hydrogen chloride HCl	7647-01-0 36.461	col gas	-114.18 1.600 g/L	-85 a,b,c,d,e	vs H_2O
942	Hydrogen cyanide HCN	74-90-8 27.026	col liq	-13.4 0.684	26 a,b,c,e	vs H_2O, EtOH; sl eth
943	Hydrogen fluoride HF	7664-39-3 20.006	col gas	-83.36 0.878 g/L	20 a,b,c,e	vs H_2O, EtOH; sl eth
944	Hydrogen iodide HI	10034-85-2 127.912	col or yel gas	-50.77 5.613 g/L	-35.55 a,b,c,d,e	vs H_2O; s os
945	Hydrogen peroxide H_2O_2	7722-84-1 34.015	col liq	-0.43 1.44	150.2 a,c,d,e	vs H_2O
946	Hydrogen selenide H_2Se	7783-07-5 80.98	col gas; flam	-65.73 3.553 g/L	-41.25 a,b,d,e	s H_2O
947	Hydrogen sulfide H_2S	7783-06-4 34.082	col gas; flam	-85.5 1.496 g/L	-59.55 a,b,c,d,e	s H_2O
948	Hydrogen disulfide H_2S_2	13465-07-1 66.148	col liq	 1.334	70.7 a,d,e	
949	Hydrogen telluride H_2Te	7783-09-7 129.62	col gas	-49 5.687 g/L	-2 a,d,e	s H_2O, EtOH, alk
950	Indium In	7440-74-6 114.818	soft wh metal	156.60 7.31	2072 a,c	s acid
951	Indium antimonide InSb	1312-41-0 236.578	blk cub cry	525 5.7747	a,c	
952	Indium arsenide InAs	1303-11-3 189.740	gray cub cry	942 5.67	a	i acids
953	Indium nitride InN	25617-98-5 128.825	hex cry	1100 6.88		
954	Indium phosphide InP	22398-80-7 145.792	blk cub cry	1062 4.81	a	sl acid
955	Indium(I) bromide InBr	14280-53-6 194.722	oran-red orth cry	290 4.96	656 a,d	reac H_2O
956	Indium(I) chloride InCl	13465-10-6 150.271	yel cub cry	211 4.19	608 a,c	reac H_2O
957	Indium(I) iodide InI	13966-94-4 241.722	orth cry	364.4 5.32	712 a,d	
958	Indium(II) bromide $InBr_2$	21264-43-7 274.626	orth cry			reac H_2O
959	Indium(II) chloride $InCl_2$	13465-11-7 185.723	col orth cry	235 3.64		reac H_2O
960	Indium(II) sulfide InS	12030-14-7 146.884	red-brn orth cry	692 5.2	a	
961	Indium(III) bromide $InBr_3$	13465-09-3 354.530	hyg yel-wh monocl cry	420 4.74	a	
962	Indium(III) chloride $InCl_3$	10025-82-8 221.176	yel monocl cry; hyg	583 4.0	a	vs H_2O; s EtOH
963	Indium(III) fluoride InF_3	7783-52-0 171.813	wh hex cry; hyg	1170 4.39	>1200	sl H_2O; s dil acid
964	Indium(III) fluoride trihydrate $InF_3 \cdot 3H_2O$	14166-78-0 225.859	wh cry	dec 100		s H_2O
965	Indium(III) hydroxide $In(OH)_3$	20661-21-6 165.840	cub cry	 4.4		

No.	Name Formula	CAS RN Mol. Wt.	Physical Form	mp/°C den/g cm⁻³	bp/°C Other Data	Solubility
966	Indium(III) iodide InI_3	13510-35-5 495.531	yel-red monocl cry; hyg	207 4.69	a	
967	Indium(III) oxide In_2O_3	1312-43-2 277.634	yel cub cry	1913 7.18	a	i H_2O; s hot acid
968	Indium(III) perchlorate octahydrate $In(ClO_4)_3 \cdot 8H_2O$	13465-15-1 557.291	wh cry	≈80	dec 200	
969	Indium(III) phosphate $InPO_4$	14693-82-4 209.789	wh orth cry	4.9		i H_2O
970	Indium(III) selenide In_2Se_3	1312-42-1 466.52	blk hex cry	660 5.8		
971	Indium(III) sulfate $In_2(SO_4)_3$	13464-82-9 517.827	hyg wh powder	3.44		s H_2O
972	Indium(III) sulfide In_2S_3	12030-24-9 325.834	oran cub cry	1050 4.45	a	
973	Indium(III) telluride In_2Te_3	1312-45-4 612.44	blk cub cry	667 5.75		
974	Iodine I_2	7553-56-2 253.809	blue-blk plates	113.7 4.933	184.4 a,b,c,d,e	sl H_2Os; s bz, EtOH, eth, ctc, chl
975	Iodic acid HIO_3	7782-68-5 175.910	col orth cry	110 dec 4.63	a	vs H_2O; i EtOH, eth
976	Periodic acid dihydrate $HIO_4 \cdot 2H_2O$	10450-60-9 227.940	monocl hyg cry	122 dec		s H_2O, EtOH; sl eth
977	Iodine tetroxide I_2O_4	12399-08-5 317.807	yel cry	dec 85 4.2		sl H_2O
978	Iodine pentoxide I_2O_5	12029-98-0 333.806	hyg wh cry	≈300 dec 4.98		s H_2O; i EtOH, eth, CS_2
979	Iodine nonaoxide I_4O_9	73560-00-6 651.613	hyg yel powder	dec 75		
980	Iodine bromide IBr	7789-33-5 206.808	blk orth cry	40 4.3	116 dec a,b	s H_2O, EtOH, eth
981	Iodine chloride ICl	7790-99-0 162.357	red cry or oily liq	27.38 3.24	100 dec a,c	reac H_2O; s EtOH
982	Iodine trichloride ICl_3	865-44-1 233.262	yel tricl cry; hyg	101 tp (16 atm) 3.2	64 sp (dec)	reac H_2O; s EtOH, bz
983	Iodine fluoride IF	13873-84-2 145.902	disproportionates at room temp		a	
984	Iodine trifluoride IF_3	22520-96-3 183.899	yel solid, stable at low temp	dec -28		
985	Iodine pentafluoride IF_5	7783-66-6 221.896	yel liq	9.43 3.19	100.5 a,d,e	reac H_2O
986	Iodine heptafluoride IF_7	16921-96-3 259.893	col liq	6.5 tp 2.67	4.8 sp e	s H_2O
987	Iridium Ir	7439-88-5 192.217	silv-wh metal; cub	2446 22.5	4428 a,c	s aqua regia
988	Iridium(III) sulfide Ir_2S_3	12136-42-4 480.632	orth cry	10.2	a	
989	Iridium(III) bromide $IrBr_3$	10049-24-8 431.929	red-brn monocl cry	6.82		i H_2O, acid, alk
990	Iridium(III) bromide tetrahydrate $IrBr_3 \cdot 4H_2O$	10049-24-8* 503.991	grn-brn cry			s H_2O; i EtOH
991	Iridium(III) chloride $IrCl_3$	10025-83-9 298.575	brn monocl cry	dec 763 5.30	a	i H_2O, acid, alk
992	Iridium(III) fluoride IrF_3	23370-59-4 249.212	blk hex cry	dec 250 ≈8.0		i H_2O, dil acid
993	Iridium(III) iodide IrI_3	7790-41-2 572.930	dark brn monocl cry	≈7.4		i H_2O, acid, bz, chl; s alk
994	Iridium(III) oxide Ir_2O_3	1312-46-5 432.432	blue-blk cry	dec 1000		i H_2O; sl hot HCl
995	Iridium(IV) oxide IrO_2	12030-49-8 224.216	brn tetr cry	1100 dec 11.7	a	
996	Iridium(IV) sulfide IrS_2	12030-51-2 256.349	orth cry	9.3	a	
997	Iridium(VI) fluoride IrF_6	7783-75-7 306.207	yel cub cry; hyg	44 4.8	53 a,c,d	reac H_2O
998	Iron Fe	7439-89-6 55.845	silv-wh or gray met	1538 7.87	2861 a,c,e	s dil acid
999	Iron arsenide $FeAs$	12044-16-5 130.767	gray orth cry	1030 7.85		
1000	Iron boride Fe_2B	12006-86-9 122.501	refr solid; tetr	1389 7.3		
1001	Iron boride FeB	12006-84-7 66.656	refr solid; orth	1650 ≈7		
1002	Iron pentacarbonyl $Fe(CO)_5$	13463-40-6 195.897	yel oily liq; flam	-20 1.490	103 a,e	i H_2O; s eth, bz, ace
1003	Iron nonacarbonyl $Fe_2(CO)_9$	15321-51-4 363.784	oran-yel cry	100 dec 2.85		
1004	Iron dodecacarbonyl $Fe_3(CO)_{12}$	12088-65-2 503.660	blk cry	140 2.00		

No.	Name / Formula	CAS RN / Mol. Wt.	Physical Form	mp/°C / den/g cm⁻³	bp/°C / Other Data	Solubility
1005	Iron hydrocarbonyl $H_2Fe(CO)_4$	17440-90-3 169.903	col liq; unstable	-70	dec	s alk
1006	Iron phosphide Fe_2P	1310-43-6 142.664	gray hex needles	1370 6.8		i H_2O, dil acid, alk
1007	Iron disulfide FeS_2	1317-66-4 119.977	blk cub cry	dec >600 5.02	a,f	i H_2O
1008	Iron silicide $FeSi$	12022-95-6 83.931	gray cub cry	1410 6.1		
1009	Iron silicide $FeSi_2$	12022-99-0 112.016	gray tetr cry	1220 4.74		
1010	Iron(II) aluminate $Fe(AlO_2)_2$	12068-49-4 173.806	blk cub cry	4.3		
1011	Iron(II) bromide $FeBr_2$	7789-46-0 215.653	yel-brn hex cry; hyg	691 4.636	dec a,c	vs H_2O, EtOH
1012	Iron(II) bromide hexahydrate $FeBr_2 \cdot 6H_2O$	13463-12-2 323.744	grn hyg cry	27 dec 4.64		s H_2O, EtOH
1013	Iron(II) carbonate $FeCO_3$	563-71-3 115.854	gray-brn hex cry	3.9	f	
1014	Iron(II) chloride $FeCl_2$	7758-94-3 126.750	wh hex cry; hyg	677 3.16	1023 a,c,e	vs H_2O, EtOH, ace; sl bz
1015	Iron(II) chloride dihydrate $FeCl_2 \cdot 2H_2O$	16399-77-2 162.781	wh-grn monocl cry	dec 120 2.39		s H_2O
1016	Iron(II) chloride tetrahydrate $FeCl_2 \cdot 4H_2O$	13478-10-9 198.812	grn monocl cry	dec 105 1.93		s H_2O, EtOH
1017	Iron(II) chromite $FeCr_2O_4$	1308-31-2 223.835	blk cub cry	5.0	f	
1018	Iron(II) fluoride FeF_2	7789-28-8 93.842	wh tetr cry	1100 4.09	a,c	sl H_2O; s dil HF; i EtOH, eth
1019	Iron(II) fluoride tetrahydrate $FeF_2 \cdot 4H_2O$	13940-89-1 165.904	col hex cry	2.20		
1020	Iron(II) hydroxide $Fe(OH)_2$	18624-44-7 89.860	wh-grn hex cry	3.4		i H_2O
1021	Iron(II) iodide FeI_2	7783-86-0 309.654	red-viol hex cry; hyg	587 5.3	a,c	s H_2O, EtOH, eth
1022	Iron(II) iodide tetrahydrate $FeI_2 \cdot 4H_2O$	7783-86-0* 381.716	blk hyg leaflets	dec 90 2.87		s H_2O, EtOH
1023	Iron(II) molybdate $FeMoO_4$	13718-70-2 215.78	brn-yel monocl cry	1115 5.6	a	i H_2O
1024	Iron(II) nitrate $Fe(NO_3)_2$	14013-86-6 179.854	grn solid			s H_2O
1025	Iron(II) nitrate hexahydrate $Fe(NO_3)_2 \cdot 6H_2O$	13476-08-9 287.946	grn solid	60 dec		vs H_2O
1026	Iron(II) oxalate dihydrate $FeC_2O_4 \cdot 2H_2O$	6047-25-2 179.895	yel cry	dec 150 2.28		sl H_2O; s acid
1027	Iron(II) oxide FeO	1345-25-1 71.844	blk cub cry	1377 6.0	a,c	i H_2O, alk; s acid
1028	Iron(II) phosphate octahydrate $Fe_3(PO_4)_2 \cdot 8H_2O$	14940-41-1 501.600	gray-blue monocl cry; hyg	2.58	f	i H_2O; s acid
1029	Iron(II) selenide $FeSe$	1310-32-3 134.81	blk hex cry	6.7		i H_2O
1030	Iron(II) orthosilicate Fe_2SiO_4	10179-73-4 203.774	brn orth cry	4.30	a,f	
1031	Iron(II) sulfate $FeSO_4$	7720-78-7 151.909	wh orth cry; hyg	3.65	a	s H_2O
1032	Iron(II) sulfate monohydrate $FeSO_4 \cdot H_2O$	17375-41-6 169.924	wh-yel monocl cry	dec 300 3.0		s H_2O
1033	Iron(II) sulfate heptahydrate $FeSO_4 \cdot 7H_2O$	7782-63-0 278.015	blue-grn monocl cry	dec ≈60 1.895	f	vs H_2O; i EtOH
1034	Iron(II) sulfide FeS	1317-37-9 87.911	col hex or tetr cry; hyg	1188 4.7	dec a,c	i H_2O; reac acid
1035	Iron(II) telluride $FeTe$	12125-63-2 183.45	tetr cry	914 6.8		
1036	Iron(II) thiocyanate trihydrate $Fe(SCN)_2 \cdot 3H_2O$	6010-09-9 226.058	grn monocl cry			s H_2O, EtOH, eth
1037	Iron(II) titanate $FeTiO_3$	12168-52-4 151.710	blk rhomb cry	≈1470 4.72	f	
1038	Iron(II) tungstate $FeWO_4$	13870-24-1 303.68	monocl cry	7.51	a,f	
1039	Iron(II,III) oxide Fe_3O_4	1317-61-9 231.533	blk cub cry or amorp powder	1597 5.17	a,c,f	i H_2O; s acid
1040	Iron(III) acetate, basic $FeOH(C_2H_3O_2)_2$	10450-55-2 190.942	brn-red amorp powder			i H_2O; s EtOH, acid
1041	Iron(III) acetylacetonate $Fe(OC(CH_3)=CHC(O)CH_3)_3$	14024-18-1 353.173	red-oran cry	179 5.24		sl H_2O; s os
1042	Iron(III) arsenate dihydrate $FeAsO_4 \cdot 2H_2O$	10102-49-5 230.795	grn-brn powder	dec 3.18		i H_2O; s dil acid
1043	Iron(III) bromide $FeBr_3$	10031-26-2 295.557	dark red hex cry; hyg	dec 4.5	a	s H_2O, EtOH, eth

No.	Name Formula	CAS RN Mol. Wt.	Physical Form	mp/°C den/g cm⁻³	bp/°C Other Data	Solubility
1044	Iron(III) chloride $FeCl_3$	7705-08-0 162.203	grn hex cry; hyg	304 2.90	≈316 a,c,e	s H_2O, EtOH, eth, ace
1045	Iron(III) chloride hexahydrate $FeCl_3 \cdot 6H_2O$	10025-77-1 270.294	yel-oran monocl cry; hyg	dec 37 1.82		s H_2O, EtOH, eth, ace
1046	Iron(III) chromate $Fe_2(CrO_4)_3$	10294-52-7 459.671	yel powder			i H_2O, EtOH; s acid
1047	Iron(III) citrate pentahydrate $FeC_6H_5O_7 \cdot 5H_2O$	3522-50-7 335.023	red-brn cry			s H_2O; i EtOH
1048	Iron(III) dichromate $Fe_2(Cr_2O_7)_3$	10294-53-8 759.654	red-brn solid			s H_2O, acid
1049	Iron(III) ferrocyanide $Fe_4[Fe(CN)_6]_3$	14038-43-8 859.234	dark blue powder	1.80		i H_2O, dil acid, os
1050	Iron(III) fluoride FeF_3	7783-50-8 112.840	grn hex cry	>1000 3.87		sl H_2O; i EtOH, eth, bz
1051	Iron(III) fluoride trihydrate $FeF_3 \cdot 3H_2O$	15469-38-2 166.886	yel-brn tetr cry	2.3		
1052	Iron(III) hydroxide $Fe(OH)_3$	1309-33-7 106.867	yel monocl cry	3.12		
1053	Iron(III) hydroxide oxide $FeO(OH)$	20344-49-4 88.852	red-brn orth cry	4.26	f	i H_2O; s acid
1054	Iron(III) nitrate nonahydrate $Fe(NO_3)_3 \cdot 9H_2O$	7782-61-8 403.997	viol-gray hyg cry	47 dec 1.68		vs H_2O, EtOH, ace
1055	Iron(III) oxalate $Fe_2(C_2O_4)_3$	19469-07-9 375.749	yel amorp powder	dec 100		s H_2O, acid; i alk
1056	Iron(III) oxide Fe_2O_3	1309-37-1 159.688	red-brn hex cry	1565 5.25	a,f	i H_2O; s acid
1057	Iron(III) phosphate dihydrate $FePO_4 \cdot 2H_2O$	10045-86-0 186.847	gray-wh orth cry	2.87	f	i H_2O; s HCl
1058	Iron(III) pyrophosphate nonahydrate $Fe_4(P_2O_7)_3 \cdot 9H_2O$	10058-44-3 907.348	yel powder			i H_2O; s acid
1059	Iron(III) hypophosphite $Fe(H_2PO_2)_3$	7783-84-8 250.811	wh-gray powder			i H_2O
1060	Iron(III) sulfate $Fe_2(SO_4)_3$	10028-22-5 399.881	gray-wh rhomb cry; hyg	3.10		sl H_2O
1061	Iron(III) sulfate nonahydrate $Fe_2(SO_4)_3 \cdot 9H_2O$	13520-56-4 562.018	yel hex cry	dec 400 2.1	f	vs H_2O
1062	Iron(III) thiocyanate monohydrate $Fe(SCN)_3 \cdot H_2O$	4119-52-2 248.111	red hyg cry	dec		s H_2O, EtOH, ace; i tol, chl
1063	Iron(III) metavanadate $Fe(VO_3)_3$	65842-03-7 352.665	gray-brn powder			i H_2O, EtOH; s acid
1064	Krypton Kr	7439-90-9 83.80	col gas	-157.36 3.677 g/L	-153.22 a,b,c,d,e	sl H_2O
1065	Krypton difluoride KrF_2	13773-81-4 121.80	col tetr cry	dec ≈25 3.24		reac H_2O
1066	Lanthanum La	7439-91-0 138.906	silv metal; hex	918 6.15	3464 a,c	s dil acid
1067	Lanthanum boride LaB_6	12008-21-8 203.772	blk cub cry; refrac	2715 4.76		
1068	Lanthanum bromide $LaBr_3$	13536-79-3 378.618	wh hex cry; hyg	788 5.1		s H_2O
1069	Lanthanum carbide LaC_2	12071-15-7 162.928	tetr cry	2360 5.29		
1070	Lanthanum carbonate octahydrate $La_2(CO_3)_3 \cdot 8H_2O$	6487-39-4 601.961	wh cry powder	2.6		i H_2O; s dil acid
1071	Lanthanum chloride $LaCl_3$	10099-58-8 245.264	wh hex cry; hyg	859 3.84	a	s H_2O
1072	Lanthanum chloride heptahydrate $LaCl_3 \cdot 7H_2O$	20211-76-1 371.371	wh tricl cry; hyg	dec 91		s H_2O, EtOH
1073	Lanthanum fluoride LaF_3	13709-38-1 195.901	wh hex cry; hyg	1493 5.9		i H_2O, acid
1074	Lanthanum hydride LaH_3	13864-01-2 141.930	blk cub cry	5.36		
1075	Lanthanum hydroxide $La(OH)_3$	14507-19-8 189.928	wh amorp solid	dec		i H_2O
1076	Lanthanum iodide LaI_3	13813-22-4 519.619	wh orth cry; hyg	778 5.6		s H_2O
1077	Lanthanum nitrate hexahydrate $La(NO_3)_3 \cdot 6H_2O$	10277-43-7 433.012	wh hyg cry	dec ≈40		vs H_2O, EtOH
1078	Lanthanum nitride LaN	25764-10-7 152.913	cub cry	6.73		
1079	Lanthanum oxide La_2O_3	1312-81-8 325.809	wh amorp powder	2305 6.51	4200 a	i H_2O; s dil acid
1080	Lanthanum silicide $LaSi_2$	12056-90-5 195.077	gray tetr cry	5.0		
1081	Lanthanum sulfate nonahydrate $La_2(SO_4)_3 \cdot 9H_2O$	10294-62-9 728.139	hex cry	2.82		sl H_2O; i EtOH
1082	Lanthanum sulfide La_2S_3	12031-49-1 374.009	red cub cry	2110 4.9		

No.	Name Formula	CAS RN Mol. Wt.	Physical Form	mp/°C den/g cm^{-3}	bp/°C Other Data	Solubility
1083	Lanthanum sulfide LaS	12031-30-0 170.972	yel cub cry	2300 5.61	a	
1084	Lawrencium Lr	22537-19-5 262	metal	1627	a	
1085	Lead Pb	7439-92-1 207.2	soft silv-gray metal; cub	327.46 11.3	1749 a,c,d,e	s conc acid
1086	Lead(II) acetate Pb(C$_2$H$_3$O$_2$)$_2$	301-04-2 325.3	wh cry	280 3.25	dec	vs H$_2$O
1087	Lead(II) acetate trihydrate Pb(C$_2$H$_3$O$_2$)$_2$·3H$_2$O	6080-56-4 427.3	col cry	75 dec 2.55		vs H$_2$O; sl EtOH
1088	Lead(II) acetate, basic Pb(C$_2$H$_3$O$_2$)$_2$·2PbO·2H$_2$O	1335-32-6 807.7	wh powder	dec		
1089	Lead(II) antimonate Pb$_3$(SbO$_4$)$_2$	13510-89-9 993.1	oran-yel powder	6.58		i H$_2$O, dil acid
1090	Lead(II) arsenate Pb$_3$(AsO$_4$)$_2$	3687-31-8 899.4	wh cry	1042 dec 5.8		i H$_2$O; s HNO$_3$
1091	Lead(II) arsenite Pb(AsO$_2$)$_2$	10031-13-7 421.0	wh powder	5.85		i H$_2$O; s dil HNO$_3$
1092	Lead(II) azide Pb(N$_3$)$_2$	13424-46-9 291.2	col orth needles; exp	exp ≈350 4.7		i H$_2$O; vs HOAc
1093	Lead(II) borate monohydrate Pb(BO$_2$)$_2$·H$_2$O	10214-39-8 310.8	wh powder	500 dec 5.6		i H$_2$O; s dil HNO$_3$
1094	Lead(II) bromate monohydrate Pb(BrO$_3$)$_2$·H$_2$O	10031-21-7 481.0	col cry	≈180 dec 5.53		sl H$_2$O
1095	Lead(II) bromide PbBr$_2$	10031-22-8 367.0	wh orth cry	371 6.69	892 a,c,d,e	sl H$_2$O; i EtOH
1096	Lead(II) butanoate Pb(C$_4$H$_7$O$_2$)$_2$	819-73-8 381.4	col solid	≈90		i H$_2$O; s dil HNO$_3$
1097	Lead(II) carbonate PbCO$_3$	598-63-0 267.2	col orth cry	dec ≈315 6.6	a,f	i H$_2$O
1098	Lead(II) carbonate, basic Pb(OH)$_2$·2PbCO$_3$	1319-46-6 775.6	wh hex cry	400 dec ≈6.5		i H$_2$O, EtOH; s acid
1099	Lead(II) chlorate Pb(ClO$_3$)$_2$	10294-47-0 374.1	col hyg cry	dec 230 3.9		vs H$_2$O, EtOH
1100	Lead(II) chloride PbCl$_2$	7758-95-4 278.1	wh orth needles or powder	501 5.98	951 a,c,d,e,f	sl H$_2$O; s alk
1101	Lead(II) chloride fluoride PbClF	13847-57-9 261.7	tetr cry	7.05		i H$_2$O
1102	Lead(II) chromate PbCrO$_4$	7758-97-6 323.2	yel-oran monocl cry	844 6.12	a,f	i H$_2$O; s akl, dil acid
1103	Lead(II) chromate(VI) oxide PbCrO$_4$·PbO	18454-12-1 546.4	red powder			i H$_2$O
1104	Lead(II) citrate trihydrate Pb$_3$(C$_6$H$_5$O$_7$)$_2$·3H$_2$O	512-26-5 1053.8	wh cry powder			s H$_2$O; sl EtOH
1105	Lead(II) cyanide Pb(CN)$_2$	592-05-2 259.2	wh-yel powder			sl H$_2$O; reac acid
1106	Lead(II) fluoride PbF$_2$	7783-46-2 245.2	wh orth cry	830 8.44	1293 a,c,d,e	sl H$_2$O
1107	Lead(II) fluoroborate Pb(BF$_4$)$_2$	13814-96-5 380.8	stable only in aq soln			s H$_2$O
1108	Lead(II) formate Pb(CHO$_2$)$_2$	811-54-1 297.2	wh prisms or needles	dec 190 4.63		sl H$_2$O; i EtOH
1109	Lead(II) hydrogen arsenate PbHAsO$_4$	7784-40-9 347.1	wh monocl cry	dec 280 5.943		i H$_2$O; s HNO$_3$, alk
1110	Lead(II) hydrogen phosphate PbHPO$_4$	15845-52-0 303.2	wh monocl cry	dec 5.66		
1111	Lead(II) hydroxide Pb(OH)$_2$	19783-14-3 241.2	wh powder	145 dec 7.59		i H$_2$O; s acid
1112	Lead(II) iodate Pb(IO$_3$)$_2$	25659-31-8 557.0	wh orth cry	6.50		i H$_2$O
1113	Lead(II) iodide PbI$_2$	10101-63-0 461.0	yel hex cry or powder	410 6.16	872 dec a,c,d,e	sl H$_2$O; i EtOH
1114	Lead(II) lactate Pb(C$_3$H$_5$O$_3$)$_2$	18917-82-3 385.3	wh cry powder			s H$_2$O, hot EtOH
1115	Lead(II) molybdate PbMoO$_4$	10190-55-3 367.1	yel tetr cry	≈1060 6.7	a,f	i H$_2$O; HNO$_3$, NaOH
1116	Lead(II) niobate Pb(NbO$_3$)$_2$	12034-88-7 489.0	rhomb or tetr cry	1343 6.6		i H$_2$O
1117	Lead(II) nitrate Pb(NO$_3$)$_2$	10099-74-8 331.2	col cub cry	470 4.53	a	vs H$_2$O; sl EtOH
1118	Lead(II) oleate Pb(C$_{18}$H$_{33}$O$_2$)$_2$	1120-46-3 770.1	wax-like solid			i H$_2$O; s EtOH, bz, eth
1119	Lead(II) oxalate PbC$_2$O$_4$	814-93-7 295.2	wh powder	dec 300 5.28	a	i H$_2$O; s dil HNO$_3$
1120	Lead(II) oxide (litharge) PbO	1317-36-8 223.2	red tetr cry	888 9.35	a,e,f	i H$_2$O, EtOH; s dil HNO$_3$
1121	Lead(II) oxide (massicot) PbO	1317-36-8 223.2	yel orth cry	stab >489 9.64	a,e,f	i H$_2$O, EtOH; s dil HNO$_3$

No.	Name / Formula	CAS RN / Mol. Wt.	Physical Form	mp/°C / den/g cm⁻³	bp/°C / Other Data	Solubility
1122	Lead(II) oxide hydrate $3PbO \cdot H_2O$	1311-11-1 / 687.6	wh powder	/ 7.41		i H_2O; s dil acid
1123	Lead(II) perchlorate $Pb(ClO_4)_2$	13453-62-8 / 406.1	wh cry			vs H_2O
1124	Lead(II) perchlorate trihydrate $Pb(ClO_4)_2 \cdot 3H_2O$	13637-76-8 / 460.1	wh cry	100 dec / 2.6		vs cold H_2O; s EtOH
1125	Lead(II) phosphate $Pb_3(PO_4)_2$	7446-27-7 / 811.5	wh hex cry	1014 / 7.01		i H_2O, EtOH
1126	Lead(II) hypophosphite $Pb(H_2PO_2)_2$	10294-58-3 / 337.2	hyg cry powder	dec		sl H_2O; i EtOH
1127	Lead(II) metasilicate $PbSiO_3$	10099-76-0 / 283.3	wh monocl cry powder	764 / 6.49	a	i H_2O, os
1128	Lead(II) orthosilicate Pb_2SiO_4	13566-17-1 / 506.5	monocl cry	743 / 7.60	a	
1129	Lead(II) hexafluorosilicate dihydrate $PbSiF_6 \cdot 2H_2O$	1310-03-8 / 385.3	col cry	dec		vs H_2O
1130	Lead(II) selenate $PbSeO_4$	7446-15-3 / 350.2	orth cry	/ 6.37	a	i H_2O; s conc acid
1131	Lead(II) selenide $PbSe$	12069-00-0 / 286.2	gray cub cry	1078 / 8.1	a	i H_2O; s HNO_3
1132	Lead(II) selenite $PbSeO_3$	7488-51-9 / 334.2	wh monocl cry	≈500 / 7.0		i H_2O
1133	Lead(II) sodium thiosulfate $Na_4Pb(S_2O_3)_3$	10101-94-7 / 635.6	wh cry			sl H_2O
1134	Lead(II) stearate $Pb(C_{18}H_{35}O_2)_2$	1072-35-1 / 774.2	wh powder	≈100 / 1.4		i H_2O, s hot EtOH
1135	Lead(II) sulfate $PbSO_4$	7446-14-2 / 303.3	orth cry	1087 / 6.29	a,f	i H_2O, acid; sl alk
1136	Lead(II) sulfide PbS	1314-87-0 / 239.3	blk powder or silv cub cry	1118 / 7.60	a,c,e,f	i H_2O; s acid
1137	Lead(II) sulfite $PbSO_3$	7446-10-8 / 287.3	wh powder	dec	a	i H_2O; s HNO_3
1138	Lead(II) tantalate $PbTa_2O_6$	12065-68-8 / 665.1	orth cry	/ 7.9		i H_2O
1139	Lead(II) telluride $PbTe$	1314-91-6 / 334.8	gray cub cry	924 / 8.164	a,f	i H_2O, acid
1140	Lead(II) thiocyanate $Pb(SCN)_2$	592-87-0 / 323.4	wh-yel powder	/ 3.82		sl H_2O
1141	Lead(II) thiosulfate PbS_2O_3	13478-50-7 / 319.3	wh cry	dec / 5.18		i H_2O; s acid
1142	Lead(II) titanate $PbTiO_3$	12060-00-3 / 303.1	yel tetr cry	/ 7.9		i H_2O; reac HCl
1143	Lead(II) tungstate (stolzite) $PbWO_4$	7759-01-5 / 455.0	yel tetr cry	1130 / 8.24	f	i H_2O; s alk
1144	Lead(II) tungstate (raspite) $PbWO_4$	7759-01-5 / 455.0	monocl cry	trans 400 / 8.46	f	i H_2O; s alk
1145	Lead(II) metavanadate $Pb(VO_3)_2$	10099-79-3 / 405.1	yel powder			i H_2O; reac HNO_3
1146	Lead(II) zirconate $PbZrO_3$	12060-01-4 / 346.4	col orth cry	≈8		i H_2O, alk; s acid
1147	Lead(II,IV) oxide Pb_2O_3	1314-27-8 / 462.4	blk monocl cry or red amorp powder	dec 530 / 10.05		i H_2O; s alk; reac conc HCl
1148	Lead(II,II,IV) oxide Pb_3O_4	1314-41-6 / 685.6	red tetr cry	830 / 8.92	a,f	i H_2O, EtOH; s hot HCl
1149	Lead(IV) acetate $Pb(C_2H_3O_2)_4$	546-67-8 / 443.4	col monocl cry	≈175 / 2.23		reac H_2O, EtOH; s bz, chl
1150	Lead(IV) bromide $PbBr_4$	13701-91-2 / 526.8	unstable liq			
1151	Lead(IV) chloride $PbCl_4$	13463-30-4 / 349.0	yel oily liq	-15	dec ≈50 / a	
1152	Lead(IV) fluoride PbF_4	7783-59-7 / 283.2	wh tetr cry; hyg	≈600 / 6.7		
1153	Lead(IV) oxide PbO_2	1309-60-0 / 239.2	red tetr cry or brn powder	290 dec / 9.64	a	
1154	Lithium Li	7439-93-2 / 6.941	soft silv-wh metal	180.5 / 0.534	1342 / a,c,e	reac H_2O
1155	Lithium acetate dihydrate $LiC_2H_3O_2 \cdot 2H_2O$	6108-17-4 / 102.017	wh rhomb cry	58 dec / 1.3		vs H_2O; s EtOH
1156	Lithium aluminum hydride $LiAlH_4$	16853-85-3 / 37.955	gray-wh monocl cry	dec >125 / 0.917	a	reac H_2O, EtOH; s eth, thf
1157	Lithium amide $LiNH_2$	7782-89-0 / 22.964	tetr cry	380 / 1.18	a	reac H_2O
1158	Lithium arsenate Li_3AsO_4	13478-14-3 / 159.743	col orth cry	/ 3.07		sl H_2O; s HOAc
1159	Lithium azide LiN_3	19597-69-4 / 48.961	hyg monocl cry; exp	/ 1.83		vs H_2O
1160	Lithium borate $LiBO_2$	13453-69-5 / 49.751	wh monocl cry; hyg	849 / 2.18	a	vs H_2O; s EtOH

No.	Name Formula	CAS RN Mol. Wt.	Physical Form	mp/°C den/g cm^{-3}	bp/°C Other Data	Solubility
1161	Lithium borohydride LiBH$_4$	16949-15-8 21.784	wh-gray orth cry or powder	268 0.66	dec 380 a	s alk, eth, thf
1162	Lithium bromide LiBr	7550-35-8 86.845	wh cub cry; hyg	552 3.464	≈1300 a,c,e	s H$_2$O, EtOH, eth
1163	Lithium carbonate Li$_2$CO$_3$	554-13-2 73.891	wh monocl cry	723 2.11	dec 1300 a,c	sl H$_2$O; s acid; i EtOH
1164	Lithium chloride LiCl	7447-41-8 42.394	wh cub cry or powder; hyg	610 2.07	1383 a,c,e	s H$_2$O, EtOH, ace, py
1165	Lithium chromate dihydrate Li$_2$CrO$_4$·2H$_2$O	7789-01-7 165.906	yel orth cry; hyg	dec 75 2.15		vs H$_2$O; s EtOH
1166	Lithium dichromate dihydrate Li$_2$Cr$_2$O$_7$·2H$_2$O	10022-48-7 265.901	yel-red hyg cry	130 dec 2.34		vs H$_2$O
1167	Lithium fluoride LiF	7789-24-4 25.939	wh cub cry or powder	848.2 2.640	1673 a,c,d,e	sl H$_2$O; s acid
1168	Lithium formate monohydrate Li(CHO$_2$)·H$_2$O	6108-23-2 69.974	col-wh cry	1.46		s H$_2$O
1169	Lithium hydride LiH	7580-67-8 7.949	gray cub cry or powder; hyg	688.7 0.78	a,c	reac H$_2$O, EtOH
1170	Lithium hydroxide LiOH	1310-65-2 23.948	col tetr cry	471.2 1.45	1626 a,c,d	s H$_2$O; sl EtOH
1171	Lithium hydroxide monohydrate LiOH·H$_2$O	1310-66-3 41.964	wh monocl cry or powder	1.51		s H$_2$O; sl EtOH
1172	Lithium iodide LiI	10377-51-2 133.845	wh cub cry; hyg	469 4.06	1171 a,c,e	vs H$_2$O
1173	Lithium iodide trihydrate LiI·3H$_2$O	7790-22-9 187.891	wh hyg cry	73 3.48		vs H$_2$O, EtOH, ace
1174	Lithium niobate LiNbO$_3$	12031-63-9 147.845	wh hex cry	≈1240 4.30		
1175	Lithium nitrate LiNO$_3$	7790-69-4 68.946	col hex cry; hyg	253 2.38	a,c	vs H$_2$O; s EtOH
1176	Lithium nitride Li$_3$N	26134-62-3 34.830	red hex cry	813 1.27		reac H$_2$O
1177	Lithium phosphate Li$_3$PO$_4$	10377-52-3 115.794	wh orth cry	1205 2.46	a	i H$_2$O
1178	Lithium oxide Li$_2$O	12057-24-8 29.881	wh cub cry	1570 2.013	a	
1179	Lithium perchlorate LiClO$_4$	7791-03-9 106.392	wh orth cry or powder	236 2.428	dec 430 a,c	s H$_2$O, EtOH, ace, eth
1180	Lithium peroxide Li$_2$O$_2$	12031-80-0 45.881	wh hex cry	2.31	a	s H$_2$O; i EtOH
1181	Lithium selenate monohydrate Li$_2$SeO$_4$·H$_2$O	7790-71-8 174.86	monocl cry	2.56		vs H$_2$O
1182	Lithium metasilicate Li$_2$SiO$_3$	10102-24-6 89.966	wh orth needles	1201 2.52	a,c	i cold H$_2$O; reac dil acid
1183	Lithium sulfate Li$_2$SO$_4$	10377-48-7 109.946	wh monocl cry; hyg	859 2.21	a,c	s H$_2$O
1184	Lithium sulfate monohydrate Li$_2$SO$_4$·H$_2$O	10102-25-7 127.961	col cry	dec 130 2.06		vs H$_2$O; sl EtOH
1185	Lithium sulfide Li$_2$S	12136-58-2 45.948	wh cub cry; hyg	1372 1.64	a	
1186	Lutetium Lu	7439-94-3 174.967	silv metal; hex	1663 9.84	3402 a,c	s dil acid
1187	Lutetium boride LuB$_4$	12688-52-7 218.211	tetr cry	2600 ≈7.0		
1188	Lutetium bromide LuBr$_3$	14456-53-2 414.679	wh hyg cry	1025		vs H$_2$O
1189	Lutetium chloride LuCl$_3$	10099-66-8 281.325	wh monocl cry; hyg	925 3.98	a	s H$_2$O
1190	Lutetium fluoride LuF$_3$	13760-81-1 231.962	orth cry	1182 8.3	2200	i H$_2$O
1191	Lutetium iodide LuI$_3$	13813-45-1 555.680	brn hex cry; hyg	1050 ≈5.6	a	vs H$_2$O
1192	Lutetium nitride LuN	12125-25-6 188.974	cub cry	11.6		
1193	Lutetium oxide Lu$_2$O$_3$	12032-20-1 397.932	wh cub cry or powder	2490 9.41	a	
1194	Lutetium sulfate octahydrate Lu$_2$(SO$_4$)$_3$·8H$_2$O	13473-77-3 782.247	wh cry			vs H$_2$O
1195	Lutetium sulfide Lu$_2$S$_3$	12163-20-1 446.132	gray rhomb cry	1750 dec 6.26		
1196	Lutetium telluride Lu$_2$Te$_3$	12163-22-3 732.73	orth cry	7.8		
1197	Magnesium Mg	7439-95-4 24.305	silv-wh metal	650 1.74	1090 a,c,e	s dil acid
1198	Magnesium acetate tetrahydrate Mg(C$_2$H$_3$O$_2$)$_2$·4H$_2$O	16674-78-5 214.455	col monocl cry; hyg	80 dec 1.45		vs H$_2$O, EtOH
1199	Magnesium amide Mg(NH$_2$)$_2$	7803-54-5 56.350	wh powder; flam	dec 1.39		reac H$_2$O

No.	Name Formula	CAS RN Mol. Wt.	Physical Form	mp/°C den/g cm^{-3}	bp/°C Other Data	Solubility
1200	Magnesium antimonide Mg_3Sb_2	12057-75-9 316.435	hex cry	1245 3.99		
1201	Magnesium boride MgB_2	12007-25-9 45.927	hex cry	dec 800 2.57		
1202	Magnesium bromate hexahydrate $Mg(BrO_3)_2 \cdot 6H_2O$	7789-36-8 388.201	col cub cry	dec 200 2.29		vs H_2O
1203	Magnesium bromide $MgBr_2$	7789-48-2 184.113	wh hex cry; hyg	711 3.72	a,c	vs H_2O
1204	Magnesium bromide hexahydrate $MgBr_2 \cdot 6H_2O$	13446-53-2 292.204	col monocl cry	dec 165 2.0		vs H_2O; s EtOH
1205	Magnesium carbonate $MgCO_3$	546-93-0 84.314	wh hex cry	990 3.05	f	i H_2O, EtOH; s acid
1206	Magnesium chlorate hexahydrate $Mg(ClO_3)_2 \cdot 6H_2O$	10326-21-3* 299.298	wh hyg cry	dec ≈35 1.80		vs H_2O; sl EtOH
1207	Magnesium chloride $MgCl_2$	7786-30-3 95.210	wh hex leaflets; hyg	714 2.325	1412 a,c,e	vs H_2O
1208	Magnesium chloride hexahydrate $MgCl_2 \cdot 6H_2O$	7791-18-6 203.301	wh hyg cry	dec ≈100 1.56		vs H_2O; s EtOH
1209	Magnesium fluoride MgF_2	7783-40-6 62.302	wh tetr cry	1263 3.148	2227 a,c,f	sl H_2O
1210	Magnesium formate dihydrate $Mg(CHO_2)_2 \cdot 2H_2O$	6150-82-9 150.371	wh cry	dec		s H_2O; i EtOH
1211	Magnesium germanide Mg_2Ge	1310-52-7 121.22	cub cry	1117 3.09		
1212	Magnesium hydride MgH_2	60616-74-2 26.321	wh tetr cry	327 1.45		reac H_2O
1213	Magnesium hydrogen phosphate trihydrate $MgHPO_4 \cdot 3H_2O$	7757-86-0 174.331	wh powder	dec 550 2.13	f	sl H_2O; s dil acid
1214	Magnesium hydroxide $Mg(OH)_2$	1309-42-8 58.320	wh hex cry	350 2.37	a,f	i H_2O; s dil acid
1215	Magnesium iodide MgI_2	10377-58-9 278.114	wh hex cry; hyg	634 4.43	a,c	vs H_2O
1216	Magnesium iodide octahydrate $MgI_2 \cdot 8H_2O$	7790-31-0 422.236	wh orth cry; hyg	41 dec 2.10		vs H_2O; s EtOH
1217	Magnesium nitrate $Mg(NO_3)_2$	10377-60-3 148.314	wh cub cry	≈2.3	a	vs H_2O
1218	Magnesium nitrate dihydrate $Mg(NO_3)_2 \cdot 2H_2O$	15750-45-5 184.345	wh cry	≈100 dec 1.45		s H_2O, EtOH
1219	Magnesium nitrate hexahydrate $Mg(NO_3)_2 \cdot 6H_2O$	13446-18-9 256.406	col monocl cry; hyg	dec ≈95 1.46		vs H_2O; s EtOH
1220	Magnesium nitride Mg_3N_2	12057-71-5 100.928	yel cub cry	dec ~1500 2.71		
1221	Magnesium oxalate dihydrate $MgC_2O_4 \cdot 2H_2O$	6150-88-5 148.355	wh powder			i H_2O, EtOH; s dil acid
1222	Magnesium oxide MgO	1309-48-4 40.304	wh cub cry	2826 3.6	3600 a,c,f	sl H_2O; i EtOH
1223	Magnesium perchlorate $Mg(ClO_4)_2$	10034-81-8 223.205	wh hyg powder	dec 250 2.2		vs H_2O
1224	Magnesium perchlorate hexahydrate $Mg(ClO_4)_2 \cdot 6H_2O$	13446-19-0 331.297	wh hyg cry	dec 190 1.98		vs H_2O; s EtOH
1225	Magnesium permanganate hexahydrate $Mg(MnO_4)_2 \cdot 6H_2O$	10377-62-5 370.268	blue-blk cry	dec 2.18		s H_2O
1226	Magnesium peroxide MgO_2	1335-26-8 56.304	wh cub cry	dec 100 ≈3.0		i H_2O; s dil acid
1227	Magnesium phosphate octahydrate $Mg_3(PO_4)_2 \cdot 8H_2O$	13446-23-6 406.980	wh monocl cry	2.17		i H_2O; s acid
1228	Magnesium pyrophosphate trihydrate $Mg_2P_2O_7 \cdot 3H_2O$	10102-34-8 276.600	wh powder	dec 100 2.56		i H_2O; s acid
1229	Magnesium phosphide Mg_3P_2	12057-74-8 134.863	yel cub cry	2.06		reac H_2O
1230	Magnesium selenate hexahydrate $MgSeO_4 \cdot 6H_2O$	13446-28-1 275.35	wh monocl cry	1.928		s H_2O
1231	Magnesium selenide $MgSe$	1313-04-8 103.27	brn cub cry	4.2		reac H_2O
1232	Magnesium selenite hexahydrate $MgSeO_3 \cdot 6H_2O$	15593-61-0 259.36	col hex cry	2.09		i H_2O; s dil acid
1233	Magnesium metasilicate $MgSiO_3$	13776-74-4 100.389	wh monocl cry	dec ≈1550 3.19	f	i H_2O; sl HF
1234	Magnesium orthosilicate Mg_2SiO_4	26686-77-1 140.694	wh orth cry	1898 3.21	a,c,f	i H_2O
1235	Magnesium hexafluorosilicate hexahydrate $MgSiF_6 \cdot 6H_2O$	60950-56-3 274.472	wh cry	120 dec 1.79		s H_2O; i EtOH
1236	Magnesium silicide Mg_2Si	22831-39-6 76.696	gray cub cry	1102 1.99		reac H_2O
1237	Magnesium stannide Mg_2Sn	1313-08-2 167.320	blue cub cry	771 3.60		s H_2O, dil HCl

No.	Name Formula	CAS RN Mol. Wt.	Physical Form	mp/°C den/g cm^{-3}	bp/°C Other Data	Solubility
1238	Magnesium sulfate $MgSO_4$	7487-88-9 120.369	col orth cry	1127 2.66	a,c	s H_2O
1239	Magnesium sulfate heptahydrate $MgSO_4 \cdot 7H_2O$	10034-99-8 246.475	col orth cry	dec 150 1.67	f	vs H_2O; sl EtOH
1240	Magnesium sulfate monohydrate $MgSO_4 \cdot H_2O$	14168-73-1 138.384	col monocl cry	dec 150 2.57	f	vs H_2O
1241	Magnesium sulfide MgS	12032-36-9 56.371	red-brn cub cry	2227 2.68	a	reac H_2O
1242	Magnesium sulfite hexahydrate $MgSO_3 \cdot 6H_2O$	13446-29-2 212.461	wh hex cry	dec 200 1.72		sl H_2O; i EtOH
1243	Magnesium sulfite trihydrate $MgSO_3 \cdot 3H_2O$	19086-20-5 158.415	col orth cry	 2.12		sl H_2O
1244	Magnesium thiosulfate hexahydrate $MgS_2O_3 \cdot 6H_2O$	13446-30-5 244.527	col cry	dec 170 1.82		vs H_2O; i EtOH
1245	Magnesium titanate $MgTiO_3$	1312-99-8 120.170	col hex cry	1565 3.85	f	
1246	Magnesium tungstate $MgWO_4$	13573-11-0 272.14	wh monocl cry	 6.89		i H_2O, EtOH
1247	Manganese Mn	7439-96-5 54.938	hard gray metal	1246 7.3	2061 a,c,e	s dil acids
1248	Manganese antimonide $MnSb$	12032-82-5 176.698	hex cry	840 6.9		
1249	Manganese antimonide Mn_2Sb	12032-97-2 231.636	tetr cry	948 7.0		
1250	Manganese boride MnB	12045-15-7 65.749	orth cry	1890 6.45		
1251	Manganese boride MnB_2	12228-50-1 76.560	hex cry	1827 5.3		
1252	Manganese boride Mn_2B	12045-16-8 120.687	red-brn tetr cry	1580 7.20		
1253	Manganese carbide Mn_3C	12266-65-8 176.825	refrac solid	1520 6.89		
1254	Manganese carbonyl $Mn_2(CO)_{10}$	10170-69-1 389.980	yel monocl cry	154 1.75		i H_2O; s os
1255	Manganese phosphide Mn_2P	12333-54-9 140.850	hex cry	1327 6.0		
1256	Manganese phosphide MnP	12032-78-9 85.912	orth cry	1147 5.49		
1257	Manganese(II) acetate tetrahydrate $Mn(C_2H_3O_2)_2 \cdot 4H_2O$	15243-27-3 245.088	red monocl cry	80 1.59		s H_2O, EtOH
1258	Manganese(II) tetraborate octahydrate $MnB_4O_7 \cdot 8H_2O$	12228-91-0 354.300	red solid			i H_2O, EtOH; s dil acid
1259	Manganese(II) bromide $MnBr_2$	13446-03-2 214.746	pink hex cry	698 4.385	a	vs H_2O
1260	Manganese(II) bromide tetrahydrate $MnBr_2 \cdot 4H_2O$	10031-20-6 286.808	red hyg cry	64 dec		vs H_2O
1261	Manganese(II) carbonate $MnCO_3$	598-62-9 114.947	pink hex cry	dec >200 3.70	f	sl H_2O; s dil acid
1262	Manganese(II) chloride $MnCl_2$	7773-01-5 125.843	pink trig cry; hyg	650 2.977	1190 a,c,e	s H_2O, py, EtOH; i eth
1263	Manganese(II) chloride tetrahydrate $MnCl_2 \cdot 4H_2O$	13446-34-9 197.905	red monocl cry; hyg	87.5 1.913		vs H_2O; s EtOH; i eth
1264	Manganese(II) dihydrogen phosphate dihydrate $Mn(H_2PO_4)_2 \cdot 2H_2O$	18718-07-5 284.944	col hyg cry			s H_2O; i EtOH
1265	Manganese(II) fluoride MnF_2	7782-64-1 92.935	red tetr cry	930 3.98		sl H_2O; i EtOH
1266	Manganese(II) hydroxide $Mn(OH)_2$	18933-05-6 88.953	pink hex cry	dec 3.26	f	i H_2O
1267	Manganese(II) iodide MnI_2	7790-33-2 308.747	wh hex cry; hyg	638 5.04		s H_2O, EtOH
1268	Manganese(II) iodide tetrahydrate $MnI_2 \cdot 4H_2O$	7790-33-2* 380.809	red cry			vs H_2O; s EtOH
1269	Manganese(II) molybdate $MnMoO_4$	14013-15-1 214.88	yel monocl cry	 4.05		
1270	Manganese(II) nitrate $Mn(NO_3)_2$	10377-93-2 178.948	col orth cry; hyg	 2.2		vs H_2O; s diox, thf
1271	Manganese(II) nitrate hexahydrate $Mn(NO_3)_2 \cdot 6H_2O$	10377-66-9 287.040	rose monocl cry	28 dec 1.8		vs H_2O, EtOH
1272	Manganese(II) nitrate tetrahydrate $Mn(NO_3)_2 \cdot 4H_2O$	20694-39-7 251.010	pink hyg cry	37.1 dec 2.13		vs H_2O; s EtOH
1273	Manganese(II) oxalate dihydrate $MnC_2O_4 \cdot 2H_2O$	6556-16-7 178.988	wh cry powder	dec 150 2.45		sl H_2O; s acid
1274	Manganese(II) oxide MnO	1344-43-0 70.937	gr cub cry or powder	1840 5.37	a,c,f	i H_2O; s acid
1275	Manganese(II) perchlorate hexahydrate $Mn(ClO_4)_2 \cdot 6H_2O$	15364-94-0 361.930	pink hex cry	 2.10		
1276	Manganese(II) pyrophosphate $Mn_2P_2O_7$	53731-35-4 283.819	wh monocl cry	1196 3.71		i H_2O

No.	Name Formula	CAS RN Mol. Wt.	Physical Form	mp/°C den/g cm^{-3}	bp/°C Other Data	Solubility
1277	Manganese(II) metasilicate $MnSiO_3$	7759-00-4 131.022	red orth cry	1291 3.48	a,f	i H_2O
1278	Manganese(II) orthosilicate Mn_2SiO_4	13568-32-6 201.960	orth cry	4.11	a	i H_2O
1279	Manganese(II) selenide MnSe	1313-22-0 133.90	gray cub cry	1460 5.45	a	i H_2O
1280	Manganese(II) sulfate $MnSO_4$	7785-87-7 151.002	wh orth cry	700 3.25	dec 850	vs H_2O
1281	Manganese(II) sulfate monohydrate $MnSO_4 \cdot H_2O$	10034-96-5 169.017	red monocl cry	2.95		vs H_2O; i EtOH
1282	Manganese(II) sulfate tetrahydrate $MnSO_4 \cdot 4H_2O$	10101-68-5 223.063	red monocl cry	38 dec 2.26		s H_2O; i EtOH
1283	Manganese(II) sulfide (α) MnS	18820-29-6 87.004	grn cub cry	1610 4.0	a,f	i H_2O; s dil acid
1284	Manganese(II) sulfide (β) MnS	18820-29-6 87.004	red cub cry	3.3	a,f	i H_2O; s dil acid
1285	Manganese(II) sulfide (γ) MnS	18820-29-6 87.004	red hex cry	$\approx$3.3	a,f	i H_2O; s dil acid
1286	Manganese(II) telluride MnTe	12032-88-1 182.54	hex cry	$\approx$1150 6.0		
1287	Manganese(II) titanate $MnTiO_3$	12032-74-5 150.803	red hex cry	1360 4.55		
1288	Manganese(II) tungstate $MnWO_4$	13918-22-4 302.78	wh monocl cry	7.2	f	
1289	Manganese(II,III) oxide Mn_3O_4	1317-35-7 228.812	brn tetr cry	1567 4.84	a,f	i H_2O; s HCl
1290	Manganese(III) fluoride MnF_3	7783-53-1 111.933	red monocl cry; hyg	dec >600 3.54		reac H_2O
1291	Manganese(III) hydroxide MnO(OH)	1332-63-4 87.945	blk monocl cry	dec 250 $\approx$4.3	f	i H_2O
1292	Manganese(III) oxide Mn_2O_3	1317-34-6 157.874	blk cub cry	dec 1080 $\approx$5.0	a	i H_2O
1293	Manganese(IV) oxide MnO_2	1313-13-9 86.937	blk tetr cry	535 dec 5.08	a,f	i H_2O, HNO_3
1294	Manganese(VII) oxide Mn_2O_7	12057-92-0 221.872	grn oil; exp	5.9 2.40	exp 95	vs H_2O
1295	Mendelevium Md	7440-11-1 258	Metal	827	a	
1296	Mercury Hg	7439-97-6 200.59	heavy silv liq	-38.83 13.5336	356.73 a,b,c,d,e	i H_2O
1297	Mercury(I) acetate $Hg_2(C_2H_3O_2)_2$	631-60-7 519.27	col scales	dec		sl H_2O; i EtOH, eth
1298	Mercury(I) bromide Hg_2Br_2	15385-58-7 560.99	wh tetr cry or powder	407 7.307	a	i H_2O, EtOH, eth
1299	Mercury(I) chlorate $Hg_2(ClO_3)_2$	10294-44-7 568.08	wh cry	dec $\approx$250 6.41		sl H_2O; s EtOH
1300	Mercury(I) chloride Hg_2Cl_2	10112-91-1 472.09	wh tetr cry	525 tp 7.16	383 sp a,f	i H_2O, EtOH, eth
1301	Mercury(I) fluoride Hg_2F_2	13967-25-4 439.18	yel cub cry	570 dec 8.73	subl	reac H_2O
1302	Mercury(I) iodide Hg_2I_2	15385-57-6 654.99	yel amorp powder	290 7.70	a,c	i H_2O, EtOH, eth
1303	Mercury(I) nitrate dihydrate $Hg_2(NO_3)_2 \cdot 2H_2O$	7782-86-7 561.22	col cry	70 dec 4.8		reac H_2O
1304	Mercury(I) oxide Hg_2O	15829-53-5 417.18	prob mixture of HgO+Hg	100 dec 9.8		i H_2O; s HNO_3
1305	Mercury(I) sulfate Hg_2SO_4	7783-36-0 497.24	wh-yel cry powder	7.56	a	i H_2O; s dil HNO_3
1306	Mercury(II) acetate $Hg(C_2H_3O_2)_2$	1600-27-7 318.68	wh-yel cry or powder	179 dec 3.28		s H_2O, EtOH
1307	Mercury(II) amide chloride $Hg(NH_2)Cl$	10124-48-8 252.07	wh solid	5.38	subl	i H_2O, EtOH; s warm acid
1308	Mercury(II) bromide $HgBr_2$	7789-47-1 360.40	wh rhomb cry or powder	236 6.05	322 a,b,c,d,e	sl H_2O, chl; s EtOH, MeOH
1309	Mercury(II) chloride $HgCl_2$	7487-94-7 271.50	wh orth cry	276 5.6	304 a,b,c,d,e	sl H_2O, bz; s EtOH, MeOH, ace, eth
1310	Mercury(II) chromate $HgCrO_4$	13444-75-2* 316.58	red monocl cry	6.06		sl H_2O
1311	Mercury(II) cyanide $Hg(CN)_2$	592-04-1 252.63	col tetr cry	dec 320 4.00		s H_2O, EtOH; sl eth
1312	Mercury(II) dichromate $HgCr_2O_7$	7789-10-8 416.58	red cry powder			i H_2O; s acid
1313	Mercury(II) fluoride HgF_2	7783-39-3 238.59	wh cub cry; hyg	dec 645 8.95		reac H_2O
1314	Mercury(II) fulminate $Hg(CNO)_2$	628-86-4 284.62	gray cry	exp 4.42		sl H_2O; s EtOH, NH_4OH
1315	Mercury(II) hydrogen arsenate $HgHAsO_4$	7784-37-4 340.52	yel powder			i H_2O; s acid

No.	Name Formula	CAS RN Mol. Wt.	Physical Form	mp/°C den/g cm⁻³	bp/°C Other Data	Solubility
1316	Mercury(II) iodate $Hg(IO_3)_2$	7783-32-6 550.40	wh powder	dec 175		i H_2O
1317	Mercury(II) iodide HgI_2	7774-29-0 454.40	red tetr cry or powder	259 6.28	354 a,b,c,d,e	i H_2O; sl EtOH, ace, eth
1318	Mercury(II) nitrate $Hg(NO_3)_2$	10045-94-0 324.60	col hyg cry	79 4.3		s H_2O; i EtOH
1319	Mercury(II) nitrate dihydrate $Hg(NO_3)_2 \cdot 2H_2O$	10045-94-0* 360.63	monocl cry	4.78		s H_2O
1320	Mercury(II) nitrate monohydrate $Hg(NO_3)_2 \cdot H_2O$	7783-34-8 342.62	wh-yel hyg cry	4.3		s H_2O, dil acid
1321	Mercury(II) oxide HgO	21908-53-2 216.59	red or yel orth cry	dec 500 11.14	a,f	i H_2O, EtOH; s dil acid
1322	Mercury(II) oxycyanide $Hg(CN)_2 \cdot HgO$	1335-31-5 469.21	wh orth cry	exp 4.44		s H_2O
1323	Mercury(II) phosphate $Hg_3(PO_4)_2$	7782-66-3 791.71	wh-yel powder			i H_2O, EtOH; s acid
1324	Mercury(II) selenide HgSe	20601-83-6 279.55	gray cub cry	8.21	subl	i H_2O
1325	Mercury(II) sulfate $HgSO_4$	7783-35-9 296.65	wh monocl cry	6.47	a	reac H_2O
1326	Mercury(II) sulfide (black) HgS	1344-48-5 232.66	blk cub cry or powder	583 7.70	a,f	i H_2O; s acid, EtOH
1327	Mercury(II) sulfide (red) HgS	1344-48-5 232.66	red hex cry	trans to blk HgO 386 8.17	a,f	i H_2O, acid; s aqua regia
1328	Mercury(II) telluride HgTe	12068-90-5 328.19	gray cub cry	673 8.63	a	
1329	Mercury(II) thiocyanate $Hg(SCN)_2$	592-85-8 316.76	monocl cry	dec ≈165 3.71		i H_2O; s dil HCl
1330	Molybdenum Mo	7439-98-7 95.94	gray-blk metal; cub	2623 10.2	4639 a,c,e	i H_2O, dil acid, alk
1331	Molybdenum boride Mo_2B	12006-99-4 202.69	refrac tetr cry	2000 9.2		
1332	Molybdenum boride Mo_2B_5	12007-97-5 245.94	refrac hex cry	1600 ≈7.2		
1333	Molybdenum carbide MoC	12011-97-1 107.95	refrac solid; cub	2577		
1334	Molybdenum carbide Mo_2C	12069-89-5 203.89	gray orth cry	2687 9.18		
1335	Molybdenum carbonyl $Mo(CO)_6$	13939-06-5 264.00	wh orth cry	dec 150 1.96	subl a,d	i H_2O; s bz; sl eth
1336	Molybdenum nitride MoN	12033-19-1 109.95	hex cry	1750 9.20		
1337	Molybdenum nitride Mo_2N	12033-31-7 205.89	gray cub cry	790 dec 9.46		
1338	Molybdenum phosphide MoP	12163-69-8 126.91	blk hex cry	7.34		
1339	Molybdenum silicide $MoSi_2$	12136-78-6 152.11	gray tetr cry	≈1900 6.2		i H_2O; s HF
1340	Molybdenum(II) bromide $MoBr_2$	13446-56-5 255.75	yel-red cry	dec 900		
1341	Molybdenum(II) chloride $MoCl_2$	13478-17-6 166.85	yel cry	dec 530		
1342	Molybdenum(II) iodide MoI_2	14055-74-4 349.75	blk hyg cry	5.278		
1343	Molybdenum(III) bromide $MoBr_3$	13446-57-6 335.65	grn hex cry	977 4.89		i H_2O
1344	Molybdenum(III) chloride $MoCl_3$	13478-18-7 202.30	dark red monocl cry	1027 3.74		i H_2O
1345	Molybdenum(III) fluoride MoF_3	20193-58-2 152.94	brn hex cry	>600 4.64		i H_2O
1346	Molybdenum(III) iodide MoI_3	14055-75-5 476.65	blk solid	927		i H_2O
1347	Molybdenum(III) oxide Mo_2O_3	1313-29-7 239.88	gray-blk powder			i H_2O; sl acid
1348	Molybdenum(IV) bromide $MoBr_4$	13520-59-7 415.56	blk cry	dec		reac H_2O
1349	Molybdenum(IV) chloride $MoCl_4$	13320-71-3 237.75	blk cry	dec >170		reac H_2O
1350	Molybdenum(IV) fluoride MoF_4	23412-45-5 171.93	grn cry	dec		reac H_2O
1351	Molybdenum(IV) oxide MoO_2	18868-43-4 127.94	brn-viol tetr cry	dec ≈1100 6.47	a	sl H_2O
1352	Molybdenum(IV) selenide $MoSe_2$	12058-18-3 253.86	gray hex cry	>1200 6.90		
1353	Molybdenum(IV) sulfide MoS_2	1317-33-5 160.07	blk powder or hex cry	2375 5.06	a,f	i H_2O; s conc acid

No.	Name Formula	CAS RN Mol. Wt.	Physical Form	mp/°C den/g cm^{-3}	bp/°C Other Data	Solubility
1354	Molybdenum(IV) telluride MoTe$_2$	12058-20-7 351.14	gray hex cry	7.7		
1355	Molybdenum(V) chloride MoCl$_5$	10241-05-1 273.20	gr-blk monocl cry; hyg	194 2.93	268 b,c,d	s EtOH, eth
1356	Molybdenum(V) fluoride MoF$_5$	13819-84-6 190.93	yel monocl cry	67 3.5	213	
1357	Molybdenum(V) oxytrichloride MoOCl$_3$	13814-74-9 218.30	blk monocl cry	297	subl	reac H$_2$O
1358	Molybdenum(VI) acid monohydrate H$_2$MoO$_4$•H$_2$O	7782-91-4 179.97	wh powder	3.1		sl H$_2$O; s alk
1359	Molybdenum(VI) fluoride MoF$_6$	7783-77-9 209.93	wh cub cry or col liq; hyg	17.5 2.54	34 a,b,c,d,e	reac H$_2$O
1360	Molybdenum(VI) oxytetrafluoride MoOF$_4$	14459-59-7 187.93	volatile solid	98		
1361	Molybdenum(VI) oxytetrachloride MoOCl$_4$	13814-75-0 253.75	grn hyg powder	101		
1362	Molybdenum(VI) dioxydichloride MoO$_2$Cl$_2$	13637-68-8 198.84	yel-oran solid	≈175 3.31		reac H$_2$O
1363	Molybdenum(VI) oxide MoO$_3$	1313-27-5 143.94	wh-yel rhomb cry	801 4.70	1155 a,c,d,e	sl H$_2$O; s conc acid
1364	Molybdenum(VI) metaphosphate Mo(PO$_3$)$_6$	133863-98-6 569.77	yel powder	3.28		i H$_2$O, acid
1365	Neodymium Nd	7440-00-8 144.24	silv metal; hex	1021 7.01	3074 a,c	
1366	Neodymium boride NdB$_6$	12008-23-0 209.11	blk cub cry	2610 4.93		
1367	Neodymium bromide NdBr$_3$	13536-80-6 383.95	viol orth cry; hyg	682 5.3	1540	s H$_2$O
1368	Neodymium chloride NdCl$_3$	10024-93-8 250.60	viol hex cry	758 4.13	1600 a	vs H$_2$O, EtOH; i eth, chl
1369	Neodymium chloride hexahydrate NdCl$_3$•6H$_2$O	13477-89-9 358.69	purp cry	dec 124 2.3		vs H$_2$O; s EtOH
1370	Neodymium fluoride NdF$_3$	13709-42-7 201.24	viol hex cry; hyg	1377 6.51	2300 a	i H$_2$O
1371	Neodymium iodide NdI$_3$	13813-24-6 524.95	grn orth cry; hyg	784 5.85		s H$_2$O
1372	Neodymium nitrate hexahydrate Nd(NO$_3$)$_3$•6H$_2$O	14517-29-4 438.35	purp hyg cry			vs H$_2$O; s EtOH, ace
1373	Neodymium nitride NdN	25764-11-8 158.25	blk cub cry	7.69		
1374	Neodymium oxide Nd$_2$O$_3$	1313-97-9 336.48	blue hex cry; hyg	2320 7.24	a	i H$_2$O; s dil acid
1375	Neodymium sulfate Nd$_2$(SO$_4$)$_3$	13477-91-3 576.67	pink needles	dec ≈700		s H$_2$O
1376	Neodymium sulfide Nd$_2$S$_3$	12035-32-4 384.68	orth cry	2207 5.46		
1377	Neodymium telluride Nd$_2$Te$_3$	12035-35-7 671.28	gray orth cry	1377 7.0		
1378	Neon Ne	7440-01-9 20.180	col gas	-248.59 0.885 g/L	-246.08 a,b,c,d,e	sl H$_2$O
1379	Neptunium Np	7439-99-8 237	silv metal	644 20.2	c	s HCl
1380	Neptunium(IV) oxide NpO$_2$	12035-79-9 269	grn cub cry	2547 11.1		
1381	Nickel Ni	7440-02-0 58.693	wh metal; cub	1455 8.90	2913 a,c,e	i H$_2$O; sl dil acid
1382	Nickel antimonide NiSb	12035-52-8 180.453	hex cry	1147 8.74	f	
1383	Nickel arsenide NiAs	27016-75-7 133.615	hex cry	967 7.77	f	
1384	Nickel boride Ni$_3$B	12007-02-2 186.891	refrac solid	1156 8.17		
1385	Nickel boride NiB	12007-00-0 69.504	grn refrac solid	1035 7.13		
1386	Nickel boride Ni$_2$B	12007-01-1 128.198	refrac solid	1125 7.90		
1387	Nickel carbonyl Ni(CO)$_4$	13463-39-3 170.735	col liq	-19.3 1.31	43 (exp ≈60) a,e	i H$_2$O; s EtOH, bz, ace, ctc
1388	Nickel phosphide Ni$_2$P	12035-64-2 148.361	hex cry	1100 7.33		
1389	Nickel silicide Ni$_2$Si	12059-14-2 145.473	orth cry	1255 7.40		
1390	Nickel silicide NiSi$_2$	12201-89-7 114.864	cub cry	993 4.83		
1391	Nickel(II) arsenate octahydrate Ni$_3$(AsO$_4$)$_2$•8H$_2$O	7784-48-7 598.040	yel-grn powder	dec 4.98		i H$_2$O; s acid
1392	Nickel(II) bromide NiBr$_2$	13462-88-9 218.501	yel hex cry; hyg	963 5.10	subl a	vs H$_2$O

No.	Name / Formula	CAS RN / Mol. Wt.	Physical Form	mp/°C / den/g cm^{-3}	bp/°C / Other Data	Solubility
1393	Nickel(II) bromide trihydrate / $NiBr_2 \cdot 3H_2O$	13462-88-9* / 272.547	yel-grn hyg cry	dec 200		vs H_2O; s EtOH, eth
1394	Nickel(II) carbonate / $NiCO_3$	3333-67-3 / 118.702	grn rhomb cry	4.39		i H_2O; s dil acid
1395	Nickel(II) chloride / $NiCl_2$	7718-54-9 / 129.598	yel hex cry; hyg	1009 / 3.51	subl / a,c,e	s H_2O, EtOH
1396	Nickel(II) chloride hexahydrate / $NiCl_2 \cdot 6H_2O$	7791-20-0 / 237.689	grn monocl cry			vs H_2O; s EtOH
1397	Nickel(II) cyanide tetrahydrate / $Ni(CN)_2 \cdot 4H_2O$	13477-95-7 / 182.790	grn plates	dec 200		i H_2O; sl dil acid; s NH_4OH
1398	Nickel(II) fluoride / NiF_2	10028-18-9 / 96.690	yel tetr cry	1474 / 4.7	a	sl H_2O; i EtOH, eth
1399	Nickel(II) hydroxide / $Ni(OH)_2$	12054-48-7 / 92.708	grn hex cry	dec 230 / 4.1	a	i H_2O
1400	Nickel(II) hydroxide monohydrate / $Ni(OH)_2 \cdot H_2O$	36897-37-7 / 110.723	grn powder			i H_2O; s dil acid
1401	Nickel(II) iodide / NiI_2	13462-90-3 / 312.502	blk hex cry; hyg	780 / 5.22	subl / a	vs H_2O
1402	Nickel(II) iodide hexahydrate / $NiI_2 \cdot 6H_2O$	7790-34-3 / 420.593	grn monocl cry; hyg			vs H_2O, EtOH
1403	Nickel(II) nitrate hexahydrate / $Ni(NO_3)_2 \cdot 6H_2O$	13478-00-7 / 290.794	grn monocl cry; hyg	dec 56 / 2.05		vs H_2O; s EtOH
1404	Nickel(II) oxide / NiO	1313-99-1 / 74.692	grn cub cry	1955 / 6.72	f	i H_2O; s acid
1405	Nickel(II) selenide / $NiSe$	1314-05-2 / 137.65	yel-grn hex cry	980 / 7.2		
1406	Nickel(II) sulfate / $NiSO_4$	7786-81-4 / 154.757	grn-yel orth cry	dec 840 / 4.01	a	vs H_2O
1407	Nickel(II) sulfate heptahydrate / $NiSO_4 \cdot 7H_2O$	10101-98-1 / 280.863	grn orth cry	1.98		s H_2O, EtOH
1408	Nickel(II) sulfate hexahydrate / $NiSO_4 \cdot 6H_2O$	10101-97-0 / 262.848	blue-grn tetr cry	dec ≈100 / 2.07		vs H_2O; sl EtOH
1409	Nickel(II) sulfide / NiS	16812-54-7 / 90.759	yel hex cry	976 / 5.5	a,c,f	i H_2O
1410	Nickel(II) titanate / $NiTiO_3$	12035-39-1 / 154.558	brn hex cry	5.0		
1411	Nickel(II,III) sulfide / Ni_3S_4	12137-12-1 / 304.344	cub cry	995 / 4.77		
1412	Nickel(III) oxide / Ni_2O_3	1314-06-3 / 165.385	gray-blk cub cry	dec ≈600	a	i H_2O; s hot acid
1413	Nickel(III) sulfide / Ni_3S_2	12035-72-2 / 240.212	hex cry	787 / 5.87		
1414	Niobium / Nb	7440-03-1 / 92.906	gray metal; cub	2477 / 8.57	4744 / a,c	i acid
1415	Niobium boride / NbB	12045-19-1 / 103.717	gray orth cry	2270 / 7.5		
1416	Niobium boride / NbB_2	12007-29-3 / 114.528	gray hex cry	2900 / 6.8		
1417	Niobium carbide / NbC	12069-94-2 / 104.917	gray cub cry	3608 / 7.82	4300	i H_2O, acid
1418	Niobium carbide / Nb_2C	12011-99-3 / 197.824	refrac hex cry	3080 / 7.8		i H_2O
1419	Niobium nitride / NbN	24621-21-4 / 106.913	gray cry; cub	2300 / 8.47		i HCl, acid
1420	Niobium phosphide / NbP	12034-66-1 / 123.880	tetr cry	6.5		
1421	Niobium silicide / $NbSi_2$	12034-80-9 / 149.077	gray hex cry	1950 / 5.7		
1422	Niobium(II) oxide / NbO	12034-57-0 / 108.905	gray cub cry	1937 / 7.30	a,c	
1423	Niobium(III) bromide / $NbBr_3$	15752-41-7 / 332.618	dark brn solid		subl	
1424	Niobium(III) chloride / $NbCl_3$	13569-59-0 / 199.264	blk solid			
1425	Niobium(III) fluoride / NbF_3	15195-53-6 / 149.901	blue cub cry	4.2		
1426	Niobium(IV) chloride / $NbCl_4$	13569-70-5 / 234.717	viol-blk monocl cry	3.2		
1427	Niobium(IV) fluoride / NbF_4	13842-88-1 / 168.900	blk tetr cry; hyg	dec >350 / 4.01		
1428	Niobium(IV) iodide / NbI_4	13870-21-8 / 600.524	gray orth cry	503 / 5.6		
1429	Niobium(IV) oxide / NbO_2	12034-59-2 / 124.905	wh tetr cry or powder	1902 / 5.9	a,c	
1430	Niobium(IV) selenide / $NbSe_2$	12034-77-4 / 250.83	gray hex cry	>1300 / 6.3		
1431	Niobium(IV) sulfide / NbS_2	12136-97-9 / 157.038	blk rhomb cry	4.4		

No.	Name Formula	CAS RN Mol. Wt.	Physical Form	mp/°C den/g cm⁻³	bp/°C Other Data	Solubility
1432	Niobium(IV) telluride $NbTe_2$	12034-83-2 348.11	hex cry	7.6		
1433	Niobium(V) bromide $NbBr_5$	13478-45-0 492.426	oran orth cry	254 4.36	360	s H_2O, EtOH
1434	Niobium(V) chloride $NbCl_5$	10026-12-7 270.170	yel monocl cry; hyg	204.7 2.78	254.0 a,b,c,d	reac H_2O; s HCl, ctc
1435	Niobium(V) fluoride NbF_5	7783-68-8 187.898	col monocl cry; hyg	80 2.70	229 a,b,c,d,e	reac H_2O; sl CS_2, chl
1436	Niobium(V) iodide NbI_5	13779-92-5 727.428	yel-blk monocl cry	dec ≈200 5.32		
1437	Niobium(V) oxide Nb_2O_5	1313-96-8 265.810	wh orth cry	1512 4.6	a,c	i H_2O; s HF
1438	Niobium(V) oxybromide $NbOBr_3$	14459-75-7 348.617	yel-brn cry	dec ≈320	subl	
1439	Niobium(V) oxychloride $NbOCl_3$	13597-20-1 215.263	wh tetr cry	3.72	subl	
1440	Niobium(V) dioxyfluoride NbO_2F	15195-33-2 143.903	wh cub cry	4.0		
1441	Nitrogen N_2	7727-37-9 28.013	col gas	-210.00 1.229 g/L	-195.79 a,b,c,d,e	sl H_2O; i EtOH
1442	Nitramide NO_2NH_2	7782-94-7 62.028	unstable wh cry	72 dec	a	s H_2O, EtOH, ace, eth; i chl
1443	Nitric acid HNO_3	7697-37-2 63.013	col liq; hyg	-41.6 1.55	83 a,c,d	vs H_2O
1444	Nitrous acid HNO_2	7782-77-6 47.014	stable only in soln		a	
1445	Nitrous oxide N_2O	10024-97-2 44.012	col gas	-90.8 1.931 g/L	-88.48 a,b,c,d,e	sl H_2O; s EtOH, eth
1446	Nitric oxide NO	10102-43-9 30.006	col gas	-163.6 1.317 g/L	-151.74 a,b,c,d,e	sl H_2O
1447	Nitrogen dioxide NO_2	10102-44-0 46.006	brn gas; equil with N_2O_4	2.019 g/L	see N_2O_4 a	reac H_2O
1448	Nitrogen trioxide N_2O_3	10544-73-7 76.011	blue solid or liq (low temp)	-101.1 1.4(2°C)	≈3 dec a	reac H_2O
1449	Nitrogen tetroxide N_2O_4	10544-72-6 92.011	col liq; equil with NO_2	-9.3 1.45(20°C)	21.15 a,b,c,d,e	reac H_2O
1450	Nitrogen pentoxide N_2O_5	10102-03-1 108.010	col hex cry	30 2.0	47 a,e	s chl; sl ctc
1451	Nitrogen tribromide NBr_3	15162-90-0 253.719	unstable solid	exp -100		
1452	Nitrogen trichloride NCl_3	10025-85-1 120.365	yel oily liq; exp	-40 1.653	71 a	i H_2O; s CS_2, bz, ctc
1453	Nitrogen trifluoride NF_3	7783-54-2 71.002	col gas	-206.79 3.116 g/L	-128.75 a,b,d,e	i H_2O
1454	Nitrogen triiodide NI_3	13444-85-4 394.720	unstable blk cry; exp			
1455	Difluoramine NHF_2	10405-27-3 53.012	col gas	2.326 g/L	-23 a,b	
1456	cis-Difluorodiazine N_2F_2	13812-43-6 66.010	col gas	<-195 2.896 g/L	-105.75 a,b	
1457	trans-Difluorodiazine N_2F_2	13776-62-0 66.010	col gas	-172 2.896 g/L	-111.45 a,b	
1458	Tetrafluorohydrazine N_2F_4	10036-47-2 104.007	col gas	-164.5 4.564 g/L	-74 a,b,d	
1459	Nitrosyl bromide NOBr	13444-87-6 109.910	red gas	-56 4.823 g/L	≈0 a	reac H_2O
1460	Nitrosyl chloride NOCl	2696-92-6 65.459	yel gas	-59.6 2.872 g/L	-5.5 a,b,d,e	reac H_2O
1461	Nitrosyl fluoride NOF	7789-25-5 49.004	col gas	-132.5 2.150 g/L	-59.9 a,d,e	
1462	Trifluoramine oxide NOF_3	13847-65-9 87.001	col gas	-161 3.818 g/L	-87.5 b	
1463	Nitryl chloride NO_2Cl	13444-90-1 81.459	col gas	-145 3.574 g/L	-15 a,d	
1464	Nitryl fluoride NO_2F	10022-50-1 65.004	col gas	-166 2.852 g/L	-72.4 a,b,d,e	reac H_2O
1465	Nitrogen selenide N_4Se_4	12033-88-4 371.87	red monocl cry; hyg	exp 4.2		i H_2O, eth, EtOH; sl bz, CS_2
1466	Nobelium No	10028-14-5 259	metal	827	a,c	
1467	Osmium Os	7440-04-2 190.23	blue-wh metal; hex	3033 22.59	5012 a,c	s aqua regia
1468	Osmium carbonyl $Os_3(CO)_{12}$	15696-40-9 906.82	yel cry	3.48		
1469	Osmium(III) bromide $OsBr_3$	59201-51-3 429.94	dark gray cry	dec 340		
1470	Osmium(III) chloride $OsCl_3$	13444-93-4 296.59	gray cub cry	dec >450	a	i H_2O; s HNO_3

No.	Name Formula	CAS RN Mol. Wt.	Physical Form	mp/°C den/g cm^{-3}	bp/°C Other Data	Solubility
1471	Osmium(IV) chloride $OsCl_4$	10026-01-4 332.04	red-blk orth cry	 4.38	450 sp	reac H_2O
1472	Osmium(IV) fluoride OsF_4	54120-05-7 266.22	yel cry	230		
1473	Osmium(IV) oxide OsO_2	12036-02-1 222.23	yel-brn tetr cry	 11.4		i H_2O, acid
1474	Osmium(V) fluoride OsF_5	31576-40-6 285.22	blue cry	70		reac H_2O
1475	Osmium(VI) fluoride OsF_6	13768-38-2 304.22	yel cub cry	33.2 4.1	46 a	reac H_2O
1476	Osmium(VIII) oxide OsO_4	20816-12-0 254.23	yel monocl cry	41 5.1	135 a,b,c,e	sl H_2O
1477	Oxygen O_2	7782-44-7 31.999	col gas	-218.79 1.404 g/L	-182.95 a,b,c,d,e	sl H_2O, EtOH, os
1478	Ozone O_3	10028-15-6 47.998	blue gas	-193 2.106 g/L	-111.35 a,b,e	sl H_2O
1479	Palladium Pd	7440-05-3 106.42	silv-wh metal; cub	1554.9 12.0	2963 a,c	s aqua regia
1480	Palladium(II) sulfide PdS	12125-22-3 138.49	gray tetr cry	 6.7	a	
1481	Palladium(II) bromide $PdBr_2$	13444-94-5 266.23	red-blk monocl cry; hyg	250 dec ≈5.2		i H_2O
1482	Palladium(II) chloride $PdCl_2$	7647-10-1 177.33	red rhomb cry; hyg	679 4.0		s H_2O, EtOH, ace
1483	Palladium(II) fluoride PdF_2	13444-96-7 144.42	viol tetr cry; hyg	952 5.76		reac H_2O
1484	Palladium(II) iodide PdI_2	7790-38-7 360.23	blk cry	360 dec 6.0		i H_2O, EtOH, eth
1485	Palladium(II) nitrate $Pd(NO_3)_2$	10102-05-3 230.43	brn hyg cry	dec		sl H_2O; s dil HNO_3
1486	Palladium(II) oxide PdO	1314-08-5 122.42	grn-blk tetr cry	750 dec 8.3	a	i H_2O, acid; sl aqua regia
1487	Phosphorus (white) P	7723-14-0 30.974	col waxlike cub cry	44.15 1.823	280.5 a,b,c,d,e	i H_2O; sl bz, EtOH, chl; s CS_2
1488	Phosphorus (red) P	7723-14-0 30.974	red-viol amorp powder	590tp 2.16	431sp a,b,c,d,e	i H_2O, os
1489	Phosphorus (black) P	7723-14-0 30.974	blk orth cry or amorp solid	610 2.69	 a,b,c,d,e	i os
1490	Phosphine PH_3	7803-51-2 33.998	col gas; flam	-133 1.492 g/L	-87.75 a,b,d,e	i H_2O; sl EtOH, eth
1491	Diphosphine P_2H_4	13445-50-6 65.980	col liq	-99	63.5 dec a,d	reac H_2O
1492	Phosphoric acid (orthophosphoric acid) H_3PO_4	7664-38-2 97.995	col visc liq	42.4	407 a,c	s H_2O, EtOH
1493	Metaphosphoric acid HPO_3	37267-86-0 79.980	gl solid; hyg			sl H_2O; s EtOH
1494	Pyrophosphoric acid $H_4P_2O_7$	2466-09-3 177.975	wh cry	71.5	a	s H_2O
1495	Hypophosphoric acid $H_4P_2O_6$	7803-60-3 161.976	col orth cry	73 dec		vs H_2O
1496	Difluorophosphoric acid HPO_2F_2	13779-41-4 101.978	col liq	≈-94 1.583	dec 110	reac H_2O
1497	Phosphorous acid H_3PO_3	13598-36-2 81.996	wh hyg cry	74.4 1.65	200 a,c	vs H_2O, EtOH
1498	Hypophosphorous acid HPH_2O_2	6303-21-5 65.997	hyg cry or col oily liq	26.5 1.49	130 a,c	vs H_2O, EtOH, eth
1499	Phosphonium iodide PH_4I	12125-09-6 161.910	col tetr cry	18.5 2.86	62.5 e	reac H_2O, EtOH
1500	Phosphorus heptasulfide P_4S_7	12037-82-0 348.357	pale yel monocl cry	308 2.19	523	sl CS_2
1501	Phosphorus nitride P_3N_5	12136-91-3 162.955	wh amorp solid	dec 800		i H_2O; s os
1502	Phosphonitrilic chloride trimer $(PNCl_2)_3$	940-71-6 347.657	wh hyg cry	128.8 1.98		reac H_2O
1503	Phosphorus(III) bromide PBr_3	7789-60-8 270.686	col liq	-40 2.8	172.95 a,b,d,e	reac H_2O, EtOH; s ace, CS_2
1504	Phosphorus(III) chloride PCl_3	7719-12-2 137.332	col liq	-112 1.574	75.95 a,b,c,d,e	reac H_2O, EtOH; s bz, chl, eth
1505	Phosphorus(III) chloride difluoride $PClF_2$	14335-40-1 104.424	col gas	 4.582 g/L	-47.25 b,d	
1506	Phosphorus(III) dichloride fluoride PCl_2F	15597-63-4 120.877	col gas	 5.304 g/L	14 b,d	
1507	Phosphorus(III) fluoride PF_3	7783-55-3 87.969	col gas	-151.5 3.860 g/L	-101.5 a,b,d,e	reac H_2O
1508	Phosphorus(III) iodide PI_3	13455-01-1 411.687	red-oran hex cry; hyg	61.5 4.18	227 dec a,d	reac H_2O; s EtOH
1509	Phosphorus(III) oxide P_2O_3	1314-24-5 109.946	col monocl cry or liq	23.8 2.13	173 e	reac H_2O

No.	Name Formula	CAS RN Mol. Wt.	Physical Form	mp/°C den/g cm^{-3}	bp/°C Other Data	Solubility
1510	Phosphorus(III) selenide P_2Se_3	1314-86-9 298.83	oran-red cry	245 1.31	≈380	reac H_2O; s bz, ctc, CS_2, ace
1511	Phosphorus(III) sulfide P_2S_3	12165-69-4 158.146	yel solid	290	490	reac H_2O; s EtOH, eth, CS_2
1512	Phosphorus(V) bromide PBr_5	7789-69-7 430.494	yel orth cry, Donnay	dec >100 3.61	a	reac H_2O, EtOH; s CS_2, ctc
1513	Phosphorus(V) chloride PCl_5	10026-13-8 208.238	wh-yel tetr cry; hyg	167 tp 2.1	160 sp a,b,e	reac H_2O; s CS_2, ctc
1514	Phosphorus(V) fluoride PF_5	7647-19-0 125.966	col gas	-93.8 5.527 g/L	-84.6 a,d	reac H_2O
1515	Phosphorus(V) oxide P_2O_5	1314-56-3 141.945	wh orth cry; hyg	562 2.30	605 c,e	reac H_2O, EtOH
1516	Phosphorus(V) oxybromide $POBr_3$	7789-59-5 286.685	faint oran plates	56 2.822	191.7 a,d	reac H_2O; s bz, eth, chl
1517	Phosphorus(V) oxychloride $POCl_3$	10025-87-3 153.331	col liq	1 1.645	105.5 a,c,d,e	reac H_2O, EtOH
1518	Phosphorus(V) oxyfluoride POF_3	13478-20-1 103.968	col gas	-39.1 tp 4.562 g/L	-39.7 sp a	reac H_2O
1519	Phosphorus(V) selenide P_2Se_5	1314-82-5 456.75	blk-purp amorp solid			reac hot H_2O, ctc; i CS_2
1520	Phosphorus(V) sulfide P_2S_5	1314-80-3 222.278	grn-yel hyg cry	285 2.03	515	reac H_2O; s CS_2
1521	Phosphorus(V) sulfide trifluoride PSF_3	2404-52-6 120.035	gas	-148.8 5.267 g/L	-52.25 b,d	
1522	Phosphorus(V) sulfide trichloride $PSCl_3$	3982-91-0 169.398	fuming liq	-36.2 1.635	125 e	reac H_2O, s bz, ctc, chl, CS_2
1523	Phosphorus(V) sulfide chloride difluoride $PSClF_2$	2524-02-9 136.490	col gas	-155 5.989 g/L	6.3 b	
1524	Platinum Pt	7440-06-4 195.08	silv-gray metal; cub	1768.4 21.5	3825 a,c,e,f	i acid; s aqua regia
1525	Hexachloroplatinic acid hexahydrate $H_2PtCl_6 \cdot 6H_2O$	16941-12-1 517.90	brn-yel hyg cry	60 2.43		vs H_2O, EtOH
1526	Platinum silicide PtSi	12137-83-6 223.17	orth cry	1229 12.4		
1527	Platinum(II) bromide $PtBr_2$	13455-12-4 354.89	red-brn powder	dec 250 6.65	a	i H_2O
1528	Platinum(II) chloride $PtCl_2$	10025-65-7 265.99	grn hex cry	dec 581 6.0	a	i H_2O, EtOH, eth; s HCl
1529	Platinum(II) iodide PtI_2	7790-39-8 448.89	blk powder	dec 325 6.4		i H_2O
1530	Platinum(II) oxide PtO	12035-82-4 211.08	blk tetr cry	dec 325 14.1		i H_2O, EtOH; s aqua regia
1531	Platinum(II) sulfide PtS	12038-20-9 227.15	tetr cry	10.25	a,f	
1532	Platinum(III) bromide $PtBr_3$	25985-07-3 434.79	grn-blk cry	dec 200	a	
1533	Platinum(III) chloride $PtCl_3$	25909-39-1 301.44	grn-blk cry	435 dec 5.26	a	
1534	Platinum(IV) bromide $PtBr_4$	68938-92-1 514.70	brn-blk cry	dec 180	a	i H_2O; sl EtOH, eth
1535	Platinum(IV) chloride $PtCl_4$	37773-49-2 336.89	red-brn cub cry	327 dec 4.30		sl H_2O
1536	Platinum(IV) chloride pentahydrate $PtCl_4 \cdot 5H_2O$	13454-96-1 426.97	red cry	2.43		s H_2O, EtOH
1537	Platinum(IV) fluoride PtF_4	13455-15-7 271.07	red cry	600		
1538	Platinum(IV) iodide PtI_4	7790-46-7 702.70	brn-blk powder	dec 130	a	s H_2O
1539	Platinum(IV) oxide PtO_2	1314-15-4 227.08	blk hex cry	450 11.8		i H_2O; s conc acid, dil alk
1540	Platinum(IV) sulfide PtS_2	12038-21-0 259.21	hex cry	7.85	a	
1541	Platinum(VI) fluoride PtF_6	13693-05-5 309.07	red cub cry	61.3 ≈4.0	69.1 a	
1542	Plutonium Pu	7440-07-5 244	silv-wh metal; monocl	640 19.7	3228 a,c	
1543	Plutonium nitride PuN	12033-54-4 258	gray cub cry	2550 14.4		
1544	Plutonium(II) oxide PuO	12035-83-5 260	cub cry	14.0		
1545	Plutonium(III) bromide $PuBr_3$	15752-46-2 484	grn orth cry	681 6.75		s H_2O
1546	Plutonium(III) chloride $PuCl_3$	13569-62-5 350	grn hex cry	760 5.71		s H_2O
1547	Plutonium(III) fluoride PuF_3	13842-83-6 301	purp hex cry	1425 9.33		i H_2O; sl acid
1548	Plutonium(III) iodide PuI_3	13813-46-2 625	grn orth cry; hyg	777 6.92		s H_2O

No.	Name Formula	CAS RN Mol. Wt.	Physical Form	mp/°C den/g cm^{-3}	bp/°C Other Data	Solubility
1549	Plutonium(III) oxide Pu_2O_3	12036-34-9 536	blk cub cry	 10.5		
1550	Plutonium(IV) fluoride PuF_4	13709-56-3 320	red-brn monocl cry	1027 7.1		
1551	Plutonium(IV) oxide PuO_2	12059-95-9 276	yel-brn cub cry	2400 11.5		
1552	Plutonium(VI) fluoride PuF_6	13693-06-6 358	red-brn orth cry	52 5.08		
1553	Polonium Po	7440-08-6 209	silv metal; cub	254 9.20	962 a	
1554	Polonium(IV) chloride $PoCl_4$	10026-02-5 351	yel hyg cry	≈300	390	s H_2O, EtOH, ace
1555	Polonium(IV) oxide PoO_2	7446-06-2 241	yel cub cry	dec 500 8.9		
1556	Potassium K	7440-09-7 39.098	soft silv-wh metal; cub	63.38 0.89	759 a,c,e	reac H_2O
1557	Potassium acetate $KC_2H_3O_2$	127-08-2 98.143	wh hyg cry	292 1.57	 a	vs H_2O; s EtOH; i eth
1558	Potassium aluminate trihydrate $K_2Al_2O_4 \cdot 3H_2O$	12003-63-3* 250.204	wh orth cry	 2.13		vs H_2O; i EtOH
1559	Potassium aluminum silicate $KAlSi_3O_8$	1327-44-2 278.332	col monocl cry	 2.56		i H_2O
1560	Potassium aluminum sulfate $KAl(SO_4)_2$	10043-67-1 258.207	wh hyg powder			sl H_2O
1561	Potassium aluminum sulfate dodecahydrate $KAl(SO_4)_2 \cdot 12H_2O$	7784-24-9 474.391	col cry	dec ≈100 1.72		sl H_2O
1562	Potassium arsenate K_3AsO_4	7784-41-0 256.215	col cry	 2.8		s H_2O
1563	Potassium azide KN_3	20762-60-1 81.118	tetr cry; exp	 2.04		vs H_2O
1564	Potassium borohydride KBH_4	13762-51-1 53.941	wh cub cry	dec ≈500 1.11	 a	s H_2O
1565	Potassium bromate $KBrO_3$	7758-01-2 167.000	wh hex cry	434 dec 3.27	 a	sl H_2O; i EtOH
1566	Potassium bromide KBr	7758-02-3 119.002	col cub cry; hyg	734 2.74	1435 a,c,e	vs H_2O; sl EtOH
1567	Potassium carbonate K_2CO_3	584-08-7 138.206	wh monocl cry; hyg	898 2.29	dec a,c	s H_2O; i EtOH
1568	Potassium carbonate sesquihydrate $K_2CO_3 \cdot 1.5H_2O$	6381-79-9 165.229	granular cry			vs H_2O
1569	Potassium chlorate $KClO_3$	3811-04-9 122.549	wh monocl cry	368 2.32	dec a	s H_2O
1570	Potassium chloride KCl	7447-40-7 74.551	wh cub cry	771 1.988	 a,c,e,f	s H_2O; i eth, ace
1571	Potassium chromate K_2CrO_4	7789-00-6 194.191	yel orth cry	975 2.73		vs H_2O
1572	Potassium cyanate KCNO	590-28-3 81.115	wh tetr cry	dec ≈700 2.05		s H_2O; sl EtOH
1573	Potassium cyanide KCN	151-50-8 65.116	wh cub cry; hyg	634 1.55	 a	vs H_2O; sl EtOH
1574	Potassium dichromate $K_2Cr_2O_7$	7778-50-9 294.185	oran-red tricl cry	398 2.68	dec ≈500	s H_2O
1575	Potassium dihydrogen arsenate KH_2AsO_4	7784-41-0 180.034	col cry	288 2.87		s H_2O; i EtOH
1576	Potassium dihydrogen phosphate KH_2PO_4	7778-77-0 136.085	wh tetr cry	253 2.34	 a	vs H_2O; sl EtOH
1577	Potassium ferricyanide $K_3Fe(CN)_6$	13746-66-2 329.246	red cry	dec 1.89		vs H_2O
1578	Potassium ferrocyanide trihydrate $K_4Fe(CN)_6 \cdot 3H_2O$	14459-95-1 422.390	yel monocl cry	dec 60 1.85		s H_2O; i EtOH, eth
1579	Potassium fluoride KF	7789-23-3 58.096	wh cub cry	858 2.48	1502 a,c,e	s H_2O
1580	Potassium fluoride dihydrate $KF \cdot 2H_2O$	13455-21-5 94.127	monocl cry	41 dec 2.5		vs H_2O
1581	Potassium fluoroborate KBF_4	14075-53-7 125.903	col orth cry	530 2.505		sl H_2O, EtOH
1582	Potassium formate $KCHO_2$	590-29-4 84.116	col hyg cry	167 1.91	 a	vs H_2O
1583	Potassium hexachloroosmate(IV) K_2OsCl_6	16871-60-6 481.14	red cub cry			vs H_2O; sl EtOH
1584	Potassium hexachloroplatinate K_2PtCl_6	16921-30-5 485.99	yel-oran cub cry	dec 250 3.50		sl H_2O; i EtOH
1585	Potassium hexacyanocobalt $K_3Co(CN)_6$	13963-58-1 332.334	yel monocl cry	dec 1.91		vs H_2O; i EtOH
1586	Potassium hexafluoromanganate(IV) K_2MnF_6	16962-31-5 247.125	yel hex cry			reac H_2O
1587	Potassium hexafluorosilicate K_2SiF_6	16871-90-2 220.273	wh cry	dec 2.27	 a	sl H_2O; i EtOH

No.	Name Formula	CAS RN Mol. Wt.	Physical Form	mp/°C den/g cm^{-3}	bp/°C Other Data	Solubility
1588	Potassium hydride KH	7693-26-7 40.106	cub cry	 1.43	 a	reac H_2O
1589	Potassium hydrogen arsenite $KAsO_2 \cdot HAsO_2$	10124-50-2 253.947	wh hyg powder			s H_2O
1590	Potassium hydrogen carbonate $KHCO_3$	298-14-6 100.115	col monocl cry	dec ≈100 2.17	 a	vs H_2O; i EtOH
1591	Potassium hydrogen fluoride KHF_2	7789-29-9 78.103	col tetr cry	238.9 2.37	 a,c	s H_2O; i EtOH
1592	Potassium hydrogen phosphate K_2HPO_4	7758-11-4 174.176	wh hyg cry	dec		vs H_2O; sl EtOH
1593	Potassium hydrogen phosphite K_2HPO_3	13492-26-7 158.177	wh hyg powder	dec		vs H_2O; i EtOH
1594	Potassium hydrogen selenite $KHSeO_3$	7782-70-9 167.06	hyg orth cry	dec >100		s H_2O; sl EtOH
1595	Potassium hydrogen sulfate $KHSO_4$	7646-93-7 136.170	wh monocl cry; hyg	≈200 2.32	 a	vs H_2O
1596	Potassium hydrogen sulfide KHS	1310-61-8 72.172	wh hex cry; hyg	≈450 1.69		s H_2O, EtOH
1597	Potassium hydrogen sulfide hemihydrate $KHS \cdot 0.5H_2O$	1310-61-8* 81.179	wh-yel hyg cry	≈175 1.7		vs H_2O, EtOH
1598	Potassium hydrogen sulfite $KHSO_3$	7773-03-7 120.170	wh cry powder	dec 190		s H_2O; i EtOH
1599	Potassium hydrogen tartrate $KHC_4H_4O_6$	868-14-4 188.178	wh cry	 1.98		sl H_2O; s acid, alk; i EtOH
1600	Potassium hydroxide KOH	1310-58-3 56.105	wh rhomb cry; hyg	406 2.044	1327 a,c,e	vs H_2O; s EtOH; s MeOH
1601	Potassium hypophosphite KH_2PO_2	7782-87-8 104.087	wh hyg cry	dec		vs H_2O; s EtOH
1602	Potassium iodate KIO_3	7758-05-6 214.001	wh monocl cry	560 dec 3.89	 a	sl H_2O
1603	Potassium iodide KI	7681-11-0 166.003	col cub cry	681 3.12	1323 a,c,e	vs H_2O; sl EtOH
1604	Potassium manganate K_2MnO_4	10294-64-1 197.133	grn cry	190 dec		s H_2O; reac HCl
1605	Potassium metabisulfite $K_2S_2O_5$	16731-55-8 222.326	wh powder	dec ≈150 2.3		vs H_2O; reac acid; i EtOH
1606	Potassium metaborate KBO_2	13709-94-9 81.908	wh hex cry	≈2.3	 a	
1607	Potassium molybdate K_2MoO_4	13446-49-6 238.14	wh hyg cry	919 2.3		vs H_2O; i EtOH
1608	Potassium niobate $KNbO_3$	12030-85-2 180.002	wh rhomb cry	≈1100 4.64		i H_2O
1609	Potassium nitrate KNO_3	7757-79-1 101.103	col rhomb cry or powder	337 2.11	dec 400 a,c	s H_2O; i EtOH
1610	Potassium nitrite KNO_2	7758-09-0 85.104	wh hyg cry	441 1.915	exp 537 a	vs H_2O; sl EtOH
1611	Potassium oxalate monohydrate $K_2C_2O_4 \cdot H_2O$	6487-48-5 184.232	col cry	dec 160 2.13		vs H_2O
1612	Potassium oxide K_2O	12136-45-7 94.196	gray cub cry	350 dec 2.35	 a	s H_2O, EtOH, eth
1613	Potassium perchlorate $KClO_4$	7778-74-7 138.549	col orth cry; hyg	525 2.52	 a	sl H_2O
1614	Potassium periodate KIO_4	7790-21-8 230.001	col tetr cry	582 3.618	exp a	sl H_2O
1615	Potassium permanganate $KMnO_4$	7722-64-7 158.034	purp orth cry	dec 2.7	 a	s H_2O; reac EtOH
1616	Potassium peroxide K_2O_2	17014-71-0 110.196	yel amorp solid	490	 a	reac H_2O
1617	Potassium persulfate $K_2S_2O_8$	7727-21-1 270.324	col cry	dec ≈100 2.48		sl H_2O
1618	Potassium phosphate K_3PO_4	7778-53-2 212.266	wh orth cry; hyg	1340 2.564	 a	vs H_2O; i EtOH
1619	Potassium pyrophosphate trihydrate $K_4P_2O_7 \cdot 3H_2O$	7320-34-5* 384.383	col hyg cry	1090 2.33		vs H_2O; i EtOH
1620	Potassium pyrosulfate $K_2S_2O_7$	7790-62-7 254.325	col needles	≈325 2.28		s H_2O
1621	Potassium selenate K_2SeO_4	7790-59-2 221.16	wh powder	 3.07		vs H_2O
1622	Potassium selenide K_2Se	1312-74-9 157.16	red cub cry; hyg	800 2.29		s H_2O
1623	Potassium stannate trihydrate $K_2SnO_3 \cdot 3H_2O$	12142-33-5* 298.951	col cry	 3.20		vs H_2O; i EtOH
1624	Potassium sulfate K_2SO_4	7778-80-5 174.261	wh orth cry	1069 2.66	 a,c,f	s H_2O; i EtOH
1625	Potassium sulfide K_2S	1312-73-8 110.263	red-yel cub cry; hyg	948 1.74	 a,c	s H_2O, EtOH; i eth
1626	Potassium sulfide pentahydrate $K_2S \cdot 5H_2O$	1312-73-8* 200.339	col rhomb cry	60		vs H_2O, EtOH; i eth

No.	Name Formula	CAS RN Mol. Wt.	Physical Form	mp/°C den/g cm^{-3}	bp/°C Other Data	Solubility
1627	Potassium sulfite dihydrate $K_2SO_3 \cdot 2H_2O$	7790-56-9 194.292	wh monocl cry	dec		s H_2O; sl EtOH; dec dil acid
1628	Potassium superoxide KO_2	12030-88-5 71.097	yel tetr cry; hyg	380 2.16	a	reac H_2O
1629	Potassium tellurate(VI) trihydrate $K_2TeO_4 \cdot 3H_2O$	15571-91-2* 323.84	wh cry powder			s H_2O
1630	Potassium tellurite K_2TeO_3	7790-58-1 253.80	wh hyg cry	dec ≈460		vs H_2O
1631	Potassium tetraborate pentahydrate $K_2B_4O_7 \cdot 5H_2O$	1332-77-0 323.513	wh cry powder			s H_2O; sl EtOH
1632	Potassium tetrachloroaurate dihydrate $KAuCl_4 \cdot 2H_2O$	13682-61-6 413.907	yel monocl cry			s H_2O, EtOH, eth
1633	Potassium tetrachloroplatinate K_2PtCl_4	10025-99-7 415.09	pink-red tetr cry	dec 500 3.38		s H_2O; i EtOH
1634	Potassium tetracyanoplatinate(II) trihydrate $K_2Pt(CN)_4 \cdot 3H_2O$	562-76-5* 431.39	col rhomb prisms			s H_2O
1635	Potassium tetraiodomercurate(II) K_2HgI_4	7783-33-7 786.40	yel hyg cry	4.29		vs H_2O; s EtOH, eth, ace
1636	Potassium thiocyanate KSCN	333-20-0 97.182	col tetr cry; hyg	173 1.88	dec 500 a	vs H_2O; s EtOH
1637	Potassium thiosulfate $K_2S_2O_3$	10294-66-3 190.327	col hyg cry			vs H_2O; i EtOH
1638	Potassium titanate K_2TiO_3	12030-97-6 174.062	wh orth cry	1515 3.1		reac H_2O
1639	Potassium triiodide monohydrate $KI_3 \cdot H_2O$	7790-42-3 437.827	brn monocl cry; hyg	dec 225 3.5		s H_2O; reac EtOH, eth
1640	Potassium thiocarbonate K_2CS_3	26750-66-3 186.406	yel-red hyg cry			vs H_2O
1641	Potassium tungstate K_2WO_4	7790-60-5 326.04	hyg cry	921 3.12		vs H_2O; i EtOH
1642	Potassium uranate $K_2U_2O_7$	7790-63-8 666.251	oran cub cry	6.12		i H_2O; s acid
1643	Praseodymium Pr	7440-10-0 140.908	silv metal; hex	931 6.77	3520 a,c	
1644	Praseodymium boride PrB_6	12008-27-4 205.774	blk cub cry	2610 4.84		
1645	Praseodymium bromide $PrBr_3$	13536-53-3 380.620	grn hex cry; hyg	693 5.28		s H_2O
1646	Praseodymium chloride $PrCl_3$	10361-79-2 247.266	grn hex needles; hyg	786 4.0	a	s H_2O, EtOH
1647	Praseodymium chloride heptahydrate $PrCl_3 \cdot 7H_2O$	10025-90-8 373.373	grn cry	dec 110		s H_2O, EtOH
1648	Praseodymium fluoride PrF_3	13709-46-1 197.903	grn hex cry	1395 6.3		
1649	Praseodymium iodide PrI_3	13813-23-5 521.621	orth hyg cry	737 ≈5.8		s H_2O
1650	Praseodymium nitrate hexahydrate $Pr(NO_3)_3 \cdot 6H_2O$	14483-17-1 435.014	ligtt grn cry			vs H_2O
1651	Praseodymium nitride PrN	25764-09-4 154.915	cub cry	7.46		
1652	Praseodymium oxide Pr_2O_3	12036-32-7 329.813	wh hex cry	2300 6.9	a	
1653	Praseodymium silicide $PrSi_2$	12066-83-0 197.079	tetr cry	1712 5.46		
1654	Praseodymium sulfide Pr_2S_3	12038-13-0 378.013	cub cry	1765 5.1		
1655	Praseodymium telluride Pr_2Te_3	12038-12-9 664.62	cub cry	1500 ≈7.0		
1656	Promethium Pm	7440-12-2 145	silv metal; hex	1042 7.26	3000 a	
1657	Protactinium Pa	7440-13-3 231.036	shiny metal; tetr or cub	1572 15.4	a,c	
1658	Protactinium(V) chloride $PaCl_5$	13760-41-3 408.300	yel monocl cry	306 3.74	a	
1659	Radium Ra	7440-14-4 226	wh metal; cub	700 5	a	
1660	Radium bromide $RaBr_2$	10031-23-9 386	wh orth cry	728 5.79		s H_2O, EtOH
1661	Radium chloride $RaCl_2$	10025-66-8 297	wh orth cry	1000 4.9		s H_2O, EtOH
1662	Radium fluoride RaF_2	20610-49-5 264	wh cub cry	6.7		
1663	Radium sulfate $RaSO_4$	7446-16-4 322	wh cry	a		i H_2O, acid
1664	Radon Rn	10043-92-2 222	col gas	-71 9.741 g/L	-61.7 a,b,e	sl H_2O
1665	Rhenium Re	7440-15-5 186.207	silv-gray metal	3186 20.8	5596 a,c	i HCl

No.	Name Formula	CAS RN Mol. Wt.	Physical Form	mp/°C den/g cm^{-3}	bp/°C Other Data	Solubility
1666	Perrhenic acid $HReO_4$	13768-11-1 251.213	exists only in soln		a	vs H_2O, os
1667	Rhenium carbonyl $Re_2(CO)_{10}$	14285-68-8 652.518	yel-wh cry	170 dec 2.87		s os
1668	Rhenium(III) bromide $ReBr_3$	13569-49-8 425.919	red-brn monocl cry	6.10	subl 500 a	s ace, MeOH, EtOH
1669	Rhenium(III) chloride $ReCl_3$	13569-63-6 292.565	red-blk hyg cry	dec 500 4.81	a	s H_2O
1670	Rhenium(III) iodide ReI_3	15622-42-1 566.920	blk solid	dec		
1671	Rhenium(IV) chloride $ReCl_4$	13569-71-6 328.018	purp-blk cry; hyg	dec 300 4.9		
1672	Rhenium(IV) fluoride ReF_4	15192-42-4 262.201	blue tetr cry	7.49	subl >300	
1673	Rhenium(IV) oxide ReO_2	12036-09-8 218.206	gray orth cry	dec 900 11.4		
1674	Rhenium(IV) sulfide ReS_2	12038-63-0 250.339	tricl cry	7.6		
1675	Rhenium(IV) telluride $ReTe_2$	12067-00-4 441.41	orth cry	8.50		
1676	Rhenium(V) bromide $ReBr_5$	30937-53-2 585.727	brn solid	110 dec		
1677	Rhenium(V) chloride $ReCl_5$	39368-69-9 363.471	brn-blk solid	220 4.9		reac H_2O
1678	Rhenium(V) fluoride ReF_5	30937-52-1 281.199	yel-grn solid	48	≈220	
1679	Rhenium(V) oxide Re_2O_5	12165-05-8 452.411	blue-blk tetr cry	≈7		
1680	Rhenium(VI) chloride $ReCl_6$	31234-26-1 398.923	red-grn solid	29		
1681	Rhenium(VI) fluoride ReF_6	10049-17-9 300.197	yel liq or cub cry	18.5 4.06(cry)	33.7	s HNO_3
1682	Rhenium(VI) oxide ReO_3	1314-28-9 234.205	redcub cry	400 dec 6.9		i H_2O, acid, alk
1683	Rhenium(VI) oxytetrachloride $ReOCl_4$	13814-76-1 344.017	brn cry	29.3	223 b	reac H_2O
1684	Rhenium(VI) oxytetrafluoride $ReOF_4$	17026-29-8 278.200	blue solid	108	171	
1685	Rhenium(VII) fluoride ReF_7	17029-21-9 319.196	yel cub cry	48.3 4.32	73.7	
1686	Rhenium(VII) oxide Re_2O_7	1314-68-7 484.410	yel hyg cry	297 6.10	360 a,b,c,e	s H_2O, EtOH, eth, diox, py
1687	Rhenium(VII) trioxychloride ReO_3Cl	7791-09-5 269.658	col liq	4.5 3.87	128	reac H_2O
1688	Rhenium(VII) trioxyfluoride ReO_3F	42246-24-2 253.203	yel solid	147	164	
1689	Rhenium(VII) dioxytrifluoride ReO_2F_3	57246-89-6 275.201	yel solid	90	185	reac H_2O
1690	Rhenium(VII) oxypentafluoride $ReOF_5$	23377-53-9 297.198	cream solid	43.8	73.0	
1691	Rhenium(VII) sulfide Re_2S_7	12038-67-4 596.876	brn-blk tetr cry	4.87		i H_2O
1692	Rhodium Rh	7440-16-6 102.906	silv-wh metal; cub	1964 12.4	3695 a,c	i acid, sl aqua regia
1693	Rhodium carbonyl chloride $[Rh(CO)_2Cl]_2$	14523-22-9 388.758	red-oran cry	124		s os
1694	Rhodium dodecacarbonyl $Rh_4(CO)_{12}$	19584-30-6 747.747	red hyg cry	2.52		reac H_2O
1695	Rhodium(III) chloride $RhCl_3$	10049-07-7 209.264	red monocl cry	5.38	717 a	i H_2O; s alk
1696	Rhodium(III) fluoride RhF_3	60804-25-3 159.901	red hex cry	5.4		
1697	Rhodium(III) iodide RhI_3	15492-38-3 483.619	blk monocl cry; hyg	6.4		
1698	Rhodium(III) oxide Rh_2O_3	12036-35-0 253.809	gray hex cry	1100 dec 8.2	a	
1699	Rhodium(III) sulfate $Rh_2(SO_4)_3$	10489-46-0 494.002	red-yel solid	dec >500		
1700	Rhodium(IV) oxide RhO_2	12137-27-8 134.905	blk tetr cry	7.2		
1701	Rhodium(VI) fluoride RhF_6	13693-07-7 216.896	blk cub cry	≈70 3.1		
1702	Rubidium Rb	7440-17-7 85.468	soft silv metal; cub	39.31 1.53	688 a,c,e	reac H_2O
1703	Rubidium aluminum sulfate $RbAl(SO_4)_2$	13530-57-9 304.577	hex cry	≈3.1		sl H_2O; i EtOH

No.	Name Formula	CAS RN Mol. Wt.	Physical Form	mp/°C den/g cm^{-3}	bp/°C Other Data	Solubility
1704	Rubidium aluminum sulfate dodecahydrate $RbAl(SO_4)_2 \cdot 12H_2O$	7784-29-4 520.761	col cub cry	≈100 dec ≈1.9		s H_2O; i EtOH
1705	Rubidium azide RbN_3	22756-36-1 127.488	tetr cry; exp	317 2.79		
1706	Rubidium bromide RbBr	7789-39-1 165.372	wh cub cry; hyg	682 3.35	1340 a,c,e	vs H_2O
1707	Rubidium carbonate Rb_2CO_3	584-09-8 230.945	col monocl cry; hyg	837 	a	vs H_2O
1708	Rubidium chloride RbCl	7791-11-9 120.921	wh cub cry; hyg	715 2.76	1390 a,c,e	s H_2O; sl EtOH
1709	Rubidium cyanide RbCN	19073-56-4 111.486	wh cub cry	 2.3		s H_2O; i EtOH, eth
1710	Rubidium fluoride RbF	13446-74-7 104.466	wh cub cry; hyg	833 3.2	1410 a,c,e	vs H_2O; i EtOH
1711	Rubidium fluoride $RbHF_2$	12280-64-7 124.473	tetr cry	188 3.3	a	
1712	Rubidium hydride RbH	13446-75-8 86.476	wh cub cry; flam	dec ≈170 2.60	a	reac H_2O
1713	Rubidium hydrogen carbonate $RbHCO_3$	19088-74-5 146.485	wh rhomb cry	dec 175 		vs H_2O
1714	Rubidium hydrogen sulfate $RbHSO_4$	15587-72-1 182.540	col monocl cry	208 2.9	a	s H_2O
1715	Rubidium hydroxide RbOH	1310-82-3 102.475	gray-wh orth cry; hyg	382 3.2	a	vs H_2O; s EtOH
1716	Rubidium iodide RbI	7790-29-6 212.372	wh cub cry	642 3.55	1300 a,c,e	s H_2O, EtOH
1717	Rubidium nitrate $RbNO_3$	13126-12-0 147.473	wh hex cry; hyg	305 3.11	a,c	vs H_2O
1718	Rubidium oxide Rb_2O	18088-11-4 186.935	yel-brn cub cry; hyg	dec 400 4.0	a	reac H_2O
1719	Rubidium perchlorate $RbClO_4$	13510-42-4 184.919	wh hyg cry	281 2.8	dec 600 a	sl H_2O
1720	Rubidium peroxide Rb_2O_2	23611-30-5 202.935	wh orth cry	 3.8	a	reac H_2O
1721	Rubidium selenide Rb_2Se	31052-43-4 249.90	wh cub cry	733 3.22		reac H_2O
1722	Rubidium sulfate Rb_2SO_4	7488-54-2 267.000	wh orth cry	1050 3.6	a	vs H_2O
1723	Rubidium sulfide Rb_2S	31083-74-6 203.002	wh cub cry	425 2.91		s H_2O
1724	Rubidium superoxide RbO_2	12137-25-6 117.467	tetr cry	412 ≈3.0	a	
1725	Ruthenium Ru	7440-18-8 101.07	silv-wh metal; hex	2334 12.1	4150 a,c	i acid, aqua regia
1726	Ruthenium dodecacarbonyl $Ru_3(CO)_{12}$	15243-33-1 639.34	oran cry	dec 150		
1727	Ruthenium(III) acetylacetonate $Ru(CH_3COCHCOCH_3)_3$	14284-93-6 398.40	red-brn cry	230		
1728	Ruthenium(III) bromide $RuBr_3$	14014-88-1 340.78	brn hex cry	dec >400 5.3	a	
1729	Ruthenium(III) chloride $RuCl_3$	10049-08-8 207.43	brn hex cry	dec >500 3.1	a	i H_2O; s EtOH
1730	Ruthenium(III) fluoride RuF_3	51621-05-7 158.07	brn rhomb cry	dec >600 5.36		
1731	Ruthenium(III) iodide RuI_3	13896-65-6 481.78	blk hex cry	 6.0	a	
1732	Ruthenium(IV) fluoride RuF_4	71500-16-8 177.06	yel cry			reac H_2O
1733	Ruthenium(IV) oxide RuO_2	12036-10-1 133.07	gray-blk tetr cry	 7.05	a	i H_2O, acid
1734	Ruthenium(V) fluoride RuF_5	14521-18-7 196.06	grn monocl cry	86.5 3.90	227	
1735	Ruthenium(VI) fluoride RuF_6	13693-08-8 215.06	dark brn orth cry	54 3.54		reac H_2O
1736	Ruthenium(VIII) oxide RuO_4	20427-56-9 165.07	yel monocl prisms	25.4 3.29	40 a	sl H_2O; vs ctc; reac EtOH
1737	Samarium Sm	7440-19-9 150.36	silv metal; rhomb	1074 7.52	1794 a,c	
1738	Samarium boride SmB_6	12008-29-6 215.23	refrac solid	2580 5.07		
1739	Samarium silicide $SmSi_2$	12300-22-0 206.53	orth cry	 5.14		
1740	Samarium(II) bromide $SmBr_2$	50801-97-3 310.17	brn cry	669		reac H_2O
1741	Samarium(II) chloride $SmCl_2$	13874-75-4 221.27	brn cry	855 3.69		reac H_2O
1742	Samarium(II) fluoride SmF_2	15192-17-3 188.36	purp cry			reac H_2O

No.	Name / Formula	CAS RN / Mol. Wt.	Physical Form	mp/°C den/g cm⁻³	bp/°C Other Data	Solubility
1743	Samarium(II) iodide / SmI_2	32248-43-4 / 404.17	grn cry	520		reac H_2O
1744	Samarium(III) bromide / $SmBr_3$	13759-87-0 / 390.07	yel cry	640		reac H_2O
1745	Samarium(III) chloride / $SmCl_3$	10361-82-7 / 256.72	yel cry	682 / 4.46	a	reac H_2O
1746	Samarium(III) chloride hexahydrate / $SmCl_3 \cdot 6H_2O$	13465-55-9 / 364.81	yel cry	dec / 2.38		reac H_2O
1747	Samarium(III) fluoride / SmF_3	13765-24-7 / 207.36	wh cry	1306	a	reac H_2O
1748	Samarium(III) iodide / SmI_3	13813-25-7 / 531.07	oran cry	850		reac H_2O
1749	Samarium(III) oxide / Sm_2O_3	12060-58-1 / 348.72	yel-wh cub cry	2335 / 7.6	a	
1750	Samarium(III) sulfate octahydrate / $Sm_2(SO_4)_3 \cdot 8H_2O$	13465-58-2 / 733.03	yel cry	/ 2.93		sl H_2O
1751	Samarium(III) sulfide / Sm_2S_3	12067-22-0 / 396.92	gray-brn cub cry	1720 / 5.87		
1752	Samarium(III) telluride / Sm_2Te_3	12040-00-5 / 683.52	orth cry	/ 7.31		
1753	Scandium / Sc	7440-20-2 / 44.956	silv metal; hex	1541 / 2.99	2836 / a,c	
1754	Scandium boride / ScB_2	12007-34-0 / 66.578	refrac solid	2250 / 3.17		
1755	Scandium bromide / $ScBr_3$	13465-59-3 / 284.668	wh hyg cry	969 / 9.33	a	s H_2O
1756	Scandium chloride / $ScCl_3$	10361-84-9 / 151.314	wh hyg cry	967 / 2.4	a	s H_2O; i EtOH
1757	Scandium fluoride / ScF_3	13709-47-2 / 101.951	wh powder	1515	a	sl H_2O
1758	Scandium oxide / Sc_2O_3	12060-08-1 / 137.910	wh cub cry	2485 / 3.864	a	s conc acid
1759	Scandium sulfide / Sc_2S_3	12166-29-9 / 186.110	yel orth cry	1775 / 2.91		
1760	Scandium telluride / Sc_2Te_3	12166-44-8 / 472.71	blk hex cry	/ 5.29		
1761	Selenium (α form) / Se	7782-49-2 / 78.96	red monocl cry	221 / 4.39	685 / a,b,c,d,e	i H_2O, EtOH; sl eth
1762	Selenium (gray) / Se	7782-49-2 / 78.96	gray metallic cry; hex	221 / 4.81	685 / a,b,c,d,e	i H_2O, CS_2
1763	Selenium (vitreous) / Se	7782-49-2 / 78.96	blk amorp solid	trans to gray Se 180 / 4.28	685 / a,b,c,d,e	i H_2O; sl CS_2
1764	Selenic acid / H_2SeO_4	7783-08-6 / 144.97	wh hyg solid	58 / 2.95	260 dec / a	vs H_2O; reac EtOH
1765	Selenous acid / H_2SeO_3	7783-00-8 / 128.97	wh hyg cry	dec 70 / 3.0		vs H_2O; s EtOH
1766	Selenium dioxide / SeO_2	7446-08-4 / 110.96	wh tetr needles or powder	340 tp / 3.95	315 sp / a,e	s H_2O, EtOH, MeOH; sl ace
1767	Selenium trioxide / SeO_3	13768-86-0 / 126.96	wh tetr cry; hyg	118 / 3.44	subl	s H_2O, os
1768	Selenium bromide / Se_2Br_2	7789-52-8 / 317.73	red liq	/ 3.60	225 dec	reac H_2O; s CS_2, chl
1769	Selenium chloride / Se_2Cl_2	10025-68-0 / 228.83	yel-brn oily liq	-85 / 2.774	130 dec	reac H_2O; s CS_2, bz, ctc, chl
1770	Selenium tetrabromide / $SeBr_4$	7789-65-3 / 398.58	oran-red cry	123		reac H_2O; s CS_2, chl
1771	Selenium tetrachloride / $SeCl_4$	10026-03-6 / 220.77	wh-yel cry	305 / 2.6	subl / e	reac H_2O
1772	Selenium tetrafluoride / SeF_4	13465-66-2 / 154.95	col liq	-10 / 2.75	106 / d	reac H_2O, vs EtOH, eth
1773	Selenium hexafluoride / SeF_6	7783-79-1 / 192.95	col gas	-34.6 tp / 8.467 g/L	-46.6 sp / a,b,e	i H_2O
1774	Selenium oxybromide / $SeOBr_2$	7789-51-7 / 254.77	red-yel solid	41.6 / 3.38	220 dec	reac H_2O; s CS_2, bz, ctc
1775	Selenium oxychloride / $SeOCl_2$	7791-23-3 / 165.86	col or yel liq	8.5 / 2.44	177 / b,e	reac H_2O; s ctc, chl, bz, tol
1776	Selenium oxyfluoride / $SeOF_2$	7783-43-9 / 132.96	col liq	15 / 2.8	125	reac H_2O
1777	Selenium dioxydifluoride / SeO_2F_2	14984-81-7 / 148.96	col gas	-99.5 / 6.536 g/L	-8.4	reac H_2O
1778	Selenium sulfide / Se_2S_6	75926-26-0 / 350.32	oran needles	121.5 / 2.44		s CS_2; sl bz
1779	Selenium sulfide / Se_4S_4	75926-28-2 / 444.10	red cry	113 dec / 3.29		s bz; sl CS_2
1780	Silicon / Si	7440-21-3 / 28.086	gray cry or brn amorp solid	1414 / 2.3290	3265 / a,c,e	i H_2O, acid; s alk
1781	Silane / SiH_4	7803-62-5 / 32.118	col gas; flam	-185 / 1.409 g/L	-111.9 / a,d,e	reac H_2O; i EtOH, bz

No.	Name Formula	CAS RN Mol. Wt.	Physical Form	mp/°C den/g cm⁻³	bp/°C Other Data	Solubility
1782	Bromosilane SiH_3Br	13465-73-1 111.014	col gas	-94 4.871 g/L	1.9 a,d,e	
1783	Bromotrichlorosilane $SiCl_3Br$	13465-74-2 214.348	col liq	-62 1.826	80.3 a	reac H_2O
1784	Chlorosilane SiH_3Cl	13465-78-6 66.563	col gas	-118 2.921 g/L	-30.4 a,d,e	
1785	Chlorotrifluorosilane SiF_3Cl	14049-36-6 120.534	col gas	-138 5.289 g/L	-70.0 b,d	reac H_2O
1786	Dibromosilane SiH_2Br_2	13768-94-0 189.910	liq	-70.1	66 a,d,e	
1787	Dichlorosilane SiH_2Cl_2	4109-96-0 101.007	col gas; flam	-122 4.432 g/L	8.3 a,d	reac H_2O
1788	Dichlorodifluorosilane $SiCl_2F_2$	18356-71-3 136.988	col gas	-44 6.011 g/L	-32 b,d,e	reac H_2O
1789	Difluorosilane SiH_2F_2	13824-36-7 68.099	col gas	-122 2.988 g/L	-77.8 d,e	
1790	Diiodosilane SiH_2I_2	13760-02-6 283.911	col liq	-1	150 e	
1791	Fluorosilane SiH_3F	13537-33-2 50.108	col gas	2.199 g/L	-98.6 a,d,e	
1792	Iodosilane SiH_3I	13598-42-0 158.014	col liq	-57	45.6 a,e	
1793	Tetrabromosilane $SiBr_4$	7789-66-4 347.702	col fuming liq	5.2 2.8	154 a,b,d	reac H_2O
1794	Tetrachlorosilane $SiCl_4$	10026-04-7 169.897	col fuming liq	-68.85 1.5	57.65 a,b,c,d,e	reac H_2O
1795	Tetrafluorosilane SiF_4	7783-61-1 104.080	col gas	-90.2 4.567 g/L	-86 a,b,e	reac H_2O
1796	Tetraiodosilane SiI_4	13465-84-4 535.704	wh powder	120.5 4.1	287.35 a,b,c,d	
1797	Tribromosilane $SiHBr_3$	7789-57-3 268.806	flam liq	-73 2.7	109 a,b,d,e	reac H_2O
1798	Tribromochlorosilane $SiBr_3Cl$	13465-76-4 303.251	col liq	-20.8 2.497	127 a	reac H_2O
1799	Trichlorosilane $SiHCl_3$	10025-78-2 135.452	fuming liq	-128.2 1.331	33 a,b,d,e	reac H_2O
1800	Trichlorofluorosilane $SiCl_3F$	14965-52-7 153.442	col gas	6.733 g/L	12.25 b,e	
1801	Trifluorosilane $SiHF_3$	13465-71-9 86.089	col gas	-131 3.778 g/L	-95 a,d,e	
1802	Triiodosilane $SiHI_3$	13465-72-0 409.807	liq	8	220 dec	
1803	Disilane Si_2H_6	1590-87-0 62.219	col gas; flam	-132.5 2.730 g/L	-14.3 a,d,e	reac H_2O, ctc, chl; s EtOH, bz
1804	Trisilane Si_3H_8	7783-26-8 92.321	flam liq	-117.4 0.725(-117°C)	52.9 a,d,e	reac H_2O
1805	Tetrasilane Si_4H_{10}	7783-29-1 122.421	col liq	-84.3 0.79(-84°C)	107.4 e	reac H_2O
1806	Disiloxane $(SiH_3)_2O$	13597-73-4 78.218	gas	-144 3.432 g/L	-15.2 e	
1807	Metasilicic acid H_2SiO_3	7699-41-4 78.100	wh amorp powder		a,f	i H_2O; s HF
1808	Orthosilicic acid H_4SiO_4	10193-36-9 96.116	exists only in soln		a	
1809	Fluorosilicic acid H_2SiF_6	16961-83-4 144.092	stable only in aq soln			s H_2O
1810	Silicon carbide SiC	409-21-2 40.097	hard grn-black hex cry	2830 3.16	a,f	i H_2O, EtOH
1811	Silicon nitride Si_3N_4	12033-89-5 140.284	gray refrac solid; hex	1900 3.17	a	
1812	Silicon monoxide SiO	10097-28-6 44.085	blk cub cry, stable >1200	2.18	a	
1813	Silicon dioxide (α quartz) SiO_2	14808-60-7 60.085	col hex cry	573 trans to quartz 2.648	2950 a,c,f	i H_2O, acid; s HF
1814	Silicon dioxide (β quartz) SiO_2	14808-60-7 60.085	col hex cry	867 trans to tridymite 2.533(600°C)	2950 a,c,f	i H_2O, acid; s HF
1815	Silicon dioxide (tridymite) SiO_2	15468-32-3 60.085	col hex cry	1470 trans to cristobalite 2.265	2950	i H_2O, acid; s HF
1816	Silicon dioxide (cristobalite) SiO_2	14464-46-1 60.085	col hex cry	1713 2.334	2950 e	i H_2O, acid; s HF
1817	Silicon dioxide (vitreous) SiO_2	60676-86-0 60.085	col amorp solid	1713 2.196	2950	i H_2O, acid; s HF
1818	Silicon monosulfide SiS	50927-81-6 60.152	yel-red hyg powder	≈900 1.85	940	reac H_2O
1819	Silicon disulfide SiS_2	13759-10-9 92.218	wh rhomb cry	1090 2.04	subl	reac H_2O, EtOH; i bz

No.	Name Formula	CAS RN Mol. Wt.	Physical Form	mp/°C den/g cm^{-3}	bp/°C Other Data	Solubility
1820	Silver Ag	7440-22-4 107.868	silv metal; cub	961.78 10.5	2162 a,c,e,f	
1821	Silver azide AgN$_3$	13863-88-2 149.888	orth cry; exp	exp ≈250 4.9		
1822	Silver subfluoride Ag$_2$F	1302-01-8 234.734	yel hex cry	dec 100 8.6		reac H$_2$O
1823	Silver(I) acetate AgC$_2$H$_3$O$_2$	563-63-3 166.913	wh needles or powder	dec 3.26		s H$_2$O
1824	Silver(I) acetylide Ag$_2$C$_2$	7659-31-6 239.758	wh powder; exp			
1825	Silver(I) acetylide AgC$_2$H	13092-75-6 132.898	wh powder; exp			
1826	Silver(I) bromate AgBrO$_3$	7783-89-3 235.770	wh tetr cry	dec 5.2	a	sl H$_2$O
1827	Silver(I) bromide AgBr	7785-23-1 187.772	yel cub cry	432 6.47	1502 a,c,d,f	i H$_2$O, acid
1828	Silver(I) carbonate Ag$_2$CO$_3$	534-16-7 275.745	yel monocl cry	218 6.077		sl H$_2$O; s acid
1829	Silver(I) chlorate AgClO$_3$	7783-92-8 191.319	wh tetr cry	230 4.430	dec 270 a	s H$_2$O; sl EtOH
1830	Silver(I) chloride AgCl	7783-90-6 143.321	wh cub cry	455 5.56	1547 a,c,d,e,f	i H$_2$O
1831	Silver(I) chromate Ag$_2$CrO$_4$	7784-01-2 331.730	brn-red monocl cry	5.625	a	i H$_2$O
1832	Silver(I) citrate Ag$_3$C$_6$H$_5$O$_7$	126-45-4 512.707	wh cry powder			i H$_2$O; s HNO$_3$
1833	Silver(I) cyanide AgCN	506-64-9 133.886	wh-gray hex cry	320 dec 3.95	a	i H$_2$O, EtOH, dil acid
1834	Silver(I) dichromate Ag$_2$Cr$_2$O$_7$	7784-02-3 431.724	red cry	4.770		sl H$_2$O
1835	Silver(I) fluoride AgF	7775-41-9 126.866	yel-brn cub cry; hyg	435 5.852	1159 a	s H$_2$O
1836	Silver(I) hydrogen fluoride AgHF$_2$	12249-52-4 146.873	hyg cry	dec		
1837	Silver(I) iodate AgIO$_3$	7783-97-3 282.770	wh orth cry	>200 5.53	a	i H$_2$O
1838	Silver(I) iodide AgI	7783-96-2 234.772	yel powder; hex	558 5.68	1506 a,c,d,e,f	i H$_2$O, acid
1839	Silver(I) lactate monohydrate AgC$_3$H$_5$O$_3$·H$_2$O	128-00-7 214.955	gray cry powder			sl H$_2$O, EtOH
1840	Silver(I) molybdate Ag$_2$MoO$_4$	13765-74-7 375.67	yel cub cry	483 6.18		sl H$_2$O
1841	Silver(I) nitrate AgNO$_3$	7761-88-8 169.873	col rhomb cry	212 4.35	dec 440 a,c	vs H$_2$O; sl EtOH, ace
1842	Silver(I) nitrite AgNO$_2$	7783-99-5 153.874	yel needles	dec 140 4.453		sl H$_2$O; i EtOH; reac acid
1843	Silver(I) oxalate Ag$_2$C$_2$O$_4$	533-51-7 303.756	wh cry powder	exp 140 5.03		i H$_2$O
1844	Silver(I) oxide Ag$_2$O	20667-12-3 231.735	brn-blk cub cry	dec ≈200 7.2	a	i H$_2$O, EtOH; s acid, alk
1845	Silver(I) perchlorate AgClO$_4$	7783-93-9 207.319	col cub cry; hyg	dec 486 2.806	a	vs H$_2$O; s bz, py, os
1846	Silver(I) perchlorate monohydrate AgClO$_4$·H$_2$O	14242-05-8 225.334	hyg wh cry	dec 43		
1847	Silver(I) permanganate AgMnO$_4$	7783-98-4 226.804	viol monocl cry	dec 4.49		sl H$_2$O; reac EtOH
1848	Silver(I) phosphate Ag$_3$PO$_4$	7784-09-0 418.576	yel powder	849 6.37		i H$_2$O; sl dil acid
1849	Silver(I) picrate monohydrate AgC$_6$H$_2$N$_3$O$_7$·H$_2$O	146-84-9 353.981	yel cry			sl H$_2$O, EtOH; i chl, eth
1850	Silver(I) selenate Ag$_2$SeO$_4$	7784-07-8 358.69	orth cry	5.72		sl H$_2$O
1851	Silver(I) selenide Ag$_2$Se	1302-09-6 294.70	gray hex needles	880 8.216		i H$_2$O
1852	Silver(I) selenite Ag$_2$SeO$_3$	7784-05-6 342.69	needles	530 5.930	dec >550	sl H$_2$O; s acid
1853	Silver(I) sulfate Ag$_2$SO$_4$	10294-26-5 311.800	col cry or powder	652 5.45	a	sl H$_2$O
1854	Silver(I) sulfide Ag$_2$S	21548-73-2 247.802	gray-blk orth powder	825 7.23	a,c,f	i H$_2$O; s acid
1855	Silver(I) telluride Ag$_2$Te	12002-99-2 343.34	blk orth cry	955 8.4	f	
1856	Silver(I) tetraiodomercurate(II) Ag$_2$HgI$_4$	7784-03-4 923.94	yel tetr cry	trans to red cub ≈40 6.1		i H$_2$O, dil acid
1857	Silver(I) thiocyanate AgSCN	14104-20-2 165.952	wh powder	dec		i H$_2$O

No.	Name Formula	CAS RN Mol. Wt.	Physical Form	mp/°C den/g cm^{-3}	bp/°C Other Data	Solubility
1858	Silver(II) oxide AgO	1301-96-8 123.867	gray powder; monocl or cub	dec >100 7.5		i H$_2$O; s alk; reac acid
1859	Silver(II) fluoride AgF$_2$	7783-95-1 145.865	wh or gray hyg cry	690 4.58	a	reac H$_2$O
1860	Sodium Na	7440-23-5 22.990	soft silv met; cub	97.80 0.97	883 a,c,e	reac H$_2$O
1861	Sodium acetate NaC$_2$H$_3$O$_2$	127-09-3 82.035	col cry	324 1.528	a	vs H$_2$O
1862	Sodium acetate trihydrate NaC$_2$H$_3$O$_2$·3H$_2$O	6131-90-4 136.080	col cry	58 dec 1.45		vs H$_2$O; sl EtOH
1863	Sodium aluminate NaAlO$_2$	1302-42-7 81.971	wh orth cry; hyg	1650 4.63		vs H$_2$O; i EtOH
1864	Sodium aluminum sulfate dodecahydrate NaAl(SO$_4$)$_2$·12H$_2$O	10102-71-3 458.283	col cry	≈60 1.61		vs H$_2$O; i EtOH
1865	Sodium amide NaNH$_2$	7782-92-5 39.013	wh-grn orth cry	210 1.39	dec 500 a	reac H$_2$O
1866	Sodium ammonium phosphate tetrahydrate NaNH$_4$HPO$_4$·4H$_2$O	13011-54-6 209.069	monocl cry	≈80 dec 1.54		s H$_2$O; i EtOH
1867	Sodium arsenite NaAsO$_2$	7784-46-5 129.911	wh-gray hyg powder	 1.87		vs H$_2$O; i EtOH
1868	Sodium azide NaN$_3$	26628-22-8 65.010	col hex cry	dec 300; exp 1.846	a	vs H$_2$O; sl EtOH; i eth
1869	Sodium borohydride NaBH$_4$	16940-66-2 37.833	wh cub cry; hyg	dec ≈400 1.07	a	vs H$_2$O; reac EtOH
1870	Sodium bromate NaBrO$_3$	7789-38-0 150.892	col cub cry	381 3.34	a,c	s H$_2$O; i EtOH
1871	Sodium bromide NaBr	7647-15-6 102.894	wh cub cry	747 3.200	1390 a,c,e	vs H$_2$O; s EtOH
1872	Sodium bromide dihydrate NaBr·2H$_2$O	13466-08-5 138.925	wh cry	36 dec 2.18		s H$_2$O; sl EtOH
1873	Sodium carbonate Na$_2$CO$_3$	497-19-8 105.989	wh hyg powder	858.1 2.54	a,c	s H$_2$O; i EtOH
1874	Sodium carbonate decahydrate Na$_2$CO$_3$·10H$_2$O	6132-02-1 286.142	col cry	34 dec 1.46		vs H$_2$O; i EtOH
1875	Sodium carbonate monohydrate Na$_2$CO$_3$·H$_2$O	5968-11-6 124.005	col orth cry	dec 100 2.25	f	vs H$_2$O; i EtOH
1876	Sodium chlorate NaClO$_3$	7775-09-9 106.441	col cub cry	248 2.5	dec >300 a,c	vs H$_2$O; sl EtOH
1877	Sodium chloride NaCl	7647-14-5 58.443	col cub cry	800.7 2.17	1465 a,c,e,f	s H$_2$O; sl EtOH
1878	Sodium chlorite NaClO$_2$	7758-19-2 90.442	wh hyg cry	dec ≈180 	a	vs H$_2$O
1879	Sodium chromate Na$_2$CrO$_4$	7775-11-3 161.974	yel orth cry	792 2.72		vs H$_2$O; sl EtOH
1880	Sodium chromate tetrahydrate Na$_2$CrO$_4$·4H$_2$O	10034-82-9 234.035	yel hyg cry	dec 		vs H$_2$O; sl EtOH
1881	Sodium citrate dihydrate Na$_3$C$_6$H$_5$O$_7$·2H$_2$O	6132-04-3 294.101	wh cry	dec 150 		vs H$_2$O; i EtOH, eth
1882	Sodium cyanate NaOCN	917-61-3 65.007	col needles	550 1.89	a	s H$_2$O; sl EtOH; i eth
1883	Sodium cyanide NaCN	143-33-9 49.008	wh cub cry; hyg	563 1.6	a,e	vs H$_2$O; sl EtOH
1884	Sodium cyanoborohydride NaBH$_3$(CN)	25895-60-7 62.843	wh hyg powder	240 dec 1.12		vs H$_2$O; s thf; sl EtOH; i bz, eth
1885	Sodium dichromate Na$_2$Cr$_2$O$_7$	10588-01-9 261.968	red hyg cry	357 	dec 400	vs H$_2$O
1886	Sodium dihydrogen phosphate monohydrate NaH$_2$PO$_4$·H$_2$O	10049-21-5 137.993	wh hyg cry	dec 100 		vs H$_2$O; i EtOH
1887	Sodium dihydrogen phosphate dihydrate NaH$_2$PO$_4$·2H$_2$O	13472-35-0 156.008	col orth cry	60 dec 1.91		vs H$_2$O; i EtOH
1888	Sodium dihydrogen pyrophosphate Na$_2$H$_2$P$_2$O$_7$	7758-16-9 221.939	wh powder	dec 220 ≈1.9		s H$_2$O
1889	Sodium dithionate dihydrate Na$_2$S$_2$O$_6$·2H$_2$O	7631-94-9* 242.139	col orth cry	dec 110 2.19		s H$_2$O; i EtOH
1890	Sodium ferricyanide monohydrate Na$_3$Fe(CN)$_6$·H$_2$O	14217-21-1* 298.935	red hyg cry			s H$_2$O; i EtOH
1891	Sodium ferrocyanide decahydrate Na$_4$Fe(CN)$_6$·10H$_2$O	13601-19-9 484.063	yel monocl cry	dec ≈50 1.46		s H$_2$O; i os
1892	Sodium fluoride NaF	7681-49-4 41.988	col cub or tetr cry	996 2.78	1704 a,c,e,f	sl H$_2$O; i EtOH
1893	Sodium fluoroborate NaBF$_4$	13755-29-8 109.795	wh orth prisms	384 2.47	a	vs H$_2$O; sl EtOH
1894	Sodium formate NaCHO$_2$	141-53-7 68.008	wh hyg cry	253 1.92	dec a	vs H$_2$O; sl EtOH
1895	Sodium hexafluoroaluminate Na$_3$AlF$_6$	13775-53-6 209.941	col monocl cry; trans cub 560	1009 2.97	f	i H$_2$O

No.	Name Formula	CAS RN Mol. Wt.	Physical Form	mp/°C den/g cm^{-3}	bp/°C Other Data	Solubility
1896	Sodium hexafluorosilicate Na_2SiF_6	16893-85-9 188.056	wh hex cry	dec 2.7	a	sl H_2O; i EtOH
1897	Sodium hydride NaH	7646-69-7 23.998	silv cub cry; flam	425 dec 1.39	a	reac H_2O, EtOH
1898	Sodium hydrogen arsenate heptahydrate $Na_2HAsO_4 \cdot 7H_2O$	10048-95-0 312.014	wh monocl cry	dec ≈50 1.87		vs H_2O; sl EtOH
1899	Sodium hydrogen carbonate $NaHCO_3$	144-55-8 84.007	wh monocl cry	dec ≈50 2.20	a	s H_2O; i EtOH
1900	Sodium hydrogen fluoride $NaHF_2$	1333-83-1 61.995	wh hex cry	dec >160 2.08	a	s H_2O
1901	Sodium hydrogen phosphate Na_2HPO_4	7558-79-4 141.959	wh hyg powder	1.7	a	s H_2O
1902	Sodium hydrogen phosphate dodecahydrate $Na_2HPO_4 \cdot 12H_2O$	10039-32-4 358.143	col cry	dec ≈35 ≈1.5		s H_2O; i EtOH
1903	Sodium hydrogen phosphate heptahydrate $Na_2HPO_4 \cdot 7H_2O$	7782-85-6 268.066	col cry	≈1.7		s H_2O; i EtOH
1904	Sodium hydrogen sulfate $NaHSO_4$	7681-38-1 120.062	wh hyg cry	≈315 2.43	a	vs H_2O
1905	Sodium hydrogen sulfate monohydrate $NaHSO_4 \cdot H_2O$	10034-88-5 138.077	wh monocl cry	2.10		vs H_2O; reac EtOH
1906	Sodium hydrogen sulfide NaHS	16721-80-5 56.064	col rhomb cry	350 1.79		s H_2O, EtOh, eth
1907	Sodium hydrogen sulfide dihydrate $NaHS \cdot 2H_2O$	16721-80-5 92.095	yel hyg needles	dec 55		vs H_2O, EtOH, eth
1908	Sodium hydrogen sulfite $NaHSO_3$	7631-90-5 104.062	wh cry	1.48		s H_2O; sl EtOH
1909	Sodium hydroxide NaOH	1310-73-2 39.997	wh orth cry; hyg	323 2.13	1388 a,c,d,e	vs H_2O; s EtOH, MeOH
1910	Sodium hypochlorite NaClO	7681-52-9 74.442	stable in aq soln	anh form exp		vs H_2O
1911	Sodium hypochlorite pentahydrate $NaOCl \cdot 5H_2O$	7681-52-9* 164.518	pale grn orth cry	18 1.6		s H_2O
1912	Sodium iodate $NaIO_3$	7681-55-2 197.892	wh orth cry	dec 4.28	a	s H_2O; i EtOH
1913	Sodium iodide NaI	7681-82-5 149.894	wh cub cry; hyg	660 3.67	1304 a,c,e	vs H_2O; s EtOH, ace
1914	Sodium bismuthate $NaBiO_3$	12232-99-4 279.968	yel-brn hyg cry			i cold H_2O, reac acid
1915	Sodium metabisulfite $Na_2S_2O_5$	7681-57-4 190.109	wh cry			vs H_2O; sl EtOH
1916	Sodium metaborate $NaBO_2$	7775-19-1 65.800	wh hex cry	966 2.46	1434 a,c	s H_2O
1917	Sodium metasilicate Na_2SiO_3˙	6834-92-0 122.064	wh amorp solid; hyg	1089 2.61	a,c	s cold H_2O; reac hot H_2O
1918	Sodium molybdate Na_2MoO_4	7631-95-0 205.92	col cub cry	687 ≈3.5	a	vs H_2O
1919	Sodium molybdate dihydrate $Na_2MoO_4 \cdot 2H_2O$	10102-40-6 241.95	cry powder	dec 100 ≈3.5		vs H_2O
1920	Sodium niobate $NaNbO_3$	12034-09-2 163.894	rhomb cry	1422 4.55		i H_2O
1921	Sodium nitrate $NaNO_3$	7631-99-4 84.995	col hex cry; hyg	307 2.26	a,c	vs H_2O; sl EtOH, MeOH
1922	Sodium nitrite $NaNO_2$	7632-00-0 68.996	wh orth cry; hyg	271 2.17	dec >320 a	vs H_2O; sl EtOH; reac acid
1923	Sodium oxalate $Na_2C_2O_4$	62-76-0 134.000	wh powder	dec ≈250 2.34	a	sl H_2O; i EtOH
1924	Sodium oxide Na_2O	1313-59-3 61.979	wh amorp powder	1132 dec 2.27	a,c	reac H_2O
1925	Sodium perchlorate $NaClO_4$	7601-89-0 122.441	wh orth cry; hyg	480 dec 2.52	a	vs H_2O
1926	Sodium perchlorate monohydrate $NaClO_4 \cdot H_2O$	7791-07-3 140.456	wh hyg cry	dec ≈130 2.02		vs H_2O
1927	Sodium periodate $NaIO_4$	7790-28-5 213.892	wh tetr cry	dec ≈300 3.86	a	s H_2O, acid
1928	Sodium periodate trihydrate $NaIO_4 \cdot 3H_2O$	13472-31-6 267.938	wh hex cry	dec 175 3.22		s H_2O
1929	Sodium permanganate trihydrate $NaMnO_4 \cdot 3H_2O$	10101-50-5* 195.972	red-blk hyg cry	dec 170 2.47		vs H_2O; reac EtOH
1930	Sodium peroxide Na_2O_2	1313-60-6 77.979	yel hyg powder	675 2.805	a	reac H_2O
1931	Sodium persulfate $Na_2S_2O_8$	7775-27-1 238.107	wh hyg cry			vs H_2O; reac EtOH
1932	Sodium phosphate dodecahydrate $Na_3PO_4 \cdot 12H_2O$	7601-54-9 380.124	col hex cry	≈75 1.62		s H_2O; i EtOH
1933	Sodium potassium tartrate tetrahydrate $NaK(C_4H_4O_6) \cdot 4H_2O$	304-59-6 282.221	wh cry	dec ≈70 1.79	anh at 130	vs H_2O; i EtOH

No.	Name Formula	CAS RN Mol. Wt.	Physical Form	mp/°C den/g cm⁻³	bp/°C Other Data	Solubility
1934	Sodium pyrophosphate $Na_4P_2O_7$	7722-88-5 265.902	col cry	988 2.53		sl H_2O
1935	Sodium selenate decahydrate $Na_2SeO_4 \cdot 10H_2O$	10102-23-5 369.09	wh cry	1.61		vs H_2O
1936	Sodium selenide Na_2Se	1313-85-5 124.94	amorp solid	>875 2.62		reac H_2O
1937	Sodium selenite Na_2SeO_3	10102-18-8 172.94	wh tetr cry			vs H_2O; i EtOH
1938	Sodium sulfate Na_2SO_4	7757-82-6 142.044	wh orth cry or powder	884 2.7	a,c,f	s H_2O; i EtOH
1939	Sodium sulfate decahydrate $Na_2SO_4 \cdot 10H_2O$	7727-73-3 322.197	col monocl cry	32 dec 1.46	f	s H_2O; i EtOH
1940	Sodium sulfide Na_2S	1313-82-2 78.046	wh cub cry; hyg	1172 1.856	a,c	s H_2O; sl EtOH; i eth
1941	Sodium sulfide nonahydrate $Na_2S \cdot 9H_2O$	1313-84-4 240.184	wh-yel hyg cry	dec ≈50 1.43		vs H_2O; sl EtOH; i eth
1942	Sodium sulfide pentahydrate $Na_2S \cdot 5H_2O$	1313-83-3 168.122	col orth cry	120 dec 1.58		vs H_2O; s EtOH; i eth
1943	Sodium sulfite Na_2SO_3	7757-83-7 126.044	wh hex cry	dec 2.63	a	s H_2O; i EtOH
1944	Sodium sulfite heptahydrate $Na_2SO_3 \cdot 7H_2O$	10102-15-5 252.151	wh monocl cry; unstable	1.56		vs H_2O; sl EtOH
1945	Sodium superoxide NaO_2	12034-12-7 54.989	yel cub cry	552 2.2	a	reac H_2O
1946	Sodium tellurate Na_2TeO_4	10102-20-2 237.58	wh powder			sl H_2O
1947	Sodium tetraborate $Na_2B_4O_7$	1330-43-4 201.220	col gl solid; hyg	743 2.4	1575 a	sl H_2O, MeOH
1948	Sodium tetraborate decahydrate $Na_2B_4O_7 \cdot 10H_2O$	1303-96-4 381.373	wh monocl cry	75 dec 1.73	f	sl H_2O; i EtOH
1949	Sodium tetraborate pentahydrate $Na_2B_4O_7 \cdot 5H_2O$	12045-88-4 291.296	hex cry	dec 1.88		sl H_2O
1950	Sodium tetraborate tetrahydrate $Na_2B_4O_7 \cdot 4H_2O$	12045-87-3 273.281	wh monocl cry	1.95	f	sl H_2O
1951	Sodium tetrachloroaluminate $NaAlCl_4$	7784-16-9 191.783	orth cry	2.01		s H_2O
1952	Sodium tetrafluoroberyllate Na_2BeF_4	13871-27-7 130.986	orth cry	575 2.47		sl H_2O
1953	Sodium thiocyanate $NaSCN$	540-72-7 81.074	col hyg cry	287		vs H_2O
1954	Sodium thiosulfate pentahydrate $Na_2S_2O_3 \cdot 5H_2O$	10102-17-7 248.186	col cry	dec ≈50 1.69		vs H_2O; i EtOH
1955	Sodium tungstate dihydrate $Na_2WO_4 \cdot 2H_2O$	10213-10-2 329.85	wh orth cry	dec 100 3.25		vs H_2O; i EtOH
1956	Strontium Sr	7440-24-6 87.62	silv-wh metal; cub	777 2.64	1382 a,c,e	reac H_2O; s EtOH
1957	Strontium bromate monohydrate $Sr(BrO_3)_2 \cdot H_2O$	14519-18-7 361.44	wh-yel hyg cry	dec 240 3.77		s H_2O
1958	Strontium bromide $SrBr_2$	10476-81-0 247.43	wh tetr cry	657 4.216	a,c	vs H_2O
1959	Strontium bromide hexahydrate $SrBr_2 \cdot 6H_2O$	7789-53-9 355.52	col hyg cry	88 dec		vs H_2O; s EtOH; i eth
1960	Strontium carbide SrC_2	12071-29-3 111.64	tetr cry	3.19		
1961	Strontium carbonate $SrCO_3$	1633-05-2 147.63	wh orth cry; hyg	1494 3.5	a,f	i H_2O; s dil acid
1962	Strontium chlorate $Sr(ClO_3)_2$	7791-10-8 254.52	col cry	120 dec 3.15		vs H_2O; sl EtOH
1963	Strontium chloride $SrCl_2$	10476-85-4 158.53	wh cub cry; hyg	874 3.052	1250 a,c	vs H_2O
1964	Strontium chloride hexahydrate $SrCl_2 \cdot 6H_2O$	10025-70-4 266.62	col hyg cry	dec 100 1.96		vs H_2O; s EtOH
1965	Strontium chromate $SrCrO_4$	7789-06-2 203.61	yel monocl cry	dec 3.9		sl H_2O; s dil acid
1966	Strontium fluoride SrF_2	7783-48-4 125.62	wh cub cry or powder	1477 4.24	2460 a,c	s H_2O, dil acid
1967	Strontium hydride SrH_2	13598-33-9 89.64	orth cry	1050 3.26	a	reac H_2O
1968	Strontium hydroxide $Sr(OH)_2$	18480-07-4 121.64	col orth cry; hyg	535 3.625	dec 710 a,c	s H_2O
1969	Strontium iodide SrI_2	10476-86-5 341.43	wh hyg cry	538 4.55	1773 dec a,c	vs H_2O
1970	Strontium iodide hexahydrate $SrI_2 \cdot 6H_2O$	73796-25-5 449.52	wh-yel hex cry; hyg	dec 120 4.4		vs H_2O; s EtOH
1971	Strontium niobate $SrNb_2O_6$	12034-89-8 369.43	monocl cry	1225 5.11		i H_2O
1972	Strontium nitrate $Sr(NO_3)_2$	10042-76-9 211.63	wh cub cry	570 2.99	645 a	vs H_2O; sl EtOH, ace

No.	Name Formula	CAS RN Mol. Wt.	Physical Form	mp/°C den/g cm^{-3}	bp/°C Other Data	Solubility
1973	Strontium nitrite $Sr(NO_2)_2$	13470-06-9 179.63	wh-yel hyg needles	dec 240 2.8	a	vs H_2O
1974	Strontium oxide SrO	1314-11-0 103.62	col cub cry	2532 5.1	a,c,e	reac H_2O
1975	Strontium peroxide SrO_2	1314-18-7 119.62	wh tetr cry; unstable	215 dec 4.78		reac H_2O
1976	Strontium phosphate $Sr_3(PO_4)_2$	7446-28-8 452.80	wh powder			i H_2O; s acid
1977	Strontium selenate $SrSeO_4$	7446-21-1 230.58	orth cry	4.25		i H_2O; s hot HCl
1978	Strontium selenide SrSe	1315-07-7 166.58	wh cub cry	1600 4.54	a	
1979	Strontium orthosilicate Sr_2SiO_4	13597-55-2 267.32	orth cry	4.5	a	
1980	Strontium silicide $SrSi_2$	12138-28-2 143.79	silv-gray cub cry	1100 3.35		
1981	Strontium sulfate $SrSO_4$	7759-02-6 183.68	wh orth cry	1607 3.96	a,f	i H_2O, EtOH; sl acid
1982	Strontium sulfide SrS	1314-96-1 119.69	gray cub cry	2227 3.70	a	sl H_2O; s acid
1983	Strontium titanate $SrTiO_3$	12060-59-2 183.49	wh cub cry	2080 5.1		i H_2O
1984	Sulfur (α form) S	7704-34-9 32.066	yel orth cry	trans to β form 94.5 2.07	444.60 a,b,c,d,e,f	i H_2O; sl EtOH, bz, eth; s CS_2
1985	Sulfur (β form) S	7704-34-9 32.066	yel monocl needles, stable 94.5-115	115.21 2.07	444.60 a,b,c,d,e,f	i H_2O; sl EtOH, bz, eth; s CS_2
1986	Sulfuric acid H_2SO_4	7664-93-9 98.080	col oily liq	10.31 1.8	337 a,c,e	vs H_2O
1987	Peroxysulfuric acid H_2SO_5	7722-86-3 114.079	wh cry; unstable	dec 45		vs H_2O
1988	Chlorosulfonic acid $SO_2(OH)Cl$	7790-94-5 116.525	col-yel liq	-80 1.75	152	reac H_2O; s py
1989	Fluorosulfonic acid $SO_2(OH)F$	7789-21-1 100.070	col liq	-89 1.726	163	reac H_2O
1990	Nitrosylsulfuric acid $HNOSO_4$	7782-78-7 127.078	prisms	dec 73		reac H_2O; s H_2SO_4
1991	Sulfurous acid H_2SO_3	7782-99-2 82.080	exists only in soln			soln of SO_2 in H_2O
1992	Sulfur dioxide SO_2	7446-09-5 64.065	col gas	-75.5 2.811 g/L	-10.05 a,b,d,e	s H_2O, EtOh, eth, chl
1993	Sulfur trioxide SO_3	7446-11-9 80.064	col liq	16.8 1.92	45 a,b,d,e	reac H_2O
1994	Sulfur bromide $SSBr_2$	13172-31-1 223.940	red oily liq	-46 2.63	dec >25 a	reac H_2O
1995	Sulfur chloride $SSCl_2$	10025-67-9 135.037	yel-red oily liq	-77 1.69	137 a,e	reac H_2O; s EtOH, bz, eth, ctc
1996	Sulfur fluoride SSF_2	16860-99-4 102.129	col gas	-164.6 4.481 g/L	-10.6	reac H_2O
1997	Sulfur fluoride FSSF	13709-35-8 102.129	col gas	-133 4.481 g/L	15	reac H_2O
1998	Sulfur dichloride SCl_2	10545-99-0 102.971	red visc liq	-122 1.62	59.6 a	reac H_2O
1999	Sulfur tetrafluoride SF_4	7783-60-0 108.060	col gas	-125 4.742 g/L	-40.45 a,b,d	reac H_2O
2000	Sulfur hexafluoride SF_6	2551-62-4 146.056	col gas	-50.7 tp 6.409 g/L	-63.8 sp a,b,c,d,e	sl H_2O; s EtOH
2001	Sulfur bromide pentafluoride SF_5Br	15607-89-3 206.962	col gas	-79 9.081 g/L	3.1	
2002	Sulfur chloride pentafluoride SF_5Cl	13780-57-9 162.511	col gas	-64 7.131 g/L	-19.05 a,b	
2003	Sulfur decafluoride S_2F_{10}	5714-22-7 254.116	liq	-52.7 2.08	30; dec 150	i H_2O
2004	Sulfuryl amide $SO_2(NH_2)_2$	7803-58-9 96.110	orth plates	93	dec 250	s H_2O, hot EtOH, ace
2005	Sulfuryl chloride SO_2Cl_2	7791-25-5 134.970	col liq	-51 1.680	69.4 a,d,e	reac H_2O; s bz, tol, eth
2006	Sulfuryl fluoride SO_2F_2	2699-79-8 102.062	col gas	-135.8 4.478 g/L	-55.4 a	sl H_2O, EtOH; s tol, ctc
2007	Thionyl bromide $SOBr_2$	507-16-4 207.873	yel liq	-50	140 e	reac H_2O
2008	Thionyl chloride $SOCl_2$	7719-09-7 118.970	yel fuming liq	-101 1.631	75.6 a,d,e	reac H_2O; s bz, ctc, chl
2009	Thionyl fluoride SOF_2	7783-42-8 86.062	col gas	-129.5 3.776 g/L	-43.8 a,d	reac H_2O; s bz, eth
2010	Sulfur fluoride hypofluorite F_5SOF	15179-32-5 162.055	col gas	-86 7.111 g/L	-35.1	
2011	Tantalum Ta	7440-25-7 180.948	gray metal; cub	3017 16.4	5458 a,c	reac HF

No.	Name / Formula	CAS RN / Mol. Wt.	Physical Form	mp/°C / den/g cm⁻³	bp/°C / Other Data	Solubility
2012	Tantalum aluminide / TaAl$_3$	12004-76-1 / 261.893	gray refrac powder	≈1400 / 7.02		i H$_2$O, acid, alk
2013	Tantalum boride / TaB	12007-07-7 / 191.759	refrac orth cry	2040 / 14.2		
2014	Tantalum boride / TaB$_2$	12007-35-1 / 202.570	blk hex cry	3140 / 11.2		i H$_2$O, acid, alk
2015	Tantalum carbide / TaC	12070-06-3 / 192.959	gold-brown powder; cub	3880 / 14.3	4780	s HF-HNO$_3$ mixture
2016	Tantalum carbide / Ta$_2$C	12070-07-4 / 373.907	refrac hex cry	3327 / 15.1		
2017	Tantalum nitride / TaN	12033-62-4 / 194.955	blk hex cry	3090 / 13.7		i H$_2$O; sl aqua regia; reac alk
2018	Tantalum silicide / TaSi$_2$	12039-79-1 / 237.119	gray powder	2200 / 9.14		
2019	Tantalum(IV) oxide / TaO$_2$	12036-14-5 / 212.947	tetr cry	/ 10.0		
2020	Tantalum(IV) selenide / TaSe$_2$	12039-55-3 / 338.87	hex cry	/ 6.7		
2021	Tantalum(IV) sulfide / TaS$_2$	12143-72-5 / 245.080	blk hex cry	>3000 / 6.86		i H$_2$O
2022	Tantalum(IV) telluride / TaTe$_2$	12067-66-2 / 436.15	monocl cry	/ 9.4		
2023	Tantalum(V) bromide / TaBr$_5$	13451-11-1 / 580.468	yel cry powder	265 / 4.99	349 / a,b,c,d	
2024	Tantalum(V) chloride / TaCl$_5$	7721-01-9 / 358.212	yel monocl cry; hyg	216 / 3.68	239.35 / a,b,c,d	reac H$_2$O; s EtOH
2025	Tantalum(V) fluoride / TaF$_5$	7783-71-3 / 275.940	wh monocl cry; hyg	95.1 / 5.0	229.2 / a,d,e	s H$_2$O, eth; sl CS$_2$, ctc
2026	Tantalum(V) iodide / TaI$_5$	14693-81-3 / 815.470	blk hex cry; hyg	496 / 5.80	543	
2027	Tantalum(V) oxide / Ta$_2$O$_5$	1314-61-0 / 441.893	wh rhomb cry or powder	1785 / 8.2	a,c	i H$_2$O, EtOH, acid; s HF
2028	Technetium / Tc	7440-26-8 / 98	hex cry	2157 / 11	4265 / a,c	
2029	Technetium(V) fluoride / TcF$_5$	31052-14-9 / 193	yel solid	50	dec	
2030	Technetium(VI) fluoride / TcF$_6$	13842-93-8 / 212	yel cub cry	37.4 / 3.0	55.3	
2031	Tellurium / Te	13494-80-9 / 127.60	gray-wh rhomb cry	449.51 / 6.24	988 / a,c,d,e	i H$_2$O, bz, CS$_2$
2032	Telluric(VI) acid / H$_6$TeO$_6$	7803-68-1 / 229.64	wh monocl cry	136 / 3.07		s H$_2$O
2033	Tellurous acid / H$_2$TeO$_3$	10049-23-7 / 177.61	wh cry	dec 40 / 3.0		sl H$_2$O; s dil acid, alk
2034	Tellurium dioxide / TeO$_2$	7446-07-3 / 159.60	wh orth cry	733 / 5.9	1245 / a	i H$_2$O; s alk, acid
2035	Tellurium trioxide / TeO$_3$	13451-18-8 / 175.60	yel-oran cry	430 / 5.07		i H$_2$O
2036	Tellurium dibromide / TeBr$_2$	7789-54-0 / 287.41	grn-brn hyg cry	210	339	reac H$_2$O; s eth; sl chl
2037	Tellurium dichloride / TeCl$_2$	10025-71-5 / 198.51	blk amorp solid; hyg	208 / 6.9	328	reac H$_2$O; i ctc
2038	Tellurium tetrabromide / TeBr$_4$	10031-27-3 / 447.22	yel-oran monocl cry	388 / 4.3	≈420 dec / a	reac H$_2$O; s eth
2039	Tellurium tetrachloride / TeCl$_4$	10026-07-0 / 269.41	wh monocl cry; hyg	224 / 3.0	387 / a,b,d,e	reac H$_2$O; s EtOH, tol
2040	Tellurium tetrafluoride / TeF$_4$	15192-26-4 / 203.59	col cry	129	dec 195	reac H$_2$O
2041	Tellurium tetraiodide / TeI$_4$	7790-48-9 / 635.22	blk orth cry	280 / 5.05		reac H$_2$O; sl ace
2042	Tellurium hexafluoride / TeF$_6$	7783-80-4 / 241.59	col gas	-37.6 tp / 10.601 g/L	-38.9 sp / a,b,e	reac H$_2$O
2043	Terbium / Tb	7440-27-9 / 158.925	silv metal; hex	1356 / 8.23	3230 / a,c	
2044	Terbium chloride / TbCl$_3$	10042-88-3 / 265.283	wh orth cry; hyg	588 / 4.35	a	s H$_2$O
2045	Terbium chloride hexahydrate / TbCl$_3$·6H$_2$O	13798-24-8 / 373.374	hyg cry	/ 4.35		vs H$_2$O
2046	Terbium iodide / TbI$_3$	13813-40-6 / 539.638	hex cry; hyg	957 / ≈5.2		s H$_2$O
2047	Terbium nitride / TbN	12033-64-6 / 172.932	cub cry	/ 9.55		
2048	Terbium oxide / Tb$_2$O$_3$	12036-41-8 / 365.849	wh cub cry	2410 / 7.91	a	
2049	Terbium silicide / TbSi$_2$	12039-80-4 / 215.096	orth cry	/ 6.66		
2050	Terbium sulfide / Tb$_2$S$_3$	12138-11-3 / 414.049	cub cry	/ 6.35		

No.	Name / Formula	CAS RN / Mol. Wt.	Physical Form	mp/°C / den/g cm^{-3}	bp/°C / Other Data	Solubility
2051	Thallium / Tl	7440-28-0 / 204.383	soft blue-wh metal	304 / 11.8	1473 / a,c,e	i H_2O; reac acid
2052	Thallium(I) acetate / $TlC_2H_3O_2$	563-68-8 / 263.428	hyg wh cry	131 / 3.68		s H_2O, EtOH
2053	Thallium(I) bromide / TlBr	7789-40-4 / 284.287	yel cub cry	460 / 7.5	819 / a,c,d,e	i H_2O
2054	Thallium(I) carbonate / Tl_2CO_3	6533-73-9 / 468.776	wh monocl cry	272 / 7.11	a,c	sl H_2O; i EtOH
2055	Thallium(I) chlorate / $TlClO_3$	13453-30-0 / 287.834	col hex cry	5.5		sl H_2O
2056	Thallium(I) chloride / TlCl	7791-12-0 / 239.836	wh cub cry	430 / 7.0	720 / a,c,d,e	sl H_2O; i EtOH
2057	Thallium(I) cyanide / TlCN	13453-34-4 / 230.401	wh hex plates	6.523		s H_2O, acid, EtOH
2058	Thallium(I) ethoxide / TlC_2H_5O	20398-06-5 / 249.444	cloudy liq	-3 / 3.49	dec 130	reac H_2O
2059	Thallium(I) fluoride / TlF	7789-27-7 / 223.381	wh orth cry	326 / 8.36	826 / a,c	s H_2O
2060	Thallium(I) formate / $TlCHO_2$	992-98-3 / 249.401	hyg col needles	101 / 4.97		vs H_2O; s MeOH
2061	Thallium(I) hexafluorophosphate / $TlPF_6$	60969-19-9 / 349.347	wh cub cry	4.6		
2062	Thallium(I) hydroxide / TlOH	12026-06-1 / 221.390	yel needles	139 dec / 7.44	a	vs H_2O
2063	Thallium(I) iodide / TlI	7790-30-9 / 331.287	yel cry powder	441.8 / 7.1	824 / a,c,d,e	i H_2O, EtOH
2064	Thallium(I) molybdate / Tl_2MoO_4	34128-09-1 / 568.71	yel-wh cub cry			i H_2O
2065	Thallium(I) nitrate / $TlNO_3$	10102-45-1 / 266.388	wh cry	206 / 5.55	450 dec / a,c	s H_2O; i EtOH
2066	Thallium(I) nitrite / $TlNO_2$	13826-63-6 / 250.389	cub cry	5.7		vs H_2O
2067	Thallium(I) oxalate / $Tl_2C_2O_4$	30737-24-7 / 496.787	wh powder	6.31		
2068	Thallium(I) oxide / Tl_2O	1314-12-1 / 424.766	blk rhomb cry; hyg	579 / 9.52	≈1080 / a	s H_2O, EtOH
2069	Thallium(I) perchlorate / $TlClO_4$	13453-40-2 / 303.834	col orth cry	4.8		s H_2O
2070	Thallium(I) selenate / Tl_2SeO_4	7446-22-2 / 551.73	orth cry	>400 / 6.875		sl H_2O; i EtOH, eth
2071	Thallium(I) selenide / Tl_2Se	15572-25-5 / 487.73	gray plates	340	a	i H_2O, acid
2072	Thallium(I) sulfate / Tl_2SO_4	7446-18-6 / 504.831	wh rhomb prisms	632 / 6.77	a,c	sl H_2O
2073	Thallium(I) sulfide / Tl_2S	1314-97-2 / 440.833	blue-blk cry	448 / 8.39	1367 / a,c,d	sl H_2O, alk; s acid
2074	Thallium(III) bromide tetrahydrate / $TlBr_3 \cdot 4H_2O$	13701-90-1 / 516.157	yel orth cry	3.65		s H_2O, EtOH
2075	Thallium(III) chloride / $TlCl_3$	13453-32-2 / 310.741	monocl cry	155 / 4.7	a	vs H_2O, EtOH, eth
2076	Thallium(III) chloride tetrahydrate / $TlCl_3 \cdot 4H_2O$	13453-32-2* / 382.803	orth cry	3.00		s H_2O
2077	Thallium(III) fluoride / TlF_3	7783-57-5 / 261.378	wh orth cry; hyg	550 dec / 8.65		reac H_2O
2078	Thallium(III) nitrate / $Tl(NO_3)_3$	13746-98-0 / 390.398	col cry			reac H_2O
2079	Thallium(III) oxide / Tl_2O_3	1314-32-5 / 456.765	brn cub cry	834 / 10.2		i H_2O; reac acid
2080	Thorium / Th	7440-29-1 / 232.038	soft gray-wh metal; cub	1750 / 11.7	4788 / a,c	s acids
2081	Thorium hydride / ThH_2	16689-88-6 / 234.054	tetr cry	9.5	a	
2082	Thorium boride / ThB_6	12229-63-9 / 296.904	refrac solid	2450 / 6.99		
2083	Thorium bromide / $ThBr_4$	13453-49-1 / 551.654	wh hyg cry	679		reac H_2O
2084	Thorium carbide / ThC	12012-16-7 / 244.049	cub cry	2500 / 10.6		reac H_2O
2085	Thorium dicarbide / ThC_2	12071-31-7 / 256.060	yel monocl cry	≈2650 / 9.0		reac H_2O
2086	Thorium chloride / $ThCl_4$	10026-08-1 / 373.849	gray-wh tetr needles; hyg	770 / 4.59	921 / a,c,d	s H_2O, EtOH
2087	Thorium fluoride / ThF_4	13709-59-6 / 308.032	wh monocl cry; hyg	1110 / 6.1	1680 / a,d	
2088	Thorium iodide / ThI_4	7790-49-0 / 739.656	wh-yel monocl cry	570	837	
2089	Thorium nitrate tetrahydrate / $Th(NO_3)_4 \cdot 4H_2O$	33088-16-3 / 552.119	wh hyg cry	dec 500		vs H_2O; s EtOH

No.	Name Formula	CAS RN Mol. Wt.	Physical Form	mp/°C den/g cm^{-3}	bp/°C Other Data	Solubility
2090	Thorium nitride ThN	12033-65-7 246.045	refrac cub cry	2820 11.6		reac H_2O
2091	Thorium oxide ThO_2	1314-20-1 264.037	wh cub cry	3390 10.0	4400 a,f	i H_2O, alk; sl acid
2092	Thorium selenide $ThSe_2$	60763-24-8 389.96	orth cry	8.5		
2093	Thorium orthosilicate $ThSiO_4$	14553-44-7 324.122	brn tetr cry	6.7	f	
2094	Thorium silicide $ThSi_2$	12067-54-8 288.209	tetr cry	1850 7.9		
2095	Thorium sulfate nonahydrate $Th(SO_4)_2 \cdot 9H_2O$	10381-37-0 586.303	wh monocl cry	dec 2.8		s H_2O
2096	Thorium sulfide ThS_2	12138-07-7 296.170	dark brn cry	1905 7.30		i H_2O; s acid
2097	Thulium Tm	7440-30-4 168.934	silv metal; hex	1545 9.32	1950 a,c	s dil acid
2098	Thulium bromide $TmBr_3$	14456-51-0 408.646	wh hyg cry	954		s H_2O
2099	Thulium chloride $TmCl_3$	13537-18-3 275.292	yel hyg cry	824	a	s H_2O
2100	Thulium chloride heptahydrate $TmCl_3 \cdot 7H_2O$	13778-39-7 401.399	hyg cry			s H_2O, EtOH
2101	Thulium fluoride TmF_3	13760-79-7 225.929	wh cry	1158		s H_2O
2102	Thulium iodide TmI_3	13813-43-9 549.647	yel hyg cry	1021		
2103	Thulium oxide Tm_2O_3	12036-44-1 385.866	grn-wh cub cry	2425 8.6	a	sl acid
2104	Tin (gray) Sn	7440-31-5 118.710	cub cry	trans to wh Sn 13.2 5.769	2602 a,c,e	
2105	Tin (white) Sn	7440-31-5 118.710	silv tetr cry	231.93 7.265	2602 a,c,e	
2106	Stannane SnH_4	2406-52-2 122.742	unstable col gas	-146 5.386 g/L	-51.8 a,d,e	
2107	Methylstannane SnH_3CH_3	1631-78-3 136.769	col gas	6.001 g/L	0	reac H_2O
2108	Tin(II) acetate $Sn(C_2H_3O_2)_2$	638-39-1 236.800	wh orth cry	183 2.31	subl	s dil HCl
2109	Tin(II) bromide $SnBr_2$	10031-24-0 278.518	yel powder	215 5.12	639 a,d	s H_2O, EtOH, eth, ace
2110	Tin(II) chloride $SnCl_2$	7772-99-8 189.615	wh orth cry	247 3.90	623 a,c,d,e	s H_2O, EtOH, ace, eth; i xyl
2111	Tin(II) chloride dihydrate $SnCl_2 \cdot 2H_2O$	10025-69-1 225.646	wh monocl cry	37 dec 2.71		s H_2O, EtOH, NaOH; vs HCl
2112	Tin(II) fluoride SnF_2	7783-47-3 156.707	wh monocl cry; hyg	213 4.57	850	s H_2O; i EtOH, eth, chl
2113	Tin(II) hexafluorozirconate $SnZrF_6$	12419-43-1 323.924	cry	4.21		s H_2O
2114	Tin(II) hydroxide $Sn(OH)_2$	12026-24-3 152.725	wh amorp solid		a	
2115	Tin(II) iodide SnI_2	10294-70-9 372.519	red-oran powder	320 5.28	714 a,d	sl H_2O; s bz, chl, CS_2
2116	Tin(II) oxalate SnC_2O_4	814-94-8 206.730	wh powder	280 dec 3.56		i H_2O; s dil HCl
2117	Tin(II) oxide SnO	21651-19-4 134.709	blue-blk tetr cry	1080 dec 6.45	a	i H_2O, EtOH; s acid
2118	Tin(II) pyrophosphate $Sn_2P_2O_7$	15578-26-4 411.363	wh amorp powder	dec 400 4.009		i H_2O; s conc acid
2119	Tin(II) selenide SnSe	1315-06-6 197.67	gray orth cry	861 6.18		i H_2O; s aqua regia
2120	Tin(II) sulfate $SnSO_4$	7488-55-3 214.774	wh orth cry	dec 378 4.15		reac H_2O
2121	Tin(II) sulfide SnS	1314-95-0 150.776	gray orth cry	880 5.08	1210 a	i H_2O; s conc acid
2122	Tin(II) tartrate $SnC_4H_4O_6$	815-85-0 266.782	wh cry powder			s H_2), dil HCl
2123	Tin(II) telluride SnTe	12040-02-7 246.31	gray cub cry	790 6.5		
2124	Tin(IV) bromide $SnBr_4$	7789-67-5 438.326	wh cry	31 3.34	205 a,b,c,d,e	vs H_2O; s EtOH
2125	Tin(IV) chloride $SnCl_4$	7646-78-8 260.521	col fuming liq	-33 2.234	114.15 a,b,c,d,e	reac H_2O; s EtOH, ctc, bz, ace
2126	Tin(IV) chloride pentahydrate $SnCl_4 \cdot 5H_2O$	10026-06-9 350.597	wh-yel cry	dec 56 2.04		vs H_2O; s EtOH
2127	Tin(IV) chromate $Sn(CrO_4)_2$	38455-77-5 350.697	brn-yel cry powder	dec		s H_2O
2128	Tin(IV) fluoride SnF_4	7783-62-2 194.704	wh tetr cry	4.78	subl 705	reac H_2O

No.	Name Formula	CAS RN Mol. Wt.	Physical Form	mp/°C den/g cm^{-3}	bp/°C Other Data	Solubility
2129	Tin(IV) iodide SnI$_4$	7790-47-8 626.328	yel-brn cub cry	143 4.46	364.35 a,b,d,e	reac H$_2$O; s EtOH, bz, chl, eth
2130	Tin(IV) oxide SnO$_2$	18282-10-5 150.709	gray tetr cry	1630 6.85	a,f	i H$_2$O, EtOH; s hot conc alk
2131	Tin(IV) selenide SnSe$_2$	20770-09-6 276.63	red-brn cry	650 ≈5.0		i H$_2$O; s alk, conc acid
2132	Tin(IV) selenite Sn(SeO$_3$)$_2$	7446-25-5 372.63	cry powder			i H$_2$O; s hot HCl
2133	Tin(IV) sulfide SnS$_2$	1315-01-1 182.842	gold-yel hex cry	dec 600 4.5		i H$_2$O; s alk, aqua regia
2134	Titanium Ti	7440-32-6 47.867	gray metal; hex	1668 4.506	3287 a,c	
2135	Titanium hydride TiH$_2$	7704-98-5 49.883	gray-blk powder	dec ≈450 3.75		i H$_2$O
2136	Titanium boride TiB$_2$	12045-63-5 69.489	gray refrac solid; hex	3225 4.38		
2137	Titanium carbide TiC	12070-08-5 59.878	cub cry	3067 4.93		i H$_2$O; s HNO$_3$
2138	Titanium nitride TiN	25583-20-4 61.874	yel-brn cub cry	3290 5.43		i H$_2$O; s aqua regia
2139	Titanium phosphide TiP	12037-65-9 78.841	gray hex cry	1990 4.08		
2140	Titanium silicide TiSi$_2$	12039-83-7 104.038	blk orth cry	1500 4.0		i H$_2$O, acid, alk; s HF
2141	Titanium(II) bromide TiBr$_2$	13783-04-5 207.675	blk powder	4.0	a	reac H$_2$O
2142	Titanium(II) chloride TiCl$_2$	10049-06-6 118.772	blk hex cryc	1035 3.13	1500 a,d	reac H$_2$O; s EtOH; i chl, eth
2143	Titanium(II) iodide TiI$_2$	13783-07-8 301.676	blk hex cry	5.02		reac H$_2$O
2144	Titanium(II) oxide TiO	12137-20-1 63.866	cub cry	1750 4.95	a	
2145	Titanium(II) sulfide TiS	12039-07-5 79.933	brn hex cry	1780 3.85		s conc acid
2146	Titanium(III) bromide TiBr$_3$	13135-31-4 287.579	blue-blk hex cry		a	s H$_2$O
2147	Titanium(III) chloride TiCl$_3$	7705-07-9 154.225	red-viol hex cry; hyg	425 dec 2.64	a,d	reac H$_2$O
2148	Titanium(III) fluoride TiF$_3$	13470-08-1 104.862	viol hex cry	1200 2.98	1400	i H$_2$O, dil acid, alk
2149	Titanium(III) oxide Ti$_2$O$_3$	1344-54-3 143.732	viol hex cry	1842 4.486	a	s hot HF
2150	Titanium(III) sulfate Ti$_2$(SO$_4$)$_3$	10343-61-0 383.925	grn cry			i H$_2$O, EtOH; s dil HCl
2151	Titanium(III) sulfide Ti$_2$S$_3$	12039-16-6 191.932	blk hex cry	3.56		
2152	Titanium(III,IV) oxide Ti$_3$O$_5$	12065-65-5 223.598	blk monocl cry	1777 4.24	a	
2153	Titanium(IV) bromide TiBr$_4$	7789-68-6 367.483	yel-oran cub cry; hyg	39 3.37	230 a,b,c,d	reac H$_2$O
2154	Titanium(IV) chloride TiCl$_4$	7550-45-0 189.678	col or yel liq	-25 1.73	136.45 a,b,c,d,e	reac H$_2$O; s EtOH
2155	Titanium(IV) fluoride TiF$_4$	7783-63-3 123.861	wh hyg powder	284 2.798	subl	reac H$_2$O; s EtOH, py
2156	Titanium(IV) iodide TiI$_4$	7720-83-4 555.485	red hyg powder	150 4.3	377 a,b,c,d	reac H$_2$O
2157	Titanium(IV) oxide TiO$_2$	13463-67-7 79.866	wh tetr cry	1843 4.23	a,f	i H$_2$O, dil acid; s conc acid
2158	Titanium(IV) oxysulfate monohydrate TiOSO$_4$·H$_2$O	13825-74-6* 177.945	col orth cry	2.71		reac H$_2$O
2159	Titanium(IV) sulfide TiS$_2$	12039-13-3 111.999	yel-brn hex cry; hyg	3.37		s H$_2$SO$_4$
2160	Tungsten W	7440-33-7 183.84	gray-wh metal; cub	3422 19.3	5555 a,c,e	
2161	Tungstic acid H$_2$WO$_4$	7783-03-1 249.85	yel amorp powder	dec 100 5.5		i H$_2$O, acid; s alk
2162	Tungsten boride W$_2$B	12007-10-2 378.49	refrac blk powder	2670 16.0		i H$_2$O
2163	Tungsten boride WB	12007-09-9 194.65	blk refrac powder	2665 15.2		i H$_2$O
2164	Tungsten boride W$_2$B$_5$	12007-98-6 421.74	refrac solid	2365 11.0		i H$_2$O
2165	Tungsten carbide W$_2$C	12070-13-2 379.69	refrac hex cry	≈2800 14.8		i H$_2$O
2166	Tungsten carbide WC	12070-12-1 195.85	gray hex cry	2785 15.6		i H$_2$O; s HNO$_3$/HF
2167	Tungsten carbonyl W(CO)$_6$	14040-11-0 351.90	wh cry	dec 170 2.65	subl	i H$_2$O; s os

No.	Name Formula	CAS RN Mol. Wt.	Physical Form	mp/°C den/g cm⁻³	bp/°C Other Data	Solubility
2168	Tungsten nitride WN₂	60922-26-1 211.85	hex cry	dec 600 7.7		
2169	Tungsten nitride W₂N	12033-72-6 381.69	gray cub cry	dec 17.8		
2170	Tungsten silicide WSi₂	12039-88-2 240.01	blue-gray tetr cry	2160 9.3		i H₂O
2171	Tungsten silicide W₅Si₃	12039-95-1 1003.46	blue-gray refrac solid	2320 14.4		
2172	Tungsten(II) bromide WBr₂	13470-10-5 343.65	yel powder	dec 400		
2173	Tungsten(II) chloride WCl₂	13470-12-7 254.75	yel solid	dec >500		s H₂O
2174	Tungsten(II) iodide WI₂	13470-17-2 437.65	oran cry	6.79		
2175	Tungsten(III) bromide WBr₃	15163-24-3 423.55	blk hex cry	dec >80		i H₂O
2176	Tungsten(III) chloride WCl₃	20193-56-0 290.20	red solid	550 dec	subl	reac H₂O
2177	Tungsten(IV) bromide WBr₄	14055-81-3 503.46	blk orth cry			reac H₂O
2178	Tungsten(IV) chloride WCl₄	13470-13-8 325.65	blk hyg powder	4.62		reac H₂O
2179	Tungsten(IV) fluoride WF₄	13766-47-7 259.83	red-brn cry	dec >800		
2180	Tungsten(IV) iodide WI₄	14055-84-6 691.46	blk powder	dec		reac H₂O; s EtOH; i eth chl
2181	Tungsten(IV) oxide WO₂	12036-22-5 215.84	blue monocl cry	dec 1500-1700 10.8	a	i H₂O, os
2182	Tungsten(IV) selenide WSe₂	12067-46-8 341.76	gray hex cry	9.2		
2183	Tungsten(IV) sulfide WS₂	12138-09-9 247.97	gray hex cry	1250 dec 7.6		i H₂O, HCl, alk
2184	Tungsten(IV) telluride WTe₂	12067-76-4 439.04	gray orth cry	1020 9.43		
2185	Tungsten(V) bromide WBr₅	13470-11-6 583.36	brn-blk hyg solid	286	333	
2186	Tungsten(V) chloride WCl₅	13470-14-9 361.10	blk hyg cry	242	286	reac H₂O
2187	Tungsten(V) fluoride WF₅	19357-83-6 278.83	yel solid	dec >80		
2188	Tungsten(V) oxytribromide WOBr₃	20213-56-3 439.55	dark brn tetr cry	≈5.9		
2189	Tungsten(V) oxytrichloride WOCl₃	14249-98-0 306.20	grn tetr cry	≈4.6		
2190	Tungsten(VI) bromide WBr₆	13701-86-5 663.26	blue-blk cry	309	a	
2191	Tungsten(VI) chloride WCl₆	13283-01-7 396.56	purp hex cry; hyg	275 3.52	346.75 a,b,c,d	s EtOH, os
2192	Tungsten(VI) dioxydibromide WO₂Br₂	13520-75-7 375.65	red cry			
2193	Tungsten(VI) dioxydichloride WO₂Cl₂	13520-76-8 286.74	yel orth cry	265 4.67		i H₂O
2194	Tungsten(VI) dioxydiiodide WO₂I₂	14447-89-3 469.65	monocl cry	6.39		
2195	Tungsten(VI) fluoride WF₆	7783-82-6 297.83	col gas	2.3 13.069 g/L	17 a,b,c,d,e	reac H₂O
2196	Tungsten(VI) oxide WO₃	1314-35-8 231.84	yel powder	1472 7.2	a,c	i H₂O; sl acid; s alk
2197	Tungsten(VI) oxytetrabromide WOBr₄	13520-77-9 519.46	red tetr cry	277 ≈5.5	327	reac H₂O
2198	Tungsten(VI) oxytetrachloride WOCl₄	13520-78-0 341.65	red hyg cry	211 11.92	227.55 b,c,d	reac H₂O; s bz, CS₂
2199	Tungsten(VI) oxytetrafluoride WOF₄	13520-79-1 275.83	wh monocl cry	106 5.07	186	reac H₂O
2200	Tungsten(VI) sulfide WS₃	12125-19-8 280.04	brn powder			sl H₂O; s alk
2201	Uranium U	7440-61-1 238.029	silv-wh orth cry	1135 19.1	4131 a,c	
2202	Uranium boride UB₂	12007-36-2 259.651	refrac solid	2430 12.7		
2203	Uranium boride UB₄	12007-84-0 281.273	refrac solid	2530 9.32		i H₂O
2204	Uranium carbide UC	12070-09-6 250.040	gray cub cry	2790		
2205	Uranium carbide UC₂	12071-33-9 262.051	gray tetr cry	2350 11.3	4370	reac H₂O; sl EtOH
2206	Uranium carbide U₂C₃	12076-62-9 512.091	gray cub cry	dec ≈1700 12.7		

No.	Name Formula	CAS RN Mol. Wt.	Physical Form	mp/°C den/g cm^{-3}	bp/°C Other Data	Solubility
2207	Uranium nitride UN	25658-43-9 252.036	gray cub cry	2805 14.3		i H_2O
2208	Uranium nitride U_2N_3	12033-83-9 518.078	cub cry	dec 11.3		
2209	Uranium(III) bromide UBr_3	13470-19-4 477.741	red hyg cry	727		s H_2O
2210	Uranium(III) chloride UCl_3	10025-93-1 344.387	grn hyg cry	837 5.51	a	vs H_2O; i bz, ctc
2211	Uranium(III) fluoride UF_3	13775-06-9 295.024	blk hex cry	dec 8.9	a	i H_2O; s acid
2212	Uranium(III) hydride UH_3	13598-56-6 241.053	gray-blk cub cry	11.1	a	
2213	Uranium(III) iodide UI_3	13775-18-3 618.742	blk hyg cry	766		s H_2O
2214	Uranium(IV) bromide UBr_4	13470-20-7 557.645	brn hyg cry	519		s H_2O, EtOH
2215	Uranium(IV) chloride UCl_4	10026-10-5 379.840	grn octahed cry	590 4.72	791 a,c	reac H_2O; s EtOH
2216	Uranium(IV) fluoride UF_4	10049-14-6 314.023	grn monocl cry	1036 6.7	1417 a	i H_2O; s conc acid, alk
2217	Uranium(IV) iodide UI_4	13470-22-9 745.647	blk hyg cry	506		s H_2O, EtOH
2218	Uranium(IV) oxide UO_2	1344-57-6 270.028	brn cub cry	2827 10.97	a,f	i H_2O, dil acid; s conc acid
2219	Uranium(IV,V) oxide U_4O_9	12037-15-9 1096.111	cub cry	11.2	a	
2220	Uranium(V) bromide UBr_5	13775-16-1 637.549	brn hyg cry			reac H_2O
2221	Uranium(V) chloride UCl_5	13470-21-8 415.293	brn hyg cry	287		reac H_2O
2222	Uranium(V) fluoride UF_5	13775-07-0 333.021	pale blue tetr cry; hyg	348 5.81		s H_2O
2223	Uranium(V,VI) oxide U_3O_8	1344-59-8 842.082	grn-blk orth cry	1300 dec 8.38	a	
2224	Uranium(VI) chloride UCl_6	13763-23-0 450.745	green hex cry	177 3.6	a	
2225	Uranium(VI) fluoride UF_6	7783-81-5 352.019	wh monocl solid	64.0 tp 5.09	56.5 sp a,b,c,e	reac H_2O; s ctc, chl
2226	Uranium(VI) oxide UO_3	1344-58-7 286.027	oran-yel cry	≈7.3	a	i H_2O; s acid
2227	Uranium(VI) oxide monohydrate $UO_3 \cdot H_2O$	12326-21-5 304.043	yel orth cry	dec 570 7.05		
2228	Uranyl chloride UO_2Cl_2	7791-26-6 340.933	yel orth cry; hyg	577		vs H_2O; s EtOH, ace; i bz
2229	Uranyl nitrate hexahydrate $UO_2(NO_3)_2 \cdot 6H_2O$	13520-83-7 502.129	yel orth cry; hyg	60 2.81	dec 118	vs H_2O; s EtOH, eth
2230	Uranyl sulfate trihydrate $UO_2SO_4 \cdot 3H_2O$	20910-28-5 420.138	yel cry	3.28		s H_2O, sl EtOH
2231	Vanadium V	7440-62-2 50.942	gray-wh metal; cub	1910 6.0	3407 a,c	i H_2O; s acid
2232	Vanadium boride VB	12045-27-1 61.753	refrac solid	2250		i H_2O
2233	Vanadium boride VB_2	12007-37-3 72.564	refrac solid	2450		
2234	Vanadium carbide VC	12070-10-9 62.953	refrac blk cry; cub	2810 5.77		i H_2O
2235	Vanadium carbide V_2C	12012-17-8 113.894	hex cry	2167		
2236	Vanadium carbonyl $V(CO)_6$	20644-87-5 219.004	blue-grn cry; flam	dec 60	subl	
2237	Vanadium nitride VN	24646-85-3 64.949	blk powder; cub	2050 6.13		i H_2O; s aqua regia
2238	Vanadium silicide VSi_2	12039-87-1 107.113	metallic prisms	4.42		s HF
2239	Vanadium silicide V_3Si	12039-76-8 180.911	cub cry	1935 5.70		
2240	Vanadium(II) bromide VBr_2	14890-41-6 210.750	oran-brn hex cry	4.58	subl 800	reac H_2O
2241	Vanadium(II) chloride VCl_2	10580-52-6 121.847	grn hex plates	3.23	subl 910	reac H_2O; s EtOH, eth
2242	Vanadium(II) fluoride VF_2	13842-80-3 88.939	blue hyg cry			reac H_2O
2243	Vanadium(II) iodide VI_2	15513-84-5 304.751	red-viol hex cry	5.44		reac H_2O
2244	Vanadium(II) oxide VO	12035-98-2 66.941	grn cry	1790 5.758	a,c	s acid
2245	Vanadium(II) sulfate heptahydrate $VSO_4 \cdot 7H_2O$	36907-42-3 273.112	viol cry			

No.	Name Formula	CAS RN Mol. Wt.	Physical Form	mp/°C den/g cm^{-3}	bp/°C Other Data	Solubility
2246	Vanadium(III) acetylacetonate $V(CHCOCHCOCH_3)_3$	13476-99-8 348.270	brn cry	≈185 ≈1.0	subl	s MeOH, ace, bz chl
2247	Vanadium(III) bromide VBr_3	13470-26-3 290.654	gray-brn hyg cry	 4.00		reac H_2O
2248	Vanadium(III) chloride VCl_3	7718-98-1 157.300	red-viol hex cry; hyg	 3.00	a	reac H_2O; s EtOH, eth
2249	Vanadium(III) fluoride VF_3	10049-12-4 107.937	yel-grn hex cry	≈1400 3.363	subl	i H_2O, EtOH
2250	Vanadium(III) fluoride trihydrate $VF_3 \cdot 3H_2O$	10049-12-4* 161.983	grn rhomb cry	dec ≈100		sl H_2O
2251	Vanadium(III) iodide VI_3	15513-94-7 431.655	brn-blk rhomb cry; hyg	 5.21		reac H_2O
2252	Vanadium(III) oxide V_2O_3	1314-34-7 149.881	blk powder	2067 4.87	a	i H_2O
2253	Vanadium(III) sulfate $V_2(SO_4)_3$	13701-70-7 390.074	yel powder	dec ≈400		sl H_2O
2254	Vanadium(III) sulfide V_2S_3	1315-03-3 198.081	grn-blk powder	dec 4.7		i H_2O, s hot HCl
2255	Vanadium(IV) bromide VBr_4	13595-30-7 370.558	unstable magenta cry	dec -23	a	
2256	Vanadium(IV) chloride VCl_4	7632-51-1 192.753	unstable red liq	-25.7 1.816	148 a,c,d	reac H_2O; s EtOH, eth
2257	Vanadium(IV) fluoride VF_4	10049-16-8 126.936	grn hyg powder	dec 325 3.15	subl a	vs H_2O
2258	Vanadium(IV) oxide VO_2	12036-21-4 82.941	blue-blk powder	1967 4.339		i H_2O; s acid, alk
2259	Vanadium(V) fluoride VF_5	7783-72-4 145.934	col liq	19.5 2.50	48.3 a,c,d	reac H_2O
2260	Vanadium(V) oxide V_2O_5	1314-62-1 181.880	yel-brn orth cry	670 3.35	1800 dec a,c	sl H_2O; s conc acid, alk; i EtOH
2261	Vanadyl bromide VOBr	13520-88-2 146.845	viol cry	dec 480		
2262	Vanadyl chloride VOCl	13520-87-1 102.394	brn orth cry	dec >700 1.72	a,b	
2263	Vanadyl dibromide $VOBr_2$	13520-89-3 226.749	yel-brn cry	dec 180		
2264	Vanadyl dichloride $VOCl_2$	10213-09-9 137.846	grn hyg cry	dec 380 2.88		reac H_2O; s EtOH
2265	Vanadyl difluoride VOF_2	13814-83-0 104.938	yel cry			
2266	Vanadyl selenite hydrate $VOSeO_3 \cdot H_2O$	133578-89-9 211.92	grn tricl plates	 3.506		
2267	Vanadyl sulfate dihydrate $VOSO_4 \cdot 2H_2O$	27774-13-6 199.036	blue cry powder			s H_2O
2268	Vanadyl tribromide $VOBr_3$	13520-90-6 306.653	deep red liq		dec 180	reac H_2O
2269	Vanadyl trichloride $VOCl_3$	7727-18-6 173.299	fuming red liq	-79 1.829	127 a,d,e	reac H_2O; s MeOH, eth, ace
2270	Vanadyl trifluoride VOF_3	13709-31-4 123.936	yel hyg powder	300 2.459	480	reac H_2O
2271	Water H_2O	7732-18-5 18.015	col liq	0.00 0.9970	100.0 a,b,c,d,e	s EtOH, MeOH, ace
2272	Xenon Xe	7440-63-3 131.29	col gas	-111.75 5.761 g/L	-108.04 a,b,c,d,e	sl H_2O
2273	Xenon trioxide XeO_3	13776-58-4 179.29	col orth cry	exp ≈25 4.55		s H_2O
2274	Xenon tetroxide XeO_4	12340-14-6 195.29	yel solid; exp	-35.9	dec ≈0	
2275	Xenon difluoride XeF_2	13709-36-9 169.29	col tetr cry	129.03 tp 4.32	114.35 sp b	sl H_2O
2276	Xenon tetrafluoride XeF_4	13709-61-0 207.28	col monocl cry	117.10 tp 4.04	115.75 sp a,b	reac H_2O
2277	Xenon hexafluoride XeF_6	13693-09-9 245.28	col monocl cry	49.5 3.56	75.6	reac H_2O
2278	Xenon dioxydifluoride XeO_2F_2	13875-06-4 201.29	col orth cry	30.8 exp 4.10		
2279	Xenon oxytetrafluoride $XeOF_4$	13774-85-1 223.28	col liq	-46.2 3.17(0°C)		reac H_2O
2280	Xenon fluoride hexafluororuthenate $XeFRuF_6$	22527-13-5 365.35	yel-grn monocl cry	110 3.78		
2281	Xenon fluoride undecafluoroantimonate $XeFSb_2F_{11}$	15364-10-0 602.79	yel monocl cry	63 3.69		
2282	Xenon fluoride hexafluoroarsenate $Xe_2F_3AsF_6$	50432-32-1 508.49	yel-grn monocl cry	99 3.62		reac H_2O
2283	Xenon fluoride hexafluoroantimonate XeF_3SbF_6	39797-63-2 424.04	yel-grn monocl cry	≈110 3.92		
2284	Xenon trifluoride undecafluoroantimonate $XeF_3Sb_2F_{11}$	35718-37-7 640.79	yel-grn tricl cry	82 3.98		

No.	Name Formula	CAS RN Mol. Wt.	Physical Form	mp/°C den/g cm^{-3}	bp/°C Other Data	Solubility
2285	Xenon pentafluoride hexafluoroarsenate XeF$_5$AsF$_6$	20328-94-3 415.19	wh monocl cry	130.5 3.51		
2286	Xenon pentafluoride hexafluororuthenate XeF$_5$RuF$_6$	39796-98-0 441.34	grn orth cry	152 3.79		
2287	Ytterbium Yb	7440-64-4 173.04	silv metal; cub	819 6.90	1196 a,c	s dil acid
2288	Ytterbium silicide YbSi$_2$	12039-89-3 229.21	hex cry	7.54		
2289	Ytterbium(II) bromide YbBr$_2$	25502-05-0 332.85	yel cry	673		reac H$_2$O
2290	Ytterbium(II) chloride YbCl$_2$	13874-77-6 243.95	grn cry	721 5.27		reac H$_2$O
2291	Ytterbium(II) iodide YbI$_2$	19357-86-9 426.85	blk cry	772		reac H$_2$O
2292	Ytterbium(III) chloride YbCl$_3$	10361-91-8 279.40	wh hyg powder	875	a	s H$_2$O
2293	Ytterbium(III) chloride hexahydrate YbCl$_3$·6H$_2$O	19423-87-1 387.49	grn hyg cry	dec 150 2.57		vs H$_2$O
2294	Ytterbium(III) fluoride YbF$_3$	13760-80-0 230.04	wh cry	1157 8.2		i H$_2$O
2295	Ytterbium(III) oxide Yb$_2$O$_3$	1314-37-0 394.08	col cub cry	2435 9.2	a	s dil acid
2296	Ytterbium(III) sulfate octahydrate Yb$_2$(SO$_4$)$_3$·8H$_2$O	10034-98-7 778.39	col cry	3.3		s H$_2$O
2297	Yttrium Y	7440-65-5 88.906	silv metal; hex	1522 4.47	3345 a,c	reac H$_2$O; s dil acid
2298	Yttrium aluminum oxide Y$_3$Al$_5$O$_{12}$	12005-21-9 593.619	grn cub cry	≈4.5		
2299	Yttrium antimonide YSb	12186-97-9 210.666	cub cry	2310 5.97		
2300	Yttrium arsenide YAs	12255-48-0 163.828	cub cry	5.59		
2301	Yttrium boride YB$_6$	12008-32-1 153.772	refrac solid	2600 3.72		
2302	Yttrium bromide YBr$_3$	13469-98-2 328.618	col hyg cry	904		vs H$_2$O
2303	Yttrium carbide YC$_2$	12071-35-1 112.928	refrac solid	≈2400 4.13		
2304	Yttrium carbonate trihydrate Y$_2$(CO$_3$)$_3$·3H$_2$O	5970-44-5 411.886	red-brn powder			i H$_2$O; s dil acid
2305	Yttrium chloride YCl$_3$	10361-92-9 195.264	wh monocl cry; hyg	721 2.61	a	vs H$_2$O
2306	Yttrium fluoride YF$_3$	13709-49-4 145.901	wh hyg powder	≈1150 4.0	a	i H$_2$O
2307	Yttrium nitrate hexahydrate Y(NO$_3$)$_3$·6H$_2$O	13494-98-9 383.012	hyg cry			vs H$_2$O
2308	Yttrium oxide Y$_2$O$_3$	1314-36-9 225.810	wh cry; cub	2439 5.03	a,c	s dil acid
2309	Yttrium phosphide YP	12294-01-8 119.880	cub cry	≈4.4		
2310	Yttrium sulfate octahydrate Y$_2$(SO$_4$)$_3$·8H$_2$O	7446-33-5 610.125	red monocl cry	2.6		s H$_2$O
2311	Yttrium sulfide Y$_2$S$_3$	12039-19-9 274.010	yel cub cry	1925 3.87		
2312	Zinc Zn	7440-66-6 65.39	blue-wh metal; hex	419.53 7.14	907 a,c,e	s acid, alk
2313	Zinc acetate dihydrate Zn(C$_2$H$_3$O$_2$)$_2$·2H$_2$O	5970-45-6 219.51	wh powder	237 dec 1.735		vs H$_2$O; s EtOH
2314	Zinc antimonide ZnSb	12039-35-9 187.15	silv-wh orth cry	565 6.33		reac H$_2$O
2315	Zinc arsenate Zn$_3$(AsO$_4$)$_2$	13464-44-3 474.01	wh powder			i H$_2$O; s acid, alk
2316	Zinc arsenate octahydrate Zn$_3$(AsO$_4$)$_2$·8H$_2$O	13464-45-4 618.13	wh monocl cry	3.33		i H$_2$O; s acid, alk
2317	Zinc arsenite Zn(AsO$_2$)$_2$	10326-24-6 279.23	col powder			i H$_2$O; s acid
2318	Zinc borate 3ZnO·2B$_2$O$_3$	27043-84-1 383.41	wh amorp powder	3.64		sl H$_2$O; s dil acid
2319	Zinc borate hemiheptahydrate 2ZnO·3B$_2$O$_3$·3.5H$_2$O	12513-27-8 434.69	wh cry	980 4.22		i H$_2$O
2320	Zinc bromate hexahydrate Zn(BrO$_3$)$_2$·6H$_2$O	13517-27-6 429.29	wh hyg solid	100 2.57		vs H$_2$O
2321	Zinc bromide ZnBr$_2$	7699-45-8 225.20	wh hex cry; hyg	394 4.5	697 a,c,d	vs H$_2$O, EtOH; s eth
2322	Zinc caprylate Zn(C$_8$H$_{15}$O$_2$)$_2$	557-09-5 351.80	wh hyg cry	136		sl H$_2$O
2323	Zinc carbonate ZnCO$_3$	3486-35-9 125.40	wh rhomb cry	dec 4.4	a,f	i H$_2$O; s dil acid, alk

No.	Name Formula	CAS RN Mol. Wt.	Physical Form	mp/°C den/g cm⁻³	bp/°C Other Data	Solubility
2324	Zinc chlorate $Zn(ClO_3)_2$	10361-95-2 232.29	yel hyg cry	dec 60 2.15		vs H_2O
2325	Zinc chloride $ZnCl_2$	7646-85-7 136.29	wh hyg cry	290 2.907	732 a,d,e	vs H_2O; s EtOH, ace
2326	Zinc chromite $ZnCr_2O_4$	12018-19-8 233.38	grn cub cry	 5.29		
2327	Zinc citrate dihydrate $Zn_3(C_6H_5O_7)_2 \cdot 2H_2O$	546-46-3 610.40	col powder			sl H_2O; s dil acid, alk
2328	Zinc cyanide $Zn(CN)_2 \cdot$	557-21-1 117.43	wh powder	 1.852		i H_2O; reac acid
2329	Zinc fluoride ZnF_2	7783-49-5 103.39	wh tetr needles; hyg	872 4.9	1500 a,d,e	sl H_2O
2330	Zinc fluoride tetrahydrate $ZnF_2 \cdot 4H_2O$	13986-18-0 175.45	wh orth cry	 2.30		sl H_2O
2331	Zinc fluoroborate hexahydrate $Zn(BF_4)_2 \cdot 6H_2O$	27860-83-9 347.09	hex cry	 2.12		vs H_2O; s EtOH
2332	Zinc formate dihydrate $Zn(CHO_2)_2 \cdot 2H_2O$	5970-62-7 191.46	wh cry	 2.207		s H_2O, i EtOH
2333	Zinc hexafluorosilicate hexahydrate $ZnSiF_6 \cdot 6H_2O$	16871-71-9 315.56	wh cry			s H_2O
2334	Zinc hydroxide $Zn(OH)_2$	20427-58-1 99.41	col orth cry	125 dec 3.05	a	sl H_2O
2335	Zinc iodate $Zn(IO_3)_2$	7790-37-6 415.20	wh cry powder			sl H_2O
2336	Zinc iodide ZnI_2	10139-47-6 319.20	wh hyg cry	446 4.74	625 a	vs H_2O; s EtOH, eth
2337	Zinc laurate $Zn(C_{12}H_{23}O_2)_2$	2452-01-9 464.02	wh powder	128		sl H_2O
2338	Zinc molybdate $ZnMoO_4$	13767-32-3 225.33	wh tetr cry	>700 4.3		i H_2O
2339	Zinc nitrate $Zn(NO_3)_2$	7779-88-6 189.40	wh powder		a	vs H_2O
2340	Zinc nitrate hexahydrate $Zn(NO_3)_2 \cdot 6H_2O$	10196-18-6 297.49	col orth cry	36 dec 2.067		vs H_2O, EtOH
2341	Zinc nitride Zn_3N_2	1313-49-1 224.18	blue-gray cub cry			
2342	Zinc nitrite $Zn(NO_2)_2$	10102-02-0 157.40	hyg solid			reac H_2O
2343	Zinc oleate $Zn(C_{18}H_{33}O_2)_2$	557-07-3 628.31	wh powder	70 dec		i H_2O; s EtOH, eth, bz
2344	Zinc oxalate dihydrate $ZnC_2O_4 \cdot 2H_2O$	547-68-2 189.44	wh powder	100 dec 2.56		i H_2O; s dil acid
2345	Zinc oxide ZnO	1314-13-2 81.39	wh powder; hex	1975 5.6	a,c,f	i H_2O; s dil acid
2346	Zinc perchlorate hexahydrate $Zn(ClO_4)_2 \cdot 6H_2O$	10025-64-6 372.38	wh cub cry; hyg	106 dec 2.2		vs H_2O; s EtOH
2347	Zinc permanganate hexahydrate $Zn(MnO_4)_2 \cdot 6H_2O$	23414-72-4 411.35	blk orth cry; hyg	 2.45		s H_2O; reac EtOH
2348	Zinc peroxide ZnO_2	1314-22-3 97.39	yel-wh powder	dec >150 1.57	exp 212	i H_2O; reac acid, EtOH, ace
2349	Zinc phosphate $Zn_3(PO_4)_2$	7779-90-0 386.11	wh monocl cry	900 4.0		i H_2O
2350	Zinc phosphate tetrahydrate $Zn_3(PO_4)_2 \cdot 4H_2O$	7543-51-3 458.17	col orth cry	 3.04	f	i H_2O, EtOH; s dil acid, alk
2351	Zinc phosphide Zn_3P_2	1314-84-7 258.12	gray tetr cry	1160 4.55		i H_2O, EtOH; reac acid; s bz
2352	Zinc pyrophosphate $Zn_2P_2O_7$	7446-26-6 304.72	wh cry powder	 3.75		i H_2O; s dil acid
2353	Zinc selenate pentahydrate $ZnSeO_4 \cdot 5H_2O$	13597-54-1 298.42	tricl cry	dec 50 2.59		s H_2O
2354	Zinc selenide $ZnSe$	1315-09-9 144.35	yel-red cub cry	>1100 5.65	subl a	i H_2O; s dil acid
2355	Zinc orthosilicate Zn_2SiO_4	13597-65-4 222.86	wh hex cry	1509 4.1	a,f	i H_2O, dil acid
2356	Zinc stearate $Zn(C_{18}H_{35}O_2)_2$	557-05-1 632.34	wh powder	130 1.095		i H_2O, EtOH, eth; s bz
2357	Zinc sulfate $ZnSO_4$	7733-02-0 161.45	col orth cry	680 dec 3.8	a	vs H_2O
2358	Zinc sulfate monohydrate $ZnSO_4 \cdot H_2O$	7446-19-7 179.47	wh monocl cry	238 dec 3.20		vs H_2O; i EtOH
2359	Zinc sulfate heptahydrate $ZnSO_4 \cdot 7H_2O$	7446-20-0 287.56	col orth cry	100 dec 1.97	f	vs H_2O; i EtOH
2360	Zinc sulfide (sphalerite) ZnS	1314-98-3 97.46	gray-wh cub cry	1700 4.04	a,f	i H_2O, EtOH; s dil acid
2361	Zinc sulfide (wurtzite) ZnS	1314-98-3 97.46	wh hex cry	1700 4.09	a,f	i H_2O; s dil acid
2362	Zinc sulfite dihydrate $ZnSO_3 \cdot 2H_2O$	123714-57-8 181.49	wh powder	200 dec		i H_2O, EtOH; reac hot H_2O

No.	Name Formula	CAS RN Mol. Wt.	Physical Form	mp/°C den/g cm^{-3}	bp/°C Other Data	Solubility
2363	Zinc telluride ZnTe	1315-11-3 192.99	red cub cry	1239 5.9		i H_2O
2364	Zinc thiocyanate Zn(SCN)$_2$	557-42-6 181.56	wh hyg cry			sl H_2O; s EtOH
2365	Zirconium Zr	7440-67-7 91.224	gray-wh metal; hex	1855 6.52	4409 a,c	s hot conc acid
2366	Zirconium boride ZrB$_2$	12045-64-6 112.846	gray refrac solid; hex	3245 6.17		
2367	Zirconium bromide ZrBr$_4$	13777-25-8 410.840	wh cub cry	450 tp 3.98	360 sp a,b,e	
2368	Zirconium carbide ZrC	12020-14-3 103.235	gray refrac solid; cub	3532 6.73		s HF
2369	Zirconium chloride ZrCl$_4$	10026-11-6 233.035	wh monocl cry; hyg	437 tp 2.80	331 sp a,b,c,e	reac H_2O; s EtOH, eth
2370	Zirconium fluoride ZrF$_4$	7783-64-4 167.218	wh monocl cry	932 tp 4.43	912 sp a,c	sl H_2O
2371	Zirconium hydride ZrH$_2$	7704-99-6 93.240	gray tetr cry	5.6	a	i H_2O
2372	Zirconium hydroxide Zr(OH)$_4$	14475-63-9 159.254	wh amorp powder	dec 3.25		i H_2O; s acid
2373	Zirconium iodide ZrI$_4$	13986-26-0 598.842	oran cub cry; hyg	499 tp 4.85	431 sp a,b,e	vs H_2O
2374	Zirconium nitrate pentahydrate Zr(NO$_3$)$_4$•5H$_2$O	13746-89-9 429.320	wh hyg cry	dec 100		vs H_2O; s EtOH
2375	Zirconium nitride ZrN	25658-42-8 105.231	yel cub cry	2960 7.09		s conc HF; sl dil acid
2376	Zirconium oxide ZrO$_2$	1314-23-4 123.223	wh amorp powder	2710 5.68	a,c,f	i H_2O; sl acid
2377	Zirconium phosphide ZrP$_2$	12037-80-8 153.172	orth cry	≈5.1		
2378	Zirconium orthosilicate ZrSiO$_4$	10101-52-7 183.308	wh tetr cry	dec 1540 4.6	a,f	i H_2O, acid
2379	Zirconium silicide ZrSi$_2$	12039-90-6 147.395	gray powder	1620 4.88		i H_2O, aqua regia; s HF
2380	Zirconium sulfate tetrahydrate Zr(SO$_4$)$_2$•4H$_2$O	7446-31-3 355.413	wh tetr cry	dec 100 2.80		vs H_2O
2381	Zirconium sulfide ZrS$_2$	12039-15-5 155.356	red-brn hex cry	1480 3.82		i H_2O
2382	Zirconyl chloride octahydrate ZrOCl$_2$•8H$_2$O	13520-92-8 322.251	tetr cry	dec 400 1.91		vs H_2O, EtOH

Acanthite	1854
Acetylenogen	425
Actinium tribromide	2
Actinium trichloride	3
Actinium trifluoride	4
Actinium triiodide	5
Adams' catalyst	1539
ADP	84
Alabandite	1283
Alamoisite	1127
Allemontite	163
Altaite	1139
Aluctyl	34
Alum	1561
Alumina	40
Aluminum ammonium chloride	150
Aluminum calcium hydride	476
Aluminum chlorohydroxide	30
Aluminum ethylate	23
Aluminum hexadecanoate	42
Aluminum iron oxide	1010
Aluminum metaphosphate	39
Aluminum orthophosphate	38
Aluminum potassium sulfate	1560
Aluminum sodium oxide	1863
Aluminum subacetate	22
Aluminum tetrahydroborate	14
Aluminum tribromide	16
Aluminum trichloride	20
Aluminum trichloride hexahydrate	21
Aluminum trifluoride	24
Aluminum triiodide	32
Aluminum triiodide hexahydrate	33
Aluminum trioleate	37
Aluminum tripalmitate	42
Aluminum tristearate	48
Aluminum yttrium aluminate	2298
Alunogenite	49
Americium tetrafluoride	60
Americium tribromide	56
Americium trichloride	57
Americium trifluoride	58
Americium triiodide	59
Aminomercuric chloride	1307
Ammonia solution	116
Ammonium alum	8
Ammonium aluminum fluoride	100
Ammonium bicarbonate	107
Ammonium bichromate	82
Ammonium bifluoride	109
Ammonium binoxalate monohydrate	110
Ammonium bisulfate	112
Ammonium bisulfide	113
Ammonium bisulfite	114
Ammonium bitartrate	115
Ammonium carbazoate	139
Ammonium diuranate	157
Ammonium heptamolybdate tetrahydrate	123
Ammonium hexacyanoferrate(II) trihydrate	90
Ammonium hexacyanoferrate(III) trihydrate	89
Ammonium hexadecanoate	130
Ammonium hydrogen malate	66
Ammonium hyposulfite	155
Ammonium molybdophosphate	137
Ammonium nitroprusside	127
Ammonium octadecanoate	143
Ammonium octanoate	69
Ammonium osmium chloride	97
Ammonium persulfate	134
Ammonium phosphate monobasic	84
Ammonium phosphate, dibasic	111
Ammonium platinic chloride	99
Ammonium platinous chloride	151
Ammonium rhodanide	154
Ammonium soap	128
Ammonium sodium phosphate tetrahydrate	1866

Ammonium sulfocarbamate	85
Ammonium tetraborate tetrahydrate	67
Ammonium tungstophosphate	138
Ammonium vanadate	122
Andalusite	46
Anglesite	1135
Anhydrite	470
Antarcticite	432
Antimonous sulfate	172
Antimony pentachloride	176
Antimony pentafluoride	177
Antimony pentasulfide	180
Antimony pentoxide	179
Antimony tetroxide	175
Antimony tribromide	164
Antimony trichloride	165
Antimony trifluoride	166
Antimony triiodide	167
Antimony trioxide	169
Antimony triselenide	171
Antimony trisulfide	173
Antlerite	746
Aragonite	426
Arcanite	1624
Argentite	1854
Arnaudon's green	597
Arsenic antimonide	163
Arsenic hydride	183
Arsenic pentachloride	200
Arsenic pentafluoride	201
Arsenic pentaselenide	203
Arsenic pentasulfide	204
Arsenic pentoxide	202
Arsenic tribromide	191
Arsenic trichloride	192
Arsenic trifluoride	193
Arsenic triiodide	194
Arsenic trioxide	195,196
Arsenic triselenide	197
Arsenic trisulfide	198
Arsenolite	195
Auric bromide	891
Auric chloride	892
Auric cyanide trihydrate	893
Auric fluoride	894
Auric hydroxide	895
Auric iodide	896
Auric oxide	897
Auric selenate	898
Auric selenide	899
Auric sulfide	900
Aurous bromide	886
Aurous chloride	887
Aurous cyanide	888
Aurous iodide	889
Aurous sulfide	890
Avogadrite	1581
Baddeleyite	2376
Barite	270
Barium boride	228
Barium diphosphate	259
Barium disilicate	263
Barium hyposulfate dihydrate	224
Barium hyposulfite	279
Barium hyposulfite monohydrate	280
Barium mercuric iodide	275
Barium metaphosphate	257
Barium metasilicate	264
Barium metatitanate	281
Barium monoxide	252
Barium phosphate, dibasic	231
Barium platinous cyanide tetrahydrate	274
Barium silicofluoride	229
Barium tungsten oxide	282
Barium white	282
Barium zirconium oxide	285

Basic antimony chloride	170
Basic lead acetate	1088
Bayerite	29
Berlinite	38
Beryl	292
Beryllia	307
Beryllium sodium fluoride	1952
Berzelianite	696
Bieberite	663
Bischofite	1208
Bismite	328
Bismuth bromine oxide	329
Bismuth iodide oxide	331
Bismuth pentafluoride	319
Bismuth peroxide	339
Bismuth subcarbonate	314
Bismuth titanium oxide	340
Bismuth tribromide	315
Bismuth trichloride	316
Bismuth trifluoride	318
Bismuth triiodide	322
Bismuthine	320
Bismuthinite	337
Black phosphorus	1489
Blue verdigris	709
Blue vitriol	745
Bobierrite	1227
Boehmite	41
Borax	1948
Borazole	351
Boric oxide	366
Boron tetrachloride	361
Boron tetrafluoride	363
Boron tribromide	358
Boron trichloride	360
Boron trifluoride	362
Boron triiodide	364
Braggite	1531
Breithauptite	1382
Bremen green	713
Bromocyanide	499
Bromogermane	854
Bromophosgene	495
Bromosilane	1782
Bromotrichlorosilane	1783
Bromyrite	1827
Brucite	1214
Brushite	444
Buckminsterfullerene	484,485
Bunsenite	1404
Burnt alum	1560
Cadmium dibromide	384
Cadmium dichloride	388
Cadmium dichloride hemipentahydrate	389
Cadmium difluoride	393
Cadmium diiodide	396
Cadmium metasilicate	397
Cadmoselite	408
Calcite	427
Calcium aluminum hydride	476
Calcium carbimide	435
Calcium dioxide	460
Calcium diphosphate	464
Calcium metasilicate	467
Calcium phosphate, dibasic	443
Calcium phosphate, monobasic, monohydrate	438
Calomel	1300
Carbon oxychloride	496
Carbon sulfotelluride	494
Carbonic acid anhydride	488
Carbonyl selenide	491
Carbonyl sulfide	492
Caro's acid	1987
Cassiopeium	1186
Cassiterite	2130

Caustic baryta	234
Caustic soda	1909
Celestite	1981
Celphos	44
Cerargyrite	1830
Ceria	524
Ceric ammonium nitrate	73
Ceric fluoride	523
Ceric oxide	524
Ceric sulfate tetrahydrate	525
Cerium hexaboride	504
Cerous ammonium sulfate tetrahydrate	72
Cerous bromide	511
Cerous bromide heptahydrate	512
Cerous carbide	513
Cerous carbonate hydrate	514
Cerous chloride	515
Cerous chloride heptahydrate	516
Cerous fluoride	517
Cerous iodide	518
Cerous iodide nonahydrate	519
Cerous nitride	506
Cerous oxide	520
Cerous sulfate octahydrate	521
Cerous sulfide	522
Cerussite	1097
Cervantite	175
Chalcanthite	745
Chalcocite	697
Chalcocyanite	744
Chalcopyrite	723
Chereul's salt	701
Chile saltpeter	1921
China clay	47
Chlorine monofluoride	554
Chlorine tetroxyfluoride	806
Chlorocyanide	500
Chlorogermane	855
Chlorosilane	1784
Chlorosulfuric acid	1988
Chlorotrifluorogermane	856
Chrome alum	599
Chrome alum ammonium	77
Chrome yellow	1102
Chromic acetate hexahydrate	583
Chromic acid	608
Chromic anhydride	608
Chromic bromide	584
Chromic bromide hexahydrate	585,586
Chromic chloride	587
Chromic chloride hexahydrate	588
Chromic fluoride	589
Chromic fluoride trihydrate	590
Chromic hydroxide trihydrate	591
Chromic iodide	592
Chromic nitrate	593
Chromic nitrate nonahydrate	594
Chromic oxide	595
Chromic phosphate	596
Chromic phosphate hemiheptahydrate	597
Chromic phosphate hexahydrate	598
Chromic sulfate	600
Chromic sulfide	601
Chromic telluride	602
Chromite	1017
Chromium diboride	601
Chromium dioxide	605
Chromium trioxide	608
Chromium(VI) oxychloride	609
Chromous acetate monohydrate	574
Chromous bromide	575
Chromous chloride	576
Chromous chloride tetrahydrate	577
Chromous fluoride	578
Chromous oxalate monohydrate	580
Chromous sulfate pentahydrate	581
Chrysoberyl	291
Cianurina	1311
Cinnabar	1327
Claudetite	196
Clausthalite	1131
Clinoenstatite	1233
Cobalt black	676
Cobalt hydroxide oxide	677
Cobalt yellow	678
Cobaltic acetate	670
Cobaltic ammonium phosphate	78
Cobaltic ammonium sulfate hexahydrate	79
Cobaltic fluoride	672
Cobaltic hexafluorosilicate hexahydrate	642
Cobaltic hydroxide	674
Cobaltic nitrate	675
Cobaltic orthotitanate	680
Cobaltic oxide	676
Cobaltic sulfide	679
Cobaltic-cobaltous oxide	669
Cobaltite	612
Cobaltous acetate	623
Cobaltous acetate tetrahydrate	624
Cobaltous arsenate octahydrate	626
Cobaltous bromate hexahydrate	627
Cobaltous bromide	628
Cobaltous bromide hexahydrate	629
Cobaltous chloride	631
Cobaltous chloride dihydrate	632
Cobaltous chloride hexahydrate	633
Cobaltous chromate	634
Cobaltous chromate(III)	635
Cobaltous cyanide	636
Cobaltous cyanide dihydrate	637
Cobaltous dipotassium sulfate hexahydrate	656
Cobaltous ferricyanide	638
Cobaltous fluoride	639
Cobaltous fluoride tetrahydrate	640
Cobaltous formate dihydrate	641
Cobaltous hydroxide	643
Cobaltous iodate	644
Cobaltous iodide	645
Cobaltous iodide hexahydrate	646
Cobaltous metatitanate	647
Cobaltous molybdate	648
Cobaltous nitrate	649
Cobaltous nitrate hexahydrate	650
Cobaltous orthosilicate	660
Cobaltous oxalate	651
Cobaltous oxalate dihydrate	652
Cobaltous oxide	653
Cobaltous perchlorate	654
Cobaltous phosphate octahydrate	655
Cobaltous selenate pentahydrate	657
Cobaltous selenide	658
Cobaltous selenite dihydrate	659
Cobaltous stannate	661
Cobaltous sulfate	662
Cobaltous sulfate monohydrate	664
Cobaltous sulfide	665
Cobaltous telluride	666
Cobaltous thiocyanate trihydrate	667
Cobaltous tungstate	668
Coccinite	1317
Coloradoite	1328
Copper hydroxide sulfate	746
Coquimbite	1061
Corundum	40
Cotunnite	1100
Covellite	747
Cristobalite	1816
Crocoite	1102
Cryolite	1895
Cryptohalite	104
Cupric acetate	702
Cupric acetate monohydrate	704
Cupric acetoarsenite	703
Cupric acetylide	705
Cupric ammonium chloride	80
Cupric arsenate	706
Cupric azide	708
Cupric borate	710
Cupric bromide	711
Cupric butyrate monohydrate	712
Cupric chlorate hexahydrate	714
Cupric chloride	715
Cupric chloride dihydrate	716
Cupric chromate	717
Cupric chromite	718
Cupric cyanide	720
Cupric dichromate dihydrate	721
Cupric ferrocyanide	722
Cupric fluoride	724
Cupric fluoride dihydrate	725
Cupric formate	726
Cupric formate tetrahydrate	727
Cupric hexafluorosilicate tetrahydrate	728
Cupric hydroxide	729
Cupric molybdate	730
Cupric nitrate	731
Cupric nitrate hexahydrate	732
Cupric nitrate trihydrate	733
Cupric oleate	734
Cupric oxalate	735
Cupric oxide	736
Cupric perchlorate	737
Cupric perchlorate hexahydrate	738
Cupric phosphate trihydrate	739
Cupric selenate pentahydrate	740
Cupric selenide	741
Cupric selenite dihydrate	742
Cupric stearate	743
Cupric sulfate	744
Cupric sulfate pentahydrate	745
Cupric sulfide	747
Cupric tartrate trihydrate	748
Cupric tungstate	750
Cupric tungstate dihydrate	751
Cuprite	695
Cuprous acetate	685
Cuprous acetylide	686
Cuprous azide	687
Cuprous bromide	688
Cuprous chloride	689
Cuprous cyanide	690
Cuprous fluoride	691
Cuprous hydride	692
Cuprous iodide	693
Cuprous oxide	695
Cuprous selenide	696
Cuprous sulfide	697
Cuprous sulfite hemihydrate	698
Cuprous telluride	699
Cuprous tetraiodomercurate	694
Cuprous thiocyanate	700
Cyanogas	436
Cyanogran	1883
Decaborane	349
Dehydrite	1223
Detia	44
Diammonium citrate	108
Diammonium hydrogen phosphite monohydrate	136
Diamond	482
Diarsane	184
Diaspore	41
Diborane	342
Diboron tetrachloride	361
Dibromogermane	857
Dibromosilane	1786
Dichlorine hexoxide	552
Dichlorine trioxide	551

Dichlorodifluorogermane	859
Dichlorodifluorosilane	1788
Dichlorogermane	858
Dichlorosilane	1787
Dicobalt octacarbonyl	618
Difluoroazane	1455
Difluorodiazine	1457
Difluorophosphoric acid	1496
Difluorosilane	1789
Digermane	866
Dihydrogen disulfide	948
Diiodine pentoxide	978
Diiodine tetroxide	977
Diiodosilane	1790
Diniobium carbide	1418
Dinitrogen pentoxide	1450
Dinitrogen tetraoxide	1449
Dinitrogen trioxide	1448
Dioxygen difluoride	804
Diphosphine	1491
Diphosphoric acid	1494
Diselenium dibromide	1768
Diselenium dichloride	1769
Disilane	1803
Disiloxane	1806
Disulfur decafluoride	2003
Disulfur dibromide	1994
Disulfur dichloride	1995
Disulfur difluoride	1996, 1997
Dysprosia	764
Enneaborane(15)	348
Enstatite	1233
Epsomite	1239
Erbia	776
Erdmann's salt	671
Eremeyevite	12
Eriochaleite	716
Eskolaite	595
Etard's salt	698
Europia	798
Europic bromide	793
Europic chloride	794
Europic chloride hexahydrate	795
Europic fluoride	796
Europic nitrate hexahydrate	797
Europic sulfate	799
Europic sulfate octahydrate	800
Europous bromide	786
Europous chloride	787
Europous fluoride	788
Europous iodide	789
Fayalite	1030
Ferberite	1038
Ferric acetylacetonate	1041
Ferric alum	88
Ferric ammonium chromate	86
Ferric ammonium oxalate trihydrate	87
Ferric arsenate dihydrate	1042
Ferric basic acetate	1040
Ferric bromide	1043
Ferric chloride	1044
Ferric chloride hexahydrate	1045
Ferric chromate	1046
Ferric citrate pentahydrate	1047
Ferric dichromate	1048
Ferric ferrocyanide	1049
Ferric ferrous oxide	1039
Ferric fluoride	1050
Ferric fluoride trihydrate	1051
Ferric hexacyanoferrate(II)	1049
Ferric hydroxide	1052
Ferric hypophosphite	1059
Ferric metavanadate	1063
Ferric nitrate nonahydrate	1054
Ferric oxalate	1055

Ferric oxide	1056
Ferric oxide hydroxide	1053
Ferric phosphate dihydrate	1057
Ferric pyrophosphate nonahydrate	1058
Ferric sulfate	1060
Ferric sulfate nonahydrate	1061
Ferric thiocyanate sesquihydrate	1062
Ferric vanadate	1063
Ferrous aluminate	1010
Ferrous ammonium sulfate	91
Ferrous bromide	1011
Ferrous bromide hexahydrate	1012
Ferrous carbonate	1013
Ferrous chloride	1014
Ferrous chloride dihydrate	1015
Ferrous chloride tetrahydrate	1016
Ferrous chromite	1017
Ferrous fluoride	1018
Ferrous fluoride tetrahydrate	1019
Ferrous hydroxide	1020
Ferrous iodide	1021
Ferrous iodide tetrahydrate	1022
Ferrous molybdate	1023
Ferrous nitrate	1024
Ferrous nitrate hexahydrate	1025
Ferrous oxalate dihydrate	1026
Ferrous oxide	1027
Ferrous phosphate octahydrate	1028
Ferrous phosphide	1006
Ferrous selenide	1029
Ferrous sulfate	1031
Ferrous sulfate heptahydrate	1033
Ferrous sulfate monohydrate	1032
Ferrous sulfide	1034
Ferrous telluride	1035
Ferrous thiocyanate trihydrate	1036
Ferrous titanate	1037
Fischer's yellow	678
Fluellite	25
Fluorine oxide	803
Fluorite	439
Fluoroboric acid	356
Fluorocyanide	501
Fluorogermane	860
Fluorosilane	1791
Fluorosulfuric acid	1989
Fluorspar	439
Forsterite	1234
Fullerene	484,485
Gadolinia	821
Galena	1136
Gallane	840
Gallium dichloride	831
Gallium tribromide	836
Gallium trichloride	837
Gallium trifluoride	838
Gallium trifluoride trihydrate	839
Gallium triiodide	842
Geikielite	1245
Gerhardite	733
Germanium dibromide	868
Germanium dichloride	869
Germanium difluoride	870
Germanium diiodide	871
Germanium dioxide	880
Germanium monoxide	872
Germanium tetrabromide	876
Germanium tetrachloride	877
Germanium tetrafluoride	878
Germanium tetrahydride	853
Germanium tetraiodide	879
Gibbsite	29
Glauber's salt	1939
Glucinium	288
Goethite	1053
Gold trichloride acid	885

Goslarite	2359
Graphite	483
Greenockite	412
Guanajuatite	334
Gummite	2227
Gunningite	2358
Gypsum	471
Hafnia	910
Hafnium orthosilicate	914
Hafnium tetrabromide	903
Hafnium tetrachloride	905
Hafnium tetrafluoride	906
Hafnium tetraiodide	908
Halite	1877
Hatchett's brown	722
Hausmannite	1289
Heazlewoodite	1413
Hematite	1056
Hercynite	1010
Hessite	1855
Hexaaminecobalt trichloride	673
Hieratite	1587
Holmia	925
Hopeite	2350
Huebnerite	1288
Hydrazine	928
Hydrazoic acid	936
Hydrobromic acid	940
Hydrocerussite	1098
Hydrochloric acid	941
Hydrocyanic acid	942
Hydrofluoric acid	943
Hydrogen azide	936
Hydrogen hexafluorosilicate	1809
Hydrogen tetrabromoaurate pentahydrate	884
Hydrogen tetracarbonylferrate(II)	1005
Hydrogen tetrachloroaurate tetrahydrate	885
Hydroiodic acid	944
Hydrophilite	430
Hydrous gold bromide	884
Hydroxylamine	937
Hypo	1954
Hypochlorous acid	547
Ilmenite	1037
Indium dibromide	958
Indium dichloride	959
Indium sesquioxide	967
Indium tribromide	961
Indium trichloride	962
Indium trifluoride	963
Indium trifluoride trihydrate	964
Indium triiodide	966
Iodine monobromide	980
Iodine monochloride	981
Iodine monofluoride	983
Iodocyanide	502
Iodogermane	861
Iodosilane	1792
Iodyrite	1838
Iridium hexafluoride	997
Iridium tribromide	989
Iridium tribromide tetrahydrate	990
Iridium trichloride	991
Iridium trifluoride	992
Iridium triiodide	993
Iridium trioxide	994
Iron orthosilicate	1030
Jeremejevite	12
Kalinite	1561
Kalium	1556
Kaolin	47
Kernite	1950
Kieserite	1240
Koettigite	2316
Lautarite	447
Lead butyrate	1096

Platinum monoxide	1530
Platinum tetrabromide	1534
Platinum tetrachloride	1535
Platinum tetrafluoride	1537
Platinum tetraiodide	1538
Platinum tribromide	1532
Platinum trichloride	1533
Plattnerite	1153
Plessy's green	597
Plutonium dioxide	1551
Plutonium hexafluoride	1552
Plutonium monoxide	1544
Plutonium sesquioxide	1549
Plutonium tetrafluoride	1550
Plutonium tribromide	1545
Plutonium trichloride	1546
Plutonium trifluoride	1547
Plutonium triiodide	1548
Polonium dioxide	1555
Polonium tetrachloride	1554
Polydymite	1411
Portlandite	445
Potassium bicarbonate	1590
Potassium bichromate	1574
Potassium bifluoride	1591
Potassium biselenite	1594
Potassium bisulfate	1595
Potassium bisulfide	1596
Potassium bisulfite	1598
Potassium bitartrate	1599
Potassium chloroplatinate	1633
Potassium cobalticyanide	1585
Potassium dioxide	1628
Potassium hexacyanoferrate(II) trihydrate	1578
Potassium hexacyanoferrate(III)	1577
Potassium hyposulfite	1637
Potassium monoxide	1612
Potassium orthodiuranate	1642
Potassium peroxydisulfate	1617
Potassium phosphate, dibasic	1592
Potassium pyrosulfite	1605
Potassium sodium tartrate	1933
Potassium trithiocarbonate	1640
Potcrate	1569
Powellite	450
Prussic acid	942
Pucherite	326
Pyrite	1007
Pyrochroite	1266
Pyrolusite	1293
Pyrophenite	1287
Pyrophosphoric acid	1494
Quartz	1813-1816
Quartz	1815
Quicklime	457
Quicksilver	1296
Quilonorm	1155
Raspite	1144
Realgar	190
Red lead	1148
Red lead chromate	1103
Red phosphorus	1488
Retgersite	1408
Rhenium dioxide	1673
Rhenium disulfide	1674
Rhenium ditelluride	1675
Rhenium heptafluoride	1685
Rhenium heptasulfide	1691
Rhenium heptoxide	1686
Rhenium hexachloride	1680
Rhenium hexafluoride	1681
Rhenium pentabromide	1676
Rhenium pentachloride	1677
Rhenium pentafluoride	1678
Rhenium pentoxide	1679

Rhenium tetrachloride	1671
Rhenium tetrafluoride	1672
Rhenium tribromide	1668
Rhenium trichloride	1669
Rhenium triiodide	1670
Rhenium trioxide	1682
Rhodium dioxide	1700
Rhodium hexafluoride	1701
Rhodium trichloride	1695
Rhodium trifluoride	1696
Rhodium triiodide	1697
Rhodium trioxide	1698
Rhodochrosite	1261
Rhodonite	1277
Rochelle salt	1933
Rock salt	1877
Rohrbach's solution	275
Rubidium alum	1704
Rubidium bicarbonate	1713
Rubidium bifluoride	1711
Rubidium bisulfate	1714
Rubidium dioxide	1724
Ruthenium dioxide	1733
Ruthenium hexafluoride	1735
Ruthenium monoxide	1718
Ruthenium pentafluoride	1734
Ruthenium tetrafluoride	1732
Ruthenium tetroxide	1736
Ruthenium tribromide	1728
Ruthenium trichloride	1729
Ruthenium trifluoride	1730
Ruthenium triiodide	1731
Rutile	2157
Sacchite	1262
Sal ammoniac	75
Sal soda	1874
Salt of tartar	1567
Saltpeter	1609
Samaria	1749
Sanbornite	263
Sassolite	352
Scandia	1758
Scheele's green	707
Scheelite	480
Schultenite	1109
Selenium hexasulfide	1778
Selenium tetrasulfide	1779
Selenium(IV) bromide	1770
Selenium(IV) oxide	1766
Sellaite	1209
Senarmontite	168
Siderite	1013
Silane	1781
Silica	1813
Silicon chloride trifluoride	1785
Silicon decahydride	1805
Silicon octahydride	1804
Silicon tetrabromide	1793
Silicon tetrachloride	1794
Silicon tetrafluoride	1795
Silicon tetraiodide	1796
Silicon(II) oxide	1812
Silver difluoride	1859
Sinjarite	431
Skutterudite	615
Smaltite	614
Smite	1868
Smithsonite	2323
Soda ash	1873
Soda lye	1909
Sodamide	1865
Sodium acid pyrophosphate	1888
Sodium alum	1864
Sodium bicarbonate	1899
Sodium bifluoride	1900
Sodium bisulfate	1904

Sodium bisulfate monohydrate	1905
Sodium bisulfide	1906
Sodium bisulfide dihydrate	1907
Sodium bisulfite	1908
Sodium borate	1947
Sodium borofluoride	1893
Sodium dioxide	1945
Sodium metaarsenite	1867
Sodium metabismuthate	1914
Sodium metaperiodate	1928
Sodium metasilicate	1917
Sodium monobasic phosphate	1886
Sodium monoxide	1924
Sodium peroxydisulfate	1931
Sodium phosphate, dibasic	1901
Sodium prussiate yellow	1891
Sodium pyrosulfite	1915
Sodium tetrafluoroborate	1893
Sphalerite	2360
Spherocobaltite	630
Stannane	2106
Stannic bromide	2124
Stannic chloride	2125
Stannic chloride pentahydrate	2126
Stannic chromate	2127
Stannic fluoride	2128
Stannic iodide	2129
Stannic oxide	2130
Stannic selenide	2131
Stannic selenite	2132
Stannic sulfide	2133
Stannous acetate	2108
Stannous bromide	2109
Stannous chloride	2110
Stannous chloride dihydrate	2111
Stannous fluoride	2112
Stannous hexafluorozirconate	2113
Stannous hydroxide	2114
Stannous iodide	2115
Stannous monoxide	2117
Stannous oxalate	2116
Stannous pyrophosphate	2118
Stannous selenide	2119
Stannous sulfate	2120
Stannous sulfide	2121
Stannous tartrate	2122
Stannous telluride	2123
Stibine	162
Stibium	161
Stibnite	173
Stilleite	2354
Stolzite	1143
Strengite	1057
Strontia	1974
Strontianite	1961
Strontium orthosilicate	1979
Sulfamide	2004
Sulfinyl dibromide	2007
Sulfinyl dichloride	2008
Sulfinyl difluoride	2009
Sulfur chloride	1998
Sulfur decafluoride	2003
Sulfur monobromide	1994
Sulfur monochloride	1995
Sulfur monofluoride	1996, 1997
Sycoporite	665
Sylvite	1570
Szmikite	1281
Szomolnokite	1032
Tantalum boride	2014
Tantalum dioxide	2019
Tantalum diselenide	2020
Tantalum disulfide	2021
Tantalum ditelluride	2022
Tantalum pentabromide	2023

Tantalum pentachloride	2024
Tantalum pentafluoride	2025
Tantalum pentaiodide	2026
Tantalum pentoxide	2027
Tarapacaite	1571
Technetium hexafluoride	2030
Technetium pentafluoride	2029
Tellurite	2034
Tellurous bromide	2036
Tellurous chloride	2037
Tenorite	736
Tephroite	1277
Tephroite	1278
Terbia	2048
Tetrabromosilane	1793
Tetrachlorodiborane(4)	361
Tetrachlorosilane	1794
Tetracobalt dodecacarbonyl	622
Tetradymite	338
Tetrafluorosilane	1795
Tetraiodine nonaoxide	979
Tetraiodosilane	1796
Tetrasilane	1805
Thallic bromide tetrahydrate	2074
Thallic chloride	2075
Thallic chloride tetrahydrate	2076
Thallic fluoride	2077
Thallic nitrate	2078
Thallic oxide	2079
Thallium sesquioxide	2079
Thallous acetate	2052
Thallous bromide	2053
Thallous carbonate	2054
Thallous chlorate	2055
Thallous chloride	2056
Thallous cyanide	2057
Thallous ethoxide	2058
Thallous fluoride	2059
Thallous formate	2060
Thallous hexafluorophosphate	2061
Thallous hydroxide	2062
Thallous iodide	2063
Thallous molybdate	2064
Thallous nitrate	2065
Thallous nitrite	2066
Thallous oxalate	2067
Thallous oxide	2068
Thallous perchlorate	2069
Thallous selenate	2070
Thallous selenide	2071
Thallous sulfate	2072
Thallous sulfide	2073
Thenard's blue	625
Thenardite	1938
Thermonatrite	1875
Thiophosphoryl chloride difluoride	1523
Thiophosphoryl trifluoride	1521
Thiothionyl bromide	1994
Thiothionyl chloride	1995
Thiothionyl fluoride	1996
Thoria	2091
Thorianite	2091
Thorite	2093
Thorium hexaboride	2082
Thorium orthosilicate	2093
Thorium tetrabromide	2083
Thorium tetrachloride	2086
Thorium tetrafluoride	2087
Thorium tetraiodide	2088
Thulia	2103
Tiemannite	1324
Tin hydride	2106
Tincal	1948
Tincalconite	1949
Titanium ammonium oxalate monohydrate	156

Titanium dibromide	2141
Titanium dichloride	2142
Titanium diiodide	2143
Titanium dioxide	2157
Titanium disilicide	2140
Titanium disulfide	2159
Titanium monosulfide	2145
Titanium monoxide	2144
Titanium pentoxide	2152
Titanium tetrabromide	2153
Titanium tetrachloride	2154
Titanium tetrafluoride	2155
Titanium tetraiodide	2156
Titanium tribromide	2146
Titanium trichloride	2147
Titanium trifluoride	2148
Titanium triiodide	2149
Titanium trisulfide	2151
Titanyl sulfate monohydrate	2158
Triazaborane	351
Tribromochlorosilane	1798
Tribromogermane	862
Tribromosilane	1797
Trichlorofluorogermane	864
Trichlorofluorosilane	1800
Trichlorogermane	863
Trichlorosilane	1799
Tridymite	1815
Trifluoroamine oxide	1462
Trifluorosilane	1801
Trigermane	867
Triiodosilane	1802
Trisilane	1804
Troilite	1034
Tschernigite	9
Tungsten dibromide	2172
Tungsten dichloride	2173
Tungsten diiodide	2174
Tungsten dinitride	2168
Tungsten dioxide	2181
Tungsten dioxydibromide	2192
Tungsten diselenide	2182
Tungsten disilicide	2170
Tungsten disulfide	2183
Tungsten ditelluride	2184
Tungsten hexabromide	2190
Tungsten hexacarbonyl	2167
Tungsten hexachloride	2191
Tungsten hexafluoride	2195
Tungsten oxybromide	2197
Tungsten oxychloride	2198
Tungsten oxyfluoride	2199
Tungsten oxyiodide	2194
Tungsten oxytribromide	2188
Tungsten oxytrichloride	2189
Tungsten pentaboride	2164
Tungsten pentabromide	2185
Tungsten pentachloride	2186
Tungsten pentafluoride	2187
Tungsten tetrabromide	2177
Tungsten tetrachloride	2178
Tungsten tetrafluoride	2179
Tungsten tetraiodide	2180
Tungsten tribromide	2175
Tungsten trichloride	2176
Tungsten trioxide	2196
Tungsten trisulfide	2200
Tungstenite	2183
Tysonite	1073
Uraninite	2218
Uranium dioxide	2218
Uranium hexachloride	2224
Uranium hexafluoride	2225
Uranium oxychloride	2228
Uranium pentabromide	2220

Uranium pentachloride	2221
Uranium pentafluoride	2222
Uranium tetrabromide	2214
Uranium tetrachloride	2215
Uranium tetrafluoride	2216
Uranium tetraiodide	2217
Uranium tribromide	2209
Uranium tricarbide	2206
Uranium trichloride	2210
Uranium trifluoride	2211
Uranium trihydride	2212
Uranium triiodide	2213
Uranium trioxide	2226
Uranyl ammonium fluoride	158
Vanadium dibromide	2240
Vanadium dichloride	2241
Vanadium difluoride	2242
Vanadium diiodide	2243
Vanadium dioxide	2258
Vanadium hexacarbonyl	2236
Vanadium monoxide	2244
Vanadium oxide bromide	2261
Vanadium oxide chloride	2262
Vanadium pentafluoride	2259
Vanadium pentoxide	2260
Vanadium tetrabromide	2255
Vanadium tetrachloride	2256
Vanadium tetrafluoride	2257
Vanadium tribromide	2247
Vanadium trichloride	2248
Vanadium trifluoride	2249
Vanadium trifluoride trihydrate	2250
Vanadium triiodide	2251
Vanadium trioxide	2252
Vanadium trisulfate	2253
Vanadium trisulfide	2254
Verdigris	704
Villiaumite	1892
Vitreous silica	1817
Vivianite	1028
Vulcanite	749
Washing soda	1874
Whewellite	456
White lead	1098
White phosphorus	1487
Whitlockite	461
Willemite	2355
Witherite	215
Wolfram	2160
Wulfenite	1115
Wulframite	415
Wurtzite	2361
Wustite	1027
YAG	2298
Yellow phosphorus	1487
Ytterbia	2295
Yttria	2308
Zinc ammonium chloride	131,152
Zinc dodecanoate	2337
Zinc metaarsenite	2317
Zinc octanoate	2322
Zinc orthoarsenate	2315
Zinc orthosilicate	2355
Zincite	2345
Zinkosite	2357
Zircon	2378
Zirconia	2376
Zirconium ammonium carbonate	160
Zirconium orthosilicate	2378
Zirconium tetrabromide	2367
Zirconium tetrachloride	2369
Zirconium tetrafluoride	2370
Zirconium tetraiodide	2373

CAS	No.	CAS	No.	CAS	No.	CAS	No.	CAS	No.	CAS	No.
10294-27-6	886	12004-76-1	2012	12019-52-2	699	12035-79-9	1380	12057-71-5	1220
10294-28-7	891	12005-21-9	2298	12019-88-4	755	12035-82-4	1530	12057-74-8	1229
10294-29-8	887	12005-67-3	61	12020-14-3	2368	12035-83-5	1544	12057-75-9	1200
10294-31-2	889	12005-69-5	357	12020-21-2	775	12035-98-2	2244	12057-92-0	1294
10294-32-3	898	12006-15-4	382	12020-28-9	777	12036-02-1	1473	12058-18-3	1352
10294-33-4	358	12006-77-8	617	12020-39-2	781	12036-09-8	1673	12058-20-7	1354
10294-34-5	360	12006-79-0	563	12020-58-5	784	12036-10-1	1733	12059-14-2	1389
10294-38-9	216	12006-84-7	1001	12020-65-4	791	12036-14-5	2019	12059-95-9	1551
10294-39-0	254	12006-86-9	1000	12020-66-5	790	12036-21-4	2258	12060-00-3	1142
10294-40-3	220	12006-99-4	1331	12020-69-8	792	12036-22-5	2181	12060-01-4	1146
10294-42-5	525	12007-00-0	1385	12021-68-0	642	12036-32-7	1652	12060-08-1	1758
10294-44-7	1299	12007-01-1	1386	12022-95-6	1008	12036-34-9	1549	12060-58-1	1749
10294-46-9	738	12007-02-2	1384	12022-99-0	1009	12036-35-0	1698	12060-59-2	1983
10294-47-0	1099	12007-07-7	2013	12023-99-3	841	12036-41-8	2048	12061-16-4	776
10294-48-1	553	12007-09-9	2163	12024-10-1	833	12036-44-1	2103	12062-24-7	728
10294-50-5	655	12007-10-2	2162	12024-11-2	832	12037-15-9	2219	12063-98-8	829
10294-52-7	1046	12007-16-8	564	12024-14-5	834	12037-65-9	2139	12064-03-8	826
10294-53-8	1048	12007-23-7	902	12024-20-3	830	12037-80-8	2377	12064-62-9	821
10294-54-9	545	12007-25-9	1201	12024-21-4	844	12037-82-0	1500	12065-10-0	873
10294-58-3	1126	12007-29-3	1416	12024-22-5	849	12038-12-9	1655	12065-11-1	881
10294-62-9	1081	12007-33-9	368	12024-24-7	846	12038-13-0	1654	12065-36-0	852
10294-64-1	1604	12007-34-0	1754	12024-27-0	850	12038-20-9	1531	12065-65-5	2152
10294-66-3	1637	12007-35-1	2014	12024-81-6	813	12038-21-0	1540	12065-68-8	1138
10294-70-9	2115	12007-36-2	2202	12025-32-0	874	12038-63-0	1674	12066-83-0	1653
10325-94-7	400	12007-37-3	2233	12025-34-2	882	12038-67-4	1691	12067-00-4	1675
10326-21-3*	1206	12007-38-4	565	12025-39-7	875	12039-07-5	2145	12067-22-0	1751
10326-24-6	2317	12007-84-0	2203	12026-06-1	2062	12039-13-3	2159	12067-46-8	2182
10326-26-8	211	12007-97-5	1332	12026-24-3	2114	12039-15-5	2381	12067-54-8	2094
10326-27-9	218	12007-98-6	2164	12027-06-4	119	12039-16-6	2151	12067-66-2	2022
10326-28-0	405	12007-99-7	422	12028-48-7	121	12039-19-9	2311	12067-76-4	2184
10340-06-4	494	12008-02-5	504	12029-81-1	924	12039-35-9	2314	12068-49-4	1010
10343-61-0	2150	12008-05-8	783	12029-98-0	978	12039-55-3	2020	12068-69-8	334
10361-37-2	217	12008-06-9	809	12030-14-7	960	12039-76-8	2239	12068-90-5	1328
10361-43-0	321	12008-19-4	347	12030-24-9	972	12039-79-1	2018	12069-00-0	1131
10361-46-3	332	12008-21-8	1067	12030-49-8	995	12039-80-4	2049	12069-32-8	359
10361-79-2	1646	12008-23-0	1366	12030-51-2	996	12039-83-7	2140	12069-69-1	713
10361-82-7	1745	12008-27-4	1644	12030-85-2	1608	12039-87-1	2238	12069-85-1	904
10361-84-9	1756	12008-29-6	1738	12030-88-5	1628	12039-88-2	2170	12069-89-5	1334
10361-91-8	2292	12008-32-1	2301	12030-97-6	1638	12039-89-3	2288	12069-94-2	1417
10361-92-9	2305	12009-14-2	245	12031-30-0	1083	12039-90-6	2379	12070-06-3	2015
10361-95-2	2324	12009-18-6*	268	12031-49-1	1082	12039-95-1	2171	12070-07-4	2016
10377-48-7	1183	12009-18-6	267	12031-63-9	1174	12040-00-5	1752	12070-08-5	2137
10377-51-2	1172	12009-21-1	285	12031-80-0	1180	12040-02-7	2123	12070-09-6	2204
10377-52-3	1177	12011-97-1	1333	12032-20-1	1193	12041-50-8	13	12070-10-9	2234
10377-58-9	1215	12011-99-3	1418	12032-36-9	1241	12042-68-1	419	12070-12-1	2166
10377-60-3	1217	12012-16-7	2084	12032-74-5	1287	12042-78-3	420	12070-13-2	2165
10377-62-5	1225	12012-17-8	2235	12032-78-9	1256	12043-29-7	52	12071-15-7	1069
10377-66-9	1271	12012-32-7	505	12032-82-5	1248	12044-16-5	999	12071-29-3	1960
10377-93-2	1270	12012-35-0	566	12032-88-1	1286	12044-42-7	614	12071-31-7	2085
10380-31-1	283	12013-10-4	621	12032-97-2	1249	12044-54-1	199	12071-33-9	2205
10381-37-0	2095	12013-55-7	469	12033-19-1	1336	12045-01-1	616	12071-35-1	2303
10402-15-0	719	12013-56-8	468	12033-31-7	1337	12045-15-7	1250	12076-62-9	2206
10405-27-3	1455	12013-57-9	475	12033-54-4	1543	12045-16-8	1252	12088-65-2	1004
10450-55-2	1040	12013-82-0	453	12033-62-4	2017	12045-19-1	1415	12115-63-8	513
10450-60-9	976	12014-14-1	414	12033-64-6	2047	12045-27-1	2232	12124-97-9	68
10476-81-0	1958	12014-28-7	406	12033-65-7	2090	12045-63-5	2136	12124-99-1	113
10476-85-4	1963	12014-29-8	381	12033-72-6	2169	12045-64-6	2366	12125-01-8	92
10476-86-5	1969	12014-82-3	510	12033-83-9	2208	12045-87-3	1950	12125-02-9	75
10489-46-0	1699	12014-85-6	507	12033-88-4	1811	12045-88-4	1949	12125-08-5	97
10534-89-1	673	12014-93-6	522	12033-89-5	1920	12046-08-1	228	12125-09-6	1499
10544-72-6	1449	12016-80-7	677	12034-09-2	1945	12047-27-7	281	12125-19-8	2200
10544-73-7	1448	12017-01-5	647	12034-12-7	1422	12047-79-9	247	12125-22-3	1480
10545-99-0	1998	12017-08-2	660	12034-57-0	1429	12048-50-9	339	12125-25-6	1192
10553-31-8	212	12017-12-8	620	12034-59-2	1420	12048-51-0	340	12125-63-2	1035
10567-69-8	238	12017-13-9	666	12034-66-1	1430	12049-50-2	479	12133-07-2	756
10580-03-7	156	12017-37-8	680	12034-77-4	1421	12052-42-5	611	12133-10-7	765
10580-52-6	2241	12018-01-8	605	12034-80-9	1420	12053-12-2	561	12134-02-0	619
10588-01-9	1885	12018-09-6	573	12034-83-2	1432	12053-13-3	571	12134-75-7	811
11121-16-7	12	12018-10-9	718	12034-88-7	1116	12053-27-9	568	12134-77-9	823
11126-81-1*	15	12018-19-8	2326	12034-89-8	1971	12053-39-3	602	12135-76-1	146
12002-03-8	703	12018-22-3	601	12035-32-4	1376	12054-48-7	1399	12136-24-2	926
12002-61-8	6	12018-34-7	582	12035-35-7	1377	12054-85-2	123	12136-42-4	988
12002-99-2	1855	12018-36-9	572	12035-39-1	1410	12055-23-1	910	12136-45-7	1612
12003-63-3*	1558	12018-61-0	544	12035-52-8	1382	12055-62-8	925	12136-58-2	1185
12004-04-5	209	12019-11-3	684	12035-64-2	1388	12056-90-5	1080	12136-78-6	1339
12004-06-7	291	12019-23-7		12035-72-2		12057-24-8	1413	12136-91-3	1501

CAS	No.	CAS	No.	CAS	No.	CAS	No.	CAS	No.
13776-58-4	2273	13940-89-1	1019	14763-77-0	720	16893-85-9	1896	20328-94-3	2285
13776-62-0	1457	13963-58-1	1585	14779-70-5	862	16903-35-8	885	20344-49-4	1053
13776-74-4	1233	13965-73-6	363	14808-60-7	1813	16919-19-0	104	20398-06-5	2058
13777-22-5	903	13966-94-4	957	14808-60-7	1814	16919-58-7	99	20405-64-5	696
13777-23-6	908	13967-25-4	1301	14871-79-5*	237	16921-30-5	1584	20427-11-6	637
13777-25-8	2367	13972-68-4	398	14884-42-5	606	16921-96-3	986	20427-56-9	1736
13778-39-7	2100	13973-87-0	373	14890-41-6	2240	16925-39-6	441	20427-58-1	2334
13779-41-4	1496	13982-53-1*	698	14933-38-1	56	16940-66-2	1869	20427-59-2	729
13779-92-5	1436	13986-18-0	2330	14940-41-1	1028	16940-92-4	96	20548-54-3	473
13780-06-8	454	13986-26-0	2373	14965-52-7	1800	16941-10-9	476	20601-83-6	1324
13780-42-2	848	14013-15-1	1269	14977-61-8	609	16941-11-0	103	20610-49-5	1662
13780-48-8	786	14013-86-6	1024	14984-81-7	1777	16941-12-1	1525	20619-16-3	872
13780-57-9	2002	14014-88-1	1728	15098-87-0	26	16949-15-8	1161	20644-87-5	2236
13782-01-9*	678	14024-18-1	1041	15162-90-0	1451	16961-83-4	1809	20661-21-6	965
13783-04-5	2141	14038-43-8	1049	15163-24-3	2175	16962-07-5	14	20665-52-5	845
13783-07-8	2143	14040-11-0	2167	15168-20-4	742	16962-31-5	1586	20667-12-3	1844
13798-24-8	2045	14049-36-6	1785	15179-32-5	2010	16962-47-3	102	20694-39-7	1272
13812-43-6	1456	14049-81-1	638	15192-17-3	1742	17014-71-0	1616	20712-42-9	403
13813-22-4	1076	14055-74-4	1342	15192-26-4	2040	17026-29-8	1684	20762-60-1	1563
13813-23-5	1649	14055-75-5	1346	15192-42-4	1672	17029-21-9	1685	20770-09-6	2131
13813-24-6	1371	14055-81-3	2177	15195-33-2	1440	17083-68-0	884	20816-12-0	1476
13813-25-7	1748	14055-84-6	2180	15195-53-6	1425	17099-70-6	27	20859-73-8	44
13813-40-6	2046	14059-33-7	326	15230-48-5	858	17125-80-3	229	20910-28-5	2230
13813-41-7	923	14075-53-7	1581	15238-00-3	645	17194-00-2	234	21041-93-0	643
13813-42-8	774	14077-39-5	788	15238-00-3*	646	17375-41-6	1032	21041-95-2	394
13813-43-9	2102	14104-20-2	1857	15243-27-3	1257	17440-85-6	295	21109-95-5	271
13813-45-1	1191	14166-78-0	964	15243-33-1	1726	17440-90-3	1005	21159-32-0	534
13813-46-2	1548	14168-73-1	1240	15244-35-6	411	17496-59-2	551	21255-83-4	372
13813-47-3	59	14188-40-0	856	15321-51-4	1003	17702-41-9	349	21264-43-7	958
13814-72-7	812	14215-29-3	383	15364-10-0	2281	17786-31-1	622	21308-80-5	371
13814-74-9	1357	14215-30-6	708	15364-94-0	1275	18088-11-4	1718	21351-79-1	538
13814-75-0	1361	14217-21-1*	1890	15385-57-6	1302	18282-10-5	2130	21548-73-2	1854
13814-76-1	1683	14221-48-8*	89	15385-58-7	1298	18283-93-7	343	21645-51-2	29
13814-81-8	701	14242-05-8	1846	15468-32-3	1815	18288-22-7	320	21651-19-4	2117
13814-83-0	2265	14249-98-0	2189	15469-38-2	1051	18356-71-3	1788	21679-31-2	560
13814-96-5	1107	14280-53-6	955	15474-63-2	763	18433-40-4	158	21908-53-2	1321
13817-37-3	640	14284-93-6	1727	15477-33-5	19	18433-84-6	345	21995-38-0*	72
13818-75-2	814	14285-68-8	1667	15492-38-3	1697	18454-12-1	1103	22015-35-6	789
13818-89-8	866	14307-33-6*	437	15513-84-5	2243	18480-07-4	1968	22205-45-4	697
13819-84-6	1356	14312-00-6	391	15513-94-7	2251	18618-55-8	516	22205-57-8	527
13820-41-2	151	14335-40-1	1505	15520-84-0	675	18624-44-7	1020	22326-55-2	235
13820-62-7	625	14336-80-2	687	15571-91-2*	1629	18718-07-5	1264	22398-80-7	954
13821-06-2*	225	14402-70-1	127	15572-25-5	2071	18721-05-6	586	22441-45-8	200
13824-36-7	1789	14405-43-7	835	15578-26-4	2118	18810-58-7	210	22520-96-3	984
13825-74-6*	2158	14447-89-3	2194	15587-72-1	1714	18820-29-6	1285	22527-13-5	2280
13825-76-8	920	14452-39-2	43	15593-61-0	1232	18820-29-6	1283	22537-19-5	1084
13825-86-0	581	14456-34-9	911	15597-63-4	1506	18820-29-6	1284	22750-54-5*	387
13826-63-6	2066	14456-48-5	759	15597-88-3	603	18855-94-2	917	22750-57-8	528
13826-83-0	93	14456-51-0	2098	15607-89-3	2001	18868-43-4	1351	22756-36-1	1705
13842-80-3	2242	14456-53-2	1188	15622-42-1	1670	18917-82-3	1114	22831-39-6	1236
13842-83-6	1547	14457-84-2	41	15696-40-9	1468	18917-91-4	34	22831-42-1	11
13842-88-1	1427	14457-87-5	511	15750-45-5	1218	18933-05-6	1266	22886-66-4	839
13842-93-8	2030	14457-87-5*	512	15752-41-7	1423	19034-13-0	659	22986-54-5	3
13843-28-2	607	14459-59-7	1360	15752-46-2	1545	19073-56-4	1709	23370-59-4	992
13845-17-5	224	14459-75-7	1438	15823-43-5	916	19086-20-5	1243	23377-53-9	1690
13847-57-9	1101	14459-95-1	1578	15829-53-5	1304	19088-74-5	1713	23412-45-5	1350
13847-65-9	1462	14464-46-1	1816	15845-52-0	1110	19139-47-0	509	23414-72-4	2347
13862-78-7	153	14475-63-9	2372	15942-63-9	184	19168-23-1	98	23611-30-5	1720
13863-41-7	374	14481-29-9*	90	15947-41-8	60	19287-45-7	342	23777-80-2	346
13863-59-7	375	14483-17-1	1650	16122-03-5*	124	19357-83-6	2187	24094-93-7	569
13863-88-2	1821	14507-19-8	1075	16399-77-2	1015	19357-86-9	2291	24304-00-5	36
13864-01-2	1074	14517-29-4	1372	16544-92-6	632	19423-81-5	816	24415-00-7	868
13870-13-8	914	14519-18-7	1957	16671-27-5	590	19423-87-1	2293	24422-20-6	864
13870-21-8	1428	14521-18-7	1734	16674-78-5	1198	19465-30-6	348	24422-21-7	859
13870-24-1	1038	14523-22-9	1693	16689-88-6	2081	19469-07-9	1055	24597-12-4	831
13871-27-7	1952	14553-44-7	2093	16721-80-5	1906	19584-30-6	1694	24613-38-5	634
13873-84-2	983	14590-13-7	78	16721-80-5	1907	19597-69-4	1159	24621-21-4	1419
13874-75-4	1741	14590-19-3	657	16731-55-8	1605	19598-90-4	819	24646-85-3	2237
13874-77-6	2290	14639-94-2	101	16774-21-3	73	19624-22-7	344	24719-19-5	626
13875-06-4	2278	14639-97-5	152	16812-54-7	1409	19783-14-3	1111	25152-52-7	10
13876-85-2	694	14639-98-6	131	16853-85-3	1156	20193-56-0	2176	25417-81-6	232
13896-65-6	1731	14691-44-2	867	16860-99-4	1996	20193-58-2	1345	25502-05-0	2289
13918-22-4	1288	14693-81-3	2026	16871-60-6	1583	20205-91-8	367	25583-20-4	2138
13931-94-7	577	14693-82-4	969	16871-71-9	2333	20211-76-1	1072	25617-97-4	828
13939-06-5	1335	14720-21-9	894	16871-90-2	1587	20213-56-3	2188	25617-98-5	953
13940-63-1	870	14721-21-2	714	16872-11-0	356	20281-00-9	542	25658-42-8	2375

25658-43-9	2207	27218-16-2	558	35112-53-9	279	51621-05-7	1730	65842-03-7	1063
25659-31-8	1112	27774-13-6	2267	35718-37-7	2284	51898-99-8	323	68016-36-4	278
25764-08-3	506	27860-83-9	2331	36377-94-3	758	52003-58-4	120	68938-92-1	1534
25764-09-4	1651	30737-24-7	2067	36897-37-7	1400	52503-64-7	709	71500-16-8	1732
25764-10-7	1078	30937-52-1	1678	36907-42-3	2245	52788-53-1	820	71626-98-7	449
25764-11-8	1373	30937-53-2	1676	37267-86-0	1493	52951-38-9	325	72520-94-6	514
25764-15-2	810	31032-13-0	896	37773-49-2	1535	53731-35-4	1276	73560-00-6	979
25817-87-2	909	31052-14-9	2029	38455-77-5	2127	54120-05-7	1472	73796-25-5	1970
25895-60-7	1884	31052-43-4	1721	39290-85-2	710	54723-94-3	137	75926-26-0	1778
25909-39-1	1533	31083-74-6	1723	39368-69-9	1677	57246-89-6	1689	75926-28-2	1779
25985-07-3	1532	31234-26-1	1680	39416-30-3	284	58724-12-2	536	84359-31-9	597
26124-86-7	243	31576-40-6	1474	39796-98-0	2286	59201-51-3	1469	84359-31-9	598
26134-62-3	1176	32248-43-4	1743	39797-63-2	2283	60616-74-2	1212	92141-86-1	540
26342-61-0	570	32287-65-3	25	42246-24-2	1688	60676-86-0	1817	97126-35-7	667
26628-22-8	1868	32823-06-6	39	42739-38-8	159	60763-24-8	2092	99685-96-8	484
26686-77-1	1234	33088-16-3	2089	50432-32-1	2282	60804-25-3	1696	115383-22-7	485
26750-66-3	1640	33689-80-4	4	50801-97-3	1740	60883-64-9	298	123714-57-8	2362
27016-73-5	613	33689-81-5	2	50813-65-5	214	60922-26-1	2168	133578-89-9	2266
27016-75-7	1383	33689-82-6	5	50927-81-6	1818	60950-56-3	1235	133863-98-6	1364
27043-84-1	2318	34128-09-1	2064	51503-61-8	136	60969-19-9	2061	134929-59-2	486
27133-66-0	258	34330-64-8	390						

* Registry number refers to the anhydrous compound.

PHYSICAL PROPERTIES OF THE RARE EARTH METALS

K.A. Gschneidner, Jr.

Table 1
Data for the Trivalent Ions of the Rare Earth Elements

Rare earth	Symbol	Atomic no.	Atomic wt.[a]	No. 4f electrons	S	L	J	Spectroscopic ground state symbol
				Electronic configuration for R^{3+}				
Scandium	Sc	21	44.955910	0	—	—	—	—
Yttrium	Y	39	88.90585	0	—	—	—	—
Lanthanum	La	57	138.9055	0	—	—	—	—
Cerium	Ce	58	140.115	1	1/2	3	5/2	$^2F_{5/2}$
Praseodymium	Pr	59	140.90765	2	1	5	4	$^3H_4$
Neodymium	Nd	60	144.24	3	3/2	6	9/2	$^4I_{9/2}$
Promethium	Pm	61	(145)	4	2	6	4	$^5I_4$
Samarium	Sm	62	150.36	5	5/2	5	5/2	$^6H_{5/2}$
Europium	Eu	63	151.965	6	3	3	0	$^7F_0$
Gadolinium	Gd	64	157.25	7	7/2	0	7/2	$^8S_{7/2}$
Terbium	Tb	65	158.92534	8	3	3	6	$^7F_6$
Dysprosium	Dy	66	162.50	9	5/2	5	15/2	$^6H_{15/2}$
Holmium	Ho	67	164.93032	10	2	6	8	$^5I_8$
Erbium	Er	68	167.26	11	3/2	6	15/2	$^4I_{15/2}$
Thulium	Tm	69	168.93421	12	1	5	6	$^3H_6$
Ytterbium	Yb	70	173.04	13	1/2	3	7/2	$^2F_{7/2}$
Lutetium	Lu	71	174.967	14	—	—	—	—

Note: For additional information, see Goldschmidt, Z.B., in *Handbook on the Physics and Chemistry of Rare Earths,* Vol. 1, Gschneidner, K.A., Jr. and Eyring, L., Eds., North-Holland Physics, Amsterdam, 1978; DeLaeter, J.R., and Heumann, K.G., *J. Phys. Chem. Ref. Data,* 20, 1313, 1991; *Pure Appl. Chem.,* 66, 2423, 1994.

[a] 1993 standard atomic weights.

Table 2
Crystallographic Data for the Rare Earth Metals at 24°C (297 K) or Below

Rare earth metal	Crystal structure[a]	a_o	b_o	c_o	Metallic radius CN = 12 (Å)	Atomic volume (cm³/mol)	Density (g/cm³)
			Lattice constants(Å)				
αSc	hcp	3.3088	—	5.2680	1.6406	15.039	2.989
αY	hcp	3.6482	—	5.7318	1.8012	19.893	4.469
αLa	dhcp	3.7740	—	12.171	1.8791	22.602	6.146
αCe[b]	fcc	4.85[b]	—	—	1.72[b]	17.2[b]	8.16[b]
βCe	dhcp	3.6810	—	11.857	1.8321	20.947	6.689
γCe[c]	fcc	5.1610	—	—	1.8247	20.696	6.770
αPr	dhcp	3.6721	—	11.8326	1.8279	20.803	6.773
αNd	dhcp	3.6582	—	11.7966	1.8214	20.583	7.008
αPm	dhcp	3.65	—	11.65	1.811	20.24	7.264
αSm	rhomb[d]	3.6290[d]	—	26.207	1.8041	20.000	7.520
Eu	bcc	4.5827	—	—	2.0418	28.979	5.244
αGd	hcp	3.6336	—	5.7810	1.8013	19.903	7.901
α'Tb[e]	ortho	3.605[e]	6.244[e]	5.706[e]	1.784[e]	19.34[e]	8.219[e]
αTb	hcp	3.6055	—	5.6966	1.7833	19.310	8.230
α'Dy[f]	ortho	3.595[f]	6.184[f]	5.678[f]	1.774[f]	19.00[f]	8.551[f]
αDy	hcp	3.5915	—	5.6501	1.7740	19.004	8.551
Ho	hcp	3.5778	—	5.6178	1.7661	18.752	8.795

Table 2
Crystallographic Data for the Rare Earth Metals at 24°C (297 K) or Below (continued)

Rare earth metal	Crystal structure[a]	Lattice constants(Å)			Metallic radius CN = 12 (Å)	Atomic volume (cm³/mol)	Density (g/cm³)
		a_o	b_o	c_o			
Er	hcp	3.5592	—	5.5850	1.7566	18.449	9.066
Tm	hcp	3.5375	—	5.5540	1.7462	18.124	9.321
αYb[g]	hcp	3.8799[g]	—	6.3859[g]	1.9451[g]	25.067[g]	6.903[g]
βYb	fcc	5.4848	—	—	1.9392	24.841	6.966
Lu	hcp	3.5052	—	5.5494	1.7349	17.779	9.841

Note: For additional information, see Gschneidner, K.A., Jr. and Calderwood, F.W., in *Handbook on the Physics and Chemistry of Rare Earths,* Vol. 8, Gschneidner, K.A., Jr. and Eyring, L., Eds., North-Holland Physics, Amsterdam, 1986; Gschneidner, K.A., Jr., Pecharsky, V.K., Cho, Jaephil and Martin, S.W., *Scripta Mater.*, 1996, to be published.

[a] hcp = hexagonal close-packed; P6₃/mmc, hP2, A3, Mg-type; dhcp = double-c hexagonal close-packed; P6₃/mmc, hP4, A3′, αLa-type; fcc = face-centered cubic; Fm3̄m, cF4, A1, Cu-type; rhomb = rhombohedral; R3̄m, hR3, αSm-type; bcc = body-centered cubic; Im3̄m, cI2, A2, W-type; ortho = orthorhombic; Cmcm, oC4, α′ Dy-type.
[b] At 77 K (–196°C).
[c] Equilibrium room temperature (standard state) phase.
[d] Rhombohedral is the primitive cell. Lattice parameters given are for the nonprimitive hexagonal cell.
[d] At 220 K (–53°C).
[f] At 86 K (–187°C).
[g] At 23°C.

Table 3
Crystallographic Data for Rare Earth Metals at High Temperature

Rare earth metal	Structure	Lattice parameter (Å)	Temp. (°C)	Metallic radius		Atomic volume (cm³/mol)	Density (g/cm³)
				CN = 8 (Å)	CN = 12 (Å)		
βSc	bcc	3.73 (est.)	1337	1.62	1.66	15.6	2.88
βY	bcc	4.10[a]	1478	1.78	1.83	20.8	4.28
βLa	fcc	5.303	325	—	1.875	22.45	6.187
γLa	bcc	4.26	887	1.84	1.90	23.3	5.97
δCe	bcc	4.12	757	1.78	1.84	21.1	6.65
βPr	bcc	4.13	821	1.79	1.84	21.2	6.64
βNd	bcc	4.13	883	1.79	1.84	21.2	6.80
βPm	bcc	4.10 (est.)	890	1.78	1.83	20.8	6.99
βSm	hcp	$a = 3.6630$ $c = 5.8448$	450[b]	—	1.8176	20.450	7.353
γSm	bcc	4.10 (est.)	922	1.77	1.82	20.8	7.25
βGd	bcc	4.06	1265	1.76	1.81	20.2	7.80
βTb	bcc	4.07[a]	1289	1.76	1.81	20.3	7.82
βDy	bcc	4.03[a]	1381	1.75	1.80	19.7	8.23
γYb	bcc	4.44	763[c]	1.92	1.98	26.4	6.57

Note: The rare earths Eu, Ho, Er, Tm, and Lu are monomorphic. For additional information, see Gschneidner, K.A., Jr. and Calderwood, F.W., in *Handbook on the Physics and Chemistry of Rare Earths,* Vol. 8, Gschneidner, K.A., Jr. and Eyring, L., Eds., North-Holland Physics, Amsterdam, 1986, 1.

[a] Determined by extrapolation to 0% solute of a vs. composition data for R-Mg alloys at 24°C and corrected for thermal expansion to temperature given.
[b] The hcp phase was stabilized by impurities and the temperature of measurement was below the equilibrium transition temperature (see Table 4).
[c] The bcc phase was stabilized by impurities and the temperature of measurement was below the equilibrium transition temperature (see Table 4).

Table 4
High Temperature Transition Temperatures and Melting Point of Rare Earth Metals

Rare earth metal	Transition I $(\alpha - \beta)$[a]		Transition II $(\beta - \gamma)$[a]		Melting point
	Temp. (°C)	Phases	Temp. (C°)	Phases	(C°)
Sc	1337	hcp ⇌ bcc	—	—	1541
Y	1478	hcp ⇌ bcc	—	—	1522
La[b]	310	dhcp → fcc	865	fcc ⇌ bcc	918
Ce[c,d]	139	dhcp → fcc (β - γ)	726	fcc ⇌ bcc (γ - δ)	798
Pr	795	dhcp ⇌ bcc	—	—	931
Nd	863	dhcp ⇌ bcc	—	—	1021
Pm	890	dhcp ⇌ bcc	—	—	1042
Sm[e]	734	rhom → hcp	922	hcp ⇌ bcc	1074
Eu	—	—	—	—	822
Gd	1235	hcp ⇌ bcc	—	—	1313
Tb	1289	hcp ⇌ bcc	—	—	1356
Dy	1381	hcp ⇌ bcc	—	—	1412
Ho	—	—	—	—	1474
Er	—	—	—	—	1529
Tm	—	—	—	—	1545
Yb	795	fcc ⇌ bcc (β - γ)	—	—	819
Lu	—	—	—	—	1663

Note: For additional information, see Gschneidner, K.A., Jr. and Calderwood, F.W., in *Handbook on the Physics and Chemistry of Rare Earths,* Vol. 8, Gschneidner, K.A., Jr. and Eyring, L., Eds., North-Holland Physics, Amsterdam, 1986; Gschneidner, K.A., Jr., Pecharsky, V.K., Cho, Jaephil and Martin, S.W., *Scripta Mater.*, 34, 1717, 1996.

[a] For all the transformations listed, unless otherwise noted.
[b] On cooling, fcc → dhcp (β → α), 260°C.
[c] The β ⇌ γ equilibrium transition temperature is 10 ± 5°C.
[d] On cooling, fcc → dhcp (γ → β), −16°C.
[e] On cooling, hcp → rhomb (β → α), 727°C.

Table 5
Low Temperature Transition Temperatures of the Rare Earth Metals

Rare earth metal	Cooling			Rare earth metal	Heating		
	Transformation	°C	K		Transformation	°C	K
Ce	γ → β[a]	−16	257	Ce	α → β	−148	125
	γ → α	−172	101		α → β + γ	−104	169
	β → α	−228	45		β → γ[a]	139	412
Tb	α → α′	−53	220	Yb	α → β	7	280
Dy	α → α′	−187	86				
Yb	β → α	−13	260				

Note: For additional information, see Beaudry, B.J. and Gschneidner, K.A., Jr., in *Handbook on the Physics and Chemistry of Rare Earths,* Vol. 1, Gschneidner, K.A., Jr. and Eyring, L., Eds., North-Holland Physics, Amsterdam, 1978, 173. Koskenmaki, D.C. and Gschneidner, K.A., Jr., 1978, in *Handbook on the Physics and Chemistry of Rare Earths,* Vol. 1, Gschneidner, K.A., Jr. and Eyring, L., Eds., North-Holland Physics, Amsterdam, 1978, 337. Gschneidner, K.A., Jr., Pecharsky, V.K., Cho, Jaephil and Martin, S.W., *Scripta Mater.*, 34, 1717, 1996.

[a] The β ⇌ γ equilibrium transition temperature is 10 ± 5°C (283 ± 5K).

Table 6
Heat Capacity, Standard Entropy, Heats of Transformation, and Fusion of the Rare Earth Metals

Rare earth metal	Heat capacity at 298 K (J/mol K)	Standard entropy $S°_{298}$ (J/mol K)	Heat of transformation (kJ/mol)				Heat of fusion (kJ/mol)
			trans. 1	ΔH_{tr}^1	trans. 2	ΔH_{tr}^2	
Sc	25.5	34.6	α ⇌ β	4.00	—	—	14.1
Y	26.5	44.4	α ⇌ β	4.99	—	—	11.4
La	27.1	56.9	α ⇌ β	0.36	β ⇌ γ	3.12	6.20
Ce	26.9	72.0	β ⇌ γ	0.05	γ ⇌ δ	2.99	5.46
Pr	27.2	73.2	α ⇌ β	3.17	—	—	6.89
Nd	27.5	71.5	α ⇌ β	3.03	—	—	7.14
Pm	27.3[a]	71.6[a]	α ⇌ β	3.0[a]	—	—	7.7[a]
Sm	29.5	69.6	α ⇌ β	0.2[a]	β ⇌ γ	3.11	8.62
Eu	27.7	77.8	—	—	—	—	9.21
Gd	37.0	68.1	α ⇌ β	3.91	—	—	10.0
Tb	28.9	73.2	α ⇌ β	5.02	—	—	10.79
Dy	27.7	75.6	α ⇌ β	4.16	—	—	11.06
Ho	27.2	75.3	—	—	—	—	17.0[a]
Er	28.1	73.2	—	—	—	—	19.9
Tm	27.0	74.0	—	—	—	—	16.8
Yb	26.7	59.9	β ⇌ γ	1.75	—	—	7.66
Lu	26.9	51.0	—	—	—	—	22[a]

Note: For additional information, see Hultgren, R., Desai, P.D., Hawkins, D.T., Gleiser, M., Kelley, K.K., and Wagman, D.D., *Selected Values of the Thermodynamic Properties of the Elements,* ASM International, Metals Park, Ohio, 1973; Wagman, D.D., Evans, W.H., Parker, V.B., Schumm, R.H., Halow, I., Bailey, S.M., Churney, K.L., and Nuttall, R.L., *The NBS Tables of Chemical Thermodynamic Properties, J. Phys. Chem. Ref. Data,* Vol. 11, Suppl 2, 1982; Amitin, E.B., Bessergenev, W.G., Kovalevskaya, Yu. A., and Paukov, I.E., *J. Chem. Thermodyn.,* 15, 181, 1983; Amitin, E.B., Bessergenev, W.G., Kovalevskaya, Yu. A., and Paukov, I.E., *J. Chem. Thermodyn.,* 15, 181, 1983.

[a] Estimated.

Table 7
Vapor Pressures, Boiling Points, and Heats of Sublimation of Rare Earth Metals

Rare earth metal	Temperature in °C[a] for a vapor pressure of				Boiling point[a] (°C)	Heat of sublimation at 25°C (kJ/mol)
	10^{-8} atm (0.001 Pa)	10^{-6} atm (0.101 Pa)	10^{-4} atm (10.1 Pa)	10^{-2} atm (1013 Pa)		
Sc	1036	1243	1533	1999	2836	377.8
Y	1222	1460	1812	2360	3345	424.7
La	1301	1566	1938	2506	3464	431.0
Ce	1290	1554	1926	2487	3443	422.6
Pr	1083	1333	1701	2305	3520	355.6
Nd	955	1175	1500	2029	3074	327.6
Pm	—	—	—	—	3000[b]	348[b]
Sm	508	642	835	1150	1794	206.7
Eu	399	515	685	964	1529	175.3
Gd	1167	1408	1760	2306	3273	397.5
Tb	1124	1354	1698	2237	3230	388.7
Dy	804	988	1252	1685	2567	290.4
Ho	845	1036	1313	1771	2700	300.8
Er	908	1113	1405	1896	2868	317.1
Tm	599	748	964	1300	1950	232.2
Yb	301	400	541	776	1196	152.1
Lu	1241	1483	1832	2387	3402	427.6

Note: For additional information, see Hultgren, R., Desai, P.D., Hawkins, D.T., Gleiser, M., Kelley, K.K., and Wagman, D.D., *Selected Values of the Thermodynamic Properties of the Elements,* ASM International, Metals Park, Ohio, 1973 and Beaudry, B.J. and Gschneidner, K.A., Jr., in *Handbook on the Physics and Chemistry of Rare Earths,* Vol. 1, Gschneidner, K.A., Jr. and Eyring, L., Eds., North-Holland Physics, Amsterdam, 1978, 173.

[a] International Temperature Scale of 1990 (ITS-90) values.
[b] Estimated.

Table 8
Magnetic Properties of the Rare Earth Metals

Rare earth metal	$\chi_A \times 10^6$ at 298 K (emu/mol)	Paramagnetic at ~298 K Theory[a]	Paramagnetic Obs.	Ferromagnetic at ~0 K Theory[b]	Ferromagnetic Obs.	Easy axis	Néel temp. T_N (K) Hex sites	Cubic sites	Curie temp. T_C (K)	θ_p (K) ∥c	⊥c	Polycryst. or avg.
αSc	295.2	—	—	—	—	—	—	—	—	—	—	—
αY	187.7	—	—	—	—	—	—	—	—	—	—	—
αLa	95.9	—	—	—	—	—	—	—	—	—	—	—
βLa	105	—	—	—	—	—	—	—	—	—	—	−50
γCe	2,270	2.54	2.52	2.14	—	—	13.7	14.4	—	—	—	−50
βCe	2,500	2.54	2.61	2.14	—	—	0.03	12.5	—	—	—	−41
αPr	5,530	3.58	3.56	3.20	2.7[c]	a	—	—	—	—	—	0
αNd	5,930	3.62	3.45	3.27	2.2[c]	b	19.9	7.5	—	0	5	3.3
αPm	—	2.68	—	2.40	—	—	—	—	—	—	—	—
αSm	1,278[d]	0.85	1.74	0.71	0.5[c]	a	109	14.0	—	—	—	100
Eu	30,900	7.94	8.48	7.0	5.9	<110>	—	90.4	—	—	—	—
αGd	185,000[e]	7.94	7.98	7.0	7.63	30° to c	—	—	293.4	317	317	317
αTb	170,000	9.72	9.77	9.0	9.34	b	230.0	—	219.5	195	239	224
α'Tb	—	—	—	—	—	—	—	—	—	—	—	—
αDy	98,000	10.64	10.83	10.0	10.33	a	180.2	—	—	121	169	153
α'Dy	—	—	—	10.0	—	—	—	—	90.5[g]	—	—	—
Ho	72,900	10.60	11.2	10.0	10.34	b	132	—	19.5	73.0	88.0	83.0
Er	48,000	9.58	9.9	9.0	9.1	30° to c	85	—	18.7	61.7	32.5	42.2
Tm	24,700	7.56	7.61	7.0	7.14	c	58	—	32.0	41.0	−17.0	2.3
βYb	67[d]	—	—	—	—	—	—	—	—	—	—	—
Lu	182.9	—	—	—	—	—	—	—	—	—	—	—

Note: For additional information, see McEwen, K.A., in *Handbook on the Physics and Chemistry of Rare Earths*, Vol. 1, Gschneidner, K.A., Jr. and Eyring, L., Eds., North-Holland Physics, Amsterdam, 1978, 411 and Legvold, S., in *Ferromagnetic Materials*, Vol. 1, Wohlfarth, E.P., Ed., North-Holland Physics, Amsterdam, 1980, 183; Pecharsky, V.K., Gschneidner, K.A., Jr. and Fort, D., *Phys. Rev. B*, 47, 5063, 1993; Pecharsky, V.K., Gschneidner, K.A., Jr. and Fort, D., 1996, to be published; Steward, A.M. and Collocott, S.J., *J. Phys.: Condens. Matter*, 1, 677, 1988.

a $g[J(J+1)]^{1/2}$.
b gJ.
c At 38 T and 4.2 K.
d At 290 K.
e At 350 K.
g On cooling T_C = 89.6 K and on warming T_C = 91.5 K.

Table 9

Room Temperature Coefficient of Thermal Expansion, Thermal Conductivity, Electrical Resistance, and Hall Coefficient

Rare earth metal	Expansion $\alpha_1 \times 10^6$ ($°C^{-1}$)			Thermal conductivity (W/cm-K)	Electrical resistance ($\mu\Omega\cdot cm$)			Hall coefficient ($R_i \times 10^{12}$) (V·cm/A·Oe)		
	α_a	α_c	α_{poly}		ρ_a	ρ_c	ρ_{poly}	R_a	R_c	R_{poly}
αSc	7.6	15.3	10.2	0.158	70.9	26.9	56.2[a]	—	—	-0.13
αY	6.0	19.7	10.6	0.172	72.5	35.5	59.6	-0.27	-1.6	—
aLa	4.5	27.2	12.1	0.134	—	—	61.5	—	—	-0.35
bCe	—	—	—	—	—	—	82.8	—	—	—
γCe	6.3	—	6.3	0.113	—	—	74.4	—	—	+1.81
αPr	4.5	11.2	6.7	0.125	—	—	70.0	—	—	+0.709
αNd	7.6	13.5	9.6	0.165	—	—	64.3	—	—	+0.971
αPm	9[b]	16[b]	11[b]	0.15[b]	—	—	75[b]	—	—	—
αSm	9.6	19.0	12.7	0.133	—	—	94.0	—	—	-0.21
Eu	35.0	—	35.0	0.139[b]	—	—	90.0	—	—	+24.4
αGd	9.1[c]	10.0[c]	9.4[c]	0.105	135.1	121.7	131.0	-10	-54	-4.48[d]
αTb	9.3	12.4	10.3	0.111	123.5	101.5	115.0	-1.0	-3.7	—
αDy	7.1	15.6	9.9	0.107	111.0	76.6	92.6	-0.3	-3.7	—
Ho	7.0	19.5	11.2	0.162	101.5	60.5	81.4	+0.2	-3.2	—
Er	7.9	20.9	12.2	0.145	94.5	60.3	86.0	+0.3	-3.6	—
Tm	8.8	22.2	13.3	0.169	88.0	47.2	67.6	—	—	-1.8
βYb	26.3	—	26.3	0.385	—	—	25.0	—	—	+3.77
Lu	4.8	20.0	9.9	0.164	76.6	34.7	58.2	+0.45	-2.6	-0.535

Note: For additional information, see Beaudry, B. J. and Gschneidner, K.A., Jr., in *Handbook on the Physics and Chemistry of Rare Earths*, Vol. 1, Gschneidner, K.A., Jr. and Eyring, L., Eds., North-Holland Physics, Amsterdam, 1978, 173 and McEwen, K.A., in *Handbook on the Physics and Chemistry of Rare Earths*, Vol. 1, Gschneidner, K.A., Jr. and Eyring, L., Eds., North-Holland Physics, Amsterdam, 1978, 411.

[a] Calculated from single crystal values.
[b] Estimated.
[c] At 100°C.
[d] At 77°C.

Table 10
Electronic Specific Heat Constant (γ), Electron-Electron (Coulomb) Coupling Constant (μ^*), Electron-Phonon Coupling Constant (λ), Debye Temperature at 0 K(θ_D), and Superconducting Transition Temperature

Rare earth metal	γ (mJ/mol·K^2)	μ^*	λ	θ_D (K) from Heat capacity	θ_D (K) from Elastic constants	Superconducting temperature (K)
αSc	10.334	0.16	0.30	345.3	—	0.050[a]
αY	7.878	0.15	0.30	244.4	258	1.3[b]
αLa	9.45	0.08	0.76	150	154	5.10
βLa	11.5	—	—	140	—	6.00
αCe	12.8	—	—	179	—	0.022[c]
αPr	20	—	1.07[d]	155[e]	153	—
αNd	f	—	0.86[d]	157[e]	163	—
αPm	—	—	—	159[e]	—	—
αSm	8.1 ± 1.5[g]	—	0.81[d]	162[e,f]	169	—
Eu	f	—	—	f	118	—
αGd	4.48	—	0.30	169	182	—
α'Tb	3.71	—	0.34[d]	169.6	177	—
α'Dy	4.9	—	0.32[d]	192	183	—
Ho	2.1	—	0.30[d]	175[e]	190	—
Er	8.7	—	0.33[d]	176.9	188	—
Tm	f	—	0.36[d]	179[e]	200	—
αYb	3.30	—	—	117.6	118	—
βYb	8.36	—	—	109	—	0.022[h]
Lu	8.194	0.14	0.31	183.2	185	—

Note: For additional information, see Sundström, L.J., in *Handbook on the Physics and Chemistry of Rare Earths,* Vol. 1, Gschneidner, K.A., Jr., and Eyring, L., Eds., North-Holland Physics, Amsterdam, 1978, 379, Scott, T., in *Handbook on the Physics and Chemistry of Rare Earths,* Vol. 1, Gschneidner, K.A., Jr. and Eyring, L., Eds., North-Holland Physics, Amsterdam, 1978, 591, Probst, C. and Wittig, J., in *Handbook on the Physics and Chemistry of Rare Earths,* Vol. 1, Gschneidner, K.A., Jr. and Eyring, L., Eds., North-Holland Physics, Amsterdam, 1978, 749, and Tsang, T.-W.E., Gschneidner, K.A., Jr., Schmidt, F.A., and Thome, D.K., *Phys. Rev.,* B, 31, 235, 1985. Collocott, S.J., Hill, R.W. and Stewart, A.M., *J. Phys. F,* 18, L223, 1988. Hill, R.W. and Gschneidner, K.A., Jr., *J. Phys. F,* 18, 2545, 1988. Skriver, H.L. and Mertig, I., *Phys. Rev. B,* 41, 6553, 1990. Collocott, S.J. and Stewart, A.M., *J. Phys.: Condens. Matter,* 4, 6743, 1992. Pecharsky, V.K., Gschneidner, K.A., Jr. and Fort, D., *Phys. Rev. B,* 47, 5063, 1993.

[a] At 18.6 GPa.
[b] At 11 GPa.
[c] At 2.2 GPa.
[d] Calculated value.
[e] Estimated.
[f] Heat capacity results have been reported, but the resultant γ and θ_D values are unreliable because of the presence of impurities and/or there was no reliable procedure or model to correct for the magnetic contribution to the heat capacity.
[g] Based on the values reported for the purer Sm sample (IV).
[h] At 4.5 GPa.

Table 11
Room Temperature Elastic Moduli and Mechanical Properties

Rare earth metal	Elastic moduli (GPa)				Mechanical properties (MPa)				Recryst. temp. (°C)
	Young's (elastic) modulus	Shear modulus	Bulk modulus	Poisson's ratio	Yield strength 0.2% offset	Ultimate tensile strength	Uniform elongation (%)	Reduction in area (%)	
Sc	74.4	29.1	56.6	0.279	173[a]	255[a]	5.0[a]	8.0[a]	550
Y	63.5	25.6	41.2	0.243	42	129	34.0	—	550
αLa	36.6	14.3	27.9	0.280	126[a]	130	7.9[a]	—	300
βCe	—	—	—	—	86	138	—	24.0	—
γCe	33.6	13.5	21.5	0.24	28	117	22.0	30.0	325
αPr	37.3	14.8	28.8	0.281	73	147	15.4	67.0	400
αNd	41.4	16.3	31.8	0.281	71	164	25.0	72.0	400
αPm	46[b]	18[b]	33[b]	0.28[b]	—	—	—	—	400[b]
αSm	49.7	19.5	37.8	0.274	68	156	17.0	29.5	440
Eu	18.2	7.9	8.3	0.152	—	—	—	—	300
αGd	54.8	21.8	37.9	0.259	15	118	37.0	56.0	500
αTb	55.7	22.1	38.7	0.261	—	—	—	—	500
αDy	61.4	24.7	40.5	0.247	43	139	30.0	30.0	550
Ho	64.8	26.3	40.2	0.231	—	—	—	—	520
Er	69.9	28.3	44.4	0.237	60	136	11.5	11.9	520
Tm	74.0	30.5	44.5	0.213	—	—	—	—	600
βYb	23.9	9.9	30.5	0.207	7	58	43.0	92.0	300
Lu	68.6	27.2	47.6	0.261	—	—	—	—	600

Note: For additional information, see Scott, T., in *Handbook on the Physics and Chemistry of Rare Earths,* Vol. 1, Gschneidner, K.A., Jr. and Eyring, L., Eds., North-Holland Physics, Amsterdam, 1978, 591.

[a] Value is questionable.
[b] Estimated.

Table 12
Liquid Metal Properties Near the Melting Point

Rare earth metal	Density (g/cm³)	Surface tension (N/m)	Viscosity (centipoise)	Heat capacity (J/mol K)	Thermal conductivity (W/cm K)	Magnetic susceptibility $\chi \times 10^4$ (emu/mol)	Electrical resistivity (μΩ·cm)	$\Delta V_{L \rightarrow S}$[a] (%)	Spectral emittance at λ = 645 nm ε (%)	Temp. range (°C)
Sc	2.80	0.954	—	44.2[b]	—	—	—	—	—	—
Y	4.24	0.871	—	43.1	—	—	—	—	36.8	1522–1647
La	5.96	0.718	2.65	34.3	0.238	1.20	133	–0.6	25.4	920–1287
Ce	6.68	0.706	3.20	37.7	0.210	9.37	130	+1.1	32.2	877–1547
Pr	6.59	0.707	2.85	43.0	0.251	17.3	139	–0.02	28.4	931–1537
Nd	6.72	0.687	—	48.8	0.195	18.7	151	–0.9	39.4	1021–1567
Pm	6.9[b]	0.680[b]	—	50[b]	—	—	160[b]	—	—	—
Sm	7.16	0.431	—	50.2[b]	—	18.3	182	–3.6	43.7	1075
Eu	4.87	0.264	—	38.1	—	97	242	–4.8	—	—
Gd	7.4	0.664	—	37.2	0.149	67	195	–2.0	34.2	1313–1600
Tb	7.65	0.669	—	46.5	—	82	193	–3.1	—	—
Dy	8.2	0.648	—	49.9	0.187	95	210	–4.5	29.7	1412–1437
Ho	8.34	0.650	—	43.9	—	88	221	–7.4	—	—
Er	8.6	0.637	—	38.7	—	69	226	–9.0	37.2	1529–1587
Tm	9.0[b]	—	—	41.4	—	41	235[b]	–6.9	—	—
Yb	6.21	0.320	2.67	36.8	—	—	113	–5.1	—	—
Lu	9.3	0.940	—	47.9[b]	—	—	224	–3.6	—	—

Note: For additional information, see Van Zytveld, J., in Handbook on the Physics and Chemistry of Rare Earths, Vol. 12, Gschneidner, K.A., Jr. and Eyring, L., Eds. North-Holland Physics, Amsterdam, 1989, 357. Stretz, L.A. and Bautista, R.G., in Temperature, Its Measurement and Control in Science and Industry, Vol. 4, part I, H.H. Plumb, Ed., Instrument Society of America, Pittsburgh, 1972, 489. King, T.S., Baria, D.N., and Bautista, R.G., Met. Trans. B, 7, 411, 1976. Baria, D.N., King, T.S., and Bautista, R.G., Met. Trans. B, 7, 577, 1976.

[a] Volume change on freezing.
[b] Estimated.

PHYSICAL PROPERTIES OF THE RARE EARTH METALS (continued)

Table 13
Ionization Potentials (Electronvolts)

Rare earth	I Neutral atom	II Singly ionized	III Doubly ionized	IV Triply ionized	V Quadruply ionized
Sc	6.56144	12.79967	24.75666	73.4894	91.65
Y	6.217	12.24	20.52	60.597	77.0
La	5.5770	11.060	19.1773	49.95	61.6
Ce	5.5387	10.85	20.198	36.758	65.55
Pr	5.464	10.55	21.624	38.98	57.53
Nd	5.5250	10.73	22.1	40.41	—
Pm	5.554	10.90	22.3	41.1	—
Sm	5.6437	11.07	23.4	41.4	—
Eu	5.6704	11.241	24.92	42.7	—
Gd	6.1500	12.09	20.63	44.0	—
Tb	5.8639	11.52	21.91	39.79	—
Dy	5.9389	11.67	22.8	41.47	—
Ho	6.0216	11.80	22.84	42.5	—
Er	6.1078	11.93	22.74	42.7	—
Tm	6.18431	12.05	23.68	42.7	—
Yb	6.25416	12.1761	25.05	43.56	—
Lu	5.42585	13.9	20.9594	45.25	66.8

Note: For references, see the table "Ionization Potentials of Atoms and Atomic Ions" in Section 10.

Table 14
Effective Ionic Radii (Å)[a]

Rare earth ion	R²⁺ CN = 6	R²⁺ CN = 8	R³⁺ CN = 6	R³⁺ CN = 8	R³⁺ CN = 12	R⁴⁺ CN = 6	R⁴⁺ CN = 8
Sc	—	—	0.745	0.87	1.116	—	—
Y	—	—	0.900	1.015	1.220	—	—
La	—	—	1.045	1.18	1.320	—	—
Ce	—	—	1.010	1.14	1.290	0.80	0.97
Pr	—	—	0.997	1.14	1.286	0.78	0.96
Nd	—	—	0.983	1.12	1.276	—	—
Pm	—	—	0.97	1.10	1.267	—	—
Sm	1.19	1.27	0.958	1.09	1.260	—	—
Eu	1.17	1.25	0.947	1.07	1.252	—	—
Gd	—	—	0.938	1.06	1.246	—	—
Tb	—	—	0.923	1.04	1.236	0.76	0.88
Dy	—	—	0.912	1.03	1.228	—	—
Ho	—	—	0.901	1.02	1.221	—	—
Er	—	—	0.890	1.00	1.214	—	—
Tm	—	—	0.880	0.99	1.207	—	—
Yb	1.00	1.07	0.868	0.98	1.199	—	—
Lu	—	—	0.861	0.97	1.194	—	—

Note: For additional information, see Shannon, R.D. and Prewitt, C.T., *Acta Cryst.*, 25, 925, 1969 and Shannon, R.D. and Prewitt, C.T., *Acta Cryst.*, 26, 1046, 1970.

[a] Radius of O^{2-} is 1.40 Å for a coordination number (CN) of 6.

MELTING, BOILING, AND CRITICAL TEMPERATURES OF THE ELEMENTS

This table gives the melting point (T_m), normal boiling point (T_b) at a pressure of 101.325 kPa (1 atmosphere), and critical temperature (T_c) for all the elements for which measurements or reliable estimates are available. Values are given in °C on the ITS-90 scale. A "t" after a value indicates a triple point, and "s" indicates sublimation temperature (i.e., vapor pressure of the solid phase reaches 101.325 kPa, or 1 atmosphere).

Name	$T_m/°C$	$T_b/°C$	$T_c/°C$	Name	$T_m/°C$	$T_b/°C$	$T_c/°C$
Actinium	1051	3198		Mercury	-38.83	356.73	1477
Aluminum	660.32	2519		Molybdenum	2623	4639	
Americium	1176	2011		Neodymium	1021	3074	
Antimony	630.63	1587		Neon	-248.59	-246.08	-228.7
Argon	-189.35	-185.85	-122.28	Neptunium	644		
Arsenic	817 t	614 s	1400	Nickel	1455	2913	
Astatine	302			Niobium	2477	4744	
Barium	727	1897		Nitrogen	-210.00	-195.79	-146.94
Berkelium	1050			Nobelium	827		
Beryllium	1287	2471		Osmium	3033	5012	
Bismuth	271.40	1564		Oxygen	-218.79	-182.95	-118.56
Boron	2075	4000		Palladium	1554.9	2963	
Bromine	-7.2	58.8	315	Phosphorus	44.15	280.5	721
Cadmium	321.07	767		Platinum	1768.4	3825	
Calcium	842	1484		Plutonium	640	3228	
Californium	900			Polonium	254	962	
Carbon	4492 t	3642 s		Potassium	63.38	759	
Cerium	798	3443		Praseodymium	931	3520	
Cesium	28.44	671		Promethium	1042	3000	
Chlorine	-101.5	-34.04	143.8	Protactinium	1572		
Chromium	1907	2671		Radium	700		
Cobalt	1495	2927		Radon	-71	-61.7	104
Copper	1084.62	2562		Rhenium	3186	5596	
Curium	1345			Rhodium	1964	3695	
Dysprosium	1412	2567		Rubidium	39.31	688	
Einsteinium	860			Ruthenium	2334	4150	
Erbium	1529	2868		Samarium	1074	1794	
Europium	822	1596		Scandium	1541	2836	
Fermium	1527			Selenium	221	685	1493
Fluorine	-219.62	-188.12	-129.02	Silicon	1414	3265	
Francium	27			Silver	961.78	2162	
Gadolinium	1313	3273		Sodium	97.80	883	
Gallium	29.76	2204		Strontium	777	1382	
Germanium	938.25	2833		Sulfur	115.21	444.60	1041
Gold	1064.18	2856		Tantalum	3017	5458	
Hafnium	2233	4603		Technetium	2157	4265	
Helium		-268.93	-267.96	Tellurium	449.51	988	
Holmium	1474	2700		Terbium	1356	3230	
Hydrogen	-259.34	-252.87	-240.18	Thallium	304	1473	
Indium	156.60	2072		Thorium	1750	4788	
Iodine	113.7	184.4	546	Thulium	1545	1950	
Iridium	2446	4428		Tin	231.93	2602	
Iron	1538	2861		Titanium	1668	3287	
Krypton	-157.36	-153.22	-63.74	Tungsten	3422	5555	
Lanthanum	918	3464		Uranium	1135	4131	
Lawrencium	1627			Vanadium	1910	3407	
Lead	327.46	1749		Xenon	-111.75	-108.04	16.58
Lithium	180.5	1342		Ytterbium	819	1196	
Lutetium	1663	3402		Yttrium	1522	3345	
Magnesium	650	1090		Zinc	419.53	907	
Manganese	1246	2061		Zirconium	1855	4409	
Mendelevium	827						

HEAT CAPACITY OF THE ELEMENTS AT 25°C

This table gives the specific heat capacity (c_p) in J/g K and the molar heat capacity (C_p) in J/mol K at a temperature of 25°C and a pressure of 100 kPa (1 bar or 0.987 standard atmospheres) for all the elements for which reliable data are available.

Name	c_p J/g K	C_p J/mol K	Name	c_p J/g K	C_p J/mol K
Actinium	0.120	27.2	Molybdenum	0.251	24.06
Aluminum	0.897	24.200	Neodymium	0.190	27.45
Antimony	0.207	25.23	Neon	1.030	20.786
Argon	0.520	20.786	Nickel	0.444	26.07
Arsenic	0.329	24.64	Niobium	0.265	24.60
Barium	0.204	28.07	Nitrogen (N_2)	1.040	29.124
Beryllium	1.825	16.443	Osmium	0.130	24.7
Bismuth	0.122	25.52	Oxygen (O_2)	0.918	29.378
Boron	1.026	11.087	Palladium	0.246	25.98
Bromine (Br_2)	0.226	36.057	Phosphorous (white)	0.769	23.824
Cadmium	0.232	26.020	Platinum	0.133	25.86
Calcium	0.647	25.929	Potassium	0.757	29.600
Carbon (graphite)	0.709	8.517	Praseodymium	0.193	27.20
Cerium	0.192	26.94	Radon	0.094	20.786
Cesium	0.242	32.210	Rhenium	0.137	25.48
Chlorine (Cl_2)	0.479	33.949	Rhodium	0.243	24.98
Chromium	0.449	23.35	Rubidium	0.363	31.060
Cobalt	0.421	24.81	Ruthenium	0.238	24.06
Copper	0.385	24.440	Samarium	0.197	29.54
Dysprosium	0.170	27.7	Scandium	0.568	25.52
Erbium	0.168	28.12	Selenium	0.321	25.363
Europium	0.182	27.66	Silicon	0.705	19.789
Fluorine (F_2)	0.824	31.304	Silver	0.235	25.350
Gadolinium	0.236	37.03	Sodium	1.228	28.230
Gallium	0.371	25.86	Strontium	0.301	26.4
Germanium	0.320	23.222	Sulfur (rhombic)	0.710	22.75
Gold	0.129	25.418	Tantalum	0.140	25.36
Hafnium	0.144	25.73	Tellurium	0.202	25.73
Helium	5.193	20.786	Terbium	0.182	28.91
Holmium	0.165	27.15	Thallium	0.129	26.32
Hydrogen (H_2)	14.304	28.836	Thorium	0.113	26.230
Indium	0.233	26.74	Thulium	0.160	27.03
Iodine (I_2)	0.145	36.888	Tin (white)	0.228	27.112
Iridium	0.131	25.10	Titanium	0.523	25.060
Iron	0.449	25.10	Tungsten	0.132	24.27
Krypton	0.248	20.786	Uranium	0.116	27.665
Lanthanum	0.195	27.11	Vanadium	0.489	24.89
Lead	0.129	26.650	Xenon	0.158	20.786
Lithium	3.582	24.860	Ytterbium	0.155	26.74
Lutetium	0.154	26.86	Yttrium	0.298	26.53
Magnesium	1.023	24.869	Zinc	0.388	25.390
Manganese	0.479	26.32	Zirconium	0.278	25.36
Mercury	0.140	27.983			

VAPOR PRESSURE OF THE METALLIC ELEMENTS

C. B. Alcock

This table gives coefficients in an equation for the vapor pressure of 65 metallic elements in both the solid and liquid state. Vapor pressures in the range 10^{-10} to 10^2 Pa (10^{-15} to 10^{-3} atm) are covered. The equation is:

for p in pascals: $\log(p/\text{Pa}) = 5.006 + A + BT^{-1} + C\log T + DT^{-3}$

for p in atmospheres: $\log(p/\text{atm}) = A + BT^{-1} + C\log T + DT^{-3}$, where T is the temperature in K

This equation reproduces the observed vapor pressures to an accuracy of ±5% or better. Further details are given in:

C. B. Alcock, V. P. Itkin, and M. K. Horrigan, *Canadian Metallurgical Quarterly*, 23, 309, 1984.

Reprinted with permission of the publisher, Pergamon Press.

Element, state	A	B	C	D	Temperature range
Li sol	5.667	−8310			298–m.p.
Li liq	5.055	−8023			m.p.–1000
Na sol	5.298	−5603			298–m.p.
Na liq	4.704	−5377			m.p.–700
K sol	4.961	−4646			298–m.p.
K liq	4.402	−4453			m.p.–600
Rb sol	4.857	−4215			298–m.p.
Rb liq	4.312	−4040			m.p.–550
Cs sol	4.711	−3999			298–m.p.
Cs liq	4.165	−3830			m.p.–550
Be sol	8.042	−17020	−0.4440		298–m.p.
Be liq	5.786	−15731			m.p.–1800
Mg sol	8.489	−7813	−0.8253		298–m.p.
Ca sol	10.127	−9517	−1.4030		298–m.p.
Sr sol	9.226	−8572	−1.1926		298–m.p.
Ba sol	12.405	−9690	−2.2890		298–m.p.
Ba liq	4.007	−8163			m.p.–1200
Al sol	9.459	−17342	−0.7927		298–m.p.
Al liq	5.911	−16211			m.p.–1800
Ga sol	6.657	−14208			298–m.p.
Ga liq	6.754	−13984	−0.3413		m.p.–1600
In sol	5.991	−12548			298–m.p.
In liq	5.374	−12276			m.p.–1500
Tl sol	5.971	−9447			298–m.p.
Tl liq	5.259	−9037			m.p.–1100
Sn sol	6.036	−15710			298–m.p.
Sn liq	5.262	−15332			m.p.–1850
Pb sol	5.643	−10143			298–m.p.
Pb liq	4.911	−9701			m.p.–1200
Sc sol	6.650	−19721	0.2885	−0.3663	298–m.p.
Sc liq	5.795	−17681			m.p.–2000
Y sol	9.735	−22306	−0.8705		298–m.p.
Y liq	5.795	−20341			m.p.–2300
La sol	7.463	−22551	−0.3142		298–m.p.
La liq	5.911	−21855			m.p.–2450
Ti sol	11.925	−24991	−1.3376		298–m.p.
Ti liq	6.358	−22747			m.p.–2400
Zr sol	10.008	−31512	−0.7890		298–m.p.
Zr liq	6.806	−30295			m.p.–2500
Hf sol	9.445	−32482	−0.6735		298–m.p.
V sol	9.744	−27132	−0.5501		298–m.p.
V liq	6.929	−25011			m.p.–2500
Nb sol	8.822	−37818	−0.2575		298–2500
Ta sol	16.807	−41346	−3.2152	0.7437	298–2500
Cr sol	6.800	−20733	0.4391	−0.4094	298–2000
Mo sol	11.529	−34626	−1.1331		298–2500

VAPOR PRESSURE OF THE METALLIC ELEMENTS (continued)

Element, state	A	B	C	D	Temperature range
W sol	2.945	−44 094	1.3677		298–2350
W sol	−54.527	−57 687	−12.2231		2200–2500
Mn sol	12.805	−15 097	−1.7896		298–m.p.
Re sol	11.543	−40 726	−1.1629		298–2500
Fe sol	7.100	−21 723	0.4536	−0.5846	298–m.p.
Fe liq	6.347	−19 574			m.p.–2100
Ru sol	9.755	−34 154	−0.4723		298–m.p.
Os sol	9.419	−41 198	−0.3896		298–2500
Co sol	10.976	−22 576	−1.0280		298–m.p.
Co liq	6.488	−20 578			m.p.–2150
Rh sol	10.168	−29 010	−0.7068		298–m.p.
Rh liq	6.802	−26 792			m.p.–2500
Ir sol	10.506	−35 099	−0.7500		298–2500
Ni sol	10.557	−22 606	−0.8717		298–m.p.
Ni liq	6.666	−20 765			m.p.–2150
Pd sol	9.502	−19 813	−0.9258		298–m.p.
Pd liq	5.426	−17 899			m.p.–2100
Pt sol	4.882	−29 387	1.1039	−0.4527	298–m.p.
Pt liq	6.386	−26 856			m.p.–2500
Cu sol	9.123	−17 748	−0.7317		298–m.p.
Cu liq	5.849	−16 415			m.p.–1850
Ag sol	9.127	−14 999	−0.7845		298–m.p.
Ag liq	5.752	−13 827			m.p.–1600
Au sol	9.152	−19 343	−0.7479		298–m.p.
Au liq	5.832	−18 024			m.p.–2050
Zn sol	6.102	−6776			298–m.p.
Zn liq	5.378	−6286			m.p.–750
Cd sol	5.939	−5799			298–m.p.
Cd liq	5.242	−5392			m.p.–650
Hg liq	5.116	−3190			298–400
Ce sol	6.139	−21 752			298–m.p.
Ce liq	5.611	−21 200			m.p.–2450
Pr sol	8.859	−18 720	−0.9512		298–m.p.
Pr liq	4.772	−17 315			m.p.–2200
Nd sol	8.996	−17 264	−0.9519		298–m.p.
Nd liq	4.912	−15 824			m.p.–2000
Sm sol	9.988	−11 034	−1.3287		298–m.p.
Eu sol	9.240	−9459	−1.1661		298–m.p.
Gd sol	8.344	−20 861	−0.5775		298–m.p.
Gd liq	5.557	−19 389			m.p.–2250
Tb sol	9.510	−20 457	−0.9247		298–m.p.
Tb liq	5.411	−18 639			m.p.–2200
Dy sol	9.579	−15 336	−1.1114		298–m.p.
Ho sol	9.785	−15 899	−1.1753		298–m.p.
Er sol	9.916	−16 642	−1.2154		298–m.p.
Er liq	4.688	−14 380			m.p.–1900
Tm sol	8.882	−12 270	−0.9564		298–1400
Yb sol	9.111	−8111	−1.0849		298–900
Lu sol	8.793	−22 423	−0.6200		298–m.p.
Lu liq	5.648	−20 302			m.p.–2350
Th sol	8.668	−31 483	−0.5288		298–m.p.
Th liq	−18.453	−24 569	6.6473		m.p.–2500
Pa sol	10.552	−34 869	−1.0075		298–m.p.
Pa liq	6.177	−32 874			m.p.–2500
U sol	0.770	−27 729	2.6982	−1.5471	298–m.p.
U liq	20.735	−28 776	−4.0962		m.p.–2500
Np sol	19.643	−24 886	−3.9991		298–m.p.
Np liq	10.076	−23 378	−1.3250		m.p.–2500
Pu sol	26.160	−19 162	−6.6675		298–600
Pu sol	18.858	−18 460	−4.4720		500–m.p.
Pu liq	3.666	−16 658			m.p.–2450
Am sol	11.311	−15 059	−1.3449		298–m.p.
Cm sol	8.369	−20 364	−0.5770		298–m.p.
Cm liq	5.223	−18 292			m.p.–2200

DENSITY OF MOLTEN ELEMENTS AND REPRESENTATIVE SALTS

This table lists the liquid density at the melting point, ρ_m, for elements that are solid at room temperature, as well as for some representative salts of these elements. Densities at higher temperatures (up to the t_{max} given in the last column) may be estimated from the equation

$$\rho(t) = \rho_m - k(t - t_m)$$

where t_m is the melting point and k is given in the fifth column of the table. If a value of t_{max} is not given, the equation should not be used to extrapolate more than about 20°C beyond the melting point.

Data for the elements were selected from the primary literature; the assistance of Gernot Lang in compiling these data is gratefully acknowledged. The molten salt data were derived from Reference 1.

REFERENCE

1. Janz, G. J., Thermodynamic and Transport Properties of Molten Salts: Correlation Equations for Critically Evaluated Density, Surface Tension, Electrical Conductance, and Viscosity Data, *J. Phys. Chem. Ref. Data*, 17, Suppl. 2, 1988.
2. Nasch, P. M., and Steinemann, S. G., *Phys. Chem. Liq.*, 29, 43, 1995.

Formula	Name	$t_m/°C$	$\rho_m/\text{g cm}^{-3}$	$k/\text{g cm}^{-3}\ °C^{-1}$	t_{max}
Ag	Silver	961.78	9.320	0.0009	1500
AgBr	Silver(I) bromide	432	5.577	0.001035	667
AgCl	Silver(I) chloride	455	4.83	0.00094	627
AgI	Silver(I) iodide	558	5.58	0.00101	802
AgNO$_3$	Silver(I) nitrate	212	3.970	0.001098	360
Ag$_2$SO$_4$	Silver(I) sulfate	652	4.84	0.001089	770
Al	Aluminum	660.32	2.375	0.000233	1340
AlBr$_3$	Aluminum bromide	97.5	2.647	0.002435	267
AlCl$_3$	Aluminum chloride	192.6	1.302	0.002711	296
AlI$_3$	Aluminum iodide	188.32	3.223	0.0025	240
As	Arsenic	817	5.22	0.000544	
Au	Gold	1064.18	17.31	0.001343	1200
B	Boron	2075	2.08		
Ba	Barium	727	3.338	0.000299	1550
BaBr$_2$	Barium bromide	857	3.991	0.000924	900
BaCl$_2$	Barium chloride	962	3.174	0.000681	1081
BaF$_2$	Barium fluoride	1368	4.14	0.000999	1727
BaI$_2$	Barium iodide	711	4.26	0.000977	975
Be	Beryllium	1287	1.690	0.00011	
BeCl$_2$	Beryllium chloride	415	1.54	0.0011	473
BeF$_2$	Beryllium fluoride	552	1.96	0.000015	850
Bi	Bismuth	271.40	10.05	0.00135	800
BiBr$_3$	Bismuth bromide	218	4.76	0.002637	927
BiCl$_3$	Bismuth chloride	230	3.916	0.0023	350
Ca	Calcium	842	1.378	0.000230	1484
CaBr$_2$	Calcium bromide	742	3.111	0.0005	791
CaCl$_2$	Calcium chloride	775	2.085	0.000422	950
CaF$_2$	Calcium fluoride	1418	2.52	0.000391	2027
CaI$_2$	Calcium iodide	783	3.443	0.000751	1028
Cd	Cadmium	321.07	7.996	0.001218	500
CdBr$_2$	Cadmium bromide	568	4.075	0.00108	720
CdCl$_2$	Cadmium chloride	564	3.392	0.00082	807
CdI$_2$	Cadmium iodide	387	4.396	0.001117	700
Ce	Cerium	799	6.55	0.000710	1460
CeCl$_3$	Cerium(III) chloride	817	3.25	0.00092	950
CeF$_3$	Cerium(III) fluoride	1430	4.659	0.000936	1927
Co	Cobalt	1495	7.75	0.00165	1580
Cr	Chromium	1907	6.3	0.0011	2100
Cs	Cesium	28.44	1.843	0.000556	510
CsBr	Cesium bromide	636	3.133	0.001223	860
CsCl	Cesium chloride	645	2.79	0.001065	906
CsF	Cesium fluoride	703	3.649	0.001282	912
CsI	Cesium iodide	621	3.197	0.001183	907

Formula	Name	$t_m/°C$	$\rho_m/g\ cm^{-3}$	$k/g\ cm^{-3}\ °C^{-1}$	t_{max}
$CsNO_3$	Cesium nitrate	414	2.820	0.001166	491
Cs_2SO_4	Cesium sulfate	1005	3.1	0.00095	1530
Cu	Copper	1084.62	8.02	0.000609	1630
CuCl	Copper(I) chloride	430	3.692	0.00076	585
Dy	Dysprosium	1411	8.37	0.00143	1540
$DyCl_3$	Dysprosium(III) chloride	680	3.62	0.00068	987
Er	Erbium	1529	8.86	0.00157	1700
Eu	Europium	822	5.13	0.0028	980
Fe	Iron	1538	6.98	0.000572	1680
$FeCl_2$	Iron(II) chloride	677	2.348	0.000555	877
Ga	Gallium	29.76	6.08	0.00062	400
$GaBr_3$	Gallium(III) bromide	121.5	3.116	0.00246	135
$GaCl_3$	Gallium(III) chloride	77.9	2.053	0.002083	141
GaI_3	Gallium(III) iodide	212	3.630	0.002377	252
Gd	Gadolinium	1314	7.4		
$GdCl_3$	Gadolinium(III) chloride	609	3.56	0.000671	1007
GdI_3	Gadolinium(III) iodide	925	4.12	0.000908	1032
Ge	Germanium	938.25	5.60	0.00055	1600
Hf	Hafnium	2233	12		
$HgBr_2$	Mercury(II) bromide	236	5.126	0.003233	319
$HgCl_2$	Mercury(II) chloride	276	4.368	0.002862	304
HgI_2	Mercury(II) iodide	259	5.222	0.003235	354
Ho	Holmium	1472	8.34		
In	Indium	156.60	7.02	0.000836	500
$InBr_3$	Indium(III) bromide	420	3.121	0.0015	528
$InCl_3$	Indium(III) chloride	583	2.140	0.0021	666
InI_3	Indium(III) iodide	207	3.820	0.0015	360
Ir	Iridium	2446	19		
K	Potassium	63.38	0.828	0.000232	500
KBr	Potassium bromide	734	2.127	0.000825	930
KCl	Potassium chloride	771	1.527	0.000583	939
KF	Potassium fluoride	858	1.910	0.000651	1037
KI	Potassium iodide	681	2.448	0.000956	904
KNO_3	Potassium nitrate	337	1.865	0.000723	457
La	Lanthanum	920	5.94	0.00061	1600
$LaBr_3$	Lanthanum bromide	788	4.933	0.000096	912
$LaCl_3$	Lanthanum chloride	859	3.209	0.000777	973
LaF_3	Lanthanum fluoride	1493	4.589	0.000682	2177
LaI_3	Lanthanum iodide	778	4.29	0.001110	907
Li	Lithium	180.5	0.512	0.00052	285
LiBr	Lithium bromide	552	2.528	0.000652	739
LiCl	Lithium chloride	610	1.502	0.000432	781
LiF	Lithium fluoride	848.2	1.81	0.000490	1047
LiI	Lithium iodide	469	3.109	0.000917	667
$LiNO_3$	Lithium nitrate	253	1.781	0.000546	441
Li_2SO_4	Lithium sulfate	859	2.003	0.000407	1214
Lu	Lutetium	1663	9.3		
Mg	Magnesium	650	1.584	0.000234	900
$MgBr_2$	Magnesium bromide	711	2.62	0.000478	935
$MgCl_2$	Magnesium chloride	714	1.68	0.000271	826
MgI_2	Magnesium iodide	634	3.05	0.000651	888
Mn	Manganese	1246	5.95	0.00105	1590
$MnCl_2$	Manganese(II) chloride	650	2.353	0.000437	850
Mo	Molybdenum	2623	9.33		
Na	Sodium	97.80	0.927	0.00023	600
NaBr	Sodium bromide	747	2.342	0.000816	945
Na_2CO_3	Sodium carbonate	858.1	1.972	0.000448	1004
NaCl	Sodium chloride	800.7	1.556	0.000543	1027
NaF	Sodium fluoride	996	1.948	0.000636	1097
NaI	Sodium iodide	660	2.742	0.000949	912

Formula	Name	$t_m/°C$	$\rho_m/g\ cm^{-3}$	$k/g\ cm^{-3}\ °C^{-1}$	t_{max}
$NaNO_3$	Sodium nitrate	307	1.90	0.000715	370
Na_2SO_4	Sodium sulfate	884	2.069	0.000483	1077
Nd	Neodymium	1016	6.89	0.00076	1350
Ni	Nickel	1455	7.81	0.000726	1700
$NiCl_2$	Nickel(II) chloride	1009	2.653	0.00066	1057
Os	Osmium	3033	20		
Pb	Lead	327.46	10.66	0.00122	700
$PbBr_2$	Lead(II) bromide	371	5.73	0.00165	600
$PbCl_2$	Lead(II) chloride	501	4.951	0.0015	710
PbI_2	Lead(II) iodide	410	5.691	0.001594	697
Pd	Palladium	1554.9	10.38	0.001169	1700
Pr	Praseodymium	931	6.50	0.00093	1460
$PrCl_3$	Praseodymium chloride	786	3.23	0.00074	977
Pt	Platinum	1768.4	19.77	0.0024	2200
Pu	Plutonium	640	16.63	0.001419	950
Rb	Rubidium	39.31	1.46	0.000451	800
$RbBr$	Rubidium bromide	682	2.715	0.001072	907
Rb_2CO_3	Rubidium carbonate	837	2.84	0.000640	1007
$RbCl$	Rubidium chloride	715	2.248	0.000883	923
RbF	Rubidium fluoride	833	2.87	0.00102	1067
RbI	Rubidium iodide	642	2.904	0.001143	902
$RbNO_3$	Rubidium nitrate	305	2.519	0.001068	417
Rb_2SO_4	Rubidium sulfate	1050	2.56	0.000665	1545
Re	Rhenium	3186	18.9		
Rh	Rhodium	1964	10.7	0.000895	2200
Ru	Ruthenium	2334	10.65		
S	Sulfur	115.21	1.819	0.00080	160
Sb	Antimony	630.63	6.53	0.00067	745
$SbCl_3$	Antimony(III) chloride	73.4	2.681	0.002293	77
$SbCl_5$	Antimony(V) chloride	4	2.37	0.001869	77
SbI_3	Antimony(III) iodide	168	4.171	0.002483	322
Sc	Scandium	1541	2.80		
Se	Selenium	221	3.99		
Si	Silicon	1414	2.57	0.000936	1500
Sm	Samarium	1072	7.16		
Sn	Tin	231.93	6.99	0.000601	1200
$SnCl_2$	Tin(II) chloride	247	3.36	0.001253	480
$SnCl_4$	Tin(IV) chloride	-33	2.37	0.002687	138
Sr	Strontium	777	6.980		
$SrBr_2$	Strontium bromide	657	3.70	0.000745	1004
$SrCl_2$	Strontium chloride	874	2.727	0.000578	1037
SrF_2	Strontium fluoride	1477	3.470	0.000751	1927
SrI_2	Strontium iodide	538	4.085	0.000885	1026
Ta	Tantalum	3017	15		
$TaCl_5$	Tantalum(V) chloride	216	2.700	0.004316	457
Tb	Terbium	1359	7.65		
Te	Tellurium	449.51	5.70	0.00035	600
$ThCl_4$	Thorium chloride	770	3.363	0.0014	847
ThF_4	Thorium fluoride	1110	6.058	0.000759	1378
Ti	Titanium	1668	4.11		
$TiCl_4$	Titanium(IV) chloride	-25	1.807	0.001735	137
Tl	Thallium	304	11.22	0.00144	600
$TlBr$	Thallium(I) bromide	460	5.98	0.001755	647
$TlCl$	Thallium(I) chloride	430	5.628	0.0018	642
TlI	Thallium(I) iodide	441.8	6.15	0.001761	737
$TlNO_3$	Thallium(I) nitrate	206	4.91	0.001873	279
Tl_2SO_4	Thallium(I) sulfate	632	5.62	0.00130	927
Tm	Thulium	1545	8.56	0.00050	1675
U	Uranium	1135	17.3		
UCl_3	Uranium(III) chloride	837	4.84	0.007943	1057

DENSITY OF MOLTEN ELEMENTS AND REPRESENTATIVE SALTS (continued)

Formula	Name	t_m/°C	ρ_m/g cm^{-3}	k/g cm^{-3} °C^{-1}	t_{max}
UCl$_4$	Uranium(IV) chloride	590	3.572	0.001945	667
UF$_4$	Uranium(IV) fluoride	1036	6.485	0.000992	1341
V	Vanadium	1910	5.5		
W	Tungsten	3422	17.6		
Y	Yttrium	1526	4.24		
YCl$_3$	Yttrium chloride	721	2.510	0.0005	845
Yb	Ytterbium	824	6.21		
Zn	Zinc	419.53	6.57	0.0011	700
ZnBr$_2$	Zinc bromide	394	3.47	0.000959	602
ZnCl$_2$	Zinc chloride	290	2.54	0.00053	557
ZnI$_2$	Zinc iodide	446	3.878	0.00136	588
ZnSO$_4$	Zinc sulfate	680	3.14	0.00047	987
Zr	Zirconium	1855	5.8		
ZrCl$_4$	Zirconium chloride	437	1.643	0.007464	492

MAGNETIC SUSCEPTIBILITY OF THE ELEMENTS AND INORGANIC COMPOUNDS

When a material is placed in a magnetic field H, a magnetization (magnetic moment per unit volume) M is induced in the material which is related to H by $M = \kappa H$, where κ is called the volume susceptibility. Since H and M have the same dimensions, κ is dimensionless. A more useful parameter is the molar susceptibility χ_m, defined by

$$\chi_m = \kappa V_m = \kappa M/\rho$$

where V_m is the molar volume of the substance, M the molar mass, and ρ the mass density. When the cgs system is used, the customary units for χ_m are $cm^3\ mol^{-1}$; the corresponding SI units are $m^3\ mol^{-1}$.

Substances that have no unpaired electron orbital or spin angular momentum generally have negative values of χ_m and are called diamagnetic. Their molar susceptibility varies only slightly with temperature. Substances with unpaired electrons, which are termed paramagnetic, have positive χ_m and show a much stronger temperature dependence, varying roughly as $1/T$. The net susceptibility of a paramagnetic substance is the sum of the paramagnetic and diamagnetic contributions, but the former almost always dominates.

This table gives values of χ_m for the elements and selected inorganic compounds. All values refer to nominal room temperature (285 to 300 K) unless otherwise indicated. When the physical state (s = solid, l = liquid, g = gas, aq = aqueous solution) is not given, the most common crystalline form is understood. An entry of "Ferro." indicates a ferromagnetic substance.

Substances are arranged in alphabetical order by the most common name, except that compounds such as hydrides, oxides, and acids are grouped with the parent element (the same ordering used in the table "Physical Constants of Inorganic Compounds").

In keeping with customary practice, the molar susceptibility is given here in units appropriate to the cgs system. These values should be multiplied by 4π to obtain values for use in SI equations (where the magnetic field strength H has units of A m^{-1}).

REFERENCES

1. *Landolt-Börnstein, Numerical Data and Functional Relationships in Science and Technology*, New Series, II/16, *Diamagnetic Susceptibility*, Springer-Verlag, Heidelberg, 1986.
2. *Landolt-Börnstein, Numerical Data and Functional Relationships in Science and Technology*, New Series, III/19, Subvolumes a to i2, *Magnetic Properties of Metals*, Springer-Verlag, Heidelberg, 1986-1992.
3. *Landolt-Börnstein, Numerical Data and Functional Relationships in Science and Technology*, New Series, II/2, II/8, II/10, II/11, and II/12a, *Coordination and Organometallic Transition Metal Compounds*, Springer-Verlag, Heidelberg, 1966-1984.
4. *Tables de Constantes et Données Numérique, Volume 7, Relaxation Paramagnetique*, Masson, Paris, 1957.

Name	Formula	$\chi_m/10^{-6}\ cm^3\ mol^{-1}$	Name	Formula	$\chi_m/10^{-6}\ cm^3\ mol^{-1}$
Aluminum	Al	+16.5	Arsenic (yellow)	As	-23.2
Aluminum trifluoride	AlF$_3$	-13.9	Arsine (g)	AsH$_3$	-35.2
Aluminum oxide	Al$_2$O$_3$	-37	Arsenic(III) bromide	AsBr$_3$	-106
Aluminum sulfate	Al$_2$(SO$_4$)$_3$	-93	Arsenic(III) chloride	AsCl$_3$	-72.5
Ammonia (g)	NH$_3$	-16.3	Arsenic(III) iodide	AsI$_3$	-142.2
Ammonia (aq)	NH$_3$	-18.3	Arsenic(III) oxide	As$_2$O$_3$	-30.34
Ammonium acetate	NH$_4$C$_2$H$_3$O$_2$	-41.1	Arsenic(III) sulfide	As$_2$S$_3$	-70
Ammonium bromide	NH$_4$Br	-47	Barium	Ba	+20.6
Ammonium carbonate	(NH$_4$)$_2$CO$_3$	-42.5	Barium bromide	BaBr$_2$	-92
Ammonium chlorate	NH$_4$ClO$_3$	-42.1	Barium bromide dihydrate	BaBr$_2$·2H$_2$O	-119.3
Ammonium chloride	NH$_4$Cl	-36.7	Barium carbonate	BaCO$_3$	-58.9
Ammonium fluoride	NH$_4$F	-23	Barium chloride	BaCl$_2$	-72.6
Ammonium iodate	NH$_4$IO$_3$	-62.3	Barium chloride dihydrate	BaCl$_2$·2H$_2$O	-100
Ammonium iodide	NH$_4$I	-66	Barium fluoride	BaF$_2$	-51
Ammonium nitrate	NH$_4$NO$_3$	-33	Barium hydroxide	Ba(OH)$_2$	-53.2
Ammonium sulfate	(NH$_4$)$_2$SO$_4$	-67	Barium iodate	Ba(IO$_3$)$_2$	-122.5
Ammonium thiocyanate	NH$_4$SCN	-48.1	Barium iodide	BaI$_2$	-124.4
Antimony	Sb	-99	Barium iodide dihydrate	BaI$_2$·2H$_2$O	-163
Stibine (g)	SbH$_3$	-34.6	Barium nitrate	Ba(NO$_3$)$_2$	-66.5
Antimony(III) bromide	SbBr$_3$	-111.4	Barium oxide	BaO	-29.1
Antimony(III) chloride	SbCl$_3$	-86.7	Barium peroxide	BaO$_2$	-40.6
Antimony(III) fluoride	SbF$_3$	-46	Barium sulfate	BaSO$_4$	-65.8
Antimony(III) iodide	SbI$_3$	-147.2	Beryllium	Be	-9.0
Antimony(III) oxide	Sb$_2$O$_3$	-69.4	Beryllium chloride	BeCl$_2$	-26.5
Antimony(III) sulfide	Sb$_2$S$_3$	-86	Beryllium hydroxide	Be(OH)$_2$	-23.1
Antimony(V) chloride	SbCl$_5$	-120.5	Beryllium oxide	BeO	-11.9
Argon (g)	Ar	-19.32	Beryllium sulfate	BeSO$_4$	-37
Arsenic (gray)	As	-5.6	Bismuth	Bi	-280.1

Name	Formula	$\chi_m/10^{-6}$ cm^3 mol^{-1}	Name	Formula	$\chi_m/10^{-6}$ cm^3 mol^{-1}
Bismuth tribromide	$BiBr_3$	-147	Cesium	Cs	+29
Bismuth trichloride	$BiCl_3$	-26.5	Cesium bromate	$CsBrO_3$	-75.1
Bismuth fluoride	BiF_3	-61.2	Cesium bromide	$CsBr$	-67.2
Bismuth hydroxide	$Bi(OH)_3$	-65.8	Cesium carbonate	Cs_2CO_3	-103.6
Bismuth triiodide	BiI_3	-200.5	Cesium chlorate	$CsClO_3$	-65
Bismuth nitrate pentahydrate	$Bi(NO_3)_3 \cdot 5H_2O$	-159	Cesium chloride	$CsCl$	-56.7
Bismuth oxide	Bi_2O_3	-83	Cesium fluoride	CsF	-44.5
Bismuth phosphate	$BiPO_4$	-77	Cesium iodide	CsI	-82.6
Bismuth sulfate	$Bi_2(SO_4)_3$	-199	Cesium superoxide	CsO_2	+1534
Bismuth sulfide	Bi_2S_3	-123	Cesium sulfate	Cs_2SO_4	-116
Boron	B	-6.7	Chlorine (l)	Cl_2	-40.4
Diborane (g)	B_2H_6	-21.0	Chlorine trifluoride (g)	ClF_3	-26.5
Boric acid (orthoboric acid)	H_3BO_3	-34.1	Chromium	Cr	+167
Boron trichloride	BCl_3	-59.9	Chromium(II) chloride	$CrCl_2$	+7230
Boron oxide	B_2O_3	-38.7	Chromium(III) chloride	$CrCl_3$	+6350
Bromine (l)	Br_2	-56.4	Chromium(III) fluoride	CrF_3	+4370
Bromine (g)	Br_2	-73.5	Chromium(III) oxide	Cr_2O_3	+1960
Bromine trifluoride	BrF_3	-33.9	Chromium(III) sulfate	$Cr_2(SO_4)_3$	+11800
Bromine pentafluoride	BrF_5	-45.1	Chromium(VI) oxide	CrO_3	+40
Cadmium	Cd	-19.7	Cobalt	Co	Ferro.
Cadmium bromide	$CdBr_2$	-87.3	Cobalt(II) bromide	$CoBr_2$	+13000
Cadmium bromide tetrahydrate	$CdBr_2 \cdot 4H_2O$	-131.5	Cobalt(II) chloride	$CoCl_2$	+12660
Cadmium carbonate	$CdCO_3$	-46.7	Cobalt(II) chloride hexahydrate	$CoCl_2 \cdot 6H_2O$	+9710
Cadmium chloride	$CdCl_2$	-68.7	Cobalt(II) cyanide	$Co(CN)_2$	+3825
Cadmium chromate	$CdCrO_4$	-16.8	Cobalt(II) fluoride	CoF_2	+9490
Cadmium cyanide	$Cd(CN)_2$	-54	Cobalt(II) iodide	CoI_2	+10760
Cadmium fluoride	CdF_2	-40.6	Cobalt(II) sulfate	$CoSO_4$	+10000
Cadmium hydroxide	$Cd(OH)_2$	-41	Cobalt(II) sulfide	CoS	+225
Cadmium iodate	$Cd(IO_3)_2$	-108.4	Cobalt(II,III) oxide	Co_3O_4	+7380
Cadmium iodide	CdI_2	-117.2	Cobalt(III) fluoride	CoF_3	+1900
Cadmium nitrate	$Cd(NO_3)_2$	-55.1	Cobalt(III) oxide	Co_2O_3	+4560
Cadmium nitrate tetrahydrate	$Cd(NO_3)_2 \cdot 4H_2O$	-140	Copper	Cu	-5.46
Cadmium oxide	CdO	-30	Copper(I) bromide	$CuBr$	-49
Cadmium sulfate	$CdSO_4$	-59.2	Copper(I) chloride	$CuCl$	-40
Cadmium sulfide	CdS	-50	Copper(I) cyanide	$CuCN$	-24
Calcium	Ca	+40	Copper(I) iodide	CuI	-63
Calcium bromide	$CaBr_2$	-73.8	Copper(I) oxide	Cu_2O	-20
Calcium carbonate	$CaCO_3$	-38.2	Copper(II) bromide	$CuBr_2$	+685
Calcium chloride	$CaCl_2$	-54.7	Copper(II) chloride	$CuCl_2$	+1080
Calcium fluoride	CaF_2	-28	Copper(II) chloride dihydrate	$CuCl_2 \cdot 2H_2O$	+1420
Calcium hydroxide	$Ca(OH)_2$	-22	Copper(II) fluoride	CuF_2	+1050
Calcium iodate	$Ca(IO_3)_2$	-101.4	Copper(II) fluoride dihydrate	$CuF_2 \cdot 2H_2O$	+1600
Calcium iodide	CaI_2	-109	Copper(II) hydroxide	$Cu(OH)_2$	+1170
Calcium oxide	CaO	-15.0	Copper(II) nitrate trihydrate	$Cu(NO_3)_2 \cdot 3H_2O$	+1570
Calcium sulfate	$CaSO_4$	-49.7	Copper(II) nitrate hexahydrate	$Cu(NO_3)_2 \cdot 6H_2O$	+1625
Calcium sulfate dihydrate	$CaSO_4 \cdot 2H_2O$	-74	Copper(II) oxide	CuO	+238
Carbon (diamond)	C	-5.9	Copper(II) sulfate	$CuSO_4$	+1330
Carbon (graphite)	C	-6.0	Copper(II) sulfate pentahydrate	$CuSO_4 \cdot 5H_2O$	+1460
Carbon monoxide (g)	CO	-9.8	Copper(II) sulfide	CuS	-2.0
Carbon dioxide (g)	CO_2	-21.0	Dysprosium (α)	Dy	+98000
Cerium (β)	Ce	+2500	Dysprosium(III) oxide	Dy_2O_3	+89600
Cerium(II) sulfide	CeS	+2110	Dysprosium(III) sulfide	Dy_2S_3	+95200
Cerium(III) chloride	$CeCl_3$	+2490	Erbium	Er	+48000
Cerium(III) fluoride	CeF_3	+2190	Erbium oxide	Er_2O_3	+73920
Cerium(III) sulfide	Ce_2S_3	+5080	Erbium sulfate octahydrate	$Er_2(SO_4)_3 \cdot 8H_2O$	+74600
Cerium(IV) oxide	CeO_2	+26	Erbium sulfide	Er_2S_3	+77200
Cerium(IV) sulfate tetrahydrate	$Ce(SO_4)_2 \cdot 4H_2O$	-97			

Name	Formula	$\chi_m/10^{-6}$ cm^3 mol^{-1}	Name	Formula	$\chi_m/10^{-6}$ cm^3 mol^{-1}
Europium	Eu	+30900	Iodic acid	HIO_3	-48
Europium(II) bromide	$EuBr_2$	+26800	Iodine pentoxide	I_2O_5	-79.4
Europium(II) chloride	$EuCl_2$	+26500	Iodine chloride	ICl	-54.6
Europium(II) fluoride	EuF_2	+23750	Iodine trichloride	ICl_3	-90.2
Europium(II) iodide	EuI_2	+26000	Iodine pentafluoride	IF_5	-58.1
Europium(II) sulfide	EuS	+23800	Iridium	Ir	+25
Europium(III) oxide	Eu_2O_3	+10100	Iridium(III) chloride	$IrCl_3$	-14.4
Europium(III) sulfate	$Eu_2(SO_4)_3$	+10400	Iridium(IV) oxide	IrO_2	+224
Gadolinium (350 K)	Gd	+185000	Iron	Fe	Ferro.
Gadolinium(III) chloride	$GdCl_3$	+27930	Iron(II) bromide	$FeBr_2$	+13600
Gadolinium(III) oxide	Gd_2O_3	+53200	Iron(II) carbonate	$FeCO_3$	+11300
Gadolinium(III) sulfate octahydrate	$Gd_2(SO_4)_3 \cdot 8H_2O$	+53280	Iron(II) chloride	$FeCl_2$	+14750
			Iron(II) chloride tetrahydrate	$FeCl_2 \cdot 4H_2O$	+12900
Gadolinium(III) sulfide	Gd_2S_3	+55500	Iron(II) fluoride	FeF_2	+9500
Gallium	Ga	-21.6	Iron(II) iodide	FeI_2	+13600
Gallium suboxide	Ga_2O	-34	Iron(II) oxide	FeO	+7200
Gallium(II) sulfide	GaS	-23	Iron(II) sulfate	$FeSO_4$	+12400
Gallium(III) chloride	$GaCl_3$	-63	Iron(II) sulfate monohydrate	$FeSO_4 \cdot H_2O$	+10500
Gallium(III) sulfide	Ga_2S_3	-80	Iron(II) sulfate heptahydrate	$FeSO_4 \cdot 7H_2O$	+11200
Germanium	Ge	-11.6	Iron(II) sulfide	FeS	+1074
Germane (g)	GeH_4	-29.7	Iron(III) chloride	$FeCl_3$	+13450
Germanium(II) oxide	GeO	-28.8	Iron(III) chloride hexahydrate	$FeCl_3 \cdot 6H_2O$	+15250
Germanium(II) sulfide	GeS	-40.9	Iron(III) fluoride	FeF_3	+13760
Germanium(IV) chloride	$GeCl_4$	-72	Iron(III) fluoride trihydrate	$FeF_3 \cdot 3H_2O$	+7870
Germanium(IV) fluoride	GeF_4	-50	Iron(III) nitrate nonahydrate	$Fe(NO_3)_3 \cdot 9H_2O$	+15200
Germanium(IV) iodide	GeI_4	-171	Krypton (g)	Kr	-29.0
Germanium(IV) oxide	GeO_2	-34.3	Lanthanum (α)	La	+95.9
Germanium(IV) sulfide	GeS_2	-53.9	Lanthanum oxide	La_2O_3	-78
Gold	Au	-28	Lanthanum sulfate nonahydrate	$La_2(SO_4)_3 \cdot 9H_2O$	-262
Gold(I) bromide	AuBr	-61			
Gold(I) chloride	AuCl	-67	Lanthanum sulfide	La_2S_3	-37
Gold(I) iodide	AuI	-91	Lead	Pb	-23
Gold(III) chloride	$AuCl_3$	-112	Lead(II) acetate	$Pb(C_2H_3O_2)_2$	-89.1
Hafnium	Hf	+71	Lead(II) bromide	$PbBr_2$	-90.6
Hafnium oxide	HfO_2	-23	Lead(II) carbonate	$PbCO_3$	-61.2
Helium (g)	He	-2.02	Lead(II) chloride	$PbCl_2$	-73.8
Holmium	Ho	+72900	Lead(II) chromate	$PbCrO_4$	-18
Holmium oxide	Ho_2O_3	+88100	Lead(II) fluoride	PbF_2	-58.1
Hydrazine (l)	N_2H_4	-201	Lead(II) iodate	$Pb(IO_3)_2$	-131
Hydrogen (l, 20.3 K)	H_2	-5.44	Lead(II) iodide	PbI_2	-126.5
Hydrogen (g)	H_2	-3.99	Lead(II) nitrate	$Pb(NO_3)_2$	-74
Hydrogen chloride (l)	HCl	-22.6	Lead(II) oxide	PbO	-42
Hydrogen chloride (aq)	HCl	-22	Lead(II) phosphate	$Pb_3(PO_4)_2$	-182
Hydrogen fluoride (l)	HF	-8.6	Lead(II) sulfate	$PbSO_4$	-69.7
Hydrogen fluoride (aq)	HF	-9.3	Lead(II) sulfide	PbS	-83.6
Hydrogen iodide (s, 195 K)	HI	-47.3	Lithium	Li	+14.2
Hydrogen iodide (l, 233 K)	HI	-48.3	Lithium bromide	LiBr	-34.3
Hydrogen iodide (aq)	HI	-50.2	Lithium carbonate	Li_2CO_3	-27
Hydrogen peroxide (l)	H_2O_2	-17.3	Lithium chloride	LiCl	-24.3
Hydrogen sulfide (g)	H_2S	-25.5	Lithium fluoride	LiF	-10.1
Indium	In	-10.2	Lithium hydride	LiH	-4.6
Indium(I) chloride	InCl	-30	Lithium hydroxide (aq)	LiOH	-12.3
Indium(II) chloride	$InCl_2$	-56	Lithium iodide	LiI	-50
Indium(II) sulfide	InS	-28	Lithium sulfate	Li_2SO_4	-41.6
Indium(III) bromide	$InBr_3$	-107	Lutetium	Lu	+182.9
Indium(III) chloride	$InCl_3$	-86	Magnesium	Mg	+13.1
Indium(III) oxide	In_2O_3	-56	Magnesium bromide	$MgBr_2$	-72
Indium(III) sulfide	In_2S_3	-98	Magnesium carbonate	$MgCO_3$	-32.4
Iodine	I_2	-90	Magnesium chloride	$MgCl_2$	-47.4

Name	Formula	$\chi_m/10^{-6}$ cm^3 mol^{-1}	Name	Formula	$\chi_m/10^{-6}$ cm^3 mol^{-1}
Magnesium fluoride	MgF_2	-22.7	Molybdenum(VI) fluoride	MoF_6	-26.0
Magnesium hydroxide	$Mg(OH)_2$	-22.1	Molybdenum(VI) oxide	MoO_3	+3
Magnesium iodide	MgI_2	-111	Neodymium (α)	Nd	+5930
Magnesium oxide	MgO	-10.2	Neodymium fluoride	NdF_3	+4980
Magnesium sulfate	$MgSO_4$	-42	Neodymium oxide	Nd_2O_3	+10200
Magnesium sulfate monohydrate	$MgSO_4 \cdot H_2O$	-61	Neodymium sulfate	$Nd_2(SO_4)_3$	+9990
			Neodymium sulfide	Nd_2S_3	+5550
Magnesium sulfate heptahydrate	$MgSO_4 \cdot 7H_2O$	-135.7	Neon (g)	Ne	-6.96
Manganese	Mn	+511	Neptunium	Np	+575
Manganese(II) bromide	$MnBr_2$	+13900	Nickel	Ni	Ferro.
Manganese(II) carbonate	$MnCO_3$	+11400	Nickel(II) bromide	$NiBr_2$	+5600
Manganese(II) chloride	$MnCl_2$	+14350	Nickel(II) chloride	$NiCl_2$	+6145
Manganese(II) chloride tetrahydrate	$MnCl_2 \cdot 4H_2O$	+14600	Nickel(II) chloride hexahydrate	$NiCl_2 \cdot 6H_2O$	+4240
Manganese(II) fluoride	MnF_2	+10700	Nickel(II) fluoride	NiF_2	+2410
Manganese(II) hydroxide	$Mn(OH)_2$	+13500	Nickel(II) hydroxide	$Ni(OH)_2$	+4500
Manganese(II) iodide	MnI_2	+14400	Nickel(II) iodide	NiI_2	+3875
Manganese(II) oxide	MnO	+4850	Nickel(II) nitrate hexahydrate	$Ni(NO_3)_2 \cdot 6H_2O$	+4300
Manganese(II) sulfate	$MnSO_4$	+13660	Nickel(II) oxide	NiO	+660
Manganese(II) sulfate monohydrate	$MnSO_4 \cdot H_2O$	+14200	Nickel(II) sulfate	$NiSO_4$	+4005
			Nickel(II) sulfide	NiS	+190
Manganese(II) sulfate tetrahydrate	$MnSO_4 \cdot 4H_2O$	+14600	Nickel(III) sulfide	Ni_3S_2	+1030
Manganese(II) sulfide (α form)	MnS	+5630	Niobium	Nb	+208
			Niobium(V) oxide	Nb_2O_5	-10
Manganese(II) sulfide (β form)	MnS	+3850	Nitrogen (g)	N_2	-12.0
			Nitric acid (l)	HNO_3	-19.9
Manganese(II,III) oxide	Mn_3O_4	+12400	Nitrous oxide (g)	N_2O	-18.9
Manganese(III) fluoride	MnF_3	+10500	Nitric oxide (s, 90 K)	NO	+19.8
Manganese(III) oxide	Mn_2O_3	+14100	Nitric oxide (l, 118 K)	NO	+114.2
Manganese(IV) oxide	MnO_2	+2280	Nitric oxide (g)	NO	+1461
Mercury (s, 234 K)	Hg	-24.1	Nitrogen dioxide (g, 408 K)	NO_2	+150
Mercury (l)	Hg	-33.5	Nitrogen trioxide (g)	N_2O_3	-16
Mercury(I) bromide	Hg_2Br_2	-105	Nitrogen tetroxide (g)	N_2O_4	-23.0
Mercury(I) chloride	Hg_2Cl_2	-120	Osmium	Os	+11
Mercury(I) fluoride	Hg_2F_2	-106	Oxygen (s, 54 K)	O_2	+10200
Mercury(I) iodide	Hg_2I_2	-166	Oxygen (l, 90 K)	O_2	+7699
Mercury(I) nitrate	$Hg_2(NO_3)_2$	-121	Oxygen (g)	O_2	+3449
Mercury(I) oxide	Hg_2O	-76.3	Ozone (l)	O_3	+6.7
Mercury(I) sulfate	Hg_2SO_4	-123	Palladium	Pd	+540
Mercury(II) bromide	$HgBr_2$	-94.2	Palladium(II) chloride	$PdCl_2$	-38
Mercury(II) chloride	$HgCl_2$	-82	Phosphorus (white)	P	-26.66
Mercury(II) cyanide	$Hg(CN)_2$	-67	Phosphorus (red)	P	-20.77
Mercury(II) fluoride	HgF_2	-57.3	Phosphine (g)	PH_3	-26.2
Mercury(II) iodide	HgI_2	-165	Phosphoric acid (aq)	H_3PO_4	-43.8
Mercury(II) nitrate	$Hg(NO_3)_2$	-74	Phosphorous acid (aq)	H_3PO_3	-42.5
Mercury(II) oxide	HgO	-46	Phosphorus(III) chloride (l)	PCl_3	-63.4
Mercury(II) sulfate	$HgSO_4$	-78.1	Platinum	Pt	+193
Mercury(II) sulfide	HgS	-55.4	Platinum(II) chloride	$PtCl_2$	-54
Mercury(II) thiocyanate	$Hg(SCN)_2$	-96.5	Platinum(III) chloride	$PtCl_3$	-66.7
Molybdenum	Mo	+72	Platinum(IV) chloride	$PtCl_4$	-93
Molybdenum(III) bromide	$MoBr_3$	+525	Platinum(IV) fluoride	PtF_4	+445
Molybdenum(III) chloride	$MoCl_3$	+43	Plutonium	Pu	+525
Molybdenum(III) oxide	Mo_2O_3	-42.0	Plutonium(IV) fluoride	PuF_4	+1760
Molybdenum(IV) bromide	$MoBr_4$	+520	Plutonium(IV) oxide	PuO_2	+730
Molybdenum(IV) chloride	$MoCl_4$	+1750	Plutonium(VI) fluoride	PuF_6	+173
Molybdenum(IV) oxide	MoO_2	+41	Potassium	K	+20.8
Molybdenum(V) chloride	$MoCl_5$	+990	Potassium bromate	$KBrO_3$	-52.6
			Potassium bromide	KBr	-49.1
			Potassium carbonate	K_2CO_3	-59

Name	Formula	$\chi_m/10^{-6}$ cm^3 mol^{-1}	Name	Formula	$\chi_m/10^{-6}$ cm^3 mol^{-1}
Potassium chlorate	$KClO_3$	-42.8	Silane (g)	SiH_4	-20.4
Potassium chloride	KCl	-38.8	Disilane (g)	Si_2H_6	-37.3
Potassium chromate	K_2CrO_4	-3.9	Tetramethylsilane (l)	$(CH_3)_4Si$	-74.80
Potassium cyanide	KCN	-37	Tetraethylsilane (l)	$(C_2H_5)_4Si$	-120.2
Potassium ferricyanide	$K_3Fe(CN)_6$	+2290	Tetrabromosilane (l)	$SiBr_4$	-126
Potassium ferrocyanide trihydrate	$K_4Fe(CN)_6 \cdot 3H_2O$	-172.3	Tetrachlorosilane (l)	$SiCl_4$	-87.5
			Silicon carbide	SiC	-12.8
Potassium fluoride	KF	-23.6	Silicon dioxide	SiO_2	-29.6
Potassium hydrogen sulfate	$KHSO_4$	-49.8	Silver	Ag	-19.5
Potassium hydroxide (aq)	KOH	-22	Silver(I) bromide	$AgBr$	-61
Potassium iodate	KIO_3	-63.1	Silver(I) carbonate	Ag_2CO_3	-80.90
Potassium iodide	KI	-63.8	Silver(I) chloride	$AgCl$	-49
Potassium nitrate	KNO_3	-33.7	Silver(I) chromate	Ag_2CrO_4	-40
Potassium nitrite	KNO_2	-23.3	Silver(I) cyanide	$AgCN$	-43.2
Potassium permanganate	$KMnO_4$	+20	Silver(I) fluoride	AgF	-36.5
Potassium sulfate	K_2SO_4	-67	Silver(I) iodide	AgI	-80
Potassium sulfide	K_2S	-60	Silver(I) nitrate	$AgNO_3$	-45.7
Potassium superoxide	KO_2	+3230	Silver(I) nitrite	$AgNO_2$	-42
Potassium thiocyanate	$KSCN$	-48	Silver(I) oxide	Ag_2O	-134
Praseodymium (α)	Pr	+5530	Silver(I) phosphate	Ag_3PO_4	-120
Praseodymium chloride	$PrCl_3$	+44.5	Silver(I) sulfate	Ag_2SO_4	-92.90
Praseodymium oxide	Pr_2O_3	+8994	Silver(I) thiocyanate	$AgSCN$	-61.8
Praseodymium sulfide	Pr_2S_3	+10770	Silver(II) oxide	AgO	-19.6
Protactinium	Pa	+277	Sodium	Na	+16
Rhenium	Re	+67	Sodium acetate	$NaC_2H_3O_2$	-37.6
Rhenium(IV) oxide	ReO_2	+44	Sodium bromate	$NaBrO_3$	-44.2
Rhenium(IV) sulfide	ReS_2	+38	Sodium bromide	$NaBr$	-41
Rhenium(V) chloride	$ReCl_5$	+1225	Sodium carbonate	Na_2CO_3	-41
Rhenium(VI) oxide	ReO_3	+16	Sodium chlorate	$NaClO_3$	-34.7
Rhenium(VII) oxide	Re_2O_7	-16	Sodium chloride	$NaCl$	-30.2
Rhodium	Rh	+102	Sodium dichromate	$Na_2Cr_2O_7$	+55
Rhodium(III) chloride	$RhCl_3$	-7.5	Sodium fluoride	NaF	-15.6
Rhodium(III) oxide	Rh_2O_3	+104	Sodium hydrogen phosphate	Na_2HPO_4	-56.6
Rubidium	Rb	+17	Sodium hydroxide (aq)	$NaOH$	-15.8
Rubidium bromide	$RbBr$	-56.4	Sodium iodate	$NaIO_3$	-53
Rubidium carbonate	Rb_2CO_3	-75.4	Sodium iodide	NaI	-57
Rubidium chloride	$RbCl$	-46	Sodium nitrate	$NaNO_3$	-25.6
Rubidium fluoride	RbF	-31.9	Sodium nitrite	$NaNO_2$	-14.5
Rubidium iodide	RbI	-72.2	Sodium oxide	Na_2O	-19.8
Rubidium nitrate	$RbNO_3$	-41	Sodium peroxide	Na_2O_2	-28.10
Rubidium sulfate	Rb_2SO_4	-88.4	Sodium sulfate	Na_2SO_4	-52
Rubidium superoxide	RbO_2	+1527	Sodium sulfate decahydrate	$Na_2SO_4 \cdot 10H_2O$	-184
Ruthenium	Ru	+39	Sodium sulfide	Na_2S	-39
Ruthenium(III) chloride	$RuCl_3$	+1998	Sodium tetraborate	$Na_2B_4O_7$	-85
Ruthenium(IV) oxide	RuO_2	+162	Strontium	Sr	+92
Samarium (α)	Sm	+1278	Strontium bromide	$SrBr_2$	-86.6
Samarium(II) bromide	$SmBr_2$	+5337	Strontium bromide hexahydrate	$SrBr_2 \cdot 6H_2O$	-160
Samarium(III) bromide	$SmBr_3$	+972			
Samarium(III) oxide	Sm_2O_3	+1988	Strontium carbonate	$SrCO_3$	-47
Samarium(III) sulfate octahydrate	$Sm_2(SO_4)_3 \cdot 8H_2O$	+1710	Strontium chlorate	$Sr(ClO_3)_2$	-73
			Strontium chloride	$SrCl_2$	-61.5
Samarium(III) sulfide	Sm_2S_3	+3300	Strontium chloride hexahydrate	$SrCl_2 \cdot 6H_2O$	-145
Scandium (α)	Sc	+295.2			
Selenium	Se	-25	Strontium chromate	$SrCrO_4$	-5.1
Selenium dioxide	SeO_2	-27.2	Strontium fluoride	SrF_2	-37.2
Selenium bromide	Se_2Br_2	-113	Strontium hydroxide	$Sr(OH)_2$	-40
Selenium chloride (l)	Se_2Cl_2	-94.8	Strontium iodate	$Sr(IO_3)_2$	-108
Selenium hexafluoride (g)	SeF_6	-51	Strontium iodide	SrI_2	-112
Silicon	Si	-3.12	Strontium nitrate	$Sr(NO_3)_2$	-57.2

Name	Formula	$\chi_m/10^{-6}$ cm^3 mol^{-1}	Name	Formula	$\chi_m/10^{-6}$ cm^3 mol^{-1}
Strontium oxide	SrO	-35	Tungsten	W	+53
Strontium peroxide	SrO$_2$	-32.3	Tungsten carbide	WC	+10
Strontium sulfate	SrSO$_4$	-57.9	Tungsten(II) chloride	WCl$_2$	-25
Sulfur (rhombic)	S	-15.5	Tungsten(IV) oxide	WO$_2$	+57
Sulfur (monoclinic)	S	-14.9	Tungsten(IV) sulfide	WS$_2$	+5850
Sulfuric acid (l)	H$_2$SO$_4$	-39	Tungsten(V) bromide	WBr$_5$	+270
Sulfur dioxide (g)	SO$_2$	-18.2	Tungsten(V) chloride	WCl$_5$	+387
Sulfur trioxide (l)	SO$_3$	-28.54	Tungsten(VI) chloride	WCl$_6$	-71
Sulfur chloride (l)	SSCl$_2$	-62.2	Tungsten(VI) fluoride (g)	WF$_6$	-53
Sulfur dichloride (l)	SCl$_2$	-49.4	Tungsten(VI) oxide	WO$_3$	-15.8
Sulfur hexafluoride (g)	SF$_6$	-44	Uranium	U	+409
Thionyl chloride (l)	SOCl$_2$	-44.3	Uranium(III) bromide	UBr$_3$	+4740
Tantalum	Ta	+154	Uranium(III) chloride	UCl$_3$	+3460
Tantalum(V) chloride	TaCl$_5$	+140	Uranium(III) hydride	UH$_3$	+6244
Tantalum(V) oxide	Ta$_2$O$_5$	-32	Uranium(III) iodide	UI$_3$	+4460
Technetium	Tc	+115	Uranium(IV) bromide	UBr$_4$	+3530
Tellurium	Te	-38	Uranium(IV) chloride	UCl$_4$	+3680
Tellurium dibromide	TeBr$_2$	-106	Uranium(IV) fluoride	UF$_4$	+3530
Tellurium dichloride	TeCl$_2$	-94	Uranium(IV) oxide	UO$_2$	+2360
Tellurium hexafluoride (g)	TeF$_6$	-66	Uranium(VI) fluoride	UF$_6$	+43
Terbium (α)	Tb	+170000	Uranium(VI) oxide	UO$_3$	+128
Terbium oxide	Tb$_2$O$_3$	+78340	Vanadium	V	+285
Thallium	Tl	-50	Vanadium(II) bromide	VBr$_2$	+3230
Thallium(I) bromate	TlBrO$_3$	-75.9	Vanadium(II) chloride	VCl$_2$	+2410
Thallium(I) bromide	TlBr	-63.9	Vanadium(III) bromide	VBr$_3$	+2910
Thallium(I) carbonate	Tl$_2$CO$_3$	-101.6	Vanadium(III) chloride	VCl$_3$	+3030
Thallium(I) chlorate	TlClO$_3$	-65.5	Vanadium(III) fluoride	VF$_3$	+2757
Thallium(I) chloride	TlCl	-57.8	Vanadium(III) oxide	V$_2$O$_3$	+1976
Thallium(I) chromate	Tl$_2$CrO$_4$	-39.3	Vanadium(III) sulfide	V$_2$S$_3$	+1560
Thallium(I) cyanide	TlCN	-49	Vanadium(IV) chloride	VCl$_4$	+1215
Thallium(I) fluoride	TlF	-44.4	Vanadium(IV) oxide	VO$_2$	+99
Thallium(I) iodate	TlIO$_3$	-86.8	Vanadium(V) oxide	V$_2$O$_5$	+128
Thallium(I) iodide	TlI	-82.2	Water (s, 273 K)	H$_2$O	-12.63
Thallium(I) nitrate	TlNO$_3$	-56.5	Water (l, 293 K)	H$_2$O	-12.96
Thallium(I) nitrite	TlNO$_2$	-50.8	Water (l, 373 K)	H$_2$O	-13.09
Thallium(I) sulfate	Tl$_2$SO$_4$	-112.6	Water (g, 373 K))	H$_2$O	-13.1
Thallium(I) sulfide	Tl$_2$S	-88.8	Xenon (g)	Xe	-45.5
Thorium	Th	+97	Ytterbium (β)	Yb	+67
Thorium(IV) oxide	ThO$_2$	-16	Yttrium (α)	Y	+187.7
Thulium	Tm	+24700	Yttrium oxide	Y$_2$O$_3$	+44.4
Thulium oxide	Tm$_2$O$_3$	+51444	Yttrium sulfide	Y$_2$S$_3$	+100
Tin (gray)	Sn	-37.4	Zinc	Zn	-9.15
Tin(II) chloride	SnCl$_2$	-69	Zinc carbonate	ZnCO$_3$	-34
Tin(II) chloride dihydrate	SnCl$_2$·2H$_2$O	-91.4	Zinc chloride	ZnCl$_2$	-55.33
Tin(II) oxide	SnO	-19	Zinc cyanide	Zn(CN)$_2$	-46
Tin(IV) bromide	SnBr$_4$	-149	Zinc fluoride	ZnF$_2$	-34.3
Tin(IV) chloride (l)	SnCl$_4$	-115	Zinc hydroxide	Zn(OH)$_2$	-67
Tin(IV) oxide	SnO$_2$	-41	Zinc iodide	ZnI$_2$	-108
Titanium	Ti	+151	Zinc oxide	ZnO	-27.2
Titanium(II) bromide	TiBr$_2$	+720	Zinc phosphate	Zn$_3$(PO$_4$)$_2$	-141
Titanium(II) chloride	TiCl$_2$	+484	Zinc sulfate	ZnSO$_4$	-47.8
Titanium(II) iodide	TiI$_2$	+1790	Zinc sulfate monohydrate	ZnSO$_4$·H$_2$O	-63
Titanium(II) sulfide	TiS	+432	Zinc sulfate heptahydrate	ZnSO$_4$·7H$_2$O	-138
Titanium(III) bromide	TiBr$_3$	+660	Zinc sulfide	ZnS	-25
Titanium(III) chloride	TiCl$_3$	+1110	Zirconium	Zr	+120
Titanium(III) fluoride	TiF$_3$	+1300	Zirconium carbide	ZrC	-26
Titanium(III) oxide	Ti$_2$O$_3$	+132	Zirconium nitrate pentahydrate	Zr(NO$_3$)$_4$·5H$_2$O	-77
Titanium(IV) chloride	TiCl$_4$	-54	Zirconium(IV) oxide	ZrO$_2$	-13.8
Titanium(IV) oxide	TiO$_2$	+5.9			

INDEX OF REFRACTION OF INORGANIC LIQUIDS

This table gives the index of refraction n of several inorganic substances in the liquid state at specified temperatures. The measurements refer to ambient atmospheric pressure except for substances whose normal boiling points are greater than the indicated temperature; in this case the pressure is the saturated vapor pressure of the substance. All values refer to a wavelength of 589 nm unless otherwise indicated. Entries are arranged in alphabetical order by chemical formula as normally written.

Data on the index of refraction at other temperatures and wavelengths may be found in Reference 1.

REFERENCES

1. Wohlfarth, C., and Wohlfarth, B., *Landolt-Börnstein, Numerical Data and Functional Relationships in Science and Technology, New Series,* III/38A, Martienssen, W., Editor, Springer-Verlag, Heidelberg, 1996.
2. Francis, A.W., *J. Chem. Eng. Data,* 5, 534, 1960.

Formula	Name	$t/°C$	n
Ar	Argon	-188	1.2312
$AsCl_3$	Arsenic(III) chloride	16	1.604
BBr_3	Boron tribromide	16	1.312
BrF_3	Bromine trifluoride	25	1.4536
BrF_5	Bromine pentafluoride	25	1.3529
Br_2	Bromine	15	1.659
COS	Carbon oxysulfide	25	1.3506
CO_2	Carbon dioxide	24	1.6630
CS_2	Carbon disulfide	20	1.62774
C_3O_2	Carbon suboxide	0	1.453
Cl_2	Chlorine	20	1.3834
CrO_2Cl_2	Chromyl chloride	23	1.524
$Fe(CO)_5$	Iron pentacarbonyl	14	1.523
$GeBr_4$	Germanium(IV) bromide	26	1.6269
$GeCl_4$	Germanium(IV) chloride	25	1.4614
HBr	Hydrogen bromide	10	1.325
HCN	Hydrogen cyanide	20	1.26136
HCl	Hydrogen chloride	18	1.3287 [a]
$HClO_4$	Perchloric acid	50	1.3819
HF	Hydrogen fluoride	25	1.1574
HI	Hydrogen iodide	16	1.466
HNO_3	Nitric acid	25	1.393
H_2	Hydrogen	-253	1.1096
H_2O	Water	20	1.33336
H_2O_2	Hydrogen peroxide	28	1.4061
H_2S	Hydrogen sulfide	-80	1.460
		20	1.3682
H_2SO_4	Sulfuric acid	20	1.4183
H_2S_2	Hydrogen disulfide	20	1.630
He	Helium	-269	1.02451 [c]
Kr	Krypton	-157	1.3032 [c]
NH_3	Ammonia	-77	1.3944 [b]
		20	1.3327
NO	Nitric oxide	-90	1.330
N_2	Nitrogen	-196	1.19876 [b]
N_2H_4	Hydrazine	22	1.470
N_2O	Nitrous oxide	25	1.238
O_2	Oxygen	-183	1.2243 [c]
PBr_3	Phosphorus(III) bromide	25	1.687
PCl_3	Phosphorus(III) chloride	21	1.5122
PH_3	Phosphine	17	1.317
P_2O_3	Phosphorus(III) oxide	27	1.540
S	Sulfur	125	1.9170
SCl_2	Sulfur dichloride	14	1.557
SF_6	Sulfur hexafluoride	25	1.167
$SOCl_2$	Thionyl chloride	10	1.527
SO_2	Sulfur dioxide	25	1.3396
SO_2Cl_2	Sulfuryl chloride	12	1.444

INDEX OF REFRACTION OF INORGANIC LIQUIDS (continued)

Formula	Name	$t/°C$	n
SO_3	Sulfur trioxide	20	1.40965
$SSCl_2$	Sulfur chloride	20	1.671
$SbCl_5$	Antimony(V) chloride	22	1.5925
$SiBr_4$	Tetrabromosilane	31	1.5685
$SiCl_4$	Tetrachlorosilane	25	1.41156
$SnBr_4$	Tin(IV) bromide	31	1.6628
$SnCl_4$	Tin(IV) chloride	25	1.5086
$TiCl_4$	Titanium(IV) chloride	18	1.6076
Xe	Xenon	-112	1.3918 [c]

[a] At 581 nm
[b] At 578 nm
[c] At 546 nm

PHYSICAL AND OPTICAL PROPERTIES OF MINERALS

The chemical formula, crystal system, density, hardness, and index of refraction of some common minerals are given in this table. Entries are arranged alphabetically by mineral name. The columns are:

- **Formula:** Chemical formula for a typical sample of the mineral. Composition often varies considerably with the origin of the sample.
- **Crystal system:** tricl = triclinic; monocl = monoclinic; orth = orthorhombic; tetr = tetragonal; hex = hexagonal; rhomb = rhombohedral; cub = cubic.
- **Density:** Typical density in g/cm^3. Individual samples may vary by a few percent.
- **Hardness:** On the Mohs' scale (range of 1 to 10, with talc = 1 and diamond = 10).
- **Index of refraction:** Values are given for the three coordinate axes in the order of least, intermediate, and greatest index. For cubic crystals there is only a single value. See Reference 1 for details on the axis systems. Variations of several percent, depending on the origin and exact composition of the sample, are common.

REFERENCES

1. Deer, W.A., Howie, R.A., and Zussman, J., *An Introduction to the Rock-Forming Minerals*, 2nd Edition, Longman Scientific & Technical, Harlow, Essex, 1992.
2. Carmichael, R.S., *Practical Handbook of Physical Properties of Rocks and Minerals*, CRC Press, Boca Raton, FL, 1989.
3. Donnay, J.D.H., and Ondik, H.M., *Crystal Data Determinative Tables, Third Edition, Volume 2, Inorganic Compounds*, Joint Committee on Powder Diffraction Standards, Swarthmore, PA, 1973.

Name	Formula	Crystal system	Density g/cm^3	Hardness	n_α	n_β	n_γ
Acanthite	Ag_2S	orth	7.2	2.3			
Actinolite	$Ca_2(Mg,Fe)_5Si_8O_{22}(OH,F)_2$	monocl	3.23	5.5	1.624	1.655	1.664
Aegirine	$NaFe(SiO_3)_2$	monocl	3.58	6	1.763	1.800	1.815
Akermanite	$Ca_2MgSi_2O_7$	tetr	2.94	5.5	1.632	1.640	
Alabandite	MnS	cub	4.0	3.8			
Albite	$NaAlSi_3O_8$	tricl	2.63	6.3	1.527	1.531	1.538
Allanite	$(Ca,Mn,Ce,La,Y,Th)_2(Fe,Ti)(Al,Fe)O \cdot OH(Si_2O_7)(SiO_4)$	monocl	3.8	5.8	1.75	1.78	1.80
Allemontite	$SbAs$	hex	6.0	3.5			
Almandine	$Fe_3Al_2Si_3O_{12}$	cub	4.32	6.8	1.830		
Altaite	$PbTe$	cub	8.16	3			
Aluminite	$Al_2(SO_4)(OH)_4 \cdot 7H_2O$	monocl	1.74	1.5	1.459	1.464	1.470
Alunite	$(K,Na)Al_3(SO_4)_2(OH)_6$	rhomb	2.8	3.8	1.572	1.592	
Alunogen	$Al_2(SO_4)_3 \cdot 18H_2O$	monocl	1.69	1.8	1.467	1.47	1.478
Amblygonite	$(Li,Na)Al(PO_4)(F,OH)$	tricl	3.1	5.8	1.591	1.604	1.613
Analcite	$NaAlSi_2O_6 \cdot H_20$	cub	2.27	5.5	1.486		
Anatase	TiO_2	tetr	4.23	5.8	2.488	2.561	
Andalusite	Al_2OSiO_4	orth	3.15	7.5	1.635	1.639	1.644
Andesine	$([NaSi]_{0.7-0.5}[CaAl]_{0.3-0.5}AlSi_2O_8$	tricl	2.67	6.3	1.550	1.553	1.557
Andorite	$PbAgSb_3S_6$	rhomb	5.35	3.3			
Andradite	$Ca_3(Fe,Ti)_2Si_3O_{12}$	cub	3.86	6.8	1.887		
Anglesite	$PbSO_4$	orth	6.29	2.8	1.877	1.883	1.894
Anhydrite	$CaSO_4$	orth	2.96	3.5	1.570	1.575	1.614
Ankerite	$Ca(Fe,Mg,Mn)(CO_3)_2$	rhomb	3.0	3.8	1.529	1.720	
Anorthite	$CaAl_2Si_2O_8$	tricl	2.76	6.3	1.577	1.585	1.590
Anorthoclase	$(Na,K)AlSi_3O_8$	tricl	2.58	6	1.523	1.528	1.529
Anthophyllite	$(Mg,Fe)_7Si_8O_{22}(OH,F)_2$	rhomb	3.21	5.8	1.645	1.658	1.668
Apatite	$Ca_5(PO_4)_3(OH,F,Cl)$	hex	3.2	5	1.645	1.648	
Apophyllite	$KFCa_4Si_8O_{20} \cdot 8H_2O$	tetr	2.35	4.8	1.535	1.536	
Aragonite	$CaCO_3$	orth	2.83	3.5	1.531	1.680	1.686
Arcanite	K_2SO_4	orth	2.66		1.494	1.494	1.497
Argentite	Ag_2S	orth	7.2	2.3			
Arsenolite	As_2O_3	cub	3.86	1.5	1.755		
Arsenopyrite	$FeAsS$	monocl	6.1	5.8			
Atacamite	$Cu_2(OH)_3Cl$	rhomb	3.76	3.3	1.831	1.861	1.880
Augelite	$Al_2(PO_4)(OH)_3$	monocl	2.70	4.8	1.574	1.576	1.588
Augite	$(Ca,Mg,Fe,Ti,Al)_2(Si,Al)_2O_6$	monocl	3.38	6	1.703	1.707	1.738
Autunite	$Ca(UO_{22})(PO_4)_2 \cdot 10H_20$	tetr	3.2	2.3	1.553	1.577	
Axinite	$(Ca,Mn,Fe)_3Al_2BO_3Si_4O_{12}(OH)$	tricl	3.31	6.8	1.684	1.691	1.694

Name	Formula	Crystal system	Density g/cm³	Hardness	Index of refraction n_α	n_β	n_γ
Azurite	$Cu_3(OH)_2(CO_3)_2$	monocl	3.77	3.8	1.730	1.758	1.838
Baddeleyite	ZrO_2	monocl	5.7	6.5	2.13	2.19	2.20
Barite	$BaSO_4$	orth	4.49	3.3	1.636	1.637	1.648
Benitoite	$BaTi(SiO_3)_3$	rhomb	3.65	6.3	1.757	1.804	
Bertrandite	$Be_4Si_2O_7(OH)_2$	rhomb	2.6	6	1.589	1.602	1.613
Beryl	$Be_3Al_2(SiO_3)_6$	hex	2.64	7.8	1.582	1.589	
Beryllonite	$NaBe(PO)_4$	monocl	2.81	5.8	1.552	1.558	1.561
Biotite	$K(Mg,Fe)_3AlSi_3O_{10}(OH,F)_2$	monocl	3.0	2.8	1.595	1.651	1.651
Bismuthinite	Bi_2S_3	orth	6.78	2			
Bixbyite	$(Mn,Fe)_2O_3$	cub	4.95	6.3			
Bloedite	$Na_2Mg(SO_4)_2 \cdot 4H_2O$	monocl	2.25	2.8	1.483	1.486	1.487
Boehmite	$AlO(OH)$	orth	3.44	3.8	1.64	1.65	1.66
Boracite	$Mg_3B_7O_{13}Cl$	rhomb	2.94	7.3	1.66	1.66	1.67
Borax	$Na_2B_4O_7 \cdot 10H_2O$	monocl	1.73	2.3	1.447	1.469	1.472
Bornite	Cu_5FeS_4	cub	5.07	3			
Boulangerite	$Pb_5Sb_4S_{11}$	monocl	6.1	2.8			
Bournonite	$PbCuSbS_3$	rhomb	5.83	2.8			
Braggite	PtS	tetr	10.2				
Braunite	$(Mn,Si)_2O_3$	tetr	4.78	6.3			
Bravoite	$(Ni,Fe)S_2$	cub	4.62	5.8			
Breithauptite	$NiSb$	hex	≈8.7	5.5			
Brochantite	$Cu_4(SO_4)(OH)_6$	monocl	3.79	3.8	1.728	1.771	1.800
Bromyrite	$AgBr$	cub	6.47	2.5	2.253		
Brookite	TiO_2	tetr	4.23	5.8	2.583	2.584	2.700
Brucite	$Mg(OH)_2$	hex	2.37	2.5	1.575	1.59	
Bunsenite	NiO	cub	6.72	5.5			
Cacoxenite	$Fe_4(PO_4)_3(OH)_3 \cdot 12H_2O$	hex	2.3	3.5	1.580	1.646	
Calcite	$CaCO_3$	hex	2.71	3	1.486	1.658	
Caledonite	$Cu_2Pb_5(SO_4)_3(CO_3)(OH)_6$	rhomb	5.76	2.8	1.818	1.866	1.909
Calomel	Hg_2Cl_2	tetr	7.16	1.5	1.973	2.656	
Cancrinite	$(Na,Ca,K)_7[Al_6Si_6O_{24}]$ $(CO_3,SO_4,Cl,OH)_2 \cdot H_2O$	hex	2.42	5.5	1.495	1.509	
Carnalite	$KMgCl_3 \cdot 6H_2O$	rhomb	1.60	2.5	1.466	1.475	1.494
Carnotite	$K_2(UO_2)_2(VO_4)_2 \cdot 3H_2O$	rhomb		1.5	1.75	1.92	1.95
Cassiterite	SnO_2	tetr	6.85	6.5	2.006	2.097	
Celestite	$SrSO_4$	orth	3.96	3.3	1.622	1.624	1.631
Celsian	$BaAl_2Si_2O_8$	monocl	3.25	6.3	1.583	1.588	1.594
Cerargyrite	$AgCl$	cub	5.56	2.5	2.071		
Cerussite	$PbCO_3$	orth	6.6	3.3	1.804	2.076	2.079
Cervantite	Sb_2O_4	orth	6.64	4.5			
Chabazite	$Ca[Al_2Si_4O_{12}] \cdot 6H_2O$	trig	2.08	4.5	1.482		
Chalcanthite	$CuSO_4 \cdot 5H_2O$	tricl	2.29	2.5	1.514	1.537	1.543
Chalcocite	Cu_2S	orth	5.6	2.8			
Chalcopyrite	$CuFeS_2$	tetr	4.2	3.8			
Chiolite	$Na_5Al_3F_{14}$	tetr	3.00	3.8	1.342	1.349	
Chlorite	$(Mg,Al,Fe)_{12}(Si,Al)_8O_{20}(OH)_{16}$	monocl	3.0	2.5	1.61	1.62	1.62
Chloritoid	$FeAl_4O_2(SiO_4)_2(OH)_4$	monocl	3.66	6.5	1.717	1.721	1.726
Chondrodite	$Mg(OH,F)_2 \cdot 2Mg_2SiO_4$	monocl	3.21	6.5	1.604	1.615	1.634
Chromite	$FeCr_2O_4$	cub	5.0	5.5	2.16		
Chrysoberyl	$BeAl_2O_4$	orth	3.65	8.5	1.746	1.748	1.756
Chrysocolla	$CuSiO_3 \cdot 2H_2O$	rhomb	2.4	2	1.575	1.597	1.598
Cinnabar	HgS	hex	8.17	2.3	2.814	3.143	
Claudetite	As_2O_3	monocl	3.74	2.5	1.87	1.92	2.01
Clinohumite	$Mg(OH,F)_2 \cdot 4Mg_2SiO_4$	monocl	3.21	6	1.633	1.647	1.668
Clinozoisite	$Ca_2Al_3Si_3O_{12}(OH)$	monocl	3.30	6.5	1.693	1.700	1.712
Cobaltite	$CoAsS$	cub	≈6.1	5.5			
Colemanite	$Ca_2B_6O_{11} \cdot 5H_2O$	monocl	2.42	4.5	1.586	1.592	1.614
Columbite	$(Fe,Mn)(Nb,Ta)_2O_6$	rhomb	5.20	6			
Connellite	$Cu_{19}(SO_4)Cl_4(OH)_{32} \cdot 3H_2O$	hex	3.36	3	1.731	1.752	

Name	Formula	Crystal system	Density g/cm^3	Hardness	Index of refraction n_α	n_β	n_γ
Copiapite	$(Fe,Mg)Fe_4(SO_4)_6(OH)_2 \cdot 20H_2O$	tricl	2.13	2.8	1.52	1.54	1.59
Coquimbite	$Fe_2(SO_4)_3 \cdot 9H_2O$	hex	2.1	2.5	1.54	1.56	
Cordierite	$Al_3(Mg,Fe)_2Si_5AlO_{18}$	rhomb	2.66	7	1.540	1.549	1.553
Corundum	Al_2O_3	hex	3.97	9	1.761	1.769	
Cotunnite	$PbCl_2$	orth	5.98	2.5	2.199	2.217	2.260
Covellite	CuS	hex	4.8	1.8			
Cristobalite	SiO_2	hex	2.33	6.5	1.484	1.487	
Crocoite	$PbCrO_4$	monocl	6.12	2.8	2.29	2.36	2.66
Cryolite	Na_3AlF_6	monocl	2.97	2.5	1.338	1.338	1.339
Cryolithionite	$Na_3Li_3Al_2F_{12}$	cub	2.77	2.8	1.340		
Cubanite	$CuFe_2S_3$	rhomb	4.11	3.5			
Cummingtonite	$(Mg,Fe)_7Si_8O_{22}(OH)_2$	monocl	3.4	5.5	1.650	1.660	1.676
Cuprite	Cu_2O	cub	6.0	3.8			
Danburite	$CaSi_2B_2O_8$	rhomb	3.0	7	1.63	1.63	1.63
Datolite	$CaBSiO_4(OH)$	monocl	2.98	5.3	1.624	1.652	1.668
Daubreelite	Cr_2FeS_4	cub	3.81				
Derbylite	$Fe_6Ti_6Sb_2O_{23}$	rhomb	4.53	5	2.45	2.45	2.51
Diamond	C	cub	3.51	10	2.418		
Diaspore	$AlO(OH)$	orth	3.4	6.8	1.694	1.715	1.741
Digenite	$Cu_{2-x}S$	cub	5.55	2.8			
Diopside	$CaMgSi_2O_6$	monocl	3.30	6	1.680	1.687	1.708
Dioptase	$CuSiO_2(OH)_2$	rhomb	3.5	5	1.65	1.70	
Dolomite	$CaMg(CO_3)_2$	rhomb	2.86	3.5	1.500	1.679	
Douglasite	$K_2FeCl_4 \cdot 2H_2O$	orth	2.16		1.488	1.500	
Dyscrasite	Ag_3Sb	rhomb	9.74	3.8			
Eddingtonite	$BaAl_2Si_3O_{10} \cdot 4H_2O$	rhomb	2.8		1.541	1.553	1.557
Eglestonite	Hg_4OCl_2	cub	8.4	2.5	2.49		
Emplectite	$CuBiS_2$	rhomb	6.38	2			
Enargite	Cu_3AsS_4	rhomb	4.5	3			
Enstatite	$MgSiO_3$	monocl	3.19	5.5	1.656	1.662	1.669
Epidote	$Ca_2Al_2(Al,Fe)OH(SiO_4)_3$	monocl	3.44	6	1.733	1.755	1.765
Epsomite	$MgSO_4 \cdot 7H_2O$	orth	1.67	2.3	1.433	1.455	1.461
Erythrite	$(Co,Ni)_3(AsO_4)_2 \cdot 8H_2O$	monocl	3.06	2	1.626	1.661	1.699
Eucairite	$CuAgSe$	orth	7.7	2.5			
Euclasite	$BeAlSiO_4(OH)$	monocl	3.1	7.5	1.651	1.655	1.671
Eudialite	$(Na,Ca,Ce)_5(Fe,Mn)(Zr,Ti)(Si_3O_9)_2$ (OH,Cl)	hex	3.0	5.5	1.623	1.600	1.615
Eulytite	$Bi_4Si_3O_{12}$	cub	6.6	4.5	2.05		
Euxenite	$(Y,Ca,Ce,U,Th)(Nb,Ta,Ti)_2O_6$	rhomb	5.5	6	2.2		
Fayalite	Fe_2SiO_4	orth	4.30	6.5	1.827	1.869	1.879
Ferberite	$FeWO_4$	monocl	7.51	4.3			
Fergussonite	$(Y,Er,Ce,Fe)(Nb,Ta,Ti)O_4$	tetr	5.7	6	2.1		
Fluorite	CaF_2	cub	3.18	4	1.434		
Forsterite	Mg_2SiO_4	orth	3.21	7	1.635	1.651	1.670
Franklinite	$ZnFe_2O_4$	cub	5.21	6	2.36		
Gahnite	$ZnAl_2O_4$	cub	4.62	7.8	1.805		
Galaxite	$MnAl_2O_4$	cub	4.04	7.8	1.92		
Galena	PbS	cub	7.60	2.5	3.91		
Galenabismuthite	$PbBi_2S_4$	rhomb	7.04	3			
Ganomalite	$(Ca,Pb)_{10}(OH,Cl)_2(Si_2O_7)_3$	hex	5.6	3.5	1.910	1.945	
Gaylussite	$Na_2Ca(CO_3)_2 \cdot 5H_2O$	monocl	1.99	2.8	1.444	1.516	1.523
Gehlenite	$Ca_2Al_2SiO_7$	tetr	3.04	5.5	1.658	1.669	
Geikielite	$MgTiO_3$	hex	3.85	5.5	1.95	2.31	
Gibbsite	$Al(OH)_3$	monocl	2.42	3	1.57	1.57	1.59
Glauberite	$Na_2Ca(SO_4)_2$	monocl	2.80	2.8	1.515	1.535	1.536
Glauconite	$(K,Na,Ca)_{1.6}(Fe,Al,Mg)_{4.0}Si_{7.3}Al_{0.7}$ $O_{20}(OH)_4$	monocl	2.7	2	1.60	1.63	1.63
Glaucophane	$Na_2Mg_3Al_2Si_8O_{22}(OH)_2$	monocl	3.19	6	1.634	1.645	1.648
Gmelinite	$(Ca,Na_2)[Al_2Si_4O_{12}] \cdot 6H_2O$	hex	2.10	4.5	1.477	1.485	

Name	Formula	Crystal system	Density g/cm³	Hard-ness	Index of refraction		
					n_α	n_β	n_γ
Goethite	$FeO(OH)$	orth	4.3	5.3	2.268	2.401	2.457
Goslarite	$ZnSO_4 \cdot 7H_2O$	orth	1.97	2.3	1.457	1.480	1.484
Greenockite	CdS	cub	4.8	3.3	2.506	2.529	
Grossularite	$Ca_3Al_2Si_3O_{12}$	cub	3.59	6.8	1.734		
Gummite	$UO_3 \cdot H_2O$	orth	7.05	3.8			
Gypsum	$CaSO_4 \cdot 2H_2O$	monocl	2.32	2	1.520	1.525	1.530
Halite	$NaCl$	cub	2.17	2	1.544		
Hambergite	$Be_2(OH)(BO_3)$	rhomb	2.36	7.5	1.56	1.59	1.63
Hanksite	$Na_{22}K(SO_4)_9(CO_3)_2Cl$	hex	2.56	3.3	1.461	1.481	
Harmotome	$Ba[Al_2Si_6O_{16}] \cdot 6H_2O$	monocl	2.44	4.5	1.506	1.507	1.511
Hausmannite	Mn_3O_4	tetr	4.84	5.5	2.15	2.46	
Haüyne	$(Na,Ca)_{4-8}Al_6Si_6O_{24}(SO_4,S)_{1-2}$	cub	2.47	5.8	1.502		
Hedenbergite	$CaFeSi_2O_6$	monocl	3.53	6	1.721	1.727	1.746
Helvite	$Mn_4Be_3Si_3O_{12}S$	cub	3.32	6	1.739		
Hematite	Fe_2O_3	hex	5.25	6	2.91	3.19	
Hemimorphite	$Zn_4Si_2O_7(OH)_2 \cdot H_2O$	rhomb	3.45	5	1.614	1.617	1.636
Hercynite	$Fe(AlO_2)_2$	cub	4.3	7.8	1.835		
Herderite	$CaBe(PO_4)(Fe,OH)$	monocl	2.98	5.3	1.592	1.612	1.621
Hessite	Ag_2Te	orth	8.4	2.5			
Heulandite	$(Ca,Na_2,K_2)[Al_2Si_7O_{18}] \cdot 6H_2O$	monocl	2.2	3.8	1.498	1.498	1.506
Hopeite	$Zn_3(PO_4)_2 \cdot 4H_2O$	orth	3.0	3.2	1.58	1.59	1.59
Hornblende	$Ca_2(Mg,Fe)_4Al(Si_7AlO_{22})(OH)_2$	monocl	3.24	5.5	1.67	1.67	1.69
Huebnerite	$MnWO_4$	monocl	7.2	4.3	2.17	2.22	2.32
Humite	$Mg(OH,F)_2 \cdot 3Mg_2SiO_4$	orth	3.3	6	1.625	1.636	1.657
Huntite	$Mg_3Ca(CO_3)_4$	trig	2.70				
Hydrogrossularite	$Ca_3Al_2Si_2O_8(SiO_4)_{1-m}(OH)_{4m}$	cub	3.4	6.8	1.70		
Hydromagnesite	$3MgCO_3 \cdot Mg(OH)_2 \cdot 3H_2O$	monocl	2.24	3.5	1.523	1.527	1.545
Illite	$KAl_4[Si_7AlO_{20}](OH)_4$	monocl	2.8	1.5	1.56	1.59	1.59
Ilmenite	$FeTiO_3$	rhomb	4.72	5.5			
Iodyrite	AgI	hex	5.68	1.5	2.21	2.22	
Jacobsite	$MnFe_2O_4$	cub	4.87	7.8	2.3		
Jadeite	$NaAlSi_2O_6$	monocl	3.34	6	1.649	1.654	1.663
Jamesonite	$Pb_4FeSb_6S_{14}$	monocl	5.63	2.5			
Jarosite	$KFe_3(SO_4)_2(OH)_6$	rhomb	3.09	3	1.715	1.820	
Kainite	$KMg(SO_4)Cl \cdot 3H_2O$	monocl	2.15	2.8	1.494	1.505	1.516
Kaliophylite	$KAlSiO_4$	hex	2.61	6	1.532	1.537	
Kaolinite	$Al_4Si_4O_{10}(OH)_8$	tricl	2.65	2.3	1.549	1.564	1.565
Kernite	$Na_2B_4O_7 \cdot 4H_2O$	monocl	1.95	2.5	1.454	1.472	1.488
Kieserite	$MgSO_4 \cdot H_2O$	monocl	2.57	3.5	1.520	1.533	1.584
Kyanite	Al_2OSiO_4	tricl	3.59	6.3	1.715	1.722	1.731
Lanarkite	$Pb_2(SO_4)O$	monocl	6.92	2.3	1.928	2.007	2.036
Lanthanite	$(La,Ce)_2(CO_3)_3 \cdot 8H_2O$	rhomb	2.72	2.8	1.52	1.587	1.613
Laumontite	$Ca_4[Al_8Si_{16}O_{48}] \cdot 16H_2O$	monocl	2.3	3.3	1.508	1.517	1.519
Laurionite	$Pb(OH)Cl$	rhomb	6.24	3.3	2.08	2.12	2.16
Lawsonite	$CaAl_2(OH)_2Si_2O_7 \cdot H_2O$	rhomb	3.08	6	1.655	1.675	1.685
Lazulite	$(Mg,Fe)Al_2(PO_4)_2(OH)_2$	monocl	3.23	5.8	1.615	1.64	1.650
Lazurite	$Na_4SSi_3Al_3O_{12}$	cub	2.42	5.3	1.500		
Leadhillite	$Pb_4(SO_4)(CO_3)_2(OH)_2$	monocl	6.55	2.8	1.87	2.00	2.01
Lepidocrocite	$FeO(OH)$	orth	4.26	5	1.94	2.20	2.51
Lepidolite	$K_2(Li,Al)_{5-6}[Si_{6-7}Al_{2-1}O_{20}](OH,F)_4$	monocl	2.85	3.3	1.536	1.565	1.566
Leucite	$KAlSi_2O_6$	tetr	2.49	5.8	1.510		
Levyne	$(Ca,Na_2)Al_2Si_4O_{12} \cdot 6H_2O$	rhomb	2.10	4.5	1.496	1.501	
Litharge	PbO	tetr	9.35	2	2.535	2.665	
Loellingite	$FeAs_2$	rhomb	7.40	5.3			
Maghemite	Fe_2O_3	cub	4.88	7.8	2.63		
Magnesite	$MgCO_3$	hex	3.05	4	1.536	1.741	
Magnetite	Fe_3O_4	cub	5.17	6	2.42		
Malachite	$Cu_2(OH)_2(CO_3)$	monocl	4.05	3.8	1.655	1.875	1.909
Manganite	$MnO(OH)$	monocl	≈4.3	4	2.25	2.25	2.53

Name	Formula	Crystal system	Density g/cm³	Hardness	n_α	n_β	n_γ
Manganosite	MnO	cub	5.37	5.5			
Marcasite	FeS_2	cub	5.02	6.3			
Marialite	$Na_4Al_3Si_9O_{24}Cl$	tetr	2.56	5.5	1.541	1.548	
Marshite	CuI	cub	5.67	2.5	2.346		
Mascagnite	$(NH_4)_2SO_4$	orth	1.77	2.3	1.520	1.523	1.533
Matlockite	$PbClF$	tetr	7.05	2.8	2.006	2.145	
Meionite	$Ca_4Al_6Si_6O_{24}CO_3$	tetr	2.78	5.5	1.559	1.595	
Melanterite	$FeSO_4 \cdot 7H_2O$	monocl	1.89	2	1.47	1.48	1.49
Melilite	$(Ca,Na)_2(Mg,Fe,Al,Si)_3O_7$	tetr	3.00	5.5	1.639	1.645	
Mellite	$Al_2C_{12}O_{12} \cdot 18H_2O$	tetr	1.64	2.3	1.511	1.539	
Mendipite	$Pb_3O_2Cl_2$	rhomb	7.24	2.5	2.24	2.27	2.31
Mesolite	$Na_2Ca_2(Al_2Si_3O_{10})_3 \cdot 8H_2O$	orth	2.26	5	1.506		
Metacinnabar	HgS	cub	7.70	3			
Microcline	$KAlSi_3O_8$	monocl	2.56	6.3	1.522	1.526	1.530
Miersite	AgI	hex	5.68	2.5	2.20		
Millerite	NiS	hex	5.5	3.3			
Mimetite	$Pb_5(AsO_4,PO_4)_3Cl$	hex	7.24	3.8	2.128	2.147	
Minium	Pb_3O_4	tetr	8.9	2.5			
Mirabilite	$Na_2SO_4 \cdot 10H_2O$	monocl	1.46	1.8	1.394	1.396	1.398
Moissanite	SiC	hex	3.16	9.5	2.648	2.691	
Molybdenite	MoS_2	hex	5.06	1.3			
Monazite	$(Ce,La,Th)PO_4$	monocl	5.2	5	1.787	1.789	1.840
Monetite	$CaHPO_4$	tricl	2.92	3.5	1.587	1.61	1.640
Monticellite	$Ca(Mg,Fe)SiO_4$	orth	3.18	5.5	1.647	1.655	1.664
Montmorillonite	$(0.5Ca,Na)_{0.7}(Al,Mg,Fe)_4$ $[(Si,Al)_8O_{20}](OH)_4 \cdot nH_2O$	monocl	2.5	1.5	1.55	1.57	1.57
Montroydite	HgO	orth	11.14	2.5	2.37	2.50	2.65
Mordenite	$(Na,K,Ca)[Al_2Si_{10}O_{24}] \cdot 7H_2O$	orth	2.13	3.5	1.478	1.480	1.482
Muscovite	$KAl_2Si_3AlO_{10}(OH,F)_2$	monocl	2.83	2.8	1.563	1.596	1.602
Nantokite	$CuCl$	cub	4.14	2.5	1.930		
Natrolite	$Na_2Al_2Si_3O_{10} \cdot 2H_2O$	orth	2.23	5	1.478	1.481	1.491
Nepheline	$Na_3KAl_4Si_4O_{16}$	hex	2.61	5.8	1.534	1.538	
Newberyite	$MgHPO_4 \cdot 3H_2O$	orth	2.13	3.3	1.514	1.517	1.533
Niccolite	$NiAs$	hex	7.77	5.3			
Norbergite	$Mg(OH,F)_2 \cdot Mg_2SiO_4$	orth	3.21	6.5	1.565	1.573	1.592
Nosean	$Na_8Al_6Si_6O_{24}SO_4$	cub	2.35	5.5	1.495		
Oldhamite	CaS	cub	2.59	4	2.137		
Oligoclase	$([NaSi]_{0.9-0.7}[CaAl]_{0.1-0.3})AlSi_2O_8$	tricl	2.64	6.3	1.539	1.543	1.547
Olivenite	$Cu_2(AsO_4)(OH)$	rhomb	4.2	3	1.77	1.80	1.85
Olivine	$(Mg,Fe)SiO_4$	rhomb	3.81	6.8	1.73	1.76	1.78
Opal	$SiO_2 \cdot nH_2O$	amorp	1.9	5	1.44		
Orpiment	As_2S_3	monocl	3.46	1.8	2.40	2.81	3.02
Orthoclase	$KAlSi_3O_8$	monocl	2.56	6	1.523	1.527	1.531
Orthopyroxene	$(Mg,Fe)SiO_3$	rhomb	3.6	5.5	1.709	1.712	1.723
Paragonite	$NaAl_2Si_3AlO_{10}(OH)_2$	monocl	2.85	2.5	1.572	1.602	1.605
Parisite	$(Ce,La,Na)FCO_3 \cdot CaCO_3$	hex	4.42	4.5	1.672	1.771	
Pectolite	$Ca_2NaH(SiO_3)_3$	tricl	2.88	4.8	1.603	1.610	1.639
Penfieldite	$Pb_4Cl_6(OH)_2$	hex	6.6		2.13	2.21	
Pentlandite	$(Fe,Ni)_9S_8$	cub	4.8	3.8			
Percylite	$PbCuCl_2(OH)_2$	cub		2.5	2.05		
Periclase	MgO	cub	3.6	5.5	1.735		
Perovskite	$CaTiO_3$	cub	3.98	5.5	2.34		
Petalite	$LiAlSi_4O_{10}$	monocl	2.42	6.5	1.506	1.511	1.519
Pharmacosiderite	$Fe_3(AsO_4)_2(OH)_3 \cdot 5H_2O$	cub	2.80	2.5	1.690		
Phenakite	Be_2SiO_4	rhomb	2.98	7.5	1.654	1.670	
Phillipsite	$K(Ca_{0.5},Na)_2[Al_3Si_5O_{16}] \cdot 6H_2O$	monocl	2.2	4.3	1.494	1.497	1.505
Phlogopite	$KMg_3AlSi_3O_{10}(OH,F)_2$	monocl	2.83	2.3	1.560	1.597	1.598
Phosgenite	$Pb_2(CO_3)Cl_2$	tetr	6.13	2.5	2.118	2.145	
Piemontite	$Ca_2(Mn,Fe,Al)_3O(Si_2O_7)(SiO_4)(OH)$	monocl	3.49	6	1.762	1.773	1.796

Name	Formula	Crystal system	Density g/cm^3	Hard-ness	Index of refraction n_α	n_β	n_γ
Pigeonite	$(Mg,Fe,Ca)(Mg,Fe)Si_2O_6$	monocl	3.38	6	1.702	1.703	1.728
Pollucite	$CsAlSi_2O_6$	tetr	2.9	6.5	1.517		
Polybasite	$(Ag,Cu)_{16}Sb_2S_{11}$	monocl	6.1	2.5			
Powellite	$Ca(Mo,W)O_4$	tetr	4.35	3.8	1.971	1.980	
Prehnite	$Ca_2Al_2Si_3O_{10}(OH)_2$	rhomb	2.93	6.3	1.622	1.628	1.648
Proustite	Ag_3AsS_3	rhomb	5.57	2.3	2.792	3.088	
Pseudobrookite	Fe_2TiO_5	rhomb	4.36	6	2.38	2.39	2.42
Psilomelane	$BaMn_9O_{16}(OH)_4$	rhomb	4.71	5.5			
Pumpellyite	$Ca_2Al_2(Al,Fe,Mg)[Si_2(O,OH)_7]$ $(SiO_4)(OH,O)_3$	monocl	3.21	5.5	1.688	1.695	1.705
Pyrargyrite	Ag_3SbS_3	rhomb	5.85	2.5	2.88	3.08	
Pyrite	FeS_2	cub	5.02	6.3			
Pyrochlore	$NaCaNb_2O_6F$	cub	5.3	5.3			
Pyrochroite	$Mn(OH)_2$	hex	3.26	2.5	1.68	1.72	
Pyrolusite	MnO_2	tetr	5.08	6.3			
Pyromorphite	$Pb_5(PO_4,AsO_4)_3Cl$	hex	7.04	3.8	2.048	2.058	
Pyrope	$Mg_3Al_2Si_3O_{12}$	cub	3.58	6.8	1.714		
Pyrophyllite	$Al_2Si_4O_{10}(OH)_2$	monocl	2.78	1.5	1.545	1.579	1.599
Pyrrhotite	Fe_7S_8	hex	4.62	4			
Quartz	SiO_2	hex	2.65	7	1.544	1.553	
Rammelsbergite	$NiAs_2$	orth	7.1	5.8			
Raspite	$PbWO_4$	monocl	8.46	2.8	1.27	1.27	1.30
Realgar	As_4S_4	monocl	3.5	1.8	2.538	2.684	2.704
Rhodochrosite	$MnCO_3$	hex	3.70	3.8	1.597	1.816	
Rhodonite	$(Mn,Fe,Ca)SiO_3$	orth	3.48	6	1.725	1.729	1.737
Riebeckite	$Na_2Fe_5(Si_8O_{22})(OH)_2$	monocl	3.3	5	1.675	1.683	1.694
Rutile	TiO_2	tetr	4.23	6.2	2.609	2.900	
Safflorite	$(Co,Fe)As_2$	rhomb	7.3	4.8			
Samarskite	(Y,Er,Ce,U,Ca,Fe,Pb,Th) $(Nb,Ta,Ti,Sn)_2O_6$	rhomb	5.69	5.5	2.200		
Sapphirine	$(Mg,Fe)_2Al_4O_6SiO_4$	monocl	3.49	7.5	1.709	1.712	1.715
Scapolite	$(Na,Ca)_4Al_3(Al,Si)_3Si_6O_{24}$ (Cl,F,OH,CO_3,SO_4)	tetr	2.64	5.5	1.551	1.573	
Scheelite	$CaWO_4$	tetr	6.06	4.8	1.920	1.936	
Scolecite	$CaAl_2Si_3O_{10}\cdot3H_2O$	monocl	2.27	5	1.510	1.518	1.519
Scorodite	$Fe(AsO_4)\cdot2H_2O$	rhomb	3.28	3.8	1.784	1.795	1.814
Sellaite	MgF_2	tetr	3.15	5	1.378	1.390	
Senarmontite	Sb_2O_3	cub	5.58	2.3	2.087		
Serpentine	$Mg_3Si_2O_5(OH)_4$	monocl	2.55	3	1.55	1.56	1.56
Siderite	$FeCO_3$	hex	3.9	4.3	1.635	1.875	
Sillimanite	Al_2OSiO_4	rhomb	3.25	7	1.658	1.660	1.660
Skutterudite	$(Co,Ni)As_3$	cub	6.8	5.8			
Smithsonite	$ZnCO_3$	rhomb	4.4	4.3	1.621	1.848	
Sodalite	$Na_8Al_6Si_6O_{24}Cl_2$	cub	2.30	5.8	1.485		
Sperrylite	$PtAs_2$	cub	10.58	6.5			
Spessartite	$Mn_3Al_2Si_3O_{12}$	cub	4.19	6.8	1.800		
Sphalerite	ZnS	cub	4.0	3.8	2.369		
Sphene	$CaTiSiO_4(O,OH,F)$	monocl	3.50	5	1.90	1.95	2.03
Spinel	$MgAl_2O_4$	cub	3.55	7.8	1.719		
Spodumene	$LiAlSi_2O_6$	monocl	3.13	6.8	1.656	1.662	1.671
Stannite	Cu_2FeSn_4	tetr	4.4	4			
Staurolite	$(Fe,Mg,Zn)_2(Al,Fe,Ti)_9O_6$ $[(Si,Al)O_4]_4(O,OH)_2$	monocl	3.79	7.5	1.743	1.747	1.755
Stercorite	$Na(NH_4)H(PO_4)\cdot4H_2O$	tricl	1.62	2	1.439	1.442	1.469
Stibiotantalite	$Sb(Ta,Nb)O_4$	rhomb	6.6	5.5	2.38	2.41	2.46
Stibnite	Sb_2S_3	orth	4.56	2			
Stilbite	$NaCa_2[Al_5Si_{13}O_{36}]\cdot14H_2O$	monocl	2.2	3.8	1.492	1.499	1.503

Name	Formula	Crystal system	Density g/cm³	Hardness	Index of refraction		
					n_α	n_β	n_γ
Stilpnomelane	$(K,Na,Ca)_{0.6}(Fe,Mg)_6Si_8Al(O,OH)_{27} \cdot 2H_2O$	monocl	2.8	3.5	1.585	1.665	1.665
Stolzite	$PbWO_4$	tetr	8.2	2.8	2.19	2.27	
Strengite	$FePO_4 \cdot 2H_2O$	orth	2.87	4	1.707	1.719	1.741
Strontianite	$SrCO_3$	orth	3.5	3.5	1.518	1.666	1.668
Struvite	$Mg(NH_4)(PO_4) \cdot 6H_2O$	rhomb	1.71	2	1.495	1.496	1.504
Sulfur	S	orth	2.07	2	1.958	2.038	2.245
Sylvanite	$(Ag,Au)Te_2$	monocl	8.16	1.8			
Sylvite	KCl	cub	1.99	2	1.490		
Talc	$Mg_3Si_4O_{10}(OH)_2$	monocl	2.71	1	1.545	1.592	1.595
Tantalite	$(Fe,Mn)(Ta,Nb)_2O_6$	rhomb	7.95	6.5	2.26	2.32	2.43
Tapiolite	$FeTa_2O_6$	tetr	7.9	6.3	2.27	2.42	
Tellurobismuthite	Bi_2Te_3	hex	7.74	1.8			
Terlinguaite	Hg_2CCl	monocl	8.73	2.5	2.35	2.64	2.66
Tetrahedrite	$(Cu,Fe)_{12}Sb_4S_{13}$	cub	4.9	3.8			
Thenardite	Na_2SO_4	orth	2.7	2.8	1.468	1.475	1.483
Thermonatrite	$Na_2CO_3 \cdot H_2O$	orth	2.25	1.3	1.420	1.506	1.524
Thomsenolite	$NaCaAlF_6 \cdot H_2O$	monocl	2.98	2	1.407	1.414	1.415
Thorianite	ThO_2	cub	10.0	6.5	2.200		
Thorite	$ThSiO_4$	tetr	6.7	4.8	1.8		
Topaz	$Al_2SiO_3(OH,F)_2$	rhomb	3.53	8	1.618	1.620	1.627
Torbernite	$Cu(UO_2)_2(PO_4)_2 \cdot 8H_2O$	tetr	3.22	2.3	1.582	1.592	
Tourmaline	$Na(Mg,Fe,Mn,Li,Al)_3Al_6Si_6O_{18}(BO_3)_3$	rhomb	3.14	7	1.62	1.65	
Tremolite	$Ca_2Mg_5Si_8O_{22}(OH,F)_2$	monocl	3.0	5.5	1.599	1.612	1.622
Trevorite	$NiFe_2O_4$	cub	5.33	7.8	2.3		
Tridymite	SiO_2	hex	2.27	7	1.475	1.476	1.479
Triphyllite-Lithiophyllite	$Li(Fe,Mn)PO_4$	rhomb	3.46	4.5	1.68	1.68	1.69
Troegerite	$(UO_2)_3(AsO_4)_2 \cdot 12H_2O$	tetr		2.5	1.59	1.630	
Troilite	FeS	hex	4.7	4			
Trona	$Na_3H(CO_3)_2 \cdot 2H_2O$	monocl	2.14	2.8	1.412	1.492	1.540
Turquois	$Cu(Al,Fe)_6(PO_4)_4(OH)_8 \cdot 4H_2O$	tricl	2.9	5.3	1.70	1.73	1.75
Ullmannite	$NiSbS$	cub	6.65	5.3			
Uraninite	UO_2	cub	11.0	5.5			
Uvarovite	$Ca_3Cr_2Si_3O_{12}$	cub	3.83	6.8	1.865		
Valentinite	Sb_2O_3	orth	5.7	2.8	2.18	2.35	2.35
Vanadinite	$Pb_5(VO_4)_3Cl$	hex	6.8	2.9	2.350	2.416	
Variseite-Strengite	$(Al,Fe)(PO_4) \cdot 2H_2O$	rhomb	2.72	4	1.635	1.654	1.668
Vaterite	$CaCO_3$	hex	2.71		1.550	1.645	
Vermiculite	$(Mg,Ca)_{0.7}(Mg,Fe,Al)_6[(Al,Si)_8O_{20}](OH)_4 \cdot 8H_2O$	monocl	2.3	1.5	1.542	1.556	1.556
Vesuvianite	$Ca_{10}(Mg,Fe)_2Al_4(Si_2O_7)_2(SiO_4)_5(OH,F)_4$	tetr	3.33	6.5	1.72	1.73	
Villiaumite	NaF	cub	2.78	2.3	1.327		
Vivianite	$Fe_3(PO_4)_2 \cdot 8H_2O$	monocl	2.58	1.8	1.598	1.629	1.652
Wagnerite	$Mg_2(PO_4)F$	monocl	3.15	5.3	1.568	1.572	1.582
Wavellite	$Al_3(OH)_3(PO_4)_2 \cdot 5H_2O$	rhomb	2.36	3.6	1.527	1.535	1.553
Whewellite	$CaC_2O_4 \cdot H_2O$	cub	2.2	2.8	1.491	1.554	1.650
Willemite	Zn_2SiO_4	hex	4.1	5.5	1.691	1.719	
Witherite	$BaCO_3$	orth	4.29	3.5	1.529	1.676	1.677
Wolframite	$(Fe,Mn)WO_4$	monocl	7.3	4.3	2.26	2.32	2.42
Wollastonite	$CaSiO_3$	monocl	2.92	4.8	1.628	1.639	1.642
Wulfenite	$PbMoO_4$	tetr	6.7	2.9	2.283	2.403	
Wurtzite	ZnS	hex	4.09	3.8	2.356	2.378	
Xenotime	YPO_4	tetr	4.8	4.5	1.721	1.816	
Zeunerite	$Cu(UO_2)_2(AsO_4)_2 \cdot 10H_2O$	tetr			1.606		
Zincite	ZnO	hex	5.6	4	2.013	2.029	
Zircon	$ZrSiO_4$	tetr	4.6	7.5	1.94	1.99	
Zoisite	$Ca_2Al_3Si_3O_{12}(OH)$	rhomb	3.26	6	1.695	1.699	1.711

CRYSTALLOGRAPHIC DATA ON MINERALS

This table contains x-ray crystallographic data on about 400 common minerals, as well as selected crystalline elements. Entries are arranged alphabetically by mineral name. The columns are:

Name: Common name of the mineral.

Formula: Chemical formula for a typical sample of the mineral. Composition often varies considerably with the origin of the sample.

Crystal system: tricl = triclinic; monocl = monoclinic; orth = orthorhombic; tetr = tetragonal; hex = hexagonal; rhomb = rhombohedral; cubic = cubic.

Structure type: Prototype for the structural arrangement of the crystallographic cell.

Z: Number of formula units per the unit cell.

a, b, c: Lengths of the cell edges in Å ($1Å = 10^{-8}$ cm).

α, β, γ : Angles between cell axes.

REFERENCES

1. Robie, R.A., Bethke, P.M., and Beardsley, K.M., *U. S. Geological Survey Bulletin 1248*, U. S. Government Printing Office, Washington, D.C.
2. Donnay, J.D.H., and Ondik, H.M., *Crystal Data Determinative Tables, Third Edition, Volume 2, Inorganic Compounds*, Joint Committee on Powder Diffraction Standards, Swarthmore, PA, 1973.
3. Deer, W.A., Howie, R.A., and Zussman, J., *An Introduction to the Rock-Forming Minerals, 2nd Edition*, Longman Scientific & Technical, Harlow, Essex, 1992.

Name	Formula	Crystal system	Structure type	Z	a/Å	b/Å	c/Å	α	β	γ
Acanthite	Ag_2S	monocl		4	4.228	6.928	7.862		99.58°	
Acmite (Aegirine)	$NaFe(SiO_3)_2$	monocl	diopside	4	9.658	8.795	5.294		107.42°	
Akermanite	$Ca_2MgSi_2O_7$	tetr	melilite	2	7.8435		5.010			
Alabandite	MnS	cubic	rock salt	4	5.223					
Almandine (Almandite)	$Fe_3Al_2Si_3O_{12}$	cubic	garnet	8	11.526					
Altaite	$PbTe$	cubic	rock salt	4	6.4606					
Aluminum	Al	cubic	copper	4	4.049					
Alunite	$KAl_3(SO_4)_2(OH)_6$	rhomb		3	6.982		17.32			
Analcite	$NaAlSi_2O_6 \cdot H_2O$	cubic		16	13.733					
Anatase	TiO_2	tetr		4	3.785		9.514			
Andalusite	Al_2OSiO_4	orth		4	7.7959	7.8983	5.5583			
Andradite	$Ca_3Fe_2Si_3O_{12}$	cubic	garnet	8	12.048					
Anglesite	$PbSO_4$	orth	barite	4	8.480	5.398	6.958			
Anhydrite	$CaSO_4$	orth	anhydrite	4	6.991	6.996	6.238			
Annite	$KFe_3[AlSi_3O_{10}](OH)_2$	monocl	1M mica	2	10.29	9.33	5.39		105.1°	
Anorthite	$CaAl_2Si_2O_8$	tricl	primitive cell	8	8.177	12.877	14.169	93.17°	115.85°	91.22°
Anthophyllite	$Mg_7Si_8O_{22}(OH)_2$	orth		4	18.61	18.01	5.24			
Antimony	Sb	rhomb	arsenic	6	4.2996		11.2516			
Aragonite	$CaCO_3$	orth	aragonite	4	5.741	7.968	4.959			
Arcanite	K_2SO_4	orth	arcanite	4	5.772	10.072	7.483			
Argentite	Ag_2S	cubic		2	4.870					
Argentopyrite	$AgFe_2S_3$	orth		4	6.64	11.47	6.45			
Arsenic	As	rhomb	arsenic	6	3.760		10.555			
Arsenolite	As_2O_3	cubic	diamond	16	11.074					
Arsenopyrite	$FeAsS$	tricl		4	5.760	5.690	5.785	90.00°	112.23°	90.00°
Azurite	$Cu_3(OH)_2(CO_3)_2$	monocl		2	5.008	5.844	10.336		92.45°	
Baddeleyite	ZrO_2	monocl	baddeleyite	4	5.1454	5.2075	5.3107		99.23°	
Banalsite	$BaNa_2Al_4Si_4O_{16}$	orth		4	8.50	9.97	16.72			
Barite	$BaSO_4$	orth	barite	4	8.878	5.450	7.152			
Berlinite	$AlPO_4$	hex	α-quartz	3	4.942		10.97			
Beryl	$Be_3Al_2(SiO_3)_6$	hex	beryl	2	9.215		9.192			
Berzelianite	Cu_2Se	cubic	pseudo-orth	4	5.85					
Bismite	Bi_2O_3	monocl		4	7.48	8.14	5.83		112.9°	
Bismuth	Bi	rhomb	arsenic	6	4.5367		11.8383			
Bismuthinite	Bi_2S_3	orth	stibnite	4	11.150	11.300	3.981			
Bixbyite	Mn_2O_3	cubic	thallium trioxide	16	9.411					
Boehmite	$AlO(OH)$	orth	lepidocrocite	4	2.868	12.227	3.700			
Borax	$Na_2B_4O_7 \cdot 10H_2O$	monocl		8	11.858	10.674	12.197		106.68°	
Bornite (metastable)	Cu_5FeS_4	cubic		2	10.94					
Breithauptite	$NiSb$	hex	niccolite	2	3.942		5.155			
Brochantite	$Cu_4SO_4(OH)_6$	monocl		4	13.066	9.85	6.022		103.27°	
Bromargyrite	$AgBr$	cubic	rock salt	4	5.7745					
Bromellite	BeO	hex	zincite	2	2.6979		4.3772			
Brookite	TiO_2	orth		8	5.456	9.182	5.143			
Brucite	$Mg(OH)_2$	hex	cadmium iodide	1	3.147		4.769			

Name	Formula	Crystal system	Structure type	Z	$a/\text{Å}$	$b/\text{Å}$	$c/\text{Å}$	α	β	γ
Bunsenite	NiO	cubic	rock salt	4	4.177					
Bustamite	$CaMn(SiO_3)_2$	tricl		6	7.736	7.157	13.824	90.52°	94.58°	103.87°
Cadmium telluride	CdTe	cubic	sphalerite	4	6.4805					
Cadmoselite	CdSe	hex	zincite	2	4.2977		7.0021			
Calcite	$CaCO_3$	rhomb	calcite	6	4.9899		17.064			
Calomel	Hg_2Cl_2	tetr		4	4.478		10.910			
Carbonate-apatite	$Ca_{10}(PO_4)_6CO_3 \cdot H_2O$	hex	apatite	1	9.436		6.883			
Cassiterite	SnO_2	tetr	rutile	2	4.738		3.188			
Cattierite	CoS_2	cubic	pyrite	4	5.5345					
Celestite	$SrSO_4$	orth	barite	4	8.359	5.352	6.866			
Celsian	$BaAl_2Si_2O_8$	monocl		8	8.627	13.045	14.408		115.20°	
Cerianite	CeO_2	cubic	fluorite	4	5.4110					
Cerussite	$PbCO_3$	orth	aragonite	4	6.152	8.436	5.195			
Cervantite	Sb_2O_4	orth		4	5.424	11.76	4.804			
Chalcanthite	$CuSO_4 \cdot 5H_2O$	tricl		2	6.1045	10.72	5.949	97.57°	107.28°	77.43°
Chalcocite	Cu_2S	orth		96	11.881	27.323	13.491			
Chalcopyrite	$CuFeS_2$	tetr		4	5.2988		10.434			
Chlorapatite	$Ca_5(PO_4)_3Cl$	hex	apatite	2	9.629		6.777			
Chlorargyrite	AgCl	cubic	rock salt	4	5.5491					
Chloritoid	$FeAl_2O_2(SiO_4)_2(OH)_4$	monocl		8	9.48	5.48	18.18		101.77°	
Chloromagnesite	$MgCl_2$	rhomb		3	3.632		17.795			
Chondrodite	$2Mg_2SiO_4 \cdot MgF_2$	monocl		2	7.89	4.743	10.29		109.03°	
Chrysoberyl	$BeAl_2O_4$	orth	olivine	4	5.4756	9.4041	4.4267			
Cinnabar	HgS	hex	cinnabar	3	4.149		9.495			
Claudetite	As_2O_3	monocl		4	5.339	12.984	4.5405		94.27°	
Clausthalite	PbSe	cubic	rock salt	4	6.1255					
Clinoenstatite	$MgSiO_3$	monocl		8	9.620	8.825	5.188		108.33°	
Clinoferrosilite	$FeSiO_3$	monocl		8	9.7085	9.0872	5.2284		108.43°	
Clinohumite	$4Mg_2SiO_4 \cdot MgF_2$	monocl		2	13.68	4.75	10.27		100.83°	
Clinozoisite	$Ca_2Al_3(SiO_4)_3OH$	monocl		2	8.887	5.581	10.14		115.93°	
Cobalt olivine	Co_2SiO_4	orth	olivine	4	4.782	10.301	6.003			
Cobalt oxide	CoO	cubic	rock salt	4	4.260					
Cobalt sulfide	CoS	cubic	sphalerite	4	5.339					
Cobalt titanate	$CoTiO_3$	rhomb	ilmenite	6	5.066		13.918			
Cobalticalcite	$CoCO_3$	rhomb	calcite	6	4.6581		14.958			
Cobaltite	CoAsS	cubic	NiSbS	4	5.60					
Coesite	SiO_2	monocl		16	7.152	12.379	7.152		120.00°	
Coffinite	$USiO_4$	tetr	zircon	4	6.995		6.263			
Colemanite	$Ca_2B_6O_{11} \cdot 5H_2O$	monocl		4	8.743	11.264	6.102		110.12°	
Coloradoite	HgTe	cubic	sphalerite	4	6.4600					
Cooperite	PtS	tetr		2	3.4699		6.1098			
Copper	Cu	cubic	face-centered cubic	4	3.6150					
Corundum	Al_2O_3	rhomb	corundum	6	4.7591		12.9894			
Cotunnite	$PbCl_2$	orth		4	4.535	7.62	9.05			
Covellite	CuS	hex		6	3.792		16.34			

Name	Formula	Crystal system	Structure type	Z	$a/\text{Å}$	$b/\text{Å}$	$c/\text{Å}$	α	β	γ
Cristobalite (α)	SiO_2	tetr		4	4.971		6.918			
Cristobalite (β)	SiO_2	cubic		8	7.1382					
Cryolite	Na_3AlF_6	monocl		2	5.40	5.60	7.776		90.18°	
Cubanite	$CuFe_2S_3$	orth		4	6.46	11.12	6.23			
Cummingtonite	$(Mg,Fe,Mn)_7(Si_4O_{11})_2(OH)_2$	monocl	tremolite	2	9.522	18.223	5.332		101.92°	
Cuprite	Cu_2O	cubic		2	4.2696					
Danburite	$CaB_2Si_2O_8$	orth		4	8.04	8.77	7.74			
Datolite	$CaBSiO_4(OH)$	monocl		4	9.62	7.60	4.84		90.15°	
Daubreeite	$FeCr_2S_4$	cubic	spinel	8	9.966					
Diamond	C	cubic	diamond	8	3.5670					
Diaspore	$AlO(OH)$	orth		4	4.401	9.421	2.845			
Dickite	$Al_2Si_2O_5(OH)_4$	monocl		4	5.150	8.940	14.736		103.58°	
Digenite	$Cu_{1.79}S$	cubic	deformed fluorite	4	5.5695					
Diopside	$CaMg(SiO_3)_2$	monocl	diopside	4	9.743	8.923	5.251		105.93°	
Dioptase	$CuSiO_2(OH)_2$	rhomb	phenacite	18	14.61		7.80			
Dolerophanite	$Cu_2O(SO_4)$	monocl		4	8.334	6.312	7.628		108.4°	
Dolomite	$CaMg(CO_3)_2$	rhomb	calcite	3	4.8079		16.010			
Dravite	$NaMg_3Al_6B_3Si_6O_{27}(OH)_4$	rhomb	tourmaline	3	15.942		7.224			
Elbaite	$NaLiAl_{7.67}B_3Si_6O_{27}(OH)_4$	rhomb	tourmaline	3	15.842		7.009			
Enargite	Cu_3AsS_4	orth		2	6.426	7.422	6.144			
Enstatite	$MgSiO_3$	orth		16	8.829	18.22	5.192			
Epidote	$Ca_2Al_2(Al,Fe)OH(SiO_4)_3$	monocl		2	8.89	5.63	10.19		115.40°	
Epsomite	$MgSO_4 \cdot 7H_2O$	orth		4	11.86	11.99	6.858			
Eskolaite	Cr_2O_3	rhomb	corundum	6	4.9607		13.599			
Eucairite	$AgCuSe$	orth		10	4.105	20.35	6.31			
Euclase	$AlBeSiO_4(OH)$	monocl		4	4.763	14.29	4.618		100.25°	
Famatimite	Cu_3SbS_4	tetr		2	5.384		10.770			
Fayalite	Fe_2SiO_4	orth	olivine	4	4.817	10.477	6.105			
Fe-Cordierite	$Fe_2Al_3(AlSi_5O_{18})$	orth	cordierite	4	9.726	17.065	9.287			
Fe-Gehlenite	$Ca_2Fe_2SiO_7$	tetr	melilite	2	7.54		4.855			
Fe-Indialite	$Fe_2Al_3(AlSi_5O_{18})$	hex	beryl	2	9.860		9.285			
Fe-Leucite	$KFeSi_2O_6$	tetr		16	13.205		13.970			
Fe-Microcline	$KFeSi_3O_8$	tricl		4	8.68	13.10	7.340	90.75°	116.05°	86.23°
Fe-Sanidine	$KFeSi_3O_8$	monocl		4	8.689	13.12	7.319		116.10°	
Fe-Skutterudite	$FeAs_{2.95}$	cubic		8	8.1814					
Ferberite	$FeWO_4$	monocl	wolframite	2	4.732	5.708	4.965		90.00°	
Ferriannite	$KFe_3[FeSi_3O_{10}](OH)_2$	monocl		2	5.430	9.404	10.341		100.07°	
Ferroselite	$FeSe_2$	orth	marcasite	2	4.801	5.778	3.587			
Ferrotremolite	$Ca_2Fe_5[Si_8O_{22}](OH)_2$	monocl	tremolite	2	9.97	18.34	5.30		104.50°	
Fluor-edenite	$NaCa_2Mg_5[AlSi_7O_{22}]F_2$	monocl	tremolite	2	9.847	18.00	5.282		104.83°	
Fluor-humite	$3Mg_2SiO_4 \cdot MgF_2$	orth		4	10.243	20.72	4.735			
Fluor-norbergite	$Mg_2SiO_4 \cdot MgF_2$	orth		4	8.727	10.271	4.709			
Fluor-phlogopite	$KMg_3[AlSi_3O_{10}]F_2$	monocl	1M mica	2	5.299	9.188	10.135		99.92°	
Fluor-richterite	$Na_2CaMg_5[Si_8O_{22}]F_2$	monocl	tremolite	2	9.823	17.96	5.268		104.33°	
Fluor-tremolite	$Ca_2Mg_5[Si_8O_{22}]F_2$	monocl	tremolite	2	9.781	18.01	5.267		104.52°	

Name	Formula	Crystal system	Structure type	Z	a/Å	b/Å	c/Å	α	β	γ
Fluorapatite	$Ca_5(PO_4)_3F$	hex	apatite	2	9.3684		6.8841			
Fluorite	CaF_2	cubic	fluorite	4	5.4638					
Forsterite	Mg_2SiO_4	orth	olivine	4	4.758	10.214	5.984			
Frohbergite	$FeTe_2$	orth	marcasite	2	5.265	6.265	3.869			
Gahnite	$ZnAl_2O_4$	cubic	spinel	8	8.0848					
Galaxite	$MnAl_2O_4$	cubic	spinel	8	8.258					
Galena	PbS	cubic	rock salt	4	5.9360					
Gallium oxide	Ga_2O_3	rhomb	corundum	6	4.9793		13.429			
Gehlenite	$Ca_2Al_2SiO_7$	tetr	melilite	2	7.690		5.0675			
Geikielite	$MgTiO_3$	rhomb	ilmenite	6	5.054		13.898			
Gerhardite	$Cu_2(NO_3)(OH)_3$	orth		4	6.075	13.812	5.592			
Gersdorffite	$NiAsS$	cubic		4	5.693					
Gibbsite	$Al(OH)_3$	monocl		8	9.719	5.0705	8.6412		94.57°	
Glauchroite	$CaMnSiO_4$	orth	olivine	4	4.944	11.19	6.529			
Glaucodot	$(Co,Fe)AsS$	orth		24	6.64	28.39	5.64			
Glaucophane I	$Na_2Mg_3Al_2[Si_8O_{22}](OH)_2$	monocl	tremolite	2	9.748	17.915	5.273		102.78°	
Glaucophane II	$Na_2Mg_3Al_2[Si_8O_{22}](OH)_2$	monocl	tremolite	2	9.663	17.696	5.277		103.67°	
Goethite	$FeO(OH)$	orth		4	4.596	9.957	3.021			
Gold	Au	cubic	face-centered cubic	4	4.0786					
Goldmanite	$Ca_3V_2Si_3O_{12}$	cubic	garnet	8	12.070					
Goslarite	$ZnSO_4·7H_2O$	orth	epsomite	4	11.779	12.050	6.822			
Graphite	C	hex	graphite	4	2.4612		6.7079			
Greenockite	CdS	hex	zincite	2	4.1354		6.7120			
Greigite	Fe_3S_4	cubic	spinel	8	9.876					
Grossularite	$Ca_3Al_2Si_3O_{12}$	cubic	garnet	8	11.851					
Grunerite	$Fe_7[Si_8O_{22}](OH)_2$	monocl	tremolite	2	9.572	18.44	5.342		101.77°	
Gudmundite	$FeSbS$	monocl		8	10.00	5.93	6.73		90.00°	
Gypsum	$CaSO_4·2H_2O$	monocl		4	5.68	15.18	6.29		113.83°	
Hafnia	HfO_2	monocl	baddeleyite	4	5.1156	5.1722	5.2948		99.18°	
Halite	$NaCl$	cubic	rock salt	4	5.6402					
Hambergite	$Be_2(OH,F)BO_3$	orth		8	9.755	12.201	4.426			
Hardystonite	$Ca_2ZnSi_2O_7$	tetr	melilite	2	7.87		5.01			
Hauerite	MnS_2	cubic	pyrite	4	6.1014					
Hausmannite	Mn_3O_4	tetr		8	8.136		9.422			
Hawleyite	CdS	cubic	sphalerite	4	5.833					
Heazelwoodite	Ni_3S_2	rhomb		3	5.746		7.134			
Hedenbergite	$CaFe(SiO_3)_2$	monocl	diopside	4	9.854	9.024	5.263		104.23°	
Hematite	Fe_2O_3	rhomb	corundum	6	5.0329		13.749			
Hemimorphite	$Zn_4(OH)_2Si_2O_7·H_2O$	orth		2	8.370	10.719	5.120			
Hercynite	$Fe(AlO_2)_2$	cubic	spinel	8	8.150					
Herzenbergite	SnS	orth	germanium sulfide	4	4.328	11.190	3.978			
Hessite	Ag_2Te	monocl		4	8.13	4.48	8.09		111.9°	
Hexahydrite	$MgSO_4·6H_2O$	monocl		8	10.110	7.212	24.41		98.30°	
High albite (Analbite)	$NaAlSi_3O_8$	tricl		4	8.160	12.870	7.106	93.54°	116.36°	90.19°
High argentite	Ag_2S	cubic		4	6.269					

Name	Formula	Crystal system	Structure type	Z	a/Å	b/Å	c/Å	α	β	γ
High bornite	Cu_5FeS_4	cubic		1	5.50					
High carnegeite	$NaAlSiO_4$	cubic		4	7.325					
High chalcocite	Cu_2S	hex		2	3.961		6.722			
High clinoenstatite	$MgSiO_3$	tricl		8	10.000	8.934	5.170	88.27°	70.03°	91.01°
High digenite	Cu_2S	cubic		4	5.725					
High germania	GeO_2	hex	α–quartz	3	4.987		5.652			
High leucite	$KAlSi_2O_6$	cubic		16	13.43					
High naumanite	Ag_2Se	cubic		2	4.993					
High sanidine	$KAlSi_3O_8$	monocl		2	8.615	13.031	7.177		115.98°	
Huebnerite	$MnWO_4$	monocl	wolframite	2	4.834	5.758	4.999		91.18°	
Huntite	$Mg_3Ca(CO_3)_4$	rhomb	calcite	3	9.498		7.816			
Hydroxylapatite	$Ca_5(PO_4)_3OH$	hex	apatite	2	9.418		6.883			
Ice	H_2O	hex		4	4.5212		7.3666			
Ilmenite	$FeTiO_3$	rhomb	ilmenite	6	5.093		14.055			
Indialite (Cordierite)	$Mg_2Al_3(AlSi_5O_{18})$	hex	beryl	2	9.7698		9.3517			
Iodargyrite	AgI	hex	zincite	2	4.5955		7.5005			
Iron (α)	Fe	cubic	body-centered cubic	2	2.8664					
Jacobsite	$MnFe_2O_4$	cubic	spinel	8	8.499					
Jadeite	$NaAl(SiO_3)_2$	monocl	diopside	4	9.409	8.564	5.220		107.50°	
Jalpaite	$Ag_{1.5}Cu_{0.45}S$	tetr		16	8.673		11.756			
Johannsenite	$CaMn(SiO_3)_2$	monocl	diopside	4	9.83	9.04	5.27		105.00°	
Kaliophilite	$KAlSiO_4$	hex		54	26.930		8.522			
Kalsilite	$KAlSiO_4$	hex		2	5.1597		8.7032			
Kaolinite	$Al_2Si_2O_5(OH)_4$	tricl		2	5.155	8.959	7.407	91.68°	104.87°	89.93°
Karelianite	V_2O_3	rhomb	corundum	6	4.952		14.002			
Keatite	SiO_2	tetr		12	7.456		8.604			
Kernite	$Na_2B_4O_7 \cdot 4H_2O$	monocl		4	7.022	9.151	15.676		108.83°	
Kerschsteinite	$CaFeSiO_4$	orth	olivine	4	4.886	11.146	6.434			
Klockmannite	$CuSe$	hex	deformed covellite	78	14.206		17.25			
Knebelite	$MnFeSiO_4$	orth	olivine	4	4.854	10.602	6.162			
Kyanite	Al_2OSiO_4	tricl		4	7.123	7.848	5.564	89.92°	101.25°	105.97°
Larnite	Ca_2SiO_4	monocl		4	5.48	6.76	9.28		94.55°	
Laurite	RuS_2	cubic	pyrite	4	5.60					
Lawrencite	$FeCl_2$	rhomb		3	3.593		17.58			
Lawsonite	$CaAl_2Si_2O_7(OH)_2 \cdot H_2O$	orth		4	8.787	5.836	13.123			
Lead	Pb	cubic	face-centered cubic	4	4.9505					
Leonhardtite	$MgSO_4 \cdot 4H_2O$	monocl		4	5.922	13.604	7.905		90.85°	
Lepidocrocite	$FeO(OH)$	orth		4	3.868	12.525	3.066			
Lepidolite	$K_2Al_3Li_2AlSi_7O_{20}(OH)_4$	monocl	2M2 mica	2	9.2	5.3	20.0		98.00°	
Leucite	$KAlSi_2O_6$	tetr		16	13.074		13.738			
Lime	CaO	cubic	rock salt	4	4.8108					
Lime olivine	Ca_2SiO_4	orth	olivine	4	5.091	11.371	6.782			
Linnaeite	Co_3S_4	cubic	spinel	8	9.401					
Litharge	PbO	tetr		2	3.9759		5.023			
Loellingite	$FeAs_2$	orth	marcasite	2	5.300	5.981	2.882			

CRYSTALLOGRAPHIC DATA ON MINERALS (continued)

Name	Formula	Crystal system	Structure type	Z	a/Å	b/Å	c/Å	α	β	γ
Low albite	$NaAlSi_3O_8$	tricl		4	8.139	12.788	7.160	94.27°	116.57°	87.68°
Low bornite	Cu_5FeS_4	tetr		16	10.94		21.88			
Low cordierite	$Mg_2Al_3(AlSi_5O_{18})$	orth		4	9.721	17.062	9.339			
Low germania	GeO_2	tetr	rutile	2	4.3963		2.8626			
Low nepheline	$NaAlSiO_4$	hex		8	9.986		8.330			
Luzonite	Cu_3AsS_4	tetr		2	5.289		10.440			
Mackinawite	FeS	tetr		2	3.675		5.030			
Magnesioriebeckite	$Na_2Mg_3Fe_2[Si_8O_{22}](OH)_2$	monocl	tremolite	2	9.733	17.946	5.299		103.30°	
Magnesite	$MgCO_3$	rhomb	calcite	6	4.6330		15.016			
Magnetite	Fe_3O_4	cubic	spinel	8	8.3940					
Malachite	$Cu_2(OH)_2CO_3$	monocl		4	9.502	11.974	3.240		98.75°	
Maldonite	Au_2Bi	cubic		8	7.958					
Manganese sulfide (γ)	MnS	hex	zincite	2	3.976		6.432			
Manganese sulfide (β)	MnS	cubic	sphalerite	4	5.611					
Manganosite	MnO	cubic	rock salt	4	4.4448					
Marcasite	FeS_2	orth	marcasite	2	4.443	5.423	3.3876			
Margarite	$CaAl_2[AlSi_2O_{10}](OH)_2$	monocl	2M mica	4	5.13	8.92	19.50		95.00°	
Marialite	$Na_4Al_3Si_9O_{24}Cl$	tetr		2	12.064		7.514			
Marshite	CuI	cubic	sphalerite	4	6.0507					
Mascagnite	$(NH_4)_2SO_4$	orth	arcanite	4	7.782	5.993	10.636			
Massicot	PbO	orth		4	5.489	4.755	5.891			
Matlockite	$PbClF$	tetr		2	4.106		7.23			
Maucherite	$Ni_{11}As_8$	tetr		4	6.870		21.81			
Meionite	$Ca_4Al_6Si_6O_{24}CO_3$	tetr		2	12.174		7.652			
Melanophlogite	SiO_2	cubic	clathrate type	46	13.402					
Melanterite	$FeSO_4 \cdot 7H_2O$	monocl		4	14.072	6.503	11.041		105.57°	
Melonite	$NiTe_2$	hex	cadmium iodide	1	3.869		5.308			
Metacinnabar	HgS	cubic	sphalerite	4	5.8517					
Miargyrite	$AgSbS_2$	monocl		8	12.862	4.111	13.220		98.63°	
Microcline	$KAlSi_3O_8$	tricl		4	8.582	12.964	7.222	90.62°	115.92°	87.68°
Miersite	AgI	cubic	sphalerite	4	6.4963					
Millerite	NiS	rhomb		9	9.616		3.152			
Minium	Pb_3O_4	tetr		4	8.815		6.565			
Minnesotaite	$Fe_3Si_4O_{10}(OH)_2$	monocl		4	5.4	9.42	19.4		100.00°	
Mirabilite	$Na_2SO_4 \cdot 10H_2O$	monocl		4	11.51	10.38	12.83		107.75°	
Mn-Indialite	$Mn_2Al_3(AlSi_5O_{18})$	hex	beryl	2	9.925		9.297			
Molybdenite	MoS_2	hex	molybdenite	2	3.1604		12.295			
Molybdenum	Mo	cubic		2	3.1653					
Molybdite	MoO_3	orth		4	3.962	13.858	3.697			
Monteponite	CdO	cubic	rock salt	4	4.6953					
Monticellite	$CaMgSiO_4$	orth	olivine	4	4.827	11.084	6.376			
Montroydite	HgO	orth		4	6.608	5.518	3.519			
Mullite (2:1)	$2Al_2O_3 \cdot SiO_2$	orth		6	7.5788	7.6909	2.8883			
Mullite (3:2)	$3Al_2O_3 \cdot 2SiO_2$	orth		3	7.557	7.6876	2.8842			
Muscovite	$KAl_2AlSi_3O_{10}(OH)_2$	monocl	2M2 mica	4	5.203	8.995	20.030		94.47°	

Name	Formula	Crystal system	Structure type	Z	a/Å	b/Å	c/Å	α	β	γ
Nacrite	$Al_2Si_2O_5(OH)_4$	monocl		4	8.909	5.146	15.697		113.70°	
Nantokite	$CuCl$	cubic	sphalerite	4	5.416					
Natroalunite	$NaAl_3(SO_4)_2(OH)_6$	rhomb		3	6.974		16.69			
Natrolite	$Na_2Al_2Si_3O_{10}·2H_2O$	orth		8	18.30	18.63	6.60			
Neighborite	$NaMgF_3$	orth	perovskite	4	5.363	7.676	5.503			
Ni-Skutterudite	$NiAs_{2.95}$	cubic		8	8.3300					
Niccolite	$NiAs$	hex	niccolite	2	3.618		5.034			
Nickel	Ni	cubic	face-centered cubic	4	3.5238					
Nickel carbonate	$NiCO_3$	rhomb	calcite	6	4.5975		14.723			
Nickel olivine	Ni_2SiO_4	orth	olivine	4	4.727	10.121	5.915			
Nickel selenide	$NiSe_2$	cubic	pyrite	4	5.9604					
Niter	KNO_3	orth	aragonite	4	6.431	9.164	5.414			
Norsethite	$BaMg(CO_3)_2$	rhomb	calcite	3	5.020		16.75			
Oldhamite	CaS	cubic	rock salt	4	5.689					
Orpiment	As_2S_3	monocl		4	11.49	9.59	4.25		90.45°	
Orthoclase	$KAlSi_3O_8$	monocl		4	8.562	12.996	7.193		116.02°	
Orthoferrosilite	$FeSiO_3$	orth	enstatite	16	9.080	18.431	5.238			
Otavite	$CdCO_3$	rhomb	calcite	6	4.9204		16.298			
Paracelsian	$BaAl_2Si_2O_8$	monocl		4	8.58	9.583	9.08		90.00°	
Paragonite	$NaAl_2AlSi_3O_{10}(OH)_2$	monocl	2M1 mica	4	5.13	8.89	19.32		95.17°	
Pararammelsbergite	$NiAs_2$	orth		8	5.75	5.82	11.428			
Paratellurite	TeO_2	tetr		4	4.810		7.613			
Parawollastonite	$CaSiO_3$	monocl		12	15.417	7.321	7.066		95.40°	
Pectolite	$Ca_2NaH(SiO_3)_3$	tricl		2	7.99	7.04	7.02	90.05°	95.27°	102.47°
Pentlandite	$Fe_{5.25}Ni_{3.75}S_8$	cubic		4	10.196					
Pentlandite	$Fe_{4.75}Ni_{5.25}S_8$	cubic		4	10.095					
Periclase	MgO	cubic	rock salt	4	4.2117					
Perovskite	$CaTiO_3$	orth	perovskite	4	5.3670	7.6438	5.4439			
Petalite	$LiAlSi_4O_{10}$	monocl		2	11.32	5.14	7.62		105.90°	
Petzite	Ag_3AuTe_2	cubic		8	10.38					
Phenacite	Be_2SiO_4	rhomb	phenacite	18	12.472		8.252			
Phlogopite	$KMg_3AlSi_3O_{10}(OH)_2$	monocl	1M mica	2	5.326	9.210	10.311		100.17°	
Picrochromite	$MgCr_2O_4$	cubic	spinel	8	8.333					
Piemontite	$Ca_2Al_{1.5}Mn_{1.5}(SiO_4)_3OH$	monocl		2	8.95	5.70	9.41		115.70°	
Platinum	Pt	cubic	face-centered cubic	4	3.9231					
Polymidite	Ni_3S_4	cubic	spinel	8	9.480					
Portlandite	$Ca(OH)_2$	hex	cadmium iodide	1	3.5933					
Powellite	$CaMoO_4$	tetr	scheelite	4	5.226		11.43			
Protoenstatite	$MgSiO_3$	orth		8	9.25	8.74	5.32			
Proustite	Ag_3AsS_3	rhomb		6	10.816		8.6948			
Pseudowollastonite	$CaSiO_3$	tricl		24	6.90	11.78	19.65	90.00°	90.80°	90.00°
Pyrargyrite	Ag_3SbS_3	rhomb		6	11.052		8.7177			
Pyrite	FeS_2	cubic	pyrite	4	5.4175					
Pyrolusite	MnO_2	tetr	rutile	2	4.388		2.865			
Pyrope	$Mg_3Al_2Si_3O_{12}$	cubic	garnet	8	11.459					

Name	Formula	Crystal system	Structure type	Z	a/Å	b/Å	c/Å	α	β	γ
Pyrophanite	$MnTiO_3$	rhomb	ilmenite	6	5.155		14.18			
Pyrophyllite	$Al_2Si_4O_{10}(OH)_2$	monocl	2M1 mica	4	5.14	8.90	18.55	84.00°	99.92°	
Pyroxmangite	$MnFe(SiO_3)_2$	tricl		7	7.56	17.45	6.67		94.30°	113.70°
Pyrrhotite	$Fe_{0.980}S$	hex	defect niccolite	2	3.446		5.848			
Pyrrhotite	$Fe_{0.885}S$	hex	defect niccolite	2	3.440		5.709			
Quartz (α)	SiO_2	hex		3	4.9136		5.4051			
Quartz (β)	SiO_2	hex		3	4.999		5.4592			
Rammelsbergite	$NiAs_2$	orth	marcasite	2	4.757	5.797	3.542			
Realgar	AsS	monocl		16	9.29	13.53	6.57		106.55°	
Retgersite	$NiSO_4 \cdot 4H_2O$	tetr		4	6.782		18.28			
Rhodochrosite	$MnCO_3$	rhomb	calcite	6	4.7771		15.664			
Rhodonite	$MnSiO_3$	tricl		10	7.682	11.818	6.707	92.36°	93.95°	105.66°
Riebeckite	$Na_2Fe_3FSi_8O_{22}(OH)_2$	monocl	tremolite	2	9.729	18.065	5.334		103.31°	
Rutile	TiO_2	tetr		2	4.5937		2.9618			
Safflorite	$Co_{0.5}Fe_{0.5}As_2$	orth	marcasite	2	5.231	5.953	2.962			
Sanmartinite	$ZnWO_4$	monocl	wolframite	2	4.691	5.720	4.925		89.36°	
Sapphirine	$Mg_2Al_4O_6SiO_4$	monocl		8	9.96	28.60	9.85		110.5°	
Scacchite	$MnCl_2$	rhomb		3	3.711		17.59			
Scheelite	$CaWO_4$	tetr	scheelite	4	5.242		11.372			
Schorl	$NaFe_3Al_6B_3Si_6O_{27}(OH)_4$	rhomb	tourmaline	3	16.032		7.149			
Selenium	Se	hex		3	4.3642		4.9588			
Selenolite	SeO_2	tetr		8	8.35		5.05			
Sellaite	MgF_2	tetr	rutile	2	4.621		3.050			
Senarmontite	Sb_2O_3	cubic	arsenic trioxide	16	11.152					
Shandite	$Ni_3Pb_2S_2$	rhomb		3	5.576		13.658			
Shortite	$Na_2Ca_2(CO_3)_3$	orth		2	4.961	11.03	7.12			
Siderite	$FeCO_3$	rhomb	calcite	6	4.6887		15.373			
Silicon	Si	cubic	diamond	8	5.4305					
Sillimanite	Al_2OSiO_4	orth		4	7.4843	7.6730	5.7711			
Silver	Ag	cubic	face-centered cubic	4	4.0862					
Silver telluride I	Ag_2Te	cubic		2	5.29					
Silver telluride II	Ag_2Te	cubic		4	6.585					
Smithsonite	$ZnCO_3$	rhomb	calcite	6	4.6528		15.025			
Soda niter	$NaNO_3$	rhomb	calcite	6	5.0696		16.829			
Sodium melilite	$NaCaAlSi_2O_7$	tetr	melilite	2	8.511		4.809			
Sperrylite	$PtAs_2$	cubic	pyrite	4	5.968					
Spessartite	$Mn_3Al_2Si_3O_{12}$	cubic	garnet	8	11.621					
Sphalerite	ZnS	cubic	sphalerite	4	5.4093					
Sphene	$CaTiSiO_5$	monocl		4	7.07	8.72	6.56		113.95°	
Spinel	$MgAl_2O_4$	cubic	spinel	8	8.080					
Spodumene	$LiAl(SiO_3)_2$	monocl	diopside	4	9.451	8.387	5.208		110.07°	
Spodumene (β)	$LiAl(SiO_3)_2$	tetr		4	7.5332		9.1540		90.00°	
Staurolite	$Fe_2Al_9Si_4O_{22}(OH)_2$	monocl		2	7.90	16.65	5.63			
Sternbergite	$AgFe_2S_3$	orth		8	11.60	12.675	6.63			
Stibnite	Sb_2S_3	orth	stibnite	4	11.229	11.310	3.8389			

Name	Formula	Crystal system	Structure type	Z	a/Å	b/Å	c/Å	α	β	γ
Stilleite	$ZnSe$	cubic	sphalerite	4	5.6685					
Stishovite	SiO_2	tetr	rutile	2	4.1790		2.6649			
Stolzite	$PbWO_4$	tetr	scheelite	4	5.4616		12.046			
Stromeyerite	$Ag_{0.93}Cu_{1.07}S$	orth		4	4.066	6.628	7.972			
Strontianite	$SrCO_3$	orth	aragonite	4	6.029	8.414	5.107			
Sulfur (monoclinic)	S	monocl	S8 ring molecules	48	11.04	10.98	10.92		96.73°	
Sulfur (orthorhombic)	S	orth	S8 ring molecules	128	10.4646	12.8660	24.4860			
Sulfur (rhombohedral)	S	rhomb	S6 ring molecules	18	10.818		4.280			
Sylvite	KCl	cubic	rock salt	4	6.2931					
Syngenite	$K_2Ca(SO_4)_2 \cdot H_2O$	monocl		2	9.775	7.156	6.251		104.00°	
Synthetic anorthite	$CaAl_2Si_2O_8$	hex		2	5.10		14.72			
Synthetic anorthite	$CaAl_2Si_2O_8$	orth		2	8.22	8.60	4.83			
Talc	$Mg_3Si_4O_{10}(OH)_2$	monocl	2M1 mica	4	5.287	9.158	18.95		99.50°	
Tantalum	Ta	cubic	tungsten	2	3.3058					
Teallite	$PbSnS_2$	orth	germanium sulfide	2	4.266	11.419	4.090			
Tellurite	TeO_2	orth	tellurite	8	5.607	12.034	5.463			
Tellurium	Te	hex	selenium	3	4.4570		5.9290			
Tellurobismuthite	Bi_2Te_3	rhomb		3	4.3835		30.487			
Tennantite	$Cu_{12}As_4S_{13}$	cubic	tetrahedrite	2	10.190					
Tenorite	CuO	monocl		4	4.684	3.425	5.129		99.47°	
Tephroite	Mn_2SiO_4	orth	olivine	4	4.871	10.636	6.232			
Tetrahedrite	$Cu_{12}Sb_4S_{13}$	cubic	tetrahedrite	2	10.327					
Thenardite	Na_2SO_4	orth	thenardite	8	5.863	12.304	9.821			
Thorianite	ThO_2	cubic	fluorite	4	5.5952					
Thorite	$ThSiO_4$	tetr	zircon	4	7.143		6.327			
Tiemannite	$HgSe$	cubic	sphalerite	4	6.0853					
Tin	Sn	tetr		4	5.8315		3.1813			
Titanium	Ti	hex		2	2.953		4.729			
Titanium(III) oxide	Ti_2O_3	rhomb	corundum	6	5.149		13.642			
Topaz	$Al_2(SiO_4)(OH)$	orth		4	8.394	8.792	4.649			
Tremolite	$Ca_2Mg_5Si_8O_{22}(OH)_2$	monocl	tremolite	2	9.840	18.052	5.275		104.70°	
Trevorite	$NiFe_2O_4$	cubic	spinel	8	8.339					
Tridymite (β)	SiO_2	hex		4	5.0463		8.2563			
Trogtalite	$CoSe_2$	cubic	pyrite	4	5.8588					
Troilite	FeS	hex	niccolite	2	3.446		5.877			
Tschermakite	$CaAl_2SiO_6$	monocl	diopside	4	9.615	8.661	5.272		106.12°	
Tungsten	W	cubic		2	3.1653					
Tungstenite	WS_2	hex	molybdenite	2	3.154		12.362			
Turquois	$CuAl_6(PO_4)_4(OH)_8 \cdot 4H_2O$	tricl		1	7.424	7.629	9.910	68.61°	69.71°	65.08°
Umangite	Cu_3Se_2	tetr		2	6.402		4.276			
Uraninite	UO_2	cubic	fluorite	4	5.4682					
Ureyite	$NaCr(SiO_3)_2$	monocl	diopside	4	9.550	8.712	5.273		107.44°	
Uvarovite	$Ca_3Cr_2Si_3O_{12}$	cubic	garnet	8	11.999					
Uvite	$CaMg_4Al_5B_3Si_6O_{27}(OH)_4$	rhomb	tourmaline	3	15.86		7.19			
Vaesite	NiS_2	cubic	pyrite	4	5.6873					

Name	Formula	Crystal system	Structure type	Z	a/Å	b/Å	c/Å	α	β	γ
Valentinite	Sb_2O_3	orth	antimony trioxide	4	4.914	12.468	5.421			
Vanthoffite	$MgSO_4 \cdot 3Na_2SO_4$	monocl		2	9.797	9.217	8.199		113.50°	
Vaterite	$CaCO_3$	hex		6	7.135		8.524			
Villiaumite	NaF	cubic	rock salt	4	4.6342					
Violarite	$FeNi_2S_4$	cubic	spinel	8	9.464		9.309			
Willemite	Zn_2SiO_4	rhomb	phenacite	18	13.94					
Witherite	$BaCO_3$	orth	aragonite	4	6.430	8.904	5.314			
Wolframite	$Fe_{0.5}Mn_{0.5}WO_4$	monocl	wolframite	2	4.782	5.731	4.982		90.57°	
Wollastonite	$CaSiO_3$	tricl		6	7.94	7.32	7.07	90.03°	95.37°	103.43°
Wulfenite	$PbMoO_4$	tetr	scheelite	4	5.435		12.110			
Wurtzite	ZnS	hex	zincite	2	3.8230		6.2565			
Wustite	$Fe_{0.953}O$	cubic	defect rock salt	4	4.3088					
Xenotime	YPO_4	tetr	zircon	4	6.885		5.982			
Zinc	Zn	hex	hexagonal close pack	2	2.665		4.947			
Zinc telluride	$ZnTe$	cubic	sphalerite	4	6.1020					
Zincite	ZnO	hex	zincite	2	3.2495		5.2069			
Zinkosite	$ZnSO_4$	orth	barite	4	8.588	6.740	4.770			
Zircon	$ZrSiO_4$	tetr	zircon	4	6.604		5.979			
Zoisite	$Ca_2Al_3(SiO_4)_3OH$	orth		4	16.15	5.581	10.06			

Section 5
Thermochemistry, Electrochemistry, and Kinetics

CODATA KEY VALUES FOR THERMODYNAMICS

The Committee on Data for Science and Technology (CODATA) has conducted a project to establish internationally agreed values for the thermodynamic properties of key chemical substances. This table presents the final results of the project. Use of these recommended, internally consistent values is encouraged in the analysis of thermodynamic measurements, data reduction, and preparation of other thermodynamic tables.

The table includes the standard enthalpy of formation at 298.15 K, the entropy at 298.15 K, and the quantity $H°$ (298.15 K)$-H°$ (0). A value of 0 in the $\Delta_f H°$ column for an element indicates the reference state for that element. The standard state pressure is 100000 Pa (1 bar). See the reference for information on the dependence of gas-phase entropy on the choice of standard state pressure.

Substances are listed in alphabetical order of their chemical formulas when written in the most common form.

The table is reprinted with permission of CODATA.

REFERENCE

Cox, J. D., Wagman, D. D., and Medvedev, V. A., *CODATA Key Values for Thermodynamics,* Hemisphere Publishing Corp., New York, 1989.

Substance	State	$\Delta_f H°$ (298.15 K) kJ·mol^{-1}	$S°$ (298.15 K) J·K^{-1}·mol^{-1}	$H°$ (298.15 K)$-H°$ (0) kJ·mol^{-1}
Ag	cr	0	42.55 ± 0.20	5.745 ± 0.020
Ag	g	284.9 ± 0.8	172.997 ± 0.004	6.197 ± 0.001
Ag$^+$	aq	105.79 ± 0.08	73.45 ± 0.40	
AgCl	cr	-127.01 ± 0.05	96.25 ± 0.20	12.033 ± 0.020
Al	cr	0	28.30 ± 0.10	4.540 ± 0.020
Al	g	330.0 ± 4.0	164.554 ± 0.004	6.919 ± 0.001
Al^{+3}	aq	-538.4 ± 1.5	-325 ± 10	
AlF$_3$	cr	-1510.4 ± 1.3	66.5 ± 0.5	11.62 ± 0.04
Al$_2$O$_3$	cr, corundum	-1675.7 ± 1.3	50.92 ± 0.10	10.016 ± 0.020
Ar	g	0	154.846 ± 0.003	6.197 ± 0.001
B	cr, rhombic	0	5.90 ± 0.08	1.222 ± 0.008
B	g	565 ± 5	153.436 ± 0.015	6.316 ± 0.002
BF$_3$	g	-1136.0 ± 0.8	254.42 ± 0.20	11.650 ± 0.020
B$_2$O$_3$	cr	-1273.5 ± 1.4	53.97 ± 0.30	9.301 ± 0.040
Be	cr	0	9.50 ± 0.08	1.950 ± 0.020
Be	g	324 ± 5	136.275 ± 0.003	6.197 ± 0.001
BeO	cr	-609.4 ± 2.5	13.77 ± 0.04	2.837 ± 0.008
Br	g	111.87 ± 0.12	175.018 ± 0.004	6.197 ± 0.001
Br$^-$	aq	-121.41 ± 0.15	82.55 ± 0.20	
Br$_2$	l	0	152.21 ± 0.30	24.52 ± 0.01
Br$_2$	g	30.91 ± 0.11	245.468 ± 0.005	9.725 ± 0.001
C	cr, graphite	0	5.74 ± 0.10	1.050 ± 0.020
C	g	716.68 ± 0.45	158.100 ± 0.003	6.536 ± 0.001
CO	g	-110.53 ± 0.17	197.660 ± 0.004	8.671 ± 0.001
CO$_2$	g	-393.51 ± 0.13	213.785 ± 0.010	9.365 ± 0.003
CO$_2$	aq, undissoc.	-413.26 ± 0.20	119.36 ± 0.60	
CO$_3^{-2}$	aq	-675.23 ± 0.25	-50.0 ± 1.0	
Ca	cr	0	41.59 ± 0.40	5.736 ± 0.040
Ca	g	177.8 ± 0.8	154.887 ± 0.004	6.197 ± 0.001
Ca^{+2}	aq	-543.0 ± 1.0	-56.2 ± 1.0	
CaO	cr	-634.92 ± 0.90	38.1 ± 0.4	6.75 ± 0.06
Cd	cr	0	51.80 ± 0.15	6.247 ± 0.015
Cd	g	111.80 ± 0.20	167.749 ± 0.004	6.197 ± 0.001
Cd^{+2}	aq	-75.92 ± 0.60	-72.8 ± 1.5	
CdO	cr	-258.35 ± 0.40	54.8 ± 1.5	8.41 ± 0.08
CdSO$_4$·8/3H$_2$O	cr	-1729.30 ± 0.80	229.65 ± 0.40	35.56 ± 0.04
Cl	g	121.301 ± 0.008	165.190 ± 0.004	6.272 ± 0.001
Cl$^-$	aq	-167.080 ± 0.10	56.60 ± 0.20	
ClO$_4^-$	aq	-128.10 ± 0.40	184.0 ± 1.5	
Cl$_2$	g	0	223.081 ± 0.010	9.181 ± 0.001
Cs	cr	0	85.23 ± 0.40	7.711 ± 0.020
Cs	g	76.5 ± 1.0	175.601 ± 0.003	6.197 ± 0.001
Cs$^+$	aq	-258.00 ± 0.50	132.1 ± 0.5	

Substance	State	$\Delta_f H°$ (298.15 K) kJ·mol^{-1}	$S°$ (298.15 K) J·K^{-1}·mol^{-1}	$H°$ (298.15 K)–$H°$ (0) kJ·mol^{-1}
Cu	cr	0	33.15 ± 0.08	5.004 ± 0.008
Cu	g	337.4 ± 1.2	166.398 ± 0.004	6.197 ± 0.001
Cu^{+2}	aq	64.9 ± 1.0	-98 ± 4	
CuSO$_4$	cr	-771.4 ± 1.2	109.2 ± 0.4	16.86 ± 0.08
F	g	79.38 ± 0.30	158.751 ± 0.004	6.518 ± 0.001
F$^-$	aq	-335.35 ± 0.65	-13.8 ± 0.8	
F$_2$	g	0	202.791 ± 0.005	8.825 ± 0.001
Ge	cr	0	31.09 ± 0.15	4.636 ± 0.020
Ge	g	372 ± 3	167.904 ± 0.005	7.398 ± 0.001
GeF$_4$	g	-1190.20 ± 0.50	301.9 ± 1.0	17.29 ± 0.10
GeO$_2$	cr, tetragonal	-580.0 ± 1.0	39.71 ± 0.15	7.230 ± 0.020
H	g	217.998 ± 0.006	114.717 ± 0.002	6.197 ± 0.001
H$^+$	aq	0	0	
HBr	g	-36.29 ± 0.16	198.700 ± 0.004	8.648 ± 0.001
HCO$_3^-$	aq	-689.93 ± 0.20	98.4 ± 0.5	
HCl	g	-92.31 ± 0.10	186.902 ± 0.005	8.640 ± 0.001
HF	g	-273.30 ± 0.70	173.779 ± 0.003	8.599 ± 0.001
HI	g	26.50 ± 0.10	206.590 ± 0.004	8.657 ± 0.001
HPO$_4^{-2}$	aq	-1299.0 ± 1.5	-33.5 ± 1.5	
HS$^-$	aq	-16.3 ± 1.5	67 ± 5	
HSO$_4^-$	aq	-886.9 ± 1.0	131.7 ± 3.0	
H$_2$	g	0	130.680 ± 0.003	8.468 ± 0.001
H$_2$O	l	-285.830 ± 0.040	69.95 ± 0.03	13.273 ± 0.020
H$_2$O	g	-241.826 ± 0.040	188.835 ± 0.010	9.905 ± 0.005
H$_2$PO$_4^-$	aq	-1302.6 ± 1.5	92.5 ± 1.5	
H$_2$S	g	-20.6 ± 0.5	205.81 ± 0.05	9.957 ± 0.010
H$_2$S	aq, undissoc.	-38.6 ± 1.5	126 ± 5	
H$_3$BO$_3$	cr	-1094.8 ± 0.8	89.95 ± 0.60	13.52 ± 0.04
H$_3$BO$_3$	aq, undissoc.	-1072.8 ± 0.8	162.4 ± 0.6	
He	g	0	126.153 ± 0.002	6.197 ± 0.001
Hg	l	0	75.90 ± 0.12	9.342 ± 0.008
Hg	g	61.38 ± 0.04	174.971 ± 0.005	6.197 ± 0.001
Hg^{+2}	aq	170.21 ± 0.20	-36.19 ± 0.80	
HgO	cr, red	-90.79 ± 0.12	70.25 ± 0.30	9.117 ± 0.025
Hg$_2^{+2}$	aq	166.87 ± 0.50	65.74 ± 0.80	
Hg$_2$Cl$_2$	cr	-265.37 ± 0.40	191.6 ± 0.8	23.35 ± 0.20
Hg$_2$SO$_4$	cr	-743.09 ± 0.40	200.70 ± 0.20	26.070 ± 0.030
I	g	106.76 ± 0.04	180.787 ± 0.004	6.197 ± 0.001
I$^-$	aq	-56.78 ± 0.05	106.45 ± 0.30	
I$_2$	cr	0	116.14 ± 0.30	13.196 ± 0.040
I$_2$	g	62.42 ± 0.08	260.687 ± 0.005	10.116 ± 0.001
K	cr	0	64.68 ± 0.20	7.088 ± 0.020
K	g	89.0 ± 0.8	160.341 ± 0.003	6.197 ± 0.001
K$^+$	aq	-252.14 ± 0.08	101.20 ± 0.20	
Kr	g	0	164.085 ± 0.003	6.197 ± 0.001
Li	cr	0	29.12 ± 0.20	4.632 ± 0.040
Li	g	159.3 ± 1.0	138.782 ± 0.010	6.197 ± 0.001
Li$^+$	aq	-278.47 ± 0.08	12.24 ± 0.15	
Mg	cr	0	32.67 ± 0.10	4.998 ± 0.030
Mg	g	147.1 ± 0.8	148.648 ± 0.003	6.197 ± 0.001
Mg^{+2}	aq	-467.0 ± 0.6	-137 ± 4	
MgF$_2$	cr	-1124.2 ± 1.2	57.2 ± 0.5	9.91 ± 0.06
MgO	cr	-601.60 ± 0.30	26.95 ± 0.15	5.160 ± 0.020
N	g	472.68 ± 0.40	153.301 ± 0.003	6.197 ± 0.001
NH$_3$	g	-45.94 ± 0.35	192.77 ± 0.05	10.043 ± 0.010
NH$_4^+$	aq	-133.26 ± 0.25	111.17 ± 0.40	
NO$_3^-$	aq	-206.85 ± 0.40	146.70 ± 0.40	

Substance	State	$\Delta_f H°$ (298.15 K) kJ·mol^{-1}	$S°$ (298.15 K) J·K^{-1}·mol^{-1}	$H°$ (298.15 K)–$H°$ (0) kJ·mol^{-1}
N_2	g	0	191.609 ± 0.004	8.670 ± 0.001
Na	cr	0	51.30 ± 0.20	6.460 ± 0.020
Na	g	107.5 ± 0.7	153.718 ± 0.003	6.197 ± 0.001
Na$^+$	aq	-240.34 ± 0.06	58.45 ± 0.15	
Ne	g	0	146.328 ± 0.003	6.197 ± 0.001
O	g	249.18 ± 0.10	161.059 ± 0.003	6.725 ± 0.001
OH$^-$	aq	-230.015 ± 0.040	-10.90 ± 0.20	
O_2	g	0	205.152 ± 0.005	8.680 ± 0.002
P	cr, white	0	41.09 ± 0.25	5.360 ± 0.015
P	g	316.5 ± 1.0	163.199 ± 0.003	6.197 ± 0.001
P_2	g	144.0 ± 2.0	218.123 ± 0.004	8.904 ± 0.001
P_4	g	58.9 ± 0.3	280.01 ± 0.50	14.10 ± 0.20
Pb	cr	0	64.80 ± 0.30	6.870 ± 0.030
Pb	g	195.2 ± 0.8	175.375 ± 0.005	6.197 ± 0.001
Pb^{+2}	aq	0.92 ± 0.25	18.5 ± 1.0	
$PbSO_4$	cr	-919.97 ± 0.40	148.50 ± 0.60	20.050 ± 0.040
Rb	cr	0	76.78 ± 0.30	7.489 ± 0.020
Rb	g	80.9 ± 0.8	170.094 ± 0.003	6.197 ± 0.001
Rb$^+$	aq	-251.12 ± 0.10	121.75 ± 0.25	
S	cr, rhombic	0	32.054 ± 0.050	4.412 ± 0.006
S	g	277.17 ± 0.15	167.829 ± 0.006	6.657 ± 0.001
SO_2	g	-296.81 ± 0.20	248.223 ± 0.050	10.549 ± 0.010
SO_4^{-2}	aq	-909.34 ± 0.40	18.50 ± 0.40	
S_2	g	128.60 ± 0.30	228.167 ± 0.010	9.132 ± 0.002
Si	cr	0	18.81 ± 0.08	3.217 ± 0.008
Si	g	450 ± 8	167.981 ± 0.004	7.550 ± 0.001
SiF_4	g	-1615.0 ± 0.8	282.76 ± 0.50	15.36 ± 0.05
SiO_2	cr, alpha quartz	-910.7 ± 1.0	41.46 ± 0.20	6.916 ± 0.020
Sn	cr, white	0	51.18 ± 0.08	6.323 ± 0.008
Sn	g	301.2 ± 1.5	168.492 ± 0.004	6.215 ± 0.001
Sn^{+2}	aq	-8.9 ± 1.0	-16.7 ± 4.0	
SnO	cr, tetragonal	-280.71 ± 0.20	57.17 ± 0.30	8.736 ± 0.020
SnO_2	cr, tetragonal	-577.63 ± 0.20	49.04 ± 0.10	8.384 ± 0.020
Th	cr	0	51.8 ± 0.5	6.35 ± 0.05
Th	g	602 ± 6	190.17 ± 0.05	6.197 ± 0.003
ThO_2	cr	-1226.4 ± 3.5	65.23 ± 0.20	10.560 ± 0.020
Ti	cr	0	30.72 ± 0.10	4.824 ± 0.015
Ti	g	473 ± 3	180.298 ± 0.010	7.539 ± 0.002
$TiCl_4$	g	-763.2 ± 3.0	353.2 ± 4.0	21.5 ± 0.5
TiO_2	cr, rutile	-944.0 ± 0.8	50.62 ± 0.30	8.68 ± 0.05
U	cr	0	50.20 ± 0.20	6.364 ± 0.020
U	g	533 ± 8	199.79 ± 0.10	6.499 ± 0.020
UO_2	cr	-1085.0 ± 1.0	77.03 ± 0.20	11.280 ± 0.020
UO_2^{+2}	aq	-1019.0 ± 1.5	-98.2 ± 3.0	
UO_3	cr, gamma	-1223.8 ± 1.2	96.11 ± 0.40	14.585 ± 0.050
U_3O_8	cr	-3574.8 ± 2.5	282.55 ± 0.50	42.74 ± 0.10
Xe	g	0	169.685 ± 0.003	6.197 ± 0.001
Zn	cr	0	41.63 ± 0.15	5.657 ± 0.020
Zn	g	130.40 ± 0.40	160.990 ± 0.004	6.197 ± 0.001
Zn^{+2}	aq	-153.39 ± 0.20	-109.8 ± 0.5	
ZnO	cr	-350.46 ± 0.27	43.65 ± 0.40	6.933 ± 0.040

STANDARD THERMODYNAMIC PROPERTIES OF CHEMICAL SUBSTANCES

This table gives the standard state chemical thermodynamic properties of about 2400 individual substances in the crystalline, liquid, and gaseous states. Substances are listed by molecular formula in a modified Hill order; all compounds not containing carbon appear first, followed by those that contain carbon. The properties tabulated are:

$\Delta_f H°$	Standard molar enthalpy (heat) of formation at 298.15 K in kJ/mol
$\Delta_f G°$	Standard molar Gibbs energy of formation at 298.15 K in kJ/mol
$S°$	Standard molar entropy at 298.15 K in J/mol K
C_p	Molar heat capacity at constant pressure at 298.15 K in J/mol K

The standard state pressure is 100 kPa (1 bar). The standard states are defined for different phases by:

- The standard state of a pure gaseous substance is that of the substance as a (hypothetical) ideal gas at the standard state pressure.
- The standard state of a pure liquid substance is that of the liquid under the standard state pressure.
- The standard state of a pure crystalline substance is that of the crystalline substance under the standard state pressure.

An entry of 0.0 for $\Delta_f H°$ for an element indicates the reference state of that element. See References 1 and 2 for further information on reference states. A blank means no value is available.

We are indebted to M.V. Korobov for providing data on fullerene compounds.

REFERENCES

1. Cox, J.D., Wagman, D.D., and Medvedev, V.A., *CODATA Key Values for Thermodynamics*, Hemisphere Publishing Corp., New York, 1989.
2. Wagman, D.D., Evans, W.H., Parker, V.B., Schumm, R.H., Halow, I., Bailey, S.M., Churney, K.L., and Nuttall, R.L., *The NBS Tables of Chemical of Chemical Thermodynamic Properties, J. Phys. Chem. Ref. Data*, Vol. 11, Suppl. 2, 1982.
3. Chase M.W., Davies, C.A., Downey, J.R., Frurip, D.J., McDonald, R.A., and Syverud, A.N., *JANAF Thermochemical Tables, Third Edition, J. Phys. Chem. Ref. Data*, Vol. 14, Suppl. 1, 1985.
4. Daubert, T.E., Danner, R.P., Sibul, H.M., and Stebbins, C.C., *Physical and Thermodynamic Properties of Pure Compounds: Data Compilation*, extant 1994 (core with 4 supplements), Taylor & Francis, Bristol, PA.
5. Pedley, J.B., Naylor, R.D., and Kirby, S.P., *Thermochemical Data of Organic Compounds, Second Edition*, Chapman & Hall, London, 1986.
6. Pedley, J.B., *Thermochemical Data and Structures of Organic Compounds*, Thermodynamic Research Center, Texas A & M University, College Station, TX, 1994.
7. Dolmalski, E.S., Evans, W.H., Hearing E.D., *Heat Capacities and Entropies of Organic Compounds in the Condensed Phase, J. Phys. Chem. Ref. Data*, Vol. 13, Suppl. 1, 1984; Vol. 19, No. 4, 881—1047, 1990.
8. Gurvich, L.V., Veyts, I.V., and Alcock, C.B., *Thermodynamic Properties of Individual Substances, Fourth Edition*, Vol. 1, Hermisphere Publishing Corp., New York, 1989.
9. Gurvich, L.V., Veyts, I.V., and Alcock, C.B., *Thermodynamic Properties of Individual Substances, Fourth Edition*, Vol. 3, CRC Press, Boca Raton, FL, 1994.

Molecular formula	Name	$\Delta_f H°$/kJ mol⁻¹			$\Delta_f G°$/kJ mol⁻¹			S°/J mol⁻¹ K⁻¹			C_P/J mol⁻¹ K⁻¹		
		Crys.	Liq.	Gas	Crys.	Liq.	Gas	Crys.	Liq.	Gas	Crys.	Liq.	Gas
Ac	Actinium	0.0		406.0			366.0	56.5		188.1	27.2		20.8
Ag	Silver	0.0		284.9			246.0	42.6		173.0	25.4		20.8
AgBr	Silver(I) bromide	-100.4			-96.9			107.1			52.4		
AgBrO₃	Silver(I) bromate	-10.5			71.3			151.9					
AgCl	Silver(I) chloride	-127.0			-109.8			96.3			50.8		
AgClO₃	Silver(I) chlorate	-30.3			64.5			142.0					
AgClO₄	Silver(I) perchlorate	-31.1											
AgF	Silver(I) fluoride	-204.6											
AgF₂	Silver(II) fluoride	-360.0											
AgI	Silver(I) iodide	-61.8			-66.2			115.5			56.8		
AgIO₃	Silver(I) iodate	-171.1			-93.7			149.4			102.9		
AgNO₃	Silver(I) nitrate	-124.4			-33.4			140.9			93.1		
Ag₂	Disilver			410.0			358.8			257.1			37.0
Ag₂CrO₄	Silver(I) chromate	-731.7			-641.8			217.6			142.3		
Ag₂O	Silver(I) oxide	-31.1			-11.2			121.3			65.9		
Ag₂O₂	Silver(II) oxide	-24.3			27.6			117.0			88.0		
Ag₂O₃	Silver(III) oxide	33.9			121.4			100.0					
Ag₂O₄S	Silver(I) sulfate	-715.9			-618.4			200.4			131.4		
Ag₂S	Silver(I) sulfide (argentite)	-32.6			-40.7			144.0			76.5		
Al	Aluminum	0.0		330.0			289.4	28.3		164.6	24.4		21.4
AlB₃H₁₂	Aluminum borohydride		-16.3	13.0		145.0	147.0		289.1	379.2		194.6	
AlBr	Aluminum monobromide			-4.0			-42.0			239.5			35.6
AlBr₃	Aluminum tribromide	-527.2		-425.1				180.2			100.6		
AlCl	Aluminum monochloride			-47.7			-74.1			228.1			35.0
AlCl₂	Aluminum dichloride			-331.0									
AlCl₃	Aluminum trichloride	-704.2		-583.2	-628.8			109.3			91.1		
AlF	Aluminum monofluoride			-258.2			-283.7			215.0			31.9
AlF₃	Aluminum trifluoride	-1510.4		-1204.6	-1431.1		-1188.2	66.5		277.1	75.1		62.6
AlF₄Na	Sodium tetrafluoroaluminate			-1869.0			-1827.5			345.7			105.9
AlH	Aluminum hydride			259.2			231.2			187.9			29.4
AlH₃	Aluminum hydride	-46.0						30.0			40.2		
AlH₄K	Potassium aluminum hydride	-183.7											
AlH₄Li	Lithium aluminum hydride	-116.3			-44.7			78.7			83.2		
AlH₄Na	Sodium aluminum hydride	-15.5											
AlI	Aluminum monoiodide			65.5									
AlI₃	Aluminum triiodide	-313.8		-207.5	-300.8			159.0			98.7		
AlN	Aluminum nitride	-318.0			-287.0			20.2			30.1		
AlO	Aluminum monoxide			91.2			65.3			218.4			30.9
AlO₄P	Aluminum phosphate	-1733.8			-1617.9			90.8			93.2		
AlP	Aluminum phosphide	-166.5											
AlS	Aluminum monosulfide			200.9			150.1			230.6			33.4
Al₂	Dialuminum			485.9			433.3			233.2			36.4
Al₂Br₆	Aluminum hexabromide			-970.7									
Al₂Cl₆	Aluminum hexachloride			-1290.8			-1220.4			490.0			

STANDARD THERMODYNAMIC PROPERTIES OF CHEMICAL SUBSTANCES (continued)

Molecular formula	Name	$\Delta_f H°$/kJ mol⁻¹ Crys.	Liq.	Gas	$\Delta_f G°$/kJ mol⁻¹ Crys.	Liq.	Gas	S°/J mol⁻¹ K⁻¹ Crys.	Liq.	Gas	C_p/J mol⁻¹ K⁻¹ Crys.	Liq.	Gas
Al₂F₆	Aluminum hexafluoride	-2628.0											
Al₂I₆	Aluminum hexaiodide			-516.7									
Al₂O	Aluminum oxide (Al₂O)			-130.0			-159.0			259.4			45.7
Al₂O₃	Aluminum oxide (corundum)	-1675.7			-1582.3			50.9			79.0		
Al₂S₃	Aluminum sulfide	-724.0						116.9			105.1		
Am	Americium	0.0											
Ar	Argon			0.0						154.8			20.8
As	Arsenic (gray)	0.0						35.1			24.6		
As	Arsenic (yellow)	14.6		302.5			261.0			174.2			20.8
AsBr₃	Arsenic(III) bromide	-197.5		-130.0			-159.0			363.9			79.2
AsCl₃	Arsenic(III) chloride		-305.0	-261.5		-259.4	-248.9		216.3	327.2			75.7
AsF₃	Arsenic(III) fluoride		-821.3	-785.8		-774.2	-770.8		181.2	289.1		126.6	65.6
AsGa	Gallium arsenide	-71.0			-67.8			64.2			46.2		
AsH₃	Arsine			66.4			68.9			222.8			38.1
AsH₃O₄	Arsenic acid	-906.3											
AsI₃	Arsenic(III) iodide	-58.2			-59.4			213.1		388.3	105.8		80.6
AsIn	Indium arsenide	-58.6			-53.6			75.7			47.8		
AsO	Arsenic monoxide			70.0									
As₂	Diarsenic			222.2			171.9			239.4			35.0
As₂O₅	Arsenic(V) oxide	-924.9			-782.3			105.4			116.5		
As₂S₃	Arsenic(III) sulfide	-169.0			-168.6			163.6			116.3		
At	Astatine	0.0											
Au	Gold	0.0		366.1			326.3	47.4		180.5	25.4		20.8
AuBr	Gold(I) bromide	-14.0											
AuBr₃	Gold(III) bromide	-53.3											
AuCl	Gold(I) chloride	-34.7											
AuCl₃	Gold(III) chloride	-117.6											
AuF₃	Gold(III) fluoride	-363.6											
AuH	Gold hydride			295.0			265.7			211.2			29.2
AuI	Gold(I) iodide	0.0											
Au₂	Digold			515.1									36.9
B	Boron (rhombic)	0.0		565.0			521.0	5.9		153.4	11.1		20.8
BBr	Bromoborane			238.1			195.4			225.0			32.9
BBr₃	Boron tribromide		-239.7	-205.6		-238.5	-232.5		229.7	324.2			67.8
BCl	Chloroborane			149.5			120.9			213.2			31.7
BClO	Chloroxyborane			-314.0									
BCl₃	Boron trichloride		-427.2	-403.8		-387.4	-388.7		206.3	290.1		106.7	62.7
BCsO₂	Cesium metaborate	-972.0			-915.0			104.4			80.6		
BF	Fluoroborane			-122.2			-149.8			200.5			29.6
BFO	Fluorooxyborane			-607.0									
BF₃	Boron trifluoride			-1136.0			-1119.4			254.4			
BF₃H₃N	Aminetrifluoroboron	-1353.9											
BF₃H₃P	Trihydro(phosphorus trifluoride)boron			-854.0									
BF₄Na	Sodium tetrafluoroborate	-1844.7			-1750.1			145.3			120.3		

Molecular formula	Name	$\Delta_f H°$/kJ mol⁻¹ Crys.	Liq.	Gas	$\Delta_f G°$/kJ mol⁻¹ Crys.	Liq.	Gas	$S°$/J mol⁻¹ K⁻¹ Crys.	Liq.	Gas	C_p/J mol⁻¹ K⁻¹ Crys.	Liq.	Gas
BH	Boron monohydride			449.6			419.6			171.9			29.2
BHO₂	Metaboric acid (monoclinic)	-794.3		-561.9	-723.4		-551.0	38.0		240.1			42.2
BH₃	Borane			100.0									
BH₃O₃	Boric acid (orthoboric acid)	-1094.3		-994.1	-968.9			90.0			86.1		
BH₄K	Potassium borohydride	-227.4			-160.3			106.3			96.1		
BH₄Li	Lithium borohydride	-190.8			-125.0			75.9			82.6		
BH₄Na	Sodium borohydride	-188.6			-123.9			101.3			86.8		
BI₃	Boron triiodide			71.1			20.7			349.2			70.8
BKO₂	Potassium metaborate	-981.6			-923.4			80.0			66.7		
BLiO₂	Lithium metaborate	-1032.2			-976.1			51.5			59.7		
BN	Boron nitride	-254.4		647.5	-228.4		614.5	14.8		212.3	19.7		29.5
BNaO₂	Sodium metaborate	-977.0			-920.7			73.5			65.9		
BO	Boron monoxide			25.0			-4.0			203.5			29.2
BO₂	Boron dioxide			-300.4			-305.9			229.6			43.0
BO₂Rb	Rubidium metaborate	-971.0			-913.0			94.3			74.1		
BS	Boron monosulfide			342.0			288.8			216.2			30.0
B₂	Diboron			830.5			774.0			201.9			30.5
B₂Cl₄	Tetrachlorodiborane		-523.0	-490.4		-464.8	-460.6		262.3	357.4		137.7	95.4
B₂F₄	Tetrafluorodiborane			-1440.1			-1410.4			317.3			79.1
B₂H₆	Diborane			35.6			86.7			232.1			56.9
B₂O₂	Diboron dioxide			-454.8			-462.3			242.5			57.3
B₂O₃	Boron oxide	-1273.5		-843.8	-1194.3		-832.0	54.0		279.8	62.8		66.9
B₂S₃	Boron sulfide	-240.6		67.0				100.0			111.7		
B₃H₆N₃	Borazine		-541.0	66.1		-392.7			199.6				
B₄H₁₀	Tetraborane												
B₄Na₂O₇	Sodium tetraborate	-3291.1			-3096.0			189.5			186.8		
B₅H₉	Pentaborane(9)		42.7	73.2		171.8	175.0		184.2	275.9		151.1	96.8
B₅H₁₁	Pentaborane(11)		73.2	103.3									
B₆H₁₀	Hexaborane		56.3	94.6									
Ba	Barium	0.0		180.0			146.0	62.5		170.2	28.1		20.8
BaBr₂	Barium bromide	-757.3			-736.8			146.0					
BaCl₂	Barium chloride	-855.0			-806.7			123.7			75.1		
BaCl₂H₄O₂	Barium chloride dihydrate	-1456.9			-1293.2			203.0					
BaF₂	Barium fluoride	-1207.1			-1156.8			96.4			71.2		
BaH₂	Barium hydride	-177.0			-138.2			63.0			46.0		
BaH₂O₂	Barium hydroxide	-944.7											
BaI₂	Barium iodide	-602.1											
BaN₂O₄	Barium nitrite	-768.2											
BaN₂O₆	Barium nitrate	-988.0			-792.6			214.0			151.4		
BaO	Barium oxide	-548.0			-520.3			72.1			47.3		
BaO₄S	Barium sulfate	-1473.2			-1362.2			132.2			101.8		
BaS	Barium sulfide	-460.0			-456.0			78.2			49.4		
Be	Beryllium	0.0		324.0			286.6	9.5		136.3	16.4		20.8
BeBr₂	Beryllium bromide	-353.5						108.0			69.4		
BeCl₂	Beryllium chloride	-490.4			-445.6			75.8			62.4		

Molecular formula	Name	$\Delta_f H°$/kJ mol⁻¹ Crys.	Liq.	Gas	$\Delta_f G°$/kJ mol⁻¹ Crys.	Liq.	Gas	$S°$/J mol⁻¹ K⁻¹ Crys.	Liq.	Gas	C_p/J mol⁻¹ K⁻¹ Crys.	Liq.	Gas
BeF_2	Beryllium fluoride	-1026.8			-979.4			53.4			51.8		
BeH_2O_2	Beryllium hydroxide	-902.5			-815.0			45.5			62.1		
BeI_2	Beryllium iodide	-192.5						121.0			71.1		
BeO	Beryllium oxide	-609.4			-580.1			13.8			25.6		
BeO_4S	Beryllium sulfate	-1205.2			-1093.8			77.9			85.7		
BeS	Beryllium sulfide	-234.3						34.0			34.0		
Bi	Bismuth	0.0		207.1			168.2	56.7		187.0	25.5		20.8
$BiClO$	Bismuth oxychloride	-366.9			-322.1			120.5					
$BiCl_3$	Bismuth trichloride	-379.1		-265.7	-315.0		-256.0	177.0		358.9	105.0		79.7
BiH_3O_3	Bismuth hydroxide	-711.3											
BiI_3	Bismuth triiodide	-175.3			-175.3								
Bi_2	Dibismuth			219.7									36.9
Bi_2O_3	Bismuth oxide	-573.9			-493.7			151.5			113.5		
$Bi_2O_{12}S_3$	Bismuth sulfate	-2544.3											
Bi_2S_3	Bismuth sulfide	-143.1			-140.6			200.4			122.2		
Bk	Berkelium	0.0											
Br	Bromine (atomic)			111.9			82.4			175.0			20.8
$BrCl$	Bromine chloride			14.6			-1.0			240.1			35.0
$BrCl_3Si$	Bromotrichlorosilane									350.1			90.9
$BrCs$	Cesium bromide	-405.8			-391.4			113.1			52.9		
$BrCu$	Copper(I) bromide	-104.6			-100.8			96.1			54.7		
BrF	Bromine fluoride			-93.8			-109.2			229.0			33.0
BrF_3	Bromine trifluoride		-300.8	-255.6		-240.5	-229.4		178.2	292.5		124.6	66.6
BrF_5	Bromine pentafluoride		-458.6	-428.9		-351.8	-350.6		225.1	320.2			99.6
$BrGe$	Germanium monobromide			235.6						274.8			37.1
$BrGeH_3$	Bromogermane												56.4
BrH	Hydrogen bromide			-36.3			-53.4			198.7			29.1
$BrHSi$	Bromosilylene												
BrH_3Si	Bromosilane			-464.4						262.4			52.8
BrH_4N	Ammonium bromide	-270.8			-175.2			113.0			96.0		
BrI	Iodine bromide			40.8			3.7			258.8			36.4
$BrIn$	Indium(I) bromide	-175.3		-56.9	-169.0		-94.3	113.0		259.5			36.7
BrK	Potassium bromide	-393.8			-380.7			95.9			52.3		
$BrKO_3$	Potassium bromate	-360.2			-271.2			149.2			105.2		
$BrKO_4$	Potassium perbromate	-287.9			-174.4			170.1			120.2		
$BrLi$	Lithium bromide	-351.2			-342.0			74.3					
$BrNO$	Nitrosyl bromide			82.2			82.4			273.7			45.5
$BrNa$	Sodium bromide	-361.1		-143.1	-349.0		-177.1	86.8		241.2	51.4		36.3
$BrNaO_3$	Sodium bromate	-334.1			-242.6			128.9					
BrO	Bromine monoxide			125.8			108.2			237.6			32.1
BrO_2	Bromine dioxide	48.5											
$BrRb$	Rubidium bromide	-394.6			-381.8			110.0			52.8		
$BrSi$	Bromosilylidyne			209.0									
$BrTl$	Thallium(I) bromide	-173.2		-37.7	-167.4			120.5					38.6
Br_2	Bromine		0.0	30.9			3.1		152.2	245.5		75.7	36.0

Molecular formula	Name	$\Delta_f H°$/kJ mol^{-1}			$\Delta_f G°$/kJ mol^{-1}			$S°$/J mol^{-1} K^{-1}			C_p/J mol^{-1} K^{-1}		
		Crys.	Liq.	Gas	Crys.	Liq.	Gas	Crys.	Liq.	Gas	Crys.	Liq.	Gas
Br₂Ca	Calcium bromide	-682.8			-663.6			130.0			76.7		
Br₂Cd	Cadmium bromide	-316.2			-296.3			137.2			79.5		
Br₂Co	Cobalt(II) bromide	-220.9											
Br₂Cr	Chromium(II) bromide	-302.1											
Br₂Cu	Copper(II) bromide	-141.8											
Br₂Fe	Iron(II) bromide	-249.8			-238.1			140.6					
Br₂H₂Si	Dibromosilane									309.7			65.5
Br₂Hg	Mercury(II) bromide	-170.7			-153.1			172.0					
Br₂Hg₂	Mercury(I) bromide	-206.9			-181.1			218.0					
Br₂Mg	Magnesium bromide	-524.3			-503.8			117.2					
Br₂Mn	Manganese(II) bromide	-384.9											
Br₂Ni	Nickel(II) bromide	-212.1											
Br₂Pb	Lead(II) bromide	-278.7			-261.9			161.5			80.1		
Br₂Pt	Platinum(II) bromide	-82.0											
Br₂S₂	Sulfur bromide		-13.0										
Br₂Se	Selenium bromide			-21.0									
Br₂Sn	Tin(II) bromide	-243.5			-697.1			135.1			75.3		
Br₂Sr	Strontium bromide	-717.6											
Br₂Ti	Titanium(II) bromide	-402.0			-312.1			138.5					
Br₂Zn	Zinc bromide	-328.7											
Br₃ClSi	Tribromochlorosilane									377.1			95.3
Br₃Fe	Iron(III) bromide	-268.2			-359.8			180.0					
Br₃Ga	Gallium(III) bromide	-386.6				-336.4	-328.5		248.1	348.6			80.8
Br₃HSi	Tribromosilane		-355.6	-317.6									
Br₃In	Indium(III) bromide	-428.9		-282.0									
Br₃OP	Phosphorus(V) oxybromide	-458.6								359.8			89.9
Br₃P	Phosphorus(III) bromide		-184.5	-139.3		-175.7	-162.8		240.2	348.1			76.0
Br₃Pt	Platinum(III) bromide	-120.9											
Br₃Re	Rhenium(III) bromide	-167.0											
Br₃Ru	Ruthenium(III) bromide	-138.0											
Br₃Sb	Antimony(III) bromide	-259.4		-194.6			-223.9	207.1		372.9			80.2
Br₃Sc	Scandium bromide	-743.1			-239.3								
Br₃Ti	Titanium(III) bromide	-548.5			-523.8			176.6	280.7	396.2	101.7		101.8
Br₄Ge	Germanium(IV) bromide		-347.7	-300.0	-787.8	-331.4	-318.0	234.0					
Br₄Pa	Protactinium(IV) bromide	-824.0											
Br₄Pt	Platinum(IV) bromide	-156.5			-350.2								
Br₄Si	Tetrabromosilane		-457.3	-415.5		-443.9	-431.8	264.4	277.8	377.9	131.5		97.1
Br₄Sn	Tin(IV) bromide	-377.4		-314.6			-331.4	243.5		411.9			103.4
Br₄Te	Tellurium tetrabromide	-190.4			-589.5								
Br₄Ti	Titanium(IV) bromide	-616.7		-549.4			-568.2			398.4			100.8
Br₄V	Vanadium(IV) bromide	-760.7											
Br₄Zr	Zirconium(IV) bromide	-269.9											
Br₅P	Phosphorus(V) bromide	-598.3		-336.8									
Br₅Ta	Tantalum(V) bromide	-348.5											
Br₆W	Tungsten(VI) bromide												

Molecular formula	Name	$\Delta_f H°$/kJ mol⁻¹			$\Delta_f G°$/kJ mol⁻¹			$S°$/J mol⁻¹ K⁻¹			C_p/J mol⁻¹ K⁻¹		
		Crys.	Liq.	Gas	Crys.	Liq.	Gas	Crys.	Liq.	Gas	Crys.	Liq.	Gas
Ca	Calcium	0.0		177.8			144.0	41.6		154.9	25.9		20.8
CaCl₂	Calcium chloride	-795.4			-748.8			108.4			72.9		
CaF₂	Calcium fluoride	-1228.0			-1175.6			68.5			67.0		
CaH₂	Calcium hydride	-181.5			-142.5			41.4			41.0		
CaH₂O₂	Calcium hydroxide	-985.2			-897.5			83.4			87.5		
CaI₂	Calcium iodide	-533.5			-528.9			142.0					
CaN₂O₆	Calcium nitrate	-938.2			-742.8			193.2			149.4		
CaO	Calcium oxide	-634.9			-603.3			38.1			42.0		
CaO₄S	Calcium sulfate	-1434.5			-1322.0			106.5			99.7		
CaS	Calcium sulfide	-482.4			-477.4			56.5			47.4		
Ca₃O₈P₂	Calcium phosphate	-4120.8			-3884.7			236.0			227.8		
Cd	Cadmium	0.0		111.8				51.8		167.7	26.0		20.8
CdCl₂	Cadmium chloride	-391.5			-343.9			115.3			74.7		
CdF₂	Cadmium fluoride	-700.4			-647.7			77.4					
CdH₂O₂	Cadmium hydroxide	-560.7			-473.6			96.0					
CdI₂	Cadmium iodide	-203.3			-201.4			161.1			80.0		
CdO	Cadmium oxide	-258.4			-228.7			54.8			43.4		
CdO₄S	Cadmium sulfate	-933.3			-822.7			123.0			99.6		
CdS	Cadmium sulfide	-161.9			-156.5			64.9					
CdTe	Cadmium telluride	-92.5			-92.0			100.0			114.6		
Ce	Cerium (γ, fcc)	0.0		423.0			385.0	72.0		191.8	26.9		23.1
CeCl₃	Cerium(III) chloride	-1060.5			-984.8			151.0			87.4		
CeI₃	Cerium(III) iodide	-669.3											
CeO₂	Cerium(IV) oxide	-1088.7			-1024.6			62.3			61.6		
CeS	Cerium(II) sulfide	-459.4			-451.5			78.2			50.0		
Ce₂O₃	Cerium(III) oxide	-1796.2			-1706.2			150.6			114.6		
Cf	Californium	0.0											
Cl	Chlorine (atomic)			121.3			105.3			165.2			21.8
ClCs	Cesium chloride	-443.0			-414.5			101.2			52.5		
ClCsO₄	Cesium perchlorate	-443.1			-314.3			175.1			108.3		
ClCu	Copper(I) chloride	-137.2			-119.9			86.2			48.5		
ClF	Chlorine fluoride			-50.3			-51.8			217.9			32.1
ClFO₃	Perchloryl fluoride			-23.8			48.2			279.0			64.9
ClF₃	Chlorine trifluoride		-189.5	-163.2			-123.0			281.6			63.9
ClF₅S	Sulfur chloride pentafluoride		-1065.7										
ClGe	Germanium monochloride			155.2			124.2			247.0			36.9
ClGeH₃	Chlorogermane									263.7			54.7
ClH	Hydrogen chloride			-92.3			-95.3			186.9			29.1
ClHO	Hypochlorous acid			-78.7			-66.1			236.7			37.2
ClHO₄	Perchloric acid		-40.6										
ClH₃Si	Chlorosilane												
ClH₄N	Ammonium chloride	-314.4			-202.9			94.6			84.1		
ClH₄NO₄	Ammonium perchlorate	-295.3			-88.8			186.2					
ClH₄P	Phosphonium chloride	-145.2											
ClI	Iodine chloride		-23.9	17.8		-13.6	-5.5		135.1	247.6			35.6

STANDARD THERMODYNAMIC PROPERTIES OF CHEMICAL SUBSTANCES (continued)

Molecular formula	Name	$\Delta_f H°$/kJ mol⁻¹ Crys.	Liq.	Gas	$\Delta_f G°$/kJ mol⁻¹ Crys.	Liq.	Gas	$S°$/J mol⁻¹ K⁻¹ Crys.	Liq.	Gas	C_p/J mol⁻¹ K⁻¹ Crys.	Liq.	Gas
ClIn	Indium(I) chloride	-186.2		-75.0									
ClK	Potassium chloride	-436.5		-214.6	-408.5		-233.3	82.6		239.1	51.3		36.5
ClKO₃	Potassium chlorate	-397.7			-296.3			143.1			100.3		
ClKO₄	Potassium perchlorate	-432.8			-303.1			151.0			112.4		
ClLi	Lithium chloride	-408.6			-384.4			59.3			48.0		
ClLiO₄	Lithium perchlorate	-381.0											
ClNO	Nitrosyl chloride			51.7			66.1			261.7			44.7
ClNO₂	Nitryl chloride			12.6			54.4			272.2			53.2
ClNa	Sodium chloride	-411.2			-384.1			72.1			50.5		
ClNaO₂	Sodium chlorite	-307.0											
ClNaO₃	Sodium chlorate	-365.8			-262.3			123.4					
ClNaO₄	Sodium perchlorate	-383.3			-254.9			142.3					
ClO	Chlorine monoxide			101.8			98.1			226.6			31.5
ClOV	Vanadyl chloride	-607.0			-556.0			75.0					
ClO₂	Chlorine dioxide			102.5			120.5			256.8			42.0
ClO₂	Chlorine superoxide (ClOO)			89.1			105.0			263.7			46.0
ClO₄Rb	Rubidium perchlorate	-437.2			-306.9			161.1					
ClRb	Rubidium chloride	-435.4			-407.8			95.9			52.4		
ClSi	Chlorosilylidyne			189.9									
ClTl	Thallium(I) chloride	-204.1		-67.8	-184.9			111.3			50.9		36.9
Cl₂	Chlorine			0.0						223.1			33.9
Cl₂Co	Cobalt(II) chloride	-312.5			-269.8			109.2			78.5		
Cl₂Cr	Chromium(II) chloride	-395.4			-356.0			115.3			71.2		
Cl₂CrO₂	Chromyl chloride		-579.5	-538.1		-510.8	-501.6		221.8	329.8			84.5
Cl₂Cu	Copper(II) chloride	-220.1			-175.7			108.1			71.9		
Cl₂Fe	Iron(II) chloride	-341.8			-302.3			118.0			76.7		
Cl₂H₂Si	Dichlorosilane									285.7			60.5
Cl₂Hg	Mercury(II) chloride	-224.3			-178.6			146.0					
Cl₂Hg₂	Mercury(I) chloride	-265.4			-210.7			191.6					
Cl₂Mg	Magnesium chloride	-641.3			-591.8			89.6			71.4		
Cl₂Mn	Manganese(II) chloride	-481.3			-440.5			118.2			72.9		
Cl₂Ni	Nickel(II) chloride	-305.3			-259.0			97.7			71.7		
Cl₂O	Chlorine oxide			80.3			97.9			266.2			45.4
Cl₂OS	Thionyl chloride		-245.6	-212.5			-198.3			309.8		121.0	66.5
Cl₂O₂S	Sulfuryl chloride		-394.1	-364.0			-320.0			311.9		134.0	77.0
Cl₂O₂U	Uranyl chloride	-1243.9			-1146.4			150.5			107.9		
Cl₂Pb	Lead(II) chloride	-359.4			-314.1			136.0					
Cl₂Pt	Platinum(II) chloride	-123.4											
Cl₂S	Sulfur dichloride		-50.0										
Cl₂S₂	Sulfur chloride		-59.4										
Cl₂Sn	Tin(II) chloride	-325.1											
Cl₂Sr	Strontium chloride	-828.9			-781.1			114.9			75.6		
Cl₂Ti	Titanium(II) chloride	-513.8			-464.4			87.4			69.8		
Cl₂Zn	Zinc chloride	-415.1		-266.1	-369.4			111.5			71.3		
Cl₂Zr	Zirconium(II) chloride	-502.0											

Molecular formula	Name	ΔfH°/kJ mol⁻¹ Crys.	Liq.	Gas	ΔfG°/kJ mol⁻¹ Crys.	Liq.	Gas	S°/J mol⁻¹ K⁻¹ Crys.	Liq.	Gas	Cp/J mol⁻¹ K⁻¹ Crys.	Liq.	Gas
Cl₃Cr	Chromium(III) chloride	-556.5			-486.1			123.0			91.8		
Cl₃Dy	Dysprosium(III) chloride	-1000.0									100.0		
Cl₃Er	Erbium chloride	-998.7											
Cl₃Eu	Europium(III) chloride	-936.0									96.7		
Cl₃Fe	Iron(II) chloride	-399.5						142.3					
Cl₃Ga	Gallium(III) chloride	-524.7						142.0					
Cl₃Gd	Gadolinium(III) chloride	-1008.0									88.0		
Cl₃HSi	Trichlorosilane		-539.3	-513.0		-482.5	-482.0		227.6	313.9			75.8
Cl₃Ho	Holmium chloride	-1005.4									88.0		
Cl₃In	Indium(III) chloride	-537.2											
Cl₃Ir	Iridium(III) chloride	-245.6		-374.0									
Cl₃La	Lanthanum chloride	-1072.2									108.8		
Cl₃Lu	Lutetium chloride	-945.6		-649.0									
Cl₃N	Nitrogen trichloride		230.0										
Cl₃Nd	Neodymium chloride	-1041.0									113.0		
Cl₃OP	Phosphorus(V) oxychloride		-597.1	-558.5		-520.8	-512.9		222.5	325.5		138.8	84.9
Cl₃OV	Vanadyl trichloride		-734.7	-695.6		-668.5	-659.3		244.3	344.3			89.9
Cl₃Os	Osmium(III) chloride	-190.4											
Cl₃P	Phosphorus(III) chloride		-319.7	-287.0		-272.3	-267.8		217.1	311.8			71.8
Cl₃Pr	Praseodymium chloride	-1056.9									100.0		
Cl₃Pt	Platinum(III) chloride	-182.0											
Cl₃Re	Rhenium(III) chloride	-264.0			-188.0			123.8			92.4		
Cl₃Rh	Rhodium(III) chloride	-299.2											
Cl₃Ru	Ruthenium(III) chloride	-205.0											
Cl₃Sb	Antimony(III) chloride	-382.2			-323.7			184.1			107.9		
Cl₃Sc	Scandium chloride	-925.1											
Cl₃Sm	Samarium(III) chloride	-1025.9											
Cl₃Tb	Terbium chloride	-997.0											
Cl₃Ti	Titanium(III) chloride	-720.9			-653.5			139.7			97.2		
Cl₃Tl	Thallium(III) chloride	-315.1											
Cl₃Tm	Thulium chloride	-986.6											
Cl₃U	Uranium(III) chloride	-866.5			-799.1			159.0			102.5		
Cl₃V	Vanadium(II) chloride	-580.7			-511.2			131.0			93.2		
Cl₃Y	Yttrium chloride	-1000.0		-750.2									
Cl₃Yb	Ytterbium(III) chloride	-959.8											
Cl₄Ge	Germanium(IV) chloride		-531.8	-495.8		-462.7	-457.3		245.6	347.7	120.5		
Cl₄Hf	Hafnium(IV) chloride	-990.4		-884.5	-901.3			190.8					
Cl₄Pa	Protactinium(IV) chloride	-1043.0			-953.0			192.0					
Cl₄Pb	Lead(IV) chloride		-329.3										
Cl₄Pt	Platinum(IV) chloride	-231.8											
Cl₄Si	Tetrachlorosilane		-687.0	-657.0		-619.8	-617.0		239.7	330.7		145.3	90.3
Cl₄Sn	Tin(IV) chloride		-511.3	-471.5		-440.1	-432.2		258.6	365.8		165.3	98.3
Cl₄Te	Tellurium tetrachloride	-326.4									138.5		
Cl₄Th	Thorium(IV) chloride	-1186.2		-964.4	-1094.1		-932.0	190.4			120.3		107.5
Cl₄Ti	Titanium(IV) chloride		-804.2	-763.2		-737.2	-726.3		252.3	353.2		145.2	95.4

Molecular formula	Name	$\Delta_f H°$/kJ mol⁻¹			$\Delta_f G°$/kJ mol⁻¹			S°/J mol⁻¹ K⁻¹			C_p/J mol⁻¹ K⁻¹		
		Crys.	Liq.	Gas	Crys.	Liq.	Gas	Crys.	Liq.	Gas	Crys.	Liq.	Gas
Cl_4U	Uranium(IV) chloride	-1019.2		-809.6	-930.0		-786.6	197.1		419.0	122.0		96.2
Cl_4V	Vanadium(IV) chloride		-569.4	-525.5		-503.7	-492.0		255.0	362.4			
Cl_4Zr	Zirconium(IV) chloride	-980.5			-889.9			181.6			119.8		
Cl_5Nb	Niobium(V) chloride	-797.5		-703.7	-683.2		-646.0	210.5		400.6	148.1		120.8
Cl_5P	Phosphorus(V) chloride	-443.5		-374.9			-305.0			364.6			112.8
Cl_5Pa	Protactinium(V) chloride	-1145.0			-1034.0			238.0					
Cl_5Ta	Tantalum(V) chloride	-859.0											
Cl_6U	Uranium(VI) chloride	-1092.0		-1013.0	-962.0		-928.0	285.8		431.0	175.7		
Cl_6W	Tungsten(VI) chloride	-602.5		-513.8									
Cm	Curium	0.0			0.0								
Co	Cobalt	0.0		424.7	0.0		380.3	30.0		179.5	24.8		23.0
CoF_2	Cobalt(II) fluoride	-692.0			-647.2			82.0			68.8		
CoH_2O_2	Cobalt(II) hydroxide	-539.7			-454.3			79.0					
CoI_2	Cobalt(II) iodide	-88.7											
CoN_2O_6	Cobalt(II) nitrate	-420.5											
CoO	Cobalt(II) oxide	-237.9			-214.2			53.0			55.2		
CoO_4S	Cobalt(II) sulfate	-888.3			-782.3			118.0					
CoS	Cobalt(II) sulfide	-82.8											
Co_2S_3	Cobalt(III) sulfide	-147.3											
Co_3O_4	Cobalt(II,III) oxide	-891.0			-774.0			102.5			123.4		
Cr	Chromium	0.0		396.6	0.0		351.8	23.8		174.5	23.4		20.8
CrF_2	Chromium(II) fluoride	-778.0											
CrF_3	Chromium(III) fluoride	-1159.0			-1088.0			93.9			78.7		
CrI_2	Chromium(II) iodide	-156.9											
CrI_3	Chromium(III) iodide	-205.0											
CrO_2	Chromium(IV) oxide	-598.0											
CrO_4Pb	Lead(II) chromate	-930.9											
Cr_2FeO_4	Chromium iron oxide	-1444.7			-1343.8			146.0			133.6		
Cr_2O_3	Chromium(III) oxide	-1139.7			-1058.1			81.2			118.7		
Cr_3O_4	Chromium(II,III) oxide	-1531.0											
Cs	Cesium	0.0		76.5	0.0		49.6	85.2		175.6	32.2		20.8
CsF	Cesium fluoride	-553.5			-525.5			92.8			51.1		
CsF_2H	Cesium hydrogen fluoride	-923.8			-858.9			135.2			87.3		
CsH	Cesium hydride	-54.2											
CsHO	Cesium hydroxide	-417.2											
$CsHO_4S$	Cesium hydrogen sulfate	-1158.1											
CsH_2N	Cesium amide	-118.4											
CsI	Cesium iodide	-346.6			-340.6			123.1			52.8		
$CsNO_3$	Cesium nitrate	-506.0			-406.5			155.2					
CsO_2	Cesium superoxide	-286.2			-308.1								
Cs_2O	Cesium oxide	-345.8						146.9			76.0		
Cs_2O_3S	Cesium sulfite	-1134.7											
Cs_2O_4S	Cesium sulfate	-1443.0			-1323.6			211.9			134.9		
Cs_2S	Cesium sulfide	-359.8											
Cu	Copper	0.0		337.4	0.0		297.7	33.2		166.4	24.4		20.8

Molecular formula	Name	$\Delta_f H°$/kJ mol⁻¹			$\Delta_f G°$/kJ mol⁻¹			S°/J mol⁻¹ K⁻¹			C_p/J mol⁻¹ K⁻¹		
		Crys.	Liq.	Gas	Crys.	Liq.	Gas	Crys.	Liq.	Gas	Crys.	Liq.	Gas
CuF_2	Copper(II) fluoride	-542.7											
CuH_2O_2	Copper(II) hydroxide	-449.8											
CuI	Copper(I) iodide	-67.8			-69.5			96.7			54.1		
CuN_2O_6	Copper(II) nitrate	-302.9											
CuO	Copper(II) oxide	-157.3			-129.7			42.6			42.3		
CuO_4S	Copper(II) sulfate	-771.4			-662.2			109.2					
CuO_4W	Copper(II) tungstate	-1105.0											
CuS	Copper(II) sulfide	-53.1			-53.6			66.5			47.8		
$CuSe$	Copper(II) selenide	-39.5											
Cu_2	Dicopper			484.2			431.9			241.6			36.6
Cu_2O	Copper(I) oxide	-168.6			-146.0			93.1			63.6		
Cu_2S	Copper(I) sulfide	-79.5			-86.2			120.9			76.3		
Dy	Dysprosium	0.0		290.4			254.4	75.6		196.6	27.7		20.8
Dy_2O_3	Dysprosium(III) oxide	-1863.1			-1771.5			149.8			116.3		
Er	Erbium	0.0		317.1			280.7	73.2		195.6	28.1		20.8
ErF_3	Erbium fluoride	-1711.0											
Er_2O_3	Erbium oxide	-1897.9			-1808.7			155.6			108.5		
Es	Einsteinium	0.0											
Eu	Europium	0.0		175.3			142.2	77.8		188.8	27.7		20.8
Eu_2O_3	Europium(III) oxide	-1651.4			-1556.8			146.0			122.2		
Eu_3O_4	Europium(II,III) oxide	-2272.0			-2142.0			205.0					
F	Fluorine (atomic)			79.4			62.3			158.8			22.7
FGa	Gallium monofluoride			-251.9									33.3
FGe	Germanium monofluoride			-33.4									34.7
$FGeH_3$	Fluorogermane									252.8			51.6
FH	Hydrogen fluoride		-299.8	-273.3			-275.4			173.8			
FH_3Si	Fluorosilane									238.4			47.4
FH_4N	Ammonium fluoride	-464.0			-348.7			72.0			65.3		
FI	Iodine monofluoride			-95.7			-118.5			236.2			33.4
FIn	Indium monofluoride			-203.4									
FK	Potassium fluoride	-567.3			-537.8			66.6			49.0		
FLi	Lithium fluoride	-616.0			-587.7			35.7			41.6		
FNO	Nitrosyl fluoride			-66.5			-51.0			248.1			41.3
FNO_2	Nitryl fluoride									260.4			49.8
FNS	Thionitrosyl fluoride (NSF)									259.8			44.1
FNa	Sodium fluoride	-576.6			-546.3			51.1			46.9		
FO	Fluorine monoxide			109.0			105.0			216.8			30.5
FRb	Rubidium fluoride	-557.7											
FSi	Fluorosilylidyne			7.1			-24.3			225.8			32.6
FTl	Thallium(I) fluoride	-324.7		-182.4									
F_2	Fluorine			0.0						202.8			31.3
F_2Fe	Iron(II) fluoride	-711.3			-668.6			87.0			68.1		
F_2HK	Potassium hydrogen fluoride	-927.7			-859.7			104.3			76.9		
F_2HN	Difluoramine									252.8			43.4
F_2HNa	Sodium hydrogen fluoride	-920.3			-852.2			90.9			75.0		

Molecular formula	Name	$\Delta_f H°$/kJ mol⁻¹			$\Delta_f G°$/kJ mol⁻¹			$S°$/J mol⁻¹ K⁻¹			C_p/J mol⁻¹ K⁻¹		
		Crys.	Liq.	Gas	Crys.	Liq.	Gas	Crys.	Liq.	Gas	Crys.	Liq.	Gas
F2HRb	Rubidium hydrogen fluoride	-922.6			-855.6			120.1			79.4		
F2Mg	Magnesium fluoride	-1124.2			-1071.1			57.2			61.6		
F2N	Difluoroamidogen			43.1			57.8			249.9			41.0
F2N2	cis-Difluorodiazine			69.5									
F2N2	trans-Difluorodiazine			82.0									
F2Ni	Nickel(II) fluoride	-651.4			-604.1			73.6			64.1		
F2O	Fluorine monoxide			24.7			41.9			247.4			43.3
F2OS	Thionyl fluoride									278.7			56.8
F2O2	Fluorine dioxide			18.0									
F2O2S	Sulfuryl fluoride									284.0			66.0
F2O2U	Uranyl fluoride	-1653.5			-1557.4			135.6			103.2		
F2Pb	Lead(II) fluoride	-664.0			-617.1			110.5					
F2Si	Difluorosilylene			-619.0			-628.0			252.7			43.9
F2Sr	Strontium fluoride	-1216.3			-1164.8			82.1			70.0		
F2Zn	Zinc fluoride	-764.4			-713.3			73.7			65.7		
F3Ga	Gallium(II) fluoride	-1163.0			-1085.3			84.0					
F3Gd	Gadolinium(III) fluoride			-1297.0									
F3HSi	Trifluorosilane									271.9			60.5
F3Ho	Holmium fluoride	-1707.0											
F3N	Nitrogen trifluoride			-132.1			-90.6			260.8			53.4
F3Nd	Neodymium fluoride	-1657.0											
F3OP	Phosphorus(V) oxyfluoride			-1254.3			-1205.8			285.4			68.8
F3P	Phosphorus(III) fluoride			-958.4			-936.9			273.1			58.7
F3Sb	Antimony(III) fluoride	-915.5											
F3Sc	Scandium fluoride	-1629.2		-1247.0	-1555.6		-1234.0	92.0		300.5			67.8
F3Sm	Samarium(II) fluoride	-1778.0											
F3Th	Thorium(III) fluoride			-1166.1			-1160.6			339.2			73.3
F3U	Uranium(III) fluoride	-1502.1		-1058.5	-1433.4		-1051.9	123.4		331.9	95.1		74.3
F3Y	Yttrium fluoride	-1718.8		-1288.7	-1644.7		-1277.8	100.0		311.8			70.3
F4Ge	Germanium(IV) fluoride			-1190.2			-1150.0			301.9			
F4Hf	Hafnium fluoride	-1930.5		-1669.8	-1830.4			113.0		301.2			79.2
F4N2	Tetrafluorohydrazine			-8.4			79.9						
F4Pb	Lead(IV) fluoride	-941.8											
F4S	Sulfur tetrafluoride			-763.2			-722.0			299.6			77.6
F4Si	Tetrafluorosilane			-1615.0			-1572.8			282.8			73.6
F4Th	Thorium(IV) fluoride	-2097.8		-1759.0	-2003.4		-1724.0	142.0		341.7	110.7		93.0
F4U	Uranium(IV) fluoride	-1914.2		-1598.7	-1823.3		-1572.7	151.7		368.0	116.0		91.2
F4V	Vanadium(IV) fluoride	-1403.3											
F4Xe	Xenon tetrafluoride	-261.5											
F4Zr	Zirconium(IV) fluoride	-1911.3			-1809.9			104.6			103.7		
F5I	Iodine pentafluoride		-864.8	-822.5			-751.7			327.7			99.2
F5Nb	Niobium(V) fluoride	-1813.8		-1739.7	-1699.0		-1673.6	160.2		321.9	134.7		97.1
F5P	Phosphorus(V) fluoride			-1594.4			-1520.7			300.8			84.8
F5Ta	Tantalum(V) fluoride	-1903.6	-1480.3			-1373.1			175.7				
F5V	Vanadium(V) fluoride			-1433.9			-1369.8			320.9			98.6

Molecular formula	Name	$\Delta_f H°$/kJ mol⁻¹ Crys.	Liq.	Gas	$\Delta_f G°$/kJ mol⁻¹ Crys.	Liq.	Gas	S°/J mol⁻¹ K⁻¹ Crys.	Liq.	Gas	C_p/J mol⁻¹ K⁻¹ Crys.	Liq.	Gas
F₆H₈N₂Si	Ammonium hexafluorosilicate	-2681.7			-2365.3			280.2			228.1		
F₆Ir	Iridium(VI) fluoride	-579.7		-544.0	-461.6		-460.0	247.7		357.8			121.1
F₆K₂Si	Potassium hexafluorosilicate	-2956.0			-2798.6			226.0					
F₆Mo	Molybdenum(VI) fluoride		-1585.5	-1557.7		-1473.0	-1472.2		259.7	350.5		169.8	120.6
F₆Na₂Si	Sodium hexafluorosilicate	-2909.6			-2754.2			207.1			187.1		
F₆Os	Osmium(VI) fluoride							246.0		358.1			120.8
F₆Pt	Platinum(VI) fluoride							235.6		348.3			122.8
F₆S	Sulfur hexafluoride			-1220.5			-1116.5			291.5			97.0
F₆Se	Selenium hexafluoride			-1117.0			-1017.0			313.9			110.5
F₆Te	Tellurium hexafluoride			-1318.0									
F₆U	Uranium(VI) fluoride	-2197.0		-2147.4	-2068.5		-2063.7	227.6		377.9	166.8		129.6
F₆W	Tungsten(VI) fluoride		-1747.7	-1721.7		-1631.4	-1632.1		251.5	341.1			119.0
Fe	Iron	0.0		416.3			370.7	27.3		180.5	25.1		25.7
FeI₂	Iron(II) iodide	-113.0											
FeI₃	Iron(III) iodide			71.0									
FeMoO₄	Iron(II) molybdate	-1075.0			-975.0			129.3			118.5		
FeO	Iron(II) oxide	-272.0											
FeO₄S	Iron(II) sulfate	-928.4			-820.8			107.5			100.6		
FeO₄W	Iron(II) tungstate	-1155.0			-1054.0			131.8			114.6		
FeS	Iron(II) sulfide	-100.0			-100.4			60.3			50.5		
FeS₂	Iron disulfide	-178.2			-166.9			52.9			62.2		
Fe₂O₃	Iron(III) oxide	-824.2			-742.2			87.4			103.9		
Fe₂O₄Si	Iron(II) orthosilicate	-1479.9			-1379.0			145.2			132.9		
Fe₃O₄	Iron(II,III) oxide	-1118.4			-1015.4			146.4			143.4		
Fm	Fermium	0.0											
Fr	Francium	0.0						95.4					
Ga	Gallium	0.0	5.6	272.0	0.0		233.7	40.8		169.0	26.1		25.3
GaH₃O₃	Gallium(III) hydroxide	-964.4			-831.3			100.0					
GaI₃	Gallium(III) iodide	-238.9						205.0			100.0		
GaN	Gallium nitride	-110.5											
GaO	Gallium monoxide			279.5			253.5			231.1			32.1
GaP	Gallium phosphide	-88.0											
GaSb	Gallium antimonide	-41.8			-38.9			76.1			48.5		
Ga₂	Digallium			438.5									
Ga₂O	Gallium suboxide	-356.0			-356.0								
Ga₂O₃	Gallium(III) oxide	-1089.1			-998.3			85.0			92.1		
Gd	Gadolinium	0.0		397.5			359.8	68.1		194.3	37.0		27.5
Gd₂O₃	Gadolinium(III) oxide	-1819.6									106.7		
Ge	Germanium	0.0		372.0	0.0		331.2	31.1		167.9	23.3		30.7
GeH₃I	Iodogermane									283.2			57.5
GeH₄	Germane			90.8			113.4			217.1			45.0
GeI₄	Germanium(IV) iodide	-141.8			-144.3			271.1		428.9			104.1
GeO	Germanium(II) oxide	-261.9		-56.9	-237.2		-106.3	50.0		224.3			30.9
GeO₂	Germanium(IV) oxide	-580.0		-46.2	-521.4		-73.2	39.7			52.1		

Molecular formula	Name	$\Delta_f H°$/kJ mol⁻¹ Crys.	Liq.	Gas	$\Delta_f G°$/kJ mol⁻¹ Crys.	Liq.	Gas	S°/J mol⁻¹ K⁻¹ Crys.	Liq.	Gas	C_p/J mol⁻¹ K⁻¹ Crys.	Liq.	Gas
GeP	Germanium phosphide	-21.0			-17.0			63.0					
GeS	Germanium(II) sulfide	-69.0		92.0	-71.5		42.0	71.0		234.0			33.7
GeTe	Germanium(II) telluride	20.0											
Ge₂	Digermanium			473.1			416.3			252.8			35.6
Ge₂H₆	Digermane		137.3	162.3									
Ge₃H₈	Trigermane		193.7	226.8									
H	Hydrogen (atomic)			218.0			203.3			114.7			20.8
HI	Hydrogen iodide			26.5			1.7			206.6			29.2
HIO₃	Iodic acid	-230.1											
HK	Potassium hydride	-57.7											
HKO	Potassium hydroxide	-424.6		-228.0	-378.7		-229.7	78.9		238.3	64.9		49.2
HKO₄S	Potassium hydrogen sulfate	-1160.6			-1031.3			138.1					
HLi	Lithium hydride	-90.5			-68.3			20.0			27.9		
HLiO	Lithium hydroxide	-484.9			-439.0			42.8			49.7		
HN	Imidogen			351.5			345.6			181.2			29.2
HNO₂	Nitrous acid			-79.5			-46.0			254.1			45.6
HNO₃	Nitric acid		-174.1	-133.9		-80.7	-73.5		155.6	266.9		109.9	54.1
HN₃	Hydrazoic acid		264.0	294.1		327.3	328.1		140.6	239.0			43.7
HNa	Sodium hydride	-56.3			-33.5			40.0			36.4		
HNaO	Sodium hydroxide	-425.6			-379.5			64.5			59.5		
HNaO₄S	Sodium hydrogen sulfate	-1125.5			-992.8			113.0					
HNa₂O₄P	Sodium hydrogen phosphate	-1748.1			-1608.2			150.5			135.3		
HO	Hydroxyl			39.0			34.2			183.7			29.9
HORb	Rubidium hydroxide	-418.2											
HOTl	Thallium(I) hydroxide	-238.9			-195.8			88.0					
HO₂	Hydroperoxy			10.5			22.6			229.0			34.9
HO₃P	Metaphosphoric acid	-948.5											
HO₄RbS	Rubidium hydrogen sulfate	-1159.0											
HO₄Re	Perrhenic acid	-762.3			-656.4			158.2					
HRb	Rubidium hydride	-52.3											
HS	Mercapto			142.7			113.3			195.7			32.3
HSi	Silylidyne			361.0									
HTa₂	Tantalum hydride	-32.6			-69.0			79.1			90.8		
H₂	Hydrogen			0.0						130.7			28.8
H₂KN	Potassium amide	-128.9											
H₂KO₄P	Potassium dihydrogen phosphate	-1568.3			-1415.9			134.9			116.6		
H₂LiN	Lithium amide	-179.5											
H₂Mg	Magnesium hydride	-75.3			-35.9			31.1			35.4		
H₂MgO₂	Magnesium hydroxide	-924.5			-833.5			63.2			77.0		
H₂N	Amidogen			184.9			194.6			195.0			33.9
H₂NNa	Sodium amide	-123.8			-64.0			76.9			66.2		
H₂NRb	Rubidium amide	-113.0											
H₂N₂O₂	Nitramide	-89.5											
H₂NiO₂	Nickel(II) hydroxide	-529.7			-447.2			88.0					

Molecular formula	Name	ΔfH°/kJ mol⁻¹ Crys.	Liq.	Gas	ΔfG°/kJ mol⁻¹ Crys.	Liq.	Gas	S°/J mol⁻¹ K⁻¹ Crys.	Liq.	Gas	Cp/J mol⁻¹ K⁻¹ Crys.	Liq.	Gas
H_2O	Water		-285.8	-241.8		-237.1	-228.6		70.0	188.8		75.3	33.6
H_2O_2	Hydrogen peroxide		-187.8	-136.3		-120.4	-105.6		109.6	232.7		89.1	43.1
H_2O_2Sn	Tin(II) hydroxide	-561.1			-491.6			155.0					
H_2O_2Sr	Strontium hydroxide	-959.0											
H_2O_2Zn	Zinc hydroxide	-641.9			-553.5			81.2					
H_2O_3Si	Metasilicic acid	-1188.7			-1092.4			134.0					
H_2O_4S	Sulfuric acid		-814.0										
H_2O_4Se	Selenic acid	-530.1				-690.0			156.9			138.9	
H_2S	Hydrogen sulfide			-20.6			-33.4			205.8			34.2
H_2S_2	Hydrogen disulfide		-18.1	15.5								84.1	51.5
H_2Se	Hydrogen selenide			29.7			15.9			219.0			34.7
H_2Sr	Strontium hydride	-180.3											
H_2Te	Hydrogen telluride			99.6									
H_2Th	Thorium hydride	-139.7			-100.0			50.7			36.7		
H_2Zr	Zirconium(II) hydride	-169.0			-128.8			35.0			31.0		
H_3ISi	Iodosilane									270.9			54.4
H_3N	Ammonia			-45.9			-16.4			192.8			35.1
H_3NO	Hydroxylamine	-114.2											
H_3O_2P	Hypophosphorous acid	-604.6	-595.4										
H_3O_3P	Phosphorous acid	-964.4											
H_3O_4P	Phosphoric acid (orthophosphoric acid)	-1284.4	-1271.7		-1124.3	-1123.6		110.5	150.8		106.1	145.0	
H_3P	Phosphine			5.4			13.4			210.2			37.1
H_3Sb	Stibine			145.1			147.8			232.8			41.1
H_3U	Uranium(III) hydride	-127.2			-72.8			63.7			49.3		
H_4IN	Ammonium iodide	-201.4			-112.5			117.0					
H_4N_2	Hydrazine		50.6	95.4		149.3	159.4		121.2	238.5		98.9	48.4
$H_4N_2O_2$	Ammonium nitrite	-256.5						151.1					
$H_4N_2O_3$	Ammonium nitrate	-365.6			-183.9			112.5			139.3		
H_4N_4	Ammonium azide	115.5			274.2			192.0					
H_4O_4Si	Orthosilicic acid	-1481.1			-1332.9								
$H_4O_7P_2$	Pyrophosphoric acid	-2241.0	-2231.7										
H_4P_2	Diphosphine		-5.0										
H_4Si	Silane			20.9			56.9			204.6			42.8
H_4Sn	Stannane			34.3			188.3			227.7			49.0
H_5NO	Ammonium hydroxide		-361.2	162.8		-254.0			165.6			154.9	
H_5NO_3S	Ammonium hydrogen sulfite	-768.6											
H_5NO_4S	Ammonium hydrogen sulfate	-1027.0											
H_6Si_2	Disilane			80.3			127.3			272.7			80.8
$H_8N_2O_4S$	Ammonium sulfate	-1180.9			-901.7			220.1			187.5		
H_8Si_3	Trisilane		92.5	120.9									
$H_9N_2O_4P$	Diammonium hydrogen phosphate	-1566.9									188.0		
$H_{12}N_3O_4P$	Ammonium phosphate	-1671.9											
He	Helium			0.0						126.2			20.8

STANDARD THERMODYNAMIC PROPERTIES OF CHEMICAL SUBSTANCES (continued)

Molecular formula	Name	$\Delta_f H°$/kJ mol⁻¹ Crys.	Liq.	Gas	$\Delta_f G°$/kJ mol⁻¹ Crys.	Liq.	Gas	$S°$/J mol⁻¹ K⁻¹ Crys.	Liq.	Gas	C_p/J mol⁻¹ K⁻¹ Crys.	Liq.	Gas
Hf	Hafnium	0.0		619.2			576.5	43.6		186.9	25.7		20.8
HfO₂	Hafnium oxide	-1144.7			-1088.2			59.3			60.3		
Hg	Mercury		0.0	61.4			31.8		75.9	175.0		28.0	20.8
HgI₂	Mercury(II) iodide	-105.4			-101.7			180.0					
HgO	Mercury(II) oxide	-90.8			-58.5			70.3			44.1		
HgO₄S	Mercury(II) sulfate	-707.5											
HgS	Mercury(II) sulfide	-58.2			-50.6			82.4			48.4		
HgTe	Mercury(II) telluride	-42.0											
Hg₂	Dimercury			108.8			68.2			288.1			37.4
Hg₂I₂	Mercury(I) iodide	-121.3			-111.0			233.5			132.0		
Hg₂O₄S	Mercury(I) sulfate	-743.1			-625.8			200.7					
Ho	Holmium	0.0		300.8			264.8	75.3		195.6	27.2		20.8
Ho₂O₃	Holmium oxide	-1880.7			-1791.1			158.2			115.0		
I	Iodine (atomic)			106.8			70.2			180.8			20.8
IIn	Indium(I) iodide	-116.3		7.5	-120.5		-37.7	130.0		267.3			36.8
IK	Potassium iodide	-327.9			-324.9			106.3			52.9		
IKO₃	Potassium iodate	-501.4			-418.4			151.5			106.5		
IKO₄	Potassium periodate	-467.2			-361.4			175.7					
ILi	Lithium iodide	-270.4			-270.3			86.8			51.0		
INa	Sodium iodide	-287.8			-286.1			98.5			52.1		
INaO₃	Sodium iodate	-481.8			-323.0			163.0			92.0		
INaO₄	Sodium periodate	-429.3											
IO	Iodine monoxide			175.1			149.8			245.5			32.9
IRb	Rubidium iodide	-333.8			-328.9			118.4			53.2		
ITl	Thallium(I) iodide	-123.8		7.1	-125.4			127.6					
I₂	Iodine (rhombic)	0.0		62.4			19.3	116.1		260.7	54.4		36.9
I₂Mg	Magnesium iodide	-364.0			-358.2			129.7					
I₂Ni	Nickel(II) iodide	-78.2											
I₂Pb	Lead(II) iodide	-175.5			-173.6			174.9			77.4		
I₂Sn	Tin(II) iodide	-143.5											
I₂Sr	Strontium iodide	-558.1											
I₂Zn	Zinc iodide	-208.0			-209.0			161.1			81.6		
I₃In	Indium(III) iodide	-238.0		-120.5						374.4			78.4
I₃La	Lanthanum iodide	-668.9											
I₃Lu	Lutetium iodide	-548.0											
I₃P	Phosphorus(III) iodide	-45.6											
I₃Ru	Ruthenium(III) iodide	-65.7											
I₃Sb	Antimony(III) iodide	-100.4											
I₄Pt	Platinum(IV) iodide	-72.8											
I₄Si	Tetraiodosilane	-189.5											
I₄Sn	Tin(IV) iodide										84.9		
I₄Ti	Titanium(IV) iodide	-375.7		-277.8	-371.5			249.4		446.1	125.7		105.4
I₄V	Vanadium(IV) iodide			-122.6									
I₄Zr	Zirconium(IV) iodide	-481.6											
In	Indium	0.0		243.3			208.7	57.8		173.8	26.7		20.8

Molecular formula	Name	$\Delta_f H°$/kJ mol⁻¹			$\Delta_f G°$/kJ mol⁻¹			$S°$/J mol⁻¹ K⁻¹			C_p/J mol⁻¹ K⁻¹		
		Crys.	Liq.	Gas	Crys.	Liq.	Gas	Crys.	Liq.	Gas	Crys.	Liq.	Gas
InO	Indium monoxide			387.0			364.4			236.5			32.6
InP	Indium phosphide	-88.7			-77.0			59.8			45.4		
InS	Indium(II) sulfide	-138.1		238.0	-131.8			67.0					
InSb	Indium antimonide	-30.5		344.3	-25.5			86.2			49.5		
In₂	Diindium			380.9									
In₂O₃	Indium(III) oxide	-925.8			-830.7			104.2			92.0		
In₂S₃	Indium(III) sulfide	-427.0			-412.5			163.6			118.0		
Ir	Iridium	0.0		665.3	0.0		617.9	35.5		193.6	25.1		20.8
IrO₂	Iridium(IV) oxide	-274.1									57.3		
IrS₂	Iridium(IV) sulfide	-138.0											
Ir₂S₃	Iridium(III) sulfide	-234.0											
K	Potassium	0.0		89.0	0.0		60.5	64.7		160.3	29.6		20.8
KMnO₄	Potassium permanganate	-837.2			-737.6			171.7			117.6		
KNO₂	Potassium nitrite	-369.8			-306.6			152.1			107.4		
KNO₃	Potassium nitrate	-494.6			-394.9			133.1			96.4		
KNa	Potassium sodium		6.3										
KO₂	Potassium superoxide	-284.9			-239.4			116.7			77.5		
K₂	Dipotassium			123.7			87.5			249.7			37.9
K₂O	Potassium oxide	-361.5			-425.1			102.1					
K₂O₂	Potassium peroxide	-494.1											
K₂O₄S	Potassium sulfate	-1437.8			-1321.4			175.6			131.5		
K₂S	Potassium sulfide	-380.7			-364.0			105.0					
K₃O₄P	Potassium phosphate	-1950.2											
Kr	Krypton	0.0		0.0						164.1			20.8
La	Lanthanum	0.0		431.0	0.0		393.6	56.9		182.4	27.1		22.8
LaS	Lanthanum sulfide	-456.0			-451.5			73.2			59.0		
La₂O₃	Lanthanum oxide	-1793.7			-1705.8			127.3			108.8		
Li	Lithium	0.0		159.3	0.0		126.6	29.1		138.8	24.8		20.8
LiNO₂	Lithium nitrite	-372.4			-302.0			96.0					
LiNO₃	Lithium nitrate	-483.1			-381.1			90.0					
Li₂	Dilithium			215.9			174.4			197.0			36.1
Li₂O	Lithium oxide	-597.9			-561.2			37.6			54.1		
Li₂O₂	Lithium peroxide	-634.3											
Li₂O₃Si	Lithium metasilicate	-1648.1			-1557.2			79.8			99.1		
Li₂O₄S	Lithium sulfate	-1436.5			-1321.7			115.1			117.6		
Li₂S	Lithium sulfide	-441.4											
Li₃O₄P	Lithium phosphate	-2095.8											
Lr	Lawrencium	0.0											
Lu	Lutetium	0.0		427.6			387.8	51.0		184.8	26.9		20.9
Lu₂O₃	Lutetium oxide	-1878.2			-1789.0			110.0			101.8		
Md	Mendelevium	0.0											
Mg	Magnesium	0.0		147.1			112.5	32.7		148.6	24.9		20.8
MgN₂O₆	Magnesium nitrate	-790.7			-589.4			164.0			141.9		
MgO	Magnesium oxide	-601.6			-569.3			27.0			37.2		
MgO₄S	Magnesium sulfate	-1284.9			-1170.6			91.6			96.5		

Molecular formula	Name	$\Delta_f H°$/kJ mol⁻¹			$\Delta_f G°$/kJ mol⁻¹			$S°$/J mol⁻¹ K⁻¹			C_p/J mol⁻¹ K⁻¹		
		Crys.	Liq.	Gas	Crys.	Liq.	Gas	Crys.	Liq.	Gas	Crys.	Liq.	Gas
MgO₄Se	Magnesium selenate	-968.5											
MgS	Magnesium sulfide	-346.0			-341.8			50.3			45.6		
Mg₂	Dimagnesium			287.7									
Mg₂O₄Si	Magnesium orthosilicate	-2174.0			-2055.1			95.1			118.5		
Mn	Manganese	0.0		280.7			238.5	32.0		173.7	26.3		20.8
MnN₂O₆	Manganese(II) nitrate	-576.3											
MnNa₂O₄	Sodium permanganate	-1156.0											
MnO	Manganese(II) oxide	-385.2			-362.9			59.7			45.4		
MnO₂	Manganese(IV) oxide	-520.0			-465.1			53.1			54.1		
MnO₃Si	Manganese(II) metasilicate	-1320.9			-1240.5			89.1			86.4		
MnS	Manganese(II) sulfide	-214.2			-218.4			78.2			50.0		
MnSe	Manganese(II) selenide	-106.7			-111.7			90.8			51.0		
Mn₂O₃	Manganese(III) oxide	-959.0			-881.1			110.5			107.7		
Mn₂O₄Si	Manganese(II) orthosilicate	-1730.5			-1632.1			163.2			129.9		
Mn₃O₄	Manganese(II,III) oxide	-1387.8			-1283.2			155.6			139.7		
Mo	Molybdenum	0.0		658.1			612.5	28.7		182.0	24.1		20.8
MoNa₂O₄	Sodium molybdate	-1468.1			-1354.3			159.7			141.7		
MoO₂	Molybdenum(IV) oxide	-588.9			-533.0			46.3			56.0		
MoO₃	Molybdenum(VI) oxide	-745.1			-668.0			77.7			75.0		
MoO₄Pb	Lead(II) molybdate	-1051.9			-951.4			166.1			119.7		
MoS₂	Molybdenum(IV) sulfide	-235.1			-225.9			62.6			63.6		
N	Nitrogen (atomic)			472.7			455.5			153.3			20.8
NNaO₂	Sodium nitrite	-358.7			-284.6			103.8					
NNaO₃	Sodium nitrate	-467.9			-367.0			116.5			92.9		
NO	Nitric oxide			91.3			87.6			210.8			29.9
NO₂	Nitrogen dioxide			33.2			51.3			240.1			37.2
NO₂Rb	Rubidium nitrite	-367.4			-306.2			172.0			102.1		
NO₃Rb	Rubidium nitrate	-495.1			-395.8			147.3			99.5		
NO₃Tl	Thallium(I) nitrate	-243.9			-152.4			160.7					
NP	Phosphorus nitride	-63.0											
N₂	Nitrogen			0.0						191.6			29.1
N₂O	Nitrous oxide			81.6			103.7			220.0			38.6
N₂O₃	Nitrogen trioxide		50.3	86.6			142.4			314.7			72.7
N₂O₄	Nitrogen tetroxide		-19.5	11.1		97.5	99.8		209.2	304.4		142.7	79.2
N₂O₄Sr	Strontium nitrite	-762.3											
N₂O₅	Nitrogen pentoxide	-43.1		13.3	113.9		117.1	178.2		355.7	143.1		95.3
N₂O₆Pb	Lead(II) nitrate	-451.9											
N₂O₆Ra	Radium nitrate	-992.0			-796.1			222.0					
N₂O₆Sr	Strontium nitrate	-978.2			-780.0			194.6			149.9		
N₂O₆Zn	Zinc nitrate	-483.7											
N₃Na	Sodium azide	21.7			93.8			96.9			76.6		
N₄Si₃	Silicon nitride	-743.5			-642.6			101.3					
Na	Sodium	0.0		107.5			77.0	51.3		153.7	28.2		20.8
NaO₂	Sodium superoxide	-260.2			-218.4			115.9			72.1		
Na₂	Disodium			142.1			103.9			230.2			37.6

Molecular formula	Name	$\Delta_f H°$/kJ mol⁻¹ Crys.	Liq.	Gas	$\Delta_f G°$/kJ mol⁻¹ Crys.	Liq.	Gas	S°/J mol⁻¹ K⁻¹ Crys.	Liq.	Gas	C_p/J mol⁻¹ K⁻¹ Crys.	Liq.	Gas
Na₂O	Sodium oxide	-414.2			-375.5			75.1			69.1		
Na₂O₂	Sodium peroxide	-510.9			-447.7			95.0			89.2		
Na₂O₃S	Sodium sulfite	-1100.8			-1012.5			145.9			120.3		
Na₂O₃Si	Sodium metasilicate	-1554.9			-1462.8			113.9			128.2		
Na₂O₄S	Sodium sulfate	-1387.1			-1270.2			149.6					
Na₂S	Sodium sulfide	-364.8			-349.8			83.7					
Nb	Niobium	0.0		725.9			681.1	36.4		186.3	24.6		30.2
NbO	Niobium(II) oxide	-405.8			-378.6			48.1			41.3		
NbO₂	Niobium(IV) oxide	-796.2			-740.5			54.5			57.5		
Nb₂O₅	Niobium(V) oxide	-1899.5			-1766.0			137.2			132.1		
Nd	Neodymium	0.0		327.6			292.4	71.5		189.4	27.5		22.1
Nd₂O₃	Neodymium oxide	-1807.9			-1720.8			158.6			111.3		
Ne	Neon			0.0						146.3			20.8
Ni	Nickel	0.0		429.7			384.5	29.9		182.2	26.1		23.4
NiO₄S	Nickel(II) sulfate	-872.9			-759.7			92.0			138.0		
NiS	Nickel(II) sulfide	-82.0			-79.5			53.0			47.1		
Ni₂O₃	Nickel(III) oxide	-489.5											
No	Nobelium	0.0		0.0									
O	Oxygen (atomic)			249.2			231.7			161.1			21.9
OP	Phosphorus monoxide			-28.5			-51.9			222.8			31.8
OPb	Lead(II) oxide (massicot)	-217.3			-187.9			68.7			45.8		
OPb	Lead(II) oxide (litharge)	-219.0			-188.9			66.5			45.8		
OPd	Palladium(II) oxide	-85.4		348.9			325.9			218.0	31.4		
ORa	Radium oxide	-523.0											
ORb₂	Rubidium oxide	-339.0											
ORh	Rhodium monoxide			385.0									
OS	Sulfur monoxide			6.3			-19.9			222.0			30.2
OSe	Selenium monoxide			53.4			26.8			234.0			31.3
OSi	Silicon monoxide			-99.6			-126.4			211.6			29.9
OSn	Tin(II) oxide	-280.7		15.1	-251.9		-8.4	57.2		232.1	44.3		31.6
OSr	Strontium oxide	-592.0			-561.9			54.4			45.0		
OTi	Titanium(II) oxide	-519.7			-495.0			50.0			40.0		
OTl₂	Thallium(I) oxide	-178.7			-147.3			126.0					
OU	Uranium(II) oxide			21.0									
OV	Vanadium(II) oxide	-431.8			-404.2			38.9			45.4		
OZn	Zinc oxide	-350.5			-320.5			43.7			40.3		
O₂	Oxygen			0.0			0.0			205.2			29.4
O₂P	Phosphorus dioxide			-279.9			-281.6			252.1			39.5
O₂Pb	Lead(IV) oxide	-277.4			-217.3			68.6			64.6		
O₂Rb	Rubidium superoxide	-278.7											
O₂Rb₂	Rubidium peroxide	-472.0											
O₂Ru	Ruthenium(IV) oxide	-305.0											
O₂S	Sulfur dioxide		-320.5	-296.8			-300.1			248.2			39.9
O₂Se	Selenium dioxide	-225.4											
O₂Si	Silicon dioxide (α-quartz)	-910.7		-322.0	-856.3			41.5			44.4		

Molecular formula	Name	$\Delta_f H°$/kJ mol^{-1}			$\Delta_f G°$/kJ mol^{-1}			$S°$/J mol^{-1} K^{-1}			C_p/J mol^{-1} K^{-1}		
		Crys.	Liq.	Gas	Crys.	Liq.	Gas	Crys.	Liq.	Gas	Crys.	Liq.	Gas
O$_2$Sn	Tin(IV) oxide	-577.6			-515.8			49.0			52.6		
O$_2$Te	Tellurium dioxide	-322.6			-270.3			79.5					
O$_2$Th	Thorium(IV) oxide	-1226.4			-1169.2			65.2			61.8		
O$_2$Ti	Titanium(IV) oxide	-944.0			-888.8			50.6			55.0		
O$_2$U	Uranium(IV) oxide	-1085.0		-465.7	-1031.8		-471.5	77.0		274.6	63.6		51.4
O$_2$W	Tungsten(IV) oxide	-589.7			-533.9			50.5			56.1		
O$_2$Zr	Zirconium(IV) oxide	-1100.6			-1042.8			50.4			56.2		
O$_3$	Ozone			142.7			163.2			238.9			39.2
O$_3$PbS	Lead(II) sulfite	-669.9											
O$_3$PbSi	Lead(II) metasilicate	-1145.7			-1062.1			109.6			90.0		
O$_3$Pr$_2$	Praseodymium oxide	-1809.6									117.4		
O$_3$Rh$_2$	Rhodium(III) oxide	-343.0									103.8		
O$_3$S	Sulfur trioxide	-454.5	-441.0	-395.7	-374.2	-373.8	-371.1	70.7	113.8	256.8			50.7
O$_3$Sc$_2$	Scandium oxide	-1908.8			-1819.4			77.0			94.2		
O$_3$SiSr	Strontium metasilicate	-1633.9			-1549.7			96.7			88.5		
O$_3$Sm$_2$	Samarium(III) oxide	-1823.0			-1734.6			151.0			114.5		
O$_3$Tb$_2$	Terbium oxide	-1865.2									115.9		
O$_3$Ti$_2$	Titanium(III) oxide	-1520.9			-1434.2			78.8			97.4		
O$_3$Tm$_2$	Thulium oxide	-1888.7			-1794.5			139.7			116.7		
O$_3$U	Uranium(VI) oxide	-1223.8			-1145.7			96.1			81.7		
O$_3$V$_2$	Vanadium(III) oxide	-1218.8			-1139.3			98.3			103.2		
O$_3$W	Tungsten(VI) oxide	-842.9			-764.0			75.9			73.8		
O$_3$Y$_2$	Yttrium oxide	-1905.3			-1816.6			99.1			102.5		
O$_3$Yb$_2$	Ytterbium(III) oxide	-1814.6			-1726.7			133.1			115.4		
O$_4$Os	Osmium(VIII) oxide	-394.1		-337.2	-304.9		-292.8	143.9		293.8			74.1
O$_4$PbS	Lead(II) sulfate	-920.0			-813.0			148.5			103.2		
O$_4$PbSe	Lead(II) selenate	-609.2			-504.9			167.8					
O$_4$Pb$_2$Si	Lead(II) orthosilicate	-1363.1			-1252.6			186.6			137.2		
O$_4$Pb$_3$	Lead(II,IV) oxide	-718.4			-601.2			211.3			146.9		
O$_4$RaS	Radium sulfate	-1471.1			-1365.6			138.0					
O$_4$Rb$_2$S	Rubidium sulfate	-1435.6			-1316.9			197.4			134.1		
O$_4$Ru	Ruthenium(VIII) oxide	-239.3			-152.2			146.4					
O$_4$SSr	Strontium sulfate	-1453.1			-1340.9			117.0					
O$_4$STl$_2$	Thallium(I) sulfate	-931.8			-830.4			230.5					
O$_4$SZn	Zinc sulfate	-982.8			-871.5			110.5			99.2		
O$_4$SiSr$_2$	Strontium orthosilicate	-2304.5			-2191.1			153.1			134.3		
O$_4$SiZn$_2$	Zinc orthosilicate	-1636.7			-1523.2			131.4			123.3		
O$_4$SiZr	Zirconium(IV) orthosilicate	-2033.4			-1919.1			84.1			98.7		
O$_5$Sb$_2$	Antimony(V) oxide	-971.9			-829.2			125.1					
O$_5$Ta$_2$	Tantalum(V) oxide	-2046.0			-1911.2			143.1			135.1		
O$_5$Ti$_3$	Titanium(III,IV) oxide	-2459.4			-2317.4			129.3			154.8		
O$_5$V$_2$	Vanadium(V) oxide	-1550.6			-1419.5			131.0			127.7		
O$_5$V$_3$	Vanadium(III,IV) oxide	-1933.0			-1803.0			163.0					
O$_7$Re$_2$	Rhenium(VII) oxide	-1240.1		-1100.0	-1066.0		-994.0	207.1		452.0	166.1		
O$_8$U$_3$	Uranium(IV,VI) oxide	-3427.1			-3242.9			250.5			215.5		

Molecular formula	Name	$\Delta_fH°$/kJ mol⁻¹ Crys.	Liq.	Gas	$\Delta_fG°$/kJ mol⁻¹ Crys.	Liq.	Gas	$S°$/J mol⁻¹ K⁻¹ Crys.	Liq.	Gas	C_p/J mol⁻¹ K⁻¹ Crys.	Liq.	Gas
O_8S_2Zr	Zirconium(IV) sulfate	-2217.1									172.0		
O_8U_3	Uranium(V,VI) oxide	-3574.8			-3369.5			282.6			238.4		
O_9U_4	Uranium(IV,V) oxide	-4510.4			-4275.1			334.1			293.3		
Os	Osmium	0.0		791.0			745.0	32.6		192.6	24.7		20.8
P	Phosphorus (white)	0.0		316.5			280.1	41.1		163.2	23.8		20.8
P	Phosphorus (red)	-17.6											
P	Phosphorus (black)	-39.3						22.8			21.2		
P_2	Diphosphorus			144.0			103.5			218.1			32.1
P_4	Tetraphosphorus			58.9			24.4			280.0			67.2
Pa	Protactinium	0.0		607.0			563.0	51.9		198.1			22.9
Pb	Lead	0.0		195.2			162.2	64.8		175.4	26.4		20.8
PbS	Lead(II) sulfide	-100.4			-98.7			91.2			49.5		
PbSe	Lead(II) selenide	-102.9			-101.7			102.5			50.2		
PbTe	Lead(II) telluride	-70.7			-69.5			110.0			50.5		
Pd	Palladium	0.0		378.2			339.7	37.6		167.1	26.0		20.8
PdS	Palladium(II) sulfide	-75.0			-67.0			46.0					
Pm	Promethium	0.0											
Po	Polonium	0.0								187.1			24.3
Pr	Praseodymium	0.0		355.6			320.9	73.2		189.8	27.2		21.4
Pt	Platinum	0.0		565.3			520.5	41.6		192.4	25.9		25.5
PtS	Platinum(II) sulfide	-81.6			-76.1			55.1			43.4		
PtS_2	Platinum(IV) sulfide	-108.8			-99.6			74.7			65.9		
Pu	Plutonium	0.0											
Ra	Radium	0.0		159.0			130.0	71.0		176.5			20.8
Rb	Rubidium	0.0		80.9			53.1	76.8		170.1	31.1		20.8
Re	Rhenium	0.0		769.9			724.6	36.9		188.9	25.5		20.8
Rh	Rhodium	0.0		556.9			510.8	31.5		185.8	25.0		21.0
Rn	Radon	0.0		0.0						176.2			20.8
Ru	Ruthenium	0.0		642.7			595.8	28.5		186.5	24.1		21.5
S	Sulfur (rhombic)	0.0		277.2			236.7	32.1		167.8	22.6		23.7
S	Sulfur (monoclinic)	0.3											
SSi	Silicon monosulfide			112.5			60.9			223.7			32.3
SSn	Tin(II) sulfide	-100.0			-98.3			77.0			49.3		
SSr	Strontium sulfide	-472.4			-467.8			68.2			48.7		
STl_2	Thallium(I) sulfide	-97.1			-93.7			151.0					
SZn	Zinc sulfide (wurtzite)	-192.6											
SZn	Zinc sulfide (sphalerite)	-206.0			-201.3			57.7			46.0		
S_2	Disulfur			128.6			79.7			228.2			32.5
Sb	Antimony	0.0		262.3			222.1	45.7		180.3	25.2		20.8
Sb_2	Diantimony			235.6			187.0			254.9			36.4
Sc	Scandium	0.0		377.8			336.0	34.6		174.8	25.5		22.1
Se	Selenium	0.0		227.1			187.0	42.4		176.7	25.4		20.8
SeSr	Strontium selenide	-385.8											
$SeTl_2$	Thallium(I) selenide	-59.0			-59.0			172.0					
SeZn	Zinc selenide	-163.0			-163.0			84.0					

Molecular formula	Name	$\Delta_f H°$/kJ mol⁻¹			$\Delta_f G°$/kJ mol⁻¹			$S°$/J mol⁻¹ K⁻¹			C_p/J mol⁻¹ K⁻¹		
		Crys.	Liq.	Gas	Crys.	Liq.	Gas	Crys.	Liq.	Gas	Crys.	Liq.	Gas
Se₂	Diselenium			146.0			96.2			252.0			35.4
Si	Silicon	0.0		450.0			405.5	18.8		168.0	20.0		22.3
Si₂	Disilicon			594.0			536.0			229.9			34.4
Sm	Samarium	0.0		206.7			172.8	69.6		183.0	29.5		30.4
Sn	Tin (white)	0.0		301.2			266.2	51.2		168.5	27.0		21.3
Sn	Tin (gray)	-2.1			0.1			44.1			25.8		
Sr	Strontium	0.0		164.4			130.9	55.0		164.6	26.8		20.8
Ta	Tantalum	0.0		782.0			739.3	41.5		185.2	25.4		20.9
Tb	Terbium	0.0		388.7			349.7	73.2		203.6	28.9		24.6
Tc	Technetium	0.0		678.0						181.1			20.8
Te	Tellurium	0.0		196.7			157.1	49.7		182.7	25.7		20.8
Te₂	Ditellurium			168.2			118.0			268.1			36.7
Th	Thorium	0.0		602.0			560.7	51.8		190.2	27.3		20.8
Ti	Titanium	0.0		473.0			428.4	30.7		180.3	25.0		24.4
Tl	Thallium	0.0		182.2			147.4	64.2		181.0	26.3		20.8
Tm	Thulium	0.0		232.2			197.5	74.0		190.1	27.0		20.8
U	Uranium	0.0		533.0			488.4	50.2		199.8	27.7		23.7
V	Vanadium	0.0		514.2			469.9	28.9		182.3	24.9		26.0
W	Tungsten	0.0		849.4			807.1	32.6		174.0	24.3		21.3
Xe	Xenon			0.0						169.7			20.8
Y	Yttrium	0.0		421.3			381.1	44.4		179.5	26.5		25.9
Yb	Ytterbium	0.0		152.3			118.4	59.9		173.1	26.7		20.8
Zn	Zinc	0.0		130.4			94.8	41.6		161.0	25.4		20.8
Zr	Zirconium	0.0		608.8			566.5	39.0		181.4	25.4		26.7
C	Carbon (graphite)	0.0		716.7			671.3	5.7		158.1	8.5		20.8
C	Carbon (diamond)	1.9			2.9			2.4			6.1		
CAgN	Silver(I) cyanide	146.0			156.9			107.2			66.7		
CAg₂O₃	Silver(I) carbonate	-505.8			-436.8			167.4			112.3		
CBaO₃	Barium carbonate	-1213.0			-1134.4			112.1			86.0		
CBeO₃	Beryllium carbonate	-1025.0						52.0			65.0		
CBrClF₂	Bromochlorodifluoro-methane									318.5			74.6
CBrCl₂F	Bromodichlorofluoro-methane									330.6			80.0
CBrCl₃	Bromotrichloromethane			-41.1									85.3
CBrF₃	Bromotrifluoromethane			-648.3						248.3			69.3
CBrN	Cyanogen bromide		140.5	186.2			165.3						46.9
CBrN₃O₆	Bromotrinitromethane		32.5	80.3									
CBr₂ClF	Dibromochlorofluoro-methane									342.8			82.4
CBr₂Cl₂	Dibromodichloromethane									347.8			87.1
CBr₂F₂	Dibromodifluoromethane									325.3			77.0
CBr₂O	Carbonyl bromide		-127.2	-96.2			-110.9			309.1			61.8
CBr₃Cl	Tribromochloromethane									357.8			89.4
CBr₃F	Tribromofluoromethane									345.9			84.4

Molecular formula	Name	$\Delta_f H°/\text{kJ mol}^{-1}$			$\Delta_f G°/\text{kJ mol}^{-1}$			$S°/\text{J mol}^{-1}\text{ K}^{-1}$			$C_p/\text{J mol}^{-1}\text{ K}^{-1}$		
		Crys.	Liq.	Gas	Crys.	Liq.	Gas	Crys.	Liq.	Gas	Crys.	Liq.	Gas
CBr4	Tetrabromomethane	29.4		83.9	47.7		67.0	212.5		358.1	144.3		91.2
CCaO3	Calcium carbonate (calcite)	-1207.6			-1129.1			91.7			83.5		
CCaO3	Calcium carbonate (aragonite)	-1207.8			-1128.2			88.0			82.3		
CCdO3	Cadmium carbonate	-750.6			-669.4			92.5					
CClFO	Carbonyl chloride fluoride												
CClF3	Chlorotrifluoromethane			-706.3						276.7			66.9
CClN	Cyanogen chloride		112.1	138.0			131.0			236.2			45.0
CClN3O6	Chlorotrinitromethane		-27.1	18.4									
CCl2F2	Dichlorodifluoromethane			-477.4			-439.4			300.8			72.3
CCl2O	Carbonyl chloride			-219.1			-204.9			283.5			57.7
CCl3	Trichloromethyl			59.0									52.4
CCl3F	Trichlorofluoromethane		-301.3	-268.3		-236.8						121.6	78.1
CCl4	Tetrachloromethane		-128.2	-95.7					225.4			130.7	83.3
CCoO3	Cobalt(II) carbonate	-713.0											
CCs2O3	Cesium carbonate	-1139.7			-1054.3			204.5			123.9		
CCuN	Copper(I) cyanide	96.2			111.3			84.5					
CFN	Cyanogen fluoride									224.7			41.8
CF2O	Carbonyl fluoride			-639.8									46.8
CF3	Trifluoromethyl			-477.0			-464.0			264.5			49.6
CF3I	Trifluoroiodomethane			-587.8						307.4			70.9
CF4	Tetrafluoromethane			-933.6						261.6			61.1
CFeO3	Iron(II) carbonate	-740.6			-666.7			92.9			82.1		
CFe3	Iron carbide	25.1			20.1			104.6			105.9		
CH	Methylidyne			595.8									
CHBrClF	Bromochlorofluoromethane									304.3			63.2
CHBrCl2	Bromodichloromethane									316.4			67.4
CHBrF2	Bromodifluoromethane			-424.9						295.1			58.7
CHBr2Cl	Chlorodibromomethane									327.7			69.2
CHBr2F	Dibromofluoromethane									316.8			65.1
CHBr3	Tribromomethane		-22.3	23.8		-5.0	8.0		220.9	330.9		130.7	71.2
CHClF2	Chlorodifluoromethane			-482.6						280.9			55.9
CHCl2F	Dichlorofluoromethane									293.1			60.9
CHCl3	Trichloromethane		-134.1	-102.7		-73.7	6.0		201.7	295.7		114.2	65.7
CHCsO3	Cesium hydrogen carbonate	-966.1											
CHFO	Formyl fluoride									246.6			39.9
CHF3	Trifluoromethane			-695.4						259.7			51.0
CHI3	Triiodomethane	-181.1		251.0						356.2			75.0
CHKO2	Potassium formate	-679.7											
CHKO3	Potassium hydrogen carbonate	-963.2			-863.5			115.5					
CHN	Hydrogen cyanide		108.9	135.1		125.0	124.7		112.8	201.8		70.6	35.9
CHNO	Isocyanic acid (HNCO)									238.0			44.9
CHNS	Isothiocyanic acid			127.6			113.0			247.8			46.9
CHN3O6	Trinitromethane		-32.8	-0.2									
CHNaO2	Sodium formate	-666.5			-599.9			103.8			82.7		

Molecular formula	Name	$\Delta_f H°$/kJ mol⁻¹ Crys.	Liq.	Gas	$\Delta_f G°$/kJ mol⁻¹ Crys.	Liq.	Gas	$S°$/J mol⁻¹ K⁻¹ Crys.	Liq.	Gas	C_p/J mol⁻¹ K⁻¹ Crys.	Liq.	Gas
$CHNaO_3$	Sodium hydrogen carbonate	-950.8			-851.0			101.7			87.6		
CHO	Oxomethyl (HCO)			43.1			28.0			224.7			34.6
CH_2	Methylene			390.4			372.9			194.9			33.8
CH_2BrCl	Bromochloromethane									287.6			52.7
CH_2BrF	Bromofluoromethane									276.3			49.2
CH_2Br_2	Dibromomethane									293.2			54.7
CH_2ClF	Chlorofluoromethane									264.4			47.0
CH_2Cl_2	Dichloromethane		-124.2	-95.4					177.8	270.2		101.2	51.0
CH_2F_2	Difluoromethane			-452.3					174.1	246.7		134.0	42.9
CH_2I_2	Diiodomethane		68.5	119.5		90.4	95.8			309.7			57.7
CH_2N_2	Diazomethane									242.9			52.5
CH_2N_2	Cyanamide	58.8											
$CH_2N_2O_4$	Dinitromethane		-104.9	-58.9									
CH_2O	Formaldehyde			-108.6			-102.5			218.8			35.4
$(CH_2O)_x$	Paraformaldehyde	-177.6											
CH_2O_2	Formic acid		-425.0	-378.7		-361.4			129.0			99.0	
CH_2S_3	Trithiocarbonic acid		24.0										
CH_3	Methyl			145.7			147.9			194.2			38.7
CH_3BO	Borane carbonyl			-111.2			-92.9			249.4			59.5
CH_3Br	Bromomethane		-59.8	-35.4			-26.3			246.4			42.4
CH_3Cl	Chloromethane			-81.9						234.6			40.8
CH_3F	Fluoromethane									222.9			37.5
CH_3I	Iodomethane		-13.6	14.4						254.1			44.1
CH_3NO	Formamide		-254.0	-193.9		-14.4			163.2			126.0	
CH_3NO_2	Nitromethane		-112.6	-74.3			-6.8		171.8	275.0		106.6	57.3
CH_3NO_2	Methyl nitrite			-66.1									
CH_3NO_3	Methyl nitrate		-156.3	-122.4		-43.4	-39.2		217.1	318.5		157.3	
CH_4	Methane			-74.6			-50.5			186.3	134.0		35.7
CH_4N_2	Ammonium cyanide	0.4											
CH_4N_2O	Urea	-333.1		-245.8									
CH_4N_2S	Thiourea	-89.1		22.9									
$CH_4N_4O_2$	Nitroguanidine	-92.4											
CH_4O	Methanol		-239.2	-201.0		-166.6	-162.3		126.8	239.9		81.1	44.1
CH_4S	Methanethiol		-46.7	-22.9		-7.7	-9.3		169.2	255.2		90.5	50.3
CH_5N	Methylamine		-47.3	-22.5		35.7	32.7		150.2	242.9		102.1	50.1
CH_5NO_3	Ammonium hydrogen carbonate	-849.4			-665.9			120.9					
CH_5N_3	Guanidine	-56.0											
CH_5N_3S	Hydrazinecarbothioamide	24.7											
$CH_5N_5O_2$	3-Amino-1-nitroguanidine	22.1											
CH_6ClN	Methylamine hydrochloride	-298.1											
CH_6N_2	Methylhydrazine		54.2	94.7		180.0	187.0		165.9	278.8		134.9	71.1
CH_6Si	Methylsilane									256.5			65.9
CHg_2O_3	Mercury(I) carbonate	-553.5			-468.1			180.0					
CIN	Cyanogen iodide	166.2		225.5	185.0		196.6	96.2		256.8			48.3

Molecular formula	Name	$\Delta_f H°$/kJ mol⁻¹			$\Delta_f G°$/kJ mol⁻¹			S°/J mol⁻¹ K⁻¹			C_p/J mol⁻¹ K⁻¹		
		Crys.	Liq.	Gas	Crys.	Liq.	Gas	Crys.	Liq.	Gas	Crys.	Liq.	Gas
CI4	Tetraiodomethane	-392.9		474.0						391.9			95.9
CKN	Potassium cyanide	-113.0			-101.9			128.5			66.3		
CKNS	Potassium thiocyanate	-200.2			-178.3			124.3			88.5		
CK2O3	Potassium carbonate	-1151.0			-1063.5			155.5			114.4		
CLi2O3	Lithium carbonate	-1215.9			-1132.1			90.4			99.1		
CMgO3	Magnesium carbonate	-1095.8			-1012.1			65.7			75.5		
CMnO3	Manganese(II) carbonate	-894.1			-816.7			85.8			81.5		
CN	Cyanide			437.6			407.5			202.6			29.2
CNNa	Sodium cyanide	-87.5			-76.4			115.6			70.4		
CNNaO	Sodium cyanate	-405.4			-358.1			96.7			86.6		
CN4O8	Tetranitromethane		38.4	88.3									
CNa2O3	Sodium carbonate	-1130.7			-1044.4			135.0			112.3		
CO	Carbon monoxide			-110.5			-137.2			197.7			29.1
COS	Carbon oxysulfide			-142.0			-169.2			231.6			41.5
CO2	Carbon dioxide			-393.5			-394.4			213.8			37.1
CO3Pb	Lead(II) carbonate	-699.1			-625.5			131.0			87.4		
CO3Rb2	Rubidium carbonate	-1136.0			-1051.0			181.3			117.6		
CO3Sr	Strontium carbonate	-1220.1			-1140.1			97.1			81.4		
CO3Tl2	Thallium(I) carbonate	-700.0			-614.6			155.2					
CO3Zn	Zinc carbonate	-812.8			-731.5			82.4			79.7		
CS	Carbon sulfide			234.0			184.0			210.6			29.8
CS2	Carbon disulfide		89.0	116.7		64.6	67.1		151.3	237.8		76.4	45.4
CSe2	Carbon diselenide		164.8										
CSi	Silicon carbide (cubic)	-65.3			-62.8			16.6			26.9		
CSi	Silicon carbide (hexagonal)	-62.8			-60.2			16.5			26.7		
C2	Dicarbon			831.9			775.9			199.4			43.2
C2BrF5	Bromopentafluoroethane			-1064.4									
C2Br2ClF3	1,2-Dibromo-1-chloro-1,2,2-trifluoroethane		-691.7	-656.6									
C2Br2F4	1,2-Dibromotetrafluoroethane		-817.7	-789.1									
C2Br4	Tetrabromoethane									387.1			102.7
C2Br6	Hexabromoethane									441.9			139.3
C2Ca	Calcium carbide	-59.8			-64.9			70.0			62.7		
C2CaN2	Calcium cyanide	-184.5											
C2CaO4	Calcium oxalate	-1360.6											
C2ClF3	Chlorotrifluoroethylene		-522.7	-505.5			-523.8			322.1			83.9
C2ClF5	Chloropentafluoroethane			-1118.8									
C2Cl2F4	1,2-Dichlorotetrafluoroethane		-960.2	-937.0									184.2
C2Cl2O2	Ethanedioyl dichloride		-367.6	-335.8									
C2Cl3F3	1,1,2-Trichlorotrifluoroethane		-745.0	-716.8					266.9			111.7	
C2Cl3N	Trichloroacetonitrile											170.1	
C2Cl4	Tetrachloroethylene		-50.6	-10.9		3.0				336.6		143.4	96.1

Molecular formula	Name	$\Delta_f H°$/kJ mol⁻¹			$\Delta_f G°$/kJ mol⁻¹			$S°$/J mol⁻¹ K⁻¹			C_p/J mol⁻¹ K⁻¹		
		Crys.	Liq.	Gas	Crys.	Liq.	Gas	Crys.	Liq.	Gas	Crys.	Liq.	Gas
C₂Cl₄F₂	1,1,1,2-Tetrachloro-2,2-difluoroethane			-489.9			-407.0			382.9			123.4
C₂Cl₄F₂	1,1,2,2-Tetrachloro-1,2-difluoroethane											173.6	
C₂Cl₄O	Trichloroacetyl chloride		-280.8	-239.8									
C₂Cl₆	Hexachloroethane	-202.8		-143.6				237.3			198.2		
C₂F₃N	Trifluoroacetonitrile			-497.9						298.1			77.9
C₂F₄	Tetrafluoroethylene			-658.9						300.1			80.5
C₂F₆	Hexafluoroethane	-820.5		-1344.2						332.3			106.7
C₂HBr	Bromoacetylene									253.7			55.7
C₂HBrClF₃	1-Bromo-2-chloro-1,1,2-trifluoroethane		-675.3	-644.8									
C₂HBrClF₃	2-Bromo-2-chloro-1,1,1-trifluoroethane		-720.0	-690.4									
C₂HCl	Chloroacetylene									242.0			54.3
C₂HClF₂	1-Chloro-2,2-difluoroethylene			-315.5			-289.1			303.0			72.1
C₂HCl₂F	1,1-Dichloro-2-fluoroethylene									313.9			76.5
C₂HCl₂F₃	2,2-Dichloro-1,1,1-trifluoroethane									352.8			102.5
C₂HCl₃	Trichloroethylene		-43.6	-9.0						324.8			80.3
C₂HCl₃O	Trichloroacetaldehyde		-234.5	-196.6					228.4			124.4	
C₂HCl₃O	Dichloroacetyl chloride		-280.4	-241.0								151.0	
C₂HCl₃O₂	Trichloroacetic acid		-187.6	-142.0									
C₂HCl₅	Pentachloroethane	-503.3											
C₂HF	Fluoroacetylene			-490.5						231.7			52.4
C₂HF₃	Trifluoroethylene												
C₂HF₃O₂	Trifluoroacetic acid		-1069.9	-1031.4									
C₂HF₅	Pentafluoroethane			-1100.4									
C₂H₂	Acetylene			227.4			209.9			200.9			44.0
C₂H₂BrF₃	2-Bromo-1,1,1-trifluoroethane			-694.5									
C₂H₂Br₂	cis-1,2-Dibromoethylene									311.3			68.8
C₂H₂Br₂	trans-1,2-Dibromoethylene									313.5			70.3
C₂H₂Br₂Cl₂	1,2-Dibromo-1,2-dichloroethane			-36.9									
C₂H₂Br₄	1,1,2,2-Tetrabromoethane									326.5		165.7	89.1
C₂H₂ClF₃	2-Chloro-1,1,1-trifluoroethane												
C₂H₂Cl₂	1,1-Dichloroethylene		-23.9	2.8		24.1	25.4		201.5	289.0		111.3	67.1
C₂H₂Cl₂	cis-1,2-Dichloroethylene		-26.4	4.6					198.4	289.6		116.4	65.1
C₂H₂Cl₂	trans-1,2-Dichloroethylene		-24.3	5.0		27.3	28.6		195.9	290.0		116.8	66.7
C₂H₂Cl₂O	Chloroacetyl chloride		-283.7										
C₂H₂Cl₂O₂	Dichloroacetic acid		-496.3	-244.8									

Molecular formula	Name	$\Delta_f H°$/kJ mol⁻¹ Crys.	Liq.	Gas	$\Delta_f G°$/kJ mol⁻¹ Crys.	Liq.	Gas	$S°$/J mol⁻¹ K⁻¹ Crys.	Liq.	Gas	C_p/J mol⁻¹ K⁻¹ Crys.	Liq.	Gas
$C_2H_2Cl_3NO$	2,2,2-Trichloroacetamide	-358.0											
$C_2H_2Cl_4$	1,1,1,2-Tetrachloroethane		-195.0	-149.2					246.9	356.0		162.3	102.7
$C_2H_2Cl_4$	1,1,2,2-Tetrachloroethane			-335.0						362.8			100.8
$C_2H_2F_2$	1,1-Difluoroethylene									266.2			60.1
$C_2H_2F_2$	cis-1,2-Difluoroethylene									268.3			58.2
$C_2H_2F_3I$	1,1,1-Trifluoro-2-iodoethane			-644.5									
$C_2H_2I_2$	cis-1,2-Diiodoethylene			-207.4									51.8
C_2H_2O	Ketene		-67.9	-47.5			-48.3			247.6			
$C_2H_2O_2$	Glyoxal			-212.0									
$C_2H_2O_4$	Oxalic acid	-821.7		-723.7									
$C_2H_2O_4Sr$	Strontium formate	-1393.3						109.8			91.0		
$C_2H_2S_2$	Thiirene			300.0			275.8			255.3			54.7
C_2H_3Br	Bromoethylene			79.2			81.8			275.8			55.5
C_2H_3BrO	Acetyl bromide		-223.5	-190.4									
C_2H_3Cl	Chloroethylene	-94.1	14.6	37.2			53.6			264.0	59.4		53.7
$C_2H_3ClF_2$	1-Chloro-1,1-difluoroethane									307.2			82.5
C_2H_3ClO	Acetyl chloride		-272.9	-242.8		-208.0	-205.8		200.8	295.1		117.0	67.8
$C_2H_3ClO_2$	Chloroacetic acid	-510.5		-435.2									
$C_2H_3Cl_2F$	1,1-Dichloro-1-fluoroethane									320.2			88.7
$C_2H_3Cl_3$	1,1,1-Trichloroethane		-177.4	-144.4					227.4	323.1		144.3	93.3
$C_2H_3Cl_3$	1,1,2-Trichloroethane		-190.8	-151.3					232.6	337.2		150.9	89.0
C_2H_3F	Fluoroethylene			-138.8									
C_2H_3FO	Acetyl fluoride		-467.2	-442.1									
$C_2H_3F_3$	1,1,1-Trifluoroethane			-744.6						279.9			78.2
$C_2H_3F_3$	1,1,2-Trifluoroethane			-730.7									
$C_2H_3F_3O$	2,2,2-Trifluoroethanol		-932.4	-888.4									
C_2H_3I	Iodoethylene		-163.5	-126.4						285.0			57.9
C_2H_3IO	Acetyl iodide												
$C_2H_3KO_2$	Potassium acetate	-723.0											
C_2H_3N	Acetonitrile		40.6	74.0		86.5	91.9		149.6	243.4		91.5	52.2
C_2H_3N	Isocyanomethane		130.8	163.5		159.5	165.7		159.0	246.9			52.9
C_2H_3NO	Methylisocyanate		-92.0										
$C_2H_3NO_3$	Oxamic acid	-661.2		-552.3									
C_2H_3NS	Methyl isothiocyanate	79.4									79.9		
$C_2H_3NaO_2$	Sodium acetate	-708.8			-607.2			123.0					
C_2H_4	Ethylene			52.4			68.4			219.3			42.9
C_2H_4BrCl	1-Bromo-2-chloroethane												
$C_2H_4Br_2$	1,1-Dibromoethane		-66.2	-37.5						327.7		130.1	80.8
$C_2H_4Br_2$	1,2-Dibromoethane		-79.2						223.3			136.0	
C_2H_4ClF	1-Chloro-1-fluoroethane			-313.4									
$C_2H_4Cl_2$	1,1-Dichloroethane		-158.4	-127.7		-73.8	-70.8		211.8	305.1		126.3	76.2
$C_2H_4Cl_2$	1,2-Dichloroethane		-166.8	-126.4						308.4		128.4	78.7

Molecular formula	Name	$\Delta_f H°$/kJ mol^{-1}			$\Delta_f G°$/kJ mol^{-1}			$S°$/J mol^{-1} K^{-1}			C_p/J mol^{-1} K^{-1}		
		Crys.	Liq.	Gas	Crys.	Liq.	Gas	Crys.	Liq.	Gas	Crys.	Liq.	Gas
C$_2$H$_4$F$_2$	1,1-Difluoroethane			-497.0						282.5			67.8
C$_2$H$_4$I$_2$	1,2-Diiodoethane	9.3		75.0									
C$_2$H$_4$N$_2$O$_2$	Oxamide	-504.4		-387.1									
C$_2$H$_4$N$_2$O$_2$	Ethanedial, dioxime	-90.5											
C$_2$H$_4$N$_2$O$_4$	1,1-Dinitroethane		-148.2										
C$_2$H$_4$N$_2$O$_4$	1,2-Dinitroethane		-165.2										
C$_2$H$_4$N$_2$S$_2$	Ethanedithioamide	-20.8		83.0									
C$_2$H$_4$N$_4$	1H-1,2,4-Triazol-3-amine	76.8											
C$_2$H$_4$O	Acetaldehyde		-192.2	-166.2		-127.6	-133.0		160.2	263.8		89.0	55.3
C$_2$H$_4$O	Ethylene oxide		-78.0	-52.6		-11.8	-13.0		153.9	242.5		88.0	47.9
C$_2$H$_4$OS	Thioacetic acid		-216.9	-175.1									
C$_2$H$_4$O$_2$	Acetic acid		-484.3	-432.2		-389.9	-374.2		159.8	283.5		123.3	63.4
C$_2$H$_4$O$_2$	Methyl formate		-386.1	-357.4						285.3		119.1	64.4
C$_2$H$_4$S	Thiirane		51.6	82.0			96.8			255.2			53.3
C$_2$H$_5$Si	Ethynylsilane									269.4			72.6
C$_2$H$_5$Br	Bromoethane		-90.5	-61.9		-25.8	-23.9		198.7	286.7		100.8	64.5
C$_2$H$_5$Cl	Chloroethane		-136.8	-112.1		-59.3	-60.4		190.8	276.0		104.3	62.8
C$_2$H$_5$ClO	Ethylene chlorohydrin		-295.4										
C$_2$H$_5$F	Fluoroethane		-40.0	-8.1		14.7	19.2		211.7	264.5		115.1	58.6
C$_2$H$_5$I	Iodoethane		91.9	126.5						306.0			66.9
C$_2$H$_5$N	Ethyleneimine			-238.3									
C$_2$H$_5$NO	Acetamide	-317.0						115.0			91.3		
C$_2$H$_5$NO	N-Methylformamide		-143.9	-102.3									
C$_2$H$_5$NO$_2$	Nitroethane			-392.1									
C$_2$H$_5$NO$_2$	Glycine	-528.5											
C$_2$H$_5$NO$_3$	2-Nitroethanol		-350.7	-154.1									
C$_2$H$_5$NO$_3$	Ethyl nitrate		-190.4										
C$_2$H$_5$NS	Thioacetamide	-71.7		11.4									
C$_2$H$_6$	Ethane			-84.0			-32.0			229.2			52.5
C$_2$H$_6$Cd	Dimethyl cadmium		63.6	101.6		139.0	146.9		201.9	303.0		132.0	
C$_2$H$_6$Hg	Dimethyl mercury		59.8	94.4		140.3	146.1		209.0	306.0			83.3
C$_2$H$_6$N$_2$O	N-Methylurea	-332.8											
C$_2$H$_6$N$_4$O$_2$	1,2-Hydrazinedicarboxamide	-498.7											
C$_2$H$_6$N$_4$O$_2$	Ethanedioic acid, dihydrazide	-295.2			-302.4			142.0					
C$_2$H$_6$O	Ethanol		-277.6	-234.8		-174.8	-167.9		160.7	281.6		112.3	65.6
C$_2$H$_6$O	Dimethyl ether		-203.3	-184.1			-112.6			266.4			64.4
C$_2$H$_6$OS	Dimethyl sulfoxide		-204.2	-151.3		-99.9			188.3			153.0	
C$_2$H$_6$O$_2$	Ethylene glycol		-460.0	-392.2					163.2	303.8		148.6	82.7
C$_2$H$_6$O$_2$S	Dimethyl sulfone	-450.1		-373.1			-272.7			310.6			100.0
C$_2$H$_6$O$_3$S	Dimethyl sulfite		-523.6	-483.4									
C$_2$H$_6$O$_4$S	Dimethyl sulfate		-735.5	-687.0									
C$_2$H$_6$S	Ethanethiol		-73.6	-46.1		-5.5	-4.8		207.0	296.2		117.9	72.7
C$_2$H$_6$S	Dimethyl sulfide		-65.3	-37.4					196.4	286.0		118.1	74.1
C$_2$H$_6$S$_2$	1,2-Ethanedithiol		-54.3	-9.7									
C$_2$H$_6$S$_2$	Dimethyl disulfide		-62.6	-24.7					235.4			146.1	

Molecular formula	Name	$\Delta_f H°$/kJ mol⁻¹ Crys.	Liq.	Gas	$\Delta_f G°$/kJ mol⁻¹ Crys.	Liq.	Gas	$S°$/J mol⁻¹ K⁻¹ Crys.	Liq.	Gas	C_P/J mol⁻¹ K⁻¹ Crys.	Liq.	Gas
C₂H₆Zn	Dimethyl zinc		23.4	53.0					201.6			129.2	
C₂H₇N	Ethylamine		-74.1	-47.5			36.3			283.8		130.0	71.5
C₂H₇N	Dimethylamine		-43.9	-18.8		70.0	68.5		182.3	273.1		137.7	70.7
C₂H₇NO	Ethanolamine											195.5	
C₂H₈ClN	Dimethylamine hydrochloride	-289.3											
C₂H₈N₂	1,2-Ethanediamine		-63.0	-18.0									
C₂H₈N₂	1,1-Dimethylhydrazine		48.9	84.1								172.6	
C₂H₈N₂	1,2-Dimethylhydrazine		52.7	92.2		206.4			198.0			164.1	
C₂H₈N₂O₄	Ammonium oxalate	-1123.0											
C₂HgO₄	Mercury(II) oxalate	-678.2											
C₂I₂	Diiodoacetylene									313.1	226.0		70.3
C₂I₄	Tetraiodoethylene	305.0											
C₂K₂O₄	Potassium oxalate	-1346.0											
C₂MgO₄	Magnesium oxalate	-1269.0											
C₂N₂	Cyanogen		285.9	306.7						241.9			56.8
C₂N₄O₆	Trinitroacetonitrile		183.7										
C₂Na₂O₄	Sodium oxalate	-1318.0											
C₂O₄Pb	Lead(II) oxalate	-851.4			-750.1			146.0			105.4		
C₃F₈	Perfluoropropane			-1783.2									
C₃H₂N₂	Malononitrile	186.4		265.5									
C₃H₂O₂	2-Propynoic acid		-193.2										
C₃H₂O₃	1,3-Dioxol-2-one		-459.9	-418.6									
C₃H₃Cl₃	1,2,3-Trichloropropene		-101.8										
C₃H₃F₃	3,3,3-Trifluoropropene			-614.2									
C₃H₃N	2-Propenenitrile		147.1	180.6									
C₃H₃NO	Oxazole		-48.0	-15.5									
C₃H₃NO	Isoxazole		42.1	78.6									
C₃H₄	Allene			190.5									
C₃H₄	Propyne			184.9									
C₃H₄	Cyclopropene			277.1									
C₃H₄Cl₂	2,3-Dichloropropene		-73.3										
C₃H₄Cl₄	1,1,1,3-Tetrachloropropane		-208.7										
C₃H₄Cl₄	1,2,2,3-Tetrachloropropane		-251.8										
C₃H₄F₄O	2,2,3,3-Tetrafluoro-1-propanol		-1114.9	-1061.3									
C₃H₄N₂	1H-Pyrazole	49.8	105.4	179.4									
C₃H₄N₂	Imidazole			132.9									
C₃H₄O₂	1,2-Propanedione		-309.1	-271.0									
C₃H₄O₂	Propenoic acid		-383.8										
C₃H₄O₂	2-Oxetanone		-329.9	-282.9									
C₃H₄O₃	Ethylene carbonate		-571.5	-508.4					175.3			145.7	
C₃H₅Br	cis-1-Bromopropene		7.9	40.8								122.1	
C₃H₅Br	3-Bromopropene		12.2	45.2								133.9	
C₃H₅BrO	Bromoacetone			-181.0									

Molecular formula	Name	$\Delta_f H°$/kJ mol^{-1} Crys.	Liq.	Gas	$\Delta_f G°$/kJ mol^{-1} Crys.	Liq.	Gas	$S°$/J mol^{-1} K^{-1} Crys.	Liq.	Gas	C_p/J mol^{-1} K^{-1} Crys.	Liq.	Gas
C₃H₅Cl	2-Chloropropene			-21.0								125.1	
C₃H₅Cl	3-Chloropropene											131.6	
C₃H₅ClO	Epichlorohydrin		-148.4	-107.8									
C₃H₅ClO₂	2-Chloropropanoic acid	-549.3	-522.5	-475.8									
C₃H₅ClO₂	3-Chloropropanoic acid												
C₃H₅ClO₂	Ethyl chloroformate		-505.3	-462.9								183.6	
C₃H₅Cl₃	1,2,3-Trichloropropane		-230.6	-182.9									
C₃H₅I	3-Iodopropene		53.7	91.5									
C₃H₅IO	Iodoacetone			-130.5									
C₃H₅IO₂	3-Iodopropanoic acid	-460.0											
C₃H₅N	Propanenitrile		15.5	51.7								119.3	
C₃H₅N	2-Propyn-1-amine		205.7										
C₃H₅N	Ethyl isocyanide		108.6	141.7									
C₃H₅NO₃	Nitroacetone		-278.6										
C₃H₅NO₄	Methyl nitroacetate		-464.0										
C₃H₅N₃O₉	Trinitroglycerol		-370.9	-270.9									
C₃H₆	Propene		4.0	20.0			104.5			237.5			55.6
C₃H₆	Cyclopropane		35.2	53.3									
C₃H₆Br₂	1,2-Dibromopropane		-113.6	-71.6								149.1	
C₃H₆Cl₂	1,2-Dichloropropane		-198.8	-162.8									
C₃H₆Cl₂	1,3-Dichloropropane		-199.9	-159.2									
C₃H₆Cl₂	2,2-Dichloropropane		-205.8	-173.2									
C₃H₆Cl₂O	2,3-Dichloro-1-propanol		-381.5	-316.3									
C₃H₆Cl₂O	1,3-Dichloro-2-propanol		-385.3	-318.4									
C₃H₆I₂	1,2-Diiodopropane			35.6									
C₃H₆I₂	1,3-Diiodopropane		-9.0										
C₃H₆N₂O₂	Propanediamide	-546.1											
C₃H₆N₂O₂	N-(Aminocarbonyl)acetamide	-544.2		-441.2									
C₃H₆N₂O₄	1,1-Dinitropropane		-163.2	-100.7									
C₃H₆N₂O₄	1,3-Dinitropropane		-207.1										
C₃H₆N₂O₄	2,2-Dinitropropane		-181.2										
C₃H₆O	Allyl alcohol		-171.8	-124.5								138.9	
C₃H₆O	Propanal		-215.6	-185.6						304.5		126.3	80.7
C₃H₆O	Acetone		-248.4	-217.1			-152.7		199.8	295.3		120.4	74.5
C₃H₆O	Methyloxirane		-123.0	-94.7					196.5	286.9			72.6
C₃H₆O	Oxetane		-110.8	-80.5									
C₃H₆O₂	Propanoic acid		-510.7	-455.7					191.0			152.8	
C₃H₆O₂	Ethyl formate											149.3	
C₃H₆O₂	Methyl acetate		-445.9	-413.3						324.4		141.9	86.0
C₃H₆O₂	1,3-Dioxolane		-333.5	-298.0								118.0	
C₃H₆O₂S	Thiolactic acid		-468.4										
C₃H₆O₃	Trioxane	-522.5		-465.9				133.0			111.4		
C₃H₆S	Thiacyclobutane		24.7	60.6			107.1		184.9	285.0			68.3
C₃H₆S	Methylthiirane		11.3	45.8									

Molecular formula	Name	$\Delta_f H°$/kJ mol⁻¹ Crys.	Liq.	Gas	$\Delta_f G°$/kJ mol⁻¹ Crys.	Liq.	Gas	S°/J mol⁻¹ K⁻¹ Crys.	Liq.	Gas	C_p/J mol⁻¹ K⁻¹ Crys.	Liq.	Gas
C₃H₆S₂	1,2-Dithiolane			0.0			47.7			313.5			86.5
C₃H₆S₂	1,3-Dithiolane			10.0			54.7			323.3			84.7
C₃H₇Br	1-Bromopropane		-121.9	-87.0									
C₃H₇Br	2-Bromopropane		-130.5	-99.4									
C₃H₇Cl	1-Chloropropane		-160.5	-131.9									
C₃H₇Cl	2-Chloropropane		-172.3	-144.9									
C₃H₇ClO₂	3-Chloro-1,2-propanediol		-525.3										
C₃H₇ClO₂	2-Chloro-1,3-propanediol		-517.5										
C₃H₇F	1-Fluoropropane			-285.9									
C₃H₇F	2-Fluoropropane			-293.5									
C₃H₇I	1-Iodopropane		-66.0	-30.0									
C₃H₇I	2-Iodopropane		-74.8	-40.3									
C₃H₇N	Allylamine		-10.0										
C₃H₇N	Cyclopropylamine		45.8	77.0					187.7			147.1	
C₃H₇NO	N,N-Dimethylformamide		-239.3	-192.4								150.6	
C₃H₇NO	Propanamide	-338.2		-259.0									
C₃H₇NO₂	1-Nitropropane		-167.2	-123.8									
C₃H₇NO₂	2-Nitropropane		-180.3	-138.9								170.3	
C₃H₇NO₂	DL-Alanine	-563.6											
C₃H₇NO₂	D-Alanine	-561.2											
C₃H₇NO₂	L-Alanine	-604.0		-465.9									
C₃H₇NO₂	β-Alanine	-558.0		-424.0									
C₃H₇NO₂	N-Methylglycine	-513.3		-367.3									
C₃H₇NO₂S	L-Cysteine	-534.1											
C₃H₇NO₃	Propyl nitrate		-214.5	-173.9									
C₃H₇NO₃	Isopropyl nitrate		-229.7	-191.0									
C₃H₇NO₃	DL-Serine	-739.0											
C₃H₇NO₃	L-Serine	-732.7											
C₃H₈	Propane		-120.9	-103.8			-23.4			270.3			73.6
C₃H₈N₂O	N-Ethylurea	-357.8											
C₃H₈N₂O	N,N-Dimethylurea	-319.1											
C₃H₈N₂O	N,N'-Dimethylurea	-312.1											
C₃H₈N₂O₃	Oxymethurea	-717.0											
C₃H₈O	1-Propanol		-302.6	-255.1					193.6	322.6		143.9	85.6
C₃H₈O	2-Propanol		-318.1	-272.6					181.1	309.2		156.5	89.3
C₃H₈O	Ethyl methyl ether			-216.4						309.2			93.3
C₃H₈O₂	1,2-Propylene glycol		-501.0	-429.8								190.8	
C₃H₈O₂	1,3-Propylene glycol		-480.8	-408.0									
C₃H₈O₂	Ethylene glycol monomethyl ether		-377.8	-348.5					244.0			171.1	
C₃H₈O₂	Dimethoxymethane											162.0	
C₃H₈O₃	Glycerol		-669.6	-577.9					206.3			218.9	
C₃H₈S	1-Propanethiol		-99.9	-67.8					242.5			144.6	
C₃H₈S	2-Propanethiol		-105.9	-76.2					233.5			145.3	
C₃H₈S	Ethyl methyl sulfide		-91.6	-59.6					239.1			144.6	

Molecular formula	Name	$\Delta_f H°$/kJ mol⁻¹			$\Delta_f G°$/kJ mol⁻¹			$S°$/J mol⁻¹ K⁻¹			C_p/J mol⁻¹ K⁻¹		
		Crys.	Liq.	Gas	Crys.	Liq.	Gas	Crys.	Liq.	Gas	Crys.	Liq.	Gas
C₃H₈S₂	1,3-Propanedithiol		-79.4	-29.8					209.4			155.6	
C₃H₉Al	Trimethyl aluminum		-136.4	-74.1		-9.9	-35.9		238.9	314.7			88.5
C₃H₉B	Trimethylborane		-143.1	-124.3		-32.1	-243.5		278.2	369.1			
C₃H₉ClSi	Trimethylchlorosilane		-382.8	-352.8		-246.4				325.4		164.1	91.2
C₃H₉N	Propylamine		-101.5	-70.1			39.9		218.3	312.2		163.8	97.5
C₃H₉N	Isopropylamine		-112.3	-83.7			32.2		208.5	287.1		137.9	91.8
C₃H₉N	Trimethylamine		-45.7	-23.6									
C₃H₁₀ClN	Propylamine hydrochloride	-354.7											
C₃H₁₀ClN	Trimethylamine hydrochloride	-282.9											
C₃H₁₀N₂	1,2-Propanediamine, (±)-		-97.8	-53.6								127.2	
C₃H₁₀Si	Trimethylsilane		-142.5							331.0			117.9
C₃H₁₂BN	Trimethylamine borane	-142.5			70.7			187.0					
C₃H₁₂BN	Aminetrimethylboron	-284.1			-79.3			218.0					
C₄F₈	Perfluorocyclobutane			-1542.6									
C₄F₁₀	Perfluorobutane												
C₄H₂N₂	trans-2-Butenedinitrile	268.2		340.2									
C₄H₂O₃	Maleic anhydride	-469.8		-398.3									
C₄H₂O₄	2-Butynedioic acid	-577.3											
C₄H₃NO₃	2-Nitrofuran	-104.1		-28.8									
C₄H₄BrNO₂	N-Bromosuccinimide	-335.9											
C₄H₄ClNO₂	N-Chlorosuccinimide	-357.9											
C₄H₄N₂	Succinonitrile	139.8	139.7	209.7					191.6			145.6	
C₄H₄N₂	Pyrazine		145.9	196.1									
C₄H₄N₂	Pyrimidine			195.7									
C₄H₄N₂	Pyridazine		224.9	278.3							120.5		
C₄H₄N₂O₂	Uracil	-429.4		-302.9									
C₄H₄N₂O₃	2,4,6-Trihydroxypyrimidine	-634.7											
C₄H₄O	Furan		-62.3	-34.8					177.0	267.2		114.8	65.4
C₄H₄O₂	Diketene		-233.1	-190.3									
C₄H₄O₃	Succinic anhydride	-608.6		-527.9									
C₄H₄O₄	Maleic acid	-789.4		-679.4				160.8			137.0		
C₄H₄O₄	Fumaric acid	-811.7		-675.8				168.0			142.0		
C₄H₄S	Thiophene		80.2	114.9			126.1		181.2	278.8		123.8	72.8
C₄H₅N	trans-2-Butenenitrile		95.1	134.3									
C₄H₅N	3-Butenenitrile		117.8	159.7					156.4			127.7	
C₄H₅N	Pyrrole		63.1	108.2									
C₄H₅N	Cyclopropanecarbonitrile		140.8	182.8									
C₄H₅NO₂	Succinimide	-459.0		-375.4									
C₄H₅NS	4-Methylthiazole		67.9	111.8									
C₄H₅N₃O	Cytosine	-221.3									132.6		
C₄H₆	1,2-Butadiene		138.6	162.3					199.0			123.6	
C₄H₆	1,3-Butadiene		88.5	110.0									
C₄H₆	1-Butyne		141.4	165.2									
C₄H₆	2-Butyne		119.1	145.7									

Molecular formula	Name	$\Delta_f H°$/kJ mol⁻¹ Crys.	Liq.	Gas	$\Delta_f G°$/kJ mol⁻¹ Crys.	Liq.	Gas	$S°$/J mol⁻¹ K⁻¹ Crys.	Liq.	Gas	C_p/J mol⁻¹ K⁻¹ Crys.	Liq.	Gas
C_4H_6	Cyclobutene			156.7									
$C_4H_6N_2O_2$	2,5-Piperazinedione	-446.5											
C_4H_6O	Divinyl ether		-39.8	-13.6									
C_4H_6O	trans-2-Butenal		-138.7	-100.6									
$C_4H_6O_2$	trans-Crotonic acid												
$C_4H_6O_2$	Methacrylic acid												
$C_4H_6O_2$	Vinyl acetate		-349.2	-314.0								161.1	
$C_4H_6O_2$	Methyl acrylate		-362.2	-333.0					239.5			158.8	
$C_4H_6O_2$	γ-Butyrolactone		-420.9	-366.5								141.4	
$C_4H_6O_3$	Acetic anhydride		-624.4	-572.5									
$C_4H_6O_3$	Propylene carbonate		-613.2	-582.5								218.6	
$C_4H_6O_4$	Succinic acid	-940.5		-823.0				167.3			153.1		
$C_4H_6O_4$	Dimethyl oxalate	-756.3		-708.9									
C_4H_6S	2,3-Dihydrothiophene		52.9	90.7			133.5			303.5			79.8
C_4H_6S	2,5-Dihydrothiophene		47.0	86.9			131.6			297.1			83.3
C_4H_7ClO	2-Chloroethyl vinyl ether		-208.1	-170.1									
$C_4H_7ClO_2$	2-Chlorobutanoic acid		-575.5										
$C_4H_7ClO_2$	3-Chlorobutanoic acid		-556.3										
$C_4H_7ClO_2$	4-Chlorobutanoic acid		-566.3										
$C_4H_7ClO_2$	Propyl chlorocarbonate		-533.4	-492.7									
C_4H_7N	Butanenitrile		-5.8	33.6									
C_4H_7N	2-Methylpropanenitrile		-13.8	23.4									
C_4H_7NO	2-Pyrrolidone		-286.2										
C_4H_7NO	2-Methyl-2-oxazoline		-169.5	-130.5									
$C_4H_7NO_4$	Iminodiacetic acid	-932.6											
$C_4H_7NO_4$	Ethyl nitroacetate		-487.1										
$C_4H_7NO_4$	L-Aspartic acid	-973.3											
$C_4H_7N_3O$	4H-Imidazol-4-one, 2-amino-1,5-dihydro-1-methyl-	-238.5											
C_4H_8	1-Butene		-20.8	0.1					227.0			118.0	
C_4H_8	cis-2-Butene		-29.8	-7.1					219.9			127.0	
C_4H_8	trans-2-Butene		-33.3	-11.4									
C_4H_8	Isobutene		-37.5	-16.9									
C_4H_8	Cyclobutane		3.7	27.7									
C_4H_8	Methylcyclopropane		1.7										
$C_4H_8Br_2$	1,2-Dibromobutane		-142.1	-91.6									
$C_4H_8Br_2$	1,3-Dibromobutane		-148.0										
$C_4H_8Br_2$	1,4-Dibromobutane		-140.3	-87.8									
$C_4H_8Br_2$	2,3-Dibromobutane		-139.6	-102.0									
$C_4H_8Br_2$	1,2-Dibromo-2-methylpropane		-156.6	-113.3									
$C_4H_8Cl_2$	1,3-Dichlorobutane		-237.3	-195.0									
$C_4H_8Cl_2$	1,4-Dichlorobutane		-229.8	-183.4									
$C_4H_8Cl_2O$	Bis(2-chloroethyl) ether											220.9	

Molecular formula	Name	$\Delta_f H°$/kJ mol^{-1}			$\Delta_f G°$/kJ mol^{-1}			$S°$/J mol^{-1} K^{-1}			C_p/J mol^{-1} K^{-1}		
		Crys.	Liq.	Gas	Crys.	Liq.	Gas	Crys.	Liq.	Gas	Crys.	Liq.	Gas
C$_4$H$_8$I$_2$	1,4-Diiodobenzene		-30.0										
C$_4$H$_8$N$_2$O$_2$	Succinamide	-581.2											
C$_4$H$_8$N$_2$O$_2$	Dimethylglyoxime	-199.7											
C$_4$H$_8$N$_2$O$_3$	L-Asparagine	-789.4											
C$_4$H$_8$N$_2$O$_3$	N-Glycylglycine	-747.7											
C$_4$H$_8$N$_2$O$_4$	1,4-Dinitrobutane		-237.5										
C$_4$H$_8$O	Ethyl vinyl ether		-167.4	-140.8									
C$_4$H$_8$O	1,2-Epoxybutane		-168.9										
C$_4$H$_8$O	Butanal		-239.2	-204.8					230.9			147.0	
C$_4$H$_8$O	Isobutanal		-247.3	-215.7					246.6	343.7		163.7	103.4
C$_4$H$_8$O	2-Butanone		-273.3	-238.5					239.1	339.9		158.7	101.7
C$_4$H$_8$O	Tetrahydrofuran		-216.2	-184.1					204.3	302.4		124.0	76.3
C$_4$H$_8$OS	S-Ethyl thioacetate		-268.2	-228.1									
C$_4$H$_8$O$_2$	Butanoic acid		-533.8	-475.9					222.2			178.6	
C$_4$H$_8$O$_2$	2-Methylpropanoic acid		-500.3	-462.7								173.0	
C$_4$H$_8$O$_2$	Propyl formate		-479.3	-443.6					257.7			170.7	
C$_4$H$_8$O$_2$	Ethyl acetate		-379.7	-340.6								143.9	
C$_4$H$_8$O$_2$	1,3-Dioxane		-353.9	-315.3					270.2			152.1	
C$_4$H$_8$O$_2$	1,4-Dioxane		-386.9	-352.0									
C$_4$H$_8$O$_2$	2-Methyl-1,3-dioxolane												
C$_4$H$_8$O$_2$S	Sulfolane						45.8					180.0	92.5
C$_4$H$_8$S	Tetrahydrothiophene		-72.9	-34.1			72.4			309.6			110.4
C$_4$H$_8$S$_2$	1,3-Dithiane			-10.0						333.5			
C$_4$H$_9$Br	1-Bromobutane		-143.8	-107.1									
C$_4$H$_9$Br	2-Bromobutane		-154.9	-120.3									
C$_4$H$_9$Br	2-Bromo-2-methylpropane		-164.4	-132.4									
C$_4$H$_9$Cl	1-Chlorobutane		-188.1	-154.4									
C$_4$H$_9$Cl	2-Chlorobutane		-192.8	-161.1									
C$_4$H$_9$Cl	1-Chloro-2-methylpropane		-191.1	-159.3									
C$_4$H$_9$Cl	2-Chloro-2-methylpropane		-211.3	-182.2									
C$_4$H$_9$ClO	2-Chloroethyl ethyl ether		-335.6	-301.3									
C$_4$H$_9$I	1-Iodo-2-methylpropane		-107.5	-72.1									
C$_4$H$_9$I	2-Iodo-2-methylpropane		5.6	41.2									
C$_4$H$_9$N	Cyclobutanamine		-41.1	-3.6									
C$_4$H$_9$N	Pyrrolidine		-346.9	-282.0					204.1			156.6	
C$_4$H$_9$NO	Butanamide	-368.6											
C$_4$H$_9$NO	N-Methylpropanamide			-282.6									
C$_4$H$_9$NO	2-Methylpropanamide			-228.0									
C$_4$H$_9$NO	N,N-Dimethylacetamide		-278.3										
C$_4$H$_9$NO	Morpholine		-192.5	-143.9								179.0	
C$_4$H$_9$NO$_2$	1-Nitrobutane		-217.2	-177.1								175.6	
C$_4$H$_9$NO$_2$	2-Nitroisobutane											164.8	
C$_4$H$_9$NO$_2$	Propyl carbamate	-552.6		-471.4									
C$_4$H$_9$NO$_2$	4-Aminobutanoic acid	-581.0		-441.0								162.3	
C$_4$H$_9$NO$_3$	3-Nitro-2-butanol		-390.0										

Molecular formula	Name	ΔfH°/kJ mol⁻¹ Crys.	ΔfH°/kJ mol⁻¹ Liq.	ΔfH°/kJ mol⁻¹ Gas	ΔfG°/kJ mol⁻¹ Crys.	ΔfG°/kJ mol⁻¹ Liq.	ΔfG°/kJ mol⁻¹ Gas	S°/J mol⁻¹ K⁻¹ Crys.	S°/J mol⁻¹ K⁻¹ Liq.	S°/J mol⁻¹ K⁻¹ Gas	Cp/J mol⁻¹ K⁻¹ Crys.	Cp/J mol⁻¹ K⁻¹ Liq.	Cp/J mol⁻¹ K⁻¹ Gas
$C_4H_9NO_3$	2-Methyl-2-nitro-1-propanol	-410.1											
$C_4H_9NO_3$	DL-Threonine	-758.8											
$C_4H_9NO_3$	L-Threonine	-807.2											
$C_4H_9N_3O_2$	Creatine	-537.2											
C_4H_{10}	Butane		-147.3	-125.7								140.9	
C_4H_{10}	Isobutane		-154.2	-134.2									
$C_4H_{10}Hg$	Diethyl mercury		30.1	75.3								182.8	
$C_4H_{10}N_2$	Piperazine	-45.6											
$C_4H_{10}N_2O$	Trimethylurea	-330.5											
$C_4H_{10}N_2O_2$	N-Nitrodiethylamine		-106.2	-53.0									
$C_4H_{10}N_2O_4$	L-Asparagine, monohydrate	-1086.6											
$C_4H_{10}O$	1-Butanol		-327.3	-274.9					225.8			177.2	
$C_4H_{10}O$	2-Butanol		-342.6	-292.8					214.9	359.5		196.9	112.7
$C_4H_{10}O$	2-Methyl-1-propanol		-334.7	-283.8					214.7			181.5	
$C_4H_{10}O$	2-Methyl-2-propanol		-359.2	-312.5					193.3	326.7		218.6	113.6
$C_4H_{10}O$	Diethyl ether		-279.5	-252.1					172.4	342.7		175.6	119.5
$C_4H_{10}O$	Methyl propyl ether		-266.0	-238.1					262.9			165.4	
$C_4H_{10}O$	Isopropyl methyl ether		-278.8	-252.0					253.8			161.9	
$C_4H_{10}OS$	Diethyl sulfoxide		-268.0	-205.6									
$C_4H_{10}O_2$	1,2-Butanediol, (±)-		-523.6										
$C_4H_{10}O_2$	1,3-Butanediol		-501.0	-433.2									
$C_4H_{10}O_2$	1,4-Butanediol		-505.3	-428.7					223.4			200.1	
$C_4H_{10}O_2$	2,3-Butanediol		-541.5	-482.3								213.0	
$C_4H_{10}O_2$	2-Methyl-1,2-propanediol		-539.7										
$C_4H_{10}O_2$	Ethylene glycol monoethyl ether		-376.6									210.8	
$C_4H_{10}O_2$	Ethylene glycol dimethyl ether		-420.6	-389.7								193.3	
$C_4H_{10}O_2$	Dimethylacetal		-293.6	-245.9									
$C_4H_{10}O_2$	tert-Butyl hydroperoxide		-628.5	-571.2									
$C_4H_{10}O_3$	Diethylene glycol		-600.7	-552.2									
$C_4H_{10}O_3S$	Diethyl sulfite		-813.2	-756.3									
$C_4H_{10}O_4S$	Sulfuric acid, diethyl ester		-124.7	-88.0								244.8	
$C_4H_{10}S$	1-Butanethiol		-131.0	-96.9								171.2	
$C_4H_{10}S$	2-Butanethiol		-132.0	-97.3									
$C_4H_{10}S$	2-Methyl-1-propanethiol		-140.5	-109.6									
$C_4H_{10}S$	2-Methyl-2-propanethiol		-119.4	-83.5									
$C_4H_{10}S$	Diethyl sulfide		-118.5	-82.2					269.3	368.1		171.4	117.0
$C_4H_{10}S$	Methyl propyl sulfide		-124.7	-90.5					272.5			171.6	
$C_4H_{10}S$	Isopropyl methyl sulfide		-105.7	-50.6					263.1			172.4	
$C_4H_{10}S_2$	1,4-Butanedithiol		-120.1	-79.4									
$C_4H_{10}S_2$	Diethyl disulfide		-127.6	-91.9					269.3			171.4	
$C_4H_{11}N$	Butylamine		-137.5	-104.6								179.2	
$C_4H_{11}N$	sec-Butylamine		-150.6	-121.0								192.1	
$C_4H_{11}N$	tert-Butylamine		-132.6	-98.7								183.2	
$C_4H_{11}N$	Isobutylamine		-103.7	-72.2								169.2	
$C_4H_{11}N$	Diethylamine												

Molecular formula	Name	$\Delta_f H°$/kJ mol⁻¹			$\Delta_f G°$/kJ mol⁻¹			$S°$/J mol⁻¹ K⁻¹			C_p/J mol⁻¹ K⁻¹		
		Crys.	Liq.	Gas	Crys.	Liq.	Gas	Crys.	Liq.	Gas	Crys.	Liq.	Gas
$C_4H_{11}NO$	*N,N*-Dimethylethanolamine		-253.7	-203.6							233.5		
$C_4H_{11}NO_2$	Diethanolamine	-493.8		-397.1									
$C_4H_{11}NO_3$	2-Amino-2-(hydroxymethyl)-1,3-propanediol	-717.8											
$C_4H_{12}BrN$	Tetramethylammonium bromide	-251.0											
$C_4H_{12}ClN$	Diethylamine hydrochloride	-358.6											
$C_4H_{12}ClN$	Tetramethylammonium chloride	-276.4											
$C_4H_{12}IN$	Tetramethylammonium iodide	-203.9											
$C_4H_{12}N_2$	2-Methyl-1,2-propanediamine		-133.9	-90.3									
$C_4H_{12}Pb$	Tetramethyl lead		97.9	135.9									
$C_4H_{12}Si$	Tetramethylsilane		-264.0	-239.1		-100.0	-99.9		277.3	359.0		204.1	143.9
$C_4H_{12}Sn$	Tetramethylstannane		-52.3	-18.8								254.0	
$C_4H_{13}N_3$	Diethylenetriamine												
C_4N_2	2-Butynedinitrile		500.4	529.2									
C_4NiO_4	Nickel carbonyl		-633.0	-602.9		-588.2	-587.2		313.4	410.6		204.6	145.2
C_5FeO_5	Iron pentacarbonyl		-774.0			-705.3			338.1			240.6	
$C_5H_2F_6O_2$	Hexafluoroacetylacetone	-2286.7											
$C_5H_3NO_5$	5-Nitro-2-furancarboxylic acid	-516.8											
$C_5H_4N_4$	1*H*-Purine	169.4											
$C_5H_4N_4O$	Hypoxanthine	-110.8						145.6			134.5		
$C_5H_4N_4O_2$	Xanthine	-379.6						161.1			151.3		
$C_5H_4N_4O_3$	Uric acid	-618.8						173.2			166.1		
$C_5H_4O_2$	Furfural		-201.6	-151.0								163.2	
$C_5H_4O_3$	2-Furancarboxylic acid	-498.4		-390.0									
$C_5H_4O_3$	3-Methyl-2,5-furandione		-504.5	-447.2									
$C_5H_5F_3O_2$	1,1,1-Trifluoro-2,4-pentanedione		-1040.2	-993.3									
C_5H_5N	Pyridine		100.2	140.4								132.7	
C_5H_5NO	1*H*-Pyrrole-2-carboxaldehyde	-106.4											
$C_5H_5N_5$	Adenine	96.9		205.7							147.0		
$C_5H_5N_5O$	Guanine	-183.9											
C_5H_6	*cis*-3-Penten-1-yne		226.5										
C_5H_6	*trans*-3-Penten-1-yne		228.2										
C_5H_6	1,3-Cyclopentadiene		105.9	134.3									
$C_5H_6N_2O_2$	Thymine	-462.8		-328.7							150.8		
$C_5H_6O_2$	Furfuryl alcohol		-276.2	-211.8								204.0	
$C_5H_6O_4$	*trans*-1-Propene-1,2-dicarboxylic acid	-824.4											
C_5H_6S	2-Methylthiophene		44.6	83.5					218.5				
C_5H_6S	3-Methylthiophene		43.1	82.5								149.8	
C_5H_7N	*trans*-3-Pentenenitrile		80.9	125.7									
C_5H_7N	Cyclobutanecarbonitrile		103.0	147.4									

Molecular formula	Name	$\Delta_f H°/\text{kJ mol}^{-1}$			$\Delta_f G°/\text{kJ mol}^{-1}$			$S°/\text{J mol}^{-1}\text{K}^{-1}$			$C_p/\text{J mol}^{-1}\text{K}^{-1}$		
		Crys.	Liq.	Gas	Crys.	Liq.	Gas	Crys.	Liq.	Gas	Crys.	Liq.	Gas
C_5H_7N	1-Methylpyrrole		62.4	103.1									
C_5H_7N	2-Methylpyrrole		23.3	74.0									
C_5H_7N	3-Methylpyrrole		20.5	70.2									
$C_5H_7NO_2$	Ethyl cyanoacetate											220.2	
C_5H_8	1,2-Pentadiene			140.7									
C_5H_8	cis-1,3-Pentadiene			81.4									
C_5H_8	trans-1,3-Pentadiene			76.1									
C_5H_8	1,4-Pentadiene			105.7									
C_5H_8	2,3-Pentadiene			133.1									
C_5H_8	3-Methyl-1,2-butadiene		101.2										
C_5H_8	2-Methyl-1,3-butadiene		48.2	75.5					229.3			152.6	
C_5H_8	Cyclopentene		4.3	34.0					201.2			122.4	
C_5H_8	Methylenecyclobutane		93.8	121.6									
C_5H_8	Spiropentane		157.5	185.2					193.7			134.5	
$C_5H_8N_4O_{12}$	Pentaerythritol tetranitrate	-538.6											
C_5H_8O	Cyclopentanone		-235.9	-192.1									
$C_5H_8O_2$	4-Pentenoic acid	-430.6											
$C_5H_8O_2$	Methyl trans-2-butenoate		-382.9	-341.9									
$C_5H_8O_2$	Methyl methacrylate												
$C_5H_8O_2$	Allyl acetate		-423.8	-382.0									
$C_5H_8O_2$	2,4-Pentanedione		-461.3	-406.5								191.2	
$C_5H_8O_2$	4-Methyl-gamma-butyrolactone		-436.7	-379.6								184.1	
$C_5H_8O_2$	Tetrahydro-2H-pyran-2-one												
$C_5H_8O_4$	Glutaric acid	-960.0											
$C_5H_9ClO_2$	Propyl chloroacetate		-515.5	-467.0									
C_5H_9N	Pentanenitrile		-33.1	10.5									
C_5H_9N	2,2-Dimethylpropanenitrile		-39.8	-2.3					232.0			179.4	
C_5H_9N	1,2,5,6-Tetrahydropyridine		33.5										
C_5H_9NO	2-Piperidinone	-306.6											
C_5H_9NO	N-Methyl-2-pyrrolidone		-262.2	-366.2								307.8	
$C_5H_9NO_2$	L-Proline	-515.2											
$C_5H_9NO_4$	D-Glutamic acid	-1005.3											
$C_5H_9NO_4$	L-Glutamic acid	-1009.7											
C_5H_{10}	1-Pentene		-46.9	-21.1					262.6			154.0	
C_5H_{10}	cis-2-Pentene		-53.7	-27.6					258.6			151.7	
C_5H_{10}	trans-2-Pentene		-58.2	-31.9					256.5			157.0	
C_5H_{10}	2-Methyl-1-butene		-61.1	-35.2					254.0			157.2	
C_5H_{10}	3-Methyl-1-butene		-51.5	-27.5					253.3			156.1	
C_5H_{10}	2-Methyl-2-butene		-68.6	-41.7					251.0			152.8	
C_5H_{10}	Cyclopentane		-105.1	-76.4					204.5			128.8	
C_5H_{10}	Methylcyclobutane		-44.5										
C_5H_{10}	Ethylcyclopropane		-24.8										
C_5H_{10}	1,1-Dimethylcyclopropane		-33.3	-8.2									
C_5H_{10}	cis-1,2-Dimethylcyclopropane		-26.3										

Molecular formula	Name	$\Delta_f H°$/kJ mol^{-1} Crys.	Liq.	Gas	$\Delta_f G°$/kJ mol^{-1} Crys.	Liq.	Gas	$S°$/J mol^{-1} K^{-1} Crys.	Liq.	Gas	C_p/J mol^{-1} K^{-1} Crys.	Liq.	Gas
C_5H_{10}	trans-1,2-Dimethylcyclopropane		-30.7										
$C_5H_{10}Br_2$	1,3-Dibromo-2-methylpropane			-137.6									
$C_5H_{10}N_2O$	N-Nitrosopiperidine		-31.1	16.6									
$C_5H_{10}N_2O_2$	N-Nitropiperidine		-93.0	-44.5									
$C_5H_{10}N_2O_3$	L-Glutamine	-826.4											
$C_5H_{10}O$	Cyclopentanol		-300.1	-242.5					206.3			184.1	
$C_5H_{10}O$	Pentanal		-267.2	-228.4									
$C_5H_{10}O$	2-Pentanone		-297.3	-258.8								184.1	
$C_5H_{10}O$	3-Pentanone		-296.5	-257.9					266.0			190.9	
$C_5H_{10}O$	3-Methyl-2-butanone		-299.5	-262.6					268.5			179.9	
$C_5H_{10}O$	3,3-Dimethyloxetane		-182.2	-148.2									
$C_5H_{10}O$	Tetrahydropyran		-258.3	-223.4									
$C_5H_{10}OS$	S-Propyl thioacetate		-294.5	-250.4									
$C_5H_{10}O_2$	Pentanoic acid		-559.4	-491.9					259.8			210.3	
$C_5H_{10}O_2$	2-Methylbutanoic acid		-554.5										
$C_5H_{10}O_2$	3-Methylbutanoic acid		-561.6	-510.0									
$C_5H_{10}O_2$	2,2-Dimethylpropanoic acid	-564.5		-491.3									
$C_5H_{10}O_2$	Butyl formate											200.2	
$C_5H_{10}O_2$	Propyl acetate											196.2	
$C_5H_{10}O_2$	Isopropyl acetate		-518.9	-481.6								199.4	
$C_5H_{10}O_2$	Ethyl propanoate		-502.7	-463.4									
$C_5H_{10}O_2$	cis-1,2-Cyclopentanediol	-485.0											
$C_5H_{10}O_2$	trans-1,2-Cyclopentanediol	-490.1											
$C_5H_{10}O_2$	4-Methyl-1,3-dioxane	-416.1		-376.9									
$C_5H_{10}O_2$	(Ethoxymethyl)oxirane		-296.5										
$C_5H_{10}O_2$	Tetrahydrofurfuryl alcohol		-435.7	-369.1									
$C_5H_{10}O_3$	Diethyl carbonate		-681.5	-637.9									
$C_5H_{10}O_3$	Ethylene glycol momomethyl ether acetate												
$C_5H_{10}O_3$	Ethyl lactate		-909.2									310.0	
$C_5H_{10}O_4$	1,2,3-Propanetriol, 1-acetate, (±)-											254.0	
$C_5H_{10}O_5$	D-Ribose	-1047.2											
$C_5H_{10}O_5$	D-Xylose	-1057.8											
$C_5H_{10}O_5$	α-D-Arabinopyranose	-1057.9											
$C_5H_{10}S$	Thiacyclohexane		-106.3	-63.5			53.1		218.2	323.0		163.3	109.7
$C_5H_{10}S$	Cyclopentanethiol		-89.5	-48.1					256.9			165.2	
$C_5H_{11}Br$	1-Bromopentane		-170.2	-128.9									
$C_5H_{11}Cl$	1-Chloropentane		-213.2	-174.9									
$C_5H_{11}Cl$	2-Chloro-2-methylbutane		-235.7	-202.2									
$C_5H_{11}Cl$	1-Chloro-3-methylbutane		-216.0	-179.7									
$C_5H_{11}Cl$	2-Chloro-3-methylbutane		-226.6	-185.1									
$C_5H_{11}N$	Cyclopentylamine		-95.1	-54.9					241.0			181.2	

Molecular formula	Name	$\Delta_f H°$/kJ mol⁻¹			$\Delta_f G°$/kJ mol⁻¹			S°/J mol⁻¹ K⁻¹			C_p/J mol⁻¹ K⁻¹		
		Crys.	Liq.	Gas	Crys.	Liq.	Gas	Crys.	Liq.	Gas	Crys.	Liq.	Gas
C₅H₁₁N	Piperidine		-86.4	-47.1					210.0			179.9	
C₅H₁₁NO	Pentanamide	-379.5		-290.2									
C₅H₁₁NO	2,2-Dimethylpropanamide	-399.7		-313.1									
C₅H₁₁NO₂	1-Nitropentane		-215.4										
C₅H₁₁NO₂	DL-Valine	-628.9											
C₅H₁₁NO₂	L-Valine	-617.9		-455.1									
C₅H₁₁NO₂	5-Aminopentanoic acid	-604.1		-460.0									
C₅H₁₁NO₂S	L-Methionine	-577.5		-413.5									
C₅H₁₁NO₄	2-Ethyl-2-nitro-1,3-propanediol	-606.4											
C₅H₁₂	Pentane		-173.5	-146.9					260.4			167.2	
C₅H₁₂	Isopentane		-178.4	-153.6								164.8	
C₅H₁₂	Neopentane		-190.2	-168.0									
C₅H₁₂N₂O	Butylurea	-419.5											
C₅H₁₂N₂O	tert-Butylurea	-417.4											
C₅H₁₂N₂O	N,N-Diethylurea	-372.2											
C₅H₁₂N₂O	Tetramethylurea	-262.2											
C₅H₁₂N₂S	Tetramethylthiourea	-38.1		44.9									
C₅H₁₂O	1-Pentanol		-351.6	-294.6								208.1	
C₅H₁₂O	2-Pentanol		-365.2	-311.0									
C₅H₁₂O	3-Pentanol		-368.9	-314.9								239.7	
C₅H₁₂O	2-Methyl-1-butanol		-356.6	-301.4									
C₅H₁₂O	3-Methyl-1-butanol		-356.4	-300.7									
C₅H₁₂O	2-Methyl-2-butanol		-379.5	-329.3								247.1	
C₅H₁₂O	3-Methyl-2-butanol		-366.6	-313.5									
C₅H₁₂O	2,2-Dimethyl-1-propanol		-399.4										
C₅H₁₂O	Butyl methyl ether		-290.6	-258.1					295.3			192.7	
C₅H₁₂O	tert-Butyl methyl ether		-313.6	-283.7					265.3			187.5	
C₅H₁₂O	Ethyl propyl ether		-303.6	-272.0					295.0			197.2	
C₅H₁₂O₂	1,5-Pentanediol		-528.8	-450.8									
C₅H₁₂O₂	2,2-Dimethyl-1,3-propanediol	-551.2		-414.7									
C₅H₁₂O₂	Diethoxymethane		-450.5										
C₅H₁₂O₂	1,1-Dimethoxypropane		-443.6										
C₅H₁₂O₂	2,2-Dimethoxypropane		-459.4	-429.9									
C₅H₁₂O₃	Diethylene glycol monomethyl ether											271.1	
C₅H₁₂O₃	2-(Hydroxymethyl)-2-methyl-1,3-propanediol	-744.6											
C₅H₁₂O₄	Pentaerythritol	-920.6		-776.7									
C₅H₁₂O₅	Xylitol	-1118.5											
C₅H₁₂S	1-Pentanethiol		-151.3	-110.0									
C₅H₁₂S	2-Methyl-1-butanethiol		-154.4	-114.9									
C₅H₁₂S	3-Methyl-1-butanethiol		-154.4	-114.9									
C₅H₁₂S	2-Methyl-2-butanethiol		-162.8	-127.1									
C₅H₁₂S	3-Methyl-2-butanethiol		-158.8	-121.3					290.1			198.1	

Molecular formula	Name	ΔfH°/kJ mol⁻¹			ΔfG°/kJ mol⁻¹			S°/J mol⁻¹ K⁻¹			Cp/J mol⁻¹ K⁻¹		
		Crys.	Liq.	Gas	Crys.	Liq.	Gas	Crys.	Liq.	Gas	Crys.	Liq.	Gas
C₅H₁₂S	2,2-Dimethyl-1-propanethiol		-165.4	-129.0					307.5			200.9	
C₅H₁₂S	Butyl methyl sulfide		-142.9	-102.4					276.1			199.9	
C₅H₁₂S	tert-Butyl methyl sulfide		-157.1	-121.3					309.5			198.4	
C₅H₁₂S	Ethyl propyl sulfide		-144.8	-104.8									
C₅H₁₂S	Ethyl isopropyl sulfide		-156.1	-118.3									
C₅H₁₃N	Pentylamine											218.0	
C₅H₁₄N₂	N,N,N',N'-Tetramethyl-methanediamine		-51.1	-18.2									
C₆ClF₅	Chloropentafluorobenzene	-858.4		-809.3									
C₆Cl₆	Hexachlorobenzene	-127.6		-35.5									
C₆F₆	Hexafluorobenzene			-955.4				260.2			201.2		
C₆F₁₀	Perfluorocyclohexene		-991.3	-1932.7									
C₆F₁₂	Perfluorocyclohexane		-1963.5	-2370.4					280.8			221.6	
C₆HF₅	Pentafluorobenzene	-852.7	-841.8	-806.5									
C₆HF₅O	Pentafluorophenol	-1024.1	-1007.7										
C₆H₂F₄	1,2,4,5-Tetrafluorobenzene		-683.8										
C₆H₃Cl₃	1,2,3-Trichlorobenzene	-70.8		3.8									
C₆H₃Cl₃	1,2,4-Trichlorobenzene		-63.1	-8.1									
C₆H₃Cl₃	1,3,5-Trichlorobenzene	-78.4		-13.4									
C₆H₃N₃O₈	2,4,6-Trinitro-1,3-benzenediol	-467.5											
C₆H₄Cl₂	o-Dichlorobenzene		-17.5	30.2								162.4	
C₆H₄Cl₂	m-Dichlorobenzene		-20.7	25.7									
C₆H₄Cl₂	p-Dichlorobenzene	-42.3		22.5				175.4			147.8		
C₆H₄F₂	o-Difluorobenzene		-330.0	-293.8					222.6			159.0	
C₆H₄F₂	m-Difluorobenzene		-343.9	-309.2					223.8			159.1	
C₆H₄F₂	p-Difluorobenzene		-342.3	-306.7								157.5	
C₆H₄N₂O₅	2,4-Dinitrophenol	-232.7		-128.1									
C₆H₄O₂	p-Benzoquinone	-185.7		-122.9									
C₆H₅Br	Bromobenzene		60.9						219.2			154.3	
C₆H₅Cl	Chlorobenzene		11.1	52.0								150.1	
C₆H₅ClO	m-Chlorophenol	-206.4	-189.3										
C₆H₅ClO	p-Chlorophenol	-197.7	-181.3										
C₆H₅F	Fluorobenzene		-150.6	-115.9					205.9			146.4	
C₆H₅I	Iodobenzene		117.2	164.9					205.4		129.0	158.7	
C₆H₅NO₂	Nitrobenzene		12.5	67.5								185.8	
C₆H₅NO₂	3-Pyridinecarboxylic acid	-344.9											
C₆H₅NO₃	2-Nitrophenol	-202.4		-221.5									
C₆H₅N₃	1H-Benzotriazole	236.5		335.5									
C₆H₅N₃O₄	2,3-Dinitroaniline	-11.7											
C₆H₅N₃O₄	2,4-Dinitroaniline	-67.8											
C₆H₅N₃O₄	2,5-Dinitroaniline	-44.3											
C₆H₅N₃O₄	2,6-Dinitroaniline	-50.6											
C₆H₅N₃O₄	3,5-Dinitroaniline	-38.9											
C₆H₆	1,5-Hexadiyne		384.2										
C₆H₆	Benzene		49.1	82.9		124.5	129.7		173.4	269.2		136.0	82.4

Molecular formula	Name	$\Delta_f H°$/kJ mol⁻¹			$\Delta_f G°$/kJ mol⁻¹			$S°$/J mol⁻¹ K⁻¹			C_p/J mol⁻¹ K⁻¹		
		Crys.	Liq.	Gas	Crys.	Liq.	Gas	Crys.	Liq.	Gas	Crys.	Liq.	Gas
C₆H₆N₂O₂	o-Nitroaniline	-26.1	-9.4	63.8							166.0		
C₆H₆N₂O₂	m-Nitroaniline	-38.3	-14.4	58.4							158.8		
C₆H₆N₂O₂	p-Nitroaniline	-42.0	-20.7	58.8							167.0		
C₆H₆O	Phenol	-165.1		-96.4				144.0			127.4		
C₆H₆O	2-Vinylfuran		-10.3	27.8									
C₆H₆O₂	p-Hydroquinone	-364.5		-265.3							136.0		
C₆H₆O₂	Pyrocatechol	-354.1		-267.5									
C₆H₆O₂	Resorcinol	-368.0		-274.7									
C₆H₆O₃	1,2,3-Benzenetriol	-551.1		-434.2									
C₆H₆O₃	1,2,4-Benzenetriol	-563.8		-444.0									
C₆H₆O₃	1,3,5-Benzenetriol	-584.6		-452.9									
C₆H₆O₃	3,4-Dimethyl-2,5-furandione	-581.4											
C₆H₆O₆	cis-1-Propene-1,2,3-tricarboxylic acid	-1224.4											
C₆H₆O₆	trans-1-Propene-1,2,3-tricarboxylic acid	-1232.7											
C₆H₆S	Benzenethiol		63.7	111.3					222.8			173.2	
C₆H₇N	Aniline		31.6	87.5			-7.0			317.9		191.9	107.9
C₆H₇N	2-Methylpyridine		56.7	99.2								158.6	
C₆H₇N	3-Methylpyridine		61.9	106.4					216.3			158.7	
C₆H₇N	4-Methylpyridine		59.2	103.8					209.1			159.0	
C₆H₇N	1-Cyclopentenecarbonitrile		111.5	156.5									
C₆H₈N₂	Adiponitrile		85.1	149.5								128.7	
C₆H₈N₂	o-Phenylenediamine	-0.3											
C₆H₈N₂	m-Phenylenediamine	-7.8											
C₆H₈N₂	p-Phenylenediamine	3.0											
C₆H₈N₂	Phenylhydrazine		141.0	202.9									
C₆H₈N₂S	Bis(2-cyanoethyl) sulfide		96.3					154.5			159.6		
C₆H₈O₄	Dimethyl maleate											263.2	
C₆H₈O₆	L-Ascorbic acid	-1164.6											
C₆H₈O₇	2-Hydroxy-1,2,3-propane-tricarboxylic acid	-1543.8											
C₆H₉Cl₃O₂	Butyl trichloroacetate		-545.8	-492.3									
C₆H₉Cl₃O₂	Isobutyl trichloroacetate		-553.4	-500.2									
C₆H₉N	Cyclopentanecarbonitrile		0.7	44.1									
C₆H₉N	2,4-Dimethylpyrrole	-422.3											
C₆H₉N	2,5-Dimethylpyrrole		-16.7	39.8									
C₆H₉NO₃	Triacetamide	-610.5		-550.1									
C₆H₉N₃O₆	Nitrilotriacetic acid	-1311.9											
C₆H₉N₃O₂	L-Histidine	-466.7											
C₆H₁₀	1,5-Hexadiene		54.1	84.2									
C₆H₁₀	3,3-Dimethyl-1-butyne		78.4										
C₆H₁₀	Cyclohexene		-38.5	-5.0					214.6			148.3	
C₆H₁₀	1-Methylcyclopentene		-36.4	-3.8									
C₆H₁₀	3-Methylcyclopentene		-23.7	7.4									

Molecular formula	Name	$\Delta_f H°$/kJ mol⁻¹ Crys.	Liq.	Gas	$\Delta_f G°$/kJ mol⁻¹ Crys.	Liq.	Gas	S°/J mol⁻¹ K⁻¹ Crys.	Liq.	Gas	C_p/J mol⁻¹ K⁻¹ Crys.	Liq.	Gas
C_6H_{10}	4-Methylcyclopentene		-17.6	14.6									
$C_6H_{10}Cl_2O_2$	Butyl dichloroacetate		-550.1	-497.8									
$C_6H_{10}O$	Cyclohexanone		-271.2	-226.1								182.2	
$C_6H_{10}O$	2-Methylcyclopentanone		-265.2										
$C_6H_{10}O$	Mesityl oxide											212.5	
$C_6H_{10}O_2$	Ethyl trans-2-butenoate		-420.0	-375.6									
$C_6H_{10}O_2$	Methyl cyclobutane-carboxylate		-395.0	-350.2									
$C_6H_{10}O_3$	Ethyl acetoacetate												
$C_6H_{10}O_3$	Propanoic anhydride		-679.1	-626.5								248.0	
$C_6H_{10}O_4$	Adipic acid	-994.3											
$C_6H_{10}O_4$	Diethyl oxalate		-805.5	-742.0									
$C_6H_{10}O_4$	Ethylene glycol diacetate											310.0	
$C_6H_{11}Cl$	Chlorocyclohexane		-207.2	-163.7									
$C_6H_{11}ClO_2$	Ethyl 4-chlorobutanoate		-566.5	-513.8									
$C_6H_{11}ClO_2$	Propyl 3-chloropropanoate		-537.6	-485.7									
$C_6H_{11}ClO_2$	Butyl chloroacetate		-538.4	-487.4									
$C_6H_{11}NO$	Caprolactam	-329.4		-239.6							156.8		
$C_6H_{11}NO$	1-Methyl-2-piperidinone		-293.0										
C_6H_{12}	1-Hexene		-74.2	-43.5					295.2			183.3	
C_6H_{12}	cis-2-Hexene		-83.9	-52.3									
C_6H_{12}	trans-2-Hexene		-85.5	-53.9									
C_6H_{12}	cis-3-Hexene		-78.9	-47.6									
C_6H_{12}	trans-3-Hexene		-86.1	-54.4									
C_6H_{12}	2-Methyl-1-pentene		-90.0	-59.4									
C_6H_{12}	3-Methyl-1-pentene		-78.2	-49.5									
C_6H_{12}	4-Methyl-1-pentene		-80.0	-51.3									
C_6H_{12}	3-Methyl-cis-2-pentene		-94.5	-62.3									
C_6H_{12}	4-Methyl-cis-2-pentene		-87.0	-57.5									
C_6H_{12}	3-Methyl-trans-2-pentene		-94.6	-63.1									
C_6H_{12}	4-Methyl-trans-2-pentene		-91.6	-61.5									
C_6H_{12}	2-Ethyl-1-butene		-87.1	-56.0									
C_6H_{12}	2,3-Dimethyl-1-butene		-93.2	-62.4									
C_6H_{12}	3,3-Dimethyl-1-butene		-87.5	-60.3									
C_6H_{12}	2,3-Dimethyl-2-butene		-101.4	-68.1									
C_6H_{12}	Cyclohexane		-156.4	-123.4					270.2			174.7	
C_6H_{12}	Methylcyclopentane		-137.9	-106.2								154.9	
C_6H_{12}	Ethylcyclobutane		-59.0	-27.5									
C_6H_{12}	1,1,2-Trimethylcyclo-propane		-96.2										
C_6H_{12}	2-Methyl-2-pentene		-98.5	-66.9									
$C_6H_{12}N_2O_4S_2$	L-Cystine	-1032.7											
$C_6H_{12}O$	Cyclohexanol		-348.2	-286.2								208.2	
$C_6H_{12}O$	cis-2-Methylcyclopentanol		-345.5										
$C_6H_{12}O$	Butyl vinyl ether		-218.8	-182.6								232.0	

Molecular formula	Name	ΔfH°/kJ mol⁻¹ Crys.	Liq.	Gas	ΔfG°/kJ mol⁻¹ Crys.	Liq.	Gas	S°/J mol⁻¹ K⁻¹ Crys.	Liq.	Gas	Cp/J mol⁻¹ K⁻¹ Crys.	Liq.	Gas
$C_6H_{12}O$	2-Hexanone		-322.0	-278.9									213.3
$C_6H_{12}O$	3-Hexanone		-320.2	-277.6					305.3				216.9
$C_6H_{12}O$	Methyl isobutyl ketone												213.3
$C_6H_{12}O$	2-Methyl-3-pentanone		-325.9	-286.0									
$C_6H_{12}O$	3,3-Dimethyl-2-butanone		-328.6	-290.6									
$C_6H_{12}O_2$	Hexanoic acid		-583.8	-511.9									
$C_6H_{12}O_2$	Butyl acetate		-529.2	-485.3								227.8	
$C_6H_{12}O_2$	Isobutyl acetate											233.8	
$C_6H_{12}O_2$	Ethyl butanoate											228.0	
$C_6H_{12}O_2$	Methyl pentanoate		-514.2	-471.1								229.3	
$C_6H_{12}O_2$	Methyl 2,2-dimethyl-propanoate		-530.0	-491.2									
$C_6H_{12}O_2$	Diacetone alcohol											257.9	
$C_6H_{12}O_3$	Ethylene glycol ethyl ether acetate											221.3	
$C_6H_{12}O_3$	Paraldehyde		-673.1	-631.7									376.0
$C_6H_{12}O_6$	β-D-Fructose	-1265.6											
$C_6H_{12}O_6$	D-Galactose	-1286.3											
$C_6H_{12}O_6$	α-D-Glucose	-1273.3											
$C_6H_{12}O_6$	D-Mannose	-1263.0											
$C_6H_{12}O_6$	L-Sorbose	-1271.5											
$C_6H_{12}S$	Cyclohexanethiol		-140.7	-96.2					255.6			192.6	
$C_6H_{12}S$	Cyclopentyl methyl sulfide		-109.8	-64.7					453.0			203.5	
$C_6H_{13}Br$	1-Bromohexane		-194.2	-148.3									
$C_6H_{13}Cl$	2-Chlorohexane		-246.1	-204.3									
$C_6H_{13}N$	Cyclohexylamine		-147.6	-104.0									
$C_6H_{13}N$	(±)-2-Methylpiperidine		-124.9	-84.4									
$C_6H_{13}NO$	Hexanamide		-397.9	-324.2									
$C_6H_{13}NO$	N-Butylacetamide		-380.9	-305.9									
$C_6H_{13}NO_2$	DL-Leucine	-640.6											
$C_6H_{13}NO_2$	D-Leucine	-637.3											
$C_6H_{13}NO_2$	L-Leucine	-637.4		-486.8									
$C_6H_{13}NO_2$	DL-Isoleucine	-635.3											
$C_6H_{13}NO_2$	L-Isoleucine	-637.8									200.1		
$C_6H_{13}NO_2$	Norleucine	-639.1											
$C_6H_{13}NO_2$	6-Aminohexanoic acid	-637.3											
C_6H_{14}	Hexane		-198.7	-166.9					290.6			195.6	
C_6H_{14}	2-Methylpentane		-204.6	-174.6					292.5			193.7	
C_6H_{14}	3-Methylpentane		-202.4	-171.9					272.5			190.7	
C_6H_{14}	2,2-Dimethylbutane		-213.8	-185.9					287.8			191.9	
C_6H_{14}	2,3-Dimethylbutane		-207.4	-178.1								189.7	
$C_6H_{14}N_2$	Azopropane		11.5	51.3									
$C_6H_{14}N_2O_2$	Lysine	-678.7						250.6					
$C_6H_{14}N_4O_2$	D-Arginine	-623.5									232.0		
$C_6H_{14}O$	1-Hexanol		-377.5	-315.9					287.4			240.4	

Molecular formula	Name	$\Delta_f H°$/kJ mol⁻¹ Crys.	Liq.	Gas	$\Delta_f G°$/kJ mol⁻¹ Crys.	Liq.	Gas	$S°$/J mol⁻¹ K⁻¹ Crys.	Liq.	Gas	C_p/J mol⁻¹ K⁻¹ Crys.	Liq.	Gas
$C_6H_{14}O$	2-Hexanol		-392.0	-333.5								286.2	
$C_6H_{14}O$	3-Hexanol		-392.4									248.0	
$C_6H_{14}O$	2-Methyl-1-pentanol											275.9	
$C_6H_{14}O$	3-Methyl-2-pentanol		-394.7									273.0	
$C_6H_{14}O$	4-Methyl-2-pentanol		-396.4										
$C_6H_{14}O$	2-Methyl-3-pentanol											293.4	
$C_6H_{14}O$	3-Methyl-3-pentanol												
$C_6H_{14}O$	Dipropyl ether		-328.8	-293.0					323.9			221.6	
$C_6H_{14}O$	Diisopropyl ether		-351.5	-319.2								216.8	
$C_6H_{14}O$	Butyl ethyl ether											159.0	
$C_6H_{14}O$	Ethyl tert-butyl ether			-313.9									
$C_6H_{14}OS$	Dipropyl sulfoxide		-329.4	-254.9									
$C_6H_{14}O_2$	1,2-Hexanediol		-577.1	-490.1									
$C_6H_{14}O_2$	1,6-Hexanediol	-569.9	-548.6	-461.2									
$C_6H_{14}O_2$	Hexylene glycol											336.0	
$C_6H_{14}O_2$	Ethylene glycol monobutyl ether											281.0	
$C_6H_{14}O_2$	1,1-Diethoxyethane		-491.4	-453.5									
$C_6H_{14}O_2$	1,2-Diethoxyethane		-451.4	-408.1								259.4	
$C_6H_{14}O_3$	Diethylene glycol monoethyl ether											301.0	
$C_6H_{14}O_3$	Diethylene glycol dimethyl ether											274.1	
$C_6H_{14}O_3$	Trimethylolpropane	-750.9	-804.3	-725.0									
$C_6H_{14}O_4$	Triethylene glycol		-859.0	-792.0									
$C_6H_{14}O_4S$	Dipropyl sulfate		-1317.0										
$C_6H_{14}O_6$	Galactitol		-1314.5										
$C_6H_{14}O_6$	D-Mannitol												
$C_6H_{14}S$	1-Hexanethiol		-175.7	-129.9									
$C_6H_{14}S$	2-Methyl-2-pentanethiol		-188.3	-148.3									
$C_6H_{14}S$	2,3-Dimethyl-2-butanethiol		-187.1	-147.9									
$C_6H_{14}S$	Diisopropyl sulfide		-181.6	-142.0					313.0			232.0	
$C_6H_{14}S$	Butyl ethyl sulfide		-172.3	-127.8									
$C_6H_{14}S$	Methyl pentyl sulfide		-167.1	-121.8									
$C_6H_{14}S_2$	Dipropyl disulfide		-171.5	-118.3									
$C_6H_{15}B$	Triethylborane		-194.6	-157.7		9.4	16.1		336.7	437.8		241.2	
$C_6H_{15}N$	Dipropylamine		-156.1	-116.0									
$C_6H_{15}N$	Diisopropylamine		-178.5	-143.8									
$C_6H_{15}N$	Triethylamine		-127.7	-92.7								219.9	
$C_6H_{15}NO$	2-Diethylaminoethanol		-305.9										
$C_6H_{15}NO_3$	Triethanolamine	-664.2		-558.3							389.0		
$C_6H_{18}N_3OP$	Hexamethylphosphoric triamide					-541.5	-534.5					321.0	
$C_6H_{16}OSi_2$	Hexamethyldisiloxane		-815.0	-777.7			-856.0		433.8	535.0		311.4	238.5
C_6MoO_6	Molybdenum hexacarbonyl	-982.8		-912.1	-877.7		-856.0	325.9		490.0	242.3		205.0

Molecular formula	Name	$\Delta_f H^\circ$/kJ mol^{-1}			$\Delta_f G^\circ$/kJ mol^{-1}			S°/J mol^{-1} K^{-1}			C_p/J mol^{-1} K^{-1}		
		Crys.	Liq.	Gas	Crys.	Liq.	Gas	Crys.	Liq.	Gas	Crys.	Liq.	Gas
C$_6$N$_4$	Tetracyanoethylene	623.8		705.0									
C$_7$F$_8$	Perfluorotoluene		-1311.1						355.5			262.3	
C$_7$F$_{14}$	Perfluoromethylcyclohexane		-2931.1	-2897.2								353.1	
C$_7$F$_{16}$	Perfluoroheptane		-3420.0	-3383.6					561.8			419.0	
C$_7$H$_3$F$_5$	2,3,4,5,6-Pentafluorotoluene		-883.8	-842.7					306.4			225.8	
C$_7$H$_4$Cl$_2$O	m-Chlorobenzoyl chloride		-189.7										
C$_7$H$_4$N$_2$O$_6$	3,5-Dinitrobenzoic acid	-409.8											
C$_7$H$_5$ClO	Benzoyl chloride		-158.0	-103.2									
C$_7$H$_5$ClO$_2$	o-Chlorobenzoic acid	-404.5		-325.0									
C$_7$H$_5$ClO$_2$	m-Chlorobenzoic acid	-424.3		-342.3									
C$_7$H$_5$ClO$_2$	p-Chlorobenzoic acid	-428.9		-341.0									
C$_7$H$_5$F$_3$	(Trifluoromethyl)benzene										163.2	188.4	
C$_7$H$_5$N	Benzonitrile		163.2	215.7					209.1			165.2	
C$_7$H$_5$NO	Benzoxazole	-24.2		44.8									
C$_7$H$_5$NO$_4$	2-Nitrobenzoic acid	-378.8											
C$_7$H$_5$NO$_4$	3-Nitrobenzoic acid	-394.7											
C$_7$H$_5$NO$_4$	4-Nitrobenzoic acid	-392.2											
C$_7$H$_6$N$_2$	1H-Benzimidazole	79.5		181.7									
C$_7$H$_6$N$_2$	1H-Indazole	151.9		243.0									
C$_7$H$_6$O	Benzaldehyde		-87.0	-36.7					221.2			172.0	
C$_7$H$_6$O$_2$	Benzoic acid	-385.2		-294.0				167.6			146.8		
C$_7$H$_6$O$_2$	3-(2-Furanyl)-2-propenal	-182.0		-105.9									
C$_7$H$_6$O$_2$	Salicylaldehyde											222.0	
C$_7$H$_6$O$_3$	Salicylic acid	-589.9		-494.8									
C$_7$H$_7$Cl	o-Chlorotoluene		-32.5	18.9								166.8	
C$_7$H$_7$Cl	(Chloromethyl)benzene		-186.9										
C$_7$H$_7$F	p-Fluorotoluene			-147.4								171.2	
C$_7$H$_7$NO	Benzamide	-202.6		-100.9									
C$_7$H$_7$NO$_2$	2-Aminobenzoic acid	-380.4		-296.0									
C$_7$H$_7$NO$_2$	Aniline-3-carboxylic acid	-389.8		-283.6									
C$_7$H$_7$NO$_2$	Aniline-4-carboxylic acid	-391.9		-296.7									
C$_7$H$_7$NO$_2$	o-Nitrotoluene		-9.7										
C$_7$H$_7$NO$_2$	m-Nitrotoluene		-31.5										
C$_7$H$_7$NO$_2$	p-Nitrotoluene	-48.1		31.0				172.3			172.3		
C$_7$H$_7$NO$_2$	(Nitromethyl)benzene		-22.8	30.7									
C$_7$H$_7$NO$_2$	Salicylaldoxime	-183.7											
C$_7$H$_8$	Toluene		12.4	50.5								157.3	
C$_7$H$_8$N$_2$O	Phenylurea	-218.6											
C$_7$H$_8$O	o-Cresol	-204.6		-128.6				165.4			154.6		
C$_7$H$_8$O	m-Cresol		-194.0	-132.3					212.6			224.9	
C$_7$H$_8$O	p-Cresol	-199.3		-125.4				167.3			150.2		
C$_7$H$_8$O	Benzyl alcohol		-160.7	-100.4					216.7			217.9	
C$_7$H$_8$O	Anisole		-114.8	-67.9									
C$_7$H$_9$N	Benzylamine		34.2	94.4									
C$_7$H$_9$N	o-Methylaniline		-6.3	56.4			167.6			351.0			130.2

Molecular formula	Name	$\Delta_f H°$/kJ mol⁻¹			$\Delta_f G°$/kJ mol⁻¹			S°/J mol⁻¹ K⁻¹			C_p/J mol⁻¹ K⁻¹		
		Crys.	Liq.	Gas	Crys.	Liq.	Gas	Crys.	Liq.	Gas	Crys.	Liq.	Gas
C₇H₉N	*m*-Methylaniline		-8.1	54.6			165.4			352.5			125.5
C₇H₉N	*p*-Methylaniline	-23.5		55.3			167.7			347.0			126.2
C₇H₉N	*N*-Methylaniline											207.1	
C₇H₉N	1-Cyclohexenecarbonitrile		48.1	101.6									
C₇H₉N	2,3-Dimethylpyridine		19.4	67.1					243.7			189.5	
C₇H₉N	2,4-Dimethylpyridine		16.1	63.6					248.5			184.8	
C₇H₉N	2,5-Dimethylpyridine		18.7	66.5					248.8			184.7	
C₇H₉N	2,6-Dimethylpyridine		12.7	58.1					244.2			185.2	
C₇H₉N	3,4-Dimethylpyridine		18.3	68.8					240.7			191.8	
C₇H₉N	3,5-Dimethylpyridine		22.5	72.0					241.7			184.5	
C₇H₁₀O₂	Ethyl 2-pentynoate		-301.8	-250.3									
C₇H₁₀O₂	Methyl 2-hexynoate		-242.7										
C₇H₁₁Cl₃O₂	3-Methylbutyl trichloroacetate		-580.9	-523.1									
C₇H₁₁N	Cyclohexanecarbonitrile		-47.2	4.8									
C₇H₁₂	Methylenecyclohexane		-61.3	-25.2									
C₇H₁₂	Vinylcyclopentane		-34.8										
C₇H₁₂	1-Ethylcyclopentene		-53.3	-19.8									
C₇H₁₂	Bicyclo[2.2.1]heptane	-95.1		-54.8									
C₇H₁₂	1-Methylbicyclo(3,1,0)hexane		-33.2	1.7									
C₇H₁₂O	2-Methylenecyclohexanol	-277.6											
C₇H₁₂O₄	Diethyl malonate											285.0	
C₇H₁₃ClO₂	Butyl 2-chloropropanoate		-571.7	-517.3									
C₇H₁₃ClO₂	Isobutyl 2-chloropropanoate		-603.1	-549.6									
C₇H₁₃ClO₂	Butyl 3-chloropropanoate		-557.9	-502.3									
C₇H₁₃ClO₂	Isobutyl 3-chloropropanoate		-572.6	-517.3									
C₇H₁₃ClO₂	Propyl 2-chlorobutanoate		-630.7	-578.4									
C₇H₁₃N	Heptanenitrile		-82.8	-31.0									
C₇H₁₄	1-Heptene		-97.9	-62.3					327.6			211.8	
C₇H₁₄	*cis*-2-Heptene		-105.1										
C₇H₁₄	*trans*-2-Heptene		-109.5										
C₇H₁₄	*cis*-3-Heptene		-104.3										
C₇H₁₄	*trans*-3-Heptene		-109.3										
C₇H₁₄	5-Methyl-1-hexene		-100.0	-65.7									
C₇H₁₄	3-Methyl-*cis*-3-hexene		-115.9	-79.4									
C₇H₁₄	3-Methyl-*trans*-3-hexene		-112.7	-76.8									
C₇H₁₄	2,4-Dimethyl-1-pentene		-117.0	-83.8									
C₇H₁₄	4,4-Dimethyl-1-pentene		-110.6	-81.6									
C₇H₁₄	2,4-Dimethyl-2-pentene		-123.1	-88.7									
C₇H₁₄	*cis*-4,4-Dimethyl-2-pentene		-105.3	-72.6									
C₇H₁₄	*trans*-4,4-Dimethyl-2-pentene		-121.7	-88.8									
C₇H₁₄	2-Ethyl-3-methyl-1-butene		-114.1	-79.5									
C₇H₁₄	2,3,3-Trimethyl-1-butene		-117.7	-85.5									
C₇H₁₄	Cycloheptane		-156.6	-118.1									
C₇H₁₄	Methylcyclohexane		-190.1	-154.7									184.8

Molecular formula	Name	$\Delta_f H°$/kJ mol^{-1}			$\Delta_f G°$/kJ mol^{-1}			$S°$/J mol^{-1} K^{-1}			C_p/J mol^{-1} K^{-1}		
		Crys.	Liq.	Gas	Crys.	Liq.	Gas	Crys.	Liq.	Gas	Crys.	Liq.	Gas
C_7H_{14}	Ethylcyclopentane		-163.4	-126.9					279.9				
C_7H_{14}	1,1-Dimethylcyclopentane		-172.1	-138.2									
C_7H_{14}	*cis*-1,2-Dimethylcyclopentane		-165.3	-129.5					269.2				
C_7H_{14}	*trans*-1,2-Dimethylcyclopentane		-171.2	-136.6									
C_7H_{14}	*cis*-1,3-Dimethylcyclopentane		-170.1	-135.8									
C_7H_{14}	*trans*-1,3-Dimethylcyclopentane		-168.1	-133.6									
C_7H_{14}	1,1,2,2-Tetramethylcyclopropane		-119.8										
$C_7H_{14}Br_2$	1,2-Dibromoheptane		-212.3	-157.9									
$C_7H_{14}O$	1-Heptanal		-311.5	-263.8					335.4			230.1	
$C_7H_{14}O$	2-Heptanone		-356.1	-313.6								232.6	
$C_7H_{14}O$	2,2-Dimethyl-3-pentanone		-352.9	-311.3									
$C_7H_{14}O$	2,4-Dimethyl-3-pentanone			-327.0					318.0			233.7	
$C_7H_{14}O$	*cis*-2-Methylcyclohexanol		-390.2										
$C_7H_{14}O$	*trans*-2-Methylcyclohexanol		-415.7	-352.5									
$C_7H_{14}O$	*cis*-3-Methylcyclohexanol		-416.1	-350.9									
$C_7H_{14}O$	*trans*-3-Methylcyclohexanol		-394.4	-329.1									
$C_7H_{14}O$	*cis*-4-Methylcyclohexanol		-413.2	-347.5									
$C_7H_{14}O$	*trans*-4-Methylcyclohexanol		-433.3	-367.2									
$C_7H_{14}O_2$	Heptanoic acid		-610.2	-536.2									
$C_7H_{14}O_2$	Pentyl acetate		-553.0	-505.9								265.4	
$C_7H_{14}O_2$	Isopentyl acetate		-571.0	-527.0								261.0	
$C_7H_{14}O_2$	Ethyl pentanoate		-577.2	-536.0								248.5	
$C_7H_{14}O_2$	Ethyl 3-methylbutanoate												
$C_7H_{14}O_2$	Ethyl 2,2-dimethylpropanoate												
$C_7H_{14}O_2$	Methyl hexanoate		-540.2	-492.2									
$C_7H_{14}O_6$	α-Methylglucoside	-1233.3											
$C_7H_{15}Br$	1-Bromoheptane		-218.4	-167.8									
C_7H_{16}	Heptane		-224.2	-187.6					323.3			224.7	
C_7H_{16}	2-Methylhexane		-229.5	-194.5								222.9	
C_7H_{16}	3-Methylhexane		-226.4	-191.3									
C_7H_{16}	3-Ethylpentane		-224.9	-189.5					314.5			219.6	
C_7H_{16}	2,2-Dimethylpentane		-238.3	-205.7					300.3			221.1	
C_7H_{16}	2,3-Dimethylpentane		-233.1	-198.7					303.2			224.2	
C_7H_{16}	2,4-Dimethylpentane		-234.6	-201.6									
C_7H_{16}	3,3-Dimethylpentane		-234.2	-201.0									
C_7H_{16}	2,2,3-Trimethylbutane		-236.5	-204.4									
$C_7H_{16}O$	1-Heptanol		-403.3	-336.5					292.2			213.5	
$C_7H_{16}O$	*tert*-Butyl isopropyl ether		-392.8	-358.1									
$C_7H_{16}O_2$	1,7-Heptanediol		-574.2									272.1	

Molecular formula	Name	$\Delta_f H°$/kJ mol⁻¹			$\Delta_f G°$/kJ mol⁻¹			$S°$/J mol⁻¹ K⁻¹			C_p/J mol⁻¹ K⁻¹		
		Crys.	Liq.	Gas	Crys.	Liq.	Gas	Crys.	Liq.	Gas	Crys.	Liq.	Gas
$C_7H_{16}O_2$	2,2-Diethoxypropane		-538.9	-506.9									
$C_7H_{16}S$	1-Heptanethiol		-200.5	-149.9									
$C_8H_4O_3$	Phthalic anhydride	-460.1		-371.4				180.0			160.0		
$C_8H_5NO_2$	1H-Indole-2,3-dione	-268.2											
$C_8H_6O_4$	Phthalic acid	-782.0		-696.3				207.9			188.1		
$C_8H_6O_4$	Isophthalic acid	-803.0		-717.9									
$C_8H_6O_4$	Terephthalic acid	-816.1											
C_8H_6S	Benzo[b]thiophene	100.6		166.3									
C_8H_7N	1H-Indole	86.6		156.5									
C_8H_8	Styrene		103.8	147.9								182.0	
C_8H_8O	Phenyl vinyl ether		-26.2	22.7									
C_8H_8O	Acetophenone		-142.5	-86.7									
$C_8H_8O_2$	o-Toluic acid	-416.5									174.9		
$C_8H_8O_2$	m-Toluic acid	-426.1									163.6		
$C_8H_8O_2$	p-Toluic acid	-429.2									169.0		
$C_8H_8O_2$	Methyl benzoate		-343.5	-287.9								221.3	
$C_8H_8O_3$	Methyl salicylate											249.0	
C_8H_{10}	1,7-Octadiyne		334.4										
C_8H_{10}	Ethylbenzene		-12.3	29.9								183.2	
C_8H_{10}	o-Xylene		-24.4	19.1								186.1	
C_8H_{10}	m-Xylene		-25.4	17.3								183.0	
C_8H_{10}	p-Xylene		-24.4	18.0								181.5	
$C_8H_{10}O$	o-Ethylphenol		-208.8	-145.2									
$C_8H_{10}O$	m-Ethylphenol		-214.3	-146.1									
$C_8H_{10}O$	p-Ethylphenol	-224.4		-144.1									
$C_8H_{10}O$	2,3-Xylenol	-241.1		-157.2									
$C_8H_{10}O$	2,4-Xylenol		-228.7	-163.8									
$C_8H_{10}O$	2,5-Xylenol	-246.6		-161.6									
$C_8H_{10}O$	2,6-Xylenol	-237.4		-162.1									
$C_8H_{10}O$	3,4-Xylenol	-242.3		-157.3									
$C_8H_{10}O$	3,5-Xylenol	-244.4		-162.4									
$C_8H_{10}O$	Phenetole		-152.6	-101.6									
$C_8H_{10}O_2$	Veratrole		-290.3	-223.3									
$C_8H_{11}N$	N-Ethylaniline		4.0	56.3									
$C_8H_{11}N$	N,N-Dimethylaniline		47.7	100.5									
$C_8H_{11}N$	2,6-Xylidine										206.9	228.5	
C_8H_{12}	1-Octen-3-yne		140.7	166.5									
C_8H_{12}	cis-1,2-Divinylcyclo-butane		124.3										
C_8H_{12}	trans-1,2-Divinylcyclo-butane		101.3	143.5									
$C_8H_{12}N_4$	2,2'-Azobis[isobutyronitrile]	228.9											
$C_8H_{12}O_2$	2,2,4,4-Tetramethyl-1,3-cyclobutanedione	-379.9		-307.6								238.9	
C_8H_{14}	Ethylidenecyclohexane		-103.5	-59.5									

Molecular formula	Name	$\Delta_f H°$/kJ mol⁻¹ Crys.	Liq.	Gas	$\Delta_f G°$/kJ mol⁻¹ Crys.	Liq.	Gas	$S°$/J mol⁻¹ K⁻¹ Crys.	Liq.	Gas	C_p/J mol⁻¹ K⁻¹ Crys.	Liq.	Gas
C_8H_{14}	Allylcyclopentane		-64.5	-24.1								283.7	
$C_8H_{14}O_3$	Butanoic anhydride												
$C_8H_{15}ClO_2$	3-Methylbutyl 2-chloropropanoate		-627.3	-575.0									
$C_8H_{15}ClO_2$	3-Methylbutyl 3-chloropropanoate		-593.4	-539.4									
$C_8H_{15}N$	Octanenitrile		-107.3	-50.5								241.0	
C_8H_{16}	1-Octene		-121.8	-81.3								239.0	
C_8H_{16}	cis-2-Octene		-135.7									239.0	
C_8H_{16}	trans-2-Octene		-135.7										
C_8H_{16}	2,2-Dimethyl-cis-3-hexene		-126.4	-89.3									
C_8H_{16}	2,2-Dimethyl-trans-3-hexene		-144.9	-107.7									
C_8H_{16}	3-Ethyl-2-methyl-1-pentene		-137.9	-100.3									
C_8H_{16}	2,4,4-Trimethyl-1-pentene		-145.9	-110.5									
C_8H_{16}	2,4,4-Trimethyl-2-pentene		-142.4	-104.9									
C_8H_{16}	Cyclooctane		-167.7	-124.4									
C_8H_{16}	Ethylcyclohexane		-212.1	-171.5					280.9			211.8	
C_8H_{16}	1,1-Dimethylcyclohexane		-218.7	-180.9					267.2			209.2	
C_8H_{16}	cis-1,2-Dimethylcyclohexane		-211.8	-172.1					274.1			210.2	
C_8H_{16}	trans-1,2-Dimethylcyclohexane		-218.2	-179.9					273.2			209.4	
C_8H_{16}	cis-1,3-Dimethylcyclohexane		-222.9	-184.6					272.6			209.4	
C_8H_{16}	trans-1,3-Dimethylcyclohexane		-215.7	-176.5					276.3			212.8	
C_8H_{16}	cis-1,4-Dimethylcyclohexane		-215.6	-176.6					271.1			212.1	
C_8H_{16}	trans-1,4-Dimethylcyclohexane		-222.4	-184.5					268.0			210.2	
C_8H_{16}	Propylcyclopentane		-188.8	-147.7					310.8			216.3	
C_8H_{16}	1-Ethyl-1-methylcyclopentane		-193.8										
C_8H_{16}	cis-1-Ethyl-2-methylcyclopentane		-190.8										
C_8H_{16}	trans-1-Ethyl-2-methylcyclopentane		-195.1	-156.2									
C_8H_{16}	cis-1-Ethyl-3-methylcyclopentane		-194.4										
C_8H_{16}	trans-1-Ethyl-3-methylcyclopentane		-196.0										
$C_8H_{16}O$	2-Ethylhexanal		-348.5	-299.6									
$C_8H_{16}O$	2-Octanone		-381.6	-338.3								273.3	
$C_8H_{16}O$	2,2,4-Trimethyl-3-pentanone		-636.0	-554.3									
$C_8H_{16}O_2$	Octanoic acid		-635.1	-559.5								297.9	
$C_8H_{16}O_2$	2-Ethylhexanoic acid												
$C_8H_{16}O_2$	Hexyl acetate		-583.0	-533.6								282.8	
$C_8H_{16}O_2$	Propyl pentanoate		-592.2	-544.9									
$C_8H_{16}O_2$	Isopropyl pentanoate												

Molecular formula	Name	$\Delta_f H°$/kJ mol⁻¹			$\Delta_f G°$/kJ mol⁻¹			$S°$/J mol⁻¹ K⁻¹			C_p/J mol⁻¹ K⁻¹		
		Crys.	Liq.	Gas	Crys.	Liq.	Gas	Crys.	Liq.	Gas	Crys.	Liq.	Gas
C₈H₁₆O₂	Methyl heptanoate		-567.1	-515.5									285.1
C₈H₁₇Br	1-Bromooctane		-245.1	-189.3									
C₈H₁₇Cl	1-Chlorooctane		-291.3	-238.9									
C₈H₁₇NO	Octanamide	-473.2		-362.7									
C₈H₁₈	Octane		-250.1	-208.5								254.6	
C₈H₁₈	2-Methylheptane		-255.0	-215.3					356.4			252.0	
C₈H₁₈	3-Methylheptane		-252.3	-212.5					362.6			250.2	
C₈H₁₈	4-Methylheptane		-251.6	-211.9								251.1	
C₈H₁₈	3-Ethylhexane		-250.4	-210.7									
C₈H₁₈	2,2-Dimethylhexane		-261.9	-224.5									
C₈H₁₈	2,3-Dimethylhexane		-252.6	-213.8									
C₈H₁₈	2,4-Dimethylhexane		-257.0	-219.2									
C₈H₁₈	2,5-Dimethylhexane		-260.4	-222.5								249.2	
C₈H₁₈	3,3-Dimethylhexane		-257.5	-219.9								246.6	
C₈H₁₈	3,4-Dimethylhexane		-251.8	-212.8									
C₈H₁₈	3-Ethyl-2-methylpentane		-249.6	-211.0									
C₈H₁₈	3-Ethyl-3-methylpentane		-252.8	-214.8									
C₈H₁₈	2,2,3-Trimethylpentane		-256.9	-220.0								239.1	
C₈H₁₈	2,2,4-Trimethylpentane		-259.2	-224.0								245.6	
C₈H₁₈	2,3,3-Trimethylpentane		-253.5	-216.3								247.3	
C₈H₁₈	2,3,4-Trimethylpentane		-255.0	-217.3									
C₈H₁₈	2,2,3,3-Tetramethylbutane	-269.0		-226.0				273.7			239.2		
C₈H₁₈N₂	Azobutane		-40.1	9.2									
C₈H₁₈O	1-Octanol		-426.5	-355.6								305.2	
C₈H₁₈O	2-Octanol											330.1	
C₈H₁₈O	2-Ethyl-1-hexanol		-432.8	-365.3								317.5	
C₈H₁₈O	Dibutyl ether		-377.9	-332.8					347.0			278.2	
C₈H₁₈O	Di-sec-butyl ether		-401.5	-360.6									
C₈H₁₈O	Di-tert-butyl ether		-399.6	-362.0								276.1	
C₈H₁₈O	tert-Butyl isobutyl ether		-409.1	-369.0									
C₈H₁₈O₂	1,8-Octanediol	-626.6											
C₈H₁₈O₂	2,5-Dimethyl-2,5-hexanediol	-681.7											
C₈H₁₈O₃	Diethylene glycol diethyl ether											341.4	
C₈H₁₈O₃S	Dibutyl sulfite		-693.1	-625.3									
C₈H₁₈O₅	Tetraethylene glycol		-981.7	-883.0					405.1			428.8	
C₈H₁₈S	Dibutyl sulfide		-220.7	-167.7								284.3	
C₈H₁₈S	Di-sec-butyl sulfide		-220.7	-167.7									
C₈H₁₈S	Di-tert-butyl sulfide		-232.6	-188.8									
C₈H₁₈S	Diisobutyl sulfide		-229.2	-180.5									
C₈H₁₈S₂	Dibutyl disulfide		-222.9	-160.6									
C₈H₁₈S₂	Di-tert-butyl disulfide		-255.2	-201.0									
C₈H₁₉N	Dibutylamine		-206.0	-156.6								292.9	
C₈H₁₉N	Diisobutylamine		-218.5	-179.2									

Molecular formula	Name	$\Delta_fH°$/kJ mol⁻¹			$\Delta_fG°$/kJ mol⁻¹			$S°$/J mol⁻¹ K⁻¹			C_p/J mol⁻¹ K⁻¹		
		Crys.	Liq.	Gas	Crys.	Liq.	Gas	Crys.	Liq.	Gas	Crys.	Liq.	Gas
C₈H₂₀BrN	Tetraethylammonium bromide	-342.7											
C₈H₂₀Pb	Tetraethyl lead		52.7	109.6					464.6			307.4	
C₈H₂₀Si	Tetraethylsilane		141.2	200.5								298.1	
C₉H₇N	Quinoline		144.3	204.6					216.0			196.2	
C₉H₇N	Isoquinoline			-25.5									
C₉H₇NO	2-Hydroxyquinoline	-144.9		163.4									
C₉H₈	Indene		110.6	150.5					215.3			186.9	
C₉H₈O₄	2-(Acetyloxy)benzoic acid	-815.6											
C₉H₁₀	Cyclopropylbenzene		100.3	60.3									
C₉H₁₀	Indan		11.5						56.0			190.2	
C₉H₁₀N₂	2,2'-Dipyrrolylmethane	126.2											
C₉H₁₀O₂	Benzyl acetate			-312.9								148.5	
C₉H₁₀O₂	Ethyl benzoate								287.8			246.0	
C₉H₁₁NO₂	L-Phenylalanine	-466.9						213.6			203.0		
C₉H₁₁NO₃	L-Tyrosine	-685.1						214.0			216.4		
C₉H₁₂	Propylbenzene		-38.3	7.9								214.7	
C₉H₁₂	Cumene		-41.1	4.0								210.7	
C₉H₁₂	o-Ethyltoluene		-46.4	1.3									
C₉H₁₂	m-Ethyltoluene		-48.7	-1.8									
C₉H₁₂	p-Ethyltoluene		-49.8	-3.2									
C₉H₁₂	1,2,3-Trimethylbenzene		-58.5	-9.5								216.4	
C₉H₁₂	1,2,4-Trimethylbenzene		-61.8	-13.8								215.0	
C₉H₁₂	Mesitylene		-63.4	-15.9					267.9			209.3	
C₉H₁₂O	2-Isopropylphenol		-233.7	-182.2									
C₉H₁₂O	3-Isopropylphenol		-252.5	-196.0									
C₉H₁₂O	4-Isopropylphenol		-265.9	-209.4									
C₉H₁₂O₂	Hydroperoxide, 1-methyl-1-phenylethyl		-148.3	-78.4									
C₉H₁₃NO₂	Ethyl 3,5-dimethylpyrrole-2-carboxylate	-474.5											
C₉H₁₃NO₂	Ethyl 2,4-dimethylpyrrole-3-carboxylate	-463.2											
C₉H₁₃NO₂	Ethyl 2,5-dimethylpyrrole-3-carboxylate	-478.7											
C₉H₁₃NO₂	Ethyl 4,5-dimethylpyrrole-3-carboxylate	-470.3											
C₉H₁₄O	Isophorone											253.5	
C₉H₁₄O₆	Triacetin		-1330.8	-1245.0					458.3			384.7	
C₉H₁₅N	3-Ethyl-2,4,5-trimethyl-pyrrole	-89.2											
C₉H₁₆O₄	Nonanedioic acid	-1054.3											
C₉H₁₇NO	2,2,6,6-Tetramethyl-4-piperidinone	-334.2		-273.4									
C₉H₁₈	Propylcyclohexane		-237.4	-192.3					311.9			242.0	

Molecular formula	Name	$\Delta_f H°$/kJ mol⁻¹			$\Delta_f G°$/kJ mol⁻¹			$S°$/J mol⁻¹ K⁻¹			C_p/J mol⁻¹ K⁻¹		
		Crys.	Liq.	Gas	Crys.	Liq.	Gas	Crys.	Liq.	Gas	Crys.	Liq.	Gas
C_9H_{18}	1,3,5-Trimethylcyclo-hexane (1′,3′,5′)			-212.1									
$C_9H_{18}O$	2-Nonanone		-397.2	-340.7					401.4			303.6	
$C_9H_{18}O$	5-Nonanone		-398.2	-344.9								297.3	
$C_9H_{18}O$	2,6-Dimethyl-4-heptanone		-408.5	-357.6								362.4	
$C_9H_{18}O_2$	Nonanoic acid	-613.3	-659.7	-577.3									
$C_9H_{18}O_2$	Butyl pentanoate		-624.2	-560.2									
$C_9H_{18}O_2$	sec-Butyl pentanoate		-620.0	-573.2									
$C_9H_{18}O_2$	Isobutyl pentanoate		-590.3	-568.6									
$C_9H_{18}O_2$	Methyl octanoate		-171.8	-533.9									
$C_9H_{19}N$	N-Butylpiperidine			-159.9									
$C_9H_{19}N$	2,2,6,6-Tetramethyl-piperidine		-206.9	-228.2									
C_9H_{20}	Nonane		-274.7									284.4	
C_9H_{20}	2,2-Dimethylheptane		-288.1										
C_9H_{20}	2,2,3-Trimethylhexane		-282.7										
C_9H_{20}	2,2,4-Trimethylhexane		-282.8										
C_9H_{20}	2,2,5-Trimethylhexane		-293.3										
C_9H_{20}	2,3,3-Trimethylhexane		-281.1										
C_9H_{20}	2,3,5-Trimethylhexane		-284.0	-242.6									
C_9H_{20}	2,4,4-Trimethylhexane		-280.2										
C_9H_{20}	3,3,4-Trimethylhexane		-277.5										
C_9H_{20}	3,3-Diethylpentane		-275.4	-233.3								278.2	
C_9H_{20}	3-Ethyl-2,2-dimethylpentane		-272.7										
C_9H_{20}	3-Ethyl-2,4-dimethylpentane		-269.7										
C_9H_{20}	2,2,3,3-Tetramethylpentane		-278.3	-237.1								271.5	
C_9H_{20}	2,2,3,4-Tetramethylpentane		-277.7	-236.9									
C_9H_{20}	2,2,4,4-Tetramethylpentane		-280.0	-241.6								266.3	
C_9H_{20}	2,3,3,4-Tetramethylpentane		-277.9	-236.1									
$C_9H_{20}N_2O$	Tetraethylurea	-403.0		-376.5									
$C_9H_{20}O$	1-Nonanol	-657.6	-453.4										
$C_9H_{20}O_2$	1,9-Nonanediol		-207.1	-161.0									
$C_9H_{21}N$	Tripropylamine	29.8											
$C_{10}H_6N_2O_4$	1,5-Dinitronaphthalene	39.7		119.8									
$C_{10}H_6N_2O_4$	1,8-Dinitronaphthalene	55.4	54.6	137.4								212.6	
$C_{10}H_7Cl$	1-Chloronaphthalene			233.8									
$C_{10}H_7Cl$	2-Chloronaphthalene	144.3	161.5	235.1									
$C_{10}H_7I$	1-Iodonaphthalene	42.6		111.2									
$C_{10}H_7I$	2-Iodonaphthalene	78.5		150.6	201.6		224.1	167.4		333.1	165.7		131.9
$C_{10}H_7NO_2$	1-Nitronaphthalene	212.3		289.1							166.9		
$C_{10}H_8$	Naphthalene	-121.5		-30.4									
$C_{10}H_8$	Azulene	67.8	-124.1	-29.9									
$C_{10}H_8O$	1-Naphthol			132.8									
$C_{10}H_8O$	2-Naphthol												
$C_{10}H_9N$	1-Naphthalenamine												

Molecular formula	Name	$\Delta_f H°$/kJ mol⁻¹			$\Delta_f G°$/kJ mol⁻¹			S°/J mol⁻¹ K⁻¹			C_p/J mol⁻¹ K⁻¹		
		Crys.	Liq.	Gas	Crys.	Liq.	Gas	Crys.	Liq.	Gas	Crys.	Liq.	Gas
$C_{10}H_9N$	2-Naphthalenamine	60.2		134.3									
$C_{10}H_{10}$	1,2-Dihydronaphthalene		71.6										
$C_{10}H_{10}$	1,4-Dihydronaphthalene		84.2										
$C_{10}H_{10}O$	1-Tetralone	-209.6											
$C_{10}H_{10}O_4$	Dimethyl phthalate											303.1	
$C_{10}H_{10}O_4$	Dimethyl isophthalate	-730.9											
$C_{10}H_{10}O_4$	Dimethyl terephthalate	-732.6									261.1		
$C_{10}H_{12}$	1,2,3,4-Tetrahydronaphthalene		-29.2	26.0									
$C_{10}H_{14}$	Butylbenzene		-63.2	-11.8								217.5	
$C_{10}H_{14}$	sec-Butylbenzene		-66.4	-18.4					321.2			243.4	
$C_{10}H_{14}$	tert-Butylbenzene		-70.7	-23.0									
$C_{10}H_{14}$	Isobutylbenzene		-69.8	-21.9									
$C_{10}H_{14}$	o-Cymene		-73.3										
$C_{10}H_{14}$	m-Cymene		-78.6										
$C_{10}H_{14}$	p-Cymene		-78.0									236.4	
$C_{10}H_{14}$	o-Diethylbenzene		-68.5										
$C_{10}H_{14}$	m-Diethylbenzene		-73.5										
$C_{10}H_{14}$	p-Diethylbenzene		-72.8										
$C_{10}H_{14}$	3-Ethyl-o-xylene		-80.5										
$C_{10}H_{14}$	4-Ethyl-1,2-dimethylbenzene		-86.0										
$C_{10}H_{14}$	2-Ethyl-1,3-dimethylbenzene		-80.1										
$C_{10}H_{14}$	2-Ethyl-1,4-dimethylbenzene		-84.8										
$C_{10}H_{14}$	1-Ethyl-2,4-dimethylbenzene		-84.1										
$C_{10}H_{14}$	1-Ethyl-3,5-dimethylbenzene		-87.8										
$C_{10}H_{14}$	1,2,4,5-Tetramethylbenzene	-119.9		-218.5				245.6			215.1		
$C_{10}H_{14}O$	Thymol	-309.7											
$C_{10}H_{16}$	cis,cis-2,6-Dimethyl-2,4,6-octatriene		-24.0										
$C_{10}H_{16}$	Dipentene		-50.8	-2.6								249.4	
$C_{10}H_{16}$	D-Limonene		-54.5									249.0	
$C_{10}H_{16}$	β-Myrcene		14.5										
$C_{10}H_{16}$	α-Pinene		-16.4	28.3									
$C_{10}H_{16}$	β-Pinene		-7.7	38.7									
$C_{10}H_{16}$	α-Terpinene			-20.6									
$C_{10}H_{16}N_2O_8$	Glycine, N,N'-1,2-ethanediylbis[N-(carboxymethyl)]-	-1759.5											
$C_{10}H_{16}O$	Camphor	-319.4		-267.5							271.2		
$C_{10}H_{18}$	1,1'-Bicyclopentyl		-178.9										
$C_{10}H_{18}$	cis-Decahydronaphthalene		-219.4	-169.2					265.0			232.0	
$C_{10}H_{18}$	trans-Decahydronaphthalene		-230.6	-182.1					264.9			228.5	
$C_{10}H_{18}O_4$	Sebacic acid	-1082.6		-921.9									
$C_{10}H_{19}N$	Decanenitrile		-158.4	-91.5									
$C_{10}H_{20}$	1-Decene		-173.8	-123.3					425.0			300.8	
$C_{10}H_{20}$	cis-1,2-Di-tert-butylethylene		-163.6										

Molecular formula	Name	$\Delta_f H°$/kJ mol⁻¹ Crys.	Liq.	Gas	$\Delta_f G°$/kJ mol⁻¹ Crys.	Liq.	Gas	$S°$/J mol⁻¹ K⁻¹ Crys.	Liq.	Gas	C_p/J mol⁻¹ K⁻¹ Crys.	Liq.	Gas
C₁₀H₂₀	Butylcyclohexane		-263.1	-213.7					345.0			271.0	
C₁₀H₂₀O₂	Decanoic acid	-713.7	-684.3	-594.9									
C₁₀H₂₀O₂	Methyl nonanoate		-616.2	-554.2									
C₁₀H₂₁NO₂	1-Nitrodecane		-351.5										
C₁₀H₂₂	Decane		-300.9	-249.5									
C₁₀H₂₂	2-Methylnonane		-309.8	-260.2					420.1			314.4	
C₁₀H₂₂	5-Methylnonane		-307.9	-258.6					423.8			313.3	
C₁₀H₂₂O	1-Decanol		-478.1	-396.6								314.4	
C₁₀H₂₂O	Dipentyl ether											370.6	
C₁₀H₂₂O	Diisopentyl ether											250.0	
C₁₀H₂₂O₂	1,10-Decanediol	-678.9										379.0	
C₁₀H₂₂O₂	Ethylene glycol dibutyl ether												
C₁₀H₂₂S	1-Decanethiol	-309.9	-276.5	-211.5					476.1			350.0	
C₁₀H₂₂S	Dipentyl sulfide		-266.4	-204.9								350.4	
C₁₀H₂₂S	Diisopentylsulfide		-281.8	-221.5									
C₁₁H₈O₂	1-Naphthalenecarboxylic acid	-333.5		-223.1									
C₁₁H₈O₂	2-Naphthoic acid	-346.1		-232.5									
C₁₁H₁₀	1-Methylnaphthalene		56.3	106.7					254.8			224.4	
C₁₁H₁₀	2-Methylnaphthalene	44.9						220.0			196.0		
C₁₁H₁₂N₂O₂	L-Tryptophan	-415.3						251.0			238.1		
C₁₁H₁₄	1,1-Dimethylindan		-53.6	-1.6									
C₁₁H₁₆	Pentamethylbenzene	-144.6		-67.2									
C₁₁H₂₀	Spiro[5.5]undecane		-244.5	-188.3									
C₁₁H₂₂	1-Undecene												
C₁₁H₂₂O₂	Methyl decanoate		-640.5	-573.8								344.9	
C₁₁H₂₄	Undecane		-327.2	-270.8								344.9	
C₁₁H₂₄O	1-Undecanol		-504.8										
C₁₂F₂₇N	Trinonafluorobutylamine											418.4	
C₁₂H₈	Acenaphthylene	186.7		259.7									
C₁₂H₈N₂	Phenazine	237.0		328.8									
C₁₂H₈O	Dibenzofuran	-5.3		83.4									
C₁₂H₈S	Dibenzothiophene	120.0		205.1									
C₁₂H₈S₂	Thianthrene	182.0		286.0									
C₁₂H₉N	Carbazole	101.7		200.7								166.4	
C₁₂H₁₀	Acenaphthene	70.3		156.0				188.9			190.4		
C₁₂H₁₀	Biphenyl	99.4		181.4				209.4			198.4		
C₁₂H₁₀N₂O	trans-Azoxybenzene	243.4		342.0									
C₁₂H₁₀N₂O	N-Nitrosodiphenylamine	227.2											
C₁₂H₁₀O	Diphenyl ether	-32.1	-14.9	52.0				233.9			216.6		
C₁₂H₁₀O₂	1-Naphthaleneacetic acid	-359.2											
C₁₂H₁₀O₂	2-Naphthaleneacetic acid	-371.9											
C₁₂H₁₁N	2-Aminobiphenyl	93.8		184.4									
C₁₂H₁₁N	p-Biphenylamine	81.0											
C₁₂H₁₁N	Diphenylamine	130.2		219.3									
C₁₂H₁₂N₂	p-Benzidine	70.7											

Molecular formula	Name	$\Delta_f H°$/kJ mol⁻¹			$\Delta_f G°$/kJ mol⁻¹			$S°$/J mol⁻¹ K⁻¹			C_p/J mol⁻¹ K⁻¹		
		Crys.	Liq.	Gas	Crys.	Liq.	Gas	Crys.	Liq.	Gas	Crys.	Liq.	Gas
C₁₂H₁₄O₄	Diethyl phthalate		-776.6	-688.4					425.1			366.1	
C₁₂H₁₆	Cyclohexylbenzene		-76.6	-16.7									
C₁₂H₁₇NO₄	Diethyl 3,5-dimethyl-pyrrole-2,4-dicarboxylate	-916.7											
C₁₂H₁₈	3,9-Dodecadiyne		197.8										
C₁₂H₁₈	5,7-Dodecadiyne		181.5										
C₁₂H₁₈	Hexamethylbenzene	-162.4		-77.4				306.3			245.6		
C₁₂H₂₂	Cyclohexylcyclohexane		-273.7	-215.7									
C₁₂H₂₂O₄	Dodecanedioic acid	-1130.0											
C₁₂H₂₂O₁₁	Sucrose	-2226.1											
C₁₂H₂₂O₁₁	β-Lactose	-2236.7											
C₁₂H₂₄	1-Dodecene		-226.2	-165.4									
C₁₂H₂₄O₂	Dodecanoic acid	-774.6	-737.9	-642.0					484.8		404.3	360.7	
C₁₂H₂₄O₂	Methyl undecanoate		-665.2	-593.8									
C₁₂H₂₄O₁₂	α-Lactose monohydrate	-2484.1											
C₁₂H₂₅Br	1-Bromododecane		-344.7	-269.9									
C₁₂H₂₅Cl	1-Chlorododecane		-392.3	-321.1									
C₁₂H₂₆	Dodecane		-350.9	-289.4								375.8	
C₁₂H₂₆O	1-Dodecanol		-528.5	-436.6								438.1	
C₁₂H₂₆O₃	Diethylene glycol dibutyl ether		-281.6									452.0	
C₁₂H₂₇N	Tributylamine	-191.5											
C₁₃H₈O₂	Xanthone	179.4											
C₁₃H₉N	Acridine	141.9		273.9									
C₁₃H₉N	Phenanthridine	150.6		240.5									
C₁₃H₉N	Benzo[f]quinoline	159.2		233.7									
C₁₃H₁₀N₂	9-Acridinamine												
C₁₃H₁₀O	Benzophenone	-34.5		54.9									
C₁₃H₁₁N	9-Methylcarbazole	105.5		201.0									
C₁₃H₁₂	Diphenylmethane	71.5	89.7	139.0				239.3					
C₁₃H₂₄O₄	Tridecanedioic acid	-1148.3											
C₁₃H₂₆	1-Tridecene												
C₁₃H₂₆O₂	Methyl dodecanoate		-693.0	-614.9								391.8	
C₁₃H₂₈	Tridecane											406.7	
C₁₃H₂₈O	1-Tridecanol	-599.4											
C₁₄H₈O₂	9,10-Anthracenedione	-188.5		-75.7									
C₁₄H₈O₂	9,10-Phenanthrenedione	-154.7		-46.6									
C₁₄H₈O₄	1,4-Dihydroxy-9,10-anthracenedione	-595.8		-471.7									
C₁₄H₁₀	Anthracene	129.2		230.9				207.5			210.5		
C₁₄H₁₀	Phenanthrene	116.2		207.5				215.1			220.6		
C₁₄H₁₀	Diphenylacetylene	312.4									225.9		
C₁₄H₁₀O₂	Benzil	-153.9		-55.5									
C₁₄H₁₂	cis-Stilbene		183.3	252.3									
C₁₄H₁₂	trans-Stilbene	136.9		236.1									

| Molecular formula | Name | $\Delta_f H°$/kJ mol^{-1} | | | $\Delta_f G°$/kJ mol^{-1} | | | $S°$/J mol^{-1} K^{-1} | | | C_p/J mol^{-1} K^{-1} | | |
		Crys.	Liq.	Gas	Crys.	Liq.	Gas	Crys.	Liq.	Gas	Crys.	Liq.	Gas
C$_{14}$H$_{14}$	1,1-Diphenylethane		48.7	142.9									
C$_{14}$H$_{14}$	1,2-Diphenylethane	51.5											
C$_{14}$H$_{23}$N$_3$O$_{10}$	Glycine, *N,N*-bis[2-[bis(carboxymethyl)amino]ethyl]	-2225.2											
C$_{14}$H$_{27}$N	Tetradecanenitrile		-260.2	-174.9									
C$_{14}$H$_{28}$O$_2$	Tetradecanoic acid	-833.5	-788.8	-693.7									
C$_{14}$H$_{28}$O$_2$	Methyl tridecanoate		-717.9	-635.3							432.0		
C$_{14}$H$_{30}$O	1-Tetradecanol	-629.6	-580.6								388.0		
C$_{15}$H$_{16}$O$_2$	2,2-Bis(4-hydroxyphenyl)propane	-368.6											
C$_{15}$H$_{30}$	Decylcyclopentane		-367.3										
C$_{15}$H$_{30}$O$_2$	Pentadecanoic acid	-861.7	-811.7	-699.0									
C$_{15}$H$_{30}$O$_2$	Methyl tetradecanoate		-743.9	-656.9							443.3		
C$_{15}$H$_{32}$O	1-Pentadecanol	-658.2											
C$_{16}$H$_{10}$	Fluoranthene	189.9		289.0				230.6			230.2		
C$_{16}$H$_{10}$	Pyrene	125.5		225.7				224.9			229.7		
C$_{16}$H$_{22}$O$_4$	Dibutyl phthalate		-842.6	-750.9									
C$_{16}$H$_{22}$O$_{11}$	α-D-Glucose pentaacetate	-2249.4											
C$_{16}$H$_{22}$O$_{11}$	β-D-Glucose pentaacetate	-2232.6											
C$_{16}$H$_{26}$	Decylbenzene		-218.3	-138.6									
C$_{16}$H$_{32}$	1-Hexadecene		-328.7	-248.4								488.9	
C$_{16}$H$_{32}$O$_2$	Hexadecanoic acid	-891.5	-838.1	-737.1				452.4	587.9		460.7		
C$_{16}$H$_{32}$O$_2$	Methyl pentadecanoate		-771.0	-680.0									
C$_{16}$H$_{33}$Br	1-Bromohexadecane		-444.5	-350.2									
C$_{16}$H$_{34}$	Hexadecane		-456.1	-374.8								501.6	
C$_{16}$H$_{34}$O	1-Hexadecanol	-686.5		-517.0							422.0		
C$_{16}$H$_{36}$IN	Tetrabutyl ammonium iodide	-498.6											
C$_{17}$H$_{34}$O$_2$	Margaric acid	-924.4	-865.6								475.7		
C$_{18}$H$_{12}$	Benz[a]anthracene	170.8		293.0									
C$_{18}$H$_{12}$	Chrysene	145.3		269.8									
C$_{18}$H$_{15}$N	Triphenylamine	234.7		326.8									
C$_{18}$H$_{34}$O$_2$	Oleic acid												
C$_{18}$H$_{34}$O$_4$	Dibutyl sebacate											577.0	
C$_{18}$H$_{36}$O$_2$	Stearic acid	-947.7	-884.7	-781.2							501.5	619.0	
C$_{18}$H$_{37}$Cl	1-Chlorooctadecane		-544.1	-446.0									
C$_{18}$H$_{38}$	Octadecane	-567.4		-414.6				480.2			485.6		
C$_{18}$H$_{39}$N	Trihexylamine		-433.0										
C$_{19}$H$_{16}$O	Triphenylmethanol	-2.5											
C$_{19}$H$_{36}$O$_2$	Methyl oleate		-734.5	-649.9									
C$_{19}$H$_{36}$O$_2$	Methyl *trans*-9-octadecenoate		-737.0										
C$_{20}$H$_{12}$	Perylene	182.8						264.6			274.9		
C$_{20}$H$_{14}$O$_4$	1,2-Benzenedicarboxylic acid, diphenyl ester	-489.2											
C$_{20}$H$_{38}$O$_2$	Ethyl *cis*-9-octadecenoate		-775.8										

Molecular formula	Name	$\Delta_f H°$/kJ mol⁻¹			$\Delta_f G°$/kJ mol⁻¹			$S°$/J mol⁻¹ K⁻¹			C_p/J mol⁻¹ K⁻¹		
		Crys.	Liq.	Gas	Crys.	Liq.	Gas	Crys.	Liq.	Gas	Crys.	Liq.	Gas
$C_{20}H_{38}O_2$	Ethyl *trans*-9-octa-decenoate		-773.3										
$C_{20}H_{40}O_2$	Arachidic acid	-1011.9	-940.0	-812.4							545.1		
$C_{22}H_{42}O_2$	Brassidic acid	-960.7											
$C_{22}H_{42}O_2$	Butyl oleate		-816.9										
$C_{22}H_{44}O_2$	Butyl stearate												
$C_{24}H_{38}O_4$	Bis(2-ethylhexyl) phthalate											704.7	
$C_{24}H_{51}N$	Trioctylamine		-585.0										
$C_{26}H_{18}$	9,10-Diphenylanthracene	308.7		465.6									
$C_{26}H_{54}$	5-Butyldocosane		-713.5	-587.6									
$C_{26}H_{54}$	11-Butyldocosane		-716.0	-593.4									
$C_{28}H_{18}$	9,9'-Bianthracene	326.2		454.3									
$C_{31}H_{64}$	11-Decylheneicosane		-848.0	-705.8									
$C_{32}H_{66}$	Dotriacontane		-968.3	-697.2									
C_{60}	Fullerene-C_{60}	2327.0		2502.0	2302.0		2442.0	426.0		544.0	520.0		512.0
C_{70}	Fullerene-C_{70}	2555.0		2755.0	2537.0		2692.0	464.0		614.0	650.0		585.0

THERMODYNAMIC PROPERTIES AS A FUNCTION OF TEMPERATURE

L. V. Gurvich, V. S. Iorish, V. S. Yungman, and O. V. Dorofeeva

The thermodynamic properties $C_p^\circ(T)$, $S^\circ(T)$, $H^\circ(T)-H^\circ(T_r)$, $-[G^\circ(T)-H^\circ(T_r)]/T$ and formation properties $\Delta_f H^\circ(T)$, $\Delta_f G^\circ(T)$, $\log K_f^\circ(T)$ are tabulated as functions of temperature in the range 298.15 to 1500 K for 80 substances in the standard state. The reference temperature, T_r, is equal to 298.15 K. The standard state pressure is taken as 1 bar (100,000 Pa). The tables are presented in the JANAF Thermochemical Tables format (Reference 2). The numerical data are extracted from IVTANTHERMO databases except for C_2H_4O, C_3H_6O, C_6H_6, C_6H_6O, $C_{10}H_8$, and CH_5N, which are based upon TRC Tables. See the references for information on standard states and other details.

REFERENCES

1. Gurvich, L. V., Veyts, I. V., and Alcock, C. B., Eds., *Thermodynamic Properties of Individual Substances, 4th ed.*, Hemisphere Publishing Corp., New York, 1989.
2. Chase, M. W., et al., *JANAF Thermochemical Tables, 3rd ed., J. Phys. Chem. Ref. Data*, 14, Suppl. 1, 1985.

Order of Listing of Tables

No.	Formula	Name	State	No.	Formula	Name	State
1	Ar	Argon	g	41	$CuCl_2$	Copper dichloride	cr, l
2	Br	Bromine	g	42	$CuCl_2$	Copper dichloride	g
3	Br_2	Dibromine	g	43	F	Fluorine	g
4	BrH	Hydrogen bromide	g	44	F_2	Difluorine	g
5	C	Carbon (graphite)	cr	45	FH	Hydrogen fluoride	g
6	C	Carbon (diamond)	cr	46	Ge	Germanium	cr, l
7	C_2	Dicarbon	g	47	Ge	Germanium	g
8	C_3	Tricarbon	g	48	GeO_2	Germanium dioxide	cr, l
9	CO	Carbon oxide	g	49	$GeCl_4$	Germanium tetrachloride	g
10	CO_2	Carbon dioxide	g	50	H	Hydrogen	g
11	CH_4	Methane	g	51	H_2	Dihydrogen	g
12	C_2H_2	Acetylene	g	52	HO	Hydroxyl	g
13	C_2H_4	Ethylene	g	53	H_2O	Water	l
14	C_2H_6	Ethane	g	54	H_2O	Water	g
15	C_3H_6	Cyclopropane	g	55	I	Iodine	g
16	C_3H_8	Propane	g	56	I_2	Diiodine	cr, l
17	C_6H_6	Benzene	l	57	I_2	Diiodine	g
18	C_6H_6	Benzene	g	58	IH	Hydrogen iodide	g
19	$C_{10}H_8$	Naphthalene	cr, l	59	K	Potassium	cr, l
20	$C_{10}H_8$	Naphthalene	g	60	K	Potassium	g
21	CH_2O	Formaldehyde	g	61	K_2O	Dipotassium oxide	cr, l
22	CH_4O	Methanol	g	62	KOH	Potassium hydroxide	cr, l
23	C_2H_4O	Acetaldehyde	g	63	KOH	Potassium hydroxide	g
24	C_2H_6O	Ethanol	g	64	KCl	Potassium chloride	cr, l
25	$C_2H_4O_2$	Acetic acid	g	65	KCl	Potassium chloride	g
26	C_3H_6O	Acetone	g	66	N_2	Dinitrogen	g
27	C_6H_6O	Phenol	g	67	NO	Nitric oxide	g
28	CF_4	Carbon tetrafluoride	g	68	NO_2	Nitrogen dioxide	g
29	CHF_3	Trifluoromethane	g	69	NH_3	Ammonia	g
30	$CClF_3$	Chlorotrifluoromethane	g	70	O	Oxygen	g
31	CCl_2F_2	Dichlorodifluoromethane	g	71	O_2	Dioxygen	g
32	$CHClF_2$	Chlorodifluoromethane	g	72	S	Sulfur	cr, l
33	CH_5N	Methylamine	g	73	S	Sulfur	g
34	Cl	Chlorine	g	74	S_2	Disulfur	g
35	Cl_2	Dichlorine	g	75	S_8	Octasulfur	g
36	ClH	Hydrogen chloride	g	76	SO_2	Sulfur dioxide	g
37	Cu	Copper	cr, l	77	Si	Silicon	cr
38	Cu	Copper	g	78	Si	Silicon	g
39	CuO	Copper oxide	cr	79	SiO_2	Silicon dioxide	cr
40	Cu_2O	Dicopper oxide	cr	80	$SiCl_4$	Silicon tetrachloride	g

T/K	C_p°	S°	-(G°-H°(T_r))/T	H°-H°(T_r)	Δ_fH°	Δ_fG°	Log K_f
	J/K·mol			kJ/mol			

1. ARGON   Ar (g)

T/K	$C_p°$	$S°$	$-(G°-H°(T_r))/T$	$H°-H°(T_r)$	$\Delta_f H°$	$\Delta_f G°$	$\text{Log } K_f$
298.15	20.786	154.845	154.845	0.000	0.000	0.000	0.000
300	20.786	154.973	154.845	0.038	0.000	0.000	0.000
400	20.786	160.953	155.660	2.117	0.000	0.000	0.000
500	20.786	165.591	157.200	4.196	0.000	0.000	0.000
600	20.786	169.381	158.924	6.274	0.000	0.000	0.000
700	20.786	172.585	160.653	8.353	0.000	0.000	0.000
800	20.786	175.361	162.322	10.431	0.000	0.000	0.000
900	20.786	177.809	163.909	12.510	0.000	0.000	0.000
1000	20.786	179.999	165.410	14.589	0.000	0.000	0.000
1100	20.786	181.980	166.828	16.667	0.000	0.000	0.000
1200	20.786	183.789	168.167	18.746	0.000	0.000	0.000
1300	20.786	185.453	169.434	20.824	0.000	0.000	0.000
1400	20.786	186.993	170.634	22.903	0.000	0.000	0.000
1500	20.786	188.427	171.773	24.982	0.000	0.000	0.000

2. BROMINE   Br (g)

T/K	$C_p°$	$S°$	$-(G°-H°(T_r))/T$	$H°-H°(T_r)$	$\Delta_f H°$	$\Delta_f G°$	$\text{Log } K_f$
298.15	20.786	175.017	175.017	0.000	111.870	82.379	−14.432
300	20.786	175.146	175.018	0.038	111.838	82.196	−14.311
400	20.787	181.126	175.833	2.117	96.677	75.460	−9.854
500	20.798	185.765	177.373	4.196	96.910	70.129	−7.326
600	20.833	189.559	179.097	6.277	97.131	64.752	−5.637
700	20.908	192.776	180.827	8.364	97.348	59.338	−4.428
800	21.027	195.575	182.499	10.461	97.568	53.893	−3.519
900	21.184	198.061	184.093	12.571	97.796	48.420	−2.810
1000	21.365	200.302	185.604	14.698	98.036	42.921	−2.242
1100	21.559	202.347	187.034	16.844	98.291	37.397	−1.776
1200	21.752	204.231	188.390	19.010	98.560	31.850	−1.386
1300	21.937	205.980	189.676	21.195	98.844	26.279	−1.056
1400	22.107	207.612	190.900	23.397	99.141	20.686	−0.772
1500	22.258	209.142	192.065	25.615	99.449	15.072	−0.525

3. DIBROMINE   Br_2 (g)

T/K	$C_p°$	$S°$	$-(G°-H°(T_r))/T$	$H°-H°(T_r)$	$\Delta_f H°$	$\Delta_f G°$	$\text{Log } K_f$
298.15	36.057	245.467	245.467	0.000	30.910	3.105	−0.544
300	36.074	245.690	245.468	0.067	30.836	2.933	−0.511
332.25	36.340	249.387	245.671	1.235		pressure = 1 bar	
400	36.729	256.169	246.892	3.711	0.000	0.000	0.000
500	37.082	264.406	249.600	7.403	0.000	0.000	0.000
600	37.305	271.188	252.650	11.123	0.000	0.000	0.000
700	37.464	276.951	255.720	14.862	0.000	0.000	0.000
800	37.590	281.962	258.694	18.615	0.000	0.000	0.000
900	37.697	286.396	261.530	22.379	0.000	0.000	0.000
1000	37.793	290.373	264.219	26.154	0.000	0.000	0.000
1100	37.883	293.979	266.763	29.938	0.000	0.000	0.000
1200	37.970	297.279	269.170	33.730	0.000	0.000	0.000
1300	38.060	300.322	271.451	37.532	0.000	0.000	0.000
1400	38.158	303.146	273.615	41.343	0.000	0.000	0.000
1500	38.264	305.782	275.673	45.164	0.000	0.000	0.000

4. HYDROGEN BROMIDE   HBr (g)

T/K	$C_p°$	$S°$	$-(G°-H°(T_r))/T$	$H°-H°(T_r)$	$\Delta_f H°$	$\Delta_f G°$	$\text{Log } K_f$
298.15	29.141	198.697	198.697	0.000	−36.290	−53.360	9.348
300	29.141	198.878	198.698	0.054	−36.333	−53.466	9.309
400	29.220	207.269	199.842	2.971	−52.109	−55.940	7.305
500	29.454	213.811	202.005	5.903	−52.484	−56.854	5.939
600	29.872	219.216	204.436	8.868	−52.844	−57.694	5.023
700	30.431	223.861	206.886	11.882	−53.168	−58.476	4.363
800	31.063	227.965	209.269	14.957	−53.446	−59.214	3.866

	J/K·mol			kJ/mol			
T/K	C_p°	S°	$-(G^\circ - H^\circ(T_r))/T$	$H^\circ - H^\circ(T_r)$	$\Delta_f H^\circ$	$\Delta_f G^\circ$	Log K_f

4. HYDROGEN BROMIDE   HBr (g) (continued)

900	31.709	231.661	211.555	18.095	−53.677	−59.921	3.478
1000	32.335	235.035	213.737	21.298	−53.864	−60.604	3.166
1100	32.919	238.145	215.816	24.561	−54.012	−61.271	2.909
1200	33.454	241.032	217.799	27.880	−54.129	−61.925	2.696
1300	33.938	243.729	219.691	31.250	−54.220	−62.571	2.514
1400	34.374	246.261	221.499	34.666	−54.291	−63.211	2.358
1500	34.766	248.646	223.230	38.123	−54.348	−63.846	2.223

5. CARBON (GRAPHITE)   C (cr; graphite)

298.15	8.536	5.740	5.740	0.000	0.000	0.000	0.000
300	8.610	5.793	5.740	0.016	0.000	0.000	0.000
400	11.974	8.757	6.122	1.054	0.000	0.000	0.000
500	14.537	11.715	6.946	2.385	0.000	0.000	0.000
600	16.607	14.555	7.979	3.945	0.000	0.000	0.000
700	18.306	17.247	9.113	5.694	0.000	0.000	0.000
800	19.699	19.785	10.290	7.596	0.000	0.000	0.000
900	20.832	22.173	11.479	9.625	0.000	0.000	0.000
1000	21.739	24.417	12.662	11.755	0.000	0.000	0.000
1100	22.452	26.524	13.827	13.966	0.000	0.000	0.000
1200	23.000	28.502	14.968	16.240	0.000	0.000	0.000
1300	23.409	30.360	16.082	18.562	0.000	0.000	0.000
1400	23.707	32.106	17.164	20.918	0.000	0.000	0.000
1500	23.919	33.749	18.216	23.300	0.000	0.000	0.000

6. CARBON (DIAMOND)   C (cr; diamond)

298.15	6.109	2.362	2.362	0.000	1.850	2.857	−0.501
300	6.201	2.400	2.362	0.011	1.846	2.863	−0.499
400	10.321	4.783	2.659	0.850	1.645	3.235	−0.422
500	13.404	7.431	3.347	2.042	1.507	3.649	−0.381
600	15.885	10.102	4.251	3.511	1.415	4.087	−0.356
700	17.930	12.709	5.274	5.205	1.361	4.537	−0.339
800	19.619	15.217	6.361	7.085	1.338	4.993	−0.326
900	21.006	17.611	7.479	9.118	1.343	5.450	−0.316
1000	22.129	19.884	8.607	11.277	1.372	5.905	−0.308
1100	23.020	22.037	9.731	13.536	1.420	6.356	−0.302
1200	23.709	24.071	10.842	15.874	1.484	6.802	−0.296
1300	24.222	25.990	11.934	18.272	1.561	7.242	−0.291
1400	24.585	27.799	13.003	20.714	1.646	7.675	−0.286
1500	24.824	29.504	14.047	23.185	1.735	8.103	−0.282

7. DICARBON   C$_2$ (g)

298.15	43.548	197.095	197.095	0.000	830.457	775.116	−135.795
300	43.575	197.365	197.096	0.081	830.506	774.772	−134.898
400	42.169	209.809	198.802	4.403	832.751	755.833	−98.700
500	39.529	218.924	201.959	8.483	834.170	736.423	−76.933
600	37.837	225.966	205.395	12.342	834.909	716.795	−62.402
700	36.984	231.726	208.758	16.078	835.148	697.085	−52.016
800	36.621	236.637	211.943	19.755	835.020	677.366	−44.227
900	36.524	240.943	214.931	23.411	834.618	657.681	−38.170
1000	36.569	244.793	217.728	27.065	834.012	638.052	−33.328
1100	36.696	248.284	220.349	30.728	833.252	618.492	−29.369
1200	36.874	251.484	222.812	34.406	832.383	599.006	−26.074
1300	37.089	254.444	225.133	38.104	831.437	579.596	−23.288
1400	37.329	257.201	227.326	41.824	830.445	560.261	−20.903
1500	37.589	259.785	229.405	45.570	829.427	540.997	−18.839

	J/K·mol			kJ/mol			
T/K	C_p°	S°	$-(G^\circ-H^\circ(T_r))/T$	$H^\circ-H^\circ(T_r)$	$\Delta_f H^\circ$	$\Delta_f G^\circ$	Log K_f

8. TRICARBON   C$_3$ (g)

298.15	42.202	237.611	237.611	0.000	839.958	774.249	−135.643
300	42.218	237.872	237.611	0.078	839.989	773.841	−134.736
400	43.383	250.164	239.280	4.354	841.149	751.592	−98.147
500	44.883	260.003	242.471	8.766	841.570	729.141	−76.172
600	46.406	268.322	246.104	13.331	841.453	706.659	−61.519
700	47.796	275.582	249.807	18.042	840.919	684.230	−51.057
800	48.997	282.045	253.440	22.884	840.053	661.901	−43.217
900	50.006	287.876	256.948	27.835	838.919	639.698	−37.127
1000	50.844	293.189	260.310	32.879	837.572	617.633	−32.261
1100	51.535	298.069	263.524	37.999	836.059	595.711	−28.288
1200	52.106	302.578	266.593	43.182	834.420	573.933	−24.982
1300	52.579	306.768	269.524	48.417	832.690	552.295	−22.191
1400	52.974	310.679	272.326	53.695	830.899	530.793	−19.804
1500	53.307	314.346	275.006	59.010	829.068	509.421	−17.739

9. CARBON OXIDE   CO (g)

298.15	29.141	197.658	197.658	0.000	−110.530	−137.168	24.031
300	29.142	197.838	197.659	0.054	−110.519	−137.333	23.912
400	29.340	206.243	198.803	2.976	−110.121	−146.341	19.110
500	29.792	212.834	200.973	5.930	−110.027	−155.412	16.236
600	30.440	218.321	203.419	8.941	−110.157	−164.480	14.319
700	31.170	223.067	205.895	12.021	−110.453	−173.513	12.948
800	31.898	227.277	208.309	15.175	−110.870	−182.494	11.915
900	32.573	231.074	210.631	18.399	−111.378	−191.417	11.109
1000	33.178	234.538	212.851	21.687	−111.952	−200.281	10.461
1100	33.709	237.726	214.969	25.032	−112.573	−209.084	9.928
1200	34.169	240.679	216.990	28.426	−113.228	−217.829	9.482
1300	34.568	243.430	218.920	31.864	−113.904	−226.518	9.101
1400	34.914	246.005	220.763	35.338	−114.594	−235.155	8.774
1500	35.213	248.424	222.527	38.845	−115.291	−243.742	8.488

10. CARBON DIOXIDE   CO$_2$ (g)

298.15	37.135	213.783	213.783	0.000	−393.510	−394.373	69.092
300	37.220	214.013	213.784	0.069	−393.511	−394.379	68.667
400	41.328	225.305	215.296	4.004	−393.586	−394.656	51.536
500	44.627	234.895	218.280	8.307	−393.672	−394.914	41.256
600	47.327	243.278	221.762	12.909	−393.791	−395.152	34.401
700	49.569	250.747	225.379	17.758	−393.946	−395.367	29.502
800	51.442	257.492	228.978	22.811	−394.133	−395.558	25.827
900	53.008	263.644	232.493	28.036	−394.343	−395.724	22.967
1000	54.320	269.299	235.895	33.404	−394.568	−395.865	20.678
1100	55.423	274.529	239.172	38.893	−394.801	−395.984	18.803
1200	56.354	279.393	242.324	44.483	−395.035	−396.081	17.241
1300	57.144	283.936	245.352	50.159	−395.265	−396.159	15.918
1400	57.818	288.196	248.261	55.908	−395.488	−396.219	14.783
1500	58.397	292.205	251.059	61.719	−395.702	−396.264	13.799

11. METHANE   CH$_4$ (g)

298.15	35.695	186.369	186.369	0.000	−74.600	−50.530	8.853
300	35.765	186.590	186.370	0.066	−74.656	−50.381	8.772
400	40.631	197.501	187.825	3.871	−77.703	−41.827	5.462
500	46.627	207.202	190.744	8.229	−80.520	−32.525	3.398
600	52.742	216.246	194.248	13.199	−82.969	−22.690	1.975
700	58.603	224.821	198.008	18.769	−85.023	−12.476	0.931
800	64.084	233.008	201.875	24.907	−86.693	−1.993	0.130

THERMODYNAMIC PROPERTIES AS A FUNCTION OF TEMPERATURE (continued)

	J/K·mol			kJ/mol			
T/K	C_p°	S°	$-(G^\circ-H^\circ(T_r))/T$	$H^\circ-H^\circ(T_r)$	$\Delta_f H^\circ$	$\Delta_f G^\circ$	Log K_f

11. METHANE   CH$_4$ (g) (continued)

900	69.137	240.852	205.773	31.571	−88.006	8.677	−0.504
1000	73.746	248.379	209.660	38.719	−88.996	19.475	−1.017
1100	77.919	255.607	213.511	46.306	−89.698	30.358	−1.442
1200	81.682	262.551	217.310	54.289	−90.145	41.294	−1.797
1300	85.067	269.225	221.048	62.630	−90.367	52.258	−2.100
1400	88.112	275.643	224.720	71.291	−90.390	63.231	−2.359
1500	90.856	281.817	228.322	80.242	−90.237	74.200	−2.584

12. ACETYLENE   C$_2$H$_2$ (g)

298.15	44.036	200.927	200.927	0.000	227.400	209.879	−36.769
300	44.174	201.199	200.927	0.082	227.397	209.770	−36.524
400	50.388	214.814	202.741	4.829	227.161	203.928	−26.630
500	54.751	226.552	206.357	10.097	226.846	198.154	−20.701
600	58.121	236.842	210.598	15.747	226.445	192.452	−16.754
700	60.970	246.021	215.014	21.704	225.968	186.823	−13.941
800	63.511	254.331	219.418	27.931	225.436	181.267	−11.835
900	65.831	261.947	223.726	34.399	224.873	175.779	−10.202
1000	67.960	268.995	227.905	41.090	224.300	170.355	−8.898
1100	69.909	275.565	231.942	47.985	223.734	164.988	−7.835
1200	71.686	281.725	235.837	55.067	223.189	159.672	−6.950
1300	73.299	287.528	239.592	62.317	222.676	154.400	−6.204
1400	74.758	293.014	243.214	69.721	222.203	149.166	−5.565
1500	76.077	298.218	246.709	77.264	221.774	143.964	−5.013

13. ETHYLENE   C$_2$H$_4$ (g)

298.15	42.883	219.316	219.316	0.000	52.400	68.358	−11.976
300	43.059	219.582	219.317	0.079	52.341	68.457	−11.919
400	53.045	233.327	221.124	4.881	49.254	74.302	−9.703
500	62.479	246.198	224.864	10.667	46.533	80.887	−8.450
600	70.673	258.332	229.441	17.335	44.221	87.982	−7.659
700	77.733	269.770	234.393	24.764	42.278	95.434	−7.121
800	83.868	280.559	239.496	32.851	40.655	103.142	−6.734
900	89.234	290.754	244.630	41.512	39.310	111.036	−6.444
1000	93.939	300.405	249.730	50.675	38.205	119.067	−6.219
1100	98.061	309.556	254.756	60.280	37.310	127.198	−6.040
1200	101.670	318.247	259.688	70.271	36.596	135.402	−5.894
1300	104.829	326.512	264.513	80.599	36.041	143.660	−5.772
1400	107.594	334.384	269.225	91.223	35.623	151.955	−5.669
1500	110.018	341.892	273.821	102.107	35.327	160.275	−5.581

14. ETHANE   C$_2$H$_6$ (g)

298.15	52.487	229.161	229.161	0.000	−84.000	−32.015	5.609
300	52.711	229.487	229.162	0.097	−84.094	−31.692	5.518
400	65.459	246.378	231.379	5.999	−88.988	−13.473	1.759
500	77.941	262.344	235.989	13.177	−93.238	5.912	−0.618
600	89.188	277.568	241.660	21.545	−96.779	26.086	−2.271
700	99.136	292.080	247.835	30.972	−99.663	46.800	−3.492
800	107.936	305.904	254.236	41.334	−101.963	67.887	−4.433
900	115.709	319.075	260.715	52.525	−103.754	89.231	−5.179
1000	122.552	331.628	267.183	64.445	−105.105	110.750	−5.785
1100	128.553	343.597	273.590	77.007	−106.082	132.385	−6.286
1200	133.804	355.012	279.904	90.131	−106.741	154.096	−6.708
1300	138.391	365.908	286.103	103.746	−107.131	175.850	−7.066
1400	142.399	376.314	292.178	117.790	−107.292	197.625	−7.373
1500	145.905	386.260	298.121	132.209	−107.260	219.404	−7.640

	J/K·mol			kJ/mol			
T/K	C_p°	S°	$-(G^\circ - H^\circ(T_r))/T$	$H^\circ - H^\circ(T_r)$	$\Delta_f H^\circ$	$\Delta_f G^\circ$	Log K_f

15. CYCLOPROPANE C_3H_6 (g)

298.15	55.571	237.488	237.488	0.000	53.300	104.514	−18.310
300	55.941	237.832	237.489	0.103	53.195	104.832	−18.253
400	76.052	256.695	239.924	6.708	47.967	122.857	−16.043
500	93.859	275.637	245.177	15.230	43.730	142.091	−14.844
600	108.542	294.092	251.801	25.374	40.405	162.089	−14.111
700	120.682	311.763	259.115	36.854	37.825	182.583	−13.624
800	130.910	328.564	266.755	49.447	35.854	203.404	−13.281
900	139.658	344.501	274.516	62.987	34.384	224.441	−13.026
1000	147.207	359.616	282.277	77.339	33.334	245.618	−12.830
1100	153.749	373.961	289.965	92.395	32.640	266.883	−12.673
1200	159.432	387.588	297.538	108.060	32.249	288.197	−12.545
1300	164.378	400.549	304.967	124.257	32.119	309.533	−12.437
1400	168.689	412.892	312.239	140.915	32.215	330.870	−12.345
1500	172.453	424.662	319.344	157.976	32.507	352.193	−12.264

16. PROPANE C_3H_8 (g)

298.15	73.597	270.313	270.313	0.000	−103.847	−23.458	4.110
300	73.931	270.769	270.314	0.136	−103.972	−22.959	3.997
400	94.014	294.739	273.447	8.517	−110.33	15.029	−0.657
500	112.591	317.768	280.025	18.872	−115.658	34.507	−3.605
600	128.700	339.753	288.162	30.955	−119.973	64.961	−5.655
700	142.674	360.668	297.039	44.540	−123.384	96.065	−7.168
800	154.766	380.528	306.245	59.427	−126.016	127.603	−8.331
900	165.352	399.381	315.555	75.444	−127.982	159.430	−9.253
1000	174.598	417.293	324.841	92.452	−129.380	191.444	−10.000
1100	182.673	434.321	334.026	110.325	−130.296	223.574	−10.617
1200	189.745	450.526	343.064	128.954	−130.802	255.770	−11.133
1300	195.853	465.961	351.929	148.241	−130.961	287.993	−11.572
1400	201.209	480.675	360.604	168.100	−130.829	320.217	−11.947
1500	205.895	494.721	369.080	188.460	−130.445	352.422	−12.272

17. BENZENE C_6H_6 (l)

298.15	135.950	173.450	173.450	0.000	49.080	124.521	−21.815
300	136.312	174.292	173.453	.252	49.077	124.989	−21.762
400	161.793	216.837	179.082	15.102	48.978	150.320	−19.630
500	207.599	257.048	190.639	33.204	50.330	175.559	−18.340

18. BENZENE C_6H_6 (g)

298.15	82.430	269.190	269.190	0.000	82.880	129.750	−22.731
300	83.020	269.700	269.190	0.153	82.780	130.040	−22.641
400	113.510	297.840	272.823	10.007	77.780	146.570	−19.140
500	139.340	326.050	280.658	22.696	73.740	164.260	−17.160
600	160.090	353.360	290.517	37.706	70.490	182.680	−15.903
700	176.790	379.330	301.360	54.579	67.910	201.590	−15.042
800	190.460	403.860	312.658	72.962	65.910	220.820	−14.418
900	201.840	426.970	324.084	92.597	64.410	240.280	−13.945
1000	211.430	448.740	335.473	113.267	63.340	259.890	−13.575
1100	219.580	469.280	346.710	134.827	62.620	277.640	−13.184
1200	226.540	488.690	357.743	157.137	62.200	299.320	−13.029
1300	232.520	507.070	368.534	180.097	62.000	319.090	−12.821
1400	237.680	524.490	379.056	203.607	61.990	338.870	−12.643
1500	242.140	541.040	389.302	227.607	62.110	358.640	−12.489

T/K	C_p°	S°	$-(G^\circ - H^\circ(T_r))/T$	$H^\circ - H^\circ(T_r)$	$\Delta_f H^\circ$	$\Delta_f G^\circ$	Log K_f
		J/K·mol			kJ/mol		

19. NAPHTHALENE $C_{10}H_8$ (cr, l)

T/K	C_p°	S°	$-(G^\circ - H^\circ(T_r))/T$	$H^\circ - H^\circ(T_r)$	$\Delta_f H^\circ$	$\Delta_f G^\circ$	Log K_f
298.15	165.720	167.390	167.390	0.000	78.530	201.585	-35.316
300	167.001	168.419	167.393	0.308	78.466	202.349	-35.232
353.43	208.722	198.948	169.833	10.290	96.099	224.543	-33.186

PHASE TRANSITION: $\Delta_{trs} H$ = 18.980 kJ/mol, $\Delta_{trs} S$ = 53.702 J/K·mol, cr–l

T/K	C_p°	S°	$-(G^\circ - H^\circ(T_r))/T$	$H^\circ - H^\circ(T_r)$	$\Delta_f H^\circ$	$\Delta_f G^\circ$	Log K_f
353.43	217.200	252.650	169.833	29.270	96.099	224.543	-33.186
400	241.577	280.916	181.124	39.917	96.067	241.475	-31.533
470	276.409	322.712	199.114	58.091	97.012	266.859	-29.658

20. NAPHTHALENE $C_{10}H_8$ (g)

T/K	C_p°	S°	$-(G^\circ - H^\circ(T_r))/T$	$H^\circ - H^\circ(T_r)$	$\Delta_f H^\circ$	$\Delta_f G^\circ$	Log K_f
298.15	131.920	333.150	333.150	0.000	150.580	224.100	-39.260
300	132.840	333.970	333.157	0.244	150.450	224.560	-39.098
400	180.070	378.800	338.950	15.940	144.190	250.270	-32.681
500	219.740	423.400	351.400	36.000	139.220	277.340	-28.973
600	251.530	466.380	367.007	59.624	135.350	305.330	-26.581
700	277.010	507.140	384.146	86.096	132.330	333.950	-24.919
800	297.730	545.520	401.935	114.868	130.050	362.920	-23.696
900	314.850	581.610	419.918	145.523	128.430	392.150	-22.759
1000	329.170	615.550	437.806	177.744	127.510	421.700	-22.027
1100	341.240	647.500	455.426	211.281	127.100	450.630	-21.398
1200	351.500	677.650	472.707	245.932	126.960	480.450	-20.913
1300	360.260	706.130	489.568	281.531	127.060	509.770	-20.482
1400	367.780	733.110	506.009	317.941	127.390	539.740	-20.137
1500	374.270	758.720	522.019	355.051	127.920	568.940	-19.812

21. FORMALDEHYDE H_2CO (g)

T/K	C_p°	S°	$-(G^\circ - H^\circ(T_r))/T$	$H^\circ - H^\circ(T_r)$	$\Delta_f H^\circ$	$\Delta_f G^\circ$	Log K_f
298.15	35.387	218.760	218.760	0.000	-108.700	-102.667	17.987
300	35.443	218.979	218.761	0.066	-108.731	-102.630	17.869
400	39.240	229.665	220.192	3.789	-110.438	-100.340	13.103
500	43.736	238.900	223.028	7.936	-112.073	-97.623	10.198
600	48.181	247.270	226.381	12.534	-113.545	-94.592	8.235
700	52.280	255.011	229.924	17.560	-114.833	-91.328	6.815
800	55.941	262.236	233.517	22.975	-115.942	-87.893	5.739
900	59.156	269.014	237.088	28.734	-116.889	-84.328	4.894
1000	61.951	275.395	240.603	34.792	-117.696	-80.666	4.213
1100	64.368	281.416	244.042	41.111	-118.382	-76.929	3.653
1200	66.453	287.108	247.396	47.655	-118.966	-73.134	3.183
1300	68.251	292.500	250.660	54.392	-119.463	-69.294	2.784
1400	69.803	297.616	253.833	61.297	-119.887	-65.418	2.441
1500	71.146	302.479	256.915	68.346	-120.249	-61.514	2.142

22. METHANOL CH_3OH (g)

T/K	C_p°	S°	$-(G^\circ - H^\circ(T_r))/T$	$H^\circ - H^\circ(T_r)$	$\Delta_f H^\circ$	$\Delta_f G^\circ$	Log K_f
298.15	44.101	239.865	239.865	0.000	-201.000	-162.298	28.434
300	44.219	240.139	239.866	0.082	-201.068	-162.057	28.216
400	51.713	253.845	241.685	4.864	-204.622	-148.509	19.393
500	59.800	266.257	245.374	10.442	-207.750	-134.109	14.010
600	67.294	277.835	249.830	16.803	-210.387	-119.125	10.371
700	73.958	288.719	254.616	23.873	-212.570	-103.737	7.741
800	79.838	298.987	259.526	31.569	-214.350	-88.063	5.750
900	85.025	308.696	264.455	39.817	-215.782	-72.188	4.190
1000	89.597	317.896	269.343	48.553	-216.916	-56.170	2.934
1100	93.624	326.629	274.158	57.718	-217.794	-40.050	1.902
1200	97.165	334.930	278.879	67.262	-218.457	-23.861	1.039
1300	100.277	342.833	283.497	77.137	-218.936	-7.624	0.306
1400	103.014	350.367	288.007	87.304	-219.261	8.644	-0.322
1500	105.422	357.558	292.405	97.729	-219.456	24.930	-0.868

T/K	$C_p°$	$S°$	$-(G°-H°(T_r))/T$	$H°-H°(T_r)$	$\Delta_f H°$	$\Delta_f G°$	Log K_f
	J/K·mol			kJ/mol			

23. ACETALDEHYDE C$_2$H$_4$O (g)

T/K	$C_p°$	$S°$	$-(G°-H°(T_r))/T$	$H°-H°(T_r)$	$\Delta_f H°$	$\Delta_f G°$	Log K_f
298.15	55.318	263.840	263.840	0.000	−166.190	−133.010	23.302
300	55.510	264.180	263.837	0.103	−166.250	−132.800	23.122
400	66.282	281.620	266.147	6.189	−169.530	−121.130	15.818
500	76.675	297.540	270.850	13.345	−172.420	−108.700	11.356
600	85.942	312.360	276.550	21.486	−174.870	−95.720	8.334
700	94.035	326.230	282.667	30.494	−176.910	−82.350	6.145
800	101.070	339.260	288.938	40.258	−178.570	−68.730	4.487
900	107.190	351.520	295.189	50.698	−179.880	−54.920	3.187
1000	112.490	363.100	301.431	61.669	−180.850	−40.930	2.138
1100	117.080	374.040	307.537	73.153	−181.560	−27.010	1.283
1200	121.060	384.400	313.512	85.065	−182.070	−12.860	0.560
1300	124.500	394.230	319.350	97.344	−182.420	1.240	−0.050
1400	127.490	403.570	325.031	109.954	−182.640	15.470	−0.577
1500	130.090	412.460	330.571	122.834	−182.750	29.580	−1.030

24. ETHANOL   C$_2$H$_5$OH (g)

T/K	$C_p°$	$S°$	$-(G°-H°(T_r))/T$	$H°-H°(T_r)$	$\Delta_f H°$	$\Delta_f G°$	Log K_f
298.15	65.652	281.622	281.622	0.000	−234.800	−167.874	29.410
300	65.926	282.029	281.623	0.122	−234.897	−167.458	29.157
400	81.169	303.076	284.390	7.474	−239.826	−144.216	18.832
500	95.400	322.750	290.115	16.318	−243.940	−119.820	12.517
600	107.656	341.257	297.112	26.487	−247.260	−94.672	8.242
700	118.129	358.659	304.674	37.790	−249.895	−69.023	5.151
800	127.171	375.038	312.456	50.065	−251.951	−43.038	2.810
900	135.049	390.482	320.276	63.185	−253.515	−16.825	0.976
1000	141.934	405.075	328.033	77.042	−254.662	9.539	−0.498
1100	147.958	418.892	335.670	91.543	−255.454	36.000	−1.709
1200	153.232	431.997	343.156	106.609	−255.947	62.520	−2.721
1300	157.849	444.448	350.473	122.168	−256.184	89.070	−3.579
1400	161.896	456.298	357.612	138.160	−256.206	115.630	−4.314
1500	165.447	467.591	364.571	154.531	−256.044	142.185	−4.951

25. ACETIC ACID   C$_2$H$_4$O$_2$ (g)

T/K	$C_p°$	$S°$	$-(G°-H°(T_r))/T$	$H°-H°(T_r)$	$\Delta_f H°$	$\Delta_f G°$	Log K_f
298.15	63.438	283.470	283.470	0.000	−432.249	−374.254	65.567
300	63.739	283.863	283.471	0.118	−432.324	−373.893	65.100
400	79.665	304.404	286.164	7.296	−436.006	−353.840	46.206
500	93.926	323.751	291.765	15.993	−438.875	−332.950	34.783
600	106.181	341.988	298.631	26.014	−440.993	−311.554	27.123
700	116.627	359.162	306.064	37.169	−442.466	−289.856	21.629
800	125.501	375.331	313.722	49.287	−443.395	−267.985	17.497
900	132.989	390.558	321.422	62.223	−443.873	−246.026	14.279
1000	139.257	404.904	329.060	75.844	−443.982	−224.034	11.702
1100	144.462	418.429	336.576	90.039	−443.798	−202.046	9.594
1200	148.760	431.189	343.933	104.707	−443.385	−180.086	7.839
1300	152.302	443.240	351.113	119.765	−442.795	−158.167	6.355
1400	155.220	454.637	358.105	135.146	−442.071	−136.299	5.085
1500	157.631	465.432	364.903	150.793	−441.247	−114.486	3.987

26. ACETONE   C$_3$H$_6$O (g)

T/K	$C_p°$	$S°$	$-(G°-H°(T_r))/T$	$H°-H°(T_r)$	$\Delta_f H°$	$\Delta_f G°$	Log K_f
298.15	74.517	295.349	295.349	0.000	−217.150	−152.716	26.757
300	74.810	295.809	295.349	0.138	−217.233	−152.339	26.521
400	91.755	319.658	298.498	8.464	−222.212	−129.913	16.962
500	107.864	341.916	304.988	18.464	−226.522	−106.315	11.107
600	122.047	362.836	312.873	29.978	−230.120	−81.923	7.133
700	134.306	382.627	321.470	42.810	−233.049	−56.986	4.252
800	144.934	401.246	330.265	56.785	−235.350	−31.673	2.068

T/K	J/K·mol			kJ/mol			
	C_p°	S°	$-(G^\circ-H^\circ(T_r))/T$	$H^\circ-H^\circ(T_r)$	$\Delta_f H^\circ$	$\Delta_f G^\circ$	Log K_f

26. ACETONE C_3H_6O (g) (continued)

T/K	C_p°	S°	$-(G^\circ-H^\circ(T_r))/T$	$H^\circ-H^\circ(T_r)$	$\Delta_f H^\circ$	$\Delta_f G^\circ$	Log K_f
900	154.097	418.860	339.141	71.747	−237.149	−6.109	0.353
1000	162.046	435.513	347.950	87.563	−238.404	19.707	−1.030
1100	168.908	451.286	356.617	104.136	−239.283	45.396	−2.157
1200	174.891	466.265	365.155	121.332	−239.827	71.463	−3.110
1300	180.079	480.491	373.513	139.072	−240.120	97.362	−3.912
1400	184.556	493.963	381.596	157.314	−240.203	123.470	−4.607
1500	188.447	506.850	389.533	175.975	−240.120	149.369	−5.202

27. PHENOL C_6H_6O (g)

T/K	C_p°	S°	$-(G^\circ-H^\circ(T_r))/T$	$H^\circ-H^\circ(T_r)$	$\Delta_f H^\circ$	$\Delta_f G^\circ$	Log K_f
298.15	103.220	314.810	314.810	0.000	−96.400	−32.630	5.720
300	103.860	315.450	314.810	0.192	−96.490	−32.230	5.610
400	135.790	349.820	319.278	12.217	−100.870	−10.180	1.330
500	161.910	383.040	328.736	27.152	−104.240	12.970	−1.360
600	182.480	414.450	340.430	44.412	−106.810	36.650	−3.190
700	198.840	443.860	353.134	63.508	−108.800	60.750	−4.530
800	212.140	471.310	366.211	84.079	−110.300	85.020	−5.550
900	223.190	496.950	379.327	105.861	−111.370	109.590	−6.360
1000	232.490	520.960	392.302	128.658	−111.990	134.280	−7.010
1100	240.410	543.500	405.033	152.314	−112.280	158.620	−7.530
1200	247.200	564.720	417.468	176.703	−112.390	183.350	−7.980
1300	253.060	584.740	429.568	201.723	−112.330	208.070	−8.360
1400	258.120	603.680	441.331	227.288	−112.120	233.050	−8.700
1500	262.520	621.650	452.767	253.325	−111.780	257.540	−8.970

28. CARBON TETRAFLUORIDE CF_4 (g)

T/K	C_p°	S°	$-(G^\circ-H^\circ(T_r))/T$	$H^\circ-H^\circ(T_r)$	$\Delta_f H^\circ$	$\Delta_f G^\circ$	Log K_f
298.15	61.050	261.455	261.455	0.000	−933.200	−888.518	155.663
300	61.284	261.833	261.456	0.113	−933.219	−888.240	154.654
400	72.399	281.057	264.001	6.822	−933.986	−873.120	114.016
500	80.713	298.153	269.155	14.499	−934.372	−857.852	89.618
600	86.783	313.434	275.284	22.890	−934.490	−842.533	73.348
700	91.212	327.162	281.732	31.801	−934.431	−827.210	61.726
800	94.479	339.566	288.199	41.094	−934.261	−811.903	53.011
900	96.929	350.842	294.542	50.670	−934.024	−796.622	46.234
1000	98.798	361.156	300.695	60.460	−933.745	−781.369	40.814
1100	100.250	370.643	306.629	70.416	−933.442	−766.146	36.381
1200	101.396	379.417	312.334	80.500	−933.125	−750.952	32.688
1300	102.314	387.571	317.811	90.687	−932.800	−735.784	29.564
1400	103.059	395.181	323.069	100.957	−932.470	−720.641	26.887
1500	103.671	402.313	328.116	111.295	−932.137	−705.522	24.568

29. TRIFLUOROMETHANE CHF_3 (g)

T/K	C_p°	S°	$-(G^\circ-H^\circ(T_r))/T$	$H^\circ-H^\circ(T_r)$	$\Delta_f H^\circ$	$\Delta_f G^\circ$	Log K_f
298.15	51.069	259.675	259.675	0.000	−696.700	−662.237	116.020
300	51.258	259.991	259.676	0.095	−696.735	−662.023	115.267
400	61.148	276.113	261.807	5.722	−698.427	−650.186	84.905
500	69.631	290.700	266.149	12.275	−699.715	−637.969	66.647
600	76.453	304.022	271.368	19.593	−700.634	−625.528	54.456
700	81.868	316.230	276.917	27.519	−701.253	−612.957	45.739
800	86.201	327.455	282.542	35.930	−701.636	−600.315	39.196
900	89.719	337.818	288.116	44.732	−701.832	−587.636	34.105
1000	92.617	347.426	293.572	53.854	−701.879	−574.944	30.032
1100	95.038	356.370	298.879	63.240	−701.805	−562.253	26.699
1200	97.084	364.730	304.022	72.849	−701.629	−549.574	23.922
1300	98.833	372.571	308.997	82.647	−701.368	−536.913	21.573
1400	100.344	379.952	313.804	92.607	−701.033	−524.274	19.561
1500	101.660	386.921	318.449	102.709	−700.635	−511.662	17.817

	J/K·mol			kJ/mol			
T/K	C_p°	S°	$-(G^\circ-H^\circ(T_r))/T$	$H^\circ-H^\circ(T_r)$	$\Delta_f H^\circ$	$\Delta_f G^\circ$	Log K_f

30. CHLOROTRIFLUOROMETHANE $\quad$ CClF$_3$ (g)

298.15	66.886	285.419	285.419	0.000	−707.800	−667.238	116.896
300	67.111	285.834	285.421	0.124	−707.810	−666.986	116.131
400	77.528	306.646	288.187	7.383	−708.153	−653.316	85.313
500	85.013	324.797	293.734	15.532	−708.170	−639.599	66.818
600	90.329	340.794	300.271	24.314	−707.975	−625.901	54.489
700	94.132	355.020	307.096	33.547	−707.654	−612.246	45.686
800	96.899	367.780	313.897	43.106	−707.264	−598.642	39.087
900	98.951	379.317	320.536	52.903	−706.837	−585.090	33.957
1000	100.507	389.827	326.947	62.880	−706.396	−571.586	29.856
1100	101.708	399.465	333.108	72.993	−705.950	−558.126	26.503
1200	102.651	408.357	339.013	83.213	−705.505	−544.707	23.710
1300	103.404	416.604	344.668	93.517	−705.064	−531.326	21.349
1400	104.012	424.290	350.084	103.889	−704.628	−517.977	19.326
1500	104.512	431.484	355.273	114.316	−704.196	−504.660	17.574

31. DICHLORODIFLUOROMETHANE $\quad$ CCl$_2$F$_2$ (g)

298.15	72.476	300.903	300.903	0.000	−486.000	−447.030	78.317
300	72.691	301.352	300.905	0.134	−486.002	−446.788	77.792
400	82.408	323.682	303.883	7.919	−485.945	−433.716	56.637
500	89.063	342.833	309.804	16.514	−485.618	−420.692	43.949
600	93.635	359.500	316.729	25.663	−485.136	−407.751	35.497
700	96.832	374.189	323.909	35.196	−484.576	−394.897	29.467
800	99.121	387.276	331.027	44.999	−483.984	−382.126	24.950
900	100.801	399.053	337.942	55.000	−483.388	−369.429	21.441
1000	102.062	409.742	344.596	65.146	−482.800	−356.799	18.637
1100	103.030	419.517	350.969	75.402	−482.226	−344.227	16.346
1200	103.786	428.515	357.061	85.745	−481.667	−331.706	14.439
1300	104.388	436.847	362.882	96.154	−481.121	−319.232	12.827
1400	104.874	444.602	368.445	106.618	−480.588	−306.799	11.447
1500	105.270	451.851	373.767	117.126	−480.065	−294.404	10.252

32. CHLORODIFLUOROMETHANE $\quad$ CHClF$_2$ (g)

298.15	55.853	280.915	280.915	0.000	−475.000	−443.845	77.759
300	56.039	281.261	280.916	0.104	−475.028	−443.652	77.246
400	65.395	298.701	283.231	6.188	−476.390	−432.978	56.540
500	73.008	314.145	287.898	13.123	−477.398	−422.001	44.086
600	78.940	328.003	293.448	20.733	−478.103	−410.851	35.767
700	83.551	340.533	299.294	28.867	−478.574	−399.603	29.818
800	87.185	351.936	305.172	37.411	−478.870	−388.299	25.353
900	90.100	362.379	310.956	46.280	−479.031	−376.967	21.878
1000	92.475	371.999	316.586	55.413	−479.090	−365.622	19.098
1100	94.433	380.908	322.033	64.761	−479.068	−354.276	16.823
1200	96.066	389.196	327.289	74.289	−478.982	−342.935	14.927
1300	97.438	396.941	332.352	83.966	−478.843	−331.603	13.324
1400	98.601	404.206	337.228	93.769	−478.661	−320.283	11.950
1500	99.593	411.044	341.923	103.681	−478.443	−308.978	10.759

33. METHYLAMINE $\quad$ CH$_5$N (g)

298.15	50.053	242.881	242.881	0.000	−22.529	32.734	−5.735
300	50.227	243.196	242.893	0.091	−22.614	33.077	−5.759
400	60.171	258.986	244.975	5.604	−26.846	52.294	−6.829
500	70.057	273.486	249.244	12.121	−30.431	72.510	−7.575
600	78.929	287.063	254.431	19.579	−33.364	93.382	−8.129
700	86.711	299.826	260.008	27.873	−35.712	114.702	−8.559
800	93.545	311.865	265.749	36.893	−37.548	136.316	−8.900

T/K	$C_p°$ (J/K·mol)	$S°$	$-(G°-H°(T_r))/T$	$H°-H°(T_r)$ (kJ/mol)	$\Delta_f H°$	$\Delta_f G°$	Log K_f

33. METHYLAMINE CH$_5$N (g) (continued)

T/K	$C_p°$	$S°$	$-(G°-H°(T_r))/T$	$H°-H°(T_r)$	$\Delta_f H°$	$\Delta_f G°$	Log K_f
900	99.573	323.239	271.511	46.555	−38.949	158.138	−9.178
1000	104.886	334.006	277.220	56.786	−39.967	180.098	−9.407
1100	109.576	344.233	282.861	67.509	−40.681	201.822	−9.584
1200	113.708	353.944	288.374	78.685	−41.136	224.240	−9.761
1300	117.341	363.190	293.775	90.239	−41.376	246.364	−9.899
1400	120.542	372.012	299.061	102.131	−41.451	268.504	−10.018
1500	123.353	380.426	304.209	114.326	−41.381	290.639	−10.121

34. CHLORINE Cl (g)

T/K	$C_p°$	$S°$	$-(G°-H°(T_r))/T$	$H°-H°(T_r)$	$\Delta_f H°$	$\Delta_f G°$	Log K_f
298.15	21.838	165.190	165.190	0.000	121.302	105.306	−18.449
300	21.852	165.325	165.190	0.040	121.311	105.207	−18.318
400	22.467	171.703	166.055	2.259	121.795	99.766	−13.028
500	22.744	176.752	167.708	4.522	122.272	94.203	−9.841
600	22.781	180.905	169.571	6.800	122.734	88.546	−7.709
700	22.692	184.411	171.448	9.074	123.172	82.813	−6.179
800	22.549	187.432	173.261	11.337	123.585	77.019	−5.029
900	22.389	190.079	174.986	13.584	123.971	71.175	−4.131
1000	22.233	192.430	176.615	15.815	124.334	65.289	−3.410
1100	22.089	194.542	178.150	18.031	124.675	59.368	−2.819
1200	21.959	196.458	179.597	20.233	124.996	53.416	−2.325
1300	21.843	198.211	180.963	22.423	125.299	47.439	−1.906
1400	21.742	199.826	182.253	24.602	125.587	41.439	−1.546
1500	21.652	201.323	183.475	26.772	125.861	35.418	−1.233

35. DICHLORINE Cl$_2$ (g)

T/K	$C_p°$	$S°$	$-(G°-H°(T_r))/T$	$H°-H°(T_r)$	$\Delta_f H°$	$\Delta_f G°$	Log K_f
298.15	33.949	223.079	223.079	0.000	0.000	0.000	0.000
300	33.981	223.290	223.080	0.063	0.000	0.000	0.000
400	35.296	233.263	224.431	3.533	0.000	0.000	0.000
500	36.064	241.229	227.021	7.104	0.000	0.000	0.000
600	36.547	247.850	229.956	10.736	0.000	0.000	0.000
700	36.874	253.510	232.926	14.408	0.000	0.000	0.000
800	37.111	258.450	235.815	18.108	0.000	0.000	0.000
900	37.294	262.832	238.578	21.829	0.000	0.000	0.000
1000	37.442	266.769	241.203	25.566	0.000	0.000	0.000
1100	37.567	270.343	243.692	29.316	0.000	0.000	0.000
1200	37.678	273.617	246.052	33.079	0.000	0.000	0.000
1300	37.778	276.637	248.290	36.851	0.000	0.000	0.000
1400	37.872	279.440	250.416	40.634	0.000	0.000	0.000
1500	37.961	282.056	252.439	44.426	0.000	0.000	0.000

36. HYDROGEN CHLORIDE HCl (g)

T/K	$C_p°$	$S°$	$-(G°-H°(T_r))/T$	$H°-H°(T_r)$	$\Delta_f H°$	$\Delta_f G°$	Log K_f
298.15	29.136	186.902	186.902	0.000	−92.310	−95.298	16.696
300	29.137	187.082	186.902	0.054	−92.314	−95.317	16.596
400	29.175	195.468	188.045	2.969	−92.587	−96.278	12.573
500	29.304	201.990	190.206	5.892	−92.911	−97.164	10.151
600	29.576	207.354	192.630	8.835	−93.249	−97.983	8.530
700	29.988	211.943	195.069	11.812	−93.577	−98.746	7.368
800	30.500	215.980	197.435	14.836	−93.879	−99.464	6.494
900	31.063	219.604	199.700	17.913	−94.149	−100.145	5.812
1000	31.639	222.907	201.858	21.049	−94.384	−100.798	5.265
1100	32.201	225.949	203.912	24.241	−94.587	−101.430	4.816
1200	32.734	228.774	205.867	27.488	−94.760	−102.044	4.442
1300	33.229	231.414	207.732	30.786	−94.908	−102.645	4.124
1400	33.684	233.893	209.513	34.132	−95.035	−103.235	3.852
1500	34.100	236.232	211.217	37.522	−95.146	−103.817	3.615

	J/K·mol			kJ/mol			
T/K	C_p°	S°	$-(G^\circ-H^\circ(T_r))/T$	$H^\circ-H^\circ(T_r)$	$\Delta_f H^\circ$	$\Delta_f G^\circ$	Log K_f

37. COPPER   Cu (cr, l)

298.15	24.440	33.150	33.150	0.000	0.000	0.000	0.000
300	24.460	33.301	33.150	0.045	0.000	0.000	0.000
400	25.339	40.467	34.122	2.538	0.000	0.000	0.000
500	25.966	46.192	35.982	5.105	0.000	0.000	0.000
600	26.479	50.973	38.093	7.728	0.000	0.000	0.000
700	26.953	55.090	40.234	10.399	0.000	0.000	0.000
800	27.448	58.721	42.322	13.119	0.000	0.000	0.000
900	28.014	61.986	44.328	15.891	0.000	0.000	0.000
1000	28.700	64.971	46.245	18.726	0.000	0.000	0.000
1100	29.553	67.745	48.075	21.637	0.000	0.000	0.000
1200	30.617	70.361	49.824	24.644	0.000	0.000	0.000
1300	31.940	72.862	51.501	27.769	0.000	0.000	0.000
1358	32.844	74.275	52.443	29.647	0.000	0.000	0.000
		PHASE TRANSITION: $\Delta_{trs} H$ = 13.141 kJ/mol, $\Delta_{trs} S$ = 9.676 J/K·mol, cr–l					
1358	32.800	83.951	52.443	42.788	0.000	0.000	0.000
1400	32.800	84.950	53.403	44.166	0.000	0.000	0.000
1500	32.800	87.213	55.583	47.446	0.000	0.000	0.000

38. COPPER   Cu (g)

298.15	20.786	166.397	166.397	0.000	337.600	297.873	−52.185
300	20.786	166.525	166.397	0.038	337.594	297.626	−51.821
400	20.786	172.505	167.213	2.117	337.179	284.364	−37.134
500	20.786	177.143	168.752	4.196	336.691	271.215	−28.333
600	20.786	180.933	170.476	6.274	336.147	258.170	−22.475
700	20.786	184.137	172.205	8.353	335.554	245.221	−18.298
800	20.786	186.913	173.874	10.431	334.913	232.359	−15.171
900	20.786	189.361	175.461	12.510	334.219	219.581	−12.744
1000	20.786	191.551	176.963	14.589	333.463	206.883	−10.806
1100	20.788	193.532	178.380	16.667	332.631	194.265	−9.225
1200	20.793	195.341	179.719	18.746	331.703	181.726	−7.910
1300	20.803	197.006	180.986	20.826	330.657	169.270	−6.801
1400	20.823	198.548	182.186	22.907	316.342	157.305	−5.869
1500	20.856	199.986	183.325	24.991	315.146	145.987	−5.084

39. COPPER OXIDE   CuO (cr)

298.15	42.300	42.740	42.740	0.000	−162.000	−134.277	23.524
300	42.417	43.002	42.741	0.078	−161.994	−134.105	23.349
400	46.783	55.878	44.467	4.564	−161.487	−124.876	16.307
500	49.190	66.596	47.852	9.372	−160.775	−115.803	12.098
600	50.827	75.717	51.755	14.377	−159.973	−106.883	9.305
700	52.099	83.651	55.757	19.526	−159.124	−98.102	7.320
800	53.178	90.680	59.691	24.791	−158.247	−89.444	5.840
900	54.144	97.000	63.491	30.158	−157.356	−80.897	4.695
1000	55.040	102.751	67.134	35.617	−156.462	−72.450	3.784
1100	55.890	108.037	70.615	41.164	−155.582	−64.091	3.043
1200	56.709	112.936	73.941	46.794	−154.733	−55.812	2.429
1300	57.507	117.507	77.118	52.505	−153.940	−47.601	1.913
1400	58.288	121.797	80.158	58.295	−166.354	−39.043	1.457
1500	59.057	125.845	83.070	64.163	−165.589	−29.975	1.044

40. DICOPPER OXIDE   Cu$_2$O (cr)

298.15	62.600	92.550	92.550	0.000	−173.100	−150.344	26.339
300	62.721	92.938	92.551	0.116	−173.102	−150.203	26.152
400	67.587	111.712	95.078	6.654	−173.036	−142.572	18.618
500	70.784	127.155	99.995	13.580	−172.772	−134.984	14.101

	J/K·mol			kJ/mol			
T/K	C_p°	S°	$-(G^\circ - H^\circ(T_r))/T$	$H^\circ - H^\circ(T_r)$	$\Delta_f H^\circ$	$\Delta_f G^\circ$	Log K_f

40. DICOPPER OXIDE Cu_2O (cr) (continued)

600	73.323	140.291	105.643	20.789	−172.389	−127.460	11.096
700	75.552	151.764	111.429	28.235	−171.914	−120.009	8.955
800	77.616	161.989	117.121	35.894	−171.363	−112.631	7.354
900	79.584	171.245	122.629	43.755	−170.750	−105.325	6.113
1000	81.492	179.729	127.920	51.809	−170.097	−98.091	5.124
1100	83.360	187.584	132.992	60.052	−169.431	−90.922	4.317
1200	85.202	194.917	137.850	68.480	−168.791	−83.814	3.648
1300	87.026	201.808	142.507	77.092	−168.223	−76.756	3.084
1400	88.836	208.324	146.978	85.885	−194.030	−68.926	2.572
1500	90.636	214.515	151.276	94.858	−193.438	−60.010	2.090

41. COPPER DICHLORIDE $CuCl_2$ (cr, l)

298.15	71.880	108.070	108.070	0.000	−218.000	−173.826	30.453
300	71.998	108.515	108.071	0.133	−217.975	−173.552	30.218
400	76.338	129.899	110.957	7.577	−216.494	−158.962	20.758
500	78.654	147.204	116.532	15.336	−214.873	−144.765	15.123
600	80.175	161.687	122.884	23.282	−213.182	−130.901	11.396
675	81.056	171.183	127.732	29.329	−211.185	−120.693	9.340

PHASE TRANSITION: $\Delta_{trs} H = 0.700$ kJ/mol, $\Delta_{trs} S = 1.037$ J/K·mol, crII–crI

675	82.400	172.220	127.732	30.029	−211.185	−120.693	9.340
700	82.400	175.216	129.375	32.089	−210.719	−117.350	8.757
800	82.400	186.219	135.808	40.329	−208.898	−104.137	6.799
871	82.400	193.226	140.207	46.179	−192.649	−94.893	5.691

PHASE TRANSITION: $\Delta_{trs} H = 15.001$ kJ/mol, $\Delta_{trs} S = 17.221$ J/K·mol, crI–l

871	100.000	210.447	140.207	61.180	−192.649	−94.893	5.691
900	100.000	213.723	142.523	64.080	−191.640	−91.655	5.319
1000	100.000	224.259	150.179	74.080	−188.212	−80.730	4.217
1100	100.000	233.790	157.353	84.080	−184.873	−70.144	3.331
1130.75	100.000	236.547	159.470	87.155	−183.867	−66.951	3.093

42. COPPER DICHLORIDE $CuCl_2$ (g)

298.15	56.814	278.418	278.418	0.000	−43.268	−49.883	8.739
300	56.869	278.769	278.419	0.105	−43.271	−49.924	8.692
400	58.992	295.456	280.679	5.911	−43.428	−52.119	6.806
500	60.111	308.752	285.010	11.871	−43.606	−54.271	5.670
600	60.761	319.774	289.911	17.918	−43.814	−56.385	4.909
700	61.168	329.173	294.865	24.015	−44.060	−58.462	4.362
800	61.439	337.360	299.677	30.147	−44.349	−60.500	3.950
900	61.630	344.608	304.274	36.301	−44.688	−62.499	3.627
1000	61.776	351.109	308.638	42.471	−45.088	−64.457	3.367
1100	61.900	357.003	312.771	48.655	−45.566	−66.372	3.152
1200	62.022	362.394	316.685	54.851	−46.139	−68.239	2.970
1300	62.159	367.364	320.395	61.060	−46.829	−70.053	2.815
1400	62.325	371.976	323.916	67.284	−60.784	−71.404	2.664
1500	62.531	376.283	327.265	73.526	−61.613	−72.133	2.512

43. FLUORINE   F (g)

298.15	22.746	158.750	158.750	0.000	79.380	62.280	−10.911
300	22.742	158.891	158.750	0.042	79.393	62.173	−10.825
400	22.432	165.394	159.639	2.302	80.043	56.332	−7.356
500	22.100	170.363	161.307	4.528	80.587	50.340	−5.259
600	21.832	174.368	163.161	6.724	81.046	44.246	−3.852
700	21.629	177.717	165.008	8.897	81.442	38.081	−2.842
800	21.475	180.595	166.780	11.052	81.792	31.862	−2.080
900	21.357	183.117	168.458	13.193	82.106	25.601	−1.486

T/K	C_p°	S°	$-(G^\circ - H^\circ(T_r))/T$	$H^\circ - H^\circ(T_r)$	$\Delta_f H^\circ$	$\Delta_f G^\circ$	Log K_f
		J/K·mol			kJ/mol		

43. FLUORINE   F (g) (continued)

T/K	C_p°	S°	$-(G^\circ - H^\circ(T_r))/T$	$H^\circ - H^\circ(T_r)$	$\Delta_f H^\circ$	$\Delta_f G^\circ$	Log K_f
1000	21.266	185.362	170.039	15.324	82.391	19.308	−1.009
1100	21.194	187.386	171.525	17.447	82.654	12.986	−0.617
1200	21.137	189.227	172.925	19.563	82.897	6.642	−0.289
1300	21.091	190.917	174.245	21.675	83.123	0.278	−0.011
1400	21.054	192.479	175.492	23.782	83.335	−6.103	0.228
1500	21.022	193.930	176.673	25.886	83.533	−12.498	0.435

44. DIFLUORINE F_2 (g)

T/K	C_p°	S°	$-(G^\circ - H^\circ(T_r))/T$	$H^\circ - H^\circ(T_r)$	$\Delta_f H^\circ$	$\Delta_f G^\circ$	Log K_f
298.15	31.304	202.790	202.790	0.000	0.000	0.000	0.000
300	31.337	202.984	202.790	0.058	0.000	0.000	0.000
400	32.995	212.233	204.040	3.277	0.000	0.000	0.000
500	34.258	219.739	206.453	6.643	0.000	0.000	0.000
600	35.171	226.070	209.208	10.117	0.000	0.000	0.000
700	35.839	231.545	212.017	13.669	0.000	0.000	0.000
800	36.343	236.365	214.765	17.279	0.000	0.000	0.000
900	36.740	240.669	217.409	20.934	0.000	0.000	0.000
1000	37.065	244.557	219.932	24.625	0.000	0.000	0.000
1100	37.342	248.103	222.334	28.346	0.000	0.000	0.000
1200	37.588	251.363	224.619	32.093	0.000	0.000	0.000
1300	37.811	254.381	226.794	35.863	0.000	0.000	0.000
1400	38.019	257.191	228.866	39.654	0.000	0.000	0.000
1500	38.214	259.820	230.843	43.466	0.000	0.000	0.000

45. HYDROGEN FLUORIDE   HF (g)

T/K	C_p°	S°	$-(G^\circ - H^\circ(T_r))/T$	$H^\circ - H^\circ(T_r)$	$\Delta_f H^\circ$	$\Delta_f G^\circ$	Log K_f
298.15	29.137	173.776	173.776	0.000	−273.300	−275.399	48.248
300	29.137	173.956	173.776	0.054	−273.302	−275.412	47.953
400	29.149	182.340	174.919	2.968	−273.450	−276.096	36.054
500	29.172	188.846	177.078	5.884	−273.679	−276.733	28.910
600	29.230	194.169	179.496	8.804	−273.961	−277.318	24.142
700	29.350	198.683	181.923	11.732	−274.277	−277.852	20.733
800	29.549	202.614	184.269	14.676	−274.614	−278.340	18.174
900	29.827	206.110	186.505	17.645	−274.961	−278.785	16.180
1000	30.169	209.270	188.626	20.644	−275.309	−279.191	14.583
1100	30.558	212.163	190.636	23.680	−275.652	−279.563	13.275
1200	30.974	214.840	192.543	26.756	−275.988	−279.904	12.184
1300	31.403	217.336	194.355	29.875	−276.315	−280.217	11.259
1400	31.831	219.679	196.081	33.037	−276.631	−280.505	10.466
1500	32.250	221.889	197.729	36.241	−276.937	−280.771	9.777

46. GERMANIUM   Ge (cr, l)

T/K	C_p°	S°	$-(G^\circ - H^\circ(T_r))/T$	$H^\circ - H^\circ(T_r)$	$\Delta_f H^\circ$	$\Delta_f G^\circ$	Log K_f
298.15	23.222	31.090	31.090	0.000	0.000	0.000	0.000
300	23.249	31.234	31.090	0.043	0.000	0.000	0.000
400	24.310	38.083	32.017	2.426	0.000	0.000	0.000
500	24.962	43.582	33.798	4.892	0.000	0.000	0.000
600	25.452	48.178	35.822	7.414	0.000	0.000	0.000
700	25.867	52.133	37.876	9.980	0.000	0.000	0.000
800	26.240	55.612	39.880	12.586	0.000	0.000	0.000
900	26.591	58.723	41.804	15.227	0.000	0.000	0.000
1000	26.926	61.542	43.639	17.903	0.000	0.000	0.000
1100	27.252	64.124	45.386	20.612	0.000	0.000	0.000
1200	27.571	66.509	47.048	23.353	0.000	0.000	0.000
1211.4	27.608	66.770	47.232	23.668	0.000	0.000	0.000
PHASE TRANSITION: $\Delta_{trs} H$ = 37.030 kJ/mol, $\Delta_{trs} S$ = 30.568 J/K·mol, cr–l							
1211.4	27.600	97.338	47.232	60.698	0.000	0.000	0.000
1300	27.600	99.286	50.714	63.143	0.000	0.000	0.000

T/K	C_p°	S°	$-(G^\circ - H^\circ(T_r))/T$	$H^\circ - H^\circ(T_r)$	$\Delta_f H^\circ$	$\Delta_f G^\circ$	Log K_f
		J/K·mol		kJ/mol			

46. GERMANIUM   Ge (cr, l) (continued)

T/K	C_p°	S°	$-(G^\circ - H^\circ(T_r))/T$	$H^\circ - H^\circ(T_r)$	$\Delta_f H^\circ$	$\Delta_f G^\circ$	Log K_f
1400	27.600	101.331	54.258	65.903	0.000	0.000	0.000
1500	27.600	103.236	57.460	68.663	0.000	0.000	0.000

47. GERMANIUM   Ge (g)

T/K	C_p°	S°	$-(G^\circ - H^\circ(T_r))/T$	$H^\circ - H^\circ(T_r)$	$\Delta_f H^\circ$	$\Delta_f G^\circ$	Log K_f
298.15	30.733	167.903	167.903	0.000	367.800	327.009	−57.290
300	30.757	168.094	167.904	0.057	367.814	326.756	−56.893
400	31.071	177.025	169.119	3.162	368.536	312.959	−40.868
500	30.360	183.893	171.415	6.239	369.147	298.991	−31.235
600	29.265	189.334	173.965	9.222	369.608	284.914	−24.804
700	28.102	193.758	176.487	12.090	369.910	270.773	−20.205
800	27.029	197.439	178.882	14.845	370.060	256.598	−16.754
900	26.108	200.567	181.122	17.501	370.073	242.414	−14.069
1000	25.349	203.277	183.205	20.072	369.969	228.234	−11.922
1100	24.741	205.664	185.141	22.575	369.763	214.069	−10.165
1200	24.264	207.795	186.941	25.025	369.471	199.928	−8.703
1300	23.898	209.722	188.621	27.432	332.088	188.521	−7.575
1400	23.624	211.483	190.192	29.807	331.704	177.492	−6.622
1500	23.426	213.105	191.666	32.159	331.296	166.491	−5.798

48. GERMANIUM DIOXIDE   GeO$_2$ (cr, l)

T/K	C_p°	S°	$-(G^\circ - H^\circ(T_r))/T$	$H^\circ - H^\circ(T_r)$	$\Delta_f H^\circ$	$\Delta_f G^\circ$	Log K_f
298.15	50.166	39.710	39.710	0.000	−580.200	−521.605	91.382
300	50.475	40.021	39.711	0.093	−580.204	−521.242	90.755
400	61.281	56.248	41.850	5.759	−579.893	−501.610	65.503
500	66.273	70.519	46.191	12.164	−579.013	−482.134	50.368
600	69.089	82.872	51.299	18.943	−577.915	−462.859	40.295
700	70.974	93.671	56.597	25.952	−576.729	−443.776	33.115
800	72.449	103.247	61.841	33.125	−575.498	−424.866	27.741
900	73.764	111.857	66.928	40.436	−574.235	−406.113	23.570
1000	75.049	119.696	71.819	47.877	−572.934	−387.502	20.241
1100	76.378	126.910	76.504	55.447	−571.582	−369.024	17.523
1200	77.796	133.616	80.987	63.155	−570.166	−350.671	15.264
1300	79.332	139.903	85.279	71.010	−605.685	−329.732	13.249
1308	79.460	140.390	85.615	71.646	−584.059	−328.034	13.100
PHASE TRANSITION:	$\Delta_{trs} H = 21.500$ kJ/mol,	$\Delta_{trs} S = 16.437$ J/K·mol, crII–crI					
1308	80.075	156.827	85.615	93.146	−584.059	−328.034	13.100
1388	81.297	161.617	89.858	99.601	−565.504	−312.415	11.757
PHASE TRANSITION:	$\Delta_{trs} H = 17.200$ kJ/mol,	$\Delta_{trs} S = 12.392$ J/K·mol, crI–l					
1388	78.500	174.009	89.858	116.801	−565.504	−312.415	11.757
1400	78.500	174.685	90.582	117.743	−565.328	−310.228	11.575
1500	78.500	180.100	96.372	125.593	−563.882	−292.057	10.170

49. GERMANIUM TETRACHLORIDE   GeCl$_4$ (g)

T/K	C_p°	S°	$-(G^\circ - H^\circ(T_r))/T$	$H^\circ - H^\circ(T_r)$	$\Delta_f H^\circ$	$\Delta_f G^\circ$	Log K_f
298.15	95.918	348.393	348.393	0.000	−500.000	−461.582	80.866
300	96.041	348.987	348.395	0.178	−499.991	−461.343	80.326
400	100.750	377.342	352.229	10.045	−499.447	−448.540	58.573
500	103.206	400.114	359.604	20.255	−498.845	−435.882	45.536
600	104.624	419.067	367.980	30.652	−498.234	−423.347	36.855
700	105.509	435.266	376.463	41.162	−497.634	−410.914	30.662
800	106.096	449.396	384.715	51.744	−497.057	−398.565	26.023
900	106.504	461.917	392.611	62.375	−496.509	−386.287	22.419
1000	106.799	473.155	400.113	73.041	−495.993	−374.068	19.539
1100	107.020	483.344	407.224	83.733	−495.512	−361.899	17.185
1200	107.189	492.664	413.961	94.444	−495.067	−349.772	15.225
1300	107.320	501.249	420.349	105.169	−531.677	−334.973	13.459
1400	107.425	509.206	426.416	115.907	−531.265	−319.857	11.934
1500	107.509	516.621	432.185	126.654	−530.861	−304.771	10.613

	J/K·mol			kJ/mol			
T/K	C_p°	S°	$-(G^\circ - H^\circ(T_r))/T$	$H^\circ - H^\circ(T_r)$	$\Delta_f H^\circ$	$\Delta_f G^\circ$	Log K_f

50. HYDROGEN   H (g)

298.15	20.786	114.716	114.716	0.000	217.998	203.276	−35.613
300	20.786	114.845	114.716	0.038	218.010	203.185	−35.377
400	20.786	120.824	115.532	2.117	218.635	198.149	−25.875
500	20.786	125.463	117.071	4.196	219.253	192.956	−20.158
600	20.786	129.252	118.795	6.274	219.867	187.639	−16.335
700	20.786	132.457	120.524	8.353	220.476	182.219	−13.597
800	20.786	135.232	122.193	10.431	221.079	176.712	−11.538
900	20.786	137.680	123.780	12.510	221.670	171.131	−9.932
1000	20.786	139.870	125.282	14.589	222.247	165.485	−8.644
1100	20.786	141.852	126.700	16.667	222.806	159.781	−7.587
1200	20.786	143.660	128.039	18.746	223.345	154.028	−6.705
1300	20.786	145.324	129.305	20.824	223.864	148.230	−5.956
1400	20.786	146.864	130.505	22.903	224.360	142.393	−5.313
1500	20.786	148.298	131.644	24.982	224.835	136.522	−4.754

51. DIHYDROGEN   H$_2$ (g)

298.15	28.836	130.680	130.680	0.000	0.000	0.000	0.000
300	28.849	130.858	130.680	0.053	0.000	0.000	0.000
400	29.181	139.217	131.818	2.960	0.000	0.000	0.000
500	29.260	145.738	133.974	5.882	0.000	0.000	0.000
600	29.327	151.078	136.393	8.811	0.000	0.000	0.000
700	29.440	155.607	138.822	11.749	0.000	0.000	0.000
800	29.623	159.549	141.172	14.702	0.000	0.000	0.000
900	29.880	163.052	143.412	17.676	0.000	0.000	0.000
1000	30.204	166.217	145.537	20.680	0.000	0.000	0.000
1100	30.580	169.113	147.550	23.719	0.000	0.000	0.000
1200	30.991	171.791	149.460	26.797	0.000	0.000	0.000
1300	31.422	174.288	151.275	29.918	0.000	0.000	0.000
1400	31.860	176.633	153.003	33.082	0.000	0.000	0.000
1500	32.296	178.846	154.653	36.290	0.000	0.000	0.000

52. HYDROXYL   OH (g)

298.15	29.886	183.737	183.737	0.000	39.349	34.631	−6.067
300	29.879	183.922	183.738	0.055	39.350	34.602	−6.025
400	29.604	192.476	184.906	3.028	39.384	33.012	−4.311
500	29.495	199.067	187.104	5.982	39.347	31.422	−3.283
600	29.513	204.445	189.560	8.931	39.252	29.845	−2.598
700	29.655	209.003	192.020	11.888	39.113	28.287	−2.111
800	29.914	212.979	194.396	14.866	38.945	26.752	−1.747
900	30.265	216.522	196.661	17.874	38.763	25.239	−1.465
1000	30.682	219.731	198.810	20.921	38.577	23.746	−1.240
1100	31.135	222.677	200.848	24.012	38.393	22.272	−1.058
1200	31.603	225.406	202.782	27.149	38.215	20.814	−0.906
1300	32.069	227.954	204.621	30.332	38.046	19.371	−0.778
1400	32.522	230.347	206.374	33.562	37.886	17.941	−0.669
1500	32.956	232.606	208.048	36.836	37.735	16.521	−0.575

53. WATER   H$_2$O (l)

298.15	75.300	69.950	69.950	0.000	−285.830	−237.141	41.546
300	75.281	70.416	69.951	0.139	−285.771	−236.839	41.237
373.21	76.079	86.896	71.715	5.666	−283.454	−225.160	31.513

54. WATER   H$_2$O (g)

298.15	33.598	188.832	188.832	0.000	−241.826	−228.582	40.046

T/K	$C_p°$	$S°$	$-(G°-H°(T_r))/T$	$H°-H°(T_r)$	$\Delta_f H°$	$\Delta_f G°$	Log K_f
		J/K·mol			**kJ/mol**		

54. WATER H$_2$O (g) (continued)

T/K	$C_p°$	$S°$	$-(G°-H°(T_r))/T$	$H°-H°(T_r)$	$\Delta_f H°$	$\Delta_f G°$	Log K_f
300	33.606	189.040	188.833	0.062	−241.844	−228.500	39.785
400	34.283	198.791	190.158	3.453	−242.845	−223.900	29.238
500	35.259	206.542	192.685	6.929	−243.822	−219.050	22.884
600	36.371	213.067	195.552	10.509	−244.751	−214.008	18.631
700	37.557	218.762	198.469	14.205	−245.620	−208.814	15.582
800	38.800	223.858	201.329	18.023	−246.424	−203.501	13.287
900	40.084	228.501	204.094	21.966	−247.158	−198.091	11.497
1000	41.385	232.792	206.752	26.040	−247.820	−192.603	10.060
1100	42.675	236.797	209.303	30.243	−248.410	−187.052	8.882
1200	43.932	240.565	211.753	34.574	−248.933	−181.450	7.898
1300	45.138	244.129	214.108	39.028	−249.392	−175.807	7.064
1400	46.281	247.516	216.374	43.599	−249.792	−170.132	6.348
1500	47.356	250.746	218.559	48.282	−250.139	−164.429	5.726

55. IODINE I (g)

T/K	$C_p°$	$S°$	$-(G°-H°(T_r))/T$	$H°-H°(T_r)$	$\Delta_f H°$	$\Delta_f G°$	Log K_f
298.15	20.786	180.787	180.787	0.000	106.760	70.172	−12.294
300	20.786	180.915	180.787	0.038	106.748	69.945	−12.178
400	20.786	186.895	181.602	2.117	97.974	58.060	−7.582
500	20.786	191.533	183.142	4.196	75.988	50.202	−5.244
600	20.786	195.323	184.866	6.274	76.190	45.025	−3.920
700	20.786	198.527	186.594	8.353	76.385	39.816	−2.971
800	20.787	201.303	188.263	10.432	76.574	34.579	−2.258
900	20.789	203.751	189.851	12.510	76.757	29.319	−1.702
1000	20.795	205.942	191.352	14.589	76.936	24.038	−1.256
1100	20.806	207.924	192.770	16.669	77.109	18.740	−0.890
1200	20.824	209.735	194.110	18.751	77.277	13.426	−0.584
1300	20.851	211.403	195.377	20.835	77.440	8.098	−0.325
1400	20.889	212.950	196.577	22.921	77.596	2.758	−0.103
1500	20.936	214.392	197.717	25.013	77.745	−2.592	0.090

56. DIIODINE I$_2$ (cr, l)

T/K	$C_p°$	$S°$	$-(G°-H°(T_r))/T$	$H°-H°(T_r)$	$\Delta_f H°$	$\Delta_f G°$	Log K_f
298.15	54.440	116.139	116.139	0.000	0.000	0.000	0.000
300	54.518	116.476	116.140	0.101	0.000	0.000	0.000
386.75	61.531	131.039	117.884	5.088	0.000	0.000	0.000
	PHASE TRANSITION: $\Delta_{trs} H$ = 15.665 kJ/mol, $\Delta_{trs} S$ = 40.504 J/K·mol, cr–l						
386.75	79.555	171.543	117.884	20.753	0.000	0.000	0.000
400	79.555	174.223	119.706	21.807	0.000	0.000	0.000
457.67	79.555	184.938	127.266	26.395	0.000	0.000	0.000

57. DIIODINE I$_2$ (g)

T/K	$C_p°$	$S°$	$-(G°-H°(T_r))/T$	$H°-H°(T_r)$	$\Delta_f H°$	$\Delta_f G°$	Log K_f
298.15	36.887	260.685	260.685	0.000	62.420	19.324	−3.385
300	36.897	260.913	260.685	0.068	62.387	19.056	−3.318
400	37.256	271.584	262.138	3.778	44.391	19.056	−0.711
457.67	37.385	276.610	263.652	5.931		5.447	
500	37.464	279.921	264.891	7.515	pressure = 1 bar		
500	37.464	279.921	264.891	7.515	0.000	0.000	0.000
600	37.613	286.765	267.983	11.269	0.000	0.000	0.000
700	37.735	292.573	271.092	15.037	0.000	0.000	0.000
800	37.847	297.619	274.099	18.816	0.000	0.000	0.000
900	37.956	302.083	276.965	22.606	0.000	0.000	0.000
1000	38.070	306.088	279.681	26.407	0.000	0.000	0.000
1100	38.196	309.722	282.249	30.220	0.000	0.000	0.000
1200	38.341	313.052	284.679	34.047	0.000	0.000	0.000
1300	38.514	316.127	286.981	37.890	0.000	0.000	0.000
1400	38.719	318.989	289.166	41.751	0.000	0.000	0.000
1500	38.959	321.668	291.245	45.635	0.000	0.000	0.000

T/K	C_p°	S°	$-(G^\circ - H^\circ(T_r))/T$	$H^\circ - H^\circ(T_r)$	$\Delta_f H^\circ$	$\Delta_f G^\circ$	Log K_f
	J/K·mol			kJ/mol			

58. HYDROGEN IODIDE   HI (g)

T/K	C_p°	S°	$-(G^\circ - H^\circ(T_r))/T$	$H^\circ - H^\circ(T_r)$	$\Delta_f H^\circ$	$\Delta_f G^\circ$	Log K_f
298.15	29.157	206.589	206.589	0.000	26.500	1.700	-0.298
300	29.158	206.769	206.589	0.054	26.477	1.546	-0.269
400	29.329	215.176	207.734	2.977	17.093	-6.289	0.821
500	29.738	221.760	209.904	5.928	-5.481	-9.946	1.039
600	30.351	227.233	212.348	8.931	-5.819	-10.806	0.941
700	31.070	231.965	214.820	12.002	-6.101	-11.614	0.867
800	31.807	236.162	217.230	15.145	-6.323	-12.386	0.809
900	32.511	239.950	219.548	18.362	-6.489	-13.133	0.762
1000	33.156	243.409	221.763	21.646	-6.608	-13.865	0.724
1100	33.735	246.597	223.878	24.991	-6.689	-14.586	0.693
1200	34.249	249.555	225.896	28.391	-6.741	-15.302	0.666
1300	34.703	252.314	227.823	31.839	-6.775	-16.014	0.643
1400	35.106	254.901	229.666	35.330	-6.797	-16.723	0.624
1500	35.463	257.336	231.430	38.858	-6.814	-17.432	0.607

59. POTASSIUM   K (cr, l)

T/K	C_p°	S°	$-(G^\circ - H^\circ(T_r))/T$	$H^\circ - H^\circ(T_r)$	$\Delta_f H^\circ$	$\Delta_f G^\circ$	Log K_f
298.15	29.600	64.680	64.680	0.000	0.000	0.000	0.000
300	29.671	64.863	64.681	0.055	0.000	0.000	0.000
336.86	32.130	68.422	64.896	1.188	0.000	0.000	0.000
PHASE TRANSITION: $\Delta_{trs} H = 2.321$ kJ/mol, $\Delta_{trs} S = 6.891$ J/K·mol, cr–l							
336.86	32.129	75.313	64.896	3.509	0.000	0.000	0.000
400	31.552	80.784	66.986	5.519	0.000	0.000	0.000
500	30.741	87.734	70.469	8.632	0.000	0.000	0.000
600	30.158	93.283	73.824	11.675	0.000	0.000	0.000
700	29.851	97.905	76.943	14.673	0.000	0.000	0.000
800	29.838	101.887	79.818	17.655	0.000	0.000	0.000
900	30.130	105.415	82.470	20.651	0.000	0.000	0.000
1000	30.730	108.618	84.927	23.691	0.000	0.000	0.000
1039.4	31.053	109.812	85.847	24.908	0.000	0.000	0.000

60. POTASSIUM   K (g)

T/K	C_p°	S°	$-(G^\circ - H^\circ(T_r))/T$	$H^\circ - H^\circ(T_r)$	$\Delta_f H^\circ$	$\Delta_f G^\circ$	Log K_f
298.15	20.786	160.340	160.340	0.000	89.000	60.479	-10.596
300	20.786	160.468	160.340	0.038	88.984	60.302	-10.499
400	20.786	166.448	161.155	2.117	85.598	51.332	-6.703
500	20.786	171.086	162.695	4.196	84.563	42.887	-4.480
600	20.786	174.876	164.419	6.274	83.599	34.643	-3.016
700	20.786	178.080	166.148	8.353	82.680	26.557	-1.982
800	20.786	180.856	167.817	10.431	81.776	18.601	-1.215
900	20.786	183.304	169.404	12.510	80.859	10.759	-0.624
1000	20.786	185.494	170.905	14.589	79.897	3.021	-0.158
1039.4	20.786	186.297	171.474	15.408		pressure = 1 bar	
1100	20.786	187.475	172.323	16.667	0.000	0.000	0.000
1200	20.786	189.284	173.662	18.746	0.000	0.000	0.000
1300	20.789	190.948	174.929	20.825	0.000	0.000	0.000
1400	20.793	192.489	176.129	22.904	0.000	0.000	0.000
1500	20.801	193.923	177.268	24.983	0.000	0.000	0.000

61. DIPOTASSIUM OXIDE K_2O (cr, l)

T/K	C_p°	S°	$-(G^\circ - H^\circ(T_r))/T$	$H^\circ - H^\circ(T_r)$	$\Delta_f H^\circ$	$\Delta_f G^\circ$	Log K_f
298.15	72.000	96.000	96.000	0.000	-361.700	-321.171	56.267
300	72.130	96.446	96.001	0.133	-361.704	-320.920	55.876
400	79.154	118.158	98.914	7.698	-366.554	-306.416	40.013
500	86.178	136.575	104.647	15.964	-366.043	-291.423	30.444
590	92.500	151.348	110.662	24.005	-364.204	-278.079	24.619
PHASE TRANSITION: $\Delta_{trs} H = 0.700$ kJ/mol, $\Delta_{trs} S = 1.186$ J/K·mol, crIII–crII							

T/K	$C_p°$	$S°$	$-(G°-H°(T_r))/T$	$H°-H°(T_r)$	$\Delta_f H°$	$\Delta_f G°$	Log K_f
	J/K·mol			**kJ/mol**			

61. DIPOTASSIUM OXIDE K_2O (cr, l) (continued)

T/K	$C_p°$	$S°$	$-(G°-H°(T_r))/T$	$H°-H°(T_r)$	$\Delta_f H°$	$\Delta_f G°$	Log K_f
590	100.000	152.534	110.662	24.705	−364.204	−278.079	24.619
600	100.000	154.215	111.374	25.705	−363.968	−276.621	24.082
645	100.000	161.447	114.618	30.205	−358.901	−270.109	21.874

PHASE TRANSITION: $\Delta_{trs} H = 4.000$ kJ/mol, $\Delta_{trs} S = 6.202$ J/K·mol, crII–crI

T/K	$C_p°$	$S°$	$-(G°-H°(T_r))/T$	$H°-H°(T_r)$	$\Delta_f H°$	$\Delta_f G°$	Log K_f
645	100.000	167.649	114.618	34.205	−358.901	−270.109	21.874
700	100.000	175.832	119.111	39.705	−357.592	−262.592	19.595
800	100.000	189.185	127.054	49.705	−355.224	−249.183	16.270
900	100.000	200.963	134.625	59.705	−352.919	−236.067	13.701
1000	100.000	211.499	141.794	69.705	−350.732	−223.202	11.659
1013	100.000	212.791	142.697	71.005	−323.459	−221.546	11.424

PHASE TRANSITION: $\Delta_{trs} H = 27.000$ kJ/mol, $\Delta_{trs} S = 26.654$ J/K·mol, crI–l

T/K	$C_p°$	$S°$	$-(G°-H°(T_r))/T$	$H°-H°(T_r)$	$\Delta_f H°$	$\Delta_f G°$	Log K_f
1013	100.000	239.444	142.697	98.005	−323.459	−221.546	11.424
1100	100.000	247.684	150.679	106.705	−479.439	−203.633	9.670
1200	100.000	256.385	159.131	116.705	−475.371	−178.740	7.780
1300	100.000	264.389	166.924	126.705	−471.321	−154.185	6.195
1400	100.000	271.800	174.154	136.705	−467.287	−129.941	4.848
1500	100.000	278.699	180.896	146.705	−463.268	−105.986	3.691

62. POTASSIUM HYDROXIDE   KOH (cr, l)

T/K	$C_p°$	$S°$	$-(G°-H°(T_r))/T$	$H°-H°(T_r)$	$\Delta_f H°$	$\Delta_f G°$	Log K_f
298.15	64.900	78.870	78.870	0.000	−424.580	−378.747	66.354
300	65.038	79.272	78.871	0.120	−424.569	−378.463	65.895
400	72.519	99.007	81.512	6.998	−426.094	−362.765	47.372
500	80.000	115.993	86.745	14.624	−424.572	−347.093	36.260
520	81.496	119.159	87.931	16.239	−417.725	−344.002	34.555

PHASE TRANSITION: $\Delta_{trs} H = 6.450$ kJ/mol, $\Delta_{trs} S = 12.404$ J/K·mol, crII–crI

T/K	$C_p°$	$S°$	$-(G°-H°(T_r))/T$	$H°-H°(T_r)$	$\Delta_f H°$	$\Delta_f G°$	Log K_f
520	79.000	131.563	87.931	22.689	−417.725	−344.002	34.555
600	79.000	142.868	94.520	29.009	−416.274	−332.766	28.969
678	79.000	152.523	100.649	35.171	−405.464	−321.998	24.807

PHASE TRANSITION: $\Delta_{trs} H = 9.400$ kJ/mol, $\Delta_{trs} S = 13.865$ J/K·mol, crI–l

T/K	$C_p°$	$S°$	$-(G°-H°(T_r))/T$	$H°-H°(T_r)$	$\Delta_f H°$	$\Delta_f G°$	Log K_f
678	83.000	166.388	100.649	44.571	−405.464	−321.998	24.807
700	83.000	169.038	102.757	46.397	−404.981	−319.297	23.826
800	83.000	180.121	111.750	54.697	−402.808	−307.206	20.058
900	83.000	189.897	119.901	62.997	−400.694	−295.383	17.143
1000	83.000	198.642	127.345	71.297	−398.668	−283.791	14.824
1100	83.000	206.553	134.192	79.597	−475.618	−267.780	12.716
1200	83.000	213.775	140.527	87.897	−472.711	−249.014	10.839
1300	83.000	220.418	146.421	96.197	−469.843	−230.490	9.261
1400	83.000	226.569	151.929	104.497	−467.011	−212.184	7.917
1500	83.000	232.296	157.098	112.797	−464.217	−194.080	6.758

63. POTASSIUM HYDROXIDE   KOH (g)

T/K	$C_p°$	$S°$	$-(G°-H°(T_r))/T$	$H°-H°(T_r)$	$\Delta_f H°$	$\Delta_f G°$	Log K_f
298.15	49.184	238.283	238.283	0.000	−227.989	−229.685	40.239
300	49.236	238.588	238.284	0.091	−228.007	−229.696	39.993
400	51.178	253.053	240.243	5.124	−231.377	−229.667	29.991
500	52.178	264.591	243.998	10.296	−232.309	−229.129	23.937
600	52.804	274.163	248.251	15.547	−233.145	−228.413	19.885
700	53.296	282.340	252.551	20.853	−233.934	−227.562	16.981
800	53.758	289.487	256.730	26.206	−234.708	−226.599	14.795
900	54.229	295.846	260.730	31.605	−235.495	−225.538	13.090
1000	54.713	301.585	264.533	37.052	−236.322	−224.388	11.721
1100	55.203	306.823	268.143	42.548	−316.077	−218.535	10.377
1200	55.686	311.647	271.570	48.092	−315.925	−209.674	9.127
1300	56.153	316.122	274.827	53.684	−315.764	−200.826	8.069
1400	56.598	320.300	277.927	59.322	−315.595	−191.991	7.163
1500	57.016	324.220	280.884	65.003	−315.420	−183.169	6.378

		J/K·mol			kJ/mol		
T/K	C_p°	S°	$-(G^\circ-H^\circ(T_r))/T$	$H^\circ-H^\circ(T_r)$	$\Delta_f H^\circ$	$\Delta_f G^\circ$	Log K_f

64. POTASSIUM CHLORIDE   KCl (cr, l)

298.15	51.300	82.570	82.570	0.000	−436.490	−408.568	71.579
300	51.333	82.887	82.571	0.095	−436.481	−408.395	71.107
400	52.977	97.886	84.605	5.312	−438.463	−398.651	52.058
500	54.448	109.867	88.498	10.685	−437.990	−388.749	40.612
600	55.885	119.921	92.919	16.201	−437.332	−378.960	32.991
700	57.425	128.649	97.413	21.865	−436.502	−369.295	27.557
800	59.205	136.430	101.812	27.694	−435.505	−359.760	23.490
900	61.361	143.523	106.058	33.719	−434.337	−350.360	20.334
1000	64.032	150.121	110.138	39.983	−432.981	−341.100	17.817
1044	65.405	152.908	111.882	42.830	−485.450	−336.720	16.847
PHASE TRANSITION: $\Delta_{trs} H = 26.320$ kJ/mol, $\Delta_{trs} S = 25.210$ J/K·mol, cr–l							
1044	72.000	178.118	111.882	69.150	−485.450	−336.720	16.847
1100	72.000	181.880	115.351	73.182	−483.633	−328.790	15.613
1200	72.000	188.145	121.160	80.382	−480.393	−314.856	13.705
1300	72.000	193.908	126.537	87.582	−477.158	−301.192	12.102
1400	72.000	199.244	131.542	94.782	−473.928	−287.778	10.737
1500	72.000	204.211	136.223	101.982	−470.704	−274.594	9.562

65. POTASSIUM CHLORIDE   KCl (g)

298.15	36.505	239.091	239.091	0.000	−214.575	−233.320	40.876
300	36.518	239.317	239.092	0.068	−214.594	−233.436	40.644
400	37.066	249.904	240.532	3.749	−218.112	−239.107	31.224
500	37.384	258.212	243.267	7.473	−219.287	−244.219	25.513
600	37.597	265.048	246.344	11.222	−220.396	−249.100	21.686
700	37.769	270.857	249.441	14.991	−221.461	−253.799	18.938
800	37.907	275.910	252.441	18.775	−222.509	−258.347	16.868
900	38.041	280.382	255.302	22.572	−223.568	−262.764	15.250
1000	38.162	284.397	258.014	26.383	−224.667	−267.061	13.950
1100	38.279	288.039	260.581	30.205	−304.696	−266.627	12.661
1200	38.401	291.375	263.010	34.039	−304.821	−263.161	11.455
1300	38.518	294.454	265.312	37.885	−304.941	−259.684	10.434
1400	38.639	297.313	267.496	41.743	−305.053	−256.199	9.559
1500	38.761	299.983	269.574	45.613	−305.159	−252.706	8.800

66. DINITROGEN   N$_2$ (g)

298.15	29.124	191.608	191.608	0.000	0.000	0.000	0.000
300	29.125	191.788	191.608	0.054	0.000	0.000	0.000
400	29.249	200.180	192.752	2.971	0.000	0.000	0.000
500	29.580	206.738	194.916	5.911	0.000	0.000	0.000
600	30.109	212.175	197.352	8.894	0.000	0.000	0.000
700	30.754	216.864	199.812	11.936	0.000	0.000	0.000
800	31.433	221.015	202.208	15.046	0.000	0.000	0.000
900	32.090	224.756	204.509	18.222	0.000	0.000	0.000
1000	32.696	228.169	206.706	21.462	0.000	0.000	0.000
1100	33.241	231.311	208.802	24.759	0.000	0.000	0.000
1200	33.723	234.224	210.801	28.108	0.000	0.000	0.000
1300	34.147	236.941	212.708	31.502	0.000	0.000	0.000
1400	34.517	239.485	214.531	34.936	0.000	0.000	0.000
1500	34.842	241.878	216.275	38.404	0.000	0.000	0.000

67. NITRIC OXIDE   NO (g)

298.15	29.862	210.745	210.745	0.000	91.277	87.590	−15.345
300	29.858	210.930	210.746	0.055	91.278	87.567	−15.247
400	29.954	219.519	211.916	3.041	91.320	86.323	−11.272
500	30.493	226.255	214.133	6.061	91.340	85.071	−8.887

	J/K·mol			kJ/mol			
T/K	C_p°	S°	$-(G^\circ - H^\circ(T_r))/T$	$H^\circ - H^\circ(T_r)$	$\Delta_f H^\circ$	$\Delta_f G^\circ$	Log K_f

67. NITRIC OXIDE   NO (g) (continued)

600	31.243	231.879	216.635	9.147	91.354	83.816	−7.297
700	32.031	236.754	219.168	12.310	91.369	82.558	−6.160
800	32.770	241.081	221.642	15.551	91.386	81.298	−5.308
900	33.425	244.979	224.022	18.862	91.405	80.036	−4.645
1000	33.990	248.531	226.298	22.233	91.426	78.772	−4.115
1100	34.473	251.794	228.469	25.657	91.445	77.505	−3.680
1200	34.883	254.811	230.540	29.125	91.464	76.237	−3.318
1300	35.234	257.618	232.516	32.632	91.481	74.967	−3.012
1400	35.533	260.240	234.404	36.170	91.495	73.697	−2.750
1500	35.792	262.700	236.209	39.737	91.506	72.425	−2.522

68. NITROGEN DIOXIDE   NO$_2$ (g)

298.15	37.178	240.166	240.166	0.000	34.193	52.316	−9.165
300	37.236	240.397	240.167	0.069	34.181	52.429	−9.129
400	40.513	251.554	241.666	3.955	33.637	58.600	−7.652
500	43.664	260.939	244.605	8.167	33.319	64.882	−6.778
600	46.383	269.147	248.026	12.673	33.174	71.211	−6.199
700	48.612	276.471	251.575	17.427	33.151	77.553	−5.787
800	50.405	283.083	255.107	22.381	33.213	83.893	−5.478
900	51.844	289.106	258.555	27.496	33.334	90.221	−5.236
1000	53.007	294.631	261.891	32.741	33.495	96.534	−5.042
1100	53.956	299.729	265.102	38.090	33.686	102.828	−4.883
1200	54.741	304.459	268.187	43.526	33.898	109.105	−4.749
1300	55.399	308.867	271.148	49.034	34.124	115.363	−4.635
1400	55.960	312.994	273.992	54.603	34.360	121.603	−4.537
1500	56.446	316.871	276.722	60.224	34.604	127.827	−4.451

69. AMMONIA   NH$_3$ (g)

298.15	35.630	192.768	192.768	0.000	−45.940	−16.407	2.874
300	35.678	192.989	192.769	0.066	−45.981	−16.223	2.825
400	38.674	203.647	194.202	3.778	−48.087	−5.980	0.781
500	41.994	212.633	197.011	7.811	−49.908	4.764	−0.498
600	45.229	220.578	200.289	12.174	−51.430	15.846	−1.379
700	48.269	227.781	203.709	16.850	−52.682	27.161	−2.027
800	51.112	234.414	207.138	21.821	−53.695	38.639	−2.523
900	53.769	240.589	210.516	27.066	−54.499	50.231	−2.915
1000	56.244	246.384	213.816	32.569	−55.122	61.903	−3.233
1100	58.535	251.854	217.027	38.309	−55.589	73.629	−3.496
1200	60.644	257.039	220.147	44.270	−55.920	85.392	−3.717
1300	62.576	261.970	223.176	50.432	−56.136	97.177	−3.905
1400	64.339	266.673	226.117	56.779	−56.251	108.975	−4.066
1500	65.945	271.168	228.971	63.295	−56.282	120.779	−4.206

70. OXYGEN   O (g)

298.15	21.911	161.058	161.058	0.000	249.180	231.743	−40.600
300	21.901	161.194	161.059	0.041	249.193	231.635	−40.331
400	21.482	167.430	161.912	2.207	249.874	225.677	−29.470
500	21.257	172.197	163.511	4.343	250.481	219.556	−22.937
600	21.124	176.060	165.290	6.462	251.019	213.319	−18.571
700	21.040	179.310	167.067	8.570	251.500	206.997	−15.446
800	20.984	182.115	168.777	10.671	251.932	200.610	−13.098
900	20.944	184.584	170.399	12.767	252.325	194.171	−11.269
1000	20.915	186.789	171.930	14.860	252.686	187.689	−9.804
1100	20.893	188.782	173.372	16.950	253.022	181.173	−8.603
1200	20.877	190.599	174.733	19.039	253.335	174.628	−7.601
1300	20.864	192.270	176.019	21.126	253.630	168.057	−6.753

	J/K·mol			kJ/mol			
T/K	C_p°	S°	$-(G^\circ-H^\circ(T_r))/T$	$H^\circ-H^\circ(T_r)$	$\Delta_f H^\circ$	$\Delta_f G^\circ$	Log K_f

70. OXYGEN   O (g) (continued)

1400	20.853	193.815	177.236	23.212	253.908	161.463	−6.024
1500	20.845	195.254	178.389	25.296	254.171	154.851	−5.392

71. DIOXYGEN O_2 (g)

298.15	29.378	205.148	205.148	0.000	0.000	0.000	0.000
300	29.387	205.330	205.148	0.054	0.000	0.000	0.000
400	30.109	213.873	206.308	3.026	0.000	0.000	0.000
500	31.094	220.695	208.525	6.085	0.000	0.000	0.000
600	32.095	226.454	211.045	9.245	0.000	0.000	0.000
700	32.987	231.470	213.612	12.500	0.000	0.000	0.000
800	33.741	235.925	216.128	15.838	0.000	0.000	0.000
900	34.365	239.937	218.554	19.244	0.000	0.000	0.000
1000	34.881	243.585	220.878	22.707	0.000	0.000	0.000
1100	35.314	246.930	223.096	26.217	0.000	0.000	0.000
1200	35.683	250.019	225.213	29.768	0.000	0.000	0.000
1300	36.006	252.888	227.233	33.352	0.000	0.000	0.000
1400	36.297	255.568	229.162	36.968	0.000	0.000	0.000
1500	36.567	258.081	231.007	40.611	0.000	0.000	0.000

72. SULFUR   S (cr, l)

298.15	22.690	32.070	32.070	0.000	0.000	0.000	0.000
300	22.737	32.210	32.070	0.042	0.000	0.000	0.000
368.3	24.237	37.030	32.554	1.649	0.000	0.000	0.000
PHASE TRANSITION: $\Delta_{trs} H = 0.401$ kJ/mol, $\Delta_{trs} S = 1.089$ J/K·mol, crII–crI							
368.3	24.773	38.119	32.553	2.050	0.000	0.000	0.000
388.36	25.180	39.444	32.875	2.551	0.000	0.000	0.000
PHASE TRANSITION: $\Delta_{trs} H = 1.722$ kJ/mol, $\Delta_{trs} S = 4.431$ J/K·mol, crI–l							
388.36	31.710	43.875	32.872	4.273	0.000	0.000	0.000
400	32.369	44.824	33.206	4.647	0.000	0.000	0.000
500	38.026	53.578	36.411	8.584	0.000	0.000	0.000
600	34.371	60.116	39.842	12.164	0.000	0.000	0.000
700	32.451	65.278	43.120	15.511	0.000	0.000	0.000
800	32.000	69.557	46.163	18.715	0.000	0.000	0.000
882.38	32.000	72.693	48.496	21.351	0.000	0.000	0.000

73. SULFUR   S (g)

298.15	23.673	167.828	167.828	0.000	277.180	236.704	−41.469
300	23.669	167.974	167.828	0.044	277.182	236.453	−41.170
400	23.233	174.730	168.752	2.391	274.924	222.962	−29.115
500	22.741	179.860	170.482	4.689	273.286	210.145	−21.953
600	22.338	183.969	172.398	6.942	271.958	197.646	−17.206
700	22.031	187.388	174.302	9.160	270.829	185.352	−13.831
800	21.800	190.314	176.125	11.351	269.816	173.210	−11.309
900	21.624	192.871	177.847	13.522	215.723	162.258	−9.417
1000	21.489	195.142	179.465	15.677	216.018	156.301	−8.164
1100	21.386	197.185	180.985	17.821	216.284	150.317	−7.138
1200	21.307	199.043	182.413	19.955	216.525	144.309	−6.282
1300	21.249	200.746	183.759	22.083	216.743	138.282	−5.556
1400	21.209	202.319	185.029	24.206	216.940	132.239	−4.934
1500	21.186	203.781	186.231	26.325	217.119	126.182	−4.394

74. DISULFUR S_2 (g)

298.15	32.505	228.165	228.165	0.000	128.600	79.696	−13.962
300	32.540	228.366	228.165	0.060	128.576	79.393	−13.823
400	34.108	237.956	229.462	3.398	122.703	63.380	−8.276
500	35.133	245.686	231.959	6.863	118.296	49.031	−5.122

	J/K·mol			kJ/mol			
T/K	C_p°	S°	$-(G^\circ - H^\circ(T_r))/T$	$H^\circ - H^\circ(T_r)$	$\Delta_f H^\circ$	$\Delta_f G^\circ$	$\text{Log } K_f$

74. DISULFUR S_2 (g) (continued)

600	35.815	252.156	234.800	10.413	114.685	35.530	−3.093
700	36.305	257.715	237.686	14.020	111.599	22.588	−1.685
800	36.697	262.589	240.501	17.671	108.841	10.060	−0.657
882.38	36.985	266.200	242.734	20.706		pressure = 1 bar	
900	37.045	266.932	243.201	21.358	0.000	0.000	0.000
1000	37.377	270.852	245.773	25.079	0.000	0.000	0.000
1100	37.704	274.430	248.218	28.833	0.000	0.000	0.000
1200	38.030	277.725	250.541	32.620	0.000	0.000	0.000
1300	38.353	280.781	252.751	36.439	0.000	0.000	0.000
1400	38.669	283.635	254.856	40.290	0.000	0.000	0.000
1500	38.976	286.314	256.865	44.173	0.000	0.000	0.000

75. OCTASULFUR S_8 (g)

298.15	156.500	432.536	432.536	0.000	101.277	48.810	−8.551
300	156.768	433.505	432.539	0.290	101.231	48.484	−8.442
400	167.125	480.190	438.834	16.542	80.642	32.003	−4.179
500	173.181	518.176	451.022	33.577	66.185	21.409	−2.237
600	177.936	550.180	464.951	51.137	55.101	13.549	−1.180
700	182.441	577.948	479.152	69.157	46.349	7.343	−0.548
800	186.764	602.596	493.071	87.620	39.177	2.263	−0.148
900	190.595	624.821	506.495	106.494	−392.062	6.554	−0.380
1000	193.618	645.067	519.355	125.712	−387.728	50.614	−2.644
1100	195.684	663.625	531.639	145.185	−383.272	94.233	−4.475
1200	196.825	680.707	543.359	164.817	−378.786	137.444	−5.983
1300	197.195	696.480	554.539	184.524	−374.356	180.283	−7.244
1400	196.988	711.089	565.206	204.237	−370.048	222.785	−8.312
1500	196.396	724.662	575.389	223.909	−365.905	264.984	−9.227

76. SULFUR DIOXIDE SO_2 (g)

298.15	39.842	248.219	248.219	0.000	−296.810	−300.090	52.574
300	39.909	248.466	248.220	0.074	−296.833	−300.110	52.253
400	43.427	260.435	249.828	4.243	−300.240	−300.935	39.298
500	46.490	270.465	252.978	8.744	−302.735	−300.831	31.427
600	48.938	279.167	256.634	13.520	−304.699	−300.258	26.139
700	50.829	286.859	260.413	18.513	−306.308	−299.386	22.340
800	52.282	293.746	264.157	23.671	−307.691	−298.302	19.477
900	53.407	299.971	267.796	28.958	−362.075	−295.987	17.178
1000	54.290	305.646	271.301	34.345	−362.012	−288.647	15.077
1100	54.993	310.855	274.664	39.810	−361.934	−281.314	13.358
1200	55.564	315.665	277.882	45.339	−361.849	−273.989	11.926
1300	56.033	320.131	280.963	50.920	−361.763	−266.671	10.715
1400	56.426	324.299	283.911	56.543	−361.680	−259.359	9.677
1500	56.759	328.203	286.735	62.203	−361.605	−252.053	8.777

77. SILICON   Si (cr)

298.15	19.789	18.810	18.810	0.000	0.000	0.000	0.000
300	19.855	18.933	18.810	0.037	0.000	0.000	0.000
400	22.301	25.023	19.624	2.160	0.000	0.000	0.000
500	23.610	30.152	21.231	4.461	0.000	0.000	0.000
600	24.472	34.537	23.092	6.867	0.000	0.000	0.000
700	25.124	38.361	25.006	9.348	0.000	0.000	0.000
800	25.662	41.752	26.891	11.888	0.000	0.000	0.000
900	26.135	44.802	28.715	14.478	0.000	0.000	0.000
1000	26.568	47.578	30.464	17.114	0.000	0.000	0.000
1100	26.974	50.130	32.138	19.791	0.000	0.000	0.000
1200	27.362	52.493	33.737	22.508	0.000	0.000	0.000

THERMODYNAMIC PROPERTIES AS A FUNCTION OF TEMPERATURE (continued)

	J/K·mol			kJ/mol			
T/K	C_p°	S°	$-(G^\circ-H^\circ(T_r))/T$	$H^\circ-H^\circ(T_r)$	$\Delta_f H^\circ$	$\Delta_f G^\circ$	Log K_f

77. SILICON   Si (cr) (continued)

1300	27.737	54.698	35.265	25.263	0.000	0.000	0.000
1400	28.103	56.767	36.728	28.055	0.000	0.000	0.000
1500	28.462	58.719	38.130	30.883	0.000	0.000	0.000

78. SILICON   Si (g)

298.15	22.251	167.980	167.980	0.000	450.000	405.525	−71.045
300	22.234	168.117	167.980	0.041	450.004	405.249	−70.559
400	21.613	174.416	168.843	2.229	450.070	390.312	−50.969
500	21.316	179.204	170.456	4.374	449.913	375.388	−39.216
600	21.153	183.074	172.246	6.497	449.630	360.508	−31.385
700	21.057	186.327	174.032	8.607	449.259	345.682	−25.795
800	21.000	189.135	175.748	10.709	448.821	330.915	−21.606
900	20.971	191.606	177.375	12.808	448.329	316.205	−18.352
1000	20.968	193.815	178.911	14.904	447.791	301.553	−15.751
1100	20.989	195.815	180.358	17.002	447.211	286.957	−13.626
1200	21.033	197.643	181.723	19.103	446.595	272.416	−11.858
1300	21.099	199.329	183.014	21.209	445.946	257.927	−10.364
1400	21.183	200.895	184.236	23.323	445.268	243.489	−9.085
1500	21.282	202.360	185.396	25.446	444.563	229.101	−7.978

79. SILICON DIOXIDE   SiO$_2$ (cr)

298.15	44.602	41.460	41.460	0.000	−910.700	−856.288	150.016
300	44.712	41.736	41.461	0.083	−910.708	−855.951	149.032
400	53.477	55.744	43.311	4.973	−910.912	−837.651	109.385
500	60.533	68.505	47.094	10.705	−910.540	−819.369	85.598
600	64.452	79.919	51.633	16.971	−909.841	−801.197	69.749
700	68.234	90.114	56.414	23.590	−908.958	−783.157	58.439
800	76.224	99.674	61.226	30.758	−907.668	−765.265	49.966
848	82.967	104.298	63.533	34.569	−906.310	−756.747	46.613

PHASE TRANSITION: $\Delta_{trs} H = 0.411$ kJ/mol, $\Delta_{trs} S = 0.484$ J/K·mol, crII–crII′

848	67.446	104.782	63.532	34.980	−906.310	−756.747	46.613
900	67.953	108.811	66.033	38.500	−905.922	−747.587	43.388
1000	68.941	116.021	70.676	45.345	−905.176	−730.034	38.133
1100	69.940	122.639	75.104	52.289	−904.420	−712.557	33.836
1200	70.947	128.768	79.323	59.333	−901.382	−695.148	30.259

PHASE TRANSITION: $\Delta_{trs} H = 2.261$ kJ/mol, $\Delta_{trs} S = 1.883$ J/K·mol, crII′–crI

1200	71.199	130.651	79.323	61.594	−901.382	−695.148	30.259
1300	71.743	136.372	83.494	68.742	−900.574	−677.994	27.242
1400	72.249	141.707	87.463	75.941	−899.782	−660.903	24.658
1500	72.739	146.709	91.248	83.191	−899.004	−643.867	22.421

80. SILICON TETRACHLORIDE   SiCl$_4$ (g)

298.15	90.404	331.446	331.446	0.000	−662.200	−622.390	109.039
300	90.562	332.006	331.448	0.167	−662.195	−622.143	108.323
400	96.893	359.019	335.088	9.572	−661.853	−608.841	79.505
500	100.449	381.058	342.147	19.456	−661.413	−595.637	62.225
600	102.587	399.576	350.216	29.616	−660.924	−582.527	50.713
700	103.954	415.500	358.432	39.948	−660.417	−569.501	42.496
800	104.875	429.445	366.455	50.392	−659.912	−556.548	36.338
900	105.523	441.837	374.155	60.914	−659.422	−543.657	31.553
1000	105.995	452.981	381.490	71.491	−658.954	−530.819	27.727
1100	106.349	463.101	388.456	82.109	−658.515	−518.027	24.599
1200	106.620	472.366	395.068	92.758	−658.107	−505.274	21.994
1300	106.834	480.909	401.347	103.431	−657.735	−492.553	19.791
1400	107.003	488.833	407.316	114.123	−657.400	−479.860	17.904
1500	107.141	496.220	413.000	124.830	−657.104	−467.189	16.269

THERMODYNAMIC PROPERTIES OF AQUEOUS SYSTEMS

This table contains standard state thermodynamic properties of ions and neutral species in aqueous solution. It includes enthalpy and Gibbs energy of formation, entropy, and heat capacity, and thus serves as a companion to the preceding table, "Standard Thermodynamic Properties of Chemical Substances". The standard state is the hypothetical ideal solution with molality $m = 1$ mol/kg (mean ionic molality $m_\pm$ in the case of a species which is assumed to dissociate at infinite dilution). Further details on conventions may be found in Reference 1.

Cations are listed by formula in the first part of the table, followed by anions and finally neutral species. All values refer to standard conditions of 25°C and 100 kPa pressure.

REFERENCES

1. Wagman, D. D., Evans, W. H., Parker, V. B., Schumm, R. H., Halow, I., Bailey, S. M., Churney, K. L., and Nuttall, R. L., *The NBS Tables of Chemical Thermodynamic Properties, J. Phys. Chem. Ref. Data*, Vol. 11, Suppl. 2, 1982.
2. Zemaitis, J. F., Clark, D. M., Rafal, M., and Scrivner, N. C., *Handbook of Aqueous Electrolyte Thermodynamics*, American Institute of Chemical Engineers, New York, 1986.

Species	$\Delta_f H°/$ kJ mol⁻¹	$\Delta_f G°/$ kJ mol⁻¹	$S°/$ J mol⁻¹K⁻¹	$C_p/$ J mol⁻¹K⁻¹	Species	$\Delta_f H°/$ kJ mol⁻¹	$\Delta_f G°/$ kJ mol⁻¹	$S°/$ J mol⁻¹K⁻¹	$C_p/$ J mol⁻¹K⁻¹
Cations					$InOH^{+2}$	-370.3	-313.0	-88.0	
					$In(OH)_2^+$	-619.0	-525.0	25.0	
Ag^+	105.6	77.1	72.7	21.8	K^+	-252.4	-283.3	102.5	21.8
Al^{+3}	-531.0	-485.0	-321.7		La^{+3}	-707.1	-683.7	-217.6	-13.0
$AlOH^{+2}$		-694.1			Li^+	-278.5	-293.3	13.4	68.6
Ba^{+2}	-537.6	-560.8	9.6		Lu^{+3}	-665.0	-628.0	-264.0	25.0
$BaOH^+$		-730.5			LuF^{+2}		-931.4		
Be^{+2}	-382.8	-379.7	-129.7		Mg^{+2}	-466.9	-454.8	-138.1	
Bi^{+3}		82.8			$MgOH^+$		-626.7		
$BiOH^{+2}$		-146.4			Mn^{+2}	-220.8	-228.1	-73.6	50.0
Ca^{+2}	-542.8	-553.6	-53.1		$MnOH^+$	-450.6	-405.0	-17.0	
$CaOH^+$		-718.4			NH_4^+	-132.5	-79.3	113.4	79.9
Cd^{+2}	-75.9	-77.6	-73.2		$N_2H_5^+$	-7.5	82.5	151.0	70.3
$CdOH^+$		-261.1			Na^+	-240.1	-261.9	59.0	46.4
Ce^{+3}	-696.2	-672.0	-205.0		Nd^{+3}	-696.2	-671.6	-206.7	-21.0
Ce^{+4}	-537.2	-503.8	-301.0		Ni^{+2}	-54.0	-45.6	-128.9	
Co^{+2}	-58.2	-54.4	-113.0		$NiOH^+$	-287.9	-227.6	-71.0	
Co^{+3}	92.0	134.0	-305.0		PH_4^+		92.1		
Cr^{+2}	-143.5				Pa^{+4}	-619.0			
Cs^+	-258.3	-292.0	133.1	-10.5	Pb^{+2}	-1.7	-24.4	10.5	
Cu^+	71.7	50.0	40.6		$PbOH^+$		-226.3		
Cu^{+2}	64.8	65.5	-99.6		Pd^{+2}	149.0	176.5	-184.0	
Dy^{+3}	-699.0	-665.0	-231.0	21.0	Po^{+2}		71.0		
Er^{+3}	-705.4	-669.1	-244.3	21.0	Po^{+4}		293.0		
Eu^{+2}	-527.0	-540.2	-8.0		Pr^{+3}	-704.6	-679.1	-209.0	-29.0
Eu^{+3}	-605.0	-574.1	-222.0	8.0	Pt^{+2}		254.8		
Fe^{+2}	-89.1	-78.9	-137.7		Ra^{+2}	-527.6	-561.5	54.0	
Fe^{+3}	-48.5	-4.7	-315.9		Rb^+	-251.2	-284.0	121.5	
$FeOH^+$	-324.7	-277.4	-29.0		Re^+		-33.0		
$FeOH^{+2}$	-290.8	-229.4	-142.0		Sc^{+3}	-614.2	-586.6	-255.0	
$Fe(OH)_2^+$		-438.0			$ScOH^{+2}$	-861.5	-801.2	-134.0	
Ga^{+2}		-88.0			Sm^{+2}		-497.5		
Ga^{+3}	-211.7	-159.0	-331.0		Sm^{+3}	-691.6	-666.6	-211.7	-21.0
$GaOH^{+2}$		-380.3			Sn^{+2}	-8.8	-27.2	-17.0	
$Ga(OH)_2^+$		-597.4			$SnOH^+$	-286.2	-254.8	50.0	
Gd^{+3}	-686.0	-661.0	-205.9		Sr^{+2}	-545.8	-559.5	-32.6	
H^+	0	0	0	0	$SrOH^+$		-721.3		
Hg^{+2}	171.1	164.4	-32.2		Tb^{+3}	-682.8	-651.9	-226.0	17.0
Hg_2^{+2}	172.4	153.5	84.5		$Te(OH)_3^+$	-608.4	-496.1	111.7	
$HgOH^+$	-84.5	-52.3	71.0		Th^{+4}	-769.0	-705.1	-422.6	
Ho^{+3}	-705.0	-673.7	-226.8	17.0	$Th(OH)^{+3}$	-1030.1	-920.5	-343.0	
In^+		-12.1			$Th(OH)_2^{+2}$	-1282.4	-1140.9	-218.0	
In^{+2}		-50.7			Tl^+	5.4	-32.4	125.5	
In^{+3}	-105.0	-98.0	-151.0		Tl^{+3}	196.6	214.6	-192.0	

Species	$\Delta_f H°/$ kJ mol^{-1}	$\Delta_f G°/$ kJ mol^{-1}	$S°/$ J mol^{-1}K^{-1}	$C_p/$ J mol^{-1}K^{-1}
TlOH^{+2}		-15.9		
Tl(OH)$_2^+$		-244.7		
Tm^{+3}	-697.9	-662.0	-243.0	25.0
U^{+3}	-489.1	-476.2	-188.0	
U^{+4}	-591.2	-531.9	-410.0	
Y^{+3}	-723.4	-693.8	-251.0	
Y$_2$(OH)$_2^{+4}$		-1780.3		
Yb^{+2}		-527.0		
Yb^{+3}	-674.5	-644.0	-238.0	25.0
Y(OH)$^{+2}$		-879.1		
Zn^{+2}	-153.9	-147.1	-112.1	46.0
ZnOH$^+$		-330.1		

Anions

Species	$\Delta_f H°/$ kJ mol^{-1}	$\Delta_f G°/$ kJ mol^{-1}	$S°/$ J mol^{-1}K^{-1}	$C_p/$ J mol^{-1}K^{-1}
AlO$_2^-$	-930.9	-830.9	-36.8	
Al(OH)$_4^-$	-1502.5	-1305.3	102.9	
AsO$_2^-$	-429.0	-350.0	40.6	
AsO$_4^{-3}$	-888.1	-648.4	-162.8	
BF$_4^-$	-1574.9	-1486.9	180.0	
BH$_4^-$	48.2	114.4	110.5	
BO$_2^-$	-772.4	-678.9	-37.2	
B$_4$O$_7^{-2}$		-2604.8		
BeO$_2^{-2}$	-790.8	-640.1	-159.0	
Br$^-$	-121.6	-104.0	82.4	-141.8
BrO$^-$	-94.1	-33.4	42.0	
BrO$_3^-$	-67.1	18.6	161.7	
BrO$_4^-$	13.0	118.1	199.6	
CHOO$^-$	-425.6	-351.0	92.0	-87.9
CH$_3$COO$^-$	-486.0	-369.3	86.6	-6.3
C$_2$O$_4^{-2}$	-825.1	-673.9	45.6	
C$_2$O$_4$H$^-$	-818.4	-698.3	149.4	
Cl$^-$	-167.2	-131.2	56.5	-136.4
ClO$^-$	-107.1	-36.8	42.0	
ClO$_2^-$	-66.5	17.2	101.3	
ClO$_3^-$	-104.0	-8.0	162.3	
ClO$_4^-$	-129.3	-8.5	182.0	
CN$^-$	150.6	172.4	94.1	
CO$_3^{-2}$	-677.1	-527.8	-56.9	
CrO$_4^{-2}$	-881.2	-727.8	50.2	
Cr$_2$O$_7^{-2}$	-1490.3	-1301.1	261.9	
F$^-$	-332.6	-278.8	-13.8	-106.7
Fe(CN)$_6^{-3}$	561.9	729.4	270.3	
Fe(CN)$_6^{-4}$	455.6	695.1	95.0	
HB$_4$O$_7^-$		-2685.1		
HCO$_3^-$	-692.0	-586.8	91.2	
HF$_2^-$	-649.9	-578.1	92.5	
HPO$_3$F$^-$		-1198.2		
HPO$_4^{-2}$	-1292.1	-1089.2	-33.5	
HP$_2$O$_7^{-3}$	-2274.8	-1972.2	46.0	
HS$^-$	-17.6	12.1	62.8	
HSO$_3^-$	-626.2	-527.7	139.7	
HSO$_4^-$	-887.3	-755.9	131.8	-84.0
HS$_2$O$_4^-$		-614.5		
HSe$^-$	15.9	44.0	79.0	
HSeO$_3^-$	-514.6	-411.5	135.1	
HSeO$_4^-$	-581.6	-452.2	149.4	
H$_2$AsO$_3^-$	-714.8	-587.1	110.5	
H$_2$AsO$_4^-$	-909.6	-753.2	117.0	

Species	$\Delta_f H°/$ kJ mol^{-1}	$\Delta_f G°/$ kJ mol^{-1}	$S°/$ J mol^{-1}K^{-1}	$C_p/$ J mol^{-1}K^{-1}
H$_2$PO$_4^-$	-1296.3	-1130.2	90.4	
H$_2$P$_2$O$_7^{-2}$	-2278.6	-2010.2	163.0	
I$^-$	-55.2	-51.6	111.3	-142.3
IO$^-$	-107.5	-38.5	-5.4	
IO$_3^-$	-221.3	-128.0	118.4	
IO$_4^-$	-151.5	-58.5	222.0	
MnO$_4^-$	-541.4	-447.2	191.2	-82.0
MnO$_4^{-2}$	-653.0	-500.7	59.0	
MoO$_4^{-2}$	-997.9	-836.3	27.2	
NO$_2^-$	-104.6	-32.2	123.0	-97.5
NO$_3^-$	-207.4	-111.3	146.4	-86.6
N$_3^-$	275.1	348.2	107.9	
OCN$^-$	-146.0	-97.4	106.7	
OH$^-$	-230.0	-157.2	-10.8	-148.5
PO$_4^{-3}$	-1277.4	-1018.7	-220.5	
P$_2$O$_7^{-4}$	-2271.1	-1919.0	-117.0	
Re$^-$	46.0	10.1	230.0	
S^{-2}	33.1	85.8	-14.6	
SCN$^-$	76.4	92.7	144.3	-40.2
SO$_3^{-2}$	-635.5	-486.5	-29.0	
SO$_4^{-2}$	-909.3	-744.5	20.1	-293.0
S$_2^{-2}$	30.1	79.5	28.5	
S$_2$O$_3^{-2}$	-652.3	-522.5	67.0	
S$_2$O$_4^{-2}$	-753.5	-600.3	92.0	
S$_2$O$_8^{-2}$	-1344.7	-1114.9	244.3	
Se^{-2}		129.3		
SeO$_3^{-2}$	-509.2	-369.8	13.0	
SeO$_4^{-2}$	-599.1	-441.3	54.0	
VO$_3^-$	-888.3	-783.6	50.0	
VO$_4^{-3}$		-899.0		
WO$_4^{-2}$	-1075.7			

Neutral species

Species	$\Delta_f H°/$ kJ mol^{-1}	$\Delta_f G°/$ kJ mol^{-1}	$S°/$ J mol^{-1}K^{-1}	$C_p/$ J mol^{-1}K^{-1}
AgBr	-16.0	-26.9	155.2	-120.1
AgCl	-61.6	-54.1	129.3	-114.6
AgF	-227.1	-201.7	59.0	-84.9
AgI	50.4	25.5	184.1	-120.5
AgNO$_3$	-101.8	-34.2	219.2	-64.9
Ag$_2$SO$_4$	-698.1	-590.3	165.7	-251.0
AlBr$_3$	-895.0	-799.0	-74.5	
AlCl$_3$	-1033.0	-879.0	-152.3	
AlF$_3$	-1531.0	-1322.0	-363.2	
AlI$_3$	-699.0	-640.0	12.1	
Al$_2$(SO$_4$)$_3$	-3791.0	-3205.0	-583.2	
BaBr$_2$	-780.7	-768.7	174.5	
BaCO$_3$	-1214.8	-1088.6	-47.3	
BaCl$_2$	-872.0	-823.2	122.6	
BaF$_2$	-1202.9	-1118.4	-18.0	
Ba(HCO$_3$)$_2$	-1921.6	-1734.3	192.0	
BaI$_2$	-648.0	-663.9	232.2	
Ba(NO$_3$)$_2$	-952.4	-783.3	302.5	
BaSO$_4$	-1446.9	-1305.3	29.7	
BeSO$_4$	-1292.0	-1124.3	-109.6	
CCl$_3$COOH	-516.3			
CHCl$_2$COOH	-512.1			
CHOOCs	-683.8	-643.0	226.0	
CHOOH	-425.6	-351.0	92.0	-87.9
CHOOK	-677.9	-634.2	192.0	-66.1

Species	$\Delta_f H°/$ kJ mol^{-1}	$\Delta_f G°/$ kJ mol^{-1}	$S°/$ J mol^{-1}K^{-1}	$C_p/$ J mol^{-1}K^{-1}	Species	$\Delta_f H°/$ kJ mol^{-1}	$\Delta_f G°/$ kJ mol^{-1}	$S°/$ J mol^{-1}K^{-1}	$C_p/$ J mol^{-1}K^{-1}
CHOONH$_4$	-558.1	-430.4	205.0	-7.9	GdCl$_3$	-1188.0	-1059.0	-36.8	-410.0
CHOONa	-665.7	-612.9	151.0	-41.4	HBr	-121.6	-104.0	82.4	-141.8
CHOORb	-676.7	-635.1	213.0		HCN	150.6	172.4	94.1	
CH$_2$ClCOOH	-501.3				HCl	-167.2	-131.2	56.5	-136.4
CH$_3$COOCs	-744.3	-661.3	219.7		HF	-332.6	-278.8	-13.8	-106.7
CH$_3$COOH	-486.0	-369.3	86.6	-6.3	HI	-55.2	-51.6	111.3	-142.3
CH$_3$COOK	-738.4	-652.6	189.1	15.5	HNO$_3$	-207.4	-111.3	146.4	-86.6
CH$_3$COONH$_4$	-618.5	-448.6	200.0	73.6	HSCN	76.4	92.7	144.3	-40.2
CH$_3$COONa	-726.1	-631.2	145.6	40.2	H$_2$SO$_4$	-909.3	-744.5	20.1	-293.0
CH$_3$COORb	-737.2	-653.3	207.9		HoCl$_3$	-1206.7	-1067.3	-57.7	-393.0
(COOH)$_2$	-825.1	-673.9	45.6		KBr	-373.9	-387.2	184.9	-120.1
(CH$_3$)$_3$N	-76.0	93.1	133.5		KCl	-419.5	-414.5	159.0	-114.6
CaBr$_2$	-785.9	-761.5	111.7		KF	-585.0	-562.1	88.7	-84.9
CaCO$_3$	-1220.0	-1081.4	-110.0		KHCO$_3$	-944.4	-870.0	193.7	
CaCl$_2$	-877.1	-816.0	59.8		KHSO$_4$	-1139.7	-1039.2	234.3	-63.0
CaF$_2$	-1208.1	-1111.2	-80.8		KI	-307.6	-334.9	213.8	-120.5
CaI$_2$	-653.2	-656.7	169.5		KNO$_3$	-459.7	-394.5	248.9	-64.9
Ca(NO$_3$)$_2$	-957.6	-776.1	239.7		K$_2$CO$_3$	-1181.9	-1094.4	148.1	
CaSO$_4$	-1452.1	-1298.1	-33.1		K$_2$S	-471.5	-480.7	190.4	
CdBr$_2$	-319.0	-285.5	91.6		K$_2$SO$_4$	-1414.0	-1311.1	225.1	-251.0
CdCl$_2$	-410.2	-340.1	39.7		K$_2$Se		-437.2		
CdF$_2$	-741.2	-635.2	-100.8		LaCl$_3$	-1208.8	-1077.3	-50.0	-423.0
CdI$_2$	-186.3	-180.8	149.4		LiBr	-400.0	-397.3	95.8	-73.2
Cd(NO$_3$)$_2$	-490.6	-300.1	219.7		LiCl	-445.6	-424.6	69.9	-67.8
CdSO$_4$	-985.2	-822.1	-53.1		LiF	-611.1	-571.9	-0.4	-38.1
CeCl$_3$	-1197.5	-1065.6	-38.0		LiI	-333.7	-344.8	124.7	-73.6
CoBr$_2$	-301.2	-262.3	50.0		LiNO$_3$	-485.9	-404.5	160.2	-18.0
CoCl$_2$	-392.5	-316.7			Li$_2$CO$_3$	-1234.1	-1114.6	-29.7	
CoI$_2$	-168.6	-157.7	109.0		Li$_2$SO$_4$	-1466.2	-1331.2	47.3	-155.6
Co(NO$_3$)$_2$	-472.8	-276.9	180.0		LuCl$_3$	-1167.0	-1021.0	-96.0	-385.0
CoSO$_4$	-967.3	-799.1	-92.0		MgBr$_2$	-709.9	-662.7	26.8	
CsBr	-379.8	-396.0	215.5		MgCl$_2$	-801.2	-717.1	-25.1	
CsCl	-425.4	-423.2	189.5	-146.9	MgI$_2$	-577.2	-558.1	84.5	
CsF	-590.9	-570.8	119.2		Mg(NO$_3$)$_2$	-881.6	-677.3	154.8	
CsHCO$_3$	-950.3	-878.8	224.3		MgSO$_4$	-1376.1	-1199.5	-118.0	
CsHSO$_4$	-1145.6	-1047.9	264.8		MnBr$_2$	-464.0			
CsI	-313.5	-343.6	244.3	-152.7	MnCl$_2$	-555.1	-490.8	38.9	-222.0
CsNO$_3$	-465.6	-403.3	279.5	-99.0	MnI$_2$	-331.0			
Cs$_2$CO$_3$	-1193.7	-1111.9	209.2		Mn(NO$_3$)$_2$	-635.5	-450.9	218.0	-121.0
Cs$_2$S	-483.7	-498.3	251.0		MnSO$_4$	-1130.1	-972.7	-53.6	-243.0
Cs$_2$SO$_4$	-1425.8	-1328.6	286.2		NH$_4$Br	-254.1	-183.3	195.8	-61.9
Cs$_2$Se		-454.8			NH$_4$BrO$_3$	-199.6	-60.7	275.1	
Cu(NO$_3$)$_2$	-350.0	-157.0	193.3		NH$_4$CN	18.0	93.0	207.5	
CuSO$_4$	-844.5	-679.0	-79.5		NH$_4$Cl	-299.7	-210.5	169.9	-56.5
DyCl$_3$	-1197.0	-1059.0	-61.9	-389.0	NH$_4$ClO$_3$	-236.5	-87.3	275.7	
ErCl$_3$	-1207.1	-1062.7	-75.3	-389.0	NH$_4$ClO$_4$	-261.8	-87.8	295.4	
EuCl$_2$	-862.0				NH$_4$F	-465.1	-358.1	99.6	-26.8
EuCl$_3$	-1106.2	-967.7	-54.0	-402.0	NH$_4$HCO$_3$	-824.5	-666.1	204.6	
FeBr$_2$	-332.2	-286.8	27.2		NH$_4$HS	-150.2	-67.2	176.1	
FeBr$_3$	-413.4	-316.7	-68.6		NH$_4$HSO$_3$	-758.7	-607.0	253.1	
FeCl$_2$	-423.4	-341.3	-24.7		NH$_4$HSO$_4$	-1019.9	-835.2	245.2	-3.8
FeCl$_3$	-550.2	-398.3	-146.4		NH$_4$HSeO$_4$	-714.2	-531.6	262.8	
FeF$_2$	-754.4	-636.5	-165.3		NH$_4$H$_2$AsO$_3$	-847.3	-666.4	223.8	
FeF$_3$	-1046.4	-840.9	-357.3		NH$_4$H$_2$AsO$_4$	-1042.1	-832.5	230.5	
FeI$_2$	-199.6	-182.1	84.9		NH$_4$H$_2$PO$_4$	-1428.8	-1209.6	203.8	
FeI$_3$	-214.2	-159.4	18.0		NH$_4$H$_3$P$_2$O$_7$	-2409.1	-2102.6	326.0	
Fe(NO$_3$)$_3$	-670.7	-338.3	123.4		NH$_4$I	-187.7	-130.9	224.7	-62.3
FeSO$_4$	-998.3	-823.4	-117.6		NH$_4$IO$_3$	-354.0	-207.4	231.8	
Fe$_2$(SO$_4$)$_3$	-2825.0	-2242.8	-571.5		NH$_4$NO$_2$	-237.2	-111.6	236.4	-17.6

Species	$\Delta_f H°/$ kJ mol^{-1}	$\Delta_f G°/$ kJ mol^{-1}	$S°/$ J mol^{-1}K^{-1}	$C_p/$ J mol^{-1}K^{-1}	Species	$\Delta_f H°/$ kJ mol^{-1}	$\Delta_f G°/$ kJ mol^{-1}	$S°/$ J mol^{-1}K^{-1}	$C_p/$ J mol^{-1}K^{-1}
NH_4NO_3	-339.9	-190.6	259.8	-6.7	$RaSO_4$	-1436.8	-1306.2	75.0	
NH_4OH	-362.5	-236.5	102.5	-68.6	$RbBr$	-372.7	-387.9	203.9	
NH_4SCN	-56.1	13.4	257.7	39.7	$RbCl$	-418.3	-415.2	178.0	
$(NH_4)_2CO_3$	-942.2	-686.4	169.9		RbF	-583.8	-562.8	107.5	
$(NH_4)_2CrO_4$	-1146.2	-886.4	277.0		$RbHCO_3$	-943.2	-870.8	212.7	
$(NH_4)_2Cr_2O_7$	-1755.2	-1459.5	488.7		$RbHSO_4$	-1138.5	-1039.9	253.1	
$(NH_4)_2HAsO_4$	-1171.4	-873.2	225.1		RbI	-306.4	-335.6	232.6	
$(NH_4)_2HPO_4$	-1557.2	-1247.8	193.3		$RbNO_3$	-458.5	-395.2	267.8	
$(NH_4)_2S$	-231.8	-72.6	212.1		Rb_2CO_3	-1179.5	-1095.8	186.2	
$(NH_4)_2SO_3$	-900.4	-645.0	197.5		Rb_2S	-469.4	-482.0	228.4	
$(NH_4)_2SO_4$	-1174.3	-903.1	246.9	-133.1	Rb_2SO_4	-1411.6	-1312.5	263.2	
$(NH_4)_2SeO_4$	-864.0	-599.8	280.7		$SmCl_3$	-1193.3	-1060.2	-42.7	-431.0
$(NH_4)_3PO_4$	-1674.9	-1256.6	117.0		$SrBr_2$	-788.9	-767.4	132.2	
$NaBr$	-361.7	-365.8	141.4	-95.4	$SrCO_3$	-1222.9	-1087.3	-89.5	
$NaCl$	-407.3	-393.1	115.5	-90.0	$SrCl_2$	-880.1	-821.9	80.3	
NaF	-572.8	-540.7	45.2	-60.2	SrI_2	-656.2	-662.6	190.0	
$NaHCO_3$	-932.1	-848.7	150.2		$Sr(NO_3)_2$	-960.5	-782.0	260.2	
$NaHSO_4$	-1127.5	-1017.8	190.8	-38.0	$SrSO_4$	-1455.1	-1304.0	-12.6	
NaI	-295.3	-313.5	170.3	-95.8	$TbCl_3$	-1184.1	-1045.5	-59.0	-393.0
$NaNO_3$	-447.5	-373.2	205.4	-40.2	$TlBr$	-116.2	-136.4	207.9	
Na_2CO_3	-1157.4	-1051.6	61.1		$TlBr_3$	-168.2	-97.1	54.0	
Na_2S	-447.3	-438.1	103.3		$TlCl$	-161.8	-163.6	182.0	
Na_2SO_4	-1389.5	-1268.4	138.1	-201.0	$TlCl_3$	-305.0	-179.0	-23.0	
Na_2Se		-394.6			TlF	-327.3	-311.2	111.7	
$NdCl_3$	-1197.9	-1065.6	-37.7	-431.0	TlI	-49.8	-84.0	236.8	
$NiBr_2$	-297.1	-253.6	36.0		$TlNO_3$	-202.0	-143.7	272.0	
$NiCl_2$	-388.3	-307.9	-15.1		Tl_2SO_4	-898.6	-809.3	271.1	
NiF_2	-719.2	-603.3	-156.5		$TmCl_3$	-1199.1	-1055.6	-75.0	-385.0
NiI_2	-164.4	-149.0	93.7		UCl_4	-1259.8	-1056.8	-184.0	
$Ni(NO_3)_2$	-468.6	-268.5	164.0		UO_2CO_3	-1696.6	-1481.5	-154.4	
$NiSO_4$	-963.2	-790.3	-108.8		$UO_2(NO_3)_2$	-1434.3	-1176.0	195.4	
$PbBr_2$	-244.8	-232.3	175.3		UO_2SO_4	-1928.8	-1698.2	-77.4	
$PbCl_2$	-336.0	-286.9	123.4		$YbCl_3$	-1176.1	-1037.6	-71.0	-385.0
PbF_2	-666.9	-582.0	-17.2		$ZnBr_2$	-397.0	-355.0	52.7	-238.0
PbI_2	-112.1	-127.6	233.0		$ZnCl_2$	-488.2	-409.5	0.8	-226.0
$Pb(NO_3)_2$	-416.3	-246.9	303.3		ZnF_2	-819.1	-704.6	-139.7	-167.0
$PrCl_3$	-1206.2	-1072.7	-42.0	-439.0	ZnI_2	-264.3	-250.2	110.5	-238.0
$RaCl_2$	-861.9	-823.8	167.0		$Zn(NO_3)_2$	-568.6	-369.6	180.7	-126.0
$Ra(NO_3)_2$	-942.2	-784.0	347.0		$ZnSO_4$	-1063.2	-891.6	-92.0	-247.0

HEAT OF COMBUSTION

The heat of combustion of a substance at 25°C can be calculated from the enthalpy of formation ($\Delta_f H°$) data in the table "Standard Thermodynamic Properties of Chemical Substances" in this Section. We can write the general combustion reaction as

$$X + O_2 \rightarrow CO_2(g) + H_2O(l) + \text{other products}$$

For a compound containing only carbon, hydrogen, and oxygen, the reaction is simply

$$C_a H_b O_c + \left(a + \frac{1}{4}b - \frac{1}{2}c \right) O_2 \rightarrow a\ CO_2(g) + \frac{1}{2}b\ H_2O(l)$$

and the standard heat of combustion $\Delta_c H°$, which is defined as the negative of the enthalpy change for the reaction (i.e., the heat released in the combustion process), is given by

$$\Delta_c H° = -a\ \Delta_f H°(CO_2, g) - \frac{1}{2}b\ \Delta_f H°(H_2O, l) + \Delta_f H°(C_a H_b O_c)$$

$$393.51\ a + 142.915\ b + \Delta_f H°(C_a H_b O_c)$$

This equation applies if the reactants start in their standard states (25°C and one atmosphere pressure) and the products return to the same conditions. The same equation applies to a compound containing another element if that element ends in its standard reference state (e.g., nitrogen, if the product is N_2); in general, however, the exact products containing the other elements must be known in order to calculate the heat of combustion.

The following table gives the standard heat of combustion calculated in this manner for a few representative substances.

Molecular formula	Name	$\Delta_c H°$/kJ mol^{-1}
Inorganic substances		
C	Carbon (graphite)	393.5
CO	Carbon monoxide (g)	283.0
H$_2$	Hydrogen (g)	285.8
H$_3$N	Ammonia (g)	382.8
H$_4$N$_2$	Hydrazine (g)	667.1
N$_2$O	Nitrous oxide (g)	82.1
Hydrocarbons		
CH$_4$	Methane (g)	890.8
C$_2$H$_2$	Acetylene (g)	1301.1
C$_2$H$_4$	Ethylene (g)	1411.2
C$_2$H$_6$	Ethane (g)	1560.7
C$_3$H$_6$	Propylene (g)	2058.0
C$_3$H$_6$	Cyclopropane (g)	2091.3
C$_3$H$_8$	Propane (g)	2219.2
C$_4$H$_6$	1,3-Butadiene (g)	2541.5
C$_4$H$_{10}$	Butane (g)	2877.6
C$_5$H$_{12}$	Pentane (l)	3509.0
C$_6$H$_6$	Benzene (l)	3267.6
C$_6$H$_{12}$	Cyclohexane (l)	3919.6
C$_6$H$_{14}$	Hexane (l)	4163.2
C$_7$H$_8$	Toluene (l)	3910.3
C$_7$H$_{16}$	Heptane (l)	4817.0
C$_{10}$H$_8$	Naphthalene (s)	5156.3
Alcohols and ethers		
CH$_4$O	Methanol (l)	726.1
C$_2$H$_6$O	Ethanol (l)	1366.8
C$_2$H$_6$O	Dimethyl ether (g)	1460.4
C$_2$H$_6$O$_2$	Ethylene glycol (l)	1189.2

Molecular formula	Name	$\Delta_c H°$/kJ mol^{-1}
C$_3$H$_8$O	1-Propanol (l)	2021.3
C$_3$H$_8$O$_3$	Glycerol (l)	1655.4
C$_4$H$_{10}$O	Diethyl ether (l)	2723.9
C$_5$H$_{12}$O	1-Pentanol (l)	3330.9
C$_6$H$_6$O	Phenol (s)	3053.5
Carbonyl compounds		
CH$_2$O	Formaldehyde (g)	570.7
C$_2$H$_2$O	Ketene (g)	1025.4
C$_2$H$_4$O	Acetaldehyde (l)	1166.9
C$_3$H$_6$O	Acetone (l)	1789.9
C$_3$H$_6$O	Propanal (l)	1822.7
C$_4$H$_8$O	2-Butanone (l)	2444.1
Acids and esters		
CH$_2$O$_2$	Formic acid (l)	254.6
C$_2$H$_4$O$_2$	Acetic acid (l)	874.2
C$_2$H$_4$O$_2$	Methyl formate (l)	972.6
C$_3$H$_6$O$_2$	Methyl acetate (l)	1592.2
C$_4$H$_8$O$_2$	Ethyl acetate (l)	2238.1
C$_7$H$_6$O$_2$	Benzoic acid (s)	3226.9
Nitrogen compounds		
CHN	Hydrogen cyanide (g)	671.5
CH$_3$NO$_2$	Nitromethane (l)	709.2
CH$_5$N	Methylamine (g)	1085.6
C$_2$H$_3$N	Acetonitrile (l)	1247.2
C$_2$H$_5$NO	Acetamide (s)	1184.6
C$_3$H$_9$N	Trimethylamine (g)	2443.1
C$_5$H$_5$N	Pyridine (l)	2782.3
C$_6$H$_7$N	Aniline (l)	3392.8

MOLAR CONDUCTIVITY OF AQUEOUS HF, HCl, HBr, AND HI

The molar conductivity Λ of an electrolyte solution is defined as the conductivity divided by amount-of-substance concentration. The customary unit is S cm²mol⁻¹ (i.e., Ω⁻¹ cm²mol⁻¹). The first part of this table gives the molar conductivity of the hydrohalogen acids at 25°C as a function of the concentration in mol/L. The second part gives the temperature dependence of Λ for HCl and HBr. More extensive tables and mathematical representations may be found in the reference.

REFERENCE

Hamer, W.J., and DeWane, H.J., *Electrolytic Conductance and the Conductances of the Hydrohalogen Acids in Water*, Natl. Stand. Ref. Data Sys.-Natl. Bur. Standards (U.S.), No. 33, 1970.

c/mol L⁻¹	HF	HCl	HBr	HI	c/mol L⁻¹	HF	HCl	HBr	HI
Inf. dil.	405.1	426.1	427.7	426.4	3.5		218.3	217.5	215.4
0.0001		424.5	425.9	424.6	4.0		200.0	199.4	195.1
0.0005		422.6	424.3	423.0	4.5		183.1	182.4	176.8
0.001		421.2	422.9	421.7	5.0		167.4	166.5	160.4
0.005	128.1	415.7	417.6	416.4	5.5		152.9	151.8	145.5
0.01	96.1	411.9	413.7	412.8	6.0		139.7	138.2	131.7
0.05	50.1	398.9	400.4	400.8	6.5		127.7	125.7	118.6
0.10	39.1	391.1	391.9	394.0	7.0		116.9	114.2	105.7
0.5	26.3	360.7	361.9	369.8	7.5		107.0	103.8	
1.0	24.3	332.2	334.5	343.9	8.0		98.2	94.4	
1.5		305.8	307.6	316.4	8.5		90.3	85.8	
2.0		281.4	281.7	288.9	9.0		83.1		
2.5		258.9	257.8	262.5	9.5		76.6		
3.0		237.6	236.8	237.9	10.0		70.7		

c/mol L⁻¹	–20°C	–10°C	0°C	10°C	20°C	30°C	40°C	50°C
					HCl			
0.5			228.7	283.0	336.4	386.8	436.9	482.4
1.0			211.7	261.6	312.2	359.0	402.9	445.3
1.5			196.2	241.5	287.5	331.1	371.6	410.8
2.0			182.0	222.7	262.9	303.3	342.4	378.2
2.5		131.7	168.5	205.1	239.8	277.0	315.2	347.6
3.0		120.8	154.6	188.5	219.3	253.3	289.3	319.0
3.5	85.5	111.3	139.6	172.2	201.6	232.9	263.9	292.1
4.0	79.3	102.7	129.2	158.1	185.6	214.2	242.2	268.2
4.5	73.7	94.9	119.5	145.4	170.6	196.6	222.5	246.7
5.0	68.5	87.8	110.3	133.5	156.6	180.2	204.1	226.5
5.5	63.6	81.1	101.7	122.5	143.6	165.0	187.1	207.7
6.0	58.9	74.9	93.7	112.3	131.5	151.0	171.3	190.3
6.5	54.4	69.1	86.2	103.0	120.4	138.2	156.9	174.3
7.0	50.2	63.7	79.3	94.4	110.2	126.4	143.3	159.7
7.5	46.3	58.6	73.0	86.5	100.9	115.7	131.6	146.2
8.0	42.7	54.0	67.1	79.4	92.4	106.1	120.6	134.0
8.5	39.4	49.8	61.7	72.9	84.7	97.3	110.7	123.0
9.0	36.4	45.9	56.8	67.1	77.8	89.4	101.7	112.9
9.5	33.6	42.3	52.3	61.8	71.5	82.3	93.6	103.9
10.0	31.2	39.1	48.2	57.0	65.8	75.9	86.3	95.7
10.5	28.9	36.1	44.5	52.7	60.7	70.1	79.6	88.4
11.0	26.8	33.4	41.1	48.8	56.1	64.9	73.6	81.7
11.5	24.9	31.0	38.0	45.3	51.9	60.1	68.0	75.6
12.0	23.1	28.7	35.3	42.0	48.0	55.6	62.8	70.0
12.5	21.4	26.7	32.7	39.0	44.4	51.4	57.9	64.8

MOLAR CONDUCTIVITY OF AQUEOUS HF, HCl, HBr, AND HI (continued)

c/mol L^{-1}	–20°C	–10°C	0°C	10°C	20°C	30°C	40°C	50°C
				HBr				
0.5			240.9	295.9	347.0	398.9	453.6	496.8
1.0			229.6	276.0	329.0	380.4	418.6	465.2
1.5			209.5	254.9	298.9	340.6	381.8	421.4
2.0		150.8	188.6	231.3	271.8	314.1	350.5	387.4
2.5		136.8	171.7	208.3	244.8	281.7	316.0	349.1
3.0		125.7	157.2	189.5	222.2	255.0	287.8	318.6
3.5		116.1	144.1	174.6	203.2	234.4	263.7	291.9
4.0	84.0	107.5	132.3	160.2	186.8	214.2	239.7	266.9
4.5	78.0	99.0	123.0	146.4	171.2	195.1	218.8	242.6
5.0	72.3	91.4	112.6	134.0	155.7	178.2	199.6	221.3
5.5	67.0	84.2	103.1	122.7	142.1	162.8	181.4	201.8
6.0	61.8	77.2	94.3	112.0	129.6	148.0	165.4	183.4
6.5	56.8	70.7	86.0	102.0	118.0	134.1	150.5	166.3
7.0	51.9	64.6	78.4	92.6	107.1	121.4	136.3	150.8

STANDARD SOLUTIONS FOR CALIBRATING CONDUCTIVITY CELLS

Grams of KCl per kg of H$_2$O	k/(S/m)			
	0°C	18°C	25°C	25°C (Ref. 2)
76.5829	6.514	9.781	11.131	
7.47458	0.7134	1.1163	1.2852	1.2854
0.745819	0.07733	0.12201	0.14083	0.14086
7.45510 (0.1 m)				1.28217
0.745510 (0.01 m)				0.14079

Data are from Reference 1 except last column.

REFERENCES

1. K. N. Marsh, Ed., *Recommended Reference Materials for the Realization of Physicochemical Properties*, Blackwell, Oxford, 1987.
2. Y. C. Wu, K. W. Pratt, and W. F. Koch, *J. Solution Chem.*, 18, 515,1989.

EQUIVALENT CONDUCTIVITY OF ELECTROLYTES IN AQUEOUS SOLUTION

Petr Vanýsek

This table gives the equivalent (molar) conductivity Λ at 25°C for some common electrolytes in aqueous solution at concentrations up to 0.1 mol/L. The units of Λ are 10^{-4} m^2 S mol^{-1}.

For very dilute solutions, the equivalent conductivity for any electrolyte of concentration c can be approximately calculated using the Debye-Hückel-Onsager equation, which can be written for a symmetrical (equal charge on cation and anion) electrolyte as

$$\Lambda = \Lambda^\circ - (A + B\Lambda^\circ)c^{1/2}$$

For a solution at 25°C and both cation and anion with charge $|1|$, the constants are $A = 60.20$ and $B = 0.229$. Λ° can be found from the next table, "Ionic Conductivity and Diffusion at Infinite Dilution". The equation is reliable for $c < 0.001$ mol/L; with higher concentration the error increases.

Compound	Infinite dilution Λ°	Concentration (mol/L) Λ						
		0.0005	0.001	0.005	0.01	0.02	0.05	0.1
AgNO$_3$	133.29	131.29	130.45	127.14	124.70	121.35	115.18	109.09
1/2BaCl$_2$	139.91	135.89	134.27	127.96	123.88	119.03	111.42	105.14
1/2CaCl$_2$	135.77	131.86	130.30	124.19	120.30	115.59	108.42	102.41
1/2Ca(OH)$_2$	258	—	—	233	226	214	—	—
1/2CuSO$_4$	133.6	121.6	115.20	94.02	83.08	72.16	59.02	50.55
HCl	425.95	422.53	421.15	415.59	411.80	407.04	398.89	391.13
KBr	151.9	149.8	148.9	146.02	143.36	140.41	135.61	131.32
KCl	149.79	147.74	146.88	143.48	141.20	138.27	133.30	128.90
KClO$_4$	139.97	138.69	137.80	134.09	131.39	127.86	121.56	115.14
1/3K$_3$Fe(CN)$_6$	174.5	166.4	163.1	150.7	—	—	—	—
1/4K$_4$Fe(CN)$_6$	184	—	167.16	146.02	134.76	122.76	107.65	97.82
KHCO$_3$	117.94	116.04	115.28	112.18	110.03	107.17	—	—
KI	150.31	148.2	143.32	144.30	142.11	139.38	134.90	131.05
KIO$_4$	127.86	125.74	124.88	121.18	118.45	114.08	106.67	98.2
KNO$_3$	144.89	142.70	141.77	138.41	132.75	132.34	126.25	120.34
KMnO$_4$	134.8	132.7	131.9	—	126.5	—	—	113
KOH	271.5	—	234	230	228	—	219	213
KReO$_4$	128.20	126.03	125.12	121.31	118.49	114.49	106.40	97.40
1/3LaCl$_3$	145.9	139.6	137.0	127.5	121.8	115.3	106.2	99.1
LiCl	114.97	113.09	112.34	109.35	107.27	104.60	100.06	95.81
LiClO$_4$	105.93	104.13	103.39	100.52	98.56	96.13	92.15	88.52
1/2MgCl$_2$	129.34	125.55	124.15	118.25	114.49	109.99	103.03	97.05
NH$_4$Cl	149.6	147.5	146.7	134.4	141.21	138.25	133.22	128.69
NaCl	126.39	124.44	123.68	120.59	118.45	115.70	111.01	106.69
NaClO$_4$	117.42	115.58	114.82	111.70	109.54	106.91	102.35	98.38
NaI	126.88	125.30	124.19	121.19	119.18	116.64	112.73	108.73
NaOOCCH$_3$	91.0	89.2	88.5	85.68	83.72	81.20	76.88	72.76
NaOH	247.7	245.5	244.6	240.7	237.9	—	—	—
Na picrate	80.45	78.7	78.6	75.7	73.7	—	66.3	61.8
1/2Na$_2$SO$_4$	129.8	125.68	124.09	117.09	112.38	106.73	97.70	89.94
1/2SrCl$_2$	135.73	131.84	130.27	124.18	120.23	115.48	108.20	102.14
1/2ZnSO$_4$	132.7	121.3	114.47	95.44	84.87	74.20	61.17	52.61

IONIC CONDUCTIVITY AND DIFFUSION AT INFINITE DILUTION
Petr Vanýsek

This table gives the molar (equivalent) conductivity λ for common ions at infinite dilution. All values refer to aqueous solutions at 25°C. It also lists the diffusion coefficient D of the ion in dilute aqueous solution, which is related to λ through the equation

$$D = \left(RT / F^2 \right)\left(\lambda / |z|\right)$$

where R is the molar gas constant, T the temperature, F the Faraday constant, and z the charge on the ion. The variation with temperature is fairly sharp; for typical ions, λ and D increase by 2 to 3% per degree as the temperature increases from 25°C.

The diffusion coefficient for a salt, D_{salt}, may be calculated from the D_+ and D_- values of the constituent ions by the relation

$$D_{salt} = \frac{\left(z_+ + |z_-|\right)D_+ D_-}{z_+ D_+ + |z_-| D_-}$$

For solutions of simple, pure electrolytes (one positive and one negative ionic species), such as NaCl or $CaCl_2$, equivalent ionic conductivity $\Lambda°$, which is the conductivity per unit concentration of charge, is defined as

$$\Lambda° = \lambda_+ + \lambda_-$$

where λ_+ and λ_- are equivalent ionic conductivities of the cation and anion.

The ions are arranged in four groups, Inorganic Cations, Inorganic Anions, Organic Cations, and Organic Anions. Within each group, entries are listed in alphabetical order either by chemical formula (for inorganic ions) or name (for organic ions).

REFERENCES

1. Gray, D. E., Ed., *American Institute of Physics Handbook*, McGraw-Hill, New York, 1972, 2—226.
2. Robinson, R. A., and Stokes, R. H., *Electrolyte Solutions*, Butterworths, London, 1959.
3. Lobo, V. M. M., and Quaresma, J. L., *Handbook of Electrolyte Solutions*, Physical Science Data Series 41, Elsevier, Amsterdam, 1989.

Ion	λ 10^{-4} m^2 S mol^{-1}	D 10^{-5} cm^2 s^{-1}	Ion	λ 10^{-4} m^2 S mol^{-1}	D 10^{-5} cm^2 s^{-1}
Inorganic Cations			$1/3La^{3+}$	69.7	0.619
			Li^+	38.66	1.029
Ag^+	61.9	1.648	$1/2Mg^{2+}$	53.0	0.706
$1/3Al^{3+}$	61	0.541	$1/2Mn^{2+}$	53.5	0.712
$1/2Ba^{2+}$	63.6	0.847	NH_4^+	73.5	1.957
$1/2Be^{2+}$	45	0.599	$N_2H_5^+$	59	1.571
$1/2Ca^{2+}$	59.47	0.792	Na^+	50.08	1.334
$1/2Cd^{2+}$	54	0.719	$1/3Nd^{3+}$	69.4	0.616
$1/3Ce^{3+}$	69.8	0.620	$1/2Ni^{2+}$	49.6	0.661
$1/2Co^{2+}$	55	0.732	$1/4[Ni_2(trien)_3]^{4+}$	52	0.346
$1/3[Co(NH_3)_6]^{3+}$	101.9	0.904	$1/2Pb^{2+}$	71	0.945
$1/3[Co(en)_3]^{3+}$	74.7	0.663	$1/3Pr^{3+}$	69.5	0.617
$1/6[Co_2(trien)_3]^{6+}$	69	0.306	$1/2Ra^{2+}$	66.8	0.889
$1/3Cr^{3+}$	67	0.595	Rb^+	77.8	2.072
Cs^+	77.2	2.056	$1/3Sc^{3+}$	64.7	0.574
$1/2Cu^{2+}$	53.6	0.714	$1/3Sm^{3+}$	68.5	0.608
D^+	249.9	6.655	$1/2Sr^{2+}$	59.4	0.791
$1/3Dy^{3+}$	65.6	0.582	Tl^+	74.7	1.989
$1/3Er^{3+}$	65.9	0.585	$1/3Tm^{3+}$	65.4	0.581
$1/3Eu^{3+}$	67.8	0.602	$1/2UO_2^{2+}$	32	0.426
$1/2Fe^{2+}$	54	0.719	$1/3Y^{3+}$	62	0.550
$1/3Fe^{3+}$	68	0.604	$1/3Yb^{3+}$	65.6	0.582
$1/3Gd^{3+}$	67.3	0.597	$1/2Zn^{2+}$	52.8	0.703
H^+	349.65	9.311			
$1/2Hg^{2+}$	68.6	0.913	**Inorganic Anions**		
$1/2Hg^{2+}$	63.6	0.847			
$1/3Ho^{3+}$	66.3	0.589	$Au(CN)_2^-$	50	1.331
K^+	73.48	1.957	$Au(CN)_4^-$	36	0.959

Ion	λ 10^{-4} m^2 S mol^{-1}	D 10^{-5} cm^2 s^{-1}	Ion	λ 10^{-4} m^2 S mol^{-1}	D 10^{-5} cm^2 s^{-1}
$B(C_6H_5)_4^-$	21	0.559	**Organic Cations**		
Br^-	78.1	2.080			
Br_3^-	43	1.145	Benzyltrimethylammonium$^+$	34.6	0.921
BrO_3^-	55.7	1.483	Isobutylammonium$^+$	38	1.012
CN^-	78	2.077	Butyltrimethylammonium$^+$	33.6	0.895
CNO^-	64.6	1.720	Decylpyridinium$^+$	29.5	0.786
$1/2CO_3^{2-}$	69.3	0.923	Decyltrimethylammonium$^+$	24.4	0.650
Cl^-	76.31	2.032	Diethylammonium$^+$	42.0	1.118
ClO_2^-	52	1.385	Dimethylammonium$^+$	51.8	1.379
ClO_3^-	64.6	1.720	Dipropylammonium$^+$	30.1	0.802
ClO_4^-	67.3	1.792	Dodecylammonium$^+$	23.8	0.634
$1/3[Co(CN)_6]^{3-}$	98.9	0.878	Dodecyltrimethylammonium$^+$	22.6	0.602
$1/2CrO_4^{2-}$	85	1.132	Ethanolammonium$^+$	42.2	1.124
F^-	55.4	1.475	Ethylammonium$^+$	47.2	1.257
$1/4[Fe(CN)_6]^{4-}$	110.4	0.735	Ethyltrimethylammonium$^+$	40.5	1.078
$1/3[Fe(CN)_6]^{3-}$	100.9	0.896	Hexadecyltrimethylammonium$^+$	20.9	0.557
$H_2AsO_4^-$	34	0.905	Hexyltrimethylammonium$^+$	29.6	0.788
HCO_3^-	44.5	1.185	Histidyl$^+$	23.0	0.612
HF_2^-	75	1.997	Hydroxyethyltrimethylarsonium$^+$	39.4	1.049
$1/2HPO_4^{2-}$	33	0.439	Methylammonium$^+$	58.7	1.563
$H_2PO_4^-$	33	0.879	Octadecylpyridinium$^+$	20	0.533
$H_2PO_2^-$	46	1.225	Octadecyltributylammonium$^+$	16.6	0.442
HS^-	65	1.731	Octadecyltriethylammonium$^+$	17.9	0.477
HSO_3^-	50	1.331	Octadecyltrimethylammonium$^+$	19.9	0.530
HSO_4^-	50	1.331	Octadecyltripropylammonium$^+$	17.2	0.458
$H_2SbO_4^-$	31	0.825	Octyltrimethylammonium$^+$	26.5	0.706
I^-	76.8	2.045	Pentylammonium$^+$	37	0.985
IO_3^-	40.5	1.078	Piperidinium$^+$	37.2	0.991
IO_4^-	54.5	1.451	Propylammonium$^+$	40.8	1.086
MnO_4^-	61.3	1.632	Pyrilammonium$^+$	24.3	0.647
MoO_4^-	74.5	1.984	Tetrabutylammonium$^+$	19.5	0.519
$N(CN)_2^-$	54.5	1.451	Tetradecyltrimethylammonium$^+$	21.5	0.573
NO_2^-	71.8	1.912	Tetraethylammonium$^+$	32.6	0.868
NO_3^-	71.42	1.902	Tetramethylammonium$^+$	44.9	1.196
$NH_2SO_3^-$	48.3	1.286	Tetraisopentylammonium$^+$	17.9	0.477
N_3^-	69	1.837	Tetrapentylammmonium$^+$	17.5	0.466
OCN^-	64.6	1.720	Tetrapropylammonium$^+$	23.4	0.623
OH^-	198	5.273	Triethylammonium$^+$	34.3	0.913
PF_6^-	56.9	1.515	Triethylsulfonium$^+$	36.1	0.961
$1/2PO_3F^{2-}$	63.3	0.843	Trimethylammonium$^+$	47.23	1.258
$1/3PO_4^{3-}$	69.0	0.612	Trimethylhexylammonium$^+$	34.6	0.921
$1/4P_2O_7^{4-}$	96	0.639	Trimethylsulfonium$^+$	51.4	1.369
$1/3P_3O_9^{3-}$	83.6	0.742	Tripropylammonium$^+$	26.1	0.695
$1/5P_3O_{10}^{5-}$	109	0.581			
ReO_4^-	54.9	1.462	**Organic Anions**		
SCN^-	66	1.758			
$1/2SO_3^{2-}$	79.9	1.064	Acetate$^-$	40.9	1.089
$1/2SO_4^{2-}$	80.0	1.065	p-Anisate$^-$	29.0	0.772
$1/2S_2O_3^{2-}$	85.0	1.132	$1/2$Azelate^{2-}	40.6	0.541
$1/2S_4O_6^{2-}$	66.5	0.885	Benzoate$^-$	32.4	0.863
$1/2S_2O_6^{2-}$	93	1.238	Bromoacetate$^-$	39.2	1.044
$1/2S_2O_8^{2-}$	86	1.145	Bromobenzoate$^-$	30	0.799
$Sb(OH)_6^-$	31.9	0.849	Butyrate$^-$	32.6	0.868
$SeCN^-$	64.7	1.723	Chloroacetate$^-$	42.2	1.124
$1/2SeO_4^{2-}$	75.7	1.008	m-Chlorobenzoate$^-$	31	0.825
$1/2WO_4^{2-}$	69	0.919	o-Chlorobenzoate$^-$	30.2	0.804

Ion	λ 10^{-4} m^2 S mol^{-1}	D 10^{-5} cm^2 s^{-1}	Ion	λ 10^{-4} m^2 S mol^{-1}	D 10^{-5} cm^2 s^{-1}
1/3Citrate^{3-}	70.2	0.623	Iodoacetate$^-$	40.6	1.081
Crotonate$^-$	33.2	0.884	Lactate$^-$	38.8	1.033
Cyanoacetate$^-$	43.4	1.156	1/2Malate^{2-}	58.8	0.783
Cyclohexane carboxylate$^-$	28.7	0.764	1/2Maleate^{2-}	61.9	0.824
1/2 1,1-Cyclopropanedicarboxylate^{2-}	53.4	0.711	1/2Malonate^{2-}	63.5	0.845
Decylsulfate$^-$	26	0.692	Methylsulfate$^-$	48.8	1.299
Dichloroacetate$^-$	38.3	1.020	Naphthylacetate$^-$	28.4	0.756
1/2Diethylbarbiturate^{2-}	26.3	0.350	1/2Oxalate^{2-}	74.11	0.987
Dihydrogencitrate$^-$	30	0.799	Octylsulfate$^-$	29	0.772
1/2Dimethylmalonate^{2-}	49.4	0.658	Phenylacetate$^-$	30.6	0.815
3,5-Dinitrobenzoate$^-$	28.3	0.754	1/2o-Phthalate^{2-}	52.3	0.696
Dodecylsulfate$^-$	24	0.639	1/2m-Phthalate^{2-}	54.7	0.728
Ethylmalonate$^-$	49.3	1.313	Picrate$^-$	30.37	0.809
Ethylsulfate$^-$	39.6	1.055	Pivalate$^-$	31.9	0.849
Fluoroacetate$^-$	44.4	1.182	Propionate$^-$	35.8	0.953
Fluorobenzoate$^-$	33	0.879	Propylsulfate$^-$	37.1	0.988
Formate$^-$	54.6	1.454	Salicylate$^-$	36	0.959
1/2Fumarate^{2-}	61.8	0.823	1/2Suberate^{2-}	36	0.479
1/2Glutarate^{2-}	52.6	0.700	1/2Succinate^{2-}	58.8	0.783
Hydrogenoxalate$^-$	40.2	1.070	1/2Tartarate^{2-}	59.6	0.794
Isovalerate$^-$	32.7	0.871	Trichloroacetate$^-$	36.6	0.975

ACTIVITY COEFFICIENTS OF ACIDS, BASES, AND SALTS
Petr Vanýsek

This table gives mean activity coefficients at 25°C for molalities in the range 0.1 to 1.0. See the following table for definitions, references, and data over a wider concentration range.

	0.1	0.2	0.3	0.4	0.5	0.6	0.7	0.8	0.9	1.0
AgNO$_3$	0.734	0.657	0.606	0.567	0.536	0.509	0.485	0.464	0.446	0.429
AlCl$_3$	0.337	0.305	0.302	0.313	0.331	0.356	0.388	0.429	0.479	0.539
Al$_2$(SO$_4$)$_3$	0.035	0.0225	0.0176	0.0153	0.0143	0.014	0.0142	0.0149	0.0159	0.0175
BaCl$_2$	0.500	0.444	0.419	0.405	0.397	0.391	0.391	0.391	0.392	0.395
BeSO$_4$	0.150	0.109	0.0885	0.0769	0.0692	0.0639	0.0600	0.0570	0.0546	0.0530
CaCl$_2$	0.518	0.472	0.455	0.448	0.448	0.453	0.460	0.470	0.484	0.500
CdCl$_2$	0.2280	0.1638	0.1329	0.1139	0.1006	0.0905	0.0827	0.0765	0.0713	0.0669
Cd(NO$_3$)$_2$	0.513	0.464	0.442	0.430	0.425	0.423	0.423	0.425	0.428	0.433
CdSO$_4$	0.150	0.103	0.0822	0.0699	0.0615	0.0553	0.0505	0.0468	0.0438	0.0415
CoCl$_2$	0.522	0.479	0.463	0.459	0.462	0.470	0.479	0.492	0.511	0.531
CrCl$_3$	0.331	0.298	0.294	0.300	0.314	0.335	0.362	0.397	0.436	0.481
Cr(NO$_3$)$_3$	0.319	0.285	0.279	0.281	0.291	0.304	0.322	0.344	0.371	0.401
Cr$_2$(SO$_4$)$_3$	0.0458	0.0300	0.0238	0.0207	0.0190	0.0182	0.0181	0.0185	0.0194	0.0208
CsBr	0.754	0.694	0.654	0.626	0.603	0.586	0.571	0.558	0.547	0.538
CsCl	0.756	0.694	0.656	0.628	0.606	0.589	0.575	0.563	0.553	0.544
CsI	0.754	0.692	0.651	0.621	0.599	0.581	0.567	0.554	0.543	0.533
CsNO$_3$	0.733	0.655	0.602	0.561	0.528	0.501	0.478	0.458	0.439	0.422
CsOH	0.795	0.761	0.744	0.739	0.739	0.742	0.748	0.754	0.762	0.771
CsOAc	0.799	0.771	0.761	0.759	0.762	0.768	0.776	0.783	0.792	0.802
Cs$_2$SO$_4$	0.456	0.382	0.338	0.311	0.291	0.274	0.262	0.251	0.242	0.235
CuCl$_2$	0.508	0.455	0.429	0.417	0.411	0.409	0.409	0.410	0.413	0.417
Cu(NO$_3$)$_2$	0.511	0.460	0.439	0.429	0.426	0.427	0.431	0.437	0.445	0.455
CuSO$_4$	0.150	0.104	0.0829	0.0704	0.0620	0.0559	0.0512	0.0475	0.0446	0.0423
FeCl$_2$	0.5185	0.473	0.454	0.448	0.450	0.454	0.463	0.473	0.488	0.506
HBr	0.805	0.782	0.777	0.781	0.789	0.801	0.815	0.832	0.850	0.871
HCl	0.796	0.767	0.756	0.755	0.757	0.763	0.772	0.783	0.795	0.809
HClO$_4$	0.803	0.778	0.768	0.766	0.769	0.776	0.785	0.795	0.808	0.823
HI	0.818	0.807	0.811	0.823	0.839	0.860	0.883	0.908	0.935	0.963

ACTIVITY COEFFICIENTS OF ACIDS, BASES, AND SALTS (continued)

	0.1	0.2	0.3	0.4	0.5	0.6	0.7	0.8	0.9	1.0
HNO_3	0.791	0.754	0.735	0.725	0.720	0.717	0.717	0.718	0.721	0.724
H_2SO_4	0.2655	0.2090	0.1826	—	0.1557	—	0.1417	—	—	0.1316
KBr	0.772	0.722	0.693	0.673	0.657	0.646	0.636	0.629	0.622	0.617
KCl	0.770	0.718	0.688	0.666	0.649	0.637	0.626	0.618	0.610	0.604
$KClO_3$	0.749	0.681	0.635	0.599	0.568	0.541	0.518	—	—	—
K_2CrO_4	0.456	0.382	0.340	0.313	0.292	0.276	0.263	0.253	0.243	0.235
KF	0.775	0.727	0.700	0.682	0.670	0.661	0.654	0.650	0.646	0.645
$K_3Fe(CN)_6$	0.268	0.212	0.184	0.167	0.155	0.146	0.140	0.135	0.131	0.128
$K_4Fe(CN)_6$	0.139	0.0993	0.0808	0.0693	0.0614	0.0556	0.0512	0.0479	0.0454	—
KH_2PO_4	0.731	0.653	0.602	0.561	0.529	0.501	0.477	0.456	0.438	0.421
KI	0.778	0.733	0.707	0.689	0.676	0.667	0.660	0.654	0.649	0.645
KNO_3	0.739	0.663	0.614	0.576	0.545	0.519	0.496	0.476	0.459	0.443
$KOAc$	0.796	0.766	0.754	0.750	0.751	0.754	0.759	0.766	0.774	0.783
KOH	0.798	0.760	0.742	0.734	0.732	0.733	0.736	0.742	0.749	0.756
$KSCN$	0.769	0.716	0.685	0.663	0.646	0.633	0.623	0.614	0.606	0.599
K_2SO_4	0.441	0.360	0.316	0.286	0.264	0.246	0.232	—	—	—
$LiBr$	0.796	0.766	0.756	0.752	0.753	0.758	0.767	0.777	0.789	0.803
$LiCl$	0.790	0.757	0.744	0.740	0.739	0.743	0.748	0.755	0.764	0.774
$LiClO_4$	0.812	0.794	0.792	0.798	0.808	0.820	0.834	0.852	0.869	0.887
LiI	0.815	0.802	0.804	0.813	0.824	0.838	0.852	0.870	0.888	0.910
$LiNO_3$	0.788	0.752	0.736	0.728	0.726	0.727	0.729	0.733	0.737	0.743
$LiOH$	0.760	0.702	0.665	0.638	0.617	0.599	0.585	0.573	0.563	0.554
$LiOAc$	0.784	0.742	0.721	0.709	0.700	0.691	0.689	0.688	0.688	0.689
Li_2SO_4	0.468	0.398	0.361	0.337	0.319	0.307	0.297	0.289	0.282	0.277
$MgCl_2$	0.529	0.489	0.477	0.475	0.481	0.491	0.506	0.522	0.544	0.570
$MgSO_4$	0.150	0.107	0.0874	0.0756	0.0675	0.0616	0.0571	0.0536	0.0508	0.0485
$MnCl_2$	0.516	0.469	0.450	0.442	0.440	0.443	0.448	0.455	0.466	0.479
$MnSO_4$	0.150	0.105	0.0848	0.0725	0.0640	0.0578	0.0530	0.0493	0.0463	0.0439
NH_4Cl	0.770	0.718	0.687	0.665	0.649	0.636	0.625	0.617	0.609	0.603
NH_4NO_3	0.740	0.677	0.636	0.606	0.582	0.562	0.545	0.530	0.516	0.504
$(NH_4)_2SO_4$	0.439	0.356	0.311	0.280	0.257	0.240	0.226	0.214	0.205	0.196
$NaBr$	0.782	0.741	0.719	0.704	0.697	0.692	0.689	0.687	0.687	0.687
$NaCl$	0.778	0.735	0.710	0.693	0.681	0.673	0.667	0.662	0.659	0.657
$NaClO_3$	0.772	0.720	0.688	0.664	0.645	0.630	0.617	0.606	0.597	0.589
$NaClO_4$	0.775	0.729	0.701	0.683	0.668	0.656	0.648	0.641	0.635	0.629
Na_2CrO_4	0.464	0.394	0.353	0.327	0.307	0.292	0.280	0.269	0.261	0.253
NaF	0.765	0.710	0.676	0.651	0.632	0.616	0.603	0.592	0.582	0.573
NaH_2PO_4	0.744	0.675	0.629	0.593	0.563	0.539	0.517	0.499	0.483	0.468
NaI	0.787	0.751	0.735	0.727	0.723	0.723	0.724	0.727	0.731	0.736
$NaNO_3$	0.762	0.703	0.666	0.638	0.617	0.599	0.583	0.570	0.558	0.548
$NaOAc$	0.791	0.757	0.744	0.737	0.735	0.736	0.740	0.745	0.752	0.757
$NaOH$	0.766	0.727	0.708	0.697	0.690	0.685	0.681	0.679	0.678	0.678
$NaSCN$	0.787	0.750	—	0.720	0.715	0.712	0.710	0.710	0.711	0.712
Na_2SO_4	0.445	0.365	0.320	0.289	0.266	0.248	0.233	0.221	0.210	0.201
$NiCl_2$	0.522	0.479	0.463	0.460	0.464	0.471	0.482	0.496	0.515	0.563
$NiSO_4$	0.150	0.105	0.0841	0.0713	0.0627	0.0562	0.0515	0.0478	0.0448	0.0425
$Pb(NO_3)_2$	0.395	0.308	0.260	0.228	0.205	0.187	0.172	0.160	0.150	0.141
$RbBr$	0.763	0.706	0.673	0.650	0.632	0.617	0.605	0.595	0.586	0.578
$RbCl$	0.764	0.709	0.675	0.652	0.634	0.620	0.608	0.599	0.590	0.583
RbI	0.762	0.705	0.671	0.647	0.629	0.614	0.602	0.591	0.583	0.575
$RbNO_3$	0.734	0.658	0.606	0.565	0.534	0.508	0.485	0.465	0.446	0.430
$RbOAc$	0.796	0.767	0.756	0.753	0.755	0.759	0.766	0.773	0.782	0.792
Rb_2SO_4	0.451	0.374	0.331	0.301	0.279	0.263	0.249	0.238	0.228	0.219
$SrCl_2$	0.511	0.462	0.442	0.433	0.430	0.431	0.434	0.441	0.449	0.461
$TlClO_4$	0.730	0.652	0.599	0.559	0.527	—	—	—	—	—
$TlNO_3$	0.702	0.606	0.545	0.500	—	—	—	—	—	—
UO_2Cl_2	0.544	0.510	0.520	0.505	0.517	0.532	0.549	0.571	0.595	0.620
UO_2SO_4	0.150	0.102	0.0807	0.0689	0.0611	0.0566	0.0515	0.0483	0.0458	0.0439
$ZnCl_2$	0.515	0.462	0.432	0.411	0.394	0.380	0.369	0.357	0.348	0.339
$Zn(NO_3)_2$	0.531	0.489	0.474	0.469	0.473	0.480	0.489	0.501	0.518	0.535
$ZnSO_4$	0.150	0.140	0.0835	0.0714	0.0630	0.0569	0.0523	0.0487	0.0458	0.0435

MEAN ACTIVITY COEFFICIENTS OF ELECTROLYTES AS A FUNCTION OF CONCENTRATION

The mean activity coefficient γ of an electrolyte X_aY_b is defined as

$$\gamma = \left(\gamma_+^a \gamma_-^b\right)^{1/(a+b)}$$

where γ_+ and γ_- are activity coefficients of the individual ions (which cannot be directly measured). This table gives the mean activity coefficients of about 100 electrolytes in aqueous solution as a function of concentration, expressed in molality terms. All values refer to a temperature of 25°C. Substances are arranged in alphabetical order by formula.

REFERENCES

1. Hamer, W. J., and Wu, Y. C., *J. Phys. Chem. Ref. Data*, 1, 1047, 1972.
2. Staples, B. R., *J. Phys. Chem. Ref. Data*, 6, 385, 1977; 10, 767, 1981; 10, 779, 1981.
3. Goldberg, R. N. et al., *J. Phys. Chem. Ref. Data*, 7, 263, 1978; 8, 923, 1979; 8, 1005, 1979; 10, 1, 1981; 10, 671, 1981.

Mean Activity Coefficient at 25°C

m/mol kg^{-1}	$AgNO_3$	$BaBr_2$	$BaCl_2$	BaI_2	$CaBr_2$	$CaCl_2$	CaI_2
0.001	0.964	0.881	0.887	0.890	0.890	0.888	0.890
0.002	0.950	0.850	0.849	0.853	0.853	0.851	0.853
0.005	0.924	0.785	0.782	0.792	0.791	0.787	0.791
0.010	0.896	0.727	0.721	0.737	0.735	0.727	0.736
0.020	0.859	0.661	0.653	0.678	0.674	0.664	0.677
0.050	0.794	0.573	0.559	0.600	0.594	0.577	0.600
0.100	0.732	0.517	0.492	0.551	0.540	0.517	0.552
0.200	0.656	0.463	0.436	0.520	0.502	0.469	0.524
0.500	0.536	0.435	0.391	0.536	0.500	0.444	0.554
1.000	0.430	0.470	0.393	0.664	0.604	0.495	0.729
2.000	0.316	0.654		1.242	1.125	0.784	
5.000	0.181				18.7	5.907	
10.000	0.108					43.1	
15.000	0.085						

m/mol kg^{-1}	$Cd(NO_2)_2$	$Cd(NO_3)_2$	$CoBr_2$	$CoCl_2$	CoI_2	$Co(NO_3)_2$	$CsBr$
0.001	0.881	0.888	0.890	0.889	0.887	0.888	0.965
0.002	0.837	0.851	0.854	0.852	0.849	0.850	0.951
0.005	0.759	0.787	0.794	0.789	0.783	0.786	0.925
0.010	0.681	0.728	0.740	0.732	0.724	0.728	0.898
0.020	0.589	0.664	0.681	0.670	0.661	0.663	0.864
0.050	0.451	0.576	0.605	0.586	0.582	0.576	0.806
0.100	0.344	0.515	0.556	0.528	0.540	0.516	0.752
0.200	0.247	0.465	0.523	0.483	0.527	0.469	0.691
0.500	0.148	0.428	0.538	0.465	0.596	0.446	0.605
1.000	0.098	0.437	0.685	0.532	0.845	0.492	0.540
2.000	0.069	0.517	1.421	0.864	2.287	0.722	0.485
5.000	0.054		13.9		55.3	3.338	0.454
10.000					196		

m/mol kg^{-1}	$CsCl$	CsF	CsI	$CsNO_3$	$CsOH$	Cs_2SO_4	$CuBr_2$
0.001	0.965	0.965	0.965	0.964	0.966	0.885	0.889
0.002	0.951	0.952	0.951	0.951	0.953	0.845	0.853
0.005	0.925	0.929	0.925	0.924	0.930	0.775	0.791
0.010	0.898	0.905	0.898	0.897	0.906	0.709	0.735
0.020	0.864	0.876	0.863	0.860	0.878	0.634	0.674
0.050	0.805	0.830	0.804	0.796	0.836	0.526	0.594
0.100	0.751	0.792	0.749	0.733	0.802	0.444	0.541

m/mol kg^{-1}	CsCl	CsF	CsI	CsNO$_3$	CsOH	Cs$_2$SO$_4$	CuBr$_2$
0.200	0.691	0.755	0.688	0.655	0.772	0.369	0.504
0.500	0.607	0.721	0.601	0.529	0.755	0.285	0.503
1.000	0.546	0.726	0.534	0.421	0.782	0.233	0.591
2.000	0.496	0.803	0.470				0.859
5.000	0.474						
10.000	0.508						

m/mol kg^{-1}	CuCl$_2$	Cu(ClO$_4$)$_2$	Cu(NO$_3$)$_2$	FeCl$_2$	HBr	HCl	HClO$_4$
0.001	0.887	0.890	0.888	0.888	0.966	0.965	0.966
0.002	0.849	0.854	0.851	0.850	0.953	0.952	0.953
0.005	0.783	0.795	0.787	0.785	0.930	0.929	0.929
0.010	0.722	0.741	0.729	0.725	0.907	0.905	0.906
0.020	0.654	0.685	0.664	0.659	0.879	0.876	0.878
0.050	0.561	0.613	0.577	0.570	0.837	0.832	0.836
0.100	0.495	0.572	0.516	0.509	0.806	0.797	0.803
0.200	0.441	0.553	0.466	0.462	0.783	0.768	0.776
0.500	0.401	0.617	0.431	0.443	0.790	0.759	0.769
1.000	0.405	0.892	0.456	0.500	0.872	0.811	0.826
2.000	0.453	2.445	0.615	0.782	1.167	1.009	1.055
5.000	0.601		2.083		3.800	2.380	3.100
10.000					33.4	10.4	30.8
15.000							323

m/mol kg^{-1}	HF	HI	HNO$_3$	H$_2$SO$_4$	KBr	KCNS	KCl
0.001	0.551	0.966	0.965	0.804	0.965	0.965	0.965
0.002	0.429	0.953	0.952	0.740	0.952	0.951	0.951
0.005	0.302	0.931	0.929	0.634	0.927	0.927	0.927
0.010	0.225	0.909	0.905	0.542	0.902	0.901	0.901
0.020	0.163	0.884	0.875	0.445	0.870	0.869	0.869
0.050	0.106	0.847	0.829	0.325	0.817	0.815	0.816
0.100	0.0766	0.823	0.792	0.251	0.771	0.768	0.768
0.200	0.0550	0.811	0.756	0.195	0.772	0.716	0.717
0.500	0.0352	0.845	0.725	0.146	0.658	0.647	0.649
1.000	0.0249	0.969	0.730	0.125	0.617	0.598	0.604
2.000	0.0175	1.363	0.788	0.119	0.593	0.556	0.573
5.000	0.0110	4.760	1.063	0.197	0.626	0.525	0.593
10.000	0.0085	49.100	1.644	0.527			
15.000	0.0077		2.212	1.077			
20.000	0.0075		2.607	1.701			

m/mol kg^{-1}	KClO$_3$	K$_2$CrO$_4$	KF	KH$_2$PO$_4$*	K$_2$HPO$_4$**	KI	KNO$_3$
0.001	0.965	0.886	0.965	0.964	0.886	0.965	0.964
0.002	0.951	0.847	0.952	0.950	0.847	0.952	0.950
0.005	0.926	0.779	0.927	0.924	0.779	0.927	0.924
0.010	0.899	0.715	0.902	0.896	0.715	0.902	0.896
0.020	0.865	0.643	0.870	0.859	0.643	0.871	0.860
0.050	0.805	0.539	0.818	0.793	0.538	0.820	0.797
0.100	0.749	0.460	0.773	0.730	0.457	0.776	0.735
0.200	0.681	0.385	0.726	0.652	0.379	0.731	0.662
0.500	0.569	0.296	0.670	0.529	0.283	0.676	0.546
1.000		0.239	0.645	0.422		0.646	0.444

m/mol kg^{-1}	KClO$_3$	K$_2$CrO$_4$	KF	KH$_2$PO$_4$*	K$_2$HPO$_4$**	KI	KNO$_3$
2.000		0.199	0.658			0.638	0.332
5.000			0.871				
10.000			1.715				
15.000			3.120				

m/mol kg^{-1}	KOH	K$_2$SO$_4$	LiBr	LiCl	LiClO$_4$	LiI	LiNO$_3$
0.001	0.965	0.885	0.965	0.965	0.966	0.966	0.965
0.002	0.952	0.844	0.952	0.952	0.953	0.953	0.952
0.005	0.927	0.772	0.929	0.928	0.931	0.930	0.928
0.010	0.902	0.704	0.905	0.904	0.908	0.908	0.904
0.020	0.871	0.625	0.877	0.874	0.882	0.882	0.874
0.050	0.821	0.511	0.832	0.827	0.843	0.843	0.827
0.100	0.779	0.424	0.797	0.789	0.815	0.817	0.788
0.200	0.740	0.343	0.767	0.756	0.795	0.802	0.753
0.500	0.710	0.251	0.754	0.739	0.806	0.824	0.726
1.000	0.733		0.803	0.775	0.887	0.912	0.743
2.000	0.860		1.012	0.924	1.161	1.197	0.837
5.000	1.697		2.696	2.000			1.298
10.000	6.110		20.0	9.600			2.500
15.000	19.9		147	30.9			3.960
20.000	46.4		486				4.970

m/mol kg^{-1}	LiOH	Li$_2$SO$_4$	MgBr$_2$	MgCl$_2$	MgI$_2$	MnBr$_2$	MnCl$_2$
0.001	0.964	0.887	0.889	0.889	0.889	0.889	0.888
0.002	0.950	0.847	0.852	0.852	0.853	0.853	0.850
0.005	0.923	0.780	0.790	0.790	0.791	0.791	0.786
0.010	0.895	0.716	0.733	0.734	0.736	0.735	0.727
0.020	0.858	0.645	0.672	0.672	0.677	0.674	0.662
0.050	0.794	0.544	0.593	0.590	0.602	0.595	0.574
0.100	0.735	0.469	0.543	0.535	0.556	0.543	0.513
0.200	0.668	0.400	0.512	0.493	0.535	0.508	0.464
0.500	0.579	0.325	0.540	0.485	0.594	0.519	0.437
1.000	0.522	0.284	0.715	0.577	0.858	0.650	0.477
2.000	0.484	0.270	1.590	1.065	2.326	1.224	0.661
5.000	0.493		36.1	14.40	109.8	6.697	1.539

m/mol kg^{-1}	Mn(ClO$_4$)$_2$	NH$_4$Cl	NH$_4$ClO$_4$	(NH$_4$)$_2$HPO$_4$**	NH$_4$NO$_3$	NaBr	NaBrO$_3$
0.001	0.892	0.965	0.964	0.882	0.964	0.965	0.965
0.002	0.858	0.952	0.950	0.839	0.951	0.952	0.951
0.005	0.801	0.927	0.924	0.763	0.925	0.928	0.926
0.010	0.752	0.901	0.895	0.688	0.897	0.903	0.900
0.020	0.700	0.869	0.859	0.600	0.862	0.873	0.867
0.050	0.637	0.816	0.794	0.469	0.801	0.824	0.811
0.100	0.604	0.769	0.734	0.367	0.744	0.783	0.759
0.200	0.596	0.718	0.663	0.273	0.678	0.742	0.698
0.500	0.686	0.649	0.560	0.171	0.582	0.697	0.605
1.000	1.030	0.603	0.479	0.114	0.502	0.687	0.528
2.000	3.072	0.569	0.399	0.074	0.419	0.730	0.449
5.000		0.563			0.303	1.083	
10.000					0.220		
15.000					0.179		
20.000					0.154		

MEAN ACTIVITY COEFFICIENTS OF ELECTROLYTES AS A FUNCTION
OF CONCENTRATION (continued)

m/mol kg^{-1}	Na$_2$CO$_3$	NaCl	NaClO$_3$	NaClO$_4$	Na$_2$CrO$_4$	NaF	Na$_2$HPO$_4$*
0.001	0.887	0.965	0.965	0.965	0.887	0.965	0.887
0.002	0.847	0.952	0.952	0.952	0.849	0.951	0.848
0.005	0.780	0.928	0.927	0.928	0.783	0.926	0.780
0.010	0.716	0.903	0.902	0.903	0.722	0.901	0.717
0.020	0.644	0.872	0.870	0.872	0.653	0.868	0.644
0.050	0.541	0.822	0.818	0.821	0.554	0.813	0.539
0.100	0.462	0.779	0.771	0.777	0.479	0.764	0.456
0.200	0.385	0.734	0.719	0.729	0.406	0.710	0.373
0.500	0.292	0.681	0.646	0.668	0.318	0.633	0.266
1.000	0.229	0.657	0.590	0.630	0.261	0.573	0.191
2.000	0.182	0.668	0.537	0.608	0.231		0.133
5.000		0.874		0.648			

m/mol kg^{-1}	NaI	NaNO$_3$	NaOH	Na$_2$SO$_3$	Na$_2$SO$_4$	Na$_2$WO$_4$	NiBr$_2$
0.001	0.965	0.965	0.965	0.887	0.886	0.886	0.889
0.002	0.952	0.951	0.952	0.847	0.846	0.846	0.853
0.005	0.928	0.926	0.927	0.779	0.777	0.777	0.791
0.010	0.904	0.900	0.902	0.716	0.712	0.712	0.735
0.020	0.874	0.866	0.870	0.644	0.637	0.638	0.675
0.050	0.827	0.810	0.819	0.540	0.529	0.534	0.596
0.100	0.789	0.759	0.775	0.462	0.446	0.457	0.546
0.200	0.753	0.701	0.731	0.386	0.366	0.388	0.514
0.500	0.722	0.617	0.685	0.296	0.268	0.320	0.535
1.000	0.734	0.550	0.674	0.237	0.204	0.291	0.692
2.000	0.823	0.480	0.714	0.196	0.155	0.291	1.476
5.000	1.402	0.388	1.076				
10.000	4.011	0.329	3.258				
15.000			9.796				
20.000			19.410				

m/mol kg^{-1}	NiCl$_2$	Ni(ClO$_4$)$_2$	Ni(NO$_3$)$_2$	Pb(ClO$_4$)$_2$	Pb(NO$_3$)$_2$	RbBr	RbCl
0.001	0.889	0.891	0.889	0.889	0.882	0.965	0.965
0.002	0.852	0.855	0.851	0.851	0.840	0.951	0.951
0.005	0.789	0.797	0.787	0.787	0.764	0.926	0.926
0.010	0.732	0.745	0.730	0.729	0.690	0.900	0.900
0.020	0.669	0.690	0.666	0.666	0.604	0.866	0.867
0.050	0.584	0.621	0.581	0.580	0.476	0.811	0.811
0.100	0.527	0.582	0.524	0.522	0.379	0.760	0.761
0.200	0.482	0.567	0.481	0.476	0.291	0.705	0.707
0.500	0.465	0.639	0.467	0.458	0.195	0.630	0.633
1.000	0.538	0.946	0.528	0.516	0.136	0.578	0.583
2.000	0.915	2.812	0.797	0.799		0.535	0.546
5.000	4.785			4.043		0.514	0.544
10.000				33.8			

m/mol kg^{-1}	RbF	RbI	RbNO$_3$	Rb$_2$SO$_4$	SrBr$_2$	SrCl$_2$	SrI$_2$
0.001	0.965	0.965	0.964	0.886	0.889	0.888	0.890
0.002	0.952	0.951	0.950	0.845	0.852	0.850	0.854
0.005	0.927	0.926	0.924	0.776	0.790	0.785	0.793
0.010	0.902	0.900	0.896	0.710	0.734	0.725	0.740
0.020	0.871	0.866	0.859	0.635	0.673	0.659	0.681

m/mol kg^{-1}	RbF	RbI	RbNO$_3$	Rb$_2$SO$_4$	SrBr$_2$	SrCl$_2$	SrI$_2$
0.050	0.821	0.810	0.795	0.526	0.591	0.569	0.606
0.100	0.780	0.759	0.733	0.443	0.535	0.506	0.557
0.200	0.739	0.703	0.657	0.365	0.492	0.455	0.526
0.500	0.701	0.627	0.536	0.274	0.476	0.421	0.542
1.000	0.697	0.574	0.430	0.217	0.545	0.451	0.686
2.000	0.724	0.532	0.320		0.921	0.650	
5.000		0.517					

m/mol kg^{-1}	UO$_2$Cl$_2$	UO$_2$(NO$_3$)$_2$	ZnBr$_2$	ZnCl$_2$	ZnI$_2$
0.001	0.888	0.888	0.890	0.887	0.893
0.002	0.851	0.849	0.854	0.847	0.859
0.005	0.787	0.784	0.794	0.781	0.804
0.010	0.729	0.726	0.741	0.719	0.757
0.020	0.666	0.663	0.683	0.652	0.708
0.050	0.583	0.583	0.606	0.561	0.644
0.100	0.529	0.535	0.553	0.499	0.601
0.200	0.493	0.509	0.515	0.447	0.574
0.500	0.501	0.532	0.516	0.384	0.635
1.000	0.601	0.673	0.558	0.330	0.836
2.000	0.948	1.223	0.578	0.283	1.062
5.000		3.020	0.788	0.342	1.546
10.000			2.317	0.876	4.698
15.000			5.381	1.914	
20.000			7.965	2.968	

*   The anion is H$_2$PO$_4^-$.

**   The anion is HPO$_4^{-2}$.

ENTHALPY OF DILUTION OF ACIDS

The quantity given in this table is $-\Delta_{dil}H$, the negative of the enthalpy (heat) of dilution to infinite dilution for aqueous solutions of several common acids; i.e., the negative of the enthalpy change when a solution of molality m at a temperature of 25°C is diluted with an infinite amount of water. The tabulated numbers thus represent the heat produced (or, if the value is negative, the heat absorbed) when the acid is diluted. The initial molality m is given in the first column. The second column gives the dilution ratio, which is the number of moles of water that must be added to one mole of the acid to produce a solution of the molality in the first column.

REFERENCE

Parker, V. B., *Thermal Properties of Aqueous Uni-Univalent Electrolytes*, Natl. Stand. Ref. Data Ser. - Natl. Bur. Stand. (U.S.) 2, U.S. Government Printing Office, 1965.

$-\Delta_{dil}H$ in kJ/mol at 25°C

m	Dil. ratio	HF	HCl	HClO$_4$	HBr	HI	HNO$_3$	CH$_2$O$_2$	C$_2$H$_4$O$_2$
55.506	1.0		45.61		48.83		19.73	0.046	2.167
20	2.775	14.88	19.87	13.81	19.92	21.71	9.498	0.038	2.075
15	3.700	14.34	15.40	7.920	14.29	14.02	6.883	0.109	1.962
10	5.551	13.87	10.24	2.013	8.694	7.615	3.933	0.205	1.824
9	6.167	13.81	9.213	1.280	7.719	6.569	3.368	0.230	1.782
8	6.938	13.77	8.201	0.611	6.786	5.607	2.791	0.255	1.724
7	7.929	13.73	7.217	0.046	5.925	4.728	2.251	0.280	1.648
6	9.251	13.69	6.268	-0.351	5.004	3.975	1.749	0.280	1.540
5.5506	10	13.66	5.841	-0.490	4.590	3.577	1.540	0.285	1.477
5	11.10	13.62	5.318	-0.628	4.113	3.197	1.310	0.289	1.393
4.5	12.33	13.58	4.899	-0.732	3.711	2.828	1.109	0.289	1.310
4	13.88	13.53	4.402	-0.787	3.330	2.460	0.958	0.289	1.218
3.5	15.86	13.47	3.958	-0.820	2.966	2.105	0.791	0.289	1.121
3	18.50	13.45	3.506	-0.782	2.611	1.787	0.665	0.289	1.025
2.5	22.20	13.43	3.063	-0.724	2.301	1.527	0.582	0.285	0.912
2	27.75	13.40	2.623	-0.623	1.996	1.318	0.527	0.276	0.803
1.5	37.00	13.36	2.167	-0.431	1.665	1.125	0.506	0.259	0.678
1	55.51	13.30	1.695	-0.201	1.314	0.933	0.506	0.226	0.544
0.5551	100	13.22	1.234	0.050	0.983	0.736	0.502	0.184	0.423
0.5	111.0	13.20	1.172	0.075	0.941	0.711	0.498	0.176	0.406
0.2	277.5	13.09	0.761	0.247	0.649	0.536	0.439	0.146	0.331
0.1	555.1	12.80	0.556	0.272	0.498	0.439	0.372	0.134	0.289
0.0925	600	12.79	0.540	0.272	0.481	0.427	0.368	0.134	0.285
0.0793	700	12.70	0.502	0.272	0.452	0.402	0.351	0.134	0.285
0.0694	800	12.61	0.473	0.268	0.427	0.385	0.339	0.130	0.280
0.0617	900	12.50	0.448	0.264	0.406	0.368	0.326	0.126	0.276
0.05551	1000	12.42	0.427	0.259	0.385	0.351	0.318	0.121	0.272
0.05	1110	12.24	0.406	0.259	0.372	0.339	0.305	0.121	0.272
0.02775	2000	11.29	0.310	0.226	0.285	0.264	0.247	0.117	0.264
0.01850	3000	10.66	0.251	0.197	0.234	0.218	0.213	0.117	0.259
0.01388	4000	10.25	0.226	0.180	0.205	0.192	0.192	0.113	0.259
0.01110	5000	9.874	0.197	0.167	0.184	0.172	0.176	0.109	0.255
0.00555	10000	8.912	0.142	0.126	0.130	0.121	0.130	0.105	0.243
0.00278	20000	7.531	0.105	0.092	0.092	0.084	0.096	0.096	0.230
0.00111	50000	5.439	0.067	0.059	0.054	0.050	0.063	0.084	0.222
0.000555	100000	3.766	0.042	0.042	0.038	0.038	0.046	0.054	0.209
0.000111	500000	1.255	0.021	0.021	0.021	0.021	0.021	0.038	0.167
0	∞	0	0	0	0	0	0	0	0

ENTHALPY OF SOLUTION OF ELECTROLYTES

This table gives the molar enthalpy (heat) of solution at infinite dilution for some common uni-univalent electrolytes. This is the enthalpy change when 1 mol of solute in its standard state is dissolved in an infinite amount of water. Values are given in kilojoules per mole at 25°C.

REFERENCE

Parker, V. B., *Thermal Properties of Uni-Univalent Electrolytes*, Natl. Stand. Ref. Data Series — Natl. Bur. Stand.(U.S.), No.2, 1965.

Solute	State	$\Delta_{sol} H°$ kJ/mol	Solute	State	$\Delta_{sol} H°$ kJ/mol	Solute	State	$\Delta_{sol} H°$ kJ/mol
HF	g	−61.50	$LiBr \cdot 2H_2O$	c	−9.41	KCl	c	17.22
HCl	g	−74.84	$LiBrO_3$	c	1.42	$KClO_3$	c	41.38
$HClO_4$	l	−88.76	LiI	c	−63.30	$KClO_4$	c	51.04
$HClO_4 \cdot H_2O$	c	−32.95	$LiI \cdot H_2O$	c	−29.66	KBr	c	19.87
HBr	g	−85.14	$LiI \cdot 2H_2O$	c	−14.77	$KBrO_3$	c	41.13
HI	g	−81.67	$LiI \cdot 3H_2O$	c	0.59	KI	c	20.33
HIO_3	c	8.79	$LiNO_2$	c	−11.00	KIO_3	c	27.74
HNO_3	l	−33.28	$LiNO_2 \cdot H_2O$	c	7.03	KNO_2	c	13.35
HCOOH	l	−0.86	$LiNO_3$	c	−2.51	KNO_3	c	34.89
CH_3COOH	l	−1.51				$KC_2H_3O_2$	c	−15.33
			NaOH	c	−44.51	KCN	c	11.72
NH_3	g	−30.50	$NaOH \cdot H_2O$	c	−21.41	KCNO	c	20.25
NH_4Cl	c	14.78	NaF	c	0.91	KCNS	c	24.23
NH_4ClO_4	c	33.47	NaCl	c	3.88	$KMnO_4$	c	43.56
NH_4Br	c	16.78	$NaClO_2$	c	0.33			
NH_4I	c	13.72	$NaClO_2 \cdot 3H_2O$	c	28.58	RbOH	c	−62.34
NH_4IO_3	c	31.80	$NaClO_3$	c	21.72	$RbOH \cdot H_2O$	c	−17.99
NH_4NO_2	c	19.25	$NaClO_4$	c	13.88	$RbOH \cdot 2H_2O$	c	0.88
NH_4NO_3	c	25.69	$NaClO_4 \cdot H_2O$	c	22.51	RbF	c	−26.11
$NH_4C_2H_3O_2$	c	−2.38	NaBr	c	−0.60	$RbF \cdot H_2O$	c	−0.42
NH_4CN	c	17.57	$NaBr \cdot 2H_2O$	c	18.64	$RbF \cdot 1.5H_2O$	c	1.34
NH_4CNS	c	22.59	$NaBrO_3$	c	26.90	RbCl	c	17.28
CH_3NH_3Cl	c	5.77	NaI	c	−7.53	$RbClO_3$	c	47.74
$(CH_3)_3NHCl$	c	1.46	$NaI \cdot 2H_2O$	c	16.13	$RbClO_4$	c	56.74
$N(CH_3)_4Cl$	c	4.08	$NaIO_3$	c	20.29	RbBr	c	21.88
$N(CH_3)_4Br$	c	24.27	$NaNO_2$	c	13.89	$RbBrO_3$	c	48.95
$N(CH_3)_4I$	c	42.07	$NaNO_3$	c	20.50	RbI	c	25.10
			$NaC_2H_3O_2$	c	−17.32	$RbNO_3$	c	36.48
$AgClO_4$	c	7.36	$NaC_2H_3O_2 \cdot 3H_2O$	c	19.66			
$AgNO_2$	c	36.94	NaCN	c	1.21	CsOH	c	−71.55
$AgNO_3$	c	22.59	$NaCN \cdot 0.5H_2O$	c	3.31	$CsOH \cdot H_2O$	c	−20.50
			$NaCN \cdot 2H_2O$	c	18.58	CsF	c	−36.86
LiOH	c	−23.56	NaCNO	c	19.20	$CsF \cdot H_2O$	c	−10.46
$LiOH \cdot H_2O$	c	−6.69	NaCNS	c	6.83	$CsF \cdot 1.5H_2O$	c	−5.44
LiF	c	4.73				CsCl	c	17.78
LiCl	c	−37.03	KOH	c	−57.61	$CsClO_4$	c	55.44
$LiCl \cdot H_2O$	c	−19.08	$KOH \cdot H_2O$	c	−14.64	CsBr	c	25.98
$LiClO_4$	c	−26.55	$KOH \cdot 1.5H_2O$	c	−10.46	$CsBrO_3$	c	50.46
$LiClO_4 \cdot 3H_2O$	c	32.61	KF	c	−17.73	CsI	c	33.35
LiBr	c	−48.83	$KF \cdot 2H_2O$	c	6.97	$CsNO_3$	c	40.00
$LiBr \cdot H_2O$	c	−23.26						

KINETIC AND PHOTOCHEMICAL DATA FOR ATMOSPHERIC CHEMISTRY

R. Atkinson, D. L. Baulch, R. A. Cox, R. F. Hampson, Jr., J. A. Kerr, and J. Troe
IUPAC Subcommittee on Gas Kinetic Data Evaluation for Atmospheric Chemistry

The International Union of Pure and Applied Chemistry (IUPAC) maintains a Subcommittee on Gas Kinetic Data evaluation for Atmospheric Chemistry which issues periodic recommendations on rate constants and related data for chemical reactions that are important in the chemistry of the atmosphere. These recommended values are based on the latest (1992) evaluation of all relevant data reported in the literature. The table below provides a summary of the recommended values and their assigned uncertainties. Full details of the data and evaluation process may be found in the references.

This table is reprinted with permission of IUPAC and of the American Institute of Physics and American Chemical Society, who are the publishers of the *Journal of Physical and Chemical Reference Data*.

REFERENCE

1. Atkinson, R., et al., *Journal of Physical and Chemical Reference Data*, 21, 1125-1568, 1992; 18, 881-1097, 1989; 13, 1259-1380, 1984; 11, 327-496, 1982; 9, 295-471, 1980.

Reaction	k_{298} cm³ molecule⁻¹ s⁻¹	$\Delta\log k_{298}$	Temp. dependence of k/cm³ molecule⁻¹ s⁻¹	Temp. range/K	$\Delta(E/R)$/ K
***O_x* Reactions**					
$O + O_2 + M \rightarrow O_3 + M$	$6.0 \times 10^{-34} [O_2]$ (k_0)	± 0.05	$6.0 \times 10^{-34}(T/300)^{-2.8}[O_2]$	100–300	$\Delta n = \pm 0.5$
	$5.6 \times 10^{-34} [N_2]$ (k_0)	± 0.05	$5.6 \times 10^{-34}(T/300)^{-2.8}[N_2]$	100–300	$\Delta n = \pm 0.5$
$O + O_2 \rightarrow O_3^*$	See previous evaluation				
$O_3^* + M \rightarrow O_3 + M$	See previous evaluation				
$O + O_3 \rightarrow 2 O_2$	8.0×10^{-15}	± 0.08	$8.0 \times 10^{-12} \exp(-2060/T)$	200–400	± 200
$O + O_3^* \rightarrow$ products	See previous evaluation				
$O(^1D) + O_2 \rightarrow O(^3P) + O_2$	4.0×10^{-11}	± 0.05	$3.2 \times 10^{-11} \exp(67/T)$	200–350	± 100
$O(^1D) + O_3 \rightarrow O_2 + 2 O(^3P)$	1.2×10^{-10}	± 0.1	2.4×10^{-10}	100–400	$\Delta\log k = \pm 0.05$
$\rightarrow 2 O_2(^3\Sigma_g^-)$	1.2×10^{-10}	± 0.1			
$O_2^* + O_3 \rightarrow O + 2 O_2$	See data sheet				
$O_2(^1\Delta_g) + M \rightarrow O_2(^3\Sigma_g^-) + M$	1.6×10^{-18} $(M = O_2)$	± 0.2	$3.0 \times 10^{-18} \exp(-200/T)$	100–450	± 200
	$\leqslant 1.4 \times 10^{-19}$ $(M = N_2)$				
	5×10^{-18} $(M = H_2O)$	± 0.3			
	$\leqslant 2 \times 10^{-20}$ $(M = CO_2)$				
$O_2(^1\Sigma_g^+) + M \rightarrow O_2(^3\Sigma_g^-) + M$	See previous evaluation				
$O_2(^1\Sigma_g^+) + O_3 \rightarrow$ products	See previous evaluation				
$O_2(^1\Sigma_g^+)^* + O_2 \rightarrow O_2(^1\Sigma_g^+) + O_2$	See previous evaluation				
$O_2 + h\nu \rightarrow$ products	See data sheets				
$O_3 + h\nu \rightarrow$ products	See data sheets				
***HO_x* Reactions**					
$H + HO_2 \rightarrow H_2 + O_2$	5.6×10^{-12}	± 0.5	5.6×10^{-12}	245–300	$\Delta\log k = \pm 0.5$
$\rightarrow 2 HO$	7.2×10^{-11}	± 0.1	7.2×10^{-11}	245–300	$\Delta\log k = \pm 0.1$
$\rightarrow H_2O + O$	2.4×10^{-12}	± 0.5	2.4×10^{-12}	245–300	$\Delta\log k = \pm 0.5$
$H + O_2 + M \rightarrow HO_2 + M$	$6.2 \times 10^{-32}[N_2]$ (k_0)	± 0.05	$6.2 \times 10^{-32}(T/300)^{-1.6}[N_2]$	200–600	$\Delta n = \pm 0.6$
	7.5×10^{-11} (k_∞)	± 0.3	7.5×10^{-11}	200–300	$\Delta n = \pm 0.6$
	$F_c = 0.55$	$\Delta F_c = \pm 0.15$	$F_c = \exp(-T/498)$	200–300	
$H + O_3 \rightarrow HO + O_2$	See previous evaluation				
$H + O_3 \rightarrow HO^* + O_2$	See previous evaluation				
$O + H_2 \rightarrow HO + H$	See previous evaluation				
$O + HO \rightarrow O_2 + H$	3.3×10^{-11}	± 0.1	$2.3 \times 10^{-11} \exp(110/T)$	220–500	± 100
$O + HO_2 \rightarrow HO + O_2$	5.8×10^{-11}	± 0.08	$2.7 \times 10^{-11} \exp(224/T)$	200–400	± 100
$O + H_2O_2 \rightarrow HO + HO_2$	1.7×10^{-15}	± 0.3	$1.4 \times 10^{-12} \exp(-2000/T)$	250–390	± 1000
$O(^1D) + H_2 \rightarrow HO + H$	1.1×10^{-10}	± 0.1	1.1×10^{-10}	200–350	± 100
$O(^1D) + H_2O \rightarrow 2 HO$	2.2×10^{-10}	± 0.1	2.2×10^{-10}	200–350	± 100
$HO + H_2 \rightarrow H_2O + H$	6.7×10^{-15}	± 0.1	$7.7 \times 10^{-12} \exp(-2100/T)$	200–450	± 200
$HO + H_2(v=1) \rightarrow H_2O + H$	See previous evaluation				
$HO + HO \rightarrow H_2O + O$	1.9×10^{-12}	± 0.15	$4.2 \times 10^{-12} \exp(-240/T)$	250–500	± 240
$HO + HO + M \rightarrow H_2O_2 + M$	$8 \times 10^{-31}[N_2]$ (k_0)	± 0.3	$8 \times 10^{-31}(T/300)^{-0.8}[N_2]$	200–300	$\Delta n = \pm 0.5$
	3×10^{-11} (k_∞)	± 0.3	3×10^{-11}	200–300	$\Delta n = \pm 0.5$
	$F_c = 0.5$		$F_c = 0.5$	200–300	
$HO + HO_2 \rightarrow H_2O + O_2$	1.1×10^{-10}	± 0.1	$4.8 \times 10^{-11} \exp(250/T)$	250–400	± 200
$HO + H_2O_2 \rightarrow H_2O + HO_2$	1.7×10^{-12}	± 0.1	$2.9 \times 10^{-12} \exp(-160/T)$	240–460	± 100
$HO + O_3 \rightarrow HO_2 + O_2$	6.7×10^{-14}	± 0.15	$1.9 \times 10^{-12} \exp(-1000/T)$	220–450	± 300
$HO^* + M \rightarrow HO + M$	See previous evaluation				
$HO^* + O_3 \rightarrow$ products	See previous evaluation				
$HO_2 + HO_2 \rightarrow H_2O_2 + O_2$	1.6×10^{-12}	± 0.15	$2.2 \times 10^{-13} \exp(600/T)$	230–420	± 200
$HO_2 + HO_2 + M \rightarrow H_2O_2 + O_2 + M$	$5.2 \times 10^{-32} [N_2]$	± 0.15	$1.9 \times 10^{-33} \exp(980/T)[N_2]$	230–420	± 300
	$4.5 \times 10^{-32} [O_2]$	± 0.15			
	See data sheets for effect of H_2O				
$HO_2 + O_3 \rightarrow HO + 2 O_2$	2.0×10^{-15}	± 0.2	$1.4 \times 10^{-14} \exp(-600/T)$	250–350	$+500$ -100
$H_2O + h\nu \rightarrow HO + H$	See data sheets				
$H_2O_2 + h\nu \rightarrow 2 HO$	See data sheets				

Reaction	k_{298} cm³ molecule⁻¹ s⁻¹		$\Delta\log k_{298}$	Temp. dependence of k/cm³ molecule⁻¹ s⁻¹	Temp. range/K	$\Delta(E/R)/$ K
NO$_x$ Reactions						
O + NO + M → NO$_2$ + M	1.0×10^{-31} [N$_2$]	(k_0)	± 0.1	$1.0 \times 10^{-31}(T/300)^{-1.6}$[N$_2$]	200–300	$\Delta n = \pm 0.3$
	3.0×10^{-11}	(k_∞)	± 0.3	$3.0 \times 10^{-11}(T/300)^{0.3}$	200–1500	$\Delta n = \pm 0.3$
	$F_c = 0.85$		$\Delta F_c = \pm 0.1$	$F_c = \exp(-T/1850)$	200–300	
O + NO$_2$ → O$_2$ + NO	9.7×10^{-12}		± 0.06	$6.5 \times 10^{-12}\exp(120/T)$	230–350	± 120
O + NO$_2$ + M → NO$_3$ + M	9.0×10^{-32} [N$_2$]	(k_0)	± 0.1	$9.0 \times 10^{-32}(T/300)^{-2.0}$[N$_2$]	200–400	$\Delta n = \pm 1$
	2.2×10^{-11}	(k_∞)	± 0.2	2.2×10^{-11}	200–400	$\Delta n = \pm 0.5$
	$F_c = 0.8$		$\Delta F_c = \pm 0.1$	$F_c = \exp(-T/1300)$	200–400	
O + NO$_3$ → O$_2$ + NO$_2$	1.7×10^{-11}		± 0.3			
O + N$_2$O$_5$ → products	See previous evaluation					
O(^{1}D) + N$_2$ → O(^{3}P) + N$_2$	2.6×10^{-11}		± 0.1	$1.8 \times 10^{-11}\exp(107/T)$	200–350	± 100
O(^{1}D) + N$_2$O → N$_2$ + O$_2$	4.4×10^{-11}		± 0.15	4.4×10^{-11}	200–350	± 100
→ 2 NO	7.2×10^{-11}		± 0.15	7.2×10^{-11}	200–350	± 100
N + HO → NO + H	See previous evaluation					
N + O$_2$ → NO + O	See previous evaluation					
N + O$_2$($^1\Delta_g$) → NO + O	See previous evaluation					
N + O$_3$ → NO + O$_2$	See previous evaluation					
N + NO → N$_2$ + O	See previous evaluation					
N + NO$_2$ → N$_2$O + O	See previous evaluation					
HO + NH$_3$ → H$_2$O + NH$_2$	1.6×10^{-13}		± 0.1	$3.5 \times 10^{-12}\exp(-925/T)$	230–450	± 200
HO + HONO → H$_2$O + NO$_2$	4.9×10^{-12}		± 0.3	$1.8 \times 10^{-11}\exp(-390/T)$	280–340	± 400
HO + HONO$_2$ → H$_2$O + NO$_3$	1.5×10^{-13}	(1 bar)	± 0.1	See data sheets		
HO + HO$_2$NO$_2$ → products	5.0×10^{-12}		± 0.2	$1.5 \times 10^{-12}\exp(360/T)$	240–340	$^{+300}_{-600}$
HO + NO + M → HONO + M	7.4×10^{-31} [N$_2$]	(k_0)	± 0.1	$7.4 \times 10^{-31}(T/300)^{-2.4}$[N$_2$]	200–300	$\Delta n = \pm 0.5$
	3.2×10^{-11}	(k_∞)	± 0.3	3.2×10^{-11}	200–400	$\Delta\log k = \pm 0.3$
	$F_c = 0.8$					
HO + NO$_2$ + M → HONO$_2$ + M	2.6×10^{-30} [N$_2$]	(k_0)	± 0.1	$2.6 \times 10^{-30}(T/300)^{-2.9}$[N$_2$]	200–300	$\Delta n = \pm 0.5$
	6.0×10^{-11}	(k_∞)	± 0.1	6.0×10^{-11}	200–300	$\Delta n = \pm 0.5$
	$F_c = 0.43$					
HO + NO$_3$ → HO$_2$ + NO$_2$	2.3×10^{-11}		± 0.2			
HO$_2$ + NO → HO + NO$_2$	8.3×10^{-12}		± 0.1	$3.7 \times 10^{-12}\exp(240/T)$	230–500	± 100
HO$_2$ + NO$_2$ + M → HO$_2$NO$_2$ + M	1.8×10^{-31} [N$_2$]	(k_0)	± 0.1	$1.8 \times 10^{-31}(T/300)^{-3.2}$[N$_2$]	200–300	$\Delta n = \pm 1$
	4.7×10^{-12}	(k_∞)	± 0.2	4.7×10^{-12}	200–300	$\Delta n = \pm 1$
	$F_c = 0.6$			$F_c = 0.6$		
HO$_2$NO$_2$ + M → HO$_2$ + NO$_2$ + M	1.3×10^{-20} [N$_2$]	(k_0/s^{-1})	± 0.3	$5 \times 10^{-6}\exp(-10000/T)$[N$_2$]	260–300	± 500
	0.34	(k_∞/s^{-1})	± 0.5	$2.6 \times 10^{15}\exp(-10900/T)$	260–300	± 500
	$F_c = 0.6$					
HO$_2$ + NO$_3$ → O$_2$ + HONO$_2$ → HO + NO$_2$ + O$_2$ }	4.3×10^{-12}		± 0.2			
NH$_2$ + HO → products	See previous evaluation					
NH$_2$ + HO$_2$ → products	See previous evaluation					
NH$_2$ + O$_2$ → products	$< 3 \times 10^{-18}$					
NH$_2$ + O$_3$ → products	1.7×10^{-13}		± 0.5	$4.9 \times 10^{-12}\exp(-1000/T)$	250–380	± 500
NH$_2$ + NO → products	1.6×10^{-11}		± 0.2	$1.6 \times 10^{-11}(T/298)^{-1.5}$	210–500	$\Delta n = \pm 0.5$
NH$_2$ + NO$_2$ → products	2.0×10^{-11}		± 0.2	$2.0 \times 10^{-11}(T/298)^{-2.0}$	250–500	$\Delta n = \pm 0.7$
2 NO + O$_2$ → 2 NO$_2$	2.0×10^{-38} (cm⁶ molecule⁻² s⁻¹)		± 0.1	$3.3 \times 10^{-39}\exp(530/T)$	273–600	± 400
NO + O$_3$ → NO$_2$ + O$_2$	1.8×10^{-14}		± 0.08	$1.8 \times 10^{-12}\exp(-1370/T)$	195–304	± 200
NO + NO$_3$ → 2 NO$_2$	2.6×10^{-11}		± 0.1	$1.8 \times 10^{-11}\exp(110/T)$	220–400	± 100
NO$_2$ + O$_3$ → NO$_3$ + O$_2$	3.2×10^{-17}		± 0.06	$1.2 \times 10^{-13}\exp(-2450/T)$	230–360	± 150
NO$_2$ + NO$_3$ + M → N$_2$O$_5$ + M	2.7×10^{-30}[N$_2$]	(k_0)	± 0.1	$2.7 \times 10^{-30}(T/300)^{-3.4}$[N$_2$]	200–300	$\Delta n = \pm 0.5$
	2.0×10^{-12}	(k_∞)	± 0.2	$2.0 \times 10^{-12}(T/300)^{0.2}$	200–500	$\Delta n = \pm 0.6$
	$F_c = 0.33$			$F_c = [\exp(-T/250) + \exp(-1050/T)]$	200–500	
N$_2$O$_5$ + M → NO$_2$ + NO$_3$ + M	1.6×10^{-19}[N$_2$]	(k_0/s^{-1})	± 0.2	$2.2 \times 10^{-3}(T/300)^{-4.4}$ $\exp(-11080/T)$[N$_2$]	220–300	$\Delta n = \pm 0.5$
	6.9×10^{-2}	(k_∞/s^{-1})	± 0.3	$9.7 \times 10^{14}(T/300)^{0.1}$ $\exp(-11080/T)$	200–300	$\Delta n = \pm 0.2$
	$F_c = 0.33$			$F_c = [\exp(-T/250) + \exp(-1050/T)]$	200–300	
N$_2$O$_5$ + H$_2$O → 2 HONO$_2$	$< 2 \times 10^{-21}$					
HONO + $h\nu$ → products	See data sheets					
HONO$_2$ + $h\nu$ → products	See data sheets					
HO$_2$NO$_2$ + $h\nu$ → products	See data sheets					
NO + $h\nu$ → products	See previous evaluation					
NO$_2$ + $h\nu$ → products	See data sheets					
NO$_3$ + $h\nu$ → products	See data sheets					
N$_2$O + $h\nu$ → products	See data sheets					
N$_2$O$_5$ + $h\nu$ → products	See data sheets					
Organic Reactions						
O + CH$_3$ → HCHO + H	1.4×10^{-10}		± 0.1	1.4×10^{-10}	200–900	± 100
O + CN → CO + N(^{2}D)	See previous evaluation					
→ CO + N(^{4}S)	See previous evaluation					
O(^{1}D) + CH$_4$ → HO + CH$_3$	1.35×10^{-10}		± 0.1	1.35×10^{-10}	200–300	± 100
→ HCHO + H$_2$	1.5×10^{-11}		± 0.1	1.5×10^{-11}	200–300	± 100
HO + CH$_4$ → H$_2$O + CH$_3$	7.0×10^{-15}		± 0.10	$3.9 \times 10^{-12}\exp(-1885/T)$	240–300	± 100
HO + C$_2$H$_2$ + M → C$_2$H$_2$OH + M	5×10^{-30}[N$_2$]	(k_0)	± 0.1	$5 \times 10^{-30}(T/300)^{-1.5}$[N$_2$]	200–300	$\Delta n = \pm 1.5$
	9.0×10^{-13}	(k_∞)	± 0.1	$9.0 \times 10^{-13}(T/300)^{2}$	200–300	$\Delta n = \pm 1$
	$F_c = 0.62$			$F_c = \exp(-T/623)$	200–300	

Reaction	k_{298} cm³ molecule⁻¹ s⁻¹		$\Delta \log k_{298}$	Temp. dependence of k/cm³ molecule⁻¹ s⁻¹	Temp. range/K	$\Delta(E/R)$/ K
HO + C₂H₄ + M → C₂H₄OH + M	$7 \times 10^{-29}[\text{N}_2]$	(k_0)	±0.3	$7 \times 10^{-29}(T/300)^{-3.1}[\text{N}_2]$	200–300	$\Delta n = \pm 2$
	9×10^{-12}	(k_∞)	±0.3	9×10^{-12}	200–300	$\Delta n = \pm 0.5$
	$F_c = 0.7$					
HO + C₂H₆ → H₂O + C₂H₅	2.5×10^{-13}		±0.10	$7.8 \times 10^{-12}\exp(-1020/T)$	240–300	±100
HO + C₃H₆ + M → C₃H₆OH + M	$8 \times 10^{-27}[\text{N}_2]$	(k_0)	±1	$8 \times 10^{-27}(T/300)^{-3.5}[\text{N}_2]$	200–300	$\Delta n = \pm 1$
	3.0×10^{-11}	(k_∞)	±0.1	3.0×10^{-11}	200–300	$\Delta n = \pm 1$
	$F_c = 0.5$					
HO + C₃H₈ → H₂O + C₃H₇	1.14×10^{-12}		±0.10	$9.8 \times 10^{-12}\exp(-640/T)$	~300	±150
HO + CO → H + CO₂	$1.5 \times 10^{-13}(1 + 0.6\ P/\text{bar})$		±0.1	$1.5 \times 10^{-13}(1 + 0.6\ P/\text{bar})$	200–300	±300
HO + HCHO → H₂O + HCO	9.6×10^{-12}		±0.10	$8.8 \times 10^{-12}\exp(25/T)$	240–300	±150
HO + CH₃CHO → H₂O + CH₃CO	1.6×10^{-11}		±0.10	$5.6 \times 10^{-12}\exp(310/T)$	240–530	±200
HO + C₂H₅CHO → products	2.0×10^{-11}		±0.15			
HO + (CHO)₂ → products	1.1×10^{-11}		±0.3			
HO + HOCH₂CHO → H₂O + HOCH₂CO	8.0×10^{-12}		±0.3			
→ H₂O + HOCHCHO	2.0×10^{-12}		±0.3			
HO + CH₃COCHO → H₂O + CH₃COCO	1.7×10^{-11}		±0.3			
HO + CH₃COCH₃ → H₂O + CH₂COCH₃	2.3×10^{-13}		±0.2	$1.7 \times 10^{-12}\exp(-600/T)$	240–440	±300
HO + CH₃OH → H₂O + CH₂OH	7.8×10^{-13}		±0.15 }	$3.3 \times 10^{-12}\exp(-380/T)$	240–300	±200
→ H₂O + CH₃O	1.4×10^{-13}		±0.15			
HO + C₂H₅OH → H₂O + CH₂CH₂OH	1.6×10^{-13}		±0.15 }	$4.1 \times 10^{-12}\exp(-70/T)$	270–340	±200
→ H₂O + CH₃CHOH	2.9×10^{-12}		±0.15			
→ H₂O + CH₃CH₂O	1.6×10^{-13}		±0.15			
HO + n-C₃H₇OH → products	5.5×10^{-12}		±0.2			
HO + i-C₃H₇OH → products	5.7×10^{-12}		±0.2	5.7×10^{-12}	240–440	±200
HO + CH₃COCH₂OH → products	3.0×10^{-12}		±0.3			
HO + CH₃OOH → H₂O + CH₂OOH	1.9×10^{-12}		±0.2	$1.0 \times 10^{-12}\exp(190/T)$	220–430	±150
→ H₂O + CH₃OO	3.6×10^{-12}		±0.2	$1.9 \times 10^{-12}\exp(190/T)$	220–430	±150
HO + HCOOH → products	4.5×10^{-13}		±0.15	4.5×10^{-13}	290–450	±250
HO + CH₃COOH → products	8×10^{-13}		±0.3			
HO + CH₃ONO₂ → products	3.5×10^{-13}	(1 bar)	±0.10	$1.0 \times 10^{-14}\exp(1060/T)$ (1 bar)	290–400	±500
HO + C₂H₅ONO₂ → products	4.9×10^{-13}	(1 bar)	±0.15	$4.4 \times 10^{-14}\exp(720/T)$ (1 bar)	290–380	±500
HO + n-C₃H₇ONO₂ → products	7.3×10^{-13}	(1 bar)	±0.15	7.3×10^{-13} (1 bar)	290–370	±500
HO + i-C₃H₇ONO₂ → products	4.9×10^{-13}	(1 bar)	±0.25			
HO + CH₃CO₃NO₂ → products	1.1×10^{-13}		±0.2	$9.5 \times 10^{-13}\exp(-650/T)$	270–300	±400
HO + HCN → products	3×10^{-14}	(1 bar)	±0.5	$1.2 \times 10^{-13}(-400/T)$ (1 bar)	290–440	±300
HO + CH₃CN → products	2.2×10^{-14}	(1 bar)	±0.15	$8.1 \times 10^{-13}\exp(-1080/T)$ (1 bar)	250–390	±200
HO₂ + CH₃O₂ → O₂ + CH₃O₂H	5.2×10^{-12}		±0.3	$3.8 \times 10^{-13}\exp(780/T)$	225–580	±500
HO₂ + HOCH₂O₂ → O₂ + HOCH₂O₂H / → O₂ + HCO₂H + H₂O	1.2×10^{-11}		±0.3	$5.6 \times 10^{-15}\exp(2300/T)$	275–335	±1500
HO₂ + C₂H₅O₂ → O₂ + C₂H₅O₂H	5.8×10^{-12}		±0.2	$6.5 \times 10^{-13}\exp(650/T)$	240–380	±200
HO₂ + CH₃CO₃ → O₂ + CH₃CO₃H	4.2×10^{-12}		±0.3	$1.3 \times 10^{-13}\exp(1040/T)$	250–370	±500
→ O₃ + CH₃CO₂H	1.0×10^{-11}		±0.3	$3.0 \times 10^{-13}\exp(1040/T)$	250–370	±500
HO₂ + HOCH₂CH₂O₂ → products	1.0×10^{-11}		±0.3			
HO₂ + HCHO → HOCH₂OO	7.9×10^{-14}		±0.3	$9.7 \times 10^{-15}\exp(625/T)$	275–333	±600
HOCH₂OO → HO₂ + HCHO	1.5×10^{2}	(k/s^{-1})	±0.3	$2.4 \times 10^{12}\exp(-7000/T)$	275–333	±2000
NO₃ + C₂H₂ → products	$<1 \times 10^{-16}$					
NO₃ + C₂H₄ → products	2.1×10^{-16}		±0.2	$3.3 \times 10^{-12}\exp(-2880/T)$	270–330	±500
NO₃ + C₃H₄ → products	9.4×10^{-15}		±0.2			
NO₃ + HCHO → HNO₃ + HCO	5.8×10^{-16}		±0.3			
NO₃ + CH₃CHO → HNO₃ + CH₃CO	2.7×10^{-15}		±0.2	$1.4 \times 10^{-12}\exp(-1860/T)$	260–370	±500
NO₃ + CH₃OH → products	2.4×10^{-16}		±0.5	$1.3 \times 10^{-12}\exp(-2560/T)$	290–480	±700
NO₃ + C₂H₅OH → products	$<2 \times 10^{-15}$					
NO₃ + i-C₃H₇OH → products	$<5 \times 10^{-15}$					
CH₃ + O₂ + M → CH₃O₂ + M	$1.0 \times 10^{-30}[\text{N}_2]$	(k_0)	±0.2	$1.0 \times 10^{-30}(T/300)^{-3.3}[\text{N}_2]$	200–300	$\Delta n = \pm 1$
	2.2×10^{-12}	(k_∞)	±0.3	$2.2 \times 10^{-12}(T/300)^{1.0}$	200–300	$\Delta n = \pm 1$
	$F_c = 0.27$					
C₂H₅ + O₂ → C₂H₄ + HO₂	3.8×10^{-15}	(1 bar air)	±0.5			
	1.9×10^{-14}	(0.133 bar air)	±0.5			
C₂H₅ + O₂ + M → C₂H₅O₂ + M	$5.9 \times 10^{-29}[\text{N}_2]$	(k_0)	±0.3	$5.9 \times 10^{-29}(T/300)^{-3.8}[\text{N}_2]$	200–300	$\Delta n = \pm 1$
	7.8×10^{-12}	(k_∞)	±0.2	7.8×10^{-12}	200–300	$\Delta \log k = \pm 0.2$
	$F_c = 0.54$			$F_c = \{0.58\exp(-T/1250)$ $+ 0.42\exp(-T/183)\}$	200–300	
n-C₃H₇ + O₂ + M → n-C₃H₇O₂ + M	8×10^{-12}	(k_∞)	±0.2	8×10^{-12}	200–300	$\Delta \log k = \pm 0.2$
i-C₃H₇ + O₂ + M → i-C₃H₇O₂ + M	1.1×10^{-11}	(k_∞)	±0.3	1.1×10^{-11}	200–300	$\Delta \log k = \pm 0.3$
CH₃COCH₂ + O₂ + M → CH₃COCH₂O₂ + M	1.5×10^{-12}	(k_∞)	±0.5			
HCO + O₂ → CO + HO₂	5.5×10^{-12}		±0.15	5.5×10^{-12}	200–400	±150
CH₃CO + O₂ + M → CH₃CO₃ + M	5.0×10^{-12}	(k_∞)	±0.5	5.0×10^{-12}	200–300	$\Delta \log k = \pm 0.5$
CH₂OH + O₂ → HCHO + HO₂	9.4×10^{-12}		±0.12			
CH₃CHOH + O₂ → CH₃CHO + HO₂	1.9×10^{-11}		±0.3			
CH₂CH₂OH + O₂ → products	3.0×10^{-12}		±0.3			
CH₃O + O₂ → HCHO + HO₂	1.9×10^{-15}		±0.2	$7.2 \times 10^{-14}\exp(-1080/T)$	298–610	±300
C₂H₅O + O₂ → CH₃CHO + HO₂	9.5×10^{-15}		±0.2	$6.0 \times 10^{-14}\exp(-550/T)$	295–425	±300
n-C₃H₇O + O₂ → C₂H₅CHO + HO₂	8×10^{-15}		±0.5			
i-C₃H₇O + O₂ → CH₃COCH₃ + HO₂	8×10^{-15}		±0.3	$1.5 \times 10^{-14}\exp(-200/T)$	290–390	±200
CH₃ + O₃ → products	2.5×10^{-12}		±0.3	$5.1 \times 10^{-12}\exp(-210/T)$	240–400	±200
CH₃O + NO + M → CH₃ONO + M	$1.6 \times 10^{-29}[\text{N}_2]$	(k_0)	±0.1	$1.6 \times 10^{-29}(T/300)^{-3.5}[\text{N}_2]$	200–400	$\Delta n = \pm 0.5$
	3.6×10^{-11}	(k_∞)	±0.5	$3.6 \times 10^{-11}(T/300)^{-0.6}$	200–400	$\Delta n = \pm 0.5$
	$F_c = 0.6$					
CH₃O + NO → HCHO + HNO	4×10^{-12}			$4 \times 10^{-12}(T/300)^{-0.7}$	200–400	$\Delta n = \pm 0.5$
C₂H₅O + NO + M → C₂H₅ONO + M	4.4×10^{-11}	(k_∞)	±0.3	4.4×10^{-11}	200–300	$\Delta n = \pm 0.5$
C₂H₅O + NO → CH₃CHO + HNO	1.3×10^{-11}					
i-C₃H₇O + NO + M → i-C₃H₇ONO + M	3.4×10^{-11}	(k_∞)	±0.3	3.4×10^{-11}	200–300	$\Delta n = \pm 0.5$

Reaction	k_{298} cm³ molecule⁻¹ s⁻¹		Δlog k_{298}	Temp. dependence of k/cm³ molecule⁻¹ s⁻¹	Temp. range/K	Δ(E/R)/ K
i-C₃H₇O + NO → CH₃COCH₃ + HNO	6.5×10^{-12}		±0.5			
CH₃O + NO₂ + M → CH₃ONO₂ + M	2.8×10^{-29}[N₂]	(k_0)	±0.3	$2.8 \times 10^{-29}(T/300)^{-4.5}$[N₂]	200–400	Δn = ±1
	2×10^{-11}	(k_∞)	±0.3	2×10^{-11}	200–400	Δn = ±0.5
	$F_c = 0.44$					
CH₃O + NO₂ → HCHO + HONO	See data sheets					
C₂H₅O + NO₂ + M → C₂H₅ONO₂ + M	2.8×10^{-11}	(k_∞)	±0.3	2.8×10^{-11}	200–300	Δn = ±0.5
C₂H₅O + NO₂ → CH₃CHO + HONO	See data sheets					
i-C₃H₇O + NO₂ + M → i-C₃H₇ONO₂ + M	3.5×10^{-11}	(k_∞)	±0.3	3.5×10^{-11}	200–300	Δn = ±0.5
i-C₃H₇O + NO₂ → CH₃COCH₃ + HONO	See data sheets					
CH₃O₂ + NO → CH₃O + NO₂	7.6×10^{-12}		±0.1	$4.2 \times 10^{-12}\exp(180/T)$	240–360	±180
C₂H₅O₂ + NO → C₂H₅O + NO₂ ·	8.9×10^{-12}		±0.3			
C₂H₅O₂ + NO (+M) → C₂H₅ONO₂ (+M)	$\leqslant 1.3 \times 10^{-13}$	(1 bar)				
n-C₃H₇O₂ + NO → n-C₃H₇O + NO₂	8.7×10^{-12}		±0.3			
n-C₃H₇O₂ + NO (+M) → n-C₃H₇ONO₂ (+M)	1.8×10^{-13}	(1 bar)	±0.5			
i-C₃H₇O₂ + NO → i-C₃H₇O + NO₂	8.5×10^{-12}		±0.5			
i-C₃H₇O₂ + NO (+M) → i-C₃H₇ONO₂ (+M)	3.7×10^{-13}	(1 bar)	±0.5			
CH₃CO₃ + NO → CH₃ + CO₂ + NO₂	2.0×10^{-11}		±0.2	2.0×10^{-11}	280–325	±600
CH₃O₂ + NO₂ + M → CH₃O₂NO₂ + M	2.5×10^{-30}[N₂]	(k_0)	±0.3	$2.5 \times 10^{-30}(T/300)^{-5.5}$[N₂]	250–350	Δn = ±1
	7.5×10^{-12}	(k_∞)	±0.3	7.5×10^{-12}	250–350	Δn = ±0.5
	$F_c = 0.4$					
CH₃O₂NO₂ + M → CH₃O₂ + NO₂ + M	6.8×10^{-19}[N₂]	(k_0/s^{-1})	±0.3	$9 \times 10^{-5}\exp(-9690/T)$[N₂]	250–300	±500
	4.5	(k_∞/s^{-1})	±0.3	$1.1 \times 10^{16}\exp(-10560/T)$	250–300	±500
	$F_c = 0.4$					
C₂H₅O₂ + NO₂ + M → C₂H₅O₂NO₂ + M	1.3×10^{-29}[N₂]	(k_0)	±0.3	$1.3 \times 10^{-29}(T/300)^{-6.2}$[N₂]	200–300	Δn = ±1
	8.8×10^{-12}	(k_∞)	±0.3	8.8×10^{-12}	200–300	Δlog k = ±0.3
	$F_c = 0.31$			$F_c = 0.31$	250–300	
C₂H₅O₂NO₂ + M → C₂H₅O₂ + NO₂ + M	1.4×10^{-17}[N₂]	(k_0/s^{-1})	±0.5	$4.8 \times 10^{-4}\exp(-9285/T)$[N₂]	250–300	±1000
	5.4	(k_∞/s^{-1})	±0.5	$8.8 \times 10^{15}\exp(-10440/T)$	250–300	±1000
	$F_c = 0.31$			$F_c = 0.31$	250–300	
CH₃CO₃ + NO₂ + M → CH₃CO₃NO₂ + M	2.7×10^{-28}[N₂]	(k_0)	±0.4	$2.7 \times 10^{-28}(T/300)^{-7.1}$[N₂]	250–300	Δn = ±2
	1.2×10^{-11}	(k_∞)	±0.2	$1.2 \times 10^{-11}(T/300)^{-0.9}$	250–300	Δn = ±1
	$F_c = 0.3$					
CH₃CO₃NO₂ + M → CH₃CO₃ + NO₂ + M	1.1×10^{-20}[N₂]	(k_0/s^{-1})	±0.4	$4.9 \times 10^{-3}\exp(-12100/T)$[N₂]	300–330	±1000
	6.1×10^{-4}	(k_∞/s^{-1})	±0.2	$4.0 \times 10^{16}\exp(-13600/T)$	280–330	±200
	$F_c = 0.3$					
CH₃O₂ + NO₃ → products	No recommendation (see data sheets)					
CH₃O₂ + CH₃O₂ → CH₃OH + HCHO + O₂ → 2 CH₃O + O₂ → CH₃OOCH₃ + O₂	3.7×10^{-13}		±0.12	$1.1 \times 10^{-13}\exp(365/T)$	200–400	±200
CH₃O₂ + CH₃CO₃ → CH₃O + CH₃CO₂ + O₂	5.5×10^{-12}		±0.5			
→ CH₃CO₂H + HCHO + O₂	5.5×10^{-12}		±0.5			
HOCH₂O₂ + HOCH₂O₂ → HCOOH + CH₂(OH)₂ + O₂	7.0×10^{-13}		±0.3	$5.7 \times 10^{-14}\exp(750/T)$	275–325	±750
→ 2 HOCH₂O + O₂	5.5×10^{-12}		±0.3			
C₂H₅O₂ + C₂H₅O₂ → C₂H₅OH + CH₃CHO + O₂ → 2 C₂H₅O + O₂ → C₂H₅OOC₂H₅ + O₂	6.8×10^{-14}		±0.12	$9.8 \times 10^{-14}\exp(-110/T)$	250–450	$^{+300}_{-100}$
CH₃CO₃ + CH₃CO₃ → 2 CH₃CO₂ + O₂	1.6×10^{-11}		±0.5	$2.8 \times 10^{-12}\exp(530/T)$	250–370	±500
HOCH₂CH₂O₂ + HOCH₂CH₂O₂ → HOCH₂CH₂OH + HOCH₂CHO + O₂	1.5×10^{-12}		±0.3			
→ 2 HOCH₂CH₂O + O₂	8.3×10^{-13}		±0.3			
n-C₃H₇O₂ + n-C₃H₇O₂ → n-C₃H₇OH + C₂H₅CHO + O₂ → 2 n-C₃H₇O + O₂	3×10^{-13}		±0.5			
i-C₃H₇O₂ + i-C₃H₇O₂ → i-C₃H₇OH + CH₃COCH₃ + O₂	4.4×10^{-16}		±0.3	$1.6 \times 10^{-12}\exp(-2200/T)$	300–400	±300
→ 2 i-C₃H₇O + O₂	5.6×10^{-16}		±0.3			
CH₃COCH₂O₂ + CH₃COCH₂O₂ → CH₃COCH₂OH + CH₃COCHO + O₂ → 2 CH₃COCH₂O + O₂	$\leqslant 1 \times 10^{-11}$					
RCHOO + H₂O → RCOOH + H₂O RCHOO + NO₂ → RCHO + NO₃ RCHOO + SO₂ → products RCHOO + HCHO → products	No recommendations (see data sheets)					
CN + O₂ → products	See previous evaluation					
O₃ + C₂H₂ → products	1×10^{-20}		±1.0			
O₃ + C₂H₄ → products	1.7×10^{-18}		±0.10	$1.2 \times 10^{-14}\exp(-2630/T)$	180–360	±100
O₃ + C₃H₆ → products	1.2×10^{-17}		±0.15	$6.5 \times 10^{-15}\exp(-1880/T)$	230–370	±400
HCHO + $h\nu$ → products	See data sheets					
CH₃CHO + $h\nu$ → products	See data sheets					
C₂H₅CHO + $h\nu$ → products	See data sheets					
(CHO)₂ + $h\nu$ → products	See data sheets					
CH₃COCHO + $h\nu$ → products	See data sheets					
CH₃COCH₃ + $h\nu$ → products	See data sheets					
CH₃OOH + $h\nu$ → products	See data sheets					
CH₃ONO₂ + $h\nu$ → products	See data sheets					
C₂H₅ONO₂ + $h\nu$ → products	See data sheets					
n-C₃H₇ONO₂ + $h\nu$ → products	See data sheets					
i-C₃H₇ONO₂ + $h\nu$ → products	See data sheets					
CH₃O₂NO₂ + $h\nu$ → products	See data sheets					
CH₃CO₃NO₂ + $h\nu$ → products	See data sheets					

Reaction	k_{298} cm^3 molecule^{-1} s^{-1}		$\Delta\log k_{298}$	Temp. dependence of k/cm^3 molecule^{-1} s^{-1}	Temp. range/K	$\Delta(E/R)$/ K
SO$_x$ Reactions						
O + H$_2$S → HO + HS	See previous evaluation					
O + CS → CO + S	2.1×10^{-11}		±0.1	$2.7 \times 10^{-10}\exp(-760/T)$	150–300	±250
O + CH$_3$SCH$_3$ → CH$_3$SO + CH$_3$	5.0×10^{-11}		±0.1	$1.3 \times 10^{-11}\exp(409/T)$	270–560	±100
O + CS$_2$ → SO + CS	3.6×10^{-12}		±0.2	$3.2 \times 10^{-11}\exp(-650/T)$	200–500	±100
O + CH$_3$SSCH$_3$ → CH$_3$SO + CH$_3$S	1.3×10^{-10}		±0.3	$5.5 \times 10^{-11}\exp(250/T)$	290–570	±100
O + OCS → SO + CO	1.2×10^{-14}		±0.2	$1.6 \times 10^{-11}\exp(-2150/T)$	220–500	±150
O + SO$_2$ + M → SO$_3$ + M	1.4×10^{-33}[N$_2$]	(k_0)	±0.3	$4.0 \times 10^{-32}\exp(-1000/T)$[N$_2$]	200–400	±200
S + O$_2$ → SO + O	2.1×10^{-12}		±0.2	2.1×10^{-12}	230–400	±200
S + O$_3$ → SO + O$_2$	1.2×10^{-11}		±0.3			
Cl + H$_2$S → HCl + HS	5.7×10^{-11}		±0.3	5.7×10^{-11}	210–350	±100
HO + H$_2$S → H$_2$O + HS	4.8×10^{-12}		±0.08	$6.3 \times 10^{-12}\exp(-80/T)$	200–300	±80
HO + SO$_2$ + M → HOSO$_2$ + M	4.0×10^{-31}[N$_2$]	(k_0)	±0.3	$4.0 \times 10^{-31}(T/300)^{-3.3}$[N$_2$]	300–400	$\Delta n = \pm 1$
	$F_c = 0.8$					
	2×10^{-12}	(k_∞)	±0.3	2×10^{-12}	200–300	$\Delta\log k = \pm 0.3$
	$F_c = 0.45$					
HOSO$_2$ + O$_2$ → HO$_2$ + SO$_3$	4.0×10^{-13}		±0.1	$1.3 \times 10^{-12}\exp(-330/T)$	290–420	±200
HO + OCS → products	2.0×10^{-15}		±0.3	$1.1 \times 10^{-13}\exp(-1200/T)$	250–500	±500
HO + CS$_2$ + M → HOCS$_2$ + M	8×10^{-31}[N$_2$]	(k_0)	±0.5	8×10^{-31}[N$_2$]	270–300	$\Delta\log k = \pm 0.5$
	8×10^{-12}	(k_∞)	±0.5	8×10^{-12}	250–300	$\Delta\log k = \pm 0.5$
	$F_c = 0.8$					
HO + CS$_2$ → HS + OCS	$<2 \times 10^{-15}$					
HOCS$_2$ + M → HO + CS$_2$ + M	4.8×10^{-14}[N$_2$]	(k_0/s^{-1})	±0.5	$1.6 \times 10^{-6}\exp(-5160/T)$[N$_2$]	250–300	±500
	4.8×10^{5}	(k_∞/s^{-1})	±0.5	$1.6 \times 10^{13}\exp(-5160/T)$	250–300	±500
	$F_c = 0.8$					
HOCS$_2$ + O$_2$ → products	2.8×10^{-14}		±0.3	2.8×10^{-14}	240–300	$\Delta\log k = \pm 0.3$
HO + CH$_3$SH → products	3.3×10^{-11}		±0.10	$9.9 \times 10^{-12}\exp(356/T)$	240–430	±100
HO + CH$_3$SCH$_3$ → H$_2$O + CH$_2$SCH$_3$	4.4×10^{-12}		±0.10	$9.6 \times 10^{-12}\exp(-234/T)$	250–400	±300
→ CH$_3$S(OH)CH$_3$	See data sheets					
HO + CH$_3$SSCH$_3$ → products	2.0×10^{-10}		±0.10	$6.0 \times 10^{-11}\exp(380/T)$	250–370	±300
HO$_2$ + SO$_2$ → products	$\leqslant 1 \times 10^{-18}$					
NO$_3$ + H$_2$S → products	$<1 \times 10^{-15}$					
NO$_3$ + CS$_2$ → products	$<1 \times 10^{-15}$					
NO$_3$ + OCS → products	$<1 \times 10^{-16}$					
NO$_3$ + SO$_2$ → products	$<1 \times 10^{-19}$					
NO$_3$ + CH$_3$SH → products	9.2×10^{-13}		±0.15	9.2×10^{-13}	250–370	±400
NO$_3$ + CH$_3$SCH$_3$ → products	1.1×10^{-12}		±0.15	$1.9 \times 10^{-13}\exp(520/T)$	250–380	±200
NO$_3$ + CH$_3$SSCH$_3$ → products	7×10^{-13}		±0.3	7×10^{-13}	300–380	±500
CH$_3$O$_2$ + SO$_2$ → CH$_3$O + SO$_3$	See previous evaluation					
→ CH$_3$O$_2$SO$_2$	See previous evaluation					
HS + O$_2$ → products	$\leqslant 4 \times 10^{-19}$					
HS + O$_3$ → HSO + O$_2$	3.7×10^{-12}		±0.2	$9.5 \times 10^{-12}\exp(-280/T)$	290–450	±250
HS + NO + M → HSNO + M	2.4×10^{-31}[N$_2$]	(k_0)	±0.3	$2.4 \times 10^{-31}(T/300)^{-2.5}$[N$_2$]	200–300	$\Delta n = \pm 1$
	2.7×10^{-11}	(k_∞)	±0.5	2.7×10^{-11}	200–300	$\Delta\log k = \pm 0.5$
	$F_c = 0.6$					
HS + NO$_2$ → HSO + NO	5.8×10^{-11}		±0.3	$2.6 \times 10^{-11}\exp(240/T)$	220–450	±200
HSO + O$_2$ → products	$\leqslant 2.0 \times 10^{-17}$					
HSO + O$_3$ → products	1.1×10^{-13}		±0.3			
HSO + NO → products	$\leqslant 1.0 \times 10^{-15}$					
HSO + NO$_2$ → products	9.6×10^{-12}		±0.3			
HSO$_2$ + O$_2$ → products	3.0×10^{-13}		±0.8			
SO + O$_2$ → SO$_2$ + O	6.7×10^{-17}		±0.15	$1.4 \times 10^{-13}\exp(-2280/T)$	230–420	±500
SO + O$_3$ → SO$_2$ + O$_2$	8.9×10^{-14}		±0.10	$4.5 \times 10^{-12}\exp(-1170/T)$	230–420	±150
SO + NO$_2$ → SO$_2$ + NO	1.4×10^{-11}		±0.1	1.4×10^{-11}	210–360	±100
SO$_3$ + H$_2$O → products	$<6 \times 10^{-15}$					
CS + O$_2$ → products	2.9×10^{-19}		±0.6			
CS + O$_3$ → OCS + O$_2$	3.0×10^{-16}		±0.5			
CS + NO$_2$ → OCS + NO	7.6×10^{-17}		±0.5			
CH$_3$S + O$_2$ → products	$<2.5 \times 10^{-18}$					
CH$_3$S + O$_3$ → products	4.1×10^{-12}		±0.5			
CH$_3$S + NO + M → CH$_3$SNO + M	3.2×10^{-29}[N$_2$]	(k_0)	±0.3	$3.2 \times 10^{-29}(T/298)^{-4}$[N$_2$]	250–450	$\Delta n = \pm 2$
	4×10^{-11}	(k_∞)	±0.5	4×10^{-11}	250–450	$\Delta\log k = \pm 0.5$
	$F_c = 0.60$			$F_c = \exp(-T/580)$		
CH$_3$S + NO$_2$ → CH$_3$SO + NO	5.6×10^{-11}		±0.2			
CH$_3$SO + O$_3$ → products	1×10^{-12}		±0.7			
CH$_3$SO + NO$_2$ → products	1.2×10^{-11}		±0.5			
O$_3$ + CH$_3$SCH$_3$ → products	$<1 \times 10^{-18}$					
OCS + $h\nu$ → products	See data sheets					
CS$_2$ + $h\nu$ → products	See data sheets					
CH$_3$SSCH$_3$ + $h\nu$ → products	See data sheets					
CH$_3$SNO + $h\nu$ → products	See data sheets					
FO$_x$ Reactions						
O + FO → O$_2$ + F	5×10^{-11}		±0.5			
O + FO$_2$ → O$_2$ + FO	5×10^{-11}		±0.7			
O(^{1}D) + HF → HO + F	1×10^{-10}		±0.5			
→ O(^{3}P) + HF						
O(^{1}D) + COF$_2$ → CO$_2$ + F$_2$	2.2×10^{-11}		±0.2			
→ O(^{3}P) + COF$_2$	5.2×10^{-11}		±0.2			
O(^{1}D) + CH$_3$F → products	1.4×10^{-10}		±0.5			
O(^{1}D) + CH$_2$F$_2$ → products	9×10^{-11}		±0.5			
O(^{1}D) + CHF$_3$ → products	8.4×10^{-12}		±0.5			
O(^{1}D) + CH$_3$CHF$_2$ → products	1×10^{-10}		±0.7			

Reaction	k_{298} cm^3 molecule^{-1} s^{-1}		$\Delta \log k_{298}$	Temp. dependence of k/cm^3 molecule^{-1} s^{-1}	Temp. range/K	$\Delta(E/R)/$ K
$O(^1D) + CH_3CF_3 \rightarrow$ products	1×10^{-10}		± 0.5			
$O(^1D) + CH_2FCF_3 \rightarrow$ products	1×10^{-10}		± 0.7			
$O(^1D) + CHF_2CF_3 \rightarrow$ products	5×10^{-11}		± 0.5			
$F + H_2 \rightarrow HF + H$	2.6×10^{-11}		± 0.1	$1.4 \times 10^{-10} \exp(-500/T)$	200–375	± 200
$F + H_2O \rightarrow HF + HO$	1.4×10^{-11}		± 0.1	1.4×10^{-11}	240–370	± 200
$F + O_2 + M \rightarrow FO_2 + M$	$3.7 \times 10^{-33}[N_2]$	(k_0)	± 0.3	$3.7 \times 10^{-33}(T/300)^{-1}[N_2]$	300–400	$\Delta n = \pm 0.5$
$FO_2 + M \rightarrow F + O_2 + M$	No recommendation (see data sheets)					
$F + O_3 \rightarrow FO + O_2$	1.3×10^{-11}		± 0.3	$2.8 \times 10^{-11} \exp(-230/T)$	250–365	± 200
$F + HONO_2 \rightarrow HF + NO_3$	2.3×10^{-11}		± 0.1	$6.0 \times 10^{-12} \exp(400/T)$	260–320	± 200
$F + NO_2 + M \rightarrow FONO + M$	See previous evaluation					
$F + CH_4 \rightarrow HF + CH_3$	8×10^{-11}		± 0.2	$3.0 \times 10^{-10} \exp(-400/T)$	250–450	± 200
$HO + CH_3F \rightarrow H_2O + CH_2F$	1.7×10^{-14}		± 0.10	$3.7 \times 10^{-12} \exp(-1600/T)$	270–340	± 300
$HO + CH_2F_2 \rightarrow H_2O + CHF_2$	1.1×10^{-14}		± 0.10	$2.0 \times 10^{-12} \exp(-1545/T)$	240–300	± 200
$HO + CHF_3 \rightarrow H_2O + CF_3$	2.4×10^{-16}		± 0.5	$1.0 \times 10^{-12} \exp(-2490/T)$	270–340	± 500
$HO + CH_3CH_2F \rightarrow$ products	2.3×10^{-14}		± 0.3			
$HO + CH_3CHF_2 \rightarrow$ products	3.6×10^{-14}		± 0.10	$1.0 \times 10^{-12} \exp(-990/T)$	240–300	± 200
$HO + CH_3CF_3 \rightarrow H_2O + CH_2CF_3$	1.3×10^{-15}		± 0.15	$1.05 \times 10^{-12} \exp(-1990/T)$	240–300	± 300
$HO + CH_2FCHF_2 \rightarrow$ products	1.8×10^{-15}		± 0.3			
$HO + CH_2FCF_3 \rightarrow H_2O + CHFCF_3$	4.9×10^{-15}		± 0.2	$8.4 \times 10^{-13} \exp(-1535/T)$	240–300	± 300
$HO + CHF_2CHF_2 \rightarrow H_2O + CF_2CHF_2$	5.7×10^{-15}		± 0.3			
$HO + CHF_2CF_3 \rightarrow H_2O + CF_2CF_3$	1.9×10^{-15}		± 0.2	$4.9 \times 10^{-13} \exp(-1655/T)$	240–300	± 300
$HO + CF_3CHO \rightarrow H_2O + CF_3CO$	1.1×10^{-12}		± 0.3			
$FO + O_3 \rightarrow$ products	No recommendation (see data sheets)					
$FO + NO \rightarrow F + NO_2$	2.6×10^{-11}		± 0.3			
$FO + NO_2 + M \rightarrow FONO_2 + M$	See previous evaluation					
$FO + FO \rightarrow$ products	1.5×10^{-11}		± 0.3			
$COF_2 + h\nu \rightarrow$ products	See data sheets					
$HCOF + h\nu \rightarrow$ products	See data sheets					
$CF_3COF + h\nu \rightarrow$ products	See data sheets					
ClO$_x$ Reactions						
$O + HCl \rightarrow HO + Cl$	See previous evaluation					
$O + HOCl \rightarrow HO + ClO$	No recommendation (see data sheets)					
$O + ClO \rightarrow Cl + O_2$	3.8×10^{-11}		± 0.1	3.8×10^{-11}	200–300	± 250
$O + OClO \rightarrow O_2 + ClO$	1.0×10^{-13}		± 0.3	$2.5 \times 10^{-12} \exp(-950/T)$	240–400	± 300
$O + OClO + M \rightarrow ClO_3 + M$	$1.8 \times 10^{-31}[N_2]$	(k_0)	± 0.3	$1.8 \times 10^{-31}(T/298)^{-1}[N_2]$	250–300	$\Delta n = \pm 0.5$
	3.1×10^{-11}	(k_∞)	± 0.3	$3.1 \times 10^{-11}(T/298)$	250–300	$\Delta n = \pm 1$
	$F_c = 0.48$					
$O + Cl_2O \rightarrow ClO + ClO$	3.5×10^{-12}		± 0.15	$2.9 \times 10^{-11} \exp(-630/T)$	235–300	± 200
$O + ClONO_2 \rightarrow$ products	2.0×10^{-13}		± 0.1	$3.0 \times 10^{-12} \exp(-800/T)$	213–295	± 200
$O(^1D) + CHF_2Cl \rightarrow$ products	9.5×10^{-11}		± 0.2	9.5×10^{-11}	175–340	$\Delta \log k = \pm 0.2$
$O(^1D) + CHFCl_2 \rightarrow$ products	1.9×10^{-10}		± 0.2	1.9×10^{-10}	175–340	$\Delta \log k = \pm 0.2$
$O(^1D) + CH_3CF_2Cl \rightarrow$ products	1.4×10^{-10}		± 0.3			
$O(^1D) + CH_3CFCl_2 \rightarrow$ products	1.5×10^{-10}		± 0.5			
$O(^1D) + CH_2ClCF_3 \rightarrow$ products	1.5×10^{-10}		± 0.3			
$O(^1D) + CH_2ClCF_2Cl \rightarrow$ products	1.6×10^{-10}		± 0.3			
$O(^1D) + CHFClCF_3 \rightarrow$ products	1.0×10^{-10}		± 0.5			
$O(^1D) + CHCl_2CF_3 \rightarrow$ products	2.2×10^{-10}		± 0.3			
$O(^1D) + CF_2Cl_2 \rightarrow$ products	1.4×10^{-10}		± 0.1			
$O(^1D) + CFCl_3 \rightarrow$ products	2.3×10^{-10}		± 0.1			
$O(^1D) + CCl_4 \rightarrow$ products	3.3×10^{-10}		± 0.1			
$O(^1D) + COFCl \rightarrow$ products	1.9×10^{-10}		± 0.3			
$O(^1D) + COCl_2 \rightarrow$ products	3.6×10^{-10}		± 0.3			
$Cl + H_2 \rightarrow HCl + H$	1.6×10^{-14}		± 0.1	$3.7 \times 10^{-11} \exp(-2300/T)$	200–300	± 200
$Cl + HO_2 \rightarrow HCl + O_2$	3.2×10^{-11}		± 0.2	$1.8 \times 10^{-11} \exp(170/T)$	250–420	± 250
$\rightarrow ClO + HO$	9.1×10^{-12}		± 0.3	$4.1 \times 10^{-11} \exp(-450/T)$	250–420	± 250
$Cl + H_2O_2 \rightarrow HCl + HO_2$	4.1×10^{-13}		± 0.2	$1.1 \times 10^{-11} \exp(-980/T)$	265–424	± 500
$Cl + O_2 + M \rightarrow ClOO + M$	$1.4 \times 10^{-33}[N_2]$	(k_0)	± 0.2	$1.4 \times 10^{-33}(T/300)^{-3.9}[N_2]$	160–300	$\Delta n = \pm 1$
	$1.6 \times 10^{-33}[O_2]$	(k_0)	± 0.2	$1.6 \times 10^{-33}(T/300)^{-2.9}[O_2]$	160–300	$\Delta n = \pm 1$
$ClOO + M \rightarrow Cl + O_2 + M$	$6.2 \times 10^{-13}[N_2]$	(k_0/s^{-1})	± 0.3	$2.8 \times 10^{-10} \exp(-1820/T)[N_2]$	200–300	± 200
$Cl + O_3 \rightarrow ClO + O_2$	1.2×10^{-11}		± 0.06	$2.9 \times 10^{-11} \exp(-260/T)$	205–298	± 100
$Cl + HONO_2 \rightarrow HCl + NO_3$	$< 2.0 \times 10^{-16}$					
$Cl + NO_3 \rightarrow ClO + NO_2$	2.6×10^{-11}		± 0.3	2.6×10^{-11}	200–300	± 400
$Cl + OClO \rightarrow ClO + ClO$	5.8×10^{-11}		± 0.1	$3.4 \times 10^{-11} \exp(160/T)$	298–450	± 200
$Cl + Cl_2O \rightarrow Cl_2 + ClO$	9.8×10^{-11}		± 0.1	9.8×10^{-11}	200–300	± 250
$Cl + Cl_2O_2 \rightarrow Cl_2 + ClOO$	1.0×10^{-10}		± 0.3	1.0×10^{-10}	230–298	± 300
$Cl + ClONO_2 \rightarrow Cl_2 + NO_3$	1.2×10^{-11}		± 0.12	$6.8 \times 10^{-12} \exp(160/T)$	219–298	± 200
$Cl + CH_4 \rightarrow HCl + CH_3$	1.0×10^{-13}		± 0.08	$9.6 \times 10^{-12} \exp(-1350/T)$	200–300	± 250
$Cl + C_2H_2 + M \rightarrow C_2H_2Cl + M$	$6 \times 10^{-30}[N_2]$	(k_0)	± 0.3	$6 \times 10^{-30}(T/300)^{-3.5}[N_2]$	200–300	$\Delta n = \pm 1$
	2.3×10^{-10}	(k_∞)	± 0.5	2.3×10^{-10}	200–300	$\Delta n = \pm 1$
	$F_c = 0.6$					
$Cl + C_2H_4 + M \rightarrow C_2H_4Cl + M$	$1.6 \times 10^{-29}[air]$	(k_0)	± 0.5	$1.6 \times 10^{-29}(T/298)^{-3.5}[air]$	250–300	$\Delta n = \pm 1$
	3×10^{-10}	(k_∞)	± 0.3	3×10^{-10}	250–300	$\Delta n = \pm 1$
	$F_c = 0.6$					
$Cl + C_2H_6 \rightarrow HCl + C_2H_5$	5.9×10^{-11}		± 0.06	$8.2 \times 10^{-11} \exp(-100/T)$	220–600	± 100
$Cl + C_3H_8 \rightarrow HCl + C_3H_7$	1.4×10^{-10}		± 0.12	$1.2 \times 10^{-10} \exp(40/T)$	220–600	± 200
$Cl + HCHO \rightarrow HCl + HCO$	7.3×10^{-11}		± 0.06	$8.2 \times 10^{-11} \exp(-34/T)$	200–500	± 100
$Cl + CH_3CHO \rightarrow HCl + CH_3CO$	7.2×10^{-11}		± 0.15	7.2×10^{-11}	210–340	± 300
$Cl + C_2H_5CHO \rightarrow$ products	1.2×10^{-10}		± 0.3			
$Cl + CH_3COCH_3 \rightarrow HCl + CH_3COCH_2$	3.5×10^{-12}		± 0.3			
$Cl + CH_3OH \rightarrow HCl + CH_2OH$	5.3×10^{-11}		± 0.15	5.3×10^{-11}	200–500	± 200
$Cl + C_2H_5OH \rightarrow$ products	9.4×10^{-11}		± 0.2			
$Cl + n\text{-}C_3H_7OH \rightarrow$ products	1.5×10^{-10}		± 0.2			
$Cl + i\text{-}C_3H_7OH \rightarrow$ products	8.4×10^{-11}		± 0.3			
$Cl + CH_3OOH \rightarrow$ products	5.9×10^{-11}		± 0.5			

Reaction	k_{298} cm^3 molecule^{-1} s^{-1}	$\Delta\log k_{298}$	Temp. dependence of k/cm^3 molecule^{-1} s^{-1}	Temp. range/K	$\Delta(E/R)/$ K
Cl + HCOOH → products	2.0×10^{-13}	± 0.2			
Cl + CH$_3$COOH → products	2.8×10^{-14}	± 0.3			
Cl + CH$_3$ONO$_2$ → products	2.4×10^{-13}	± 0.3			
Cl + C$_2$H$_5$ONO$_2$ → products	4.7×10^{-12}	± 0.2			
Cl + n-C$_3$H$_7$ONO$_2$ → products	2.7×10^{-11}	± 0.2			
Cl + i-C$_3$H$_7$ONO$_2$ → products	5.8×10^{-12}	± 0.3			
Cl + CH$_3$CO$_3$NO$_2$ → products	$<2 \times 10^{-14}$				
Cl + CH$_3$CN → products	$\leqslant 2 \times 10^{-15}$				
Cl + HCOCl → HCl + ClCO	7.8×10^{-13}	± 0.15	$1.2 \times 10^{-11}\exp(-815/T)$	265–325	± 300
Cl + CH$_3$Cl → HCl + CH$_2$Cl	4.9×10^{-13}	± 0.15	$3.3 \times 10^{-11}\exp(-1250/T)$	233–322	± 300
Cl + CH$_2$Cl$_2$ → HCl + CHCl$_2$	4.1×10^{-13}	± 0.25	$8.7 \times 10^{-12}\exp(-910/T)$	270–330	± 400
Cl + CHCl$_3$ → HCl + CCl$_3$	7.6×10^{-14}	± 0.3	$4.9 \times 10^{-12}\exp(-1240/T)$	240–330	± 400
Cl + CH$_3$CCl$_3$ → HCl + CH$_2$CCl$_3$	$<4 \times 10^{-14}$				
HO + HCl → H$_2$O + Cl	8.1×10^{-13}	± 0.1	$2.4 \times 10^{-12}\exp(-330/T)$	200–300	± 150
HO + HOCl → ClO + H$_2$O	5.0×10^{-13}	± 0.5	$3.0 \times 10^{-12}\exp(-500/T)$	200–300	± 500
HO + ClO → HO$_2$ + Cl} → HCl + O$_2$}	1.7×10^{-11}	± 0.2	$1.1 \times 10^{-11}\exp(120/T)$	200–373	± 150
HO + OClO → HOCl + O$_2$	7.0×10^{-12}	± 0.3	$4.5 \times 10^{-13}\exp(800/T)$	290–480	± 200
HO + ClNO$_2$ → HOCl + NO$_2$	3.5×10^{-14}	± 0.3			
HO + ClONO$_2$ → products	3.9×10^{-13}	± 0.2	$1.2 \times 10^{-12}\exp(-330/T)$	246–387	± 200
HO + CH$_3$Cl → H$_2$O + CH$_2$Cl	4.3×10^{-14}	± 0.10	$1.8 \times 10^{-12}\exp(-1115/T)$	240–300	± 200
HO + CH$_2$FCl → H$_2$O + CHFCl	4.4×10^{-14}	± 0.10	$2.0 \times 10^{-11}\exp(-1135/T)$	240–300	± 200
HO + CHF$_2$Cl → H$_2$O + CF$_2$Cl	4.6×10^{-15}	± 0.10	$7.8 \times 10^{-13}\exp(-1530/T)$	240–300	± 200
HO + CHFCl$_2$ → H$_2$O + CFCl$_2$	3.0×10^{-14}	± 0.10	$8.8 \times 10^{-13}\exp(-1010/T)$	240–300	± 200
HO + CH$_2$Cl$_2$ → H$_2$O + CHCl$_2$	1.4×10^{-13}	± 0.10	$4.4 \times 10^{-12}\exp(-1030/T)$	240–300	± 250
HO + CHCl$_3$ → H$_2$O + CCl$_3$	1.0×10^{-13}	± 0.10	$3.3 \times 10^{-12}\exp(-1030/T)$	240–300	± 100
HO + CFCl$_3$ → HOCl + CFCl$_2$	$<5 \times 10^{-18}$		$<1 \times 10^{-12}\exp(-3650/T)$	250–480	
HO + CF$_2$Cl$_2$ → HOCl + CF$_2$Cl	$<7 \times 10^{-18}$		$<1 \times 10^{-12}\exp(-3540/T)$	250–478	
HO + CCl$_4$ → HOCl + CCl$_3$	$<5 \times 10^{-16}$		$<1 \times 10^{-12}\exp(-2260/T)$	250–300	
HO + C$_2$HCl$_3$ → products	2.2×10^{-12}	± 0.10	$5.0 \times 10^{-13}\exp(445/T)$	230–420	± 200
HO + C$_2$Cl$_4$ → products	1.7×10^{-13}	± 0.10	$9.4 \times 10^{-12}\exp(-1200/T)$	300–420	± 200
HO + CH$_2$CF$_2$Cl → H$_2$O + CH$_2$CF$_2$Cl	3.0×10^{-15}	± 0.10	$9.2 \times 10^{-13}\exp(-1705/T)$	240–300	± 200
HO + CH$_3$CFCl$_2$ → H$_2$O + CH$_2$CFCl$_2$	5.9×10^{-15}	± 0.2	$7.0 \times 10^{-13}\exp(-1425/T)$	240–300	± 300
HO + CH$_3$CCl$_3$ → H$_2$O + CH$_2$CCl$_3$	9.5×10^{-15}	± 0.10	$1.2 \times 10^{-12}\exp(-1440/T)$	240–300	± 200
HO + CH$_2$ClCF$_3$ → H$_2$O + CHClCF$_3$	1.3×10^{-14}	± 0.2	$5.2 \times 10^{-13}\exp(-1100/T)$	260–380	± 250
HO + CH$_2$ClCF$_2$Cl → H$_2$O + CHClCF$_2$Cl	1.6×10^{-14}	± 0.3	$3.2 \times 10^{-12}\exp(-1580/T)$	250–350	± 500
HO + CHFClCF$_3$ → H$_2$O + CFClCF$_3$	9.5×10^{-15}	± 0.10	$5.4 \times 10^{-13}\exp(-1205/T)$	240–300	± 200
HO + CHCl$_2$CF$_3$ → H$_2$O + CCl$_2$CF$_3$	3.6×10^{-14}	± 0.15	$5.5 \times 10^{-13}\exp(-815/T)$	240–300	± 200
HO + CHCl$_2$CF$_2$CF$_3$ → H$_2$O + CCl$_2$CF$_2$CF$_3$	2.5×10^{-14}	± 0.15	$1.1 \times 10^{-12}\exp(-1130/T)$	270–400	± 300
HO + CHFClCF$_2$CF$_2$Cl → H$_2$O + CFClCF$_2$CF$_2$Cl	8.9×10^{-15}	± 0.10	$5.5 \times 10^{-13}\exp(-1230/T)$	290–400	± 300
HO + CH$_3$CF$_2$CFCl$_2$ → H$_2$O + CH$_2$CF$_2$CFCl$_2$	2.4×10^{-15}	± 0.3	$7.0 \times 10^{-13}\exp(-1690/T)$	290–370	± 300
HO + HCOCl → H$_2$O + ClCO	$<5 \times 10^{-13}$				
HO + COCl$_2$ → products	$<5 \times 10^{-15}$				
HO + CH$_2$ClCHO → products	3.0×10^{-12}	± 0.3			
HO + CHCl$_2$CHO → products	2.4×10^{-12}	± 0.3			
HO + CCl$_3$CHO → H$_2$O + CCl$_3$CO	1.4×10^{-12}	± 0.3			
HO + CH$_3$COCl → H$_2$O + CH$_2$COCl	9×10^{-15}	± 1.0			
HO + CHF$_2$OCHClCF$_3$ → products	2.1×10^{-14}	± 0.5			
HO + CHF$_2$OCF$_2$CHFCl → products	1.6×10^{-15}	± 0.5	$6.1 \times 10^{-13}\exp(-1080/T)$	300–430	± 500
NO$_3$ + C$_2$HCl$_3$ → products	2.9×10^{-16}	± 0.3			
NO$_3$ + C$_2$Cl$_4$ → products	$<1 \times 10^{-16}$				
ClO + HO$_2$ → HOCl + O$_2$} → HCl + O$_3$}	5.0×10^{-12}	± 0.15	$4.6 \times 10^{-13}\exp(710/T)$	200–300	± 300
ClO + O$_2$($^1\Delta_g$) → sym-ClO$_3$	See previous evaluation				
ClO + O$_3$ → ClOO + O$_2$ → OClO + O$_2$	$<1.5 \times 10^{-17}$ $<1 \times 10^{-18}$				
ClO + NO → Cl + NO$_2$	1.7×10^{-11}	± 0.1	$6.2 \times 10^{-12}\exp(294/T)$	202–415	± 100
ClO + NO$_2$ + M → ClONO$_2$ + M	1.6×10^{-31}[N$_2$] $\quad (k_0)$	± 0.1	$1.6 \times 10^{-31}(T/300)^{-3.4}$[N$_2$]	200–300	$\Delta n = \pm 1$
	2×10^{-11} $\quad (k_\infty)$	± 0.3	2×10^{-11}	200–300	$\Delta\log k = \pm 0.3$
	$F_c = 0.5$		$F_c = \exp(-T/430)$	200–300	
ClO + NO$_3$ → ClOO + NO$_2$} → OClO + NO$_2$}	4.0×10^{-13}	± 0.3			
ClO + HCHO → products	See previous evaluation				
ClO + ClO → Cl + ClOO	3.4×10^{-15}	± 0.3			
→ Cl + OClO	1.7×10^{-15}	± 0.3			
→ Cl$_2$ + O$_2$	4.9×10^{-15}	± 0.3			
ClO + ClO + M → Cl$_2$O$_2$ + M	1.7×10^{-32}[N$_2$] $\quad (k_0)$	± 0.1	$1.7 \times 10^{-32}(T/300)^{-4}$[N$_2$]	200–260	$\Delta n = \pm 1.5$
	5.4×10^{-12} $\quad (k_\infty)$	± 0.3	5.4×10^{-12}	200–300	$\Delta\log k = \pm 0.3$
	$F_c = 0.6$				
Cl$_2$O$_2$ + M → ClO + ClO + M	2.7×10^{-18}[N$_2$] $\quad (k_0/\text{s}^{-1})$	± 0.3	$1.35 \times 10^{-5}(T/300)^{-5}$ $\exp(-8720/T)$[N$_2$]	200–300	± 900
	8.7×10^2 $\quad (k_\infty/\text{s}^{-1})$	± 0.3	$1.8 \times 10^{15}\exp(-8450/T)$	200–300	± 900
	$F_c = 0.6$				
ClO + OClO + M → Cl$_2$O$_3$ + M	2.8×10^{-31}[N$_2$] $\quad (k_0; 226\text{ K})$	± 0.5 (226 K)			
Cl$_2$O$_3$ + M → ClO + OClO + M	2.8×10^{-18}[N$_2$] $\quad (k_0/\text{s}^{-1}; 226\text{ K})$	± 0.5 (226 K)			
ClO + CH$_3$O$_2$ → ClOO + CH$_3$O	$<4 \times 10^{-12}$ $\quad$ (200 K)				
→ OClO + CH$_3$O	$<1 \times 10^{-15}$ $\quad$ (200 K)				
OClO + O$_3$ → ClO$_3$ + O$_2$	3.0×10^{-19}	± 0.4	$2.1 \times 10^{-12}\exp(-4700/T)$	262–298	± 1000
OClO + NO → NO$_2$ + ClO	3.4×10^{-13}	± 0.3			
Cl$_2$O$_2$ + O$_3$ → ClO + ClOO + O$_2$	$<1 \times 10^{-19}$ $\quad$ (200 K)				

Reaction	k_{298} cm^3 molecule^{-1} s^{-1}		$\Delta\log k_{298}$	Temp. dependence of k/cm^3 molecule^{-1} s^{-1}	Temp. range/K	$\Delta(E/R)/$ K
$CF_3 + O_2 + M \rightarrow CF_3O_2 + M$	$1.9 \times 10^{-29}[N_2]$	(k_0)	± 0.2	$1.9 \times 10^{-29}(T/300)^{-4.7}[N_2]$	200–300	$\Delta n = \pm 1$
	1.0×10^{-11}	(k_∞)	± 0.3	1.0×10^{-11}	200–400	$\Delta\log k = \pm 0.3$
	$F_c = 0.6$					
$CF_2Cl + O_2 + M \rightarrow CF_2ClO_2 + M$	$1.4 \times 10^{-29}[N_2]$	(k_0)	± 0.5	$1.4 \times 10^{-29}(T/300)^{-5}[N_2]$	200–300	$\Delta n = \pm 2$
	9×10^{-12}	(k_∞)	± 0.5	9×10^{-12}	200–300	$\Delta\log k = \pm 0.5$
	$F_c = 0.6$					
$CFCl_2 + O_2 + M \rightarrow CFCl_2O_2 + M$	$5.5 \times 10^{-30}[N_2]$	(k_0)	± 0.3	$5.5 \times 10^{-30}(T/300)^{-6}[N_2]$	200–300	$\Delta n = \pm 2$
	9×10^{-12}	(k_∞)	± 0.5	9×10^{-12}	200–300	$\Delta n = \pm 1$
	$F_c = 0.6$					
$CCl_3 + O_2 + M \rightarrow CCl_3O_2 + M$	$1.6 \times 10^{-30}[N_2]$	(k_0)	± 0.3	$1.6 \times 10^{-30}(T/300)^{-6}[N_2]$	200–300	$\Delta n = \pm 2$
	3.6×10^{-12}	(k_∞)	± 0.5	3.6×10^{-12}	200–300	$\Delta\log k = \pm 0.5$
	$F_c = 0.6$					
$CF_3O \rightarrow COF_2 + F$	$< 10^{-5}$					
$CF_2ClO \rightarrow COF_2 + Cl$	7×10^5	(k/s^{-1})	± 1.0	$3 \times 10^{13}\exp(-5250/T)$	220–300	± 1000
$CFCl_2O \rightarrow COFCl + Cl$	7×10^5	(k/s^{-1})	± 1.0	$3 \times 10^{13}\exp(-5250/T)$	220–300	± 1000
$CCl_3O \rightarrow COCl_2 + Cl$	8×10^6	(k/s^{-1})	± 1.0	$4 \times 10^{13}\exp(-4600/T)$	220–300	± 1000
$CF_3O_2 + NO \rightarrow CF_3O + NO_2$	1.6×10^{-11}		± 0.2	$1.6 \times 10^{-11}(T/300)^{-1.2}$	230–430	$\Delta\log k = \pm 0.2$
$CF_2ClO_2 + NO \rightarrow CF_2ClO + NO_2$	1.6×10^{-11}		± 0.3	$1.6 \times 10^{-11}(T/300)^{-1.5}$	230–430	$\Delta\log k = \pm 0.3$
$CFCl_2O_2 + NO \rightarrow CFCl_2O + NO_2$	1.5×10^{-11}		± 0.2	$1.5 \times 10^{-11}(T/300)^{-1.3}$	230–430	$\Delta\log k = \pm 0.2$
$CCl_3O_2 + NO \rightarrow CCl_3O + NO_2$	1.8×10^{-11}		± 0.2	$1.8 \times 10^{-11}(T/300)^{-1.0}$	230–430	$\Delta\log k = \pm 0.2$
$CF_3O_2 + NO_2 + M \rightarrow CF_3O_2NO_2 + M$	$4.5 \times 10^{-29}[N_2]$	(k_0)	± 0.3	$4.5 \times 10^{-29}(T/300)^{-6.4}[N_2]$	220–300	$\Delta n = \pm 1$
	7.5×10^{-12}	(k_∞)	± 0.5	7.5×10^{-12}	200–300	$\Delta\log k = \pm 0.5$
	$F_c = 0.28$					
$CF_3O_2NO_2 + M \rightarrow CF_3O_2 + NO_2 + M$	$3.6 \times 10^{-19}[N_2]$	(k_0/s^{-1})	± 0.4	$5 \times 10^{-1}(T/300)^{-6}$ $\exp(-12460/T)[N_2]$	233–373	± 500
	5.6×10^{-2}	(k_∞/s^{-1})	± 0.5	$1.2 \times 10^{17}\exp(-12580/T)$	233–373	± 500
	$F_c = 0.28$				220–300	
$CF_2ClO_2 + NO_2 + M \rightarrow CF_2ClO_2NO_2 + M$	$1.4 \times 10^{-28}[N_2]$	(k_0)	± 0.5	$1.4 \times 10^{-28}(T/300)^{-6.4}[N_2]$	200–300	$\Delta n = \pm 2$
	7.5×10^{-12}	(k_∞)	± 0.3	7.5×10^{-12}	200–300	$\Delta\log k = \pm 0.3$
	$F_c = 0.26$				220–300	
$CF_2ClO_2NO_2 + M \rightarrow CF_2ClO_2 + NO_2 + M$	$9.0 \times 10^{-19}[N_2]$	(k_0/s^{-1})	± 0.3	$1.8 \times 10^{-3}\exp(-10500/T)[N_2]$	260–300	± 500
	5.4×10^{-2}	(k_∞/s^{-1})	± 0.3	$1.6 \times 10^{16}\exp(-11990/T)$	260–300	± 500
	$F_c = 0.26$				250–300	
$CFCl_2O_2 + NO_2 + M \rightarrow CFCl_2O_2NO_2 + M$	$1.7 \times 10^{-28}[N_2]$	(k_0)	± 0.3	$1.7 \times 10^{-28}(T/300)^{-6.7}[N_2]$	230–300	$\Delta n = \pm 2$
	7.5×10^{-12}	(k_∞)	± 0.3	7.5×10^{-12}	250–300	$\Delta\log k = \pm 0.3$
	$F_c = 0.23$				230–300	
$CFCl_2O_2NO_2 + M \rightarrow CFCl_2O_2 + NO_2 + M$	$1.5 \times 10^{-18}[N_2]$	(k_0/s^{-1})	± 0.3	$1.0 \times 10^{-2}\exp(-10860/T)$	250–300	± 500
	9.6×10^{-2}	(k_∞/s^{-1})	± 0.3	$6.6 \times 10^{16}\exp(-12240/T)$	250–300	± 500
	$F_c = 0.23$				250–300	
$CCl_3O_2 + NO_2 + M \rightarrow CCl_3O_2NO_2 + M$	$3.2 \times 10^{-28}[N_2]$	(k_0)	± 0.5	$3.2 \times 10^{-28}(T/300)^{-7.7}[N_2]$	230–300	$\Delta n = \pm 3$
	7.5×10^{-12}	(k_∞)	± 0.3	7.5×10^{-12}	250–300	$\Delta\log k = \pm 0.3$
	$F_c = 0.21$				250–300	
$CCl_3O_2NO_2 + M \rightarrow CCl_3O_2 + NO_2 + M$	$7.6 \times 10^{-18}[N_2]$	(k_0/s^{-1})	± 0.3	$6.3 \times 10^{-3}\exp(-10235/T)[N_2]$	250–300	± 500
	0.29	(k_∞/s^{-1})	± 0.3	$4.8 \times 10^{16}\exp(-11820/T)$	250–300	± 500
	$F_c = 0.20$				250–300	
$O_3 + C_2HCl_3 \rightarrow$ products	$< 5 \times 10^{-20}$					
$O_3 + C_2Cl_4 \rightarrow$ products	$< 10^{-21}$					
$HCl + h\nu \rightarrow$ products	See data sheets					
$HOCl + h\nu \rightarrow$ products	See data sheets					
$OCIO + h\nu \rightarrow$ products	See data sheets					
$Cl_2O + h\nu \rightarrow$ products	See data sheets					
$Cl_2O_2 + h\nu \rightarrow$ products	See data sheets					
$Cl_2O_3 + h\nu \rightarrow$ products	See data sheets					
$ClNO + h\nu \rightarrow$ products	See data sheets					
$ClONO + h\nu \rightarrow$ products	See data sheets					
$ClNO_2 + h\nu \rightarrow$ products	See data sheets					
$ClONO_2 + h\nu \rightarrow$ products	See data sheets					
$Cl_2 + h\nu \rightarrow$ products	See data sheets					
$CH_3Cl + h\nu \rightarrow$ products	See data sheets					
$CHF_2Cl + h\nu \rightarrow$ products	See data sheets					
$CF_2Cl_2 + h\nu \rightarrow$ products	See data sheets					
$CFCl_3 + h\nu \rightarrow$ products	See data sheets					
$CCl_4 + h\nu \rightarrow$ products	See data sheets					
$CH_3CF_2Cl + h\nu \rightarrow$ products	See data sheets					
$CH_3CFCl_2 + h\nu \rightarrow$ products	See data sheets					
$CH_3CCl_3 + h\nu \rightarrow$ products	See data sheets					
$CF_3CHFCl + h\nu \rightarrow$ products	See data sheets					
$CF_3CHCl_2 + h\nu \rightarrow$ products	See data sheets					
$CF_2ClCFCl_2 + h\nu \rightarrow$ products	See data sheets					
$CF_2ClCF_2Cl + h\nu \rightarrow$ products	See data sheets					
$CF_3CF_2Cl + h\nu \rightarrow$ products	See data sheets					
$CF_3CF_2CHCl_2 + h\nu \rightarrow$ products	See data sheets					
$CF_2ClCF_2CHFCl + h\nu \rightarrow$ products	See data sheets					
$HCOCl + h\nu \rightarrow$ products	See data sheets					
$COFCl + h\nu \rightarrow$ products	See data sheets					
$COCl_2 + h\nu \rightarrow$ products	See data sheets					
$CCl_3CHO + h\nu \rightarrow$ products	See data sheets					
$CF_3COCl + h\nu \rightarrow$ products	See data sheets					

BrO$_x$ Reactions

Reaction	k_{298}		$\Delta\log k_{298}$
$O + HBr \rightarrow HO + Br$	See previous evaluation		
$O + Br_2 \rightarrow BrO + Br$	See previous evaluation		
$O + BrO \rightarrow O_2 + Br$	3×10^{-11}		± 0.5

Reaction	k_{298} cm^3 molecule^{-1} s^{-1}		$\Delta\log k_{298}$	Temp. dependence of k/cm^3 molecule^{-1} s^{-1}	Temp. range/K	$\Delta(E/R)$/K
Br + HO$_2$ → HBr + O$_2$	2.0×10^{-12}		±0.3	$1.4 \times 10^{-11}\exp(-590/T)$	260–390	±200
Br + H$_2$O$_2$ → HBr + HO$_2$ ⎫	$<5 \times 10^{-16}$					
→ HOBr + HO ⎭						
Br + O$_3$ → BrO + O$_2$	1.2×10^{-12}		±0.08	$1.7 \times 10^{-11}\exp(-800/T)$	195–392	±200
Br + NO$_2$ + M → BrNO$_2$ + M	4.2×10^{-31} [N$_2$]	(k_0)	±0.3	$4.2 \times 10^{-31}(T/300)^{-2.4}$[N$_2$]	200–300	$\Delta n = \pm 1$
	2.7×10^{-11}	(k_∞)	±0.4	2.7×10^{-11}	200–300	$\Delta\log k = \pm 0.4$
	$F_c = 0.55$					
Br + OClO → BrO + ClO	3.4×10^{-13}		±0.3	$2.6 \times 10^{-11}\exp(-1300/T)$	200–450	±300
Br + Cl$_2$O → BrCl + ClO	3.8×10^{-11}		±0.3	$2.1 \times 10^{-11}\exp(-520/T)$	220–298	±300
Br + Cl$_2$O$_2$ → BrCl + ClOO	3.0×10^{-12}		±0.3			
Br + HCHO → HBr + HCO	1.0×10^{-12}		±0.15	$1.7 \times 10^{-11}\exp(-800/T)$	223–480	±250
Br + CH$_3$CHO → HBr + CH$_3$CO	3.9×10^{-12}		±0.2	$1.3 \times 10^{-11}\exp(-360/T)$	250–400	±250
HO + HBr → H$_2$O + Br	1.1×10^{-11}		±0.1	1.1×10^{-11}	249–416	±250
HO + Br$_2$ → HOBr + Br	4.5×10^{-11}		±0.15	$1.2 \times 10^{-11}\exp(400/T)$	260–360	±400
HO + CH$_3$Br → H$_2$O + CH$_2$Br	3.0×10^{-14}		±0.10	$1.9 \times 10^{-12}\exp(-1240/T)$	240–300	±200
HO + CHF$_2$Br → H$_2$O + CF$_2$Br	9.5×10^{-15}		±0.2	$7.7 \times 10^{-13}\exp(-1310/T)$	240–300	±200
HO + CF$_3$Br → products	$<1 \times 10^{-16}$					
HO + CF$_2$ClBr → products	$<1 \times 10^{-16}$					
HO + CF$_2$Br$_2$ → products	$<5 \times 10^{-16}$					
HO + CF$_3$CHFBr → H$_2$O + CF$_3$CFBr	1.7×10^{-14}		±0.3	$1.1 \times 10^{-12}\exp(-1250/T)$	270–430	±500
HO + CF$_3$CHClBr → H$_2$O + CF$_3$CClBr	5.8×10^{-14}		±0.3			
HO + CF$_2$BrCF$_2$Br → products	$<1.3 \times 10^{-16}$					
BrO + HO$_2$ → HOBr + O$_2$ ⎫	3.3×10^{-11}		±0.5	$6.2 \times 10^{-12}\exp(500/T)$	200–300	±500
→ HBr + O$_3$ ⎭						
BrO + O$_3$ → Br + 2O$_2$	$<5 \times 10^{-15}$					
BrO + NO → Br + NO$_2$	2.1×10^{-11}		±0.1	$8.7 \times 10^{-12}\exp(260/T)$	224–425	±100
BrO + NO$_2$ + M → BrONO$_2$ + M	4.7×10^{-31} [N$_2$]	(k_0)	±0.1	$4.7 \times 10^{-31}(T/300)^{-3.1}$[N$_2$]	200–300	$\Delta n = \pm 1$
	1.7×10^{-11}	(k_∞)	±0.1	$1.7 \times 10^{-11}(T/298)^{-0.6}$	200–300	$\Delta n = \pm 1$
	$F_c = 0.40$			$F_c = \exp(-T/327)$	200–300	
BrO + ClO → Br + OClO	6.8×10^{-12}		±0.1	$1.6 \times 10^{-12}\exp(430/T)$	220–400	±200
→ Br + ClOO	6.1×10^{-12}		±0.1	$2.9 \times 10^{-12}\exp(220/T)$	220–400	±200
→ BrCl + O$_2$	1.0×10^{-12}		±0.1	$5.8 \times 10^{-13}\exp(170/T)$	220–400	±200
BrO + BrO → 2Br + O$_2$	2.1×10^{-12}		±0.1 ⎫	$1.1 \times 10^{-12}\exp(250/T)$	223–400	±200
→ Br$_2$ + O$_2$	4.1×10^{-13}		±0.2 ⎭			
HOBr + $h\nu$ → products	See data sheets					
BrO + $h\nu$ → products	See data sheets					
BrONO$_2$ + $h\nu$ → products	See data sheets					
CH$_3$Br + $h\nu$ → products	See data sheets					
CF$_3$Br + $h\nu$ → products	See data sheets					
CF$_2$ClBr + $h\nu$ → products	See data sheets					
CF$_2$Br$_2$ + $h\nu$ → products	See data sheets					
CHBr$_3$ + $h\nu$ → products	See data sheets					
CF$_2$BrCF$_2$Br + $h\nu$ → products	See data sheets					

IO$_x$ Reactions

Reaction	k_{298} cm^3 molecule^{-1} s^{-1}		$\Delta\log k_{298}$	Temp. dependence of k/cm^3 molecule^{-1} s^{-1}	Temp. range/K	$\Delta(E/R)$/K
O + I$_2$ → IO + I	1.4×10^{-10}		±0.3	1.4×10^{-10}	200–400	±250
O + IO → O$_2$ + I	3×10^{-11}		±0.5			
I + HO$_2$ → HI + O$_2$	3.8×10^{-13}		±0.3	$1.5 \times 10^{-11}\exp(-1090/T)$	250–350	±500
I + O$_3$ → IO + O$_2$	1.0×10^{-12}		±0.2	$2.0 \times 10^{-11}\exp(-890/T)$	200–350	±300
I + NO + M → INO + M	1.8×10^{-32} [N$_2$]	(k_0)	±0.1	$1.8 \times 10^{-32}(T/300)^{-1.0}$[N$_2$]	200–400	$\Delta n = \pm 0.5$
	1.7×10^{-11}	(k_∞)	±0.3	1.7×10^{-11}	200–400	$\Delta n = \pm 0.5$
	$F_c = 0.75$			$F_c = [\exp(-T/1040) +$ $\exp(-4160/T)]$	200–400	
I + NO$_2$ + M → INO$_2$ + M	3.0×10^{-31} [N$_2$]	(k_0)	±0.2	$3.0 \times 10^{-31}(T/300)^{-1}$[N$_2$]	200–400	$\Delta n = \pm 1$
	6.6×10^{-11}	(k_∞)	±0.3	6.6×10^{-11}	200–400	$\Delta\log k = \pm 0.3$
	$F_c = 0.63$			$F_c = [\exp(-T/650) +$ $\exp(-2600/T)]$	200–400	
HO + HI → H$_2$O + I	3.0×10^{-11}		±0.3			
HO + I$_2$ → HOI + I	1.8×10^{-10}		±0.3			
HO + CH$_3$I → H$_2$O + CH$_2$I	7.2×10^{-14}		±0.5	$3.1 \times 10^{-12}\exp(-1120/T)$	270–430	±500
NO$_3$ + HI → HNO$_3$ + I	No recommendation (see data sheets)					
IO + HO$_2$ → HOI + O$_2$	6.4×10^{-11}		±0.3			
IO + IO → products	5.2×10^{-11}		±0.3	$1.7 \times 10^{-12}\exp(1020/T)$	250–373	±500
IO + NO → I + NO$_2$	2.2×10^{-11}		±0.3	$7.3 \times 10^{-12}\exp(330/T)$	200–400	±150
IO + NO$_2$ + M → IONO$_2$ + M	7.7×10^{-31} [N$_2$]	(k_0)	±0.3	$7.7 \times 10^{-31}(T/300)^{-5}$[N$_2$]	250–350	$\Delta n = \pm 2$
	1.6×10^{-11}	(k_∞)	±0.3	1.6×10^{-11}	250–350	$\Delta\log k = \pm 0.3$
	$F_c = 0.4$					
IO + CH$_3$SCH$_3$ → products	1.2×10^{-14}		±0.3			
INO + INO → I$_2$ + 2 NO	1.3×10^{-14}		±0.4	$8.4 \times 10^{-11}\exp(-2620/T)$	300–450	±600
INO$_2$ + INO$_2$ → I$_2$ + 2 NO$_2$	4.7×10^{-15}		±0.5	$2.9 \times 10^{-11}\exp(-2600/T)$	298–400	±1000
HOI + $h\nu$ → products	See data sheets					
IO + $h\nu$ → products	See data sheets					
INO + $h\nu$ → products	See data sheets					
INO$_2$ + $h\nu$ → products	See data sheets					
IONO$_2$ + $h\nu$ → products	See data sheets					

*No data sheet or recommendation presented in this article. See our earlier evaluation; J. Phys. Chem. Ref. Data 18, 881 (1989) for our most recent recommendation.

KINETIC DATA FOR COMBUSTION MODELLING

D. L. Baulch, C. J. Cobos, R. A. Cox, C. Esser, P. Frank, Th. Just, J. A. Kerr, M. J. Pilling, J. Troe, R. W. Walker, and J. Warnatz

The following tables present evaluated rate constants and other chemical kinetic data required for modelling the combustion of hydrocarbons. The compilation was prepared as part of the project "Kinetics and Mechanisms of Chemical Processes in Combustion", which is one of the projects in the third European Community Energy Research and Development Program. The tables are reprinted from the *Journal of Physical and Chemical Reference Data* by permission of the authors and the American Institute of Physics.

Table 1 lists all the reactions studied and gives the recommended rate constant *k* for every bimolecular reaction, as well as the applicable temperature range and the associated error limits. Where more than one set of products is possible, rate constants or branching ratios are given for all channels considered feasible. The data for decomposition reactions and combination reactions are given in Tables 2 and 3, respectively. The reference includes a detailed data sheet for each reaction listed here, covering the thermodynamic data, kinetic measurements, and reliability assessments.

REFERENCE

Baulch, D. L., et al., *J. Phys. Chem. Ref. Data*, 21, 411-734, 1992.

Table 1
BIMOLECULAR REACTIONS

Reaction	k/cm^3 molecule^{-1}s^{-1}	Temp/K	Error limits ($\Delta \log k$)
O Atom Reactions			
$O + H_2 \rightarrow OH + H$	$8.5 \times 10^{-20} T^{2.67} \exp(-3160/T)$	300–2500	± 0.5 at 300 K falling to ± 0.2 for $T > 500$ K
$O + OH \rightarrow O_2 + H$	$2.0 \times 10^{-11} \exp(112/T)$ $2.4 \times 10^{-11} \exp(-353/T)$	220–500 1000–2000	± 0.2 ± 0.1
$O + HO_2 \rightarrow OH + O_2$	5.3×10^{-11}	300–1000	± 0.3 at 300 K rising to ± 0.5 at 1000 K.
$O + H_2O_2 \rightarrow OH + HO_2$	$1.1 \times 10^{-12} \exp(-2000/T)$	300–500	± 0.3
$O + NH_3 \rightarrow OH + NH_2$	$1.6 \times 10^{-11} \exp(-3670/T)$	500–2500	± 0.5
$O + CH \rightarrow CO + H$ $\rightarrow CHO^+ + e$	6.6×10^{-11} $4.2 \times 10^{-13} \exp(-850/T)$	300–2000 300–2500	± 0.5 ± 0.5
$O + {}^3CH_2 \rightarrow CO + 2H$ $\rightarrow CO + H_2$	2×10^{-10} $k_1/k = 0.6 \pm 0.3$ over whole range	300–2500	± 0.2 at 300 K rising to ± 0.7 at 2500 K.
$O + CH_3 \rightarrow HCHO + H$	1.4×10^{-10}	300–2500	± 0.2
$O + CH_4 \rightarrow OH + CH_3$	$1.5 \times 10^{-15} T^{1.56} \exp(-4270/T)$	300–2500	± 0.3 at 300 K falling to ± 0.15 at 2500 K.
$O + CHO \rightarrow OH + CO$ $\rightarrow CO_2 + H$	5.0×10^{-11} 5.0×10^{-11}	300–2500 300–2500	± 0.3 ± 0.3
$O + HCHO \rightarrow OH + CHO$	$6.9 \times 10^{-13} T^{0.57} \exp(-1390/T)$	250–2200	± 0.1 at 250 K rising to ± 0.3 at 2200 K.
$O + CH_3O \rightarrow O_2 + CH_3$ $\rightarrow OH + HCHO$	2.5×10^{-11} $k_2/k = (0.12 \pm 0.1)$ at 300 K	300–1000	± 0.3 at 300 K rising to ± 0.7 at 1000 K.
$O + CN \rightarrow CO + N(^4S)$ $\rightarrow CO + N(^2D)$	1.7×10^{-11}	300–5000	± 0.2 at 300 K rising to $\pm 0.6 \times 5000$ K.
$O + NCO \rightarrow NO + CO$ $\rightarrow O_2 + CN$	7.0×10^{-11}	1450–2600	± 0.8

Reaction	k/cm^3 molecule^{-1}s^{-1}	Temp/K	Error limits ($\Delta \log k$)
O + HCN → NCO + H → CO + NH → OH + CN	$2.3 \times 10^{-18} \, T^{2.1} \exp(-3075/T)$	450–2500	± 0.2 at 450 K rising to ±0.3 at 2500 K.
O + CH$_3$OOH → OH + CH$_2$COOH → OH + CH$_3$O$_2$	$6.9 \times 10^{-13} \, T^{0.57} \exp(-1390/T)$ [estimate]	250–2200	± 0.1 at 250 K rising to ± 0.3 at 2200 K.
O + C$_2$H → CO + CH	1.7×10^{-11}	300–2500	± 1.0
O + C$_2$H$_2$ → CO + ^{3}CH$_2$ → CHCO + H	$3.6 \times 10^{-20} \, T^{2.8} \exp(-250/T)$ $k_1/k = 0.5 \pm 0.3$ over whole range.	300–2500	± 0.2
O + C$_2$H$_3$ → OH + C$_2$H$_2$ → CO + CH$_3$ → HCO + CH$_2$	5×10^{-11}	300–2000	± 0.5
O + C$_2$H$_4$ → CH$_2$CHO + H → HCO + CH$_3$ → HCHO + CH$_2$ → CH$_2$CO + H$_2$	$5.75 \times 10^{-18} \, T^{2.08}$ $k_1/k = 0.35 \pm 0.05$ at p > 3 Torr $k_2/k = 0.6 \pm 0.10$	300–2000 over whole temperature range	± 0.1 for $T < 1000$ K rising to ± 0.3 at 2000 K.
O + C$_2$H$_5$ → CH$_3$CHO + H → HCHO + CH$_3$	1.1×10^{-10} $k_2/k = 0.17 \pm 0.2$ at 300 K	300–2500	± 0.3 from 300 to 1000 K ± 0.5 from 1000 to 2500 K
O + C$_2$H$_6$ → OH + C$_2$H$_5$	$1.66 \times 10^{-15} \, T^{1.5} \exp(-2920/T)$	300–1200	± 0.3 at 300 K falling to ± 0.15 at 1200 K.
O + CHCO → 2CO + H	1.6×10^{-10}	300–2500	± 0.3
O + CH$_2$CO → CH$_2$O + CO → HCO + H + CO → HCO + HCO	$3.8 \times 10^{-12} \exp(-680/T)$	230–500	± 0.3
O + CH$_3$CHO → OH + CH$_3$CO → OH + CH$_2$CHO	$9.7 \times 10^{-12} \exp(-910/T)$	300–1500	± 0.05 at 300 K rising to ± 0.5 at 1500 K.
O + C$_2$H$_5$OOH → OH + C$_2$H$_4$OOH → OH + C$_2$H$_5$OO	$6.9 \times 10^{-13} \, T^{0.57} \exp(-1390/T)$ [estimate]	250–2200	± 0.1 at 150 K rising to ± 0.3 at 2200 K.
O + C$_6$H$_6$ → OH + C$_6$H$_5$ → C$_6$H$_5$OH	$1.2 \times 10^{-22} \, T^{3.7} \exp(-570/T)$	300–1000	± 0.5
O + C$_6$H$_5$CH$_2$ → HCO + C$_6$H$_6$ → C$_6$H$_5$CH + H → CH$_2$O + C$_6$H$_5$	5.5×10^{-10} No recommendation	300	±0.3
O + C$_6$H$_5$CH$_3$ → products	$5.3 \times 10^{-15} \, T^{1.21} \exp(-1260/T)$	300–2800	± 0.1 at 300 K rising to ± 0.4 at 2800 K
O + p-C$_6$H$_4$(CH$_3$)$_2$ → products	$2.6 \times 10^{-11} \exp(-1409/T)$	300–600	± 0.3
O + C$_6$H$_5$C$_2$H$_5$ → products	1.0×10^{-13}	298	± 0.3

O$_2$ Reactions

Reaction	k/cm^3 molecule^{-1}s^{-1}	Temp/K	Error limits ($\Delta \log k$)
O$_2$ + CH$_4$ → HO$_2$ + CH$_3$	$6.6 \times 10^{-11} \exp(-28630/T)$	500–2000	± 0.5 at 500 K rising to ± 1.0 at 2000 K.

Reaction	$k/cm^3 \, molecule^{-1}s^{-1}$	Temp/K	Error limits ($\Delta \log k$)
$O_2 + C_2H_6 \rightarrow HO_2 + C_2H_5$	$1.0 \times 10^{-10} \exp(-26100/T)$	500–2000	± 0.5 at 500 K rising to ± 1.0 at 2000 K
$O_2 + HCHO \rightarrow HO_2 + HCO$	$1.0 \times 10^{-10} \exp(-20460/T)$	700–1000	± 0.5
$O_2 + CH_3CHO \rightarrow HO_2 + CH_3CO$	$5.0 \times 10^{-11} \exp(-19700/T)$	600–1100	± 0.5 at 600 K rising to ± 1.0 at 1100 K.
H Atom Reactions			
$H + O_2 \rightarrow OH + O$	$3.3 \times 10^{-10} \exp(-8460/T)$	300–2500	± 0.1 at 300 K rising to ± 0.2 at 2500 K.
$H + O_2 + Ar \rightarrow HO_2 + Ar$	See Table 3		
$H + O_2 + H_2 \rightarrow HO_2 + H_2$	See Table 3		
$H + O_2 + N_2 \rightarrow HO_2 + N_2$	See Table 3		
$H + O_2 + H_2O \rightarrow HO_2 + H_2O$	See Table 3		
$H + H + Ar \rightarrow H_2 + Ar$	See Table 3		
$H + H + H_2 \rightarrow H_2 + H_2$	See Table 3		
$H + OH + H_2O \rightarrow H_2O + H_2O$	See Table 3		
$H + OH + Ar \rightarrow H_2O + Ar$	See Table 3		
$H + HO + N_2 \rightarrow H_2O + N_2$	See Table 3		
$H + HO_2 \rightarrow H_2 + O_2$	$7.1 \times 10^{-11} \exp(-710/T)$	300–1000	± 0.3
$\rightarrow 2 \, OH$	$2.8 \times 10^{-10} \exp(-440/T)$	300–1000	± 0.3
$\rightarrow H_2O + O$	$5.0 \times 10^{-11} \exp(-866/T)$	300–1000	± 0.3
$H + H_2O \rightarrow OH + H_2$	$7.5 \times 10^{-16} \, T^{1.6} \exp(-9270/T)$	300–2500	±0.2
$H + H_2O_2 \rightarrow H_2 + HO_2$	$2.8 \times 10^{-12} \exp(-1890/T)$	300–1000	± 0.3
$\rightarrow OH + H_2O$	$1.7 \times 10^{-11} \exp(-1800/T)$	300–1000	± 0.3
$H + NH \rightarrow H_2 + N$	1.7×10^{-11}	1500–2500	± 1.0
$H + NH_2 \rightarrow H_2 + NH$	1.0×10^{-11}	2000–3000	± 1.0
$H + {}^3CH_2 \rightarrow H_2 + CH$	$1.0 \times 10^{-11} \exp(900/T)$	300–3000	± 0.7
$H + CH_3 \rightarrow H_2 + {}^1CH_2$	$1.0 \times 10^{-10} \exp(-7600/T)$	300–2500	± 1.0
$\rightarrow CH_4$	See Table 3		
$H + CH_4 \rightarrow H_2 + CH_3$	$2.2 \times 10^{-20} \, T^{3.0} \exp(-4045/T)$	300–2500	± 0.2
$H + CHO \rightarrow H_2 + CO$	1.5×10^{-10}	300–2500	± 0.3
$H + HCHO \rightarrow H_2 + HCO$	$3.8 \times 10^{-14} \, T^{1.05} \exp(-1650/T)$	300–2200	± 0.1 at 300 K rising to ± 0.5 at 2200 K
$H + CH_3O \rightarrow H_2 + HCHO$	3.0×10^{-11}	300–1000	± 0.5
$H + HNCO \rightarrow NH_2 + CO$	No recommendation		
$\rightarrow H_2 + NCO$	$3.4 \times 10^{-10} \, T^{-0.27} \exp(-10190/T)$	500–1000	± 1.0
$H + NCO \rightarrow NH + CO$ $\rightarrow HCN + O$	8.7×10^{-11}	1400–1500	± 0.5

Reaction	k/cm^3 molecule^{-1}s^{-1}	Temp/K	Error limits ($\Delta \log k$)
H + C$_2$H$_2$ → H$_2$ + C$_2$H → C$_2$H$_3$	$1.0 \times 10^{-10} \exp(-14000/T)$ See Table 3	1000–3000	± 1.0
H + C$_2$H$_3$ → H$_2$ + C$_2$H$_2$ → C$_2$H$_4$	2.0×10^{-11} See Table 3	300–2500	± 0.5
H + C$_2$H$_4$ → C$_2$H$_3$ + H$_2$ → C$_2$H$_5$	$9.0 \times 10^{-10} \exp(-7500/T)$ See Table 3	700–2000	± 0.5
H + C$_2$H$_5$ → 2CH$_3$ → C$_2$H$_6$	6.0×10^{-11} See Table 3	300–2000	± 0.3
H + C$_2$H$_6$ → H$_2$ + C$_2$H$_5$	$2.4 \times 10^{-15} T^{1.5} \exp(-3730/T)$	300–2000	± 0.15 at 300 K rising to ± 0.3 at 2000 K
H + CHCO → CH$_2$ + CO → H$_2$ + C$_2$O → HCCOH]	2.5×10^{-10}	300–2500	± 0.4
H + CH$_2$CO → CH$_3$ + CO → CH$_2$CHO	$3.0 \times 10^{-11} \exp(-1700/T)$ k_2/k very small	200–2000	± 0.5 at 200 K rising to ± 1.0 at 2000 K.
H + CH$_3$CHO → H$_2$ + CH$_3$CO → H$_2$ + CH$_2$CHO]	$6.8 \times 10^{-15} T^{1.16} \exp(-1210/T)$	300–2000	± 0.1 at 300 rising to ± 0.4 at 2000 K.
H + C$_6$H$_5$ + M → C$_6$H$_6$ + M	See Table 3		
H + C$_6$H$_6$ → H$_2$ + C$_6$H$_5$ → C$_6$H$_7$	No recommendation See Table 3		
H + C$_6$H$_5$O + M → C$_6$H$_5$OH + M	See Table 3		
H + C$_6$H$_5$OH → C$_6$H$_5$O + H$_2$ → C$_6$H$_6$ + OH	$1.9 \times 10^{-10} \exp(-6240/T)$ $3.7 \times 10^{-11} \exp(-3990/T)$	1000–1150 1000–1150	± 0.3 ± 0.3
H + C$_6$H$_5$CH$_2$ + M → C$_6$H$_5$CH$_3$ + M	See Table 3		
H + C$_6$H$_5$CH$_3$ → H$_2$ + C$_6$H$_5$CH$_2$ → H$_2$ + C$_6$H$_4$CH$_3$ → C$_6$H$_6$ + CH$_3$ → C$_6$H$_6$CH$_3$	$6.6 \times 10^{-22} T^{3.44} \exp(-1570/T)$ No recommendation No recommendation See Table 3	600–2800	± 0.3 at 600 K rising to ± 0.5 at 2800 K.
H + p-C$_6$H$_4$(CH$_3$)$_2$ → products	5.8×10^{-13}	298	± 0.1
H + C$_6$H$_5$C$_2$H$_5$ → H$_2$ + C$_6$H$_5$C$_2$H$_4$ → C$_6$H$_6$C$_2$H$_5$	2.4×10^{-12} See Table 3	773	± 0.1

H$_2$ Reactions

Reaction	k/cm^3 molecule^{-1}s^{-1}	Temp/K	Error limits ($\Delta \log k$)
H$_2$ + Ar → 2H + Ar	See Table 2		
H$_2$ + H$_2$ → 2H + H$_2$	See Table 2		

OH Radical Reactions

Reaction	k/cm^3 molecule^{-1}s^{-1}	Temp/K	Error limits ($\Delta \log k$)
OH + H$_2$ → H$_2$O + H	$1.7 \times 10^{-16} T^{1.6} \exp(-1660/T)$	300–2500	± 0.1 at 300 K rising to ± 0.3 at 2500 K
OH + OH → H$_2$O + O	$2.5 \times 10^{-15} T^{1.14} \exp(-50/T)$	250–2500	± 0.2
OH + OH + M → H$_2$O$_2$ + M	See Table 3		

Reaction	k/cm^3 molecule^{-1}s^{-1}	Temp/K	Error limits ($\Delta \log k$)
OH + HO$_2$ → H$_2$O + O$_2$	4.8×10^{-11} exp(250/T)	300–2000	± 0.2 at 300 K rising to ± 0.5 at 2000 K.
OH + H$_2$O$_2$ → H$_2$O + HO$_2$	1.3×10^{-11} exp($-670/T$)	300–1000	± 0.2
OH + NH → NO + H$_2$ → H$_2$O + N	8.0×10^{-11}	300–1000	± 0.5
OH + NH$_2$ → O + NH$_3$ → H$_2$O + NH	3.3×10^{-14} $T^{0.405}$ exp($-250/T$) No recommendation	500–2500	± 0.5
OH + CO → H + CO$_2$	1.05×10^{-17} $T^{1.5}$ exp(250/T)	300–2000	± 0.2 at 300 K rising to ± 0.5 at 2000 K.
OH + CH$_3$ → H + CH$_2$OH → H + CH$_3$O → H$_2$O + ^{1}CH$_2$ → CH$_3$OH	6.0×10^{-11} See Table 3	300–2000	± 0.7
OH + CH$_4$ → H$_2$O + CH$_3$	2.6×10^{-17} $T^{1.83}$ exp($-1400/T$)	250–2500	± 0.07 at 250 K rising to ± 0.15 at 1200 K.
OH + CHO → H$_2$O + CO	1.7×10^{-10}	300–2500	± 0.3
OH + HCHO → H$_2$O + CHO	5.7×10^{-15} $T^{1.18}$ exp(225/T)	300–3000	± 0.1 at 300 K rising to ± 0.7 at 3000 K.
OH + CN → O + HCN → NCO + H	1.0×10^{-10}	1500–3000	± 0.5
OH + HCN → H$_2$O + CN → HOCN + H → HNCO + H	1.5×10^{-11} exp($-5400/T$) No recommendation	1500–2500	± 0.5
OH + CH$_3$OOH → H$_2$O + CH$_3$OO	1.2×10^{-12} exp(130/T)	300–1000	± 0.2 at 300 K rising to ± 0.4 at 1000 K
→ H$_2$O + CH$_2$OOH	1.8×10^{-12} exp(220/T)	300–1000	± 0.1 at 300 K rising to ±0.3 at 1000 K.
OH + C$_2$H$_2$ → H$_2$O + C$_2$H → H + CH$_2$CO → C$_2$H$_2$OH	1.0×10^{-10} exp($-6500/T$) See Table 3	1000–2000	± 1.0
OH + C$_2$H$_4$ → H$_2$O + C$_2$H$_3$	3.4×10^{-11} exp($-2990/T$)	650–1500	± 0.5
OH + C$_2$H$_6$ → H$_2$O + C$_2$H$_5$	1.2×10^{-17} $T^{2.0}$ exp($-435/T$)	250–2000	± 0.07 at 250 K rising to ± 0.15 at 2000 K.
OH + CH$_2$CO → CH$_2$OH + CO → H$_2$CO + HCO	1.7×10^{-11}	300–2000	± 1.0
OH + CH$_3$CHO → H$_2$O + CH$_3$CO → H$_2$O + CH$_2$CHO	3.9×10^{-14} $T^{0.73}$ exp(560/T)	250–1200	± 0.1 at 250 K rising to ± 0.3 at 1200 K.
OH + C$_2$H$_5$OOH → H$_2$O + C$_2$H$_5$OO → H$_2$O + C$_2$H$_4$OOH	3.0×10^{-12} exp(190/T) [estimate]	250–1000	± 0.3 at 250 K rising to ± 0.7 at 1000 K

Reaction	k/cm^3 molecule^{-1}s^{-1}	Temp/K	Error limits ($\Delta \log k$)
$OH + C_6H_6 \rightarrow H_2O + C_6H_5$	$2.7 \times 10^{-16} T^{1.42} \exp(-730/T)$	400–1500	± 0.3
$\rightarrow H + C_6H_5OH$	$2.2 \times 10^{-11} \exp(-5330/T)$	1000–1150	± 0.3
$\rightarrow C_6H_6OH$	See Table 3		
$OH + C_6H_5OH \rightarrow C_6H_5(OH)_2$	See Table 3		
$\rightarrow H_2O + C_6H_5O$			
$\rightarrow H_2O + C_6H_4OH$]	1.0×10^{-11}	1000–1150	± 0.5
$OH + C_6H_5CH_3 \rightarrow H_2O + C_6H_5CH_2$	$8.6 \cdot 10^{-15} T \exp(-1440/T)$ See Table 3	400–1200	± 0.5 at 400 K reducing to ±0.3 at 1200 K.
$OH + p\text{-}C_6H_4(CH_3)_2 \rightarrow C_6H_4CH_2CH_3 + H_2O$	$6.4 \times 10^{-11} \exp(-1440/T)$	500–960	±0.1
$\rightarrow p\text{-}C_6H_4(CH_3)_2OH$	See Table 3		
$OH + C_6H_5C_2H_5 \rightarrow HOC_6H_5C_2H_5$	See Table 3		
$\rightarrow H_2O + C_6H_5C_2H_4$	8.7×10^{-12}	773	± 0.1
$\rightarrow H_2O + C_6H_4C_2H_5$]			

H_2O Reactions

$H_2O + M \rightarrow H + OH + M$	See Table 2		

HO_2 Radical Reactions

Reaction	k/cm^3 molecule^{-1}s^{-1}	Temp/K	Error limits
$HO_2 + HO_2 \rightarrow H_2O_2 + O_2$	$3.1 \times 10^{-12} \exp(-775/T)$	550–1250	± 0.15 at 550 K rising to ± 0.3 at 1250 K.
$HO_2 + NH_2 \rightarrow NH_3 + O_2$			
$\rightarrow HNO + H_2O$]	2.6×10^{-11}	300–400	± 0.4
$HO_2 + CH_3 \rightarrow OH + CH_3O$	3×10^{-11}	300–2500	± 0.7
$\rightarrow O_2 + CH_4$	No recommendation		
$HO_2 + CH_4 \rightarrow H_2O_2 + CH_3$	$1.5 \times 10^{-11} \exp(-12400/T)$	600–1000	± 0.2 at 600 K rising to ± 0.3 at 1000 K.
$HO_2 + HCHO \rightarrow H_2O_2 + CHO$	$5.0 \times 10^{-12} \exp(-6580/T)$	600–1000	± 0.5
$HO_2 + C_2H_4 \rightarrow OH + C_2H_4O$	$3.7 \times 10^{-12} \exp(-8650/T)$	600–900	± 0.15 at 600 K rising to ± 0.25 at 900 K.
$HO_2 + C_2H_6 \rightarrow H_2O_2 + C_2H_5$	$2.2 \times 10^{-11} \exp(-10300/T)$	500–1000	± 0.2 at 500 K rising to ± 0.3 at 1000 K.
$HO_2 + CH_3CHO \rightarrow H_2O_2 + CH_3CO$	$5.0 \times 10^{-12} \exp(-6000/T)$	900–1200	± 0.7

H_2O_2 Reactions

$H_2O_2 + M \rightarrow 2OH + M$	See Table 2		

N Atom Reactions

Reaction	k/cm^3 molecule^{-1}s^{-1}	Temp/K	Error limits
$N + CN \rightarrow N_2 + C$	3×10^{-10}	300–2500	± 1.0
$N + NCO \rightarrow NO + CN$	No recommendation		
$\rightarrow N_2 + CO$	3.3×10^{-11}	1700	± 0.5

Reaction	k/cm^3 molecule^{-1}s^{-1}	Temp/K	Error limits ($\Delta \log k$)
NH Radical Reactions			
NH + O$_2$ → NO + OH → NO$_2$ + H → HNO + O	1.26×10^{-13} exp($-770/T$)	270–550	± 0.2 at 270 K rising to ± 0.5 at 550 K.
NH + NO → N$_2$O + H → HN$_2$ + O → N$_2$ + OH	5.0×10^{-11}	270–380	± 0.2
NH$_2$ Radical Reactions			
NH$_2$ + O$_2$ → products	$<3 \times 10^{-18}$	298	
NH$_2$ + NO → N$_2$ + H$_2$O → N$_2$ + H + OH → N$_2$H + OH → N$_2$O + H$_2$	1.8×10^{-12} exp($650/T$) $(k_2 + k_3)/k \approx 0.12$ at 298 K.	220–2000	± 0.5
^{1}C$_2$ and ^{3}C$_2$ Radical Reactions	See data sheets.		
CH Radical Reactions			
CH + O$_2$ → CHO + O → CO + OH	5.5×10^{-11}	300–2000	± 0.3 at 300 K rising to ± 0.5 at 2000 K.
CH + H$_2$ → CH$_2$ + H → CH$_3$	2.4×10^{-10} exp($-1760/T$)	300–1000	± 0.3
CH + H$_2$O → products	9.5×10^{-12} exp($380/T$)	300–1000	± 1.0
CH + CO → products	4.6×10^{-13} exp($860/T$)	300–1000	± 1.0
CH + CO$_2$ → products	5.7×10^{-12} exp($-345/T$)	300–1000	± 1.0
CH + CH$_4$ → products	5.0×10^{-11} exp($200/T$)	200–700	± 1.0
CH + C$_2$H$_2$ → products	3.5×10^{-10} exp($61/T$)	200–700	± 1.0
CH + C$_2$H$_4$ → products	2.2×10^{-10} exp($173/T$)	200–700	± 1.0
CH + C$_2$H$_6$ → products	1.8×10^{-10} exp($132/T$)	200–700	± 1.0
CH + C$_3$H$_8$ → products	1.9×10^{-10} exp($240/T$)	300–700	± 1.0
CH + n-C$_4$H$_{10}$ → products	4.4×10^{-10} exp($28/T$)	250–700	± 1.0
CH + i-C$_4$H$_{10}$ → products	2.0×10^{-10} exp($240/T$)	300–700	± 1.0
CH + neo-C$_5$H$_{12}$ → products	1.6×10^{-10} exp($340/T$)	300–700	± 1.0
CH + CH$_3$C$_2$H → products	No recommendation		
CH + CH$_2$O → products	1.6×10^{-10} exp($260/T$)	300–700	± 1.0

Reaction	k/cm^3 molecule^{-1}s^{-1}	Temp/K	Error limits ($\Delta \log k$)
$^3CH_2$ Radical Reactions			
$^3CH_2 + O_2 \rightarrow CO + H + OH$ $\rightarrow CO_2 + H + H$ $\rightarrow CO + H_2O$ $\rightarrow CO_2 + H_2$ $\rightarrow HCHO + O$	$4.1 \times 10^{-11} \exp(-750/T)$	300–1000	± 0.3 at 300 K rising to ± 0.5 at 1000 K.
$^3CH_2 + {}^3CH_2 \rightarrow C_2H_2 + H_2$ $\rightarrow C_2H_2 + 2H$	$2.0 \times 10^{-10} \exp(-400/T)$ $k_2/k = 0.9 \pm 0.1$ over range 300–3000 K.	300–3000	± 0.5
$^3CH_2 + CH_3 \rightarrow C_2H_4 + H$	7.0×10^{-11}	300–3000	± 0.3 at 300 K rising to ± 0.5 at 3000 K.
$^3CH_2 + C_2H_2 \rightarrow C_3H_4$	See Table 3		
$^3CH_2 + C_2H_4 \rightarrow C_3H_6$ $\rightarrow c\text{-}C_3H_6$ $\rightarrow CH_2CHCH_2 + H$	See Table 3		
$^1CH_2$ Radical Reactions			
$^1CH_2 + Ar \rightarrow {}^3CH_2 + Ar$	6.0×10^{-12}	300–2000	± 0.3
$^1CH_2 + N_2 \rightarrow {}^3CH_2 + N_2$	1.0×10^{-11}	300–2000	± 0.3
$^1CH_2 + CH_4 \rightarrow {}^3CH_2 + CH_4$	1.2×10^{-11}	300–2000	± 0.4
$^1CH_2 + C_2H_2 \rightarrow {}^3CH_2 + C_2H_2$	8.0×10^{-11}	300–2000	± 0.4
$^1CH_2 + C_2H_4 \rightarrow {}^3CH_2 + C_2H_4$	2.3×10^{-11}	300–2000	± 0.4
$^1CH_2 + C_2H_6 \rightarrow {}^3CH_2 + C_2H_6$	3.6×10^{-11}	300–2000	± 0.4
$^1CH_2 + O_2 \rightarrow CO + H + OH$ $\rightarrow CO_2 + H_2$ $\rightarrow CO + H_2O$ $\rightarrow {}^3CH_2 + O_2$	5.2×10^{-11}	300–1000	± 0.3 at 300 K rising to ± 0.5 at 1000 K
$^1CH_2 + H_2 \rightarrow CH_3 + H$	1.2×10^{-10}	300–1000	± 0.1 at 300 K rising to ± 0.3 at 1000 K
$^1CH_2 + C_2H_2 \rightarrow CH_2CCH_2$ $\rightarrow CH_3CCH$ $\rightarrow CH_2CCH + H$ $\rightarrow {}^3CH_2 + C_2H_2$	See Table 3 See earlier entry		
$^1CH_2 + C_2H_4 \rightarrow C_3H_6$ $\rightarrow {}^3CH_2 + C_2H_4$	See Table 3 See earlier entry		
CH_3 Radical Reactions			
$CH_3 + M \rightarrow CH_2 + H + M$	See Table 2		
$CH_3 + O_2 \rightarrow CH_3O + O$ $\rightarrow HCHO + OH$ $\rightarrow CH_3O_2$	$2.2 \times 10^{-10} \exp(-15800/T)$ $5.5 \times 10^{-13} \exp(-4500/T)$ See Table 3	300–2500 1000–2500	± 0.5 ± 0.5

Reaction	k/cm^3 molecule^{-1}s^{-1}	Temp/K	Error limits ($\Delta \log k$)
$CH_3 + H_2 \rightarrow CH_4 + H$	$1.14 \times 10^{-20} T^{2.74} \exp(-4740/T)$	300–2500	$\pm$ 0.15 in the range 300–700 K. $\pm$ 0.3 in the range 700–2500 K.
$CH_3 + CH_3 \rightarrow C_2H_5 + H$ $\rightarrow C_2H_4 + H_2$ $\rightarrow C_2H_6$	$5 \times 10^{-11} \exp(-6800/T)$ No recommendation (see data sheets) See Table 3	1300–2500	$\pm$ 0.6
$CH_3 + HCHO \rightarrow CH_4 + HCO$	$6.8 \times 10^{-12} \exp(-4450/T)$	300–1000	$\pm$ 0.3
$CH_3 + C_2H_2 + M \rightarrow C_3H_5 + M$ $\rightarrow CH_4 + C_2H$	See Table 3 No recommendation		
$CH_3 + C_2H_4 \rightarrow CH_4 + C_2H_3$ $\rightarrow n\text{-}C_3H_7$	$6.9 \times 10^{-12} \exp(-5600/T)$ See Table 3	400–3000	$\pm$ 0.5
$CH_3 + C_2H_5 \rightarrow CH_4 + C_2H_4$ $\rightarrow C_3H_8$	1.9×10^{-12} See Table 3	300–800	$\pm$ 0.4
$CH_3 + C_2H_6 \rightarrow CH_4 + C_2H_5$	$2.5 \times 10^{-31} T^{6.0} \exp(-3043/T)$	300–1500	$\pm$ 0.1 at 300 K rising to $\pm$ 0.2 at 1500 K.
$CH_3 + CH_3CHO \rightarrow CH_4 + CH_3CO$ $\rightarrow CH_4 + CH_2CHO$	$3.3 \times 10^{-30} T^{5.64} \exp(-1240/T)$ No recommendation (see data sheets)	300–1250	$\pm$ 0.3

CH$_4$ Reactions

$CH_4 + M \rightarrow CH_3 + H + M$	See Table 2		

CHO Radical Reactions

$CHO + O_2 \rightarrow CO + HO_2$ $\rightarrow OH + CO_2$ $\rightarrow HCO_3$	5.0×10^{-12}	300–2500	$\pm$ 0.3
$CHO + CHO \rightarrow HCHO + CO$	5.0×10^{-11}	300	$\pm$ 0.3

HCHO Reactions

$HCHO + M \rightarrow H + CHO + M$ $\rightarrow H_2 + CO + M$	See Table 2		

CH$_2$OH Reactions

$CH_2OH + O_2 \rightarrow CH_2O + HO_2$	$2.6 \times 10^{-9} T^{-1.0} +$ $1.2 \times 10^{-10} \exp(-1800/T)$	300–1200	$\pm$ 0.1 at 300 K rising to $\pm$ 0.3 at 1200 K.

CH$_3$O Radical Reactions

$CH_3O + M \rightarrow HCHO + H + M$	See Table 2		
$CH_3O + O_2 \rightarrow HCHO + HO_2$	$6.7 \times 10^{-14} \exp(-1070/T)$	300–1000	$\pm$ 0.2 at 500 K rising to $\pm$ 0.3 at 300 K and 1000 K.

Reaction	k/cm^3 molecule^{-1}s^{-1}	Temp/K	Error limits ($\Delta \log k$)
CH$_3$OOH Reactions			
$CH_3OOH + M \rightarrow CH_3O + OH + M$	See Table 2		
CN Radical Reactions			
$CN + O_2 \rightarrow NCO + O$	$1.1 \times 10^{-11} \exp(205/T)$	300–2500	$\pm$ 0.25 at 300 K rising to $\pm$ 0.5 at 2500 K.
$CN + H_2O \rightarrow HCN + OH$ $\rightarrow HOCN + H$]	$1.3 \times 10^{-11} \exp(-3750/T)$	500–3000	$\pm$ 0.3 at 500 K rising to $\pm$ 0.5 at 3000 K.
$CN + CH_4 \rightarrow HCN + CH_3$	$1.5 \times 10^{-11} \exp(-940/T)$	260–400	$\pm$ 0.3
NCO Radical Reactions			
$NCO + M \rightarrow N + CO + M$	See Table 2		
$NCO + NO \rightarrow N_2O + CO$ $\rightarrow N_2 + CO_2$ $\rightarrow N_2 + CO + O$]	$1.7 \times 10^{-11} \exp(200/T)$	300–600	$\pm$ 0.5
C$_2$H Radical Reactions			
$C_2H + O_2 \rightarrow CO_2 + CH$ $\rightarrow 2CO + H$ $\rightarrow C_2HO + O$ $\rightarrow CO + HCO$]	3.0×10^{-11}	300	$\pm$ 0.5
$C_2H + H_2 \rightarrow C_2H_2 + H$	$2.5 \times 10^{-11} \exp(-1560/T)$	300–2500	$\pm$ 0.3 at 300 K rising to $\pm$ 0.7 at 2500 K
$C_2H + C_2H_2 \rightarrow C_4H_2 + H$	5.0×10^{-11}	300–2700	$\pm$ 0.3
$C_2H + CH_4 \rightarrow$ products	2.0×10^{-12}	298	$\pm$ 1
$C_2H + C_2H_6 \rightarrow$ products	No recommendation		
C$_2$H$_3$ Radical Reactions			
$C_2H_3 + M \rightarrow C_2H_2 + H + M$	See Table 2		
$C_2H_3 + O_2 \rightarrow HCHO + CHO$	9.0×10^{-12}	300–2000	$\pm$ 0.3 at 300 K rising to $\pm$ 0.5 at 2000 K
C$_2$H$_5$ Radical Reactions			
$C_2H_5 + O_2 \rightarrow C_2H_4 + HO_2$	$1.7 \times 10^{-14} \exp(1100/T)$	600–1200	$\pm$ 0.3
$C_2H_5 + C_2H_5 \rightarrow C_2H_6 + C_2H_4$ $\rightarrow n\text{-}C_4H_{10}$	2.4×10^{-12} See Table 3	300–1200	$\pm$ 0.4
C$_2$H$_6$ Reactions			
$C_2H_6 + M \rightarrow CH_3 + CH_3 + M$	See Table 2		

Reaction	k/cm^3 molecule^{-1}s^{-1}	Temp/K	Error limits ($\Delta \log k$)
CHCO Reactions			
CHCO + O$_2$ → CO$_2$ + HCO → 2CO + OH → C$_2$O + HO$_2$ → CHO$_2$CO	2.7×10^{-12} exp(430/T) M = He, 2 Torr	300–550	± 0.7
CH$_2$CHO Radical Reactions			
CH$_2$CHO + O$_2$ → HO$_2$ + CH$_2$CHO → HCHO + CO + OH → O$_2$CH$_2$CHO	$k_\infty = 2.6 \times 10^{-13}$ $k_2 = 3.0 \times 10^{-14}$	250–1000 300	± 0.2 ± 0.3
CH$_3$CO Radical Reactions			
CH$_3$CO + O$_2$ + M → CH$_3$CO$_3$ + M	See Table 3		
CH$_3$CHO Reactions			
CH$_3$CHO + M → CH$_3$ + HCO + M	See Table 2		
C$_2$H$_5$O Reactions			
C$_2$H$_5$O + M → HCHO + CH$_3$ + M → CH$_3$CHO + H + M	See Table 2		
C$_2$H$_5$O + O$_2$ → CH$_3$CHO + HO$_2$	1.0×10^{-13} exp(−830/T)	300–1000	± 0.3 at 300 K rising to ± 0.5 at 1000 K
C$_2$H$_5$OOH Reactions			
C$_2$H$_5$OOH + M → C$_2$H$_5$O + OH + M	See Table 2		
C$_6$H$_5$ Radical Reactions			
C$_6$H$_5$ + M → C$_2$H$_2$ + C$_4$H$_3$ + M → C$_2$H$_3$ + C$_4$H$_2$ + M → linear-C$_6$H$_5$ + M	See Table 2		
C$_6$H$_6$ Reactions			
C$_6$H$_6$ + M → C$_6$H$_5$ + H + M → C$_4$H$_4$ + C$_2$H$_2$ + M	See Table 2		
C$_6$H$_5$O Radical Reactions			
C$_6$H$_5$O + M → C$_5$H$_5$ + CO + M	See Table 2		
C$_6$H$_5$CH$_2$ Radical Reactions			
C$_6$H$_5$CH$_2$ + M → C$_3$H$_3$ + 2C$_2$H$_2$ + M → C$_4$H$_4$ + C$_3$H$_3$ + M → C$_5$H$_5$ + C$_2$H$_2$ + M → C$_7$H$_7$(BCH) + M	See Table 2		

Reaction	$k/\text{cm}^3 \text{ molecule}^{-1}\text{s}^{-1}$	Temp/K	Error limits ($\Delta \log k$)

$C_6H_5CH_3$ Reactions

$\left.\begin{array}{l} C_6H_5CH_3 + M \rightarrow C_6H_5CH_2 + H + M \\ \qquad\qquad \rightarrow C_6H_5 + CH_3 + M \end{array}\right\}$ See Table 2

p-$C_6H_4(CH_3)_2$ Reactions

p-$C_6H_4(CH_3)_2 + M \rightarrow C_6H_4CH_2CH_3 + H + M$ See Table 2

$C_6H_5C_2H_5$ Reactions

$\left.\begin{array}{l} C_6H_5C_2H_5 + M \rightarrow C_6H_5CH_2 + CH + M \\ \qquad\qquad \rightarrow C_6H_6 + C_2H_4 + M \\ \qquad\qquad \rightarrow C_6H_5CHCH_2 + H_2 + M \\ \qquad\qquad \rightarrow C_6H_5 + C_2H_5 + M \\ \qquad\qquad \rightarrow C_6H_5CHCH_3 + H + M \end{array}\right\}$ See Table 2

Table 2
DECOMPOSITION REACTIONS

Reaction	k_∞/s^{-1} $k_0/\text{cm}^3 \text{ molecule}^{-1}\text{ s}^{-1}$ F_c $k/\text{s}^{-1} = \dfrac{k_0 k_\infty [M]}{k_0[M] + k_\infty} F$	Temp/K	Error limits ($\Delta \log k$)
$H_2 + Ar \rightarrow 2H + Ar$	$k_0 = 3.7 \times 10^{-10} \exp(-48350/T)$	2500–8000	± 0.3
$H_2 + H_2 \rightarrow 2H + H_2$	$k_0 = 1.5 \times 10^{-9} \exp(-48350/T)$	2500–8000	± 0.5
$H_2O + N_2 \rightarrow H + OH + N_2$	$k_0 = 5.8 \times 10^{-9} \exp(-52920/T)$	2000–6000	± 0.5
$H_2O_2 + M \rightarrow 2\,OH + M$	$k_0(\text{Ar}) = 3 \times 10^{-8} \exp(-21600/T)$ $k_0(\text{N}_2) = 2 \times 10^{-7} \exp(-22900/T)$ $k_\infty = 3 \times 10^{14} \exp(-24400/T)$ $F_c(\text{Ar}) = 0.5$	1000–1500 700–1500 1000–1500 700–1500	± 0.2 ± 0.2 ± 0.5 $\Delta F_c = \pm 0.1$
$CH_3 + M \rightarrow CH_2 + H + M$	$k_0 = 1.7 \times 10^{-8} \exp(-45600/T)$	1500–3000	± 0.5
$CH_4 + M \rightarrow CH_3 + H + M$	$k_0(\text{Ar}) = 1.2 \times 10^{-6} \exp(-47000/T)$ $k_0(\text{CH}_4) = 1.4 \times 10^{-5} \exp(-48100/T)$ $k_\infty = 2.4 \times 10^{16} \exp(-52800/T)$ $F_c(\text{Ar}) = \exp(-0.45 - T/3231)$ $F_c(\text{CH}_4) = \exp(-0.37 - T/2210)$	1000–3000 1000–2000 1000–3000 1000–3000 1000–2000	± 0.3 ± 0.3 ± 0.5 $\Delta F_c = \pm 0.1$ $\Delta F_c = \pm 0.1$
$\left.\begin{array}{l} HCHO + M \rightarrow H + CHO + M \\ \qquad\qquad \rightarrow H_2 + CO + M \end{array}\right\}$	$k_0(1) = 2.1 \times 10^{-8} \exp(-39200/T)$ $k_1/k_2 = 0.5$ at 2200 K	1500–2500	± 0.3
$CH_3O + M \rightarrow HCHO + H + M$	$k_0 = 3.16 \times 10^2 T^{-2.7} \exp(-15400/T)$ [estimate]	300–1000	± 1.0
$CH_3OOH + M \rightarrow CH_3O + OH + M$	$k_\infty = 4 \times 10^{15} \exp(-21600/T)$	400–1000	± 0.5 at 600 K rising to ± 1.0 at 400 and 1000 K
$NCO + Ar \rightarrow N + CO + Ar$	$k_0 = 1.7 \times 10^{-9} \exp(-23500/T)$	1450–2600	± 0.4

Table 2
DECOMPOSITION REACTIONS (continued)

Reaction	k_∞/s^{-1} k_0/cm^3 molecule^{-1} s^{-1} F_c $k/s^{-1} = \dfrac{k_0\, k_\infty\, [M]}{k_0[M] + k_\infty} F$	Temp/K	Error limits ($\Delta \log k$)
$C_2H_3 + M \rightarrow C_2H_2 + H + M$	$k_0 = 6.9 \times 10^{17}\, T^{-7.5} \exp(-22900/T)$ $k_\infty = 2 \times 10^{14} \exp(-20000/T)$ $F_c = 0.35$	500–2500 500–2500 500–2500	± 0.5 ± 0.5 $\Delta F_c = \pm 0.1$
$C_2H_6 + M \rightarrow 2CH_3 + M$	$k_0(Ar) = 1.1 \times 10^{25}\, T^{-8.24} \exp(-47090/T)$ $k_0(C_2H_6) = 4.5 \times 10^{-2} \exp(-41930/T)$ $k_\infty = 1.8 \times 10^{21}\, T^{-1.24} \exp(-45700/T)$ $F_c(Ar) = 0.38 \exp(-T/73) + 0.62 \exp(-T/1180)$ $F_c(C_2H_6) = 0.54 \exp(-T/1250)$	300–2000 800–1000 300–2000 300–2000 800–1000	± 0.5 ± 0.5 ± 0.3 $\Delta F_c = \pm 0.1$ $\Delta F_c = \pm 0.1$
$CH_3CHO + M \rightarrow CH_3 + CHO + M$	$k(1\text{ atm.}) = 7 \times 10^{15} \exp(-41100/T)$ (pressure dependent region)	750–1200	± 0.4
$C_2H_5O + M \rightarrow HCHO + CH_3 + M$	$k_\infty = 8 \times 10^{13} \exp(-10830/T)$ [estimate]	300–600	± 1.0
$C_2H_5OOH + M \rightarrow C_2H_5O + OH + M$	$k_\infty\ 4 \times 10^{15} \exp(-21600/T)$	400–1000	± 1.0
$C_6H_5 + M \rightarrow C_2H_2 + C_4H_3 + M$ $ \rightarrow C_2H_3 + C_4H_2 + M$	No recommendation		
$ \rightarrow \text{linear-}CH_5 + M$	$4.0 \times 10^{13} \exp(-36700/T)$	1450–1900	± 0.4
$C_6H_6 + M \rightarrow C_6H_5 + H + M$ $ \rightarrow C_4H_4 + H_2 + M$	$9.0 \times 10^{15} \exp(-54060/T)$	1200–2500	± 0.4 at 1200 K reducing to ± 0.3 at 2500 K
$C_6H_5O + M \rightarrow C_5H_5 + CO + M$	$2.5 \times 10^{11} \exp(-22100/T)$	1000–1580	± 0.2
$C_6H_5CH_2 + M \rightarrow C_3H_3 + 2C_2H_2 + M$ $ \rightarrow C_4H_4 + C_3H_3 + M$ $ \rightarrow C_5H_5 + C_2H_2 + M$ $ \rightarrow C_7H_7\ (BCH) + M$	$5.1 \times 10^{13} \exp(-36370/T)$	1350–1900	± 0.3 at 1350 K rising to ± 0.5 1900 K
$C_6H_5CH_3 + M \rightarrow C_6H_5CH_2 + H + M$ $ \rightarrow C_6H_5 + CH_3 + M$	$3.1 \times 10^{15} \exp(-44890/T)$ No recommendation	920–2200	± 0.3 at 900 K rising to ± 0.5 at 2200 K
$p\text{-}C_6H_4(CH_3)_2 + M \rightarrow p\text{-}C_6H_4CH_2CH_3 + H + M$	$4.0 \times 10^{15} \exp(-42600/T)$	1400–1800	± 0.5
$C_6H_5C_2H_5 + M \rightarrow C_6H_5CH_2 + CH + M$ $ \rightarrow C_6H_6 + C_2H_4 + M$ $ \rightarrow C_6H_5CHCH_2 + H_2 + M$ $ \rightarrow C_6H_5 + C_2H_5 + M$ $ \rightarrow C_6H_5CHCH_3 + H + M$	$6.1 \times 10^{15} \exp(-37800/T)$ No recommendations	770–1800	± 0.1 at 770 K rising to ± 0.4 at

Table 3
COMBINATION REACTIONS

Reaction	k_∞/cm^3 molecule^{-1}s^{-1} k_0/cm^6 molecule^{-2} s^{-1} F_c k/cm^3 molecule^{-1}s^{-1} $= \dfrac{k_0\, k_\infty\, [M]}{k_0[M] + k_\infty} F$	Temp/K	Error limits ($\Delta \log k$)
$H + O_2 + Ar \rightarrow HO_2 + Ar$	$k_0 = 1.7 \times 10^{-30}\, T^{-0.8}$	300–2000	± 0.5

Reaction	k_∞/cm^3 molecule^{-1}s^{-1} k_0/cm^6 molecule^{-2} s^{-1} F_c k/cm^3 molecule^{-1}s^{-1} = $\dfrac{k_0 \, k_\infty \, [M]}{k_0[M] + k_\infty} F$	Temp/K	Error limits ($\Delta \log k$)
$H + O_2 + H_2 \rightarrow HO_2 + H_2$	$k_0 = 5.8 \times 10^{-30} T^{-0.8}$	300–2000	± 0.5
$H + O_2 + N_2 \rightarrow HO_2 + N_2$	$k_0 = 3.9 \times 10^{-30} T^{-0.8}$	300–2000	± 0.5
$H + O_2 + H_2O \rightarrow HO_2 + H_2O$	$k_0 = 4.3 \times 10^{-30} T^{-0.8}$	300–2000	± 0.5
$H + H + Ar \rightarrow H_2 + Ar$	$k_0 = 1.8 \times 10^{-30} T^{-1.0}$	300–2500	± 0.5
$H + H + H_2 \rightarrow H_2 + H_2$	$k_0 = 2.7 \times 10^{-31} T^{-0.6}$	100–5000	± 0.5
$H + OH + H_2O \rightarrow H_2O + H_2O$	$k_0 = 3.9 \times 10^{-25} T^{-2.0}$	300–3000	± 0.3
$H + OH + Ar \rightarrow H_2O + Ar$	$k_0 = 2.3 \times 10^{-26} T^{-2.0}$	300–3000	± 0.3
$H + OH + N_2 \rightarrow H_2O + N_2$	$k_0 = 6.1 \times 10^{-26} T^{-2.0}$	300–3000	± 0.3
$H + CH_3 + M \rightarrow CH_4 + M$	$k_0(He) = 6.2 \times 10^{-29} (T/3000)^{-1.8}$ $k_0(Ar) = 6 \times 10^{-29} (T/300)^{-1.8}$ $k_0(C_2H_6) = 3 \times 10^{-28} (T/300)^{-1.8}$ $k_\infty = 3.5 \times 10^{-10}$ $F_c(He,Ar) = \exp(-0.45 - T/3231)$ $F_c(C_2H_6) = \exp(-0.34 - T/3053)$	300–1000 300–1000 300–1000 300–1000 300–1000 300–1000	± 0.3 ± 0.5 ± 0.5 ± 0.3 $\Delta F_c = \pm 0.1$ $\Delta F_c = \pm 0.1$
$H + C_2H_2 + He \rightarrow C_2H_3 + He$	$k_\infty = 1.4 \times 10^{-11} \exp(-1300/T)$ $k_0 = 3.3 \times 10^{-30} \exp(-740/T)$ $F_c = 0.44$	200–400 200–400 200–400	± 0.3 ± 0.5 $\Delta F_c = \pm 0.1$
$H + C_2H_3 + M \rightarrow C_2H_4 + M$	No recommendation		
$H + C_2H_4 + M \rightarrow C_2H_5 + M$	No recommendation		
$H + C_2H_5 + M \rightarrow C_2H_6 + M$	No recommendation		
$H + C_6H_5 + M \rightarrow C_6H_6 + M$	$k_\infty = 1.3 \times 10^{-10}$	1400–1700	± 0.5
$H + C_6H_6 + M \rightarrow C_6H_7 + M$	$k_\infty = 6.7 \times 10^{-11} \exp(-2170/T)$	300–1000	± 0.2
$H + C_6H_5O + M \rightarrow C_6H_5OH + M$	$k_\infty = 4.2 \times 10^{-10}$	1000	± 0.3
$H + C_6H_5CH_2 + M \rightarrow C_6H_5CH_3 + M$	$k_\infty = 5.5 \times 10^{-10}$	300–2000	± 0.2 at 300 K rising to ± 0.7 at 2000 K.
$H + C_6H_5CH_3 + M \rightarrow C_6H_6CH_3 + M$	$k_\infty = 1.2 \times 10^{-13}$	298	± 0.2
$H + C_6H_5C_2H_5 + M \rightarrow C_6H_6C_2H_5 + M$	$k_\infty = 3.3 \times 10^{-13}$	298	± 0.1
$OH + OH + M \rightarrow H_2O_2 + M$	$k_0(N_2) = 8 \times 10^{-31} (T/300)^{-0.76}$ $k_0(H_2O) = 4 \times 10^{-30}$ $k_\infty = 1.5 \times 10^{-11} (T/300)^{-0.37}$ $F_c(N_2) = 0.5$	250–1400 300–400 200–1500 200–1500	± 0.4 ± 0.5 $\Delta F_c = \pm 0.2$
$OH + CH_3 + M \rightarrow CH_3OH + M$	No data available for this channel (See Table 1)		

Reaction	k_∞/cm^3 molecule^{-1}s^{-1} k_0/cm^6 molecule^{-2} s^{-1} F_c k/cm^3 molecule^{-1}s^{-1} $= \dfrac{k_0 \, k_\infty \, [M]}{k_0[M] + k_\infty} F$	Temp/K	Error limits ($\Delta \log k$)
$OH + C_2H_2 + M \rightarrow C_2H_2OH + M$	See data sheet		
$OH + C_6H_6 + M \rightarrow C_6H_6OH + M$	$k_\infty = 3.8 \times 10^{-12} \exp(-340/T)$	240–340	± 0.2
$OH + C_6H_5OH + M \rightarrow C_6H_5(OH)_2 + M$	$k_\infty = 2.8 \times 10^{-11}$	298	± 0.1
$OH + C_6H_5CH_3 + M \rightarrow HOC_6H_5CH_3 + M$	$k_\infty = 3.8 \times 10^{-12} \exp(180/T)$	200–300	± 0.4
$OH + C_6H_4(CH_3)_2 + M \rightarrow C_6H_4(CH_3)_2OH + M$	$k_\infty = 1.4 \times 10^{-11}$	300–320	± 0.1
$OH + C_6H_5C_2H_5 + M \rightarrow HOC_6H_5C_2H_5 + M$	7.5×10^{-12} at p $\leqslant$ 1 atm.	298	± 0.1
$^3CH_2 + C_2H_2 + M \rightarrow C_3H_4 + M$	$2.0 \times 10^{-11} \exp(-3330/T)$ at $p = \leqslant$ 10 Torr.	300–1000	± 0.3
$^3CH_2 + C_2H_4 + M \rightarrow C_3H_6 + M$ $\rightarrow c\text{-}C_3H_6 + M$ $\rightarrow C_3H_5 + H + M$	$5.3 \times 10^{-12} \exp(-2660/T)$	300–1000	± 0.2 at 300 K rising to ± 0.3 at 1000 K
$^1CH_2 + C_2H_2 + M \rightarrow CH_2CCH_2 + M$ $\rightarrow CH_3CCH + M$ $\rightarrow CH_2CCH + H + M$	3.7×10^{-10} independent of p	300–1000	± 0.3 at 300 K rising to ± 0.7 at 1000 K.
$^1CH_2 + C_2H_4 + M \rightarrow C_3H_6$	1.1×10^{-10} independent of p	300–1000	± 0.2 at 300 K rising to ± 0.5 at 1000 K.
$CH_3 + O_2 + M \rightarrow CH_3O_2 + M$	$k_0(Ar) = 1.5 \times 10^{-22} T^{-3.3}$ $k_0(N_2) = 1.6 \times 10^{-22} T^{-3.3}$ $k_\infty = 1.3 \times 10^{-15} T^{-1.2}$ $F_c = 0.466 - 1.30 \times 10^{-4} T$	300–800 300–800 300–800 300–800	± 0.3 ± 0.3 ± 0.3
$CH_3 + CH_3 + Ar \rightarrow C_2H_6 + Ar$	$k_\infty = 6 \times 10^{-11}$	300–2000	± 0.05 at 300 K rising to ± 0.3 at 2000 K
	$k_0 = 3.5 \times 10^{-7} T^{-7.0} \exp(-1390/T)$ $F_c = 0.38 \exp(-T/73)$ $+ 0.62\exp(-T/1180)$	300–2000 300–2000	± 0.3 $\Delta F_c = \pm$ 0.1
$CH_3 + C_2H_2 + M \rightarrow C_3H_5 + M$	$k_\infty = 1 \times 10^{-12} \exp(-3900/T)$	300–600	± 0.5
$CH_3 + C_2H_4 + M \rightarrow n\text{-}C_3H_7 + M$	$3.5 \times 10^{-13} \exp(-3700/T)$	300–600	± 0.3
$CH_3 + C_2H_5 + M \rightarrow C_3H_8 + M$	$k_\infty = 4.7 \times 10^{-11}$	300–800	± 0.3
$C_2H_5 + C_2H_5 + M \rightarrow n\text{-}C_4H_{10} + M$	$k_\infty = 1.9 \times 10^{-11}$	300–1200	± 0.3
$CH_3CO + O_2 + M \rightarrow CH_3CO_3 + M$	2×10^{-12} for $p = 1$–4 Torr.	300	± 0.3

Section 6
Fluid Properties

THERMODYNAMIC PROPERTIES OF AIR

These tables summarize the thermodynamic properties of air in the liquid and gaseous states, as well as along the saturation line. In the table for the saturation state, P(boil) is the bubble point temperature (i.e., the pressure at which boiling begins as the temperature of the liquid is raised), and P(con) is the dew point temperature (pressure at which condensation begins as the temperature of the gas is lowered). The other properties tabulated are density (ρ), enthalpy (H), entropy (S), and isobaric heat capacity (C_p). More detailed tables may be found in the references.

REFERENCES

1. Vasserman, A.A., and Rabinovich, V.A., *Thermophysical Properties of Liquid Air and its Components*, Izdatel'stvo Komiteta, Standartov, Moscow, 1968.
2. Vasserman, A.A., et al., *Thermophysical Properties of Air and Air Components,* Izdatel'stvo Nauka, Moscow, 1966.

Properties in the saturation state:

T K	P(boil) bar	P(con) bar	ρ (liq) g/cm^3	ρ (gas) g/L
65	0.1468	0.0861	0.939	0.464
70	0.3234	0.2052	0.917	1.033
75	0.6366	0.4321	0.894	2.048
80	1.146	0.8245	0.871	3.709
85	1.921	1.453	0.845	6.258
90	3.036	2.397	0.819	9.980
95	4.574	3.748	0.792	15.21
100	6.621	5.599	0.763	22.39
110	12.59	11.22	0.699	45.15
120	21.61	20.14	0.622	87.34
130	34.16	33.32	0.487	184.33
132.55	37.69	37.69	0.313	312.89

Properties of liquid air:

P bar	T K	ρ g/cm^3	H J/g	S J/g K	C_p J/g K
1	75	0.8935	−131.7	2.918	1.843
5	75	0.8942	−131.4	2.916	1.840
5	80	0.8718	−122.3	3.031	1.868
5	85	0.8482	−112.9	3.143	1.901
5	90	0.8230	−103.3	3.250	1.941
5	95	0.7962	−93.5	3.356	1.991
10	75	0.8952	−131.1	2.913	1.836
10	80	0.8729	−122.0	3.028	1.863
10	90	0.8245	−103.1	3.246	1.932
10	100	0.7695	−83.2	3.452	2.041
50	75	0.9025	−128.2	2.892	1.806
50	100	0.7859	−81.8	3.415	1.939
50	125	0.6222	−28.3	3.889	2.614
50	150	0.1879	91.9	4.764	2.721
100	75	0.9111	−124.5	2.867	1.774
100	100	0.8033	−79.4	3.376	1.852
100	125	0.6746	−31.4	3.805	2.062
100	150	0.4871	32.8	4.271	2.832

Properties of air in the gaseous state:

P bar	T K	ρ g/L	H J/g	S J/g K	C_p J/g K
1	100	3.556	98.3	5.759	1.032
1	200	1.746	199.7	6.463	1.007
1	300	1.161	300.3	6.871	1.007

THERMODYNAMIC PROPERTIES OF AIR (continued)

P bar	T K	ρ g/L	H J/g	S J/g K	C_p J/g K
1	500	0.696	503.4	7.389	1.030
1	1000	0.348	1046.6	8.138	1.141
10	200	17.835	195.2	5.766	1.049
10	300	11.643	298.3	6.204	1.021
10	500	6.944	502.9	6.727	1.034
10	1000	3.471	1047.2	7.477	1.142
100	200	213.950	148.8	4.949	1.650
100	300	116.945	279.9	5.486	1.158
100	500	66.934	499.0	6.048	1.073
100	1000	33.613	1052.4	6.812	1.151

PROPERTIES OF WATER IN THE RANGE 0 — 100 °C

This table summarizes the best available values of the density, specific heat capacity at constant pressure (C_p), vapor pressure, viscosity, thermal conductivity, dielectric constant, and surface tension for liquid water in the range 0 — 100 °C. All values (except vapor pressure) refer to a pressure of 100 kPa (1 bar). The temperature scale is IPTS-68.

t °C	Density g/cm³	C_p J/g K	Vap. pres. kPa	Visc. µPa s	Ther. cond. mW/K m	Diel. const.	Surf. ten. mN/m
0	0.99984	4.2176	0.6113	1793	561.0	87.90	75.64
10	0.99970	4.1921	1.2281	1307	580.0	83.96	74.23
20	0.99821	4.1818	2.3388	1002	598.4	80.20	72.75
30	0.99565	4.1784	4.2455	797.7	615.4	76.60	71.20
40	0.99222	4.1785	7.3814	653.2	630.5	73.17	69.60
50	0.98803	4.1806	12.344	547.0	643.5	69.88	67.94
60	0.98320	4.1843	19.932	466.5	654.3	66.73	66.24
70	0.97778	4.1895	31.176	404.0	663.1	63.73	64.47
80	0.97182	4.1963	47.373	354.4	670.0	60.86	62.67
90	0.96535	4.2050	70.117	314.5	675.3	58.12	60.82
100	0.95840	4.2159	101.325	281.8	679.1	55.51	58.91
Ref.	1—3	2	1, 3	3	3	4	5

REFERENCES

1. L. Harr, J. S. Gallagher, and G. S. Kell, *NBS/NRC Steam Tables*, Hemisphere Publishing Corp., 1984.
2. K. N. Marsh, Ed., *Recommended Reference Materials for the Realization of Physicochemical Properties*, Blackwell Scientific Publications, Oxford, 1987.
3. J. V. Sengers and J. T. R. Watson, Improved international formulations for the viscosity and thermal conductivity of water substance, *J. Phys. Chem. Ref. Data*, 15, 1291, 1986.
4. D. G. Archer and P. Wang, The dielectric constant of water and Debye-Hückel limiting law slopes, *J. Phys. Chem. Ref. Data*, 19, 371, 1990.
5. N. B. Vargaftik, et al., International tables of the surface tension of water, *J. Phys. Chem. Ref. Data*, 12, 817, 1983.

ENTHALPY OF VAPORIZATION OF WATER

The enthalpy (heat) of vaporization of water is tabulated as a function of temperature on the IPTS-68 scale.

REFERENCE

Marsh, K. N., Ed., *Recommended Reference Materials for the Realization of Physicochemical Properties*, Blackwell, Oxford, 1987.

t °C	$\Delta_{vap}H$ kJ/mol	t °C	$\Delta_{vap}H$ kJ/mol
0	45.054	200	34.962
25	43.990	220	33.468
40	43.350	240	31.809
60	42.482	260	29.930
80	41.585	280	27.795
100	40.657	300	25.300
120	39.684	320	22.297
140	38.643	340	18.502
160	37.518	360	12.966
180	36.304	374	2.066

FIXED POINT PROPERTIES OF H$_2$O AND D$_2$O

Temperatures are given on the IPTS-68 scale.

REFERENCES

1. Haar, L., Gallagher, J.S., and Kell, G.S., *NBS/NRC Steam Tables*, Hemisphere Publishing Corp., 1984.
2. Levelt Sengers, J.M.H., Straub, J., Watanabe, K., and Hill, P.G., Assessment of critical parameter values for H$_2$O and D$_2$O, *J. Phys. Chem. Ref. Data,* 14, 193, 1985.
3. Kestin, J. et. al., Thermophysical properties of fluid D$_2$O, *J. Phys. Chem. Ref. Data,* 13, 601, 1984.
4. Kestin, J. et. al., Thermophysical properties of fluid H$_2$O, *J. Phys. Chem. Ref. Data,* 13, 175, 1984.
5. Hill, P.G., MacMillan, R.D.C., and Lee, V., A fundamental equation of state for heavy water, *J. Phys. Chem. Ref. Data*, 11, 1, 1982.

	Unit	H$_2$O	D$_2$O
Molar mass	g/mol	18.01528	20.02748
Melting point(101.325 kPa)	°C	0.00	3.82
Boiling point(101.325 kPa)	°C	100.00	101.42
Triple point temperature	°C	0.01	3.82
Triple point pressure	Pa	611.73	661
Triple point density(l)	g/cm^3	0.99978	1.1055
Triple point density(g)	mg/L	4.885	5.75
Critical temperature	°C	373.99	370.74
Critical pressure	MPa	22.064	21.671
Critical density	g/cm^3	0.322	0.356
Critical specific volume	cm^3/g	3.11	2.81
Maximum density(saturated liquid)	g/cm^3	0.99995	1.1053
Temperature of maximum density	°C	4.0	11.2

THERMAL CONDUCTIVITY OF SATURATED H$_2$O AND D$_2$O

This table gives the thermal conductivity λ for water (H$_2$O or D$_2$O) in equilibrium with its vapor. Values for the liquid (λ_l) and vapor (λ_v) are listed, as well as the vapor pressure.

REFERENCES

1. Sengers, J.V. and Watson, J.T.R., Improved international formulations for the viscosity and thermal conductivity of water substance, *J. Phys. Chem. Ref. Data*, 15, 1291, 1986.
2. Matsunaga, N. and Nagashima, A., Transport properties of liquid and gaseous D$_2$O over a wide range of temperature and pressure, *J. Phys. Chem. Ref. Data*, 12, 933, 1983.

t/°C	H$_2$O			D$_2$O		
	P/kPa	λ_l/(mW/K m)	λ_v/(mW/K m)	P/kPa	λ_l/(mW/K m)	λ_v/(mW/K m)
0	0.6	561.0	16.49			
10	1.2	580.0	17.21	1.0	575	17.0
20	2.3	598.4	17.95	2.0	589	17.8
30	4.2	615.4	18.70	3.7	600	18.5
40	7.4	630.5	19.48	6.5	610	19.3
50	12.3	643.5	20.28	11.1	618	20.2
60	19.9	654.3	21.10	18.2	625	21.0
70	31.2	663.1	21.96	28.8	629	21.9
80	47.4	670.0	22.86	44.2	633	22.8
90	70.1	675.3	23.80	66.1	635	23.8
100	101.3	679.1	24.79	96.2	636	24.8
150	476	682.1	30.77	465	625	30.8
200	1555	663.4	39.10	1546	592	39.0
250	3978	621.4	51.18	3995	541	52.0
300	8593	547.7	71.78	8688	473	75.2
350	16530	447.6	134.59	16820	391	143.0

STANDARD DENSITY OF WATER

This table gives the density ρ of standard mean ocean water (SMOW), free from dissolved salts and gases, at a pressure of 101325 Pa. SMOW is a standard water sample of high purity and known isotopic composition. Methods of correcting for different isotopic compositions are discussed in the reference. The table below is reprinted with the permission of IUPAC. Note that the temperature scale is IPTS-68.

REFERENCE

Marsh, K. N., Ed., *Recommended Reference Materials for the Realization of Physicochemical Properties,* Blackwell Scientific Publications, Oxford, 1987.

$$\rho/\text{kg m}^{-3}$$

$t_{68}/°C$	0.0	0.1	0.2	0.3	0.4	0.5	0.6	0.7	0.8	0.9
0	999.8426	8493	8558	8622	8683	8743	8801	8857	8912	8964
1	999.9015	9065	9112	9158	9202	9244	9284	9323	9360	9395
2	999.9429	9461	9491	9519	9546	9571	9595	9616	9636	9655
3	999.9672	9687	9700	9712	9722	9731	9738	9743	9747	9749
4	999.9750	9748	9746	9742	9736	9728	9719	9709	9696	9683
5	999.9668	9651	9632	9612	9591	9568	9544	9518	9490	9461
6	999.9430	9398	9365	9330	9293	9255	9216	9175	9132	9088
7	999.9043	8996	8948	8898	8847	8794	8740	8684	8627	8569
8	999.8509	8448	8385	8321	8256	8189	8121	8051	7980	7908
9	999.7834	7759	7682	7604	7525	7444	7362	7279	7194	7108
10	999.7021	6932	6842	6751	6658	6564	6468	6372	6274	6174
11	999.6074	5972	5869	5764	5658	5551	5443	5333	5222	5110
12	999.4996	4882	4766	4648	4530	4410	4289	4167	4043	3918
13	999.3792	3665	3536	3407	3276	3143	3010	2875	2740	2602
14	999.2464	2325	2184	2042	1899	1755	1609	1463	1315	1166
15	999.1016	0864	0712	0558	0403	0247	0090	9932*	9772*	9612*
16	998.9450	9287	9123	8957	8791	8623	8455	8285	8114	7942
17	998.7769	7595	7419	7243	7065	6886	6706	6525	6343	6160
18	998.5976	5790	5604	5416	5228	5038	4847	4655	4462	4268
19	998.4073	3877	3680	3481	3282	3081	2880	2677	2474	2269
20	998.2063	1856	1649	1440	1230	1019	0807	0594	0380	0164
21	997.9948	9731	9513	9294	9073	8852	8630	8406	8182	7957
22	997.7730	7503	7275	7045	6815	6584	6351	6118	5883	5648
23	997.5412	5174	4936	4697	4456	4215	3973	3730	3485	3240
24	997.2994	2747	2499	2250	2000	1749	1497	1244	0990	0735
25	997.0480	0223	9965*	9707*	9447*	9186*	8925*	8663*	8399*	8135*
26	996.7870	7604	7337	7069	6800	6530	6259	5987	5714	5441
27	996.5166	4891	4615	4337	4059	3780	3500	3219	2938	2655
28	996.2371	2087	1801	1515	1228	0940	0651	0361	0070	9778*
29	995.9486	9192	8898	8603	8306	8009	7712	7413	7113	6813
30	995.6511	6209	5906	5602	5297	4991	4685	4377	4069	3760
31	995.3450	3139	2827	2514	2201	1887	1572	1255	0939	0621
32	995.0302	9983*	9663*	9342*	9020*	8697*	8373*	8049*	7724*	7397*
33	994.7071	6743	6414	6085	5755	5423	5092	4759	4425	4091
34	994.3756	3420	3083	2745	2407	2068	1728	1387	1045	0703
35	994.0359	0015	9671*	9325*	8978*	8631*	8283*	7934*	7585*	7234*
36	993.6883	6531	6178	5825	5470	5115	4759	4403	4045	3687
37	993.3328	2968	2607	2246	1884	1521	1157	0793	0428	0062
38	992.9695	9328	8960	8591	8221	7850	7479	7107	6735	6361
39	992.5987	5612	5236	4860	4483	4105	3726	3347	2966	2586
40	992.2204									

* The leading figure decreases by 1.

VOLUMETRIC PROPERTIES OF AQUEOUS SODIUM CHLORIDE SOLUTIONS

This table gives the following properties of aqueous solutions of NaCl as a function of temperature and concentration:

Specific volume v (reciprocal of density) in cm^3/g
Isothermal compressibility $\kappa_T = -(1/v)(\partial v/\partial P)_T$ in GPa^{-1}
Cubic expansion coefficient $\alpha_v = (1/v)(\partial v/\partial T)_P$ in kK^{-1}

All data refer to a pressure of 100 kPa (1 bar). The reference gives properties over a wider range of temperature and pressure.

REFERENCE

Rogers, P. S. Z., and Pitzer, K. S., *J. Phys. Chem. Ref. Data*, 11, 15, 1982.

Molality in mol/kg

$T/°C$	0.100	0.250	0.500	0.750	1.000	2.000	3.000	4.000	5.000

Specific volume v in cm^3/g

$T/°C$	0.100	0.250	0.500	0.750	1.000	2.000	3.000	4.000	5.000
0	0.995732	0.989259	0.978889	0.968991	0.959525	0.925426	0.896292	0.870996	0.848646
10	0.995998	0.989781	0.979804	0.970256	0.961101	0.927905	0.899262	0.874201	0.851958
20	0.997620	0.991564	0.981833	0.972505	0.963544	0.930909	0.902565	0.877643	0.855469
25	0.998834	0.992832	0.983185	0.973932	0.965038	0.932590	0.904339	0.879457	0.857301
30	1.000279	0.994319	0.984735	0.975539	0.966694	0.934382	0.906194	0.881334	0.859185
40	1.003796	0.997883	0.988374	0.979243	0.970455	0.938287	0.910145	0.885276	0.863108
50	1.008064	1.002161	0.992668	0.983551	0.974772	0.942603	0.914411	0.889473	0.867241
60	1.0130	1.0071	0.9976	0.9885	0.9797	0.9474	0.9191	0.8940	0.8716
70	1.0186	1.0127	1.0031	0.9939	0.9851	0.9526	0.9240	0.8987	0.8762
80	1.0249	1.0188	1.0092	0.9999	0.9909	0.9581	0.9293	0.9037	0.8809
90	1.0317	1.0256	1.0157	1.0063	0.9972	0.9640	0.9348	0.9089	0.8858
100	1.0391	1.0329	1.0228	1.0133	1.0040	0.9703	0.9406	0.9144	0.8910

Compressibility κ_T in GPa^{-1}

$T/°C$	0.100	0.250	0.500	0.750	1.000	2.000	3.000	4.000	5.000
0	0.503	0.492	0.475	0.459	0.443	0.389	0.346	0.315	0.294
10	0.472	0.463	0.449	0.436	0.423	0.377	0.341	0.313	0.294
20	0.453	0.446	0.433	0.422	0.411	0.371	0.338	0.313	0.294
25	0.447	0.440	0.428	0.417	0.407	0.369	0.337	0.313	0.294
30	0.443	0.436	0.425	0.414	0.404	0.367	0.337	0.313	0.294
40	0.438	0.432	0.421	0.411	0.401	0.367	0.338	0.315	0.296
50	0.438	0.431	0.421	0.411	0.402	0.369	0.340	0.317	0.299
60	0.44	0.44	0.43	0.42	0.41	0.38	0.35	0.32	0.30
70	0.45	0.44	0.43	0.42	0.42	0.38	0.36	0.33	0.31
80	0.46	0.45	0.44	0.43	0.43	0.39	0.37	0.34	0.32
90	0.47	0.47	0.46	0.45	0.44	0.41	0.38	0.35	0.33
100	0.49	0.48	0.47	0.46	0.45	0.42	0.39	0.37	0.34

Cubic expansion coefficient α_v in kK^{-1}

$T/°C$	0.100	0.250	0.500	0.750	1.000	2.000	3.000	4.000
0	-0.058	-0.026	0.024	0.069	0.110	0.237	0.313	0.355
10	0.102	0.123	0.156	0.186	0.213	0.297	0.349	0.380
20	0.218	0.232	0.254	0.274	0.292	0.349	0.384	0.406
25	0.267	0.278	0.296	0.312	0.327	0.373	0.401	0.420
30	0.311	0.320	0.334	0.347	0.359	0.395	0.418	0.433
40	0.389	0.394	0.402	0.410	0.417	0.438	0.451	0.460
50	0.458	0.460	0.464	0.467	0.470	0.479	0.484	0.486
60	0.52	0.52	0.52	0.52	0.52	0.52	0.52	0.52
70	0.58	0.58	0.58	0.57	0.57	0.56	0.55	0.54
80	0.64	0.63	0.63	0.62	0.61	0.60	0.58	0.56
90	0.69	0.68	0.67	0.67	0.66	0.63	0.61	0.59
100	0.74	0.73	0.72	0.71	0.70	0.66	0.64	0.61

DENSITY OF D$_2$O

Density of liquid D$_2$O in g/cm^3 at a pressure of 100 kPa (1 bar).

REFERENCE

Kirillin, V.A., Ed., *Heavy Water: Thermophysical Properties*, Gosudarstvennoe Energeticheskoe Izdatel'stvo, Moscow, 1963.

$t/°C$	3.8	5	10	15	20	25	30
Density	1.1053	1.1055	1.1057	1.1056	1.105	1.1044	1.1034

$t/°C$	35	40	45	50	55	60	65
Density	1.1019	1.1001	1.0979	1.0957	1.0931	1.0905	1.0875

$t/°C$	70	75	80	85	90	95	100
Density	1.0847	1.0815	1.0783	1.0748	1.0712	1.0673	1.0635

VAPOR PRESSURE OF ICE

The values of the vapor (sublimation) pressure of ice in this table were calculated from the equation recommended by the International Association for the Properties of Steam (IAPS) in 1993. Temperature values correspond to the ITS-90 temperature scale. The uncertainty in the pressure is estimated to be 0.1% for $t > -25°C$ and 0.5% for $t < -25°C$. The first entry in the table is the triple point of water.

REFERENCE

Wagner, W., Saul, A., and Pruss, A., *J. Phys. Chem. Ref. Data*, 23, 515, 1994.

$t/°C$	p/Pa	$t/°C$	p/Pa	$t/°C$	p/Pa
0.01	611.657	−16	150.68	−33	27.71
0	611.15	−17	137.25	−34	24.90
−1	562.67	−18	124.92	−35	22.35
−2	517.72	−19	113.62	−36	20.04
−3	476.06	−20	103.26	−37	17.96
−4	437.47	−21	93.77	−38	16.07
−5	401.76	−22	85.10	−39	14.37
−6	368.73	−23	77.16	−40	12.84
−7	338.19	−24	69.91	−45	7.202
−8	309.98	−25	63.29	−50	3.936
−9	283.94	−26	57.25	−55	2.093
−10	259.90	−27	51.74	−60	1.080
−11	237.74	−28	46.73	−65	0.540
−12	217.32	−29	42.16	−70	0.261
−13	198.52	−30	38.01	−75	0.122
−14	181.22	−31	34.24	−80	0.055
−15	165.30	−32	30.82		

VAPOR PRESSURE OF WATER FROM 0 TO 370° C

This table gives the vapor pressure of water at intervals of 1° C from the melting point to the critical point.

REFERENCE

Haar, L., Gallagher, J.S., and Kell, G.S., *NBS/NRC Steam Tables*, Hemisphere Publishing Corp., New York, 1984.

t/°C	P/kPa	t/°C	P/kPa	t/°C	P/kPa	t/°C	P/kPa
0	0.61129	53	14.303	106	125.03	159	602.11
1	0.65716	54	15.012	107	129.39	160	617.66
2	0.70605	55	15.752	108	133.88	161	633.53
3	0.75813	56	16.522	109	138.50	162	649.73
4	0.81359	57	17.324	110	143.24	163	666.25
5	0.87260	58	18.159	111	148.12	164	683.10
6	0.93537	59	19.028	112	153.13	165	700.29
7	1.0021	60	19.932	113	158.29	166	717.83
8	1.0730	61	20.873	114	163.58	167	735.70
9	1.1482	62	21.851	115	169.02	168	753.94
10	1.2281	63	22.868	116	174.61	169	772.52
11	1.3129	64	23.925	117	180.34	170	791.47
12	1.4027	65	25.022	118	186.23	171	810.78
13	1.4979	66	26.163	119	192.28	172	830.47
14	1.5988	67	27.347	120	198.48	173	850.53
15	1.7056	68	28.576	121	204.85	174	870.98
16	1.8185	69	29.852	122	211.38	175	891.80
17	1.9380	70	31.176	123	218.09	176	913.03
18	2.0644	71	32.549	124	224.96	177	934.64
19	2.1978	72	33.972	125	232.01	178	956.66
20	2.3388	73	35.448	126	239.24	179	979.09
21	2.4877	74	36.978	127	246.66	180	1001.9
22	2.6447	75	38.563	128	254.25	181	1025.2
23	2.8104	76	40.205	129	262.04	182	1048.9
24	2.9850	77	41.905	130	270.02	183	1073.0
25	3.1690	78	43.665	131	278.20	184	1097.5
26	3.3629	79	45.487	132	286.57	185	1122.5
27	3.5670	80	47.373	133	295.15	186	1147.9
28	3.7818	81	49.324	134	303.93	187	1173.8
29	4.0078	82	51.342	135	312.93	188	1200.1
30	4.2455	83	53.428	136	322.14	189	1226.9
31	4.4953	84	55.585	137	331.57	190	1254.2
32	4.7578	85	57.815	138	341.22	191	1281.9
33	5.0335	86	60.119	139	351.09	192	1310.1
34	5.3229	87	62.499	140	361.19	193	1338.8
35	5.6267	88	64.958	141	371.53	194	1368.0
36	5.9453	89	67.496	142	382.11	195	1397.6
37	6.2795	90	70.117	143	392.92	196	1427.8
38	6.6298	91	72.823	144	403.98	197	1458.5
39	6.9969	92	75.614	145	415.29	198	1489.7
40	7.3814	93	78.494	146	426.85	199	1521.4
41	7.7840	94	81.465	147	438.67	200	1553.6
42	8.2054	95	84.529	148	450.75	201	1586.4
43	8.6463	96	87.688	149	463.10	202	1619.7
44	9.1075	97	90.945	150	475.72	203	1653.6
45	9.5898	98	94.301	151	488.61	204	1688.0
46	10.094	99	97.759	152	501.78	205	1722.9
47	10.620	100	101.32	153	515.23	206	1758.4
48	11.171	101	104.99	154	528.96	207	1794.5
49	11.745	102	108.77	155	542.99	208	1831.1
50	12.344	103	112.66	156	557.32	209	1868.4
51	12.970	104	116.67	157	571.94	210	1906.2
52	13.623	105	120.79	158	586.87	211	1944.6

$t/°C$	P/kPa	$t/°C$	P/kPa	$t/°C$	P/kPa	$t/°C$	P/kPa
212	1983.6	253	4178.9	294	7881.3	335	13701
213	2023.2	254	4249.1	295	7995.2	336	13876
214	2063.4	255	4320.2	296	8110.3	337	14053
215	2104.2	256	4392.2	297	8226.8	338	14232
216	2145.7	257	4465.1	298	8344.5	339	14412
217	2187.8	258	4539.0	299	8463.5	340	14594
218	2230.5	259	4613.7	300	8583.8	341	14778
219	2273.8	260	4689.4	301	8705.4	342	14964
220	2317.8	261	4766.1	302	8828.3	343	15152
221	2362.5	262	4843.7	303	8952.6	344	15342
222	2407.8	263	4922.3	304	9078.2	345	15533
223	2453.8	264	5001.8	305	9205.1	346	15727
224	2500.5	265	5082.3	306	9333.4	347	15922
225	2547.9	266	5163.8	307	9463.1	348	16120
226	2595.9	267	5246.3	308	9594.2	349	16320
227	2644.6	268	5329.8	309	9726.7	350	16521
228	2694.1	269	5414.3	310	9860.5	351	16725
229	2744.2	270	5499.9	311	9995.8	352	16931
230	2795.1	271	5586.4	312	10133	353	17138
231	2846.7	272	5674.0	313	10271	354	17348
232	2899.0	273	5762.7	314	10410	355	17561
233	2952.1	274	5852.4	315	10551	356	17775
234	3005.9	275	5943.1	316	10694	357	17992
235	3060.4	276	6035.0	317	10838	358	18211
236	3115.7	277	6127.9	318	10984	359	18432
237	3171.8	278	6221.9	319	11131	360	18655
238	3228.6	279	6317.0	320	11279	361	18881
239	3286.3	280	6413.2	321	11429	362	19110
240	3344.7	281	6510.5	322	11581	363	19340
241	3403.9	282	6608.9	323	11734	364	19574
242	3463.9	283	6708.5	324	11889	365	19809
243	3524.7	284	6809.2	325	12046	366	20048
244	3586.3	285	6911.1	326	12204	367	20289
245	3648.8	286	7014.1	327	12364	368	20533
246	3712.1	287	7118.3	328	12525	369	20780
247	3776.2	288	7223.7	329	12688	370	21030
248	3841.2	289	7330.2	330	12852	371	21283
249	3907.0	290	7438.0	331	13019	372	21539
250	3973.6	291	7547.0	332	13187	373	21799
251	4041.2	292	7657.2	333	13357	373.98	22055
252	4109.6	293	7768.6	334	13528		

BOILING POINT OF WATER AT VARIOUS PRESSURES

Data are based on the equation of state recommended by the International Association for the Properties of Steam in 1984, as presented in Haar, Gallagher, and Kell, *NBS-NRC Steam Tables* (Hemisphere Publishing Corp., New York, 1984). The temperature scale is IPTS-68.
Note that: 1 mbar = 100 Pa = 0.000986923 atmos = 0.750062 mmHg.

P/mbar	T/°C	P/mbar	T/°C	P/mbar	T/°C	P/mbar	T/°C
50	32.88	915	97.17	1013.25	100.00	1200	104.81
100	45.82	920	97.32	1015	100.05	1250	105.99
150	53.98	925	97.47	1020	100.19	1300	107.14
200	60.07	930	97.62	1025	100.32	1350	108.25
250	64.98	935	97.76	1030	100.46	1400	109.32
300	69.11	940	97.91	1035	100.60	1450	110.36
350	72.70	945	98.06	1040	100.73	1500	111.38
400	75.88	950	98.21	1045	100.87	1550	112.37
450	78.74	955	98.35	1050	101.00	1600	113.33
500	81.34	960	98.50	1055	101.14	1650	114.26
550	83.73	965	98.64	1060	101.27	1700	115.18
600	85.95	970	98.78	1065	101.40	1750	116.07
650	88.02	975	98.93	1070	101.54	1800	116.94
700	89.96	980	99.07	1075	101.67	1850	117.79
750	91.78	985	99.21	1080	101.80	1900	118.63
800	93.51	990	99.35	1085	101.93	1950	119.44
850	95.15	995	99.49	1090	102.06	2000	120.24
900	96.71	1000	99.63	1095	102.19	2050	121.02
905	96.87	1005	99.77	1100	102.32	2100	121.79
910	97.02	1010	99.91	1150	103.59	2150	122.54

MELTING POINT OF ICE AS A FUNCTION OF PRESSURE

This table gives values of the melting temperature of ice at various pressures, as calculated from the equation for the ice I - liquid water phase boundary recommended by the International Association for the Properties of Steam (IAPS). Temperatures are on the ITS-90 scale. See the Reference for information on forms of ice that exist at higher pressures. The transition points for transformations of the various forms of ice (in each case in equilibrium with liquid water) are:

ice I - ice III	209.9 MPa	−21.985°C
ice III - ice V	350.1	−16.986
ice V - ice VI	632.4	0.16
ice VI - ice VII	2216	82

REFERENCE

Wagner, W., Saul, A., and Pruss, A., *J. Phys. Chem. Ref. Data*, 23, 515, 1994.

p/MPa	t/°C	p/MPa	t/°C	p/MPa	t/°C
0.1	0.00	40	-3.15	130	-12.07
1	-0.06	50	-4.02	140	-13.22
2	-0.14	60	-4.91	150	-14.40
3	-0.21	70	-5.83	160	-15.62
4	-0.29	80	-6.79	170	-16.85
5	-0.36	90	-7.78	180	-18.11
10	-0.74	100	-8.80	190	-19.39
20	-1.52	110	-9.86	200	-20.69
30	-2.32	120	-10.95	210	-22.00

STEAM TABLES

This table gives properties of compressed water and superheated steam at selected pressures and temperatures. It was generated from the formulation approved by the International Association for the Properties of Steam in 1984. The properties tabulated are density (ρ), enthalpy (H), entropy (S), and isobaric specific heat capacity (C_p).

The reference state for these tables is the liquid at the triple point, at which the internal energy and entropy are taken as zero. A duplicate entry in the temperature column indicates a phase transition (liquid-vapor) at that temperature; property values are then given for both phases.

REFERENCE

Haar, L., Gallagher, J.S., and Kell, G.S., *NBS/NRC Steam Tables*, Hemisphere Press, New York, 1984

P MPa	T K	ρ kg/m³	H J/g	S J/g K	C_p J/g K
0.1	273.15	999.83	0.06	−0.00015	4.2282
0.1	300	996.57	112.58	0.3928	4.1831
0.1	372.78	958.66	417.51	1.3027	4.2166
0.1	372.78	0.59021	2675.1	7.3589	2.0427
0.1	373.15	0.58958	2675.9	7.3609	2.0418
0.1	400	0.54765	2730.0	7.5010	1.9973
0.1	500	0.43517	2927.9	7.9427	1.9816
0.1	600	0.36186	3128.2	8.3076	2.0272
0.1	700	0.30988	3333.8	8.6245	2.0869
0.1	800	0.27102	3545.8	8.9074	2.1526
0.1	900	0.24085	3764.4	9.1649	2.2216
0.1	1000	0.21673	3990.1	9.4026	2.2921
0.1	1100	0.19701	4222.8	9.6244	2.3621
0.1	1200	0.18058	4462.5	9.8329	2.4302
0.1	1300	0.16668	4708.8	10.030	2.4954
0.1	1400	0.15477	4961.4	10.217	2.5568
0.1	1500	0.14445	5220.0	10.396	2.6143
0.1	1600	0.13542	5484.2	10.566	2.6676
0.1	1700	0.12745	5753.4	10.729	2.7166
0.1	1800	0.12037	6027.3	10.886	2.7617
0.1	1900	0.11403	6305.6	11.036	2.8031
0.1	2000	0.10833	6587.9	11.181	2.8412
1	273.15	1000.30	0.98	−0.0001	4.2233
1	300	996.97	113.41	0.3926	4.1806
1	373.15	958.81	419.74	1.3062	4.2150
1	400	937.92	533.47	1.6005	4.2593
1	453.07	887.15	762.88	2.1388	4.4030
1	453.07	5.1445	2777.7	6.5859	2.5569
1	500	4.5348	2890.0	6.8220	2.2760
1	600	3.6877	3108.0	7.2199	2.1332
1	700	3.1307	3321.1	7.5483	2.1386
1	800	2.7266	3537.0	7.8365	2.1821
1	900	2.4175	3758.0	8.0968	2.2402
1	1000	2.1724	3985.2	8.3361	2.3046
1	1100	1.9730	4219.0	8.5588	2.3710
1	1200	1.8074	4459.4	8.7680	2.4368
1	1300	1.6677	4706.3	8.9656	2.5004
1	1400	1.5481	4959.3	9.1531	2.5608
1	1500	1.4445	5218.3	9.3317	2.6175
1	1600	1.3540	5482.7	9.5024	2.6702
1	1700	1.2742	5752.2	9.6657	2.7188
1	1800	1.2034	6026.3	9.8224	2.7636
1	1900	1.1399	6304.8	9.9729	2.8047
1	2000	1.0829	6587.2	10.1180	2.8426

P MPa	T K	ρ kg/m³	H J/g	S J/g K	C_p J/g K
10	273.15	1004.80	10.10	0.0004	4.1765
10	300	1001.00	121.66	0.3900	4.1564
10	373.15	962.98	426.52	1.2992	4.1948
10	400	942.47	539.67	1.5921	4.2360
10	500	838.17	977.04	2.5666	4.5903
10	584.18	688.63	1407.3	3.3591	6.1244
10	584.18	55.477	2724.5	5.6139	6.8973
10	600	49.830	2817.8	5.7716	5.1447
10	700	35.392	3176.1	6.3279	2.8769
10	800	29.128	3442.6	6.6844	2.5324
10	900	25.132	3690.6	6.9764	2.4463
10	1000	22.243	3934.5	7.2334	2.4395
10	1100	20.015	4179.5	7.4670	2.4654
10	1200	18.227	4428.0	7.6832	2.5059
10	1300	16.751	4680.9	7.8856	2.5529
10	1400	15.508	4938.6	8.0766	2.6018
10	1500	14.443	5201.3	8.2577	2.6503
10	1600	13.520	5468.6	8.4303	2.6969
10	1700	12.710	5740.6	8.5951	2.7410
10	1800	11.994	6016.8	8.7530	2.7823
10	1900	11.356	6296.9	8.9044	2.8206
10	2000	10.784	6580.8	9.0500	2.8563
100	273.15	1045.30	95.40	−0.0085	3.9092
100	300	1037.20	201.35	0.3614	3.9846
100	373.15	999.70	495.00	1.2371	4.0392
100	400	981.74	603.77	1.5186	4.0633
100	500	899.23	1015.3	2.4365	4.1854
100	600	791.46	1447.6	3.2237	4.5025
100	700	651.37	1925.2	3.9588	5.0954
100	800	482.17	2466.3	4.6806	5.6041
100	900	343.48	2997.0	5.3065	4.8280
100	1000	265.84	3431.3	5.7649	3.9368
100	1100	220.96	3798.6	6.1154	3.4632
100	1200	191.59	4131.4	6.4051	3.2174
100	1300	170.56	4445.8	6.6568	3.0848
100	1400	154.56	4750.3	6.8825	3.0129
100	1500	141.85	5049.5	7.0890	2.9759
100	1600	131.43	5346.2	7.2804	2.9600
100	1700	122.68	5641.9	7.4597	2.9570
100	1800	115.18	5937.8	7.6289	2.9620
100	1900	108.67	6234.5	7.7893	2.9720
100	2000	102.94	6532.4	7.9420	2.9852

PERMITTIVITY (DIELECTRIC CONSTANT) OF WATER AT VARIOUS FREQUENCIES

The permittivity of liquid water in the radiofrequency and microwave regions can be represented by the Debye equation (References 1 and 2):

$$\varepsilon' = \varepsilon_\infty + \frac{\varepsilon_s - \varepsilon_\infty}{1 + \omega^2 \tau^2}$$

$$\varepsilon'' = \frac{(\varepsilon_s - \varepsilon_\infty)\omega\tau}{1 + \omega^2 \tau^2}$$

where $\varepsilon = \varepsilon' + i\,\varepsilon''$ is the (complex) relative permittivity (i.e., the absolute permittivity divided by the permittivity of free space $\varepsilon_0 = 8.854 \cdot 10^{-12}$ F m^{-1}). Here ε_s is the static permittivity (see Reference 3 and the table "Properties of Water in the Range 0—100°C" in this Section); ε_∞ is a parameter describing the permittivity in the high frequency limit; τ is the relaxation time for molecular orientation; and $\omega = 2\pi f$ is the angular frequency. The values in this table have been calculated from parameters given in Reference 2:

	0°C	25°C	50°C
ε_∞	5.7	5.2	4.0
τ/ps	17.67	8.27	4.75

Other useful quantities that can be calculated from the values in the table are the loss tangent:

$$\tan\delta = \varepsilon''/\varepsilon'$$

and the absorption coefficient α which describes the power attentuation per unit length ($P = P_0\, e^{-\alpha l}$):

$$\alpha = \frac{\pi f\, \varepsilon''}{c\sqrt{\varepsilon'}}$$

and c is the speed of light. The last equation is valid when $\varepsilon''/\varepsilon' \ll 1$.

REFERENCES

1. Fernendez, D.P., Mulev, Y., Goodwin, A.R.H., and Levelt Sengers, J.M.H., *J. Phys. Chem. Ref. Data*, 24, 33, 1995.
2. Kaatze, U., *J. Chem. Eng. Data*, 34, 371, 1989.
3. Archer, D.G., and Wang, P., *J. Phys. Chem. Ref. Data*, 12, 817, 1983.

Frequency	0°C ε'	0°C ε''	25°C ε'	25°C ε''	50°C ε'	50°C ε''
0	87.90	0.00	78.36	0.00	69.88	0.00
1 kHz	87.90	0.00	78.36	0.00	69.88	0.00
1 MHz	87.90	0.01	78.36	0.00	69.88	0.00
10 MHz	87.90	0.09	78.36	0.04	69.88	0.02
100 MHz	87.89	0.91	78.36	0.38	69.88	0.20
200 MHz	87.86	1.82	78.35	0.76	69.88	0.39
500 MHz	87.65	4.55	78.31	1.90	69.87	0.98
1 GHz	86.90	9.01	78.16	3.79	69.82	1.96
2 GHz	84.04	17.39	77.58	7.52	69.65	3.92
3 GHz	79.69	24.64	76.62	11.13	69.36	5.85
4 GHz	74.36	30.49	75.33	14.58	68.95	7.75
5 GHz	68.54	34.88	73.73	17.81	68.45	9.62
10 GHz	42.52	40.88	62.81	29.93	64.49	18.05
20 GHz	19.56	30.78	40.37	36.55	52.57	28.99
30 GHz	12.50	22.64	26.53	33.25	40.57	32.74
40 GHz	9.67	17.62	18.95	28.58	31.17	32.43
50 GHz	8.28	14.34	14.64	24.53	24.42	30.47

THERMOPHYSICAL PROPERTIES OF FLUIDS

These tables give thermodynamic and transport properties of some important fluids, as generated from the equations of state presented in the references below. The properties tabulated are density (ρ), energy (E), enthalpy (H), entropy (S), isochoric heat capacity (C_v), isobaric heat capacity (C_p), speed of sound (v_s), viscosity (η), thermal conductivity (λ), and dielectric constant (D). All extensive properties are given on a molar basis. Not all properties are included for every substance. The references should be consulted for information on the uncertainties and the reference states for E, H, and S.

Values are given as a function of temperature for several isobars. The phase can be determined by noting the sharp decrease in density between two successive temperature entries; all lines above this point refer to the liquid phase, and all lines below refer to the gas phase. If there is no sharp discontinuity in density, all data in the table refer to the supercritical region (i.e., the isobar is above the critical pressure).

REFERENCES

1. Younglove, B.A., *Thermophysical Properties of Fluids. Part I, J. Phys. Chem. Ref. Data,* 11, Suppl. 1, 1982.
2. Younglove, B.A., and Ely, J.F., *Thermophysical Properties of Fluids. Part II, J. Phys. Chem. Ref. Data,* 16, 577, 1987.
3. McCarty, R.D., *Thermodynamic Properties of Helium, J. Phys. Chem. Ref. Data,* 2, 923, 1973.

Nitrogen (N₂)

T K	ρ mol/L	E J/mol	H J/mol	S J/mol K	C_v J/mol K	C_p J/mol K	η μPa s	λ mW/m K	D
$P = 0.1$ MPa (1 bar)									
70	30.017	−3828	−3824	73.8	28.5	57.2	203.9	143.5	1.45269
77.25	28.881	−3411	−3407	79.5	27.8	57.8	152.2	133.8	1.43386
77.25	0.163	1546	2161	151.6	21.6	31.4	5.3	7.6	1.00215
100	0.123	2041	2856	159.5	21.1	30.0	6.8	9.6	1.00162
200	0.060	4140	5800	179.9	20.8	29.2	12.9	18.4	1.00079
300	0.040	6223	8717	191.8	20.8	29.2	18.0	25.8	1.00053
400	0.030	8308	11635	200.2	20.9	29.2	22.2	32.3	1.00040
500	0.024	10414	14573	206.7	21.2	29.6	26.1	38.5	1.00032
600	0.020	12563	17554	212.2	21.8	30.1	29.5	44.5	1.00026
700	0.017	14770	20593	216.8	22.4	30.7	32.8	50.5	1.00023
800	0.015	17044	23698	221.0	23.1	31.4	35.8	56.3	1.00020
900	0.013	19383	26869	224.7	23.7	32.0	38.7	62.0	1.00017
1000	0.012	21786	30103	228.1	24.3	32.6	41.5	67.7	1.00016
1500	0.008	34530	47004	241.8	26.4	34.7	54.0	93.3	1.00010
$P = 1$ MPa									
70	30.070	−3838	−3805	73.6	28.9	56.9	205.9	144.1	1.45355
80	28.504	−3267	−3232	81.3	27.8	57.7	139.5	130.7	1.42760
90	26.721	−2685	−2648	88.2	26.7	59.4	100.1	115.3	1.39824
100	24.634	−2073	−2032	94.6	26.2	64.4	73.1	98.5	1.36417
103.75	23.727	−1828	−1786	97.1	26.2	67.8	64.8	91.8	1.34947
103.75	1.472	1788	2467	138.1	24.1	45.0	7.6	12.5	1.01954
200	0.614	4048	5675	160.3	21.0	30.4	13.2	19.3	1.00812
300	0.402	6171	8661	172.5	20.9	29.6	18.1	26.3	1.00529
400	0.300	8273	11609	180.9	20.9	29.5	22.4	32.7	1.00395
500	0.240	10389	14563	187.5	21.3	29.7	26.1	38.8	1.00315
600	0.200	12544	17554	193.0	21.8	30.2	29.6	44.8	1.00262
700	0.171	14756	20600	197.7	22.4	30.8	32.8	50.7	1.00224
800	0.150	17032	23709	201.8	23.1	31.4	35.9	56.5	1.00196
900	0.133	19374	26884	205.6	23.7	32.1	38.8	62.2	1.00174
1000	0.120	21778	30121	209.0	24.3	32.7	41.5	67.8	1.00157
1500	0.080	34527	47029	222.7	26.4	34.8	54.0	93.4	1.00104
$P = 10$ MPa									
65.32	31.120	−4176	−3855	68.6	31.8	53.8	275.7	153.8	1.47067
100	26.201	−2328	−1946	92.0	27.4	56.3	90.2	112.3	1.38942
200	7.117	3037	4442	136.4	22.7	45.5	17.6	30.4	1.09698
300	3.989	5667	8174	151.7	21.4	33.4	20.1	31.9	1.05347
400	2.898	7941	11392	161.0	21.3	31.3	23.7	36.7	1.03860
500	2.302	10148	14492	167.9	21.5	30.8	27.1	42.0	1.03055

T K	ρ mol/L	E J/mol	H J/mol	S J/mol K	C_v J/mol K	C_p J/mol K	η μPa s	λ mW/m K	D
600	1.918	12361	17575	173.5	21.9	30.9	30.4	47.4	1.02538
700	1.647	14613	20683	178.3	22.5	31.3	33.5	53.0	1.02175
800	1.445	16919	23837	182.5	23.2	31.8	36.4	58.6	1.01904
900	1.288	19283	27046	186.3	23.8	32.4	39.3	64.1	1.01694
1000	1.162	21705	30308	189.8	24.4	32.9	42.0	69.6	1.01526
1500	0.783	34504	47283	203.5	26.5	34.8	54.3	94.7	1.01020

Oxygen (O₂)

T K	ρ mol/L	E J/mol	H J/mol	S J/mol K	C_v J/mol K	C_p J/mol K	η μPa s	λ mW/m K	D
$P = 0.1$ MPa (1 bar)									
60	40.049	−5883	−5880	72.4	34.9	53.4	425.2	188.2	1.55619
80	37.204	−4814	−4812	87.7	31.0	53.6	251.7	166.1	1.51114
100	0.123	2029	2840	172.9	21.4	30.5	7.5	9.3	1.00146
120	0.102	2458	3442	178.4	21.0	29.8	9.0	11.2	1.00121
140	0.087	2881	4035	182.9	20.9	29.5	10.5	13.1	1.00103
160	0.076	3301	4624	186.9	20.9	29.4	11.9	15.0	1.00090
180	0.067	3720	5210	190.3	20.8	29.3	13.3	16.7	1.00080
200	0.060	4138	5796	193.4	20.8	29.3	14.6	18.4	1.00072
220	0.055	4556	6381	196.2	20.8	29.3	15.9	20.1	1.00065
240	0.050	4974	6966	198.8	20.9	29.3	17.2	21.7	1.00060
260	0.046	5393	7552	201.1	20.9	29.3	18.4	23.2	1.00055
280	0.043	5812	8138	203.3	21.0	29.4	19.5	24.8	1.00051
300	0.040	6234	8726	205.3	21.1	29.4	20.6	26.3	1.00048
320	0.038	6657	9316	207.2	21.2	29.5	21.7	27.8	1.00045
340	0.035	7082	9908	209.0	21.3	29.7	22.8	29.3	1.00042
360	0.033	7510	10503	210.7	21.5	29.8	23.8	30.8	1.00040
380	0.032	7941	11100	212.3	21.6	30.0	24.8	32.2	1.00038
$P = 1$ MPa									
60	40.084	−5887	−5863	72.3	34.9	53.3	428.5	188.4	1.55674
80	37.254	−4822	−4795	87.6	31.0	53.5	253.8	166.4	1.51192
100	34.153	−3741	−3712	99.7	28.5	55.2	155.6	137.9	1.46381
120	1.198	2163	2997	156.7	24.0	40.6	9.4	13.9	1.01429
140	0.950	2683	3735	162.4	22.2	34.4	10.8	14.9	1.01133
160	0.802	3151	4398	166.8	21.5	32.2	12.2	16.3	1.00955
180	0.698	3598	5030	170.5	21.2	31.2	13.5	17.7	1.00831
200	0.620	4035	5647	173.8	21.1	30.6	14.8	19.3	1.00738
220	0.559	4466	6255	176.7	21.0	30.3	16.1	20.8	1.00665
240	0.509	4894	6858	179.3	21.0	30.1	17.3	22.3	1.00606
260	0.468	5321	7458	181.7	21.0	29.9	18.5	23.8	1.00556
280	0.433	5748	8056	183.9	21.1	29.9	19.6	25.2	1.00515
300	0.403	6174	8654	186.0	21.1	29.9	20.7	26.7	1.00479
320	0.377	6602	9252	187.9	21.2	29.9	21.8	28.2	1.00448
340	0.355	7032	9851	189.7	21.4	30.0	22.8	29.6	1.00421
360	0.335	7463	10452	191.4	21.5	30.1	23.9	31.1	1.00397
380	0.317	7898	11056	193.1	21.7	30.2	24.9	32.6	1.00376
$P = 10$ MPa									
60	40.419	−5931	−5684	71.5	35.1	53.0	461.8	189.9	1.56210
80	37.727	−4893	−4628	86.7	31.6	52.7	274.4	168.6	1.51936
100	34.881	−3856	−3570	98.5	29.1	53.4	171.0	141.2	1.47500
120	31.721	−2796	−2481	108.4	27.3	55.9	113.0	115.1	1.42677
140	27.890	−1662	−1304	117.5	26.2	62.9	76.3	91.8	1.36972
160	22.379	−322	125	127.0	26.1	84.8	48.6	71.2	1.29037
180	13.232	1489	2245	139.5	26.6	105.9	26.2	46.8	1.16560
200	8.666	2681	3835	147.9	24.0	60.6	21.2	34.0	1.10650

T K	ρ mol/L	E J/mol	H J/mol	S J/mol K	C_v J/mol K	C_p J/mol K	η μPa s	λ mW/m K	D
220	6.868	3424	4880	152.9	22.6	46.4	20.5	30.8	1.08380
240	5.836	4029	5742	156.6	22.0	40.6	20.8	30.1	1.07090
260	5.134	4573	6521	159.7	21.8	37.6	21.4	30.2	1.06219
280	4.613	5086	7254	162.5	21.6	35.8	22.1	30.8	1.05575
300	4.205	5581	7959	164.9	21.6	34.7	22.9	31.6	1.05073
320	3.874	6063	8645	167.1	21.7	33.9	23.7	32.6	1.04667
340	3.598	6538	9318	169.1	21.8	33.4	24.6	33.7	1.04329
360	3.363	7009	9982	171.0	21.9	33.0	25.4	34.9	1.04043
380	3.161	7477	10641	172.8	22.0	32.8	26.3	36.1	1.03796

Hydrogen (H_2)

T K	ρ mol/L	E J/mol	H J/mol	S J/mol K	C_v J/mol K	C_p J/mol K	v_s m/s	D
P = 0.1 MPa (1 bar)								
15	37.738	−605	−603	11.2	9.7	14.4	1319	1.24827
20	35.278	−524	−521	15.8	11.3	19.1	1111	1.23093
40	0.305	491	818	75.6	12.5	21.3	521	1.00186
60	0.201	748	1244	84.3	13.1	21.6	636	1.00122
80	0.151	1030	1694	90.7	15.3	23.7	714	1.00091
100	0.120	1370	2202	96.4	18.7	27.1	773	1.00073
120	0.100	1777	2776	101.6	21.8	30.2	827	1.00061
140	0.086	2237	3401	106.4	23.8	32.2	883	1.00052
160	0.075	2723	4054	110.8	24.6	33.0	940	1.00046
180	0.067	3216	4714	114.7	24.6	32.9	998	1.00041
200	0.060	3703	5367	118.1	24.1	32.4	1054	1.00037
220	0.055	4179	6009	121.2	23.4	31.8	1110	1.00033
240	0.050	4641	6638	123.9	22.8	31.2	1163	1.00030
260	0.046	5093	7256	126.4	22.3	30.6	1214	1.00028
280	0.043	5535	7865	128.6	21.9	30.2	1263	1.00026
300	0.040	5970	8466	130.7	21.6	29.9	1310	1.00024
400	0.030	8093	11421	139.2	21.0	29.3	1518	1.00018
P = 1 MPa								
15	38.109	−609	−583	10.9	10.1	14.1	1315	1.25089
20	35.852	−532	−504	15.5	11.4	18.4	1155	1.23496
40	3.608	399	676	54.1	12.9	28.4	498	1.02209
60	2.098	697	1173	64.3	13.2	23.5	635	1.01280
80	1.523	994	1651	71.1	15.4	24.7	719	1.00928
100	1.204	1343	2174	77.0	18.8	27.7	779	1.00733
120	0.999	1756	2758	82.3	21.9	30.6	835	1.00608
140	0.854	2219	3390	87.1	23.9	32.5	891	1.00520
160	0.747	2709	4048	91.5	24.7	33.2	949	1.00454
180	0.663	3204	4712	95.4	24.6	33.1	1006	1.00404
200	0.597	3693	5368	98.9	24.1	32.5	1063	1.00363
220	0.543	4170	6012	102.0	23.5	31.9	1118	1.00330
240	0.498	4634	6643	104.7	22.9	31.2	1171	1.00303
260	0.460	5087	7263	107.2	22.3	30.7	1222	1.00279
280	0.427	5530	7873	109.5	21.9	30.3	1271	1.00259
300	0.399	5966	8475	111.5	21.6	30.0	1317	1.00242
400	0.299	8091	11433	120.1	21.0	29.4	1525	1.00182
P = 10 MPa								
20	39.669	−568	−316	13.0	10.9	15.0	1458	1.26198
40	31.344	−209	110	27.3	13.2	27.0	1171	1.20354
60	21.273	255	725	39.7	13.8	32.5	931	1.13527
80	14.830	686	1360	48.8	15.9	31.1	886	1.09303
100	11.417	1110	1986	55.8	19.3	31.9	904	1.07109

T K	ρ mol/L	E J/mol	H J/mol	S J/mol K	C_v J/mol K	C_p J/mol K	v_s m/s	D
120	9.357	1571	2640	61.8	22.4	33.5	941	1.05801
140	7.969	2068	3323	67.0	24.3	34.6	989	1.04925
160	6.963	2583	4020	71.7	25.0	34.9	1042	1.04294
180	6.195	3099	4713	75.7	24.9	34.4	1096	1.03814
200	5.588	3604	5393	79.3	24.4	33.6	1150	1.03436
220	5.094	4094	6057	82.5	23.7	32.8	1203	1.03129
240	4.683	4569	6704	85.3	23.1	32.0	1254	1.02874
260	4.336	5030	7336	87.8	22.6	31.3	1302	1.02659
280	4.038	5481	7958	90.1	22.1	30.8	1349	1.02475
300	3.780	5924	8570	92.3	21.8	30.4	1394	1.02315
400	2.869	8073	11559	100.9	21.2	29.6	1592	1.01753

Helium (He-4)

T K	ρ mol/L	E J/mol	H J/mol	S J/mol K	C_v J/mol K	C_p J/mol K	v_s m/s	η μPa s	D
P = 0.1 MPa (1 bar)									
3	35.794	−39	−36	9.8	7.6	9.4	222	3.85	1.05646
4	32.477	−27	−24	13.3	9.1	16.3	185	3.33	1.05114
5	2.935	52	86	39.1	12.7	27.1	120	1.39	1.00456
10	1.238	120	201	55.2	12.5	21.7	185	2.26	1.00192
20	0.602	247	413	69.9	12.5	21.0	264	3.58	1.00093
50	0.240	623	1039	89.0	12.5	20.8	417	6.36	1.00037
100	0.120	1247	2079	103.4	12.5	20.8	589	9.78	1.00019
200	0.060	2494	4158	117.8	12.5	20.8	833	15.14	1.00009
300	0.040	3741	6237	126.3	12.5	20.8	1020	19.93	1.00006
400	0.030	4988	8315	132.3	12.5	20.8	1177	24.29	1.00005
500	0.024	6236	10394	136.9	12.5	20.8	1316	28.36	1.00004
600	0.020	7483	12472	140.7	12.5	20.8	1441	32.22	1.00003
700	0.017	8730	14551	143.9	12.5	20.8	1557	35.89	1.00003
800	0.015	9977	16630	146.7	12.5	20.8	1664	39.43	1.00002
900	0.013	11224	18708	149.1	12.5	20.8	1765	42.85	1.00002
1000	0.012	12471	20787	151.3	12.5	20.8	1861	46.16	1.00002
1500	0.008	18707	31179	159.7	12.5	20.8	2279	61.55	1.00001
P = 1 MPa									
3	39.703	−42	−16	8.6	7.1	7.8	300	5.63	1.06274
4	38.210	−34	−7	11.2	8.3	10.9	290	5.01	1.06034
5	35.818	−22	6	14.0	9.7	15.1	269	4.38	1.05650
10	15.378	78	143	32.2	12.3	30.5	198	3.07	1.02402
20	6.067	228	393	49.8	12.6	22.9	274	3.94	1.00943
50	2.353	617	1042	69.8	12.5	21.1	428	6.53	1.00365
100	1.186	1245	2089	84.3	12.5	20.9	597	9.89	1.00184
200	0.597	2495	4170	98.7	12.5	20.8	838	15.21	1.00093
300	0.399	3742	6249	107.1	12.5	20.8	1024	19.96	1.00062
400	0.300	4990	8327	113.1	12.5	20.8	1180	24.32	1.00046
500	0.240	6237	10406	117.8	12.5	20.8	1319	28.38	1.00037
600	0.200	7485	12484	121.5	12.5	20.8	1444	32.23	1.00031
700	0.172	8732	14562	124.7	12.5	20.8	1559	35.91	1.00027
800	0.150	9979	16641	127.5	12.5	20.8	1666	39.44	1.00023
900	0.133	11227	18719	130.0	12.5	20.8	1767	42.86	1.00021
1000	0.120	12474	20798	132.2	12.5	20.8	1862	46.17	1.00019
1500	0.080	18710	31190	140.6	12.5	20.8	2280	61.55	1.00012
P = 10 MPa									
4	51.978	−24	169	6.7	6.0	7.3	586	24.27	1.08262
5	51.118	−18	177	8.5	7.9	9.3	576	18.16	1.08122
10	46.872	23	236	16.6	11.0	14.5	546	9.31	1.07432

T K	ρ mol/L	E J/mol	H J/mol	S J/mol K	C_v J/mol K	C_p J/mol K	v_s m/s	η μPa s	D
20	37.092	154	423	29.5	12.6	20.7	498	6.99	1.05854
50	19.192	572	1093	49.9	12.9	22.4	541	8.07	1.03003
100	10.525	1231	2181	65.0	12.8	21.3	674	10.93	1.01640
200	5.605	2500	4284	79.6	12.6	20.9	889	15.82	1.00871
300	3.829	3755	6367	88.0	12.6	20.8	1063	20.25	1.00595
400	2.908	5006	8445	94.0	12.6	20.8	1212	24.54	1.00452
500	2.344	6256	10522	98.6	12.5	20.8	1346	28.56	1.00364
600	1.963	7505	12599	102.4	12.5	20.8	1467	32.38	1.00305
700	1.689	8754	14676	105.6	12.5	20.8	1580	36.04	1.00262
800	1.481	10003	16753	108.4	12.5	20.8	1685	39.56	1.00230
900	1.320	11252	18830	110.9	12.5	20.8	1784	42.96	1.00205
1000	1.189	12500	20907	113.0	12.5	20.8	1877	46.26	1.00185
1500	0.797	18742	31294	121.5	12.5	20.8	2289	61.62	1.00124

Argon (Ar)

T K	ρ mol/L	E J/mol	H J/mol	S J/mol K	C_v J/mol K	C_p J/mol K	v_s m/s	η μPa s	λ mW/m K
P = 0.1 MPa (1 bar)									
85	35.243	−4811	−4808	53.6	23.1	44.7	820	278.8	132.4
90	0.138	1077	1802	129.4	13.1	22.5	174	7.5	6.0
100	0.123	1211	2024	131.8	12.9	21.9	184	8.2	6.6
120	0.102	1471	2456	135.7	12.6	21.4	203	9.8	7.8
140	0.087	1727	2881	139.0	12.6	21.1	220	11.4	9.0
160	0.076	1980	3302	141.8	12.5	21.0	235	13.0	10.2
180	0.067	2232	3722	144.3	12.5	21.0	250	14.5	11.4
200	0.060	2483	4141	146.5	12.5	20.9	263	16.0	12.5
220	0.055	2734	4559	148.5	12.5	20.9	276	17.5	13.7
240	0.050	2984	4976	150.3	12.5	20.9	289	18.9	14.8
260	0.046	3234	5394	152.0	12.5	20.9	300	20.3	15.8
280	0.043	3484	5811	153.5	12.5	20.8	312	21.6	16.9
300	0.040	3734	6227	155.0	12.5	20.8	323	22.9	17.9
320	0.038	3984	6644	156.3	12.5	20.8	333	24.2	18.9
340	0.035	4234	7060	157.6	12.5	20.8	344	25.4	19.9
360	0.033	4484	7477	158.7	12.5	20.8	354	26.6	20.8
380	0.032	4734	7893	159.9	12.5	20.8	363	27.8	21.7
P = 1 MPa									
85	35.307	−4820	−4792	53.5	23.1	44.6	823	281.3	133.0
90	34.542	−4598	−4569	56.1	21.6	44.7	808	242.7	124.2
100	32.909	−4145	−4115	60.9	19.9	46.2	753	185.0	109.2
120	1.181	1210	2057	114.3	14.7	30.1	189	10.3	9.3
140	0.945	1544	2603	118.5	13.5	25.4	212	11.8	10.1
160	0.799	1838	3089	121.8	13.0	23.6	231	13.3	11.1
180	0.697	2116	3551	124.5	12.8	22.7	247	14.8	12.1
200	0.619	2384	3999	126.9	12.7	22.2	262	16.3	13.2
220	0.559	2648	4438	128.9	12.6	21.8	275	17.7	14.2
240	0.509	2908	4873	130.8	12.6	21.6	288	19.1	15.3
260	0.468	3167	5304	132.6	12.6	21.5	301	20.4	16.3
280	0.433	3423	5732	134.2	12.6	21.4	312	21.8	17.3
300	0.403	3679	6159	135.6	12.5	21.3	324	23.1	18.3
320	0.377	3934	6583	137.0	12.5	21.2	334	24.3	19.2
340	0.355	4188	7007	138.3	12.5	21.2	345	25.5	20.2
360	0.335	4441	7429	139.5	12.5	21.1	355	26.7	21.1
380	0.317	4694	7851	140.6	12.5	21.1	365	27.9	22.0
P = 10 MPa									
90	35.208	−4694	−4410	55.0	21.9	43.2	846	265.2	129.5

T K	ρ mol/L	E J/mol	H J/mol	S J/mol K	C_v J/mol K	C_p J/mol K	v_s m/s	η μPa s	λ mW/m K
100	33.744	−4271	−3974	59.6	20.4	44.0	800	205.0	115.1
120	30.525	−3396	−3069	67.8	18.8	46.9	672	131.2	92.1
140	26.609	−2447	−2072	75.5	17.6	54.1	526	85.9	71.7
160	20.816	−1279	−799	83.9	17.4	78.6	357	51.3	52.8
180	12.296	228	1042	94.8	17.3	83.6	257	27.8	32.0
200	8.442	1118	2302	101.4	15.3	48.6	268	23.3	23.6
220	6.776	1661	3137	105.4	14.2	36.8	284	22.8	21.6
240	5.787	2087	3815	108.4	13.7	31.6	300	23.2	21.3
260	5.105	2458	4416	110.8	13.4	28.8	314	23.9	21.4
280	4.596	2798	4974	112.9	13.2	27.1	327	24.8	21.8
300	4.195	3119	5503	114.7	13.1	25.9	339	25.7	22.3
320	3.869	3427	6012	116.3	13.0	25.0	350	26.7	22.9
340	3.596	3726	6506	117.8	13.0	24.4	361	27.7	23.5
360	3.364	4017	6989	119.2	12.9	23.9	372	28.7	24.2
380	3.164	4303	7464	120.5	12.9	23.5	381	29.7	24.9

Methane (CH₄)

T K	ρ mol/L	E J/mol	H J/mol	S J/mol K	C_v J/mol K	C_p J/mol K	η μPa s	λ mW/m K	D
P = 0.1 MPa (1 bar)									
100	27.370	−5258	−5254	73.0	33.4	54.1	156.3	208.1	1.65504
125	0.099	3026	4039	156.5	25.4	34.6	5.0	13.4	1.00193
150	0.081	3667	4896	162.7	25.2	34.0	5.9	16.2	1.00159
175	0.069	4301	5743	168.0	25.2	33.8	6.9	19.1	1.00136
200	0.061	4935	6587	172.5	25.3	33.8	7.8	21.9	1.00119
225	0.054	5571	7434	176.5	25.5	34.0	8.7	24.8	1.00105
250	0.048	6216	8288	180.1	26.0	34.4	9.6	27.8	1.00095
275	0.044	6875	9156	183.4	26.6	35.0	10.4	30.9	1.00086
300	0.040	7552	10042	186.4	27.5	35.9	11.2	34.1	1.00079
325	0.037	8252	10951	189.4	28.5	36.9	12.0	37.6	1.00073
350	0.034	8979	11887	192.1	29.7	38.0	12.8	41.2	1.00068
375	0.032	9737	12853	194.8	30.9	39.3	13.5	45.1	1.00063
400	0.030	10528	13852	197.4	32.3	40.7	14.3	49.1	1.00059
425	0.028	11354	14886	199.9	33.7	42.1	15.0	53.3	1.00056
450	0.027	12215	15956	202.3	35.2	43.5	15.7	57.6	1.00053
500	0.024	14047	18204	207.1	38.0	46.4	17.0	66.5	1.00047
600	0.020	18111	23101	216.0	42.9	51.3	19.4	84.1	1.00039
P = 1 MPa									
100	27.413	−5268	−5231	72.9	33.4	54.0	158.1	208.9	1.65617
125	25.137	−3882	−3842	85.3	32.4	57.4	89.2	168.2	1.59261
150	0.969	3282	4315	140.9	27.9	45.2	6.2	18.4	1.01911
175	0.765	4041	5348	147.3	26.4	38.9	7.1	20.6	1.01507
200	0.644	4736	6289	152.3	25.9	36.8	8.0	23.1	1.01268
225	0.560	5410	7197	156.6	25.9	36.0	8.9	25.8	1.01102
250	0.497	6081	8093	160.4	26.2	35.8	9.7	28.7	1.00979
275	0.448	6758	8991	163.8	26.8	36.1	10.6	31.7	1.00882
300	0.408	7449	9901	167.0	27.6	36.7	11.4	34.9	1.00803
325	0.375	8160	10829	169.9	28.6	37.6	12.1	38.3	1.00738
350	0.347	8897	11781	172.8	29.7	38.6	12.9	41.9	1.00683
375	0.323	9662	12760	175.5	31.0	39.8	13.6	45.7	1.00636
400	0.302	10460	13770	178.1	32.4	41.1	14.4	49.6	1.00595
425	0.284	11291	14814	180.6	33.8	42.4	15.1	53.8	1.00559
450	0.268	12157	15892	183.1	35.2	43.8	15.7	58.1	1.00527
500	0.241	13997	18153	187.8	38.1	46.6	17.0	66.9	1.00474
600	0.200	18073	23070	196.8	43.0	51.4	19.5	84.5	1.00394

T K	ρ mol/L	E J/mol	H J/mol	S J/mol K	C_v J/mol K	C_p J/mol K	η μPa s	λ mW/m K	D

P = 10 MPa

T K	ρ mol/L	E J/mol	H J/mol	S J/mol K	C_v J/mol K	C_p J/mol K	η μPa s	λ mW/m K	D
100	27.815	−5362	−5003	72.0	33.8	53.2	175.4	217	1.66668
125	25.754	−4036	−3648	84.1	32.7	55.3	100.4	178.8	1.60895
150	23.441	−2655	−2229	94.4	31.4	58.6	65.7	144.6	1.54553
175	20.613	−1175	−689	103.9	30.3	65.5	44.9	113.4	1.47021
200	16.602	542	1144	113.6	30.1	84.7	29.4	85.8	1.36789
225	10.547	2680	3628	125.3	30.8	102.2	17.6	61.0	1.22352
250	7.013	4289	5714	134.1	29.3	67.4	14.3	47.6	1.14481
275	5.530	5387	7195	139.8	28.7	53.4	13.8	44.1	1.11297
300	4.685	6320	8454	144.2	28.9	48.0	13.9	44.6	1.09513
325	4.115	7192	9622	147.9	29.6	45.8	14.3	46.6	1.08322
350	3.695	8047	10753	151.3	30.5	44.9	14.7	49.2	1.07450
375	3.366	8903	11874	154.4	31.7	44.8	15.2	52.3	1.06773
400	3.101	9774	12999	157.3	32.9	45.2	15.8	55.7	1.06227
425	2.880	10666	14138	160.0	34.3	46.0	16.3	59.4	1.05775
450	2.692	11584	15298	162.7	35.7	46.9	16.9	63.3	1.05392
500	2.389	13507	17692	167.7	38.5	48.9	18.0	71.6	1.04775
600	1.963	17700	22795	177.0	43.3	52.9	20.2	88.3	1.03911

Ethane (C₂H₆)

T K	ρ mol/L	E J/mol	H J/mol	S J/mol K	C_v J/mol K	C_p J/mol K	v_s m/s	D

P = 0.1 MPa (1 bar)

T K	ρ mol/L	E J/mol	H J/mol	S J/mol K	C_v J/mol K	C_p J/mol K	v_s m/s	D
95	21.50	−14555	−14550	80.2	47.2	68.7	1970	1.93480
100	21.32	−14210	−14205	83.8	47.1	69.3	1943	1.92500
125	20.41	−12468	−12463	99.3	45.0	69.8	1775	1.87634
150	19.47	−10717	−10712	112.1	43.4	70.4	1587	1.82726
175	18.49	−8938	−8933	123.1	42.7	72.1	1396	1.77671
200	0.062	5503	7123	210.1	34.5	43.8	258	1.00208
225	0.054	6401	8238	215.4	36.5	45.5	273	1.00183
250	0.049	7349	9401	220.3	38.9	47.7	287	1.00164
275	0.044	8360	10624	224.9	41.6	50.2	300	1.00148
300	0.040	9439	11914	229.4	44.5	53.1	312	1.00136
325	0.037	10592	13278	233.8	47.6	56.1	324	1.00125
350	0.035	11823	14719	238.1	50.7	59.2	335	1.00116
375	0.032	13133	16240	242.3	54.0	62.4	345	1.00108
400	0.030	14525	17841	246.4	57.2	65.6	355	1.00101
450	0.027	17548	21282	254.5	63.6	72.0	375	1.00090
500	0.024	20883	25035	262.4	69.7	78.1	393	1.00081
600	0.020	28429	33415	277.6	80.9	89.3	428	1.00067

P = 1 MPa

T K	ρ mol/L	E J/mol	H J/mol	S J/mol K	C_v J/mol K	C_p J/mol K	v_s m/s	D
95	21.514	−14562	−14515	80.2	47.3	68.7	1972	1.93537
100	21.334	−14217	−14170	83.7	47.2	69.3	1946	1.92560
125	20.427	−12478	−12429	99.2	45.0	69.8	1778	1.87709
150	19.494	−10731	−10679	112.0	43.4	70.3	1592	1.82823
175	18.515	−8957	−8903	123.0	42.7	72.0	1402	1.77800
200	17.464	−7127	−7070	132.7	42.9	74.9	1209	1.72513
225	16.288	−5199	−5137	141.8	43.8	80.2	1008	1.66733
250	0.564	6762	8534	198.7	41.6	57.5	260	1.01909
275	0.489	7902	9949	204.1	43.2	56.2	280	1.01650
300	0.435	9063	11363	209.0	45.5	57.2	297	1.01467
325	0.393	10273	12815	213.7	48.3	59.1	311	1.01327
350	0.360	11546	14321	218.1	51.3	61.5	325	1.01214
375	0.333	12889	15893	222.5	54.4	64.2	337	1.01121
400	0.310	14306	17534	226.7	57.5	67.1	349	1.01043

THERMOPHYSICAL PROPERTIES OF FLUIDS (continued)

T K	ρ mol/L	E J/mol	H J/mol	S J/mol K	C_v J/mol K	C_p J/mol K	v_s m/s	D
450	0.272	17367	21038	234.9	63.8	73.0	370	1.00917
500	0.244	20730	24836	242.9	69.9	78.9	390	1.00819
600	0.201	28313	33278	258.3	81.0	89.8	427	1.00677
P = 10 MPa								
95	21.624	−14626	−14163	79.5	47.4	68.5	2000	1.94104
100	21.448	−14286	−13819	83.0	47.4	69.1	1974	1.93146
125	20.570	−12572	−12086	98.5	45.5	69.3	1814	1.88436
150	19.678	−10858	−10350	111.1	43.9	69.6	1637	1.83753
175	18.758	−9130	−8596	121.9	43.3	70.8	1459	1.79010
200	17.793	−7363	−6801	131.5	43.5	73.0	1284	1.74134
225	16.760	−5535	−4938	140.3	44.3	76.4	1110	1.69017
250	15.620	−3609	−2969	148.6	45.8	81.5	935	1.63488
275	14.301	−1539	−839	156.7	47.9	89.4	758	1.57249
300	12.666	757	1547	165.0	50.8	102.7	577	1.49740
325	10.398	3443	4404	174.1	54.7	129.1	399	1.39745
350	7.292	6643	8015	184.8	58.8	150.1	290	1.26832
375	5.182	9419	11349	194.1	60.0	115.7	289	1.18570
400	4.182	11577	13968	200.8	61.4	96.9	310	1.14797
450	3.204	15379	18500	211.5	65.8	87.5	347	1.11193
500	2.677	19135	22870	220.7	71.2	88.0	378	1.09288
600	2.076	27160	31978	237.3	81.8	94.7	427	1.07142

Propane (C₃H₈)

T K	ρ mol/L	E J/mol	H J/mol	S J/mol K	C_v J/mol K	C_p J/mol K	v_s m/s	D
P = 0.1 MPa (1 bar)								
90	16.526	−21486	−21426	87.3	59.2	84.5	2126	2.07988
100	16.295	−20639	−20577	96.2	59.6	85.2	2041	2.05806
125	15.726	−18495	−18432	115.4	59.2	86.5	1856	2.00674
150	15.156	−16319	−16253	131.3	58.9	88.0	1685	1.95796
175	14.577	−14096	−14028	145.0	59.5	90.3	1521	1.91036
200	13.982	−11806	−11735	157.3	61.0	93.5	1359	1.86300
225	13.339	−9395	−9387	168.5	63.4	97.9	1197	1.81487
250	0.050	9194	11213	257.6	57.2	66.8	228	1.00238
275	0.045	10691	12930	264.1	61.6	70.7	239	1.00215
300	0.041	12297	14752	270.5	66.2	75.1	249	1.00195
325	0.037	14019	16689	276.7	71.1	79.8	259	1.00179
350	0.035	15862	18744	282.8	76.0	84.6	269	1.00166
375	0.032	17827	20921	288.8	80.9	89.5	278	1.00154
400	0.030	19912	23217	294.7	85.7	94.3	286	1.00144
450	0.027	24441	28166	306.4	95.2	103.6	303	1.00128
500	0.024	29428	33573	317.7	104.1	112.6	318	1.00115
600	0.020	40677	45658	339.7	120.4	128.8	347	1.00095
P = 1 MPa								
90	16.526	−21486	−21426	87.2	59.3	84.5	2128	2.08034
100	16.295	−20639	−20577	96.2	59.7	85.2	2043	2.05856
125	15.726	−18495	−18432	115.3	59.2	86.4	1859	2.00736
150	15.156	−16319	−16253	131.2	59.0	88.0	1690	1.95873
175	14.577	−14096	−14028	144.9	59.6	90.2	1526	1.91132
200	13.982	−11806	−11735	157.2	61.1	93.4	1365	1.86421
225	13.361	−9424	−9349	168.4	63.4	97.7	1205	1.81642
250	12.696	−6919	−6840	179.0	66.4	103.3	1045	1.76672
275	11.962	−4252	−4169	189.1	70.0	110.8	881	1.71316
300	11.102	−1360	−1270	199.2	74.1	121.9	708	1.65216
325	0.428	13278	15614	255.2	74.1	89.6	233	1.02067

T K	ρ mol/L	E J/mol	H J/mol	S J/mol K	C_v J/mol K	C_p J/mol K	v_S m/s	D
350	0.383	15259	17869	261.9	78.0	91.2	248	1.01846
375	0.349	17318	20183	268.3	82.2	94.2	261	1.01678
400	0.322	19472	22582	274.4	86.7	97.8	272	1.01544
450	0.279	24092	27672	286.4	95.7	105.9	293	1.01337
500	0.248	29137	33172	298.0	104.4	114.1	312	1.01184
600	0.203	40455	45374	320.2	120.5	129.7	344	1.00968
P = 10 MPa								
90	16.590	−21553	−20951	86.5	59.9	84.4	2146	2.08489
100	16.364	−20714	−20103	95.4	60.1	85.1	2068	2.06350
125	15.810	−18595	−17962	114.5	59.6	86.1	1895	2.01342
150	15.259	−16448	−15793	130.3	59.3	87.5	1733	1.96617
175	14.705	−14261	−13581	144.0	59.9	89.5	1577	1.92048
200	14.141	−12016	−11309	156.1	61.4	92.4	1425	1.87557
225	13.562	−9692	−8955	167.2	63.7	96.1	1277	1.83076
250	12.960	−7268	−6496	177.5	66.7	100.7	1133	1.78529
275	12.322	−4721	−3909	187.4	70.2	106.4	991	1.73826
300	11.631	−2027	−1167	196.9	74.1	113.2	851	1.68849
325	10.860	843	1764	206.3	78.4	121.5	715	1.63437
350	9.973	3924	4927	215.7	82.9	132.0	582	1.57361
375	8.905	7270	8393	225.2	87.7	146.1	455	1.50271
400	7.561	10957	12279	235.3	93.0	165.7	339	1.41671
450	4.614	18845	21013	255.8	101.8	167.8	249	1.24060
500	3.241	25567	28652	272.0	107.8	142.7	276	1.16439
600	2.242	38131	42591	297.4	121.7	140.5	332	1.11122

VIRIAL COEFFICIENTS OF SELECTED GASES

Henry V. Kehiaian

This table gives second virial coefficients of about 110 inorganic and organic gases as a function of temperature. Selected data from the literature have been fitted by least squares to the equation

$$B/cm^3\,mol^{-1} = \sum_{i=1}^{n} a(i)\left[(T_o/T)-1\right]^{i-1}$$

where $T_o = 298.15$ K. The table gives the coefficients $a(i)$ and values of B at fixed temperature increments, as calculated from this smoothing equation.

The equation may be used with the tabulated coefficients for interpolation within the indicated temperature range. It should not be used for extrapolation beyond this range.

Compounds are listed in the modified Hill order (see Introduction), with carbon-containing compounds following those compounds not containing carbon.

A useful compilation of virial coefficient data from the literature may be found in:

J. H. Dymond and E. B. Smith, *The Virial Coefficients of Pure Gases and Mixtures, A Critical Compilation*, Oxford University Press, Oxford, 1980.

Compounds Not Containing Carbon

Mol. form.	Name	T/K	B/cm³ mol⁻¹
Ar	Argon	100	−184
		120	−131
		140	−98
	$a(1) = -16$	160	−76
	$a(2) = -60$	80	−60
	$a(3) = -10$	200	−48
		300	−16
		400	−1
		500	7
		600	12
		700	15
		800	18
		900	20
		1000	22
BF₃	Boron trifluoride	200	−338
		240	−202
		280	−129
	$a(1) = -106$	320	−85
	$a(2) = -330$	360	−56
	$a(3) = -251$	400	−37
	$a(4) = -80$	440	−23
ClH	Hydrogen chloride	190	−451
		230	−269
		270	−181
	$a(1) = -144$	310	−132
	$a(2) = -325$	350	−102
	$a(3) = -277$	390	−81
	$a(4) = -170$	430	−66
		470	−54
Cl₂	Chlorine	210	−508
		220	−483

Mol. form.	Name		T/K	B/cm^3 mol^{-1}
			230	−457
	$a(1) = -303$		240	−432
	$a(2) = -555$		250	−407
	$a(3) = 9$		260	−383
	$a(4) = 329$		270	−360
	$a(5) = 68$		280	−339
			290	−318
			300	−299
			350	−221
			400	−166
			450	−126
			500	−97
			600	−59
			700	−36
			800	−22
			900	−12
F$_2$	Fluorine		80	−386
			110	−171
			140	−113
	$a(1) = -25$		170	−73
	$a(2) = 21$		200	−47
	$a(3) = -185$		230	−32
	$a(4) = 113$		260	−25
F$_4$Si	Silicon tetrafluoride		210	−268
			240	−213
			270	−170
	$a(1) = -138$		300	−136
	$a(2) = -312$		330	−108
			360	−84
			390	−64
			420	−47
			450	−32
F$_5$I	Iodine pentafluoride		320	−2540
			330	−2344
			340	−2172
	$a(1) = -3077$		350	−2021
	$a(2) = -8474$		360	−1890
	$a(3) = -9116$		370	−1775
			380	−1674
			390	−1587
			400	−1510
			410	−1443
F$_5$P	Phosphorus pentafluoride		320	−162
			340	−143
			360	−127
	$a(1) = -186$		380	−112
	$a(2) = -345$		400	−98
			420	−86
			440	−75
			460	−64
F$_6$Mo	Molybdenum hexafluoride		300	−896
			310	−810
			320	−737

Mol. form.	Name	T/K	$B/cm^3\ mol^{-1}$
	a(1) = –914	330	–677
	a(2) = –2922	340	–627
	a(3) = –4778	350	–586
		360	–553
		370	–527
		380	–506
		390	–491
F_6S	Sulfur hexafluoride	200	–685
		250	–416
		300	–275
	a(1) = –279	350	–190
	a(2) = –647	400	–135
	a(3) = –335	450	–96
	a(4) = –72	500	–68
F_6U	Uranium hexafluoride	320	–1030
		340	–905
		360	–805
	a(1) = –1204	380	–724
	a(2) = –2690	400	–658
	a(3) = –2144	420	–604
		440	–560
F_6W	Tungsten hexafluoride	320	–641
		340	–578
		360	–523
	a(1) = –719	380	–473
	a(2) = –1143	400	–428
		420	–387
		440	–350
		460	–317
H_2	Hydrogen	15	–230
		20	–151
		25	–108
	a(1) = 15.4	30	–82
	a(2) = –9.0	35	–64
	a(3) = –0.2	40	–52
		45	–42
		50	–35
		60	–24
		70	–16
		80	–11
		90	–7
		100	–3
		200	11
		300	15
		400	18
H_2O	Water	300	–1126
		320	–850
		340	–660
	a(1) = –1158	360	–526
	a(2) = –5157	380	–428
	a(3) = –10301	400	–356
	a(4) = –10597	420	–301
	a(5) = –4415	440	–258

Mol. form.	Name		T/K	B/cm^3 mol^{-1}
			460	–224
			480	–197
			500	–175
			600	–104
			700	–67
			800	–44
			900	–30
			1000	–20
			1100	–14
			1200	–11
H_3N	Ammonia		290	–302
			300	–265
			310	–236
	$a(1) = -271$		320	–213
	$a(2) = -1022$		330	–194
	$a(3) = -2715$		340	–179
	$a(4) = -4189$		350	–166
			360	–154
			370	–144
			380	–135
			400	–118
			420	–101
H_3P	Phosphine		190	–457
			200	–404
			210	–364
	$a(1) = -146$		220	–332
	$a(2) = -733$		230	–305
	$a(3) = 1022$		240	–281
	$a(4) = -1220$		250	–258
			260	–235
			270	–213
			280	–190
			290	–166
He	Helium		2	–172
			6	–48
			10	–24
	$a(1) = 12$		14	–13
	$a(2) = -1$		18	–7
			22	–3
			26	–1
			30	1
			50	6
			70	8
			90	10
			110	10
			150	11
			250	12
			650	13
			700	13
Kr	Krypton		110	–363
			120	–307
			130	–263
	$a(1) = -51$		140	–229
	$a(2) = -118$		150	–201

Mol. form.	Name	T/K	B/cm³ mol⁻¹

Mol. form.	Name	T/K	$B/\text{cm}^3\ \text{mol}^{-1}$
	a(3) = −29	160	−178
	a(4) = −5	170	−159
		180	−143
		190	−129
		200	−117
		250	−75
		300	−51
		400	−23
		500	−8
		600	2
		700	8
NO	Nitric oxide	120	−232
		130	−176
		140	−138
	a(1) = −12	150	−113
	a(2) = −119	160	−96
	a(3) = 89	170	−83
	a(4) = −73	180	−73
		190	−65
		200	−58
		210	−52
		230	−42
		250	−32
		270	−24
N_2	Nitrogen	75	−274
		100	−161
		125	−104
	a(1) = −4	150	−71
	a(2) = −56	175	−49
	a(3) = −12	200	−34
		225	−24
		250	−15
		300	−4
		400	9
		500	16
		600	21
		700	24
N_2O	Nitrous oxide	240	−219
		260	−181
		280	−151
	a(1) = −130	300	−128
	a(2) = −307	320	−110
	a(3) = −248	340	−96
		360	−85
		380	−76
		400	−68
Ne	Neon	60	−25
		80	−13
		100	−6
	a(1) = 10.8	120	−1
	a(2) = −7.5	140	2
	a(3) = 0.4	160	4
		180	6
		200	7

Mol. form.	Name	T/K	$B/cm^3\ mol^{-1}$
		300	11
		400	13
		500	14
		600	15
O_2	Oxygen	90	−241
		110	−161
		130	−117
	a(1) = −16	150	−88
	a(2) = −62	170	−69
	a(3) = −8	190	−55
	a(4) = −3	210	−44
		230	−36
		250	−29
		270	−23
		290	−18
		310	−14
		330	−10
		350	−7
		400	−1
O_2S	Sulfur dioxide	290	−465
		320	−354
		350	−276
	a(1) = −430	380	−221
	a(2) = −1193	410	−181
	a(3) = −1029	440	−153
		470	−132
Xe	Xenon	160	−421
		170	−377
		180	−340
	a(1) = −130	190	−307
	a(2) = −262	200	−280
	a(3) = −87	210	−255
		220	−234
		230	−215
		240	−199
		250	−184
		300	−129
		350	−93
		400	−69
		500	−39
		600	−21
		650	−14

Compounds Containing Carbon

Mol. form.	Name	T/K	$B/cm^3\ mol^{-1}$
$CClF_3$	Chlorotrifluoromethane	240	−369
		290	−237
		340	−165
	a(1) = −223	390	−119
	a(2) = −504	440	−86
	a(3) = −340	490	−60
	a(4) = −291	540	−39

Mol. form.	Name		T/K	$B/cm^3\ mol^{-1}$
CCl_2F_2	Dichlorodifluoromethane		250	−769
			280	−570
			310	−441
	a(1) = −486		340	−353
	a(2) = −1217		370	−289
	a(3) = −1188		400	−241
	a(4) = −698		430	−204
			460	−174
CCl_3F	Trichlorofluoromethane		240	−1140
			280	−879
			320	−689
	a(1) = −786		360	−545
	a(2) = −1428		400	−431
	a(3) = −142		440	−340
			480	−265
CCl_4	Tetrachloromethane		320	−1345
			340	−1171
			360	−1040
	a(1) = −1600		380	−942
	a(2) = −4059		400	−868
	a(3) = −4653		420	−814
CF_4	Tetrafluoromethane		250	−137
			300	−87
			350	−55
	a(1) = −88		400	−32
	a(2) = −238		450	−16
	a(3) = −70		500	−4
			600	14
			700	25
			800	33
$CHClF_2$	Chlorodifluoromethane		300	−343
			325	−298
			350	−257
	a(1) = −347		375	−221
	a(2) = −575		400	−188
	a(3) = 187		425	−158
$CHCl_2F$	Dichlorofluoromethane		250	−728
			275	−634
			300	−557
	a(1) = −562		325	−491
	a(2) = −862		350	−434
			375	−385
			400	−343
			425	−305
			450	−271
$CHCl_3$	Trichloromethane		320	−1001
			330	−926
			340	−858
	a(1) = −1193		350	−797
	a(2) = −2936		360	−740
	a(3) = −1751		370	−689
			380	−642

Mol. form.	Name	T/K	$B/cm^3\ mol^{-1}$
		390	−599
		400	−559
CHF$_3$	Trifluoromethane	200	−433
		220	−350
		240	−288
	a(1) = −177	260	−241
	a(2) = −399	280	−204
	a(3) = −250	300	−174
		320	−151
		340	−132
		360	−116
		380	−103
		400	−91
CH$_2$Cl$_2$	Dichloromethane	320	−706
		330	−634
		340	−574
	a(1) = −913	350	−524
	a(2) = −3371	360	−482
	a(3) = −5013	370	−447
		380	−420
		400	−380
		420	−357
CH$_2$F$_2$	Difluoromethane	280	−375
		290	−343
		300	−316
	a(1) = −321	310	−294
	a(2) = −754	320	−275
	a(3) = −1300	330	−260
		340	−248
		350	−238
CH$_3$Br	Bromomethane	280	−645
		290	−596
		300	−551
	a(1) = −559	310	−509
	a(2) = −1324	320	−469
		340	−396
		360	−332
		380	−274
CH$_3$Cl	Chloromethane	280	−466
		300	−402
		320	−348
	a(1) = −407	340	−304
	a(2) = −887	360	−266
	a(3) = −385	380	−234
		400	−206
		420	−182
		440	−161
		460	−142
		480	−126
		500	−112
		600	−58
CH$_3$F	Fluoromethane	280	−244

Mol. form.	Name		T/K	B/cm^3 mol^{-1}
			300	−205
			320	−174
	a(1) = −209		340	−150
	a(2) = −525		360	−129
	a(3) = −365		380	−112
			400	−99
			420	−87
CH$_3$I	Iodomethane		310	−725
			320	−646
			330	−582
	a(1) = −844		340	−531
	a(2) = −3353		350	−492
	a(3) = −6590		360	−462
			370	−441
			380	−427
CH$_4$	Methane		110	−328
			120	−276
			130	−237
	a(1) = −43		140	−206
	a(2) = −114		150	−181
	a(3) = −19		160	−160
	a(4) = −7		170	−143
			180	−128
			190	−116
			200	−105
			250	−66
			300	−43
			350	−27
			400	−16
			500	0
			600	10
CH$_4$O	Methanol		320	−1431
			330	−1299
			340	−1174
	a(1) = −1752		350	−1056
	a(2) = −4694		360	−945
			370	−840
			380	−741
			390	−646
			400	−557
CH$_5$N	Methylamine		300	−451
			325	−367
			350	−304
	a(1) = −459		375	−257
	a(2) = −1191		400	−220
	a(3) = −995		425	−192
			450	−170
			500	−140
			550	−122
CO	Carbon monoxide		210	−36
			240	−24
			270	−15
	a(1) = −9		300	−8

Mol. form.	Name	T/K	$B/cm^3\,mol^{-1}$
	$a(2) = -58$	330	−3
	$a(3) = -18$	360	1
		420	7
		480	11
CO_2	Carbon dioxide	220	−244
		240	−204
		260	−172
	$a(1) = -127$	280	−146
	$a(2) = -288$	300	−126
	$a(3) = -118$	320	−108
		340	−94
		360	−81
		380	−71
		400	−62
		500	−30
		600	−13
		700	−1
		800	7
		900	12
		1000	16
		1100	19
CS_2	Carbon disulfide	280	−932
		310	−740
		340	−603
	$a(1) = -807$	370	−504
	$a(2) = -1829$	400	−431
	$a(3) = -1371$	430	−375
$C_2Cl_2F_4$	1,2-Dichloro-1,1,2,2-tetrafluoroethane	300	−801
		320	−695
		340	−608
	$a(1) = -812$	360	−536
	$a(2) = -1773$	380	−475
	$a(3) = -963$	400	−423
		420	−379
		440	−341
		460	−307
		480	−279
		500	−253
$C_2Cl_3F_3$	1,1,2-Trichloro-1,2,2-trifluoroethane	290	−1041
		310	−943
		330	−856
	$a(1) = -999$	350	−780
	$a(2) = -1479$	370	−712
		390	−651
		410	−596
		430	−546
		450	−500
C_2H_2	Ethyne	200	−573
		210	−500
		220	−440
	$a(1) = -216$	230	−390
	$a(2) = -375$	240	−349
	$a(3) = -716$	250	−315

Mol. form.	Name		T/K	B/cm^3 mol^{-1}
			260	−287
			270	−263
C$_2$H$_3$N	Ethanenitrile		330	−3468
			340	−2971
			350	−2563
	a(1) = −5840		360	−2233
	a(2) = −29175		370	−1970
	a(3) = −47611		380	−1765
			390	−1610
			400	−1499
			410	−1425
C$_2$H$_4$	Ethene		240	−218
			270	−172
			300	−139
	a(1) = −140		330	−113
	a(2) = −296		360	−92
	a(3) = −101		390	−76
			420	−63
			450	−52
C$_2$H$_4$Cl$_2$	1,2-Dichloroethane		370	−812
			390	−716
			410	−635
	a(1) = −1362		430	−566
	a(2) = −3240		450	−508
	a(3) = −2100		470	−458
			490	−416
			510	−379
			530	−347
			550	−319
			570	−295
C$_2$H$_4$O	Ethanal		290	−1352
			320	−927
			350	−654
	a(1) = −1217		380	−482
	a(2) = −4647		410	−375
	a(3) = −5725		440	−314
			470	−283
C$_2$H$_4$O$_2$	Methyl methanoate		320	−821
			330	−744
			340	−677
	a(1) = −1035		350	−620
	a(2) = −3425		360	−571
	a(3) = −4203		370	−528
			380	−492
			390	−461
			400	−435
C$_2$H$_5$Cl	Chloroethane		320	−634
			360	−450
			400	−330
	a(1) = −777		440	−249
	a(2) = −2205		480	−195
	a(3) = −1764		520	−157

Mol. form.	Name	T/K	$B/cm^3\ mol^{-1}$
		560	−131
		600	−114
C_2H_6	Ethane	200	−409
		220	−337
		240	−284
	a(1) = −184	260	−242
	a(2) = −376	280	−209
	a(3) = −143	300	−181
	a(4) = −54	320	−159
		340	−140
		360	−123
		380	−109
		400	−96
		500	−52
		600	−24
C_2H_6O	Ethanol	320	−2710
		330	−2135
		340	−1676
	a(1) = −4475	350	−1317
	a(2) = −29719	360	−1043
	a(3) = −56716	370	−843
		380	−705
		390	−622
C_2H_6O	Dimethyl ether	275	−536
		280	−517
		285	−499
	a(1) = −455	290	−482
	a(2) = −965	295	−465
		300	−449
		305	−433
		310	−418
C_2H_7N	Dimethylamine	310	−606
		320	−563
		330	−523
	a(1) = −662	340	−487
	a(2) = −1504	350	−454
	a(3) = −667	360	−423
		370	−395
		380	−369
		390	−345
		400	−322
C_2H_7N	Ethylamine	300	−773
		310	−710
		320	−654
	a(1) = −785	330	−604
	a(2) = −2012	340	−558
	a(3) = −1397	350	−517
		360	−480
		370	−447
		380	−416
		390	−389
		400	−363
C_3H_6	Cyclopropane	300	−383

Mol. form.	Name		T/K	B/cm³ mol⁻¹

Mol. form.	Name		T/K	$B/\text{cm}^3\,\text{mol}^{-1}$
			310	−356
			320	−332
	a(1) = −388		330	−310
	a(2) = −861		340	−290
	a(3) = −538		350	−272
			360	−256
			370	−241
			380	−227
			390	−215
			400	−204
C_3H_6	Propene		280	−395
			300	−342
			320	−299
	a(1) = −347		340	−262
	a(2) = −727		360	−232
	a(3) = −325		380	−205
			400	−183
			420	−163
			440	−146
			460	−131
			480	−118
			500	−106
C_3H_6O	2-Propanone		300	−1996
			320	−1522
			340	−1198
	a(1) = −2051		360	−971
	a(2) = −8903		380	−806
	a(3) = −18056		400	−683
	a(4) = −16448		420	−586
			440	−506
			460	−437
			480	−375
C_3H_6O	Ethyl methanoate		330	−1003
			340	−916
			350	−839
	a(1) = −1371		360	−771
	a(2) = −4231		370	−712
	a(3) = −4312		380	−660
			390	−614
C_3H_6O	Methyl ethanoate		320	−1320
			330	−1186
			340	−1074
	a(1) = −1709		350	−980
	a(2) = −6348		360	−903
	a(3) = −9650		370	−840
			380	−789
			390	−749
C_3H_7Cl	1-Chloropropane		310	−1001
			340	−772
			370	−614
	a(1) = −1121		400	−501
	a(2) = −3271		430	−417
	a(3) = −3786		460	−352
	a(4) = −1974		490	−302

Mol. form.	Name		T/K	B/cm^3 mol^{-1}
			520	−261
			550	−227
			580	−198
C_3H_8	Propane		240	−641
			260	−527
			280	−444
	$a(1) = -386$		300	−381
	$a(2) = -844$		320	−331
	$a(3) = -720$		340	−292
	$a(4) = -574$		360	−259
			380	−232
			400	−208
			440	−169
			480	−138
			520	−112
			560	−90
C_3H_8O	1-Propanol		380	−873
			385	−826
			390	−783
	$a(1) = -2690$		395	−744
	$a(2) = -12040$		400	−709
	$a(3) = -16738$		405	−679
			410	−651
			415	−627
			420	−606
C_3H_8O	2-Propanol		380	−821
			385	−766
			390	−717
	$a(1) = -3165$		395	−674
	$a(2) = -16092$		400	−636
	$a(3) = -24197$		405	−604
			410	−576
			415	−552
			420	−533
C_3H_9N	Trimethylamine		310	−675
			320	−628
			330	−585
	$a(1) = -737$		340	−547
	$a(2) = -1669$		350	−512
	$a(3) = -986$		360	−480
			370	−450
C_4H_8	1-Butene		300	−624
			320	−539
			340	−470
	$a(1) = -633$		360	−413
	$a(2) = -1442$		380	−366
	$a(3) = -932$		400	−327
			420	−294
C_4H_8O	2-Butanone		310	−2056
			320	−1878
			330	−1712
	$a(1) = -2282$		340	−1555

Mol. form.	Name		T/K	$B/\text{cm}^3 \text{ mol}^{-1}$
	a(2) = −5907		350	−1407
			360	−1267
			370	−1135
$C_4H_8O_2$	Propyl methanoate		330	−1496
			340	−1354
			350	−1231
	a(1) = −2118		360	−1126
	a(2) = −7299		370	−1035
	a(3) = −8851		380	−957
			390	−890
			400	−834
$C_4H_8O_2$	Ethyl ethanoate		330	−1543
			340	−1385
			350	−1254
	a(1) = −2272		360	−1144
	a(2) = −8818		370	−1055
	a(3) = −13130		380	−982
			390	−923
			400	−878
$C_4H_8O_2$	Methyl propanoate		330	−1588
			340	−1444
			350	−1319
	a(1) = −2216		360	−1211
	a(2) = −7339		370	−1117
	a(3) = −8658		380	−1037
			390	−968
			400	−908
C_4H_9Cl	1-Chlorobutane		330	−1224
			370	−898
			410	−691
	a(1) = −1643		450	−551
	a(2) = −4897		490	−449
	a(3) = −6178		530	−371
	a(4) = −3718		570	−309
C_4H_{10}	Butane		250	−1170
			280	−863
			310	−668
	a(1) = −735		340	−536
	a(2) = −1835		370	−442
	a(3) = −1922		400	−371
	a(4) = −1330		430	−315
			460	−270
			490	−232
			520	−199
			550	−171
C_4H_{10}	2-Methylpropane		270	−900
			300	−697
			330	−553
	a(1) = −707		360	−450
	a(2) = −1719		390	−374
	a(3) = −1282		420	−317
			450	−273

Mol. form.	Name	T/K	$B/cm^3\ mol^{-1}$
		480	−240
		510	−215
$C_4H_{10}O$	1-Butanol	350	−1693
		360	−1544
		370	−1402
	$a(1) = −2629$	380	−1268
	$a(2) = −6315$	390	−1141
		400	−1021
		420	−796
		440	−593
$C_4H_{10}O$	2-Methyl-1-propanol	390	−1076
		400	−979
		410	−887
	$a(1) = −2269$	420	−800
	$a(2) = −5065$	430	−716
		440	−636
$C_4H_{10}O$	2-Butanol	380	−1110
		390	−1005
		400	−906
	$a(1) = −2232$	410	−811
	$a(2) = −5209$	420	−721
$C_4H_{10}O$	2-Methyl-2-propanol	380	−924
		390	−827
		400	−736
	$a(1) = −1952$	410	−649
	$a(2) = −4775$	420	−567
$C_4H_{10}O$	Diethyl ether	280	−1550
		300	−1199
		320	−954
	$a(1) = −1226$	340	−776
	$a(2) = −4458$	360	−638
	$a(3) = −7746$	380	−525
	$a(4) = −10005$	400	−428
		420	−340
$C_4H_{11}N$	Diethylamine	320	−1228
		330	−1134
		340	−1056
	$a(1) = −1522$	350	−988
	$a(2) = −5204$	360	−926
	$a(3) = −15047$	370	−868
	$a(4) = −28835$	380	−812
		390	−755
		400	−697
C_5H_5N	Pyridine	350	−1257
		360	−1176
		370	−1099
	$a(1) = −1765$	380	−1026
	$a(2) = −3431$	390	−957
		400	−892
		420	−770
		440	−659

Mol. form.	Name	T/K	B/cm³ mol⁻¹

Mol. form.	Name	T/K	$B/\text{cm}^3\,\text{mol}^{-1}$
C_5H_{10}	Cyclopentane	300	−1049
		305	−1015
		310	−981
	a(1) = −1062	315	−949
	a(2) = −2116	320	−918
C_5H_{10}	1-Pentene	310	−966
		320	−898
		330	−836
	a(1) = −1055	340	−780
	a(2) = −2377	350	−729
	a(3) = −1189	360	−681
		370	−638
		380	−598
		390	−561
		400	−527
		410	−495
$C_5H_{10}O$	2-Pentanone	330	−2850
		340	−2420
		350	−2076
	a(1) = −4962	360	−1804
	a(2) = −26372	370	−1595
	a(3) = −46537	380	−1440
		390	−1332
C_5H_{12}	Pentane	300	−1234
		310	−1130
		320	−1038
	a(1) = −1254	330	−957
	a(2) = −3345	340	−884
	a(3) = −2726	350	−818
		400	−579
		450	−436
		500	−348
		550	−294
C_5H_{12}	2-Methylbutane	280	−1263
		290	−1166
		300	−1079
	a(1) = −1095	310	−1001
	a(2) = −2503	320	−931
	a(3) = −1534	330	−867
		340	−810
		350	−757
		400	−557
		450	−424
C_5H_{12}	2,2-Dimethylpropane	300	−916
		310	−843
		320	−780
	a(1) = −931	330	−724
	a(2) = −2387	340	−674
	a(3) = −2641	350	−629
	a(4) = −1810	360	−590
		370	−554
		380	−521
		390	−492

Mol. form.	Name		T/K	$B/\text{cm}^3\ \text{mol}^{-1}$
			400	−464
			450	−357
			500	−279
			550	−218
C_6H_6	Benzene		290	−1588
			300	−1454
			310	−1335
		$a(1) = -1477$	320	−1231
		$a(2) = -3851$	330	−1139
		$a(3) = -3683$	340	−1056
		$a(4) = -1423$	350	−983
			400	−712
			450	−542
			500	−429
			550	−349
			600	−291
C_6H_7N	2-Methylpyridine		360	−1656
			370	−1523
			380	−1404
		$a(1) = -2940$	390	−1297
		$a(2) = -8813$	400	−1202
		$a(3) = -7809$	410	−1117
			420	−1040
			430	−972
C_6H_7N	3-Methylpyridine		380	−1819
			390	−1612
			400	−1448
		$a(1) = -6304$	410	−1322
		$a(2) = -30415$	420	−1230
		$a(3) = -44549$	430	−1166
C_6H_7N	4-Methylpyridine		380	−1787
			390	−1578
			400	−1417
		$a(1) = -6553$	410	−1297
		$a(2) = -32873$	420	−1214
		$a(3) = -49874$	430	−1163
C_6H_{12}	Cyclohexane		300	−1698
			320	−1391
			340	−1170
		$a(1) = -1733$	360	−1007
		$a(2) = -5618$	380	−883
		$a(3) = -9486$	400	−786
		$a(4) = -7936$	420	−707
			440	−641
			460	−584
			480	−534
			500	−488
			520	−446
			540	−406
			560	−368
C_6H_{12}	Methylcyclopentane		305	−1447
			315	−1357

Mol. form.	Name		T/K	$B/cm^3 \, mol^{-1}$
		a(1) = −1512	325	−1272
		a(2) = −2910	335	−1192
			345	−1117
C_6H_{14}	Hexane		300	−1920
			310	−1724
			320	−1561
		a(1) = −1961	330	−1424
		a(2) = −6691	340	−1309
		a(3) = −13167	350	−1209
		a(4) = −15273	360	−1123
			370	−1046
			380	−978
			390	−916
			400	−859
			410	−806
			430	−707
			450	−616
$C_6H_{15}N$	Triethylamine		330	−1562
			340	−1444
			350	−1340
		a(1) = −2061	360	−1249
		a(2) = −5735	370	−1169
		a(3) = −5899	380	−1099
			390	−1037
			400	−983
C_7H_8	Toluene		350	−1641
			360	−1511
			370	−1394
		a(1) = −2620	380	−1289
		a(2) = −7548	390	−1195
		a(3) = −6349	400	−1110
			410	−1034
			420	−965
			430	−903
C_7H_{14}	1-Heptene		340	−1781
			350	−1651
			360	−1532
		a(1) = −2491	370	−1424
		a(2) = −6230	380	−1324
		a(3) = −3780	390	−1233
			400	−1150
			410	−1073
C_7H_{16}	Heptane		300	−2782
			320	−2297
			340	−1928
		a(1) = −2834	360	−1641
		a(2) = −8523	380	−1415
		a(3) = −10068	400	−1233
		a(4) = −5051	420	−1085
			440	−963
			460	−862
			480	−775
			500	−702

Mol. form.	Name		T/K	$B/cm^3\ mol^{-1}$
			540	−583
			580	−490
			620	−416
			660	−355
			700	−304
C_8H_{10}	1,2-Dimethylbenzene		380	−2046
			390	−1848
			400	−1681
	$a(1) = -5632$		410	−1543
	$a(2) = -22873$		420	−1428
	$a(3) = -28900$		430	−1335
			440	−1261
C_8H_{10}	1,3-Dimethylbenzene		380	−2082
			390	−1865
			400	−1679
	$a(1) = -5808$		410	−1521
	$a(2) = -23244$		420	−1388
	$a(3) = -27607$		430	−1276
			440	−1184
C_8H_{10}	1,4-Dimethylbenzene		380	−2043
			390	−1851
			400	−1680
	$a(1) = -4921$		410	−1529
	$a(2) = -16843$		420	−1395
	$a(3) = -16159$		430	−1276
			440	−1171
C_8H_{16}	1-Octene		360	−2147
			370	−2000
			380	−1861
	$a(1) = -3273$		390	−1729
	$a(2) = -6557$		400	−1604
			410	−1485
C_8H_{18}	Octane		300	−4042
			350	−2511
			400	−1704
	$a(1) = -4123$		450	−1234
	$a(2) = -13120$		500	−936
	$a(3) = -16408$		550	−732
	$a(4) = -8580$		600	−583
			650	−468
			700	−375

VAN DER WAALS CONSTANTS FOR GASES

The van der Waals equation of state for a real gas is

$$(P + n^2a/V^2)(V - nb) = nRT$$

where P is the pressure, V the volume, T the temperature, n the amount of substance (in moles), and R the gas constant. The van der Waals constants a and b are characteristic of the substance and are independent of temperature. They are related to the critical temperature and pressure, T_c and P_c, by

$$a = 27R^2T_c^2/64P_c \qquad b = RT_c/8P_c$$

This table gives values of a and b for some common gases. Most of the values have been calculated from the critical temperature and pressure values given in the table "Critical Constants" in this section. Van der Waals constants for other gases may easily be calculated from the data in that table.

To convert the van der Waals constants to SI units, note that 1 bar L²/mol² = 0.1 Pa m⁶/mol² and 1 L/mol = 0.001 m³/mol.

REFERENCE

Reid, R.C, Prausnitz, J. M., and Poling, B.E., *The Properties of Gases and Liquids, Fourth Edition*, McGraw-Hill, New York, 1987.

Substance	a bar L²/mol²	b L/mol	Substance	a bar L²/mol²	b L/mol
Acetic acid	17.71	0.1065	Hydrogen sulfide	4.544	0.0434
Acetone	16.02	0.1124	Isobutane	13.32	0.1164
Acetylene	4.516	0.0522	Krypton	5.193	0.0106
Ammonia	4.225	0.0371	Methane	2.303	0.0431
Aniline	29.14	0.1486	Methanol	9.476	0.0659
Argon	1.355	0.0320	Methylamine	7.106	0.0588
Benzene	18.82	0.1193	Neon	0.208	0.0167
Bromine	9.75	0.0591	Neopentane	17.17	0.1411
Butane	13.89	0.1164	Nitric oxide	1.46	0.0289
1-Butanol	20.94	0.1326	Nitrogen	1.370	0.0387
2-Butanone	19.97	0.1326	Nitrogen dioxide	5.36	0.0443
Carbon dioxide	3.658	0.0429	Nitrogen trifluoride	3.58	0.0545
Carbon disulfide	11.25	0.0726	Nitrous oxide	3.852	0.0444
Carbon monoxide	1.472	0.0395	Octane	37.88	0.2374
Chlorine	6.343	0.0542	1-Octanol	44.71	0.2442
Chlorobenzene	25.80	0.1454	Oxygen	1.382	0.0319
Chloroethane	11.66	0.0903	Ozone	3.570	0.0487
Chloromethane	7.566	0.0648	Pentane	19.09	0.1449
Cyclohexane	21.92	0.1411	1-Pentanol	25.88	0.1568
Cyclopropane	8.34	0.0747	Phenol	22.93	0.1177
Decane	52.74	0.3043	Propane	9.39	0.0905
1-Decanol	59.51	0.3086	1-Propanol	16.26	0.1079
Diethyl ether	17.46	0.1333	2-Propanol	15.82	0.1109
Dimethyl ether	8.690	0.0774	Propene	8.442	0.0824
Dodecane	69.38	0.3758	Pyridine	19.77	0.1137
1-Dodecanol	75.70	0.3750	Pyrrole	18.82	0.1049
Ethane	5.580	0.0651	Silane	4.38	0.0579
Ethanol	12.56	0.0871	Sulfur dioxide	6.865	0.0568
Ethylene	4.612	0.0582	Sulfur hexafluoride	7.857	0.0879
Fluorine	1.171	0.0290	Tetrachloromethane	20.01	0.1281
Furan	12.74	0.0926	Tetrachlorosilane	20.96	0.1470
Helium	0.0346	0.0238	Tetrafluoroethylene	6.954	0.0809
Heptane	31.06	0.2049	Tetrafluoromethane	4.040	0.0633
1-Heptanol	38.17	0.2150	Tetrafluorosilane	5.259	0.0724
Hexane	24.84	0.1744	Tetrahydrofuran	16.39	0.1082
1-Hexanol	31.79	0.1856	Thiophene	17.21	0.1058
Hydrazine	8.46	0.0462	Toluene	24.86	0.1497
Hydrogen	0.2452	0.0265	1,1,1-Trichloroethane	20.15	0.1317
Hydrogen bromide	4.500	0.0442	Trichloromethane	15.34	0.1019
Hydrogen chloride	3.700	0.0406	Trifluoromethane	5.378	0.0640
Hydrogen cyanide	11.29	0.0881	Trimethylamine	13.37	0.1101
Hydrogen fluoride	9.565	0.0739	Water	5.537	0.0305
Hydrogen iodide	6.309	0.0530	Xenon	4.192	0.0516

MEAN FREE PATH AND RELATED PROPERTIES OF GASES

In the simplest version of the kinetic theory of gases, molecules are treated as hard spheres of diameter d which make binary collisions only. In this approximation the mean distance traveled by a molecule between successive collisions, the mean free path l, is related to the collision diameter by:

$$l = \frac{kT}{\pi\sqrt{2}Pd^2}$$

where P is the pressure, T the absolute temperature, and k the Boltzmann constant. At standard conditions ($P = 100\,000$ Pa and $T = 298.15$ K) this relation becomes:

$$l = \frac{9.27 \cdot 10^{-27}}{d^2}$$

where l and d are in meters.

Using the same model and the same standard pressure, the collision diameter can be calculated from the viscosity η by the kinetic theory relation:

$$\eta = \frac{2.67 \cdot 10^{-20}(MT)^{1/2}}{d^2}$$

where η is in units of μPa s and M is the molar mass in g/mol. Kinetic theory also gives a relation for the mean velocity $\bar{v}$ of molecules of mass m:

$$\bar{v} = \left(\frac{8kT}{\pi m}\right)^{1/2} = 145.5(T/M)^{1/2} \text{ m/s}$$

Finally, the mean time τ between collisions can be calculated from the relation $\tau\bar{v} = l$.

The table below gives values of l, $\bar{v}$, and τ for some common gases at 25°C and atmospheric pressure, as well as the value of d, all calculated from measured gas viscosities (see References 2 and 3 and the table "Viscosity of Gases" in this section). It is seen from the above equations that the mean free path varies directly with T and inversely with P, while the mean velocity varies as the square root of T and, in this approximation, is independent of P.

A more accurate model, in which molecular interactions are described by a Lennard-Jones potential, gives mean free path values about 5% lower than this table (see Reference 4).

REFERENCES

1. Reid, R.C., Prausnitz, J.M., and Poling, B.E., *The Properties of Gases and Liquids, Fourth Edition*, McGraw-Hill, New York, 1987.
2. Lide, D.R., and Kehiaian, H.V., *CRC Handbook of Thermophysical and Thermochemical Data*, CRC Press, Boca Raton, FL, 1994.
3. Vargaftik, N.B., *Tables of Thermophysical Properties of Liquids and Gases, Second Edition*, John Wiley, New York, 1975.
4. Kaye, G.W.C., and Laby, T.H., *Tables of Physical and Chemical Constants, 15th Edition*, Longman, London, 1986.

Gas	d	l	$\bar{v}$	τ
Air	$3.66 \cdot 10^{-10}$ m	$6.91 \cdot 10^{-8}$ m	467 m/s	148 ps
Ar	3.58	7.22	397	182
CO_2	4.53	4.51	379	119
H_2	2.71	12.6	1769	71
He	2.15	20.0	1256	159
Kr	4.08	5.58	274	203
N_2	3.70	6.76	475	142
NH_3	4.32	4.97	609	82
Ne	2.54	14.3	559	256
O_2	3.55	7.36	444	166
Xe	4.78	4.05	219	185

INFLUENCE OF PRESSURE ON FREEZING POINTS

This table illustrates the variation of the freezing point of representative types of liquids with pressure. Substances are listed in alphabetical order. Note that 1 MPa = 0.01 kbar = 9.87 atm.

REFERENCES

1. Isaacs, N.S., *Liquid Phase High Pressure Chemistry*, John Wiley, New York, 1981.
2. Merrill, L., *J. Phys. Chem. Ref. Data*, 6, 1205, 1977; 11, 1005, 1982.

Substance	Molecular formula	Freezing point in °C at: 0.1 MPa	100 MPa	1000 MPa
Acetic acid	$C_2H_4O_2$	16.6	37	
Acetophenone	C_8H_8O	20.0	41.2	
Aniline	C_6H_7N	−6.0	13.5	140
Benzene	C_6H_6	5.5	33.4	
Benzonitrile	C_7H_5N	−12.8	7.6	
Benzyl alcohol	C_7H_8O	−15.2	0.2	
Bromobenzene	C_6H_5Br	−30.6	−12	108
Bromoethane	C_2H_5Br	−118.6	−108	
1-Bromonaphthalene	$C_{10}H_7Br$	−1.8	6.1	
1-Bromopropane	C_3H_7Br	−110	−98	
p-Bromotoluene	C_7H_7Br	28.0	56.7	
Butanoic acid	$C_4H_8O_2$	−5.7	13.8	
1-Butanol	$C_4H_{10}O$	−89.8	−77.2	
Carbon disulfide	CS_2	−111.5	−98	
Chlorobenzene	C_6H_5Cl	−45.2	−28	84
p-Chlorotoluene	C_7H_7Cl	6.9	33.1	
o-Cresol	C_7H_8O	29.8	47.7	
m-Cresol	C_7H_8O	11.8	25.6	
p-Cresol	C_7H_8O	35.8	56.2	
Cyclohexane	C_6H_{12}	6.6	32.5	
Cyclohexanol	$C_6H_{12}O$	25.5	62.3	
1,2-Dibromoethane	$C_2H_4Br_2$	9.9	34.0	
p-Dichlorobenzene	$C_6H_4Cl_2$	52.7	79.1	
Dichloromethane	CH_2Cl_2	−95.1	−83	
N,N-Dimethylaniline	$C_8H_{11}N$	2.5	26.3	
1,4-Dioxane	$C_4H_8O_2$	11	23	
Ethanol	C_2H_6O	−114.1	−108	
Formamide	CH_3NO	−15.5	10.8	
Formic acid	CH_2O_2	8.3	20.6	
Furan	C_4H_4O	−85.6	−73	
Hexamethyldisiloxane	$C_6H_{18}OSi_2$	−66	−37	
Menthol	$C_{10}H_{20}O$	42	60	
Methyl benzoate	$C_8H_8O_2$	−15	31.8	
2-Methyl-2-butanol	$C_5H_{12}O$	−8.8	13.4	
2-Methyl-2-propanol	$C_4H_{10}O$	25.4	58.1	
Naphthalene	$C_{10}H_8$	78.2	115.7	
Nitrobenzene	$C_6H_5NO_2$	5.7	13.5	
m-Nitrotoluene	$C_7H_7NO_2$	15.5	40.6	
Pentachloroethane	C_2HCl_5	−29.0	−6.3	
Potassium	K	63.7	78	170
Potassium chloride	ClK	771		945
Propanoic acid	$C_3H_6O_2$	−20.7	−1.2	
Silver chloride	AgCl	455		545
Sodium	Na	97.8	106	167
Sodium chloride	ClNa	800.7		997
Sodium fluoride	FNa	996		1115
Tetrachloromethane	CCl_4	−23.0	14.2	
Tribromomethane	$CHBr_3$	8.1	31.5	
Trichloromethane	$CHCl_3$	−63.6	−45.2	
Water	H_2O	0.0	−9.0	
o-Xylene	C_8H_{10}	−25.2	−3.5	
m-Xylene	C_8H_{10}	−47.8	−25.2	
p-Xylene	C_8H_{10}	13.2	46.0	

CRITICAL CONSTANTS

Douglas Ambrose

This table gives the liquid - gas critical constants and normal boiling points of about 700 inorganic and organic substances. The properties tabulated are:

T_b: Normal boiling point in K at a pressure of 101.325 kPa (1 atmosphere); an "s" following the value indicates a sublimation point (temperature at which the solid is in equilibrium with the gas at a pressure of 101.325 kPa)

T_c: Critical temperature in K

P_c: Critical pressure in MPa

V_c: Critical molar volume in cm^3/mol

The number of digits given for T_b, T_c, and P_c indicates the estimated accuracy of these quantities; however, values of T_c greater than 750 K may be in error by 10 K or more. Although most V_c values are given to three figures, they cannot be assumed accurate to better than a few percent.

Many of the critical constants in this table are taken from reviews produced by the IUPAC Commission on Thermodynamics (References 2-6) and from the publications of the Design Institute for Physical Property Data (References 8-9). Compounds are listed by molecular formula in modified Hill order, with compounds not containing carbon preceding those that do contain carbon.

An asterisk following an entry indicates the value was obtained by extrapolation.

REFERENCES

1. Ambrose, D., "Vapor-Liquid Constants of Fluids", in Stevenson, R. M., and Malanowski, S., *Handbook of the Thermodynamics of Organic Compounds*, Elsevier, New York, 1987.
2. Ambrose, D., and Tsonopoulos, C., *J. Chem. Eng. Data*, 40, 531, 1995.
3. Tsonopoulos, C., and Ambrose, D., *J. Chem. Eng. Data*, 40, 547, 1995.
4. Gude, M., and Teja, A. S., *J. Chem. Eng. Data*, 40, 1025, 1995.
5. Daubert, T. E., *J. Chem. Eng. Data*, 41, 365, 1966.
6. Tsonopoulos, C., and Ambrose, D., *J. Chem. Eng. Data*, 41, 645, 1996.
7. Das, A., Frenkel, M., Gadalla, N. A. M., Kudchadker, S., Marsh, K. N., Rodgers, A. S., and Wilhoit, R. C., *J. Phys. Chem. Ref. Data*, 22, 659, 1993.
8. Wilson, L. C., Wilson, H. L., Wilding, W. V., and Wilson, G. M., *J. Chem. Eng. Data*, 41, 1252, 1997.
9. Daubert, T. E., Danner, R. P., Sibul, H. M., and Stebbins, C. C., *Physical and Thermodynamic Properties of Pure Compounds: Data Compilation*, extant 1996 (core with 6 supplements), Taylor & Francis, Bristol, PA.
10. Reid, R. C., Prausnitz, J. M., and Poling, B. E., *The Properties of Gases and Liquids, Fourth Edition*, McGraw-Hill, New York, 1987.
11. Vargaftik, N.B., *Tables of Thermophysical Properties of Liquids and Gases, Second Edition*, John Wiley, New York, 1975.

Mol. Form.	Name	T_b/K	T_c/K	P_c/MPa	V_c/cm^3 mol^{-1}
AlBr$_3$	Aluminum tribromide	528	763	2.89	310
AlCl$_3$	Aluminum trichloride	453 s	620	2.63	257
AlI$_3$	Aluminum triiodide	655	983		408
Ar	Argon	87.30	150.87	4.898	75
As	Arsenic	876 s	1673		35
AsCl$_3$	Arsenic(III) chloride	403	654		252
AsH$_3$	Arsine	210.7	373.1		
BBr$_3$	Boron tribromide	364	581		272
BCl$_3$	Boron trichloride	285.80	455	3.87	239
BF$_3$	Boron trifluoride	172	260.8	4.98	115
BI$_3$	Boron triiodide	483	773		356
B$_2$H$_6$	Diborane	180.7	289.8	4.05	
BiBr$_3$	Bismuth tribromide	726	1220		301
BiCl$_3$	Bismuth trichloride	720	1179	12.0	261
BrH	Hydrogen bromide	206.77	363.2	8.55	
BrI	Iodine bromide	389	719		139
Br$_2$	Bromine	332.0	588	10.34	127
Br$_2$Hg	Mercury(II) bromide	595	1012		
Br$_3$Ga	Gallium(III) bromide	552	806.7		303
Br$_3$HSi	Tribromosilane	382	610.0		305
Br$_3$P	Phosphorus(III) bromide	446.10	711		300
Br$_3$Sb	Antimony(III) bromide	553	904		300
Br$_4$Ge	Germanium(IV) bromide	459.50	718		392
Br$_4$Hf	Hafnium(IV) bromide	596 s	746		415

Mol. Form.	Name	T_b/K	T_c/K	P_c/MPa	V_c/cm^3 mol^{-1}
Br$_4$Si	Tetrabromosilane	427	663		382
Br$_4$Sn	Tin(IV) bromide	478	744		417
Br$_4$Ti	Titanium(IV) bromide	503	795.7		391
Br$_4$Zr	Zirconium(IV) bromide	633 s	805		424
Br$_5$Ta	Tantalum(V) bromide	622	974		461
ClFO$_3$	Perchloryl fluoride	226.40	368.4	5.37	161
ClF$_2$N	Nitrogen chloride difluoride	207	337.5	5.15	
ClF$_2$P	Phosphorus(III) chloride difluoride	225.90	362.4	4.52	
ClF$_2$PS	Phosphorus(V) sulfide chloride difluoride	279.5	439.2	4.14	
ClF$_3$Si	Chlorotrifluorosilane	203.1	307.7	3.46	
ClF$_5$	Chlorine pentafluoride	260.0	416	5.27	233
ClF$_5$S	Sulfur chloride pentafluoride	254.10	390.9		
ClH	Hydrogen chloride	188	324.7	8.31	81
ClH$_4$N	Ammonium chloride	611 s	1155	163.5	
ClH$_4$P	Phosphonium chloride	246 s	322.3	7.37	
ClNO	Nitrosyl chloride	267.7	440		
ClOV	Vanadyl chloride	400	636		171
Cl$_2$	Chlorine	239.11	416.9	7.991	123
Cl$_2$FP	Phosphorus(III) dichloride fluoride	287	463.0	4.96	
Cl$_2$F$_2$Si	Dichlorodifluorosilane	241	369.0	3.5	
Cl$_2$Hg	Mercury(II) chloride	577	973		174
Cl$_2$OSe	Selenium oxychloride	450	730	7.09	235
Cl$_3$FSi	Trichlorofluorosilane	285.40	438.6	3.58	
Cl$_3$Ga	Gallium(III) chloride	474	694		263
Cl$_3$HSi	Trichlorosilane	306	479		268
Cl$_3$P	Phosphorus(III) chloride	349.10	563		264
Cl$_3$Sb	Antimony(III) chloride	493.5	794		272
Cl$_4$Ge	Germanium(IV) chloride	359.70	553.2	3.861	330
Cl$_4$Hf	Hafnium(IV) chloride	590 s	725.7	5.42	314
Cl$_4$ORe	Rhenium(VI) oxytetrachloride	496	781		362
Cl$_4$OW	Tungsten(VI) oxytetrachloride	500.70	782		338
Cl$_4$Si	Tetrachlorosilane	330.80	508.1	3.593	326
Cl$_4$Sn	Tin(IV) chloride	387.30	591.9	3.75	351
Cl$_4$Te	Tellurium tetrachloride	660	1002	8.56	310
Cl$_4$Ti	Titanium(IV) chloride	409.60	638	4.66	339
Cl$_4$Zr	Zirconium(IV) chloride	604 s	778	5.77	319
Cl$_5$Mo	Molybdenum(V) chloride	541	850		369
Cl$_5$Nb	Niobium(V) chloride	527.2	803.5	4.88	397
Cl$_5$P	Phosphorus(V) chloride	433 s	646		
Cl$_5$Ta	Tantalum(V) chloride	512.50	767		402
Cl$_6$W	Tungsten(VI) chloride	619.90	923		422
Cs	Cesium	944	1938	9.4	341
FH	Hydrogen fluoride	293	461	6.48	69
FNO$_2$	Nitryl fluoride	200.7	349.5		
F$_2$	Fluorine	85.03	144.13	5.172	66
F$_2$HN	Difluoramine	250	403		
F$_2$N$_2$	cis-Difluorodiazine	167.40	272	7.09	
F$_2$N$_2$	trans-Difluorodiazine	161.70	260	5.57	
F$_2$O	Fluorine monoxide	128.40	215		
F$_2$Xe	Xenon difluoride	387.50 s	631	9.32	148
F$_3$N	Nitrogen trifluoride	144.40	234.0	4.46	126
F$_3$NO	Trifluoramine oxide	185.6	303	6.43	147
F$_3$P	Phosphorus(III) fluoride	171.6	271.2	4.33	
F$_3$PS	Phosphorus(V) sulfide trifluoride	220.90	346.0	3.82	
F$_4$N$_2$	Tetrafluorohydrazine	199	309	3.75	
F$_4$S	Sulfur tetrafluoride	232.70	364		
F$_4$Si	Tetrafluorosilane	187	259.0	3.72	
F$_4$Xe	Xenon tetrafluoride	388.90 s	612	7.04	188
F$_5$Nb	Niobium(V) fluoride	502	737	6.28	155

CRITICAL CONSTANTS (continued)

Mol. Form.	Name	T_b/K	T_c/K	P_c/MPa	V_c/cm^3 mol^{-1}
F_6Mo	Molybdenum(VI) fluoride	307.2	473	4.75	226
F_6S	Sulfur hexafluoride	209.4 s	318.69	3.77	199
F_6Se	Selenium hexafluoride	226.6 s	345.5		
F_6Te	Tellurium hexafluoride	234.3 s	356		
F_6U	Uranium(VI) fluoride	329.7 s	505.8	4.66	250
F_6W	Tungsten(VI) fluoride	290.3	444	4.34	233
GaI_3	Gallium(III) iodide	613	951	395	
GeH_4	Germane	185.0	312.2	4.95	147
GeI_4	Germanium(IV) iodide	650	973		500
HI	Hydrogen iodide	237.60	424.0	8.31	
H_2	Hydrogen	20.28	32.97	1.293	65
H_2O	Water	373.15	647.14	22.06	56
H_2O_2	Hydrogen peroxide	423.4	728*	22*	
H_2S	Hydrogen sulfide	213.60	373.2	8.94	99
H_2Se	Hydrogen selenide	231.90	411	8.92	
H_3N	Ammonia	239.82	405.5	11.35	72
H_3P	Phosphine	185.40	324.5	6.54	
H_4N_2	Hydrazine	386.70	653	14.7	
He	Helium	4.22	5.19	0.227	57
HfI_4	Hafnium iodide	667 s	916		528
Hg	Mercury	629.88	1750	172.00	43
HgI_2	Mercury(II) iodide	627	1072		
I_2	Iodine	457.6	819		155
I_3Sb	Antimony(III) iodide	674	1102		
I_4Si	Tetraiodosilane	560.50	944		558
I_4Sn	Tin(IV) iodide	637.50	968		531
I_4Ti	Titanium(IV) iodide	650	1040		505
I_4Zr	Zirconium(IV) iodide	704 s	960		530
K	Potassium	1032	2223*	16*	209*
Kr	Krypton	119.93	209.41	5.50	91
Li	Lithium	1615	3223*	67*	66*
NO	Nitric oxide	121.41	180	6.48	58
N_2	Nitrogen	77.36	126.21	3.39	90
N_2O	Nitrous oxide	184.67	309.57	7.255	97
N_2O_4	Nitrogen tetroxide	294.30	431	10.1	167
Na	Sodium	1156	2573*	35*	116*
Ne	Neon	27.07	44.4	2.76	42
O_2	Oxygen	90.20	154.59	5.043	73
O_2S	Sulfur dioxide	263.10	430.8	7.884	122
O_3	Ozone	161.80	261.1	5.57	89
O_3S	Sulfur trioxide	318	491.0	8.2	127
O_4Os	Osmium(VIII) oxide	408	678		
O_7Re_2	Rhenium(VII) oxide	633	942		334
P	Phosphorus	553.7	994		
Rb	Rubidium	961	2093*	16*	247*
Rn	Radon	211.5	377	6.28	
S	Sulfur	717.75	1314	20.7	
Se	Selenium	958	1766	27.2	
Xe	Xenon	165.04	289.73	5.84	118
$CBrClF_2$	Bromochlorodifluoromethane	269.5	426.88	4.254	246
$CBrF_3$	Bromotrifluoromethane	215.3	340.2	3.97	196
CBr_2F_2	Dibromodifluoromethane	298	471.3		
$CClF_3$	Chlorotrifluoromethane	191.7	302	3.870	180
CCl_2F_2	Dichlorodifluoromethane	243.3	384.95	4.136	217
CCl_2O	Carbonyl chloride	281	455	5.67	190
CCl_3F	Trichlorofluoromethane	296.8	471.1	4.47	247
CCl_4	Tetrachloromethane	350.0	556.6	4.516	276
CF_4	Tetrafluoromethane	145.1	227.6	3.74	140
$CHClF_2$	Chlorodifluoromethane	232.5	369.5	5.035	164

Mol. Form.	Name	T_b/K	T_c/K	P_c/MPa	V_c/cm^3 mol^{-1}
CHCl$_2$F	Dichlorofluoromethane	282.0	451.58	5.18	196
CHCl$_3$	Trichloromethane	334.32	536.4	5.47	239
CHF$_3$	Trifluoromethane	191.0	299.3	4.858	133
CHN	Hydrogen cyanide	299	456.7	5.39	139
CH$_2$Cl$_2$	Dichloromethane	313	510	6.10	
CH$_2$F$_2$	Difluoromethane	221.6	351.6	5.830	121
CH$_2$O$_2$	Formic acid	374	588		
CH$_3$Cl	Chloromethane	249.06	416.25	6.679	139
CH$_3$Cl$_3$Si	Methyltrichlorosilane	338.8	517	3.28	348
CH$_3$F	Fluoromethane	194.7	317.8	5.88	113
CH$_3$I	Iodomethane	315.70	528		
CH$_3$NO$_2$	Nitromethane	374.34	588	5.87	173
CH$_4$	Methane	111.8	190.56	4.599	98.60
CH$_4$O	Methanol	337.8	512.5	8.084	117
CH$_4$S	Methanethiol	279.0	470.0	7.23	145
CH$_5$ClSi	Chloromethylsilane	280	517.8		
CH$_5$N	Methylamine	266.83	430.7	7.614	
CH$_6$N$_2$	Methylhydrazine	360.7	567	8.24	271
CH$_6$Si	Methylsilane	215.7	352.5		
CO	Carbon monoxide	81.6	132.91	3.499	93
COS	Carbon oxysulfide	223	375	5.88	137
CO$_2$	Carbon dioxide	194.8 s	304.13	7.375	94
CS$_2$	Carbon disulfide	319	552	7.90	173
C$_2$Br$_2$ClF$_3$	1,2-Dibromo-1-chloro-1,2,2-trifluoroethane	366	560.7	3.61	368
C$_2$Br$_2$F$_4$	1,2-Dibromotetrafluoroethane	320.50	487.8	3.393	341
C$_2$ClF$_3$	Chlorotrifluoroethylene	245.3	379	4.05	212
C$_2$ClF$_5$	Chloropentafluoroethane	235.20	353.2	3.229	252
C$_2$Cl$_2$F$_4$	1,1-Dichlorotetrafluoroethane	277	418.6	3.30	294
C$_2$Cl$_2$F$_4$	1,2-Dichlorotetrafluoroethane	277.0	418.78	3.252	297
C$_2$Cl$_3$F$_3$	1,1,2-Trichloro-1,2,2-trifluoroethane	320.8	487.3	3.42	325
C$_2$Cl$_4$	Tetrachloroethylene	394.5	620.2		
C$_2$Cl$_4$F$_2$	1,1,2,2-Tetrachloro-1,2-difluoroethane	366.0	551		
C$_2$F$_3$N	Trifluoroacetonitrile	204.3	311.11	3.618	202
C$_2$F$_4$	Tetrafluoroethylene	197.2	306.5	3.94	172
C$_2$F$_6$	Hexafluoroethane	195.0	293		222
C$_2$HClF$_2$	1-Chloro-2,2-difluoroethylene	254.7	400.6	4.46	197
C$_2$HClF$_4$	1-Chloro-1,1,2,2-tetrafluoroethane	263	399.9	3.72	244
C$_2$HClF$_4$	1-Chloro-1,2,2,2-tetrafluoroethane	261	395.4	3.66	
C$_2$HCl$_2$F$_3$	1,2-Dichloro-1,1,2-trifluoroethane	301	461.6		278
C$_2$HCl$_2$F$_3$	2,2-Dichloro-1,1,1-trifluoroethane	300.97	456.83	3.661	278
C$_2$HCl$_3$	Trichloroethylene	360.36	544.2	5.02	
C$_2$HF$_3$O$_2$	Trifluoroacetic acid	346	491.3	3.258	204
C$_2$HF$_5$	Pentafluoroethane	224.7	339.41	3.639	209.7
C$_2$HF$_5$O	Trifluoromethyl difluoromethyl ether	237	354.49	3.33	
C$_2$H$_2$	Acetylene	188.4 s	308.3	6.138	112.2
C$_2$H$_2$Cl$_2$	cis-1,2-Dichloroethylene	333.3	544.2		
C$_2$H$_2$Cl$_2$	trans-1,2-Dichloroethylene	321.8	516.5	5.51	
C$_2$H$_2$Cl$_4$	1,1,2,2-Tetrachloroethane	419.7	661.15		
C$_2$H$_2$F$_2$	1,1-Difluoroethylene	187.5	302.9	4.46	154
C$_2$H$_2$F$_4$	1,1,1,2-Tetrafluoroethane	247.07	374.18	4.065	198
C$_2$H$_2$F$_4$	1,1,2,2-Tetrafluoroethane	253.2	391.8		191
C$_2$H$_3$ClF$_2$	1-Chloro-1,1-difluoroethane	263.5	410.34	4.048	225
C$_2$H$_3$Cl$_2$F	1,1-Dichloro-1-fluoroethane	305.2	477.5	4.194	255
C$_2$H$_3$Cl$_3$	1,1,1-Trichloroethane	347.24	545	4.30	
C$_2$H$_3$F	Fluoroethylene	201	327.9	5.24	144
C$_2$H$_3$F$_3$	1,1,1-Trifluoroethane	225.90	346.04	3.775	194
C$_2$H$_3$F$_3$O	Methyl trifluoromethyl ether	249	378.0	3.680	228
C$_2$H$_3$N	Acetonitrile	354.80	545.5	4.85	173
C$_2$H$_4$	Ethylene	169.5	282.34	5.041	131.1

Mol. Form.	Name	T_b/K	T_c/K	P_c/MPa	V_c/cm^3 mol^{-1}
$C_2H_4Br_2$	1,2-Dibromoethane	404.8	583.0	7.2	
$C_2H_4Cl_2$	1,1-Dichloroethane	330.5	523	5.07	236
$C_2H_4Cl_2$	1,2-Dichloroethane	356.7	561	5.4	225
$C_2H_4F_2$	1,1-Difluoroethane	248.20	386.7	4.50	181
C_2H_4O	Acetaldehyde	293.3	466		154
C_2H_4O	Ethylene oxide	283.8	469	7.19	140
$C_2H_4O_2$	Acetic acid	391.0	592.71	5.786	171
$C_2H_4O_2$	Methyl formate	304.8	487.2	5.998	172
C_2H_5Br	Bromoethane	311.7	503.9	6.23	215
C_2H_5Cl	Chloroethane	285.5	460.4	5.3	
C_2H_5F	Fluoroethane	235.6	375.31	5.028	
C_2H_6	Ethane	184.5	305.32	4.872	145.5
$C_2H_6Cl_2Si$	Dichlorodimethylsilane	343.5	520.4	3.49	350
C_2H_6O	Ethanol	351.44	514.0	6.137	168
C_2H_6O	Dimethyl ether	248.3	400.0	5.37	190
$C_2H_6O_2$	Ethylene glycol	470.5	718		
C_2H_6S	Ethanethiol	308.3	499	5.49	207
C_2H_6S	Dimethyl sulfide	310.48	503.0	5.53	201
C_2H_7N	Ethylamine	289.7	456	5.62	182
C_2H_7N	Dimethylamine	280.03	437.22	5.340	
$C_2H_8N_2$	1,2-Ethanediamine	390	613.1	6.707	
C_2N_2	Cyanogen	252.1	400	5.98	
C_3ClF_5O	Chloropentafluoroacetone	281	410.6	2.878	
C_3F_6O	Perfluoroacetone	245.7	357.14	2.84	329
C_3F_6O	Perfluorooxetane	244	361.8	3.094	272
C_3F_8	Perfluoropropane	236.6	345.1	2.680	299
$C_3F_8O_2$	Perfluorodimethoxymethane	263	372.3	2.333	363
C_3HF_7O	Trifluoromethyl 1,1,2,2-tetrafluoroethyl ether	270	387.7	2.640	337
$C_3H_2F_6$	1,1,1,2,3,3-Hexafluoropropane	279.3	412.38	3.412	269
$C_3H_2F_6O$	1,2,2,2-Tetrafluoroethyl difluoromethyl ether	296.50	428.95	3.050	315
$C_3H_3F_3$	3,3,3-Trifluoropropene	256	376.2	3.80	211
$C_3H_3F_5$	1,1,1,2,2-Pentafluoropropane	255.7	380.11	3.137	273
$C_3H_3F_5O$	Perfluoroethyl methyl ether	278.74	406.80	2.887	301
C_3H_3N	Acrylonitrile	350.5	540	4.660	
C_3H_3NO	Isoxazole	368	552.0		
C_3H_4	Allene	238.7	394	5.25	
C_3H_4	Propyne	250.0	402.4	5.63	163.5
C_3H_5Cl	3-Chloropropene	318.3	514		
$C_3H_5F_3O$	2,2,2-Trifluoroethyl methyl ether	304.77	448.98	3.513	277
C_3H_5N	Propanenitrile	370.29	561.3	4.26	229
C_3H_6	Propene	225.46	364.9	4.60	184.6
C_3H_6	Cyclopropane	240.34	398.0	5.54	162
C_3H_6O	Allyl alcohol	370.2	545.1		
C_3H_6O	Propanal	321	504.4	5.27	204
C_3H_6O	Acetone	329.20	508.1	4.700	209
C_3H_6O	Methyloxirane	308	482.2	4.92	186
$C_3H_6O_2$	Propanoic acid	414.30	604	4.53	222
$C_3H_6O_2$	Ethyl formate	327.5	508.5	4.74	229
$C_3H_6O_2$	Methyl acetate	330.02	506.55	4.75	228
C_3H_7Cl	1-Chloropropane	319.7	503	4.58	
C_3H_7NO	N,N-Dimethylformamide	426	649.6		262
C_3H_8	Propane	231.1	369.83	4.248	200
C_3H_8O	1-Propanol	370.3	536.8	5.169	218
C_3H_8O	2-Propanol	355.5	508.3	4.764	222
C_3H_8O	Ethyl methyl ether	280.5	437.8	4.40	221
$C_3H_8O_2$	Ethylene glycol monomethyl ether	397.3	597.6	5.285	263
C_3H_8S	1-Propanethiol	341.0	536.6		286
C_3H_8S	Ethyl methyl sulfide	339.8	533	4.26	
$C_3H_9BO_3$	Trimethyl borate	340.7	501.7	3.59	

Mol. Form.	Name	T_b/K	T_c/K	P_c/MPa	V_c/cm^3 mol^{-1}
C_3H_9ClSi	Trimethylchlorosilane	333	497.8	3.20	366
C_3H_9N	Propylamine	320.37	497.0	4.72	
C_3H_9N	Isopropylamine	304.91	471.8	4.54	221
C_3H_9N	Trimethylamine	276.02	432.79	4.087	254
$C_4Br_2F_8$	1,4-Dibromooctafluorobutane	370	532.5	2.39	
C_4F_8	Perfluorocyclobutane	267.3	388.46	2.784	324
C_4F_{10}	Perfluorobutane	271.3	386.4	2.323	378
C_4F_{10}	Perfluoroisobutane	273	395.4		
$C_4H_2F_8O$	Perfluoroethyl 2,2,2-trifluoroethyl ether	301.04	421.68	2.330	409
$C_4H_3F_7O$	Perfluoropropyl methyl ether	307.38	437.70	2.481	377
$C_4H_3F_7O$	Perfluoroisopropyl methyl ether	302.56	433.30	2.553	369
C_4H_4O	Furan	304.7	490.2	5.50	218
C_4H_4S	Thiophene	357.2	579.4	5.69	219
$C_4H_5F_5O$	Perfluoroethyl ethyl ether	301.26	431.23	2.533	366
C_4H_5N	Pyrrole	402.94	639.7	6.34	200
C_4H_6	1,3-Butadiene	268.74	425	4.32	221
C_4H_6	1-Butyne	281.23	440	4.60	208
C_4H_6	2-Butyne	300.0	488.7		
$C_4H_6O_2$	γ-Butyrolactone	477	731.0	5.131	
$C_4H_6O_3$	Acetic anhydride	412.7	606	4.0	
C_4H_7N	Butanenitrile	390.8	585.4	3.88	
C_4H_8	1-Butene	266.89	419.5	4.02	240.8
C_4H_8	cis-2-Butene	276.86	435.5	4.21	233.8
C_4H_8	trans-2-Butene	274.03	428.6	4.10	237.7
C_4H_8	Isobutene	266.3	417.9	4.000	238.8
C_4H_8	Cyclobutane	285.8	460.0	4.98	210
C_4H_8O	Ethyl vinyl ether	308.7	475	4.07	
C_4H_8O	Butanal	348.0	537.2	4.32	258
C_4H_8O	2-Butanone	352.74	536.78	4.207	267
C_4H_8O	Tetrahydrofuran	338	540.1	5.19	224
C_4H_8OS	Ethyl thioacetate	389.5	590.55	4.075	319
$C_4H_8O_2$	Butanoic acid	436.90	624	4.03	290
$C_4H_8O_2$	2-Methylpropanoic acid	427.60	605	3.7	292
$C_4H_8O_2$	Propyl formate	354.0	538.0	4.06	285
$C_4H_8O_2$	Ethyl acetate	350.26	523.3	3.882	286
$C_4H_8O_2$	Methyl propanoate	353.0	530.6	4.004	282
$C_4H_8O_2$	1,4-Dioxane	374.7	587	5.21	238
C_4H_8S	Tetrahydrothiophene	394.2	632.0		
C_4H_9N	Pyrrolidine	359.71	568	6.00	238
C_4H_{10}	Butane	272.7	425.12	3.796	255
C_4H_{10}	Isobutane	261.42	407.8	3.640	259
$C_4H_{10}O$	1-Butanol	390.88	563.0	4.414	274
$C_4H_{10}O$	2-Butanol	372.66	536.2	4.202	269
$C_4H_{10}O$	2-Methyl-1-propanol	381.04	547.8	4.295	274
$C_4H_{10}O$	2-Methyl-2-propanol	355.5	506.2	3.972	275
$C_4H_{10}O$	Diethyl ether	307.7	466.74	3.638	280
$C_4H_{10}O$	Methyl propyl ether	312.3	476.25	3.801	
$C_4H_{10}O$	Isopropyl methyl ether	303.92	464.48	3.762	
$C_4H_{10}O_2$	1,2-Butanediol	463.7	680	5.21	303
$C_4H_{10}O_2$	1,3-Butanediol	480.7	676	4.02	305
$C_4H_{10}O_2$	1,4-Butanediol	508	727	6.22	
$C_4H_{10}O_2$	Ethylene glycol dimethyl ether	358	536	3.87	271
$C_4H_{10}O_2$	Propylene glycol monomethyl ether	392	579.8	4.113	
$C_4H_{10}S$	1-Butanethiol	371.7	570.1		324
$C_4H_{10}S$	Diethyl sulfide	365.3	557	3.96	318
$C_4H_{11}N$	Butylamine	350.15	531.9	4.25	277
$C_4H_{11}N$	sec-Butylamine	335.88	514.3	4.20	278
$C_4H_{11}N$	tert-Butylamine	317.19	483.9	3.84	292
$C_4H_{11}N$	Isobutylamine	340.90	519	4.07	278

Mol. Form.	Name	T_b/K	T_c/K	P_c/MPa	V_c/cm^3 mol^{-1}
$C_4H_{11}N$	Diethylamine	328.7	499.99	3.758	
$C_4H_{12}Si$	Tetramethylsilane	299.8	448.64	2.821	362
$C_4H_{12}Sn$	Tetramethylstannane	351	521.8	2.981	
C_5F_{12}	Perfluoropentane	302.3	420.59	2.045	473
$C_5H_2F_6O_2$	Hexafluoroacetylacetone	327.30	485.1	2.767	
$C_5H_4O_2$	Furfural	434.9	670	5.89	
C_5H_5N	Pyridine	388.38	620.0	5.67	243
$C_5H_6N_2$	2-Methylpyrazine	410	634.3	5.01	283
C_5H_6O	2-Methylfuran	338	527	4.72	247
C_5H_7N	1-Methylpyrrole	385.96	596.0	4.86	271
C_5H_7N	2-Methylpyrrole	420.7	654	5.08	266
C_5H_7N	3-Methylpyrrole	416.0	647	5.08	266
C_5H_8	1-Pentyne	313.3	493.5		
C_5H_8	Cyclopentene	317.3	506.5	4.80	245
C_5H_8O	Cyclopentanone	403.72	624.5	4.60	
C_5H_8O	3,4-Dihydro-2*H*-pyran	359	561.7	4.56	268
C_5H_9N	Pentanenitrile	414.5	610.3	3.58	
C_5H_9NO	*N*-Methyl-2-pyrrolidone	475	721.8		311
C_5H_{10}	1-Pentene	303.11	464.8	3.56	298.4
C_5H_{10}	cis-2-Pentene	310.08	475	3.69	
C_5H_{10}	trans-2-Pentene	309.49	471	3.52	
C_5H_{10}	2-Methyl-1-butene	304.3	470	3.8	
C_5H_{10}	3-Methyl-1-butene	293.3	452.7	3.53	304.9
C_5H_{10}	2-Methyl-2-butene	311.71	470	3.42	
C_5H_{10}	Cyclopentane	322.5	511.7	4.51	259
$C_5H_{10}O$	Cyclopentanol	413.57	619.5	4.9	
$C_5H_{10}O$	Allyl ethyl ether	340.8	518		
$C_5H_{10}O$	Pentanal	376	566.1	3.97	313
$C_5H_{10}O$	2-Pentanone	375.41	561.08	3.694	301
$C_5H_{10}O$	3-Pentanone	375.11	561.46	3.729	336
$C_5H_{10}O$	3-Methyl-2-butanone	367.48	553.4	3.85	310
$C_5H_{10}O$	Tetrahydropyran	361	572.2	4.77	263
$C_5H_{10}O$	2-Methyltetrahydrofuran	351	537	3.76	267
$C_5H_{10}O_2$	Pentanoic acid	459.3	643	3.58	340
$C_5H_{10}O_2$	3-Methylbutanoic acid	449.7	629	3.40	
$C_5H_{10}O_2$	Isobutyl formate	371.3	551	3.88	352
$C_5H_{10}O_2$	Propyl acetate	374.69	549.7	3.36	345
$C_5H_{10}O_2$	Isopropyl acetate	361.8	531		
$C_5H_{10}O_2$	Ethyl propanoate	372.3	546.0	3.362	345
$C_5H_{10}O_2$	Methyl butanoate	376.0	554.4	3.47	340
$C_5H_{10}O_2$	Methyl isobutanoate	365.7	540.8	3.43	339
$C_5H_{11}N$	Piperidine	379.37	594	4.94	288
C_5H_{12}	Pentane	309.21	469.7	3.370	311
C_5H_{12}	Isopentane	301.03	460.4	3.38	306
C_5H_{12}	Neopentane	282.63	433.8	3.196	307
$C_5H_{12}O$	1-Pentanol	411.13	588.1	3.897	326
$C_5H_{12}O$	2-Pentanol	392.5	560.3	3.675	329
$C_5H_{12}O$	3-Pentanol	389.40	559.6		325
$C_5H_{12}O$	2-Methyl-1-butanol	400.7	575.4	3.94	
$C_5H_{12}O$	3-Methyl-1-butanol	404.3	577.2	3.93	
$C_5H_{12}O$	2-Methyl-2-butanol	375.5	543.7	3.71	
$C_5H_{12}O$	3-Methyl-2-butanol	386.0	556.1	3.87	
$C_5H_{12}O$	Butyl methyl ether	343.31	512.78	3.371	329
$C_5H_{12}O$	Methyl *tert*-butyl ether	328.3	497.1	3.430	
$C_5H_{12}O$	Ethyl propyl ether	336.36	500.23	3.370	339
$C_5H_{12}O_2$	Ethylene glycol monopropyl ether	423.0	614.6		364
$C_5H_{12}O_3$	Diethylene glycol monomethyl ether	466	672	3.67	
$C_5H_{12}S$	3-Methyl-1-butanethiol	393	604		
C_6BrF_5	Bromopentafluorobenzene	410	601	3.0	

Mol. Form.	Name	T_b/K	T_c/K	P_c/MPa	$V_c/cm^3\ mol^{-1}$
C_6ClF_5	Chloropentafluorobenzene	391.11	570.81	3.238	376
$C_6Cl_2F_4$	1,2-Dichlorotetrafluorobenzene	430.9	626	5.32	
$C_6Cl_3F_3$	1,3,5-Trichloro-2,4,6-trifluorobenzene	471.6	684.8	3.27	448
C_6F_6	Hexafluorobenzene	353.41	516.73	3.273	335
C_6F_{10}	Perfluorocyclohexene	325.2	461.8		
C_6F_{12}	Perfluoro-1-hexene	330.2	454.4		
C_6F_{12}	Perfluorocyclohexane	326.0	457.2	2.43	
C_6F_{14}	Perfluorohexane	329.8	448.77	1.868	606
C_6F_{14}	Perfluoro-2-methylpentane	330.8	455.3	1.923	532
C_6F_{14}	Perfluoro-3-methylpentane	331.5	450	1.69	
C_6F_{14}	Perfluoro-2,3-dimethylbutane	333.0	463	1.87	525
C_6HF_5	Pentafluorobenzene	358.89	530.97	3.531	324
C_6HF_5O	Pentafluorophenol	418.8	609	4.0	348
C_6HF_{11}	Undecafluorocyclohexane	335.2	477.7		
$C_6H_2F_4$	1,2,3,4-Tetrafluorobenzene	367.5	550.83	3.791	313
$C_6H_2F_4$	1,2,3,5-Tetrafluorobenzene	357.5	535.25	3.747	
$C_6H_2F_4$	1,2,4,5-Tetrafluorobenzene	363.3	543.35	3.801	
$C_6H_4F_2$	p-Difluorobenzene	362	556	4.40	
C_6H_5Br	Bromobenzene	429.21	670	4.52	324
C_6H_5Cl	Chlorobenzene	404.87	632.4	4.52	308
C_6H_5F	Fluorobenzene	357.88	560.09	4.551	269
C_6H_5I	Iodobenzene	461.6	721	4.52	351
C_6H_6	Benzene	353.24	562.05	4.895	256
C_6H_6O	Phenol	455.02	694.2	6.13	
C_6H_7N	Aniline	457.32	699	4.89	287
C_6H_7N	2-Methylpyridine	402.53	621.0	4.60	292
C_6H_7N	3-Methylpyridine	417.29	645.0	4.65	288
C_6H_7N	4-Methylpyridine	418.51	645.7	4.70	292
C_6H_{10}	1,5-Hexadiene	332.5	508		
C_6H_{10}	Cyclohexene	356.13	560.4		
$C_6H_{10}O$	Cyclohexanone	428.58	653.0	4.0	
$C_6H_{10}O_2$	Ethyl trans-2-butenoate	411	599		
$C_6H_{10}S$	Diallylsulfide	411.8	653		
$C_6H_{11}N$	Hexanenitrile	436.80	633.8	3.30	
C_6H_{12}	1-Hexene	336.63	504.0	3.21	355.1
C_6H_{12}	Cyclohexane	353.88	553.8	4.08	308
C_6H_{12}	Methylcyclopentane	345.0	532.7	3.79	318
$C_6H_{12}O$	Hexanal	404	591	3.46	
$C_6H_{12}O$	2-Hexanone	400.8	587.0	3.32	
$C_6H_{12}O$	3-Hexanone	396.7	582.82	3.320	
$C_6H_{12}O$	4-Methyl-2-pentanone	389.7	571	3.27	
$C_6H_{12}O$	Cyclohexanol	433.99	647.1	4.401	
$C_6H_{12}O_2$	Hexanoic acid	478.4	662	3.20	
$C_6H_{12}O_2$	Pentyl formate	403.5	576	3.46	
$C_6H_{12}O_2$	Isopentyl formate	396.7	578		
$C_6H_{12}O_2$	Butyl acetate	399.3	579		
$C_6H_{12}O_2$	Isobutyl acetate	389.7	561	3.16	
$C_6H_{12}O_2$	Propyl propanoate	395.7	578		
$C_6H_{12}O_2$	Ethyl butanoate	394.7	566	3.06	421
$C_6H_{12}O_2$	Ethyl 2-methylpropanoate	383.3	553	3.07	421
$C_6H_{12}O_2$	Methyl pentanoate	400.5	567	3.19	
$C_6H_{12}O_3$	Ethylene glycol monoethyl ether acetate	429.6	607.3	3.166	443
$C_6H_{12}O_3$	Ethylene glycol monoethyl ether acetate	429.6	611	3.18	441
$C_6H_{12}O_3$	Paraldehyde	397.5	563		
C_6H_{14}	Hexane	341.88	507.6	3.025	368
C_6H_{14}	2-Methylpentane	333.41	497.7	3.04	368
C_6H_{14}	3-Methylpentane	336.42	504.6	3.12	368
C_6H_{14}	2,2-Dimethylbutane	322.88	489.0	3.10	358
C_6H_{14}	2,3-Dimethylbutane	331.08	500.0	3.15	361

Mol. Form.	Name	T_b/K	T_c/K	P_c/MPa	$V_c/cm^3\,mol^{-1}$
$C_6H_{14}O$	1-Hexanol	430.8	610.3	3.417	387
$C_6H_{14}O$	2-Hexanol	413	585.9	3.31	384
$C_6H_{14}O$	3-Hexanol	408	582.4	3.36	383
$C_6H_{14}O$	2-Methyl-1-pentanol	422	604.4	3.45	
$C_6H_{14}O$	4-Methyl-1-pentanol	425.1	603.5		
$C_6H_{14}O$	2-Methyl-2-pentanol	394.3	559.5		
$C_6H_{14}O$	4-Methyl-2-pentanol	404.8	574.4		
$C_6H_{14}O$	2-Methyl-3-pentanol	399.7	576.0	3.46	
$C_6H_{14}O$	3-Methyl-3-pentanol	395.5	575.6	3.52	
$C_6H_{14}O$	Dipropyl ether	363.23	530.6	3.028	
$C_6H_{14}O$	Diisopropyl ether	341.66	500.32	2.832	386
$C_6H_{14}O$	Methyl pentyl ether	372	546.53	3.042	391
$C_6H_{14}O_2$	Ethylene glycol monobutyl ether	441.6	633.9		424
$C_6H_{14}O_2$	1,1-Diethoxyethane	375.40	527		
$C_6H_{14}O_3$	Diethylene glycol monoethyl ether	469	670	3.167	
$C_6H_{15}N$	Dipropylamine	382.5	555.8	3.63	
$C_6H_{15}N$	Diisopropylamine	357.0	523.1	3.02	
$C_6H_{15}N$	Triethylamine	362	535.6	3.032	389
C_7F_8	Perfluorotoluene	377.7	534.47	2.705	428
C_7F_{14}	Perfluoro-1-heptene	354.2	478.2		
C_7F_{14}	Perfluoromethylcyclohexane	349.5	485.91	2.019	570
C_7F_{16}	Perfluoroheptane	355.7	474.8	1.62	664
C_7HF_{15}	1H-Pentadecafluoroheptane	369.2	495.8		
$C_7H_3F_5$	2,3,4,5,6-Pentafluorotoluene	390.7	566.52	3.126	384
C_7H_5N	Benzonitrile	464.3	699.4	4.21	
C_7H_6O	Benzaldehyde	452.2	695	4.65	
C_7H_8	Toluene	383.78	591.75	4.108	316
C_7H_8O	o-Cresol	464.19	697.6	5.01	
C_7H_8O	m-Cresol	475.42	705.8	4.56	309
C_7H_8O	p-Cresol	475.13	704.6	5.15	
C_7H_8O	Benzyl alcohol	478.46	715	4.3	
C_7H_8O	Anisole	426.9	646.1	4.222	341
C_7H_9N	o-Methylaniline	473.5	707	4.37	
C_7H_9N	m-Methylaniline	476.5	707	4.28	
C_7H_9N	p-Methylaniline	473.6	706	4.58	
C_7H_9N	N-Methylaniline	469.4	701	5.20	
C_7H_9N	2,3-Dimethylpyridine	434.27	655.4	4.10	356
C_7H_9N	2,4-Dimethylpyridine	431.53	647	3.95	361
C_7H_9N	2,5-Dimethylpyridine	430.13	645	3.85	361
C_7H_9N	2,6-Dimethylpyridine	417.16	624	3.85	361
C_7H_9N	3,4-Dimethylpyridine	452.25	684	4.20	355
C_7H_9N	3,5-Dimethylpyridine	444.99	668	4.05	361
C_7H_{14}	1-Heptene	366.79	537.3	2.92	409
C_7H_{14}	Cycloheptane	391.5	604.2	3.82	353
C_7H_{14}	Methylcyclohexane	374.08	572.1	3.48	369
C_7H_{14}	Ethylcyclopentane	376.7	569.5	3.40	375
C_7H_{14}	1,1-Dimethylcyclopentane	360.7	547	3.45	
C_7H_{14}	cis-1,2-Dimethylcyclopentane	372.7	565	3.45	
C_7H_{14}	$trans$-1,2-Dimethylcyclopentane	365.0	553	3.45	
C_7H_{14}	cis-1,3-Dimethylcyclopentane	364.0	551	3.45	
C_7H_{14}	$trans$-1,3-Dimethylcyclopentane	364.8	553	3.45	
$C_7H_{14}O$	2-Heptanone	424.20	611.5	3.436	
$C_7H_{14}O_2$	Heptanoic acid	495.4	679	2.90	
$C_7H_{14}O_2$	Pentyl acetate	422.4	600	2.68	470
$C_7H_{14}O_2$	Isopentyl acetate	415.7	599		
$C_7H_{14}O_2$	Isobutyl propanoate	410	592		
$C_7H_{14}O_2$	Propyl butanoate	416.2	600		
$C_7H_{14}O_2$	Propyl isobutanoate	408.5	589		
$C_7H_{14}O_2$	Ethyl pentanoate	419.3	570		

Mol. Form.	Name	T_b/K	T_c/K	P_c/MPa	$V_c/cm^3\ mol^{-1}$
$C_7H_{14}O_2$	Ethyl 3-methylbutanoate	408.2	588		
C_7H_{16}	Heptane	371.7	540.2	2.74	428
C_7H_{16}	2-Methylhexane	363.19	530.4	2.74	421
C_7H_{16}	3-Methylhexane	365	535.2	2.81	404
C_7H_{16}	3-Ethylpentane	366.7	540.6	2.89	416
C_7H_{16}	2,2-Dimethylpentane	352.3	520.5	2.77	416
C_7H_{16}	2,3-Dimethylpentane	362.93	537.3	2.91	393
C_7H_{16}	2,4-Dimethylpentane	353.64	519.8	2.74	418
C_7H_{16}	3,3-Dimethylpentane	359.21	536.4	2.95	414
C_7H_{16}	2,2,3-Trimethylbutane	354.01	531.1	2.95	398
$C_7H_{16}O$	1-Heptanol	449.60	632.6	3.058	435
$C_7H_{16}O$	2-Heptanol	432	608.3	3.021	442
$C_7H_{16}O$	3-Heptanol	430	605.4		434
$C_7H_{16}O$	4-Heptanol	429	602.6		432
$C_8F_{16}O$	Perfluoro-2-butyltetrahydrofuran	375.8	500.2	1.607	588
C_8F_{18}	Perfluorooctane	379.0	502	1.66	
C_8H_7N	4-Methylbenzonitrile	490.2	723		
C_8H_7N	Indole	526.7	794	4.8	356
C_8H_8O	Acetophenone	475	713	4.40	380
$C_8H_8O_3$	Methyl salicylate	496.1	709		
C_8H_{10}	Ethylbenzene	409.34	617.15	3.609	374
C_8H_{10}	o-Xylene	417.7	630.3	3.732	370
C_8H_{10}	m-Xylene	412.27	617.0	3.541	375
C_8H_{10}	p-Xylene	411.52	616.2	3.511	378
$C_8H_{10}O$	o-Ethylphenol	477.7	703.0		
$C_8H_{10}O$	m-Ethylphenol	491.6	718.8		
$C_8H_{10}O$	p-Ethylphenol	491.1	716.4		
$C_8H_{10}O$	2,3-Xylenol	490.1	722.8		
$C_8H_{10}O$	2,4-Xylenol	484.13	707.6		
$C_8H_{10}O$	2,5-Xylenol	484.3	706.9		
$C_8H_{10}O$	2,6-Xylenol	474.22	701.0		
$C_8H_{10}O$	3,4-Xylenol	500	729.8		
$C_8H_{10}O$	3,5-Xylenol	494.89	715.6		
$C_8H_{10}O$	Phenetole	442.96	647	3.42	
$C_8H_{11}N$	N-Ethylaniline	476.2	698		
$C_8H_{11}N$	N,N-Dimethylaniline	467.30	687	3.63	
$C_8H_{14}O_4$	Diethyl succinate	490.9	663		
$C_8H_{15}N$	Octanenitrile	478.40	674.4	2.85	
C_8H_{16}	1-Octene	394.44	567.0	2.68	468
C_8H_{16}	Cyclooctane	422	647.2	3.56	410
C_8H_{16}	Ethylcyclohexane	405.0	609	3.04	
C_8H_{16}	cis-1,2-Dimethylcyclohexane	403.0	606	2.95	
C_8H_{16}	trans-1,2-Dimethylcyclohexane	396.7	596	2.94	
C_8H_{16}	cis-1,3-Dimethylcyclohexane	393.3	591	2.94	
C_8H_{16}	trans-1,3-Dimethylcyclohexane	397.7	598	2.94	
C_8H_{16}	trans-1,4-Dimethylcyclohexane	392.5	587.7		
$C_8H_{16}O_2$	Octanoic acid	512	695	2.64	
$C_8H_{16}O_2$	2-Ethylhexanoic acid	501	674.6	2.778	528
$C_8H_{16}O_2$	Isopentyl propanoate	433.4	611		
$C_8H_{16}O_2$	Isobutyl butanoate	430.1	611		
$C_8H_{16}O_2$	Isobutyl isobutanoate	421.8	602		
$C_8H_{16}O_2$	Propyl 3-methylbutanoate	429.1	609		
$C_8H_{16}O_4$	Diethylene glycol monoethyl ether acetate	491.7	663	2.73	
C_8H_{18}	Octane	398.82	568.7	2.49	492
C_8H_{18}	2-Methylheptane	390.81	559.7	2.50	488
C_8H_{18}	3-Methylheptane	392.0	563.6	2.55	464
C_8H_{18}	4-Methylheptane	390.87	561.7	2.54	476
C_8H_{18}	3-Ethylhexane	391.8	565.5	2.61	455
C_8H_{18}	2,2-Dimethylhexane	380.01	549.8	2.53	478

Mol. Form.	Name	T_b/K	T_c/K	P_c/MPa	V_c/cm^3 mol^{-1}
C$_8$H$_{18}$	2,3-Dimethylhexane	388.77	563.5	2.63	468
C$_8$H$_{18}$	2,4-Dimethylhexane	382.7	553.5	2.56	472
C$_8$H$_{18}$	2,5-Dimethylhexane	382.27	550.0	2.49	482
C$_8$H$_{18}$	3,3-Dimethylhexane	385.12	562.0	2.65	443
C$_8$H$_{18}$	3,4-Dimethylhexane	390.88	568.8	2.69	466
C$_8$H$_{18}$	3-Ethyl-2-methylpentane	388.81	567.1	2.70	442
C$_8$H$_{18}$	3-Ethyl-3-methylpentane	391.42	576.5	2.81	455
C$_8$H$_{18}$	2,2,3-Trimethylpentane	383	563.5	2.73	436
C$_8$H$_{18}$	2,2,4-Trimethylpentane	372.37	543.8	2.57	468
C$_8$H$_{18}$	2,3,3-Trimethylpentane	388.0	573.5	2.82	455
C$_8$H$_{18}$	2,3,4-Trimethylpentane	386.7	566.4	2.73	460
C$_8$H$_{18}$	2,2,3,3-Tetramethylbutane	379.60	567.8	2.87	461
C$_8$H$_{18}$O	1-Octanol	468.31	652.5	2.777	497
C$_8$H$_{18}$O	2-Octanol	453	629.6	2.754	519
C$_8$H$_{18}$O	3-Octanol	444	628.5		515
C$_8$H$_{18}$O	4-Octanol	449.5	625.1		515
C$_8$H$_{18}$O	4-Methyl-3-heptanol	443	623.5		
C$_8$H$_{18}$O	5-Methyl-3-heptanol	445	621.2		
C$_8$H$_{18}$O	2-Ethyl-1-hexanol	457.8	640.6	2.8	
C$_8$H$_{18}$O	Dibutyl ether	413.43	584.1	3.01	
C$_8$H$_{18}$O	Di-tert-butyl ether	380.38	550		
C$_8$H$_{19}$N	Dibutylamine	432.8	607.5	3.11	
C$_8$H$_{19}$N	Diisobutylamine	412.8	584.4	3.20	
C$_8$H$_{20}$Si	Tetraethylsilane	427.9	603.7	2.602	
C$_9$F$_{20}$	Perfluorononane	398.5	524	1.56	
C$_9$H$_7$N	Quinoline	510.31	782	4.86	371
C$_9$H$_7$N	Isoquinoline	516.37	803	5.10	374
C$_9$H$_{10}$	Indan	451.12	684.9	3.95	
C$_9$H$_{12}$	Propylbenzene	432.39	638.35	3.200	440
C$_9$H$_{12}$	Isopropylbenzene	425.56	631.0	3.209	
C$_9$H$_{12}$	o-Ethyltoluene	438.4	651	3.38	
C$_9$H$_{12}$	m-Ethyltoluene	434.5	637	3.25	
C$_9$H$_{12}$	p-Ethyltoluene	435	640.2	3.23	
C$_9$H$_{12}$	1,2,3-Trimethylbenzene	449.27	664.5	3.454	
C$_9$H$_{12}$	1,2,4-Trimethylbenzene	442.53	649.1	3.232	
C$_9$H$_{12}$	1,3,5-Trimethylbenzene	437.89	637.3	3.127	
C$_9$H$_{13}$N	2-Methyl-N,N-dimethylaniline	467.3	668	3.12	
C$_9$H$_{18}$	1-Nonene	420.1	594.0		526
C$_9$H$_{18}$	Cyclononane	451.6	682	3.34	
C$_9$H$_{18}$	1α,3α,5α-1,3,5-Trimethylcyclohexane	413.7	602.2		
C$_9$H$_{18}$O	2-Nonanone	468.5	651.9	2.482	
C$_9$H$_{18}$O	5-Nonanone	461.60	640		
C$_9$H$_{18}$O$_2$	Nonanoic acid	527.7	711	2.40	
C$_9$H$_{18}$O$_2$	Isopentyl butanoate	452	619		
C$_9$H$_{18}$O$_2$	Isobutyl 3-methylbutanoate	441.7	621		
C$_9$H$_{20}$	Nonane	423.97	594.6	2.29	555
C$_9$H$_{20}$	2-Methyloctane	416.4	582.8	2.31	
C$_9$H$_{20}$	2,2-Dimethylheptane	405.8	576.7	2.35	
C$_9$H$_{20}$	2,2,5-Trimethylhexane	397.24	569.8		
C$_9$H$_{20}$	2,2,3,3-Tetramethylpentane	413.4	607.5	2.74	
C$_9$H$_{20}$	2,2,3,4-Tetramethylpentane	406.2	592.6	2.60	
C$_9$H$_{20}$	2,2,4,4-Tetramethylpentane	395.44	574.6	2.49	
C$_9$H$_{20}$	2,3,3,4-Tetramethylpentane	414.7	607.5	2.72	
C$_9$H$_{20}$O	1-Nonanol	486.52	670.7	2.528	572
C$_9$H$_{20}$O	2-Nonanol	466.7	649.6	2.53	575
C$_9$H$_{20}$O	3-Nonanol	468	648.0		577
C$_9$H$_{20}$O	4-Nonanol	465.7	645.1		575
C$_{10}$F$_8$	Perfluoronaphthalene	482	673.1		
C$_{10}$F$_{18}$	Perfluorodecalin	415	566	1.52	

Mol. Form.	Name	T_b/K	T_c/K	P_c/MPa	V_c/cm^3 mol^{-1}
$C_{10}F_{22}$	Perfluorodecane	417.4	542	1.45	
$C_{10}H_8$	Naphthalene	491.1	748.4	4.05	407
$C_{10}H_9N$	1-Naphthylamine	573.8	850	5.0	438
$C_{10}H_9N$	2-Naphthylamine	579.3	850	4.9	438
$C_{10}H_{12}$	1,2,3,4-Tetrahydronaphthalene	480.8	720	3.65	408
$C_{10}H_{14}$	Butylbenzene	456.46	660.5	2.89	497
$C_{10}H_{14}$	Isobutylbenzene	445.94	650	3.05	
$C_{10}H_{14}$	1-Isopropyl-4-methylbenzene	450.3	652	2.8	
$C_{10}H_{14}$	p-Diethylbenzene	456.9	657.9	2.803	
$C_{10}H_{14}$	1,2,4,5-Tetramethylbenzene	470.0	676	2.9	
$C_{10}H_{14}O$	Thymol	505.7	698		
$C_{10}H_{16}$	d-Limonene	451	653		470
$C_{10}H_{16}$	α-Pinene	429	644		454
$C_{10}H_{16}$	(+)-3-Carene	444	658		487
$C_{10}H_{18}$	1,3-Decadiene	442	615		
$C_{10}H_{18}$	cis-Decahydronaphthalene	469.0	702.3	3.20	
$C_{10}H_{18}$	trans-Decahydronaphthalene	460.5	687.1		
$C_{10}H_{20}$	1-Decene	443.7	617	2.22	584
$C_{10}H_{20}O$	Decanal	481.7	674.2		
$C_{10}H_{20}O$	Menthol	489	694		
$C_{10}H_{20}O_2$	Decanoic acid	541.8	726	2.23	
$C_{10}H_{20}O_2$	Ethyl octanoate	481.7	659		
$C_{10}H_{20}O_4$	Diethylene glycol monobutyl ether acetate	518	681	3.15	
$C_{10}H_{22}$	Decane	447.30	617.7	2.11	624
$C_{10}H_{22}$	3,3,5-Trimethylheptane	428.9	609.5	2.32	
$C_{10}H_{22}$	2,2,3,3-Tetramethylhexane	433.5	623.0	2.51	
$C_{10}H_{22}$	2,2,5,5-Tetramethylhexane	410.6	581.4	2.19	
$C_{10}H_{22}O$	1-Decanol	504.3	687.3	2.315	649
$C_{10}H_{22}O$	2-Decanol	484	668.6		646
$C_{10}H_{22}O$	3-Decanol		666.1		643
$C_{10}H_{22}O$	4-Decanol	483.7	663.7		643
$C_{10}H_{22}O$	5-Decanol	474	663.2		646
$C_{10}H_{22}S$	Diisopentylsulfide	484	664		
$C_{11}H_{10}$	1-Methylnaphthalene	517.8	772	3.60	
$C_{11}H_{10}$	2-Methylnaphthalene	514.3	761		
$C_{11}H_{22}O_2$	Ethyl nonanoate	500.2	674		
$C_{11}H_{24}$	Undecane	469.1	639	1.98	689
$C_{11}H_{24}O$	1-Undecanol	516	703.6	2.147	718
$C_{12}H_8$	Acenaphthylene	553	792	3.20	
$C_{12}H_9N$	Carbazole	627.84	901.8	3.13	454
$C_{12}H_{10}$	Biphenyl	529.3	773	3.38	497
$C_{12}H_{10}O$	Diphenyl ether	531.2	766.8		
$C_{12}H_{12}$	2,7-Dimethylnaphthalene	538	775	3.23	601
$C_{12}H_{18}$	Hexamethylbenzene	536.6	758		
$C_{12}H_{24}$	1-Dodecene	487.0	658	1.93	
$C_{12}H_{26}$	Dodecane	489.47	658	1.82	754
$C_{12}H_{26}O$	1-Dodecanol	532	719.4	1.994	
$C_{13}H_9N$	Acridine	618.01	891.1	3.21	548
$C_{13}H_9N$	Phenanthridine	622.0	895	3.6	548
$C_{13}H_{11}N$	9-Methylcarbazole	616.79	890	3.38	572
$C_{13}H_{12}$	Diphenylmethane	538.2	760	2.71	563
$C_{13}H_{26}O_2$	Methyl dodecanoate	540	712		
$C_{13}H_{28}$	Tridecane	508.62	675	1.68	823
$C_{13}H_{28}O$	1-Tridecanol	546.8	734	1.935	
$C_{14}H_{10}$	Anthracene	613.1	869.3		554
$C_{14}H_{10}$	Phenanthrene	613	869		
$C_{14}H_{30}$	Tetradecane	526.73	693	1.57	894
$C_{14}H_{30}O$	1-Tetradecanol	562	747	1.81	

Mol. Form.	Name	T_b/K	T_c/K	P_c/MPa	V_c/cm^3 mol^{-1}
$C_{15}H_{32}$	Pentadecane	543.8	708	1.48	966
$C_{16}H_{34}$	Hexadecane	560.01	723	1.40	1034
$C_{16}H_{34}$	2,2,4,4,6,8,8-Heptamethylnonane	519.5	692		
$C_{16}H_{34}O$	1-Hexadecanol	607	770	1.61	
$C_{17}H_{36}$	Heptadecane	575.2	736	1.34	1103
$C_{17}H_{36}O$	1-Heptadecanol	606	780	1.50	
$C_{18}H_{14}$	o-Terphenyl	605	857	2.99	731
$C_{18}H_{14}$	m-Terphenyl	636	883	2.48	724
$C_{18}H_{14}$	p-Terphenyl	649	908	2.99	729
$C_{18}H_{38}$	Octadecane	589.5	747	1.29	1189
$C_{18}H_{38}O$	1-Octadecanol	616.7	790	1.44	
$C_{19}H_{40}$	Nonadecane	603.1	755	1.16	
$C_{20}H_{42}$	Eicosane	616	768	1.07	
$C_{20}H_{42}O$	1-Eicosanol	582	809	1.30	
$C_{21}H_{44}$	Heneicosane	629.7	778	1.03	
$C_{22}H_{46}$	Docosane	641.8	786	0.98	
$C_{23}H_{48}$	Tricosane	653	790	0.92	
$C_{24}H_{50}$	Tetracosane	664.5	800	0.87	

SUBLIMATION PRESSURE OF SOLIDS

This table gives the sublimation pressure of some representative solids as a function of temperature. The last entry for each substance is the solid-liquid-gas triple point. Substances are listed by molecular formula in the Hill order. Note that the pressure is given in kilopascals (kPa), except for water and naphthalene, for which the pressure unit is pascals. For conversion, 1 kPa = 7.506 Torr = 0.0098692 atm.

Ar	T/K	55	60	65	70	75	80	83.81	
Argon	P/kPa	0.2	0.8	2.8	7.7	18.7	40.7	68.8	
BrH	T/K	135	140	150	160	170	180	185.1	
Hydrogen bromide	P/kPa	0.1	0.3	1.1	3.3	8.7	20.1	27.4	
ClH	T/K	120	130	140	150	155	159.0		
Hydrogen chloride	P/kPa	0.1	0.5	1.9	5.8	9.5	13.5		
F_4Si	T/K	130	140	150	160	170	175	180	186.3
Silicon tetrafluoride	P/kPa	0.2	0.9	3.9	14.0	43.8	74.2	122.4	220.8
F_6S	T/K	150	165	180	190	200	210	220	223.1
Sulfur hexafluoride	P/kPa	0.4	2.6	11.3	25.9	54.5	106.1	195.1	232.7
HI	T/K	160	170	180	190	200	210	220	222.4
Hydrogen iodide	P/kPa	0.2	0.8	2.2	5.3	11.7	23.6	44.1	49.3
H_2O	T/K	190	210	225	240	250	260	270	273.16
Water	P/Pa	0.032	0.702	4.942	27.28	76.04	195.8	470.1	611.66
H_2S	T/K	140	150	160	165	170	175	180	187.6
Hydrogen sulfide	P/kPa	0.2	0.6	1.9	3.2	5.2	8.3	12.7	22.7
H_3N	T/K	160	170	180	190	195	195.4		
Ammonia	P/kPa	0.1	0.4	1.2	3.5	5.8	6.12		
Kr	T/K	80	90	95	100	105	110	115.8	
Krypton	P/kPa	0.4	2.7	6.0	12.1	22.8	40.4	73.1	
NO	T/K	85	90	95	100	105	109.5		
Nitric oxide	P/kPa	0.1	0.4	1.3	3.8	10.0	21.9		
Xe	T/K	110	120	130	140	150	155	160	161.4
Xenon	P/kPa	0.3	1.5	4.9	14.0	34.2	51.1	74.2	81.7
CO	T/K	50	55	60	65	68.13			
Carbon monoxide	P/kPa	0.1	0.6	2.6	8.2	15.4			
CO_2	T/K	130	140	155	170	185	194.7	205	216.58
Carbon dioxide	P/kPa	0.032	0.187	1.674	9.987	44.02	101.3	227.1	518.0
CHN	T/K	200	210	220	230	240	250	255	259.83
Hydrogen cyanide	P/kPa	0.2	0.4	1.0	2.2	4.8	9.7	13.6	18.62
CH_4	T/K	65	70	75	80	85	90.69		
Methane	P/kPa	0.1	0.3	0.8	2.1	4.9	11.70		
C_2H_2	T/K	130	140	150	160	170	180	190	192.4
Acetylene	P/kPa	0.2	0.7	2.6	7.8	20.6	49.0	106.3	126.0
$C_2H_4O_2$	T/K	250	260	270	280	289.7			
Acetic acid	P/kPa	0.092	0.199	0.406	0.79	1.29			
C_5H_{12}	T/K	200	210	220	230	240	250	255	256.58
Neopentane	P/kPa	0.7	1.6	3.6	7.3	13.9	24.8	32.4	35.8
$C_{10}H_8$	T/K	250	270	280	290	300	310	330	353.43
Naphthalene	P/Pa	0.036	0.514	1.662	4.918	13.43	34.15	182.9	999.6

VAPOR PRESSURE

This table gives vapor pressure data for about 1800 inorganic and organic substances. In order to accommodate elements and compounds ranging from refractory to highly volatile in a single table, the temperature at which the vapor pressure reaches specified pressure values is listed. The pressure values run in decade steps from 1 Pa (about 7.5 µm Hg) to 100 kPa (about 750 mm Hg). All temperatures are given in °C.

The data used in preparing the table came from a large number of sources; the main references used for each substance are indicated in the last column. Since the data were refit in most cases, values appearing in this table may not be identical with values in the source cited. The temperature entry in the 100 kPa column is close to, but not identical with, the normal boiling point (which is defined as the temperature at which the vapor pressure reaches 101.325 kPa). Although some temperatures are quoted to 0.1°C, uncertainties of several degrees should generally be assumed. Values followed by an "e" were obtained by extrapolating (usually with an Antoine equation) beyond the region for which experimental measurements were available and are thus subject to even greater uncertainty.

Compounds are listed by molecular formula following the Hill convention. Substances not containing carbon are listed first, followed by those that contain carbon. To locate an organic compound by name or CAS Registry Number when the molecular formula is not known, use the table *Physical Constants of Organic Compounds* in Section 3 and its indexes to determine the molecular formula. The indexes to *Physical Constants of Inorganic Compounds* in Section 4 can be used in a similar way.

More extensive and detailed vapor pressure data on selected important substances appear in other tables in this section of the *Handbook*. These substances are flagged by a symbol following the name as follows:

*   See *Vapor Pressure of Fluids below 300 K*
**   See *IUPAC Recommended Data for Vapor Pressure Calibration*
***   See *Vapor Pressure of Ice* and *Vapor Pressure of Water from 0 to 370°C*

The following notations appear after individual temperature entries:

s — Indicates the substance is a solid at this temperature.
e — Indicates an extrapolation beyond the region where experimental measurements exist.
i — Indicates the value was calculated from ideal gas thermodynamic functions, such as those in the *JANAF Thermochemical Tables* (see Reference 8).

REFERENCES

1. Lide, D.R., and Kehiaian, H.V., *CRC Handbook of Thermophysical and Thermochemical Data*, CRC Press, Boca Raton, FL, 1994.
2. Stull, D., in *American Institute of Physics Handbook, Third Edition*, Gray, D.E., Ed., McGraw Hill, New York, 1972.
3. Hultgren, R., Desai, P.D., Hawkins, D.T., Gleiser, M., Kelley, K.K., and Wagman, D.D., *Selected Values of Thermodynamic Properties of the Elements*, American Society for Metals, Metals Park, OH, 1973.
4. Stull, D., *Ind. Eng. Chem.*, 39, 517, 1947.
5. *TRCVP, Vapor Pressure Database, Version 2.2P*, Thermodynamic Research Center, Texas A&M University, College Station, TX.
6. *TRC Thermodynamic Tables,* Thermodynamic Research Center, Texas A&M University, College Station, TX.
7. Ohe, S., *Computer Aided Data Book of Vapor Pressure*, Data Book Publishing Co., Tokyo, 1976.
8. Chase, M.W., Davies, C.A., Downey, J.R., Frurip, D.J., McDonald, R.A., and Syverud, A.N., *JANAF Thermochemical Tables, Third Edition, J. Phys. Chem. Ref. Data*, Vol. 14, Suppl. 1, 1985.
9. Barin, I., *Thermochemical Data of Pure Substances*, VCH Publishers, New York, 1993.
10. Jacobsen, R.T., et. al, *International Thermodynamic Tables of the Fluid State, No. 10. Ethylene*, Blackwell Scientific Publications, Oxford, 1988.
11. Wakeham, W.A., *International Thermodynamic Tables of the Fluid State, No. 12. Methanol*, Blackwell Scientific Publications, Oxford, 1993.
12. Janz, G.J., *Molten Salts Handbook*, Academic Press, New York, 1967.
13. Ohse, R.W. *Handbook of Thermodynamic and Transport Properties of Alkali Metals*, Blackwell Scientific Publications, Oxford, 1994.
14. Gschneidner, K.A., in *CRC Handbook of Chemistry and Physics, 77th Edition*, p. 4-112, CRC Press, Boca Raton, FL, 1996.
15. Leider, H.R., Krikorian, O.H., and Young, D.A., *Carbon*, 11, 555, 1973.
16. Ruzicka, K., and Majer, V., *J. Phys. Chem. Ref. Data*, 23, 1, 1994.
17. Tillner-Roth, R., and Baehr, H.D., *J. Phys. Chem. Ref. Data*, 23, 657, 1994.
18. Younglove, B.A., and McLinden, M.O., *J. Phys. Chem. Ref. Data*, 23, 731, 1994.
19. Outcalt, S.L., and McLinden, M.O., *J. Phys. Chem. Ref. Data*, 25, 605, 1996.
20. Weber, L.A., and Defibaugh, D.R., *J. Chem. Eng. Data*, 41, 382, 1996.
21. Rodrigues, M.F., and Bernardo-Gil, M.G., *J. Chem. Eng. Data*, 41, 581, 1996.
22. Piacente, V., Gigli, G., Scardala, P., and Giustini, A., *J. Phys. Chem.*, 100, 9815, 1996.
23. Barton, J.L., and Bloom, H., *J. Phys. Chem.*, 60, 1413, 1956.
24. Sense, K.A., Alexander, C.A., Bowman, R.E., and Filbert, R.B., *J. Phys. Chem.*, 61, 337, 1957.
25. Ewing, C.T., and Stern, K.H., *J. Phys. Chem.* 78, 1998, 1974.
26. Cady, G.H., and Hargreaves, G.B., *J. Chem. Soc.*, 1961, 1563; 1961, 1568.
27. Skudlarski, K., Dudek, J., and Kapala, J., *J. Chem. Thermodynamics*, 19, 857, 1987.
28. Wagner, W., and de Reuck, K.M., *International Thermodynamic Tables of the Fluid State, No. 9. Oxygen*, Blackwell Scientific Publications, Oxford, 1987.

29. Marsh, K.N., Editor, *Recommended Reference Materials for the Realization of Physicochemical Properties*, Blackwell Scientific Publications, Oxford, 1987.

30. Alcock, C.B., Itkin, V.P., and Horrigan, M.K., *Canadian Metallurgical Quarterly*, 23, 309, 1984.

31. Stewart, R.B., and Jacobsen, R.T., *J. Phys. Chem. Ref. Data*, 18, 639, 1989.

32. Sifner, O., and Klomfar, J., *J. Phys. Chem. Ref. Data*, 23, 63, 1994.

33. Bah, A., and Dupont-Pavlovsky, N., *J. Chem. Eng. Data*, 40, 869, 1995.

34. Behrens, R.G., and Rosenblatt, G., *J. Chem. Thermodynamics*, 4, 175, 1972.

35. Behrens, R.G., and Rosenblatt, G., *J. Chem. Thermodynamics*, 5, 173, 1973.

36. Haar, L., Gallagher, J.S., and Kell, G.S., *NBS/NRC Steam Tables*, Hemisphere Publishing Corp., New York, 1984.

37. Wagner, W., Saul, A., and Pruss, A., *J. Phys. Chem. Ref. Data*, 23, 515. 1994.

38. Behrens, R.G., Lemons, R.S., and Rosenblatt, G., *J. Chem. Thermodynamics*, 6, 457, 1974.

39. Boublik, T., Fried, V., and Hala, E., *The Vapor Pressure of Pure Substances, Second Edition*, Elsevier, Amsterdam, 1984.

40. Goodwin, R.D., *J. Phys. Chem. Ref. Data*, 14, 849, 1985.

41. Younglove, B.A., and Ely, J.F., *J. Phys. Chem. Ref. Data*, 16, 577, 1987.

		Temperature in °C for the indicated pressure						
Mol. Form.	Name	1 Pa	10 Pa	100 Pa	1 kPa	10 kPa	100 kPa	Ref.

Substances not containing carbon:

Mol. Form.	Name	1 Pa	10 Pa	100 Pa	1 kPa	10 kPa	100 kPa	Ref.
Ag	Silver	1010	1140	1302	1509	1782	2160	2
AgBr	Silver(I) bromide	569 i	656 i	765 i	905 i	1093 i	1359 i	9
AgCl	Silver(I) chloride	670	769	873	1052	1264	1561	4
AgI	Silver(I) iodide	594	686	803	959	1177	1503	4
Al	Aluminum	1209	1359	1544	1781	2091	2517	2
AlB$_3$H$_{12}$	Aluminum borohydride				-46.8	-9.4	45.5	4
AlCl$_3$	Aluminum trichloride	58.4 s	76.5 s	97.1 s	120.7 s	148.2 s	180.5 s	4
AlF$_3$	Aluminum trifluoride	744 s	819 s	906 s	1008 s	1130 s	1276 s	8
AlI$_3$	Aluminum triiodide				218	285	385	4
Al$_2$O$_3$	Aluminum oxide			2122	2351	2629	2975	4
Ar	Argon*		-226.4 s	-220.3 s	-212.4 s	-201.7 s	-186.0	1,5,31
As	Arsenic	280 s	323 s	373 s	433 s	508 s	601 s	3
AsCl$_3$	Arsenic(III) chloride			-8 e	21.3	63.1	129.4	1
AsF$_3$	Arsenic(III) fluoride				8.1	56.0		4
AsI$_3$	Arsenic(III) iodide				187	261	367 e	7
As$_2$O$_3$	Arsenic(III) oxide (arsenolite)	133.7 s	163.0 s	196.8 s	236.2 s	283.0		34
At	Astatine	88 s	119 s	156 s	202 s	258 s	334	2
Au	Gold	1373	1541	1748	2008	2347	2805	2
B	Boron	2075	2289	2549	2868	3272	3799	2
BBr$_3$	Boron tribromide			-45 e	-15 e	27.5	90.4	1
BCl$_3$	Boron trichloride*			-94.0	-70.5	-37.4	12.3	4
BF$_3$	Boron trifluoride*	-173.9 s	-166.0 s	-156.0 s	-143.0 s	-125.9	-101.1	4
B$_2$F$_4$	Tetrafluorodiborane						-34	1
B$_2$H$_6$	Diborane			-162 e	-147.0	-125.8	-92.6	1
B$_5$H$_9$	Pentaborane(9)				-34.8	3.8	57.6	4
Ba	Barium	638 s	765	912	1115	1413	1897	9
Be	Beryllium	1189 s	1335	1518	1750	2054	2469	2
BeBr$_2$	Beryllium bromide	203 s	240 s	283 s	335 s	397 s	473 s	4
BeCl$_2$	Beryllium chloride	196 s	237 s	284 s	339 s	402 s	487	4
BeF$_2$	Beryllium fluoride		686 e	767 e	869	999	1172 e	7
BeI$_2$	Beryllium iodide	188 s	229 s	276 s	333 s	402 s	487	4
Bi	Bismuth	668	768	892	1052	1265	1562	2
BiBr$_3$	Bismuth tribromide			217 s	273 i	348 i	455 i	4,9
BiCl$_3$	Bismuth trichloride				248.9	328.6	438.7	1,4
BrCs	Cesium bromide	531 s	601 s	701 i	834 i	1019 i	1293 e	9
BrH	Hydrogen bromide*		-153.3 s	-140.4 s	-123.8 s	-101.5 s	-67.0	5
BrH$_3$Si	Bromosilane				-81.0	-47.3	2.2	4
BrH$_4$N	Ammonium bromide	121 s	154 s	195 s	246 s	310.4 s	395.1 s	5
BrK	Potassium bromide	597 s	674 s	773				25
BrLi	Lithium bromide		630	733	868	1049	1308	4
BrNa	Sodium bromide			791	931	1120	1389	4
BrRb	Rubidium bromide			766	903	1087	1350	4
BrTl	Thallium(I) bromide				509	635	817	4
Br$_2$	Bromine*	-87.7 s	-71.8 s	-52.7 s	-29.3 s	2.5	58.4	1
Br$_2$Cd	Cadmium bromide	373 s	435 s	509 s				27
Br$_2$Hg	Mercury(II) bromide	71 s	98 s	132 s	174 s	227 s	318	4
Br$_2$OS	Thionyl bromide	-49 e	-29 e	-5 e	27.8	72.9	139.6	5

Mol. Form.	Name	Temperature in °C for the indicated pressure						Ref.
		1 Pa	10 Pa	100 Pa	1 kPa	10 kPa	100 kPa	
Br_2Pb	Lead(II) bromide	374	431	502	597	726	914	4
Br_2S_2	Sulfur bromide	-7 e	15 e	42 e	78.4	128.1	200.9	5
Br_3In	Indium(III) bromide			304.6 s	328.7 s	364.8 s		1
Br_3OP	Phosphorus(V) oxybromide				64 e	115.5	191.4	5
Br_3P	Phosphorus(III) bromide		-23 e	5 e	42.3	94.6	172.6	5
Br_3Sb	Antimony(III) bromide				136.5	196.9	286.5	1
Br_4Ge	Germanium(IV) bromide				51	105	188	4
Br_4Sn	Tin(IV) bromide				67	122	204	4
Br_4Zr	Zirconium(IV) bromide	136 s	167 s	203 s	245 s	295 s	356 s	4
Br_5P	Phosphorus(V) bromide		-19 s	4 s	31 s	65.5 s	110.1	5
Ca	Calcium	591 s	683 s	798 s	954	1170	1482	2
Cd	Cadmium	257 s	310 s	381	472	594	767	2
$CdCl_2$	Cadmium chloride	412 s	471 s	541 s	634	768	959	23, 27
CdF_2	Cadmium fluoride				1257	1461	1742	4
CdI_2	Cadmium iodide	296 s	344 s	406	498	622	795	4,27
CdO	Cadmium oxide	770 s	866 s	983 s	1128 s	1314 s	1558 s	4
Ce	Cerium	1719	1921	2169	2481	2886	3432	14
$ClCs$	Cesium chloride			730	864	1043	1297	4
$ClCu$	Copper(I) chloride		459	543	675	914	1477	4
ClF	Chlorine fluoride*				-144.4	-122.6	-90.2	5
ClF_2P	Phosphorus(III) chloride difluoride				-119.5	-91.1	-47.6	5
ClF_3	Chlorine trifluoride				-63.7	-33.0	11.4	5
ClF_5	Chlorine pentafluoride				-88 e	-59	-14	7
ClH	Hydrogen chloride*				-138.2 s	-118.0	-85.2	1,5
$ClHO_3S$	Chlorosulfonic acid	-40 e	-20 e	5 e	38.7	85.0	153.6	5
ClH_4N	Ammonium chloride	91 s	121 s	159 s	204.7 s	263.1 s	339.5 s	5
ClK	Potassium chloride	625 s	704 s	804	945	1137	1411	23,25
$ClLi$	Lithium chloride		649 i	761 i	905 i	1101 i	1381 i	8
$ClNO$	Nitrosyl chloride		-116 s	-100 s	-78.7 s	-50.2	-5.7	5
$ClNO_2$	Nitryl chloride	-121 e	-113 e	-102 e	-86.1	-60.9	-15.7	5
$ClNa$	Sodium chloride	653 s	733 s	835	987	1182	1461	23,25
ClO_2	Chlorine dioxide*					-34.3	10.5	5
$ClRb$	Rubidium chloride			777	916	1105	1379	4
$ClTl$	Thallium(I) chloride				504	626	806	4
Cl_2	Chlorine*	-145 s	-133.7 s	-120.2 s	-103.6 s	-76.1	-34.2	1
Cl_2Co	Cobalt(II) chloride					818	1048	4
Cl_2FP	Phosphorus(III) dichloride fluoride				-71.1	-37.4	13.5	5
Cl_2F_3P	Phosphorus(V) dichloride trifluoride		-120 e	-101 e	-77.1	-44.3	3 e	7
Cl_2Fe	Iron(II) chloride				685	821	1025	4
Cl_2Hg	Mercury(II) chloride	64.4 s	94.7 s	130.8 s	174.5 s	228.5 s	304.0	4
Cl_2Mg	Magnesium chloride			762	908	1111	1414	4
Cl_2Mn	Manganese(II) chloride				760	933	1189	4
Cl_2Ni	Nickel(II) chloride	534 s	592 s	662 s	747 s	852 s	985 s	4
Cl_2OS	Thionyl chloride	-99 e	-81 e	-58 e	-27.1	14.6	75.2	5
Cl_2O_2S	Sulfuryl chloride				-27 e	11.8	69.0	5
Cl_2Pb	Lead(II) chloride			541 e	637	765	949	23
Cl_2S	Sulfur dichloride	-76 e	-61 e	-41 e	-16.7	15.3	58.7	5
Cl_2S_2	Sulfur chloride	-55 e	-36 e	-12 e	21.0	67.2	137.1	5
Cl_2Sn	Tin(II) chloride		253	308	381	479	622	4
Cl_2Zn	Zinc chloride	305 i	356 i	419 i	497 i	596 i	726 i	4,9,12
Cl_3Fe	Iron(III) chloride	118 s	153 s	190 s	229 s	268 s	319	4
Cl_3HSi	Trichlorosilane			-81 e	-56 e	-21 e	31.6	7
Cl_3N	Nitrogen trichloride				-25 e	13.2	70.6	5
Cl_3OP	Phosphorus(V) oxychloride					39.9	105.0	5
Cl_3P	Phosphorus(III) chloride	-93 e	-77 e	-55 e	-26.0	14.5	75.7	5
Cl_4Po	Polonium(IV) chloride					300.6	389.4	5
Cl_4Se	Selenium tetrachloride	23 s	45 s	71 s	102 s	141.4 s	191.1 s	5
Cl_4Si	Tetrachlorosilane*				-39 e	0 e	57.3	1
Cl_4Te	Tellurium tetrachloride				237 e	299.4	387.8	5
Cl_4Zr	Zirconium(IV) chloride	117 s	146 s	181 s	222 s	272 s	336 s	9
Cl_5P	Phosphorus(V) chloride	-2 s	19 s	44 s	74 s	111.4 s	158.9 s	5
Co	Cobalt	1517	1687	1892	2150	2482	2925	2
Cr	Chromium	1383 s	1534 s	1718 s	1950	2257	2669	2
Cs	Cesium	144.5	195.6	260.9	350.0	477.1	667.0	13,30

Mol. Form.	Name	1 Pa	10 Pa	100 Pa	1 kPa	10 kPa	100 kPa	Ref.
					Temperature in °C for the indicated pressure			
CsF	Cesium fluoride				825	999	1249	4
CsI	Cesium iodide	523 s	595 s	692	854	1029	1278	4,25
Cu	Copper	1236	1388	1577	1816	2131	2563	2
CuI	Copper(I) iodide				636	864	1331	4
Dy	Dysprosium	1105 s	1250 s	1431 i	1681 i	2031 i	2558 i	3
Er	Erbium	1231 s	1390 s	1612 i	1890 i	2279 i	2859 i	3
Eu	Europium	590 s	684 s	799 s	961	1179	1523	14
FH	Hydrogen fluoride*				-71.1	-33.7	19.2	1,5
FHO$_3$S	Fluorosulfonic acid	-14 e	4 e	28 e	59.1	101.3	162.2	5
FK	Potassium fluoride			869	1017	1216	1499	4
FLi	Lithium fluoride	801 s	896	1024	1188	1395	1672	4,12,25
FNO	Nitrosyl fluoride			-131 e	-116.1	-94.3	-60.1	5
FNO$_2$	Nitryl fluoride		-156 e	-144 e	-128.1	-106.0	-72.6	5
FNO$_3$	Fluorine nitrate	-160 e	-149 e	-135 e	-115.1	-87.4	-45.0	5
FNa	Sodium fluoride		920 s	1058	1218	1426	1702	4,12,24
FRb	Rubidium fluoride			910	1001	1145	1409	4,12
F$_2$	Fluorine*	-235 s	-229.5 s	-222.9 s	-214.8	-204.3	-188.3	1,5
F$_2$O	Fluorine monoxide*	-211.7	-204.7	-195.9	-184.2	-168.2	-144.9	5
F$_2$OS	Thionyl fluoride			-124 e	-106.5	-81.5	-44.1	5
F$_2$O$_2$Re	Rhenium(VI) dioxydifluoride				89.2	131.9	185 e	26
F$_2$Pb	Lead(II) fluoride				865	1054	1292	4
F$_2$Xe	Xenon difluoride			2.9 s	31.8 s	67.9 s	114 s	1,5
F$_2$Zn	Zinc fluoride	731 s	813 s	911 i	1048 i	1237 i	1503 i	9
F$_3$N	Nitrogen trifluoride*	-201 e	-194 e	-185 e	-172.8	-155.5	-129.2	5
F$_3$OP	Phosphorus(V) oxyfluoride	-124 s	-113 s	-100 s	-83.7 s	-64.1 s	-39.7 s	5
F$_3$P	Phosphorus(III) fluoride*				-152 e	-132.6	-101.4	5
F$_4$MoO	Molybdenum(VI) oxytetrafluoride	-21 s	3 s	33 s	69.3 s	117.3	184.1	26
F$_4$ORe	Rhenium(VI) oxytetrafluoride	5 s	26 s	50.7 s	80.1 s	117.1	171.2	26
F$_4$OW	Tungsten(VI) oxytetrafluoride	2 s	25 s	52.1 s	84.3 s	126.7	185.4	26
F$_4$S	Sulfur tetrafluoride				-110.0	-82.1	-40.3	5
F$_4$Se	Selenium tetrafluoride				13.6	51.6	104.7	5
F$_4$Si	Tetrafluorosilane*	-166 s	-157 s	-145.6 s	-132.3 s	-115.7 s	-94.9 s	4,7
F$_5$Mo	Molybdenum(V) fluoride				86.6	140.3	213 e	26
F$_5$Nb	Niobium(V) fluoride				80	140	224	4
F$_5$ORe	Rhenium(VII) oxypentafluoride	-103 s	-84 s	-59 s	-28 s	13.7 s	72.8	26
F$_5$Os	Osmium(V) fluoride			74.1	113.2	162.3	226 e	26
F$_5$P	Phosphorus(V) fluoride	-157 s	-148 s	-137 s	-124.5 s	-108.6 s	-84.8 s	5
F$_5$Re	Rhenium(V) fluoride			58.8	99.5	152 e	221 e	26
F$_5$Ta	Tantalum(V) fluoride					119	229	4
F$_6$Ir	Iridium(VI) fluoride	-88 s	-71 s	-51 s	-27 s	3.8 s	53.1	26
F$_6$Mo	Molybdenum(VI) fluoride	-98 s	-82 s	-64 s	-41.2 s	-13.4 s	33.5	26
F$_6$Os	Osmium(VI) fluoride	-89 s	-73 s	-54 s	-30.6 s	-1.7 s	47.4	26
F$_6$Re	Rhenium(VI) fluoride	-97 s	-82 s	-63 s	-40.2 s	-11.9 s	33.4	26
F$_6$S	Sulfur hexafluoride*	-158 s	-147 s	-133.6 s	-116.6 s	-94.4 s	-64.1 s	5
F$_6$Se	Selenium hexafluoride	-143 s	-132 s	-118 s	-100.7 s	-77.8 s	-46.5 s	5
F$_6$Te	Tellurium hexafluoride	-142 s	-130 s	-115 s	-96 s	-71.8 s	-39.1 s	5
F$_6$W	Tungsten(VI) fluoride	-107 s	-92 s	-74 s	-52.1 s	-24.8 s	16.9	26
F$_{10}$S$_2$	Sulfur decafluoride					-22.0	28.5	5
Fe	Iron	1455 s	1617	1818	2073	2406	2859	2
Fr	Francium	131 e	181 e	246 e	335 e	465 e	673 e	2
Ga	Gallium	1037	1175	1347	1565	1852	2245	2
Gd	Gadolinium	1563 i	1755 i	1994 i	2300 i	2703 i	3262 i	3
Ge	Germanium	1371	1541	1750	2014	2360	2831	2
HI	Hydrogen iodide*	-146 s	-135.2 s	-120.8 s	-101.9 s	-75.9 s	-35.9	5
HKO	Potassium hydroxide	520 e	601 e	704	842	1035	1325	4
HNO$_3$	Nitric acid			-37 e	-9 e	28.4	82.2	5
HN$_3$	Hydrazoic acid			-79 e	-54 e	-18.0	35.7	5
HNaO	Sodium hydroxide	513	605	722	874	1080	1377	4
H$_2$	Hydrogen*					-258.6	-252.8	1
H$_2$I$_2$Si	Diiodosilane				11.8	70.5	149.4	4
H$_2$O	Water***	-60.7 s	-42.2 s	-20.3 s	7.0	45.8	99.6	36,37
H$_2$O$_2$	Hydrogen peroxide			13 e	45 e	89.0	149.8	5
H$_2$O$_4$S	Sulfuric acid	72	103	140	187	248	330	4
H$_2$S	Hydrogen sulfide*		-149	-136	-118.9 s	-95.9 s	-60.5	1,5
H$_2$S$_2$	Hydrogen disulfide				-27 e	12.2	70.7	5
H$_2$Se	Hydrogen selenide	-145 s	-134 s	-120 s	-102.8 s	-78.9 s	-41.5	5

Mol. Form.	Name	1 Pa	10 Pa	100 Pa	1 kPa	10 kPa	100 kPa	Ref.
				Temperature in °C for the indicated pressure				
H_2Te	Hydrogen telluride					-46.6	-2.3	5
H_3ISi	Iodosilane				-47.7	-10.1	45.2	4
H_3N	Ammonia*	-139 s	-127 s	-112 s	-94.5 s	-71.3	-33.6	1,5,6
H_3NO	Hydroxylamine				43.7	73.3	109.8	4
H_3P	Phosphine*	-182 s	-173 s	-161 s	-145 s	-122.7	-88.0	5
H_4IN	Ammonium iodide	125 s	159 s	201 s	253 s	318.4 s	405.2 s	5
H_4N_2	Hydrazine				14.7	55.6	113 e	5
H_4Si	Silane*			-181	-165.4	-143.7	-111.8	4
He	Helium*					-270.6	-268.9	2
Hf	Hafnium	2416	2681	3004	3406	3921	4603	9
Hg	Mercury**	42.0	76.6	120.0	175.6	250.3	355.9	29,30
HgI_2	Mercury(II) iodide	85.1 s	115.6 s	152.4 s	197.8 s	255.1 s	353.6	4
Ho	Holmium	1159 s	1311 s	1502 i	1767 i	2137 i	2691 i	3
IK	Potassium iodide			731	866	1052	1322	4
ILi	Lithium iodide	545	619	710	824	972	1170	4
INa	Sodium iodide			753	883	1058	1301	4
IRb	Rubidium iodide			733	866	1045	1302	4
ITl	Thallium(I) iodide				520	644	821	4
I_2	Iodine (rhombic)	-12.8 s	9.3 s	35.9 s	68.7 s	108 s	184.0	1,2
I_2Pb	Lead(II) iodide			470	558	682	869	4
I_2Zn	Zinc iodide	301 s	351 s	409 s	488 i	598 i	750 i	9
I_3Sb	Antimony(III) iodide				214.9	292.0	401.2	4
I_4Sn	Tin(IV) iodide				167.1	242.7	347.7	4
I_4Zr	Zirconium(IV) iodide	187 s	220 s	259 s	305 s	361 s	430 s	4
In	Indium	923	1052	1212	1417	1689	2067	2
Ir	Iridium	2440 s	2684	2979	3341	3796	4386	2
K	Potassium	200.2	256.5	328	424	559	756.2	13,30
Kr	Krypton*	-214.0 s	-208.0 s	-199.4 s	-188.9 s	-174.6 s	-153.6	5
La	Lanthanum	1732 i	1935 i	2185 i	2499 i	2905 i	3453 i	3
Li	Lithium	524.3	612.3	722.1	871.2	1064.3	1337.1	13,30
Lu	Lutetium	1633 s	1829.8	2072.8	2380 i	2799 i	3390 i	3
Mg	Magnesium	428 s	500 s	588 s	698	859	1088	2
Mn	Manganese	955 s	1074 s	1220 s	1418	1682	2060	2
Mo	Molybdenum	2469 s	2721	3039	3434	3939	4606	2
MoO_3	Molybdenum(VI) oxide				801	935	1151	4
NO	Nitric oxide*	-201 s	-195 s	-188 s	-179.3 s	-168.1 s	-151.9	5
N_2	Nitrogen*	-236 s	-232 s	-226.8 s	-220.2 s	-211.1 s	-195.9	1,5
N_2O	Nitrous oxide*	-167 s	-157 s	-145.4 s	-131.1 s	-112.9 s	-88.7	5
N_2O_4	Nitrogen tetroxide	-92 s	-78 s	-61 s	-41.1 s	-16.6 s	28.7	5
N_2O_5	Nitrogen pentoxide	-71 s	-56 s	-40 s	-19.9 s	3.9 s	33.2	5
Na	Sodium	280.6	344.2	424.3	529	673	880.2	13,30
Nb	Niobium	2669	2934	3251	3637	4120	4740	2
Nd	Neodymium	1322.3	1501.2	1725.3	2023 i	2442 i	3063 i	3
Ne	Neon*	-261 s	-260 s	-258 s	-255 s	-252 s	-246.1	2
Ni	Nickel	1510	1677	1881	2137	2468	2911	2
OPb	Lead(II) oxide	724	816	928	1065	1241	1471	4
OSr	Strontium oxide	1789 s	1903 s	2047 s	2235 s	2488 s		4
O_2	Oxygen*				-211.9	-200.5	-183.1	1,28
O_2S	Sulfur dioxide*			-98 s	-80 s	-52.2	-10.3	1,5
O_2Se	Selenium dioxide	124.5 s	153.9 s	188 s	228 s	275 s	315 s	38
O_2Si	Silicon dioxide	1966 i	2149 i	2368 i				8
O_3	Ozone*	-189 e	-182 e	-172 e	-158 e	-139.7	-111.5	5
O_3P_2	Phosphorus(III) oxide				47.3	100.3	172.8	4
O_3S	Sulfur trioxide				-20 s	6.6 s	44.5	5
O_3Sb_2	Antimony(III) oxide (valentinite)	426.1 s	478 s	539 s	610 s	907	1420	4,35
O_5P_2	Phosphorus(V) oxide	285 s	328 s	377.5 s	434.4 s	500.5 s	591	4
O_7Re_2	Rhenium(VII) oxide	147 s	176 s	208 s	244 s	284 s	362	4
Os	Osmium	2887 s	3150	3478	3875	4365	4983	2
P	Phosphorus (white)	6 s	34 s	69	115	180	276	3,9
P	Phosphorus (red)	182 s	216 s	256 s	303 s	362 s	431 s	2,3
Pb	Lead	705	815	956	1139	1387	1754	2
PbS	Lead(II) sulfide	656 s	741 s	838 s	953 s	1088 s	1280	4
Pd	Palladium	1448 s	1624	1844	2122	2480	2961	2
Po	Polonium				573 e	730.2	963.3	5
Pr	Praseodymium	1497.7	1699.4	1954 i	2298 i	2781 i	3506 i	3
Pt	Platinum	2057	2277 e	2542	2870	3283	3821	2

Mol. Form.	Name	Temperature in °C for the indicated pressure						Ref.
		1 Pa	10 Pa	100 Pa	1 kPa	10 kPa	100 kPa	
Pu	Plutonium	1483	1680	1925	2238	2653	3226	2
Ra	Radium	546 s	633 s	764	936	1173	1526	2
Rb	Rubidium	160.4	212.5	278.9	368	496.1	685.3	13,30
Re	Rhenium	3030 s	3341	3736	4227	4854	5681	2
Rh	Rhodium	2015	2223	2476	2790	3132	3724	2
Rn	Radon*	-163 s	-152 s	-139 s	-121.4 s	-97.6 s	-62.3	5
Ru	Ruthenium	2315 s	2538	2814	3151	3572	4115	2
S	Sulfur	102 s	135	176	235	318	444	3
Sb	Antimony	534 s	603 s	738	946	1218	1585	2,3
Sc	Scandium	1372 s	1531 s	1733 i	1993 i	2340 i	2828 i	3
Se	Selenium	227	279	344	431	540	685	3
Si	Silicon	1635	1829	2066	2363	2748	3264	3
Sm	Samarium	728 s	833 s	967 s	1148 i	1402 i	1788 i	2
Sn	Tin	1224	1384	1582	1834	2165	2620	3
Sr	Strontium	523 s	609 s	717 s	866	1072	1373	2
Ta	Tantalum	3024	3324	3684	4122	4666	5361	2
Tb	Terbium	1516.1	1706.1	1928 i	2232 i	2640 i	3218 i	3
Tc	Technetium	2454 e	2725 e	3051 e	3453 e	3961 e	4621 e	2
Te	Tellurium			502 e	615 e	768.8	992.4	5
Th	Thorium	2360	2634	2975	3410	3986	4782	2
Ti	Titanium	1709	1898	2130 e	2419	2791	3285	2
Tl	Thallium	609	704	824	979	1188	1485	2
Tm	Thulium	844 s	962 s	1108 s	1297 s	1548 i	1944 i	3
U	Uranium	2052	2291	2586	2961	3454	4129	2
V	Vanadium	1828 s	2016	2250	2541	2914	3406	2
W	Tungsten	3204 s	3500	3864	4306	4854	5550	2
Xe	Xenon*	-190 s	-181 s	-170 s	-155.8 s	-136.6 s	-108.4	5,32
Y	Yttrium	1610.1	1802.3	2047 i	2354 i	2763 i	3334 i	3
Yb	Ytterbium	463 s	540 s	637 s	774 s	993 i	1192 i	3
Zn	Zinc	337 s	397 s	477	579	717	912 e	2
Zr	Zirconium	2366	2618	2924	3302	3780	4405	2

Substances containing carbon:

Mol. Form.	Name	1 Pa	10 Pa	100 Pa	1 kPa	10 kPa	100 kPa	Ref.
C	Carbon (graphite)		2566 s	2775 s	3016 s	3299 s	3635 s	15
$CBrClF_2$	Bromochloro-difluoromethane	-136 e	-123 e	-106 e	-83.4	-51.8	-4.3	1
$CBrCl_3$	Bromotrichloromethane				-6 e	38.9	104.4	5
$CBrF_3$	Bromotrifluoromethane*	-168 e	-156 e	-142 e	-122.8	-96.6	-58.1	5
CBrN	Cyanogen bromide				-13 s	17.7 s	61.0	1
CBr_2F_2	Dibromodifluoromethane		-110 e	-91 e	-66 e	-30 e	22.5	1
CBr_4	Tetrabromomethane			25.6 s	65.8 s	111.6	188.9	5
$CClF_3$	Chlorotrifluoromethane	-176 e	-167 e	-155 e	-139 e	-116 e	-81.7	5
CClN	Cyanogen chloride		-94.6 s	-78.1 s	-57 s	-29 s	13.0	5
CCl_2F_2	Dichlorodifluoromethane*	-150 e	-138 e	-122 e	-101.8	-73.1	-30.0	5
CCl_2O	Carbonyl chloride	-127 e	-113 e	-96 e	-73 e	-40.6	7.2	5
CCl_3F	Trichlorofluoromethane*		-107 e	-89 e	-63 e	-28.5	23.3	1,5
CCl_3NO_2	Trichloronitromethane		-59 e	-30 e	4.4	47.8	112.0	5
CCl_4	Tetrachloromethane*	-79.4 s	-70.8 s	-53.5 s	-24.4 s	15.8	76.2	1,5
CFN	Cyanogen fluoride		-135 s	-121.2 s	-104.1 s	-82.8 s	-46.2	1,5
CF_4	Tetrafluoromethane*	-199.9 s	-193 s	-183.9 s	-171.6	-153.9	-128.3	1,5
$CHBrF_2$	Bromodifluoromethane		-128 s	-111.4 s	-89.7 s	-59.7 s	-16 s	5
$CHBr_3$	Tribromomethane				30.5	78.3	148.8	1
$CHClF_2$	Chlorodifluoromethane*	-152 e	-141 e	-126 e	-107.1	-80.5	-41.1	5
$CHCl_2F$	Dichlorofluoromethane	-76 e	-70 e	-61 e	-49 e	-28.7	8.6	1
$CHCl_3$	Trichloromethane*			-61 e	-34 e	4.3	60.8	1
CHF_3	Trifluoromethane*			-152 e	-136 e	-114.4	-82.3	1
CHI_3	Triiodomethane	51.1 s	82.7 s	121 e			218.0	5
CHN	Hydrogen cyanide*			-77 s	-52.6 s	-22.7 s	25.4	1,5
CHNO	Cyanic acid			-81.1	-56.8	-23.9	23 e	5
CH_2BrCl	Bromochloromethane	-83 e	-69 e	-50 e	-25 e	11.4	67.7	1
CH_2Br_2	Dibromomethane			-37 e	-7 e	35.2	96.5	5
CH_2ClF	Chlorofluoromethane		-124 e	-108 e	-86.2	-55.7	-9.4	5
CH_2Cl_2	Dichloromethane*		-92 e	-73 e	-48 e	-12.5	39.3	1
CH_2F_2	Difluoromethane*	-156.7	-145.8	-131.9	-113.6	-88.6	-51.9	1
CH_2I_2	Diiodomethane			17 e	55 e	106.1	181.6	5
CH_2O	Formaldehyde*			-91 e	-61.7	-19.3		1

Mol. Form.	Name	Temperature in °C for the indicated pressure						Ref.
		1 Pa	10 Pa	100 Pa	1 kPa	10 kPa	100 kPa	
CH_2O_2	Formic acid	-56 s	-40.4 s	-22.3 s	-0.8 s	37.0	100.2	1,5
CH_3AsF_2	Methyldifluoroarsine				-15 e	22.1	76.1	5
CH_3BO	Borane carbonyl			-124	-99	-64	-64	4
CH_3Br	Bromomethane				-77 e	-44.3	3.3	1
CH_3Cl	Chloromethane*	-140.2 s	-128.6 s	-114.7 s	-96 e	-67.1	-24.4	1,33
CH_3Cl_3Si	Methyltrichlorosilane		-83 e	-61 e	-33 e	7 e	65.7	1
CH_3F	Fluoromethane*				-130 e	-111 e	-78.6	1
CH_3I	Iodomethane				-49 e	-12.4	42.1	1
CH_3NO	Formamide		22 e	53 e	93 e	145.0	218 e	5
CH_3NO_2	Nitromethane				-2 e	40 e	100.8	1
CH_3NO_3	Methyl nitrate		-75 e	-55 e	-27 e	9.8	63 e	5
CH_4	Methane*	-220 s	-214.2 s	-206.8 s	-197 s	-183.6 s	-161.7	5,41
CH_4Cl_2Si	Dichloromethylsilane			-77 e	-51 e	-14 e	40.5	1
CH_4O	Methanol*	-87 e	-69 e	-47.5	-20.4	15.2	64.2	11
CH_4S	Methanethiol			-115 e	-74 e	-41.7	5.7	1
CH_5ClSi	Chloromethylsilane	-129 e	-115 e	-97.9	-74.4	-41.5	8.3	5
CH_5N	Methylamine				-76.7	-48.1	-6.6	1
CH_6N_2	Methylhydrazine			-31 e	-4.7	32.9	91 e	1
CH_6OSi	Methyl silyl ether				-90.2	-61.8	-18 e	1
CH_6Si	Methylsilane			-144 e	-124.6	-97.5	-57.5	5
CIN	Cyanogen iodide						153.8	5
$CNNa$	Sodium cyanide		672 e	798	961	1182	1497	4
CN_4O_8	Tetranitromethane				18.0	61.8	124 e	5
CO	Carbon monoxide*			-223 s	-216.5 s	-207.2 s	-191.7	40
COS	Carbon oxysulfide*			-136 e	-117 e	-90.0	-50.4	1
$COSe$	Carbon oxyselenide			-120	-98	-67	-22	4
CO_2	Carbon dioxide*	-159.1 s	-148.9 s	-136.7 s	-121.6 s	-103.1 s	-78.6 s	5
CS_2	Carbon disulfide		-96 e	-76 e	-49 e	-10.9	45.9	1
CSe_2	Carbon diselenide			-24 e	9.4	56.2	127 e	1
$C_2Br_2ClF_3$	1,2-Dibromo-1-chloro-1,2,2-trifluoroethane						92.3	5
$C_2Br_2F_4$	1,2-Dibromotetrafluoroethane		-97	-75 e	-46	-7.2	47.1	5
C_2Br_4	Tetrabromoethylene		-54.5 s	-31.7 s	-3.5 s	32.2 s	226.0	5
C_2ClF_3	Chlorotrifluoroethylene	-146 e	-134 e	-119 e	-99 e	-71 e	-28.4	1
C_2ClF_5	Chloropentafluoroethane					-80.3	-39.4	1
$C_2Cl_2F_4$	1,1-Dichlorotetrafluoroethane					-45.4	2.7	5
$C_2Cl_2F_4$	1,2-Dichlorotetrafluoroethane				-76.8	-44.9	3.2	5
$C_2Cl_3F_3$	1,1,1-Trichlorotrifluoroethane						45.6	1,5
$C_2Cl_3F_3$	1,1,2-Trichlorotrifluoroethane					-8.2	47.3	1,5
C_2Cl_3N	Trichloroacetonitrile				-16 e	25.3	85.1	1
C_2Cl_4	Tetrachloroethylene			-22 e	10 e	54.4	120.7	1
$C_2Cl_4F_2$	1,1,1,2-Tetrachloro-2,2-difluoroethane				-7 e	31.0	91.1	5
$C_2Cl_4F_2$	1,1,2,2-Tetrachloro-1,2-difluoroethane					32.3	92.5	1
C_2Cl_4O	Trichloroacetyl chloride			-25 e	7 e	51.7	117.8	1,5
C_2Cl_6	Hexachloroethane	-7.6 s	9.9 s	33.6 s	67.7 s	116.9 s	184.2 s	5
C_2F_3N	Trifluoroacetonitrile				-126.1	-102.5	-67.8	1
C_2F_4	Tetrafluoroethylene				-132.3	-109.7	-75.8	1
$C_2F_4N_2O_4$	1,1,2,2-Tetrafluoro-1,2-dinitroethane				-30 e	6.4	59.5	5
C_2F_6	Hexafluoroethane**			-155.2 s	-137.5 s	-113.4 s	-78.4 s	1,5
$C_2HBrClF_3$	2-Bromo-2-chloro-1,1,1-trifluoroethane				-41.4	-4.8	49.8	1
C_2HBr_3O	Tribromoacetaldehyde			15.0	52.7	103.0	173.5	5
C_2HClF_4	1-Chloro-1,1,2,2-tetrafluoroethane			-110 e	-87.6	-57.0	-12.1	5
$C_2HCl_2F_3$	2,2-Dichloro-1,1,1-trifluoroethane		-101.0	-82.2	-57.4	-23.3	26.7	18
C_2HCl_3	Trichloroethylene	-74 e	-59 e	-39 e	-12 e	26.7	86.8	1
C_2HCl_3O	Trichloroacetaldehyde			-41.6	-9.8	33.8	97.4	5
$C_2HCl_3O_2$	Trichloroacetic acid				83.8	130.0	197.2	1,5
C_2HCl_5	Pentachloroethane		-23 e	3 e	37.4	86.0	159.4	1
$C_2HF_3O_2$	Trifluoroacetic acid					16.8	71.4	1,5
C_2HF_5O	Trifluoromethyl difluoromethyl ether	-147 e	-136 e	-121 e	-102 e	-75.0	-35.4	20
C_2H_2	Acetylene*			-146.6 s	-130.7 s	-110.6 s	-84.8 s	5

Mol. Form.	Name	Temperature in °C for the indicated pressure						Ref.
		1 Pa	10 Pa	100 Pa	1 kPa	10 kPa	100 kPa	
$C_2H_2Br_2$	cis-1,2-Dibromoethylene		-45 e	-21 e	10 e	52.2	114.8	1
$C_2H_2Br_2$	trans-1,2-Dibromoethylene				-4 e	42.2	107.4	5
$C_2H_2Br_2Cl_2$	1,2-Dibromo-1,1-dichloroethane					103.6	177.8	5
$C_2H_2Br_2Cl_2$	1,2-Dibromo-1,2-dichloroethane		-11 e	22 e	64.1	119 e	193 e	5
$C_2H_2Br_4$	1,1,2,2-Tetrabromoethane	14 e	38 e	69 e	109 e	163.7	242.9	5
$C_2H_2Cl_2$	1,1-Dichloroethylene	-116 e	-101 e	-82 e	-57 e	-21.4	31.2	1
$C_2H_2Cl_2$	cis-1,2-Dichloroethylene			-62 e	-34 e	3.8	60.3	1
$C_2H_2Cl_2$	trans-1,2-Dichloroethylene				-44 e	-7.5	47.3	1
$C_2H_2Cl_2F_2$	1,2-Dichloro-1,1-difluoroethane	-101 e	-87 e	-68 e	-42.2	-6.8	46.3	5
$C_2H_2Cl_2O$	Chloroacetyl chloride			-23.7	5.6	46.1	105.6	5
$C_2H_2Cl_4$	1,1,1,2-Tetrachloroethane	-58 e	-40 e	-15 e	17 e	62.2	129.7	1
$C_2H_2Cl_4$	1,1,2,2-Tetrachloroethane		-22 e	1 e	32.4	76.9	144.7	1
$C_2H_2F_4$	1,1,1,2-Tetrafluoroethane				-94.3	-66.8	-26.4	17
$C_2H_2F_4$	1,1,2,2-Tetrafluoroethane				-96.0	-66.9	-23.3	5
C_2H_2O	Ketene		-151 e	-135 e	-115 e	-88.2	-50.0	1
C_2H_3Br	Bromoethylene	-124 e	-110 e	-92 e	-68 e	-34.5	15.4	5
C_2H_3BrO	Acetyl bromide	-78 e	-65 e	-49 e	-25 e	13.9	84 e	5
$C_2H_3Br_3$	1,1,2-Tribromoethane	-18 e	4 e	32 e	68 e	117.1	188.4	5
C_2H_3Cl	Chloroethylene	-139 e	-127 e	-110 e	-89 e	-59.0	-14.1	1
$C_2H_3ClF_2$	1-Chloro-1,1-difluoroethane		-123 e	-107 e	-85.3	-55.4	-10.5	5
C_2H_3ClO	Acetyl chloride	-100 e	-85 e	-66 e	-40 e	-3.6	50.4	1
$C_2H_3ClO_2$	Chloroacetic acid				78.4	123.9	188.9	1
$C_2H_3Cl_2F$	1,1-Dichloro-1-fluoroethane		-101 e	-83 e	-57.9	-22.7	31.4	5
$C_2H_3Cl_2F$	1,2-Dichloro-1-fluoroethane			-50 e	-23.8	14.1	73.4	5
$C_2H_3Cl_3$	1,1,1-Trichloroethane				-25.3	14.2	73.7	5
$C_2H_3Cl_3$	1,1,2-Trichloroethane			-23 e	7 e	49.9	113.4	1
C_2H_3F	Fluoroethylene				-153.3	-135.2	-109.9	-72.2	5
C_2H_3FO	Acetyl fluoride					-64.1	17.0	5
$C_2H_3F_3$	1,1,1-Trifluoroethane				-113 e	-86.6	-47.8	1
$C_2H_3F_3O$	2,2,2-Trifluoroethanol			-33 e	-8 e	26.0	74 e	5
C_2H_3I	Iodoethylene				-41 e	-3 e	55.6	5
C_2H_3IO	Acetyl iodide				-0.6	47 e	107.0	5
C_2H_3N	Acetonitrile				-20 e	21.4	81.2	1
C_2H_3NO	Methylisocyanate				-43.5	-10.2	38.8	1
C_2H_3NS	Methyl thiocyanate			-18.4	16.2	63.5	132.5	5
C_2H_4	Ethylene*				-155.6	-135.1	-104.0	1,10
C_2H_4BrCl	1-Bromo-2-chloroethane				-0.4	41.7	105.7	6
$C_2H_4Br_2$	1,1-Dibromoethane		-49 e	-26 e	5 e	46.4	107.6	5
$C_2H_4Br_2$	1,2-Dibromoethane				18 e	62.2	130.9	1
C_2H_4ClF	1-Chloro-1-fluoroethane				-69.9	-36.1	15.8	5
$C_2H_4Cl_2$	1,1-Dichloroethane		-84 e	-64 e	-36.7	1.0	56.9	1
$C_2H_4Cl_2$	1,2-Dichloroethane				-16.4	23.7	83.1	1
$C_2H_4F_2$	1,1-Difluoroethane				-115.2	-94.6	-66.1	-24.3	19
$C_2H_4N_2O_6$	Ethylene glycol dinitrate	4 e	25.6	51.0	81 e	117 e	162 e	5
C_2H_4O	Acetaldehyde		-105 e	-87 e	-62.8	-29.4	20.0	5
C_2H_4O	Ethylene oxide		-111 e	-93 e	-70 e	-37.0	10.2	1
$C_2H_4O_2$	Acetic acid	-42.8 s	-26.7 s	-8 s	14.2 s	55.9	117.5	1,5
$C_2H_4O_2$	Methyl formate		-95 e	-76 e	-51.8	-18.1	31.4	5
$C_2H_4O_3$	Peroxyacetic acid				14.4	55.3	109.7	5
$C_2H_4O_3$	Glycolic acid						99.9	5
$C_2H_5AsF_2$	Ethyldifluoroarsine			-36 e	-6.0	35.0	93.1	5
C_2H_5Br	Bromoethane	-111 e	-96 e	-77 e	-51.3	-15.5	38.0	5
C_2H_5Cl	Chloroethane	-126 e	-112 e	-94 e	-70 e	-37.0	12.0	1
C_2H_5ClO	2-Chloroethanol	-61 e	-39 e	-12 e	23 e	67.1	127.3	5
C_2H_5ClO	Chloromethyl methyl ether	-96 e	-80 e	-59 e	-32 e	6 e	61 e	5
$C_2H_5Cl_3OSi$	Trichloroethoxysilane	-78 e	-60 e	-36.0	-4.6	38.7	102.0	5
$C_2H_5Cl_3Si$	Trichloroethylsilane	-79 e	-61 e	-38 e	-8 e	34.9	98.7	5
C_2H_5F	Fluoroethane		-142 e	-127 e	-106.3	-78.7	-37.9	1
C_2H_5FO	2-Fluoroethanol			-22 e	8.3	47.5	99 e	5
C_2H_5I	Iodoethane	-94 e	-78 e	-56 e	-27.9	11.9	71.9	5
C_2H_5N	Ethyleneimine		-74 e	-55 e	-30 e	4.1	55 e	5
C_2H_5NO	Acetamide	16.7 s	39.1 s	65.2 s	102.8	150.8	218.2	5
C_2H_5NO	N-Methylformamide		13 e	41 e	78 e	127.9	199.1	1
$C_2H_5NO_2$	Nitroethane	-61 e	-44 e	-21 e	8.3	50.1	113.5	5
$C_2H_5NO_3$	Ethyl nitrate	-81 e	-63 e	-41 e	-12 e	28.2	87 e	1
C_2H_6	Ethane*	-183.3 s	-173.2	-161.3	-145.3	-122.8	-88.8	41
$C_2H_6Cl_2Si$	Dichlorodimethylsilane					11.1	70.1	5

Mol. Form.	Name	\multicolumn{6}{c}{Temperature in °C for the indicated pressure}	Ref.
		1 Pa	10 Pa	100 Pa	1 kPa	10 kPa	100 kPa	
C_2H_6Hg	Dimethyl mercury				-13.5	29.0	92.1	5
$C_2H_6N_2O$	N-Nitrosodimethylamine				30.7	80.5	149.8	5
C_2H_6O	Ethanol	-73 e	-56 e	-34 e	-7 e	29.2	78.0	1,5
C_2H_6O	Dimethyl ether*		-135 e	-118 e	-96.8	-67.6	-25.1	1,5
C_2H_6OS	Dimethyl sulfoxide			27.4	65.0	115.9	188.6	1
$C_2H_6O_2$	Ethylene glycol	2 e	24 e	51.1	86.1	132.5	196.9	1
$C_2H_6O_2$	Ethyl hydroperoxide	-70 e	-49 e	-25 e	6.8	47.0	101 e	5
$C_2H_6O_2S$	Dimethyl sulfone				109 e	166.8	248.9	5
C_2H_6S	Ethanethiol	-112 e	-97 e	-78 e	-53 e	-18 e	34.7	1
C_2H_6S	Dimethyl sulfide		-96 e	-77 e	-51.2	-16.0	37.0	1,5
$C_2H_6S_2$	Dimethyl disulfide	-71 e	-53 e	-29 e	1.7	45.0	109.3	5
$C_2H_7BO_2$	Dimethoxyborane	-116 e	-101.9	-83.5	-59.2	-25.4	25 e	5
C_2H_7N	Ethylamine			-71 e	-53 e	-27 e	16.4	1
C_2H_7N	Dimethylamine			-88 e	-66.9	-37.2	6.6	1
C_2H_7NO	Ethanolamine		11 e	35 e	66.2	109.0	170.6	1
$C_2H_8N_2$	1,2-Ethanediamine				17.0	57.5	116.6	1,5
$C_2H_8N_2$	1,1-Dimethylhydrazine			-52 e	-25.6	10.5	63 e	5
$C_2H_8N_2$	1,2-Dimethylhydrazine		-49 e	-33 e	-9 e	26.4	88 e	1
C_2N_2	Cyanogen	-127 s	-114.1 s	-98.5 s	-79.2 s	-54.9 s	-21.4	5
C_3ClF_5O	Chloropentafluoroacetone	-122 e	-109 e	-93 e	-71 e	-39.4	7.4	5
C_3Cl_6	Hexachloropropene	-12 e	11 e	40 e	79 e	132.8	213.6	5
C_3F_6	Perfluoropropene	-150 e	-138 e	-122 e	-101 e	-72 e	-30.6	5
C_3F_6O	Perfluoroacetone			-113 e	-94 e	-67.8	-27.6	5
C_3F_8	Perfluoropropane		-139 e	-124 e	-105 e	-77.5	-37.0	1
C_3HN	Cyanoacetylene			-58.7 s	-35.6 s	-7 s	42.0	5
$C_3H_2F_6O$	1,1,1,3,3,3-Hexafluoro-2-propanol					12.7	57.1	5
$C_3H_3F_5$	1,1,1,2,2-Pentafluoropropane					-60 e	-17.9	5
C_3H_3N	2-Propenenitrile		-72 e	-50 e	-22 e	17.7	77.0	1
C_3H_3NS	Thiazole					54.4	117.8	5
C_3H_4	Allene*		-129 e	-118 e	-101.4	-76.7	-34.7	5
C_3H_4	Propyne				-94 e	-65.3	-23.2	1
$C_3H_4ClF_3$	3-Chloro-1,1,1-trifluoropropane	-102 e	-87 e	-68 e	-43 e	-8 e	45.3	5
$C_3H_4Cl_2O$	1,1-Dichloroacetone				1 e	47.8	118.0	5
$C_3H_4Cl_2O_2$	Methyl dichloroacetate	-44 e	-25 e	0 e	33 e	77.7	142.3	5
$C_3H_4Cl_4$	1,1,1,2-Tetrachloropropane	-48 e	-28 e	-2 e	32 e	79.1	149.5	5
$C_3H_4F_4O$	2,2,3,3-Tetrafluoro-1-propanol			-10 e	17 e	53.9	107.2	5
C_3H_4O	Acrolein		-87 e	-67 e	-40 e	-3.0	52.8	1
$C_3H_4O_2$	Propenoic acid				35 e	78.0	140.7	1
$C_3H_4O_2$	Vinyl formate			-58 e	-34 e	-1.6	46.2	1
$C_3H_4O_2$	2-Oxetanone		-21 e	8 e	45.5	93.8	159.3	5
$C_3H_4O_3$	Ethylene carbonate	12.7 s	37 e				247	5
C_3H_5Br	cis-1-Bromopropene	-100 e	-84 e	-64 e	-37 e	1.0	57.4	5
C_3H_5Br	2-Bromopropene	-112 e	-95 e	-75 e	-47 e	-9 e	48.0	5
C_3H_5Br	3-Bromopropene	-98 e	-80 e	-58 e	-28 e	12 e	69.6	5
C_3H_5Cl	cis-1-Chloropropene	-114 e	-100 e	-81 e	-55 e	-20.1	32.4	5
C_3H_5Cl	trans-1-Chloropropene		-97 e	-77 e	-52 e	-16.2	37.0	5
C_3H_5Cl	2-Chloropropene	-120 e	-106 e	-87 e	-63 e	-28.7	22.3	5
C_3H_5Cl	3-Chloropropene	-107 e	-92 e	-72.4	-46.3	-9.8	44.6	5
C_3H_5ClO	Epichlorohydrin			-21 e	11 e	53.8	115.5	5
$C_3H_5ClO_2$	Methyl chloroacetate		-28 e	-5 e	25 e	66.9	129.1	5
$C_3H_5Cl_3$	1,1,3-Trichloropropane	-51 e	-31 e	-5 e	28 e	75.3	145.1	5
$C_3H_5Cl_3$	1,2,3-Trichloropropane			2 e	37 e	84.9	156.3	5
$C_3H_5Cl_3Si$	Trichloro-2-propenylsilane					53.0	116.5	5
C_3H_5I	3-Iodopropene	-80 e	-62 e	-39 e	-8 e	36 e	101.5	5
C_3H_5N	Propanenitrile	-69.4	-55.3	-36.0	-7.9	35.2	97.4	1,5
C_3H_5NO	Acrylamide			109.6	161 e			5
C_3H_5NO	3-Hydroxypropanenitrile	-11 e	18 e	53 e	96.1	150.3	220.8	5
C_3H_5NS	Ethyl thiocyanate	-39 e	-20 e	4 e	35 e	79.1	143.4	5
C_3H_5NS	Ethyl isothiocyanate				17.4	66 e	136 e	5
$C_3H_5N_3O_9$	Trinitroglycerol	48.6	75.7	118 e	191 e	353 e	1007 e	5
C_3H_6	Propene*	-160.6	-149.0	-134.3	-114.9	-88.2	-47.9	1,5
C_3H_6	Cyclopropane			-124 e	-104 e	-75.7	-33.1	1
C_3H_6BrCl	1-Bromo-3-chloropropane	-51 e	-31 e	-6 e	28 e	74.1	142.9	5
$C_3H_6Br_2$	1,2-Dibromopropane	-46 e	-26 e	-2 e	31 e	75.3	139.5	5

Mol. Form.	Name	Temperature in °C for the indicated pressure						Ref.
		1 Pa	10 Pa	100 Pa	1 kPa	10 kPa	100 kPa	
$C_3H_6Br_2$	1,3-Dibromopropane	-30 e	-9 e	17 e	52 e	98.7	166.8	5
$C_3H_6Cl_2$	1,1-Dichloropropane				-14 e	27.0	87.7	5
$C_3H_6Cl_2$	1,2-Dichloropropane	-78 e	-61 e	-38.1	-8.1	33.7	95.9	5
$C_3H_6Cl_2$	1,3-Dichloropropane	-65 e	-46 e	-22 e	10 e	54.0	119.9	5
$C_3H_6Cl_2$	2,2-Dichloropropane				-28 e	10.8	68.9	5
$C_3H_6Cl_2O$	1,3-Dichloro-2-propanol			21.8	59.0	107.6	173.9	5
$C_3H_6N_2O_4$	1,1-Dinitropropane	-9 e	12 e	39 e	73.2	120 e	187 e	5
C_3H_6O	Allyl alcohol	-63 e	-48 e	-21.9	6.8	44.5	96.2	5
C_3H_6O	Methyl vinyl ether			-114 e	-89 e	-52.7	4.6	1
C_3H_6O	Propanal			-69 e	-42 e	-6 e	47.7	1
C_3H_6O	Acetone	-95	-81.8	-62.8	-35.6	1.3	55.7	1,5
C_3H_6O	Methyloxirane	-109 e	-95 e	-76 e	-51.5	-17.2	33.9	5
$C_3H_6O_2$	Propanoic acid			0 e	35.1	79.9	140.8	1,5
$C_3H_6O_2$	Ethyl formate		-80 e	-61 e	-35 e	1 e	54.0	1
$C_3H_6O_2$	Methyl acetate	-95 e	-79 e	-59 e	-33 e	3.3	56.6	1
$C_3H_6O_2$	1,3-Dioxolane		-72 e	-50 e	-22 e	17.0	75.3	1
$C_3H_6O_3$	1,3,5-Trioxane				53 e		113.7	1
C_3H_6S	Thiacyclobutane		-62 e	-40 e	-9 e	32.5	94.5	5
C_3H_7Br	1-Bromopropane	-95 e	-78 e	-57 e	-28 e	11.6	70.6	1
C_3H_7Br	2-Bromopropane		-84 e	-65 e	-39.6	-1.7	59.1	1,5
C_3H_7Cl	1-Chloropropane	-106 e	-90 e	-71 e	-44.5	-8.1	46.2	1
C_3H_7Cl	2-Chloropropane		-91 e	-74 e	-51.1	-17.8	35.4	1,5
C_3H_7ClO	2-Chloro-1-propanol			23 e	63.8		125.7	5
C_3H_7F	1-Fluoropropane	-133 e	-120 e	-103 e	-80.7	-49.4	-2.8	5
C_3H_7I	1-Iodopropane	-78 e	-60 e	-37 e	-6 e	36.9	102.0	5
C_3H_7I	2-Iodopropane	-89 e	-71 e	-47 e	-16.3	26.5	89.2	5
C_3H_7N	Allylamine		-88 e	-65 e	-37 e	0.4	52 e	5
C_3H_7NO	N,N-Dimethylformamide	-39 e	-20 e	5 e	38.0	83.9	152.6	1
C_3H_7NO	N-Methylacetamide	-13.3 s	13 s	43 e	83.8	136.1	206.3	5
$C_3H_7NO_2$	1-Nitropropane	-56 e	-37 e	-13 e	20 e	64.8	130.8	1
$C_3H_7NO_2$	2-Nitropropane		-48 e	-22 e	10.7	55.6	119.8	1
$C_3H_7NO_3$	Propyl nitrate			-23.9	6.1	48.1	111 e	5
C_3H_8	Propane*	-156.9	-145.6	-130.9	-111.4	-83.8	-42.3	1,41
C_3H_8O	1-Propanol	-54 e	-38 e	-16 e	10 e	47 e	96.9	1,5
C_3H_8O	2-Propanol	-65 e	-49 e	-28 e	-1.3	33.6	82.0	1,5
C_3H_8O	Ethyl methyl ether	-98 e	-89 e	-77 e	-60 e	-34.8	7.0	5
$C_3H_8O_2$	1,2-Propylene glycol	-11 e	13 e	42 e	78 e	125.0	187.2	5
$C_3H_8O_2$	1,3-Propylene glycol	4 e	30 e	62 e	101 e	149.9	214.0	5
$C_3H_8O_2$	Ethylene glycol monomethyl ether	-57 e	-37 e	-12 e	21 e	63.8	124.3	1
$C_3H_8O_2$	Dimethoxymethane	-93 e	-81 e	-64 e	-42 e	-9.3	41.7	5
$C_3H_8O_3$	Glycerol	96 e	113 e	136 e	168 e	213.4	287 e	1
C_3H_8S	1-Propanethiol	-94 e	-78 e	-57 e	-29.1	9.6	67.4	1,5
C_3H_8S	2-Propanethiol	-102 e	-87 e	-67 e	-41 e	-3 e	52.2	1
C_3H_8S	Ethyl methyl sulfide	-94 e	-78 e	-57 e	-29.7	8.8	66.3	1
$C_3H_8S_2$	1,3-Propanedithiol	-53 e	-28 e	3 e	43 e	97 e	172.4	5
C_3H_9As	Trimethylarsine		-74 e	-45 e	-5.4	52.0		5
$C_3H_9BO_3$	Trimethyl borate			-14 e	15.6	67.9		5
C_3H_9BS	Methyl dimethylthioborane		-62 e	-30.4	11.4	70.7		5
C_3H_9ClSi	Trimethylchlorosilane		-37.8	0.4	57.3			5
C_3H_9N	Propylamine		-81 e	-63 e	-38.3	-4.1	46.9	1,5
C_3H_9N	Isopropylamine		-91 e	-74 e	-50.4	-17.6	31.5	1,5
C_3H_9N	Trimethylamine		-114 e	-97 e	-75.0	-43.8	2.6	1,5
C_3H_9NO	1-Amino-2-propanol			18 e	53.2	98.2	157.9	5
$C_3H_9O_4P$	Trimethyl phosphate	-31 e	-7 e	23.6	62.8	116.0	192.0	5
C_3H_9P	Trimethylphosphine			-81 e	-53 e	-15.0	37.1	5
C_3H_9Sb	Trimethylstibine			-56 e	-23.8	19 e	80 e	5
$C_3H_{10}N_2$	1,2-Propanediamine		-35.4	-12.0	18.8	61 e	119 e	5
C_3N_2O	Carbonyl dicyanide			-21.7	15.3	65.2		5
C_4Cl_6	Hexachloro-1,3-butadiene	-1 e	22 e	50 e	86.7	137.0	209.7	5
$C_4F_6O_3$	Trifluoroacetic acid anhydride			-63 e	-39 e	-7.1	38.8	5
C_4F_8	Perfluorocyclobutane					-6.2		1
C_4F_{10}	Perfluorobutane		-122 e	-105 e	-82 e	-49.8	-2.5	1,5
$C_4H_2Cl_2O_2$	trans-2-Butenedioyl dichloride			8.0	45.6	94.3	159.8	5
$C_4H_2Cl_2S$	2,5-Dichlorothiophene			-20 e	22 e	81.4	171 e	5

Mol. Form.	Name	Temperature in °C for the indicated pressure						Ref.
		1 Pa	10 Pa	100 Pa	1 kPa	10 kPa	100 kPa	
$C_4H_2O_3$	Maleic anhydride				73.7	127.9	201.7	5
C_4H_3ClS	2-Chlorothiophene		-62 e	-35 e	2 e	51.8	123 e	5
C_4H_3IS	2-Iodothiophene			-25 e	23 e	94.9	181.0	5
C_4H_4	1-Buten-3-yne			-96.1	-73.4	-41.8	4.9	5
$C_4H_4N_2$	Succinonitrile	24.8 s					266.0	5
C_4H_4O	Furan			-78 e	-54 e	-20 e	31.0	1
$C_4H_4O_2$	Diketene				19.3	63.3	126 e	5
$C_4H_4O_3$	Succinic anhydride				121 e	180.8	260.8	5
$C_4H_4O_4$	Fumaric acid	123.9 s	150 s	180 s				5
C_4H_4S	Thiophene				-17 e	23.7	83.7	5
C_4H_5Cl	2-Chloro-1,3-butadiene	-113 e	-95 e	-71 e	-41 e	0.3	59.0	5
C_4H_5ClO	2-Methyl-2-propenoyl chloride		-57 e	-35 e	-5 e	36.4	98.2	5
$C_4H_5Cl_3O_2$	Ethyl trichloroacetate			15.3	51.9	100.1	166.6	5
C_4H_5N	3-Butenenitrile	-67 e	-48 e	-23.1	9.3	53.7	118.4	5
C_4H_5N	Methylacrylonitrile				-12 e	29.0	89.8	5
C_4H_5N	Pyrrole			-8 e	24 e	66.7	129.4	1
$C_4H_5NO_2$	Methyl cyanoacetate	-3 e	19 e	48 e	84 e	134.0	204.6	5
C_4H_5NS	Allyl isothiocyanate	-45 e	-27 e	-3 e	32.1	89 e	198 e	5
C_4H_5NS	4-Methylthiazole						67.0	5
C_4H_6	1,2-Butadiene	-132 e	-117 e	-98 e	-72.8	-38.9	10.5	5
C_4H_6	1,3-Butadiene*			-106 e	-83 e	-51.9	-4.7	1
C_4H_6	1-Butyne	-125 e	-111 e	-94 e	-71.2	-39.4	7.8	1
C_4H_6	2-Butyne		-89.2 s	-73.8 s	-53.5 s	-23.9	26.6	5
$C_4H_6Cl_2O_2$	Ethyl dichloroacetate			2.6	40.1	89.1	156.3	5
C_4H_6O	Divinyl ether		-99 e	-80 e	-56 e	-22.1	28.0	5
C_4H_6O	trans-2-Butenal	-74 e	-56 e	-33 e	-3 e	39.7	102.4	5
C_4H_6O	3-Buten-2-one					21 e	81.0	5
C_4H_6O	Cyclobutanone			-34 e	-4 e	37.1	97 e	5
$C_4H_6O_2$	cis-Crotonic acid			30 e	63 e	106.7	168.9	5
$C_4H_6O_2$	trans-Crotonic acid				74 e	120.8	184.9	5
$C_4H_6O_2$	3-Butenoic acid	-19 e	2 e	27 e	61 e	105.6	168.6	5
$C_4H_6O_2$	Methacrylic acid			22 e	56 e	99.9	161.5	5
$C_4H_6O_2$	Vinyl acetate	-88 e	-71 e	-50 e	-22 e	16.2	72.2	1
$C_4H_6O_2$	Methyl acrylate		-71 e	-48 e	-18 e	22 e	79.9	5
$C_4H_6O_2$	2,3-Butanedione					30.7	84.8	5
$C_4H_6O_2$	gamma-Butyrolactone		-17 e	24 e	72 e	130.2	203 e	5
$C_4H_6O_3$	Acetic anhydride	-44 e	-25 e	-1 e	31 e	75.1	139.7	1
$C_4H_6O_3$	Propylene carbonate	-40 e	-5 e	43 e	112 e	220 e	410 e	5
$C_4H_6O_4$	Dimethyl oxalate				50.5	98.1	163.0	5
C_4H_7Br	trans-1-Bromo-1-butene	-87 e	-68 e	-43.3	-11.4	31.9	94.4	5
C_4H_7Br	2-Bromo-1-butene	-87 e	-70 e	-48 e	-20 e	20.7	80.6	5
C_4H_7Br	cis-2-Bromo-2-butene	-90 e	-72 e	-49.0	-18.5	23.5	85.2	5
C_4H_7Br	trans-2-Bromo-2-butene	-86 e	-67 e	-43.4	-12.0	31.0	93.5	5
$C_4H_7Br_3$	1,2,3-Tribromobutane	0 e	23 e	53 e	91 e	143.7	219.5	5
$C_4H_7Br_3$	1,2,4-Tribromobutane	-3 e	20 e	49 e	87 e	139.4	214.5	5
C_4H_7Cl	3-Chloro-1-butene			-64 e	-36 e	4 e	63.6	5
C_4H_7Cl	cis-2-Chloro-2-butene	-100 e	-83 e	-62 e	-34 e	6 e	66.4	5
C_4H_7Cl	trans-2-Chloro-2-butene	-102 e	-86 e	-65 e	-37 e	3 e	62.2	5
C_4H_7Cl	3-Chloro-2-methylpropene		-75 e	-54 e	-25 e	13.8	71.5	5
$C_4H_7ClO_2$	Ethyl chloroacetate			-2.6	32.6	79.1	143.8	5
C_4H_7N	Butanenitrile	-67 e	-48 e	-24 e	8 e	52.3	117.2	1
C_4H_8	1-Butene	-139.0	-125.2	-107.8	-85.3	-53.7	-6.6	1,5
C_4H_8	cis-2-Butene	-131.2	-117.4	-99.8	-76.7	-44.8	3.4	1,5
C_4H_8	trans-2-Butene			-102 e	-80 e	-47.6	0.6	1
C_4H_8	Isobutene	-139.1	-125.5	-108.2	-85.5	-54.5	-7.3	1,5
C_4H_8	Cyclobutane				-71.8	-38.1	12.1	5
C_4H_8	Methylcyclopropane	-130 e	-116 e	-99.3	-76.3	-44.2	4.2	5
$C_4H_8Br_2$	1,2-Dibromobutane	-54 e	-30 e	0.4	39.6	92.1	166.1	5
$C_4H_8Br_2$	1,4-Dibromobutane	-13 e	9 e	37 e	74 e	124.0	196.5	5
$C_4H_8Cl_2$	1,1-Dichlorobutane			-25 e	6 e	49.3	113.4	5
$C_4H_8Cl_2$	1,2-Dichlorobutane			-28.4	5.8	53.1	123.1	5
$C_4H_8Cl_2$	1,4-Dichlorobutane		-26 e	0 e	35 e	82.4	153.4	5
$C_4H_8Cl_2$	2,2-Dichlorobutane		-58 e	-35 e	-5 e	37.8	102.1	5
$C_4H_8Cl_2O$	Bis(2-chloroethyl) ether	-32 e	-9 e	19.8	56.9	106.9	177.9	5
C_4H_8O	Ethyl vinyl ether		-102 e	-81 e	-53.1	-16.5	34.7	5
C_4H_8O	1,2-Epoxybutane	-135 e	-114 e	-87 e	-53 e	-5.5	62.1	5

Mol. Form.	Name	\multicolumn{6}{c}{Temperature in °C for the indicated pressure}	Ref.
		1 Pa	10 Pa	100 Pa	1 kPa	10 kPa	100 kPa	
C_4H_8O	Butanal	-88 e	-72 e	-50 e	-22 e	16.6	74.5	1,5
C_4H_8O	Isobutanal		-56 e	-29 e	8 e	63.8		1
C_4H_8O	2-Butanone	-85 e	-68 e	-46 e	-18.1	21.2	79.2	1
C_4H_8O	Tetrahydrofuran	-94 e	-78 e	-57.3	-29.8	9 e	65.6	1
$C_4H_8O_2$	Butanoic acid			12.9	52.2	101.4	163.3	1,5
$C_4H_8O_2$	2-Methylpropanoic acid	-30.1	-8.2	18.1	50.5	92.9	154.0	5
$C_4H_8O_2$	Propyl formate	-78 e	-62 e	-42 e	-15.1	23.0	80.4	1,5
$C_4H_8O_2$	Isopropyl formate	-80 e	-65 e	-47 e	-22.2	13.2	67.7	5
$C_4H_8O_2$	Ethyl acetate	-83 e	-66 e	-45 e	-18 e	20.4	76.8	1
$C_4H_8O_2$	Methyl propanoate	-80 e	-64 e	-43 e	-15.8	22.2	79.0	1
$C_4H_8O_2$	cis-2-Butene-1,4-diol	17 e	44 e	77 e	117.4	168.5	234.9	5
$C_4H_8O_2$	1,3-Dioxane			-37 e	-3 e	43.4	106.0	5
$C_4H_8O_2$	1,4-Dioxane					39.6	101.0	1
$C_4H_8O_2S$	Sulfolane		49 e	87 e	135 e	198.0	283.5	5
C_4H_8S	Tetrahydrothiophene	-66 e	-47 e	-23 e	9.4	54.1	120.5	1
C_4H_9Br	1-Bromobutane	-68.4	-53.9	-34.1	-5.4	37.6	101.1	1,5
C_4H_9Br	2-Bromobutane	-86 e	-68 e	-46 e	-16 e	26.6	90.7	5
C_4H_9Br	1-Bromo-2-methylpropane	-85 e	-68 e	-46 e	-16 e	26.8	91.1	5
C_4H_9Br	2-Bromo-2-methylpropane					11.7	72.4	1,5
C_4H_9Cl	1-Chlorobutane	-87 e	-71 e	-49 e	-21 e	18.4	78.1	1
C_4H_9Cl	2-Chlorobutane	-96 e	-80 e	-59 e	-31.0	8.5	67.9	1
C_4H_9Cl	1-Chloro-2-methylpropane	-94 e	-78 e	-56.6	-28.7	10.2	68.5	5
C_4H_9Cl	2-Chloro-2-methylpropane					-4.2	50.3	5
$C_4H_9Cl_3Si$	Butyltrichlorosilane					77.2	148.4	5
C_4H_9F	1-Fluorobutane	-114 e	-99 e	-80 e	-55 e	-20.0	32.1	5
C_4H_9F	2-Fluorobutane	-117 e	-103 e	-85 e	-60.7	-26.7	24.7	5
C_4H_9I	1-Iodobutane	-62 e	-43 e	-19 e	14 e	60.5	130.0	5
C_4H_9I	2-Iodobutane	-70 e	-51 e	-27 e	5 e	50 e	119.5	5
C_4H_9I	1-Iodo-2-methylpropane		-47 e	-21.4	12.0	56.8	120.0	5
C_4H_9I	2-Iodo-2-methylpropane	-75.1 s	-58.8 s	-39.5 s	-5.2	41 e	100.0	5
C_4H_9N	Pyrrolidine		-59 e	-38 e	-10 e	28.5	86.2	1
C_4H_9NO	N-Methylpropanamide				81.1	105 e		5
C_4H_9NO	N,N-Dimethylacetamide	-8 e	8 e	28.0	56.4	98.2	165.7	5
C_4H_9NO	2-Butanone oxime		-18 e	7 e	38.9	81.9	142.9	5
C_4H_9NO	Morpholine				21 e	64.5	128.5	1
$C_4H_9NO_3$	Isobutyl nitrate			-18 e	15.1	59.2	123.0	5
C_4H_{10}	Butane*	-134.3	-121.0	-103.9	-81.1	-49.1	-0.8	1,41
C_4H_{10}	Isobutane*		-129.0	-113.0	-90.9	-59.4	-12.0	1,41
$C_4H_{10}O$	1-Butanol	-37 e	-20 e	0 e	28 e	64 e	117.4	1
$C_4H_{10}O$	2-Butanol	-50 e	-34 e	-14 e	12.6	48.2	99.2	1,5
$C_4H_{10}O$	2-Methyl-1-propanol	-39 e	-24 e	-5 e	20.9	56.0	107.6	1,5
$C_4H_{10}O$	2-Methyl-2-propanol					34.4	82.1	1,5
$C_4H_{10}O$	Diethyl ether	-111 e	-96 e	-77 e	-52.6	-17.8	34.1	1
$C_4H_{10}O$	Methyl propyl ether				-40 e	-11.3	38.7	5
$C_4H_{10}O$	Isopropyl methyl ether				-56 e	-21.2	30.4	5
$C_4H_{10}O_2$	1,3-Butanediol	-4 e	23 e	55 e	94 e	142.9	206.1	5
$C_4H_{10}O_2$	1,4-Butanediol		45 e	77 e	116 e	164.7	227.6	5
$C_4H_{10}O_2$	2,3-Butanediol		15 e	43 e	77 e	121.2	180.3	5
$C_4H_{10}O_2$	Ethylene glycol monoethyl ether	-49 e	-29 e	-3 e	30 e	73.6	135.3	1
$C_4H_{10}O_2$	Ethylene glycol dimethyl ether			-44 e	-15 e	25.2	85.2	1
$C_4H_{10}O_2$	Dimethylacetal	-89 e	-74 e	-55 e	-29 e	7.7	64.1	5
$C_4H_{10}O_2$	Diethylperoxide			-39 e	3.6	65.0		5
$C_4H_{10}O_2S$	Bis(2-hydroxyethyl) sulfide			31 e	114.2		282.0	5
$C_4H_{10}O_3$	Diethylene glycol	35 e	58 e	86 e	123 e	173.6	245.2	1
$C_4H_{10}O_4S$	Diethyl sulfate		3 e	36 e	79 e	134 e	208.3	5
$C_4H_{10}S$	1-Butanethiol	-77 e	-59 e	-37 e	-6 e	35.4	98.0	5
$C_4H_{10}S$	2-Butanethiol	-86 e	-69 e	-47 e	-17 e	23.4	84.5	5
$C_4H_{10}S$	2-Methyl-1-propanethiol		-66 e	-44 e	-15 e	26.5	88.1	5
$C_4H_{10}S$	2-Methyl-2-propanethiol					5.8	63.8	5
$C_4H_{10}S$	Diethyl sulfide	-80 e	-62 e	-40 e	-10.8	30.3	91.7	1
$C_4H_{10}S$	Methyl propyl sulfide	-78 e	-61 e	-38 e	-8 e	33.1	95.1	5
$C_4H_{10}S$	Isopropyl methyl sulfide	-85 e	-68 e	-46 e	-17 e	23.4	84.3	5
$C_4H_{10}S_2$	1,4-Butanedithiol	-17 e	5 e	32 e	69.1	119.9	195.1	5
$C_4H_{10}S_2$	Diethyl disulfide	-46 e	-26 e	0 e	35 e	82.4	153.5	5

Mol. Form.	Name	Temperature in °C for the indicated pressure						Ref.
		1 Pa	10 Pa	100 Pa	1 kPa	10 kPa	100 kPa	
$C_4H_{11}N$	Butylamine			-46 e	-18.1	20.0	75.9	5
$C_4H_{11}N$	sec-Butylamine			-55 e	-29.1	7.5	62.3	5
$C_4H_{11}N$	tert-Butylamine			-67 e	-42.4	-8.1	43.7	5
$C_4H_{11}N$	Isobutylamine	-85 e	-70 e	-50 e	-24.5	12.0	67.3	5
$C_4H_{11}N$	Diethylamine			-46 e	-26 e	5 e	55.2	1
$C_4H_{11}NO$	N,N-Dimethylethanolamine	-52 e	-31 e	-6 e	27 e	70.9	133 e	5
$C_4H_{11}NO_2$	Diethanolamine	53 e	77 e	107 e	146 e	197.3	268 e	5
$C_4H_{12}BN$	(Dimethylamino)dimethyl-borane		-81 e	-60.1	-31.9	7.0	64.2	5
$C_4H_{12}Cl_2OSi_2$	1,3-Dichloro-1,1,3,3-tetramethyldisiloxane		-33 e	-9 e	23.8	69.1	136.5	5
$C_4H_{12}O_4Si$	Tetramethyl silicate				14.4	59.3	119.7	5
$C_4H_{12}Si$	Tetramethylsilane			-83 e	-59 e	-25 e	26.7	5
$C_4H_{12}Sn$	Tetramethylstannane			-55.0	-25.6	16.6	77.7	5
$C_4H_{13}N_3$	Diethylenetriamine	-10 e	13 e	43 e	80 e	129.6	198 e	5
C_4NiO_4	Nickel carbonyl					-12	42	4
C_5F_{12}	Perfluoropentane				-54.7	-20.9	28.6	5
C_5FeO_5	Iron pentacarbonyl				0	44	105	4
C_5H_4ClN	2-Chloropyridine			7.4	45.8	97.3	169.9	5
$C_5H_4O_2$	Furfural	-26 e	-8 e	16 e	47 e	92.4	161.4	1
C_5H_5N	Pyridine			-23 e	8 e	51.0	114.9	1
C_5H_6	1,3-Cyclopentadiene			-77 e	-51 e	-14 e	39.8	5
$C_5H_6N_2$	Pentanedinitrile	24.1	52 e	85 e	126 e	178 e	245 e	5
C_5H_6O	2-Methylfuran			-66 e	-35 e	6 e	64.5	1
$C_5H_6O_2$	Furfuryl alcohol	-30 e	-5 e	25 e	62.6	109.3	169.7	5
C_5H_6S	2-Methylthiophene		-58 e	-32 e	2 e	47.9	112.2	1
C_5H_6S	3-Methylthiophene		-53 e	-28 e	6 e	50.6	115.1	1
C_5H_7N	1-Methylpyrrole				8 e	49.9	112.3	5
$C_5H_7NO_2$	Ethyl cyanoacetate	16 e	39 e	67.0	102.1	146.7	205.6	5
C_5H_8	1,2-Pentadiene	-109 e	-93 e	-73 e	-46.1	-9.7	44.5	5
C_5H_8	cis-1,3-Pentadiene	-109 e	-93 e	-73 e	-47.0	-10.5	43.7	1,5
C_5H_8	trans-1,3-Pentadiene			-75 e	-49.0	-13 e	42 e	1
C_5H_8	1,4-Pentadiene	-120 e	-105 e	-86 e	-60.9	-26.2	25.6	5
C_5H_8	2,3-Pentadiene	-106 e	-90 e	-70 e	-42.9	-6.3	47.9	5
C_5H_8	3-Methyl-1,2-butadiene	-111 e	-95 e	-75 e	-49.2	-13.1	40.4	5
C_5H_8	2-Methyl-1,3-butadiene	-115 e	-100 e	-81 e	-55.4	-19.7	33.7	1,5
C_5H_8	1-Pentyne			-75 e	-49.1	-13.5	39.9	5
C_5H_8	2-Pentyne	-100 e	-85 e	-65 e	-37.9	-0.5	55.7	5
C_5H_8	3-Methyl-1-butyne			-82 e	-57.5	-23.1	28.6	5
C_5H_8	Cyclopentene	-109 e	-94 e	-74 e	-48 e	-11.1	43.8	5
C_5H_8	Spiropentane	-110 e	-95 e	-76 e	-51 e	-15 e	38.6	5
C_5H_8O	3-Methyl-3-buten-2-one			-35 e	-5 e	36.0	97.3	5
C_5H_8O	Cyclopropyl methyl ketone		-57 e	-31 e	3 e	49 e	112 e	5
C_5H_8O	Cyclopentanone		-39 e	-14 e	19 e	64 e	130.3	1
C_5H_8O	3,4-Dihydro-2H-pyran			-22 e	22.0	84.9		5
$C_5H_8O_2$	4-Pentenoic acid	0 e	19 e	44 e	77 e	122.0	187.5	5
$C_5H_8O_2$	Vinyl propanoate				31.2	94 e		5
$C_5H_8O_2$	Ethyl acrylate		-55 e	-32.7	-2.8	38.5	99.2	5
$C_5H_8O_2$	Methyl methacrylate			-31 e	-1 e	39.7	100.0	1
$C_5H_8O_2$	2,4-Pentanedione			-5 e	24.7	67.8	137.4	1
$C_5H_8O_2$	Tetrahydro-2H-pyran-2-one			5 e	35.1	74.4	128.3	207.0	5
$C_5H_8O_3$	Methyl acetoacetate				50.1	101.1	171.3	5
$C_5H_8O_4$	Glutaric acid		121 e	153.2	191.9	240.3	302.5	5
$C_5H_8O_4$	Dimethyl malonate	-22 e	1 e	30.0	66.7	114.7	180.2	5
$C_5H_9ClO_2$	Ethyl 2-chloropropanoate			1.4	36.4	82.5	146.0	5
$C_5H_9ClO_2$	Isopropyl chloroacetate			-2 e	35.0	83.3	148.1	5
C_5H_9N	Pentanenitrile	-54 e	-34 e	-8 e	26 e	72.2	140.9	1
C_5H_9N	2,2-Dimethylpropanenitrile					41.1	104.8	5
C_5H_9NO	N-Methyl-2-pyrrolidone	1 e	24 e	53.1	92.3	147.2	229 e	5
C_5H_{10}	1-Pentene	-118.9	-103.4	-84.0	-58.8	-23.3	29.6	1,5
C_5H_{10}	cis-2-Pentene	-113.8	-98.1	-78.4	-52.7	-16.8	36.6	1,5
C_5H_{10}	trans-2-Pentene	-114.5	-98.9	-79.1	-53.3	-17.5	36.0	1,5
C_5H_{10}	2-Methyl-1-butene	-117.7	-102.2	-82.7	-57.2	-21.9	30.8	1,5
C_5H_{10}	3-Methyl-1-butene	-125.0	-110.1	-91.2	-66.7	-32.1	19.7	1,5
C_5H_{10}	2-Methyl-2-butene	-113.4	-97.6	-77.7	-51.6	-15.8	38.2	1,5

		Temperature in °C for the indicated pressure						
Mol. Form.	Name	1 Pa	10 Pa	100 Pa	1 kPa	10 kPa	100 kPa	Ref.
C_5H_{10}	Cyclopentane			-77.0	-45.4	-7.1	48.8	5
C_5H_{10}	Ethylcyclopropane	-118 e	-102 e	-83 e	-57 e	-20 e	35.5	5
C_5H_{10}	cis-1,2-Dimethylcyclo-propane	-118 e	-103 e	-83 e	-57 e	-20 e	36.6	5
C_5H_{10}	trans-1,2-Dimethylcyclo-propane	-122 e	-108 e	-89 e	-63 e	-27 e	27.8	5
$C_5H_{10}Br_2$	1,5-Dibromopentane	1 e	25 e	54 e	93 e	145.6	221.8	5
$C_5H_{10}Cl_2$	1,2-Dichloropentane				30 e	77.4	147.8	5
$C_5H_{10}Cl_2$	1,5-Dichloropentane	-31 e	-10 e	17 e	54 e	104.1	178.9	5
$C_5H_{10}N_2$	3-(Dimethylamino)-propanenitrile				51.1	101.8	171.4	5
$C_5H_{10}O$	Cyclopentanol		-13 e	11.5	42.2	82.5	140.0	5
$C_5H_{10}O$	Allyl ethyl ether			-56 e	-28.7	9.8	67.2	5
$C_5H_{10}O$	Pentanal	-71 e	-53 e	-31 e	-1 e	40.8	102.6	5
$C_5H_{10}O$	2-Pentanone				-1 e	40.3	101.9	1,5
$C_5H_{10}O$	3-Pentanone			-31 e	-1 e	40 e	101.6	1
$C_5H_{10}O$	3-Methyl-2-butanone	-69 e	-54 e	-34 e	-6.9	32.2	94.0	1,5
$C_5H_{10}O$	Tetrahydropyran				-15 e	26.0	88 e	5
$C_5H_{10}O$	2-Methyltetrahydrofuran				-20 e	19.7	79.8	5
$C_5H_{10}O_2$	Pentanoic acid	-7.4	15.3	42.7	76.3	122.1	185.7	5
$C_5H_{10}O_2$	2-Methylbutanoic acid	-10 e	10 e	36 e	69 e	112.8	175.2	5
$C_5H_{10}O_2$	3-Methylbutanoic acid	-15.8	4 e	30.0	64.7	110.6	176.1	5
$C_5H_{10}O_2$	Butyl formate			-29 e	2 e	44.4	105.7	5
$C_5H_{10}O_2$	Isobutyl formate	-69 e	-53 e	-31 e	-3 e	37.4	97.6	5
$C_5H_{10}O_2$	Propyl acetate	-69 e	-51 e	-29 e	0 e	40.9	101.2	1
$C_5H_{10}O_2$	Isopropyl acetate			-61 e	-40 e	-11 e	29.8	88.2	5
$C_5H_{10}O_2$	Ethyl propanoate	-69 e	-52 e	-30 e	-1 e	38.9	98.7	1
$C_5H_{10}O_2$	Methyl butanoate	-68 e	-50 e	-28 e	0.9	41.7	102.3	5
$C_5H_{10}O_2$	Methyl isobutanoate	-83 e	-65 e	-41 e	-11 e	31 e	92.1	5
$C_5H_{10}O_2$	Tetrahydrofurfuryl alcohol	-40 e	-16 e	15 e	55 e	106 e	176.8	5
$C_5H_{10}O_3$	Diethyl carbonate		-42 e	-17 e	17 e	61.6	125.9	5
$C_5H_{10}O_3$	Ethylene glycol monomethyl ether acetate	-47 e	-26 e	0 e	34 e	79.4	144.1	5
$C_5H_{10}S$	Thiacyclohexane				24 e	71.1	141.2	5
$C_5H_{10}S$	Cyclopentanethiol				18 e	64 e	131.7	5
$C_5H_{11}Br$	1-Bromopentane	-60 e	-41 e	-16 e	16 e	61.5	129.1	5
$C_5H_{11}Br$	2-Bromopentane	-69 e	-51 e	-27 e	5 e	49.7	116.9	5
$C_5H_{11}Br$	3-Bromopentane	-68 e	-50 e	-26 e	6 e	50.8	118.1	5
$C_5H_{11}Br$	1-Bromo-3-methylbutane	-67 e	-49 e	-25 e	8 e	52.4	119.9	5
$C_5H_{11}Cl$	1-Chloropentane	-73 e	-55 e	-32 e	-1 e	42.5	107.9	5
$C_5H_{11}Cl$	2-Chloropentane	-80 e	-62 e	-39 e	-9 e	33.2	96.1	5
$C_5H_{11}Cl$	3-Chloropentane	-77 e	-60 e	-37 e	-7 e	34.9	97.3	5
$C_5H_{11}Cl$	2-Chloro-2-methylbutane				-52 e	-21 e	21.8	85.2	5
$C_5H_{11}Cl$	1-Chloro-2,2-dimethyl-propane				-17 e	23.5	83.9	5
$C_5H_{11}F$	1-Fluoropentane	-97 e	-80 e	-60 e	-32 e	5.7	62.4	5
$C_5H_{11}I$	1-Iodopentane	-47 e	-27 e	-1 e	34 e	83.0	156.5	5
$C_5H_{11}I$	1-Iodo-3-methylbutane		-34 e	-6.6	28.8	77.3	147.8	5
$C_5H_{11}N$	Cyclopentylamine	-66 e	-48 e	-26 e	4 e	45.8	108 e	5
$C_5H_{11}N$	Piperidine				2 e	43.3	105.8	5
$C_5H_{11}N$	N-Methylpyrrolidine				-23 e	18.5	78 e	5
$C_5H_{11}NO_3$	3-Methylbutyl nitrate		-26 e	1.0	35.5	81.7	147.0	5
C_5H_{12}	Pentane**	-115.5	-99.8	-80.0	-54.0	-18.1	35.7	16
C_5H_{12}	Isopentane	-119 e	-105 e	-86 e	-61 e	-26 e	27.5	1
C_5H_{12}	Neopentane*		-107.5 s	-90.8 s	-68.8 s	-38.5 s	9.2	1,5
$C_5H_{12}N_2O$	Tetramethylurea			20.7	58.0	106.7	179.5	5
$C_5H_{12}O$	1-Pentanol	-27 e	-10 e	12 e	41 e	79.8	137.4	5
$C_5H_{12}O$	2-Pentanol	-35 e	-19 e	1 e	28.0	64.9	118.7	1
$C_5H_{12}O$	3-Pentanol	-41 e	-25 e	-4 e	24 e	61.1	114.9	5
$C_5H_{12}O$	2-Methyl-1-butanol	-27 e	-11 e	9 e	36.2	73.4	128.3	1
$C_5H_{12}O$	3-Methyl-1-butanol	-22 e	-7 e	13 e	39.1	75.7	130.1	5
$C_5H_{12}O$	2-Methyl-2-butanol			-5 e	17.7	50.6	101.7	1,5
$C_5H_{12}O$	3-Methyl-2-butanol			-3 e	22.7	58.2	111.1	5
$C_5H_{12}O$	2,2-Dimethyl-1-propanol					59.2	112.7	5
$C_5H_{12}O$	Butyl methyl ether			-54 e	-27 e	12 e	69.8	1
$C_5H_{12}O$	Methyl tert-butyl ether			-66 e	-39 e	-2 e	54.8	1
$C_5H_{12}O$	Ethyl propyl ether	-92 e	-77 e	-57 e	-30.5	6.7	63.4	1,5

Mol. Form.	Name	Temperature in °C for the indicated pressure						Ref.
		1 Pa	10 Pa	100 Pa	1 kPa	10 kPa	100 kPa	
$C_5H_{12}O_2$	1,5-Pentanediol	25 e	52 e	85 e	125 e	175.1	238.9	5
$C_5H_{12}O_2$	Ethylene glycol monopropyl ether				40 e	85.6	149.3	5
$C_5H_{12}O_2$	Diethoxymethane		-65 e	-43 e	-14 e	27.3	87.7	5
$C_5H_{12}O_3$	Diethylene glycol monomethyl ether		12 e	40 e	76 e	124.2	193.7	1
$C_5H_{12}S$	1-Pentanethiol	-60 e	-41 e	-17 e	15 e	60 e	126.2	1
$C_5H_{12}S$	2-Pentanethiol	-70 e	-52 e	-28 e	3 e	46.6	111.9	5
$C_5H_{12}S$	3-Pentanethiol	-70 e	-51 e	-28 e	4 e	47.7	113.4	5
$C_5H_{12}S$	2-Methyl-1-butanethiol				8.0	52.3	118.5	5
$C_5H_{12}S$	3-Methyl-1-butanethiol				7.8	51.9	117.9	5
$C_5H_{12}S$	2-Methyl-2-butanethiol				-8.0	34.6	98.7	5
$C_5H_{12}S$	Butyl methyl sulfide		-43 e	-19 e	13 e	57 e	123.0	1
$C_5H_{12}S$	tert-Butyl methyl sulfide				-7.8	34.7	98.4	5
$C_5H_{12}S$	Ethyl propyl sulfide	-64 e	-46 e	-23 e	9 e	52.7	118.0	5
$C_5H_{12}S$	Ethyl isopropyl sulfide	-72 e	-54 e	-31 e	0 e	42.7	106.9	5
$C_5H_{13}N$	Pentylamine		-52 e	-29 e	1 e	42.8	104.0	5
C_6BrF_5	Bromopentafluorobenzene			-10 e	23 e	68 e	136.0	5
C_6ClF_5	Chloropentafluorobenzene		-44 e	-21 e	11 e	53.8	117.6	1
$C_6Cl_3F_3$	1,3,5-Trichloro-2,4,6-trifluorobenzene	-19 e	4 e	32 e	70 e	121.7	197.9	1
C_6F_6	Hexafluorobenzene		-56.9 s	-36 s	-11.5 s	22.6	79.9	1,5
C_6F_{12}	Perfluorocyclohexane				-46.2 s	-7.6 s	48.9 s	5
C_6F_{14}	Perfluorohexane		-75 e	-57 e	-32 e	2.8	56.8	5
C_6F_{14}	Perfluoro-2-methylpentane				-33 e	2.9	57.1	5
C_6F_{14}	Perfluoro-3-methylpentane	-95 e	-80 e	-60 e	-34 e	2.8	57.9	5
C_6F_{14}	Perfluoro-2,3-dimethylbutane					4.3	59.3	5
C_6HF_5	Pentafluorobenzene			-41 e	-13 e	27 e	85.3	5
C_6HF_5O	Pentafluorophenol				39 e	82 e	145.2	5
$C_6H_2F_4$	1,2,3,4-Tetrafluorobenzene			-36 e	-7 e	33.8	94.0	1
$C_6H_2F_4$	1,2,3,5-Tetrafluorobenzene			-43 e	-14 e	25.5	84.1	1
$C_6H_2F_4$	1,2,4,5-Tetrafluorobenzene					30.7	89.9	1
$C_6H_3Cl_3O$	2,4,6-Trichlorophenol			71.8	114.0	169.5	245.7	5
$C_6H_3F_3$	1,3,5-Trifluorobenzene					18.2	75.0	5
$C_6H_4Br_2$	m-Dibromobenzene	-7 e	16 e	44 e	83 e	137.0	218.2	5
$C_6H_4ClNO_2$	1-Chloro-4-nitrobenzene	15.4 s	35.8 s		97 e	156.0	238 e	5
$C_6H_4Cl_2$	o-Dichlorobenzene		-13 e	16.3	53.9	104.6	180.0	1,5
$C_6H_4Cl_2$	m-Dichlorobenzene		-22 e	8.0	46.7	97.8	172.5	1,5
$C_6H_4Cl_2$	p-Dichlorobenzene	-45.5 s	-21.8 s	8 s	46.7 s	99.0	173.6	1,5
$C_6H_4O_2$	p-Benzoquinone	-4.1 s	17.8 s	43.5 s	74.3 s	111.6 s		5
$C_6H_5AsCl_2$	Dichlorophenylarsine	6.9	35.2	70 e	113 e	170 e	245 e	5
C_6H_5Br	Bromobenzene		-25 e	1 e	34.9	83.1	155.4	1
C_6H_5Cl	Chlorobenzene		-43 e	-17 e	16.8	62.9	131.3	1,5
C_6H_5ClO	o-Chlorophenol				45.8	97.9	173.9	5
C_6H_5ClO	m-Chlorophenol			39.7	80.2	135.1	213.4	5
C_6H_5ClO	p-Chlorophenol			45.0	86.5	142.0	219.9	5
$C_6H_5Cl_3Si$	Trichlorophenylsilane			33 e	70.2	122.6	201 e	5
C_6H_5F	Fluorobenzene				-16.9	24.2	84.4	1
C_6H_5I	Iodobenzene	-30 e	-7 e	20.9	58.5	110.6	187.8	1
$C_6H_5NO_2$	Nitrobenzene		10 e	40 e	78 e	132 e	210.3	1
$C_6H_5NO_3$	p-Nitrophenol	72.6 s	97.4 s					5
C_6H_6	1,5-Hexadien-3-yne	-82 e	-66 e	-44.3	-16.0	23.7	83.6	5
C_6H_6	Benzene**		-40 s	-15.1 s	20.0	79.7		1,5
C_6H_6ClN	o-Chloroaniline		10 e	39.0	75.2	131.4	208.3	5
C_6H_6ClN	m-Chloroaniline	-5 e	19.7	49.4	94.2	162 e	1069 e	5
$C_6H_6N_2O_2$	p-Nitroaniline	87.8 s			192.0	252.6	331.2	5
C_6H_6O	Phenol	-9.7 s	9.6 s	34.1 s	68.9	113.7	181.4	1,5
$C_6H_6O_3$	1,2,3-Benzenetriol				162.0	222.8	308.3	5
C_6H_6S	Benzenethiol		-15 e	12 e	47 e	96.0	168.6	5
C_6H_7N	Aniline		-2.5	26.7	63.5	112.5	183.5	1,5
C_6H_7N	2-Methylpyridine	-56.5	-37.8	-13.9	18.3	62.9	129.0	1,5
C_6H_7N	3-Methylpyridine			-5 e	28.8	75.2	143.7	1
C_6H_7N	4-Methylpyridine	-58.2 s	-43.1 s	-3.9 s	29.6	76.1	144.9	1,5
C_6H_8	cis-1,3,5-Hexatriene					21 e	78 e	5
C_6H_8	1,3-Cyclohexadiene	-88 e	-71 e	-50 e	-21 e	19 e	79.9	5
C_6H_8	1,4-Cyclohexadiene				-15 e	27.3	85.0	5
$C_6H_8N_2$	Adiponitrile	30 e	61 e	100 e	148.6	211.8	297 e	5

Mol. Form.	Name	Temperature in °C for the indicated pressure						Ref.
		1 Pa	10 Pa	100 Pa	1 kPa	10 kPa	100 kPa	
$C_6H_8N_2$	m-Phenylenediamine			94.5	140.2	200.8	285.0	5
$C_6H_8N_2$	Phenylhydrazine		38 e	69 e	109 e	163.9	242.5	5
$C_6H_8O_4$	Dimethyl maleate		5 e	36 e	76 e	127.3	197 e	5
C_6H_8S	2,5-Dimethylthiophene		-43 e	-16 e	20 e	67.5	134.8	5
C_6H_{10}	trans-1,3-Hexadiene	-86 e	-70 e	-51 e	-24 e	14 e	72 e	5
C_6H_{10}	trans-1,4-Hexadiene	-98 e	-81 e	-60 e	-33 e	7 e	65 e	5
C_6H_{10}	1,5-Hexadiene	-99 e	-84 e	-64 e	-37 e	0.9	59.2	5
C_6H_{10}	cis,cis-2,4-Hexadiene					18 e	79.6	5
C_6H_{10}	trans,cis-2,4-Hexadiene	-89 e	-73 e	-52 e	-23 e	18 e	79.6	5
C_6H_{10}	trans,trans-2,4-Hexadiene				-23 e	18 e	79.6	5
C_6H_{10}	trans-2-Methyl-1,3-pentadiene	-92 e	-75 e	-54 e	-26 e	14 e	75.6	5
C_6H_{10}	2,3-Dimethyl-1,3-butadiene			-59 e	-30 e	9.7	68.1	5
C_6H_{10}	1-Hexyne	-91 e	-75 e	-54 e	-26 e	12.8	71.0	5
C_6H_{10}	2-Hexyne	-84 e	-67 e	-46 e	-17 e	23.6	84.1	5
C_6H_{10}	3-Hexyne	-86 e	-69 e	-48 e	-19.1	21.0	81.0	1,5
C_6H_{10}	4-Methyl-1-pentyne	-97 e	-81 e	-61 e	-34 e	4.1	60.7	5
C_6H_{10}	4-Methyl-2-pentyne	-91 e	-74 e	-54 e	-26 e	13.8	72.7	5
C_6H_{10}	Cyclohexene	-87 e	-70 e	-49 e	-19 e	21 e	82.6	1
$C_6H_{10}Cl_2$	1,1-Dichlorocyclohexane	-39 e	-19 e	8 e	43 e	93.5	170.5	5
$C_6H_{10}Cl_2$	cis-1,2-Dichlorocyclohexane			27 e	69 e	125.7	206.2	5
$C_6H_{10}O$	4-Methyl-4-penten-2-one	-59 e		-17 e	14 e	57.0	121.0	5
$C_6H_{10}O$	Cyclohexanone		-25 e	1 e	36 e	84 e	155.2	1
$C_6H_{10}O$	Mesityl oxide	-56 e	-37 e	-13 e	19 e	63.5	129.3	5
$C_6H_{10}O_2$	Vinyl butanoate					53 e	114.5	5
$C_6H_{10}O_2$	Ethyl methacrylate				8 e	53.2	116.8	5
$C_6H_{10}O_2$	Allyl glycidyl ether				40.1	85.7	152.8	5
$C_6H_{10}O_3$	Ethyl acetoacetate	-25 e	-3 e	25.7	62.3	111.3	180.2	5
$C_6H_{10}O_3$	Propanoic anhydride	-32 e	-15 e	6 e	36 e	77.6	142.9	5
$C_6H_{10}O_4$	Diethyl oxalate	-5 e	18 e	44.9	79.4	124.3	185.2	5
$C_6H_{10}O_4$	Dimethyl succinate			30 e	70.4	123.3	195.4	5
$C_6H_{10}O_4$	Ethylene glycol diacetate	-17 e	6 e	35.0	71.9	121.1	190.0	5
$C_6H_{10}S$	Diallylsulfide	-58 e	-38 e	-12.4	21.7	68.8	138.1	5
$C_6H_{11}Cl$	Chlorocyclohexane		-35 e	-9 e	25 e	71.6	142.1	5
$C_6H_{11}N$	Hexanenitrile	-40 e	-19 e	8 e	43 e	91.5	163.2	1,5
$C_6H_{11}N$	4-Methylpentanenitrile		-50 e	-20 e	20 e	75.2	155.2	5
$C_6H_{11}NO$	Caprolactam	36.8 s	58.9 s	86.6 s			270	5
C_6H_{12}	1-Hexene	-99.8	-82.8	-61.4	-33.7	5.2	63.1	1,5
C_6H_{12}	cis-2-Hexene	-97 e	-80 e	-58 e	-30 e	9.9	68.5	5
C_6H_{12}	trans-2-Hexene	-94 e	-78 e	-57 e	-30 e	9.3	67.5	5
C_6H_{12}	cis-3-Hexene	-96 e	-79 e	-59 e	-30.8	7.9	66.0	5
C_6H_{12}	trans-3-Hexene	-95 e	-79 e	-58 e	-30.0	8.8	66.7	5
C_6H_{12}	2-Methyl-1-pentene	-98 e	-82 e	-62 e	-34.2	4.1	61.7	5
C_6H_{12}	3-Methyl-1-pentene	-104 e	-88 e	-68 e	-41.5	-3.6	53.8	5
C_6H_{12}	4-Methyl-1-pentene	-105 e	-89 e	-69 e	-41.6	-3.6	53.5	5
C_6H_{12}	2-Methyl-2-pentene	-95 e	-78 e	-58 e	-30 e	9.0	66.9	5
C_6H_{12}	3-Methyl-cis-2-pentene	-95 e	-79 e	-58 e	-30 e	8.9	67.3	5
C_6H_{12}	3-Methyl-trans-2-pentene	-93 e	-77 e	-55 e	-27.4	11.7	70.0	5
C_6H_{12}	4-Methyl-cis-2-pentene	-102 e	-86 e	-66 e	-38.7	-0.9	56.0	5
C_6H_{12}	4-Methyl-trans-2-pentene	-100 e	-84 e	-64 e	-36.8	1.2	58.2	5
C_6H_{12}	2-Ethyl-1-butene	-98 e	-81 e	-60 e	-32 e	6.6	64.3	5
C_6H_{12}	2,3-Dimethyl-1-butene	-103 e	-87 e	-67 e	-39.9	-1.9	55.2	5
C_6H_{12}	3,3-Dimethyl-1-butene	-110 e	-95 e	-76 e	-50.8	-14.5	40.8	5
C_6H_{12}	2,3-Dimethyl-2-butene		-75 e	-54 e	-25 e	14 e	72.9	1
C_6H_{12}	Cyclohexane	-85.6 s	-68.9 s	-47.6 s	-19.8 s	19.3	80.4	1,5
C_6H_{12}	Methylcyclopentane	-97 e	-80 e	-58 e	-28.8	11.6	71.4	1,5
C_6H_{12}	Ethylcyclobutane	-99 e	-82 e	-61 e	-32 e	9 e	70.2	5
C_6H_{12}	Isopropylcyclopropane	-104 e	-88 e	-68 e	-40 e	-1 e	57.9	5
C_6H_{12}	1-Ethyl-1-methylcyclopropane	-105 e	-89 e	-69 e	-41 e	-3 e	56.3	5
C_6H_{12}	1,1,2-Trimethylcyclopropane	-109 e	-94 e	-73 e	-46 e	-7 e	52.0	5
$C_6H_{12}Cl_2$	1,2-Dichlorohexane				49 e	98.1	171.7	5
$C_6H_{12}Cl_2O$	2,2'-Dichlorodiisopropyl ether		-1 e	27.3	63.4	112.3	182.1	5
$C_6H_{12}O$	Butyl vinyl ether	-87 e	-67 e	-42 e	-9.3	33.6	93.2	5
$C_6H_{12}O$	Isobutyl vinyl ether	-87 e	-68 e	-44 e	-13 e	26.5	80.7	5
$C_6H_{12}O$	Hexanal	-56 e	-37 e	-13 e	19 e	62.6	127.8	5
$C_6H_{12}O$	2-Hexanone	-43 e	-21 e	4.2	34.5	61.9	127.2	1,5
$C_6H_{12}O$	3-Hexanone		-40 e	-16 e	15 e	58.5	123.1	1
$C_6H_{12}O$	3-Methyl-2-pentanone				8.5	52.7	117.0	5

Mol. Form.	Name	Temperature in °C for the indicated pressure						Ref.
		1 Pa	10 Pa	100 Pa	1 kPa	10 kPa	100 kPa	
$C_6H_{12}O$	4-Methyl-2-pentanone	-61 e	-43 e	-21 e	9 e	51.5	116.1	5
$C_6H_{12}O$	2-Methyl-3-pentanone					50.2	113.0	5
$C_6H_{12}O$	3,3-Dimethyl-2-butanone			-30 e	0 e	42.5	105.7	1
$C_6H_{12}O$	Cyclohexanol			34 e	61 e	99.2	160.7	1
$C_6H_{12}O_2$	Hexanoic acid		33 e	59 e	93 e	139.3	204.5	1
$C_6H_{12}O_2$	4-Methylpentanoic acid	36 e	49 e	67.1	92.9	133.6	206.8	5
$C_6H_{12}O_2$	Diethylacetic acid	-9 e	16 e	46 e	83 e	130.7	192.5	5
$C_6H_{12}O_2$	Isopentyl formate	-60 e	-41 e	-17 e	15 e	59.1	124 e	5
$C_6H_{12}O_2$	Butyl acetate	-63 e	-43 e	-19 e	14 e	61.0	125.6	1,5
$C_6H_{12}O_2$	Isobutyl acetate	-63 e	-45 e	-21 e	10 e	53.4	116 e	5
$C_6H_{12}O_2$	Propyl propanoate	-62 e	-42 e	-18 e	14 e	58.3	122.0	5
$C_6H_{12}O_2$	Ethyl butanoate	-49 e	-34 e	-14 e	14.3	55.2	121.1	5
$C_6H_{12}O_2$	Ethyl 2-methylpropanoate	-65 e	-47 e	-24.6	5.4	47.3	109.8	5
$C_6H_{12}O_2$	Methyl pentanoate				19.2	63.7	127.4	5
$C_6H_{12}O_2$	Methyl isopentanoate					53.3	116.3	5
$C_6H_{12}O_2$	Diacetone alcohol	-41 e	-17 e	13 e	50.1	98.5	164 e	5
$C_6H_{12}O_3$	Ethylene glycol monoethyl ether acetate	-25 e	-8 e	14 e	44.6	88.0	155.6	5
$C_6H_{12}O_3$	Paraldehyde				17 e	62.2	124 e	5
$C_6H_{12}S$	Cyclohexanethiol					84.8	158.3	5
$C_6H_{12}S$	cis-Tetrahydro-2,5-dimethylthiophene	-53 e	-34 e	-8 e	25 e	72.0	142.1	5
$C_6H_{12}S$	Tetrahydro-3-methyl-2H-thiopyran	-48 e	-27 e	0 e	35 e	84.1	157.5	5
$C_6H_{13}Br$	1-Bromohexane	-45 e	-25 e	2 e	36 e	83.7	154.8	5
$C_6H_{13}Cl$	1-Chlorohexane	-55 e	-36 e	-11 e	21 e	66.7	134.6	5
$C_6H_{13}F$	1-Fluorohexane	-80 e	-62 e	-40 e	-11 e	30.4	91.1	5
$C_6H_{13}I$	1-Iodohexane	-33 e	-11 e	16 e	53 e	104.0	180.8	5
$C_6H_{13}N$	Cyclohexylamine			-9 e	22 e	66.6	133.5	1
C_6H_{14}	Hexane	-96.4 s	-79.2	-57.6	-29.3	9.8	68.3	16
C_6H_{14}	2-Methylpentane	-100 e	-84 e	-64 e	-36 e	2 e	59.9	1
C_6H_{14}	3-Methylpentane	-99 e	-83 e	-62 e	-34.3	4.6	62.9	1
C_6H_{14}	2,2-Dimethylbutane		-90 e	-71.5	-45.5	-7.7	49.4	1
C_6H_{14}	2,3-Dimethylbutane	-103 e	-87 e	-66 e	-39.0	-0.4	57.6	1
$C_6H_{14}O$	1-Hexanol		5 e	28 e	56.8	97.3	157.1	1
$C_6H_{14}O$	2-Hexanol	-28 e	-10 e	12 e	41.4	81.5	139.6	1
$C_6H_{14}O$	3-Hexanol	-43 e	-23 e	1 e	33 e	75.4	135.1	1
$C_6H_{14}O$	2-Methyl-1-pentanol			14 e	45.9	88.3	147.6	5
$C_6H_{14}O$	4-Methyl-1-pentanol			24 e	53 e	92.4	151.4	5
$C_6H_{14}O$	2-Methyl-2-pentanol	-29 e	-15 e	3 e	27.1	63.0	120.9	5
$C_6H_{14}O$	3-Methyl-2-pentanol				36.5	76.1	133.8	5
$C_6H_{14}O$	4-Methyl-2-pentanol	-43 e	-24 e	0 e	30 e	71.9	131.3	5
$C_6H_{14}O$	2-Methyl-3-pentanol					29.8	68.8	126.0	5
$C_6H_{14}O$	3-Methyl-3-pentanol		-23 e	-4 e	22.9	61.1	121.1	5
$C_6H_{14}O$	2-Ethyl-1-butanol		-5 e	17 e	46 e	85.7	146.1	5
$C_6H_{14}O$	3,3-Dimethyl-1-butanol	-37 e	-16 e	9 e	42 e	84.3	142.5	5
$C_6H_{14}O$	2,3-Dimethyl-2-butanol			-5 e	23 e	61.3	118.2	5
$C_6H_{14}O$	Dipropyl ether	-80 e	-63 e	-41 e	-12 e	28.8	89.7	1
$C_6H_{14}O$	Diisopropyl ether		-76 e	-55 e	-28 e	11 e	68.1	1
$C_6H_{14}O$	Butyl ethyl ether	-78 e	-61 e	-39 e	-10 e	31.0	91.9	1
$C_6H_{14}O$	tert-Butyl ethyl ether	-90 e	-74 e	-53 e	-24.6	14.4	72.6	5
$C_6H_{14}O_2$	2-Methyl-2,4-pentanediol	-8 e	17 e	48 e	86 e	134.4	197.5	5
$C_6H_{14}O_2$	Ethylene glycol monobutyl ether	-31 e	-8 e	20 e	55 e	103.2	170.2	5
$C_6H_{14}O_2$	1,1-Diethoxyethane	-68 e	-49 e	-26 e	3.7	44.2	101.9	5
$C_6H_{14}O_2$	Ethylene glycol diethyl ether		-59 e	-35.3	-2.8	44.4	118.8	5
$C_6H_{14}O_3$	1,2,6-Hexanetriol	92 e	114.8	146.0	191 e			5
$C_6H_{14}O_3$	Dipropylene glycol				110 e	162.6	231.4	5
$C_6H_{14}O_3$	Diethylene glycol monoethyl ether			40 e	80.3	132.4	201.4	5
$C_6H_{14}O_3$	Diethylene glycol dimethyl ether	-42 e	-20 e	8.3	44.3	92.3	159.4	5
$C_6H_{14}O_3$	Trimethylolpropane	73 e	98 e	128 e	167.8	220.5	295 e	5
$C_6H_{14}O_4$	Triethylene glycol	44 e	74 e	109.0	152.6	207.2	277.9	5
$C_6H_{14}S$	1-Hexanethiol	-45 e	-25 e	1 e	35 e	81.7	152.2	5
$C_6H_{14}S$	2-Hexanethiol	-50 e	-32 e	-8 e	25 e	69.9	138.4	5
$C_6H_{14}S$	Dipropyl sulfide	-50 e	-30 e	-6 e	28 e	73.6	142.4	5

		Temperature in °C for the indicated pressure						
Mol. Form.	Name	1 Pa	10 Pa	100 Pa	1 kPa	10 kPa	100 kPa	Ref.
C₆H₁₄S	Diisopropyl sulfide	-65 e	-47 e	-23 e	9 e	53.1	119.6	5
C₆H₁₄S	Isopropyl propyl sulfide				18.5	63.8	131.6	5
C₆H₁₄S	Butyl ethyl sulfide	-49 e	-30 e	-5 e	29 e	74.8	143.8	5
C₆H₁₅N	Hexylamine			-10 e	22 e	66.0	130.6	5
C₆H₁₅N	Butylethylamine				6.1	47.7	107.0	5
C₆H₁₅N	Dipropylamine		-48 e	-25 e	6 e	47.5	108.8	5
C₆H₁₅N	Diisopropylamine			-47 e	-17.5	23.5	84.0	5
C₆H₁₅N	Triethylamine	-58 e	-45 e	-29 e	-5 e	29.9	88.5	1
C₆H₁₅NO	2-Diethylaminoethanol					97 e	160.6	5
C₆H₁₅NO₃	Triethanolamine	75 e	108 e	148 e	196 e	256.7	334 e	5
C₆H₁₅O₄P	Triethyl phosphate			34	76	132	211	4
C₆H₁₆N₂	Hexamethylenediamine				76.0	128.2	199.0	5
C₆H₁₆O₂Si	Diethoxydimethylsilane	-62 e	-44 e	-21.2	9.1	51.0	113.0	5
C₆H₁₈Cl₂O₂Si₃	1,5-Dichloro-1,1,3,3,5,5-hexamethyltrisiloxane	-29 e	-7 e	22.2	59.7	110.5	183.4	5
C₆H₁₈OSi₂	Hexamethyldisiloxane		-56 e	-34 e	-5 e	37.1	100.1	5
C₆MoO₆	Molybdenum hexacarbonyl		17.4 s	42.8 s	73.1 s	109.9 s	155.4 s	5
C₇F₁₄	Perfluoromethylcyclohexane				-21 e	18 e	75.9	1
C₇F₁₆	Perfluoroheptane		-62 e	-41 e	-14 e	24.7	82.1	1
C₇HF₁₅	1H-Pentadecafluoroheptane				-7 e	35.9	96.0	5
C₇H₃ClF₃NO₂	1-Chloro-2-nitro-4-(trifluoromethyl)benzene	3 e	26 e	55 e	92.8	145.2	222.0	5
C₇H₃F₅	2,3,4,5,6-Pentafluorotoluene			-20 e	11 e	53.6	117.0	5
C₇H₄ClF₃	1-Chloro-2-(trifluoromethyl)benzene				1 e	34.5	81.8	151.8	5
C₇H₄ClF₃	1-Chloro-3-(trifluoromethyl)benzene	-53 e	-34 e	-9 e	24.2	69.8	137.2	5
C₇H₄ClF₃	1-Chloro-4-(trifluoromethyl)benzene			-9 e	24.2	70.4	138.1	5
C₇H₄Cl₂O	o-Chlorobenzoyl chloride				93 e	149 e	237.0	5
C₇H₄Cl₂O	m-Chlorobenzoyl chloride				87.8	147 e	225.0	5
C₇H₄F₃NO₂	1-Nitro-3-(trifluoromethyl)benzene		11 e	39 e	76.2	127.3	202.2	5
C₇H₄F₄	1-Fluoro-4-(trifluoromethyl)benzene			-38 e	-6 e	38.6	102.3	5
C₇H₅BrO	Benzoyl bromide	-15 e	11 e	42.6	83.9	139.5	218.0	5
C₇H₅ClO	Benzoyl chloride			27.5	67.0	120.4	196.7	5
C₇H₅Cl₃	(Trichloromethyl)benzene		9 e	40.6	81.5	136.2	213.0	5
C₇H₅F₃	(Trifluoromethyl)benzene				-3 e	39 e	101.6	5
C₇H₅N	Benzonitrile		-6 e	23.9	63.1	115.7	190.0	5
C₇H₅NS	Phenyl isothiocyanate				79.4	105 e	117 e	5
C₇H₆Cl₂	2,4-Dichlorotoluene		6 e	33 e	68.3	119.5	199.1	5
C₇H₆Cl₂	3,4-Dichlorotoluene	-13 e	9 e	38 e	76 e	129.3	208.4	5
C₇H₆Cl₂	(Dichloromethyl)benzene			31	72	130	213	4
C₇H₆O	Benzaldehyde		-9 e	19 e	54.6	104.6	178.3	1
C₇H₆O₂	Salicylaldehyde		-1 e	29 e	68 e	120.7	196.2	5
C₇H₇Br	o-Bromotoluene		-10 e	17 e	54 e	104.8	181.1	5
C₇H₇Br	m-Bromotoluene	-34 e	-11 e	19.4	58.1	109.9	183.1	5
C₇H₇Br	p-Bromotoluene				57 e	107.8	183.8	5
C₇H₇Br	(Bromomethyl)benzene			25.4	66.8	121.7	198.3	5
C₇H₇Cl	o-Chlorotoluene		-24 e	3 e	38 e	86.3	158.7	1,5
C₇H₇Cl	m-Chlorotoluene	-41 e	-21 e	6 e	41 e	89 e	161.8	5
C₇H₇Cl	p-Chlorotoluene				40 e	88.9	161.5	1,5
C₇H₇Cl	(Chloromethyl)benzene	-34 e	-11 e	17.7	55.4	106.3	178.9	5
C₇H₇ClO	1-Chloro-2-methoxy-benzene	-22 e	2 e	33 e	72 e	125.2	201 e	5
C₇H₇F	o-Fluorotoluene		-50 e	-26 e	5 e	49.0	113.9	5
C₇H₇F	m-Fluorotoluene	-67 e	-48 e	-25 e	7 e	51.0	116.1	5
C₇H₇F	p-Fluorotoluene		-48 e	-24 e	7 e	51 e	116.2	5
C₇H₇NO₂	o-Nitrotoluene	23 e	40 e	62 e	94 e	141.9	221.9	5
C₇H₇NO₂	m-Nitrotoluene			45 e	89.7	148.7	231.3	5
C₇H₇NO₃	2-Nitroanisole	15 e	45 e	82 e	129 e	189.4	271.8	5
C₇H₈	Toluene	-78.1	-57.1	-31.3	1.5	45.2	110.1	5
C₇H₈	Bicyclo[2.2.1]hepta-2,5-diene				-15 e	27.4	91 e	5
C₇H₈Cl₂Si	Dichloromethylphenylsilane			32.4	71.8	126.0	205.0	5
C₇H₈O	o-Cresol	-6.4 s	12.8 s	40.2	72.3	120.3	190.5	1,5
C₇H₈O	m-Cresol	20.8	33.6	52.4	82.6	130.6	201.8	1,5

Mol. Form.	Name	\multicolumn{7}{c}{Temperature in °C for the indicated pressure}	Ref.
		1 Pa	10 Pa	100 Pa	1 kPa	10 kPa	100 kPa	
C7H8O	p-Cresol	-0.2 s	20.7 s	52.7	83.1	130.7	201.5	1,5
C7H8O	Benzyl alcohol	8 e	28 e	54 e	88 e	134.7	204.9	1
C7H8O	Anisole		-21 e	4 e	38 e	84 e	153.2	1,5
C7H8S	3-Methylbenzenethiol		0 e	29 e	66 e	117.9	194.6	5
C7H9N	Benzylamine			25.6	62.6	112.7	183.9	5
C7H9N	o-Methylaniline	1.0	18.8	42.6	76.1	125.6	199.9	1,5
C7H9N	m-Methylaniline	3.8	22.0	46.2	80.1	128.8	202.9	1,5
C7H9N	p-Methylaniline				77.1	126.2	199.9	5
C7H9N	N-Methylaniline	-16 e	6 e	34 e	70.3	121.1	195.8	1
C7H9N	2-Ethylpyridine	-46 e	-26 e	-1 e	33 e	79.3	149.0	5
C7H9N	3-Ethylpyridine	-38 e	-17 e	9 e	44 e	92.7	166.5	5
C7H9N	4-Ethylpyridine	-35 e	-15 e	11 e	46 e	94.4	168.6	5
C7H9N	2,3-Dimethylpyridine				42 e	89.9	160.6	5
C7H9N	2,4-Dimethylpyridine		-25 e	3.7	40.0	87.5	157.9	1,5
C7H9N	2,5-Dimethylpyridine			4 e	39 e	86.2	156.6	1
C7H9N	2,6-Dimethylpyridine			-3 e	29.9	75.8	143.6	1
C7H9N	3,4-Dimethylpyridine		-9 e	19 e	55 e	104.8	178.6	5
C7H9N	3,5-Dimethylpyridine			11 e	48 e	98 e	171.5	1
C7H10N2	Toluene-2,4-diamine			100.4	145.3	202.9	279.5	5
C7H12	1-Heptyne	-75 e	-57 e	-35 e	-5 e	37.1	99.5	5
C7H12	2-Heptyne		-51 e	-27 e	4 e	46.9	111.5	5
C7H12	3-Heptyne	-71 e	-53 e	-31 e	0 e	42.7	106.4	5
C7H12	5-Methyl-1-hexyne	-80 e	-62 e	-40 e	-11 e	30.1	91.4	5
C7H12	5-Methyl-2-hexyne	-75 e	-57 e	-34 e	-4 e	38.6	102.0	5
C7H12	2-Methyl-3-hexyne	-78 e	-61 e	-39 e	-9 e	32.6	94.8	5
C7H12	4,4-Dimethyl-1-pentyne		-73 e	-52 e	-24 e	15.9	75.6	5
C7H12	4,4-Dimethyl-2-pentyne		-70 e	-48 e	-19 e	21.4	82.6	5
C7H12	Bicyclo[4.1.0]heptane					49.9	116.3	5
C7H12	Cycloheptene			-30.0	3.4	47.5	108 e	5
C7H12	1-Methylbicyclo(3,1,0)hexane					29.8	92.6	5
C7H12	Methylenecyclohexane	-76 e	-58 e	-35 e	-5 e	38 e	103.0	5
C7H12	1-Methylcyclohexene	-72 e	-53 e	-30 e	1 e	45 e	109.8	5
C7H12	4-Methylcyclohexene	-76 e	-59 e	-36 e	-5 e	37.9	102.3	5
C7H12	1-Ethylcyclopentene	-75 e	-57 e	-34 e	-3 e	40.7	105.8	5
C7H12	1,2-Dimethylcyclopentene	-75 e	-57 e	-34 e	-3 e	40.2	105.3	5
C7H12	1,5-Dimethylcyclopentene	-77 e	-59 e	-36 e	-5.5	37.3	101.5	5
C7H12O	Cycloheptanone			18 e	53.7	104.0	178.7	5
C7H12O2	Butyl acrylate	-52 e	-31 e	-4.5	30.4	78.0	146.9	5
C7H12O2	Propyl methacrylate				26 e	73.8	139.7	5
C7H12O3	Ethyl levulinate		17 e	45.3	82.6	133.2	205.7	5
C7H12O4	Diethyl malonate	-23 e	4 e	36.0	76.4	128.5	198.3	5
C7H12O4	Dimethyl glutarate	-11 e	15 e	47 e	87.7	139.8	209.5	5
C7H13ClO	Heptanoyl chloride	-17 e	4 e	29.4	59.7	96.9	144.0	5
C7H14	1-Heptene	-82.1	-63.8	-40.6	-10.7	31.1	93.2	1,5
C7H14	cis-2-Heptene	-79 e	-61 e	-38 e	-8 e	34.3	98.0	5
C7H14	trans-2-Heptene	-79 e	-61 e	-39 e	-8 e	34.0	97.5	5
C7H14	cis-3-Heptene	-80 e	-62 e	-40 e	-10 e	32.3	95.3	5
C7H14	trans-3-Heptene	-80 e	-62 e	-40 e	-10 e	32.2	95.2	5
C7H14	2-Methyl-1-hexene	-81 e	-64 e	-42 e	-12 e	29.3	91.6	5
C7H14	4-Methyl-1-hexene	-84 e	-67 e	-45 e	-16 e	25.3	86.3	5
C7H14	2-Methyl-2-hexene	-80 e	-63 e	-40 e	-10 e	32.0	95.0	5
C7H14	cis-3-Methyl-2-hexene	-79 e	-62 e	-39 e	-9 e	33.4	96.8	5
C7H14	trans-4-Methyl-2-hexene	-83 e	-66 e	-44 e	-15 e	25.9	87.1	5
C7H14	trans-5-Methyl-2-hexene	-83 e	-66 e	-44 e	-15 e	26.3	87.7	5
C7H14	trans-2-Methyl-3-hexene	-84 e	-67 e	-45 e	-16 e	24.6	85.5	5
C7H14	3-Ethyl-1-pentene	-85 e	-68 e	-46 e	-17 e	23.2	83.7	5
C7H14	2,3-Dimethyl-1-pentene	-85 e	-68 e	-46 e	-17 e	23.4	83.8	5
C7H14	2,4-Dimethyl-1-pentene	-88 e	-71 e	-50 e	-21 e	20.0	81.2	5
C7H14	3,3-Dimethyl-1-pentene	-87 e	-71 e	-50 e	-21 e	18.1	77.1	5
C7H14	4,4-Dimethyl-1-pentene	-94 e	-78 e	-57 e	-28 e	11.5	72.1	5
C7H14	2,3-Dimethyl-2-pentene	-79 e	-62 e	-39 e	-9 e	33.5	96.9	5
C7H14	2,4-Dimethyl-2-pentene	-84 e	-68 e	-46 e	-18 e	22.6	82.9	5
C7H14	cis-3,4-Dimethyl-2-pentene	-83 e	-65 e	-43 e	-14 e	27.2	88.8	5
C7H14	trans-3,4-Dimethyl-2-pentene	-82 e	-64 e	-42 e	-13 e	29.0	91.1	5
C7H14	cis-4,4-Dimethyl-2-pentene	-90 e	-73 e	-51 e	-22 e	18.6	80.0	5
C7H14	trans-4,4-Dimethyl-2-pentene	-90 e	-73 e	-52 e	-23 e	16.6	76.3	5
C7H14	2,3,3-Trimethyl-1-butene	-91 e	-75 e	-53 e	-24.2	16.3	77.5	5
C7H14	Cycloheptane				6 e	51.1	118.4	1

Mol. Form.	Name	Temperature in °C for the indicated pressure						Ref.
		1 Pa	10 Pa	100 Pa	1 kPa	10 kPa	100 kPa	
C_7H_{14}	Methylcyclohexane	-79 e	-62 e	-39 e	-7.9	35.5	100.5	1
C_7H_{14}	Ethylcyclopentane	-76 e	-59 e	-35 e	-5 e	38.4	103.0	5
C_7H_{14}	1,1-Dimethylcyclopentane		-69 e	-47 e	-17 e	24.8	87.4	5
C_7H_{14}	cis-1,2-Dimethylcyclopentane			-38 e	-8 e	34.9	99.0	5
C_7H_{14}	trans-1,2-Dimethylcyclopentane	-83 e	-66 e	-43 e	-13 e	28.4	91.4	5
C_7H_{14}	cis-1,3-Dimethylcyclopentane	-84 e	-66 e	-44 e	-14 e	28.2	91.1	5
C_7H_{14}	trans-1,3-Dimethylcyclopentane	-84 e	-67 e	-44 e	-14 e	27.4	90.3	5
$C_7H_{14}O$	1-Heptanal	-41 e	-21 e	4 e	37 e	83.7	152.3	5
$C_7H_{14}O$	2-Heptanone		-22 e	3 e	36 e	82.2	150.6	1
$C_7H_{14}O$	3-Heptanone		-28 e	0 e	36 e	83.2	147.0	5
$C_7H_{14}O$	4-Heptanone	-27 e	-6 e	18.8	50.2	90.3	143.4	5
$C_7H_{14}O$	5-Methyl-2-hexanone		-27 e	-2 e	31.0	76.6	144.4	5
$C_7H_{14}O$	2,4-Dimethyl-3-pentanone	-61 e	-42 e	-18 e	14 e	58.5	124.8	1
$C_7H_{14}O_2$	Heptanoic acid	24 e	46 e	72 e	107 e	154.6	222.6	5
$C_7H_{14}O_2$	Pentyl acetate	-58 e	-39 e	-14 e	20 e	70.1	149 e	5
$C_7H_{14}O_2$	Isopentyl acetate	-51 e	-30 e	-4 e	30.3	76.2	141.4	5
$C_7H_{14}O_2$	Isobutyl propanoate	-35 e	-19 e	2 e	31 e	72.0	136.1	5
$C_7H_{14}O_2$	Propyl butanoate	-35 e	-19 e	3 e	32.0	74.9	142.8	5
$C_7H_{14}O_2$	Propyl isobutanoate		-28 e	-5.7	24.5	67.5	133.3	5
$C_7H_{14}O_2$	Isopropyl isobutanoate		-44 e	-19.7	12.2	56.0	120.1	5
$C_7H_{14}O_2$	Ethyl 3-methylbutanoate	-57 e	-36 e	-10 e	23.9	69.5	134.4	5
$C_7H_{14}O_2$	Methyl hexanoate	-47 e	-26 e	2 e	36.6	83.3	149 e	5
$C_7H_{14}O_2$	4-Methoxy-4-methyl-2-pentanone				43 e	89.8	160 e	5
$C_7H_{15}Br$	1-Bromoheptane	-30 e	-9 e	18 e	54 e	104.4	178.4	5
$C_7H_{15}Cl$	1-Chloroheptane	-39 e	-19 e	7 e	41 e	88.6	159.9	5
$C_7H_{15}F$	1-Fluoroheptane	-64 e	-45 e	-22 e	10 e	53.3	117.4	5
$C_7H_{15}I$	1-Iodoheptane	-19 e	3 e	32 e	71 e	123.8	203.4	5
C_7H_{16}	Heptane	-78.6	-60.2	-37.0	-6.6	35.4	98.0	16
C_7H_{16}	2-Methylhexane	-82 e	-65 e	-43 e	-13 e	27.8	89.7	1
C_7H_{16}	3-Methylhexane	-81 e	-64 e	-42 e	-12 e	29.2	91.5	1
C_7H_{16}	3-Ethylpentane	-81 e	-63 e	-41 e	-11 e	30.5	93.1	1
C_7H_{16}	2,2-Dimethylpentane	-90 e	-73 e	-52 e	-22.9	17.6	78.8	1
C_7H_{16}	2,3-Dimethylpentane	-87 e	-68.4	-45.3	-14.9	26.8	89.3	5
C_7H_{16}	2,4-Dimethylpentane	-89 e	-72 e	-50 e	-21.3	19.2	80.1	1
C_7H_{16}	3,3-Dimethylpentane	-88 e	-71 e	-49 e	-18.8	22.9	85.6	1
C_7H_{16}	2,2,3-Trimethylbutane				-23.2	18.1	80.4	5
$C_7H_{16}O$	1-Heptanol		17 e	40 e	70.1	112.5	176 e	1
$C_7H_{16}O$	2-Heptanol	-9 e	7 e	27 e	55.0	95.2	158.7	5
$C_7H_{16}O$	3-Heptanol	-8 e	7 e	27 e	54.5	93.9	156.3	5
$C_7H_{16}O$	4-Heptanol	-16 e	1 e	22 e	51 e	91.9	154.6	5
$C_7H_{16}O$	2,2-Dimethyl-3-pentanol			9 e	35 e	73.1	135.5	5
$C_7H_{16}S$	1-Heptanethiol	-30 e	-9 e	18 e	53 e	102.7	176.4	5
$C_7H_{17}N$	Heptylamine			5 e	39 e	86.7	156.4	5
$C_7H_{18}N_2$	N,N-Diethyl-1,3-propanediamine			50.1	99.9	167.7	5
C_8F_{18}	Perfluorooctane			5 e	45.0	105.6	5
$C_8H_4O_3$	Phthalic anhydride	48.2 s	72.4 s			192.7	284.2	5
C_8H_6O	Benzofuran		-16 e	12 e	47.9	97.7	170.7	5
C_8H_7Cl	o-Chlorostyrene	-33 e	-10 e	20 e	58 e	110.8	188 e	5
C_8H_7N	2-Methylbenzonitrile		1 e	32.1	72.2	126.6	204.7	5
C_8H_7N	4-Methylbenzonitrile			40.1	78.7	134.3	221.3	5
C_8H_7N	Benzeneacetonitrile	-3 e	23 e	55.3	97.4	153.7	233.1	5
C_8H_7N	Indole	20.6 s	44.5 s				254.0	5
$C_8H_7NO_4$	Methyl 2-nitrobenzoate	17 e	49 e	89 e	140 e	208 e	302 e	5
C_8H_8	Styrene		-31 e	-5 e	28.6	75.4	144.7	1
C_8H_8	1,3,5,7-Cyclooctatetraene				24.3	71.0	140.1	5
C_8H_8O	Acetophenone			36 e	73 e	125.3	201.5	5
$C_8H_8O_2$	Phenyl acetate		3 e	33.1	72.2	123.9	195.5	5
$C_8H_8O_2$	Methyl benzoate		-1 e	29 e	68 e	121.2	198.9	5
$C_8H_8O_2$	4-Methoxybenzaldehyde	9 e	35 e	68.1	110.8	167.9	248.5	5
$C_8H_8O_3$	Methyl salicylate	-1 e	22 e	51 e	88.8	141.8	219.9	5
C_8H_9Cl	1-Chloro-2-ethylbenzene	-30 e	-9 e	18 e	54 e	103.7	177.9	5
C_8H_9Cl	1-Chloro-4-ethylbenzene	-27 e	-6 e	22 e	58 e	108.7	183.9	5
$C_8H_9NO_2$	1-Ethyl-4-nitrobenzene	10 e	36 e	69 e	111.6	168 e	245 e	5
C_8H_{10}	Ethylbenzene	-56.2	-36.8	-12.0	21.1	67.1	135.7	1
C_8H_{10}	o-Xylene			-7 e	27 e	74.2	143.9	1
C_8H_{10}	m-Xylene		-35 e	-10 e	23.4	69.8	138.7	1
C_8H_{10}	p-Xylene			22.4	68.9	137.9	1

Mol. Form.	Name	Temperature in °C for the indicated pressure						Ref.
		1 Pa	10 Pa	100 Pa	1 kPa	10 kPa	100 kPa	
$C_8H_{10}O$	o-Ethylphenol		16.9	44.5	81.1	130.9	204.0	5
$C_8H_{10}O$	m-Ethylphenol	5.6	29.2	57.5	91.9	144.8	217.9	5
$C_8H_{10}O$	p-Ethylphenol			60 e	95.5	144.6	217.5	5
$C_8H_{10}O$	2,3-Xylenol	14.3 s	34.3 s	57.2 s	91.4	141.7	216.4	1,5
$C_8H_{10}O$	2,4-Xylenol			50.2	85.5	137.2	210.5	1,5
$C_8H_{10}O$	2,5-Xylenol	13.4 s	33.2 s	55.9 s	87.4	137.0	210.6	5
$C_8H_{10}O$	2,6-Xylenol	-3.1 s	16.7 s	39.6 s	75.3	125.9	200.6	1,5
$C_8H_{10}O$	3,4-Xylenol	19.7 s	40.2 s	63.7 s	102.1	152.3	226.4	1,5
$C_8H_{10}O$	3,5-Xylenol	16.5 s	37.2 s	61.1 s	98.0	147.9	221.3	1,5
$C_8H_{10}O$	Benzeneethanol	2 e	25 e	54 e	92 e	143.6	217.7	5
$C_8H_{10}O$	Phenetole		-9 e	17 e	51 e	99 e	169.3	5
$C_8H_{10}O_2$	2-Phenoxyethanol	21 e	46 e	75.9	115.4	168.7	244.8	5
$C_8H_{10}O_2$	1,3-Dimethoxybenzene	18 e	34 e	56 e	86.7	135.5	223 e	5
$C_8H_{11}N$	p-Ethylaniline	-2 e	21 e	49 e	87 e	139.4	216.7	5
$C_8H_{11}N$	N-Ethylaniline	-15 e	8 e	38 e	76.4	128.8	204.2	5
$C_8H_{11}N$	N,N-Dimethylaniline			28 e	66 e	118.1	193.6	1
$C_8H_{11}N$	2,4-Xylidine	-2 e	21 e	51 e	88 e	139.1	210.9	5
$C_8H_{11}N$	2,6-Xylidine			37 e	80 e	137.7	217.7	5
$C_8H_{11}N$	5-Ethyl-2-picoline	-33 e	-9.3	20 e			178.0	5
$C_8H_{11}NO$	o-Phenetidine	0 e	27 e	60 e	102.2	156.0	228.1	5
C_8H_{12}	1,5-Cyclooctadiene		-37 e	-8 e	30 e	80.2	150 e	5
C_8H_{12}	4-Vinylcyclohexene	-62 e	-43 e	-19 e	14.1	59.9	129 e	5
$C_8H_{12}O_4$	Diethyl maleate	-6 e	20 e	52.2	93.5	148.4	224.8	5
C_8H_{14}	2,5-Dimethyl-1,5-hexadiene	-38 e	-26 e	-10 e	14 e	50.8	115.1	5
C_8H_{14}	1-Octyne	-59 e	-40 e	-16 e	16 e	60.3	125.8	1
C_8H_{14}	2-Octyne	-52 e	-33 e	-8 e	25 e	70.6	137.8	1
C_8H_{14}	3-Octyne	-55 e	-35 e	-11 e	22 e	66.8	132.8	1
C_8H_{14}	4-Octyne	-56 e	-36 e	-12 e	21 e	65.6	131.4	1
C_8H_{14}	1-Ethylcyclohexene	-55 e	-35 e	-11 e	22 e	68 e	136.5	5
$C_8H_{14}O_2$	Cyclohexyl acetate					103.1	172.9	5
$C_8H_{14}O_2$	Butyl methacrylate				47 e	93.3	159.0	5
$C_8H_{14}O_3$	Butanoic anhydride	-28 e	-2 e	30 e	71 e	123.8	196.5	5
$C_8H_{14}O_4$	Ethyl succinate	-6 e	20 e	51.0	91.1	143.7	216.1	5
$C_8H_{14}O_4$	Dipropyl oxalate	-4 e	20 e	49.9	88.6	140.4	213.0	5
$C_8H_{14}O_4$	Dimethyl adipate		28 e	61 e	103 e	156.1	227.3	5
$C_8H_{15}Br$	(2-Bromoethyl)cyclohexane	-14 e	8 e	36.9	75.3	129.7	212.5	5
$C_8H_{15}ClO$	Octanoyl chloride	1 e	22 e	46 e	74.7	109 e	150 e	5
$C_8H_{15}N$	Octanenitrile	-15 e	8 e	37 e	75 e	127.7	204.4	5
C_8H_{16}	1-Octene	-65.7	-46.1	-21.4	10.5	54.9	120.9	1,5
C_8H_{16}	cis-2-Octene	-59 e	-41 e	-17 e	15 e	59 e	125.2	5
C_8H_{16}	trans-2-Octene	-59 e	-41 e	-17 e	14 e	59 e	124.5	5
C_8H_{16}	cis-3-Octene	-65 e	-46 e	-22 e	10 e	55.1	122.4	5
C_8H_{16}	trans-3-Octene	-61 e	-43 e	-19 e	13 e	57 e	122.8	5
C_8H_{16}	cis-4-Octene	-63 e	-44 e	-20 e	11 e	56 e	122.1	5
C_8H_{16}	trans-4-Octene	-65 e	-46 e	-22 e	10 e	54.6	121.8	5
C_8H_{16}	2-Methyl-1-heptene	-66 e	-48 e	-24 e	8 e	52.3	118.7	5
C_8H_{16}	2,2-Dimethyl-cis-3-hexene	-74 e	-56 e	-33 e	-3 e	40.1	105.0	5
C_8H_{16}	2,3-Dimethyl-2-hexene	-65 e	-47 e	-23 e	10 e	54.3	121.3	5
C_8H_{16}	2,3,3-Trimethyl-1-pentene		-53 e	-30 e	1 e	43.8	107.9	5
C_8H_{16}	2,4,4-Trimethyl-1-pentene	-79 e	-61 e	-38 e	-7 e	36.2	101.0	5
C_8H_{16}	2,3,4-Trimethyl-2-pentene	-68 e	-49 e	-26 e	6 e	50.0	115.8	5
C_8H_{16}	2,4,4-Trimethyl-2-pentene	-73 e	-56 e	-33 e	-2 e	40.4	104.5	5
C_8H_{16}	Cyclooctane				30 e	78 e	150.7	1
C_8H_{16}	Ethylcyclohexane	-61 e	-42 e	-17 e	15.8	61.9	131.3	5
C_8H_{16}	1,1-Dimethylcyclohexane			-27 e	5 e	50.6	119.1	5
C_8H_{16}	cis-1,2-Dimethylcyclohexane		-44 e	-20 e	14 e	59.7	129.2	5
C_8H_{16}	trans-1,2-Dimethylcyclohexane	-68 e	-49 e	-25 e	8 e	53.9	122.9	5
C_8H_{16}	cis-1,3-Dimethylcyclohexane	-68 e	-48 e	-23 e	10 e	55.6	123.1	5
C_8H_{16}	trans-1,3-Dimethylcyclohexane	-62 e	-45 e	-23 e	8 e	51.5	120.9	5
C_8H_{16}	cis-1,4-Dimethylcyclohexane	-66 e	-47 e	-23 e	10 e	55.3	123.8	5
C_8H_{16}	trans-1,4-Dimethylcyclohexane			-27 e	5 e	50.6	118.9	5
C_8H_{16}	Propylcyclopentane	-60 e	-41 e	-16 e	16.5	62.1	130.5	5
C_8H_{16}	Isopropylcyclopentane	-65 e	-46 e	-21 e	12 e	57.3	125.9	5
C_8H_{16}	1-Ethyl-1-methylcyclopentane	-67 e	-49 e	-24 e	8 e	53.2	121.0	5
C_8H_{16}	cis-1-Ethyl-2-methylcyclopentane	-63 e	-44 e	-19 e	13.3	59.1	127.6	5
C_8H_{16}	1,1,2-Trimethylcyclopentane				2 e	46.2	113.2	5
C_8H_{16}	1,1,3-Trimethylcyclopentane	-77 e	-59 e	-36 e	-5 e	38.7	104.4	5

Mol. Form.	Name	Temperature in °C for the indicated pressure						Ref.
		1 Pa	10 Pa	100 Pa	1 kPa	10 kPa	100 kPa	
C_8H_{16}	1′,2′,4a-1,2,4-Trimethylcyclopentane	-70 e	-52 e	-28 e	4 e	48.9	116.2	5
C_8H_{16}	1′,2a,4′-1,2,4-Trimethylcyclopentane	-74 e	-56 e	-33 e	-1 e	42.8	108.8	5
$C_8H_{16}O$	1-Propylcyclopentanol	9 e	24 e	43 e	69.0	108.4	173.5	5
$C_8H_{16}O$	Octanal			6 e	45.7	97.8	170.2	5
$C_8H_{16}O$	2-Octanone		-3 e	23 e	57 e	103.8	172.1	5
$C_8H_{16}O$	3-Octanone			8 e	47.7	97 e	161 e	5
$C_8H_{16}O$	2,2,4-Trimethyl-3-pentanone			11.3	42.1	81.7	134.6	5
$C_8H_{16}O_2$	Octanoic acid	37 e	58 e	85 e	120 e	165.5	238.4	1,5
$C_8H_{16}O_2$	2-Ethylhexanoic acid				108 e	159.6	226.6	5
$C_8H_{16}O_2$	Hexyl acetate	-37 e	-13 e	16 e	52.8	100.4	164 e	5
$C_8H_{16}O_2$	Isopentyl propanoate			3.1	40.7	90.6	159.8	5
$C_8H_{16}O_2$	Isobutyl isobutanoate	-47 e	-26 e	0.4	34.8	81.1	147.0	5
$C_8H_{16}O_2$	Propyl 3-methylbutanoate			1.8	38.9	87.9	155.6	5
$C_8H_{16}O_2$	Ethyl hexanoate	-31 e	-9 e	18.7	53.9	100.7	166.2	5
$C_8H_{16}O_2$	Methyl heptanoate	-30 e	-9 e	19 e	54.2	102.4	172 e	5
$C_8H_{16}O_4$	Diethylene glycol monoethyl ether acetate	-16 e	10.6	43.9	86.2	141.3	216.6	5
$C_8H_{17}Br$	1-Bromooctane	-17 e	6 e	34 e	72 e	123.8	200.3	5
$C_8H_{17}Cl$	1-Chlorooctane	-25 e	-4 e	23 e	59 e	108.8	182.9	5
$C_8H_{17}Cl$	3-(Chloromethyl)heptane					100.3	172.4	5
$C_8H_{17}F$	1-Fluorooctane				29 e	74.6	141.8	5
$C_8H_{17}I$	1-Iodooctane	-6 e	18 e	48 e	87 e	142.5	224.5	5
C_8H_{18}	Octane		-42.6	-17.9	14.4	58.9	125.3	16
C_8H_{18}	2-Methylheptane	-69 e	-49.1	-24.5	7.6	51.6	117.2	1,5
C_8H_{18}	3-Methylheptane	-67 e	-48.1	-23.6	8.5	52.7	118.5	1,5
C_8H_{18}	4-Methylheptane	-65 e	-47 e	-24 e	7.8	51.6	117.2	5
C_8H_{18}	3-Ethylhexane				8 e	52.1	118.1	5
C_8H_{18}	2,2-Dimethylhexane	-73 e	-55 e	-32 e	-1.5	41.6	106.4	5
C_8H_{18}	2,3-Dimethylhexane				5 e	49.2	115.1	5
C_8H_{18}	2,4-Dimethylhexane				0.6	43.9	109.0	5
C_8H_{18}	2,5-Dimethylhexane	-71 e	-53 e	-30 e	0.7	43.8	108.6	5
C_8H_{18}	3,3-Dimethylhexane	-72 e	-54 e	-30 e	1.4	45.4	111.5	5
C_8H_{18}	3,4-Dimethylhexane				7 e	50.9	117.3	5
C_8H_{18}	3-Ethyl-2-methylpentane	-69 e	-50 e	-27 e	5 e	48.9	115.2	5
C_8H_{18}	3-Ethyl-3-methylpentane	-70 e	-51 e	-27 e	5 e	50.2	117.8	5
C_8H_{18}	2,2,3-Trimethylpentane	-74 e	-56 e	-32 e	-0.8	43.1	109.4	5
C_8H_{18}	2,2,4-Trimethylpentane	-81.9	-63.4	-39.8	-8.9	34.0	98.8	5
C_8H_{18}	2,3,3-Trimethylpentane	-72 e	-54 e	-30 e	2.1	46.9	114.3	5
C_8H_{18}	2,3,4-Trimethylpentane	-74 e	-54.5	-30.0	2.2	46.7	113.1	1,5
C_8H_{18}	2,2,3,3-Tetramethylbutane	-62.5 s	-44 s	-20.9 s	8.9 s	48.8 s	105.8	5
$C_8H_{18}O$	1-Octanol	12 e	30 e	53 e	84 e	128.2	194.8	1,39
$C_8H_{18}O$	2-Octanol			40 e	69.9	112.5	179.4	1,39
$C_8H_{18}O$	3-Octanol	12 e	24 e	40 e	64 e	102.8	174.1	1
$C_8H_{18}O$	4-Octanol			40 e	66.9	107.3	176.0	1,39
$C_8H_{18}O$	4-Methyl-3-heptanol	-52 e	-28 e	1 e	39 e	87.6	155.0	5
$C_8H_{18}O$	5-Methyl-3-heptanol	-35 e	-16 e	8 e	40 e	84.8	153.0	5
$C_8H_{18}O$	4-Methyl-4-heptanol	-17 e	1 e	24 e	55 e	97.2	160.7	5
$C_8H_{18}O$	2-Ethyl-1-hexanol			45 e	75 e	118.3	184.2	1
$C_8H_{18}O$	2-Ethyl-2-hexanol	-13 e	4 e	26 e	55 e	96.3	160.3	5
$C_8H_{18}O$	2,4,4-Trimethyl-2-pentanol		-7 e	13 e	40 e	79.8	146.1	5
$C_8H_{18}O$	2,2,4-Trimethyl-3-pentanol	-2 e	9 e	24 e	47 e	82.6	150.4	5
$C_8H_{18}O$	Dibutyl ether	-55 e	-35 e	-8 e	26 e	73.0	141.2	5
$C_8H_{18}O$	Di-sec-butyl ether			-19 e	12.1	55.4	120.6	5
$C_8H_{18}O$	Di-tert-butyl ether			-33 e	-2 e	41.7	106.8	1
$C_8H_{18}O_2$	Ethylene glycol monohexyl ether	-13 e	14 e	46 e	86 e	137.7	206.9	5
$C_8H_{18}O_2$	1,2-Dipropoxyethane			-44.2	-2.0	63.6	179.2	5
$C_8H_{18}O_2$	Di-tert-butyl peroxide			-26 e	4.3	46.6	110.5	5
$C_8H_{18}O_3$	Diethylene glycol monobutyl ether	14 e	37 e	66.8	104.9	153 e	230.4	5
$C_8H_{18}O_3$	Diethylene glycol diethyl ether	-32 e	-7 e	25 e	64.9	117.1	189 e	5
$C_8H_{18}O_5$	Tetraethylene glycol	89 e	117 e	151.1	192.2	242.9	307.3	5
$C_8H_{18}S$	1-Octanethiol	-15 e	6 e	34 e	71 e	122.1	198.5	5
$C_8H_{18}S$	Dibutyl sulfide	-22 e	0 e	27 e	63 e	113.5	188.4	5

Mol. Form.	Name	\multicolumn{6}{c}{Temperature in °C for the indicated pressure}	Ref.
		1 Pa	10 Pa	100 Pa	1 kPa	10 kPa	100 kPa	
$C_8H_{19}N$	Dibutylamine	-37 e	-16 e	10 e	44 e	90.8	159.1	5
$C_8H_{19}N$	Diisobutylamine	-57 e	-36 e	-9.0	25.5	72.2	139.0	5
$C_8H_{20}O_4Si$	Ethyl silicate	-77 e	-52 e	-21 e	21.6	80.5	164.1	5
$C_8H_{20}Si$	Tetraethylsilane			-6.5	30.5	80.6	152.6	5
C_9F_{20}	Perfluorononane					40 e	114.7	5
$C_9H_6N_2O_2$	Toluene-2,4-diisocyanate		39 e	72 e	113.9	169.7	247 e	5
C_9H_7N	Quinoline	-1.3	23.7	55.4	96.8	153.4	236.5	1,5
C_9H_7N	Isoquinoline		30.2	60.7	101.3	157.9	242.7	1,5
C_9H_8	Indene			12 e	53.0	106.8	181.0	5
C_9H_{10}	cis-1-Propenylbenzene	-38 e	-15.4	13.3	51.4	103.7	178.4	5
C_9H_{10}	trans-1-Propenylbenzene		-16 e	13.3	51.6	103.7	178.4	5
C_9H_{10}	Isopropenylbenzene			3.2	41.5	92.8	164.9	5
C_9H_{10}	Indan	-33 e	-12 e	16 e	52 e	102.3	177.5	1
$C_9H_{10}O$	2,4-Dimethylbenzaldehyde	-3 e	23 e	54 e	93.2	144.6	214.5	5
$C_9H_{10}O_2$	Ethyl benzoate	-18 e	8 e	39 e	80.1	135.1	212.8	5
$C_9H_{10}O_2$	Benzyl acetate	-11 e	15 e	46.6	86.9	139.5	211 e	5
$C_9H_{11}Br$	1-Bromo-4-isopropylbenzene	-8 e	15 e	45 e	84 e	138.1	218.5	5
$C_9H_{11}Cl$	1-Chloro-2-isopropylbenzene	-23 e	-1 e	27 e	64 e	114.6	190.5	5
$C_9H_{11}Cl$	1-Chloro-4-isopropylbenzene		3 e	31 e	69 e	120.5	197.8	5
C_9H_{12}	Propylbenzene	-43 e	-23 e	4 e	38 e	86.7	158.8	1
C_9H_{12}	Isopropylbenzene	-46 e	-26 e	-1 e	33 e	80.9	152.0	1
C_9H_{12}	o-Ethyltoluene	-40 e	-19 e	8 e	43 e	92.1	164.7	5
C_9H_{12}	m-Ethyltoluene	-42 e	-21 e	5 e	40.4	88.9	160.8	5
C_9H_{12}	p-Ethyltoluene	-41 e	-21 e	6 e	41 e	89.2	161.5	5
C_9H_{12}	1,2,3-Trimethylbenzene		-12 e	15 e	52 e	101.5	175.6	1
C_9H_{12}	1,2,4-Trimethylbenzene	-37 e	-16 e	11 e	47 e	95.9	168.9	1
C_9H_{12}	1,3,5-Trimethylbenzene	-39 e	-18 e	9 e	43.7	92.4	164.3	1
$C_9H_{12}O$	Benzyl ethyl ether		-10 e	20.4	59.3	111.3	184.5	5
$C_9H_{12}O$	Phenyl propyl ether		-10 e	21 e	61 e	113.9	189.3	5
$C_9H_{12}O$	Phenyl isopropyl ether	-20 e	-1 e	23 e	56 e	103.7	176.9	5
$C_9H_{13}N$	2,4,6-Trimethylaniline	12 e	36 e	66 e	104.1	154.9	226 e	5
$C_9H_{13}N$	N,N-Dimethyl-o-toluidine	-25 e	-3 e	24.4	60.6	110.7	184.5	5
$C_9H_{13}N$	Amphetamine			33 e	70.1	118 e	202.0	5
$C_9H_{14}O$	Isophorone		1 e	33.1	75.1	132.4	215.1	5
$C_9H_{14}O_6$	Triacetin	37.6	62 e	90 e	124 e	165 e	214 e	5
$C_9H_{16}O_4$	Diethyl glutarate	-1 e	26 e	60.2	103.3	159.6	236.5	5
$C_9H_{17}N$	Nonanenitrile	-3 e	21 e	50.9	90.7	145.4	225.1	5
C_9H_{18}	1-Nonene	-50.1	-29.4	-3.3	30.4	77.1	146.4	1,5
C_9H_{18}	2-Methyl-1-octene	-53 e	-34 e	-9 e	25 e	72 e	144.1	5
C_9H_{18}	Butylcyclopentane	-45 e	-24 e	1 e	36 e	84 e	156.1	5
C_9H_{18}	Propylcyclohexane	-46 e	-26 e	0 e	35.1	83.6	156.2	5
C_9H_{18}	Isopropylcyclohexane	-48 e	-28 e	-2 e	33 e	81.3	154.0	5
C_9H_{18}	trans-1-Ethyl-4-methylcyclo-hexane	-53 e	-33 e	-8 e	25 e	71.8	141.5	5
C_9H_{18}	1,1,2-Trimethylcyclohexane			-12 e	23 e	71.5	145.5	5
C_9H_{18}	1,1,3-Trimethylcyclohexane	-60 e	-41 e	-16 e	18 e	65.2	136.1	5
C_9H_{18}	1′,2a,4a-1,2,4-Trimethylcyclo-hexane	-71 e	-50 e	-22 e	15 e	65.7	140.7	5
C_9H_{18}	1′,3′,5′-1,3,5-Trimethylcyclo-hexane	-72 e	-50 e	-22 e	14 e	65.1	140.0	5
C_9H_{18}	Isobutylcyclopentane	-105 e	-88 e	-64 e	-28 e	31 e	147.0	5
C_9H_{18}	cis-1-Methyl-2-propylcyclo-pentane	-52 e	-33 e	-7 e	28 e	77 e	152.0	5
C_9H_{18}	trans-1-Methyl-2-propylcyclo-pentane	-56 e	-36 e	-11 e	23 e	72 e	145.8	5
C_9H_{18}	1,1,3,3-Tetramethylcyclo-pentane	-72 e	-54 e	-30 e	2 e	47 e	117.4	5
$C_9H_{18}O$	Nonanal		-3 e	27.4	65.5	115.6	184.6	5
$C_9H_{18}O$	2-Nonanone		8 e	35 e	71 e	121.0	194.0	5
$C_9H_{18}O$	5-Nonanone			-1 e	39.1	94 e	188 e	5
$C_9H_{18}O$	2,6-Dimethyl-4-heptanone	-32 e	-12 e	14 e	48 e	96.2	167.7	5
$C_9H_{18}O_2$	Nonanoic acid	48 e	69 e	97 e	133 e	182.7	255.1	5
$C_9H_{18}O_2$	Heptyl acetate	-16 e	6 e	34 e	70 e	119.9	191.9	5
$C_9H_{18}O_2$	Isopentyl butanoate				55 e	105.6	178.4	5
$C_9H_{18}O_2$	Isobutyl 3-methylbutanoate			11.3	48.3	97.9	168.3	5
$C_9H_{18}O_2$	Propyl hexanoate	-26 e	-2 e	28 e	65.1	113.4	178 e	5
$C_9H_{18}O_2$	Methyl octanoate	-26 e	-9 e	13 e	40 e	76 e	127.9	5

Mol. Form.	Name	Temperature in °C for the indicated pressure						Ref.
		1 Pa	10 Pa	100 Pa	1 kPa	10 kPa	100 kPa	
$C_9H_{19}Cl$	1-Chlorononane	-11 e	11 e	39 e	76 e	127.8	204.7	5
C_9H_{20}	Nonane	-46.8	-26.0	0.0	34.0	80.8	150.3	16
C_9H_{20}	2-Methyloctane	-49 e	-30 e	-5 e	28 e	73.9	142.8	5
C_9H_{20}	3-Methyloctane	-49 e	-29 e	-5 e	29 e	74.7	143.7	5
C_9H_{20}	4-Methyloctane	-50 e	-30 e	-6 e	27 e	73.2	141.9	5
C_9H_{20}	2,2-Dimethylheptane	-58 e	-39 e	-15 e	18 e	63.6	132.3	5
C_9H_{20}	2,3-Dimethylheptane	-53 e	-33 e	-9 e	25 e	70.8	140.0	5
C_9H_{20}	2,6-Dimethylheptane	-55 e	-36 e	-12 e	21 e	66.4	134.7	5
C_9H_{20}	3-Ethyl-4-methylhexane			-9 e	24 e	70.6	139.9	5
C_9H_{20}	2,2,4-Trimethylhexane	-66.1	-46.4	-21.3	11.8	57.7	126.0	5
C_9H_{20}	2,2,5-Trimethylhexane	-65.1	-45.8	-21.2	11.2	56.2	123.7	1,5
C_9H_{20}	2,3,3-Trimethylhexane	-58 e	-38 e	-13 e	20 e	66.7	137.2	5
C_9H_{20}	2,3,5-Trimethylhexane	-60 e	-41 e	-16 e	17 e	62.3	130.9	5
C_9H_{20}	2,4,4-Trimethylhexane	-62 e	-43 e	-18 e	15 e	61.0	130.2	5
C_9H_{20}	3,3,4-Trimethylhexane	-53 e	-33 e	-7 e	28 e	76.3	148.9	5
C_9H_{20}	3,3-Diethylpentane			-9 e	26 e	73.7	145.7	1
C_9H_{20}	3-Ethyl-2,4-dimethylpentane	-58 e	-38 e	-13 e	20 e	66.7	136.2	5
C_9H_{20}	2,2,3,3-Tetramethylpentane				21 e	68.5	139.8	1
C_9H_{20}	2,2,3,4-Tetramethylpentane	-61 e	-42 e	-17 e	16 e	62.5	132.6	1
C_9H_{20}	2,2,4,4-Tetramethylpentane		-49 e	-25 e	8 e	53.2	121.8	1
C_9H_{20}	2,3,3,4-Tetramethylpentane	-57 e	-37 e	-12 e	22 e	69.7	141.1	1
$C_9H_{20}O$	1-Nonanol		40 e	64 e	96.9	141.0	213.0	5,39
$C_9H_{20}O$	3-Nonanol		24 e	47 e	78 e	123.0	194.2	5
$C_9H_{20}O$	4-Nonanol			45 e	76.4	121.3	192.0	5
$C_9H_{20}O$	5-Nonanol	13 e	31 e	54 e	84.5	128.1	194.7	5
$C_9H_{20}O$	2,2,4,4-Tetramethyl-3-pentanol				58	100	167	5
$C_9H_{20}S$	1-Nonanethiol	-2 e	21 e	49 e	87 e	140.4	219.2	5
$C_9H_{21}BO_3$	Triisopropyl borate				73.1	139.0		5
$C_9H_{21}N$	Nonylamine		9 e	37 e	75 e	126.2	202.1	5
$C_9H_{21}N$	Tripropylamine	-39 e	-18 e	8 e	42 e	88.2	156.0	5
$C_{10}F_8$	Perfluoronaphthalene	5.2 s	25.1 s	48.1 s				5
$C_{10}F_{22}$	Perfluorodecane					52 e	132.9	5
$C_{10}H_7Br$	1-Bromonaphthalene	17 e	45 e	80.3	126.7	189.8	280.5	5
$C_{10}H_7Cl$	1-Chloronaphthalene	14 e	39 e	70.5	112.8	171.6	258.6	5
$C_{10}H_8$	Naphthalene**	3.2 s	24.1 s	49.3 s	80.7	135.6	217.5	1,5
$C_{10}H_8$	Azulene	24.1 s	46 s	71.5 s	103.3	162.6	244.0	5
$C_{10}H_8O$	1-Naphthol				137.2	196.7	281.8	5
$C_{10}H_8O$	2-Naphthol				140.7	200.5	286.8	5
$C_{10}H_9N$	1-Naphthalenamine		62 e	99.0	146.9	210.7	300.1	5
$C_{10}H_9N$	2-Naphthalenamine	36.3 s	65.9 s	103 s	150.9	215.1	305.5	5
$C_{10}H_9N$	2-Methylquinoline	5.3	31.9	63.8	102.9	165.8	247.2	5
$C_{10}H_9N$	4-Methylquinoline	29 e	54 e	85 e	127 e	183.0	265.1	5
$C_{10}H_9N$	6-Methylquinoline	27 e	51 e	81 e	122 e	179.2	264.5	5
$C_{10}H_9N$	8-Methylquinoline	15 e	40 e	70 e	111 e	166.1	247.3	5
$C_{10}H_{10}$	m-Divinylbenzene	-29 e	-4 e	27.1	67.6	122.1	199 e	5
$C_{10}H_{10}O_4$	Dimethyl phthalate	27 e	56 e	92.7	137.8	195.8	272.7	5
$C_{10}H_{10}O_4$	Dimethyl isophthalate			85 e	129.5	189.2	273 e	5
$C_{10}H_{10}O_4$	Dimethyl terephthalate	56.6 s	79.4 s	106.1 s	137.9 s	197.9	282 e	5
$C_{10}H_{12}$	1,2,3,4-Tetrahydronaphthalene	-21 e	3 e	33.2	74.1	127.4	207.8	5
$C_{10}H_{12}$	2-Ethylstyrene	-31 e	-8 e	21 e	60 e	111.7	187 e	5
$C_{10}H_{12}$	3-Ethylstyrene	-28 e	-5.3	24.1	62.6	116 e	193 e	5
$C_{10}H_{12}$	4-Ethylstyrene	-31 e	-8.2	21.3	60.5	115 e	196 e	5
$C_{10}H_{12}O$	Estragole			48.5	88.0	140.7	214.6	5
$C_{10}H_{12}O$	4-Isopropylbenzaldehyde			54.1	96.0	152.2	231.5	5
$C_{10}H_{12}O_2$	4-Allyl-2-methoxyphenol	9 e	37 e	72 e	115.9	173.8	252.9	5
$C_{10}H_{12}O_2$	2-Phenylethyl acetate	-4 e	22 e	54 e	96 e	152.3	232.0	5
$C_{10}H_{12}O_2$	Propyl benzoate	-8 e	18 e	50.2	92.3	149.2	230.5	5
$C_{10}H_{12}O_2$	Ethyl phenylacetate	-9 e	19 e	52 e	95 e	150.2	225 e	5
$C_{10}H_{12}O_2$	Isoeugenol				125 e	185.3	267.1	5
$C_{10}H_{14}$	Butylbenzene	-28 e	-7 e	21 e	56.9	107.6	182.8	1,5
$C_{10}H_{14}$	sec-Butylbenzene	-35 e	-14 e	13 e	48 e	98.3	172.8	5
$C_{10}H_{14}$	tert-Butylbenzene	-37 e	-16 e	10 e	46 e	94.9	168.6	5
$C_{10}H_{14}$	Isobutylbenzene	-36 e	-15 e	12 e	47.9	97.8	172.3	5
$C_{10}H_{14}$	o-Cymene	-39 e	-16 e	13 e	51 e	103.1	177.8	5
$C_{10}H_{14}$	m-Cymene	-34 e	-13 e	14 e	50 e	99.9	174.6	5
$C_{10}H_{14}$	p-Cymene	-33 e	-12 e	16 e	52 e	102.2	176.6	5
$C_{10}H_{14}$	o-Diethylbenzene	-28 e	-6 e	21 e	58 e	107.9	182.9	5

Mol. Form.	Name	1 Pa	10 Pa	100 Pa	1 kPa	10 kPa	100 kPa	Ref.
				Temperature in °C for the indicated pressure				
$C_{10}H_{14}$	m-Diethylbenzene	-28 e	-7 e	20 e	56 e	106.2	180.6	5
$C_{10}H_{14}$	p-Diethylbenzene	-28 e	-6 e	21 e	57 e	108.1	183.3	5
$C_{10}H_{14}$	3-Ethyl-1,2-dimethylbenzene	-22 e	0 e	28 e	66 e	117.2	193.4	5
$C_{10}H_{14}$	4-Ethyl-1,2-dimethylbenzene	-24 e	-2 e	26 e	63 e	113.6	189.2	5
$C_{10}H_{14}$	2-Ethyl-1,3-dimethylbenzene		-2 e	26 e	63 e	113.7	189.5	5
$C_{10}H_{14}$	2-Ethyl-1,4-dimethylbenzene	-27 e	-5 e	23 e	60 e	110.6	186.4	5
$C_{10}H_{14}$	1-Ethyl-2,4-dimethylbenzene	-25 e	-4 e	24 e	61 e	112.2	187.9	5
$C_{10}H_{14}$	1-Ethyl-3,5-dimethylbenzene	-28 e	-6 e	21 e	58 e	108.3	183.2	5
$C_{10}H_{14}$	1-Methyl-2-propylbenzene	-27 e	-6 e	22 e	58.2	108.9	184.3	5
$C_{10}H_{14}$	1-Methyl-3-propylbenzene	-29 e	-8 e	20 e	56.1	106.5	181.3	5
$C_{10}H_{14}$	1-Methyl-4-propylbenzene	-29 e	-7 e	20 e	56.6	107.4	182.8	5
$C_{10}H_{14}$	1,2,3,4-Tetramethylbenzene		7 e	36 e	74 e	126.6	204.5	5
$C_{10}H_{14}$	1,2,3,5-Tetramethylbenzene	-19 e	3 e	32 e	69 e	120.9	197.5	5
$C_{10}H_{14}$	1,2,4,5-Tetramethylbenzene					119.9	196.3	5
$C_{10}H_{14}O$	2-Butylphenol	7 e	31 e	61 e	101 e	155.2	234.4	5
$C_{10}H_{14}O$	Butyl phenyl ether	-16 e	8 e	38 e	77 e	131.3	209.7	5
$C_{10}H_{14}O$	Thymol	18.9 s	37.9 s	59.5	101.2	155.0	230.4	5
$C_{10}H_{15}N$	2-Methyl-5-isopropylaniline	19 e	43 e	72 e	107.4	150 e	204 e	5
$C_{10}H_{15}N$	N-Butylaniline	11 e	35 e	66 e	106 e	160.9	241.0	5
$C_{10}H_{15}N$	N,N-Diethylaniline	-11 e	14 e	44.3	84.2	138.4	216.3	5
$C_{10}H_{16}$	Dipentene	-42 e	-19 e	10.6	48.7	100.2	173.9	5
$C_{10}H_{16}$	d-Limonene	-45 e	-21 e	9.1	48.0	100.4	174.5	5
$C_{10}H_{16}$	l-Limonene	-33 e	-12 e	16 e	52.0	102.3	177.0	21
$C_{10}H_{16}$	β-Myrcene			9.4	47.3	98.3	171.0	5
$C_{10}H_{16}$	α-Pinene	-48 e	-27 e	-1 e	33.6	82.2	155.1	21
$C_{10}H_{16}$	β-Pinene	-43 e	-22 e	5.0	40.6	90.5	165.5	21
$C_{10}H_{16}$	Camphene					90.7	160.1	4
$C_{10}H_{16}$	Terpinolene			26.5	64.9	115.4	184.6	5
$C_{10}H_{16}$	β-Phellandrene			16 e	53.2	104 e	171.0	5
$C_{10}H_{16}O$	(+)-Camphor	-15.8 s	10 s	41.5 s	80.8 s	131.4 s	207.6	5
$C_{10}H_{16}O$	Pulegone	37 e	49.1	66.4	92.2	135.1	220.2	5
$C_{10}H_{18}$	1-Decyne	-34 e	-13 e	14 e	51 e	100.3	173.5	5
$C_{10}H_{18}$	cis-Decahydronaphthalene	-26 e	-4 e	24 e	62.4	115.5	195.3	1
$C_{10}H_{18}$	trans-Decahydronaphthalene		-10 e	18 e	55.3	107.9	186.8	1
$C_{10}H_{18}O$	α-Terpineol			48	89	142	217	4
$C_{10}H_{18}O$	Eucalyptol			10.6	48.5	100.3	175.4	5
$C_{10}H_{18}O$	trans-Geraniol	4 e	31 e	63.2	104.3	157.7	229.6	5
$C_{10}H_{18}O_4$	Sebacic acid	125.9 s						5
$C_{10}H_{18}O_4$	Dipropyl succinate	11 e	38 e	72.1	115.4	172.3	250.4	5
$C_{10}H_{18}O_4$	Diethyl adipate	4 e	35 e	72 e	116.6	171.2	239.5	5
$C_{10}H_{19}N$	Decanenitrile	13 e	36 e	66 e	105.8	160.6	241.6	5
$C_{10}H_{20}$	1-Decene	-35.5	-13.7	13.7	49.0	97.9	170.1	1,5
$C_{10}H_{20}$	Cyclodecane			29 e	68 e	121.3	201.8	1
$C_{10}H_{20}$	Butylcyclohexane	-31 e	-9 e	18 e	54 e	104.7	180.4	5
$C_{10}H_{20}$	Isobutylcyclohexane	-37 e	-16 e	10 e	46 e	95.9	170.8	5
$C_{10}H_{20}$	tert-Butylcyclohexane	-39 e	-18 e	9 e	45 e	95.3	171.1	5
$C_{10}H_{20}O$	Decanal		16 e	47.2	86.3	137.7	208.0	5
$C_{10}H_{20}O_2$	Decanoic acid	58 e	80 e	108 e	145 e	195.2	269.5	5
$C_{10}H_{20}O_2$	Octyl acetate	-26 e	-3 e	27 e	66.3	120.0	198.2	5
$C_{10}H_{20}O_2$	2-Ethylhexyl acetate	-11 e	5 e	26 e	57.6	107.1	197.2	5
$C_{10}H_{20}O_2$	Isopentyl isopentanoate			22 e	62.8	116.9	193.6	5
$C_{10}H_{20}O_2$	Ethyl octanoate	-17 e	9 e	41 e	81.4	133.2	203 e	5
$C_{10}H_{20}O_4$	Diethylene glycol monobutyl ether acetate	6 e	34 e	69 e	112.6	169.2	245.4	5
$C_{10}H_{21}Br$	1-Bromodecane	9 e	33 e	63 e	104 e	159.2	240.0	5
$C_{10}H_{21}Cl$	1-Chlorodecane	2 e	25 e	54 e	92 e	145.7	225.3	5
$C_{10}H_{21}F$	1-Fluorodecane	-22 e	0 e	27 e	64 e	113.3	185.7	5
$C_{10}H_{22}$	Decane		-10.6	16.7	52.3	101.1	173.7	16
$C_{10}H_{22}$	2-Methylnonane	-34 e	-14 e	12 e	47 e	94.8	166.5	5
$C_{10}H_{22}$	3-Methylnonane	-34 e	-14 e	12 e	47 e	95.1	167.3	5
$C_{10}H_{22}$	4-Methylnonane	-36 e	-16 e	10 e	45 e	93.1	165.2	5
$C_{10}H_{22}$	5-Methylnonane	-36 e	-16 e	10 e	45 e	92.6	164.6	5
$C_{10}H_{22}$	2,4-Dimethyloctane				38 e	84.9	155.4	5
$C_{10}H_{22}$	2,7-Dimethyloctane	-39 e	-19 e	7 e	41 e	88.4	159.4	5
$C_{10}H_{22}$	2,2,6-Trimethylheptane	-46 e	-27 e	-2 e	32 e	78.5	148.4	5
$C_{10}H_{22}$	3,3,5-Trimethylheptane			0 e	35 e	82.7	155.2	5
$C_{10}H_{22}$	2,2,3,3-Tetramethylhexane	-46 e	-25 e	1 e	36 e	85.6	159.8	5

		Temperature in °C for the indicated pressure						
Mol. Form.	Name	1 Pa	10 Pa	100 Pa	1 kPa	10 kPa	100 kPa	Ref.
$C_{10}H_{22}$	2,2,5,5-Tetramethylhexane			-10 e	22 e	68.3	137.0	5
$C_{10}H_{22}$	2,4-Dimethyl-3-isopropylpentane	-46 e	-26 e	0 e	35 e	83.2	156.5	5
$C_{10}H_{22}$	2,2,3,3,4-Pentamethylpentane		-24 e	3 e	39 e	89.1	165.5	5
$C_{10}H_{22}$	2,2,3,4,4-Pentamethylpentane		-29 e	-3 e	33 e	82.8	158.7	5
$C_{10}H_{22}O$	1-Decanol	30 e	50 e	75 e	109 e	157.3	230.6	1,39
$C_{10}H_{22}O$	4-Decanol	18 e	37 e	61 e	93 e	139 e	210 e	5
$C_{10}H_{22}O$	Dipentyl ether	-31 e	-8 e	22 e	60 e	111.6	186.2	5
$C_{10}H_{22}O$	Diisopentyl ether			14.0	51.5	101.8	172.8	5
$C_{10}H_{22}O_2$	Ethylene glycol dibutyl ether	0 e	20 e	44 e	78.4	127.1	202.9	5
$C_{10}H_{22}O_5$	Tetraethylene glycol dimethyl ether				138 e	200.9	275.3	5
$C_{10}H_{22}S$	1-Decanethiol	11 e	34 e	64 e	103 e	157.5	238.6	5
$C_{10}H_{22}S$	Diisopentylsulfide			7 e	82 e	118 e	139 e	5
$C_{10}H_{23}N$	Dipentylamine				77 e	127.7	202.0	5
$C_{10}H_{30}O_3Si_4$	Decamethyltetrasiloxane	-31 e	-6 e	26 e	66.8	118.8	193.9	5
$C_{10}H_{30}O_5Si_5$	Decamethylcyclopentasiloxane	-2 e	19 e	46 e	82 e	132.9	210.4	5
$C_{11}H_8O_2$	1-Naphthalenecarboxylic acid				191.9	239.3	299.6	5
$C_{11}H_8O_2$	2-Naphthalenecarboxylic acid				197.9	246.0	308.1	5
$C_{11}H_{10}$	1-Methylnaphthalene	5 e	29 e	60 e	102 e	159.1	244.1	1
$C_{11}H_{10}$	2-Methylnaphthalene			57 e	99 e	156.0	240.5	1
$C_{11}H_{12}O_2$	Ethyl trans-cinnamate			79	125	187	271	4
$C_{11}H_{12}O_3$	Myristicin	23 e	53 e	88.9	135.2	196.0	279.4	5
$C_{11}H_{14}$	4-Isopropylstyrene	-25 e	-1 e	30.2	70.3	124.5	202.1	5
$C_{11}H_{14}$	1,2,3,4-Tetrahydro-5-methylnaphthalene	9 e	31 e	60 e	99 e	153.1	233.8	5
$C_{11}H_{14}$	1,2,3,4-Tetrahydro-6-methylnaphthalene	17 e	36 e	62 e	97 e	147.8	228.5	5
$C_{11}H_{14}O_2$	Butyl benzoate	6 e	34 e	67.9	110.3	165 e	237 e	5
$C_{11}H_{16}$	Pentylbenzene	-14 e	8 e	37 e	74 e	126.7	204.9	5
$C_{11}H_{16}$	p-tert-Butyltoluene	-24 e	-2 e	27 e	64.1	115.5	190.8	5
$C_{11}H_{16}$	1,3-Diethyl-5-methylbenzene	-26 e	-1 e	29.5	69.5	123.5	200.2	5
$C_{11}H_{16}$	2-Ethyl-1,3,5-trimethylbenzene		6 e	36 e	75.7	129.6	207.6	5
$C_{11}H_{16}$	1-Ethyl-2,4,5-trimethylbenzene	-13 e	11 e	40 e	79.4	132.1	207.7	5
$C_{11}H_{20}$	1-Undecyne	-22 e	0 e	29 e	67 e	118.5	194.5	5
$C_{11}H_{20}$	2-Undecyne	-17 e	6 e	35 e	74 e	127.4	205.4	5
$C_{11}H_{20}O_2$	10-Undecenoic acid	35 e	67 e	105 e	150.0	205.4	274.5	5
$C_{11}H_{20}O_4$	Ethyl diethylmalonate			74 e	105 e	149.4	219 e	5
$C_{11}H_{21}N$	Undecanenitrile			78.6	120.3	177.3	259.9	5
$C_{11}H_{22}$	1-Undecene	-21.6	1.2	29.7	66.4	117.1	192.2	5
$C_{11}H_{22}$	cis-2-Undecene	-14 e	7 e	34 e	70.2	120.6	196 e	5
$C_{11}H_{22}$	trans-2-Undecene	-14 e	7 e	33 e	69.3	119.6	195 e	5
$C_{11}H_{22}$	cis-4-Undecene	-19 e	3 e	30 e	66.6	117.1	192 e	5
$C_{11}H_{22}$	trans-4-Undecene	-17 e	4 e	31 e	67.1	117.4	193 e	5
$C_{11}H_{22}$	cis-5-Undecene	-19 e	2 e	30 e	66.2	116.7	191 e	5
$C_{11}H_{22}$	trans-5-Undecene	-18 e	3 e	31 e	67.0	117.4	192 e	5
$C_{11}H_{22}$	Pentylcyclohexane	-17 e	6 e	34 e	72 e	124.2	202.7	5
$C_{11}H_{22}$	Hexylcyclopentane	-15 e	7 e	36 e	73 e	125.0	202.5	5
$C_{11}H_{22}O$	2-Undecanone	17 e	37 e	64.3	103.0	153.6	232.6	1,5
$C_{11}H_{22}O$	6-Undecanone		28 e	57 e	95 e	148.4	226.9	1
$C_{11}H_{22}O_2$	Undecanoic acid	68 e	90 e	118 e	156 e	207.2	283.6	5
$C_{11}H_{22}O_2$	Heptyl butanoate	2 e	29 e	62 e	102.6	155.1	224.7	5
$C_{11}H_{22}O_2$	Propyl octanoate	-2 e	23 e	55 e	94.0	145.2	215 e	5
$C_{11}H_{22}O_2$	Methyl decanoate	10 e	33 e	62 e	100.9	154.0	232 e	5
$C_{11}H_{24}$	Undecane	-18.4	4.3	32.6	69.5	120.2	195.4	16
$C_{11}H_{24}$	2-Methyldecane	-20 e	1 e	28 e	64 e	114.0	188.7	5
$C_{11}H_{24}$	3-Methyldecane	-35 e	-10 e	22 e	61.9	115.6	190.4	5
$C_{11}H_{24}$	4-Methyldecane	-38 e	-12 e	20 e	60.8	113.9	186.4	5
$C_{11}H_{24}$	2,4,7-Trimethyloctane				43 e	94 e	170.4	5
$C_{11}H_{24}O$	1-Undecanol	52.2	80.0	82 e	118 e	167.6	244.1	5
$C_{11}H_{24}S$	1-Undecanethiol	23 e	47 e	77 e	118 e	173.6	256.8	5
$C_{12}F_{27}N$	Trinonafluorobutylamine			3 e	29.0	63.3	109.9	176.8	5
$C_{12}H_8$	Acenaphthylene	24 s	49.8 s	80.6 s				5
$C_{12}H_9N$	Carbazole					254.7	354.0	5
$C_{12}H_{10}$	Acenaphthene				126.2	187 e	276 e	1
$C_{12}H_{10}$	Biphenyl			69.0	111.1	169.5	254.7	1
$C_{12}H_{10}N_2$	Azobenzene			98.1	144.8	206.7	292.7	4
$C_{12}H_{10}O$	Diphenyl ether		44 e	75 e	116 e	173 e	257.4	5

Mol. Form.	Name	1 Pa	10 Pa	100 Pa	1 kPa	10 kPa	100 kPa	Ref.
					Temperature in °C for the indicated pressure			
$C_{12}H_{10}O$	1-Acetonaphthone	37 e	69 e	107.0	154.6	215.2	294.9	5
$C_{12}H_{10}O$	2-Acetonaphthone	48.3 s		118.7	163.0	221.1	300.3	5
$C_{12}H_{10}S$	Diphenyl sulfide	20 e	51 e	88.7	137.5	202.2	291.8	5
$C_{12}H_{11}N$	Diphenylamine	48 s		102.8	150.5	213.7	301.4	5
$C_{12}H_{12}$	1-Ethylnaphthalene	16 e	41 e	72 e	114 e	171.8	257.7	5
$C_{12}H_{12}$	2-Ethylnaphthalene	14 e	39 e	71 e	113 e	171.2	257.3	5
$C_{12}H_{12}$	1,2-Dimethylnaphthalene	26 e	51 e	82 e	123 e	180.5	265.7	5
$C_{12}H_{12}$	2,7-Dimethylnaphthalene	31.5 s	53.1 s	78.8 s	115.9	175 e	260 e	5
$C_{12}H_{14}O_4$	Diethyl phthalate	12 e	51 e	96 e	150.5	215.9	296.2	5
$C_{12}H_{16}$	p-Isopropenylisopropylbenzene	-11 e	15 e	46 e	87 e	142.4	221 e	5
$C_{12}H_{16}$	Cyclohexylbenzene		28 e	58 e	98 e	154.7	239.5	5
$C_{12}H_{16}O_2$	3-Methylbutyl benzoate			66 e	115.0	177.7	261.4	5
$C_{12}H_{18}$	Hexylbenzene	-2 e	22 e	51 e	90 e	144.5	225.5	5
$C_{12}H_{18}$	1,2-Diisopropylbenzene	-14 e	9 e	37 e	74 e	125.9	203.2	5
$C_{12}H_{18}$	1,3-Diisopropylbenzene	-14 e	8 e	36 e	74 e	125.5	202.6	5
$C_{12}H_{18}$	1,4-Diisopropylbenzene	-6 e	18 e	49 e	90 e	148.8	238 e	5
$C_{12}H_{18}$	Hexamethylbenzene	46.3 s	72.5 s	81.7 s	121.8 s	178.3	263.7	5
$C_{12}H_{18}$	1,5,9-Cyclododecatriene	-14 e	11 e	44 e	87 e	145.0	229.8	5
$C_{12}H_{20}O_2$	Geranyl acetate			67.7	110.8	166.9	242.9	5
$C_{12}H_{20}O_4$	Dibutyl maleate	12.3	50.4	94.0	144.2	203 e	272 e	5
$C_{12}H_{22}$	1-Dodecyne	-11 e	13 e	43 e	82 e	135.8	214.4	5
$C_{12}H_{22}$	Cyclohexylcyclohexane		20 e	53.1	96.0	154.1	237.2	5
$C_{12}H_{22}O_2$	Methyl 10-undecenoate	10 e	38 e	73 e	116 e	172.2	247.1	5
$C_{12}H_{22}O_4$	Dimethyl sebacate		53 e	97	150	214	293	4
$C_{12}H_{23}N$	Dodecanenitrile	36 e	60 e	92 e	133 e	190.5	275.5	5
$C_{12}H_{24}$	1-Dodecene	-8.3	15.2	44.8	82.9	135.4	212.8	5
$C_{12}H_{24}$	Hexylcyclohexane	-3 e	20 e	50 e	89 e	143.1	224.2	5
$C_{12}H_{24}$	Heptylcyclopentane	-1 e	22 e	51 e	90 e	143.5	223.5	5
$C_{12}H_{24}O$	Dodecanal			70 e	116.2	175.9	256.6	5
$C_{12}H_{24}O_2$	Dodecanoic acid	78 e	100 e	128 e	166 e	219.1	298.1	5
$C_{12}H_{24}O_2$	Decyl acetate	12 e	40 e	74 e	115.1	168.1	238 e	5
$C_{12}H_{24}O_2$	Ethyl decanoate	8 e	35 e	69 e	111.8	166.1	238 e	5
$C_{12}H_{25}Br$	1-Bromododecane	31 e	57 e	90 e	132 e	190.8	275.3	5
$C_{12}H_{25}Cl$	1-Chlorododecane	27 e	51 e	81 e	122 e	178.7	262.6	5
$C_{12}H_{26}$	Dodecane	-5.4	18.2	47.6	85.8	138.2	215.8	16
$C_{12}H_{26}O$	1-Dodecanol				133 e	185.0	264.1	1
$C_{12}H_{26}O_3$	Diethylene glycol dibutyl ether	5 e	34.4	70.2	115.3	174.1	253.8	5
$C_{12}H_{27}N$	Tributylamine	-26 e	1 e	35 e	77.7	134.5	213.4	5
$C_{12}H_{27}N$	Triisobutylamine		1 e	28.9	64.9	112.5	178.5	5
$C_{12}H_{27}O_4P$	Tributyl phosphate					205 e	288.3	5
$C_{12}H_{36}O_6Si_6$	Dodecamethylcyclohexasiloxane	18 e	41 e	69 e	108 e	162.2	244.7	5
$C_{13}H_9N$	Acridine			124.4	176.2	246.0	345.4	5
$C_{13}H_9N$	Phenanthridine	79 s						5
$C_{13}H_{10}$	Fluorene	48.4 s			137.4	205.4	295 e	5
$C_{13}H_{10}O_2$	Phenyl benzoate			102.3	151.4	217.9	313.3	5
$C_{13}H_{10}O_3$	Phenyl salicylate				166.0	224.8	312.4	5
$C_{13}H_{12}$	Diphenylmethane		45 e	77 e	119.3	177.7	263.6	1,5
$C_{13}H_{13}N$	Methyldiphenylamine	35 e	63 e	98.4	143.1	201.6	281.6	5
$C_{13}H_{14}$	1-Isopropylnaphthalene	27 e	51 e	82 e	123.2	180.8	267.3	5
$C_{13}H_{20}$	Heptylbenzene	12 e	36 e	66 e	107 e	162.7	246.2	5
$C_{13}H_{24}O_2$	Ethyl 10-undecenoate	32 e	55 e	86 e	125.2	179.5	258.4	5
$C_{13}H_{26}$	1-Tridecene	4.1	28.5	59.0	98.3	152.5	232.3	5
$C_{13}H_{26}$	Heptylcyclohexane	11 e	34 e	65 e	105 e	160.9	244.3	5
$C_{13}H_{26}$	Octylcyclopentane	13 e	36 e	66 e	106 e	160.9	243.1	5
$C_{13}H_{26}O_2$	Tridecanoic acid	87 e	109 e	138 e	176 e	230.3	311.5	5
$C_{13}H_{26}O_2$	Methyl dodecanoate	38 e	61 e	90 e	130 e	184.9	269 e	5
$C_{13}H_{28}$	Tridecane	7.2	31.5	61.8	101.1	155.1	234.9	16
$C_{13}H_{28}O$	1-Tridecanol	71.6	101.0	103 e	140 e	192.3	273.1	5
$C_{14}H_{10}$	Anthracene	89.2 s	125.9 s	151.5 s	165 s	238.8	340.2	1,5
$C_{14}H_{10}$	Phenanthrene	53 s	83 s	120.8	170.4	238.4	337.7	5
$C_{14}H_{10}O_2$	Benzil			123	175	246	346	4
$C_{14}H_{12}$	cis-Stilbene	26 e	54 e	88 e	130.4	183 e	253 e	5
$C_{14}H_{12}$	trans-Stilbene				155.6	218.1	305.8	5
$C_{14}H_{12}O_2$	Benzoin				181	248	342	4
$C_{14}H_{14}$	1,1-Diphenylethane	19 e	47 e	82.0	125.3	181 e	254 e	5
$C_{14}H_{15}N$	Dibenzylamine	48 e	77 e	113.1	158.9	218.5	299.4	5
$C_{14}H_{16}$	1-Butylnaphthalene	67 e	82 e	103 e	135 e	186.7	288.6	5

Mol. Form.	Name	Temperature in °C for the indicated pressure						Ref.
		1 Pa	10 Pa	100 Pa	1 kPa	10 kPa	100 kPa	
$C_{14}H_{16}$	2-Butylnaphthalene	44 e	67 e	98 e	139 e	197.5	287.4	5
$C_{14}H_{22}$	Octylbenzene	20.1	46.2	79.1	121.9	178.1	263.8	5
$C_{14}H_{26}O_4$	Diethyl sebacate		83 e	120	166	225	305	4
$C_{14}H_{27}N$	Tetradecanenitrile	52 e	79 e	114.0	159.0	219.7	306.3	5
$C_{14}H_{28}$	1-Tetradecene	16.1	41.3	72.7	113.2	168.7	250.6	5
$C_{14}H_{28}$	Octylcyclohexane	16.9	44.3	77.8	120.0	177.6	263.2	5
$C_{14}H_{28}$	Nonylcyclopentane	25 e	49 e	80 e	120 e	177.2	261.5	5
$C_{14}H_{28}O_2$	Tetradecanoic acid	96 e	118 e	147 e	186 e	241.3	325.6	5
$C_{14}H_{30}$	Tetradecane	19.1	44.1	75.3	115.7	171.1	253.0	16
$C_{14}H_{30}O$	1-Tetradecanol	80.0	110.5	149.6	152	205.3	286.7	5
$C_{14}H_{31}N$	Tetradecylamine			104 e	147 e	206.1	290.9	5
$C_{14}H_{42}O_5Si_6$	Tetradecamethylhexasiloxane	6 e	36 e	72 e	117 e	176.0	259.1	5
$C_{15}H_{18}$	1-Pentylnaphthalene	34 e	62 e	96 e	141.3	202.2	289 e	5
$C_{15}H_{24}$	Nonylbenzene	33.0	58.9	92.0	135.4	193.7	281.4	5
$C_{15}H_{30}$	Nonylcyclohexane	35 e	60 e	92 e	134 e	193.4	280.9	5
$C_{15}H_{30}$	Decylcyclopentane	37 e	61 e	93 e	134 e	192.5	278.8	5
$C_{15}H_{30}O_2$	Methyl tetradecanoate		75 e	110	155	214	295	4
$C_{15}H_{32}$	Pentadecane	30.5	56.1	88.1	129.6	186.3	270.1	16
$C_{16}H_{22}O_4$	Dibutyl phthalate		104.0	142.7	191.5	254.5	339.4	4
$C_{16}H_{32}$	1-Hexadecene	38.4	65.0	98.1	140.5	198.8	284.3	5
$C_{16}H_{32}O_2$	Hexadecanoic acid		136 e	165 e	205 e	261.9	350.2	5
$C_{16}H_{34}$	Hexadecane	41.1	67.4	100.3	142.7	200.7	286.3	16
$C_{16}H_{34}O$	1-Hexadecanol	99.5	130.6	171.9	175 e	229.0	311.7	5
$C_{16}H_{35}N$	Hexadecylamine	63 e	91 e	126 e	171 e	232.6	320.5	5
$C_{17}H_{10}O$	Benzanthrone		184 e	229.3	290.3	377.2	511 e	5
$C_{17}H_{34}O_2$	Methyl hexadecanoate	65 e	93	129	177			4
$C_{17}H_{36}$	Heptadecane	51.5	78.5	112.0	155.3	214.5	302 e	16
$C_{17}H_{36}O$	1-Heptadecanol	94 e	117 e	146 e	185 e	240.1	323.3	5
$C_{18}H_{14}$	o-Terphenyl	66 e	94 e	129 e	176 e	241.3	336.3	5
$C_{18}H_{14}$	m-Terphenyl	87 e	118 e	156 e	206.6	275.3	374.6	5
$C_{18}H_{14}$	p-Terphenyl	127.1 s	154.7 s		217.2	284.0	383.0	5
$C_{18}H_{30}$	Hexaethylbenzene				144.1	206.8	297.5	5
$C_{18}H_{34}O_2$	Oleic acid	94 e	126 e	165.5	214.5	277.0	359.7	5
$C_{18}H_{34}O_2$	Elaidic acid		124 e	166	216	280	361	4
$C_{18}H_{36}O$	Stearaldehyde			142 e	186 e	246.9	336.7	5
$C_{18}H_{36}O_2$	Stearic acid		153 e	183 e	223 e	281.6	374.5	5
$C_{18}H_{38}$	Octadecane	61.5	89.0	123.1	167.3	227.6	316 e	16
$C_{18}H_{38}O$	1-Octadecanol	106 e	130 e	160 e	200.5	257.3	343.0	5
$C_{19}H_{16}$	Triphenylmethane	81 s		112 e	175 e	254.6	360.0	5
$C_{19}H_{36}O_2$	Methyl oleate	85 e	114 e	149.7	195.6	256 e	340 e	5
$C_{19}H_{40}$	Nonadecane	71.1	99.1	133.8	178.8	240.1	330 e	16
$C_{20}H_{42}$	Eicosane	80.4	108.9	144.2	189.8	252.1	344 e	16
$C_{20}H_{42}O$	1-Eicosanol	119 e	143 e	173 e	213 e	270.0	355.1	5
$C_{20}H_{60}O_8Si_9$	Eicosamethylnonasiloxane			141 e	183.1	236.7	307.1	5
$C_{21}H_{21}O_4P$	Tri-o-cresyl phosphate	119.0	156.1	201.0	256.3	326.3	418 e	5
$C_{21}H_{21}O_4P$	Tri-m-cresyl phosphate	147.8	177.3	211.4	251.3	298 e	355 e	5
$C_{21}H_{21}O_4P$	Tri-p-cresyl phosphate	140.6	174 e	214 e	262 e	320 e	392 e	5
$C_{21}H_{44}$	Heneicosane	82.3	113.5	152.2	201.6	263.8	355.9	5
$C_{22}H_{42}O_2$	Brassidic acid	134 e	166 e	203.6	249.8	307.6	382.0	5
$C_{22}H_{42}O_2$	Erucic acid	126 e	160 e	199.4	247.4	306.5	381.1	5
$C_{22}H_{42}O_2$	Butyl oleate	95.5	124.2	158 e	198 e	245 e	304 e	5
$C_{22}H_{44}O_2$	Behenic acid	145.4	176.5	213.7	259.3	316.2	390 e	5
$C_{22}H_{44}O_2$	Butyl stearate	99.6	128 e	162 e	201 e	249 e	307 e	5
$C_{22}H_{46}$	Docosane	83.5	115.0	154.0	203.6	274.8	368.0	5
$C_{23}H_{48}$	Tricosane	102.9	135.1	174.8	221 e	285.3	379.5	5
$C_{24}H_{38}O_4$	Dioctyl phthalate	130 e	163.7	203.8	252 e	311 e	385 e	5
$C_{24}H_{38}O_4$	Bis(2-ethylhexyl) phthalate	122.0	153.2	189.2	231.3	281.1	341.1	5
$C_{24}H_{50}$	Tetracosane	115.0	148.1	188.5	239.1	295.4	390.6	5
$C_{25}H_{52}$	Pentacosane	119.7	152.7	193.2	244.4	305.0	401.1	5
$C_{26}H_{54}$	Hexacosane	125.1	158.8	200.1	252.1	314.3	411.3	5
$C_{27}H_{56}$	Heptacosane	136.7	168.8	206.5	255.8	323.3	421.2	5
$C_{28}H_{58}$	Octacosane	136.5	169.8	210.9	263.1	332.0	430.6	5
$C_{29}H_{60}$	Nonacosane	148.2	182.8	221.2	271.5	340.2	439.7	5
$C_{30}H_{62}$	Squalane	66 e	84 e	105.8	131.9	163.7	203.2	5
C_{70}	Carbon (fullerene-C_{70})	598 s	662 s					22

VAPOR PRESSURE OF FLUIDS AT TEMPERATURES BELOW 300 K

This table gives vapor pressures of 67 important fluids in the temperature range 2 to 300 K. Helium (^{4}He), hydrogen (H_2), and neon (Ne) are covered on this page. The remaining fluids are listed on subsequent pages by molecular formula in the Hill order (see Introduction). The data have been taken from evaluated sources; references are listed at the end of the table.

Pressures are given in kilopascals (kPa). Note that:

1 kPa = 7.50062 Torr
100 kPa = 1 bar
101.325 kPa = 1 atmos

s following an entry indicates that the compound is solid at that temperature.

Helium		Hydrogen		Neon	
T/K	*P*/kPa	*T*/K	*P*/kPa	*T*/K	*P*/kPa
2.2	5.3	14.0	7.90	25.0	51.3
2.3	6.7	14.5	10.38	26.0	71.8
2.4	8.3	15.0	13.43	27.0	98.5
2.5	10.2	15.5	17.12	28.0	132.1
2.6	12.4	16.0	21.53	29.0	173.5
2.7	14.8	16.5	26.74	30.0	223.8
2.8	17.5	17.0	32.84	31.0	284.0
2.9	20.6	17.5	39.92	32.0	355.2
3.0	24.0	18.0	48.08	33.0	438.6
3.1	27.8	18.5	57.39	34.0	535.2
3.2	32.0	19.0	67.96	35.0	646.2
3.3	36.5	19.5	79.89	36.0	772.8
3.4	41.5	20.0	93.26	37.0	916.4
3.5	47.0	20.5	108.2	38.0	1078
3.6	52.9	21.0	124.7	39.0	1260
3.7	59.3	21.5	143.1	40.0	1462
3.8	66.1	22.0	163.2	41.0	1688
3.9	73.5	22.5	185.3	42.0	1939
4.0	81.5	23.0	209.4	43.0	2216
4.1	90.0	23.5	235.7	44.0	2522
4.2	99.0	24.0	264.2		
4.3	108.7	24.5	295.1		
4.4	119.0	25.0	328.5		
4.5	129.9	25.5	364.3		
4.6	141.6	26.0	402.9		
4.7	153.9	26.5	444.3		
4.8	167.0	27.0	488.5		
4.9	180.8	27.5	535.7		
5.0	195.4	28.0	586.1		
5.1	210.9	28.5	639.7		
		29.0	696.7		
		29.5	757.3		
		30.0	821.4		
		30.5	889.5		
		31.0	961.5		
		31.5	1038.0		
		32.0	1119.0		
		32.5	1204.0		
Ref.	17,18		1		13

T/K	Ar Argon		BCl_3 Boron trichloride	BF_3 Boron trifluoride	BrH Hydrogen bromide		Br_2 Bromine		ClF Chlorine fluoride	ClH Hydrogen chloride	
50	0.1	s									
55	0.2	s									
60	0.8	s									
65	2.8	s									
70	7.7	s									
75	18.7	s									
80	40.7	s									
85	79.0										
90	134										
95	213										
100	324										
105	473										
110	666										
115	910								0.1		
120	1214								0.3	0.1	s
125	1584								0.6	0.3	s
130	2027								1.2	0.5	s
135	2553				0.1	s			2.1	1.0	s
140	3170				0.3	s			3.6	1.9	s
145	3892			7.7	0.6	s			6.0	3.4	s
150	4736			13.4	1.1	s			9.5	5.8	s
155				22.3	1.9	s			14.6	9.5	s
160				35.2	3.3	s			21.8	14.7	
165				53.7	5.4	s			31.7	22.0	
170				79.1	8.7	s			44.8	31.9	
175				113	13.4	s			62.0	45.1	
180			0.1	157	20.1	s			84.2	62.5	
185			0.2	214	29.5	s			112	84.7	
190			0.3	285	37.9				147	113	
195			0.5	372	51.8				190	148	
200			0.8	479	69.5				242	190	
205			1.2	608	91.8				304	242	
210			1.8	762	119				378	304	
215			2.6	944	153				464	377	
220			3.8	1160	194		0.1	s	564	463	
225			5.2	1413	242		0.2	s	680	563	
230			7.2	1709	299		0.3	s	812	678	
235			9.7	2056	366		0.4	s	961	811	
240			12.9	2460	443		0.7	s	1130	961	
245			17.0	2913	532		1.1	s	1319	1132	
250			22.0	3481	633		1.7	s	1529	1325	
255			28.1	4123	748		2.6	s	1762	1542	
260			35.6	4874	878		3.8	s	2019	1784	
265			44.5		1023		5.5	s	2301	2054	
270			55.1		1185		7.3		2608	2354	
275			67.6		1364		9.5		2941	2686	
280			82.2		1562		12.3		3303	3053	
285			99.1		1780		15.6		3693	3457	
290			119		2018		19.7		4111	3901	
295			141		2278		24.6		4560	4388	
300			166		2561		30.5		5039	4921	
Ref.	8,15		12	12	12		12		12	12	

T/K	ClO_2 Chlorine dioxide	Cl_2 Chlorine	Cl_4Si Silicon tetrachloride	FH Hydrogen fluoride	F_2 Fluorine	F_2O Difluorine oxide	F_3N Nitrogen trifluoride
50							
55					0.4		
60					1.5		
65					4.8		
70					12.3		
75					27.6	0.1	
80					55.3	0.2	
85					101	0.5	0.1
90					172	1.2	0.2
95					276	2.6	0.4
100					420	5.3	0.9
105					615	10.1	2.0
110					870	18.0	4.0
115					1196	30.5	7.3
120					1605	49.3	12.8
125					2108	76.7	21.1
130					2721	115	33.5
135					3458	168	51.1
140					4339	237	75.4
145						328	108
150						444	150
155						588	205
160						766	273
165						981	357
170						1238	459
175		1.8				1541	581
180		2.8				1895	726
185		4.2				2303	896
190		6.1		0.3		2771	1092
195	0.1	8.7		0.5		3302	1319
200	0.3	12.3		0.8		3899	1578
205	0.5	16.9		1.2		4567	1871
210	0.9	22.9	0.1	1.7		5308	2203
215	1.4	30.5	0.2	2.3			2577
220	2.3	40.1	0.3	3.2			2995
225	3.5	51.9	0.5	4.4			3464
230	5.3	66.4	0.7	5.9			3991
235	7.6	84.0	1.0	7.9			
240	10.8	105	1.5	10.3			
245	14.9	130	2.0	13.4			
250	20.1	160	2.8	17.2			
255	26.6	194	3.8	21.8			
260	34.6	234	5.0	27.4			
265	44.4	280	6.6	34.2			
270	56.1	332	8.6	42.2			
275	69.9	392	11.1	51.8			
280	86.2	459	14.2	63.1			
285	105	535	17.9	76.3			
290	127	619	22.3	91.7			
295	151	714	27.7	110			
300	179	818	34.0	130			
Ref.	12	5	12	12	12	12	1

T/K	F_3P Phosphorous trifluoride	F_4Si Silicon tetrafluoride		F_6S Sulfur hexafluoride		HI Hydrogen iodide		H_2S Hydrogen sulfide		H_3N Ammonia		H_3P Phosphine
50												
55												
60												
65												
70												
75												
80												
85												
90												
95												
100												
105	0.1											
110	0.2											0.1
115	0.5											0.2
120	1.0											0.4
125	1.9	0.1	s									0.7
130	3.5	0.2	s									1.3
135	5.9	0.4	s					0.1	s			2.3
140	9.5	0.9	s	0.1	s			0.2	s			3.9
145	14.9	1.9	s	0.2	s			0.3	s			6.2
150	22.5	3.8	s	0.4	s			0.6	s			9.6
155	33.1	7.5	s	0.8	s	0.1	s	1.1	s			14.5
160	47.3	14.0	s	1.5	s	0.2	s	1.9	s	0.1	s	21.1
165	66.0	25.2	s	2.6	s	0.4	s	3.2	s	0.2	s	30.0
170	90.1	43.8	s	4.4	s	0.8	s	5.2	s	0.3	s	41.6
175	121	74.2	s	7.1	s	1.3	s	8.3	s	0.6	s	56.6
180	159	122	s	11.3	s	2.2	s	12.7	s	1.2	s	75.6
185	206	197	s	17.3	s	3.4	s	18.9	s	2.1	s	99.2
190	262	280		25.9	s	5.3	s	26.6		3.5	s	128
195	330	376		38.0	s	8.0	s	36.7		5.8	s	163
200	410	488		54.4	s	11.7	s	49.8		8.7		205
205	503	618		76.6	s	16.8	s	66.4		12.6		254
210	611	766		106	s	23.6	s	87.1		17.9		312
215	736	932		145	s	32.5	s	113		24.9		379
220	877	1117		195	s	44.0	s	144		34.1		456
225	1037	1324		249		56.2		182		45.9		544
230	1217	1555		305		71.4		227		60.8		644
235	1418	1816		371		89.7		281		79.6		756
240	1640	2111		448		112		344		103		881
245	1885	2449		536		137		416		131		1019
250	2154	2841		636		168		500		165		1172
255	2448	3301		750		203		597		207		1341
260	2767			878		244		706		256		1525
265	3112			1021		290		830		313		1725
270				1181		343		969		381		1942
275				1358		404		1124		460		2176
280				1554		472		1297		552		2428
285				1768		548		1488		655		2699
290				2003		633		1698		774		2987
295				2258		727		1929		909		3295
300				2534		831		2181		1062		3621
Ref.	12	12		12,15		12		12,15		11		12

T/K	H₄Si Silane	Kr Krypton		NO Nitric oxide		N₂ Nitrogen		N₂O Nitrous oxide	O₂ Oxygen	O₂S Sulfur dioxide
50						0.4	s			
55						1.8	s		0.2	
60						6.3	s		0.7	
65						17.4			2.3	
70						38.6			6.3	
75		0.1	s			76.1			14.5	
80		0.4	s			137			30.1	
85		1.1	s	0.1	s	229			56.8	
90		2.7	s	0.4	s	361			99.3	
95	0.1	6.0	s	1.3	s	541			163	
100	0.2	12.1	s	3.8	s	779			254	
105	0.4	22.8	s	10.0	s	1084			379	
110	1.0	40.4	s	23.5		1467			543	
115	1.9	68.0	s	46.8		1939		0.1	756	
120	3.5	103		86.5		2513		0.1	1022	
125	6.1	150		151		3209		0.3	1351	
130	10.0	211		248				0.7	1749	
135	15.8	290		391				1.3	2225	
140	24.1	388		592				2.5	2788	
145	35.3	509		867				4.3	3448	
150	50.3	655		1231				7.1	4219	
155	69.8	830		1703				11.4		
160	94.6	1037		2302				17.6		
165	126	1278		3050				26.4		
170	164	1557		3971				38.5		0.1
175	210	1877		5089				54.7		0.2
180	265	2241		6433				75.9		0.3
185	331	2655						103		0.5
190	408	3120						138		0.8
195	498	3641						181		1.3
200	602	4223						234		2.0
205	722	4870						298		3.0
210	859							374		4.4
215	1017							465		6.3
220	1196							571		9.0
225	1398							694		12.6
230	1628							835		17.3
235	1888							996		23.3
240	2180							1179		31.1
245	2509							1385		40.9
250	2880							1615		53.2
255	3296							1870		68.3
260	3763							2152		86.7
265	4288							2462		109
270								2802		136
275								3172		168
280								3573		205
285								4006		249
290								4473		300
295								4973		359
300								5508		426
Ref.	12	13, 15		12, 15		1		12	3	12

T/K	O_3 Ozone	Rn Radon	Xe Xenon		$CBrF_3$ Bromotri-fluoromethane	$CClF_3$ Chlorotri-fluoromethane	CCl_2F_2 Dichlorodi-fluoromethane	CCl_3F Trichloro-fluoromethane
50								
55								
60								
65								
70								
75								
80								
85								
90								
95								
100	0.1		0.1	s				
105	0.2		0.1	s				
110	0.4		0.3	s				
115	1.0		0.7	s		0.1		
120	2.0		1.5	s		0.2		
125	3.8		2.7	s		0.3		
130	6.8	0.1	4.9	s		0.6		
135	11.5	0.3	8.5	s	0.1	1.1		
140	18.7	0.5	14.0	s	0.3	2.0		
145	29.1	0.9	22.2	s	0.5	3.3		
150	43.7	1.5	34.2	s	0.9	5.3		
155	63.6	2.4	51.1	s	1.5	8.3	0.1	
160	89.9	3.8	74.2	s	2.5	12.6	0.3	
165	124	5.8	101		3.9	18.6	0.5	
170	168	8.6	134		5.9	26.8	0.8	
175	222	12.5	173		8.8	37.6	1.3	
180	289	17.7	222		12.8	51.7	2.1	
185	367	24.5	280		18.1	69.7	3.2	
190	468	33.2	348		25.1	92.3	4.8	0.2
195	584	44.4	428		34.1	120	6.9	0.3
200	721	58.2	521		45.6	155	9.9	0.4
205	881	75.3	628		60.0	196	13.7	0.6
210	1068	96	750		77.8	246	18.8	1.0
215	1285	121	889		99.5	304	25.2	1.4
220	1536	151	1045		126	372	33.3	2.0
225	1824	185	1220		157	451	43.3	2.9
230	2155		1416		194	542	55.5	4.1
235	2534		1633		237	646	70.4	5.6
240	2968		1872		287	763	88.1	7.6
245	3464		2136		344	896	109	10.1
250	4031		2425		410	1044	134	13.3
255	4678		2742		485	1210	163	17.2
260	5417		3087		570	1394	196	22.1
265			3462		665	1598	234	28.0
270			3869		771	1823	278	35.1
275			4310		889	2071	327	43.7
280			4786		1021	2343	383	53.8
285			5299		1166	2641	445	65.7
290					1325	2968	515	79.6
295					1501	3325	593	95.6
300					1692	3716	679	114.1
Ref.	12	15	12,13		12	12	12	12

VAPOR PRESSURE OF FLUIDS AT TEMPERATURES BELOW 300 K (continued)

T/K	CCl₄ Tetrachloro-methane	CF₄ Tetrafluoro-methane	CO Carbon monoxide	COS Carbon oxysulfide	CO₂ Carbon dioxide	CHClF₂ Chlorodifluo-methane	CHCl₃ Trichloro-methane
50			0.1 s				
55			0.6 s				
60			2.6 s				
65			8.2 s				
70			21.0				
75			44.4				
80			83.7				
85			147				
90		0.1	239				
95		0.3	371				
100		0.8	545				
105		1.7	771				
110		3.4	1067				
115		6.5	1428				
120		11.5	1877				
125		19.3	2400				
130		30.8	3064				
135		47.4			0.1		
140		70.2		0.1	0.2		
145		101		0.2	0.4		
150		141		0.4	0.8	0.1	
155		191		0.8	1.7	0.3	
160		254		1.3	3.1	0.5	
165		332		2.2	5.7	0.8	
170		425		3.4	9.9	1.4	
175		537		5.2	16.8	2.3	
180		669		7.8	27.6	3.6	
185		824		11.3	44.0	5.5	
190		1005		15.9	68.4	8.1	
195		1216		22.1	104	11.8	
200		1460		30.0	155	16.7	
205		1743		40.1	227	23.1	
210		2073		52.7	327	31.5	
215		2457		68.2	465	42.1	0.1
220		2907		87.2	600	55.3	0.2
225		3438		110	735	71.7	0.3
230				137	894	91.6	0.4
235				169	1075	116	0.7
240				207	1283	144	1.0
245				250	1519	178	1.4
250				301	1786	218	2.0
255	1.5			358	2085	264	2.7
260	2.1			423	2419	317	3.7
265	2.8			497	2790	377	5.0
270	3.7			580	3203	446	6.6
275	4.9			673	3658	525	8.7
280	6.4			777	4160	613	11.3
285	8.2			892	4712	711	14.4
290	10.5			1019	5315	821	18.3
295	13.2			1159	5984	944	22.9
300	16.5			1313	6710	1080	28.5
Ref.	12	12	9	12	6	12	12

T/K	CHF_3 Trifluoro-methane	CHN Hydrogen cyanide		CH_2Cl_2 Dichloro-methane	CH_2F_2 Difluoro methane	CH_2O Formaldehyde	CH_3Cl Chloromethane	CH_3F Fluoromethane
50								
55								
60								
65								
70								
75								
80								
85								
90								
95								
100								
105								
110								
115								
120	0.1							
125	0.2							
130	0.4							
135	0.7							0.6
140	1.4				0.1			1.2
145	2.5				0.2			2.1
150	4.3				0.3			3.6
155	7.1				0.6			5.9
160	11.1				1.0			9.3
165	17.0				1.7			14.1
170	25.3				2.8			20.9
175	36.5				4.4			29.9
180	51.4				6.8			42.0
185	70.9				10.2	1.3	2.1	57.6
190	95.8				14.8	2.0	3.1	77.4
195	127				21.2	3.0	4.6	102
200	166	0.1	s	0.1	29.5	4.4	6.7	133
205	214	0.2	s	0.2	40.5	6.4	9.5	171
210	271	0.4	s	0.3	54.5	9.1	13.1	216
215	340	0.6	s	0.4	72.1	12.7	17.9	270
220	421	1	s	0.6	94.1	17.4	24.0	333
225	516	1.5	s	0.9	121	23.4	31.8	408
230	626	2.2	s	1.4	154	31.0	41.4	495
235	754	3.3	s	2.0	193	40.6	53.3	595
240	900	4.7	s	2.8	240	52.5	67.7	711
245	1067	6.8	s	3.8	295	67.0	85.1	843
250	1257	9.7	s	5.3	360	84.6	106	993
255	1472	13.6	s	7.1	434	106	131	1163
260	1713	18.8		9.5	521	131	159	1355
265	1984	24.1		12.4	620	161	193	1571
270	2287	30.5		16.1	732	196	232	1813
275	2624	38.3		20.7	860	236	277	2084
280	3000	47.7		26.3	1004	283	327	2387
285	3418	58.8		33.0	1165	337	385	2724
290	3881	72.1		41.1	1346	399	450	3099
295	4393	87.6		50.8	1547	470	524	3516
300		105.9		62.1	1770	549	606	3978
Ref.	12	12,16		12	12	12	12	12

T/K	CH$_4$ Methane	CH$_4$O Methanol	C$_2$H$_2$ Acetylene	C$_2$H$_4$ Ethylene	C$_2$H$_6$ Ethane	C$_2$H$_6$O Dimethyl ether	C$_3$H$_4$ Propadiene
50							
55							
60							
65	0.1						
70	0.3						
75	0.8						
80	2.1						
85	4.9						
90	10.6						
95	20.0						
100	34.5						
105	57.0						
110	88.4			0.3			
115	133			0.8	0.1		
120	192			1.4	0.4		
125	269			2.7	0.7		
130	368		0.1 s	4.5	1.3		
135	491		0.3 s	7.7	2.2		
140	642		0.7 s	11.9	3.8		
145	824		1.3 s	18.3	6.0		
150	1041		2.6 s	27.5	9.7		0.1
155	1297		4.6 s	39.9	15.0	0.1	0.2
160	1594		7.8 s	56.4	21.5	0.2	0.3
165	1937		12.8 s	77.9	31.0	0.3	0.6
170	2331		20.6 s	105	42.9	0.5	1.0
175	2779		32.2 s	140	59.0	0.9	1.7
180	3288		49.0 s	182	78.7	1.4	2.7
185	3865		72.9 s	234	104	2.1	4.1
190	4520		106 s	296	135	3.2	6.1
195			146	369	172	4.7	8.9
200			190	456	217	6.8	12.5
205			244	557	271	9.6	17.4
210			309	673	334	13.3	23.7
215			385	806	407	18.1	31.6
220			475	958	492	24.3	41.4
225			579	1128	590	32.1	53.5
230		0.1	699	1321	700	41.9	68.2
235		0.2	837	1535	826	53.9	85.8
240		0.4	993	1774	967	68.6	107
245		0.5	1170	2039	1125	86.3	131
250		0.8	1370	2331	1301	108	160
255		1.2	1593	2652	1496	133	193
260		1.7	1843	3005	1712	162	230
265		2.4	2121	3391	1949	197	273
270		3.3	2429	3813	2210	237	322
275		4.5	2771	4275	2495	283	376
280		6.2	3150		2806	335	438
285		8.3	3567		3146	395	506
290		11	4028		3515	463	582
295		14.4	4535		3917	538	666
300		18.7	5093		4355	623	759
Ref.	2,16	12	12,16	4	2	12	12

T/K	C_3H_6 Propylene	C_3H_8 Propane	C_4H_6 Buta-1,3-diene	C_4H_{10} n-Butane	C_4H_{10} Isobutane	C_5H_{12} n-Pentane	C_5H_{12} Neopentane	
50								
55								
60								
65								
70								
75								
80								
85								
90								
95								
100								
105								
110								
115								
120								
125								
130								
135								
140	0.1							
145	0.2							
150	0.4							
155	0.7							
160	1.2	0.8			0.1			
165	2.0	1.4			0.1			
170	3.1	2.2	0.1	0.1	0.3			
175	4.7	3.3	0.2	0.2	0.4			
180	7.0	5.0	0.4	0.3	0.7			
185	10.1	7.3	0.6	0.5	1.1		0.1	s
190	14.2	10.5	1.0	0.8	1.7		0.2	s
195	19.7	15.0	1.5	1.3	2.5		0.4	s
200	26.9	20.1	2.3	1.9	3.7		0.7	s
205	35.9	27.0	3.4	2.8	5.3		1.1	s
210	47.3	36.0	4.8	4.0	7.4		1.6	s
215	61.3	47.0	6.7	5.7	10.2		2.4	s
220	78.5	60.0	9.2	7.8	13.8	1.0	3.6	s
225	99.2	77.0	12.5	10.6	18.3	1.5	5.2	s
230	124	97.0	16.7	14.1	24.0	2.1	7.3	s
235	153	120	21.9	18.5	31.1	3.0	10.2	s
240	188	148	28.4	24.1	39.8	4.2	13.9	s
245	228	180	36.3	30.9	50.3	5.7	18.7	s
250	274	218	46.0	39.1	62.9	7.6	24.8	s
255	327	261	57.6	49.1	77.8	10.0	32.4	s
260	387	311	71.3	61.0	95.4	13.0	41.6	
265	456	367	87.6	75.0	116	16.6	51.4	
270	533	431	107	91.5	140	21.1	63.0	
275	619	502	129	111	167	26.6	76.6	
280	715	582	154	133	198	33.1	92.3	
285	822	671	184	159	234	40.8	111	
290	940	769	217	188	274	50.0	131	
295	1069	878	255	221	319	60.7	155	
300	1212	998	297	258	370	73.2	182	
Ref.	7	2	12	2	2	14	12,16	

VAPOR PRESSURE OF FLUIDS AT TEMPERATURES BELOW 300 K (continued)

REFERENCES

1. B. A. Younglove, Thermophysical properties of fluids. I. Ethylene, parahydrogen, nitrogen trifluoride, and oxygen, *J. Phys. Chem. Ref. Data*, 11, Supp. 1, 1982.
2. B. A. Younglove and J. F. Ely, Thermophysical properties of fluids. II. Methane, ethane, propane, isobutane, and normal butane, *J. Phys. Chem. Ref. Data*, 16, 577, 1987.
3. W. Wagner, et al., *International Tables for the Fluid State: Oxygen*, Blackwell Scientific Publications, Oxford, 1987.
4. R. T. Jacobsen, et al., *International Tables for the Fluid State: Ethylene*, Blackwell Scientific Publications, Oxford, 1988.
5. S. Angus, et al., *International Tables for the Fluid State: Chlorine*, Pergamon Press, Oxford, 1985.
6. S. Angus, et al., *International Tables for the Fluid State: Carbon Dioxide*, Pergamon Press, Oxford, 1976.
7. S. Angus, et al., *International Tables for the Fluid State: Propylene*, Pergamon Press, Oxford, 1980.
8. R. B. Stewart and R. T. Jacobsen, Thermophysical properties of argon, *J. Phys. Chem. Ref. Data*, 18, 639, 1989.
9. R. D. Goodwin, Carbon monoxide thermophysical properties, *J. Phys. Chem. Ref. Data*, 14, 849, 1985.
10. R. D. Goodwin, Methanol thermophysical properties, *J. Phys. Chem. Ref. Data*, 16, 799, 1987.
11. L. Haar, Thermodynamic properties of ammonia, *J. Phys. Chem. Ref. Data*, 7, 635, 1978.
12. DIPPR Data Compilation of Pure Compound Properties, Design Institute for Physical Properties Data, American Institute of Chemical Engineers, 1987.
13. V. A. Rabinovich, et al., *Thermophysical Properties of Neon, Argon, Krypton, and Xenon*, Hemisphere Publishing Corp., New York, 1987.
14. K. N. Marsh, *Recommended Reference Methods for the Realization of Physicochemical Properties*, Blackwell Scientific Publications, Oxford, 1987.
15. TRC Thermodynamic Tables: Non-Hydrocarbons, Thermodynamic Research Center, Texas A & M University, College Station, Texas, 1985.
16. R. M. Stevenson and S. Malanowski, *Handbook of the Thermodynamics of Organic Compounds*, Elsevier, New York, 1987.
17. S. Angus and K. M. de Reuck, *International Tables of the Fluid State: Helium-4*, Pergamon Press, Oxford, 1977.
18. R. D. McCarty, *J. Phys. Chem. Ref. Data*, 2, 923, 1973.

IUPAC RECOMMENDED DATA FOR VAPOR PRESSURE CALIBRATION

Vapor pressures are given in kPa (1 kPa = 0.0098692 atmos = 7.5006 Torr). Reprinted with permission if IUPAC.

REFERENCE

Marsh, K.N., Ed., *Recommended Reference Materials for the Realization of Physicochemical Properties*, Blackwell Scientific Publications, Oxford (1987).

T/K	$CO_2(s)$	$H_2O(s)$	$C_{10}H_8(s)$	n-C_5H_{12}	C_6H_6	C_6F_6	H_2O	Hg
180	27.62							
190	68.44							
200	155.11	0.0002						
210	327.17	0.0007						
220		0.0026						
230		0.0089						
240		0.0273						
250		0.0760		7.60				
260		0.1958	0.0001	12.98				
270		0.4701	0.0005	21.15			0.485	
280			0.0017	33.11	5.148	4.322	0.991	
290			0.0049	50.01	8.606	7.463	1.919	
300			0.0134	73.17	13.816	12.328	3.535	
310			0.0341	104.07	21.389	19.576	6.228	
320			0.0814	144.3	32.054	30.009	10.540	
330			0.1829	195.7	46.656	44.578	17.202	
340			0.3899	260.1	66.152	64.380	27.167	
350			0.7920	339.4	91.609	90.664	41.647	
360				435.9	124.192	124.816	62.139	
370				551.5	165.2	168.4	90.453	
380				688.8	215.9	223.0	128.74	
390				850.2	277.7	290.4	179.48	
400				1038	353.2	372.6	245.54	0.138
410				1256	441.0	471.5	330.15	0.215
420				1507	545.5	589.3	436.89	0.329
430				1793	667.6	728.3	569.73	0.493
440				2120	808.8	890.9	732.99	0.724
450				2490	971.1	1080	931.36	1.045
460				2910	1156	1297	1169.9	1.485
470					1366	1547	1453.9	2.078
480					1602	1833	1789.0	2.866
490					1868	2159	2181.4	3.899
500					2164	2530	2637.3	5.239
510					2494	2954	3163.3	6.955
520					2861		3766.4	9.131
530					3267		4453.9	11.861
540					3717		5233.5	15.256
550					4216		6113.4	19.438
560					4770		7102.0	25.547
570							8208.6	30.74
580							9443.0	38.19
590							10816	47.09
600							12339	57.64

ENTHALPY OF VAPORIZATION

The molar enthalpy (heat) of vaporization $\Delta_{vap}H$, which is defined as the enthalpy change in the conversion of one mole of liquid to gas at constant temperature, is tabulated here for approximately 850 inorganic and organic compounds. Values are given, when available, both at the normal boiling point t_b, referred to a pressure of 101.325 kPa (760 mmHg), and at 25°C. Substances are listed by molecular formula in the modified Hill order (see Preface).

The values in this table were measured either by calorimetric techniques or by application of the Claperyon equation to the variation of vapor pressure with temperature. See Reference 1 for a discussion of the accuracy of different experimental techniques and for methods of estimating enthalpy of vaporization at other temperatures.

REFERENCES

1. Majer, V. and Svoboda, V., *Enthalpies of Vaporization of Organic Compounds,* Blackwell Scientific Publications, Oxford, 1985.
2. Chase, M. W., Davies, C. A., Downey, J. R., Frurip, D. J., McDonald, R. A., and Syverud, A. N., *JANAF Thermochemical Tables, Third Edition, J. Phys. Chem. Ref. Data,* Vol. 14, Suppl. 1, 1985.
3. *Landolt-Börnstein, Numerical Data and Functional Relationships in Science and Technology, Sixth Edition,* II/4, *Caloric Quantities of State,* Springer-Verlag, Heidelberg, 1961.
4. Daubert, T. E., Danner, R. P., Sibul, H. M., and Stebbins, C. C., *Physical and Thermodynamic Properties of Pure Compounds: Data Compilation,* extant 1994 (core with 4 supplements), Taylor & Francis, Bristol, PA.
5. Ruzicka, K. and Majer, V., "Simultaneous Treatment of Vapor Pressures and Related Thermal Data Between the Triple and Normal Boiling Temperatures for *n*-Alkanes C_5 - C_{20}", *J. Phys. Chem. Ref. Data,* 23, 1, 1994.
6. Verevkin, S. P., "Thermochemistry of Amines: Experimental Standard Molar Enthalpies of Formation of Some Aliphatic and Aromatic Amines", *J. Chem. Thermodynamics,* 29, 891, 1997.
7. Cady, G. H. and Hargreaves, G. B., "The Vapor Pressure of Some Heavy Transition Metal Hexafluorides", *J. Chem. Soc.,* 1961, 1563; 1961, 1578.
8. Steele, W. V., Chirico, R. D., Knipmeyer, S. E., and Nguyen, A., *J. Chem. Eng. Data,* 41, 1255, 1996.

Mol. Form.	Name	$t_b/°C$	$\Delta_{vap}H(t_b)$ kJ/mol	$\Delta_{vap}H(25°C)$ kJ/mol
AgBr	Silver(I) bromide	1502	198	
AgCl	Silver(I) chloride	1547	199	
AgI	Silver(I) iodide	1506	143.9	
Al	Aluminum	2519	294	
AlB_3H_{12}	Aluminum borohydride	44.5	30	
$AlBr_3$	Aluminum tribromide	255	23.5	
AlI_3	Aluminum triiodide	382	32.2	
Ar	Argon	-185.85	6.43	
$AsBr_3$	Arsenic(III) bromide	221	41.8	
$AsCl_3$	Arsenic(III) chloride	130	35.01	
AsF_3	Arsenic(III) fluoride	57.8	29.7	
AsF_5	Arsenic(V) fluoride	-52.8	20.8	
AsH_3	Arsine	-62.5	16.69	
AsI_3	Arsenic(III) iodide	424	59.3	
Au	Gold	2856	324	
B	Boron	4000	480	
BBr_3	Boron tribromide	91	30.5	
BCl_3	Boron trichloride	12.65	23.77	23.1
BF_3	Boron trifluoride	-101	19.33	
BI_3	Boron triiodide	210	40.5	
B_2F_4	Tetrafluorodiborane	-34	28	
B_2H_6	Diborane	-92.4	14.28	
B_4H_{10}	Tetraborane	18	27.1	
B_5H_{11}	Pentaborane(11)	63	31.8	
Ba	Barium	1897	140	
$BeCl_2$	Beryllium chloride	482	105	
BeI_2	Beryllium iodide	487	70.5	
Bi	Bismuth	1564	151	
$BiBr_3$	Bismuth tribromide	453	75.4	
$BiCl_3$	Bismuth trichloride	447	72.61	
BrF	Bromine fluoride	20	25.1	
BrF_3	Bromine trifluoride	125.8	47.57	

Mol. Form.	Name	$t_b/°C$	$\Delta_{vap}H(t_b)$ kJ/mol	$\Delta_{vap}H(25°C)$ kJ/mol
BrF_5	Bromine pentafluoride	40.76	30.6	
BrH	Hydrogen bromide	-66.38		12.69
BrH_3Si	Bromosilane	1.9	24.4	
BrIn	Indium(I) bromide	656	92	
BrTl	Thallium(I) bromide	819	99.56	
Br_2	Bromine	58.8	29.96	30.91
Br_2Cd	Cadmium bromide	844	115	
Br_2H_2Si	Dibromosilane	66	31	
Br_2Hg	Mercury(II) bromide	322	58.89	
Br_2Pb	Lead(II) bromide	892	133	
Br_2Sn	Tin(II) bromide	639	102	
Br_2Zn	Zinc bromide	697	118	
Br_3Ga	Gallium(III) bromide	279	38.9	
Br_3HSi	Tribromosilane	109	34.8	
Br_3OP	Phosphorus(V) oxybromide	191.7	38	
Br_3P	Phosphorus(III) bromide	172.95	38.8	
Br_3Sb	Antimony(III) bromide	280	59	
Br_4Ge	Germanium(IV) bromide	186.35	41.4	
Br_4Si	Tetrabromosilane	154	37.9	
Br_4Sn	Tin(IV) bromide	205	43.5	
Br_4Ti	Titanium(IV) bromide	230	44.37	
Br_5Ta	Tantalum(V) bromide	349	62.3	
Cd	Cadmium	767	99.87	
$CdCl_2$	Cadmium chloride	960	124.3	
CdF_2	Cadmium fluoride	1748	214	
CdI_2	Cadmium iodide	742	115	
ClF	Chlorine fluoride	-101.1	24	
$ClFO_3$	Perchloryl fluoride	-46.75	19.33	
ClF_2P	Phosphorus(III) chloride difluoride	-47.25	17.6	
ClF_3	Chlorine trifluoride	11.75	27.53	
ClF_3Si	Chlorotrifluorosilane	-70.0	18.7	
ClH	Hydrogen chloride	-85	16.15	9.08
ClH_3Si	Chlorosilane	-30.4	21	
ClNO	Nitrosyl chloride	-5.5	25.78	
$ClNO_2$	Nitryl chloride	-15	25.7	
ClO_2	Chlorine dioxide	11	30	
ClTl	Thallium(I) chloride	720	102.2	
Cl_2	Chlorine	-34.04	20.41	17.65
Cl_2Cr	Chromium(II) chloride	1300	197	
Cl_2CrO_2	Chromyl chloride	117	35.1	
Cl_2FP	Phosphorus(III) dichloride fluoride	14	24.9	
Cl_2F_2Si	Dichlorodifluorosilane	-32	21.2	
Cl_2H_2Si	Dichlorosilane	8.3	25	24.2
Cl_2Hg	Mercury(II) chloride	304	58.9	
Cl_2O	Chlorine monoxide	2.2	25.9	
Cl_2OS	Thionyl chloride	75.6	31.7	31
Cl_2O_2S	Sulfuryl chloride	69.4	31.4	30.1
Cl_2Pb	Lead(II) chloride	951	127	
Cl_2Sn	Tin(II) chloride	623	86.8	
Cl_2Ti	Titanium(II) chloride	1500	232	
Cl_2Zn	Zinc chloride	732	126	
Cl_3Ga	Gallium(III) chloride	201	23.9	
Cl_3HSi	Trichlorosilane	33		25.7
Cl_3OP	Phosphorus(V) oxychloride	105.5	34.35	38.6
Cl_3OV	Vanadyl trichloride	127	36.78	
Cl_3P	Phosphorus(III) chloride	75.95	30.5	32.1
Cl_3Sb	Antimony(III) chloride	220.3	45.19	
Cl_3Ti	Titanium(III) chloride	960	124	

Mol. Form.	Name	t_b/°C	$\Delta_{vap}H(t_b)$ kJ/mol	$\Delta_{vap}H(25°C)$ kJ/mol
Cl_4Ge	Germanium(IV) chloride	86.55	27.9	
Cl_4OW	Tungsten(VI) oxytetrachloride	227.55	67.8	
Cl_4Si	Tetrachlorosilane	57.65	28.7	29.7
Cl_4Sn	Tin(IV) chloride	114.15	34.9	
Cl_4Te	Tellurium tetrachloride	387	77	
Cl_4Th	Thorium(IV) chloride	921	146.4	
Cl_4Ti	Titanium(IV) chloride	136.45	36.2	
Cl_4V	Vanadium(IV) chloride	148	41.4	42.5
Cl_5Mo	Molybdenum(V) chloride	268	62.8	
Cl_5Nb	Niobium(V) chloride	254.0	52.7	
Cl_5Ta	Tantalum(V) chloride	239.35	54.8	
Cl_6W	Tungsten(VI) chloride	346.75	52.7	
FH_3Si	Fluorosilane	-98.6	18.8	
FLi	Lithium fluoride	1673	147	
FNO	Nitrosyl fluoride	-59.9	19.28	
FNO_2	Nitryl fluoride	-72.4	18.05	
FNS	Thionitrosyl fluoride (NSF)	4.8	22.2	
F_2	Fluorine	-188.12	6.62	
F_2H_2Si	Difluorosilane	-77.8	16.3	
F_2O	Fluorine monoxide	-144.75	11.09	
F_2OS	Thionyl fluoride	-43.8	21.8	
F_2O_2	Fluorine dioxide	-57	19.1	
F_2Pb	Lead(II) fluoride	1293	160.4	
F_2Zn	Zinc fluoride	1500	190.1	
F_3HSi	Trifluorosilane	-95	16.2	
F_3N	Nitrogen trifluoride	-128.75	11.56	
F_3O_2Re	Rhenium(VII) dioxytrifluoride	185.4	65.7	
F_3P	Phosphorus(III) fluoride	-101.5	16.5	
F_3PS	Phosphorus(V) sulfide trifluoride	-52.25	19.6	
F_4MoO	Molybdenum(VI) oxytetrafluoride	186.0	50.6	
F_4N_2	Tetrafluorohydrazine	-74	13.27	
F_4ORe	Rhenium(VI) oxytetrafluoride	171.7	61.0	
F_4OW	Tungsten(VI) oxytetrafluoride	185.9	59.5	
F_4S	Sulfur tetrafluoride	-40.45	26.44	
F_4Se	Selenium tetrafluoride	106	47.2	
F_4Th	Thorium(IV) fluoride	1680	258	
F_5I	Iodine pentafluoride	100.5	41.3	
F_5Mo	Molybdenum(V) fluoride	213.6	51.8	
F_5Nb	Niobium(V) fluoride	229	52.3	
F_5Os	Osmium(V) fluoride	225.9	65.6	
F_5P	Phosphorus(V) fluoride	-84.6	17.2	
F_5Re	Rhenium(V) fluoride	221.3	58.1	
F_5Ta	Tantalum(V) fluoride	229.2	56.9	
F_5V	Vanadium(V) fluoride	48.3	44.52	
F_6Ir	Iridium(VI) fluoride	53.6	30.9	
F_6Mo	Molybdenum(VI) fluoride	34.0	29.0	
F_6Os	Osmium(VI) fluoride	47.5	28.1	
F_6Re	Rhenium(VI) fluoride	33.8	28.7	
F_6S	Sulfur hexafluoride			8.99
F_6W	Tungsten(VI) fluoride	17.1	26.5	
Ga	Gallium	2204	254	
GaI_3	Gallium(III) iodide	340	56.5	
Ge	Germanium	2833	334	
GeH_4	Germane	-88.1	14.06	
Ge_2H_6	Digermane	30.8	25.1	
Ge_3H_8	Trigermane	110.5	32.2	
HI	Hydrogen iodide	-35.55	19.76	17.36
$HLiO$	Lithium hydroxide	1626	188	

Mol. Form.	Name	$t_b/°C$	$\Delta_{vap}H(t_b)$ kJ/mol	$\Delta_{vap}H(25°C)$ kJ/mol
HNO_3	Nitric acid	83		39.1
HN_3	Hydrazoic acid	35.7	30.5	
$HNaO$	Sodium hydroxide	1388	175	
H_2	Hydrogen	-252.87	0.90	
H_2O	Water	100.0	40.65	43.98
H_2O_2	Hydrogen peroxide	150.2		51.6
H_2S	Hydrogen sulfide	-59.55	18.67	14.08
H_2S_2	Hydrogen disulfide	70.7		33.78
H_2Se	Hydrogen selenide	-41.25	19.7	
H_2Te	Hydrogen telluride	-2	19.2	
H_3N	Ammonia	-33.33	23.33	19.86
H_3P	Phosphine	-87.75	14.6	
H_3Sb	Stibine	-17	21.3	
H_4N_2	Hydrazine	113.55	41.8	44.7
H_4P_2	Diphosphine	63.5	28.8	
H_4Si	Silane	-111.9	12.1	
H_4Sn	Stannane	-51.8	19.05	
H_6Si_2	Disilane	-14.3	21.2	
H_8Si_3	Trisilane	52.9	28.5	
He	Helium	-268.93	0.08	
Hg	Mercury	356.73	59.11	
HgI_2	Mercury(II) iodide	354	59.2	
IIn	Indium(I) iodide	712	90.8	
ITl	Thallium(I) iodide	824	104.7	
I_2	Iodine	184.4	41.57	
I_2Pb	Lead(II) iodide	872	104	
I_2Sn	Tin(II) iodide	714	105	
I_3P	Phosphorus(III) iodide	227	43.9	
I_3Sb	Antimony(III) iodide	401	68.6	
I_4Si	Tetraiodosilane	287.35	50.2	
I_4Sn	Tin(IV) iodide	364.35	56.9	
I_4Ti	Titanium(IV) iodide	377	58.4	
Kr	Krypton	-153.22	9.08	
MoO_3	Molybdenum(VI) oxide	1155	138	
NO	Nitric oxide	-151.74	13.83	
N_2	Nitrogen	-195.79	5.57	
N_2O	Nitrous oxide	-88.48	16.53	
N_2O_4	Nitrogen tetroxide	21.15	38.12	
Ne	Neon	-246.08	1.71	
O_2	Oxygen	-182.95	6.82	
O_2S	Sulfur dioxide	-10.05	24.94	22.92
O_3S	Sulfur trioxide	45	40.69	43.14
P	Phosphorus	280.5	12.4	14.2
Pb	Lead	1749	179.5	
S	Sulfur	444.60	45	
STl_2	Thallium(I) sulfide	1367	154	
Se	Selenium	685	95.48	
Te	Tellurium	988	114.1	
Xe	Xenon	-108.11	12.57	
$CClF_3$	Chlorotrifluoromethane	-81.4	15.8	
CCl_2F_2	Dichlorodifluoromethane	-29.8	20.1	
CCl_3F	Trichlorofluoromethane	23.7	25.1	
CCl_4	Tetrachloromethane	76.8	29.82	32.43
$CHBr_3$	Tribromomethane	149.1	39.66	46.05
$CHClF_2$	Chlorodifluoromethane	-40.7	20.2	
$CHCl_2F$	Dichlorofluoromethane	8.9	25.2	
$CHCl_3$	Trichloromethane	61.17	29.24	31.28
CH_2BrCl	Bromochloromethane	68.0	30.0	

Mol. Form.	Name	$t_b/°C$	$\Delta_{vap}H(t_b)$ kJ/mol	$\Delta_{vap}H(25°C)$ kJ/mol
CH_2Br_2	Dibromomethane	97	32.92	36.97
CH_2Cl_2	Dichloromethane	40	28.06	28.82
CH_2I_2	Diiodomethane	182	42.5	
CH_2O_2	Formic acid	101	22.69	20.10
CH_3Br	Bromomethane	3.5	23.91	22.81
CH_3Cl	Chloromethane	-24.09	21.40	18.92
CH_3I	Iodomethane	42.55	27.34	27.97
CH_3NO	Formamide	220		60.15
CH_3NO_2	Nitromethane	101.19	33.99	38.27
CH_4	Methane	-161.48	8.19	
CH_4O	Methanol	64.6	35.21	37.43
CH_5N	Methylamine	-6.32	25.60	23.37
CH_6N_2	Methylhydrazine	87.5	36.12	40.37
CN_4O_8	Tetranitromethane	126.1	40.74	49.93
CO	Carbon monoxide	-191.5	6.04	
CS_2	Carbon disulfide	46	26.74	27.51
$C_2Br_2ClF_3$	1,2-Dibromo-1-chloro-1,2,2-trifluoroethane	93	31.17	35.04
$C_2Br_2F_4$	1,2-Dibromotetrafluoroethane	47.35	27.03	28.39
C_2ClF_5	Chloropentafluoroethane	-37.95	19.41	
$C_2Cl_2F_4$	1,2-Dichlorotetrafluoroethane	3.8	23.3	
$C_2Cl_3F_3$	1,1,1-Trichlorotrifluoroethane	46.1	26.85	28.08
$C_2Cl_3F_3$	1,1,2-Trichloro-1,2,2-trifluoroethane	47.7	27.04	28.40
C_2Cl_4	Tetrachloroethylene	121.3	34.68	39.68
C_2F_6	Hexafluoroethane	-78.1	16.15	
$C_2HBrClF_3$	2-Bromo-2-chloro-1,1,1-trifluoroethane	50.2	28.08	29.61
C_2HCl_3	Trichloroethylene	87.21	31.40	34.54
C_2HCl_5	Pentachloroethane	159.8	36.9	
$C_2HF_3O_2$	Trifluoroacetic acid	73	33.3	
$C_2H_2Br_4$	1,1,2,2-Tetrabromoethane	243.5	48.7	
$C_2H_2Cl_2$	1,1-Dichloroethylene	31.6	26.14	26.48
$C_2H_2Cl_2$	cis-1,2-Dichloroethylene	60.1	30.2	
$C_2H_2Cl_2$	trans-1,2-Dichloroethylene	48.7	28.9	
$C_2H_2Cl_4$	1,1,2,2-Tetrachloroethane	146.5	37.64	45.71
C_2H_3Br	Bromoethylene	15.8	23.4	
C_2H_3Cl	Chloroethylene	-13.3	20.8	
$C_2H_3Cl_2F$	1,1-Dichloro-1-fluoroethane	32.0	26.06	26.48
$C_2H_3Cl_3$	1,1,1-Trichloroethane	74.09	29.86	32.50
$C_2H_3Cl_3$	1,1,2-Trichloroethane	113.8	34.82	40.24
$C_2H_3F_3$	1,1,1-Trifluoroethane	-47.25	18.99	
C_2H_3N	Acetonitrile	81.65	29.75	32.94
C_2H_4	Ethylene	-103.77	13.53	
$C_2H_4Br_2$	1,2-Dibromoethane	131.6	34.77	41.73
$C_2H_4Cl_2$	1,1-Dichloroethane	57.4	28.85	30.62
$C_2H_4Cl_2$	1,2-Dichloroethane	83.5	31.98	35.16
$C_2H_4F_2$	1,1-Difluoroethane	-24.95	21.56	19.08
C_2H_4O	Acetaldehyde	20.1	25.76	25.47
C_2H_4O	Ethylene oxide	10.6	25.54	24.75
$C_2H_4O_2$	Acetic acid	117.9	23.70	23.36
$C_2H_4O_2$	Methyl formate	31.7	27.92	28.35
C_2H_5Br	Bromoethane	38.5	27.04	28.03
C_2H_5Cl	Chloroethane	12.3	24.65	
C_2H_5ClO	2-Chloroethanol	128.6	41.4	
C_2H_5I	Iodoethane	72.5	29.44	31.93
C_2H_5NO	N-Methylformamide	199.51		56.19
$C_2H_5NO_2$	Nitroethane	114.0	38.0	
C_2H_6	Ethane	-88.6	14.69	5.16
C_2H_6O	Ethanol	78.29	38.56	42.32
C_2H_6O	Dimethyl ether	-24.8	21.51	18.51

Mol. Form.	Name	t_b/°C	$\Delta_{vap}H(t_b)$ kJ/mol	$\Delta_{vap}H(25°C)$ kJ/mol
C$_2$H$_6$OS	Dimethyl sulfoxide	189	43.1	
C$_2$H$_6$O$_2$	Ethylene glycol	197.3	50.5	
C$_2$H$_6$S	Ethanethiol	35.1	26.79	27.30
C$_2$H$_6$S	Dimethyl sulfide	37.33	27.0	27.65
C$_2$H$_6$S$_2$	1,2-Ethanedithiol	146.1	37.93	44.68
C$_2$H$_6$S$_2$	Dimethyl disulfide	109.8	33.78	37.86
C$_2$H$_7$N	Dimethylamine	6.88	26.40	25.05
C$_2$H$_7$NO	Ethanolamine	171	49.83	
C$_2$H$_8$N$_2$	1,2-Ethanediamine	117	37.98	44.98
C$_2$H$_8$N$_2$	1,1-Dimethylhydrazine	63.9	32.55	35.0
C$_2$N$_2$	Cyanogen	-21.1	23.33	19.75
C$_3$Cl$_2$F$_6$	1,2-Dichlorohexafluoropropane	34.1	26.28	26.93
C$_3$H$_3$Cl$_3$O$_2$	Methyl trichloroacetate	153.8		48.33
C$_3$H$_3$N	Acrylonitrile	77.3	32.6	
C$_3$H$_4$Cl$_2$O$_2$	Methyl dichloroacetate	142.9	39.28	47.72
C$_3$H$_4$O	Acrolein	52.6	28.3	
C$_3$H$_4$O$_2$	2-Oxetanone	162		47.03
C$_3$H$_5$Br	3-Bromopropene	70.1	30.24	32.73
C$_3$H$_5$Cl	3-Chloropropene	45.1	29.0	
C$_3$H$_5$ClO$_2$	Methyl chloroacetate	129.5	39.23	46.73
C$_3$H$_5$Cl$_3$	1,2,3-Trichloropropane	157	37.1	
C$_3$H$_5$N	Propanenitrile	97.14	31.81	36.03
C$_3$H$_6$	Propene	-47.69	18.42	14.24
C$_3$H$_6$	Cyclopropane	-32.81	20.05	16.93
C$_3$H$_6$Br$_2$	1,2-Dibromopropane	141.9	35.61	41.67
C$_3$H$_6$Br$_2$	1,3-Dibromopropane	167.3		47.45
C$_3$H$_6$Cl$_2$	1,3-Dichloropropane	120.9	35.18	40.75
C$_3$H$_6$O	Allyl alcohol	97.0	40.0	
C$_3$H$_6$O	Propanal	48	28.31	29.62
C$_3$H$_6$O	Acetone	56.05	29.10	30.99
C$_3$H$_6$O	Methyloxirane	35	27.35	27.89
C$_3$H$_6$O	Oxetane	47.6	28.67	29.85
C$_3$H$_6$O$_2$	Propanoic acid	141.15		32.14
C$_3$H$_6$O$_2$	Ethyl formate	54.4	29.91	31.96
C$_3$H$_6$O$_2$	Methyl acetate	56.87	30.32	32.29
C$_3$H$_6$S	Thiacyclobutane	95	32.32	35.97
C$_3$H$_7$Br	1-Bromopropane	71.1	29.84	32.01
C$_3$H$_7$Br	2-Bromopropane	59.5	28.33	30.17
C$_3$H$_7$Cl	1-Chloropropane	46.5	27.18	28.35
C$_3$H$_7$Cl	2-Chloropropane	35.7	26.30	26.90
C$_3$H$_7$I	1-Iodopropane	102.6	32.08	36.25
C$_3$H$_7$I	2-Iodopropane	89.5	30.68	34.06
C$_3$H$_7$NO	N-Ethylformamide	198		58.44
C$_3$H$_7$NO	N,N-Dimethylformamide	153		46.89
C$_3$H$_7$NO$_2$	1-Nitropropane	131.1	38.5	
C$_3$H$_7$NO$_2$	2-Nitropropane	120.2	36.8	
C$_3$H$_8$	Propane	-42.1	19.04	14.79
C$_3$H$_8$O	1-Propanol	97.2	41.44	47.45
C$_3$H$_8$O	2-Propanol	82.3	39.85	45.39
C$_3$H$_8$O$_2$	1,2-Propylene glycol	187.6	52.4	
C$_3$H$_8$O$_2$	1,3-Propylene glycol	214.4	57.9	
C$_3$H$_8$O$_2$	Ethylene glycol monomethyl ether	124.1	37.54	45.17
C$_3$H$_8$O$_3$	Glycerol	290	61.0	
C$_3$H$_8$S	1-Propanethiol	67.8	29.54	31.89
C$_3$H$_8$S	2-Propanethiol	52.6	27.91	29.45
C$_3$H$_8$S	Ethyl methyl sulfide	66.7	29.53	31.85
C$_3$H$_8$S$_2$	1,3-Propanedithiol	172.9		49.66
C$_3$H$_9$N	Propylamine	47.22	29.55	31.27

Mol. Form.	Name	$t_b/°C$	$\Delta_{vap}H(t_b)$ kJ/mol	$\Delta_{vap}H(25°C)$ kJ/mol
C_3H_9N	Isopropylamine	31.76	27.83	28.36
C_3H_9N	Trimethylamine	2.87	22.94	21.66
$C_3H_{10}N_2$	1,3-Propanediamine	139.8	40.85	50.16
C_4F_8	Perfluorocyclobutane	-5.9	23.2	
C_4F_{10}	Perfluorobutane	-1.9	22.9	
$C_4H_4N_2$	Succinonitrile	266	48.5	
$C_4H_4N_2$	Pyrimidine	123.8	43.09	49.79
$C_4H_4N_2$	Pyridazine	208		53.47
C_4H_4O	Furan	31.5	27.10	27.45
$C_4H_4O_2$	Diketene	126.1	36.80	42.89
C_4H_4S	Thiophene	84.0	31.48	34.70
$C_4H_5Cl_3O_2$	Ethyl trichloroacetate	167.5		50.97
C_4H_5N	2-Methylacrylonitrile	90.3	31.8	
C_4H_5N	Pyrrole	129.79	38.75	45.09
C_4H_5N	Cyclopropanecarbonitrile	135.1	35.55	41.94
$C_4H_5NO_2$	Methyl cyanoacetate	200.5	48.2	
C_4H_5NS	4-Methylthiazole	133.3	37.58	43.85
C_4H_6	1,2-Butadiene	10.9	24.02	23.21
C_4H_6	1,3-Butadiene	-4.41	22.47	20.86
C_4H_6	1-Butyne	8.08	24.52	23.35
$C_4H_6Cl_2O_2$	Ethyl dichloroacetate	155		50.60
$C_4H_6O_2$	Vinyl acetate	72.5	34.6	
$C_4H_6O_2$	Methyl acrylate	80.7	33.1	
$C_4H_6O_2$	γ-Butyrolactone	204	52.2	
$C_4H_6O_3$	Acetic anhydride	139.5	38.2	
C_4H_6S	2,3-Dihydrothiophene	112.1	33.24	37.74
C_4H_6S	2,5-Dihydrothiophene	122.4	34.83	39.95
$C_4H_7ClO_2$	Ethyl chloroacetate	144.3	40.43	49.47
C_4H_7N	Butanenitrile	117.6	33.68	39.33
C_4H_7N	2-Methylpropanenitrile	103.9	32.39	37.13
C_4H_8	1-Butene	-6.26	22.07	20.22
C_4H_8	cis-2-Butene	3.71	23.34	22.16
C_4H_8	trans-2-Butene	0.88	22.72	21.40
C_4H_8	Cyclobutane	12.6	24.19	23.51
$C_4H_8Br_2$	1,4-Dibromobutane	197		53.09
$C_4H_8Cl_2$	1,2-Dichlorobutane	124.1	33.90	39.58
$C_4H_8Cl_2$	1,4-Dichlorobutane	161		46.36
$C_4H_8Cl_2O$	Bis(2-chloroethyl) ether	178.5	45.2	
C_4H_8O	Ethyl vinyl ether	35.5	26.2	
C_4H_8O	1,2-Epoxybutane	63.4	30.3	
C_4H_8O	Butanal	74.8	31.5	
C_4H_8O	2-Butanone	79.59	31.30	34.79
C_4H_8O	Tetrahydrofuran	65	29.81	31.99
$C_4H_8O_2$	Butanoic acid	163.75		40.45
$C_4H_8O_2$	2-Methylpropanoic acid	154.45		35.30
$C_4H_8O_2$	Propyl formate	80.9	33.61	37.53
$C_4H_8O_2$	Ethyl acetate	77.11	31.94	35.60
$C_4H_8O_2$	Methyl propanoate	79.8	32.24	35.85
$C_4H_8O_2$	1,3-Dioxane	106.1	34.37	39.09
$C_4H_8O_2$	1,4-Dioxane	101.5	34.16	38.60
C_4H_8S	Tetrahydrothiophene	121.0	34.66	39.43
C_4H_9Br	1-Bromobutane	101.6	32.51	36.64
C_4H_9Br	2-Bromobutane	91.3	30.77	34.41
C_4H_9Br	1-Bromo-2-methylpropane	91.1	31.33	34.82
C_4H_9Br	2-Bromo-2-methylpropane	73.3	29.23	31.81
C_4H_9Cl	1-Chlorobutane	78.6	30.39	33.51
C_4H_9Cl	2-Chlorobutane	68.2	29.17	31.53
C_4H_9Cl	1-Chloro-2-methylpropane	68.5	29.22	31.67

Mol. Form.	Name	$t_b/°C$	$\Delta_{vap}H(t_b)$ kJ/mol	$\Delta_{vap}H(25°C)$ kJ/mol
C_4H_9Cl	2-Chloro-2-methylpropane	50.9	27.55	28.98
C_4H_9I	1-Iodobutane	130.6	34.66	40.63
C_4H_9I	2-Iodobutane	120.1	33.27	38.46
C_4H_9I	1-Iodo-2-methylpropane	121.1	33.54	38.83
C_4H_9I	2-Iodo-2-methylpropane	100.1	31.43	35.41
C_4H_9N	Pyrrolidine	86.56	33.01	37.52
C_4H_9NO	N-Ethylacetamide	205		64.89
C_4H_9NO	N,N-Dimethylacetamide	165		50.24
C_4H_9NO	Morpholine	128	37.1	
C_4H_{10}	Butane	-0.5	22.44	21.02
C_4H_{10}	Isobutane	-11.73	21.30	19.23
$C_4H_{10}O$	1-Butanol	117.73	43.29	52.35
$C_4H_{10}O$	2-Butanol	99.51	40.75	49.72
$C_4H_{10}O$	2-Methyl-1-propanol	107.89	41.82	50.82
$C_4H_{10}O$	2-Methyl-2-propanol	82.4	39.07	46.69
$C_4H_{10}O$	Diethyl ether	34.5	26.52	27.10
$C_4H_{10}O$	Methyl propyl ether	39.1	26.75	27.60
$C_4H_{10}O$	Isopropyl methyl ether	30.77	26.05	26.41
$C_4H_{10}O_2$	1,2-Butanediol	190.5	52.84	71.55
$C_4H_{10}O_2$	1,3-Butanediol	207.5	54.31	74.46
$C_4H_{10}O_2$	Ethylene glycol monoethyl ether	135	39.22	48.21
$C_4H_{10}O_2$	Ethylene glycol dimethyl ether	85	32.42	36.39
$C_4H_{10}O_3$	Diethylene glycol	245.8	52.3	
$C_4H_{10}S$	1-Butanethiol	98.5	32.23	36.63
$C_4H_{10}S$	2-Butanethiol	85	30.59	33.99
$C_4H_{10}S$	2-Methyl-1-propanethiol	88.5	31.01	34.63
$C_4H_{10}S$	2-Methyl-2-propanethiol	64.3	28.45	30.78
$C_4H_{10}S$	Diethyl sulfide	92.1	31.77	35.80
$C_4H_{10}S$	Methyl propyl sulfide	95.6	32.08	36.24
$C_4H_{10}S$	Isopropyl methyl sulfide	84.8	30.71	34.15
$C_4H_{10}S_2$	1,4-Butanedithiol	195.5		55.10
$C_4H_{10}S_2$	Diethyl disulfide	154.1	37.58	45.18
$C_4H_{11}N$	Butylamine	77.00	31.81	35.72
$C_4H_{11}N$	sec-Butylamine	62.73	29.92	32.85
$C_4H_{11}N$	tert-Butylamine	44.04	28.27	29.64
$C_4H_{11}N$	Isobutylamine	67.75	30.61	33.85
$C_4H_{11}N$	Diethylamine	55.5	29.06	31.31
$C_4H_{11}N$	Isopropylmethylamine	50.4	28.71	30.69
$C_4H_{11}NO$	2-Amino-2-methyl-1-propanol	165.5	50.6	
$C_4H_{11}NO_2$	Diethanolamine	268.8	65.2	
$C_5H_2F_6O_2$	Hexafluoroacetylacetone	54.15	27.05	30.58
$C_5H_4O_2$	Furfural	161.7	43.2	
C_5H_5N	Pyridine	115.23	35.09	40.21
$C_5H_6O_2$	Furfuryl alcohol	171	53.6	
C_5H_6S	2-Methylthiophene	112.6	33.90	38.87
C_5H_6S	3-Methylthiophene	115.5	34.24	39.43
C_5H_7N	trans-3-Pentenenitrile	142.6	37.09	44.77
C_5H_7N	Cyclobutanecarbonitrile	149.6	36.88	44.34
C_5H_8	Spiropentane	39	26.76	27.49
C_5H_8O	Cyclopropyl methyl ketone	111.3	34.07	39.41
C_5H_8O	Cyclopentanone	130.57	36.35	42.72
$C_5H_8O_2$	Methyl cyclopropanecarboxylate	114.9	35.25	41.27
$C_5H_8O_2$	Allyl acetate	103.5	36.3	
$C_5H_8O_2$	Ethyl acrylate	99.4	34.7	
$C_5H_8O_2$	Methyl methacrylate	100.5	36.0	
$C_5H_8O_2$	2,4-Pentanedione	138	34.30	41.77
C_5H_9N	Pentanenitrile	141.3	36.09	43.60
C_5H_9N	3-Methylbutanenitrile	127.5	35.10	41.64

Mol. Form.	Name	t_b/°C	$\Delta_{vap}H(t_b)$ kJ/mol	$\Delta_{vap}H(25°C)$ kJ/mol
C_5H_9N	2,2-Dimethylpropanenitrile	106.1	32.40	37.35
C_5H_{10}	1-Pentene	29.96	25.20	25.47
C_5H_{10}	cis-2-Pentene	36.93		26.86
C_5H_{10}	trans-2-Pentene	36.34		26.76
C_5H_{10}	2-Methyl-1-butene	31.2	25.50	25.92
C_5H_{10}	3-Methyl-1-butene	20.1		23.77
C_5H_{10}	2-Methyl-2-butene	38.56	26.31	27.06
C_5H_{10}	Cyclopentane	49.3	27.30	28.52
$C_5H_{10}Cl_2$	1,2-Dichloropentane	148.3	36.45	43.89
$C_5H_{10}Cl_2$	1,5-Dichloropentane	179		50.71
$C_5H_{10}O$	Cyclopentanol	140.42		57.05
$C_5H_{10}O$	2-Pentanone	102.26	33.44	38.40
$C_5H_{10}O$	3-Pentanone	101.96	33.45	38.52
$C_5H_{10}O$	3-Methyl-2-butanone	94.33	32.35	36.78
$C_5H_{10}O$	3,3-Dimethyloxetane	80.6	30.85	33.94
$C_5H_{10}O$	Tetrahydropyran	88	31.17	34.58
$C_5H_{10}O_2$	Pentanoic acid	186.1	44.1	
$C_5H_{10}O_2$	2-Methylbutanoic acid	177		46.91
$C_5H_{10}O_2$	Butyl formate	106.1	36.58	41.11
$C_5H_{10}O_2$	Isobutyl formate	98.2	33.6	
$C_5H_{10}O_2$	Propyl acetate	101.54	33.92	39.72
$C_5H_{10}O_2$	Isopropyl acetate	88.6	32.93	37.20
$C_5H_{10}O_2$	Ethyl propanoate	99.1	33.88	39.21
$C_5H_{10}O_2$	Methyl butanoate	102.8	33.79	39.28
$C_5H_{10}O_2$	Methyl isobutanoate	92.5	32.61	37.32
$C_5H_{10}O_2$	Tetrahydrofurfuryl alcohol	178	45.2	
$C_5H_{10}O_3$	Diethyl carbonate	126		43.60
$C_5H_{10}O_3$	Ethylene glycol monomethyl ether acetate	143	43.9	
$C_5H_{10}S$	Thiacyclohexane	141.8	35.96	42.58
$C_5H_{10}S$	Cyclopentanethiol	132.1	35.32	41.42
$C_5H_{11}Br$	1-Bromopentane	129.8	35.01	41.28
$C_5H_{11}Cl$	1-Chloropentane	107.8	33.15	38.24
$C_5H_{11}Cl$	2-Chloropentane	97.0	31.79	36.03
$C_5H_{11}Cl$	1-Chloro-3-methylbutane	98.9	32.02	36.24
$C_5H_{11}I$	1-Iodopentane	155		45.27
$C_5H_{11}N$	Piperidine	106.22		39.29
C_5H_{12}	Pentane	36.06	25.79	26.43
C_5H_{12}	Isopentane	27.88	24.69	24.85
C_5H_{12}	Neopentane	9.48	22.74	21.84
$C_5H_{12}O$	1-Pentanol	137.98	44.36	57.02
$C_5H_{12}O$	2-Pentanol	119.3	41.40	54.21
$C_5H_{12}O$	3-Pentanol	116.25		54.0
$C_5H_{12}O$	2-Methyl-1-butanol	128		55.16
$C_5H_{12}O$	3-Methyl-1-butanol	131.1	44.07	55.61
$C_5H_{12}O$	2-Methyl-2-butanol	102.4	39.04	50.10
$C_5H_{12}O$	3-Methyl-2-butanol	112.9		53.0
$C_5H_{12}O$	Butyl methyl ether	70.16	29.55	32.37
$C_5H_{12}O$	sec-Butyl methyl ether	59.1	28.09	30.23
$C_5H_{12}O$	Methyl tert-butyl ether	55.2	27.94	29.82
$C_5H_{12}O$	Isobutyl methyl ether	58.6	28.02	30.13
$C_5H_{12}O$	Ethyl propyl ether	63.21	28.94	31.43
$C_5H_{12}O$	Ethyl isopropyl ether	54.1	28.21	30.08
$C_5H_{12}O_2$	1-Ethoxy-2-methoxyethane	102.1	34.33	39.83
$C_5H_{12}O_2$	1,5-Pentanediol	239	60.7	
$C_5H_{12}O_2$	Ethylene glycol monopropyl ether	149.8	41.40	52.12
$C_5H_{12}O_2$	Diethoxymethane	88	31.33	35.65
$C_5H_{12}O_3$	Diethylene glycol monomethyl ether	193	46.6	
$C_5H_{12}S$	1-Pentanethiol	126.6	34.88	41.24

Mol. Form.	Name	$t_b/°C$	$\Delta_{vap}H(t_b)$ kJ/mol	$\Delta_{vap}H(25°C)$ kJ/mol
$C_5H_{12}S$	2-Methyl-1-butanethiol	119.1	33.79	39.45
$C_5H_{12}S$	2-Methyl-2-butanethiol	99.1	31.37	35.67
$C_5H_{12}S$	Butyl methyl sulfide	123.5	34.47	40.46
$C_5H_{12}S$	*tert*-Butyl methyl sulfide	99	31.47	35.84
$C_5H_{12}S$	Ethyl propyl sulfide	118.6	34.24	39.97
$C_5H_{12}S$	Ethyl isopropyl sulfide	107.5	32.74	37.78
$C_5H_{13}N$	Pentylamine	104.3	34.01	40.08
$C_5H_{13}N$	Ethylisopropylamine	69.6	29.94	33.13
C_6ClF_5	Chloropentafluorobenzene	117.96	34.76	41.07
C_6F_6	Hexafluorobenzene	80.26	31.66	35.71
C_6HF_5	Pentafluorobenzene	85.74	32.15	36.27
$C_6H_4Cl_2$	*o*-Dichlorobenzene	180	39.66	50.21
$C_6H_4Cl_2$	*m*-Dichlorobenzene	173	38.62	48.58
$C_6H_4Cl_2$	*p*-Dichlorobenzene	174	38.79	49.0
$C_6H_4F_2$	*o*-Difluorobenzene	94	32.21	36.18
$C_6H_4F_2$	*m*-Difluorobenzene	82.6	31.10	34.59
$C_6H_4F_2$	*p*-Difluorobenzene	89	31.77	35.54
C_6H_5Br	Bromobenzene	156.06		44.54
C_6H_5Cl	Chlorobenzene	131.72	35.19	40.97
C_6H_5F	Fluorobenzene	84.73	31.19	34.58
C_6H_5I	Iodobenzene	188.4	39.5	
$C_6H_5NO_2$	Nitrobenzene	210.8		55.01
C_6H_6	Benzene	80.09	30.72	33.83
C_6H_6ClN	*o*-Chloroaniline	208.8	44.4	
C_6H_6O	Phenol	181.87	45.69	57.82
C_6H_6S	Benzenethiol	169.1	39.93	47.56
C_6H_7N	Aniline	184.17	42.44	55.83
C_6H_7N	2-Methylpyridine	129.38	36.17	42.48
C_6H_7N	3-Methylpyridine	144.14	37.35	44.44
C_6H_7N	4-Methylpyridine	145.36	37.51	44.56
C_6H_7N	1-Cyclopentenecarbonitrile			44.98
C_6H_9N	Cyclopentanecarbonitrile			43.43
$C_6H_9NO_3$	Triacetamide			60.41
C_6H_{10}	Cyclohexene	82.98	30.46	33.47
$C_6H_{10}O$	Cyclohexanone	155.43		45.06
$C_6H_{10}O$	Mesityl oxide	130	36.1	
$C_6H_{10}O_2$	Methyl cyclobutanecarboxylate	135.5	37.13	44.72
$C_6H_{10}O_3$	Propanoic anhydride	170	41.7	
$C_6H_{10}O_4$	Diethyl oxalate	185.7	42.0	
$C_6H_{10}O_4$	Ethylene glycol diacetate	190		61.44
$C_6H_{11}N$	Hexanenitrile	163.65		47.91
C_6H_{12}	1-Hexene	63.48		30.61
C_6H_{12}	*cis*-2-Hexene	68.8		32.19
C_6H_{12}	*trans*-2-Hexene	67.9		31.60
C_6H_{12}	*cis*-3-Hexene	66.4		31.23
C_6H_{12}	*trans*-3-Hexene	67.1		31.55
C_6H_{12}	2-Methyl-1-pentene	62.1		30.48
C_6H_{12}	3-Methyl-1-pentene	54.2		28.62
C_6H_{12}	4-Methyl-1-pentene	53.9		28.71
C_6H_{12}	2-Methyl-2-pentene	67.3		31.60
C_6H_{12}	3-Methyl-*cis*-2-pentene	67.7		32.09
C_6H_{12}	3-Methyl-*trans*-2-pentene	70.4		31.35
C_6H_{12}	4-Methyl-*cis*-2-pentene	56.3		29.48
C_6H_{12}	4-Methyl-*trans*-2-pentene	58.6		29.97
C_6H_{12}	2-Ethyl-1-butene	64.7		31.13
C_6H_{12}	2,3-Dimethyl-1-butene	55.6		29.18
C_6H_{12}	3,3-Dimethyl-1-butene	41.2		26.61
C_6H_{12}	2,3-Dimethyl-2-butene	73.3	29.64	32.51

Mol. Form.	Name	$t_b/°C$	$\Delta_{vap}H(t_b)$ kJ/mol	$\Delta_{vap}H(25°C)$ kJ/mol
C_6H_{12}	Cyclohexane	80.73	29.97	33.01
C_6H_{12}	Methylcyclopentane	71.8	29.08	31.64
C_6H_{12}	Ethylcyclobutane	70.8	28.67	31.24
$C_6H_{12}Cl_2$	1,2-Dichlorohexane	173		48.16
$C_6H_{12}O$	Butyl vinyl ether	94	31.58	36.17
$C_6H_{12}O$	2-Hexanone	127.6	36.35	43.14
$C_6H_{12}O$	3-Hexanone	123.5	35.36	42.47
$C_6H_{12}O$	3-Methyl-2-pentanone	117.5	34.16	40.53
$C_6H_{12}O$	4-Methyl-2-pentanone	116.5	34.49	40.61
$C_6H_{12}O$	2-Methyl-3-pentanone	113.5	33.84	39.79
$C_6H_{12}O$	3,3-Dimethyl-2-butanone	106.1	33.39	37.91
$C_6H_{12}O$	Cyclohexanol	160.84		62.01
$C_6H_{12}O_2$	Butyl acetate	126.1	36.28	43.86
$C_6H_{12}O_2$	*tert*-Butyl acetate	95.1	33.07	38.03
$C_6H_{12}O_2$	Isobutyl acetate	116.5	35.9	
$C_6H_{12}O_2$	Propyl propanoate	122.5	35.54	43.45
$C_6H_{12}O_2$	Ethyl butanoate	121.5	35.47	42.68
$C_6H_{12}O_2$	Ethyl 2-methylpropanoate	110.1	33.67	39.83
$C_6H_{12}O_2$	Methyl pentanoate	127.4	35.36	43.10
$C_6H_{12}O_2$	Methyl 2,2-dimethylpropanoate	101.1	33.42	38.76
$C_6H_{12}O_3$	Ethylene glycol monoethyl ether acetate	156.4	40.76	52.61
$C_6H_{12}S$	Cyclohexanethiol	158.9	37.06	44.57
$C_6H_{13}Br$	1-Bromohexane	155.3		45.89
$C_6H_{13}Cl$	1-Chlorohexane	135	35.67	42.83
$C_6H_{13}I$	1-Iodohexane	181		49.75
$C_6H_{13}N$	Cyclohexylamine	134	36.14	43.67
C_6H_{14}	Hexane	68.73	28.85	31.56
C_6H_{14}	2-Methylpentane	60.26	27.79	29.89
C_6H_{14}	3-Methylpentane	63.27	28.06	30.28
C_6H_{14}	2,2-Dimethylbutane	49.73	26.31	27.68
C_6H_{14}	2,3-Dimethylbutane	57.93	27.38	29.12
$C_6H_{14}N_2$	Azopropane	114		39.88
$C_6H_{14}O$	1-Hexanol	157.6	44.50	61.61
$C_6H_{14}O$	2-Hexanol	140	41.01	58.46
$C_6H_{14}O$	2-Methyl-1-pentanol	149	50.2	
$C_6H_{14}O$	4-Methyl-1-pentanol	151.9	44.46	60.47
$C_6H_{14}O$	2-Methyl-2-pentanol	121.1	39.59	54.77
$C_6H_{14}O$	4-Methyl-2-pentanol	131.6	44.2	
$C_6H_{14}O$	2-Ethyl-1-butanol	147	43.2	
$C_6H_{14}O$	Dipropyl ether	90.08	31.31	35.69
$C_6H_{14}O$	Diisopropyl ether	68.51	29.10	32.12
$C_6H_{14}O$	Butyl ethyl ether	92.3	31.63	36.32
$C_6H_{14}O$	Methyl pentyl ether	99	32.02	36.85
$C_6H_{14}O_2$	2-Methyl-2,4-pentanediol	197.1	57.3	
$C_6H_{14}O_2$	Ethylene glycol monobutyl ether	168.4		56.59
$C_6H_{14}O_2$	1,1-Diethoxyethane	102.25	36.28	43.20
$C_6H_{14}O_2$	Ethylene glycol diethyl ether	119.4	36.28	43.20
$C_6H_{14}O_3$	Bis(ethoxymethyl) ether	140.6	36.17	44.69
$C_6H_{14}O_3$	Diethylene glycol monoethyl ether	196	47.5	
$C_6H_{14}O_3$	Diethylene glycol dimethyl ether	162	36.17	44.69
$C_6H_{14}O_4$	Triethylene glycol	285	71.4	
$C_6H_{14}S$	Dipropyl sulfide	142.9	36.60	44.21
$C_6H_{14}S$	Diisopropyl sulfide	120.1	33.80	39.60
$C_6H_{14}S$	Isopropyl propyl sulfide	132.1	35.11	41.78
$C_6H_{14}S$	Butyl ethyl sulfide	144.3	37.01	44.51
$C_6H_{14}S$	Methyl pentyl sulfide	145.1	37.41	45.24
$C_6H_{15}N$	Hexylamine	132.8	36.54	45.10
$C_6H_{15}N$	Butylethylamine	107.5	33.97	40.15

Mol. Form.	Name	t_b/°C	$\Delta_{vap}H(t_b)$ kJ/mol	$\Delta_{vap}H(25°C)$ kJ/mol
$C_6H_{15}N$	Dipropylamine	109.3	33.47	40.04
$C_6H_{15}N$	Diisopropylamine	83.9	30.40	34.61
$C_6H_{15}N$	Isopropylpropylamine	96.9	32.14	37.23
$C_6H_{15}N$	Triethylamine	89	31.01	34.84
C_6MoO_6	Molybdenum hexacarbonyl	701	72.51	
$C_7H_3F_5$	2,3,4,5,6-Pentafluorotoluene	117.5	34.75	41.12
$C_7H_5F_3$	(Trifluoromethyl)benzene	102.1	32.63	37.60
C_7H_5N	Benzonitrile	191.1	45.9	
C_7H_6O	Benzaldehyde	179.0	42.5	
$C_7H_6O_2$	Salicylaldehyde	197	38.2	
C_7H_7Cl	o-Chlorotoluene	159.0	37.5	
C_7H_7Cl	p-Chlorotoluene	162.4	38.7	
C_7H_7F	o-Fluorotoluene	115	35.4	
C_7H_7F	p-Fluorotoluene	116.6	34.08	39.42
C_7H_8	Toluene	110.63	33.18	38.01
C_7H_8O	o-Cresol	191.04	45.19	
C_7H_8O	m-Cresol	202.27	47.40	61.71
C_7H_8O	p-Cresol	201.98	47.45	
C_7H_8O	Benzyl alcohol	205.31	50.48	
C_7H_8O	Anisole	153.7	38.97	46.90
C_7H_9N	Benzylamine	185		60.16
C_7H_9N	o-Methylaniline	200.3	44.6	
C_7H_9N	m-Methylaniline	203.3	44.9	
C_7H_9N	p-Methylaniline	200.4	44.3	
C_7H_9N	1-Cyclohexenecarbonitrile			53.55
C_7H_9N	2,3-Dimethylpyridine	161.12	39.08	47.82
C_7H_9N	2,4-Dimethylpyridine	158.38	38.53	47.49
C_7H_9N	2,5-Dimethylpyridine	156.98	38.68	47.04
C_7H_9N	2,6-Dimethylpyridine	144.01	37.46	45.34
C_7H_9N	3,4-Dimethylpyridine	179.10	39.99	50.50
C_7H_9N	3,5-Dimethylpyridine	171.84	39.46	49.33
$C_7H_{10}O$	Dicyclopropyl ketone	161		53.70
$C_7H_{11}N$	Cyclohexanecarbonitrile			51.92
C_7H_{12}	1-Methylbicyclo(3,1,0)hexane	93.1	31.07	34.77
$C_7H_{12}O_4$	Diethyl malonate	200	54.8	
C_7H_{14}	1-Heptene	93.64		35.49
C_7H_{14}	cis-2-Heptene	98.4		36.26
C_7H_{14}	trans-2-Heptene	98		36.27
C_7H_{14}	cis-3-Heptene	95.8		35.81
C_7H_{14}	trans-3-Heptene	95.7		35.84
C_7H_{14}	cis-3-Methyl-3-hexene	95.4		36.31
C_7H_{14}	trans-3-Methyl-3-hexene	93.5		35.70
C_7H_{14}	2,4-Dimethyl-1-pentene	81.6		33.03
C_7H_{14}	4,4-Dimethyl-1-pentene	72.5		31.13
C_7H_{14}	2,4-Dimethyl-2-pentene	83.4		34.19
C_7H_{14}	cis-4,4-Dimethyl-2-pentene	80.4		32.56
C_7H_{14}	trans-4,4-Dimethyl-2-pentene	76.7		32.81
C_7H_{14}	2-Ethyl-3-methyl-1-butene	89		34.35
C_7H_{14}	2,3,3-Trimethyl-1-butene	77.9		32.09
C_7H_{14}	Methylcyclohexane	100.93	31.27	35.36
C_7H_{14}	Ethylcyclopentane	103.5	31.96	36.40
C_7H_{14}	cis-1,3-Dimethylcyclopentane	90.8	30.40	34.20
$C_7H_{14}O$	2-Heptanone	151.05		47.24
$C_7H_{14}O$	2,2-Dimethyl-3-pentanone	125.6	36.09	42.34
$C_7H_{14}O$	2,4-Dimethyl-3-pentanone	125.4	34.64	41.51
$C_7H_{14}O$	1-Methylcyclohexanol	155	79.0	
$C_7H_{14}O$	cis-2-Methylcyclohexanol	165	48.5	
$C_7H_{14}O$	trans-2-Methylcyclohexanol	167.5	53.0	

Mol. Form.	Name	t_b/°C	$\Delta_{vap}H(t_b)$ kJ/mol	$\Delta_{vap}H(25°C)$ kJ/mol
$C_7H_{14}O_2$	Pentyl acetate	149.2	38.42	48.56
$C_7H_{14}O_2$	Isopentyl acetate	142.5	37.5	
$C_7H_{14}O_2$	Ethyl pentanoate	146.1	36.96	47.01
$C_7H_{14}O_2$	Ethyl 3-methylbutanoate	135.0	37.0	
$C_7H_{14}O_2$	Ethyl 2,2-dimethylpropanoate	118.4	34.51	41.25
$C_7H_{14}O_2$	Methyl hexanoate	149.5	38.55	48.04
$C_7H_{15}Br$	1-Bromoheptane	179		50.60
$C_7H_{15}Cl$	1-Chloroheptane	159		47.66
C_7H_{16}	Heptane	98.5	31.77	36.57
C_7H_{16}	2-Methylhexane	90.04	30.62	34.87
C_7H_{16}	3-Methylhexane	92	30.9	
C_7H_{16}	3-Ethylpentane	93.5	31.12	35.22
C_7H_{16}	2,2-Dimethylpentane	79.2	29.23	32.42
C_7H_{16}	2,3-Dimethylpentane	89.78	30.46	34.26
C_7H_{16}	2,4-Dimethylpentane	80.49	29.55	32.88
C_7H_{16}	3,3-Dimethylpentane	86.06	29.62	33.03
C_7H_{16}	2,2,3-Trimethylbutane	80.86	28.90	32.05
$C_7H_{16}O$	Hexyl methyl ether	126.1	34.93	42.07
$C_7H_{16}O$	1-Heptanol	176.45		66.81
$C_7H_{16}O$	3-Heptanol	157	42.5	
$C_7H_{16}O$	Butyl propyl ether	118.1	33.72	40.22
$C_7H_{16}O$	Ethyl pentyl ether	117.6	34.41	41.01
$C_7H_{17}N$	Heptylamine	156		49.96
C_8F_{18}	Perfluorooctane	105.9	33.38	41.13
C_8H_8	Styrene	145	38.7	
C_8H_8O	Acetophenone	202	43.98	55.40
$C_8H_8O_2$	Methyl benzoate	199		55.57
$C_8H_8O_3$	Methyl salicylate	222.9	46.7	
C_8H_{10}	Ethylbenzene	136.19	35.57	42.24
C_8H_{10}	o-Xylene	144.5	36.24	43.43
C_8H_{10}	m-Xylene	139.12	35.66	42.65
C_8H_{10}	p-Xylene	138.37	35.67	42.40
$C_8H_{10}O$	2,4-Xylenol	210.98		64.96
$C_8H_{10}O$	2,5-Xylenol	211.1	46.9	
$C_8H_{10}O$	2,6-Xylenol	201.07		75.31
$C_8H_{10}O$	3,4-Xylenol	227		85.03
$C_8H_{10}O$	3,5-Xylenol	221.74		82.01
$C_8H_{10}O$	Phenetole	169.81		51.04
$C_8H_{11}N$	N-Ethylaniline	203.0		58.3
$C_8H_{11}N$	N,N-Dimethylaniline	194.15		52.83
$C_8H_{11}N$	2,4-Dimethylaniline	214		61.3
$C_8H_{11}N$	2,5-Dimethylaniline	214		61.7
$C_8H_{11}N$	2,3,6-Trimethylpyridine	171.6	39.95	50.61
$C_8H_{11}N$	2,4,6-Trimethylpyridine	170.6	39.87	50.33
C_8H_{14}	1-Octyne	126.3	35.83	42.30
C_8H_{14}	2-Octyne	137.6	37.26	44.49
C_8H_{14}	3-Octyne	133.1	36.94	43.92
C_8H_{14}	4-Octyne	131.6	36.0	42.73
$C_8H_{14}O_3$	Butanoic anhydride	200	50.0	
$C_8H_{15}N$	Octanenitrile	205.25		56.80
C_8H_{16}	1-Octene	121.29	34.07	40.34
C_8H_{16}	cis-2,2-Dimethyl-3-hexene	105.5		36.86
C_8H_{16}	trans-2,2-Dimethyl-3-hexene	100.8		37.03
C_8H_{16}	3-Ethyl-2-methyl-1-pentene	109.5		37.27
C_8H_{16}	2,4,4-Trimethyl-1-pentene	101.4		35.59
C_8H_{16}	2,4,4-Trimethyl-2-pentene	104.9		37.23
C_8H_{16}	Ethylcyclohexane	131.9	34.04	40.56
C_8H_{16}	1,1-Dimethylcyclohexane	119.6	32.51	37.92

Mol. Form.	Name	t_b/°C	$\Delta_{vap}H(t_b)$ kJ/mol	$\Delta_{vap}H(25°C)$ kJ/mol
C_8H_{16}	cis-1,2-Dimethylcyclohexane	129.8	33.47	39.70
C_8H_{16}	trans-1,2-Dimethylcyclohexane	123.5	32.96	38.36
C_8H_{16}	cis-1,3-Dimethylcyclohexane	120.1	32.91	38.26
C_8H_{16}	trans-1,3-Dimethylcyclohexane	124.5	33.39	39.16
C_8H_{16}	cis-1,4-Dimethylcyclohexane	124.4	33.28	39.02
C_8H_{16}	trans-1,4-Dimethylcyclohexane	119.4	32.56	37.90
C_8H_{16}	Propylcyclopentane	131	34.70	41.08
C_8H_{16}	Isopropylcyclopentane	126.5	33.56	39.44
C_8H_{16}	1-Ethyl-1-methylcyclopentane	121.6	33.20	38.85
$C_8H_{16}O$	2,2,4-Trimethyl-3-pentanone	135.1	35.64	43.30
$C_8H_{16}O_2$	Octanoic acid	239	58.5	
$C_8H_{16}O_2$	2-Ethylhexanoic acid	228		75.60
$C_8H_{16}O_2$	Isobutyl isobutanoate	148.6	38.2	
$C_8H_{16}O_2$	Ethyl hexanoate	167		51.72
$C_8H_{16}O_2$	Methyl heptanoate	174		51.62
$C_8H_{17}Br$	1-Bromooctane	200		55.77
$C_8H_{17}Cl$	1-Chlorooctane	181.5		52.42
$C_8H_{17}F$	1-Fluorooctane	142.4	40.43	49.65
C_8H_{18}	Octane	125.67	34.41	41.49
C_8H_{18}	2-Methylheptane	117.66	33.26	39.67
C_8H_{18}	3-Methylheptane	118.9	33.66	39.83
C_8H_{18}	4-Methylheptane	117.72	33.35	39.69
C_8H_{18}	3-Ethylhexane	118.6	33.59	39.64
C_8H_{18}	2,2-Dimethylhexane	106.86	32.07	37.28
C_8H_{18}	2,3-Dimethylhexane	115.62	33.17	38.78
C_8H_{18}	2,4-Dimethylhexane	109.5	32.51	37.76
C_8H_{18}	2,5-Dimethylhexane	109.12	32.54	37.85
C_8H_{18}	3,3-Dimethylhexane	111.97	32.31	37.53
C_8H_{18}	3,4-Dimethylhexane	117.73	33.24	38.97
C_8H_{18}	3-Ethyl-2-methylpentane	115.66	32.93	38.52
C_8H_{18}	3-Ethyl-3-methylpentane	118.27	32.78	37.99
C_8H_{18}	2,2,3-Trimethylpentane	110	31.94	36.91
C_8H_{18}	2,2,4-Trimethylpentane	99.22	30.79	35.14
C_8H_{18}	2,3,3-Trimethylpentane	114.8	32.12	37.27
C_8H_{18}	2,3,4-Trimethylpentane	113.5	32.36	37.75
C_8H_{18}	2,2,3,3-Tetramethylbutane	106.45		42.90
$C_8H_{18}N_2$	Azobutane			49.31
$C_8H_{18}O$	1-Octanol	195.16		70.98
$C_8H_{18}O$	2-Octanol	180	44.4	
$C_8H_{18}O$	2-Ethyl-1-hexanol	184.6	54.2	
$C_8H_{18}O$	Dibutyl ether	140.28	36.49	44.97
$C_8H_{18}O$	Di-sec-butyl ether	121.1	34.06	40.84
$C_8H_{18}O$	Di-tert-butyl ether	107.23	32.15	37.61
$C_8H_{18}O_2$	1,2-Dipropoxyethane			50.62
$C_8H_{18}O_3$	Diethylene glycol diethyl ether	188		58.40
$C_8H_{18}S$	Dibutyl sulfide	185		52.96
$C_8H_{18}S$	Di-tert-butyl sulfide	149.1	33.26	43.76
$C_8H_{18}S$	Diisobutyl sulfide	171		48.71
$C_8H_{19}N$	Dibutylamine	159.6	38.44	49.45
$C_8H_{19}N$	2-Ethylhexylamine	169.2	40.0	
C_9H_7N	Quinoline	237.16	49.7	59.30
C_9H_7N	Isoquinoline	243.22	49.0	60.26
C_9H_{10}	Cyclopropylbenzene	173.6		50.22
C_9H_{10}	Indan	177.97	39.63	48.79
$C_9H_{10}O_2$	Benzyl acetate	213	49.4	
C_9H_{12}	Propylbenzene	159.24		46.22
C_9H_{12}	Isopropylbenzene	152.41		45.13
C_9H_{12}	1,2,3-Trimethylbenzene	176.12		49.05

Mol. Form.	Name	t_b/°C	$\Delta_{vap}H(t_b)$ kJ/mol	$\Delta_{vap}H(25°C)$ kJ/mol
C_9H_{12}	1,2,4-Trimethylbenzene	169.38		47.93
C_9H_{12}	1,3,5-Trimethylbenzene	164.74		47.50
$C_9H_{14}O_6$	Triacetin	259		85.74
C_9H_{18}	Butylcyclopentane	156.6	36.16	45.89
C_9H_{18}	Propylcyclohexane	156.7		45.08
C_9H_{18}	Isopropylcyclohexane	154.8		44.02
$C_9H_{18}O$	2-Nonanone	195.3		56.44
$C_9H_{18}O$	5-Nonanone	188.45		53.30
$C_9H_{18}O$	2,6-Dimethyl-4-heptanone	169.4		50.92
$C_9H_{18}O_2$	Methyl octanoate	192.9		56.41
C_9H_{20}	Nonane	150.82	37.18	46.55
C_9H_{20}	2,2,5-Trimethylhexane	124.09	33.65	40.16
C_9H_{20}	2,3,5-Trimethylhexane	131.4	34.43	41.41
C_9H_{20}	3,3-Diethylpentane	146.3	34.61	42.0
C_9H_{20}	2,2,4,4-Tetramethylpentane	122.29	32.51	38.49
$C_9H_{20}O$	1-Nonanol	213.37		76.86
$C_{10}H_7Br$	1-Bromonaphthalene	281	39.3	
$C_{10}H_7Cl$	1-Chloronaphthalene	259	52.1	
$C_{10}H_8$	Naphthalene	217.9	43.2	
$C_{10}H_9N$	2-Methylquinoline	246.5		66.1
$C_{10}H_9N$	4-Methylquinoline	262		67.6
$C_{10}H_9N$	6-Methylquinoline	258.6		67.7
$C_{10}H_9N$	8-Methylquinoline	247.5		65.7
$C_{10}H_{12}$	1,2,3,4-Tetrahydronaphthalene	207.6	43.9	
$C_{10}H_{14}$	Butylbenzene	183.31	38.87	51.36
$C_{10}H_{14}$	sec-Butylbenzene	173.3		47.98
$C_{10}H_{14}$	tert-Butylbenzene	169.1		47.71
$C_{10}H_{14}$	Isobutylbenzene	172.79		47.86
$C_{10}H_{14}$	1-Isopropyl-4-methylbenzene	177.1	38.2	
$C_{10}H_{16}O$	(+)-Camphor	207.4	59.5	
$C_{10}H_{18}$	cis-Decahydronaphthalene	195.8	41.0	
$C_{10}H_{18}$	trans-Decahydronaphthalene	187.3	40.2	
$C_{10}H_{19}N$	Decanenitrile	243		66.84
$C_{10}H_{20}$	1-Decene	170.5		50.43
$C_{10}H_{20}$	Butylcyclohexane	180.9		49.36
$C_{10}H_{20}O_2$	2-Ethylhexyl acetate	199	43.5	
$C_{10}H_{20}O_2$	Isopentyl isopentanoate	190.4	45.9	
$C_{10}H_{22}$	Decane	174.15	39.58	51.42
$C_{10}H_{22}$	2-Methylnonane	167.1	38.23	49.63
$C_{10}H_{22}$	3-Methylnonane	167.9	38.26	49.71
$C_{10}H_{22}$	5-Methylnonane	165.1	38.14	49.34
$C_{10}H_{22}$	2,4-Dimethyloctane	156	36.47	47.13
$C_{10}H_{22}O$	1-Decanol	231.1		81.50
$C_{10}H_{22}O$	Diisopentyl ether	172.5	35.1	
$C_{10}H_{22}S$	1-Decanethiol	240.6		65.48
$C_{11}H_{10}$	1-Methylnaphthalene	244.7	45.5	
$C_{11}H_{21}N$	Undecanenitrile	253		71.14
$C_{11}H_{22}$	Pentylcyclohexane	203.7		53.88
$C_{11}H_{24}$	Undecane	195.9	41.91	56.58
$C_{11}H_{24}$	2-Methyldecane	189.3	40.25	54.28
$C_{11}H_{24}$	4-Methyldecane	187	40.70	53.76
$C_{11}H_{24}$	2,4,7-Trimethyloctane	168.1	38.22	49.91
$C_{12}F_{27}N$	Tris(perfluorobutyl)amine	178	46.4	
$C_{12}H_{10}O$	Diphenyl ether	258.0	48.2	
$C_{12}H_{16}$	Cyclohexylbenzene	240.1		59.94
$C_{12}H_{22}$	Cyclohexylcyclohexane	238		57.98
$C_{12}H_{23}N$	Dodecanenitrile	277		76.12
$C_{12}H_{24}$	1-Dodecene	213.8		60.78

Mol. Form.	Name	$t_b/°C$	$\Delta_{vap}H(t_b)$ kJ/mol	$\Delta_{vap}H(25°C)$ kJ/mol
$C_{12}H_{26}$	2,2,4,6,6-Pentamethylheptane	177.8		48.97
$C_{12}H_{26}$	Dodecane	216.32	44.09	61.52
$C_{12}H_{26}O$	1-Dodecanol	259		91.96
$C_{12}H_{27}BO_3$	Tributyl borate	234	56.1	
$C_{12}H_{27}N$	Tributylamine	216.5	46.9	
$C_{13}H_{13}N$	N-Benzylaniline	306.5		79.6
$C_{13}H_{26}O_2$	Methyl dodecanoate	267		77.17
$C_{13}H_{28}$	Tridecane	235.47	46.20	66.68
$C_{14}H_{10}$	Phenanthrene	340		75.50
$C_{14}H_{12}O_2$	Benzyl benzoate	323.5	53.6	
$C_{14}H_{27}N$	Tetradecanenitrile			85.29
$C_{14}H_{30}$	Tetradecane	253.58	48.16	71.73
$C_{14}H_{30}O$	1-Tetradecanol	289		102.20
$C_{15}H_{32}$	Pentadecane	270.6	50.08	76.77
$C_{16}H_{22}O_4$	Dibutyl phthalate	340	79.2	
$C_{16}H_{32}$	1-Hexadecene	284.9		80.25
$C_{16}H_{34}$	Hexadecane	286.86	51.84	81.35
$C_{17}H_{36}$	Heptadecane	302.0	53.58	86.47
$C_{18}H_{34}O_2$	Oleic acid	360	67.4	
$C_{18}H_{38}$	Octadecane	316.3	55.23	91.44
$C_{19}H_{40}$	Nonadecane	329.9	56.93	96.4
$C_{20}H_{42}$	Eicosane	343	58.49	101.81

ENTHALPY OF FUSION

This table lists the molar enthalpy (heat) of fusion of about 600 inorganic and organic compounds. Values of $\Delta_{fus}H$ are given in kJ/mol at the normal melting point t_m, except for substances whose triple-point pressures are greater than normal atmospheric pressure (i.e., the sublimation pressure of the solid exceeds 1 atm at temperatures for which the liquid phase does not exist). In such cases the entries are designated by "t" and refer to the triple-point temperature. All values refer to the enthalpy change at equilibrium between the liquid phase and the most stable solid phase at the transition temperature T_m. Substances are listed by molecular formula in the Hill order, with compounds not containing carbon preceding those that do contain carbon.

REFERENCES

1. Chase, M.W., Davies, C.A., Downey, J.R., Frurip, D.J., McDonald, R.A., and Syverud, A.N., *JANAF Thermochemical Tables, Third Edition, J. Phys. Chem. Ref. Data*, Vol. 14, Suppl. 1, 1985.
2. Daubert, T.E., Danner, R.P., Sibul, H.M., and Stebbins, C.C., *Physical and Thermodynamic Properties of Pure Compounds: Data Compilation*, extant 1994 (core with 4 supplements), Taylor & Francis, Bristol, PA.
3. Janz, G.J., et al., *Physical Properties Data Compilations Relevant to Energy Storage. II. Molten Salts*, Natl. Stand. Ref. Data Sys.- Natl. Bur. Standards (U.S.), No. 61, Part 2, 1979.
4. Dinsdale, A.T., *CALPHAD*, 15, 317, 1991.
5. *Landolt-Börnstein, Numerical Values and Functions for Physics, Chemistry, Astronomy, Geophysics, and Technology, Sixth Edition*, Vol. 2, Part 4, Springer-Verlag, Heidelberg, 1961.
6. Gurvich, L.V., Veyts, I.V., and Alcock, C.B., *Thermodynamic Properties of Individual Substances, Fourth Edition*, Vol. 3, CRC Press, Boca Raton, FL, 1994.
7. *Physical Constants of Hydrocarbon and Non-Hydrocarbon Compounds, 2nd Edition*, ASTM Data Series DS 4B, ASTM, Philadelphia, 1988.
8. Trotman-Dickenson, A.F., Executive Editor, *Comprehensive Inorganic Chemistry*, Vol. 1-5, Pergamon Press, Oxford, 1973.

Molecular formula	Name	$t_m/°C$	$\Delta_{fus}H/kJ\ mol^{-1}$
AgBr	Silver(I) bromide	432	9.12
AgCl	Silver(I) chloride	455	13.2
AgI	Silver(I) iodide	558	9.41
AgNO$_3$	Silver(I) nitrate	212	11.5
Ag$_2$S	Silver(I) sulfide	825	14.1
Al	Aluminum	660.32	10.71
AlBr$_3$	Aluminum tribromide	97.5	11.25
AlCl$_3$	Aluminum trichloride	192.6	35.4
AlF$_3$	Aluminum trifluoride	2250 t	98
AlI$_3$	Aluminum triiodide	188.32	15.9
Al$_2$O$_3$	Aluminum oxide (corundum)	2054	111.4
Al$_2$S$_3$	Aluminum sulfide	1100	55
Am	Americium	1176	14.39
Ar	Argon	-189.3	1.12
As	Arsenic	817 t	24.44
AsBr$_3$	Arsenic(III) bromide	31.1	11.7
AsCl$_3$	Arsenic(III) chloride	-16	10.1
AsF$_3$	Arsenic(III) fluoride	-5.9	10.4
Au	Gold	1064.18	12.55
B	Boron	2075	50.2
BCl$_3$	Boron trichloride	-107	2.10
BF$_3$	Boron trifluoride	-126.8	4.20
BHO$_2$	Metaboric acid	236	14.3
BH$_3$O$_3$	Boric acid	170.9	22.3
BN	Boron nitride	2967	81
BNaO$_2$	Sodium metaborate	966	36.2
B$_2$O$_3$	Boron oxide	450	24.56
Ba	Barium	727	7.12
BaBr$_2$	Barium bromide	857	32.2
BaCl$_2$	Barium chloride	962	15.85
BaF$_2$	Barium fluoride	1368	17.8
BaH$_2$	Barium hydride	1200	25
BaH$_2$O$_2$	Barium hydroxide	408	16
BaI$_2$	Barium iodide	711	26.5
BaO	Barium oxide	1973	46

Molecular formula	Name	$t_m/°C$	$\Delta_{fus}H/kJ$ mol^{-1}
BaO_4S	Barium sulfate	1580	40
BaS	Barium sulfide	2230	63
Be	Beryllium	1287	7.895
$BeBr_2$	Beryllium bromide	508	18
$BeCl_2$	Beryllium chloride	415	8.66
BeF_2	Beryllium fluoride	552	4.77
BeI_2	Beryllium iodide	470	18
BeO	Beryllium oxide	2578	86
BeO_4S	Beryllium sulfate	1127	6
Bi	Bismuth	271.40	11.30
$BiCl_3$	Bismuth trichloride	230	10.9
BrF_5	Bromine pentafluoride	-60.5	5.67
BrH	Hydrogen bromide	-86.81	2.41
BrIn	Indium(I) bromide	290	15
BrK	Potassium bromide	734	25.5
BrLi	Lithium bromide	552	17.6
BrNa	Sodium bromide	747	26.11
$BrNaO_3$	Sodium bromate	381	28.11
BrRb	Rubidium bromide	682	15.5
BrTl	Thallium(I) bromide	460	16.4
Br_2	Bromine	-7.2	10.57
Br_2Ca	Calcium bromide	742	29.1
Br_2Cd	Cadmium bromide	568	20.9
Br_2Fe	Iron(II) bromide	691	50.2
Br_2Hg	Mercury(II) bromide	236	17.9
Br_2Mg	Magnesium bromide	711	39.3
Br_2Pb	Lead(II) bromide	371	16.44
Br_2Sr	Strontium bromide	657	10.1
Br_2Zn	Zinc bromide	394	16.7
Br_3Ga	Gallium(III) bromide	121.5	12.1
Br_3In	Indium(III) bromide	420	26
Br_3Pu	Plutonium(III) bromide	681	55.2
Br_3U	Uranium(III) bromide	727	43.9
Br_4Sn	Tin(IV) bromide	31	12
Br_4Th	Thorium bromide	679	66.9
Br_4Ti	Titanium(IV) bromide	39	12.9
Br_4U	Uranium(IV) bromide	519	55.2
Br_5Ta	Tantalum(V) bromide	265	45.6
Ca	Calcium	842	8.54
$CaCl_2$	Calcium chloride	775	28.05
CaF_2	Calcium fluoride	1418	30
CaH_2	Calcium hydride	1000	6.7
CaI_2	Calcium iodide	783	41.8
CaO	Calcium oxide	2899	80
CaO_4S	Calcium sulfate	1460	28
CaS	Calcium sulfide	2525	70
Cd	Cadmium	321.07	6.19
$CdCl_2$	Cadmium chloride	564	48.58
CdF_2	Cadmium fluoride	1110	22.6
CdI_2	Cadmium iodide	387	15.3
Ce	Cerium	798	5.46
$CeCl_3$	Cerium(III) chloride	817	54.4
ClCs	Cesium chloride	645	15.9
ClCu	Copper(I) chloride	430	10.2
ClH	Hydrogen chloride	-114.1	2.00
ClI	Iodine chloride	27.38	11.6
ClIn	Indium(I) chloride	211	21.3
ClK	Potassium chloride	771	26.53
ClLi	Lithium chloride	610	19.9

Molecular formula	Name	$t_m/°C$	$\Delta_{fus}H/kJ$ mol^{-1}
$ClLiO_4$	Lithium perchlorate	236	29
$ClNa$	Sodium chloride	800.7	28.16
$ClNaO_3$	Sodium chlorate	248	22.1
$ClRb$	Rubidium chloride	715	18.4
$ClTl$	Thallium(I) chloride	430	15.56
Cl_2	Chlorine	-101.5	6.40
Cl_2Co	Cobalt(II) chloride	740	45
Cl_2Cr	Chromium(II) chloride	814	32.2
Cl_2Cu	Copper(II) chloride	630	20.4
Cl_2Fe	Iron(II) chloride	677	43.01
Cl_2Hg	Mercury(II) chloride	276	19.41
Cl_2Mg	Magnesium chloride	714	43.1
Cl_2Mn	Manganese(II) chloride	650	30.7
Cl_2Ni	Nickel(II) chloride	1009	71.2
Cl_2Pb	Lead(II) chloride	501	21.9
Cl_2Sn	Tin(II) chloride	247	12.8
Cl_2Sr	Strontium chloride	874	17.5
Cl_3Fe	Iron(III) chloride	304	43.1
Cl_3Ga	Gallium(III) chloride	77.9	11.13
Cl_3In	Indium(III) chloride	583	27
Cl_3La	Lanthanum chloride	859	43.1
Cl_3OP	Phosphorus(V) oxychloride	1	13.1
Cl_3P	Phosphorus(III) chloride	-112	7.10
Cl_3Sb	Antimony(III) chloride	73.4	12.7
Cl_4OW	Tungsten(VI) oxytetrachloride	211	45
Cl_4Si	Tetrachlorosilane	-68.85	7.60
Cl_4Sn	Tin(IV) chloride	-33	9.20
Cl_4Th	Thorium(IV) chloride	770	40.2
Cl_4Ti	Titanium(IV) chloride	-25	9.97
Cl_4U	Uranium(IV) chloride	590	45
Cl_4V	Vanadium(IV) chloride	-25.7	2.30
Cl_4Zr	Zirconium(IV) chloride	437 t	50
Cl_5Mo	Molybdenum(V) chloride	194	19
Cl_5Nb	Niobium(V) chloride	204.7	38.3
Cl_5Ta	Tantalum(V) chloride	216	41.6
Cl_6W	Tungsten(VI) chloride	275	6.60
Co	Cobalt	1495	16.2
CoF_2	Cobalt(II) fluoride	1127	59
Cr	Chromium	1907	21.0
Cr_2O_3	Chromium(III) oxide	2330	130
Cs	Cesium	28.44	2.09
CsF	Cesium fluoride	703	21.7
Cs_2O_4S	Cesium sulfate	1005	35.7
Cu	Copper	1084.62	13.26
CuF_2	Copper(II) fluoride	836	55
CuO	Copper(II) oxide	1446	11.8
Dy	Dysprosium	1412	11.06
Er	Erbium	1529	19.9
Eu	Europium	822	9.21
FH	Hydrogen fluoride	-83.36	4.58
FK	Potassium fluoride	858	27.2
FLi	Lithium fluoride	848.2	27.09
FNa	Sodium fluoride	996	33.35
FRb	Rubidium fluoride	833	17.3
FTl	Thallium(I) fluoride	326	13.87
F_2	Fluorine	-219.6	0.51
F_2Fe	Iron(II) fluoride	1100	52
F_2HK	Potassium hydrogen fluoride	238.9	6.62
F_2Mg	Magnesium fluoride	1263	58.5

Molecular formula	Name	$t_m/°C$	$\Delta_{fus}H/kJ$ mol^{-1}
F_2Pb	Lead(II) fluoride	830	14.7
F_2Sr	Strontium fluoride	1477	28.5
F_3In	Indium(III) fluoride	1170	64
F_3Pu	Plutonium(III) fluoride	1396	59.8
F_4Pu	Plutonium(IV) fluoride	1027	65.3
F_4Th	Thorium(IV) fluoride	1110	44.0
F_4U	Uranium(IV) fluoride	1036	42.7
F_4Zr	Zirconium(IV) fluoride	932 t	64.2
F_5Nb	Niobium(V) fluoride	80	12.2
F_5V	Vanadium(V) fluoride	19.5	49.96
F_6Ir	Iridium(VI) fluoride	44	8.40
F_6Mo	Molybdenum(VI) fluoride	17.5	4.33
F_6Pu	Plutonium(VI) fluoride	52	17.6
F_6S	Sulfur hexafluoride	-50.7 t	5.02
F_6U	Uranium(VI) fluoride	64.0	19.19
F_6W	Tungsten(VI) fluoride	2.3	4.10
Fe	Iron	1538	13.81
FeI_2	Iron(II) iodide	587	45
FeO	Iron(II) oxide	1377	24
FeS	Iron(II) sulfide	1188	31.5
Fe_3O_4	Iron(II,III) oxide	1597	138
Ga	Gallium	29.76	5.59
GaI_3	Gallium(III) iodide	212	12.9
$GaSb$	Gallium antimonide	712	25.1
Ga_2O_3	Gallium(III) oxide	1807	100
Gd	Gadolinium	1313	10.0
Ge	Germanium	938.25	36.94
HI	Hydrogen iodide	-50.77	2.87
HKO	Potassium hydroxide	406	7.9
HLi	Lithium hydride	688.7	22.59
$HLiO$	Lithium hydroxide	471.2	20.88
HNO_3	Nitric acid	-41.6	10.5
$HNaO$	Sodium hydroxide	323	6.60
H_2	Hydrogen	-259.3	0.12
H_2Mg	Magnesium hydride	327	14
H_2O	Water	0.00	6.01
H_2O_2	Hydrogen peroxide	-0.43	12.50
H_2O_2Sr	Strontium hydroxide	535	23
H_2O_4S	Sulfuric acid	10.31	10.71
H_2S	Hydrogen sulfide	-85.5	23.8
H_2Sr	Strontium hydride	1050	23
H_3N	Ammonia	-77.74	5.66
H_3O_2P	Hypophosphorous acid	26.5	9.7
H_3O_3P	Phosphorous acid	74.4	12.8
H_3O_4P	Phosphoric acid	42.4	13.4
H_4IN	Ammonium iodide	551	21
H_4N_2	Hydrazine	1.4	12.6
$H_4N_2O_3$	Ammonium nitrate	210	6.40
Hf	Hafnium	2233	27.2
Hg	Mercury	-38.83	2.29
HgI_2	Mercury(II) iodide	259	18.9
Hg_2I_2	Mercury(I) iodide	290	27
Ho	Holmium	1474	17.0 (est.)
IIn	Indium(I) iodide	364.4	17.26
IK	Potassium iodide	681	24
ILi	Lithium iodide	469	14.6
INa	Sodium iodide	660	23.6
IRb	Rubidium iodide	642	12.5
ITl	Thallium(I) iodide	441.8	14.73

Molecular formula	Name	$t_m/°C$	$\Delta_{fus}H/kJ$ mol^{-1}
I_2	Iodine (rhombic)	113.7	15.52
I_2Mg	Magnesium iodide	634	26
I_2Pb	Lead(II) iodide	410	23.4
I_2Sr	Strontium iodide	538	19.7
I_3In	Indium(III) iodide	207	18.48
I_4Si	Tetraiodosilane	120.5	19.7
I_4Th	Thorium iodide	570	61.4
I_4Ti	Titanium(IV) iodide	150	19.8
I_4U	Uranium(IV) iodide	506	70.7
In	Indium	156.60	3.28
InSb	Indium antimonide	525	25.5
In_2O_3	Indium(III) oxide	1913	105
Ir	Iridium	2446	41.12
K	Potassium	63.38	2.33
KNO_3	Potassium nitrate	337	10.1
K_2O_4S	Potassium sulfate	1069	36.4
K_2S	Potassium sulfide	948	16.15
Kr	Krypton	-157.3	1.37
La	Lanthanum	918	6.20
Li	Lithium	180.5	3.00
$LiNO_3$	Lithium nitrate	253	24.9
Li_2O_3Si	Lithium metasilicate	1201	28
Li_2O_4S	Lithium sulfate	859	7.50
Lu	Lutetium	1663	22 (est.)
Mg	Magnesium	650	8.48
MgO	Magnesium oxide	2826	77
MgO_4S	Magnesium sulfate	1127	14.6
MgS	Magnesium sulfide	2227	63
Mg_2O_4Si	Magnesium orthosilicate	1898	71
Mn	Manganese	1246	12.91
MnO	Manganese(II) oxide	1840	54.4
Mo	Molybdenum	2623	37.48
MoO_3	Molybdenum(VI) oxide	801	48
$NNaO_3$	Sodium nitrate	307	15
NO	Nitric oxide	-163.6	2.30
NO_3Rb	Rubidium nitrate	305	5.60
NO_3Tl	Thallium(I) nitrate	206	9.6
N_2	Nitrogen	-210.0	0.71
N_2O	Nitrous oxide	-90.8	6.54
N_2O_4	Nitrogen tetroxide	-9.3	14.65
Na	Sodium	97.80	2.60
Na_2O	Sodium oxide	1132	48
Na_2O_3Si	Sodium metasilicate	1089	52
Na_2O_4S	Sodium sulfate	884	23.6
Na_2S	Sodium sulfide	1172	19
Nb	Niobium	2477	30
NbO	Niobium(II) oxide	1937	85
NbO_2	Niobium(IV) oxide	1902	92
Nb_2O_5	Niobium(V) oxide	1512	104.3
Nd	Neodymium	1021	7.14
Ne	Neon	-248.5	0.34
Ni	Nickel	1455	17.48
NiS	Nickel(II) sulfide	976	30.1
Np	Neptunium	644	3.20
OSr	Strontium oxide	2532	81
OTl_2	Thallium(I) oxide	579	30.3
OV	Vanadium(II) oxide	1790	63
OZn	Zinc oxide	1975	52.3
O_2	Oxygen	-218.7	0.44

Molecular formula	Name	$t_m/°C$	$\Delta_{fus}H/kJ$ mol^{-1}
O_2Si	Silicon dioxide (cristobalite)	1713	8.51
O_2Zr	Zirconium(IV) oxide	2710	87
O_3S	Sulfur trioxide	16.8	8.60
O_3Tl_2	Thallium(III) oxide	834	53
O_3W	Tungsten(VI) oxide	1472	73
O_3Y_2	Yttrium oxide	2439	105
O_4Os	Osmium(VIII) oxide	41	9.8
O_4SSr	Strontium sulfate	1607	36
O_4STl_2	Thallium(I) sulfate	632	23
O_5P_2	Phosphorus(V) oxide	562	27.2
O_5Ta_2	Tantalum(V) oxide	1785	120
O_5V_2	Vanadium(V) oxide	670	64.5
O_7Re_2	Rhenium(VII) oxide	297	64.2
Os	Osmium	3033	57.85
P	Phosphorus (white)	44.15	0.66
Pa	Protactinium	1572	12.34
Pb	Lead	327.46	4.77
PbS	Lead(II) sulfide	1118	19
Pd	Palladium	1554.9	16.74
Pr	Praseodymium	931	6.89
Pt	Platinum	1768.4	22.17
Pu	Plutonium	640	2.82
Rb	Rubidium	39.31	2.19
Re	Rhenium	3186	60.43
Rh	Rhodium	1964	26.59
Ru	Ruthenium	2334	38.59
S	Sulfur (monoclinic)	115.21	1.72
SSr	Strontium sulfide	2227	63
STl_2	Thallium(I) sulfide	448	12
Sb	Antimony	630.63	19.87
Sc	Scandium	1541	14.1
Se	Selenium	221	6.69
Si	Silicon	1414	50.21
Sm	Samarium	1074	8.62
Sn	Tin (white)	231.93	7.03
Sr	Strontium	777	7.43
Ta	Tantalum	3017	36.57
Tb	Terbium	1356	10.79
Tc	Technetium	2157	33.29
Te	Tellurium	449.51	17.49
Th	Thorium	1750	13.81
Ti	Titanium	1668	14.15
Tl	Thallium	304	4.14
Tm	Thulium	1545	16.84
U	Uranium	1135	9.14
V	Vanadium	1910	21.5
W	Tungsten	3422	52.31
Xe	Xenon	-111.7	1.81
Y	Yttrium	1522	11.4
Yb	Ytterbium	819	7.66
Zn	Zinc	419.53	7.32
Zr	Zirconium	1855	21.00
C	Carbon (graphite)	4492 t	117
$CBaO_3$	Barium carbonate	1555	40
$CBrCl_3$	Bromotrichloromethane	-5.7	2.54
$CCaO_3$	Calcium carbonate	1330	36
CCl_2O	Carbonyl chloride	-127.9	5.74
CCl_4	Tetrachloromethane	-23	3.28
$CHCl_3$	Trichloromethane	-63.6	8.80

Molecular formula	Name	$t_m/°C$	$\Delta_{fus}H/kJ$ mol^{-1}
CHN	Hydrogen cyanide	-13.4	8.41
CH_2Cl_2	Dichloromethane	-95.14	6
CH_2N_2	Cyanamide	44	8.76
CH_2O_2	Formic acid	8.3	12.72
CH_3Br	Bromomethane	-93.7	5.98
CH_4	Methane	-182.4	0.94
CH_4O	Methanol	-97.68	3.18
CH_4S	Methanethiol	-123	5.91
CH_5N	Methylamine	-93.44	6.13
CK_2O_3	Potassium carbonate	898	27.6
CLi_2O_3	Lithium carbonate	723	41
$CMgO_3$	Magnesium carbonate	990	59
CNa_2O_3	Sodium carbonate	858.1	29.7
CO	Carbon monoxide	-205	0.83
CO_2	Carbon dioxide	-56.57 t	9.02
CO_3Sr	Strontium carbonate	1494	40
CO_3Tl_2	Thallium(I) carbonate	272	18.4
CS_2	Carbon disulfide	-111.5	4.40
$C_2Br_2F_4$	1,2-Dibromotetrafluoroethane	-110.4	7.04
$C_2Cl_2F_4$	1,2-Dichlorotetrafluoroethane	-94	1.51
$C_2Cl_3F_3$	1,1,2-Trichlorotrifluoroethane	-35	2.47
$C_2Cl_4F_2$	1,1,2,2-Tetrachloro-1,2-difluoroethane	26	3.7
$C_2HCl_3O_2$	Trichloroacetic acid	57.5	5.88
$C_2H_3ClO_2$	Chloroacetic acid	63	12.28
$C_2H_3Cl_3$	1,1,1-Trichloroethane	-30.4	2.73
$C_2H_3Cl_3$	1,1,2-Trichloroethane	-36.6	11.54
$C_2H_3F_3$	1,1,1-Trifluoroethane	-111.3	6.19
C_2H_4	Ethylene	-169	3.35
$C_2H_4Br_2$	1,2-Dibromoethane	9.93	10.84
$C_2H_4Cl_2$	1,2-Dichloroethane	-35.5	8.83
$C_2H_4O_2$	Acetic acid	16.6	11.54
C_2H_5Cl	Chloroethane	-138.7	4.45
C_2H_6	Ethane	-182.8	2.86
C_2H_6O	Ethanol	-114.1	5.02
C_2H_6O	Dimethyl ether	-141.5	4.94
$C_2H_6O_2$	Ethylene glycol	-13	11.23
C_2H_6S	Ethanethiol	-147.8	4.98
C_2H_6S	Dimethyl sulfide	-98.3	7.99
$C_2H_6S_2$	Dimethyl disulfide	-85	9.19
C_2H_7N	Dimethylamine	-92.2	5.94
$C_2H_8N_2$	1,2-Ethanediamine	11.1	22.58
C_3H_3N	2-Propenenitrile	-83.5	6.23
$C_3H_4O_2$	Propenoic acid	12.3	11.16
$C_3H_5N_3O_9$	Trinitroglycerol	13.5	21.87
C_3H_6	Propene	-185.2	3.00
C_3H_6	Cyclopropane	-127.4	5.44
$C_3H_6Br_2$	1,3-Dibromopropane	-34.2	13.6
$C_3H_6Cl_2$	(±)-1,2-Dichloropropane	-100.5	6.4
C_3H_6O	Acetone	-94.8	5.69
$C_3H_6O_3$	Trioxane	60.2	15.11
C_3H_7Cl	2-Chloropropane	-117.2	7.39
C_3H_7N	Cyclopropylamine	-35.4	13.18
C_3H_8	Propane	-187.6	3.53
C_3H_8O	1-Propanol	-126.1	5.2
C_3H_8O	2-Propanol	-89.5	5.37
$C_3H_8O_3$	Glycerol	18.2	8.48
C_3H_9N	Propylamine	-83	10.97
C_3H_9N	Isopropylamine	-95.14	7.33
C_3H_9N	Trimethylamine	-117.1	6.55

Molecular formula	Name	$t_m/°C$	$\Delta_{fus}H/\text{kJ mol}^{-1}$
$C_4H_4N_2$	Succinonitrile	54.5	3.92
$C_4H_4O_3$	Succinic anhydride	119	20.41
C_4H_4S	Thiophene	-39.4	4.97
C_4H_5N	Pyrrole	-23.42	7.91
C_4H_6	1,3-Butadiene	-108.9	7.98
C_4H_6	2-Butyne	-32.36	9.23
$C_4H_6O_2$	*cis*-Crotonic acid	15	12.57
$C_4H_6O_2$	*trans*-Crotonic acid	72	9.12
$C_4H_6O_2$	γ-Butyrolactone	-43.37	9.57
$C_4H_6O_4$	Succinic acid	188	32.95
$C_4H_6O_4$	Dimethyl oxalate	54.35	21.07
C_4H_8	*cis*-2-Butene	-138.9	7.58
C_4H_8	Isobutene	-140.4	5.93
C_4H_8O	2-Butanone	-86.67	8.44
C_4H_8O	Tetrahydrofuran	-108.3	8.54
$C_4H_8O_2$	Butanoic acid	-5.7	11.08
$C_4H_8O_2$	Ethyl acetate	-83.6	10.48
$C_4H_8O_2$	1,4-Dioxane	11.8	12.85
C_4H_9Br	(±)-2-Bromobutane	-111.9	6.89
C_4H_9N	Pyrrolidine	-57.84	8.58
C_4H_{10}	Butane	-138.2	4.66
C_4H_{10}	Isobutane	-138.3	4.54
$C_4H_{10}O$	1-Butanol	-89.8	9.28
$C_4H_{10}O$	2-Methyl-2-propanol	25.4	6.79
$C_4H_{10}O$	Diethyl ether	-116.3	7.27
$C_4H_{11}N$	*tert*-Butylamine	-66.95	0.88
$C_4H_{12}Si$	Tetramethylsilane	-99.04	6.88
C_5H_5N	Pyridine	-41.66	8.28
C_5H_8	1,4-Pentadiene	-148.8	6.14
C_5H_8	2-Methyl-1,3-butadiene	-145.9	4.79
C_5H_8	Cyclopentene	-135.1	3.36
C_5H_8	Spiropentane	-134.6	5.76
$C_5H_8O_3$	4-Oxopentanoic acid	33	9.22
$C_5H_8O_4$	Glutaric acid	97.8	20.9
C_5H_{10}	1-Pentene	-165.2	5.81
C_5H_{10}	*cis*-2-Pentene	-151.4	7.12
C_5H_{10}	*trans*-2-Pentene	-140.2	8.36
C_5H_{10}	Cyclopentane	-93.8	0.61
$C_5H_{10}O$	2-Pentanone	-76.9	10.63
$C_5H_{10}O$	3-Pentanone	-39	11.59
$C_5H_{11}N$	Cyclopentylamine	-82.7	8.31
$C_5H_{11}N$	Piperidine	-11.03	14.85
C_5H_{12}	Pentane	-129.7	8.42
C_5H_{12}	Isopentane	-159.9	5.15
C_5H_{12}	Neopentane	-16.6	3.10
$C_5H_{12}O$	1-Pentanol	-78.9	9.83
C_6Cl_6	Hexachlorobenzene	231.8	23.85
C_6F_6	Hexafluorobenzene	5.35	11.59
C_6HF_5	Pentafluorobenzene	-47.3	10.85
C_6HF_5O	Pentafluorophenol	32.8	12.85
$C_6H_3Cl_3$	1,3,5-Trichlorobenzene	63.5	18.2
$C_6H_4ClNO_2$	1-Chloro-3-nitrobenzene	44.4	19.37
$C_6H_4ClNO_2$	1-Chloro-4-nitrobenzene	83.5	20.77
$C_6H_4Cl_2$	*o*-Dichlorobenzene	-16.7	12.93
$C_6H_4Cl_2$	*m*-Dichlorobenzene	-24.8	12.64
$C_6H_4Cl_2$	*p*-Dichlorobenzene	52.7	17.15
$C_6H_4F_2$	*m*-Difluorobenzene	-69.09	8.58
$C_6H_4O_2$	*p*-Benzoquinone	115.7	18.53
C_6H_5Br	Bromobenzene	-30.6	10.62

Molecular formula	Name	$t_m/°C$	$\Delta_{fus}H/kJ\ mol^{-1}$
C_6H_5Cl	Chlorobenzene	-45.2	9.61
C_6H_5ClO	o-Chlorophenol	9.8	12.52
C_6H_5ClO	m-Chlorophenol	32.6	14.91
C_6H_5ClO	p-Chlorophenol	42.7	14.07
C_6H_5F	Fluorobenzene	-42.21	11.31
C_6H_5I	Iodobenzene	-31.3	9.76
$C_6H_5NO_2$	Nitrobenzene	5.7	11.59
$C_6H_5NO_3$	2-Nitrophenol	44.8	17.44
$C_6H_5NO_3$	3-Nitrophenol	96.8	19.2
$C_6H_5NO_3$	4-Nitrophenol	113.8	18.25
C_6H_6	Benzene	5.53	9.95
$C_6H_6N_2O_2$	o-Nitroaniline	71.2	16.11
$C_6H_6N_2O_2$	m-Nitroaniline	114	23.68
$C_6H_6N_2O_2$	p-Nitroaniline	147	21.1
C_6H_6O	Phenol	40.9	11.29
$C_6H_6O_2$	p-Hydroquinone	172.3	27.11
C_6H_6S	Benzenethiol	-14.9	11.48
C_6H_7N	Aniline	-6.02	10.56
C_6H_7N	2-Methylpyridine	-66.68	9.72
C_6H_7N	3-Methylpyridine	-18.14	14.18
$C_6H_8N_2$	Phenylhydrazine	19.6	16.43
C_6H_{10}	Cyclohexene	-103.5	3.29
$C_6H_{10}O_4$	Adipic acid	153.2	34.85
C_6H_{12}	1-Hexene	-139.7	9.35
C_6H_{12}	2,3-Dimethyl-2-butene	-74.6	5.46
C_6H_{12}	Cyclohexane	6.6	2.63
C_6H_{12}	Methylcyclopentane	-142.5	6.93
$C_6H_{12}O$	Cyclohexanol	25.46	1.76
$C_6H_{12}O$	2-Hexanone	-55.5	14.9
$C_6H_{12}O$	3-Hexanone	-55.5	13.49
C_6H_{14}	Hexane	-95.3	13.08
C_6H_{14}	2-Methylpentane	-153.7	6.27
C_6H_{14}	2,2-Dimethylbutane	-99	0.58
C_6H_{14}	2,3-Dimethylbutane	-128.8	0.80
$C_6H_{14}O$	Dipropyl ether	-126.1	8.83
$C_6H_{14}O$	Diisopropyl ether	-86.8	11.03
C_7F_8	Perfluorotoluene	-65.6	11.58
$C_7H_3F_5$	2,3,4,5,6-Pentafluorotoluene	-29.8	12.99
$C_7H_5ClO_2$	o-Chlorobenzoic acid	140.2	25.73
$C_7H_6O_2$	Benzoic acid	122.4	18.06
C_7H_7NO	Benzamide	129.1	18.49
$C_7H_7NO_2$	p-Nitrotoluene	51.6	16.81
C_7H_8	Toluene	-94.99	6.85
C_7H_8O	o-Cresol	29.8	13.94
C_7H_8O	m-Cresol	11.8	9.41
C_7H_8O	p-Cresol	35.5	11.89
C_7H_8O	Benzyl alcohol	-15.2	8.97
C_7H_9N	p-Methylaniline	43.75	18.22
C_7H_{14}	1-Heptene	-119.7	12.66
C_7H_{14}	Cycloheptane	-8.03	1.88
C_7H_{14}	Methylcyclohexane	-126.6	6.75
C_7H_{16}	Heptane	-90.6	14.16
C_7H_{16}	2-Methylhexane	-118.2	8.87
C_7H_{16}	3-Ethylpentane	-118.6	9.55
C_7H_{16}	2,2-Dimethylpentane	-123.8	5.86
C_7H_{16}	2,4-Dimethylpentane	-119.9	6.69
C_7H_{16}	3,3-Dimethylpentane	-134.9	7.07
C_7H_{16}	2,2,3-Trimethylbutane	-25	2.20
$C_8H_8O_2$	o-Toluic acid	103.7	20.17

Molecular formula	Name	$t_m/°C$	$\Delta_{fus}H/\text{kJ mol}^{-1}$
$C_8H_8O_2$	m-Toluic acid	108.75	15.72
$C_8H_8O_2$	p-Toluic acid	179.6	22.73
$C_8H_8O_2$	Benzeneacetic acid	76.7	14.49
C_8H_{10}	o-Xylene	-25.2	13.61
C_8H_{10}	m-Xylene	-47.8	11.55
C_8H_{10}	p-Xylene	13.2	16.81
$C_8H_{10}O$	2,3-Xylenol	72.8	21.02
$C_8H_{10}O$	2,5-Xylenol	74.8	23.38
$C_8H_{10}O$	2,6-Xylenol	45.7	18.9
$C_8H_{10}O$	3,4-Xylenol	60.8	18.13
$C_8H_{10}O$	3,5-Xylenol	63.6	18
C_8H_{16}	Cyclooctane	14.83	2.41
C_8H_{16}	Ethylcyclohexane	-111.3	8.33
C_8H_{16}	1,1-Dimethylcyclohexane	-33.3	2.06
C_8H_{16}	cis-1,2-Dimethylcyclohexane	-49.9	1.64
C_8H_{16}	trans-1,2-Dimethylcyclohexane	-90	10.49
C_8H_{16}	cis-1,3-Dimethylcyclohexane	-75.6	10.82
C_8H_{16}	trans-1,3-Dimethylcyclohexane	-90.1	9.86
C_8H_{16}	cis-1,4-Dimethylcyclohexane	-87.4	9.31
C_8H_{16}	trans-1,4-Dimethylcyclohexane	-36.9	12.33
$C_8H_{16}O_2$	Octanoic acid	16.3	21.36
C_8H_{18}	Octane	-56.8	20.65
C_8H_{18}	3-Methylheptane	-120	11.38
C_8H_{18}	4-Methylheptane	-121	10.84
C_8H_{18}	2,2,4-Trimethylpentane	-107.3	9.04
C_9H_7N	Quinoline	-14.78	10.66
C_9H_7N	Isoquinoline	26.47	13.54
C_9H_{12}	Propylbenzene	-99.56	9.27
C_9H_{12}	1,2,3-Trimethylbenzene	-25.4	8.37
C_9H_{12}	1,2,4-Trimethylbenzene	-43.8	3.76
C_9H_{12}	1,3,5-Trimethylbenzene	-44.7	9.51
C_9H_{18}	Propylcyclohexane	-94.9	10.37
$C_9H_{18}O$	5-Nonanone	-5.9	24.93
$C_9H_{18}O_2$	Nonanoic acid	12.35	20.28
C_9H_{20}	Nonane	-53.5	15.47
C_9H_{20}	3,3-Diethylpentane	-33.1	10.09
C_9H_{20}	2,2,3,3-Tetramethylpentane	-9.8	2.33
C_9H_{20}	2,2,4,4-Tetramethylpentane	-66.54	9.75
$C_{10}H_7Br$	1-Bromonaphthalene	-0.9	15.16
$C_{10}H_7Cl$	1-Chloronaphthalene	-2.5	12.9
$C_{10}H_8$	Naphthalene	80.2	19.06
$C_{10}H_8O$	1-Naphthol	95	23.33
$C_{10}H_8O$	2-Naphthol	123	17.51
$C_{10}H_{14}$	Butylbenzene	-87.9	11.22
$C_{10}H_{14}$	p-Cymene	-68.9	9.6
$C_{10}H_{14}$	1,2,4,5-Tetramethylbenzene	79.3	21
$C_{10}H_{14}O$	Thymol	51.5	17.27
$C_{10}H_{18}O_4$	Sebacic acid	130.8	40.8
$C_{10}H_{20}$	Butylcyclohexane	-74.73	14.16
$C_{10}H_{20}O_2$	Decanoic acid	31.99	28.02
$C_{10}H_{22}$	Decane	-29.7	28.78
$C_{11}H_{10}$	2-Methylnaphthalene	34.4	11.97
$C_{11}H_{24}$	Undecane	-25.6	22.32
$C_{12}H_9N$	Carbazole	246.2	26.9
$C_{12}H_{10}$	Acenaphthene	93.4	21.54
$C_{12}H_{10}$	Biphenyl	69	18.6
$C_{12}H_{10}N_2$	Azobenzene	68.5	22.04
$C_{12}H_{10}N_2O$	trans-Azoxybenzene	36	17.93
$C_{12}H_{10}O$	Diphenyl ether	26.87	17.22

Molecular formula	Name	$t_m/°C$	$\Delta_{fus}H/kJ$ mol^{-1}
$C_{12}H_{11}N$	Diphenylamine	52.98	17.86
$C_{12}H_{16}$	Cyclohexylbenzene	7.3	15.3
$C_{12}H_{24}O_2$	Dodecanoic acid	43.22	36.64
$C_{12}H_{26}$	Dodecane	-9.6	36.58
$C_{13}H_{10}$	9H-Fluorene	114.8	19.58
$C_{13}H_{10}O$	Benzophenone	47.88	18.19
$C_{14}H_{10}$	Anthracene	215.0	28.83
$C_{14}H_{10}$	Phenanthrene	99.24	16.46
$C_{14}H_{10}O_2$	Benzil	94.86	23.54
$C_{14}H_{12}$	trans-Stilbene	123	27.4
$C_{14}H_{12}O_2$	α-Phenylbenzeneacetic acid	147	31.27
$C_{14}H_{28}O_2$	Tetradecanoic acid	53.96	45.38
$C_{16}H_{10}$	Fluoranthene	107.8	18.87
$C_{16}H_{10}$	Pyrene	151.2	17.11
$C_{16}H_{32}O_2$	Hexadecanoic acid	61.82	42.04
$C_{16}H_{34}O$	1-Hexadecanol	49.3	34.29
$C_{18}H_{12}$	Chrysene	258.2	26.15
$C_{18}H_{14}$	p-Terphenyl	210.1	35.5
$C_{18}H_{36}O_2$	Stearic acid	68.82	56.59
$C_{18}H_{38}$	Octadecane	28.2	61.39
$C_{19}H_{40}$	Nonadecane	32.1	45.82
$C_{20}H_{12}$	Perylene	274	31.75
$C_{20}H_{42}$	Eicosane	36.8	69.88
$C_{24}H_{12}$	Coronene	437.3	19.2

PRESSURE AND TEMPERATURE DEPENDENCE OF LIQUID DENSITY

This table gives data on the variation of the density of some common liquids with pressure and temperature. The pressure dependence is described to first order by the isothermal compressibility coefficient κ defined as

$$\kappa = -(1/V)\,(\partial V/\partial P)_T$$

where V is the volume, and the temperature dependence by the cubic expansion coefficient α,

$$\alpha = (1/V)\,(\partial V/\partial T)_P$$

Substances are listed by molecular formula in the Hill order. More precise data on the variation of density with temperature over a wide temperature range can be found in Reference 1.

REFERENCES

1. Lide, D. R., and Kehiaian, H. V., *CRC Handbook of Thermophysical and Thermochemical Data*, CRC Press, Boca Raton, FL, 1994.
2. Le Neindre, B., *Effets des Hautes et Très Hautes Pressions*, in *Techniques de l'Ingénieur*, Paris, 1991.
3. *Landolt-Börnstein, Numerical Data and Functional Relationships in Science and Technology*, New Series, IV/4, *High-pressure Properties of Matter*, Springer-Verlag, Heidelberg, 1980.
4. Riddick, J.A., Bunger, W.B., and Sakano, T.K., *Organic Solvents, Fourth Edition*, John Wiley & Sons, New York, 1986.
5. Isaacs, N. S., *Liquid Phase High Pressure Chemistry*, John Wiley, New York, 1981.

Molecular formula	Name	Isothermal Compressibility		Cubic Thermal Expansion	
		$t/°C$	$\kappa \times 10^4/\text{MPa}^{-1}$	$t/°C$	$\alpha \times 10^3/°C^{-1}$
Cl_3P	Phosphorus trichloride	20	9.45	20	1.9
H_2O	Water	20	4.591	20	0.206
		25	4.524	25	0.256
		30	4.475	30	0.302
Hg	Mercury	20	0.401	20	1.811
CCl_4	Tetrachloromethane	20	10.50	20	1.14
		40	12.20	40	1.21
		70	15.6	70	1.33
$CHBr_3$	Tribromomethane	50	8.76	25	0.91
$CHCl_3$	Trichloromethane	20	9.96	20	1.21
		50	12.9	50	1.33
CH_2Br_2	Dibromomethane	27	6.85		
CH_2Cl_2	Dichloromethane	25	10.3	25	1.39
CH_3I	Iodomethane	27	10.3	25	1.26
CH_4O	Methanol	20	12.14	20	1.49
		40	13.83	40	1.59
CS_2	Carbon disulfide	20	9.38	20	1.12
		40	10.6	35	1.16
C_2Cl_4	Tetrachloroethylene	25	7.56	25	1.02
C_2HCl_3	Trichloroethylene	25	8.57	25	1.17
$C_2H_2Cl_2$	trans-1,2-Dichloroethylene	25	11.2	25	1.36
$C_2H_4Cl_2$	1,1-Dichloroethane	20	7.97	25	0.93
$C_2H_4Cl_2$	1,2-Dichloroethane	30	8.46	20	1.14
$C_2H_4O_2$	Acetic acid	20	9.08	20	1.08
		80	13.7	80	1.38
C_2H_5Br	Bromoethane	20	11.53	20	1.31
C_2H_5I	Iodoethane	20	9.82	25	1.17
C_2H_6O	Ethanol	20	11.19	20	1.40
		70	15.93	70	1.67
$C_2H_6O_2$	Ethylene glycol	20	3.64	20	0.626
C_3H_6O	Acetone	20	12.62	20	1.46
		40	15.6	40	1.57
C_3H_7Br	1-Bromopropane	0	10.22	25	1.2
C_3H_7Cl	1-Chloropropane	0	12.09	20	1.4
C_3H_7I	1-Iodopropane	0	10.22	25	1.09
C_3H_8O	1-Propanol	0	8.43	0	1.22
C_3H_8O	2-Propanol	40	13.32	40	1.55
$C_3H_8O_2$	1,2-Propanediol	0	4.45	20	0.695

Molecular formula	Name	Isothermal Compressibility		Cubic Thermal Expansion	
		$t/°C$	$\kappa \times 10^4/\text{MPa}^{-1}$	$t/°C$	$\alpha \times 10^3/°C^{-1}$
$C_3H_8O_2$	1,3-Propanediol	0	4.09	20	0.61
$C_3H_8O_3$	Glycerol	0	2.54	20	0.520
$C_4H_8O_2$	Ethyl acetate	20	11.32	20	1.35
		60	16.2	60	1.54
C_4H_9Br	1-Bromobutane	25	10.26	20	1.13
C_4H_9I	1-Iodobutane	0	7.73	25	1.02
$C_4H_{10}O$	1-Butanol	0	8.10	0	1.12
$C_4H_{10}O$	Diethyl ether	20	18.65	20	1.65
		30	20.85	30	1.72
$C_4H_{10}O_3$	Diethylene glycol	0	3.34	20	0.635
C_5H_{10}	Cyclopentane	20	13.31	20	1.35
$C_5H_{11}Br$	1-Bromopentane	0	8.42	25	1.04
$C_5H_{11}I$	1-Iodopentane	0	7.56		
C_5H_{12}	Pentane	25	21.80	25	1.64
$C_5H_{12}O$	1-Pentanol	0	7.71	0	1.02
C_6H_5Br	Bromobenzene	20	6.46	20	0.86
C_6H_5Cl	Chlorobenzene	20	7.45	20	0.94
$C_6H_5NO_2$	Nitrobenzene	20	4.93	25	0.833
C_6H_6	Benzene	25	9.66	25	1.14
		45	11.28	45	1.21
C_6H_6O	Phenol	60	6.05	60	0.82
C_6H_7N	Aniline	20	4.53	20	0.81
		80	6.32	80	0.91
C_6H_{12}	Cyclohexane	20	11.30	20	1.15
		60	15.2	60	1.29
C_6H_{14}	Hexane	25	16.69	25	1.41
		45	20.27	45	1.52
C_6H_{14}	2-Methylpentane	0	13.97	25	1.43
		0	14.57	25	1.40
C_6H_{14}	2,3-Dimethylbutane	20	17.97	25	1.39
$C_6H_{14}O$	1-Hexanol	25	8.24	25	1.03
$C_6H_{15}NO_3$	Triethanolamine	0	3.61	55	0.53
C_7H_8	Toluene	20	8.96	20	1.05
		50	11.0	50	1.13
C_7H_8O	Anisole	20	6.60	20	0.951
C_7H_{14}	Cycloheptane	20	9.22		
C_7H_{16}	Heptane	25	14.38	25	1.26
C_8H_{10}	o-Xylene	25	8.10	25	0.96
C_8H_{10}	m-Xylene	20	8.46	20	0.99
C_8H_{10}	p-Xylene	25	8.59	25	1.00
C_8H_{16}	Cyclooctane	20	8.03		
C_8H_{18}	Octane	25	12.82	25	1.16
		45	15.06	45	1.23
$C_8H_{18}O$	1-Octanol	25	7.64	25	0.827
C_9H_{12}	Mesitylene	25	8.14	25	0.94
$C_9H_{14}O_6$	Triacetin	0	4.49	25	0.94
C_9H_{20}	Nonane	25	11.75	25	1.08
$C_{10}H_{22}$	Decane	25	10.94	25	1.02
$C_{11}H_{24}$	Undecane	25	10.31	25	0.97
$C_{12}H_{26}$	Dodecane	25	9.88	25	0.93
$C_{13}H_{28}$	Tridecane	25	9.48	25	0.90
$C_{14}H_{30}$	Tetradecane	25	9.10	25	0.87
$C_{15}H_{32}$	Pentadecane	25	8.82		
$C_{16}H_{22}O_4$	Butyl phthalate	0	5.0	25	0.86
$C_{16}H_{34}$	Hexadecane	25	8.57		
		45	9.78		
$C_{19}H_{36}O_2$	Methyl oleate	0	6.18	60	0.85

PROPERTIES OF CRYOGENIC FLUIDS

This table gives physical and thermodynamic properties of eight cryogenic fluids. The properties are:

M	Molar mass in grams per mole
T_t	Triple point temperature in kelvins
P_t	Triple point pressure in kilopascals
$\rho_t(l)$	Liquid density at the triple point in grams per milliliter
$\Delta_{fus}H$ @ T_t	Enthalpy of fusion at the triple point in joules per gram
T_b	Normal boiling point in kelvins at a pressure of 101325 pascals (760 mmHg)
$\Delta_{vap}H$ @ T_b	Enthalpy of vaporization at the normal boiling point in joules per gram
$\rho(l)$ @ T_b	Liquid density at the normal boiling point in grams per milliliter
$\rho(g)$ @ T_b	Vapor density at the normal boiling point in grams per liter
$C_p(l)$ @ T_b	Liquid heat capacity at constant pressure at the normal boiling point in joules per gram kelvin
$C_p(g)$ @ T_b	Vapor heat capacity at constant pressure at the normal boiling point in joules per gram kelvin
T_c	Critical temperature in kelvins
P_c	Critical pressure in megapascals
ρ_c	Critical density in grams per milliliter

In the case of air, the value given for the triple point temperature is the incipient solidification temperature, and the normal boiling point value is the incipient boiling (bubble) point. See Reference 3 for more details.

Property	Units	Air	N_2	O_2	H_2	He	Ne	Ar	Kr	Xe	CH_4
M	g/mol	28.96	28.014	31.999	2.0159	4.0026	20.180	39.948	83.800	131.290	16.043
T_t	K	59.75	63.15	54.3584	13.8		24.5561	83.8058	115.8	161.4	90.694
P_t	kPa		12.463	0.14633	7.042		50	68.95	72.92	81.55	11.696
$\rho_t(l)$	g/mL	0.959	0.870	1.306	0.0770		1.251	1.417	2.449	2.978	0.4515
$\Delta_{fus}H$ @ T_t	J/g		25.3	13.7	59.5		16.8	28.0	16.3	13.8	58.41
T_b	K	78.67	77.35	90.188	20.28	4.2221	27.07	87.293	119.92	165.10	111.668
$\Delta_{vap}H$ @ T_b	J/g	198.7	198.8	213.1	445	20.7	84.8	161.0	108.4	96.1	510.83
$\rho(l)$ @ T_b	g/mL	0.8754	0.807	1.141	0.0708	0.124901	1.204	1.396	2.418	2.953	0.4224
$\rho(g)$ @ T_b	g/L	3.199	4.622	4.467	1.3390	16.89	9.51	5.79	8.94		1.816
$C_p(l)$ @ T_b	J/g K	1.865	2.042	1.699	9.668	4.545	1.877	1.078	0.533	0.340	3.481
$C_p(g)$ @ T_b	J/g K		1.341	0.980	12.24	9.78		0.570	0.248	0.158	2.218
T_c	K	132.5	126.20	154.581	32.98	5.1953	44.40	150.663	209.40	289.73	190.56
P_c	MPa	3.766	3.390	5.043	1.293	0.227460	2.760	4.860	5.500	5.840	4.592
ρ_c	g/mL	0.316	0.313	0.436	0.031	0.06964	0.484	0.531	0.919	1.110	0.1627

REFERENCES

1. Younglove, B. A., J. Phys. Chem. Ref. Data, 11, Suppl. 1, 1982.
2. Daubert, T. E., Danner, R. P., Sibul, H. M., and Stebbins, C. C., Physical and Thermodynamic Properties of Pure Compounds: Data Compilation, extant 1994 (core with 4 supplements), Taylor & Francis, Bristol, PA (also available as database).
3. Sytchev, V. V., et al., Thermodynamic Properties of Air, Hemisphere Publishing, New York, 1987.
4. Jacobsen, R. T., Stewart, R. B., and Jahangiri, M., J. Phys. Chem. Ref. Data, 15, 735, 1986. [Nitrogen]
5. Stewart, R. B., Jacobsen, R. T., and Wagner, W., J. Phys. Chem. Ref. Data, 20, 917, 1991. [Oxygen]
6. McCarty, R. D., J. Phys. Chem. Ref. Data, 2, 923, 1973. [Helium] Also, Donnelly, R. J., private communication.
7. Stewart, R. B. and Jacobsen, R. T., J. Phys. Chem. Ref. Data, 18, 639, 1989. [Argon]
8. Setzmann, U. and Wagner, W., J. Phys. Chem. Ref. Data, 20, 1061, 1991. [Methane]
9. Vargaftik, N. B., Thermophysical Properties of Liquids and Gases, 2nd ed., John Wiley, New York, 1975.

HALOCARBON REFRIGERANTS

Halogen derivatives of the lower aliphatic hydrocarbons, especially fluorocarbons and chlorocarbons, are widely used as refrigerants, solvents, and cleaning agents. This table lists the most important halocarbons in current use or being considered as possible substitutes for those compounds that are potential environmental hazards. The code number for each compound appears in the first column; this number is frequently used in expressions like R 11 or CFC 11 or HFC 134.

In addition to name, molecular formula, CAS Registry Number, and molecular weight, the table lists the following properties:

- t_m: normal melting point
- t_b: normal boiling point (at 101.324 kPa)
- t_c: critical temperature
- TLV: Threshold Limit Value, which is the maximum safe concentration in air in the workplace, expressed as the time-weighted average (TWA) in parts per million by volume over an 8-hr workday and 40-hr workweek (see Reference 3). An * following the TLV indicates that the substance is a confirmed or suspected human carcinogen.

REFERENCES

1. *1989 ASHRAE Handbook: Fundamentals*, American Society of Heating, Refrigerating, and Air Conditioning Engineers, Atlanta, GA, 1989.
2. Platzer, B., Polt, A., and Mauer, G., *Thermophysical Properties of Refrigerants*, Springer, Berlin, 1990.
3. 1991-1992 Threshold Limit Values and Biological Exposure Indexes, American Conference of Government Industrial Hygienists, Cincinnati, OH, 1991.

Code no.	Name	Molecular formula	CAS reg. no.	Molecular weight	t_m/°C	t_b/°C	t_c/°C	TLV ppm
10	Tetrachloromethane	CCl$_4$	56-23-5	153.822	-23	76.8	283.5	5*
11	Trichlorofluoromethane	CCl$_3$F	75-69-4	137.368	-111.11	23.75	198.1	1000
12	Dichlorodifluoromethane	CCl$_2$F$_2$	75-71-8	120.913	-158	-29.8	111.80	1000
12B1	Bromochlorodifluoromethane	CBrClF$_2$	353-59-3	165.365	-159.5	-3.72	153.73	
13	Chlorotrifluoromethane	CClF$_3$	75-72-9	104.459	-181	-81.4	29	
13B1	Bromotrifluoromethane	CBrF$_3$	75-63-8	148.910	-172	-57.89	67.1	
14	Tetrafluoromethane	CF$_4$	75-73-0	88.005	-183.59	-128.02	-45.5	
20	Trichloromethane	CHCl$_3$	67-66-3	119.377	-63.6	61.17	263.3	10*
21	Dichlorofluoromethane	CHCl$_2$F	75-43-4	102.923	-135	8.95	178.43	10
22	Chlorodifluoromethane	CHClF$_2$	75-45-6	86.468	-157.42	-40.75	96.2	1000
23	Trifluoromethane	CHF$_3$	75-46-7	70.014	-155.18	-82.1	26.2	
30	Dichloromethane	CH$_2$Cl$_2$	75-09-2	84.932	-95.14	40	237	50*
31	Chlorofluoromethane	CH$_2$ClF	593-70-4	68.478	-133	-9.1		
32	Difluoromethane	CH$_2$F$_2$	75-10-5	52.024	-136	-51.69	78.5	
40	Chloromethane	CH$_3$Cl	74-87-3	50.488	-97.7	-24.09	143.10	50*
41	Fluoromethane	CH$_3$F	593-53-3	34.033	-141.8	-78.41	44.7	1*
110	Hexachloroethane	C$_2$Cl$_6$	67-72-1	236.738	187	187		
111	Pentachlorofluoroethane	C$_2$Cl$_5$F	354-56-3	220.284				
112	Tetrachloro-1,2-difluoroethane	C$_2$Cl$_4$F$_2$	76-12-0	203.830	26	93	278	500
112a	1,1,1,2-Tetrachloro-2,2-difluoroethane	C$_2$Cl$_4$F$_2$	76-11-9	203.830	40.6	91.5		
113	1,1,2-Trichlorotrifluoroethane	C$_2$Cl$_3$F$_3$	76-13-1	187.375	-35	47.7	214.2	1000
113a	1,1,1-Trichlorotrifluoroethane	C$_2$Cl$_3$F$_3$	354-58-5	187.375	14.2	46.1	145.63	
114	1,2-Dichlorotetrafluoroethane	C$_2$Cl$_2$F$_4$	76-14-2	170.921	-94	3.8	145.5	1000
114a	1,1-Dichlorotetrafluoroethane	C$_2$Cl$_2$F$_4$	374-07-2	170.921	-56.6	4		

HALOCARBON REFRIGERANTS (continued)

Code no.	Name	Molecular formula	CAS reg. no.	Molecular weight	t_m/°C	t_b/°C	t_c/°C	TLV ppm
114B2	1,2-Dibromotetrafluoroethane	$C_2Br_2F_4$	124-73-2	259.824	-110.4	47.35	214.7	
115	Chloropentafluoroethane	C_2ClF_5	76-15-3	154.467	-99.44	-37.95	80.1	1000
116	Hexafluoroethane	C_2F_6	76-16-4	138.012	-100.7	-78.1	20	
120	Pentachloroethane	C_2HCl_5	76-01-7	202.293	-29	159.88		
121	1,1,2,2-Tetrachloro-1-fluoroethane	C_2HCl_4F	354-14-3	185.839	-82.6	116.6		
121a	1,1,1,2-Tetrachloro-2-fluoroethane	C_2HCl_4F	354-11-0	185.840	-95.35	116.5		
122	2,2,2-Trichloro-1,1-difluoroethane	$C_2HCl_3F_2$	354-21-2	169.385	-140	71.9		
122a	1,1,2-Trichloro-1,2-difluoroethane	$C_2HCl_3F_2$	354-15-4	169.385		72.5		
122b	1,1,1-Trichloro-2,2-difluoroethane	$C_2HCl_3F_2$	354-12-1	169.385		73		
123	2,2-Dichloro-1,1,1-trifluoroethane	$C_2HCl_2F_3$	306-83-2	152.931	-107	27.1		
123a	1,2-Dichloro-1,2,2-trifluoroethane	$C_2HCl_2F_3$	354-23-4	152.931		28.2		
124	2-Chloro-1,1,1,2-tetrafluoroethane	C_2HClF_4	2837-89-0	136.476	-117	-12	126.8	
124a	1-Chloro-1,1,2,2-tetrafluoroethane	C_2HClF_4	354-25-6	136.476		-10.2		
125	Pentafluoroethane	C_2HF_5	354-33-6	120.022	-103	-48.5		
130	1,1,2,2-Tetrachloroethane	$C_2H_2Cl_4$	79-34-5	167.849	-43.8	146.5	388.00	1*
131	1,1,2-Trichloro-2-fluoroethane	$C_2H_2Cl_3F$	359-28-4	151.394		102.5		
132	1,2-Dichloro-1,2-difluoroethane	$C_2H_2Cl_2F_2$	431-06-1	134.940	-155	59		
132b	1,2-Dichloro-1,1-difluoroethane	$C_2H_2Cl_2F_2$	1649-08-7	134.940				
133	1-Chloro-1,2,2-trifluoroethane	$C_2H_2ClF_3$	431-07-2	118.486	-105.5	17		
133a	2-Chloro-1,1,1-trifluoroethane	$C_2H_2ClF_3$	75-88-7	118.486		6.1		
133b	1-Chloro-1,1,2-trifluoroethane	$C_2H_2ClF_3$	421-04-5	118.486		12		
134	1,1,2,2-Tetrafluoroethane	$C_2H_2F_4$	359-35-3	102.032	-89	-19.9	119	
134a	1,1,1,2-Tetrafluoroethane	$C_2H_2F_4$	811-97-2	102.032	-101	-26.5	101.2	
140	1,1,2-Trichloroethane	$C_2H_3Cl_3$	79-00-5	133.404	-36.6	113.8		10
140a	1,1,1-Trichloroethane	$C_2H_3Cl_3$	71-55-6	133.404	-30.4	74.09	272	350
141	1,2-Dichloro-1-fluoroethane	$C_2H_3Cl_2F$	430-57-9	116.950		75.7		
141b	1,1-Dichloro-1-fluoroethane	$C_2H_3Cl_2F$	1717-00-6	116.950	-103.5	32.11		
142	2-Chloro-1,1-difluoroethane	$C_2H_3ClF_2$	338-65-8	100.495		35.1		
142b	1-Chloro-1,1-difluoroethane	$C_2H_3ClF_2$	75-68-3	100.495	-130.8	-9.75	137.14	
143	1,1,2-Trifluoroethane	$C_2H_3F_3$	430-66-0	84.041	-84	5		
143a	1,1,1-Trifluoroethane	$C_2H_3F_3$	420-46-2	84.041	-111.3	-47.55	73.2	
150	1,2-Dichloroethane	$C_2H_4Cl_2$	107-06-2	98.959	-35.5	83.5	288	10*
150a	1,1-Dichloroethane	$C_2H_4Cl_2$	75-34-3	98.959	-96.96	57.4	250	100
151	1-Chloro-2-fluoroethane	C_2H_4ClF	762-50-5	82.505		53.2		
151a	1-Chloro-1-fluoroethane	C_2H_4ClF	1615-75-4	82.505		16.1		
152	1,2-Difluoroethane	$C_2H_4F_2$	624-72-6	66.051		30.7		
152a	1,1-Difluoroethane	$C_2H_4F_2$	75-37-6	66.051	-117	-24.95	113.6	
160	Chloroethane	C_2H_5Cl	75-00-3	64.514	-138.7	13.1	187.3	1000
161	Fluoroethane	C_2H_5F	353-36-6	48.060	-143.2	-37.65	102.16	
218	Perfluoropropane	C_3F_8	76-19-7	188.020	-147.69	-36.65	72.0	
1112a	1,1-Dichloro-2,2-difluoroethylene	$C_2Cl_2F_2$	79-35-6	132.924				
1113	Chlorotrifluoroethylene	C_2ClF_3	79-38-9	116.470	-158	-27.85	106	

HALOCARBON REFRIGERANTS (continued)

Code no.	Name	Molecular formula	CAS reg. no.	Molecular weight	$t_m/°C$	$t_b/°C$	$t_c/°C$	TLV ppm
1114	Tetrafluoroethylene	C_2F_4	116-14-3	100.016	-142.5	-75.95	33.4	
1120	Trichloroethylene	C_2HCl_3	79-01-6	131.388	-84.75	87.21	271.1	50*
1130	cis-1,2-Dichloroethylene	$C_2H_2Cl_2$	156-59-2	96.943	-80	60.19	271.1	200
1130	trans-1,2-Dichloroethylene	$C_2H_2Cl_2$	156-60-5	96.943	-49.8	48.73	243.4	200
1132a	1,1-Difluoroethylene	$C_2H_2F_2$	75-38-7	64.035	-144	-85.7	29.8	
1140	Chloroethylene	C_2H_3Cl	75-01-4	62.499	-153.79	-13.37		5*
1141	Fluoroethylene	C_2H_3F	75-02-5	46.044	-160.5	-72	54.8	
C316	1,2-Dichlorohexafluorocyclobutane	$C_4Cl_2F_6$	356-18-3	232.940				
C317	Chloroheptafluorocyclobutane	C_4ClF_7	377-41-3	216.486				
C318	Perfluorocyclobutane	C_4F_8	115-25-3	200.031	-40.19	-5.99	115.31	

* Confirmed or suspected human carcinogen.

DENSITY AND SPECIFIC VOLUME OF MERCURY

The data in this table have been adjusted to the ITS-90 temperature scale. The uncertainty in density values is 0.0003 g/mL between −20 and −10°C; 0.0001 or less between −10 and 200°C; and 0.0002 between 200 and 300°C.

REFERENCE

Ambrose, D., *Metrologia*, 27, 245, 1990.

t/°C	ρ/(g/mL)	v/(mL/kg)	t/°C	ρ/(g/mL)	v/(mL/kg)	t/°C	ρ/(g/mL)	v/(mL/kg)
−20	13.64461	73.2890	27	13.52869	73.9170	74	13.41423	74.5477
−19	13.64212	73.3024	28	13.52624	73.9304	75	13.41181	74.5612
−18	13.63964	73.3157	29	13.52379	73.9438	76	13.40939	74.5746
−17	13.63716	73.3291	30	13.52134	73.9572	77	13.40697	74.5881
−16	13.63468	73.3424	31	13.51889	73.9705	78	13.40455	74.6016
−15	13.63220	73.3558	32	13.51645	73.9839	79	13.40213	74.6150
−14	13.62972	73.3691	33	13.51400	73.9973	80	13.39971	74.6285
−13	13.62724	73.3824	34	13.51156	74.0107	81	13.39729	74.6420
−12	13.62476	73.3958	35	13.50911	74.0241	82	13.39487	74.6554
−11	13.62228	73.4091	36	13.50667	74.0375	83	13.39245	74.6689
−10	13.61981	73.4225	37	13.50422	74.0509	84	13.39003	74.6824
−9	13.61733	73.4358	38	13.50178	74.0643	85	13.38762	74.6959
−8	13.61485	73.4492	39	13.49934	74.0777	86	13.38520	74.7094
−7	13.61238	73.4625	40	13.49690	74.0911	87	13.38278	74.7229
−6	13.60991	73.4759	41	13.49446	74.1045	88	13.38037	74.7364
−5	13.60743	73.4892	42	13.49202	74.1179	89	13.37795	74.7498
−4	13.60496	73.5026	43	13.48958	74.1313	90	13.37554	74.7633
−3	13.60249	73.5160	44	13.48714	74.1447	91	13.37313	74.7768
−2	13.60002	73.5293	45	13.48470	74.1581	92	13.37071	74.7903
−1	13.59755	73.5427	46	13.48226	74.1715	93	13.36830	74.8038
0	13.59508	73.5560	47	13.47982	74.1850	94	13.36589	74.8173
1	13.59261	73.5694	48	13.47739	74.1984	95	13.36347	74.8308
2	13.59014	73.5827	49	13.47495	74.2118	96	13.36106	74.8443
3	13.58768	73.5961	50	13.47251	74.2252	97	13.35865	74.8579
4	13.58521	73.6095	51	13.47008	74.2386	98	13.35624	74.8714
5	13.58275	73.6228	52	13.46765	74.2520	99	13.35383	74.8849
6	13.58028	73.6362	53	13.46521	74.2655	100	13.35142	74.8984
7	13.57782	73.6495	54	13.46278	74.2789	110	13.3273	75.0337
8	13.57535	73.6629	55	13.46035	74.2923	120	13.3033	75.1693
9	13.57289	73.6763	56	13.45791	74.3057	130	13.2793	75.3052
10	13.57043	73.6896	57	13.45548	74.3192	140	13.2553	75.4413
11	13.56797	73.7030	58	13.45305	74.3326	150	13.2314	75.5778
12	13.56551	73.7164	59	13.45062	74.3460	160	13.2075	75.7147
13	13.56305	73.7297	60	13.44819	74.3594	170	13.1836	75.8519
14	13.56059	73.7431	61	13.44576	74.3729	180	13.1597	75.9895
15	13.55813	73.7565	62	13.44333	74.3863	190	13.1359	76.1274
16	13.55567	73.7698	63	13.44090	74.3998	200	13.1120	76.2659
17	13.55322	73.7832	64	13.43848	74.4132	210	13.0882	76.4047
18	13.55076	73.7966	65	13.43605	74.4266	220	13.0644	76.5440
19	13.54831	73.8100	66	13.43362	74.4401	230	13.0406	76.6838
20	13.54585	73.8233	67	13.43120	74.4535	240	13.0167	76.8241
21	13.54340	73.8367	68	13.42877	74.4670	250	12.9929	76.9650
22	13.54094	73.8501	69	13.42635	74.4804	260	12.9691	77.1064
23	13.53849	73.8635	70	13.42392	74.4939	270	12.9453	77.2484
24	13.53604	73.8769	71	13.42150	74.5073	280	12.9214	77.3909
25	13.53359	73.8902	72	13.41908	74.5208	290	12.8975	77.5341
26	13.53114	73.9036	73	13.41665	74.5342	300	12.8736	77.6779

THERMAL PROPERTIES OF MERCURY
Lev R. Fokin

The first of these tables gives the molar heat capacity at constant pressure of liquid and gaseous mercury as a function of temperature. To convert to specific heat in units of J/g K, divide these values by 200.59, the atomic weight of mercury.

REFERENCE

Douglas, T. B., Ball, A. T., and Ginnings, D. C., *J. Res. Natl. Bur. Stands.*, 46, 334, 1951.

t/°C	C_p/(J/mol K) Liquid	Gas	t/°C	C_p/(J/mol K) Liquid	Gas	t/°C	C_p/(J/mol K) Liquid	Gas
−38.84	28.2746	20.786	140	27.3675	20.786	340	27.1500	20.836
−20	28.1466	20.786	160	27.3090	20.786	356.73	27.1677	20.849
0	28.0190	20.786	180	27.2588	20.790	360	27.1709	20.853
20	27.9002	20.786	200	27.2169	20.790	380	27.1981	20.870
25	27.8717	20.786	220	27.1834	20.794	400	27.2324	20.891
40	27.7897	20.786	240	27.1583	20.794	420	27.2738	20.916
60	27.6880	20.786	260	27.1412	20.799	440	27.3207	20.941
80	27.5952	20.786	280	27.1320	20.807	460	27.3742	20.974
100	27.5106	20.786	300	27.1303	20.815	480	27.4332	21.008
120	27.4349	20.786	320	27.1366	20.824	500	27.4985	21.046

The second table gives the molar heat capacity of solid mercury in its rhombohedral (α-mercury) form.

REFERENCES

1. Busey and Giaque, *J. Am. Chem. Soc.,* 75, 806, 1953.
2. Amitin, Lebedeva, and Paukov, *Rus. J. Phys. Chem.,* 2666, 1979.

t/°C	C_p/J mol⁻¹	t/°C	C_p/J mol⁻¹	t/°C	C_p/J mol⁻¹	t/°C	C_p/J mol⁻¹
−268.99	0.99*	−248.15	12.74	−193.15	23.16	−113.15	26.15
−268.99	0.97**	−243.15	14.78	−183.15	23.76	−93.15	26.69
−268.15	1.6	−233.15	17.90	−173.15	24.24	−73.15	27.28
−263.15	4.6	−223.15	19.94	−153.15	25.00	−53.15	27.96
−258.15	7.6	−213.15	21.40	−133.15	25.61	−38.87	28.5
−253.15	10.33	−203.15	22.42				

*   Superconducting state
**   Normal state

The final table gives the cubic thermal expansion coefficient α, the isothermal compressibility coefficient κ_T, and the speed of sound U for liquid mercury as a function of temperature. These properties are defined as follows:

$$\alpha = \frac{1}{v}\left(\frac{\partial v}{\partial T}\right)_p \qquad \kappa_T = -\frac{1}{v}\left(\frac{\partial v}{\partial P}\right)_T \qquad U^2 = \left(\frac{\partial P}{\partial \rho}\right)_s \qquad \rho = v^{-1}$$

where v is the specific volume (given in the table on the preceding page).

REFERENCE

Vukalovich, M. P., et al., *Thermophysical Properties of Mercury,* Moscow Standard Press, 1971.

t/°C	$\alpha \times 10^4$/K⁻¹	$\kappa_T \times 10^6$/bar⁻¹ At 1 bar	At 1000 bar	U/m s⁻¹	t/°C	$\alpha \times 10^4$/K⁻¹	$\kappa_T \times 10^6$/bar⁻¹ At 1 bar	At 1000 bar	U/m s⁻¹
−20	1.818	3.83		1470	120	1.8058	4.513	4.33	1404.7
0	1.8144	3.918	3.78	1460.8	140	1.8074	4.622		1395.4
20	1.8110	4.013	3.87	1451.4	160	1.8100	4.731	4.53	1386.1
40	1.8083	4.109	3.96	1442.0	180	1.8136	4.844		1376.7
60	1.8064	4.207		1432.7	200	1.818	4.96		1367
80	1.8053	4.308	4.14	1423.4	250	1.834	5.26		1344
100	1.8051	4.410		1414.1	300	1.856	5.59		1321

SURFACE TENSION OF COMMON LIQUIDS

The surface tension γ of about 200 liquids is tabulated here as a function of temperature. Values of γ are given in units of millinewtons per meter (mN/m), which is equivalent to dyn/cm in cgs units. The values refer to a nominal pressure of one atmosphere (about 100 kPa) except in cases where the indicated temperature is above the normal boiling point of the substance; in those cases, the applicable pressure is the saturation vapor pressure at the temperature in question.

The uncertainty of the values is 0.1 to 0.2 mN/m or less in most cases. Values at temperatures between the points tabulated can be obtained by linear interpolation to a good approximation.

Substances are listed by molecular formula in the modified Hill order, with substances not containing carbon appearing before those that do contain carbon. A more extensive compilation of surface tension may be found in the Reference.

REFERENCE

Jasper, J. J., *J. Phys. Chem. Ref. Data*, 1, 841, 1972.

Mol. form.	Name	γ in mN/m				
		10°C	25°C	50°C	75°C	100°C
Br_2	Bromine	43.68	40.95	36.40		
Cl_2O_2S	Sulfuryl chloride		28.78			
Cl_3OP	Phosphoryl chloride		32.03	28.85	25.66	
Cl_3P	Phosphorus trichloride		27.98	24.81		
Cl_4Si	Silicon tetrachloride	19.78	18.29	15.80		
H_2O	Water	74.23	71.99	67.94	63.57	58.91
H_4N_2	Hydrazine		66.39			
Hg	Mercury	488.55	485.48	480.36	475.23	470.11
CCl_4	Tetrachloromethane		26.43	23.37	20.31	17.25
CS_2	Carbon disulfide	33.81	31.58	27.87		
$CHBr_3$	Tribromomethane		44.87	41.60	38.33	
$CHCl_3$	Trichloromethane		26.67	23.44	20.20	
CH_2Br_2	Dibromomethane		39.05	35.33	31.61	
CH_2Cl_2	Dichloromethane		27.20			
CH_2O_2	Formic acid		37.13	34.38	31.64	
CH_3I	Iodomethane	32.19	30.34			
CH_3NO	Formamide		57.03	54.92	52.82	50.71
CH_3NO_2	Nitromethane	39.04	36.53	32.33		
CH_4O	Methanol	23.23	22.07	20.14		
CH_5N	Methylamine		19.15			
C_2HCl_5	Pentachloroethane		34.15	31.20	28.26	
$C_2HF_3O_2$	Trifluoroacetic acid		13.53	11.42		
$C_2H_2Cl_4$	1,1,2,2-Tetrachloroethane		35.58	32.41	29.24	26.07
$C_2H_3Cl_3$	1,1,1-Trichloroethane		25.18	22.07		
$C_2H_3Cl_3$	1,1,2-Trichloroethane		34.02	30.65	27.27	23.89
C_2H_3N	Acetonitrile		28.66	25.51		
$C_2H_4Br_2$	1,2-Dibromoethane		39.55	36.25	32.95	
$C_2H_4Cl_2$	1,1-Dichloroethane		24.07			
$C_2H_4Cl_2$	1,2-Dichloroethane		31.86	28.29	24.72	
C_2H_4O	Acetaldehyde	22.54	20.50	17.10		
$C_2H_4O_2$	Acetic acid		27.10	24.61	22.13	
$C_2H_4O_2$	Methyl formate	26.72	24.36	20.43	16.50	12.57
C_2H_5Br	Bromoethane	25.36	23.62			
C_2H_5I	Iodoethane	30.38	28.46	25.24		
$C_2H_5NO_2$	Nitroethane	34.02	32.13	29.00		
C_2H_6O	Ethanol	23.22	21.97	19.89		
C_2H_6OS	Dimethyl sulfoxide		42.92	40.06		
$C_2H_6O_2$	Ethylene glycol		47.99	45.76	43.54	41.31
C_2H_6S	Dimethyl sulfide	25.27	24.06			
C_2H_6S	Ethanethiol		23.08			
$C_2H_6S_2$	Dimethyl disulfide		33.39	30.04		
C_2H_7N	Dimethylamine		26.34			
C_2H_7N	Ethylamine		19.20			

Mol. form.	Name	γ in mN/m				
		10°C	25°C	50°C	75°C	100°C
C₂H₇NO	Ethanolamine		48.32	45.53	42.73	
C₃H₅Br	3-Bromopropene		26.31	23.17		
C₃H₅Cl	3-Chloropropene		23.14			
C₃H₅ClO	Epichlorohydrin	38.40	36.36	32.96	29.56	26.16
C₃H₅N	Propanenitrile		26.75	23.87		
C₃H₆Cl₂	1,2-Dichloropropane		28.32	25.22	22.12	
C₃H₆O	Acetone		23.46	20.66		
C₃H₆O	Allyl alcohol	26.63	25.28	23.02	20.77	
C₃H₆O₂	Ethyl formate	25.16	23.18			
C₃H₆O₂	Methyl acetate	26.66	24.73	21.51		
C₃H₆O₂	Propanoic acid		26.20	23.72	21.23	
C₃H₇Br	1-Bromopropane	27.08	25.26	22.21		
C₃H₇Br	2-Bromopropane	25.03	23.25	20.30		
C₃H₇Cl	1-Chloropropane	23.16	21.30			
C₃H₇Cl	2-Chloropropane	20.49	19.16			
C₃H₇NO₂	2-Nitropropane	31.02	29.29	26.39		
C₃H₈O	1-Propanol	24.48	23.32	21.38	19.43	
C₃H₈O	2-Propanol	22.11	20.93	18.96	16.98	
C₃H₈O₂	2-Methoxyethanol	32.32	30.84	28.38	25.92	23.46
C₃H₈S	1-Propanethiol		24.20	21.02		
C₃H₈S	2-Propanethiol		21.33	18.39		
C₃H₉N	Propylamine		21.75			
C₃H₉N	Trimethylamine		13.41			
C₄H₄N₂	Pyridazine	49.51	47.96	45.37	42.78	40.19
C₄H₄N₂	Pyrimidine		30.33	27.80	25.28	22.75
C₄H₄S	Thiophene		30.68	27.36		
C₄H₅N	Pyrrole	38.71	37.06	34.31		
C₄H₆O₃	Acetic anhydride	34.08	31.93	28.34	24.75	21.16
C₄H₇N	Butanenitrile		26.92	24.33	21.73	
C₄H₈O	2-Butanone		23.97	21.16		
C₄H₈O₂	1,4-Dioxane		32.75	29.28	25.80	22.32
C₄H₈O₂	Ethyl acetate	25.13	23.39	20.49	17.58	14.68
C₄H₈O₂	Methyl propanoate	26.32	24.44	21.29		
C₄H₈O₂	Butanoic acid		26.05	23.75	21.45	
C₄H₉Br	1-Bromobutane	27.58	25.90	23.08	20.27	17.45
C₄H₉Cl	1-Chlorobutane	24.85	23.18	20.39		
C₄H₉I	1-Iodobutane	29.79	28.24	25.67	23.09	20.51
C₄H₉N	Pyrrolidine	30.58	29.23	26.98		
C₄H₁₀O	1-Butanol	26.28	24.93	22.69	20.44	18.20
C₄H₁₀O	2-Butanol	23.74	22.54	20.56	18.57	16.58
C₄H₁₀O	2-Methyl-2-propanol		19.96	17.71		
C₄H₁₀O	Diethyl ether		16.65			
C₄H₁₀O₂	2-Ethoxyethanol		28.35	26.11	23.86	21.62
C₄H₁₀O₃	Diethylene glycol		44.77	42.57	40.37	38.17
C₄H₁₀S	Diethyl sulfide	26.22	24.57	21.80		
C₄H₁₁N	Butylamine		23.44	20.63		
C₄H₁₁N	Isobutylamine		21.75	19.02		
C₄H₁₁N	tert-Butylamine		16.87			
C₄H₁₁N	Diethylamine		19.85			
C₅H₄O₂	Furfural	45.08	43.09	39.78	36.46	33.14
C₅H₅N	Pyridine		36.56	33.29	30.03	
C₅H₈	Cyclopentene	24.45	22.20			
C₅H₈O	Cyclopentanone	34.45	32.80	30.05	27.30	24.55
C₅H₁₀	1-Pentene	17.10	15.45			
C₅H₁₀	2-Methyl-2-butene	18.61	17.15			
C₅H₁₀	Cyclopentane	24.07	21.88	18.22		
C₅H₁₀O	2-Pentanone		23.25	21.62		

SURFACE TENSION OF COMMON LIQUIDS (continued)

Mol. form.	Name	γ in mN/m				
		10°C	25°C	50°C	75°C	100°C
$C_5H_{10}O$	3-Pentanone		24.74	22.13		
$C_5H_{10}O$	Pentanal	26.95	25.44	22.91		
$C_5H_{10}O_2$	Butyl formate	26.05	24.52	21.95	19.39	16.82
$C_5H_{10}O_2$	Propyl acetate	25.48	23.80	21.00	18.20	15.40
$C_5H_{10}O_2$	Isopropyl acetate	23.37	21.76	19.08	16.40	
$C_5H_{10}O_2$	Ethyl propanoate	25.55	23.80	20.88	17.96	
$C_5H_{10}O_2$	Methyl butanoate	26.34	24.62	21.76	18.89	16.03
$C_5H_{11}Cl$	1-Chloropentane	26.01	24.40	21.71	19.02	16.33
$C_5H_{11}N$	Piperidine	30.64	28.91	26.03	23.14	20.26
C_5H_{12}	Pentane	17.15	15.49			
$C_5H_{12}O$	1-Pentanol	26.67	25.36	23.17	20.99	18.80
$C_5H_{12}O$	2-Pentanol	24.96	23.45	20.94	18.43	15.92
$C_5H_{12}O$	3-Methyl-1-butanol	24.94	23.71	21.66	19.61	17.56
$C_5H_{13}N$	Pentylamine		24.69	22.14	19.58	
$C_6H_4Cl_2$	m-Dichlorobenzene	37.15	35.43	32.57	29.70	26.83
C_6H_5Br	Bromobenzene	36.98	35.24	32.34	29.44	26.54
C_6H_5Cl	Chlorobenzene	34.78	32.99	30.02	27.04	24.06
C_6H_5ClO	o-Chlorophenol		39.70	36.89	34.09	31.28
C_6H_5ClO	m-Chlorophenol		41.18	38.66	36.13	33.61
C_6H_5F	Fluorobenzene	28.47	26.66	23.65	20.64	
C_6H_5I	Iodobenzene	40.40	38.71	35.91	33.10	30.29
$C_6H_5NO_2$	Nitrobenzene			40.56	37.66	34.77
C_6H_6	Benzene		28.22	25.00	21.77	
C_6H_6O	Phenol			38.20	35.53	32.86
C_6H_7N	Aniline		42.12	39.41	36.69	
C_6H_7N	2-Methylpyridine		33.00	29.90	26.79	
$C_6H_8N_2$	Adiponitrile		45.45	43.02	40.58	
C_6H_{10}	Cyclohexene	28.01	26.17	23.12		
$C_6H_{10}O$	Cyclohexanone	36.43	34.57	31.46	28.36	25.25
$C_6H_{11}N$	Hexanenitrile		27.37	25.11	22.84	
C_6H_{12}	Cyclohexane	26.43	24.65	21.68		
C_6H_{12}	Methylcyclopentane	23.47	21.72	18.82		
C_6H_{12}	1-Hexene	19.44	17.90	15.33		
$C_6H_{12}O$	Cyclohexanol		32.92	30.50	28.09	25.67
$C_6H_{12}O$	2-Hexanone		25.45	22.72		
$C_6H_{12}O_2$	Butyl acetate	26.48	24.88	22.21	19.54	16.87
$C_6H_{12}O_2$	Isobutyl acetate	24.58	23.06	20.53	17.99	15.46
$C_6H_{12}O_2$	Ethyl butanoate	25.51	23.94	21.33	18.71	16.10
$C_6H_{12}O_3$	Paraldehyde	27.22	25.63	22.97	20.32	17.66
$C_6H_{13}Cl$	1-Chlorohexane	27.28	25.73	23.13	20.54	17.94
$C_6H_{13}N$	Cyclohexylamine		31.22	28.25	25.28	
C_6H_{14}	Hexane	19.42	17.89	15.33		
C_6H_{14}	2-Methylpentane	18.37	16.88	14.39		
C_6H_{14}	3-Methylpentane	19.20	17.61	14.96		
$C_6H_{14}O$	Diisopropyl ether		17.27	14.65		
$C_6H_{14}O$	1-Hexanol		25.81	23.81	21.80	19.80
$C_6H_{14}O_2$	1,1-Diethoxyethane		20.89	18.31	15.74	
$C_6H_{14}O_2$	2-Butoxyethanol	27.36	26.14	24.10	22.06	20.02
$C_6H_{15}N$	Triethylamine		20.22	17.74		
$C_6H_{15}N$	Dipropylamine		22.31	19.75	17.20	
$C_6H_{15}N$	Diisopropylamine		19.14	16.45		
C_7H_5N	Benzonitrile		38.79	35.90	33.00	
C_7H_6O	Benzaldehyde	39.63	38.00	35.27	32.55	29.82
C_7H_8	Toluene	29.71	27.93	24.96	21.98	19.01
C_7H_8O	o-Cresol		36.90	34.38	31.85	29.32
C_7H_8O	m-Cresol		35.69	33.38	31.07	28.76
C_7H_8O	Benzyl alcohol				27.89	24.44

Mol. form.	Name	γ in mN/m				
		10°C	25°C	50°C	75°C	100°C
C_7H_8O	Anisole		35.10	32.09	29.08	
C_7H_9N	N-Methylaniline		36.90	34.47	32.05	
C_7H_9N	2,3-Dimethylpyridine		32.71	30.04	27.36	
C_7H_9N	Benzylamine		39.30	36.27	33.23	
C_7H_{14}	Methylcyclohexane	24.98	23.29	20.46		
C_7H_{14}	1-Heptene	21.29	19.80	17.33	14.85	
$C_7H_{14}O$	2-Heptanone		26.12	23.48		
$C_7H_{14}O_2$	Pentyl acetate	26.67	25.17	22.69	20.20	17.72
$C_7H_{14}O_2$	Heptanoic acid		27.76	25.64		
C_7H_{16}	Heptane	21.12	19.65	17.20	14.75	
C_7H_{16}	3-Methylhexane	20.76	19.31	16.88	14.46	
C_8H_8O	Acetophenone		39.04	36.15	33.27	
$C_8H_8O_2$	Methyl benzoate		37.17	34.25	31.32	
$C_8H_8O_3$	Methyl salicylate	40.98	39.22	36.28	33.35	30.41
C_8H_{10}	Ethylbenzene	30.39	28.75	26.01	23.28	20.54
C_8H_{10}	o-Xylene	31.41	29.76	27.01	24.25	21.50
C_8H_{10}	m-Xylene	30.13	28.47	25.71	22.95	20.19
C_8H_{10}	p-Xylene		28.01	25.32	22.64	19.95
$C_8H_{10}O$	Phenetole		32.41	29.65	26.89	
$C_8H_{11}N$	N,N-Dimethylaniline		35.52	32.90	30.27	
$C_8H_{11}N$	N-Ethylaniline		36.33	33.65	30.98	
C_8H_{16}	Ethylcyclohexane	26.73	25.15	22.51		
C_8H_{18}	Octane	22.57	21.14	18.77	16.39	14.01
C_8H_{18}	2,5-Dimethylhexane	20.77	19.40	17.12	14.84	12.56
$C_8H_{18}O$	1-Octanol	28.30	27.10	25.12		
$C_8H_{19}N$	Dibutylamine		24.12	21.74	19.36	
$C_8H_{19}N$	Diisobutylamine		21.72	19.44	17.16	
C_9H_7N	Quinoline	44.19	42.59	39.94	37.28	34.62
C_9H_{12}	Cumene	29.27	27.69	25.05	22.42	19.78
C_9H_{12}	1,2,4-Trimethylbenzene	30.74	29.20	26.64	24.07	21.51
C_9H_{12}	Mesitylene	28.89	27.55	25.31	23.07	20.82
$C_9H_{18}O$	5-Nonanone		26.28	23.85		
C_9H_{20}	Nonane	23.79	22.38	20.05	17.71	15.37
$C_9H_{20}O$	1-Nonanol	29.03	27.89	26.00	24.10	22.20
$C_{10}H_{12}$	1,2,3,4-Tetrahydronaphthalene		33.17	30.78	28.40	
$C_{10}H_{22}$	Decane	24.75	23.37	21.07	18.77	16.47
$C_{10}H_{22}O$	1-Decanol	29.61	28.51	26.68	24.85	23.02
$C_{11}H_{24}$	Undecane	25.56	24.21	21.96	19.70	17.45
$C_{12}H_{10}O$	Diphenyl ether		26.75	24.80		
$C_{12}H_{27}N$	Tributylamine		24.39	22.32	20.24	
$C_{13}H_{28}$	Tridecane	26.86	25.55	23.37	21.19	19.01
$C_{14}H_{12}O_2$	Benzyl benzoate	44.47	42.82	40.06	37.31	34.55
$C_{14}H_{30}$	Tetradecane	27.43	26.13	23.96	21.78	19.61
$C_{16}H_{34}$	Hexadecane		27.05	24.91	22.78	20.64
$C_{18}H_{38}$	Octadecane		27.87	25.77	23.66	21.55

PERMITTIVITY (DIELECTRIC CONSTANT) OF LIQUIDS

Christian Wohlfarth

The permittivity of a substance (often called the dielectric constant) is the ratio of the electric displacement D to the electric field strength E when an external field is applied to the substance. The quantity tabulated here is the relative permittivity, which is the ratio of the actual permittivity to the permittivity of a vacuum; it is a dimensionless number.

The table gives the static relative permittivity ε_r, i.e., the relative permittivity measured in static fields or at low frequencies where no relaxation effects occur. The fourth column of the table lists the value of ε_r at the temperature specified in the third column, usually 293.15 or 298.15 K. Otherwise, the temperature closest to 293.15 K was chosen, or (as it is the case for many of the substances included here) ε_r is given at the only temperature for which data are available.

The static permittivity refers to nominal atmospheric pressure as long as the corresponding temperature is below the normal boiling point. Otherwise, at temperatures above the normal boiling point, the pressure is understood to be the saturated vapor pressure of the substance considered.

For substances where information on the temperature dependence of the permittivity is available, the table gives the coefficients of a simple polynomial fitting of permittivity to temperature with an equation of the form

$$\varepsilon_r(T) = a + bT + cT^2 + dT^3$$

where T is the absolute temperature in K. Since the parameter d was used in only a few cases where the quadratic fit was not satisfactory, only a, b, and c are listed as columns in the table, while the d values are given at the end of this introduction. For all other substances, $d = 0$. The temperature range of the fit is given in the last column. The coefficients of the fitting equation can be used to calculate dielectric constants within the fitted temperature range but should not be used for extrapolation outside this range. The user who needs dielectric constant data with more accuracy than can be provided by this equation is referred to Reference 1, which gives the original data together with their literature source.

Substances are listed by molecular formula in modified Hill order, with substances not containing carbon preceding those that do contain carbon.

* Indicates that the isomer was not specified in the original reference.
** Indicates a compound for which the cubic term is needed:

Ethanol	$d = -0.15512E-05$
N-Methylacetamide	$d = -0.12998E-04$
1,2-Propylene glycol	$d = -0.32544E-05$
1-Butanol	$d = -0.48841E-06$
2-Butanol	$d = -0.89512E-06$
2-Methyl-1-propanol	$d = -0.45229E-06$
2-Methyl-2-propanol	$d = -0.25968E-05$
N-Butylacetamide	$d = -0.48716E-05$

REFERENCES

1. Wohlfarth, Ch., "Static Dielectric Constants of Pure Liquids and Binary Liquid Mixtures", *Landolt-Börnstein, Numerical Data and Functional Relationships in Science and Technology*, New Series, Editor in Chief, O. Madelung, Group IV, Macroscopic and Technical Properties of Matter, Volume 6, Springer-Verlag, Berlin, Heidelberg, New York, 1991.
2. Marsh, K. N., Ed., *Recommended Reference Materials for the Realization of Physicochemical Properties*, Blackwell Scientific Publications, Oxford, 1987.

Mol. Form.	Name	T/K	ε	a	b	c	Range/K
$AlBr_3$	Aluminum tribromide	373.2	3.38				
Ar	Argon	140.00	1.3247	0.12408E+01	0.68755E-02	-0.45344E-04	87-149
AsH_3	Arsine	200.9	2.40	0.37674E+01	-0.97454E-02	0.14537E-04	157-201
BBr_3	Boron tribromide	273.2	2.58				
B_2H_6	Diborane	180.66	1.8725	0.23848E+01	-0.29501E-02	0.64189E-06	108-181
B_5H_9	Pentaborane(9)	298.2	21.1	0.40952E+03	-0.24414E+01	0.38225E-02	226-298
BrF_3	Bromine trifluoride	298.2	106.8				
BrF_5	Bromine pentafluoride	297.7	7.91	0.11428E+02	-0.11822E-01		262-298
BrH	Hydrogen bromide	186.8	8.23				
BrNO	Nitrosyl bromide	288.4	13.4				
Br_2	Bromine	297.9	3.1484	0.32701E+01	-0.12535E-03		273-327
Br_2OS	Thionyl bromide	293.2	9.06				
Br_3OV	Vanadyl tribromide	298.2	3.6	0.61112E+01	-0.84211E-02		203-298
Br_4Ge	Germanium(IV) bromide	299.9	2.955	0.34450E+01	-0.16083E-02		300-316
Br_4Sn	Tin(IV) bromide	303.45	3.169	0.50001E+01	-0.60383E-02		304-316
$ClFO_3$	Perchloryl fluoride	150.2	2.194	0.23808E+01	-0.38629E-03	-0.57143E-05	125-150
ClF_3	Chlorine trifluoride	293.2	4.394	0.96716E+01	-0.18000E-01		273-313

Mol. Form.	Name	T/K	ε	a	b	c	Range/K
ClF_5	Chlorine pentafluoride	193.2	4.28	0.78192E+01	-0.20860E-01	0.13132E-04	193-256
ClH	Hydrogen chloride	158.9	14.3	0.47316E+02	-0.28455E+00	0.48650E-03	159-258
$ClNO$	Nitrosyl chloride	285.2	18.2				
Cl_2	Chlorine	208.0	2.147	0.29440E+01	-0.44649E-02	0.30388E-05	208-240
Cl_2F_3P	Phosphorus(V) dichloride trifluoride	228.63	2.8129	0.46501E+01	-0.80358E-02		172-229
Cl_2OS	Thionyl chloride	298.2	8.675				
Cl_2OSe	Selenium oxychloride	293.2	46.2				
Cl_2O_2S	Sulfuryl chloride	293.2	9.1				
Cl_2S	Sulfur dichloride	298.2	2.915				
Cl_2S_2	Sulfur chloride	288.2	4.79				
Cl_3F_2P	Phosphorus(V) trichloride difluoride	268.0	2.3752	0.28905E+01	-0.19228E-02		215-268
Cl_3OP	Phosphorus(V) oxychloride	293.2	14.1				
Cl_3OV	Vanadyl trichloride	298.2	3.4				
Cl_3P	Phosphorus(III) chloride	290.2	3.498	0.59098E+01	-0.83322E-02		290-333
Cl_3PS	Phosphorus(V) sulfide trichloride	298.2	4.94				
Cl_4FP	Phosphorus(V) tetrachloride fluoride	272.64	2.6499	0.33503E+01	-0.29651E-02		244-273
Cl_4Ge	Germanium(IV) chloride	273.2	2.463	-0.55078E+01	0.64881E-01	-0.13091E-03	246-273
Cl_4Pb	Lead(IV) chloride	293.2	2.78				
Cl_4Si	Tetrachlorosilane	273.2	2.248	0.58041E+01	-0.27129E-01	0.51678E-04	207-273
Cl_4Sn	Tin(IV) chloride	273.2	3.014	0.43951E+01	-0.48805E-02		234-273
Cl_4Ti	Titanium(IV) chloride	257.4	2.843	0.33668E+01	-0.19675E-02		237-257
Cl_4V	Vanadium(IV) chloride	298.2	3.05				
Cl_5P	Phosphorus(V) chloride	433.2	2.85				
Cl_5Sb	Antimony(V) chloride	293.0	3.222	0.45413E+01	-0.45078E-02		276-320
FH	Hydrogen fluoride	273.2	83.6	0.50352E+03	-0.19297E+01	0.14372E-02	200-273
F_2	Fluorine	53.48	1.4913	0.14144E+01	0.26387E-02	-0.28356E-04	54-144
F_5I	Iodine pentafluoride	293.2	37.13	0.95184E+02	-0.19800E+00		273-313
F_6S	Sulfur hexafluoride	223.2	1.81				
F_6Xe	Xenon hexafluoride	398.2	4.10				
F_7I	Iodine heptafluoride	298.2	1.75				
$F_{10}S_2$	Sulfur decafluoride	293.2	2.0202				
HI	Hydrogen iodide	220.2	3.87	0.51557E+03	-0.44552E+01	0.96795E-02	220-236
H_2	Hydrogen	13.52	1.2792	0.13327E+01	-0.51946E-02		14-19
H_2O	Water	293.2	80.100	0.24921E+03	-0.79069E+00	0.72997E-03	273-372
H_2O_2	Hydrogen peroxide	290.2	74.6	0.48511E+03	-0.23145E+01	0.31020E-02	233-303
H_2S	Hydrogen sulfide	283.2	5.93	0.14736E+02	-0.33675E-01	0.96740E-05	212-363
H_3N	Ammonia	293.2	16.61	0.66756E+02	-0.24696E+00	0.25913E-03	238-323
H_4N_2	Hydrazine	298.2	51.7	0.22061E+03	-0.89633E+00	0.11066E-02	278-323
He	Helium	2.055	1.0555	0.10640E+01	-0.35584E-02		2-4
I_2	Iodine	391.25	11.08	0.64730E+02	-0.29266E+00	0.39759E-03	391-441
Kr	Krypton	119.80	1.664				
Mn_2O_7	Manganese(VII) oxide	293.2	3.28	0.37655E+01	-0.16463E-02		283-312
NO	Nitric oxide		1.997				
N_2	Nitrogen	63.15	1.4680	0.12550E+01	0.67949E-02	-0.56704E-04	63-126
N_2O_3	Nitrogen trioxide	203.2	31.13	0.92287E+02	-0.43306E+00	0.65000E-03	203-243
N_2O_4	Nitrogen tetroxide	293.2	2.44	0.28212E+01	-0.13000E-02		253-293
Ne	Neon	26.11	1.1907	0.12667E+01	-0.29064E-02		26-29
O_2	Oxygen	54.478	1.5684	0.15434E+01	0.14615E-02	-0.21964E-04	55-154
O_2S	Sulfur dioxide	298.2	16.3	0.52045E+02	-0.16125E+00	0.11042E-03	213-449
O_3	Ozone	90.2	4.75	0.86344E+01	-0.54807E-01	0.12596E-03	90-185
O_3S	Sulfur trioxide	291.2	3.11				
P	Phosphorus	307.2	4.096	0.79018E+00	0.23911E-01	-0.42826E-04	307-358
S	Sulfur	407.2	3.4991	0.51651E+01	-0.77381E-02	0.89120E-05	407-479
Se	Selenium	510.65	5.44	0.67569E+01	-0.25829E-02		511-575
Xe	Xenon	161.35	1.880				
$CBrClF_2$	Bromochlorodifluoromethane	123.2	3.920	0.52442E+01	-0.11000E-01		123-223
$CBrCl_3$	Bromotrichloromethane	293.2	2.405	0.29249E+01	-0.17650E-02		273-333
$CBrF_3$	Bromotrifluoromethane	123.2	3.730	0.54154E+01	-0.13680E-01		123-173
CBr_2Cl_2	Dibromodichloromethane	298.2	2.542	0.32330E+01	-0.23162E-02		298-333

Mol. Form.	Name	T/K	ε	a	b	c	Range/K
CBr_2F_2	Dibromodifluoromethane	273.2	2.939	0.67296E+01	-0.22133E-01	0.30213E-04	139-273
CBr_3Cl	Tribromochloromethane	333.2	2.601				
CBr_3F	Tribromofluoromethane	293.2	3.00	0.53203E+01	-0.11061E-01	0.10688E-04	206-323
CBr_3NO_2	Tribromonitromethane	298.2	9.034	0.16079E+02	-0.23630E-01		298-328
$CClF_3$	Chlorotrifluoromethane	123.2	3.010	0.43677E+01	-0.11020E-01		123-173
CCl_2F_2	Dichlorodifluoromethane	123.2	3.500	0.46984E+01	-0.97600E-02		123-223
CCl_2O	Carbonyl chloride	295.2	4.30				
CCl_3D	Trichloromethane-d	298.2	4.67				
CCl_3F	Trichlorofluoromethane	293.2	3.00	0.53203E+01	-0.11061E-01	0.10688E-04	206-323
CCl_3NO_2	Trichloronitromethane	293.2	7.319	0.14403E+02	-0.24178E-01		276-333
CCl_4	Tetrachloromethane	293.2	2.2379	0.28280E+01	-0.20339E-02	0.71795E-07	283-333
CF_4	Tetrafluoromethane	126.3	1.685	0.20350E+01	-0.27616E-02		126-142
$CHBr_3$	Tribromomethane	283.2	4.404	0.71707E+01	-0.98000E-02		283-343
$CHCl_3$	Trichloromethane	293.2	4.8069	0.15115E+02	-0.51830E-01	0.56803E-04	218-323
CHF_3	Trifluoromethane	294.0	5.2	0.11442E+03	-0.75600E+00	0.13562E-02	130-263
CHN	Hydrogen cyanide	293.2	114.9	0.37331E+04	-0.23180E+02	0.36963E-01	258-299
CH_2Br_2	Dibromomethane	283.2	7.77	0.18060E+02	-0.36333E-01		283-313
CH_2Cl_2	Dichloromethane	298.0	8.93	0.40452E+02	-0.17748E+00	0.23942E-03	184-306
CH_2F_2	Difluoromethane	152.2	53.74	0.19428E+03	-0.12939E+01	0.24280E-02	152-224
CH_2I_2	Diiodomethane	298.2	5.32				
CH_2O_2	Formic acid	298.2	51.1	0.14040E+03	-0.24673E+00	-0.17151E-03	287-358
CH_3Br	Bromomethane	275.7	9.71	0.40580E+02	-0.18418E+00	0.26219E-03	195-276
CH_3Cl	Chloromethane	295.2	10.0	0.42775E+02	-0.16175E+00	0.17108E-03	190-392
CH_3ClO_2S	Methanesulfonyl chloride	293.2	34.0	0.10384E+03	-0.33838E+00	0.34156E-03	293-373
CH_3DO	Methan-d_1-ol	297.5	31.68	0.20839E+03	-0.10318E+01	0.14740E-02	176-298
CH_3F	Fluoromethane	131.0	51.0	0.11338E+03	-0.63979E+00	0.96983E-03	150-299
CH_3I	Iodomethane	293.2	6.97	0.24264E+02	-0.93914E-01	0.11926E-03	223-303
CH_3NO	Formamide	293.2	111.0	0.26076E+03	-0.61145E+00	0.34296E-03	278-333
CH_3NO_2	Nitromethane	293.2	37.27	0.11227E+03	-0.35591E+00	0.34206E-03	288-343
CH_3NO_2	Methyl nitrite	200.0	20.77	0.11071E+03	-0.73428E+00	0.14054E-02	110-260
CH_3NO_3	Methyl nitrate	293.2	23.9				
CH_4	Methane	91.0	1.6761	0.15996E+01	0.27434E-02	-0.22086E-04	91-184
CH_4O	Methanol	293.2	33.0	0.19341E+03	-0.92211E+00	0.12839E-02	177-293
CH_5N	Methylamine	215.2	16.7	0.34398E+02	-0.73630E-01	-0.41279E-04	198-258
CN_4O_8	Tetranitromethane	293.2	2.317				
COS	Carbon oxysulfide	185.0	4.47	0.84702E+01	-0.21488E-01		143-185
$COSe$	Carbon oxyselenide	283.2	3.47	0.48740E+01	-0.49425E-02		219-283
CO_2	Carbon dioxide	295.0	1.4492	0.79062E+00	0.10639E-01	-0.28510E-04	220-300
CS_2	Carbon disulfide	293.2	2.6320	0.45024E+01	-0.12054E-01	0.19147E-04	154-319
$C_2Br_2F_4$	1,2-Dibromotetrafluoroethane	298.2	2.34				
$C_2Cl_2F_4$	1,2-Dichlorotetrafluoroethane	273.2	2.4842	0.36663E+01	-0.42271E-02	-0.36255E-06	193-273
$C_2Cl_2O_2$	Oxalyl chloride	294.35	3.470				
C_2Cl_3N	Trichloroacetonitrile	292.2	7.85				
C_2Cl_4	Tetrachloroethylene	303.2	2.268				
$C_2Cl_4F_2$	1,1,2,2-Tetrachloro-1,2-difluoroethane	308.2	2.52				
C_2HBr_3O	Tribromoacetaldehyde	293.2	7.6				
C_2HCl_3	Trichloroethylene	301.5	3.390	0.58319E+01	-0.80828E-02		302-338
$C_2HCl_3F_2$	1,2,2-Trichloro-1,1-difluoroethane	303.2	4.01	0.75423E+01	-0.11667E-01		303-333
C_2HCl_3O	Trichloroacetaldehyde	298.2	6.8				
$C_2HCl_3O_2$	Trichloroacetic acid	333.2	4.34	0.13412E+01	0.90000E-02	-0.24130E-14	333-393
C_2HCl_5	Pentachloroethane	298.2	3.716	0.65972E+01	-0.96800E-02		298-338
$C_2HF_3O_2$	Trifluoroacetic acid	293.2	8.42	0.21652E+02	-0.68146E-01	0.78571E-04	263-323
C_2H_2	Acetylene	195.0	2.4841				
$C_2H_2Br_2$	cis-1,2-Dibromoethylene	298.2	7.08				
$C_2H_2Br_2$	trans-1,2-Dibromoethylene	298.2	2.88				
$C_2H_2Br_4$	1,1,2,2-Tetrabromoethane	303.2	6.72	0.16246E+02	-0.31500E-01		303-333
$C_2H_2Cl_2$	1,1-Dichloroethylene	293.2	4.60				
$C_2H_2Cl_2$	cis-1,2-Dichloroethylene	298.2	9.20				
$C_2H_2Cl_2$	trans-1,2-Dichloroethylene	293.2	2.14				

Mol. Form.	Name	T/K	ε	a	b	c	Range/K
$C_2H_2Cl_2O_2$	Dichloroacetic acid	293.2	8.33	0.11014E+02	-0.10859E-01	0.49242E-05	284-363
$C_2H_2Cl_4$	1,1,1,2-Tetrachloroethane	207.2	9.22	0.19606E+02	-0.49847E-01		207-233
$C_2H_2Cl_4$	1,1,2,2-Tetrachloroethane	293.2	8.50				
$C_2H_2I_2$	cis-1,2-Diiodoethylene	345.65	4.46				
C_2H_3ClO	Acetyl chloride	295.2	15.8				
$C_2H_3ClO_2$	Chloroacetic acid	338.2	12.35	0.17310E+02	-0.14674E-01		338-393
$C_2H_3Cl_2NO_2$	1,1-Dichloro-1-nitroethane	303.2	16.3	0.37576E+02	-0.70400E-01		303-333
$C_2H_3Cl_3$	1,1,1-Trichloroethane	293.2	7.243	0.27705E+02	-0.10621E+00	0.12424E-03	258-318
$C_2H_3Cl_3$	1,1,2-Trichloroethane	298.2	7.1937	0.17147E+02	-0.33371E-01		288-318
$C_2H_3F_3O$	2,2,2-Trifluoroethanol	293.2	27.68	0.90593E+02	-0.21421E+00		293-318
C_2H_3N	Acetonitrile	293.2	36.64	0.29724E+03	-0.15508E+01	0.22591E-02	288-333
C_2H_3NO	Methyl isocyanate	288.7	21.75				
C_2H_4	Ethylene	270.0	1.4833	0.13546E+01	0.62614E-02	-0.21374E-04	200-270
C_2H_4BrCl	1-Bromo-2-chloroethane	283.2	7.41	0.19493E+02	-0.59054E-01	0.58036E-04	263-363
$C_2H_4Br_2$	1,2-Dibromoethane	293.2	4.9612	0.67142E+01	-0.59800E-02		293-313
$C_2H_4Cl_2$	1,1-Dichloroethane	298.2	10.10	0.24429E+02	-0.48000E-01		288-318
$C_2H_4Cl_2$	1,2-Dichloroethane	293.2	10.42	0.24404E+02	-0.47892E-01		293-343
$C_2H_4Cl_2O$	Bis(chloromethyl) ether	293.2	3.51				
$C_2H_4N_2O_6$	Ethylene glycol dinitrate	293.2	28.26				
C_2H_4O	Acetaldehyde	291.2	21.0				
C_2H_4O	Ethylene oxide	293.2	12.42	0.52661E+02	-0.21337E+00	0.25947E-03	293-243
C_2H_4OS	Thioacetic acid	298.2	14.30				
$C_2H_4O_2$	Acetic acid	293.2	6.20	-0.15731E+02	0.12662E+00	-0.17738E-03	293-363
$C_2H_4O_2$	Methyl formate	288.2	9.20	0.19699E+02	-0.36429E-01		288-302
$C_2H_4O_3S$	Ethylene glycol sulfite	298.2	39.6	0.85483E+02	-0.15400E+00		298-328
C_2H_5Br	Bromoethane	298.2	9.01	0.28473E+02	-0.85495E-01	0.67971E-04	243-308
C_2H_5Cl	Chloroethane	293.2	9.45	0.60693E+02	-0.31290E+00	0.47154E-03	237-293
C_2H_5ClO	2-Chloroethanol	293.2	25.80	0.11155E+03	-0.30149E+00		140-175
C_2H_5I	Iodoethane	293.2	7.82	0.25598E+02	-0.94367E-01	0.11424E-03	183-343
C_2H_5N	Ethyleneimine	298.2	18.3	0.61405E+02	-0.14474E+00		273-298
C_2H_5NO	Acetamide	363.7	67.6	-0.20055E+03	0.15515E+01	-0.22392E-02	364-448
C_2H_5NO	N-Methylformamide	293.2	189.0	0.10383E+04	-0.43165E+01	0.48398E-02	276-353
C_2H_5NO	Acetaldoxime	298.2	4.70				
$C_2H_5NO_2$	Nitroethane	288.2	29.11	0.57406E+02	-0.97657E-01		276-333
$C_2H_5NO_2$	Methyl carbamate	328.2	18.48	0.36773E+02	-0.55700E-01		328-368
$C_2H_5NO_3$	Ethyl nitrate	293.2	19.7				
C_2H_6	Ethane	95.0	1.9356	0.20185E+01	-0.51493E-03	-0.48148E-05	95-295
C_2H_6O	Ethanol	293.2	25.3	0.15145E+03	-0.87020E+00	0.19570E-02	163-523
C_2H_6O	Dimethyl ether	258.0	6.18	0.22389E+02	-0.86524E-01	0.91291E-04	155-258
C_2H_6OS	Dimethyl sulfoxide	293.2	47.24	0.38478E+02	0.16939E+00	-0.47423E-03	288-343
$C_2H_6O_2$	Ethylene glycol	293.2	41.4	0.14355E+03	-0.48573E+00	0.46703E-03	293-423
$C_2H_6O_2S$	Dimethyl sulfone	383.2	47.39	0.10830E+03	-0.15900E+00		383-398
$C_2H_6O_4S$	Dimethyl sulfate	298.2	55.0				
C_2H_6S	Ethanethiol	298.2	6.667				
C_2H_6S	Dimethyl sulfide	294.2	6.70				
$C_2H_6S_2$	1,2-Ethanedithiol	293.2	7.26	0.11228E+02	-0.13500E-01		293-333
$C_2H_6S_2$	Dimethyl disulfide	298.2	9.6	0.19109E+02	-0.32000E-01		298-323
C_2H_7N	Ethylamine	273.2	8.7	0.30163E+02	-0.79000E-01		233-273
C_2H_7NO	Ethanolamine	293.2	31.94	0.14890E+03	-0.62491E+00	0.77143E-03	253-293
$C_2H_8N_2$	1,2-Ethanediamine	293.2	13.82	0.48922E+02	-0.17021E+00	0.17262E-03	273-333
C_3Cl_6O	Hexachloroacetone	291.9	3.925	0.76423E+01	-0.15838E-01	0.10618E-04	269-303
C_3F_6O	Perfluoroacetone	202.2	2.104	0.34809E+01	-0.92883E-02	0.12282E-04	151-238
C_3HN	Cyanoacetylene	291.9	72.3	0.91803E+03	-0.49149E+01	0.69104E-02	281-314
$C_3H_2F_6O$	1,1,1,3,3,3-Hexafluoro-2-propanol	293.2	16.70				
$C_3H_3ClO_3$	4-Chloro-1,3-dioxolan-2-one	313.2	62.0				
C_3H_3N	Acrylonitrile	293.2	33.0	0.11109E+03	-0.36806E+00	0.34879E-03	233-413
$C_3H_3NO_2$	Cyanoacetic acid	277.2	33.4				
C_3H_4	Allene	269.0	2.025	0.26049E+01	-0.44147E-03	-0.63420E-05	156-269
C_3H_4	Propyne	246.0	3.218	0.60871E+01	-0.11730E-01		185-246

Mol. Form.	Name	T/K	ε	a	b	c	Range/K
$C_3H_4ClF_3$	3-Chloro-1,1,1-trifluoropropane	295.2	7.32	0.22361E+02	-0.68840E-01	0.60594E-04	275-313
C_3H_4ClNO	2-Chloroethyl isocyanate	288.2	29.1	0.64311E+02	-0.12217E+00		288-403
$C_3H_4Cl_2O$	1,1-Dichloroacetone	293.2	14.6				
$C_3H_4F_4O$	2,2,3,3-Tetrafluoro-1-propanol	298.2	21.03				
C_3H_4O	Propargyl alcohol	293.2	20.8	0.99895E+02	-0.38911E+00	0.40776E-03	213-293
$C_3H_4O_3$	Ethylene carbonate	313.2	89.78	0.20746E+03	-0.37610E+00		313-343
C_3H_5Br	3-Bromopropene	293.2	7.0				
$C_3H_5BrO_2$	2-Bromopropanoic acid	294.2	11.0				
$C_3H_5Br_3$	1,2,3-Tribromopropane	303.2	6.00	0.11024E+02	-0.16596E-01		303-358
C_3H_5Cl	2-Chloropropene	299.25	8.92				
C_3H_5Cl	3-Chloropropene	293.2	8.2				
$C_3H_5ClN_2O_6$	3-Chloro-1,2-propanediol dinitrate	293.2	17.50				
C_3H_5ClO	Epichlorohydrin	293.2	22.6				
$C_3H_5ClO_2$	Ethyl chloroformate	308.7	9.736	0.15356E+02	-0.18250E-01		309-349
$C_3H_5ClO_2$	Methyl chloroacetate	293.2	12.0				
$C_3H_5Cl_3$	1,2,3-Trichloropropane	293.2	7.5				
C_3H_5I	3-Iodopropene	292.2	6.1				
C_3H_5N	Propanenitrile	293.2	29.7	0.82222E+02	-0.22937E+00	0.17424E-03	213-473
C_3H_5NO	Ethyl isocyanate	293.2	19.7				
C_3H_5NS	Ethyl isothiocyanate	293.2	19.6				
$C_3H_5N_3O_9$	Trinitroglycerol	293.2	19.25				
C_3H_6	Propene	220.0	2.1365	0.29623E+01	-0.37564E-02		220-250
$C_3H_6Br_2$	1,2-Dibromopropane	283.2	4.60	0.54973E+01	-0.31695E-02		283-333
$C_3H_6Br_2$	1,3-Dibromopropane	293.2	9.482	0.29193E+02	-0.94450E-01	0.92800E-04	293-368
$C_3H_6ClNO_2$	2-Chloro-2-nitropropane	250.4	31.90				
$C_3H_6Cl_2$	1,2-Dichloropropane	293.2	8.37	0.18915E+02	-0.35907E-01		281-323
$C_3H_6Cl_2$	1,3-Dichloropropane	303.2	10.27	0.21609E+02	-0.37333E-01		303-333
$C_3H_6Cl_2$	2,2-Dichloropropane	293.2	11.37	0.32421E+02	-0.72188E-01		245-293
$C_3H_6N_2O_4$	2,2-Dinitropropane	325.1	42.4				
C_3H_6O	Allyl alcohol	293.2	19.7	0.62714E+02	-0.14771E+00	0.37879E-05	213-303
C_3H_6O	Propanal	290.2	18.5				
C_3H_6O	Acetone	293.2	21.01	0.88157E+02	-0.34300E+00	0.38925E-03	273-323
$C_3H_6O_2$	Propanoic acid	298.2	3.44	0.18793E+01	0.46841E-02	0.19983E-05	289-408
$C_3H_6O_2$	Ethyl formate	288.2	8.57	0.15884E+02	-0.25333E-01		288-318
$C_3H_6O_2$	Methyl acetate	288.2	7.07	0.13190E+02	-0.21226E-01		276-318
$C_3H_6O_3$	3-Hydroxypropanoic acid	296.2	30.0				
$C_3H_6O_3$	Dimethyl carbonate	298.2	3.087				
$C_3H_6O_3$	1,3,5-Trioxane	338.2	15.55				
C_3H_7Br	1-Bromopropane	293.2	8.09	0.17769E+02	-0.32599E-01		274-328
C_3H_7Br	2-Bromopropane	293.2	9.46	0.26195E+02	-0.72995E-01	0.55454E-04	186-328
C_3H_7Cl	1-Chloropropane	293.2	8.588	0.21214E+02	-0.43130E-01		273-313
C_3H_7ClO	3-Chloro-1-propanol	215.2	36.0	0.12436E+03	-0.60841E+00	0.92060E-03	145-215
C_3H_7ClO	1-Chloro-2-propanol	153.2	59.0	-0.19169E+02	0.13605E+01	-0.55567E-02	153-177
$C_3H_7ClO_2$	3-Chloro-1,2-propanediol	293.2	31.0				
C_3H_7I	1-Iodopropane	293.2	7.07	0.13744E+02	-0.22745E-01		293-323
C_3H_7I	2-Iodopropane	298.2	8.19				
C_3H_7NO	N-Ethylformamide	298.2	102.7	0.64764E+03	-0.28499E+01	0.34286E-02	298-338
C_3H_7NO	N,N-Dimethylformamide	293.2	38.25	0.15364E+03	-0.60367E+00	0.71505E-03	213-353
C_3H_7NO	N-Methylacetamide	303.2	179.0	0.15975E+04	-0.90451E+01	0.18345E-01	303-473
$C_3H_7NO_2$	1-Nitropropane	288.2	24.70	0.94999E+02	-0.38358E+00	0.48480E-03	276-333
$C_3H_7NO_2$	2-Nitropropane	288.2	26.74	0.60138E+02	-0.11566E+00		276-303
$C_3H_7NO_2$	Propyl nitrite	250.0	12.35	0.70552E+02	-0.40362E+00	0.66687E-03	110-310
$C_3H_7NO_2$	Isopropyl nitrite	260.0	13.92	0.74578E+02	-0.38283E+00	0.57071E-03	150-300
$C_3H_7NO_2$	Ethyl carbamate	328.2	14.14	0.32431E+02	-0.65097E-01	0.28571E-04	328-368
C_3H_8	Propane	293.19	1.6678	0.22883E+01	-0.23276E-02	0.84710E-06	90-300
C_3H_8O	1-Propanol	293.2	20.8	0.98045E+02	-0.36860E+00	0.36422E-03	193-493
C_3H_8O	2-Propanol	293.2	20.18	0.10416E+03	-0.41011E+00	0.42049E-03	193-493
$C_3H_8O_2$	1,2-Propylene glycol	303.2	27.5	0.24546E+03	-0.15738E+01	0.38068E-02	193-403
$C_3H_8O_2$	1,3-Propylene glycol	293.2	35.1	0.11365E+03	-0.36680E+00	0.33766E-03	288-328

Mol. Form.	Name	T/K	ε	a	b	c	Range/K
C$_3$H$_8$O$_2$	Ethylene glycol monomethyl ether	298.2	17.2	0.11803E+03	-0.58000E+00	0.81001E-03	254-318
C$_3$H$_8$O$_2$	Dimethoxymethane	293.2	2.644	0.25877E+01	-0.93019E-03	0.38472E-05	171-293
C$_3$H$_8$O$_3$	Glycerol	293.2	46.53	0.77503E+02	-0.37984E-01	-0.23107E-03	288-343
C$_3$H$_8$S	1-Propanethiol	288.2	5.937	0.11602E+02	-0.19580E-01		273-318
C$_3$H$_8$S	2-Propanethiol	298.2	5.952				
C$_3$H$_8$S$_2$	1,2-Propanedithiol	293.2	7.24	0.14667E+02	-0.32660E-01	0.25000E-04	293-333
C$_3$H$_8$S$_2$	1,3-Propanedithiol	303.2	8.11	0.66607E+01	0.31310E-01	-0.87500E-04	303-343
C$_3$H$_9$BO$_3$	Trimethyl borate	293.2	2.2762				
C$_3$H$_9$ClSi	Trimethylchlorosilane	273.2	10.21	-0.19492E+02	0.29806E+00	-0.69284E-03	223-273
C$_3$H$_9$N	Propylamine	296.2	5.08	0.17719E+02	-0.59022E-01	0.54780E-04	204-296
C$_3$H$_9$N	Isopropylamine	293.2	5.6268	0.40429E+02	-0.21441E+00	0.32634E-03	213-298
C$_3$H$_9$N	Trimethylamine	298.2	2.440	0.39745E+01	-0.51331E-02		273-298
C$_3$H$_9$O$_4$P	Trimethyl phosphate	293.2	20.6				
C$_4$Cl$_6$	Hexachloro-1,3-butadiene	293.2	2.55				
C$_4$Cl$_6$O$_3$	Trichloroacetic anhydride	298.2	5.0				
C$_4$F$_6$O$_3$	Trifluoroacetic acid anhydride	298.2	2.7				
C$_4$H$_2$Cl$_4$O$_3$	Dichloroacetic anhydride	298.2	15.8				
C$_4$H$_2$O$_3$	Maleic anhydride	326.2	52.75				
C$_4$H$_3$F$_7$O	2,2,3,3,4,4,4-Heptafluoro-1-butanol	298.2	14.4				
C$_4$H$_4$N$_2$	Succinonitrile	298.2	62.6	0.17724E+03	-0.54654E+00	0.54046E-03	236-351
C$_4$H$_4$N$_2$	Pyrazine	323.2	2.80				
C$_4$H$_4$O	Furan	277.1	2.88	0.13636E+01	0.12864E-01	-0.22701E-04	188-277
C$_4$H$_4$S	Thiophene	293.2	2.739	0.32941E+01	-0.19019E-02		253-293
C$_4$H$_5$Cl	2-Chloro-1,3-butadiene	293.2	4.914				
C$_4$H$_5$Cl$_3$O$_2$	Ethyl trichloroacetate	293.2	8.428				
C$_4$H$_5$N	Pyrrole	293.0	8.00	0.12672E+02	-0.14075E-01	-0.62671E-05	293-357
C$_4$H$_5$NO	Allyl isocynate	288.2	15.15	0.34299E+02	-0.66444E-01		288-333
C$_4$H$_6$	1,3-Butadiene	265.0	2.050	0.27674E+01	-0.26738E-02		185-265
C$_4$H$_6$O	Divinyl ether	288.2	3.94				
C$_4$H$_6$O	Ethoxyacetylene	298.2	8.05				
C$_4$H$_6$O	Cyclobutanone	298.2	14.27	0.43974E+02	-0.15712E+00	0.19264E-03	220-317
C$_4$H$_6$O$_2$	Methyl acrylate	303.2	7.03	0.11968E+02	-0.16500E-01		303-333
C$_4$H$_6$O$_2$	2,3-Butanedione	298.2	4.04	0.46907E+01	-0.22302E-02		278-348
C$_4$H$_6$O$_2$	γ-Butyrolactone	293.2	39.0				
C$_4$H$_6$O$_3$	Acetic anhydride	293.2	22.45				
C$_4$H$_6$O$_3$	Propylene carbonate	293.0	66.14	0.15940E+03	-0.39530E+00	0.26284E-03	273-333
C$_4$H$_7$Br	cis-2-Bromo-2-butene	293.2	5.38				
C$_4$H$_7$Br	trans-2-Bromo-2-butene	293.2	6.76				
C$_4$H$_7$BrO$_2$	2-Bromobutanoic acid	293.2	7.2				
C$_4$H$_7$BrO$_2$	Ethyl bromoacetate	303.2	9.75	0.15627E+02	-0.19600E-01		303-333
C$_4$H$_7$BrO$_2$	Methyl 3-bromopropanoate	303.2	5.81	0.36001E+01	0.72500E-02		303-343
C$_4$H$_7$ClO$_2$	Propyl chlorocarbonate	293.2	11.2				
C$_4$H$_7$ClO$_2$	Methyl 2-chloropropanoate	303.2	11.45	0.22449E+02	-0.36250E-01		303-343
C$_4$H$_7$N	Butanenitrile	293.2	24.83	0.53884E+02	-0.99257E-01		293-333
C$_4$H$_7$N	2-Methylpropanenitrile	293.2	24.42	0.52554E+02	-0.96000E-01		293-313
C$_4$H$_7$NO	2-Pyrrolidone	298.2	28.18	0.11054E+03	-0.47945E+00	0.68182E-03	298-338
C$_4$H$_8$	1-Butene	220.0	2.2195	0.29354E+01	-0.32580E-02		220-250
C$_4$H$_8$	cis-2-Butene	296.0	1.960	0.28802E+01	-0.31064E-02		197-296
C$_4$H$_8$	Isobutene	288.7	2.1225	0.33701E+01	-0.43295E-02		220-289
C$_4$H$_8$Br$_2$	1,2-Dibromobutane	293.2	4.74	0.11199E+03	-0.63334E+00	0.91250E-03	293-333
C$_4$H$_8$Br$_2$	1,3-Dibromobutane	293.2	9.14	0.34031E+02	-0.13254E+00	0.16250E-03	293-333
C$_4$H$_8$Br$_2$	1,4-Dibromobutane	303.2	8.68	0.20944E+02	-0.55620E-01	0.50000E-04	303-333
C$_4$H$_8$Br$_2$	2,3-Dibromobutane	298.2	6.245	0.23849E+02	-0.96300E-01	0.12500E-03	293-333
C$_4$H$_8$Br$_2$	1,2-Dibromo-2-methylpropane	293.2	4.1				
C$_4$H$_8$Cl$_2$	1,2-Dichlorobutane	293.2	7.74	0.31925E+02	-0.13232E+00	0.17007E-03	293-356
C$_4$H$_8$Cl$_2$	1,4-Dichlorobutane	308.2	9.30	0.59766E+01	0.49300E-01	-0.12500E-03	308-338
C$_4$H$_8$Cl$_2$	1,2-Dichloro-2-methylpropane	296.0	7.15	0.39429E+02	-0.20028E+00	0.30917E-03	165-296
C$_4$H$_8$Cl$_2$O	Bis(2-chloroethyl) ether	293.2	21.20				
C$_4$H$_8$O	Butanal	298.2	13.45				

Mol. Form.	Name	T/K	ε	a	b	c	Range/K
C₄H₈O	2-Butanone	293.2	18.56	0.15457E+02	0.90152E-01	-0.27100E-03	293-333
C₄H₈O	Tetrahydrofuran	295.2	7.52	0.30739E+02	-0.12946E+00	0.17195E-03	224-295
C₄H₈O₂	Butanoic acid	287.2	2.98	0.15010E+01	0.50046E-02		287-403
C₄H₈O₂	2-Methylpropanoic acid	293.2	2.58				
C₄H₈O₂	Propyl formate	303.2	6.92				
C₄H₈O₂	Ethyl acetate	293.2	6.0814	0.15646E+02	-0.44066E-01	0.39137E-04	293-433
C₄H₈O₂	Methyl propanoate	293.2	6.200	0.12798E+02	-0.22540E-01		293-333
C₄H₈O₂	1,4-Dioxane	293.2	2.2189	0.27299E+01	-0.17440E-02		293-313
C₄H₈O₃	2-Hydroxybutanoic acid	296.2	37.7				
C₄H₈O₃	3-Hydroxybutanoic acid	296.2	31.5				
C₄H₈O₃	Ethyl methyl carbonate	293.2	2.985				
C₄H₈O₃	Ethylene glycol monoacetate	303.2	12.95				
C₄H₉Br	1-Bromobutane	283.2	7.315	0.22542E+02	-0.79306E-01	0.89867E-04	183-363
C₄H₉Br	2-Bromobutane	298.2	8.64	0.18461E+02	-0.32933E-01		274-328
C₄H₉Br	1-Bromo-2-methylpropane	273.2	7.70	0.37558E+02	-0.20571E+00	0.35496E-03	112-273
C₄H₉Br	2-Bromo-2-methylpropane	293.0	10.98	0.35085E+02	-0.14075E+00	0.19960E-03	258-293
C₄H₉Cl	1-Chlorobutane	293.2	7.276	0.13565E+02	-0.10161E-01	-0.38750E-04	273-323
C₄H₉Cl	2-Chlorobutane	293.2	8.564	0.30376E+02	-0.11377E+00	0.13429E-03	273-323
C₄H₉Cl	1-Chloro-2-methylpropane	293.2	7.027	0.14945E+02	-0.33747E-01	0.23036E-04	273-323
C₄H₉Cl	2-Chloro-2-methylpropane	293.2	9.663	0.35077E+02	-0.12867E+00	0.14304E-03	273-323
C₄H₉I	1-Iodobutane	293.2	6.27	0.16493E+02	-0.50262E-01	0.52485E-04	293-323
C₄H₉I	2-Iodobutane	293.2	7.873	0.10883E+02	-0.14680E-02	-0.30000E-04	293-323
C₄H₉I	2-Iodo-2-methylpropane	283.2	6.65	0.76780E+01	0.69900E-02	-0.37500E-04	283-323
C₄H₉N	Pyrrolidine	293.0	8.30	0.38191E+02	-0.15462E+00	0.17941E-03	274-333
C₄H₉NO	N-Methylpropanamide	293.2	170.0				
C₄H₉NO	N-Ethylacetamide	293.2	135.0	0.74494E+03	-0.31400E+01	0.36131E-02	213-353
C₄H₉NO	N,N-Dimethylacetamide	294.2	38.85	0.15420E+03	-0.57506E+00	0.61911E-03	294-433
C₄H₉NO	2-Butanone oxime	293.2	3.4				
C₄H₉NO	Morpholine	298.2	7.42				
C₄H₉NO₂	tert-Butyl nitrite	298.2	11.47				
C₄H₉NO₂	Propyl carbamate	338.2	12.06	0.24356E+02	-0.36400E-01		338-378
C₄H₉NO₂	Ethyl-N-methyl carbamate	298.2	21.10	0.11477E+03	-0.47568E+00	0.54127E-03	298-373
C₄H₉NO₂	N-Acetylethanolamine	298.2	96.6	0.37016E+03	-0.13113E+01	0.13214E-02	298-348
C₄H₉NO₃	Butyl nitrate	293.2	13.10				
C₄H₁₀	Butane	295.0	1.7697	0.22379E+01	-0.13884E-02	-0.66711E-06	135-303
C₄H₁₀	Isobutane	295.0	1.7518	0.23295E+01	-0.19953E-02	0.14197E-06	115-303
C₄H₁₀O	1-Butanol	293.2	17.84	0.10578E+03	-0.50587E+00	0.84733E-03	193-553
C₄H₁₀O	2-Butanol	293.2	17.26	0.13850E+03	-0.75146E+00	0.14086E-02	172-533
C₄H₁₀O	2-Methyl-1-propanol	293.2	17.93	0.10762E+03	-0.51398E+00	0.83702E-03	173-533
C₄H₁₀O	2-Methyl-2-propanol	298.2	12.47	0.22541E+03	-0.14990E+01	0.34050E-02	298-503
C₄H₁₀O	Diethyl ether	293.2	4.2666	0.79725E+01	-0.12519E-01		283-301
C₄H₁₀O₂	1,2-Butanediol	298.2	22.4	0.63702E+02	-0.13807E+00		278-323
C₄H₁₀O₂	1,3-Butanediol	298.2	28.8	0.72883E+02	-0.14770E+00		278-323
C₄H₁₀O₂	1,4-Butanediol	298.2	31.9	0.13079E+03	-0.46985E+00	0.46320E-03	288-328
C₄H₁₀O₂	Ethylene glycol monoethyl ether	298.2	13.38				
C₄H₁₀O₂	Ethylene glycol dimethyl ether	296.7	7.30	0.48832E+02	-0.24218E+00	0.34413E-03	256-318
C₄H₁₀O₂S	Bis(2-hydroxyethyl) sulfide	293.2	28.61	0.13128E+03	-0.52719E+00	0.60465E-03	253-333
C₄H₁₀O₃	Diethylene glycol	293.2	31.82	0.13973E+03	-0.54725E+00	0.61149E-03	288-343
C₄H₁₀O₃S	Diethyl sulfite	293.2	15.6				
C₄H₁₀O₄	1,2,3,4-Butanetetrol	393.2	28.2				
C₄H₁₀O₄S	Diethyl sulfate	293.2	29.2				
C₄H₁₀S	1-Butanethiol	288.2	5.204	0.11201E+02	-0.20767E-01		273-318
C₄H₁₀S	2-Butanethiol	288.2	5.645	0.10866E+02	-0.17993E-01		273-318
C₄H₁₀S	2-Methyl-1-propanethiol	298.2	4.961				
C₄H₁₀S	2-Methyl-2-propanethiol	293.2	5.475	0.10597E+02	-0.17500E-01		283-313
C₄H₁₀S	Diethyl sulfide	298.2	5.723				
C₄H₁₁N	Butylamine	293.2	4.71	0.13322E+02	-0.44176E-01	0.50250E-04	223-333
C₄H₁₁N	Diethylamine	293.2	3.680	0.26462E+02	-0.13750E+00	0.20373E-03	243-323
C₄H₁₁NO₂	Diethanolamine	293.2	25.75	0.73435E+02	-0.21377E+00	0.17500E-03	273-323

Mol. Form.	Name	T/K	ε	a	b	c	Range/K
$C_4H_{12}O_2Si$	Dimethoxydimethylsilane	298.2	3.663				
$C_4H_{12}O_3Si$	Trimethoxymethylsilane	298.2	4.9				
$C_4H_{12}O_4Si$	Tetramethyl silicate	293.2	6.0				
$C_4H_{12}Si$	Diethylsilane	293.2	2.544				
$C_4H_{12}Si$	Tetramethylsilane	293.2	1.921				
$C_4H_{13}N_3$	Diethylenetriamine	293.2	12.62	0.57840E+02	-0.23873E+00	0.28841E-03	213-333
C_5FeO_5	Iron pentacarbonyl	293.2	2.602				
C_5H_4BrN	2-Bromopyridine	298.2	23.18	0.73391E+02	-0.23678E+00	0.22930E-03	298-398
C_5H_4ClN	2-Chloropyridine	298.2	27.32	0.98702E+02	-0.34237E+00	0.34502E-03	298-398
$C_5H_4F_8O$	2,2,3,3,4,4,5,5-Octafluoro-1-pentanol	298.2	15.30				
$C_5H_4O_2$	Furfural	293.2	42.1				
C_5H_5N	Pyridine	293.2	13.260	0.43991E+02	-0.15150E+00	0.15925E-03	293-323
C_5H_5NO	Pyridine-1-oxide	343.0	35.94	0.20878E+02	0.16450E+00	-0.35269E-03	343-398
C_5H_6O	2-Methylfuran	293.2	2.76				
$C_5H_6O_2$	Furfuryl alcohol	298.2	16.85				
$C_5H_7Cl_3O_2$	Propyl trichloroacetate	298.2	8.32				
$C_5H_7NO_2$	Ethyl cyanoacetate	263.2	31.62				
C_5H_8	1,3-Pentadiene*	298.2	2.319				
C_5H_8	1,4-Pentadiene	294.0	2.054	0.29994E+01	-0.34578E-02	0.85300E-06	178-294
C_5H_8	2-Methyl-1,3-butadiene	293.2	2.098	0.28170E+01	-0.23147E-02	-0.43975E-06	198-293
C_5H_8	Cyclopentene	295.0	2.083	0.28177E+01	-0.27597E-02	0.89346E-06	171-319
C_5H_8O	Cyclopentanone	298.2	13.58	0.24083E+02	-0.30286E-01	-0.16802E-04	219-298
$C_5H_8O_2$	Ethyl acrylate	303.2	6.05	0.47827E+02	-0.24394E+00	0.35000E-03	303-343
$C_5H_8O_2$	Methyl trans-2-butenoate	293.2	6.6645				
$C_5H_8O_2$	Methyl methacrylate	303.2	6.32	0.32098E+02	-0.14568E+00	0.20000E-03	303-343
$C_5H_8O_2$	2,4-Pentanedione	303.2	26.524				
$C_5H_8O_4$	Dimethyl malonate	293.2	9.82	0.26470E+02	-0.76656E-01	0.67888E-04	293-433
$C_5H_9BrO_2$	Ethyl 2-bromopropanoate	293.2	9.4				
$C_5H_9ClO_2$	Isobutyl chlorocarbonate	293.2	9.1				
$C_5H_9ClO_2$	Ethyl 2-chloropropanoate	303.2	11.95	0.25965E+02	-0.46250E-01		303-343
$C_5H_9ClO_2$	Ethyl 3-chloropropanoate	303.2	10.19	0.21951E+02	-0.38750E-01		303-343
$C_5H_9ClO_2$	Methyl 4-chlorobutanoate	303.2	9.51	0.17127E+02	-0.25000E-01		303-343
C_5H_9N	Pentanenitrile	293.2	20.04	0.55793E+02	-0.15750E+00	0.12432E-03	183-333
C_5H_9N	2,2-Dimethylpropanenitrile	293.2	21.1	0.58418E+02	-0.16884E+00	0.14131E-03	293-453
C_5H_9NO	Isobutyl isocyanate	293.2	11.638	0.38026E+02	-0.12714E+00	0.12679E-03	293-353
C_5H_9NO	N-Methyl-2-pyrrolidone	293.2	32.55				
C_5H_{10}	1-Pentene	293.2	2.011	-0.11438E+01	0.25420E-01	-0.50000E-04	273-293
C_5H_{10}	2-Methyl-1-butene	293.2	2.180				
C_5H_{10}	2-Methyl-2-butene	296.0	1.979	0.26064E+01	-0.19578E-02	-0.53908E-06	225-296
C_5H_{10}	Cyclopentane	293.2	1.9687	0.24287E+01	-0.15304E-02	-0.13095E-06	278-313
C_5H_{10}	Ethylcyclopropane	293.2	1.933				
$C_5H_{10}Br_2$	1,2-Dibromopentane	298.2	4.39				
$C_5H_{10}Br_2$	1,4-Dibromopentane	293.2	9.05	0.26443E+02	-0.88640E-01	0.10000E-03	293-333
$C_5H_{10}Br_2$	1,5-Dibromopentane	303.2	9.14	0.38192E+02	-0.15648E+00	0.20000E-03	303-333
$C_5H_{10}Cl_2$	1,2-Dichloropentane	293.2	6.89	0.19016E+02	-0.57954E-01	0.56801E-04	293-356
$C_5H_{10}Cl_2$	1,5-Dichloropentane	298.2	9.92				
$C_5H_{10}O$	Cyclopentanol	288.2	18.5	0.10565E+03	-0.44244E+00	0.48657E-03	258-323
$C_5H_{10}O$	Pentanal	293.2	10.00				
$C_5H_{10}O$	2,2-Dimethylpropanal	293.2	9.051	0.18645E+02	-0.32395E-01	-0.16157E-05	280-333
$C_5H_{10}O$	2-Pentanone	293.2	15.45	0.40893E+02	-0.10423E+00	0.60557E-04	204-353
$C_5H_{10}O$	3-Pentanone	293.2	17.00	0.12690E+02	0.95177E-01	-0.27321E-03	233-353
$C_5H_{10}O$	3-Methyl-2-butanone	293.2	10.37	0.30695E+02	-0.10962E+00	0.13810E-03	293-328
$C_5H_{10}O$	Tetrahydropyran	293.2	5.66	0.19793E+02	-0.76071E-01	0.94852E-04	234-333
$C_5H_{10}O$	2-Methyltetrahydrofuran	298.2	6.97				
$C_5H_{10}O_2$	Pentanoic acid	294.4	2.661	0.33491E+01	-0.75156E-02	0.17820E-04	250-344
$C_5H_{10}O_2$	Butyl formate	303.2	6.10	0.21532E+02	-0.84106E-01	0.10952E-03	288-323
$C_5H_{10}O_2$	Isobutyl formate	293.2	6.41				
$C_5H_{10}O_2$	Propyl acetate	293.2	5.62	0.17677E+02	-0.61404E-01	0.69196E-04	253-353
$C_5H_{10}O_2$	Ethyl propanoate	293.2	5.76				

Mol. Form.	Name	T/K	ε	a	b	c	Range/K
$C_5H_{10}O_2$	Methyl butanoate	301.2	5.48	0.38604E+02	-0.19171E+00	0.27128E-03	301-343
$C_5H_{10}O_2$	Tetrahydrofurfuryl alcohol	303.2	13.48				
$C_5H_{10}O_2S$	3-Methyl sulfolane	298.2	29.4	0.53158E+02	-0.93730E-01	0.47275E-04	298-398
$C_5H_{10}O_3$	Diethyl carbonate	297.2	2.820				
$C_5H_{10}O_3$	Ethyl lactate	303.2	15.4	0.31225E+02	-0.43531E-01	-0.28571E-04	273-373
$C_5H_{10}O_4$	1,2,3-Propanetriol-1-acetate	242.2	38.57	0.10653E+03	-0.26439E+00	-0.62371E-04	215-242
$C_5H_{11}Br$	2-Bromo-2-methylbutane	298.2	9.21				
$C_5H_{11}Br$	1-Bromopentane	299.2	6.31	0.20954E+02	-0.78743E-01	0.98908E-04	183-328
$C_5H_{11}Br$	3-Bromopentane	298.2	8.37				
$C_5H_{11}Br$	1-Bromo-3-methylbutane	291.5	6.33	0.27743E+02	-0.13927E+00	0.22627E-03	123-292
$C_5H_{11}Cl$	1-Chloropentane	293.2	6.654	0.18626E+02	-0.54719E-01	0.47143E-04	273-323
$C_5H_{11}Cl$	1-Chloro-3-methylbutane	292.0	6.10	0.22228E+02	-0.93189E-01	0.12991E-03	171-297
$C_5H_{11}Cl$	2-Chloro-2-methylbutane	222.75	12.31	0.55104E+02	-0.29866E+00	0.47840E-03	201-223
$C_5H_{11}F$	1-Fluoropentane	293.2	3.931				
$C_5H_{11}I$	1-Iodopentane	293.2	5.78	0.15753E+02	-0.50543E-01	0.56401E-04	293-323
$C_5H_{11}I$	3-Iodopentane	293.2	7.432				
$C_5H_{11}I$	1-Iodo-3-methylbutane	292.2	5.6				
$C_5H_{11}I$	2-Iodo-2-methylbutane	293.2	8.192				
$C_5H_{11}N$	Piperidine	293.0	4.33	0.82317E+01	-0.11229E-01	-0.71429E-05	293-333
$C_5H_{11}N$	N-Methylpyrrolidine	298.2	32.2				
$C_5H_{11}NO$	2,2-Dimethylpropanamide	298.2	20.13	0.10400E+03	-0.46017E+00	0.60000E-03	298-328
$C_5H_{11}NO$	N,N-Diethylformamide	293.2	29.6				
$C_5H_{11}NO$	2-Pentanone oxime	293.2	3.3				
$C_5H_{11}NO_2$	Pentyl nitrite	298.2	7.21				
C_5H_{12}	Pentane	293.2	1.8371				
C_5H_{12}	Isopentane	293.2	1.845	0.22384E+01	-0.12985E-02	-0.16182E-06	143-293
C_5H_{12}	Neopentane	296.0	1.769	0.10949E+02	-0.63057E-01	0.10835E-03	251-296
$C_5H_{12}N_2O$	Tetramethylurea	293.2	23.10				
$C_5H_{12}O$	1-Pentanol	298.2	15.13	0.73397E+02	-0.28165E+00	0.28427E-03	213-513
$C_5H_{12}O$	2-Pentanol	298.2	13.71	0.16437E+03	-0.86506E+00	0.11955E-02	273-323
$C_5H_{12}O$	3-Pentanol	298.2	13.35	0.12838E+03	-0.60980E+00	0.75000E-03	288-318
$C_5H_{12}O$	2-Methyl-1-butanol	298.2	15.63	0.14020E+02	0.13948E+00	-0.45000E-03	288-318
$C_5H_{12}O$	3-Methyl-1-butanol	293.2	15.63	0.79733E+02	-0.31272E+00	0.32014E-03	173-513
$C_5H_{12}O$	2-Methyl-2-butanol	298.2	5.78	0.11662E+03	-0.69756E+00	0.10920E-02	268-318
$C_5H_{12}O$	3-Methyl-2-butanol	298.2	12.1				
$C_5H_{12}O$	2,2-Dimethyl-1-propanol	333.2	8.35	0.92350E+02	-0.41870E+00	0.50000E-03	333-373
$C_5H_{12}O_2$	1,2-Pentanediol	296.8	17.31	0.18436E+03	-0.10682E+01	0.17037E-02	197-297
$C_5H_{12}O_2$	1,4-Pentanediol	295.7	26.74	0.13568E+03	-0.59198E+00	0.75398E-03	193-318
$C_5H_{12}O_2$	1,5-Pentanediol	293.2	26.2	0.11858E+03	-0.45920E+00	0.49341E-03	243-343
$C_5H_{12}O_2$	2,3-Pentanediol	296.9	17.37	0.95876E+02	-0.46463E+00	0.67434E-03	238-297
$C_5H_{12}O_2$	2,4-Pentanediol	294.2	24.69	0.11914E+03	-0.52569E+00	0.69607E-03	224-294
$C_5H_{12}O_2$	Diethoxymethane	293.2	2.527	0.25294E+01	0.73988E-04	-0.28331E-06	227-293
$C_5H_{12}O_4$	Tetramethoxymethane	293.2	2.40				
$C_5H_{12}O_5$	Xylitol	293.2	40.0				
$C_5H_{12}S$	1-Pentanethiol	293.2	4.847	0.71131E+01	-0.30228E-02	-0.16414E-04	273-333
$C_5H_{12}S$	2-Methyl-2-butanethiol	293.2	5.087	0.15116E+02	-0.50700E-01	0.56250E-04	273-333
$C_5H_{12}S_4$	Tetrakis(methylthio)methane	343.2	2.818				
$C_5H_{13}N$	Pentylamine	293.2	4.27	0.11274E+02	-0.34965E-01	0.37706E-04	223-353
$C_5H_{13}N_3$	1,1,3,3-Tetramethylguanidine	298.2	11.5				
$C_5H_{14}OSi$	Ethoxytrimethylsilane	298.2	3.013				
C_6F_6	Hexafluorobenzene	298.2	2.029	0.24041E+01	-0.83086E-03	-0.14286E-05	298-338
C_6F_{14}	Perfluorohexane	298.2	1.76				
$C_6H_3N_3O_7$	2,4,6-Trinitrophenol	294.2	4.0				
C_6H_4BrF	1-Bromo-2-fluorobenzene	298.2	4.72				
C_6H_4BrF	1-Bromo-3-fluorobenzene	298.2	4.85				
C_6H_4BrF	1-Bromo-4-fluorobenzene	298.2	2.60				
$C_6H_4BrNO_2$	1-Bromo-3-nitrobenzene	328.2	20.2	0.81413E+02	-0.27645E+00	0.27367E-03	328-413
$C_6H_4Br_2$	o-Dibromobenzene	293.2	7.86	-0.81849E-02	0.62671E-01	-0.12222E-03	293-353
$C_6H_4Br_2$	m-Dibromobenzene	293.2	4.81	0.93214E+01	-0.20273E-01	0.16667E-04	293-353

Mol. Form.	Name	T/K	ε	a	b	c	Range/K
$C_6H_4Br_2$	p-Dibromobenzene	368.2	2.57				
C_6H_4ClF	1-Chloro-2-fluorobenzene	298.2	6.10				
C_6H_4ClF	1-Chloro-3-fluorobenzene	298.2	4.96				
C_6H_4ClF	1-Chloro-4-fluorobenzene	298.2	3.34				
$C_6H_4ClNO_2$	1-Chloro-2-nitrobenzene	323.2	37.7	0.16800E+03	-0.59708E+00	0.59957E-03	323-436
$C_6H_4ClNO_2$	1-Chloro-3-nitrobenzene	323.2	20.9	0.77193E+02	-0.25118E+00	0.23798E-03	323-433
$C_6H_4ClNO_2$	1-Chloro-4-nitrobenzene	393.2	8.09				
$C_6H_4Cl_2$	o-Dichlorobenzene	293.2	10.12	0.13629E+02	0.10622E-02	-0.44444E-04	293-353
$C_6H_4Cl_2$	m-Dichlorobenzene	293.2	5.02	0.77565E+01	-0.93333E-02	-0.26880E-14	293-353
$C_6H_4Cl_2$	p-Dichlorobenzene	328.2	2.3943	0.26999E+01	-0.35325E-03	-0.17619E-05	328-363
C_6H_4FI	1-Fluoro-2-iodobenzene	298.2	8.22				
C_6H_4FI	1-Fluoro-4-iodobenzene	298.2	3.12				
$C_6H_4F_2$	o-Difluorobenzene	301.2	13.38	0.59107E+02	-0.23611E+00	0.27987E-03	273-323
$C_6H_4F_2$	m-Difluorobenzene	301.2	5.01	0.14448E+02	-0.46982E-01	0.51948E-04	273-323
$C_6H_4I_2$	o-Diiodobenzene	323.2	5.41	0.31150E+02	-0.14428E+00	0.20000E-03	323-353
$C_6H_4I_2$	m-Diiodobenzene	323.2	4.11				
$C_6H_4I_2$	p-Diiodobenzene	393.2	2.88				
$C_6H_4N_2$	2-Pyridinecarbonitrile	303.2	93.77	0.45596E+03	-0.17746E+01	0.19105E-02	303-398
$C_6H_4N_2$	3-Pyridinecarbonitrile	323.2	20.54	0.60484E+02	-0.17280E+00	0.15218E-03	323-398
$C_6H_4N_2$	4-Pyridinecarbonitrile	353.2	5.23	0.12533E+02	-0.30115E-01	0.26674E-04	353-398
$C_6H_4N_2O_4$	1,3-Dinitrobenzene	365.2	22.9	0.10406E+03	-0.34133E+00	0.32609E-03	365-413
C_6H_5Br	Bromobenzene	293.2	5.45	0.94100E+01	-0.12537E-01	-0.31127E-05	234-333
C_6H_5Cl	Chlorobenzene	293.2	5.6895	0.19471E+02	-0.70786E-01	0.82466E-04	293-430
C_6H_5ClO	o-Chlorophenol	296.2	7.40	0.29755E+02	-0.11256E+00	0.12390E-03	296-448
C_6H_5ClO	m-Chlorophenol	293.2	6.255				
C_6H_5ClO	p-Chlorophenol	314.2	11.18	0.31997E+02	-0.94241E-01	0.88392E-04	314-453
$C_6H_5ClO_2S$	Benzenesulfonyl chloride	323.2	28.90	0.83886E+02	-0.23405E+00	0.19713E-03	323-473
C_6H_5ClS	4-Chlorobenzenethiol	338.2	3.59				
C_6H_5F	Fluorobenzene	293.2	5.465				
C_6H_5I	Iodobenzene	293.2	4.59	0.89442E+01	-0.20008E-01	0.17641E-04	243-323
C_6H_5NOS	N-Sulfinylaniline	298.2	6.97				
$C_6H_5NO_2$	Nitrobenzene	293.0	35.6	0.11212E+03	-0.35211E+00	0.31128E-03	279-533
$C_6H_5NO_3$	o-Nitrophenol	323.2	16.50	0.33827E+02	-0.62123E-01	0.26774E-04	323-453
$C_6H_5NO_3$	m-Nitrophenol	373.2	35.45	0.18967E+03	-0.66144E+00	0.66532E-03	373-458
$C_6H_5NO_3$	p-Nitrophenol	393.2	42.20	0.22901E+03	-0.74264E+00	0.68006E-03	393-463
C_6H_6	Benzene	293.2	2.2825	0.26706E+01	-0.91648E-03	-0.14257E-05	293-513
C_6H_6BrN	m-Bromoaniline	293.2	13.0				
C_6H_6ClN	o-Chloroaniline	293.2	13.40				
C_6H_6ClN	m-Chloroaniline	293.2	13.3				
$C_6H_6N_2O_2$	o-Nitroaniline	353.0	47.3	0.18900E+03	-0.56977E+00	0.47484E-03	353-468
$C_6H_6N_2O_2$	m-Nitroaniline	398.0	35.6	0.20352E+03	-0.66582E+00	0.61310E-03	398-468
$C_6H_6N_2O_2$	p-Nitroaniline	428.0	78.5	0.48673E+03	-0.15040E+01	0.12857E-02	428-468
C_6H_6O	Phenol	303.2	12.40	0.63391E+02	-0.24988E+00	0.26930E-03	303-433
$C_6H_6O_2$	Pyrocatechol	388.2	17.57	0.74930E+02	-0.22142E+00	0.18919E-03	388-463
$C_6H_6O_2$	Resorcinol	393.2	13.55	0.30252E+02	-0.56443E-01	0.35578E-04	393-463
C_6H_6S	Benzenethiol	303.2	4.26	0.57155E+01	-0.70336E-02	0.73617E-05	303-358
C_6H_7N	Aniline	293.2	7.06	0.89534E+01	0.38990E-02	-0.36310E-04	293-413
C_6H_7N	2-Methylpyridine	293.2	10.18	0.34560E+02	-0.11980E+00	0.12500E-03	293-333
C_6H_7N	3-Methylpyridine	303.0	11.10	0.19643E+03	-0.11167E+01	0.16667E-02	303-333
C_6H_7N	4-Methylpyridine	293.0	12.2	0.33765E+02	-0.10113E+00	0.93860E-04	274-333
C_6H_7NO	2-Methylpyridine-1-oxide	323.2	36.4	0.11705E+03	-0.35301E+00	0.32000E-03	323-398
C_6H_7NO	3-Methylpyridine-1-oxide	318.2	28.26	0.59851E+02	-0.12682E+00	0.86622E-04	318-398
C_6H_8	1,3-Cyclohexadiene	184.2	2.68				
C_6H_8	1,4-Cyclohexadiene	296.0	2.211	0.27459E+01	-0.16975E-02	-0.36461E-06	232-356
$C_6H_8N_2$	Phenylhydrazine	293.2	7.15				
$C_6H_8N_2$	2,5-Dimethylpyrazine	293.2	2.436				
$C_6H_8N_2$	2,6-Dimethylpyrazine	308.2	2.653				
$C_6H_8O_2$	1,4-Cyclohexanedione	351.2	4.40				
$C_6H_9Cl_3O_2$	Butyl trichloroacetate	293.2	7.480				

Mol. Form.	Name	T/K	ε	a	b	c	Range/K
$C_6H_9Cl_3O_2$	Isobutyl trichloroacetate	293.2	7.667				
C_6H_9N	Cyclopentanecarbonitrile	293.2	22.68	0.69830E+02	-0.25303E+00	0.31491E-03	201-293
C_6H_{10}	1,5-Hexadiene	294.0	2.125	0.30014E+01	-0.28668E-02	-0.31026E-06	151-294
C_6H_{10}	cis,cis-2,4-Hexadiene	297.0	2.163	0.27284E+01	-0.17178E-02	-0.62926E-06	234-351
C_6H_{10}	trans,trans-2,4-Hexadiene	297.0	2.123	0.26774E+01	-0.16977E-02	-0.55637E-06	232-353
C_6H_{10}	2-Methyl-1,3-pentadiene*	298.2	2.422				
C_6H_{10}	3-Methyl-1,3-pentadiene	298.2	2.426				
C_6H_{10}	4-Methyl-1,3-pentadiene	293.2	2.599	0.51328E+01	-0.12774E-01	0.14215E-04	198-323
C_6H_{10}	2,3-Dimethyl-1,3-butadiene	293.2	2.102	0.26258E+01	-0.17990E-02	0.12035E-06	223-323
C_6H_{10}	1-Hexyne	296.0	2.621	0.58591E+01	-0.17099E-01	0.20856E-04	184-296
C_6H_{10}	Cyclohexene	293.2	2.2176	0.30598E+01	-0.39841E-02	0.37554E-05	141-313
$C_6H_{10}O$	Butoxyacetylene	298.2	6.62				
$C_6H_{10}O$	Cyclohexanone	293.0	16.1	0.41577E+02	-0.11463E+00	0.92454E-04	253-423
$C_6H_{10}O$	Mesityl oxide	273.2	15.6				
$C_6H_{10}O_2$	Ethyl 2-butenoate	293.2	5.4				
$C_6H_{10}O_2$	Ethyl methacrylate	303.2	5.68	0.40962E+02	-0.20520E+00	0.29286E-03	303-343
$C_6H_{10}O_3$	Ethyl acetoacetate	293.2	14.0				
$C_6H_{10}O_3$	Propanoic anhydride	293.2	18.30				
$C_6H_{10}O_4$	Monomethyl glutarate	293.2	8.37	0.16779E+02	-0.39839E-01	0.38095E-04	293-363
$C_6H_{10}O_4$	Diethyl oxalate	293.2	8.266	0.21938E+02	-0.66226E-01	0.66800E-04	293-368
$C_6H_{10}O_4$	Dimethyl succinate	293.2	7.19	0.13551E+02	-0.23109E-01	0.55440E-05	293-433
$C_6H_{10}O_4$	Ethylene glycol diacetate	290.2	7.7	0.25093E+02	-0.95171E-01	0.12224E-03	223-290
$C_6H_{11}Br$	Bromocyclohexane	303.2	8.0026				
$C_6H_{11}BrO_2$	Ethyl 2-bromobutanoate	303.2	8.57	0.49005E+02	-0.23193E+00	0.32500E-03	303-333
$C_6H_{11}BrO_2$	Ethyl 2-bromo-2-methylpropanoate	303.2	8.55	0.77044E+02	-0.40784E+00	0.60000E-03	303-333
$C_6H_{11}Cl$	Chlorocyclohexane	303.2	7.9505				
$C_6H_{11}N$	Hexanenitrile	298.2	17.26				
$C_6H_{11}N$	4-Methylpentanenitrile	295.2	17.5				
$C_6H_{11}NO$	Cyclohexanone oxime	362.2	3.04				
C_6H_{12}	1-Hexene	294.0	2.077	0.31476E+01	-0.50003E-02	0.46673E-05	149-294
C_6H_{12}	trans-2-Hexene	295.0	1.978	0.24338E+01	-0.11323E-02	-0.13720E-05	157-295
C_6H_{12}	cis-3-Hexene	296.0	2.069	0.30691E+01	-0.45458E-02	0.39898E-05	155-296
C_6H_{12}	trans-3-Hexene	293.2	1.954				
C_6H_{12}	Cyclohexane	293.2	2.0243	0.24293E+01	-0.12095E-02	-0.58741E-06	283-333
C_6H_{12}	Methylcyclopentane	293.2	1.9853	0.21587E+01	-0.22450E-03	-0.12500E-05	293-323
C_6H_{12}	Ethylcyclobutane	293.2	1.965				
$C_6H_{12}Br_2$	1,6-Dibromohexane	298.2	8.52	-0.55185E+01	0.11746E+00	-0.23658E-03	274-328
$C_6H_{12}Br_2$	3,4-Dibromohexane	298.2	6.732				
$C_6H_{12}Cl_2$	1,6-Dichlorohexane	308.2	8.60	0.11277E+02	0.67200E-02	-0.50000E-04	308-338
$C_6H_{12}O$	1-Methylcyclopentanol	310.1	7.11	0.75444E+02	-0.36617E+00	0.47021E-03	310-333
$C_6H_{12}O$	Isobutyl vinyl ether	293.2	3.34	0.48060E+01	-0.50000E-02	-0.41495E-14	293-323
$C_6H_{12}O$	2-Hexanone	293.2	14.56	0.70378E+02	-0.29385E+00	0.35289E-03	243-293
$C_6H_{12}O$	4-Methyl-2-pentanone	293.2	13.11	0.36341E+02	-0.97119E-01	0.61896E-04	204-373
$C_6H_{12}O$	3,3-Dimethyl-2-butanone	293.2	12.73	0.66857E+02	-0.28552E+00	0.34422E-03	243-293
$C_6H_{12}O$	Cyclohexanol	293.2	16.40	0.10173E+03	-0.43072E+00	0.47926E-03	293-423
$C_6H_{12}O_2$	Hexanoic acid	298.2	2.600	0.21730E+01	0.14840E-02	-0.16526E-06	298-433
$C_6H_{12}O_2$	2-Ethylbutanoic acid	296.2	2.72				
$C_6H_{12}O_2$	tert-Butylacetic acid	296.2	2.85				
$C_6H_{12}O_2$	Pentyl formate	292.2	5.7				
$C_6H_{12}O_2$	Isopentyl formate	288.2	5.44	0.29257E+02	-0.14028E+00	0.20000E-03	288-323
$C_6H_{12}O_2$	Butyl acetate	293.2	5.07	0.13825E+02	-0.43994E-01	0.48214E-04	253-353
$C_6H_{12}O_2$	sec-Butyl acetate	293.2	5.135	0.12427E+02	-0.32035E-01	0.24286E-04	273-323
$C_6H_{12}O_2$	tert-Butyl acetate	293.2	5.672	0.55435E+02	-0.30494E+00	0.46107E-03	273-323
$C_6H_{12}O_2$	Isobutyl acetate	293.2	5.068	0.14323E+02	-0.46048E-01	0.49286E-04	273-323
$C_6H_{12}O_2$	Propyl propanoate	293.2	5.249				
$C_6H_{12}O_2$	Ethyl butanoate	301.2	5.18	0.48698E+02	-0.25660E+00	0.37237E-03	301-343
$C_6H_{12}O_2$	Methyl pentanoate	293.2	4.992				
$C_6H_{12}O_2$	Diacetone alcohol	298.2	18.2				
$C_6H_{12}O_3$	Ethylene glycol monoethyl ether acetate	303.2	7.567	0.23290E+02	-0.71566E-01	0.65000E-04	303-323

Mol. Form.	Name	T/K	ε	a	b	c	Range/K
$C_6H_{12}S$	Cyclohexanethiol	298.2	5.420				
$C_6H_{13}Br$	1-Bromohexane	298.2	5.82	0.15233E+02	-0.44385E-01	0.43039E-04	274-328
$C_6H_{13}Cl$	1-Chlorohexane	293.2	6.104	0.15994E+02	-0.43647E-01	0.33393E-04	273-323
$C_6H_{13}ClO$	6-Chloro-1-hexanol	242.2	21.6	-0.73364E+01	0.46377E+00	-0.14202E-02	195-242
$C_6H_{13}I$	1-Iodohexane	293.3	5.35	0.16685E+02	-0.61309E-01	0.77262E-04	293-323
$C_6H_{13}N$	Cyclohexylamine	293.2	4.547				
$C_6H_{13}NO$	N-Propylpropanamide	298.2	118.1	0.58846E+03	-0.22012E+01	0.20870E-02	298-328
$C_6H_{13}NO$	N-Butylacetamide	293.2	104.0	0.70739E+03	-0.37369E+01	0.71585E-02	253-493
$C_6H_{13}NO$	N,N-Diethylacetamide	293.2	32.1				
C_6H_{14}	Hexane	293.2	1.8865	0.19768E+01	0.70933E-03	-0.34470E-05	293-473
C_6H_{14}	2-Methylpentane	293.2	1.886	0.20745E+01	0.50871E-03	-0.39286E-05	273-323
C_6H_{14}	3-Methylpentane	293.2	1.886	0.24739E+01	-0.23190E-02	0.10714E-05	273-323
C_6H_{14}	2,2-Dimethylbutane	293.2	1.869	0.22740E+01	-0.96229E-03	-0.14286E-05	273-313
C_6H_{14}	2,3-Dimethylbutane	293.2	1.889	0.24305E+01	-0.20081E-02	0.53571E-06	273-323
$C_6H_{14}O$	1-Hexanol	293.2	13.03	0.62744E+02	-0.24214E+00	0.24704E-03	233-513
$C_6H_{14}O$	2-Hexanol	298.2	11.06				
$C_6H_{14}O$	3-Hexanol	298.2	9.66				
$C_6H_{14}O$	3-Methyl-1-pentanol	298.2	15.2				
$C_6H_{14}O$	3-Methyl-3-pentanol	293.2	4.322				
$C_6H_{14}O$	2-Ethyl-1-butanol	362.2	6.19				
$C_6H_{14}O$	2,2-Dimethyl-1-butanol	293.2	10.5	0.14054E+03	-0.72925E+00	0.97821E-03	243-393
$C_6H_{14}O$	Dipropyl ether	297.0	3.38	0.14600E+02	-0.72670E-01	0.11742E-03	161-297
$C_6H_{14}O$	Diisopropyl ether	303.2	3.805				
$C_6H_{14}OS$	Dipropyl sulfoxide	303.2	30.37	0.84868E+02	-0.23486E+00	0.18198E-03	303-373
$C_6H_{14}O_2$	2-Methyl-2,4-pentanediol	293.2	25.86	0.14531E+03	-0.65285E+00	0.83503E-03	203-333
$C_6H_{14}O_2$	Ethylene glycol diethyl ether	293.2	3.90	0.99099E+01	-0.33403E-01	0.44048E-04	223-303
$C_6H_{14}O_2S$	Dipropyl sulfone	303.2	32.62	0.70195E+02	-0.15008E+00	0.86506E-04	303-398
$C_6H_{14}O_3$	1,2,6-Hexanetriol	285.3	31.5	0.26127E+03	-0.14552E+01	0.22765E-02	261-285
$C_6H_{14}O_3$	Diethylene glycol dimethyl ether	298.2	7.23	0.28291E+02	-0.11236E+00	0.14000E-03	298-333
$C_6H_{14}O_4$	Triethylene glycol	293.2	23.69	0.91845E+02	-0.33827E+00	0.36062E-03	253-333
$C_6H_{14}O_6$	D-Glucitol	353.2	35.5				
$C_6H_{14}O_6$	D-Mannitol	443.2	24.6				
$C_6H_{14}S$	1-Hexanethiol	293.2	4.436	0.11774E+02	-0.37298E-01	0.41875E-04	273-333
$C_6H_{15}B$	Triethylborane	293.2	1.974				
$C_6H_{15}N$	Hexylamine	293.2	4.08	0.80244E+01	-0.16627E-01	0.10874E-04	253-373
$C_6H_{15}N$	Dipropylamine	293.2	2.923	0.11376E+02	-0.49796E-01	0.71792E-04	243-323
$C_6H_{15}N$	Triethylamine	293.2	2.418	0.29205E+01	-0.14007E-02	-0.13469E-05	233-323
$C_6H_{15}OP$	Triethylphosphine oxide	323.2	35.5				
$C_6H_{15}O_4P$	Triethyl phosphate	298.2	13.20	0.61230E+02	-0.26047E+00	0.33333E-03	298-333
$C_6H_{15}PS$	Triethylphosphine sulfide	371.2	39.0				
$C_6H_{16}O_2Si$	Diethoxydimethylsilane	298.2	3.216				
$C_6H_{16}Si$	Triethylsilane	293.2	2.323				
$C_6H_{18}N_3OP$	Hexamethylphosphoric triamide	293.2	31.3	0.95666E+02	-0.29769E+00	0.26407E-03	283-363
$C_6H_{18}N_4$	N,N'-Bis(2-aminoethyl)-1,2-ethanediamine	293.2	10.76	0.50699E+02	-0.21730E+00	0.27582E-03	213-333
$C_6H_{18}OSi_2$	Hexamethyldisiloxane	293.2	2.179	0.34537E+01	-0.61530E-02	0.61544E-05	213-313
$C_6H_{18}O_3Si_3$	Hexamethylcyclotrisiloxane	343.2	2.139				
$C_6H_{19}NSi_2$	Hexamethyldisilazane	294.2	2.273	0.23358E+01	0.16127E-02	-0.62078E-05	294-333
C_7F_{14}	Perfluoromethylcyclohexane	298.2	1.82				
C_7F_{16}	Perfluoroheptane	289.2	1.847				
$C_7H_3Cl_5$	2,3,4,5,6-Pentachlorotoluene	293.2	4.8				
C_7H_4ClNO	4-Chlorophenyl isocyanate	288.2	3.177	0.40896E+01	-0.31667E-02		288-348
C_7H_5BrO	Benzoyl bromide	293.2	21.33	0.84231E+02	-0.31089E+00	0.32857E-03	283-313
C_7H_5ClO	Benzoyl chloride	293.2	23.0				
C_7H_5FO	Benzoyl fluoride	293.2	22.7				
$C_7H_5F_3$	(Trifluoromethyl)benzene	298.2	9.22				
C_7H_5N	Benzonitrile	293.2	25.9	0.57605E+02	-0.13354E+00	0.87767E-04	273-453
C_7H_5NO	Phenyl isocyanate	293.2	8.940	0.17541E+02	-0.29790E-01	0.15476E-05	293-353
$C_7H_6ClNO_2$	4-Chloro-3-nitrotoluene	301.2	28.07				

Mol. Form.	Name	T/K	ε	a	b	c	Range/K
$C_7H_6Cl_2$	2,4-Dichlorotoluene	301.2	5.68				
$C_7H_6Cl_2$	2,6-Dichlorotoluene	301.2	3.36				
$C_7H_6Cl_2$	3,4-Dichlorotoluene	301.2	9.39				
$C_7H_6Cl_2$	(Dichloromethyl)benzene	293.2	6.9				
C_7H_6O	Benzaldehyde	293.2	17.85	0.35046E+02	-0.61271E-01	0.16222E-04	301-346
$C_7H_6O_2$	Salicylaldehyde	293.2	18.35	0.51315E+02	-0.15379E+00	0.14111E-03	289-453
C_7H_7Br	o-Bromotoluene	293.2	4.641	0.10229E+02	-0.25050E-01	0.20357E-04	273-323
C_7H_7Br	m-Bromotoluene	293.2	5.566	0.11522E+02	-0.24946E-01	0.15714E-04	273-323
C_7H_7Br	p-Bromotoluene	293.2	5.503	0.10014E+02	-0.13918E-01	-0.50000E-05	273-293
C_7H_7Br	(Bromomethyl)benzene	293.2	6.658	0.18482E+02	-0.57207E-01	0.57321E-04	273-323
C_7H_7BrO	o-Bromoanisole	303.2	8.96	0.12023E+02	-0.59116E-02	-0.13787E-04	303-358
C_7H_7BrO	p-Bromoanisole	303.2	7.40	0.74367E+01	0.12648E-01	-0.42128E-04	303-358
C_7H_7Cl	o-Chlorotoluene	293.2	4.721	0.11507E+02	-0.31148E-01	0.27143E-04	273-323
C_7H_7Cl	m-Chlorotoluene	293.2	5.763	0.13921E+02	-0.37186E-01	0.31786E-04	273-323
C_7H_7Cl	p-Chlorotoluene	293.2	6.25	0.20265E+01	0.40060E-01	-0.87500E-04	293-333
C_7H_7Cl	(Chloromethyl)benzene	293.2	6.854	0.17108E+02	-0.45285E-01	0.35000E-04	273-323
C_7H_7ClO	p-Chloroanisole	293.2	7.84	0.64019E+01	0.30560E-01	-0.87500E-04	293-333
$C_7H_7ClO_2S$	p-Toluenesulfonyl chloride	343.2	22.6				
$C_7H_7ClO_3S$	4-Methoxybenzenesulfonyl chloride	314.2	27.2				
C_7H_7F	o-Fluorotoluene	298.2	4.23				
C_7H_7F	m-Fluorotoluene	298.2	5.41				
C_7H_7F	p-Fluorotoluene	298.2	5.88				
C_7H_7I	p-Iodotoluene	308.2	4.4				
C_7H_7N	2-Vinylpyridine	293.2	9.126				
C_7H_7N	4-Vinylpyridine	293.2	10.50				
$C_7H_7NO_2$	Benzyl nitrite	298.2	7.78				
$C_7H_7NO_2$	o-Nitrotoluene	293.0	26.26	0.10420E+03	-0.41726E+00	0.51607E-03	273-323
$C_7H_7NO_2$	m-Nitrotoluene	303.2	24.95	0.62492E+02	-0.16235E+00	0.12844E-03	303-403
$C_7H_7NO_2$	p-Nitrotoluene	331.2	22.2				
$C_7H_7NO_2S$	4-Nitrothioanisole	346.0	21.7				
$C_7H_7NO_3$	2-Nitroanisole	293.2	45.75	0.16684E+03	-0.58196E+00	0.57382E-03	293-423
$C_7H_7NO_3$	3-Nitroanisole	318.2	25.7	0.65402E+02	-0.16460E+00	0.12560E-03	318-443
$C_7H_7NO_3$	4-Nitroanisole	338.2	26.95	0.59811E+02	-0.10955E+00	0.36042E-04	338-443
C_7H_8	Toluene	296.35	2.379	0.32584E+01	-0.34410E-02	0.15937E-05	207-316
C_7H_8O	o-Cresol	298.2	6.76	0.21633E+02	-0.71069E-01	0.70590E-04	298-453
C_7H_8O	m-Cresol	298.2	12.44	0.81716E+02	-0.35039E+00	0.39878E-03	274-463
C_7H_8O	p-Cresol	298.2	13.05	0.70253E+02	-0.28870E+00	0.31979E-03	298-453
C_7H_8O	Benzyl alcohol	303.2	11.916	0.13661E+03	-0.72127E+00	0.10225E-02	303-333
C_7H_8O	Anisole	294.2	4.30	0.10887E+02	-0.32372E-01	0.33629E-04	294-413
$C_7H_8O_2$	2-Methoxyphenol	298.2	11.95	0.31751E+02	-0.88173E-01	0.72953E-04	291-448
$C_7H_8O_2$	3-Methoxyphenol	298.2	11.59	0.37279E+02	-0.12113E+00	0.11698E-03	298-433
$C_7H_8O_2$	4-Methoxyphenol	333.7	11.05	0.39483E+02	-0.12142E+00	0.10841E-03	334-453
$C_7H_8O_2S$	Ethyl thiophene-2-carboxylate	293.2	6.18				
$C_7H_8O_2S$	Methyl phenyl sulfone	373.2	37.9				
C_7H_8S	Benzenemethanethiol	298.2	4.705	0.16628E+02	-0.68276E-01	0.94636E-04	298-358
C_7H_8S	4-Methylbenzenethiol	323.2	4.74	0.87052E+01	-0.15347E-01	0.95238E-05	323-358
C_7H_8S	(Methylthio)benzene	303.2	4.88	0.21841E+02	-0.97630E-01	0.13750E-03	303-343
C_7H_9N	Benzylamine	293.2	5.18				
C_7H_9N	o-Methylaniline	298.2	6.138	0.10988E+02	-0.18976E-01	0.91958E-05	298-398
C_7H_9N	m-Methylaniline	298.2	5.816	0.13477E+02	-0.35551E-01	0.33135E-04	298-398
C_7H_9N	p-Methylaniline	333.2	5.058	0.78897E+01	-0.10196E-01	0.51190E-05	333-403
C_7H_9N	N-Methylaniline	293.2	5.96				
C_7H_9N	2-Ethylpyridine	293.2	8.33	0.36397E+02	-0.15070E+00	0.18750E-03	293-333
C_7H_9N	4-Ethylpyridine	293.2	10.98	-0.73831E+01	0.14326E+00	-0.27500E-03	293-333
C_7H_9N	2,4-Dimethylpyridine	293.2	9.60	0.25895E+02	-0.73900E-01	0.62500E-04	293-333
C_7H_9N	2,6-Dimethylpyridine	293.2	7.33	0.17714E+02	-0.39080E-01	0.12500E-04	293-333
C_7H_9NO	2,6-Dimethylpyridine-1-oxide	298.2	46.11	0.22765E+03	-0.90760E+00	0.10011E-02	298-398
C_7H_9NO	o-Methoxyaniline	303.2	5.230	0.79911E+01	-0.92183E-02	0.37879E-06	303-393
C_7H_9NO	m-Methoxyaniline	298.2	8.76	0.28179E+02	-0.97840E-01	0.11027E-03	289-393

Mol. Form.	Name	T/K	ε	a	b	c	Range/K
C_7H_9NO	p-Methoxyaniline	333.2	7.85	0.30149E+02	-0.10523E+00	0.11467E-03	333-453
$C_7H_{10}N_2$	1-Methyl-1-phenylhydrazine	292.2	7.3				
$C_7H_{11}Cl_3O_2$	Isopentyl trichloroacetate	293.2	7.287				
C_7H_{12}	1,6-Heptadiene	293.0	2.161	0.30815E+01	-0.36095E-02	0.16354E-05	184-293
C_7H_{12}	Cycloheptene	295.0	2.265	0.32309E+01	-0.42373E-02	0.32572E-05	227-363
$C_7H_{12}O$	Cycloheptanone	298.2	13.16	0.17511E+03	-0.11221E+01	0.19417E-02	258-298
$C_7H_{12}O$	2-Methylcyclohexanone	293.2	14.0				
$C_7H_{12}O$	3-Methylcyclohexanone	293.2	12.4				
$C_7H_{12}O$	4-Methylcyclohexanone	293.2	12.35				
$C_7H_{12}O_2$	Cyclohexanecarboxylic acid	304.2	2.67				
$C_7H_{12}O_2$	Cyclohexyl formate	293.2	6.47				
$C_7H_{12}O_2$	Butyl acrylate	301.2	5.25	0.38296E+02	-0.19109E+00	0.27006E-03	301-343
$C_7H_{12}O_4$	Monomethyl adipate	293.2	6.69	0.11962E+02	-0.23973E-01	0.20608E-04	293-433
$C_7H_{12}O_4$	Diethyl malonate	304.2	7.550	0.14809E+02	-0.31207E-01	0.24066E-04	304-393
$C_7H_{12}O_4$	Dimethyl glutarate	293.2	7.87	0.20697E+02	-0.57794E-01	0.48405E-04	293-433
$C_7H_{12}O_5$	1,2,3-Propanetriol-1,3-diacetate	288.2	9.80	0.28321E+02	-0.89073E-01	0.86891E-04	258-374
C_7H_{14}	1-Heptene	293.2	2.092	0.21755E+01	0.13896E-02	-0.57049E-05	273-323
C_7H_{14}	2-Methyl-2-hexene	293.2	2.962				
C_7H_{14}	3-Ethyl-2-pentene	293.2	2.051				
C_7H_{14}	Cycloheptane	293.2	2.0784	0.25136E+01	-0.15089E-02	0.84915E-07	278-333
C_7H_{14}	Methylcyclohexane	293.2	2.024				
$C_7H_{14}Br_2$	1,2-Dibromoheptane	298.2	3.77				
$C_7H_{14}Br_2$	2,3-Dibromoheptane	298.2	5.08				
$C_7H_{14}Br_2$	3,4-Dibromoheptane	298.2	4.70				
$C_7H_{14}Cl_2$	1,7-Dichloroheptane	298.2	8.34				
$C_7H_{14}O$	1-Heptanal	295.2	9.07				
$C_7H_{14}O$	2-Heptanone	293.2	11.95	0.38348E+02	-0.12531E+00	0.12005E-03	253-413
$C_7H_{14}O$	3-Heptanone	293.2	12.7				
$C_7H_{14}O$	4-Heptanone	293.2	12.60	0.41520E+02	-0.13839E+00	0.13497E-03	253-393
$C_7H_{14}O$	5-Methyl-2-hexanone	293.2	13.53	0.52353E+02	-0.17695E+00	0.15195E-03	293-333
$C_7H_{14}O$	Cyclohexanemethanol	333.2	9.70	0.10164E+03	-0.45839E+00	0.54762E-03	333-368
$C_7H_{14}O$	2-Methylcyclohexanol*	293.2	9.375	0.17315E+03	-0.98794E+00	0.14634E-02	273-323
$C_7H_{14}O$	3-Methylcyclohexanol*	293.2	13.79	0.65896E+02	-0.21954E+00	0.14107E-03	273-323
$C_7H_{14}O$	4-Methylcyclohexanol*	293.2	13.45	0.65021E+02	-0.22896E+00	0.17946E-03	273-323
$C_7H_{14}O_2$	Heptanoic acid	288.2	3.04	0.36423E+01	-0.31996E-02	0.39362E-05	288-423
$C_7H_{14}O_2$	Pentyl acetate	293.2	4.79	0.12091E+02	-0.36536E-01	0.39732E-04	253-353
$C_7H_{14}O_2$	Isopentyl acetate	293.2	4.72				
$C_7H_{14}O_2$	Butyl propanoate	293.2	4.838				
$C_7H_{14}O_2$	Propyl butanoate	293.2	4.3				
$C_7H_{14}O_2$	Ethyl pentanoate	291.2	4.71				
$C_7H_{14}O_2$	Ethyl 3-methylbutanoate	293.2	4.71				
$C_7H_{14}O_2$	Methyl hexanoate	293.2	4.615				
$C_7H_{15}Br$	1-Bromoheptane	303.2	5.255	0.15289E+02	-0.50621E-01	0.57753E-04	203-343
$C_7H_{15}Br$	2-Bromoheptane	295.2	6.46				
$C_7H_{15}Br$	4-Bromoheptane	295.2	6.81				
$C_7H_{15}Cl$	1-Chloroheptane	293.2	5.521	0.14279E+02	-0.39431E-01	0.32321E-04	273-323
$C_7H_{15}Cl$	2-Chloroheptane	295.2	6.52				
$C_7H_{15}Cl$	3-Chloroheptane	295.2	6.70				
$C_7H_{15}Cl$	4-Chloroheptane	295.2	6.54				
$C_7H_{15}I$	1-Iodoheptane	298.2	4.92	0.11856E+02	-0.33493E-01	0.34368E-04	294-323
$C_7H_{15}I$	3-Iodoheptane	295.2	6.39				
C_7H_{16}	Heptane	293.2	1.9209	0.24740E+01	-0.22577E-02	0.12428E-05	273-373
C_7H_{16}	2-Methylhexane	293.2	1.9221	0.24759E+01	-0.22535E-02	0.12500E-05	293-323
C_7H_{16}	3-Methylhexane	293.2	1.920	0.27089E+01	-0.37908E-02	0.37500E-05	273-323
C_7H_{16}	3-Ethylpentane	293.2	1.942	0.23771E+01	-0.15140E-02	0.10093E-06	163-363
C_7H_{16}	2,2-Dimethylpentane	293.2	1.915	0.23414E+01	-0.14362E-02	-0.51322E-07	153-353
C_7H_{16}	2,3-Dimethylpentane	293.2	1.929	0.25637E+01	-0.26328E-02	0.16071E-05	273-323
C_7H_{16}	2,4-Dimethylpentane	293.2	1.902	0.23979E+01	-0.17436E-02	0.17857E-06	273-323
C_7H_{16}	3,3-Dimethylpentane	291.3	1.9419	0.24007E+01	-0.16802E-02	0.36069E-06	291-322

Mol. Form.	Name	T/K	ε	a	b	c	Range/K
C_7H_{16}	2,2,3-Trimethylbutane	293.2	1.930				
$C_7H_{16}O$	1-Heptanol	293.2	11.75	0.60662E+02	-0.24049E+00	0.25155E-03	239-513
$C_7H_{16}O$	2-Heptanol	293.7	9.72	0.10050E+03	-0.49793E+00	0.64504E-03	207-365
$C_7H_{16}O$	3-Heptanol	296.1	7.07	0.19586E+03	-0.11465E+01	0.17175E-02	248-349
$C_7H_{16}O$	4-Heptanol	296.2	6.18	0.28995E+03	-0.18499E+01	0.30109E-02	270-301
$C_7H_{16}O$	2-Methyl-2-hexanol	297.0	3.257				
$C_7H_{16}O$	3-Methyl-2-hexanol	297.2	4.990	0.59724E+02	-0.32417E+00	0.47058E-03	244-372
$C_7H_{16}O$	3-Methyl-3-hexanol	298.2	3.248				
$C_7H_{16}O$	3-Ethyl-3-pentanol	293.2	3.158				
$C_7H_{16}O$	2,2-Dimethyl-1-pentanol	293.2	6.020	0.37318E+02	-0.17095E+00	0.22022E-03	283-393
$C_7H_{16}O$	Ethyl pentyl ether	296.2	3.6				
$C_7H_{16}O$	Ethyl isopentyl ether	293.2	3.955	0.66541E+01	-0.55450E-02	-0.12500E-04	293-323
$C_7H_{16}O_3$	Triethoxymethane	293.2	4.779				
$C_7H_{16}S$	1-Heptanethiol	293.2	4.194	0.71333E+01	-0.97320E-02	-0.12500E-05	273-333
$C_7H_{17}N$	Heptylamine	293.2	3.81	0.87794E+01	-0.24363E-01	0.25325E-04	253-373
$C_7H_{18}O_3Si$	Triethoxymethylsilane	298.2	3.845				
$C_8H_4F_6$	1,3-Bis(trifluoromethyl)benzene	303.2	5.98				
C_8H_6	Phenylacetylene	298.2	2.98				
$C_8H_6Cl_2$	2,5-Dichlorostyrene	298.2	2.58				
$C_8H_6Cl_4$	1,2,3,4-Tetrachloro-5,6-dimethylbenzene	293.2	8.0				
$C_8H_6Cl_4$	1,2,3,5-Tetrachloro-4,6-dimethylbenzene	293.2	5.4				
C_8H_6O	Phenoxyacetylene	298.2	4.76				
C_8H_7N	Benzeneacetonitrile	299.2	17.87	0.82175E+02	-0.37416E+00	0.53220E-03	299-343
$C_8H_7NO_2$	4-Methoxyphenyl isocyanate	333.2	10.26	0.20780E+02	-0.31571E-01		333-403
$C_8H_7NO_4$	Methyl 2-nitrobenzoate	300.1	27.76				
C_8H_8	Styrene	293.2	2.4737	0.44473E+01	-0.11422E-01	0.16000E-04	293-313
C_8H_8O	Acetophenone	298.2	17.44	0.26099E+02	0.64048E-02	-0.11905E-03	298-333
$C_8H_8O_2$	Benzeneacetic acid	353.2	3.47	0.24104E+01	0.30000E-02		353-393
$C_8H_8O_2$	Benzyl formate	303.2	6.34	0.26162E+02	-0.11026E+00	0.14787E-03	303-358
$C_8H_8O_2$	Phenyl acetate	298.2	5.403	0.11327E+02	-0.26707E-01	0.22938E-04	298-404
$C_8H_8O_2$	Methyl benzoate	302.7	6.642	0.17486E+02	-0.51027E-01	0.50222E-04	303-393
$C_8H_8O_2$	(Hydroxyacetyl)benzene	298.2	21.33	0.42286E+02	-0.69215E-01	-0.35714E-05	298-368
$C_8H_8O_2$	4-Methoxybenzaldehyde	303.2	22.00				
$C_8H_8O_3$	Methyl salicylate	314.4	8.80	0.20501E+02	-0.39045E-01	0.68298E-05	223-398
C_8H_9Br	1-Bromo-2-ethylbenzene	298.2	5.55				
C_8H_9Br	1-Bromo-3-ethylbenzene	298.2	5.56				
C_8H_9Br	1-Bromo-4-ethylbenzene	298.2	5.42				
C_8H_9BrO	1-Bromo-2-ethoxybenzene	313.2	7.04	0.23146E+02	-0.75753E-01	0.77778E-04	313-358
C_8H_9Cl	1-Chloro-2-ethylbenzene	298.2	4.36				
C_8H_9Cl	1-Chloro-3-ethylbenzene	298.2	5.18				
C_8H_9Cl	1-Chloro-4-ethylbenzene	298.2	5.16				
$C_8H_9NO_2$	1-Ethyl-2-nitrobenzene	273.4	21.9				
$C_8H_9NO_2$	Methyl 2-aminobenzoate	298.2	21.9				
$C_8H_9NO_2$	Ethyl 4-pyridinecarboxylate	293.2	8.95				
C_8H_{10}	Ethylbenzene	293.2	2.4463	0.35969E+01	-0.53169E-02	0.47500E-05	293-323
C_8H_{10}	o-Xylene	293.2	2.562	0.36163E+01	-0.40177E-02	0.14286E-05	273-323
C_8H_{10}	m-Xylene	293.2	2.359	0.28421E+01	-0.10191E-02	-0.21429E-05	273-323
C_8H_{10}	p-Xylene	293.2	2.2735	0.23140E+01	0.97221E-03	-0.37500E-05	293-363
$C_8H_{10}O$	2,3-Xylenol	343.2	4.81	0.14399E+02	-0.41438E-01	0.39244E-04	343-433
$C_8H_{10}O$	2,4-Xylenol	303.2	5.060	0.22125E+02	-0.85543E-01	0.96548E-04	303-363
$C_8H_{10}O$	2,5-Xylenol	338.2	5.36	0.18049E+02	-0.54991E-01	0.51656E-04	338-455
$C_8H_{10}O$	2,6-Xylenol	313.2	4.90	0.12284E+02	-0.32996E-01	0.29867E-04	313-453
$C_8H_{10}O$	3,4-Xylenol	333.2	9.02	0.54423E+02	-0.21153E+00	0.22508E-03	333-453
$C_8H_{10}O$	3,5-Xylenol	323.2	9.06	0.54251E+02	-0.21647E+00	0.23542E-03	323-453
$C_8H_{10}O$	Benzeneethanol	293.2	12.31	0.12170E+03	-0.63124E+00	0.87776E-03	278-333
$C_8H_{10}O$	1-Phenylethanol	293.2	8.77	0.32971E+02	-0.12042E+00	0.12809E-03	293-423
$C_8H_{10}O$	Phenetole	293.2	4.216	-0.15043E+02	0.13752E+00	-0.24500E-03	293-313

Mol. Form.	Name	T/K	ε	a	b	c	Range/K
$C_8H_{10}O$	2-Methylanisole	293.2	3.502	0.50825E+01	-0.62297E-02	0.28571E-05	293-333
$C_8H_{10}O$	3-Methylanisole	293.2	3.967	0.12830E+02	-0.49701E-01	0.66429E-04	293-333
$C_8H_{10}O$	4-Methylanisole	293.2	3.914	0.86608E+01	-0.23510E-01	0.25000E-04	293-333
$C_8H_{10}O_2$	1,2-Dimethoxybenzene	293.2	4.45	0.74604E+01	-0.13445E-01	0.10737E-04	293-443
$C_8H_{10}O_2$	1,3-Dimethoxybenzene	298.2	5.363	0.11911E+02	-0.30804E-01	0.29643E-04	298-358
$C_8H_{10}O_2$	1,4-Dimethoxybenzene	333.7	5.60	0.11289E+02	-0.20765E-01	0.11987E-04	334-463
$C_8H_{10}O_2S$	Ethyl phenyl sulfone	348.2	39.0				
$C_8H_{10}S$	(Ethylthio)benzene	298.2	4.95				
$C_8H_{11}N$	p-Ethylaniline	298.2	4.84				
$C_8H_{11}N$	N-Ethylaniline	293.2	5.87				
$C_8H_{11}N$	N,N-Dimethylaniline	298.2	4.90	0.84052E+01	-0.13549E-01	0.62835E-05	289-453
$C_8H_{11}N$	2,4,6-Trimethylpyridine	298.2	7.807	0.20990E+02	-0.57419E-01	0.44286E-04	298-358
$C_8H_{11}NO$	4-Ethoxyaniline	298.2	7.43				
$C_8H_{12}N_2O_2$	Hexamethylene diisocyanate	288.2	14.41	0.26715E+02	-0.42696E-01		288-403
$C_8H_{12}O_4$	Diethyl maleate	298.2	7.560	0.13953E+02	-0.21969E-01	0.17817E-05	298-343
$C_8H_{12}O_4$	Diethyl fumarate	296.2	6.56				
C_8H_{14}	1,7-Octadiene	293.0	2.186	0.28376E+01	-0.17442E-02	-0.16141E-05	214-293
C_8H_{14}	cis-Cyclooctene	296.0	2.306	0.31115E+01	-0.32058E-02	0.16713E-05	269-406
C_8H_{14}	1,2-Dimethylcyclohexene	296.0	2.144	0.26443E+01	-0.17973E-02	0.35815E-06	211-374
C_8H_{14}	1,3-Dimethylcyclohexene	296.0	2.182	0.29951E+01	-0.34615E-02	0.24026E-05	213-373
$C_8H_{14}O_2$	Methyl cyclohexanecarboxylate	293.2	4.87				
$C_8H_{14}O_2$	Cyclohexyl acetate	293.2	5.08				
$C_8H_{14}O_3$	Butanoic anhydride	293.2	12.8				
$C_8H_{14}O_3$	2-Methylpropanoic anhydride	292.2	13.6				
$C_8H_{14}O_4$	Diisopropyl oxalate	293.2	6.403	0.10709E+02	-0.16328E-01	0.56000E-05	293-368
$C_8H_{14}O_4$	Diethyl succinate	293.2	6.098	0.80213E+01	0.11810E-02	-0.26400E-04	293-343
$C_8H_{14}O_4$	Dimethyl adipate	293.2	6.84	0.11739E+02	-0.17281E-01	0.11447E-05	293-433
$C_8H_{15}N$	Octanenitrile	293.2	13.90				
C_8H_{16}	1-Octene	293.2	2.113	0.24348E+01	0.34200E-03	-0.50000E-05	273-323
C_8H_{16}	cis-3-Octene	298.2	2.062				
C_8H_{16}	trans-3-Octene	298.2	2.002				
C_8H_{16}	cis-4-Octene	298.2	2.053				
C_8H_{16}	trans-4-Octene	298.2	2.004				
C_8H_{16}	3-Methyl-2-heptene*	293.2	2.436				
C_8H_{16}	2,5-Dimethyl-2-hexene	293.2	2.431				
C_8H_{16}	2,4,4-Trimethyl-1-pentene	298.2	2.0908				
C_8H_{16}	Cyclooctane	295.0	2.116	0.25036E+01	-0.12460E-02	-0.23175E-06	295-411
$C_8H_{16}Br_2$	1,8-Dibromooctane	298.2	7.43	0.94117E+00	0.61520E-01	-0.13333E-03	298-328
$C_8H_{16}Cl_2$	1,8-Dichlorooctane	298.2	7.64				
$C_8H_{16}O$	2-Octanone	293.2	9.51	-0.16219E+02	0.18799E+00	-0.34156E-03	293-333
$C_8H_{16}O$	3-Octanone	303.2	10.50				
$C_8H_{16}O_2$	Octanoic acid	288.2	2.85	0.29391E+01	-0.38721E-03		288-423
$C_8H_{16}O_2$	2-Ethylhexanoic acid	296.2	2.64				
$C_8H_{16}O_2$	Hexyl acetate	293.2	4.42				
$C_8H_{16}O_2$	Pentyl propanoate	293.2	4.552				
$C_8H_{16}O_2$	Isopentyl propanoate	273.2	5.21	0.17665E+02	-0.71718E-01	0.95635E-04	273-373
$C_8H_{16}O_2$	Butyl butanoate	298.2	4.39	0.79684E+01	-0.12000E-01	0.15266E-13	298-318
$C_8H_{16}O_2$	Propyl pentanoate	292.2	4.0				
$C_8H_{16}O_2$	Ethyl hexanoate	293.2	4.45	0.11007E+02	-0.32800E-01	0.35714E-04	253-353
$C_8H_{16}O_2$	Methyl heptanoate	293.2	4.355				
$C_8H_{16}O_3$	Isopentyl lactate	273.2	11.2	0.48649E+02	-0.21253E+00	0.27619E-03	273-373
$C_8H_{17}Br$	1-Bromooctane	293.2	5.0957	0.12404E+02	-0.35050E-01	0.34542E-04	283-353
$C_8H_{17}Br$	2-Bromooctane	293.2	5.44				
$C_8H_{17}Cl$	1-Chlorooctane	298.2	5.05	0.11346E+02	-0.25120E-01	0.13450E-04	274-328
$C_8H_{17}Cl$	2-Chlorooctane	293.2	5.42				
$C_8H_{17}F$	1-Fluorooctane	293.2	3.89				
$C_8H_{17}I$	1-Iodooctane	293.2	4.67	0.12452E+02	-0.41229E-01	0.50108E-04	233-313
$C_8H_{17}NO_2$	1-Nitrooctane	293.2	11.46				
C_8H_{18}	Octane	293.2	1.948	0.22590E+01	-0.84212E-03	-0.75758E-06	233-393

Mol. Form.	Name	T/K	ε	a	b	c	Range/K
C$_8$H$_{18}$	2-Methylheptane	293.2	1.9519				
C$_8$H$_{18}$	3-Ethylhexane	293.2	1.9617				
C$_8$H$_{18}$	2,2-Dimethylhexane	293.2	1.9498				
C$_8$H$_{18}$	2,5-Dimethylhexane	293.95	1.9619	0.25821E+01	-0.26804E-02	0.19404E-05	294-324
C$_8$H$_{18}$	3,3-Dimethylhexane	293.2	1.9645				
C$_8$H$_{18}$	3,4-Dimethylhexane	292.1	1.9814	0.26849E+01	-0.33712E-02	0.32949E-05	292-324
C$_8$H$_{18}$	3-Ethyl-3-methylpentane	291.49	1.9869	0.25983E+01	-0.28027E-02	0.24195E-05	292-324
C$_8$H$_{18}$	2,2,3-Trimethylpentane	293.2	1.960				
C$_8$H$_{18}$	2,2,4-Trimethylpentane	293.2	1.943	0.23677E+01	-0.14768E-02	0.94261E-07	173-373
C$_8$H$_{18}$	2,3,3-Trimethylpentane	293.2	1.9780				
C$_8$H$_{18}$	2,3,4-Trimethylpentane	293.2	1.9738				
C$_8$H$_{18}$O	1-Octanol	293.2	10.30	0.51647E+02	-0.20371E+00	0.21320E-03	258-513
C$_8$H$_{18}$O	2-Octanol	293.2	8.13	0.63760E+02	-0.27643E+00	0.31075E-03	213-513
C$_8$H$_{18}$O	3-Octanol	293.2	5.55	0.12505E+03	-0.70646E+00	0.10245E-02	223-383
C$_8$H$_{18}$O	4-Octanol	293.2	4.48	0.51049E+02	-0.26664E+00	0.37280E-03	243-403
C$_8$H$_{18}$O	2-Methyl-1-heptanol	293.1	5.16	0.61698E+02	-0.33647E+00	0.49066E-03	236-328
C$_8$H$_{18}$O	3-Methyl-1-heptanol	290.3	2.884	0.84687E+01	-0.33712E-01	0.49793E-04	241-316
C$_8$H$_{18}$O	4-Methyl-1-heptanol	290.6	4.63	0.48612E+02	-0.26773E+00	0.39972E-03	237-332
C$_8$H$_{18}$O	5-Methyl-1-heptanol	290.4	7.68	0.54581E+02	-0.24772E+00	0.29734E-03	235-328
C$_8$H$_{18}$O	6-Methyl-1-heptanol	290.3	10.54	0.57997E+02	-0.23517E+00	0.24663E-03	265-328
C$_8$H$_{18}$O	2-Methyl-2-heptanol	292.2	3.43				
C$_8$H$_{18}$O	3-Methyl-2-heptanol	289.6	7.47	0.39178E+02	-0.17976E+00	0.24218E-03	229-329
C$_8$H$_{18}$O	4-Methyl-2-heptanol	290.0	3.59	0.39715E+02	-0.23115E+00	0.36771E-03	240-333
C$_8$H$_{18}$O	5-Methyl-2-heptanol	278.5	7.5	0.68568E+02	-0.40706E+00	0.67433E-03	230-279
C$_8$H$_{18}$O	6-Methyl-2-heptanol	290.1	6.41	0.77520E+02	-0.41724E+00	0.59448E-03	239-329
C$_8$H$_{18}$O	2-Methyl-3-heptanol	293.2	3.260	-0.59739E+01	0.56700E-01	-0.83125E-04	343-403
C$_8$H$_{18}$O	3-Methyl-3-heptanol	293.2	3.013	-0.38440E+01	0.42327E-01	-0.61250E-04	343-403
C$_8$H$_{18}$O	4-Methyl-3-heptanol	293.2	3.312	-0.48003E+01	0.50740E-01	-0.75000E-04	343-403
C$_8$H$_{18}$O	5-Methyl-3-heptanol	293.2	3.832	0.61967E+01	-0.63750E-02		343-383
C$_8$H$_{18}$O	6-Methyl-3-heptanol	293.2	4.992	0.23037E+02	-0.98029E-01	0.12479E-03	283-383
C$_8$H$_{18}$O	2-Methyl-4-heptanol	296.3	3.338	0.42102E+00	0.10427E-01	-0.20438E-05	230-333
C$_8$H$_{18}$O	3-Methyl-4-heptanol	290.0	7.46	0.33354E+02	-0.14077E+00	0.17750E-03	230-330
C$_8$H$_{18}$O	4-Methyl-4-heptanol	296.2	2.902				
C$_8$H$_{18}$O	2-Ethyl-1-hexanol	298.2	7.58	0.86074E+02	-0.42636E+00	0.55078E-03	208-318
C$_8$H$_{18}$O	2,2-Dimethyl-1-hexanol	293.2	4.50	0.91244E+01	-0.21785E-01	0.21018E-04	283-393
C$_8$H$_{18}$O	Dibutyl ether	293.2	3.0830	0.65383E+01	-0.16172E-01	0.14969E-04	293-314
C$_8$H$_{18}$OS	Dibutyl sulfoxide	313.2	24.73	0.67156E+02	-0.16448E+00	0.92275E-04	313-393
C$_8$H$_{18}$O$_2$	2-Ethyl-1,3-hexanediol	293.2	18.73	0.57919E+02	-0.17128E+00	0.12949E-03	233-333
C$_8$H$_{18}$O$_2$S	Dibutyl sulfone	323.2	25.72	0.66248E+02	-0.16417E+00	0.12001E-03	323-398
C$_8$H$_{18}$O$_4$	Triethylene glycol dimethyl ether	298.2	7.62				
C$_8$H$_{18}$O$_5$	Tetraethylene glycol	293.2	20.44	0.83547E+02	-0.31691E+00	0.34689E-03	253-333
C$_8$H$_{18}$S	1-Octanethiol	293.2	3.949	0.63667E+01	-0.87920E-02	0.18750E-05	273-333
C$_8$H$_{18}$S	Dibutyl sulfide	298.2	4.29				
C$_8$H$_{19}$N	Octylamine	293.2	3.58	0.77931E+01	-0.20015E-01	0.19347E-04	273-373
C$_8$H$_{19}$N	Dibutylamine	293.2	2.765	0.52504E+01	-0.10538E-01	0.71485E-05	243-323
C$_8$H$_{20}$O$_4$Si	Ethyl silicate	293.2	2.50				
C$_8$H$_{20}$Si	Tetraethylsilane	293.2	2.090				
C$_8$H$_{20}$Sn	Tetraethylstannane	293.2	2.241				
C$_8$H$_{23}$N$_5$	Tetraethylenepentamine	293.2	9.40	0.40553E+02	-0.16681E+00	0.20659E-03	213-333
C$_8$H$_{24}$O$_4$Si$_4$	Octamethylcyclotetrasiloxane	296.2	2.390	0.36286E+01	-0.56885E-02	0.50874E-05	296-333
C$_9$H$_6$N$_2$O$_2$	Toluene-2,4-diisocyanate	293.2	8.433	0.22174E+02	-0.66982E-01	0.68571E-04	293-353
C$_9$H$_6$O$_2$	2H-1-Benzopyran-2-one	343.2	34.04	0.11311E+03	-0.33804E+00	0.31324E-03	343-423
C$_9$H$_7$N	Quinoline	293.2	9.16	0.33432E+02	-0.13497E+00	0.17788E-03	258-323
C$_9$H$_7$N	Isoquinoline	298.2	11.0	0.14412E+03	-0.79935E+00	0.11839E-02	298-323
C$_9$H$_8$O	Cinnamaldehyde	305.8	17.72	0.41837E+02	-0.11060E+00	0.10401E-03	306-354
C$_9$H$_8$O$_4$	2-(Acetyloxy)benzoic acid	333.2	6.55	0.69994E+01	-0.14553E-02		333-416
C$_9$H$_{10}$	1-Propenylbenzene	293.2	2.73				
C$_9$H$_{10}$	Allylbenzene	293.2	2.63				
C$_9$H$_{10}$	Isopropenylbenzene	293.2	2.28				

Mol. Form.	Name	T/K	ε	a	b	c	Range/K
$C_9H_{10}OS$	4-Acetylthioanisole	355.2	11.34				
$C_9H_{10}O_2$	Ethyl benzoate	293.2	6.20	0.18216E+02	-0.62361E-01	0.72884E-04	288-343
$C_9H_{10}O_2$	Methyl 4-methylbenzoate	306.2	4.3				
$C_9H_{10}O_2$	Benzyl acetate	303.2	5.34	0.11727E+02	-0.30869E-01	0.32340E-04	303-358
$C_9H_{10}O_2$	Phenyl propanoate	293.2	4.77				
$C_9H_{10}O_2$	4-Acetylanisole	313.2	17.3				
$C_9H_{10}O_3$	Ethyl salicylate	308.2	8.48	0.18910E+02	-0.35623E-01	0.46529E-05	225-321
$C_9H_{10}O_3$	Methyl 2-methoxybenzoate	294.2	7.7				
$C_9H_{11}Br$	(3-Bromopropyl)benzene	302.2	5.41	0.11360E+02	-0.27471E-01	0.25775E-04	302-358
$C_9H_{11}NO$	N-Ethylbenzamide	352.7	42.6	-0.20109E+03	0.17866E+01	-0.31065E-02	353-389
$C_9H_{11}NO$	N,N-Dimethylbenzamide	318.2	20.77	0.76725E+02	-0.26908E+00	0.29409E-03	318-443
$C_9H_{11}NO_2$	Ethyl 2-aminobenzoate	298.2	4.14				
C_9H_{12}	Propylbenzene	293.2	2.370	0.26933E+01	0.21679E-03	-0.44643E-05	273-323
C_9H_{12}	Isopropylbenzene	293.2	2.381	0.31149E+01	-0.30801E-02	0.19643E-05	273-323
C_9H_{12}	o-Ethyltoluene	293.2	2.595				
C_9H_{12}	m-Ethyltoluene	293.2	2.365				
C_9H_{12}	p-Ethyltoluene	293.2	2.265				
C_9H_{12}	1,2,3-Trimethylbenzene	293.2	2.656	0.76006E+01	-0.29118E-01	0.41786E-04	273-323
C_9H_{12}	1,2,4-Trimethylbenzene	293.2	2.377	0.31517E+01	-0.30634E-02	0.14286E-05	273-323
C_9H_{12}	1,3,5-Trimethylbenzene	293.2	2.279	0.38998E+01	-0.88072E-02	0.11149E-04	288-358
$C_9H_{12}O$	Benzenepropanol	293.2	11.97	0.94482E+02	-0.45540E+00	0.59307E-03	213-303
$C_9H_{12}O$	α-Ethylbenzenemethanol	293.2	6.68	0.44520E+02	-0.21505E+00	0.29443E-03	233-373
$C_9H_{12}O$	α,α-Dimethylbenzenemethanol	303.2	5.61	0.57072E+01	0.86568E-02	-0.29580E-04	303-373
$C_9H_{12}O$	1-Phenyl-2-propanol	293.2	9.35	0.10762E+03	-0.56026E+00	0.76915E-03	233-373
$C_9H_{12}O$	Benzyl ethyl ether	298.2	3.90				
$C_9H_{12}O$	2,6-Dimethylanisole	293.2	3.780	0.76700E+01	-0.18298E-01	0.17143E-04	293-333
$C_9H_{12}O$	3,5-Dimethylanisole	293.2	3.711	0.54981E+01	-0.56651E-02	-0.14286E-05	293-333
$C_9H_{12}O_2S$	Butyl thiophene-2-carboxylate	293.2	6.40				
$C_9H_{12}S$	Benzenepropanethiol	303.2	4.36	0.82411E+01	-0.15034E-01	0.73617E-05	303-358
$C_9H_{13}N$	Benzylethylamine	293.2	4.3				
$C_9H_{13}N$	N-Propylaniline	293.2	5.48				
$C_9H_{13}N$	2-Methyl-N,N-dimethylaniline	293.2	3.4				
$C_9H_{13}N$	4-Methyl-N,N-dimethylaniline	293.2	3.9				
$C_9H_{14}OSi$	Trimethylphenoxysilane	298.2	3.3953				
$C_9H_{14}O_6$	Triacetin	293.6	7.11	0.17819E+02	-0.53656E-01	0.57759E-04	219-304
$C_9H_{14}Si$	Trimethylphenylsilane	298.2	2.3533	0.21463E+01	0.32711E-02	-0.86264E-05	288-323
$C_9H_{16}O_2$	2-Nonenoic acid	296.2	2.5				
$C_9H_{16}O_2$	Cyclohexyl propanoate	293.2	4.82				
$C_9H_{16}O_2$	Ethyl cyclohexanecarboxylate	293.2	4.64				
$C_9H_{16}O_4$	Diethyl glutarate	303.2	6.659				
$C_9H_{17}N$	Nonanenitrile	293.2	12.08				
C_9H_{18}	1-Nonene	293.2	2.180	0.22710E+01	0.15797E-02	-0.64286E-05	273-323
$C_9H_{18}Br_2$	1,9-Dibromononane	293.2	7.153	0.18931E+02	-0.57764E-01	0.60000E-04	293-343
$C_9H_{18}O$	2-Nonanone	295.2	9.14				
$C_9H_{18}O$	5-Nonanone	293.2	10.6				
$C_9H_{18}O$	Di-$tert$-butyl ketone	287.65	10.0				
$C_9H_{18}O$	2,6-Dimethyl-4-heptanone	293.2	9.91	0.33178E+02	-0.11290E+00	0.11454E-03	273-393
$C_9H_{18}O_2$	Nonanoic acid	294.9	2.475	0.25039E+01	0.67274E-03	-0.24180E-05	295-365
$C_9H_{18}O_2$	2-Methyloctanoic acid	293.2	2.39				
$C_9H_{18}O_2$	2-Ethylheptanoic acid	293.2	1.98				
$C_9H_{18}O_2$	Heptyl acetate	293.2	4.2				
$C_9H_{18}O_2$	Pentyl butanoate	301.2	4.08	0.59029E+01	-0.49905E-02	-0.34292E-05	301-343
$C_9H_{18}O_2$	Isopentyl butanoate	293.2	4.0				
$C_9H_{18}O_2$	Isobutyl pentanoate	292.2	3.8				
$C_9H_{18}O_2$	Methyl octanoate	293.2	4.101				
$C_9H_{19}Br$	1-Bromononane	298.2	4.74	0.79870E+01	-0.10488E-01	-0.13450E-05	274-328
$C_9H_{19}Cl$	1-Chlorononane	293.2	4.803	0.95528E+01	-0.16200E-01	-0.16365E-13	293-323
$C_9H_{19}NO$	N,N-Dibutylformamide	293.2	18.4				
C_9H_{20}	Nonane	293.2	1.9722	0.23894E+01	-0.14830E-02	0.14881E-06	253-393

Mol. Form.	Name	T/K	ε	a	b	c	Range/K
C_9H_{20}	2-Methyloctane	293.2	1.967				
C_9H_{20}	4-Methyloctane	293.2	1.967				
C_9H_{20}	2,4-Dimethylheptane	293.2	1.89				
C_9H_{20}	2,5-Dimethylheptane	293.2	1.89				
C_9H_{20}	2,6-Dimethylheptane	293.2	1.987				
$C_9H_{20}N_2O$	Tetraethylurea	296.8	14.29	0.52820E+02	-0.18790E+00	0.19580E-03	205-411
$C_9H_{20}O$	1-Nonanol	293.2	8.83	0.97467E+02	-0.51103E+00	0.71429E-03	288-343
$C_9H_{20}O$	2-Nonanol	298.2	6.66	0.10136E+03	-0.55612E+00	0.80000E-03	288-308
$C_9H_{20}O$	3-Nonanol	298.2	4.49	0.55214E+02	-0.31920E+00	0.50000E-03	288-308
$C_9H_{20}O$	4-Nonanol	298.2	3.69	0.27954E+01	0.30000E-02	-0.52375E-13	288-308
$C_9H_{20}O$	5-Nonanol	298.2	3.54	-0.25463E+01	0.35320E-01	-0.50000E-04	288-308
$C_9H_{21}B$	Tripropylborane	293.2	2.026				
$C_9H_{21}N$	Nonylamine	293.2	3.42	0.53575E+01	-0.71982E-02	0.19481E-05	293-373
$C_9H_{21}N$	Tripropylamine	293.2	2.380	0.33380E+01	-0.86332E-02	0.18322E-04	243-293
$C_9H_{21}O_4P$	Tripropyl phosphate	293.2	10.93	0.33166E+02	-0.10514E+00	0.10000E-03	293-373
$C_{10}H_7Br$	1-Bromonaphthalene	298.2	4.768	0.10561E+02	-0.27671E-01	0.27655E-04	293-323
$C_{10}H_7Cl$	1-Chloronaphthalene	298.2	5.04	0.84861E+01	-0.12357E-01	0.26899E-05	274-328
$C_{10}H_7NO_2$	1-Nitronaphthalene	333.2	19.68	0.36267E+02	-0.41283E-01	-0.25595E-04	333-403
$C_{10}H_8$	Naphthalene	363.2	2.54				
$C_{10}H_8O$	1-Naphthol	373.0	5.03	0.16489E+02	-0.46700E-01	0.42857E-04	373-453
$C_{10}H_8O$	2-Naphthol	413.0	4.95	0.92865E+01	-0.10500E-01	0.42501E-15	413-453
$C_{10}H_9N$	1-Naphthylamine	333.2	5.20	0.10577E+02	-0.22114E-01	0.17857E-04	333-453
$C_{10}H_9N$	2-Naphthylamine	393.0	5.26	0.19722E+02	-0.60679E-01	0.60714E-04	393-473
$C_{10}H_9N$	2-Methylquinoline	293.2	7.24	0.11688E+02	-0.78400E-02	-0.25000E-04	293-333
$C_{10}H_9N$	4-Methylquinoline	293.2	9.31	0.17788E+02	-0.32580E-01	0.12500E-04	293-333
$C_{10}H_9N$	6-Methylquinoline	293.2	8.48	0.21696E+02	-0.63400E-01	0.62500E-04	293-333
$C_{10}H_9N$	8-Methylquinoline	293.2	6.58	0.19356E+02	-0.61900E-01	0.62500E-04	293-333
$C_{10}H_{10}O_4$	Methyl 2-(acetyloxy)benzoate	328.9	5.31	0.19579E+02	-0.69970E-01	0.80889E-04	329-371
$C_{10}H_{10}O_4$	Dimethyl phthalate	293.2	8.66				
$C_{10}H_{12}$	1,2,3,4-Tetrahydronaphthalene	298.2	2.771	0.29172E+01	0.12832E-02	-0.59453E-05	298-343
$C_{10}H_{12}$	4-Ethylstyrene	298.2	3.350				
$C_{10}H_{12}$	Dicyclopentadiene	313.2	2.43	0.30564E+01	-0.20000E-02	0.82443E-15	313-373
$C_{10}H_{12}O$	Tetrahydro-2-naphthol*	293.2	11.70	0.98978E+02	-0.48267E+00	0.63008E-03	293-363
$C_{10}H_{12}O$	4-Isopropylbenzaldehyde	288.2	10.68				
$C_{10}H_{12}O_2$	4-Allyl-2-methoxyphenol	293.2	9.55	0.52377E+02	-0.24380E+00	0.33333E-03	273-323
$C_{10}H_{12}O_2$	2-Phenylethyl acetate	297.2	4.93				
$C_{10}H_{12}O_2$	Benzyl propanoate	303.0	5.11	0.42301E+01	0.13962E-01	-0.36426E-04	303-358
$C_{10}H_{12}O_2$	Phenyl butanoate	293.2	4.48				
$C_{10}H_{12}O_2$	Propyl benzoate	303.2	5.78	0.10927E+02	-0.20535E-01	0.11745E-04	303-358
$C_{10}H_{12}O_2$	Ethyl phenylacetate	293.2	5.320				
$C_{10}H_{14}$	Butylbenzene	293.2	2.359				
$C_{10}H_{14}$	sec-Butylbenzene	293.2	2.357	0.28348E+01	-0.68586E-03	-0.32143E-05	273-323
$C_{10}H_{14}$	tert-Butylbenzene	293.2	2.359	0.27924E+01	-0.38350E-03	-0.37500E-05	273-323
$C_{10}H_{14}$	Isobutylbenzene	293.2	2.318	0.28055E+01	-0.92614E-03	-0.25000E-05	273-323
$C_{10}H_{14}$	1-Isopropyl-4-methylbenzene	298.2	2.2322	0.25266E+01	-0.25121E-03	-0.24867E-05	277-333
$C_{10}H_{14}$	o-Diethylbenzene	293.2	2.594				
$C_{10}H_{14}$	m-Diethylbenzene	293.2	2.369				
$C_{10}H_{14}$	p-Diethylbenzene	293.2	2.259				
$C_{10}H_{14}$	1-Ethyl-3,5-dimethylbenzene	293.2	2.275				
$C_{10}H_{14}$	1,2,3,4-Tetramethylbenzene	296.0	2.538	0.33822E+01	-0.33630E-02	0.17475E-05	273-412
$C_{10}H_{14}$	1,2,4,5-Tetramethylbenzene	356.0	2.223	0.26834E+01	-0.10327E-02	-0.73533E-06	356-430
$C_{10}H_{14}N_2$	L-Nicotine	293.2	8.937	0.21347E+02	-0.57177E-01	0.50655E-04	293-363
$C_{10}H_{14}O$	1-Phenyl-2-methyl-2-propanol	298.2	5.71	0.21922E+02	-0.84231E-01	0.99475E-04	298-423
$C_{10}H_{14}O$	Butyl phenyl ether	293.2	3.734				
$C_{10}H_{14}O$	Thymol	333.2	4.259				
$C_{10}H_{15}N$	N,N-Diethylaniline	303.2	5.15	0.50773E+01	0.15399E-01	-0.50000E-04	303-328
$C_{10}H_{16}$	γ-Terpinene	298.2	2.2738				
$C_{10}H_{16}$	d-Limonene	298.2	2.3746				
$C_{10}H_{16}$	l-Limonene	298.2	2.3738				

Mol. Form.	Name	T/K	ε	a	b	c	Range/K
$C_{10}H_{16}$	Terpinolene	298.2	2.2918				
$C_{10}H_{16}$	α-Pinene	298.2	2.1787				
$C_{10}H_{16}$	β-Pinene	298.2	2.4970				
$C_{10}H_{16}$	α-Terpinene	298.2	2.4526				
$C_{10}H_{16}$	β-Myrcene	298.2	2.3				
$C_{10}H_{16}O$	Carvenone	293.2	18.8				
$C_{10}H_{16}O$	d-Fenchone	294.2	12.8				
$C_{10}H_{17}Cl$	2-Chlorobornane	368.2	5.21				
$C_{10}H_{18}$	Pinane	298.2	2.1456				
$C_{10}H_{18}$	cis-Decahydronaphthalene	293.2	2.219	0.25410E+01	-0.11420E-02	0.15092E-06	293-373
$C_{10}H_{18}$	trans-Decahydronaphthalene	293.2	2.184	0.26615E+01	-0.21241E-02	0.16864E-05	293-373
$C_{10}H_{18}O$	Eucalyptol	298.2	4.57				
$C_{10}H_{18}O_2$	Cyclohexyl butanoate	293.2	4.58				
$C_{10}H_{18}O_4$	Diethyl adipate	293.2	6.109	0.14824E+02	-0.40749E-01	0.37600E-04	293-343
$C_{10}H_{20}$	1-Decene	293.2	2.136	0.19091E+01	0.33442E-02	-0.87500E-05	273-323
$C_{10}H_{20}$	cis-5-Decene	298.2	2.071				
$C_{10}H_{20}$	trans-5-Decene	298.2	2.030				
$C_{10}H_{20}$	5-Methyl-4-nonene	293.2	2.175				
$C_{10}H_{20}$	2,4,6-Trimethyl-3-heptene	293.2	2.293				
$C_{10}H_{20}Br_2$	1,10-Dibromodecane	303.2	6.56	0.17350E+02	-0.50328E-01	0.48633E-04	303-368
$C_{10}H_{20}Cl_2$	1,10-Dichlorodecane	308.2	6.68	-0.57423E+01	0.94220E-01	-0.17500E-03	308-338
$C_{10}H_{20}O$	2-Decanone	287.2	8.3				
$C_{10}H_{20}O$	Menthol	309.3	3.90	0.68202E+01	-0.15894E-01	0.20837E-04	309-358
$C_{10}H_{20}O_2$	2,2-Dimethyloctanoic acid	296.2	2.8				
$C_{10}H_{20}O_2$	Octyl acetate	288.2	4.18	-0.34691E+01	0.58106E-01	-0.10952E-03	288-323
$C_{10}H_{20}O_2$	2-Methylheptyl acetate	288.2	4.27	0.23285E+02	-0.11538E+00	0.17143E-03	288-323
$C_{10}H_{20}O_2$	Pentyl pentanoate	305.6	4.076	0.77641E+01	-0.14335E-01	0.73740E-05	306-393
$C_{10}H_{20}O_2$	Isopentyl pentanoate	292.2	3.6				
$C_{10}H_{20}O_2$	Isopentyl isopentanoate	288.2	4.39	0.14698E+02	-0.57726E-01	0.76190E-04	288-323
$C_{10}H_{20}O_2$	Methyl nonanoate	293.2	3.943				
$C_{10}H_{21}Br$	1-Bromodecane	298.2	4.44	0.11202E+02	-0.33491E-01	0.36314E-04	274-328
$C_{10}H_{21}Cl$	1-Chlorodecane	293.2	4.581	0.68741E+01	-0.12210E-02	-0.22500E-04	293-323
$C_{10}H_{21}NO$	N,N-Dibutylacetamide	293.2	19.1				
$C_{10}H_{22}$	Decane	293.2	1.9853	0.24054E+01	-0.15445E-02	0.44643E-06	253-393
$C_{10}H_{22}$	2,7-Dimethyloctane	293.2	1.98				
$C_{10}H_{22}$	4-Propylheptane	293.2	1.9955				
$C_{10}H_{22}O$	1-Decanol	293.2	7.93	0.47195E+02	-0.20740E+00	0.24942E-03	293-343
$C_{10}H_{22}O$	2-Decanol	298.2	5.82	0.13621E+03	-0.81000E+00	0.12500E-02	288-308
$C_{10}H_{22}O$	3-Decanol	298.2	4.05	0.52090E+02	-0.31020E+00	0.50000E-03	288-308
$C_{10}H_{22}O$	4-Decanol	298.2	3.42	-0.11260E+02	0.93960E-01	-0.15000E-03	288-308
$C_{10}H_{22}O$	5-Decanol	298.2	3.24	-0.25832E+01	0.31456E-01	-0.40000E-04	288-308
$C_{10}H_{22}O$	2,2-Dimethyl-1-octanol	293.2	7.86	0.69536E+02	-0.34596E+00	0.46250E-03	293-333
$C_{10}H_{22}O$	Dipentyl ether	298.2	2.798				
$C_{10}H_{22}O$	Diisopentyl ether	293.2	2.817	0.44690E+01	-0.63710E-02	0.25000E-05	293-323
$C_{10}H_{22}OS$	Dipentyl sulfoxide	348.2	18.8				
$C_{10}H_{22}O_5$	Tetraethylene glycol dimethyl ether	298.2	7.68				
$C_{10}H_{22}S$	Dipentyl sulfide	298.2	3.826				
$C_{10}H_{23}N$	Decylamine	293.2	3.31	0.61497E+01	-0.12801E-01	0.10606E-04	293-373
$C_{10}H_{30}O_3Si_4$	Decamethyltetrasiloxane	293.2	2.370				
$C_{10}H_{30}O_5Si_5$	Decamethylcyclopentasiloxane	293.2	2.50				
$C_{11}H_{10}$	1-Methylnaphthalene	293.2	2.915	0.45126E+01	-0.76480E-02	0.75000E-05	293-333
$C_{11}H_{10}$	2-Methylnaphthalene	313.2	2.747				
$C_{11}H_{10}O$	1-Methoxynaphthalene	293.2	4.020	0.71885E+01	-0.14838E-01	0.13750E-04	293-333
$C_{11}H_{10}O$	2-Methoxynaphthalene	353.2	3.563	0.56702E+01	-0.69754E-02	0.28571E-05	353-373
$C_{11}H_{12}O_2$	Ethyl trans-cinnamate	293.2	5.63				
$C_{11}H_{12}O_3$	Ethyl benzoylacetate	303.2	13.50	0.93644E+01	0.74280E-01	-0.20000E-03	303-323
$C_{11}H_{14}O_2$	Benzyl butanoate	301.2	4.55				
$C_{11}H_{14}O_2$	Phenyl pentanoate	293.2	4.30				
$C_{11}H_{14}O_2$	Butyl benzoate	303.2	5.52	0.77854E+01	-0.34972E-02	-0.13149E-04	303-358

Mol. Form.	Name	T/K	ε	a	b	c	Range/K
$C_{11}H_{14}O_2$	Isobutyl benzoate	291.2	5.39				
$C_{11}H_{16}$	1,3-Diethyl-5-methylbenzene	293.2	2.264				
$C_{11}H_{16}$	Pentamethylbenzene	334.0	2.358	0.30196E+01	-0.22619E-02	0.83831E-06	334-413
$C_{11}H_{22}$	1-Undecene	293.2	2.137	0.22132E+01	0.13121E-02	-0.53571E-05	273-323
$C_{11}H_{22}O$	2-Undecanone	285.3	8.3				
$C_{11}H_{22}O_2$	Nonyl acetate	293.2	3.87				
$C_{11}H_{22}O_2$	Pentyl hexanoate	288.2	4.22	0.83503E+01	-0.18449E-01	0.14286E-04	288-323
$C_{11}H_{23}Br$	1-Bromoundecane	272.6	4.61				
$C_{11}H_{24}$	Undecane	293.2	1.9972	0.23637E+01	-0.12500E-02	-0.85869E-16	283-363
$C_{11}H_{24}O$	1-Undecanol	313.2	5.98				
$C_{11}H_{25}N$	Undecylamine	293.2	3.25	0.54945E+01	-0.96161E-02	0.66017E-05	293-373
$C_{12}F_{27}N$	Tris(perfluorobutyl)amine	293.2	2.15				
$C_{12}H_8O$	Dibenzofuran	373.2	3.00				
$C_{12}H_{10}$	Biphenyl	348.2	2.53	0.26869E+01	0.63072E-03	-0.30995E-05	348-428
$C_{12}H_{10}N_2O$	trans-Azoxybenzene	311.2	5.2				
$C_{12}H_{10}O$	Diphenyl ether	283.2	3.726				
$C_{12}H_{10}O$	2-Acetonaphthone	333.2	13.03	0.14538E+03	-0.73040E+00	0.10000E-02	333-363
$C_{12}H_{10}OS$	Diphenyl sulfoxide	344.7	16.6				
$C_{12}H_{10}O_2S$	Diphenyl sulfone	406.2	21.1				
$C_{12}H_{10}S$	Diphenyl sulfide	298.2	5.43				
$C_{12}H_{11}N$	Diphenylamine	323.2	3.73				
$C_{12}H_{11}NO$	N-1-Naphthylenylacetamide	433.2	24.3	0.84739E+02	-0.12391E+00	-0.35714E-04	433-533
$C_{12}H_{12}$	1,6-Dimethylnaphthalene	293.2	2.7250				
$C_{12}H_{12}O$	1-Ethoxynaphthalene	292.2	3.3				
$C_{12}H_{14}O_2$	Propyl cinnamate	293.2	5.45				
$C_{12}H_{14}O_4$	Diethyl phthalate	293.2	7.86				
$C_{12}H_{16}O$	2-Cyclohexylphenol	328.2	3.97				
$C_{12}H_{16}O$	4-Cyclohexylphenol	404.2	4.42				
$C_{12}H_{16}O_2$	Pentyl benzoate	293.2	5.07				
$C_{12}H_{16}O_3$	Pentyl salicylate	301.2	6.25				
$C_{12}H_{16}O_3$	Isopentyl salicylate	293.12	7.26	0.13129E+02	-0.19190E-01	-0.36060E-05	225-397
$C_{12}H_{17}NO$	N-Butyl-N-phenylacetamide	298.2	11.66				
$C_{12}H_{18}$	Hexylbenzene	293.2	2.3				
$C_{12}H_{18}$	1,3,5-Triethylbenzene	293.2	2.256				
$C_{12}H_{18}$	Hexamethylbenzene	449.0	2.172	0.35710E+01	-0.46912E-02	0.35088E-05	449-489
$C_{12}H_{20}O_2$	l-Bornyl acetate	303.2	4.46	0.60791E+01	0.98200E-02	-0.50000E-04	303-323
$C_{12}H_{22}O$	Dicyclohexyl ether	293.2	3.45	0.95324E+01	-0.31740E-01	0.37500E-04	293-333
$C_{12}H_{22}O$	Cyclododecanone	303.2	11.4	0.39327E+02	-0.13248E+00	0.13298E-03	303-423
$C_{12}H_{22}O_6$	Dibutyl tartrate	314.2	9.4				
$C_{12}H_{24}$	1-Dodecene	293.2	2.152	0.22581E+01	0.11106E-02	-0.50000E-05	273-323
$C_{12}H_{24}O_2$	Decyl acetate	293.2	3.75				
$C_{12}H_{24}O_2$	Ethyl decanoate	293.2	3.75	0.70969E+01	-0.15080E-01	0.12500E-04	293-353
$C_{12}H_{24}O_2$	Methyl undecanoate	293.2	3.671				
$C_{12}H_{25}Br$	1-Bromododecane	298.2	4.07	0.86103E+01	-0.20891E-01	0.18994E-04	274-328
$C_{12}H_{25}Cl$	1-Chlorododecane	298.2	4.17	0.10002E+02	-0.27798E-01	0.27559E-04	274-328
$C_{12}H_{25}I$	1-Iodododecane	298.2	3.91	0.34641E+01	0.97404E-02	-0.27602E-04	293-323
$C_{12}H_{26}$	Dodecane	293.2	2.0120	0.23697E+01	-0.12200E-02	-0.36375E-16	283-363
$C_{12}H_{26}O$	1-Dodecanol	303.2	5.82	0.18518E+02	-0.44859E-01	0.99900E-05	303-358
$C_{12}H_{26}O$	2-Butyl-1-octanol	363.2	3.28				
$C_{12}H_{27}BO_3$	Tributyl borate	293.2	2.23				
$C_{12}H_{27}N$	Dodecylamine	303.2	3.07	0.27999E+01	0.44810E-02	-0.11905E-04	303-373
$C_{12}H_{27}N$	Tributylamine	293.2	2.340	0.19846E+01	0.28108E-02	-0.54545E-05	233-293
$C_{12}H_{27}O_4P$	Tributyl phosphate	293.2	8.34	0.26304E+02	-0.88480E-01	0.92857E-04	293-373
$C_{12}H_{28}O_4Si$	Tetrapropoxysilane	298.2	3.21				
$C_{12}H_{28}Sn$	Tetrapropylstannane	293.2	2.267				
$C_{12}H_{30}OSi_2$	Hexaethyldisiloxane	298.2	2.259	0.36559E+01	-0.72406E-02	0.85714E-05	298-333
$C_{13}H_{10}O$	Benzophenone	300.2	12.62	0.34130E+02	-0.10249E+00	0.10268E-03	300-420
$C_{13}H_{10}O_3$	Phenyl salicylate	290.2	6.92	0.26545E+02	-0.11180E+00	0.15220E-03	290-358
$C_{13}H_{12}$	Diphenylmethane	303.2	2.540	0.30638E+01	-0.17286E-02		303-333

Mol. Form.	Name	T/K	ε	a	b	c	Range/K
C$_{13}$H$_{12}$O	Benzyl phenyl ether	313.2	3.748				
C$_{13}$H$_{18}$O$_2$	Hexyl benzoate	293.2	4.80				
C$_{13}$H$_{20}$	Heptylbenzene	293.2	2.26				
C$_{13}$H$_{20}$O	α-Ionone*	292.4	10.78				
C$_{13}$H$_{20}$O	β-Ionone*	297.65	11.66				
C$_{13}$H$_{24}$O$_4$	Diethyl nonanedioate	303.2	5.133				
C$_{13}$H$_{26}$	1-Tridecene	293.2	2.139	0.14154E+01	0.66514E-02	-0.14286E-04	273-323
C$_{13}$H$_{26}$O	7-Tridecanone	303.2	7.6				
C$_{13}$H$_{26}$O$_2$	Ethyl undecanoate	293.2	3.55				
C$_{13}$H$_{26}$O$_2$	Methyl dodecanoate	293.2	3.539				
C$_{13}$H$_{27}$Br	1-Bromotridecane	281.15	4.19				
C$_{13}$H$_{28}$	Tridecane	293.2	2.0213	0.23731E+01	-0.12000E-02	-0.21841E-15	283-363
C$_{13}$H$_{28}$	5-Butylnonane	293.2	2.0319				
C$_{13}$H$_{28}$O	1-Tridecanol	333.2	4.02				
C$_{14}$H$_{10}$	Anthracene	502.0	2.649	0.20571E+02	-0.69169E-01	0.66667E-04	502-516
C$_{14}$H$_{10}$	Phenanthrene	383.2	2.72				
C$_{14}$H$_{10}$O$_2$	Benzil	368.2	13.04	-0.23599E+02	0.22715E+00	-0.34667E-03	368-393
C$_{14}$H$_{12}$O$_2$	Benzyl benzoate	303.2	5.26	0.76856E+01	-0.80000E-02	-0.80361E-15	303-358
C$_{14}$H$_{12}$O$_3$	Benzyl salicylate	301.2	4.12				
C$_{14}$H$_{14}$	1,2-Diphenylethane	331.2	2.47	0.31178E+01	-0.21572E-02	0.59800E-06	331-451
C$_{14}$H$_{14}$O	Dibenzyl ether	293.2	3.821	0.80154E+01	-0.20536E-01	0.21250E-04	293-333
C$_{14}$H$_{15}$N	Dibenzylamine	293.2	3.446				
C$_{14}$H$_{16}$O$_2$Si	Dimethyldiphenoxysilane	298.2	3.500	0.51669E+01	-0.77001E-02	0.70156E-05	283-353
C$_{14}$H$_{18}$O$_2$	Pentyl cinnamate	293.2	4.89				
C$_{14}$H$_{22}$	Octylbenzene	293.2	2.26				
C$_{14}$H$_{26}$O$_4$	Diisobutyl adipate	293.2	5.19				
C$_{14}$H$_{26}$O$_4$	Diethyl sebacate	303.2	4.995	0.39143E+02	-0.20965E+00	0.32000E-03	303-313
C$_{14}$H$_{28}$O$_2$	Dodecyl acetate	293.2	3.6				
C$_{14}$H$_{28}$O$_2$	Ethyl laurate	273.2	3.94				
C$_{14}$H$_{28}$O$_2$	Methyl tridecanoate	293.2	3.442				
C$_{14}$H$_{29}$Br	1-Bromotetradecane	293.2	3.84	0.10058E+02	-0.33905E-01	0.43528E-04	274-328
C$_{14}$H$_{30}$	Tetradecane	293.2	2.0343	0.23832E+01	-0.11900E-02	-0.51229E-16	283-363
C$_{14}$H$_{30}$O	1-Tetradecanol	318.2	4.42	0.12272E+02	-0.24667E-01	-0.13168E-13	318-358
C$_{14}$H$_{31}$N	Tetradecylamine	312.55	2.90				
C$_{15}$H$_{12}$O$_4$	Phenyl 2-(acetyloxy)benzoate	384.2	4.33				
C$_{15}$H$_{26}$O$_6$	Tributyrin	282.8	5.72	0.13152E+02	-0.36684E-01	0.36795E-04	199-283
C$_{15}$H$_{30}$O$_2$	Methyl tetradecanoate	293.2	3.352				
C$_{15}$H$_{31}$Br	1-Bromopentadecane	293.35	3.88				
C$_{15}$H$_{32}$	Pentadecane	293.2	2.0391	0.23792E+01	-0.11600E-02	-0.71069E-16	283-363
C$_{15}$H$_{32}$O	1-Pentadecanol	333.2	3.70				
C$_{15}$H$_{33}$N	Pentadecylamine	313.25	2.85				
C$_{15}$H$_{33}$N	Triisopentylamine	294.2	2.29				
C$_{16}$H$_{22}$O$_4$	Dibutyl phthalate	293.2	6.58	0.12444E+02	-0.20000E-01		293-333
C$_{16}$H$_{32}$O$_2$	Hexadecanoic acid	338.2	2.417				
C$_{16}$H$_{32}$O$_2$	Ethyl myristate	293.2	3.50	0.52642E+01	-0.60000E-02	-0.47358E-15	293-353
C$_{16}$H$_{32}$O$_2$	Methyl pentadecanoate	293.2	3.296				
C$_{16}$H$_{33}$Br	1-Bromohexadecane	298.2	3.68	0.58668E+01	-0.73333E-02	-0.52666E-14	298-328
C$_{16}$H$_{33}$I	1-Iodohexadecane	293.2	3.57	0.79531E+01	-0.22859E-01	0.26955E-04	293-323
C$_{16}$H$_{34}$	Hexadecane	293.2	2.0460	0.23861E+01	-0.11600E-02	0.25555E-15	293-363
C$_{16}$H$_{34}$O	1-Hexadecanol	333.2	3.69	0.85935E+01	-0.14714E-01	-0.45533E-13	333-363
C$_{16}$H$_{35}$N	Hexadecylamine	328.35	2.71				
C$_{16}$H$_{36}$Sn	Tetrabutylstannane	293.2	9.74	0.56115E+02	-0.24812E+00	0.30682E-03	293-313
C$_{17}$H$_{12}$O$_3$	2-Naphthyl salicylate	293.0	6.30	0.11229E+02	-0.18857E-01	0.70332E-05	293-353
C$_{17}$H$_{34}$O	9-Heptadecanone	328.2	5.43	0.44176E+02	-0.21183E+00	0.28571E-03	328-363
C$_{17}$H$_{34}$O$_2$	Methyl palmitate	313.2	3.124				
C$_{17}$H$_{36}$	Heptadecane	293.2	2.0578	0.23627E+01	-0.10400E-02	-0.10397E-12	293-308
C$_{17}$H$_{36}$O	1-Heptadecanol	333.2	3.41				
C$_{18}$H$_{26}$O$_4$	Dipentyl phthalate	293.2	6.00				
C$_{18}$H$_{28}$O$_2$	Phenyl laurate	293.2	3.28				

Mol. Form.	Name	T/K	ε	a	b	c	Range/K
$C_{18}H_{30}O_2$	Linolenic acid	293.2	2.825	0.33867E+01	-0.19181E-02		274-368
$C_{18}H_{30}O_4$	Dicyclohexyl adipate	308.2	4.84				
$C_{18}H_{32}O_2$	Linoleic acid	293.2	2.754	0.32073E+01	-0.15477E-02		275-368
$C_{18}H_{34}O_2$	Oleic acid	293.2	2.336	0.25385E+01	-0.69448E-03		275-368
$C_{18}H_{34}O_4$	Dibutyl sebacate	293.2	4.54				
$C_{18}H_{36}O_2$	Stearic acid	293.2	2.314	0.27159E+01	-0.13300E-02		293-373
$C_{18}H_{36}O_2$	Hexadecyl acetate	308.2	3.19	0.47310E+01	-0.50000E-02	0.41338E-14	308-348
$C_{18}H_{36}O_2$	Ethyl palmitate	303.2	3.07	0.57938E+01	-0.12294E-01	0.10919E-04	303-455
$C_{18}H_{36}O_2$	Methyl heptadecanoate	313.2	3.07				
$C_{18}H_{37}Br$	1-Bromooctadecane	303.35	3.53	0.46790E+01	-0.30355E-02	-0.24798E-05	303-332
$C_{18}H_{38}O$	1-Octadecanol	333.2	3.38	0.73784E+01	-0.12000E-01	-0.22871E-13	333-363
$C_{18}H_{39}BO_3$	Trihexyl borate	293.2	2.22				
$C_{18}H_{39}N$	Octadecylamine	326.35	2.67				
$C_{19}H_{16}$	Triphenylmethane	367.2	2.46	0.40201E+01	-0.66507E-02	0.65329E-05	367-448
$C_{19}H_{18}O_3Si$	Methyltriphenoxysilane	298.2	3.628				
$C_{19}H_{32}O_2$	Methyl linolenate	293.2	3.355				
$C_{19}H_{34}O_2$	Methyl linoleate	293.2	3.466				
$C_{19}H_{36}O_2$	Methyl oleate	293.2	3.211				
$C_{19}H_{38}O$	10-Nonadecanone	353.2	5.37				
$C_{19}H_{38}O_2$	Methyl stearate	313.2	3.021				
$C_{19}H_{40}$	Nonadecane	293.2	2.0706				
$C_{20}H_{30}O_4$	Dihexyl phthalate	293.2	5.62				
$C_{20}H_{38}O_2$	Ethyl oleate	301.2	3.17	0.57033E+01	-0.11223E-01	0.93447E-05	301-423
$C_{20}H_{40}O_2$	Octadecyl acetate	308.2	3.07	0.44569E+01	-0.45000E-02	0.33923E-14	308-348
$C_{20}H_{40}O_2$	Ethyl stearate	313.2	2.958	0.70930E+01	-0.19081E-01	0.19555E-04	331-440
$C_{20}H_{40}O_2$	Methyl nonadecanoate	313.2	2.982				
$C_{20}H_{42}O$	1-Eicosanol	338.2	3.13	0.21700E+01	0.12497E-01	-0.28571E-04	338-363
$C_{20}H_{42}O$	Didecyl ether	293.2	2.644	0.41465E+01	-0.62240E-02	0.37500E-05	293-333
$C_{20}H_{60}O_8Si_9$	Eicosamethylnonasiloxane	293.2	2.645	0.57840E+01	-0.16568E-01	0.20000E-04	293-323
$C_{21}H_{21}O_4P$	Tricresyl phosphate*	298.2	6.7				
$C_{21}H_{38}O_6$	1,2,3-Propanetriyl hexanoate	293.2	4.476				
$C_{22}H_{42}O_2$	Butyl oleate	298.2	4.00				
$C_{22}H_{44}O_2$	Butyl stearate	298.2	3.120	0.73894E+02	-0.46261E+00	0.75500E-03	298-343
$C_{22}H_{46}$	Docosane	293.2	2.0840				
$C_{22}H_{46}O$	1-Docosanol	348.2	2.94	0.82062E+01	-0.25069E-01	0.28571E-04	348-373
$C_{24}H_{20}O_4Si$	Tetraphenoxysilane	333.2	3.4915				
$C_{24}H_{38}O_4$	Dioctyl phthalate	293.2	5.22				
$C_{26}H_{50}O_4$	Dioctyl sebacate	299.2	4.01				
$C_{27}H_{50}O_6$	1,2,3-Propanetriyl octanoate	293.2	3.931				
$C_{30}H_{58}O_4$	Ethylene glycol ditetradecanoate	343.2	2.98				
$C_{30}H_{62}$	Triacontane	373.2	1.9112				
$C_{30}H_{62}$	2,6,10,15,19,23-Hexamethyltetracosane	373.2	1.9106				
$C_{34}H_{66}O_4$	Ethylene glycol dipalmitate	348.2	2.89				
$C_{34}H_{68}O_2$	Hexadecyl stearate	333.2	2.61				
$C_{38}H_{74}O_4$	Ethylene glycol distearate	353.2	2.79				
$C_{39}H_{74}O_6$	Glycerol trilaurate	313.2	3.287				
$C_{51}H_{98}O_6$	Glycerol tripalmitate	328.2	2.901	-0.29131E+01	0.32206E-01	-0.44154E-04	328-393
$C_{57}H_{104}O_6$	Glycerol trioleate	293.2	3.109				
$C_{57}H_{104}O_6$	Glycerol trielaidate	313.2	2.980				
$C_{57}H_{110}O_6$	Glycerol tristearate	353.2	2.740				

PERMITTIVITY (DIELECTRIC CONSTANT) OF GASES

This table gives the permittivity ε (often called the dielectric constant) of some common gases at a temperature of 20°C and pressure of one atmosphere (101.325 kPa). Values of the permanent dipole moment μ in Debye Units (1 D = 3.33564 × 10⁻³⁰ C m) are also included.

The density dependence of the permittivity is given by the equation

$$\frac{\varepsilon - 1}{\varepsilon + 2} = \rho_m \left(\frac{4\pi N \alpha}{3} + \frac{4\pi N \mu^2}{9kT} \right)$$

where ρ_m is the molar density, N is Avogadro's number, k is the Boltzmann constant, T is the temperature, and α is the molecular polarizability. Therefore, in regions where the gas can be considered ideal, $\varepsilon - 1$ is approximately proportional to the pressure at constant temperature. For nonpolar gases ($\mu = 0$), $\varepsilon - 1$ is inversely proportional to temperature at constant pressure.

The number of significant figures indicates the accuracy of the values given. The values of ε for air, Ar, H_2, He, N_2, O_2, and CO_2 are recommended as reference values; these are accurate to 1 ppm or better.

The second part of the table gives the permittivity of water vapor in equilibrium with liquid water as a function of temperature (derived from Reference 4).

REFERENCE

1. A. A. Maryott and F. Buckley, *Table of Dielectric Constants and Electric Dipole Moments of Substances in the Gaseous State*, National Bureau of Standards Circular 537, 1953.
2. B. A. Younglove, *J. Phys. Chem. Ref. Data*, 11, Suppl. 1, 1982; 16, 577, 1987 (for data on N_2, H_2, O_2, and hydrocarbons over a range of pressure and temperature).
3. Landolt-Börnstein, *Numerical Data and Functional Relationships in Science and Technology*, New Series, Group IV, Vol. 4, Springer-Verlag, Heidelberg, 1980 (for data at high pressures).
4. G. Birnbaum and S. K. Chatterjee, *J. Appl. Phys.*, 23, 220, 1952 (for data on water vapor).

Mol. form.	Name	ε	μ/D
	Compounds not containing carbon		
	Air (dry, CO_2 free)	1.0005364	
Ar	Argon	1.0005172	0
BF_3	Boron trifluoride	1.0011	0
BrH	Hydrogen bromide	1.00279	0.827
ClH	Hydrogen chloride	1.00390	1.109
F_3N	Nitrogen trifluoride	1.0013	0.235
F_6S	Sulfur hexafluoride	1.00200	0
HI	Hydrogen iodide	1.00214	0.448
H_2	Hydrogen	1.0002538	0
H_2S	Hydrogen sulfide	1.00344	0.97
H_3N	Ammonia	1.00622	1.471
He	Helium	1.0000650	0
Kr	Krypton	1.00078	0
NO	Nitric oxide	1.00060	0.159
N_2	Nitrogen	1.0005480	0
N_2O	Nitrous oxide	1.00104	0.161
Ne	Neon	1.00013	0
O_2	Oxygen	1.0004947	0
O_2S	Sulfur dioxide	1.00825	1.633
O_3	Ozone	1.0017	0.534
Xe	Xenon	1.00126	0
	Compounds containing carbon		
CF_4	Tetrafluoromethane	1.00121	0
CO	Carbon monoxide	1.00262	0.110
CO_2	Carbon dioxide	1.000922	0

Mol. form.	Name	ε	μ/D
CH_3Br	Bromomethane	1.01028	1.822
CH_3Cl	Chloromethane	1.01080	1.892
CH_3F	Fluoromethane	1.00973	1.858
CH_3I	Iodomethane	1.00914	1.62
CH_4	Methane	1.00081	0
C_2H_2	Acetylene	1.00124	0
C_2H_3Cl	Chloroethylene	1.0075	1.45
C_2H_4	Ethylene	1.00134	0
C_2H_5Cl	Chloroethane	1.01325	2.05
C_2H_6	Ethane	1.00140	0
C_2H_6O	Dimethyl ether	1.0062	1.30
C_3H_6	Propene	1.00228	0.366
C_3H_6	Cyclopropane	1.00178	0
C_3H_8	Propane	1.00200	0.084
C_4H_{10}	Butane	1.00258	0
C_4H_{10}	Isobutane	1.00260	0.132

PERMITTIVITY OF SATURATED WATER VAPOR

$t/°C$	ε	$t/°C$	ε
0	1.00007	60	1.00144
10	1.00012	70	1.00213
20	1.00022	80	1.00305
30	1.00037	90	1.00428
40	1.00060	100	1.00587
50	1.00095		

AZEOTROPIC DATA FOR BINARY MIXTURES

Liquid mixtures having an extremum (maximum or minimum) vapor pressure at constant temperature, as a function of composition, are called azeotropic mixtures, or simply azeotropes. Mixtures that do not show a maximum or minimum are called zeotropic. Azeotropes in which the pressure is a maximum are often called positive azeotropes, while pressure-minimum azeotropes are called negative azeotropes. The coordinates of an azeotropic point are the azeotropic temperature t_{az}, pressure P_{az}, and liquid-phase composition, usually expressed as mole fractions. At the azeotropic point, the vapor-phase composition is the same as the liquid-phase composition.

This table gives azeotropic data for a number of binary mixtures at normal atmospheric pressure ($P_{az} = 101.3$ kPa). Component 1 of each mixture is given in bold face. The temperature t_{az} and mole fraction x_1 of component 1 are listed for each choice of component 2.

The components are arranged in a modified Hill order, with substances that do not contain carbon preceding those that do contain carbon.

REFERENCES

1. Lide, D.R., and Kehiaian, H.V., *CRC Handbook of Thermophysical and Thermochemical Data*, CRC Press, Boca Raton, FL, 1994.
2. Horsley, L.H., *Azeotropic Data, III*, American Chemical Society, Washington, D.C., 1973.

Molecular formula	Name	$t_{az}/°C$	x_1
Water H$_2$O			
CHCl$_3$	Trichloromethane	56.1	0.160
CH$_2$O$_2$	Formic acid	107.2	0.427
CH$_3$NO$_2$	Nitromethane	83.6	0.511
CS$_2$	Carbon disulfide	42.6	0.109
C$_2$H$_3$N	Acetonitrile	76.5	0.307
C$_2$H$_5$NO$_2$	Nitroethane	87.2	0.624
C$_2$H$_6$O	Ethanol	78.2	0.096
C$_4$H$_8$O$_2$	Ethyl acetate	70.4	0.312
C$_4$H$_{10}$O	1-Butanol	92.7	0.753
C$_4$H$_{10}$O	2-Butanol	87	0.601
C$_5$H$_5$N	Pyridine	93.6	0.755
C$_5$H$_{11}$N	Piperidine	92.8	0.718
C$_5$H$_{12}$	Pentane	34.6	0.054
C$_6$H$_5$Cl	Chlorobenzene	90.2	0.712
C$_6$H$_6$	Benzene	69.3	0.295
C$_6$H$_6$O	Phenol	99.5	0.981
C$_6$H$_{10}$	Cyclohexene	70.8	0.308
C$_6$H$_{12}$	Cyclohexane	69.5	0.300
C$_6$H$_{14}$	Hexane	61.6	0.221
C$_7$H$_8$	Toluene	84.1	0.444
C$_7$H$_{16}$	Heptane	79.2	0.452
C$_8$H$_{10}$	1,3-Dimethylbenzene	92	0.767
C$_8$H$_{10}$	Ethylbenzene	92	0.744
C$_8$H$_{18}$	Octane	89.6	0.673
C$_8$H$_{18}$O	Dibutyl ether	92.9	0.781
C$_9$H$_{20}$	Nonane	94.8	0.970
C$_{12}$H$_{27}$N	Tributylamine	99.7	0.976
Tetrachloromethane CCl$_4$			
CH$_2$O$_2$	Formic acid	66.7	0.569
CH$_3$NO$_2$	Nitromethane	71.3	0.660
CH$_4$O	Methanol	55.7	0.445
C$_2$H$_3$N	Acetonitrile	65.1	0.566
C$_2$H$_6$O	Ethanol	65.0	0.615
C$_3$H$_6$O	Acetone	56.1	0.047
C$_3$H$_8$O	1-Propanol	73.4	0.820
C$_4$H$_{10}$O	1-Butanol	76.6	0.951
Formic acid CH$_2$O$_2$			
CS$_2$	Carbon disulfide	42.6	0.253
Nitromethane CH$_3$NO$_2$			
CS$_2$	Carbon disulfide	41.2	0.845
Methanol CH$_4$O			
C$_3$H$_6$O	Acetone	55.5	0.198

Molecular formula	Name	$t_{az}/°C$	x_1
$C_3H_6O_2$	Methyl acetate	53.5	0.352
C_5H_{10}	Cyclopentane	38.8	0.263
C_5H_{12}	Pentane	30.9	0.145
$C_5H_{12}O$	*tert*-Butyl methyl ether	51.3	0.315
C_6H_6	Benzene	57.5	0.610
C_6H_{12}	Cyclohexane	53.9	0.601
C_7H_8	Toluene	63.5	0.883
C_7H_{16}	Heptane	59.1	0.769
C_8H_{18}	Octane	62.8	0.881
C_9H_{20}	Nonane	64.1	0.953
Carbon disulfide CS$_2$			
C_2H_6O	Ethanol	42.6	0.860
C_3H_6O	Acetone	39.3	0.608
C_3H_8O	1-Propanol	45.7	0.931
$C_4H_8O_2$	Ethyl acetate	46.1	0.974
Acetonitrile C$_2$H$_3$N			
C_2H_6O	Ethanol	72.5	0.469
C_7H_8	Toluene	81.4	0.900
Acetic acid C$_2$H$_4$O$_2$			
$C_4H_8O_2$	1,4-Dioxane	119.5	0.831
C_5H_5N	Pyridine	138.1	0.579
C_6H_6	Benzene	80.1	0.026
C_6H_{12}	Cyclohexane	78.8	0.130
C_6H_{14}	Hexane	68.3	0.084
$C_6H_{15}N$	Triethylamine	163	0.774
C_7H_8	Toluene	100.7	0.375
C_7H_{16}	Heptane	91.7	0.451
C_8H_{10}	Ethylbenzene	114.7	0.774
C_8H_{18}	Octane	105.7	0.688
C_9H_{20}	Nonane	112.9	0.826
Iodoethane C$_2$H$_5$I			
C_6H_{14}	Hexane	64.7	0.420
Ethanol C$_2$H$_6$O			
C_5H_{10}	Cyclopentane	44.7	0.110
C_5H_{12}	Pentane	34.3	0.076
C_6H_6	Benzene	67.9	0.440
C_6H_{12}	Cyclohexane	64.8	0.430
C_6H_{14}	Hexane	58.7	0.332
C_7H_8	Toluene	76.7	0.810
C_8H_{18}	Octane	77	0.898
Ethylene glycol C$_2$H$_6$O$_2$			
C_7H_8	Toluene	110.1	0.034
C_7H_{16}	Heptane	97.9	0.048
$C_8H_{18}O$	Dibutyl ether	139.5	0.125
$C_{10}H_{22}$	Decane	161	0.406
Dimethyl sulfide C$_2$H$_6$S			
C_5H_{12}	Pentane	31.8	0.503
1,2-Ethanediamine C$_2$H$_8$N$_2$			
C_7H_8	Toluene	104	0.406
Propanenitrile C$_3$H$_5$N			
C_6H_{14}	Hexane	63.5	0.134
Acetone C$_3$H$_6$O			
$C_3H_6O_2$	Methyl acetate	55.8	0.544
C_5H_{10}	Cyclopentane	41	0.404
C_6H_{12}	Cyclohexane	53	0.751
Ethyl formate C$_3$H$_6$O$_2$			
C_5H_{12}	Pentane	32.5	0.294
Methyl acetate C$_3$H$_6$O$_2$			
C_6H_{12}	Cyclohexane	55.5	0.801

Molecular formula	Name	$t_{az}/^\circ C$	x_1
C_6H_{14}	Hexane	51.8	0.642
Propanoic acid $C_3H_6O_2$			
C_5H_5N	Pyridine	148.6	0.686
C_7H_{16}	Heptane	97.8	0.027
C_9H_{12}	Propylbenzene	139.5	0.830
1-Nitropropane $C_3H_7NO_2$			
C_3H_8O	1-Propanol	97.0	0.061
C_7H_{16}	Heptane	96.6	0.149
1-Propanol C_3H_8O			
$C_4H_8O_2$	1,4-Dioxane	95.3	0.642
C_6H_6	Benzene	77.1	0.209
C_6H_{12}	Cyclohexane	74.7	0.241
C_7H_{16}	Heptane	84.6	0.470
2-Propanol C_3H_8O			
$C_4H_{11}N$	Butylamine	74.7	0.646
C_5H_{12}	Pentane	35.5	0.071
C_6H_{12}	Cyclohexane	69.4	0.397
C_7H_8	Toluene	80.6	0.773
Ethyl methyl sulfide C_3H_8S			
C_6H_{12}	Methylcyclopentane	65.6	0.664
C_6H_{14}	2,2-Dimethylpentane	66.4	0.908
1-Propanethiol C_3H_8S			
C_6H_{12}	Cyclohexane	67.8	0.978
C_6H_{14}	Hexane	64.4	0.557
$C_6H_{14}O$	Diisopropyl ether	65.9	0.714
Thiophene C_4H_4S			
C_6H_{12}	Cyclohexane	77.9	0.412
C_6H_{14}	Hexane	68.5	0.114
Butanal C_4H_8O			
C_6H_{14}	Hexane	60	0.296
2-Butanone C_4H_8O			
C_4H_9Cl	1-Chlorobutane	77	0.440
$C_4H_{11}N$	Butylamine	74	0.353
C_6H_6	Benzene	78.3	0.460
C_6H_{12}	Cyclohexane	71.8	0.438
C_7H_{16}	Heptane	77	0.764
Butanoic acid $C_4H_8O_2$			
C_5H_5N	Pyridine	163.2	0.912
C_6H_5Cl	Chlorobenzene	131.8	0.035
C_8H_{10}	1,2-Dimethylbenzene	143	0.118
1,4-Dioxane $C_4H_8O_2$			
C_4H_9Br	1-Bromobutane	98	0.580
Ethyl acetate $C_4H_8O_2$			
C_6H_{14}	Hexane	65.2	0.394
Methyl propanoate $C_4H_8O_2$			
C_4H_9Cl	1-Chlorobutane	76.8	0.392
Propyl formate $C_4H_8O_2$			
C_4H_9Cl	1-Chlorobutane	76.1	0.392
C_6H_6	Benzene	78.5	0.440
C_6H_{12}	Cyclohexane	75	0.469
1-Butanol $C_4H_{10}O$			
C_5H_5N	Pyridine	118.6	0.704
C_6H_5Cl	Chlorobenzene	115.3	0.659
C_6H_{10}	Cyclohexene	82	0.055
C_7H_8	Toluene	105.5	0.324
C_7H_{16}	Heptane	93.9	0.229
C_8H_{10}	1,2-Dimethylbenzene	116.8	0.811
$C_8H_{18}O$	Dibutyl ether	117.7	0.892

Molecular formula	Name	$t_{az}/°C$	x_1
	2-Butanol $C_4H_{10}O$		
C_6H_6	Benzene	78.5	0.161
C_7H_{16}	Heptane	88.1	0.439
	Diethyl ether $C_4H_{10}O$		
C_5H_{12}	Pentane	33.7	0.553
	***tert*-Butyl alcohol** $C_4H_{10}O$		
C_6H_6	Benzene	74.0	0.378
C_7H_{16}	Heptane	78	0.688
	Methyl propyl ether $C_4H_{10}O$		
C_5H_{12}	Pentane	35.6	0.215
	2-Ethoxyethanol $C_4H_{10}O_2$		
C_7H_{16}	Heptane	96.5	0.153
C_9H_{12}	Propylbenzene	134.6	0.842
	2-Furaldehyde $C_5H_4O_2$		
C_7H_{16}	Heptane	98.3	0.055
C_9H_{12}	Propylbenzene	151.4	0.475
	Pyridine C_5H_5N		
C_7H_8	Toluene	110.1	0.249
	Benzene C_6H_6		
C_6H_{10}	Cyclohexene	78.9	0.635
C_6H_{12}	Cyclohexane	77.6	0.538
	Phenol C_6H_6O		
C_6H_7N	2-Methylpyridine	185.5	0.752
C_7H_9N	2,4-Dimethylpyridine	193.4	0.601
C_9H_{12}	1,3,5-Trimethylbenzene	163.5	0.253
$C_{10}H_{22}$	Decane	168	0.449
	Aniline C_6H_7N		
C_9H_{12}	1,3,5-Trimethylbenzene	164.4	0.150
$C_{10}H_{22}O$	Dipentyl ether	177.5	0.675
$C_{12}H_{26}$	Dodecane	180.4	0.821
	2-Methylpyridine C_6H_7N		
C_8H_{18}	Octane	121.1	0.470
	Cyclohexanol $C_6H_{12}O$		
C_8H_{10}	1,2-Dimethylbenzene	143	0.147

VISCOSITY OF GASES

The following table gives the viscosity of some common gases as a function of temperature. Unless otherwise noted, the viscosity values refer to a pressure of 100 kPa (1 bar). The notation P=0 indicates the low pressure limiting value is given. The difference between the viscosity at 100 kPa and the limiting value is generally less than 1%. Viscosity is given in units of μPa s; note that 1μPa s = 10^{-5} poise. Substances are listed in the modified Hill order (see Introduction).

| | | Viscosity in micropascal seconds (μPa s) | | | | | | |
		100 K	200 K	300 K	400 K	500 K	600 K	Ref.
	Air	7.1	13.3	18.6	23.1	27.1	30.8	1
Ar	Argon	8.0	15.9	22.9	28.8	34.2	39.0	2,8
BF_3	Boron trifluoride		12.3	17.1	21.7	26.1	30.2	13
ClH	Hydrogen chloride			14.6	19.7	24.3		13
F_6S	Sulfur hexafluoride							
	(P=0)			15.3	19.8	23.9	27.7	10
H_2	Hydrogen (P=0)	4.2	6.8	9.0	10.9	12.7	14.4	4
D_2	Deuterium (P=0)	5.9	9.6	12.6	15.4	17.9	20.3	11
H_2O	Water			10.0	13.3	17.3	21.4	6
D_2O	Deuterium oxide			11.1	13.7	17.7	22.0	7
He	Helium (P=0)	9.7	15.3	20.0	24.4	28.4	32.3	8
Kr	Krypton (P=0)	8.8	17.1	25.6	33.1	39.8	45.9	8
NO	Nitric oxide		13.8	19.2	23.8	28.0	31.9	13
N_2	Nitrogen (P=0)		12.9	17.9	22.2	26.1	29.6	12
N_2O	Nitrous oxide		10.0	15.0	19.4	23.6	27.4	13
Ne	Neon (P=0)	14.4	24.3	32.1	38.9	45.0	50.8	8
O_2	Oxygen (P=0)	7.5	14.6	20.8	26.1	30.8	35.1	12
O_2S	Sulfur dioxide		8.6	12.9	17.5	21.7		13
Xe	Xenon (P=0)	8.3	15.4	23.2	30.7	37.6	44.0	8
CO	Carbon monoxide	6.7	12.9	17.8	22.1	25.8	29.1	13
CO_2	Carbon dioxide		10.0	15.0	19.7	24.0	28.0	9,10
$CHCl_3$	Chloroform			10.2	13.7	16.9	20.1	13
CH_4	Methane		7.7	11.2	14.3	17.0	19.4	10
CH_4O	Methanol				13.2	16.5	19.6	13
C_2H_2	Acetylene			10.4	13.5	16.5		13
C_2H_4	Ethylene		7.0	10.4	13.6	16.5	19.1	3
C_2H_6	Ethane		6.4	9.5	12.3	14.9	17.3	5
C_2H_6O	Ethanol				11.6	14.5	17.0	13
C_3H_8	Propane			8.3	10.9	13.4	15.8	5
C_4H_{10}	n-Butane			7.5	10.0	12.3	14.6	5
C_4H_{10}	Isobutane			7.6	10.0	12.3	14.6	5
$C_4H_{10}O$	Diethyl ether			7.6	10.1	12.4		13
C_5H_{12}	n-Pentane			6.7	9.2	11.4	13.4	13
C_6H_{14}	n-Hexane				8.6	10.8	12.8	13

REFERENCES

1. K. Kadoya, N. Matsunaga, and A. Nagashima, Viscosity and thermal conductivity of dry air in the gaseous phase, *J. Phys. Chem. Ref. Data*, 14, 947, 1985.
2. B. A. Younglove and H. J. M. Hanley, The viscosity and thermal conductivity coefficients of gaseous and liquid argon, *J. Phys. Chem. Ref. Data*, 15, 1323, 1986.
3. P. M. Holland, B. E. Eaton, and H. J. M. Hanley, A Correlation of the viscosity and thermal conductivity data of gaseous and liquid ethylene, *J. Phys. Chem. Ref. Data*, 12, 917, 1983.
4. M. J. Assael, S. Mixafendi, and W. A. Wakeham, The viscosity and thermal conductivity of normal hydrogen in the limit zero density, *J. Phys. Chem. Ref. Data*, 15, 1315, 1986.
5. B. A. Younglove and J. F. Ely, Thermophysical properties of fluids. II. Methane, ethane, propane, isobutane, and normal butane, *J. Phys. Chem. Ref. Data*, 16, 577, 1987.

6. J. V. Sengers and J. T. R. Watson, Improved international formulations for the viscosity and thermal conductivity of water substance, *J. Phys. Chem. Ref. Data*, 15, 1291, 1986.

7. N. Matsunaga and A. Nagashima, Transport properties of liquid and gaseous D_2O over a wide range of temperature and pressure, *J. Phys. Chem. Ref. Data*, 12, 933, 1983.

8. J. Kestin, et al., Equilibrium and transport properties of the noble gases and their mixtures at low density, *J. Phys. Chem. Ref. Data*, 13, 299, 1984.

9. V. Vescovic, et al., The transport properties of carbon dioxide, *J. Phys. Chem. Ref. Data*, 19, 1990.

10. R. D. Trengove and W. A. Wakeham, The viscosity of carbon dioxide, methane, and sulfur hexafluoride in the limit of zero density, *J. Phys. Chem. Ref. Data*, 16, 175, 1987.

11. M. J. Assael, S. Mixafendi, and W. A. Wakeham, The viscosity of normal deuterium in the limit of zero density, *J. Phys. Chem. Ref. Data*, 16, 189, 1987.

12. W. A. Cole and W. A. Wakeham, The viscosity of nitrogen, oxygen, and their binary mixtures in the limit of zero density, *J. Phys. Chem. Ref. Data*, 14, 209, 1985.

13. C. Y. Ho, Ed., *Properties of Inorganic and Organic Fluids, CINDAS Data Series on Materials Properties*, Vol. V-1, Hemisphere Publishing Corp., New York, 1988.

VISCOSITY OF LIQUIDS

The absolute viscosity of some common liquids at temperatures between –25 and 100°C is given in this table. Values were derived by fitting experimental data to suitable expressions for the temperature dependence. The substances are arranged by molecular formula in the modified Hill order (see Preface). All values are given in units of millipascal seconds (mPa s); this unit is identical to centipoise (cp).

Viscosity values correspond to a nominal pressure of 1 atmosphere. If a value is given at a temperature above the normal boiling point, the applicable pressure is understood to be the vapor pressure of the liquid at that temperature. A few values are given at a temperature slightly below the normal freezing point; these refer to the supercooled liquid.

The accuracy ranges from 1% in the best cases to 5 to 10% in the worst cases. Additional significant figures are included in the table to facilitate interpolation.

REFERENCES

1. Viswanath, D. S. and Natarajan, G., *Data Book on the Viscosity of Liquids*, Hemisphere Publishing Corp., New York, 1989.
2. Daubert, T. E., Danner, R. P., Sibul, H. M., and Stebbins, C. C., *Physical and Thermodynamic Properties of Pure Compounds: Data Compilation*, extant 1994 (core with 4 supplements), Taylor & Francis, Bristol, PA (also available as database).
3. Ho, C. Y., Ed., *CINDAS Data Series on Material Properties*, Vol. V-1, *Properties of Inorganic and Organic Fluids*, Hemisphere Publishing Corp., New York, 1988.
4. Stephan, K. and Lucas, K., *Viscosity of Dense Fluids*, Plenum Press, New York, 1979.
5. Vargaftik, N. B., *Tables of Thermophysical Properties of Liquids and Gases*, 2nd ed., John Wiley, New York, 1975.

Molecular formula	Name	Viscosity in mPa s					
		–25°C	0°C	25°C	50°C	75°C	100°C
Compounds not containing carbon							
Br_2	Bromine		1.252	0.944	0.746		
Cl_3HSi	Trichlorosilane		0.415	0.326			
Cl_3P	Phosphorous trichloride	0.870	0.662	0.529	0.439		
Cl_4Si	Tetrachlorosilane			99.4	96.2		
H_2O	Water		1.793	0.890	0.547	0.378	0.282
H_4N_2	Hydrazine			0.876	0.628	0.480	0.384
Hg	Mercury			1.526	1.402	1.312	1.245
NO_2	Nitrogen dioxide		0.532	0.402			
Compounds containing carbon							
CCl_3F	Trichlorofluoromethane	0.740	0.539	0.421			
CCl_4	Tetrachloromethane		1.321	0.908	0.656	0.494	
CS_2	Carbon disulfide		0.429	0.352			
$CHBr_3$	Tribromomethane			1.857	1.367	1.029	
$CHCl_3$	Trichloromethane	0.988	0.706	0.537	0.427		
CHN	Hydrogen cyanide		0.235	0.183			
CH_2Br_2	Dibromomethane	1.948	1.320	0.980	0.779	0.652	
CH_2Cl_2	Dichloromethane	0.727	0.533	0.413			
CH_2O_2	Formic acid			1.607	1.030	0.724	0.545
CH_3I	Iodomethane		0.594	0.469			
CH_3NO	Formamide		7.114	3.343	1.833		
CH_3NO_2	Nitromethane	1.311	0.875	0.630	0.481	0.383	0.317
CH_4O	Methanol	1.258	0.793	0.544			
CH_5N	Methylamine	0.319	0.231				
$C_2Cl_3F_3$	1,1,2-Trichlorotrifluoro-ethane	1.465	0.945	0.656	0.481		
C_2Cl_4	Tetrachloroethylene		1.114	0.844	0.663	0.535	0.442
C_2HCl_3	Trichloroethylene		0.703	0.545	0.444	0.376	
C_2HCl_5	Pentachloroethane		3.761	2.254	1.491	1.061	
$C_2HF_3O_2$	Trifluoroacetic acid			0.808	0.571		
$C_2H_2Cl_2$	*cis*-1,2-Dichloroethylene	0.786	0.575	0.445			
$C_2H_2Cl_2$	*trans*-1,2-Dichloroethylene	0.522	0.398	0.317	0.261		
$C_2H_2Cl_4$	1,1,1,2-Tetrachloroethane	3.660	2.200	1.437	1.006	0.741	0.570
$C_2H_3ClF_2$	1-Chloro-1,1-difluoro-ethane	0.477	0.376				

Molecular formula	Name	Viscosity in mPa s					
		−25°C	0°C	25°C	50°C	75°C	100°C
C₂H₃ClO	Acetyl chloride			0.368	0.294		
C₂H₃Cl₃	1,1,1-Trichloroethane	1.847	1.161	0.793	0.578	0.428	
C₂H₃N	Acetonitrile		0.400	0.369	0.284	0.234	
C₂H₄Br₂	1,2-Dibromoethane			1.595	1.116	0.837	0.661
C₂H₄Cl₂	1,1-Dichloroethane			0.464	0.362		
C₂H₄Cl₂	1,2-Dichloroethane		1.125	0.779	0.576	0.447	
C₂H₄O₂	Acetic acid			1.056	0.786	0.599	0.464
C₂H₄O₂	Methyl formate		0.424	0.325			
C₂H₅Br	Bromoethane	0.635	0.477	0.374			
C₂H₅Cl	Chloroethane	0.416	0.319				
C₂H₅I	Iodoethane		0.723	0.556	0.444	0.365	
C₂H₅NO	N-Methylformamide		2.549	1.678	1.155	0.824	0.606
C₂H₅NO₂	Nitroethane	1.354	0.940	0.688	0.526	0.415	0.337
C₂H₆O	Ethanol	3.262	1.786	1.074	0.694	0.476	
C₂H₆OS	Dimethyl sulfoxide			1.987	1.290		
C₂H₆O₂	Ethylene glycol			16.1	6.554	3.340	1.975
C₂H₆S	Dimethyl sulfide		0.356	0.284			
C₂H₆S	Ethanethiol		0.364	0.287			
C₂H₇N	Dimethylamine	0.300	0.232				
C₂H₇NO	Ethanolamine			21.1	8.560	3.935	1.998
C₃H₅Br	3-Bromopropene		0.620	0.471	0.373		
C₃H₅Cl	3-Chloropropene		0.408	0.314			
C₃H₅ClO	Epichlorohydrin	2.492	1.570	1.073	0.781	0.597	0.474
C₃H₅N	Propanenitrile			0.294	0.240	0.202	
C₃H₆O	Acetone	0.540	0.395	0.306	0.247		
C₃H₆O	Allyl alcohol			1.218	0.759	0.505	
C₃H₆O	Propanal			0.321	0.249		
C₃H₆O₂	Ethyl formate		0.506	0.380	0.300		
C₃H₆O₂	Methyl acetate		0.477	0.364	0.284		
C₃H₆O₂	Propanoic acid		1.499	1.030	0.749	0.569	0.449
C₃H₇Br	1-Bromopropane		0.645	0.489	0.387		
C₃H₇Br	2-Bromopropane		0.612	0.458	0.359		
C₃H₇Cl	1-Chloropropane		0.436	0.334			
C₃H₇Cl	2-Chloropropane		0.401	0.303			
C₃H₇I	1-Iodopropane		0.970	0.703	0.541	0.436	0.363
C₃H₇I	2-Iodopropane		0.883	0.653	0.506	0.407	
C₃H₇NO	N,N-Dimethylformamide		1.176	0.794	0.624		
C₃H₇NO₂	1-Nitropropane	1.851	1.160	0.798	0.589	0.460	0.374
C₃H₈O	1-Propanol	8.645	3.815	1.945	1.107	0.685	
C₃H₈O	2-Propanol		4.619	2.038	1.028	0.576	
C₃H₈O₂	1,2-Propylene glycol		248	40.4	11.3	4.770	2.750
C₃H₈O₃	Glycerol			934	152	39.8	14.8
C₃H₈S	1-Propanethiol		0.503	0.385			
C₃H₈S	2-Propanethiol		0.477	0.357	0.280		
C₃H₉N	Propylamine			0.376			
C₃H₉N	Isopropylamine		0.454	0.325			
C₄H₄O	Furan	0.661	0.475	0.361			
C₄H₅N	Pyrrole		2.085	1.225	0.828	0.612	
C₄H₆O₃	Acetic anhydride		1.241	0.843	0.614	0.472	0.377
C₄H₇N	Butanenitrile			0.553	0.418	0.330	0.268
C₄H₈O	2-Butanone	0.720	0.533	0.405	0.315	0.249	
C₄H₈O	Tetrahydrofuran	0.849	0.605	0.456	0.359		
C₄H₈O₂	1,4-Dioxane			1.177	0.787	0.569	
C₄H₈O₂	Ethyl acetate		0.578	0.423	0.325	0.259	
C₄H₈O₂	Methyl propionate		0.581	0.431	0.333	0.266	
C₄H₈O₂	Propyl formate		0.669	0.485	0.370	0.293	
C₄H₈O₂	Butanoic acid		2.215	1.426	0.982	0.714	0.542

Molecular formula	Name	Viscosity in mPa s					
		−25°C	0°C	25°C	50°C	75°C	100°C
$C_4H_8O_2$	2-Methylpropanoic acid		1.857	1.226	0.863	0.639	0.492
$C_4H_8O_2S$	Sulfolane				6.280	3.818	2.559
C_4H_8S	Tetrahydrothiophene			0.973	0.912		
C_4H_9Br	1-Bromobutane		0.815	0.606	0.471	0.379	
C_4H_9Cl	1-Chlorobutane		0.556	0.422	0.329	0.261	
C_4H_9N	Pyrrolidine	1.914	1.071	0.704	0.512		
C_4H_9NO	N,N-Dimethylacetamide			1.956	1.279	0.896	0.661
C_4H_9NO	Morpholine			2.021	1.247	0.850	0.627
$C_4H_{10}O$	1-Butanol	12.19	5.185	2.544	1.394	0.833	0.533
$C_4H_{10}O$	2-Butanol			3.096	1.332	0.698	0.419
$C_4H_{10}O$	2-Methyl-2-propanol			4.312	1.421	0.678	
$C_4H_{10}O$	Diethyl ether		0.283	0.224			
$C_4H_{10}O_3$	Diethylene glycol			30.200	11.130	4.917	2.505
$C_4H_{10}S$	Diethyl sulfide		0.558	0.422	0.331	0.267	
$C_4H_{11}N$	Butylamine		0.830	0.574	0.409	0.298	
$C_4H_{11}N$	Isobutylamine		0.770	0.571	0.367		
$C_4H_{11}N$	Diethylamine			0.319	0.239		
$C_4H_{11}NO_2$	Diethanolamine				109.5	28.7	9.100
$C_5H_4O_2$	Furfural		2.501	1.587	1.143	0.906	0.772
C_5H_5N	Pyridine		1.361	0.879	0.637	0.497	0.409
C_5H_{10}	1-Pentene	0.313	0.241	0.195			
C_5H_{10}	2-Methyl-2-butene		0.255	0.203			
C_5H_{10}	Cyclopentane		0.555	0.413	0.321		
$C_5H_{10}O$	Mesityl oxide	1.291	0.838	0.602	0.465	0.381	0.326
$C_5H_{10}O$	2-Pentanone		0.641	0.470	0.362	0.289	0.238
$C_5H_{10}O$	3-Pentanone		0.592	0.444	0.345	0.276	0.227
$C_5H_{10}O_2$	Butyl formate		0.937	0.644	0.472	0.362	0.289
$C_5H_{10}O_2$	Propyl acetate		0.768	0.544	0.406	0.316	0.255
$C_5H_{10}O_2$	Ethyl propanoate		0.691	0.501	0.380	0.299	0.242
$C_5H_{10}O_2$	Methyl butanoate		0.759	0.541	0.406	0.318	0.257
$C_5H_{10}O_2$	Methyl isobutanoate		0.672	0.488	0.373	0.296	
$C_5H_{11}N$	Piperidine			1.573	0.958	0.649	0.474
C_5H_{12}	Pentane	0.351	0.274	0.224			
C_5H_{12}	Isopentane	0.376	0.277	0.214			
$C_5H_{12}O$	1-Pentanol	25.4	8.512	3.619	1.820	1.035	0.646
$C_5H_{12}O$	2-Pentanol			3.470	1.447	0.761	0.465
$C_5H_{12}O$	3-Pentanol			4.149	1.473	0.727	0.436
$C_5H_{12}O$	2-Methyl-1-butanol			4.453	1.963	1.031	0.612
$C_5H_{12}O$	3-Methyl-1-butanol		8.627	3.692	1.842	1.031	0.631
$C_5H_{13}N$	Pentylamine		1.030	0.702	0.493	0.356	
C_6F_6	Hexafluorobenzene			2.789	1.730	1.151	
$C_6H_4Cl_2$	o-Dichlorobenzene		1.958	1.324	0.962	0.739	0.593
$C_6H_4Cl_2$	m-Dichlorobenzene		1.492	1.044	0.787	0.628	0.525
C_6H_5Br	Bromobenzene		1.560	1.074	0.798	0.627	0.512
C_6H_5Cl	Chlorobenzene	1.703	1.058	0.753	0.575	0.456	0.369
C_6H_5ClO	o-Chlorophenol			3.589	1.835	1.131	0.786
C_6H_5ClO	m-Chlorophenol				4.041		
C_6H_5F	Fluorobenzene		0.749	0.550	0.423	0.338	
C_6H_5I	Iodobenzene		2.354	1.554	1.117	0.854	0.683
$C_6H_5NO_2$	Nitrobenzene		3.036	1.863	1.262	0.918	0.704
C_6H_6	Benzene			0.604	0.436	0.335	
C_6H_6ClN	o-Chloroaniline			3.316	1.913	1.248	0.887
C_6H_6O	Phenol				3.437	1.784	1.099
C_6H_7N	Aniline			3.847	2.029	1.247	0.850
$C_6H_8N_2$	Phenylhydrazine			13.0	4.553	1.850	0.848
C_6H_{10}	Cyclohexene		0.882	0.625	0.467	0.364	
$C_6H_{10}O$	Cyclohexanone			2.017	1.321	0.919	0.671

Molecular formula	Name	Viscosity in mPa s					
		–25°C	0°C	25°C	50°C	75°C	100°C
$C_6H_{11}N$	Hexanenitrile			0.912	0.650	0.488	0.382
C_6H_{12}	Cyclohexane			0.894	0.615	0.447	
C_6H_{12}	Methylcyclopentane	0.927	0.653	0.479	0.364		
C_6H_{12}	1-Hexene	0.441	0.326	0.252	0.202		
$C_6H_{12}O$	Cyclohexanol			57.5	12.3	4.274	1.982
$C_6H_{12}O$	2-Hexanone	1.300	0.840	0.583	0.429	0.329	0.262
$C_6H_{12}O$	4-Methyl-2-pentanone			0.545	0.406		
$C_6H_{12}O_2$	Butyl acetate		1.002	0.685	0.500	0.383	0.305
$C_6H_{12}O_2$	Isobutyl acetate			0.676	0.493	0.370	0.286
$C_6H_{12}O_2$	Ethyl butanoate			0.639	0.453		
$C_6H_{12}O_2$	Diacetone alcohol	28.7	6.621	2.798	1.829	1.648	
$C_6H_{12}O_3$	Paraldehyde			1.079	0.692	0.485	0.362
$C_6H_{13}N$	Cyclohexylamine			1.944	1.169	0.782	0.565
C_6H_{14}	Hexane		0.405	0.300	0.240		
C_6H_{14}	2-Methylpentane		0.372	0.286	0.226		
C_6H_{14}	3-Methylpentane		0.395	0.306			
$C_6H_{14}O$	Dipropyl ether		0.542	0.396	0.304	0.242	
$C_6H_{14}O$	1-Hexanol			4.578	2.271	1.270	0.781
$C_6H_{15}N$	Triethylamine		0.455	0.347	0.273	0.221	
$C_6H_{15}N$	Dipropylamine		0.751	0.517	0.377	0.288	0.228
$C_6H_{15}N$	Diisopropylamine			0.393	0.300	0.237	
$C_6H_{15}NO_3$	Triethanolamine			609	114	31.5	11.7
C_7H_5N	Benzonitrile			1.267	0.883	0.662	0.524
C_7H_7Cl	o-Chlorotoluene		1.390	0.964	0.710	0.547	0.437
C_7H_7Cl	m-Chlorotoluene		1.165	0.823	0.616	0.482	0.391
C_7H_7Cl	p-Chlorotoluene			0.837	0.621	0.483	0.390
C_7H_8	Toluene	1.165	0.778	0.560	0.424	0.333	0.270
C_7H_8O	o-Cresol				3.035	1.562	0.961
C_7H_8O	m-Cresol			12.9	4.417	2.093	1.207
C_7H_8O	Benzyl alcohol			5.474	2.760	1.618	1.055
C_7H_8O	Anisole			1.056	0.747	0.554	0.427
C_7H_9N	N-Methylaniline		4.120	2.042	1.222	0.825	0.606
C_7H_9N	o-Methyl aniline		10.3	3.823	1.936	1.198	0.839
C_7H_9N	m-Methyl aniline		8.180	3.306	1.679	1.014	0.699
C_7H_9N	Benzylamine			1.624	1.080	0.769	0.577
C_7H_{14}	Methylcyclohexane		0.991	0.679	0.501	0.390	0.316
C_7H_{14}	1-Heptene		0.441	0.340	0.273	0.226	
$C_7H_{14}O$	2-Heptanone			0.714	0.407	0.297	
$C_7H_{14}O_2$	Heptanoic acid			3.840	2.282	1.488	1.041
C_7H_{16}	Heptane	0.757	0.523	0.387	0.301	0.243	
C_7H_{16}	3-Methylhexane			0.350			
$C_7H_{16}O$	1-Heptanol			5.810	2.603	1.389	0.849
$C_7H_{16}O$	2-Heptanol			3.955	1.799	0.987	0.615
$C_7H_{16}O$	3-Heptanol				1.957	0.976	0.584
$C_7H_{16}O$	4-Heptanol			4.207	1.695	0.882	0.539
$C_7H_{17}N$	Heptylamine			1.314	0.865	0.600	0.434
C_8H_8	Styrene		1.050	0.695	0.507	0.390	0.310
C_8H_8O	Acetophenone			1.681			0.634
$C_8H_8O_2$	Methyl benzoate			1.857			
$C_8H_8O_3$	Methyl salicylate					1.102	0.815
C_8H_{10}	Ethylbenzene		0.872	0.631	0.482	0.380	0.304
C_8H_{10}	o-Xylene		1.084	0.760	0.561	0.432	0.345
C_8H_{10}	m-Xylene		0.795	0.581	0.445	0.353	0.289
C_8H_{10}	p-Xylene			0.603	0.457	0.359	0.290
$C_8H_{10}O$	Phenetole			1.197	0.817	0.594	0.453
$C_8H_{11}N$	N,N-Dimethylaniline		1.996	1.300	0.911	0.675	0.523
$C_8H_{11}N$	N-Ethylaniline		3.981	2.047	1.231	0.825	0.596

Molecular formula	Name	Viscosity in mPa s					
		−25°C	0°C	25°C	50°C	75°C	100°C
C_8H_{16}	Ethylcyclohexane		1.139	0.784	0.579		
$C_8H_{16}O_2$	Octanoic acid			5.020	2.656	1.654	1.147
C_8H_{18}	Octane		0.700	0.508	0.385	0.302	0.243
$C_8H_{18}O$	1-Octanol			7.288	3.232	1.681	0.991
$C_8H_{18}O$	4-Methyl-3-heptanol		1.904	1.085	0.702	0.497	0.375
$C_8H_{18}O$	5-Methyl-3-heptanol		2.052	1.178	0.762	0.536	0.401
$C_8H_{18}O$	2-Ethyl-1-hexanol		20.7	6.271	2.631	1.360	0.810
$C_8H_{18}O$	Dibutyl ether	1.417	0.918	0.637	0.466	0.356	0.281
$C_8H_{19}N$	Dibutylamine		1.509	0.918	0.619	0.449	0.345
$C_8H_{19}N$	Diisobutylamine		1.115	0.723	0.511	0.384	0.303
C_9H_7N	Quinoline			3.337	1.892	1.201	0.833
C_9H_{10}	Indane		2.230	1.357	0.931	0.692	0.545
C_9H_{12}	Cumene		1.075	0.737	0.547		
$C_9H_{14}O$	Isophorone		4.201	2.329	1.415	0.923	0.638
$C_9H_{18}O$	5-Nonanone			1.199	0.834	0.619	0.484
$C_9H_{18}O_2$	Nonanoic acid			7.011	3.712	2.234	1.475
C_9H_{20}	Nonane		0.964	0.665	0.488	0.375	0.300
$C_9H_{20}O$	1-Nonanol			9.123	4.032		
$C_{10}H_{10}O_4$	Dimethyl phthalate		63.2	14.4	5.309	2.824	1.980
$C_{10}H_{14}$	Butylbenzene			0.950	0.683	0.515	
$C_{10}H_{18}$	cis-Decahydronaphthalene	12.8	5.645	3.042	1.875	1.271	0.924
$C_{10}H_{18}$	trans-Decahydronaphthalene	6.192	3.243	1.948	1.289	0.917	0.689
$C_{10}H_{20}O_2$	Decanoic acid				4.327	2.651	
$C_{10}H_{22}$	Decane	2.188	1.277	0.838	0.598	0.453	0.359
$C_{10}H_{22}O$	1-Decanol			10.9	4.590		
$C_{11}H_{24}$	Undecane		1.707	1.098	0.763	0.562	0.433
$C_{12}H_{10}O$	Diphenyl ether				2.130	1.407	1.023
$C_{12}H_{26}$	Dodecane		2.277	1.383	0.930	0.673	0.514
$C_{13}H_{12}$	Diphenylmethane					1.265	0.929
$C_{13}H_{28}$	Tridecane		2.909	1.724	1.129	0.796	0.594
$C_{14}H_{30}$	Tetradecane			2.128	1.376	0.953	0.697
$C_{16}H_{22}O_4$	Dibutyl phthalate	483	66.4	16.6	6.470	3.495	2.425
$C_{16}H_{34}$	Hexadecane			3.032	1.879	1.260	0.899
$C_{18}H_{38}$	Octadecane				2.487	1.609	1.132

THERMAL CONDUCTIVITY OF GASES

This table gives the thermal conductivity of several gases as a function of temperature. Unless otherwise noted, the values refer to a pressure of 100 kPa (1 bar) or to the saturation vapor pressure if that is less than 100 kPa. The notation $P = 0$ indicates the low pressure limiting value is given. In general, the $P = 0$ and $P = 100$ kPa values differ by less than 1%. Units are milliwatts per meter kelvin. Substances are listed in the modified Hill order.

MF	Name	Thermal conductivity in mW/m K						Ref.
		100 K	200 K	300 K	400 K	500 K	600 K	
	Air	9.4	18.4	26.2	33.3	39.7	45.7	1
Ar	Argon	6.2	12.4	17.9	22.6	26.8	30.6	2,8
BF_3	Boron trifluoride			19.0	24.6			11
ClH	Hydrogen chloride		9.2	14.5	19.5	24.0	28.1	11
F_6S	Sulfur hexafluoride ($P = 0$)			13.0	20.6	27.5	33.8	16
H_2	Hydrogen ($P = 0$)	68.6	131.7	186.9	230.4			4
H_2O	Water			18.7	27.1	35.7	47.1	6
	Deuterium oxide				27.0	36.5	47.6	7
H_2S	Hydrogen sulfide			14.6	20.5	26.4	32.4	11
H_3N	Ammonia			24.4	37.4	51.6	66.8	11
He	Helium ($P = 0$)	75.5	119.3	156.7	190.6	222.3	252.4	8
Kr	Krypton ($P = 0$)	3.3	6.4	9.5	12.3	14.8	17.1	8
NO	Nitric oxide		17.8	25.9	33.1	39.6	46.2	11
N_2	Nitrogen	9.8	18.7	26.0	32.3	38.3	44.0	12
N_2O	Nitrous oxide		9.8	17.4	26.0	34.1	41.8	11
Ne	Neon ($P = 0$)	22.3	37.6	49.8	60.3	69.9	78.7	8
O_2	Oxygen	9.3	18.4	26.3	33.7	41.0	48.1	10
O_2S	Sulfur dioxide			9.6	14.3	20.0	25.6	11
Xe	Xenon ($P = 0$)	2.0	3.6	5.5	7.3	8.9	10.4	8
CCl_2F_2	Dichlorodifluoromethane			9.9	15.0	20.1	25.2	13
CF_4	Tetrafluoromethane ($P = 0$)			16.0	24.1	32.2	39.9	16
CO	Carbon monoxide ($P = 0$)			25.0	32.3	39.2	45.7	14
CO_2	Carbon dioxide		9.6	16.8	25.1	33.5	41.6	9
$CHCl_3$	Trichloromethane			7.5	11.1	15.1		11
CH_4	Methane		22.5	34.1	49.1	66.5	84.1	5,15
CH_4O	Methanol				26.2	38.6	53.0	11
$C_2Cl_2F_4$	1,2-Dichlorotetrafluoro-ethane			10.25	15.7	21.1		13
$C_2Cl_3F_3$	1,1,2-Trichlorotrifluoro-ethane			9.0	13.6	18.3		13
C_2H_2	Acetylene			21.4	33.3	45.4	56.8	11
C_2H_4	Ethylene		11.1	20.5	34.6	49.9	68.6	3
C_2H_6	Ethane		11.0	21.3	35.4	52.2	70.5	5
C_2H_6O	Ethanol			14.4	25.8	38.4	53.2	11
C_3H_6O	Acetone			11.5	20.2	30.6	42.7	11
C_3H_8	Propane			18.0	30.6	45.5	61.9	5
C_4F_8	Perfluorocyclobutane			12.5	19.5			13
C_4H_{10}	Butane			16.4	28.4	43.0	59.1	5
C_4H_{10}	Isobutane			16.1	27.9	42.1	57.6	5
$C_4H_{10}O$	Diethyl ether			15.1	25.0	37.1		11
C_5H_{12}	Pentane			14.4	24.9	37.8	52.7	11
C_6H_{14}	Hexane				23.4	35.4	48.7	11

REFERENCES

1. Kadoya, K. Matsunaga, N., and Nagashima, A., Viscosity and thermal conductivity of dry air in the gaseous phase, *J. Phys. Chem. Ref. Data*, 14, 947, 1985.
2. Younglove, B. A. and Hanley, H. J. M., The viscosity and thermal conductivity coefficients of gaseous and liquid argon, *J. Phys. Chem. Ref. Data*, 15, 1323, 1986.
3. Holland, P. M., Eaton, B. E., and Hanley, H. J. M., A correlation of the viscosity and thermal conductivity data of gaseous and liquid ethylene, *J. Phys. Chem. Ref. Data*, 12, 917, 1983.

4. Assael, M. J., Mixafendi, S., and Wakeham, W. A., The viscosity and thermal conductivity of normal hydrogen in the limit of zero density, *J. Phys. Chem. Ref. Data*, 15, 1315, 1986.

5. Younglove, B. A. and Ely, J. F., Thermophysical properties of fluids. II. Methane, ethane, propane, isobutane, and normal butane, *J. Phys. Chem. Ref. Data*, 16, 577, 1987.

6. Sengers, J. V. and Watson, J. T. R., Improved international formulations for the viscosity and thermal conductivity of water substance, *J. Phys. Chem. Ref. Data*, 15, 1291, 1986.

7. Matsunaga, N. and Nagashima, A., Transport properties of liquid and gaseous D_2O over a wide range of temperature and pressure, *J. Phys. Chem. Ref. Data*, 12, 933, 1983.

8. Kestin, J. et al., Equilibrium and transport properties of the noble gases and their mixtures at low density, *J. Phys. Chem. Ref. Data*, 13, 229, 1984.

9. Vescovic, V. et al., The transport properties of carbon dioxide, *J. Phys. Chem. Ref. Data*, 19, 1990.

10. Younglove, B. A., Thermophysical properties of fluids. I. Argon, ethylene, parahydrogen, nitrogen, nitrogen trifluoride, and oxygen, *J. Phys. Chem. Ref. Data*, 11, Suppl. 1, 1982.

11. Ho, C. Y., Ed., *Properties of Inorganic and Organic Fluids, CINDAS Data Series on Materials Properties*, Volume V-1, Hemisphere Publishing Corp., New York, 1988.

12. Stephen, K., Krauss, R., and Laesecke, A., Viscosity and thermal conductivity of nitrogen for a wide range of fluid states, *J. Phys. Chem. Ref. Data*, 16, 993, 1987.

13. Krauss, R. and Stephan, K., Thermal conductivity of refrigerants in a wide range of temperature and pressure, *J. Phys. Chem. Ref. Data*, 18, 43, 1989.

14. Millat, J. and Wakeham, W. A., The thermal conductivity of nitrogen and carbon monoxide in the limit of zero density, *J. Phys. Chem. Ref. Data*, 18, 565, 1989.

15. Friend, D. G., Ely, J. F., and Ingham, H., Thermophysical properties of methane, *J. Phys. Chem. Ref. Data*, 18, 583, 1989.

16. Uribe, F. J., Mason, E. A., and Kestin, J., Thermal conductivity of nine polyatomic gases at low density, *J. Phys. Chem. Ref. Data*, 19, 1123, 1990.

THERMAL CONDUCTIVITY OF LIQUIDS

This table gives the thermal conductivity of some common liquids at temperatures between -25 and 100°C. All values are given in units of watts per meter kelvin (W/m K). Values refer to nominal atmospheric pressure (about 100 kPa); when an entry is given at a temperature above the normal boiling point of the substance, the pressure is understood to be the saturation vapor pressure at that temperature.

Substances are arranged by molecular formula in the modified Hill order, with compounds not containing carbon preceding those that do contain carbon.

The values for water, benzene, toluene, heptane, and dimethyl phthalate are particularly well determined and can be used for calibration purposes.

REFERENCES

1. Daubert, T. E., Danner, R. P., Sibul, H. M., and Stebbins, C. C., *Physical and Thermodynamic Properties of Pure Compounds: Data Compilation*, extant 1994 (core with 4 supplements), Taylor & Francis, Bristol, PA (also available as database).
2. Marsh, K. N., Ed., *Recommended Reference Materials for the Realization of Physicochemical Properties*, Blackwell Scientific Publications, Oxford, 1987.

Molecular formula	Name	Thermal conductivity in W/m K					
		−25°C	0°C	25°C	50°C	75°C	100°C
Cl_4Si	Silicon tetrachloride			0.099	0.096		
H_2O	Water		0.5610	0.6071	0.6435	0.6668	0.6791
Hg	Mercury	7.25	7.77	8.25	8.68	9.07	9.43
CCl_4	Tetrachloromethane		0.104	0.099	0.093	0.088	
CS_2	Carbon disulfide		0.154	0.149			
$CHCl_3$	Trichloromethane	0.127	0.122	0.117	0.112	0.107	0.102
CH_2Br_2	Dibromomethane	0.120	0.114	0.108	0.103	0.097	
CH_4O	Methanol	0.214	0.207	0.200	0.193		
C_2Cl_4	Tetrachloroethylene		0.117	0.110	0.104	0.097	0.091
C_2HCl_3	Trichloroethylene	0.133	0.124	0.116	0.108	0.100	
$C_2H_3Cl_3$	1,1,1-Trichloroethane		0.106	0.101	0.096		
C_2H_3N	Acetonitrile	0.208	0.198	0.188	0.178	0.168	
$C_2H_4O_2$	Acetic acid			0.158	0.153	0.149	0.144
C_2H_5Cl	Chloroethane	0.145	0.132	0.119	0.106	0.093	
C_2H_5NO	N-Methylformamide			0.203	0.201	0.199	0.196
C_2H_6O	Ethanol		0.176	0.169	0.162		
$C_2H_6O_2$	Ethylene glycol		0.256	0.256	0.256	0.256	0.256
C_2H_7NO	Ethanolamine			0.299	0.286	0.274	0.261
C_3H_5ClO	Epichlorohydrin	0.142	0.137	0.131	0.125	0.119	0.114
C_3H_6O	Acetone		0.169	0.161			
$C_3H_6O_2$	Methyl acetate	0.174	0.164	0.153	0.143	0.133	0.122
C_3H_7NO	N,N-Dimethylformamide			0.184	0.178	0.171	0.165
C_3H_8O	1-Propanol	0.162	0.158	0.154	0.149	0.145	0.141
C_3H_8O	2-Propanol	0.146	0.141	0.135	0.129	0.124	0.118
$C_3H_8O_2$	1,2-Propanediol		0.202	0.200	0.199	0.198	0.197
$C_3H_8O_3$	Glycerol			0.292	0.295	0.297	0.300
C_3H_9N	Trimethylamine	0.143	0.133				
C_4H_4O	Furan	0.142	0.134	0.126			
C_4H_4S	Thiophene			0.199	0.195	0.191	0.186
C_4H_6	2-Butyne	0.137	0.129	0.121			
C_4H_8O	2-Butanone	0.158	0.151	0.145	0.139	0.133	
C_4H_8O	Tetrahydrofuran	0.132	0.126	0.120	0.114		
$C_4H_8O_2$	1,4-Dioxane			0.159	0.147	0.135	0.123
$C_4H_8O_2$	Ethyl acetate	0.162	0.153	0.144	0.135	0.126	
$C_4H_{10}O$	1-Butanol		0.158	0.154	0.149		
$C_4H_{10}O$	Diethyl ether	0.150	0.140	0.130	0.120	0.110	0.100
C_5H_5N	Pyridine		0.169	0.165	0.161	0.158	
C_5H_8	Cyclopentene	0.143	0.136	0.129			
C_5H_{10}	1-Pentene	0.131	0.124	0.116			
C_5H_{10}	Cyclopentane	0.140	0.133	0.126			
C_5H_{12}	Pentane	0.132	0.122	0.113	0.103	0.095	0.087
$C_5H_{12}O$	1-Pentanol		0.157	0.153	0.149	0.145	
C_6H_5Cl	Chlorobenzene	0.136	0.131	0.127	0.122	0.117	0.112

Molecular formula	Name	Thermal conductivity in W/m K					
		−25°C	0°C	25°C	50°C	75°C	100°C
C_6H_6	Benzene			0.1411	0.1329	0.1247	
C_6H_6O	Phenol				0.156	0.153	0.151
C_6H_{10}	Cyclohexene	0.142	0.136	0.130	0.124	0.118	
$C_6H_{10}O$	Mesityl oxide	0.170	0.163	0.156	0.149	0.142	0.134
C_6H_{12}	Cyclohexane			0.123	0.117	0.111	
C_6H_{12}	1-Hexene	0.137	0.129	0.121	0.113		
$C_6H_{12}O$	Cyclohexanol			0.134	0.131		
$C_6H_{12}O$	2-Hexanone	0.151	0.145	0.139	0.133	0.127	0.121
C_6H_{14}	Hexane	0.137	0.128	0.120	0.111	0.102	0.093
$C_6H_{14}O$	1-Hexanol	0.159	0.154	0.150	0.145	0.141	0.137
C_7H_6O	Benzaldehyde			0.151	0.141	0.131	0.121
C_7H_8	Toluene	0.1461	0.1386	0.1311	0.1236	0.1161	
C_7H_8O	Anisole	0.170	0.163	0.156	0.150	0.143	0.136
C_7H_{16}	Heptane	0.1378	0.1303	0.1228	0.1152	0.1077	
$C_7H_{16}O$	1-Heptanol		0.166	0.159	0.153	0.147	0.141
C_8H_8	Styrene	0.148	0.142	0.137	0.131	0.126	0.120
C_8H_{10}	Ethylbenzene			0.130	0.124	0.118	0.112
C_8H_{10}	o-Xylene			0.131	0.126	0.120	0.114
C_8H_{10}	m-Xylene			0.130	0.124	0.118	0.113
C_8H_{10}	p-Xylene			0.130	0.124	0.118	0.112
C_8H_{18}	Octane	0.143	0.135	0.128	0.120	0.113	0.106
$C_8H_{18}O$	1-Octanol		0.168	0.161	0.154	0.147	0.141
C_9H_{12}	Cumene			0.128	0.120	0.112	0.107
C_9H_{12}	Mesitylene	0.147	0.141	0.136	0.130	0.124	0.118
C_9H_{20}	Nonane	0.144	0.138	0.131	0.124	0.118	0.111
$C_9H_{20}O$	1-Nonanol		0.166	0.161	0.155	0.149	0.143
$C_{10}H_{10}O_4$	Dimethyl phthalate		0.1501	0.1473	0.1443	0.1409	0.1373
$C_{10}H_{14}$	p-Cymene	0.132	0.127	0.122	0.117	0.112	0.107
$C_{10}H_{22}$	Decane	0.144	0.138	0.132	0.126	0.119	0.113
$C_{10}H_{22}O$	1-Decanol			0.162	0.156	0.150	0.145
$C_{11}H_{24}$	Undecane			0.140	0.135	0.129	0.123
$C_{12}H_{10}O$	Diphenyl ether			0.139		0.135	0.131
$C_{12}H_{26}$	Dodecane		0.157	0.152	0.146	0.140	0.135
$C_{12}H_{26}O$	1-Dodecanol			0.146	0.142	0.139	0.135
$C_{13}H_{28}$	Tridecane			0.137	0.132	0.127	0.122
$C_{14}H_{30}$	Tetradecane			0.136	0.131	0.126	0.121
$C_{14}H_{30}O$	1-Tetradecanol				0.167	0.162	0.157
$C_{16}H_{22}O_4$	Dibutyl phthalate	0.144	0.140	0.136	0.133	0.129	0.125
$C_{16}H_{34}$	Hexadecane			0.140	0.135	0.130	0.125
$C_{18}H_{38}$	Octadecane				0.146	0.142	0.137

DIFFUSION IN GASES

This table gives binary diffusion coefficients D_{12} for a number of common gases as a function of temperature. Values refer to atmospheric pressure. The diffusion coefficient is inversely proportional to pressure as long as the gas is in a regime where binary collisions dominate. See Reference 1 for a discussion of the dependence of D_{12} on temperature and composition.

The first part of the table gives data for several gases in the presence of a large excess of air. The remainder applies to equimolar mixtures of gases. Each gas pair is ordered alphabetically according to the most common way of writing the formula. The listing of pairs then follows alphabetical order by the first constituent.

REFERENCES

1. Marrero, T. R., and Mason, E. A., *J. Phys. Chem. Ref. Data*, 1, 1, 1972.
2. Kestin, J., et al., *J. Phys. Chem. Ref. Data*, 13, 229, 1984.

$D_{12}/\text{cm}^2\,\text{s}^{-1}$ for $p = 101.325$ kPa and the Specified T/K

System	200	273.15	293.15	373.15	473.15	573.15	673.15
Large Excess of Air							
Ar-air		0.167	0.148	0.289	0.437	0.612	0.810
CH_4-air			0.106	0.321	0.485	0.678	0.899
CO-air			0.208	0.315	0.475	0.662	0.875
CO_2-air			0.160	0.252	0.390	0.549	0.728
H_2-air		0.668	0.627	1.153	1.747	2.444	3.238
H_2O-air			0.242	0.399	0.638	0.873	1.135
He-air		0.617	0.580	1.057	1.594	2.221	2.933
SF_6-air				0.150	0.233	0.329	0.438
Equimolar Mixture							
Ar-CH_4				0.306	0.467	0.657	0.876
Ar-CO		0.168	0.187	0.290	0.439	0.615	0.815
Ar-CO_2		0.129	0.078	0.235	0.365	0.517	0.689
Ar-H_2		0.698	0.794	1.228	1.876	2.634	3.496
Ar-He	0.381	0.645	0.726	1.088	1.617	2.226	2.911
Ar-Kr	0.064	0.117	0.134	0.210	0.323	0.456	0.605
Ar-N_2		0.168	0.190	0.290	0.439	0.615	0.815
Ar-Ne	0.160	0.277	0.313	0.475	0.710	0.979	1.283
Ar-O_2		0.166	0.189	0.285	0.430	0.600	0.793
Ar-SF_6				0.128	0.202	0.290	0.389
Ar-Xe	0.052	0.095	0.108	0.171	0.264	0.374	0.498
CH_4-H_2			0.782	1.084	1.648	2.311	3.070
CH_4-He			0.723	0.992	1.502	2.101	2.784
CH_4-N_2			0.220	0.317	0.480	0.671	0.890
CH_4-O_2			0.210	0.341	0.523	0.736	0.978
CH_4-SF_6				0.167	0.257	0.363	0.482
CO-CO_2			0.162	0.250	0.384		
CO-H_2	0.408	0.686	0.772	1.162	1.743	2.423	3.196
CO-He	0.365	0.619	0.698	1.052	1.577	2.188	2.882
CO-Kr		0.131	0.147	0.227	0.346	0.485	0.645
CO-N_2	0.133	0.208	0.231	0.336	0.491	0.673	0.878
CO-O_2			0.202	0.307	0.462	0.643	0.849
CO-SF_6				0.144	0.226	0.323	0.432
CO_2-C_3H_8			0.084	0.133	0.209		
CO_2-H_2	0.315	0.552	0.412	0.964	1.470	2.066	2.745
CO_2-H_2O			0.162	0.292	0.496	0.741	1.021
CO_2-He	0.300	0.513	0.400	0.878	1.321		
CO_2-N_2			0.160	0.253	0.392	0.553	0.733
CO_2-N_2O	0.055	0.099	0.113	0.177	0.276		
CO_2-Ne	0.131	0.227	0.199	0.395	0.603	0.847	

System	200	273.15	293.15	373.15	473.15	573.15	673.15
CO_2-O_2			0.159	0.248	0.380	0.535	0.710
CO_2-SF_6				0.099	0.155		
D_2-H_2	0.631	1.079	1.219	1.846	2.778	3.866	5.103
H_2-He	0.775	1.320	1.490	2.255	3.394	4.726	6.242
H_2-Kr	0.340	0.601	0.682	1.053	1.607	2.258	2.999
H_2-N_2	0.408	0.686	0.772	1.162	1.743	2.423	3.196
H_2-Ne	0.572	0.982	0.317	1.684	2.541	3.541	4.677
H_2-O_2		0.692	0.756	1.188	1.792	2.497	3.299
H_2-SF_6			0.208	0.649	0.998	1.400	1.851
H_2-Xe		0.513	0.122	0.890	1.349	1.885	2.493
H_2O-N_2			0.242	0.399			
H_2O-O_2			0.244	0.403	0.645	0.882	1.147
He-Kr	0.330	0.559	0.629	0.942	1.404	1.942	2.550
He-N_2	0.365	0.619	0.698	1.052	1.577	2.188	2.882
He-Ne	0.563	0.948	1.066	1.592	2.362	3.254	4.262
He-O_2		0.641	0.697	1.092	1.640	2.276	2.996
He-SF_6			1.109	0.592	0.871	1.190	1.545
He-Xe	0.282	0.478	0.538	0.807	1.201	1.655	2.168
Kr-N_2		0.131	0.149	0.227	0.346	0.485	0.645
Kr-Ne	0.131	0.228	0.258	0.392	0.587	0.812	1.063
Kr-Xe	0.035	0.064	0.073	0.116	0.181	0.257	0.344
N_2-Ne			0.258	0.483	0.731	1.021	1.351
N_2-O_2			0.202	0.307	0.462	0.643	0.849
N_2-SF_6				0.148	0.231	0.328	0.436
N_2-Xe		0.107	0.123	0.188	0.287	0.404	0.539
Ne-Xe	0.111	0.193	0.219	0.332	0.498	0.688	0.901
O_2-SF_6			0.097	0.154	0.238	0.334	0.441

DIFFUSION COEFFICIENTS IN LIQUIDS AT INFINITE DILUTION

This table lists diffusion coefficients D_{AB} at infinite dilution for some binary liquid mixtures. Although values are given to two decimal places, measurements in the literature are often in poor agreement. Therefore most values in the table cannot be relied upon to better than 10%.

Solvents are listed in alphabetical order, as are the solutes within each solvent group.

REFERENCE

Landolt-Börnstein, *Numerical Data and Functional Relationships in Science and Technology*, Sixth Edition, Vol. II/5a, 1969.

Solute	Solvent	$t/°C$	D_{AB} 10^{-5} cm^2 s^{-1}	Solute	Solvent	$t/°C$	D_{AB} 10^{-5} cm^2 s^{-1}
Acetic acid	Acetone	25	3.31	Acetone	Tetrachloromethane	25	1.75
Benzoic acid	Acetone	25	2.62	Benzene	Tetrachloromethane	25	1.42
Formic acid	Acetone	25	3.77	Cyclohexane	Tetrachloromethane	25	1.30
Nitrobenzene	Acetone	20	2.94	Ethanol	Tetrachloromethane	25	1.90
Tetrachloromethane	Acetone	25	3.29	Iodine	Tetrachloromethane	30	1.63
Trichloromethane	Acetone	25	3.64	Trichloromethane	Tetrachloromethane	25	1.66
Water	Acetone	25	4.56	Acetic acid	Toluene	25	2.26
Acetic acid	Benzene	25	2.09	Benzene	Toluene	25	2.54
Aniline	Benzene	25	1.96	Benzoic acid	Toluene	25	1.49
Benzoic acid	Benzene	25	1.38	Cyclohexane	Toluene	25	2.42
Bromobenzene	Benzene	8	1.45	Formic acid	Toluene	25	2.65
2-Butanone	Benzene	30	2.09	Water	Toluene	25	6.19
Chloroethylene	Benzene	8	1.77	Acetone	Trichloromethane	25	2.55
Cyclohexane	Benzene	25	2.25	Benzene	Trichloromethane	25	2.89
Ethanol	Benzene	25	3.02	2-Butanone	Trichloromethane	25	2.13
Formic acid	Benzene	25	2.28	Butyl acetate	Trichloromethane	25	1.71
Heptane	Benzene	25	1.78	Diethyl ether	Trichloromethane	25	2.15
Methanol	Benzene	25	3.80	Ethanol	Trichloromethane	15	2.20
Toluene	Benzene	25	1.85	Ethyl acetate	Trichloromethane	25	2.02
1,2,4-Trichlorobenzene	Benzene	8	1.34	Acetic acid	Water	25	1.29
Trichloromethane	Benzene	25	2.26	Acetone	Water	25	1.28
Adipic acid	1-Butanol	30	0.40	Acetonitrile	Water	15	1.26
Benzene	1-Butanol	25	1.00	Alanine	Water	25	0.91
Biphenyl	1-Butanol	25	0.63	Allyl alcohol	Water	15	0.90
Butyric acid	1-Butanol	30	0.51	Aniline	Water	20	0.92
p-Dichlorobenzene	1-Butanol	25	0.82	Arabinose	Water	20	0.69
Methanol	1-Butanol	30	0.59	Benzene	Water	20	1.02
Oleic acid	1-Butanol	30	0.25	1-Butanol	Water	25	0.56
Propane	1-Butanol	25	1.57	Caprolactam	Water	25	0.87
Water	1-Butanol	25	0.56	Chloroethylene	Water	25	1.34
Benzene	Cyclohexane	25	1.41	Cyclohexane	Water	20	0.84
Tetrachloromethane	Cyclohexane	25	1.49	Diethylamine	Water	20	0.97
Toluene	Cyclohexane	25	1.57	Ethanol	Water	25	1.24
Allyl alcohol	Ethanol	20	0.98	Ethanolamine	Water	25	1.08
Benzene	Ethanol	25	1.81	Ethyl acetate	Water	20	1.00
Iodine	Ethanol	25	1.32	Ethylbenzene	Water	20	0.81
Iodobenzene	Ethanol	20	1.00	Ethylene glycol	Water	25	1.16
3-Methyl-1-butanol	Ethanol	20	0.81	Glucose	Water	25	0.67
Pyridine	Ethanol	20	1.10	Glycerol	Water	25	1.06
Tetrachloromethane	Ethanol	25	1.50	Glycine	Water	25	1.05
Water	Ethanol	25	1.24	Lactose	Water	15	0.38
Acetic acid	Ethyl acetate	20	2.18	Maltose	Water	15	0.38
Acetone	Ethyl acetate	20	3.18	Mannitol	Water	15	0.50
2-Butanone	Ethyl acetate	30	2.93	Methane	Water	25	1.49
Ethyl benzoate	Ethyl acetate	20	1.85	Methanol	Water	15	1.28
Nitrobenzene	Ethyl acetate	20	2.25	3-Methyl-1-butanol	Water	10	0.69
Water	Ethyl acetate	25	3.20	Methylcyclopentane	Water	20	0.85
Benzene	Heptane	25	3.91	Phenol	Water	20	0.89
Toluene	Heptane	25	3.72	1-Propanol	Water	15	0.87
Bromobenzene	Hexane	8	2.60	Propene	Water	25	1.44
2-Butanone	Hexane	30	3.74	Pyridine	Water	25	0.58
Dodecane	Hexane	25	2.73	Raffinose	Water	15	0.33
Iodine	Hexane	25	4.45	Sucrose	Water	25	0.52
Methane	Hexane	25	0.09	Toluene	Water	20	0.85
Propane	Hexane	25	4.87	Urea	Water	25	1.38
Tetrachloromethane	Hexane	25	3.70	Urethane	Water	15	0.80
Toluene	Hexane	25	4.21				

Section 7
Biochemistry

PROPERTIES OF AMINO ACIDS

This table gives selected properties of some important amino acids and closely related compounds. The first part of the table lists the 20 "standard" amino acids that are the basic constituents of proteins (structures of these amino acids may be found in the following table). The second part includes other amino acids and related compounds of biochemical importance. Within each part of the table the compounds are listed by name in alphabetical order.

Symbol — Three-letter symbol for the standard amino acids
M_r — Molecular weight
t_m — Melting point
pK_a, pK_b, pK_c, pK_d — Negative of the logarithm of the acid dissociation constants for the COOH and NH_2 groups (and, in some cases, other groups) in the molecule (at 25°C)
pI — pH at the isoelectric point
S — Solubility in water at 25°C in units of grams of compound per kilogram of water; when quantitative data are not available, the notations sl.s. (for slightly soluble) and v.s. (for very soluble) are used.

Data on the enthalpy of formation of many of these compounds are included in the table "Standard Thermodynamic Properties of Chemical Substances" in Section 5 of this Handbook. Absorption spectra and optical rotation data can be found in Reference 3. Partial molar volume and other thermodynamic properties, including solubility as a function of temperature, are given in References 3 and 5. Most of the pK values come from Reference 7.

REFERENCES

1. Dawson, R. M. C., Elliott, D. C., Elliott, W. H., and Jones, K. M., *Data for Biochemical Research*, 3rd ed., Clarendon Press, Oxford, 1986.
2. Budavari, S., Ed., *The Merck Index, Twelfth Edition*, Merck & Co., Rahway, NJ, 1996.
3. Sober, H. A., Ed., *CRC Handbook of Biochemistry. Selected Data for Molecular Biology*, CRC Press, Boca Raton, FL, 1968.
4. Voet, D. and Voet, J. G., *Biochemistry, Second Edition*, John Wiley & Sons, New York, 1995.
5. Hinz, H. J., Ed., *Thermodynamic Data for Biochemistry and Biotechnology*, Springer-Verlag, Heidelberg, 1986.
6. Fasman, G. D., Ed., *Practical Handbook of Biochemistry and Molecular Biology*, CRC Press, Boca Raton, FL, 1989.
7. Smith, R. M., and Martell, A. E., *NIST Standard Reference Database 46: Critically Selected Stability Constants of Metal Complexes Database, Version 3.0*, National Institute of Standards and Technology, Gaithersburg, MD, 1997.
8. Jin, Z. and Chao, K. C., *J. Chem. Eng. Data*, 37, 199, 1992.

The standard amino acids:

Symbol	Name	Mol. form.	M_r	t_m/°C	pK_a	pK_b	pK_c	pI	S/g kg^{-1}
Ala	Alanine	$C_3H_7NO_2$	89.09	297	2.33	9.71		6.00	165.0
Arg	Arginine	$C_6H_{14}N_4O_2$	174.20	244	2.03	9.00	12.10	10.76	182.6
Asn	Asparagine	$C_4H_8N_2O_3$	132.12	235	2.16	8.73		5.41	25.1
Asp	Aspartic acid	$C_4H_7NO_4$	133.10	270	1.95	9.66	3.71	2.77	4.95
Cys	Cysteine	$C_3H_7NO_2S$	121.16	240	1.91	10.28	8.14	5.07	v.s.
Glu	Glutamic acid	$C_5H_9NO_4$	147.13	160	2.16	9.58	4.15	3.22	8.61
Gln	Glutamine	$C_5H_{10}N_2O_3$	146.15	185	2.18	9.00		5.65	42
Gly	Glycine	$C_2H_5NO_2$	75.07	290	2.34	9.58		5.97	250.9
His	Histidine	$C_6H_9N_3O_2$	155.16	287	1.70	9.09	6.04	7.59	43.5
Ile	Isoleucine	$C_6H_{13}NO_2$	131.17	284	2.26	9.60		6.02	34.2
Leu	Leucine	$C_6H_{13}NO_2$	131.17	293	2.32	9.58		5.98	22.0
Lys	Lysine	$C_6H_{14}N_2O_2$	146.19	224	2.15	9.16	10.67	9.74	5.8
Met	Methionine	$C_5H_{11}NO_2S$	149.21	281	2.16	9.08		5.74	56
Phe	Phenylalanine	$C_9H_{11}NO_2$	165.19	283	2.18	9.09		5.48	27.9
Pro	Proline	$C_5H_9NO_2$	115.13	221	1.95	10.47		6.30	1623
Ser	Serine	$C_3H_7NO_3$	105.09	228	2.13	9.05		5.68	421.7
Thr	Threonine	$C_4H_9NO_3$	119.12	256	2.20	8.96		5.60	98.1
Trp	Tryptophan	$C_{11}H_{12}N_2O_2$	204.23	289	2.38	9.34		5.89	13.2
Tyr	Tyrosine	$C_9H_{11}NO_3$	181.19	343	2.24	9.04	10.10	5.66	0.46
Val	Valine	$C_5H_{11}NO_2$	117.15	315	2.27	9.52		5.96	88.5

Other amino acids and related compounds:

Name	Mol. form.	M_r	$t_m/°C$	pK_a	pK_b	pK_c	pK_d	S/g kg
N-Acetylglutamic acid	$C_7H_{11}NO_5$	189.17	199					
N6-Acetyl-L-lysine	$C_8H_{16}N_2O_3$	188.23	265	2.12	9.51			
β-Alanine	$C_3H_7NO_2$	89.09	200	3.51	10.08			891
DL-2-Aminobutanoic acid	$C_4H_9NO_2$	103.12	304	2.30	9.63			210
DL-3-Aminobutanoic acid	$C_4H_9NO_2$	103.12	194	3.43	10.05			1250
4-Aminobutanoic acid	$C_4H_9NO_2$	103.12	203	4.02	10.35			v.s.
5-Aminolevulinic acid	$C_5H_9NO_3$	131.13	118	4.05	8.90			
L-3-Amino-2-methylpropanoic acid	$C_4H_9NO_2$	103.12	183					
Azaserine	$C_5H_7N_3O_4$	173.1?	150		8.55			v.s.
L-γ-Carboxyglutamic acid	$C_6H_9NO_6$	191.14	167	1.70	9.90	4.75	3.20	
Carnosine	$C_9H_{14}N_4O_3$	226.24	260	2.51	9.35	6.76		322
Citrulline	$C_6H_{13}N_3O_3$	175.19	222	2.32	9.30			
Creatine	$C_4H_9N_3O_2$	131.13	303	2.63	14.30			16
L-Cysteic acid	$C_3H_7NO_5S$	169.16	260	1.89	8.70	1.30		v.s.
L-Cystine	$C_6H_{12}N_2O_4S_2$	240.30	260	1.50	8.80	2.05	8.03	0.11
L-3,5-Diiodotyrosine	$C_9H_9I_2NO_3$	432.98	213	2.12	9.10	6.16		0.62
Dopamine	$C_8H_{11}NO_2$	153.18			10.36	8.88		
L-Ethionine	$C_6H_{13}NO_2S$	163.24	273	2.18	9.05	13.10		
Glycocyamine	$C_3H_7N_3O_2$	117.11	282	2.82				5
N-Glycylglycine	$C_4H_8N_2O_3$	132.12	263	3.13	8.10			
Histamine	$C_5H_9N_3$	111.15	83		9.83	6.11		v.s.
Homocysteine	$C_4H_9NO_2S$	135.19	232	2.15	8.57	10.38		
Homocystine	$C_8H_{16}N_2O_4S_2$	268.36	264	1.59	9.44	2.54	8.52	0.2
L-Homoserine	$C_4H_9NO_3$	119.12	203	2.27	9.28			1100
5-Hydroxylysine	$C_6H_{14}N_2O_3$	162.19		2.13	8.85	9.83		
trans-4-Hydroxyproline	$C_5H_9NO_3$	131.13	274	1.82	9.47			361
L-3-Iodotyrosine	$C_9H_{10}INO_3$	307.09	205	2.20	9.10	8.70		sl.s.
L-Kynurenine	$C_{10}H_{12}N_2O_3$	208.22	185					sl.s.
L-Lanthionine	$C_6H_{12}N_2O_4S$	208.24	295					1.5
Levodopa	$C_9H_{11}NO_4$	197.19	277	2.20	8.75	9.81	13.40	1650
2-Methylalanine	$C_4H_9NO_2$	103.12	335	2.36	10.21			137
L-1-Methylhistidine	$C_7H_{11}N_3O_2$	169.18	250	1.69	8.85	6.48		
L-Norleucine	$C_6H_{13}NO_2$	131.17	301	2.31	9.68			15
L-Norvaline	$C_5H_{11}NO_2$	117.15	305	2.31	9.65			107
L-Ornithine	$C_5H_{12}N_2O_2$	132.16	140	1.94	8.78	10.52		v.s.
O-Phosphoserine	$C_3H_8NO_6P$	185.07	166	2.14	9.80	5.70		
L-Pyroglutamic acid	$C_5H_7NO_3$	129.12	162	3.32				
Sarcosine	$C_3H_7NO_2$	89.09	212	2.18	9.97			428
L-Thyroxine	$C_{15}H_{11}I_4NO_4$	776.87	235	2.20	10.01	6.45		sl.s.

STRUCTURES OF COMMON AMINO ACIDS

Alanine
(Ala, A)

$$H_3C-\overset{\overset{NH_2}{|}}{\underset{\underset{H}{|}}{C}}-COOH$$

Leucine
(Leu, L)

$$\overset{CH_3}{\underset{CH_3}{>}}HC-CH_2-\overset{\overset{NH_2}{|}}{\underset{\underset{H}{|}}{C}}-COOH$$

Arginine
(Arg, R)

$$H_2N-\overset{\overset{NH}{||}}{C}-\overset{\overset{H}{|}}{N}-CH_2-CH_2-CH_2-\overset{\overset{NH_2}{|}}{\underset{\underset{H}{|}}{C}}-COOH$$

Lysine
(Lys, K)

$$H_2N-CH_2-CH_2-CH_2-CH_2-\overset{\overset{NH_2}{|}}{\underset{\underset{H}{|}}{C}}-COOH$$

Asparagine
(Asn, N)

$$H_2N-\overset{\overset{O}{||}}{C}-CH_2-\overset{\overset{NH_2}{|}}{\underset{\underset{H}{|}}{C}}-COOH$$

Methionine
(Met, M)

$$CH_3-S-CH_2-CH_2-\overset{\overset{NH_2}{|}}{\underset{\underset{H}{|}}{C}}-COOH$$

Aspartic acid
(Asp, D)

$$HOOC-CH_2-\overset{\overset{NH_2}{|}}{\underset{\underset{H}{|}}{C}}-COOH$$

Phenylalanine
(Phe, F)

$$C_6H_5-CH_2-\overset{\overset{NH_2}{|}}{\underset{\underset{H}{|}}{C}}-COOH$$

Cysteine
(Cys, C)

$$HS-CH_2-\overset{\overset{NH_2}{|}}{\underset{\underset{H}{|}}{C}}-COOH$$

Proline
(Pro, P)

$$\begin{array}{c} CH_2-CH_2 \\ | \qquad\quad | \\ CH_2-N \quad CH-COOH \\ | \\ H \end{array}$$

Glutamine
(Gln, Q)

$$H_2N-\overset{\overset{O}{||}}{C}-CH_2-CH_2-\overset{\overset{NH_2}{|}}{\underset{\underset{H}{|}}{C}}-COOH$$

Serine
(Ser, S)

$$HO-CH_2-\overset{\overset{NH_2}{|}}{\underset{\underset{H}{|}}{C}}-COOH$$

Glutamic acid
(Glu, E)

$$HOOC-CH_2-CH_2-\overset{\overset{NH_2}{|}}{\underset{\underset{H}{|}}{C}}-COOH$$

Threonine
(Thr, T)

$$CH_3-\overset{\overset{}{|}}{\underset{\underset{OH}{|}}{CH}}-\overset{\overset{NH_2}{|}}{\underset{\underset{H}{|}}{C}}-COOH$$

Glycine
(Gly, G)

$$H-\overset{\overset{NH_2}{|}}{\underset{\underset{H}{|}}{C}}-COOH$$

Tryptophan
(Trp, W)

$$CH_2-\overset{\overset{NH_2}{|}}{\underset{\underset{H}{|}}{C}}-COOH$$

Histidine
(His, H)

$$CH_2-\overset{\overset{NH_2}{|}}{\underset{\underset{H}{|}}{C}}-COOH$$

Tyrosine
(Tyr, Y)

$$HO-C_6H_4-CH_2-\overset{\overset{NH_2}{|}}{\underset{\underset{H}{|}}{C}}-COOH$$

Isoleucine
(Ile, I)

$$CH_3-CH_2-\overset{\overset{}{|}}{\underset{\underset{CH_3}{|}}{HC}}-\overset{\overset{NH_2}{|}}{\underset{\underset{H}{|}}{C}}-COOH$$

Valine
(Val, V)

$$\overset{CH_3}{\underset{CH_3}{>}}HC-\overset{\overset{NH_2}{|}}{\underset{\underset{H}{|}}{C}}-COOH$$

PROPERTIES OF PURINE AND PYRIMIDINE BASES

This table lists some of the important purine and pyrimidine bases that occur in nucleic acids. The pK_a values (negative logarithm of the acid dissociation constant) are given for each ionization stage. The last column gives the aqueous solubility S at the indicated temperature in units of grams per 100 grams of solution.

The numbering system in the rings is:

Purine   Pyrimidine

REFERENCES

1. R. M. C. Dawson, et al., *Data for Biochemical Research*, 3rd Ed., Clarendon Press, Oxford, 1986.
2. S. Budavari, Ed., *The Merck Index*, 11th Ed., Merk and Co., Rahway, N.J., 1989.

Common name	Systematic name	Mol. form.	Mol. wt.	pK_a values			S/mass % (temp.)
Pyrimidines							
Cytosine	4-Amino-2-hydroxypyrimidine	$C_4H_5N_3O$	111.10	4.5	12.2		0.76 (25°C)
5-Methylcytosine	4-Amino-2-hydroxy-5-methylpyrimidine	$C_5H_7N_3O$	125.13	4.6	12.4		0.45 (25°C)
5-Hydroxymethyl-cytosine	4-Amino-2-hydroxy-5-hydroxy-methylpyrimidine	$C_5H_7N_3O_2$	141.13	4.3	13		
Uracil	2,4-Dihydroxypyrimidine	$C_4H_4N_2O_2$	112.09	0.5	9.5	>13	0.36 (25°C)
Thymine	5-Methyluracil	$C_5H_6N_2O_2$	126.11	9.9	>13		0.4 (25°C)
Orotic acid	Uracil-6-carboxylic acid	$C_5H_4N_2O_4$	156.10	2.4	9.5	>13	0.18 (18°C)
Purines							
Adenine	6-Aminopurine	$C_5H_5N_5$	135.14	<1	4.1	9.8	0.09 (25°C)
Guanine	2-Amino-6-hydroxypurine	$C_5H_5N_5O$	151.13	3.3	9.2	12.3	0.004 (40°C)
7-Methylguanine	7-Methyl-2-amino-6-hydroxypurine	$C_6H_7N_5O$	165.16	3.5	9.9		
Isoguanine	6-Amino-2-hydroxypurine	$C_5H_5N_5O$	151.13	4.5	9.0		0.006 (25°C)
Xanthine	2,6-Dioxopurine	$C_5H_4N_4O_2$	152.11	0.8	7.4	11.1	0.05 (20°C)
Hypoxanthine	6-Hydroxypurine	$C_5H_4N_4O$	136.11	2.0	8.9	12.1	0.07 (19°C)
Uric acid	2,6,8-Trihydroxypurine	$C_5H_4N_4O_3$	168.11	5.4	11.3		0.002 (20°C)

THE GENETIC CODE

This table gives the correspondence between a messenger RNA codon and the amino acid which it specifies. The symbols for bases in the codon are:

U: uracil
C: cytosine
A: adenine
G: guanine

The amino acid symbols are given in the table entitled "Structures of Common Amino Acids". A chain-initiating codon is indicated by **init** and a chain-terminating codon by **term**.

Example: UCA codes for **Ser**, UAC codes for **Tyr**, etc.

First position	Second position				Third position
	U	C	A	G	
U	Phe	Ser	Tyr	Cys	U
	Phe	Ser	Tyr	Cys	C
	Leu	Ser	**term**	**term**	A
	Leu	Ser	**term**	Trp	G
C	Leu	Pro	His	Arg	U
	Leu	Pro	His	Arg	C
	Leu	Pro	Gln	Arg	A
	Leu	Pro	Gln	Arg	G
A	Ile	Thr	Asn	Ser	U
	Ile	Thr	Asn	Ser	C
	Ile	Thr	Lys	Arg	A
	Met (**init**)	Thr	Lys	Arg	G
G	Val	Ala	Asp	Gly	U
	Val	Ala	Asp	Gly	C
	Val	Ala	Glu	Gly	A
	Val (**init**)	Ala	Glu	Gly	G

PROPERTIES OF FATTY ACIDS

This table gives the systematic names and selected properties of some of the more important fatty acids of five or more carbon atoms. Compounds are listed first by degree of saturation and, secondly, by number of carbon atoms. The following data are included:

M_r: Molecular weight

t_m: Melting point in °C

S: Aqueous solubility at 20°C in units of grams of solute per 100 grams of water

REFERENCES

1. Dawson, R. M. C., Elliott, D. C., Elliott, W. H., and Jones, K. M., *Data for Biochemical Research*, Third Edition, Clarendon Press, Oxford, 1986.
2. Fasman, G. D., Ed., *Practical Handbook of Biochemistry and Molecular Biology*, CRC Press, Boca Raton, FL, 1989.

Common name	Systematic name	Mol. form.	M_r	t_m/°C	S
Saturated					
Valeric acid	Pentanoic acid	$C_5H_{10}O_2$	102.13	-34	2.5
Isovaleric acid	3-Methylbutanoic acid	$C_5H_{10}O_2$	102.13	-29.3	4.3
Caproic acid	Hexanoic acid	$C_6H_{12}O_2$	116.16	-3	0.967
Enanthic acid	Heptanoic acid	$C_7H_{14}O_2$	130.19	-7.5	0.24
Caprylic acid	Octanoic acid	$C_8H_{16}O_2$	144.21	16.3	0.080
Pelargonic acid	Nonanoic acid	$C_9H_{18}O_2$	158.24	12.3	0.0284
Capric acid	Decanoic acid	$C_{10}H_{20}O_2$	172.27	31.9	0.015
Lauric acid	Dodecanoic acid	$C_{12}H_{24}O_2$	200.32	43.2	0.0055
Tridecylic acid	Tridecanoic acid	$C_{13}H_{26}O_2$	214.35	41.5	0.0033
Myristic acid	Tetradecanoic acid	$C_{14}H_{28}O_2$	228.38	53.9	0.0020
Pentadecylic acid	Pentadecanoic acid	$C_{15}H_{30}O_2$	242.40	52.3	0.0012
Palmitic acid	Hexadecanoic acid	$C_{16}H_{32}O_2$	256.43	63.1	0.00072
Margaric acid	Heptadecanoic acid	$C_{17}H_{34}O_2$	270.46	61.3	0.00042
Stearic acid	Octadecanoic acid	$C_{18}H_{36}O_2$	284.48	69.6	0.00029
Arachidic acid	Eicosanoic acid	$C_{20}H_{40}O_2$	312.54	76.5	
Phytanic acid	3,7,11,15-Tetramethylhexadecanoic acid	$C_{20}H_{40}O_2$	312.54	-65	
Behenic acid	Docosanoic acid	$C_{22}H_{44}O_2$	340.59	81.5	
Lignoceric acid	Tetracosanoic acid	$C_{24}H_{48}O_2$	368.64	87.5	
Cerotic acid	Hexacosanoic acid	$C_{26}H_{52}O_2$	396.70	88.5	
Montanic acid	Octacosanoic acid	$C_{28}H_{56}O_2$	424.75	90.9	
Monounsaturated					
Caproleic acid	9-Decenoic acid	$C_{10}H_{18}O_2$	170.25	26.5	
Palmitoleic acid	cis-9-Hexadecenoic acid	$C_{16}H_{30}O_2$	254.41	-0.1	
Oleic acid	cis-9-Octadecenoic acid	$C_{18}H_{34}O_2$	282.47	13.4	
Elaidic acid	trans-9-Octadecenoic acid	$C_{18}H_{34}O_2$	282.47	45	
Vaccenic acid	trans-11-Octadecenoic acid	$C_{18}H_{34}O_2$	282.47	44	
Erucic acid	cis-13-Docosenoic acid	$C_{22}H_{42}O_2$	338.57	34.7	
Brassidic acid	trans-13-Docosenoic acid	$C_{22}H_{42}O_2$	338.57	61.9	
Nervonic acid	cis-15-Tetracosenoic acid	$C_{24}H_{46}O_2$	366.63	43	
Diunsaturated					
Linoleic acid	cis,cis-9,12-Octadecadienoic acid	$C_{18}H_{32}O_2$	280.45	-12	
Triunsaturated					
cis-Eleostearic acid	trans,cis,trans-9,11,13-Octadecatrienoic acid	$C_{18}H_{30}O_2$	278.44	49	
trans-Eleostearic acid	trans,trans,trans-9,11,13-Octadecatrienoic acid	$C_{18}H_{30}O_2$	278.44	71.5	
Linolenic acid	cis,cis,cis-9,12,15-Octadecatrienoic acid	$C_{18}H_{30}O_2$	278.44	-11	
Tetraunsaturated					
Arachidonic acid	5,8,11,14-Eicosatetraenoic acid, (all-trans)	$C_{20}H_{32}O_2$	304.47	-49.5	

CARBOHYDRATE NAMES AND SYMBOLS

The following table lists the systematic names and symbols for selected carbohydrates and some of their derivatives. The symbols for monosaccharide residues and derivatives are recommended by IUPAC for use in describing the structures of oligosaccharide chains. A more complete list can be found in the reference.

REFERENCE

McNaught, A. D., *Pure Appl. Chem.*, 68, 1919-2008, 1996.

Common Name	Symbol	Systematic Name
Abequose	Abe	3,6-Dideoxy-D-*xylo*-hexose
N-Acetyl-2-deoxyneur-2-enaminic acid	Neu2en5Ac	
N-Acetylgalactosamine	GalNAc	
N-Acetylglucosamine	GlcNAc	
N-Acetylneuraminic acid	Neu5Ac	
Allose	All	*allo*-Hexose
Altrose	Alt	*altro*-Hexose
Apiose	Api	3-*C*-(Hydroxymethyl)-*glycero*-tetrose
Arabinitol	Ara-ol	Arabinitol
Arabinose	Ara	*arabino*-Pentose
Arcanose		2,6-Dideoxy-3-*C*-methyl-3-*O*-methyl-*xylo*-hexose
Ascarylose		3,6-Dideoxy-L-*arabino*-hexose
Boivinose		2,6-Dideoxy-D-gulose
Chalcose		4,6-Dideoxy-3-*O*-methyl-D-*xylo*-hexose
Cladinose		2,6-Dideoxy-3-*C*-methyl-3-*O*-methyl-L-*ribo*-hexose
Colitose		3,6-Dideoxy-L-*xylo*-hexose
Cymarose		6-Deoxy-3-*O*-methyl-*ribo*-hexose
3-Deoxy-D-*manno*-oct-2-ulosonic acid	Kdo	
2-Deoxyribose	dRib	2-Deoxy-*erythro*-pentose
2,3-Diamino-2,3-dideoxy-D-glucose	GlcN3N	
Diginose		2,6-Dideoxy-3-*O*-methyl-*lyxo*-hexose
Digitalose		6-Deoxy-3-*O*-methyl-D-galactose
Digitoxose		2,6-Dideoxy-D-*ribo*-hexose
3,4-Di-*O*-methylrhamnose	Rha3,4Me$_2$	
Ethyl glucopyranuronate	Glc*p*A6Et	
Evalose		6-Deoxy-3-*C*-methyl-D-mannose
Fructose	Fru	*arabino*-Hex-2-ulose
Fucitol	Fuc-ol	6-Deoxy-D-galactitol
Fucose	Fuc	6-Deoxygalactose
β-D-Galactopyranose 4-sulfate	β-D-Gal*p*4S	
Galactosamine	GalN	2-Amino-2-deoxygalactose
Galactose	Gal	*galacto*-Hexose
Glucitol	Glc-ol	
Glucosamine	GlcN	2-Amino-2-deoxyglucose
Glucose	Glc	*gluco*-Hexose
Glucuronic acid	GlcA	
N-Glycoloylneuraminic acid	Neu5Gc	
Gulose	Gul	*gulo*-Hexose
Hamamelose		2-*C*-(Hydroxymethyl)-D-ribose
Idose	Ido	*ido*-Hexose
Iduronic acid	IdoA	
Lactose	Lac	β-D-Galactopyranosyl-(1→4)-D-glucose
Lyxose	Lyx	*lyxo*-Pentose
Maltose		α-D-Glucopyranosyl-(1→4)-D-glucose
Mannose	Man	*manno*-Hexose
2-*C*-Methylxylose	Xyl2*C*Me	
Muramic acid	Mur	2-Amino-3-*O*-[(R)-1-carboxyethyl]-2-deoxy-D-glucose
Mycarose		2,6-Dideoxy-3-*C*-methyl-L-*ribo*-hexose
Mycinose		6-Deoxy-2,3-di-*O*-methyl-D-allose
Neuraminic acid	Neu	5-Amino-3,5-dideoxy-D-*glycero*-D-*galacto*-non-2-ulosonic acid

Common Name	Symbol	Systematic Name
Panose		α-D-Glucopyranosyl-(1→6)-α-D-glucopyranosyl-(1→ 4)-D-glucose
Paratose		3,6-Dideoxy-D-*ribo*-hexose
Primeverose		β-D-Xylopyranosyl-(1→6)-D-glucose
Psicose	Psi	*ribo*-Hex-2-ulose
Quinovose	Qui	6-Deoxyglucose
Raffinose		β-D-Fructofuranosyl-α-D-galactopyranosyl-(1→6)-α-D-glucopyranoside
Rhamnose	Rha	6-Deoxymannose
Rhodinose		2,3,6-Trideoxy-L-*threo*-hexose
Ribose	Rib	*ribo*-Pentose
Ribose 5-phosphate	Rib5*P*	
Ribulose	Ribulo (Rul)	*erythro*-Pent-2-ulose
Rutinose		α-L-Rhamnopyranosyl-(1→6)-D-glucose
Sarmentose		2,6-Dideoxy-3-*O*-methyl-D-*xylo*-hexose
Sedoheptulose		D-*altro*-Hept-2-ulose
Sorbose	Sor	*xylo*-Hex-2-ulose
Streptose		5-Deoxy-3-*C*-formyl-L-lyxose
Sucrose		β-D-Fructofuranosyl-α-D-glucopyranoside
Tagatose	Tag	*lyxo*-Hex-2-ulose
Talose	Tal	*talo*-Hexose
Turanose		α-D-Glucopyranosyl-(1→3)-D-fructose
Tyvelose	Tyv	3,6-Dideoxy-D-*arabino*-hexose
Xylose	Xyl	*xylo*-Pentose
Xylulose	Xylulo (Xul)	*threo*-Pent-2-ulose

BIOLOGICAL BUFFERS

This table of frequently used buffers gives the pK_a value at 25°C and the useful pH range of each buffer. The buffers are listed in order of increasing pH.

The table is reprinted with permission of Sigma Chemical Company, St. Louis, Mo.

Acronym	Name	Mol. wt.	pK_a	Useful pH range
MES	2-(N-Morpholino)ethanesulfonic acid	195.2	6.1	5.5—6.7
BIS TRIS	Bis(2-hydroxyethyl)iminotris(hydroxymethyl)methane	209.2	6.5	5.8—7.2
ADA	N-(2-Acetamido)-2-iminodiacetic acid	190.2	6.6	6.0—7.2
ACES	2-[(2-Amino-2-oxoethyl)amino]ethanesulfonic acid	182.2	6.8	6.1—7.5
PIPES	Piperazine-N,N′-bis(2-ethanesulfonic acid)	302.4	6.8	6.1—7.5
MOPSO	3-(N-Morpholino)-2-hydroxypropanesulfonic acid	225.3	6.9	6.2—7.6
BIS TRIS PROPANE	1,3-Bis[tris(hydroxymethyl)methylamino]propane	282.3	6.8[a]	6.3—9.5
BES	N,N-Bis(2-hydroxyethyl)-2-aminoethanesulfonic acid	213.2	7.1	6.4—7.8
MOPS	3-(N-Morpholino)propanesulfonic acid	209.3	7.2	6.5—7.9
HEPES	N-(2-Hydroxyethyl)piperazine-N′-(2-ethanesulfonic acid)	238.3	7.5	6.8—8.2
TES	N-Tris(hydroxymethyl)methyl-2-aminoethanesulfonic acid	229.2	7.5	6.8—8.2
DIPSO	3-[N,N-Bis(2-hydroxyethyl)amino]-2-hydroxypropanesulfonic acid	243.3	7.6	7.0—8.2
TAPSO	3-[N-Tris(hydroxymethyl)methylamino)-2-hydroxypropanesulfonic acid	259.3	7.6	7.0—8.2
TRIZMA	Tris(hydroxymethyl)aminomethane	121.1	8.1	7.0—9.1
HEPPSO	N-(2-hydroxyethyl)piperazine-N′-(2-hydroxypropanesulfonic acid)	268.3	7.8	7.1—8.5
POPSO	Piperazine-N,N′-bis(2-hydroxypropanesulfonic acid)	362.4	7.8	7.2—8.5
EPPS	N-(2-Hydroxyethyl)piperazine-N′-(3-propanesulfonic acid)	252.3	8.0	7.3—8.7
TEA	Triethanolamine	149.2	7.8	7.3—8.3
TRICINE	N-Tris(hydroxymethyl)methylglycine	179.2	8.1	7.4—8.8
BICINE	N,N-Bis(2-hydroxyethyl)glycine	163.2	8.3	7.6—9.0
TAPS	N-Tris(hydroxymethyl)methyl-3-aminopropanesulfonic acid	243.3	8.4	7.7—9.1
AMPSO	3-[(1,1-Dimethyl-2-hydroxyethyl)amino]-2-hydroxypropanesulfonic acid	227.3	9.0	8.3—9.7
CHES	2-(N-Cyclohexylamino)ethanesulfonic acid	207.3	9.3	8.6—10.0
CAPSO	3-(Cyclohexylamino)-2-hydroxy-1-propanesulfonic acid	237.3	9.6	8.9—10.3
AMP	2-Amino-2-methyl-1-propanol	89.1	9.7	9.0—10.5
CAPS	3-(Cyclohexylamino)-1-propanesulfonic acid	221.3	10.4	9.7—11.1

[a] $pK_a = 9.0$ for the second dissociation stage.

TYPICAL pH VALUES OF BIOLOGICAL MATERIALS AND FOODS

This table gives typical pH ranges for various biological fluids and common foods. All values refer to 25°C.

Biological Materials

Blood, human	7.35-7.45
Blood, dog	6.9-7.2
Spinal fluid, human	7.3-7.5
Saliva, human	6.5-7.5
Gastric contents, human	1.0-3.0
Duodenal contents, human	4.8-8.2
Feces, human	4.6-8.4
Urine, human	4.8-8.4
Milk, human	6.6-7.6
Bile, human	6.8-7.0

Foods

Apples	2.9-3.3
Apricots	3.6-4.0
Asparagus	5.4-5.8
Bananas	4.5-4.7
Beans	5.0-6.0
Beers	4.0-5.0
Beets	4.9-5.5
Blackberries	3.2-3.6
Bread, white	5.0-6.0
Butter	6.1-6.4
Cabbage	5.2-5.4
Carrots	4.9-5.3
Cheese	4.8-6.4
Cherries	3.2-4.0
Cider	2.9-3.3
Corn	6.0-6.5
Crackers	6.5-8.5
Dates	6.2-6.4
Eggs, fresh white	7.6-8.0
Flour, wheat	5.5-6.5
Gooseberries	2.8-3.0
Grapefruit	3.0-3.3
Grapes	3.5-4.5
Hominy (lye)	6.8-8.0
Jams, fruit	3.5-4.0
Jellies, fruit	2.8-3.4
Lemons	2.2-2.4
Limes	1.8-2.0
Maple syrup	6.5-7.0
Milk, cows	6.3-6.6
Olives	3.6-3.8
Oranges	3.0-4.0
Oysters	6.1-6.6
Peaches	3.4-3.6
Pears	3.6-4.0
Peas	5.8-6.4
Pickles, dill	3.2-3.6
Pickles, sour	3.0-3.4
Pimento	4.6-5.2
Plums	2.8-3.0
Potatoes	5.6-6.0
Pumpkin	4.8-5.2
Raspberries	3.2-3.6
Rhubarb	3.1-3.2
Salmon	6.1-6.3
Sauerkraut	3.4-3.6
Shrimp	6.8-7.0
Soft drinks	2.0-4.0
Spinach	5.1-5.7
Squash	5.0-5.4
Strawberries	3.0-3.5
Sweet potatoes	5.3-5.6
Tomatoes	4.0-4.4
Tuna	5.9-6.1
Turnips	5.2-5.6
Vinegar	2.4-3.4
Water, drinking	6.5-8.0
Wines	2.8-3.8

CHEMICAL COMPOSITION OF THE HUMAN BODY

The elemental composition of the "standard man" of mass 70 kg is given below.

REFERENCES

1. Padikal, T.N., and Fivozinsky, S.P., *Medical Physics Data Book, National Bureau of Standards Handbook 138*, U. S. Government Printing Office, Washington, DC, 1981.
2. Snyde, W.S., et al., *Reference Man: Anatomical, Physiological, and Metabolic Characteristics*, Pergamon, New York, 1975.

Element	Amount (g)	Percent of total body mass
Oxygen	43,000	61
Carbon	16,000	23
Hydrogen	7000	10
Nitrogen	1800	2.6
Calcium	1000	1.4
Phosphorus	780	1.1
Sulfur	140	0.20
Potassium	140	0.20
Sodium	100	0.14
Chlorine	95	0.12
Magnesium	19	0.027
Silicon	18	0.026
Iron	4.2	0.006
Fluorine	2.6	0.0037
Zinc	2.3	0.0033
Rubidium	0.32	0.00046
Strontium	0.32	0.00046
Bromine	0.20	0.00029
Lead	0.12	0.00017
Copper	0.072	0.00010
Aluminum	0.061	0.00009
Cadmium	0.050	0.00007
Boron	<0.048	0.00007
Barium	0.022	0.00003
Tin	<0.017	0.00002
Manganese	0.012	0.00002
Iodine	0.013	0.00002
Nickel	0.010	0.00001
Gold	<0.010	0.00001
Molybdenum	<0.0093	0.00001
Chromium	<0.0018	0.000003
Cesium	0.0015	0.000002
Cobalt	0.0015	0.000002
Uranium	0.00009	0.0000001
Beryllium	0.000036	
Radium	$3.1 \cdot 10^{-11}$	

Section 8
Analytical Chemistry

PREPARATION OF SPECIAL ANALYTICAL REAGENTS

Aluminon (qualitative test for aluminum). Aluminon is a trade name for the ammonium salt of aurintricarboxylic acid. Dissolve 1 g of the salt in 1 L of distilled water. Shake the solution well to insure thorough mixing.

Bang's reagent (for glucose estimation). Dissolve 100 g of K_2CO_3, 66 g of KCl and 160 g of $KHCO_3$ in the order given in about 700 mL of water at 30°C. Add 4.4 g of $CuSO_4$ and dilute to 1 L after the CO_2 is evolved. This solution should be shaken only in such a manner as not to allow entry of air. After 24 hours 300 mL are diluted to 1 L with saturated KCl solution, shaken gently and used after 24 hours; 50 mL is equivalent to 10 mg glucose.

Barfoed's reagent (test for glucose). See Cupric acetate.

Baudisch's reagent. See Cupferron.

Benedict's solution (qualitative reagent for glucose). With the aid of heat, dissolve 173 g of sodium citrate and 100 g of Na_2CO_3 in 800 mL of water. Filter, if necessary, and dilute to 850 mL. Dissolve 17.3 g of $CuSO_4 \cdot 5H_2O$ in 100 mL of water. Pour the latter solution, with constant stirring, into the carbonate-citrate solution, and dilute to 1 L.

Benzidine hydrochloride solution (for sulfite determination). Make a paste of 8 g of benzidine hydrochloride ($C_{12}H_8(NH_3)_2 \cdot 2HCl$) and 20 mL of water, add 20 mL of HCl (sp. gr. 1.12) and dilute to 1 L with water. Each mL of this solution is equivalent to 0.00357 g of H_2SO_4.

Bertrand's reagent (glucose estimation). Consists of the following solutions:
1. Dissolve 200 g of Rochelle salt and 150 g of NaOH in sufficient water to make 1 L of solution.
2. Dissolve 40 g of $CuSO_4$ in enough water to make 1 L of solution.
3. Dissolve 50 g of $Fe_2(SO_4)_3$ and 200 g of H_2SO_4 (sp. gr. 1.84) in sufficient water to make 1 L of solution.
4. Dissolve 5 g of $KMnO_4$ in sufficient water to make 1 L of solution.

Bial's reagent (for pentose). Dissolve 1 g of orcinol (5-methyl-1,3-benzenediol) in 500 mL of 30% HCl to which 30 drops of a 10% solution of $FeCl_3$ has been added.

Boutron — Boudet soap solution:
1. Dissolve 100 g of pure castile soap in about 2.5 L of 56% ethanol.
2. Dissolve 0.59 g of $Ba(NO_3)_2$ in 1 L of water.

Adjust the castile soap solution so that 2.4 mL of it will give a permanent lather with 40 mL of solution (b). When adjusted, 2.4 mL of soap solution is equivalent to 220 parts per million of hardness (as $CaCO_3$) for a 40 mL sample. See also Soap solution.

Brucke's reagent (protein precipitation). See Potassium iodide-mercuric iodide.

Clarke's soap solution (estimation of hardness in water).
1. Dissolve 100 g of pure powdered castile soap in 1 L of 80% ethanol and allow to stand over night.
2. Prepare a solution of $CaCl_2$ by dissolving 0.5 g of $CaCO_3$ in HCl (sp. gr. 1.19), neutralize with NH_4OH and make slightly alkaline to litmus, and dilute to 500 mL. One mL is equivalent to 1 mg of $CaCO_3$.

Titrate (1) against (2) and dilute (1) with 80% ethanol until 1 mL of the resulting solution is equivalent to 1 mL of (2) after making allowance for the lather factor (the amount of standard soap solution required to produce a permanent lather in 50 mL of distilled water). One mL of the adjusted solution after subtracting the lather factor is equivalent to 1 mg of $CaCO_3$. See also Soap solution.

Cobalticyanide paper (Rinnmann's test for Zn). Dissolve 4 g of $K_3Co(CN)_6$ and 1 g of $KClO_3$ in 100 mL of water. Soak filter paper in solution and dry at 100°C. Apply drop of zinc solution and burn in an evaporating dish. A green disk is obtained if zinc is present.

Cochineal. Extract 1 g of cochineal for 4 days with 20 mL of alcohol and 60 mL of distilled water. Filter.

Congo red. Dissolve 0.5 g of congo red in 90 mL of distilled water and 10 mL of alcohol.

Cupferron (Baudisch's reagent for iron analysis). Dissolve 6 g of the ammonium salt of *N*-hydroxy-*N*-nitrosoaniline (cupferron) in 100 mL of H_2O. Reagent good for 1 week only and must be kept in the dark.

Cupric acetate (Barfoed's reagent for reducing monosaccharides). Dissolve 66 g of cupric acetate and 10 mL of glacial acetic acid in water and dilute to 1 L.

Cupric oxide, ammoniacal; Schweitzer's reagent (dissolves cotton, linen, and silk, but not wool).
1. Dissolve 5 g of cupric sulfate in 100 mL of boiling water, and add sodium hydroxide until precipitation is complete. Wash the precipitate well, and dissolve it in a minimum quantity of ammonium hydroxide.
2. Bubble a slow stream of air through 300 mL of strong ammonium hydroxide containing 50 g of fine copper turnings. Continue for 1 hour.

Cupric sulfate in glycerin-potassium hydroxide (reagent for silk). Dissolve 10 g of cupric sulfate, $CuSO_4 \cdot 5H_2O$, in 100 mL of water and add 5 g of glycerol. Add KOH solution slowly until a deep blue solution is obtained.

Cupron (precipitates copper). Dissolve 5 g of benzoinoxime in 100 mL of 95% ethanol.

Cuprous chloride, acidic (reagent for CO in gas analysis).
1. Cover the bottom of a 2-L flask with a layer of cupric oxide about 0.5 inch deep, suspend a coil of copper wire so as to reach from the bottom to the top of the solution, and fill the flask with hydrochloric acid (sp. gr. 1.10). Shake occasionally. When the solution becomes nearly colorless, transfer to reagent bottles, which should also contain copper wire. The stock bottle may be refilled with dilute hydrochloric acid until either the cupric oxide or the copper wire is used up. Copper sulfate may be substituted for copper oxide in the above procedure.
2. Dissolve 340 g of $CuCl_2 \cdot 2H_2O$ in 600 mL of conc. HCl and reduce the cupric chloride by adding 190 mL of a saturated solution of stannous chloride or until the solution is colorless. The stannous chloride is prepared by treating 300 g of metallic tin in a 500 mL flask with conc. HCl until no more tin goes into solution.
3. (Winkler method). Add a mixture of 86 g of CuO and 17 g of finely divided metallic Cu, made by the reduction of CuO with hydrogen, to a solution of HCl, made by diluting 650 mL of conc. HCl with 325 mL of water. After the mixture has been added slowly and with frequent stirring, a spiral of copper wire is suspended in the bottle, reaching all the way to the bottom. Shake occasionally, and when the solution becomes colorless, it is ready for use.

Cuprous chloride, ammoniacal (reagent for CO in gas analysis).

    1. The acid solution of cuprous chloride as prepared above is neutralized with ammonium hydroxide until an ammonia odor persists. An excess of metallic copper must be kept in the solution.

    2. Pour 800 mL of acidic cuprous chloride, prepared by the Winkler method, into about 4 L of water. Transfer the precipitate to a 250 mL graduate. After several hours, siphon off the liquid above the 50 mL mark and refill with 7.5% NH_4OH solution which may be prepared by diluting 50 mL of conc. NH_4OH with 150 mL of water. The solution is well shaken and allowed to stand for several hours. It should have a faint odor of ammonia.

Dichlorofluorescein indicator. Dissolve 1 g in 1 L of 70% alcohol or 1 g of the sodium salt in 1 L of water.

Dimethyglyoxime, 0.01 N. Dissolve 0.6 g of dimethylglyoxime (2,3-butanedione oxime) in 500 mL of 95% ethanol. This is an especially sensitive test for nickel, a very definite crimson color being produced.

Diphenylamine (reagent for rayon). Dissolve 0.2 g in 100 mL of concentrated sulfuric acid.

Diphenylamine sulfonate (for titration of iron with $K_2Cr_2O_7$). Dissolve 0.32 g of the barium salt of diphenylamine sulfonic acid in 100 mL of water, add 0.5 g of sodium sulfate and filter off the precipitate of $BaSO_4$.

Diphenylcarbazide. Dissolve 0.2 g of diphenylcarbazide in 10 mL of glacial acetic acid and dilute to 100 mL with 95% ethanol.

Esbach's reagent (estimation of protein). To a water solution of 10 g of picric acid and 20 g of citric acid, add sufficient water to make 1 L of solution.

Eschka's compound. Two parts of calcined ("light") magnesia are thoroughly mixed with 1 part of anhydrous sodium carbonate.

Fehling's solution (reagent for reducing sugars.)

    1. Copper sulfate solution. Dissolve 34.66 g of $CuSO_4 \cdot 5H_2O$ in water and dilute to 500 mL.

    2. Alkaline tartrate solution. Dissolve 173 g of potassium sodium tartrate (Rochelle salt, $KNaC_4H_4O_6 \cdot 4H_2O$) and 50 g of NaOH in water and dilute when cold to 500 mL.

    Mix equal volumes of the two solutions at the time of using.

Ferric-alum indicator. Dissolve 140 g of ferric ammonium sulfate crystals in 400 mL of hot water. When cool, filter, and make up to a volume of 500 mL with dilute nitric acid.

Folin's mixture (for uric acid). To 650 mL of water add 500 g of $(NH_4)_2SO_4$, 5 g of uranium acetate, and 6 g of glacial acetic acid. Dilute to 1 L.

Formaldehyde — sulfuric acid (Marquis' reagent for alkaloids). Add 10 mL of formaldehyde solution to 50 mL of sulfuric acid.

Froehde's reagent. See Sulfomolybdic acid.

Fuchsin (reagent for linen). Dissolve 1 g of fuchsin in 100 mL of alcohol.

Fuchsin — sulfurous acid (Schiff's reagent for aldehydes). Dissolve 0.5 g of fuchsin and 9 g of sodium bisulfite in 500 mL of water, and add 10 mL of HCl. Keep in well-stoppered bottles and protect from light.

Gunzberg's reagent (detection of HCl in gastric juice). Prepare as needed a solution containing 4 g of phloroglucinol (1,3,5-benzenetriol) and 2 g of vanillin in 100 mL of absolute ethanol.

Hager's reagent. See Picric acid.

Hanus solution (for iodine number). Dissolve 13.2 g of resublimed iodine in 1 L of glacial acetic acid which will pass the dichromate test for reducible matter. Add sufficient bromine to double the halogen content, determined by titration (3 mL is about the proper amount). The iodine may be dissolved by the aid of heat, but the solution should be cold when the bromine is added.

Iodine, tincture of. To 50 mL of water add 70 g of I_2 and 50 g of KI. Dilute to 1 L with alcohol.

Iodo-potassium iodide (Wagner's reagent for alkaloids). Dissolve 2 g of iodine and 6 g of KI in 100 mL of water.

Litmus (indicator). Extract litmus powder three times with boiling alcohol, each treatment consuming an hour. Reject the alcoholic extract. Treat residue with an equal weight of cold water and filter; then exhaust with five times its weight of boiling water, cool and filter. Combine the aqueous extracts.

Magnesia mixture (reagent for phosphates and arsenates). Dissolve 55 g of magnesium chloride and 105 g of ammonium chloride in water, barely acidify with hydrochloric acid, and dilute to 1 L. The ammonium hydroxide may be omitted until just previous to use. The reagent, if completely mixed and stored for any period of time, becomes turbid

Magnesium uranyl acetate. Dissolve 100 g of $UO_2(C_2H_3O_2)_2 \cdot 2H_2O$ in 60 mL of glacial acetic acid and dilute to 500 mL. Dissolve 330 g of $Mg(C_2H_3O_2)_2 \cdot 4H_2O$ in 60 mL of glacial acetic acid and dilute to 200 mL. Heat solutions to the boiling point until clear, pour the magnesium solution into the uranyl solution, cool and dilute to 1 L. Let stand over night and filter if necessary.

Marme's reagent. See Potassium-cadmium iodide.

Marquis' reagent. See Formaldehyde-sulfuric acid.

Mayer's reagent (white precipitate with most alkaloids in slightly acid solutions). Dissolve 1.358 g of $HgCl_2$ in 60 mL of water and pour into a solution of 5 g of KI in 10 mL of H_2O. Add sufficient water to make 100 mL.

Methyl orange indicator. Dissolve 1 g of methyl orange in 1 L of water. Filter, if necessary.

Methyl orange, modified. Dissolve 2 g of methyl orange and 2.8 g of xylene cyanole FF in 1 L of 50% alcohol.

Methyl red indicator. Dissolve 1 g of methyl red in 600 mL of alcohol and dilute with 400 mL of water.

Methyl red, modified. Dissolve 0.50 g of methyl red and 1.25 g of xylene cyanole FF in 1 L of 90% alcohol. Or, dissolve 1.25 g of methyl red and 0.825 g of methylene blue in 1 L of 90% alcohol.

Millon's reagent (for albumins and phenols). Dissolve 1 part of mercury in 1 part of cold fuming nitric acid. Dilute with twice the volume of water and decant the clear solution after several hours.

Molisch's reagent. See 1-Naphthol.

1-Naphthol (Molisch's reagent for wool). Dissolve 15 g of 1-naphthol in 100 mL of alcohol or chloroform.

Nessler's reagent (for ammonia). Dissolve 50 g of KI in the smallest possible quantity of cold water (50 mL). Add a saturated solution of mercuric chloride (about 22 g in 350 mL of water will be needed) until an excess is indicated by the formation of a precipitate. Then add 200 mL of 5 N NaOH and dilute to 1 L. Let settle, and draw off the clear liquid.

Nickel oxide, ammoniacal (reagent for silk). Dissolve 5 g of nickel sulfate in 100 mL of water, and add sodium hydroxide solution until nickel hydroxide is completely precipitated. Wash the precipitate well and dissolve in 25 mL of concentrated ammonium hydroxide and 25 mL of water.

Nitron (detection of nitrate radical). Dissolve 10 g of nitron (1,4-diphenyl-3-(phenylamino)-1,2,4-triazolium hydroxide) in 5 mL of glacial acetic acid and 95 mL of water. The solution may be filtered with slight suction through an alumdum crucible and kept in a dark bottle.

1-Nitroso-2-naphthol. Make a saturated solution in 50% acetic acid (1 part of glacial acetic acid with 1 part of water). Does not keep well.

Nylander's solution (carbohydrates). Dissolve 20 g of bismuth subnitrate and 40 g of Rochelle salt in 1 L of 8% NaOH solution. Cool and filter.

Obermayer's reagent (for indoxyl in urine). Dissolve 4 g of $FeCl_3$ in 1 L of HCl (sp. gr. 1.19).

Oxine. Dissolve 14 g of 8-hydroxyquinoline in 30 mL of glacial acetic acid. Warm slightly, if necessary. Dilute to 1 L.

Oxygen absorbent. Dissolve 300 g of ammonium chloride in 1 L of water and add 1 L of concentrated ammonium hydroxide solution. Shake the solution thoroughly. For use as an oxygen absorbent, a bottle half full of copper turnings is filled nearly full with the NH_4Cl-NH_4OH solution and the gas passed through.

Pasteur's salt solution. To 1 L of distilled water add 2.5 g of potassium phosphate, 0.25 g of calcium phosphate, 0.25 g of magnesium sulfate, and 12.00 g of ammonium tartrate.

Pavy's solution (glucose reagent). To 120 mL of Fehling's solution, add 300 mL of NH_4OH (sp. gr. 0.88) and dilute to 1 L with water.

Phenanthroline ferrous ion indicator. Dissolve 1.485 g of 1,10-phenanthroline monohydrate in 100 mL of 0.025 M ferrous sulfate solution.

Phenolphthalein. Dissolve 1 g of phenolphthalein in 50 mL of alcohol and add 50 mL of water.

Phenolsulfonic acid (determination of nitrogen as nitrate). Dissolve 25 g of phenol in 150 mL of conc. H_2SO_4, add 75 mL of fuming H_2SO_4 (15% SO_3), stir well and heat for 2 hours at 100°C.

Phloroglucinol solution (pentosans). Make a 3% phloroglucinol (1,3,5-benzenetriol) solution in alcohol. Keep in a dark bottle.

Phosphomolybdic acid (Sonnenschein's reagent for alkaloids).

    1. Prepare ammonium phosphomolybdate and after washing with water, boil with nitric acid and expel NH_3; evaporate to dryness and dissolve in 2 M nitric acid.

    2. Dissolve ammonium molybdate in HNO_3 and treat with phosphoric acid. Filter, wash the precipitate, and boil with aqua regia until the ammonium salt is decomposed. Evaporate to dryness. The residue dissolved in 10 % HNO_3 constitutes Sonnenschein's reagent.

Phosphoric acid — sulfuric acid mixture. Dilute 150 mL of conc. H_2SO_4 and 100 mL of conc. H_3PO_4 (85%) with water to a volume of 1 L.

Phosphotungstic acid (Schcibicr's reagent for alkaloids).

    1. Dissolve 20 g of sodium tungstate and 15 g of sodium phosphate in 100 mL of water containing a little nitric acid.

    2. The reagent is a 10% solution of phosphotungstic acid in water. Thc.phosphotungstic acid is prepared by evaporating a mixture of 10 g of sodium tungstate dissolved in 5 g of phosphoric acid (sp. gr. 1.13) and enough boiling water to effect solution. Crystals of phosphotungstic acid separate.

Picric acid (Hager's reagent for alkaloids, wool and silk). Dissolve 1 g of picric acid in 100 mL of water.

Potassium antimonate (reagent for sodium). Boil 22 g of potassium antimonate with 1 L of water until nearly all of the salt has dissolved, cool quickly, and add 35 mL of 10% potassium hydroxide. Filter after standing overnight.

Potassium-cadmium iodide (Marme's reagent for alkaloids). Add 2 g of CdI_2 to a boiling solution of 4 g of KI in 12 mL of water, and then mix with 12 mL of saturated KI solution.

Potassium hydroxide (for CO_2 absorption). Dissolve 360 g of KOH in water and dilute to 1 L.

Potassium iodide — mercuric iodide (Brucke's reagent for proteins). Dissolve 50 g of KI in 500 mL of water, and saturate with mercuric iodide (about 120 g). Dilute to 1 L.

Potassium pyrogallate (for oxygen absorption). For mixtures of gases containing less than 28% oxygen, add 100 mL of KOH solution (50 g of KOH to 100 mL of water) to 5 g of pyrogallol. For mixtures containing more than 28% oxygen the KOH solution should contain 120 g of KOH to 100 mL of water.

Pyrogallol, alkaline.

    1. Dissolve 75 g of pyrogallic acid in 75 mL of water.

    2. Dissolve 500 g of KOH in 250 mL of water. When cool, adjust until sp. gr. is 1.55.

    For use, add 270 mL of solution (2) to 30 mL of solution (1).

Rosolic acid (indicator). Dissolve 1 g of rosolic acid in 10 mL of alcohol and add 100 mL of water.

Scheibler's reagent. See Phosphotungstic acid.

Schiff's reagent. See Fuchsin-sulfurous acid.

Schweitzer's reagent. See Cupric oxide, ammoniacal.

Soap solution (reagent for hardness in water). Dissolve 100 g of dry castile soap in 1 L of 80% alcohol (5 parts alcohol to 1 part water). Allow to stand several days and dilute with 70% to 80% alcohol until 6.4 mL produces a permanent lather with 20 mL of standard calcium solution. The latter solution is made by dissolving 0.2 g of $CaCO_3$ in a small amount of dilute HCl, evaporating to dryness and making up to 1 L.

Sodium bismuthate (oxidation of manganese). Heat 20 parts of NaOH nearly to redness in an iron or nickel crucible and add slowly 10 parts of basic bismuth nitrate which has been previously dried. Add 2 parts of sodium peroxide, and pour the brownish-yellow fused mass onto an iron plate to cool. When cold, break up in a mortar, extract with water, and collect on an asbestos filter.

Sodium hydroxide (for CO_2 absorption). Dissolve 330 g of NaOH in water and dilute to 1 L.

Sodium nitroprusside (reagent for hydrogen sulfide and wool). Use a freshly prepared solution of 1 g of sodium nitroferricyanide in 10 mL of water.

Sodium oxalate (primary standard). Dissolve 30 g of the commercial salt in 1 L of water, make slightly alkaline with sodium hydroxide, and let stand until perfectly clear. Filter and evaporate the filtrate to 100 mL. Cool and filter. Pulverize the residue and wash it several times with small volumes of water. The procedure is repeated until the mother liquor is free from sulfate and is neutral to phenolphthalein.

Sodium plumbite (reagent for wool). Dissolve 5 g of sodium hydroxide in 100 mL of water. Add 5 g of litharge (PbO) and boil until dissolved.

Sodium polysulfide. Dissolve 480 g of $Na_2S \cdot 9H_2O$ in 500 mL of water, add 40 g of NaOH and 18 g of sulfur. Stir thoroughly and dilute to 1 L with water.

Sonnenschein's reagent. See Phosphomolybdic acid.

Starch solution.

1. Make a paste with 2 g of soluble starch and 0.01 g of HgI_2 with a small amount of water. Add the mixture slowly to 1 L of boiling water and boil for a few minutes. Keep in a glass stoppered bottle. If other than soluble starch is used, the solution will not clear on boiling; it should be allowed to stand and the clear liquid decanted.

2. A solution of starch which keeps indefinitely is made as follows: Mix 500 mL of saturated NaCl solution (filtered), 80 mL of glacial acetic acid, 20 mL of water and 3 g of starch. Bring slowly to a boil and boil for 2 minutes.

3. Make a paste with 1 g of soluble starch and 5 mg of HgI_2, using as little cold water as possible. Then pour about 200 mL of boiling water on the paste and stir immediately. This will give a clear solution if the paste is prepared correctly and the water actually boiling. Cool and add 4 g of KI. Starch solution decomposes on standing due to bacterial action, but this solution will keep well if stored under a layer of toluene.

Stoke's reagent. Dissolve 30 g of $FeSO_4$ and 20 g of tartaric acid in water and dilute to 1 L. Just before using, add concentrated NH_4OH until the precipitate first formed is redissolved.

Sulfanilic acid (reagent for nitrites). Dissolve 0.5 g of sulfanilic acid in a mixture of 15 mL of glacial acetic acid and 135 mL of recently boiled water.

Sulfomolybdic acid (Froehde's reagent for alkaloids and glucosides). Dissolve 10 g of molybdic acid or sodium molybdate in 100 mL of conc. H_2SO_4.

Tannic acid (reagent for albumin, alkaloids, and gelatin). Dissolve 10 g of tannic acid in 10 mL of alcohol and dilute with water to 100 mL.

Titration mixture. (residual chlorine in water analyasis). Prepare 1 L of dilute HCl (100 mL of HCl (sp. gr. 1.19) in sufficient water to make 1 L). Dissolve 1 g of o-tolidine in 100 mL of the dilute HCl and dilute to 1 L with dilute HCl solution.

Trinitrophenol solution. See Picric acid.

Turmeric tincture (reagent for borates). Digest ground turmeric root with several quantities of water which are discarded. Dry the residue and digest it several days with six times its weight of alcohol. Filter.

Uffelmann's reagent (turns yellow in presence of lactic acid). To a 2% solution of pure phenol in water, add a water solution of $FeCl_3$ until the phenol solution becomes violet in color.

Wagner's reagent. See Iodo-potassium iodide.

Wagner's solution (used in phosphate rock analysis to prevent precipitation of iron and aluminum). Dissolve 25 g of citric acid and 1 g of salicylic acid in water and dilute to 1 L. Use 50 mL of the reagent.

Wij's iodine monochloride solution (for iodine number). Dissolve 13 g of resublimed iodine in 1 L of glacial acetic acid which will pass the dichromate test for reducible matter. Set aside 25 mL of this solution. Pass into the remainder of the solution dry chlorine gas (dried and washed by passing through H_2SO_4 (sp. gr. 1.84)) until the characteristic color of free iodine has been discharged. Now add the iodine solution which was reserved, until all free chlorine has been destroyed. A slight excess of iodine does little or no harm, but an excess of chlorine must be avoided. Preserve in well stoppered, amber colored bottles. Avoid use of solutions which have been prepared for more than 30 days.

Wij's special solution (for iodine number). To 200 mL of glacial acetic acid that will pass the dichromate test for reducible matter, add 12 g of dichloramine T (N,N-dichloro-4-methyl-benzenesulfonamide), and 16.6 g of dry KI (in small quantities with continual shaking until all the KI has dissolved). Make up to 1 L with the same quality of acetic acid used above and preserve in a dark colored bottle.

Zimmermann-Reinhardt reagent (determination of iron). Dissolve 70 g of $MnSO_4 \cdot 4H_2O$ in 500 mL of water, add 125 mL of conc. H_2SO_4 and 125 mL of 85% H_3PO_4, and dilute to 1 L.

Zinc chloride solution, basic (reagent for silk). Dissolve 1000 g of zinc chloride in 850 mL of water, and add 40 g of zinc oxide. Heat until solution is complete.

Zinc uranyl acetate (reagent for sodium). Dissolve 10 g of $UO_2(C_2H_3O_2)_2 \cdot 2H_2O$ in 6 g of 30% acetic acid with heat, if necessary, and dilute to 50 mL. Dissolve 30 g of $Zn(C_2H_3O_2)_2 \cdot H_2O$ in 3 g of 30% acetic acid and dilute to 50 mL. Mix the two solutions, add 50 mg of NaCl, allow to stand overnight and filter.

STANDARD SOLUTIONS OF ACIDS, BASES, AND SALTS

For each compound listed, the last column of this table gives the mass in grams which is contained in 1 liter of a solution whose amount-of-substance concentration divided by the equivalence factor of the compound equals 0.1 mol/L. In the older literature such a solution is often referred to as a "decinormal solution" (0.1 N).

REFERENCE

Compendium of Analytical Nomenclature (IUPAC), Pergamon Press, Oxford, 1978.

Name	Formula	Atomic or molecular weight	Equivalence factor	Mass in grams
Acetic acid	$HC_2H_3O_2$	60.0530	1	6.0053
Ammonia	NH_3	17.0306	1	1.7031
Ammonium ion	NH_4^+	18.0386	1	1.8039
Ammonium chloride	NH_4Cl	53.4916	1	5.3492
Ammonium sulfate	$(NH_4)_2SO_4$	132.1388	1/2	6.6069
Ammonium thiocyanate	NH_4CNS	76.1204	1	7.6120
Barium	Ba	137.34	1/2	6.867
Barium carbonate	$BaCO_3$	197.3494	1/2	9.8675
Barium chloride hydrate	$BaCl_2 \cdot 2H_2O$	244.2767	1/2	12.2138
Barium hydroxide	$Ba(OH)_2$	171.3547	1/2	8.5677
Barium oxide	BaO	153.3394	1/2	7.6670
Bromine	Be	79.909	1	7.9909
Calcium	Ca	40.08	1/2	2.004
Calcium carbonate	$CaCO_3$	100.0894	1/2	5.0045
Calcium chloride	$CaCl_2$	110.9860	1/2	5.5493
Calcium chloride hydrate	$CaCl_2 \cdot 6H_2O$	219.0150	1/2	10.9508
Calcium hydroxide	$Ca(OH)_2$	74.0947	1/2	3.7047
Calcium oxide	CaO	56.0794	1/2	2.8040
Chlorine	Cl	35.453	1	3.5453
Citric acid	$C_6H_8O_7 \cdot H_2O$	210.1418	1/3	7.0047
Cobalt	Co	58.9332	1/2	2.9466
Copper	Cu	63.54	1/2	3.177
Copper oxide (cupric)	CuO	79.5394	1/2	3.9770
Copper sulfate hydrate	$CuSO_4 \cdot 5H_2O$	249.6783	1/2	12.4839
Hydrochloric acid	HCl	36.4610	1	3.6461
Hydrocyanic acid	HCN	27.0258	1	2.7026
Iodine	I	126.9044	1	12.6904
Lactic acid	$C_3H_6O_3$	90.0795	1	9.0080
Malic acid	$C_4H_6O_5$	134.0894	1/2	6.7045
Magnesium	Mg	24.312	1/2	1.2156
Magnesium carbonate	$MgCO_3$	84.3214	1/2	4.2161
Magnesium chloride	$MgCl_2$	95.2180	1/2	4.7609
Magnesium chloride hydrate	$MgCl_2 \cdot 6H_2O$	203.2370	1/2	10.1623
Magnesium oxide	MgO	40.3114	1/2	2.0156
Manganese	Mn	54.938	1/2	2.7469
Manganese sulfate	$MnSO_4$	150.9996	1/2	7.5500
Mercuric chloride	$HgCl_2$	271.4960	1/2	13.5748
Nickel	Ni	58.71	1/2	2.9356
Nitric acid	HNO_3	63.0129	1	6.3013
Oxalic acid	$H_2C_2O_4$	90.0358	1/2	4.5018
Oxalic acid hydrate	$H_2C_2O_4 \cdot 2H_2O$	126.0665	1/2	6.3033
Oxalic acid anhydride	C_2O_3	72.0205	1/2	3.6010
Phosphoric acid	H_3PO_4	97.9953	1/3	3.2665
Potassium	K	39.102	1	3.9102
Potassium bicarbonate	$KHCO_3$	100.1193	1	10.0119
Potassium carbonate	K_2CO_3	138.2134	1/2	6.9106
Potassium chloride	KCl	74.5550	1	7.4555

Name	Formula	Atomic or molecular weight	Equivalence factor	Mass in grams
Potassium cyanide	KCN	65.1199	1	6.5120
Potassium hydroxide	KOH	56.1094	1	5.6109
Potassium oxide	K_2O	94.2034	1/2	4.7102
Potassium tartrate	$K_2H_4C_4O_6$	226.2769	1/2	11.3139
Silver	Ag	107.87	1	10.787
Silver nitrate	$AgNO_3$	169.8749	1	16.9875
Sodium	Na	22.9898	1	2.2990
Sodium bicarbonate	$NaHCO_3$	84.0071	1	8.4007
Sodium carbonate	Na_2CO_3	105.9890	1/2	5.2995
Sodium chloride	NaCl	58.4428	1	5.8443
Sodium hydroxide	NaOH	39.9972	1	3.9997
Sodium oxide	Na_2O	61.9790	1/2	3.0990
Sodium sulfide	Na_2S	78.0436	1/2	3.9022
Succinic acid	$H_2C_4H_4O_4$	118.0900	1/2	5.9045
Sulfuric acid	H_2SO_4	98.0775	1/2	4.9039
Tartaric acid	$C_4H_6O_6$	150.0888	1/2	7.5044
Zinc	Zn	65.37	1/2	3.269
Zinc sulfate hydrate	$ZnSO_4 \cdot 7H_2O$	287.5390	1/2	14.3769

STANDARD SOLUTIONS OF OXIDATION AND REDUCTION REAGENTS

For each reagent listed, the last column of this table gives the mass in grams which is contained in a solution whose amount-of-substance concentration divided by the equivalence factor of the compound equals 0.1 mol/L. The equivalence factor given refers to the most common reactions of the reagent. In the older literature such a solution is often called a "decinormal solution" (0.1 N).

REFERENCE

Compendium of Analytical Nomenclature (IUPAC), Pergamon Press, Oxford, 1978.

Name	Formula	Atomic or molecular weight	Equivalence factor	Mass in grams
Antimony	Sb	121.75	1/2	6.0875
Arsenic	As	74.9216	1/2	3.7461
Arsenic trisulfide	As_2S_3	246.0352	1/4	6.1509
Arsenous oxide	As_2O_3	197.8414	1/4	4.9460
Barium peroxide	BaO_2	169.3388	1/2	8.4669
Barium peroxide hydrate	$BaO_2 \cdot 8H_2O$	313.4615	1/2	15.6730
Calcium	Ca	40.08	1/2	2.004
Calcium carbonate	$CaCO_3$	100.0894	1/2	5.0045
Calcium hypochlorite	$Ca(OCl)_2$	142.9848	1/4	3.5746
Calcium oxide	CaO	56.0794	1/2	2.8040
Chlorine	Cl	35.453	1	3.5453
Chromium trioxide	CrO_3	99.9942	1/3	3.3331
Ferrous ammonium sulfate	$FeSO_4(NH_4)SO_4 \cdot 6H_2O$	392.0764	1	39.2076
Hydroferrocyanic acid	$H_4Fe(CN)_6$	215.9860	1	21.5986
Hydrogen peroxide	H_2O_2	34.0147	1/2	1.7007
Hydrogen sulfide	H_2S	34.0799	1/2	1.7040
Iodine	I	126.9044	1	12.6904
Iron	Fe	55.847	1	5.5847
Iron oxide (ferrous)	FeO	71.8464	1	7.1846
Iron oxide (ferric)	Fe_2O_3	159.6922	1/2	7.9846
Lead peroxide	PbO_2	239.1888	1/2	11.9594
Manganese dioxide	MnO_2	86.9368	1/2	4.3468
Nitric acid	HNO_3	63.0129	1/3	2.1004
Nitrogen trioxide	N_2O_3	76.0116	1/4	1.9002
Nitrogen pentoxide	N_2O_5	108.0104	1/6	1.8001
Oxalic acid	$C_2H_2O_4$	90.0358	1/2	4.5018
Oxalic acid hydrate	$C_2H_2O_4 \cdot 2H_2O$	126.0665	1/2	6.3033
Oxygen	O	15.9994	1/2	0.8000
Potassium dichromate	$K_2Cr_2O_7$	294.1918	1/6	4.9032
Potassium chlorate	$KClO_3$	122.5532	1/6	2.0425
Potassium chromate	K_2CrO_4	194.1076	1/3	6.4733
Potassium ferrocyanide	$K_4Fe(CN)_6$	368.3621	1	36.8362
Potassium ferrocyanide hydrate	$K_4Fe(CN)_6 \cdot 3H_2O$	422.4081	1	42.2408
Potassium iodide	KI	166.0064	1	16.6006
Potassium nitrate	KNO_3	101.1069	1/3	3.3702
Potassium perchlorate	$KClO_4$	138.5526	1/8	1.7319
Potassium permanganate	$KMnO_4$	158.0376	1/5	3.1608
Sodium chlorate	$NaClO_3$	106.4410	1/6	1.7740
Sodium nitrate	$NaNO_3$	84.9947	1/3	2.8332
Sodium thiosulfate hydrate	$Na_2S_2O_3 \cdot 5H_2O$	248.1825	1	24.8183
Stannous chloride	$SnCl_2$	189.5960	1/2	9.4798
Stannous oxide	SnO	134.6894	1/2	6.7345
Sulfur dioxide	SO_2	64.0628	1/2	3.2031
Tin	Sn	118.69	1/2	5.935

ORGANIC ANALYTICAL REAGENTS FOR THE DETERMINATION OF INORGANIC SUBSTANCES
G. Ackermann, L. Sommer, and D. Thorburn Burns

Determination	Reagents	Ref.
Aluminium	Alizarin Red S	Onishi, Part II a, p 28. (5), Snell, *Metals I*, p 587. (7)
	Aluminon	Fries/Getrost, p 16. (2), Onishi, IIa, p 21. (5), Snell, *Metals I*, p 590. (7)
	Aluminon + Cetyltrimethylammonium bromide	Huaxue Shiji, *8*, 85, (1986)
	Chrome Azurol S	Onishi, Part IIa, p 26. (5), Snell, *Metals I*, p 605. (7)
	Chrome Azurol S + Cetyltrimethylammonium bromide	Marczenko, p 133 (3), Snell, *Metals I*, p 606. (7)
	Chromazol KS + Cetylpyridinium bromide	*Analyst*, *107*, 428, (1982).
	Eriochrome Cyanine R	Fries/Getrost, p 19 (2), Onishi, Part IIa p 25. (5), Snell, *Metals I*, p 611. (7)
	Eriochrome Cyanine R + Cetyltrimethylammonium bromide	Snell, *Metals I*, p 613. (7), *Analyst*, *107*, 1431, (1982).
	8-Hydroxyquinoline	Fries/Getrost, p 22 (2), Marczenko, p 131 (3), Onishi, Part IIa, p 31. (5), Snell, *Metals I* , p 622 (7)
Ammonia	Phenol + Sodium hypochlorite	Boltz, p 210 (1), Marczenko, p 413 (3), Snell, *Nonmetals*, p 604 (9)
Antimony	Brilliant Green	Onishi, Part IIa, p 102. (5), Snell, *Metals I*, p 384. (7)
	Bromopyrogallol Red	*Talanta*, *13*, 507, (1966).
	Rhodamine B	Fries/Getrost, p 32, (2), Marczenko, p 141. (3), Onishi, Part IIa, p 93. (5), Snell, *Metals I*, p 404. (7)
	Silver diethyldithiocarbamate	Fries/Getrost, p 36. (2)
Arsenic	Silver diethyldithiocarbamate	Fries/Getrost, p 41. (2), Marczenko, p 153. (3), Onishi, Part IIa, p 153. (5), Snell, *Metals I*, p 370. (7)
Barium	Sulfonazo III	Fries/Getrost, p 46. (2), Snell, *Metals II*, p 1782. (8), Onishi, Part IIa, p 202. (5)
Beryllium	Beryllon II	Snell, *Metals I*, p 667. (7)
	Chrome Azurol S	Marczenko, p 163. (3), Snell, *Metals I*, p 672. (7)
	Chrome Azurol S + Cetyltrimethylammonium bromide	Marczenko, p 164. (3), Snell, *Metals I*, p 673. (7)
	Eriochrome Cyanine R	Snell, *Metals I*, p 675. (7), *Talanta*, *31*, 249, (1984).
	Eriochrome Cyanine R + Cetyltrimethylammonium bromide	Zh, *Anal. Khim.*, *33*, 1298, (1978).
Bismuth	Dithizone	Onishi, Part IIa, p 262. (5), Snell, *Metals I*, p 303. (7)
	Pyrocatechol Violet	Fres. Z. *Anal. Chem.*, *186*, 418, (1962).
	Pyrocatechol Violet + Cetyltrimethylammonium bromide	Zh. *Anal. Khim.*, *38*, 216, (1983).
	Thiourea	Onishi, Part IIa, p 260. (5), Snell, *Metals I*, p 317. (7)
	Xylenol Orange	Friez/Getrost, p 57. (2), Marczenko, p 172. (3), Snell, *Metals I*, p 320. (7)
Boron	Azomethine H	Snell, Nonmetals, p 165. (9)
	Carminic acid	Boltz, p 14. (1), Fries/Getrost, p 65. (2), Snell, Nonmetals, p 170. (9), Williams, p 35. (11)
	Curcumin	Boltz, p 8. (1), Fries/Getrost, p 68. (2), Marczenko, p 180. (3), Snell, Nonmetals, p 180. (9), Fres. Z. A*nal. Chem.*, *323*, 266, (1986).

Determination	Reagents	Ref.
	Methylene Blue	Boltz, p 21. (1), Marczenko, p 183. (3), Snell, Nonmetals, p 205. (9), *Talanta, 31*, 547, (1984).
Bromide	Fluorescein	Boltz, p 48. (1), Snell, Nonmetals, p 276., Fres. Z. *Anal. Chem., 301*, 28 (1980).
	Phenol Red	Boltz, p 44. (1), Marczenko, p 190. (3), Snell, Nonmetals, p 28. (9)
Cadmium	2-(5-Bromo-2-pyridylazo)-5-diethylaminophenol	Marczenko, p 197. (3)
	Cadion	Onishi, Part IIa, p 323. (5)
	Dithizone	Fries/Getrost, p 78. (2), Onishi, Part IIa, p 315. (5), Snell, *Metals I*, p 279. (7), West, p 25. (10).
Calcium	4–(2-Pyridylazo)resorcinol	Fres. Z. *Anal. Chem., 310*, 51, (1982).
	Chlorophosphonazo III	Marczenko, p 207. (3), Snell, *Metals II*, p 1744. (8)
	Glyoxal-bis(2-hydroxyanil)	Fries/Getrost, p 86. (2), Onishi, Part IIa, p 352. (5), Snell, *Metals I*, p 1762. (8)
	Murexide	Onishi, Part IIa, p 357. (5), Snell, *Metals II*, p 1769. (8)
	Phthalein Purple	*Anal. Chim. Acta, 34*, 71 (1966).
Cerium	*N*-benzoyl-*N*-phenylhydroxylamine	*Anal. Chim. Acta, 48*, 155, (1969).
	8-Hydroxyquinoline	Fries/Getrost, p 93. (2), Marczenko, p 220. (3), Onishi, Part IIa, p 383. (7)
Chlorine	N,N-Diethyl-1,4-phenylenediamine	Boltz, p 92. (1), Fries/Getrost, p 101. (2), Snell, Nonmetals, p 225. (9), *Analyst, 90*, 187, (1965).
Chromium	1,5-Diphenylcarbazide	Fries/Getrost, p 105. (2), Onishi, Part IIa, p 412. (5), Snell, *Metals I*, p 714. (7), West, p 12. (10)
	4-(2-Pyridylazo)resorcinol	Snell, *Metals I*, p 736. (7), West, p 17. (10)
	4-(2-Pyridylazo)resorcinol + Tetradecyldimethyl-benzylammonium chloride	West, p 17. (10), *Anal. Chim. Acta, 67*, 297, (1973).
	4-(2-Pyridylazo)resorcinol + Hydrogen peroxide	Fres. Z. *Anal. Chem., 304*, 382, (1980).
Cobalt	Nitroso-R salt	Fries/Getrost, p 118. (2), Onishi, Part IIa, p 454. (5), Snell, *Metals I*, p 953. (7)
	1-Nitroso-2-naphthol	Fries/Getrost, p 111. (2), Marczenko, p 246. (3), Snell, *Metals I*, p 947. (5)
	2-Nitroso-1-naphthol	Fries/Getrost, p 113. (2), Onishi, Part IIa, p 459. (5), Snell, *Metals I*, p 949. (7), West, p 45. (10)
	4-(2-Pyridylazo)resorcinol	Snell, *Metals I*, p 969. (7), West, p 44. (10)
	4-(2-Pyridylazo)resorcinol + Diphenylguanidine	Zh. *Anal. Khim., 35*, 1306, (1980).
Copper	Bathocuproine	Fries/Getrost, p 135. (2), Snell, *Metals I*, p 148. (7)
	Bathocuproine disulfonic acid	Fries/Getrost, p 137. (2), West, p 52. (10)
	Dithizone	Marczenko, p 258. (3), Onishi, Part IIa, p 529. (5), Snell, *Metals I*, p 199. (7)
	Neocuproine	Snell, *Metals I*, p 217. (5), West, p 51. (10)
	Cuprizone	Onishi, Part IIa, p 534. (5), Snell, *Metals I*, p 157. (7), West, p 53. (10)
	4-(2-pyridylazo)resorcinol + Tetradecyldimethyl-benzylammonium chloride	*Anal. Chim. Acta, 138*, 321, (1982).
Cyanide	Barbituric Acid + Pyridine	Fries/Getrost, p 153. (2), Snell, Nonmetals, p 653. (9)
	Barbituric Acid + Pyridine-4-carboxylic acid	*Anal. Chim. Acta, 99*, 197, (1978).

Determination	Reagents	Ref.
Fluoride	Alizarin Fluorine blue + Lanthanum(III) ion	Boltz, p 129. (1), Fries/Getrost, p 158. (2), Snell, *Nonmetals,* p 333. (9), Williams, p 354. (11)
	Eriochrome Cyanine R + Zirconium(IV) ion	Boltz, p 119. (1), Snell, *Nonmetals,* p 359. (2), Williams, p 357. (10)
Gallium	Pyrocatechol violet + Diphenylguanidine	Snell, *Metals I,* p 500. (7)
	8-Hydroxyquinoline	Onishi Pt IIa, p 582. (5), Snell, *Metals I,* p 505. (7)
	1-(2-Pyridylazo)-2-naphthol	Snell, *Metals I,* p 512. (7)
	4-(2-Pyridylazo)resorcinol	Snell, *Metals I,* p 513. (7)
	Rhodamine B	Marczenko, p 284. (3), Onishi, Part IIa, p 578. (5), Snell, *Metals I,* p 515. (7)
	Xylenol Orange	Fries/Getrost, p 166. (2), Snell, *Metals I,* p 523. (7)
	Xylenol Orange + 8-Hydroxyquinoline	*Zh. Anal. Khim.,* 26, 75, (1971).
Germanium	Brilliant Green + Molybdate	Snell, *Metals I,* p 562. (7)
	Phenylfluorone	Fries/Getrost, p 168. (2), Marczenko, p 292. (3), Onishi, Part IIa, p 607. (5), Snell, *Metals I,* p 570. (7)
Gold	5-(4-Diethylaminobenzylidene) rhodanine	Fries/Getrost, p 173. (2), Onishi, Part IIa, p 631. (5), Snell, *Metals II,* p 1516. (8)
	Rhodamine B	Fries/Getrost, p 175. (2), Marczenko, p 301. (3), Onishi, Part IIa, p 637. (5), Snell, *Metals II,* p 513. (8)
Hafnium	Arsenazo III	Snell, *Metals II,* p 1184. (8), *Talanta,* 19, 807, (1972).
Indium	Bromopyrogallol Red	Snell, *Metals I,* p 469. (7)
	Chrome Azurol S	Snell, *Metals I,* p 474. (7)
	Chrome Azurol S + Cetyltrimethylammonium bromide	*Anal. Chim. Acta,* 67, 107, (1973).
	Dithizone	Fries/Getrost, p 179. (2), Onishi, Part IIa, p 672. (5), Snell, *Metals I,* p 474. (7)
	8-Hydroxyquinoline	Onishi, Part IIa, p 670. (5), Snell, *Metals I,* p 475. (7)
	1-(2-Pyridylazo)-2-naphthol	Snell, *Metals I,* p 480. (7)
	4-(2-Pyridylazo)resorcinol	Marczenko, p 309. (3), Snell, Metals I, p 480. (7)
Iodide	Neocuproine + Copper(II)	*Anal. Chim. Acta,* 69, 321, (1974).
Iodine	Starch	Boltz, p 162. (1), Marczenko, p 316. (3), Snell, *Nonmetals,* p 307. (9)
Iridium	Rhodamine 6G + Tin(II)	Marczenko, p 323. (3)
	N,N-Dimethyl-4-nitrosoaniline	*Anal. Chem.,* 27, 1776, (1955).
Iron	Bathophenanthroline	Fries/Getrost, p 189. (2), Onishi, Part IIa, p 729. (5), Snell, *Metals I,* p 763. (7)
	Bathophenanthroline disulfonic acid	Fries/Getrost, p 191. (2), Snell, *Metals I,* p 772. (7)
	2,2′-Bipyridyl	Snell, *Metals I,* p 750. (7)
	Chrome Azurol S + Cetyltrimethylammonium bromide	Snell, *Metals I,* p 757. (7), *Coll. Czech. Chem. Comm.,* 45, 2656, (1980).
	1,10-Phenanthroline	Fries/Getrost, p 199. (2), Marczenko, p 331. (3), Onishi, Part IIa, p 725. (5), Snell, *Metals I,* p 795. (7)
	1,10-Phenanthroline + Bromothymol Blue	*Zh. Anal. Khim.,* 25, 1348, (1970).
	Ferrozine	Onishi, Part IIa, p 730. (5), Snell, *Metals I,* p 783. (7)
Lanthanum	Arsenazo III	Marczenko, p 468. (3), Snell, *Metals II,* p 1910. (8)

Determination	Reagents	Ref.
Lead	Dithizone	Fries/Getrost, p 207. (2), Onishi, Part IIa, p 824. (5), Snell, *Metals I*, p 2. (7), West, p 34. (10)
	Sodium diethyldithiocarbamate	Fries/Getrost, p 214. (2), Snell, *Metals I*, p 27. (7)
	4-(2-Pyridylazo)resorcinol	Fries/Getrost, p 220. (2), Marczenko, p 347. (3), Snell, *Metals I*, p 34. (7)
Lithium	Thoron	Onishi, Part IIa, p 863. (5), Snell, *Metals II*, p 1726. (8), *Talanta, 30*, 587, (1983).
Magnesium	Eriochrome Black T	Fries/Getrost, p 226. (2), Marczenko, p 355. (3), Onishi, Part IIb, p 13. (6), Snell, *Metals II*, p 1932. (8)
	8-Hydroxyquinoline	Onishi, Part IIb, p 11. (6), Snell, *Metals II*, p 1938. (8)
	8-Hydroxyquinoline + Butylamine	Fries/Getrost, p 228. (2), Snell, *Metals II*, p 1938. (8)
	Titan Yellow	Fries/Getrost, p 234. (2), Marczenko, p 352. (3), Snell, *Metals II*, p 1945. (8)
	Xylidyl Blue	Fries/Getrost, p 231. (2), Onishi, Part IIb, p 14. (6), Snell, *Metals II*, p 1950. (8)
Manganese	Formaldoxime	Fries/Getrost, p 236. (2), Marczenko, p 364. (3), Onishi Part IIb, p 38. (6), Snell, *Metals II*, 1010. (8)
Mercury	Dithizone	Fries/Getrost, p 243. (2), Marczenko, p 373. (3), Onishi, Part IIb, p 66. (6), Snell, *Metals I*, p 107. (7), West, p 29. (10)
	Michler's thioketone	Marczenko, p 375. (3), Snell, *Metals I*, p 126. (7)
	Xylenol Orange	*Talanta, 16*, 1023, (1969)
Molybdenum	Bromopyrogallol Red + Cetylpyridium chloride	West, p 58. (10)
	Phenylfluorone	Snell, *Metals II*, p 1311., *Microchem. J., 31*, 56, (1985).
	Toluene-3,4-dithiol	Fries/Getrost, p 251. (2), Marczenko, p 384. (3), Onishi, Part IIb, p 96. (6), Snell, *Metals II*, p 1301. (8)
Nickel	2-(5-Bromo-2-pyridylazo)-5-diethylaminophenol	Marczenko, p 397. (3), *Talanta 28*, 189, (1981).
	Dimethylglyoxime	Fries/Getrost, p 263. (2), Marczenko, p 393. (3), Onishi, Part IIb, p 125. (6), Snell, *Metals I*, p 887. (7)
	Dimethylglyoxime + Oxidant	Fries/Getrost, p 263. (2), Onishi, Part IIb, p 125. (6), Snell, *Metals I*, p 887. (7)
	2,2'-Furildioxime	Marczenko, p 396. (3), Snell, *Metals I*, p 904. (7)
	2-(2-Pyridylazo)-2-naphthol	Snell, *Metals I*, p 910. (7)
	4-(2-Pyridylazo)resorcinol	Snell, *Metals I*, p 911. (7), West, p 39. (10), *Anal. Chim. Acta, 82*, 431, (1976).
Niobium	N-Benzoyl-N-phenylhydroxylamine	Snell, *Metals II*, p 1425. (8)
	Pyrocatechol + EDTA or 2,2'Bipyridyl or 1-(2-thenoyl)-3,3,3,-trifluoroacetone	Snell, *Metals II*, p 1427. (8)
	Bromopyrogallol red	Marczenko, p 407. (3), Snell, *Metals II*, p 1426. (8)
	Bromopyrogallol red + Cetylpyridinium chloride	*Talanta, 32*, 189, (1985).
	4-(2-Pyridylazo)resorcinol	Fries/Getrost, p 274. (2), Marczenko, p 406. (3), Onishi, Part IIb, p 160. (7), Snell, *Metals II*, p 1447. (8)

Determination	Reagents	Ref.
	Sulfochlorophenol S	Onishi, Part IIb, p 161. (7), Snell, *Metals II,* p 1430. (8)
	Xylenol Orange	Onishi, Part IIb, p 164. (7)
Nitrate	Brucine	Boltz, p 227. (1), Fries/Getrost, p 280. (2), Snell, *Nonmetals,* p 546. (9)
	Chromotropic acid	Boltz, p 229. (1), Fries/Getrost, p 281. (2), Snell, *Nonmetals,* p 548. (9), Williams, p 132. (11), *Fres. Z. Anal. Chem., 320,* 490, (1985).
	Sulfanilamide + *N*-(1-Naphthyl)ethylenediamine dihydrochloride	Fries/Getrost, p 279. (2), Snell, *Nonmetals,* p 559. (9)
Nitrite	Sulfanilamide + *N*-(1-Naphthyl)ethylenediamine dihydrochlorine	Boltz, p 241. (1), Snell, N*onmetals, p* 585. (8), *Analyst, 109,* 1281, (1984).
	Sulfanilic acid + 1-Naphthylamine	Boltz, p 237. (1), Fries/Getrost, p 285. (2), Marczenko, p 419. (3), Snell, *Nonmetals,* p 586. (9)
Osmium	1,5-Diphenylcarbazide	Marczenko, p 428. (3)
Palladium	2-(5 Bromo-2-pyridylazo)-5-diethylaminophenol	*Talanta, 33,* 939, (1986).
	Dithizone	Marczenko, p 440. (3), Onishi, Part IIb, p 227. (6), Snell, *Metals II,* p 1577. (8)
	2-Nitroso-1-naphthol	Fries/Getrost, p 294. (2), Onishi, Part IIb, p 226. (6), Snell, *Metals II,* p 1581. (8)
	4-(2-Pyridylazo)resorcinol	Snell, *Metals II,* p 1583. (8) *Analyst, 107,* 708, (1982).
Phosphate	Rhodamine B + Molybdate	Snell, *Nonmetals,* p 103. (9)
	Malachite Green + Molybdate	Snell, *Nonmetals,* p 12. (9), Analyst, *108,* 361, (1983).
Platinum	Sulfochlorophenolazorhodamine	Onishi, Part IIb, p 253. (6), *Talanta, 34,* 87, (1987).
	Dithizone	Fries/Getrost, p 300. (2), Onishi, Part IIb, p 253. (6), Snell, *Metals II,* p 1534. (8)
	2-Mercaptobenzothiazole	Fries/Getrost, p 302. (2), *Zh. Anal. Khim., 24,* 1172, (1969).
Rare Earths	Arsenazo I	Marczenko, p 470. (3), Onishi, Part IIa, p 785. (5), Snell, *Metals II,* p 1857. (8)
	Arsenazo III	Fries/Getrost, p 309. (2), Marczenko, p 468. (3), Onishi, Part IIa, p 786. (5), Snell, *Metals II,* p 1862. (8)
	Xylenol Orange	Onishi, Part IIa, p 787. (5), Snell, *Metals II,* p 1874. (8)
Rhenium	2,2′-Furildioxime	Fries/Getrost, p 310. (2), Marczenko, p 481. (3), Onishi, Part IIb, p 288. (6), Snell, *Metals II,* p 1659. (8)
Rhodium	1-(2-Pyridylazo)-2-naphthol	Fries/Getrost, p 311. (2), Snell, *Metals II,* p 1553. (8)
Ruthenium	1,10-Phenanthroline	Onishi, Part IIb, p 331. (6), Snell, *Metals II,* p 1623. (8)
	Thiourea	Fries/Getrost, p 318. (2), Onishi, Part IIb, 329. (6), Snell, *Metals II,* p 1626. (8)
	1,4-Diphenylthiosemicarbazide	Marczenko, p 493. (3), Onishi, Part IIb, p 330. (8)
Scandium	Alizarin red S	Fries/Getrost, p 319. (2), Onishi, Part IIb, p 360. (6), Snell, *Metals I,* p 536. (7)
	Arsenazo III	Onishi, Part IIb, p 359. (6), Snell, *Metals I,* p 539. (7)

Determination	Reagents	Ref.
	Chrome Azurol S	Snell, Metals I, p 551. (7), *Anal. Chim. Acta, 159*, 309, (1984).
	Xylenol Orange	Marczenko, p 501. (3), Onishi, Part IIb, p 357. (6), Snell, *Metals I*, p 547. (7)
Selenium	3,3'-Diaminobenzidine	Boltz, p 391. (1), Fries/Getrost, p 323. (2), Marczenko, p 508. (3), Snell, *Nonmetals*, p 490. (9), West, p 4. (10).
	2,3-Diaminonaphthaline	Snell, *Nonmetals*, p 501. (9)
Silver	Dithizone	Fries/Getrost, p 328. (2), Marczenko, p 524. (3), Onishi, Part IIb, p 379. (6), Snell, *Metals I*, p 82. (7)
Sulfate	Eosin + 1,10-Phenanthroline	Snell, *Metals I*, p 93. (7)
Sulfide	Methylthymol blue + Barium (II)	Snell, *Nonmetals*, p 457. (9)
	N,N,-Dimethyl-1,4-phenylenediamine	Boltz, p 483. (1), Fries/Getrost, p 344. (2), Snell, *Nonmetals*, p 400. (9), Williams, p 578. (11)
Sulfite	Pararosaniline + Formaldehyde	Boltz, p 478. (1), Marczenko, p 540. (3), Snell, *Nonmetals*, p 430. (9), Williams, p 591. (11)
Tantalum	Methyl Violet	Marczenko, p 551. (3), Snell, *Metals II*, p 1485. (8)
	4-(2-Pyridylazo)resorcinol	Snell, *Metals II*, p 1488. (8)
	Phenylfluorone	Onishi, Part IIb, p 166. (6), Snell, *Metals II*, p 1486. (8)
Tellurium	Diethyldithiocarbamate	Boltz, p 402. (1), Fries/Getrost, p 348. (2), Snell, *Nonmetals*, p 533. (9), Williams, p 220. (10)
	Bismuthiol II	Boltz, p 401. (1), Marczenko, p 557. (3), Snell, *Nonmetals*, p 524. (9)
Thallium	Brilliant green	Fries/Getrost, p 352. (2), Marczenko, p 567. (3), Onishi, Part IIb, p 426. (6), Snell, *Metals I*, p 45. (7)
	Dithizone	Fries/Getrost, p 355. (2), Onishi, Part IIb, p 426. (6), Snell, *Metals I*, p 54. (7)
	Rhodamine B	Fries/Getrost, p 354. (2), Marczenko, p 566. (3), Onishi, Part IIb, p 424. (6), Snell, *Metals I*, p 63. (7)
Thorium	Arsenazo III	Fries/Getrost, p 360. (2), Marczenko, p 575. (3), Onishi, Part IIb, p 460. (6), Snell, *Metals II*, p 1820. (8)
	Thoron	Marczenko, p 574. (3), Onishi, Part IIb, p 463. (6), Snell, *Metals I*, p 1835. (7)
	Xylenol Orange	Snell, *Metals I*, p 1852. (7)
Tin	Xylenol Orange + Cetyltrimethylammonium bromide	*Talanta, 26*, 499, (1979).
	Pyrocatechol violet (and + Cetyltrimethylammonium bromide)	Marczenko, p 585. (3), Onishi, Part IIb, p 501. (6), Snell, *Metals I, p* 422. (7)
	Gallein	Onishi, Part IIb, p 507, 510. (6), Snell, *Metals I*, p 432. (7)
	Phenylfluorone	Fries/Getrost, p 368. (2), Marczenko, p 582. (3), Onishi, Part IIb, p 497. (6), Snell, *Metals I*, p 444. (7)
	Toluene-3,4-dithiol + Dispersant	Fries/Getrost, p 366. (2), Onishi, Part IIb, p 502. (6), Snell, *Metals I*, p 427. (7)
Titanium	Chromotropic acid	Marczenko, p 593. (3), Onishi, Part IIb, p 551. (6), Snell, *Metals II*, p 1080. (8)

Determination	Reagents	Ref.
	Diantipyrinylmethane	Onishi, Part IIb, p 545. (6), Snell, *Metals II,* 1085. (8)
	Tiron	Fries/Getrost, p 376. (2), Onishi, Part IIb, p 549. (6), Snell, *Metals II,* p 1114. (8)
Tungsten	Pyrocatechol Violet	Snell, Metals II, p 1265. (8)
	Tetraphenylarsonium chloride + Thiocyanate	Onishi, Part IIb, p 596. (6), Snell, *Metals II,* p 1278. (8)
	Toluene-3,5-dithiol	Marczenko, p 605. (3), Onishi, Part IIb, p 590. (6), Snell, *Metals II,* p 1267. (8)
Uranium	Arsenazo III	Marczenko, p 611. (3), Onishi, Part IIb, p 627. (6), Snell, *Metals II,* p 1356. (8)
	2-(5-Bromo-2-pyridylazo)diethylaminophenol	Fries/Getrost, p 388. (2), Onishi, Part IIb, p 625. (6)
	Chlorophosphonazo III	Snell, Metals II, p 1367. (8), *Fres. Z. Anal. Chem., 306,* 110, (1981).
	1-(2-Pyridylazo)-2-naphthol	Fries/Getrost, p 386. (2), Onishi, Part IIb, p 625. (6), Snell, *Metals II,* p 1387. (8)
Vanadium	N-Benzoyl-N-phenylhydroxylamine	Fries/Getrost, p 395. (2), Marczenko, p 625. (3), Snell, *Metals II,* p 1196. (8)
	8-Hydroxyquinoline	Marczenko, p 623. (3), Snell, *Metals II,* p 1209. (8)
	4-(2-pyridylazo)resorcinol	Fries/Getrost, p 404. (23), Marczenko, p 628. (3), Onishi, Part IIb, p 625. (6), Snell, *Metals II,* p 1226. (8)
Yttrium	Alizarin Red S	Fries/Getrost, p 406. (2), Onishi, Part IIa, p 784. (5), Snell, *Metals II,* p 1919. (8)
	Arsenazo III	Marczenko, p 468. (3), Onishi, Part IIa, p 786. (5), Snell, *Metals II,* p 1921. (8)
	Xylenol Orange	Fries/Getrost, p 406. (2), Onishi, Part IIa, p 787. (5), Snell, *Metals II,* p 1923. (8)
Zinc	Dithizone	Fries/Getrost, p 408. (2), Marczenko, p 637. (3), Onishi, Part IIb, p 708. (6), Snell, *Metals II,* p 1042. (8)
	1-(2-Pyridylazo)-2-naphthol	Marczenko, p 639. (3), Onishi, Part IIb, p 719. (6), Snell, *Metals II,* p 1056. (8)
	Xylenol Orange	Fries/Getrost, p 417. (2), Snell, *Metals II,* p 1062. (8), Talanta, *26,* 693, (1979).
	Zircon	Fries/Getrost, p 412. (2), Onishi, Part IIb, p 719. (6), Snell, *Metals II,* p 1063. (8), West, p 23. (10)
Zirconium	Alizarin Red S	Fries/Getrost, p 421. (2), Marczenko, p 647. (3), Onishi, Part IIb, p 763. (6), Snell, *Metals II,* p 1136. (8)
	Arsenazo III	Fries/Getrost, p 421. (2), Onishi, Part IIb, p 770. (6), Snell, *Metals II,* p 1143. (8)
	Pyrocatechol Violet	Onishi, Part IIb, p 771. (6), Snell, *Metals II,* p 1149. (8)
	Morin	Fries/Getrost, p 424. (2), Onishi, Part IIb, p 765. (6), Snell, *Metals II,* p 1158. (8)
	Xylenol Orange	Fries/Getrost, p 419. (2), Marczenko, p 648. (3), Onishi, Part IIb, p 767. (6), Snell, *Metals II,* p 1167. (8)

ORGANIC ANALYTICAL REAGENTS FOR THE DETERMINATION OF INORGANIC SUBSTANCES (continued)

REVIEWS

Sommer, L, Ackermann, G., Thorburn Burns, D., and Savvin, S. B., *Pure and Applied Chem.*, 62, 2147, 1990.

Sommer, L., Ackermann, G., and Thorburn Burns, D., *Pure and Applied Chem.*, 62, 2323, 1990)

Sommer, L., Komarek, J., and Thorburn Burns, D., *Pure and Applied Chem.*, 64, 213, 1992.

Savvin, S. B., *Crit. Rev. Anal. Chem.*, 8, 55, 1979.

MONOGRAPHS

1. Boltz, D. F., and Howell, J. A., *Colorimetric Determination of Nonmetals,* 2nd ed, Wiley, New York, 1978.
2. Fries, J. and Getrost, H., *Organic Reagents for Trace Analysis,* E Merck, Darmstadt, 1977.
3. Marczenko, Z., *Separation and Spectrophotometric Determination of Elements,* Ellis Horwood, Chichester, 1986.
4. Sandell, E. B. and Onishi, H., *Photometric Determination of Traces of Metals. General Aspects, Part I,* 4th ed, J. Wiley, New York, 1978.
5. Onishi, H., *Photometric Determination of Traces of Metals. Part IIa: Individual Metals, Aluminium to Lithium,* 4th ed, J. Wiley, New York, 1986.
6. Onishi, H., *Photometric Determination of Traces of Metals. Part IIb: Individual Metals, Magnesium to Zinc,* 4th ed, J. Wiley, New York, 1989.
7. Snell, F. D., *Photometric and Fluorimetric Methods of Analysis, Metals Part 1,* J. Wiley, New York, 1978.
8. Snell, F. D., *Photometric and Fluorimetric Methods of Analysis, Metals Part 2,* J. Wiley, New York, 1978.
9. Snell, F. D., *Photometric and Fluorimetric Methods of Analysis, Nonmetals,* J. Wiley, New York, 1981.
10. West, T. S. and Nürnberg, H. W., Eds., *The Determination of Trace Metals in Natural Waters,* Blackwell, Oxford, 1988.
11. Williams, W. J.,, *Handbook of Anion Determination,* Butterworth, London, 1979.
12. Townshend, A., Burns, D. T., Guilbault, G. G., Lobinski, R., Mavenzenko, Z., Newman, E., and Onishi, H., *Dictionary of Analytical Reagents*, Chapman & Hall, London, 1993.

ACID-BASE INDICATORS

A. K. Covington

The first part of this table lists some common acid-base indicators in alphabetical order along with the approximate pH range(s) at which a color change occurs. Following this is a table of the same indicators ordered by pH range, which includes the nature of the color change, instructions on preparation of the indicator solution, and the acid dissociation constant pK, when available. The color code is:

C = colorless   A = amber   B/G = blue-green   Pk = pink   Y = yellow   V = violet   R = red   B = blue

P = purple   O = orange

REFERENCE

Bishop, E., Ed., *Indicators*, Pergamon, Oxford, 1972.

Indicator	pH Range
Alizarin	5.6-7.2; 11.0-12.4
Alizarin Red S	4.6-6.0
Alizarin Yellow R	10.1-12.0
Benzopurpurine 4B	2.2-4.2
4,4'-Bis(2-amino-1-naphthylazo)-2,2'-stilbenedisulfonic acid	3.0-4.0
4,4'-Bis(4-amino-1-naphthylazo)-2,2'-stilbenedisulfonic acid	8.0-9.0
Brilliant Yellow	6.6-7.8
Bromocresol Green	3.8-5.4
Bromocresol Purple	5.2-6.8
Bromophenol Blue	3.0-4.6
Bromothymol Blue	6.0-7.6
Chlorophenol Red	5.2-6.8
Clayton Yellow	12.2-13.2
Congo Red	3.0-5.0
o-Cresolphthalein	8.2-9.8
Cresol Red	0.0-1.0; 7.0-8.8
Crystal Violet	0.0-1.8
Curcumin (Turmaric)	7.4-8.6
p-(2,4-Dihydroxyphenylazo) benzenesulfonic acid, sodium salt	11.4-12.6
p-Dimethylaminoazobenzene	2.8-4.4
4-(4-Dimethylamino-1-naphylazo)-3-methoxybenzenesulfonic acid	3.5-4.8
2-(p-Dimethylamino-phenylazo)pyridine	0.2-1.8; 4.4-5.6
N,N-Dimethyl-p-(m-tolylazo)aniline	2.6-4.8
2,4-Dinitrophenol	2.0-4.7
2-(2,4 Dinitrophenylazo)-1-naphthol-3,6-disulfonic acid, disodium salt	6.0-7.0
6,8-Dinitro-2,4-(1H)quinazolinedione	6.4-8.0

Indicator	pH Range
Erythrosin, disodium salt	2.2-3.6
4-(p-Ethoxyphenylazo)-m-phenylene-diamine monohydrochloride	4.4-5.8
Ethyl bis(2,4-dimethylphenyl) ethanoate	8.4-9.6
Ethyl Orange	3.4-4.8
Ethyl Red	4.0-5.8
Ethyl Violet	0.0-2.4
5,5'-Indigodisulfonic acid, disodium salt	11.4-13.0
Malachite Green	0.2-1.8
Metacresol Purple	1.2-2.8; 7.4-9.0
Metanil Yellow	1.2-2.4
Methyl Green	0.2-1.8
Methyl Orange	3.2-4.4
Methyl Red	4.8-6.0
Methyl Violet	0.0-1.6
p-Naphtholbenzein	8.2-10.0
Neutral Red	6.8-8.0
p-Nitrophenol	5.4-6.6
m-Nitrophenol	6.8-8.6
Orange IV	1.4-2.8
Paramethyl Red	1.0-3.0
Phenolphthalein	8.2-10.0
Phenol Red	6.6-8.0
4-Phenylazodiphenylamine	1.2-2.6
4-Phenylazo-1-naphthylamine	4.0-5.6
Propyl Red	4.8-6.6
Quinaldine Red	1.4-3.2
Resazurin	3.8-6.4
Resorcin Blue	4.4-6.2
Tetrabromophenolphthalein ethyl ester, potassium salt	3.0-4.2
Thymol Blue	1.2-2.8; 8.0-9.6
Thymolphthalein	9.4-10.6
4-o-Tolylazo-o-toluidine	1.4-2.8
1,3,5-Trinitrobenzene	12.0-14.0
2,4,6-Trinitrotoluene	11.5-13.0
Turmaric	7.4-8.6

pH range	Color change	Indicator	pK	Preparation
0.0-1.0	R-Y	Cresol Red		0.1 g in 26.2 mL 0.01 M NaOH + 223.8 mL water
0.0-1.6	Y-B	Methyl Violet		0.01-0.05% in water
0.0-1.8	Y-B	Crystal Violet		0.02% in water
0.0-2.4	Y-B	Ethyl Violet		0.1 g in 50 mL 50% v/v methanol-water
0.2-1.8	Y-B/G	Malachite Green	1.3	water
0.2-1.8	Y-B	Methyl Green		0.1% in water
0.2-1.8	Y-R	2-(p-Dimethylaminophenylazo)pyridine		0.1% in ethanol
1.0-3.0	R-Y	Paramethyl Red		ethanol
1.2-2.4	R-Y	Metanil Yellow		0.01% in water
1.2-2.6	R-Y	4-Phenylazodiphenylamine		0.01 g in 1 mL 1 M HCl + 50 mL ethanol + 49 mL water
1.2-2.8	R-Y	Thymol Blue	1.65	0.1 g in 21.5 mL 0.01 M NaOH + 228.5 mL water
1.2-2.8	R-Y	Metacresol Purple	1.51	0.1 g in 26.2 mL 0.01 M NaOH + 223.8 mL water
1.4-2.8	R-Y	Orange IV		0.01% in water
1.4-2.8	O-Y	4-o-Tolylazo-o-toluidine		water
1.4-3.2	C-R	Quinaldine Red	2.63	1% in ethanol
2.0-4.7	C-Y	2,4-Dinitrophenol	3.96	sat. solution in water
2.2-3.6	O-R	Erythrosin, disodium salt		0.1% in water
2.2-4.2	V-R	Benzopurpurine 4B		0.1% in water
2.6-4.8	R-Y	N,N-Dimethyl-p-(m-tolylazo)aniline		0.1% in water
2.8-4.4	R-Y	p-Dimethylaminoazobenzene		0.1 g in 100 mL 90% v/v ethanol-water
3.0-4.0	P-R	4,4'-Bis(2-amino-1-naphthylazo)-2,2'-stilbenedisulfonic acid		0.1 g in 5.9 mL 0.05 M NaOH + 94.1 mL water
3.0-4.2	Y-B	Tetrabromophenolphthalein ethyl ester, potassium salt		0.1% in ethanol
3.0-4.6	Y-B	Bromophenol Blue	4.10	0.1 g in 14.9 mL 0.01 M NaOH + 235.1 mL water
3.0-5.0	B-R	Congo Red		0.1% in water
3.2-4.4	R-Y	Methyl Orange	3.46	0.1% in water
3.4-4.8	R-Y	Ethyl Orange	4.34	0.05-0.2% in water or aqueous ethanol
3.5-4.8	V-Y	4-(4-Dimethylamino-1-naphylazo)-3-methoxybenzenesulfonic acid		0.1% in 60% ethanol-water
3.8-5.4	Y-B	Bromocresol Green	4.90	0.1 g in 14.3 mL 0.01 M NaOH + 235.7 mL water
3.8-6.4	O-V	Resazurin		water
4.0-5.6	R-Y	4-Phenylazo-1-naphthylamine		0.1% in ethanol
4.0-5.8	C-R	Ethyl Red	5.42	0.1 g in 100 mL 50% v/v methanol-water
4.4-5.6	R-Y	2-(p-Dimethylaminophenylazo)pyridine		0.1% in ethanol
4.4-5.8	O-Y	4-(p-Ethoxyphenylazo)-m-phenylene-diamine monohydrochloride		0.1% in water
4.4-6.2	R-B	Resorcin Blue		0.2% in ethanol
4.6-6.0	Y-R	Alizarin Red S		water
4.8-6.0	R-Y	Methyl Red	5.00	0.02 g in 100 mL 60% v/v ethanol-water
4.8-6.6	R-Y	Propyl Red	5.48	ethanol
5.2-6.8	Y-P	Bromocresol Purple	6.40	0.1 g in 18.5 mL 0.01 M NaOH + 231.5 mL water
5.2-6.8	Y-R	Chlorophenol Red	6.25	0.1 g in 23.6 mL 0.01 M NaOH + 226.4 mL water
5.4-6.6	C-Y	p-Nitrophenol	7.15	0.1% in water
5.6-7.2	Y-R	Alizarin		0.1% in methanol
6.0-7.0	Y-B	2-(2,4-Dinitrophenylazo)-1-naphthol-3,6-disulfonic acid, disodium salt		0.1% in water
6.0-7.6	Y-B	Bromothymol Blue	7.30	0.1 g in 16 mL 0.01 M NaOH + 234 mL water
6.4-8.0	C-Y	6,8-Dinitro-2,4-(1H)quinazolinedione		25 g in 115 mL 1 M NaOH + 50 mL water at 100°C
6.6-7.8	Y-R	Brilliant Yellow		1% in water
6.6-8.0	Y-R	Phenol Red	8.00	0.1 g in 28.2 mL 0.01 M NaOH + 221.8 mL water
6.8-8.0	R-A	Neutral Red		0.01 g in 100 mL 50% v/v ethanol-water
6.8-8.6	C-Y	m-Nitrophenol	8.28	0.3% in water
7.0-8.8	Y-R	Cresol Red	8.46	0.1 g in 26.2 mL 0.01 M NaOH + 223.8 mL water
7.4-8.6	Y-R	Turmaric (Curcumin)		ethanol
7.4-9.0	Y-P	Metacresol Purple	8.3	0.1 g in 26.2 mL 0.01 M NaOH + 223.8 mL water

pH range	Color change	Indicator	pK	Preparation
8.0-9.0	B-R	4,4'-Bis(4-amino-1-naphthylazo)-2,2'-stilbenedisulfonic acid		0.1 g in 5.9 mL 0.05 M NaOH + 94.1 mL water
8.0-9.6	Y-B	Thymol Blue	9.20	0.1 g in 21.5 mL 0.01 M NaOH + 228.5 mL water
8.2-10.0	O-B	p-Naphtholbenzein		1% in dil. alkali
8.2-10.0	C-Pk	Phenolphthalein	9.5	0.5 g in 100 mL 50% v/v ethanol-water
8.2-9.8	C-R	o-Cresolphthalein		0.04% in ethanol
8.4-9.6	C-B	Ethyl bis(2,4-dimethylphenyl)ethanoate		sat. solution in 50% acetone-ethanol
9.4-10.6	C-B	Thymolphthalein		0.04 g in 100 mL 50% v/v ethanol-water
10.1-12.0	Y-R	Alizarin Yellow R		0.01% in water
11.0-12.4	R-P	Alizarin		0.1% in methanol
11.4-12.6	Y-O	p-(2,4-Dihydroxyphenylazo)benzenesulfonic acid, sodium salt		0.1% in water
11.4-13.0	B-Y	5,5'-Indigodisulfonic acid, disodium salt	water	
11.5-13.0	C-O	2,4,6-Trinitrotoluene		0.1-0.5% in ethanol
12.0-14.0	C-O	1,3,5-Trinitrobenzene		0.1-0.5% in ethanol
12.2-13.2	Y-A	Clayton Yellow		0.1% in water

FLUORESCENT INDICATORS

Jack DeMent

Fluorescent indicators are substances which show definite changes in fluorescence with change in pH. Some fluorescent materials are not suitable for indicators since their change in fluorescence is too gradual. Fluorescent indicators find greatest utility in the titration of opaque, highly turbid or deeply colored solutions. A long wavelength ultraviolet ("black light") lamp in a dimly lighted room provides the best environment for titrations involving fluorescent indicators, although bright daylight is sometimes sufficient to evoke a response in the bright green, yellow and orange fluorescent indicators. Titrations are carried out in non-fluorescent glassware. One should check the glassware prior to use to make certain that it does not fluoresce due to the wavelengths of light involved in the titration. The meniscus of the liquid in the burette can be followed when a few particles of an insoluble fluorescent solid are dropped onto its surface.

In this table the indicators are arranged by approximate pH range covered. In the case of some of the dyestuffs the end point may vary slightly with the source or manufacturer.

pH 0 to 2

Indicator	C.I.	From pH	To pH
Benzoflavine	—	0.3, yellow fl.	1.7, green fl.
3,6-Dioxyphthalimide	—	0, blue fl.	2.4, green fl.
Eosine YS	768	0, yellow colored	3.0, yellow fl.
Erythrosine	772	0, yellow colored	3.6, yellow fl.
Esculin	—	1.5, colorless	2, blue fl.
4-Ethoxyacridone	—	1.2, green fl.	3.2, blue fl.
3,6-Tetramethyldiaminooxanthone	—	1.2, green fl.	3.4, blue fl.

pH 2 to 4

Indicator	C.I.	From pH	To pH
Chromotropic acid	—	3.5, colorless	4.5, blue fl.
Fluorescein	766	4, colorless	4.5, green fl.
Magdala Red	—	3.0, purple colored	4.0, fl.
α-Naphthylamine	—	3.4, colorless	4.8, blue fl.
β-Naphthylamine	—	2.8, colorless	4.4, violet fl.
Phloxine	774	3.4, colorless	5.0, bright yellow fl.
Salicylic acid	—	2.5, colorless	3.5, blue fl.

pH 4 to 6

Indicator	C.I.	From pH	To pH
Acridine	788	4.9, green fl.	5.1, violet colored
Dichlorofluorescein	—	4.0, colorless	5.0, green fl.
3,6-Dioxyxanthone	—	5.4, colorless	7.6, blue-violet fl.
Erythrosine	772	4.0, colorless	4.5, yellow-green fl.
β-Methylesculetin	—	4.0, colorless	6.2, blue fl.
Neville-Winther acid	—	6.0, colorless	6.5, blue fl.
Resorufin	—	4.4, yellow fl.	6.4, weak orange fl.
Quininic acid	—	4.0, yellow colored	5.0, blue fl.
Quinine [first end point]	—	5.0, blue fl.	6.1, violet fl.

pH 6 to 8

Indicator	C.I.	From pH	To pH
Acid R Phosphine	—	(claimed for range pH 6.0–7.0)	
Brilliant Diazol Yellow	—	6.5, colorless	7.5, violet fl.
Cleves acid	—	6.5, colorless	7.5, green fl.
Coumaric acid	—	7.2, colorless	9.0, green fl.
3,6-Dioxyphthalic dinitrile	—	5.8, blue fl.	8.2, green fl.
Magnesium 8-hydroxyquinolinate	—	6.5, colorless	7.5, golden fl.
β-Methylumbelliferone	—	7.0, colorless	7.5, blue fl.
1-Naphthol-4-sulfonic acid	—	6.0, colorless	6.5, blue fl.
Orcinaurine	—	6.5, colorless	8.0, green fl.
Patent Phosphine	789	(for the range pH 6.0–7.0, green-yellow fl.)	
Thioflavine	816	(for the region pH 6.5–7.0, yellow fl.)	
Umbelliferone	—	6.5, colorless	7.6, blue fl.

pH 8 to 10

Indicator	C.I.	From pH	To pH
Acridine Orange	788	8.4, orange colored	10.4, green fl.
Ethoxyphenylnaphthostilbazonium chloride	—	9, green fl.	11, non-fl.
G Salt	—	9.0, dull blue fl.	9.5, bright blue fl.
Naphthazol derivatives	—	8.2, colorless	10.0, yellow or green fl.
α-Naphthionic acid	—	9, blue fl.	11, green fl.
2-Naphthol-3,6-disulfonic acid	—	9.5, dark blue fl.	Light blue fl. at higher pH
β-Naphthol	—	8.6, colorless	Blue fl. at higher pH
α-Naphtholsulfonic acid	—	8.0, dark blue fl.	9.0, bright violet fl.
1,4-Naphtholsulfonic acid	—	8.2, dark blue fl.	Light blue fl. at higher pH
Orcinsulfonphthalein	—	8.6, yellow colored	10.0 fl.
Quinine [second end point]	—	9.5, violet fl.	10.0, colorless
R-Salt	—	9.0, dull blue fl.	9.5, bright blue fl.
Sodium 1-naphthol-2-sulfonate	—	9.0, dark blue fl.	10.0, bright violet fl.

pH 10 to 12

Indicator	C.I.	From pH	To pH
Coumarin	—	9.8, deep green fl.	12, light green fl.
Eosine BN	771	10.5, colorless	14.0, yellow fl.
Papaverine (permanganate oxidized)	—	9.5, yellow fl.	11.0, blue fl.
Schaffers Salt	—	5.0, violet fl.	11.0, green-blue fl.
SS-Acid (sodium salt)	—	10.0, violet fl.	12.0, yellow colored

pH 12 to 14

Indicator	C.I.	From pH	To pH
Cotarnine	—	12.0, yellow fl.	13.0, white fl.
α-Naphthionic acid	—	12, blue fl.	13, green fl.
β-Naphthionic acid	—	12, blue fl.	13, violet fl.

CONVERSION FORMULAS FOR CONCENTRATION OF SOLUTIONS

A = Weight per cent of solute	G = Molality
B = Molecular weight of solvent	M = Molarity
E = Molecular weight of solute	N = Mole fraction
F = Grams of solute per liter of solution	R = Density of solution in grams per milliliter

Concentration of solute—SOUGHT	Concentration of solute—GIVEN				
	A	N	G	M	F
A	—	$\dfrac{100N \times E}{N \times E + (1-N)B}$	$\dfrac{100G \times E}{1000 + G \times E}$	$\dfrac{M \times E}{10R}$	$\dfrac{F}{10R}$
N	$\dfrac{\dfrac{A}{E}}{\dfrac{A}{E} + \dfrac{100-A}{B}}$	—	$\dfrac{B \times G}{B \times G + 1000}$	$\dfrac{B \times M}{M(B-E) + 1000R}$	$\dfrac{B \times F}{F(B-E) + 1000R \times E}$
G	$\dfrac{1000A}{E(100-A)}$	$\dfrac{1000N}{B - N \times B}$	—	$\dfrac{1000M}{1000R - (M \times E)}$	$\dfrac{1000F}{E(1000R - F)}$
M	$\dfrac{10R \times A}{E}$	$\dfrac{1000R \times N}{N \times E + (1-N)B}$	$\dfrac{1000R \times G}{1000 + E \times G}$	—	$\dfrac{F}{E}$
F	$10AR$	$\dfrac{1000R \times N \times E}{N \times E + (1-N)B}$	$\dfrac{1000R \times G \times E}{1000 + G \times E}$	$M \times E$	—

ELECTROCHEMICAL SERIES

Petr Vanýsek

There are three tables for this electrochemical series. Each table lists standard reduction potentials, $E°$ values, at 298.15 K (25°C), and at a pressure of 101.325 kPa (1 atm). Table 1 is an alphabetical listing of the elements, according to the symbol of the elements. Thus, data for silver (Ag) precedes those for aluminum (Al). Table 2 lists only those reduction reactions which have $E°$ values positive in respect to the standard hydrogen electrode. In Table 2, the reactions are listed in the order of increasing positive potential, and they range from 0.0000 V to + 3.4 V. Table 3 lists only those reduction potentials which have $E°$ negative with respect to the standard hydrogen electrode. In Table 3, the reactions are listed in the order of decreasing potential and range from 0.0000 V to –4.10 V. The reliability of the potentials is not the same for all the data. Typically, the values with fewer significant figures have lower reliability. The values of reduction potentials, in particular those of less common reactions, are not definite; they are subject to occasional revisions.

Abbreviations: ac = acetate; bipy = 2,2′-dipyridine, or bipyridine; en = ethylenediamine; phen = 1,10-phenanthroline.

REFERENCES

1. G. Milazzo, S. Caroli, and V. K. Sharma, *Tables of Standard Electrode Potentials*, Wiley, Chichester, 1978.
2. A. J. Bard, R. Parsons, and J. Jordan, *Standard Potentials in Aqueous Solutions*, Marcel Dekker, New York, 1985.
3. S. G. Bratsch, *J. Phys. Chem. Ref. Data*, 18, 1—21, 1989.

TABLE 1
Alphabetical Listing

Reaction	$E°$/V	Reaction	$E°$/V
$Ac^{3+} + 3\,e \rightleftharpoons Ac$	−2.20	$Al(OH)_4^- + 3\,e \rightleftharpoons Al + 4\,OH^-$	−2.328
$Ag^+ + e \rightleftharpoons Ag$	0.7996	$H_2AlO_3^- + H_2O + 3\,e \rightleftharpoons Al + 4\,OH^-$	−2.33
$Ag^{2+} + e \rightleftharpoons Ag^+$	1.980	$AlF_6^{3-} + 3\,e \rightleftharpoons Al + 6\,F^-$	−2.069
$Ag(ac) + e \rightleftharpoons Ag + (ac)^-$	0.643	$Am^{4+} + e \rightleftharpoons Am^{3+}$	2.60
$AgBr + e \rightleftharpoons Ag + Br^-$	0.07133	$Am^{2+} + 2\,e \rightleftharpoons Am$	−1.9
$AgBrO_3 + e \rightleftharpoons Ag + BrO_3^-$	0.546	$Am^{3+} + 3\,e \rightleftharpoons Am$	−2.048
$Ag_2C_2O_4 + 2\,e \rightleftharpoons 2\,Ag + C_2O_4^{2-}$	0.4647	$Am^{3+} + e \rightleftharpoons Am^{2+}$	−2.3
$AgCl + e \rightleftharpoons Ag + Cl^-$	0.22233	$As + 3\,H^+ + 3\,e \rightleftharpoons AsH_3$	−0.608
$AgCN + e \rightleftharpoons Ag + CN^-$	−0.017	$As_2O_3 + 6\,H^+ + 6\,e \rightleftharpoons 2\,As + 3\,H_2O$	0.234
$Ag_2CO_3 + 2\,e \rightleftharpoons 2\,Ag + CO_3^{2-}$	0.47	$HAsO_2 + 3\,H^+ + 3\,e \rightleftharpoons As + 2\,H_2O$	0.248
$Ag_2CrO_4 + 2\,e \rightleftharpoons 2\,Ag + CrO_4^{2-}$	0.4470	$AsO_2^- + 2\,H_2O + 3\,e \rightleftharpoons As + 4\,OH^-$	−0.68
$AgF + e \rightleftharpoons Ag + F^-$	0.779	$H_3AsO_4 + 2\,H^+ + 2\,e^- \rightleftharpoons HAsO_2 + 2\,H_2O$	0.560
$Ag_4[Fe(CN)_6] + 4\,e \rightleftharpoons 4\,Ag + [Fe(CN)_6]^{4-}$	0.1478	$AsO_4^{3-} + 2\,H_2O + 2\,e \rightleftharpoons AsO_2^- + 4\,OH^-$	−0.71
$AgI + e \rightleftharpoons Ag + I^-$	−0.15224	$At_2 + 2\,e \rightleftharpoons 2\,At^-$	0.3
$AgIO_3 + e \rightleftharpoons Ag + IO_3^-$	0.354	$Au^+ + e \rightleftharpoons Au$	1.692
$Ag_2MoO_4 + 2\,e \rightleftharpoons 2\,Ag + MoO_4^{2-}$	0.4573	$Au^{3+} + 2\,e \rightleftharpoons Au^+$	1.401
$AgNO_2 + e \rightleftharpoons Ag + 2\,NO_2^-$	0.564	$Au^{3+} + 3\,e \rightleftharpoons Au$	1.498
$Ag_2O + H_2O + 2\,e \rightleftharpoons 2\,Ag + 2\,OH^-$	0.342	$Au^{2+} + e^- \rightleftharpoons Au^+$	1.8
$Ag_2O_3 + H_2O + 2\,e \rightleftharpoons 2\,AgO + 2\,OH^-$	0.739	$AuOH^{2+} + H^+ + 2\,e \rightleftharpoons Au^+ + H_2O$	1.32
$Ag^{3+} + 2\,e \rightleftharpoons Ag^+$	1.9	$AuBr_2^- + e \rightleftharpoons Au + 2\,Br^-$	0.959
$Ag^{3+} + e \rightleftharpoons Ag^{2+}$	1.8	$AuBr_4^- + 3\,e \rightleftharpoons Au + 4\,Br^-$	0.854
$Ag_2O_2 + 4\,H^+ + e \rightleftharpoons 2\,Ag + 2\,H_2O$	1.802	$AuCl_4^- + 3\,e \rightleftharpoons Au + 4\,Cl^-$	1.002
$2\,AgO + H_2O + 2\,e \rightleftharpoons Ag_2O + 2\,OH^-$	0.607	$Au(OH)_3 + 3\,H^+ + 3\,e \rightleftharpoons Au + 3\,H_2O$	1.45
$AgOCN + e \rightleftharpoons Ag + OCN^-$	0.41	$H_2BO_3^- + 5\,H_2O + 8\,e \rightleftharpoons BH_4^- + 8\,OH^-$	−1.24
$Ag_2S + 2\,e \rightleftharpoons 2\,Ag + S^{2-}$	−0.691	$H_2BO_3^- + H_2O + 3\,e \rightleftharpoons B + 4\,OH^-$	−1.79
$Ag_2S + 2\,H^+ + 2\,e \rightleftharpoons 2\,Ag + H_2S$	−0.0366	$H_3BO_3 + 3\,H^+ + 3\,e \rightleftharpoons B + 3\,H_2O$	−0.8698
$AgSCN + e \rightleftharpoons Ag + SCN^-$	0.08951	$B(OH)_3 + 7\,H^+ + 8\,e \rightleftharpoons BH_4^- + 3\,H_2O$	−0.481
$Ag_2SeO_3 + 2\,e \rightleftharpoons 2\,Ag + SeO_4^{2-}$	0.3629	$Ba^{2+} + 2\,e \rightleftharpoons Ba$	−2.912
$Ag_2SO_4 + 2\,e \rightleftharpoons 2\,Ag + SO_4^{2-}$	0.654	$Ba^{2+} + 2\,e \rightleftharpoons Ba(Hg)$	−1.570
$Ag_2WO_4 + 2\,e \rightleftharpoons 2\,Ag + WO_4^{2-}$	0.4660	$Ba(OH)_2 + 2\,e \rightleftharpoons Ba + 2\,OH^-$	−2.99
$Al^{3+} + 3\,e \rightleftharpoons Al$	−1.662	$Be^{2+} + 2\,e \rightleftharpoons Be$	−1.847
$Al(OH)_3 + 3\,e \rightleftharpoons Al + 3\,OH^-$	−2.31	$Be_2O_3^{2-} + 3\,H_2O + 4\,e \rightleftharpoons 2\,Be + 6\,OH^-$	−2.63

Reaction	$E°/V$	Reaction	$E°/V$
p–benzoquinone $+ 2 H^+ + 2 e \rightleftharpoons$ hydroquinone	0.6992	$HClO_2 + 3 H^+ + 4 e \rightleftharpoons Cl^- + 2 H_2O$	1.570
$Bi^+ + e \rightleftharpoons Bi$	0.5	$ClO_2^- + H_2O + 2 e \rightleftharpoons ClO^- + 2 OH^-$	0.66
$Bi^{3+} + 3 e \rightleftharpoons Bi$	0.308	$ClO_2^- + 2 H_2O + 4 e \rightleftharpoons Cl^- + 4 OH^-$	0.76
$Bi^{3+} + 2 e \rightleftharpoons Bi^+$	0.2	$ClO_2(aq) + e \rightleftharpoons ClO_2^-$	0.954
$Bi + 3 H^+ + 3 e \rightleftharpoons BiH_3$	−0.8	$ClO_3^- + 2 H^+ + e \rightleftharpoons ClO_2 + H_2O$	1.152
$BiCl_4^- + 3 e \rightleftharpoons Bi + 4 Cl^-$	0.16	$ClO_3^- + 3 H^+ + 2 e \rightleftharpoons HClO_2 + H_2O$	1.214
$Bi_2O_3 + 3 H_2O + 6 e \rightleftharpoons 2 Bi + 6 OH^-$	−0.46	$ClO_3^- + 6 H^+ + 5 e \rightleftharpoons 1/2 Cl_2 + 3 H_2O$	1.47
$Bi_2O_4 + 4 H^+ + 2 e \rightleftharpoons 2 BiO^+ + 2 H_2O$	1.593	$ClO_3^- + 6 H^+ + 6 e \rightleftharpoons Cl^- + 3 H_2O$	1.451
$BiO^+ + 2 H^+ + 3 e \rightleftharpoons Bi + H_2O$	0.320	$ClO_3^- + H_2O + 2 e \rightleftharpoons ClO_2^- + 2 OH^-$	0.33
$BiOCl + 2 H^+ + 3 e \rightleftharpoons Bi + Cl^- + H_2O$	0.1583	$ClO_3^- + 3 H_2O + 6 e \rightleftharpoons Cl^- + 6 OH^-$	0.62
$Bk^{4+} + e \rightleftharpoons Bk^{3+}$	1.67	$ClO_4^- + 2 H^+ + 2 e \rightleftharpoons ClO_3^- \cdot H_2O$	1.189
$Bk^{2+} + 2 e \rightleftharpoons Bk$	−1.6	$ClO_4^- + 8 H^+ + 7 e \rightleftharpoons 1/2 Cl_2 + 4 H_2O$	1.39
$Bk^{3+} + e \rightleftharpoons Bk^{2+}$	−2.8	$ClO_4^- + 8 H^+ + 8 e \rightleftharpoons Cl^- + 4 H_2O$	1.389
$Br_2(aq) + 2 e \rightleftharpoons 2 Br^-$	1.0873	$ClO_4^- + H_2O + 2 e \rightleftharpoons ClO_3^- + 2 OH^-$	0.36
$Br_2(l) + 2 e \rightleftharpoons 2 Br^-$	1.066	$Cm^{4+} + e \rightleftharpoons Cm^{3+}$	3.0
$HBrO + H^+ + 2 e \rightleftharpoons Br^- + H_2O$	1.331	$Cm^{3+} + 3 e \rightleftharpoons Cm$	−2.04
$HBrO + H^+ + e \rightleftharpoons 1/2 Br_2(aq) + H_2O$	1.574	$Co^{2+} + 2 e \rightleftharpoons Co$	−0.28
$HBrO + H^+ + e \rightleftharpoons 1/2 Br_2(l) + H_2O$	1.596	$Co^{3+} + e \rightleftharpoons Co^{2+}$	1.92
$BrO^- + H_2O + 2 e \rightleftharpoons Br^- + 2 OH^-$	0.761	$[Co(NH_3)_6]^{3+} + e \rightleftharpoons [Co(NH_3)_6]^{2+}$	0.108
$BrO_3^- + 6 H^+ + 5 e \rightleftharpoons 1/2 Br_2 + 3 H_2O$	1.482	$Co(OH)_2 + 2 e \rightleftharpoons Co + 2 OH^-$	−0.73
$BrO_3^- + 6 H^+ + 6 e \rightleftharpoons Br^- + 3 H_2O$	1.423	$Co(OH)_3 + e \rightleftharpoons Co(OH)_2 + OH^-$	0.17
$BrO_3^- + 3 H_2O + 6 e \rightleftharpoons Br^- + 6 OH^-$	0.61	$Cr^{2+} + 2 e \rightleftharpoons Cr$	−0.913
$(CN)_2 + 2 H^+ + 2 e \rightleftharpoons 2 HCN$	0.373	$Cr^{3+} + e \rightleftharpoons Cr^{2+}$	−0.407
$2 HCNO + 2 H^+ + 2 e \rightleftharpoons (CN)_2 + 2 H_2O$	0.330	$Cr^{3+} + 3 e \rightleftharpoons Cr$	−0.744
$(CNS)_2 + 2 e \rightleftharpoons 2 CNS^-$	0.77	$Cr_2O_7^{2-} + 14 H^+ + 6 e \rightleftharpoons 2 Cr^{3+} + 7 H_2O$	1.232
$CO_2 + 2 H^+ + 2 e \rightleftharpoons HCOOH$	−0.199	$CrO_2^- + 2 H_2O + 3 e \rightleftharpoons Cr + 4 OH^-$	−1.2
$Ca^+ + e \rightleftharpoons Ca$	−3.80	$HCrO_4^- + 7 H^+ + 3 e \rightleftharpoons Cr^{3+} + 4 H_2O$	1.350
$Ca^{2+} + 2 e \rightleftharpoons Ca$	−2.868	$CrO_2 + 4 H^+ + e \rightleftharpoons Cr^{3+} + 2 H_2O$	1.48
$Ca(OH)_2 + 2 e \rightleftharpoons Ca + 2 OH^-$	−3.02	$Cr(V) + e \rightleftharpoons Cr(IV)$	1.34
Calomel electrode, 1 molal KCl	0.2800	$CrO_4^{2-} + 4 H_2O + 3 e \rightleftharpoons Cr(OH)_3 + 5 OH^-$	−0.13
Calomel electrode, 1 molar KCl (NCE)	0.2801	$Cr(OH)_3 + 3 e \rightleftharpoons Cr + 3 OH^-$	−1.48
Calomel electrode, 0.1 molar KCl	0.3337	$Cs^+ + e \rightleftharpoons Cs$	−3.026
Calomel electrode, saturated KCl (SCE)	0.2412	$Cu^+ + e \rightleftharpoons Cu$	0.521
Calomel electrode, saturated NaCl (SSCE)	0.2360	$Cu^{2+} + e \rightleftharpoons Cu^+$	0.153
$Cd^{2+} + 2 e \rightleftharpoons Cd$	−0.4030	$Cu^{2+} + 2 e \rightleftharpoons Cu$	0.3419
$Cd^{2+} + 2 e \rightleftharpoons Cd(Hg)$	−0.3521	$Cu^{2+} + 2 e \rightleftharpoons Cu(Hg)$	0.345
$Cd(OH)_2 + 2 e \rightleftharpoons Cd(Hg) + 2 OH^-$	−0.809	$Cu^{3+} + e \rightleftharpoons Cu^{2+}$	2.4
$CdSO_4 + 2 e \rightleftharpoons Cd + SO_4^{2-}$	−0.246	$Cu_2O_3 + 6 H^+ + 2 e \rightleftharpoons 2 Cu^{2+} + 3 H_2O$	2.0
$Cd(OH)_4^{2-} + 2 e \rightleftharpoons Cd + 4 OH^-$	−0.658	$Cu^{2+} + 2 CN^- + e \rightleftharpoons [Cu(CN)_2]^-$	1.103
$CdO + H_2O + 2 e \rightleftharpoons Cd + 2 OH^-$	−0.783	$CuI_2^- + e \rightleftharpoons Cu + 2 I^-$	0.00
$Ce^{3+} + 3 e \rightleftharpoons Ce$	−2.336	$Cu_2O + H_2O + 2 e \rightleftharpoons 2 Cu + 2 OH^-$	−0.360
$Ce^{3+} + 3 e \rightleftharpoons Ce(Hg)$	−1.4373	$Cu(OH)_2 + 2 e \rightleftharpoons Cu + 2 OH^-$	−0.222
$Ce^{4+} + e \rightleftharpoons Ce^{3+}$	1.72	$2 Cu(OH)_2 + 2 e \rightleftharpoons Cu_2O + 2 OH^- + H_2O$	−0.080
$CeOH^{3+} + H^+ + e \rightleftharpoons Ce^{3+} + H_2O$	1.715	$2 D^+ + 2 e \rightleftharpoons D_2$	−0.013
$Cf^{4+} + e \rightleftharpoons Cf^{3+}$	3.3	$Dy^{2+} + 2 e \rightleftharpoons Dy$	−2.2
$Cf^{3+} + e \rightleftharpoons Cf^{2+}$	−1.6	$Dy^{3+} + 3 e \rightleftharpoons Dy$	−2.295
$Cf^{3+} + 3 e \rightleftharpoons Cf$	−1.94	$Dy^{3+} + e \rightleftharpoons Dy^{2+}$	−2.6
$Cf^{2+} + 2 e \rightleftharpoons Cf$	−2.12	$Er^{2+} + 2 e \rightleftharpoons Er$	−2.0
$Cl_2(g) + 2 e \rightleftharpoons 2 Cl^-$	1.35827	$Er^{3+} + 3 e \rightleftharpoons Er$	−2.331
$HClO + H^+ + e \rightleftharpoons 1/2 Cl_2 + H_2O$	1.611	$Er^{3+} + e \rightleftharpoons Er^{2+}$	−3.0
$HClO + H^+ + 2 e \rightleftharpoons Cl^- + H_2O$	1.482	$Es^{3+} + e \rightleftharpoons Es^{2+}$	−1.3
$ClO^- + H_2O + 2 e \rightleftharpoons Cl^- + 2 OH^-$	0.81	$Es^{3+} + 3 e \rightleftharpoons Es$	−1.91
$ClO_2 + H^+ + e \rightleftharpoons HClO_2$	1.277	$Es^{2+} + 2 e \rightleftharpoons Es$	−2.23
$HClO_2 + 2 H^+ + 2 e \rightleftharpoons HClO + H_2O$	1.645	$Eu^{2+} + 2 e \rightleftharpoons Eu$	−2.812
$HClO_2 + 3 H^+ + 3 e \rightleftharpoons 1/2 Cl_2 + 2 H_2O$	1.628	$Eu^{3+} + 3 e \rightleftharpoons Eu$	−1.991

Reaction	$E°/V$
$Eu^{3+} + e \rightleftharpoons Eu^{2+}$	−0.36
$F_2 + 2\,H^+ + 2\,e \rightleftharpoons 2\,HF$	3.053
$F_2 + 2\,e \rightleftharpoons 2\,F^-$	2.866
$F_2O + 2\,H^+ + 4\,e \rightleftharpoons H_2O + 2\,F^-$	2.153
$Fe^{2+} + 2\,e \rightleftharpoons Fe$	−0.447
$Fe^{3+} + 3\,e \rightleftharpoons Fe$	−0.037
$Fe^{3+} + e \rightleftharpoons Fe^{2+}$	0.771
$2\,HFeO_4^- + 8\,H^+ + 6\,e \rightleftharpoons Fe_2O_3 + 5\,H_2O$	2.09
$HFeO_4^- + 4\,H^+ + 3\,e \rightleftharpoons FeOOH + 2\,H_2O$	2.08
$HFeO_4^- + 7\,H^+ + 3\,e \rightleftharpoons Fe^{3+} + 4\,H_2O$	2.07
$Fe_2O_3 + 4\,H^+ + 2\,e \rightleftharpoons 2\,FeOH^+ + H_2O$	0.16
$[Fe(CN)_6]^{3-} + e \rightleftharpoons [Fe(CN)_6]^{4-}$	0.358
$FeO_4^{2-} + 8\,H^+ + 3\,e \rightleftharpoons Fe^{3+} + 4\,H_2O$	2.20
$[Fe(bipy)_2]^{3+} + e \rightleftharpoons Fe(bipy)_2]^{2+}$	0.78
$[Fe(bipy)_3]^{3+} + e \rightleftharpoons Fe(bipy)_3]^{2+}$	1.03
$Fe(OH)_3 + e \rightleftharpoons Fe(OH)_2 + OH^-$	−0.56
$[Fe(phen)_3]^{3+} + e \rightleftharpoons [Fe(phen)_3]^{2+}$	1.147
$[Fe(phen)_3]^{3+} + e \rightleftharpoons [Fe(phen)_3]^{2+} (1\ molar\ H_2SO_4)$	1.06
$[Ferricinium]^+ + e \rightleftharpoons ferrocene$	0.400
$Fm^{3+} + e \rightleftharpoons Fm^{2+}$	−1.1
$Fm^{3+} + 3\,e \rightleftharpoons Fm$	−1.89
$Fm^{2+} + 2\,e \rightleftharpoons Fm$	−2.30
$Fr^+ + e \rightleftharpoons Fr$	−2.9
$Ga^{3+} + 3\,e \rightleftharpoons Ga$	−0.549
$Ga^+ + e \rightleftharpoons Ga$	−0.2
$GaOH^{2+} + H^+ + 3\,e \rightleftharpoons Ga + H_2O$	−0.498
$H_2GaO_3^- + H_2O + 3\,e \rightleftharpoons Ga + 4\,OH^-$	−1.219
$Gd^{3+} + 3\,e \rightleftharpoons Gd$	−2.279
$Ge^{2+} + 2\,e \rightleftharpoons Ge$	0.24
$Ge^{4+} + 4\,e \rightleftharpoons Ge$	0.124
$Ge^{4+} + 2\,e \rightleftharpoons Ge^{2+}$	0.00
$GeO_2 + 2\,H^+ + 2\,e \rightleftharpoons GeO + H_2O$	−0.118
$H_2GeO_3 + 4\,H^+ + 4\,e \rightleftharpoons Ge + 3\,H_2O$	−0.182
$2\,H^+ + 2\,e \rightleftharpoons H_2$	0.00000
$H_2 + 2\,e \rightleftharpoons 2\,H^-$	−2.23
$HO_2 + H^+ + e \rightleftharpoons H_2O_2$	1.495
$2\,H_2O + 2\,e \rightleftharpoons H_2 + 2\,OH^-$	−0.8277
$H_2O_2 + 2\,H^+ + 2\,e \rightleftharpoons 2\,H_2O$	1.776
$Hf^{4+} + 4\,e \rightleftharpoons Hf$	−1.55
$HfO^{2+} + 2\,H^+ + 4\,e \rightleftharpoons Hf + H_2O$	−1.724
$HfO_2 + 4\,H^+ + 4\,e \rightleftharpoons Hf + 2\,H_2O$	−1.505
$HfO(OH)_2 + H_2O + 4\,e \rightleftharpoons Hf + 4\,OH^-$	−2.50
$Hg^{2+} + 2\,e \rightleftharpoons Hg$	0.851
$2\,Hg^{2+} + 2\,e \rightleftharpoons Hg_2^{2+}$	0.920
$Hg_2^{2+} + 2\,e \rightleftharpoons 2\,Hg$	0.7973
$Hg_2(ac)_2 + 2\,e \rightleftharpoons 2\,Hg + 2(ac)^-$	0.51163
$Hg_2Br_2 + 2\,e \rightleftharpoons 2\,Hg + 2\,Br^-$	0.13923
$Hg_2Cl_2 + 2\,e \rightleftharpoons 2\,Hg + 2\,Cl^-$	0.26808
$Hg_2HPO_4 + 2\,e \rightleftharpoons 2\,Hg + HPO_4^{2-}$	0.6359
$Hg_2I_2 + 2\,e \rightleftharpoons 2\,Hg + 2\,I^-$	−0.0405
$Hg_2O + H_2O + 2\,e \rightleftharpoons 2\,Hg + 2\,OH^-$	0.123
$HgO + H_2O + 2\,e \rightleftharpoons Hg + 2\,OH^-$	0.0977
$Hg(OH)_2 + 2\,H^+ + 2\,e \rightleftharpoons Hg + 2\,H_2O$	1.034
$Hg_2SO_4 + 2\,e \rightleftharpoons 2\,Hg + SO_4^{2-}$	0.6125
$Ho^{2+} + 2\,e \rightleftharpoons Ho$	−2.1

Reaction	$E°/V$
$Ho^{3+} + 3\,e \rightleftharpoons Ho$	−2.33
$Ho^{3+} + e \rightleftharpoons Ho^{2+}$	−2.8
$I_2 + 2\,e \rightleftharpoons 2\,I^-$	0.5355
$I_3^- + 2\,e \rightleftharpoons 3\,I^-$	0.536
$H_3IO_6^{2-} + 2\,e \rightleftharpoons IO_3^- + 3\,OH^-$	0.7
$H_5IO_6 + H^+ + 2\,e \rightleftharpoons IO_3^- + 3\,H_2O$	1.601
$2\,HIO + 2\,H^+ + 2\,e \rightleftharpoons I_2 + 2\,H_2O$	1.439
$HIO + H^+ + 2\,e \rightleftharpoons I^- + H_2O$	0.987
$IO^- + H_2O + 2\,e \rightleftharpoons I^- + 2\,OH^-$	0.485
$2\,IO_3^- + 12\,H^+ + 10\,e \rightleftharpoons I_2 + 6\,H_2O$	1.195
$IO_3^- + 6\,H^+ + 6\,e \rightleftharpoons I^- + 3\,H_2O$	1.085
$IO_3^- + 2\,H_2O + 4\,e \rightleftharpoons IO^- + 4\,OH^-$	0.15
$IO_3^- + 3\,H_2O + 6\,e \rightleftharpoons IO^- + 6\,OH^-$	0.26
$In^+ + e \rightleftharpoons In$	−0.14
$In^{2+} + e \rightleftharpoons In^+$	−0.40
$In^{3+} + e \rightleftharpoons In^{2+}$	−0.49
$In^{3+} + 2\,e \rightleftharpoons In^+$	−0.443
$In^{3+} + 3\,e \rightleftharpoons In$	−0.3382
$In(OH)_3 + 3\,e \rightleftharpoons In + 3\,OH^-$	−0.99
$In(OH)_4^- + 3\,e \rightleftharpoons In + 4\,OH^-$	−1.007
$In_2O_3 + 3\,H_2O + 6\,e \rightleftharpoons 2\,In + 6\,OH^-$	−1.034
$Ir^{3+} + 3\,e \rightleftharpoons Ir$	1.156
$[IrCl_6]^{2-} + e \rightleftharpoons [IrCl_6]^{3-}$	0.8665
$[IrCl_6]^{3-} + 3\,e \rightleftharpoons Ir + 6\,Cl^-$	0.77
$Ir_2O_3 + 3\,H_2O + 6\,e \rightleftharpoons 2\,Ir + 6\,OH^-$	0.098
$K^+ + e \rightleftharpoons K$	−2.931
$La^{3+} + 3\,e \rightleftharpoons La$	−2.379
$La(OH)_3 + 3\,e \rightleftharpoons La + 3\,OH^-$	−2.90
$Li^+ + e \rightleftharpoons Li$	−3.0401
$Lr^{3+} + 3\,e \rightleftharpoons Lr$	−1.96
$Lu^{3+} + 3\,e \rightleftharpoons Lu$	−2.28
$Md^{3+} + e \rightleftharpoons Md^{2+}$	−0.1
$Md^{3+} + 3\,e \rightleftharpoons Md$	−1.65
$Md^{2+} + 2\,e \rightleftharpoons Md$	−2.40
$Mg^+ + e \rightleftharpoons Mg$	−2.70
$Mg^{2+} + 2\,e \rightleftharpoons Mg$	−2.372
$Mg(OH)_2 + 2\,e \rightleftharpoons Mg + 2\,OH^-$	−2.690
$Mn^{2+} + 2\,e \rightleftharpoons Mn$	−1.185
$Mn^{3+} + 3\,e \rightleftharpoons Mn^{2+}$	1.5415
$MnO_2 + 4\,H^+ + 2\,e \rightleftharpoons Mn^{2+} + 2\,H_2O$	1.224
$MnO_4^- + e \rightleftharpoons MnO_4^{2-}$	0.558
$MnO_4^- + 4\,H^+ + 3\,e \rightleftharpoons MnO_2 + 2\,H_2O$	1.679
$MnO_4^- + 8\,H^+ + 5\,e \rightleftharpoons Mn^{2+} + 4\,H_2O$	1.507
$MnO_4^- + 2\,H_2O + 3\,e \rightleftharpoons MnO_2 + 4\,OH^-$	0.595
$MnO_4^{2-} + 2\,H_2O + 2\,e \rightleftharpoons MnO_2 + 4\,OH^-$	0.60
$Mn(OH)_2 + 2\,e \rightleftharpoons Mn + 2\,OH^-$	−1.56
$Mn(OH)_3 + e \rightleftharpoons Mn(OH)_2 + OH^-$	0.15
$Mn_2O_3 + 6\,H^+ + e \rightleftharpoons 2\,Mn^{2+} + 3\,H_2O$	1.485
$Mo^{3+} + 3\,e \rightleftharpoons Mo$	−0.200
$MoO_2 + 4\,H^+ + 4\,e \rightleftharpoons Mo + 4\,H_2O$	−0.152
$H_3Mo_7O_{24}^{3-} + 45\,H^+ + 42\,e \rightleftharpoons 7\,Mo + 24\,H_2O$	0.082
$MoO_3 + 6\,H^+ + 6\,e \rightleftharpoons Mo + 3\,H_2O$	0.075
$N_2 + 2\,H_2O + 6\,H^+ + 6\,e \rightleftharpoons 2\,NH_4OH$	0.092
$3\,N_2 + 2\,H^+ + 2\,e \rightleftharpoons 2\,HN_3$	−3.09
$N_5^+ + 3\,H^+ + 2\,e \rightleftharpoons 2\,NH_4^+$	1.275

Reaction	$E°/V$	Reaction	$E°/V$
$N_2O + 2H^+ + 2e \rightleftharpoons N_2 + H_2O$	1.766	$H_2P_2^- + e \rightleftharpoons P + 2OH^-$	−1.82
$H_2N_2O_2 + 2H^+ + 2e \rightleftharpoons N_2 + 2H_2O$	2.65	$H_3PO_2 + H^+ + e \rightleftharpoons P + 2H_2O$	−0.508
$N_2O_4 + 2e \rightleftharpoons 2NO_2^-$	0.867	$H_3PO_3 + 2H^+ + 2e \rightleftharpoons H_3PO_2 + H_2O$	−0.499
$N_2O_4 + 2H^+ + 2e \rightleftharpoons 2HNO_2$	1.065	$H_3PO_3 + 3H^+ + 3e \rightleftharpoons P + 3H_2O$	−0.454
$N_2O_4 + 4H^+ + 4e \rightleftharpoons 2NO + 2H_2O$	1.035	$HPO_3^{2-} + 2H_2O + 2e \rightleftharpoons H_2PO_2^- + 3OH^-$	−1.65
$2NH_3OH^+ + H^+ + 2e \rightleftharpoons N_2H_5^+ + 2H_2O$	1.42	$HPO_3^{2-} + 2H_2O + 3e \rightleftharpoons P + 5OH^-$	−1.71
$2NO + 2H^+ + 2e \rightleftharpoons N_2O + H_2O$	1.591	$H_3PO_4 + 2H^+ + 2e \rightleftharpoons H_3PO_3 + H_2O$	−0.276
$2NO + H_2O + 2e \rightleftharpoons N_2O + 2OH^-$	0.76	$PO_4^{3-} + 2H_2O + 2e \rightleftharpoons HPO_3^{2-} + 3OH^-$	−1.05
$HNO_2 + H^+ + e \rightleftharpoons NO + H_2O$	0.983	$Pa^{3+} + 3e \rightleftharpoons Pa$	−1.34
$2HNO_2 + 4H^+ + 4e \rightleftharpoons H_2N_2O_2 + 2H_2O$	0.86	$Pa^{4+} + 4e \rightleftharpoons Pa$	−1.49
$2HNO_2 + 4H^+ + 4e \rightleftharpoons N_2O + 3H_2O$	1.297	$Pa^{4+} + e \rightleftharpoons Pa^{3+}$	−1.9
$NO_2^- + H_2O + e \rightleftharpoons NO + 2OH^-$	−0.46	$Pb^{2+} + 2e \rightleftharpoons Pb$	−0.1262
$2NO_2^- + 2H_2O + 4e \rightleftharpoons N_2O_2^{2-} + 4OH^-$	−0.18	$Pb^{2+} + 2e \rightleftharpoons Pb(Hg)$	−0.1205
$2NO_2^- + 3H_2O + 4e \rightleftharpoons N_2O + 6OH^-$	0.15	$PbBr_2 + 2e \rightleftharpoons Pb + 2Br^-$	−0.284
$NO_3^- + 3H^+ + 2e \rightleftharpoons HNO_2 + H_2O$	0.934	$PbCl_2 + 2e \rightleftharpoons Pb + 2Cl^-$	−0.2675
$NO_3^- + 4H^+ + 3e \rightleftharpoons NO + 2H_2O$	0.957	$PbF_2 + 2e \rightleftharpoons Pb + 2F^-$	−0.3444
$2NO_3^- + 4H^+ + 2e \rightleftharpoons N_2O_4 + 2H_2O$	0.803	$PbHPO_4 + 2e \rightleftharpoons Pb + HPO_4^{2-}$	−0.465
$NO_3^- + H_2O + 2e \rightleftharpoons NO_2^- + 2OH^-$	0.01	$PbI_2 + 2e \rightleftharpoons Pb + 2I^-$	−0.365
$2NO_3^- + 2H_2O + 2e \rightleftharpoons N_2O_4 + 4OH^-$	−0.85	$PbO + H_2O + 2e \rightleftharpoons Pb + 2OH^-$	−0.580
$Na^+ + e \rightleftharpoons Na$	−2.71	$PbO_2 + 4H^+ + 2e \rightleftharpoons Pb^{2+} + 2H_2O$	1.455
$Nb^{3+} + 3e \rightleftharpoons Nb$	−1.099	$HPbO_2^- + H_2O + 2e \rightleftharpoons Pb + 3OH^-$	−0.537
$NbO_2 + 2H^+ + 2e \rightleftharpoons NbO + H_2O$	−0.646	$PbO_2 + H_2O + 2e \rightleftharpoons PbO + 2OH^-$	0.247
$NbO_2 + 4H^+ + 4e \rightleftharpoons Nb + 2H_2O$	−0.690	$PbO_2 + SO_4^{2-} + 4H^+ + 2e \rightleftharpoons PbSO_4 + 2H_2O$	1.6913
$NbO + 2H^+ + 2e \rightleftharpoons Nb + H_2O$	−0.733	$PbSO_4 + 2e \rightleftharpoons Pb + SO_4^{2-}$	−0.3588
$Nb_2O_5 + 10H^+ + 10e \rightleftharpoons 2Nb + 5H_2O$	−0.644	$PbSO_4 + 2e \rightleftharpoons Pb(Hg) + SO_4^{2-}$	−0.3505
$Nd^{3+} + 3e \rightleftharpoons Nd$	−2.323	$Pd^{2+} + 2e \rightleftharpoons Pd$	0.951
$Nd^{2+} + 2e \rightleftharpoons Nd$	−2.1	$[PdCl_4]^{2-} + 2e \rightleftharpoons Pd + 4Cl^-$	0.591
$Nd^{3+} + e \rightleftharpoons Nd^{2+}$	−2.7	$[PdCl_6]^{2-} + 2e \rightleftharpoons [PdCl_4]^{2-} + 2Cl^-$	1.288
$Ni^{2+} + 2e \rightleftharpoons Ni$	−0.257	$Pd(OH)_2 + 2e \rightleftharpoons Pd + 2OH^-$	0.07
$Ni(OH)_2 + 2e \rightleftharpoons Ni + 2OH^-$	−0.72	$Pm^{2+} + 2e \rightleftharpoons Pm$	−2.2
$NiO_2 + 4H^+ + 2e \rightleftharpoons Ni^{2+} + 2H_2O$	1.678	$Pm^{3+} + 3e \rightleftharpoons Pm$	−2.30
$NiO_2 + 2H_2O + 2e \rightleftharpoons Ni(OH)_2 + 2OH^-$	−0.490	$Pm^{3+} + e \rightleftharpoons Pm^{2+}$	−2.6
$No^{3+} + e \rightleftharpoons No^{2+}$	1.4	$Po^{2+} + 2e \rightleftharpoons Po^{2+}$	0.9
$No^{3+} + 3e \rightleftharpoons No$	−1.20	$Po^{4+} + 4e \rightleftharpoons Po$	0.76
$No^{2+} + 2e \rightleftharpoons No$	−2.50	$Pr^{4+} + e \rightleftharpoons Pr^{3+}$	3.2
$Np^{3+} + 3e \rightleftharpoons Np$	−1.856	$Pr^{2+} + 2e \rightleftharpoons Pr$	−2.0
$Np^{4+} + e \rightleftharpoons Np^{3+}$	0.147	$Pr^{3+} + 3e \rightleftharpoons Pr$	−2.353
$NpO_2 + H_2O + H^+ + e \rightleftharpoons Np(OH)_3$	−0.962	$Pr^{3+} + e \rightleftharpoons Pr^{2+}$	−3.1
$O_2 + 2H^+ + 2e \rightleftharpoons H_2O_2$	0.695	$Pt^{2+} + 2e \rightleftharpoons Pt$	1.18
$O_2 + 4H^+ + 4e \rightleftharpoons 2H_2O$	1.229	$[PtCl_4]^{2-} + 2e \rightleftharpoons Pt + 4Cl^-$	0.755
$O_2 + H_2O + 2e \rightleftharpoons HO_2^- + OH^-$	−0.076	$[PtCl_6]^{2-} + 2e \rightleftharpoons [PtCl_4]^{2-} + 2Cl^-$	0.68
$O_2 + 2H_2O + 2e \rightleftharpoons H_2O_2 + 2OH^-$	−0.146	$Pt(OH)_2 + 2e \rightleftharpoons Pt + 2OH^-$	0.14
$O_2 + 2H_2O + 4e \rightleftharpoons 4OH^-$	0.401	$PtO_3 + 2H^+ + 2e \rightleftharpoons PtO_2 + H_2O$	1.7
$O_3 + 2H^+ + 2e \rightleftharpoons O_2 + H_2O$	2.076	$PtO_3 + 4H^+ + 2e \rightleftharpoons Pt(OH)_2^{2+} + H_2O$	1.5
$O_3 + H_2O + 2e \rightleftharpoons O_2 + 2OH^-$	1.24	$PtOH^+ + H^+ + 2e \rightleftharpoons Pt + H_2O$	1.2
$O(g) + 2H^+ + 2e \rightleftharpoons H_2O$	2.421	$PtO_2 + 2H^+ + 2e \rightleftharpoons PtO + H_2O$	1.01
$OH + e \rightleftharpoons OH^-$	2.02	$PtO_2 + 4H^+ + 4e \rightleftharpoons Pt + 2H_2O$	1.00
$HO_2^- + H_2O + 2e \rightleftharpoons 3OH^-$	0.878	$Pu^{3+} + 3e \rightleftharpoons Pu$	−2.031
$OsO_4 + 8H^+ + 8e \rightleftharpoons Os + 4H_2O$	0.838	$Pu^{4+} + e \rightleftharpoons Pu^{3+}$	1.006
$OsO_4 + 4H^+ + 4e \rightleftharpoons OsO_2 + 2H_2O$	1.02	$Pu^{5+} + e \rightleftharpoons Pu^{4+}$	1.099
$[Os(bipy)_2]^{3+} + e \rightleftharpoons [Os(bipy)_2]^{2+}$	0.81	$PuO_2(OH)_2 + 2H^+ + 2e \rightleftharpoons Pu(OH)_4$	1.325
$[Os(bipy)_3]^{3+} + e \rightleftharpoons [Os(bipy)_3]^{2+}$	0.80	$PuO_2(OH)_2 + H^+ + e \rightleftharpoons PuO_2OH + H_2O$	1.062
$P(red) + 3H^+ + 3e \rightleftharpoons PH_3(g)$	−0.111	$Ra^{2+} + 2e \rightleftharpoons Ra$	−2.8
$P(white) + 3H^+ + 3e \rightleftharpoons PH_3(g)$	−0.063	$Rb^+ + e \rightleftharpoons Rb$	−2.98
$P + 3H_2O + 3e \rightleftharpoons PH_3(g) + 3OH^-$	−0.87	$Re^{3+} + 3e \rightleftharpoons Re$	0.300

Reaction	$E°/V$	Reaction	$E°/V$
$ReO_4^- + 4 H^+ + 3 e \rightleftharpoons ReO_2 + 2 H_2O$	0.510	$SiO_2 \text{ (quartz)} + 4 H^+ + 4 e \rightleftharpoons Si + 2 H_2O$	0.857
$ReO_2 + 4 H^+ + 4 e \rightleftharpoons Re + 2 H_2O$	0.2513	$SiO_3^{2-} + 3 H_2O + 4 e \rightleftharpoons Si + 6 OH^-$	−1.697
$ReO_4^- + 2 H^+ + e \rightleftharpoons ReO_3 + H_2O$	0.768	$Sm^{3+} + e \rightleftharpoons Sm^{2+}$	−1.55
$ReO_4^- + 4 H_2O + 7 e \rightleftharpoons Re + 8 OH^-$	−0.584	$Sm^{3+} + 3 e \rightleftharpoons Sm$	−2.304
$ReO_4^- + 8 H^+ + 7 e \rightleftharpoons Re + 4 H_2O$	0.368	$Sm^{2+} + 2 e \rightleftharpoons Sm$	−2.68
$Rh^+ + e \rightleftharpoons Rh$	0.600	$Sn^{2+} + 2 e \rightleftharpoons Sn$	−0.1375
$Rh^+ + 2e \rightleftharpoons Rh$	0.600	$Sn^{4+} + 2 e \rightleftharpoons Sn^{2+}$	0.151
$Rh^{3+} + 3 e \rightleftharpoons Rh$	0.758	$Sn(OH)_3^+ + 3 H^+ + 2 e \rightleftharpoons Sn^{2+} + 3 H_2O$	0.142
$[RhCl_6]^{3-} + 3 e \rightleftharpoons Rh + 6 Cl^-$	0.431	$SnO_2 + 4 H^+ + 2 e^- \rightleftharpoons Sn^{2+} + 2 H_2O$	−0.094
$RhOH^{2+} + H^+ + 3 e \rightleftharpoons Rh + H_2O$	0.83	$SnO_2 + 4 H^+ + 4 e \rightleftharpoons Sn + 2 H_2O$	−0.117
$Ru^{2+} + 2 e \rightleftharpoons Ru$	0.455	$SnO_2 + 3 H^+ + 2 e \rightleftharpoons SnOH^+ + H_2O$	−0.194
$Ru^{3+} + e \rightleftharpoons Ru^{2+}$	0.2487	$SnO_2 + 2 H_2O + 4 e \rightleftharpoons Sn + 4 OH^-$	−0.945
$RuO_2 + 4 H^+ + 2 e \rightleftharpoons Ru^{2+} + 2 H_2O$	1.120	$HSnO_2^- + H_2O + 2 e \rightleftharpoons Sn + 3 OH^-$	−0.909
$RuO_4^- + e \rightleftharpoons RuO_4^{2-}$	0.59	$Sn(OH)_6^{2-} + 2 e \rightleftharpoons HSnO_2^- + 3 OH^- + H_2O$	−0.93
$RuO_4 + e \rightleftharpoons RuO_4^-$	1.00	$Sr^+ + e \rightleftharpoons Sr$	−4.10
$RuO_4 + 6 H^+ + 4 e \rightleftharpoons Ru(OH)_2^{2+} + 2 H_2O$	1.40	$Sr^{2+} + 2 e \rightleftharpoons Sr$	−2.899
$RuO_4 + 8 H^+ + 8 e \rightleftharpoons Ru + 4 H_2O$	1.038	$Sr^{2+} + 2 e \rightleftharpoons Sr(Hg)$	−1.793
$[Ru(bipy)_3]^{3+} + e^- \rightleftharpoons [Ru(bipy)_3]^{2+}$	1.24	$Sr(OH)_2 + 2 e \rightleftharpoons Sr + 2 OH^-$	−2.88
$[Ru(H_2O)_6]^{3+} + e^- \rightleftharpoons [Ru(H_2O)_6]^{2+}$	0.23	$Ta_2O_5 + 10 H^+ + 10 e \rightleftharpoons 2 Ta + 5 H_2O$	−0.750
$[Ru(NH_3)_6]^{3+} + e^- \rightleftharpoons [Ru(NH_3)_6]^{2+}$	0.10	$Ta^{3+} + 3 e \rightleftharpoons Ta$	−0.6
$[Ru(en)_3]^{3+} + e^- \rightleftharpoons [Ru(en)_3]^{2+}$	0.210	$Tc^{2+} + 2 e \rightleftharpoons Tc$	0.400
$[Ru(CN)_6]^{3-} + e^- \rightleftharpoons [Ru(CN)_6]^{4-}$	0.86	$TcO_4^- + 4 H^+ + 3 e \rightleftharpoons TcO_2 + 2 H_2O$	0.782
$S + 2 e \rightleftharpoons S^{2-}$	−0.47627	$Tc^{3+} + e \rightleftharpoons Tc^{2+}$	0.3
$S + 2 H^+ + 2 e \rightleftharpoons H_2S(aq)$	0.142	$TcO_4^- + 8 H^+ + 7 e \rightleftharpoons Tc + 4 H_2O$	0.472
$S + H_2O + 2 e \rightleftharpoons SH^- + OH^-$	−0.478	$Tb^{4+} + e \rightleftharpoons Tb^{3+}$	3.1
$2 S + 2 e \rightleftharpoons S_2^{2-}$	−0.42836	$Tb^{3+} + 3 e \rightleftharpoons Tb$	−2.28
$S_2O_6^{2-} + 4 H^+ + 2 e \rightleftharpoons 2 H_2SO_3$	0.564	$Te + 2 e \rightleftharpoons Te^{2-}$	−1.143
$S_2O_8^{2-} + 2 e \rightleftharpoons 2 SO_4^-$	2.010	$Te + 2 H^+ + 2 e \rightleftharpoons H_2Te$	−0.793
$S_2O_8^{2-} + 2 H^+ + 2 e \rightleftharpoons 2 HSO_4^-$	2.123	$Te^{4+} + 4 e \rightleftharpoons Te$	0.568
$S_4O_6^{2-} + 2 e \rightleftharpoons 2 S_2O_3^{2-}$	0.08	$TeO_2 + 4 H^+ + 4 e \rightleftharpoons Te + 2 H_2O$	0.593
$2 H_2SO_3 + H^+ + 2 e \rightleftharpoons HS_2O_4^- + 2 H_2O$	−0.056	$TeO_3^{2-} + 3 H_2O + 4 e \rightleftharpoons Te + 6 OH^-$	−0.57
$H_2SO_3 + 4 H^+ + 4 e \rightleftharpoons S + 3 H_2O$	0.449	$TeO_4^- + 8 H^+ + 7 e \rightleftharpoons Te + 4 H_2O$	0.472
$2 SO_3^{2-} + 2 H_2O + 2 e \rightleftharpoons S_2O_4^{2-} + 4 OH^-$	−1.12	$H_6TeO_6 + 2 H^+ + 2 e \rightleftharpoons TeO_2 + 4 H_2O$	1.02
$2 SO_3^{2-} + 3 H_2O + 4 e \rightleftharpoons S_2O_3^{2-} + 6 OH^-$	−0.571	$Th^{4+} + 4 e \rightleftharpoons Th$	−1.899
$SO_4^{2-} + 4 H^+ + 2 e \rightleftharpoons H_2SO_3 + H_2O$	0.172	$ThO_2 + 4 H^+ + 4 e \rightleftharpoons Th + 2 H_2O$	−1.789
$2 SO_4^{2-} + 4 H^+ + 2 e \rightleftharpoons S_2O_6^{2-} + H_2O$	−0.22	$Th(OH)_4 + 4 e \rightleftharpoons Th + 4 OH^-$	−2.48
$SO_4^{2-} + H_2O + 2 e \rightleftharpoons SO_3^{2-} + 2 OH^-$	−0.93	$Ti^{2+} + 2 e \rightleftharpoons Ti$	−1.630
$Sb + 3 H^+ + 3 e \rightleftharpoons SbH_3$	−0.510	$Ti^{3+} + e \rightleftharpoons Ti^{2+}$	−0.9
$Sb_2O_3 + 6 H^+ + 6 e \rightleftharpoons 2 Sb + 3 H_2O$	0.152	$TiO_2 + 4 H^+ + 2 e \rightleftharpoons Ti^{2+} + 2 H_2O$	−0.502
$Sb_2O_5 \text{ (senarmontite)} + 4 H^+ + 4 e \rightleftharpoons Sb_2O_3 + 2 H_2O$	0.671	$Ti^{3+} + 3 e \rightleftharpoons Ti$	−1.37
$Sb_2O_5 \text{ (valentinite)} + 4 H^+ + 4 e \rightleftharpoons Sb_2O_3 + 2 H_2O$	0.649	$TiOH^{3+} + H^+ + e \rightleftharpoons Ti^{3+} + H_2O$	−0.055
$Sb_2O_5 + 6 H^+ + 4 e \rightleftharpoons 2 SbO^+ + 3 H_2O$	0.581	$Tl^+ + e \rightleftharpoons Tl$	−0.336
$SbO^+ + 2 H^+ + 3 e \rightleftharpoons Sb + 2 H_2O$	0.212	$Tl^+ + e \rightleftharpoons Tl(Hg)$	−0.3338
$SbO_2^- + 2 H_2O + 3 e \rightleftharpoons Sb + 4 OH^-$	−0.66	$Tl^{3+} + 2 e \rightleftharpoons Tl^+$	1.252
$SbO_3^- + H_2O + 2 e \rightleftharpoons SbO_2^- + 2 OH^-$	−0.59	$Tl^{3+} + 3 e \rightleftharpoons Tl$	0.741
$Sc^{3+} + 3 e \rightleftharpoons Sc$	−2.077	$TlBr + e \rightleftharpoons Tl + Br^-$	−0.658
$Se + 2 e \rightleftharpoons Se^{2-}$	−0.924	$TlCl + e \rightleftharpoons Tl + Cl^-$	−0.5568
$Se + 2 H^+ + 2 e \rightleftharpoons H_2Se(aq)$	−0.399	$TlI + e \rightleftharpoons Tl + I^-$	−0.752
$H_2SeO_3 + 4 H^+ + 4 e \rightleftharpoons Se + 3 H_2O$	0.74	$Tl_2O_3 + 3 H_2O + 4 e \rightleftharpoons 2 Tl^+ + 6 OH^-$	0.02
$Se + 2 H^+ + 2 e \rightleftharpoons H_2Se$	−0.082	$TlOH + e \rightleftharpoons Tl + OH^-$	−0.34
$SeO_3^{2-} + 3 H_2O + 4 e \rightleftharpoons Se + 6 OH^-$	−0.366	$Tl(OH)_3 + 2 e \rightleftharpoons TlOH + 2 OH^-$	−0.05
$SeO_4^{2-} + 4 H^+ + 2 e \rightleftharpoons H_2SeO_3 + H_2O$	1.151	$Tl_2SO_4 + 2 e \rightleftharpoons Tl + SO_4^{2-}$	−0.4360
$SeO_4^{2-} + H_2O + 2 e \rightleftharpoons SeO_3^{2-} + 2 OH^-$	0.05	$Tm^{3+} + e \rightleftharpoons Tm^{2+}$	−2.2
$SiF_6^{2-} + 4 e \rightleftharpoons Si + 6 F^-$	−1.24	$Tm^{3+} + 3 e \rightleftharpoons Tm$	−2.319
$SiO + 2 H^+ + 2 e \rightleftharpoons Si + H_2O$	−0.8	$Tm^{2+} + 2 e \rightleftharpoons Tm$	−2.4

TABLE 1
Alphabetical Listing (continued)

Reaction	$E°/V$	Reaction	$E°/V$
$U^{3+} + 3e \rightleftharpoons U$	−1.798	$2 WO_3 + 2 H^+ + 2e \rightleftharpoons W_2O_5 + H_2O$	−0.029
$U^{4+} + e \rightleftharpoons U^{3+}$	−0.607	$H_4XeO_6 + 2 H^+ + 2e \rightleftharpoons XeO_3 + 3 H_2O$	2.42
$UO_2^+ + 4 H^+ + e \rightleftharpoons U^{4+} + 2 H_2O$	0.612	$XeO_3 + 6 H^+ + 6e \rightleftharpoons Xe + 3 H_2O$	2.10
$UO_2^{2+} + e \rightleftharpoons UO_2^+$	0.062	$XeF + e \rightleftharpoons Xe + F^-$	3.4
$UO_2^{2+} + 4 H^+ + 2e \rightleftharpoons U^{4+} + 2 H_2O$	0.327	$Y^{3+} + 3e \rightleftharpoons Y$	−2.372
$UO_2^{2+} + 4 H^+ + 6e \rightleftharpoons U + 2 H_2O$	−1.444	$Yb^{3+} + e \rightleftharpoons Yb^{2+}$	−1.05
$V^{2+} + 2e \rightleftharpoons V$	−1.175	$Yb^{3+} + 3e \rightleftharpoons Yb$	−2.19
$V^{3+} + e \rightleftharpoons V^{2+}$	−0.255	$Yb^{2+} + 2e \rightleftharpoons Yb$	−2.76
$VO^{2+} + 2 H^+ + e \rightleftharpoons V^{3+} + H_2O$	0.337	$Zn^{2+} + 2e \rightleftharpoons Zn$	−0.7618
$VO_2^+ + 2 H^+ + e \rightleftharpoons VO^{2+} + H_2O$	0.991	$Zn^{2+} + 2e \rightleftharpoons Zn(Hg)$	−0.7628
$V_2O_5 + 6 H^+ + 2e \rightleftharpoons 2 VO^{2+} + 3 H_2O$	0.957	$ZnO_2^{2-} + 2 H_2O + 2e \rightleftharpoons Zn + 4 OH^-$	−1.215
$V_2O_5 + 10 H^+ + 10e \rightleftharpoons 2 V + 5 H_2O$	−0.242	$ZnSO_4 \cdot 7 H_2O + 2e = Zn(Hg) + SO_4^{2-} + 7 H_2O$	−0.7993
$V(OH)_4^+ + 2 H^+ + e \rightleftharpoons VO^{2+} + 3 H_2O$	1.00	(Saturated $ZnSO_4$)	
$V(OH)_4^+ + 4 H^+ + 5e \rightleftharpoons V + 4 H_2O$	−0.254	$ZnOH^+ + H^+ + 2e \rightleftharpoons Zn + H_2O$	−0.497
$[V(phen)_3]^{3+} + e \rightleftharpoons [V(phen)_3]^{2+}$	0.14	$Zn(OH)_4^{2-} + 2e \rightleftharpoons Zn + 4 OH^-$	−1.199
$W^{3+} + 3e \rightleftharpoons W$	0.1	$Zn(OH)_2 + 2e \rightleftharpoons Zn + 2 OH^-$	−1.249
$W_2O_5 + 2 H^+ + 2e \rightleftharpoons 2 WO_2 + H_2O$	−0.031	$ZnO + H_2O + 2e \rightleftharpoons Zn + 2 OH^-$	−1.260
$WO_2 + 4 H^+ + 4e \rightleftharpoons W + 2 H_2O$	−0.119	$ZrO_2 + 4 H^+ + 4e \rightleftharpoons Zr + 2 H_2O$	−1.553
$WO_3 + 6 H^+ + 6e \rightleftharpoons W + 3 H_2O$	−0.090	$ZrO(OH)_2 + H_2O + 4e \rightleftharpoons Zr + 4 OH^-$	−2.36
$WO_3 + 2 H^+ + 2e \rightleftharpoons WO_2 + H_2O$	0.036	$Zr^{4+} + 4e \rightleftharpoons Zr$	−1.45

TABLE 2
Reduction Reactions Having $E°$ Values More Positive than that of the Standard Hydrogen Electrode

Reaction	$E°/V$	Reaction	$E°/V$
$2 H^+ + 2e \rightleftharpoons H_2$	0.00000	$Sn(OH)_3^+ + 3 H^+ + 2e \rightleftharpoons Sn^{2+} + 3 H_2O$	0.142
$CuI_2^- + e \rightleftharpoons Cu + 2 I^-$	0.00	$Np^{4+} + e \rightleftharpoons Np^{3+}$	0.147
$Ge^{4+} + 2e \rightleftharpoons Ge^{2+}$	0.00	$Ag_4[Fe(CN)_6] + 4e \rightleftharpoons 4 Ag + [Fe(CN)_6]^{4-}$	0.1478
$NO_3^- + H_2O + 2e \rightleftharpoons NO_2^- + 2 OH^-$	0.01	$IO_3^- + 2 H_2O + 4e \rightleftharpoons IO^- + 4 OH^-$	0.15
$Tl_2O_3 + 3 H_2O + 4e \rightleftharpoons 2 Tl^+ + 6 OH^-$	0.02	$Mn(OH)_3 + e \rightleftharpoons Mn(OH)_2 + OH^-$	0.15
$SeO_4^{2-} + H_2O + 2e \rightleftharpoons SeO_3^{2-} + 2 OH^-$	0.05	$2 NO_2^- + 3 H_2O + 4e \rightleftharpoons N_2O + 6 OH^-$	0.15
$WO_3 + 2 H^+ + 2e \rightleftharpoons WO_2 + H_2O$	0.036	$Sn^{4+} + 2e \rightleftharpoons Sn^{2+}$	0.151
$UO_2^{2+} + e = UO_2^+$	0.062	$Sb_2O_3 + 6 H^+ + 6e \rightleftharpoons 2 Sb + 3 H_2O$	0.152
$Pd(OH)_2 + 2e \rightleftharpoons Pd + 2 OH^-$	0.07	$Cu^{2+} + e \rightleftharpoons Cu^+$	0.153
$AgBr + e \rightleftharpoons Ag + Br^-$	0.07133	$BiOCl + 2 H^+ + 3e \rightleftharpoons Bi + Cl^- + H_2O$	0.1583
$MoO_3 + 6 H^+ + 6e \rightleftharpoons Mo + 3 H_2O$	0.075	$BiCl_4^- + 3e \rightleftharpoons Bi + 4 Cl^-$	0.16
$S_4O_6^{2-} + 2e \rightleftharpoons 2 S_2O_3^{2-}$	0.08	$Fe_2O_3 + 4 H^+ + 2e \rightleftharpoons 2 FeOH^+ + H_2O$	0.16
$H_3Mo_7O_{24}^{3-} + 45 H^+ + 42e \rightleftharpoons 7 Mo + 24 H_2O$	0.082	$Co(OH)_3 + e \rightleftharpoons Co(OH)_2 + OH^-$	0.17
$AgSCN + e \rightleftharpoons Ag + SCN^-$	0.8951	$SO_4^{2-} + 4 H^+ + 2e \rightleftharpoons H_2SO_3 + H_2O$	0.172
$N_2 + 2 H_2O + 6 H^+ + 6e \rightleftharpoons 2 NH_4OH$	0.092	$Bi^{3+} + 2e \rightleftharpoons Bi^+$	0.2
$HgO + H_2O + 2e \rightleftharpoons Hg + 2 OH^-$	0.0977	$[Ru(en)_3]^{3+} + e \rightleftharpoons [Ru(en)_3]^{2+}$	0.210
$Ir_2O_3 + 3 H_2O + 6e \rightleftharpoons 2 Ir + 6 OH^-$	0.098	$SbO^+ + 2 H^+ + 3e \rightleftharpoons Sb + 2 H_2O$	0.212
$2 NO + 2e \rightleftharpoons N_2O_2^{2-}$	0.10	$AgCl + e \rightleftharpoons Ag + Cl^-$	0.22233
$[Ru(NH_3)_6]^{3+} + e \rightleftharpoons [Ru(NH_3)_6]^{2+}$	0.10	$[Ru(H_2O)_6]^{3+} + e \rightleftharpoons [Ru(H_2O)_6]^{2+}$	0.23
$W^{3+} + 3e \rightleftharpoons W$	0.1	$As_2O_3 + 6 H^+ + 6e \rightleftharpoons 2 As + 3 H_2O$	0.234
$[Co(NH_3)_6]^{3+} + e \rightleftharpoons [Co(NH_3)_6]^{2+}$	0.108	Calomel electrode, saturated NaCl (SSCE)	0.2360
$Hg_2O + H_2O + 2e \rightleftharpoons 2 Hg + 2 OH^-$	0.123	$Ge^{2+} + 2e \rightleftharpoons Ge$	0.24
$Ge^{4+} + 4e \rightleftharpoons Ge$	0.124	$Ru^{3+} + e \rightleftharpoons Ru^{2+}$	0.24
$Hg_2Br_2 + 2e \rightleftharpoons 2 Hg + 2 Br^-$	0.13923	Calomel electrode, saturated KCl	0.2412
$Pt(OH)_2 + 2e \rightleftharpoons Pt + 2 OH^-$	0.14	$PbO_2 + H_2O + 2e \rightleftharpoons PbO + 2 OH^-$	0.247
$[V(phen)_3]^{3+} + e \rightleftharpoons [V(phen)_3]^{2+}$	0.14	$HAsO_2 + 3 H^+ + 3e \rightleftharpoons As + 2 H_2O$	0.248
$S + 2 H^+ + 2e \rightleftharpoons H_2S(aq)$	0.142	$Ru^{3+} + e \rightleftharpoons Ru^{2+}$	0.2487

TABLE 2
Reduction Reactions Having $E°$ Values More Positive than that of the Standard Hydrogen Electrode
(continued)

Reaction	$E°$/V	Reaction	$E°$/V
$ReO_2 + 4 H^+ + 4 e \rightleftharpoons Re + 2 H_2O$	0.2513	$[PdCl_4]^{2-} + 2 e \rightleftharpoons Pd + 4 Cl^-$	0.591
$IO_3^- + 3 H_2O + 6 e \rightleftharpoons I^- + OH^-$	0.26	$TeO_2 + 4 H^+ + 4 e \rightleftharpoons Te + 2 H_2O$	0.593
$Hg_2Cl_2 + 2 e \rightleftharpoons 2 Hg + 2 Cl^-$	0.26808	$MnO_4^- + 2 H_2O + 3 e \rightleftharpoons MnO_2 + 4 OH^-$	0.595
Calomel electrode, 1 molal KCl	0.2800	$Rh^{2+} + 2 e \rightleftharpoons Rh$	0.600
Calomel electrode, 1 molar KCl (NCE)	0.2801	$Rh^+ + e \rightleftharpoons Rh$	0.600
$At_2 + 2 e \rightleftharpoons 2 At^-$	0.3	$MnO_4^{2-} + 2 H_2O + 2 e \rightleftharpoons MnO_2 + 4 OH^-$	0.60
$Re^{3+} + 3 e \rightleftharpoons Re$	0.300	$2 AgO + H_2O + 2 e \rightleftharpoons Ag_2O + 2 OH^-$	0.607
$Tc^{3+} + e \rightleftharpoons Tc^{2+}$	0.3	$BrO_3^- + 3 H_2O + 6 e \rightleftharpoons Br^- + 6 OH^-$	0.61
$Bi^{3+} + 3 e \rightleftharpoons Bi$	0.308	$UO_2^+ + 4 H^+ + e \rightleftharpoons U^{4+} + 2 H_2O$	0.612
$BiO^+ + 2 H^+ + 3 e \rightleftharpoons Bi + H_2O$	0.320	$Hg_2SO_4 + 2 e \rightleftharpoons 2 Hg + SO_4^{2-}$	0.6125
$UO_2^{2+} + 4 H^+ + 2 e \rightleftharpoons U^{4+} + 2 H_2O$	0.327	$ClO_3^- + 3 H_2O + 6 e \rightleftharpoons Cl^- + 6 OH^-$	0.62
$ClO_3^- + H_2O + 2 e \rightleftharpoons ClO_2^- + 2 OH^-$	0.33	$Hg_2HPO_4 + 2 e \rightleftharpoons 2 Hg + HPO_4^{2-}$	0.6359
$2 HCNO + 2 H^+ + 2 e \rightleftharpoons (CN)_2 + 2 H_2O$	0.330	$Ag(ac) + e \rightleftharpoons Ag + (ac)^-$	0.643
Calomel electrode, 0.1 molar KCl	0.3337	$Sb_2O_5(valentinite) + 4 H^+ + 4 e \rightleftharpoons Sb_2O_3 + 2 H_2O$	0.649
$VO^{2+} + 2 H^+ + e \rightleftharpoons V^{3+} + H_2O$	0.337	$Ag_2SO_4 + 2 e \rightleftharpoons 2 Ag + SO_4^{2-}$	0.654
$Cu^{2+} + 2 e \rightleftharpoons Cu$	0.3419	$ClO_2^- + H_2O + 2 e \rightleftharpoons ClO^- + 2 OH^-$	0.66
$Ag_2O + H_2O + 2 e \rightleftharpoons 2 Ag + 2 OH^-$	0.342	$Sb_2O_5(senarmontite) + 4 H^+ + 4 e \rightleftharpoons Sb_2O_5 + 2 H_2O$	0.671
$Cu^{2+} + 2 e \rightleftharpoons Cu(Hg)$	0.345	$[PtCl_6]^{2-} + 2 e \rightleftharpoons [PtCl_4]^{2-} + 2 Cl^-$	0.68
$AgIO_3 + e \rightleftharpoons Ag + IO_3^-$	0.354	$O_2 + 2 H^+ + 2 e \rightleftharpoons H_2O_2$	0.695
$[Fe(CN)_6]^{3-} + e \rightleftharpoons [Fe(CN)_6]^{4-}$	0.358	p-benzoquinone $+ 2 H^+ + 2 e \rightleftharpoons$ hydroquinone	0.6992
$ClO_4^- + H_2O + 2 e \rightleftharpoons ClO_3^- + 2 OH^-$	0.36	$H_3IO_6^{2-} + 2 e \rightleftharpoons IO_3^- + 3 OH^-$	0.7
$Ag_2SeO_3 + 2 e \rightleftharpoons 2 Ag + SeO_3^{2-}$	0.3629	$Ag_2O_3 + H_2O + 2 e \rightleftharpoons 2 AgO + 2 OH^-$	0.739
$ReO_4^- + 8 H^+ + 7 e \rightleftharpoons Re + 4 H_2O$	0.368	$Tl^{3+} + 3 e \rightleftharpoons Tl$	0.741
$(CN)_2 + 2 H^+ + 2 e \rightleftharpoons 2 HCN$	0.373	$[PtCl_4]^{2-} + 2 e \rightleftharpoons Pt + 4 Cl^-$	0.755
$[Ferricinium]^+ + e \rightleftharpoons$ ferrocene	0.400	$Rh^{3+} + 3 e \rightleftharpoons Rh$	0.758
$Tc^{2+} + 2 e \rightleftharpoons Tc$	0.400	$ClO_2^- + 2 H_2O + 4 e \rightleftharpoons Cl^- + 4 OH^-$	0.76
$O_2 + 2 H_2O + 4 e \rightleftharpoons 4 OH^-$	0.401	$2 NO + H_2O + 2 e \rightleftharpoons N_2O + 2 OH^-$	0.76
$AgOCN + e \rightleftharpoons Ag + OCN^-$	0.41	$Po^{4+} + 4 e \rightleftharpoons Po$	0.76
$[RhCl_6]^{3-} + 3 e \rightleftharpoons Rh + 6 Cl^-$	0.431	$BrO^- + H_2O + 2 e \rightleftharpoons Br^- + 2 OH^-$	0.761
$Ag_2CrO_4 + 2 e \rightleftharpoons 2 Ag + CrO_4^{2-}$	0.4470	$ReO_4^- + 2 H^+ + e \rightleftharpoons ReO_3 + H_2O$	0.768
$H_2SO_3 + 4 H^+ + 4 e \rightleftharpoons S + 3 H_2O$	0.449	$(CNS)_2 + 2 e \rightleftharpoons 2 CNS^-$	0.77
$Ru^{2+} + 2 e \rightleftharpoons Ru$	0.455	$[IrCl_6]^{3-} + 3 e \rightleftharpoons Ir + 6 Cl^-$	0.77
$Ag_2MoO_4 + 2 e \rightleftharpoons 2 Ag + MoO_4^{2-}$	0.4573	$Fe^{3+} + e \rightleftharpoons Fe^{2+}$	0.771
$Ag_2C_2O_4 + 2 e \rightleftharpoons 2 Ag + C_2O_4^{2-}$	0.4647	$AgF + e \rightleftharpoons Ag + F^-$	0.779
$Ag_2WO_4 + 2 e \rightleftharpoons 2 Ag + WO_4^{2-}$	0.4660	$[Fe(bipy)_2]^{3+} + e \rightleftharpoons [Fe(bipy)_2]^{2+}$	0.78
$Ag_2CO_3 + 2 e \rightleftharpoons 2 Ag + CO_3^{2-}$	0.47	$TcO_4^- + 4 H^+ + 3 e \rightleftharpoons TcO_2 + 2 H_2O$	0.782
$TcO_4^- + 8 H^+ + 7 e \rightleftharpoons Tc + 4 H_2O$	0.472	$Hg_2^{2+} + 2 e \rightleftharpoons 2 Hg$	0.7973
$TeO_4^- + 8 H^+ + 7 e \rightleftharpoons Te + 4 H_2O$	0.472	$Ag^+ + e \rightleftharpoons Ag$	0.7996
$IO^- + H_2O + 2 e \rightleftharpoons I^- + 2 OH^-$	0.485	$[Os(bipy)_3]^{3+} + e \rightleftharpoons [Os(bipy)_3]^{2+}$	0.80
$NiO_2 + 2 H_2O + 2 e \rightleftharpoons Ni(OH)_2 + 2 OH^-$	0.490	$2 NO_3^- + 4 H^+ + 2 e \rightleftharpoons N_2O_4 + 2 H_2O$	0.803
$Bi^+ + e \rightleftharpoons Bi$	0.5	$[Os(bipy)_2]^{3+} + e \rightleftharpoons [Os(bipy)_2]^{2+}$	0.81
$ReO_4^- + 4 H^+ + 3 e \rightleftharpoons ReO_2 + 2 H_2O$	0.510	$RhOH^{2+} + H + 3 e \rightleftharpoons Rh + H_2O$	0.83
$Hg_2(ac)_2 + 2 e \rightleftharpoons 2 Hg + 2(ac)^-$	0.51163	$OsO_4 + 8 H^+ + 8 e \rightleftharpoons Os + 4 H_2O$	0.838
$Cu^+ + e \rightleftharpoons Cu$	0.521	$ClO^- + H_2O + 2 e \rightleftharpoons Cl^- + 2 OH^-$	0.841
$I_2 + 2 e \rightleftharpoons 2 I^-$	0.5355	$Hg^{2+} + 2 e \rightleftharpoons Hg$	0.851
$I_3^- + 2 e \rightleftharpoons 3 I^-$	0.536	$AuBr_4^- + 3 e \rightleftharpoons Au + 4 Br^-$	0.854
$AgBrO_3 + e \rightleftharpoons Ag + BrO_3^-$	0.546	$SiO_2(quartz) + 4 H^+ + 4 e \rightleftharpoons Si + 2 H_2O$	0.857
$MnO_4^- + e \rightleftharpoons MnO_2^-$	0.558	$2 HNO_2 + 4 H^+ + 4 e \rightleftharpoons H_2N_2O_2 + H_2O$	0.86
$H_3AsO_4 + 2 H^+ + 2 e \rightleftharpoons HAsO_2 + 2 H_2O$	0.560	$[Ru(CN)_6]^{3-} + e^- \rightleftharpoons [Ru(CN)_6]^{4-}$	0.86
$S_2O_6^{2-} + 4 H^+ + 2 e \rightleftharpoons 2 H_2SO_3$	0.564	$[IrCl_6]^{2-} + e \rightleftharpoons [IrCl_6]^{3-}$	0.8665
$AgNO_2 + e \rightleftharpoons Ag + NO_2^-$	0.564	$N_2O_4 + 2 e \rightleftharpoons 2 NO_2^-$	0.867
$Te^{4+} + 4 e \rightleftharpoons Te$	0.568	$HO_2^- + H_2O + 2 e \rightleftharpoons 3 OH^-$	0.878
$Sb_2O_5 + 6 H^+ + 4 e \rightleftharpoons 2 SbO^+ + 3 H_2O$	0.581	$Po^{4+} + 2 e \rightleftharpoons Po^{2+}$	0.9
$RuO_4^- + e \rightleftharpoons RuO_4^{2-}$	0.59	$2 Hg^+ + 2 e \rightleftharpoons Hg_2^{2+}$	0.920

TABLE 2
Reduction Reactions Having $E°$ Values More Positive than that of the Standard Hydrogen Electrode
(continued)

Reaction	$E°/V$	Reaction	$E°/V$
$NO_3^- + 3 H^+ + 2 e \rightleftharpoons HNO_2 + H_2O$	0.934	$Cl_2(g) + 2 e \rightleftharpoons 2 Cl^-$	1.35827
$Pd^{2+} + 2 e \rightleftharpoons Pd$	0.951	$ClO_4^- + 8 H^+ + 8 e \rightleftharpoons Cl^- + 4 H_2O$	1.389
$ClO_2(aq) + e \rightleftharpoons ClO_2^-$	0.954	$ClO_4^- + 8 H^+ + 7 e \rightleftharpoons 1/2 Cl_2 + 4 H_2O$	1.39
$NO_3^- + 4 H^+ + 3 e \rightleftharpoons NO + 2 H_2O$	0.957	$No^{3+} + e \rightleftharpoons No^{2+}$	1.4
$V_2O_5 + 6 H^+ + 2 e \rightleftharpoons 2 VO^{2+} + 3 H_2O$	0.957	$RuO_4 + 6 H^+ + 4 e \rightleftharpoons Ru(OH)_2^{2+} + 2 H_2O$	1.40
$AuBr_2^- + e \rightleftharpoons Au + 2 Br^-$	0.959	$Au^{3+} + 2 e \rightleftharpoons Au^+$	1.401
$HNO_2 + H^+ + e \rightleftharpoons NO + H_2O$	0.983	$2 NH_3OH^+ + H^+ + 2 e \rightleftharpoons N_2H_5^+ + 2 H_2O$	1.42
$HIO + H^+ + 2 e \rightleftharpoons I^- + H_2O$	0.987	$BrO_3^- + 6 H^+ + 6 e \rightleftharpoons Br^- + 3 H_2O$	1.423
$VO_2^+ + 2 H^+ + e \rightleftharpoons VO^{2+} + H_2O$	0.991	$2 HIO + 2 H^+ + 2 e \rightleftharpoons I_2 + 2 H_2O$	1.439
$PtO_2 + 4 H^+ + 4 e \rightleftharpoons Pt + 2 H_2O$	1.00	$Au(OH)_3 + 3 H^+ + 3 e \rightleftharpoons Au + 3 H_2O$	1.45
$RuO_4 + e \rightleftharpoons RuO_4^-$	1.00	$3 IO_3^- + 6 H^+ + 6 e \rightleftharpoons Cl^- + 3 H_2O$	1.451
$V(OH)_4^+ + 2 H^+ + e \rightleftharpoons VO^{2+} + 3 H_2O$	1.00	$PbO_2 + 4 H^+ + 2 e \rightleftharpoons Pb^{2+} + 2 H_2O$	1.455
$AuCl_4^- + 3 e \rightleftharpoons Au + 4 Cl^-$	1.002	$ClO_3^- + 6 H^+ + 5 e \rightleftharpoons 1/2 Cl_2 + 3 H_2O$	1.47
$Pu^{4+} + e \rightleftharpoons Pu^{3+}$	1.006	$CrO_2 + 4 H^+ + e \rightleftharpoons Cr^{3+} + 2 H_2O$	1.48
$PtO_2 + 2 H^+ + 2 e \rightleftharpoons PtO + H_2O$	1.01	$BrO_3^- + 6 H^+ + 5 e \rightleftharpoons 1/2 Br_2 + 3 H_2O$	1.482
$OsO_4 + 4 H + 4 e \rightleftharpoons OsO_2 + 2 H_2O$	1.02	$HClO + H^+ + 2 e \rightleftharpoons Cl^- + H_2O$	1.482
$H_6TeO_6 + 2 H^+ + 2 e \rightleftharpoons TeO_2 + 4 H_2O$	1.02	$Mn_2O_3 + 6 H^+ + e \rightleftharpoons 2 Mn^{2+} + 3 H_2O$	1.485
$[Fe(bipy)_3]^{3+} + e \rightleftharpoons [Fe(bipy)_3]^{2+}$	1.03	$HO_2 + H^+ + e \rightleftharpoons H_2O_2$	1.495
$Hg(OH)_2 + 2 H^+ + 2 e \rightleftharpoons Hg + 2 H_2O$	1.034	$Au^{3+} + 3 e \rightleftharpoons Au$	1.498
$N_2O_4 + 4 H^+ + 4 e \rightleftharpoons 2 NO + 2 H_2O$	1.035	$PtO_3 + 4 H^+ + 2 e \rightleftharpoons Pt(OH)_2^{2+} + H_2O$	1.5
$RuO_4 + 8 H^+ + 8 e \rightleftharpoons Ru + 4 H_2O$	1.038	$MnO_4^- + 8 H^+ + 5 e \rightleftharpoons Mn^{2+} + 4 H_2O$	1.507
$[Fe(phen)_3]^{3+} + e \rightleftharpoons [Fe(phen)_3]^{2+}$ (1 molar H_2SO_4)	1.06	$Mn^{3+} + e \rightleftharpoons Mn^{2-}$	1.5415
$PuO_2(OH)_2 + H^+ + e \rightleftharpoons PuO_2OH + H_2O$	1.062	$HClO_2 + 3 H^+ + 4 e \rightleftharpoons Cl^- + 2 H_2O$	1.570
$N_2O_4 + 2 H^+ + 2 e \rightleftharpoons 2 HNO_2$	1.065	$HBrO + H^+ + e \rightleftharpoons 1/2 Br_2(aq) + H_2O$	1.574
$Br_2(l) + 2 e \rightleftharpoons 2 Br^-$	1.066	$2 NO + 2 H^+ + 2 e \rightleftharpoons N_2O + H_2O$	1.591
$IO_3^- + 6 H^+ + 6 e \rightleftharpoons I^- + 3 H_2O$	1.085	$Bi_2O_4 + 4 H^+ + 2 e \rightleftharpoons 2 BiO^+ + 2 H_2O$	1.593
$Br_2(aq) + 2 e \rightleftharpoons 2 Br^-$	1.0873	$HBrO + H^+ + e \rightleftharpoons 1/2 Br_2(l) + H_2O$	1.596
$Pu^{5+} + e \rightleftharpoons Pu^{4+}$	1.099	$H_5IO_6 + H^+ + 2 e \rightleftharpoons IO_3^- + 3 H_2O$	1.601
$Cu^{2+} + 2 CN^- + e \rightleftharpoons [Cu(CN)_2]^-$	1.103	$HClO + H^+ + e \rightleftharpoons 1/2 Cl_2 + H_2O$	1.611
$RuO_2 + 4 H^+ + 2 e \rightleftharpoons Ru^{2+} + 2 H_2O$	1.120	$HClO_2 + 3 H^+ + 3 e \rightleftharpoons 1/2 Cl_2 + 2 H_2O$	1.628
$[Fe(phen)_3]^{3+} + e \rightleftharpoons [Fe(phen)_3]^{2+}$	1.147	$HClO_2 + 2 H^+ + 2 e \rightleftharpoons HClO + H_2O$	1.645
$SeO_4^{2-} + 4 H^+ + 2 e \rightleftharpoons H_2SeO_3 + H_2O$	1.151	$Bk^{4+} + e \rightleftharpoons Bk^{3+}$	1.67
$ClO_3^- + 2 H^+ + e \rightleftharpoons ClO_2 + H_2O$	1.152	$NiO_2 + 4 H^+ + 2 e \rightleftharpoons Ni^{2+} + 2 H_2O$	1.678
$Ir^{3+} + 3 e \rightleftharpoons Ir$	1.156	$MnO_4^- + 4 H^+ + 3 e \rightleftharpoons MnO_2 + 2 H_2O$	1.679
$Pt^{2+} + 2 e \rightleftharpoons Pt$	1.18	$PbO_2 + SO_4^{2-} + 4 H^+ + 2 e \rightleftharpoons PbSO_4 + 2 H_2O$	1.6913
$ClO_4^- + 2 H^+ + 2 e \rightleftharpoons ClO_3^- + H_2O$	1.189	$Au^+ + e \rightleftharpoons Au$	1.692
$2 IO_3^- + 12 H^+ + 10 e \rightleftharpoons I_2 + 6 H_2O$	1.195	$PtO_3 + 2 H^+ + 2 e \rightleftharpoons PtO_2 + H_2O$	1.7
$PtOH^+ + H^+ + 2 e \rightleftharpoons Pt + H_2O$	1.2	$CeOH^{3+} + H^+ + e \rightleftharpoons Ce^{3+} + H_2O$	1.715
$ClO_3^- + 3 H^+ + 2 e \rightleftharpoons HClO_2 + H_2O$	1.214	$Ce^{4+} + e \rightleftharpoons Ce^{3+}$	1.72
$MnO_2 + 4 H^+ + 2 e \rightleftharpoons Mn^{2+} + 2 H_2O$	1.224	$N_2O + 2 H^+ + 2 e \rightleftharpoons N_2 + H_2O$	1.766
$O_2 + 4 H^+ + 4 e \rightleftharpoons 2 H_2O$	1.229	$H_2O_2 + 2 H^+ + 2 e \rightleftharpoons 2 H_2O$	1.776
$Cr_2O_7^{2-} + 14 H^+ + 6 e \rightleftharpoons 2 Cr^{3+} + 7 H_2O$	1.232	$Ag^{3+} + e \rightleftharpoons Ag^{2+}$	1.8
$O_3 + H_2O + 2 e \rightleftharpoons O_2 + 2 OH^-$	1.24	$Au^{2+} + e^- \rightleftharpoons Au^+$	1.8
$[Ru(bipy)_3]^{3+} + e \rightleftharpoons [Ru(bipy)_3]^{2+}$	1.24	$Ag_2O_2 + 4 H^+ + e \rightleftharpoons 2 Ag + 2 H_2O$	1.802
$Tl^{3+} + 2 e \rightleftharpoons Tl^+$	1.252	$Co^{3+} + e \rightleftharpoons Co^{2-}$ (2 molar H_2SO_4)	1.83
$N_2H_5^+ + 3 H^+ + 2 e \rightleftharpoons 2 NH_4^+$	1.275	$Ag^{3+} + 2 e \rightleftharpoons Ag^+$	1.9
$ClO_2 + H^+ + e \rightleftharpoons HClO_2$	1.277	$Co^{3+} + e \rightleftharpoons Co^{2+}$	1.92
$[PdCl_6]^{2-} + 2 e \rightleftharpoons [PdCl_4]^{2-} + 2 Cl^-$	1.288	$Ag^{2+} + e \rightleftharpoons Ag^+$	1.980
$2 HNO_2 + 4 H^+ + 4 e \rightleftharpoons N_2O + 3 H_2O$	1.297	$Cu_2O_3 + 6 H^+ + 2 e \rightleftharpoons 2 Cu^{2+} + 3 H_2O$	2.0
$AuOH^{2+} + H^+ + 2 e \rightleftharpoons Au^+ + H_2O$	1.32	$S_2O_8^{2-} + 2 e \rightleftharpoons 2 SO_4^{2-}$	2.010
$PuO_2(OH)_2 + 2 H^- + 2 e \rightleftharpoons Pu(OH)_4$	1.325	$OH + e \rightleftharpoons OH^-$	2.02
$HBrO + H^+ + 2 e \rightleftharpoons Br^- + H_2O$	1.331	$HFeO_4^- + 7 H^+ + 3 e \rightleftharpoons Fe^{3+} + 4 H_2O$	2.07
$Cr(V) + e \rightleftharpoons Cr(IV)$	1.34	$O_3 + 2 H^+ + 2 e \rightleftharpoons O_2 + H_2O$	2.076
$HCrO_4^- + 7 H^+ + 3 e \rightleftharpoons Cr^{3+} + 4 H_2O$	1.350	$HFeO_4^- + 4 H^+ + 3 e \rightleftharpoons FeOOH + 2 H_2O$	2.08

TABLE 2
Reduction Reactions Having $E°$ Values More Positive than that of the Standard Hydrogen Electrode (continued)

Reaction	$E°/V$	Reaction	$E°/V$
$2 HFeO_4^- + 8 H^+ + 6 e \rightleftharpoons Fe_2O_3 + 5 H_2O$	2.09	$H_2N_2O_2 + 2 H^+ + 2 e \rightleftharpoons N_2 + 2 H_2O$	2.65
$XeO_3 + 6 H^+ + 6 e \rightleftharpoons Xe + 3 H_2O$	2.10	$F_2 + 2 e \rightleftharpoons 2 F^-$	2.866
$S_2O_8^{2-} + 2 H^+ + 2 e \rightleftharpoons 2 HSO_4^-$	2.123	$Cm^{4+} + e \rightleftharpoons Cm^{3+}$	3.0
$F_2O + 2 H^+ + 4 e \rightleftharpoons H_2O + 2 F^-$	2.153	$F_2 + 2 H^+ + 2 e \rightleftharpoons 2 HF$	3.053
$FeO_4^{2-} + 8 H^+ + 3 e \rightleftharpoons Fe^{3+} + 4 H_2O$	2.20	$Tb^{4+} + e \rightleftharpoons Tb^{3+}$	3.1
$Cu^{3+} + e \rightleftharpoons Cu^{2+}$	2.4	$Pr^{4+} + e \rightleftharpoons Pr^{3+}$	3.2
$H_4XeO_6 + 2 H^+ + 2 e \rightleftharpoons XeO_3 + 3 H_2O$	2.42	$Cf^{4+} + e \rightleftharpoons Cf^{3+}$	3.3
$O(g) + 2 H^+ + 2 e \rightleftharpoons H_2O$	2.421	$XeF + e \rightleftharpoons Xe + F^-$	3.4
$Am^{4+} + e \rightleftharpoons Am^{3+}$	2.60		

TABLE 3
Reduction Reactions Having $E°$ Values More Negative than that of the Standard Hydrogen Electrode

Reaction	$E°/V$	Reaction	$E°/V$
$2 H^+ + 2 e \rightleftharpoons H_2$	0.00000	$Cu(OH)_2 + 2 e \rightleftharpoons Cu + 2 OH^-$	-0.222
$2 D^+ + 2 e \rightleftharpoons D_2$	-0.013	$V_2O_5 + 10 H^+ + 10 e \rightleftharpoons 2 V + 5 H_2O$	-0.242
$AgCN + e \rightleftharpoons Ag + CN^-$	-0.017	$CdSO_4 + 2 e \rightleftharpoons Cd + SO_4^{2-}$	-0.246
$2 WO_3 + 2 H^+ + 2 e \rightleftharpoons W_2O_5 + H_2O$	-0.029	$V(OH)_4^+ + 4 H^+ + 5 e \rightleftharpoons V + 4 H_2O$	-0.254
$W_2O_5 + 2 H^+ + 2 e \rightleftharpoons 2 WO_2 + H_2O$	-0.031	$V^{3+} + e \rightleftharpoons V^{2+}$	-0.255
$Ag_2S + 2 H^+ + 2 e \rightleftharpoons 2 Ag + H_2S$	-0.0366	$Ni^{2+} + 2 e \rightleftharpoons Ni$	-0.257
$Fe^{3+} + 3 e \rightleftharpoons Fe$	-0.037	$PbCl_2 + 2 e \rightleftharpoons Pb + 2 Cl^-$	-0.2675
$Hg_2I_2 + 2 e \rightleftharpoons 2 Hg + 2 I^-$	-0.0405	$H_3PO_4 + 2 H^+ + 2 e \rightleftharpoons H_3PO_3 + H_2O$	-0.276
$Tl(OH)_3 + 2 e \rightleftharpoons TlOH + 2 OH^-$	-0.05	$Co^{2+} + 2 e \rightleftharpoons Co$	-0.28
$TiOH^{3+} + H^+ + e \rightleftharpoons Ti^{3+} + H_2O$	-0.055	$PbBr_2 + 2 e \rightleftharpoons Pb + 2 Br^-$	-0.284
$2 H_2SO_3 + H^+ + 2 e \rightleftharpoons HS_2O_4^- + 2 H_2O$	-0.056	$Tl^+ + e \rightleftharpoons Tl(Hg)$	-0.3338
$P(white) + 3 H^+ + 3 e \rightleftharpoons PH_3(g)$	-0.063	$Tl^+ + e \rightleftharpoons Tl$	-0.336
$O_2 + H_2O + 2 e \rightleftharpoons HO_2^- + OH^-$	-0.076	$In^{3+} + 3 e \rightleftharpoons In$	-0.3382
$2 Cu(OH)_2 + 2 e \rightleftharpoons Cu_2O + 2 OH^- + H_2O$	-0.080	$TlOH + e \rightleftharpoons Tl + OH^-$	-0.34
$Se + 2 H^+ + 2 e \rightleftharpoons H_2Se$	-0.082	$PbF_2 + 2 e \rightleftharpoons Pb + 2 F^-$	-0.3444
$WO_3 + 6 H^+ + 6 e \rightleftharpoons W + 3 H_2O$	-0.090	$PbSO_4 + 2 e \rightleftharpoons Pb(Hg) + SO_4^{2-}$	-0.3505
$SnO_2 + 4 H^+ + 2 e \rightleftharpoons Sn^{2+} + 2 H_2O$	-0.094	$Cd^{2+} + 2 e \rightleftharpoons Cd(Hg)$	-0.3521
$Md^{3+} + e \rightleftharpoons Md^{2+}$	-0.1	$PbSO_4 + 2 e \rightleftharpoons Pb + SO_4^{2-}$	-0.3588
$P(red) + 3 H^+ + 3 e \rightleftharpoons PH_3(g)$	-0.111	$Cu_2O + H_2O + 2 e \rightleftharpoons 2 Cu + 2 OH^-$	-0.360
$SnO_2 + 4 H^+ + 4 e \rightleftharpoons Sn + 2 H_2O$	-0.117	$Eu^{3+} + e \rightleftharpoons Eu^{2+}$	-0.36
$GeO_2 + 2 H^+ + 2 e \rightleftharpoons GeO + H_2O$	-0.118	$PbI_2 + 2 e \rightleftharpoons Pb + 2 I^-$	-0.365
$WO_2 + 4 H^+ + 4 e \rightleftharpoons W + 2 H_2O$	-0.119	$SeO_3^{2-} + 3 H_2O + 4 e \rightleftharpoons Se + 6 OH^-$	-0.366
$Pb^{2+} + 2 e \rightleftharpoons Pb(Hg)$	-0.1205	$Se + 2 H^+ + 2 e \rightleftharpoons H_2Se(aq)$	-0.399
$Pb^{2+} + 2 e \rightleftharpoons Pb$	-0.1262	$In^{2+} + e \rightleftharpoons In^+$	-0.40
$CrO_4^{2-} + 4 H_2O + 3 e \rightleftharpoons Cr(OH)_3 + 5 OH^-$	-0.13	$Cd^{2+} + 2 e \rightleftharpoons Cd$	-0.4030
$Sn^{2-} + 2 e \rightleftharpoons Sn$	-0.1375	$Cr^{3+} + e \rightleftharpoons Cr^{2+}$	-0.407
$In^+ + e \rightleftharpoons In$	-0.14	$2 S + 2 e \rightleftharpoons S_2^{2-}$	-0.42836
$O_2 + 2 H_2O + 2 e \rightleftharpoons H_2O_2 + 2 OH^-$	-0.146	$Tl_2SO_4 + 2 e \rightleftharpoons Tl + SO_4^{2-}$	-0.4360
$MoO_2 + 4 H^+ + 4 e \rightleftharpoons Mo + 4 H_2O$	-0.152	$In^{3+} + 2 e \rightleftharpoons In^+$	-0.443
$AgI + e \rightleftharpoons Ag + I^-$	-0.15224	$Fe^{2+} + 2 e \rightleftharpoons Fe$	-0.447
$2 NO_2^- + 2 H_2O + 4 e \rightleftharpoons N_2O_2^{2-} + 4 OH^-$	-0.18	$H_3PO_3 + 3 H^+ + 3 e \rightleftharpoons P + 3 H_2O$	-0.454
$H_2GeO_3 + 4 H^+ + 4 e \rightleftharpoons Ge + 3 H_2O$	-0.182	$Bi_2O_3 + 3 H_2O + 6 e \rightleftharpoons 2 Bi + 6 OH^-$	-0.46
$SnO_2 + 3 H^+ + 2 e \rightleftharpoons SnOH^+ + H_2O$	-0.194	$NO_2^- + H_2O + e \rightleftharpoons NO + 2 OH$	-0.46
$CO_2 + 2 H^+ + 2 e \rightleftharpoons HCOOH$	-0.199	$PbHPO_4 + 2 e \rightleftharpoons Pb + HPO_4^{2-}$	-0.465
$Mo^{3+} + 3 e \rightleftharpoons Mo$	-0.200	$S + 2 e \rightleftharpoons S^{2-}$	-0.47627
$Ga^+ + e \rightleftharpoons Ga$	-0.2	$S + H_2O + 2 e \rightleftharpoons HS^- + OH^-$	-0.478
$2 SO_2^{2-} + 4 H^+ + 2 e \rightleftharpoons S_2O_6^{2-} + H_2O$	-0.22	$B(OH)_3 + 7 H^+ + 8 e \rightleftharpoons BH_4^- + 3 H_2O$	-0.481

TABLE 3
Reduction Reactions Having $E°$ Values More Negative than that of the Standard Hydrogen Electrode
(continued)

Reaction	$E°/V$	Reaction	$E°/V$
$In^{3+} + e \rightleftharpoons In^{2+}$	-0.49	$SnO_2 + 2 H_2O + 4 e \rightleftharpoons Sn + 4 OH^-$	-0.945
$ZnOH^+ + H^+ + 2 e \rightleftharpoons Zn + H_2O$	-0.497	$In(OH)_3 + 3 e \rightleftharpoons In + 3 OH^-$	-0.99
$GaOH^{2+} + H^+ + 3 e \rightleftharpoons Ga + H_2O$	-0.498	$NpO_2 + H_2O + H^+ + e \rightleftharpoons Np(OH)_3$	-0.962
$H_3PO_3 + 2 H^+ + 2 e \rightleftharpoons H_3PO_2 + H_2O$	-0.499	$In(OH)_4^- + 3 e \rightleftharpoons In + 4 OH^-$	-1.007
$TiO_2 + 4 H^+ + 2 e \rightleftharpoons Ti^{2+} + 2 H_2O$	-0.502	$In_2O_3 + 3 H_2O + 6 e \rightleftharpoons 2 In + 6 OH^-$	-1.034
$H_3PO_2 + H^+ + e \rightleftharpoons P + 2 H_2O$	-0.508	$PO_4^{3-} + 2 H_2O + 2 e \rightleftharpoons HPO_3^{2-} + 3 OH^-$	-1.05
$Sb + 3 H^+ + 3 e \rightleftharpoons SbH_3$	-0.510	$Yb^{3+} + e \rightleftharpoons Yb^{2+}$	-1.05
$HPbO_2^- + H_2O + 2 e \rightleftharpoons Pb + 3 OH^-$	-0.537	$Nb^{3+} + 3 e \rightleftharpoons Nb$	-1.099
$Ga^{3+} + 3 e \rightleftharpoons Ga$	-0.549	$Fm^{3+} + e \rightleftharpoons Fm^{2+}$	-1.1
$TlCl + e \rightleftharpoons Tl + Cl^-$	-0.5568	$2 SO_3^{2-} + 2 H_2O + 2 e \rightleftharpoons S_2O_4^{2-} + 4 OH^-$	-1.12
$Fe(OH)_3 + e \rightleftharpoons Fe(OH)_2 + OH^-$	-0.56	$Te + 2 e \rightleftharpoons Te^{2-}$	-1.143
$TeO_3^{2-} + 3 H_2O + 4 e \rightleftharpoons Te + 6 OH^-$	-0.57	$V^{2+} + 2 e \rightleftharpoons V$	-1.175
$2 SO_3^{2-} + 3 H_2O + 4 e \rightleftharpoons S_2O_3^{2-} + 6 OH^-$	-0.571	$Mn^{2+} + 2 e \rightleftharpoons Mn$	-1.185
$PbO + H_2O + 2 e \rightleftharpoons Pb + 2 OH^-$	-0.580	$Zn(OH)_4^{2-} + 2 e \rightleftharpoons Zn + 40 H^-$	-1.199
$ReO_2^- + 4 H_2O + 7 e \rightleftharpoons Re + 8 OH^-$	-0.584	$CrO_2 + 2 H_2O + 3 e \rightleftharpoons Cr + 4 OH^-$	-1.2
$SbO_3^- + H_2O + 2 e \rightleftharpoons SbO_2^- + 2 OH^-$	-0.59	$No^{3+} + 3 e \rightleftharpoons No$	-1.20
$Ta^{3+} + 3 e \rightleftharpoons Ta$	-0.6	$ZnO_2^- + 2 H_2O + 2 e \rightleftharpoons Zn + 4 OH^-$	-1.215
$U^{4+} + e \rightleftharpoons U^{3+}$	-0.607	$H_2GaO_3^- + H_2O + 3 e \rightleftharpoons Ga + 4 OH^-$	-1.219
$As + 3 H^+ + 3 e \rightleftharpoons AsH_3$	-0.608	$H_2BO_3^- + 5 H_2O + 8 e \rightleftharpoons BH_4^- + 8 OH^-$	-1.24
$Nb_2O_5 + 10 H^+ + 10 e \rightleftharpoons 2 Nb + 5 H_2O$	-0.644	$SiF_6^{2-} + 4 e \rightleftharpoons Si + 6 F^-$	-1.24
$NbO_2 + 2 H^+ + 2 e \rightleftharpoons NbO + H_2O$	-0.646	$Zn(OH)_2 + 2 e \rightleftharpoons Zn + 2 OH^-$	-1.249
$Cd(OH)_4^{2-} + 2 e \rightleftharpoons Cd + 4 OH^-$	-0.658	$ZnO + H_2O + 2 e \rightleftharpoons Zn + 2 OH^-$	-1.260
$TlBr + e \rightleftharpoons Tl + Br^-$	-0.658	$Es^{3+} + e \rightleftharpoons Es^{2+}$	-1.3
$SbO_2^- + 2 H_2O + 3 e \rightleftharpoons Sb + 4 OH^-$	-0.66	$Pa^{3+} + 3 e \rightleftharpoons Pa$	-1.34
$AsO_2^- + 2 H_2O + 3 e \rightleftharpoons As + 4 OH^-$	-0.68	$Ti^{3+} + 3 e \rightleftharpoons Ti$	-1.37
$NbO_2 + 4 H^+ + 4 e \rightleftharpoons Nb + 2 H_2O$	-0.690	$Ce^{3+} + 3 e \rightleftharpoons Ce(Hg)$	-1.4373
$Ag_2S + 2 e \rightleftharpoons 2 Ag + S^{2-}$	-0.691	$UO_2^{2+} + 4 H^+ + 6 e \rightleftharpoons U + 2 H_2O$	-1.444
$AsO_4^{3-} + 2 H_2O + 2 e \rightleftharpoons AsO_2^- + 4 OH^-$	-0.71	$Zr^{4+} + 4 e \rightleftharpoons Zr$	-1.45
$Ni(OH)_2 + 2 e \rightleftharpoons Ni + 2 OH^-$	-0.72	$Cr(OH)_3 + 3 e \rightleftharpoons Cr + 3 OH^-$	-1.48
$Co(OH)_2 + 2 e \rightleftharpoons Co + 2 OH^-$	-0.73	$Pa^{4+} + 4 e \rightleftharpoons Pa$	-1.49
$NbO + 2 H^+ + 2 e \rightleftharpoons Nb + H_2O$	-0.733	$HfO_2 + 4 H^+ + 4 e \rightleftharpoons Hf + 2 H_2O$	-1.505
$H_2SeO_3 + 4 H^+ + 4 e \rightleftharpoons Se + 3 H_2O$	-0.74	$Hf^{4+} + 4 e \rightleftharpoons Hf$	-1.55
$Cr^{3+} + 3 e \rightleftharpoons Cr$	-0.744	$Sm^{3+} + e \rightleftharpoons Sm^{2+}$	-1.55
$Ta_2O_5 + 10 H^+ + 10 e \rightleftharpoons 2 Ta + 5 H_2O$	-0.750	$ZrO_2 + 4 H^+ + 4 e \rightleftharpoons Zr + 2 H_2O$	-1.553
$TlI + e \rightleftharpoons Tl + I^-$	-0.752	$Mn(OH)_2 + 2 e \rightleftharpoons Mn + 2 OH^-$	-1.56
$Zn^{2+} + 2 e \rightleftharpoons Zn$	-0.7618	$Ba^{2+} + 2 e \rightleftharpoons Ba(Hg)$	-1.570
$Zn^{2+} + 2 e \rightleftharpoons Zn(Hg)$	-0.7628	$Bk^{2+} + 2 e \rightleftharpoons Bk$	-1.6
$CdO + H_2O + 2 e \rightleftharpoons Cd + 2 OH^-$	-0.783	$Cf^{3+} + e \rightleftharpoons Cf^{2+}$	-1.6
$Te + 2 H^+ + 2 e \rightleftharpoons H_2Te$	-0.793	$Ti^{2+} + 2 e \rightleftharpoons Ti$	-1.630
$ZnSO_4 \cdot 7H_2O + 2 e \rightleftharpoons Zn(Hg) + SO_4^{2-} + 7 H_2O$	-0.7993	$Md^{3+} + 3 e \rightleftharpoons Md$	-1.65
(Saturated $ZnSO_4$)		$HPO_3^{2-} + 2 H_2O + 2 e \rightleftharpoons H_2PO_2^- + 3 OH^-$	-1.65
$Bi + 3 H^+ + 3 e \rightleftharpoons BiH_3$	-0.8	$Al^{3+} + 3 e \rightleftharpoons Al$	-1.662
$SiO + 2 H^+ + 2 e \rightleftharpoons Si + H_2O$	-0.8	$SiO_3^{2-} + H_2O + 4 e \rightleftharpoons Si + 6 OH^-$	-1.697
$Cd(OH)_2 + 2 e \rightleftharpoons Cd(Hg) + 2 OH^-$	-0.809	$HPO_3^{2-} + 2 H_2O + 3 e \rightleftharpoons P + 5 OH^-$	-1.71
$2 H_2O + 2 e \rightleftharpoons H_2 + 2 OH^-$	-0.8277	$HfO^{2+} + 2 H^+ + 4 e \rightleftharpoons Hf + H_2O$	-1.724
$2 NO_3^- + 2 H_2O + 2 e \rightleftharpoons N_2O_4 + 4 OH^-$	-0.85	$ThO_2 + 4 H^+ + 4 e \rightleftharpoons Th + 2 H_2O$	-1.789
$H_3BO_3 + 3 H^+ + 3 e \rightleftharpoons B + 3 H_2O$	-0.8698	$H_2BO_3^- + H_2O + 3 e \rightleftharpoons B + 4 OH^-$	-1.79
$P + 3 H_2O + 3 e \rightleftharpoons PH_3(g) + 3 OH^-$	-0.87	$Sr^{2+} + 2 e \rightleftharpoons Sr(Hg)$	-1.793
$Ti^{3+} + e \rightleftharpoons Ti^{2+}$	-0.9	$U^{3+} + 3 e \rightleftharpoons U$	-1.798
$HSnO_2^- + H_2O + 2 e \rightleftharpoons Sn + 3 OH^-$	-0.909	$H_2PO_2^- + e \rightleftharpoons P + 2 OH^-$	-1.82
$Cr^{2+} + 2 e \rightleftharpoons Cr$	-0.913	$Be^{2+} + 2 e \rightleftharpoons Be$	-1.847
$Se + 2 e \rightleftharpoons Se^{2-}$	-0.924	$Np^{3+} + 3 e \rightleftharpoons Np$	-1.856
$SO_4^{2-} + H_2O + 2 e \rightleftharpoons SO_3^{2-} + 2 OH^-$	-0.93	$Fm^{3+} + 3 e \rightleftharpoons Fm$	-1.89
$Sn(OH)_6^{2-} + 2 e \rightleftharpoons HSnO_2^- + 3 OH^- + H_2O$	-0.93	$Th^{4+} + 4 e \rightleftharpoons Th$	-1.899

TABLE 3
Reduction Reactions Having $E°$ Values More Negative than that of the Standard Hydrogen Electrode
(continued)

Reaction	$E°/V$	Reaction	$E°/V$
$Am^{2+} + 2e \rightleftharpoons Am$	−1.9	$ZrO(OH)_2 + H_2O + 4e \rightleftharpoons Zr + 4OH^-$	−2.36
$Pa^{4+} + e \rightleftharpoons Pa^{3+}$	−1.9	$Mg^{2+} + 2e \rightleftharpoons Mg$	−2.372
$Es^{3+} + 3e \rightleftharpoons Es$	−1.91	$Y^{3+} + 3e \rightleftharpoons Y$	−2.372
$Cf^{3+} + 3e \rightleftharpoons Cf$	−1.94	$La^{3+} + 3e \rightleftharpoons La$	−2.379
$Lr^{3+} + 3e \rightleftharpoons Lr$	−1.96	$Tm^{2+} + 2e \rightleftharpoons Tm$	−2.4
$Eu^{3+} + 3e \rightleftharpoons Eu$	−1.991	$Md^{2+} + 2e \rightleftharpoons Md$	−2.40
$Er^{2+} + 2e \rightleftharpoons Er$	−2.0	$Th(OH)_4 + 4e \rightleftharpoons Th + 4OH^-$	−2.48
$Pr^{2+} + 2e \rightleftharpoons Pr$	−2.0	$HfO(OH)_2 + H_2O + 4e \rightleftharpoons Hf + 4OH^-$	−2.50
$Pu^{3+} + 3e \rightleftharpoons Pu$	−2.031	$No^{2+} + 2e \rightleftharpoons No$	−2.50
$Cm^{3+} + 3e \rightleftharpoons Cm$	−2.04	$Dy^{3+} + e \rightleftharpoons Dy^{2+}$	−2.6
$Am^{3+} + 3e \rightleftharpoons Am$	−2.048	$Pm^{3+} + e \rightleftharpoons Pm^{2+}$	−2.6
$AlF_6^{3-} + 3e \rightleftharpoons Al + 6F^-$	−2.069	$Be_2O_3^{2-} + 3H_2O + 4e \rightleftharpoons 2Be + 6OH^-$	−2.63
$Sc^{3+} + 3e \rightleftharpoons Sc$	−2.077	$Sm^{2+} + 2e \rightleftharpoons Sm$	−2.68
$Ho^{2+} + 2e \rightleftharpoons Ho$	−2.1	$Mg(OH)_2 + 2e \rightleftharpoons Mg + 2OH^-$	−2.690
$Nd^{2+} + 2e \rightleftharpoons Nd$	−2.1	$Nd^{3+} + e \rightleftharpoons Nd^{2+}$	−2.7
$Cf^{2+} + 2e \rightleftharpoons Cf$	−2.12	$Mg^+ + e \rightleftharpoons Mg$	−2.70
$Yb^{3+} + 3e \rightleftharpoons Yb$	−2.19	$Na^+ + e \rightleftharpoons Na$	−2.71
$Ac^{3+} + 3e \rightleftharpoons Ac$	−2.20	$Yb^{2+} + 2e \rightleftharpoons Yb$	−2.76
$Dy^{2+} + 2e \rightleftharpoons Dy$	−2.2	$Bk^{3+} + e \rightleftharpoons Bk^{2+}$	−2.8
$Tm^{3+} + e \rightleftharpoons Tm^{2+}$	−2.2	$Ho^{3+} + e \rightleftharpoons Ho^{2+}$	−2.8
$Pm^{2+} + 2e \rightleftharpoons Pm$	−2.2	$Ra^{2+} + 2e \rightleftharpoons Ra$	−2.8
$Es^{2+} + 2e \rightleftharpoons Es$	−2.23	$Eu^{2+} + 2e \rightleftharpoons Eu$	−2.812
$H_2 + 2e \rightleftharpoons 2H^-$	−2.23	$Ca^{2+} + 2e \rightleftharpoons Ca$	−2.868
$Gd^{3+} + 3e \rightleftharpoons Gd$	−2.279	$Sr(OH)_2 + 2e \rightleftharpoons Sr + 2OH^-$	−2.88
$Tb^{3+} + 3e \rightleftharpoons Tb$	−2.28	$Sr^{2+} + 2e \rightleftharpoons Sr$	−2.89
$Lu^{3+} + 3e \rightleftharpoons Lu$	−2.28	$Fr^+ + e \rightleftharpoons Fr$	−2.9
$Dy^{3+} + 3e \rightleftharpoons Dy$	−2.295	$La(OH)_3 + 3e \rightleftharpoons La + 3OH^-$	−2.90
$Am^{3+} + e \rightleftharpoons Am^{2+}$	−2.3	$Ba^{2+} + 2e \rightleftharpoons Ba$	−2.912
$Fm^{2+} + 2e \rightleftharpoons Fm$	−2.30	$K^+ + e \rightleftharpoons K$	−2.931
$Pm^{3+} + 3e \rightleftharpoons Pm$	−2.30	$Rb^+ + e \rightleftharpoons Rb$	−2.98
$Sm^{3+} + 3e \rightleftharpoons Sm$	−2.304	$Ba(OH)_2 + 2e \rightleftharpoons Ba + 2OH^-$	−2.99
$Al(OH)_3 + 3e \rightleftharpoons Al + 3OH^-$	−2.31	$Er^{3+} + e \rightleftharpoons Er^{2+}$	−3.0
$Tm^{3+} + 3e \rightleftharpoons Tm$	−2.319	$Ca(OH)_2 + 2e \rightleftharpoons Ca + 2OH^-$	−3.02
$Nd^{3+} + 3e \rightleftharpoons Nd$	−2.323	$Cs^+ + e \rightleftharpoons Cs$	−3.026
$Al(OH)^- + 3e \rightleftharpoons Al + 4OH^-$	−2.328	$Li^+ + e \rightleftharpoons Li$	−3.0401
$H_2AlO_3^- + H_2O + 3e \rightleftharpoons Al + 4OH^-$	−2.33	$3N_2 + 2H^+ + 2e \rightleftharpoons 2HN_3$	−3.09
$Ho^{3+} + 3e \rightleftharpoons Ho$	−2.33	$Pr^{3+} + e \rightleftharpoons Pr^{2+}$	−3.1
$Er^{3+} + 3e \rightleftharpoons Er$	−2.331	$Ca^+ + e \rightleftharpoons Ca$	−3.80
$Ce^{3+} + 3e \rightleftharpoons Ce$	−2.336	$Sr^+ + e \rightleftharpoons Sr$	−4.10
$Pr^{3+} + 3e \rightleftharpoons Pr$	−2.353		

REDUCTION AND OXIDATION POTENTIALS FOR CERTAIN ION RADICALS

Petr Vanýsek

There are two tables for ion radicals. The first table lists reduction potentials for organic compounds which produce anion radicals during reduction, a process described as $A + e^- \rightleftharpoons A^-$. The second table lists oxidation potentials for organic compounds which produce cation radicals during oxidation, a process described as $A \rightleftharpoons A^+ + e^-$. To obtain reduction potential for a reverse reaction, the sign for the potential is changed.

Unlike the table of the Electrochemical Series, which lists *standard* potentials, values for radicals are experimental values with experimental conditions given in the second column. Since the measurements leading to potentials for ion radicals are very dependent on conditions, an attempt to report standard potentials for radicals would serve no useful purpose. For the same reason, the potentials are also reported as experimental values, usually a half-wave potential ($E_{1/2}$ in polarography) or a peak potential (E_p in cyclic voltammetry). Unless otherwise stated, the values are reported vs. SCE (saturated calomel electrode). To obtain a value vs. normal hydrogen electrode, 0.241 V has to be added to the SCE values. All the ion radicals chosen for inclusion in the tables result from electrochemically reversible reactions. More detailed data on ion radicals can be found in the *Encyclopedia of Electrochemistry of Elements*, (A. J. Bard, Ed.), Vol. XI and XII in particular, Marcel Dekker, New York, 1978.

Abbreviations are: CV — cyclic voltammetry; DMF — *N,N*-Dimethylformamide; E swp — potential sweep; E° — standard potential; E_p — peak potential; $E_{p/2}$ — half-peak potential; $E_{1/2}$ — half wave potential; M — mol/L; MeCN — acetonitrile; pol — polarography; rot Pt dsk — rotated Pt disk; SCE — saturated calomel electrode; TBABF$_4$ — tetrabutylammonium tetrafluoroborate; TBAI — tetrabutylammonium iodide; TBAP — tetrabutylammonium perchlorate; TEABr — tetraethylammonium bromide; TEAP — tetraethylammonium perchlorate; THF — tetrahydrofuran; TPACF$_3$SO$_3$ — tetrapropylammonium trifluoromethanesulfite; TPAP — tetrapropylammonium perchlorate; and wr — wire.

Reduction Potentials (Products are Anion Radicals)

Substance	Conditions/electrode/technique	Potential V (vs. SCE)
Acetone	DMF, 0.1 M TEABr/Hg/pol	$E_{1/2} = -2.84$
1-Naphthyphenylacetylene	DMF, 0.03 M TBAI/Hg/pol	$E_{1/2} = -1.91$
1-Naphthalenecarboxaldehyde	-/Hg/pol	$E_{1/2} = -0.91$
2-Naphthalenecarboxaldehyde	-/Hg/pol	$E_{1/2} = -0.96$
2-Phenanthrenecarboxaldehyde	-/Hg/pol	$E_{1/2} = -1.00$
3-Phenanthrenecarboxaldehyde	-/Hg/pol	$E_{1/2} = -0.94$
9-Phenanthrenecarboxaldehyde	-/Hg/pol	$E_{1/2} = -0.83$
1-Anthracenecarboxaldehyde	-/Hg/pol	$E_{1/2} = -0.75$
1-Pyrenecarboxaldehyde	-/Hg/pol	$E_{1/2} = -0.76$
2-Pyrenecarboxaldehyde	-/Hg/pol	$E_{1/2} = -1.00$
Anthracene	DMF, 0.1 M TBAP/Pt dsk/CV	$E_p = -2.00$
	DMF, 0.5 M TBABF$_4$/Hg/CV	$E_{1/2} = -1.93$
	MeCN, 0.1 M TEAP/Hg/CV	$E_{1/2} = -2.07$
	DMF, 0.1 M TBAI/Hg/pol	$E_{1/2} = -1.92$
9,10-Dimethylanthracene	DMF, 0.1 M TBAP/Pt/CV	$E_p = -2.08$
	MeCN, 0.1 M TBAP/Pt/CV	$E_p = -2.10$
1-Phenylanthracene	DMF, 0.5 M TBABF$_2$/Hg/CV	$E_{1/2} = -1.91$
	DMF, 0.1 M TBAI/Hg/pol	$E_{1/2} = -1.878$
2-Phenylanthracene	DMF, 0.1 M TBAI/Hg/pol	$E_{1/2} = -1.875$
8-Phenylanthracene	DMF, 0.5 M TBABF$_4$/Hg/CV	$E_{1/2} = -1.91$
9-Phenylanthracene	DMF, 0.5 M TBABF$_4$/Hg/CV	$E_{1/2} = -1.93$
	DMF, 0.1 M TBAI/Hg/pol	$E_{1/2} = -1.863$
1,8-Diphenylanthracene	DMF, 0.5 M TBABF$_4$/Hg/CV	$E_{1/2} = -1.88$
1,9-Diphenylanthracene	DMF, 0.1 M TBAI/Hg/pol	$E_{1/2} = -1.846$
1,10-Diphenylanthracene	DMF, 0.1 M TBAI/Hg/pol	$E_{1/2} = -1.786$
8,9-Diphenylanthracene	DMF, 0.5 M TBABF$_4$/Hg/CV	$E_{1/2} = -1.90$
9,10-Diphenylanthracene	MeCN, 0.1 M TBAP/rot Pt/E swp	$E_{1/2} = -1.83$
	DMF, 0.1 M TBAI/Hg/pol	$E_{1/2} = -1.835$
1,8,9-Triphenylanthracene	DMF, 0.5 M TBABF$_4$/Hg/CV	$E_{1/2} = -1.85$
1,8,10-Triphenylanthracene	DMF, 0.5 M TBABF$_4$/Hg/CV	$E_{1/2} = -1.81$
9,10-Dibiphenylanthracene	MeCN, 0.1 M TBAP/rot Pt/E swp	$E_{1/2} = -1.94$
Benz(a)anthracene	MeCN, 0.1 M TEAP/Hg/CV	$E_{1/2} = -2.11$
	MeCN, 0.1 M TEAP/Hg/pol	$E_{1/2} = -2.40^a$
Azulene	DMF, 0.1 M TBAI/Hg/pol	$E_{1/2} = -1.10^c$
Annulene	DMF, 0.5 M TBAP 0°C/Hg/pol	$E_{1/2} = -1.23$
Benzaldehyde	DMF, 0.1 M TBAP/Hg/pol	$E_{1/2} = -1.67$
Benzil	DMSO, 0.1 M TBAP/Hg/pol	$E_{1/2} = -1.04$

Reduction Potentials (Products are Anion Radicals) (continued)

Substance	Conditions/electrode/technique	Potential V (vs. SCE)
Benzophenone	-/Hg/pol	$E_{1/2} = -1.80$
	DMF/Pt dsk/CV	$E° = -1.72$
Chrysene	MeCN, 0.1 M TEAP/Hg/pol	$E_{1/2} = -2.73^a$
Fluoranthrene	DMF, 0.1 M TBAP/Pt dsk/CV	$E_p = -1.76$
Cyclohexanone	DMF, 0.1 M TEABr/Hg/pol	$E_{1/2} = -2.79$
5,5-Dimethyl-3-phenyl-2-cyclohexen-1-one	DMF, 0.5 M/Hg/pol	$E_{1/2} = -1.71$
1,2,3-Indanetrione hydrate (ninhydrin)	DMF, 0.2 M NaNO$_3$/Hg/pol	$E_{1/2} = -0.039$
Naphthacene	DMF, 0.1 M TBAI/Hg/pol	$E_{1/2} = -1.53$
Naphthalene	DMF, 0.1 M TBAP/Pt dsk/CV	$E_p = -2.55$
	DMF, 0.5 M TBABF$_4$/Hg/CV	$E_{1/2} = -2.56$
	DMF, MeCN, 0.1 M TEAP/Hg/CV	$E_{1/2} = -2.63$
	DMF, 0.1 M TBAI/Hg/pol	$E_{1/2} = -2.50$
1-Phenylnaphthalene	DMF, 0.5 M TBABF$_4$/Hg/CV	$E_{1/2} = -2.36$
1,2-Diphenylnaphthalene	DMF, 0.5 M TBABF$_4$/Hg/CV	$E_{1/2} = -2.25$
Cyclopentanone	DMF, 0.1 M TEABr/Hg/pol	$E_{1/2} = -2.82$
Phenanthrene	MeCN, 0.1 M TBAP/Pt wr/CV	$E_{1/2} = -2.47$
	MeCN, 0.1 M TEAP/Hg/pol	$E_{1/2} = -2.88^a$
Pentacene	THF, 0.1 M TBAP/rot Pt dsk/E swp	$E_{1/2} = -1.40$
Perylene	MeCN, 0.1 M TEAP/Hg/CV	$E_{1/2} = -1.73$
1,3-Diphenyl-1,3-propanedione	DMSO, 0.2 M TBAP/Hg/pol	$E_{1/2} = -1.42$
2,2-Dimethyl-1,3-diphenyl-1,3 propanedione	DMSO, TBAP/Hg/CV	$E_{1/2} = -1.80$
Pyrene	DMF, 0.1 M TBAP/Pt/CV	$E_p = -2.14$
	MeCN, 0.1 M TEAP/Hg/pol	$E_{1/2} = -2.49^a$
Diphenylsulfone	DMF, TEABr	$E_{1/2} = -2.16$
Triphenylene	MeCN, 0.1 M TEAP/Hg/pol	$E_{1/2} = -2.87^a$
9,10-Anthraquinone	DMF, 0.5 M TBAP, 20°/Pt dsk/CV	$E_{1/2} = -1.01$
1,4-Benzoquinone	MeCN, 0.1 M TEAP/Pt/CV	$E_p = -0.54$
1,4-Naphthohydroquinone, dipotassium salt	DMF, 0.5 M TBAP, 20°/Pt dsk/CV	$E_{1/2} = -1.55$
Rubrene	DMF, 0.1 M TBAP/Pt dsk/CV	$E_p = -1.48$
	DMF, 0.1 M TBAI/Hg/pol	$E_{1/2} = -1.410$
Benzocyclooctatetraene	THF, 0.1 M TBAP/Hg/pol	$E_{1/2} = -2.13$
sym-Dibenzocyclooctatetraene	THF, 0.1 M TBAP/Hg/pol	$E_{1/2} = -2.29$
Ubiquinone-6	MeCN, 0.1 M TEAP/Pt/CV	$E_p = -1.05^e$
(9-Phenyl-fluorenyl)$^+$	10.2 M H$_2$SO$_4$/Hg/CV	$E_p = -0.01^b$
(Triphenylcyclopropenyl)$^+$	MeCN, 0.1 M TEAP/Hg/CV	$E_p = -1.87$
(Triphenylmethyl)$^+$	MeCN, 0.1 M TBAP/Hg/pol	$E_{1/2} = 0.27$
	H$_2$SO$_4$, 10.2 M/Hg/CV	$E_p = -0.58^b$
(Tribiphenylmethyl)$^+$	MeCN, 0.1 M TBAP/Hg/pol	$E_{1/2} = 0.19$
(Tri-4-t-butyl-5-phenylmethyl)$^+$	MeCN, 0.1 M TBAP/Hg/pol	$E_{1/2} = 0.13$
(Tri-4-isopropylphenylmethyl)$^+$	MeCN, 0.1 M TBAP/Hg/pol	$E_{1/2} = 0.07$
(Tri-4-methylphenylmethyl)$^+$	MeCN, 0.1 M TBAP/Hg/pol	$E_{1/2} = 0.05$
(Tri-4-cyclopropylphenylmethyl)$^+$	MeCN, 0.1 M TBAP/Hg/pol	$E_{1/2} = 0.01$
(Tropylium)$^+$	MeCN, 0.1 M TBAP/Hg/pol	$E_{1/2} = -0.17$
	DMF, 0.15 M TBAI/Hg/pol	$E_{1/2} = -1.55$
	DMF, 0.15 M TBAI/Hg/pol	$E_{1/2} = -1.55$
	DMF, 0.15 M TBAI/Hg/pol	$E_{1/2} = -1.57$
	DMF, 0.15 M TBAI/Hg/pol	$E_{1/2} = -1.60$
	DMF, 0.15 M TBAI/Hg/pol	$E_{1/2} = -1.87$
	DMF, 0.15 M TBAI/Hg/pol	$E_{1/2} = -1.96$
	DMF, 0.15 M TBAI/Hg/pol	$E_{1/2} = -2.05$

Oxidation Potentials (Products are Cation Radicals)

Substance	Conditions/electrode/technique	Potential V (vs. SCE)
Anthracene	CH_2Cl_2, 0.2 M TBABF$_4$, $-70°C$/Pt dsk/CV	$E_p = +0.73^d$
9,10-Dimethylanthracene	MeCN, 0.1 M LiClO$_4$/Pt wr/CV	$E_p = +1.0$
9,10-Dipropylanthracene	MeCN, 0.1 M TEAP/Pt/CV	$E_p = +1.08$
1,8-Diphenylanthracene	CH_2Cl_2, 0.2 M TPrACF$_3$SO$_3$/rot Pt wr/E swp	$E_{1/2} = +1.34$
8,9-Diphenylanthracene	CH_2Cl_2, 0.2 M TPrACF$_3$SO$_3$/rot Pt wr/E swp	$E_{1/2} = +1.30$
9,10-Diphenylanthracene	MeCN/Pt/CV	$E_p = +1.22$
Perylene	MeCN, 0.1 M TBAP/Pt/CV	$E_p = +1.34$
Pyrene	DMF, 0.1 M TBAP/Pt dsk/CV	$E_p = +1.25$
Rubrene	DMF, 0.1 M TBAP/Pt dsk/CV	$E_p = +1.10$
Tetracene	CH_2Cl_2, 0.2 M TBABF$_4$, $-70°C$/Pt wr/CV	$E_p = +0.35^d$
1,4-Dithiabenzene	MeCN, 0.1 M TEAP/Pt dsk/rot	$E_{1/2} = +0.69$
1,4-Dithianaphthalene	MeCN, 0.1 M TEAP/Pt dsk/rot	$E_{1/2} = +0.80$
Thianthrene	0.1 M TPAP/Pt/CV	$E_{1/2} = +1.28$

[a] vs 0.01 M Ag/AgClO$_4$
[b] vs. Hg/Hg$_2$SO$_4$, 17 M H$_2$SO$_4$
[c] vs Hg pool
[d] vs Ag/saturated AgNO$_3$
[e] vs Ag/0.01 M Ag$^+$

pH SCALE FOR AQUEOUS SOLUTIONS

A. K. Covington

The pH value is the negative decadic logarithm of the (relative) ion activity of the hydrogen ion in the solution.

$$pH = -\log a_H \tag{1}$$

This is only a notional definition since Equation 1 involves a single ion activity, which is immeasurable, and has to be attained through a nonthermodynamic assumption such as that described in Equation 5 below. In terms of substance concentration, molarity, Equation 1 may be rewritten

$$pH = -\log (c_H y_H / c^o) \tag{2}$$

where c^o is an arbitrary constant representing the standard state condition and equal to 1 mol dm^{-3}, c_H is the concentration of hydrogen ion and y_H is the single ion activity of the hydrogen ion. In terms of molality, Equation 1 may be rewritten

$$pH = -\log (m_H \gamma_H / m^o) \tag{3}$$

where m^o is an arbitrary constant representing the standard state condition and equal to 1 mol kg^{-1}, m_H is the concentration of hydrogen ion and γ_H is the single ion activity of the hydrogen ion. For most purposes the difference between these two scales can be ignored for dilute aqueous solutions; the difference is 0.001 at 25° C and 0.02 at 100° C. Arising from the nonexperimental determinability of single ion activities, the definition and determination of pH have an operational basis, and depend on the assignment of pH values to a standard solution (or solutions) together with the determination of pH difference by a cell with liquid junction called the operational cell.

The Operational Definition of pH Difference[1,2]

The electromotive force, EMF, $E(X)$ of the cell with liquid junction:

$$\text{Reference electrode} \mid \text{KCl (aq., concentrated)} \parallel \text{Solution X} \mid H_2 \mid Pt \tag{I}$$

is measured, and likewise that, $E(S)$, of the cell:

$$\text{Reference electrode} \mid \text{KCl (aq., concentrated)} \parallel \text{Solution S} \mid H_2 \mid Pt \tag{II}$$

The temperature of both cells (I and II) must be equal and uniform throughout, and the hydrogen gas pressures identical. The two bridge solutions may be any molality of KCl not less than 3.5 mol kg^{-1} provided they are the same.

The pH of the solution X, pH(X), is then related to the assigned pH of the solution S, pH(S) by the definition:

$$pH(X) = pH(S) + [E(S) - E(X)] / [(RT/F) \ln 10] \tag{4}$$

where R is the gas constant, T the thermodynamic temperature, F the Faraday constant. The quantity $k = (RT/F) \ln 10$ is called the slope factor whose values are given as a function of temperature in Table 1. As a consequence of this definition any difference in liquid junction potential between cells I and II is subsumed into the value of pH(X).

The pH Scale

The pH scale at a particular temperature is defined by Equation 4 as a straight line, on the plot of pH against $E(X)$, having a slope of k drawn through the pH value assigned to the Reference Value Standard (RVS) solution (as given in Table 2) and the value of $E(S)$ for cell II when it contains the Reference Value Standard solution. The solution chosen for the RVS is 0.05 mol kg^{-1} aqueous potassium hydrogen phthalate. The procedure by which pH(RVS) values have been assigned to the Reference Value Standard (RVS) is the cell III without transference:[1,3]

$$Pt(Pd) \mid H_2 (g, p=1 \text{ atm} = 101\,325 \text{ Pa}) \mid RVS, Cl^- \mid AgCl \mid Ag \tag{III}$$

The palladised-platinum hydrogen electrode is used to reduce the catalytised chemical reduction of the phthalate by hydrogen gas. The calculation involves a non-thermodynamic assumption, the Bates-Guggenheim Convention, for the single ion activity of the chloride ion[1,2] as

$$\log (\gamma_{Cl})^o = -A(I/m^o)^{1/2} / [1 + 1.5 (I/m^o)^{1/2}] \tag{5}$$

where I is the ionic strength = $(1/2)\Sigma m_i z_i^2 = 0.0534$ mol kg^{-1} for the RVS solution and A is a known function of temperature (Table 1).

To prepare the RVS solution, dry the sample at 110° C for 2 h before use. The water should have a conductivity of less than 0.1 mS m^{-1}. The required solution contains 10.211 g kg^{-1} water. It can be prepared on a volume basis by dissolving 10.138 g potassium hydrogen phthalate in water and making up to 1 L at 20° C. This solution is 0.04964 mol/L with a density of 1.00300 g/L at 20° C.

Primary Standards[2]

pH values may be assigned by the cell without transference method (cell III) to six other buffer solutions which meet certain criteria of reproducibility of preparation and properties. These solutions are called primary pH standards (PS), and details and the pH(PS) values assigned to them are given in Table 3. When these PS solutions are used in the operational cell I, the experimental value of the slope will not be in accord with the slope factor values of Table 1, and, moreover, the experimental value could change if additional primary solutions were to be defined. Hence the pH value determined for an unknown solution can be slightly dependent (±0.02) on the choice of primary standard.[2,4,5] Some useful data for standard buffers are given in Table 5.

Operational Standards[2,6]

Operational standards (OS) are also defined which are traceable to the Reference Value Standard (RVS). Values are assigned by means of the operational cells I and II where the liquid junctions are the free diffusion type reproducibly formed in 1 mm vertical capillary tubes. These operational standards are not restricted in number provided certain preparation criteria are met, and pH(OS) values for 16 solutions are given in Table 4.[2] These OS represent an alternative procedure and are in no way to be regarded as inferior to the primary standards. As a consequence of their definition, all pH(OS) values fall on the line with slope given by the slope factor value for the appropriate temperature in Table 1. Any difference in liquid junction potential between the solutions of cells I and II and KCl is subsumed into the assigned value of pH(OS).

Measurement of pH. Choice of Standard Reference Solution

1a. If pH is not required to better than ±0.05 any standard reference solution may be selected.

1b. If pH is required to ±0.002 and interpretation in terms of hydrogen ion concentration or activity is desired, choose a standard reference solution, pH(PS) or pH(OS), to match X as closely as possible in terms of pH, composition and ionic strength.

2. Alternatively, a bracketting procedure may be adopted whereby two standard reference solutions are chosen whose pH values, pH(S1), pH(S2) are on either side of pH(X). Then if the corresponding potential difference measurements are $E(S1)$, $E(S2)$, $E(X)$, then pH(X) is obtained from

$$pH(X) = pH(S1) + [E(X) - E(S1)]/\ \%k$$

where $\%k = 100[E(S2) - E(S1)]/[pH(S2) - pH(S1)]$ is the apparent percentage slope. This procedure is very easily done on some pH meters simply by adjusting downwards the slope factor control with the electrodes in S2. The purpose of the bracketting procedure is to compensate for deficiencies in the electrodes and measuring system.

Information to be Given about the Measurement of pH(X)

The standard solutions selected for calibration of the pH meter system should be reported with the measurement as follows,

1. System calibrated with pH(RVS) = ... at ...K.
2. System calibrated with two primary standards pH(PS1) = ... and pH(PS2) = ... at ... K.
3. System calibrated with two operational standards pH(OS1) =..... and pH(OS2) = ... at .... K.

Interpretation of pH(X) in Terms of Hydrogen Ion Concentration

The operationally defined pH has no simple interpretation in terms of hydrogen ion concentration but the mean ionic activity coefficient of a typical 1:1 electrolyte can be substituted into equation 2 or 3 to obtain hydrogen ion concentration subject to an uncertainty of 3.9% in concentration corresponding to 0.02 in pH.

REFERENCES

1. R. G. Bates, *Measurement of pH. Theory and Practice*, 2nd ed., John Wiley & Sons,, New York, 1973.
2. A. K. Covington, R. G. Bates, and R. A. Durst, *Pure Appl. Chem.*, 57, 531, 1985.
3. H. P. Butikofer and A. K. Covington, *Anal. Chim. Acta*, 108, 179, 1979.
4. R. G. Bates, *Crit. Rev. Anal. Chem.*, 10, 247, 1981.
5. A. K. Covington, *Anal. Chim. Acta*, 127, 1, 1981.
6. A. K. Covington and M. J. Rebelo, *Anal. Chim. Acta.*, 200, 245, 1987.
7. V. E. Bower and R. G. Bates, *J. Res. Natl. Bur. Stand.*, 39, 263, 1954.

TABLE 1
Standard EMF, Slope Factor and Debye-Huckel Constant A
(Unit Weight of Solvent) as Functions of Temperature

Temperature/°C	$E°$/mV[7]	Slope Factor k/mV	A[1]
0	236.55	54.199	0.4918
5	234.13	55.191	0.4952
10	231.42	56.183	0.4988
15	228.57	57.175	0.5026
20	225.57	58.167	0.5066
25	222.34	59.159	0.5108
30	219.04	60.152	0.5150
35	215.65	61.144	0.5196
40	212.08	62.136	0.5242
45	208.35	63.128	0.5291
50	204.49	64.120	0.5341
55	200.56	65.112	0.5393
60	196.49	66.104	0.5448
70	187.82	68.088	0.5562
80	178.73	70.073	0.5685
90	169.52	72.057	0.5817
95	165.11	73.049	0.5886

TABLE 2
Values of pH(RVS) for the Reference Value Standard of 0.05 mol kg^{-1} Potassium Hydrogen Phthalate at Various Temperatures

t/°C	pH(RVS)	t/°C	pH(RVS)	t/°C	pH(RVS)
0	4.000	35	4.018	65	4.097
5	3.998	37	4.022	70	4.116
10	3.997	40	4.027	75	4.137
15	3.998	45	4.038	80	4.159
20	4.001	50	4.050	85	4.183
25	4.005	55	4.064	90	4.21
30	4.011	60	4.080	95	4.24

pH SCALE FOR AQUEOUS SOLUTIONS (continued)

TABLE 3
Values of pH(PS) for Primary Standard Reference Solutions

Primary ref. standard	t/°C																		
	0	5	10	15	20	25	30	35	37	40	50	60	70	80	90	95
Saturated (at 25° C) Potassium hydrogen tartrate	—	—	—	—	—	3.557	3.552	3.549	3.548	3.547	3.549	3.560	3.580	3.610	3.650	3.674
0.1 mol/kg Potassium dihydrogen citrate	3.863	3.840	3.820	3.802	3.788	3.776	3.766	3.759	3.756	3.754	3.749	—	—	—	—	—
0.025 mol/kg Disodium hydrogen phosphate +0.025 mol/kg Potassium dihydrogen phosphate	6.984	6.951	6.923	6.900	6.881	6.865	6.853	6.844	6.841	6.838	6.833	6.836	6.845	6.859	6.876	6.886
0.03043 mol/kg Disodium hydrogen phosphate +0.008695 mol/kg Potassium dihydrogen phosphate	7.534	7.500	7.472	7.448	7.429	7.413	7.400	7.389	7.386	7.380	7.367	—	—	—	—	—
0.01 mol/kg Disodium tetraborate	9.464	9.395	9.332	9.276	9.225	9.180	9.139	9.102	9.088	9.068	9.011	8.962	8.921	8.884	8.850	8.833
0.025 mol/kg Sodium hydrogen carbonate +0.025 mol/kg sodium carbonate	10.317	10.245	10.179	10.118	10.062	10.012	9.966	9.926	9.910	9.889	9.828	—	—	—	—	—

Note: Based on an uncertainty of ±0.2 mV in determined (E-E^0), the uncertainty is ±0.003 in pH in the range 0—50° C.

pH SCALE FOR AQUEOUS SOLUTIONS (continued)

TABLE 4

pH (OS) Values for Operational Reference Solutions

Operational standard ref. solution	0	5	10	15	20	25	30	37	40	50	60	70	80	90	95
0.1 mol/kg Potassium tetroxalate[a]	—	—	—	—	1.475	1.479	1.483	1.490	1.493	1.503	1.513	1.52	1.53	1.53	1.53
0.05 mol/kg potassium tetroxalate[a]	—	—	1.638	1.642	1.644	1.646	1.648	1.649	1.650	1.653	1.660	1.671	1.689	1.72	1.73
0.05 mol/kg sodium hydrogen diglycolate[b]	—	3.466	3.470	3.476	3.484	3.492	3.502	3.519	3.527	3.558	3.595	—	—	—	3.649
Saturated (at 25° C) potassium hydrogen tartrate	—	—	—	—	—	3.556	3.549	3.544	3.542	3.544	3.553	3.570	3.596	3.627	3.649
0.05 mol/kg Potassium hydrogen phthalate (RVS)	4.000	3.998	3.997	3.998	4.000	4.005	4.011	4.022	4.027	4.050	4.080	4.115	4.159	4.21	4.24
0.1 mol/dm³ Acetic acid + 0.1 mol/dm³ sodium acetate	4.664	4.657	4.652	4.647	4.645	4.644	4.643	4.647	4.650	4.663	4.684	4.713	4.75	4.80	4.83
0.01 mol/dm³ Acetic acid + 0.1 mol/dm³ sodium acetate	4.729	4.722	4.717	4.714	4.712	4.713	4.715	4.722	4.726	4.743	4.768	4.800	4.839	4.88	4.91
0.02 mol/kg Piperazine phosphate[c]	—	6.477	6.419	6.364	6.310	6.259	6.209	6.143	6.116	6.030	5.952	—	—	—	—
0.025 mol/kg Disodium hydrogen phosphate + 0.025 mol/kg potassium dihydrogen phosphate	6.961	6.935	6.912	6.891	6.873	6.857	6.843	6.828	6.823	6.814	6.817	6.830	6.85	6.90	6.92
0.03043 mol/kg Disodium hydrogen phosphate + 0.008695 mol/kg potassium disodium phosphate	7.506	7.482	7.460	7.441	7.423	7.406	7.390	7.369	—	—	—	—	—	—	—
0.04 mol/kg Disodium hydrogen phosphate + 0.01 mol/kg potassium dihydrogen phosphate	—	7.512	7.488	7.466	7.445	7.428	7.414	7.404	—	—	—	—	—	—	—
0.05 mol/kg Tris hydrochloride + 0.01667 mol/kg Tris[d]	8.399	8.238	8.083	7.933	7.788	7.648	7.513	7.332	7.257	7.018	6.794	—	—	—	—
0.05 mol/kg Disodium tetraborate ($Na_2B_4O_7$)	9.475	9.409	9.347	9.288	9.233	9.182	9.134	9.074	9.051	8.983	8.932	8.898	8.88	8.84	8.89
0.01 mol/kg Disodium tetraborate ($Na_2B_4O_7$)	9.451	9.388	9.329	9.275	9.225	9.179	9.138	9.086	9.066	9.009	8.965	8.932	8.91	8.90	8.89

$t/°C$

TABLE 4

pH(OS) Values for Operational Standard Reference Solutions (continued)

Operational standard ref. solution	0	5	10	15	20	25	30	37	40	50	60	70	80	90	95
								$t/°C$							
0.025 mol/kg Sodium hydrogen carbonate + 0.025 mol/kg sodium carbonate	10.273	10.212	10.154	10.098	10.045	9.995	9.948	9.889	9.866	9.800	9.753	9.728	9.725	9.75	9.77
Saturated (at 20° C) calcium hydroxide	13.360	13.159	12.965	12.780	12.602	12.431	12.267	12.049	11.959	11.678	11.423	11.192	10.984	10.80	10.71

Note: Uncertainty is ±0.003 in pH between 0 and 60° C rising to ±0.01 above 70° C.

a Potassium trihydrogen dioxalate ($KH_3C_4O_8$).
b Sodium hydrogen 2,2'-oxydiethanoate.
c $C_4H_{10}N_2 \cdot H_3PO_4$.
d 2-Amino-2-(hydroxymethyl)-1,3 propanediol or tris(hydroxymethyl)aminomethane.

TABLE 5

Useful Data on Some Standard Buffer Solutions

	Molecular formula	Molality (mol/kg)	Relative molar mass	Density at 20° C (g/cm³)	Molarity at 20° C (mol/L)	Mass of 1 L at 20° C (g)	Mass tolerance for ±0.001 pH[a] (g)	Mass tolerance expressed as a percentage (%)
Potassium tetraoxalate	$KH_3C_4O_8 \cdot 2H_2O$	0.1	254.1913	1.0091	0.09875	25.1017	0.07	0.27
Potassium tetraoxalate	$KH_3C_4O_8 \cdot 2H_2O$	0.05	254.1913	1.0038	0.04965	12.6202	0.034	0.26
Disodium hydrogen orthophosphate	Na_2HPO_4	0.025	141.9588	1.0038	0.02492	3.5379	0.02	0.56
Potassium dihydrogen orthophosphate	KH_2PO_4	0.025	136.0852			3.3912	0.02	0.58
Disodium tetraborate	$Na_2B_4O_7 \cdot 10H_2O$	0.05	381.367	1.0075	0.04985	19.0117	0.9	4.73
Disodium tetraborate	$Na_2B_4O_7 \cdot 10H_2O$	0.01	381.367	1.0001	0.009981	3.8064	0.19	0.49
Sodium carbonate	Na_2CO_3	0.025	105.9887	1.0021	0.02494	2.6428	0.017	0.064
Sodium hydrogen carbonate	$NaHCO_3$	0.025	84.0069			2.0947	0.013	0.62

a Calculated from known dilution value of solution.

PRACTICAL pH MEASUREMENTS ON NATURAL WATERS

A. K. Covington and W. Davison

(1) Dilute solutions and freshwater including 'acid-rain' samples ($I < 0.02$ mol kg^{-1})

Major problems could be encountered due to errors associated with the liquid junction. It is recommended that either a free diffusion junction is used or it is verified that the junction is working correctly using dilute solutions as follows. For commercial electrodes calibrated with IUPAC aqueous RVS or PS standards, the pH(X) of dilute solutions should be within ±0.02 of those given in Table 1. The difference in determined pH(X) between a stirred and unstirred dilute solution should be < 0.02. The characteristics of glass electrodes are such that below pH 5 the readings should be stable within 2 min, but for pH 5 to 8, 8 or so minutes may be necessary to attain stability. Interpretation of pH(X) measured in this way in terms of activity of hydrogen ion, a_{H^+} is subject[1] to an uncertainty of ±0.02 in pH.

(2) Seawater

Measurements made by calibration of electrodes with IUPAC aqueous RVS or PS standards to obtain pH(X) are perfectly valid. However, the interpretation of pH(X) in terms of the activity of hydrogen ion is complicated by the non zero residual liquid junction potential as well as by systematic differences between electrode pairs, principally attributable to the reference electrode. For 35‰ salinity seawater ($S = 0.035$) a_{H^+} calculated from pH(X) is typically 12% too low. Special seawater pH scales have been devised to overcome this problem:

(i) The total hydrogen ion scale, pH$_T$, is defined in terms of the sum of free and complexed (total) hydrogen ion concentrations, where

$$^T C_H = [H^+] + [HSO_4^-] + [HF].$$

$$\text{So, } pH_T = -\log {}^T C_H$$

Calibration of the electrodes with a buffer having a composition similar to that of seawater, to which pH$_T$ has been assigned, results in values of pHT(X) (Tables 2, 3) which are accurately interpretable in terms of $^T C_H$.

(ii) The free hydrogen ion scale, pH$_F$, is defined, and fully interpretable, in terms of the concentration of free hydrogen ions.

$$pH_F = -\log [H^+]$$

Values of pH$_F$ as a function of temperature have been assigned to the same set of pH$_T$ seawater buffers, and so alternatively can be used for calibration (Tables 2, 3) [2,3]

(3) Estuarine water

Prescriptions for seawater scale buffers are available for a range of salinities. Reliable estuarine pH measurements can be made by calibrating with a buffer of the same salinity as the sample. However, these buffers are difficult to prepare and their use presumes prior knowledge of salinity of the sample. Interpretable measurements of estuarine pH can be made by calibration with IUPAC aqueous RVS or PS standards if the electrode pair is additionally calibrated using a 20‰ salinity seawater buffer.[4] The difference between the assigned pH$_{SWS}$ of the seawater buffer and its measured pH(X) value using RVS or PS standards is

$$\Delta pH = pH_{SWS} - pH(X)$$

Values of ΔpH should be in the range of 0.08 to 0.18. It empirically corrects for differences between the two pH scales and for measurement errors associated with the electrode pair. The pH(X) of samples measured using IUPAC aqueous buffers, can be converted to pH$_T$ or pH$_F$ using the appropriate measured ΔpH:

$$pH_T = pH(X) - \Delta pH$$
$$\text{or } pH_F = pH(X) - \Delta pH$$

This simple procedure is appropriate to pH measurement at salinities from 2‰ to 35‰. For salinities lower than 2‰ the procedures for freshwaters should be adopted.

REFERENCES

1. Davison, W. and Harbinson, T. R., *Analyst*, 113, 709, 1988.
2. Culberson, C. H., in *Marine Electrochemistry*, Whitfield, M. and Jagner, D., Eds., Wiley, 1981.
3. Millero, F. J., *Limnol. Oceanogr.*, 31, 839, 1986.
4. Covington, A. K., Whalley, P. D., Davison, W., and Whitfield, M., in *The Determination of Trace Metals in Natural Waters*, West, T. S. and Nurnberg, H. W., Eds., Blackwell, Oxford, 1988.
5. Koch, W. F., Marinenko, G., and Paule, R. C., *J. Res. NBS*, 91, 33, 1986.

Table 1
pH of Dilute Solutions at 25°C, Degassed and Equilibrated with Air, Suitable as Quality Control Standards

	Ionic strength mmol kg^{-1}	Concentration(x) mmol kg^{-1}	pH $p_{CO_2} = 0$	pH $p_{CO_2} =$ air
Potassium hydrogen phthalate	10.7	10	4.12	4.12
	1.1	1	4.33	4.33
xKH$_2$PO$_4$ + xNa$_2$HPO$_4$	9.9	2.5	7.07	7.05
xKH$_2$PO$_4$ + 3.5xNa$_2$HPO$_4$	10	0.87	7.61	7.58
Na$_2$B$_4$O$_7 \cdot$ 10H$_2$O	10	5	9.20	—
HCl	0.1	0.1	4.03	4.03
SRM2694-I[a]	—	—	4.30	—
SRM2694-II[a]	—	—	3.59	—

Note: The pH of solutions near to pH 4 is virtually independent of temperature over the range of 5 to 30°C.

[a] Simulated rainwater samples are available (Reference 5) from NIST containing sulfate, nitrate, chloride, fluoride, sodium, potassium, calcium and magnesium

Table 2
Composition of Seawater Buffer of Salinity $S = 35‰$ at 25°C
(Reference 3)

Solute	mol dm^{-3}	mol kg^{-1}	g kg^{-1}	g dm^{-3}
NaCl	0.3666	0.3493	20.416	20.946
Na$_2$SO$_4$	0.02926	0.02788	3.96	4.063
KCl	0.01058	0.01008	0.752	0.772
CaCl$_2$	0.01077	0.01026	1.139	1.169
MgCl$_2$	0.05518	0.05258	5.006	5.139
Tris	0.06	0.05717	6.926	7.106
Tris · HCl	0.06	0.05717	9.010	9.244

Tris = tris(hydroxymethyl)aminomethane (HOCH$_2$)$_3$CNH$_2$.
A 20‰ buffer is made by diluting the 35‰ in the ratio 20:35.

Table 3
Assigned Values of 20‰ and 35‰ Buffers on Free and Total
Hydrogen Ion Scales. Calculated from Equations Provided by
Millero (Reference 3)

Temp (°C)	pH$_T$ $S = 20‰$	pH$_T$ $S = 35‰$	pH$_F$ $S = 20‰$	pH$_F$ $S = 35‰$
5	8.683	8.718	8.759	8.81
10	8.513	8.542	8.597	8.647
15	8.351	8.374	8.442	8.491
20	8.195	8.212	8.292	8.341
25	8.045	8.057	8.149	8.197
30	7.901	7.908	8.011	8.059
35	7.762	7.764	7.879	7.926

BUFFER SOLUTIONS GIVING ROUND VALUES OF pH AT 25°C

A pH	A x	B pH	B x	C pH	C x	D pH	D x	E pH	E x
1.00	67.0	2.20	49.5	4.10	1.3	5.80	3.6	7.00	46.6
1.10	52.8	2.30	45.8	4.20	3.0	5.90	4.6	7.10	45.7
1.20	42.5	2.40	42.2	4.30	4.7	6.00	5.6	7.20	44.7
1.30	33.6	2.50	38.8	4.40	6.6	6.10	6.8	7.30	43.4
1.40	26.6	2.60	35.4	4.50	8.7	6.20	8.1	7.40	42.0
1.50	20.7	2.70	32.1	4.60	11.1	6.30	9.7	7.50	40.3
1.60	16.2	2.80	28.9	4.70	13.6	6.40	11.6	7.60	38.5
1.70	13.0	2.90	25.7	4.80	16.5	6.50	13.9	7.70	36.6
1.80	10.2	3.00	22.3	4.90	19.4	6.60	16.4	7.80	34.5
1.90	8.1	3.10	18.8	5.00	22.6	6.70	19.3	7.90	32.0
2.00	6.5	3.20	15.7	5.10	25.5	6.80	22.4	8.00	29.2
2.10	5.10	3.30	12.9	5.20	28.8	6.90	25.9	8.10	26.2
2.20	3.9	3.40	10.4	5.30	31.6	7.00	29.1	8.20	22.9
		3.50	8.2	5.40	34.1	7.10	32.1	8.30	19.9
		3.60	6.3	5.50	36.6	7.20	34.7	8.40	17.2
		3.70	4.5	5.60	38.8	7.30	37.0	8.50	14.7
		3.80	2.9	5.70	40.6	7.40	39.1	8.60	12.2
		3.90	1.4	5.80	42.3	7.50	40.9	8.70	10.3
		4.00	0.1	5.90	43.7	7.60	42.4	8.80	8.5
						7.70	43.5	8.90	7.0
						7.80	44.5	9.00	5.7
						7.90	45.3		
						8.00	46.1		

F pH	F x	G pH	G x	H pH	H x	I pH	I x	J pH	J x
8.00	20.5	9.20	0.9	9.60	5.0	10.90	3.3	12.00	6.0
8.10	19.7	9.30	3.6	9.70	6.2	11.00	4.1	12.10	8.0
8.20	18.8	9.40	6.2	9.80	7.6	11.10	5.1	12.20	10.2
8.30	17.7	9.50	8.8	9.90	9.1	11.20	6.3	12.30	12.8
8.40	16.6	9.60	11.1	10.00	10.7	11.30	7.6	12.40	16.2
8.50	15.2	9.70	13.1	10.10	12.2	11.40	9.1	12.50	20.4
8.60	13.5	9.80	15.0	10.20	13.8	11.50	11.1	12.60	25.6
8.70	11.6	9.90	16.7	10.30	15.2	11.60	13.5	12.70	32.2
8.80	9.6	10.00	18.3	10.40	16.5	11.70	16.2	12.80	41.2
8.90	7.1	10.10	19.5	10.50	17.8	11.80	19.4	12.90	53.0
9.00	4.6	10.20	20.5	10.60	19.1	11.90	23.0	13.00	66.0
9.10	2.0	10.30	21.3	10.70	20.2	12.00	26.9		
		10.40	22.1	10.80	21.2				
		10.50	22.7	10.90	22.0				
		10.60	23.3	11.00	22.7				
		10.70	23.8						
		10.80	24.25						

A. 25 ml of 0.2 molar KCl + x ml of 0.2 molar HCl.
B. 50 ml of 0.1 molar potassium hydrogen phthalate + x ml of 0.1 molar HCl.
C. 50 ml of 0.1 molar potassium hydrogen phthalate + x ml of 0.1 molar NaOH.
D. 50 ml of 0.1 molar potassium dihydrogen phosphate + x ml of 0.1 molar NaOH.
E. 50 ml of 0.1 molar tris(hydroxymethyl)aminomethane + x ml of 0.1 M HCl.
F. 50 ml of 0.025 molar borax + x ml of 0.1 molar HCl.
G. 50 ml of 0.025 molar borax + x ml of 0.1 molar NaOH.
H. 50 ml of 0.05 molar sodium bicarbonate + x ml of 0.1 molar NaOH.
I. 50 ml of 0.05 molar disodium hydrogen phosphate + x ml of 0.1 molar NaOH.
J. 25 ml of 0.2 molar KCl + x ml of 0.2 molar NaOH.

Final volume of mixtures = 100 ml.

REFERENCES

1. Bower, V.E., and Bates, R.G., *J. Res. Natl. Bur. Stand.*, 55, 197, 1955 (A–D).
2. Bates, R.G., and Bower, V.E., *Anal. Chem.*, 28, 1322, 1956 (E–J).

DISSOCIATION CONSTANTS OF INORGANIC ACIDS AND BASES

The data in this table are presented as values of pK_a, defined as the negative logarithm of the acid dissociation constant K_a for the reaction

$$BH \rightleftharpoons B^- + H^+$$

Thus $pK_a = -\log K_a$, and the hydrogen ion concentration $[H^+]$ can be calculated from

$$K_a = \frac{\left[H^+\right]\left[B^-\right]}{[BH]}$$

In the case of bases, the entry in the table is for the conjugate acid; e.g., ammonium ion for ammonia. The OH^- concentration in the system

$$NH_3 + H_2O \rightleftharpoons NH_4^+ + OH^-$$

can be calculated from the equation

$$K_b = K_{water} / K_a = \frac{\left[OH^-\right]\left[NH_4^+\right]}{\left[NH_3\right]}$$

where $K_{water} = 1.01 \times 10^{-14}$ at 25 °C. Note that $pK_a + pK_b = pK_{water}$.

All values refer to dilute aqueous solutions at zero ionic strength at the temperature indicated. The table is arranged alphabetically by compound name.

REFERENCE

1. Perrin, D. D., *Ionization Constants of Inorganic Acids and Bases in Aqueous Solution, Second Edition*, Pergamon, Oxford, 1982.

Name	Formula	Step	$t/°C$	pK_a
Aluminum(III) ion	Al^{+3}		25	5.0
Ammonia	NH_3		25	9.25
Arsenic acid	H_3AsO_4	1	25	2.26
		2	25	6.76
		3	25	11.29
Arsenious acid	H_2AsO_3		25	9.29
Barium(II) ion	Ba^{+2}		25	13.4
Boric acid	H_3BO_3	1	20	9.27
		2	20	>14
Calcium(II) ion	Ca^{+2}		25	12.6
Carbonic acid	H_2CO_3	1	25	6.35
		2	25	10.33
Chlorous acid	$HClO_2$		25	1.94
Chromic acid	H_2CrO_4	1	25	0.74
		2	25	6.49
Cyanic acid	$HCNO$		25	3.46
Germanic acid	H_2GeO_3	1	25	9.01
		2	25	12.3
Hydrazine	N_2H_4		25	8.1
Hydrazoic acid	HN_3		25	4.6
Hydrocyanic acid	HCN		25	9.21
Hydrofluoric acid	HF		25	3.20
Hydrogen peroxide	H_2O_2		25	11.62
Hydrogen selenide	H_2Se	1	25	3.89
		2	25	11.0
Hydrogen sulfide	H_2S	1	25	7.05
		2	25	19
Hydrogen telluride	H_2Te	1	18	2.6
		2	25	11
Hydroxylamine	NH_2OH		25	5.94
Hypobromous acid	$HBrO$		25	8.55

Name	Formula	Step	$t/°C$	pK_a
Hypochlorous acid	HClO		25	7.40
Hypoiodous acid	HIO		25	10.5
Iodic acid	HIO_3		25	0.78
Lithium ion	Li^+		25	13.8
Magnesium(II) ion	Mg^{+2}		25	11.4
Nitrous acid	HNO_2		25	3.25
Perchloric acid	$HClO_4$		20	-1.6
Periodic acid	HIO_4		25	1.64
Phosphoric acid	H_3PO_4	1	25	2.16
		2	25	7.21
		3	25	12.32
Phosphorous acid	H_3PO_3	1	20	1.3
		2	20	6.70
Pyrophosphoric acid	$H_4P_2O_7$	1	25	0.91
		2	25	2.10
		3	25	6.70
		4	25	9.32
Selenic acid	H_2SeO_4	2	25	1.7
Selenious acid	H_2SeO_3	1	25	2.62
		2	25	8.32
Silicic acid	H_4SiO_4	1	30	9.9
		2	30	11.8
		3	30	12
		4	30	12
Sodium ion	Na^+		25	14.8
Strontium(II) ion	Sr^{+2}		25	13.2
Sulfamic acid	NH_2SO_3H		25	1.05
Sulfuric acid	H_2SO_4	2	25	1.99
Sulfurous acid	H_2SO_3	1	25	1.85
		2	25	7.2
Telluric acid	H_2TeO_4	1	18	7.68
		2	18	11.0
Tellurous acid	H_2TeO_3	1	25	6.27
		2	25	8.43
Tetrafluoroboric acid	HBF_4		25	0.5
Thiocyanic acid	HSCN		25	-1.8
Water	H_2O		25	13.995

DISSOCIATION CONSTANTS OF ORGANIC ACIDS AND BASES

This table lists the acid-base dissociation constants of over 600 organic compounds, including many amino acids. All data apply to dilute aqueous solutions and are presented in the form of pK_a, which is the negative of the logarithm of the acid dissociation constant K_a. See the preceding table, "Dissociation Constants of Inorganic Acids and Bases", for further details on notation.

Compounds are listed by molecular formula in the Hill order.

REFERENCES

1. Perrin, D. D., *Dissociation Constants of Organic Bases in Aqueous Solution*, Butterworths, London, 1965; Supplement, 1972.
2. Serjeant, E. P., and Dempsey, B., *Ionization Constants of Organic Acids in Aqueous Solution*, Pergamon, Oxford, 1979.
3. Albert, A., "Ionization Constants of Heterocyclic Substances", in *Physical Methods in Heterocyclic Chemistry*, Katritzky, A. R., Ed., Academic Press, New York, 1963.
4. Sober, H. A., Ed., *CRC Handbook of Biochemistry*, CRC Press, Cleveland, Ohio, 1968.
5. Perrin, D. D., Dempsey, B., and Serjeant, E. P., *pK_a Prediction for Organic Acids and Bases*, Chapman & Hall, London, 1981.
6. Dawson, R. M. C., Elliot, D. C., Elliot, W. H., and Jones, K. M., *Data for Biochemical Research*, Oxford Science Publications, Oxford, 1986.

Molecular formula	Name	Step	T/°C	pK_a
CH_2O_2	Formic acid		20	3.75
CH_4N_2O	Urea		21	0.10
CH_5N	Methylamine		25	10.63
$C_2HCl_3O_2$	Trichloroacetic acid		25	0.70
$C_2H_2Cl_2O_2$	Dichloroacetic acid		25	1.48
$C_2H_2O_3$	Glyoxylic acid		25	3.18
$C_2H_2O_4$	Oxalic acid	1	25	1.23
		2	25	4.19
$C_2H_3BrO_2$	Bromoacetic acid		25	2.69
$C_2H_3ClO_2$	Chloroacetic acid		25	2.85
$C_2H_3IO_2$	Iodoacetic acid		25	3.12
C_2H_4OS	Thioacetic acid		25	3.33
$C_2H_4O_2$	Acetic acid		25	4.76
$C_2H_4O_3$	Glycolic acid		25	3.83
C_2H_5N	Ethyleneimine		25	8.01
C_2H_5NO	Acetamide		25	0.63
$C_2H_5NO_2$	Glycine	1	25	2.35
		2	25	9.78
$C_2H_6O_2$	Ethylene glycol		25	14.22
$C_2H_7AsO_2$	Cacodylic acid	1	25	1.57
		2	25	6.27
C_2H_7N	Dimethylamine		25	10.68
C_2H_7N	Ethylamine		25	10.70
C_2H_7NO	Ethanolamine		25	9.50
$C_2H_7NO_3S$	Taurine	1	25	1.5
		2	25	9.061
C_2H_7NS	Cysteamine	1	25	8.35
		2	25	10.81
$C_2H_8N_2$	1,2-Ethanediamine	1	0	10.712
		2	0	7.564
$C_3H_3NO_2$	Cyanoacetic acid		25	2.45
C_3H_3NS	Thiazole		20	2.44
$C_3H_4N_2$	1H-Imidazole		25	6.953
$C_3H_4N_2S$	2-Thiazolamine		20	5.36
$C_3H_4O_2$	Acrylic acid		25	4.25
$C_3H_4O_3$	Pyruvic acid		25	2.39
$C_3H_4O_4$	Malonic acid	1	25	2.83
		2	25	5.69
$C_3H_5ClO_2$	2-Chloropropanoic acid		25	2.83
$C_3H_5ClO_2$	3-Chloropropanoic acid		25	3.98
$C_3H_6N_6$	Melamine		25	5.00
$C_3H_6O_2$	Propanoic acid		25	4.86
$C_3H_6O_3$	3-Hydroxypropanoic acid		25	4.51
$C_3H_6O_3$	Lactic acid		100	3.08
$C_3H_6O_4$	Glyceric acid		25	3.52

Molecular formula	Name	Step	$T/°C$	pK_a
C_3H_7N	Azetidine		25	11.29
$C_3H_7NO_2$	L-Alanine	1	25	2.34
		2	25	9.87
$C_3H_7NO_2$	β-Alanine	1	25	3.55
		2	25	10.24
$C_3H_7NO_2$	Methylglycine	1	25	2.21
		2	25	10.12
$C_3H_7NO_2S$	Cysteine	1	25	1.92
		2	25	8.37
		3	25	10.70
$C_3H_7NO_3$	Serine	1	25	2.19
		2	25	9.21
$C_3H_7NO_5S$	l-Cysteic acid	1	25	1.3
		2	25	1.9
		3	25	8.70
$C_3H_7N_3O_2$	Glycocyamine		25	2.82
$C_3H_8O_3$	Glycerol		25	14.15
C_3H_9N	Propylamine		20	10.60
C_3H_9N	Trimethylamine		25	9.80
C_3H_9NO	1-Amino-2-methoxyethane		10	9.89
C_3H_9NO	Trimethylamine oxide		25	4.65
$C_3H_{10}N_2$	1,2-Propanediamine	1	25	9.82
		2	25	6.61
$C_3H_{10}N_2$	1,3-Propanediamine	1	10	10.94
		2	10	9.03
$C_3H_{11}N_3$	1,2,3-Triaminopropane	1	20	9.59
		2	20	7.95
$C_4H_4N_2$	Pyrazine		27	0.65
$C_4H_4N_2$	Pyridazine		20	2.24
$C_4H_4N_2O_3$	Barbituric acid		25	4.01
$C_4H_4N_2O_5$	Alloxanic acid		25	6.64
$C_4H_4N_4O_2$	5-Nitropyrimidinamine		20	0.35
$C_4H_4O_4$	trans-Fumaric acid	1	18	3.03
		2	18	4.44
$C_4H_4O_4$	Maleic acid	1	25	1.83
		2	25	6.07
$C_4H_4O_5$	Oxaloacetic acid	1	25	2.22
		2	25	3.89
		3	25	13.03
$C_4H_5N_3$	2-Pyrimidinamine		20	3.45
$C_4H_6N_2$	1-Methylimidazol		25	6.95
$C_4H_6N_4O_3$	Allantoin		25	8.96
$C_4H_6O_2$	3-Butenoic acid		25	4.34
$C_4H_6O_2$	trans-Crotonic acid		25	4.69
$C_4H_6O_3$	Acetoacetic acid		18	3.58
$C_4H_6O_3$	2-Oxobutanoic acid		25	2.50
$C_4H_6O_4$	Methylmalonic acid	1	25	3.07
		2	25	5.76
$C_4H_6O_4$	Succinic acid	1	25	4.16
		2	25	5.61
$C_4H_6O_5$	Malic acid	1	25	3.40
		2	25	5.11
$C_4H_6O_6$	α-Tartaric acid	1	25	2.98
		2	25	4.34
$C_4H_6O_6$	meso-Tartaric acid	1	25	3.22
		2	25	4.82
$C_4H_6O_8$	Dihydroxytartaric acid		25	1.92
$C_4H_7ClO_2$	2-Chlorobutanoic acid		RT	2.86
$C_4H_7ClO_2$	3-Chlorobutanoic acid		RT	4.05
$C_4H_7ClO_2$	4-Chlorobutanoic acid		RT	4.52

Molecular formula	Name	Step	T/°C	pK_a
$C_4H_7NO_2$	4-Cyanobutanoic acid		25	2.42
$C_4H_7NO_3$	N-Acetylglycine		25	3.669
$C_4H_7NO_4$	Aspartic acid	1	25	1.99
		2	25	3.90
		3	25	9.90
$C_4H_7N_3O$	Creatinine	1	25	4.83
		2		9.2
$C_4H_8N_2O_3$	Asparagine	1	20	2.17
		2	20	8.80
$C_4H_8N_2O_3$	N-Glycylglycine		25	3.139
$C_4H_8O_2$	Butanoic acid		25	4.83
$C_4H_8O_2$	2-Methylpropanoic acid		25	4.88
$C_4H_8O_3$	3-Hydroxybutanoic acid		25	4.70
$C_4H_8O_3$	4-Hydroxybutanoic acid		25	4.72
C_4H_9N	Pyrrolidine		25	11.27
C_4H_9NO	Morpholine		25	8.33
$C_4H_9NO_2$	2-Aminobutanoic acid	1	25	2.29
		2	25	9.83
$C_4H_9NO_2$	4-Aminobutanoic acid	1	25	4.031
		2	25	10.556
$C_4H_9NO_2$	N,N-Dimethylglycine		25	9.89
$C_4H_9NO_2$	2-Methylalanine	1	25	2.36
		2	25	10.21
$C_4H_9NO_2S$	Homocysteine	1	25	2.22
		2	25	8.87
		3	25	10.86
$C_4H_9NO_3$	Homoserine	1	25	2.71
		2	25	9.62
$C_4H_9NO_3$	DL-Methoxyalanine		25	2.037
$C_4H_9NO_3$	Threonine	1	25	2.09
		2	25	9.10
$C_4H_9N_3O_2$	Creatine	1	25	2.63
		2	25	14.3
$C_4H_{10}N_2$	Piperazine	1	23	9.83
		2	23	5.56
$C_4H_{10}N_2O_2$	2,4-Diaminobutanoic acid	1	25	1.85
		2	25	8.24
		3	25	10.44
$C_4H_{11}N$	Butylamine		20	10.77
$C_4H_{11}N$	sec-Butylamine		25	10.56
$C_4H_{11}N$	tert-Butylamine		25	10.68
$C_4H_{11}N$	Diethylamine		40	11.02
$C_4H_{12}N_2$	1,4-Butanediamine	1	20	10.80
		2	20	9.35
$C_4H_{12}O_2$	1,2-Dimethylaminoethane	1	25	10.40
		2	25	8.26
C_5H_4BrN	3-Bromopyridine		25	2.84
C_5H_4ClN	3-Chloropyridine		25	2.84
$C_5H_4N_4$	Purine	1	20	2.30
		2	20	8.96
$C_5H_4N_4O_3$	Uric acid		12	3.89
C_5H_5N	Pyridine		25	5.25
C_5H_5NO	4-Pyridinol	1	20	3.20
		2	20	11.12
C_5H_5NO	2(1H)-Pyridinone	1	20	0.75
		2	20	11.65
$C_5H_5N_5$	1H-Purin-6-amine	1	25	4.12
		2	25	9.83
$C_5H_6N_2$	2-Methylpyrazine		27	1.45
$C_5H_6N_2$	2-Pyridinamine		20	6.82

Molecular formula	Name	Step	$T/°C$	pK_a
$C_5H_6N_2$	4-Pyridinamine		25	9.114
$C_5H_6O_4$	1,1-Cyclopropanedicarboxylic acid	1	25	1.82
		2	25	7.43
$C_5H_6O_4$	Itaconic acid	1	25	3.85
		2	25	5.45
$C_5H_6O_4$	Mesaconic acid	1	25	3.09
		2	25	4.75
$C_5H_6O_5$	2-Oxoglutaric acid	1	25	2.47
		2	25	4.68
$C_5H_7NO_3$	L-2-Pyrrolidone-5-carboxylic aci		25	3.32
$C_5H_7N_3$	Methylaminopyrazine		25	3.39
$C_5H_7N_3$	2,5-Pyridinediamine		20	6.48
$C_5H_8N_2$	2,4-Dimethylimidazol		25	8.36
$C_5H_8O_4$	Dimethylmalonic acid		25	3.15
$C_5H_8O_4$	Glutaric acid	1	25	4.31
		2	25	5.41
$C_5H_8O_4$	Methylsuccinic acid	1	25	4.13
		2	25	5.64
$C_5H_9NO_2$	Proline	1	25	1.952
		2	25	10.64
$C_5H_9NO_3$	5-Aminolevulinic acid	1	25	4.05
		2	25	8.90
$C_5H_9NO_3$	trans-4-Hydroxyproline	1	25	1.818
		2	25	9.662
$C_5H_9NO_3$	8-Hydroxypurine	1	20	2.56
		2	20	8.26
$C_5H_9NO_4$	L-Glutamic acid	1	25	2.13
		2	25	4.31
$C_5H_9N_3$	Histamine	1	25	6.04
		2	25	9.75
$C_5H_{10}N_2O_2$	Diaminopimelic acid	1	25	1.8
		2	25	2.2
		3	25	8.8
		4	25	9.9
$C_5H_{10}N_2O_3$	L-Glutamine	1	25	2.17
		2	25	9.13
$C_5H_{10}N_2O_3$	Glycylalanine		25	3.15
$C_5H_{10}N_2O_4$	Glycylserine	1	25	2.98
		2	25	8.38
$C_5H_{10}O_2$	2-Methylbutanoic acid		25	4.80
$C_5H_{10}O_2$	3-Methylbutanoic acid		25	4.77
$C_5H_{10}O_2$	Pentanoic acid		25	4.84
$C_5H_{10}O_2$	Trimethylacetic acid		25	5.03
$C_5H_{11}N$	N-Methylpyrrolidine		25	10.32
$C_5H_{11}N$	Piperidine		25	11.123
$C_5H_{11}NO_2$	5-Aminopentanoic acid	1	25	4.27
		2	25	10.766
$C_5H_{11}NO_2$	Betaine		0	1.83
$C_5H_{11}NO_2$	Norvaline	1	25	2.32
		2	25	9.81
$C_5H_{11}NO_2$	N-Propylglycine	1	25	2.35
		2	25	10.19
$C_5H_{11}NO_2$	Valine	1	25	2.29
		2	25	9.74
$C_5H_{11}NO_2S$	Methionine	1	25	2.13
		2	25	9.27
$C_5H_{12}N_2O_2$	Ornithine	1	25	1.705
		2	25	8.69
		3	25	10.76

Molecular formula	Name	Step	$T/°C$	pK_a
$C_5H_{13}N$	1-Amino-2,2-dimethylpropane		25	10.15
$C_5H_{13}N$	Diethylmethylamine		25	10.35
$C_5H_{13}N$	3-Methyl-1-butanamine		25	10.60
$C_5H_{13}N$	2-Methyl-2-butanamine		19	10.85
$C_5H_{13}N$	3-Pentanamine		17	10.59
$C_5H_{13}N$	Pentylamine		25	10.63
$C_5H_{14}NO$	Choline		25	13.9
$C_5H_{14}N_2$	Cadaverine	1	25	10.05
		2	25	10.93
$C_6H_3ClN_4$	6-Chloropteridine		20	3.68
$C_6H_3N_3O_7$	Picric acid		25	0.38
$C_6H_4Cl_2O$	2,3-Dichlorophenol		25	7.44
$C_6H_4N_2O_3$	5-Hydroxylysine		25	2.13
$C_6H_4N_2O_5$	2,4-Dinitrophenol		15	3.96
$C_6H_4N_2O_5$	3,6-Dinitrophenol		15	5.15
$C_6H_4N_4$	Pteridine		20	4.05
$C_6H_5Br_2N$	3,5-Dibromoaniline		25	2.34
C_6H_5ClO	2-Chlorophenol		25	8.49
C_6H_5ClO	3-Chlorophenol		25	8.85
C_6H_5ClO	4-Chlorophenol		25	9.18
$C_6H_5Cl_2N$	2,4-Dichloroaniline		22	2.05
C_6H_5NO	2-Pyridinecarboxaldehyde		20	3.80
$C_6H_5NO_2$	Nitrobenzene		0	3.98
$C_6H_5NO_2$	Picolinic acid	1	25	1.07
		2	25	5.25
$C_6H_5NO_2$	3-Pyridinecarboxylic acid		25	4.85
$C_6H_5NO_2$	4-Pyridinecarboxylic acid		25	4.96
$C_6H_5NO_3$	2-Nitrophenol		25	7.17
$C_6H_5NO_3$	3-Nitrophenol		25	8.28
$C_6H_5NO_3$	4-Nitrophenol		25	7.15
$C_6H_5N_5O$	2-Amino-4-hydroxypteridine	1	20	2.27
		2	20	7.96
$C_6H_5N_5O_2$	Xanthopterin	2	20	6.59
		3	20	9.31
C_6H_6BrN	2-Bromoaniline		25	2.53
C_6H_6BrN	3-Bromoaniline		25	3.58
C_6H_6BrN	4-Bromoaniline		25	3.86
C_6H_6ClN	2-Chloroaniline		25	2.65
C_6H_6ClN	3-Chloroaniline		25	3.46
C_6H_6ClN	4-Chloroaniline		25	4.15
C_6H_6FN	2-Fluoroaniline		25	3.20
C_6H_6FN	3-Fluoroaniline		25	3.50
C_6H_6FN	4-Fluoroaniline		25	4.65
C_6H_6IN	2-Iodoaniline		25	2.60
$C_6H_6N_2O$	2-Pyridinecarboxaldehyde, oxime	1	20	3.59
		2	20	10.18
$C_6H_6N_2O_2$	2-Nitroaniline		25	-0.26
$C_6H_6N_2O_2$	3-Nitroaniline		25	2.466
$C_6H_6N_2O_2$	4-Nitroaniline		25	1.0
C_6H_6O	Phenol		20	9.89
$C_6H_6O_2$	Hydroquinone		20	10.35
$C_6H_6O_2$	Pyrocatcchol		20	9.85
$C_6H_6O_2$	Resorcinol		25	9.81
$C_6H_6O_3S$	Benzenesulfonic acid		25	0.70
$C_6H_6O_6$	cis-Aconitic acid		25	1.95
$C_6H_6O_6$	trans-Aconitic acid	1	25	2.80
		2	25	4.46
C_6H_7N	Aniline		25	4.63
C_6H_7N	2-Methylpyridine		20	5.97

Molecular formula	Name	Step	T/°C	pK$_a$
C$_6$H$_7$N	3-Methylpyridine		20	5.68
C$_6$H$_7$N	4-Methylpyridine		20	6.02
C$_6$H$_7$NO	Methoxypyridine		25	6.47
C$_6$H$_7$NO$_3$S	o-Aminobenzenesulfonic acid		25	2.48
C$_6$H$_7$NO$_3$S	m-Aminobenzenesulfonic acid		25	3.73
C$_6$H$_7$NO$_3$S	p-Aminobenzenesulfonic acid		25	3.24
C$_6$H$_8$N$_2$	N-Methylpyridinamine		20	9.65
C$_6$H$_8$O$_6$	Ascorbic acid	1	24	4.10
		2	16	11.79
C$_6$H$_8$O$_7$	Citric acid	1	20	3.14
		2	20	4.77
		3	20	6.39
C$_6$H$_8$O$_7$	Isocitric acid	1	25	3.29
		2	25	4.71
		3	25	6.40
C$_6$H$_9$NO$_6$	γ-Carboxyglutamic acid	1	25	1.7
		2	25	3.2
		3	25	4.75
		4	25	9.9
C$_6$H$_9$N$_3$	4,6-Dimethylpyrimidinamine		20	4.82
C$_6$H$_9$N$_3$O$_2$	Histidine	1	25	1.80
		2	25	6.04
		3	25	9.33
C$_6$H$_{10}$O$_3$	2-Oxo-3-methylpentanoic acid		25	2.3
C$_6$H$_{10}$O$_4$	Adipic acid	1	25	4.43
		2	25	5.41
C$_6$H$_{10}$O$_4$	3-Methylglutaric acid		25	4.24
C$_6$H$_{11}$NO$_2$	l-Pipecolic acid	1	25	2.28
		2	25	10.72
C$_6$H$_{11}$NO$_3$	Adipamic acid		25	4.63
C$_6$H$_{11}$NO$_4$	2-Aminoadipic acid	1	25	2.14
		2	25	4.21
		3	25	9.77
C$_6$H$_{11}$N$_3$O$_4$	Glycylasparagine	1	25	2.942
		2	18	8.44
C$_6$H$_{11}$N$_3$O$_4$	N-(N-Glycylglycyl)glycine	1	25	3.225
		2	25	8.09
C$_6$H$_{12}$N$_2$O$_4$S$_2$	Cystine	1	35	2.1
		2	35	8.0
C$_6$H$_{12}$O$_2$	Hexanoic acid		25	4.85
C$_6$H$_{12}$O$_2$	4-Methylpentanoic acid		18	4.84
C$_6$H$_{13}$N	Cyclohexylamine		24	10.66
C$_6$H$_{13}$N	1,2-Dimethylpyrrolidine		26	10.20
C$_6$H$_{13}$N	1-Methylpiperidine		25	10.08
C$_6$H$_{13}$NO$_2$	6-Aminohexanoic acid	1	25	4.373
		2	25	10.804
C$_6$H$_{13}$NO$_2$	Isoleucine	1	25	2.32
		2	25	9.76
C$_6$H$_{13}$NO$_2$	L-Leucine	1	25	2.328
		2	25	9.744
C$_6$H$_{13}$NO$_2$	Norleucine	1	25	2.335
		2	25	9.83
C$_6$H$_{13}$N$_3$O$_3$	Citrulline	1	25	2.43
		2	25	9.69
C$_6$H$_{14}$N$_2$	cis-1,2-Cyclohexanediamine	1	20	9.93
		2	20	6.13
C$_6$H$_{14}$N$_2$	trans-1,2-Cyclohexanediamine	1	20	9.94
		2	20	6.47
C$_6$H$_{14}$N$_2$	2,5-Dimethylpiperazine	1	25	9.66
		2	25	5.20

Molecular formula	Name	Step	$T/°C$	pK_a
$C_6H_{14}N_2O_2$	Lysine	1	25	2.16
		2	25	9.06
		3	25	10.54
$C_6H_{14}N_4O_2$	Arginine	1	25	1.82
		2	25	8.99
		3	25	12.48
$C_6H_{15}N$	3-Amino-3-methylpentane		16	11.01
$C_6H_{15}N$	Diisopropylamine		25	11.05
$C_6H_{15}N$	Hexylamine		25	10.56
$C_6H_{15}N$	Triethylamine		25	10.75
$C_6H_{16}N_2$	Hexamethylenediamine	1	0	11.857
		2	0	10.762
$C_7H_5BrO_2$	2-Bromobenzoic acid		25	2.84
$C_7H_5BrO_2$	3-Bromobenzoic acid		25	3.86
$C_7H_5ClO_2$	2-Chlorobenzoic acid		25	2.92
$C_7H_5ClO_2$	3-Chlorobenzoic acid		25	3.82
$C_7H_5ClO_2$	4-Chlorobenzoic acid		25	3.98
$C_7H_5IO_2$	2-Iodobenzoic acid		25	2.85
$C_7H_5IO_2$	3-Iodobenzoic acid		25	3.80
$C_7H_5NO_3S$	Saccharin		18	11.68
$C_7H_5NO_4$	Dinicotinic acid		25	2.80
$C_7H_5NO_4$	Dipicolinic acid	1	25	2.16
		2	25	4.76
$C_7H_5NO_4$	Lutidinic acid		25	2.15
$C_7H_5NO_4$	2-Nitrobenzoic acid		18	2.16
$C_7H_5NO_4$	3-Nitrobenzoic acid		25	3.47
$C_7H_5NO_4$	4-Nitrobenzoic acid		25	3.41
$C_7H_5NO_4$	Quinolinic acid	1	25	2.43
		2	25	4.78
$C_7H_6N_2$	Benzimidazole		25	5.532
$C_7H_6N_4O$	6-Hydroxy-4-methylpteridine	1	20	4.08
		2	20	6.41
$C_7H_6O_2$	Benzoic acid		25	4.19
$C_7H_6O_3$	o-Hydroxybenzoic acid	1	19	2.97
		2	18	13.40
$C_7H_6O_3$	m-Hydroxybenzoic acid	1	19	4.06
		2	19	9.92
$C_7H_6O_3$	p-Hydroxybenzoic acid	1	19	4.48
		2	19	9.32
$C_7H_6O_4$	2,5-Dihydroxybenzoic acid	1	25	2.97
$C_7H_6O_4$	3,4-Dihydroxybenzoic acid	1	25	4.48
		2	25	8.83
		3	25	12.6
$C_7H_6O_4$	3,5-Dihydroxybenzoic acid	1	25	4.04
$C_7H_6O_4$	Dihydroxymalic acid		25	1.92
$C_7H_6O_5$	Gallic acid		25	4.41
$C_7H_6O_5$	2,4,6-Trihydroxybenzoic acid		25	1.68
$C_7H_7NO_2$	2-Aminobenzoic acid	1	25	2.108
		2	25	4.946
$C_7H_7NO_2$	3-Aminobenzoic acid	2	25	4.78
$C_7H_7NO_2$	4-Aminobenzoic acid	1	25	2.501
		2	25	4.874
$C_7H_8N_4O_2$	Theobromine		18	7.89
C_7H_8O	o-Cresol		25	10.20
C_7H_8O	m-Cresol		25	10.01
C_7H_8O	p-Cresol		25	10.17
C_7H_9N	Benzylamine		25	9.33
C_7H_9N	2,3-Dimethylpyridine		25	6.57
C_7H_9N	2,4-Dimethylpyridine		25	6.99

Molecular formula	Name	Step	$T/°C$	pK_a
C_7H_9N	2,5-Dimethylpyridine		25	6.40
C_7H_9N	2,6-Dimethylpyridine		25	6.65
C_7H_9N	3,4-Dimethylpyridine		25	6.46
C_7H_9N	3,5-Dimethylpyridine		25	6.15
C_7H_9N	2-Ethylpyridine		25	5.89
C_7H_9N	N-Methylaniline		25	4.84
C_7H_9N	o-Methylaniline		25	4.44
C_7H_9N	m-Methylaniline		25	4.73
C_7H_9N	p-Methylaniline		25	5.08
C_7H_9NO	o-Anisidine		25	4.52
C_7H_9NO	m-Anisidine		25	4.23
C_7H_9NO	p-Anisidine		25	5.34
C_7H_9NS	4-Methylthioaniline		25	4.35
$C_7H_9N_5$	2-Dimethylaminopurine	1	20	4.00
		2	20	10.24
$C_7H_{11}N_3O_2$	l-1-Methylhistidine	1	25	1.69
		2	25	6.48
		3	25	8.85
$C_7H_{11}N_3O_2$	l-3-Methylhistidine	1	25	1.92
		2	25	6.56
		3	25	8.73
$C_7H_{12}O_2$	Cyclohexanecarboxylic acid		25	4.90
$C_7H_{12}O_4$	Heptanedioic acid	1	25	4.71
		2	25	5.58
$C_7H_{13}NO$	3-Acetylpiperidine		25	3.18
$C_7H_{13}NO_4$	α-Ethylglutamic acid	1	25	3.846
		2	25	7.838
$C_7H_{14}O_2$	Heptanoic acid		25	4.89
$C_7H_{15}N$	1,2-Dimethylpiperidine		25	10.22
$C_7H_{15}N$	1-Ethylpiperidine		23	10.45
$C_7H_{15}NO_3$	Carnitine		25	3.80
$C_7H_{17}N$	2-Heptanamine		19	10.7
$C_7H_{17}N$	Heptylamine		25	10.67
$C_8H_6N_2$	Cinnoline		20	2.37
$C_8H_6N_2$	Quinazoline		20	3.43
$C_8H_6N_2$	Quinoxaline		20	0.56
$C_8H_6N_2O$	5-Hydoxyquinazoline	1	20	3.62
		2	20	7.41
$C_8H_6O_4$	o-Phthalic acid	1	25	2.89
		2	25	5.51
$C_8H_6O_4$	m-Phthalic acid	1	25	3.54
		2	18	4.60
$C_8H_6O_4$	p-Phthalic acid	1	25	3.51
		2	16	4.82
$C_8H_6O_4$	Terephthalic acid		25	3.51
$C_8H_7ClO_2$	2-Chlorophenylacetic acid		25	4.07
$C_8H_7ClO_2$	3-Chlorophenylacetic acid		25	4.14
$C_8H_7ClO_2$	4-Chlorophenylacetic acid		25	4.19
$C_8H_7ClO_3$	2-Chlorophenoxyacetic acid		25	3.05
$C_8H_7ClO_3$	3-Chlorophenoxyacetic acid		25	3.10
$C_8H_7NO_4$	2-Nitrophenylacetic acid		25	4.00
$C_8H_7NO_4$	3-Nitrophenylacetic acid		25	3.97
$C_8H_7NO_4$	4-Nitrophenylacetic acid		25	3.85
$C_8H_8N_2$	2-Methylbenzimidazole		25	6.19
$C_8H_8O_2$	Phenylacetic acid		18	4.28
$C_8H_8O_2$	o-Toluic acid		25	3.91
$C_8H_8O_2$	m-Toluic acid		25	4.27
$C_8H_8O_2$	p-Toluic acid		25	4.36
$C_8H_8O_3$	DL-Mandelic acid		25	3.85

Molecular formula	Name	Step	$T/^\circ C$	pK_a
$C_8H_8O_4$	Homogentisic acid		25	4.40
$C_8H_9NO_2$	2-(Methylamino)benzoic acid		25	5.34
$C_8H_9NO_2$	3-(Methylamino)benzoic acid		25	5.10
$C_8H_9NO_2$	4-(Methylamino)benzoic acid		25	5.04
$C_8H_9NO_2$	Phenylglycine	1	25	1.83
		2	25	4.39
$C_8H_{10}BrN$	4-Bromo-N,N-dimethylaniline		25	4.23
$C_8H_{10}ClN$	3-Chloro-N,N-dimethylaniline		20	3.83
$C_8H_{10}ClN$	4-Chloro-N,N-dimethylaniline		20	4.39
$C_8H_{10}N_2O_2$	N,N-Dimethyl-3-nitroaniline		25	2.62
$C_8H_{11}N$	N,N-Dimethylaniline		25	5.15
$C_8H_{11}N$	N-Ethylaniline		24	5.12
$C_8H_{11}N$	Phenylethylamine		25	9.84
$C_8H_{11}N$	2,4,6-Trimethylpyridine		25	7.43
$C_8H_{11}NO$	o-Phenetidine		28	4.43
$C_8H_{11}NO$	m-Phenetidine		25	4.18
$C_8H_{11}NO$	p-Phenetidine		28	5.20
$C_8H_{11}NO$	Tyramine	1	25	9.74
		2	25	10.52
$C_8H_{11}NO_2$	Dopamine	1	25	8.9
		2	25	10.6
$C_8H_{11}NO_3$	Noradrenaline	1	25	8.64
		2	25	9.70
$C_8H_{12}N_2O_3$	Veronal		25	7.43
$C_8H_{14}O_4$	Octanedioic acid		25	4.52
$C_8H_{16}N_2O_3$	N-Glycylleucine		25	3.18
$C_8H_{16}N_2O_3$	N-Leucylglycine	1	25	3.25
		2	25	8.2
$C_8H_{16}N_2O_4S_2$	Homocystine	1	25	1.59
		2	25	2.54
		3	25	8.52
		4	25	9.44
$C_8H_{16}O_2$	Octanoic acid		25	4.89
$C_8H_{17}N$	2,2,4-Trimethylpiperidine		30	11.04
$C_8H_{19}N$	Dibutylamine		21	11.25
$C_8H_{19}N$	N-Methyl-2-heptanamine		17	10.99
$C_8H_{19}N$	Octylamine		25	10.65
C_9H_6BrN	3-Bromoquinoline		25	2.69
$C_9H_7ClO_2$	o-Chlorocinnamic acid		25	4.23
$C_9H_7ClO_2$	m-Chlorocinnamic acid		25	4.29
$C_9H_7ClO_2$	p-Chlorocinnamic acid		25	4.41
C_9H_7N	Isoquinoline		20	5.42
C_9H_7N	Quinoline		20	4.90
C_9H_7NO	7-Isoquinolinol	1	20	5.68
		2	20	8.90
C_9H_7NO	3-Quinolinol	1	20	4.28
		2	20	8.08
C_9H_7NO	8-Quinolinol	1	20	5.017
		2	25	9.812
$C_9H_7NO_3$	o-Cyanophenoxyacetic acid		25	2.98
$C_9H_7NO_3$	m-Cyanophenoxyacetic acid		25	3.03
$C_9H_7NO_3$	p-Cyanophenoxyacetic acid		25	2.93
$C_9H_8N_2$	1-Isoquinolinamine		20	7.59
$C_9H_8N_2$	3-Quinolinamine		20	4.91
$C_9H_8O_2$	cis-Cinnamic acid		25	3.89
$C_9H_8O_2$	trans-Cinnamic acid		25	4.44
$C_9H_9ClO_2$	3-(2-Chlorophenyl)propanoic acid		25	4.58
$C_9H_9ClO_2$	3-(3-Chlorophenyl)propanoic acid		25	4.59
$C_9H_9ClO_2$	3-(4-Chlorophenyl)propanoic acid		25	4.61

Molecular formula	Name	Step	$T/°C$	pK_a
$C_9H_9I_2NO_3$	3,5-Diiodotyrosine	1	25	2.12
		2	25	5.32
		3	25	9.48
$C_9H_9NO_3$	Hippuric acid		25	3.62
$C_9H_9NO_4$	3-(2-Nitrophenyl)propanoic acid		25	4.50
$C_9H_9NO_4$	3-(4-Nitrophenyl)propanoic acid		25	4.47
$C_9H_{10}INO_3$	3-Iodotyrosine	1	25	2.2
		2	25	8.7
		3	25	9.1
$C_9H_{10}N_2$	2-Ethylbenzimidazole		25	6.18
$C_9H_{10}O_2$	Mesitylenic acid		25	4.32
$C_9H_{10}O_2$	α-Phenylpropanoic acid		25	4.64
$C_9H_{10}O_2$	β-Phenylpropanoic acid		25	4.37
$C_9H_{11}N$	N-Allylaniline		25	4.17
$C_9H_{11}N$	1-Aminoindane		22	9.21
$C_9H_{11}NO_2$	Phenylalanine	1	25	2.20
		2	25	9.31
$C_9H_{11}NO_3$	Tyrosine	1	25	2.20
		2	25	9.21
		3	25	10.46
$C_9H_{11}NO_4$	L-3,4-Dihydroxyphenylalanine	1	25	2.32
		2	25	8.72
		3	25	9.96
		4	25	11.79
$C_9H_{12}N_2O_2$	Tyrosineamide		25	7.33
$C_9H_{13}NO_3$	D-Adrenaline	1	25	8.66
		2	25	9.95
$C_9H_{14}N_4O_3$	N-β-Alanylhistidine	1	20	2.73
		2	20	6.87
		3	20	9.73
$C_9H_{14}N_4O_3$	L-Carnosine	1	25	2.62
		2	25	6.66
		3	25	9.24
$C_9H_{18}O_2$	Nonanic acid		25	4.96
$C_9H_{19}N$	1-Butylpiperidine		23	10.47
$C_9H_{19}N$	2,2,6,6-Tetramethylpiperidine		25	11.07
$C_9H_{21}N$	Nonylamine		25	10.64
$C_{10}H_7NO_2$	8-Quinolinecarboxylic acid		25	1.82
$C_{10}H_8O$	1-Naphthol		25	9.34
$C_{10}H_8O$	2-Naphthol		25	9.51
$C_{10}H_9N$	2-Methylquinoline		20	5.83
$C_{10}H_9N$	4-Methylquinoline		20	5.67
$C_{10}H_9N$	5-Methylquinoline		20	5.20
$C_{10}H_9N$	α-Naphthylamine		25	3.92
$C_{10}H_9N$	β-Naphthylamine		25	4.16
$C_{10}H_9NO$	1-Amino-6-hydroxynaphthalene		25	3.97
$C_{10}H_9NO$	6-Methoxyquinoline		20	5.03
$C_{10}H_{10}O_2$	o-Methylcinnamic acid		25	4.50
$C_{10}H_{10}O_2$	m-Methylcinnamic acid		25	4.44
$C_{10}H_{10}O_2$	p-Methylcinnamic acid		25	4.56
$C_{10}H_{12}N_2$	Tryptamine		25	10.2
$C_{10}H_{12}N_2O$	5-Hydroxytryptamine	1	25	9.8
		2	25	11.1
$C_{10}H_{12}O_2$	4-Phenylbutanoic acid		25	4.76
$C_{10}H_{12}O_3$	2-(m-Anisyl)propanoic acid		25	4.65
$C_{10}H_{12}O_3$	2-(p-Anisyl)propanoic acid		25	4.69
$C_{10}H_{12}O_3$	2-(o-Anisyl)propanonic acid		25	4.80
$C_{10}H_{14}N_2$	Nicotine	1	25	8.02
		2	25	3.12

Molecular formula	Name	Step	$T/°C$	pK_a
$C_{10}H_{15}N$	N,N-Diethylaniline		22	6.61
$C_{10}H_{15}NO$	d-Ephedrine		10	10.139
$C_{10}H_{15}NO$	l-Ephedrine		10	9.958
$C_{10}H_{17}N_3O_6S$	l-Glutathione	1	25	2.12
		2	25	3.59
		3	25	8.75
		4	25	9.65
$C_{10}H_{17}N_5O_6$	Tetraglycylglycine	1	20	3.10
		2	20	8.02
$C_{10}H_{18}N_4O_5$	L-Argininosuccinic acid	1	25	1.62
		2	25	2.70
		3	25	4.26
		4	25	9.58
$C_{10}H_{19}N$	Bornylamine		25	10.17
$C_{10}H_{19}N$	Neobornylamine		25	10.01
$C_{10}H_{21}N$	Butylcyclohexylamine		25	11.23
$C_{10}H_{23}N$	Decylamine		25	10.64
$C_{11}H_8N_2$	Perimidine		20	6.35
$C_{11}H_8O_2$	1-Naphthoic acid		25	3.70
$C_{11}H_8O_2$	2-Naphthoic acid		25	4.17
$C_{11}H_{11}N$	Methyl-1-naphthylamine		27	3.67
$C_{11}H_{12}N_2O_2$	Tryptophan	1	25	2.46
		2	25	9.41
$C_{11}H_{16}N_2O_2$	Pilocarpine		30	6.87
$C_{11}H_{25}N$	Undecylamine		25	10.63
$C_{12}H_8N_2$	1,10-Phenanthroline		25	4.84
$C_{12}H_{11}N$	2-Aminobiphenyl		22	3.82
$C_{12}H_{11}N$	2-Benzylpyridine		25	5.13
$C_{12}H_{11}N$	Diphenylamine		25	0.79
$C_{12}H_{11}N_3$	4-Aminoazobenzene		25	2.82
$C_{12}H_{12}N_2$	p-Benzidine	1	30	4.66
		2	30	3.57
$C_{12}H_{13}N$	N,N-Dimethyl-1-naphthylamine		25	4.83
$C_{12}H_{13}N$	N,N-Dimethyl-2-naphthylamine		25	4.566
$C_{12}H_{27}N$	Dodecylamine		25	10.63
$C_{13}H_9N$	Acridine		20	5.58
$C_{13}H_9N$	Phenanthridine		20	5.58
$C_{13}H_{10}N_2$	2-Phenylbenzimidazole	1	25	5.23
		2	25	11.91
$C_{13}H_{10}O_2$	2-Phenylbenzoic acid		25	3.46
$C_{13}H_{12}N_2O$	4-(p-Aminobenzoyl)aniline	1	25	2.93
$C_{13}H_{13}N$	4-Benzylaniline		25	2.17
$C_{13}H_{29}N$	Tridecylamine		25	10.63
$C_{14}H_{12}O_2$	Diphenylacetic acid		25	3.94
$C_{14}H_{15}N_3$	4-Dimethylaminoazobenzene		25	3.226
$C_{14}H_{31}N$	Tetradecylamine		25	10.62
$C_{15}H_{11}I_4NO_4$	L-Thyroxine	1	25	2.2
		2	25	6.45
		3	25	10.1
$C_{15}H_{33}N$	Pentadecylamine		25	10.61
$C_{16}H_{35}N$	Hexadecylamine		25	10.63
$C_{17}H_{19}NO_3$	Morphine		25	8.21
$C_{18}H_{21}NO_3$	Codeine		25	8.21
$C_{18}H_{39}N$	Octadecylamine		25	10.60
$C_{20}H_{21}NO_4$	Papaverine		25	6.40
$C_{20}H_{24}N_2O_2$	Quinine	1	25	8.52
		2	25	4.13
$C_{21}H_{22}N_2O_2$	Strychnine		25	8.26
$C_{23}H_{26}N_2O_4$	Brucine	1	25	8.28

CONCENTRATIVE PROPERTIES OF AQUEOUS SOLUTIONS:
DENSITY, REFRACTIVE INDEX, FREEZING POINT DEPRESSION, AND VISCOSITY

This table gives properties of aqueous solutions of 66 substances as a function of concentration. All data refer to a temperature of 20°C. The properties are:

Mass %:   Mass of solute divided by total mass of solution, expressed as percent.
m Molality (moles of solute per kg of water).
c Molarity (moles of solute per liter of solution).
ρ Density of solution in g/cm^3.
n Index of refraction, relative to air, at a wavelength of 589 nm (sodium D line); the index of pure water at 20°C is 1.3330.
Δ Freezing point depression in °C relative to pure water.
η Absolute (dynamic) viscosity in mPa s (equal to centipoise, cP); the viscosity of pure water at 20°C is 1.002 mPa s.

Density data for aqueous solutions over a wider range of temperatures and pressures (and for other compounds) may be found in Reference 2. Solutes are listed in the following order:

Acetic acid	Lithium chloride	2-Propanol
Acetone	Magnesium chloride	Silver nitrate
Ammonia	Magnesium sulfate	Sodium acetate
Ammonium chloride	Maltose	Sodium bicarbonate
Ammonium sulfate	Manganese(II) sulfate	Sodium bromide
Barium chloride	D-Mannitol	Sodium carbonate
Calcium chloride	Methanol	Sodium chloride
Cesium chloride	Nitric acid	Sodium citrate
Citric acid	Oxalic acid	Sodium hydroxide
Copper sulfate	Phosphoric acid	Sodium nitrate
Disodium ethylenediamine	Potassium bicarbonate	Sodium phosphate
tetraacetate (EDTA sodium)	Potassium bromide	Sodium hydrogen phosphate
Ethanol	Potassium carbonate	Sodium dihydrogen phosphate
Ethylene glycol	Potassium chloride	Sodium sulfate
Ferric chloride	Potassium hydroxide	Sodium thiosulfate
Formic acid	Potassium iodide	Strontium chloride
D-Fructose	Potassium nitrate	Sucrose
D-Glucose	Potassium permanganate	Sulfuric acid
Glycerol	Potassium hydrogen phosphate	Trichloroacetic acid
Hydrochloric acid	Potassium dihydrogen phosphate	Tris(hydroxymethyl)methylamine
Lactic acid	Potassium sulfate	Urea
Lactose	1-Propanol	Zinc sulfate

REFERENCES

1. Wolf, A. V., Aqueous Solutions and Body Fluids, Hoeber, 1966.
2. Söhnel, O., and Novotny, P., *Densities of Aqueous Solutions of Inorganic Substances*, Elsevier, Amsterdam, 1985.

Solute	Mass %	m/mol kg^{-1}	c/mol L^{-1}	ρ/g cm^{-3}	n	Δ/°C	η/mPa s
Acetic acid	0.5	0.084	0.083	0.9989	1.3334	0.16	1.012
CH_3COOH	1.0	0.168	0.166	0.9996	1.3337	0.32	1.022
	2.0	0.340	0.333	1.0011	1.3345	0.63	1.042
	3.0	0.515	0.501	1.0025	1.3352	0.94	1.063
	4.0	0.694	0.669	1.0038	1.3359	1.26	1.084
	5.0	0.876	0.837	1.0052	1.3366	1.58	1.105
	6.0	1.063	1.006	1.0066	1.3373	1.90	1.125
	7.0	1.253	1.175	1.0080	1.3381	2.23	1.143
	8.0	1.448	1.345	1.0093	1.3388	2.56	1.162
	9.0	1.647	1.515	1.0107	1.3395	2.89	1.186
	10.0	1.850	1.685	1.0121	1.3402	3.23	1.210
	12.0	2.271	2.028	1.0147	1.3416	3.91	1.253
	14.0	2.711	2.372	1.0174	1.3430	4.61	1.298
	16.0	3.172	2.718	1.0200	1.3444	5.33	1.341
	18.0	3.655	3.065	1.0225	1.3458	6.06	1.380
	20.0	4.163	3.414	1.0250	1.3472	6.81	1.431

Solute	Mass %	m/mol kg^{-1}	c/mol L^{-1}	ρ/g cm^{-3}	n	$\Delta/°C$	η/mPa s
	22.0	4.697	3.764	1.0275	1.3485	7.57	1.478
	24.0	5.259	4.116	1.0299	1.3498	8.36	1.525
	26.0	5.851	4.470	1.0323	1.3512	9.17	1.572
	28.0	6.476	4.824	1.0346	1.3525	10.00	1.613
	30.0	7.137	5.180	1.0369	1.3537	10.84	1.669
	32.0	7.837	5.537	1.0391	1.3550	11.70	1.715
	34.0	8.579	5.896	1.0413	1.3562	12.55	1.762
	36.0	9.367	6.255	1.0434	1.3574	13.38	1.812
	38.0	10.207	6.615	1.0454	1.3586		1.852
	40.0	11.102	6.977	1.0474	1.3598		1.912
	50.0	16.653	8.794	1.0562	1.3653		2.158
	60.0	24.979	10.620	1.0629	1.3700		2.409
	70.0	38.857	12.441	1.0673	1.3738		2.629
	80.0	66.611	14.228	1.0680	1.3767		2.720
	90.0	149.875	15.953	1.0644	1.3771		2.386
	92.0	191.507	16.284	1.0629	1.3766		2.240
	94.0	260.894	16.602	1.0606	1.3759		2.036
	96.0	399.667	16.911	1.0578	1.3748		1.813
	98.0	815.987	17.198	1.0538	1.3734		1.535
	100.0		17.447	1.0477	1.3716		1.223
Acetone	0.5	0.087	0.086	0.9975	1.3334	0.16	1.013
$(CH_3)_2CO$	1.0	0.174	0.172	0.9968	1.3337	0.32	1.024
	2.0	0.351	0.343	0.9954	1.3344	0.65	1.047
	3.0	0.533	0.513	0.9940	1.3352	0.97	1.072
	4.0	0.717	0.684	0.9926	1.3359	1.30	1.099
	5.0	0.906	0.853	0.9912	1.3366	1.63	1.125
	6.0	1.099	1.023	0.9899	1.3373	1.96	1.150
	7.0	1.296	1.191	0.9886	1.3381	2.29	1.174
	8.0	1.497	1.360	0.9874	1.3388	2.62	1.198
	9.0	1.703	1.528	0.9861	1.3395	2.95	1.221
	10.0	1.913	1.696	0.9849	1.3402	3.29	1.244
Ammonia	0.5	0.295	0.292	0.9960	1.3332	0.55	1.009
NH_3	1.0	0.593	0.584	0.9938	1.3335	1.14	1.015
	2.0	1.198	1.162	0.9895	1.3339	2.32	1.029
	3.0	1.816	1.736	0.9853	1.3344	3.53	1.043
	4.0	2.447	2.304	0.9811	1.3349	4.78	1.057
	5.0	3.090	2.868	0.9770	1.3354	6.08	1.071
	6.0	3.748	3.428	0.9730	1.3359	7.43	1.085
	7.0	4.420	3.983	0.9690	1.3365	8.95	1.099
	8.0	5.106	4.533	0.9651	1.3370	10.34	1.113
	9.0	5.807	5.080	0.9613	1.3376	11.90	1.127
	10.0	6.524	5.622	0.9575	1.3381	13.55	1.141
	12.0	8.007	6.695	0.9502	1.3393	17.13	1.169
	14.0	9.558	7.753	0.9431	1.3404	21.13	1.195
	16.0	11.184	8.794	0.9361	1.3416	25.63	1.218
	18.0	12.889	9.823	0.9294	1.3428	30.70	1.237
	20.0	14.679	10.837	0.9228	1.3440	36.42	1.254
	22.0	16.561	11.838	0.9164	1.3453	43.36	1.268
	24.0	18.542	12.826	0.9102	1.3465	51.38	1.280
	26.0	20.630	13.801	0.9040	1.3477	60.77	1.288
	28.0	22.834	14.764	0.8980	1.3490	71.66	
	30.0	25.164	15.713	0.8920	1.3502	84.06	
Ammonium	0.5	0.094	0.093	0.9998	1.3340	0.32	0.999
chloride	1.0	0.189	0.187	1.0014	1.3349	0.64	0.996
NH_4Cl	2.0	0.382	0.376	1.0045	1.3369	1.27	0.992

Solute	Mass %	m/mol kg^{-1}	c/mol L^{-1}	ρ/g cm^{-3}	n	$\Delta/°C$	η/mPa s
	3.0	0.578	0.565	1.0076	1.3388	1.91	0.988
	4.0	0.779	0.756	1.0107	1.3407	2.57	0.985
	5.0	0.984	0.948	1.0138	1.3426	3.25	0.982
	6.0	1.193	1.141	1.0168	1.3445	3.94	0.979
	7.0	1.407	1.335	1.0198	1.3464	4.66	0.976
	8.0	1.626	1.529	1.0227	1.3483	5.40	0.974
	9.0	1.849	1.726	1.0257	1.3502	6.16	0.972
	10.0	2.077	1.923	1.0286	1.3521	6.95	0.970
	12.0	2.549	2.320	1.0344	1.3559	8.60	0.969
	14.0	3.043	2.722	1.0401	1.3596		0.969
	16.0	3.561	3.128	1.0457	1.3634		0.971
	18.0	4.104	3.537	1.0512	1.3671		0.973
	20.0	4.674	3.951	1.0567	1.3708		0.978
	22.0	5.273	4.368	1.0621	1.3745		0.986
	24.0	5.903	4.789	1.0674	1.3782		0.996
Ammonium	0.5	0.038	0.038	1.0012	1.3338	0.17	1.008
sulfate	1.0	0.076	0.076	1.0042	1.3346	0.33	1.014
$(NH_4)_2SO_4$	2.0	0.154	0.153	1.0101	1.3363	0.63	1.027
	3.0	0.234	0.231	1.0160	1.3379	0.92	1.041
	4.0	0.315	0.309	1.0220	1.3395	1.21	1.057
	5.0	0.398	0.389	1.0279	1.3411	1.49	1.073
	6.0	0.483	0.469	1.0338	1.3428	1.77	1.090
	7.0	0.570	0.551	1.0397	1.3444	2.05	1.108
	8.0	0.658	0.633	1.0456	1.3460	2.33	1.127
	9.0	0.748	0.716	1.0515	1.3476	2.61	1.147
	10.0	0.841	0.800	1.0574	1.3492	2.89	1.168
	12.0	1.032	0.971	1.0691	1.3523	3.47	1.210
	14.0	1.232	1.145	1.0808	1.3555	4.07	1.256
	16.0	1.441	1.323	1.0924	1.3586	4.69	1.305
	18.0	1.661	1.504	1.1039	1.3616		1.359
	20.0	1.892	1.688	1.1154	1.3647		1.421
	22.0	2.134	1.876	1.1269	1.3677		1.490
	24.0	2.390	2.067	1.1383	1.3707		1.566
	26.0	2.659	2.262	1.1496	1.3737		1.650
	28.0	2.943	2.460	1.1609	1.3766		1.743
	30.0	3.243	2.661	1.1721	1.3795		1.847
	32.0	3.561	2.866	1.1833	1.3824		1.961
	34.0	3.898	3.073	1.1945	1.3853		2.086
	36.0	4.257	3.284	1.2056	1.3881		2.222
	38.0	4.638	3.499	1.2166	1.3909		2.371
	40.0	5.045	3.716	1.2277	1.3938		2.530
Barium	0.5	0.024	0.024	1.0026	1.3337	0.12	1.009
chloride	1.0	0.049	0.048	1.0070	1.3345	0.23	1.016
$BaCl_2$	2.0	0.098	0.098	1.0159	1.3360	0.46	1.026
	3.0	0.149	0.148	1.0249	1.3375	0.69	1.037
	4.0	0.200	0.199	1.0341	1.3391	0.93	1.049
	5.0	0.253	0.251	1.0434	1.3406	1.18	1.062
	6.0	0.307	0.303	1.0528	1.3422	1.44	1.075
	7.0	0.361	0.357	1.0624	1.3438	1.70	1.087
	8.0	0.418	0.412	1.0721	1.3454	1.98	1.101
	9.0	0.475	0.468	1.0820	1.3470	2.27	1.114
	10.0	0.534	0.524	1.0921	1.3487	2.58	1.129
	12.0	0.655	0.641	1.1128	1.3520	3.22	1.161
	14.0	0.782	0.763	1.1342	1.3555	3.92	1.195
	16.0	0.915	0.889	1.1564	1.3591	4.69	1.234

Solute	Mass %	m/mol kg^{-1}	c/mol L^{-1}	ρ/g cm^{-3}	n	Δ/°C	η/mPa s
	18.0	1.054	1.019	1.1793	1.3627		1.277
	20.0	1.201	1.156	1.2031	1.3664		1.325
	22.0	1.355	1.297	1.2277	1.3703		1.378
	24.0	1.517	1.444	1.2531	1.3741		1.437
	26.0	1.687	1.597	1.2793	1.3781		1.503
Calcium	0.5	0.045	0.045	1.0024	1.3342	0.22	1.015
chloride	1.0	0.091	0.091	1.0065	1.3354	0.44	1.028
CaCl$_2$	2.0	0.184	0.183	1.0148	1.3378	0.88	1.050
	3.0	0.279	0.277	1.0232	1.3402	1.33	1.078
	4.0	0.375	0.372	1.0316	1.3426	1.82	1.110
	5.0	0.474	0.469	1.0401	1.3451	2.35	1.143
	6.0	0.575	0.567	1.0486	1.3475	2.93	1.175
	7.0	0.678	0.667	1.0572	1.3500	3.57	1.208
	8.0	0.784	0.768	1.0659	1.3525	4.28	1.242
	9.0	0.891	0.872	1.0747	1.3549	5.04	1.279
	10.0	1.001	0.976	1.0835	1.3575	5.86	1.319
	12.0	1.229	1.191	1.1014	1.3625	7.70	1.408
	14.0	1.467	1.413	1.1198	1.3677	9.83	1.508
	16.0	1.716	1.641	1.1386	1.3730	12.28	1.625
	18.0	1.978	1.878	1.1579	1.3784	15.11	1.764
	20.0	2.253	2.122	1.1775	1.3839	18.30	1.930
	22.0	2.541	2.374	1.1976	1.3895	21.70	2.127
	24.0	2.845	2.634	1.2180	1.3951	25.30	2.356
	26.0	3.166	2.902	1.2388	1.4008	29.70	2.645
	28.0	3.504	3.179	1.2600	1.4066	34.70	3.000
	30.0	3.862	3.464	1.2816	1.4124	41.00	3.467
	32.0	4.240	3.759	1.3036	1.4183	49.70	4.035
	34.0	4.642	4.062	1.3260	1.4242		4.820
	36.0	5.068	4.375	1.3488	1.4301		5.807
	38.0	5.522	4.698	1.3720	1.4361		7.321
	40.0	6.007	5.030	1.3957	1.4420		8.997
Cesium	0.5	0.030	0.030	1.0020	1.3334	0.10	1.000
chloride	1.0	0.060	0.060	1.0058	1.3337	0.20	0.997
CsCl	2.0	0.121	0.120	1.0135	1.3345	0.40	0.992
	3.0	0.184	0.182	1.0214	1.3353	0.61	0.988
	4.0	0.247	0.245	1.0293	1.3361	0.81	0.984
	5.0	0.313	0.308	1.0374	1.3369	1.02	0.980
	6.0	0.379	0.373	1.0456	1.3377	1.22	0.977
	7.0	0.447	0.438	1.0540	1.3386	1.43	0.974
	8.0	0.516	0.505	1.0625	1.3394	1.64	0.971
	9.0	0.587	0.573	1.0711	1.3403	1.85	0.969
	10.0	0.660	0.641	1.0798	1.3412	2.06	0.966
	12.0	0.810	0.782	1.0978	1.3430	2.51	0.961
	14.0	0.967	0.928	1.1163	1.3448	2.97	0.955
	16.0	1.131	1.079	1.1355	1.3468	3.46	0.950
	18.0	1.304	1.235	1.1552	1.3487	3.96	0.945
	20.0	1.485	1.397	1.1756	1.3507	4.49	0.939
	22.0	1.675	1.564	1.1967	1.3528		0.934
	24.0	1.876	1.737	1.2185	1.3550		0.930
	26.0	2.087	1.917	1.2411	1.3572		0.926
	28.0	2.310	2.103	1.2644	1.3594		0.924
	30.0	2.546	2.296	1.2885	1.3617		0.922
	32.0	2.795	2.497	1.3135	1.3641		0.922
	34.0	3.060	2.705	1.3393	1.3666		0.924
	36.0	3.341	2.921	1.3661	1.3691		0.926

Solute	Mass %	m/mol kg^{-1}	c/mol L^{-1}	ρ/g cm^{-3}	n	Δ/°C	η/mPa s
	38.0	3.640	3.146	1.3938	1.3717		0.930
	40.0	3.960	3.380	1.4226	1.3744		0.934
	42.0	4.301	3.624	1.4525	1.3771		0.940
	44.0	4.667	3.877	1.4835	1.3800		0.947
	46.0	5.060	4.142	1.5158	1.3829		0.956
	48.0	5.483	4.418	1.5495	1.3860		0.967
	50.0	5.940	4.706	1.5846	1.3892		0.981
	60.0	8.910	6.368	1.7868	1.4076		1.120
	64.0	10.560	7.163	1.8842	1.4167		1.238
Citric acid	0.5	0.026	0.026	1.0002	1.3336	0.05	1.013
(HO)C(COOH)$_3$	1.0	0.053	0.052	1.0022	1.3343	0.11	1.024
	2.0	0.106	0.105	1.0063	1.3356	0.21	1.048
	3.0	0.161	0.158	1.0105	1.3368	0.32	1.073
	4.0	0.217	0.211	1.0147	1.3381	0.43	1.098
	5.0	0.274	0.265	1.0189	1.3394	0.54	1.125
	6.0	0.332	0.320	1.0232	1.3407	0.65	1.153
	7.0	0.392	0.374	1.0274	1.3420	0.76	1.183
	8.0	0.453	0.430	1.0316	1.3433	0.88	1.214
	9.0	0.515	0.485	1.0359	1.3446	1.00	1.247
	10.0	0.578	0.541	1.0402	1.3459	1.12	1.283
	12.0	0.710	0.655	1.0490	1.3486	1.38	1.357
	14.0	0.847	0.771	1.0580	1.3514	1.66	1.436
	16.0	0.991	0.889	1.0672	1.3541	1.95	1.525
	18.0	1.143	1.008	1.0764	1.3569	2.26	1.625
	20.0	1.301	1.130	1.0858	1.3598	2.57	1.740
	22.0	1.468	1.254	1.0953	1.3626	2.88	1.872
	24.0	1.644	1.380	1.1049	1.3655	3.21	2.017
	26.0	1.829	1.508	1.1147	1.3684	3.55	2.178
	28.0	2.024	1.639	1.1246	1.3714	3.89	2.356
	30.0	2.231	1.772	1.1346	1.3744	4.25	2.549
Copper	0.5	0.031	0.031	1.0033	1.3339	0.08	1.017
sulfate	1.0	0.063	0.063	1.0085	1.3348	0.14	1.036
CuSO$_4$	2.0	0.128	0.128	1.0190	1.3367	0.26	1.084
	3.0	0.194	0.194	1.0296	1.3386	0.37	1.129
	4.0	0.261	0.261	1.0403	1.3405	0.48	1.173
	5.0	0.330	0.329	1.0511	1.3424	0.59	1.221
	6.0	0.400	0.399	1.0620	1.3443	0.70	1.276
	7.0	0.472	0.471	1.0730	1.3462	0.82	1.336
	8.0	0.545	0.543	1.0842	1.3481	0.93	1.400
	9.0	0.620	0.618	1.0955	1.3501	1.05	1.469
	10.0	0.696	0.694	1.1070	1.3520	1.18	1.543
	12.0	0.854	0.850	1.1304	1.3560	1.45	1.701
	14.0	1.020	1.013	1.1545	1.3601	1.75	1.889
	16.0	1.193	1.182	1.1796	1.3644		2.136
	18.0	1.375	1.360	1.2059	1.3689		2.449
Disodium	0.5	0.015	0.015	1.0009	1.3339	0.07	1.017
ethylenediamine	1.0	0.030	0.030	1.0036	1.3348	0.14	1.032
tetraacetate	1.5	0.045	0.045	1.0062	1.3356	0.21	1.046
(EDTA sodium)	2.0	0.061	0.060	1.0089	1.3365	0.27	1.062
Na$_2$C$_{10}$H$_{14}$N$_2$O$_8$	2.5	0.076	0.075	1.0115	1.3374	0.33	1.077
	3.0	0.092	0.090	1.0142	1.3383	0.40	1.093
	3.5	0.108	0.106	1.0169	1.3392	0.46	1.109
	4.0	0.124	0.121	1.0196	1.3400	0.52	1.125
	4.5	0.140	0.137	1.0223	1.3409	0.58	1.142

Solute	Mass %	m/mol kg^{-1}	c/mol L^{-1}	ρ/g cm^{-3}	n	$\Delta/°C$	η/mPa s
	5.0	0.157	0.152	1.0250	1.3418	0.65	1.160
	5.5	0.173	0.168	1.0277	1.3427	0.71	1.178
	6.0	0.190	0.184	1.0305	1.3436	0.77	1.197
Ethanol	0.5	0.109	0.108	0.9973	1.3333	0.20	1.023
CH$_3$CH$_2$OH	1.0	0.219	0.216	0.9963	1.3336	0.40	1.046
	2.0	0.443	0.432	0.9945	1.3342	0.81	1.095
	3.0	0.671	0.646	0.9927	1.3348	1.23	1.140
	4.0	0.904	0.860	0.9910	1.3354	1.65	1.183
	5.0	1.142	1.074	0.9893	1.3360	2.09	1.228
	6.0	1.385	1.286	0.9878	1.3367	2.54	1.279
	7.0	1.634	1.498	0.9862	1.3374	2.99	1.331
	8.0	1.887	1.710	0.9847	1.3381	3.47	1.385
	9.0	2.147	1.921	0.9833	1.3388	3.96	1.442
	10.0	2.412	2.131	0.9819	1.3395	4.47	1.501
	12.0	2.960	2.551	0.9792	1.3410	5.56	1.627
	14.0	3.534	2.967	0.9765	1.3425	6.73	1.761
	16.0	4.134	3.382	0.9739	1.3440	8.01	1.890
	18.0	4.765	3.795	0.9713	1.3455	9.40	2.019
	20.0	5.427	4.205	0.9687	1.3469	10.92	2.142
	22.0	6.122	4.613	0.9660	1.3484	12.60	2.259
	24.0	6.855	5.018	0.9632	1.3498	14.47	2.370
	26.0	7.626	5.419	0.9602	1.3511	16.41	2.476
	28.0	8.441	5.817	0.9571	1.3524	18.43	2.581
	30.0	9.303	6.212	0.9539	1.3535	20.47	2.667
	32.0	10.215	6.601	0.9504	1.3546	22.44	2.726
	34.0	11.182	6.987	0.9468	1.3557	24.27	2.768
	36.0	12.210	7.370	0.9431	1.3566	25.98	2.803
	38.0	13.304	7.747	0.9392	1.3575	27.62	2.829
	40.0	14.471	8.120	0.9352	1.3583	29.26	2.846
	42.0	15.718	8.488	0.9311	1.3590	30.98	2.852
	44.0	17.055	8.853	0.9269	1.3598	32.68	2.850
	46.0	18.490	9.213	0.9227	1.3604	34.36	2.843
	48.0	20.036	9.568	0.9183	1.3610	36.04	2.832
	50.0	21.706	9.919	0.9139	1.3616	37.67	2.813
	60.0	32.559	11.605	0.8911	1.3638	44.93	2.547
	70.0	50.648	13.183	0.8676	1.3652		2.214
	80.0	86.824	14.649	0.8436	1.3658		1.881
	90.0	195.355	15.980	0.8180	1.3650		1.542
	92.0	249.620	16.225	0.8125	1.3646		1.475
	94.0	340.062	16.466	0.8070	1.3642		1.407
	96.0	520.946	16.697	0.8013	1.3636		1.342
	98.0		16.920	0.7954	1.3630		1.273
	100.0		17.133	0.7893	1.3614		1.203
Ethylene	0.5	0.081	0.080	0.9988	1.3335	0.15	1.010
glycol	1.0	0.163	0.161	0.9995	1.3339	0.30	1.020
(CH$_2$OH)$_2$	2.0	0.329	0.322	1.0007	1.3348	0.61	1.048
	3.0	0.498	0.484	1.0019	1.3358	0.92	1.074
	4.0	0.671	0.646	1.0032	1.3367	1.24	1.099
	5.0	0.848	0.809	1.0044	1.3377	1.58	1.125
	6.0	1.028	0.972	1.0057	1.3386	1.91	1.153
	7.0	1.213	1.136	1.0070	1.3396	2.26	1.182
	8.0	1.401	1.299	1.0082	1.3405	2.62	1.212
	9.0	1.593	1.464	1.0095	1.3415	2.99	1.243
	10.0	1.790	1.628	1.0108	1.3425	3.37	1.277
	12.0	2.197	1.959	1.0134	1.3444	4.16	1.348

CONCENTRATIVE PROPERTIES OF AQUEOUS SOLUTIONS:
DENSITY, REFRACTIVE INDEX, FREEZING POINT DEPRESSION, AND VISCOSITY (continued)

Solute	Mass %	m/mol kg^{-1}	c/mol L^{-1}	ρ/g cm^{-3}	n	Δ/°C	η/mPa s
	14.0	2.623	2.292	1.0161	1.3464	5.01	1.424
	16.0	3.069	2.626	1.0188	1.3484	5.91	1.500
	18.0	3.537	2.962	1.0214	1.3503	6.89	1.578
	20.0	4.028	3.300	1.0241	1.3523	7.93	1.661
	24.0	5.088	3.981	1.0296	1.3564	10.28	1.843
	28.0	6.265	4.669	1.0350	1.3605	13.03	2.047
	32.0	7.582	5.364	1.0405	1.3646	16.23	2.280
	36.0	9.062	6.067	1.0460	1.3687	19.82	2.537
	40.0	10.741	6.776	1.0514	1.3728	23.84	2.832
	44.0	12.659	7.491	1.0567	1.3769	28.32	3.166
	48.0	14.872	8.212	1.0619	1.3811	33.30	3.544
	52.0	17.453	8.939	1.0670	1.3851	38.81	3.981
	56.0	20.505	9.671	1.0719	1.3892	44.83	4.475
	60.0	24.166	10.406	1.0765	1.3931	51.23	5.026
Ferric	0.5	0.031	0.031	1.0025	1.3344	0.21	1.024
chloride	1.0	0.062	0.062	1.0068	1.3358	0.39	1.047
FeCl$_3$	2.0	0.126	0.125	1.0153	1.3386	0.75	1.093
	3.0	0.191	0.189	1.0238	1.3413	1.15	1.139
	4.0	0.257	0.255	1.0323	1.3441	1.56	1.187
	5.0	0.324	0.321	1.0408	1.3468	2.00	1.238
	6.0	0.394	0.388	1.0493	1.3496	2.48	1.292
	7.0	0.464	0.457	1.0580	1.3524	2.99	1.350
	8.0	0.536	0.526	1.0668	1.3552	3.57	1.412
	9.0	0.610	0.597	1.0760	1.3581	4.19	1.480
	10.0	0.685	0.669	1.0853	1.3611	4.85	1.553
	12.0	0.841	0.817	1.1040	1.3670	6.38	1.707
	14.0	1.004	0.969	1.1228	1.3730	8.22	1.879
	16.0	1.174	1.126	1.1420		10.45	2.080
	18.0	1.353	1.289	1.1615		13.08	2.311
	20.0	1.541	1.457	1.1816		16.14	2.570
	24.0	1.947	1.810	1.2234		23.79	3.178
	28.0	2.398	2.189	1.2679		33.61	4.038
	32.0	2.901	2.595	1.3153		49.16	5.274
	36.0	3.468	3.030	1.3654			7.130
	40.0	4.110	3.496	1.4176			9.674
Formic acid	0.5	0.109	0.109	0.9994	1.3333	0.21	1.006
HCOOH	1.0	0.219	0.217	1.0006	1.3336	0.42	1.011
	2.0	0.443	0.436	1.0029	1.3342	0.82	1.017
	3.0	0.672	0.655	1.0053	1.3348	1.24	1.195
	4.0	0.905	0.876	1.0077	1.3354	1.67	1.032
	5.0	1.143	1.097	1.0102	1.3359	2.10	1.039
	6.0	1.387	1.320	1.0126	1.3365	2.53	1.046
	7.0	1.635	1.544	1.0150	1.3371	2.97	1.052
	8.0	1.889	1.768	1.0175	1.3376	3.40	1.058
	9.0	2.149	1.994	1.0199	1.3382	3.84	1.064
	10.0	2.414	2.221	1.0224	1.3387	4.27	1.070
	12.0	2.962	2.678	1.0273	1.3397	5.19	1.082
	14.0	3.537	3.139	1.0322	1.3408	6.11	1.094
	16.0	4.138	3.605	1.0371	1.3418	7.06	1.106
	18.0	4.769	4.074	1.0419	1.3428	8.08	1.119
	20.0	5.431	4.548	1.0467	1.3437	9.11	1.132
	28.0	8.449	6.481	1.0654	1.3475	13.10	1.179
	36.0	12.220	8.477	1.0839	1.3511	17.65	1.227
	44.0	17.070	10.529	1.1015	1.3547	22.93	1.281
	52.0	23.535	12.633	1.1183	1.3581	29.69	1.340

Solute	Mass %	m/mol kg^{-1}	c/mol L^{-1}	ρ/g cm^{-3}	n	Δ/°C	η/mPa s
	60.0	32.587	14.813	1.1364	1.3612	38.26	1.410
	68.0	46.166	17.054	1.1544	1.3641		1.490
D-Fructose	0.5	0.028	0.028	1.0002	1.3337	0.05	1.015
$C_6H_{12}O_6$	1.0	0.056	0.056	1.0021	1.3344	0.10	1.028
	2.0	0.113	0.112	1.0061	1.3358	0.21	1.054
	3.0	0.172	0.168	1.0101	1.3373	0.32	1.080
	4.0	0.231	0.225	1.0140	1.3387	0.43	1.106
	5.0	0.292	0.283	1.0181	1.3402	0.54	1.134
	6.0	0.354	0.340	1.0221	1.3417	0.66	1.165
	7.0	0.418	0.399	1.0262	1.3431	0.78	1.198
	8.0	0.483	0.458	1.0303	1.3446	0.90	1.232
	9.0	0.549	0.517	1.0344	1.3461	1.03	1.270
	10.0	0.617	0.576	1.0385	1.3476	1.16	1.309
	12.0	0.757	0.697	1.0469	1.3507	1.43	1.391
	14.0	0.904	0.820	1.0554	1.3538	1.71	1.483
	16.0	1.057	0.945	1.0640	1.3569	2.01	1.587
	18.0	1.218	1.072	1.0728	1.3601	2.32	1.703
	20.0	1.388	1.201	1.0816	1.3634	2.64	1.837
	22.0	1.566	1.332	1.0906	1.3667	3.05	1.986
	24.0	1.753	1.465	1.0996	1.3700	3.43	2.154
	26.0	1.950	1.600	1.1089	1.3734	3.82	2.348
	28.0	2.159	1.738	1.1182	1.3768	4.20	2.562
	30.0	2.379	1.878	1.1276	1.3803		2.817
	32.0	2.612	2.020	1.1372	1.3839		3.112
	34.0	2.859	2.164	1.1469	1.3874		3.462
	36.0	3.122	2.312	1.1568	1.3911		3.899
	38.0	3.402	2.461	1.1668	1.3948		4.418
	40.0	3.700	2.613	1.1769	1.3985		5.046
	42.0	4.019	2.767	1.1871	1.4023		5.773
	44.0	4.361	2.925	1.1975	1.4062		6.644
	46.0	4.728	3.084	1.2080	1.4101		7.753
	48.0	5.124	3.247	1.2187	1.4141		9.060
D-Glucose	0.5	0.028	0.028	1.0001	1.3337	0.05	1.010
$C_6H_{12}O_6$	1.0	0.056	0.056	1.0020	1.3344	0.11	1.021
	2.0	0.113	0.112	1.0058	1.3358	0.21	1.052
	3.0	0.172	0.168	1.0097	1.3373	0.32	1.083
	4.0	0.231	0.225	1.0136	1.3387	0.43	1.113
	5.0	0.292	0.282	1.0175	1.3402	0.55	1.145
	6.0	0.354	0.340	1.0214	1.3417	0.67	1.179
	7.0	0.418	0.398	1.0254	1.3432	0.79	1.214
	8.0	0.483	0.457	1.0294	1.3447	0.91	1.250
	9.0	0.549	0.516	1.0334	1.3462	1.04	1.289
	10.0	0.617	0.576	1.0375	1.3477	1.17	1.330
	12.0	0.757	0.697	1.0457	1.3508	1.44	1.416
	14.0	0.904	0.819	1.0540	1.3539	1.73	1.512
	16.0	1.057	0.944	1.0624	1.3571	2.03	1.625
	18.0	1.218	1.070	1.0710	1.3603	2.35	1.757
	20.0	1.388	1.199	1.0797	1.3635	2.70	1.904
	22.0	1.566	1.329	1.0884	1.3668	3.07	2.063
	24.0	1.753	1.462	1.0973	1.3702	3.48	2.242
	26.0	1.950	1.597	1.1063	1.3736	3.90	2.458
	28.0	2.159	1.734	1.1154	1.3770	4.34	2.707
	30.0	2.379	1.873	1.1246	1.3805	4.79	2.998
	32.0	2.612	2.014	1.1340	1.3840		3.324
	34.0	2.859	2.158	1.1434	1.3876		3.704

Solute	Mass %	m/mol kg^{-1}	c/mol L^{-1}	ρ/g cm^{-3}	n	$\Delta/°C$	η/mPa s
	36.0	3.122	2.304	1.1529	1.3912		4.193
	38.0	3.402	2.452	1.1626	1.3949		4.786
	40.0	3.700	2.603	1.1724	1.3986		5.493
	42.0	4.019	2.756	1.1823	1.4024		6.288
	44.0	4.361	2.912	1.1924	1.4062		7.235
	46.0	4.728	3.071	1.2026	1.4101		8.454
	48.0	5.124	3.232	1.2130	1.4141		9.883
	50.0	5.551	3.396	1.2235	1.4181		11.884
	52.0	6.013	3.562	1.2342	1.4222		14.489
	54.0	6.516	3.732	1.2451	1.4263		17.916
	56.0	7.064	3.905	1.2562	1.4306		22.886
	58.0	7.665	4.081	1.2676	1.4349		29.389
	60.0	8.326	4.261	1.2793	1.4394		37.445
Glycerol	0.5	0.055	0.054	0.9994	1.3336	0.07	1.011
$CH_2OHCHOHCH_2OH$	1.0	0.110	0.109	1.0005	1.3342	0.18	1.022
	2.0	0.222	0.218	1.0028	1.3353	0.41	1.048
	3.0	0.336	0.327	1.0051	1.3365	0.63	1.074
	4.0	0.452	0.438	1.0074	1.3376	0.85	1.100
	5.0	0.572	0.548	1.0097	1.3388	1.08	1.127
	6.0	0.693	0.659	1.0120	1.3400	1.32	1.157
	7.0	0.817	0.771	1.0144	1.3412	1.56	1.188
	8.0	0.944	0.883	1.0167	1.3424	1.81	1.220
	9.0	1.074	0.996	1.0191	1.3436	2.06	1.256
	10.0	1.207	1.109	1.0215	1.3448	2.32	1.291
	12.0	1.481	1.337	1.0262	1.3472	2.88	1.365
	14.0	1.768	1.568	1.0311	1.3496	3.47	1.445
	16.0	2.068	1.800	1.0360	1.3521	4.09	1.533
	18.0	2.384	2.035	1.0409	1.3547	4.76	1.630
	20.0	2.715	2.271	1.0459	1.3572	5.46	1.737
	24.0	3.429	2.752	1.0561	1.3624	7.01	1.988
	28.0	4.223	3.242	1.0664	1.3676	8.77	2.279
	32.0	5.110	3.742	1.0770	1.3730	10.74	2.637
	36.0	6.108	4.252	1.0876	1.3785	12.96	3.088
	40.0	7.239	4.771	1.0984	1.3841	15.50	3.653
	44.0	8.532	5.300	1.1092	1.3897		4.443
	48.0	10.024	5.838	1.1200	1.3954		5.413
	52.0	11.764	6.385	1.1308	1.4011		6.666
	56.0	13.820	6.944	1.1419	1.4069		8.349
	60.0	16.288	7.512	1.1530	1.4129		10.681
	64.0	19.305	8.092	1.1643	1.4189		13.657
	68.0	23.075	8.680	1.1755	1.4249		18.457
	72.0	27.923	9.277	1.1866	1.4310		27.625
	76.0	34.387	9.884	1.1976	1.4370		40.571
	80.0	43.436	10.498	1.2085	1.4431		59.900
	84.0	57.009	11.121	1.2192	1.4492		84.338
	88.0	79.632	11.753	1.2299	1.4553		147.494
	92.0	124.878	12.392	1.2404	1.4613		384.467
	96.0	260.615	13.039	1.2508	1.4674		780.458
	100.0		13.694	1.2611	1.4735		
Hydrochloric	0.5	0.138	0.137	1.0007	1.3341	0.49	1.008
acid	1.0	0.277	0.275	1.0031	1.3353	0.99	1.015
HCl	2.0	0.560	0.553	1.0081	1.3376	2.08	1.029
	3.0	0.848	0.833	1.0130	1.3399	3.28	1.044
	4.0	1.143	1.117	1.0179	1.3422	4.58	1.059
	5.0	1.444	1.403	1.0228	1.3445	5.98	1.075

Solute	Mass %	m/mol kg^{-1}	c/mol L^{-1}	ρ/g cm^{-3}	n	Δ/°C	η/mPa s
	6.0	1.751	1.691	1.0278	1.3468	7.52	1.091
	7.0	2.064	1.983	1.0327	1.3491	9.22	1.108
	8.0	2.385	2.277	1.0377	1.3515	11.10	1.125
	9.0	2.713	2.574	1.0426	1.3538	13.15	1.143
	10.0	3.047	2.873	1.0476	1.3561	15.40	1.161
	12.0	3.740	3.481	1.0576	1.3607	20.51	1.199
	14.0	4.465	4.099	1.0676	1.3653		1.239
	16.0	5.224	4.729	1.0777	1.3700		1.282
	18.0	6.020	5.370	1.0878	1.3746		1.326
	20.0	6.857	6.023	1.0980	1.3792		1.374
	22.0	7.736	6.687	1.1083	1.3838		1.426
	24.0	8.661	7.362	1.1185	1.3884		1.483
	26.0	9.636	8.049	1.1288	1.3930		1.547
	28.0	10.666	8.748	1.1391	1.3976		1.620
	30.0	11.754	9.456	1.1492	1.4020		1.705
	32.0	12.907	10.175	1.1594	1.4066		1.799
	34.0	14.129	10.904	1.1693	1.4112		1.900
	36.0	15.427	11.642	1.1791	1.4158		2.002
	38.0	16.810	12.388	1.1886	1.4204		2.105
	40.0	18.284	13.140	1.1977	1.4250		
Lactic acid	0.5	0.056	0.055	0.9992	1.3335	0.10	1.014
CH$_3$CHOHCOOH	1.0	0.112	0.111	1.0002	1.3340	0.19	1.027
	2.0	0.227	0.223	1.0023	1.3350	0.38	1.056
	3.0	0.343	0.334	1.0043	1.3360	0.57	1.084
	4.0	0.463	0.447	1.0065	1.3370	0.76	1.110
	5.0	0.584	0.560	1.0086	1.3380	0.95	1.138
	6.0	0.709	0.673	1.0108	1.3390	1.16	1.167
	7.0	0.836	0.787	1.0131	1.3400	1.36	1.198
	8.0	0.965	0.902	1.0153	1.3410	1.57	1.229
	9.0	1.098	1.017	1.0176	1.3420	1.79	1.262
	10.0	1.233	1.132	1.0199	1.3430	2.02	1.296
	12.0	1.514	1.365	1.0246	1.3450	2.49	1.366
	14.0	1.807	1.600	1.0294	1.3470	2.99	1.441
	16.0	2.115	1.837	1.0342	1.3491	3.48	1.522
	18.0	2.437	2.076	1.0390	1.3511	3.96	1.607
	20.0	2.775	2.318	1.0439	1.3532	4.44	1.699
	24.0	3.506	2.807	1.0536	1.3573		1.902
	28.0	4.317	3.305	1.0632	1.3615		2.136
	32.0	5.224	3.811	1.0728	1.3657		2.414
	36.0	6.244	4.325	1.0822	1.3700		2.730
	40.0	7.401	4.847	1.0915	1.3743		3.114
	44.0	8.722	5.377	1.1008	1.3786		3.566
	48.0	10.247	5.917	1.1105	1.3828		4.106
	52.0	12.026	6.466	1.1201	1.3871		4.789
	56.0	14.129	7.023	1.1297	1.3914		5.579
	60.0	16.652	7.588	1.1392	1.3958		6.679
	64.0	19.736	8.161	1.1486	1.4001		8.024
	68.0	23.590	8.741	1.1579	1.4045		9.863
	72.0	28.546	9.328	1.1670	1.4088		12.866
	76.0	35.154	9.922	1.1760	1.4131		16.974
	80.0	44.405	10.522	1.1848	1.4173		22.164
Lactose	0.5	0.015	0.015	1.0002	1.3337	0.03	1.013
C$_{12}$H$_{22}$O$_{11}$	1.0	0.030	0.029	1.0021	1.3345	0.06	1.026
	2.0	0.060	0.059	1.0061	1.3359	0.11	1.058
	3.0	0.090	0.089	1.0102	1.3375	0.17	1.089
	4.0	0.122	0.119	1.0143	1.3390	0.23	1.120

Solute	Mass %	m/mol kg^{-1}	c/mol L^{-1}	ρ/g cm^{-3}	n	Δ/°C	η/mPa s
	5.0	0.154	0.149	1.0184	1.3406	0.29	1.154
	6.0	0.186	0.179	1.0225	1.3421	0.35	1.191
	7.0	0.220	0.210	1.0267	1.3437	0.42	1.232
	8.0	0.254	0.241	1.0308	1.3453	0.50	1.276
	9.0	0.289	0.272	1.0349	1.3468		1.321
	10.0	0.325	0.304	1.0390	1.3484		1.370
	12.0	0.398	0.367	1.0473	1.3515		1.476
	14.0	0.476	0.432	1.0558	1.3548		1.593
	16.0	0.556	0.498	1.0648	1.3582		1.724
	18.0	0.641	0.565	1.0746	1.3619		1.869
Lithium	0.5	0.119	0.118	1.0012	1.3341	0.42	1.019
chloride	1.0	0.238	0.237	1.0041	1.3351	0.84	1.037
LiCl	2.0	0.481	0.476	1.0099	1.3373	1.72	1.072
	3.0	0.730	0.719	1.0157	1.3394	2.68	1.108
	4.0	0.983	0.964	1.0215	1.3415	3.73	1.146
	5.0	1.241	1.211	1.0272	1.3436	4.86	1.185
	6.0	1.506	1.462	1.0330	1.3457	6.14	1.226
	7.0	1.775	1.715	1.0387	1.3478	7.56	1.269
	8.0	2.051	1.971	1.0444	1.3499	9.11	1.313
	9.0	2.333	2.230	1.0502	1.3520	10.79	1.360
	10.0	2.621	2.491	1.0560	1.3541	12.61	1.411
	12.0	3.217	3.022	1.0675	1.3583	16.59	1.522
	14.0	3.840	3.564	1.0792	1.3625	21.04	1.647
	16.0	4.493	4.118	1.0910	1.3668		1.787
	18.0	5.178	4.683	1.1029	1.3711		1.942
	20.0	5.897	5.260	1.1150	1.3755		2.128
	22.0	6.653	5.851	1.1274	1.3799		2.341
	24.0	7.449	6.453	1.1399	1.3844		2.600
	26.0	8.288	7.069	1.1527	1.3890		2.925
	28.0	9.173	7.700	1.1658	1.3936		3.318
	30.0	10.109	8.344	1.1791	1.3983		3.785
Magnesium	0.5	0.053	0.053	1.0022	1.3343	0.26	1.024
chloride	1.0	0.106	0.106	1.0062	1.3356	0.52	1.046
MgCl$_2$	2.0	0.214	0.213	1.0144	1.3381	1.06	1.091
	3.0	0.325	0.322	1.0226	1.3406	1.65	1.139
	4.0	0.438	0.433	1.0309	1.3432	2.30	1.188
	5.0	0.553	0.546	1.0394	1.3457	3.01	1.241
	6.0	0.670	0.660	1.0479	1.3483		1.298
	7.0	0.791	0.777	1.0564	1.3508		1.358
	8.0	0.913	0.895	1.0651	1.3534		1.423
	9.0	1.039	1.015	1.0738	1.3560		1.493
	10.0	1.167	1.137	1.0826	1.3587		1.570
	12.0	1.432	1.387	1.1005	1.3641		1.745
	14.0	1.710	1.645	1.1189	1.3695		1.956
	16.0	2.001	1.911	1.1372	1.3749		2.207
	18.0	2.306	2.184	1.1553	1.3804		2.507
	20.0	2.626	2.467	1.1742	1.3859		2.867
	22.0	2.962	2.758	1.1938	1.3915		3.323
	24.0	3.317	3.060	1.2140	1.3972		3.917
	26.0	3.690	3.371	1.2346	1.4030		4.694
	28.0	4.085	3.692	1.2555	1.4089		5.709
	30.0	4.501	4.022	1.2763	1.4148		7.017
Magnesium	0.5	0.042	0.042	1.0033	1.3340	0.10	1.027
sulfate	1.0	0.084	0.084	1.0084	1.3350	0.19	1.054
MgSO$_4$	2.0	0.170	0.169	1.0186	1.3371	0.36	1.112

Solute	Mass %	m/mol kg^{-1}	c/mol L^{-1}	ρ/g cm^{-3}	n	Δ/°C	η/mPa s
	3.0	0.257	0.256	1.0289	1.3391	0.52	1.177
	4.0	0.346	0.345	1.0392	1.3411	0.69	1.249
	5.0	0.437	0.436	1.0497	1.3431	0.87	1.328
	6.0	0.530	0.528	1.0602	1.3451	1.05	1.411
	7.0	0.625	0.623	1.0708	1.3471	1.24	1.498
	8.0	0.722	0.719	1.0816	1.3492	1.43	1.593
	9.0	0.822	0.817	1.0924	1.3512	1.64	1.702
	10.0	0.923	0.917	1.1034	1.3532	1.85	1.829
	12.0	1.133	1.122	1.1257	1.3572	2.31	2.104
	14.0	1.352	1.336	1.1484	1.3613	2.86	2.412
	16.0	1.582	1.557	1.1717	1.3654	3.67	2.809
	18.0	1.824	1.788	1.1955	1.3694		3.360
	20.0	2.077	2.027	1.2198	1.3735		4.147
	22.0	2.343	2.275	1.2447	1.3776		5.199
	24.0	2.624	2.532	1.2701	1.3817		6.498
	26.0	2.919	2.800	1.2961	1.3858		8.066
Maltose	0.5	0.015	0.015	1.0003	1.3337	0.03	1.016
$C_{12}H_{22}O_{11}$	1.0	0.030	0.029	1.0023	1.3345	0.06	1.030
	2.0	0.060	0.059	1.0063	1.3359	0.11	1.060
	3.0	0.090	0.089	1.0104	1.3374	0.17	1.092
	4.0	0.122	0.119	1.0144	1.3389	0.23	1.126
	5.0	0.154	0.149	1.0184	1.3404	0.29	1.162
	6.0	0.186	0.179	1.0224	1.3420	0.35	1.200
	7.0	0.220	0.210	1.0265	1.3435	0.42	1.239
	8.0	0.254	0.241	1.0305	1.3450	0.48	1.281
	9.0	0.289	0.272	1.0345	1.3466	0.55	1.325
	10.0	0.325	0.303	1.0385	1.3482	0.62	1.372
	12.0	0.398	0.367	1.0465	1.3513	0.77	1.474
	14.0	0.476	0.431	1.0545	1.3546	0.92	1.588
	16.0	0.556	0.497	1.0629	1.3578	1.08	1.715
	18.0	0.641	0.564	1.0716	1.3612	1.25	1.859
	20.0	0.730	0.631	1.0801	1.3644	1.43	2.021
	22.0	0.824	0.700	1.0894	1.3678	1.64	2.216
	24.0	0.923	0.770	1.0984	1.3714	1.85	2.463
	26.0	1.026	0.842	1.1080	1.3749	2.08	2.753
	28.0	1.136	0.914	1.1171	1.3785	2.34	3.066
	30.0	1.252	0.988	1.1269	1.3821	2.62	3.427
	40.0	1.948	1.375	1.1769	1.4013	4.41	6.926
	50.0	2.921	1.797	1.2304	1.4217		17.786
	52.0	3.165	1.886	1.2416	1.4260		22.034
	54.0	3.429	1.976	1.2528	1.4308		28.757
	56.0	3.718	2.068	1.2638	1.4350		38.226
	58.0	4.034	2.159	1.2740	1.4394		49.298
	60.0	4.382	2.253	1.2855	1.4440		
Manganese(II)	1.0	0.067	0.067	1.0080	1.3348	0.16	1.046
sulfate	2.0	0.135	0.135	1.0178	1.3366	0.31	1.090
$MnSO_4$	3.0	0.205	0.204	1.0277	1.3384	0.44	1.137
	4.0	0.276	0.275	1.0378	1.3402	0.57	1.187
	5.0	0.349	0.347	1.0480	1.3420	0.70	1.242
	6.0	0.423	0.421	1.0583	1.3438	0.84	1.301
	7.0	0.498	0.495	1.0688	1.3457	0.98	1.363
	8.0	0.576	0.572	1.0794	1.3475	1.12	1.431
	9.0	0.655	0.650	1.0902	1.3494	1.28	1.505
	10.0	0.736	0.729	1.1012	1.3513	1.44	1.587
	12.0	0.903	0.893	1.1236	1.3551	1.80	1.779

Solute	Mass %	m/mol kg^{-1}	c/mol L^{-1}	ρ/g cm^{-3}	n	Δ/°C	η/mPa s
	14.0	1.078	1.063	1.1467	1.3589	2.21	2.005
	16.0	1.261	1.240	1.1705	1.3629	2.67	2.272
	18.0	1.454	1.424	1.1950	1.3668	3.19	2.580
	20.0	1.656	1.616	1.2203	1.3708	3.80	2.938
D-Mannitol	0.5	0.028	0.027	1.0000	1.3337	0.05	1.019
$CH_2(CHOH)_4CH_2OH$	1.0	0.055	0.055	1.0017	1.3345	0.10	1.032
	2.0	0.112	0.110	1.0053	1.3359	0.21	1.057
	3.0	0.170	0.166	1.0088	1.3374	0.32	1.081
	4.0	0.229	0.222	1.0124	1.3389	0.43	1.107
	5.0	0.289	0.279	1.0159	1.3403	0.54	1.135
	6.0	0.350	0.336	1.0195	1.3418	0.66	1.166
	7.0	0.413	0.393	1.0230	1.3433	0.77	1.200
	8.0	0.477	0.451	1.0266	1.3447	0.90	1.236
	9.0	0.543	0.509	1.0302	1.3462	1.02	1.275
	10.0	0.610	0.567	1.0338	1.3477	1.15	1.314
	11.0	0.678	0.626	1.0375	1.3491	1.28	1.355
	12.0	0.749	0.686	1.0412	1.3506	1.41	1.398
	13.0	0.820	0.746	1.0450	1.3521	1.55	1.443
	14.0	0.894	0.806	1.0489	1.3536	1.69	1.489
	15.0	0.969	0.867	1.0529	1.3552	1.84	1.537
Methanol	0.5	0.157	0.156	0.9973	1.3331	0.28	1.022
CH_3OH	1.0	0.315	0.311	0.9964	1.3332	0.56	1.040
	2.0	0.637	0.621	0.9947	1.3334	1.14	1.070
	3.0	0.965	0.930	0.9930	1.3336	1.75	1.100
	4.0	1.300	1.238	0.9913	1.3339	2.37	1.131
	5.0	1.643	1.544	0.9896	1.3341	3.02	1.163
	6.0	1.992	1.850	0.9880	1.3343	3.71	1.196
	7.0	2.349	2.155	0.9864	1.3346	4.41	1.229
	8.0	2.714	2.459	0.9848	1.3348	5.13	1.264
	9.0	3.087	2.762	0.9832	1.3351	5.85	1.297
	10.0	3.468	3.064	0.9816	1.3354	6.60	1.329
	12.0	4.256	3.665	0.9785	1.3359	8.14	1.389
	14.0	5.081	4.262	0.9755	1.3365	9.72	1.446
	16.0	5.945	4.856	0.9725	1.3370	11.36	1.501
	18.0	6.851	5.447	0.9695	1.3376	13.13	1.554
	20.0	7.803	6.034	0.9666	1.3381	15.02	1.604
	22.0	8.803	6.616	0.9636	1.3387	16.98	1.652
	24.0	9.856	7.196	0.9606	1.3392	19.04	1.697
	26.0	10.966	7.771	0.9576	1.3397	21.23	1.735
	28.0	12.138	8.341	0.9545	1.3402	23.59	1.769
	30.0	13.376	8.908	0.9514	1.3407	25.91	1.795
	32.0	14.688	9.470	0.9482	1.3411	28.15	1.814
	34.0	16.078	10.028	0.9450	1.3415	30.48	1.827
	36.0	17.556	10.580	0.9416	1.3419	32.97	1.835
	38.0	19.129	11.127	0.9382	1.3422	35.60	1.839
	40.0	20.807	11.669	0.9347	1.3425	38.60	1.837
	50.0	31.211	14.288	0.9156	1.3431	54.50	1.761
	60.0	46.816	16.749	0.8944	1.3426	74.50	1.600
	70.0	72.826	19.040	0.8715	1.3411		1.368
	80.0	124.844	21.144	0.8468	1.3385		1.128
	90.0	280.899	23.045	0.8204	1.3348		0.861
	100.0		24.710	0.7917	1.3290		0.586
Nitric acid	0.5	0.080	0.079	1.0009	1.3336	0.28	1.004
HNO_3	1.0	0.160	0.159	1.0037	1.3343	0.56	1.005
	2.0	0.324	0.320	1.0091	1.3356	1.12	1.007

Solute	Mass %	m/mol kg^{-1}	c/mol L^{-1}	ρ/g cm^{-3}	n	Δ/°C	η/mPa s
	3.0	0.491	0.483	1.0146	1.3368	1.70	1.010
	4.0	0.661	0.648	1.0202	1.3381	2.32	1.014
	5.0	0.835	0.814	1.0257	1.3394	2.96	1.018
	6.0	1.013	0.982	1.0314	1.3407	3.63	1.022
	7.0	1.194	1.152	1.0370	1.3421	4.33	1.027
	8.0	1.380	1.324	1.0427	1.3434	5.05	1.032
	9.0	1.570	1.498	1.0485	1.3447	5.81	1.038
	10.0	1.763	1.673	1.0543	1.3460	6.60	1.044
	12.0	2.164	2.030	1.0660	1.3487	8.27	1.058
	14.0	2.583	2.395	1.0780	1.3514	10.08	1.075
	16.0	3.023	2.768	1.0901	1.3541	12.04	1.094
	18.0	3.484	3.149	1.1025	1.3569	14.16	1.116
	20.0	3.967	3.539	1.1150	1.3596		1.141
	22.0	4.476	3.937	1.1277	1.3624		1.169
	24.0	5.011	4.344	1.1406	1.3652		1.199
	26.0	5.576	4.760	1.1536	1.3680		1.233
	28.0	6.172	5.185	1.1668	1.3708		1.271
	30.0	6.801	5.618	1.1801	1.3736		1.311
	32.0	7.468	6.060	1.1934	1.3763		1.354
	34.0	8.175	6.512	1.2068	1.3790		1.400
	36.0	8.927	6.971	1.2202	1.3817		1.450
	38.0	9.727	7.439	1.2335	1.3842		1.504
	40.0	10.580	7.913	1.2466	1.3867		1.561
Oxalic acid	0.5	0.056	0.056	1.0006	1.3336	0.16	1.013
(COOH)$_2$	1.0	0.112	0.111	1.0030	1.3342	0.30	1.023
	1.5	0.169	0.167	1.0054	1.3347	0.44	1.033
	2.0	0.227	0.224	1.0079	1.3353	0.57	1.044
	2.5	0.285	0.281	1.0103	1.3359	0.71	1.055
	3.0	0.343	0.337	1.0126	1.3364	0.84	1.065
	3.5	0.403	0.395	1.0150	1.3370	0.97	1.076
	4.0	0.463	0.452	1.0174	1.3375	1.09	1.086
	4.5	0.523	0.510	1.0197	1.3381		1.097
	5.0	0.585	0.568	1.0220	1.3386		1.108
	6.0	0.709	0.684	1.0265	1.3397		1.129
	7.0	0.836	0.802	1.0310	1.3407		1.150
	8.0	0.966	0.920	1.0355	1.3418		1.172
Phosphoric	0.5	0.051	0.051	1.0010	1.3335	0.12	1.010
acid	1.0	0.103	0.102	1.0038	1.3340	0.24	1.020
H$_3$PO$_4$	2.0	0.208	0.206	1.0092	1.3349	0.46	1.050
	3.0	0.316	0.311	1.0146	1.3358	0.69	1.079
	4.0	0.425	0.416	1.0200	1.3367	0.93	1.108
	5.0	0.537	0.523	1.0254	1.3376	1.16	1.138
	6.0	0.651	0.631	1.0309	1.3385	1.38	1.169
	7.0	0.768	0.740	1.0363	1.3394	1.62	1.200
	8.0	0.887	0.850	1.0418	1.3403	1.88	1.232
	9.0	1.009	0.962	1.0474	1.3413	2.16	1.267
	10.0	1.134	1.075	1.0531	1.3422	2.45	1.303
	12.0	1.392	1.304	1.0647	1.3441	3.01	1.382
	14.0	1.661	1.538	1.0765	1.3460	3.76	1.469
	16.0	1.944	1.777	1.0885	1.3480	4.45	1.565
	18.0	2.240	2.022	1.1009	1.3500	5.25	1.671
	20.0	2.551	2.273	1.1135	1.3520	6.23	1.788
	22.0	2.878	2.529	1.1263	1.3540	7.38	1.914
	24.0	3.223	2.791	1.1395	1.3561	8.69	2.049
	26.0	3.585	3.059	1.1528	1.3582	10.12	2.198

Solute	Mass %	m/mol kg^{-1}	c/mol L^{-1}	ρ/g cm^{-3}	n	Δ/°C	η/mPa s
	28.0	3.968	3.333	1.1665	1.3604	11.64	2.365
	30.0	4.373	3.614	1.1804	1.3625	13.23	2.553
	32.0	4.802	3.901	1.1945	1.3647	14.94	2.766
	34.0	5.257	4.194	1.2089	1.3669	16.81	3.001
	36.0	5.740	4.495	1.2236	1.3691	18.85	3.260
	38.0	6.254	4.803	1.2385	1.3713	21.09	3.544
	40.0	6.803	5.117	1.2536	1.3735	23.58	3.856
Potassium	0.5	0.050	0.050	1.0014	1.3335	0.18	1.009
bicarbonate	1.0	0.101	0.100	1.0046	1.3341	0.34	1.015
KHCO$_3$	2.0	0.204	0.202	1.0114	1.3353	0.67	1.027
	3.0	0.309	0.305	1.0181	1.3365	0.98	1.040
	4.0	0.416	0.409	1.0247	1.3376	1.29	1.053
	5.0	0.526	0.515	1.0310	1.3386	1.60	1.067
	6.0	0.638	0.622	1.0379	1.3397	1.91	1.081
	7.0	0.752	0.730	1.0446	1.3409	2.22	1.096
	8.0	0.869	0.840	1.0514	1.3419	2.53	1.112
	9.0	0.988	0.951	1.0581	1.3430	2.84	1.128
	10.0	1.110	1.064	1.0650	1.3441	3.16	1.145
	12.0	1.362	1.293	1.0788	1.3462	3.79	1.183
	14.0	1.626	1.528	1.0929	1.3484	4.41	1.224
	16.0	1.903	1.770	1.1073	1.3506		1.270
	18.0	2.193	2.017	1.1221	1.3528		1.319
	20.0	2.497	2.272	1.1372	1.3550		1.373
	22.0	2.817	2.533	1.1527	1.3572		1.432
	24.0	3.154	2.801	1.1685	1.3595		1.497
Potassium	0.5	0.042	0.042	1.0018	1.3336	0.15	1.000
bromide	1.0	0.085	0.084	1.0054	1.3342	0.29	0.998
KBr	2.0	0.171	0.170	1.0127	1.3354	0.59	0.994
	3.0	0.260	0.257	1.0200	1.3366	0.88	0.990
	4.0	0.350	0.345	1.0275	1.3379	1.18	0.985
	5.0	0.442	0.435	1.0350	1.3391	1.48	0.981
	6.0	0.536	0.526	1.0426	1.3403	1.78	0.977
	7.0	0.633	0.618	1.0503	1.3416	2.10	0.974
	8.0	0.731	0.711	1.0581	1.3429	2.42	0.970
	9.0	0.831	0.806	1.0660	1.3441	2.74	0.967
	10.0	0.934	0.903	1.0740	1.3454	3.07	0.964
	12.0	1.146	1.099	1.0903	1.3481	3.76	0.958
	14.0	1.368	1.302	1.1070	1.3507	4.49	0.953
	16.0	1.601	1.512	1.1242	1.3535	5.25	0.949
	18.0	1.845	1.727	1.1419	1.3562	6.04	0.946
	20.0	2.101	1.950	1.1601	1.3591	6.88	0.944
	22.0	2.370	2.179	1.1788	1.3620	7.76	0.943
	24.0	2.654	2.416	1.1980	1.3650	8.70	0.943
	26.0	2.952	2.661	1.2179	1.3680	9.68	0.944
	28.0	3.268	2.914	1.2383	1.3711	10.72	0.947
	30.0	3.601	3.175	1.2593	1.3743	11.82	0.952
	32.0	3.954	3.445	1.2810	1.3776	12.98	0.959
	34.0	4.329	3.724	1.3033	1.3809		0.968
	36.0	4.727	4.012	1.3263	1.3843		0.979
	38.0	5.150	4.311	1.3501	1.3878		0.993
	40.0	5.602	4.620	1.3746	1.3914		1.010
Potassium	0.5	0.036	0.036	1.0027	1.3339	0.18	1.013
carbonate	1.0	0.073	0.073	1.0072	1.3347	0.34	1.025
K$_2$CO$_3$	2.0	0.148	0.147	1.0163	1.3365	0.66	1.048

Solute	Mass %	m/mol kg^{-1}	c/mol L^{-1}	ρ/g cm^{-3}	n	Δ/°C	η/mPa s
	3.0	0.224	0.223	1.0254	1.3382	0.99	1.071
	4.0	0.301	0.299	1.0345	1.3399	1.32	1.094
	5.0	0.381	0.378	1.0437	1.3416	1.67	1.119
	6.0	0.462	0.457	1.0529	1.3433	2.03	1.146
	7.0	0.545	0.538	1.0622	1.3450	2.40	1.174
	8.0	0.629	0.620	1.0715	1.3467	2.77	1.204
	9.0	0.716	0.704	1.0809	1.3484	3.17	1.235
	10.0	0.804	0.789	1.0904	1.3501	3.57	1.269
	12.0	0.987	0.963	1.1095	1.3535	4.45	1.339
	14.0	1.178	1.144	1.1291	1.3569	5.39	1.414
	16.0	1.378	1.330	1.1490	1.3603	6.42	1.497
	18.0	1.588	1.523	1.1692	1.3637	7.55	1.594
	20.0	1.809	1.722	1.1898	1.3671	8.82	1.707
	24.0	2.285	2.139	1.2320	1.3739	11.96	1.978
	28.0	2.814	2.584	1.2755	1.3807	16.01	2.331
	32.0	3.405	3.057	1.3204	1.3874	21.46	2.834
	36.0	4.070	3.559	1.3665	1.3940	28.58	3.503
	40.0	4.824	4.093	1.4142	1.4006	37.55	4.360
	50.0	7.236	5.573	1.5404	1.4168		9.369
Potassium	0.5	0.067	0.067	1.0014	1.3337	0.23	1.000
chloride	1.0	0.135	0.135	1.0046	1.3343	0.46	0.999
KCl	2.0	0.274	0.271	1.0110	1.3357	0.92	0.999
	3.0	0.415	0.409	1.0174	1.3371	1.38	0.998
	4.0	0.559	0.549	1.0239	1.3384	1.85	0.997
	5.0	0.706	0.691	1.0304	1.3398	2.32	0.996
	6.0	0.856	0.835	1.0369	1.3411	2.80	0.994
	7.0	1.010	0.980	1.0434	1.3425	3.29	0.992
	8.0	1.166	1.127	1.0500	1.3438	3.80	0.990
	9.0	1.327	1.276	1.0566	1.3452	4.30	0.989
	10.0	1.490	1.426	1.0633	1.3466	4.81	0.988
	12.0	1.829	1.733	1.0768	1.3493	5.88	0.990
	14.0	2.184	2.048	1.0905	1.3521		0.994
	16.0	2.555	2.370	1.1043	1.3549		0.999
	18.0	2.944	2.701	1.1185	1.3577		1.004
	20.0	3.353	3.039	1.1328	1.3606		1.012
	22.0	3.783	3.386	1.1474	1.3635		1.024
	24.0	4.236	3.742	1.1623	1.3665		1.040
Potassium	0.5	0.090	0.089	1.0025	1.3340	0.30	1.010
hydroxide	1.0	0.180	0.179	1.0068	1.3350	0.61	1.019
KOH	2.0	0.364	0.362	1.0155	1.3369	1.24	1.038
	3.0	0.551	0.548	1.0242	1.3388	1.89	1.058
	4.0	0.743	0.736	1.0330	1.3408	2.57	1.079
	5.0	0.938	0.929	1.0419	1.3427	3.36	1.102
	6.0	1.138	1.124	1.0509	1.3445	4.14	1.126
	7.0	1.342	1.322	1.0599	1.3464	4.92	1.151
	8.0	1.550	1.524	1.0690	1.3483		1.177
	9.0	1.763	1.729	1.0781	1.3502		1.205
	10.0	1.980	1.938	1.0873	1.3520		1.233
	12.0	2.431	2.365	1.1059	1.3558		1.294
	14.0	2.902	2.806	1.1246	1.3595		1.361
	16.0	3.395	3.261	1.1435	1.3632		1.436
	18.0	3.913	3.730	1.1626	1.3670		1.521
	20.0	4.456	4.213	1.1818	1.3707		1.619
	22.0	5.027	4.711	1.2014	1.3744		1.732
	24.0	5.629	5.223	1.2210	1.3781		1.861

Solute	Mass %	m/mol kg^{-1}	c/mol L^{-1}	ρ/g cm^{-3}	n	Δ/°C	η/mPa s
	26.0	6.262	5.750	1.2408	1.3818		2.006
	28.0	6.931	6.293	1.2609	1.3854		2.170
	30.0	7.639	6.851	1.2813	1.3889		2.357
	40.0	11.882	9.896	1.3881	1.4068		3.879
	50.0	17.824	13.389	1.5024	1.4247		7.892
Potassium	0.5	0.030	0.030	1.0019	1.3337	0.11	0.999
iodide	1.0	0.061	0.061	1.0056	1.3343	0.22	0.997
KI	2.0	0.123	0.122	1.0131	1.3357	0.43	0.991
	3.0	0.186	0.184	1.0206	1.3370	0.64	0.986
	4.0	0.251	0.248	1.0282	1.3384	0.86	0.981
	5.0	0.317	0.312	1.0360	1.3397	1.08	0.976
	6.0	0.385	0.377	1.0438	1.3411	1.30	0.969
	7.0	0.453	0.443	1.0517	1.3425	1.53	0.963
	8.0	0.524	0.511	1.0598	1.3440	1.77	0.957
	9.0	0.596	0.579	1.0679	1.3454	2.01	0.951
	10.0	0.669	0.648	1.0762	1.3469	2.26	0.946
	12.0	0.821	0.790	1.0931	1.3498	2.77	0.937
	14.0	0.981	0.937	1.1105	1.3529	3.30	0.929
	16.0	1.147	1.088	1.1284	1.3560	3.87	0.921
	18.0	1.322	1.244	1.1469	1.3593	4.46	0.915
	20.0	1.506	1.405	1.1659	1.3626	5.09	0.910
	22.0	1.699	1.571	1.1856	1.3661	5.76	0.905
	24.0	1.902	1.744	1.2060	1.3696	6.46	0.901
	26.0	2.117	1.922	1.2270	1.3733	7.21	0.898
	28.0	2.343	2.106	1.2487	1.3771	8.01	0.895
	30.0	2.582	2.297	1.2712	1.3810	8.86	0.892
	32.0	2.835	2.495	1.2944	1.3851	9.76	0.891
	34.0	3.103	2.700	1.3185	1.3893	10.72	0.890
	36.0	3.388	2.913	1.3434	1.3936	11.73	0.890
	38.0	3.692	3.134	1.3692	1.3981	12.81	0.893
	40.0	4.016	3.364	1.3959	1.4027	13.97	0.897
Potassium	0.5	0.050	0.050	1.0014	1.3335	0.17	0.999
nitrate	1.0	0.100	0.099	1.0045	1.3339	0.33	0.996
KNO$_3$	2.0	0.202	0.200	1.0108	1.3349	0.64	0.990
	3.0	0.306	0.302	1.0171	1.3358	0.94	0.986
	4.0	0.412	0.405	1.0234	1.3368	1.22	0.983
	5.0	0.521	0.509	1.0298	1.3377	1.50	0.980
	6.0	0.631	0.615	1.0363	1.3386	1.76	0.977
	7.0	0.744	0.722	1.0428	1.3396	2.02	0.975
	8.0	0.860	0.830	1.0494	1.3405	2.27	0.973
	9.0	0.978	0.940	1.0560	1.3415	2.52	0.971
	10.0	1.099	1.051	1.0627	1.3425	2.75	0.970
	12.0	1.349	1.277	1.0762	1.3444		0.970
	14.0	1.610	1.509	1.0899	1.3463		0.972
	16.0	1.884	1.747	1.1039	1.3482		0.976
	18.0	2.171	1.991	1.1181	1.3502		0.982
	20.0	2.473	2.240	1.1326	1.3521		0.990
	22.0	2.790	2.497	1.1473	1.3541		0.999
	24.0	3.123	2.759	1.1623	1.3561		1.010
Potassium	0.5	0.032	0.032	1.0017		0.11	1.001
permanganate	1.0	0.064	0.064	1.0051		0.22	1.000
KMnO$_4$	1.5	0.096	0.096	1.0085		0.32	0.999
	2.0	0.129	0.128	1.0118		0.43	0.998
	3.0	0.196	0.193	1.0186			0.995

Solute	Mass %	m/mol kg^{-1}	c/mol L^{-1}	ρ/g cm^{-3}	n	Δ/°C	η/mPa s
	4.0	0.264	0.260	1.0254			0.992
	5.0	0.333	0.327	1.0322			0.989
	6.0	0.404	0.394	1.0390			0.985
Potassium	0.5	0.029	0.029	1.0025	1.3338	0.13	1.013
hydrogen	1.0	0.058	0.058	1.0068	1.3345	0.25	1.023
phosphate	1.5	0.087	0.087	1.0110	1.3353	0.37	1.034
K_2HPO_4	2.0	0.117	0.117	1.0153	1.3361	0.49	1.046
	2.5	0.147	0.146	1.0195	1.3368	0.61	1.057
	3.0	0.178	0.176	1.0238	1.3376	0.73	1.069
	3.5	0.208	0.207	1.0281	1.3384	0.86	1.081
	4.0	0.239	0.237	1.0324	1.3392	0.97	1.094
	4.5	0.271	0.268	1.0368	1.3399	1.10	1.107
	5.0	0.302	0.299	1.0412	1.3407	1.22	1.120
	6.0	0.366	0.362	1.0500	1.3422	1.46	1.147
	7.0	0.432	0.426	1.0590	1.3438	1.70	1.177
	8.0	0.499	0.491	1.0680	1.3453	1.95	1.209
Potassium	0.5	0.037	0.037	1.0018	1.3336	0.13	1.010
dihydrogen	1.0	0.074	0.074	1.0053	1.3342	0.25	1.019
phosphate	1.5	0.112	0.111	1.0089	1.3348	0.37	1.028
KH_2PO_4	2.0	0.150	0.149	1.0125	1.3354	0.49	1.038
	3.0	0.227	0.225	1.0197	1.3365	0.72	1.060
	4.0	0.306	0.302	1.0269	1.3377	0.96	1.083
	5.0	0.387	0.380	1.0342	1.3388	1.19	1.108
	6.0	0.469	0.459	1.0414	1.3400	1.41	1.133
	7.0	0.553	0.539	1.0486	1.3411	1.63	1.160
	8.0	0.639	0.621	1.0558	1.3422	1.84	1.187
	9.0	0.727	0.703	1.0630	1.3434	2.04	1.215
	10.0	0.816	0.786	1.0703	1.3445	2.23	1.245
Potassium	0.5	0.029	0.029	1.0022	1.3336	0.14	1.006
sulfate	1.0	0.058	0.058	1.0062	1.3343	0.26	1.011
K_2SO_4	2.0	0.117	0.116	1.0143	1.3355	0.50	1.021
	3.0	0.177	0.176	1.0224	1.3368	0.73	1.033
	4.0	0.239	0.237	1.0306	1.3380	0.95	1.045
	5.0	0.302	0.298	1.0388	1.3393	1.17	1.058
	6.0	0.366	0.360	1.0470	1.3405		1.072
	7.0	0.432	0.424	1.0553	1.3417		1.087
	8.0	0.499	0.488	1.0637	1.3428		1.102
	9.0	0.568	0.554	1.0721	1.3440		1.117
	10.0	0.638	0.620	1.0806	1.3452		1.132
1-Propanol	1.0	0.168	0.166	0.9963	1.3339	0.31	1.051
$CH_3CH_2CH_2OH$	2.0	0.340	0.331	0.9946	1.3348	0.61	1.100
	3.0	0.515	0.496	0.9928	1.3357	0.93	1.152
	4.0	0.693	0.660	0.9911	1.3366	1.24	1.208
	5.0	0.876	0.823	0.9896	1.3376	1.57	1.267
	6.0	1.062	0.987	0.9882	1.3385	1.91	1.325
	7.0	1.252	1.149	0.9868	1.3394	2.26	1.387
	8.0	1.447	1.312	0.9855	1.3404	2.61	1.449
	9.0	1.646	1.474	0.9842	1.3414	2.99	1.514
	10.0	1.849	1.635	0.9829	1.3423	3.36	1.577
	12.0	2.269	1.958	0.9804	1.3442	4.09	1.710
	14.0	2.709	2.278	0.9779	1.3460	4.91	1.849
	16.0	3.169	2.595	0.9749	1.3477	5.78	1.986
	18.0	3.652	2.911	0.9719	1.3494	6.67	2.106

Solute	Mass %	m/mol kg^{-1}	c/mol L^{-1}	ρ/g cm^{-3}	n	$\Delta/^\circ$C	η/mPa s
	20.0	4.160	3.223	0.9686	1.3510	7.76	2.218
	24.0	5.254	3.838	0.9612	1.3539	9.12	2.432
	28.0	6.471	4.441	0.9533	1.3566	10.17	2.612
	32.0	7.830	5.033	0.9452	1.3592	10.66	2.765
	36.0	9.359	5.613	0.9370	1.3614		2.900
	40.0	11.093	6.182	0.9288	1.3635		3.010
	60.0	24.958	8.860	0.8875	1.3734		3.186
	80.0	66.556	11.275	0.8470	1.3812		2.822
	100.0		13.368	0.8034	1.3852		2.227
2-Propanol	1.0	0.168	0.166	0.9960	1.3338	0.30	1.056
$CH_3CHOHCH_3$	2.0	0.340	0.331	0.9939	1.3346	0.60	1.112
	3.0	0.515	0.495	0.9920	1.3355	0.93	1.166
	4.0	0.693	0.659	0.9902	1.3364	1.26	1.225
	5.0	0.876	0.822	0.9884	1.3373	1.61	1.287
	6.0	1.062	0.985	0.9871	1.3382	1.96	1.352
	7.0	1.252	1.148	0.9855	1.3392	2.32	1.417
	8.0	1.447	1.310	0.9843	1.3400	2.68	1.485
	9.0	1.646	1.472	0.9831	1.3410	3.06	1.553
	10.0	1.849	1.633	0.9816	1.3420	3.48	1.629
	12.0	2.269	1.955	0.9793	1.3439	4.43	1.794
	14.0	2.709	2.276	0.9772	1.3459	5.29	1.970
	16.0	3.169	2.596	0.9751	1.3478	6.36	2.160
	18.0	3.652	2.913	0.9725	1.3496	7.40	2.352
	20.0	4.160	3.227	0.9696	1.3514	8.52	2.550
	40.0	11.093	6.191	0.9302	1.3642		
	60.0	24.958	8.809	0.8824	1.3717		
	80.0	66.556	11.103	0.8341	1.3765		
	100.0		13.058	0.7848	1.3742		
Silver	0.5	0.030	0.030	1.0027	1.3336	0.10	1.003
nitrate	1.0	0.059	0.059	1.0070	1.3342	0.20	1.005
$AgNO_3$	2.0	0.120	0.120	1.0154	1.3352	0.40	1.009
	3.0	0.182	0.181	1.0239	1.3363	0.59	1.013
	4.0	0.245	0.243	1.0327	1.3374	0.78	1.016
	5.0	0.310	0.307	1.0417	1.3385	0.96	1.020
	6.0	0.376	0.371	1.0506	1.3396	1.15	1.024
	7.0	0.443	0.437	1.0597	1.3407	1.33	1.027
	8.0	0.512	0.503	1.0690	1.3419	1.51	1.031
	9.0	0.582	0.571	1.0785	1.3431	1.69	1.035
	10.0	0.654	0.641	1.0882	1.3443	1.87	1.039
	12.0	0.803	0.783	1.1079	1.3467	2.21	1.049
	14.0	0.958	0.930	1.1284	1.3493	2.55	1.060
	16.0	1.121	1.083	1.1496	1.3519	2.86	1.072
	18.0	1.292	1.241	1.1715	1.3546		1.086
	20.0	1.472	1.406	1.1942	1.3574		1.101
	22.0	1.660	1.577	1.2177	1.3602		1.117
	24.0	1.859	1.755	1.2420	1.3632		1.135
	26.0	2.068	1.940	1.2672	1.3662		1.154
	28.0	2.289	2.132	1.2933	1.3694		1.176
	30.0	2.523	2.332	1.3204	1.3726		1.200
	32.0	2.770	2.541	1.3487	1.3760		1.227
	34.0	3.033	2.758	1.3780	1.3795		1.257
	36.0	3.311	2.985	1.4087	1.3832		1.290
	38.0	3.608	3.223	1.4407	1.3871		1.326
	40.0	3.925	3.472	1.4743	1.3911		1.366

Solute	Mass %	m/mol kg^{-1}	c/mol L^{-1}	ρ/g cm^{-3}	n	Δ/°C	η/mPa s
Sodium	0.5	0.061	0.061	1.0008	1.3337	0.22	1.021
acetate	1.0	0.123	0.122	1.0034	1.3344	0.43	1.040
CH$_3$COONa	2.0	0.249	0.246	1.0085	1.3358	0.88	1.080
	3.0	0.377	0.371	1.0135	1.3372	1.34	1.124
	4.0	0.508	0.497	1.0184	1.3386	1.82	1.171
	5.0	0.642	0.624	1.0234	1.3400	2.32	1.222
	6.0	0.778	0.752	1.0283	1.3414	2.85	1.278
	7.0	0.918	0.882	1.0334	1.3428	3.40	1.337
	8.0	1.060	1.013	1.0386	1.3442	3.98	1.401
	9.0	1.206	1.145	1.0440	1.3456	4.57	1.468
	10.0	1.354	1.279	1.0495	1.3470		1.539
	12.0	1.662	1.552	1.0607	1.3498		1.688
	14.0	1.984	1.829	1.0718	1.3526		1.855
	16.0	2.322	2.112	1.0830	1.3554		2.054
	18.0	2.676	2.400	1.0940	1.3583		2.284
	20.0	3.047	2.694	1.1050	1.3611		2.567
	22.0	3.438	2.993	1.1159	1.3639		2.948
	24.0	3.849	3.297	1.1268	1.3666		3.400
	26.0	4.283	3.606	1.1377	1.3693		3.877
	28.0	4.741	3.921	1.1488	1.3720		4.388
	30.0	5.224	4.243	1.1602	1.3748		4.940
Sodium	0.5	0.060	0.060	1.0018	1.3337	0.20	1.015
bicarbonate	1.0	0.120	0.120	1.0054	1.3344	0.40	1.028
NaHCO$_3$	1.5	0.181	0.180	1.0089	1.3351	0.59	1.042
	2.0	0.243	0.241	1.0125	1.3357	0.78	1.057
	2.5	0.305	0.302	1.0160	1.3364	0.98	1.071
	3.0	0.368	0.364	1.0196	1.3370	1.16	1.086
	3.5	0.432	0.426	1.0231	1.3377	1.35	1.102
	4.0	0.496	0.489	1.0266	1.3383	1.54	1.118
	4.5	0.561	0.552	1.0301	1.3390	1.72	1.134
	5.0	0.627	0.615	1.0337	1.3396	1.90	1.151
	5.5	0.693	0.679	1.0372	1.3403	2.08	1.168
	6.0	0.760	0.743	1.0408	1.3409	2.26	1.185
Sodium	0.5	0.049	0.049	1.0021	1.3337	0.17	1.004
bromide	1.0	0.098	0.098	1.0060	1.3344	0.34	1.007
NaBr	2.0	0.198	0.197	1.0139	1.3358	0.69	1.012
	3.0	0.301	0.298	1.0218	1.3372	1.04	1.017
	4.0	0.405	0.400	1.0298	1.3386	1.39	1.022
	5.0	0.512	0.504	1.0380	1.3401	1.76	1.028
	6.0	0.620	0.610	1.0462	1.3415	2.14	1.034
	7.0	0.732	0.717	1.0546	1.3430	2.53	1.040
	8.0	0.845	0.826	1.0630	1.3445	2.93	1.046
	9.0	0.961	0.937	1.0716	1.3460	3.34	1.053
	10.0	1.080	1.050	1.0803	1.3475	3.77	1.060
	12.0	1.325	1.281	1.0981	1.3506	4.67	1.077
	14.0	1.582	1.519	1.1164	1.3538	5.65	1.096
	16.0	1.851	1.765	1.1352	1.3570	6.74	1.119
	18.0	2.133	2.020	1.1546	1.3604		1.144
	20.0	2.430	2.283	1.1745	1.3638		1.174
	22.0	2.741	2.555	1.1951	1.3673		1.207
	24.0	3.069	2.837	1.2163	1.3708		1.244
	26.0	3.415	3.129	1.2382	1.3745		1.287
	28.0	3.780	3.431	1.2608	1.3783		1.336
	30.0	4.165	3.744	1.2842	1.3822		1.395
	32.0	4.574	4.069	1.3083	1.3862		1.465

Solute	Mass %	m/mol kg^{-1}	c/mol L^{-1}	ρ/g cm^{-3}	n	Δ/°C	η/mPa s
	34.0	5.007	4.406	1.3333	1.3903		1.546
	36.0	5.467	4.755	1.3592	1.3946		1.639
	38.0	5.957	5.119	1.3860	1.3990		1.745
	40.0	6.479	5.496	1.4138	1.4035		1.866
Sodium	0.5	0.047	0.047	1.0034	1.3341	0.22	1.025
carbonate	1.0	0.095	0.095	1.0086	1.3352	0.43	1.049
Na$_2$CO$_3$	2.0	0.193	0.192	1.0190	1.3375	0.75	1.102
	3.0	0.292	0.291	1.0294	1.3397	1.08	1.159
	4.0	0.393	0.392	1.0398	1.3419	1.42	1.222
	5.0	0.497	0.495	1.0502	1.3440	1.77	1.292
	6.0	0.602	0.600	1.0606	1.3462	2.13	1.367
	7.0	0.710	0.707	1.0711	1.3483		1.448
	8.0	0.820	0.816	1.0816	1.3504		1.538
	9.0	0.933	0.927	1.0922	1.3525		1.638
	10.0	1.048	1.041	1.1029	1.3547		1.754
	11.0	1.166	1.156	1.1136	1.3568		1.884
	12.0	1.287	1.273	1.1244	1.3589		2.028
	13.0	1.410	1.392	1.1353	1.3610		2.186
	14.0	1.536	1.514	1.1463	1.3631		2.361
	15.0	1.665	1.638	1.1574	1.3652		2.551
Sodium	0.5	0.086	0.086	1.0018	1.3339	0.30	1.011
chloride	1.0	0.173	0.172	1.0053	1.3347	0.59	1.020
NaCl	2.0	0.349	0.346	1.0125	1.3365	1.19	1.036
	3.0	0.529	0.523	1.0196	1.3383	1.79	1.052
	4.0	0.713	0.703	1.0268	1.3400	2.41	1.068
	5.0	0.901	0.885	1.0340	1.3418	3.05	1.085
	6.0	1.092	1.069	1.0413	1.3435	3.70	1.104
	7.0	1.288	1.256	1.0486	1.3453	4.38	1.124
	8.0	1.488	1.445	1.0559	1.3470	5.08	1.145
	9.0	1.692	1.637	1.0633	1.3488	5.81	1.168
	10.0	1.901	1.832	1.0707	1.3505	6.56	1.193
	12.0	2.333	2.229	1.0857	1.3541	8.18	1.250
	14.0	2.785	2.637	1.1008	1.3576	9.94	1.317
	16.0	3.259	3.056	1.1162	1.3612	11.89	1.388
	18.0	3.756	3.486	1.1319	1.3648	14.04	1.463
	20.0	4.278	3.928	1.1478	1.3684	16.46	1.557
	22.0	4.826	4.382	1.1640	1.3721	19.18	1.676
	24.0	5.403	4.847	1.1804	1.3757		1.821
	26.0	6.012	5.326	1.1972	1.3795		1.990
Sodium	1.0	0.039	0.039	1.0049	1.3348	0.20	1.043
citrate	2.0	0.079	0.078	1.0120	1.3366	0.39	1.081
(HO)C(COONa)$_3$	3.0	0.120	0.118	1.0186	1.3383	0.59	1.122
	4.0	0.161	0.159	1.0260	1.3401	0.79	1.166
	5.0	0.204	0.200	1.0331	1.3419	0.97	1.210
	6.0	0.247	0.242	1.0405	1.3437	1.17	1.263
	7.0	0.292	0.284	1.0482	1.3455	1.36	1.314
	8.0	0.337	0.327	1.0557	1.3473	1.57	1.371
	9.0	0.383	0.371	1.0632	1.3491	1.77	1.427
	10.0	0.431	0.415	1.0708	1.3509	1.96	1.499
	12.0	0.528	0.505	1.0861	1.3546	2.38	1.649
	14.0	0.631	0.598	1.1019	1.3583	2.82	1.832
	16.0	0.738	0.693	1.1173	1.3618	3.27	2.045
	18.0	0.851	0.790	1.1327	1.3656	3.82	2.290
	20.0	0.969	0.891	1.1492	1.3693	4.39	2.596

Solute	Mass %	m/mol kg^{-1}	c/mol L^{-1}	ρ/g cm^{-3}	n	$\Delta/°C$	η/mPa s
	24.0	1.224	1.099	1.1813	1.3767		3.409
	28.0	1.507	1.318	1.2151	1.3845		4.586
	32.0	1.823	1.548	1.2487	1.3923		6.541
	36.0	2.180	1.792	1.2843	1.4001		9.788
Sodium	0.5	0.126	0.125	1.0039	1.3344	0.43	1.027
hydroxide	1.0	0.253	0.252	1.0095	1.3358	0.86	1.054
NaOH	2.0	0.510	0.510	1.0207	1.3386	1.74	1.112
	3.0	0.773	0.774	1.0318	1.3414	2.64	1.176
	4.0	1.042	1.043	1.0428	1.3441	3.59	1.248
	5.0	1.316	1.317	1.0538	1.3467	4.57	1.329
	6.0	1.596	1.597	1.0648	1.3494	5.60	1.416
	7.0	1.882	1.883	1.0758	1.3520	6.69	1.510
	8.0	2.174	2.174	1.0869	1.3546	7.87	1.616
	9.0	2.473	2.470	1.0979	1.3572	9.12	1.737
	10.0	2.778	2.772	1.1089	1.3597	10.47	1.882
	12.0	3.409	3.393	1.1309	1.3648	13.42	2.201
	14.0	4.070	4.036	1.1530	1.3697	16.76	2.568
	15.0	4.412	4.365	1.1640	1.3722		2.789
	16.0	4.762	4.701	1.1751	1.3746		3.043
	18.0	5.488	5.387	1.1971	1.3793		3.698
	20.0	6.250	6.096	1.2192	1.3840		4.619
	22.0	7.052	6.827	1.2412	1.3885		5.765
	24.0	7.895	7.579	1.2631	1.3929		7.100
	26.0	8.784	8.352	1.2848	1.3971		8.744
	28.0	9.723	9.145	1.3064	1.4012		10.832
	30.0	10.715	9.958	1.3277	1.4051		13.517
	32.0	11.766	10.791	1.3488	1.4088		16.844
	34.0	12.880	11.643	1.3697	1.4123		20.751
	36.0	14.064	12.512	1.3901	1.4156		25.290
	38.0	15.324	13.398	1.4102	1.4186		30.461
	40.0	16.668	14.300	1.4299	1.4215		36.312
Sodium	0.5	0.059	0.059	1.0016	1.3336	0.20	1.004
nitrate	1.0	0.119	0.118	1.0050	1.3341	0.40	1.007
NaNO$_3$	2.0	0.240	0.238	1.0117	1.3353	0.79	1.012
	3.0	0.364	0.359	1.0185	1.3364	1.18	1.018
	4.0	0.490	0.483	1.0254	1.3375	1.56	1.025
	5.0	0.619	0.607	1.0322	1.3387	1.94	1.032
	6.0	0.751	0.734	1.0392	1.3398	2.32	1.040
	7.0	0.886	0.862	1.0462	1.3409	2.70	1.049
	8.0	1.023	0.991	1.0532	1.3421	3.08	1.059
	9.0	1.164	1.123	1.0603	1.3432	3.46	1.069
	10.0	1.307	1.256	1.0674	1.3443	3.84	1.081
	12.0	1.604	1.527	1.0819	1.3466	4.60	1.107
	14.0	1.915	1.806	1.0967	1.3489	5.37	1.138
	18.0	2.583	2.387	1.1272	1.3536	6.98	1.215
	20.0	2.941	2.689	1.1429	1.3559	7.81	1.263
	30.0	5.042	4.326	1.2256	1.3678		1.609
	40.0	7.844	6.200	1.3175	1.3802		2.226
Sodium	0.5	0.031	0.031	1.0042	1.3343	0.19	1.033
phosphate	1.0	0.062	0.062	1.0100	1.3356	0.37	1.064
Na$_3$PO$_4$	1.5	0.093	0.093	1.0158	1.3369	0.53	1.094
	2.0	0.124	0.125	1.0216	1.3381	0.67	1.126
	2.5	0.156	0.157	1.0275	1.3394	0.79	1.161
	3.0	0.189	0.189	1.0335	1.3406		1.198

Solute	Mass %	m/mol kg^{-1}	c/mol L^{-1}	ρ/g cm^{-3}	n	Δ/°C	η/mPa s
	3.5	0.221	0.222	1.0395	1.3419		1.238
	4.0	0.254	0.255	1.0456	1.3432		1.281
	4.5	0.287	0.289	1.0517	1.3444		1.327
	5.0	0.321	0.323	1.0579	1.3457		1.375
	5.5	0.355	0.357	1.0642	1.3470		1.426
	6.0	0.389	0.392	1.0705	1.3482		1.480
	6.5	0.424	0.427	1.0768	1.3495		1.538
	7.0	0.459	0.462	1.0832	1.3507		1.598
	7.5	0.495	0.498	1.0896	1.3519		1.662
	8.0	0.530	0.535	1.0961	1.3532		1.729
Sodium hydrogen	0.5	0.035	0.035	1.0032	1.3340	0.17	1.021
phosphate	1.0	0.071	0.071	1.0082	1.3349	0.32	1.042
Na_2HPO_4	1.5	0.107	0.107	1.0131	1.3358	0.46	1.064
	2.0	0.144	0.143	1.0180	1.3368		1.088
	2.5	0.181	0.180	1.0229	1.3377		1.113
	3.0	0.218	0.217	1.0279	1.3386		1.138
	3.5	0.255	0.255	1.0328	1.3396		1.165
	4.0	0.293	0.292	1.0378	1.3405		1.193
	4.5	0.332	0.331	1.0428	1.3414		1.223
	5.0	0.371	0.369	1.0478	1.3424		1.254
	5.5	0.410	0.408	1.0528	1.3433		1.286
Sodium	0.5	0.042	0.042	1.0019	1.3336	0.14	1.018
dihydrogen	1.0	0.084	0.084	1.0056	1.3343	0.28	1.035
phosphate	1.5	0.127	0.126	1.0094	1.3349	0.42	1.051
NaH_2PO_4	2.0	0.170	0.169	1.0131	1.3356	0.56	1.068
	2.5	0.214	0.212	1.0168	1.3362	0.70	1.085
	3.0	0.258	0.255	1.0206	1.3369	0.84	1.103
	3.5	0.302	0.299	1.0244	1.3375	0.98	1.121
	4.0	0.347	0.343	1.0281	1.3382	1.12	1.140
	4.5	0.393	0.387	1.0319	1.3388	1.25	1.160
	5.0	0.439	0.432	1.0358	1.3395	1.39	1.180
	6.0	0.532	0.522	1.0434	1.3408	1.65	1.223
	7.0	0.627	0.613	1.0511	1.3421	1.89	1.270
	8.0	0.725	0.706	1.0589	1.3434	2.12	1.319
	9.0	0.824	0.800	1.0668	1.3447	2.35	1.371
	10.0	0.926	0.896	1.0747	1.3460	2.58	1.428
	12.0	1.137	1.091	1.0907	1.3486	3.06	1.552
	14.0	1.357	1.292	1.1070	1.3512	3.53	1.694
	16.0	1.588	1.499	1.1236	1.3538	4.03	1.861
	18.0	1.830	1.711	1.1404	1.3565	4.55	2.050
	20.0	2.084	1.930	1.1576	1.3592	5.10	2.283
	22.0	2.351	2.155	1.1752	1.3618		2.550
	24.0	2.632	2.387	1.1931	1.3646		2.850
	26.0	2.929	2.625	1.2113	1.3673		3.214
	28.0	3.242	2.870	1.2299	1.3700		3.682
	30.0	3.572	3.123	1.2488	1.3728		4.300
	32.0	3.923	3.383	1.2682	1.3756		5.079
	34.0	4.294	3.650	1.2879	1.3784		6.008
	36.0	4.689	3.925	1.3080	1.3812		7.098
	38.0	5.109	4.208	1.3285	1.3840		8.363
	40.0	5.557	4.499	1.3493	1.3869		9.814
Sodium	0.5	0.035	0.035	1.0027	1.3338	0.17	1.013
sulfate	1.0	0.071	0.071	1.0071	1.3345	0.32	1.026
Na_2SO_4	2.0	0.144	0.143	1.0161	1.3360	0.61	1.058

Solute	Mass %	m/mol kg^{-1}	c/mol L^{-1}	ρ/g cm^{-3}	n	$\Delta/°C$	η/mPa s
	3.0	0.218	0.217	1.0252	1.3376	0.87	1.091
	4.0	0.293	0.291	1.0343	1.3391	1.13	1.126
	5.0	0.371	0.367	1.0436	1.3406	1.36	1.163
	6.0	0.449	0.445	1.0526	1.3420	1.56	1.202
	7.0	0.530	0.523	1.0619	1.3435		1.244
	8.0	0.612	0.603	1.0713	1.3449		1.289
	9.0	0.696	0.685	1.0808	1.3464		1.337
	10.0	0.782	0.768	1.0905	1.3479		1.390
	12.0	0.960	0.938	1.1101	1.3509		1.508
	14.0	1.146	1.114	1.1301	1.3539		1.646
	16.0	1.341	1.296	1.1503	1.3567		1.812
	18.0	1.545	1.483	1.1705	1.3595		2.005
	20.0	1.760	1.677	1.1907	1.3620		2.227
	22.0	1.986	1.875	1.2106	1.3643		2.481
Sodium	0.5	0.032	0.032	1.0024	1.3340	0.14	1.012
thiosulfate	1.0	0.064	0.064	1.0065	1.3351	0.28	1.023
$Na_2S_2O_3$	2.0	0.129	0.128	1.0148	1.3371	0.57	1.044
	3.0	0.196	0.194	1.0231	1.3392	0.84	1.066
	4.0	0.264	0.261	1.0315	1.3413	1.09	1.090
	5.0	0.333	0.329	1.0399	1.3434	1.34	1.115
	6.0	0.404	0.398	1.0483	1.3454	1.59	1.141
	7.0	0.476	0.468	1.0568	1.3475	1.83	1.169
	8.0	0.550	0.539	1.0654	1.3496	2.06	1.199
	9.0	0.626	0.611	1.0740	1.3517	2.30	1.231
	10.0	0.703	0.685	1.0827	1.3538	2.55	1.267
	12.0	0.862	0.835	1.1003	1.3581	3.06	1.345
	14.0	1.030	0.990	1.1182	1.3624	3.60	1.435
	16.0	1.205	1.150	1.1365	1.3667	4.17	1.537
	18.0	1.388	1.315	1.1551	1.3711	4.76	1.657
	20.0	1.581	1.485	1.1740	1.3756	5.37	1.798
	30.0	2.711	2.417	1.2739	1.3987		2.903
	40.0	4.216	3.498	1.3827	1.4229		5.758
Strontium	0.5	0.032	0.032	1.0027	1.3339	0.16	1.012
chloride	1.0	0.064	0.064	1.0071	1.3348	0.31	1.021
$SrCl_2$	2.0	0.129	0.128	1.0161	1.3366	0.62	1.039
	3.0	0.195	0.194	1.0252	1.3384	0.93	1.057
	4.0	0.263	0.261	1.0344	1.3402	1.26	1.076
	5.0	0.332	0.329	1.0437	1.3421	1.61	1.096
	6.0	0.403	0.399	1.0532	1.3440	1.98	1.116
	7.0	0.475	0.469	1.0628	1.3459	2.38	1.136
	8.0	0.549	0.541	1.0726	1.3478	2.80	1.157
	9.0	0.624	0.615	1.0825	1.3498	3.25	1.180
	10.0	0.701	0.689	1.0925	1.3518	3.74	1.204
	12.0	0.860	0.843	1.1131	1.3558	4.81	1.258
	14.0	1.027	1.002	1.1342	1.3599	6.03	1.317
	16.0	1.202	1.167	1.1558	1.3641	7.41	1.383
	18.0	1.385	1.338	1.1780	1.3684	8.98	1.460
	20.0	1.577	1.515	1.2008	1.3728	10.74	1.549
	22.0	1.779	1.699	1.2241	1.3772	12.74	1.650
	24.0	1.992	1.890	1.2481	1.3817	14.99	1.765
	26.0	2.216	2.087	1.2728	1.3864		1.897
	28.0	2.453	2.293	1.2983	1.3911		2.056
	30.0	2.703	2.507	1.3248	1.3961		2.245
	32.0	2.968	2.730	1.3523	1.4013		2.527

CONCENTRATIVE PROPERTIES OF AQUEOUS SOLUTIONS:
DENSITY, REFRACTIVE INDEX, FREEZING POINT DEPRESSION, AND VISCOSITY (continued)

Solute	Mass %	m/mol kg^{-1}	c/mol L^{-1}	ρ/g cm^{-3}	n	Δ/°C	η/mPa s
	34.0	3.250	2.962	1.3811	1.4067		2.846
	36.0	3.548	3.205	1.4114	1.4124		3.206
Sucrose	0.5	0.015	0.015	1.0002	1.3337	0.03	1.015
$C_{12}H_{22}O_{11}$	1.0	0.030	0.029	1.0021	1.3344	0.06	1.028
	2.0	0.060	0.059	1.0060	1.3359	0.11	1.055
	3.0	0.090	0.089	1.0099	1.3373	0.17	1.084
	4.0	0.122	0.118	1.0139	1.3388	0.23	1.114
	5.0	0.154	0.149	1.0178	1.3403	0.29	1.146
	6.0	0.186	0.179	1.0218	1.3418	0.35	1.179
	7.0	0.220	0.210	1.0259	1.3433	0.42	1.215
	8.0	0.254	0.241	1.0299	1.3448	0.49	1.254
	9.0	0.289	0.272	1.0340	1.3463	0.55	1.294
	10.0	0.325	0.303	1.0381	1.3478	0.63	1.336
	12.0	0.398	0.367	1.0465	1.3509	0.77	1.429
	14.0	0.476	0.431	1.0549	1.3541	0.93	1.534
	16.0	0.556	0.497	1.0635	1.3573	1.10	1.653
	18.0	0.641	0.564	1.0722	1.3606	1.27	1.790
	20.0	0.730	0.632	1.0810	1.3639	1.47	1.945
	22.0	0.824	0.700	1.0899	1.3672	1.67	2.124
	24.0	0.923	0.771	1.0990	1.3706	1.89	2.331
	26.0	1.026	0.842	1.1082	1.3741	2.12	2.573
	28.0	1.136	0.914	1.1175	1.3776	2.37	2.855
	30.0	1.252	0.988	1.1270	1.3812	2.64	3.187
	32.0	1.375	1.063	1.1366	1.3848	2.94	3.762
	34.0	1.505	1.139	1.1464	1.3885	3.27	4.052
	36.0	1.643	1.216	1.1562	1.3922	3.63	4.621
	38.0	1.791	1.295	1.1663	1.3960	4.02	5.315
	40.0	1.948	1.375	1.1765	1.3999	4.45	6.162
	42.0	2.116	1.456	1.1868	1.4038	4.93	7.234
	44.0	2.295	1.539	1.1972	1.4078		8.596
	46.0	2.489	1.623	1.2079	1.4118		10.301
	48.0	2.697	1.709	1.2186	1.4159		12.515
	50.0	2.921	1.796	1.2295	1.4201		15.431
	60.0	4.382	2.255	1.2864	1.4419		58.487
	70.0	6.817	2.755	1.3472	1.4654		481.561
	80.0	11.686	3.299	1.4117	1.4906		
Sulfuric acid	0.5	0.051	0.051	1.0016	1.3336	0.21	1.010
H_2SO_4	1.0	0.103	0.102	1.0049	1.3342	0.42	1.019
	2.0	0.208	0.206	1.0116	1.3355	0.80	1.036
	3.0	0.315	0.311	1.0183	1.3367	1.17	1.059
	4.0	0.425	0.418	1.0250	1.3379	1.60	1.085
	5.0	0.537	0.526	1.0318	1.3391	2.05	1.112
	6.0	0.651	0.635	1.0385	1.3403	2.50	1.136
	7.0	0.767	0.746	1.0453	1.3415	2.95	1.159
	8.0	0.887	0.858	1.0522	1.3427	3.49	1.182
	9.0	1.008	0.972	1.0591	1.3439	4.08	1.206
	10.0	1.133	1.087	1.0661	1.3451	4.64	1.230
	12.0	1.390	1.322	1.0802	1.3475	5.93	1.282
	14.0	1.660	1.563	1.0947	1.3500	7.49	1.337
	16.0	1.942	1.810	1.1094	1.3525	9.26	1.399
	18.0	2.238	2.064	1.1245	1.3551	11.29	1.470
	20.0	2.549	2.324	1.1398	1.3576	13.64	1.546
	22.0	2.876	2.592	1.1554	1.3602	16.48	1.624
	24.0	3.220	2.866	1.1714	1.3628	19.85	1.706
	26.0	3.582	3.147	1.1872	1.3653	24.29	1.797
	28.0	3.965	3.435	1.2031	1.3677	29.65	1.894

Solute	Mass %	m/mol kg^{-1}	c/mol L^{-1}	ρ/g cm^{-3}	n	$\Delta/°C$	η/mPa s
	30.0	4.370	3.729	1.2191	1.3701	36.21	2.001
	32.0	4.798	4.030	1.2353	1.3725	44.76	2.122
	34.0	5.252	4.339	1.2518	1.3749	55.28	2.255
	36.0	5.735	4.656	1.2685	1.3773		2.392
	38.0	6.249	4.981	1.2855	1.3797		2.533
	40.0	6.797	5.313	1.3028	1.3821		2.690
	42.0	7.383	5.655	1.3205	1.3846		2.872
	44.0	8.011	6.005	1.3386	1.3870		3.073
	46.0	8.685	6.364	1.3570	1.3895		3.299
	48.0	9.411	6.734	1.3759	1.3920		3.546
	50.0	10.196	7.113	1.3952	1.3945		3.826
	52.0	11.045	7.502	1.4149	1.3971		4.142
	54.0	11.969	7.901	1.4351	1.3997		4.499
	56.0	12.976	8.312	1.4558	1.4024		4.906
	58.0	14.080	8.734	1.4770	1.4050		5.354
	60.0	15.294	9.168	1.4987	1.4077		5.917
	70.0	23.790	11.494	1.6105			
	80.0	40.783	14.088	1.7272			
	90.0	91.762	16.649	1.8144			
	92.0	117.251	17.109	1.8240			
	94.0	159.734	17.550	1.8312			
	96.0	244.698	17.966	1.8355			
	98.0	499.592	18.346	1.8361			
	100.0		18.663	1.8305			
Trichloroacetic acid CCl_3COOH	0.5	0.031	0.031	1.0008	1.3337	0.11	1.011
	1.0	0.062	0.061	1.0034	1.3343	0.21	1.021
	2.0	0.125	0.123	1.0083	1.3356	0.42	1.044
	3.0	0.189	0.186	1.0133	1.3369	0.64	1.069
	4.0	0.255	0.249	1.0182	1.3381	0.86	1.096
	5.0	0.322	0.313	1.0230	1.3394	1.08	1.123
	6.0	0.391	0.377	1.0279	1.3406	1.30	1.150
	7.0	0.461	0.442	1.0328	1.3418	1.53	1.177
	8.0	0.532	0.508	1.0378	1.3431	1.76	1.204
	9.0	0.605	0.574	1.0428	1.3444	1.99	1.233
	10.0	0.680	0.641	1.0479	1.3456	2.23	1.263
	12.0	0.835	0.777	1.0583	1.3483	2.73	1.326
	14.0	0.996	0.916	1.0692	1.3510	3.26	1.393
	16.0	1.166	1.058	1.0806	1.3539	3.82	1.462
	18.0	1.343	1.203	1.0921	1.3568		1.533
	20.0	1.530	1.351	1.1035	1.3597		1.608
	24.0	1.933	1.654	1.1260	1.3652		1.768
	28.0	2.380	1.968	1.1485	1.3705		1.935
	32.0	2.880	2.294	1.1713	1.3759		2.118
	36.0	3.443	2.632	1.1947	1.3813		2.320
	40.0	4.080	2.984	1.2188	1.3868		1.543
	44.0	4.809	3.349	1.2435	1.3923		2.797
	48.0	5.650	3.726	1.2682	1.3977		3.076
Tris (hydroxymethyl)-methylamine $H_2NC(CH_2OH)_3$	0.5	0.041	0.041	0.9994	1.3337	0.08	1.014
	1.0	0.083	0.083	1.0006	1.3344	0.16	1.027
	2.0	0.168	0.166	1.0030	1.3359	0.31	1.054
	3.0	0.255	0.249	1.0054	1.3374	0.47	1.083
	4.0	0.344	0.333	1.0078	1.3388	0.64	1.115
	5.0	0.434	0.417	1.0103	1.3403	0.80	1.148
	6.0	0.527	0.502	1.0128	1.3418	0.97	1.182

Solute	Mass %	m/mol kg^{-1}	c/mol L^{-1}	ρ/g cm^{-3}	n	Δ/°C	η/mPa s
	7.0	0.621	0.587	1.0153	1.3433	1.15	1.218
	8.0	0.718	0.672	1.0179	1.3448	1.33	1.256
	9.0	0.816	0.758	1.0204	1.3463	1.51	1.295
	10.0	0.917	0.844	1.0230	1.3478	1.70	1.337
	12.0	1.126	1.019	1.0282	1.3508	2.08	1.427
	14.0	1.344	1.194	1.0335	1.3539	2.47	1.527
	16.0	1.572	1.372	1.0389	1.3570	2.90	1.642
	18.0	1.812	1.552	1.0443	1.3601	3.36	1.772
	20.0	2.064	1.733	1.0498	1.3633	3.85	1.920
	30.0	3.538	2.670	1.0781	1.3797		2.998
	40.0	5.503	3.657	1.1076	1.3970		5.208
Urea	0.5	0.084	0.083	0.9995	1.3337	0.16	1.007
$(NH_2)_2CO$	1.0	0.168	0.167	1.0007	1.3344	0.31	1.010
	2.0	0.340	0.334	1.0033	1.3358	0.62	1.012
	3.0	0.515	0.502	1.0058	1.3372	0.93	1.017
	4.0	0.694	0.672	1.0085	1.3387	1.24	1.025
	5.0	0.876	0.842	1.0111	1.3401	1.55	1.033
	6.0	1.063	1.013	1.0138	1.3416	1.88	1.041
	7.0	1.253	1.185	1.0165	1.3431	2.22	1.049
	8.0	1.448	1.358	1.0192	1.3446	2.56	1.057
	9.0	1.647	1.531	1.0220	1.3461	2.91	1.065
	10.0	1.850	1.706	1.0248	1.3476	3.26	1.074
	12.0	2.270	2.059	1.0304	1.3506	3.95	1.091
	14.0	2.710	2.415	1.0360	1.3537	4.66	1.109
	16.0	3.171	2.775	1.0417	1.3568	5.40	1.130
	18.0	3.655	3.139	1.0473	1.3599	6.19	1.153
	20.0	4.163	3.506	1.0530	1.3629	7.00	1.178
	22.0	4.696	3.878	1.0586	1.3661	7.81	1.205
	24.0	5.258	4.253	1.0643	1.3692	8.64	1.235
	26.0	5.850	4.632	1.0699	1.3723	9.52	1.266
	28.0	6.475	5.014	1.0756	1.3754	10.45	1.298
	30.0	7.136	5.401	1.0812	1.3785	11.40	1.332
	32.0	7.835	5.791	1.0869	1.3817	12.34	1.371
	34.0	8.577	6.185	1.0926	1.3848	13.27	1.413
	36.0	9.366	6.584	1.0984	1.3881	14.20	1.459
	38.0	10.205	6.988	1.1044	1.3913	15.11	1.509
	40.0	11.100	7.397	1.1106	1.3947	15.99	1.565
	42.0	12.057	7.812	1.1171	1.3982	16.83	1.629
	44.0	13.082	8.234	1.1239	1.4018	17.62	1.700
	46.0	14.183	8.665	1.1313	1.4056		1.780
Zinc sulfate	0.5	0.031	0.031	1.0034	1.3339	0.08	1.021
$ZnSO_4$	1.0	0.063	0.062	1.0085	1.3348	0.15	1.040
	2.0	0.126	0.126	1.0190	1.3366	0.28	1.081
	3.0	0.192	0.191	1.0296	1.3384	0.41	1.126
	4.0	0.258	0.258	1.0403	1.3403	0.53	1.175
	5.0	0.326	0.326	1.0511	1.3421	0.65	1.227
	6.0	0.395	0.395	1.0620	1.3439	0.77	1.283
	7.0	0.466	0.465	1.0730	1.3457	0.89	1.341
	8.0	0.539	0.537	1.0842	1.3475	1.01	1.403
	9.0	0.613	0.611	1.0956	1.3494	1.14	1.470
	10.0	0.688	0.686	1.1071	1.3513	1.27	1.545
	12.0	0.845	0.840	1.1308	1.3551	1.55	1.716
	14.0	1.008	1.002	1.1553	1.3590	1.89	1.918
	16.0	1.180	1.170	1.1806	1.3630	2.31	2.152

ION PRODUCT OF WATER SUBSTANCE
William L. Marshall and E. U. Franck

Pressure (bars)	Temperature (°C)								
	0	25	50	75	100	150	200	250	300
Saturated vapor	14.938	13.995	13.275	12.712	12.265	11.638	11.289	11.191	11.406
250	14.83	13.90	13.19	12.63	12.18	11.54	11.16	11.01	11.14
500	14.72	13.82	13.11	12.55	12.10	11.45	11.05	10.85	10.86
750	14.62	13.73	13.04	12.48	12.03	11.36	10.95	10.72	10.66
1,000	14.53	13.66	12.96	12.41	11.96	11.29	10.86	10.60	10.50
1,500	14.34	13.53	12.85	12.29	11.84	11.16	10.71	10.43	10.26
2,000	14.21	13.40	12.73	12.18	11.72	11.04	10.57	10.27	10.08
2,500	14.08	13.28	12.62	12.07	11.61	10.92	10.45	10.12	9.91
3,000	13.97	13.18	12.53	11.98	11.53	10.83	10.34	9.99	9.76
3,500	13.87	13.09	12.44	11.90	11.44	10.74	10.24	9.88	9.63
4,000	13.77	13.00	12.35	11.82	11.37	10.66	10.16	9.79	9.52
5,000	13.60	12.83	12.19	11.66	11.22	10.52	10.00	9.62	9.34
6,000	13.44	12.68	12.05	11.53	11.09	10.39	9.87	9.48	9.18
7,000	13.31	12.55	11.93	11.41	10.97	10.27	9.75	9.35	9.04
8,000	13.18	12.43	11.82	11.30	10.86	10.17	9.64	9.24	8.93
9,000	13.04	12.31	11.71	11.20	10.77	10.07	9.54	9.13	8.82
10,000	12.91	12.21	11.62	11.11	10.68	9.98	9.45	9.04	8.71

Pressure (bars)	Temperature (°C)								
	350	400	450	500	600	700	800	900	1000
Saturated vapor	12.30	—	—	—	—	—	—	—	—
250	11.77	19.43	21.59	22.40	23.27	23.81	24.23	24.59	24.93
500	11.14	11.88	13.74	16.13	18.30	19.29	19.92	20.39	20.80
750	10.79	11.17	11.89	13.01	15.25	16.55	17.35	17.93	18.39
1,000	10.54	10.77	11.19	11.81	13.40	14.70	15.58	16.22	16.72
1,500	10.22	10.29	10.48	10.77	11.59	12.50	13.30	13.97	14.50
2,000	9.98	9.98	10.07	10.23	10.73	11.36	11.98	12.54	12.97
2,500	9.79	9.74	9.77	9.86	10.18	10.63	11.11	11.59	12.02
3,000	9.61	9.54	9.53	9.57	9.78	10.11	10.49	10.89	11.24
3,500	9.47	9.37	9.33	9.34	9.48	9.71	10.02	10.35	10.62
4,000	9.34	9.22	9.16	9.15	9.23	9.41	9.65	9.93	10.13
5,000	9.13	8.99	8.90	8.85	8.85	8.95	9.11	9.30	9.42
6,000	8.96	8.80	8.69	8.62	8.57	8.61	8.72	8.86	8.97
7,000	8.81	8.64	8.51	8.42	8.34	8.34	8.40	8.51	8.64
8,000	8.68	8.50	8.36	8.25	8.13	8.10	8.13	8.21	8.38
9,000	8.57	8.37	8.22	8.10	7.95	7.89	7.89	7.95	8.12
10,000	8.46	8.25	8.09	7.96	7.78	7.70	7.68	7.70	7.85

Data in this table were calculated from the equation, $\log_{10} K_w^* = A + B/T + C/T^2 + D/T^3 + (E + F/T + G/T^2) \log_{10} \rho_w^*$, where $K_w^* = K_w/(\text{mol kg}^{-1})$, and $\rho_w^* = \rho_w/(\text{g cm}^{-3})$. The parameters are:

$A = -4.098$

$B = -3245.2$ K

$C = +2.2362 \times 10^5$ K^2

$D = -3.984 \times 10^7$ K^3

$E = +13.957$

$F = 1262.3$ K

$G = +8.5641 \times 10^5$ K^2

Reprinted with permission from W. L. Marshall and E. U. Franck, *J. Phys. Chem. Ref. Data*, 10, 295, 1981.

IONIZATION CONSTANT OF NORMAL AND HEAVY WATER

This table gives the ionization constant in molality terms for H_2O and D_2O at temperatures from 0 to 100°C at the saturated vapor pressure. The quantity tabulated is $-\log K_W$, where K_W is defined by

$$K_W = m_+ \times m_-$$

and m_+ and m_- are the molalities, in mol/kg of water, for H^+ and OH^-, respectively.

REFERENCES

1. W.L. Marshall and E.U. Franck, *J. Phys. Chem. Ref. Data*, 10, 295, 1981.
2. R.E. Mesmer and D.L. Herting, *J. Solution Chem.*, 7, 901, 1978.

	$-\log K_W$	
$t/°C$	H_2O	D_2O
0	14.938	15.972
5	14.727	15.743
10	14.528	15.527
15	14.340	15.324
20	14.163	15.132
25	13.995	14.951
30	13.836	14.779
35	13.685	14.616
40	13.542	14.462
45	13.405	14.316
50	13.275	14.176
55	13.152	14.044
60	13.034	13.918
65	12.921	13.798
70	12.814	13.683
75	12.712	13.574
80	12.613	13.470
85	12.520	13.371
90	12.428	13.276
95	12.345	13.186
100	12.265	13.099

SOLUBILITY OF SELECTED GASES IN WATER

L. H. Gevantman

The values in this table are taken almost exclusively from the International Union of Pure and Applied Chemistry "Solubility Data Series". Unless noted, they comprise evaluated data fitted to a smoothing equation. The data at each temperature are then derived from the smoothing equation which expresses the mole fraction solubility X_1 of the gas in solution as:

$$\ln X_1 = A + B/T^* + C \ln T^*$$

where

$$T^* = T/100 \text{ K}$$

All values refer to a partial pressure of the gas of 101.325 kPa (one atmosphere).

The equation constants, the standard deviation for $\ln X_1$ (except where noted), and the temperature range over which the equation applies are given in the column headed Equation constants. There are two exceptions. The equation for methane has an added term, DT^*. The equation for H_2Se and H_2S takes the form,

$$\ln X_1 = A + B/T + C \ln T + DT$$

where T is the temperature in kelvin.

Solubilities given for those gases which react with water, namely ozone, nitrogen oxides, chlorine and its oxides, carbon dioxide, hydrogen sulfide, hydrogen selenide and sulfur dioxide, are recorded as bulk solubilities; i.e., all chemical species of the gas and its reaction products with water are included.

Gas	T/K	Solubility (X_1)	Equation constants	Ref.
Hydrogen (H_2) $M_r = 2.01588$	288.15 293.15 298.15 303.15 308.15	1.510×10^{-5} 1.455×10^{-5} 1.411×10^{-5} 1.377×10^{-5} 1.350×10^{-5}	$A = -48.1611$ $B = 55.2845$ $C = 16.8893$ Std. dev. $= \pm 0.54\%$ Temp.range $= 273.15—353.15$	1
Deuterium (D_2) $M_r = 4.0282$	283.15 288.15 293.15 298.15 303.15	$1.675 \times 10^{-5} \pm 0.57\%$ $1.595 \times 10^{-5} \pm 0.57\%$ $1.512 \times 10^{-5} \pm 0.78\%$ $1.460 \times 10^{-5} \pm 0.52\%$ $1.395 \times 10^{-5} \pm 0.37\%$	Averaged experimental values Temp. range $= 278.15—303.15$	1
Helium (He) $A_r = 4.0026$	288.15 293.15 298.15 303.15 308.15	7.123×10^{-6} 7.044×10^{-6} 6.997×10^{-6} 6.978×10^{-6} 6.987×10^{-6}	$A = -41.4611$ $B = 42.5962$ $C = 14.0094$ Std. dev. $= \pm 0.54\%$ Temp.range $= 273.15—348.15$	2
Neon (Ne) $A_r = 20.1797$	288.15 293.15 298.15 303.15 308.15	8.702×10^{-6} 8.395×10^{-6} 8.152×10^{-6} 7.966×10^{-6} 7.829×10^{-6}	$A = -52.8573$ $B = 61.0494$ $C = 18.9157$ Std. dev. $= \pm 0.47\%$ Temp.range $= 273.15—348.15$	2
Argon (Ar) $A_r = 39.948$	288.15 293.15 298.15 303.15 308.15	3.025×10^{-5} 2.748×10^{-5} 2.519×10^{-5} 2.328×10^{-5} 2.169×10^{-5}	$A = -57.6661$ $B = 74.7627$ $C = 20.1398$ Std. dev. $= \pm 0.26\%$ Temp.range $= 273.15—348.15$	3
Krypton (Kr) $A_r = 83.80$	288.15 293.15 298.15	5.696×10^{-5} 5.041×10^{-5} 4.512×10^{-5}	$A = -66.9928$ $B = 91.0166$ $C = 24.2207$	4

SOLUBILITY OF SELECTED GASES IN WATER (continued)

Gas	T/K	Solubility (X_1)	Equation constants	Ref.
	303.15	4.079×10^{-5}	Std. dev. = ±0.32%	
	308.15	3.725×10^{-5}	Temp.range = 273.15—353.15	
Xenon (Xe)	288.15	10.519×10^{-5}	$A = -74.7398$	4
$A_r = 131.29$	293.15	9.051×10^{-5}	$B = 105.210$	
	298.15	7.890×10^{-5}	$C = 27.4664$	
	303.15	6.961×10^{-5}	Std. dev. = ±0.35%	
	308.15	6.212×10^{-5}	Temp.range = 273.15—348.15	
Radon-222(^{222}Rn)	288.15	2.299×10^{-4}	$A = -90.5481$	
$A_r = 222$	293.15	1.945×10^{-4}	$B = 130.026$	
	298.15	1.671×10^{-4}	$C = 35.0047$	
	303.15	1.457×10^{-4}	Std. dev. = ±1.02%	
	308.15	1.288×10^{-4}	Temp.range = 273.15—373.15	
Oxygen (O_2)	288.15	2.756×10^{-5}	$A = -66.7354$	5
$M_r = 31.9988$	293.15	2.501×10^{-5}	$B = 87.4755$	
	298.15	2.293×10^{-5}	$C = 24.4526$	
	303.15	2.122×10^{-5}	Std. dev. = ±0.36%	
	308.15	1.982×10^{-5}	Temp.range = 273.15—348.15	
Ozone (O_3)	293.15	1.885×10^{-6} ± 10% pH = 7.0	Experimental value derived from Henry's Law Constant	5
$M_r = 47.9982$				
Nitrogen (N_2)	288.15	1.386×10^{-5}	$A = -67.3877$	6
$M_r = 28.0134$	293.15	1.274×10^{-5}	$B = 86.3213$	
	298.15	1.183×10^{-5}	$C = 24.7981$	
	303.15	1.108×10^{-5}	Std. dev. = ±0.72%	
	308.15	1.047×10^{-5}	Temp.range = 273.15—348.15	
Nitrous oxide (N_2O)	288.15	5.948×10^{-4}	$A = -60.7467$	7
$M_r = 44.0129$	293.15	5.068×10^{-4}	$B = 88.8280$	
	298.15	4.367×10^{-4}	$C = 21.2531$	
	303.15	3.805×10^{-4}	Std. dev. = ±1.2%	
	308.15	3.348×10^{-4}	Temp.range = 273.15—313.15	
Nitric oxide (NO)	288.15	4.163×10^{-5}	$A = -62.8086$	7
$M_r = 30.0061$	293.15	3.786×10^{-5}	$B = 82.3420$	
	298.15	3.477×10^{-5}	$C = 22.8155$	
	303.15	3.222×10^{-5}	Std. dev. = ±0.76%	
	308.15	3.012×10^{-5}	Temp.range = 273.15—358.15	
Carbon monoxide (CO)	288.15	2.095×10^{-5}	Derived from Henry's	8
$M_r = 28.0104$	293.15	1.918×10^{-5}	Law Constant Equation	
	298.15	1.774×10^{-5}	Std. dev. = ±0.043%	
	303.15	1.657×10^{-5}	Temp.range = 273.15—328.15	
	308.15	1.562×10^{-5}		
Carbon dioxide (CO_2)	288.15	8.21×10^{-4}	Derived from Henry's	9
$M_r = 44.0098$	293.15	7.07×10^{-4}	Law Constant Equation	
	298.15	6.15×10^{-4}	Std. dev. = ±1.1%	
	303.15	5.41×10^{-4}	Temp.range = 273.15—353.15	
	308.15	4.80×10^{-4}		
Hydrogen selenide (H_2Se)	288.15	1.80×10^{-3}	$A = 9.15$	10
$M_r = 80.976$	298.15	1.49×10^{-3}	$B = 974$	
	308.15	1.24×10^{-3}	$C = -3.542$	
			$D = 0.0042$	

Gas	T/K	Solubility (X_1)	Equation constants	Ref.
			Std. dev. = $\pm 2.3 \times 10^{-5}$	
			Temp. range = 288.15—343.15	
Hydrogen sulfide (H_2S)	288.15	2.335×10^{-3}	$A = -24.912$	10
$M_r = 34.082$	293.15	2.075×10^{-3}	$B = 3477$	
	298.15	1.85×10^{-3}	$C = 0.3993$	
	303.15	1.66×10^{-3}	$D = 0.0157$	
	308.15	1.51×10^{-3}	Std. dev. = $\pm 6.5 \times 10^{-5}$	
			Temp. range = 283.15—603.15	
Sulfur dioxide (SO_2)	288.15	3.45×10^{-2}	$A = -25.2629$	11
$M_r = 64.0648$	293.15	2.90×10^{-2}	$B = 45.7552$	
	298.15	2.46×10^{-2}	$C = 5.6855$	
	303.15	2.10×10^{-2}	Std. dev. = $\pm 1.8\%$	
	308.15	1.80×10^{-2}	Temp.range = 278.15—328.15	
Chlorine (Cl_2)	283.15	$2.48 \times 10^{-3} \pm 2\%$	Experimental data	11
$M_r = 70.9054$	293.15	$1.88 \times 10^{-3} \pm 2\%$	Temp.range = 283.15—333.15	
	303.15	$1.50 \times 10^{-3} \pm 2\%$		
	313.15	$1.23 \times 10^{-3} \pm 2\%$		
Chlorine monoxide (Cl_2O)	273.15	$5.25 \times 10^{-1} \pm 1\%$	Experimental data	11
$M_r = 86.9048$	276.61	$4.54 \times 10^{-1} \pm 1\%$	Temp. range = 273.15—293.15	
	283.15	$4.273 \times 10^{-1} \pm 1\%$		
	293.15	$3.353 \times 10^{-1} \pm 1\%$		
Chlorine dioxide (ClO_2)	288.15	2.67×10^{-2}	$A = 7.9163$	11
$M_r = 67.4515$	293.15	2.20×10^{-2}	$B = 0.4791$	
	298.15	1.823×10^{-2}	$C = 11.0593$	
	303.15	1.513×10^{-2}	Std. dev. = $\pm 4.6\%$	
	308.15	1.259×10^{-2}	Temp.range = 283.15—333.15	
Methane (CH_4)	288.15	3.122×10^{-5}	$A = -115.6477$	12
$M_r = 16.0428$	293.15	2.806×10^{-5}	$B = 155.5756$	
	298.15	2.552×10^{-5}	$C = 65.2553$	
	303.15	2.346×10^{-5}	$D = -6.6170$	
	308.15	2.180×10^{-5}	Std. dev. = $\pm 0.056\%$	
			Temp.range = 273.15—328.15	
Ethane (C_2H_6)	288.15	4.556×10^{-5}	$A = -90.8225$	13
$M_r = 30.0696$	293.15	3.907×10^{-5}	$B = 126.9559$	
	298.15	3.401×10^{-5}	$C = 34.7413$	
	303.15	3.002×10^{-5}	Std. dev. = $\pm 0.13\%$	
	308.15	2.686×10^{-5}	Temp.range = 273.15—323.15	
Propane (C_3H_8)	288.15	3.813×10^{-5}	$A = -102.044$	14
$M_r = 44.097$	293.15	3.200×10^{-5}	$B = 144.345$	
	298.15	2.732×10^{-5}	$C = 39.4740$	
	303.15	2.370×10^{-5}	Std. dev. = $\pm 0.012\%$	
	308.15	2.088×10^{-5}	Temp.range = 273.15—347.15	
Butane (C_4H_{10})	288.15	3.274×10^{-5}	$A = -102.029$	14
$M_r = 58.123$	293.15	2.687×10^{-5}	$B = 146.040$	
	298.15	2.244×10^{-5}	$C = 38.7599$	
	303.15	1.906×10^{-5}	Std. dev. = $\pm 0.026\%$	
	308.15	1.645×10^{-5}	Temp.range = 273.15—349.15	
2-Methyl propane (Isobutane)	288.15	2.333×10^{-5}	$A = -129.714$	14

Gas	T/K	Solubility (X_1)	Equation constants	Ref.
(C_4H_{10})	293.15	1.947×10^{-5}	$B = 183.044$	
$M_r = 58.123$	298.15	1.659×10^{-5}	$C = 53.4651$	
	303.15	1.443×10^{-5}	Std. dev. = $\pm 0.034\%$	
	308.15	1.278×10^{-5}	Temp. range = 278.15—318.15	

REFERENCES

1. C. L. Young, Ed., *IUPAC Solubility Data Series*, Vol. 5/6, Hydrogen and Deuterium, Pergamon Press, Oxford, England, 1981.
2. H. L. Clever, Ed., *IUPAC Solubility Data Series*, Vol. 1, Helium and Neon, Pergamon Press, Oxford, England, 1979.
3. H. L. Clever, Ed., *IUPAC Solubility Data Series*, Vol. 4, Argon, Pergamon Press, Oxford, England, 1980.
4. H. L. Clever, Ed., *IUPAC Solubility Data Series*, Vol. 2, Krypton, Xenon and Radon, Pergamon Press, Oxford, England, 1979.
5. R. Battino, Ed., *IUPAC Solubility Data Series*, Vol. 7, Oxygen and Ozone, Pergamon Press, Oxford, England, 1981.
6. R. Battino, Ed., *IUPAC Solubility Data Series*, Vol. 10, Nitrogen and Air, Pergamon Press, Oxford, England, 1982.
7. C. L. Young, Ed., *IUPAC Solubility Data Series*, Vol. 8, Oxides of Nitrogen, Pergamon Press, Oxford, England, 1981.
8. R. W. Cargill, Ed., *IUPAC Solubility Data Series*, Vol. 43, Carbon Monoxide, Pergamon Press, Oxford, England, 1990.
9. R. Crovetto, Evaluation of Solubility Data for the System CO_2-H_2O, *J. Phys. Chem. Ref. Data*, 20, 575, 1991.
10. P. G. T. Fogg and C. L. Young, Eds., *IUPAC Solubility Data Series*, Vol. 32, Hydrogen Sulfide, Deuterium Sulfide, and Hydrogen Selenide, Pergamon Press, Oxford, England, 1988.
11. C. L. Young, Ed., *IUPAC Solubility Data Series*, Vol. 12, Sulfur Dioxide, Chlorine, Fluorine and Chlorine Oxides, Pergamon Press, Oxford, England, 1983.
12. H. L. Clever and C. L. Young, Eds., *IUPAC Solubility Data Series*, Vol. 27/28, Methane, Pergamon Press, Oxford, England, 1987.
13. W. Hayduk, Ed., *IUPAC Solubility Data Series*, Vol. 9, Ethane, Pergamon Press, Oxford, England, 1982.
14. W. Hayduk, Ed., *IUPAC Solubility Data Series*, Vol. 24, Propane, Butane and 2-Methylpropane, Pergamon Press, Oxford, England, 1986.

SOLUBILITY OF CARBON DIOXIDE IN WATER AT VARIOUS TEMPERATURES AND PRESSURES

The solubility of CO_2 in water, expressed as mole fraction of CO_2 in the liquid phase, is given for pressures up to atmospheric and temperatures of 0 to 100°C. Note that 1 standard atmosphere equals 101.325 kPa. The references give data over a wider range of temperature and pressure. The estimated accuracy is about 2%.

REFERENCES

1. Carroll, J. J., Slupsky, J. D., and Mather, A. E., *J. Phys. Chem. Ref. Data*, 20, 1201, 1991.
2. Fernandez-Prini, R. and Crovetto, R., *J. Phys. Chem. Ref. Data*, 18, 1231, 1989.
3. Crovetto, R., *J. Phys. Chem. Ref. Data*, 20, 575, 1991

	$1000 \times$ mole fraction of CO_2 in liquid phase						
	Partial pressure of CO_2 in kPa						
$t/°C$	5	10	20	30	40	50	100
0	0.067	0.135	0.269	0.404	0.538	0.671	1.337
5	0.056	0.113	0.226	0.338	0.451	0.564	1.123
10	0.048	0.096	0.191	0.287	0.382	0.477	0.950
15	0.041	0.082	0.164	0.245	0.327	0.409	0.814
20	0.035	0.071	0.141	0.212	0.283	0.353	0.704
25	0.031	0.062	0.123	0.185	0.247	0.308	0.614
30	0.027	0.054	0.109	0.163	0.218	0.271	0.541
35	0.024	0.048	0.097	0.145	0.193	0.242	0.481
40	0.022	0.043	0.087	0.130	0.173	0.216	0.431
45	0.020	0.039	0.078	0.117	0.156	0.196	0.389
50	0.018	0.036	0.071	0.107	0.142	0.178	0.354
55	0.016	0.033	0.065	0.098	0.131	0.163	0.325
60	0.015	0.030	0.060	0.090	0.121	0.150	0.300
65	0.014	0.028	0.056	0.084	0.112	0.140	0.279
70	0.013	0.026	0.052	0.079	0.105	0.131	0.261
75	0.012	0.025	0.049	0.074	0.099	0.123	0.245
80	0.012	0.023	0.047	0.070	0.093	0.116	0.232
85	0.011	0.022	0.044	0.067	0.089	0.111	0.221
90	0.011	0.021	0.042	0.064	0.085	0.106	0.211
95	0.010	0.020	0.041	0.061	0.082	0.102	0.203
100	0.010	0.020	0.039	0.059	0.079	0.098	0.196

AQUEOUS SOLUBILITY AND HENRY'S LAW CONSTANTS OF ORGANIC COMPOUNDS

The solubility in water of about 580 organic compounds, including many compounds of environmental interest, is tabulated here. Values are given at 25°C or at the nearest temperature to this where data are available. In some cases solubility values are given at other temperatures as well.

Solubility of solids is defined as the concentration of the compound in a solution that is in equilibrium with the solid phase at the specified temperature and one atmosphere pressure. For liquids whose water mixtures separate into two phases, the solubility given here is the concentration of the compound in the water-rich phase at equilibrium. In the case of gases (i.e., compounds whose vapor pressure at the specified temperature exceeds one atmosphere) the solubility is defined here as the concentration in the water phase when the partial pressure of the compound above the solution is 101.325 kPa (1 atm). Values for gases are marked with an asterisk.

All solubility values are expressed as mass percent of solute, $S = 100 \, w_2$, where the mass fraction w_2 is given by $w_2 = m_2/(m_1 + m_2)$. In these equations m_2 is the mass of solute and m_1 the mass of water. This quantity is related to other common measures of solubility as follows:

Molality: $m_2 = 1000w_2/M_2(1-w_2)$

Mole fraction: $x_2 = (w_2/M_2)/\{(w_2/M_2) + (1-w_2)/M_1\}$

Mass of solute per 100 g of H_2O: $r_2 = 100w_2/(1-w_2)$

Here M_2 is the molar mass of the solute and $M_1 = 18.015$ g/mol is the molar mass of water. For small values of S the amount of substance concentration c_2 in moles per liter is approximately $10S/M_2$.

Data have been selected from evaluated sources wherever possible, in particular the *IUPAC Solubility Data Series* (References 1-4). The primary source for each value is listed in the column following the solubility values. The user is cautioned that wide variations of data are found in the literature for the lower solubility compounds.

The table also contains values of the Henry's Law constant k_H, which provides a measure of the partition of a substance between the atmosphere and the aqueous phase. Here k_H is defined as the limit of p_2/c_2 as the concentration approaches zero, where p_2 is the partial pressure of the solute above the solution and c_2 is the concentration in the solution at equilibrium (other formulations of Henry's Law are often used; see Reference 5). The values of k_H listed here are based on direct experimental measurement whenever available, but many of them are simply calculated as the ratio of the pure compound vapor pressure to the solubility. This approximation is reliable only for compounds of very low solubility. In fact, values of k_H found in the literature frequently differ by a factor of two or three, and variations over an order of magnitude are not unusual (Reference 5). Therefore the data given here should be taken only as a rough indication of the true Henry's Law constant, which is difficult to measure precisely.

All values of k_H refer to 25°C. If the vapor pressure of the compound at 25°C is greater than one atmosphere, it can be assumed that the k_H value has been calculated as $101.325/c_2$ kPa m^3/mol. The source of the Henry's Law data is given in the last column. The air-water partition coefficient (i.e., ratio of air concentration to water concentration when both are expressed in the same units) is equal to k_H/RT or $k_H/2.48$ in the units used here.

Compounds are listed by molecular formula following the Hill convention. To locate a compound by name or CAS Registry Number when the molecular formula is not known, use the table "Physical Constants of Organic Compounds" in Section 3 and its indexes to determine the molecular formula.

* Indicates a value of S for a gas at a partial pressure of 101.325 kPa (1 atm) in equilibrium with the solution.

REFERENCES

1. *Solubility Data Series, International Union of Pure and Applied Chemistry, Vol. 15,* Pergamon Press, Oxford, 1982.
2. *Solubility Data Series, International Union of Pure and Applied Chemistry, Vol. 20,* Pergamon Press, Oxford, 1985.
3. *Solubility Data Series, International Union of Pure and Applied Chemistry, Vol. 37,* Pergamon Press, Oxford, 1988.
4. *Solubility Data Series, International Union of Pure and Applied Chemistry, Vol. 38,* Pergamon Press, Oxford, 1988.
5. Mackay, D., and Shiu, W.Y., *J. Phys. Chem. Ref. Data,* 10, 1175, 1981.
6. Pearlman, R.S., and Yalkowsky, S.H., *J. Phys. Chem. Ref. Data,* 13, 975, 1984.
7. Shiu, W.Y., and Mackay, D., *J. Phys. Chem. Ref. Data,* 15, 911, 1986.
8. Varhanickova, D., Lee, S.C., Shiu, W. Y., and Mackay, D., *J. Chem. Eng. Data,* 40, 620, 1995.
9. Miller, M.M., Ghodbane, S., Wasik, S. P., Tewari, Y.B., and Martire, D.E., *J. Chem. Eng. Data,* 29, 184, 1984.
10. Riddick, J.A., Bunger, W.B., and Sakano, T.K., *Organic Solvents, Fourth Edition,* John Wiley & Sons, New York, 1986.
11. Mackay, D., Shiu, W.Y., and Ma, K.C., *Illustrated Handbook of Physical-Chemical Properties and Environmental Fate for Organic Chemicals,* Vol. I, Lewis Publishers/CRC Press, Boca Raton, FL, 1992.
12. Mackay, D., Shiu, W.Y., and Ma, K.C., *Illustrated Handbook of Physical-Chemical Properties and Environmental Fate for Organic Chemicals,* Vol. II, Lewis Publishers/CRC Press, Boca Raton, FL, 1992.
13. Mackay, D., Shiu, W.Y., and Ma, K.C., *Illustrated Handbook of Physical-Chemical Properties and Environmental Fate for Organic Chemicals,* Vol. III, Lewis Publishers/CRC Press, Boca Raton, FL, 1993.
14. Horvath, A.L., *Halogenated Hydrocarbons,* Marcel Dekker, New York, 1982.
15. Howard, P.H., *Handbook of Environmental Fate and Exposure Data for Organic Chemicals, Vol. I,* Lewis Publishers/CRC Press, Boca Raton, FL, 1989.
16. Howard, P.H., *Handbook of Environmental Fate and Exposure Data for Organic Chemicals, Vol. II,* Lewis Publishers/CRC Press, Boca Raton, FL, 1990.
17. Banergee, S., Yalkowsky, S.H., and Valvani, S.C., *Environ. Sci. Technol.,* 14, 1227, 1980.
18. Gevantman, L., in *CRC Handbook of Chemistry and Physics, 77th Edition,* p. 6-3, CRC Press, Boca Raton, FL, 1996.
19. Wilhelm, E., Battino, R., and Wilcock, R.J., *Chem. Rev.,* 77, 219, 1977.
20. Stephenson, R.M., *J. Chem. Eng. Data,* 37, 80, 1992.
21. Stephenson, R.M., Stuart, J., and Tabak, M., *J. Chem. Eng. Data,* 29, 287, 1984.

Mol. Form.	Name	$t/°C$	S/mass %	Ref.	$k_H/kPa\ m^3\ mol^{-1}$	Ref.
$CBrF_3$	Bromotrifluoromethane	25	0.032*	14		
CBr_3F	Tribromofluoromethane	25	0.040	14		
CBr_4	Tetrabromomethane	30	0.024	14		
$CClF_3$	Chlorotrifluoromethane	25	0.009*	10	6.9	13
CCl_2F_2	Dichlorodifluoromethane	20	0.028*	5	41	13
CCl_3F	Trichlorofluoromethane	20	0.11	5	10.2	13
CCl_4	Tetrachloromethane	25	0.065	20	2.99	13
CF_4	Tetrafluoromethane	25	0.00187*	19		
$CHBr_3$	Tribromomethane	25	0.30	5	0.047	13
$CHClF_2$	Chlorodifluoromethane	25	0.30*	10	3.0	13
$CHCl_2F$	Dichlorofluoromethane	25	0.95*	10		
$CHCl_3$	Trichloromethane	25	0.80	20	0.43	13
CHF_3	Trifluoromethane	25	0.09*	14		
CHI_3	Triiodomethane	25	0.012	14		
CH_2BrCl	Bromochloromethane	25	1.7	10	0.18	13
CH_2Br_2	Dibromomethane	25	1.14	14	0.086	13
CH_2ClF	Chlorofluoromethane	25	1.05*	14		
CH_2Cl_2	Dichloromethane	25	1.73	20	0.30	13
CH_2I_2	Diiodomethane	30	0.124	10	0.032	13
CH_3Br	Bromomethane	20	1.80*	5	0.63	13
CH_3Cl	Chloromethane	25	0.535*	5	0.98	13
CH_3F	Fluoromethane	30	0.177*	5		
CH_3I	Iodomethane	20	1.4	10	0.54	13
CH_3NO_2	Nitromethane	25	11.1	10		
CH_4	Methane	25	0.00227*	18	67.4	5
CO	Carbon monoxide	25	0.00276*	18		
CO_2	Carbon dioxide	25	0.1501	18		
CS_2	Carbon disulfide	20	0.210	10		
C_2ClF_5	Chloropentafluoroethane	25	0.006*	10	260	13
$C_2Cl_2F_4$	1,2-Dichlorotetrafluoroethane	25	0.013*	10	127	13
$C_2Cl_3F_3$	1,1,2-Trichlorotrifluoroethane	25	0.017	10	32	13
C_2Cl_4	Tetrachloroethylene	25	0.026	20	1.73	13
$C_2Cl_4F_2$	1,1,2,2-Tetrachloro-1,2-difluoroethane	25	0.012	10		
C_2Cl_6	Hexachloroethane	25	0.0050	5	0.85	13
C_2F_4	Tetrafluoroethylene	25	0.0158*	19		
C_2HCl_3	Trichloroethylene	25	0.11	5	1.03	13
C_2HCl_5	Pentachloroethane	25	0.048	5	0.25	13
C_2H_2	Acetylene	25	0.1081*	19		
$C_2H_2Br_4$	1,1,2,2-Tetrabromoethane	30	0.0651	10		
$C_2H_2Cl_2$	1,1-Dichloroethylene	25	0.040	5	2.62	13
$C_2H_2Cl_2$	cis-1,2-Dichloroethylene	25	0.35	5	0.46	13
$C_2H_2Cl_2$	trans-1,2-Dichloroethylene	25	0.63	5	0.96	13
$C_2H_2Cl_4$	1,1,1,2-Tetrachloroethane	25	0.11	5	0.24	13
$C_2H_2Cl_4$	1,1,2,2-Tetrachloroethane	25	0.30	5	0.026	13
C_2H_3Cl	Chloroethylene	25	0.27*	5	2.68	13
$C_2H_3Cl_3$	1,1,1-Trichloroethane	25	0.072	5	1.76	13
$C_2H_3Cl_3$	1,1,2-Trichloroethane	25	0.442	5	0.092	13
C_2H_4	Ethylene	25	0.01336*	19	21.7	5
C_2H_4BrCl	1-Bromo-2-chloroethane	30	0.683	10		
$C_2H_4Br_2$	1,2-Dibromoethane	25	0.17	5	0.066	13
$C_2H_4Cl_2$	1,1-Dichloroethane	25	0.51	5	0.63	13
$C_2H_4Cl_2$	1,2-Dichloroethane	25	0.75	20	0.14	13
$C_2H_4O_2$	Methyl formate	25	23	10		
C_2H_5Br	Bromoethane	20	0.91	10	1.23	13
C_2H_5Cl	Chloroethane	20	0.57*	5	1.02	13
C_2H_5F	Fluoroethane	25	0.216*	14		
C_2H_5I	Iodoethane	30	3.88	10	0.52	13
C_2H_5NO	Acetamide	20	40.8	10		
$C_2H_5NO_2$	Nitroethane	25	4.68	10		
C_2H_6	Ethane	25	0.00568*	18	50.6	5

Mol. Form.	Name	$t/°C$	S/mass %	Ref.	k_H/kPa m^3 mol^{-1}	Ref.
C$_2$H$_6$O	Dimethyl ether	24	35.3*	10	0.077	13
C$_2$H$_6$OS	Dimethyl sulfoxide	25	25.3	10		
C$_2$H$_6$S	Dimethyl sulfide	25	2	10		
C$_3$Cl$_6$	Hexachloropropene	25	0.00170	14		
C$_3$F$_6$	Perfluoropropene	25	0.0194*	14		
C$_3$F$_8$	Perfluoropropane	15	0.0015*	14		
C$_3$H$_3$N	2-Propenenitrile	20	7.35	10		
C$_3$H$_4$	Propyne	25	0.364*	5	1.11	5
C$_3$H$_4$Cl$_2$	cis-1,3-Dichloropropene	25	0.27	5	0.24	5
C$_3$H$_4$Cl$_2$	trans-1,3-Dichloropropene	20	0.28	5	0.18	5
C$_3$H$_4$Cl$_2$	2,3-Dichloropropene	25	0.215	5	0.36	5
C$_3$H$_4$O	Acrolein	20	20.8	10		
C$_3$H$_5$Cl	3-Chloropropene	25	0.33	5	1.10	5
C$_3$H$_5$ClO	Epichlorohydrin	20	6.58	10	0.003	13
C$_3$H$_5$Cl$_3$	1,2,3-Trichloropropane	20	0.19	10	0.038	13
C$_3$H$_5$N	Propanenitrile	25	10.3	10		
C$_3$H$_6$	Propene	25	0.0200*	5	21.3	5
C$_3$H$_6$	Cyclopropane	25	0.0484*	19		
C$_3$H$_6$Br$_2$	1,2-Dibromopropane	25	0.143	10		
C$_3$H$_6$Cl$_2$	1,2-Dichloropropane	25	0.247	10	0.29	13
C$_3$H$_6$N$_6$O$_6$	Hexahydro-1,3,5-trinitro-1,3,5-triazine	25	0.0060	17		
C$_3$H$_6$O	Propanal	25	30.6	10		
C$_3$H$_6$O	Methyloxirane	20	40.5	10	0.0087	13
C$_3$H$_6$O$_2$	Ethyl formate	25	11.8	10		
C$_3$H$_6$O$_2$	Methyl acetate	20	24.5	10		
C$_3$H$_7$Br	1-Bromopropane	30	0.230	10	3.8	13
C$_3$H$_7$Br	2-Bromopropane	18	0.286	10	1.27	13
C$_3$H$_7$Cl	1-Chloropropane	20	0.271	10	1.41	13
C$_3$H$_7$Cl	2-Chloropropane	12	0.342	10		
C$_3$H$_7$F	1-Fluoropropane	14	0.386*	14		
C$_3$H$_7$F	2-Fluoropropane	15	0.366	14		
C$_3$H$_7$I	1-Iodopropane	30	0.104	10	0.93	13
C$_3$H$_7$I	2-Iodopropane	20	0.140	10		
C$_3$H$_7$NO$_2$	1-Nitropropane	25	1.50	10		
C$_3$H$_7$NO$_2$	2-Nitropropane	25	1.71	10		
C$_3$H$_8$	Propane	25	0.00669*	18	71.6	5
C$_3$H$_8$O$_2$	Dimethoxymethane	16	24.4	10		
C$_4$F$_8$	Perfluorocyclobutane	21	0.014*	14		
C$_4$H$_4$N$_2$	Succinonitrile	25	11.5	10		
C$_4$H$_4$O	Furan	25	1	10	0.54	13
C$_4$H$_5$N	Methylacrylonitrile	20	2.57	10		
C$_4$H$_5$N	Pyrrole	25	4.5	10		
C$_4$H$_6$	1,3-Butadiene	25	0.0735*	5	20.7	13
C$_4$H$_6$	1-Butyne	25	0.287*	5	1.91	5
C$_4$H$_6$O	trans-2-Butenal	20	15.6	10		
C$_4$H$_6$O$_2$	Methacrylic acid	20	8.9	10		
C$_4$H$_6$O$_2$	trans-Crotonic acid	25	9	10		
C$_4$H$_6$O$_2$	Vinyl acetate	20	2.0	10		
C$_4$H$_6$O$_2$	Methyl acrylate	25	4.94	10		
C$_4$H$_7$Cl	1-Chloro-2-methylpropene	25	0.916	5	0.12	5
C$_4$H$_7$N	Butanenitrile	20	3.3	10		
C$_4$H$_8$	1-Butene	25	0.0222*	5	25.6	13
C$_4$H$_8$	Isobutene	25	0.0263*	5	21.6	13
C$_4$H$_8$Cl$_2$O	Bis(2-chloroethyl) ether	25	1.03	20	0.003	13
C$_4$H$_8$O	cis-Crotonyl alcohol	20	16.6	10		
C$_4$H$_8$O	Ethyl vinyl ether	20	0.9	10		
C$_4$H$_8$O	Butanal	25	7.1	10		
C$_4$H$_8$O	Isobutanal	20	9.1	10		
C$_4$H$_8$O	2-Butanone	25	25.9	20		
C$_4$H$_8$O$_2$	2-Methylpropanoic acid	20	22.8	10		

Mol. Form.	Name	$t/°C$	S/mass %	Ref.	k_H/kPa m^3 mol^{-1}	Ref.
$C_4H_8O_2$	Propyl formate	22	2.05	10		
$C_4H_8O_2$	Ethyl acetate	25	8.08	10		
C_4H_9Br	1-Bromobutane	30	0.0608	10	1.2	13
C_4H_9Cl	1-Chlorobutane	20	0.11	10	1.54	13
C_4H_9Cl	1-Chloro-2-methylpropane	12	0.092	10		
C_4H_9I	1-Iodobutane	17	0.012	10	1.87	13
C_4H_{10}	Butane	25	0.00724*	18	95.9	5
C_4H_{10}	Isobutane	25	0.00535*	18	120	5
$C_4H_{10}O$	Diethyl ether	25	6.04	10	0.088	13
$C_4H_{10}O$	1-Butanol	0	10.4	1		
		25	7.4	1		
		50	6.4	1		
$C_4H_{10}O$	2-Butanol	10	23.9	1		
		25	18.1	1		
		50	14.0	1		
$C_4H_{10}O$	2-Methyl-1-propanol	0	11.5	1		
		25	8.1	1		
		50	6.5	1		
$C_4H_{10}S$	1-Butanethiol	20	0.0597	10		
$C_4H_{11}NO_2$	Diethanolamine	20	95.4	10		
$C_4H_{12}Si$	Tetramethylsilane	25	0.00196	10		
$C_5H_4O_2$	Furfural	20	8.2	10		
C_5H_6	1,3-Cyclopentadiene	25	0.068	3		
$C_5H_7NO_2$	Ethyl cyanoacetate	20	25.9	10		
C_5H_8	1,4-Pentadiene	25	0.056	3	12	5
C_5H_8	2-Methyl-1,3-butadiene	25	0.061	3	7.78	5
		50	0.076*	3		
C_5H_8	1-Pentyne	25	0.157	3	2.5	5
C_5H_8	Cyclopentene	25	0.054	3	6.56	13
$C_5H_8O_2$	Ethyl acrylate	25	1.50	10		
$C_5H_8O_2$	Methyl methacrylate	20	1.56	10		
$C_5H_8O_2$	2,4-Pentanedione	20	16.6	10		
C_5H_{10}	1-Pentene	25	0.0148	3	40.3	5
C_5H_{10}	cis-2-Pentene	25	0.0203	3	22.8	5
C_5H_{10}	3-Methyl-1-butene	25	0.013*	3	54.7	5
C_5H_{10}	2-Methyl-2-butene	25	0.041	3		
C_5H_{10}	Cyclopentane	25	0.0157	3	19.1	13
$C_5H_{10}O$	2-Pentanone	25	5.5	20		
$C_5H_{10}O$	3-Pentanone	25	4.72	20		
$C_5H_{10}O$	2-Methyltetrahydrofuran	25	13.9	10	0.67	13
$C_5H_{10}O_2$	Pentanoic acid	20	2.4	10		
$C_5H_{10}O_2$	3-Methylbutanoic acid	21	4.1	10		
$C_5H_{10}O_2$	Isobutyl formate	22	1.0	10		
$C_5H_{10}O_2$	Propyl acetate	20	2.3	10		
$C_5H_{10}O_2$	Isopropyl acetate	20	2.9	10		
$C_5H_{10}O_2$	Ethyl propanoate	20	1.92	10		
$C_5H_{11}Br$	1-Bromopentane	25	0.0127	10		
$C_5H_{11}Cl$	1-Chloropentane	25	0.02	10	2.37	13
C_5H_{12}	Pentane	25	0.0041	3	128	13
C_5H_{12}	Isopentane	25	0.00485	3	479	13
C_5H_{12}	Neopentane	25	0.00332*	3	220	13
$C_5H_{12}O$	1-Pentanol	0	3.1	1		
		25	2.20	1		
		50	1.8	1		
$C_5H_{12}O$	2-Pentanol	25	4.3	21		
$C_5H_{12}O$	3-Pentanol	25	5.6	21		
$C_5H_{12}O$	2-Methyl-1-butanol	25	3.0	3		
$C_5H_{12}O$	3-Methyl-1-butanol	25	2.7	1		
$C_5H_{12}O$	2-Methyl-2-butanol	25	11.0	1		
$C_5H_{12}O$	3-Methyl-2-butanol	25	5.6	1		

Mol. Form.	Name	$t/°C$	S/mass %	Ref.	k_H/kPa m^3 mol^{-1}	Ref.
$C_5H_{12}O$	2,2-Dimethyl-1-propanol	25	3.5	1		
$C_5H_{12}O$	Methyl tert-butyl ether	25	3.62	20	0.070	13
C_6Cl_6	Hexachlorobenzene	25	0.0000005	2	0.131	11
C_6HCl_5	Pentachlorobenzene	25	0.000055	2	0.085	11
C_6HCl_5O	Pentachlorophenol	25	0.0010	2		
$C_6H_2Br_4$	1,2,4,5-Tetrabromobenzene	25	0.00000434	2		
$C_6H_2Cl_4$	1,2,3,4-Tetrachlorobenzene	25	0.0000433	2	0.144	11
$C_6H_2Cl_4$	1,2,3,5-Tetrachlorobenzene	25	0.000346	2	0.59	11
$C_6H_2Cl_4$	1,2,4,5-Tetrachlorobenzene	25	0.0000606	2	0.122	11
$C_6H_2Cl_4O$	2,3,4,6-Tetrachlorophenol	25	1.8	2		
$C_6H_2Cl_4O_2$	3,4,5,6-Tetrachloro-1,2-benzenediol	25	0.071	8		
$C_6H_3Br_3$	1,2,4-Tribromobenzene	25	0.0010	2		
$C_6H_3Br_3$	1,3,5-Tribromobenzene	25	0.0000789	2		
$C_6H_3Br_3O$	2,4,6-Tribromophenol	15	0.0007	2		
$C_6H_3Cl_3$	1,2,3-Trichlorobenzene	25	0.00309	2	0.242	11
$C_6H_3Cl_3$	1,2,4-Trichlorobenzene	25	0.00379	2	0.277	11
$C_6H_3Cl_3$	1,3,5-Trichlorobenzene	25	0.000655	2	1.1	11
$C_6H_3Cl_3O$	2,4,5-Trichlorophenol	25	0.1	2		
$C_6H_3Cl_3O$	2,4,6-Trichlorophenol	25	0.04	2		
$C_6H_3Cl_3O_2$	3,4,5-Trichloro-1,2-benzenediol	25	0.051	8		
C_6H_4BrCl	1-Bromo-2-chlorobenzene	25	0.0124	2		
C_6H_4BrCl	1-Bromo-3-chlorobenzene	25	0.0118	2		
C_6H_4BrCl	1-Bromo-4-chlorobenzene	25	0.00442	2		
C_6H_4BrI	1-Bromo-4-iodobenzene	25	0.000794	2		
$C_6H_4Br_2$	o-Dibromobenzene	25	0.00748	2		
$C_6H_4Br_2$	m-Dibromobenzene	25	0.0064	2		
$C_6H_4Br_2$	p-Dibromobenzene	25	0.0020	2		
$C_6H_4Br_2O$	2,4-Dibromophenol	25	0.2	2		
C_6H_4ClI	1-Chloro-2-iodobenzene	25	0.00689	2		
C_6H_4ClI	1-Chloro-3-iodobenzene	25	0.00674	2		
C_6H_4ClI	1-Chloro-4-iodobenzene	25	0.00311	2		
$C_6H_4Cl_2$	o-Dichlorobenzene	0	0.0142	2		
		25	0.0147	2	0.244	11
		50	0.0212	2		
$C_6H_4Cl_2$	m-Dichlorobenzene	10	0.0103	2		
		25	0.0106	2	0.376	11
		50	0.0165	2		
$C_6H_4Cl_2$	p-Dichlorobenzene	10	0.00512	2		
		25	0.00829	2	0.160	11
		50	0.0167	2		
$C_6H_4Cl_2O$	2,4-Dichlorophenol	20	0.45	2		
$C_6H_4Cl_2O_2$	3,5-Dichloro-1,2-benzenediol	25	0.78	8		
$C_6H_4Cl_2O_2$	4,5-Dichloro-1,2-benzenediol	25	1.19	8		
$C_6H_4F_2$	o-Difluorobenzene	25	0.114	2		
$C_6H_4F_2$	m-Difluorobenzene	25	0.114	2		
$C_6H_4F_2$	p-Difluorobenzene	25	0.122	2		
$C_6H_4I_2$	o-Diiodobenzene	25	0.00192	2		
$C_6H_4I_2$	m-Diiodobenzene	25	0.000185	2		
$C_6H_4I_2$	p-Diiodobenzene	25	0.000893	2		
C_6H_5Br	Bromobenzene	10	0.0387	2		
		25	0.0445	2	0.21	5
		40	0.0516	2		
C_6H_5BrO	p-Bromophenol	25	1.86	2		
C_6H_5Cl	Chlorobenzene	10	0.0387	2		
		25	0.0495	2	0.38	11
		50	0.0882	2		
C_6H_5ClO	o-Chlorophenol	25	2.0	2		
C_6H_5ClO	m-Chlorophenol	25	2.2	2		
C_6H_5ClO	p-Chlorophenol	25	2.7	2		
C_6H_5F	Fluorobenzene	27	0.154	2	0.70	11

Mol. Form.	Name	$t/°C$	S/mass %	Ref.	k_H/kPa m^3 mol^{-1}	Ref.
C_6H_5I	Iodobenzene	10	0.0193	2		
		25	0.0226	2	0.078	11
		45	0.0279	2		
$C_6H_5NO_2$	Nitrobenzene	25	0.21	17		
C_6H_6	Benzene	10	0.178	3		
		25	0.177	3	0.557	11
		50	0.208	3		
C_6H_6ClN	o-Chloroaniline	25	0.876	10		
C_6H_6O	Phenol	25	8.66	10		
C_6H_7N	Aniline	25	3.38	10	14	15
C_6H_8	1,4-Cyclohexadiene	25	0.08	3	1.03	13
$C_6H_8N_2$	Adiponitrile	20	0.80	16		
$C_6H_8O_4$	Dimethyl maleate	25	8.0	10		
C_6H_{10}	1,5-Hexadiene	25	0.017	3		
C_6H_{10}	1-Hexyne	25	0.036	3	4.14	13
C_6H_{10}	Cyclohexene	25	0.016	3	4.57	13
$C_6H_{10}O$	Cyclohexanone	25	8.8	20		
$C_6H_{10}O$	Mesityl oxide	20	2.89	10		
$C_6H_{10}O_3$	Ethyl acetoacetate	25	12	10		
$C_6H_{11}NO$	Caprolactam	25	84.0	10		
C_6H_{12}	1-Hexene	25	0.0053	3	41.8	5
C_6H_{12}	trans-2-Hexene	25	0.0067	3		
C_6H_{12}	2-Methyl-1-pentene	25	0.0078	3	28.1	5
C_6H_{12}	4-Methyl-1-pentene	25	0.0048	3	63.2	5
C_6H_{12}	2,3-Dimethyl-1-butene	30	0.046	3		
C_6H_{12}	Cyclohexane	25	0.0058	3	19.4	13
C_6H_{12}	Methylcyclopentane	25	0.0043	3	36.7	5
$C_6H_{12}O$	1-Hexen-3-ol	25	2.52	1		
$C_6H_{12}O$	4-Hexen-2-ol	25	3.81	1		
$C_6H_{12}O$	Cyclohexanol	10	4.62	1		
		25	3.8	1		
		40	3.30	1		
$C_6H_{12}O$	Butyl vinyl ether	20	0.3	10		
$C_6H_{12}O$	2-Hexanone	20	1.75	10		
$C_6H_{12}O$	4-Methyl-2-pentanone	25	1.7	10		
$C_6H_{12}O_2$	Hexanoic acid	20	0.958	10		
$C_6H_{12}O_2$	Butyl acetate	20	0.68	10		
$C_6H_{12}O_2$	Isobutyl acetate	20	0.63	10		
$C_6H_{12}O_2$	sec-Butyl acetate	20	0.62	10		
$C_6H_{12}O_2$	Ethyl butanoate	20	0.49	10		
C_6H_{14}	Hexane	25	0.0011	3	183	13
		60	0.00136	3		
C_6H_{14}	2-Methylpentane	25	0.00137	3	176	13
C_6H_{14}	3-Methylpentane	25	0.00129	3	170	13
C_6H_{14}	2,2-Dimethylbutane	25	0.0021	3	199	13
C_6H_{14}	2,3-Dimethylbutane	25	0.0021	3	144	13
$C_6H_{14}O$	1-Hexanol	0	0.79	1		
		25	0.60	1		
		50	0.51	1		
$C_6H_{14}O$	2-Hexanol	25	1.4	1		
$C_6H_{14}O$	3-Hexanol	25	1.6	1		
$C_6H_{14}O$	Dipropyl ether	25	0.49	10	0.26	13
$C_6H_{14}O$	2-Methyl-1-pentanol	25	0.81	1		
$C_6H_{14}O$	4-Methyl-1-pentanol	25	0.76	1		
$C_6H_{14}O$	2-Methyl-2-pentanol	25	3.2	1		
$C_6H_{14}O$	3-Methyl-2-pentanol	25	1.9	1		
$C_6H_{14}O$	4-Methyl-2-pentanol	27	1.5	1		
$C_6H_{14}O$	2-Methyl-3-pentanol	25	2.0	1		
$C_6H_{14}O$	3-Methyl-3-pentanol	25	4.3	1		
$C_6H_{14}O$	2-Ethyl-1-butanol	25	1.0	1		

Mol. Form.	Name	$t/°C$	S/mass %	Ref.	k_H/kPa m^3 mol^{-1}	Ref.
$C_6H_{14}O$	2,2-Dimethyl-1-butanol	25	0.8	1		
$C_6H_{14}O$	Diisopropyl ether	20	1.2	10	0.26	13
$C_6H_{14}O$	2,3-Dimethyl-2-butanol	25	4.2	1		
$C_6H_{14}O$	3,3-Dimethyl-2-butanol	25	2.4	1		
$C_6H_{14}O_2$	1,1-Diethoxyethane	25	5	10		
$C_6H_{15}N$	Dipropylamine	20	2.5	10		
$C_6H_{15}N$	Triethylamine	20	5.5	10		
$C_7H_4Cl_4O$	2,3,4,6-Tetrachloro-5-methylphenol	25	0.00061	2		
$C_7H_5Cl_3$	(Trichloromethyl)benzene	5	0.0053	10		
$C_7H_5Cl_3O$	2,4,6-Trichloro-3-methylphenol	25	0.0112	2		
C_7H_5N	Benzonitrile	25	0.2	10		
C_7H_5NO	Benzoxazole	20	0.834	6		
$C_7H_6Cl_2$	(Dichloromethyl)benzene	30	0.025	10		
$C_7H_6Cl_2O$	2,6-Dichloro-4-methylphenol	25	0.0673	2		
$C_7H_6Cl_2O$	2,4-Dichloro-6-methylphenol	25	0.0283	2		
$C_7H_6N_2$	1H-Benzimidazole	20	0.201	6		
$C_7H_6N_2$	1H-Indazole	20	0.0827	6		
C_7H_6O	Benzaldehyde	20	0.3	10		
$C_7H_6O_2$	Salicylaldehyde	86	1.68	10		
C_7H_7Br	p-Bromotoluene	25	0.011	10		
C_7H_7Cl	(Chloromethyl)benzene	20	0.0493	10		
C_7H_7ClO	4-Chloro-2-methylphenol	25	0.68	2		
C_7H_7ClO	4-Chloro-3-methylphenol	25	0.40	2		
C_7H_7ClO	2-Chloro-6-methylphenol	25	0.36	2		
$C_7H_7NO_3$	2-Nitroanisole	30	0.619	10		
C_7H_8	Toluene	5	0.063	3		
		25	0.053	3	0.680	13
C_7H_8	1,3,5-Cycloheptatriene	25	0.064	3	0.47	13
C_7H_8	1,6-Heptadiyne	25	0.125	3		
C_7H_8O	Benzyl alcohol	20	0.08	10		
C_7H_8O	o-Cresol	40	3.08	10		
C_7H_8O	m-Cresol	40	2.51	10		
C_7H_8O	p-Cresol	40	2.26	10		
C_7H_8O	Anisole	25	0.19	20	0.025	13
C_7H_9N	o-Methylaniline	20	1.66	10		
C_7H_9N	p-Methylaniline	21	7.35	10		
C_7H_{12}	1-Heptyne	25	0.0094	3	4.47	13
C_7H_{12}	Cycloheptene	25	0.0066	3	4.9	13
C_7H_{12}	1-Methylcyclohexene	25	0.0052	3		
C_7H_{14}	1-Heptene	25	0.032	3	40.3	13
C_7H_{14}	$trans$-2-Heptene	25	0.015	3	42.2	13
C_7H_{14}	Cycloheptane	25	0.0030	3	9.59	13
C_7H_{14}	Methylcyclohexane	25	0.00151	3	43.3	13
		50	0.0019	3		
C_7H_{14}	Ethylcyclopentane	20	0.012	3		
$C_7H_{14}O$	2-Heptanone	25	0.43	10		
$C_7H_{14}O$	3-Heptanone	20	1.43	10		
$C_7H_{14}O$	2,4-Dimethyl-3-pentanone	20	0.59	10		
$C_7H_{14}O_2$	Pentyl acetate	20	0.17	10		
$C_7H_{14}O_2$	Isopentyl acetate	20	0.2	10		
$C_7H_{14}O_2$	Ethyl 3-methylbutanoate	20	0.2	10		
C_7H_{16}	Heptane	0	0.0003	3		
		25	0.00024	3	209	13
		40	0.00025	3		
C_7H_{16}	2-Methylhexane	25	0.00025	3	346	5
C_7H_{16}	3-Methylhexane	25	0.00026	3	249	13
C_7H_{16}	2,2-Dimethylpentane	25	0.00044	3	318	5
C_7H_{16}	2,3-Dimethylpentane	25	0.00052	3	175	5
C_7H_{16}	2,4-Dimethylpentane	25	0.00042	3	323	13
C_7H_{16}	3,3-Dimethylpentane	25	0.00059	3	186	5

Mol. Form.	Name	$t/°C$	S/mass %	Ref.	k_H/kPa m^3 mol^{-1}	Ref.
$C_7H_{16}O$	1-Heptanol	10	0.25	1		
		25	0.174	1		
		50	0.12	1		
$C_7H_{16}O$	2-Heptanol	30	0.33	1		
$C_7H_{16}O$	3-Heptanol	25	0.43	1		
$C_7H_{16}O$	4-Heptanol	25	0.47	1		
$C_7H_{16}O$	2-Methyl-2-hexanol	25	1.0	1		
$C_7H_{16}O$	5-Methyl-2-hexanol	25	0.49	1		
$C_7H_{16}O$	3-Methyl-3-hexanol	25	1.2	1		
$C_7H_{16}O$	2,3-Dimethyl-2-pentanol	25	1.5	1		
$C_7H_{16}O$	2,4-Dimethyl-2-pentanol	25	1.3	1		
$C_7H_{16}O$	3-Ethyl-3-pentanol	25	1.7	1		
$C_7H_{16}O$	2,2-Dimethyl-3-pentanol	25	0.82	1		
$C_7H_{16}O$	2,3-Dimethyl-3-pentanol	25	1.6	1		
$C_7H_{16}O$	2,4-Dimethyl-3-pentanol	25	0.70	1		
$C_7H_{16}O$	2,3,3-Trimethyl-2-butanol	40	2.2	1		
$C_8H_4F_6$	1,3-Bis(trifluoromethyl)benzene	25	0.0041	2		
$C_8H_6N_2$	Quinoxaline	50	54	6		
C_8H_6S	Benzo[b]thiophene	20	0.0130	6		
$C_8H_7ClO_3$	3-Chloro-4-hydroxy-5-methoxybenzaldehyde	25	0.093	8		
$C_8H_7ClO_3$	2-Chloro-4-hydroxy-5-methoxybenzaldehyde	25	0.013	8		
$C_8H_7Cl_3O$	2,4,6-Trichloro-3,5-dimethylphenol	25	0.00050	2		
C_8H_7N	Indole	20	0.187	6		
C_8H_8	Styrene	25	0.025	4	0.3	15
		50	0.046	4	0.30	13
$C_8H_8N_2$	2-Methyl-1H-benzimidazole	20	0.145	6		
C_8H_8O	Acetophenone	25	0.55			
$C_8H_8O_2$	Methyl benzoate	20	0.21	10		
$C_8H_8O_3$	Methyl salicylate	30	0.74	10		
$C_8H_8O_3$	4-Hydroxy-3-methoxybenzaldehyde	25	0.247	8		
C_8H_9ClO	4-Chloro-2,5-dimethylphenol	25	0.89	2		
C_8H_9ClO	4-Chloro-2,6-dimethylphenol	25	0.52	2		
C_8H_9ClO	4-Chloro-3,5-dimethylphenol	25	0.34	2		
C_8H_{10}	Ethylbenzene	0	0.020	4		
		25	0.0169	4	0.887	11
		40	0.0200	4		
C_8H_{10}	o-Xylene	25	0.0173	4	0.565	13
		45	0.021	4		
C_8H_{10}	m-Xylene	0	0.0203	4		
		25	0.016	4	0.730	13
		40	0.022	4		
C_8H_{10}	p-Xylene	0	0.0160	4		
		25	0.018	4	0.578	13
		40	0.022	4		
$C_8H_{10}O$	Phenetole	25	0.12	10		
$C_8H_{10}O$	2,4-Xylenol	25	0.787	10		
$C_8H_{10}O$	3,5-Xylenol	29	0.62	10		
C_8H_{12}	4-Vinylcyclohexene	25	0.005	4		
C_8H_{14}	1-Octyne	25	0.0024	4	7.87	13
C_8H_{16}	1-Octene	25	0.00027	4	96.3	13
C_8H_{16}	Cyclooctane	25	0.00079	4	10.7	13
C_8H_{16}	Ethylcyclohexane	40	0.00066	4		
C_8H_{16}	cis-1,2-Dimethylcyclohexane	25	0.00060	4	36	5
C_8H_{16}	trans-1,4-Dimethylcyclohexane	25	0.000384	4	88.2	5
C_8H_{16}	Propylcyclopentane	25	0.00020	4	90.2	5
C_8H_{16}	1,1,3-Trimethylcyclopentane	25	0.00037	4	159	5
$C_8H_{16}O$	2-Octanone	25	0.113	10		
$C_8H_{16}O_2$	Octanoic acid	25	0.0798	10		

Mol. Form.	Name	$t/°C$	S/mass %	Ref.	k_H/kPa m³ mol⁻¹	Ref.
$C_8H_{16}O_2$	Hexyl acetate	20	0.02	10		
$C_8H_{16}O_2$	sec-Hexyl acetate	20	0.13	10		
$C_8H_{16}O_2$	Isobutyl isobutanoate	20	0.5	10		
$C_8H_{17}Cl$	3-(Chloromethyl)heptane	20	0.01	10		
C_8H_{18}	Octane	25	0.000071	4	311	13
		50	0.00010	4		
C_8H_{18}	3-Methylheptane	25	0.000079	4	376	5
C_8H_{18}	2,2,4-Trimethylpentane	25	0.00022	4	307	13
C_8H_{18}	2,3,4-Trimethylpentane	25	0.00018	4	206	13
$C_8H_{18}O$	Dibutyl ether	20	0.03	10	0.48	13
$C_8H_{18}O$	1-Octanol	25	0.054	1		
$C_8H_{18}O$	2-Octanol	25	0.4	1		
$C_8H_{18}O$	2-Methyl-2-heptanol	30	0.25	1		
$C_8H_{18}O$	2-Ethyl-1-hexanol	25	0.01	1		
$C_8H_{19}N$	Dibutylamine	20	0.47	10		
$C_8H_{19}N$	2-Ethylhexylamine	20	0.25	10		
$C_8H_{20}Si$	Tetraethylsilane	25	0.0000325	10		
C_9H_7N	Quinoline	20	0.633	6		
C_9H_7N	Isoquinoline	20	0.452	6		
C_9H_9N	3-Methyl-1H-indole	20	0.050	6		
C_9H_{10}	Indan	25	0.010	4		
$C_9H_{10}O_2$	Ethyl benzoate	25	0.083	20		
C_9H_{12}	1,8-Nonadiyne	25	0.0125	4		
C_9H_{12}	Propylbenzene	25	0.0055	4	1.04	11
C_9H_{12}	Isopropylbenzene	25	0.0056	4	1.47	11
C_9H_{12}	o-Ethyltoluene	25	0.0093	5	0.529	13
C_9H_{12}	p-Ethyltoluene	25	0.0094	5	0.500	13
C_9H_{12}	1,2,3-Trimethylbenzene	25	0.0069	4	0.343	11
C_9H_{12}	1,2,4-Trimethylbenzene	25	0.0056	4	0.569	11
C_9H_{12}	1,3,5-Trimethylbenzene	25	0.00489	4	0.781	11
$C_9H_{14}O_6$	Triacetin	25	5.8	10		
C_9H_{16}	1-Nonyne	25	0.00072	4		
C_9H_{18}	1,1,3-Trimethylcyclohexane	25	0.000177	4	105	13
$C_9H_{18}O$	Diisobutyl ketone	25	0.043	10		
$C_9H_{18}O_2$	Nonanoic acid	20	0.0284	10		
C_9H_{20}	Nonane	25	0.000017	4	333	13
		50	0.000022	4		
C_9H_{20}	4-Methyloctane	25	0.0000115	4	1000	5
C_9H_{20}	2,2,5-Trimethylhexane	25	0.00008	4	246	13
$C_9H_{20}O$	3,5-Dimethyl-4-heptanol	15	0.072	1		
$C_9H_{20}O$	1-Nonanol	25	0.014	1		
$C_9H_{20}O$	2-Nonanol	15	0.026	1		
$C_9H_{20}O$	3-Nonanol	15	0.032	1		
$C_9H_{20}O$	4-Nonanol	15	0.0026	1		
$C_9H_{20}O$	5-Nonanol	15	0.0032	1		
$C_{10}H_7Cl$	1-Chloronaphthalene	25	0.00224	5		
$C_{10}H_7Cl$	2-Chloronaphthalene	25	0.00117	5		
$C_{10}H_8$	Naphthalene	10	0.0019	4		
		25	0.0031	4	0.0430	5
		50	0.0082	4		
$C_{10}H_9N$	3-Methylisoquinoline	20	0.092	6		
$C_{10}H_{10}O_4$	Dimethyl phthalate	25	0.40	15		
$C_{10}H_{14}$	Butylbenzene	25	0.0015	4	1.33	11
$C_{10}H_{14}$	sec-Butylbenzene	25	0.0014	4	1.89	11
$C_{10}H_{14}$	tert-Butylbenzene	25	0.0032	4	1.28	11
$C_{10}H_{14}$	Isobutylbenzene	25	0.0010	4	3.32	11
$C_{10}H_{14}$	p-Cymene	25	0.00234	4	0.80	5
$C_{10}H_{14}$	1,2,4,5-Tetramethylbenzene	25	0.000348	4	2.55	11
$C_{10}H_{16}$	d-Limonene	0	0.00097	4		
		25	0.00138	4		

Mol. Form.	Name	$t/°C$	S/mass %	Ref.	k_H/kPa m^3 mol^{-1}	Ref.
$C_{10}H_{16}O$	Camphor	20	0.01	10		
$C_{10}H_{18}$	trans-Decahydronaphthalene	25	0.000089	4	3	13
$C_{10}H_{20}$	Pentylcyclopentane	25	0.0000115	4	185	5
$C_{10}H_{20}$	1-Decene	25	0.00057	4		
$C_{10}H_{22}$	Decane	0	0.0000015	4	479	13
$C_{10}H_{22}O$	Diisopentyl ether	20	0.02	10		
$C_{10}H_{22}O$	1-Decanol	25	0.0037	1		
$C_{11}H_{10}$	1-Methylnaphthalene	25	0.0028	4	0.0450	12
$C_{11}H_{10}$	2-Methylnaphthalene	25	0.0025	4	0.051	12
$C_{11}H_{16}$	Pentylbenzene	25	0.00105	5	1.69	11
$C_{12}Cl_{10}$	Decachlorobiphenyl	25	0.00000000012	7	0.0208	7
$C_{12}HCl_9$	2,2',3,3',4,5,5',6,6'-Nonachlorobiphenyl	25	0.0000000018	7		
$C_{12}H_2Cl_8$	2,2',3,3'5,5',6,6'-Octachlorobiphenyl	25	0.0000003	7	0.0381	7
$Cl_2H_3Cl_7$	2,2',3,3',4,4',6-Heptachlorobiphenyl	25	0.0000002	7	0.0054	7
$C_{12}H_4Cl_6$	2,2',3,3',4,4'-Hexachlorobiphenyl	25	0.00000006	7	0.0119	7
$C_{12}H_4Cl_6$	2,2',4,4',6,6'-Hexachlorobiphenyl	25	0.00000007	7	0.818	7
$C_{12}H_4Cl_6$	2,2',3,3',6,6'-Hexachlorobiphenyl	25	0.00000008	7		
$C_{12}H_5Cl_5$	2,3,4,5,6-Pentachlorobiphenyl	25	0.0000008	7		
$C_{12}H_5Cl_5$	2,2',4,5,5'-Pentachlorobiphenyl	25	0.000001	7	0.0355	7
$C_{12}H_6Cl_4$	2,3,4,5-Tetrachlorobiphenyl	25	0.000002	7		
$C_{12}H_6Cl_4$	2,2',4',5-Tetrachlorobiphenyl	25	0.0000016	9		
$C_{12}H_7Cl_3$	2,4,5-Trichlorobiphenyl	25	0.000014	7	0.0243	7
$C_{12}H_7Cl_3$	2,4,6-Trichlorobiphenyl	25	0.00002	7	0.0495	7
$C_{12}H_8Cl_2$	2,5-Dichlorobiphenyl	25	0.0002	7	0.0201	7
$C_{12}H_8Cl_2$	2,6-Dichlorobiphenyl	25	0.00014	7		
$C_{12}H_8O$	Dibenzofuran	25	0.000656	6	0.011	12
$C_{12}H_8S$	Dibenzothiophene	25	0.000103	6		
$C_{12}H_9Cl$	2-Chlorobiphenyl	25	0.00055	7	0.0701	7
$C_{12}H_9N$	Carbazole	22	0.000120	6		
$C_{12}H_{10}$	Acenaphthene	0	0.00015	4		
		25	0.00038	4	0.012	12
		50	0.00092	4		
$C_{12}H_{10}$	Biphenyl	0	0.000272	4		
		25	0.00072	4	0.028	5
		50	0.0022	4		
$C_{12}H_{10}N_2O$	N-Nitrosodiphenylamine	25	0.0035	17		
$C_{12}H_{10}O$	Diphenyl ether	25	0.00180	6	0.027	13
$C_{12}H_{12}$	1-Ethylnaphthalene	25	0.00101	4	0.039	12
$C_{12}H_{12}$	2-Ethylnaphthalene	25	0.00080	4	0.078	12
$C_{12}H_{12}$	1,3-Dimethylnaphthalene	25	0.0008	4		
$C_{12}H_{12}$	1,4-Dimethylnaphthalene	25	0.00114	4		
$C_{12}H_{12}$	1,5-Dimethylnaphthalene	25	0.00031	4		
$C_{12}H_{12}$	2,3-Dimethylnaphthalene	25	0.00025	4		
$C_{12}H_{12}$	2,6-Dimethylnaphthalene	25	0.00017	4		
$C_{12}H_{18}$	Hexylbenzene	25	0.00021	4		
$C_{12}H_{26}$	Dodecane	25	0.00000037	4	750	5
$C_{12}H_{26}O$	1-Dodecanol	25	0.0004	1		
$C_{12}H_{27}O_4P$	Tributyl phosphate	25	0.039	10		
$C_{13}H_9N$	Acridine	25	0.00466	6		
$C_{13}H_9N$	Benzo[f]quinoline	25	0.0079	6		
$C_{13}H_{10}$	9H-Fluorene	0	0.00007	4		
		25	0.00019	4	0.0079	12
		50	0.00063	4		
$C_{13}H_{12}$	Diphenylmethane	25	0.000141	4	0.001	12
$C_{13}H_{14}$	1,4,5-Trimethylnaphthalene	25	0.00021	4		
$C_{14}H_{10}$	Anthracene	0	0.0000022	4		
		25	0.0000062	4	0.0040	12
$C_{14}H_{10}$	Phenanthrene	10	0.000050	4		
		25	0.00011	4	0.0032	12
		50	0.00041	4		

Mol. Form.	Name	$t/°C$	S/mass %	Ref.	k_H/kPa m^3 mol^{-1}	Ref.
$C_{14}H_{12}$	*trans*-Stilbene	25	0.000029	4	0.040	12
$C_{14}H_{14}$	1,2-Diphenylethane	25	0.00044	6	0.017	12
$C_{14}H_{14}O$	Dibenzyl ether	35	0.0040	10		
$C_{14}H_{30}$	Tetradecane	25	0.000012	5		
$C_{14}H_{30}O$	1-Tetradecanol	25	0.000031	1		
$C_{15}H_{12}$	2-Methylanthracene	25	0.0000030	4		
$C_{15}H_{12}$	9-Methylanthracene	25	0.000026	4		
$C_{15}H_{12}$	1-Methylphenanthrene	25	0.0000269	4		
$C_{15}H_{32}O$	1-Pentadecanol	25	0.000010	1		
$C_{16}H_{10}$	Fluoranthene	25	0.000024	4	0.0010	12
$C_{16}H_{10}$	Pyrene	25	0.0000132	4	0.00092	12
		50	0.00009	4		
$C_{16}H_{14}$	9,10-Dimethylanthracene	25	0.0000056	4		
$C_{16}H_{22}O_4$	Dibutyl phthalate	25	0.00112	15		
$C_{16}H_{34}O$	1-Hexadecanol	25	0.000003	1		
$C_{17}H_{12}$	11H-Benzo[a]fluorene	25	0.0000045	4		
$C_{17}H_{12}$	11H-Benzo[b]fluorene	25	0.0000002	4		
$C_{18}H_{12}$	Benz[a]anthracene	25	0.0000011	4	0.00058	12
$C_{18}H_{12}$	Chrysene	25	0.00000019	4	0.006	12
$C_{18}H_{12}$	Naphthacene	25	0.00000006	4	0.000004	12
$C_{18}H_{12}$	Triphenylene	25	0.0000041	4	0.00001	12
$C_{18}H_{12}N_2$	2,2'-Biquinoline	24	0.000102	6		
$C_{18}H_{34}O_4$	Dibutyl sebacate	20	0.004	10		
$C_{19}H_{14}$	5-Methylchrysene	27	0.0000062	4		
$C_{19}H_{14}$	9-Methylbenz[a]anthracene	27	0.0000066	4		
$C_{19}H_{14}$	10-Methylbenz[a]anthracene	25	0.0000055	4		
$C_{20}H_{12}$	Benzo[a]pyrene	25	0.00000038	4	0.000046	12
$C_{20}H_{12}$	Perylene	25	0.00000004	4	0.000003	12
$C_{20}H_{12}$	Benzo[e]pyrene	20	0.00000046	4	0.00002	12
$C_{20}H_{13}N$	13H-Dibenzo[a,i]carbazole	24	0.00000104	6		
$C_{20}H_{14}$	1,2-Dihydrobenz[j]aceanthrylene	25	0.00000036	6		
$C_{20}H_{42}$	Eicosane	25	0.00000019	4		
$C_{21}H_{13}N$	Dibenz[a,j]acridine	25	0.000016	6		
$C_{21}H_{16}$	1,2-Dihydro-3-methylbenz[j]aceanthrylene	25	0.00000022	6		
$C_{22}H_{12}$	Benzo[ghi]perylene	25	0.000000026	4	0.000075	12
$C_{22}H_{14}$	Picene	25	0.00000025	4		
$C_{22}H_{14}$	Benzo[b]triphenylene	25	0.0000027	4		
$C_{22}H_{14}$	Dibenz[a,h]anthracene	25	0.00000006	4		
$C_{22}H_{14}$	Dibenz[a,j]anthracene	25	0.0000012	4		
$C_{22}H_{44}O_2$	Butyl stearate	25	0.2	10		
$C_{24}H_{12}$	Coronene	25	0.000000014	4		

AQUEOUS SOLUBILITY OF INORGANIC COMPOUNDS AT VARIOUS TEMPERATURES

The solubility of over 300 common inorganic compounds in water is tabulated here as a function of temperature. Solubility is defined as the concentration of the compound in a solution that is in equilibrium with a solid phase at the specified temperature. In this table the solid phase is generally the most stable crystalline phase at the temperature in question. An asterisk * on solubility values in adjacent columns indicates that the solid phase changes between those two temperatures (usually from one hydrated phase to another or from a hydrate to the anhydrous solid). In such cases the slope of the solubility vs. temperature curve may show a discontinuity.

All solubility values are expressed as mass percent of solute, $100 \cdot w_2$, where

$$w_2 = m_2/(m_1 + m_2)$$

and m_2 is the mass of solute and m_1 the mass of water. This quantity is related to other common measures of solubility as follows:

Molality: $m_2 = 1000 w_2/M_2(1-w_2)$
Mole fraction: $x_2 = (w_2/M_2)/\{(w_2/M_2) + (1-w_2)/M_1\}$
Mass of solute per 100 g of H_2O: $r_2 = 100 w_2/(1-w_2)$

Here M_2 is the molar mass of the solute and $M_1 = 18.015$ g/mol is the molar mass of water.

The data in the table have been derived from the references indicated; in many cases the data have been refitted or interpolated in order to present solubility at rounded values of temperature. Where available, values were taken from the IUPAC *Solubility Data Series* (Reference 1) or the related papers in the *Journal of Physical and Chemical Reference Data* (References 2 to 5), which present carefully evaluated data.

The solubility of sparingly soluble compounds that do not appear in this table may be calculated from the data in the table "Solubility Product Constants". Solubility of inorganic gases may be found in the table "Solubility of Selected Gases in Water".

Compounds are listed alphabetically by chemical formula in the most commonly used form (e.g., NaCl, NH_4NO_3, etc.).

REFERENCES

1. *Solubility Data Series*, International Union of Pure and Applied Chemistry. Volumes 1 to 53 were published by Pergamon Press, Oxford, from 1979 to 1994; subsequent volumes were published by Oxford University Press, Oxford. The number following the colon is the volume number in the series.
2. Clever, H.L., and Johnston, F.J., *J. Phys. Chem. Ref. Data*, 9, 751, 1980.
3. Marcus, Y., *J. Phys. Chem. Ref. Data*, 9, 1307, 1980.
4. Clever, H.L., Johnson, S.A., and Derrick, M.E., *J. Phys. Chem. Ref. Data*, 14, 631, 1985.
5. Clever, H.L., Johnson, S.A., and Derrick, M.E., *J. Phys. Chem. Ref. Data*, 21, 941, 1992.
6. Söhnel, O., and Novotny, P., *Densities of Aqueous Solutions of Inorganic Substances*, Elsevier, Amsterdam, 1985.
7. Krumgalz, B.S., *Mineral Solubility in Water at Various Temperatures*, Israel Oceanographic and Limnological Research Ltd., Haifa, 1994.
8. Potter, R.W., and Clynne, M.A., *J. Research U.S. Geological Survey*, 6, 701, 1978; Clynne, M.A., and Potter, R.W., *J. Chem. Eng. Data*, 24, 338, 1979.
9. Marshal, W.L., and Slusher, R., *J. Phys. Chem.*, 70, 4015, 1966; Knacke, O., and Gans, W., *Zeit. Phys. Chem.*, NF, 104, 41, 1977.
10. Stephen, H., and Stephen, T., *Solubilities of Inorganic and Organic Compounds, Vol. 1*, Macmillan, New York, 1963.

AQUEOUS SOLUBILITY OF INORGANIC COMPOUNDS AT VARIOUS TEMPERATURES (continued)

Compound	0°C	10°C	20°C	25°C	30°C	40°C	50°C	60°C	70°C	80°C	90°C	100°C	Ref.
$AgBrO_3$				0.193							1.32		7
$AgClO_2$	0.17	0.31	0.47	0.55	0.64	0.82	1.02	1.22	1.44	1.66	1.88	2.11	7
$AgClO_3$				15									7
$AgClO_4$	81.6	83.0	84.2	84.8	85.3	86.3	86.9	87.5	87.9	88.3	88.6	88.8	6
$AgNO_2$	0.155			0.413									7
$AgNO_3$	55.9	62.3	67.8	70.1	72.3	76.1	79.2	81.7	83.8	85.4	86.7	87.8	6
Ag_2SO_4	0.56	0.67	0.78	0.83	0.88	0.97	1.05	1.13	1.20	1.26	1.32	1.39	7
$AlCl_3$	30.84	30.91	31.03	31.10	31.18	31.37	31.60	31.87	32.17	32.51	32.90	33.32	7
$Al(ClO_4)_3$	54.9										64.4		6
AlF_3	0.25	0.34	0.44	0.50	0.56	0.68	0.81	0.96	1.11	1.28	1.45	1.64	7
$Al(NO_3)_3$	37.0	38.2	39.9	40.8	42.0	44.5	47.3	50.4	53.8*			61.5*	6
$Al_2(SO_4)_3$	27.5			27.8	28.2	29.2	30.7	32.6	34.9	37.6	40.7	44.2	7
As_2O_3	1.19	1.48	1.80	2.01	2.27	2.86	3.43	4.11	4.89	5.77	6.72	7.71	10
$BaBr_2$	47.6	48.5	49.5	50.0	50.4	51.4	52.5	53.5	54.5	55.5	56.6	57.6	6
$Ba(BrO_3)_2$	0.285	0.442	0.656	0.788	0.935	1.30	1.74	2.27	2.90	3.61	4.40	5.25	1:14
$Ba(C_2H_3O_2)_2$	37.0			44.2									7
$BaCl_2$	23.30	24.88	26.33	27.03	27.70	29.00	30.27	31.53	32.81	34.14	35.54	37.05	8
$Ba(ClO_2)_2$	30.5			31.3								44.7	7
$Ba(ClO_3)_2$	16.90	21.23	23.66	27.50	29.43	33.16	36.69	40.05	43.04	45.90	48.70	51.17	1:14
$Ba(ClO_4)_2$	67.30	70.96	74.30	75.75	77.05	79.23	80.92	82.21	83.16	83.88	84.43	84.90	7
BaF_2	0.158	0.158	0.158	0.161									7
BaI_2	62.5	64.7	67.3	68.8	69.1	69.5	70.1	70.7	71.3	72.0	72.7	73.4	7
$Ba(IO_3)_2$	0.0182	0.0262	0.0342	0.0396	0.045*	0.058*	0.073	0.090	0.109	0.131	0.156	0.182	6
$Ba(NO_2)_2$	31.1	36.6	41.8	44.3	46.8	51.6	56.2	60.5	64.6	68.5	72.1	75.6	1:14
$Ba(NO_3)_2$	4.7	6.3	8.2	9.3	10.2	12.4	14.7	17.0	19.3	21.5	23.5	25.5	10
$Ba(OH)_2$	1.67			4.68	8.4	19	33	52	74	101			6
BaS	2.79	4.78	6.97	8.21	9.58	12.67	16.18	20.05	24.19	28.55	33.04	37.61	7
$Ba(SCN)_2$				62.6									7
$BaSO_3$				0.0011									1:26
$BeCl_2$	40.5			41.7									7
$Be(ClO_4)_2$				59.5									7
$BeSO_4$	26.69	27.58	28.61	29.22	29.90	31.51	33.39	35.50	37.78	40.21	42.72	45.28	7
$CaBr_2$	55	56	59	61	63	68	71	73					10
$CaCl_2$	36.70	39.19	42.13	44.83*	49.12*	52.85*	56.05*	56.73	57.44	58.21	59.04	59.94	8
$Ca(ClO_3)_2$	63.2	64.2	65.5	66.3	67.2								1:14
$Ca(ClO_4)_2$				65.3	69.0	69.0	71.0	73.2	75.5*	77.4*	77.7	78.0	7
CaF_2	0.0013			0.0016									10
CaI_2	64.6	66.0	67.6	68.3		70.8	72.4	74.0	76.0	78.0	79.6	81.0	7
$Ca(IO_3)_2$	0.082	0.155	0.243	0.305	0.384*	0.517*	0.590	0.652	0.811*	0.665*	0.668		1:14
$Ca(NO_2)_2$	38.6	39.5	44.5	48.6									7
$Ca(NO_3)_2$	50.1	53.1	56.7	59.0	60.9	65.4	77.8	78.1	78.2	78.3	78.4	78.5	6
$CaSO_3$				0.0054	0.0049	0.0041	0.0035	0.0030	0.0026	0.0023	0.0020	0.0019	1:26
$CaSO_4$	0.174	0.191	0.202	0.205	0.208	0.210	0.207	0.201	0.193	0.184	0.173	0.163	9

Compound	0°C	10°C	20°C	25°C	30°C	40°C	50°C	60°C	70°C	80°C	90°C	100°C	Ref.
CdBr$_2$	36.0	43.0	49.9	53.4	56.4	60.3*	60.3*	60.5	60.7	60.9	61.3	61.6	6
CdC$_2$O$_4$				0.0060									5
CdCl$_2$	47.2	50.1	53.2	54.6	56.3*	57.3*	57.5	57.8	58.1	58.51	58.98	59.5	6
Cd(ClO$_4$)$_2$				58.7								66.9	7
CdF$_2$		5.82	4.65	4.18	3.76								5
CdI$_2$	44.1	44.9	45.8	46.3	46.8	47.9	49.0	50.2	51.5	52.7	54.1	55.4	6
Cd(IO$_3$)$_2$				0.091									5
Cd(NO$_3$)$_2$	55.4	57.1	59.6	61.0	62.8	66.5	70.6	86.1	86.5	86.8	87.1	87.4	6
CdSO$_4$	43.1	43.1	43.2	43.4	43.6	44.1	43.5	42.5	41.4	40.2	38.5	36.7	6
CdSeO$_4$	42.04	40.59	39.02	38.18	37.29	35.35	33.15	30.65	27.84	24.69	21.24	17.49	5
Ce(NO$_3$)$_3$	57.99	59.80	61.89	63.05	64.31*	67.0*	68.6	71.1*	74.9*	79.2	80.9	83.1	1:13
CoCl$_2$	30.30	32.60	34.87	35.99	37.10	39.27	41.38	43.46	45.50	47.51	49.51	51.50	7
Co(ClO$_4$)$_2$	50.0			53.0									7
CoF$_2$				1.4									7
CoI$_2$	58.00	61.78	65.35	66.99	68.51	71.17	73.41	75.29	76.89	78.28	79.52	80.70	7
Co(NO$_2$)$_2$	0.076			0.49									7
Co(NO$_3$)$_2$	45.5	47.0	49.4	50.8	52.4	56.0	60.1	62.6	64.9	67.7			6
CoSO$_4$	19.9	23.0	26.1	27.7	29.2	32.3	34.4	35.9	35.5	33.2	30.6	27.8	6
Co(SCN)$_2$				50.7									7
CrO$_3$	62.2	62.3	62.6	62.8	63.0	63.5	64.1	64.7	65.5	66.2	67.1	67.9	6
CsBr				55.2									7
CsBrO$_3$	1.16	1.93	3.01	3.69	4.46	6.32	8.60	11.32	14.45	17.96	21.83	25.98	1:30
CsCl	61.83	63.48	64.96	65.64	66.29	67.50	68.60	69.61	70.54	71.40	72.21	72.96	1:47
CsClO$_3$	2.40	3.87	5.94	7.22	8.69	12.15	16.33	21.14	26.45	32.10	37.89	43.42	1:30
CsClO$_4$	0.79	1.01	1.51	1.96	2.57	4.28	6.55	9.29	12.41	15.80	19.39	23.07	7
CsI	30.9	37.2	43.2	45.9	48.6	53.3	57.3	60.7	63.6	65.9	67.7	69.2	6
CsIO$_3$	1.08	1.58	2.21	2.59	3.02	3.96	5.06	6.29	7.70	9.20	10.79	12.45	1:30
CsNO$_3$	8.46	13.0	18.6	21.8	25.1	32.0	39.0	45.7	51.9	57.3	62.1	66.2	6
CsOH					75								7
Cs$_2$SO$_4$	62.6	63.4	64.1	64.5	64.8	65.5	66.1	66.7	67.3	67.8	68.3	68.8	6
CuBr$_2$				55.8									7
CuCl$_2$	40.8	41.7	42.6	43.1	43.7	44.8	46.0	47.2	48.5	49.9	51.3	52.7	6
Cu(ClO$_4$)$_2$	54.3				59.3								7
CuF$_2$				0.075									7
Cu(NO$_3$)$_2$	45.2	49.8	56.3	59.2	61.1	62.0	63.1	64.5	65.9	67.5	69.2	71.0	6
CuSO$_4$	12.4	14.4	16.7	18.0	19.3	22.2	25.4	28.8	32.4	36.3	40.3	43.5	6
CuSeO$_4$	10.6			16.0									7
Dy(NO$_3$)$_3$	58.79	59.99	61.49	62.35	63.29	65.43	68.04	71.58	77.75				1:13
Er(NO$_3$)$_3$	61.58	63.15	64.84	65.75	66.69	68.70	70.96	73.64					1:13
Eu(NO$_3$)$_3$	55.2	56.7	58.5	59.4	60.4	62.5	64.6	66.2					1:13
FeBr$_2$				54.6								64.8*	7
FeCl$_2$	33.2*			39.4*								48.7*	7
FeCl$_3$	42.7	44.9	47.9	47.7	51.6	74.8	76.7	84.6	84.3	84.3	84.4	84.7	6

AQUEOUS SOLUBILITY OF INORGANIC COMPOUNDS AT VARIOUS TEMPERATURES (continued)

Compound	0°C	10°C	20°C	25°C	30°C	40°C	50°C	60°C	70°C	80°C	90°C	100°C	Ref.
$Fe(ClO_4)_2$	63.39												7
FeF_3				5.59									7
$Fe(NO_3)_3$	40.15			46.57									7
$Fe(NO_3)_2$	41.44			46.67									7
$FeSO_4$	13.5	17.0	20.8	22.8	24.8	28.8	32.8	35.5	33.6	30.4	27.1	24.0	6
$Gd(NO_3)_3$	56.3	57.7	59.2	60.1	61.0	62.9	65.2	67.9	71.5				1:13
HIO_3	73.45	74.10	74.98	75.48	76.03	77.20	78.46	79.78	81.13	82.48	83.82	85.14	1:30
H_3BO_3	2.61	3.57	4.77	5.48	6.27	8.10	10.3	12.9	15.9	19.3	23.1	27.3	6
$HgBr_2$	0.26	0.37	0.52	0.61	0.72	0.96	1.26	1.63	2.08	2.61	3.23	3.95	4
$Hg(CN)_2$	6.57	7.83	9.33	10.2	11.1	13.1	15.5	18.2	21.2	24.6	28.3	32.3	6
$HgCl_2$	4.24	5.05	6.17	6.81	7.62	9.53	12.02	15.18	19.16	24.06	29.90	36.62	4
HgI_2			0.0041	0.0055	0.0072	0.0122	0.0199						4
$Hg(SCN)_2$				0.070									4
Hg_2Cl_2				0.0004									3
$Hg_2(ClO_4)_2$	73.8			79.8*								85.3*	7
Hg_2SO_4	0.038	0.043	0.048	0.051	0.054	0.059	0.065	0.070	0.076	0.082	0.088	0.093	4
$Ho(NO_3)_3$				63.8									1:13
KBF_4	0.28	0.34	0.45	0.55	0.75	1.38	2.09	2.82	3.58	4.34	5.12	5.90	10
KBr	35.0	37.3	39.4	40.4	41.4	43.2	44.8	46.2	47.6	48.8	49.8	50.8	6
$KBrO_3$	2.97	4.48	6.42	7.55	8.79	11.57	14.71	18.14	21.79	25.57	29.42	33.28	1:30
$KC_2H_3O_2$	68.40	70.29	72.09	72.92	73.70	75.08	76.27	77.31	78.22	79.04	79.80	80.55	6
KCl	21.74	23.61	25.39	26.22	27.04	28.59	30.04	31.40	32.66	33.86	34.99	36.05	1:47
$KClO_3$	3.03	4.67	6.74	7.93	9.21	12.06	15.26	18.78	22.65	26.88	31.53	36.65	1:30
$KClO_4$	0.70	1.10	1.67	2.04	2.47	3.54	4.94	6.74	8.99	11.71	14.94	18.67	6
KF	30.90	39.8	47.3	50.41	53.2			60.0					7
$KHCO_3$	18.62	21.73	24.92	26.6	28.13	31.32	34.46	37.51	40.45				6
$KHSO_4$	27.1	29.7	32.3	33.6	35.0	37.8	40.5	43.4	46.2	49.02	51.82	54.6	6
KH_2PO_4	11.74	14.91	18.25	19.97	21.77	25.28	28.95	32.76	36.75	40.96	45.41	50.12	1:31
KI	56.0	57.6	59.0	59.7	60.4	61.6	62.8	63.8	64.8	65.7	66.6	67.4	6
KIO_3	4.53	5.96	7.57	8.44	9.34	11.09	13.22	15.29	17.41	19.58	21.78	24.03	1:30
KIO_4	0.16	0.22	0.37	0.51	0.70	1.24	1.96	2.83	3.82	4.89	6.02	7.17	7
$KMnO_4$	2.74	4.12	5.96	7.06	8.28	11.11	14.42	18.16					6
KNO_2	73.7	74.6	75.3	75.7	76.0	76.7	77.4	78.0	78.5	79.1	79.6	80.1	6
KNO_3	12.0	17.6	24.2	27.7	31.3	38.6	45.7	52.2	58.0	63.0	67.3	70.8	6
KOH	48.7	50.8	53.2	54.7	56.1	57.9	58.6	59.5	60.6	61.8	63.1	64.6	6
$KSCN$	63.8	66.4	69.1	70.4	71.6	74.1	76.5	78.9	81.1	83.3	85.3	87.3	6
K_2CO_3	51.3	51.7	52.3	52.7	53.1	54.0	54.9	56.0	57.2	58.4	59.6	61.0	6
K_2CrO_4	37.1	38.1	38.9	39.4	39.8	40.5	41.3	41.9	42.6	43.2	43.8	44.3	6
$K_2Cr_2O_7$	4.30	7.12	10.9	13.1	15.5	20.8	26.3	31.7	36.9	41.5	45.5	48.9	6
K_2HAsO_4	48.5*			63.6*		67.7*		72.7*				79.8*	7
K_2HPO_4	57.0	59.1	61.5	62.7	64.1								1:31
K_2MoO_4				64.7							66.5		7
K_2SO_3	51.30	51.39	51.49	51.55	51.62	51.76	51.93	52.11	52.32	52.54	52.79	53.06	1:26

AQUEOUS SOLUBILITY OF INORGANIC COMPOUNDS AT VARIOUS TEMPERATURES (continued)

Compound	0°C	10°C	20°C	25°C	30°C	40°C	50°C	60°C	70°C	80°C	90°C	100°C	Ref.
K_2SO_4	7.11	8.46	9.95	10.7	11.4	12.9	14.2	15.5	16.7	17.7	18.6	19.3	6
$K_2S_2O_3$	49.0*			62.3*							75.7*		7
$K_2S_2O_5$	22.1	26.7	31.1	33.1	35.2	39.0	42.6	46.0	49.1	52.0	54.6		1:26
K_2SeO_3	68.4*			68.5*								68.5*	7
K_2SeO_4	52.70	52.93	53.17	53.30	53.43	53.70	53.99	54.30	54.61	54.94	55.26	55.60	7
K_3AsO_4	51.5*			55.6*								73*	7
$K_3Fe(CN)_6$	23.9	27.6	31.1	32.8	34.3	37.2	39.6	41.7	43.5	45.0	46.1	47.0	6
K_3PO_4	44.3			51.4									7
$K_4Fe(CN)_6$	12.5	17.3	22.0	23.9	25.6	29.2	32.5	35.5	38.2	40.6	41.4	43.1	6
$LaCl_3$	49.0	48.5	48.6	48.9	49.3	50.5	52.1	54.0	56.3	58.9	61.7		6
$La(NO_3)_3$	55.0	56.9	58.9	60.0	61.1	63.6	66.3	69.9*	74.1*				1:13
$LiBr$	58.4	60.1	62.7	64.4	65.9	67.8	68.3	69.0	69.8	70.7	71.7	72.8	6
$LiBrO_3$	61.03	62.62	64.44	65.44	66.51	68.90	71.68*	73.24*	74.43	75.66	76.93	78.32	1:30
$LiC_2H_3O_2$	23.76	26.49	29.42	31.02	32.72	36.48	40.65	45.15	49.93	54.91	60.04	65.26	7
$LiCl$	40.45	42.46*	45.29*	45.81	46.25	47.30	48.47	49.78	51.27	52.98	54.98*	56.34*	1:47
$LiClO_3$	73.2	75.6*	80.8*	82.1	83.4	85.9*	87.1*	88.2	89.6	91.3	93.4	95.7	1:30
$LiClO_4$	30.1	32.6	35.5	37.0	38.6	41.9	45.5	49.2	53.2	57.2	61.3	71.4	6
LiF	0.120	0.126	0.131	0.134									7
LiH_2PO_4	55.8												7
LiI	59.4	60.5	61.7	62.3	63.0	64.3	65.8	67.3	68.8	81.3	81.7	82.6	6
$LiIO_3$				43.8									1:30
$LiNO_2$	41	45	49	51	53	56	60	63	66	68			10
$LiNO_3$	34.8	37.6	42.7	50.5	57.9	60.1	62.2	64.0	65.7	67.2	68.5	69.7	6
$LiOH$	10.8	10.8	11.0	11.1	11.3	11.7	12.2	12.7	13.4	14.2	15.1	16.1	6
$LiSCN$				54.5									7
Li_2CO_3	1.54	1.43	1.33	1.28	1.24	1.15	1.07	0.99	0.92	0.85	0.78	0.72	7
$Li_2C_2O_4$				5.87									7
Li_2HPO_3	9.07	8.40	7.77	7.47	7.18	6.64	6.16	5.71	5.30	4.91	4.53	4.16	7
Li_2SO_4	26.3	25.9	25.6	25.5	25.3	25.0	24.8	24.5	24.3	24.0	23.8	23.6	6
Li_3PO_4				0.027									1:31
$Lu(NO_3)_3$				71.1									1:13
$MgBr_2$	49.3	49.8	50.3	50.6	50.9	51.5	52.1	52.8	53.5	54.2	55.0	55.7	6
$Mg(BrO_3)_2$	43.0	45.2	48.0	49.4	51.0	54.3	57.9	61.6	65.3	69.0*	70.9*	71.7	1:14
$Mg(C_2H_3O_2)_2$	36.18	37.55	38.92	39.61									7
MgC_2O_4				0.038									7
$MgCl_2$	33.96	34.85	35.58	35.90	36.20	36.77	37.34	37.97	38.71	39.62	40.75	42.15	8
$Mg(ClO_3)_2$	53.35	54.40	56.81	58.66	60.91*	65.46*	67.33	69.27	71.01	72.44	73.48		1:14
$Mg(ClO_4)_2$	47.8	48.7	49.6	50.1	50.5	51.3	52.1						6
$MgCrO_4$	32.06*			35.39*						67.0			7
$MgCr_2O_7$				58.9									7
MgF_2				0.013									7
MgI_2	54.7	56.1	58.2	59.4	60.8	63.9	65.0	65.0	65.0	65.0	65.1	65.2	6
$Mg(IO_3)_2$	3.19*	6.70*	7.92	8.52	9.11	10.45	11.99	13.7	15.6	17.6	19.6		1:14

AQUEOUS SOLUBILITY OF INORGANIC COMPOUNDS AT VARIOUS TEMPERATURES (continued)

Compound	0°C	10°C	20°C	25°C	30°C	40°C	50°C	60°C	70°C	80°C	90°C	100°C	Ref.
$Mg(NO_2)_2$				47									7
$Mg(NO_3)_2$	38.4	39.5	40.8	41.6	42.4	44.1	45.9	47.9	50.0	52.2	70.6	72.0	6
$MgSO_3$	0.32	0.37	0.46	0.52	0.61	0.87*	0.85*	0.76	0.69	0.64	0.62	0.60	1:26
$MgSO_4$	18.2	21.7	25.1	26.3	28.2	30.9	33.4	35.6	36.9	35.9	34.7	33.3	6
MgS_2O_3	30.7			34.1									7
$MgSeO_4$	31.4*			35.7*								47*	7
$MnBr_2$	56.00	57.72	59.39	60.19	60.96	62.41	63.75	65.01	66.19	67.32	68.42	69.50	7
$MnCl_2$	38.7	40.6	42.5	43.6	44.7	47.0	49.4	54.1	54.7	55.2	55.7	56.1	6
MnF_2	0.80*			1.01*							0.34	0.48	7
$Mn(IO_3)_2$				0.27									7
$Mn(NO_3)_2$	50.5			61.7									7
$MnSO_4$	34.6	37.3	38.6	38.9	38.9	37.7	36.3	34.6	32.8	30.8	28.8	26.7	6
NH_4Br	37.5	40.2	42.7	43.9	45.1	47.3	49.4	51.3	53.0	54.6	56.1	57.4	7
NH_4Cl	22.92	25.12	27.27	28.34	29.39	31.46	33.50	35.49	37.46	39.40	41.33	43.24	1:47
NH_4ClO_4	10.8	14.1	17.8	19.7	21.7	25.8	29.8	33.6	37.3	40.7	43.8	46.6	6
NH_4F	41.7	43.2	44.7	45.5	46.3	47.8	49.3	50.9	52.5	54.1			7
NH_4HCO_3	10.6	13.7	17.6	19.9	22.4	27.9	34.2	41.4	49.3	58.1	67.6	78.0	7
$NH_4H_2AsO_4$	25.2	29.0	32.7	34.5	36.3	39.7	43.1	46.2	49.3	52.2	55.0		7
$NH_4H_2PO_4$	17.8	22.0	26.4	28.8	31.2	36.2	41.6	47.2	53.0	59.2	65.7	72.4	7
NH_4I	60.7	62.1	63.4	64.0	64.6	65.8	66.8	67.8	68.7	69.6	70.4	71.1	6
NH_4IO_3				3.70	4.20	5.64	7.63						1:30
NH_4NO_2	55.7	59.0	64.9	68.8	70.3	74.3	77.7	80.8	83.4	85.8	88.2	90.3	7
NH_4NO_3	54.0	60.1	65.5	68.0					81.1				6
NH_4SCN				64.4									7
$(NH_4)_2C_2O_4$	2.31	3.11	4.25	4.94	5.73	7.56	9.73	12.2	15.1	18.3	21.8	25.7	7
$(NH_4)_2HPO_4$	36.4	38.2	40.0	41.0	42.0	44.1	46.2	48.5	50.9	53.3	55.9	58.6	7
$(NH_4)_2S_2O_5$	65.5	67.9	69.8	70.5	71.3	72.3	72.9	73.1					1:26
$(NH_4)_2S_2O_8$	37.00	40.45	43.84	45.49	47.11	50.25	53.28	56.23	59.13	62.00			7
$(NH_4)_2SO_3$	32.2	34.9	37.7	39.1	40.6	43.7	47.0	50.6	54.5	58.9			1:26
$(NH_4)_2SO_4$	41.3	42.1	42.9	43.3	43.8	44.7	45.6	46.6	47.5	48.5	49.5	50.5	6
$(NH_4)_2SeO_3$	49.0		53.4	54.7	56.0	58.9	62.0	65.4	69.1				7
$(NH_4)_2SeO_4$				54.02									7
$(NH_4)_3PO_4$				15.5									7
$NaBr$	44.4	45.9	47.7	48.6	49.6	51.6	53.7	54.1	54.3	54.5	54.7	54.9	6
$NaBrO_3$	20.0	23.22	26.65	28.28	29.86	32.83	35.55	38.05	40.37	42.52			1:30
$NaCHO_2$	30.8	37.9	45.7	48.7	50.6	52.0	53.5	55.0					6
$NaC_2H_3O_2$	26.5	28.8	31.8	33.5	35.5	39.9	45.1	58.3	59.3	60.5	61.7	62.9	6
$NaCl$	26.28	26.32	26.41	26.45	26.52	26.67	26.84	27.03	27.25	27.50	27.78	28.05	1:47
$NaClO$	22.7			44.4									7
$NaClO_2$				97.0*				95.3*					7
$NaClO_3$	44.27	46.67	49.3	50.1	51.2	53.6	55.5	57.0	58.5	60.5	63.3	67.1	1:30
$NaClO_4$	61.9	64.1	66.2	67.2	68.3	70.4	72.5	74.1	74.7	75.4	76.1	76.7	6
NaF	3.52	3.72	3.89	3.97	4.05	4.20	4.34	4.46	4.57	4.66	4.75	4.82	6

AQUEOUS SOLUBILITY OF INORGANIC COMPOUNDS AT VARIOUS TEMPERATURES (continued)

Compound	0°C	10°C	20°C	25°C	30°C	40°C	50°C	60°C	70°C	80°C	90°C	100°C	Ref.
$NaHCO_3$	6.48	7.59	8.73	9.32	9.91	11.13	12.40	13.70	15.02	16.37	17.73	19.10	7
$NaHSO_4$				22.2								33.3	10
NaH_2PO_4	36.54	41.07	46.00	48.68	51.54	57.89*	61.7*	62.3*	65.9	68.7			1:31
NaI	61.2	62.4	63.9	64.8	65.7	67.7	69.8	72.0	74.7	74.8	74.9	75.1	6
$NaIO_3$	2.43	4.40	7.78*	8.65*	9.60	11.67	13.99	16.52	19.25*	21.1*	22.9	24.7	1:30
$NaIO_4$				12.62									7
$NaNO_2$	41.9	43.4	45.1	45.9	46.8	48.7	50.7	52.8	55.0	57.2	59.5	61.8	6
$NaNO_3$	42.2	44.4	46.6	47.7	48.8	51.0	53.2	55.3	57.5	59.6	61.7	63.8	6
$NaOH$	30	39	46	50	53	58	63	67	71	74	76	79	10
$NaSCN$		52.9	57.1	60.2	62.7	63.5	64.2	65.0	65.9	66.9	67.9	69.0	6
$Na_2B_4O_7$	1.23	1.71	2.50	3.07	3.82	6.02	9.7	14.9	17.1	19.9	23.5	28.0	6
Na_2CO_3	6.44	10.8	17.9	23.5	28.7	32.8	32.2	31.7	31.3	31.1	30.9	30.9	6
$Na_2C_2O_4$	2.62	2.95	3.30	3.48	3.65	4.00	4.36	4.71	5.06	5.41	5.75	6.08	6
Na_2CrO_4	22.6	32.3	44.6	46.7	46.9	48.9	51.0	53.4	55.3	55.5	55.8	56.1	6
$Na_2Cr_2O_7$	62.1	63.1	64.4	65.2	66.1	68.0	70.1	72.3	74.6	77.0	79.6	80.7	6
Na_2HAsO_4	5.6*			29.3*	36.9*							67*	7
Na_2HPO_4	1.66	4.19	7.51	10.55	16.34*	35.17*	44.64*	45.20	46.81	48.78	50.52	51.53	1:31
Na_2MoO_4	30.6	38.8	39.4	39.4	39.8	40.3	41.7	41.7	42.6	43.5	44.5	45.5	6
Na_2S	11.1	13.2	15.7	17.1	18.6	22.1	26.7	28.1	30.2	33.0	36.4	41.0	6
Na_2SO_3	12.0	16.1	20.9	23.5	26.3*	27.3*	25.9	24.8	23.7	22.8	22.1	21.5	1:26
Na_2SO_4			16.13	21.94	29.22*	32.35*	31.55	30.90	30.39	30.02	29.79	29.67	8
$Na_2S_2O_3$	33.1	36.3	40.6	43.3	45.9	52.0	62.3	65.7	68.8	69.4	70.1	71.0	1:26
$Na_2S_2O_5$		38.4	39.5	40.0	40.6	41.8	43.0	44.2	45.5	46.8	48.1	49.5	7
Na_2SeO_3				47.3*								45*	7
Na_2SeO_4	11.7			36.9*								42.1*	6
Na_2WO_4	41.6	41.9	42.3	42.6	42.9	43.6	44.4	45.3	46.2	47.3	48.4	49.5	6
Na_3PO_4	4.28	7.30	10.8	12.6	14.1	16.6	22.9	28.4	32.4	37.6	40.4	43.5	6
$Na_2P_2O_7$	2.23	3.28	4.81	6.62	7.00	10.10	14.38	20.07	27.31	36.03	32.37	30.67	6
$NdCl_3$	49.0	49.3	49.7	50.0	50.4	51.2	52.2	53.3	54.5	55.8	57.1	58.5	1:13
$Nd(NO_3)_3$	55.76	57.49	59.37	60.38	61.43	63.69	66.27	69.47					6
$NiCl_2$	34.7	36.1	38.5	40.3	41.7	42.1	43.2	45.0	46.1	46.2	46.4	46.6	7
$Ni(ClO_4)_2$	51.1			52.8									7
NiF_2				2.50							2.52		7
NiI_2	55.40	57.68	59.78	60.69	61.50	62.80	63.73	64.38	64.80	65.09	65.30		7
$Ni(NO_3)_2$	44.1	46.0	48.4	49.8	51.3	54.6	58.3	61.0	63.1	65.6	67.9	69.0	6
$NiSO_4$	21.4	24.4	27.4	28.8	30.3*	32.0*	34.1	35.8	37.7	39.9	42.3	44.8	6
$Ni(SCN)_2$				35.48									7
$NiSeO_4$	21.6		26.2*									45.6*	7
$PbBr_2$	0.449	0.620	0.841	0.966	1.118	1.46	1.89						2
$PbCl_2$	0.66	0.81	0.98	1.07	1.17	1.39	1.64	1.93	2.24	2.60	2.99	3.42	2
$Pb(ClO_4)_2$				81.5									7
PbF_2	0.041	0.0603	0.0649	0.0670	0.0693								2
PbI_2		0.052	0.067	0.076	0.086	0.112	0.144	0.187	0.243	0.315			2

AQUEOUS SOLUBILITY OF INORGANIC COMPOUNDS AT VARIOUS TEMPERATURES (continued)

Compound	0°C	10°C	20°C	25°C	30°C	40°C	50°C	60°C	70°C	80°C	90°C	100°C	Ref.
Pb(IO$_3$)$_2$				0.0025									7
Pb(NO$_3$)$_2$	28.46	32.13	35.67	37.38	39.05	42.22	45.17	47.90	50.42	52.72	54.82	56.75	2
PbSO$_4$	0.0033	0.0038	0.0042	0.0044	0.0047	0.0052	0.0058						2
PrCl$_3$	48.0	48.1	48.6	49.0	49.5	50.8	52.3	54.1	56.1	58.3			6
Pr(NO$_3$)$_3$	57.50	59.20	61.16	62.24	63.40*	65.7*	67.8	70.2	73.4				1:13
RbBr	47.4	50.1	52.6	53.8	54.9	57.0	58.8	60.6	62.1	63.5	64.8	65.9	6
RbBrO$_3$	0.97	1.55	2.36	2.87	3.45	4.87	6.64	8.78	11.29	14.15	17.32	20.76	1:30
RbCl	43.58	45.65	47.53	48.42	49.27	50.86	52.34	53.67	54.92	56.08	57.16	58.15	1:47
RbClO$_3$	2.10	3.38	5.14	6.22	7.45	10.35	13.85	17.93	22.53	27.57	32.96	38.60	1:30
RbClO$_4$	1			1.5								17	7
RbF			75										7
RbHCO$_3$			53.7										7
RbI	55.8	58.6	61.1	62.3	63.4	65.4	67.2	68.8	70.3	71.6	72.7	73.8	6
RbIO$_3$	1.09	1.53	2.07	2.38	2.74	3.52	4.41	5.42	6.52	7.74	9.00	10.36	1:30
RbNO$_3$	16.4	25.0	34.6	39.4	44.2	53.1	60.8	67.2	72.2	76.1	79.0	81.2	6
RbOH				63.4	63.4								7
Rb$_2$CrO$_4$	38.27			43.26									7
Rb$_2$SO$_4$	27.3	30.0	32.5	33.7	34.8	36.9	38.7	40.3	41.8	43.0	44.1	44.9	6
SbCl$_3$	85.7			90.8									7
SbF$_3$	79.4			83.1									7
Sc(NO$_3$)$_3$	57.0	59.3	61.6	62.8	63.9	66.2	68.5	68.1*					1:13
Sm(NO$_3$)$_3$	54.83	56.33	58.08	59.05	60.08	62.38	65.05*		70.8	74.2			1:13
SmCl$_3$		48.0	48.2	48.4	48.6	49.2	50.0						6
SnCl$_2$	46	64											7
SnI$_2$			0.97									3.87	7
SrBr$_2$	46.0	48.3	50.6	51.7	52.9	55.2	57.6	59.9	62.3	64.6	66.8	69.0	6
Sr(BrO$_3$)$_2$	18.53	22.00	25.39	27.02	28.59	31.55	34.21	36.57	38.64*	40.2*	40.8	41.0	1:14
SrCl$_2$	31.94	32.93	34.43	35.37	36.43	38.93	41.94	45.44*	46.81*	47.69	48.70	49.87	8
Sr(ClO$_2$)$_2$	13.0	13.6	14.1	14.3	14.5	14.9	15.3	15.6	15.9				1:14
Sr(ClO$_3$)$_2$	63.29	63.42	63.64	63.77	63.93	64.29	64.70	65.16	65.65	66.18	66.74	67.31	1:14
Sr(ClO$_4$)$_2$	70.04*			75.35*		78.44*							7
SrF$_2$	0.011			0.021									7
SrI$_2$	62.5	62.8	63.5	63.9	64.5	65.8	67.3	69.0	70.8	72.7	74.7	79.2	6
Sr(IO$_3$)$_2$	0.102	0.126	0.152	0.165	0.179	0.206	0.233	0.259	0.284	0.307	0.328	0.346	1:14
Sr(MnO$_4$)$_2$	2.5												7
Sr(NO$_2$)$_2$					41.9	44.3						58.6	7
Sr(NO$_3$)$_2$	28.2	34.6	41.0	44.5	47.0	47.4	47.9	48.4	48.9	49.5	50.1	50.7	6
Sr(OH)$_2$	0.9			2.2									7
SrSO$_3$				0.0015									1:26
SrSO$_4$				0.0135									7
SrS$_2$O$_3$	8.8	13.2	17.7	20.0	22.2	26.8							7
Tb(NO$_3$)$_3$			60.6	61.02									7
Tl$_2$SO$_4$	2.65	3.56	4.61	5.19	5.80	7.09	8.46	9.89	11.33	12.77	14.18	15.53	1:13

AQUEOUS SOLUBILITY OF INORGANIC COMPOUNDS AT VARIOUS TEMPERATURES (continued)

Compound	0°C	10°C	20°C	25°C	30°C	40°C	50°C	60°C	70°C	80°C	90°C	100°C	Ref.
Tm(NO$_3$)$_3$				67.9									1:13
UO$_2$(NO$_3$)$_2$	49.52	51.82	54.42	55.85	57.55	61.59	67.07						1:55
Y(NO$_3$)$_3$	55.57	56.93	58.75	59.86	61.11*	63.3*	64.9	67.9	72.5				1:13
Yb(NO$_3$)$_3$				70.5									1:13
ZnBr$_2$	79.3	80.1	81.8	83.0	84.1	85.6	85.8	86.1	86.3	86.6	86.8	87.1	6
ZnC$_2$O$_4$		0.0010	0.0019	0.0026									5
ZnCl$_2$		76.6	79.0	80.3	81.4	81.8	82.4	83.0	83.7	84.4	85.2	86.0	6
Zn(ClO$_4$)$_2$	44.29*			46.27*			48.70						7
ZnF$_2$				1.53									5
ZnI$_2$	81.1	81.2	81.3	81.4	81.5	81.7	82.0	82.3	82.6	83.0	83.3	83.7	6
Zn(IO$_3$)$_2$			0.58	0.64	0.69	0.77	0.82						5
Zn(NO$_3$)$_2$	47.8	50.8	54.4	54.6	58.5	79.1	80.1	87.5	89.9				6
ZnSO$_3$			0.1786	0.1790	0.1794	0.1803	0.1812						5
ZnSO$_4$	29.1	32.0	35.0	36.6	38.2	41.3	43.0	42.1	41.0	39.9	38.8	37.6	6
ZnSeO$_4$	33.06	34.98	37.38	38.79	40.34								5

SOLUBILITY PRODUCT CONSTANTS

The solubility product constant K_{sp} is a useful parameter for calculating the aqueous solubility of sparingly soluble compounds under various conditions. It may be determined by direct measurement or calculated from the standard Gibbs energies of formation $\Delta_f G^\circ$ of the species involved at their standard states. Thus if $K_{sp} = [M^+]^m [A^-]^n$ is the equilibrium constant for the reaction

$$M_m A_n(s) \rightleftharpoons mM^+(aq) + nA^-(aq),$$

where $M_m A_n$ is the slightly soluble substance and M^+ and A^- are the ions produced in solution by the dissociation of $M_m A_n$, then the Gibbs energy change is

$$\Delta G^\circ = m \, \Delta_f G^\circ \, (M^+, aq) + n \, \Delta_f G^\circ \, (A^-, aq) - \Delta_f G^\circ \, (M_m A_n, s)$$

The solubility product constant is calculated from the equation

$$\ln K_{sp} = -\Delta G^\circ / RT$$

The first table below gives selected values of K_{sp} at 25°C. Many of these have been calculated from standard state thermodynamic data in References 1 and 2; other values are taken from publications of the IUPAC Solubility Data Project (References 3 to 7).

The above formulation is not convenient for treating sulfides because the S^{-2} ion is usually not present in significant concentrations (see Reference 8). This is due to the hydrolysis reaction

$$S^{-2} + H_2O \rightleftharpoons HS^- + OH^-$$

which is strongly shifted to the right except in very basic solutions. Furthermore, the equilibrium constant for this reaction, which depends on the second ionization constant of H_2S, is poorly known. Therefore it is more useful in the case of sulfides to define a different solubility product K_{spa} based on the reaction

$$M_m S_n(s) + 2H^+ \rightleftharpoons mM^+ + nH_2S \, (aq)$$

Values of K_{spa}, taken from Reference 8, are given for several sulfides in the auxiliary table following the main table. Additional discussion of sulfide equilibria may be found in References 7 and 9.

REFERENCES

1. Wagman, D.D., Evans, W.H., Parker, V.B., Schumm, R.H., Halow, I., Bailey, S.M., Churney, K.L., and Nuttall, R L., *The NBS Tables of Chemical Thermodynamic Properties, J. Phys. Chem. Ref. Data*, Vol. 11, Suppl. 2, 1982.
2. Garvin, D., Parker, V.B., and White, H.J., *CODATA Thermodynamic Tables*, Hemisphere, New York, 1987.
3. *Solubility Data Series* (53 Volumes), International Union of Pure and Applied Chemistry, Pergamon Press, Oxford, 1979—1992.
4. Clever, H.L., and Johnston, F.J., *J. Phys. Chem. Ref. Data*, 9, 751, 1980.
5. Marcus, Y., *J. Phys. Chem. Ref. Data*, 9, 1307, 1980.
6. Clever, H.L., Johnson, S.A., and Derrick, M.E., *J. Phys. Chem. Ref. Data*, 14, 631, 1985.
7. Clever, H.L., Johnson, S.A., and Derrick, M.E., *J. Phys. Chem. Ref. Data*, 21, 941, 1992.
8. Myers, R.J., *J. Chem. Educ.*, 63, 687, 1986.
9. Licht, S., *J. Electrochem. Soc.*, 135, 2971, 1988.

Compound	Formula	K_{sp}
Aluminum phosphate	$AlPO_4$	$9.84 \cdot 10^{-21}$
Barium bromate	$Ba(BrO_3)_2$	$2.43 \cdot 10^{-4}$
Barium carbonate	$BaCO_3$	$2.58 \cdot 10^{-9}$
Barium chromate	$BaCrO_4$	$1.17 \cdot 10^{-10}$
Barium fluoride	BaF_2	$1.84 \cdot 10^{-7}$
Barium hydroxide octahydrate	$Ba(OH)_2 \cdot 8H_2O$	$2.55 \cdot 10^{-4}$
Barium iodate	$Ba(IO_3)_2$	$4.01 \cdot 10^{-9}$
Barium iodate monohydrate	$Ba(IO_3)_2 \cdot H_2O$	$1.67 \cdot 10^{-9}$
Barium molybdate	$BaMoO_4$	$3.54 \cdot 10^{-8}$
Barium nitrate	$Ba(NO_3)_2$	$4.64 \cdot 10^{-3}$
Barium selenate	$BaSeO_4$	$3.40 \cdot 10^{-8}$
Barium sulfate	$BaSO_4$	$1.08 \cdot 10^{-10}$
Barium sulfite	$BaSO_3$	$5.0 \cdot 10^{-10}$
Beryllium hydroxide	$Be(OH)_2$	$6.92 \cdot 10^{-22}$
Bismuth arsenate	$BiAsO_4$	$4.43 \cdot 10^{-10}$

Compound	Formula	K_{sp}
Bismuth iodide	BiI_3	$7.71 \cdot 10^{-19}$
Cadmium arsenate	$Cd_3(AsO_4)_2$	$2.2 \cdot 10^{-33}$
Cadmium carbonate	$CdCO_3$	$1.0 \cdot 10^{-12}$
Cadmium fluoride	CdF_2	$6.44 \cdot 10^{-3}$
Cadmium hydroxide	$Cd(OH)_2$	$7.2 \cdot 10^{-15}$
Cadmium iodate	$Cd(IO_3)_2$	$2.5 \cdot 10^{-8}$
Cadmium oxalate trihydrate	$CdC_2O_4 \cdot 3H_2O$	$1.42 \cdot 10^{-8}$
Cadmium phosphate	$Cd_3(PO_4)_2$	$2.53 \cdot 10^{-33}$
Calcium carbonate (calcite)	$CaCO_3$	$3.36 \cdot 10^{-9}$
Calcium fluoride	CaF_2	$3.45 \cdot 10^{-11}$
Calcium hydroxide	$Ca(OH)_2$	$5.02 \cdot 10^{-6}$
Calcium iodate	$Ca(IO_3)_2$	$6.47 \cdot 10^{-6}$
Calcium iodate hexahydrate	$Ca(IO_3)_2 \cdot 6H_2O$	$7.10 \cdot 10^{-7}$
Calcium molybdate	$CaMoO_4$	$1.46 \cdot 10^{-8}$
Calcium oxalate monohydrate	$CaC_2O_4 \cdot H_2O$	$2.32 \cdot 10^{-9}$
Calcium phosphate	$Ca_3(PO_4)_2$	$2.07 \cdot 10^{-33}$
Calcium sulfate	$CaSO_4$	$4.93 \cdot 10^{-5}$
Calcium sulfate dihydrate	$CaSO_4 \cdot 2H_2O$	$3.14 \cdot 10^{-5}$
Calcium sulfite hemihydrate	$CaSO_3 \cdot 0.5H_2O$	$3.1 \cdot 10^{-7}$
Cesium perchlorate	$CsClO_4$	$3.95 \cdot 10^{-3}$
Cesium periodate	$CsIO_4$	$5.16 \cdot 10^{-6}$
Cobalt(II) arsenate	$Co_3(AsO_4)_2$	$6.80 \cdot 10^{-29}$
Cobalt(II) hydroxide (blue)	$Co(OH)_2$	$5.92 \cdot 10^{-15}$
Cobalt(II) iodate dihydrate	$Co(IO_3)_2 \cdot 2H_2O$	$1.21 \cdot 10^{-2}$
Cobalt(II) phosphate	$Co_3(PO_4)_2$	$2.05 \cdot 10^{-35}$
Copper(I) bromide	$CuBr$	$6.27 \cdot 10^{-9}$
Copper(I) chloride	$CuCl$	$1.72 \cdot 10^{-7}$
Copper(I) cyanide	$CuCN$	$3.47 \cdot 10^{-20}$
Copper(I) iodide	CuI	$1.27 \cdot 10^{-12}$
Copper(I) thiocyanate	$CuSCN$	$1.77 \cdot 10^{-13}$
Copper(II) arsenate	$Cu_3(AsO_4)_2$	$7.95 \cdot 10^{-36}$
Copper(II) iodate monohydrate	$Cu(IO_3)_2 \cdot H_2O$	$6.94 \cdot 10^{-8}$
Copper(II) oxalate	CuC_2O_4	$4.43 \cdot 10^{-10}$
Copper(II) phosphate	$Cu_3(PO_4)_2$	$1.40 \cdot 10^{-37}$
Europium(III) hydroxide	$Eu(OH)_3$	$9.38 \cdot 10^{-27}$
Gallium(III) hydroxide	$Ga(OH)_3$	$7.28 \cdot 10^{-36}$
Iron(II) carbonate	$FeCO_3$	$3.13 \cdot 10^{-11}$
Iron(II) fluoride	FeF_2	$2.36 \cdot 10^{-6}$
Iron(II) hydroxide	$Fe(OH)_2$	$4.87 \cdot 10^{-17}$
Iron(III) hydroxide	$Fe(OH)_3$	$2.79 \cdot 10^{-39}$
Iron(III) phosphate dihydrate	$FePO_4 \cdot 2H_2O$	$9.91 \cdot 10^{-16}$
Lanthanum iodate	$La(IO_3)_3$	$7.50 \cdot 10^{-12}$
Lead(II) bromide	$PbBr_2$	$6.60 \cdot 10^{-6}$
Lead(II) carbonate	$PbCO_3$	$7.40 \cdot 10^{-14}$
Lead(II) chloride	$PbCl_2$	$1.70 \cdot 10^{-5}$
Lead(II) fluoride	PbF_2	$3.3 \cdot 10^{-8}$
Lead(II) hydroxide	$Pb(OH)_2$	$1.43 \cdot 10^{-20}$
Lead(II) iodate	$Pb(IO_3)_2$	$3.69 \cdot 10^{-13}$
Lead(II) iodide	PbI_2	$9.8 \cdot 10^{-9}$
Lead(II) selenate	$PbSeO_4$	$1.37 \cdot 10^{-7}$
Lead(II) sulfate	$PbSO_4$	$2.53 \cdot 10^{-8}$
Lithium carbonate	Li_2CO_3	$8.15 \cdot 10^{-4}$
Lithium fluoride	LiF	$1.84 \cdot 10^{-3}$
Lithium phosphate	Li_3PO_4	$2.37 \cdot 10^{-11}$
Magnesium carbonate	$MgCO_3$	$6.82 \cdot 10^{-6}$
Magnesium carbonate trihydrate	$MgCO_3 \cdot 3H_2O$	$2.38 \cdot 10^{-6}$
Magnesium carbonate pentahydrate	$MgCO_3 \cdot 5H_2O$	$3.79 \cdot 10^{-6}$
Magnesium fluoride	MgF_2	$5.16 \cdot 10^{-11}$
Magnesium hydroxide	$Mg(OH)_2$	$5.61 \cdot 10^{-12}$
Magnesium oxalate dihydrate	$MgC_2O_4 \cdot 2H_2O$	$4.83 \cdot 10^{-6}$

SOLUBILITY PRODUCT CONSTANTS (continued)

Compound	Formula	K_{sp}
Magnesium phosphate	$Mg_3(PO_4)_2$	$1.04 \cdot 10^{-24}$
Manganese(II) carbonate	$MnCO_3$	$2.24 \cdot 10^{-11}$
Manganese(II) iodate	$Mn(IO_3)_2$	$4.37 \cdot 10^{-7}$
Manganese(II) oxalate dihydrate	$MnC_2O_4 \cdot 2H_2O$	$1.70 \cdot 10^{-7}$
Mercury(I) bromide	Hg_2Br_2	$6.40 \cdot 10^{-23}$
Mercury(I) carbonate	Hg_2CO_3	$3.6 \cdot 10^{-17}$
Mercury(I) chloride	Hg_2Cl_2	$1.43 \cdot 10^{-18}$
Mercury(I) fluoride	Hg_2F_2	$3.10 \cdot 10^{-6}$
Mercury(I) iodide	Hg_2I_2	$5.2 \cdot 10^{-29}$
Mercury(I) oxalate	$Hg_2C_2O_4$	$1.75 \cdot 10^{-13}$
Mercury(I) sulfate	Hg_2SO_4	$6.5 \cdot 10^{-7}$
Mercury(I) thiocyanate	$Hg_2(SCN)_2$	$3.2 \cdot 10^{-20}$
Mercury(II) bromide	$HgBr_2$	$6.2 \cdot 10^{-20}$
Mercury(II) iodide	HgI_2	$2.9 \cdot 10^{-29}$
Neodymium carbonate	$Nd_2(CO_3)_3$	$1.08 \cdot 10^{-33}$
Nickel(II) carbonate	$NiCO_3$	$1.42 \cdot 10^{-7}$
Nickel(II) hydroxide	$Ni(OH)_2$	$5.48 \cdot 10^{-16}$
Nickel(II) iodate	$Ni(IO_3)_2$	$4.71 \cdot 10^{-5}$
Nickel(II) phosphate	$Ni_3(PO_4)_2$	$4.74 \cdot 10^{-32}$
Palladium(II) thiocyanate	$Pd(SCN)_2$	$4.39 \cdot 10^{-23}$
Potassium hexachloroplatinate	K_2PtCl_6	$7.48 \cdot 10^{-6}$
Potassium perchlorate	$KClO_4$	$1.05 \cdot 10^{-2}$
Potassium periodate	KIO_4	$3.71 \cdot 10^{-4}$
Praseodymium hydroxide	$Pr(OH)_3$	$3.39 \cdot 10^{-24}$
Radium iodate	$Ra(IO_3)_2$	$1.16 \cdot 10^{-9}$
Radium sulfate	$RaSO_4$	$3.66 \cdot 10^{-11}$
Rubidium perchlorate	$RbClO_4$	$3.00 \cdot 10^{-3}$
Scandium fluoride	ScF_3	$5.81 \cdot 10^{-24}$
Scandium hydroxide	$Sc(OH)_3$	$2.22 \cdot 10^{-31}$
Silver(I) acetate	$AgCH_3COO$	$1.94 \cdot 10^{-3}$
Silver(I) arsenate	Ag_3AsO_4	$1.03 \cdot 10^{-22}$
Silver(I) bromate	$AgBrO_3$	$5.38 \cdot 10^{-5}$
Silver(I) bromide	$AgBr$	$5.35 \cdot 10^{-13}$
Silver(I) carbonate	Ag_2CO_3	$8.46 \cdot 10^{-12}$
Silver(I) chloride	$AgCl$	$1.77 \cdot 10^{-10}$
Silver(I) chromate	Ag_2CrO_4	$1.12 \cdot 10^{-12}$
Silver(I) cyanide	$AgCN$	$5.97 \cdot 10^{-17}$
Silver(I) iodate	$AgIO_3$	$3.17 \cdot 10^{-8}$
Silver(I) iodide	AgI	$8.52 \cdot 10^{-17}$
Silver(I) oxalate	$Ag_2C_2O_4$	$5.40 \cdot 10^{-12}$
Silver(I) phosphate	Ag_3PO_4	$8.89 \cdot 10^{-17}$
Silver(I) sulfate	Ag_2SO_4	$1.20 \cdot 10^{-5}$
Silver(I) sulfite	Ag_2SO_3	$1.50 \cdot 10^{-14}$
Silver(I) thiocyanate	$AgSCN$	$1.03 \cdot 10^{-12}$
Strontium arsenate	$Sr_3(AsO_4)_2$	$4.29 \cdot 10^{-19}$
Strontium carbonate	$SrCO_3$	$5.60 \cdot 10^{-10}$
Strontium fluoride	SrF_2	$4.33 \cdot 10^{-9}$
Strontium iodate	$Sr(IO_3)_2$	$1.14 \cdot 10^{-7}$
Strontium iodate monohydrate	$Sr(IO_3)_2 \cdot H_2O$	$3.77 \cdot 10^{-7}$
Strontium iodate hexahydrate	$Sr(IO_3)_2 \cdot 6H_2O$	$4.55 \cdot 10^{-7}$
Strontium sulfate	$SrSO_4$	$3.44 \cdot 10^{-7}$
Thallium(I) bromate	$TlBrO_3$	$1.10 \cdot 10^{-4}$
Thallium(I) bromide	$TlBr$	$3.71 \cdot 10^{-6}$
Thallium(I) chloride	$TlCl$	$1.86 \cdot 10^{-4}$
Thallium(I) chromate	Tl_2CrO_4	$8.67 \cdot 10^{-13}$
Thallium(I) hydroxide	$Tl(OH)_3$	$1.68 \cdot 10^{-44}$
Thallium(I) iodate	$TlIO_3$	$3.12 \cdot 10^{-6}$
Thallium(I) iodide	TlI	$5.54 \cdot 10^{-8}$
Thallium(I) thiocyanate	$TlSCN$	$1.57 \cdot 10^{-4}$
Tin(II) hydroxide	$Sn(OH)_2$	$5.45 \cdot 10^{-27}$

SOLUBILITY PRODUCT CONSTANTS (continued)

Compound	Formula	K_{sp}
Yttrium carbonate	$Y_2(CO_3)_3$	$1.03 \cdot 10^{-31}$
Yttrium fluoride	YF_3	$8.62 \cdot 10^{-21}$
Yttrium hydroxide	$Y(OH)_3$	$1.00 \cdot 10^{-22}$
Yttrium iodate	$Y(IO_3)_3$	$1.12 \cdot 10^{-10}$
Zinc arsenate	$Zn_3(AsO_4)_2$	$2.8 \cdot 10^{-28}$
Zinc carbonate	$ZnCO_3$	$1.46 \cdot 10^{-10}$
Zinc carbonate monohydrate	$ZnCO_3 \cdot H_2O$	$5.42 \cdot 10^{-11}$
Zinc fluoride	ZnF_2	$3.04 \cdot 10^{-2}$
Zinc hydroxide	$Zn(OH)_2$	$3 \cdot 10^{-17}$
Zinc iodate dihydrate	$Zn(IO_3)_2 \cdot 2H_2O$	$4.1 \cdot 10^{-6}$
Zinc oxalate dihydrate	$ZnC_2O_4 \cdot 2H_2O$	$1.38 \cdot 10^{-9}$
Zinc selenide	$ZnSe$	$3.6 \cdot 10^{-26}$
Zinc selenite monohydrate	$ZnSeO_3 \cdot H_2O$	$1.59 \cdot 10^{-7}$

Sulfides

Compound	Formula	K_{spa}
Cadmium sulfide	CdS	$8 \cdot 10^{-7}$
Copper(II) sulfide	CuS	$6 \cdot 10^{-16}$
Iron(II) sulfide	FeS	$6 \cdot 10^{2}$
Lead(II) sulfide	PbS	$3 \cdot 10^{-7}$
Manganese(II) sulfide (green)	MnS	$3 \cdot 10^{7}$
Mercury(II) sulfide (red)	HgS	$4 \cdot 10^{-33}$
Mercury(II) sulfide (black)	HgS	$2 \cdot 10^{-32}$
Silver(I) sulfide	Ag_2S	$6 \cdot 10^{-30}$
Tin(II) sulfide	SnS	$1 \cdot 10^{-5}$
Zinc sulfide (sphalerite)	ZnS	$2 \cdot 10^{-4}$
Zinc sulfide (wurtzite)	ZnS	$3 \cdot 10^{-2}$

SOLUBILITY CHART

Abbreviations: **W**, soluble in water; **A**, insoluble in water but soluble in acids; **w**, sparingly soluble in water but soluble in acids; **a**, insoluble in water and acids; **I**, insoluble in water and only sparingly soluble in acids; **a**, insoluble in water and acids; **d**, decomposes in water. * Indicates two modifications of the salt

No.		Al	NH₄	Sb	Ba	Bi	Cd	Ca	Cr	Co	Cu	Au (I)	Au (II)	H	Fe (II)	Fe (III)
1	Acetate —(C₂H₃O₂)	W	W		W	W	W	W	W	W	W	W	W	W, C₂H₄O₂	W, Fe(—)₂	W, Fe₂(—)₆
2	Arsenate —(AsO₄)	Al(—)₃	NH₄(—)	A	Ba₃(—)₂	Bi(—)₃	Cd(—)₂	Ca(—)₂	Cr(—)₃	Co(—)₂	Cu(—)₂			H₃AsO₄	A	A
3	Arsenite —(AsO₃)	Al(—)	NH₄AsO₂	Sb(—)	Ba₃(—)₂	Bi(—)	Cd₃(—)₂	Ca₃(—)₂		Co₃(—)₂	Cu₃(—)₂				Fe₃(—)₂	Fe(—)
4	Benzoate —(C₇H₅O₂)	W	NH₄(—)	A	W	A	W	Ca(—)₂		Co₃H₆(—)₄	w			C₇H₆O₂	W	A
5	Bromide	AlBr₂	NH₄Br	d, SbBr₃	BaBr₂	BiBr₃	CdBr₂	CaBr₂	W(I)*, CrBr₃	CoBr₂	CuBr₂	AuBr	AuBr₃	HBr	FeBr₂	FeBr₃
6	Carbonate —(CO₃)	W	(NH₄)₂CO₃		BaCO₃	d	CdCO₃	CaCO₃	CrCO₃	CoCO₃					FeCO₃	
7	Chlorate —(ClO₃)	Al(—)₃	NH₄(—)		Ba(—)₂	Bi(—)₃	Cd(—)₂	Ca(—)₂	I	Co(—)₂	Cu(—)₂			HClO₃	Fe(—)₂	Fe(—)
8	Chloride	AlCl₃	NH₄Cl	SbCl₃	BaCl₂	BiCl₃	CdCl₂	CaCl₂	CrCl₃	CoCl₂	CuCl₂	AuCl	AuCl₃	HCl	FeCl₂	FeCl₃
9	Chromate —(CrO₄)	W	(NH₄)₂(—)		Ba(—)		Cd(—)	Ca(—)	A	Co(—)	A				w	Fe₂(—)₃
10	Citrate —(C₆H₅O₇)	Al(—)	NH₄(—)		Ba₃(—)₂	Bi(—)₃	Cd₃(—)₂	Ca₃(—)₂	Cr(—)	Co₃(—)₂	Cu(—)₂			C₆H₈O₇	Fe(—)₂	Fe(—)
11	Cyanide		NH₄CN		Ba(CN)₂	Bi(CN)₃	Cd(CN)₂	Ca(CN)₂	Cr(CN)	Co(CN)₂	Cu(CN)₂	AuCN	Au(CN)₃	HCN	Fe(CN)₂	Fe(—)
12	Ferricy'de —(Fe(CN)₆)	w	(NH₄)₃(—)		Ba₃(—)₂	Bi(—)	Cd₃(—)₂	Ca₃(—)₂	A	Co₃(—)₂	A			H₃(—)	I	a
13	Ferrocy'de —(Fe(CN)₆)	Al₄(—)₃	(NH₄)₄(—)		Ba₂(—)	Bi₄(—)₃	Cd₂(—)	Ca₂(—)	Cr₄(—)₃	Co₂(—)	Cu₂(—)			H₄(—)	Fe₂(—)	Fe₄(—)₃
14	Fluoride	AlF₃	NH₄F	SbF₃	BaF₂	BiF₃	CdF₂	CaF₂	W(a)*, CrF₃	CoF₂	CuF₂			HF	FeF₂	FeF₃
15	Formate —(CHO₂)	W	NH₄(—)		Ba(—)₂	Bi(—)₃	Cd(—)₂	Ca(—)₂	Cr(—)₃	Co(—)₂	Cu(—)₂			CH₂O₂	W	Fe(—)₃
16	Hydroxide	Al(OH)₃	NH₄OH	w, Sb₂O₃	Ba(OH)₂	Bi(OH)₃	Cd(OH)₂	Ca(OH)₂	Cr(OH)₃	Co(OH)₂	Cu(OH)₂	AuOH	Au(OH)₃	W	Fe(OH)₂	Fe(OH)₃
17	Iodide	AlI₃	NH₄I	d, SbI₃	BaI₂	BiI₃	CdI₂	CaI₂	CrI₃	CoI₂	CuI	AuI	AuI₃	HI	FeI₂	FeI₃
18	Nitrate	Al(NO₃)₃	NH₄NO₃	Sb(—)	Ba(NO₃)₂	Bi(NO₃)₃	Cd(NO₃)₂	Ca(NO₃)₂	Cr(NO₃)₃	Co(NO₃)₂	Cu(NO₃)₂		HNO₃	HNO₃	Fe(NO₃)₂	Fe(NO₃)₃
19	Oxalate —(C₂O₄)	Al₂(—)₃	(NH₄)₂(—)	A	Ba(—)	Bi₂(—)₃	Cd(—)	Ca(—)	Cr₂(—)₃	Co(—)	Cu(—)			C₂H₂O₄	Fe(—)	Fe₂(—)₃
20	Oxide	Al₂O₃		Sb₂O₃	BaO	Bi₂O₃	CdO	CaO	Cr₂O₃	CoO	CuO	Au₂O	Au₂O₃	H₂O₂	FeO	Fe₂O₃
21	Phosphate	AlPO₄	NH₄H₂PO₄		Ba₃(PO₄)₂	BiPO₄	Cd₃(PO₄)₂	Ca₃(PO₄)₂	Cr₂(PO₄)₂	Co₃(PO₄)₂	Cu₃(PO₄)₂		H₃PO₄	H₃PO₄	Fe₃(PO₄)₂	FePO₄
22	Silicate, —(SiO₃)	I	W		Ba(—)		Cd(—)	Ca(—)		Co₂SiO₄	Cu(—)			H₂SiO₃	W	w
23	Sulfate	Al₂(SO₄)₃	(NH₄)₂SO₄	A, Sb₂(SO₄)₃	BaSO₄	d, Bi₂(SO₄)₃	CdSO₄	CaSO₄	W(I)*, Cr₂(SO₄)₃	CoSO₄	CuSO₄	I, Au₂SO₄	I	H₂SO₄	FeSO₄	Fe₂(SO₄)₃
24	Sulfide	Al₂S₃	(NH₄)₂S	Sb₂S₃	BaS	Bi₂S₃	CdS	CaS	Cr₂S₃	CoS	CuS	Au₂S	Au₂S₃	H₂S	FeS	Fe₂S₃
25	Tartrate —(C₄H₄O₆)	Al₂(—)₃	(NH₄)₂(—)	Sb₂(—)₃	Ba(—)	Bi₂(—)₃	Cd(—)	Ca(—)	Cr₂(—)₃	Co(—)	Cu(—)			C₄H₆O₆	Fe(—)	Fe₂(—)₃

SOLUBILITY CHART (continued)

No.	Ion	Pb	NH_4	Mg	Sb	Mn	Ba	Hg(I)	Hg(II)	Cd	Ni	Ca	K	Cr	Pt	Ag	Co	Cu	Na	Au(I)	Sn(IV)	Au(II)	Sn(II)	H	Sr	Zn	Fe(II)	Fe(III)
26	Thiocyanate		W NH_4CNS				W $Ba(CNS)_2$					W $Ca(CNS)_2$					W $Co(CNS)_2$	d $CuCNS$						W $CNSH$			W $Fe(CNS)_2$	W $Fe(CNS)_3$
1	Acetate —$(C_2H_3O_2)$	W $Pb(-)_2$	W	W $Mg(-)_2$		W $Mn(-)_2$		W $Hg(-)$	W $Hg(-)_2$	W $Cd(-)_2$	W $Ni(-)_2$		W $K(-)$	W		W $Ag(-)$		d	W $Na(-)$		$Sn(-)_4$		d $Sn(-)_2$		W $Sr(-)_2$	W $Zn(-)_2$	W	W
2	Arsenate —(AsO_4)	A $Pb(-)_2$	A	A $Mg_3(-)$		w $MnH(-)$		A $Hg(-)$	A $Hg_3(-)_2$	A $Cd_3(-)_2$	A $Ni_3(-)_2$		W $K_3(-)$			A $Ag_3(-)$			W		$Sn_3(-)_2$		$Sn_3(-)_2$		w	A $Zn_3(-)_2$		
3	Arsenite —(AsO_3)	$PbH(-)$	A	$MgH(-)$		A $MnH(-)$		A $Hg_3(-)$	A $Hg_3(-)_2$	$Cd_3(-)_2$	A $Ni_3H_4(-)_4$		W K_3AsO_3			A $Ag_3(-)$			W $Na_3(-)$				A $Sn_3(-)_2$		A $Sr_3(-)_2$	A $Zn_3(-)_2$		
4	Benzoate —$(C_7H_5O_2)$	w	W	W $Mg_3(-)_2$		W $Mn(-)_2$		A $Hg_3(-)$	W $Hg(-)_2$		W $Ni(-)_2$		W $K(-)$			w $Ag(-)$			W $Na_3H(-)$						W	W	W
5	Bromide	A $PbBr_2$	W	$MgBr_2$		W $MnBr_2$		A $HgBr$	W $HgBr_2$	$CdBr_2$	W $NiBr_2$		W KBr		w $PtBr_4$	A $AgBr$			W $NaBr$		W $SnBr_4$		W $SnBr_2$	W	W $SrBr_2$	W $ZnBr_2$	W	W
6	Carbonate	A $PbCO_3$	W	W $MgCO_3$		W $MnCO_3$		Hg_2CO_3	W		W $NiCO_3$		W K_2CO_3			A Ag_2CO_3			W Na_2CO_3						W $SrCO_3$	W $ZnCO_3$		
7	Chlorate —(ClO_3)	W $Pb(-)_2$	W	$Mg(-)_2$		W $Mn(-)_2$		W $Hg(-)$	W $Hg(-)_2$		W $Ni(-)_2$		W $K(-)$			W $Ag(-)$			W $Na_3(-)$				W $Sn(-)_2$		W $Sr(-)_2$	W $Zn(-)_2$	W	A
8	Chloride	W $PbCl_2$	W	W $MgCl_2$		W $MnCl_2$		a $HgCl$	W $HgCl_2$	$CdCl_2$	W $NiCl_2$		W KCl		W $PtCl_4$	A $AgCl$			W $NaCl$		W $SnCl_4$		W $SnCl_2$	W	W $SrCl_2$	W $ZnCl_2$	W	W
9	Chromate —(CrO_4)	A $PbCrO_4$	W	W $Mg(-)$				A $Hg_2(-)$	W $Hg(-)$		W $Ni(-)$		W $K_2(-)$			A $Ag_2(-)$			W $Na_2(-)$		W		A		W $Sr(-)_2$	W $Zn(-)_2$		
10	Citrate —$(C_6H_5O_7)$	W $Pb(-)$	W	$Mg_3(-)_2$		W $MnH(-)$					W $Ni(-)$		W $K_3(-)$						W $Na_3(-)$				$Sn(-)_2$		W $Sr_3(-)_2$	$Zn_3(-)_2$		
11	Cyanide	$Pb(CN)_2$	W	$Mg_3(CN)_2$				$HgCN$	W $Hg(CN)_2$		a $Ni(CN)_2$		W KCN		I $Pt(CN)_2$	A $AgCN$			W $NaCN$						$Sr(CN)_2$	W $Zn(CN)_2$		
12	Ferricy'de —$Fe(CN)_6$	$Pb_3(-)_2$	W	$Mg_3(-)_2$					A $Hg_3(-)_2$		I $Ni_3(-)_2$		W $K_3(-)$			I $Ag_3(-)$			W $Na_3(-)$				A $Sn_3(-)_2$		$Sr_3(-)_2$	W $Zn_3(-)_2$		
13	Ferrocy'de —$Fe(CN)_6$	a $Pb_2(-)$	W	$Mg_2(-)$					I $Hg_2(-)$		I $Ni_2(-)$		W $K_4(-)$			I $Ag_4(-)$			W $Na_4(-)$				a $Sn_2(-)_2$		W $Sr_2(-)$	I $Zn_2(-)$		
14	Fluoride	W PbF_2	W	MgF_2		A MnF_2		d HgF	d HgF_2		W NiF_2		W KF		W PtF_4	W AgF			W NaF		W SnF_4		W SnF_2	W HgF	A SrF_2	W ZnF_2	W	W
15	Formate —(CHO_2)	W $Pb(-)_2$	W	$Mg(-)_2$		W $Mn(-)_2$		$Hg(-)$	W $Hg(-)_2$		W $Ni(-)_2$		W $K(-)$			W $Ag(-)$			W $Na(-)$				$Sn(-)_2$		$Sr(-)_2$	W $Zn(-)_2$	W	W
16	Hydroxide	w $Pb(OH)_2$	W	$Mg(OH)_2$		W $Mn(OH)_2$		$Hg(OH)_2$	A $Hg(OH)_2$		W $Ni(OH)_2$		W KOH		A $Pt(OH)_4$	A $Ag(-)$			W $NaOH$		w $Sn(OH)_4$		w $Sn(OH)_2$		W $Sr(OH)_2$	W $Zn(OH)_2$	W	W
17	Iodide	w PbI_2	W	MgI_2		W MnI_2		A HgI	W HgI_2		W NiI_2		W KI		I PtI_2	I AgI			W NaI		d SnI_4		d SnI_2	W HgI	W SrI_2	W ZnI_2	W	A
18	Nitrate	W $Pb(NO_3)_2$	W	$Mg(NO_3)_2$		W $Mn(NO_3)_2$		W $HgNO_3$	W $Hg(NO_3)_2$		W $Ni(NO_3)_2$		W KNO_3		W $Pt(NO_3)_4$	W $AgNO_3$			W $NaNO_3$				W $Sn(NO_3)_2$	W $HgNO_3$	W $Sr(NO_3)_2$	W $Zn(NO_3)_2$	W	W
19	Oxalate —(C_2O_4)	A $Pb(-)$	W	$Mg(-)$		W $Mn(-)$		a $Hg_2(-)$	W $Hg(-)$		W $Ni(-)$		W $K_2(-)$		A $Pt(-)$	a $Ag_2(-)$			d $Na_2(-)$		A SnO_2		A $Sn(-)$		W $Sr(-)$	W $Zn(-)$	A	A
20	Oxide	PbO		MgO		MnO		Hg_2O	HgO		NiO		K_2O		PtO	Ag_2O			Na_2O		SnO_2		SnO		SrO	ZnO		ZnO
21	Phosphate	A $Pb_3(PO_4)_2$	W	$Mg_3(PO_4)_2$		I $Mn_3(PO_4)_2$		Hg_3PO_4	$Hg_3(PO_4)_2$		W $Ni_3(PO_4)_2$		W K_3PO_4			A Ag_3PO_4			W Na_3PO_4				$Sn_3(PO_4)_2$		W $Sr_3(PO_4)_2$	W $Zn_3(PO_4)_2$		
22	Silicate —(SiO_3)	$Pb(-)$	W	$Mg(-)$		$Mn(-)$							W $K_2(-)$						W $Na_2(-)$				$Sn(-)$		$Sr(-)$	$Zn(-)$		
23	Sulfate	A $PbSO_4$	W	W $MgSO_4$		W $MnSO_4$		w Hg_2SO_4	d $HgSO_4$	$CdSO_4$	W $NiSO_4$		W K_2SO_4		W $Pt(SO_4)_2$	w Ag_2SO_4			W Na_2SO_4		W $Sn(SO_4)_2$		W $SnSO_4$		W $SrSO_4$	W $ZnSO_4$	W	W

8-116

SOLUBILITY CHART (continued)

No.		Pb	Mg	Mn	Hg (I)	Hg (II)	Ni	K	Pt	Ag	Na	Sn (IV)	Sn (II)	Sr	Zn
24	Sulfide	A PbS	d MgS	A MnS	I Hg_2S	I HgS	A NiS	W K_2S	I PtS	A Ag_2S	W Na_2S	A SnS_2	A SnS	W SrS	A ZnS
25	Tartrate —$(C_4H_4O_6)$	A $Pb(—)$	W $Mg(—)$	w $Mn(—)$	I $Hg_2(—)$	I	A $Ni(—)$	W $K_2(—)$		A $Ag_2(—)$	W $Na_2(—)$	A	A $Sn(—)$	W $Sr(—)$	A $Zn(—)$
26	Thiocy'te	w $Pb(CNS)_2$	W $Mg(CNS)_2$	W $Mn(CNS)_2$	A $HgCNS$	w $Hg(CNS)_2$		W $KCNS$		I $AgCNS$	W $NaCNS$			W $Sr(CNS)_2$	W $Zn(CNS)_2$

REDUCTION OF WEIGHINGS IN AIR TO VACUO

When the mass M of a body is determined in air, a correction is necessary for the buoyancy of the air. The corrected mass is given by $M + kM/1000$, where k is a function of the material used for the weights, given by

$$k = 1000\rho_{air}(1/\rho_{body} - 1/\rho_{weight})$$

and ρ is density. The table below is computed for an air density of 0.0012 g/cm^3 and for densities of three common weights: platinum-iridium (21.6 g/cm^3), brass (8.5 g/cm^3), and aluminum or quartz (2.65 g/cm^3).

REFERENCES

1. Kaye, G. W. C., and Laby, T. H., *Tables of Physical and Chemical Constants, 16th Edition*, pp. 25-28, Longman, London, 1995.
2. Giacomo, P., *Metrologia* 18, 33, 1982.
3. Davis, R. S., *Metrologia* 29, 67, 1992.

Density of body (g/cm³)	Pt-Ir	Brass	Quartz or Al	Density of body (g/cm³)	Pt-Ir	Brass	Quartz or Al
	Value of *k* for weights of:				Value of *k* for weights of:		
0.5	2.34	2.26	1.95	1.8	0.61	0.53	0.21
0.6	1.94	1.86	1.55	1.9	0.58	0.49	0.18
0.7	1.66	1.57	1.26	2.0	0.54	0.46	0.15
0.8	1.44	1.36	1.05	2.5	0.42	0.34	0.03
0.9	1.28	1.19	0.88	3.0	0.34	0.26	-0.05
1.0	1.14	1.06	0.75	4.0	0.24	0.16	-0.15
1.1	1.04	0.95	0.64	6.0	0.14	0.06	-0.25
1.2	0.94	0.86	0.55	8.0	0.09	0.01	-0.30
1.3	0.87	0.78	0.47	10.0	0.06	-0.02	-0.33
1.4	0.80	0.72	0.40	15.0	0.02	-0.06	-0.37
1.5	0.74	0.66	0.35	20.0	0.00	-0.08	-0.39
1.6	0.69	0.61	0.30	22.0	0.00	-0.09	-0.40
1.7	0.65	0.56	0.25				

For a more accurate calculation, use the following values of the density of air (assuming 50% relative humidity and 0.04% CO_2):

P/kPa	Air temperature 10°C	20°C	30°C
85	0.001043	0.001005	0.000968
90	0.001105	0.001065	0.001025
95	0.001166	0.001124	0.001083
100	0.001228	0.001184	0.001140
105	0.001290	0.001243	0.001198

Formulas for calculating the density of air over more extended ranges of temperature, pressure, and humidity may be found in the references.

VOLUME OF ONE GRAM OF WATER

The following table, which is designed for gravimetric calibration of volumetric apparatus, gives the specific volume of water at standard atmospheric pressure as a function of temperature.

REFERENCE

Marsh, K. N., Editor, *Recommended Reference Materials for the Realization of Physicochemical Properties*, pp. 25-27, Blackwell Scientific Publications, Oxford, 1987.

t/°C	Volume of 1 g H₂O in cm³	t/°C	Volume of 1 g H₂O in cm³	t/°C	Volume of 1 g H₂O in cm³
10	1.0002980	17	1.0012246	24	1.0027079
11	1.0003928	18	1.0014044	25	1.0029607
12	1.0005007	19	1.0015952	26	1.0032234
13	1.0006212	20	1.0017969	27	1.0034956
14	1.0007542	21	1.0020092	28	1.0037771
15	1.0008992	22	1.0022320	29	1.0040679
16	1.0010561	23	1.0024649	30	1.0043679

PROPERTIES OF CARRIER GASES FOR GAS CHROMATOGRAPHY

The following is a list of carrier gases sometimes used in gas chromatography, with properties relevant to the design of chromatographic systems. All data refer to normal atmospheric pressure (101.325 kPa).

M_r : Molecular weight (relative molar mass)
ρ_{25} : Density at 25°C in g/L
λ : Thermal conductivity in mW/m °C
η : Viscosity in μPa s (equal to 10^{-3} cp)
c_p : Specific heat at 25°C in J/g °C

REFERENCES

1. Lide, D. R., and Kehiaian, H. V., *CRC Handbook of Thermophysical and Thermochemical Data*, CRC Press, Boca Raton, FL, 1994.
2. Bruno, T. J., and Svoronos, P. D. N., *CRC Handbook of Basic Tables for Chemical Analysis*, CRC Press, Boca Raton, FL, 1989

Gas	M_r	ρ_{25} g L^{-1}	At 25°C λ mW/m °C	At 25°C η μPa s	At 250°C λ mW/m °C	At 250°C η μPa s	c_p(25 °C) J/g °C
Hydrogen	2.016	0.0824	185.9	8.9	280	13.1	14.3
Helium	4.003	0.1636	154.6	19.9	230	29.5	5.20
Argon	39.95	1.6329	17.8	22.7	27.7	35.3	0.521
Nitrogen	28.01	1.1449	25.9	17.9	39.6	26.8	1.039
Oxygen	32.00	1.3080	26.2	20.7	42.6	31.8	0.919
Carbon monoxide	28.01	1.1449	24.8	17.8	40.7	26.5	1.039
Carbon dioxide	44.01	1.7989	16.7	14.9	35.5	24.9	0.843
Sulfur hexafluoride	146.05	5.9696	13.1	28.1	15.3	24.8	0.664
Methane	16.04	0.6556	34.5	11.1	75.0	17.6	2.23
Ethane	30.07	1.2291	20.9	9.4	57.7	15.5	1.75
Ethylene	28.05	1.1465	20.5	10.3	53.8	17.2	1.53
Propane	44.10	1.8025	17.9	8.3	49.2	14.0	1.67

SOLVENTS FOR ULTRAVIOLET SPECTROPHOTOMETRY

This table lists some solvents commonly used for sample preparation for ultraviolet spectrophotometry. The properties given are:

λ_c: cutoff wavelength, below which the solvent absorption becomes excessive.

ε: dielectric constant (relative permittivity); the temperature in °C is given as a superscript.

t_b: normal boiling point.

REFERENCES

1. Bruno, T. J., and Svoronos, P. D. N., *CRC Handbook of Basic Tables for Chemical Analysis*, CRC Press, Boca Raton, FL, 1989.
2. *Landolt-Börnstein, Numerical Data and Functional Relationships in Science and Technology, New Series*, IV/6, *Static Dielectric Constants of Pure Liquids and Binary Liquid Mixtures*, Springer-Verlag, Heidelberg, 1991.

Name	λ_c/nm	ε	t_b/°C
Acetic acid	260	6.20^{20}	117.9
Acetone	330	21.01^{20}	56.0
Acetonitrile	190	36.64^{20}	81.6
Benzene	280	2.28^{20}	80.0
2-Butanol	260	17.26^{20}	99.5
Butyl acetate	254	5.07^{20}	126.1
Carbon disulfide	380	2.63^{20}	46
Carbon tetrachloride	265	2.24^{20}	76.8
1-Chlorobutane	220	7.28^{20}	78.6
Chloroform	245	4.81^{20}	61.1
Cyclohexane	210	2.02^{20}	80.7
1,2-Dichloroethane	226	10.42^{20}	83.5
Dichloromethane	235	8.93^{25}	40
Diethyl ether	218	4.27^{20}	34.5
N,N-Dimethylacetamide	268	38.85^{21}	165
N,N-Dimethylformamide	270	38.25^{20}	153
Dimethyl sulfoxide	265	47.24^{20}	189
1,4-Dioxane	215	2.22^{20}	101.5
Ethanol	210	25.3^{20}	78.2
Ethyl acetate	255	6.08^{20}	77.1
Ethylene glycol dimethyl ether	240	7.30^{24}	85
Ethylene glycol monoethyl ether	210	13.38^{25}	135
Ethylene glycol monomethyl ether	210	17.2^{25}	124.1
Glycerol	207	46.53^{20}	290
Heptane	197	1.92^{20}	98.5
Hexadecane	200	2.05^{20}	286.8
Hexane	210	1.89^{20}	68.7
Methanol	210	33.0^{20}	64.6
Methylcyclohexane	210	2.02^{20}	100.9
Methyl ethyl ketone	330	18.56^{20}	79.5
Methyl isobutyl ketone	335	13.11^{20}	116.5
2-Methyl-1-propanol	230	17.93^{20}	107.8
N-Methyl-2-pyrrolidone	285	32.55^{20}	202
Nitromethane	380	37.27^{20}	101.1
Pentane	210	1.84^{20}	36.0
Pentyl acetate	212	4.79^{20}	149.2
1-Propanol	210	20.8^{20}	97.2
2-Propanol	210	20.18^{20}	82.3
Pyridine	330	13.26^{20}	115.2
Tetrachloroethylene	290	2.27^{30}	121.3
Tetrahydrofuran	220	7.52^{22}	65
Toluene	286	2.38^{23}	110.6
1,1,2-Trichloro-1,2,2-trifluoroethane	231	2.41^{25}	47.7
2,2,4-Trimethylpentane	215	1.94^{20}	99.2
Water	191	80.10^{20}	100.0
o-Xylene	290	2.56^{20}	144.5
m-Xylene	290	2.36^{20}	139.1
p-Xylene	290	2.27^{20}	138.3

^{13}C CHEMICAL SHIFTS OF USEFUL NMR SOLVENTS

The following table gives the expected carbon-13 chemical shifts, relative to tetramethylsilane, for various useful NMR solvents. In some solvents, slight changes can occur with change of concentration.[2,3]

REFERENCES

1. Bruno, T. J. and Svoronos, P. D. N., *CRC Handbook of Basic Tables for Chemical Analysis*, CRC Press, Boca Raton, FL, 1989.
2. Silverstein, R. M., Bassler, G. C., and Morrill, T. C., *Spectrometric Identification of Organic Compounds*, John Wiley & Sons, Now York, 1981.
3. Rahman, A. U., *Nuclear Magnetic Resonance. Basic Principles*, Springer-Verlag, New York, 1986.
4. Pretsch, E., Clerc, T., Seibl, J., and Simon, W., *Spectral Data for Structure Determination of Organic Compounds, Second Edition*, Springer-Verlag, Heidelberg, 1989.

Solvent	Formula	Chemical shift (ppm)
Acetic acid-d_4	CD_3COOD	20.0 (CD_3) 205.8 (C=O)
Acetone	$(CH_3)_2C=O$	30.7 (CH_3) 206.7 (C=O)
Acetone-d_6	$(CD_3)_2C=O$	29.2 (CD_3) 204.1 (C=O)
Acetonitrile-d_3	$CD_3C\equiv N$	1.3 (CD_3) 117.1 ($C\equiv N$)
Benzene	C_6H_6	128.5
Benzene-d_6	C_6D_6	128.4
Carbon disulfide	CS_2	192.3
Carbon tetrachloride	CCl_4	96.0
Chloroform	$CHCl_3$	77.2
Chloroform-d_3	$CDCl_3$	77.05
Cyclohexane-d_{12}	C_6D_{12}	27.5
Dichloromethane-d_2	CD_2Cl_2	53.6
Dimethylformamide-d_7	$(CD_3)_2NCDO$	31 (CD_3) 36 (CD_3) 162.4 (C=O)
Dimethylsulfoxide-d_6	$(CD_3)_2S=O$	39.6
Dioxane-d_8	$C_4D_3O_2$	67.4
Formic acid-d_2	DCOOD	165.5
Methanol-d_4	CD_3OD	49.3
Nitromethane-d_3	CD_3NO_2	57.3
Pyridine	C_5H_5N	123.6 (C_3) 135.7 (C_4) 149.8 (C_2)
Pyridine-d_5	C_5D_5N	123.9 (C_3) 135.9 (C_4) 150.2 (C_2)
1,1,2,2-Tetrachloroethane-d_2	$CDCl_2CDCl_2$	75.5
Tetrahydrofuran-d_8	C_4D_8O	25.8 (C_2) 67.9 (C_1)
Trichlorofluoromethane	$CFCl_3$	117.6

MASS SPECTRAL PEAKS OF COMMON ORGANIC SOLVENTS

The strongest peaks in the mass spectra of 200 important solvents are listed in this table. The *e/m* value for each peak is followed by the relative intensity in parentheses, with the strongest peak assigned an intensity of 100. The peaks for each compound are listed in order of decreasing intensity. Solvents are listed in alphabetical order by common name.

Data on the physical properties of the same compounds may be found in Section 15 in the table "Properties of Common Laboratory Solvents".

REFERENCES

1. NIST/EPA/NIH Mass Spectral Database, National Institute of Standards and Technology, Gaithersburg, MD, 20899.
2. Lide, D.R., and Milne, G.W.A., Editors, *Handbook of Data on Organic Compounds, Third Edition*, CRC Press, Boca Raton, FL, 1994. (Also available as a CD ROM database.)

MASS SPECTRAL PEAKS OF COMMON ORGANIC SOLVENTS (continued)

Compound	e/m (intensity)									
Acetal (1,1-Diethoxyethane)	44(100)	43(92)	29(77)	31(76)	45(74)	27(52)	72(48)	73(23)	28(17)	46(15)
Acetic acid	43(100)	45(87)	60(57)	15(42)	42(14)	29(13)	14(13)	28(7)	18(6)	16(6)
Acetone	43(100)	15(34)	58(23)	27(9)	14(9)	42(8)	26(7)	29(5)	28(5)	39(4)
Acetonitrile	41(100)	40(46)	39(13)	14(9)	38(6)	28(4)	26(4)	25(3)	42(2)	27(2)
Acetylacetone	43(100)	85(31)	100(20)	27(12)	42(10)	29(10)	41(7)	39(7)	31(5)	26(5)
Acrylonitrile	53(100)	26(85)	52(79)	51(34)	27(13)	50(8)	25(7)	38(5)	54(3)	37(3)
Adiponitrile	41(100)	68(50)	54(42)	40(21)	55(20)	27(17)	39(16)	28(13)	52(7)	42(6)
Allyl alcohol	57(100)	31(34)	29(32)	28(31)	58(25)	39(22)	27(20)	30(16)	32(14)	26(11)
Allylamine	30(100)	56(80)	28(76)	57(33)	39(21)	29(20)	27(18)	26(13)	41(8)	18(8)
2-Aminoisobutanol	58(100)	41(18)	18(17)	42(13)	28(11)	56(10)	30(10)	29(8)	43(6)	59(5)
Benzal chloride	125(100)	127(32)	160(14)	89(13)	162(9)	63(9)	126(8)	62(7)	105(5)	39(5)
Benzaldehyde	51(100)	77(81)	50(55)	106(44)	105(43)	52(26)	78(16)	39(13)	27(10)	74(8)
Benzene	78(100)	77(20)	52(19)	51(17)	50(15)	39(12)	79(6)	76(5)	74(4)	38(4)
Benzonitrile	103(100)	76(34)	50(13)	104(9)	75(7)	51(7)	77(5)	52(4)	39(4)	74(3)
Benzyl chloride	91(100)	126(20)	65(14)	92(9)	39(9)	63(8)	128(6)	45(6)	89(5)	125(3)
Bromochloromethane	49(100)	130(67)	128(52)	51(31)	93(23)	81(20)	79(20)	95(17)	132(16)	47(8)
Bromoform (Tribromomethane)	173(100)	171(50)	175(49)	93(22)	91(22)	79(18)	81(17)	94(13)	92(13)	254(11)
Butyl acetate	43(100)	56(34)	41(17)	27(16)	29(15)	73(11)	61(10)	28(7)	55(6)	39(6)
Butyl alcohol	31(100)	56(81)	41(62)	43(60)	27(50)	42(31)	29(31)	28(17)	39(16)	55(12)
sec-Butyl alcohol	45(100)	31(22)	27(22)	59(20)	29(18)	43(13)	41(12)	44(8)	18(8)	28(5)
tert-Butyl alcohol	59(100)	31(33)	41(22)	43(18)	29(13)	27(11)	57(10)	42(4)	60(3)	28(3)
Butylamine	30(100)	73(10)	28(5)	41(3)	27(3)	18(3)	44(2)	42(2)	31(2)	29(2)
tert-Butylamine	58(100)	41(21)	42(15)	18(9)	30(8)	15(8)	39(7)	57(6)	28(6)	59(4)
Butyl methyl ketone	43(100)	58(60)	57(17)	100(16)	29(15)	41(13)	85(8)	27(8)	71(7)	59(5)
p-tert-Butyltoluene	133(100)	105(38)	41(23)	148(18)	93(16)	91(14)	115(13)	134(11)	39(11)	116(10)
γ-Butyrolactone	28(100)	42(74)	29(48)	27(33)	41(27)	56(25)	86(24)	26(18)	85(10)	39(10)
Caprolactam	55(100)	113(87)	30(81)	56(66)	84(60)	85(57)	42(51)	41(33)	28(26)	43(17)
Carbon disulfide	76(100)	32(22)	44(17)	78(9)	38(6)	28(5)	77(3)	64(1)	46(1)	39(1)
Carbon tetrachloride	117(100)	119(98)	121(31)	82(24)	47(23)	84(16)	35(14)	49(8)	28(8)	36(6)
1-Chloro-1,1-difluoroethane	65(100)	45(31)	85(14)	31(10)	64(8)	44(7)	35(6)	49(8)	26(6)	81(4)
Chlorobenzene	112(100)	77(63)	114(33)	51(29)	50(14)	75(8)	113(7)	78(5)	87(5)	28(4)
Chloroform	83(100)	85(64)	47(35)	35(19)	48(16)	49(12)	87(10)	76(5)	50(5)	84(4)
Chloropentafluoroethane	85(100)	31(38)	87(32)	50(17)	35(8)	119(6)	37(6)	100(3)	66(4)	47(3)
Cumene (Isopropylbenzene)	105(100)	120(25)	77(13)	51(12)	79(10)	106(9)	39(9)	27(8)	103(6)	91(5)
Cyclohexane	56(100)	84(71)	41(70)	27(37)	55(36)	39(35)	42(30)	69(23)	28(18)	43(14)
Cyclohexanol	57(100)	44(68)	41(68)	39(51)	32(40)	43(38)	31(32)	42(22)	67(18)	82(16)
Cyclohexanone	55(100)	42(85)	41(34)	27(33)	98(31)	39(27)	69(26)	70(20)	43(14)	28(14)
Cyclohexylamine	56(100)	43(23)	28(17)	99(10)	70(8)	57(6)	30(6)	93(5)	54(4)	41(4)
Cyclopentane	42(100)	70(30)	55(29)	41(29)	39(22)	27(15)	40(7)	29(5)	28(4)	43(3)
Cyclopentanone	55(100)	28(50)	84(42)	41(38)	56(29)	27(24)	39(19)	42(15)	26(9)	29(7)
p-Cymene (1-Methyl-4-isopropyl-benzene)	119(100)	91(42)	134(33)	41(20)	117(18)	65(18)	77(17)	27(16)	120(15)	
cis-Decalin	67(100)	81(87)	41(81)	138(67)	96(62)	82(62)	39(50)	55(45)	27(44)	95(42)
trans-Decalin	41(100)	68(91)	67(88)	82(67)	27(65)	96(61)	95(55)	138(51)	81(51)	29(51)

MASS SPECTRAL PEAKS OF COMMON ORGANIC SOLVENTS (continued)

e/m (intensity)

Compound	e/m (intensity)
Diacetone alcohol	43(100), 59(41), 58(17), 101(10), 41(9), 31(9), 83(6), 56(6), 55(6), 29(6)
1,2-Dibromoethane	27(100), 107(77), 109(72), 26(24), 28(10), 81(5), 79(5), 25(5), 95(4), 93(4)
Dibromofluoromethane	111(100), 113(98), 192(29), 43(16), 41(16), 190(15), 194(14), 81(9), 79(9), 122(7)
Dibromomethane	174(100), 93(96), 95(84), 172(53), 176(50), 91(11), 81(9), 79(9), 94(5), 65(5)
1,2-Dibromotetrafluoroethane	179(100), 181(97), 129(34), 131(33), 100(17), 73(15), 31(13), 260(12), 69(7), 262(6)
Dibutylamine	86(100), 72(52), 30(48), 44(40), 29(31), 57(24), 41(21), 73(15), 28(15), 43(13)
o-Dichlorobenzene	146(100), 148(64), 111(38), 75(23), 113(12), 74(12), 50(11), 150(10), 73(9), 147(7)
1,1-Dichloroethane (Ethylidene dichloride)	63(100), 27(71), 65(31), 26(19), 85(7), 83(7), 61(7), 35(6), 98(5), 62(5)
1,2-Dichloroethane (Ethylene dichloride)	62(100), 27(91), 49(40), 64(32), 26(31), 63(19), 98(14), 51(13), 61(12), 100(9)
1,1-Dichloroethylene	61(100), 96(61), 98(38), 63(32), 26(16), 60(15), 62(7), 25(7), 100(6), 35(6)
cis-1,2-Dichloroethylene	61(100), 96(73), 98(47), 63(32), 26(30), 60(21), 25(13), 35(12), 62(9), 100(8)
trans-1,2-Dichloroethylene	61(100), 96(67), 98(43), 26(34), 63(32), 60(24), 25(15), 62(10), 100(7), 47(7)
Dichloroethyl ether	93(100), 63(74), 27(38), 95(32), 65(24), 31(9), 49(4), 28(4), 94(3), 62(3)
Dichloromethane (Methylene chloride)	49(100), 84(64), 86(39), 51(31), 47(14), 48(8), 88(6), 50(3), 85(2), 83(2)
1,2-Dichloropropane	63(100), 62(71), 27(57), 41(49), 39(32), 65(31), 76(27), 64(25), 49(13), 77(12)
1,2-Dichlorotetrafluoroethane	85(100), 135(52), 87(33), 137(17), 101(9), 31(9), 103(6), 100(6), 50(5), 69(4)
Diethanolamine	30(100), 74(82), 28(77), 56(69), 18(50), 42(46), 29(36), 27(34), 45(30), 43(19)
Diethylamine	30(100), 58(81), 44(28), 73(18), 29(18), 28(17), 72(12), 42(11), 27(11), 59(4)
Diethyl carbonate	29(100), 45(70), 31(53), 27(39), 91(24), 28(15), 63(11), 26(10), 30(6), 43(5)
Diethylene glycol	45(100), 75(23), 31(20), 44(16), 27(14), 76(12), 29(12), 43(11), 42(9), 41(4)
Diethylene glycol dimethyl ether (Diglyme)	59(100), 58(43), 31(34), 29(32), 45(28), 28(19), 89(15), 43(9), 27(5), 60(4)
Diethylene glycol monoethyl ether (Carbitol)	45(100), 59(56), 72(37), 60(14), 31(13), 75(11), 44(9), 104(8), 103(7)
Diethylene glycol monoethyl ether acetate	43(100), 29(51), 31(42), 45(40), 59(24), 72(18), 44(10), 73(9), 42(9), 30(6)
Diethylene glycol monomethyl ether	45(100), 31(42), 59(41), 29(38), 28(32), 58(21), 43(14), 27(13), 44(11), 32(10)
Diethylenetriamine	44(100), 73(59), 30(35), 19(18), 56(16), 28(16), 27(15), 45(14), 42(11), 43(8)
Diethyl ether	31(100), 29(63), 59(40), 27(35), 45(33), 74(23), 15(17), 43(9), 28(9), 26(9)
Diisobutyl ketone (Isovalerone)	57(100), 85(82), 41(46), 43(39), 58(33), 39(22), 27(16), 142(11)
Diisopropyl ether	45(100), 43(39), 87(15), 41(12), 59(10), 27(8), 39(4), 31(3)
N,N-Dimethylacetamide	44(100), 87(69), 43(46), 45(23), 42(19), 72(15), 15(11), 88(4)
Dimethylamine	44(100), 45(81), 18(32), 28(30), 43(19), 42(15), 15(11), 30(8), 41(5), 27(5)
Dimethyl disulfide	94(100), 45(63), 79(59), 46(38), 47(26), 48(14), 61(13), 64(11), 96(9)
N,N-Dimethylformamide	73(100), 44(86), 42(36), 30(22), 28(20), 43(7), 72(6), 58(5), 74(4)
Dimethyl sulfoxide	63(100), 78(70), 15(40), 45(35), 29(16), 61(13), 46(12), 31(11), 47(10)
1,4-Dioxane	28(100), 29(37), 88(31), 58(24), 31(17), 27(15), 30(13), 43(11), 26(9)
1,3-Dioxolane	73(100), 29(56), 44(53), 45(28), 43(20), 27(13), 31(7), 74(5), 42(3)
Dipentene	68(100), 93(50), 67(44), 94(22), 39(22), 53(18), 107(18), 92(18), 136(16), 79(16)
Epichlorohydrin	57(100), 27(39), 29(32), 49(25), 31(22), 62(18), 28(16), 92(1)

MASS SPECTRAL PEAKS OF COMMON ORGANIC SOLVENTS (continued)

Compound	e/m (intensity)									
Ethanolamine (Glycinol)	30(100)	18(30)	28(15)	42(7)	31(6)	17(6)	61(5)	15(5)	43(3)	29(3)
Ethyl acetate	43(100)	29(46)	27(33)	45(32)	61(28)	28(25)	42(18)	73(11)	88(10)	70(10)
Ethyl acetoacetate	43(100)	29(24)	88(18)	28(16)	85(14)	27(12)	42(11)	60(9)	130(6)	45(6)
Ethyl alcohol	31(100)	45(44)	46(18)	27(18)	29(15)	43(14)	30(6)	42(3)	19(3)	14(3)
Ethylamine	30(100)	28(32)	44(20)	45(19)	27(13)	15(10)	42(9)	29(8)	41(5)	40(5)
Ethylbenzene	91(100)	106(31)	51(14)	39(10)	77(8)	65(8)	105(7)	92(7)	78(7)	27(6)
Ethyl bromide (Bromoethane)	108(100)	110(97)	29(62)	27(51)	28(35)	26(14)	93(6)	32(6)	95(5)	81(5)
Ethyl chloride (Cloroethane)	64(100)	28(91)	29(84)	27(75)	66(32)	26(28)	49(25)	51(8)	63(6)	65(4)
Ethylene carbonate	29(100)	44(62)	43(54)	88(40)	30(16)	28(11)	45(7)	58(6)	42(6)	73(4)
Ethylenediamine (1,2-Ethane-diamine)	30(100)	18(13)	42(6)	43(5)	27(5)	44(4)	29(4)	17(4)	15(4)	41(3)
Ethylene glycol	31(100)	33(35)	29(13)	32(11)	43(6)	27(5)	28(4)	62(3)	30(3)	44(2)
Ethylene glycol diethyl ether	31(100)	59(71)	29(58)	45(43)	27(33)	74(27)	43(15)	15(14)	28(12)	44(10)
Ethylene glycol dimethyl ether	45(100)	60(13)	29(13)	90(7)	58(6)	31(5)	28(5)	43(4)	59(3)	46(2)
Ethylene glycol monobutyl ether	57(100)	45(38)	29(35)	41(31)	87(16)	27(12)	56(11)	31(9)	75(7)	28(7)
Ethylene glycol monoethyl ether (Cellosolve)	31(100)	29(52)	59(50)	27(27)	45(26)	72(14)	43(14)	15(14)	28(8)	26(6)
Ethylene glycol monoethyl ether acetate	43(100)	31(34)	59(31)	72(28)	44(25)	29(24)	45(12)	27(11)	15(11)	87(7)
Ethylene glycol monomethyl ether	45(100)	31(15)	29(14)	28(11)	47(9)	76(6)	43(6)	58(4)	46(4)	27(4)
Ethylene glycol monomethyl ether acetate	43(100)	45(48)	58(42)	29(10)	42(4)	31(4)	73(3)	27(3)	59(2)	26(2)
Ethyl formate	31(100)	28(73)	27(51)	29(38)	45(34)	26(17)	74(11)	43(9)	47(8)	56(4)
Furan	68(100)	39(64)	40(9)	38(9)	42(6)	29(6)	37(5)	69(4)	34(2)	67(1)
Furfural	39(100)	96(55)	95(52)	38(38)	29(35)	37(29)	40(11)	97(9)	50(7)	42(7)
Furfuryl alcohol	98(100)	41(65)	39(59)	81(55)	53(53)	97(51)	42(49)	69(39)	70(36)	29(28)
Glycerol	61(100)	43(90)	31(57)	44(54)	29(38)	18(32)	27(12)	42(11)	60(10)	45(10)
Heptane	43(100)	41(56)	29(49)	57(47)	27(46)	71(45)	56(27)	42(11)	39(23)	70(18)
1-Heptanol	41(100)	70(87)	56(86)	31(78)	43(72)	29(70)	55(67)	27(65)	42(54)	69(41)
Hexane	57(100)	43(78)	41(77)	29(61)	27(57)	56(45)	42(39)	39(27)	28(16)	86(14)
1-Hexanol (Caproyl alcohol)	56(100)	43(78)	31(74)	41(71)	27(64)	55(58)	42(39)	42(53)	39(37)	69(27)
Hexylene glycol	59(100)	43(61)	56(25)	45(17)	41(16)	57(13)	42(13)	85(11)	61(10)	31(10)
Hexyl methyl ketone	43(100)	58(79)	41(56)	59(52)	71(49)	27(46)	29(36)	39(27)	57(18)	55(17)
Isobutyl acetate	43(100)	56(26)	73(15)	41(10)	29(5)	71(3)	57(3)	39(3)	27(3)	86(2)
Isobutyl alcohol	43(100)	33(73)	31(72)	41(66)	42(60)	27(43)	29(18)	39(17)	28(8)	74(6)
Isobutylamine	30(100)	28(9)	41(6)	73(5)	27(5)	39(4)	29(3)	15(3)	58(2)	56(2)
Isopentyl acetate	43(100)	70(49)	55(38)	61(15)	42(15)	41(14)	27(12)	87(11)	29(10)	73(9)
Isophorone	82(100)	39(20)	138(17)	54(13)	27(12)	41(10)	53(8)	83(7)	29(7)	55(6)
Isopropyl acetate	43(100)	61(17)	41(14)	87(9)	59(8)	27(8)	42(7)	39(4)	45(3)	44(2)
Isopropyl alcohol	45(100)	43(19)	27(17)	29(12)	41(7)	31(6)	19(6)	42(5)	44(4)	59(3)
Isoquinoline	129(100)	102(26)	51(20)	128(18)	50(11)	130(10)	75(10)	76(9)	103(8)	74(7)
d-Limonene (Citrene)	68(100)	93(50)	67(49)	41(22)	94(21)	79(21)	39(21)	136(20)	53(19)	121(16)
2,6-Lutidine (2,6-Dimethyl-pyridine)	107(100)	39(39)	106(29)	66(22)	92(18)	65(18)	38(12)	27(11)	79(9)	63(9)

MASS SPECTRAL PEAKS OF COMMON ORGANIC SOLVENTS (continued)

Compound	e/m (intensity)									
Mesitylene (1,3,5-Trimethyl-benzene)	105(100)	120(64)	119(15)	77(13)	39(11)	106(9)	91(9)	51(8)	27(7)	121(6)
Mesityl oxide	55(100)	83(89)	43(73)	29(42)	98(36)	39(32)	27(28)	53(11)	41(10)	56(5)
Methyl acetate	43(100)	74(52)	28(38)	42(19)	59(17)	44(8)	32(8)	29(6)	31(4)	75(2)
Methylal (Dimethoxymethane)	45(100)	75(61)	29(59)	31(13)	30(6)	15(6)	47(5)	76(2)	46(2)	44(2)
Methyl alcohol	31(100)	29(72)	32(67)	15(42)	28(12)	14(10)	30(9)	13(6)	12(3)	16(2)
Methylamine	30(100)	31(87)	28(56)	29(19)	32(15)	15(12)	27(9)			
Methyl benzoate	105(100)	77(81)	51(45)	136(24)	50(18)	106(8)	78(6)	28(6)	39(5)	27(5)
Methylcyclohexane	83(100)	55(82)	41(60)	98(44)	42(35)	56(30)	27(29)	39(27)	69(23)	70(22)
Methyl ethyl ketone	43(100)	72(24)	29(19)	27(12)	57(7)	42(5)	26(4)	28(3)	44(2)	39(2)
N-Methylformamide	59(100)	30(54)	28(34)	29(13)	58(8)	15(7)	60(3)	41(3)	27(3)	31(2)
Methyl formate	31(100)	29(63)	32(34)	60(28)	30(7)	28(7)	44(2)	18(2)	61(1)	59(1)
Methyl iodide (Iodomethane)	142(100)	127(38)	141(14)	15(13)	139(5)	140(4)	128(3)	14(1)	13(1)	71(0)
Methyl isobutyl ketone	43(100)	58(84)	29(65)	41(56)	57(44)	27(42)	39(31)	85(19)	100(14)	42(14)
Methyl isopentyl ketone	43(100)	58(34)	27(14)	41(13)	15(13)	57(11)	39(9)	71(8)	59(8)	29(8)
2-Methylpentane	43(100)	42(53)	41(35)	27(31)	71(29)	39(20)	29(18)	57(11)	15(10)	70(7)
4-Methyl-2-pentanol	45(100)	43(47)	69(30)	41(27)	27(19)	39(13)	29(12)	87(11)	84(10)	57(10)
Methyl pentyl ketone	43(100)	58(60)	71(14)	41(11)	27(11)	59(9)	39(8)	29(8)	42(5)	114(4)
Methyl propyl ketone	43(100)	41(17)	86(12)	42(12)	27(11)	39(8)	71(7)	58(7)	45(7)	44(3)
N-Methyl-2-pyrrolidone	99(100)	44(89)	98(80)	42(60)	41(38)	43(17)	28(17)	71(13)	39(11)	70(10)
Morpholine	57(100)	29(100)	87(69)	28(69)	30(38)	56(33)	86(28)	31(28)	27(12)	15(7)
Nitrobenzene	77(100)	51(59)	123(42)	50(25)	30(15)	65(14)	39(10)	93(9)	74(7)	78(6)
Nitroethane	29(100)	30(12)	28(11)	26(9)	27(8)	43(5)	41(5)	14(5)	15(3)	46(2)
Nitromethane	30(100)	61(64)	46(39)	28(30)	45(8)	27(8)	44(7)	29(7)	60(5)	43(4)
1-Nitropropane	43(100)	27(93)	41(90)	39(34)	30(25)	44(20)	42(20)	26(20)	28(15)	54(12)
2-Nitropropane	43(100)	41(73)	27(71)	39(30)	30(18)	15(11)	42(9)	28(8)	26(8)	38(6)
Octane	43(100)	57(30)	85(25)	41(25)	71(19)	29(17)	56(14)	70(10)	42(10)	27(10)
1-Octanol	41(100)	56(85)	43(82)	55(81)	31(69)	27(69)	29(68)	42(62)	70(53)	69(48)
Pentachloroethane	167(100)	165(91)	117(90)	119(89)	83(58)	169(54)	130(43)	132(42)	60(40)	85(37)
Pentamethylene glycol (1,5-Pentanediol)	31(100)	56(85)	41(67)	57(59)	55(51)	44(45)	29(37)	43(31)	43(29)	27(26)
Pentane	43(100)	42(55)	41(45)	27(42)	29(26)	39(19)	57(13)	28(9)	15(9)	72(8)
1-Pentanol (Amyl alcohol)	42(100)	70(72)	55(65)	41(56)	31(47)	29(41)	27(26)	57(22)	28(22)	43(21)
Pentyl acetate (Amyl acetate)	43(100)	70(90)	42(52)	28(51)	61(50)	73(21)	27(26)	41(20)	29(14)	69(11)
2-Picoline (2-Methylpyridine)	93(100)	66(41)	39(31)	92(20)	78(19)	65(16)	73(21)	38(13)	50(12)	52(11)
α-Pinene	93(100)	92(30)	39(24)	41(23)	77(22)	91(21)	27(21)	79(18)	121(13)	53(10)
β-Pinene	93(100)	41(64)	69(47)	39(33)	27(31)	79(20)	77(18)	53(14)	94(13)	91(13)
Piperidine (Hexahydropyridine)	84(100)	85(53)	56(46)	57(43)	28(41)	29(37)	44(34)	42(30)	30(30)	43(25)
Propanenitrile	28(100)	54(63)	26(20)	27(17)	52(11)	55(10)	51(9)	15(9)	53(7)	25(7)
Propyl acetate	43(100)	61(19)	31(18)	27(15)	42(11)	73(9)	41(9)	29(9)	59(5)	39(5)
Propyl alcohol	31(100)	27(19)	29(18)	59(11)	42(9)	60(7)	41(7)	28(7)	43(3)	32(3)
Propylamine	30(100)	28(13)	59(8)	27(7)	41(5)	42(3)	39(3)	29(3)	26(3)	18(3)
Propylbenzene	91(100)	120(21)	92(10)	38(10)	65(9)	78(6)	51(6)	27(5)	63(4)	105(3)
1,2-Propylene glycol	45(100)	18(46)	29(21)	43(19)	31(18)	27(17)	28(11)	19(8)	44(6)	61(5)

MASS SPECTRAL PEAKS OF COMMON ORGANIC SOLVENTS (continued)

Compound	e/m (intensity)
Pseudocumene (1,2,4-Trimethylbenzene)	105(100), 120(56), 119(17), 77(15), 39(15), 51(11), 91(10), 27(10), 106(9), 79(7)
Pyridine	79(100), 52(62), 51(31), 50(19), 78(11), 53(7), 39(7), 80(6), 27(3), 77(2)
Pyrrole	67(100), 41(58), 39(58), 40(51), 28(42), 38(20), 37(12), 66(7), 68(5), 27(3)
Pyrrolidine	43(100), 28(52), 70(33), 71(26), 42(22), 41(20), 27(16), 39(15), 29(10), 30(9)
2-Pyrrolidone	85(100), 42(43), 41(36), 28(33), 30(29), 56(16), 84(14), 40(12), 27(12), 29(9)
Quinoline	129(100), 51(28), 76(25), 128(24), 44(24), 50(20), 32(19), 75(18), 74(12), 103(11)
Styrene	104(100), 103(41), 78(32), 51(28), 77(23), 105(12), 50(12), 52(11), 39(11), 102(10)
Sulfolane	41(100), 28(94), 56(82), 55(72), 120(37), 27(32), 39(19), 29(17), 26(11), 48(5)
α-Terpinene	121(100), 93(85), 136(43), 91(40), 77(34), 39(33), 27(33), 79(27), 41(26), 43(18)
1,1,1,2-Tetrachloro-2,2-difluoroethane	167(100), 169(96), 117(85), 119(82), 171(31), 85(29), 121(26), 82(14), 47(14), 101(13)
Tetrachloro-1,2-difluoroethane	101(100), 103(64), 167(54), 169(52), 117(19), 119(18), 171(17), 105(11), 31(11), 132(9)
1,1,1,2-Tetrachloroethane	131(100), 133(96), 117(76), 119(73), 95(34), 135(31), 121(23), 97(23), 61(19), 60(18)
1,1,2,2-Tetrachloroethane	83(100), 85(63), 95(11), 87(10), 168(8), 133(8), 131(8), 96(8), 61(8), 60(8)
Tetrachloroethylene	166(100), 164(82), 131(71), 129(71), 168(45), 94(38), 47(31), 133(20), 59(17), 28(5)
Tetraethylene glycol	45(100), 89(10), 44(8), 43(6), 31(6), 29(6), 27(6), 101(5), 75(5), 15(10)
Tetrahydrofuran	42(100), 41(52), 27(33), 72(29), 71(27), 39(24), 43(22), 29(22), 40(13), 15(10)
1,2,3,4-Tetrahydronaphthalene	104(100), 132(53), 91(43), 51(17), 39(17), 131(15), 117(15), 115(14), 78(13), 77(13)
Tetrahydropyran	41(100), 28(64), 56(57), 45(57), 29(51), 27(49), 85(47), 86(42), 39(28), 55(23)
Tetramethylsilane	73(100), 43(14), 45(12), 74(8), 29(7), 15(5), 75(4), 44(4), 42(4), 31(4)
Toluene	91(100), 92(73), 39(20), 65(14), 63(11), 51(11), 50(7), 27(6), 93(5), 90(5)
o-Toluidine	106(100), 107(83), 77(17), 79(13), 39(12), 53(10), 52(10), 54(9), 51(9), 28(9)
Triacetin	43(100), 103(44), 145(34), 116(17), 115(13), 44(10), 86(9), 73(7), 30(5), 42(7)
Tributylamine	142(100), 100(19), 143(11), 29(8), 185(7), 57(6), 44(6), 30(5), 41(6), 86(4)
1,1,1-Trichloroethane	97(100), 99(64), 61(58), 26(31), 27(24), 117(19), 63(19), 35(17), 62(11), 86(4)
1,1,2-Trichloroethane	97(100), 83(95), 99(62), 85(60), 61(58), 26(23), 96(21), 27(17), 98(15), 62(11)
Trichloroethylene	95(100), 130(90), 132(85), 60(65), 97(64), 35(40), 134(27), 47(26), 59(13), 37(4)
Trichlorofluoromethane	101(100), 103(66), 66(13), 105(11), 35(11), 47(9), 31(8), 82(4), 37(4), 68(4)
1,1,2-Trichlorotrifluoroethane	101(100), 151(68), 103(64), 85(45), 31(45), 153(44), 35(20), 66(19), 47(18), 87(14)
Triethanolamine	118(100), 56(69), 45(60), 42(56), 44(27), 43(25), 41(14), 116(8), 57(8), 86(7)
Triethylamine	86(100), 30(68), 58(37), 28(24), 29(23), 27(19), 44(18), 56(8), 116(8), 15(7)
Triethylene glycol	45(100), 58(11), 89(9), 31(8), 29(8), 75(7), 44(7), 43(7), 28(5), 27(7)
Triethyl phosphate	99(100), 81(71), 155(56), 82(45), 45(45), 109(44), 127(41), 43(24), 111(14), 29(8)
Trimethylamine	58(100), 59(47), 30(29), 42(26), 44(17), 15(14), 28(10), 18(10), 57(7), 27(7)
Trimethylene glycol (1,3-Propanediol)	28(100), 58(93), 31(76), 57(70), 29(40), 27(26), 45(24), 43(23), 30(17), 19(18)
Trimethyl phosphate	110(100), 109(35), 79(34), 95(25), 80(23), 15(20), 140(18), 47(10), 139(5), 31(7)
Veratrole	138(100), 95(65), 77(48), 123(44), 52(42), 41(33), 65(30), 51(29), 63(17), 39(19)
o-Xylene	91(100), 106(40), 39(21), 105(17), 51(17), 77(15), 27(12), 65(10), 79(8), 92(8)
m-Xylene	91(100), 106(65), 105(29), 39(18), 51(15), 77(14), 27(10), 92(8), 78(8), 65(7)
p-Xylene	91(100), 106(62), 105(30), 51(16), 39(16), 77(13), 27(11), 92(7), 65(7), 78(7)

Section 9
Molecular Structure and Spectroscopy

BOND LENGTHS IN CRYSTALLINE ORGANIC COMPOUNDS

The following table gives average interatomic distances for bonds between the elements H, B, C, N, O, F, Si, P, S, Cl, As, Se, Br, Te, and I as determined from X-ray and neutron diffraction measurements on organic crystals. The table has been derived from an analysis of high-precision structure data on about 10,000 crystals contained in the 1985 version of the Cambridge Structural Database, which is maintained by the Cambridge Crystallographic Data Center. The explanation of the columns is:

Column 1: Specification of elements in the bond, with coordination number given in parentheses, and bond type (single, double, etc.). For carbon, the hybridization state is given.

Column 2: Substructure in which the bond is found. The target bond is set in boldface. Where X is not specified, it denotes any element type. C# indicates any sp^3 carbon atom, and C* denotes an sp^3 carbon whose bonds, in addition to those specified in the linear formulation, are to C and H atoms only.

Column 3: d is the unweighted mean in Å units of all the values for that bond length found in the sample.

Column 4: m is the median in Å units of all values.

Column 5: σ is the standard deviation in the sample.

Column 6: q_1 is the lower quartile for the sample (i.e., 25% of values are less than q_1 and 75% exceed it).

Column 7: q_2 is the upper quartile for the sample.

Column 8: n is number of observations in the sample.

Column 9: Notes refer to the footnotes in Appendix 1.

References to special cases are given in a shorthand form and listed in Appendix 2. Further information on the method of analysis of the data may be found in the reference cited below.

The table is reprinted with permission of the authors, the Royal Society of Chemistry, and the International Union of Crystallography.

REFERENCE

Frank H. Allen, Olga Kennard, David G. Watson, Lee Brammer, A. Guy Orpen, and Robin Taylor, *J. Chem. Soc. Perkin Trans. II*, S1—S19, 1987.

Bond	Substructure	d	m	σ	q_1	q_u	n	Note
As(3)–As(3)	X_2–As–As–X_2	2.459	2.457	0.011	2.456	2.466	8	
As–B	see CUDLOC (2.065), CUDLUI (2.041)							
As–Br	see CODDEE, CODDII (2.346—3.203)							
As(4)–C	X_3–As–CH_3	1.903	1.907	0.016	1.893	1.916	12	
	$(X)_2(C,O,S=)$As–Csp^3	1.927	1.929	0.017	1.921	1.937	16	
	As–Car in Ph_4As^+	1.905	1.909	0.012	1.897	1.912	108	
	$(X)_2(C,O,S=)$As–Car	1.922	1.927	0.016	1.908	1.934	36	
As(3)–C	X_2–As–Csp^3	1.963	1.965	0.017	1.948	1.978	6	
	X_2–As–Car	1.956	1.956	0.015	1.944	1.964	41	
As(3)–Cl	X_2–As–Cl	2.268	2.256	0.039	2.247	2.281	10	
As(6)–F	in AsF_6^-	1.678	1.676	0.020	1.659	1.695	36	
As(3)–I	see OPIMAS (2.579, 2.590)							
As(3)–N(3)	X_2–As–N–X_2	1.858	1.858	0.029	1.839	1.873	19	
As(4)=N(2)	see TPASSN (1.837)							
As(4)–O	$(X)_2(O=)$As–OH	1.710	1.712	0.017	1.695	1.726	6	
As(3)–O	see ASAZOC, PHASOC01 (1.787—1.845)							
As(4)=O	X_3–As=O	1.661	1.661	0.016	1.652	1.667	9	
As(3)=P(3)	see BELNIP (2.350, 2.362)							†
As(3)–P(3)	see BUTHAZ10 (2.124)							†
As(3)–S	X_2–As–S	2.275	2.266	0.032	2.247	2.298	14	
As(4)=S	X_3–As=S	2.083	2.082	0.004	2.080	2.086	9	
As(3)–Se(2)	see COSDIX, ESEARS (2.355—2.401)							†
As(3)–Si(4)	see BICGEZ, MESIAD (2.351—2.365)							†
As(3)–Te(2)	see ETEARS (2.571, 2.576)							†
B(n)–B(n)	$n = 5$—7 in boron cages	1.775	1.773	0.031	1.763	1.786	688	
B(4)–B(4)	see CETTAW (2.041)							
B(4)–B(3)	see COFVOI (1.698)							
B(3)–B(3)	X_2–B–B–X_2	1.701	1.700	0.014	1.691	1.712	8	
B(6)–Br		1.967	1.971	0.014	1.954	1.979	7	†
B(4)–Br		2.017	2.008	0.031	1.990	2.044	15	†
B(n)–C	$n = 5$—7: B–C in cages	1.716	1.717	0.020	1.707	1.728	96	
	$n = 3$—4: B–Csp^3 not cages	1.597	1.599	0.022	1.585	1.611	29	1
	$n = 4$: B–Car	1.606	1.607	0.012	1.596	1.615	41	
	$n = 4$: B–Car in Ph_4B^-	1.643	1.643	0.006	1.641	1.645	16	
B(n)–C	$n = 3$: B–Car	1.556	1.552	0.015	1.546	1.566	24	
B(n)–Cl	B(5)–Cl and B(3)–Cl	1.751	1.751	0.011	1.743	1.761	14	
	B(4)–Cl	1.833	1.833	0.013	1.821	1.843	22	
B(4)–F	B–F (B neutral)	1.366	1.368	0.017	1.356	1.375	25	
	B^-–F in BF_4^-	1.365	1.372	0.029	1.352	1.390	84	
B(4)–I	see TMPBTI (2.220, 2.253)							
B(4)–N(3)	X_3–B–N(=C)(X)	1.611	1.617	0.013	1.601	1.625	8	
	in pyrazaboles	1.549	1.552	0.015	1.536	1.560	10	

Bond	Substructure	d	m	σ	q_1	q_u	n	Note
B(3)–N(3)	X_2–B–N–C_2: all coplanar	1.404	1.404	0.014	1.389	1.408	40	2
	for τ(BN) > 30° see BOGSUL, BUSHAY, CILRUK (1.434—1.530)							
	S_2–B–N–X_2	1.447	1.443	0.013	1.435	1.470	14	
B(4)–O	B^-–O in BO_4	1.468	1.468	0.022	1.453	1.479	24	
	for neutral B–O see Note 3							3
B(3)–O(2)	X_2–B–O–X	1.367	1.367	0.024	1.349	1.382	35	
B(n)–P	n = 4: B–P	1.922	1.927	0.027	1.900	1.954	10	
	n = 3: see BUPSIB10 (1.892, 1.893)							
B(4)–S	B(4)–S(3)	1.930	1.927	0.009	1.925	1.934	10	
	B(4)–S(2)	1.896	1.896	0.004	1.893	1.899	6	
B(3)–S	N–B–S_2	1.806	1.806	0.010	1.799	1.816	28	
	(=X–)(N–)B–S	1.851	1.854	0.013	1.842	1.859	10	
Br–Br	see BEPZEB, TPASTB	2.542	2.548	0.015	2.526	2.551	4	
Br–C	Br–C*	1.966	1.967	0.029	1.951	1.983	100	4
	Br–Csp^3 (cyclopropane)	1.910	1.910	0.010	1.900	1.914	8	
	Br–Csp^2	1.883	1.881	0.015	1.874	1.894	31	4
	Br–Car (mono-Br + m,p-Br_2)	1.899	1.899	0.012	1.892	1.906	119	4
	Br–Car (o-Br_2)	1.875	1.872	0.011	1.864	1.884	8	4
$^-$Br(2)–Cl	see TEACBR (2.362—2.402)							†
Br–I	see DTHIBR10 (2.646), TPHOSI (2.695)							
Br–N	see NBBZAM (1.843)							
Br–O	see CIYFOF	1.581	1.581	0.007	1.574	1.587	4	
Br–P	see CISTED (2.366)							
Br–S(2)	see BEMLIO (2.206)							†
Br–S(3)	see CIWYIQ (2.435, 2.453)							†
Br–S(3)$^+$	see THINBR (2.321)							†
Br–Se	see CIFZUM (2.508, 2.619)							
Br–Si	see BIZJAV (2.284)							
Br–Te	In Br_6Te^{2-} see CUGBAH (2.692—2.716)							
	Br–Te(4) see BETUTE10 (3.079, 3.015)							
	Br–Te(3) see BTUPTE (2.835)							
Csp^3–Csp^3	C#–CH_2–CH_3	1.513	1.514	0.014	1.507	1.523	192	
	(C#)$_2$–CH–CH_3	1.524	1.526	0.015	1.518	1.534	226	
	(C#)$_3$–C–CH_3	1.534	1.534	0.011	1.527	1.541	825	
	C#–CH_2–CH_2–C#	1.524	1.524	0.014	1.516	1.532	2 459	
	(C#)$_2$–CH–CH_2–C#	1.531	1.531	0.012	1.524	1.538	1 217	
	(C#)$_3$–C–CH_2–C#	1.538	1.539	0.010	1.533	1.544	330	
	(C#)$_2$–CH–CH–(C#)$_2$	1.542	1.542	0.011	1.536	1.549	321	
	(C#)$_3$–C–CH–(C#)$_2$	1.556	1.556	0.011	1.549	1.562	215	
	(C#)$_3$–C–C–(C#)$_3$	1.588	1.580	0.025	1.566	1.610	21	
	C*–C* (overall)	1.530	1.530	0.015	1.521	1.539	5 777	5,6
	in cyclopropane (any subst.)	1.510	1.509	0.026	1.497	1.523	888	7
	in cyclobutane (any subst.)	1.554	1.553	0.021	1.540	1.567	679	8
	in cyclopentane (C,H-subst.)	1.543	1.543	0.018	1.532	1.554	1 641	
	in cyclohexane (C,H-subst.)	1.535	1.535	0.016	1.525	1.545	2 814	
	cyclopropyl-C* (exocyclic)	1.518	1.518	0.019	1.505	1.531	366	7
	cyclobutyl-C* (exocyclic)	1.529	1.529	0.016	1.519	1.539	376	8
	cyclopentyl-C* (exocyclic)	1.540	1.541	0.017	1.527	1.549	956	
	cyclohexyl-C* (exocyclic)	1.539	1.538	0.016	1.529	1.549	2 682	
	in cyclobutene (any subst.)	1.573	1.574	0.017	1.566	1.586	25	8
	in cyclopentene (C,H-subst.)	1.541	1.539	0.015	1.532	1.549	208	
	in cyclohexene (C,H-subst.)	1.541	1.541	0.020	1.528	1.554	586	
	in oxirane (epoxide)	1.466	1.466	0.015	1.458	1.474	249	9
	in aziridine	1.480	1.481	0.021	1.465	1.496	67	9
	in oxetane	1.541	1.541	0.019	1.527	1.557	16	
	in azetidine	1.548	1.543	0.018	1.536	1.558	22	
	oxiranyl-C* (exocyclic)	1.509	1.507	0.018	1.497	1.519	333	9
	aziridinyl-C* (exocyclic)	1.512	1.512	0.018	1.496	1.526	13	9
Csp^3–Csp^2	CH_3–C=C	1.503	1.504	0.011	1.497	1.509	215	
	C#–CH_2–C=C	1.502	1.502	0.013	1.494	1.510	483	
	(C#)$_2$–CH–C=C	1.510	1.510	0.014	1.501	1.518	564	
	(C#)$_3$–C–C=C	1.522	1.522	0.016	1.511	1.533	193	
Csp^3–Csp^2	C*–C=C (overall)	1.507	1.507	0.015	1.499	1.517	1 456	5
	C*–C=C (endocyclic)							
	in cyclopropene	1.509	1.508	0.016	1.500	1.516	20	10
	in cyclobutene	1.513	1.512	0.018	1.500	1.525	50	8
	in cyclopentene	1.512	1.512	0.014	1.502	1.521	208	
	in cyclohexene	1.506	1.505	0.016	1.495	1.516	391	
	in cyclopentadiene	1.502	1.503	0.019	1.490	1.515	18	
	in cyclohexa-1,3-diene	1.504	1.504	0.017	1.491	1.517	56	
	C*–C=C (exocyclic):							
	cyclopropenyl-C*	1.478	1.475	0.012	1.470	1.485	7	10
	cyclobutenyl-C*	1.489	1.483	0.015	1.479	1.496	11	8

Bond	Substructure	d	m	σ	q_1	q_u	n	Note
	cyclopentenyl-C*	1.504	1.506	0.012	1.495	1.512	115	
	cyclohexenyl-C*	1.511	1.511	0.013	1.502	1.519	292	
	C*–CH=O in aldehydes	1.510	1.510	0.008	1.501	1.518	7	
	(C*)$_2$–C=O							
	in ketones	1.511	1.511	0.015	1.501	1.521	952	11
	in cyclobutanone	1.529	1.530	0.016	1.514	1.545	18	
	in cyclopentanone	1.514	1.514	0.016	1.505	1.523	312	
	acyclic and 6+ rings	1.509	1.509	0.016	1.499	1.519	626	
	C*–COOH in carboxylic acids	1.502	1.502	0.014	1.495	1.510	176	
	C*–COO$^-$ in carboxylate anions	1.520	1.521	0.011	1.516	1.528	57	
	C*–C(=O)(–OC*)							
	in acyclic esters	1.497	1.496	0.018	1.484	1.509	553	12
	in β-lactones	1.519	1.519	0.020	1.500	1.538	4	13
	in γ-lactones	1.512	1.512	0.015	1.501	1.521	110	12
	in δ-lactones	1.504	1.502	0.013	1.495	1.517	27	12
	cyclopropyl (C)–C=O in ketones, acids							
	and esters	1.486	1.485	0.018	1.474	1.497	105	7
	C*–C(=O)(–NH$_2$) in acyclic amides	1.514	1.512	0.016	1.506	1.526	32	14
	C*–C(=O)(–NHC*) in acyclic amides	1.506	1.505	0.012	1.498	1.515	78	14
	C*–C(=O)[–N(C*)$_2$] in acyclic amides	1.505	1.505	0.011	1.496	1.517	15	14
Csp^3–Car	CH$_3$–Car	1.506	1.507	0.011	1.501	1.513	454	
	C#–CH$_2$–Car	1.510	1.510	0.009	1.505	1.516	674	
	(C#)$_2$–CH–Car	1.515	1.515	0.011	1.508	1.522	363	
	(C#)$_3$–C–Car	1.527	1.530	0.016	1.517	1.539	308	
	C*–Car (overall)	1.513	1.513	0.014	1.505	1.521	1 813	
	cyclopropyl (C)–Car	1.490	1.490	0.015	1.479	1.503	90	7
Csp^3–Csp^1	C*–C≡C	1.466	1.465	0.010	1.460	1.469	21	15
	C#–C≡C	1.472	1.472	0.012	1.464	1.481	88	15
	C*–C≡N	1.470	1.469	0.013	1.463	1.479	106	7b
	cyclopropyl (C)–C≡N	1.444	1.447	0.010	1.436	1.451	38	7
Csp^2–Csp^2	C=C–C=C							
	(conjugated)	1.455	1.455	0.011	1.447	1.463	30	16,18
	(unconjugated)	1.478	1.476	0.012	1.470	1.479	8	17,18
	(overall)	1.460	1.460	0.015	1.450	1.470	38	
	C=C–C=C–C=C	1.443	1.445	0.013	1.431	1.454	29	18
	C=C–C=C (endocyclic in TCNQ)	1.432	1.433	0.012	1.424	1.441	280	19
	C=C–C(=O)(–C*)							
	(conjugated)	1.464	1.462	0.018	1.453	1.476	211	16,18
	(unconjugated)	1.484	1.486	0.017	1.475	1.497	14	17,18
	(overall)	1.465	1.462	0.018	1.453	1.478	226	
	C=C–C(=O)–C=C							
	in benzoquinone (C,H-subst. only)	1.478	1.476	0.011	1.469	1.488	28	
	in benzoquinone (any subst.)	1.478	1.478	0.031	1.464	1.498	172	
	non-quinonoid	1.456	1.455	0.012	1.447	1.464	28	
	C=C–COOH	1.475	1.476	0.015	1.461	1.488	22	
	C=C–COOC*	1.488	1.489	0.014	1.478	1.497	113	
	C=C–COO$^-$	1.502	1.499	0.017	1.488	1.510	11	
	HOOC–COOH	1.538	1.537	0.007	1.535	1.541	9	
	HOOC–COO$^-$	1.549	1.552	0.009	1.546	1.553	13	
	$^-$OOC–COO$^-$	1.564	1.559	0.022	1.554	1.568	9	
	formal Csp^2–Csp^2 single bond in selected							
	non-fused heterocycles:							
	in 1H-pyrrole (C3–C4)	1.412	1.410	0.016	1.401	1.427	29	
	in furan (C3–C4)	1.423	1.423	0.016	1.412	1.433	62	
	in thiophene (C3–C4)	1.424	1.425	0.015	1.415	1.433	40	
	in pyrazole (C3–C4)	1.410	1.412	0.016	1.400	1.418	20	
	in isoxazole (C3–C4)	1.425	1.425	0.016	1.413	1.438	9	
	in furazan (C3–C4)	1.428	1.427	0.007	1.422	1.435	6	
	in furoxan (C3–C4)	1.417	1.417	0.006	1.412	1.422	14	
Csp^2–Car	C=C–Car							
	(conjugated)	1.470	1.470	0.015	1.463	1.480	37	16,18
Csp^2–Car		1.488	1.490	0.012	1.480	1.496	87	17,18
	(overall)	1.483	1.483	0.015	1.472	1.494	124	
	cyclopropenyl (C=C)–Car	1.447	1.448	0.006	1.441	1.452	8	10
	Car–C(=O)–C*	1.488	1.489	0.016	1.478	1.500	84	
	Car–C(=O)–Car	1.480	1.481	0.017	1.468	1.494	58	
	Car–COOH	1.484	1.485	0.014	1.474	1.491	75	
	Car–C(=O)(–OC*)	1.487	1.487	0.012	1.480	1.494	218	
	Car–COO$^-$	1.504	1.509	0.014	1.495	1.512	26	
	Car–C(=O)–NH$_2$	1.500	1.503	0.020	1.498	1.510	19	
	Car–C=N–C#							
	(conjugated)	1.476	1.478	0.014	1.466	1.486	27	16
	(unconjugated)	1.491	1.490	0.008	1.485	1.496	48	17
	(overall)	1.485	1.487	0.013	1.481	1.493	75	

Bond	Substructure	d	m	σ	q_l	q_u	n	Note
	in indole (C3–C3a)	1.434	1.434	0.011	1.428	1.439	40	
Csp^2–Csp^1	C=C–C≡C	1.431	1.427	0.014	1.425	1.441	11	7b
	C=C–C≡N in TCNQ	1.427	1.427	0.010	1.420	1.433	280	19
Car–Car	in biphenyls (*ortho* subst. all H)	1.487	1.488	0.007	1.484	1.493	30	
	(≥1 non-H *ortho*-subst.)	1.490	1.491	0.010	1.486	1.495	212	
Car–Csp^1	Car–C≡C	1.434	1.436	0.006	1.430	1.437	37	
	Car–C≡N	1.443	1.444	0.008	1.436	1.448	31	
Csp^1–Csp^1	C≡C–C≡C	1.377	1.378	0.012	1.374	1.384	21	
Csp^2=Csp^2	C*–CH=CH₂	1.299	1.300	0.027	1.280	1.311	42	
	(C*)₂=C=CH₂	1.321	1.321	0.013	1.313	1.328	77	
	C*–CH=CH–C*							
	(cis)	1.317	1.318	0.013	1.310	1.323	106	
	(trans)	1.312	1.311	0.011	1.304	1.320	19	
	(overall)	1.316	1.317	0.015	1.309	1.323	127	
	(C*)₂=C=CH–C*	1.326	1.328	0.011	1.319	1.334	168	
	(C*)₂=C=C–(C*)₂	1.331	1.330	0.009	1.326	1.334	89	
	(C*,H)₂=C=C–(C*,H)₂ (overall)	1.322	1.323	0.014	1.315	1.331	493	5
	in cyclopropene (any subst.)	1.294	1.288	0.017	1.284	1.302	10	10
	in cyclobutene (any subst.)	1.335	1.335	0.019	1.324	1.347	25	8
	in cyclopentene (C,H-subst.)	1.323	1.324	0.013	1.314	1.331	104	
	in cyclohexene (C,H-subst.)	1.326	1.325	0.012	1.318	1.334	196	
	C=C=C (allenes, any subst.)	1.307	1.307	0.005	1.303	1.310	18	
	C=C–C=C (C,H subst., conjugated)	1.330	1.330	0.014	1.322	1.338	76	16
	C=C–C=C–C=C (C,H subst., conjugated)	1.345	1.345	0.012	1.337	1.350	58	16
	C=C–Car (C,H subst., conjugated)	1.339	1.340	0.011	1.334	1.346	124	16
	C=C in cyclopenta-1,3-diene (any subst.)	1.341	1.341	0.017	1.328	1.356	18	
	C=C in cyclohexa-1,3-diene (any subst.)	1.332	1.332	0.013	1.323	1.341	56	
	in C=C–C=O							
	(C,H subst., conjugated)	1.340	1.340	0.013	1.332	1.348	211	16,18
	(C,H subst., unconjugated)	1.331	1.330	0.008	1.326	1.339	14	17,18
	(C,H subst., overall)	1.340	1.339	0.013	1.332	1.348	226	
	in cyclohexa-2,5-dien-1-ones	1.329	1.327	0.011	1.321	1.335	28	
	in *p*-benzoquinones							
	(C*,H subst.)	1.333	1.337	0.011	1.325	1.338	14	
	(any subst.)	1.349	1.339	0.030	1.330	1.364	86	
	in TCNQ							
	(endocyclic)	1.352	1.353	0.010	1.345	1.358	142	19
	(exocyclic)	1.392	1.391	0.017	1.379	1.405	139	19
	C=C–OH in enol tautomers	1.362	1.360	0.020	1.349	1.370	54	
	in heterocycles (any subst.):							
	1*H*-pyrrole (C2–C3, C4–C5)	1.375	1.377	0.018	1.361	1.388	58	
	furan (C2–C3, C4–C5)	1.341	1.342	0.021	1.329	1.351	125	
	thiophene (C2–C3, C4–C5)	1.362	1.359	0.025	1.346	1.377	60	
	pyrazole (C4–C5)	1.369	1.372	0.019	1.362	1.383	20	
	imidazole (C4–C5)	1.360	1.361	0.014	1.352	1.367	44	
	isoxazole (C4–C5)	1.341	1.336	0.012	1.331	1.355	9	
	indole (C2–C3)	1.364	1.363	0.012	1.355	1.371	40	
$Car \simeq Car$	in phenyl rings with C*,H subst. only							
	H–C≃C–H	1.380	1.381	0.013	1.372	1.388	2 191	
	C*–C≃C–H	1.387	1.388	0.010	1.382	1.393	891	
	C*–C≃C–C*	1.397	1.397	0.009	1.392	1.403	182	
	C≃C (overall)	1.384	1.384	0.013	1.375	1.391	3 264	
	F–C≃C–F	1.372	1.374	0.011	1.366	1.380	84	4
	Cl–C≃C–Cl	1.388	1.389	0.014	1.380	1.398	152	4
	in naphthalene (D_{2h}, any subst.)							
	C1–C2	1.364	1.364	0.014	1.356	1.373	440	
	C2–C3	1.406	1.406	0.014	1.397	1.415	218	
	C1–C8a	1.420	1.419	0.012	1.412	1.426	440	
	C4a–C8a	1.422	1.424	0.011	1.417	1.429	109	
$Car \simeq Car$	in anthracene (D_{2h}, any subst.)							
	C1–C2	1.356	1.356	0.009	1.350	1.360	56	
	C2–C3	1.410	1.410	0.010	1.401	1.416	34	
	C1–C9a	1.430	1.430	0.006	1.426	1.434	56	
	C4a–C9a	1.435	1.436	0.007	1.429	1.440	34	
	C9–C9a	1.400	1.402	0.009	1.395	1.406	68	
	in pyridine (C,H subst.)	1.379	1.381	0.012	1.371	1.387	276	20
	(any subst.)	1.380	1.380	0.015	1.371	1.389	537	20
	in pyridinium cation							
	(N⁺–H; C,H subst. on C)							
	C2–C3	1.373	1.375	0.012	1.368	1.380	30	
	C3–C4	1.379	1.380	0.011	1.371	1.388	30	
	(N⁺–X; C,H subst. on C)							
	C2–C3	1.373	1.372	0.019	1.362	1.382	151	
	C3–C4	1.383	1.385	0.019	1.372	1.394	151	

Bond	Substructure	d	m	σ	q_1	q_u	n	Note
	in pyrazine (H subst. on C)	1.379	1.377	0.010	1.370	1.388	10	
	(any subst. on C)	1.405	1.405	0.024	1.388	1.420	60	
	in pyrimidine (C,H subst. on C)	1.387	1.389	0.018	1.379	1.400	28	
$Csp^1{\equiv}Csp^1$	X–C≡C–X	1.183	1.183	0.014	1.174	1.193	119	15
	C,H–C≡C–C,H	1.181	1.181	0.014	1.173	1.192	104	15
	in C≡C–C(sp^2,ar)	1.189	1.193	0.010	1.181	1.195	38	15
	in C≡C–C≡C	1.192	1.192	0.010	1.187	1.197	42	15
	in CH≡C–C#	1.174	1.174	0.011	1.167	1.180	42	15
Csp^3–Cl	Omitting 1,2-dichlorides:							
	C–CH$_2$–Cl	1.790	1.790	0.007	1.783	1.795	13	4
	C$_2$–CH–Cl	1.803	1.802	0.003	1.800	1.807	8	4
	C$_3$–C–Cl	1.849	1.856	0.011	1.837	1.858	5	4
	X–CH$_2$–Cl (X = C,H,N,O)	1.790	1.791	0.011	1.783	1.797	37	4
	X$_2$–CH–Cl (X = C,H,N,O)	1.805	1.803	0.014	1.800	1.812	26	4
	X$_3$–C–Cl (X = C,H,N,O)	1.843	1.838	0.014	1.835	1.858	7	4
	X$_2$–C–Cl$_2$ (X = C,H,N,O)	1.779	1.776	0.015	1.769	1.790	18	4
	X–C–Cl$_3$ (X = C,H,N,O)	1.768	1.765	0.011	1.761	1.776	33	4
	Cl–CH(–C)–CH(–C)–Cl	1.793	1.793	0.013	1.786	1.800	66	4
	Cl–C(–C$_2$)–C(–C$_2$)–Cl	1.762	1.760	0.010	1.757	1.765	54	4
	cyclopropyl–Cl	1.755	1.756	0.011	1.749	1.763	64	
Csp^2–Cl	C=C–Cl (C,H,N,O subst. on C)	1.734	1.729	0.019	1.719	1.748	63	4
	C=C–Cl$_2$ (C,H,N,O subst. on C)	1.720	1.716	0.013	1.708	1.729	20	4
	Cl–C=C–Cl	1.713	1.711	0.011	1.705	1.720	80	4
Car–Cl	Car–Cl (mono-Cl + m,p-Cl$_2$)	1.739	1.741	0.010	1.734	1.745	340	4
	Car–Cl (o-Cl$_2$)	1.720	1.720	0.010	1.713	1.717	364	4
Csp^1–Cl	see HCLENE10 (1.634, 1.646)							
Csp^3–F	Omitting 1,2-difluorides							
	C–CH$_2$–F and C$_2$-CH–F	1.399	1.399	0.017	1.389	1.408	25	4
	C$_3$–C–F	1.428	1.431	0.009	1.421	1.435	11	4
	(C*,H)$_2$–C–F$_2$	1.349	1.347	0.012	1.342	1.356	58	4
	C*–C–F$_3$	1.336	1.334	0.007	1.330	1.344	12	4
	F–C*–C*–F	1.371	1.374	0.007	1.362	1.375	26	4
	X$_3$–C–F (X = C,H,N,O)	1.386	1.389	0.033	1.373	1.408	70	4
	X$_2$–C–F$_2$ (X = C,H,N,O)	1.351	1.349	0.013	1.342	1.356	58	4
	X–C–F$_3$ (X = C,H,N,O)	1.322	1.323	0.015	1.314	1.332	309	4
	F–C(–X)$_2$–C(–X)$_2$–F (X = C,H,N,O)	1.373	1.374	0.009	1.362	1.377	30	4
	F–C(–X)$_2$–NO$_2$ (X = any subst.)	1.320	1.319	0.009	1.312	1.327	18	
Csp^2–F	C=C–F (C,H,N,O subst. on C)	1.340	1.340	0.013	1.334	1.346	34	4
Car–F	Car–F (mono-F + m,p-F$_2$)	1.363	1.362	0.008	1.357	1.368	38	4
	Car–F (o-F$_2$)	1.340	1.340	0.009	1.336	1.344	167	4
Csp^3–H	C–C–H$_3$ (methyl)	1.059	1.061	0.030	1.039	1.083	83	21
	C$_2$–C–H$_2$ (primary)	1.092	1.095	0.013	1.088	1.099	100	21
	C$_3$–C–H (secondary)	1.099	1.097	0.004	1.095	1.103	14	21
	C$_{2,3}$–C–H (primary and secondary)	1.093	1.095	0.012	1.089	1.100	118	21
	X–C–H$_3$ (methyl)	1.066	1.074	0.028	1.049	1.087	160	21
	X$_2$–C–H$_2$ (primary)	1.092	1.095	0.012	1.088	1.099	230	21
	X$_3$–C–H (secondary)	1.099	1.099	0.007	1.095	1.103	117	21
	X$_{2,3}$–C–H (primary and secondary)	1.094	1.096	0.011	1.091	1.100	348	21
Csp^2–H	C–C=C–H	1.077	1.079	0.012	1.074	1.085	14	21
Car–H	Car–H	1.083	1.083	0.011	1.080	1.087	218	21
Csp^3–I	C*–I	2.162	2.159	0.015	2.149	2.179	15	4
Car–I	Car–I	2.095	2.095	0.015	2.089	2.104	51	4
Csp^3–N(4)	C*–NH$_3^+$	1.488	1.488	0.013	1.482	1.495	298	
	(C*)$_2$–NH$_2^+$	1.494	1.493	0.016	1.484	1.503	249	
	(C*)$_3$–NH$^+$	1.502	1.502	0.015	1.491	1.512	509	
	(C*)$_4$–N$^+$	1.510	1.509	0.020	1.496	1.523	319	
	C*–N$^+$ (overall)	1.499	1.498	0.018	1.488	1.510	1 370	
Csp^3–N(3)	C*–N$^+$ in N-subst. pyridinium	1.485	1.484	0.009	1.477	1.490	32	
	C*–NH$_2$ (Nsp^3: pyramidal)	1.469	1.470	0.010	1.462	1.474	19	22
	(C*)$_2$–NH (Nsp^3: pyramidal)	1.469	1.467	0.012	1.461	1.477	152	5,22
	(C*)$_3$–N (Nsp^3: pyramidal)	1.469	1.468	0.014	1.460	1.476	1 042	5,22
	C*–Nsp^3 (overall)	1.469	1.468	0.014	1.460	1.476	1 201	
	Csp^3–Nsp^3							
	in aziridine	1.472	1.471	0.016	1.464	1.482	134	
	in azetidine	1.484	1.481	0.018	1.472	1.495	21	
	in tetrahydropyrrole	1.475	1.473	0.016	1.464	1.483	66	
	in piperidine	1.473	1.473	0.013	1.460	1.479	240	
	Csp^3–Nsp^2 (N planar) in:							23
	acyclic amides C*–NH–C=O	1.454	1.451	0.011	1.446	1.461	78	14
	β-lactams C*–N(–X)–C=O (endo)	1.464	1.465	0.012	1.458	1.475	23	13
	γ-lactams							
	C*–NH–C=O (endo)	1.457	1.458	0.011	1.449	1.465	20	13
	C*–N(–C*)–C=O (endo)	1.462	1.461	0.010	1.453	1.466	15	13
	C*–N(–C*)–C=O (exo)	1.458	1.456	0.014	1.448	1.465	15	13

BOND LENGTHS IN CRYSTALLINE ORGANIC COMPOUNDS (continued)

Bond	Substructure	d	m	σ	q_1	q_u	n	Note
	δ-lactams							
	C*–NH–C=O (endo)	1.478	1.472	0.016	1.467	1.491	6	14
	C*–N(–C*)–C=O (endo)	1.479	1.476	0.007	1.475	1.482	15	14
	C*–N(–C*)–C=O (exo)	1.468	1.471	0.009	1.462	1.477	15	14
	nitro compounds (1,2-dinitro omitted):							
	C–CH$_2$–NO$_2$	1.485	1.483	0.020	1.478	1.502	8	
	C$_2$–CH–NO$_2$	1.509	1.509	0.011	1.502	1.511	12	
	C$_3$–C–NO$_2$	1.533	1.533	0.013	1.530	1.539	17	
	C$_2$–C–(NO$_2$)$_2$	1.537	1.536	0.016	1.525	1.550	19	
	1,2-dinitro: NO$_2$–C*–C*–NO$_2$	1.552	1.550	0.023	1.536	1.572	32	
Csp^3–N(2)	C#–N=N	1.493	1.493	0.020	1.477	1.506	54	
	C*–N=C–Car	1.465	1.468	0.011	1.461	1.472	75	
Csp^2–N(3)	C=C–NH$_2$ Nsp^2 planar	1.336	1.344	0.017	1.317	1.348	10	23
	C=C–NH–C# Nsp^2 planar	1.339	1.340	0.016	1.327	1.351	17	23
	C=C–N–(C#)$_2$							
	Nsp^2 planar	1.355	1.358	0.014	1.341	1.363	22	23
	Nsp^3 pyramidal	1.416	1.418	0.018	1.397	1.432	18	22
	Csp^2–Nsp^2 (N planar) in:							23
	acyclic amides							
	NH$_2$–C=O	1.325	1.323	0.009	1.318	1.331	32	14
	C*–NH–C=O	1.334	1.333	0.011	1.326	1.343	78	14
	(C*)$_2$–N–C=O	1.346	1.342	0.011	1.339	1.356	5	14
	β-lactams C*–NH–C=O	1.385	1.388	0.019	1.374	1.396	23	13
	γ-lactams							
	C*–NH–C=O	1.331	1.331	0.011	1.326	1.337	20	13
	C*–N(–C*)–C=O	1.347	1.344	0.014	1.335	1.359	15	13
	δ-lactams							
	C*–NH–C=O	1.334	1.334	0.006	1.330	1.339	6	14
	C*–N(–C*)–C=O	1.352	1.353	0.010	1.344	1.356	15	14
	peptides C#–N(–X)–C(–C#)(=O)	1.333	1.334	0.013	1.326	1.340	380	24
	ureas							
	(NH$_2$)$_2$–C=O	1.334	1.334	0.008	1.329	1.339	48	25,26
	(C#–NH)$_2$–C=O	1.347	1.345	0.010	1.341	1.354	26	25
	[(C#)$_n$–N]$_2$–C=O	1.363	1.359	0.014	1.354	1.370	40	25,27
	thioureas							
	(X$_2$N)$_2$–C=S	1.346	1.343	0.023	1.328	1.361	192	
	imides							
	[C#–C(=O)]$_2$–NH	1.376	1.377	0.012	1.369	1.383	64	
	[C#–C(=O)]$_2$–N–C#	1.389	1.383	0.017	1.376	1.404	38	
	[Csp^2–C(=O)]$_2$–N–C#	1.396	1.396	0.010	1.389	1.403	46	
	[Csp^2–C(=O)]$_2$–N–Csp^2	1.409	1.406	0.020	1.391	1.419	28	
	guanidinium [C–(NH$_2$)$_3$]$^+$ (unsubst.)	1.321	1.320	0.008	1.314	1.327	39	
	(any subst.)	1.328	1.325	0.015	1.317	1.333	140	
	in heterocyclic systems (any subst.)							
	1H-pyrrole (N1–C2, N1–C5)	1.372	1.374	0.016	1.363	1.384	58	
	indole (N1–C2)	1.370	1.370	0.012	1.364	1.377	40	
	pyrazole (N1–C5)	1.357	1.359	0.012	1.347	1.365	20	
	imidazole (N1–C2)	1.349	1.349	0.018	1.338	1.358	44	
	imidazole (N1–C5)	1.370	1.370	0.010	1.365	1.377	44	
Csp^2–N(2)	in imidazole (N3–C4)	1.376	1.377	0.011	1.369	1.384	44	
Car–N(4)	Car–N$^+$–(C,H)$_3$	1.465	1.466	0.007	1.461	1.470	23	
Car–N(3)	Car–NH$_2$							
	(Nsp^2: planar)	1.355	1.360	0.020	1.340	1.372	33	23
	(Nsp^3: pyramidal)	1.394	1.396	0.011	1.385	1.403	25	22
	(overall)	1.375	1.377	0.025	1.363	1.394	98	28
Car–N(3)	Car–NH–C#							
	(Nsp^2: planar)	1.353	1.353	0.007	1.347	1.359	16	23
	(Nsp^3: pyramidal)	1.419	1.423	0.017	1.412	1.432	8	22
	(overall)	1.380	1.364	0.032	1.353	1.412	31	28
	Car–N–(C#)$_2$							
	(Nsp^2: planar)	1.371	1.370	0.016	1.363	1.382	41	23
	(Nsp^3: pyramidal)	1.426	1.425	0.011	1.421	1.431	22	22
	(overall)	1.390	1.385	0.030	1.366	1.420	69	28
	in indole (N1–C7a)	1.372	1.372	0.007	1.367	1.376	40	
	Car–NO$_2$	1.468	1.469	0.014	1.460	1.476	556	
Car–N(2)	Car–N=N	1.431	1.435	0.020	1.422	1.442	26	
Csp^2=N(3)	in furoxan ($^+$N2=C3)	1.316	1.316	0.009	1.311	1.324	14	
Csp^2=N(2)	Car–C=N–C#	1.279	1.279	0.008	1.275	1.285	75	
	(C,H)$_2$–C=N–OH in oximes	1.281	1.280	0.013	1.273	1.288	67	
	S–C=N–X	1.302	1.302	0.021	1.285	1.319	36	
	in pyrazole (N2=C3)	1.329	1.331	0.014	1.315	1.339	20	
	in imidazole (C2=N3)	1.313	1.314	0.011	1.307	1.319	44	
	in isoxazole (N2=C3)	1.314	1.315	0.009	1.305	1.320	9	

Bond	Substructure	d	m	σ	q_1	q_u	n	Note
	in furazan (N2=C3, C4=N5)	1.298	1.299	0.006	1.294	1.303	12	
	in furoxan (C4=N5)	1.304	1.306	0.008	1.300	1.308	14	
$Car \simeq N(3)$	$C \simeq N^+$–H (pyrimidinium)	1.335	1.334	0.015	1.325	1.342	30	
	$C \simeq N^+$–C* (pyrimidinium)	1.346	1.346	0.010	1.340	1.352	64	
	$C \simeq N^+$–O⁻ (pyrimidinium)	1.362	1.359	0.013	1.353	1.369	56	
$Car \simeq N(2)$	$C \simeq N$ (pyridine)	1.337	1.338	0.012	1.330	1.344	269	
	$C \simeq N$ (pyrazine)	1.336	1.335	0.022	1.319	1.347	120	
	$C \simeq N \simeq C$ (pyrimidine)	1.339	1.338	0.015	1.333	1.342	28	
	$N \simeq C \simeq N$ (pyrimidine)	1.333	1.335	0.013	1.326	1.337	28	
	$C \simeq N$ (pyrimidine) (overall)	1.336	1.337	0.014	1.331	1.339	56	
	in any 6-membered N-containing aromatic ring:							
	H–C$\simeq$N$\simeq$C–H	1.334	1.334	0.014	1.327	1.341	146	
	H–C$\simeq$N$\simeq$C–C*	1.339	1.341	0.013	1.336	1.345	38	
	C*–C$\simeq$N$\simeq$C–C*	1.345	1.345	0.008	1.342	1.348	24	
	C$\simeq$N$\simeq$C (overall)	1.336	1.337	0.014	1.329	1.344	204	
$Csp^1 \equiv N(2)$	X–S–N$\equiv$C⁻ (isothiocyanide)	1.144	1.147	0.006	1.140	1.148	6	
$Csp^1 \equiv N(1)$	C*–C$\equiv$N	1.136	1.137	0.010	1.131	1.142	140	
	C=C–C$\equiv$N in TCNQ	1.144	1.144	0.008	1.139	1.149	284	19
	Car–C$\equiv$N	1.138	1.138	0.007	1.133	1.143	31	
	X–C$\equiv$N	1.144	1.141	0.012	1.138	1.151	10	
	(S–C$\equiv$N)⁻	1.155	1.156	0.012	1.147	1.165	14	
Csp^3–O(2)	in alcohols							
	CH₃–OH	1.413	1.414	0.018	1.395	1.425	17	
	C–CH₂–OH	1.426	1.426	0.011	1.420	1.431	75	
	C₂–CH–OH	1.432	1.431	0.011	1.425	1.439	266	
	C₃–C–OH	1.440	1.440	0.012	1.432	1.449	106	
	C*–OH (overall)	1.432	1.431	0.013	1.424	1.441	464	
	in dialkyl ethers							29
	CH₃–O–C*	1.416	1.418	0.016	1.405	1.426	110	
	C–CH₂–O–C*	1.426	1.424	0.011	1.418	1.435	34	
	C₂–CH–O–C*	1.429	1.430	0.010	1.420	1.437	53	
	C₃–C–O–C*	1.452	1.450	0.011	1.445	1.458	39	
	C*–O–C* (overall)	1.426	1.425	0.019	1.414	1.437	236	5
	in aryl alkyl ethers							29
	CH₃–O–Car	1.424	1.424	0.012	1.417	1.431	616	
	C–CH₂–O–Car	1.431	1.430	0.013	1.422	1.438	188	
	C₂–CH–O–Car	1.447	1.446	0.020	1.435	1.466	58	
	C₃–C–O–Car	1.470	1.469	0.018	1.456	1.483	55	
	C*–O–Car (overall)	1.429	1.427	0.018	1.419	1.436	917	
	in alkyl esters of carboxylic acids							12,29
	CH₃–O–C(=O)–C*	1.448	1.449	0.010	1.442	1.455	200	
	C–CH₂–O–C(=O)–C*	1.452	1.453	0.009	1.445	1.458	32	
	C₂–CH–O–C(=O)–C*	1.460	1.460	0.010	1.454	1.465	78	
	C₃–C–O–C(=O)–C*	1.477	1.475	0.008	1.472	1.484	6	
	C*–O–C(=O)–C* (overall)	1.450	1.451	0.014	1.442	1.459	314	
	in alkyl esters of α,β-unsaturated acids:							
	C*–O–C(=O)–C=C (overall)	1.453	1.452	0.013	1.444	1.459	112	
	in alkyl esters of benzoic acid							
	C*–O–C(=O)–C(phenyl) (overall)	1.454	1.454	0.012	1.446	1.463	219	
	in ring systems							
	oxirane (epoxides) (any subst.)	1.446	1.446	0.014	1.438	1.456	498	9
	oxetane (any subst.)	1.463	1.460	0.015	1.451	1.474	16	
	tetrahydrofuran (C,H subst.)	1.442	1.441	0.017	1.430	1.451	154	
Csp^3–O(2)	tetrahydropyran (C,H subst.)	1.441	1.442	0.015	1.431	1.451	22	
	β-lactones: C*–O–C(=O)	1.492	1.494	0.010	1.481	1.501	4	16
	γ-lactones: C*–O–C(=O)	1.464	1.464	0.012	1.455	1.473	110	12
	δ-lactones: C*–O–C(=O)	1.461	1.464	0.017	1.452	1.473	27	12
	O–C–O system in *gem*-diols, and pyranose and furanose sugars:							30,31
	HO–C*–OH	1.397	1.401	0.012	1.388	1.405	18	
	C₅–O₅–C₁–O₁H in pyranoses							
	O₁ axial (α):							
	C₅–O₅	1.439	1.440	0.008	1.432	1.445	29	
	O₅–C₁	1.427	1.426	0.012	1.421	1.432	29	
	C₁–O₁	1.403	1.400	0.012	1.391	1.412	29	
	O₁ equatorial (β):							
	C₅–O₅	1.435	1.436	0.008	1.429	1.440	17	
	O₅–C₁	1.430	1.431	0.010	1.424	1.436	17	
	C₁–O₁	1.393	1.393	0.007	1.386	1.399	17	
	α + β (overall):							
	C₅–O₅	1.439	1.440	0.008	1.432	1.446	60	
	O₅–C₁	1.430	1.429	0.012	1.421	1.436	60	
	C₁–O₁	1.401	1.399	0.011	1.392	1.407	60	

Bond	Substructure	d	m	σ	q_1	q_u	n	Note
	$C_4-O_4-C_1-O_1H$ in furanoses (overall values)							
	C_4-O_4	1.442	1.446	0.012	1.436	1.449	18	
	O_4-C_1	1.432	1.432	0.012	1.421	1.443	18	
	C_1-O_1	1.404	1.405	0.013	1.397	1.409	18	
	$C_5-O_5-C_1-O_1-C^*$ in pyranoses							
	O_1 axial (α):							
	C_5-O_5	1.439	1.438	0.010	1.433	1.446	67	
	O_5-C_1	1.417	1.417	0.009	1.410	1.424	67	
	C_1-O_1	1.409	1.409	0.014	1.401	1.417	67	
	O_1-C^*	1.435	1.435	0.013	1.427	1.443	67	
	O_1 equatorial (β):							
	C_5-O_5	1.434	1.435	0.006	1.429	1.439	39	
	O_5-C_1	1.424	1.424	0.008	1.418	1.431	39	
	C_1-O_1	1.390	1.390	0.011	1.381	1.400	39	
	O_1-C^*	1.437	1.438	0.013	1.428	1.445	39	
	$\alpha + \beta$ (overall):							
	C_5-O_5	1.436	1.436	0.009	1.431	1.442	126	
	O_5-C_1	1.419	1.419	0.011	1.412	1.426	126	
	C_1-O_1	1.402	1.403	0.016	1.391	1.413	126	
	O_1-C^*	1.436	1.436	0.013	1.428	1.445	126	
	$C_4-O_4-C_1-O_1-C^*$ in furanoses (overall values)							
	C_4-O_4	1.443	1.445	0.013	1.429	1.453	23	
	O_4-C_1	1.421	1.418	0.012	1.413	1.431	23	
	C_1-O_1	1.410	1.409	0.014	1.401	1.420	23	
	O_1-C^*	1.439	1.437	0.014	1.429	1.449	23	
	Miscellaneous:							
	$C\#-O-SiX_3$	1.416	1.416	0.017	1.405	1.428	29	
	C^*-O-SO_2-C	1.465	1.461	0.014	1.454	1.475	33	
$Csp^2-O(2)$	in enols: $C=C-OH$	1.333	1.331	0.017	1.324	1.342	53	
	in enol esters: $C=C-O-C^*$	1.354	1.353	0.016	1.341	1.363	40	
	in acids:							
	$C^*-C(=O)-OH$	1.308	1.311	0.019	1.298	1.320	174	
	$C=C-C(=O)-OH$	1.293	1.295	0.019	1.279	1.307	22	
	$Car-C(=O)-OH$	1.305	1.311	0.020	1.291	1.317	75	
	in esters:							
	$C^*-C(=O)-O-C^*$	1.336	1.337	0.014	1.328	1.346	551	12,29
	$C=C-C(=O)-O-C^*$	1.332	1.331	0.011	1.324	1.339	112	
	$Car-C(=O)-O-C^*$	1.337	1.335	0.013	1.329	1.344	219	12
	$C^*-C(=O)-O-C=C$	1.362	1.359	0.018	1.351	1.374	26	
	$C^*-C(=O)-O-C=C$	1.407	1.405	0.017	1.394	1.420	26	
	$C^*-C(=O)-O-Car$	1.360	1.359	0.011	1.355	1.367	40	12
	in anhydrides: $O=C-O-C=O$	1.386	1.386	0.011	1.379	1.393	70	
	in ring systems:							
	furan (O1–C2, O1–C5)	1.368	1.369	0.015	1.359	1.377	125	
	isoxazole (O1–C5)	1.354	1.354	0.010	1.345	1.360	9	
	β-lactones: $C^*-C(=O)-O-C^*$	1.359	1.359	0.013	1.348	1.371	4	13
	γ-lactones: $C^*-C(=O)-O-C^*$	1.350	1.349	0.012	1.342	1.359	110	12
	δ-lactones: $C^*-C(=O)-O-C^*$	1.339	1.339	0.016	1.332	1.347	27	12
$Car-O(2)$	in phenols: $Car-OH$	1.362	1.364	0.015	1.353	1.373	551	
	in aryl alkyl ethers: $Car-O-C^*$	1.370	1.370	0.011	1.363	1.377	920	29,32
$Car-O(2)$	in diaryl ethers: $Car-O-Car$	1.384	1.381	0.014	1.375	1.391	132	
	in esters: $Car-O-C(=O)-C^*$	1.401	1.401	0.010	1.394	1.408	40	12
$Csp^2=O(1)$	in aldehydes and ketones:							
	$C^*-CH=O$	1.192	1.192	0.005	1.188	1.197	7	
	$(C^*)_2-C=O$	1.210	1.210	0.008	1.206	1.215	474	5
	$(C\#)_2-C=O$							
	in cyclobutanones	1.198	1.198	0.007	1.194	1.204	12	
	in cyclopentanones	1.208	1.208	0.007	1.203	1.212	155	
	in cyclohexanones	1.211	1.211	0.009	1.207	1.216	312	
	$C=C-C=O$	1.222	1.222	0.010	1.216	1.229	225	
	$(C=C)_2-C=O$	1.233	1.229	0.010	1.226	1.242	28	
	$Car-C=O$	1.221	1.218	0.014	1.212	1.229	85	
	$(Car)_2-C=O$	1.230	1.226	0.015	1.220	1.238	66	
	$C=O$ in benzoquinones	1.222	1.220	0.013	1.211	1.231	86	
	delocalized double bonds in carboxylate anions:							
	$H-C\simeq O_2^-$ (formate)	1.242	1.243	0.012	1.234	1.252	24	
	$C^*-C\simeq O_2^-$	1.254	1.253	0.010	1.247	1.261	114	
	$C=C-C\simeq O_2^-$	1.250	1.248	0.017	1.238	1.261	52	
	$Car-C\simeq O_2^-$	1.255	1.253	0.010	1.249	1.262	22	
	$HOOC-C\simeq O_2^-$ (hydrogen oxalate)	1.243	1.247	0.015	1.232	1.256	26	
	$^-O_2\simeq C-C\simeq O_2^-$ (oxalate)	1.251	1.251	0.007	1.248	1.254	18	

Bond	Substructure	d	m	σ	q_1	q_u	n	Note
	in carboxylic acids (X–COOH)							
	C*–C(=O)–OH	1.214	1.214	0.019	1.203	1.224	175	
	C=C–C(=O)–OH	1.229	1.226	0.017	1.218	1.237	22	
	Car–C(=O)–OH	1.226	1.223	0.020	1.211	1.241	75	
	in esters:							
	C*–C(=O)–O–C*	1.196	1.196	0.010	1.190	1.202	551	12
	C=C–C(=O)–O–C*	1.199	1.198	0.009	1.193	1.203	113	
	Car–C(=O)–O–C*	1.202	1.201	0.009	1.196	1.207	218	12
	C*–C(=O)–O–C=C	1.190	1.190	0.014	1.184	1.198	26	
	C*–C(=O)–O–Car	1.187	1.188	0.011	1.181	1.195	40	12
	in anhydrides: O=C–O–C=O	1.187	1.187	0.010	1.184	1.193	70	
	in β-lactones: C*–C(=O)–O–C*	1.193	1.193	0.006	1.187	1.198	4	13
	γ-lactones: C*–C(=O)–O–C*	1.201	1.202	0.009	1.196	1.206	109	12
	δ-lactones: C*–C(=O)–O–C*	1.205	1.207	0.008	1.201	1.209	27	12
	in amides:							
	NH$_2$–C(–C*)=O	1.234	1.233	0.012	1.225	1.243	32	14
	(C*–)(C*,H–)N–C(–C*)=O	1.231	1.231	0.012	1.224	1.238	378	14
	β-lactams: C*–NH–C=O	1.198	1.200	0.012	1.193	1.204	23	13
	γ-lactams:							
	C*–NH–C=O	1.235	1.235	0.008	1.232	1.240	20	13
	C*–N(–C*)–C=O	1.225	1.226	0.011	1.217	1.233	15	13
	δ-lactams:							
	C*–NH–C=O	1.240	1.241	0.003	1.237	1.243	6	14
	C*–N(–C*)–C=O	1.233	1.233	0.007	1.229	1.239	15	14
	in ureas:							
	(NH$_2$)$_2$–C=O	1.256	1.256	0.007	1.249	1.261	24	25,26
	(C#–NH)$_2$–C=O	1.241	1.237	0.011	1.235	1.245	13	25
	[(C#)$_n$–N]$_2$–C=O	1.230	1.230	0.007	1.224	1.234	20	25,27
Csp^3–P(4)	C$_3$–P$^+$–C*	1.800	1.802	0.015	1.790	1.812	35	33
	C$_2$–P(=O)–CH$_3$	1.791	1.790	0.006	1.786	1.795	10	
	C$_2$–P(=O)–CH$_2$–C	1.806	1.806	0.009	1.801	1.813	45	
	C$_2$–P(=O)–CH–C$_2$	1.821	1.821	0.009	1.815	1.828	15	
	C$_2$–P(=O)–C–C$_3$	1.841	1.842	0.008	1.835	1.847	14	
	C$_2$–P(=O)–C* (overall)	1.813	1.811	0.017	1.800	1.822	84	
Csp^3–P(3)	C$_2$–P–C*	1.855	1.857	0.019	1.840	1.870	23	
Car–P(4)	C$_3$–P$^+$–Car	1.793	1.792	0.011	1.786	1.800	276	
	C$_2$–P(=O)–Car	1.801	1.802	0.011	1.796	1.807	98	
	Ph$_3$–P=N$^+$=P–Ph$_3$	1.795	1.795	0.008	1.789	1.800	197	
Car–P(3)	C$_2$–P–Car	1.836	1.837	0.010	1.830	1.844	102	
	(N$\simeq$)$_2$P–Car (P $\simeq$ N aromatic)	1.795	1.793	0.011	1.788	1.803	43	
Csp^3–S(4)	C*–SO$_2$–C (C* = CH$_3$ excluded)	1.786	1.782	0.018	1.774	1.797	75	
	C*–SO$_2$–C (overall)	1.779	1.778	0.020	1.764	1.790	94	
	C*–SO$_2$–O–X	1.745	1.744	0.009	1.738	1.754	7	34
	C*–SO$_2$–N–X$_2$	1.758	1.756	0.018	1.746	1.773	17	34
Csp^3–S(3)	C*–S(=O)–C (C* = CH$_3$ excluded)	1.818	1.814	0.024	1.802	1.829	69	
	C*–S(=O)–C (overall)	1.809	1.806	0.025	1.793	1.820	88	
	CH$_3$–S$^+$–X$_2$	1.786	1.787	0.007	1.779	1.792	21	
	C*–S$^+$–X$_2$ (C* = CH$_3$ excluded)	1.823	1.820	0.016	1.812	1.834	18	
	C*–S$^+$–X$_2$ (overall)	1.804	1.794	0.025	1.788	1.820	41	
Csp^3–S(2)	C*–SH	1.808	1.805	0.010	1.800	1.819	6	
	CH$_3$–S–C*	1.789	1.787	0.008	1.784	1.794	9	
Csp^3–S(2)	C–CH$_2$–S–C*	1.817	1.816	0.013	1.808	1.824	92	
	C$_2$–CH–S–C*	1.819	1.819	0.011	1.811	1.825	32	
	C$_3$–C–S–C*	1.856	1.860	0.011	1.854	1.863	26	
	C*–S–C* (overall)	1.819	1.817	0.019	1.809	1.827	242	
	in thiirane	1.834	1.835	0.025	1.810	1.858	4	9
	in thietane: see ZCMXSP (1.817, 1.844)							
	in tetrahydrothiophene	1.827	1.826	0.018	1.811	1.837	20	
	in tetrahydrothiopyran	1.823	1.821	0.014	1.812	1.832	24	
	C–CH$_2$–S–S–X	1.823	1.820	0.014	1.813	1.832	41	
	C$_3$–C–S–S–X	1.863	1.865	0.015	1.848	1.878	11	
	C*–S–S–X (overall)	1.833	1.828	0.022	1.818	1.848	59	
Csp^2–S(2)	C=C–S–C*	1.751	1.755	0.017	1.740	1.764	61	
	C=C–S–C=C (in tetrathiafulvalene)	1.741	1.741	0.011	1.733	1.750	88	
	C=C–S–C=C (in thiophene)	1.712	1.712	0.013	1.703	1.722	60	
	O=C–S–C#	1.762	1.759	0.018	1.747	1.778	20	
Car–S(4)	Car–SO$_2$–C	1.763	1.764	0.009	1.756	1.769	96	
	Car–SO$_2$–O–X	1.752	1.750	0.008	1.749	1.756	27	
	Car–SO$_2$–N–X$_2$	1.758	1.759	0.013	1.749	1.765	106	35
Car–S(3)	Car–S(=O)–C	1.790	1.790	0.010	1.783	1.798	41	
	Car–S$^+$–X$_2$	1.778	1.779	0.010	1.771	1.787	10	
Car–S(2)	Car–S–C*	1.773	1.774	0.009	1.765	1.779	44	
	Car–S–Car	1.768	1.767	0.010	1.762	1.774	158	
	Car–S–Car (in phenothiazine)	1.764	1.764	0.008	1.760	1.769	48	

Bond	Substructure	d	m	σ	q_1	q_u	n	Note
	Car–S–S–X	1.777	1.777	0.012	1.767	1.785	47	
Csp^1–S(2)	N≡C–S–X	1.679	1.683	0.026	1.645	1.698	10	
Csp^1–S(1)	(N≡C–S)⁻	1.630	1.630	0.014	1.619	1.641	14	
Csp^2=S(1)	(C*)$_2$–C=S: see IPMUDS (1.599)							
	(Car)$_2$–C=S: see CELDOM (1.611)							
	(X)$_2$–C=S (X = C,N,O,S)	1.671	1.675	0.024	1.656	1.689	245	
	X$_2$N–C(=S)–S–X	1.660	1.660	0.016	1.648	1.674	38	
	(X$_2$N)$_2$–C=S (thioureas)	1.681	1.684	0.020	1.669	1.693	96	
	N–C($\simeq$S)$_2$	1.720	1.721	0.012	1.709	1.731	20	
Csp^3–Se	C#–Se	1.970	1.967	0.032	1.948	1.998	21	
Csp^2–Se(2)	C=C–Se–C=C (in tetraselenafulvalene)	1.893	1.895	0.013	1.882	1.902	32	
Car–Se(3)	Ph$_3$–Se⁺	1.930	1.929	0.006	1.924	1.936	13	
Csp^3–Si(5)	C#–Si⁻–X$_4$	1.874	1.876	0.015	1.859	1.884	9	
Csp^3–Si(4)	CH$_3$–Si–X$_3$	1.857	1.857	0.018	1.848	1.869	552	
	C*–Si–X$_3$ (C* = CH$_3$ excluded)	1.888	1.887	0.023	1.872	1.905	124	
	C*–Si–X$_3$ (overall)	1.863	1.861	0.024	1.850	1.875	681	
Car–Si(4)	Car–Si–X$_3$	1.868	1.868	0.014	1.857	1.878	178	
Csp^1–Si(4)	C≡C–Si–X$_3$	1.837	1.840	0.012	1.824	1.849	8	
Csp^3–Te	C#–Te	2.158	2.159	0.030	2.128	2.177	13	
Car–Te	Car–Te	2.116	2.115	0.020	2.104	2.130	72	
Csp^2=Te	see CEDCUJ (2.044)							
Cl–Cl	see PHASCL (2.306, 2.227)							
Cl–I	see CMBIDZ (2.563), HXPASC (2.541, 2.513), METAMM (2.552), BQUINI (2.416, 2.718)							
Cl–N	see BECTAE (1.743—1.757), BOGPOC (1.705)							
Cl–O(1)	in ClO$_4^-$	1.414	1.419	0.026	1.403	1.431	252	
Cl–P	(N$\simeq$)$_2$P–Cl (N$\simeq$P aromatic)	1.997	1.994	0.015	1.989	2.004	46	
	Cl–P (overall)	2.008	2.001	0.035	1.986	2.028	111	
Cl–S	Cl–S (overall)	2.072	2.079	0.023	2.047	2.091	6	
	see also longer bonds in CILSAR (2.283), BIHXIZ (2.357), CANLUY (2.749)							
Cl–Se	see BIRGUE10, BIRHAL10, CTCNSE (2.234—2.851)							
Cl–Si(4)	Cl–Si–X$_3$ (monochloro)	2.072	2.075	0.009	2.066	2.078	5	
	Cl$_2$–Si–X$_2$ and Cl$_3$–Si–X	2.020	2.012	0.015	2.007	2.036	5	
Cl–Te	Cl–Te in range 2.34—2.60	2.520	2.515	0.034	2.493	2.537	22	36
	see also longer bonds in BARRIV, BOJPUL, CETUTE, EPHTEA, OPNTEC10 (2.73—2.94)							
F–N(3)	F–N–C$_2$ and F$_2$–N–C	1.406	1.404	0.016	1.395	1.416	9	
F–P(6)	in hexafluorophosphate, PF$_6^-$	1.579	1.587	0.025	1.563	1.598	72	
F–P(3)	(N$\simeq$)$_2$P–F (N$\simeq$P aromatic)	1.495	1.497	0.016	1.481	1.510	10	
F–S	43 observations in range 1.409—1.770 in a wide variety of environments; F–S(6) in F$_2$–SO$_2$–C$_2$ (see FPSULF10, BETJOZ)	1.640	1.646	0.011	1.626	1.649	6	
	F–S(4) in F$_2$–S(=O)–N (see BUDTEZ)	1.527	1.528	0.004	1.524	1.530	24	37
F–Si(6)	in SiF$_6^{2-}$	1.694	1.701	0.013	1.677	1.703	6	
F–Si(5)	F–Si⁻–X$_4$	1.636	1.639	0.035	1.602	1.657	10	
F–Si(4)	F–Si–X$_3$	1.588	1.587	0.014	1.581	1.599	24	
F–Te	see CUCPIZ (F–Te(6) = 1.942, 1.937), FPHTEL (F–Te(4) = 2.006)							
H–N(4)	X$_3$–N⁺–H	1.033	1.036	0.022	1.026	1.045	87	21
H–N(3)	X$_2$–N–H	1.009	1.010	0.019	0.997	1.023	95	21
H–O(2)	in alcohols C*–O–H	0.967	0.969	0.010	0.959	0.974	63	21
	C#–O–H	0.967	0.970	0.010	0.959	0.974	73	21
	in acids O=C–O–H	1.015	1.017	0.017	1.001	1.031	16	21,38
I–I	in I$_3^-$	2.917	2.918	0.011	2.907	2.927	6	
I–N	see BZPRIB, CMBIDZ, HMTITI, HMTNTI, IFORAM, IODMAM (2.042—2.475)							
I–O	X–I–O (see BZPRIB, CAJMAB, IBZDAC11)	2.144	2.144	0.028	2.127	2.164	6	
	for IO$_6^-$ see BOVMEE (1.829—1.912)							
I–P(3)	see CEHKAB (2.490—2.493)							†
I–S	see DTHIBR10 (2.687), ISUREA10 (2.629), BZTPPI (3.251)							
I–Te(4)	I–Te–X$_3$	2.926	2.928	0.026	2.902	2.944	8	
N(4)–N(3)	X$_3$–N⁺–N⁰–X$_2$ (N⁰ planar)	1.414	1.414	0.005	1.412	1.418	13	
N(3)–N(3)	(C)(C,H)–N$_a$–N$_b$–(C)(C,H)							5,39
	N$_a$, N$_b$ pyramidal	1.454	1.452	0.021	1.444	1.457	44	40
	N$_a$ pyramidal, N$_b$ planar	1.420	1.420	0.015	1.407	1.433	68	40
	N$_a$, N$_b$ planar	1.401	1.401	0.018	1.384	1.418	40	40
	overall	1.425	1.425	0.027	1.407	1.443	139	
N(3)–N(2)	in pyrazole (N1–N2)	1.366	1.366	0.019	1.350	1.375	20	
	in pyridazinium (N1⁺ $\simeq$ N2)	1.350	1.349	0.010	1.345	1.361	7	

Bond	Substructure	d	m	σ	q_1	q_u	n	Note
$N(2) \simeq N(2)$	$N \simeq N$ (aromatic) in pyridazine							
	with C,H as *ortho* substituents	1.304	1.300	0.019	1.287	1.326	6	
	with N,Cl as *ortho* substituents	1.368	1.373	0.011	1.362	1.375	9	
$N(2)=N(2)$	$C\#-N=N-C\#$							
	cis	1.245	1.244	0.009	1.239	1.252	21	
	trans	1.222	1.222	0.006	1.218	1.227	6	
	(overall)	1.240	1.241	0.012	1.230	1.251	27	
	$Car-N=N-Car$	1.255	1.253	0.016	1.247	1.262	13	
	$X-N=N=N$ (azides)	1.216	1.226	0.028	1.202	1.237	19	
$N(2)=N(1)$	$X-N=N=N$ (azides)	1.124	1.128	0.015	1.114	1.137	19	
$N(3)-O(2)$	$(C,H)_2-N-OH$ (Nsp^2: planar)	1.396	1.394	0.012	1.390	1.401	28	
	$C_2-N-O-C$							
	(Nsp^3: pyramidal)	1.463	1.465	0.012	1.457	1.468	22	
	(Nsp^2: planar)	1.397	1.394	0.011	1.388	1.409	12	
	in furoxan (N2-O1)	1.438	1.436	0.009	1.430	1.447	14	
$N(3)-O(1)$	$(C\simeq)_2N^+-O^-$ in pyridine *N*-oxides	1.304	1.299	0.015	1.291	1.316	11	
	in furoxan ($^+$N2-O6$^-$)	1.234	1.234	0.008	1.228	1.240	14	
$N(2)-O(2)$	in oximes							
	$(C\#)_2-C=N-OH$	1.416	1.418	0.006	1.416	1.420	7	
	$(H)(Csp^2)-C=N-OH$	1.390	1.390	0.011	1.380	1.401	20	
	$(C\#)(Csp^2)-C=N-OH$	1.402	1.403	0.010	1.393	1.410	18	
	$(Csp^2)_2-C=N-OH$	1.378	1.377	0.017	1.365	1.393	16	
	$(C,H)_2-C=N-OH$ (overall)	1.394	1.395	0.018	1.379	1.408	67	
	in furazan (O1-N2, O1-N5)	1.385	1.383	0.013	1.378	1.392	12	
	in furoxan (O1-N5)	1.380	1.380	0.011	1.370	1.388	14	
	in isoxazole (O1-N2)	1.425	1.425	0.010	1.417	1.434	9	
$N(3)=O(1)$	in nitrate ions NO_3^-	1.239	1.240	0.020	1.227	1.251	105	
	in nitro groups							
	C^*-NO_2	1.212	1.214	0.012	1.206	1.221	84	
	$C\#-NO_2$	1.210	1.210	0.011	1.203	1.218	251	
	$Car-NO_2$	1.217	1.218	0.011	1.211	1.215	1 116	
	$C-NO_2$ (overall)	1.218	1.219	0.013	1.210	1.226	1 733	
$N(3)-P(4)$	$X_2-P(=X)-NX_2$							
	Nsp^2: planar	1.652	1.651	0.024	1.634	1.670	205	
	Nsp^3: pyramidal	1.683	1.683	0.005	1.680	1.686	6	
	(overall)	1.662	1.662	0.029	1.639	1.682	358	
	subsets of this group are:							
	$O_2-P(=S)-NX_2$	1.628	1.624	0.015	1.615	1.634	9	
	$C-P(=S)-(NX_2)_2$	1.691	1.694	0.018	1.678	1.703	28	
	$O-P(=S)-(NX_2)_2$	1.652	1.654	0.014	1.642	1.664	28	
	$P(=O)-(NX_2)_3$	1.663	1.668	0.026	1.640	1.679	78	
$N(3)-P(3)$	$-NX-P(-X)-NX-P(-X)-$ (P_2N_2 ring)	1.730	1.721	0.017	1.716	1.748	20	
	$-NX-P(=S)-NX-P(=S)-$ (P_2N_2 ring)	1.697	1.697	0.015	1 690	1.703	44	
	in *P*-substituted phosphazenes:							
	$(N\simeq)_2P-N$ (amino)	1.637	1.638	0.014	1.625	1.651	16	
	(aziridinyl)	1.672	1.674	0.010	1.665	1.676	15	
$N(2)-P(4)$	$Ph_3-P=N^+=P-Ph_3$	1.571	1.573	0.013	1.563	1.580	66	
$N(2)-P(3)$	$Ph_3-P=N-C,S$	1.599	1.597	0.018	1.580	1.615	7	
$N(2) \simeq P(3)$	$N \simeq P$ aromatic							
	in phosphazenes	1.582	1.582	0.019	1.571	1.594	126	
	in $P\simeq N\simeq S$	1.604	1.606	0.009	1.594	1.612	36	
$N(3)-S(4)$	$C-SO_2-NH_2$	1.600	1.601	0.012	1.591	1.610	14	35
	$C-SO_2-NH-C\#$	1.633	1.633	0.019	1.615	1.652	47	35
	$C-SO_2-N-C(\#)_2$	1.642	1.641	0.024	1.623	1.659	38	35
$N(3)-S(2)$	$C-S-NX_2$ Nsp^2: planar	1.710	1.707	0.019	1.698	1.722	22	23
	(for Nsp^3 pyramidal see MODIAZ: 1.765)							
	$X-S-NX_2$ Nsp^2: planar	1.707	1.705	0.012	1.699	1.715	30	23
$N(2)-S(2)$	$C=N-S-X$	1.656	1.663	0.027	1.632	1.677	36	
$N(2) \simeq S(2)$	$N \simeq S$ aromatic in $P \simeq N \simeq S$	1.560	1.558	0.011	1.554	1.563	37	
$N(2)=S(2)$	$N=S$ in $N=S=N$ and $N=S=S$	1.541	1.546	0.022	1.521	1.558	37	
$N(3)-Se$	see COJCUZ (1.830), DSEMOR10 (1.846, 1.852), MORTRS10 (1.841)							
$N(2)-Se$	see SEBZQI (1.805), NAPSEZ10 (1.809, 1.820)							
$N(2)-Se$	see CISMUM (1.790, 1.791)							
$N(3)-Si(5)$	see DMESIP01, BOJLER, CASSAQ, CASYOK, CECXEN, CINTEY, CIPBUY, FMESIB, MNPSIL, PNPOSI (1.973—2.344)							
$N(3)-Si(4)$	$X_3-Si-NX_2$ (overall)	1.748	1.746	0.022	1.735	1.757	170	
	subsets of this group are:							
	$X_3-Si-NHX$	1.714	1.719	0.014	1.702	1.727	16	
	$X_3-Si-NX-Si-X_3$ acyclic	1.743	1.744	0.016	1.731	1.755	45	
	N-Si-N in 4-membered rings	1.742	1.742	0.009	1.735	1.748	53	
	N-Si-N in 5-membered rings	1.741	1.742	0.019	1.726	1.749	33	
$N(2)-Si(4)$	$X_3-Si-N^--Si-X_3$	1.711	1.712	0.019	1.693	1.729	15	

Bond	Substructure	d	m	σ	q_l	q_u	n	Note
N–Te	see ACLTEP (2.402), BIBLAZ (1.980), CESSAU (2.023)							
O(2)–O(2)	C*–O–O–C*,H							
	$\tau(OO) = 70$—$85°$	1.464	1.464	0.009	1.458	1.472	12	
	$\tau(OO)$ ca. $180°$	1.482	1.480	0.005	1.478	1.486	5	
	overall	1.469	1.471	0.012	1.461	1.478	17	
	O=C–O–O–C=O see ACBZPO01 (1.446), CEYLUN (1.452), CIMHIP (1.454)							
	Si–O–O–Si	1.496	1.499	0.005	1.490	1.499	10	
O(2)–P(5)	X–P–(OX)$_4$							41
	trigonal bipyramidal:							
	axial	1.689	1.685	0.024	1.675	1.712	20	
	equatorial	1.619	1.622	0.024	1.604	1.628	20	
	square pyramidal	1.662	1.661	0.020	1.649	1.673	28	
O(2)–P(4)	C–O–P($\simeq$O)$_3{}^{2-}$	1.621	1.622	0.007	1.615	1.628	12	
	(H–O)$_2$–P($\simeq$O)$^-$	1.560	1.561	0.009	1.555	1.566	16	
	(C–O)$_2$–P($\simeq$O)$_2{}^-$	1.608	1.607	0.013	1.599	1.615	16	
	(C#–O)$_3$–P=O	1.558	1.554	0.011	1.550	1.564	30	
	(Car–O)$_3$–P=O	1.587	1.588	0.014	1.572	1.599	19	
	X–O–P(=O)–(C,N)$_2$	1.590	1.585	0.016	1.577	1.601	33	
	(X–O)$_2$–P(=O)–(C,N)	1.571	1.572	0.013	1.563	1.579	70	
O(2)–P(3)	(N$\simeq$)$_2$P–O–C (N $\simeq$ P aromatic)	1.573	1.573	0.011	1.563	1.584	16	
O(1)=P(4)	C–O–P($\simeq$O)$_3{}^{2-}$ (delocalized)	1.513	1.512	0.008	1.508	1.518	42	
	(H–O)$_2$–P($\simeq$O)$_2{}^-$ (delocalized)	1.503	1.503	0.005	1.499	1.508	16	
	(C–O)$_2$–P($\simeq$O)$_2{}^-$ (delocalized)	1.483	1.485	0.008	1.474	1.490	16	
	(C–O)$_3$–P=O	1.449	1.448	0.007	1.446	1.452	18	
	C$_3$–P=O	1.489	1.486	0.010	1.481	1.496	72	
	N$_3$–P=O	1.461	1.462	0.014	1.449	1.470	26	
	(C)$_2$(N)–P=O	1.487	1.489	0.007	1.479	1.493	5	
	(C,N)$_2$(O)–P=O	1.467	1.465	0.007	1.462	1.472	33	
	(C,N)(O)$_2$–P=O	1.457	1.458	0.009	1.454	1.462	35	
O(2)–S(4)	C–O–SO$_2$–C	1.577	1.576	0.015	1.566	1.584	41	
	C–O–SO$_2$–CH$_3$	1.569	1.569	0.013	1.556	1.582	7	
	C–O–SO$_2$–Car	1.580	1.578	0.015	1.571	1.588	27	
O(1)=S(4)	C–SO$_2$–C	1.436	1.437	0.010	1.431	1.442	316	42
	X–SO$_2$–NX$_2$	1.428	1.428	0.010	1.422	1.434	326	
	C–SO$_2$–N–(C,H)$_2$	1.430	1.430	0.009	1.425	1.435	206	
	C–SO$_2$–O–C	1.423	1.423	0.008	1.418	1.428	82	
	in SO$_4{}^{2-}$	1.472	1.473	0.013	1.463	1.481	104	
O(1)=S(3)	C–S(=O)–C	1.497	1.498	0.013	1.489	1.505	90	5
O–Se	see BAPPAJ, BIRGUE10, BIRHAL10, CXMSEO, DGLYSE, SPSEBU (1.597 for O=Se to 1.974 for O–Se)							
O(2)–Si(5)	(X–O)$_3$–Si–(N)(C)	1.663	1.658	0.023	1.650	1.665	21	
O(2)–Si(4)	X$_3$–Si–O–X (overall)	1.631	1.630	0.022	1.617	1.646	191	
O(2)–Si(4)	subsets of this group are:							
	X$_3$–Si–O–C#	1.645	1.647	0.012	1.634	1.652	29	
	X$_3$–Si–O–Si–X$_3$	1.622	1.625	0.014	1.614	1.631	70	
	X$_3$–Si–O–O–Si–X$_3$	1.680	1.676	0.008	1.673	1.688	10	
O(2)–Te(6)	(X–O)$_6$–Te	1.927	1.927	0.020	1.908	1.942	16	
O(2)–Te(4)	(X–O)$_2$–Te–X$_2$	2.133	2.136	0.054	2.078	2.177	12	
P(4)–P(4)	X$_3$–P–P–X$_3$	2.256	2.259	0.025	2.243	2.277	6	
P(4)–P(3)	see CECHEX (2.197), COZPIQ (2.249)							
P(3)–P(3)	X$_2$–P–P–X$_2$	2.214	2.210	0.022	2.200	2.224	41	
P(4)=P(4)	see BUTSUE (2.054)							
P(3)=P(3)	see BALXOB (2.034)							
P(4)=S(1)	C$_3$–P=S	1.954	1.952	0.005	1.950	1.957	13	
	(N,O)$_2$(C)–P=S	1.922	1.924	0.014	1.913	1.927	26	
	(N,O)$_3$–P=S	1.913	1.914	0.014	1.906	1.921	50	
P(4)=Se(1)	X$_3$–P=Se	2.093	2.099	0.019	2.075	2.108	12	
P(3)–Si(4)	X$_2$–P–Si–X$_3$: 3- and 4-rings	2.264	2.260	0.019	2.249	2.283	22	
	excluded (see BOPFER, BOPFIV, CASTOF10, COZVIW: 2.201—2.317)							
P(4)=Te(1)	see MOPHTE (2.356), TTEBPZ (2.327)							
S(2)–S(2)	C–S–S–C							
	$\tau(SS) = 75$—$105°$	2.031	2.029	0.015	2.021	2.038	46	
	$\tau(SS) = 0$—$20°$	2.070	2.068	0.022	2.057	2.077	28	
	(overall)	2.048	2.045	0.026	2.028	2.068	99	
	in polysulphide chain–S–S–S–	2.051	2.050	0.022	2.037	2.065	126	
S(2)–S(1)	X–N=S–S	1.897	1.896	0.012	1.887	1.908	5	
S–Se(4)	see BUWZUO (2.264, 2.269)							
S–Se(2)	X–Se–S (any)	2.193	2.195	0.015	2.174	2.207	9	
S(2)–Si(4)	X$_3$–Si–S–X	2.145	2.138	0.020	2.130	2.158	19	

BOND LENGTHS IN CRYSTALLINE ORGANIC COMPOUNDS (continued)

Bond	Substructure	d	m	σ	q_l	q_u	n	Note
S(2)–Te	X–S–Te (any)	2.405	2.406	0.022	2.383	2.424	10	
	X=S–Te (any)	2.682	2.686	0.035	2.673	2.694	28	
Se(2)–Se(2)	X–Se–Se–X	2.340	2.340	0.024	2.315	2.361	15	
Se(2)–Te(2)	see BAWFUA, BAWGAH (2.524—2.561)							†
Si(4)–Si(4)	X_3–Si–Si–X_3 3-membered rings excluded: see CIHRAM (2.511)	2.359	2.359	0.012	2.349	2.366	42	
Te–Te	see CAHJOK (2.751, 2.704)							

Appendix 1. (Footnotes to Table)

1. Sample dominated by B–CH_3. For longer bonds in B^-–CH_3 see LITMEB10 [B(4)–CH_3 = 1.621—1.644Å].
2. $p(\pi)$–$p(\pi)$ Bonding with Bsp^2 and Nsp^2 coplanar ($\tau BN = 0 \pm 15°$) predominates. See G. Schmidt, R. Boese, and D. Bläser, *Z. Naturforsch.*, 1982, **37b**, 1230.
3. 84 observations range from 1.38 to 1.61 Å and individual values depend on substituents on B and O. For a discussion of borinic acid adducts see S. J. Rettig and J. Trotter, *Can. J. Chem.*, 1982, **60**, 2957.
4. See M. Kaftory in 'The Chemistry of Functional Groups. Supplement D: The Chemistry of Halides, Pseudohalides, and Azides' eds. S. Patai and Z. Rappoport, Wiley: New York, 1983, Part 2, ch. 24.
5. Bonds which are endocyclic or exocyclic to any 3- or 4-membered rings have been omitted from all averages in this section.
6. The overall average given here is for Csp^3–Csp^3 bonds which carry only C or H substituents. The value cited reflects the relative abundance of each 'substitution' group. The 'mean of means' for the 9 subgroups is 1.538 (σ = 0.022) Å.
7. See F. H. Allen, (a) *Acta Crystallogr.*, 1980, **B36**, 81; (b) 1981, **B37**, 890.
8. See F. H. Allen, *Acta Crystallogr.*, 1984, **B40**, 64.
9. See F. H. Allen, *Tetrahedron*, 1982, **38**, 2843.
10. See F. H. Allen, *Tetrahedron*, 1982, **38**, 645.
11. Cyclopropanones and cyclobutanones excluded.
12. See W. B. Schweizer and J. D. Dunitz, *Helv. Chim. Acta*, 1982, **65**, 1547.
13. See L. Norskov-Lauritsen, H.-B. Bürgi, P. Hoffmann, and H. R. Schmidt, *Helv. Chim. Acta*, 1985, **68**, 76.
14. See P. Chakrabarti and J. D. Dunitz, *Helv. Chim. Acta*, 1982, **65**, 1555.
15. See J. L. Hencher in 'The Chemistry of the C≡C Triple Bond,' ed. S. Patai, Wiley, New York, 1978, ch. 2.
16. Conjugated: torsion angle about central C–C single bond is 0 ± 20° (*cis*) or 180 ± 20° (*trans*).
17. Unconjugated: torsion angle about central C–C single bond is 20—160°.
18. Other conjugative substituents excluded.
19. TCNQ is tetracyanoquinodimethane.
20. No difference detected between C2≃C3 and C3≃C4 bonds.
21. Derived from neutron diffraction results only.
22. Nsp^3: pyramidal; mean valence angle at N is in range 108—114°.
23. Nsp^2: planar; mean valence angle at N is ⩾117.5°.
24. Cyclic and acyclic peptides.
25. See R. H. Blessing, *J. Am. Chem. Soc.*, 1983, **105**, 2776.
26. See L. Lebioda, *Acta Crystallogr.*, 1980, **B36**, 271.
27. n = 3 or 4, *i.e.* tri- or tetra-substituted ureas.
28. Overall value also includes structures with mean valence angle at N in the range 115—118°.
29. See F. H. Allen and A. J. Kirby, *J. Am. Chem. Soc.*, 1984, **106**, 6197.
30. See A. J. Kirby, 'The Anomeric Effect and Related Stereoelectronic Effects at Oxygen,' Springer, Berlin, 1983.
31. See B. Fuchs, L. Schleifer, and E. Tartakovsky, *Nouv. J. Chim.*, 1984, **8**, 275.
32. See S. C. Nyburg and C. H. Faerman, *J. Mol. Struct.*, 1986, **140**, 347.
33. Sample dominated by P–CH_3 and P–CH_2–C.
34. Sample dominated by C* = methyl.
35. See A. Kalman, M. Czugler, and G. Argay, *Acta Crystallogr.*, 1981, **B37**, 868.
36. Bimodal distribution resolved into 22 'short' bonds and 5 longer outliers.
37. All 24 observations come from BUDTEZ.
38. 'Long' O–H bonds in centrosymmetric O – – – H – – – O H-bonded dimers are excluded.
39. N–N bond length also dependent on torsion angle about N–N bond and on nature of substituent C atoms; these effects are ignored here.
40. N pyramidal has average angle at N in range 100—113.5°; N planar has average angle of ⩾117.5°.
41. See R. R. Holmes and J. A. Deiters, *J. Amer. Chem. Soc.*, 1977, **99**, 3318.
42. No detectable variation in S=O bond length with type of C-substituent.

Appendix 2.

Short-form references to individual CSD entries cited by reference code in the Table. A full list of CSD bibliographic entries is given in SUP 56701.

ACBZPO01	*J. Am. Chem. Soc.*, 1975, **97**, 6729.	BIBLAZ	*Zh. Strukt. Khim.*, 1981, **22**, 118.
ACLTEP	*J. Organomet. Chem.*, 1980, **184**, 417.	BICGEZ	*Z. Anorg. Allg. Chem.*, 1982, **486**, 90.
ASAZOC	*Dokl. Akad. Nauk SSSR*, 1979, **249**, 120.	BIHXIZ	*J. Chem. Soc., Chem. Commun.*, 1982, 982.
BALXOB	*J. Am. Chem. Soc.*, 1981, **103**, 4587.	BIRGUE10	*Z. Naturforsch., Teil B*, 1983, **38**, 20.
BAPPAJ	*Inorg. Chem.*, 1981, **20**, 3071.	BIRHAL10	*Z. Naturforsch., Teil B*, 1982, **37**, 1410.
BARRIV	*Acta Chem. Scand., Ser. A*, 1981, **35**, 443.	BIZJAV	*J. Organomet. Chem.*, 1982, **238**, C1.
BAWFUA	*Cryst. Struct. Commun.*, 1981, **10**, 1345.	BOGPOC	*Z. Naturforsch., Teil B*, 1982, **37**, 1402.
BAWGAH	*Cryst. Struct. Commun.*, 1981, **10**, 1353.	BOGSUL	*Z. Naturforsch., Teil B*, 1982, **37**, 1230.
BECTAE	*J. Org. Chem.*, 1981, **46**, 5048, 1981.	BOJLER	*Z. Anorg. Allg. Chem.*, 1982, **493**, 53.
BELNIP	*Z. Naturforsch., Teil B*, 1982, **37**, 299.	BOJPUL	*Acta Chem. Scand., Ser. A*, 1982, **36**, 829.
BEMLIO	*Chem. Ber.*, 1982, **115**, 1126.	BOPFER	*Chem. Ber.*, 1983, **116**, 146.
BEPZEB	*Cryst. Struct. Commun.*, 1982, **11**, 175.	BOPFIV	*Chem. Ber.*, 1983, **116**, 146.
BETJOZ	*J. Am. Chem. Soc.*, 1982, **104**, 1683.	BOVMEE	*Acta Crystallogr., Sect. B*, 1982, **38**, 1048.
BETUTE10	*Acta Chem. Scand., Ser. A*, 1976, **30**, 719.	BQUINI	*Acta Crystallogr., Sect. B*, 1979, **35**, 1930.

BTUPTE	*Acta Chem. Scand., Ser. A*, 1975, **29**, 738.		CUGBAH	*Acta Crystallogr., Sect. C*, 1985, **41**, 476.
BUDTEZ	*Z. Naturforsch., Teil B*, 1983, **38**, 454.		CXMSEO	*Acta Crystallogr., Sect. B*, 1973, **29**, 595.
BUPSIB10	*Z. Anorg. Allg. Chem.*, 1981, **474**, 31.		DGLYSE	*Acta Crystallogr., Sect. B*, 1975, **31**, 1785.
BUSHAY	*Z. Naturforsch., Teil. B*, 1983, **38**, 692.		DMESIP01	*Acta Crystallogr., Sect. C*, 1984, **40**, 895.
BUTHAZ10	*Inorg. Chem.*, 1984, **23**, 2582.		DSEMOR10	*J. Chem. Soc., Dalton Trans.*, 1980, 628.
BUTSUE	*J. Chem. Soc., Chem. Commun.*, 1983, 862.		DTHIBR10	*Inorg. Chem.*, 1971, **10**, 697.
BUWZUO	*Acta Chem. Scand., Ser A*, 1983, **37**, 219.		EPHTEA	*Inorg. Chem.*, 1980, **19**, 2487.
BZPRIB	*Z. Naturforsch., Teil B*, 1981, **36**, 922.		ESEARS	*J. Chem. Soc. C*, 1971, 1511.
BZTPPI	*Inorg. Chem.*, 1978, **17**, 894.		ETEARS	*J. Chem. Soc. C*, 1971, 1511.
CAHJOK	*Inorg. Chem.*, 1983, **22**, 1809.		FMESIB	*J. Organomet. Chem.*, 1980, **197**, 275.
CAJMAB	*Chem. Z*, 1983, **107**, 169.		FPHTEL	*J. Chem. Soc., Dalton Trans.*, 1980, 2306.
CANLUY	*Tetrahedron Lett.*, 1983, **24**, 4337.		FPSULF10	*J. Am. Chem. Soc.*, 1982, **104**, 1683.
CASSAQ	*J. Struct. Chem.*, 1983, **2**, 101.		HCLENE10	*Acta Crystallogr., Sect. B*, 1982, **38**, 3139.
CASTOF10	*Acta Crystallogr., Sect. C*, 1984, **40**, 1879.		HMTITI	*Acta Crystallogr., Sect. B*, 1975, **31**, 1505.
CASYOK	*J. Struct. Chem.*, 1983, **2**, 107.		HMTNTI	*Z. Anorg. Allg. Chem.*, 1974, **409**, 237.
CECHEX	*Z. Anorg. Allg. Chem.*, 1984, **508**, 61.		HXPASC	*J. Chem. Soc., Dalton Trans.*, 1975, 1381.
CECXEN	*J. Struct. Chem.*, 1983, **2**, 207.		IBZDAC11	*J. Chem. Soc., Dalton Trans.*, 1979, 854.
CEDCUJ	*J. Org. Chem.*, 1983, **48**, 5149.		IFORAM	*Monatsh. Chem.*, 1974, **105**, 621.
CEHKAB	*Z. Naturforsch., Teil B*, 1984, **39**, 139.		IODMAM	*Acta Crystallogr., Sect. B*, 1977, **33**, 3209.
CELDOM	*Acta Crystallogr., Sect. C*, 1984, **40**, 556.		IPMUDS	*Acta Crystallogr., Sect. B*, 1973, **29**, 2128.
CESSAU	*Acta Crystallogr., Sect. C*, 1984, **40**, 653.		ISUREA10	*Acta Crystallogr., Sect. B*, 1972, **28**, 643.
CETTAW	*Chem. Ber.*, 1984, **117**, 1089.		LITMEB10	*J. Am. Chem. Soc.*, 1975, **97**, 6401.
CETUTE	*Acta Chem. Scand., Ser A*, 1975, **29**, 763.		MESIAD	*Z. Naturforsch., Teil B*, 1980, **35**, 789.
CEYLUN	*Izv. Akad. Nauk SSSR, Ser. Khim.*, 1983, 2744.		METAMM	*Acta Crystallogr.*, 1964, **17**, 1336.
CIFZUM	*Acta Chem. Scand., Ser A*, 1984, **38**, 289.		MNPSIL	*J. Am. Chem. Soc.*, 1969, **91**, 4134.
CIHRAM	*Angew. Chem., Int. Ed. Engl.*, 1984, **23**, 302.		MODIAZ	*J. Heterocycl. Chem.*, 1980, **17**, 1217.
CILRUK	*J. Chem. Soc., Chem. Commun.*, 1984, 1023.		MOPHTE	*Acta Chem. Scand., Ser. A*, 1980, **34**, 333.
CILSAR	*J. Chem. Soc., Chem. Commun.*, 1984, 1021.		MORTRS10	*J. Chem. Soc., Dalton Trans.*, 1980, 628.
CIMHIP	*Acta Crystallogr., C*, 1984, **40**, 1458.		NAPSEZ10	*J. Am. Chem. Soc.*, 1980, **102**, 5070.
CINTEY	*Dokl. Akad. Nauk SSSR*, 1984, **274**, 615.		NBBZAM	*Z. Naturforsch., Teil B*, 1977, **32**, 1416.
CIPBUY	*J. Struct. Chem.*, 1983, **2**, 281.		OPIMAS	*Aust. J. Chem.*, 1977, **30**, 2417.
CISMUM	*Z. Naturforsch., Teil B*, 1984, **39**, 485.		OPNTEC10	*J. Chem. Soc., Dalton Trans.*, 1982, 251.
CISTED	*Z. Anorg. Allg. Chem.*, 1984, **511**, 95.		PHASCL	*Acta Crystallogr., Sect. B*, 1981, **37**, 1357.
CIWYIQ	*Inorg. Chem.*, 1984, **23**, 1946.		PHASOC01	*Aust. J. Chem.*, 1975, **28**, 15.
CIYFOF	*Inorg. Chem.*, 1984, **23**, 1790.		PNPOSI	*J. Am. Chem. Soc.*, 1968, **90**, 5102.
CMBIDZ	*J. Org. Chem.*, 1979, **44**, 1447.		SEBZQI	*J. Chem. Soc., Chem. Commun.*, 1977, 325.
CODDEE	*Z. Naturforsch., Teil B*, 1984, **39**, 1257.		SPSEBU	*Acta Chem. Scand., Ser. A*, 1979, **33**, 403.
CODDII	*Z. Naturforsch., Teil B*, 1984, **39**, 1257.		TEACBR	*Cryst. Struct. Commun.*, 1974, **3**, 753.
COFVOI	*Z. Naturforsch., Teil B*, 1984, **39**, 1027.		THINBR	*J. Am. Chem. Soc.*, 1970, **92**, 4002.
COJCUZ	*Chem. Ber.*, 1984, **117**, 2686.		TMPBTI	*Acta Crystallogr., Sect. B*, 1975, **31**, 1116.
COSDIX	*Z. Naturforsch., Teil B*, 1984, **39**, 1344.		TPASSN	*J. Chem. Soc., Dalton Trans.*, 1977, 514.
COZPIQ	*Chem. Ber.*, 1984, **117**, 2063.		TPASTB	*Cryst. Struct. Commun.*, 1976, **5**, 39.
COZVIW	*Z. Anorg. Allg. Chem.*, 1984, **515**, 7.		TPHOSI	*Z. Naturforsch., Teil B*, 1979, **34**, 1064.
CTCNSE	*J. Am. Chem. Soc.*, 1980, **102**, 5430.		TTEBPZ	*Z. Naturforsch., Teil B*, 1979, **34**, 256.
CUCPIZ	*J. Am. Chem. Soc.*, 1984, **106**, 7529.		ZCMXSP	*Cryst. Struct. Commun.*, 1977, **6**, 93.
CUDLOC	*J. Cryst. Spectrosc.*, 1985, **15**, 53.			
CUDLUI	*J. Cryst. Spectrosc.*, 1985, **15**, 53.			

BOND LENGTHS AND ANGLES IN GAS-PHASE MOLECULES

This table is reprinted from *Kagaku Benran, 3rd Edition*, Vol. II, pp. 649—661 (1984), with permission of the publisher, Maruzen Company, LTD. (Copyright 1984 by the Chemical Society of Japan). Translation was carried out by Kozo Kuchitsu.

Internuclear distances and bond angles are represented in units of Å (1 Å = 10^{-10} m) and degrees, respectively. The same but inequivalent atoms are discriminated by subscripts a, b, etc. In some molecules ax for axial and eq for equatorial are also used. All measurements were made in the gas phase. The methods used are abbreviated as follows. UV: ultraviolet (including visible) spectroscopy; IR: infrared spectroscopy; R: Raman spectroscopy; MW: microwave spectroscopy; ED: electron diffraction; NMR: nuclear magnetic resonance; LMR: laser magnetic resonance; EPR: electron paramagnetic resonance; MBE: molecular beam electric resonance. If two methods were used jointly for structure determination, they are listed together, as (ED, MW). If the numerical values listed refer to the equilibrium values, they are specified by r_e and θ_e. In other cases the listed values represent various average values in vibrational states; it is frequently the case that they represent the r_s structure derived from several isotopic species for MW or the r_g structure (i.e., the average internuclear distances at thermal equilibrium) for ED. These internuclear distances for the same atom pair with different definitions may sometimes differ as much as 0.01 Å. Appropriate comments are made on the symmetry and conformation in the equilibrium structure.

In general, the numerical values listed in the following tables contain uncertainties in the last digits. However, for certain molecules such as diatomic molecules, with experimental uncertainties of the order of 10^{-5} Å or smaller, numerical values are listed to four decimal places.

REFERENCES

1. L. E. Sutton, ed., *Tables of Interatomic Distances and Configuration in Molecules and Ions*, The Chemical Society Special Publication, No. 11, 18, The Chemical Society (London) (1958, 1965).
2. K.-H. Hellwege, ed., *Landolt-Börnstein Numerical Data and Functional Relations in Science and Technology*, New Series, II/7, J. H. Callomon, E. Hirota, K. Kuchitsu, W. J. Lafferty, A. G. Maki, C. S. Pote, with assistance of I. Buck and B. Starck, *Structure Data of Free Polyatomic Molecules*, Springer-Verlag (1976).
3. K. P. Huber and G. Herzberg, *Molecular Spectra and Molecular Structure IV. Constants of Diatomic Molecules*, Van Nostrand Reinhold Co., London (1979).
4. B. Starck, *Microwave Catalogue and Supplements*.
5. B. Starck, *Electron Diffraction Catalogue and Supplements*.

STRUCTURES OF ELEMENTS AND INORGANIC COMPOUNDS
Compounds are Arranged in Alphabetical Order by their Chemical Formulas
(Lengths in Å and Angles in Degrees)

Compound	Structure					Method
AgBr	Ag—Br (r_e)	2.3931				MW
AgCl	Ag—Cl (r_e)	2.2808				MW
AgF	Ag—F (r_e)	1.9832				MW
AgH	Ag—H (r_e)	1.617				UV
AgI	Ag—I (r_e)	2.5446				MW
AgO	Ag—O (r_e)	2.0030				UV
AlBr	Al—Br (r_e)	2.295				UV
AlCl	Al—Cl (r_e)	2.1301				MW
AlF	Al—F (r_e)	1.6544				MW
AlH	Al—H (r_e)	1.6482				UV
AlI	Al—I (r_e)	2.5371				MW
AlO	Al—O (r_e)	1.6176				UV
Al_2Br_6			Al—Br_a	2.22		ED
			Al—Br_b	2.38		
			$\angle Br_bAlBr_b$	82		
			$\angle Br_aAlBr_a$	118		
			(D_{2h})			
Al_2Cl_6			Al—Cl_a	2.04		ED
			Al—Cl_b	2.24		
			$\angle Cl_bAlCl_b$	87		
			$\angle Cl_aAlCl_a$	122		
			(D_{2h})			
$AsBr_3$	As—Br	2.324	$\angle BrAsBr$	99.6		ED
$AsCl_3$	As—Cl	2.165	$\angle ClAsCl$	98.6		ED, MW
AsF_3	As—F	1.710	$\angle FAsF$	95.9		ED, MW

Compound	Structure							Method
AsF_5	F_b-As structure with F_a, F_b	As—F_a	1.711		As—F_b (D_{3h})	1.656		
AsH_3	As—H (r_e)	1.511			∠HAsH (θ_e)	92.1		MW, IR
AsI_3	As—I	2.557			∠IAsI	100.2		ED
AuH	Au—H (r_e)	1.5237						UV
BBr_3	B—Br	1.893			(D_{3h})			ED
BCl_3	B—Cl	1.742			(D_{3h})			ED
BF	B—F (r_e)	1.2626						UV
BF_2H	B—H	1.189		B—F	1.311	∠FBF	118.3	MW
BF_2OH	B—F	1.32		B—O	1.34	O—H	0.941	MW
	∠FBF	118		∠FBO	123	∠BOH	114.1	
BF_3	B—F	1.313			(D_{3h})			ED, IR
BH	B—H (r_e)	1.2325						UV
BH_3PH_3	B—P	1.937		B—H	1.212	P—H	1.399	MW
	∠PBH	103.6		∠BPH	116.9	∠HBH	114.6	
	∠HPH	101.3		staggered form				
BI_3	B—I	2.118			(D_{3h})			ED
BN	B—N (r_e)	1.281						UV
BO	B—O (r_e)	1.2045						EPR
BO_2	B—O	1.265			linear			UV
BS	B—S	1.6091						UV
B_2H_6	H_a–H_b–H_a B–B structure			B—H_a	1.19			IR, ED
				B—H_b	1.33			
				B···B	1.77			
				∠H_aBH_a	122			
				∠H_bBH_b	97			
$B_3H_3O_3$	B—O	1.376		∠BOB≅∠OBO	120			ED
$B_3H_6N_3$	B—N	1.435		B—H	1.26	N—H	1.05	ED
	∠NBN	118		∠BNB	121	(C_2)		
BaH	Ba—H (r_e)	2.2318						UV
BaO	Ba—O (r_e)	1.9397						MW
BaS	Ba—S (r_e)	2.5074						MBE
BeF	Be—F (r_e)	1.3609						UV
BeH	Be—H (r_e)	1.3431						UV
BeO	Be—O (r_e)	1.3308						UV
BiBr	Bi—Br (r_e)	2.6095						MW
$BiBr_3$	Bi—Br	2.63			∠BrBiBr	90	(C_{3v})	ED
BiCl	Bi—Cl (r_e)	2.4716						MW
$BiCl_3$	Bi—Cl	2.423			∠ClBiCl	100	(C_{3v})	ED
BiF	Bi—F (r_e)	2.0516						MW
BiH	Bi—H (r_e)	1.805						UV
BiI	Bi—I (r_e)	2.8005						MW
BiO	Bi—O (r_e)	1.934						UV
BrCN	C—N (r_e)	1.157			C—Br (r_e)	1.790		IR
BrCl	Br—Cl (r_e)	2.1361						MW
BrF	Br—F (r_e)	1.7590						MW
BrF_3	F_a—Br—F_a with F_b	Br—F_a	1.810		Br—F_b	1.721		MW
		∠F_aBrF_b	86.2		(C_{2v})			
BrF_5	Br—F (average)	1.753						ED, MW
	(Br—F_{eq}) – (Br—F_{ax}) = 0.069							
	∠F_{ax}BrF_{eq}	85.1		(C_{4v})				
BrO	Br—O (r_e)	1.7172						MW
Br_2	Br—Br (r_e)	2.2811						R
CBr_4	C—Br	1.935			(T_d)			ED
CCl	C—Cl	1.6512						UV

Compound	Structure						Method
$CClF_3$	C—Cl	1.752	C—F	1.325	∠FCF	108.6	ED, MW
CCl_3F	C—Cl	1.754	C—F	1.362	∠ClCCl	111	MW
					(C_{3v})		
CCl_4	C—Cl		1.767	(T_d)			ED
CF	C—F (r_e)		1.2718				EPR
CF_3I	C—I	2.138	C—F	1.330	∠FCF	108.1	ED, MW
CF_4	C—F		1.323	(T_d)			ED
CH	C—H (r_e)		1.1199				UV
CI_4	C—I		2.15	(T_d)			ED
CN	C—N (r_e)		1.1718				MW
CO	C—O (r_e)		1.1283				MW
$COBr_2$	C—O		1.178		C—Br	1.923	ED, MW
	∠BrCBr		112.3				
COClF	C—F	1.334	C—O	1.173	C—Cl	1.725	ED, MW
	∠FCCl	108.8	∠ClCO	127.5			
$COCl_2$	C—O		1.179		C—Cl	1.742	ED, MW
	∠ClCCl		111.8				
COF_2	C—F		1.3157		C—O	1.172	ED, MW
	∠FCF		107.71				
CO_2	C—O (r_e)		1.1600				IR
CP	C—P (r_e)		1.562				UV
CS	C—S (r_e)		1.5349				MW
CS_2	C—S (r_e)		1.5526				IR
C_2	C—C (r_e)		1.2425				UV
C_3O_2	C—O		1.163		C—C	1.289	ED
	linear (large-amplitude bending vibration)						
CaH	Ca—H (r_e)		2.002				UV
CaO	Ca—O (r_e)		1.8221				UV
CaS	Ca—S (r_e)		2.3178				UV
CdH	Cd—H (r_e)		1.781				EPR
$CdBr_2$	Cd—Br		2.35	linear			ED
$CdCl_2$	Cd—Cl		2.24	linear			ED
CdI_2	Cd—I		2.56	linear			ED
ClCN	C—Cl (r_e)		1.629		C—N (r_e)	1.160	MW
ClF	Cl—F (r_e)		1.6283				MW
ClF_3	F$_a$—Cl—F$_a$			Cl—F$_a$	1.698	Cl—F$_b$ 1.598	MW
				∠F$_a$ClF$_b$	87.5	(C_{2v})	
	F$_b$						
ClO	Cl—O (r_e)		1.5696				MW, UV
ClOH	O—Cl		1.690	O—H	0.975	∠HOCl 102.5	MW, IR
ClO_2	Cl—O		1.470		∠OClO	117.38	MW
$ClO_3(OH)$	O$_a$—Cl		1.407		O$_b$—Cl	1.639	ED
				∠O$_a$ClO$_a$	114.3	∠O$_a$ClO$_b$ 104.1	

Compound	Structure						Method
Cl_2	Cl—Cl (r_e)		1.9878				UV
Cl_2O	Cl—O		1.6959		∠ClOCl	110.89	MW
CoH	Co—H (r_e)		1.542				UV
$Cr(CO)_6$	C—O		1.16		Cr—C	1.92	ED
	∠CrCO		180				
CrO	Cr—O (r_e)		1.615				UV
CsBr	Cs—Br (r_e)		3.0723				MW
CsCl	Cs—Cl (r_e)		2.9063				MW
CsF	Cs—F (r_e)		2.3454				MW
CsH	Cs—H (r_e)		2.4938				UV
CsI	Cs—I (r_e)		3.3152				MW

Compound	Structure						Method
CsOH	Cs—O (r_e)	2.395		O—H (r_e)	0.97		MW
CuBr	Cu—Br (r_e)	2.1734					MW
CuCl	Cu—Cl (r_e)	2.0512					MW
CuF	Cu—F (r_e)	1.7449					MW
CuH	Cu—H (r_e)	1.4626					UV
CuI	Cu—I (r_e)	2.3383					MW
FCN	C—F	1.262		C—N	1.159		MW
FOH	O—H	0.96	O—F	1.442	∠HOF	97.2	MW
F_2	F—F (r_e)	1.4119					R
$Fe(CO)_5$	Fe—C (average)	1.821					ED
	(Fe—C)$_{eq}$ – (Fe—C)$_{ax}$	0.020					
	C—O (average)	1.153	(D_{3h})				
GaBr	Ga—Br (r_e)	2.3525					MW
GaCl	Ga—Cl (r_e)	2.2017					MW
GaF	Ga—F (r_e)	1.7744					MW
GaF_3	Ga—F	1.88	(D_{3h})				ED
GaI	Ga—I (r_e)	2.5747					MW
GaI_3	Ga—I	2.458	(D_{3h})				ED
GdI_3	Gd—I	2.841	∠IGdI	108	(C_{3v})		ED
$GeBrH_3$	Ge—H	1.526	Ge—Br	2.299			MW, IR
	∠HGeH	106.2					
$GeBr_4$	Ge—Br	2.272	(T_d)				ED
$GeClH_3$	Ge—H	1.537	Ge—Cl	2.150			IR, MW
	∠HGeH	111.0					
$GeCl_2$	Ge—Cl	2.183	∠ClGeCl	100.3			ED
$GeCl_4$	Ge—Cl	2.113	(T_d)				ED
$GeFH_3$	Ge—H	1.522	Ge—F	1.732			MW, IR
	∠HGeH	113.0					
GeF_2	Ge—F (r_e)	1.7321	∠FGeF (θ_e)	97.17			MW
GeH	Ge—H (r_e)	1.5880					UV
GeH_4	Ge—H	1.5251	(T_d)				IR, R
GeO	Ge—O (r_e)	1.6246					MW
GeS	Ge—S (r_e)	2.0121					MW
GeSe	Ge—Se (r_e)	2.1346					MW
GeTe	Ge—Te (r_e)	2.3402					MW
Ge_2H_6	Ge—H	1.541	Ge—Ge	2.403			ED
	∠HGeH	106.4	∠GeGeH	112.5			
HBr	H—Br (r_e)	1.4145					MW
HCN	C—H (r_e)	1.0655	C—N (r_e)	1.1532			MW, IR
			linear				
HCNO	H—C	1.027	C—N	1.161	N—O	1.207	MW
					linear		
HCl	H-Cl (r_e)	1.2746					MW
HF	H—F (r_e)	0.9169					MW
HI	H—I (r_e)	1.6090					MW
HNCO	N—H	0.986	N—C	1.209	C—O	1.166	MW
	∠HNC	128.0					
HNCS	N—H	0.989	N—C	1.216	C—S	1.561	MW
	∠HNC	135.0	∠NCS	180			
HNO	N—H	1.063	N—O	1.212	∠HNO	108.6	UV

HNO$_2$ ╲N—O$_b$H

		s-trans conformer		*s-cis* conformer			MW
		O$_b$—H	0.958	0.98			
		N—O$_b$	1.432	1.39			
		N—O$_a$	1.170	1.19			
		∠O$_a$NO$_b$	110.7	114			
		∠NO$_b$H	102.1	104			

HNO$_3$

		O$_c$—H	0.96	N—O$_c$	1.41		MW
		N—O$_a$	1.20	N—O$_b$	1.21		
		∠HO$_c$N	102.2	∠O$_c$NO$_a$	113.9		
		∠O$_c$NO$_b$	115.9	planar			

O$_a$ ╲╲O$_b$

Compound	Structure						Method

Compound									Method
HNSO	N—H	1.029		N—S	1.512		S—O	1.451	MW
	∠HNS	115.8		∠NSO	120.4				
				planar					
H$_2$	H—H (r_e)	0.7414							UV
H$_2$O	O—H (r_e)	0.9575			∠HOH (θ_e)	104.51			MW, IR
H$_2$O$_2$	O—O	1.475			∠OOH	94.8			IR
	dihedral angle of internal rotation				119.8	(C$_2$)			
H$_2$S	H—S (r_e)	1.3356			∠HSH (θ_e)	92.12			MW, IR
H$_2$SO$_4$			O—H	0.97		S—O$_a$	1.574		MW
			S—O$_c$	1.422		∠H$_a$O$_a$S	108.5		
			∠O$_a$SO$_b$	101.3		∠O$_c$SO$_d$	123.3		
			∠O$_a$SO$_c$	108.6		∠O$_a$SO$_d$	106.4		
		dihedral angle between the H$_a$O$_a$S and O$_a$SO$_c$ planes				20.8			
		dihedral angle between the H$_a$O$_a$S and O$_a$SO$_b$ planes				90.9			
		dihedral angle between the H$_a$SO$_b$ and O$_c$SO$_d$ planes				88.4	(C$_2$)		
H$_2$S$_2$	S—S	2.055		S—H	1.327		∠SSH	91.3	ED, MW
	dihedral angle of internal rotation		90.6	(C$_2$)					
HfCl$_4$	Hf—Cl	2.33		(T$_d$)					ED
HgCl$_2$	Hg—Cl	2.252		linear					ED
HgH	Hg—H (r_e)	1.7404							UV
HgI$_2$	Hg—I	2.553		linear					ED
IBr	I—Br (r_e)	2.4691							MW
ICN	C—I	1.995			C—N	1.159			MW
ICl	I—Cl (r_e)	2.3210							MW
IF$_5$	I—F (average)	1.860		(I—F)$_{eq}$ − (I—F)$_{ax}$	0.03				ED, MW
	∠F$_{ax}$IF$_{eq}$	82.1		(C$_{4v}$)					
IO	I—O (r_e)	1.8676							MW
I$_2$	I—I (r_e)	2.6663							R
InBr	In—Br (r_e)	2.5432							MW
InCl	In—Cl (r_e)	2.4012							MW
InF	In—F (r_e)	1.9854							MW
InH	In—H (r_e)	1.8376							UV
InI	In—I (r_e)	2.7537							MW
IrF$_6$	Ir—F	1.830		(O$_h$)					ED
KBr	K—Br (r_e)	2.8208							MW
KCl	K—Cl (r_e)	2.6667							MW
KF	K—F (r_e)	2.1716							MW
KH	K—H (r_e)	2.244							UV
KI	K—I (r_e)	3.0478							MW
KOH	O—H	0.91		K—O	2.212 linear				MW
K$_2$	K—K (r_e)	3.9051							UV
KrF$_2$	Kr—F	1.89		linear					ED
LiBr	Li—Br (r_e)	2.1704							MW
LiCl	Li—Cl (r_e)	2.0207							MW
LiF	Li—F (r_e)	1.5639							MW
LiH	Li—H (r_e)	1.5949							MW
LiI	Li—I (r_e)	2.3919							MW
Li$_2$	Li—Li (r_e)	2.6729							UV
Li$_2$Cl$_2$				Li—Cl	2.23				ED
				Cl—Cl	3.61				
				∠ClLiCl	108				
LuCl$_3$	Lu—Cl	2.417		∠ClLuCl	112	(C$_{3v}$)			ED
MgF	Mg—F (r_e)	1.7500							UV
MgH	Mg—H (r_e)	1.7297							UV
MgO	Mg—O (r_e)	1.749							UV
MnH	Mn—H (r_e)	1.7308							UV
Mo(CO)$_6$	Mo—C	2.063		C—O	1.145		(O$_h$)		ED
MoCl$_4$O	Mo—Cl	2.279		Mo—O	1.658				ED
	∠ClMoCl	87.2		(C$_{4v}$)					

Compound	Structure						Method
MoF_6	Mo—F	1.820		(O_h)			ED
$NClH_2$	N—H	1.017		N—Cl	1.748		MW, IR
	∠HNCl	103.7		∠HNH	107		
NCl_3	N—Cl	1.759		∠ClNCl	107.1		ED
NF_2	N—F	1.3528		∠FNF	103.18		MW
NH_2	N—H	1.024		∠HNH	103.3		UV
NH_2CN	N—H	1.00		N_a—C	1.35		MW

$$\begin{array}{c} H \\ \diagdown \\ N_a\!-\!C\!\equiv\!N_b \\ \diagup \\ H \end{array}$$

			C—N_b	1.160	∠HNH	114	
			angle between the NH_2 plane and the N—C bond			142	
NH_2NO_2	N—N	1.427		N—H	1.005		MW
	∠HNH	115.2		∠ONO	130.1		
	dihedral angle between the NH_2 and NNO_2 planes				128.2		
NH_3	N—H (r_e)	1.012		∠HNH (θ_e)	106.7		IR
NH_4Cl	N—H	1.22		N—Cl	2.54	(C_{3v})	ED
NF_2CN	F_2N_b—C≡N_a			C—N_a	1.158	C—N_b 1.386	MW
	N_b—F	1.399		∠N_aCN_b	174		
	∠CN_bF	105.4		∠FN_bF	102.8		
NH	N—H (r_e)	1.0362					LMR
NH_2OH	N—H	1.02		N—O	1.453	O—H 0.962	MW
	∠HNH	107		∠HNO	103.3	∠NOH 101.4	
	The bisector of H—N—H angle is *trans* to the O—H bond						
NO	N—O (r_e)	1.1506					IR
NOCl	N—Cl	1.975		N—O	1.14	∠ONCl 113	MW
NOF	O—N	1.136		N—F	1.512	∠FNO 110.1	MW
NO_2	N—O		1.193	∠ONO	134.1		MW
NO_2Cl	N—Cl		1.840	N—O	1.202		MW
	∠ONO		130.6	(C_{2v})			
NO_2F	N—O		1.1798	N—F	1.467		MW
	∠ONO		136	(C_{2v})			
NS	N—S (r_e)		1.4940				IR
N_2	N—N (r_e)		1.0977				UV
N_2H_4	N—H		1.021	N—N	1.449		ED, MW
	∠HNH		106.6 (assumed)	∠NNH_a	112		
	∠NNH_b 106		dihedral angle of internal rotation		91		
	H_a: the H atom closer to the C_2 axis, H_b: the H atom farther from the C_2 axis						
N_2O	N—N (r_e)		1.1284	N—O (r_e)	1.1841		MW, IR
N_2O_3				N_a—N_b 1.864		N_a—O_a 1.142	MW

$$\begin{array}{c} O_a O_b \\ \diagdown \diagup \\ N_a\!-\!N_b \\ \diagdown \\ O_c \end{array}$$

				N_b—O_b 1.202		N_b—O_c 1.217	
				∠$O_aN_aN_b$ 105.05			
				∠$N_aN_bO_b$ 112.72			
				∠$N_aN_bO_c$ 117.47			
N_2O_4				N—N 1.782		N—O 1.190	ED

$$\begin{array}{c} O O \\ \diagdown \diagup \\ N\!-\!N \\ \diagup \diagdown \\ O O \end{array}$$

				∠ONO 135.4		(D_{2h})	
NaBr	Na—Br (r_e)	2.5020					MW
NaCl	Na—Cl (r_e)	2.3609					MW
NaF	Na—F (r_e)	1.9260					MW
NaH	Na—H (r_e)	1.8873					UV
NaI	Na—I (r_e)	2.7115					MW
Na_2	Na—Na (r_e)	3.0789					UV
$NbCl_5$	Nb—Cl_{eq} 2.241			Nb—Cl_{ax} 2.338 (D_{3h})		ED	
NbO	Nb—O (r_e)	1.691					UV
$Ni(CO)_4$	Ni—C	1.838		C—O	1.141	(T_d)	ED
NiH	Ni—H (r_e)	1.476					UV
NpF_6	Np—F	1.981		(O_h)			ED
OCS	C—O (r_e)	1.1578		C—S (r_e)	1.5601		MW

Compound	Structure						Method
OCSe	C—O	1.159		C—Se	1.709		MW
OF	O—F (r_e)	1.3579					LMR
OF$_2$	O—F (r_e)	1.4053		∠FOF (θ_e)	103.07	(C_{2v})	MW
O(SiH$_3$)$_2$	Si—H	1.486		Si—O	1.634		ED
	∠SiOSi	144.1					
O$_2$	O—O (r_e)	1.2074					MW
O$_2$F$_2$	O—O	1.217		F—O	1.575		MW
	∠OOF	109.5	dihedral angle of internal rotation	87.5	(C_2)		
O$_3$	O—O (r_e)	1.2716		∠OOO (θ_e)	117.47	(C_{2v})	MW
OsF$_6$	Os—F	1.831		(O_h)			ED
OsO$_4$	Os—O	1.712		(T_d)			ED
PBr$_3$	P—Br	2.220		∠BrPBr	101.0		ED
PCl$_3$	P—Cl	2.039		∠ClPCl	100.27		ED
PCl$_5$			P—Cl$_a$	2.124	P—Cl$_b$	2.020	ED
					(D_{3h})		

(PCl$_5$ structure diagram: Cl$_b$—P bonded to Cl$_a$, Cl$_b$, Cl$_a$, Cl$_b$)

Compound	Structure						Method
PF	P—F (r_e)	1.5896					UV
PF$_3$	P—F	1.570		∠FPF	97.8		ED, MW
PF$_5$	P—F$_{ax}$	1.577	P—F$_{eq}$	1.534	(D_{3h})		ED
PH	P—H (r_e)	1.4223					LMR
PH$_2$	P—H	1.418		∠HPH	91.70		UV
PH$_3$	P—H	1.4200		∠HPH	93.345		MW
PN	N—P (r_e)	1.4909					MW
PO	O—P (r_e)	1.4759					UV
POCl$_3$	P—O	1.449		P—Cl	1.993		ED
	∠ClPCl	103.3					
POF$_3$	P—O	1.436	P—F	1.524	∠FPF	101.3	ED, MW
P$_2$	P—P (r_e)	1.8931					UV
P$_2$F$_4$	P—F	1.587		P—P	2.281		ED
	∠PPF	95.4		∠FPF	99.1		
	The two PF$_2$ planes are *trans* to each other (the *gauche* conformer is less than 10%)						
P$_4$	P—P	2.21		(T_d)			ED
P$_4$O$_6$	P—O	1.638	∠POP	126.4	(T_d)		ED
PbH	Pb—H (r_e)	1.839					UV
PbO	Pb—O (r_e)	1.9218					MW
PbS	Pb—S (r_e)	2.2869					MW
PbSe	Pb—Se (r_e)	2.4022					MW
PbTe	Pb—Te (r_e)	2.5950					MW
PrI$_3$	Pr—I	2.904	∠IPrI	113	(C_{3v})		ED
PtO	Pt—O (r_e)	1.7273					UV
PuF$_6$	Pu—F	1.971		(O_h)			ED
RbBr	Rb—Br (r_e)	2.9447					MW
RbCl	Rb—Cl (r_e)	2.7869					MW
RbF	Rb—F (r_e)	2.2703					MW
RbH	Rb—H (r_e)	2.367					UV
RbI	Rb—I (r_e)	3.1768					MW
RbOH	Rb—O	2.301	O—H	0.957	linear		MW
ReClO$_3$	Re—O	1.702		Re—Cl	2.229		MW
	∠ClReO	109.4		(C_{3v})			
ReF$_6$	Re—F	1.832		(O_h)			ED
RuO$_4$	Ru—O	1.706		(T_d)			ED
SCSe	C—Se	1.693		C—S	1.553		MW
SCTe	C—S	1.557		C—Te	1.904		MW
SCl$_2$	S—Cl	2.006	∠ClSCl	103.0	(C_{2v})		ED
SF	S—F (r_e)	1.6006					MW
SF$_2$	S—F	1.5921		∠FSF	98.20		MW
SF$_6$	S—F	1.561		(O_h)			ED

Compound	Structure							Method
SO	S—O (r_e)	1.4811						MW
SOCl$_2$	S—O	1.44		S—Cl	2.072			MW
	∠ClSCl	97.2		∠OSCl	108.0			
SOF$_2$	S—O	1.420		S—F	1.583			ED
	∠OSF	106.2		∠FSF	92.2			
SOF$_4$	F$_b$ $\quad$ F$_b$		S—O	1.403		S—F$_a$	1.575	ED
	F$_a$—S—F$_a$		S—F$_b$	1.552		∠OSF$_a$	90.7	
	O		∠OSF$_b$	124.9		∠F$_a$SF$_b$	89.6	
			∠F$_b$SF$_b$	110.2		(C$_{2v}$)		
SO$_2$	S—O (r_e)	1.4308		∠OSO (θ_e)		119.329		MW
SO$_2$Cl$_2$	S—O	1.404		S—Cl	2.011		∠OSO	123.5	ED
	∠ClSCl	100.0		(C$_{2v}$)				
SO$_2$F$_2$	S—O	1.397		S—F	1.530		∠OSO	123	ED
	∠FSF	97		(C$_{2v}$)				
SO$_3$	S—O	1.4198		(D$_{3h}$)				IR
S(SiH$_3$)$_2$	Si—H	1.494		Si—S	2.136		∠SiSSi	97.4	ED
S$_2$	S—S (r_e)	1.8892						R
S$_2$Br$_2$	S—Br	2.24		S—S		1.98		ED
	∠SSBr	105	dihedral angle of internal rotation			83.5		
S$_2$Cl$_2$	S—Cl	2.057		S—S		1.931		ED
	∠SSCl	108.2	dihedral angle of internal rotation			84.1	(C$_2$)	
S$_2$O$_2$	S—O	1.458		S—S	2.025		∠OSS	112.8	MW
						planar *cis* form		
S$_8$				S—S	2.07			ED
				∠SSS	105			
				(D$_{4d}$)				
SbCl$_3$	Sb—Cl	2.333		∠ClSbCl		97.2		ED
SbH$_3$	Sb—H	1.704		∠HSbH		91.6		MW
SeF	Se—F	1.742						MW
SeF$_6$	Se—F	1.69		(O$_h$)				ED
SeO	Se—O (r_e)	1.6393						MW
SeOF$_2$	Se—O	1.576		Se—F		1.730		MW
	∠OSeF	104.82		∠FSeF		92.22		
SeO$_2$	Se—O (r_e)	1.6076		∠OSeO (θ_e)		113.83		MW
SeO$_3$	Se—O	1.69		(D$_{3h}$)				ED
Se$_2$	Se—Se (r_e)	2.1660						UV
Se$_6$	Se—Se	2.34		∠SeSeSe		102		ED
			six-membered ring with chair conformation					
SiBrF$_3$	Si—F	1.560		Si—Br		2.153		MW
	∠FSiBr	108.5		(C$_{3v}$)				
SiBrH$_3$	Si—H	1.485		Si—Br		2.210		MW
	∠HSiBr	107.8		(C$_{3v}$)				
SiClH$_3$	Si—H	1.482		Si—Cl		2.048		MW
	∠HSiCl	107.9		(C$_{3v}$)				
SiCl$_4$	Si—Cl	2.019		(T$_4$)				ED
SiF	Si—F	1.6008						UV
SiFH$_3$	Si—H	1.484		Si—F		1.593		MW, IR
	∠HSiH	110.63		(C$_{3v}$)				
SiF$_2$	Si—F (r_e)	1.590		∠FSiF (θ_e)		100.8		MW
SiF$_3$H	Si—H (r_e)	1.4468		Si—F (r_e)		1.5624		MW
	∠HSiF (θ_e)	110.64						
SiF$_4$	Si—F	1.553		(T$_d$)				ED
SiH	Si—H (r_e)	1.5201						UV
SiH$_3$I	Si—H	1.485		Si—I		2.437		MW
	∠HSH	107.8						
SiH$_4$	Si—H	1.4798		(T$_d$)				IR
SiN	N—Si (r_e)	1.572						UV
SiO	Si—O (r_e)	1.5097						MW

Compound	Structure					Method
SiS	Si—S (r_e)	1.9293				MW
SiSe	Se—Si (r_e)	2.0583				MW
Si$_2$	Si—Si (r_e)	2.246				UV
Si$_2$Cl$_6$	Si—Si	2.32	Si—Cl	2.009		ED
	∠ClSiCl	109.7				
Si$_2$F$_6$	Si—Si	2.317	Si—F	1.564		ED
	∠FSiF	108.6				
Si$_2$H$_6$	Si—H	1.492	Si—Si	2.331		ED
	∠SiSiH	110.3	∠HSiH	108.6		
			staggered form (assumed)			
SnCl$_4$	Sn—Cl	2.280	(T$_d$)			ED
SnH	Sn—H (r_e)	1.7815				UV
SnH$_4$	Sn—H	1.711	(T$_d$)			R, IR
SnO	Sn—O	1.8325				MW
SnS	S—Sn (r_e)	2.2090				MW
SnSe	Se—Sn (r_e)	2.3256				MW
SnTe	Sn—Te (r_e)	2.5228				MW
SrH	Sr—H (r_e)	2.1455				UV
SrO	Sr—O (r_e)	1.9198				MW
SrS	S—Sr (r_e)	2.4405				UV
TaCl$_5$	Ta—Cl$_{eq}$	2.227	Ta—Cl$_{ax}$	2.369	(D$_{3h}$)	ED
TaO	Ta—O (r_e)	1.6875				UV
TeF$_6$	Te—F	1.815	(O$_h$)			ED
Te$_2$	Te—Te (r_e)	2.5574				UV
ThCl$_4$	Th—Cl	2.58	(T$_d$)			ED
ThF$_4$	Th—F	2.14	(T$_d$)			ED
TlBr	Tl—Br (r_e)	2.6182				MW
TlCl	Tl—Cl (r_e)	2.4848				MW
TlF	Tl—F (r_e)	2.0844				MW
TlH	Tl—H (r_e)	1.870				UV
TlI	Tl—I (r_e)	2.8137				MW
TiBr$_4$	Ti—Br	2.339	(T$_d$)			ED
TiCl$_4$	Ti—Cl	2.170	(T$_d$)			ED
TiO	Ti—O (r_e)	1.620				UV
TiS	Ti—S (r_e)	2.0825				UV
UF$_6$	U—F	1.996	(O$_h$)			ED
V(CO)$_6$	V—C	2.015	C—O	1.138		ED
	(O$_h$, involving dynamic Jahn-Teller effect)					
VCl$_3$O	V—O	1.570	V—Cl	2.142		ED, MW
	∠ClVCl	111.3				
VCl$_4$	V—Cl	2.138	(T$_d$, involving dynamic Jahn-Teller effect)			ED
VF$_5$	V—F (average)	1.71				ED
VO	V—O (r_e)	1.5893				UV
W(CO)$_6$	W—C	2.059	C—O	1.149	(O$_h$)	ED
WClF$_5$			W—Cl	2.251		MW
			W—F (average)	1.836		
			∠F$_a$WF$_b$	88.7		
WF$_4$O	W—O	1.666	W—F	1.847		ED
	∠FWF	86.2	(C$_{4v}$)			
WF$_6$	W—F	1.832	(O$_h$)			ED
XeF$_2$	Xe—F	1.977	linear			IR
XeF$_4$	Xe—F	1.94	(D$_{4h}$)			ED
XeF$_6$	Xe—F	1.890	(large-amplitude bending vibration around the O$_h$ structure)			ED
XeO$_4$	Xe—O	1.736	(T$_d$)			ED
ZnH	Zn—H (r_e)	1.5949				UV
ZrCl$_4$	Zr—Cl	2.32	(T$_d$)			ED
ZrF$_4$	Zr—F	1.902	(T$_d$)			ED
ZrO	Zr—O (r_e)	1.7116				UV

For WClF$_5$ the structure is drawn as:

```
      Cl   F_b
       |  /
 F_b—W—F_b
      /|
   F_b  F_a
```

BOND LENGTHS AND ANGLES IN GAS-PHASE MOLECULES (continued)

STRUCTURES OF ORGANIC MOLECULES
Compounds are Arranged in Alphabetical Order by Chemical Name; Cross References are Given for Common Synonyms (Lengths in Å and Angles in Degrees)

Compound	Structure				Method
Acetaldehyde			C_a—O	1.210	ED, MW
			C_b—H	1.107	
			C_a—H	1.128	
	C_a—C_b	1.515	$\angle C_b C_a O$	124.1	
	$\angle HC_b H$	109.8	$\angle C_b C_a H$	115.3	
Acetamide	C—O	1.220	C—N	1.380	ED
CH_3CONH_2	C—C	1.519	N—H	1.022	
	C—H	1.124	$\angle NCO$	122.0	
	$\angle CCN$	115.1			
Acetic acid			C—C	1.520	ED
			C—O_a	1.214	
			C—O_b	1.364	
	C—H	1.10	$\angle CCO_a$	126.6	
	$\angle CCO_b$	110.6			
Acetone	C—C	1.520	C—O	1.213	ED, MW
$(CH_3)_2CO$	C—H	1.103	$\angle CCC$	116.0	
	$\angle HCH$	108.5	symmetry axis of each methyl group is tilted 2° from the C—C bond		
Acetonitrile	C—H	1.107	C—C	1.468	ED, MW
CH_3CN	C—N	1.159	$\angle CCH$	109.7	
Acetonitrile oxide	C—C	1.442	C—N	1.169	MW
CH_3CNO	N—O	1.217	(C_{3v})		
Acetyl chloride	C—H	1.105	C—O	1.187	ED, MW
CH_3COCl	C—C	1.506	C—Cl	1.798	
	$\angle HCH$	108.6	$\angle OCCl$	121.2	
	$\angle CCCl$	111.6			
Acetyl cyanide → Pyruvonitrile					
Acetylene	C—H (r_e)	1.060	C—C (r_e)	1.203	IR
HC≡CH					
Acrolein → Acrylaldehyde					
Acrylaldehyde			C_b—C_c	1.484	ED, MW
			C_a—C_b	1.345	
			C_c—O	1.217	
			C_a—H	1.10	
			C_c—H	1.13	
	$\angle C_a C_b C_c$	120.3	$\angle C_b C_c O$	123.3	
	$\angle HC_c C_b$	114	other CCH angles (average)	122	
	planar s-trans form				
Acrylonitrile			C_a—C_b	1.343	ED, MW
			C_b—C_c	1.438	
			C_c—N	1.167	
			C_a—H	1.114	
			$\angle C_a C_b C_c$	121.7	
	$\angle C_b C_c N$	178	$\angle HCC$	120	
Acryloyl chloride	C—H	1.086 (assumed)	C_b—C_c	1.48	MW
	C_c—Cl	1.82	C_a—C_b	1.35	
			C_c—O	1.19	
			$\angle C_a C_b H$	120 (assumed)	
			$\angle C_b C_a H$	121.5 (assumed)	
			$\angle C_a C_b C_c$	123	
			$\angle C_b C_c Cl$	116	
			$\angle C_b C_c O$	127	

Compound	Structure					Method

Allene $CH_2=C=CH_2$

C—C	1.3084		C—H	1.087	IR
∠HCH	118.2				

Allyl chloride

Cl

		cis conformer	C—Cl	1.811	MW
			∠CCCl	115.2	
		skew conformer	C—Cl	1.809	
∠CCCl	109.6	CCCCl	dihedral angle of internal rotation 122.4		

Aniline $C_6H_5NH_2$

C—C	1.392		C—N	1.431	MW
N—H	0.998		∠HNH	113.9	
dihedral angle between the NH_2 plane and the N—C bond			140.6		

Azetidine

CH_2—CH_2
| |
CH_2—NH

			C—N	1.482	ED
			C—C	1.553	
			C—H	1.107	
N—H	1.03		∠CNC	92.2	
∠CCC	86.9		∠CCN	85.8	
dihedral angle between the CCC and CNC planes			147		

Aziridine

			N—H	1.016	MW
			N—C	1.475	
			C—C	1.481	
			C—H	1.084	
			∠CNC	60.3	
			∠H$_a$NC	109.3	
∠H$_b$CH$_c$	115.7		∠H$_b$CC	117.8	
∠H$_b$CN	118.3		∠H$_c$CC	119.3	
∠H$_c$CN	114.3				

Azomethane $CH_3N=NCH_3$

C—N	1.482		N—N	1.247	ED
∠CNN	112.3		*trans* conformer		

Benzene C_6H_6

C—C	1.399		C—H	1.101	ED, IR

p-Benzoquinone

			C$_a$—O	1.225	ED
			C$_b$—C$_b$	1.344	
			C$_a$—C$_b$	1.481	
			∠C$_b$C$_a$C$_b$	118.1	

Biacetyl $CH_3COCOCH_3$

C—O	1.215		C—C (average)	1.524	ED
C—H	1.108		∠CCO	119.5	
∠CCC	116.2		*trans* conformer		

Bicyclo[1.1.0]butane

			C$_a$—C$_a$	1.497	MW
			C$_a$—C$_b$	1.498	
			C$_a$—H$_a$	1.071	
			C$_b$—H$_b$, C$_b$—H$_c$	1.093	
			∠H$_b$C$_b$H$_c$	115.6	
∠C$_b$C$_a$H$_a$	130.4		∠C$_a$C$_b$C$_a$	60.0	
∠C$_a$C$_a$H$_a$	128.4	dihedral angle between the two C$_a$C$_a$C$_b$ planes	121.7		

Bicyclo[2.2.1]hepta-2,5-diene

			C$_a$—C$_b$	1.535	ED
			C$_b$—C$_b$	1.343	
			C$_a$—C$_c$	1.573	
			C—H	1.12	
			∠C$_a$C$_c$C$_a$	94	
dihedral angle between the two C$_a$C$_b$C$_b$C$_a$ planes			115.6		
			(C_{2v})		

Bicyclo[2.2.1]heptane C_7H_{12}

See the preceding molecule for the labels of the C atoms

					ED
C$_a$—C$_b$	1.54		C$_b$—C$_b$	1.56	
C$_a$—C$_c$	1.56		C—C (average)	1.549	
–C$_a$C$_c$C$_a$	93.1	dihedral angle between the two C$_a$C$_b$C$_b$C$_a$ planes	113.1		

Compound	Structure			Method

Bicyclo[2.2.0]hexa-2,5-diene

C_b—C_b	1.345	ED
C_a—C_a	1.574	
C_a—C_b	1.524	

dihedral angle between the two $C_aC_bC_bC_a$ planes 117.3

Bicyclo[2.2.2]octane

$HC_a(C_bH_2C_bH_2)_3C_aH$

C_b—C_b	1.55	C_a—C_b	1.54	ED
		$\angle C_aC_bC_b$	109.7	

C—C (average) 1.542

large-amplitude torsional motion about the D_{3h} symmetry axis

Bicyclo[1.1.1]pentane
C_5H_8

C—C	1.557	$\angle$CCC	74.2	ED

Bicyclo[2.1.0]pentane

C_a—C_a	1.536	MW
C_b—C_b	1.565	
C_a—C_b	1.528	
C_a—C_c	1.507	

Dihedral angle between the $C_aC_aC_bC_b$ and $C_aC_aC_c$ planes 112.7

Biphenyl

C—C (intra-ring)	1.396	ED
(inter-ring)	1.49	

torsional dihedral angle
between the two rings ~40

4,4'-Bipyridyl

C—C, C—N (intra-ring)

	1.375	ED
C—C (inter-ring)	1.465	

torsional dihedral angle
between the two rings ~37

Bis (cyclopentadienyl) beryllium
$(C_5H_5)_2Be$

Be—(cyclopentadienyl plane) 1.470, 1.92 ED
C—C 1.423 (C_{5v}) (The Be atom has two equilibrium positions)

Bis (cyclopentadienyl) iron → Ferrocene

Bis (cyclopentadienyl) lead
$(C_5H_5)_2Pb$

C—C	1.430	Pb—C	2.79	ED

dihedral angle between the two C_5H_5 planes 40~50 (The two rings are not parallel.)

Bis (cyclopentadienyl) manganese
$(C_5H_5)_2Mn$

Mn—C	2.383	C—C	1.429	(D_{5h})	ED

Bis (cyclopentadienyl) nickel
$(C_5H_5)_2Ni$

Ni—C	2.196	C—C	1.430	ED
			(D_{5h})	

Bis (cyclopentadienyl) ruthenium
$(C_5H_5)_2Ru$

C—C	1.439	Ru—C	2.196	ED

Bis (cyclopentadienyl) tin
$(C_5H_5)_2Sn$

C—C	1.431	Sn—C	2.71	ED
C—H	1.14	(D_{5h})		

Bis (trifluoromethyl) peroxide
CF_3OOCF_3

O—O	1.42	C—O	1.399	ED
C—F	1.320	$\angle$COO	107	
$\angle$FCF	109.0	COOC dihedral angle of internal		
		rotation	123	

Borine carbonyl
BH_3CO

B—H	1.194	B—C	1.540	MW
C—O	1.131	$\angle$HBH	113.9	
$\angle$BCO	180	(C_{3v})		

Bromobenzene

C—H	1.072	MW
C_c—C_d	1.401	
C_b—C_c	1.375	
C—Br	1.85	
C_a—C_b	1.42	
$\angle C_bC_aC_b$	117.4	

Bromoform
$CHBr_3$

C—Br	1.924	C—H	1.11	ED, MW
$\angle$BrCBr	111.7	(C_{3v})		

Bromoiodoacetylene
$IC\equiv CBr$

C—I	1.972	C—C	1.206	ED
C—Br	1.795			

Compound	Structure					Method

1,3-Butadiene

C_aH_2
C_bH-C_bH
C_aH_2

		C_b-C_b	1.467		ED
		C_a-C_b	1.349		
		C—H (average)	1.108		
		∠CCC	124.4		
∠C_bC_aH	120.9	*anti* conformer	(C_{2h})		

1,3-Butadiyne

$HC_a≡C_b—C_b≡C_aH$		C—H	1.09		ED
$C_a—C_b$	1.218	$C_b—C_b$	1.384		
		linear			

Butane

$CH_3CH_2CH_2CH_3$

C—C	1.531	C—H	1.117		ED
∠CCC	113.8	∠CCH	111.0		
trans conformer	54%	dihedral angle for the *gauche* conformer	65		

2-Butanone → Ethyl methyl ketone

Butatriene

$H_2C_a=C_b=C_b=C_aH_2$		C—H	1.08		ED
$C_a—C_b$	1.32	$C_b—C_b$	1.28	(D_{2h})	

2-Butene

$C_aH_3—C_bH=C_bH—C_aH_3$		ED			
$C_a—C_b$	*cis* conformer	1.506	*trans* conformer 1.508		
$C_b—C_b$	1.346	1.347			
∠$C_aC_bC_b$	125.4	123.8			

3-Buten-1-yne → Vinylacetylene

tert-Butyl chloride

$(CH_3)_3CCl$

C—H	1.102	C—C	1.528		ED, MW
C—Cl	1.828	∠CCCl	107.3		
∠CCH	110.8	∠CCC	111.6		

tert-Butyl cyanide → Pivalonitrile

2-Butyne

$C_aH_3—C_b≡C_b—C_aH_3$	C—H	1.116		ED	
$C_b—C_b$	1.214	$C_a—C_b$	1.468		
∠C_bC_aH	110.7				

Carbon C₂

C—C (r_e)	1.3119				UV

Carbon C₃

C—C	1.277	linear			UV

Carbon suboxide → Tricarbon dioxide

Carbon tetrabromide

CBr_4

C—Br	1.935	(T_d)			ED

Carbon tetrachloride

CCl_4

C—Cl	1.767	(T_d)			ED

Carbon tetrafluoride

CF_4

C—F	1.323	(T_d)			ED

Carbon tetraiodide

CI_4

C—I	2.15	(T_d)			ED

Carbonyl cyanide

$CO(CN)_2$

C—O	1.209	C—C	1.466		ED, MW
C—N	1.153	∠CCC	115		
∠CCN	180				

Chloroacetylene

HC≡CCl

C—H	1.0550	C—C	1.2033		MW
C—Cl	1.6368				

Chlorobenzene

C_6H_5Cl

C—C	1.400	C—Cl	1.737		ED
C—H	1.083	∠CC(Cl)C	121.7		
∠CC(H)C	120				

Chlorobromoacetylene

ClC≡CBr

Cl—C	1.636	C—C	1.206		ED
C—Br	1.784				

Chlorocyanoacetylene

ClC≡CCN

C—Cl	1.624	C—C	1.205		ED
C—CN	1.362	C—N	1.160		

Chloroethane → Ethyl chloride

2-Chloroethanol

$ClCH_2CH_2OH$

C—O	1.413	C—C	1.519		ED
C—Cl	1.801	C—H	1.093		
O—H	1.033	∠CCCl	110.7	∠CCO	113.8
fraction of the *gauche* conformer at 37°C is 92 ~ 94%,					
dihedral angle of internal rotation		62.4			

Chloroethylene → Vinyl chloride

Chloroform

$CHCl_3$

C—H	1.100	C—Cl	1.758		MW
∠ClCCl	111.3	(C_{3v})			

Compound	Structure				Method
Chloroiodoacetylene	C—Cl	1.63	C—I	1.99	MW
ClC≡CI	C—C	1.209 (assumed)			
Chloromethane → Methyl chloride					
3-Chloropropene → Allyl chloride					
Cyanamide	N_a—C	1.346	C—N_b	1.160	MW
$H_2N_aCN_b$	N—H	1.00	∠HNH	114	
	dihedral angle between the NH_2 plane and the N—C bond			142	
Cyanoacetylene	C_b—H	1.058	C_a—C_b	1.205	MW
H—C_b≡C_a—C_c≡N	C_a—C_c	1.378	C_c—N	1.159	
Cyanocyclopropane	C—C (ring)	1.513	C—C_a	1.472	MW
$C_3H_5C_aN$	C—H	1.107	C_a—N	1.157	
	∠HCH	114.6	∠C_aCH	119.6	
Cyanogen	C—N	1.163	C—C	1.393	ED
(CN)$_2$				linear	
Cyclobutane	C—H	1.113	C—C	1.555	ED
(CH$_2$)$_4$	dihedral angle between the two CCC planes 145				

Cyclobutanone — MW

C_cH_2—C_aH_2 ... C_a=O, C_bH_2 (ring structure)

C_a—C_b	1.527	
C_b—C_c	1.556	
∠$C_bC_aC_b$	93.1	
∠$C_aC_bC_c$	88.0	

Cyclobutene — MW

H_2C_a C_aH_2 / HC_b C_bH (ring structure)

C_b—C_b	1.342		C_a—C_a	1.566	
C_a—C_b	1.517		C_a—H	1.094	
C_b—H	1.083				
∠$C_aC_bC_b$	94.2		∠C_bC_bH	133.5	
∠C_aC_aH	114.5		∠$C_aC_aC_b$	85.8	
∠HC_aH	109.2		dihedral angle between the CH_2 plane and the C_a—C_a bond 135.8		

Cyclohexane	C—C	1.536	C—H	1.119	ED
C_6H_{12}	∠CCC	111.3	chair form		

Cyclohexene — ED

HC_a=C_aH / H_2C_b C_bH_2 / C_cH_2—C_cH_2 (ring structure)

			C_a—C_a	1.334	
			C_a—C_b	1.50	
			C_b—C_c	1.52	
			C_c—C_c	1.54	
∠$C_aC_aC_b$	123.4		∠$C_aC_bC_c$	112.0	
∠$C_bC_cC_c$	110.9		(C$_2$)	half-chair form	

Cyclooctatetraene — ED

(ring structure with b and a labels)

C_a—C_b	1.476	
C—H	1.100	
C_a—C_a, C_b—C_b	1.340	
∠$C_bC_aC_a$, ∠$C_aC_bC_b$	126.1	

dihedral angle between the $C_aC_aC_aC_a$ and $C_aC_bC_bC_a$ planes   136.9
tub form   (D$_{2d}$)

1,3-Cyclopentadiene — MW

C_aH_2 / HC_b C_bH / HC_c—C_cH (ring structure)

C_a—C_b	1.509	
C_b—C_c	1.342	
C_c—C_c	1.469	
∠$C_aC_bC_c$	109.3	

∠$C_bC_cC_c$	109.4	∠$C_bC_aC_b$	102.8	

Cyclopentadienylindium — ED

In / HC CH / HC CH / CH (ring structure)

In—C	2.621	
C—C	1.426	
(C$_{5v}$)		

Cyclopentane	C—H	1.114	C—C	1.546	ED
(CH$_2$)$_5$	∠CCH	111.7			

(The out-of-plane vibration of the C atoms is essentially free pseudorotation; average value of the displacements of the C atoms from the molecular plane 0.43)

Compound	Structure						Method

Cyclopentene

C_a—C_b	1.546	ED
C_b—C_c	1.519	
C_c—C_c	1.342	
∠$C_bC_aC_b$	104.0	

∠$C_bC_cC_c$ 110.0   ∠$C_aC_bC_c$ 103.0

dihedral angle between the $C_bC_aC_b$ and $C_bC_cC_cC_b$ planes 151.2

Cyclopropane
$(CH_2)_3$

C—C	1.512	C—H	1.083	R
∠HCH	114.0			

Cyclopropanone

C—H	1.086		C_a—C_b	1.475	MW
C_b—C_b	1.575		C_a—O	1.191	
∠$C_aC_bC_b$	57.7				

∠HC_bH 114   dihedral angle between the CH_2 plane and the C_b—C_b bond 151

Cyclopropene

C_b—C_b	1.304	ED
C_a—C_b	1.519	
C_b—H	1.077	
∠C_bC_bH	133	

C_a—H 1.112
∠HC_aH 118

Decalin
$C_{10}H_{18}$

C—C (average)	1.530	C—H (average)	1.113	ED
∠CCC (average)	111.4			

Dewer benzene → Bicyclo[2.2.0] hexa-2,5-diene

Diacetylene → 1,3-Butadiyne

1,4-Diazabicyclo[2.2.2]octane

C—N	1.472	ED
C—C	1.562	
∠NCC	110.2	
∠CNC	108.7	

large-amplitude torsional motion about the D_{3h} symmetry axis

2,3-Diaza-1,3-butadiene → Formalde-hyde azine

Diazirine

C—H	1.09	MW
C—N	1.482	
N—N	1.228	
∠HCH	117	

Diazoacetonitrile

C_b—N_b	1.280	MW
N_b—N_c	1.132	
C_a—N_a	1.165	
C—H	1.082	
C_a—C_b	1.424	

∠C_aC_bH 117   ∠$C_aC_bN_b$ 119.5

Diazomethane
CH_2N_2

C—H	1.075	C—N	1.32	MW, IR
N—N	1.12	∠HCH	126.0 linear	

1,2-Dibromoethane
CH_2BrCH_2Br

C—C	1.506	C—Br	1.950	ED
C—H	1.108	∠CCBr	109.5	

∠CCH 110   fraction of the *trans* conformer at 25°C 95%

Dibromomethane
CH_2Br_2

C—H	1.08	C—Br	1.924	ED
∠HCBr	109	∠BrCBr	113.2	

2,2′-Dichlorobiphenyl
C_6H_4Cl—C_6H_4Cl

C—C	1.398	C—C inter-ring	1.495	ED
C—Cl	1.732	C—H	1.10	
∠CCCl	121.4	∠CCH	126	

dihedral angle between the two aromatic rings 74 (defined to be 0 for that of the *cis* conformer)

***trans*-1,4-Dichlorocyclohexane**
$C_6H_{10}Cl_2$

C—H	1.102	C—Cl	1.810	ED
C—C	1.530	∠CCC	111.5	
∠CCCl (*ee*)	108.6	∠CCCl (*aa*)	110.6	
∠HCCl (*ee*)	111.5	∠HCCl (*aa*)	107.6	

ee 49%   *aa* 51%   e: equatorial, a: axial

Compound	Structure				Method
1,1-Dichloroethane	C—Cl	1.766	C—C	1.540	MW
CHCl$_2$CH$_3$	∠ClCCl	112.0	∠CCCl	111.0	
1,2-Dichloroethane	C—C	1.531	C—Cl	1.790	ED
CH$_2$ClCH$_2$Cl	C—H	1.11	∠CCCl	109.0	
	∠CCH	113			
	fraction of the *trans* conformer at room temperature 73%, that of the				
	gauche conformer 27%				
1,1-Dichloroethylene	C—C	1.32 (assumed)	C—Cl	1.73	MW
CH$_2$=CCl$_2$	∠ClCC	123	(C$_{2v}$)		
cis-1,2-Dichloroethylene	C—Cl	1.718	C—C	1.354	ED
CHCl=CHCl	∠ClCC	123.8			
Dichloromethane	C—H (r_e)	1.087	C—Cl (r_e)	1.765	MW, IR
CH$_2$Cl$_2$	∠HCH (θ_e)	111.5	∠ClCCl (θ_e)	112.0	
1,1-Difluoroethane	C—C	1.498	C—H (average)	1.081	ED
CH$_3$CHF$_2$	C—F	1.364	∠CCH (average)	111.0	
	∠CCF 110.7	dihedral angle between the two CCF planes 118.9			
1,2-Difluoroethane	C—F	1.389	C—C	1.503	ED
CH$_2$FCH$_2$F	C—H	1.103	∠CCF	110.3	
	∠CCH 111	dihedral angle of internal rotation 109			
	fraction of the *gauche* conformer at 22°C 94%				
1,1-Difluoroethane	C—C	1.340	C—F	1.315	ED, MW
CH$_2$=CF$_2$	C—H	1.091	∠CCF	124.7	
	∠CCH	119.0			
cis-1,2-Difluoroethylene	C—C	1.33	C—F	1.342	ED, MW
CHF=CHF	C—H	1.099	∠CCF	122.0	
	∠CCH	124.1			
Difluoromethane	C—H	1.093	C—F	1.357	MW
CH$_2$F$_2$	∠HCH	113.7	∠FCF	108.3	
Dimethoxymethane			C$_a$—O	1.432	ED
			C$_b$—O	1.382	
			C—H (average)	1.108	
	∠COC	114.6	∠OCO	114.3	
	∠OCH	110.3			
Dimethylacetylene → 2-Butyne					
Dimethylamine	C—H	1.106	N—H	1.00	ED
(CH)$_2$NH	C—N	1.455	∠CNC	111.8	
	∠CNH	107	∠NCH	112	
	∠HCH	107			
Dimethylberyllium	Be—C	1.698	C—H	1.127	ED
(CH$_3$)$_2$Be	∠BeCH	113.9	CBeC linear		
Dimethylcadmium	C—Cd	2.112	∠HCH	108.4	R
(CH$_3$)$_2$Cd					
Dimethyl carbonate			C$_b$—O$_b$	1.209	ED
			C$_b$—O$_a$	1.34	
			C$_a$—O$_a$	1.42	
	∠O$_a$C$_b$O$_a$	107	∠C$_b$O$_a$C$_a$	114.5	
Dimethylcyanamide	C$_b$—N$_b$	1.161	C$_b$—N$_a$	1.338	ED
(C$_a$H$_3$)$_2$N$_a$—C$_b$≡N$_b$	C$_a$—N$_a$	1.463	∠C$_a$NC$_a$	115.5	
	∠C$_a$NC$_b$	116.0			
1,2-Dimethyldiborane			B—B	1.799	ED
			B—C	1.580	
			B—H$_b$	1.358 (*cis*), 1.365 (*trans*)	
	B—H$_t$	1.24			
	∠BBC	122.6 (*cis*), 121.8 (*trans*)			
Dimethyl diselenide	C—H	1.13	C—Se	1.95	ED
(CH$_3$)$_2$Se$_2$	Se—Se	2.326	∠CSeSe	98.9	
	∠HCSe 108	dihedral angle between the CSeSe and SeSeC planes 88			

Compound	Structure					Method
Dimethyl disulfide	C—S	1.816		S—S	2.029	ED
$(CH_3)_2S_2$	C—H	1.105		∠SSC	103.2	
	∠SCH	111.3	CSSC dihedral angle of internal rotation		85	
S,S'-Dimethyl dithiocarbonate	$C_aH_3SC_bSC_aH_3$			C_b—O	1.206	ED
	‖			C_b—S	1.777	
	O			C_a—S	1.802	
	∠OCS	124.9		∠CSC	99.3	
				syn-syn conformer		
Dimethyl ether	C—O	1.416		C—H	1.121	ED
$(CH_3)_2O$	∠COC	112		∠HCH	108	
Dimethylglyoxal → Biacetyl						
N,N'-Dimethylhydrazine	N—N	1.42		C—N	1.46	ED
$CH_3NH—NHCH_3$	N—H	1.03		C—H	1.12	
	∠NNC	112	CNNC dihedral angle of internal rotation		90	
Dimethylmercury	C—Hg	2.083		C—H	1.160 (assumed)	ED
$(CH_3)_2Hg$	Hg···H	2.71				
Dimethylphosphine	C—P	1.848		P—H	1.419	MW
$(CH_3)_2PH$	∠CPC	99.7		∠CPH	97.0	
Dimethyl selenide	C—H	1.093		Se—C	1.943	MW
$(CH_3)_2Se$	∠CSeC	96.2		∠SeCH	108.7	
	∠HCH	110.3				
Dimethyl sulfide	C—S	1.807		C—H	1.116	ED, MW
$(CH_3)_2S$	∠CSC	99.05		∠HCH	109.3	
Dimethyl sulfone	C—H	1.114		S—O	1.435	ED
$(CH_3)_2SO_2$	S—C	1.771		∠CSC	102	
	∠OSO	121				
Dimethyl sulfoxide	C—H	1.081		C—S	1.799	MW
$(CH_3)_2SO$	S—O	1.485		∠CSC	96.6	
	∠CSO	106.7		∠HCH	110.3	
	dihedral angle between the SCC plane and the S—O bond				115.5	
Dimethylzinc	Zn—C	1.929		∠HCH	107.7	R
$(CH_3)_2Zn$						
1,4-Dioxane	C—C	1.523		C—O	1.423	ED
	C—H	1.112		∠COC	112.45	
	∠CCO	109.2	chair form			
Ethanal → Acetaldehyde						
Ethane	C—C	1.5351		C—H	1.0940	MW
C_2H_6	∠CCH	111.17	staggered conformation			
Ethanethiol	$C_bH_3—C_aH_2—SH$			C_a—H	1.090	MW
	C_b—H	1.093		C_a—C_b	1.530	
	C_a—S	1.829		S—H	1.350	
	∠C_bC_aH	109.6		∠C_aC_bH	109.7	
	∠C_bC_aS	108.3		∠C_aSH	96.4	
Ethanol	$C_bH_3C_aH_2OH$			C—C	1.512	MW
	C—O	1.431		O—H	0.971	
	C_a—H	1.10		C_b—H	1.09	
	∠CCO	107.8		∠COH	105	
	∠C_bC_aH	111		∠C_aC_bH	110	
	staggered conformation					
Ethyl chloride	Cl			C—C	1.528	ED, MW
				C—Cl	1.802	
				C—H	1.103	
	H_b H_a			C_a—H_a=C_b—H_b (assumed)		
				∠CCCl	110.7	
	∠$H_bC_bH_b$	109.8		∠$H_aC_aH_a$	109.2	
	∠$C_bC_aH_a$	110.6				
Ethylene	C—H	1.087		C—C	1.339	MW
$CH_2=CH_2$	∠CCH	121.3				

Compound	Structure				Method

Ethylenediamine
$H_2NCH_2CH_2NH_2$

C—N	1.469	C—C	1.545		ED
C—H	1.11	∠CCN	110.2		

gauche conformer   dihedral angle between the NCC and CCN planes   64

Ethylene dibromide → 1,2-Dibromoethane
Ethylene dichloride → 1,2-Dichloroethane
Ethyleneimine → Aziridine

Ethylene oxide

CH_2
|
O
|
CH_2

C—C	1.466	C—H	1.085		MW
C—O	1.431	∠HCH	116.6		

dihedral angle between the NH_2 plane and the N—C bond   158.0

Ethylene sulfide → Thiirane

Ethyl methyl ether
$C_2H_5OCH_3$

C—O (average)	1.418	C—C	1.520		ED
C—H (average)	1.118	∠COC	111.9		
∠OCC	109.4	∠HCH	109.0		

fraction of the *trans* conformer at 20°C   80%

Ethyl methyl ketone

O
||
C_aH_3 — C_c
     C_bH_2 C_dH_3

		C—C (average)	1.518		ED
		C_c—O	1.219		
		C—H (average)	1.102		
∠C_bC_cO, ∠C_dC_cO	121.9	∠$C_aC_bC_c$	113.5		
		trans conformer	95%		

Ethyl methyl sulfide
$C_2H_5SCH_3$

C—S (average)	1.813	C—C	1.536		ED
C—H	1.111	∠CSC	97		
∠SCC	114.0	∠HCH	110		

fraction of the *gauche* conformer at 20°C   75%

Ferrocene
$(C_5H_5)_2Fe$

C—C	1.440	C—H	1.104		ED
Fe—C	2.064	(D_{5h})			

Fluoroform
CHF_3

C—H	1.098	C—F	1.332		MW
∠FCF	108.8	(C_{3v})			

Formaldehyde
H_2CO

C—H	1.116	C—O	1.208		MW
∠HCH	116.5				

Formaldehyde azine

$H_2C=N—N=CH_2$		N—N	1.418		ED
C—N	1.277	C—H	1.094		
∠CNN	111.4	∠HCN	120.7		

fraction of the *trans* conformer at –30°C   91%

Formaldehyde dimethylacetal → Dimethoxymethane

Formaldoxime

H_a OH_c
      C=N
H_b

		C—H_a	1.085		MW
		C—H_b	1.086		
		C—N	1.276		
N—O	1.408	O—H_c	0.956		
∠H_bCN	115.6	∠CNO	110.2		
∠H_aCN	121.8	∠NOH_c	102.7		

Formamide

H_c O
      N—C
H_b H_a

		C—H_a	1.125		ED, MW
		N—H	1.027		
		C—N	1.368		
C—O	1.212	∠NCO	125.0		
∠CNH (average)	119.2				

Formic acid

          O_a
H—C
          O_b—H

		C—O_a	1.202		MW
		C—O_b	1.343		
O_b—H	0.972	C—H	1.097		
∠HCO_a	124.1	∠O_aCO_b	124.9		
∠CO_bH	106.3	planar			

Compound	Structure			Method

Formic acid dimer

....

.... //

$O_a \cdots O_b$	2.703		ED
$C-O_a$	1.220		
$C-O_b$	1.323		
$\angle O_a C O_b$	126.2		
$\angle C O_a O_b$	108.5		

Formyl radical

\

•

| $C-H$ | 1.110 | $C-O$ 1.1712 | MW |
| $\angle HCO$ | 127.43 | | |

Fulvene

$C_d H_2$

C_a

$HC_b \qquad C_b H$

$HC_c-C_c H$

C_a-C_d	1.349	MW
C_a-C_b	1.470	
C_b-C_c	1.355	
C_c-C_c	1.476	
C_b-H	1.078	
C_c-H	1.080	

C_d-H	1.13		$\angle C_b C_a C_b$	106.6
$\angle C_a C_b C_c$	107.7		$\angle C_b C_c C_c$	109
$\angle C_a C_b H$	124.7		$\angle C_b C_c H$	126.4
$\angle H C_d H$	117			

2-Furaldehyde

H   H

$C_c \quad C_b$

H C_d

O_a

C_a-C_e	1.458	MW
C_e-O_b	1.250	
C_e-H	1.088	
$\angle C_e C_a C_b$	133.9	

| $\angle C_a C_e H$ | 116.9 | $\angle C_a C_e O$ | 121.6 |

trans conformer (with respect to the O_a and O_b atoms)

Furan

H_a O H_a

$C_a \qquad C_a$

C_b-C_b

$H_b \qquad H_b$

C_b-C_b	1.431	MW
C_a-C_b	1.361	
C_a-O	1.362	
C_a-H_a	1.075	
C_b-H_b	1.077	

$\angle C_a C_b C_b$	106.1	$\angle C_a O C_a$	106.6
$\angle C_b C_a O$	110.7	$\angle O C_a H_a$	115.9
$\angle C_b C_b H_b$	128.0		

Furfural → 2-Furaldehyde

Glycolaldehyde

$H_c \qquad O_b$

C_b

$H_b \diagup C_a \diagdown H_a$

$H_b \qquad O_a$

C_b-O_b	1.209	MW
C_a-O_a	1.437	
C_a-C_b	1.499	
O_a-H_a	1.051	
C_b-H_c	1.102	
C_a-H_b	1.093	

$\angle C_a C_b O_b$	122.7	$\angle C_a C_b H_c$	115.3
$\angle C_b C_a O_a$	111.5	$\angle C_a O_a H_a$	101.6
$\angle C_b C_a H_b$	109.2	$\angle H_b C_a H_b$	107.6
$\angle H_b C_a O_a$	109.7		

Glyoxal
CHOCHO

$C-C$	1.526	$C-O$	1.212	ED, UV
$C-H$	1.132	$\angle CCO$	121.2	
$\angle HCO$	112	*trans* conformer	(C_{2h} (assumed))	

Hexachloroethane
Cl₃CCCl₃

| $C-C$ | 1.56 | $C-Cl$ | 1.769 | ED |
| $\angle CCCl$ | 110.0 | | | |

2,4-Hexadiyne

$C_a H_3-C_b \equiv C_c-C_c \equiv C_b-C_a H_3$

				ED
C_a-C_b	1.450	C_b-C_c	1.208	
C_c-C_c	1.377	C_a-H	1.09	

Hexafluoroethane
F₃CCF₃

| $C-C$ | 1.545 | $C-F$ | 1.326 | ED |
| $\angle CCF$ | 109.8 | staggered conformation | | |

Hexafluoropropene
CF₂=CFCF₃

average value of the C=C and C—F distances   1.329   ED

$C-C$	1.513	$\angle CCC$	127.8	
$\angle FCC$ (CF₂)	124	$\angle FCC$ (CF)	120	
$\angle FCC$ (CF₃)	110			

Compound	Structure				Method
1,3,5-Hexatriene	$H_2C_a=C_bH-C_cH=C_cH-C_bH=C_aH_2$				ED
	C_a-C_b	1.337	C_b-C_c	1.458	
	C_c-C_c	1.368	$\angle C_aC_bC_c$	121.7	
	$\angle C_bC_cC_c$	124.4			
Iminocyanide radical	N—H	1.034	N···N	2.470	UV
$H\dot{N}CN$	$\angle HNC$	116.5	$\angle NCN$	~180	
Iodocyanoacetylene	$I-C_a$	1.985	C_a-C_b	1.207	MW
$I-C_a\equiv C_b-C_c\equiv N$	C_b-C_c	1.370	C_c-N	1.160	
Isobutane	C_a-H	1.122	C_b-H	1.113	ED, MW
$(C_bH_3)_3C_aH$	C_a-C_b	1.535	$\angle C_bC_aC_b$	110.8	
	$\angle C_aC_bH$	111.4			
Isobutylene → 2-Methylpropene					
Ketene	C—C	1.317	C—O	1.161	MW
$CH_2=C=O$	C—H	1.080	$\angle HCH$	123.0	
Malononitrile	C—H	1.091	C—C	1.480	MW
$C_aH_2(C_bN)_2$	C—N	1.147	$\angle CCC$	110.4	
	$\angle HCH$	108.4	$\angle CCN$	176.6	
	(The two N atoms are bent away from each other in the plane of $C_b-C_a-C_b$)				
Methane	C—H (r_e)	1.0870	(T_d)		MW
CH_4					
Methanethiol	C—H	1.09	C—S	1.819	MW
CH_3SH	S—H	1.34	$\angle HSC$	96.5	
	$\angle HCH$	109.8			
	angle between the CH_3 symmetry axis and the C—S bond 2.2.				
	(The axis of the CH_3 group is tilted away from the H atom with respect to the C—S bond.)				
Methanol	C—H	1.0936	C—O	1.4246	MW
CH_3OH	O—H	0.9451	$\angle HCH$	108.63	
	$\angle COH$	108.53			
	angle between the CH_3 symmetry axis and the C—O bond				
	(The axis of the CH_3 group is tilted away from the H atom				
	with respect to the C—O bond.)			3.27	
Methyl radical	C—H	1.08	planar		UV
$\cdot CH_3$					
N-Methylacetamide			C_a-C_b	1.520	ED
			$N-C_c$	1.469	
			C—H	1.107	
	C_b-N	1.386	C_b-O	1.225	
	$\angle C_bNC_c$	119.7			
	$\angle NC_bO$	121.8			
	$\angle C_aC_bN$	114.1			
Methylacetylene → Propyne					
Methylal → Dimethoxymethane					
Methylamine	N—H	1.010	C—N	1.471	MW
CH_3NH_2	C—H	1.099	$\angle NHN$	107.1	
	$\angle HNC$	110.3	$\angle HCH$	108.0	
	dihedral angle between the CH_3 symmetry axis and the				
	C—N bond (The axis of the CH_3 group is tilted away				
	from the NH_2 group with respect to the C—N bond.)			2.9	
Methyl azide	CH_3		C—H	1.09	ED
	$N_a-N_b-N_c$		$C-N_a$	1.468	
			N_a-N_b	1.216	
	N_b-N_c	1.113	$\angle CN_aN_b$	116.8	
	NNN linear				
Methyl bromide	C—H (r_e)	1.086	C—Br (r_e)	1.933	MW, IR
CH_3Br	$\angle HCH$ (θ_e)	111.2	(C_{3v})		
Methyl chloride	C—H	1.090	C—Cl	1.785	MW, IR
CH_3Cl	$\angle HCH$	110.8			

N-Methylacetamide structure:

$$H_3C_a \diagdown \qquad H$$
$$\qquad C_b-N$$
$$O \diagup \qquad \diagdown C_cH_3$$

Compound	Structure					Method

Methyldiazirine

CH_3CH (with N=N diazirine ring)

C—N	1.481	C—C	1.501	MW
N—N	1.235	∠NCN	49.3	

dihedral angle between the CNN plane and the C—C bond   122.3

Methylene
:CH_2

C—H	1.078	∠HCH	130	LMR

Methylenecyclopropane

C_cH_2 — $C_b = C_aH_2$, C_cH_2

		$C_a—C_b$	1.332	MW
		$C_b—C_c$	1.457	
		$C_c—C_c$	1.542	
$C_c—H$	1.09	∠$C_cC_bC_c$	63.9	
∠HC_aH	114.3	∠HC_cH	113.5	

dihedral angle between the C_cH_2 plane and the $C_c—C_c$ bond   150.8

3-Methyleneoxetane

O with C_cH_2, C_cH_2, $C_b = C_aH_2$

		$C_b—C_c$	1.52	MW
		$C_c—O$	1.45	
		$C_a—C_b$	1.33	
C—H	1.09 (assumed)	∠$C_cC_bC_c$	87	
∠HC_cH	114 (assumed)	∠HC_aH	120 (assumed)	

Methyl fluoride
CH_3F

C—H (r_e)	1.095			MW, IR
C—F (r_e)	1.382	∠HCH (θ_e)	110.45 (C_{3v})	

Methyl formate

C_aH_3, $O_a—C_b$, O_b, H_b

		$C_a—H$	1.08	ED
		$C_b—O_b$	1.206	
		C—O (average)	1.393	
		$C_b—H$	1.101 (assumed)	
∠$O_aC_bO_b$	127	∠COC	114	
∠O_aC_aH	110			

Methylgermane
CH_3GeH_3

C—H	1.083	Ge—H	1.529	MW
C—Ge	1.945	∠HCH	108.4	
∠HGeH	109.3			

Methyl hypochlorite
CH_3OCl

C—H	1.103	O—Cl	1.674	MW
O—C	1.389	∠HCH	109.6	
∠COCl	112.8			

Methylidyne radical
:ĊH

C—H (r_e)	1.1198	UV

Methylidyne phosphide
HCP

H—C (r_e)	1.0692	C—P (r_e)	1.5398	MW

Methyl iodide
CH_3I

C—H (r_e)	1.084	C—I (r_e)	2.132	MW, IR
∠HCH (θ_e)	111.2	(C_{3v})		

Methyl isocyanide
C_aH_3—N≡C_b

$C_a—H$	1.102	$C_a—N$	1.424	MW
N—C_b	1.166	∠NC_aH	109.12	

Methylketene

C_cH_3, $C_b = C_a = O$, H

		O—C_a	1.171	MW
		$C_b—C_c$	1.518	
		$C_c—H$	1.10	
$C_a—C_b$	1.306	$C_b—H$	1.083	
∠OC_aC_b	180.5	∠$C_aC_bC_c$	122.6	
∠C_aC_bH	113.7	∠C_cC_bH	123.7	
∠HCH	109.2			

Methylmercury chloride
CH_3HgCl

Hg—Cl	2.282	C—H	1.15	MW,
Hg—C	1.99	(C_{3v})		NMR

Methyl nitrate

H_a, H_a, O_a, C, N, H_b, O, O_b

		$C—H_a$	1.10	MW
		$C—H_b$	1.09	
		C—O	1.437	
		O—N	1.402	
N—O_a	1.205	N—O_b	1.208	
∠OCH_a	110	∠OCH_b	103	
∠CON	112.7	∠ONO_a	118.1	
∠ONO_b	112.4			

Compound	Structure					Method

Methylphosphine
CH_3PH_2

C—P	1.858	C—H	1.094		ED

2-Methylpropane → Isobutane

2-Methylpropene

$$C_aH_3 \quad \underset{C_b=C_c}{\overset{H_c}{\diagdown}} \quad C_aH_3 \quad \diagdown H_c$$

		C_a—H	1.119	ED, MW
		C_c—H_c	1.10	
		C_a—C_b	1.508	
		C_b—C_c	1.342	
∠HC_aC_b (average)	111.4	∠$H_cC_cH_c$	118.5	
∠$C_aC_bC_a$	115.6	∠$C_aC_bC_c$	122.2	
∠HC_aH	107.9	∠C_bC_cH	121	

Methylsilane
CH_3SiH_3

C—H	1.093	C—Si	1.867	MW
Si—H	1.485	∠HCH	107.7	
∠HSiH	108.3	(C_{3v})		

Methylstannane
CH_3SnH_3

C—Sn	2.143	Sn—H	1.700	MW
		(C_{3v})		

Methyl thiocyanate

$$C_aH_3 \\ \underset{}{S}—C_b—N$$

		S—C_b	1.684	S—C_a	1.824	MW
		C_b—N	1.170	C—H	1.081	
∠C_aSC_b	99.0	∠HCH	110.6	∠HCS	108.3	

Naphthalene

C_a—C_b	1.37	ED
C_b—C_b	1.41	
C_a—C_c	1.42	
C_c—C_c	1.42	
C—C (average)	1.40	
∠$C_aC_cC_c$	119.4	

Neopentane
$C(CH_3)_4$

C—C	1.537	C—H	1.114	ED
∠CCH	112			

Nickelocene → Bis (cyclopentadienyl) nickel

Nitromethane
CH_3NO_2

C—H	1.088 (assumed)	C—N	1.489	MW
N—O	1.224	∠NCH	107	
∠ONO	125.3			

***N*-Nitrosodimethylamine**
$(CH_3)_2NNO$

N—O	1.235	N—N	1.344	ED
C—N	1.461	∠ONN	113.6	
∠CNC	123.2	∠CNN	116.4	

Nitrosomethane
CH_3NO

C—N	1.49	N—O	1.22	MW
C—H	1.084	∠CNO	112.6	
∠NCH	109.0			

Norbornane → Bicyclo[2.2.1]heptane

Norbornadiene → Bicyclo[2.2.1]hepta-2,5-diene

1,2,5-Oxadiazole

O—N	1.380	∠NON	110.4	MW
C—N	1.300	∠ONC	105.8	
C—C	1.421	∠CCN	109.0	
C—H	1.076	∠CCH	130.2	
∠NCH	120.9	planar		

1,3,4-Oxadiazole

O—C	1.348	∠COC	102.0	MW
C—N	1.297	∠OCN	113.4	
N—N	1.399	∠CNN	105.6	
C—H	1.075	∠OCH	118.1	
∠NCH	128.5	planar		

Oxalic acid

C—C	1.544	ED
C—O_a	1.205	
C—O_b	1.336	
O_b—H	1.05	
∠CCO_a	123.1	
∠O_aCO_b	125.0	
∠CO_bH	104	

Compound	Structure					Method

Oxalyl dichloride

	C—O	1.182	ED
	C—C	1.534	
	C—Cl	1.744	
∠CCO 124.2	∠CCCl	111.7	

fraction of the *trans* conformer at 0°C 68%, that of the *gauche* conformer 32%

Oxetane

C—O	1.448	MW
C—C	1.546	
C—H (average)	1.090	
∠COC	92	
∠CCC 85	∠OCC	92
∠HCH (average) 109.9		

Oxirane → Ethylene oxide

Phenol

C—C (average)	1.397	MW
C_b—H	1.084	
C_c—H	1.076	
C_d—H	1.082	
C_a—O	1.364	
O—H	0.956	
∠COH	109.0	

Phosphirane

C—P	1.867	P—H	1.43	MW
C—C	1.502	C—H	1.09	
∠CPC	47.4	∠HCH	114.4	
∠HPC 95.2	∠CCH 118			

dihedral angle between the PCC plane and the PH bond 95.7

Piperazine

C—C	1.540	ED
C—N	1.467	
C—H	1.110	
∠CNC 109.0	∠CCN	110.4
	(C_{2h})	

Pivalonitrile
$(C_cH_3)_3C_b$—C_a≡N

C_a—C_b	1.495	C_a—N	1.159	MW
C_b—C_c	1.536	∠$C_cC_bC_c$	110.5

Propadiene → Allene

Propane C_3H_8

C—C	1.532	C—H	1.107	ED
∠CCC	112	∠HCH	107

Propenal → Acrylaldehyde

Propene

C_a—H_a	1.104	ED, MW
C_a—C_b	1.341	
C_c—H_d	1.117	
C_b—C_c	1.506	
∠$C_bC_aH_{a,b,c}$	121.3	
∠$C_bC_cH_d$ 110.7	∠$C_aC_bC_c$	124.3

1-Propenyl chloride
CH_3—C_bH=C_aH—Cl C_a—Cl   1.728   MW
∠C_bC_aCl 121.9   *trans* conformer

Propiolaldehyde
H_aC_a≡C_b—C_cH_cO

		C_a—H_a	1.085	ED, MW
C_a—C_b	1.211	C_b—C_c	1.453
C_c—H_c	1.130	C_c—O	1.214
∠C_bC_cO	124.2	∠$C_bC_cH_c$	113.7
∠$C_aC_bC_c$	178.6	planar	

Propylene → Propene

Propylene oxide
$C_aH_3C_bH$——C_cH_2
O

C_a—C_b	1.51	MW
∠$C_aC_bC_c$	121.0	

dihedral angle between the C_bC_cO plane and the C_aC_b bond 123.8

Compound	Structure						Method

Propynal → Propiolaldehyde

Propyne

H_3C_c—C_b≡C_aH

C_c—C_b	1.459	C_c—H	1.105	MW
C_a—H	1.056	C_b—C_a	1.206	
		∠HC_cC_b	110.2	

Pyrazine

N   N
═

C—C	1.339	C—N	1.403	ED
C—H	1.115	∠CCN	115.6	
∠CCH	123.9			

Pyridazine

H H
C_b=C_b
HC$_a$ C_aH
N—N

N—C_a	1.341		ED, MW
C_a—C_b	1.393		
N—N	1.330		
C_b—C_b	1.375		

∠NCC	123.7	∠NNC	119.3

Pyridine

H_c
H_b C_c H_b
C_b C_b
C_a C_a
H_a N H_a

N—C_a	1.340	MW
C_b—C_c	1.394	
C_b—H_b	1.081	
C_a—C_b	1.395	
C_a—H_a	1.084	
C_c—H_c	1.077	

∠C_aNC_a	116.8	∠NC_aC_b	123.9
∠$C_aC_bC_c$	118.5	∠$C_bC_cC_b$	118.3
∠NC_aH_a	115.9	∠$C_cC_bH_b$	121.3

Pyrimidine

═   N

N

N—C	1.340	C—C	1.393	ED
∠NCN	127.6	∠CNC	115.5	
(C_{2v} assumed)				

Pyrrole

H
N
H_a—C_a C_a—H_a
H_b—C_b—C_b—H_b

N—Ca	1.370		MW
C_b—C_b	1.417		
C_a—C_b	1.382		
N—H	0.996		
C_a—H_a	1.076		

C_b—H_b	1.077	∠C_aNC_a	109.8
∠NC_aC_b	107.7	∠$C_aC_bC_b$	107.4
∠NC_aH_a	121.5	∠C_bC_bH	127.1

Pyruvonitrile

O
C_aH_3—C_b
C_c≡N

C—H	1.12		ED, MW
C—N	1.17		
C—O	1.208		
C_b—C_c	1.477		

C_a—C_b	1.518	∠HCH	109.2
∠C_aC_bO	124.5	∠$C_aC_bC_c$	114.2
∠CCN	179		

Ruthenocene → Bis (cyclopentadienyl) ruthenium

Silacyclobutane

CH_2—CH_2
CH_2—SiH_2

Si—C	1.892		ED
C—C	1.600		
Si—H	1.47		

C—H	1.14	∠CSiC	80.7
∠SiCC	84.8	∠CCC	99.8
dihedral angle between the CCC and CSiC planes			146

Spiropentane

H_2C_b C_bH_2
C_a
H_2C_b C_bH_2

C_b—C_b	1.52		ED
C_a—C_b	1.47		
C—H	1.09		
∠$C_bC_aC_b$	62		

∠HCH	118	(D_{2d})	

Compound	Structure					Method

Succinonitrile

CH₂CN
|
CH₂CN

C—C	1.561	C—C(N)	1.465		ED
C—N	1.161	C—H	1.09		
∠CCC	110.4				

fraction of the *anti* conformer at 170°C 74%,
dihedral angle of CCCC for the *gauche* conformer   75

Tetrachloroethylene
$CCl_2=CCl_2$

C—Cl	1.718	C—C	1.354	ED
∠ClCCl	115.7			

Tetracyanoethylene
$(CN)_2C=C(CN)_2$

C—N	1.162	C—C	1.435	ED
C=C	1.357	∠CC=C	121.1	

Tetrafluoro-1,3-dithietane

F₂C, CF₂ with S atoms (ring structure)

	C—S	1.785	ED
	C—F	1.314	
	∠CSC	83.2	
∠FCS	113.7	(D₂ₕ assumed)	

Tetrafluoroethylene
$CF_2=CF_2$

C—C	1.31	C—F	1.319	ED
∠CCF	123.8	(D₂ₕ assumed)		

Tetrahydrofuran

CH₂CH₂
|
    O
|
CH₂CH₂

C—H	1.115	C—O	1.428	ED
C—C	1.536			

The skeletal bending vibration of the molecular plane is essentially free
pseudorotation

Tetrahydropyran

(ring structure with H₂C, CH₂ and O)

C—O	1.420		ED
C—C	1.531		
C—H	1.116		
∠COC	111.5		
∠OCC	111.8		

∠CCC (C)	108	∠CCC (O)	111

chair form

Tetrahydrothiophene

CH₂CH₂
|
    S
|
CH₂CH₂

C—S	1.839	C—H	1.120
C—C	1.536	∠CSC	93.4
∠SCC	106.1	∠CCC	105.0

Tetramethylgermane
$(CH_3)_4Ge$

Ge—C	1.945	C—H	1.12	ED
∠GeCH	108	(Tₐ excluding the H atoms)		

Tetramethyllead
$(CH_3)_4Pb$

Pb—C	2.238	(Tₐ excluding the H atoms)	ED

Tetramethylsilane
$(CH_3)_4Si$

C—H	1.115	C—Si	1.875	ED
∠HCH	109.8	(Tₐ excluding the H atoms)		

Tetramethylstannane
$(CH_3)_4Sn$

C—Sn	2.144		ED
C—H	1.12	(Tₐ excluding the H atoms)	

1,2,5-Thiadiazole

(ring: S, N, N, HC—CH)

S—N	1.631	∠NSN	99.6		MW
C—N	1.328	∠CCN	113.8		
C—C	1.420	∠CCH	126.2		
C—H	1.079	planar			

1,3,4-Thiadiazole

(ring: S, HC, CH, N—N)

S—C	1.721	∠CSC	86.4	MW
N—N	1.371	∠SCN	114.6	
C—N	1.302	∠CCN	112.2	
C—H	1.08	∠SCH	121.9	
∠NCH	123.5	planar		

Thietane

—   S

		C—S	1.847	ED, MW
		C—C	1.549	
		C—H (average)	1.100	
∠CSC	76.8	∠HCH (average)	112	

dihedral angle between the CCC and CSC planes   154

Thiirane

H₂C
| S
H₂C

C—C	1.484	∠HCH	116	MW
C—H	1.083	∠CSC	48.3	
C—S	1.815	∠CCS	65.9	

dihedral angle between the CH₂ plane and the
C—C bond   152

Compound	Structure				Method
Thioformaldehyde	C—S	1.611	C—H	1.093	MW
CH₂S	∠HCH	116.9			
Thioformamide			N—Hₐ	1.002	MW
			N—H_b	1.007	
			C—N	1.358	
	C—S	1.626	C—H_c	1.10	
	∠HₐNH_b	121.7	∠HₐNC	117.9	
	∠H_bNC	120.4	∠NCS	125.3	
	∠NCH_c	108	∠SCH_c	127	
Thiolane → Tetrahydrothiophene					
Thiophene			Cₐ—Hₐ	1.078	MW
			C_b—H_b	1.081	
	C_b—		Cₐ—S	1.714	
			Cₐ—C_b	1.370	
			C_b—C_b	1.423	
			∠CₐSCₐ	92.2	
	∠SCₐC_b	115.5	∠CₐC_bC_b	112.5	
	∠SCₐHₐ	119.9	∠C_bC_bH_b	124.3	
Toluene	CH₃	C—C (ring)	1.399	C—CH₃ 1.524	ED
	//	C—H (average)	1.11		
		the difference between the C—H(CH₃) and			
		C—H(ring): about 0.01			
1,1,1-Tribromoethane	C—Br	1.93	C—H	1.095 (assumed)	MW
CH₃CBr₃	C—C	1.51 (assumed)	∠CCBr	108	
	∠BrCBr	111	∠CCH	109.0 (assumed)	
Tribromomethane → Bromoform					
Tri-*tert*-butyl methane	Cₐ—C_b	1.611	C—H	1.111	ED
HCₐ[C_b(C_cH₃)₃]₃	C_b—C_c	1.548	∠CₐC_bC_c	113.0	
Tricarbon dioxide	C—O	1.163	C—C	1.289	ED
OCCCO	linear (with a large-amplitude bending vibration)				
Trichloroacetonitrile	C—N	1.165	C—C	1.460	ED
CCl₃CN	C—Cl	1.763	∠ClCCl	110.0	
1,1,1-Trichloroethane	C—H	1.090	C—C	1.541	MW
CH₃CCl₃	C—Cl	1.771	∠HCH	110.0	
	∠CCH	108.9	∠ClCCl	109.4	
	∠CCCl	109.6			
Trichloro(methyl)germane	Ge—Cl	2.132	Ge—C	1.89	ED, MW
CH₃GeCl₃	C—H	1.103 (assumed)	∠ClGeCl	106.4	
	∠GeCH	110.5 (assumed)			
Trichloro(methyl)silane	C—Si	1.876	Si—Cl	2.021	MW
CH₃SiCl₃			(C₃ᵥ)		
Trichloro(methyl)stannane	Sn—Cl	2.304	Sn—C	2.10	ED
CH₃SnCl₃	C—H	1.100	∠CSnCl	113.9	
	∠ClSnCl	104.7	∠SnCH	108	
Triethylenediamine → 1,4-Diazabicyclo [2.2.2]octane					
Trifluoroacetic acid	//Oₐ		C—F	1.325	ED
			C—C	1.546	
			C—Oₐ	1.192	
	O_bH				
	C—O_b	1.35	O—H	0.96 (assumed)	
	∠CCOₐ	126.8	∠CCO_b	111.1	
	∠CCF	109.5			
1,1,1-Trifluoroethane	C—C	1.494	C—F	1.340	ED
CH₃CF₃	C—H	1.081	∠CCF	119.2	
	∠CCH	112			
Trifluoromethane → Fluoroform					

Compound	Structure				Method
1,1,1-Trifluoro-2,2,2-trichloroethane	C—C	1.54	C—F	1.33	MW
CF_3CCl_3	C—Cl	1.77	∠CCF	110	
	∠CCCl	109.6	staggered conformation		
Trimethylaluminium	C—H	1.113	Al—C	1.957	ED
$(CH_3)_3Al$	∠AlCH	111.7	∠CAlC	120	
Trimethylamine	C—N	1.458	C—H	1.100	ED
$(CH_3)_3N$	∠CNC	110.9	∠HCH	110	
Trimethylarsine	C—As	1.979	∠CAsC	98.8	ED
$(CH_3)_3As$	∠AsCH	111.4			
Trimethylbismuth	Bi—C	2.263	C—H	1.07	ED
$(CH_3)_3Bi$	∠CBiC	97.1			
Trimethylborane	C—B	1.578	C—H	1.114	ED
$(CH_3)_3B$	∠CBC	120.0	∠BCH	112.5	
Trimethyleneimine → Azetidine					
Trimethylphosphine	C—P	1.847	C—H	1.091	ED
$(CH_3)_3P$	∠CPC	98.6	∠PCH	110.7	
1,3,5-Trioxane			C—O	1.422	MW
			∠OCO	112.2	
			∠COC	110.3	

1,3,5-Trioxane structure:

```
        O
       / \
  H2C      CH2
   |        |
   O        O
    \      /
       C
       H2
```

Compound	Structure				Method
Triphenylamine	C—C	1.392	C—N	1.42	ED
$(C_6H_5)_3N$	∠CNC	116	(C_3)		

torsional dihedral angle of the two phenyl rings 47° (defined to be 0
when the symmetry axis is contained in the phenyl planes)

Tropone		C_a—O	1.23	ED
		C_a—C_b	1.45	
		C_b—C_c	1.36	
		C_c—C_d	1.46	
		C_d—C_d	1.34	
		∠$C_bC_aC_b$	122	
		∠$C_aC_bC_c$	133	

Tropone structure:

```
            O
            ||
            Ca
          /    \
     HCb        CbH
      ||         ||
     HCc        CcH
          \    /
        Cd=Cd
        H    H
```

∠$C_bC_cC_d$	126	∠$C_cC_dC_d$	130	
		(C_{2v})		

Vinylacetylene		C_b—C_c	1.434	ED, MW
		C_a—C_b	1.344	
		C_c—C_d	1.215	
		C_a—H_a	1.11	
		C_d—H_d	1.09	
		∠$C_aC_bC_c$	123.1	

Vinylacetylene structure:

```
 Ha          Hc
   \         /
    Ca = Cb
   /         \
 Hb           Cc ≡ Cd
                      \
                       Hd
```

∠$C_bC_cC_d$	178	∠$H_aC_aC_b$	119	
∠$H_bC_aC_b$	122	∠$H_cC_bC_a$	122	
∠$C_cC_dH_d$	182			

Vinyl chloride		C—C	1.342	ED, MW
		C—Cl	1.730	
		C—H	1.09	

Vinyl chloride structure:

```
 Hc          Cl
   \         /
    C  =  C
   /         \
 Hb           Ha
```

∠CCCl	122.5	∠CCH_a	124	
∠CCH_b	120	∠CCH_c	121.1	

DIPOLE MOMENTS OF MOLECULES IN THE GAS PHASE

David R. Lide

This table gives values of the electric dipole moment for about 500 gas-phase molecules. When available, values determined by microwave spectroscopy, molecular beam electric resonance, and other spectroscopic techniques were selected; otherwise, the values come from measurements of the dielectric constant in the gas phase. Compounds are listed by molecular formula in the modified Hill order; compounds not containing carbon are listed first, followed by compounds containing carbon.

The dipole moment μ is given in debye units (D). The conversion factor to SI units is $1\,D = 3.33564 \times 10^{-30}\,C\,m$. The column headed Acc. indicates the estimated accuracy of the value through the following codes:

A $\pm 1\%$ for $\mu > 1.0\,D$; $\pm 0.01\,D$ for $\mu < 1.0\,D$
B $\pm 2\%$ for $\mu > 1.0\,D$; $\pm 0.02\,D$ for $\mu < 1.0\,D$
C $\pm 5\%$ for $\mu > 1.0\,D$; $\pm 0.05\,D$ for $\mu < 1.0\,D$
D $\pm 10\%$ for $\mu > 1.0\,D$; $\pm 0.10\,D$ for $\mu < 1.0\,D$
Q Questionable value
i Indicates the value may be ambiguous because of the presence of rotational isomers

Dipole moments of individual rotational isomers are given when they have been measured. The isomers are designated as *gauche*, *trans-trans*, etc. The meaning of these designations can be found in the references. Other information, such as the components of the dipole moment in the molecular framework and the variation with vibrational state and isotopic species, may also be found in the references.

REFERENCES

1. R. D. Nelson, D. R. Lide, and A. A. Maryott, *Selected Values of Electric Dipole Moments for Molecules in the Gas Phase*, Natl. Stand. Ref. Data Ser. — Natl. Bur. Stnds. 10, 1967.
2. K. H. Hellwege, Ed., *Landolt-Börnstein, Numerical Data and Functional Relationships in Science and Technology*, Group II, Vol. 6, Molecular Constants, Springer-Verlag, Heidelberg, 1974.
3. K. H. Hellwege, Ed., *Landolt-Börnstein, Numerical Data and Functional Relationships in Science and Technology*, Group II, Vol. 14, Subvolume a, Springer-Verlag, Heidelberg, 1982.
4. K. H. Hellwege, Ed., *Landolt-Börnstein, Numerical Data and Functional Relationships in Science and Technology*, Group II, Vol. 14, Subvolume b, Springer-Verlag, Heidelberg, 1983.
5. R. Takashi, et. al., *J. Mol. Spec.*, 138, 450, 1989.
6. M. Hayashi and T. Inagusa, *J. Mol. Spec.*, 138, 135, 1989.
7. F. Scappini, A. C. Fantoni, and W. Caminati, *J. Mol. Spec.*, 120, 101, 1986.

Mol. form.	Name	μ/D	Acc.	Ref.
Compounds not containing carbon				
AgCl	Silver chloride	5.70	C	2
AgF	Silver fluoride	6.22	C	2
AgI	Silver iodide	5.10	C	4
AlF	Aluminum fluoride (AlF)	1.53	D	1
$AsCl_3$	Arsenic trichloride	1.59	C	1
AsF_3	Arsenic trifluoride	2.59	B	1
AsH_3	Arsine	0.20	C	1
BF	Fluoroborane (BF)	0.5	Q	2
B_4H_{10}	Tetraborane	0.486	A	3
B_5H_9	Pentaborane	2.13	B	1
B_6H_{10}	Hexaborane	2.50	B	3
BaO	Barium oxide	7.954	A	2
BaS	Barium sulfide	10.86	A	2
BrCl	Bromine chloride	0.519	A	3
BrF	Bromine fluoride	1.42	A	3
BrF_5	Bromine pentafluoride	1.51	D	1
BrH	Hydrogen bromide	0.827	A	3
BrH_3Si	Bromosilane	1.319	A	3
BrI	Iodine bromide	0.726	A	4

Mol. form.	Name	μ/D	Acc.	Ref.
BrK	Potassium bromide	10.628	A	2
BrLi	Lithium bromide	7.268	A	2
BrNO	Nitrosyl bromide	1.8	Q	1
BrNa	Sodium bromide	9.118	A	2
BrO	Bromine oxide (BrO)	1.76	B	2
BrTl	Thallium bromide (TlBr)	4.49	A	2
CaCl	Calcium monochloride	3.6	Q	4
ClCs	Cesium chloride	10.387	A	2
ClF	Chlorine fluoride	0.888	A	2
ClFO$_3$	Perchloryl fluoride	0.023	A	3
ClF$_3$	Chlorine trifluoride	0.6	D	1
ClGeH$_3$	Chlorogermane	2.13	A	1
ClH	Hydrogen chloride	1.109	A	3
ClHO	Hypochlorous acid (HOCl)	1.3	Q	2
ClH$_3$Si	Chlorosilane	1.31	A	1
ClI	Iodine chloride	1.24	B	2
ClIn	Indium chloride	3.79	C	2
ClK	Potassium chloride	10.269	A	2
ClLi	Lithium chloride	7.129	A	2
ClNO$_2$	Nitryl chloride	0.53	A	1
ClNS	Thionitrosyl chloride (NSCl)	1.87	A	2
ClNa	Sodium chloride	9.001	A	2
ClO	Chlorine oxide (ClO)	1.239	A	2
ClRb	Rubidium chloride	10.510	A	2
ClTl	Thallium chloride (TlCl)	4.543	A	2
Cl$_2$H$_2$Si	Dichlorosilane	1.17	B	1
Cl$_2$OS	Thionyl chloride	1.45	B	1
Cl$_2$O$_2$S	Sulfuryl chloride	1.81	B	1
Cl$_2$S	Sulfur dichloride	0.36	A	3
Cl$_3$FSi	Trichlorofluorosilane	0.49	A	2
Cl$_3$HSi	Trichlorosilane	0.86	A	2
Cl$_3$N	Nitrogen trichloride	0.39	A	3
Cl$_3$OP	Phosphorous oxychloride	2.54	B	2
Cl$_3$P	Phosphorous trichloride	0.56	B	2
CsF	Cesium fluoride	7.884	A	2
CsNa	Cesium sodium (CsNa)	4.75	C	2
CuF	Copper fluoride (CuF)	5.77	C	2
FGa	Gallium fluoride (GaF)	2.45	B	2
FGeH$_3$	Fluorogermane	2.33	C	2
FH	Hydrogen fluoride	1.826	A	2
FHO	Hypofluorous acid (HOF)	2.23	C	3
FH$_3$Si	Fluorosilane	1.298	A	3
FI	Iodine fluoride	1.95	A	3
FIn	Indium fluoride (InF)	3.40	B	2
FK	Potassium fluoride	8.585	A	2
FLi	Lithium fluoride	6.326	A	3
FNO	Nitrosyl fluoride	1.730	A	3
FNO$_2$	Nitryl flouride	0.466	A	2
FNS	Thionitrosyl fluoride (NSF)	1.902	A	2
FNa	Sodium fluoride	8.156	A	2
FRb	Rubidium fluoride	8.546	A	2
FS	Sulfur fluoride (SF)	0.794	B	3
FTl	Thallium fuoride (TlF)	4.228	A	2
F$_2$Ge	Germanium fluoride (GeF$_2$)	2.61	A	2
F$_2$HN	Difluoramine	1.92	A	1
F$_2$H$_2$Si	Difluorosilane	1.55	A	1

Mol. form.	Name	μ/D	Acc.	Ref.
F_2N_2	*cis*-Difluorodiazine	0.16	A	1
F_2O	Oxygen difluoride	0.297	A	1
F_2OS	Thionyl fluoride	1.63	A	1
F_2O_2	Dioxygen difluoride (FOOF)	1.44	C	1
F_2O_2S	Sufuryl fluoride	1.12	B	1
F_2S	Sulfur difluoride	1.05	C	2
F_2Si	Difluorosilylene (SiF$_2$)	1.23	B	2
F_3HSi	Trifluorosilane	1.27	B	1
$F_3H_3Si_2$	1,1,1-Trifluorodisilane	2.03	C	3
F_3N	Nitrogen trifluoride	0.235	A	1
F_3NO	Trifluoramine oxide	0.039	A	3
F_3OP	Phosphorous oxyfluoride	1.868	A	3
F_3P	Phosphorous trifluoride	1.03	A	1
F_3PS	Thiophosphoryl trifluoride	0.64	B	1
F_4N_2	Tetrafluorohydrazine (*gauche*)	0.26	B	1
F_4S	Sulfur tetrafluoride	0.632	A	1
F_4Se	Selenium tetrafluoride	1.78	C	2
F_5I	Iodine pentafluoride	2.18	C	1
GeO	Germanium oxide (GeO)	3.282	A	2
GeS	Germanium sulfide (GeS)	2.00	C	2
HI	Hydrogen iodide	0.448	A	2
HLi	Lithium hydride	5.884	A	2
HLiO	Lithium hydroxide	4.754	A	3
HN	Imidogen (NH)	1.39	C	3
HNO	Nitrosyl hydride (HNO)	1.62	B	3
HNO_2	Nitrous acid (HONO) (*cis*)	1.423	A	2
	Nitrous acid (HONO) (*trans*)	1.855	A	2
HNO_3	Nitric acid	2.17	A	1
HN_3	Hydrazoic acid	1.70	C	3
HO	Hydroxyl (OH)	1.668	A	3
HS	Mercapto (SH)	0.758	A	3
H_2O	Water	1.854	A	3
H_2O_2	Hydrogen peroxide	1.573	A	3
H_2S	Hydrogen sulfide	0.97	A	1
H_3N	Ammonia	1.471	A	3
H_3NO	Hydroxylamine (NH$_2$OH)	0.59	C	2
H_3P	Phosphine	0.574	A	3
H_3Sb	Stibine	0.12	C	1
H_4N_2	Hydrazine	1.75	C	1
H_6OSi_2	Disiloxane	0.24	B	1
ILi	Lithium iodide	7.428	A	2
INa	Sodium iodide	9.236	A	2
IO	Iodine oxide (IO)	2.45	B	2
ITl	Thallium iodide (TlI)	4.61	B	2
KLi	Lithium potassium (LiK)	3.45	C	2
KNa	Potassium sodium (KNa)	2.73	C	3
LiNa	Lithium sodium (LiNa)	0.463	A	2
LiO	Lithium oxide (LiO)	6.84	A	2
LiRb	Lithium rubidium (LiRb)	4.00	C	2
NO	Nitric oxide	0.159	A	2
NO_2	Nitrogen dioxide	0.316	A	1
NS	Nitrogen sulfide (NS)	1.81	A	2
NP	Phosphorous nitride (PN)	2.747	A	2
N_2O	Nitrous oxide	0.161	A	3
N_2O_3	Nitrogen trioxide	2.122	A	2
NaRb	Rubidium sodium (RbNa)	3.1	D	2

Mol. form.	Name	μ/D	Acc.	Ref.
PbO	Lead oxide (PbO)	4.64	D	2
OS	Sulfur oxide (SO)	1.55	A	1
OS_2	Sulfur oxide (SSO)	1.47	B	1
OSi	Silicon monoxide (SiO)	3.098	A	2
OSn	Tin oxide (SnO)	4.32	C	2
OSr	Strontium oxide	8.900	A	2
O_2S	Sulfur dioxide	1.633	A	3
O_2Se	Selenium dioxide	2.62	B	2
O_3	Ozone	0.534	A	3
PbS	Lead sulfide (PbS)	3.59	C	2
SSi	Silicon monosulfide (SiS)	1.73	C	2
SSn	Tin sulfide (SnS)	3.18	C	2

Compounds containing carbon

Mol. form.	Name	μ/D	Acc.	Ref.
$CBrF_3$	Bromotrifluoromethane	0.65	C	1
CBr_2F_2	Dibromodifluoromethane	0.66	C	1
$CClF_3$	Chlorotrifluoromethane	0.50	A	1
CClN	Cyanogen chloride	2.833	A	3
CCl_2F_2	Dichlorodifluoromethane	0.51	C	1
CCl_2O	Phosgene	1.17	A	1
CCl_3F	Trichlorofluoromethane	0.46	B	2
CF	Fluoromethylidene	0.645	A	3
CFN	Cyanogen fluoride	2.120	A	3
CF_2	Difluoromethylene	0.47	B	3
CF_2O	Carbonyl fluoride	0.95	A	1
CF_3I	Trifluoroiodomethane	1.048	A	3
CO	Carbon monoxide	0.110	A	3
COS	Carbon oxysulfide	0.715	A	3
COSe	Carbon oxyselenide	0.73	B	1
CS	Carbon sulfide	1.958	A	2
CSe	Carbon selenide	1.99	B	3
CH	Methylidyne	1.46	Q	2
$CHBr_3$	Tribromomethane (bromoform)	0.99	B	1
$CHClF_2$	Chlorodifluoromethane	1.42	B	1
$CHCl_2F$	Dichlorofluoromethane	1.29	B	1
$CHCl_3$	Trichloromethane (chloroform)	1.04	B	2
CHF_3	Trifluoromethane (fluoroform)	1.651	A	3
CHN	Hydrogen cyanide	2.984	A	3
CHN	Hydrogen isocyanide (HNC)	3.05	C	3
CHNO	Fulminic acid (HCNO)	3.099	A	5
CHNO	Isocyanic acid (HNCO)	1.6	Q	2
CH_2Br_2	Dibromomethane	1.43	B	1
CH_2ClF	Chlorofluoromethane	1.82	B	1
CH_2Cl_2	Dichloromethane	1.60	B	1
CH_2F_2	Difluoromethane	1.978	A	3
CH_2N_2	Cyanamide	4.27	C	1
CH_2N_2	Diazomethane	1.50	A	1
CH_2O	Formaldehyde	2.332	A	3
CH_2O_2	Formic acid	1.41	A	1
CH_3BO	Carbonyltrihydroboron (BH_3CO)	1.70	B	3
CH_3Br	Bromomethane (methyl bromide)	1.822	A	3
CH_3Cl	Chloromethane (methyl chloride)	1.892	A	2
CH_3F	Fluoromethane (methyl fluoride)	1.858	A	3
CH_3I	Iodomethane (methyl iodide)	1.62	B	1
CH_3NO	Formamide	3.73	B	1

Mol. form.	Name	μ/D	Acc.	Ref.
CH_3NO_2	Nitromethane	3.46	A	1
CH_3N_3	Methyl azide	2.17	B	2
CH_4O	Methanol	1.70	A	1
CH_4S	Methanethiol (methyl mercaptan)	1.52	C	1
CH_5N	Methylamine	1.31	B	1
CH_6OSi	Methoxysilane	1.15	B	2
CH_6Si	Methylsilane	0.735	A	1
C_2ClF_3	Chlorotrifluoroethylene	0.40	D	1
C_2ClF_5	Chloropentafluoroethane	0.52	C	1
$C_2Cl_2F_4$	1,2-Dichlorotetrafluoroethane	0.5	Q	1
C_2F_3N	Trifluoroacetonitrile	1.262	A	3
C_2HBr	Bromoacetylene	0.234	A	3
C_2HCl	Chloroacetylene	0.44	A	1
C_2HCl_5	Pentachloroethane	0.92	C	1
C_2HF	Fluoroacetylene	0.721	A	3
C_2HF_3	Trifluoroethylene	1.32	B	2
$C_2HF_3O_2$	Trifluoroacetic acid	2.28	D	1
$C_2H_2Cl_2$	1,1-Dichloroethylene	1.34	A	1
$C_2H_2Cl_2$	cis-1,2-Dichloroethylene	1.90	B	1
$C_2H_2Cl_2O$	Chloroacetyl chloride	2.23	Ci	1
$C_2H_2Cl_4$	1,1,2,2-Tetrachloroethane	1.32	Ci	1
$C_2H_2F_2$	1,1-Difluoroethylene	1.368	A	3
$C_2H_2F_2$	cis-1,2-Difluoroethylene	2.42	A	1
C_2H_2O	Ketene	1.422	A	3
$C_2H_2O_2$	Glyoxal (cis)	4.8	C	2
C_2H_3Br	Bromoethylene (vinyl bromide)	1.42	B	1
C_2H_3Cl	Chloroethylene (vinyl chloride)	1.45	B	1
$C_2H_3ClF_2$	1-Chloro-1,1-difluoroethane	2.14	B	1
C_2H_3ClO	Acetyl chloride	2.72	C	1
$C_2H_3Cl_3$	1,1,1-Trichloroethane	1.755	A	2
C_2H_3F	Fluoroethylene (vinyl fluoride)	1.468	A	6
C_2H_3FO	Acetyl fluoride	2.96	A	1
$C_2H_3F_3$	1,1,1-Trifluoroethane	2.347	A	3
C_2H_3N	Acetonitrile	3.924	A	3
C_2H_3NO	Methylisocyanate	2.8	Q	1
C_2H_4ClF	1-Chloro-1-fluoroethane	2.068	A	3
$C_2H_4Cl_2$	1,1-Dichloroethane	2.06	B	1
$C_2H_4F_2$	1,1-Difluoroethane	2.27	B	1
$C_2H_4F_2$	1,2-Difluoroethane (gauche)	2.67	C	2
C_2H_4O	Acetaldehyde	2.750	A	3
C_2H_4O	Ethylene oxide (oxirane)	1.89	A	1
$C_2H_4O_2$	Acetic acid	1.70	B	2
$C_2H_4O_2$	Methyl formate	1.77	B	1
$C_2H_4O_2$	Glycolaldehyde (hydroxyacetylene)	2.73	B	2
C_2H_5Br	Bromoethane (ethyl bromide)	2.03	A	1
C_2H_5Cl	Chloroethane (ethyl chloride)	2.05	A	1
C_2H_5ClO	2-Chloroethanol	1.78	Ci	1
C_2H_5F	Fluoroethane (ethyl fluoride)	1.94	B	1
C_2H_5I	Iodoethane (ethyl iodide)	1.91	B	1
C_2H_5N	Ethyleneimine	1.90	A	1
C_2H_5NO	Acetamide	3.76	Bi	1
C_2H_5NO	N-Methylformamide	3.83	Bi	1
$C_2H_5NO_2$	Nitroethane	3.23	A	2
C_2H_6O	Dimethyl ether	1.30	A	1
C_2H_6O	Ethanol (average)	1.69	Bi	1
	Ethanol (trans)	1.44	B	2

Mol. form.	Name	μ/D	Acc.	Ref.
	Ethanol (*gauche*)	1.68	B	3
C_2H_6OS	Dimethyl sulfoxide	3.96	A	1
$C_2H_6O_2$	Ethylene glycol	2.28	Ci	1
C_2H_6S	Dimethyl sulfide	1.554	A	3
C_2H_6S	Ethanethiol (*trans*)	1.58	C	3
	Ethanethiol (*gauche*)	1.61	C	3
C_2H_7N	Dimethylamine	1.01	B	2
C_2H_7N	Ethylamine	1.22	Ci	1
$C_2H_8N_2$	1,2-Ethanediamine	1.99	Ci	1
C_3HN	Cyanoacetylene	3.724	A	3
C_3H_2O	Propynal (propargyl aldehyde)	2.47	B	1
$C_3H_3Cl_2F$	1,1-Dichloro-2-fluoropropene	2.43	A	3
$C_3H_3F_3$	3,3,3-Trifluoropropene	2.45	B	1
C_3H_3N	Propenenitrile (acrylonitrile)	3.87	B	1
C_3H_3NO	Oxazole	1.503	B	3
C_3H_3NO	Isoxazole	2.90	B	2
C_3H_4	Propyne (methylacetylene)	0.784	A	3
C_3H_4	Cyclopropene	0.45	A	1
$C_3H_4F_2$	1,1-Difluoro-1-propene	0.889	A	2
$C_3H_4N_2$	Imidazole	3.8	D	2
C_3H_4O	Propenal (*trans*)	3.12	B	1
C_3H_4O	Propargyl alcohol	1.13	C	2
C_3H_4O	Cyclopropanone	2.67	C	2
$C_3H_4O_2$	Vinyl formate	1.49	A	1
$C_3H_4O_2$	2-Oxetanone (beta-propiolactone)	4.18	A	1
$C_3H_4O_2$	3-Oxetanone	0.887	A	2
C_3H_5Br	3-Bromopropene (allyl bromide)	1.9	Q	1
C_3H_5Cl	*cis*-1-Chloropropene	1.67	C	1
C_3H_5Cl	*trans*-1-Chloropropene	1.97	C	1
C_3H_5Cl	2-Chloropropene	1.647	A	3
C_3H_5Cl	3-Chloropropene (allyl chloride)	1.94	Ci	1
C_3H_5F	*cis*-1-Fluoropropene	1.46	B	1
C_3H_5F	*trans*-1-Fluoropropene	1.9	Q	1
C_3H_5F	2-Fluoropropene	1.61	B	1
C_3H_5F	3-Fluoropropene (*cis*)	1.76	A	1
	3-Fluoropropene (*gauche*)	1.94	A	1
C_3H_5N	Propanenitrile	4.05	A	3
C_3H_6	Propene (propylene)	0.366	A	1
$C_3H_6Cl_2$	1,3-Dichloropropane	2.08	Bi	1
C_3H_6O	Acetone	2.88	A	1
C_3H_6O	Allyl alcohol (*average*)	1.60	Ci	1
	Allyl alcohol (*gauche*)	1.55	C	3
C_3H_6O	Propanal (*cis*)	2.52	B	1
C_3H_6O	Methyloxirane	2.01	A	1
C_3H_6O	Oxetane (trimethylene oxide)	1.94	A	1
C_3H_6O	Methyl vinyl ether (*cis*)	0.96	B	2
$C_3H_6O_2$	Ethyl formate (*trans*)	1.98	A	2
	Ethyl formate (*gauche*)	1.81	A	2
$C_3H_6O_2$	Methyl acetate	1.72	Ci	1
$C_3H_6O_2$	Propanoic acid (*average*)	1.75	Ci	1
	Propanoic acid (*cis*)	1.46	C	2
$C_3H_6O_2$	1,3-Dioxolane	1.19	C	3
$C_3H_6O_3$	Trioxane	2.08	A	1
C_3H_6S	Thiacyclobutane (trimethylene sulfide)	1.85	C	1
C_3H_7Br	1-Bromopropane (propyl bromide)	2.18	Ci	1
C_3H_7Br	2-Bromopropane (isopropyl bromide)	2.21	C	1

Mol. form.	Name	μ/D	Acc.	Ref.
C_3H_7Cl	1-Chloropropane (propyl chloride)	2.05	Bi	1
C_3H_7Cl	2-Chloropropane (isopropyl chloride)	2.17	C	1
C_3H_7F	1-Fluoropropane (*trans*)	2.05	B	1
	1-Fluoropropane (*gauche*)	1.90	C	1
C_3H_7F	2-Fluoropropane	1.96	B	3
C_3H_7I	1-Iodopropane (propyl iodide)	2.04	Ci	1
C_3H_7N	Allylamine	1.2	Q	1
C_3H_7N	Cyclopropylamine	1.19	A	2
C_3H_7N	Propyleneimine (*cis*)	1.77	C	2
	Propyleneimine (*trans*)	1.57	B	2
C_3H_7NO	N,N-Dimethylformamide	3.82	Bi	1
$C_3H_7NO_2$	1-Nitropropane	3.66	Bi	1
$C_3H_7NO_2$	2-Nitropropane	3.73	B	1
C_3H_8	Propane	0.084	A	1
C_3H_8O	1-Propanol (*trans*)	1.55	B	2
	1-Propanol (*gauche*)	1.58	B	2
C_3H_8O	2-Propanol (*trans*)	1.58	B	2
C_3H_8O	Ethyl methyl ether (*trans*)	1.17	B	3
$C_3H_8O_2$	2-Methoxyethanol (*gauche*)	2.36	B	2
C_3H_8S	Ethyl methyl sulfide (*trans*)	1.56	B	3
C_3H_8S	1-Propanethiol (*trans*)	1.60	C	3
	1-Propanethiol (*gauche*)	1.683	A	3
C_3H_8S	2-Propanethiol (*trans*)	1.61	B	3
	2-Propanethiol (*gauche*)	1.53	B	3
C_3H_9N	Propylamine	1.17	Ci	1
C_3H_9N	Isopropylamine	1.19	C	3
C_3H_9N	Trimethylamine	0.612	A	1
C_4H_4	1-Buten-3-yne (vinylacetylene)	0.22	B	3
$C_4H_4N_2$	Pyridazine	4.22	A	2
$C_4H_4N_2$	Pyrimidine	2.334	A	2
C_4H_4O	Furan	0.66	A	1
$C_4H_4O_2$	Diketene	3.53	B	1
C_4H_4S	Thiophene	0.55	A	2
C_4H_5N	Pyrrole	1.74	B	2
C_4H_5N	Methylacrylonitrile	3.69	C	1
C_4H_6	1,2-Butadiene (methylallene)	0.403	A	1
C_4H_6	1-Butyne (ethylacetylene)	0.80	C	1
C_4H_6	Cyclobutene	0.132	A	1
C_4H_6O	*trans*-2-Butenal (*trans*-crotonaldehyde)	3.67	Bi	1
C_4H_6O	2-Methylpropenal (methacrolein)	2.68	Ci	1
C_4H_6O	Divinyl ether	0.78	C	2
C_4H_6O	Cyclobutanone	2.89	A	2
C_4H_6O	2,3-Dihydrofuran	1.32	B	2
$C_4H_6O_2$	*gamma*-Butyrolactone	4.27	A	3
$C_4H_6O_3$	Acetic anhydride	2.8	Q	1
C_4H_6S	2,5-Dihydrothiophene	1.75	A	3
C_4H_7N	Butanenitrile	4.07	Bi	1
C_4H_7N	2-Methylpropanenitrile (isobutyronitrile)	4.29	B	3
C_4H_8	1-Butene (*cis*)	0.438	A	2
	1-Butene (*skew*)	0.359	A	2
C_4H_8	*cis*-2-Butene	0.253	A	2
C_4H_8	Isobutene	0.503	A	1
C_4H_8	Methylcyclopropane	0.139	A	2
$C_4H_8Cl_2$	1,4-Dichlorobutane	2.22	Ci	1
C_4H_8O	Butanal (butyraldehyde)	2.72	Bi	1
C_4H_8O	2-Butanone (methyl ethyl ketone)	2.78	A	2

Mol. form.	Name	μ/D	Acc.	Ref.
C_4H_8O	Tetrahydrofuran	1.75	B	2
$C_4H_8O_2$	1,3-Dioxane	2.06	B	2
$C_4H_8O_2$	Ethyl acetate	1.78	Ci	1
C_4H_9Br	1-Bromobutane (butyl bromide)	2.08	Ci	1
C_4H_9Br	2-Bromobutane (sec-butyl bromide)	2.23	Ci	1
C_4H_9Cl	1-Chlorobutane (butyl chloride)	2.05	Bi	1
C_4H_9Cl	1-Chloro-2-methylpropane (isobutyl chloride)	2.00	Ci	1
C_4H_9Cl	2-Chlorobutane (sec-butyl chloride)	2.04	Ci	1
C_4H_9Cl	2-Chloro-2-methylpropane (tert-butyl chloride)	2.13	B	1
C_4H_9I	2-Iodobutane (sec-butyl iodide)	2.12	Ci	1
C_4H_9NO	Morpholine	1.55	B	3
C_4H_{10}	Isobutane	0.132	A	1
$C_4H_{10}O$	1-Butanol	1.66	Bi	1
$C_4H_{10}O$	2-Methyl-1-propanol (isobutyl alcohol)	1.64	C	1
$C_4H_{10}O$	Diethyl ether	1.15	Bi	1
$C_4H_{10}O$	Methyl propyl ether (trans-trans)	1.107	A	3
$C_4H_{10}S$	2-Methyl-2-propanethiol (tert-butyl mercaptan)	1.66	B	3
$C_4H_{10}S$	Diethyl sulfide	1.54	Ci	1
$C_4H_{11}N$	Butylamine	1.0	Q	1
$C_4H_{11}N$	Diethylamine	0.92	Ci	1
C_5H_5N	Pyridine	2.215	A	3
C_5H_6	1-Penten-3-yne	0.66	B	2
C_5H_6	cis-3-Penten-1-yne	0.78	B	2
C_5H_6	trans-3-Penten-1-yne	1.06	C	2
C_5H_6	2-Methyl-1-buten-3-yne	0.513	B	2
C_5H_6	1,3-Cyclopentadiene	0.419	A	1
C_5H_6O	2-Methylfuran	0.65	C	2
C_5H_6O	3-Methylfuran	1.03	B	2
C_5H_6S	2-Methylthiophene	0.674	A	2
C_5H_6S	3-Methylthiophene	0.95	C	2
C_5H_7N	Cyclobutanecarbonitrile	4.11	A	3
C_5H_8	2-Methyl-1,3-butadiene (isoprene)	0.25	A	1
C_5H_8	cis-1,3-Pentadiene	0.500	B	2
C_5H_8	trans-1,3-Pentadiene	0.585	A	2
C_5H_8	Cyclopentene	0.20	B	1
C_5H_8	1-Pentyne (trans)	0.842	A	2
	1-Pentyne (gauche)	0.77	C	2
C_5H_8O	Cyclopentanone	3.3	Q	1
C_5H_8O	Cyclopropyl methyl ketone	2.62	D	2
C_5H_9N	Pentanenitrile	4.12	Bi	1
C_5H_9N	2,2-Dimethylpropanenitrile (tert-butyl cyanide)	3.95	A	1
C_5H_{10}	1-Pentene	0.5	Q	1
C_5H_{10}	3-Methyl-1-butene (trans)	0.320	A	3
	3-Methyl-1-butene (gauche)	0.398	A	3
$C_5H_{10}O$	Tetrahydropyran (chair form)	1.74	B	2
$C_5H_{10}O$	2,2-Dimethylpropanal (pivaldehyde)	2.66	B	2
$C_5H_{10}O_3$	Diethyl carbonate	1.10	Ci	1
$C_5H_{10}S$	Thiacyclohexane	1.781	A	3
$C_5H_{11}Br$	1-Bromopentane (amyl bromide)	2.20	Ci	1
$C_5H_{11}Cl$	1-Chloropentane (amyl chloride)	2.16	Ci	1
$C_5H_{11}N$	Piperidine (equatorial)	0.82	B	3
	Piperidine (axial)	1.19	B	3
C_5H_{12}	Isopentane	0.13	C	1
$C_6H_4ClNO_2$	o-Chloronitrobenzene	4.64	B	1
$C_6H_4ClNO_2$	m-Chloronitrobenzene	3.73	B	1
$C_6H_4ClNO_2$	p-Chloronitrobenzene	2.83	B	1

Mol. form.	Name	μ/D	Acc.	Ref.
$C_6H_4Cl_2$	o-Dichlorobenzene	2.50	B	1
$C_6H_4Cl_2$	m-Dichlorobenzene	1.72	C	1
$C_6H_4FNO_2$	p-Fluoronitrobenzene	2.87	B	1
$C_6H_4F_2$	o-Difluorobenzene	2.46	B	2
$C_6H_4F_2$	m-Difluorobenzene	1.51	A	2
C_6H_5Br	Bromobenzene	1.70	B	1
C_6H_5Cl	Chlorobenzene	1.69	B	1
C_6H_5ClO	p-Chlorophenol	2.11	C	1
C_6H_5F	Fluorobenzene	1.60	C	1
C_6H_5I	Iodobenzene	1.70	C	1
$C_6H_5NO_2$	Nitrobenzene	4.22	B	1
C_6H_6	Fulvene	0.424	A	2
C_6H_6O	Phenol	1.224	A	3
C_6H_7N	Aniline	1.13	B	3
C_6H_7N	2-Methylpyridine	1.85	B	2
C_6H_7N	4-Methylpyridine	2.70	A	2
C_6H_{10}	Cyclohexene (half-chair form)	0.332	A	2
C_6H_{10}	1-Hexyne (butylacetylene)	0.83	Ci	1
C_6H_{10}	3,3-Dimethyl-1-butyne (tert-butylacetylene)	0.661	A	1
$C_6H_{10}O$	Cyclohexanone	2.87	B	2
$C_6H_{11}Cl$	Chlorocyclohexane (equatorial)	2.44	C	7
	Chlorocyclohexane (axial)	1.91	B	7
$C_6H_{11}F$	Fluorocyclohexane (equatorial)	2.11	B	2
	Fluorocyclohexane (axial)	1.81	B	2
$C_6H_{12}O_2$	Pentyl formate	1.90	Ci	1
$C_6H_{12}O_3$	Paraldehyde	1.43	C	1
$C_6H_{14}O$	Dipropyl ether	1.21	Ci	1
$C_6H_{14}O$	Diisopropyl ether	1.13	Di	1
$C_6H_{15}N$	Triethylamine	0.66	Ci	1
$C_7H_5F_3$	(Trifluoromethyl)benzene	2.86	B	1
C_7H_5N	Benzonitrile	4.18	B	1
C_7H_7Cl	o-Chlorotoluene	1.56	C	1
C_7H_7Cl	p-Chlorotoluene	2.21	B	1
C_7H_7F	o-Fluorotoluene	1.37	C	1
C_7H_7F	m-Fluorotoluene	1.82	B	2
C_7H_7F	p-Fluorotoluene	2.00	C	1
C_7H_8	Toluene	0.375	A	3
C_7H_8O	Benzyl alcohol	1.71	C	1
C_7H_8O	Anisole (methyl phenyl ether)	1.38	C	1
$C_7H_{14}O_2$	Pentyl acetate	1.75	Ci	1
$C_7H_{15}Br$	1-Bromoheptane	2.16	Ci	1
C_8H_8O	Acetophenone (methyl phenyl ketone)	3.02	B	1
C_8H_{10}	Ethylbenzene	0.59	C	1
C_8H_{10}	o-Xylene	0.640	A	2
$C_8H_{10}O$	Phenetole (ethyl phenyl ether)	1.45	Di	1
$C_8H_{11}N$	N,N-Dimethylaniline	1.68	D	1
$C_8H_{18}O$	Dibutyl ether	1.17	Ci	1
C_9H_7N	Quinoline	2.29	C	1
C_9H_7N	Isoquinoline	2.73	C	1
$C_9H_{10}O_2$	Ethyl benzoate	2.00	Ci	1
C_9H_{12}	Cumene (isopropylbenzene)	0.79	Q	1
$C_{10}H_8$	Azulene	0.80	B	1
$C_{10}H_{14}$	tert-Butylbenzene	0.83	Q	1
$C_{12}H_{10}$	Acenaphthene	0.85	Q	1
$C_{12}H_{10}O$	Diphenyl ether	1.3	Q	1

STRENGTHS OF CHEMICAL BONDS*
J. A. Kerr

The strength of a chemical bond, $D°(R-X)$, often known as the bond dissociation energy, is defined as the standard enthalpy change of the reaction in which the bond is broken: $RX \rightarrow R + X$. It is given by the thermochemical equation, $D°(R-X) = \Delta_f H°(R) + \Delta_f H°(X) - \Delta_f H°(RX)$. Some authors list bond strengths at a temperature of absolute zero but here the values at 298 K are given because more thermodynamic data are available for this temperature. Bond strengths or bond dissociation energies are not equal to, and may differ considerably from, mean bond energies determined solely from thermochemical data on atoms and molecules.

Table 1
BOND STRENGTHS IN DIATOMIC MOLECULES

These have usually been measured spectroscopically or by mass spectrometric analysis of hot gases effusing a Knudsen cell. Excellent accounts of these and other methods are given in (i) *Dissociation Energies and Spectra of Diatomic Molecules*, by A. G. Gaydon, 3rd. ed., Chapman & Hall, London, 1968 and (ii) "Mass Spectrometric Determination of Bond Energies of High-Temperature Molecules", K. A. Gingerich, *Chimia*, 26, 619, 1972. The errors quoted in the table are those given in the original paper or review article. The references have been chosen primarily as a key to the literature. It should not be assumed that the author referred to was responsible for the value quoted, as the reference may be to a review article.

Bond strengths reported at a temperature of absolute zero, $D°_0$, have been converted to $D°_{298}$ by the use of enthalpy functions taken mainly from the JANAF Thermochemical Tables, Third Edition, *J. Phys. Chem. Ref. Data*, 14, Suppl. 1, 1985, wherever possible. For most bonds, however, where these data are not available, the conversion has been made by the approximate relation:

$$D°_{298} = D°_0 + (3/2)RT$$

The list below does not include the increasing number of bond strengths of diatomic molecules now being calculated by *ab initio* methods. The Table has been arranged in an alphabetical order of the atoms.

Molecule	$D°_{298}$/kJ mol^{-1}	Ref.	Molecule	$D°_{298}$/kJ mol^{-1}	Ref.	Molecule	$D°_{298}$/kJ mol^{-1}	Ref.
Ag-Ag	160.3 ± 3.4	296	Al-Au	325.9 ± 6.3	116	As-F	410	192
Ag-Al	183.7 ± 9.2	71	Al-Br	429 ± 6	176	As-Ga	209.6 ± 1.2	75
Ag-Au	202.9 ± 9.2	3	Al-Cl	511.3 ± 0.8	294	As-H	274.0 ± 2.9	26
Ag-Bi	193 ± 42	230	Al-Cu	227.1 ± 9.8	19	As-I	296.6 ± 28.0	308
Ag-Br	293 ± 29	112	Al-D	290.8	230	As-In	201	281
Ag-Cl	341.4	151	Al-F	663.6 ± 6.3	72	As-N	489 ± 2	292
Ag-Cu	174.1 ± 9.2	128	Al-H	284.9 ± 6.3	72	As-O	481 ± 8	245
Ag-D	226.8	192	Al-I	369.9 ± 2.1	250	As-P	433.5 ± 12.6	129
Ag-Dy	130 ± 19	190	Al-Kr	5.89 ± 0.81	39	As-S	379.5 ± 6.3	245
Ag-Eu	129.7 ± 12.6	58	Al-Li	175.7 ± 14.6	152	As-Sb	330.5 ± 5.4	87
Ag-F	354.4 ± 16.3	112	Al-N	297 ± 96	112	As-Se	96	272
Ag-Ga	180 ± 15	40	Al-Ni	225 ± 5	18	As-Tl	198.3 ± 14.6	284
Ag-Ge	174.5 ± 20.9	127	Al-O	511 ± 3	49,66	At-At	~80	81
Ag-H	215.1 ± 8	202	Al-P	216.7 ± 12.6	72	Au-Au	224.7 ± 1.5	31,234
Ag-Ho	123.4 ± 16.7	54	Al-Pd	254.4 ± 12.1	56	Au-B	367.8 ± 10.5	137
Ag-I	234 ± 29	112	Al-S	373.6 ± 7.9	352	Au-Ba	254.8 ± 10.0	127
Ag-In	166.5 ± 4.9	11	Al-Sb	216.3 ± 5.9	277	Au-Be	285 ± 8	112
Ag-Li	177.4 ± 6.3	260	Al-Se	337.6 ± 10.0	352	Au-Bi	297 ± 8.4	127
Ag-Mn	100 ± 21	230	Al-Si	229.3 ± 30.1	44	Au-Ca	243	127
Ag-Na	138.1 ± 8.4	275,282	Al-Te	267.8 ± 10.0	352	Au-Ce	339 ± 21	127
Ag-Nd	<209	206	Al-U	326 ± 29	115	Au-Cl	343 ± 9.6	112
Ag-O	220.1 ± 20.9	271	Al-Xe	7.43 ± 0.69	39	Au-Co	222 ± 17	127
Ag-S	217.1	329	Ar-Ar	4.73 ± 0.04	172	Au-Cr	213 ± 17	127
Ag-Se	202.5	329	Ar-He	3.89	230	Au-Cs	255 ± 3.3	37
Ag-Si	177.8 ± 10.0	303	Ar-Hg	6.15	230	Au-Cu	228.0 ± 5.0	31,127
Ag-Sn	136.0 ± 20.9	2	Ar-I	10.0	36	Au-D	318.4	192
Ag-Te	195.8	329	Ar-K	4.2	192	Au-Dy	259 ± 21	190
Al-Al	133 ± 6	110	As-As	382.0 ± 10.5	231	Au-Eu	241.0 ± 10.5	58
Al-Ar	5.34 ± 0.78	39	As-Cl	448	72	Au-Fe	187.0 ± 16.7	205
Al-As	202.9 ± 7.1	278,286	As-D	270.3	192	Au-Ga	234 ± 38	127

* Revised to December 1995.

Table 1
BOND STRENGTHS IN DIATOMIC MOLECULES (continued)

Molecule	$D°_{298}$/kJ mol^{-1}	Ref.	Molecule	$D°_{298}$/kJ mol^{-1}	Ref.	Molecule	$D°_{298}$/kJ mol^{-1}	Ref.
Au-Ge	274.1 ± 5.0	127	Ba-Br	362.8 ± 8.4	97,186,214	Br-Li	418.8 ± 4	341
Au-H	292.0 ± 8	202	Ba-Cl	436.0 ± 8.4	184,186	Br-Mg	≤327.2	192
Au-Ho	267.4 ± 16.7	54,234	Ba-D	≤193.7	192	Br-Mn	314.2 ± 9.6	112
Au-In	286.0 ± 5.7	11	Ba-F	587.0 ± 6.7	94,183	Br-N	276 ± 21	112
Au-La	336.4 ± 20.9	119	Ba-H	176 ± 14.6	112	Br-Na	367.4 ± 0.8	341,354
Au-Li	284.5 ± 6.7	260	Ba-I	320.8 ± 6.3	178	Br-Ni	360 ± 13	112
Au-Lu	332.2 ± 16.7	124	Ba-O	561.9 ± 13.4	271	Br-O	235.1 ± 0.4	72
Au-Mg	243 ± 42	230	Ba-Pd	221.8 ± 5.0	117	Br-Pb	247 ± 38	112
Au-Mn	185.4 ± 12.6	322	Ba-Rh	259.4 ± 25.1	117	Br-Rb	380.7 ± 4	341
Au-Na	215.1 ± 12.6	282	Ba-S	400.0 ± 18.8	62	Br-Sb	314 ± 59	112
Au-Nd	299.2 ± 20.9	119	Be-Be	59	32,79	Br-Sc	444 ± 63	230
Au-Ni	255 ± 21	127	Be-Br	381 ± 84	230	Br-Se	297 ± 84	230
Au-O	221.8 ± 20.9	271	Be-Cl	388.3 ± 9.2	101,181,358	Br-Si	367.8 ± 10.0	99
Au-Pb	130 ± 42	230	Be-D	203.05	192	Br-Sn	≥552	270
Au-Pd	155 ± 21	127	Be-F	577 ± 42	72,101	Br-Sr	333.0 ± 9.2	186
Au-Pr	310 ± 21	127	Be-H	200.0 ± 1.3	59	Br-Th	364	177
Au-Rb	243 ± 2.9	37	Be-O	434.7 ± 13.4	271	Br-Ti	439	230
Au-Rh	231.0 ± 29	58	Be-S	372 ± 59	112	Br-Tl	333.9 ± 1.7	25
Au-S	418 ± 25	123	Bi-Bi	200.4 ± 7.5	290,307	Br-U	377.4 ± 6.3	243
Au-Sc	280.3 ± 16.7	120	Bi-Br	267.4 ± 4.2	68	Br-V	439 ± 42	230
Au-Se	243.1	329	Bi-Cl	301 ± 4	70	Br-W	329.3	208
Au-Si	305.4 ± 5.9	131	Bi-D	283.7	249	Br-Y	485 ± 84	230
Au-Sn	254.8 ± 7.1	217	Bi-F	367 ± 13	375	Br-Zn	142 ± 29	230
Au-Sr	264 ± 42	230	Bi-Ga	159 ± 17	280	C-C	607 ± 21	72
Au-Tb	289.5 ± 33.5	144,234	Bi-H	≤283.3	249	C-Ce	444 ± 13	220
Au-Te	317.6	329	Bi-I	218.0 ± 4.6	69	C-Cl	397 ± 29	267
Au-U	318 ± 29	115	Bi-In	153.6 ± 1.7	304	C-D	341.4	192
Au-V	240.6 ± 12.1	165	Bi-Li	154.0 ± 5.0	262,288	C-F	552	182
Au-Y	307.1 ± 8.4	170	Bi-O	337.2 ± 12.6	271	C-Ge	460 ± 21	112
B-B	297 ± 21	72	Bi-P	280 ± 13	129	C-H	338.4 ± 1.2	192
B-Br	396	29	Bi-Pb	141.8 ± 14.6	307	C-Hf	540 ± 25	336
B-C	448 ± 29	230	Bi-S	315.5 ± 4.6	351	C-I	209 ± 21	112
B-Ce	305 ± 21	230	Bi-Sb	251 ± 4	228	C-Ir	632 ± 4	164
B-Cl	536	17	Bi-Se	280.3 ± 5.9	351	C-La	462 ± 20	274
B-D	341.0 ± 6.3	230	Bi-Sn	210.0 ± 8.4	127	C-Mo	481 ± 15.9	160
B-F	757	240	Bi-Te	232.2 ± 11.3	351	C-N	754.3 ± 10	38
B-H	340	287	Bi-Tl	121 ± 13	76	C-Nb	569 ± 13.0	160
B-I	220.5 ± 0.8	297	Br-Br	192.807	1	C-O	1076.5 ± 0.4	72
B-Ir	514.2 ± 17.2	357	Br-C	280 ± 21	112	C-Os	≥594	118
B-La	339 ± 63	230	Br-Ca	310.9 ± 9.2	302	C-P	513.4 ± 8	330
B-N	389 ± 21	72	Br-Cd	159 ± 96	112	C-Pt	598 ± 5.9	164,355
B-O	808.8 ± 20.9	271	Br-Cl	217.53 ± 0.29	50	C-Rh	580.0 ± 3.8	312
B-P	346.9 ± 16.7	139	Br-Co	331 ± 42	230	C-Ru	616.2 ± 10.5	313
B-Pd	329.3 ± 20.9	357	Br-Cr	328.0 ± 24.3	112	C-S	714.1 ± 1.2	63,333
B-Pt	477.8 ± 16.7	251	Br-Cs	389.1 ± 4	269,341	C-Sc	≤444 ± 21	140
B-Rh	475.7 ± 20.9	357	Br-Cu	331 ± 25	112	C-Se	590.4 ± 5.9	323
B-Ru	446.9 ± 20.9	357	Br-D	370.74	192	C-Si	451.5	83,363
B-S	580.7 ± 9.2	350	Br-F	280 ± 12	197	C-Tc	565 ± 29	305
B-Sc	276 ± 63	230	Br-Fe	247 ± 96	112	C-Th	453 ± 17	159,336
B-Se	461.9 ± 14.6	350	Br-Ga	444 ± 17	72	C-Ti	423 ± 29	155,336
B-Si	288.3	363	Br-Ge	255 ± 29	112	C-U	454.8 ± 15.1	158,162
B-Te	354.4 ± 20.1	350	Br-H	366.35	192	C-V	427 ± 23.8	160
B-Th	297	132	Br-Hg	72.8 ± 4	72	C-Y	418 ± 14	318
B-Ti	276 ± 63	230	Br-I	179.1 ± 0.4	112,291	C-Zr	561 ± 25	336
B-U	322 ± 33	230	Br-In	414 ± 21	72	Ca-Ca	≤46	371,373
B-Y	293 ± 63	230	Br-K	379.9 ± 0.8	341,354	Ca-Cl	409 ± 9	252

Table 1
BOND STRENGTHS IN DIATOMIC MOLECULES (continued)

Molecule	$D°_{298}$/kJ mol⁻¹	Ref.	Molecule	$D°_{298}$/kJ mol⁻¹	Ref.	Molecule	$D°_{298}$/kJ mol⁻¹	Ref.
Ca-D	≤169.9	192	Cl-Rb	427.6 ± 8	342	Cu-Ho	142 ± 21	190
Ca-F	527 ± 21	94,180	Cl-S	277.0	209	Cu-I	197 ± 21	112
Ca-H	167.8	112	Cl-Sb	360 ± 50	112	Cu-In	187.4 ± 7.9	11
Ca-I	284.7 ± 8.4	178	Cl-Sc	331	362	Cu-Li	192.9 ± 8.8	260
Ca-Li	84.9 ± 8.4	371	Cl-Se	322	230	Cu-Mn	158.6 ± 17	207
Ca-O	402.1 ± 16.7	271	Cl-Si	406	361	Cu-Na	176.1 ± 16.7	283
Ca-S	337.6 ± 18.8	62,192	Cl-Sm	≥423 ± 13	374	Cu-Ni	205.9 ± 17	207
Cd-Cd	7.36	235	Cl-Sn	414 ± 17	230	Cu-O	269.0 ± 20.9	271
Cd-Cl	208.4	192	Cl-Sr	406 ± 13	184,186	Cu-S	276	329
Cd-F	305 ± 21	28	Cl-Ta	544	20	Cu-Se	251	329
Cd-H	69.0 ± 0.4	112	Cl-Th	489	244	Cu-Si	221.3 ± 6.3	303
Cd-I	97.23	200	Cl-Ti	494	230	Cu-Sn	169.5 ± 6.7	2,221
Cd-In	138	230	Cl-Tl	372.8 ± 2.1	25	Cu-Tb	193 ± 19	190
Cd-O	235.6 ± 83.7	149,271	Cl-U	452 ± 8	242	Cu-Te	278.7	1
Cd-S	208.4 ± 20.9	149	Cl-V	477 ± 63	230	D-D	443.533	192
Cd-Se	127.6 ± 25.1	149	Cl-W	423 ± 42	230	D-F	576.6	192
Cd-Te	100.0 ± 15.1	149	Cl-Xe	6.7	192	D-Ga	<272.8	237
Ce-Ce	245.2	119	Cl-Y	527 ± 84	230	D-Ge	≤322	192
Ce-F	582 ± 42	230	Cl-Yb	~322	107	D-H	439.433	192
Ce-Ir	586	141	Cl-Zn	228.9 ± 19.7	65	D-Hg	42.05	192
Ce-N	519 ± 21	138	Cm-O	736	321	D-In	246.0	192
Ce-O	795 ± 8	88	Co-Co	167 ± 25	203	D-Li	240.1892 ± 0.0046	194,339
Ce-Os	506 ± 33	118	Co-Cu	167 ± 17	127	D-Mg	135.1	192
Ce-Pd	322.2	55	Co-F	435 ± 63	230	D-Ni	≤302.9	192
Ce-Pt	556	141	Co-Ge	234 ± 21	127	D-Pt	≤350.2	192
Ce-Rh	548	141	Co-H	226 ± 42	347	D-S	351	192
Ce-Ru	531 ± 25	118	Co-I	285 ± 21	230	D-Si	302.5	192
Ce-S	569	21	Co-O	384.5 ± 13.4	271	D-Sr	≥275.7	192
Ce-Se	494.5 ± 14.6	255	Co-S	331	329	D-Zn	88.7	192
Ce-Te	189.4 ± 12.8	236	Co-Si	276 ± 17	356	Dy-F	531	381
Cl-Cl	242.580 ± 0.004	192	Cr-Cr	142.9 ± 5.4	188	Dy-O	607 ± 17	88
Cl-Co	389	238	Cr-Cu	155 ± 21	207	Dy-S	414 ± 42	230
Cl-Cr	366.1 ± 24.3	112	Cr-F	444.8 ± 19.7	211	Dy-Se	322 ± 42	230
Cl-Cs	448 ± 8	269,342	Cr-Ge	154 ± 7	189	Dy-Te	234 ± 42	230
Cl-Cu	382.8 ± 4.6	154	Cr-H	190.3 ± 7.0	46	Er-F	565 ± 17	381
Cl-D	436.47	192	Cr-I	287.0 ± 24.3	112	Er-O	615 ± 13	88
Cl-Eu	~326	107	Cr-N	377.8 ± 18.8	144,334	Er-S	418 ± 42	230
Cl-F	256.23	192,264	Cr-O	429.3 ± 29.3	271	Er-Se	326 ± 42	230
Cl-Fe	~351	112	Cr-Pb	105 ± 2	189	Er-Te	238 ± 42	230
Cl-Ga	481 ± 13	72	Cr-S	331	85	Eu-Eu	33.5 ± 17	58
Cl-Ge	~431	192	Cr-Sn	141 ± 3	189	Eu-F	544	226
Cl-H	431.62	192	Cs-Cs	43.919 ± 0.010	368	Eu-Li	66.9 ± 2.9	259
Cl-Hg	100 ± 8	112	Cs-F	519 ± 8	269	Eu-O	479 ± 10	88
Cl-I	211.3 ± 0.4	112	Cs-H	175.364	376	Eu-Rh	233.9 ± 33	58
Cl-In	439 ± 8	72	Cs-Hg	8	192	Eu-S	362.3 ± 13.0	255,327
Cl-K	433.0 ± 8	342	Cs-I	337.2 ± 2.1	269,340	Eu-Se	301 ± 14.6	22,171,255
Cl-Li	469 ± 13	72	Cs-Na	63.2 ± 1.3	78	Eu-Te	243 ± 14.6	22,255
Cl-Mg	327.6 ± 2.1	98,184,358	Cs-O	295.8 ± 62.8	271	F-F	158.78	192
Cl-Mn	360.7 ± 9.6	112	Cs-Rb	49.57 ± 0.01	167	F-Ga	577 ± 14.6	254
Cl-N	333.9 ± 9.6	47	Cu-Cu	176.52 ± 2.38	127,306	F-Gd	590.4 ± 27.2	378
Cl-Na	412.1 ± 8	342	Cu-D	270.3	192	F-Ge	485 ± 21	91
Cl-Ni	372 ± 21	112	Cu-Dy	142 ± 21	190	F-H	569.87 ± 0.06	377
Cl-O	269.1	192,301	Cu-F	413.4 ± 13	92	F-Hf	650 ± 15	14
Cl-P	289 ± 42	230	Cu-Ga	215.9 ± 15.1	40	F-Hg	~180	192
Cl-Pb	301 ± 29	112	Cu-Ge	208.8 ± 21	257	F-Ho	540	381
Cl-Ra	343 ± 75	112	Cu-H	277.8	202,300	F-I	≤271.5	6,30,51,67

Table 1
BOND STRENGTHS IN DIATOMIC MOLECULES (continued)

Molecule	$D°_{298}$/kJ mol^{-1}	Ref.	Molecule	$D°_{298}$/kJ mol^{-1}	Ref.	Molecule	$D°_{298}$/kJ mol^{-1}	Ref.
F-In	506 ± 14.6	254	Gd-O	719 ± 10	88	Hg-O	220.9 ± 33.1	149
F-K	497.5 ± 2.5	15	Gd-S	526.8 ± 10.5	109,325	Hg-Rb	8.4	192
F-La	598 ± 42	230	Gd-Se	431 ± 14.6	22	Hg-S	217.1 ± 22.2	149
F-Li	577 ± 21	72	Gd-Te	343 ± 14.6	22	Hg-Se	144.3 ± 30.1	149
F-Lu	333.5	195	Ge-Ge	263.6 ± 7.1	222	Hg-Te	≤142	230
F-Mg	461.9 ± 5.0	94,183	Ge-H	≤321.7	227	Hg-Tl	4	174
F-Mn	423.4 ± 14.6	212	Ge-Ni	290.3 ± 10.9	315	Ho-Ho	84 ± 17	54
F-Mo	464.8	185	Ge-O	659.4 ± 12.6	271	Ho-O	611 ± 17	88
F-N	343	192	Ge-Pd	254.7 ± 10.5	314	Ho-S	428.4 ± 14.6	325
F-Na	519	192	Ge-S	551.0 ± 2.5	80,86	Ho-Se	335 ± 17	22
F-Nd	545.2 ± 12.6	379	Ge-Sc	271.0 ± 11	218	Ho-Te	259 ± 17	22
F-Ni	430 ± 20	77	Ge-Se	489 ± 4	266	I-I	151.088	192,348
F-O	222 ± 17	52	Ge-Si	301 ± 21	112	I-In	331	360
F-P	439 ± 96	112	Ge-Te	456 ± 13	106	I-K	325.1 ± 0.8	340,354
F-Pb	356 ± 8	383	Ge-Y	279.8 ± 11.4	219	I-Li	345.2 ± 4.2	340
F-Pm	540 ± 42	230	H-H	435.990	192	I-Mg	~285	23
F-Pr	582 ± 46	230	H-Hg	39.844	192	I-Mn	282.8 ± 9.6	112
F-Pu	538.5 ± 29	213	H-I	298.407	192	I-N	159 ± 17	230
F-Rb	494 ± 21	72	H-In	243.1	192	I-Na	304.2 ± 2.1	340,354
F-Ru	402	175	H-K	174.576	193,376	I-Ni	293 ± 21	112
F-S	342.7 ± 5.0	27,187,215	H-Li	238.049 ± 0.004	367	I-O	249.0 ± 1.3	301
F-Sb	439 ± 96	112	H-Mg	126.4 ± 2.9	12,13,93	I-Pb	193 ± 4	320
F-Sc	589.1 ± 13	380	H-Mn	234 ± 29	112	I-Rb	318.8 ± 2.1	340
F-Se	339 ± 42	230	H-N	≤339	192	I-Si	293	192
F-Si	552.7 ± 2.1	102	H-Na	185.69 ± 0.25	258,298	I-Sn	234 ± 42	230
F-Sm	565	226	H-Ni	252.3 ± 8	202	I-Sr	269.9 ± 5.9	223
F-Sn	466.5 ± 13	383	H-O	427.6	192	I-Te	192 ± 42	230
F-Sr	541.8 ± 6.7	94,183	H-P	297	192	I-Ti	310 ± 42	230
F-Ta	573 ± 13	239	H-Pb	≤157	192	I-Tl	272 ± 8	23
F-Tb	561 ± 42	230	H-Pd	234 ± 25	347	I-Zn	108.29	200
F-Th	652	246	H-Pt	≤335	192	I-Zr	305	225
F-Ti	569 ± 33	382	H-Rb	167 ± 21	112	In-In	100 ± 8	230
F-Tl	445.2 ± 19.2	25	H-Rh	247 ± 21	347	In-Li	92.5 ± 14.6	152
F-Tm	510	226	H-Ru	234 ± 21	347	In-O	<320.1 ± 41.8	271
F-U	659.0 ± 10.5	148,241	H-S	344.3 ± 12.1	198	In-P	197.9 ± 8.4	278
F-V	590 ± 63	230	H-Sc	~180	309	In-S	289 ± 17	61
F-W	548 ± 63	230	H-Se	314.47 ± 0.96	114	In-Sb	151.9 ± 10.5	73
F-Xe	15.77	299,346	H-Si	≤299.2	192	In-Se	247 ± 17	61
F-Y	605.0 ± 20.9	380	H-Sn	264 ± 17	112	In-Te	218 ± 17	61
F-Yb	≥521.3 ± 9.6	16,107,374	H-Sr	163 ± 8	112	Ir-La	577 ± 13	169
F-Zn	368 ± 63	230	H-Te	268 ± 2.1	111	Ir-O	414.6 ± 42.3	271
F-Zr	616 ± 15	14	H-Ti	204.6 ± 8.8	45	Ir-Si	462.8 ± 20.9	357
Fe-Fe	75 ± 17	310	H-Tl	188 ± 8	112	Ir-Th	573	125
Fe-Ge	210.9 ± 29	204	H-V	208.7 ± 7.0	46	Ir-Ti	422 ± 13	273
Fe-H	180 ± 25	347	H-Yb	159 ± 38	112	Ir-Y	456.1 ± 16.7	170
Fe-O	390.4 ± 17.2	271	H-Zn	85.8 ± 2.1	112	K-K	54.63 ± 0.02	5,248
Fe-S	322	85	He-He	3.8	192	K-Kr	4.6	192
Fe-Si	297 ± 25	356	He-Hg	6.61	230	K-Li	82.0 ± 4.2	96,385
Ga-Ga	112.1 ± 7	317	Hf-C	548 ± 63	230	K-Na	65.994 ± 0.008	33,385
Ga-H	<274.1	237	Hf-N	536 ± 29	144,229	K-O	277.8 ± 20.9	271
Ga-I	339 ± 9.6	112	Hf-O	801.7 ± 13.4	271	K-Xe	5.0	192
Ga-Li	133.1 ± 14.6	152	Hg-Hg		190,191	Kr-Kr	5.23	42,192
Ga-O	353.5 ± 41.8	271	Hg-I	34.69 ± 0.96	365	Kr-O	<8	230
Ga-P	229.7 ± 12.6	122	Hg-K	8.24 ± 0.21	230	Kr-Xe	5.505 ± 0.002	7
Ga-Sb	192.0 ± 12.6	279	Hg-Li	13.8	192	La-La	247 ± 21	362
Ga-Te	251 ± 25	353	Hg-Na	9.2	192,386	La-N	519 ± 42	230

Table 1
BOND STRENGTHS IN DIATOMIC MOLECULES (continued)

Molecule	D°_{298}/kJ mol^{-1}	Ref.	Molecule	D°_{298}/kJ mol^{-1}	Ref.	Molecule	D°_{298}/kJ mol^{-1}	Ref.
La-O	799 ± 4	88	Nd-O	703 ± 13	88	P-Th	550.2 ± 42	127
La-Pt	502 ± 21	256	Nd-S	471.5	22	P-Tl	209 ± 13	277
La-Rh	527 ± 17	57	Nd-Se	385 ± 17	22,147,255	P-U	297 ± 21	230
La-S	573.2 ± 1.7	199,338	Nd-Te	305 ± 17	22	P-W	305 ± 4	142
La-Se	477 ± 17	22,255	Ne-Ne	3.93	344	Pb-Pb	86.6 ± 0.8	130,288
La-Te	381 ± 17	22,145	Ni-Ni	203.26 ± 0.96	253	Pb-S	346.0 ± 1.7	351
La-Y	202.1	362	Ni-O	382.0 ± 16.7	271	Pb-Sb	161.5 ± 10.5	384
Li-Li	110.21 ± 4	359,372	Ni-Pt	273.7 ± 0.3	345	Pb-Se	302.9 ± 4	351
Li-Mg	67.4 ± 6.3	370	Ni-S	344.3	85	Pb-Te	251 ± 13	351
Li-Na	87.181 ± 0.001	95,104	Ni-Si	318 ± 17	356	Pd-Pd	100 ± 15	311
Li-O	333.5 ± 8.4	271	Ni-V	206.3 ± 0.2	332	Pd-Si	261 ± 12	316
Li-Pb	78.7 ± 7.9	262	Np-O	718.4 ± 41.8	271	Pd-Y	238 ± 17	295
Li-S	312.5 ± 7.5	216	O-O	498.36 ± 0.17	35,192	Pm-S	423 ± 63	230
Li-Sb	172.8 ± 10.0	263	O-Os	575	179	Pm-Se	339 ± 63	230
Li-Sm	49.0 ± 4.2	259	O-P	599.1 ± 12.6	271	Pm-Te	255 ± 63	230
Li-Tm	69.0 ± 3.3	259	O-Pa	788.3 ± 17.2	224	Po-Po	187.0	192
Li-Yb	37.2 ± 2.9	259	O-Pb	382.0 ± 12.6	271	Pr-S	492.5 ± 4.6	105
Lu-Lu	142 ± 33	230	O-Pd	380.7 ± 83.7	271	Pr-Se	446.4 ± 23.0	146,255
Lu-O	678 ± 8	88	O-Pm	674 ± 63	230	Pr-Te	326 ± 42	230
Lu-Pt	402 ± 33	133	O-Pr	753 ± 13	88	Pt-Pt	357.3 ± 15.1	163
Lu-S	507.1 ± 14.6	108,325	O-Pt	391.6 ± 41.8	271	Pt-Si	501.2 ± 18.0	357
Lu-Se	418 ± 17	22	O-Pu	715.9 ± 33.9	271	Pt-Th	552	125
Lu-Te	326 ± 17	22	O-Rb	255 ± 84	34	Pt-Ti	397 ± 13	166
Mg-Mg	8.552 ± 0.004	247,371	O-Re	626.8 ± 83.7	271	Pt-Y	474.0 ± 12.1	163
Mg-O	363.2 ± 12.6	271	O-Rh	405.0 ± 41.8	271	Rb-Rb	48.898 ± 0.005	4
Mg-S	234	62	O-Ru	528.4 ± 41.8	271	Rh-Rh	285.3 ± 20.9	53,285
Mn-Mn	25.9	206	O-S	521.7 ± 4.2	271	Rh-Sc	443.9 ± 10.5	168
Mn-O	402.9 ± 41.8	210,271	O-Sb	434.3 ± 41.8	271	Rh-Si	395.0 ± 18.0	357
Mn-S	301 ± 17	369	O-Sc	681.6 ± 11.3	271	Rh-Th	515 ± 21	121
Mn-Se	239.3 ± 9.2	331	O-Se	464.8 ± 21.3	271,324	Rh-Ti	390.8 ± 14.6	53
Mo-Mo	406 ± 21	161	O-Si	799.6 ± 13.4	271	Rh-U	519 ± 17	121
Mo-Nb	456 ± 25	156	O-Sm	565 ± 13	88	Rh-V	364 ± 29	127
Mo-O	560.2 ± 20.9	271	O-Sn	531.8 ± 12.6	271	Rh-Y	445.2 ± 10.5	168
N-N	945.33 ± 0.59	192	O-Sr	425.5 ± 16.7	271	Ru-Si	397.1 ± 20.9	357
N-O	630.57 ± 0.13	192	O-Ta	799.1 ± 12.6	271	Ru-Th	591.6 ± 42	126
N-P	617.1 ± 20.9	64,143	O-Tb	711 ± 13	88	Ru-V	414 ± 29	127
N-Pu	473 ± 63	230	O-Te	376.1 ± 20.9	271	S-S	425.30	192
N-S	464 ± 21	230	O-Th	878.6 ± 12.1	271	S-Sb	378.7	103
N-Sb	301 ± 50	112	O-Ti	672.4 ± 9.2	271	S-Sc	477 ± 13	338,349
N-Sc	469 ± 84	230	O-Tm	502 ± 13	88	S-Se	371.1 ± 6.7	82
N-Se	370 ± 11	333	O-U	759.4 ± 13.4	271	S-Si	623	192
N-Si	470 ± 15	293	O-V	626.8 ± 18.8	10,271	S-Sm	389	105
N-Ta	611 ± 84	230	O-W	672.0 ± 41.8	271	S-Sn	464 ± 3.3	80
N-Th	577.4 ± 33.1	136,144	O-Xe	36.4	230	S-Sr	339	41
N-Ti	476.1 ± 33.1	144,335	O-Y	719.6 ± 11.3	196,271	S-Tb	515 ± 42	230
N-U	531.4 ± 2.1	134	O-Yb	397 ± 17	88	S-Te	339 ± 21	80
N-V	477.4 ± 17.2	100,144	O-Zn	159 ± 4	48	S-Ti	418 ± 3	89,276
N-Xe	23.0	173	O-Zr	776.1 ± 13.4	271	S-Tm	368 ± 42	230
N-Y	481 ± 63	230	P-P	489.5 ± 10.5	143	S-U	522.6 ± 9.6	338
N-Zr	564.8 ± 25.1	135,144	P-Pt	≤416.7 ± 17	328	S-V	450	90,192
Na-Na	73.60 ± 0.25	364	P-Rh	353.1 ± 17	328	S-Y	528.4 ± 10.5	337
Na-O	256.1 ± 16.7	271	P-S	444 ± 8	84	S-Yb	167	230
Na-Rb	63.25	366	P-Sb	356.9	233	S-Zn	205 ± 13	74,149
Nb-Nb	510 ± 10.0	157	P-Se	363.6 ± 10.0	84	S-Zr	575.3 ± 16.7	338
Nb-O	771.5 ± 25.1	271	P-Si	363.6	326	Sb-Sb	299.2 ± 6.3	73,232
Nd-Nd	<163	230	P-Te	297.9 ± 10.0	84	Sb-Te	277.4 ± 3.8	289,343

Table 1
BOND STRENGTHS IN DIATOMIC MOLECULES (continued)

Molecule	$D°_{298}$/kJ mol^{-1}	Ref.	Molecule	$D°_{298}$/kJ mol^{-1}	Ref.	Molecule	$D°_{298}$/kJ mol^{-1}	Ref.
Sb-Tl	126.8 ± 10.5	9,277	Se-V	347 ± 21	230	Te-Tm	276 ± 42	230
Sc-Sc	162.8 ± 21	128	Se-Y	435 ± 13	230	Te-Y	339 ± 13	230
Sc-Se	385 ± 17	230	Se-Zn	170.7 ± 25.9	74,149	Te-Zn	117.6 ± 18.0	149
Sc-Si	228.7 ± 14	218	Si-Si	326.8 ± 10.0	44	Th-Th	≤289	132
Sc-Te	289 ± 17	230	Si-Te	452 ± 8	192,265	Ti-Ti	141.4 ± 21	201
Se-Se	332.6 ± 0.4	82,351	Si-Y	258.8 ± 17.3	219	Tl-Tl	64.4 ± 17	8
Se-Si	548	192	Sm-Te	272.4 ± 14.6	255	U-U	222 ± 21	230
Se-Sm	331.0 ± 14.6	255	Sn-Sn	187.1 ± 0.3	268	V-V	269.2 ± 0.2	332
Se-Sn	401.2 ± 5.9	60	Sn-Te	359.8	192	Xe-Xe	6.53 ± 0.29	43
Se-Sr	~285	24	Sr-Sr	15.5 ± 0.4	113	Y-Y	159 ± 21	230
Se-Tb	423 ± 42	230	Tb-Tb	131.4 ± 25.1	234	Yb-Yb	20.5 ± 17	153
Se-Te	291.6 ± 4	80,82,150	Tb-Te	339 ± 42	230	Zn-Zn	29	319
Se-Ti	381 ± 42	230	Te-Te	259.8 ± 5.0	261			
Se-Tm	276 ± 42	230	Te-Ti	289 ± 17	230			

REFERENCES

1. Abbasov, A. S., Azizov, T. Kh., Alleva, N. A., Aliev, I. Ya, Mustafaev, F. M., and Mamedov, A. N., *Zh. Fiz. Khim.*, 50, 2172, 1976.
2. Ackerman, M., Drowart, J., Stafford, F. E., and Verhaegen, G., *J. Chem. Phys.*, 36, 1557, 1962.
3. Ackerman, M., Stafford, F. E., and Drowart, J., *J. Chem. Phys.*, 33, 1784, 1960.
4. Amiot, C., *J. Chem. Phys.*, 93, 8591, 1990.
5. Amiot, C., *J. Mol. Spectrosc.*, 147, 370, 1991.
6. Appelman, E. H. and Clyne, M. A. A., *J. Chem. Soc. Faraday Trans. 1*, 71, 2072, 1975.
7. Balakrishnan, A., Jones, W. J., Mahajan, C. G., and Stoicheff, B. P., *Chem. Phys. Lett.*, 155, 43, 1989.
8. Balducci, G. and Piacente, V., *J. Chem. Soc. Chem. Commun.*, 1287, 1980.
9. Balducci, G., Ferro, D., and Piacente, V., *High Temp. Sci.*, 14, 207, 1981.
10. Balducci, G., Gigli, G., and Guido, M., *J. Chem. Phys.*, 79, 5616, 1983.
11. Balducci, G., Nunzio, P. E., Gigli, G., and Guido, M., *J. Chem. Phys.*, 90, 406, 1989.
12. Balfour, W. J. and Cartwright, H. M., *Astron. Astrophys. Suppl. Ser.*, 26, 389, 1976.
13. Balfour, W. J. and Lingren, B., *Can. J. Chem.*, 56, 767, 1978.
14. Barkovskii, N. V., Tsirel'nikov, V. I., Emel'yanov, A. M., and Khodeev, Yu. S., *Teplofiz. Vys. Temp.*, 29, 474, 1991.
15. Barrow, R. F. and Caunt, A. D., *Proc. R. Soc. London Ser. A*, 219, 120, 1953.
16. Barrow, R. F. and Chojnicki, A. H., *J. Chem. Soc. Faraday Trans. 2*, 71, 728, 1975.
17. Barrow, R. F., *Trans. Faraday Soc.*, 56, 952, 1960.
18. Behm, J. M., Arrington, C. A., and Morse, M. D., *J. Chem. Phys.*, 99, 6409, 1993.
19. Behm, J. M., Arrington, C. A., Langenberg, J. D., and Morse, M. D., *J. Chem. Phys.*, 99, 6394, 1993.
20. Behrens, R. G. and Feber, R. C., *J. Less-Common Met.*, 75, 281, 1980.
21. Bergman, C. and Gingerich, K. A., *J. Phys. Chem.*, 76, 2332, 1972.
22. Bergman, C., Coppens P., Drowart, J., and Smoes, S., *Trans. Faraday Soc.*, 66, 800, 1970.
23. Berkowitz, J. and Chupka, W. A., *J. Chem. Phys.*, 45, 1287, 1966.
24. Berkowitz, J. and Chupka, W. A., *J. Chem. Phys.*, 45, 4289, 1966.
25. Berkowitz, J. and Walter, T., *J. Chem. Phys.*, 49, 1184, 1968.
26. Berkowitz, J., *J. Chem. Phys.*, 89, 7065, 1988.
27. Berneike ,W., Kreuttle, U., and Neuert, H., *Chem. Phys. Lett.*, 76, 525, 1980.
28. Besenbruch, G., Kana'an, A. S., and Margrave, J. L., *J. Phys. Chem.*, 69, 3174, 1965.
29. Bharate, N. S., Bhartiya, J. B., and Behere, S. H., *Proc. Indian Natl. Sci. Acad. Part A*, 57, 419, 1991.
30. Birks, J. W., Gabelnick, S. D., and Johnston, H. S., *J. Mol. Spectrosc.*, 57, 23, 1975.
31. Bishea, G. A. and Morse, M. D., *Chem. Phys. Lett.*, 171, 430, 1990.
32. Bondybey, V. E., *Chem. Phys. Lett.*, 109, 436, 1984.
33. Breford, E. J. and Engelke, F., *J. Chem. Phys.*, 71, 1994, 1979.
34. Brewer, L. and Rosenblatt, G. M., *Adv. High Temp. Sci.*, 2, 1, 1969.
35. Brix, P. and Herzberg, G., *J. Chem. Phys.*, 21, 2240, 1953.
36. Burns, G., LeRoy, L. J., Morris, D. J., and Blake, J. A., *Proc. R. Soc. London Ser. A*, 316, 81, 1970.
37. Busse, V. B. and Weil, K. G., *Ber. Bunsenges. Phys. Chem.*, 85, 309, 1981.
38. Calculated from $D_f H∞$(CN) = 435.1 ±10 kJ mol^{-1} (Table 4), but see also Costes, M., Naulin, C., and Dorthe, G., *Astron. Astronphys.*, 232, 270, 1990.
39. Callender, C. L., Mitchell, S. A., and Hackett, P. A., *J. Chem. Phys.*, 90, 5252, 1989.
40. Carbonel, M., Bergman, C., and Laffite, M., *Colloq. Int. Cent. Nat. Rech. Sci.*, 201, 311, 1972.

41. Cater, E. D. and Johnson, E. W., *J. Chem. Phys.*, 47, 5353, 1967.
42. Chashchina, G. I. and Shreider, E. Ya., *Zh. Prikl. Spektrosk.*, 21, 696, 1974.
43. Chashchina, G. I. and Shreider, E. Ya., *Zh. Prikl. Spektrosk.*, 25, 163, 1976.
44. Chatillon, C., Allibert, M., and Pattoret, A., *C. R. Acad. Sci. Ser. C*, 280, 1505, 1975.
45. Chen, Y. M., Clemmer, D. E., and Armentrout, P. B., *J. Chem. Phys.*, 95, 1228, 1991.
46. Chen, Y.-M., Clemmer, D. E., and Armentrout, P. B., *J. Chem. Phys.*, 98, 4929, 1993.
47. Clarke, T. C. and Clyne, M. A. A., *Trans. Faraday Soc.*, 66, 877, 1970.
48. Clemmer, D. E., Daltaska, N. F., and Armentrout, P. B., *J. Chem. Phys.*, 95, 7263, 1991.
49. Clemmer, D. E., Weber, M. E., and Armentrout, P. B., *J. Phys. Chem.*, 96, 10888, 1992.
50. Clyne, M. A. A. and McDermid, I. S., *Faraday Discuss. Chem. Soc.*, 67, 316, 1979.
51. Clyne, M. A. A. and McDermid, I. S., *J. Chem. Soc. Faraday Trans. 2*, 72, 2252, 1976.
52. Clyne, M. A. A. and Watson, R. T., *Chem. Phys. Lett.*, 12, 344, 1971.
53. Cocke, D. L. and Gingerich, K. A., *J. Chem. Phys.*, 60, 1958, 1974.
54. Cocke, D. L. and Gingerich, K. A., *J. Phys. Chem.*, 75, 3264, 1971.
55. Cocke, D. L. and Gingerich, K. A., *J. Phys. Chem.*, 76, 2332, 1972.
56. Cocke, D. L., Gingerich, K. A., and Chang, C. A., *J. Chem. Soc. Faraday Trans. 1*, 72, 268, 1976.
57. Cocke, D. L., Gingerich, K. A., and Kordis, J., *High Temp. Sci.*, 5, 474, 1973.
58. Cocke, D. L., Gingerich, K. A., and Kordis, J., *High Temp. Sci.*, 7, 61, 1975.
59. Colin, R. and De Greef, D., *Can. J. Phys.*, 53, 2142, 1975.
60. Colin, R. and Drowart, J., *Trans. Faraday Soc.*, 60, 673, 1964.
61. Colin, R. and Drowart, J., *Trans. Faraday Soc.*, 64, 2611, 1968.
62. Colin, R., Goldfinger, P., and Jeunehomme, M., *Trans. Faraday Soc.*, 60, 306, 1964.
63. Coppens, P., Reynaert, J. C., and Drowart, J., *J. Chem. Soc. Faraday Trans. 2*, 75, 292, 1979.
64. Coquart, B. and Prudhomme, J. C., *J. Mol. Spectrosc.*, 87, 75, 1981.
65. Corbett, J. D. and Lynde, R. A., *Inorg. Chem.*, 6, 2199, 1967.
66. Costes, M., Naulin, C., Dorthe, G., Vaucamps, C., Nouchi, G., *Faraday Discuss. Chem. Soc.*, 84, 75, 1987.
67. Coxon, J. A., *Chem. Phys. Lett.*, 33, 136, 1975.
68. Cubicciotti, D., *Inorg. Chem.*, 7, 208, 1968.
69. Cubicciotti, D., *Inorg. Chem.*, 7, 211, 1968.
70. Cubicciotti, D., *J. Phys. Chem.*, 71, 3066, 1967.
71. Cuthill, A. M., Fabian, D. J., and Shu-Shou-Shen, S., *J. Phys. Chem.*, 77, 2008, 1973.
72. Darwent, B. de B., *Bond Dissociation Energies in Simple Molecules*, NSRDS-NBS-31, National Bureau of Standards, Washington, D. C., 1970.
73. De Maria, G., Drowart, J., and Inghram, M. G., *J. Chem. Phys.*, 31, 1076, 1959.
74. De Maria, G., Goldfinger, P., Malaspina, L., and Piacente, V., *Trans. Faraday Soc.*, 61, 2146, 1965.
75. De Maria, G., Malaspina, L., and Piacente, V., *J. Chem. Phys.*, 52, 1019, 1970.
76. De Maria, G., Malaspina, L., and Piacente, V., *J. Chem. Phys.*, 56, 1978, 1972.
77. Devore, T. C., McQuaid, M., and Gole, J. L., *High Temp. Sci.*, 30, 83, 1990.
78. Diemer, U., Weickenmeier, H., Wahl, M., and Demtroeder, W., *Chem. Phys. Lett.*, 104, 489, 1984.
79. Drowart, J. and Goldfinger, P., *Angew. Chem.*, 6, 581, 1967.
80. Drowart, J. and Goldfinger, P., *Q. Rev. (London)*, 20, 545, 1966.
81. Drowart, J. and Honig, R. E., *J. Phys. Chem.*, 61, 980, 1957.
82. Drowart, J. and Smoes, S., *J. Chem. Soc. Faraday Trans. 2*, 73, 1755, 1977.
83. Drowart, J., De Maria, G., and Inghram, M. G., *J. Chem. Phys.*, 29, 1015, 1958.
84. Drowart, J., Myers, C. E., Szwarc, R., Vander Auwera-Mahieu, A., and Uy, O. M., *High Temp. Sci.*, 5, 482, 1973.
85. Drowart, J., Pattoret, A., and Smoes, S., *Proc. Br. Ceramic Soc.*, No. 8, 67, 1967.
86. Drowart, J., Smets, J., Reynaert, J. C., and Coppens, P., *Adv. Mass Spectrom.*, 7A, 647, 1978.
87. Drowart, J., Smoes, S, and Vander Auwera-Mahieu, A., *J. Chem. Thermodyn.*, 10, 453, 1978.
88. Dulick, M., Murad, E., and Barrow, R. F., *J. Chem. Phys.*, 85, 385, 1986.
89. Edwards, J. G., Franklin, H. F., and Gilles, P. W., *J. Chem. Phys.*, 54, 545, 1971.
90. Edwards, J. G., *J. Chem. Phys.*, 96, 866, 1992
91. Ehlert, T. C. and Margrave, J. L., *J. Chem. Phys.*, 41, 1066, 1964.
92. Ehlert, T. C. and Wang, J. S., *J. Phys. Chem.*, 81, 2069, 1977.
93. Ehlert, T. C., Hilmer, R. M., and Beauchamp, E. A., *J. Inorg. Nucl. Chem.*, 30, 3112, 1968.
94. Engelke, F., *Chem. Phys.*, 39, 279, 1979.
95. Engelke, F., Ennen, G., and Meiwes, K. H., *Chem. Phys.*, 66, 391, 1982.
96. Engelke, F., Hage, H., and Sprick, U., *Chem. Phys.*, 88, 443, 1984.
97. Estler, C. and Zare, R. N., *Chem. Phys.*, 28, 253, 1978.
98. Farber, M. and Srivastava, R. D., *Chem. Phys. Lett.*, 42, 567, 1976.

99. Farber, M. and Srivastava, R. D., *High Temp. Sci.*, 12, 21, 1980.
100. Farber, M. and Srivastava, R. D., *J. Chem. Soc. Faraday Trans. 1,* 69, 390, 1973.
101. Farber, M. and Srivastava, R. D., *J. Chem. Soc. Faraday Trans. 1,* 70, 1581, 1974.
102. Farber, M. and Srivastava, R. D., *J. Chem. Soc. Faraday Trans. 1,* 74, 1089, 1978.
103. Faure, F. M., Mitchell, M. J., and Bartlett, R. W., *High Temp. Sci.,* 4, 181, 1972.
104. Fellows, C. E., *J. Chem. Phys.,* 94, 5855, 1991.
105. Fenochka, B. V. and Gorkienko, S. P., *Zh. Fiz. Khim*, 47, 2445, 1973.
106. Ferro, D., Piacente, V., and Bardi, G., *High Temp. Sci.,* 21, 69, 1986.
107. Filippenko, N. V., Morozov, E. V., Giricheva, N. L., and Krasnev, K. S., *Izv. Vyssh. Ucheb. Zaved Khim.Technol.*, 15, 1416, 1972.
108. Franzen, H. and Hariharan, A. V., *J. Chem. Phys.*, 70, 4907, 1979.
109. Fries, J. A. and Cater, E. D., *J. Chem. Phys.*, 68, 3978, 1978.
110. Fu, Z., Lemire, G. W., Bishea, G. A., and Morse, M. D., *J. Chem. Phys.*, 93, 8420, 1990.
111. Gal, J. F., Maria, P. C., and Decouzon, M., *Int. J. Mass Spectrom. Ion Processes,* 93, 87, 1989.
112. Gaydon, A. G., *Dissociation Energies and Spectra of Diatomic Molecules, 3rd ed.*, Chapman & Hall, London, 1968.
113. Gerber, G. and Moeller, R., *Contrib. Symp. At. Surf. Phys.*, 168, 1982.
114. Gibson, S. T., Greene, J. P., and Berkowitz, J., *J. Chem. Phys.*, 85, 4815, 1986.
115. Gingerich, K. A. and Blue, G. D., *J. Chem. Phys.*, 47, 5447, 1967.
116. Gingerich, K. A. and Blue, G. D., *J. Chem. Phys.*, 59, 186, 1973.
117. Gingerich, K. A. and Choudary, U. V., *J. Chem. Phys.*, 68, 3265, 1978.
118. Gingerich, K. A. and Cocke, D. L., *Inorg. Chim. Acta*, 28, L171, 1978.
119. Gingerich, K. A. and Finkbeiner, H. C., *J. Chem. Phys.*, 54, 2621, 1971.
120. Gingerich, K. A. and Finkbeiner, H. C., *Proc. 9th Rare Earth Res. Conf.*, 2, 795, 1971.
121. Gingerich, K. A. and Gupta, S. K., *J. Chem. Phys.*, 69, 505, 1978.
122. Gingerich, K. A. and Piacente, V., *J. Chem. Phys.*, 54, 2498, 1971.
123. Gingerich, K. A., *Chem. Commun.*, 580, 1970.
124. Gingerich, K. A., *Chem. Phys. Lett.*, 13, 262, 1972.
125. Gingerich, K. A., *Chem. Phys. Lett.*, 23, 270, 1973.
126. Gingerich, K. A., *Chem. Phys. Lett.,* 25, 523, 1974.
127. Gingerich, K. A., *Chem. Soc. Faraday, Symp.*, No.14, 109, 1980.
128. Gingerich, K. A., *Chimia*, 26, 619, 1972.
129. Gingerich, K. A., Cocke, D. L., and Kordis, J., *J. Phys. Chem.*, 78, 603, 1974.
130. Gingerich, K. A., Cocke, D. L., and Miller, F., *J. Chem. Phys.*, 64, 4027, 1976.
131. Gingerich, K. A., Haque, R., and Kingcade, J. E., *Thermochim. Acta*, 30, 61, 1979.
132. Gingerich, K. A., *High Temp. Sci.*, 1, 258, 1969.
133. Gingerich, K. A., *High Temp. Sci.*, 3, 415, 1971.
134. Gingerich, K. A., *J. Chem. Phys.,* 47, 2192, 1967.
135. Gingerich, K. A., *J. Chem. Phys.,* 49, 14, 1968.
136. Gingerich, K. A., *J. Chem. Phys.*, 49, 19, 1968.
137. Gingerich, K. A., *J. Chem. Phys.*, 54, 2646, 1971.
138. Gingerich, K. A., *J. Chem. Phys.*, 54, 3720, 1971.
139. Gingerich, K. A., *J. Chem. Phys.*, 56, 4239, 1972.
140. Gingerich, K. A., *J. Chem. Phys.*, 74, 6407, 1981.
141. Gingerich, K. A., *J. Chem. Soc. Faraday Trans. 2,* 70, 471, 1974.
142. Gingerich, K. A., *J. Phys. Chem.*, 68, 768, 1964.
143. Gingerich, K. A., *J. Phys. Chem.*, 73, 2734, 1969.
144. Gingerich, K. A., *NBS Spec. Publ. (U. S.),* 561, 289, 1979.
145. Gordienko, S. P. and Fenochka, B. V., *Izv. Akad. Nauk. SSSR Neorg. Mater.*, 18, 1811, 1982.
146. Gordienko, S. P., Fenochka, B. V., Viksman, G. Sh., Klockkova, L. A., and Mikhlina, T. M., *Izv. Akad. Nauk. SSSR Neorg. Mater.*, 18, 18, 1982.
147. Gordienko, S. P., *Izv. Akad. Nauk. SSSR Neorg. Mater.*, 20, 1472, 1984.
148. Gorokhov, L. N., Smirnov, V. K., and Khodeev, Yu. S., *Zh. Fiz. Khim.*, 58, 1603, 1984.
149. Grade, M. and Hirschwald, W., *Ber. Bunsenges. Phys. Chem.*, 86, 899, 1982.
150. Grade, M., Wienecke, J., Rosinger, W., and Hirschwald, W., *Ber. Bunsenges. Phys. Chem.*, 87, 355, 1983.
151. Graeber, P. and Weil, K. G., *Ber. Bunsenges. Phys. Chem.*, 76, 417, 1972.
152. Guggi, D. J., Neubert, A., and Zmbov, K. F., *Conf. Int. Thermodyn. Chim. [C. R.] 4th*, 3, 124, 1975.
153. Guido, M. and Balducci, G., *J. Chem. Phys.,* 57, 5611, 1972.
154. Guido, M., Gigli, G., and Balducci, G., *J. Chem. Phys.*, 57, 3731, 1972.
155. Gupta, S. K. and Gingerich, K. A., *High Temp. - High Pressures*, 12, 273, 1980.
156. Gupta, S. K. and Gingerich, K. A., *J. Chem. Phys.*, 69, 4318, 1978.

157. Gupta, S. K. and Gingerich, K. A., *J. Chem. Phys.*, 70, 5350, 1979.
158. Gupta, S. K. and Gingerich, K. A., *J. Chem. Phys.*, 71, 3072, 1979.
159. Gupta, S. K. and Gingerich, K. A., *J. Chem. Phys.*, 72, 2795, 1980.
160. Gupta, S. K. and Gingerich, K. A., *J. Chem. Phys.*, 74, 3584, 1981.
161. Gupta, S. K., Atkins, R. M., and Gingerich, K. A., *Inorg. Chem.,* 17, 3211, 1978.
162. Gupta, S. K., Kingcade, J. E., and Gingerich, K. A., *Adv. Mass Spectrom.,* 8A, 445, 1980.
163. Gupta, S. K., Nappi, B. M., and Gingerich, K. A., *Inorg. Chem.*, 20, 966, 1981.
164. Gupta, S. K., Nappi, B. M., and Gingerich, K. A., *J. Phys. Chem.*, 85, 971, 1981.
165. Gupta, S. K., Pelino, M., and Gingerich, K. A., *J. Chem. Phys.,* 70, 2044, 1979.
166. Gupta, S. K., Pelino, M., and Gingerich, K. A., *J. Phys. Chem.*, 83, 2335, 1979.
167. Gustavsson, T., Amiot, C., Verges, J., *Mol. Phys.*, 64, 279, 1988.
168. Haque, R. and Gingerich, K. A., *J. Chem. Thermodyn.,* 12, 439, 1980.
169. Haque, R., Pelino, M., and Gingerich, K. A., *J. Chem. Phys.*, 71, 2929, 1979.
170. Haque, R., Pelino, M., and Gingerich, K. A., *J. Chem. Phys.*, 73, 4045, 1980.
171. Hariharan, A. V. and Eick, H. A., *J. Chem. Thermodyn.*, 6, 373, 1974.
172. Herman, P. R., La Rocque, P. E., and Stoicheff, B., *J. Chem. Phys.,* 89, 4535, 1988.
173. Herman, R. and Herman, L., *J. Phys. Radium.*, 24, 73, 1963.
174. Herzberg, G., *Molecular Spectra and Molecular Structure. I. Spectra of Diatomic Molecules*, 2nd ed.,Van Nostrand, New York, 1950.
175. Hildenbrand, D. L. and Lau, K. H., *J. Chem. Phys.*, 89, 5825, 1988.
176. Hildenbrand, D. L. and Lau, K. H., *J. Chem. Phys.*, 91, 4909, 1989.
177. Hildenbrand, D. L. and Lau, K. H., *J. Chem. Phys.*, 93, 5983, 1990.
178. Hildenbrand, D. L. and Lau, K. H., *J. Chem. Phys.*, 96, 3830, 1992.
179. Hildenbrand, D. L. and Lau, K. H., *J. Phys. Chem.*, 96, 2325, 1992.
180. Hildenbrand, D. L. and Murad, E., *J. Chem. Phys.*, 44, 1524, 1966.
181. Hildenbrand, D. L. and Theard, L. P., *J. Chem. Phys.*, 50, 5350, 1969.
182. Hildenbrand, D. L., *Chem. Phys. Lett.*, 32, 523, 1975.
183. Hildenbrand, D. L., *J. Chem. Phys.*, 48, 3657, 1968.
184. Hildenbrand, D. L., *J. Chem. Phys.*, 52, 5751, 1970.
185. Hildenbrand, D. L., *J. Chem. Phys.*, 65, 614, 1976.
186. Hildenbrand, D. L., *J. Chem. Phys.*, 66, 3526, 1977.
187. Hildenbrand, D. L., *J. Phys. Chem.*, 77, 897, 1973.
188. Hilpert, K. and Ruthardt, K., *Ber. Bunsenges. Phys. Chem.*, 91, 724, 1987.
189. Hilpert, K. and Ruthardt, K., *Ber. Bunsenges. Phys. Chem.*, 93, 1070, 1989.
190. Hilpert, K., *Ber. Kernforschungsanlage Juelich,* JUEL-1744, 272, 1981.
191. Hilpert, K., *J. Chem. Phys.*, 77, 1425, 1982.
192. Huber, K. P. and Herzberg, G., *Molecular Spectra and Molecular Structure Constants of Diatomic Molecules*, Van Nostrand, New York, 1979.
193. Hussein, K., Effantin, C., D'Incan, J., Verges, J., and Barrow, R. F., *Chem. Phys. Lett.*, 124, 105, 1986.
194. Ihle, H. R. and Wu, C. H., *J. Chem. Phys.*, 63, 1605, 1975.
195. Ishwar, N. B., Varma, M. P., and Jha, B. L., *Acta Phys. Pol. A*, A61, 503, 1982.
196. Ishwar, N. B., Varma, M. P., and Jha, B. L., *Indian J. Pure Appl. Phys.*, 20, 992, 1982.
197. Jeyagopal, T., Rajavel, S. R. K., Ramakrishnan, M., and Rajamanickam, N., *Acta Phys. Hung.*, 68, 145, 1990.
198. Johns, J. W. C. and Ramsey, D. A., *Can. J. Phys.*, 39, 210, 1961.
199. Jones, R. W. and Gole, J. L., *Chem. Phys.,* 20, 311, 1977.
200. Jordan, K. J., Lipson, R. H., McDonald, N. A., and Le Roy, R. J., *J. Phys. Chem.*, 96, 4778, 1992.
201. Kant, A. and Lin, S.-S., *J. Chem. Phys.*, 51, 1644, 1969.
202. Kant, A. and Moon, K. A., *High Temp. Sci.*, 11, 55, 1979.
203. Kant, A. and Strauss, B. H., *J. Chem. Phys.*, 41, 3806, 1964.
204. Kant, A. and Strauss, B., *J. Chem. Phys.*, 49, 3579, 1968.
205. Kant, A., *J. Chem. Phys.*, 49, 5144, 1968.
206. Kant, A., Lin, S.-S, and Strauss, B., *J. Chem. Phys.*, 49, 1983, 1968.
207. Kant, A., Strauss, B., and Lin, S.-S., *J. Chem. Phys.*, 52, 2384, 1970.
208. Kaposi, O., *Magy. Kem. Foly.*, 83, 356, 1977.
209. Kaufel, R., Vahl, G., Nunkwitz, R., and Baumgaertel, H., *Z. Anorg. Allg. Chem.*, 481, 207, 1981.
210. Kazenas ,E., Tagirov, V. K., and Zviadadze, G. N., *Izv. Akad. Nauk. SSSR Met.*, 58, 1984.
211. Kent, R. A. and Margrave, J. L., *J. Am. Chem. Soc.*, 87, 3582, 1965.
212. Kent, R. A., Ehlert, T. C., and Margrave, J. L., *J. Am. Chem. Soc.*, 86, 5090, 1964.
213. Kent, R. A., *J. Am. Chem. Soc.*, 90, 5657, 1968.
214. Khitrov, A. N., Ryabova, V. G., and Gurvich, L. V., *Teplofiz. Vys. Tempo.*, 11, 1126, 1973.
215. Kiang, T. and Zare, R. N., *J. Am. Chem. Soc.*, 102, 4024, 1980.

216. Kimura, H., Asano, M., and Kubo, K., *J. Nucl. Mater.*, 97, 259, 1981.
217. Kingcade, J. E. Jr. and Gingerich, K. A., *J. Chem. Phys.*, 84, 3432, 1986.
218. Kingcade, J. E. Jr. and Gingerich, K. A., *J. Chem. Soc. Faraday Trans. 2*, 85, 195, 1989.
219. Kingcade, J. E. Jr., and Gingerich, K. A., *J. Chem. Phys.*, 84, 4574, 1986.
220. Kingcade, J. E., Cocke, D. L., and Gingerich, K. A., *High Temp. Sci.*, 16, 89, 1983,
221. Kingcade, J. E., Dufner, D. C., Gupta, S. K., and Gingerich, K. A., *High Temp. Sci.*, 10, 213, 1978.
222. Kingcade, J. E., Nagarathna, H. M., Shim, I., and Gingerich, K. A., *J. Phys. Chem.*, 90, 2830, 1986.
223. Kleinschmidt, P. D. and Hildenbrand, D. L., *J. Chem. Phys.*, 68, 2819, 1978.
224. Kleinschmidt, P. D. and Ward, J. W., *J. Less-Common. Met.*, 121, 61, 1986.
225. Kleinschmidt, P. D., Cubicciotti, D., and Hildenbrand, D. L., *J. Electrochem. Soc.*, 125, 1543, 1978; *Proc. Electrochem. Soc.,* 78, 217, 1978.
226. Kleinschmidt, P. D., Lau, K. H., and Hildenbrand, D. L., *J. Chem. Phys.*, 74, 653, 1981.
227. Klynning, L. and Lindgren, B., *Arkiv. Fysik.*, 32, 575, 1966.
228. Kohl, F. J. and Carlson, K. D., *J. Am. Chem. Soc.*, 90, 4814, 1968.
229. Kohl, F. J. and Stearns, C. A., *J. Phys. Chem.*, 78, 273, 1974.
230. Kondratiev, V. N., *Bond Dissociation Energies, Ionization Potentials and Electron Affinities*, Mauka Publishing House, Moscow, 1974.
231. Kordis, J. and Gingerich, K. A., *J. Chem. Eng. Data,* 18, 135, 1973.
232. Kordis, J. and Gingerich, K. A., *J. Chem. Phys.*, 58, 5141, 1973.
233. Kordis, J. and Gingerich, K. A., *J. Phys. Chem.,* 76, 2336, 1972.
234. Kordis, J., Gingerich, K. A., and Seyse, R. J., *J. Chem. Phys.*, 61, 5114, 1974.
235. Kowalski, A., Czaikowski, M., and Breckenridge, W. H., *Chem. Phys. Lett.*, 119, 368, 1985.
236. Koyama, T. and Yamawaki, M., *J. Nucl. Mater.*, 152, 30, 1988.
237. Kronekvist, M., Lagerqvist, A., and Neuhaus, H., *J. Mol. Spectrosc.*, 39, 516, 1971.
238. Kulkarni, M. P. and Dadape, V. V., *High Temp. Sci.*, 3, 277, 1971.
239. Lau, K. H. and Hildenbrand, D. L., *J. Chem. Phys.*, 71, 1572, 1979.
240. Lau, K. H. and Hildenbrand, D. L., *J. Chem. Phys.*, 72, 4928, 1980.
241. Lau, K. H. and Hildenbrand, D. L., *J. Chem. Phys.*, 76, 2646, 1982.
242. Lau, K. H. and Hildenbrand, D. L., *J. Chem. Phys.*, 80, 1312, 1984.
243. Lau, K. H. and Hildenbrand, D. L., *J. Chem. Phys.*, 86, 2949, 1987.
244. Lau, K. H. and Hildenbrand, D. L., *J. Chem. Phys.,* 92, 6124, 1990.
245. Lau, K. H., Brittain, R. D., and Hildenbrand, D. L., *Chem. Phys. Lett.*, 81, 227, 1981; *J. Phys. Chem.*, 86, 4429, 1982.
246. Lau, K. H., Brittain, R. D., and Hildenbrand, D. L., *J. Chem. Phys.*, 90, 1158, 1989.
247. Li, K. C. and Stwalley, W. C., *J. Chem. Phys.*, 59, 4423, 1973.
248. Li, L., Lyyra, A. M., Luh, W. T., and Stwalley, W. C., *J. Chem. Phys.*, 93, 8452, 1990.
249. Lindgren, B. and Nilsson, Ch., *J. Mol. Spectrosc.*, 55, 407, 1975.
250. Martin, E. and Barrow, R. F., *Phys. Scr.,* 17, 501, 1978.
251. McIntyre, N. S., Vander Auwera-Mahieu, A., and Drowart, J., *Trans. Faraday Soc.*, 64, 3006, 1968.
252. Menendez, M., Garay, M., Verdasco, E., and Gonzalez, U. A., *J. Chem. Phys.,* 99, 2760, 1993.
253. Morse, M. D., Hansen, G. P., Langridge-Smith, P. R. R., Zheng, L. S., Geusic, M. E., Michalopoulos, D. L., and Smalley, R. E., *J. Chem. Phys.*, 80, 5400, 1984.
254. Murad, E., Hildenbrand, D. L., and Main, R. P., *J. Chem. Phys.*, 45, 263, 1966.
255. Nagai, S., Shinmei, M., and Yokokawa, T., *J. Inorg. Nucl. Chem.*, 36, 1904, 1974.
256. Nappi, B. M. and Gingerich, K. A., *Inorg. Chem.*, 20, 522, 1981.
257. Neckel, A. and Sodeck, G., *Monatsch. Chem.*, 103, 367, 1972.
258. Nedelec, O. and Giroud, M., *J. Chem. Phys.*, 79, 2121, 1983.
259. Neubert, A. and Zmbov, K. F., *Chem. Phys.*, 76, 469, 1983.
260. Neubert, A. and Zmbov, K. F., *J. Chem. Soc. Faraday Trans. 1*, 70, 2219, 1974.
261. Neubert, A., *High Temp. Sci.*, 10, 213, 1978.
262. Neubert, A., Ihle, H. R., and Gingerich, K. A., *J. Chem. Phys.*, 73, 1406, 1980.
263. Neubert, A., Zmbov, K. F., Gingerich, K. A., and Ihle, H. R., *J. Chem. Phys.*, 77, 5218, 1982.
264. Nordine, P. C., *J. Chem. Phys.*, 61, 224, 1974.
265. O'Hare, P. A. G., *J. Phys. Chem. Ref. Data*, 22, 1455, 1993.
266. O'Hare, P. A. G., Susman, S., and Volin, K.J., *J. Chem. Thermodyn.*, 21, 827, 1989.
267. Ovcharenko, I. E., Ya, Kuzyankow, Y., and Tatevaskii, V. M., *Opt. Spectrosk.*, 19, 528, 1965.
268. Pak, K., Cai, M. F., Dzugan, T. P., and Bondybey, V. E., *Faraday Discuss. Chem. Soc.*, 86, 153, 1988.
269. Parks, E. K. and Wexler, S., *J. Phys. Chem.*, 88, 4492, 1984.
270. Parr, T. P., Behrens, R., Freedman, A., and Heron, R. R., *Chem. Phys. Lett.,* 56, 71, 1978.
271. Pedley, J. B. and Marshall, E. M., *J. Phys. Chem. Ref. Data*, 12, 967, 1984.
272. Pelevin, O. V., Mil'vidskii, M. G., Belyaev, A. I., and Khotin, B. A., *Izv. Akad. Nauk. SSSR Neorg. Mater.*, 2, 924, 1966.
273. Pelino, M. and Gingerich, K. A., *J. Chem. Phys.*, 90, 1286, 1989.

274. Pelino, M., and Gingerich, K. A., *J. Chem. Phys.*, 93, 1581, 1989.
275. Pelino, M., Piacente, V., and Ascenzo, G., *Thermochim. Acta,* 31, 383, 1979.
276. Pelino, M., Viswanadham, P., and Edwards, J. G., *J. Phys. Chem.*, 83, 2964, 1979.
277. Piacente, V. and Balducci, G., *Adv. Mass Spectrom.*, 7A, 626, 1978.
278. Piacente, V. and Balducci, G., *Dyn. Mass Spectrom.*, 4, 295, 1976.
279. Piacente, V. and Balducci, G., *High Temp. Sci.*, 6, 254, 1974.
280. Piacente, V. and Desideri, A., *J. Chem. Phys.*, 57, 2213, 1972.
281. Piacente, V. and Gigli, R., *J. Chem. Phys.*, 77, 4790, 1982.
282. Piacente, V. and Gingerich, K. A., *High Temp. Sci.*, 9, 189, 1977.
283. Piacente, V. and Gingerich, K. A., *Z. Naturforsch. Teil A*, 28, 316, 1973.
284. Piacente, V. and Malaspina, L., *J. Chem. Phys.*, 56, 1780, 1972.
285. Piacente, V., Balducci, G., and Bardi, G. , J. *Less-Commun. Metals.*, 37, 123, 1974.
286. Piacente, V., *J. Chem. Phys.*, 70, 5911, 1979.
287. Pianalto, F. S., O'Brien, L. C., Keller, P. C., and Bernath, P. F., *J. Mol. Spectrosc.*, 129, 348, 1988.
288. Pitzer, K. S., *J. Chem. Phys.*, 74, 3078, 1981.
289. Porter, R. F. and Spencer, C. W. J., *J. Chem. Phys.*, 32, 943, 1960.
290. Prasad, R., Venugopal, V., and Sood, D. D., *J. Chem. Thermodyn.*, 9, 593, 1977.
291. Rajamanickam, N., Palaniselvam, K., Rajavel, S. R. K., Rajesh, M., and Sureshkumar, G., *Acta Phys. Hung.*, 70, 141, 1991.
292. Rajamanickam, N., Senthilkumar, R. N., Ganesan, S., Gopalakrishnan, N., Rajkumer, J., Jegadesan, V., and Dandapani, C., *Acta Phys. Hung.*, 70, 71, 1991.
293. Rajamanickan, N., *Acta Ciencia Indica Phys.*, 14, 18, 1988.
294. Ram, R. S., Rai, S. B., Ram, R. S., Upadhya, K. N., *J. Chim. Phys. Phys-Chim. Biol.*, 76, 560, 1979.
295. Ramakrishnan, E. S., Shim, I., and Gingerich, K. A., *J. Chem. Soc. Faraday Trans. 2*, 80, 395, 1984.
296. Ran, Q., Schmude, R. W., Gingerich, K. A., Wilhite, D. W., and Kincade, J. E., *J. Phys. Chem.*, 97, 8535, 1993.
297. Rao, P. S. and Rao, T. V. R., *J. Quant. Spectrosc. Radiat. Transfer*, 27, 207, 1982.
298. Rao, S. P. and Rao, T. V. R., *Acta Ciencia Indica Phys.*, 7, 58, 1981.
299. Rao, T. V. R., Reddy, R. R., and Rao, P. S., *Indian J. Pure Appl. Phys.*, 19, 1219, 1981.
300. Rao, V. M., Rao, M. L. P., and Rao, P. T., *J. Quant. Spectrosc. Radiat. Transfer*, 25, 547, 1981.
301. Reddy, R. R., Rao, T. V. R., and Reddy, A. S. R., *Indian J. Pure Appl. Phys.*, 27, 243, 1989.
302. Reddy, R. R., Reddy, A. S. R., and Rao, T. V. R., *Acta Phys. Slovaca*, 36, 273, 1986.
303. Riekert, G., Lamparter, P., and Steeb, S., *Z. Metallkd.*, 72, 765, 1981.
304. Riekert, G., Rainer-Harbach, G., Lamparter, P., and Steeb, S., *Z. Metallkd.*, 76, 406, 1981.
305. Rinehart, G. H. and Behrens, R. G., *J. Phys. Chem.*, 83, 2052, 1979.
306. Rohlfing, E. A. and Valentine, J. J., *J. Chem. Phys.*, 84, 6560, 1986.
307. Rovner, L., Drowart, A., and Drowart, J., *Trans. Faraday Soc.*, 63, 2910, 1967.
308. Rusin, A. D., Zhukov, E., Agamirova, L. M., and Kalinnikov, V. T., *Zh. Neorg. Khim.*, 24, 1457, 1979.
309. Scott, P. R. and Richards, W. G., *J. Phys. B*, 7, 1679, 1974.
310. Shim, I. and Gingerich, K. A., *J. Chem. Phys.*, 77, 2490, 1982.
311. Shim, I. and Gingerich, K. A., *J. Chem. Phys.*, 80, 5107, 1984.
312. Shim, I., and Gingerich, K. A., *J. Chem. Phys.*, 81, 5937, 1984.
313. Shim, I., Finkbeiner, H. C., and Gingerich, K. A., *J. Phys. Chem.*, 91, 3171, 1987.
314. Shim, I., Kingcade, J. E. Jr., and Gingerich, K. A., *J. Chem. Phys.*, 85, 6629, 1986.
315. Shim, I., Kingcade, J. E. Jr., and Gingerich, K. A., *J. Chem. Phys.*, 89, 3104, 1988.
316. Shim, I., Kingcade, J. E. Jr., and Gingerich, K. A., *Z. Phys. D - Atoms Molecules and Clusters*, 7, 261, 1987.
317. Shim, I., Mandix, K., and Gingerich, K. A., *J. Phys. Chem.*, 95, 5435, 1991.
318. Shim, I., Pelino, M., and Gingerich, K. A., *J. Chem. Phys.*, 97, 9240, 1992.
319. Siegel, B., *Q. Rev. (London)*, 19, 77, 1965.
320. Simons, J. W., Oldenberg, R. C., and Baughaim, S. L., *J. Phys. Chem.*, 91, 3840, 1987.
321. Smith, P. K. and Peterson, D. E., *J .Chem. Phys.*, 52, 4963, 1970.
322. Smoes, S. and Drowart, J., *Chem. Commun.*, p.534, 1968.
323. Smoes, S. and Drowart, J., *J. Chem. Soc. Faraday Trans. 2*, 73, 1746, 1977.
324. Smoes, S. and Drowart, J., *J. Chem. Soc. Faraday Trans. 2*, 80, 1171, 1984.
325. Smoes, S., Coppens, P., Bergman, C., and Drowart, J., *Trans. Faraday Soc.*, 65, 682, 1969.
326. Smoes, S., Depiere, D., and Drowart, J., *Rev. Int. Hautes Temp. Refractaires Paris*, 9, 171, 1972.
327. Smoes, S., Drowart, J., Welter, J. M., *J. Chem. Thermodyn.*, 9, 275, 1977; *Adv. Mass Spectrom.*, 7A, 622, 1978.
328. Smoes, S., Huguet, R., and Drowart, J., *Z. Naturforsch. Teil A*, 26, 1934, 1971.
329. Smoes, S., Mandy, F., Vander Auwera-Mahieu, A., and Drowart, J., *Bull. Soc. Chim. Belg.*, 81, 45, 1972.
330. Smoes, S., Myers, C. E., and Drowart, J., *Chem. Phys. Lett.*, 8, 10, 1971.
331. Smoes, S., Pattje, W. R., and Drowart, J., *High Temp. Sci.*, 10, 109, 1978.

332. Spain, E. M. and Morse, M. D., *Int. J. Mass Spectrom. Ion Processes*, 102, 183, 1990.
333. Sreedhara Murthy, N., *Indian J. Phys.*, 62B, 92, 1988.
334. Srivastara, R. D. and Farber, M., *High Temp. Sci.*, 5, 489, 1973.
335. Stearns, C. A. and Kohl, F. J., *High Temp. Sci.*, 2, 146, 1970.
336. Stearns, C. A. and Kohl, F. J., *High Temp. Sci.*, 6, 284, 1974.
337. Steiger, R. A. and Cater, E. D., *High Temp. Sci.*, 7, 204, 1975.
338. Steiger, R. P. and Cater, E. D., *High Temp. Sci.*, 7, 288, 1975.
339. Stwalley, W. C., Way, K. R., and Velasco, R., *J. Chem. Phys.*, 60, 3611, 1974.
340. Su, T.-M. R. and Riley, S. J., *J. Chem. Phys.*, 71, 3194, 1979.
341. Su, T.-M. R. and Riley, S. J., *J. Chem. Phys.*, 72, 1614, 1980.
342. Su, T.-M. R. and Riley, S. J., *J. Chem. Phys.*, 72, 6632, 1980.
343. Sullivan, C. L., Zehe, M. J., and Carlson, K. D., *High Temp. Sci.*, 6, 80, 1974.
344. Tanaka, Y., Yushina, K., and Freeman, D. E., *J. Chem. Phys.*, 59, 564, 1973.
345. Taylor, S., Spain, E. M., and Morse, M. D., *J. Chem. Phys.*, 92, 2698, 1990.
346. Tellinghuisen, J., Tisone, G. C., Hoffmann, J. M., and Hays, A. K., *J. Chem. Phys.*, 64, 4796, 1976.
347. Tolbert, M. A. and Beauchamp, J. L., *J. Phys. Chem.*, 90, 5015, 1986.
348. Tromp, J. W., LeRoy, R. J., Gerstenkorn, S., and Luc, P., *J. Mol. Spectrosc.*, 100, 82, 1983.
349. Tuenge, R. T., Laabs, F., and Franzen, H. F., *J. Chem. Phys.*, 65, 2400, 1976.
350. Uy, O. M. and Drowart, J., *High Temp. Sci.*, 2, 293, 1970.
351. Uy, O. M. and Drowart, J., *Trans. Faraday Soc.*, 65, 3221, 1969.
352. Uy, O. M. and Drowart, J., *Trans. Faraday Soc.*, 67, 1293, 1971.
353. Uy, O. M., Muenow, D. W., Ficalora, P, J., and Margrave, J. L., *Trans. Faraday Soc.*, 64, 2998, 1968.
354. Van Veen, N. J. A., DeVries, M., and DeVries, A. E., *Chem. Phys. Lett.*, 64, 213, 1979.
355. Vander Auwera-Mahieu, A. and Drowart, J., *Chem. Phys. Lett.*, 1, 311, 1967.
356. Vander Auwera-Mahieu, A., McIntyre, N. S., and Drowart, J., *Chem. Phys. Lett.*, 4, 198, 1969.
357. Vander Auwera-Mahieu, A., Peeters, R., McIntyre, N. S., and Drowart, J., *Trans. Faraday Soc.*, 66, 809, 1970.
358. Varma, M. P., Ishwar, N. B., and Jha, B. L., *Indian J. Pure Appl. Phys.*, 20, 828, 1982.
359. Velasco, R., Ottinger, C., and Zare, R. N., *J. Chem. Phys.*, 51, 5522, 1969.
360. Vempati, S. N. and Jones, W. E., *J. Mol. Spectrosc.*, 127, 232, 1988.
361. Venkataramanaiah, M. and Lakshman, S. V. J., *J. Quant. Spectrosc. Radiat. Transfer*, 26, 11, 1981.
362. Verhaegen, G., Smoes, S., and Drowart, J., *J. Chem. Phys.*, 40, 239, 1964.
363. Verhaegen, G., Stafford, F. E., and Drowart, J., *J. Chem. Phys.*, 40, 1622, 1964.
364. Verma, K. K., Vu, T. H., and Stwalley, W. C., *J. Mol. Spectrosc.*, 85, 131, 1980.
365. Viswanathan, K. S. and Tellinghuisen, J., *J. Mol. Spectrosc.*, 98, 185, 1983.
366. Wang, Y. C., Kajitani, M., Kasahara, S., Bata, M., Ishikawa, K., and Kato, H., *J. Chem. Phys.*, 95, 6229, 1991.
367. Way, K. R. and Stwalley, W. C., *J. Chem. Phys.*, 59, 5298, 1973.
368. Weickenmeier, W., Diemer, U., Wahl, M., Raab, M., Demtroeder, W., and Mueller, W., *J. Phys.Chem.*, 82, 5354, 1985.
369. Wiedemeier, H. and Gilles, P. W., *J. Chem. Phys.*, 42, 2765, 1965.
370. Wu, C. H. and Ihle, H. R., *Adv. Mass Spectrom.*, 8A, 374, 1980.
371. Wu, C. H., Ihle, H. R., and Gingerich, K. A., *Int. J. Mass Spectrom. Ion Phys.*, 47, 235, 1983.
372. Wu, C. H., *J. Chem. Phys.*, 65, 3181, 1976; 65, 2040, 1976.
373. Wyss, J. C., *J. Chem. Phys.*, 71, 2949, 1979.
374. Yokozeki, A. and Menzinger, M., *Chem. Phys.*, 14, 427, 1976.
375. Yoo, R. K., Ruscic, B., and Berkowitz, J., *Chem. Phys.*, 166, 215, 1992.
376. Zemke, W. T. and Stwalley, W. C., *Chem. Phys. Lett.*, 143, 84, 1988.
377. Zemke, W. T., Stwalley, W. C., Langhoff, S. R., Valderrama, G. L., and Berry, M. J., *J. Chem. Phys.*, 95, 7846, 1991.
378. Zmbov, K .F. and Margrave, J. L., *J. Inorg. Nucl. Chem.*, 29, 59, 1967.
379. Zmbov, K. F. and Margrave, J. L., *J. Chem. Phys.*, 45, 3167, 1966.
380. Zmbov, K. F. and Margrave, J. L., *J. Chem. Phys.*, 47, 3122, 1967.
381. Zmbov, K. F. and Margrave, J. L., *J. Phys. Chem.*, 70, 3379, 1966.
382. Zmbov, K. F. and Margrave, J. L., *J. Phys. Chem.*, 71, 2893, 1967.
383. Zmbov, K. F., Hastie, J. W., and Margrave, J. L., *Trans. Faraday Soc.*, 64, 861, 1968.
384. Zmbov, K. F., Neubert, A., and Ihle, H. R., *Z. Naturforsch., A*, 36A, 914, 1981.
385. Zmbov, K. F., Wu, C. H., and Ihle, H. R., *J. Chem. Phys.*, 67, 4603, 1977.
386. Zollweg, R. J., *Contrib. Pap. Int. Conf. Phenom. Ioniz. Gases.* 11th, 402, 1973.

Table 2
ENTHALPY OF FORMATION OF GASEOUS ATOMS FROM ELEMENTS IN THEIR STANDARD STATES

For elements that are diatomic gases in their standard states these are readily obtained from the bond strength. For elements that are crystalline in their standard states they are derived from vapor pressure data.

Atom	$\Delta_f H°_{298}$/kJ mol^{-1}	Ref.	Atom	$\Delta_f H°_{298}$/kJ mol^{-1}	Ref.	Atom	$\Delta_f H°_{298}$/kJ mol^{-1}	Ref.
			Hf	619 ± 4	1	Re	774 ± 6.3	1
Ag	284.9 ± 0.8	2	Hg	61.38 ± 0.04	2	Rh	556 ± 4	1
Al	330.0 ± 4.0	2	I	106.76 ± 0.04	2	Ru	650.6 ± 6.3	1
As	302.5 ± 13	1	In	243 ± 4	1	S	277.17 ± 0.15	2
Au	368.2 ± 2.1	1	Ir	669 ± 4	1	Sb	264.4 ± 2.5	1
B	565 ± 5	2	K	89.0 ± 0.8	2	Sc	377.8 ± 4	1
Ba	177.8 ± 4	1	Li	159.3 ± 1.0	2	Se	227.2 ± 4	1
Be	324 ± 5	2	Mg	147.1 ± 0.8	2	Si	450 ± 8	2
Bi	209.6 ± 2.1	1	Mn	283.3 ± 4	1	Sn	301.2 ± 1.5	2
Br	111.87 ± 0.12	2	Mo	658.1 ± 2.1	1	Sr	163.6 ± 2.1	1
C	716.68 ± 0.45	2	N	472.68 ± 0.40	2	Ta	782.0 ± 2.5	1
Ca	177.8 ± 0.8	2	Na	107.5 ± 0.7	2	Te	196.6 ± 2.1	1
Cd	111.80 ± 0.20	2	Nb	721.3 ± 4	1	Th	602 ± 6	2
Ce	423 ± 13	1	Ni	430.1 ± 2.1	1	Ti	473 ± 3	2
Cl	121.301 ± 0.008	2	O	249.18 ± 0.10	2	Tl	182.21 ± 0.4	1
Co	428.4 ± 4	1	Os	787 ± 6.3	1	U	533 ± 8	2
Cr	397 ± 4	1	P	316.5 ± 1.0	2	V	514.2 ± 1.3	1
Cs	76.5 ± 1.0	2	Pb	195.2 ± 0.8	2	W	849.8 ± 4	1
Cu	337.4 ± 1.2	2	Pd	376.6 ± 2.1	1	Y	424.7 ± 2.1	1
Er	317.1 ± 4	1	Pt	565.7 ± 1.3	1	Yb	152.09 ± 0.8	1
F	79.38 ± 0.30	2	Pu	364.4 ± 17	1	Zn	130.40 ± 0.40	2
Ge	372 ± 3	2	Rb	80.9 ± 0.8	2	Zr	608.8 ± 4	1
H	217.998 ± 0.006	2						

REFERENCES

1. Brewer, L. and Rosenblatt, G. M., *Adv. High Temp. Chem.*, 2, 1, 1969.
2. Cox, J. D., Wagman, D. D., and Medvedev, V. A., Eds., *CODATA Key Values for Thermodynamics*, Hemisphere Publishing Corporation, New York, 1989.

Table 3
BOND STRENGTHS IN POLYATOMIC MOLECULES

The values below refer to a temperature of 298 K and have mostly been determined by kinetic methods (see (i) S. W. Benson, *J. Chem. Educ.*, 42, 502, 1965, (ii) J. A. Kerr, *Chem. Rev.*, 66, 465, 1966 and (iii) D. F. McMillen and D. M. Golden, *Ann. Rev. Phys. Chem.*, 33, 493, 1982, for a full description of the methods). An increasing number of bond strengths are being determined from gas-phase acidity cycles and from photoionization mass spectrometry (see J. Berkowitz, G. B. Ellison and D. Gutman, *J. Phys. Chem.*, 98, 2744, 1994).

Bond strengths in polyatomic molecules are notoriously difficult to measure accurately since the mechanisms of the kinetic systems involved in many of the measurements are seldom straightforward. Thus much controversy has taken place in the literature over the past 15 years concerning C-H bond strengths in simple alkanes, although these now appear to have been resolved. Other examples illustrating the difficulties involved are concerned with the C-H bond strengths in ethene and methanol or the corresponding enthalpies of formation of the vinyl and hydroxymethyl radicals and changes to the recommendations could well arise.

Some of the bond strengths have been calculated from the enthalpies of formation of the species involved according to the equations:

$$D°(R-X) = \Delta_f H°(R) + \Delta_f H°(X) - \Delta_f H°(RX)$$
$$D°(R-R) = 2 \Delta_f H°(R) - \Delta_f H°(RR)$$

The enthalpies of formation of the radicals are taken from Table 4 and for the molecules from the appropriate References following Table 3.

An attempt has been made to list all the important values obtained by methods that are considered to be valid. The references are intended to serve as a guide to the literature.

Table 3
BOND STRENGTHS IN POLYATOMIC MOLECULES (continued)

Bond	$D°_{298}$/kJ mol^{-1}	Ref.	Bond	$D°_{298}$/kJ mol^{-1}	Ref.
H-CH	424.0 ± 4.2	1	H-C(CH$_3$)$_2$CN	361.9 ± 8.4	57
H-CH$_2$	462.0 ± 4.0	1	H-CH$_2$NH$_2$	390.4 ± 8.4	57
H-CH$_3$	438.9 ± 0.4	14	H-CH$_2$NHCH$_3$	364 ± 8	57
H-CCH	556.1 ± 2.9	3,34	H-CH$_2$N(CH$_3$)$_2$	351 ± 8	57
H-CHCH$_2$	465.3 ± 3.4	34,74	H-CHO	368.5 ± 1.0	21
H-C$_2$H$_5$	423.0 ± 1.6	14	H-CHCO	440.6 ± 8.8	14
H-Cycloprop-2-en-1-yl	379.1 ± 17	57	H-COCH$_3$	373.8 ± 1.5	60
H-CH$_2$CCH	374.0 ± 8	57	H-COCHCH$_2$	364.4 ± 4.2	57
H-CH$_2$CHCH$_2$	361.9 ± 8.8	14	H-COC$_2$H$_5$	365.7 ± 4	57
H-Cyclopropyl	444.8 ± 1.3	57	H-COC$_6$H$_5$	363.6 ± 4	57
H-n-C$_3$H$_7$	420.0 ± 2.5	20,67	H-COCF$_3$	380.7 ± 8	57
H-i-C$_3$H$_7$	412.5 ± 1.6	14	H-CH$_2$CHO	394.6 ± 9.2	14
H-CH$_2$CCCH$_3$	364.8 ± 8	57	H-CH$_2$COCH$_3$	411.3 ± 7.5	57
H-CH(CH$_3$)CCH	347.7 ± 9.2	57	H-CH(CH$_3$)COCH$_3$	386.2 ± 5.9	57
H-Cyclobutyl	403.8 ± 4	57	H-CH$_2$OCH$_3$	389 ± 4	57
H-Cyclopropylmethyl	407.5 ± 6.7	57	H-CH(CH$_3$)OC$_2$H$_5$	383.7 ± 1.7	49
H-CH(CH$_3$)CHCH$_2$	345.2 ± 5.4	57	H-Tetrahydrofuran-2-yl	385 ± 4	57
H-CH$_2$CHCHCH$_3$	358.2 ± 6.3	57	H-2-Furylmethyl	361.9 ± 8	57
H-CH$_2$C(CH$_3$)CH$_2$	358.2 ± 4	81,85	H-CH$_2$OH	401.92 ± 0.63	14
H-s-C$_4$H$_9$	410.9 ± 2.1	14	H-CH(CH$_3$)OH	389 ± 4	57
H-t-C$_4$H$_9$	403.8 ± 1.6	14	H-CH(OH)CHCH$_2$	341.4 ± 7.5	57
H-Cyclopenta-1,3-dien-5-yl	297.5 ± 6.3	57	H-C(CH$_3$)$_2$OH	381 ± 4	57
H-Spiropentyl	413.4 ± 4	57	H-CH$_2$OCOC$_6$H$_5$	419.2 ± 5.4	57
H-Cyclopent-1-en-3-yl	344.3 ± 4	57	H-COOCH$_3$	387.9 ± 4	57
H-CH$_2$CHCHCHCH$_2$	347 ± 13	57	H-CH$_2$F	423.8 ± 4	70
H-CH(C$_2$H$_3$)$_2$	319.7	57,82	H-CHF$_2$	431.8 ± 4	70
H-CH(CH$_3$)CCCH$_3$	365.3 ± 11.3	57	H-CF$_3$	446.4 ± 4	57
H-C(CH$_3$)$_2$CCH	338.9 ± 9.6	57	H-CHFCl	421.7 ± 5.4	87
H-C(CH$_3$)$_2$CHCH$_2$	323.0 ± 6.3	57	H-CF$_2$Cl	421.3 ± 8.3	58
H-Cyclopentyl	403.5 ± 2.5	20,67	H-CHFCl$_2$	413.8 ± 5.0	87
H-CH$_2$C(CH$_3$)$_3$	418 ± 8	57	H-CH$_2$Cl	421.7 ± 4.2	87
H-C(CH$_3$)$_2$C$_2$H$_5$	404.0 ± 6.3	1,67,79	H-CHCl$_2$	411.7 ± 5.0	87
H-C$_6$H$_5$	465.3 ± 3.4	14	H-CCl$_3$	392.5 ± 2.5	44
H-Cyclohexa-1,3-dien-5-yl	305 ± 21	57	H-CH$_2$Br	425.1 ± 4.2	87
H-Cyclohexa-1,4-dien-3-yl	305.4 ± 8.4	42	H-CHBr$_2$	417.2 ± 7.5	87
H-Cyclohexyl	399.6 ± 4	57	H-CBr$_3$	401.7 ± 6.7	57
H-C(CH$_3$)$_2$CCCH$_3$	344.3 ± 11.3	57	H-CH$_2$I	431 ± 8	57
H-CH$_2$C(CH$_3$)C(CH$_3$)$_2$	326.4 ± 4.6	57	H-CHI$_2$	431 ± 8	57
H-C(CH$_3$)$_2$C(CH$_3$)CH$_2$	319.2 ± 4.6	57	H-CHCF$_2$	448 ± 8	80
H-CH$_2$C$_6$H$_5$	370.3 ± 6.3	14	H-CFCHF	448 ± 8	80
H-Cyclohepta-1,3,5-trien-7-yl	305.4 ± 8	57	H-CFCF$_2$	452 ± 8	80
H-Norbornyl	404.6 ± 10.5	57	H-CH$_2$CF$_3$	446.4 ± 4.6	57
H-Cycloheptyl	387.0 ± 4	57	H-CF$_2$CH$_3$	416.3 ± 10.5	57
H-CH(CH$_3$)C$_6$H$_5$	357.3 ± 6.3	57	H-C$_2$F$_5$	429.7 ± 2.1	57
H-Inden-1-yl	351 ± 13	57	H-CFCFCl	444 ± 8	80
H-C(CH$_3$)$_2$C$_6$H$_5$	353.1 ± 6.3	57	H-CHClCF$_3$	425.9 ± 6.3	57
H-1-Naphthylmethyl	356.1 ± 6.3	57	H-CClCFCl	439 ± 8	80
H-CH(C$_6$H$_5$)$_2$	340.6	73	H-CClCH$_2$	>433.5	74
H-9,10-Dihydroanthracen-9-yl	315.1 ± 6.3	57	H-CClCHCl	435 ± 8	80
H-C(CH$_3$)(C$_6$H$_5$)$_2$	339 ± 8	57	H-CCl$_2$CHCl$_2$	393 ± 8	57
H-9-Anthracenylmethyl	342.3 ± 6.3	57	H-C$_2$Cl$_5$	397 ± 8	57
H-9-Phenanthrenylmethyl	356.1 ± 6.3	57	H-CClBrCF$_3$	404.2 ± 6.3	57
H-CN	527.6 ± 1.7	14	H-n-C$_3$F$_7$	435 ± 8	57
H-CH$_2$CN	392.9 ± 8.4	14	H-i-C$_3$F$_7$	433.5 ± 2.5	35
H-CH$_2$NC	380.7 ± 8.8	14	H-CHClCHCH$_2$	370.7 ± 5.9	57
H-CH(CH$_3$)CN	376.1 ± 9.6	57	H-C$_6$F$_5$	476.6	57

Table 3
BOND STRENGTHS IN POLYATOMIC MOLECULES (continued)

Bond	D°_{298}/kJ mol^{-1}	Ref.	Bond	D°_{298}/kJ mol^{-1}	Ref.
H-CH$_2$Si(CH$_3$)$_3$	415.1 ± 4	89	CH$_3$-CH$_3$	376.0 ± 2.1	1,67,76
H-CSH	399.6 ± 5.0	14	CH$_3$-CH$_2$CCH	318.0 ± 8	57
H-CH$_2$SH	392.9 ± 8.4	14	CH$_3$-CH$_2$CCCH$_3$	308.4 ± 6.3	57
H-CH$_2$SCH$_3$	384.9 ± 5.9	45	CH$_3$-CH(CH$_3$)CCH	305.4	57
H-NH$_2$	452.7 ± 1.3	14	CH$_3$-C(CH$_3$)CCH$_2$	320.1 ± 9.2	57
H-NHCH$_3$	418.4 ± 10.5	57	CH$_3$-CH$_2$CHCHCH$_3$	305.0 ± 3.3	57
H-N(CH$_3$)$_2$	382.8 ± 8	57	CH$_3$-CH$_2$C(CH$_3$)CH$_2$	301.2 ± 3.3	81
H-NHC$_6$H$_5$	368.2 ± 8	57	CH$_3$-t-C$_4$H$_9$	425.9 ± 8	57
H-N(CH$_3$)C$_6$H$_5$	366.1 ± 8	57	CH$_3$-CH(CH$_3$)CCCH$_3$	320.9 ± 6.3	57
H-NO	≤207.1	23	CH$_3$-C(CH$_3$)$_2$CCH	295.8 ± 6.3	57
H-NO$_2$	327.6 ± 2.1	57	n-C$_3$H$_7$-CH$_2$CCH	306.3 ± 6.3	57
H-NF$_2$	316.7 ± 10.5	57	CH$_3$-C(CH$_3$)$_2$CHCH$_2$	284.9 ± 6.3	57
H-NHNH$_2$	366.1	41	n-C$_3$H$_7$-CH$_2$CHCH$_2$	295.8	86
H-NH$_3$	385 ± 21	57	CH$_3$-C$_6$H$_5$	317.1 ± 6.3	57
H-OH	498 ± 4	57	CH$_3$-C(CH$_3$)$_2$CCCH$_3$	303.3 ± 6.3	57
H-OCH$_3$	436.0 ± 3.8	14	CHCCH$_2$-s-C$_4$H$_9$	300.0 ± 6.3	57
H-OC$_2$H$_5$	437.7 ± 3.4	14	CH$_3$-CH$_2$C$_6$H$_5$	332.2 ± 4	57
H-OC(CH$_3$)$_3$	439.7 ± 4	57	CH$_3$-CH(CH$_3$)C$_6$H$_5$	312.1 ± 6.3	57
H-OCH$_2$C(CH$_3$)$_3$	428.0 ± 6.3	57	C$_2$H$_5$-CH$_2$C$_6$H$_5$	294.1 ± 4	57
H-OC$_6$H$_5$	361.9 ± 8	57	CH$_3$-1-Naphthylmethyl	305.0 ± 6.3	57
H-O$_2$H	369.0 ± 4.2	78	CH$_3$-C(CH$_3$)$_2$C$_6$H$_5$	308.4 ± 6.3	57
H-O$_2$CH$_3$	370.3 ± 2.1	50	CHCCH$_2$-CH$_2$C$_6$H$_5$	256.9 ± 8	57
H-O$_2$C(CH$_3$)$_3$	374.0 ± 0.8	43	n-C$_3$H$_7$-CH$_2$C$_6$H$_5$	292.9 ± 4	57
H-OCOCH$_3$	442.7 ± 8	57	CH$_3$-9-Anthracenylmethyl	282.8 ± 6.3	57
H-OCOC$_2$H$_5$	445.2 ± 8	57	CH$_3$-9-Phenanthrenylmethyl	305.0 ± 6.3	57
H-OCO-n-C$_3$H$_7$	443.1 ± 8	57	CH$_3$-CH(C$_6$H$_5$)$_2$	301 ± 8	57
H-ONO	327.6 ± 2.1	13	CH$_3$-C(CH$_3$)(C$_6$H$_5$)$_2$	289 ± 8	57
H-ONO$_2$	423.4 ± 2.1	13	CH$_3$-CN	509.6 ± 8	57
H-SiH	351	57	C$_2$H-CN	602 ± 4	63
H-SiH$_2$	268	57	C$_2$H$_5$-CH$_2$NH$_2$	332.2 ± 8	57
H-SiH$_3$	384.1 ± 2.0	75	CH$_3$-CH$_2$CN	336.4 ± 4	83
H-SiH$_2$CH$_3$	374.9	89	C$_2$H$_5$-CH$_2$CN	321.7 ± 7.1	57
H-SiH(CH$_3$)$_2$	374.0	89	CH$_3$-CH(CH$_3$)CN	329.7 ± 8	57
H-Si(CH$_3$)$_3$	377.8	57,89	C$_2$H$_5$-CH$_2$CN	321.7 ± 7.1	57
D-Si(CH$_3$)$_3$	389 ± 7.1	33	CH$_3$-C(CH$_3$)$_2$CN	312.5 ± 6.7	57
H-SiH$_2$C$_6$H$_5$	369.0	57,89	CH$_3$-C(CH$_3$)(CN)C$_6$H$_5$	250.6	57
H-SiF$_3$	418.8	57,89	C$_6$H$_5$CH$_2$-CH$_2$NH$_2$	284.5 ± 8	57
H-SiCl$_3$	382.0	57,89	C$_6$H$_5$CH$_2$-C$_5$H$_4$N	362.8	73
H-Si$_2$H$_5$	361.1	57	CN-CN	536 ± 4	29
H-Si(CH$_3$)$_2$Si(CH$_3$)$_3$	356.9 ± 8.4	42	CH$_3$-2-Furylmethyl	314 ± 8	57
H-Si(Si(CH$_3$)$_3$)$_3$	330.5 ± 8.4	42	C$_6$H$_5$CH$_2$-COCH$_2$C$_6$H$_5$	273.6 ± 8	57
H-PH$_2$	351.0 ± 2.1	14	CH$_3$CO-COCH$_3$	282.0 ± 9.6	57
H-SH	381.6 ± 2.9	59	C$_6$H$_5$CH$_2$-COOH	280	57
H-SCH$_3$	365.3 ± 2.5	59	C$_6$H$_5$CO-COC$_6$H$_5$	277.8	57
H-SC$_6$H$_5$	348.5 ± 8	57	(C$_6$H$_5$)$_2$CH-COOH	248.5 ± 13	57
H-SO	172.8	90	CF$_3$-COC$_6$H$_5$	308.8 ± 8	57
H-GeH$_3$	349.0 ± 8	14	CF$_2$=CF$_2$	319.2 ± 13	92
H-GeH$_2$I	331 ± 8	61	CH$_2$F-CH$_2$F	368 ± 8	46
H-Ge(CH$_3$)$_3$	339 ± 8	32	CH$_3$-CF$_3$	423.4 ± 4.6	72
H-AsH$_2$	319.2 ± 0.8	14	CF$_3$-CF$_3$	413.0 ± 10.5	57
H-SeH$_2$	334.93 ± 0.75	14	C$_6$F$_5$-C$_6$F$_5$	487.9 ± 24.7	71
H-Sn(n-C$_4$H$_9$)$_3$	308.4 ± 8.4	17	CH$_3$-BF$_2$	~473	57
H-SbH$_2$	288.3 ± 2.1	14	C$_6$H$_5$-BCl$_2$	~510	57
H-TeH	277.0 ± 5.0	14	CH$_2$CHCH$_2$-Si(CH$_3$)$_3$	293	57
HCĴCH	965 ± 8	1,22,67	s-C$_4$H$_9$-Si(CH$_3$)$_3$	414	57
H$_2$C=CH$_2$	733 ± 8	1,67	CH$_3$-NHC$_6$H$_5$	298.7 ± 8	57

Bond	$D°_{298}$/kJ mol^{-1}	Ref.	Bond	$D°_{298}$/kJ mol^{-1}	Ref.
$C_6H_5CH_2-NH_2$	297.5 ± 4	57	$F-CH_3$	472	1,55
$CH_3-N(CH_3)C_6H_5$	296.2 ± 8	57	$F-CN$	469.9 ± 5.0	57
$C_6H_5CH_2-NHCH_3$	287.4 ± 8	57	$F-COF$	535 ± 12	16
$C_6H_5CH_2-N(CH_3)_2$	259.8 ± 8	57	$F-CHFCl$	465.3 ± 9.6	87
$CH_2=N_2$	<175	52	$F-CF_2Cl$	490 ± 25	38
$CH_3-N_2CH_3$	219.7	11	$F-CFCl_2$	462.3 ± 10.0	87
$C_2H_5-N_2C_2H_5$	209.2	11	$F-CF_2CH_3$	522.2 ± 8	57
$i-C_3H_7-N_2i-C_3H_7$	198.7	11	$F-C_2F_5$	530.5 ± 7.5	57
$n-C_4H_9-N_2n-C_4H_9$	209.2	11	$Cl-CN$	421.7 ± 5.0	57
$i-C_4H_9-N_2i-C_4H_9$	205.0	11	$Cl-COC_6H_5$	310 ± 13	57
$s-C_4H_9-N_2s-C_4H_9$	195.4	11	$Cl-CSCl$	265.3 ± 2.1	64
$t-C_4H_9-N_2t-C_4H_9$	182.0	11	$Cl-CF_3$	360.2 ± 3.3	26
$C_6H_5CH_2-N_2CH_2C_6H_5$	157.3	11	$Cl-CHFCl$	354.4 ± 11.7	87
$CF_3-N_2CF_3$	231.0	11	$Cl-CF_2Cl$	346.0 ± 13.4	87
CH_3-NO	167.4 ± 3.3	57	$Cl-CFCl_2$	305 ± 8	37
$i-C_3H_7-NO$	152.7 ± 13	57	$Cl-CH_2Cl$	350.2 ± 0.8	87
$t-C_4H_9-NO$	165.3 ± 6.3	57	$Cl-CHCl_2$	338.5 ± 4.2	87
C_6H_5-NO	212.5 ± 4	57	$Cl-CCl_3$	305.9 ± 7.5	57
$NC-NO$	120.5 ± 10.5	40	$Cl-C_2F_5$	346.0 ± 7.1	26
CF_3-NO	179.1 ± 8	57	$Cl-CF_2CF_2Cl$	326 ± 8	57
C_6F_5-NO	208.4 ± 4	57	$Cl-SiCl_3$	464	89
CCl_3-NO	134 ± 13	57	$Br-CH_3$	292.9 ± 5.0	36
$t-C_4H_9-NOt-C_4H_9$	121	19	$Br-C_6H_5$	336.8 ± 8	57
CH_3-NO_2	254.4	57	$Br-CN$	367.4 ± 5.0	57
$CH_2C(CH_3)-NO_2$	245.2	57	$Br-CH_2COCH_3$	261.5	91
$i-C_3H_7-NO_2$	246.9	57	$Br-COC_6H_5$	268.6	11
$t-C_4H_9-NO_2$	244.8	57	$Br-CHF_2$	289 ± 8	57
$C_6H_5-NO_2$	298.3 ± 4	57	$Br-CF_3$	295.4 ± 13	57
$C(NO_2)_3-NO_2$	169.5 ± 4	57	$Br-CF_2CH_3$	287.0 ± 5.4	69
$CH_3-OC(CH_3)CH_2$	277.4	91	$Br-C_2F_5$	287.4 ± 6.3	57
$CH_3-OC_6H_5$	238 ± 8	66	$Br-n-C_3F_7$	278.2 ± 10.5	57
$CH_3-OCH_2C_6H_5$	280.3	25	$Br-i-C_3F_7$	274.1 ± 4.6	57
$C_2H_5-OC_6H_5$	264 ± 6.3	57	$Br-CH_2C_6F_5$	225 ± 6	48
$CH_2CHCH_2-OC_6H_5$	208.4 ± 8	57	$Br-CHClCF_3$	274.9 ± 6.3	57
$O=CO$	532.2 ± 0.4	28	$Br-CCl_3$	231.4 ± 4	57
CH_3-O_2	135.6 ± 2.9	56	$Br-CClBrCF_3$	251.0 ± 6.3	57
$C_2H_5-O_2$	147.2 ± 6.3	56	$Br-CH_2Br$	296.7 ± 1.3	87
$CH_2CHCH_2-O_2$	76.2 ± 2.1	56	$Br-CHBr_2$	292.0 ± 8	87
$i-C_3H_7-O_2$	157.9 ± 7.5	56	$Br-CBr_3$	235.1 ± 7.5	57
$t-C_4H_9-O_2$	153.6 ± 7.9	56	$Br-NO_2$	82.0 ± 7.1	51
$C_6H_5CH_2-O_2CCH_3$	280 ± 8	57	$Br-NF_2$	≤222	24
$C_6H_5CH_2-O_2CC_6H_5$	289	11	$I-CHCH_2$	259.0 ± 4.2	18
$CH_3-O_2SCH_3$	279.5	57	$I-n-C_4H_9$	205.0 ± 4	57
$CH_2CHCH_2-O_2SCH_3$	207.5	57	$I-Norbornyl$	261.5 ± 10.5	62
$C_6H_5CH_2-O_2SCH_3$	221.3	57	$I-CN$	305 ± 4	29
$CF_3-O_2CF_3$	361.5	9	$I-CF_3$	223.8 ± 2.9	2
CH_2Cl-O_2	120.9	56	$I-CF_2CH_3$	218.0 ± 4.2	57
$CHCl_2-O_2$	105.9	56	$I-CH_2CF_3$	235.6 ± 4	57
CCl_3-O_2	82.4 ± 3.8	56	$I-C_2F_5$	218.8 ± 2.9	2
CH_3-SH	312.5 ± 4.2	59	$I-n-C_3F_7$	208.4 ± 4.2	57
$t-C_4H_9-SH$	286.2 ± 6.3	57	$I-i-C_3F_7$	215.1 ± 2.9	2
C_6H_5-SH	361.9 ± 8	57	$I-n-C_4F_9$	205.0 ± 4.2	65
CH_3-SCH_3	307.9 ± 3.3	59	$I-C(CF_3)_3$	206	31
$CH_3-SC_6H_5$	290.4 ± 8	57	$I-C_6H_5$	273.6 ± 8	57
$C_6H_5CH_2-SCH_3$	256.9 ± 8	57	$I-C_6F_5$	~277.0	57
$S-CS$	430.5 ± 13	57	$C_5H_5-FeC_5H_5$	381 ± 13	53

Table 3
BOND STRENGTHS IN POLYATOMIC MOLECULES (continued)

Bond	D°_{298}/kJ mol^{-1}	Ref.	Bond	D°_{298}/kJ mol^{-1}	Ref.
CH_3-$ZnCH_3$	285 ± 17	57	Cl-NF_2	~134	1,68
C_2H_5-ZnC_2H_5	238 ± 17	57	I-NO	77.8 ± 0.4	39
CH_3-$Ga(CH_3)_2$	264 ± 17	57	I-NO_2	76.6 ± 4	88
C_2H_5-$Ga(C_2H_5)_2$	209 ± 17	57	HO-OH	213 ± 4	57
CH_3-$Ge(CH_3)_3$	347 ± 17	57	HO-$OCH_2C(CH_3)_3$	193.7 ± 7.9	57
CH_3-$As(CH_3)_2$	280 ± 17	57	CH_3O-OCH_3	157.3 ± 8	57
CH_3-$CdCH_3$	251 ± 17	57	C_2H_5O-OC_2H_5	158.6 ± 4	57
CH_3-$In(CH_3)_2$	205 ± 17	57	n-C_3H_7O-On-C_3H_7	155.2 ± 4	57
CH_3-$Sn(CH_3)_3$	297 ± 17	57	i-C_3H_7O-Oi-C_3H_7	157.7 ± 4	57
C_2H_5-$Sn(C_2H_5)_3$	264 ± 17	57	s-C_4H_9O-Os-C_4H_9	152.3 ± 4	57
CH_3-$Sb(CH_3)_2$	255 ± 17	57	t-C_4H_9O-Ot-C_4H_9	159.0 ± 4	57
C_2H_5-$Sb(C_2H_5)_2$	243 ± 17	57	$C_2H_5C(CH_3)_2O$-$OC(CH_3)_2C_2H_5$	164.4 ± 4	57
CH_3-$HgCH_3$	255 ± 17	57	$(CH_3)_3CCH_2O$-$OCH_2C(CH_3)_3$	152.3 ± 4	57
C_2H_5-HgC_2H_5	205 ± 17	57	CF_3O-OCF_3	193.3	57
CH_3-$Tl(CH_3)_2$	167 ± 17	57	$(CF_3)_3CO$-$OC(CF_3)_3$	148.5 ± 4.6	57
CH_3-$Pb(CH_3)_3$	238 ± 17	57	t-C_4H_9O-$OSi(CH_3)_3$	197	57
C_2H_5-$Pb(C_2H_5)_3$	230 ± 17	57	SF_5O-OSF_5	155.6	57
CH_3-$Bi(CH_3)_2$	218 ± 17	57	t-C_4H_9O-$OGe(C_2H_5)_3$	192	57
CO-$Cr(CO)_5$	155 ± 8	54	t-C_4H_9O-$OSn(C_2H_5)_3$	192	57
CO-$Fe(CO)_4$	172 ± 8	54	$FClO_2$-O	244.3	11
CO-$Mo(CO)_5$	167 ± 8	54	CF_3O-O_2CF_3	126.8 ± 8	57
CO-$W(CO)_5$	192 ± 8	54	SF_5O-O_2SF_5	126.8	57
BH_3-BH_3	146	11	CH_3CO_2-O_2CCH_3	127.2 ± 8	57
NH_2-NH_2	275.3	57	$C_2H_5CO_2$-$O_2CC_2H_5$	127.2 ± 8	57
NH_2-$NHCH_3$	268.2 ± 8	57	n-$C_3H_7CO_2$-O_2Cn-C_3H_7	127.2 ± 8	57
NH_2-$N(CH_3)_2$	246.9 ± 8	57	O-SO	552 ± 8	28
NH_2-NHC_6H_5	218.8 ± 8	57	F-OCF_3	182.0 ± 2.1	27
ON-NO_2	40.6 ± 2.1	57	HO-Cl	251 ± 13	47
O_2N-NO_2	56.9	57	O-ClO	247 ± 13	28
NF_2-NF_2	88 ± 4	57	HO-Br	234 ± 13	47
O-N_2	167	1,12	HO-I	234 ± 13	47
O-NO	305	1,12	O=PF_3	544 ± 21	47
O-NO_2	208.7 ± 1.1	30	O=PCl_3	510 ± 21	47
HO-NO	206.3	57	O=PBr_3	498 ± 21	47
HO-NO_2	206.7	57	HO-$Si(CH_3)_3$	536	57
HO_2-NO_2	96 ± 8	57	HS-SH	276 ± 8	57
CH_3O-NO	174.9 ± 3.8	8,10	CH_3S-SCH_3	272.8 ± 3.8	59
C_2H_5O-NO	175.7 ± 5.4	7,10	F-SF_5	420 ± 10	84
CH_3COO_2-NO_2	118.8 ± 3.0	15	I-SH	206.7 ± 8	57
n-C_3H_7O-NO	167.8 ± 7.5	10	I-SO	180	57
i-C_3H_7O-NO	171.5 ± 5.4	6,10	I-SCH_3	206.3 ± 7.1	77
n-C_4H_9O-NO	177.8 ± 6.3	10	I-$Si(CH_3)_3$	322	89
i-C_4H_9O-NO	175.7 ± 6.3	10	H_3Si-SiH_3	310	57,89
s-C_4H_9O-NO	173.6 ± 3.3	4,10	$(CH_3)_3Si$-$Si(CH_3)_3$	336.8	57,89
t-C_4H_9O-NO	171.1 ± 3.3	5,10	$(C_6H_5)_3Si$-$Si(C_6H_5)_3$	368 ± 29	57,89
HO-$NCHCH_3$	207.9	11			

REFERENCES

1. A value calculated from one of the thermochemical equations above, taking enthalpy data from Table 4.
2. Ahonkhai, S. I. and Whittle, E., *Int. J. Chem. Kinet.*, 16, 543, 1984.
3. Baldwin, D. P., Buntine, M. A., and Chandler, D.W., *J. Chem. Phys.*, 93, 6578, 1990.
4. Batt, L. and McCulloch, R. D., *Int. J. Chem. Kinet.*, 8, 911, 1976.
5. Batt, L. and Milne, R. T. , *Int. J. Chem. Kinet.*, 8, 59, 1976.
6. Batt, L. and Milne, R. T. , *Int. J. Chem. Kinet.*, 9, 141, 1977.
7. Batt, L. and Milne, R. T. , *Int. J. Chem. Kinet.*, 9, 549, 1977.
8. Batt, L. and Milne, R. T., and McCulloch, R. D., *Int. J. Chem. Kinet.*, 9, 567, 1977.

Table 3
BOND STRENGTHS IN POLYATOMIC MOLECULES (continued)

9. Batt, L. and Walsh, R., *Int. J. Chem. Kinet.*, 15, 605, 1983.
10. Batt, L., Christie, K., Milne, R. T., and Summers, A. J., *Int. J. Chem. Kinet.*, 6, 877, 1974.
11. Benson, S. W. and O'Neal, H. E., *Kinetic data on Gas Phase Unimolecular Reactions*, National Bureau of Standards, NSRDS-NBS, Washington, D. C., 21, 1970.
12. Benson, S. W., *J. Chem. Educ.*, 42, 502, 1965.
13. Benson, S. W., *Thermochemical Kinetics, 2nd ed.*, John Wiley & Sons, New York, 1976.
14. Berkowitz, J., Ellison, G. B., and Gutman, D., *J. Phys. Chem.*, 98, 2744, 1994.
15. Bridier, I., Caralp, F., Loirat, H., Lesclaux, R., Veyret, B., Becker, K. H., Reimer, A., and Zabel, F., *J. Phys. Chem.*, 95, 3594, 1991.
16. Buckley, T. J., Johnson, R. D., Huie, R. E., Zhang, Z., Kuo, S. C., and Klemm, B., *J. Phys. Chem.*, 99, 4879, 1995
17. Burkey, T. J., Majewski, M., and Griller, D., *J. Am. Chem. Soc.*, 108, 2218, 1986.
18. Cao, J. R., Zhang, J. M., Zhong, X., Huang, Y. H., Fang, W. Q., Wu, X. J., and Zhu, Q. H., *Chem. Phys.*, 138, 377, 1989.
19. Carmichael, P. J., Gowenlock, B. G., and Johnson, C. A. F., *J. Chem. Soc. Perkin Trans. 2*, 1853, 1973.
20. Castelhano, A. L. and Griller, D., *J. Am. Chem. Soc.*, 104, 3655, 1982.
21. Chuang, M.-C., Foltz, M. F., and Moore, C. B., *J. Chem. Phys.*, 87, 3855, 1987.
22. Chupka, W. A. and Liftshitz, C., *J. Chem. Phys.*, 48, 1109, 1968.
23. Clement, M. J. Y. and Ramsay, D. A., *Can. J. Phys.*, 39, 205, 1961.
24. Clyne, M. A. A. and Connor, J., *J. Chem. Soc. Faraday Trans. 1*, 68, 1220, 1972.
25. Colussi, A. J., Zabel, F., and Benson, S. W., *Int. J. Chem. Kinet.*, 9, 161, 1977.
26. Coomber, J. W. and Whittle, E., *Trans. Faraday Soc.*, 63, 2656, 1967.
27. Czarnarski, J., Castellano, E., and Schumacher, H. J., *Chem. Comm.*, p. 1255, 1968.
28. Darwent, D. deB., *Bond Dissociation Energies in Simple Molecules*, National Bureau of Standards, NSRDS-NBS, 31 Washington, D. C., 1970.
29. Davis, D. D. and Okabe, H., *J. Chem. Phys.*, 49, 5526, 1968.
30. Davis, H. F., Kim, B., Johnston, H. S., and Lee, Y. T., *J. Chem. Phys.*, 97, 2172, 1993.
31. Dobychin, S. L., Mashendzhinov, V. I., Mishin, V. I., Semenov, V. N., and Shpak, V. S., *Doklady Akademii Nauk SSSR*, 312, 1166 1991.
32. Doncaster, A. M. and Walsh, R., *J. Phys. Chem.*, 83, 578, 1979.
33. Ellul, E., Potzinger, P., Reimann, B., and Camilleri, P., *Ber. Bunsenges. Phys. Chem.*, 85, 407, 1981.
34. Ervin, K. M., Gronert, S., Barlow, S. E., Gilles, M. K., Harrison, A. G., Bierbaum, V. M., DePuy, C. H., Lineberger, W. C., and Ellison, G. B., *J. Am. Chem. Soc.*, 112, 5750, 1990.
35. Evans, B. S., Weeks, I., and Whittle, E., *J. Chem. Soc. Faraday Trans. 1*, 79, 1471, 1983.
36. Ferguson, K. C., Okafo, E. N., and Whittle, E., *J. Chem. Soc. Faraday Trans. 1*, 69, 295, 1973.
37. Foon, R. and Tait, K. B., *J. Chem. Soc. Faraday Trans. 1*, 68, 104, 1972.
38. Foon, R. and Tait, K. B., *J. Chem. Soc. Faraday Trans. 1*, 68, 1121, 1972.
39. Forte, E., Hippler, H., and van den Bergh, H., *Int. J. Chem. Kinet.*, 13, 1227, 1981.
40. Gowenlock, B. G., Jonhson, C. A. F., Keary, C. M., and Pfaf, J., *J. Chem. Soc. Perkin Trans. 2*, 71, 351, 1975.
41. Grela, M. A. and Colussi, A. J., *Int. J. Chem. Kinet.*, 20, 713, 1988.
42. Griller, D. and Wayner, D. D. M., *Pure Appl. Chem.*, 61, 717, 1989.
43. Heneghan, S. P. and Benson, S. W., *Int. J. Chem. Kinet.*, 15, 815, 1983.
44. Hudgens, J. W., Johnson, R. D., Timonen, R. S., Seetula, J. A., and Gutman, D., *J. Phys. Chem.*, 95, 4400, 1991.
45. Jefferson, A., Nicovich, J. M., and Wine, P. H., *J. Phys. Chem.*, 98, 7128, 1994.
46. Kerr, J. A. and Timlin, D. M., *Int. J. Chem. Kinet.*, 3, 427, 1971.
47. Kerr, J. A., *Chem. Rev.*, 66, 465, 1966.
48. Kominar, R. J., Krech, M. J., and Price, S. J. W., *Can. J. Chem.*, 58, 1906, 1980.
49. Kondo, O. and Benson, S. W., *Int. J. Chem. Kinet.*, 16, 949, 1984.
50. Kondo, O. and Benson, S. W., *J. Phys. Chem.*, 88, 6675, 1984.
51. Kreutter, K. D., Nicovich, J. M., and Wine, P. H., *J. Phys. Chem.*, 95, 4020, 1991.
52. Laufer, A. H. and Okabe, H., *J. Am. Chem. Soc.*, 93, 4137, 1971.
53. Lewis, K. E. and Smith, G. P., *J. Am. Chem. Soc.*, 106, 4650, 1984.
54. Lewis, K. E., Golden, D. M., and Smith, G. P., *J. Am. Chem. Soc.*, 106, 3905, 1984.
55. Lias, S. G., Bartmess, J. E., Liebman, J. F., Holmes, J. L., Levin, R. D., and Mallard, W. G., *J. Phys. Chem. Ref. Data 17*, Suppl. 1, 1988.
56. Lightfoot, P. D., Cox, R. A., Crowley, J. N., Destriau, M., Hayman G. D., Jenkin, M. E., Mootrgat, G. K., and Zabel, F., *Atmos. Environ.*, 26A, 1805, 1992.
57. McMillen, D. F. and Golden, D. M., *Ann. Rev. Phys. Chem.*, 33, 493, 1982.
58. Miyokawa, K. and Tschuikow-Roux, E., *J. Phys. Chem.*, 96, 7328, 1992.
59. Nicovich, J. M., Kreutter, K. D., van Dijk, C. A., and Wine, P. H., *J. Phys. Chem.*, 96, 2518, 1992.
60. Niiranen, J. T., Gutman, D., and Krasnoperov, L. N., *J. Phys. Chem.*, 96, 5881, 1992.
61. Noble, P. N. and Walsh, R., *Int. J. Chem. Kinet.*, 15, 561, 1983.
62. O'Neal, H. E., Bagg, J. W., and Richardson, W. H., *Int. J. Chem. Kinet.*, 2, 493, 1970.

Table 3
BOND STRENGTHS IN POLYATOMIC MOLECULES (continued)

63. Okabe, H. and Dibeler, V. H., *J. Chem. Phys.*, 59, 2430, 1973.
64. Okabe, H., *J. Chem. Phys.*, 66, 2058, 1977.
65. Okafo, E. N. and Whittle, E., *Int. J. Chem. Kinet.*, 7, 287, 1975.
66. Paul, S. and Back, M. H., *Can J. Chem.*, 53, 3330, 1975.
67. Pedley, J. B. and Rylance, J., "Sussex - N.P.L. Computer Analysed Thermochemical Data; Organic and Organometallic Compounds", University of Sussex, 1977.
68. Petry, R. C., *J. Am. Chem. Soc.*, 89, 4600, 1967.
69. Pickard, J. M. and Rodgers, A. S., *Int. J. Chem. Kinet.*, 9, 759, 1977.
70. Pickard, J. M. and Rodgers, A. S., *Int. J. Chem. Kinet.*, 15, 569, 1983.
71. Price, S. J. W. and Sapiano, H. J., *Can. J. Chem.*, 57, 1468, 1979.
72. Rogers, A. S. and Ford, W. G. F., *Int. J. Chem. Kinet.*, 5, 965, 1973.
73. Rossi, M., McMillen, D. F., and Golden, D. M., *J. Phys. Chem.*, 88, 5031, 1984.
74. Russell, J. J., Senkan, S. M., Seetula, J. A., and Gutman, D., *J. Phys. Chem.*, 93, 5184, 1989.
75. Seetula, J. A., Feng, Y., Gutman, D., Seakins, P. W., and Pilling, M. J., *J. Phys. Chem.*, 95, 1658, 1991.
76. Seetula, J. A., Russell, J. J., and Gutman, D., *J. Am. Chem. Soc.*, 112, 1347, 1990.
77. Shum, L. G. S. and Benson, S. W., *Int. J. Chem.*, 15, 433, 1983.
78. Shum, L. G. S. and Benson, S. W., *J. Phys. Chem.*, 87, 3479, 1983.
79. Stein, S. E., SRD Thermochemical Database, 25. N.I.S.T. Structures and Properties Database and Estimation Program, U.S. Department of Commerce, 1992.
80. Steinkruger, F. J. and Rowland, F. S., *J. Phys. Chem.*, 85, 136, 1981.
81. Trenwith, A. B. and Wrigley, S. P., *J. Chem. Soc. Faraday Trans. 1*, 73, 817, 1977.
82. Trenwith, A. B., *J. Chem. Soc. Faraday Trans. 1*, 78, 3131, 1982.
83. Trenwith, A. B., *J. Chem. Soc. Faraday Trans. 1*, 79, 2755, 1983.
84. Tsang, W. and Herron, J. T., *J. Chem. Phys.*, 96, 4272, 1992.
85. Tsang, W., *Int. J. Chem. Kinet.*, 5, 929, 1973.
86. Tsang, W., *Int. J. Chem. Kinet.*, 10, 1119, 1978.
87. Tschuikow-Roux, E. and Paddison, S., *Int. J. Chem. Kinet.*, 19, 15, 1987.
88. van den Bergh, H. and Troe, J., *J. Chem. Phys.*, 64, 736, 1976.
89. Walsh, R., *Acc. Chem. Res.*, 14, 246, 1981.
90. White, J. N. and Gardiner, W. C., *Chem. Phys. Lett.*, 58, 470, 1978.
91. Zabel, F., Benson, S. W., and Golden, D. M., *Int. J. Chem. Kinet.*, 10, 295, 1978.
92. Zmbov, K. F., Uy, O. M., and Margrave, J. L., *J. Am. Chem. Soc.*, 90, 5090, 1968.

Table 4
ENTHALPIES OF FORMATION OF FREE RADICALS

The enthalpies of formation of the free radicals are related to the corresponding bond strengths by the equations

$$D°(R\text{-}X) = \Delta_f H°(R) + \Delta_f H°(X) - \Delta_f H°(RX)$$

or

$$D°(R\text{-}R) = 2\Delta_f H°(R) + \Delta_f H°(RR)$$

For an excellent review of the methods of determining the enthalpies of formation of free radicals the reader is referred to "Thermochemistry of Free Radicals" by H. E. O'Neal and S. W. Benson in *Free Radicals*, Kochi, J. K., Ed., John Wiley & Sons, New York, 1973, 275.

Radical	$\Delta_f H°_{298}$/kJ mol^{-1}	Ref.	Radical	$\Delta_f H°_{298}$/kJ mol^{-1}	Ref.
CH	596.4 ± 1.2	5	C_2H_5	120.9 ± 1.6	9
CH_2(triplet)	390.4 ± 4	13,23	Cycloprop-2-en-1-yl	439.7 ± 17.2	34
CH_2(singlet)	428.3 ± 4	13	$CH≡CCH_2$	340.6 ± 8.4	34
CH_3	146.4 ± 0.4	9	$CH_2=CHCH_2$	170.7 ± 8.8	9
CH≡C	566.1 ± 2.9	4,19	Cyclopropyl	279.9 ± 1.1	34
$CH_2=CH$	300.0 ± 3.4	19,51	$n\text{-}C_3H_7$	97.5 ± 2.5	14

Radical	$\Delta_f H°_{298}/\text{kJ mol}^{-1}$	Ref.	Radical	$\Delta_f H°_{298}/\text{kJ mol}^{-1}$	Ref.
i-C$_3$H$_7$	90.0 ± 1.6	9	CH$_3$COCHCH$_3$	-70.3 ± 7.1	34
CH$_3$C≡CCH$_2$	293.7	34	CH$_3$OCH$_2$	-11.7 ± 5.0	34
CH$_2$=CHCHCH$_3$	125.5 ± 6.3	34	C$_2$H$_5$OCHCH$_3$	-84.5	28
CH≡CCHCH$_3$	295.0 ± 9.2	34	Tetrahydrofuran-2-yl	-18.0 ± 6.3	34
Cyclobutyl	214.2 ± 4.2	34	CH$_2$OH	-17.07 ± 3.3	9
Cyclopropylmethyl	213.8 ± 6.7	34	CH$_3$CHOH	-63.6 ± 4.2	34
CH$_2$=C(CH$_3$)CH$_2$	127.2 ± 5.4	34	CH$_2$=CHCHOH	0.0	34
CH$_3$CH=CHCH$_2$	125.5 ± 6.3	34	(CH$_3$)$_2$COH	-111.3 ± 4.6	34
s-C$_4$H$_9$	67.4 ± 2.1	9	COOH	-223.0	34
t-C$_4$H$_9$	51.5 ± 1.6	52,54	COOCH$_3$	-169.0 ± 4	34
Cyclopenta-1,3-dien-5-yl	242.3 ± 6.3	34	C$_6$H$_5$COOCH$_2$	-69.9 ± 8.4	34
Spiropentyl	380.7 ± 4	34	CHF	163 ± 13	48
Cyclopent-1-en-3-yl	160.7 ± 4	34	CH$_2$F	-31.8 ± 8.4	45
CH$_2$=CHCH=CHCH$_2$	205 ± 13	34	FCO	-152.1 ± 12	12
(C$_2$H$_3$)$_2$CH	205 ± 13	34	CHF$_2$	-238.9 ± 4	45
CH$_3$C≡CCHCH$_3$	272.8 ± 9.6	34	CF$_2$	-194.1 ± 9.2	43
CH≡CC(CH$_3$)$_2$	257.3 ± 8.4	34	CF$_3$	-467.4 ± 15.1	34
CH$_2$=CHC(CH$_3$)$_2$	77.4 ± 6.3	34	CH$_2$Cl	121.8 ± 4.2	59
Cyclopentyl	107.1 ± 2.5	14	CHFCl	-60.7 ± 10.0	59
(CH$_3$)$_3$CCH$_2$	36.4 ± 8	34	CF$_2$Cl	-279.1 ± 8.3	35
C$_2$H$_5$C(CH$_3$)$_2$	32.2 ± 6.3	57	ClCO	-21.8	40
C$_6$H$_5$	330.1 ± 3.4	9	CHCl$_2$	98.3 ± 5.0	59
Cyclohexa-1,3-dien-5-yl	197 ± 21	34	CFCl$_2$	-89.1 ± 10.0	59
Cyclohexyl	58.2 ± 4	34	CCl$_2$	239	42
CH$_3$C≡CC(CH$_3$)$_2$	221.8 ± 9.6	34	CCl$_3$	71.1 ± 2.5	25
(CH$_3$)$_2$C=C(CH$_3$)CH$_2$	39.8 ± 6.3	34	CH$_2$Br	169.0 ± 4.2	59
CH$_2$=C(CH$_3$)C(CH$_3$)$_2$	37.7 ± 6.3	34	CHBr$_2$	188.3 ± 9.2	59
C$_6$H$_5$CH$_2$	202.5 ± 6.3	9	CBr$_3$	207.1 ± 8	59
Cyclohepta-1,3,5-trien-7-yl	271.1 ± 8	34	CH$_2$I	230.1 ± 6.7	34
CH$_3$CH$_2$CH$_2$C(CH$_3$)$_2$	3.4 ± 8.4	55	CHI$_2$	333.9 ± 9.2	34
Norbornyl	136.4 ± 10.5	34	CH$_3$CF$_2$	-302.5 ± 8	34
Cycloheptyl	51.1 ± 4	34	CF$_3$CH$_2$	-517.1 ± 5.0	34
C$_6$H$_5$CHCH$_3$	169.0	34	C$_2$F$_5$	-892.9 ± 4	34
C$_6$H$_5$C(CH$_3$)$_2$	134.7	34	CCl=CH$_2$	>251	51
1-Naphthylmethyl	252.7	34	CHCl$_2$CCl$_2$	23.4 ± 8	34
(C$_6$H$_5$)$_2$CH	289	50	CF$_2$ClCF$_2$	-686 ± 17	34
9,10-Dihydroanthracen-9-yl	256.9 ± 6.3	34	C$_2$Cl$_5$	35.2 ± 8	34
9-Anthracenylmethyl	337.6	34	C$_6$F$_5$	-547.7 ± 8	34
9-Phenanthrenylmethyl	311.3	34	(CH$_3$)$_3$SiCH$_2$	-34.7	34
CH$_2$CN	243.1 ± 11.3	9	CS	278.5 ± 3.8	47
CH$_2$NC	326.4 ± 11.3	9	HCS	300.4 ± 8.4	9
CH$_3$CHCN	209.2 ± 9.6	34	CH$_2$SH	151.9 ± 8.4	9
(CH$_3$)$_2$CCN	166.5 ± 8.4	34	CH$_3$SCH$_2$	136.8 ± 5.9	26
C$_6$H$_5$C(CH$_3$)CN	248.5	34	NH	352.3 ± 9.6	46
CH$_2$NH$_2$	149.4 ± 8	34	NH$_2$	168.7 ± 1.3	9
CH$_3$NHCH$_2$	126 ± 8	34	NF$_2$	34 ± 4	34
(CH$_3$)$_2$NCH$_2$	109 ± 8	34	N$_2$H$_3$	243.5	22
CN	441.4 ± 4.6	9	N$_3$	469 ± 21	34
CHO	41.8 ± 0.8	16	CH$_2$NH	104.6 ± 13	22
CHCO	175.3 ± 8.4	9	CH$_3$NH	177.4 ± 8	34
CH$_3$CO	-10.0 ± 1.2	41	(CH$_3$)$_2$N	145.2 ± 8	34
CH$_2$=CHCO	72.4	34	C$_6$H$_5$NH	237.2 ± 8	34
C$_2$H$_5$CO	-42.7 ± 4.2	34	C$_6$H$_5$NCH$_3$	233.5 ± 8	34
C$_6$H$_5$CO	109.2 ± 8	34	NCO	159	15
CH$_2$CHO	10.5 ± 9.2	9	CH$_3$N$_2$	215.5 ± 7.5	2
CH$_3$COCH$_2$	-23.9 ± 10.9	34	C$_2$H$_5$N$_2$	187.4 ± 10.5	2

Table 4
ENTHALPIES OF FORMATION OF FREE RADICALS (continued)

Radical	$\Delta_f H°_{298}$/kJ mol^{-1}	Ref.	Radical	$\Delta_f H°_{298}$/kJ mol^{-1}	Ref.
i-$C_3H_7N_2$	158.6 ± 9.2	2	SiH_3	200.5 ± 2.5	53
OH	39.3	34	CH_3Si	310	62
CH_3O	17.2 ± 3.8	9	CH_3SiH	213.0 ± 14.6	60
C_2H_5O	-15.5 ± 3.4	9	CH_3SiH_2	152.7	34,62
n-C_3H_7O	-41.4	34	$(CH_3)_2Si$	109	62
i-C_3H_7O	-52.3	34	$(CH_3)_2SiH$	59.8	34,62
n-C_4H_9O	-62.8	34	$(CH_3)_3Si$	-3.3	34,62
s-C_4H_9O	-69.5 ± 3.3	34	$C_6H_5Si(CH_3)_2$	163	42
t-C_4H_9O	-90.8	34	$(C_6H_5)_2SiCH_3$	326	42
C_6H_5O	47.7	34	$(C_6H_5)_3Si$	486.2	42
CF_3O	-655.6	7	SiF	-19.3	34,62
FO	109 ± 4	8	SiF_2	-587.9	34,62
ClO	101.6 ± 0.1	1	SiF_3	-1025	34,62
BrO	125	61	SiCl	195.8	34,62
IO	107	6,49	$SiCl_2$	-163.6	34,62
HO_2	14.6	56	$SiCl_3$	-318	34,62
CH_3O_2	10.4 ± 3.1	29	SiH_3SiH	269.9 ± 14.6	60
$C_2H_5O_2$	-28.7 ± 6.5	29	Si_2H_5	223.0	34,62
$CH_2=CHCH_2O_2$	87.9 ± 5.5	29	PH_2	138.5 ± 2.5	9
i-$C_3H_7O_2$	-68.9 ± 8.1	29	HS	143.0 ± 2.8	39
t-$C_4H_9O_2$	-105.0 ± 8.1	29	CH_3S	124.6 ± 1.8	39
$HOCH_2O_2$	-162.1 ± 2.1	29	C_6H_5S	229.7 ± 8	34
CF_3O_2	-614.0	29	SO	5.0	15
CF_2ClO_2	-406.5	29	HSO	-4	30
$CFCl_2O_2$	-213.7	29	HSO_2	-222	10
CH_2ClO_2	9.2	29	CH_3SO_2	-239.3	42
$CHCl_2O_2$	1.6	29	$HOSO_2$	-385	32
CCl_3O_2	-11.3 ± 4.6	29	SO_3	-395.7	61
CH_3CHClO_2	-54.7 ± 3.7	27	SF_4	-746 ± 12	58
CH_3CO_2	-207.5 ± 4	34	SF_5	-879.9 ± 20	58
CH_3COO_2	-172 ± 20	11	CH_3S_2	68.6 ± 8	24
$C_2H_5CO_2$	-228.5 ± 4	34	$C_2H_5S_2$	43.5 ± 8	24
n-$C_3H_7CO_2$	-249.4 ± 4	34	i-$C_3H_7S_2$	13.8 ± 8	24
FO_2	26.1	31,44	t-$C_4H_9S_2$	-19.3 ± 8	24
ClO_2	97.5	3,38	$HOCS_2$	110.5	36
OClO	95.6	20,37	GeH_3	222 ± 8	9
NO_3	73.7 ± 1.4	18	AsH_2	167.8 ± 1.3	9
sym-ClO_3	232.6	17	HSe	144.8 ± 2.1	9
SiH	377	34,62	SbH_2	215.5 ± 2.5	9
SiH_2	269.0 ± 1.3	21,33	HTe	158.6 ± 5.0	9

REFERENCES

1. Abramowitz, S. and Chase, M. W., *Pure Appl. Chem.*, 63, 1448, 1991.
2. Acs, G. and Peter, A., *Int. J. Chem. Kinet.*, 19, 929, 1987.
3. Baer, S., Hippler, H., Rabu, R., Siefke, M., Seitzinger, N., and Troe, J., *J. Chem. Phys.*, 95, 6463, 1991.
4. Baldwin, D. P., Buntine, M. A., and Chandler, D. W., *J. Chem. Phys.*, 93, 6578, 1990.
5. Based on D(C-H), see Table 1 and $D_f H°$(C) and $D_f H°$(H), see Table 2.
6. Based on D(I-O), see Table 1 and $D_f H°$(I) and $D_f H°$(O) see Table 2.
7. Batt, L. and Walsh, R., *Int. J. Chem. Kinet.*, 14, 933, 1982.
8. Benson, S. W., *Thermochemical Kinetics, 2nd ed.*, John Wiley & Sons, New York, 1976.
9. Berkowitz, J., Ellison, G. B., and Gutman, D., *J. Phys. Chem.*, 98, 2744, 1994.
10. Boyd, R. J., Gupta, A., Langler, R. F., Lownie, S. P., and Pincock, J. A., *Can. J. Chem.*, 58, 331, 1980.
11. Bridier, I., Caralp, F., Loirat, H., Lesclaux, R., Veyret, B., Becker, K. H., Reimer, A., and Zabel, F., *J. Phys. Chem.*, 95, 3594, 1991.
12. Buckley, T. J., Johnson, R. D., Huie, R. E., Zhang, Z., Kuo, S. C., and Klemm, B., *J. Phys. Chem.*, 99, 4879, 1995.
13. Bunker, P. R. and Sears, T. J., *J. Chem. Phys.*, 83, 4866, 1985.

14. Castelhano, A. L. and Griller, D., *J. Am. Chem. Soc.*, 104, 3655, 1982; value adjusted to $D_f H°(CH_3) = 146$ kJ mol[-1] and error limits assigned here.
15. Chase, M. W. Jr., Davies, C. A., Downey, J.R. Jr., Frurip, D. J., McDonald, R. A., and Syverud, A. N., *J. Phys. Chem. Ref. Data,* 14, Suppl. 1, 1985.
16. Chuang, M.-C., Foltz, M. F., and Moore, C. B., *J. Chem. Phys.*, 87, 3855, 1987.
17. Colussi, A. J., *J. Phys. Chem.*, 94, 8922, 1990.
18. Davis, H. F., Kim, B., Johnston, H. S., and Lee, Y. T., *J. Phys. Chem.,* 97, 2172, 1993.
19. Ervin, K. M., Gronert, S., Barlow, S. E., Gilles, M. K., Harrison, A. G., Bierbaum, V. M., DePuy, C. H., Lineberger, W. C., and Ellison, G. B., *J. Am. Chem. Soc.*, 112, 5750, 1990.
20. Flesch, R., E. Rühl, K. Hottmann, and H. Baumgartel, *J. Phys. Chem.*, 97, 837 (1993).
21. Frey, H. M., Walsh, R., and Watts, I. M., *J. Chem. Soc. Chem. Comm.*, 1189, 1986.
22. Grela, M. A. and Colussi, A. J., *Int. J. Chem. Kinet.*, 20, 713, 1988.
23. Gurvich, l.V., Veyts, I.V., Alcock, C.B., *Thermodynamic Properties of Individual Substances, 4th ed.*, Hemisphere, New York, 1991, Vol. 2.
24. Howari, J. A. Griller, D., and Lossing, F. P., *J. Am. Chem. Soc.*, 108, 3273, 1986.
25. Hudgens, J. W., Johnson, R. D., Timonen, R. S., Seetula, J. A., and Gutman, D., *J. Phys. Chem.*, 95, 4400, 1991.
26. Jefferson, A., Nicovich, J. M., and Wine, P. H., *J. Phys. Chem.*, 98, 7128, 1994.
27. Knyazev, V. D., Bencsura, A., Dubinsky, I. A., Gutman, D., Melius, C. F., and Senkan, S. M., *J. Phys. Chem.*, 99, 230, 1995.
28. Kondo, O. and Benson, S. W., *Int. J. Chem. Kinet.*, 16, 949, 1984.
29. Lightfoot, P. D., Cox, R. A., Crowley, J. N., Destriau, M., Hayman, G.D., Jenkin, M. E., Moortgat, G. K., and Zabel, F., *Atmos Environ.*, 26A, 1805, 1992.
30. Lovejoy, E. R., Wang, N. S., and Howard, C. J., *J. Phys. Chem.,* 91, 5749, 1987.
31. Lyman, J. L. and Holland, R., *J. Phys. Chem.*, 92, 7232, 1988.
32. Margitan, J. J., *J. Phys. Chem.*, 88, 3314, 1984.
33. Martin, J. G., Ring, M. A., and O'Neal, H. E., *Int. J. Chem. Kinet.*, 19, 715, 1987.
34. McMillen, D. F. and Golden, D. M., *Ann. Rev. Phys. Chem.*, 33, 493, 1982.
35. Miyokawa, K. and Tschuikow-Roux, E., *J. Phys. Chem.,* 96, 7328, 1992.
36. Murrells, T. P., Lovejoy, E.R., and Ravishankara, A. R., *J. Phys. Chem.*, 80, 4065, 1984.
37. Nickolaisen, S. L., R. R. Friedl, and S. P. Sander, *J. Phys. Chem.*, 98, 155 (1994).
38. Nicovich, J. M., Kreutter, K. D., Shockelford, C. J., and Wine, P. H., *Chem. Phys. Lett.*, 179, 367, 1991.
39. Nicovich, J. M., Kreutter, K. D., van Dijk, C. A., and Wine, P. H., *J. Phys. Chem.*, 96, 2518, 1992.
40. Nicovich, J.M., Kreutter, K. D., and Wine, P.H., *J. Chem. Phys.*, 92, 3539 (1990).
41. Niiranen, J. T., Gutman, D., and Krasnoperov, L. N., *J. Phys. Chem.*, 96, 5881, 1992.
42. O'Neal, H. E. and Benson, S. W., in *Free Radicals*, Kochi, J. K., Ed., John Wiley & Sons, New York, 1973, 275.
43. Okafo, E. N. and Whittle, E., *J. Chem. Soc. Faraday Soc.*, 1, 70, 1366, 1974.
44. Pagsberg, P., Ratajczak, E., Sillesen, A., and Jodkowski, J. T., *Chem. Phys. Lett.*, 141, 88, 1987.
45. Pickard, J. M. and Rodgers, A. S., *Int. J. Chem. Kinet.*, 15, 569, 1983.
46. Piper, L. G., *J. Chem. Phys.*, 70, 3417, 1979.
47. Prinslow, D. A. and Armentrout, P. B., *J. Chem. Phys.*, 94, 3563, 1991.
48. Pritchard, G. O., Nilson, W. B., and Kirtman, B., *Int. J. Chem. Kinet.*, 16, 1637, 1984.
49. Reddy, R. R., Rao, T. V. R., and Reddy, A. S. R., *Indian J. Pure Appl. Phys.,* 27, 243, 1989.
50. Rossi, M., McMillen, D. F., and Golden, D. M., *J. Phys. Chem.*, 88, 5031, 1984.
51. Russell, J. R., Senkan, S. M., Seetula, J. A., and Gutman, D., *J. Phys. Chem.*, 93, 5184, 1989.
52. Seakins, P. W., Pilling, M. J., Niiranen, J. T., Gutman, D., and Krasnoperov, L. N., *J. Phys. Chem.*, 96, 9847, 1992.
53. Seetula, J. A., Feng, Y., Gutman, D., Seakins, P. W., and Pilling, M. J., *J. Phys. Chem.,* 95, 1658, 1991.
54. Seetula, J. A., Russell, J. A., and Gutman, D., *J. Am. Chem. Soc.*, 112, 1347, 1990.
55. Seres, L., Gorgenyi, M., and Farkas, J., *Int. J. Chem. Kinet.*, 15, 1133, 1983.
56. Shum, L. G. S. and Benson, S. W., *J. Phys. Chem.,* 87, 3479, 1983.
57. Stein, S. E., SRD Thermochemical Database, 25. N.I.S.T. Structures and Properties Database and Estimation Program, U.S. Department of Commerce, 1992.
58. Tsang, W. and Herron, J. T., *J. Chem. Phys.*, 96, 4272, 1992.
59. Tschuikow-Roux, E. and Paddison, S., *Int. J. Chem. Kinet.*, 19, 15, 1987.
60. Vanderwielen, A. J., Ring, M. A., and O'Neal, H. E., *J. Am. Chem. Soc.*, 97, 993, 1975.
61. Wagman, D. D., Evans, W. H., Parker, V. B., Schumm, R. H., Halow, I., Bailey, S. M., Churney, K. L., and Nuttall, R.L., *J. Phys. Chem. Ref. Data*, 11, Suppl. 2, 1978.
62. Walsh, R., *Acc. Chem. Res.*, 14, 246, 1981.

Table 5
BOND STRENGTHS OF SOME ORGANIC MOLECULES

Bond strengths at 298 K expressed in kJ/mol for some organic molecules of the general formula R-X are presented below. Some are experimental values taken from the preceding tables; the remainder are calculated from the enthalpies of formation of the radicals, listed in Table 4 above, and the enthalpies of formation of the parent compounds from sources indicated by the references below. The table also includes bond strengths for the inorganic molecules, hydrogen, the hydrogen halides, water and ammonia.

	H	F	Cl	Br	I	OH	NH2	CH3O	CH3	CH3CO	NO	CF3	CCl3
H	435.990	570.3	431.62	366.35	298.407	498	449	436	439	374	207	446	393
CH3	439	472	349c	293	239c	387i	354i	348i	376i	339i	167	423	362d
C2H5	423	463f	353d	294d	235d	395i	354i	355i	372i	338i	—	—	—
i-C3H7	413	463f	356i	300i	238i	402i	359i	360i	371i	328i	153	—	—
t-C4H9	404	—	355i	296i	230i	403i	358i	353i	365i	317i	165	—	—
C6H5	465	524i	399i	337i	274i	465i	427i	415i	425i	391i	213	—	—
C6H5CH2	370	—	303i	248i	207i	340i	—	—	332	274i	—	—	—
CCl3	393	435a	288c	225h	167e	—	—	—	362d	—	134	335d	292d
CF3	446	545c	360c	295c	224	—	—	—	427d	—	179	413	335d
C2F5	430	531d	346j	287g	219	—	—	—	—	—	—	—	—
CH3CO	374	499i	340i	278i	209i	447i	399i	403	339i	282	—	—	—
CN	528	470	422	367	305	—	—	—	—	—	121	—	—
C6F5	477	477b	385b	337	277b	448b	—	—	444b	—	208b	—	—

REFERENCES

a   Chen, S. S., Wilhoit, R. C., and Zwolinski, B. J., *J. Phys. Chem. Ref. Data*, 5, 571, 1976.

b   Choo, K. Y., Mendenhall, G. D., Golden, D.M., and Benson, S. W., *Int. J. Chem. Kinet.*, 6, 813, 1974.

c   Kolesov, V. P., *Russ. Chem. Rev.*, 47, 599, 1978.

d   Kolesov, V. P. and Papina, T. S., *Russ. Chem. Rev.*, 52, 425, 1983.

e   Kudchadker, S. A. and Kudchadker, A. P., *J. Phys. Chem. Ref. Data*, 1, 1285, 1978.

f   Lias, S. G., Bartmess, J. E., Liebman, J. F., Holmes, J. L., Levin, R. D., and Mallard, W. G., *J. Phys. Chem. Ref. Data*, 17, Suppl. No. 1, 1988.

g   McMillen, D. F. and Golden, D. M., *Ann. Rev. Phys. Chem.*, 33, 493, 1982.

h   Mendenhall, G. D., Golden, D. M., and Benson, S. W., *J. Phys. Chem.*, 77, 2707, 1973.

i   Pedley, J. B. and Rylance, J., "Sussex.N.P.L. Computer Analysed Thermochemical Data: Organic and Organometallic Compounds", University of Sussex, 1977.

j   Wu, E.-C. and Rodgers, A. S., *J. Am. Chem. Soc.*, 98, 6112, 1976.

ELECTRONEGATIVITY

Electronegativity is a parameter originally introduced by Pauling which describes, on a relative basis, the tendency of an atom in a molecule to attract bonding electrons. While electronegativity is not a precisely defined molecular property, the electronegativity difference between two atoms provides a useful measure of the polarity and ionic character of the bond between them. This table gives the electronegativity X, on the Pauling scale, for the most common oxidation state. Other scales are described in the references.

REFERENCES

1. Pauling, L., *The Nature of the Chemical Bond, Third Edition*, Cornell University Press, Ithaca, New York, 1960.
2. Allen, L.C., *J. Am. Chem. Soc.*, 111, 9003, 1989.
3. Allred, A.L., *J. Inorg. Nucl. Chem.*, 17, 215, 1961.

Z	Symbol	X	Z	Symbol	X	Z	Symbol	X
1	H	2.20	33	As	2.18	65	Tb	—
2	He	—	34	Se	2.55	66	Dy	1.22
3	Li	0.98	35	Br	2.96	67	Ho	1.23
4	Be	1.57	36	Kr	—	68	Er	1.24
5	B	2.04	37	Rb	0.82	69	Tm	1.25
6	C	2.55	38	Sr	0.95	70	Yb	—
7	N	3.04	39	Y	1.22	71	Lu	1.0
8	O	3.44	40	Zr	1.33	72	Hf	1.3
9	F	3.98	41	Nb	1.6	73	Ta	1.5
10	Ne	—	42	Mo	2.16	74	W	1.7
11	Na	0.93	43	Tc	2.10	75	Re	1.9
12	Mg	1.31	44	Ru	2.2	76	Os	2.2
13	Al	1.61	45	Rh	2.28	77	Ir	2.2
14	Si	1.90	46	Pd	2.20	78	Pt	2.2
15	P	2.19	47	Ag	1.93	79	Au	2.4
16	S	2.58	48	Cd	1.69	80	Hg	1.9
17	Cl	3.16	49	In	1.78	81	Tl	1.8
18	Ar	—	50	Sn	1.96	82	Pb	1.8
19	K	0.82	51	Sb	2.05	83	Bi	1.9
20	Ca	1.00	52	Te	2.1	84	Po	2.0
21	Sc	1.36	53	I	2.66	85	At	2.2
22	Ti	1.54	54	Xe	2.60	86	Rn	—
23	V	1.63	55	Cs	0.79	87	Fr	0.7
24	Cr	1.66	56	Ba	0.89	88	Ra	0.9
25	Mn	1.55	57	La	1.10	89	Ac	1.1
26	Fe	1.83	58	Ce	1.12	90	Th	1.3
27	Co	1.88	59	Pr	1.13	91	Pa	1.5
28	Ni	1.91	60	Nd	1.14	92	U	1.7
29	Cu	1.90	61	Pm	—	93	Np	1.3
30	Zn	1.65	62	Sm	1.17	94	Pu	1.3
31	Ga	1.81	63	Eu	—			
32	Ge	2.01	64	Gd	1.20			

FORCE CONSTANTS FOR BOND STRETCHING

Representative force constants (*f*) for stretching of chemical bonds are listed in this table. Except where noted, all force constants are derived from values of the harmonic vibrational frequencies ω_e. Values derived from the observed vibrational fundamentals ν, which are noted by a, are lower than the harmonic force constants, typically by 2 to 3% in the case of heavy atoms (often by 5 to 10% if one of the atoms is hydrogen). Values are given in the SI unit newton per centimeter (N/cm), which is identical to the commonly used cgs unit mdyn/Å.

REFERENCES

1. Huber, K. P., and Herzberg, G., *Molecular Spectra and Molecular Structure. IV. Constants of Diatomic Molecules*, Van Nostrand Reinhold, New York, 1979.
2. Shimanouchi, T., *The Molecular Force Field*, in Eyring, H., Henderson, D., and Yost, W., Eds., *Physical Chemistry: An Advanced Treatise*, Vol. IV, Academic Press, New York, 1970.
3. Tasumi, M., and Nakata, M., *Pure and Appl. Chem.*, 57, 121—147, 1985.

Bond	Molecule	*f*/(N/cm)	Note	Bond	Molecule	*f*/(N/cm)	Note
H-H	H_2	5.75			OCS	7.44	
Be-H	BeH	2.27		C-N	CN	16.29	
B-H	BH	3.05			HCN	18.78	
C-H	CH	4.48			CH_3CN	18.33	
	CH_4	5.44	b		CH_3NH_2	5.12	a,c
	C_2H_6	4.83	a,b,c	C-P	CP	7.83	
	CH_3CN	5.33	b	Si-Si	Si_2	2.15	
	CH_3Cl	5.02	a,b,c	Si-O	SiO	9.24	
	$CCl_2=CH_2$	5.57	b	Si-F	SiF	4.90	
	HCN	6.22		Si-Cl	SiCl	2.63	
N-H	NH	5.97		N-N	N_2	22.95	
O-H	OH	7.80			N_2O	18.72	
	H_2O	8.45		N-O	NO	15.95	
P-H	PH	3.22			N_2O	11.70	
S-H	SH	4.23		P-P	P_2	5.56	
	H_2S	4.28		P-O	PO	9.45	
F-H	HF	9.66		O-O	O_2	11.77	
Cl-H	HCl	5.16			O_3	5.74	a
Br-H	HBr	4.12		S-O	SO	8.30	
I-H	HI	3.14			SO_2	10.33	a
Li-H	LiH	1.03		S-S	S_2	4.96	
Na-H	NaH	0.78		F-F	F_2	4.70	
K-H	KH	0.56		Cl-F	ClF	4.48	
Rb-H	RbH	0.52		Br-F	BrF	4.06	
Cs-H	CsH	0.47		Cl-Cl	Cl_2	3.23	
C-C	C_2	12.16		Br-Cl	BrCl	2.82	
	$CCl_2=CH_2$	8.43		Br-Br	Br_2	2.46	
	C_2H_6	4.50	a,c	I-I	I_2	1.72	
	CH_3CN	5.16		Li-Li	Li_2	0.26	
C-F	CF	7.42		Li-Na	LiNa	0.21	
	CH_3F	5.71	a,c	Na-Na	Na_2	0.17	
C-Cl	CCl	3.95		Li-F	LiF	2.50	
	CH_3Cl	3.44	a,c	Li-Cl	LiCl	1.43	
	$CCl_2=CH_2$	4.02	b	Li-Br	LiBr	1.20	
C-Br	CH_3Br	2.89	a,c	Li-I	LiI	0.97	
C-I	CH_3I	2.34	a,c	Na-F	NaF	1.76	
C-O	CO	19.02		Na-Cl	NaCl	1.09	
	CO_2	16.00		Na-Br	NaBr	0.94	
	OCS	16.14		Na-I	NaI	0.76	
	CH_3OH	5.42	a,c	Be-O	BeO	7.51	
C-S	CS	8.49		Mg-O	MgO	3.48	
	CS_2	7.88		Ca-O	CaO	3.61	

a Derived from fundamental frequency, without anharmonicity correction.
b Average of symmetric and antisymmetric (or degenerate) modes.
c Calculated from Local Symmetry Force Field (see Reference 2).

FUNDAMENTAL VIBRATIONAL FREQUENCIES OF SMALL MOLECULES

This table lists the fundamental vibrational frequencies of selected three-, four-, and five-atom molecules. Both stable molecules and transient free radicals are included. The data have been taken from evaluated sources. In general, the selected values are based on gas-phase infrared, Raman, or ultraviolet spectra; when these were not available, liquid-phase or matrix-isolation spectra were used.

Molecules are grouped by structural type. Within each group, related molecules appear together for convenient comparison.

The vibrational modes are described by their approximate character in terms of stretching, bending, deformation, etc. However, it should be emphasized that most such descriptions are only approximate, and that the true normal mode usually involves a mixture of motions. Abbreviations are:

sym.	symmetric
antisym.	antisymmetric
str.	stretch
deform.	deformation
scis.	scissors
rock.	rocking
deg.	degenerate

In the case of free radicals, strong interactions may exist between the electronic and bending vibrational motions. Details can be found in References 3 and 4. The references should be consulted for information on the accuracy of the data and for data on other molecules not listed here.

All fundamental frequencies (more precisely, wavenumbers) are given in units of cm^{-1}.

XY_2 Molecules
Point groups $D_{\infty h}$(linear) and C_{2v}(bent)

Molecule	Structure	Sym. str.	Bend	Antisym. str.
CO_2	Linear	1333	667	2349
CS_2	Linear	658	397	1535
C_3	Linear	1224	63	2040
CNC	Linear		321	1453
NCN	Linear	1197	423	1476
BO_2	Linear	1056	447	1278
BS_2	Linear	510	120	1015
KrF_2	Linear	449	233	590
XeF_2	Linear	515	213	555
$XeCl_2$	Linear	316		481
H_2O	Bent	3657	1595	3756
D_2O	Bent	2671	1178	2788
F_2O	Bent	928	461	831
Cl_2O	Bent	639	296	686
O_3	Bent	1103	701	1042
H_2S	Bent	2615	1183	2626
D_2S	Bent	1896	855	1999
SF_2	Bent	838	357	813
SCl_2	Bent	525	208	535
SO_2	Bent	1151	518	1362
H_2Se	Bent	2345	1034	2358
D_2Se	Bent	1630	745	1696
NH_2	Bent	3219	1497	3301
NO_2	Bent	1318	750	1618
NF_2	Bent	1075	573	942
ClO_2	Bent	945	445	1111
CH_2	Bent		963	
CD_2	Bent		752	
CF_2	Bent	1225	667	1114
CCl_2	Bent	721	333	748
CBr_2	Bent	595	196	641
SiH_2	Bent	2032	990	2022
SiD_2	Bent	1472	729	1468
SiF_2	Bent	855	345	870

Molecule	Structure	Sym. str.	Bend	Antisym. str.
$SiCl_2$	Bent	515		505
$SiBr_2$	Bent	403		400
GeH_2	Bent	1887	920	1864
$GeCl_2$	Bent	399	159	374
SnF_2	Bent	593	197	571
$SnCl_2$	Bent	352	120	334
$SnBr_2$	Bent	244	80	231
PbF_2	Bent	531	165	507
$PbCl_2$	Bent	314	99	299
ClF_2	Bent	500		576

XYZ Molecules
Point Groups $C_{\infty v}$ (linear) and C_s(bent)

Molecule	Structure	XY str.	Bend	YZ str.
HCN	Linear	3311	712	2097
DCN	Linear	2630	569	1925
FCN	Linear	1077	451	2323
ClCN	Linear	744	378	2216
BrCN	Linear	575	342	2198
ICN	Linear	486	305	2188
CCN	Linear	1060	230	1917
CCO	Linear	1063	379	1967
HCO	Bent	2485	1081	1868
HCC	Linear	3612		1848
OCS	Linear	2062	520	859
NCO	Linear	1270	535	1921
NNO	Linear	2224	589	1285
HNB	Linear	3675		2035
HNC	Linear	3653		2032
HNSi	Linear	3583	523	1198
HBO	Linear		754	1817
FBO	Linear		500	2075
ClBO	Linear	676	404	1958
BrBO	Linear	535	374	1937
FNO	Bent	766	520	1844
ClNO	Bent	596	332	1800
BrNO	Bent	542	266	1799
HNF	Bent		1419	1000
HNO	Bent	2684	1501	1565
HPO	Bent	2095	983	1179
HOF	Bent	3537	886	1393
HOCl	Bent	3609	1242	725
HOO	Bent	3436	1392	1098
FOO	Bent	579	376	1490
ClOO	Bent	407	373	1443
BrOO	Bent			1487
HSO	Bent		1063	1009
NSF	Bent	1372	366	640
NSCl	Bent	1325	273	414
HCF	Bent		1407	1181
HCCl	Bent		1201	815
HSiF	Bent	1913	860	834
HSiCl	Bent		808	522
HSiBr	Bent	1548	774	408

Symmetric XY$_3$ Molecules
Point Groups D$_{3h}$ (planar) and C$_{3v}$ (pyramidal)

Molecule	Structure	Sym. str.	Sym. deform.	Deg. str.	Deg. deform.
NH$_3$	Pyram.	3337	950	3444	1627
ND$_3$	Pyram.	2420	748	2564	1191
PH$_3$	Pyram.	2323	992	2328	1118
AsH$_3$	Pyram.	2116	906	2123	1003
SbH$_3$	Pyram.	1891	782	1894	831
NF$_3$	Pyram.	1032	647	907	492
PF$_3$	Pyram.	892	487	860	344
AsF$_3$	Pyram.	741	337	702	262
PCl$_3$	Pyram.	504	252	482	198
PI$_3$	Pyram.	303	111	325	79
AsI$_3$	Pyram.	219	94	224	71
AlCl$_3$	Pyram.	375	183	595	150
SO$_3$	Planar	1065	498	1391	530
BF$_3$	Planar	888	691	1449	480
BH$_3$	Planar		1125	2808	1640
CH$_3$	Planar		606	3161	1396
CD$_3$	Planar		453	2369	1029
CF$_3$	Pyram.	1090	701	1260	510
SiF$_3$	Pyram.	830	427	937	290

Linear XYYX Molecules
Point Group D$_{\infty h}$

Molecule	Sym. XY str.	Antisym. XY str.	YY str.	Bend	Bend
C$_2$H$_2$	3374	3289	1974	612	730
C$_2$D$_2$	2701	2439	1762	505	537
C$_2$N$_2$	2330	2158	851	507	233

Planar X$_2$YZ Molecules
Point Group C$_{2v}$

Molecule	Sym.XY str.	YZ str.	YX$_2$ scis.	Antisym. XY str.	YX$_2$ rock	YX$_2$ wag
H$_2$CO	2783	1746	1500	2843	1249	1167
D$_2$CO	2056	1700	1106	2160	990	938
F$_2$CO	965	1928	584	1249	626	774
Cl$_2$CO	567	1827	285	849	440	580
O$_2$NF	1310	822	568	1792	560	742
O$_2$NCl	1286	793	370	1685	408	652

Tetrahedral XY$_4$ Molecules
Point Group T$_d$

Molecule	Sym. str.	Deg. deform.(e)	Deg. str.(f)	Deg. deform.(f)
CH$_4$	2917	1534	3019	1306
CD$_4$	2109	1092	2259	996
CF$_4$	909	435	1281	632
CCl$_4$	459	217	776	314

Molecule	Sym. str.	Deg. deform.(e)	Deg. str.(f)	Deg. deform.(f)
CBr_4	267	122	672	182
CI_4	178	90	555	125
SiH_4	2187	975	2191	914
SiD_4	1558	700	1597	681
SiF_4	800	268	1032	389
$SiCl_4$	424	150	621	221
GeH_4	2106	931	2114	819
GeD_4	1504	665	1522	596
$GeCl_4$	396	134	453	172
$SnCl_4$	366	104	403	134
$TiCl_4$	389	114	498	136
$ZrCl_4$	377	98	418	113
$HfCl_4$	382	102	390	112
RuO_4	885	322	921	336
OsO_4	965	333	960	329

REFERENCES

1. T. Shimanouchi, Tables of Molecular Vibrational Frequencies, Consolidated Volume I, Natl. Stand. Ref. Data Ser. Natl. Bur. Stand. (U.S.), 39, 1972.
2. T. Shimanouchi, Tables of Molecular Vibrational Frequencies, Consolidated Volume II, *J. Phys. Chem. Ref. Data*, 6, 993, 1977.
3. G. Herzberg, *Electronic Spectra and Electronic Structure of Polyatomic Molecules*, D. Van Nostrand Co., Princeton, 1966.
4. M. E. Jacox, Ground state vibrational energy levels of polyatomic transient molecules, *J. Phys. Chem. Ref. Data*, 13, 945, 1984.

SPECTROSCOPIC CONSTANTS OF DIATOMIC MOLECULES

This table lists the leading spectroscopic constants and equilibrium internuclear distance r_e in the ground electronic state for selected diatomic molecules. The constants are those describing the vibrational and rotational energy through the expressions:

$$E_{vib}/hc = \omega_e(v+1/2) - \omega_e x_e(v+1/2)^2 + \cdots$$

$$E_{rot}/hc = B_v J(J+1) - D_v[J(J+1)]^2 + \cdots$$

where

$$B_v = B_e - \alpha_e(v+1/2) + \cdots$$

$$D_v = D_e + \cdots$$

Here v and J are the vibrational and rotational quantum numbers, respectively, h is Planck's constant, and c is the speed of light. In this customary formulation the constants ω_e, B_e, etc. have dimensions of inverse length; in this table they are given in units of cm^{-1}.

Users should note that higher order terms in the above energy expressions are required for very precise calculations; constants for many of these terms can be found in the references. Also, if the ground electronic state is not $^1\Sigma$, additional terms are needed to account for the interaction between electronic and pure rotational angular momentum. For some molecules in the table the data have been analyzed in terms of the Dunham series expansion:

$$E/hc = \Sigma_{lm} Y_{lm}(v+1/2)^l J^m(J+1)^m$$

In such cases it has been assumed that $Y_{10} = \omega_e$, $Y_{01} = B_e$, etc., although in the highest approximations these identities are not precisely correct. Some of the values of r_e in the table have been corrected for breakdown of the Born-Oppenheimer approximation, which can affect the last decimal place. Because of differences in the method of data analysis and limitations in the model, care should be taken in comparing r_e values for different molecules to a precision beyond 0.001 Å.

Molecules are listed in alphabetical order by formula as written in the most common form. In most cases this form places the more electropositive element first, but there are exceptions such as OH, NH, CH, etc.

* Indicates a value for the interval between $v = 0$ and $v = 1$ states instead of a value of ω_e.

REFERENCES

1. Huber, K. P., and Herzberg, G., *Molecular Spectra and Molecular Structure IV. Constants of Diatomic Molecules*, Van Nostrand Reinhold, New York, 1979.
2. Lovas, F. J., and Tiemann, E., *J. Phys. Chem. Ref. Data*, 3, 609, 1974.
3. *Landolt-Börnstein, Numerical Data and Functional Relationships in Science and Technology, New Series*, II/6 (1974), II/14a (1982), II/14b (1983), II/19a (1992), II/19d-1 (1995), *Molecular Constants*, Springer-Verlag, Heidelberg.

Molecule	State	ω_e cm^{-1}	$\omega_e x_e$ cm^{-1}	B_e cm^{-1}	α_e cm^{-1}	D_e 10^{-6}cm^{-1}	r_e Å
^{107}Ag^{79}Br	$^1\Sigma^+$	249.57	0.63	0.064833	0.0002361	0.0175	2.39311
^{107}Ag^{35}Cl	$^1\Sigma^+$	343.49	1.17	0.12298388	0.00059541	0.06305	2.28079
^{107}Ag^{19}F	$^1\Sigma^+$	513.45	2.59	0.2657020	0.0019206	0.284	1.98318
^{107}Ag^{1}H	$^1\Sigma^+$	1759.9	34.06	6.449	0.201	344	1.618
^{107}Ag^{2}H	$^1\Sigma^+$	1250.70	17.17	3.2572	0.0722	85.9	1.6180
^{107}Ag^{127}I	$^1\Sigma^+$	206.50	0.46	0.04486821	0.0001414	0.00847	2.54463
^{107}Ag^{16}O	$^2\Pi_{1/2}$	490.2	3.1	0.3020	0.0025	0.45	2.003
^{27}Al$_2$	$^3\Sigma_g^-$	350.01	2.02	0.2054	0.0012	0.31	2.466
^{27}Al^{79}Br	$^1\Sigma^+$	378.0	1.28	0.15919713	0.00086045	0.11285	2.29481
^{27}Al^{35}Cl	$^1\Sigma^+$	481.30	1.95	0.24393012	0.00161113	0.2503	2.13011
^{27}Al^{19}F	$^1\Sigma^+$	802.3	4.77	0.5524798	0.0049841	1.0464	1.65437
^{27}Al^{1}H	$^1\Sigma^+$	1682.56	29.09	6.3907	0.1858	356.5	1.6478
^{27}Al^{2}H	$^1\Sigma^+$	1211.95	15.14	3.3186	0.0697	97	1.6463
^{27}Al^{127}I	$^1\Sigma^+$	316.1	1.0	0.11769985	0.00055859		2.53710
^{27}Al^{16}O	$^2\Sigma^+$	979.23	6.97	0.6414	0.0058	1.08	1.6179
^{27}Al^{32}S	$^2\Sigma^+$	617.1	3.33	0.2799	0.0018	0.22	2.029
^{75}As$_2$	$^1\Sigma_g^+$	429.55	1.12	0.10179	0.000333		2.1026
^{75}As^{1}H	$^3\Sigma^-$	2130*		7.3067	0.2117	327	1.52315
^{75}As^{2}H	$^3\Sigma^-$	1484*		3.6688		90	1.5306
^{75}As^{14}N	$^1\Sigma^+$	1068.54	5.41	0.54551	0.003366	0.53	1.6184
^{75}As^{16}O	$^2\Pi_{1/2}$	967.08	4.85	0.48482	0.003299	0.49	1.6236

Molecule	State	ω_e cm^{-1}	$\omega_e x_e$ cm^{-1}	B_e cm^{-1}	α_e cm^{-1}	D_e 10^{-6}cm^{-1}	r_e Å
$^{197}Au_2$	$^1\Sigma_g^+$	190.9	0.42	0.028013	0.0000723	0.00250	2.4719
$^{197}Au^1H$	$^1\Sigma^+$	2305.01	43.12	7.2401	0.2136	279	1.5239
$^{197}Au^2H$	$^1\Sigma^+$	1634.98	21.65	3.6415	0.07614	70.9	1.5238
$^{11}B_2$	$^3\Sigma_g^-$	1051.3	9.35	1.212	0.014		1.590
$^{11}B^{79}Br$	$^1\Sigma^+$	684.31	3.52	0.4894	0.0035	1.00	1.888
$^{11}B^{35}Cl$	$^1\Sigma^+$	840.29	5.49	0.684282	0.006812	1.84	1.71528
$^{11}B^{19}F$	$^1\Sigma^+$	1402.1	11.8	1.516950	0.019056	7.105	1.26267
$^{11}B^1H$	$^1\Sigma^+$	2366.9	49.40	12.021	0.412	1242	1.2324
$^{11}B^2H$	$^1\Sigma^+$	1703.3	28	6.54	0.17	400	1.2324
$^{11}B^{14}N$	$^3\Pi$	1514.6	12.3	1.666	0.025	8.1	1.281
$^{11}B^{16}O$	$^2\Sigma^+$	1885.69	11.81	1.7820	0.0166	6.32	1.2045
$^{11}B^{32}S$	$^2\Sigma^+$	1180.17	6.31	0.7949	0.0061	1.40	1.6092
$^{138}Ba^{79}Br$	$^2\Sigma^+$	193.77	0.41	0.0415082	0.0001219	0.00762	2.84449
$^{138}Ba^{35}Cl$	$^2\Sigma^+$	279.92	0.82	0.08396717	0.00033429	0.03022	2.68276
$^{138}Ba^{19}F$	$^2\Sigma^+$	468.9	1.79	0.2159	0.0012	0.175	2.163
$^{138}Ba^1H$	$^2\Sigma^+$	1168.31	14.50	3.38285	0.06599	112.67	2.23175
$^{138}Ba^2H$	$^2\Sigma^+$	829.77	7.32	1.7071	0.02363	28.77	2.2304
$^{138}Ba^{127}I$	$^2\Sigma^+$	152.14	0.27	0.02680587	0.00006634	0.00333	3.08476
$^{138}Ba^{16}O$	$^1\Sigma^+$	669.76	2.03	0.3126140	0.0013921	0.2724	1.93969
$^{138}Ba^{32}S$	$^1\Sigma^+$	379.42	0.88	0.10331	0.0003188	0.0306	2.5074
$^9Be^{19}F$	$^2\Sigma^+$	1247.36	9.12	1.4889	0.0176	8.28	1.3610
$^9Be^1H$	$^2\Sigma^+$	2060.78	36.31	10.3164	0.3030	1022.1	1.3426
$^9Be^2H$	$^2\Sigma^+$	1530.32	20.71	5.6872	0.1225	313.8	1.3419
$^9Be^{16}O$	$^1\Sigma^+$	1487.32	11.83	1.6510	0.0190	8.20	1.3309
$^9Be^{32}S$	$^1\Sigma^+$	997.94	6.14	0.79059	0.00664	2.00	1.7415
$^{209}Bi_2$	$^1\Sigma_g^+$	172.71	0.34	0.022781	0.000055	0.00150	2.6596
$^{209}Bi^1H$	$^3\Sigma^-$	1635.73	31.6	5.137	0.148	183	1.805
$^{209}Bi^2H$	$^3\Sigma^-$	1173.32	16.1	2.592	0.054	50.6	1.804
$^{79}Br_2$	$^1\Sigma_g^+$	325.32	1.08	0.082107	0.0003187	0.02092	2.2811
$^{79}Br^{35}Cl$	$^1\Sigma^+$	444.28	1.84	0.152470	0.000770	0.07183	2.13607
$^{79}Br^{19}F$	$^1\Sigma^+$	670.75	4.05	0.35584	0.00261	0.401	1.75894
$^{79}Br^{16}O$	$^2\Pi_{3/2}$	779	6.8	0.429598	0.003639	0.523	1.717
$^{12}C_2$	$^1\Sigma_g^+$	1854.71	13.34	1.8198	0.0177	6.92	1.2425
$^{12}C^{35}Cl$	$^2\Pi_{1/2}$	866.72*	6.2	0.6936	0.00672	1.9	1.6450
$^{12}C^{19}F$	$^2\Pi_{1/2}$	1308.1	11.10	1.4172	0.0184	6.5	1.2718
$^{12}C^1H$	$^2\Pi_{1/2}$	2858.5	63.0	14.457	0.534	1450	1.1199
$^{12}C^2H$	$^2\Pi_{1/2}$	2099.8	34.02	7.806	0.208	420	1.1190
$^{12}C^{14}N$	$^2\Sigma^+$	2068.59	13.09	1.8997830	0.0173717	6.4034	1.17181
$^{12}C^{16}O$	$^1\Sigma^+$	2169.81	13.29	1.93128075	0.01750390	6.1216	1.12823
$^{12}C^{31}P$	$^2\Sigma^+$	1239.67	6.86	0.7986	0.00597	1.33	1.562
$^{12}C^{32}S$	$^1\Sigma^+$	1285.15	6.50	0.8200434	0.0059182	1.336	1.53482
$^{12}C^{80}Se$	$^1\Sigma^+$	1035.36	4.86	0.5750	0.00379	0.71	1.67609
$^{40}Ca^{35}Cl$	$^2\Sigma^+$	367.53	1.31	0.1522302	0.0007990	0.1029	2.43676
$^{40}Ca^{19}F$	$^2\Sigma^+$	581.1	2.74	0.339	0.0026	0.45	1.967
$^{40}Ca^1H$	$^2\Sigma^+$	1298.34	19.10	4.2766	0.0970	183.7	2.0025
$^{40}Ca^2H$	$^2\Sigma^+$	910*		2.1769	0.035	47.9	2.002
$^{40}Ca^{127}I$	$^2\Sigma^+$	238.70	0.63	0.0693263	0.0002634	0.0234	2.82859
$^{40}Ca^{16}O$	$^1\Sigma^+$	732.03	4.83	0.444441	0.003282	0.6541	1.8221
$^{40}Ca^{32}S$	$^1\Sigma^+$	462.23	1.78	0.1766757	0.0008270	0.1032	2.31775
$^{114}Cd^1H$	$^2\Sigma^+$	1337.1*		5.323		314	1.781
$^{114}Cd^2H$	$^2\Sigma^+$			2.704		76	1.775
$^{35}Cl_2$	$^1\Sigma_g^+$	559.7	2.68	0.2440	0.0015	0.186	1.988
$^{35}Cl^{19}F$	$^1\Sigma^+$	786.15	6.16	0.516479	0.004358	0.88	1.62831
$^{35}Cl^{16}O$	$^2\Pi_{3/2}$	853.8	5.5	0.62345	0.0058	1.33	1.56963
$^{52}Cr^1H$	$^6\Sigma^+$	1581*	32	6.220	0.179	347	1.656
$^{52}Cr^2H$	$^6\Sigma^+$	1182*		3.14		88.8	1.664
$^{52}Cr^{16}O$	$^5\Pi$	898.4	6.8	0.5231	0.0070		1.615
$^{133}Cs_2$	$^1\Sigma_g^+$	42.02	0.08	0.0127	0.0000264	0.00464	4.47

SPECTROSCOPIC CONSTANTS OF DIATOMIC MOLECULES (continued)

Molecule	State	ω_e cm^{-1}	$\omega_e x_e$ cm^{-1}	B_e cm^{-1}	α_e cm^{-1}	D_e 10^{-6}cm^{-1}	r_e Å
^{133}Cs^{79}Br	$^1\Sigma+$	149.66	0.37	0.03606925	0.00012401	0.00838	3.07225
^{133}Cs^{35}Cl	$^1\Sigma+$	214.17	0.73	0.07209149	0.00033756	0.03268	2.90627
^{133}Cs^{19}F	$^1\Sigma+$	352.56	1.62	0.18436969	0.0011756	0.20168	1.34535
^{133}Cs^{1}H	$^1\Sigma+$	891.0	12.9	2.7099	0.0579	113	2.4938
^{133}Cs^{2}H	$^1\Sigma+$	619.1*		1.354		20	2.505
^{133}Cs^{127}I	$^1\Sigma+$	119.18	0.25	0.02362736	0.00006826	0.00371	3.31519
^{133}Cs^{16}O	$^2\Sigma+$	357.5*		0.223073	0.001303	0.348	2.3007
^{63}Cu$_2$	$^1\Sigma_g+$	264.55	1.02	0.10874	0.000614	0.0716	2.2197
^{63}Cu^{79}Br	$^1\Sigma+$	314.8	0.96	0.10192625	0.00045214	0.04274	2.17344
^{65}Cu^{35}Cl	$^1\Sigma+$	415.29	1.58	0.17628802	0.00099647	0.12706	2.05118
^{63}Cu^{19}F	$^1\Sigma+$	622.7	3.95	0.3794029	0.0032298	0.563	1.74493
^{63}Cu^{1}H	$^1\Sigma+$	1941.26	37.51	7.9441	0.2563	520	1.46263
^{63}Cu^{2}H	$^1\Sigma+$	1384.14	18.97	4.0381	0.0917	136.2	1.4626
^{63}Cu^{127}I	$^1\Sigma+$	264.5	0.60	0.07328742	0.00028390	0.02244	2.33832
^{63}Cu^{16}O	$^2\Pi_{3/2}$	640.17	4.43	0.44454	0.00456	0.85	1.7244
^{63}Cu^{32}S	$^2\Pi_{3/2}$	415.0	1.75	0.1891		0.18	2.051
^{19}F$_2$	$^1\Sigma_g+$	916.64	11.24	0.89019	0.013847	3.3	1.41193
^{56}Fe^{16}O	$^5\Delta$	965*		0.650		0.72	1.444
^{69}Ga^{81}Br	$^1\Sigma+$	263.0	0.81	0.081839	0.0003207	0.032	2.35248
^{69}Ga^{35}Cl	$^1\Sigma+$	365.67	1.25	0.1499046	0.0007936	0.1008	2.20169
^{69}Ga^{19}F	$^1\Sigma+$	622.2	3.2	0.3595161	0.0028642	0.50	1.77437
^{69}Ga^{1}H	$^1\Sigma+$	1604.52	28.77	6.137	0.181	342	1.663
^{69}Ga^{2}H	$^1\Sigma+$			3.083	0.06	84	1.663
^{69}Ga^{127}I	$^1\Sigma+$	216.38	0.47	0.0569359	0.0001897	0.015770	2.57464
^{69}Ga^{16}O	$^2\Sigma$	767.5	6.24	0.4271		0.37	1.744
^{74}Ge^{79}Br	$^2\Pi_{1/2}$	295	0.7				
^{74}Ge^{35}Cl	$^2\Pi_{1/2}$	407.6	1.36				
^{72}Ge^{1}H	$^2\Pi_{1/2}$	1833.77	37	6.726	0.192	326	1.5880
^{72}Ge^{2}H	$^2\Pi_{1/2}$	1320.09	19	3.415	0.070	83.2	1.5874
^{74}Ge^{16}O	$^1\Sigma+$	986.49	4.47	0.4856981	0.0030787	0.4709	1.62464
^{74}Ge^{32}S	$^1\Sigma+$	575.8	1.80	0.18656576	0.00074910	0.07883	2.01209
^{74}Ge^{80}Se	$^1\Sigma+$	408.7	1.36	0.09634051	0.00028904	0.02207	2.13463
^{74}Ge^{130}Te	$^1\Sigma+$	323.9	0.75	0.06533821	0.00017246	0.012	2.34017
^{1}H$_2$	$^1\Sigma_g+$	4401.21	121.34	60.853	3.062	47100	0.74144
^{2}H$_2$	$^1\Sigma_g+$	3115.50	61.82	30.444	1.0786	11410	0.74152
^{3}H$_2$	$^1\Sigma_g+$	2546.5	41.23	20.335	0.5887		0.74142
^{1}H^{81}Br	$^1\Sigma+$	2648.97	45.22	8.46488	0.23328	345.8	1.41444
^{2}H^{81}Br	$^1\Sigma+$	1884.75	22.72	4.245596	0.084	88.32	1.4145
^{1}H^{35}Cl	$^1\Sigma+$	2990.95	52.82	10.59342	0.30718	531.94	1.27455
^{2}H^{35}Cl	$^1\Sigma+$	2145.16	27.18	5.448796	0.113292	140	1.27458
^{1}H^{19}F	$^1\Sigma+$	4138.32	89.88	20.9557	0.798	2151	0.91681
^{2}H^{19}F	$^1\Sigma+$	2998.19	45.76	11.0102	0.3017	594	0.91694
^{1}H^{127}I	$^1\Sigma+$	2309.01	39.64	6.4263650	0.1689	206.9	1.60916
^{202}Hg^{1}H	$^2\Sigma+$	1203.24*		5.3888		395.3	1.7662
^{202}Hg^{2}H	$^2\Sigma+$	896.12*		2.739		91	1.757
^{127}I$_2$	$^1\Sigma_g+$	214.50	0.61	0.03737	0.000114	0.0043	2.666
^{127}I^{79}Br	$^1\Sigma+$	268.64	0.81	0.0568325	0.0001969	0.0102	2.46899
^{127}I^{35}Cl	$^1\Sigma+$	384.29	1.50	0.1141587	0.0005354	0.0403	2.32088
^{127}I^{19}F	$^1\Sigma+$	610.24	3.12	0.2797111	0.0018738	0.2356	1.90976
^{127}I^{16}O	$^2\Pi_{3/2}$	681.5	4.3	0.34026	0.00270	0.36	1.8676
^{115}In^{81}Br	$^1\Sigma+$	221.0	0.65	0.05489468	0.00018672	0.01350	2.54315
^{115}In^{35}Cl	$^1\Sigma+$	317.39	1.03	0.1090583	0.0005177	0.0515	2.40117
^{115}In^{19}F	$^1\Sigma+$	535.4	2.6	0.2623241	0.0018798	0.252	1.98540
^{115}In^{1}H	$^1\Sigma+$	1476.0	25.61	4.995	0.143	223	1.8380
^{115}In^{2}H	$^1\Sigma+$	1048.2	12.4	2.523	0.051	58	1.837
^{115}In^{127}I	$^1\Sigma+$	177.08	0.34	0.03686702	0.00010411	0.00639	2.75364
^{39}K$_2$	$^1\Sigma_g+$	92.02	0.28	0.056743	0.000165	0.0863	3.9051
^{39}K^{79}Br	$^1\Sigma+$	213	0.80	0.08122109	0.00040481	0.04462	2.82078

Molecule	State	ω_e cm^{-1}	$\omega_e x_e$ cm^{-1}	B_e cm^{-1}	α_e cm^{-1}	D_e 10^{-6}cm^{-1}	r_e Å
$^{39}K^{35}Cl$	$^1\Sigma^+$	281	1.30	0.1286348	0.0007899	0.1087	2.66665
$^{39}K^{19}F$	$^1\Sigma^+$	426.26	2.45	0.27993741	0.00233492	0.4829	2.17146
$^{39}K^1H$	$^1\Sigma^+$	983.6	14.3	3.416400	0.085313	163.55	2.243
$^{39}K^2H$	$^1\Sigma^+$	707	7.7	1.754	0.0318	50	2.240
$^{39}K^{127}I$	$^1\Sigma^+$	186.53	0.57	0.06087473	0.00026776	0.02593	3.04784
$^{139}La^{16}O$	$^2\Sigma^+$	812.8	2.22	0.35252001	0.00142365	0.2626	1.82591
$^7Li_2$	$^1\Sigma_g^+$	351.43	2.61	0.67264	0.00704	9.87	2.6729
$^7Li^{79}Br$	$^1\Sigma^+$	563.2	3.5	0.555399	0.005644	2.159	2.17043
$^7Li^{35}Cl$	$^1\Sigma^+$	642.95	4.47	0.7065225	0.0080102	3.409	2.02067
$^7Li^{19}F$	$^1\Sigma^+$	910.57	8.21	1.3452583	0.0202887	11.745	1.56386
$^7Li^1H$	$^1\Sigma^+$	1405.65	23.20	7.51373	0.21665	862	1.59490
$^7Li^2H$	$^1\Sigma^+$	1054.80	12.94	4.23310	0.09155	276	1.5941
$^7Li^{127}I$	$^1\Sigma^+$	496.85	2.85	0.4431766	0.0040862	1.4104	2.39192
$^7Li^{16}O$	$^2\Pi$	814.62	7.78	1.212830	0.017899	0.1079	1.68822
$^{24}Mg_2$	$^1\Sigma_g^+$	51.12	1.64	0.09287	0.00378	1.22	3.891
$^{24}Mg^{35}Cl$	$^2\Sigma^+$	462.12*	2.1	0.2456154	0.0016204	0.2723	2.19639
$^{24}Mg^{19}F$	$^2\Sigma^+$	711.69*	4.9	0.51922	0.00470	1.080	1.7500
$^{24}Mg^1H$	$^2\Sigma^+$	1495.20	31.89	5.8257	0.1859	344	1.7297
$^{24}Mg^2H$	$^2\Sigma^+$	1077.9	16.1	3.0306	0.06289	92	1.7302
$^{24}Mg^{16}O$	$^1\Sigma^+$	784.78	5.26	0.57470436	0.00532377	1.2328	1.74838
$^{55}Mn^1H$	$^7\Sigma$	1548.0	28.8	5.6841	0.1570	303.9	1.7311
$^{55}Mn^2H$	$^7\Sigma$	1103	13.9	2.8957	0.051	79.5	1.7310
$^{14}N_2$	$^1\Sigma_g^+$	2358.57	14.32	1.99824	0.017318	5.76	1.09769
$^{14}N^{79}Br$	$^3\Sigma^-$	691.75	4.72	0.444	0.0040		1.79
$^{14}N^{35}Cl$	$^3\Sigma^-$	827.96	5.30	0.649770	0.006414	1.598	1.61071
$^{14}N^{19}F$	$^3\Sigma^-$	1141.37	8.99	1.2057	0.01492	5.39	1.3170
$^{14}N^1H$	$^3\Sigma^-$	3282.3	78.4	16.6993	0.6490	1709.7	1.0362
$^{14}N^2H$	$^3\Sigma^-$	2398	42	8.7913	0.2531	490.4	1.0361
$^{14}N^{16}O$	$^2\Pi_{1/2}$	1904.20	14.07	1.67195	0.0171	0.5	1.15077
$^{14}N^{32}S$	$^2\Pi_{1/2}$	1218.7	7.28	0.769602	0.0064	1.2	1.4940
$^{23}Na_2$	$^1\Sigma_g^+$	159.13	0.72	0.154707	0.008736	0.581	3.0789
$^{23}Na^{79}Br$	$^1\Sigma^+$	302	1.5	0.1512533	0.0009410	0.1554	2.50204
$^{23}Na^{35}Cl$	$^1\Sigma^+$	366	2.05	0.2180631	0.0016248	0.3120	2.36080
$^{23}Na^{19}F$	$^1\Sigma^+$	535.66	3.57	0.4369011	0.0045580	1.163	1.92595
$^{23}Na^1H$	$^1\Sigma^+$	1172.2	19.72	4.9033634	0.1370919	343.40	1.88654
$^{23}Na^2H$	$^1\Sigma^+$	826.1*		2.557089	0.051600	93.46	1.88654
$^{23}Na^{127}I$	$^1\Sigma^+$	258	1.1	0.1178056	0.0006478	0.0973	2.71145
$^{23}Na^{16}O$	$^2\Pi$	492.3		0.424630	0.004506	1.2638	2.05155
$^{93}Nb^{16}O$	$^4\Sigma^-$	989.0	3.8	0.4321	0.0021	0.22	1.691
$^{58}Ni^1H$	$^2\Delta_{5/2}$	1926.6	38	7.700	0.23	481	1.476
$^{58}Ni^2H$	$^2\Delta_{5/2}$	1390.1	19	3.992	0.092	130	1.465
$^{16}O_2$	$^3\Sigma_g^-$	1580.19	11.98	1.44563	0.0159	4.839	1.20752
$^{16}O^1H$	$^2\Pi_{3/2}$	3737.76	84.88	18.911	0.7242	1938	0.96966
$^{16}O^2H$	$^2\Pi_{3/2}$	2720.24	44.05	10.021	0.276	537.4	0.9698
$^{31}P_2$	$^1\Sigma_g^+$	780.77	2.84	0.30362	0.00149	0.188	1.8934
$^{31}P^{35}Cl$	$^3\Sigma^-$	551.38	2.23	0.2528748	0.0015119	0.2124	2.01461
$^{31}P^{19}F$	$^3\Sigma^-$	846.75	4.49	0.5665	0.00456		1.58938
$^{31}P^1H$	$^3\Sigma^-$	2365.2	44.5	8.5371	0.2514	436	1.42140
$^{31}P^2H$	$^3\Sigma^-$	1699.2	23.0	4.4081	0.0928	116	1.4220
$^{31}P^{14}N$	$^1\Sigma^+$	1337.24	6.98	0.7864854	0.0055364	1.091	1.49087
$^{31}P^{16}O$	$^2\Pi_{1/2}$	1233.34	6.56	0.7337	0.0055	1.3	1.4759
$^{208}Pb_2$		110.5	0.35				
$^{208}Pb^{79}Br$	$^2\Pi_{1/2}$	207.5	0.50				
$^{208}Pb^{35}Cl$	$^2\Pi_{1/2}$	303.9	0.88				
$^{208}Pb^{19}F$	$^2\Pi_{1/2}$	502.73	2.28	0.22875	0.001473	0.183	2.0575
$^{208}Pb^1H$	$^2\Pi_{1/2}$	1564.1	29.75	4.971	0.144	201	1.839
$^{208}Pb^{16}O$	$^1\Sigma^+$	720.96	3.52	0.30730373	0.00190977	0.2138	1.92181
$^{208}Pb^{32}S$	$^1\Sigma^+$	429.17	1.26	0.11632307	0.00043510	0.03418	2.28678

Molecule	State	ω_e cm⁻¹	$\omega_e x_e$ cm⁻¹	B_e cm⁻¹	α_e cm⁻¹	D_e 10⁻⁶cm⁻¹	r_e Å
^{208}Pb^{80}Se	$^1\Sigma^+$	277.6	0.51	0.05059953	0.00012993	0.0070	2.40218
^{208}Pb^{130}Te	$^1\Sigma^+$	212.0	0.43	0.03130774	0.00006743	0.0027	2.59492
^{195}Pt^{12}C	$^1\Sigma^+$	1051.13	4.86	0.53044	0.003273	0.546	1.6767
^{195}Pt^{1}H	$^2\Delta_{5/2}$	2294.68*	46	7.1963	0.1996	261	1.52852
^{195}Pt^{2}H	$^2\Delta_{5/2}$	1644.3*	23	3.640	0.071	66	1.524
^{85}Rb^{79}Br	$^1\Sigma^+$	169.46	0.46	0.04752798	0.00018596	0.01496	2.94474
^{85}Rb^{35}Cl	$^1\Sigma^+$	228	0.92	0.0876404	0.0004537	0.04947	2.78673
^{85}Rb^{19}F	$^1\Sigma^+$	376	1.9	0.2106640	0.0015228	0.2684	2.27033
^{85}Rb^{1}H	$^1\Sigma^+$	936.9	14.21	3.020	0.072	123	2.367
^{85}Rb^{127}I	$^1\Sigma^+$	138.51	0.33	0.03283293	0.00010946	0.00738	3.17688
^{85}Rb^{16}O	$^2\Sigma^+$	388.4*		0.246481	0.002174	0.397	2.25420
^{32}S$_2$	$^3\Sigma_g^-$	725.65	2.84	0.2955	0.001570	0.19	1.8892
^{32}S^{19}F	$^2\Pi_{3/2}$			0.552174			1.60058
^{32}S^{1}H	$^2\Pi_{3/2}$	2711.6	59.9	9.5995	0.2785	480.6	1.34066
^{32}S^{2}H	$^2\Pi_{3/2}$	1885	31	4.95130	0.10308	130	1.34049
^{32}S^{16}O	$^3\Sigma^-$	1149.2	5.6	0.7208171	0.005737	1.134	1.48109
^{121}Sb^{35}Cl	$^3\Sigma^-$	374.7	0.6				
^{121}Sb^{19}F	$^3\Sigma^-$	605.0	2.6	0.2792	0.0020	0.23	1.918
^{121}Sb^{1}H	$^3\Sigma^-$			5.684		240	1.723
^{121}Sb^{2}H	$^3\Sigma^-$			2.8782		45	1.7194
^{121}Sb^{14}N	$^1\Sigma^+$	942.0	5.6				
^{121}Sb^{16}O	$^2\Pi_{1/2}$	816	4.2	0.3580	0.0022	0.270	1.826
^{45}Sc^{19}F	$^1\Sigma^+$	735.6	3.8	0.3950	0.00266		1.788
^{80}Se$_2$	$^3\Sigma_g^-$	385.30	0.96	0.08992	0.000288	0.024	2.166
^{80}Se^{1}H	$^2\Pi_{3/2}$	2400*		8.02	0.23	330	1.48
^{80}Se^{2}H	$^2\Pi_{3/2}$	1708*		3.94			1.48
^{80}Se^{16}O	$^3\Sigma^-$	914.69	4.52	0.4655	0.00323	0.5	1.648
^{28}Si$_2$	$^3\Sigma_g^-$	510.98	2.02	0.2390	0.0014	0.21	2.246
^{28}Si^{35}Cl	$^2\Pi_{1/2}$	535.60	2.17	0.2561	0.0016	0.25	2.058
^{28}Si^{19}F	$^2\Pi_{1/2}$	857.19	4.73	0.5812	0.00494	1.07	1.6011
^{28}Si^{1}H	$^2\Pi_{1/2}$	2041.80	35.51	7.4996	0.2190	397	1.5201
^{28}Si^{2}H	$^2\Pi_{1/2}$	1469.32	18.23	3.8840	0.0781	105.4	1.5199
^{28}Si^{14}N	$^2\Sigma^+$	1151.4	6.47	0.7311	0.00565	1.2	1.572
^{28}Si^{16}O	$^1\Sigma^+$	1241.54	5.97	0.7267521	0.0050379	0.9923	1.50975
^{28}Si^{32}S	$^1\Sigma^+$	749.64	2.58	0.30352788	0.00147308	0.201	1.92926
^{28}Si^{80}Se	$^1\Sigma^+$	580.0	1.78	0.1920117	0.0007767	0.0842	2.05832
^{120}Sn^{79}Br	$^2\Pi_{1/2}$	247.2	0.6				
^{120}Sn^{35}Cl	$^2\Pi_{1/2}$	351.1	1.06	0.1117	0.0004		2.361
^{118}Sn^{19}F	$^2\Pi_{1/2}$	577.6	2.69	0.2727	0.0014	0.26	1.944
^{120}Sn^{1}H	$^2\Pi_{1/2}$			5.31488		207.5	1.78146
^{120}Sn^{2}H	$^2\Pi_{1/2}$	1188.0*		2.6950	0.049	53.4	1.7770
^{120}Sn^{127}I	$^2\Pi_{1/2}$	199.0	0.6				
^{120}Sn^{16}O	$^1\Sigma^+$	822.13	3.72	0.35571998	0.00214432	0.26638	1.83251
^{120}Sn^{32}S	$^1\Sigma^+$	487.26	1.36	0.13686139	0.00050563	0.0424	2.20898
^{120}Sn^{80}Se	$^1\Sigma^+$	331.2	0.74	0.0649978	0.0001705	0.011	2.32557
^{120}Sn^{130}Te	$^1\Sigma^+$	259.5	0.50	0.04247917	0.00009543	0.0055	2.52280
^{88}Sr^{79}Br	$^2\Sigma^+$	216.60	0.52	0.0541847	0.0001827	0.01356	2.73522
^{88}Sr^{35}Cl	$^2\Sigma^+$	302.3	0.95				
^{88}Sr^{19}F	$^2\Sigma^+$	502.4	2.3	0.2505346	0.0015513	0.2498	2.07537
^{88}Sr^{1}H	$^2\Sigma^+$	1206.2	17.0	3.6751	0.0814	135	2.1456
^{88}Sr^{2}H	$^2\Sigma^+$	841	8.6	1.8609	0.0292	34.7	2.1449
^{88}Sr^{127}I	$^2\Sigma^+$	173.77	0.35	0.0367097	0.0001060	0.00655	2.94364
^{88}Sr^{16}O	$^1\Sigma^+$	653.5	3.96	0.33798	0.00219	0.36	1.91983
^{181}Ta^{16}O	$^2\Delta_{3/2}$	1028.69	3.51	0.40284	0.00182	0.2450	1.68746
^{130}Te$_2$	$^3\Sigma_g^-$	247.07	0.51	0.039681	0.000106	0.0044	2.5574
^{130}Te^{1}H	$^2\Pi_{3/2}$			5.56			1.74
^{130}Te^{16}O	0^+	797.11	4.00	0.3554	0.00237	0.27	1.825
^{232}Th^{16}O	$^1\Sigma^+$	895.77	2.39	0.332644	0.001302	0.1833	1.84032

Molecule	State	ω_e cm^{-1}	$\omega_e x_e$ cm^{-1}	B_e cm^{-1}	α_e cm^{-1}	D_e 10^{-6}cm^{-1}	r_e Å
^{48}Ti^{16}O	$^3\Delta_1$	1009.02	4.50	0.53541	0.00301	0.603	1.6202
^{205}Tl^{81}Br	$^1\Sigma^+$	192.10	0.39	0.0423899	0.0001276	0.0083	1.61817
^{205}Tl^{35}Cl	$^1\Sigma^+$	284.71	0.86	0.09139702	0.00039784	0.0377	2.48483
^{205}Tl^{19}F	$^1\Sigma^+$	476.86	2.24	0.22315014	0.00150380	0.1955	2.08439
^{205}Tl^{1}H	$^1\Sigma^+$	1390.7	22.7	4.806	0.154	254	1.870
^{205}Tl^{2}H	$^1\Sigma^+$	987.7	12.04	2.419	0.057	60	1.869
^{205}Tl^{127}I	$^1\Sigma^+$	150*		0.0271676	0.0000664	0.0036	2.81361
^{51}V^{16}O	$^4\Sigma^-$	1011.3	4.86	0.54825	0.00352	0.6	1.5893
^{89}Y^{35}Cl	$^1\Sigma$	380.7	1.3	0.1160	0.0003	0.09	2.41
^{89}Y^{19}F	$^1\Sigma^+$	631.29	2.50	0.29042	0.00163	0.237	1.9257
^{89}Y^{16}O	$^2\Sigma^+$	861.0	2.9	0.3881	0.0018	0.32	1.790
^{174}Yb^{1}H	$^2\Sigma^+$	1249.54	21.06	3.9931	0.0957	161.8	2.0526
^{174}Yb^{2}H	$^2\Sigma^+$	886.6	10.57	2.01162	0.03425	41.60	2.0516
^{64}Zn^{35}Cl	$^2\Sigma$	390.5	1.6				
^{64}Zn^{19}F	$^2\Sigma$	628	3.5				
^{64}Zn^{1}H	$^2\Sigma^+$	1607.6	55.14	6.6794	0.2500	466	1.5949
^{64}Zn^{2}H	$^2\Sigma^+$	1072	28	3.350		124	1.6054
^{64}Zn^{127}I	$^2\Sigma$	223.4	0.6				
^{90}Zr^{16}O	$^1\Sigma^+$	969.8	4.9	0.42263	0.0023	0.319	1.7116

INFRARED CORRELATION CHARTS

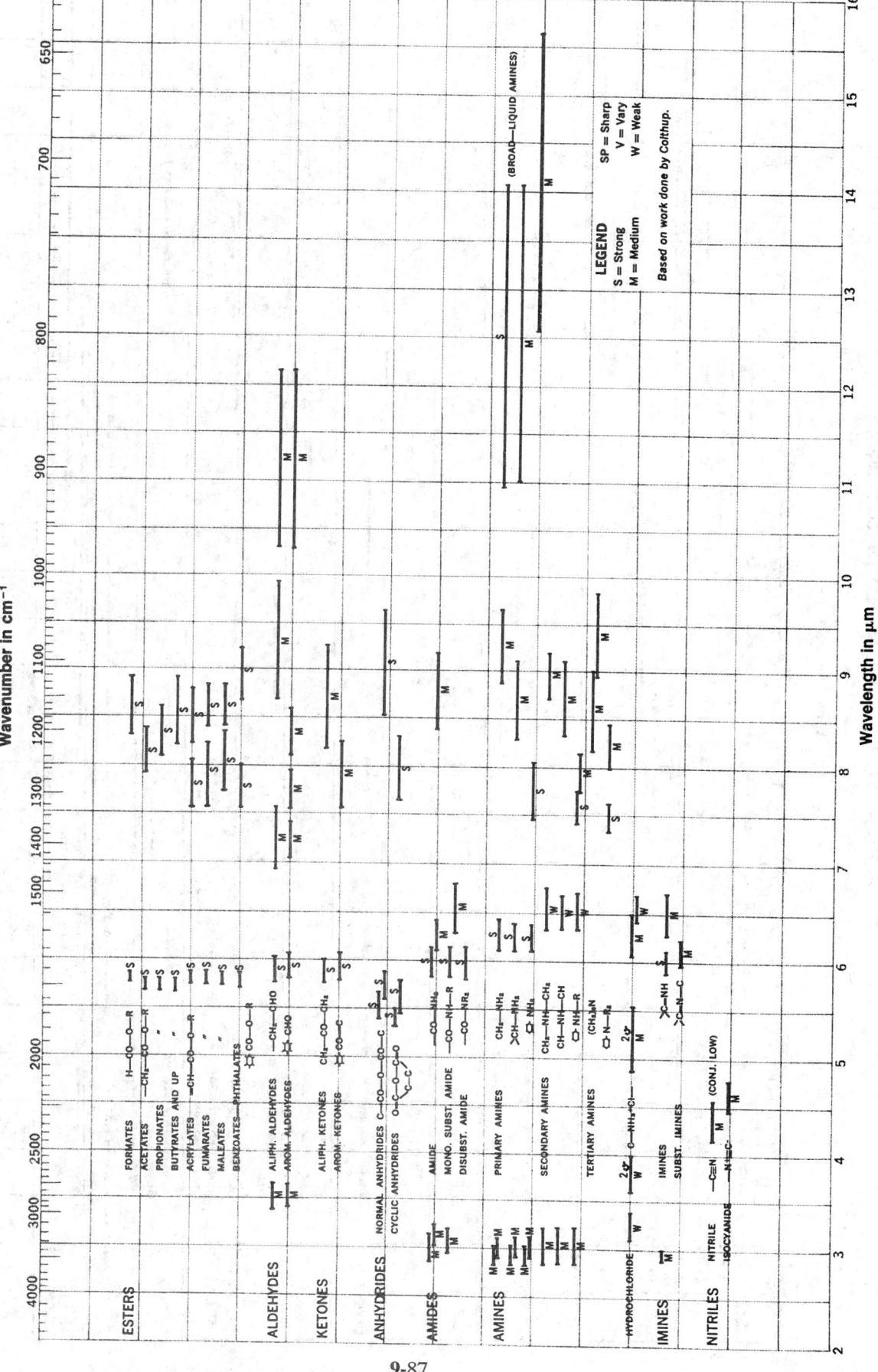

INFRARED CORRELATION CHARTS (continued)

9-87

INFRARED CORRELATION CHARTS (continued)

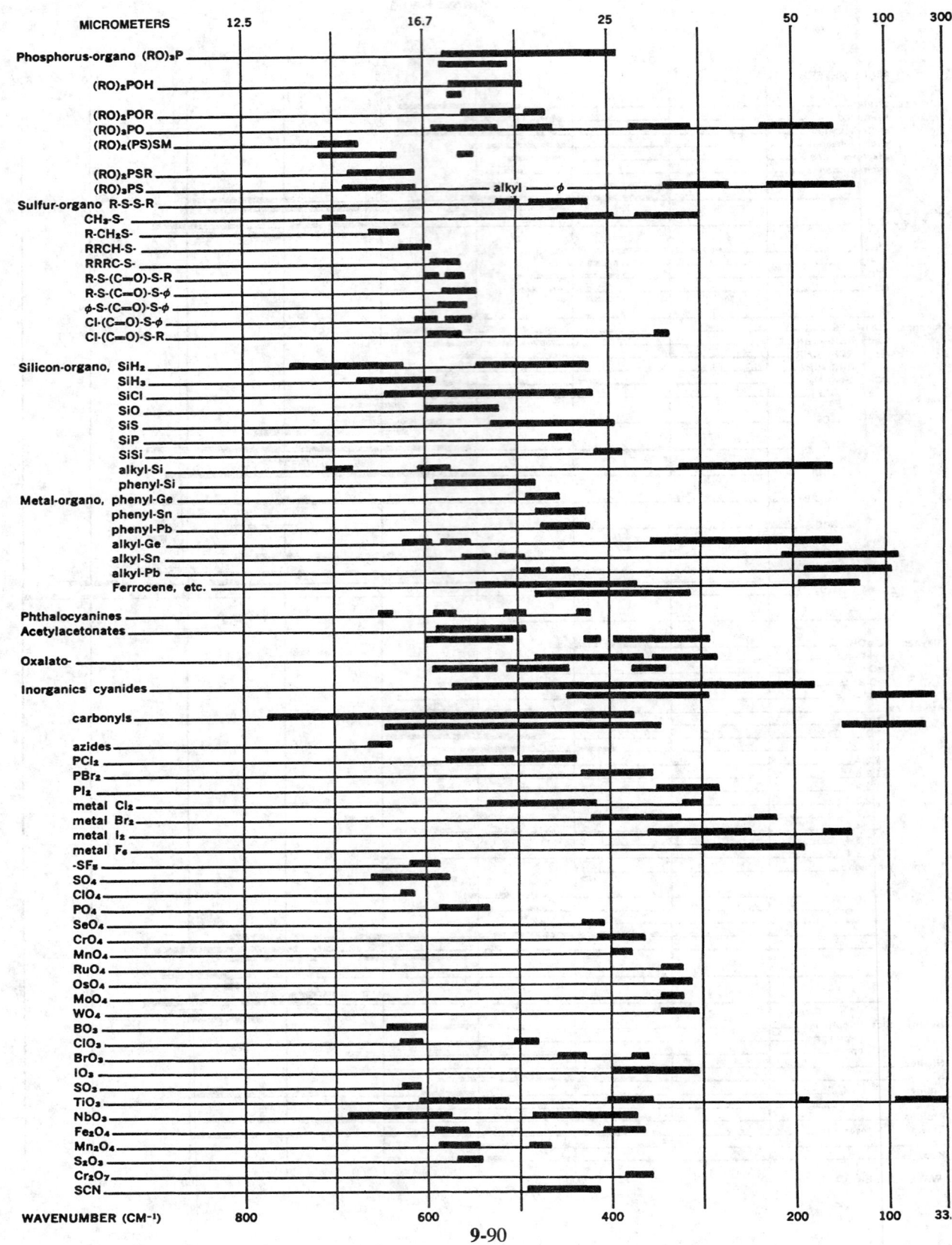

NUCLEAR SPINS, MOMENTS, AND OTHER DATA RELATED TO NMR SPECTROSCOPY

This table presents the following data relevant to nuclear magnetic resonance spectroscopy:

Z: Atomic number
Isotope: Element symbol and mass number
Abundance: Natural abundance of the isotope in percent. An * indicates a radioactive nuclide; if no value is given, the nuclide is not present in nature or its abundance is highly variable.
I: Nuclear spin
ν: Resonant frequency in megahertz for an applied field H_o of 1 tesla (in cgs units, 10 kilogauss)
Relative sensitivity: Sensitivity relative to ^{1}H (=1) assuming an equal number of nuclei and constant temperature. Values were calculated from the expressions:

   For constant H_o: $0.0076508(\mu/\mu_N)^3(I+1)/I^2$
   For constant ν: $0.23871(\mu/\mu_N)(I+1)$

μ: Nuclear magnetic moment in units of the nuclear magneton μ_N
Q: Nuclear quadrupole moment in units of femtometers squared (1 fm^2 = 10^{-2} barn)

The table includes all stable nuclides of non-zero spin for which spin and magnetic moment values have been measured, as well as selected radioactive nuclides of current or potential interest. At least one isotope is included for each element through $Z = 95$ for which data are available. See Reference 2 for a complete listing of spins and moments.
   The assistance of P. Pyykkö in providing recent data on nuclear quadrupole moments is gratefully acknowledged.

REFERENCES

1. Raghavan, P., "Table of Nuclear Moments", *At. Data Nuc. Data Tables*, 42, 189, 1989.
2. Holden, N. E., "Table of the Isotopes", in Lide, D. R., Ed., *CRC Handbook of Chemistry and Physics*, 74th Ed., CRC Press, Boca Raton, FL, 1993.
3. Pyykkö, P., *Z. Naturforsch.*, 47a, 189, 1992.

Z	Isotope	Abundance %	I	ν/MHz for H_o = 1 T	Relative sensitivity Const. H_o	Const. ν	μ/μ$_N$	Q/fm^2
1	^{1}n	*	1/2	29.1639	0.32139	0.6850	-1.91304275	
1	^{1}H	99.985	1/2	42.5764	1.00000	1.0000	+2.7928474	
1	^{2}H	0.015	1	6.53573	0.00965	0.4093	+0.8574382	+0.2860
1	^{3}H	*	1/2	45.4137	1.21354	1.0666	+2.9789625	
2	^{3}He	0.0001	1/2	32.4352	0.44212	0.7618	-2.1276248	
3	^{6}Li	7.5	1	6.2660	0.00850	0.3925	+0.8220467	-0.082
3	^{7}Li	92.5	3/2	16.5478	0.29355	1.9433	+3.256427	-4.01
4	^{9}Be	100	3/2	5.986	0.01389	0.7029	-1.1779	+5.288
5	^{10}B	19.9	3	4.5751	0.01985	1.7193	+1.800645	+8.459
5	^{11}B	80.1	3/2	13.6626	0.16522	1.6045	+2.688649	+4.059
6	^{13}C	1.10	1/2	10.7081	0.01591	0.2515	+0.7024118	
7	^{14}N	99.634	1	3.0776	0.00101	0.1928	+0.4037610	+2.02
7	^{15}N	0.366	1/2	4.3172	0.00104	0.1014	-0.2831888	
8	^{17}O	0.038	5/2	5.7741	0.02910	1.5822	-1.89379	-2.558
9	^{19}F	100	1/2	40.0765	0.83400	0.9413	+2.628868	
10	^{21}Ne	0.27	3/2	3.3630	0.00246	0.3949	-0.661797	+10.155
11	^{23}Na	100	3/2	11.2686	0.09270	1.3233	+2.217522	+10.89
12	^{25}Mg	10.00	5/2	2.6082	0.00268	0.7147	-0.85545	+19.94
13	^{27}Al	100	5/2	11.1028	0.20689	3.0424	+3.641507	+14.03
14	^{29}Si	4.67	1/2	8.4653	0.00786	0.1988	-0.55529	
15	^{31}P	100	1/2	17.2510	0.06652	0.4052	+1.13160	
16	^{33}S	0.75	3/2	3.2716	0.00227	0.3842	+0.6438212	-6.78
17	^{35}Cl	75.77	3/2	4.1764	0.00472	0.4905	+0.8218743	-8.165
17	^{37}Cl	24.23	3/2	3.4764	0.00272	0.4083	+0.6841236	-6.435
18	^{37}Ar	*	3/2	5.818	0.01276	0.6833	+1.145	
18	^{39}Ar	*	7/2	2.8	0.00617	1.3964	-1.3	
19	^{39}K	93.2581	3/2	1.9893	0.00051	0.2336	+0.3914662	+6.01
19	^{40}K	0.0117	4	2.4737	0.00523	1.5493	-1.298100	-7.49
19	^{41}K	6.7302	3/2	1.0919	0.00008	0.1282	+0.2148701	+7.33

Z	Isotope	Abundance %	I	ν/MHz for $H_0 = 1$ T	Relative sensitivity Const. H_0	Const. ν	μ/μ_N	Q/fm^2
20	^{43}Ca	0.135	7/2	2.8688	0.00642	1.4150	-1.31726	-4.08
21	^{45}Sc	100	7/2	10.3588	0.30244	5.1093	+4.756487	-22
22	^{47}Ti	7.3	5/2	2.4040	0.00210	0.6587	-0.78848	+29
22	^{49}Ti	5.5	7/2	2.4047	0.00378	1.1861	-1.10417	+24
23	^{50}V	0.250	6	4.2504	0.05571	5.5904	+3.345689	+21
23	^{51}V	99.750	7/2	11.2130	0.38360	5.5306	+5.1487057	-5.2
24	^{53}Cr	9.501	3/2	2.4114	0.00091	0.2832	-0.47454	-15
25	^{55}Mn	100	5/2	10.5760	0.17881	2.8980	+3.46872	+33
26	^{57}Fe	2.1	1/2	1.3815	0.00003	0.0324	+0.0906230	
27	^{59}Co	100	7/2	10.077	0.27841	4.9702	+4.627	+42
28	^{61}Ni	1.140	3/2	3.8113	0.00359	0.4476	-0.75002	+16.2
29	^{63}Cu	69.17	3/2	11.2979	0.09342	1.3268	+2.22329	-22.0
29	^{65}Cu	30.83	3/2	12.1027	0.11484	1.4213	+2.38167	-20.4
30	^{67}Zn	4.1	5/2	2.6693	0.00287	0.7314	+0.875479	+15.0
31	^{69}Ga	60.108	3/2	10.2475	0.06971	1.2034	+2.01659	+17.0
31	^{71}Ga	39.892	3/2	13.0204	0.14300	1.5291	+2.56227	+10.0
32	^{73}Ge	7.73	9/2	1.4897	0.00141	1.1546	-0.8794677	-17.3
33	^{75}As	100	3/2	7.3148	0.02536	0.8590	+1.439475	+31.4
34	^{77}Se	7.63	1/2	8.1566	0.00703	0.1916	+0.535042	
35	^{79}Br	50.69	3/2	10.7039	0.07945	1.2570	+2.106400	+33.1
35	^{81}Br	49.31	3/2	11.5381	0.09951	1.3550	+2.270562	+27.6
36	^{83}Kr	11.5	9/2	1.6442	0.00190	1.2744	-0.970669	+25.3
37	^{85}Rb	72.165	5/2	4.1253	0.01061	1.1304	+1.35303	+27.4
37	^{87}Rb	27.835	3/2	13.9807	0.17703	1.6418	+2.75124	+13.2
38	^{87}Sr	7.00	9/2	1.8524	0.00272	1.4358	-1.093603	+33.5
39	^{89}Y	100	1/2	2.0949	0.00012	0.0492	-0.1374154	
40	^{91}Zr	11.22	5/2	3.9747	0.00949	1.0891	-1.30362	-20.6
41	^{93}Nb	100	9/2	10.4520	0.48821	8.1011	+6.1705	-32
42	^{95}Mo	15.92	5/2	2.7874	0.00327	0.7638	-0.9142	-2.2
42	^{97}Mo	9.55	5/2	2.8462	0.00349	0.7799	-0.9335	+25.5
43	^{99}Tc	*	9/2	9.63	0.38174	7.4633	+5.6847	-12.9
44	^{99}Ru	12.7	5/2	1.9553	0.00113	0.5358	-0.6413	+7.9
44	^{101}Ru	17.0	5/2	2.192	0.00159	0.6005	-0.7188	+45.7
45	^{103}Rh	100	1/2	1.3476	0.00003	0.0317	-0.08840	
46	^{105}Pd	22.33	5/2	1.957	0.00113	0.5364	-0.642	+66.0
47	^{107}Ag	51.839	1/2	1.7330	0.00007	0.0407	-0.1136796	
47	^{109}Ag	48.161	1/2	1.9924	0.00010	0.0468	-0.1306906	
48	^{111}Cd	12.80	1/2	9.0689	0.00966	0.2130	-0.5948861	
48	^{113}Cd	12.22	1/2	9.4868	0.01106	0.2228	-0.6223009	
49	^{113}In	4.3	9/2	9.3652	0.35121	7.2588	+5.5289	+79.9
49	^{115}In	95.7	9/2	9.3854	0.35348	7.2744	+5.5408	+81
50	^{115}Sn	0.34	1/2	14.0074	0.03561	0.3290	-0.91883	
50	^{117}Sn	7.68	1/2	15.2609	0.04605	0.3584	-1.00104	
50	^{119}Sn	8.59	1/2	15.9656	0.05273	0.3750	-1.04728	
51	^{121}Sb	57.36	5/2	10.2549	0.16302	2.8100	+3.3634	-36
51	^{123}Sb	42.64	7/2	5.5530	0.04659	2.7389	+2.5498	-49
52	^{123}Te	0.908	1/2	11.2346	0.01837	0.2639	-0.7369478	
52	^{125}Te	7.139	1/2	13.5451	0.03220	0.3181	-0.8885051	
53	^{127}I	100	5/2	8.5776	0.09540	2.3504	+2.813273	-78.9
54	^{129}Xe	26.4	1/2	11.8601	0.02162	0.2786	-0.7779763	
54	^{131}Xe	21.2	3/2	3.5158	0.00282	0.4129	+0.6918619	-12.0
55	^{133}Cs	100	7/2	5.6232	0.04838	2.7735	+2.582025	-0.37
56	^{135}Ba	6.592	3/2	4.2581	0.00500	0.5001	+0.837943	+16.0
56	^{137}Ba	11.23	3/2	4.7633	0.00700	0.5594	+0.937365	+24.5
57	^{138}La	* 0.0902	5	5.6614	0.09404	5.3188	+3.713646	+45
57	^{139}La	99.9098	7/2	6.0610	0.06058	2.9895	+2.7830455	+20
58	^{137}Ce	*	3/2	4.62	0.00641	0.5431	0.91	

Z	Isotope	Abundance %	I	v/MHz for $H_o = 1$ T	Relative sensitivity Const. H_o	Const. v	μ/μ_N	Q/fm²
58	^{139}Ce	*	3/2	4.62	0.00641	0.5431	0.91	
58	^{141}Ce	*	7/2	2.37	0.00364	1.1708	1.09	
59	^{141}Pr	100	5/2	13.0355	0.33483	3.5720	+4.2754	-5.9
60	^{143}Nd	12.18	7/2	2.319	0.00339	1.1440	-1.065	-63
60	^{145}Nd	8.30	7/2	1.429	0.00079	0.7047	-0.656	-33
61	^{143}Pm	*	5/2	11.6	0.23510	3.1748	3.8	
61	^{147}Pm	*	7/2	5.7	0.04940	2.7928	+2.6	+70
62	^{147}Sm	15.0	7/2	1.7747	0.00152	0.8753	-0.8149	-26
62	^{149}Sm	13.8	7/2	1.4631	0.00085	0.7216	-0.6718	+9.4
63	^{151}Eu	47.8	5/2	10.5854	0.17929	2.9006	+3.4718	+90.3
63	^{153}Eu	52.2	5/2	4.6744	0.01544	1.2809	+1.5331	+241
64	^{155}Gd	14.80	3/2	1.317	0.00015	0.1546	-0.2591	+127
64	^{157}Gd	15.65	3/2	1.727	0.00033	0.2028	-0.3399	+135
65	^{159}Tb	100	3/2	10.23	0.06945	1.2019	+2.014	+143.2
66	^{161}Dy	18.9	5/2	1.4653	0.00048	0.4015	-0.4806	+247
66	^{163}Dy	24.9	5/2	2.0507	0.00130	0.5619	+0.6726	+265
67	^{165}Ho	100	7/2	9.0881	0.20423	4.4825	+4.173	+358
68	^{167}Er	22.95	7/2	1.2281	0.00050	0.6057	-0.5639	+357
69	^{169}Tm	100	1/2	3.531	0.00057	0.0829	-0.2316	
70	^{171}Yb	14.3	1/2	7.5259	0.00552	0.1768	+0.49367	
70	^{173}Yb	16.12	5/2	2.0730	0.00135	0.5680	-0.67989	+280
71	^{175}Lu	97.41	7/2	4.8624	0.03128	2.3983	+2.2327	+349
71	^{176}Lu	* 2.59	7	3.451	0.03975	6.0516	+3.169	+497
72	^{177}Hf	18.606	7/2	1.7281	0.00140	0.8524	+0.7935	+336
72	^{179}Hf	13.629	9/2	1.0856	0.00055	0.8414	-0.6409	+379
73	^{180}Ta	0.012	9	4.04	0.10251	11.3862	4.77	
73	^{181}Ta	99.988	7/2	5.1625	0.03744	2.5463	+2.3705	+317
74	^{183}W	14.3	1/2	1.7956	0.00008	0.0422	+0.1177847	
75	^{185}Re	37.40	5/2	9.717	0.13870	2.6627	+3.1871	+218
75	^{187}Re	*62.60	5/2	9.817	0.14300	2.6900	+3.2197	+207
76	^{187}Os	1.6	1/2	0.9856	0.00001	0.0231	+0.06465189	
76	^{189}Os	16.1	3/2	3.3535	0.00244	0.3938	+0.659933	+85.6
77	^{191}Ir	37.3	3/2	0.766	0.00003	0.0899	+0.1507	+81.6
77	^{193}Ir	62.7	3/2	0.832	0.00004	0.0977	+0.1637	+75.1
78	^{195}Pt	33.8	1/2	9.2920	0.01039	0.2182	+0.60952	
79	^{197}Au	100	3/2	0.7406	0.00003	0.0870	+0.145746	+54.7
80	^{199}Hg	16.87	1/2	7.7121	0.00594	0.1811	+0.5058855	
80	^{201}Hg	13.18	3/2	2.8468	0.00149	0.3343	-0.5602257	+38.6
81	^{203}Tl	29.524	1/2	24.7310	0.19598	0.5809	+1.6222579	
81	^{205}Tl	70.476	1/2	24.9742	0.20182	0.5866	+1.6382146	
82	^{207}Pb	22.1	1/2	9.0338	0.00955	0.2122	+0.59258	
83	^{209}Bi	100	9/2	6.9628	0.14433	5.3967	+4.1106	-50
84	^{209}Po	*	1/2	11.7	0.02096	0.2757	+0.77	
86	^{211}Rn	*	1/2	9.16	0.00997	0.2152	+0.601	
87	^{223}Fr	*	3/2	5.94	0.01362	0.6982	+1.17	+117
88	^{223}Ra	*	3/2	1.3746	0.00017	0.1614	+0.2705	+119
88	^{225}Ra	*	1/2	11.187	0.01814	0.2627	-0.7338	
89	^{227}Ac	*	3/2	5.6	0.01131	0.6564	+1.1	+170
90	^{229}Th	*	5/2	1.40	0.00042	0.3843	+0.46	+430
91	^{231}Pa	*	3/2	10.2	0.06903	1.1995	2.01	-172
92	^{235}U	* 0.7200	7/2	0.83	0.00015	0.4082	-0.38	+493.6
93	^{237}Np	*	5/2	9.57	0.13264	2.6234	+3.14	+388.6
94	^{239}Pu	*	1/2	3.09	0.00038	0.0727	+0.203	
95	^{243}Am	*	5/2	4.91	0.01788	1.3451	+1.61	+421

PROTON NMR CHEMICAL SHIFTS FOR CHARACTERISTIC ORGANIC STRUCTURES

The chart below summarizes the range of chemical shifts for protons in several classes or organic compounds and substituent groups. The chemical shifts δ are given in parts per million relative to tetramethylsilane.

REFERENCE

Mohacsi, E., *J. Chem. Edu.*, 41, 38, 1964 (with permission).

^{13}C-NMR ABSORPTIONS OF MAJOR FUNCTIONAL GROUPS

The table below lists the ^{13}C chemical shifts in parts per million (ppm), in descending order with the corresponding functional groups. A series of typical simple compounds for every family is added to illustrate the correlations above. The shifts for the carbons of interest are given in parentheses — either for each carbon as it appears from left to right in the formula, or by italics.[1-3]

REFERENCES

1. **Yoder, C. H. and Schaeffer, C. D., Jr.,** *Introduction to Multinuclear NMR: Theory and Application,* Benjamin/Cummings, Menlo Park, CA, 1987.
2. **Silverstein, R. M., Bassler, G. C., and Morrill, T. C.,** *Spectrometric Identification of Organic Compounds,* John Wiley & Sons, New York, 1981.
3. **Brown, D. W.,** A short set of ^{13}NMR Correlation Tables, *J. Chem. Educ.,* 62, 209, 1985.

^{13}C-NMR ABSORPTIONS OF MAJOR FUNCTIONAL GROUPS

δ (ppm)	Group	Family	Example (δ of italicized carbon)	
220—165	>C=O	Ketones	$(CH_3)_2CO$	(206.0)
			$(CH_3)_2CHCOCH_3$	(212.1)
		Aldehydes	CH_3CHO	(199.7)
		α,β-Unsaturated carbonyls	$CH_3CH=CHCHO$	(192.4)
			$CH_2=CHCOCH_3$	(169.9)
		Carboxylic acids	HCO_2H	(166.0)
			CH_3CO_2H	(178.1)
		Amides	$HCONH_2$	(165.0)
			CH_3CONH_2	(172.7)
		Esters	$CH_3CO_2CH_2CH_3$	(170.3)
			$CH_2=CHCO_2CH_3$	(165.5)
140—120	>C=C<	Aromatic	C_6H_6	(128.5)
		Alkenes	$CH_2=CH_2$	(123.2)
			$CH_2=CHCH_3$	(115.9,136.2)
			$CH_2=CHCH_2Cl$	(117.5.133.7)
			$CH_3CH=CHCH_2CH_3$	(132.7)
125—115	–C≡N	Nitriles	$CH_3–C≡N$	(117.7)
80—70	–C≡C–	Alkynes	$HC≡CH$	(71.9)
			$CH_3C≡CH_3$	(73.9)
70—45	–C–O	Esters	$CH_3OOCH_2CH_3$	(57.6,67.9)
		Alcohols	$HOCH_3$	(49.0)
			$HOCH_2CH_3$	(57.0)
40—20	–C–NH$_2$	Amines	CH_3NH_2	(26.9)
			$CH_3CH_2NH_2$	(35.9)
30—15	–S–CH$_3$	Sulfides (thioethers)	$C_6H_5–S–CH_3$	15.6
30—(−2.3)	–C–H	Alkanes, cycloalkanes	CH_4	(−2.3)
			CH_3CH_3	(5.7)
			$CH_3CH_2CH_3$	(15.8,16.3)
			$CH_3CH_2CH_2CH_3$	(13.4,25.2)
			$CH_3CH_2CH_2CH_2CH_3$ (13.9,22.8,34.7)	
			Cyclohexane	(26.9)

Reprinted from T. J. Bruno and P. D. N. Svoronos, *CRC Handbook of Basic Tables for Chemical Analysis*, CRC Press, Boca Raton, FL, 1989, pg. 329.

Section 10
Atomic, Molecular, and Optical Physics

LINE SPECTRA OF THE ELEMENTS

Joseph Reader and Charles H. Corliss

The original tables from which this table was derived were prepared under the auspices of the Committee on Line Spectra of the Elements of the National Academy of Sciences-National Research Council. The table contains the outstanding spectral lines of neutral (I) and singly ionized (II) atoms of the elements from hydrogen through plutonium ($Z = 1$-94); selected strong lines from doubly ionized (III), triply ionized (IV), and quadruply ionized (V) atoms are also included. Listed are lines that appear in emission from the vacuum ultraviolet to the far infrared. These lines were selected from much larger lists in such a way as to include the stronger observed lines in each spectral region. A more extensive list may be found in Reference 1.

The data were compiled by the following contributors.

J. G. Conway — Lawrence Berkeley Laboratory	L. J. Radziemski — Los Alamos Scientific Laboratory
C. H. Corliss — National Bureau of Standards	J. Reader — National Bureau of Standards
R. D. Cowan — Los Alamos Scientific Laboratory	C. J. Sansonetti — National Bureau of Standards
C. R. Cowley — University of Michigan	G. V. Shalimoff — Lawrence Berkeley Laboratory
Henry M. and Hannah Crosswhite — Argonne National Laboratory	R. W. Stanley — Purdue University
S. P. Davis — University of California, Berkeley	J. O. Stoner, Jr. — University of Arizona
V. Kaufman — National Bureau of Standards	H. H. Stroke — New York University
R. L. Kelly — Naval Postgraduate School	D. R. Wood — Wright State University
J. F. Kielkopf — University of Louisville	E. F. Worden — Lawrence Livermore Laboratory
W. C. Martin — National Bureau of Standards	J. J. Wynne — International Business Machines Corporation
T. K. McCubbin — Pennsylvania State University	R. Zalubas — National Bureau of Standards

All wavelengths are given in Ångstrom units (10^{-10} m). Below 2000 Å the wavelengths are in vacuum; above 2000 Å the wavelengths are in air. Wavelengths given to three decimal places have an uncertainty of less than 0.001 Å and are therefore suitable for calibration purposes. In the air region, the elements used most commonly for calibration are Ne, Ar, Kr, Fe, Th, and Hg; in the vacuum region, the most common are C, N, O, Si, Cu.

All data refer to natural isotopic abundance of the elements except that Kr I and Kr II lines below 11,000 Å given to three decimal places are for ^{86}Kr. A separate table for ^{198}Hg contains accurately known wavelengths that are frequently used for calibration.

A large number of the lines for neutral and singly ionized atoms were extracted from the National Bureau of Standards (NBS) *Tables of Spectral-Line Intensities* (Reference 2). The intensities of these lines represent quantitative estimates of relative line strengths that take account of varying detection sensitivity at different wavelengths. They are on a linear scale. For nearly all of the other lines the intensities represent qualitative estimates of the relative strengths of lines not greatly separated in wavelength. Because different observers frequently use different scales for their intensity estimates, these intensities are useful only as a rough indication of the appearance of a spectrum. In some cases the intensity scale is not intended to be linear. In the first and second spectra the intensities of the lines of the singly ionized atom (II) relative to those of the neutral atom (I) should be used with caution, inasmuch as the concentration of ions in a light source depends greatly on the excitation conditions.

Descriptive symbols that follow the wavelength have the following meanings:

c — complex
d — line consists of two unresolved lines
h — hazy
l — shaded to longer wavelengths
s — shaded to shorter wavelengths
p — perturbed by a close line
r — easily reversed
w — wide

The table is arranged alphabetically by element name (not symbol); for each element the lines are listed by wavelength. References to the sources of data for each element are given at the end of the table, starting on page **10**-82.

GENERAL REFERENCES

1. Reader, J., Corliss, C. H., Wiese, W. L., and Martin, G. A., *Tables of Line Spectra of the Elements, Part 1. Wavelengths and Intensities*, Nat. Stand. Ref. Data Sys.- Nat. Bur. Standards (U.S.), No. 68, 1980.
2. Meggers, W. F., Corliss, C. H., and Scribner, B. F., *Tables of Spectral Line Intensities, Part 1. Arranged by Elements*, Nat. Bur. Stand. (U.S.), Monograph 145, 1975.
3. Fuhr, J. R., Martin, W. C., Musgrove, A., Sugar, J., and Wiese, W. L., "NIST Atomic Spectroscopic Database" ver. 1.1, January 1996. *NIST Physical Reference Data*, National Institute of Standards and Technology, Gaithersburg, MD. Available at the WWW address: http://physics.nist.gov/PhysRefData/contents.html

Intensity	Wavelength/Å	Intensity	Wavelength/Å	Intensity	Wavelength/Å	Intensity	Wavelength/Å
	Actinium	100	1611.814 III	110	2637.70 II	360	6696.02 I
	Ac Z = 89	800	1611.874 III	150	2652.48 I	230	6698.67 I
2000 h	2952.55 III	150	1625.627 II	200	2660.39 I	110	7361.57 I
2000 h	3392.78 III	800	1639.06 IV	160	2669.17 II	140	7362.30 I
3000	3487.59 III	100	1644.235 II	650	2816.19 II	230	7835.31 I
2000 s	3863.12 II	100	1644.809 II	150	3041.28 II	290	7836.13 I
3000 s	4088.44 II	1000	1670.787 II	360	3050.07 I	110	8075.35 I
3000 s	4168.40 II	100	1686.250 II	450	3057.14 I	290	8640.70 II
100	4179.98 I	800	1719.440 II	150	3074.64 II	360	8772.87 I
20	4183.12 I	500	1721.244 II	4500 r	3082.153 I	450	8773.90 I
20	4194.40 I	900	1721.271 II	7200 r	3092.710 I	110	8828.91 I
20 l	4384.53 I	500	1724.952 II	1800 r	3092.839 I	180	8841.28 I
20	4396.71 I	900	1724.984 II	150	3428.92 II	140	8923.56 I
2000 h	4413.09 III	350	1760.104 II	150	3443.64 I	150	9290.65 II
20	4462.73 I	300	1761.975 II	900	3492.23 IV	110	9290.75 II
3000 h	4569.87 III	290	1763.00 I	800	3508.46 IV	150	10076.29 II
1000	5910.85 II	500	1763.869 II	450	3586.56 II	110	10768.36 I
20	6359.86 I	700	1763.952 II	360	3587.07 II	140	10782.04 I
20 l	6691.27 I	450	1765.64 I	290	3587.45 II	110	10872.98 I
	Aluminum	300	1765.815 II	870	3601.63 II	230	10891.73 I
	Al Z = 13	450	1766.38 I	220	3651.06 II	450	11253.19 I
900	125.53 V	400	1767.731 II	110	3651.10 II	570	11254.88 I
800	126.07 V	450	1769.14 I	150	3654.98 II	570	13123.41 I
800	130.41 V	1000	1818.56 IV	290	3655.00 II	450	13150.76 I
1000	130.85 V	600	1828.588 II	450	3900.68 II	230	16718.96 I
900	131.00 V	400	1832.837 II	4500 r	3944.006 I	300	16750.56 I
900	131.44 V	250	1834.808 II	9000 r	3961.520 I	140	16763.36 I
800	160.07 IV	1000	1854.716 III	110	3995.86 II	300	21093.04 I
1000	278.69 V	300	1855.929 II	290	4226.81 II	360	21163.75 I
900	281.39 V	700	1858.026 II	870	4529.19 III		**Antimony**
70	486.884 III	120	1859.980 II	150	4585.82 II		**Sb Z = 51**
30	486.912 III	1000	1862.311 II	110	4588.19 II	15	722.86 III
250	511.138 III	600	1862.790 II	550	4666.80 II	15	732.33 III
150	511.191 III	200	1929.978 II	110	4898.76 II		861.5 IV
500	560.317 III	150	1931.048 II	110	4902.77 II	4	876.84 II
200	560.433 III	200	1932.377 II	150	5280.21 II	4	921.07 II
100	670.068 III	400	1934.503 II	290	5283.77 II	6	983.57 II
200	671.118 III	150	1934.713 II	150	5285.85 II	15	999.62 III
500	695.829 III	300	1935.840 III	110	5312.32 II	6	1001.13 II
400	696.217 III	200	1935.949 III	220	5316.07 II	6	1009.43 II
200	725.683 III	150	1936.907 II	150	5371.84 II	40	1011.94 III
300	726.915 III	220	1939.261 II	180	5557.06 I	6	1052.21 II
400	855.034 III	700	1990.531 II	110	5557.95 I	8	1056.27 II
500	856.746 III	150	2016.052 II	450	5593.23 II	8	1057.32 II
400	892.024 III	150	2016.234 II	1200	5696.60 III	40	1065.90 III
50	893.887 III	100	2016.368 II	1000	5722.73 III	6	1073.81 II
450	893.897 III	200	2074.008 II	110	5853.62 II	30	1075.82 III
800	1042.17 IV	700	2094.264 II	220	5971.94 II		1087.6 IV
50	1191.812 II	150	2094.744 II	290	6001.76 II	8	1104.32 V
900	1237.19 IV	300	2094.791 II	220	6001.88 II	30	1151.49 III
900	1257.62 IV	100	2095.104 II	450	6006.42 II	40	1157.74 III
800	1264.18 IV	200	2095.141 II	150	6061.11 II		1199.1 IV
1000	1272.76 IV	400	2269.10 I	290	6068.43 II	50	1205.20 III
150	1350.18 II	120	2269.22 I	110	6068.53 II	50	1210.64 III
800	1384.13 III	140	2321.56 I	450	6073.23 II	12	1226.00 V
800	1447.51 IV	460	2367.05 I	110	6181.57 II	6	1230.30 II
800	1494.79 IV	110	2367.61 I	150	6181.68 II	8	1274.98 II
1000	1526.14 V	110	2368.11 I	290	6182.28 II	20	1306.69 III
800	1537.54 IV	180	2369.30 I	220	6182.45 II	8	1327.40 II
800	1539.830 II	140	2370.22 I	450 h	6183.42 II	6	1358.04 II
1000	1557.25 IV	160	2372.07 I	450	6201.52 II	8	1384.70 II
100	1569.385 II	850	2373.12 I	360	6201.70 II	20	1404.18 III
900	1582.04 IV	170	2373.35 I	290	6226.18 II	6	1407.83 II
800	1584.46 IV	110	2373.57 I	360	6231.78 II	8	1436.49 II
125	1596.059 II	240	2567.98 I	450	6243.36 II	20 r	1486.57 I
700	1605.766 III	480	2575.10 I	450	6335.74 II	40 h	1491.36 I

Intensity	Wavelength/Å		Intensity	Wavelength/Å		Intensity	Wavelength/Å		Intensity	Wavelength/Å	
	1499.2	IV	400	2478.32	I	30 h	7648.28	I	200	666.011	II
12	1505.70	V	150	2480.44	I	80	7844.44	I	1000	670.946	II
50 r	1512.57	I	100	2510.54	I	200	7924.65	I	3000	671.851	II
12	1524.47	V	2000 r	2528.52	I	60	8411.69	I	70	676.242	II
120 r	1532.74	I	15	2528.54	II	150	8572.64	I	30	677.952	II
80 r	1535.06	I	10	2567.75	II	100	8619.55	I	30	679.218	II
6	1565.51	II	150	2574.06	I	400	9518.68	I	200	679.401	II
8	1576.11	II	15	2590.13	III	400	9949.14	I	10	683.28	IV
7	1581.36	II	1500 r	2598.05	I	200	10078.49	I	7	688.39	IV
80 r	1599.96	I	500 r	2598.09	I	300	10261.01	I	12 p	689.01	IV
10	1606.98	II	300 r	2612.31	I	200	10585.60	I	6	699.41	IV
200 w	1612.8	I	12	2617.17	III	1000	10677.41	I	8	700.28	IV
100 w	1623.3	I	200 r	2652.60	I	800	10741.94	I	3	705.35	V
20	1657.04	II	20	2669.39	III	80	10794.11	I	5	709.20	V
100 w	1662.6	I	300 r	2670.64	I	600	10839.73	I	4	715.60	V
15	1673.89	III	200 r	2682.76	I	200	10868.58	I	3	715.65	V
15	1711.84	III	120	2692.25	I	400	10879.55	I	200	718.090	II
80 r	1716.93	I	150 r	2718.90	I	300	11012.79	I	3000	723.361	II
150 r	1717.45	I	400 r	2769.95	I	150	11266.23	I	2	725.11	V
150 r	1723.43	I	1000 r	2877.92	I	5	12116.06	I	500	725.548	II
15	1725.33	III	15	2980.96	II		**Argon**		70	730.930	II
100 r	1736.19	I	500 r	3029.83	I		**Ar Z = 18**		200	740.269	II
100 h	1765.76	I	600 r	3232.52	I	3	336.56	V	200	744.925	II
100 r	1780.87	I	20	3241.28	II	3	337.56	V	70	745.322	II
100 r	1788.24	I	700 r	3267.51	I	6	338.00	V	4	754.20	IV
150	1800.18	I	15	3498.46	II	2	338.43	V	5	761.47	IV
50 r	1810.50	I	25	3637.80	II	2	339.01	V	12	769.15	III
80 r	1814.20	I	250	3637.83	I	3	339.89	V	5	800.57	IV
100	1829.50	I	20	3722.78	II	3	350.88	V	10	801.09	IV
50 r	1868.17	I	200 r	3722.79	I	4	396.87	IV	10	801.41	IV
300 r	1871.15	I	20	3850.22	II	4	398.55	IV	5	801.91	IV
150 r	1882.56	I	200	4033.55	I	2	436.67	V	20	802.859	I
100	1927.08	I	20	4033.56	II	5	446.00	V	100	806.471	I
200 r	1950.39	I	20	4133.63	II	8	446.95	V	60	806.869	I
60 r	2029.49	I	15	4140.54	II	4	447.53	V	30	807.218	I
70 r	2039.77	I	15	4195.17	II	18	449.06	V	40	807.653	I
150 r	2049.57	I	20	4219.07	II	4	449.49	V	50	809.927	I
1000 r	2068.33	I	20	4314.32	II	3	458.12	V	120	816.232	I
100	2079.56	I	15	4514.50	II	2	458.98	V	70	816.464	I
50 r	2098.41	I	30	4596.90	II	6 p	461.23	V	80	820.124	I
80 r	2118.48	I	20	4599.09	II	3	462.42	V	4	822.16	V
100 r	2127.39	I	15	4604.77	II	7	463.94	V	120	825.346	I
50 r	2137.05	I	30	4647.32	II	30	487.227	II	120	826.365	I
100 r	2139.69	I	20	4675.74	II	50	490.650	II	5	827.05	V
10	2141.80	II	40	4711.26	II	30	490.701	II	3	827.35	V
50 r	2141.83	I	20	4757.81	II	30	519.327	II	150	834.392	I
100 r	2144.86	I	20	4765.36	II	3	522.09	V	4 p	834.88	V
1500 r	2175.81	I	30	4784.03	II	5	524.19	V	100	835.002	I
250 r	2179.19	I	20	4802.01	II	6	527.69	V	2	836.13	V
200 r	2201.32	I	20	4832.82	II	30	542.912	II	15	840.03	IV
300 r	2208.45	I	20	4877.24	II	200	543.203	II	100	842.805	I
150 r	2220.73	I	15	4947.40	II	70	547.461	II	20	843.77	IV
100	2221.98	I	15	5044.56	II	2	554.50	V	25	850.60	IV
120 r	2224.93	I	20	5238.94	II	70	556.817	II	180	866.800	I
300 r	2262.51	I	20	5354.24	II	5	558.48	V	150	869.754	I
120	2288.98	I	40 h	5556.10	I	70	573.362	II	10	871.10	III
150 r	2293.44	I	100 l	5632.02	I	30	576.736	II	9	875.53	III
300 r	2306.46	I	30	5639.75	II	70	580.263	II	180 r	876.058	I
2500 r	2311.47	I	60 h	5830.34	I	30	583.437	II	12	878.73	III
150	2315.89	I	100	6005.21	II	70	597.700	II	8	879.62	III
400 h	2373.67	I	20	6053.41	II	30	602.858	II	180 r	879.947	I
300 h	2383.64	I	30	6079.80	II	30	612.372	II	9	883.18	III
100	2395.22	I	50	6130.04	II	6	623.77	IV	10	887.40	III
150	2422.13	I	20	6154.94	II	3	635.12	V	150	894.310	I
250	2426.35	I	20	6611.49	I	500	661.867	II	5	900.36	IV
400 r	2445.51	I	30	6647.44	II	30	664.562	II	9	901.17	IV

Intensity	Wavelength/Å		Intensity	Wavelength/Å		Intensity	Wavelength/Å		Intensity	Wavelength/Å	
1000	919.781	II	15	2640.34	IV	8	3511.12	III	100	4228.158	II
1000	932.054	II	10	2654.63	III	70	3514.388	II	100	4237.220	II
1000 r	1048.220	I	8	2674.02	III	70	3545.596	II	25	4251.185	I
500 r	1066.660	I	9	2678.38	III	70	3545.845	II	200	4259.362	I
7	1669.67	III	9	2682.63	IV	7	3554.306	I	100	4266.286	I
7	1673.42	III	10	2724.84	III	100	3559.508	II	70	4266.527	II
7	1675.48	III	14	2757.92	IV	100	3561.030	II	150	4272.169	I
9	1914.40	III	7	2762.23	III	70	3576.616	II	550	4277.528	II
7	1915.56	III	10	2776.26	IV	25	3581.608	II	20	4282.898	II
10	2125.16	III	12	2784.47	III	50	3582.355	II	100	4300.101	I
15	2133.87	III	14	2788.96	IV	70	3588.441	II	25	4300.650	II
10	2138.59	III	7	2797.11	IV	7	3606.522	I	70	4309.239	II
10	2148.73	III	16	2809.44	IV	25	3622.138	II	200	4331.200	II
15	2166.19	III	10	2830.25	IV	20	3639.833	II	50	4332.030	II
10	2168.26	III	7	2842.88	III	35	3718.206	II	100	4333.561	I
20	2170.23	III	8	2855.29	III	70	3729.309	II	50	4335.338	I
25	2177.22	III	6	2874.40	IV	50	3737.889	II	25	4345.168	I
8	2184.06	III	9	2884.12	III	150	3765.270	II	800	4348.064	II
10	2188.22	III	25	2891.612	II	50	3766.119	II	50	4352.205	II
15	2192.06	III	12	2913.00	IV	20	3770.369	I	25	4362.066	II
7	2248.73	III	11	2926.33	IV	20	3770.520	II	50	4367.832	II
10	2279.10	III	200	2942.893	II	25	3780.840	II	200	4370.753	II
7	2281.22	III	100	2979.050	II	20	3795.37	III	70	4371.329	II
7	2282.21	III	10	3010.02	III	25	3803.172	II	50	4375.954	II
12	2293.03	III	12	3024.05	III	50	3809.456	II	150	4379.667	II
4	2299.72	IV	50	3033.508	II	7	3834.679	I	50	4385.057	II
10	2300.85	III	6	3037.98	IV	70	3850.581	II	70	4400.097	II
15	2302.17	III	12	3054.82	III	10	3858.32	III	200	4400.986	II
9	2317.00	III	10	3064.77	III	35	3868.528	II	400	4426.001	II
15	2317.47	III	8	3077.40	IV	7	3907.84	III	150	4430.189	II
12	2318.04	III	10	3078.15	III	35	3925.719	II	50	4430.996	II
10	2319.13	III	50	3093.402	II	50	3928.623	II	50	4433.838	II
10	2319.37	III	7	3110.41	III	25	3932.547	II	20	4439.461	II
9	2345.17	III	7	3127.90	III	70	3946.097	II	35	4448.879	II
7	2351.67	III	8	3200.37	I	7	3947.505	I	100	4474.759	II
9	2360.26	III	20	3243.689	II	35	3948.979	I	200	4481.811	II
10	2395.63	III	25	3285.85	III	8	3960.53	III	100	4510.733	I
12	2399.15	III	25	3293.640	II	20	3979.356	II	20	4522.323	I
10	2413.20	III	20	3301.88	II	35	3994.792	II	20	4530.552	II
7	2415.61	III	20	3307.228	II	50	4013.857	II	400	4545.052	II
10	2418.82	III	15	3311.25	III	6	4023.60	III	20	4564.405	II
5	2420.456	II	7	3319.34	I	50	4033.809	II	400	4579.350	II
12	2423.52	III	7	3323.59	III	20	4035.460	II	400	4589.898	II
12	2423.93	III	25	3336.13	III	150	4042.894	II	15	4596.097	I
7	2443.69	III	20	3344.72	III	50	4044.418	I	550	4609.567	II
8	2447.71	IV	25	3350.924	II	100	4052.921	II	7	4628.441	I
8	2472.95	III	15	3358.49	III	200	4072.005	II	35	4637.233	II
7	2476.10	III	7	3361.28	III	70	4072.385	II	400	4657.901	II
12	2488.86	III	7	3373.47	I	25	4076.628	II	15	4702.316	I
12	2513.28	IV	25	3376.436	II	35	4079.574	II	20	4721.591	II
10	2516.789	II	25	3388.531	II	25	4082.387	II	550	4726.868	II
6	2518.40	IV	15	3391.85	II	150	4103.912	II	50	4732.053	II
9	2525.69	IV	7	3393.73	I	300	4131.724	II	300	4735.906	II
10	2534.709	II	7	3417.49	III	5	4146.70	III	800	4764.865	II
15	2562.087	II	9	3424.25	III	35	4156.086	II	550	4806.020	II
12	2562.17	IV	8	3438.04	III	400	4158.590	I	150	4847.810	II
10	2568.07	IV	7	3461.07	I	50	4164.180	I	50	4865.910	II
7	2569.53	IV	9	3471.32	III	35	4179.297	II	800	4879.864	II
12	2599.47	IV	70	3476.747	II	50	4181.884	I	70	4889.042	II
10	2608.06	IV	20	3478.232	II	100	4190.713	I	20	4904.752	II
7	2608.44	IV	20	3480.55	III	50	4191.029	I	35	4933.209	II
12	2615.68	IV	50	3491.244	II	200	4198.317	I	200	4965.080	II
6	2619.98	IV	100	3491.536	II	400	4200.674	I	50	5009.334	II
12	2621.36	IV	12	3499.67	III	25	4218.665	II	70	5017.163	II
12	2624.92	IV	15	3503.58	III	25	4222.637	II	70	5062.037	II
7	2631.90	III	70	3509.778	II	25	4226.988	II	20	5090.495	II

Intensity	Wavelength/Å		Intensity	Wavelength/Å		Intensity	Wavelength/Å		Intensity	Wavelength/Å	
100	5141.783	II	7	6960.250	I	11	11078.869	I	800	1211.17	II
70	5145.308	II	10000	6965.431	I	30	11106.46	I	800	1218.10	II
5	5151.391	I	150	7030.251	I	12	11441.832	I	340	1223.15	II
15	5162.285	I	10000	7067.218	I	400	11488.109	I	760	1241.31	II
25	5165.773	II	100	7068.736	I	200	11668.710	I	965	1243.08	II
20	5187.746	I	25	7107.478	I	12	11719.488	I	870	1245.67	II
20	5216.814	II	25	7125.820	I	200	12112.326	I	800	1258.58	II
7	5221.271	I	1000	7147.042	I	50	12139.738	I	965	1263.77	II
5	5421.352	I	15	7158.839	I	50	12343.393	I	800	1266.34	II
10	5451.652	I	70	7206.980	I	200	12402.827	I	800	1267.59	II
25	5495.874	I	15	7265.172	I	200	12439.321	I	715	1280.99	II
5	5506.113	I	7	7270.664	I	100	12456.12	I	715	1287.54	II
25	5558.702	I	2000	7272.936	I	200	12487.663	I	715	1305.70	II
10	5572.541	I	35	7311.716	I	150	12702.281	I	340	1307.74	II
35	5606.733	I	25	7316.005	I	30	12733.418	I	760	1333.15	II
20	5650.704	I	5	7350.814	I	12	12746.232	I	965	1341.55	II
10	5739.520	I	70	7353.293	I	200	12802.739	I	760	1355.93	II
5	5834.263	I	200	7372.118	I	50	12933.195	I	965	1369.77	II
10	5860.310	I	20	7380.426	II	500	12956.659	I	800	1373.65	II
15	5882.624	I	10000	7383.980	I	200	13008.264	I	1000	1375.07	II
25	5888.584	I	20	7392.980	I	200	13213.99	I	760	1375.78	II
50	5912.085	I	15	7412.337	I	200	13228.107	I	800	1394.64	II
15	5928.813	I	10	7425.294	I	100	13230.90	I	800	1400.31	II
5	5942.669	I	25	7435.368	I	500	13272.64	I	500	1448.59	II
7	5987.302	I	10	7436.297	I	1000	13313.210	I	500	1558.88	II
5	5998.999	I	20000	7503.869	I	1000	13367.111	I	500	1570.99	II
5	6025.150	I	15000	7514.652	I	30	13499.41	I	100 r	1593.60	I
70	6032.127	I	25000	7635.106	I	1000	13504.191	I	500	1660.55	II
35	6043.223	I	15000	7723.761	I	11	13573.617	I	340	1860.34	II
10	6052.723	I	10000	7724.207	I	30	13599.333	I	1000 r	1890.42	I
20	6059.372	I	10	7891.075	I	400	13622.659	I	500	1912.94	II
7	6098.803	I	20000	7948.176	I	200	13678.550	I	800 r	1937.59	I
10	6105.635	I	20000	8006.157	I	1000	13718.577	I	585 r	1972.62	I
100	6114.923	II	25000	8014.786	I	10	13825.715	I	170 r	1990.35	I
10	6145.441	I	7	8053.308	I	10	13907.478	I	100 r	1991.13	I
7	6170.174	I	20000	8103.693	I	200	14093.640	I	100 r	1995.43	I
150	6172.278	II	35000	8115.311	I	100	15046.50	I	230 r	2003.34	I
10	6173.096	I	10000	8264.522	I	25	15172.69	I	100 r	2009.19	I
10	6212.503	I	20	8392.27	I	10	15329.34	I	200	2263.2	IV
5	6215.938	I	15000	8408.210	I	30	15989.49	I	350 r	2288.12	I
25	6243.120	II	20000	8424.648	I	30	16519.86	I	200	2301.0	IV
7	6296.872	I	15000	8521.442	I	500	16940.58	I	350 r	2349.84	I
15	6307.657	I	7	8605.776	I	12	18427.76	I	100 r	2370.77	I
7	6369.575	I	4500	8667.944	I	50	20616.23	I	135 r	2381.18	I
20	6384.717	I	20	8771.860	II	30	20986.11	I	250	2417.5	IV
70	6416.307	I	180	8849.91	I	20	23133.20	I	250	2454.0	IV
25	6483.082	II	20	9075.394	I	20	23966.52	I	170 r	2456.53	I
15	6538.112	I	35000	9122.967	I		**Arsenic**		200	2461.4	IV
15	6604.853	I	550	9194.638	I		**As Z = 33**		340	2602.00	II
25	6638.221	II	15000	9224.499	I	510	871.7	III	170 r	2780.22	I
20	6639.740	II	400	9291.531	I	325	889.0	III	300	2830.359	II
50	6643.698	II	1600	9354.220	I	325	927.5	III	300	2831.164	II
5	6660.676	I	25000	9657.786	I	325	937.2	III	100 r	2860.44	I
5	6664.051	I	4500	9784.503	I	325	953.6	III	300	2884.406	II
25	6666.359	II	180	10052.06	I	325	963.8	III	80	2926.3	III
100	6677.282	I	30	10332.72	I	250	987.7	V	615	2959.572	II
35	6684.293	II	100	10467.177	II	340	1021.96	II	300	3003.819	II
150	6752.834	I	1600	10470.054	I	250	1029.5	V	300	3116.516	II
5	6756.163	I	13	10478.034	I	340	1082.35	II	340	3842.60	II
15	6766.612	I	180	10506.50	I	500	1139.40	II	325	3922.6	III
20	6861.269	II	200	10673.565	I	615	1149.31	II	715	4190.082	II
150	6871.289	I	11	10681.773	I	555	1181.51	II	615	4197.40	II
5	6879.582	I	7	10683.034	II	555	1189.87	II	615	4242.982	II
10	6888.174	I	30	10733.87	I	615	1196.38	II	500	4315.657	II
50	6937.664	I	30	10759.16	I	615	1196.56	II	500	4323.867	II
7	6951.478	I	7	10812.896	II	340	1207.44	II	500	4336.64	II

Intensity	Wavelength/Å		Intensity	Wavelength/Å		Intensity	Wavelength/Å		Intensity	Wavelength/Å	
500	4352.145	II		1771.03	II	100	3576.28	II	400	4899.97	II
425	4352.864	II		1786.93	II	30	3577.62	I	15	4902.90	I
375	4371.17	II	100	1904.15	II	80 h	3579.67	I	20000	4934.09	II
615	4427.106	II	500	1924.70	II	200	3596.57	II	8	4947.35	I
615	4431.562	II		1985.60	II	40	3630.64	I	1000	4957.15	II
715	4458.469	II	300	1999.54	II	40 h	3636.83	I	300	4997.81	II
340	4461.075	II	10	2001.30	III	20 h	3688.47	I	1000	5013.00	II
715	4466.348	II		2009.20	II	400	3735.75	II	20 h	5159.94	I
500	4474.46	II	400	2023.95	II	200	3816.69	II	20	5267.03	I
800	4494.230	II		2052.68	II	200	3842.80	II	800	5361.35	II
850	4507.659	II		2054.57	II	100	3854.76	II	1000	5391.60	II
615	4539.74	II	500	2214.7	II	20	3889.33	I	200	5421.05	II
715	4543.483	II	800	2245.61	II	1400 l	3891.78	II	100	5424.55	I
615	4602.427	II	1000	2254.73	II	20	3892.65	I	200	5428.79	II
340	4629.787	II	1400	2304.24	II	40	3909.91	I	300	5480.30	II
340	4707.586	II	60	2331.10	III	500	3914.73	II	200	5519.05	I
340	4730.67	II	2000	2335.27	II	25	3926.85	III	1000 r	5535.48	I
340	4888.557	II	190	2347.58	II	50	3935.72	I	20 h	5620.40	I
340	5105.58	II	40	2512.28	III	20	3937.87	I	10	5680.18	I
500	5107.55	II	40	2523.83	III	200	3939.67	II	400	5777.62	I
425	5231.38	II	60	2528.51	III	500	3949.51	II	800	5784.18	II
500	5331.23	II	50	2559.54	III	25	3993.06	III	100	5800.23	I
340	5497.727	II	8 h	2596.64	I	80	3993.40	I	20	5805.69	I
425	5558.09	II	100	2634.78	II	30	3995.66	I	150	5826.28	I
425	5651.32	II	40	2681.89	III	300	4036.26	II	2800	5853.68	II
425	6110.07	II	8	2702.63	I	200	4083.77	II	15	5907.64	I
500	6170.27	II	18	2771.36	II	30 h	4084.86	I	100	5971.70	I
300	6511.74	II	15	2785.28	I	1500 h	4130.66	II	800	5981.25	II
300	7092.27	II	100 r	3071.58	I	20	4132.43	I	100	5997.09	I
300	7102.72	II	40	3079.14	III	200	4166.00	II	300	5999.85	II
340	7990.53	II	10 h	3108.21	I	500	4216.04	II	100	6019.47	I
300	8174.51	II	8	3132.60	I	800	4267.95	II	200	6063.12	I
200	9300.61	I	8 h	3135.72	I	100	4283.10	I	300	6110.78	I
230	9597.95	I	10	3137.70	I	300	4287.80	II	400	6135.83	II
290	9626.70	I	10	3155.34	I	200	4297.60	II	20000	6141.72	II
230	9833.76	I	10	3155.67	I	800	4309.32	II	150	6341.68	I
170	9915.71	I	12	3158.05	I	20 h	4323.00	I	500	6378.91	II
290	9923.05	I	12 h	3158.54	I	600	4325.73	II	10	6383.76	III
290	10024.04	I	25	3165.60	I	200	4326.74	II	90	6450.85	I
170	10614.07	I	15 h	3173.69	I	300	4329.62	II	150	6482.91	I
Astatine			30	3183.16	I	80	4350.33	I	12000	6496.90	II
At Z = 85			15	3183.96	I	60	4402.54	I	300	6498.76	I
8	2162.25	I	10	3193.91	I	400	4405.23	II	150	6527.31	I
10	2244.01	I	25 h	3203.70	I	40	4431.89	I	3000	6595.33	I
Barium			30	3221.63	I	60 h	4488.98	I	150	6654.10	I
Ba Z = 56			40	3222.19	I	50 h	4493.64	I	1500	6675.27	I
14	555.48	III	50	3261.96	I	40	4505.92	I	1800	6693.84	I
14	587.57	III	60 r	3262.34	I	200	4509.63	II	1000	6769.62	II
18	647.27	III	40	3281.50	I	60 h	4523.17	I	600	6865.69	I
300	719.86	V	15	3281.77	I	130	4524.93	II	300 h	6867.85	I
150	721.85	V	50	3322.80	I	65000	4554.03	II	1000	6874.09	II
1000	766.87	V	80 h	3356.80	I	40	4573.85	I	6000	7059.94	I
40000	794.89	IV	50	3368.18	III	80	4579.64	I	2400 hs	7120.33	I
300	877.41	V	60 r	3377.08	I	30	4599.75	I	600	7195.24	I
50000	923.74	IV	20	3377.39	I	20 h	4619.92	I	600 hl	7228.84	I
200	946.26	V	70 r	3420.32	I	25 h	4628.33	I	3000	7280.30	I
200	1486.72	II	25	3421.01	I	300	4644.10	II	1200	7392.41	I
400	1504.01	II	30 h	3421.48	I	30	4673.62	I	300	7417.53	I
300	1554.38	II	40	3463.74	I	35	4691.62	I	900 hl	7459.78	I
200	1572.73	II	200 r	3501.11	I	20	4700.43	I	600	7488.08	I
	1573.92	II	80 h	3524.97	I	800	4708.94	II	450 hl	7636.90	I
	1630.40	II	30 h	3531.35	I	40	4726.44	I	600 hl	7642.91	I
100	1674.51	II	80 h	3544.66	I	800	4843.46	II	1800	7672.09	I
400	1694.37	II	20 h	3547.68	I	300	4847.14	II	1200	7780.48	I
	1697.16	II	100	3552.45	II	200	4850.84	II	180 h	7839.57	I
	1761.75	II	200	3567.73	II	30 h	4877.65	I	1500	7905.75	I

Intensity	Wavelength/Å	Intensity	Wavelength/Å	Intensity	Wavelength/Å	Intensity	Wavelength/Å
600	7911.34 I		93.93 II		1909.0 II	10	3046.52 II
900 h	8210.24 I		94.78 II		1912. I	30	3046.69 II
8	8308.69 III		95.76 II	3	1917.03 III		3090.3 I
1800 h	8559.97 I		96.29 I		1919. I	10	3110.81 I
100	8710.74 II		97.24 I	5	1929.67 I	10	3110.92 I
100	8737.71 II		97.44 I	10	1943.68 I	20	3110.99 I
300 h	8799.76 I		97.86 I	60 h	1954.97 III		3120. I
300	8860.98 I		97.97 I		1956. I	480	3130.42 II
450	8914.99 I		98.12 I	50	1964.59 I	320	3131.07 II
300	9219.69 I		98.37 I	5	1985.13 I		3136. I
300	9308.08 I		98.66 I		1997.95 I		3150. I
300 h	9324.58 I		98.94 I		1997.98 I		3160.6 I
1500	9370.06 I		99.19 I	60	1998.01 I		3163. I
300	9455.92 I	100	100.25 III		2033.25 I		3168. I
8	9521.76 III		100.86 I		2033.28 I		3180.7 II
450	9589.37 I		101.20 I		2033.38 I		3187. I
900	9608.88 I		102.13 I	50	2055.90 I	20	3193.81 I
300 h	9645.72 I		102.49 II	100	2056.01 I	20	3197.10 II
1500 hl	9830.37 I		104.40 II	75 h	2076.94 III	30	3197.15 II
900	10001.08 I		104.67 I	60 h	2080.38 III	20	3208.60 I
600	10032.10 I		105.80 I	25	2118.56 III		3220. I
1200 h	10233.23 I		107.26 I	15 h	2122.27 III	60	3229.63 I
300	10471.26 I		107.38 I	10	2125.57 I	2	3233.52 I
120 hl	10791.25 I	3	509.99 III	20	2125.68 I	10	3241.62 II
180 hl	11012.69 I	2	549.31 III	15 h	2127.20 III	30	3241.83 II
150 h	11114.42 I	6	582.08 III	5	2137.25 III	15	3269.02 I
240	11303.04 I	4	661.32 III	25	2145. I	100	3274.58 II
120 h	11697.45 I	8	675.59 III	55	2174.99 I	30	3274.67 II
120	13207.30 I		714.0 II	55	2175.10 I	30	3282.91 I
120	13810.50 I	4	725.59 III	5	2191.57 III	30	3321.01 I
120	14077.90 I	5	725.71 II		2273.5 II	30	3321.09 I
120	15000.40 I	5	743.58 II		2324.6 II	220	3321.34 I
120	20712.00 I	7	746.23 III		2337.0 I	20	3345.43 I
150	25515.70 I	2	767.75 III	950	2348.61 I	60	3367.63 I
150	29223.90 I	8	775.37 II	20	2350.66 I		3405.6 II
Beryllium		20	842.06 II	60	2350.71 I	5	3451.37 I
Be Z = 4			865.3 II	200	2350.83 I	300	3455.18 I
	58.13 IV	2	925.25 II	2	2413.34 II	20	3476.56 I
	58.57 IV	10	943.56 II	16	2413.46 II	300	3515.54 I
	59.32 IV	10	973.27 II	20	2453.84 II	10	3555. I
	60.74 IV		981.4 II		2480.6 I	100	3720.36 III
	64.06 IV		1020.1 II	35	2494.54 I		3720.92 III
	75.93 IV	8	1026.93 II	35	2494.58 I		3722.98 III
1 h	76.10 III	5	1036.32 II	100	2494.73 I	100	3736.30 I
2	76.48 III	15	1048.23 II	16	2507.43 II	700	3813.45 I
3	78.53 III	1	1114.69 III	5	2617.99 II	40	3865.13 I
4	78.66 III	20	1143.03 II	20	2618.13 II	80	3865.42 I
1 h	78.92 III		1155.9 II	100	2650.45 I	1	3865.51 I
5	81.89 III	60	1197.19 II	60	2650.55 I	6	3865.72 I
10	82.38 III	2	1213.12 III	200	2650.62 I	100	3866.03 I
	82.58 II	1	1214.32 III	60	2650.69 I	90 h	4249.14 III
20	83.20 III	2	1362.25 III	100	2650.76 I	100	4253.05 I
	83.66 II	1	1401.52 III	5	2697.46 II	60	4253.76 I
30	84.76 III	10	1421.26 III	20	2697.58 II	300	4360.66 II
50	88.31 III	5	1422.86 III	20	2728.88 II	500	4360.99 II
	89.16 I		1426.12 I	30	2738.05 I	400	4407.94 I
	89.80 II	1	1435.17 III		2764.2 II	2	4485.52 III
	90.04 II	2	1440.77 III	20	2898.13 I	100 h	4487.30 III
	90.21 I		1491.76 I	10	2898.19 I	1	4495.09 III
	90.67 I	20	1512.30 II	20	2898.25 I	140 h	4497.8 III
	91.06 II	60	1512.43 II	30	2986.06 I		4526.6 I
	91.36 II	100	1661.49 I	10	2986.42 I		4548. I
	91.74 II	2 h	1754.69 III	60	3019.33 I	12	4572.66 I
	92.19 I	15	1776.12 II	30	3019.49 I	700	4673.33 II
	92.61 II	20	1776.34 II	30	3019.53 I	1000	4673.42 II
	93.14 II		1907. I	20	3019.60 I	6	4709.37 I
	93.42 II						

Line Spectra of the Elements (continued): Beryllium—Bismuth

Intensity	Wavelength/Å	Intensity	Wavelength/Å	Intensity	Wavelength/Å	Intensity	Wavelength/Å
200	4828.16 II	30	12098.18 II	35	1538.06 II	2800	2989.03 I
40	4849.16 I	100	14643.92 I	20	1563.67 II	700	2993.34 I
2 h	4858.22 II	60	14644.75 I	40	1573.70 II	100	3012. IV
80	5087.75 I	200	16157.72 I	60	1591.79 II	2400	3024.64 I
8	5218.12 II	80	17855.38 I	25	1601.58 II	60	3034.87 I
20	5218.33 II	120	17856.63 I	60 h	1606.40 III	100	3042. IV
3	5255.86 II	100	18143.54 I	40	1609.70 II	9000 c	3067.72 I
64	5270.28 II	160	31775.05 I	40	1611.38 II	140	3076.66 I
500	5270.81 II	200	31778.70 I	20	1652.81 II	35	3115.0 III
20	5403.04 II		**Bismuth**	20	1749.29 II	100	3239. IV
20	5410.21 II		**Bi Z = 83**	80	1777.11 II	550 c	3397.21 I
	5558.	6	420.7 IV	60	1787.47 II	10	3430.83 II
140 h	6142.01 III	6	431.2 IV	70	1791.93 II	12	3431.23 II
10	6229.11 I	2	488.39 V	70	1823.80 II	40 h	3451.0 III
16	6279.43 II	3	563.62 V	100	1902.41 II	40	3473.8 III
30	6279.73 II	5	670.76 III	9000	1954.53 I	35	3485.5 III
30	6473.54 I	6	686.88 V	7000	1960.13 I	500 c	3510.85 I
60	6547.89 II	5	730.71 V	25	1989.35 II	380 c	3596.11 I
60	6558.36 II	10	738.17 V	7000	2021.21 I	45	3613.4 III
30	6564.52 I	4	775.16 III	9000	2061.70 I	100	3643. IV
2 h	6636.44 II	6	790.5 IV	45 h	2068.9 II	12	3654.2 II
1	6756.72 II	6	790.6 IV	4600	2110.26 I	100	3682. IV
2	6757.13 II	8	792.5 IV	2500	2133.63 I	50	3695.32 III
30	6786.56 I	10	820.3 IV	15	2143.40 II	50	3695.68 III
1 h	6884.22 I	9	822.9 IV	15	2143.46 II	100	3734. IV
6 h	6884.44 I	12	824.9 IV	60	2186.9 II	70 h	3792.5 II
100	6982.75 I	15 d	864.45 V	40 h	2214.0 II	12	3811.1 II
6 h	7154.40 I	15	872.6 IV	360	2228.25 I	20	3815.8 II
40 h	7154.65 I	12	923.9 IV	1700	2230.61 I	10	3845.8 II
100	7209.13 I	15	943.3 IV	340	2276.58 I	30	3863.9 II
3	7401.20 II	25	1039.99 III	100	2311. IV	100	3868. IV
2	7401.43 II	50 h	1045.76 III	100	2326. IV	40 h	4079.1 II
10	7551.90 I	30	1051.81 III	16	2368.12 II	10	4097.2 II
10 h	7618.68 I	15	1058.88 II	12	2368.25 II	140	4121.53 I
20 h	7618.88 I	20	1085.47 II	100	2376. IV	140	4121.86 I
60	8090.06 I	10	1099.20 II	190	2400.88 I	75 h	4259.4 II
5 h	8158.99 I	24	1103.4 IV	75 h	2414.6 III	25	4272.0 II
10 h	8159.24 I	20	1139.01 III	10	2501.0 II	70 h	4301.7 II
4	8254.07 I	50	1224.64 III	25	2515.69 I	12 h	4339.8 II
10 h	8287.07 I	10	1225.43 II	70	2524.49 I	25 h	4340.5 II
30	8547.36 I	15	1232.78 II	20 h	2544.5 II	12 h	4379.4 II
60	8547.67 I	10	1241.05 II	700	2627.91 I	25 h	4476.8 II
300	8801.37 I	10	1265.35 II	100	2629. IV	60 h	4705.3 II
6	8882.18 I	15	1283.73 II	100	2677. IV	600 c	4722.52 I
40	9190.45 I	10	1306.18 II	12	2693.0 II	30	4730.3 II
20 h	9243.92 I	60	1317.0 IV	280 c	2696.76 I	20	4749.7 II
1 h	9343.89 II	20	1325.46 II	20	2713.3 II	40 h	4797.4 III
40	9392.74 I	40	1326.84 III	140 d	2730.50 I	12	4908.2 II
2	9476.43 II	20	1329.47 II	100	2767. IV	10	4916.6 II
16	9477.03 II	60	1346.12 III	100	2772. IV	12	4969.7 II
20	9847.32 I	20	1350.07 II	360	2780.52 I	20	4993.6 II
10 h	9895.63 I	25	1372.61 II	100	2786. IV	45 h	5079.3 III
20 h	9895.96 I	15	1376.02 II	15	2803.42 II	10	5091.6 II
80	9939.78 I	20	1393.92 II	11	2803.70 II	50 h	5124.3 II
16	10095.52 II	35	1423.33 III	12	2805.3 II	60 h	5144.3 II
20	10095.73 II	35	1423.52 III	140 c	2809.62 I	20	5201.5 II
60	10119.92 II	45	1436.83 II	100	2842. IV	75 h	5209.2 II
80	10331.03 I	25	1447.94 II	80 h	2855.6 III	40 h	5270.3 II
30	11066.46 I	50	1455.11 II	4000	2897.98 I	10	5397.8 II
	11173. II	60 h	1461.00 III	100	2924. IV	10 c	5552.35 I
1	11173.73 II	25	1462.14 II	100	2933. IV	3	5599.41 I
120	11496.39 I	35	1486.93 II	100	2936. IV	20	5655.2 II
2 h	11625.16 II	20	1502.50 II	15	2936.7 II	40 h	5719.2 II
	11659. II	40	1520.57 II	3200	2938.30 I	6	5742.55 I
2	11660.25 II	40	1533.17 II	20	2950.4 II	12	5818.3 II
100	12095.36 II	30	1536.77 II	12	2963.4 II	20	5860.2 II

Intensity	Wavelength/Å		Intensity	Wavelength/Å		Intensity	Wavelength/Å		Intensity	Wavelength/Å	
20	5973.0	II	160	677.14	III	40	4242.98	III	7500	1232.43	I
15	6059.1	II	40	693.95	III	70	4243.61	III	1200	1243.90	I
15	6128.0	II	40	731.36	II	110	4472.10	III	1500	1251.66	I
6	6134.82	I	40	731.44	II	110	4472.85	II	1000	1255.80	I
3	6475.73	I		749.74	V	220	4487.05	III	1500	1259.20	I
3	6476.24	I	40	758.48	III	360	4497.73	III	1200	1261.66	I
15	6497.7	II	70	758.67	III	70	4784.21	II	1200	1266.20	I
10	6577.2	II	110	882.54	II	110	4940.38	II	1000	1279.48	I
40 h	6600.2	II	110	882.68	II	110	6080.44	II	1000	1286.26	I
50 h	6808.6	II	40	984.67	II	70	6285.47	II	3000	1309.91	I
4 h	6991.12	I	110	1081.88	II	70	7030.20	II	3000	1316.74	I
12	7033.	II	110	1082.07	II	40	7031.90	II	1000	1317.37	I
2	7036.15	I	70	1112.2	IV	110	7835.25	III	2000	1317.70	I
10 h	7381.	II	450	1168.9	IV	70	7841.41	III	12000	1384.60	I
2	7502.33	I	70	1170.9	IV	20	8667.22	I	3000	1449.90	I
10 h	7637.	II	110	1230.16	II	70	8668.57	I	50000	1488.45	I
10	7750.	II	220	1362.46	II	800	11660.04	I	30000	1531.74	I
3	7838.70	I	70	1600.46	I	570	11662.47	I	25000	1540.65	I
2	7840.33	I	120	1600.73	I	125	15629.08	I	30000	1574.84	I
20	7965.	II	160	1623.58	II	200	16240.38	I	20000	1576.39	I
40	8008.	III	110	1623.77	II	250	16244.67	I	25000	1582.31	I
12 h	8050.	II	220	1624.02	II	235	18994.33	I	75000	1633.40	I
50	8070.	III	70	1624.16	II		**Bromine**		1000	2133.79	IV
15	8328.	II	160	1624.34	II		**Br Z = 35**		1000	2145.02	IV
15	8388.	II	100	1663.04	I	700	379.73	IV	1000	2257.21	IV
30	8532.	II	150	1666.87	I	700	400.37	IV	1000	2272.73	IV
2	8544.54	I	200	1667.29	I	800	482.11	V	1000	2307.40	IV
1	8579.74	I	150	1817.86	I	900	531.97	V	1000	2408.16	IV
25	8653.	II	200	1818.37	I	1000	545.43	IV	1000	2411.58	IV
2	8754.88	I	300	1825.91	I	1000	547.90	V	700	2491.14	IV
3	8761.54	I	300	1826.41	I	1000	559.76	IV	1000	2581.19	IV
25	8863.	II	110	1842.81	II	1000	569.19	IV	600	2661.40	IV
2	8907.81	I	20	1953.83	III	1000	576.59	IV	1000	2842.88	IV
2000 d	9657.04	I	550	2065.78	III	1000	585.10	IV	1100 h	2907.71	IV
40	9827.78	I	250	2066.38	I	1000	586.71	IV	500 h	2972.26	II
20	10104.5	I	250	2066.65	I	1000	597.51	IV	500	3041.18	IV
15	10138.8	I	100	2066.93	I	1000	600.09	IV	500	3074.42	III
20	10300.6	I	300	2067.19	I	1000	601.27	IV	500	3349.64	III
20	10536.19	I	450	2067.23	III	1000	607.03	IV	500	3380.56	IV
50	11072.44	I	160	2077.09	III	1000	617.85	IV	500	3540.16	III
1500 d	11710.37	I	500	2088.91	I	1000	619.87	IV	500	3562.43	III
40	11999.49	I	500	2089.57	I	1000	630.14	IV	1200	3815.65	I
200	12165.08	I	70	2220.30	II	1000	642.23	IV	1500	3992.36	I
200	12690.04	I	40	2234.09	III	1000	661.53	IV	1000	4223.89	II
100	12817.8	I	70	2234.59	III	1000	683.51	IV	2000	4365.14	I
200	14330.5	I	40	2323.03	II	1000	697.72	IV	1000	4365.60	II
50	16001.5	I	40	2328.67	II	1000	715.39	IV	1500	4425.14	I
60	22551.6	I	40	2393.20	II	1000	731.00	IV	10000	4441.74	I
	Boron		220	2395.05	II	1000	800.12	IV	10000	4472.61	I
	B Z = 5		40	2459.69	II	700	812.95	V	20000	4477.72	I
	41.00	V	40	2459.90	II	1000	813.66	IV	1000	4490.42	I
30	48.59	V	1000	2496.77	I	1000	850.81	V	3000	4513.44	I
10	52.68	IV	1000	2497.73	I	1000	889.23	II	15000	4525.59	I
30	60.31	IV	70	2524.7	IV	1000	948.97	II	3000	4575.74	I
	194.37	V	160	2530.3	IV	1000	1015.54	II	2500	4614.58	I
	262.37	V	450	2821.68	IV	1000	1049.00	II	2500	4752.28	I
160	344.0	IV	70	2824.57	IV	1000	1069.15	V	4000	4780.31	I
450	385.0	IV	285	2825.85	IV	900	1112.13	V	1600	4785.19	I
40	411.80	III	160	2918.08	II	1000	1143.56	V	4000	4979.76	I
285	418.7	IV	110	3032.26	II	1000	1189.28	I	1200	5395.48	I
20	510.77	III	70	3179.33	II	1000	1189.50	I	1200	5466.22	I
40	510.85	III	110	3323.18	II	1000	1210.73	I	1800	5852.08	I
	512.53	V	110	3323.60	II	1000	1221.13	I	1600	5940.48	I
150	518.24	III	450	3451.29	II	1000	1223.24	I	2400	6122.14	I
75	518.27	III	285	4121.93	II	1200	1224.41	I	40000	6148.60	I
110	677.00	III	110	4194.79	II	1200	1226.90	I	2000	6177.39	I

Intensity	Wavelength/Å		Intensity	Wavelength/Å		Intensity	Wavelength/Å		Intensity	Wavelength/Å	
1500	6335.48	I	9000	8964.00	I	60	567.01	IV	50	2628.979	I
60000	6350.73	I	30000	9166.06	I	150	1118.16	IV	40	2632.190	I
2500	6410.32	I	15000	9173.63	I	100	1164.65	IV	75	2639.420	I
1800	6483.56	I	20000	9178.16	I	100	1183.40	IV	40	2659.23	II
1000	6514.62	I	40000	9265.42	I	100	1256.00	II	50 h	2660.325	I
20000	6544.57	I	15000	9320.86	I	150	1296.43	II	25	2668.20	II
1500	6548.09	I	6000	9793.48	I	100	1326.50	II	50	2672.62	II
50000 c	6559.80	I	10000	9896.40	I	60	1370.48	IV	100	2677.540	I
1000	6571.31	I	3000	10140.08	I	150	1370.91	II	25	2677.748	I
1800	6579.14	I	6000	10237.74	I	60	1418.89	IV	50	2707.00	II
20000	6582.17	I	1000	10299.62	I	200	1514.26	II	75	2712.505	I
1500	6620.47	I	1500	10377.65	I	50	1545.17	III	50	2733.820	I
50000 c	6631.62	I	30000	10457.96	I	200	1571.58	II	1000	2748.54	II
20000	6682.28	I	1000	10742.14	I	100	1668.60	II	100 h	2763.894	I
10000	6692.13	I	3000	10755.92	I	50	1702.47	II	50 h	2764.230	I
8000	6728.28	I	1700	13217.17	I	40	1707.16	III	50	2774.958	I
2000	6760.06	I	1800	14354.57	I	50	1724.41	II	30	2823.19	II
2000	6779.48	I	1250	14888.70	I	40	1747.67	III	200	2836.900	I
2200	6786.74	I	1800	16731.19	I	40	1773.06	III	25	2856.46	II
6500	6790.04	I	1200	18568.31	I	100	1785.84	II	100	2868.180	I
1600 c	6791.48	I	3500	19733.62	I	75	1793.40	III	200 r	2880.767	I
1800	6861.15	I	1000	20281.73	I	40	1823.41	III	50 r	2881.224	I
10000	7005.19	I	1000	20624.67	I	100	1827.70	II	200	2914.67	II
2000	7260.45	I	1200	21787.24	I	50	1844.66	III	50	2927.87	II
10000	7348.51	I	4000	22865.65	I	40	1851.13	III	200	2929.27	II
40000	7512.96	I	1000	23513.15	I	40	1855.85	III	1000 r	2980.620	I
1600	7591.61	I	500	28346.50	I	200	1856.67	III	200 r	2981.362	I
1800	7595.07	I	500	30380.85	I	150	1874.08	III	50	2981.845	I
2000	7616.41	I	600	31630.13	I	300	1922.23	II	50	3030.60	II
30000	7803.02	I	150	38345.75	I	100	1943.54	II	150	3080.822	I
1200	7827.23	I	120	39964.36	I	40	1965.54	II	25	3081.48	II
2500 s	7881.45	I		**Cadmium**		30	1986.89	II	30	3082.593	I
2500	7881.57	I		**Cd Z = 48**		200	1995.43	II	100	3092.34	II
2500	7925.81	I	50	427.01	IV	100	2007.49	II	200	3133.167	I
30000 c	7938.68	I	50	447.85	IV	50	2032.45	II	50	3146.79	II
3000	7947.94	I	60	480.90	IV	75	2036.23	II	150	3250.33	II
3000	7950.18	I	70	493.00	IV	40	2039.83	III	300	3252.524	I
8000	7978.44	I	70	495.13	IV	50	2045.61	III	300	3261.055	I
10000	7978.57	I	70	498.14	IV	75	2087.91	III	50	3343.21	II
30000	7989.94	I	70	498.53	IV	150	2096.00	II	50	3385.49	II
2000	8026.35	I	80	504.09	IV	50	2111.60	III	30	3388.88	II
2500	8026.54	I	70	504.20	IV	1000 r	2144.41	II	800	3403.652	I
30000	8131.52	I	70	504.50	IV	50	2155.06	II	50	3417.49	II
1000 c	8152.65	I	80	506.31	IV	100	2187.79	II	50	3442.42	II
10000	8153.75	I	60	508.01	IV	1000	2194.56	II	100	3464.43	II
25000	8154.00	I	50	508.95	IV	1000	2265.02	II	1000	3466.200	I
5000	8246.86	I	70	509.55	IV	1500 r	2288.022	I	800	3467.655	I
15000	8264.96	I	70	511.40	IV	1000	2312.77	II	25	3483.08	II
75000 c	8272.44	I	80	513.00	IV	200	2321.07	II	150	3495.44	II
20000	8334.70	I	70	514.50	IV	40	2376.82	II	25	3499.952	I
10000	8343.70	I	60	519.42	IV	50	2418.69	II	100	3524.11	II
1200	8384.04	I	80	524.41	IV	50	2469.73	II	100	3535.69	II
40000	8446.55	I	70	524.47	IV	40	2487.93	II	1000	3610.508	I
4000	8477.45	I	70	525.10	IV	40	2495.58	II	800	3612.873	I
1500	8513.38	I	60	525.19	IV	50	2509.11	II	60	3614.453	I
1000	8557.73	I	70	527.07	IV	30	2516.22	II	20	3649.558	I
1000	8566.28	I	80	531.09	IV	25 h	2525.196	I	10	3981.926	I
20000	8638.66	I	80	531.51	IV	50	2544.613	I	100	4029.12	II
4000	8698.53	I	70	534.29	IV	50	2551.98	II	200	4134.77	II
10000 c	8793.47	I	70	536.77	IV	25	2553.465	I	50	4141.49	II
15000	8819.96	I	60	540.90	IV	3	2565.789	I	100	4285.08	II
25000	8825.22	I	70	541.74	IV	500	2572.93	II	8	4306.672	I
4000	8888.98	I	80	542.60	IV	50	2580.106	I	100	4412.41	II
30000	8897.62	I	80	546.55	IV	30	2592.026	I	3	4412.989	I
6000	8932.40	I	60	553.06	IV	25 h	2602.048	I	1000	4415.63	II
1800	8949.39	I	80	554.05	IV				30	4440.45	II

Intensity	Wavelength/Å	Intensity	Wavelength/Å	Intensity	Wavelength/Å	Intensity	Wavelength/Å
8	4662.352 I	400	637.93 V	20	4456.61 I	20	8020.50 II
200	4678.149 I	300	643.12 V	20	4472.04 II	70	8133.05 II
30	4744.69 II	400	646.57 V	20	4489.18 II	100	8201.72 II
300	4799.912 I	750	656.00 IV	19	4499.88 III	110	8248.80 II
50	4881.72 II	300	656.76 V	23	4526.94 I	70	8254.73 II
50	5025.50 II	500	669.70 IV	22	4578.55 I	130	8498.02 II
1000 h	5085.822 I	24	1341.89 II	23	4581.40 I	170	8542.09 II
6	5154.660 I	12	1342.54 II	23	4581.47 I	160	8662.14 II
100	5268.01 II	20	1433.75 II	24	4585.87 I	100	8912.07 II
100	5271.60 II	20	1545.29 III	24	4585.96 I	110	8927.36 II
1000	5337.48 II	60	1649.86 II	20	4685.27 I	110	9213.90 II
1000	5378.13 II	20	1807.34 II	30	4716.74 II	90	9312.00 II
200	5381.89 II	40	1814.50 II	40	4721.03 II	100	9319.56 II
40	5843.30 II	40	1838.01 II	40	4799.97 II	110	9320.65 II
50	5880.22 II	60	1840.06 II	25	4878.13 I	25	9416.97 I
300	6099.142 I	20	1843.09 II	70	5001.48 II	100	9567.97 II
100	6111.49 I	40	1850.69 II	80	5019.97 II	110	9599.24 II
100	6325.166 I	17	2123.03 III	40	5021.14 II	80	9601.82 II
30	6330.013 I	16	2152.43 III	23	5041.62 I	80	9854.74 II
400	6354.72 II	16	2687.76 III	25	5188.85 I	110	9890.63 II
500	6359.98 II	19	2881.78 III	22	5261.71 I	90	9931.39 II
2000	6438.470 I	21	2899.79 III	23	5262.24 I	100	10223.04 II
400	6464.94 II	19	2924.33 III	22	5264.24 I	20	10343.81 I
25	6567.65 II	20	2988.63 III	24	5265.56 I	20	11838.99 II
500	6725.78 II	10	3006.86 I	25	5270.27 I	25	12816.04 I
100	6759.19 II	15	3028.59 III	60	5285.27 II	24	12823.86 I
30	6778.116 I	3	3055.32 I	70	5307.22 II	25	12909.10 I
50	7237.01 II	19	3119.67 III	50	5339.19 II	30	13033.57 I
100	7284.38 II	170	3158.87 II	27	5349.47 I	21	13086.44 I
1000	7345.670 I	180	3179.33 II	23	5512.98 I	24	13134.95 I
50	8066.99 II	150	3181.28 II	25	5581.97 I	20	16150.77 I
5	8200.309 I	20	3316.51 II	27	5588.76 I	22	16157.36 I
20	9289. I	12	3361.92 I	24	5590.12 I	21	16197.04 I
15	11652. I	19	3372.67 III	26	5594.47 I	20	18925.47 I
35	14487. I	20	3461.87 II	25	5598.49 I	24	18970.14 I
80	15708. I	13	3487.60 I	24	5601.29 I	30	19046.14 I
55 d	19120. I	18	3537.77 III	24	5602.85 I	48	19309.20 I
25	24371. I	20	3644.41 I	30	5857.45 I	49	19452.99 I
35	25448. I	30	3683.70 II	27	6102.72 I	47	19505.72 I
	Calcium	40	3694.11 II	29	6122.22 I	50	19776.79 I
	Ca Z = 20	170	3706.03 II	22	6161.29 I	35	19853.10 I
250	190.46 V	180	3736.90 II	30	6162.17 I	34	19862.22 I
250	196.97 V	20	3755.67 II	22	6163.76 I	23	19917.19 I
300	199.55 V	30	3758.39 II	24	6166.44 I	24	19933.70 I
250	200.51 V	230	3933.66 II	26	6169.06 I	25	22624.93 I
265	257.98 V	220	3968.47 II	28	6169.56 I	30	22651.23 I
400	267.77 V	50	4097.10 II	35	6439.07 I		**Carbon**
300	270.31 V	60	4109.82 II	30	6449.81 I		**C Z = 6**
400	280.99 V	30	4110.28 II	22	6455.60 I	110	34.973 V
300	284.98 V	40	4206.18 II	80	6456.87 II	450	40.268 V
450 c	286.96 V	50	4220.07 II	34	6462.57 I	110	227.19 V
500	322.17 V	50	4226.73 I	29	6471.66 I	250	244.91 IV
300	323.22 V	24	4283.01 I	32	6493.78 I	160	248.66 V
300	330.94 V	22	4289.36 I	28	6499.65 I	160	248.74 V
300	334.55 V	22	4298.99 I	23	6572.78 I	200	289.14 IV
250 c	342.45 IV	25	4302.53 I	30	6717.69 I	250	289.23 IV
250	343.93 IV	20	4302.81 III	33	7148.15 I	570	312.42 IV
450	352.92 V	23	4307.74 I	31	7202.19 I	500	312.46 IV
250	377.18 V	22	4318.65 I	33	7326.15 I	250	371.69 III
200	387.08 V	20	4355.08 I	30	7575.81 II	250	371.75 III
750	425.00 V	19	4399.59 III	60	7581.11 II	150	371.78 III
600	434.57 IV	25	4425.44 I	80	7601.30 II	650	384.03 IV
250	437.77 IV	26	4434.96 I	20	7602.32 II	700	384.18 IV
750	443.82 IV	25	4435.69 I	40	7820.78 II	500	386.203 III
500	450.57 IV	30	4454.78 I	60	7843.38 II	400	419.52 IV
500	558.60 V	28	4455.89 I	20	8017.50 II	500	419.71 IV

Intensity	Wavelength/Å		Intensity	Wavelength/Å		Intensity	Wavelength/Å		Intensity	Wavelength/Å	
200	450.734	III	900	1550.774	IV	300	5380.34	I	12	16559.66	I
400	459.46	III	150	1560.310	I	250	5648.07	II	50	16890.38	I
500	459.52	III	400	1560.683	I	350	5662.47	II	10	17338.56	I
570	459.63	III	400	1560.708	I	450	5695.92	III	11	17448.60	I
250	511.522	III	100	1561.341	I	250	5801.33	IV	13	18139.80	I
250	535.288	III	400	1561.438	I	200	5811.98	IV	23	19721.99	I
300	538.080	III	150	1656.266	I	150	5826.42	III		Cerium	
350	538.149	III	120	1656.928	I	570	5889.77	II		Ce Z = 58	
400	538.312	III	300	1657.008	I	350	5891.59	II	300	399.36	V
350	574.281	III	120	1657.380	I	200	6001.13	I	200	482.96	V
9	595.022	II	120	1657.907	I	250	6006.03	I	40	741.79	IV
30	687.053	II	150	1658.122	I	110	6007.18	I	30	754.60	IV
50	687.345	II	500	1751.823	I	150	6010.68	I	75	1332.16	IV
10	858.092	II	1000	1930.905	I	300	6013.22	I	75	1372.72	IV
20	858.559	II	250	2162.94	III	250	6014.84	I	100	2000.42	IV
30	903.624	II	40	2270.91	V	800	6578.05	II	100	2009.94	IV
60	903.962	II	5	2277.25	V	570	6582.88	II	10000	2318.64	III
150	904.142	II	20	2277.92	V	200	6587.61	I	10000	2372.34	III
30	904.480	II	800	2296.87	III	150	6744.38	III	10000	2380.12	III
800	977.03	III	800	2478.56	I	250	6783.90	II	10000	2431.45	III
9	1009.86	II	250	2509.12	II	150 h	7037.25	III	15000	2439.80	III
10	1010.08	II	350	2512.06	II	250	7113.18	I	10000	2454.32	III
10	1010.37	II	200 l	2524.41	IV	250	7115.19	I	10000	2469.95	III
80	1036.337	II	300 s	2529.98	IV	250	7115.63	II	10000	2483.82	III
150	1037.018	II	250 h	2574.83	II	200	7116.99	I	10000	2497.50	III
150	1157.910	I	150	2697.75	III	350	7119.90	II	20000	2531.99	III
150	1158.019	I	110 l	2724.85	III	800	7231.32	II	10000	2603.59	III
150	1158.035	I	150 l	2725.30	III	1000	7236.42	II	340	2651.01	II
370	1174.93	III	150 l	2725.90	III	150	7612.65	III	270	2830.90	II
350	1175.26	III	350 l	2741.28	II	90 w	7726.2	IV	250	2874.14	II
330	1175.59	III	250	2746.49	II	200	7860.89	I	10000	2923.81	III
500	1175.71	III	1000	2836.71	II	200	8058.62	I	10000	2931.54	III
350	1175.99	III	800	2837.60	II	300 h	8196.48	III	400	2976.91	II
370	1176.37	III	200	2982.11	III	150	8332.99	III	10000	3022.75	III
150	1188.992	I	800 h	2992.62	II	520	8335.15	I	50000	3031.58	III
150	1189.447	I	350	3876.19	II	300	8500.32	III	95000	3055.59	III
200	1189.631	I	350	3876.41	II	250	9061.43	I	20000	3056.56	III
300	1193.009	I	350	3876.66	II	200	9062.47	I	40000	3057.23	III
300	1193.031	I	570	3918.98	II	200	9078.28	I	20000	3057.58	III
300	1193.240	I	800	3920.69	II	250	9088.51	I	680	3063.01	II
300	1193.264	I	150	4056.06	III	450	9094.83	I	40000	3085.10	III
100	1193.393	I	200	4067.94	III	300	9111.80	I	20000	3106.98	III
150	1193.649	I	250	4068.91	III	800	9405.73	I	30000	3110.53	III
150	1193.679	I	250	4070.26	III	150	9603.03	I	30000	3121.56	III
100	1194.064	I	250	4074.52	II	250	9620.80	I	20000	3141.29	III
100	1194.488	I	350 l	4075.85	II	300	9658.44	I	20000	3143.96	III
100	1261.552	I	150	4162.86	III	200	10683.08	I	20000	3147.06	III
250	1277.245	I	250 h	4186.90	III	300	10691.25	I	710	3194.83	II
250	1277.282	I	800	4267.00	II	12	11619.29	I	990	3201.71	II
300	1277.513	I	1000	4267.26	II	23	11628.83	I	710	3218.94	II
300	1277.550	I	200	4325.56	III	13	11658.85	I	880	3221.17	II
200	1280.333	I	600	4647.42	III	47	11659.68	I	710	3227.11	II
100	1311.363	I	520	4650.25	III	24	11669.63	I	20000	3228.57	III
9	1323.951	II	375	4651.47	III	85	11748.22	I	710	3234.16	II
120	1329.578	I	200 w	4658.30	IV	142	11753.32	I	990	3272.25	II
120	1329.600	I	200	4665.86	III	114	11754.76	I	20000	3353.29	III
150	1334.532	II	200	4771.75	I	11	11777.54	I	10000	3395.77	III
300	1335.708	II	200	4932.05	I	17	11892.91	I	30000	3427.36	III
100	1354.288	I	5	4943.88	V	30	11895.75	I	40000	3443.63	III
150	1355.84	I	5	4944.56	V	26	12614.10	I	30000	3454.39	III
120	1364.164	I	200	5052.17	I	20	13502.27	I	40000	3459.39	III
100	1459.032	I	350	5132.94	II	38	14399.65	I	60000	3470.92	III
200	1463.336	I	350	5133.28	II	16	14403.25	I	710	3485.05	II
120	1467.402	I	350	5143.49	II	61	14420.12	I	50000	3497.81	III
150	1481.764	I	570	5145.16	II	12	14429.03	I	60000	3504.64	III
1000	1548.202	IV	400	5151.09	II	13	14442.24	I	770	3539.08	II

Intensity	Wavelength/Å		Intensity	Wavelength/Å		Intensity	Wavelength/Å		Intensity	Wavelength/Å	
50000	3544.07	III	770	4227.75	II	35	6343.95	II	330	2076.43	III
1200	3560.80	II	980	4239.92	II	35	6371.11	II	540	2077.30	III
1000	3577.45	II	1100	4248.68	II	28	6386.84	I	410	2088.68	III
1800	3655.85	II	2000	4289.94	II	23	6393.02	II	210	2101.63	III
880	3660.64	II	1500	4296.67	II	35	6430.07	I	200	2141.47	III
880	3667.98	II	770	4300.33	II	23	6436.40	I	1000	2316.88	III
1000	3709.29	II	770	4306.72	II	35	6458.03	I	230	2325.95	III
1000	3709.93	II	980	4337.77	II	28	6467.39	I	390	2340.49	III
1400	3716.37	II	700	4349.79	II	35	6473.72	I	1600	2455.81	III
800	3728.42	II	910	4364.66	II	23	6513.59	II	1600	2477.57	III
860	3786.63	II	910	4382.17	II	45	6555.65	I	890	2485.45	III
2500	3801.52	II	700	4386.84	II	23	6579.10	I	410	2495.07	III
800	3803.09	II	1700	4391.66	II	22	6612.06	I	1400	2525.67	III
1000	3808.11	II	980	4418.78	II	30	6628.93	I	430	2573.05	III
1100	3838.54	II	770	4449.34	II	22	6652.72	II	16000	2596.86	III
860	3848.59	II	2400	4460.21	II	26	6700.66	I	390	2610.12	III
860	3853.15	II	1400	4471.24	II	35	6704.27	I	6200	2630.51	III
1200	3854.18	II	700	4479.36	II	30	6774.28	II	370	2700.32	III
1200	3854.31	II	700	4483.90	II	35	6775.59	I	710	2701.20	III
1100	3878.36	II	840	4486.91	II	30	6924.81	I	390	2776.44	III
1500	3882.45	II	770	4523.08	II	30	6986.02	I	270	2810.87	III
1000	3889.98	II	840	4527.35	II	35	7061.75	II	630	2845.70	III
770	3907.29	II	840	4528.47	II	35	7086.35	II	3100	2859.32	III
980	3912.44	II	840	4539.75	II	22	7238.36	II	200	2893.85	III
770	3918.28	II	2100	4562.36	II	25	7252.75	I	180	2921.13	III
770	3931.09	II	1100	4572.28	II	25	7329.91	I	3200	2976.86	III
770	3940.34	II	840	4593.93	II	25	7397.77	I	210	3001.28	III
2000	3942.15	II	1700	4628.16	II	25	7616.11	II	1700	3066.59	III
2700	3942.75	II	310	4737.28	II	25	7689.17	II	1100 c	3149.36	III
770	3943.89	II	470	5079.68	II	22	7844.94	II	1400	3152.36	III
3100	3952.54	II	280	5159.69	I	22	7857.54	II	8400	3268.32	III
980	3956.28	II	280	5161.48	I	30	8025.56	II	1300	3315.51	III
770	3960.91	II	370	5187.46	II	25	8772.14	II	550	3340.60	III
770	3967.05	II	260	5223.46	I	30	8891.20	II	430	3344.02	III
770	3978.65	II	260	5245.92	I		**Cesium**		1200	3349.46	III
770	3984.68	II	340	5274.23	II		**Cs Z = 55**		400	3463.45	III
700	3992.39	II	450	5353.53	II	10000	614.01	III	580	3476.83	III
910	3993.82	II	300	5393.40	II	2000	638.17	III	480	3559.82	III
2800	3999.24	II	280	5409.23	II	2500	666.25	III	7200	3597.45	III
910	4003.77	II	260	5512.08	II	5000	691.60	III	1300	3608.31	III
2700	4012.39	II	300	5696.99	I	3500	703.89	III	2300	3618.19	III
910	4014.90	II	370	5699.23	I	15000	718.14	II	300 c	3641.34	III
840	4024.49	II	240	5719.03	I	20000	721.79	III	520	3651.08	III
840	4028.41	II	230	5940.86	I	20000	722.20	III	4800	3661.40	III
840	4031.34	II	55	6001.90	I	5000	731.56	III	640	3699.50	III
2100	4040.76	II	55	6005.86	I	12000	740.29	III	430	3837.46	III
910	4042.58	II	55	6006.82	I	15000	808.76	II	2100 c	3876.15	I
700	4053.51	II	75	6013.42	I	15000	813.84	II	2900	3888.37	III
1100	4071.81	II	110	6024.20	I	7500	830.39	III	600 c	3888.61	I
1800	4073.48	II	10000	6032.54	III	35000	901.27	II	2700	3925.60	III
1500	4075.71	II	110	6043.39	II	15000	920.35	III	680 c	4001.70	III
1500	4075.85	II	55	6047.40	I	40000	926.66	II	3100	4006.55	III
910	4083.23	II	10000	6060.91	III	25000 c	1054.79	III	420	4006.78	III
770	4118.14	II	45	6098.34	II	17 c	1673.99	III	520	4043.42	III
980	4123.87	II	45	6123.67	I	12	1705.25	III	14000	4264.70	II
980	4127.37	II	35	6143.36	II	10	1801.83	III	18000 w	4277.13	II
2700	4133.80	II	35	6186.17	I	20 c	1822.40	III	370	4403.86	III
2000	4137.65	II	35	6208.98	I	11	1823.93	III	1200	4410.22	III
770	4142.40	II	35	6228.94	I	12	1824.70	III	940	4425.68	III
980	4149.94	II	23	6232.45	II	12	1841.80	III	530	4471.48	III
1400	4151.97	II	28	6237.45	I	25	1915.50	III	12000	4501.55	II
1300	4165.61	II	45	6272.05	II	25 c	1923.29	III	1200	4506.72	III
3500	4186.60	II	35	6295.58	I	12	1961.33	III	590	4522.86	III
840	4198.72	II	28	6299.51	II	17	1996.56	III	20000	4526.74	II
910	4202.94	II	23	6300.21	I	710	2035.11	III	1000 c	4555.28	I
1500	4222.60	II	35	6310.01	I	120	2056.43	III	460 c	4593.17	I

Intensity	Wavelength/Å		Intensity	Wavelength/Å		Intensity	Wavelength/Å		Intensity	Wavelength/Å	
99900	4603.79	II	18000	9172.32	I	1000	659.811	II	25000	1389.693	I
420 h	4620.61	III	5200	9208.53	I	1300	661.841	II	20000	1389.957	I
210	4665.52	III	19000	10024.36	I	2000	663.074	II	12000	1396.527	I
25000	4830.19	II	4800	10123.41	I	1500	682.053	II	500	1441.470	II
140	4851.59	III	26000	10123.60	I	1500	687.656	II	500	1528.569	II
19000	4870.04	II	2900	13424.31	I	1500	693.594	II	500	1542.942	II
37000	4952.85	II	38000 c	13588.29	I	2000	725.271	II	500	1558.144	II
370	5035.72	III	8400	13602.56	I	2500	728.951	II	500	1565.050	II
27000	5043.80	II	5700	13758.81	I	2000	777.562	II	600	1822.50	III
75000	5227.04	II	55000 c	14694.91	I	5000	787.580	II	500	1828.40	III
29000	5249.38	II	820	16535.63	I	5000	788.740	II	500	1857.488	II
11000	5274.05	II	1500	17012.32	I	5000	793.342	II	500	1901.61	III
10000 c	5349.13	II	760	20138.47	I	500	834.84	IV	500	1983.61	III
22000	5370.99	II	880	22811.86	I	500	834.97	IV	450 h	1997.370	II
230	5380.79	III	1100	23037.98	I	6000	839.297	II	450	2032.116	II
60 c	5465.94	I	3900	23344.47	I	8000	839.599	II	350 h	2088.583	II
37	5502.88	I	4400	24251.21	I	600	840.93	IV	350 h	2091.458	II
39000	5563.02	II	850	24374.96	I	5000	851.691	II	700	2253.07	III
100	5635.21	I	890 d	25763.51	I	2000	888.026	II	500	2268.95	III
210 c	5664.02	I	500	25764.73	I	2000	893.549	II	500	2278.34	III
27	5745.72	I	680 c	29310.06	I	2000	961.499	II	700	2283.93	III
24000	5831.14	II	2800	30103.27	I	500	973.21	IV	600	2323.50	III
59 c	5838.83	I	610 c	30953.06	I	600	977.56	IV	500	2336.45	III
300	5845.14	I	1100	34900.13	I	40	978.284	I	600	2340.64	III
51000	5925.63	II	190	36131.00	I	700	984.95	IV	600	2359.67	III
140	5950.14	III	2 c	39177.28	I	25	998.372	I	600	2370.37	III
110	5979.97	III	2 d	39421.25	I	25	998.432	I	700	2416.42	III
640 c	6010.49	I	1	39424.11	I	75	1002.346	I	600	2447.14	III
86	6034.09	I	**Chlorine**			500	1005.28	III	600	2448.58	III
150	6043.99	III	**Cl Z = 17**			600	1008.78	III	500	2486.91	III
870	6079.86	III	500	392.43	V	150	1013.664	I	500	2532.48	III
9800	6128.61	II	800	486.17	IV	700	1015.02	III	600	2580.67	III
330	6150.42	III	800	534.73	IV	90	1025.553	I	500	2603.59	III
1000	6213.10	I	700	535.67	IV	6000	1063.831	II	500	2632.67	III
170	6217.60	I	600	536.15	IV	3000	1067.945	II	500	2633.18	III
450	6242.96	III	900	537.61	IV	9000	1071.036	II	600	2665.54	III
320 c	6354.55	I	500	538.03	V	6000	1071.767	II	700	2710.37	III
510	6456.33	III	600	538.12	IV	5000	1075.230	II	500	2724.03	IV
8300	6495.53	II	800	542.23	V	5000	1079.080	II	500	2751.23	IV
10000 w	6536.44	II	600	542.30	V	200	1084.667	I	700	2782.47	IV
490	6586.51	I	1000	545.11	V	200	1085.171	I	600	2965.56	III
97	6628.66	II	600	546.33	V	250	1085.304	I	500	3063.13	IV
8800	6646.57	II	1000	547.63	V	400	1088.06	I	600	3076.68	IV
3300 c	6723.28	I	500	549.22	IV	350	1090.271	I	600	3104.46	III
9600	6724.47	II	700	552.02	IV	250	1090.982	I	800	3139.34	III
400	6753.12	III	600	553.30	IV	250	1092.437	I	900	3191.45	III
200	6824.65	I	700	554.62	IV	400	1094.769	I	700	3289.80	III
300	6870.45	I	600	556.23	III	350	1095.148	I	700	3320.57	III
37000	6955.50	II	700	556.61	III	350	1095.662	I	800	3329.06	III
4800	6973.30	I	700	557.12	III	400	1095.797	I	900	3340.42	III
16000	6979.67	II	350	559.305	II	250	1096.810	I	800	3392.89	III
980	6983.49	I	700	561.53	III	300	1097.369	I	800	3393.45	III
13000 w	7149.54	II	700	561.68	III	200	1098.068	I	900	3530.03	III
1900 c	7219.60	III	700	561.74	III	200	1099.523	I	800	3560.68	III
790	7228.53	I	400	571.904	II	500	1107.528	I	900	3602.10	III
130	7279.90	I	800	574.406	II	800	1139.214	II	800	3612.85	III
1100	7279.96	I	500	601.50	IV	800	1167.148	I	700	3622.69	III
2600 c	7608.90	I	500	604.59	IV	3000	1179.293	I	700	3656.95	III
3300	7943.88	I	500	606.35	III	1200	1188.774	I	700	3670.28	III
22000	7997.44	II	700	618.057	II	900	1201.353	I	700	3682.05	III
3500	8015.73	I	600	619.982	II	3000	1335.726	I	600	3705.45	III
510	8078.94	I	800	620.298	II	10000	1347.240	I	600	3707.34	III
4500	8079.04	I	700	626.735	II	5000	1351.657	I	800	3720.45	III
59000 c	8521.13	I	800	635.881	II	12000	1363.447	I	800	3748.81	III
15000 c	8761.41	I	1000	636.626	II	2500	1373.116	I	500	3779.35	III
61000 c	8943.47	I	1000	650.894	II	20000	1379.528	I	10000	3850.99	II

Intensity	Wavelength/Å		Intensity	Wavelength/Å		Intensity	Wavelength/Å		Intensity	Wavelength/Å	
25000	3860.83	II	600	7980.60	I	259	16198.5	I	280	2668.71	II
500	3925.87	III	2900	7997.85	I	717	19755.3	I	350	2671.81	II
700	3991.50	III	2200	8015.61	I	100	24470.0	I	280	2672.83	II
600	4018.50	III	1100	8023.33	I		39716.0	I	1800	2677.16	II
600	4059.07	III	400	8051.07	I		40085.5	I	320	2678.79	II
500	4104.23	III	1700	8084.51	I		40089.5	I	230	2687.09	II
500	4106.83	III	2200	8085.56	I		40532.2	I	280	2691.04	II
10000 h	4132.50	II	3000	8086.67	I		**Chromium**		180	2698.41	II
500	4608.21	III	1300	8087.73	I		**Cr Z = 24**		180	2698.69	II
40	4623.938	I	2500	8194.42	I	100	438.62	V	110	2701.99	I
50	4654.040	I	2200	8199.13	I	100	464.02	V	140	2712.31	II
80	4661.208	I	2200	8200.21	I	100	620.66	IV	170	2722.75	II
45	4691.523	I	800	8203.78	I	100	629.26	IV	420 h	2726.51	I
40	4721.255	I	18000	8212.04	I	80	630.30	IV	280 h	2731.91	I
45	4740.729	I	3000	8220.45	I	100	666.55	IV	170 h	2736.47	I
13000	4781.32	II	20000	8221.74	I	100	693.92	IV	250	2743.64	II
99000	4794.55	II	18000	8333.31	I	60	1030.47	III	110 h	2748.29	I
29000	4810.06	II	99900	8375.94	I	100	1033.69	III	330	2748.98	II
16000	4819.47	II	400	8406.199	I	100	1036.03	III	390	2750.73	II
81000	4896.77	II	15000	8428.25	I	80	1055.89	IV	280	2751.87	II
47000	4904.78	II	2200	8467.34	I	80	1068.41	III	110 h	2752.88	I
26000	4917.73	II	2200	8550.44	I	100	1116.48	V	150	2757.10	I
10000	4995.48	II	20000	8575.24	I	150	1121.07	V	350	2757.72	II
26000	5078.26	II	750	8578.02	I	150	1127.63	V	750	2762.59	II
30	5099.789	I	75000	8585.97	I	100	1263.50	V	750	2766.54	II
56000	5217.94	II	450	8628.54	I	100	1417.42	IV	250 h	2769.92	I
23000	5221.36	II	300	8641.71	I	150	1465.86	V	610	2780.70	I
15000	5392.12	II	3500	8686.26	I	150	1497.97	V	180	2822.37	II
99000	5423.23	II	2200	8912.92	I	170	1519.03	V	180	2830.47	II
10000	5423.51	II	3000	8948.06	I	220	1579.70	V	2500	2835.63	II
19000	5443.37	II	2000	9038.982	I	170	1591.72	V	110	2840.02	II
10000	5444.21	II	2500	9045.43	I	150	1603.19	V	1700	2843.25	II
40	5532.162	I	1000	9069.656	I	120	1672.66	IV	1200	2849.84	II
50 d	5796.305	I	2000	9073.17	I	120	1758.51	IV	120	2851.36	II
45	5799.914	I	7500	9121.15	I	140	1802.72	IV	880	2855.68	II
30	5856.742	I	3000	9191.731	I	130	1812.41	IV	610	2858.91	II
50	6019.812	I	500	9197.596	I	200	1837.44	V	440	2860.93	II
200	6140.245	I	4000	9288.86	I	140	1873.89	IV	790	2862.57	II
160	6194.757	I	1500	9393.862	I	140	1967.18	IV	750	2865.11	II
150	6434.833	I	3500	9452.10	I	120	1972.07	IV	610	2866.74	II
300	6932.903	I	500	9486.964	I	19000	2055.52	II	480	2867.65	II
300	6981.886	I	1000	9584.801	I	14000	2061.49	II	210	2870.44	II
600	7086.814	I	3500	9592.22	I	8900	2065.42	II	110	2871.63	I
7500	7256.62	I	250	9632.509	I	200	2226.72	III	160	2873.48	II
5000	7414.11	I	1000	9702.439	I	200	2235.91	III	320	2875.99	II
550	7462.370	I	250	9744.426	I	150	2237.59	III	230	2876.24	II
550	7489.47	I	200	9807.057	I	150	2244.10	III	180	2877.98	II
700	7492.118	I	400	9875.970	I	150	2284.44	III	120	2879.27	I
11000	7547.072	I	331	10392.549	I	150	2324.88	III	170	2887.00	I
2300	7672.42	I	300	11123.05	I	130	2383.33	I	700	2889.29	I
450	7702.828	I	269	11409.69	I	140	2408.62	I	370	2893.25	I
7000	7717.581	I	1000	11436.33	I	170	2496.31	I	190	2894.17	I
10000	7744.97	I	350	13243.8	I	110	2502.53	I	210	2896.75	I
2200	7769.16	I	310	13296.0	I	190	2504.31	I	180	2905.49	I
650	7771.09	I	550	13346.8	I	110	2516.92	I	260	2909.05	I
2200	7821.36	I	525	13821.7	I	390	2519.52	I	260	2910.90	I
1700	7830.75	I	294	14931.7	I	190	2527.12	I	250	2911.14	I
3000	7878.22	I	269	15108.0	I	160	2549.54	I	480	2967.64	I
2300	7899.31	I	381	15465.1	I	130	2560.69	I	480	2971.11	I
1800	7915.08	I	1094	15520.3	I	150	2571.74	I	210	2971.91	II
3000	7924.645	I	1487	15730.1	I	100	2577.65	I	480	2975.48	I
2100	7933.89	I	2780	15869.7	I	380	2591.85	I	190	2979.74	II
1700	7935.012	I	277	15883.3	I	250	2653.59	II	350	2980.79	I
650	7952.52	I	342	15928.9	I	250	2658.59	II	110	2985.32	II
1500	7974.72	I	735	15960.0	I	320	2663.42	II	480	2985.85	I
1300	7976.97	I	283	15970.5	I	440	2666.02	II	1500	2986.00	I

Intensity	Wavelength/Å		Intensity	Wavelength/Å		Intensity	Wavelength/Å		Intensity	Wavelength/Å	
2100	2986.47	I	130	3573.64	I	190	3984.34	I	60	5013.32	I
660	2988.65	I	330 h	3574.80	I	160	3989.99	I	70	5166.23	I
160	2989.19	II	19000	3578.69	I	960	3991.12	I	70	5184.59	I
480	2991.89	I	160 h	3584.33	I	160	3991.67	I	70	5192.00	I
230	2994.07	I	130	3585.30	II	190	3992.84	I	85	5196.44	I
300	2995.10	I	17000	3593.49	I	160	4001.44	I	5300	5204.52	I
700	2996.58	I	350	3601.67	I	120	4012.47	II	8400	5206.04	I
210	2998.79	I	13000	3605.33	I	120	4026.17	I	11000	5208.44	I
1100	3000.89	I	130	3632.84	I	190	4039.10	I	85	5224.94	I
750	3005.06	I	350	3636.59	I	160	4048.78	I	290	5247.56	I
140	3013.03	I	630	3639.80	I	120	4058.77	I	530	5264.15	I
710	3013.71	I	220	3641.83	I	140	4126.52	I	180	5265.72	I
710	3014.76	I	220	3649.00	I	120	4153.82	I	95 h	5275.17	I
1400	3014.92	I	170	3653.91	I	140	4163.62	I	70 h	5276.03	I
710	3015.19	I	220	3656.26	I	170	4174.80	I	340	5296.69	I
2800	3017.57	I	130	3663.21	I	170	4179.26	I	70 h	5297.36	I
430	3018.50	I	120	3685.55	I	110	4209.37	I	660	5298.27	I
240	3018.82	I	130	3686.80	I	20000	4254.35	I	85	5300.75	I
430	3020.67	I	130	3687.25	I	110	4263.14	I	340 h	5328.34	I
2800	3021.56	I	130	3730.81	I	16000	4274.80	I	70 h	5329.17	I
1100	3024.35	I	150	3732.03	I	10000	4289.72	I	780	5345.81	I
170	3029.16	I	480	3743.58	I	780	4337.57	I	380	5348.32	I
710	3030.24	I	570	3743.88	I	1100	4339.45	I	40	5400.61	I
140	3031.35	I	340	3749.00	I	380	4339.72	I	1400	5409.79	I
390	3034.19	I	230	3757.66	I	1900	4344.51	I	24	5628.64	I
550	3037.04	I	260	3768.24	I	380	4351.05	I	7	5642.36	I
550	3040.85	I	130	3791.38	I	2300	4351.77	I	24	5664.04	I
110	3050.14	II	130	3792.14	I	570	4359.63	I	24	5694.73	I
710	3053.88	I	120	3793.29	I	530	4371.28	I	40	5698.33	I
240	3118.65	II	130	3793.88	I	110	4374.16	I	24	5702.31	I
430	3120.37	II	140	3797.13	I	530	4384.98	I	24	5712.78	I
470	3124.94	II	200	3797.72	I	110	4458.54	I	24 h	5783.11	I
120	3128.70	II	530	3804.80	I	660	4496.86	I	30 h	5783.93	I
590	3132.06	II	110	3806.83	I	380	4526.47	I	24 h	5785.00	I
140	3136.68	II	110	3807.93	I	380	4530.74	I	19 h	5785.82	I
140	3147.23	II	180	3815.43	I	240	4535.72	I	60 h	5787.99	I
100	3155.15	I	180	3819.56	I	240	4540.50	I	180 h	5791.00	I
100	3163.76	I	130	3826.42	I	240	4540.72	I	35	6330.10	I
240	3180.70	II	130	3830.03	I	140	4544.62	I	22	6362.87	I
220	3197.08	II	380	3841.28	I	600	4545.96	I	19	6661.08	I
170	3209.18	II	190	3848.98	I	120	4565.51	I	21 h	6883.03	I
140	3217.40	II	140	3849.36	I	120	4571.68	I	27 h	6924.13	I
120	3245.54	I	290	3850.04	I	360	4580.06	I	30 h	6978.48	I
130	3251.84	I	140	3852.22	I	360	4591.39	I	85	7355.90	I
130	3257.82	I	190	3854.22	I	480	4600.75	I	130	7400.21	I
130	3339.80	II	110	3855.29	I	240	4613.37	I	150	7462.31	I
110	3342.59	II	140	3855.57	I	600	4616.14	I	40	8947.15	I
170	3358.50	II	260	3857.63	I	550	4626.19	I	19	8976.83	I
160	3360.30	II	660	3883.29	I	1600	4646.17	I		**Cobalt**	
430	3368.05	II	570	3885.22	I	570	4651.28	I		**Co Z = 27**	
140	3382.68	II	380	3886.79	I	840	4652.16	I	20	355.52	V
170	3403.32	II	260	3894.04	I	240 d	4698.46	I	18	355.88	V
360	3408.76	II	360	3902.92	I	190	4708.04	I	12	356.06	V
210	3421.21	II	960	3908.76	I	240	4718.43	I	66	609.16	IV
270	3422.74	II	120 hd	3911.82	I	120	4730.71	I	70	609.21	IV
140	3433.31	II	120	3915.84	I	140	4737.35	I	64	609.28	IV
270	3433.60	I	190	3916.24	I	340	4756.11	I	10	1018.36	V
160	3436.19	I	1900	3919.16	I	190	4789.32	I	10	1021.14	V
140	3441.44	I	600	3921.02	I	120	4801.03	I	15	1231.73	V
170	3445.62	I	600	3928.64	I	110	4829.38	I	50	1277.01	V
170	3447.43	I	410	3941.49	I	140	4870.80	I	80	1299.58	II
190	3453.33	I	1900	3963.69	I	130	4887.01	I	80	1306.95	II
130	3455.60	I	120	3969.06	I	260	4922.27	I	50	1345.67	V
100	3460.43	I	1600	3969.75	I	110	4936.33	I	1000	1696.01	III
120	3550.64	I	1600	3976.66	I	70	4942.50	I	800	1697.99	III
130	3566.16	I	960	3983.91	I	110	4954.81	I	1000	1707.35	III

Intensity	Wavelength/Å		Intensity	Wavelength/Å		Intensity	Wavelength/Å		Intensity	Wavelength/Å	
5000	1760.35	III	200	2291.98	II	200	2546.74	II	8800	3443.64	I
5000	1773.57	III	300 d	2293.38	II	340	2548.34	I	50	3446.39	II
2000	1780.05	III	300	2301.40	II	310	2553.37	I	4100	3449.17	I
3000	1782.97	III	800 d	2307.85	II	310	2555.07	I	2100	3449.44	I
1000	1787.08	III	2600	2309.02	I	300	2559.41	II	21000	3453.50	I
1000	1789.07	III	500	2311.60	II	200	2560.03	II	1000	3455.23	I
1000	1823.08	III	500	2314.05	II	960	2562.15	II	5100	3462.80	I
2000	1830.09	III	300	2314.96	II	500	2564.04	II	5100	3465.80	I
2000	1831.44	III	200 p	2317.06	II	1100	2567.35	I	8000	3474.02	I
5000	1835.00	III	2400	2323.14	I	960	2574.35	I	1900	3483.41	I
1500	1842.34	I	300 p	2324.31	II	800	2580.32	II	4800	3489.40	I
1800	1847.89	I	200 d	2326.11	II	300 d	2582.22	II	2400	3495.69	I
1800	1852.71	I	500	2326.47	II	500	2587.22	II	50	3501.72	II
2400	1855.05	I	1400	2335.99	I	500	2587.52	II	9600	3502.28	I
2000	1863.83	III	1600	2338.67	I	200	2588.91	II	7000	3506.32	I
1500	1878.28	I	200	2347.39	II	100 p	2605.71	II	50	3507.77	II
1800	1936.58	I	1600	2352.85	I	100	2612.50	II	2900	3509.84	I
1500	1946.79	I	200 d	2353.41	II	100	2614.36	II	1400	3510.43	I
1500	1951.90	I	2000	2353.42	I	100 p	2628.77	II	4800	3512.64	I
1800	1954.22	I	500	2363.80	II	100	2632.26	II	3800	3513.48	I
1800	1955.17	I	400	2378.62	II	100	2636.07	II	4800	3518.35	I
1500	1958.55	I	1400	2380.48	I	310	2646.42	I	1300	3520.08	I
1500	1961.59	I	200	2381.76	II	770	2648.64	I	2700	3521.57	I
1500 h	1968.69	I	300 p	2383.45	II	100	2653.72	II	3800	3523.43	I
1500 h	1968.93	I	1400	2384.86	I	100	2663.53	II	60	3523.51	II
3000	1970.71	I	200	2386.36	II	200	2666.73	II	6400	3526.85	I
1800 h	1971.16	I	500	2388.92	II	100	2675.85	II	2700	3529.03	I
1800 h	1972.52	I	200	2397.38	II	100	2684.42	II	7300	3529.81	I
1500	1973.85	I	1100 d	2402.06	I	100	2702.02	II	1900	3533.36	I
1800	1976.97	I	200 p	2404.16	I	200	2706.62	II	50	3545.03	II
2400 h	1980.89	I	5300	2407.25	I	200	2707.35	II	1100	3560.89	I
1500	1989.80	I	5300	2411.62	I	190	2715.99	I	80	3561.07	II
1800	1990.34	I	1600	2412.76	I	100	2727.78	II	8800	3569.38	I
1500 l	1998.49	I	4800	2414.46	I	80	2734.54	II	50	3574.95	II
1500	2002.32	I	4800	2415.30	I	190	2745.10	I	1600	3574.96	I
900	2008.04	I	300	2417.65	II	100	2753.22	II	60	3575.32	II
50	2011.51	II	4100	2424.93	I	190	2764.19	I	2500	3575.36	I
1200 h	2014.58	I	3300	2432.21	I	100	2766.70	II	60	3577.96	II
900	2016.17	I	2900	2436.66	I	100	2774.97	II	1000	3585.16	I
50	2022.35	II	2400	2439.05	I	100	2791.00	II	6700	3587.19	I
50	2027.04	II	200	2442.63	II	100	2793.73	II	1900	3594.87	I
900	2031.96	I	200 d	2446.03	II	150	2815.56	I	1600	3602.08	I
1500	2039.95	I	200 p	2447.69	II	80	2835.63	II	100	3621.21	II
1200	2041.11	I	200	2450.00	II	80	2847.35	II	1000	3627.81	I
50	2065.54	II	200	2464.20	II	80	2871.22	II	80	3643.61	II
1500 h	2077.76	I	200	2486.44	II	190	2886.44	I	60	3681.35	II
900	2085.67	I	200	2498.82	II	100	2918.38	II	1100	3745.50	I
900	2087.55	I	570	2504.52	I	100	2930.24	II	1400	3842.05	I
900	2089.35	I	500	2506.46	II	100	2954.73	II	6900	3845.47	I
900	2093.40	I	360	2506.88	I	690	2987.16	I	5500	3873.12	I
900	2094.86	I	200	2511.16	II	690	2989.59	I	2800	3873.96	II
900	2095.77	I	860	2517.87	I	60	3022.59	II	7900	3894.08	I
1200	2097.51	I	500	2519.82	II	3100	3044.00	I	1500	3935.97	I
1500	2104.73	I	4300	2521.36	I	1700	3061.82	I	80 h	3963.10	I
1500	2106.80	I	200 h	2524.65	II	80	3387.70	II	6000	3995.31	I
900	2108.98	I	300	2524.97	II	1100	3388.17	I	970	3997.91	I
900 s	2117.68	I	500	2528.62	II	2200	3395.38	I	350	4020.90	I
900	2137.78	I	2900	2528.97	I	11000	3405.12	I	370	4045.39	I
900	2138.97	I	200 p	2530.09	II	4500	3409.18	I	350	4066.37	I
900	2163.03	I	720	2530.13	I	6700	3412.34	I	830	4092.39	I
1100	2174.60	I	860	2532.18	I	2200	3412.63	I	550	4110.54	I
200	2193.60	II	200 d	2533.82	II	2700	3417.16	I	2800	4118.77	I
200	2256.73	II	2900	2535.96	I	50	3423.84	II	4400	4121.32	I
150	2260.00	II	860	2536.49	I	2500	3431.58	I	90	4190.71	I
200	2283.52	II	300	2541.94	II	4500	3433.04	I	90	4469.56	I
1000	2286.15	II	1700	2544.25	I	1600	3442.93	I	690	4530.96	I

Intensity	Wavelength/Å		Intensity	Wavelength/Å		Intensity	Wavelength/Å		Intensity	Wavelength/Å	
90	4549.66	I	300	1004.055	II	250	1450.304	II	400	1606.834	II
140	4565.59	I	300	1008.569	II	200	1452.294	II	250	1608.639	II
190	4581.60	I	300	1008.728	II	300	1458.002	II	150	1610.296	II
120	4629.38	I	300	1010.269	II	250	1459.412	II	200	1617.915	II
85	4663.41	I	250	1012.597	II	200	1463.752	II	600	1621.426	II
110	4792.86	I	500	1018.707	II	400	1463.838	II	400	1622.428	II
100	4840.27	I	500	1027.831	II	200	1466.070	II	250	1630.268	II
150	4867.88	I	250	1028.328	II	400	1470.697	II	100	1636.605	II
80 h	4964.18	II	200	1030.263	II	200	1472.395	II	1000 r	1642.21	III
50	5212.71	I	600	1036.470	II	250	1473.978	II	250	1649.458	II
50	5230.22	I	600	1039.348	II	200	1474.935	II	30 r	1655.32	I
50	5247.93	I	600	1039.582	II	150	1476.059	II	200	1656.322	II
50	5342.71	I	800	1044.519	II	300 r	1481.23	III	200	1660.001	II
50	5352.05	I	800	1044.744	II	200	1481.544	II	300	1663.002	II
	Copper		500	1049.755	II	200	1485.328	II	100	1672.776	II
	Cu Z = 29		600	1054.690	II	750	1488.831	II	30	1688.09	I
80	685.141	II	400	1055.797	II	300	1492.834	II	30	1691.08	I
100	709.313	II	600	1056.955	II	250	1493.366	II	30 r	1703.84	I
100	718.179	II	400	1058.799	II	250	1495.430	II	50 r	1713.36	I
150	724.489	II	600	1059.096	II	350	1496.687	II	150	1717.721	II
200	735.520	II	600	1060.634	II	150	1503.368	II	50 r	1725.66	I
250	736.032	II	600	1063.005	II	250	1504.757	II	100	1736.551	II
80	779.295	II	200	1065.782	II	200	1505.388	II	50 r	1741.57	I
100	797.455	II	200	1066.134	II	300	1508.632	II	150	1753.281	II
150	810.998	II	500	1069.195	II	350	1510.506	II	200 r	1774.82	I
200	813.883	II	300	1073.745	II	200	1512.465	II	100 r	1825.35	I
300	826.996	II	200	1088.395	II	200	1513.366	II	250	1929.751	II
150	848.808	II	300	1094.402	II	500	1514.492	II	250	1944.597	II
250	851.303	II	250	1097.053	II	200	1517.631	II	100	1946.493	II
250	858.487	II	150	1119.947	II	500	1519.492	II	200	1957.518	II
400	861.994	II	200	1142.640	II	600	1519.837	II	150	1970.495	II
400	865.390	II	300	1144.856	II	200	1520.540	II	150	1977.027	II
250	869.336	II	100	1250.048	II	200	1524.860	II	500	1979.956	II
150	873.263	II	150	1265.506	II	150	1525.764	II	300	1989.855	II
200	876.723	II	300	1275.572	II	500	1531.856	II	250	1999.698	II
250	877.012	II	150	1282.455	II	300	1532.131	II	270	2035.854	II
200	877.555	II	150	1287.468	II	250	1533.986	II	250	2037.127	II
500	878.699	II	150	1298.395	II	250	1535.002	II	350	2043.802	II
100	884.133	II	300	1308.297	II	500	1537.559	II	300	2054.980	II
250	885.847	II	300	1314.337	II	200	1540.239	II	100	2078.663	II
600	886.943	II	100	1320.686	II	300	1540.389	II	110	2098.398	II
600	890.567	II	100	1326.395	II	300	1540.588	II	320	2104.797	II
500	892.414	II	150	1350.594	II	750	1541.703	II	300	2112.100	II
800	893.678	II	250	1351.837	II	400	1544.677	II	320	2117.310	II
400	894.227	II	150	1355.305	II	100	1547.958	II	350	2122.980	II
600	896.759	II	300	1358.773	II	300	1550.653	II	350	2126.044	II
400	896.976	II	200	1359.009	II	300	1551.389	II	420	2134.341	II
600	901.073	II	200	1362.600	II	500	1552.646	II	900	2135.981	II
400	906.113	II	250	1367.951	II	250	1553.896	II	400	2148.984	II
800	914.213	II	200	1371.840	II	400	1555.134	II	150	2161.320	II
600	922.019	II	300 r	1376.79	III	500	1555.703	II	1300 r	2165.09	I
500	924.239	II	200 r	1377.49	III	300	1558.345	II	250	2174.982	II
400	935.232	II	100	1393.128	II	400	1565.924	II	1600 r	2178.94	I
600	935.898	II	100	1398.642	II	400	1566.415	II	700	2179.410	II
600	943.335	II	150	1402.777	II	100	1569.416	II	1700 r	2181.72	I
600	945.525	II	150	1407.169	II	300	1579.492	II	700	2189.630	II
500	945.965	II	100	1414.898	II	300	1580.626	II	900	2192.268	II
200	954.383	II	250	1418.426	II	400	1581.995	II	400	2195.683	II
250	956.290	II	250	1421.759	II	500	1583.682	II	1700 r	2199.58	I
400	958.154	II	200	1427.829	II	400	1590.165	II	1300 r	2199.75	I
200	960.414	II	400	1430.243	II	600	1593.556	II	100	2200.509	II
250	968.042	II	250	1434.904	II	500 r	1593.75	III	200	2209.806	II
200	974.759	II	150	1436.236	II	400	1598.402	II	750	2210.268	II
250	977.567	II	150	1442.139	II	400	1602.388	II	1600 r	2214.58	I
100	987.657	II	200	1445.984	II	200	1604.848	II	250	2215.106	II
250	992.953	II	200	1449.058	II	300	1605.281	II	1000 r	2215.65	I

Line Spectra of the Elements (continued): Copper

Intensity	Wavelength/Å		Intensity	Wavelength/Å		Intensity	Wavelength/Å		Intensity	Wavelength/Å	
750	2218.108	II	10000 r	3273.96	I	100	4758.433	II	550	6305.972	II
2100 r	2225.70	I	1400 h	3282.72	I	400	4812.948	II	400	6312.492	II
150	2226.780	II	400	3290.418	II	120	4851.262	II	120	6326.466	II
1600 r	2227.78	I	1500 h	3290.54	I	300	4854.988	II	400	6373.268	II
350	2228.868	II	110	3300.881	I	100	4873.304	II	750	6377.840	II
2500 r	2230.08	I	250	3301.229	II	150	4901.427	II	400	6403.384	II
1100 r	2238.45	I	2500 h	3307.95	I	1000	4909.734	II	850	6423.884	II
900	2242.618	II	200	3316.276	II	500	4918.376	II	200	6442.965	II
2300 r	2244.26	I	1500	3337.84	I	200	4926.424	II	750	6448.559	II
1000	2247.002	II	150	3338.648	II	900	4931.698	II	170	6466.246	II
1300 r	2260.53	I	200	3365.648	II	120	4943.026	II	950	6470.168	II
2200 r	2263.08	I	450	3370.454	II	700	4953.724	II	750	6481.437	II
150	2263.786	II	300	3374.952	II	500	4985.506	II	400	6484.421	II
200	2276.258	II	200	3380.712	II	400	5006.801	II	220	6517.317	II
100	2286.645	II	100	3384.945	II	350	5009.851	II	400	6530.083	II
2500 r	2293.84	I	1250 h	3483.76	I	400	5012.620	II	120	6551.286	II
170	2294.368	II	1250	3524.23	I	350	5021.279	II	200	6577.080	II
1000	2303.12	I	2000	3530.38	I	200	5039.016	II	750	6624.292	II
150	2369.890	II	1400	3599.13	I	300	5047.348	II	800	6641.396	II
2500 r	2392.63	I	1400	3602.03	I	900	5051.793	II	450	6660.962	II
120	2403.337	II	1000	3686.555	II	400	5058.910	II	100	6770.362	II
1500	2406.66	I	150	3786.270	II	500	5065.459	II	300	6806.216	II
1000 r	2441.64	I	170	3797.849	II	450	5067.094	II	400	6809.647	II
100	2485.792	II	100	3818.879	II	350	5072.302	II	320	6823.202	II
2000 r	2492.15	I	140	3826.921	II	450	5088.277	II	250	6844.157	II
150	2506.273	II	160	3864.137	II	420	5093.816	II	320	6868.791	II
120	2526.593	II	280	3884.131	II	350	5100.067	II	270	6872.231	II
300	2544.805	II	150	3892.924	II	1500	5105.54	I	270	6879.404	II
100	2571.756	II	170	3903.177	II	250	5124.476	II	220	6937.553	II
150	2590.529	II	140	3920.654	II	2000	5153.24	I	150	6952.871	II
200	2600.270	II	120	3933.268	II	100	5158.093	II	150	6977.572	II
2500 r	2618.37	I	120	3987.024	II	100	5183.367	II	200	7022.860	II
200	2666.291	II	150	3993.302	II	2500	5218.20	I	300	7194.896	II
750	2689.300	II	140	4003.476	II	100	5269.991	II	400	7326.008	II
700	2700.962	II	1250	4022.63	I	100	5276.525	II	300	7331.694	II
650	2703.184	II	100	4032.647	II	1650	5292.52	I	250	7382.277	II
700	2713.508	II	600	4043.484	II	100	5368.383	II	1000	7404.354	II
650	2718.778	II	500	4043.751	II	1500	5700.24	I	270	7434.156	II
300	2721.677	II	2000	4062.64	I	1500	5782.13	I	500	7562.015	II
120	2737.342	II	120	4068.106	II	150	5805.989	II	700	7652.333	II
270	2745.271	II	500	4131.363	II	100	5833.515	II	1000	7664.648	II
2500 r	2766.37	I	200	4143.017	II	200	5897.971	II	150	7681.788	II
800	2769.669	II	300	4153.623	II	120	5937.577	II	450	7744.097	II
200	2791.795	II	500	4161.140	II	400	5941.196	II	800	7778.738	II
170	2799.528	II	370	4164.284	II	100	5993.260	II	750	7805.184	II
100	2810.804	II	400	4171.851	II	650	6000.120	II	1500	7807.659	II
1250 r	2824.37	I	500	4179.512	II	100	6023.264	II	1000	7825.654	II
350	2837.368	II	500	4211.866	II	250	6072.218	II	350	7860.577	II
100	2857.748	II	320	4230.449	II	150	6080.343	II	300	7890.567	II
600	2877.100	II	200	4255.635	II	150	6099.990	II	700	7902.553	II
270	2884.196	II	950	4275.11	I	160	6107.412	II	1500	7933.13	I
2500 r	2961.16	I	300	4279.962	II	300	6114.493	II	400	7944.438	II
100	2986.335	II	500	4292.470	II	600	6150.384	II	400	7972.033	II
2000	2997.36	I	400	4365.370	II	750	6154.222	II	1200	7988.163	II
2000	3010.84	I	100	4444.831	II	500	6172.037	II	2000	8092.63	I
2500	3036.10	I	400	4506.002	II	550	6186.884	II	500	8277.560	II
2500	3063.41	I	150	4516.049	II	400	6188.676	II	800	8283.160	II
1400	3073.80	I	150	4541.032	II	300	6198.092	II	250	8503.396	II
1500	3093.99	I	500	4555.920	II	470	6204.261	II	750	8511.061	II
1250	3099.93	I	100	4596.906	II	450	6208.457	II	200	8609.134	II
2000	3108.60	I	120	4649.271	II	750	6216.939	II	500	9813.213	II
1400 h	3126.11	I	2000	4651.12	I	700	6219.844	II	250	9827.978	II
1500	3194.10	I	120	4661.363	II	500	6261.848	II	200	9830.798	II
1400	3208.23	I	320	4671.702	II	1000	6273.349	II	600	9861.280	II
1500 h	3243.16	I	300	4673.577	II	350	6288.696	II	600	9864.137	II
10000 r	3247.54	I	450	4681.994	II	900	6301.009	II	200	9883.969	II

Intensity	Wavelength/Å		Intensity	Wavelength/Å		Intensity	Wavelength/Å		Intensity	Wavelength/Å	
550	9916.419	II	560	3517.26	II	580	3806.27	II	4400	4218.09	I
500	9917.954	II	4400	3523.98	II	470	3812.27	I	4400	4221.11	I
550	9925.594	II	22000	3531.70	II	470	3813.67	II	2700	4225.16	I
450	9938.998	II	4400	3534.96	II	1400	3816.76	II	1000	4308.63	II
500	9960.354	II	5500	3536.02	II	700	3825.68	II	540	4409.38	II
450	10006.588	II	4400	3538.52	II	2300	3836.50	II	740	4449.70	II
550	10022.969	II	1700	3542.33	II	1400	3841.31	II	420	4577.78	I
550	10038.093	II	1400	3546.83	II	420	3846.34	II	2100	4589.36	I
650	10054.938	II	4400	3550.22	II	420	3847.02	I	990	4612.26	I
450	10080.354	II	2200	3551.62	II	1200	3853.03	II	170	4731.84	II
	Dysprosium		440 h	3558.23	II	420	3858.40	I	120 h	4775.79	II
	Dy Z = 66		440	3559.30	II	560	3868.45	II	480	4957.34	II
260	2356.91	II	2200	3563.15	II	1600	3868.81	I	70	5022.12	I
240	2410.01	II	560	3563.69	II	820	3869.86	II	160	5042.63	I
260	2439.84	II	780	3573.83	II	7000	3872.11	II	95	5070.68	I
220	2585.30	I	1400	3574.15	II	1200	3873.99	II	120	5077.67	I
440	2634.80	II	4400	3576.24	II	470	3879.11	II	80	5090.38	II
220	2755.75	II	1700	3576.87	II	5800	3898.53	II	80	5110.32	I
300	2816.39	II	830	3577.98	II	540	3914.87	II	130 h	5120.04	I
390	2913.95	II	440	3580.04	II	540	3915.59	II	190	5139.60	II
610	3038.28	II	3300	3585.06	II	540 d	3917.29	I	110	5169.69	II
830	3135.38	II	1400	3585.78	II	420	3927.86	I	80	5185.30	I
500	3141.14	II	560	3586.11	II	540	3930.14	I	290	5192.86	II
1200	3156.52	II	1100	3591.41	II	2100	3931.52	II	95	5197.66	II
670	3162.83	II	560	3591.81	II	10000	3944.68	II	70	5259.88	I
1000	3169.99	II	560	3592.11	II	800	3957.79	II	130	5260.56	I
470	3215.19	II	1800	3595.04	II	14000	3968.39	II	65	5267.11	I
830	3216.63	II	560	3600.38	II	2700	3978.57	II	55	5282.07	I
490	3235.89	II	1800	3606.12	II	1400	3981.92	II	160	5301.58	I
490	3245.12	II	440	3618.51	II	1600	3983.65	II	65	5340.30	I
1200	3251.27	II	560	3620.16	II	800	3984.21	II	85	5389.58	II
890	3280.09	II	470	3624.27	II	540	3991.32	II	80	5419.13	I
490	3282.77	II	1100	3629.42	II	1600	3996.69	II	70	5423.32	I
1100	3308.88	II	4000	3630.24	II	8000	4000.45	II	95	5451.11	I
780	3316.32	II	440	3632.78	II	420	4005.84	II	65	5547.27	I
1000	3319.88	II	1100	3640.25	II	540	4011.29	II	100	5639.50	I
780	3341.00	II	11000	3645.40	II	540	4013.82	I	55 h	5645.99	I
510	3353.58	II	1000	3648.78	II	540	4014.70	II	80	5652.01	I
510	3368.11	II	700	3664.62	II	420	4027.78	II	70 h	5718.46	I
5300	3385.02	II	990	3672.30	II	520 d	4028.32	II	55	5745.53	I
610	3388.85	II	420	3672.70	II	520	4032.47	II	55 h	5868.11	II
3800	3393.57	II	1400	3674.08	II	420	4033.65	II	70	5945.80	I
1300	3396.16	II	2200	3676.59	II	420	4036.32	II	120	5974.49	I
5300	3407.80	II	640	3678.51	I	12000	4045.97	I	140	5988.56	I
1300	3413.78	II	820	3684.85	I	1600	4050.56	II	140	6088.26	I
530	3414.82	II	1300	3685.78	I	520	4055.14	II	100	6168.43	I
780	3419.63	II	4700	3694.81	II	2500	4073.12	II	270	6259.09	I
530	3425.06	II	990	3698.21	II	7400	4077.96	II	160	6579.37	I
1900	3434.37	II	540	3701.63	II	3900	4103.30	II	75	6667.86	I
560	3440.93	II	440	3707.57	II	860	4103.87	I	180	6835.42	I
1300	3441.45	II	440	3708.22	II	1500	4111.34	II	80	6852.96	I
3800	3445.57	II	420	3710.07	II	490	4124.63	II	65	6899.32	II
830	3446.99	II	1600	3724.45	II	990	4129.42	II	55	7426.86	II
2700	3454.32	II	930	3739.34	I	1200	4143.10	II	55	7543.73	I
1300	3456.56	II	1200	3747.82	II	990	4146.06	I	80	7662.36	I
4400	3460.97	II	1400	3753.51	II	5700	4167.97	I	100	8201.57	II
720	3468.43	II	1400	3753.75	II	930	4183.72	I	45	8791.39	II
560	3471.14	II	1200	3757.05	I	12000	4186.82	I		**Erbium**	
560 d	3471.53	II	4700	3757.37	II	2200	4191.64	I		**Er Z = 68**	
1300	3477.07	II	640	3767.63	I	6800	4194.84	I	600	2277.65	III
4400	3494.49	II	640	3773.05	I	800	4198.02	I	290	2586.73	II
560	3496.34	II	420	3781.47	I	680	4201.30	I	490	2670.26	II
830	3498.71	II	3300	3786.18	II	680	4202.24	I	500	2739.27	III
830	3504.53	II	1600	3788.44	II	16000	4211.72	I	610	2755.63	II
830	3505.45	II	700	3791.87	II	1800	4213.18	I	1000	2904.47	II
1300	3506.81	II	510	3804.14	II	3700	4215.16	I	1500	2910.36	II

Intensity	Wavelength/Å		Intensity	Wavelength/Å		Intensity	Wavelength/Å		Intensity	Wavelength/Å	
1500	2964.52	II	3200	3973.58	I	55	6326.13	I	120	3058.98	I
1200	3002.41	II	1400	3974.72	II	55	6492.35	I	220	3077.36	II
1000	3055.10	III	810	3977.02	I	60	6583.48	I	120	3097.45	II
1000	3070.40	III	1100	3982.33	I	70	6601.11	I	320	3106.18	I
610	3073.34	II	810	3987.66	I	70	6759.87	I	950	3111.43	I
720	3082.08	II	14000	4007.96	I	35	6790.92	I	120	3130.73	II
610	3084.02	II	1100	4012.58	I	70	6848.10	I	50 c	3171.00	III
770	3122.72	II	3000	4020.51	I	55	6865.13	I	50 c	3183.78	III
1500	3166.25	III	1000	4046.96	I	55	7459.55	I	420	3210.57	I
870	3181.92	II	940	4055.47	II	120	7469.51	I	1000	3212.81	I
870	3220.73	II	690	4059.78	II	35	7680.01	I	420	3213.75	I
610	3223.31	II	3500	4087.63	I	35	7797.47	I	150	3272.77	II
2300	3230.58	II	1100	4098.10	I	35	7921.85	I	210	3277.78	II
2700	3264.78	II	6900	4151.11	I	30	7937.84	I	150	3301.95	II
720	3279.33	II	1000	4190.70	I	35	8312.82	I	140	3308.02	II
720	3280.22	II	1400	4218.43	I	55	8409.90	I	140	3313.33	II
2000	3301.23	III	690	4286.56	I	9	8866.84	II	950	3334.33	I
2300	3312.42	II	40000	4290.06	III	**Europium**			110	3350.40	I
770	3323.19	II	20000	4386.86	III	**Eu Z = 63**			140	3369.06	II
770	3332.70	II	810	4409.34	I	30	2124.69	III	190	3391.99	II
1300	3346.04	II	1000	4606.61	I	200	2350.51	III	280	3396.58	II
1400	3364.08	II	570	4675.62	II	4000	2375.46	III	150	3425.02	II
1400 d	3368.02	II	15000	4735.56	III	100 d	2435.14	III	150	3441.00	II
7700	3372.71	II	2000	4783.12	III	1000	2444.38	III	130	3461.38	II
970	3374.17	II	250	5007.25	I	4000	2445.99	III	470 cw	3521.09	II
1700	3385.08	II	200	5035.94	I	2000	2513.76	III	150	3542.15	II
2300	3392.00	II	210	5042.05	II	200	2522.14	III	180	3552.52	II
770	3441.13	II	120	5124.56	I	160	2564.17	II	150	3603.20	II
970	3471.71	II	130	5127.41	II	110	2568.17	II	6400	3688.42	II
610	3479.41	II	120	5131.53	I	230	2577.14	II	20000 cw	3724.94	II
970	3485.85	II	130	5133.83	II	1000	2638.77	II	350	3741.31	II
6700	3499.10	II	170	5164.77	II	380	2641.27	II	260	3761.12	II
610	3502.78	I	130	5172.78	I	640	2668.34	II	39000 cw	3819.67	II
610	3524.91	II	160	5188.90	II	110	2673.42	II	140	3844.23	II
820	3549.84	II	150	5206.52	I	250	2678.29	II	190	3865.57	I
1500	3558.02	I	140	5255.93	II	250	2685.66	II	150	3884.75	I
1000	3559.90	II	80	5272.91	I	550	2692.03	II	28000 cw	3907.10	II
920	3570.75	II	90	5348.06	I	700	2701.14	II	32000 cw	3930.48	II
1000	3580.52	II	60	5414.63	II	800	2701.90	II	30000 cw	3971.96	II
610	3590.76	I	180	5456.62	I	240	2705.28	II	180	4011.69	II
610	3599.50	II	90	5468.32	I	180	2709.99	I	150	4017.58	II
1000	3599.83	II	80	5485.97	II	700	2716.98	II	120	4039.19	I
3100	3616.56	II	80	5593.46	I	4200	2727.78	II	120	4085.38	II
720	3628.04	I	60	5611.82	I	160	2740.62	II	33000 cw	4129.70	II
1000	3633.54	II	70	5622.01	I	120	2744.26	II	60000 cw	4205.05	II
1600	3638.68	I	80	5626.53	II	480	2781.89	II	150	4298.73	I
900	3645.94	II	90	5640.36	I	1900	2802.84	II	240	4355.09	II
7900	3692.65	II	70	5664.95	I	220	2811.75	II	14000 cw	4435.56	II
1300	3729.52	II	70	5719.55	I	3400	2813.94	II	3000	4522.57	II
900	3742.64	II	100	5739.19	I	550	2816.18	II	11000	4594.03	I
900	3747.43	I	290	5762.80	I	2000	2820.78	II	9800	4627.22	I
1800	3786.84	II	70	5784.66	I	400 cw	2828.72	II	8300	4661.88	I
1600	3810.33	I	70	5800.79	I	260	2859.67	II	110	4867.62	I
4000	3816.78	III	430	5826.79	I	280	2862.57	II	150	4907.18	I
3600	3830.48	II	100	5850.07	I	200	2892.54	I	180	4911.40	I
680	3855.90	I	120	5855.31	I	140	2893.03	I	180	5013.17	I
7500	3862.85	I	140	5872.35	I	360	2893.83	I	170	5022.91	I
1500	3880.61	II	120	5881.14	I	3200	2906.68	II	110	5029.54	I
1200	3882.89	II	8000	5903.30	III	160	2908.99	I	170	5114.37	I
4200	3892.68	I	70	6022.56	I	850	2925.04	II	170	5129.10	I
5200	3896.23	II	70	6061.25	I	200 cw	2952.68	II	210	5133.52	I
11000	3906.31	II	60	6076.45	II	260	2960.21	II	270	5160.07	I
3200	3937.01	I	360	6221.02	I	300	2991.33	II	210	5166.70	I
2100	3938.63	II	55	6262.56	I	100 c	3023.93	III	200	5199.85	I
3200	3944.42	I	60	6268.87	I	200 c	3026.79	III	110	5200.96	I
2700	3973.04	I	130	6308.77	I	320 cw	3054.94	II	120	5206.44	I

Intensity	Wavelength/Å		Intensity	Wavelength/Å		Intensity	Wavelength/Å		Intensity	Wavelength/Å	
750	5215.10	I	35	7887.99	I	750	955.55	I	150	2835.63	III
300	5223.49	I	24 cw	8209.80	I	500	958.52	I	150	2860.33	III
120	5239.24	I	21 cw	8642.67	I	20	972.40	I	120	2862.86	III
200	5266.40	I	18	8870.30	I	350	973.90	I	140	2887.58	III
390	5271.96	I		**Fluorine**		100	976.22	I	150	2889.45	III
110	5272.48	I		**F Z = 9**		40	976.51	I	120	2905.30	III
150	5282.82	I	50	148.00	V	100	977.75	I	140	2913.29	III
120	5291.26	I	50	163.56	V	60	1082.31	V	160	2916.34	III
120	5294.64	I	90	165.98	V	70	1088.39	V	140	2932.49	III
540	5357.61	I	100	166.18	V	80	1219.03	III	140	2994.28	III
120	5361.61	I	50	186.84	V	80	1266.87	III	120 h	2997.21	III
110	5376.94	I	60	190.57	V	90	1267.71	III	130	2997.53	III
120	5392.94	I	70	190.84	V	70	1297.54	III	120	2999.47	III
450	5402.77	I	50	196.39	IV	70	1359.92	III	130	3039.25	III
380	5451.51	I	60	196.45	IV	110	1498.93	III	120	3039.75	III
260	5452.94	I	70	200.09	IV	120	1502.01	III	160	3042.80	III
120	5488.65	I	80	201.16	IV	110	1504.18	III	150	3049.14	III
120	5510.52	I	90	208.25	IV	140	1504.79	III	140	3113.62	III
200	5547.44	I	90	240.08	IV	130	1506.30	III	160	3115.70	III
150	5570.33	I	100	251.03	IV	110	1506.77	III	180	3121.54	III
200	5577.14	I	140	419.65	IV	100	1553.02	III	140	3124.79	III
120	5580.03	I	150	420.05	IV	110	1557.59	III	140	3134.23	III
210	5645.80	I	160	420.73	IV	100	1563.73	III	140	3146.99	III
330	5765.20	I	100	429.51	III	100	1565.54	III	180	3174.17	III
180	5783.69	I	110	430.15	III	100	1623.40	III	170	3174.76	III
170	5818.74	II	150	430.76	IV	100	1650.76	III	120	3214.00	III
600 cw	5830.98	I	90	464.29	III	130	1670.39	III	200	3501.45	II
330	5966.07	II	120	465.98	V	140	1677.40	III	200	3501.57	II
480 cw	5967.10	I	130	490.57	IV	100	1716.99	III	200	3502.96	II
170	5972.75	I	160	491.00	IV	120	1770.09	III	6	3594.10	I
240	5992.83	I	50	497.38	IV	150	1770.67	III	12	3668.17	I
110	6012.56	I	60	497.83	IV	110	1772.93	III	270	3847.09	II
420	6018.15	I	70	498.80	IV	140	1773.36	III	260	3849.99	II
170	6029.00	I	90	506.16	V	160	1791.65	III	250	3851.67	II
420	6049.51	II	100	508.08	V	110	1803.03	III	5	3898.48	I
140	6057.36	I	120	508.39	III	170	1805.90	III	8	3930.69	I
240	6083.84	I	60	514.08	V	110	1839.30	III	5	3934.26	I
240	6099.35	I	90	525.29	V	120	1839.97	III	5	3948.56	I
120	6118.78	I	100	526.30	V	110	1840.14	III	240	4024.73	II
330	6173.05	II	120	567.69	III	100	2027.44	III	220	4025.01	II
110	6178.76	I	110	567.75	III	120	2030.32	III	230	4025.49	II
260 cw	6188.13	I	140	570.64	IV	120	2217.17	III	200	4103.51	II
140	6195.07	I	140	571.30	IV	50	2298.29	III	200	4246.23	II
240	6262.25	I	150	571.39	IV	40	2451.58	IV	200	4299.17	II
170	6299.77	I	160	572.66	IV	120	2452.07	III	140 h	4420.30	III
230	6303.41	II	90	605.67	II	50	2456.92	IV	120 h	4427.35	III
120 cw	6350.04	I	100	606.80	II	130	2464.85	III	120 h	4432.32	III
120 cw	6400.93	I	90	630.20	III	130	2470.29	III	140 h	4479.99	III
180	6410.04	I	100	647.77	V	120	2478.73	III	6	4960.65	I
140	6411.32	I	110	647.87	V	150	2484.37	III	150	5012.54	III
830	6437.64	II	130	654.03	V	120	2542.77	III	160	5110.99	III
120	6457.96	I	120	656.12	III	120	2580.04	III	15	5230.41	I
1400	6645.11	II	130	656.87	III	130	2583.81	III	12	5279.01	I
50	6666.35	III	110	657.23	V	120	2593.23	III	18	5540.52	I
140	6802.72	I	140	657.33	V	130	2595.53	III	12	5552.43	I
360	6864.54	I	140	658.33	III	140	2599.28	III	10	5577.33	I
120	7040.20	I	130	676.12	IV	130	2625.01	III	20	5624.06	I
330	7077.10	II	150	677.15	IV	140	2629.70	III	12	5626.93	I
570	7194.81	II	130	677.22	IV	120	2656.44	III	15	5659.15	I
570	7217.55	II	150	678.99	IV	130	2755.55	III	40	5667.53	I
540	7301.17	II	160	679.21	IV	160	2759.63	III	90	5671.67	I
720	7370.22	II	60	757.04	V	120	2788.15	III	18	5689.14	I
300	7426.57	II	150	806.96	I	160	2811.45	III	25	5700.82	I
160	7583.91	I	125	809.60	I	40	2820.74	IV	25	5707.31	I
60 cw	7742.57	I	500	951.87	I	50	2826.13	IV	140	5753.17	III
70	7746.19	I	1000	954.83	I	140	2833.99	III	120	5761.20	III

Intensity	Wavelength/Å		Intensity	Wavelength/Å		Intensity	Wavelength/Å		Intensity	Wavelength/Å	
12	5950.15	I	1500	8274.62	I	2200	3481.28	II	1100	4306.34	I
25	5959.19	I	2000	8298.58	I	1700	3481.80	II	1800	4313.84	I
70	5965.28	I	600	8302.40	I	1700	3494.40	II	2600 d	4325.57	II
50	5994.43	I	900	8807.58	I	1400	3505.51	II	1900	4327.12	I
150	6015.83	I	1000	8900.92	I	1100	3512.50	II	1000	4344.30	II
80	6038.04	I	300	8912.78	I	4300	3545.80	II	2200	4346.46	I
900	6047.54	I	350	9025.49	I	3900	3549.36	II	1400	4401.86	I
100	6080.11	I	400	9042.10	I	1400	3557.05	II	1400	4422.41	I
150	6091.82	III	350	9178.68	I	5400	3584.96	II	1100	4430.63	I
140	6125.50	III	200	9433.67	I	1100	3592.71	II	1100	4519.66	I
800	6149.76	I	25	9505.30	I	1100	3604.87	I	910	4537.81	I
400	6210.87	I	12	9662.04	I	6100	3646.19	I	520	4614.50	I
130	6233.57	III	25	9734.34	I	3900	3654.62	II	700	4694.33	I
13000	6239.65	I	15	9822.11	I	3100	3656.15	II	410	4743.65	I
10000	6348.51	I	12	9902.65	I	1400	3662.26	II	470	4767.24	I
140	6363.05	III	80 h	10047.98	II	2700	3664.60	II	300	4784.62	I
8000	6413.65	I	15	10285.45	I	2000	3671.20	II	320	4821.69	I
450	6569.69	I	20	10862.31	I	2000	3684.13	I	280	4934.12	I
300	6580.39	I		**Francium**		3100	3687.74	II	750	5015.04	I
400	6650.41	I		**Fr Z = 87**		2000	3697.73	II	75	5039.09	I
1800	6690.48	I		7177.	I	1300	3699.73	II	5000	5091.70	III
400	6708.28	I		**Gadolinium**		2700	3712.70	II	130	5098.38	II
7000	6773.98	I		**Gd Z = 64**		2000	3713.57	I	180	5108.91	II
1500	6795.53	I	1200	1007.24	IV	1400	3716.36	II	120	5125.56	II
9000	6834.26	I	1200	1063.84	IV	2000	3717.48	I	860	5155.84	I
50000	6856.03	I	1600	1476.98	IV	1800 d	3719.45	II	190	5176.28	II
8000	6870.22	I	1500	1705.03	IV	1500	3730.84	II	410	5197.77	I
15000	6902.48	I	1600	1706.01	IV	4500	3743.47	II	280	5219.40	I
6000	6909.82	I	2000	1736.24	IV	1400	3758.31	II	130	5233.93	I
4000	6966.35	I	1500	1815.32	IV	8700	3768.39	II	320	5251.18	I
45000	7037.47	I	2200	1975.24	III	1400	3770.69	II	120	5252.14	II
30000	7127.89	I	3400	2018.07	III	2900	3783.05	I	140	5255.80	I
15000	7202.36	I	2800	2359.31	III	5100	3796.37	II	280	5283.08	I
1000	7309.03	I	1400	2397.87	IV	3700	3813.97	I	280	5301.67	I
15000	7311.02	I	2800	2697.39	III	3300	3850.69	II	220	5302.76	I
700	7314.30	I	2800	2703.28	III	5100	3850.97	II	280	5307.30	I
5000	7331.96	I	2700	2727.89	III	4300	3852.45	II	130	5321.50	I
120	7336.77	III	9000	2904.73	III	1600	3866.99	I	280	5321.78	I
130	7354.94	III	9500	2955.53	III	1500	3894.70	II	110	5327.32	I
10000	7398.69	I	1200	2999.04	III	2200	3916.51	II	170	5333.30	I
4000	7425.65	I	2100	3010.13	II	1200	3934.79	I	300	5343.00	I
2200	7482.72	I	1900	3027.60	II	1400	3945.54	I	200	5348.67	I
2500	7489.16	I	2100	3032.84	II	1200	3957.67	II	300	5350.38	I
900	7514.92	I	1600	3034.05	II	1100	4023.14	I	240	5353.26	I
5000	7552.24	I	2100	3081.99	II	1100	4028.15	I	3000	5365.96	III
5000	7573.38	I	3500	3100.50	II	1400	4037.33	II	150	5370.63	I
7000	7607.17	I	930	3145.10	II	1600	4045.01	I	4000	5553.30	III
18000	7754.70	I	980	3156.53	II	1300	4049.43	II	3000	5587.88	III
15000	7800.21	I	980	3161.37	II	2200	4049.86	II	190	5617.91	I
300	7879.18	I	4000	3176.66	III	2600	4053.64	I	110	5632.25	I
500	7898.59	I	1400	3331.38	II	2600	4058.22	I	260	5643.24	I
350	7936.31	I	1100	3336.18	II	1900	4063.39	I	3000	5658.98	III
300	7956.32	I	5400	3350.47	II	1300	4078.44	II	390	5696.22	I
80	8016.01	II	4300	3358.62	II	2800	4078.70	I	120	5733.86	II
1000	8040.93	I	5400	3362.23	II	1500	4085.56	II	240	5791.38	I
900	8075.52	I	1100	3392.53	II	1100	4092.71	I	220	5851.63	I
350	8077.52	I	1100 d	3407.56	II	2600	4098.61	II	280	5856.22	I
350	8126.56	I	6900	3422.47	II	2200	4130.37	II	110	5904.56	I
600	8129.26	I	1700	3439.21	II	1100	4132.28	II	170	5911.45	II
300	8159.51	I	2700	3439.99	II	2400	4175.54	I	85	5930.29	I
600	8179.34	I	1400	3450.38	II	2400	4184.25	II	85	5936.84	I
300	8191.24	I	1100	3451.23	II	2200	4190.78	I	65	5937.71	I
350	8208.63	I	2700	3463.98	II	1300	4212.00	II	110 h	5988.02	I
2500	8214.73	I	1700	3467.27	II	4800	4225.85	I	430	6114.07	I
3000	8230.77	I	1700	3468.99	II	1700	4251.73	II	75	6305.15	II
500	8232.19	I	1400	3473.22	II	1600	4262.09	I			

Intensity	Wavelength/Å		Intensity	Wavelength/Å		Intensity	Wavelength/Å		Intensity	Wavelength/Å	
40	6331.35	I	80	1054.56	V	76	1351.06	IV	60	746.88	V
40	6380.95	II	90	1058.12	V	70	1353.92	III	60	760.05	V
40 h	6538.15	I	80	1066.69	V	74	1364.63	IV	10	862.234	II
55	6564.78	I	60	1069.60	V	60	1395.54	IV	15	875.493	II
50	6634.36	II	80	1073.77	V	77	1402.55	IV	15	905.977	II
35	6681.23	II	90	1078.83	V	70	1405.32	IV	20	920.554	II
85	6730.73	I	110	1079.60	V	73	1465.87	IV	300	971.35	V
50	6752.67	II	60	1080.99	V	90	1495.07	III	300	990.66	V
26	6786.33	II	80	1085.00	III	50	1534.46	III	50	999.101	II
100	6828.25	I	250	1085.01	V	10	1813.98	II	300	1004.38	V
100	6916.57	I	80	1087.37	V	15	1845.30	II	100	1016.638	II
50	6985.89	II	90	1091.71	V	20	2091.34	II	900	1045.71	V
75	6991.92	I	100	1094.36	V	90	2417.70	III	700	1072.66	V
60	6996.76	II	80	1095.10	V	90	2423.98	III	100	1075.072	II
45	7006.16	II	160	1102.83	V	15	2424.36	III	300	1085.51	II
35	7122.57	I	140	1103.03	V	10	2632.66	I	40	1088.45	III
170	7168.37	I	60	1105.61	III	10	2665.05	I	800	1089.49	V
28	7189.57	II	75	1105.62	V	20	2700.47	II	200	1098.71	II
28	7262.66	I	70	1106.17	V	15	2780.15	II	500	1106.74	II
35	7441.85	I	80	1118.34	V	50	3521.77	III	1000	1116.94	V
40	7464.36	I	120	1126.40	V	80	3581.19	III	500	1120.46	II
55	7562.97	I	130	1128.10	V	100	3589.34	III	700	1163.39	V
80	7733.50	I	120	1128.53	V	10	3731.10	III	200	1164.27	II
35	7846.35	II	100	1129.94	V	10	3806.60	III	500	1181.19	II
35	7856.93	I	130	1136.07	V	10	4032.99	I	500	1181.65	II
25	7930.25	II	67	1137.06	IV	10	4172.04	I	200	1188.73	II
18	8146.15	I	130	1150.23	V	15	4251.16	II	20	1188.99	IV
21	8668.63	I	90	1150.27	III	10 h	4254.04	II	300	1191.26	II
21 h	8832.06	II	70	1156.10	IV	10	4255.77	II	700	1222.30	V
14 h	8849.14	I	120	1156.51	V	40	4262.00	II	20	1229.81	IV
18 h	8867.31	I	35	1157.74	V	100	4380.69	III	500	1237.059	II
5000	14332.88	III	70	1163.60	IV	150	4381.76	III	500	1261.905	II
Gallium			25	1169.40	V	100	4863.00	III	100	1264.710	II
Ga Z = 31			75	1170.58	IV	150	4993.78	III	200	1401.24	II
14	294.53	IV	48	1171.71	IV	10	5808.28	III	200	1538.091	II
61	295.67	II	40	1178.95	V	20	5848.25	III	500	1576.855	II
30	298.44	V	68	1185.23	IV	15	5993.51	III	75	1581.070	II
30	300.01	V	40	1186.06	IV	10	6334.2	II	100	1602.486	II
30	302.86	V	73	1190.89	IV	2000	6396.56	I	3 r	1615.57	I
41	304.99	IV	73	1193.02	IV	1000	6413.44	I	2 r	1624.130	I
30	307.03	V	75	1195.02	IV	10 h	7251.4	I	2 r	1630.173	I
30	308.26	V	69	1201.54	IV	20 h	7403.0	I	3 r	1636.31	I
30	311.79	V	72	1206.89	IV	30 h	7464.0	I	4 r	1639.730	I
30	313.68	V	80	1213.17	V	10 h	7620.5	I	2	1647.531	I
40	319.41	V	63	1216.15	IV	50 h	7734.77	I	200	1649.194	II
40	322.31	V	50	1228.03	IV	100 h	7800.01	I	2	1651.528	I
50	322.99	V	60	1236.38	IV	15 h	8002.55	I	4 r	1651.955	I
30	323.10	V	60	1238.59	IV	20 h	8074.25	I	3	1661.345	I
40	324.25	V	75	1245.53	IV	100 h	8311.86	I	4 r	1663.539	I
40	324.95	V	83	1258.77	IV	200 h	8386.49	I	10 h	1665.275	I
40	326.14	V	81	1264.66	IV	200 h	9492.92	I	4	1667.802	I
30	326.77	V	82	1267.15	IV	200 h	9493.12	I	3 r	1670.608	I
30	328.65	V	90	1267.16	III	300 h	9589.36	I	100 r	1691.090	I
25	423.18	IV	81	1279.24	IV	100 h	10905.95	I	200 r	1716.784	I
16	439.92	IV	15	1283.64	V	400	11949.12	I	100 h	1739.102	I
50	620.00	III	80	1285.33	IV	200	12109.78	I	100	1742.195	I
40	622.01	III	80	1293.46	III	60	14996.64	I	50	1746.065	I
90	806.51	III	60	1295.36	III	60	22016.81	I	200	1750.043	I
90	817.30	III	82	1295.86	IV	70	22568.71	I	100	1758.279	I
50	828.70	III	83	1299.46	IV	**Germanium**			100 h	1764.185	I
20	878.17	V	82	1303.53	IV	**Ge Z = 32**			100 h	1765.284	I
40	973.21	V	80	1309.68	IV	700	294.51	V	50 h	1766.433	I
40	989.75	V	80	1314.82	IV	1000	295.64	V	200	1774.176	I
90	1014.47	V	60	1323.15	III	200	304.98	V	200	1785.046	I
90	1019.71	V	85	1338.09	IV	20	621.52	V	100 h	1793.071	I
120	1050.48	V	77	1347.03	IV	50	724.21	V	75 h	1801.432	I

Intensity	Wavelength/Å		Intensity	Wavelength/Å		Intensity	Wavelength/Å		Intensity	Wavelength/Å	
200 h	1841.328	I	1200	2651.172	I	4	9492.559	I	300	1428.93	III
200 h	1842.410	I	500	2651.568	I	7	9625.664	I	80	1429.19	I
100 h	1844.410	I	500	2691.341	I	5	10039.436	I	250	1430.06	III
100 h	1845.872	I	850	2709.624	I	4	10200.952	I	275	1433.37	III
100 h	1846.958	I	400	2729.78	II	10	10382.427	I	50	1435.79	I
200	1853.134	I	40	2740.426	I	10	10404.913	I	250	1435.81	III
500 r	1860.086	I	650	2754.588	I	8	10734.068	I	300	1439.12	III
100	1865.052	I	70	2793.925	I	8	10947.416	I	200	1441.21	III
300 r	1874.256	I	80	2829.008	I	10	11125.130	I	150	1446.37	III
100	1895.197	I	1000	2831.843	II	230	11252.83	I	250	1448.42	III
500 r	1904.702	I	1000	2845.527	II	600	11714.76	I	250	1454.95	III
50 h	1908.434	I	750	3039.067	I	1300	12069.20	I	100	1464.72	III
30	1912.409	I	600	3067.021	I	1050	12391.58	I	150	1471.28	III
300 r	1917.592	I	20	3124.816	I	235	13107.61	I	100	1474.73	III
100 h	1923.467	I	35	3211.86	III	470	14822.38	I	150	1481.10	III
500 r	1929.826	I	40	3255.05	III	150	16759.79	I	100	1481.76	I
10 h	1934.048	I	110	3269.489	I	135	17214.34	I	300	1487.15	III
100 r	1937.483	I	40	3434.03	III	70	18811.86	I	250	1487.91	III
500	1938.008	II	300	3499.21	II	62	19279.24	I	200	1489.47	III
100 r	1938.300	I	60	3554.19	IV	28	20673.64	I	250	1500.37	III
500	1938.891	II	50	3676.65	IV		**Gold**		200	1502.47	III
30 s	1944.116	I	200	4178.96	III		**Au Z = 79**		200	1503.74	III
200	1944.731	I	70	4226.562	I	100	843.44	III	100	1542.00	III
200	1955.115	I	200	4260.85	III	100	845.14	III	100	1548.50	III
500	1962.013	I	150	4291.71	III	200	945.10	III	200	1567.54	III
30 h	1963.373	I	10	4685.829	I	100	1040.63	III	200	1574.85	III
30	1965.383	I	1000	4741.806	II	80	1044.49	III	200	1579.44	III
200	1970.880	I	1000	4814.608	II	80	1046.81	III	150	1584.10	III
200	1979.274	II	50	4824.097	II	100 h	1239.96	III	200	1587.16	I
300 h	1987.849	I	100	5131.752	II	100	1278.51	III	200	1589.56	III
300	1988.267	I	200	5178.648	II	100	1314.84	III	150	1593.41	III
500 r	1998.887	I	3	5194.583	I	100 h	1328.37	I	70	1598.24	I
200	2011.29	I	6	5265.892	I	200	1336.72	III	200	1600.51	III
1700	2019.068	I	6	5513.263	I	180	1341.68	III	250	1617.16	III
2400 r	2041.712	I	8	5564.741	I	100	1348.89	III	100	1617.78	III
1600 r	2043.770	I	8	5607.010	I	150	1350.32	III	500	1621.93	III
420	2054.461	I	6	5616.135	I	150	1355.61	III	100	1624.34	I
220 h	2057.238	I	7	5621.426	I	150	1356.13	III	300 d	1629.13	III
750 r	2065.215	I	6	5664.226	I	50	1363.98	I	250	1638.88	III
2600 r	2068.656	I	5	5664.842	I	500	1365.40	III	50	1639.90	I
420	2086.021	I	9	5691.954	I	200	1367.17	III	100	1644.17	III
2000 r	2094.258	I	6	5701.776	I	60	1368.62	I	150	1646.67	I
25	2104.45	III	5	5717.877	I	70	1374.82	I	250	1652.74	III
240	2105.824	I	6	5801.029	I	50	1375.76	I	250	1664.77	III
95 h	2124.744	I	9	5802.093	I	180	1377.73	III	100	1665.76	I
50 h	2186.451	I	1000	5893.389	II	150	1378.69	III	100	1668.11	III
340 r	2198.714	I	500	6021.041	II	150	1379.98	III	125	1673.93	III
15	2220.375	I	75	6283.452	II	125	1380.53	III	1000	1693.94	III
18	2256.001	I	100	6336.377	II	200	1381.36	III	150	1697.09	III
18	2314.201	I	100	6484.181	II	80	1382.75	I	200	1698.98	III
24	2327.918	I	6	6557.488	I	50	1385.33	I	200	1699.34	I
15	2359.233	I	50	7049.369	II	300	1385.79	II	200	1700.00	III
20	2379.144	I	30	7145.390	II	100	1389.41	III	200	1702.25	III
10	2389.472	I	5	7353.334	I	180	1391.46	III	100	1707.53	III
15	2397.885	I	7	7384.208	I	60	1392.27	I	250	1710.16	III
130	2417.367	I	10	7833.575	I	180	1396.00	III	200	1715.69	III
30	2436.412	I	10	8031.039	I	50	1402.12	I	100	1716.71	III
30	2488.25	IV	6	8044.165	I	100	1402.91	III	300	1717.83	III
90	2497.962	I	10	8256.013	I	70	1407.38	I	500	1727.31	III
500	2500.54	II	10	8482.21	I	100	1408.45	I	100 d	1733.17	III
70	2533.230	I	9	8700.60	I	225	1409.50	III	300	1738.48	III
20	2542.44	IV	5	9068.785	I	250	1413.80	III	150	1744.39	III
3	2556.298	I	5	9095.957	I	100	1414.27	III	500	1746.10	III
28	2589.188	I	6	9398.868	I	100	1417.09	III	500	1756.92	III
500	2592.534	I	20	9474.993	II	125	1417.39	III	500	1761.95	III
8	2644.184	I	20	9475.645	II	150	1427.42	III	300	1767.42	III

Intensity	Wavelength/Å		Intensity	Wavelength/Å		Intensity	Wavelength/Å		Intensity	Wavelength/Å	
100	1774.42	III	100	3122.50	II	270	921.67	V	450	2393.36	II
800	1775.17	III	1600	3122.78	I	245	951.62	V	670	2393.83	II
200	1776.40	III	100	3194.72	I	180	960.12	V	540	2405.42	II
100	1780.57	III	100	3227.99	III	180	964.74	V	370	2410.14	II
60	1783.22	II	300	3230.63	I	160	971.51	V	320	2417.69	II
300	1786.11	III	300	3308.30	I	160	1092.76	V	390	2447.25	II
150	1792.65	III	300	3309.64	I	160	1201.76	V	450	2460.49	II
500	1793.76	III	100	3309.86	III	270	1232.03	V	400	2461.74	III
200	1801.98	III	100	3320.12	I	200	1233.59	V	430	2464.19	II
400	1805.24	III	100	3355.15	I	160	1237.42	V	210	2469.18	II
100	1809.81	III	100 h	3391.31	I	160	1239.53	V	2000	2495.16	III
400	1821.17	III	100	3395.40	I	160	1244.46	V	290	2496.99	II
400	1844.89	III	100	3467.21	I	440	1396.66	V	580	2512.69	II
150	1848.83	III	300	3557.36	I	270	1400.09	V	580	2513.03	II
500	1861.80	III	300	3586.73	I	160	1401.70	V	1000	2515.16	III
150	1871.92	III	100	3611.57	I	370	1407.17	V	890	2516.88	II
100	1879.83	I	100 h	3631.31	I	370	1408.38	V	340	2531.19	II
150	1918.28	III	300	3637.90	I	270	1412.28	V	200	2537.33	II
100	1932.04	III	100 h	3645.02	I	270	1413.51	V	320	2551.40	II
100	1935.42	III	100	3650.74	I	160	1421.96	V	400 h	2560.74	III
200	1948.79	III	100	3709.62	I	220	1422.53	V	250	2563.61	II
100	1958.47	III	100	3796.01	I	370	1433.43	V	300 h	2567.46	III
400	1989.63	III	100	3874.73	I	370	1437.27	V	890	2571.67	II
150	1996.85	III	100 h	3892.26	I	500	1437.73	V	320	2573.90	II
11000	2012.00	I	400	3897.86	I	370	1445.40	V	320	2576.82	II
2600	2021.38	I	300	3909.38	I	270	1457.91	V	300	2578.14	II
150	2082.09	II	100	3927.69	I	100	1717.21	IV	320	2582.54	II
300	2083.09	III	400	4040.93	I	200	1719.32	V	390	2606.37	II
60	2110.68	II	700	4065.07	I	550	1729.08	V	450	2607.03	II
100	2159.08	III	100	4084.10	I	750	1731.83	V	230	2613.60	II
80	2167.33	III	100	4241.80	I	750	1733.96	V	450	2622.74	II
200	2172.20	III	200	4315.11	I	440	1741.74	V	160	2637.00	I
100	2184.11	III	120 h	4437.27	I	1000	1749.11	V	1100	2638.71	II
500	2188.97	III	250	4488.25	I	1000	1750.19	V	160	2641.41	II
70	2248.56	II	900 h	4607.51	I	500	1760.89	V	160	2642.75	I
80	2263.62	II	100 h	4620.56	I	370	1765.62	V	670	2647.29	II
300	2322.27	III	500	4792.58	I	270	1774.02	V	160	2657.84	II
180	2352.65	I	100	4811.60	I	6200	2012.78	II	210	2661.88	II
100	2382.40	III	100	5147.44	I	8500	2028.18	II	290	2683.35	II
120	2387.75	I	300	5230.26	I	300	2070.94	III	200	2687.22	III
150	2402.71	III	100 h	5261.76	I	1200	2096.18	II	670	2705.61	I
150	2405.12	III	100	5655.77	I	200 h	2099.30	III	210	2712.42	II
2600	2427.95	I	100 h	5721.36	I	200 h	2110.31	III	250	2718.59	I
60	2533.52	II	300	5837.37	I	200	2155.66	III	710	2738.76	II
250	2641.48	I	100 h	5862.93	I	200	2183.50	III	200	2743.64	I
3400	2675.95	I	300 h	5956.96	I	200	2195.44	III	360	2751.81	II
1100	2748.25	I	600	6278.17	I	540	2210.82	II	500	2753.60	III
100	2780.82	I	100	6562.68	I	200	2234.59	III	450	2761.63	I
1000	2802.04	II	600	7510.73	I	320	2254.01	II	160	2766.96	I
300	2819.79	II	10	8145.06	I	160	2255.15	II	170	2773.02	I
100	2822.55	II	10	9254.28	I	250	2266.83	II	980	2773.36	II
100 h	2825.44	II		**Hafnium**		620	2277.16	II	180	2774.02	II
300	2837.85	II		**Hf Z = 72**		200 h	2313.44	III	390	2779.37	I
100	2846.92	II	220	545.41	V	230	2321.14	II	230	2808.00	II
100	2856.74	II	180	600.00	V	580	2322.47	II	230	2813.86	II
300	2883.45	I	200	618.27	IV	300	2323.25	II	170	2814.48	II
300	2891.96	I	200	644.54	IV	300	2324.89	II	230	2817.68	I
100	2893.25	II	400	647.39	IV	200	2332.97	II	200	2819.74	I
100	2907.04	II	600	665.65	IV	300	2336.47	III	1200	2820.22	II
300	2913.52	II	200	673.49	IV	200	2337.33	II	490	2822.68	II
300	2918.24	II	270	867.25	V	230	2343.32	II	180	2833.28	I
100	2954.22	II	180	875.88	V	320	2347.44	II	410	2845.83	I
100 h	2990.27	II	180	885.58	V	540	2351.22	II	270	2849.21	II
300	2994.80	II	180	896.14	V	250	2380.30	II	270	2850.96	II
320	3029.20	I	180	901.54	V	250	2383.540	III	180	2851.21	II
300	3065.42	I	180	919.10	V	170	2393.18	II	180	2860.56	I

Intensity	Wavelength/Å		Intensity	Wavelength/Å		Intensity	Wavelength/Å		Intensity	Wavelength/Å	
760	2861.01	II	270	3255.28	II	1400	3777.64	I	85	6789.27	I
760	2861.70	II	180	3273.66	II	1400	3785.46	I	160	6818.94	I
2100	2866.37	I	200 h	3279.67	III	650	3793.37	II	160	7063.83	I
210	2887.14	I	270	3279.98	II	850 d	3800.38	I	570	7131.81	I
800	2889.62	I	160	3291.05	I	320	3811.78	I	650	7237.10	I
1800	2898.26	I	210	3306.12	I	1300	3820.73	I	410	7240.87	I
1200	2904.41	I	340	3310.27	I	280	3830.02	I	360	7624.40	I
890	2904.75	I	670	3312.86	I	800	3849.18	I	110	7740.17	I
2000	2916.48	I	180	3317.99	II	600	3858.31	I	310	7845.35	I
580	2918.58	I	890	3332.73	I	230	3860.91	I	130	7920.71	I
320	2919.59	II	370	3352.06	I	200	3872.55	II	250	7994.73	I
180	2924.62	I	230	3358.91	I	160	3877.10	II	130	8204.58	I
490	2929.63	II	180	3360.06	I	380	3880.82	II	150	8546.48	I
450	2929.90	I	180	3378.93	I	200	3882.52	I	160	8640.06	I
710	2937.80	II	230	3384.70	II	200	3889.23	I	40	8711.24	I
2000	2940.77	I	170	3386.21	I	200	3889.33	I	65	9004.73	I
160	2944.71	I	800	3389.83	II	620	3899.94	I	**Helium**		
1200	2950.68	I	230	3392.81	I	620	3918.09	II	**He Z = 2**		
1100	2954.20	I	230	3394.59	II	200	3923.90	II	15	231.454	II
540	2958.02	I	230	3397.26	I	320	3931.38	I	20	232.584	II
1400	2964.88	I	230	3397.60	I	410	3951.83	I	30	234.347	II
620	2966.93	I	2300	3399.80	II	160	3968.01	I	50	237.331	II
710	2968.81	II	170	3400.21	I	200	3973.48	I	100	243.027	II
890	2975.88	II	180	3402.51	I	180	4032.27	I	300	256.317	II
1100	2980.81	I	230	3410.17	II	230	4062.84	I	1000	303.780	II
210	2982.72	I	230	3417.34	I	180	4083.35	I	500	303.786	II
170	3000.10	II	410	3419.18	I	540	4093.16	II	10	320.293	I
800	3005.56	I	200	3428.37	I	1100	4174.34	I	2	505.500	I
1100	3012.90	II	250	3438.24	II	160	4206.58	II	3	505.684	I
540	3016.78	I	710	3472.40	I	190	4209.70	I	4	505.912	I
1100	3016.94	II	200	3478.99	II	170	4228.08	I	5	506.200	I
980	3018.31	I	480	3479.28	II	170	4232.44	II	7	506.570	I
1200	3020.53	I	250	3495.75	II	170	4260.98	I	10	507.058	I
410	3031.16	II	250	3497.16	I	200	4263.39	I	15	507.718	I
710	3050.76	I	980	3497.49	I	170	4272.85	II	20	508.643	I
1100	3057.02	I	1200	3505.23	II	320	4294.79	I	25	509.998	I
850	3067.41	I	980	3523.02	I	160	4330.27	I	35	512.098	I
2100	3072.88	I	980	3535.54	II	180	4336.66	II	50	515.616	I
170	3074.10	I	760	3536.62	I	250	4356.33	I	100	522.213	I
250	3074.79	I	180	3548.81	I	180	4370.97	II	400	537.030	I
430	3080.84	I	540	3552.70	II	160	4417.91	I	1000	584.334	I
200	3096.76	I	1300	3561.66	II	200	4438.04	I	50	591.412	I
340	3101.40	II	270	3567.36	I	250	4565.94	I	5	958.70	II
710	3109.12	II	1100	3569.04	II	500 d	4598.80	I	6	972.11	II
710	3131.81	I	210	3597.42	II	230	4620.86	I	8	992.36	II
850	3134.72	II	540	3599.87	I	210	4655.19	I	15	1025.27	II
170	3139.65	II	800	3616.89	I	120	4699.01	I	30	1084.94	II
220	3145.32	II	320	3630.87	II	160	4782.74	I	35	1215.09	II
220	3148.41	I	800	3644.36	II	310	4800.50	I	50	1215.17	II
450	3156.63	I	320	3649.10	I	130	4859.24	I	120	1640.34	II
270	3159.82	I	200	3651.84	I	120	4975.25	I	180	1640.47	II
710	3162.61	II	220	3665.35	II	95	5018.20	I	7	2385.40	II
450	3168.39	I	200	3672.27	I	95	5047.45	I	9	2511.20	II
890	3172.94	I	480	3675.74	I	75	5170.18	I	50	2577.6	I
450	3176.86	II	2200	3682.24	I	230	5181.86	I	1	2723.19	I
220	3181.01	I	280	3696.51	I	110	5243.99	I	12	2733.30	II
360	3193.53	II	240	3699.72	II	120	5294.87	I	2	2763.80	I
670	3194.19	II	340	3701.15	II	110	5354.73	I	10	2818.2	I
200	3196.93	I	1000	3717.80	I	110	5373.86	I	4	2829.08	I
310	3206.11	I	650	3719.28	II	75	5452.92	I	10	2945.11	I
180	3210.98	I	160	3729.10	I	230	5550.60	I	40	3013.7	I
180	3217.30	II	460	3733.79	I	230	5552.12	I	20	3187.74	I
180	3220.61	II	160	3737.88	II	95	5613.27	I	3	3202.96	II
360	3247.66	I	400	3746.80	I	160	5719.18	I	15	3203.10	II
220	3249.53	I	170	3766.92	II	95	6098.67	I	1	3354.55	I
890	3253.70	II	200	3768.25	I	95	6185.13	I	2	3447.59	I

Intensity	Wavelength/Å	Intensity	Wavelength/Å	Intensity	Wavelength/Å	Intensity	Wavelength/Å
1	3587.27 I	200	17002.47 I	410	3580.75 II	220	4979.97 I
3	3613.64 I	1	18555.55 I	410	3581.83 II	90	4995.05 I
2	3634.23 I	6	18636.8 II	630 c	3592.23 II	130	5042.37 I
3	3705.00 I	500	18685.34 I	1100 cw	3598.77 II	80	5093.07 I
1	3732.86 I	200	18697.23 I	540 c	3600.95 II	140	5127.81 I
10	3819.607 I	100	19089.38 I	410	3618.43 I	130	5142.59 II
1	3819.76 I	20	19543.08 I	430 c	3626.69 II	110	5143.22 II
500	3888.65 I	1000	20581.30 I	490	3627.25 II	160	5149.59 II
20	3964.729 I	80	21120.07 I	430 c	3631.76 II	90 c	5167.88 I
1	4009.27 I	10	21121.43 I	430 c	3638.30 II	130 c	5182.11 I
50	4026.191 I	20	21132.03 I	1600 c	3662.29 I	90	5190.11 II
5	4026.36 I	3	30908.5 II	1400	3667.97 I	65	5251.82 I
12	4120.82 I	4	40478.90 I	720	3679.19 I	90	5301.25 I
2	4120.99 I	**Holmium**	**Ho Z = 67**	670	3679.70 I	80	5330.11 I
3	4143.76 I	170	2502.91 II	720	3682.65 I	90	5359.99 I
10	4387.929 I	170	2533.80 I	580	3690.65 I	100	5407.08 I
3	4437.55 I	190	2605.86 II	410	3700.04 I	70	5566.52 I
200	4471.479 I	270	2750.35 II	490 c	3702.35 II	65	5627.60 I
25	4471.68 I	270	2769.89 II	430	3712.88 I	140	5659.58 I
6	4685.4 II	300	2824.20 II	450	3720.72 I	140 c	5691.47 I
30	4685.7 II	270 c	2831.69 II	1100	3731.40 I	140 c	5696.57 I
30	4713.146 I	270	2849.10 II	810	3736.35 I	140 c	5860.28 I
4	4713.38 I	360	2880.26 II	3200 cw	3748.17 II	70 c	5882.99 I
20	4921.931 I	460	2880.98 II	8900 c	3796.75 II	70	5921.76 I
100	5015.678 I	570 c	2909.41 II	8900 c	3810.73 II	70 cw	5948.03 I
10	5047.74 I	410 c	2979.63 II	410 c	3835.35 II	70	5972.76 I
5	5411.52 II	410	2987.64 II	1300 cw	3837.51 II	90	5973.52 I
500	5875.62 I	480 c	3049.38 II	410 c	3842.05 II	230 c	5982.90 I
100	5875.97 I	410 c	3054.00 II	1100	3843.86 II	120	6081.79 I
8	6560.10 II	500 c	3057.45 II	490 c	3846.73 II	70	6208.65 I
100	6678.15 I	500 c	3082.34 II	1800 c	3854.07 II	70 c	6305.36 I
3	6867.48 I	910	3084.36 II	2700 c	3861.68 II	70	6550.97 I
200	7065.19 I	430 c	3086.54 II	3000 c	3888.96 II	260	6604.94 I
30	7065.71 I	760	3118.50 II	13000 c	3891.02 II	120	6628.99 I
50	7281.35 I	580 c	3166.62 II	1300 cw	3905.68 II	55 cw	6694.32 I
1	7816.15 I	810	3173.78 II	580	3955.73 I	55 c	6785.43 I
2	8361.69 I	810 c	3181.50 II	490	3959.68 I	40 cw	6939.49 I
2	9063.27 I	980 c	3281.97 II	2700	4040.81 I	45 cw	6950.39 I
2	9210.34 I	630 c	3337.23 II	5400 c	4045.44 II	140	7555.09 I
10	9463.61 I	980 c	3343.58 II	8100	4053.93 I	40 c	7628.42 I
4	9516.60 I	8100 c	3398.98 II	1700	4065.09 II	50 c	7693.15 I
3	9526.17 I	810 c	3410.26 II	720	4068.05 I	60 cw	7815.48 I
1	9529.27 I	1400 c	3414.90 II	8900	4103.84 I	60	7894.64 I
1	9603.42 I	5400	3416.46 II	2900	4108.62 I	50	8512.94 I
3	9702.60 I	1200	3421.63 II	1500	4120.20 I	40	8670.19 I
6	10027.73 I	2000 c	3425.34 II	1300	4125.65 I	90	8915.98 II
2	10031.16 I	2000 c	3428.13 II	4300	4127.16 I	**Hydrogen**	**H Z = 1**
15	10123.6 II	3200	3453.14 II	1500	4136.22 I	15	926.226 I
1	10138.50 I	16000 c	3456.00 II	980 cw	4152.61 II	20	930.748 I
10	10311.23 I	1600	3461.97 II	8100	4163.03 I	30	937.803 I
2	10311.54 I	810 c	3473.91 II	2500	4173.23 I	50	949.743 I
3	10667.65 I	5400 c	3474.26 II	540	4194.35 I	100	972.537 I
300	10829.09 I	6300	3484.84 II	2000	4227.04 I	300	1025.722 I
1000	10830.25 I	2500 c	3494.76 II	1300 cw	4254.43 I	1000	1215.668 I
2000	10830.34 I	810 c	3498.88 II	490	4264.05 I	500	1215.674 I
9	10913.05 I	810	3510.73 I	1300	4350.73 I	5	3835.384 I
3	10917.10 I	4100 c	3515.59 II	300	4477.64 II	6	3889.049 I
4	11626.4 II	410 c	3519.94 II	290	4629.10 II	8	3970.072 I
30	11969.12 I	630	3540.76 II	80	4701.69 II	15	4101.74 I
20	12527.52 I	1600	3546.05 II	130	4709.84 II	30	4340.47 I
50	12784.99 I	1100 c	3556.78 II	130 c	4717.52 I	80	4861.33 I
20	12790.57 I	410	3560.15 II	290	4742.04 II	120	6562.72 I
7	12845.96 I	410 c	3573.24 II	100 c	4757.01 I	180	6562.852 I
10	12968.45 I	630 c	3574.80 II	65	4782.92 I	5	9545.97 I
2	12984.89 I	810	3579.12 I	290	4939.01 I	7	10049.4 I
12	15083.64 I			250 c	4967.21 II		

Intensity	Wavelength/Å		Intensity	Wavelength/Å		Intensity	Wavelength/Å		Intensity	Wavelength/Å	
12	10938.1	I	40	2281.64	II	250 w	3962.35	II	180 w	6115.9	II
20	12818.1	I	100 c	2306.05	II	120 c	4004.66	II	230 w	6128.7	II
40	18751.0	I	90 d	2313.21	II	140 d	4013.92	II	240 w	6129.4	II
5	21655.3	I	70 d	2327.95	II	100	4023.77	III	320 w	6132.1	II
8	26251.5	I	80 h	2334.57	II	150	4032.32	III	150 c	6140.0	II
15	40511.6	I	110 d	2382.63	II	410 w	4056.94	II	90	6143.23	II
4	46525.1	I	70 h	2427.20	II	100	4071.57	III	140 c	6148.10	II
6	74578.	I	100	2447.90	II	100	4072.93	III	190 w	6149.5	II
	Indium		110 d	2488.62	II	17000	4101.76	I	80	6161.15	II
	In Z = 49		90	2488.95	II	140 c	4205.14	II	180 w	6162.45	II
17	378.61	V	80	2498.59	II	100 d	4213.04	II	200	6197.72	III
17	386.21	V	100	2499.60	II	110 d	4219.66	II	100 c	6224.28	II
14	388.91	V	90 d	2500.99	II	100	4252.68	III	280 w	6228.3	II
25	393.89	V	110 d	2512.31	II	150 d	4372.87	II	140 w	6231.1	II
25	400.57	V	100	2521.37	I	150 c	4500.78	II	270 w	6304.8	II
25	402.39	V	100	2527.41	III	18000	4511.31	I	290 w	6362.3	II
622	472.71	IV	160 d	2554.44	II	110 c	4549.01	II	300 w	6469.0	II
689	479.39	IV	1100	2560.15	I	140 c	4570.85	II	210 c	6541.20	II
709	498.62	IV	200	2601.76	I	180 w	4578.02	II	190 c	6751.88	II
10	882.24	III	90 d	2654.70	II	180 w	4578.40	II	180 c	6765.9	II
10	890.84	III	100 d	2662.63	II	140 c	4616.08	II	100 c	6783.72	II
10	915.87	III	140 d	2668.65	II	170 c	4617.17	II	320 w	6891.5	II
85	954.67	IV	140 d	2674.56	II	250 c	4620.14	II	380 w	7182.9	II
87	973.50	IV	80	2683.12	II	150 w	4620.70	II	180 c	7255.0	II
86	991.60	IV	1600	2710.26	I	170 c	4627.30	II	210 c	7276.5	II
89	1024.68	IV	300	2713.94	I	140 c	4637.04	II	180 c	7303.4	II
85	1024.79	IV	80	2726.15	III	380 c	4638.16	II	320 c	7350.6	II
88	1031.45	IV	130 d	2749.75	II	220 c	4644.58	II	100 c	7632.7	II
82	1031.98	IV	700	2753.88	I	360 c	4655.62	II	100 c	7682.9	II
80	1054.43	IV	90 d	2818.97	II	320 w	4656.74	II	210 c	7740.7	II
84	1063.03	IV	180 c	2836.92	I	190 c	4681.11	II	100 c	7776.96	II
83	1068.25	IV	80	2865.68	II	450 w	4684.8	II	180 c	7789.0	II
82	1069.82	IV	120 d	2890.18	II	90 d	4907.06	II	90 w	7840.9	II
86	1077.64	IV	1100	2932.63	I	150 c	4973.77	II	240 c	8227.0	II
90	1082.10	IV	100	2941.05	II	80 h	5109.36	II	100 w	8813.5	II
83	1086.33	IV	100	2982.80	III	100 w	5115.14	II	80 c	8832.6	II
82	1096.81	IV	110 c	2999.40	II	140 c	5117.40	II	120 c	9197.7	II
84	1097.18	IV	100	3008.08	III	270 c	5120.80	II	120 c	9202.0	II
85	1116.10	IV	8000	3039.36	I	200 w	5121.75	II	220 w	9213.0	II
80	1124.06	IV	110 d	3099.80	II	80 d	5129.85	II	160 d	9241.1	II
90	1131.46	IV	180 c	3101.8	II	240 c	5175.42	II	100	9977.86	I
85	1144.43	IV	130 c	3138.60	II	140 c	5184.44	II	200	10257.03	I
80	1145.41	IV	80 c	3142.75	II	200	5248.77	III	100 h	10744.31	I
89	1146.62	IV	130 d	3146.70	II	150 c	5309.45	II	20	11334.72	I
83	1154.11	IV	150	3155.77	II	80	5411.41	II	20	11731.48	I
84	1154.60	IV	100 c	3158.40	II	140 c	5418.45	II	10	12912.59	I
90	1157.71	IV	90 c	3176.30	II	220 w	5436.70	II	9	13429.96	I
90	1157.82	IV	90 d	3198.11	II	130 c	5497.50	II	7	14719.08	I
85	1159.78	IV	13000	3256.09	I	140 c	5507.08	II	6	22291.06	I
88	1176.50	IV	3000	3258.56	I	320 c	5513.00	II	7	23879.13	I
85	1191.58	IV	90 c	3338.50	II	250 w	5523.28	II		**Iodine**	
83	1204.87	IV	100 c	3404.28	II	130 c	5536.50	II		**I Z = 53**	
90	1206.55	IV	110 d	3438.40	II	190 w	5555.45	II	30	363.78	V
88	1221.50	IV	180 c	3693.91	II	240 c	5576.90	II	36	380.74	V
85	1221.90	IV	95 c	3708.13	II	200 w	5636.70	II	45	565.53	V
85	1233.58	IV	380 w	3716.14	II	100	5645.15	III	50	607.57	V
87	1235.84	IV	120 c	3718.30	II	160 c	5708.50	II	6	612.46	IV
90	1373.20	IV	160 c	3718.72	II	100 c	5721.80	II	6	666.81	III
88	1398.77	IV	160 c	3723.40	II	100	5819.50	III	8	705.11	III
81	1412.09	IV	170 w	3795.21	II	210 c	5853.15	II	7	784.64	III
100	1625.42	III	230 c	3799.21	II	490 w	5903.4	II	7	784.80	III
100	1748.83	III	250 c	3834.65	II	260 w	5915.4	II	8	795.52	III
30	1842.41	III	200 c	3842.18	II	120 c	5918.78	II	7	919.28	IV
40	1850.30	III	100	3852.82	III	130 c	6062.9	II	10000	1034.66	II
30	2154.08	III	100	3889.78	II	250 c	6095.95	II	8	1094.20	III
50	2205.28	II	100 c	3902.07	II	210 c	6108.66	II	10000	1139.80	II

Intensity	Wavelength/Å		Intensity	Wavelength/Å		Intensity	Wavelength/Å		Intensity	Wavelength/Å	
10000	1160.56	II	8	2519.74	IV	2000 c	7700.20	I	10	30361.93	I
20000	1166.48	II	8	2545.67	IV	600	7897.98	I	8	30383.88	I
10000	1178.65	II	7	2545.71	III	500	7969.48	I	10	34295.73	I
15000	1187.34	II	8	2652.23	IV	1000	8003.63	I	9	34513.11	I
10000	1190.85	II	5000	3078.75	II	99000	8043.74	I	3	40228.54	I
200	1218.41	I	200	4102.23	I	300 d	8065.70	I	2	41633.80	I
20000	1220.89	II	200	4129.21	I	1000	8090.76	I		**Iridium**	
600	1224.05	I	100 d	4134.15	I	800 c	8169.38	I		**Ir Z = 77**	
600	1224.08	I	500 d	4321.84	I	500 d	8222.57	I	9900	2010.65	I
500	1228.89	I	250	4763.31	I	4000	8240.05	I	8700	2022.35	I
20000	1234.06	II	1000	4862.32	I	10000 c	8393.30	I	15000	2033.57	I
600	1251.34	I	200	4916.94	I	1000	8486.11	I	6200	2052.22	I
8	1252.35	III	10000	5119.29	I	1500 c	8664.95	I	5000	2060.64	I
2500	1259.15	I	3000 c	5161.20	II	500 c	8700.80	I	3700	2083.22	I
3000	1259.51	I	1000	5234.57	I	250 d	8748.22	I	3100	2085.74	I
800	1261.27	I	3000 c	5245.71	II	1000	8853.24	I	17000	2088.82	I
600	1267.57	I	10000	5338.22	II	2000	8853.80	I	14000	2092.63	I
600	1267.60	I	5000 c	5345.15	II	3000	8857.50	I	2700	2112.68	I
1500	1275.26	I	600 c	5427.06	I	1000 d	8898.50	I	1800	2119.54	I
3000	1289.40	I	3000	5435.83	II	400	8964.69	I	2000	2125.44	I
10000	1300.34	I	2000 c	5464.62	II	400	8993.13	I	4500	2126.81	II
3000	1302.98	I	10000	5625.69	II	5000	9022.40	I	2000	2127.52	I
3000	1313.95	I	2000 c	5690.91	II	15000	9058.33	I	4500	2127.94	I
3000	1317.54	I	4000 c	5710.53	II	1000	9098.86	I	3700	2148.22	I
2000	1330.19	I	1000 d	5764.33	I	12000	9113.91	I	2500	2150.54	I
20000	1336.52	II	2000	5894.03	I	600	9128.03	I	3500	2152.68	II
5000	1355.10	I	5000	5950.25	II	600	9227.74	I	2900	2155.81	I
3000	1357.97	I	300	5984.86	I	1000	9335.05	I	7900	2158.05	I
5000	1360.97	I	2000 d	6024.08	I	4000	9426.71	I	2100	2162.88	I
3000	1361.11	I	2000 c	6074.98	II	3000	9427.15	I	5800	2169.42	II
2500	1367.71	I	1000	6082.43	I	2000	9598.22	I	4500	2175.24	I
2500	1368.22	I	2000 c	6127.49	II	2000	9649.61	I	2700	2178.17	I
4000	1383.23	I	800	6191.88	I	3000 d	9653.06	I	1600	2187.43	II
3000	1390.75	I	500	6213.10	I	5000	9731.73	I	1100	2190.38	II
2000	1392.90	I	800	6244.48	I	500	10003.05	I	740	2191.64	I
2000	1400.01	I	1000	6293.98	I	750	10131.16	I	910	2208.09	II
8000	1425.49	I	500	6313.13	I	1000	10238.82	I	1300	2220.37	I
5000	1446.26	I	800	6330.37	I	400	10375.20	I	790	2221.07	II
5000	1453.18	I	400	6333.50	I	400	10391.74	I	2500	2242.68	II
5000	1457.39	I	2000	6337.85	I	5000	10466.54	I	620	2245.76	II
5000	1457.47	I	1000	6339.44	I	400	11236.56	I	2100	2253.38	I
10000	1457.98	I	500	6359.16	I	350	11558.46	I	2100	2255.10	I
2500	1458.79	I	1000	6566.49	I	320	11778.34	I	1400	2255.81	I
4000	1459.15	I	2000	6583.75	I	450	11996.86	I	350	2258.51	I
2500	1465.83	I	1000	6585.27	I	300	12033.69	I	1400	2258.86	I
1000	1485.92	I	5000	6619.66	I	150	12304.58	I	830	2264.61	I
5000	1492.89	I	500	6661.11	I	60	13149.16	I	1100	2266.33	I
5000	1507.04	I	500 c	6697.29	I	140	13958.27	I	1000	2268.90	I
5000	1514.68	I	400	6732.03	I	200	14287.02	I	660	2280.00	I
15000	1518.05	I	4000	6812.57	II	100	14460.00	I	950	2281.02	II
2500	1526.45	I	500	6989.78	I	225	15032.57	I	660	2281.91	I
5000	1593.58	I	500	7120.05	I	105	15528.65	I	330	2284.60	I
5000	1617.60	I	1200	7122.05	I	150	16037.33	I	330	2295.08	I
2500	1640.78	I	2000	7142.06	I	15	18275.71	I	790	2298.05	I
15000	1702.07	I	1000	7164.79	I	20	18348.52	I	460	2299.53	I
12000	1782.76	I	400 d	7191.66	I	15	18982.41	I	910	2300.50	I
5000	1799.09	I	700	7227.30	I	35	19070.17	I	2700	2304.22	I
75000	1830.38	I	1000	7236.78	I	110	19105.12	I	410	2305.47	I
15000	1844.45	I	500	7237.84	I	50	19370.02	I	210	2307.27	I
2000	2061.63	I	5000	7402.06	I	10	20648.69	I	910	2308.93	I
7	2361.13	IV	1000	7410.50	I	220	22183.03	I	460	2315.38	I
6	2372.45	IV	500	7416.48	I	150	22226.53	I	410	2321.45	I
7	2376.46	IV	5000	7468.99	I	30	22309.21	I	410	2321.58	I
8	2387.11	IV	500 c	7490.52	I	32	24420.82	I	210	2327.98	I
9	2426.10	IV	2000	7554.18	I	12	27365.42	I	540	2333.30	I
8	2475.35	IV	500 d	7556.65	I	9	27573.05	I	740	2333.84	I

Intensity	Wavelength/Å		Intensity	Wavelength/Å		Intensity	Wavelength/Å		Intensity	Wavelength/Å	
580	2334.50	I	330	2617.78	I	3400	3133.32	I	45	5625.55	I
1600	2343.18	I	210	2619.88	I	490	3168.88	I	35	5894.06	I
740	2343.61	I	250	2625.32	I	370	3177.58	I	20	6110.67	I
580	2355.00	I	700	2634.17	I	370	3198.92	I	12	6288.28	I
230	2357.53	II	250	2639.42	I	610	3212.12	I	10	6686.08	I
410	2358.16	I	3500	2639.71	I	370	3219.51	I	6	7834.32	I
500	2360.73	I	210	2644.19	I	5100	3220.78	I		**Iron**	
2500	2363.04	I	1800	2661.98	I	300	3229.28	I		**Fe Z = 26**	
370	2368.04	II	350	2662.63	I	470	3241.52	I	350	386.16	V
3500	2372.77	I	2700	2664.79	I	200	3262.01	I	350	386.88	V
290	2375.09	II	520	2669.91	I	390	3266.44	I	400	387.20	V
250	2377.28	I	520	2671.84	I	200	3322.60	I	400	387.50	V
250	2377.98	I	330	2673.61	I	560	3368.48	I	400	387.76	V
500	2379.38	I	270	2692.34	I	660	3437.02	I	400	387.78	V
540	2381.62	I	3000	2694.23	I	410	3448.97	I	400	395.90	V
210	2383.17	I	330	2772.46	I	3200	3513.64	I	400	404.62	V
1300	2386.89	I	250	2775.55	I	220	3515.95	I	400	405.50	V
2500	2390.62	I	520	2781.29	I	410	3522.03	I	800	407.42	V
2700	2391.18	I	330	2785.22	I	320	3558.99	I	600	407.44	V
230	2407.59	I	540	2797.35	I	1200	3573.72	I	400	407.49	V
290	2409.37	I	1600	2797.70	I	320	3594.39	I	500	407.75	V
290	2410.17	I	380	2798.18	I	220	3609.77	I	400	409.71	V
290	2410.73	I	410	2800.82	I	660	3628.67	I	400	410.20	V
540	2413.31	I	680	2823.18	I	220	3636.20	I	600	411.55	V
370	2415.86	I	1200	2824.45	I	300	3661.71	I	700	417.39	V
620	2418.11	I	820	2836.40	I	300	3664.62	I	700	418.04	V
210	2424.89	I	1100	2839.16	I	320	3674.98	I	500	418.47	V
370	2424.99	I	820	2840.22	I	200	3687.08	I	700	421.06	V
290	2425.66	I	3800	2849.72	I	200	3731.36	II	500	421.78	V
540	2427.61	I	380	2875.60	I	530	3747.20	I	500	422.31	V
540	2431.24	I	380	2875.98	I	3100	3800.12	I	500	426.06	V
1300	2431.94	I	270	2877.68	I	230	3817.24	I	500	426.11	V
270	2435.14	I	820	2882.64	I	480	3902.51	I	350	426.97	V
250	2445.34	I	650	2897.15	I	480	3915.38	I	17	525.69	IV
250	2447.76	I	260	2901.95	I	400	3934.84	I	15	526.29	IV
910	2452.81	I	260	2904.80	I	590	3976.31	I	13	526.63	IV
1300	2455.61	I	200	2907.24	I	460	3992.12	I	14	536.61	IV
230	2455.87	I	440	2916.36	I	350	4033.76	I	15	537.10	IV
210	2457.03	I	230	2918.57	I	370	4069.92	I	13	537.26	IV
210	2457.23	I	4400	2924.79	I	150	4070.68	I	14	537.79	IV
870	2467.30	I	1200	2934.64	I	100	4092.61	I	13	537.94	IV
3300	2475.12	I	880	2936.68	I	140	4115.78	I	13	552.14	IV
210	2478.11	I	250	2938.47	I	90	4172.56	I	14	607.53	IV
2100	2481.18	I	2700	2943.15	I	260	4268.10	I	13	608.80	IV
620	2493.08	I	230	2946.97	I	220	4311.50	I	10	813.38	III
210	2496.27	I	200	2949.76	I	160	4399.47	I	10	844.28	III
250	2502.63	I	1200	2951.22	I	65	4403.78	I	10 p	861.83	III
4100	2502.98	I	200	2974.95	I	110	4426.27	I	10	891.17	III
210	2513.71	I	440	2980.65	I	55	4478.48	I	10	950.33	III
990	2533.13	I	300	2996.08	I	55	4545.68	I	10	981.37	III
1100	2534.46	I	220	3002.25	I	30	4548.48	I	10 w	983.88	III
580	2537.22	I	600	3003.63	I	35	4568.09	I	12	1055.27	II
580	2542.02	I	270	3017.31	I	75	4616.39	I	15	1068.36	II
7900	2543.97	I	380	3029.36	I	26	4656.18	I	15	1071.60	II
790	2546.03	I	330	3039.26	I	50	4728.86	I	15	1096.89	II
210	2551.40	I	300	3047.16	I	26	4756.46	I	12	1099.12	II
210	2555.35	I	300	3049.44	I	65	4778.16	I	18	1112.09	II
910	2564.18	I	300	3057.28	I	30	4795.67	I	12	1121.99	II
210	2569.88	I	1600	3068.89	I	50	4938.09	I	12	1122.86	II
230	2572.70	I	320	3083.22	I	26	4970.48	I	12	1128.07	II
740	2577.26	I	240	3086.44	I	25	4999.74	I	12	1130.43	II
740	2592.06	I	390	3088.04	I	25	5002.74	I	15	1133.41	II
740	2599.04	I	510	3100.29	I	30	5014.98	I	12	1133.68	II
700	2608.25	I	510	3100.45	I	30	5123.66	I	12	1138.64	II
1800	2611.30	I	340	3120.76	I	35	5364.32	I	12	1142.33	II
210	2614.98	I	200	3121.78	I	75	5449.50	I	12	1143.23	II

Intensity	Wavelength/Å		Intensity	Wavelength/Å		Intensity	Wavelength/Å		Intensity	Wavelength/Å	
18	1144.95	II	13	1606.98	IV	18	1788.07	II	300	2178.118	I
12	1147.41	II	17	1609.10	IV	13	1792.10	IV	12	2180.41	III
15	1148.29	II	14	1609.83	IV	13	1796.93	IV	250	2186.486	I
12	1151.16	II	13	1610.47	IV	13	1827.98	IV	60	2186.892	I
12	1267.44	II	13	1611.20	IV	10	1869.83	III	120	2187.195	I
12	1272.00	II	13	1613.64	IV	12	1877.99	III	250	2191.839	I
400	1317.86	V	15	1614.02	IV	10	1882.05	III	150	2196.043	I
400	1323.27	V	13	1614.64	IV	12	1886.76	III	80	2200.390	I
400	1330.40	V	13	1615.00	IV	13	1890.67	III	80	2200.724	I
400	1359.01	V	16	1616.68	IV	11	1893.98	III	15	2208.41	II
600	1361.82	V	14	1617.68	IV	20	1895.46	III	10 p	2208.85	III
700	1373.59	V	12	1618.47	II	10 s	1907.58	III	20	2213.65	II
600	1373.67	V	14	1619.02	IV	19	1914.06	III	12	2218.26	III
500	1376.34	V	13	1621.16	IV	15	1915.08	III	20	2220.38	II
500	1378.56	V	14	1621.57	IV	15	1922.79	III	10	2221.83	III
800	1387.94	V	13	1623.38	IV	10 p	1926.01	III	10	2229.27	III
400	1397.97	V	13	1623.53	IV	18	1926.30	III	10	2232.43	III
600	1400.24	V	15	1626.47	IV	15	1930.39	III	10	2232.69	III
800	1402.39	V	14	1626.90	IV	14	1931.51	III	10	2235.91	III
400	1406.67	V	13	1628.54	IV	30	1934.538	I	10	2238.16	III
500	1406.82	V	13	1630.18	IV	25	1937.269	I	12 p	2241.54	III
400	1407.25	V	17	1631.08	IV	14	1937.34	III	50	2250.790	I
600	1409.45	V	15	1631.12	II	10 l	1938.90	III	60	2251.874	I
400	1415.20	V	14	1632.40	IV	14 s	1943.48	III	300	2259.511	I
600	1420.46	V	13	1634.01	IV	12	1945.34	III	12	2261.59	III
800	1430.57	V	18	1635.40	II	50	1946.988	I	60	2264.389	I
13	1431.43	IV	15	1636.32	II	10	1950.33	III	80	2267.085	I
800	1440.53	V	15	1639.40	II	12	1951.01	III	10	2267.42	III
400	1442.22	V	14	1639.40	IV	25	1951.571	I	80	2267.469	I
800	1446.62	V	16	1640.04	IV	30	1952.59	I	50	2270.862	I
700	1448.85	V	14	1640.16	IV	11	1952.65	III	150	2272.070	I
400	1449.93	V	12	1641.76	II	30	1953.005	I	150	2276.026	I
700	1456.16	V	15	1641.87	IV	13	1953.32	III	80	2279.937	I
500	1459.83	V	15	1647.09	IV	10	1953.49	III	150	2284.086	I
400	1460.73	V	12	1647.16	II	10 w	1954.22	III	150	2287.250	I
500	1462.63	V	15	1651.58	IV	60	1957.823	I	300	2292.524	I
700	1464.68	V	15	1652.90	IV	11	1958.58	III	10	2293.06	III
500	1465.38	V	13	1653.41	IV	60	1960.144	I	15	2295.86	III
400	1466.65	V	13	1656.11	IV	13	1960.32	III	200	2297.787	I
500	1469.00	V	15	1656.65	IV	30	1961.25	I	600	2298.169	I
500	1479.47	V	14	1660.10	IV	50	1962.111	I	80	2299.220	I
10 h	1505.17	III	13	1662.32	IV	12	1963.11	II	300	2300.142	I
13	1526.60	IV	13	1662.52	IV	15	1987.50	III	50	2301.684	I
13	1530.26	IV	13	1663.54	IV	14	1991.61	III	100	2303.424	I
14	1532.63	IV	13	1668.09	IV	13	1994.07	III	150	2303.581	I
13	1532.91	IV	12	1670.74	II	12	1995.56	III	120	2308.999	I
15	1533.86	IV	14	1671.04	IV	12	1996.42	III	150	2313.104	I
13	1533.95	IV	13	1673.68	IV	10	2061.55	III	10 p	2317.70	III
14	1536.58	IV	14	1675.66	IV	12	2068.24	III	10	2319.22	III
10 h	1538.63	III	13	1681.36	IV	14	2078.99	III	200	2320.358	I
13	1542.16	IV	15	1687.69	IV	100	2084.122	I	10 p	2321.71	III
14	1542.70	IV	15	1698.88	IV	10	2084.35	III	10	2326.95	III
12 h	1550.20	III	12	1702.04	II	12	2090.14	III	100	2327.40	II
13	1566.26	IV	13	1709.81	IV	15	2097.48	III	100	2331.31	II
14	1568.27	IV	15	1711.41	IV	12	2097.69	III	300	2332.80	II
13	1591.51	IV	14	1712.76	IV	12	2103.80	III	10 p	2336.77	III
13	1592.05	IV	14	1717.90	IV	10	2107.32	III	200	2338.01	II
13	1598.01	IV	14	1718.16	IV	15	2151.78	III	10	2338.96	III
13	1600.58	IV	14	1719.46	IV	12	2157.71	III	600	2343.49	II
10 h	1601.21	III	14	1722.71	IV	50	2157.794	I	80	2343.96	II
13	1601.67	IV	14	1724.06	IV	12	2158.47	III	150	2344.28	II
13	1603.18	IV	16	1725.63	IV	10	2161.27	III	200	2348.11	II
13	1603.73	IV	13	1761.08	IV	40	2166.773	I	250	2348.30	II
13	1604.88	IV	12	1761.38	II	12	2166.95	III	200	2359.12	II
13	1605.68	IV	20	1785.26	II	12	2171.04	III	150	2360.00	II
15	1605.97	IV	20	1786.74	II	15	2174.66	III	120	2360.29	II

Line Spectra of the Elements (continued): Iron

Intensity	Wavelength/Å		Intensity	Wavelength/Å		Intensity	Wavelength/Å		Intensity	Wavelength/Å	
200	2364.83	II	50	2485.990	I	80	2549.39	II	250	2742.254	I
80	2365.76	II	800	2486.373	I	600	2549.613	I	800	2742.405	I
80	2368.59	II	100	2486.691	I	400	2562.53	II	200	2743.20	II
80	2369.456	I	100	2487.066	I	200	2563.48	II	150	2743.565	I
80	2369.95	II	120	2487.370	I	150	2574.36	II	200	2744.068	I
120	2371.430	I	4000	2488.143	I	300	2576.691	I	80	2744.527	I
300	2373.624	I	100	2488.945	I	100	2582.58	II	300	2746.48	II
150	2373.74	II	80	2489.48	II	1500	2584.54	I	100	2749.32	II
120	2374.518	I	1000	2489.750	I	650	2585.88	II	500	2749.48	II
120	2376.43	II	50	2489.913	I	90	2591.54	II	1200	2750.140	I
80	2379.27	II	3000	2490.644	I	90	2593.73	II	80	2753.29	II
120	2380.76	II	100	2490.71	II	650	2598.37	II	150	2754.032	I
150	2381.835	I	2000	2491.155	I	2000	2599.40	II	100	2754.426	I
1000	2382.04	II	100	2491.40	II	300	2599.57	I	800	2755.73	II
300	2388.63	II	100	2493.18	II	60	2605.657	I	250	2756.328	I
200	2389.973	I	500	2493.26	II	300	2606.51	II	100	2757.316	I
1000	2395.62	II	60	2494.000	I	800	2606.827	I	120	2761.780	I
300	2399.24	II	50	2494.251	I	650	2607.09	II	150	2761.81	II
800	2404.88	II	100	2495.87	I	600	2611.87	II	150	2762.026	I
250	2406.66	II	600	2496.533	I	320	2613.82	II	120	2762.772	I
80	2406.97	II	150	2498.90	I	320	2617.62	II	120	2763.109	I
300	2410.52	II	1000	2501.132	I	250	2618.018	I	80	2766.910	I
200	2411.07	II	50	2501.693	I	90	2620.41	II	250	2767.522	I
150	2413.31	II	80	2506.09	II	400	2623.53	I	300	2772.07	I
80	2417.87	II	500	2507.900	I	200	2625.67	II	600	2778.220	I
60	2420.396	I	50	2508.753	I	150	2628.29	II	3000	2788.10	I
60	2423.089	I	1000	2510.835	I	250	2631.05	II	200	2797.78	I
150	2424.14	II	120	2511.76	I	250	2631.32	II	400	2804.521	I
120	2428.36	II	80	2512.275	I	100	2632.237	I	1500	2806.98	I
120	2430.08	II	400	2512.365	I	300	2635.809	I	10 p	2813.24	III
80	2432.26	II	80	2516.570	I	50	2641.646	I	2500	2813.287	I
60	2438.182	I	300	2517.661	I	200	2643.998	I	300	2823.276	I
150	2439.30	II	800	2518.102	I	300	2666.812	I	600	2825.56	I
150	2439.74	I	150	2519.629	I	60	2666.965	I	50	2825.687	I
100	2442.57	I	50	2522.480	I	600	2679.062	I	120	2828.808	I
250	2443.872	I	4000	2522.849	I	500	2684.75	II	1500	2832.436	I
100	2444.51	II	200	2523.66	I	400	2689.212	I	120	2835.950	I
50	2445.212	I	500	2524.293	I	10 h	2695.13	III	200	2838.119	I
100	2445.57	II	100	2525.02	I	200	2699.106	I	200	2843.631	I
60	2447.709	I	200	2525.39	II	80	2706.012	I	1000	2843.977	I
100	2453.476	I	300	2526.29	II	400	2706.582	I	100	2845.594	I
1500	2457.598	I	2000	2527.435	I	60	2708.571	I	800	2851.797	I
150	2458.78	II	800	2529.135	I	200	2711.655	I	50	2869.307	I
80	2461.28	II	250	2529.55	II	80	2714.41	II	50	2872.334	I
100	2461.86	II	150	2529.836	I	50	2716.257	I	80	2874.172	I
100	2462.181	I	200	2530.687	I	50	2717.786	I	50	2894.504	I
1500	2462.647	I	120	2533.63	II	250	2718.436	I	12	2904.43	III
50	2463.730	I	100	2534.42	II	4000	2719.027	I	10	2907.50	III
800	2465.149	I	120	2535.49	II	100	2719.420	I	12	2907.70	III
60	2467.732	I	400	2535.607	I	50	2720.197	I	120	2912.157	I
600	2468.879	I	200	2536.792	I	1500	2720.903	I	120	2929.007	I
80	2470.67	II	200	2536.80	II	400	2723.578	I	1200	2936.903	I
80	2470.965	I	100	2538.80	II	150	2724.953	I	60	2941.343	I
800	2472.336	I	100	2538.91	II	50	2726.235	I	1000	2947.876	I
1000	2472.895	I	150	2538.99	II	80	2727.54	II	600	2953.940	I
200	2473.16	I	50	2539.357	I	200	2728.020	I	250	2957.364	I
600	2474.814	I	200	2540.66	II	50	2728.820	I	150	2965.254	I
60	2476.657	I	600	2540.972	I	80	2728.90	II	1500	2966.898	I
120	2479.480	I	80	2541.10	II	1000	2733.581	I	120	2969.36	I
1200	2479.776	I	300	2542.10	I	60	2734.005	I	800	2970.099	I
100	2480.16	II	250	2543.92	I	50	2734.268	I	1200	2973.132	I
80	2482.12	II	150	2544.70	I	500	2735.475	I	500	2973.235	I
100	2482.66	II	800	2545.978	I	50	2735.612	I	600	2981.445	I
10000	2483.271	I	80	2546.67	II	500	2737.310	I	1000	2983.570	I
300	2483.533	I	100	2548.74	II	120	2737.83	I	1000	2994.427	I
1000	2484.185	I	80	2549.08	II	400	2739.55	II	250	2994.502	I

Intensity	Wavelength/Å		Intensity	Wavelength/Å		Intensity	Wavelength/Å		Intensity	Wavelength/Å	
500	2999.512	I	11	3276.08	III	200	3605.454	I	100	3786.68	I
120	3000.451	I	150	3286.75	I	500	3606.680	I	250	3787.880	I
800	3000.948	I	10	3288.81	III	1500	3608.859	I	250	3790.092	I
12	3001.62	III	120	3305.97	I	250	3610.16	I	150	3794.34	I
60	3001.655	I	200	3306.343	I	60	3612.068	I	400	3795.002	I
12 h	3007.28	III	400	3355.227	I	150	3617.788	I	120	3797.518	I
200	3007.282	I	80	3355.517	I	1500	3618.768	I	250	3798.511	I
500	3008.14	I	60	3369.546	I	200	3621.462	I	400	3799.547	I
120	3009.569	I	120	3370.783	I	150	3622.004	I	200	3805.345	I
15	3013.17	III	50	3378.678	I	150	3623.19	I	80	3806.696	I
60	3017.627	I	50	3380.110	I	100	3631.096	I	600	3812.964	I
60	3018.983	I	60	3383.978	I	1200	3631.463	I	60	3813.059	I
500	3020.491	I	50	3392.304	I	60	3632.041	I	1500	3815.840	I
1500	3020.639	I	150	3392.651	I	100	3638.298	I	2500	3820.425	I
600	3021.073	I	150	3399.333	I	200	3640.389	I	150	3821.179	I
500	3024.032	I	80	3404.353	I	80	3643.717	I	80	3824.306	I
150	3025.638	I	500	3407.458	I	1500	3647.842	I	2500	3824.444	I
500	3025.842	I	250	3413.131	I	250	3649.506	I	1500	3825.880	I
80	3030.148	I	60	3424.284	I	80	3650.279	I	1200	3827.823	I
60	3031.214	I	500	3427.119	I	200	3651.467	I	1000	3834.222	I
60	3034.484	I	60	3428.748	I	120	3670.024	I	120	3839.257	I
800	3037.389	I	6000	3440.606	I	150	3670.089	I	500	3840.437	I
80	3041.637	I	2500	3440.989	I	100	3676.311	I	800	3841.047	I
800	3047.604	I	1000	3443.876	I	150	3677.629	I	120	3843.256	I
600	3057.446	I	200	3445.149	I	1500	3679.913	I	80	3846.800	I
1000	3059.086	I	1200	3465.860	I	200	3682.242	I	200	3849.96	I
250	3067.244	I	2000	3475.450	I	120	3683.054	I	120	3850.817	I
120	3075.719	I	500	3476.702	I	150	3684.107	I	2500	3856.372	I
120	3091.577	I	2500	3490.574	I	120	3685.998	I	150	3859.212	I
80	3098.189	I	500	3497.840	I	500	3687.456	I	10000	3859.911	I
100	3099.895	I	10	3501.76	III	120	3689.477	I	150	3865.523	I
100	3099.968	I	250	3513.817	I	150	3694.008	I	60	3867.215	I
60	3100.303	I	300	3521.261	I	120	3695.051	I	250	3872.501	I
100	3100.665	I	400	3526.040	I	150	3701.086	I	150	3873.761	I
10 p	3136.43	III	100	3526.166	I	80	3704.462	I	250	3878.018	I
10	3174.09	III	60	3526.237	I	1200	3705.566	I	2000	3878.573	I
80	3175.445	I	60	3526.381	I	60	3707.041	I	4000	3886.282	I
10	3175.99	III	60	3526.467	I	150	3707.821	I	200	3887.048	I
10	3178.01	III	100	3533.199	I	300	3707.919	I	300	3888.513	I
150	3184.895	I	200	3536.556	I	600	3709.246	I	800	3895.656	I
250	3191.659	I	300	3541.083	I	120	3716.442	I	1200	3899.707	I
500	3193.226	I	250	3542.075	I	8000	3719.935	I	400	3902.945	I
800	3193.299	I	80	3553.739	I	1500	3722.563	I	250	3906.479	I
200	3196.928	I	400	3554.925	I	120	3724.377	I	80	3916.731	I
80	3199.500	I	200	3556.878	I	60	3725.491	I	600	3920.258	I
50	3205.398	I	400	3558.515	I	60	3727.093	I	1200	3922.911	I
100	3211.88	I	1000	3565.379	I	500	3727.619	I	1200	3927.920	I
200	3214.011	I	1200	3570.097	I	150	3732.396	I	2000	3930.296	I
200	3214.396	I	800	3570.25	I	1200	3733.317	I	60	3948.774	I
60	3215.938	I	120	3571.996	II	5000	3734.864	I	60	3949.953	I
50	3217.377	I	100	3573.393	I	120	3735.324	I	50	3951.164	I
80	3219.583	I	60	3573.829	I	6000	3737.131	I	50	3952.601	I
60	3219.766	II	60	3573.888	I	100	3738.306	I	16	3954.33	III
300	3222.045	I	4000	3581.19	I	400	3743.362	I	60	3956.454	I
600	3225.78	I	150	3582.199	I	6000	3745.561	I	250	3956.68	I
80	3227.796	I	150	3584.660	I	1200	3745.899	I	60	3966.614	I
50	3233.967	I	120	3584.929	I	3000	3748.262	I	11	3968.72	III
120	3234.613	I	300	3585.319	I	80	3748.964	I	100	3969.257	I
300	3236.222	I	150	3585.705	I	3000	3749.485	I	80	3977.741	I
100	3239.433	I	10	3586.04	III	1500	3758.232	I	10 w	3979.42	III
80	3244.187	I	200	3586.103	I	400	3760.05	I	40	3981.771	I
80	3246.005	I	400	3586.984	I	1500	3763.788	I	50	3983.956	I
80	3265.046	I	100	3594.633	I	400	3765.54	I	60	3994.114	I
50	3265.617	I	11	3600.94	III	600	3767.191	I	200	3997.392	I
13	3266.88	III	150	3603.204	I	60	3776.452	I	40	3998.053	I
50	3271.000	I	11	3603.88	III	250	3785.95	I	400	4005.241	I

Intensity	Wavelength/Å		Intensity	Wavelength/Å		Intensity	Wavelength/Å		Intensity	Wavelength/Å	
60	4009.713	I	12 h	4273.40	III	150	5216.274	I	16	6032.59	III
100	4021.867	I	12	4279.72	III	60	5226.862	I	13	6036.56	III
10	4035.42	III	1200	4282.402	I	1000	5227.150	I	11	6048.72	III
50	4040.638	I	14 h	4286.16	III	250	5232.939	I	11	6054.18	III
4000	4045.813	I	80	4291.462	I	10	5235.66	III	40	6065.482	I
11	4053.11	III	16 h	4296.85	III	18	5243.31	III	30	6102.159	I
1500	4063.594	I	250	4299.234	I	13 l	5260.34	III	40	6136.614	I
50	4066.975	I	18 h	4304.78	III	100	5266.555	I	40	6137.694	I
50	4067.977	I	1200	4307.901	I	1200	5269.537	I	40	6191.558	I
1200	4071.737	I	20 h	4310.36	III	800	5270.357	I	30	6213.429	I
40	4076.629	I	150	4315.084	I	14	5272.98	III	30	6219.279	I
12	4081.00	III	1500	4325.761	I	15	5276.48	III	40	6230.726	I
40	4100.737	I	80	4352.734	I	30	5281.789	I	20	6246.317	I
40	4107.489	I	80	4369.771	I	16	5282.30	III	80	6247.56	II
150	4118.544	I	11 h	4372.31	III	60	5283.621	I	30	6252.554	I
10	4120.90	III	14 h	4372.53	III	12	5284.83	III	20	6393.602	I
11	4122.02	III	18 h	4372.81	III	11	5298.12	III	30	6399.999	I
11	4122.78	III	800	4375.929	I	12	5299.93	III	20	6411.647	I
40	4127.608	I	3000	4383.544	I	25	5302.299	I	20	6421.349	I
400	4132.058	I	1200	4404.750	I	14 w	5302.60	III	30	6430.844	I
80	4134.676	I	300	4415.122	I	10	5306.76	III	200	6456.38	II
40	4136.997	I	12	4419.60	III	10	5322.74	III	60	6494.981	I
15	4137.76	III	600	4427.299	I	150	5324.178	I	20	6546.239	I
13	4139.35	III	400	4461.652	I	800	5328.038	I	20	6592.913	I
200	4143.415	I	120	4466.551	I	300	5328.531	I	40	6677.989	I
800	4143.869	I	80	4476.017	I	100	5332.899	I	25	7164.443	I
40	4153.898	I	80	4482.169	I	80	5339.928	I	80	7187.313	I
50	4154.500	I	200	4482.252	I	500	5341.023	I	30	7207.381	I
60	4156.799	I	50	4489.739	I	11	5346.88	III	30	7445.746	I
18	4164.73	III	50	4528.613	I	12	5353.77	III	40	7495.059	I
13	4166.84	III	30	4647.433	I	12	5363.76	III	60	7511.045	I
50	4172.744	I	30	4736.771	I	10	5368.06	III	80	7937.131	I
13	4174.26	III	50	4859.741	I	400	5371.489	I	60	7945.984	I
60	4174.912	I	120	4871.317	I	11 l	5375.47	III	80	7998.939	I
50	4175.635	I	60	4872.136	I	40	5393.167	I	60	8046.047	I
50	4177.593	I	30	4878.208	I	300	5397.127	I	50	8085.176	I
120	4181.754	I	100	4890.754	I	250	5405.774	I	150	8220.41	I
50	4184.891	I	250	4891.492	I	250	5429.695	I	120	8327.053	I
120	4187.038	I	30	4903.309	I	100	5434.523	I	20	8331.908	I
120	4187.795	I	150	4918.992	I	200	5446.871	I	120	8387.770	I
80	4191.430	I	500	4920.502	I	120	5455.609	I	30	8468.404	I
40	4195.329	I	1500	4957.597	I	25	5497.516	I	15	8514.069	I
150	4198.304	I	80	5001.862	I	20	5501.464	I	60	8661.898	I
40	4199.095	I	30	5005.711	I	30	5506.778	I	150	8688.621	I
300	4202.029	I	100	5006.117	I	30	5569.618	I	52	11422.32	I
40	4203.984	I	60	5012.067	I	60	5572.841	I	87	11439.12	I
80	4206.696	I	30	5014.941	I	120	5586.755	I	91	11593.59	I
80	4210.343	I	150	5041.755	I	200	5615.644	I	255	11607.57	I
400	4216.183	I	30	5049.819	I	20	5624.541	I	160	11638.26	I
100	4219.360	I	30	5051.634	I	50	5662.515	I	230	11689.98	I
50	4222.212	I	25	5074.748	I	11	5719.88	III	160	11783.26	I
11	4222.27	III	150	5110.357	I	10	5756.38	III	580	11882.84	I
50	4225.956	I	40	5139.251	I	20	5762.990	I	225	11884.08	I
200	4227.423	I	100	5139.462	I	18	5833.93	III	1030	11973.05	I
100	4233.602	I	25	5151.910	I	10	5854.62	III	96	14400.56	I
13	4235.56	III	12	5156.12	III	30	5862.353	I	72	14512.23	I
250	4235.936	I	80	5166.281	I	15	5891.91	III	50	14555.06	I
50	4238.809	I	2500	5167.487	I	30	5914.114	I	40	14826.43	I
12	4243.75	III	80	5168.897	I	10 p	5920.13	III	94	15294.58	I
50	4247.425	I	500	5171.595	I	18 p	5929.69	III	41	15769.42	I
200	4250.118	I	50	5191.454	I	10	5952.31	III	105	18856.65	I
300	4250.787	I	80	5192.343	I	14	5953.62	III		**Krypton**	
40	4258.315	I	200	5194.941	I	12	5979.32	III		**Kr Z = 36**	
800	4260.473	I	10	5199.08	III	30	5986.956	I	30	467.35	III
250	4271.153	I	30	5204.582	I	12 h	5989.08	III	150	472.16	V
1200	4271.759	I	25	5215.179	I	18	5999.54	III	100	484.39	V

Intensity	Wavelength/Å		Intensity	Wavelength/Å		Intensity	Wavelength/Å		Intensity	Wavelength/Å	
250	496.25	V	50	837.66	III	6	2774.70	IV	40 h	3868.70	III
120	500.77	V	22	842.04	IV	3	2829.60	IV	150 h	3875.44	II
200	507.20	V	100	844.06	II	100	2833.00	II	150	3906.177	II
30	540.86	III	50	854.73	III	3	2836.08	IV	200	3920.081	II
60	548.04	V	60	862.58	III	30	2841.00	III	5	3934.29	IV
30	565.64	III	60	864.82	II	30	2851.16	III	100	3994.840	II
30	569.16	III	60	868.87	II	5	2853.0	IV	100 h	3997.793	II
30	571.98	III	40	870.84	III	3	2859.3	IV	300	4057.037	II
30	579.83	III	50	876.08	III	50	2870.61	III	300	4065.128	II
30	585.14	III	200	884.14	III	100	2892.18	II	50	4067.37	III
30	585.96	III	1000	886.30	II	30	2909.17	III	500	4088.337	II
30	593.70	III	400	891.01	II	50	2952.56	III	250	4098.729	II
30	594.10	III	75	897.81	III	60	2992.22	III	100	4109.248	II
30	596.41	III	200	911.39	II	50	3022.30	III	40	4131.33	III
40	600.17	III	2000	917.43	II	80	3024.45	III	250	4145.122	II
30	603.67	III	50	945.44	I	50	3046.93	III	40	4154.46	III
50	605.86	III	50	946.54	I	30	3056.72	III	150	4250.580	II
35	606.47	III	20	951.06	I	60	3063.13	III	1000	4273.969	I
50	611.12	III	50	953.40	I	40	3097.16	III	100	4282.967	I
35	616.72	III	50	963.37	I	60	3112.25	III	600	4292.923	II
40	621.45	III	2000	964.97	II	30	3120.61	III	200	4300.49	II
45	622.80	III	50	987.29	III	100	3124.39	III	500 h	4317.81	II
50	625.02	III	100	1001.06	I	60	3141.35	III	400	4318.551	I
30	625.76	III	100	1003.55	I	3	3142.01	IV	1000	4319.579	I
45	628.59	III	100	1030.02	I	100	3189.11	III	150 h	4322.98	II
50	630.04	III	30	1158.74	III	80	3191.21	III	100	4351.359	I
35	633.09	III	200	1164.87	I	6	3224.99	IV	3000	4355.477	II
120	637.87	V	650	1235.84	I	40	3239.52	III	500	4362.641	I
50	639.98	III	6	1638.82	III	40	3240.44	III	200	4369.69	II
60	646.41	III	6	1914.09	III	300	3245.69	III	800	4376.121	I
50	651.20	III	3	2237.34	IV	3	3261.70	IV	300 h	4386.54	II
50	659.72	III	6	2291.26	IV	150	3264.81	III	200	4399.965	I
30	664.86	III	3	2329.3	IV	100	3268.48	III	100	4425.189	I
40	672.34	III	4	2336.75	IV	30	3271.65	III	500	4431.685	II
35	672.85	III	4	2348.27	IV	30	3285.89	III	600	4436.812	II
35	676.57	III	3	2358.5	IV	30	3304.75	III	600	4453.917	I
35	680.13	III	3	2388.05	IV	50	3311.47	III	800	4463.689	I
35	683.68	III	40	2393.94	III	200	3325.75	III	800	4475.014	II
45	686.25	III	4	2416.9	IV	60	3330.76	III	400 h	4489.88	II
45	687.98	III	3	2428.04	IV	50	3342.48	III	600	4502.353	I
	690.86	V	5	2442.68	IV	100	3351.93	III	400 h	4523.14	II
	691.75	V	4	2451.7	IV	40	3374.96	III	200 h	4556.61	II
45	691.93	III	6	2459.74	IV	100	3439.46	III	800	4577.209	II
50	695.61	III	100 h	2464.77	II	70	3474.65	III	300	4582.978	II
30	698.05	III	5	2474.06	II	100	3488.59	III	150 h	4592.80	II
50	708.36	III	60	2492.48	II	200	3507.42	III	500	4615.292	II
600	708.85	V	40	2494.01	III	100	3564.23	III	1000	4619.166	II
50	714.00	III	4	2517.0	IV	100 h	3607.88	II	800	4633.885	II
100 p	722.04	III	5	2518.02	IV	200	3631.889	II	2000	4658.876	II
60	729.40	II	6	2519.38	IV	30	3641.34	III	500	4680.406	II
30	746.70	III	5	2524.5	IV	250	3653.928	II	100	4691.301	II
200	761.18	II	5	2546.0	IV	80	3665.324	I	200	4694.360	II
100	763.98	II	6	2547.0	IV	150	3669.01	II	3000	4739.002	II
60	766.20	II	4	2558.08	IV	100	3679.559	I	300	4762.435	II
200	771.03	II	30	2563.25	III	80	3686.182	II	1000	4765.744	II
60 p	773.69	II	3	2586.9	IV	30	3690.65	III	300	4811.76	II
200	782.10	II	5	2606.17	IV	300 h	3718.02	II	300	4825.18	II
100	783.72	II	10	2609.5	IV	200	3718.595	II	800	4832.077	II
60	785.97	III	8	2615.3	IV	150	3721.350	II	700	4846.612	II
	793.44	IV	7	2621.11	IV	200	3741.638	II	150	4857.20	II
	794.11	IV	60	2639.76	III	150	3744.80	II	300	4945.59	II
7	805.76	IV	30	2680.32	III	80	3754.245	II	20 h	5016.45	III
	810.70	V	40	2681.19	III	500	3778.089	II	200	5022.40	II
18	816.82	IV	80 h	2712.40	II	500	3783.095	II	250	5086.52	II
60	818.15	II	3	2730.55	IV	3	3809.30	IV	400 h	5125.73	II
60	830.38	II	8	2748.18	IV	5	3860.58	IV	500	5208.32	II

Intensity	Wavelength/Å		Intensity	Wavelength/Å		Intensity	Wavelength/Å		Intensity	Wavelength/Å	
200	5308.66	II	100	9362.082	I	140	20446.971	I	30000 c	2962.58	IV
500	5333.41	II	200 h	9402.82	II	600	21165.471	I	70000 w	3009.51	IV
200	5468.17	II	200 h	9470.93	II	1800	21902.513	I	90000 c	3056.68	IV
10	5501.43	III	500	9577.52	II	120	22485.775	I	1000	3171.63	III
500	5562.224	I	500 h	9605.80	II	180	23340.416	I	1500	3171.74	III
2000	5570.288	I	400 h	9619.61	II	120	24260.506	I	510	3245.13	II
80	5580.386	I	200	9663.34	II	180	24292.221	I	550	3265.67	II
100	5649.561	I	200 h	9711.60	II	600	25233.820	I	800	3303.11	II
400	5681.89	II	2000	9751.758	I	180	28610.55	I	1500	3337.49	II
200 h	5690.35	II	500	9803.14	II	1000	28655.72	I	870	3344.56	II
100	5832.855	I	500	9856.314	I	150	28769.71	I	1500	3380.91	II
3000	5870.914	I	1000	10221.46	II	140	28822.49	I	320	3628.83	II
200	5992.22	II	100	11187.108	I	300	29236.69	I	1000	3645.42	II
60	5993.849	I	200	11257.711	I	300	30663.54	I	550	3713.54	II
10 h	6037.17	III	150	11259.126	I	300	30979.16	I	2400	3759.08	II
60	6056.125	I	500	11457.481	I	500	39300.6	I	3700	3790.83	II
10 h	6078.38	III	150	11792.425	I	1100	39486.52	I	3900	3794.78	II
10	6310.22	III	1500	11819.377	I	220	39557.25	I	600	3840.72	II
300	6420.18	II	600	11997.105	I	100	39572.60	I	1600	3849.02	II
100	6421.026	I	160	12077.224	I	1400	39588.4	I	3400	3871.64	II
200	6456.288	I	100	12861.892	I	1100	39589.6	I	1700	3886.37	II
150	6570.07	II	1100	13177.412	I	500	39954.8	I	1300	3916.05	II
60	6699.228	I	1000	13622.415	I	300	39966.6	I	1100	3921.54	II
100	6904.678	I	2400	13634.220	I	1300	40306.1	I	2200	3929.22	II
250	7213.13	II	800	13658.394	I	250	40685.16	I	9000	3949.10	II
100	7224.104	I	200	13711.036	I		**Lanthanum**		4400	3988.52	II
80	7287.258	I	600	13738.851	I		**La Z = 57**		3600	3995.75	II
400	7289.78	II	150	13974.027	I	100	344.12	IV	2800	4031.69	II
400	7407.02	II	550	14045.657	I	400	390.72	V	3000	4042.91	II
60	7425.541	I	140	14104.298	I	1000	432.11	V	850	4067.39	II
200	7435.78	II	180	14402.22	I	2500	435.28	V	2800	4077.35	II
100	7486.862	I	2000	14426.793	I	10000	463.14	IV	5500	4086.72	II
300	7524.46	II	100	14517.84	I	5000	482.16	V	4400	4123.23	II
1000	7587.411	I	1600	14734.436	I	7000	498.08	V	550	4141.74	II
2000	7601.544	I	550	14762.672	I	15000	499.54	IV	1100	4151.97	II
150	7641.16	II	450	14765.472	I	10000	503.58	V	1500	4196.55	II
1000	7685.244	I	400	14961.894	I	12000	526.76	V	1600	4238.38	II
1200	7694.538	I	120	15005.307	I	10000	531.07	V	480	4269.50	II
250	7735.69	II	140	15209.526	I	15000	533.23	V	600	4286.97	II
150	7746.827	I	1700	15239.615	I	8000	547.44	V	600	4296.05	II
800	7854.821	I	130	15326.480	I	40000	552.02	IV	440	4322.51	II
200	7913.423	I	1500	15334.958	I	5000	600.24	V	4600	4333.74	II
180	7928.597	I	700	15372.037	I	30000	631.26	IV	550	4354.40	II
200	7933.22	II	200	15474.026	I	400	796.99	III	2000	4429.90	II
120	7973.62	II	180	15681.02	I	2000	870.40	III	850	4522.37	II
100	7982.401	I	120	15820.09	I	1000	882.34	III	420	4526.12	II
1500	8059.503	I	200	16726.513	I	400	942.86	III	400	4558.46	II
4000	8104.364	I	2000	16785.128	I	50000	1081.61	III	400	4574.88	II
6000	8112.899	I	1000	16853.488	I	95000	1099.73	III	410	4613.39	II
60	8132.967	I	2400	16890.441	I	2000	1255.63	III	410	4619.88	II
3000	8190.054	I	1600	16896.753	I	10000	1349.18	III	540	4655.50	II
200	8202.72	II	1800	16935.806	I	25000	1368.04	IV	360	4662.51	II
80	8218.365	I	600	17098.771	I	20000	1463.47	IV	230	4692.50	II
3000	8263.240	I	700	17367.606	I	15000	1507.87	IV	230	4728.42	II
100	8272.353	I	120	17404.443	I	10000	1523.79	III	500	4740.28	II
1500	8281.050	I	150	17616.854	I	4000	1808.66	IV	390	4743.09	II
5000	8298.107	I	650	17842.737	I	5000	1902.97	IV	320	4748.73	II
100	8412.430	I	700	18002.229	I	4000 c	2197.45	IV	320	4860.91	II
3000	8508.870	I	2600	18167.315	I	770	2256.76	II	850	4899.92	II
150	8764.110	I	100	18399.786	I	25000 w	2417.58	IV	1000	4920.98	II
6000	8776.748	I	150	18580.896	I	50000	2532.75	IV	1000	4921.79	II
2000	8928.692	I	300	18696.294	I	45000	2582.05	IV	370	4949.77	I
500	9238.48	II	170	18785.460	I	95000 w	2597.50	IV	340	4970.39	II
500 hl	9293.82	II	200	18797.703	I	70000 w	2662.75	IV	370	4986.83	II
200 h	9320.99	II	140	20209.878	I	420	2808.39	II	720	4999.47	II
300	9361.95	II	300	20423.964	I	50000 w	2848.30	IV	210	5050.57	I

Intensity	Wavelength/Å		Intensity	Wavelength/Å		Intensity	Wavelength/Å		Intensity	Wavelength/Å	
470	5114.56	II	10	827.41	IV	10	1348.37	II	4	2986.876	II
470	5122.99	II	12	832.60	IV	16	1388.94	IV	10	3043.85	III
450	5145.42	I	12	845.94	IV	18	1400.26	IV	150	3118.894	I
290	5158.69	I	18	857.64	IV	10	1404.34	IV	10	3137.81	III
580	5177.31	I	16	862.33	IV	10	1433.96	II	10	3176.50	III
850	5183.42	II	20	863.97	V	14	1535.71	IV	600	3220.528	I
260	5188.22	II	14	870.44	IV	20	1553.1	III	100	3229.613	I
720	5211.86	I	6	873.71	II	10	1671.53	II	400	3240.186	I
520	5234.27	I	12	879.96	IV	10	1682.15	II	200	3262.355	I
340	5253.46	I	18	883.90	V	20	1726.75	II	35000	3572.729	I
370	5271.19	I	14	884.96	IV	10	1796.670	II	50000 r	3639.568	I
370	5301.98	II	14	884.99	IV	10	1822.050	II	20000	3671.491	I
180	5303.55	II	14	888.37	V	10	1904.77	I	70000 r	3683.462	I
500	5455.15	I	8	889.68	II	7	1921.471	II	10	3713.982	II
470	5501.34	I	16	890.72	IV	12	1959.34	IV	25000	3739.935	I
240	5648.25	I	14	894.40	V	16	1973.16	V	12	3854.08	III
180	5740.66	I	12	896.08	V	10	1998.83	V	15000	4019.632	I
370	5769.34	I	12	908.51	IV	5 r	2022.02	I	95000	4057.807	I
320	5789.24	I	14	915.71	V	10	2042.58	IV	14000	4062.136	I
450	5791.34	I	12	917.90	IV	12	2049.34	IV	10	4157.814	I
140	5821.99	I	12	918.09	V	8 r	2053.28	I	10000	4168.033	I
320	5930.62	I	12	920.28	V	12	2079.22	IV	8	4272.66	III
720	6249.93	I	12	920.66	V	6	2111.758	I	200	4340.413	I
260 d	6262.30	II	10	922.12	IV	10	2115.066	I	10	4496.15	IV
450	6394.23	I	12	922.49	IV	15	2154.01	IV	6	4499.34	III
250	6455.99	I	10	927.64	IV	500 r	2170.00	I	16	4534.60	IV
180	6709.50	I	14	932.20	IV	7	2175.580	I	7	4571.21	III
110	7045.96	I	12	954.35	V	12	2177.46	IV	10	4579.051	II
160	7066.23	II	10	967.23	II	7	2187.888	I	6	4761.12	III
50	7161.25	I	10	986.71	II	8	2189.603	I	1000	5005.416	I
110 w	7282.34	II	10	995.89	II	10	2203.534	II	100	5006.572	I
110 w	7334.18	I	10	1016.61	II	20	2237.425	I	50	5089.484	I
75 cw	7483.50	II	14	1028.61	IV	20	2246.86	I	10	5107.242	I
50	7498.83	I	20	1032.05	IV	25	2246.89	I	2000	5201.437	I
85	7539.23	I	16	1041.24	IV	20	2259.01	V	10	5372.099	II
40	7964.83	I	18	1044.14	IV	150	2332.418	I	40	5692.346	I
75	8086.05	I	12	1048.9	III	16	2359.53	IV	200	5895.624	I
85	8324.69	I	10	1049.82	II	180	2388.797	I	2000	6001.862	I
95	8346.53	I	10	1050.77	II	550 r	2393.792	I	500	6011.667	I
65	8545.44	I	10	1051.26	II	140	2399.597	I	500	6059.356	I
300	8583.45	III	15	1056.53	IV	320 r	2401.940	I	40	6081.409	II
40	8674.43	I	10	1060.66	II	320 r	2411.734	I	50	6110.520	I
35	8825.82	I	12	1072.09	IV	16	2417.61	IV	100	6235.266	I
120	9184.38	III	18	1080.81	IV	15	2424.81	V	50 c	6660.20	II
100	9212.63	III	20	1084.17	IV	150 r	2443.829	I	20000	7228.965	I
140	10284.79	III	10	1088.86	V	160 r	2446.181	I	10	7346.676	I
Lead			10	1103.94	II	130 r	2476.378	I	20	7809.259	I
Pb Z = 82			10	1108.43	II	80 r	2577.260	I	5	7896.737	I
10	496.38	IV	10	1109.84	II	500 r	2613.655	I	10	8168.001	I
12	499.94	IV	20	1116.08	IV	900 r	2614.175	I	6	8191.886	I
14	529.78	IV	10	1119.57	II	160	2628.262	I	5	8217.711	I
20	570.16	IV	10	1121.36	II	4	2634.256	II	40	8272.690	I
10	648.50	IV	10	1133.14	II	10	2657.094	I	20	8409.384	I
20	703.73	V	18	1137.84	IV	700	2663.154	I	10	8478.492	I
12	749.46	V	14	1144.93	IV	10	2697.541	I	5	8722.810	I
10	752.52	V	12	1157.88	V	25000 r	2801.995	I	10	8857.457	I
10	761.09	IV	14	1185.43	V	100	2822.58	I	8	9293.476	I
18	767.45	V	20	1189.95	IV	14000 r	2823.189	I	15	9438.05	I
18	769.49	V	10	1203.63	II	35000 r	2833.053	I	15	9604.297	I
14	771.42	V	10	1231.20	II	6	2840.557	II	15	9674.351	I
14	782.79	V	11	1233.50	V	14000 r	2873.311	I	200	10290.458	I
15	797.02	V	10	1291.10	IV	3	2914.442	II	100	10498.965	I
18	802.07	IV	20	1313.05	IV	15	2966.460	I	50	10649.249	I
12	802.82	V	10	1331.65	II	15	2972.991	I	15	10886.688	I
18	809.63	V	10	1335.20	II	15	2980.157	I	40	10969.53	I
10	812.59	IV	12	1343.06	IV					13512.6	I

Intensity	Wavelength/Å	Intensity	Wavelength/Å	Intensity	Wavelength/Å	Intensity	Wavelength/Å
	14743.0 I		2164. II		2846. I		5037.92 II
	15349.6 I		2173.4 I		2868. I		5271. I
	39039.4 I		2183. II		2895. I		5315. I
	Lithium		2214. II	2	2934.02 II		5395. I
	Li Z = 3		2222. II	2	2934.07 II		5440. I
	102.9 III		2237. II	5	2934.12 II	600 c	5483.55 II
	103.4 III	h	2249.21 II	1	2934.25 II	600 c	5485.65 II
	104.1 III		2286.82 II		2968. I	320	6103.54 I
	105.5 III		2302.57 II	3	3029.12 II	320	6103.65 I
	108.0 III		2303.33 II	3	3029.14 II	3600	6707.76 I
	113.9 III		2304.59 I		3144. I	3600	6707.91 I
	125.5 II		2304.92 I	3	3155.31 II	48	8126.23 I
	135.0 III		2305.36 I	4	3155.33 II	48	8126.45 I
	136.5 II		2305.83 I	1	3196.26 II		8517.37 II
	140.5 II		2306.29 I	9	3196.29 II		9581.42 II
	167.21 II		2306.82 I	4	3196.36 II		10120. II
	168.74 II		2307.44 I	5	3199.33 II		12232. I
	171.58 II		2308.97 I	2	3199.43 II		12782. I
	178.02 II		2309.88 I	17	3232.66 I		13566. I
	199.28 II		2310.94 I		3249.87 II		17552. I
	207.5 II		2312.11 I		3306.28 II		18697. I
	456. II		2313.49 I		3488. I		19290. I
	483. II		2315.08 I		3579.8 I		24467. I
	540. II		2316.95 I		3618. I		40475. I
	540.0 III		2319.18 I		3662. I		**Lutetium**
	729. II		2321.88 I		3684.32 II		**Lu Z = 71**
	729.1 III		2325.11 I	1	3714.00 II	100	563.72 V
	800. II		2329.02 I	5	3714.16 II	500	810.73 III
	820. II		2329.84 II	6 d	3714.27 II	2000	832.28 III
	861. II		2333.94 I	8	3714.29 II	100	861.92 V
	905.5 II	3	2336.88 II	7 d	3714.40 II	400	876.80 IV
	917.5 II	5	2336.91 II	10	3714.41 II	100	880.32 V
	936. II	2	2337.00 II	1	3714.51 II	100	891.81 V
	945. II		2340.15 I		3714.58 II	100	914.72 V
	965. II		2348.22 I	3	3718.7 I	400	1001.18 III
	972. II		2358.93 I	6	3794.72 I	100	1272.42 IV
	988. II		2373.54 I	20	3915.30 I	800	1333.79 IV
	1018. II		2381.54 II	20	3915.35 I	400	1429.38 IV
	1032. II		2383.20 II	10	3985.48 I	200	1441.76 V
	1036. II	1	2394.39 I	10	3985.54 I	200	1453.35 V
	1093. II		2402.33 II	40	4132.56 I	200	1468.99 V
	1103. II		2410.84 II	40	4132.62 I	400	1472.12 V
	1109. II		2425.43 I		4196. I	200	1473.71 V
	1116. II	3	2429.81 II	20	4273.07 I	200	1485.58 V
	1132.1 II		2460.2 I	20	4273.13 I	400	1511.26 IV
	1141. II	10	2475.06 I	5	4325.42 II	600	1772.57 IV
	1166.4 II		2506.94 II	5	4325.47 II	100 c	1786.25 V
	1198.09 II		2508.78 II	1	4325.54 II	1000	1854.57 III
	1215. II		2518. I		4516.45 II	1500	2065.35 III
	1238. II		2539.49 II	13	4602.83 I	1500 c	2070.56 III
	1253.8 II	24	2551.7 II	13	4602.89 I	600 c	2086.47 IV
	1420.89 II		2559. II		4607.34 II	1000 c	2104.41 IV
	1424. II	15	2562.31 I		4671.51 II	1000 c	2108.31 IV
3	1492.93 II		2605.08 II	6	4671.65 II	1700 h	2195.54 II
5	1492.97 II	2	2657.29 II	2	4671.70 II	1000	2236.14 III
1	1493.04 II	3	2657.30 II	3	4678.06 II	2000	2236.22 III
	1555. II		2674.46 II	1	4678.29 II	95	2276.94 II
3	1653.08 II		2728.24 II		4760. I	190	2297.41 II
5	1653.13 II	5	2728.29 II		4763. II	1300	2392.19 II
1	1653.21 II	2	2728.32 II		4788.36 II	120	2399.14 II
	1681.66 II	3	2730.47 II		4843.0 II	80	2419.21 II
	1755.33 II	1	2730.55 II	4	4881.32 II	130	2459.64 II
	2009. II	5	2741.20 I	4	4881.39 II	370	2536.95 II
	2039. I		2766.99 II	1	4881.49 II	930	2571.23 II
	2068. II		2790.31 II	8	4971.66 I	1700	2578.79 II
	2131. II		2801. I	8	4971.75 I	4500 c	2603.35 III

Intensity	Wavelength/Å	
1800	2613.40	II
18000	2615.42	II
1800	2619.26	II
2700	2657.80	II
570 h	2685.08	I
4200	2701.71	II
180 d	2719.09	I
480 h	2728.95	I
3600	2754.17	II
750 h	2765.74	I
2000	2772.55	III
2700	2796.63	II
270 c	2834.35	II
330 h	2845.13	I
3000	2847.51	II
570 h	2885.14	I
6300	2894.84	II
4500	2900.30	II
300	2903.05	I
9000	2911.39	II
270 h	2949.73	I
1200	2951.69	II
4200	2963.32	II
2400	2969.82	II
1800	2989.27	I
3000	3020.54	II
2100	3056.72	II
1000	3057.86	III
7500	3077.60	II
390	3080.11	I
5100 h	3081.47	I
3000	3118.43	I
2400	3171.36	I
260	3191.80	II
1400	3198.12	II
4800	3254.31	II
3800	3278.97	I
7600	3281.74	I
6200	3312.11	I
7600	3359.56	I
6200	3376.50	I
950	3385.50	I
160 h	3391.55	I
1400	3396.82	I
4100	3397.07	II
4800	3472.48	II
8300 c	3507.39	II
1600	3508.42	I
4800	3554.43	II
4800	3567.84	I
340	3596.34	I
800	3623.99	II
680	3636.25	I
2600	3647.77	I
110	3756.70	I
110	3756.79	I
150	3800.67	I
2700	3841.18	I
530	3876.65	II
50	3918.86	I
480	3968.46	I
670	4054.45	I
310	4122.49	I
3100	4124.73	I
150 c	4131.79	I
460	4154.08	I

Intensity	Wavelength/Å	
1600	4184.25	II
150	4277.50	I
250	4281.03	I
330 d	4295.97	I
150	4309.57	I
190 c	4430.48	I
190	4450.81	I
3300	4518.57	I
100 h	4648.21	I
1000	4658.02	I
85 h	4659.03	I
150	4785.42	II
85	4815.05	I
460	4904.88	I
180	4942.34	I
800	4994.13	II
800	5001.14	I
140	5134.05	I
2700	5135.09	I
170	5196.61	I
500	5402.57	I
140 c	5421.90	I
100	5437.88	I
2100	5476.69	II
550	5736.55	I
80	5800.59	I
690 cw	5983.9	II
140	5997.13	I
1400	6004.52	I
440	6055.03	I
150	6159.94	II
600	6198.13	III
160	6199.66	II
2100	6221.87	II
80	6235.36	II
160	6242.34	II
70 h	6345.35	I
1100	6463.12	II
29	6477.67	I
55 c	6523.18	I
35 cw	6611.28	II
23 c	6677.14	I
30 c	6793.77	I
45	6917.31	I
23	7031.24	I
45	7125.84	I
14 ch	7237.98	I
11 c	7441.52	I
9 c	8178.16	I
17	8382.08	I
35	8459.19	II
10 d	8478.50	I
29 c	8508.08	I
35 c	8610.98	I

Magnesium
Mg Z = 12

Intensity	Wavelength/Å	
400	146.95	IV
20	186.51	III
20	187.20	III
10	188.53	III
100	231.73	III
80	234.26	III
35	276.58	V
4000	320.99	IV
3000	323.31	IV
30	353.09	V

Intensity	Wavelength/Å	
150	857.29	IV
50	919.03	IV
250	1037.41	IV
300	1210.99	IV
300	1342.19	IV
800	1346.57	IV
300	1346.68	IV
600	1352.05	IV
900	1384.46	IV
500	1385.77	IV
800	1387.53	IV
300	1404.68	IV
1000	1409.36	IV
500	1437.53	IV
1000	1437.64	IV
300	1447.42	IV
300	1459.54	IV
400	1459.62	IV
400	1481.51	IV
350	1490.45	IV
300	1495.50	IV
300	1607.11	IV
5	1668.43	I
500	1683.02	IV
10	1683.41	I
400	1698.81	IV
15	1707.06	I
40	1734.84	II
50	1737.62	II
20	1747.80	I
40	1750.65	II
50	1753.46	II
30	1827.93	I
300	1844.17	IV
9	2025.82	I
25	2064.90	III
20	2091.96	III
20	2177.70	III
3	2329.58	II
20	2395.15	III
6	2449.57	II
1	2557.23	I
1	2560.94	I
1	2562.26	I
1	2564.94	I
1	2570.91	I
1	2572.25	I
2	2574.94	I
1	2577.89	I
1	2580.59	I
2	2585.56	I
3	2588.28	I
1	2591.89	I
1	2593.23	I
2	2595.97	I
2	2602.50	I
4	2603.85	I
5	2606.62	I
1	2613.36	I
2	2614.73	I
3	2617.51	I
3	2628.66	I
6	2630.05	I
8	2632.87	I
2	2644.80	I

Intensity	Wavelength/Å	
3	2646.21	I
4	2649.06	I
8	2660.76	II
8	2660.82	II
6	2668.12	I
8	2669.55	I
10	2672.46	I
3	2693.72	I
5	2695.18	I
6	2698.14	I
8	2731.99	I
10	2733.49	I
12	2736.53	I
5	2765.22	I
7	2768.34	I
38	2776.69	I
32	2778.27	I
90	2779.83	I
8	2781.29	I
32	2781.42	I
36	2782.97	I
1000	2795.53	II
600	2802.70	II
3	2809.76	I
2	2811.11	I
1	2811.78	I
12	2846.72	I
12	2846.75	I
14	2848.34	I
14	2848.42	I
16	2851.65	I
16	2851.66	I
6000	2852.13	I
2	2902.92	I
4	2906.36	I
3	2915.45	I
10	2936.74	I
12	2938.47	I
2	2942.00	I
13	2942.00	I
20	3091.08	I
22	3092.99	I
14	3096.90	I
9	3104.71	II
8	3104.81	II
6	3168.98	II
6	3172.71	II
7	3175.78	II
2	3197.62	I
17	3329.93	I
6	3332.15	I
9	3336.68	I
7	3535.04	I
8	3538.86	II
7	3549.52	II
8	3553.37	II
140	3829.30	I
300	3832.30	I
500	3838.29	I
8	3848.24	II
7	3850.40	II
3	3878.31	I
3	3895.57	I
4	3903.86	I
6	3938.40	I
8	3986.75	I

Intensity	Wavelength/Å		Intensity	Wavelength/Å		Intensity	Wavelength/Å		Intensity	Wavelength/Å	
10	4057.50	I	17	9429.81	I	80	1795.65	IV	50	2427.72	II
15	4167.27	I	19	9432.76	I	80	1795.79	IV	30	2427.94	II
20	4351.91	I	20	9438.78	I	30	1853.27	II	30	2437.37	II
9	4384.64	II	12	9631.89	II	20	1857.92	II	20	2437.84	II
10	4390.59	II	11	9632.43	II	50	1902.95	II	30	2452.49	II
8	4428.00	II	15	9953.20	I	20	1907.84	II	50	2499.00	II
9	4433.99	II	15	9983.20	I	75	1910.25	IV	30	2507.60	II
14	4481.16	II	17	9986.47	I	30	1911.41	II	20	2516.60	II
13	4481.33	II	18	9993.21	I	20 d	1914.68	II	30	2516.74	II
28	4571.10	I	14	10092.16	II	100	1915.10	II	20	2521.66	II
10	4730.03	I	35	10811.08	I	20	1918.64	II	20	2530.72	II
7	4851.10	II	11	10914.23	II	30	1919.64	II	20	2531.80	II
75	5167.33	I	10	10951.78	II	80	1921.25	II	50	2532.78	II
220	5172.68	I	25	10953.32	I	20	1923.07	II	50	2533.33	II
400	5183.61	I	27	10957.30	I	20	1923.34	II	30	2534.10	II
8	5264.21	II	28	10965.45	I	30	1925.52	II	80	2534.22	II
7	5264.37	II	15	11032.10	I	50	1926.59	II	100	2535.66	II
9	5401.54	II	14	11033.66	I	30	1931.40	II	30	2535.98	II
6	5528.41	I	45	11828.18	I	500	1941.28	III	100	2537.92	II
30	5711.09	I	30	12083.66	I	800	1943.21	III	50	2541.11	II
10	6318.72	I	28	14877.62	I	20	1945.15	II	80	2542.92	II
9	6319.24	I	35	15024.99	I	20	1947.93	II	50	2543.45	II
7	6319.49	I	30	15040.24	I	20	1950.14	II	100	2548.75	II
10	6346.74	II	25	15047.70	I	500	1952.36	III	50	2551.85	II
9	6346.96	II	10	15765.84	I	1000	1952.52	III	30	2553.27	II
11	6545.97	II	30	17108.66	I	30	1953.23	II	75	2556.57	II
7	6781.45	II	5	26392.90	I	20 d	1954.81	II	30	2556.89	II
8	6787.85	II	**Manganese**			30	1959.25	II	95	2558.59	II
7	6812.86	II	**Mn Z = 25**			20	1969.24	II	30	2559.41	II
8	6819.27	II	600	410.30	V	500	1978.95	III	150	2563.65	II
10	7193.17	I	600	410.60	V	30	1994.23	II	30	2565.22	II
10	7291.06	I	600	415.62	V	9700	1996.06	I	580	2572.76	I
12	7387.69	I	650	415.98	V	14000	1999.51	I	480	2575.51	I
20	7657.60	I	600	428.59	V	18000	2003.85	I	12000	2576.10	II
19	7659.15	I	600	435.67	V	1000 w	2027.83	III	550	2584.31	I
17	7659.90	I	1000	441.72	V	500 w	2028.14	III	30	2588.97	II
15	7691.55	I	850	442.49	V	50	2037.31	II	45	2589.71	II
12	7877.05	II	60	579.79	IV	40	2037.64	II	250	2592.94	I
13	7896.37	II	60	581.44	IV	40	2039.97	II	6200	2593.73	II
10	8098.72	I	60	581.65	IV	500	2049.68	III	250	2595.76	I
9	8115.22	II	60	585.21	IV	500	2066.38	III	95	2598.90	II
8	8120.43	II	90	1242.25	IV	1000	2069.02	III	30	2602.72	II
10	8209.84	I	90	1244.50	IV	30	2076.21	II	45	2603.72	II
20	8213.03	I	95	1251.93	IV	900	2077.38	III	4300	2605.69	II
10	8213.99	II	95	1257.28	IV	800	2084.23	III	190	2610.20	II
11	8234.64	II	90	1264.41	IV	600	2090.05	III	500	2618.14	II
10	8310.26	I	500	1283.58	III	1500	2092.16	I	140	2622.90	I
15	8346.12	I	400	1287.59	III	500	2094.78	III	150	2624.04	I
10	8710.18	I	300	1291.62	III	20	2097.46	II	40	2624.80	II
12	8712.69	I	1000	1360.72	III	500	2097.93	III	200	2625.58	II
13	8717.83	I	800	1365.20	III	500	2099.97	III	190	2632.35	II
10	8734.99	II	500 h	1609.17	III	20	2102.50	II	130	2638.17	II
17	8736.02	I	1000	1614.14	III	1700	2109.58	I	80	2639.84	II
11	8745.66	II	2000	1620.60	III	30	2113.96	II	27	2650.99	II
14	8806.76	I	500	1633.80	III	1000	2169.78	III	60	2655.91	II
10	8824.32	II	80 b	1667.00	IV	700	2174.15	III	30	2666.77	II
11	8835.08	II	80	1698.30	IV	900	2176.87	III	30	2667.03	II
20	8923.57	I	20	1726.47	II	800	2181.86	III	110	2672.59	II
10	8997.16	I	30	1732.70	II	800	2184.87	III	55	2673.37	II
14	9218.25	II	50	1733.55	II	290	2208.81	I	55	2674.43	II
13	9244.27	II	40	1734.49	II	540	2213.85	I	45	2680.34	II
12	9246.50	I	30	1737.93	II	900	2220.55	III	30	2680.68	II
30	9255.78	I	20	1740.16	II	770	2221.84	I	30	2681.25	II
10	9327.54	II	20	1742.00	II	1000	2227.42	III	55	2684.55	II
10	9340.54	II	85 b	1742.10	IV	20	2373.36	II	55	2685.94	II
25	9414.96	I	85 d	1766.27	IV	20	2427.38	II	110	2688.25	II

Intensity	Wavelength/Å		Intensity	Wavelength/Å		Intensity	Wavelength/Å		Intensity	Wavelength/Å	
27	2693.19	II	200	3330.78	II	730	4063.53	I	40	5481.40	I
55	2695.36	II	720	3441.99	II	290	4070.28	I	30	5505.87	I
27	2698.97	II	50	3460.03	II	730	4079.24	I	50	5516.77	I
85	2701.00	II	360	3460.33	II	730	4079.42	I	40	5537.76	I
50	2701.17	II	360 h	3474.04	II	1100	4082.94	I	21	5551.98	I
160	2701.70	II		3474.13	II	1100	4083.63	I	200	5946.65	III
100	2703.98	II	290	3482.91	II	200	4110.90	I	140	6013.50	I
130	2705.74	II	180	3488.68	II	150	4131.12	I	200	6016.64	I
80	2707.53	II	140	3495.84	II	120	4135.04	I	290	6021.80	I
110	2708.45	II	50	3496.81	II	150	4176.60	I	200	6231.21	III
45	2709.96	II	100	3497.54	II	120	4189.99	I	17	6440.97	I
80	2710.33	II	360	3531.85	I	370	4235.14	I	24	6491.71	I
110	2711.58	II	1100	3532.12	I	510	4235.29	I	14 h	6942.52	I
30	2716.80	II	1300	3547.80	I	190	4239.72	I	12	6989.96	I
30	2717.53	II	1100	3548.03	I	290	4257.66	I	14	7069.84	I
30	2719.01	II	390	3548.20	I	290	4265.92	I	12	7184.25	I
50	2719.74	II	2200	3569.49	I	270	4281.10	I	24 h	7283.82	I
30	2722.10	II	720	3569.80	I	50	4323.63	II	35 h	7302.89	I
30	2724.46	II	1400	3577.88	I	350	4414.88	I	50	7326.51	I
55	2728.61	II	720	3586.54	I	210	4436.35	I	12	7680.20	I
6200	2794.82	I	290	3595.12	I	800	4451.59	I	12 h	8672.06	I
5100	2798.27	I	150	3601.72	III	160	4453.00	I	12 h	8701.05	I
220	2799.84	I	420	3607.54	I	130	4455.01	I	17 h	8703.76	I
3700	2801.06	I	420	3608.49	I	160	4455.32	I	30 h	8740.93	I
110	2809.11	I	360	3610.30	I	110	4455.82	I	**Mercury 198**		
60	2815.02	II	290	3619.28	I	210	4457.55	I	**Hg Z = 80**		
30	2816.33	II	220	3623.79	I	270	4458.26	I	80	1250.564	I
60	2870.08	II	140	3629.74	I	150	4461.08	I	8	1259.242	I
30	2872.94	II	100	3660.40	I	510	4462.02	I	100	1268.825	I
80	2879.49	II	280	3693.67	I	290	4464.68	I	5	1307.751	I
70	2886.68	II	180	3696.57	I	200	4470.14	I	20	1402.619	I
160	2889.58	II	210	3706.08	I	130	4472.79	I	10	1435.503	I
55	2892.39	II	130	3718.93	I	170	4490.08	I	1000	1849.492	I
50	2898.70	II	130	3731.93	I	240	4498.90	I	60	2262.210	II
80	2900.16	II	260	3790.22	I	240	4502.22	I	20	2302.065	I
140 h	2914.60	I	110	3800.55	I	160	4709.72	I	20	2345.440	I
190 h	2925.57	I	3200	3806.72	I	180	4727.48	I	100	2378.325	I
1100	2933.06	II	700	3809.59	I	130	4739.11	I	20	2380.004	I
1500	2939.30	II	2100	3823.51	I	1000	4754.04	I	40	2399.349	I
250 h	2940.39	I	390	3823.89	I	180	4761.53	I	20	2399.729	I
1900	2949.20	II	200	3829.68	I	750	4762.38	I	20	2446.900	I
30	3019.92	II	480	3833.86	I	300	4765.86	I	15	2464.064	I
55	3031.06	II	1300	3834.36	I	500	4766.43	I	40	2481.999	I
30	3035.35	II	350	3839.78	I	940	4783.42	I	30	2482.713	I
330	3044.57	I	670	3841.08	I	1000	4823.52	I	40	2483.821	I
120	3045.59	I	350	3843.98	I	19	5004.91	I	90	2534.769	I
200	3047.04	I	120	3926.47	I	30	5074.79	I	15000	2536.506	I
30	3050.65	II	130	3982.58	I	200	5079.20	III	25	2563.861	I
250	3054.36	I	150	3985.24	I	150	5100.03	III	25	2576.290	I
140	3062.12	I	190	3986.83	I	60	5117.94	I	250	2652.043	I
170	3066.02	I	150	3987.10	I	50	5150.89	I	400	2653.683	I
170	3070.27	I	1500	4018.10	I	50	5196.59	I	100	2655.130	I
160	3073.13	I	150	4026.44	I	85	5255.32	I	50	2698.831	I
140 h	3178.50	I	27000	4030.76	I	160	5341.06	I	80	2752.783	I
220	3212.88	I	19000	4033.07	I	19	5349.88	I	20	2759.710	III
1000	3228.09	I	11000	4034.49	I	95	5377.63	I	40	2803.471	I
300	3230.72	I	1500	4035.73	I	95	5394.67	I	30	2804.438	I
850	3236.78	I	5600	4041.36	I	50	5399.49	I	750	2847.675	II
330	3243.78	I	210 d	4045.13	I	95	5407.42	I	50	2856.939	I
650	3248.52	I	1100	4048.76	I	35	5413.69	I	150	2893.598	I
100	3251.14	I	150	4055.21	I	85	5420.36	I	150	2916.227	II
310	3252.95	I	1900	4055.54	I	35	5432.55	I	60	2925.413	I
310	3256.14	I	210	4057.95	I	150	5454.07	III	1200	2967.283	I
220	3258.41	I	1100	4058.93	I	12	5457.47	I	300	3021.500	I
180	3260.23	I	150	4059.39	I	60	5470.64	I	120	3023.476	I
180	3264.71	I	730	4061.74	I	200	5474.68	III	30	3025.608	I

Intensity	Wavelength/Å		Intensity	Wavelength/Å		Intensity	Wavelength/Å		Intensity	Wavelength/Å	
50	3027.490	I	9	1681.40	III	3	2670.49	III	100	4398.62	II
400	3125.670	I	100	1702.73	II	5	2674.91	I	15	4470.58	III
320	3131.551	I	100	1707.40	II	50	2698.83	I	12	4552.84	III
320	3131.842	I	120	1727.18	II	50	2699.38	I	90	4660.28	II
80	3341.481	I	250	1732.14	II	80	2705.36	II	50	4797.01	III
2800	3650.157	I	15	1759.75	III	70	2724.43	III	80	4855.72	II
300	3654.839	I	20	1775.68	I	80	2752.78	I	10	4869.85	III
80	3662.883	I	40	1783.70	II	20	2759.71	I	5	4883.00	I
240	3663.281	I	30	1796.22	II	6	2769.22	III	5	4889.91	I
30	3701.432	I	200	1796.90	II	40	2803.46	II	80	4916.07	I
35	3704.170	I	60	1798.74	II	30	2804.43	I	5	4970.37	I
30	3801.660	I	30	1803.89	II	2	2805.34	I	80	4973.57	III
20	3901.867	I	40	1808.29	II	2	2806.77	I	5	4980.64	I
60	3906.372	I	400	1820.34	II	150	2814.93	II	20	5102.70	I
200	3983.839	II	5	1832.74	I	3	2844.76	I	40	5120.64	I
1800	4046.572	I	1000	1849.50	I	750	2847.68	II	100	5128.45	II
150	4077.838	I	160	1869.23	II	50	2856.94	I	20	5137.94	I
40	4108.057	I	300	1870.55	II	150	2893.60	I	30	5210.82	III
250	4339.224	I	200	1875.54	II	150	2916.27	II	20	5290.74	I
400	4347.496	I	1	1894.77	III	60	2925.41	I	5	5316.78	I
4000	4358.337	I	20	1900.28	I	150	2935.94	II	60	5354.05	I
80	4916.068	I	30	1927.60	II	400	2947.08	II	30	5384.63	I
1100	5460.753	I	300	1942.27	II	1200	2967.28	I	1100	5460.74	I
160	5675.922	I	100	1972.94	II	300	3021.50	I	30	5549.63	I
240	5769.598	I	200	1973.89	II	120	3023.47	I	160	5675.86	I
280	5790.663	I	150	1987.98	II	30	3025.61	I	6	5695.71	III
20	6072.713	I	90	2026.97	II	50	3027.49	I	240	5769.60	I
30	6234.402	I	90	2052.93	II	15	3090.05	III	100	5789.66	I
160	6716.429	I	70	2148.00	II	400	3125.67	I	280	5790.66	I
250	6907.461	I	5	2247.55	I	320	3131.55	I	140	5803.78	I
240	11287.407	I	60	2262.23	II	320	3131.84	I	60	5859.25	I
	Mercury		20	2302.06	I	400	3208.20	II	60	5871.73	II
	Hg Z = 80		7	2314.15	III	400	3264.06	II	20	5871.98	I
3	621.44	III	15	2323.20	I	5	3283.02	III	20	6072.72	I
2	679.68	III	5	2340.57	I	12	3312.28	III	1000	6149.50	II
2	878.59	III	20	2345.43	I	80	3341.48	I	25	6220.35	III
1	886.48	III	20	2352.48	I	100	3385.25	II	30	6234.40	I
400	893.08	II	100	2378.32	I	8	3389.01	III	35	6418.98	III
300	915.83	II	20	2380.00	I	5	3450.77	III	40	6501.38	III
150	923.39	II	4	2380.55	III	400	3451.69	II	80	6521.13	II
200	940.80	II	40	2399.38	I	3	3500.35	III	10	6584.26	III
100	962.74	II	20	2399.73	I	4	3538.88	III	6	6610.12	III
50	969.13	II	10	2400.49	I	200	3549.42	II	30	6709.29	III
1	988.89	III	60	2407.35	II	5	3557.24	III	160	6716.43	I
2	1009.29	III	50	2414.13	II	2800	3650.15	I	250	6907.52	I
5	1068.03	III	8	2431.65	III	300	3654.84	I	250	7081.90	I
800	1099.26	II	5	2441.06	I	80	3662.88	I	200	7091.86	I
2	1161.95	III	20	2446.90	I	240	3663.28	I	40	7346.37	II
80	1250.58	I	15	2464.06	I	30	3701.44	I	100	7485.87	II
8	1259.24	I	5	2480.56	III	35	3704.17	I	12	7517.46	III
100	1268.82	I	40	2482.00	I	30	3801.66	I	20	7728.82	I
5	1307.75	I	30	2482.72	I	15	3803.51	III	7	7808.10	III
300	1307.93	II	40	2483.82	I	100	3806.38	II	100	7944.66	II
400	1321.71	II	7	2484.50	III	20	3901.87	I	25	7946.75	III
400	1331.74	II	90	2534.77	I	60	3906.37	I	50	7984.51	III
80	1350.07	II	15000	2536.52	I	100	3918.92	II	5	8151.64	III
200	1361.27	II	25	2563.86	I	200	3983.96	I	2000	10139.75	I
20	1402.62	I	25	2576.29	I	1800	4046.56	I	240	11287.40	I
200	1414.43	II	5	2578.91	I	150	4077.83	I	120	13209.95	I
10	1435.51	I	2	2612.92	III	40	4108.05	I	140	13426.57	I
15	1619.46	II	4	2617.97	III	70	4122.07	III	60	13468.38	I
120	1623.95	II	15	2625.19	I	10	4140.34	III	80	13505.58	I
20	1628.25	II	5	2639.78	I	100	4216.74	III	500	13570.21	I
150	1649.94	II	250	2652.04	I	250	4339.22	I	450	13673.51	I
50	1653.64	II	400	2653.69	I	400	4347.49	I	200	13950.55	I
200	1672.41	II	100	2655.13	I	4000	4358.33	I	500	15295.82	I

Intensity	Wavelength/Å		Intensity	Wavelength/Å		Intensity	Wavelength/Å		Intensity	Wavelength/Å	
100	16881.48	I	200	2506.19	III	290	2903.07	II	1300	3344.75	I
400	16920.16	I	440	2538.46	II	80	2907.12	II	95	3346.40	II
300	16942.00	I	330	2542.67	II	600	2909.12	II	1600	3358.12	I
500	17072.79	I	80	2558.88	II	1100	2911.92	II	950	3363.78	I
400	17109.93	I	85	2564.34	II	120	2918.83	II	950	3379.97	I
20	17116.75	I	250	2593.70	II	1300	2923.39	II	1900	3384.62	I
20	17198.67	I	250	2602.80	II	140	2924.32	II	130	3395.36	II
20	17213.20	I	400	2616.78	I	1100	2930.50	II	640	3404.34	I
70	17329.41	I	440	2629.85	I	800	2934.30	II	1300	3405.94	I
30	17436.18	I	330	2636.67	II	95	2940.10	II	640	3437.22	I
50	18130.38	I	720	2638.76	II	110	2941.22	II	130	3446.08	II
40	19700.17	I	410	2640.99	I	150	2944.82	II	3200	3447.12	I
	22493.28	I	600	2644.35	II	140	2946.69	II	640	3449.07	I
250	23253.07	I	370	2646.49	II	95	2947.28	II	950	3456.39	I
	32148.06	I	640	2649.46	I	125	2947.32	III	640	3460.78	I
	36303.03	I	480	2653.35	II	95	2955.84	II	800	3504.41	I
Molybdenum			560 h	2655.03	I	240	2956.06	II	560	3508.12	I
Mo Z = 42			640	2660.58	II	70	2956.90	II	480	3521.41	I
50	867.92	IV	720	2672.84	II	95	2960.24	II	640	3537.28	I
100	884.19	IV	250	2673.27	II	250	2963.79	II	520	3558.10	I
60	886.05	IV	1000	2679.85	I	210	2965.27	II	400	3563.14	I
50	891.74	IV	95	2681.36	II	70	2971.91	II	1400	3581.89	I
100	1169.33	III	640	2683.23	II	250	2972.61	II	1400	3624.46	I
100	1254.93	III	880	2684.14	II	80	2975.40	II	1000	3635.43	I
100	1258.52	III	560	2687.99	II	95	2992.84	II	400	3657.35	I
100	1262.21	III	480	2701.42	II	95	3027.77	II	540	3664.81	I
100	1263.74	III	190	2713.51	II	100	3060.78	II	590	3672.82	I
100	1274.37	III	290	2717.35	II	800	3064.28	I	1300	3680.60	I
100	1276.40	III	85	2726.97	II	250	3065.04	II	65	3688.31	II
200	1277.40	III	140	2729.68	II	800	3074.37	I	180	3692.64	II
200	1277.58	III	80	2730.20	II	85	3077.66	II	1400	3694.94	I
200	1278.40	III	330	2732.88	II	800	3085.62	I	500	3727.69	I
150	1281.90	III	160	2736.96	II	270	3087.62	II	80	3744.37	II
150	1283.60	III	80 h	2737.88	II	190	3092.07	II	29000	3798.25	I
100	1854.73	III	290	2746.30	II	560	3094.66	I	520	3826.70	I
80	1926.26	IV	110	2756.07	II	560	3101.34	I	940	3828.87	I
100	1929.24	IV	220	2763.62	II	1400	3112.12	I	1700	3833.75	I
80	1971.06	IV	240	2769.76	II	290	3122.00	II	29000	3864.11	I
70	2010.92	IV	160	2773.78	II	14000	3132.59	I	580	3869.08	I
19000	2015.11	II	190	2774.39	II	110	3138.72	II	580	3886.82	I
40000	2020.30	II	1700	2775.40	II	220	3152.82	II	19000	3902.96	I
21000	2038.44	II	65	2777.86	II	55	3155.64	II	65	3941.48	II
17000	2045.98	II	880	2780.04	II	6000	3158.16	I	1400	4062.08	I
50	2060.38	IV	400	2784.99	II	8700	3170.35	I	2300	4069.88	I
4800	2081.68	II	100	2807.74	III	95	3172.03	II	1300	4081.44	I
2400	2089.52	II	400	2807.76	II	160	3172.74	II	940	4084.38	I
2200	2092.50	II	1700	2816.15	II	120 d	3187.59	II	730	4107.47	I
4000	2093.11	II	220	2817.44	II	7600	3193.97	I	630	4120.10	I
2700	2100.84	II	80	2827.74	II	880	3205.88	I	2900	4143.55	I
1500	2104.29	II	80	2834.39	II	3000	3208.83	I	480	4185.82	I
1400	2108.02	II	80	2835.33	II	560	3215.07	I	2500	4188.32	I
100	2184.37	III	160	2842.15	II	880	3228.22	I	1500	4232.59	I
100	2211.02	III	1700	2848.23	II	600	3229.79	I	890	4276.91	I
400	2269.69	II	370	2853.23	II	1100	3233.14	I	1200	4277.24	I
150	2269.71	III	370	2863.81	II	950	3237.08	I	1400	4288.64	I
200	2294.97	III	220	2866.69	II	65	3240.71	II	680	4292.13	I
160	2304.25	II	1700	2871.51	II	950	3256.21	I	890	4293.21	I
160	2306.97	II	85	2872.88	II	480	3264.40	I	840	4326.14	I
150	2330.93	III	220	2879.05	II	800	3270.90	I	1900	4381.64	I
110	2332.12	II	65	2888.15	II	200	3271.69	III	2500	4411.57	I
190	2341.59	II	1300	2890.99	II	1100	3289.02	I	990	4434.95	I
100	2359.76	III	95	2891.28	II	950	3290.82	I	480	4457.36	I
110	2389.20	II	190	2892.81	II	190	3292.31	II	630	4474.56	I
140	2403.61	II	950	2894.45	II	100	3313.62	II	400	4536.80	I
120	2413.01	II	140	2897.63	II	190	3320.90	II	460	4626.47	I
85	2498.28	II	70	2900.80	II	640	3323.95	I	640	4707.26	I

Intensity	Wavelength/Å		Intensity	Wavelength/Å		Intensity	Wavelength/Å		Intensity	Wavelength/Å	
700	4731.44	I	50	5869.33	I	780	3723.50	II	2000	3951.16	II
770	4760.19	I	820	5888.33	I	410	3724.87	II	810	3952.20	II
410	4819.25	I	50 h	5893.38	I	710	3728.13	II	590	3958.00	II
410	4830.51	I	160 h	5928.88	I	470	3730.58	II	510	3962.21	II
180	5014.60	I	35	6025.49	I	1000 d	3735.54	II	1400	3963.12	II
80	5029.00	I	1300	6030.66	I	440	3737.10	II	1100	3973.30	II
65	5030.78	I	40	6101.87	I	1000	3738.06	II	740	3973.69	II
100	5047.71	I	40	6357.22	I	580	3752.49	II	740	3976.85	II
50	5055.00	I	35	6401.07	I	510	3757.82	II	740	3979.49	II
200	5059.88	I	100	6424.37	I	930	3758.95	II	470	3986.25	II
100	5080.02	I	230	6619.13	I	930	3763.47	II	1400	3990.10	II
100	5096.65	I	50	6650.38	I	510	3769.65	II	1000	3991.74	II
130	5097.52	I	110	6733.98	I	1400	3775.50	II	1100	3994.68	II
130	5109.71	I	50	6746.27	I	710	3779.47	II	410	4000.50	II
80	5114.97	I	35	6753.97	I	580	3780.40	II	540	4004.02	II
150	5145.38	I	40	6838.88	I	510	3781.32	II	410	4007.43	II
110	5147.39	I	35	6914.01	I	2400	3784.25	II	3700	4012.25	II
80	5163.19	I	110	7109.87	I	370	3801.12	II	540	4012.70	II
100	5167.76	I	150	7242.50	I	1200	3803.47	II	1000	4020.87	II
160 d	5171.08	I	40	7245.85	I	2500	3805.36	II	1000	4021.34	II
230 h	5172.94	I	40	7391.36	I	470	3807.23	II	1000	4021.78	II
160 h	5174.18	I	140	7485.74	I	540	3808.77	II	1200	4023.00	II
110	5200.17	I	27	7720.77	I	440	3809.06	II	410	4030.47	II
50	5200.74	I	40 h	8328.44	I	580	3810.49	II	1200	4031.82	II
50	5211.86	I	45 h	8389.32	I	710	3814.73	II	3000	4040.80	II
80	5219.40	I	45 h	8483.39	I	410	3822.47	II	410	4043.59	II
65	5231.06	I		Neodymium		1200	3826.42	II	410	4048.81	II
100	5234.26	I		Nd Z = 60		540	3828.85	II	850	4051.15	II
460 h	5238.20	I	75	2764.98	I	440	3829.16	II	850	4059.96	II
230 h	5240.88	I	80	2993.20	II	510	3830.47	II	4700	4061.09	II
110 h	5242.81	I	95	3007.97	II	740	3836.54	II	1100	4069.28	II
100	5245.51	I	95	3014.19	II	1700	3838.98	II	710	4075.12	II
150	5259.04	I	95	3018.35	II	410 d	3841.82	II	470	4075.28	II
65	5261.14	I	140	3056.71	II	1700 d	3848.24	II	470	4080.23	II
65	5279.65	I	130	3069.73	II	1500	3848.52	II	1400	4109.08	II
210	5280.86	I	160	3075.38	II	470	3850.22	II	2500	4109.46	II
55	5292.08	I	240	3092.92	II	2400 d	3851.66	II	510	4110.48	II
55	5295.47	I	260	3115.18	II	3700 d	3863.33	II	410	4123.88	II
55	5313.89	I	290	3133.60	II	850	3869.07	II	470	4133.36	II
80	5354.88	I	220	3134.90	II	470	3875.87	II	510	4135.33	II
65	5356.48	I	170	3141.46	II	1100	3878.58	II	3000	4156.08	II
560 hl	5360.56	I	170	3142.44	II	1000	3879.55	II	510	4156.26	II
110 hl	5364.28	I	150	3203.47	II	780	3880.38	II	410	4168.00	II
65	5394.52	I	220	3259.24	II	1200	3880.78	II	810	4175.61	II
50	5400.47	I	220	3265.12	II	540	3887.87	II	2400	4177.32	II
55	5435.68	I	320	3275.22	II	1300	3889.93	II	640	4179.59	II
65	5437.75	I	290	3285.10	II	1300	3890.58	II	470	4205.60	II
50	5501.54	I	410	3328.28	II	1300	3890.94	II	470	4211.29	II
7800	5506.49	I	320	3353.59	II	580	3891.51	II	440	4227.73	II
5200	5533.05	I	410	3560.75	II	470	3892.06	II	1300	4232.38	II
50	5543.12	I	470	3587.51	II	810	3894.63	II	2000	4247.38	II
55	5556.28	I	370	3615.82	II	440	3897.63	II	850	4252.44	II
2500	5570.45	I	410	3653.15	II	2000	3900.21	II	410	4261.84	II
100	5610.93	I	470	3662.26	II	1300	3901.84	II	470	4282.44	II
330	5632.47	I	540	3665.18	II	1700	3905.89	II	710	4284.52	II
50	5634.86	I	540	3672.36	II	510	3907.84	II	5400	4303.58	II
230	5650.13	I	580	3673.54	II	2000	3911.16	II	470	4314.52	II
55	5674.47	I	1200	3685.80	II	850	3912.23	II	1100	4325.76	II
460	5689.14	I	440	3687.30	II	440	3915.13	II	510	4327.93	II
80	5705.72	I	410	3689.69	II	610	3915.95	II	540	4338.70	II
210	5722.74	I	410	3697.56	II	1100	3920.96	II	680	4351.29	II
620	5751.40	I	470	3713.70	II	510	3927.10	II	850	4358.17	II
520	5791.85	I	640 d	3714.73	II	610	3934.82	II	470 d	4374.93	II
55 h	5849.73	I	470	3715.68	II	410	3936.11	II	710	4385.66	II
50 h	5851.52	I	410	3718.54	II	510	3938.86	II	540	4400.83	II
520	5858.27	I	410	3721.35	II	2000	3941.51	II	510	4411.06	II

Intensity	Wavelength/Å	Intensity	Wavelength/Å	Intensity	Wavelength/Å	Intensity	Wavelength/Å
580	4446.39 II	55	6310.49 I	100	208.48 IV	80	541.13 IV
1400	4451.57 II	65	6385.20 II	100	208.73 IV	100	542.07 IV
740	4462.99 II	45	6630.14 I	80	208.90 IV	150	543.89 IV
410	4501.82 II	45	6650.57 II	150	212.56 IV	400	568.42 V
250	4516.36 II	40	6740.11 II	140	223.24 IV	250	569.76 V
340	4541.27 II	40	6900.43 II	120	223.60 IV	500	569.83 V
340	4542.61 II	35	7037.30 II	140	234.32 IV	250	572.11 V
340	4563.22 II	40	7066.89 II	120	234.70 IV	800	572.34 V
300	4621.94 I	29	7129.35 II	20	251.14 III	35	587.213 I
510	4634.24 I	24	7189.42 II	20	251.56 III	35	589.179 I
340	4641.10 I	20	7192.01 II	20	251.73 III	35	589.911 I
250	4645.77 II	15	7236.54 II	40	267.06 III	70	591.830 I
300	4649.67 I	12	7316.81 II	40	267.52 III	100	595.920 I
310	4683.45 I	10	7406.62 II	20	267.71 III	75	598.706 I
470	4706.54 II	10	7418.18 II	40	283.18 III	35	598.891 I
240	4719.02 I	12	7511.16 II	160	283.21 III	70	600.036 I
240	4811.34 II	17	7513.73 II	110	283.69 III	170	602.726 I
350	4825.48 II	12	7528.99 II	40	283.89 III	170	615.628 I
280	4859.02 II	10	7538.26 II	220	301.12 III	170	618.672 I
350	4883.81 I	12	7696.56 II	220	313.05 III	120	619.102 I
220	4890.70 II	10	7750.95 II	220	313.68 III	200	626.823 I
240	4891.07 I	10	7808.47 II	40	313.95 III	200	629.739 I
280	4896.93 I	12	7863.04 II	90	352.956 I	1000	735.896 I
210	4901.84 I	12	7917.01 II	60	354.962 I	400	743.720 I
330	4920.68 II	12	7958.95 I	50	357.83 IV	60	993.88 I
470	4924.53 I	12	7965.73 II	400	357.96 V	70	1068.65 I
260	4944.83 I	15	7982.09 II	500	358.47 V	90	1131.72 I
290	4954.78 I	12	7982.68 II	200	358.72 IV	100	1131.85 II
290	4959.13 II	12	8000.76 II	500	359.38 V	90	1229.83 I
250	4989.94 II	10	8120.93 II	90	361.433 II	20	1255.03 III
360	5076.59 II	12	8122.07 II	60	362.455 II	110	1255.68 III
360	5092.80 II	12	8141.75 II	1000	365.59 V	160	1257.19 III
360	5107.59 II	12	8143.27 II	220	379.31 III	90	1418.38 I
340	5123.79 II	10	8231.52 II	125	387.14 IV	90	1428.58 I
680	5130.60 II	10	8307.72 II	100	388.22 IV	90	1436.09 I
500	5191.45 II	12	8346.36 II	150	405.854 II	120	1681.68 II
630	5192.62 II	17	8839.10 II	120	407.138 II	180	1688.36 II
330	5200.12 II	**Neon**		800	416.20 V	100	1888.11 II
310	5212.37 II	**Ne Z = 10**		150	421.61 IV	100	1889.71 II
450	5234.20 II	66	119.01 V	200	445.040 II	200	1907.49 II
250	5239.79 II	200	122.52 V	300	446.256 II	500	1916.08 II
720	5249.59 II	66	125.12 V	250	446.590 II	300	1930.03 II
360	5255.51 II	45	131.99 V	180	447.815 II	200	1938.83 II
590	5273.43 II	50	132.04 V	150	454.654 II	100 c	1945.46 II
680	5293.17 II	150	140.76 V	200	455.274 II	80	2007.01 II
220	5311.46 II	150	140.79 V	10	456.275 II	65	2018.44 IV
500	5319.82 II	100	142.44 V	120	456.348 II	110	2022.19 IV
290	5361.47 II	100	142.50 V	90	456.896 II	80	2025.56 II
160	5431.53 II	150	142.72 V	1000	460.728 II	150	2085.47 II
240	5594.43 II	100	143.27 V	500	462.391 II	200	2086.96 III
220	5620.54 I	150	143.34 V	140	469.77 IV	300	2089.43 III
140 d	5675.97 I	150	147.13 V	200	469.82 IV	240	2092.44 III
220	5688.53 II	66	151.23 V	180	469.87 IV	400	2095.54 III
130	5702.24 II	120	151.42 V	140	469.92 IV	180	2096.11 II
160	5708.28 II	15	151.82 IV	250	480.41 V	120	2096.25 II
100	5729.29 I	15	152.23 IV	150	481.28 V	200	2161.22 III
160	5804.02 II	45	154.50 IV	250	481.36 V	300	2163.77 III
80	5811.57 II	15	158.65 IV	500	482.99 V	200	2180.89 III
70	5825.87 II	15	158.82 IV	285	488.10 III	30	2203.88 IV
80	5842.39 II	100	164.02 V	220	488.87 III	200	2209.35 III
55	5858.91 I	100	164.14 V	450	489.50 III	200	2211.85 III
45	6007.67 I	80	172.62 IV	70	489.64 III	240	2213.76 III
45	6034.24 II	500	173.93 V	220	490.31 III	300	2216.07 III
55	6066.03 I	80	177.16 IV	360	491.05 III	10	2220.81 IV
45	6178.59 I	150	186.58 IV	120	521.74 IV	75	2227.42 V
45	6223.39 I	100	194.28 IV	140	521.82 IV	110	2232.41 V

Line Spectra of the Elements (continued): Neon

Intensity	Wavelength/Å		Intensity	Wavelength/Å		Intensity	Wavelength/Å		Intensity	Wavelength/Å	
65	2245.48	V	300	3028.86	II	40	3369.908	I	150 p	4430.90	II
250	2258.02	IV	100	3030.79	II	100	3371.80	II	150 p	4430.94	II
65	2259.57	V	120	3034.46	II	500	3378.22	II	120	4457.05	II
175	2262.08	IV	100	3035.92	II	150	3388.42	II	100	4522.72	II
240	2263.21	III	100	3037.72	II	120	3388.94	II	10	4537.754	I
65	2263.39	V	100	3039.59	II	300	3392.80	II	10	4540.380	I
110	2264.54	IV	100	3044.09	II	100	3404.82	II	100	4569.06	II
200	2264.91	III	100	3045.56	II	120	3406.95	II	15	4704.395	I
250	2265.71	V	120	3047.56	II	100	3413.15	II	12	4708.862	I
550	2285.79	IV	100	3054.34	II	120	3416.91	II	10	4710.067	I
30	2293.14	IV	100	3054.68	II	120	3417.69	II	10	4712.066	I
250	2293.49	IV	100	3059.11	II	50	3417.904	I	15	4715.347	I
250	2350.84	IV	100	3062.49	II	15	3418.006	I	10	4752.732	I
450	2352.52	IV	100	3063.30	II	120	3428.69	II	12	4788.927	I
700	2357.96	IV	100	3070.89	II	60	3447.703	I	10	4790.22	I
250	2362.68	IV	100	3071.53	II	50	3454.195	I	10	4827.344	I
250	2363.28	IV	100	3075.73	II	100	3456.61	II	10	4884.917	I
110	2365.49	IV	120	3088.17	II	100	3459.32	II	4	5005.159	I
350	2372.16	IV	100	3092.09	II	25	3460.524	I	10	5037.751	I
65	2384.20	IV	120	3092.90	II	30	3464.339	I	10	5144.938	I
350	2384.95	IV	100	3094.01	II	30	3466.579	I	25	5330.778	I
300	2412.73	III	100	3095.10	II	60	3472.571	I	20	5341.094	I
240	2412.94	III	100	3097.13	II	150	3479.52	II	8	5343.283	I
200	2413.78	III	100	3117.98	II	200	3480.72	II	60	5400.562	I
200	2473.40	III	120	3118.16	II	200	3481.93	II	5	5562.766	I
80 p	2562.12	II	10	3126.199	I	25	3498.064	I	10	5656.659	I
90 w	2567.12	II	300	3141.33	II	30	3501.216	I	5	5719.225	I
800	2590.04	III	100	3143.72	II	25	3515.191	I	12	5748.298	I
600	2593.60	III	100 p	3148.68	II	150	3520.472	I	80	5764.419	I
400	2595.68	III	100	3164.43	II	120	3542.85	II	12	5804.450	I
300	2610.03	III	100	3165.65	II	120	3557.80	II	40	5820.156	I
240	2613.41	III	100	3188.74	II	100	3561.20	II	500	5852.488	I
200	2615.87	III	120	3194.58	II	100	3574.18	II	100	5872.828	I
80	2623.11	II	500	3198.59	II	200	3574.61	II	100	5881.895	I
80	2629.89	II	60	3208.96	II	50	3593.526	I	60	5902.462	I
90 w	2636.07	II	120	3209.36	II	30	3593.640	I	60	5906.429	I
80	2638.29	II	120	3213.74	II	15	3600.169	I	100	5944.834	I
200	2638.70	III	150	3214.33	II	20	3633.665	I	100	5965.471	I
200	2641.07	III	150	3218.19	II	150	3643.93	II	100	5974.627	I
80	2644.10	II	120	3224.82	II	200	3664.07	II	120	5975.534	I
600	2677.90	III	120	3229.57	II	20	3682.243	I	80	5987.907	I
500	2678.64	III	200	3230.07	II	12	3685.736	I	100	6029.997	I
80	2762.92	II	120	3230.42	II	200	3694.21	II	100	6074.338	I
90	2792.02	II	120	3232.02	II	10	3701.225	I	80	6096.163	I
80	2794.22	II	150	3232.37	II	150	3709.62	II	60	6128.450	I
100	2809.48	II	100	3243.40	II	250	3713.08	II	100	6143.063	I
80	2906.59	II	100	3244.10	II	250	3727.11	II	120	6163.594	I
80	2906.82	II	100	3248.34	II	800	3766.26	II	250	6182.146	I
90	2910.06	II	100	3250.36	II	1000	3777.13	II	150	6217.281	I
90	2910.41	II	150	3297.73	II	100	3818.43	II	150	6266.495	I
80	2911.14	II	150	3309.74	II	120	3829.75	II	60	6304.789	I
80	2915.12	II	300	3319.72	II	150	4219.74	II	100	6334.428	I
80	2925.62	II	1000	3323.74	II	100	4233.85	II	120	6382.992	I
80 w	2932.10	II	150	3327.15	II	120	4250.65	II	200	6402.246	I
80	2940.65	II	100	3329.16	II	120	4369.86	II	150	6506.528	I
90	2946.04	II	200	3334.84	II	70	4379.40	II	60	6532.882	I
150	2955.72	II	150	3344.40	II	150	4379.55	II	150	6598.953	I
150	2963.24	II	300	3345.45	II	100	4385.06	II	70	6652.093	I
150	2967.18	II	150	3345.83	II	200	4391.99	II	90	6678.276	I
100	2973.10	II	200	3355.02	II	150	4397.99	II	20	6717.043	I
15	2974.72	I	120	3357.82	II	150	4409.30	II	100	6929.467	I
100	2979.46	II	200	3360.60	II	100	4413.22	II	90	7024.050	I
12	2982.67	I	120	3362.16	II	100	4421.39	II	100	7032.413	I
150	3001.67	II	100	3362.71	II	100 p	4428.52	II	50	7051.292	I
120 p	3017.31	II	120	3367.22	II	100 p	4428.63	II	80	7059.107	I
300	3027.02	II	12	3369.808	I				100	7173.938	I

Neon (continued)

Intensity	Wavelength/Å	
150	7213.20	II
150	7235.19	II
100	7245.167	I
150	7343.94	II
40	7472.439	I
90	7488.871	I
100	7492.10	II
150	7522.82	II
80	7535.774	I
60	7544.044	I
100	7724.628	I
120	7740.74	II
300	7839.055	I
120	7926.20	II
400	7927.118	I
700	7936.996	I
2000	7943.181	I
2000	8082.458	I
100	8084.34	II
1000	8118.549	I
600	8128.911	I
3000	8136.406	I
2500	8259.379	I
100	8264.81	II
2500	8266.077	I
800	8267.117	I
6000	8300.326	I
100	8315.00	II
1500	8365.749	I
100	8372.11	II
8000	8377.606	I
1000	8417.159	I
4000	8418.427	I
1500	8463.358	I
800	8484.444	I
5000	8495.360	I
600	8544.696	I
1000	8571.352	I
4000	8591.259	I
6000	8634.647	I
3000	8647.041	I
15000	8654.383	I
4000	8655.522	I
100	8668.26	II
5000	8679.492	I
5000	8681.921	I
2000	8704.112	I
4000	8771.656	I
12000	8780.621	I
10000	8783.753	I
500	8830.907	I
7000	8853.867	I
1000	8865.306	I
1000	8865.755	I
3000	8919.501	I
2000	8988.57	I
100	9079.46	II
6000	9148.67	I
6000	9201.76	I
4000	9220.06	I
2000	9221.58	I
2000	9226.69	I
1000	9275.52	I
200	9287.56	II
6000	9300.85	I
1500	9310.58	I
3000	9313.97	I
6000	9326.51	I
2000	9373.31	I
5000	9425.38	I
3000	9459.21	I
5000	9486.68	I
5000	9534.16	I
3000	9547.40	I
120	9577.01	II
1000	9665.42	I
100	9808.86	II
800	10295.42	I
2000	10562.41	I
1500	10798.07	I
2000	10844.48	I
3000	11143.020	I
3500	11177.528	I
1600	11390.434	I
1100	11409.134	I
3000	11522.746	I
1500	11525.020	I
950	11536.344	I
500	11601.537	I
1200	11614.081	I
300	11688.002	I
2000	11766.792	I
1500	11789.044	I
500	11789.889	I
1000	11984.912	I
3000	12066.334	I
800	12459.389	I
1000	12689.201	I
1100	12912.014	I
700	13219.241	I
800	15230.714	I
400	17161.930	I
400	18035.80	I
1000	18083.21	I
350	18221.11	I
250	18227.02	I
2500	18276.68	I
2000	18282.62	I
1200	18303.97	I
250	18359.12	I
1200	18384.85	I
2000	18389.95	I
1000	18402.84	I
1200	18422.39	I
300	18458.65	I
400	18475.79	I
900	18591.55	I
1600	18597.70	I
350	18618.96	I
550	18625.16	I
1200	21041.295	I
750	21708.145	I
300	22247.35	I
350	22428.13	I
2250	22530.40	I
400	22661.81	I
600	23100.51	I
1000	23260.30	I
1050	23373.00	I
850	23565.36	I
3500	23636.52	I
300	23701.64	I
1100	23709.2	I
1800	23951.42	I
600	23956.46	I
1000	23978.12	I
200	24098.54	I
500	24161.42	I
600	24249.64	I
1500	24365.05	I
800	24371.60	I
400	24447.85	I
700	24459.4	I
300	24776.46	I
550	24928.88	I
250	25161.69	I
650	25524.37	I
125	28386.21	I
150	30200.	I
250	33173.09	I
450	33352.35	I
1300	33901.	I
2200	33912.10	I
600	34131.31	I
100	34471.44	I
120	35834.78	I

Neptunium
Np Z = 93

Intensity	Wavelength/Å	
300	3481.93	I
300 h	3501.50	I
300 l	3986.89	I
300 s	5044.66	I
300 l	5601.70	I
300 l	5652.75	I
300 l	5784.39	I
300 l	5878.04	I
300 s	6011.22	I
300	6056.09	I
300 s	6073.90	I
300 s	6080.05	I
300 l	6120.49	I
300	6188.59	I
300 l	6200.00	I
300 s	6215.90	I
300 s	6317.84	I
300 l	6341.38	I
300 l	6566.11	I
300 l	6720.68	I
300 s	6751.32	I
300 s	6795.21	I
300 l	6802.62	I
300 l	6805.81	I
300 s	6816.44	I
300 l	6865.45	I
300 s	6907.13	I
300 h	6912.91	I
1000 s	6930.31	I
300 l	6963.63	I
3000 s	6972.09	I
300	7014.02	I
300 l	7018.91	I
300 s	7039.14	I
300 s	7080.01	I
300 l	7174.83	I
300 l	7184.93	I
300 l	7284.28	I
300 l	7292.29	I
300 l	7332.52	I
300 s	7370.60	I
300 l	7381.03	I
300 l	7381.65	I
300 l	7402.70	I
300 s	7512.22	I
300 l	7515.15	I
300 l	7546.05	I
300 l	7624.83	I
300	7626.85	I
300 s	7681.01	I
300 s	7685.25	I
1000 l	7735.14	I
300 l	7761.61	I
1000 l	7765.75	I
300 s	7776.07	I
300	7787.46	I
1000 l	7791.38	I
300 l	7851.44	I
300 l	7887.88	I
300 l	7901.71	I
300 l	7975.98	I
300 h	8080.32	I
300 s	8124.59	I
300	8155.11	I
300 l	8167.42	I
300 l	8183.06	I
300 l	8188.61	I
300 l	8247.82	I
300 l	8287.11	I
300 s	8287.75	I
300 l	8306.22	I
300 s	8313.66	I
1000 l	8339.12	I
300	8356.79	I
300 l	8367.11	I
3000	8372.88	I
3000	8529.96	I
1000 s	8696.23	I
1000 s	8906.02	I
1000	8942.70	I
1000 s	9004.75	I
1000 l	9006.31	I
10000 l	9016.18	I
3000 l	9141.30	I
3000 s	9379.33	I
3000 l	9468.66	I
3000 s	9679.13	I
3000 l	9930.55	I
10000 l	10091.99	I
10000 s	10817.45	I
10000 l	11695.15	I
10000 l	11776.64	I
10000 l	12148.18	I
10000 s	12377.42	I
10000 l	12407.99	I
10000 l	13834.33	I

Nickel
Ni Z = 28

Intensity	Wavelength/Å	
55	315.24	V
56	315.71	V
72	354.18	V
76	354.42	V
68	354.49	V
500	630.71	III
500	676.94	III
300	713.33	III

Intensity	Wavelength/Å		Intensity	Wavelength/Å		Intensity	Wavelength/Å		Intensity	Wavelength/Å	
300	713.38	III	2000	2174.67	II	660	3500.85	I	13	7797.59	I
500	718.48	III	1500	2175.15	II	2600	3510.34	I	1000	8096.75	II
300	722.09	III	2500	2185.50	II	6600	3515.05	I	700	8121.48	II
500	729.82	III	3000	2192.09	II	660	3519.77	I	9	8862.55	I
400	731.70	III	5000	2205.55	II	8200	3524.54	I	500 w	9900.92	II
300	732.16	III	4000	2206.72	II	5000	3566.37	I		**Niobium**	
300	747.99	III	6000	2216.48	II	990	3571.87	I		**Nb Z = 41**	
300	750.05	III	1000	2264.46	II	1300	3597.70	I	80	464.55	V
300	757.80	III	2000	2270.21	II	1300	3610.46	I	80	468.32	V
400	770.22	III	1600	2289.98	I	530	3612.74	I	80	763.77	V
500	778.81	III	630	2300.78	I	6600	3619.39	I	80	774.02	V
300	788.04	III	1000	2303.00	II	200	3664.10	I	60	993.54	IV
500	811.57	III	2000	2310.96	I	130	3669.24	I	400	1005.72	IV
500	826.14	III	1700	2312.34	I	180	3670.43	I	500	1007.05	IV
500	842.14	III	1400	2313.66	I	260	3674.15	I	500	1010.19	IV
400	845.24	III	1400	2313.98	I	160	3688.42	I	100	1116.08	IV
300	847.43	III	1000	2316.04	II	80	3693.93	I	150	1120.02	IV
300	860.64	III	1400	2317.16	I	120	3722.48	I	100	1258.87	V
300	862.88	III	2600	2320.03	I	150	3736.81	I	60	1314.56	III
300	863.22	III	1900	2321.38	I	60	3739.23	I	80	1445.43	III
300	867.51	III	1400	2325.79	I	600	3775.57	I	80	1445.98	III
300	973.79	III	940	2329.96	I	700	3783.53	I	80	1447.09	III
400	979.59	III	1200	2345.54	I	700	3807.14	I	100	1456.68	III
500	1317.22	II	400	2347.52	I	110	3831.69	I	80	1484.73	III
76	1398.19	IV	1000	2375.42	II	1200	3858.30	I	100	1495.94	III
74	1411.45	IV	240	2386.58	I	110	3973.56	I	80	1498.02	III
70	1438.82	IV	1000	2394.52	II	110	4401.55	I	80	1499.45	III
73	1449.01	IV	2000	2416.13	II	85	4459.04	I	100	1501.99	III
76	1452.22	IV	240	2419.31	I	55	4470.48	I	60	1502.30	IV
73	1482.25	IV	160	2472.06	I	65	4605.00	I	80	1513.81	III
72	1489.83	IV	150	2798.65	I	75	4648.66	I	60	1524.36	IV
75	1525.31	IV	250	2821.29	I	110	4714.42	I	100	1524.91	III
74	1527.68	IV	500	2943.91	I	45	4786.54	I	100	1590.21	III
74	1527.80	IV	570	2981.65	I	45	4855.41	I	80	1598.86	III
76	1534.71	IV	500	2992.60	I	40	4904.41	I	80	1604.72	III
73	1537.25	IV	1000	2994.46	I	45	4980.16	I	80	1639.51	III
75	1543.41	IV	4000	3002.49	I	45	4984.13	I	100	1682.77	III
74	1546.23	IV	2200	3003.63	I	50	5017.59	I	100	1705.44	III
300	1604.54	III	3700	3012.00	I	100	5035.37	I	100	1707.14	III
300	1652.87	III	1700	3037.94	I	100	5080.52	I	100	1758.33	V
400	1687.90	III	3500	3050.82	I	65	5081.11	I	100	1877.34	V
1000	1692.51	III	1500	3054.32	I	40 h	5146.48	I	100	1892.92	III
800	1709.90	III	1900	3057.64	I	40 h	5155.76	I	60	1922.41	IV
650	1715.30	III	500	3064.62	I	180	5476.91	I	100	1938.84	III
500	1719.46	III	2600	3101.55	I	23	5709.56	I	60	1978.22	IV
400	1722.28	III	1300	3101.88	I	16	5754.68	I	3300	2029.32	II
500	1738.25	III	2900	3134.11	I	10	5857.76	I	65	2032.53	IV
300	1739.78	III	1100	3232.96	I	10	5892.88	I	3000	2032.99	II
1000	1741.55	II	600	3243.06	I	10	6108.12	I	2000	2109.42	II
300	1741.96	III	660	3315.66	I	10	6176.81	I	1700	2125.21	II
550	1747.01	III	2000	3331.88	II	10	6191.18	I	1100	2126.54	II
300	1752.43	III	2900	3369.57	I	13	6256.36	I	80 h	2130.24	III
400	1753.01	III	3300	3380.57	I	16	6643.64	I	1500	2131.18	II
800	1764.69	III	1300	3391.05	I	22	6767.77	I	80	2273.92	III
500	1767.94	III	3300	3392.99	I	10	6914.56	I	100	2275.23	III
2000	1769.64	III	8200	3414.76	I	26	7122.20	I	80	2279.36	III
400	1776.07	III	1600	3423.71	I	16	7393.60	I	100	2281.51	III
300	1807.24	III	2600	3433.56	I	16	7409.35	I	80	2284.40	III
300	1819.28	III	990	3437.28	I	23	7422.28	I	100	2290.36	III
800	1823.06	III	4800	3446.26	I	13	7522.76	I	370	2295.68	II
400	1830.01	III	1300	3452.89	I	19	7555.60	I	280	2302.08	II
650	1847.28	III	5000	3458.47	I	23	7617.00	I	100	2313.30	III
800	1854.15	III	5000	3461.65	I	16	7714.32	I	100	2338.09	III
300	1858.75	III	1600	3472.54	I	19	7727.61	I	80	2344.12	III
1000	2165.55	II	550	3483.77	I	19	7748.89	I	90	2349.21	III
2000	2169.10	II	5500	3492.96	I	10	7788.94	I	80	2355.54	III

Intensity	Wavelength/Å		Intensity	Wavelength/Å		Intensity	Wavelength/Å		Intensity	Wavelength/Å	
100	2362.06	III	470	2716.62	II	390	3215.60	II	2700	3726.24	I
80	2362.50	III	470	2721.98	II	800	3225.48	II	2700	3739.80	I
80	2365.70	III	310	2733.26	II	140	3229.56	II	670	3740.73	II
100	2372.73	III	110	2737.09	II	400	3236.40	II	1700	3742.39	I
170	2376.40	II	240	2768.13	II	200	3247.47	II	530	3763.49	I
110	2387.09	II	310	2773.20	I	120	3248.94	II	350	3765.08	I
100	2387.41	III	270	2780.24	II	320	3254.07	II	530	3771.85	I
140	2387.52	III	110	2793.05	II	230	3260.56	II	870	3781.01	I
80	2388.23	III	190	2827.08	II	160	3263.37	II	1700	3787.06	I
45	2388.27	II	250	2841.15	II	200	3283.46	II	1300	3790.15	I
160	2398.48	II	280	2842.65	II	160	3292.02	II	3500	3791.21	I
80	2404.89	III	160	2846.28	II	320	3296.01	I	2700	3798.12	I
55	2405.34	II	240	2861.09	II	400	3312.60	I	2700	3802.92	I
55	2405.85	II	100	2865.61	II	120	3319.58	II	670	3803.88	I
140	2412.46	II	500	2868.52	II	130	3341.60	II	530	3804.74	I
100	2413.94	III	270	2875.39	II	1300	3341.97	I	670	3810.49	I
160	2416.99	II	270	2876.95	II	1300	3343.71	I	530	3811.03	I
140	2418.69	II	530	2877.03	II	1700	3349.06	I	530	3815.51	I
100	2421.91	III	100	2880.72	II	420	3349.52	I	210	3818.86	II
75	2433.80	II	570	2883.18	II	340	3354.74	I	670	3824.88	I
40	2435.95	II	280	2888.83	II	1700	3358.42	I	350	3835.18	I
45	2437.42	II	470	2897.81	II	130	3365.58	II	350	3863.38	I
40	2442.14	II	400	2899.24	II	340	3366.96	I	530	3877.56	I
28	2442.68	II	470	2908.24	II	130	3369.16	II	870	3878.82	I
65	2451.87	II	670	2910.59	II	350	3374.92	I	670	3883.14	I
65	2453.95	II	470	2911.74	II	170	3386.24	II	1100	3885.44	I
100	2456.99	III	1100	2927.81	II	350	3392.34	I	670	3885.68	I
55	2458.09	II	110	2931.47	II	230	3408.68	II	580	3891.30	I
65	2462.89	I	870	2941.54	II	180	3409.19	II	670	3914.70	I
80	2468.72	III	110 h	2945.88	II	230	3412.94	II	530	3920.20	I
80	2475.87	III	110	2946.12	II	230	3425.42	II	670	3937.44	I
110	2477.38	II	110	2946.90	II	230	3426.57	II	520	3943.67	I
65	2478.29	II	1100	2950.88	II	180	3432.70	II	910 d	3966.09	I
65	2479.94	II	400	2972.57	II	180	3440.59	II	1100	4032.52	I
35	2483.88	II	320	2974.10	II	200	3479.56	II	16000 c	4058.94	I
100	2499.73	III	210	2977.68	II	100	3484.05	II	350	4060.79	I
110	2511.00	II	200	2982.11	II	500	3498.63	I	12000	4079.73	I
110	2521.40	II	330	2990.26	II	460	3507.96	I	440	4100.40	I
390	2544.80	II	470	2994.73	II	200	3510.26	II	6700	4100.92	I
100	2545.64	III	80	3001.84	III	200	3515.42	II	310	4116.90	I
110	2551.38	II	140	3024.74	II	200	3517.67	II	5300	4123.81	I
130	2556.94	II	350	3028.44	II	2000	3535.30	I	670	4129.43	I
80	2557.94	III	300	3032.77	II	1300	3537.48	I	770	4129.93	I
130	2562.41	II	100	3044.76	II	250	3540.96	II	2300	4137.10	I
110	2571.33	II	100	3055.52	II	500	3544.02	I	440	4139.44	I
390	2583.99	II	220	3064.53	II	300	3550.45	I	2700	4139.71	I
390	2590.94	II	110	3069.68	II	1000	3554.66	I	350	4143.21	I
80	2598.86	III	100	3070.90	II	630	3563.50	I	870	4150.12	I
80	2633.17	III	110	3071.56	II	630	3563.62	I	4400	4152.58	I
200	2642.24	II	100	3073.24	II	1500	3575.85	I	870	4163.47	I
320	2646.26	II	400	3076.87	II	5000	3580.27	I	4400	4163.66	I
330	2647.50	I	110	3080.35	II	500	3584.97	I	4000	4164.66	I
330	2654.45	I	1800	3094.18	II	750	3589.11	I	3500	4168.13	I
310	2656.08	II	140	3099.19	II	500	3589.36	I	310	4184.44	I
80	2657.99	III	270	3127.53	II	500	3593.97	I	1200	4190.88	I
110	2665.25	II	1500	3130.79	II	500	3602.56	I	870	4192.07	I
110	2666.59	II	80	3142.26	III	300	3619.51	II	870	4195.09	I
110	2667.30	II	390	3145.40	II	420	3649.85	II	1300	4195.66	I
400	2671.93	II	1200	3163.40	II	400	3651.19	II	310	4198.51	I
200	2673.57	II	150	3175.78	II	200	3659.61	II	350	4201.52	I
200	2675.94	II	390	3180.29	II	630	3660.37	I	870	4205.31	I
160	2691.77	II	300	3191.10	II	900	3664.70	I	350	4214.73	I
1000	2697.06	II	150	3191.43	II	1500	3697.85	I	420	4217.94	I
320	2698.86	II	1000	3194.98	II	330	3711.34	I	420	4229.15	I
320	2702.20	II	120	3203.35	II	3300	3713.01	I	770	4262.05	I
150	2702.52	II	300	3206.34	II	480	3716.99	I	420	4266.02	I

Intensity	Wavelength/Å		Intensity	Wavelength/Å		Intensity	Wavelength/Å		Intensity	Wavelength/Å	
400	4286.99	I	210 cw	6660.84	I	600 w	297.7	IV	550	916.012	II
580	4299.60	I	150 cw	6677.33	I	700	297.82	IV	650	916.701	II
580	4300.99	I	130 c	6723.62	I	650	300.32	IV	520	921.992	IV
390	4311.27	I	85	6828.11	I	90	303.123	IV	500	922.519	IV
350	4326.33	I	85	6990.32	I	500	303.28	IV	480	923.057	IV
390	4331.37	I	190 c	7046.81	I	150	314.715	III	520	924.283	IV
330	4410.21	I	130	7159.43	I	200	314.850	III	90	953.415	I
150	4503.04	I	190 cw	7372.50	I	90	314.877	III	100	953.655	I
530	4523.41	I	65	7515.93	I	150	315.053	IV	130	953.970	I
480	4546.82	I	170 c	7574.58	I	120	322.503	IV	1000	955.335	IV
370	4564.53	I	75 c	7726.68	I	150	322.570	IV	130	963.990	I
720	4573.08	I	35	7885.31	I	200	322.724	IV	115	964.626	I
480	4581.62	I	40	8135.20	I	120	323.175	IV	70	965.041	I
1200	4606.77	I	29 cw	8320.93	I	600	323.26	III	650	979.842	III
170	4616.17	I	29	8346.08	I	300	335.050	IV	700	979.919	III
450	4630.11	I	35	8905.78	I	500	338.35	III	900	989.790	III
450	4648.95	I		**Nitrogen**		500	340.20	III	700	991.514	III
450	4663.83	I		**N Z = 7**		500 w	351.93	IV	1000	991.579	III
340	4666.24	I	400	181.75	IV	500	351.98	III	150 w	1036.16	IV
240	4667.22	I	52	186.069	V	700	353.06	IV	90	1067.614	I
580	4672.09	I	62	186.153	V	120	362.833	III	60	1068.612	I
530	4675.37	I	400	191.7	IV	150	362.881	III	90	1078.71	IV
320	4685.14	I	400	192.9	IV	150	362.946	III	450	1083.990	II
130 c	4706.14	I	500	196.87	IV	90	362.985	III	600	1084.580	II
260	4708.29	I	500	197.23	IV	300	374.204	III	430	1085.546	II
150	4713.50	I	500	202.60	IV	350	374.441	III	650	1085.701	II
220 c	4749.70	I	500	205.94	IV	500	387.48	III	175	1097.237	I
130 c	4967.78	I	500	205.97	IV	500	420.77	IV	115	1098.095	I
190	4988.97	I	500	206.03	IV	250	451.869	III	115	1098.260	I
230	5017.75	I	90	209.303	V	300	452.226	III	105	1100.360	I
150	5026.36	I	500	217.20	IV	650	463.74	IV	40	1100.465	I
210	5039.04	I	500 d	217.90	IV	285	644.634	II	90	1101.291	I
170	5058.01	I	500 d	223.4	IV	360	644.837	II	360	1134.165	I
130	5065.25	I	800 w	225.12	IV	450	645.178	II	385	1134.415	I
750	5078.96	I	800	225.21	IV	140	647.50	I	410	1134.980	I
420	5095.30	I	600 w	234.12	IV	360	660.286	II	105	1143.65	I
170	5100.16	I	600 w	234.20	IV	170	671.016	II	130	1163.884	I
170	5120.30	I	600 w	234.25	IV	285	671.386	II	60	1164.206	I
210	5134.75	I	550	236.07	IV	150	671.630	II	105	1164.325	I
250	5160.33	I	500	237.99	IV	160	671.773	II	270	1167.448	I
250	5164.38	I	500 w	238.7	IV	170	672.001	II	105	1168.334	I
230	5180.31	I	600	238.80	IV	500	684.996	III	60	1168.417	I
190	5189.20	I	500 w	239.62	IV	570	685.513	III	195	1168.536	I
170	5193.08	I	900	247.20	IV	650	685.816	III	230	1176.510	I
150	5195.84	I	90	247.561	V	500	686.335	III	105	1176.630	I
150	5232.81	I	120	247.706	V	350	692.70	I	195	1177.695	I
150 d	5251.62	I	500 w	248.43	IV	90	713.518	V	500	1183.031	III
270	5271.53	I	500 w	248.46	IV	150	713.860	V	570	1184.550	III
130 c	5276.20	I	500 w	248.48	IV	285	746.984	II	90	1188.01	IV
250	5318.60	I	500	257.95	III	150	748.195	V	410	1199.550	I
460	5344.17	I	650	258.50	III	200	748.291	V	385	1200.223	I
340	5350.74	I	700	259.19	III	500	763.336	III	360	1200.710	I
110	5437.27	I	800	260.09	III	570	764.359	III	175	1225.026	I
85	5551.35	I	600	260.45	IV	570	765.148	IV	160	1225.37	I
170	5642.11	I	800	261.28	III	250	771.544	III	130	1228.41	I
130	5664.71	I	500	262.91	III	300	771.901	III	160	1228.79	I
170	5665.63	I	500	265.23	III	350	772.385	III	1000	1238.821	V
130	5729.19	I	500	265.27	III	200	772.891	III	900	1242.804	V
110	5760.34	I	150	266.196	V	150	772.975	III	360	1243.179	I
110	5819.43	I	200	266.379	V	650	775.965	II	315	1243.306	I
130 d	5838.64	I	500	268.70	III	90	885.67	I	290	1310.540	I
190 cw	5900.62	I	650	270.99	IV	90	909.697	I	250	1310.95	I
150	5983.22	I	250	283.42	IV	80	910.278	I	230	1319.00	I
75	6221.96	I	300	283.48	IV	40	910.645	I	315	1319.68	I
85 c	6430.46	I	350	283.58	IV	450	915.612	II	115	1326.57	I
65	6544.61	I	600	285.56	IV	450	915.962	II	115	1327.92	I

Intensity	Wavelength/Å		Intensity	Wavelength/Å		Intensity	Wavelength/Å		Intensity	Wavelength/Å	
150	1387.371	III	160	2885.27	II	160	4950.23	I	160	7406.12	I
360	1411.94	I	90 1	2974.52	V	350	4963.98	I	265	7406.24	I
700	1492.625	I	150 w	2980.78	V	285	4987.37	II	685	7423.64	I
490	1492.820	I	250 w	2981.31	V	450	4994.36	II	785	7442.29	I
640	1494.675	I	60 w	2998.43	V	650	5001.48	II	900	7468.31	I
90	1549.336	V	220	3006.83	II	360	5002.70	II	185	7608.80	I
200 1	1616.33	V	90	3078.25	IV	870	5005.15	II	60 w	7618.46	V
350 1	1619.69	V	120	3367.34	III	550	5007.32	II	450	7762.24	II
1000	1718.55	IV	360	3437.15	II	450	5010.62	II	400	8184.87	I
250	1729.945	III	90	3463.37	IV	360	5016.39	II	400	8188.02	I
775	1742.729	I	570	3478.71	IV	360	5025.66	II	250	8200.36	I
700	1745.252	I	500	3482.99	IV	550	5045.10	II	300	8210.72	I
570	1747.848	III	400	3484.96	IV	185	5281.20	I	570	8216.34	I
350	1751.218	III	90	3747.54	IV	140	5292.68	I	400	8223.14	I
650	1751.657	III	90	3754.67	III	90	5314.35	III	400	8242.39	I
150	1804.486	III	120	3771.05	III	200	5320.82	III	550	8438.74	II
200	1805.669	III	285	3838.37	II	150	5327.18	III	500	8567.74	I
150	1846.42	III	360	3919.00	II	450	5495.67	II	570	8594.00	I
90 w	1860.37	V	90	3938.52	III	285	5535.36	II	650	8629.24	I
350	1885.06	III	450	3955.85	II	650	5666.63	II	500	8655.89	I
400	1885.22	III	1000	3995.00	II	550	5676.02	II	220	8676.08	II
200	1907.99	III	150	3998.63	III	870	5679.56	II	700	8680.28	I
150	1919.55	III	200	4003.58	III	450	5686.21	II	650	8683.40	I
150	1919.77	III	360	4035.08	II	450	5710.77	II	500	8686.15	I
300	1920.65	III	550	4041.31	II	285	5747.30	II	110	8687.43	II
150	1920.84	III	360	4043.53	II	700	5752.50	I	110 h	8699.00	II
200	1921.30	III	150	4057.76	IV	240	5764.75	I	500	8703.25	I
200	2064.01	III	250	4097.33	II	265	5829.54	I	160 h	8710.54	II
250	2064.42	III	140	4099.94	I	235	5854.04	I	570	8711.70	I
120	2068.68	III	200	4103.43	III	360	5927.81	II	500	8718.83	I
90	2071.09	III	185	4109.95	I	550	5931.78	II	250	8728.89	I
90	2080.34	IV	285	4176.16	II	285	5940.24	II	200	8747.36	I
160	2095.53	II	120	4195.76	II	650	5941.65	II	500	9386.80	I
70	2096.20	II	150	4200.10	III	285	5952.39	II	570	9392.79	I
110	2096.86	II	285	4227.74	II	160	5999.43	I	250	9460.68	I
90	2117.59	III	285	4236.91	II	210	6008.47	I	200	9863.33	I
90	2121.50	III	220	4237.05	II	285	6167.76	II	160 h	9865.41	II
110	2130.18	III	450	4241.78	II	360	6379.62	II	110 h	9868.21	II
160	2142.78	II	90	4332.91	III	150	6380.77	IV	160 h	9887.39	II
90	2147.31	III	120	4345.68	III	185	6411.65	I	220 h	9891.09	II
200	2188.20	III	300	4379.11	III	210	6420.64	I	160 h	9961.86	II
150	2188.38	III	285	4432.74	II	210	6423.02	I	220 h	9969.34	II
160	2206.09	II	650	4447.03	II	210	6428.32	I	285 h	10023.27	II
160	2286.69	II	90	4510.91	II	185	6437.68	I	220 h	10035.45	II
110	2288.44	II	120	4514.86	III	235	6440.94	I	220 h	10065.15	II
220	2316.49	II	360	4530.41	II	90	6454.11	III	160 h	10070.12	II
160	2316.69	II	550	4601.48	II	185	6457.90	I	250	10105.13	I
285	2317.05	II	350	4603.73	V	120	6467.02	III	300	10108.89	I
90 w	2318.09	IV	90	4606.33	IV	300	6468.44	I	350	10112.48	I
160	2461.27	II	450	4607.16	II	265	6481.71	I	400	10114.64	I
150	2477.69	IV	360	4613.87	II	750	6482.05	II	110 h	10126.27	II
110	2496.83	II	250	4619.98	V	360	6482.70	I	250	10539.57	I
70	2496.97	II	450	4621.39	II	300	6483.75	I	200	12074.51	I
110	2520.22	II	870	4630.54	II	325	6484.80	I	380	12186.82	I
160	2520.79	II	90	4634.14	III	160	6491.22	I	225	12288.97	I
220	2522.23	II	120	4640.64	III	210	6499.54	I	290	12328.76	I
110	2590.94	II	550	4643.08	II	185	6506.31	I	310	12381.65	I
250	2645.65	IV	285	4788.13	II	750	6610.56	II	180	12438.40	I
300	2646.18	IV	450	4803.29	II	185	6622.54	I	510	12461.25	I
350	2646.96	IV	180	4847.38	I	185	6636.94	I	920	12469.62	I
250 w	2682.18	III	90	4858.82	III	235	6644.96	I	500	13429.61	I
90	2689.20	III	150	4867.15	III	185	6646.50	I	840	13581.33	I
160	2709.84	II	285	4895.11	II	235	6653.46	I	180	13587.73	I
110	2799.22	II	160	4914.94	I	210	6656.51	I	180	13602.27	I
110	2823.64	II	210	4935.12	I	185	6722.62	I	290	13624.18	I
60 1	2859.16	V	200 w	4944.56	V	210	7398.64	I	250	14757.07	I

Intensity	Wavelength/Å		Intensity	Wavelength/Å		Intensity	Wavelength/Å		Intensity	Wavelength/Å	
100	14868.87	I	5100	2838.63	I	28	5417.51	I	250	215.245	V
160	14966.60	I	2300	2844.40	I	55	5443.31	I	250	216.018	V
180	15582.27	I	1500	2850.76	I	22	5446.93	I	520	220.352	V
120 s	17516.58	I	1500	2860.96	I	22	5457.30	I	80	227.372	V
100 l	17584.86	I	9600	2909.06	I	28	5470.00	I	80	227.469	V
100	17878.26	I	2100	2912.33	I	22	5509.33	I	150	227.511	V
	Osmium		2100	2919.79	I	270	5523.53	I	80	227.549	V
	Os Z = 76		1100 h	2948.23	I	22	5546.82	I	80	227.634	V
9600	2001.45	I	1400	2949.53	I	80	5584.44	I	80	227.689	V
13000	2003.73	I	4400	3018.04	I	35	5620.08	I	150	231.823	V
17000	2010.15	I	1100	3030.70	I	22	5642.56	I	140	233.46	IV
29000	2018.14	I	2900	3040.90	I	28	5645.25	I	150	233.50	IV
14000	2022.76	I	120	3042.74	II	28	5680.88	I	110	233.52	IV
14000	2028.23	I	8600	3058.66	I	170	5721.93	I	200	233.56	IV
18000	2034.44	I	1100	3077.72	I	22	5765.05	I	110	233.60	IV
26000	2045.36	I	3100	3156.25	I	170	5780.82	I	90	238.36	IV
8600	2058.69	I	180	3173.93	II	40	5800.60	I	180	238.57	IV
13000	2061.69	I	150	3213.31	II	110	5857.76	I	110	248.459	V
7800	2067.21	II	1900	3232.06	I	28	5860.64	I	110	252.56	IV
4200	2070.67	II	3100	3262.29	I	65	5996.00	I	110	252.95	IV
7200	2076.95	I	3100	3267.94	I	35	6227.70	I	150	253.08	IV
14000	2079.97	I	1200	3290.26	I	22	6269.41	I	300	260.39	IV
2900	2082.54	I	7600	3301.56	I	22	6403.15	I	250	260.56	IV
2900	2089.03	I	960	3336.15	I	27	6729.56	I	80 d	264.34	III
2900	2089.21	I	960	3370.59	I	22	7145.54	I	110	264.48	III
6000	2097.60	I	620	3387.84	I	26	7602.95	I	110	266.97	III
5300	2100.63	I	620	3401.86	I	7	8041.29	I	150	266.98	III
2100	2117.66	I	620	3504.66	I		**Oxygen**		150	267.03	III
4800	2117.96	I	1200	3528.60	I		**O Z = 8**		150	277.38	III
5300	2137.11	I	1200	3560.86	I	80	124.616	V	300	279.63	IV
2600	2154.59	I	620	3598.11	I	110	135.523	V	375	279.94	IV
1300	2157.84	I	95	3604.48	II	80	138.109	V	110	285.71	IV
1200	2158.53	I	480	3670.89	I	110	139.029	V	150	285.84	IV
3100	2166.90	I	3700	3752.52	I	80	151.447	V	110	286.448	V
1100	2167.75	I	2100	3782.20	I	110	151.477	V	80	295.62	III
2100	2171.65	I	730	3876.77	I	150	151.546	V	110	295.66	III
1100	2234.61	I	1000	3963.63	I	80	164.574	V	120	295.72	III
1300	2252.15	I	730	3977.23	I	110	164.657	V	150	303.41	III
2000	2255.85	II	960	4066.69	I	80	164.709	V	150	303.46	III
1400	2264.60	I	1200	4112.02	I	80	166.235	V	140	303.52	III
1400	2282.26	II	2500	4135.78	I	150	167.99	V	160	303.62	III
500	2367.35	II	1200	4173.23	I	110	170.219	V	160	303.69	III
2600	2377.03	I	1200	4211.86	I	450	172.169	V	250	303.80	III
1700	2387.29	I	4900	4260.85	I	250	185.745	V	200	305.60	III
1100	2395.88	I	560	4293.95	I	375	192.751	V	250	305.66	III
200	2423.07	II	560	4311.40	I	450	192.799	V	190	305.70	III
1400	2424.97	I	4900	4420.47	I	520	192.906	V	300	305.77	III
110	2454.91	II	540	4550.41	I	80	193.003	V	190	305.84	III
1800	2461.42	I	670	4793.99	I	200	194.593	V	200	306.62	IV
110	2468.90	II	55	5031.83	I	150	195.86	IV	150	306.88	IV
530	2486.24	II	45	5039.12	I	200	196.01	IV	450	320.979	III
4500	2488.55	I	35	5072.88	I	80	202.161	V	300	328.45	III
2600	2498.41	I	35	5074.77	I	80	202.224	V	250	328.74	III
2400	2513.25	I	35	5079.09	I	80	202.283	V	300	345.31	III
780	2538.00	II	90	5103.50	I	80	202.334	V	110	355.14	III
1000	2542.51	I	55	5110.81	I	150	202.393	V	90	355.33	III
1000	2590.76	I	140	5149.74	I	110	203.78	V	80	355.47	III
1800	2613.06	I	40	5193.52	I	150	203.82	V	200	359.02	III
3800	2637.13	I	270	5202.63	I	100	203.85	V	190	359.22	III
1900	2644.11	I	35	5203.23	I	200	203.89	V	150	359.38	III
1900	2658.60	I	45	5255.82	I	100	203.94	V	210	373.80	III
2100	2689.82	I	55	5265.15	I	110	207.18	IV	200	374.00	III
3000	2714.64	I	40	5298.78	I	150	207.24	IV	300	374.08	III
1300	2720.04	I	110	5376.79	I	300	207.794	V	190	374.16	III
960	2770.71	I	120	5416.34	I	150	215.040	V	200	374.33	III
2800	2806.91	I	45	5416.69	I	200	215.103	V	210	374.44	III

Intensity	Wavelength/Å		Intensity	Wavelength/Å		Intensity	Wavelength/Å		Intensity	Wavelength/Å	
450	395.558	III	80	805.810	I	30 d	2283.42	II	160	3305.15	II
300	434.98	III	240	832.762	II	30 d	2284.89	II	160	3306.60	II
800	507.391	III	600	832.927	III	110	2293.32	II	80	3312.30	III
900	507.683	III	450	833.332	II	200	2300.35	II	110	3340.74	III
1000	508.182	III	780	833.742	III	30 d	2313.05	II	230	3348.08	IV
1000	525.795	III	600	834.467	II	30 d	2316.12	II	270	3349.11	IV
250	537.83	II	600	835.096	III	30 d	2316.79	II	160	3354.27	IV
300	538.26	II	800	835.292	III	50 d	2319.68	II	200	3375.40	IV
220	539.09	II	40	877.879	I	30 d	2322.15	II	220	3377.20	II
200	539.55	II	130	921.296	IV	30 d	2339.31	II	130	3378.06	IV
150	539.85	II	160	921.366	IV	200 d	2390.44	III	360	3381.20	IV
700	553.330	IV	80	922.008	I	80	2394.33	III	360	3385.52	IV
775	554.075	IV	200	923.367	IV	110	2411.60	II	285	3390.25	II
850	554.514	IV	130	923.433	IV	80	2422.84	III	270	3396.79	IV
700	555.261	IV	90	935.193	I	80	2425.55	II	360	3403.52	IV
700	597.818	III	40	948.686	I	250	2433.56	II	220	3407.38	II
1000	599.598	III	90	971.738	I	80 d	2436.06	II	230	3409.66	IV
580	608.398	IV	40	976.448	I	80 d	2438.83	III	160	3409.84	II
110	609.70	III	160	988.773	I	80	2444.26	II	410	3411.69	IV
640	609.829	IV	40	990.204	I	300	2445.55	II	230	3413.64	IV
160	610.04	III	250	1025.762	I	200	2449.372	IV	80	3444.10	III
200	610.75	III	90	1027.431	I	200	2450.040	IV	80	3455.12	III
100	610.85	III	160	1039.230	I	200	2454.99	III	285	3470.81	II
270	616.952	IV	60	1040.942	I	200	2493.44	IV	200	3489.83	IV
150	617.005	IV	40	1152.152	I	200	2493.77	IV	160	3492.24	IV
200	617.036	IV	900	1302.168	I	200	2507.73	IV	230	3560.39	IV
520	624.617	IV	600	1304.858	I	230	2509.19	IV	270	3563.33	IV
580	625.130	IV	300	1306.029	I	200	2517.2	IV	80	3698.70	III
640	625.852	IV	200	1338.612	IV	200	2558.06	III	80	3702.75	III
1000	629.730	V	130	1342.992	IV	80	2687.53	III	80	3703.37	III
150	644.148	II	230	1343.512	IV	110	2695.49	III	110	3707.24	III
200	672.95	II	640	1371.292	V	300	2733.34	II	220	3712.75	II
150	673.77	II	160	1476.89	III	110	2747.46	III	110	3715.08	III
230	681.272	V	160 w	1506.72	V	1000	2781.01	V	315 w	3725.93	IV
70	685.544	I	285	1590.01	III	920	2786.99	V	285	3727.33	II
800	702.332	III	160	1591.33	III	775	2789.85	V	360	3729.03	IV
800	702.822	III	315 w	1643.68	V	160	2836.26	IV	410	3736.85	IV
900	702.899	III	160	1707.996	V	160	2921.45	IV	160	3739.92	II
1000	703.850	III	220	1760.12	III	200	2941.33	V	110	3744.00	III
900	718.484	II	110	1760.42	III	210	2941.65	V	230	3744.89	IV
600	718.562	II	220	1763.22	III	80	2959.68	III	360	3749.49	II
70	744.794	I	220	1764.48	III	265	2972.29	I	150	3754.67	III
700	758.678	V	750	1767.78	III	250	2983.78	III	80	3757.21	III
640	759.441	V	550	1768.24	III	80	3017.63	III	250	3759.87	III
580	760.228	V	360	1771.67	III	80	3023.45	III	110	3791.26	III
775	760.445	V	110	1773.00	III	80	3043.02	III	160	3803.14	III
640	761.128	V	110	1773.85	III	200	3047.13	III	120	3823.41	I
700	762.003	V	220	1779.16	III	110	3059.30	III	450	3911.96	II
70	770.793	I	160	1781.03	III	460	3063.42	IV	160	3919.29	II
90	771.056	I	160	1784.85	III	410	3071.61	IV	185	3947.29	I
520	774.518	V	220	1789.66	III	80	3121.71	III	160	3947.48	I
70	775.321	I	110	1848.26	III	160	3122.62	II	140	3947.59	I
200	779.734	IV	110	1856.62	III	220	3129.44	II	220	3954.37	II
315	779.821	IV	285	1872.78	III	110	3132.86	III	100	3954.61	I
360	779.912	IV	285	1872.87	III	450	3134.82	II	200	3961.59	III
200	779.997	IV	285	1874.94	III	285	3138.44	II	450	3973.26	II
640	787.711	IV	160	1920.04	III	160	3144.66	V	220	3982.20	II
520	790.109	IV	110	1920.75	III	160	3209.66	IV	160	4069.90	II
700	790.199	IV	110	1921.52	III	80	3238.57	III	285	4072.16	II
70	791.973	I	220	1923.49	III	200	3260.98	III	450	4075.87	II
300	796.66	II	110	1923.82	III	300	3265.46	III	80 d	4083.91	II
200	802.200	IV	110	1926.94	III	80	3267.31	III	50 d	4087.14	II
160	802.255	IV	360	2013.27	III	220	3270.98	II	150 d	4089.27	II
90	804.267	I	160	2026.96	III	220	3273.52	II	110	4097.24	II
70	804.848	I	220	2045.67	III	220	3277.69	II	220	4105.00	II
70	805.295	I	160	2052.74	III	360	3287.59	II	285	4119.22	II

Intensity	Wavelength/Å		Intensity	Wavelength/Å	
100	4123.99	V	130	6500.24	V
160	4132.81	II	80	6604.91	I
50	4146.06	II	100	6653.83	I
220	4153.30	II	360	7001.92	I
285	4185.46	II	450	7002.23	I
450	4189.79	II	210	7156.70	I
80	4233.27	I	400	7254.15	I
50 d	4253.74	II	450	7254.45	I
50 d	4253.98	II	320	7254.53	I
50 d	4275.47	II	210	7476.44	I
50 d	4303.78	II	100	7477.24	I
285	4317.14	II	120	7479.08	I
160	4336.86	II	120	7480.67	I
220	4345.56	II	100	7706.75	I
285	4349.43	II	870	7771.94	I
220	4366.90	II	810	7774.17	I
100	4368.25	I	750	7775.39	I
220	4395.95	II	80	7886.27	I
450	4414.91	II	100	7943.15	I
285	4416.98	II	100	7947.17	I
160	4448.21	II	235	7947.55	I
160	4452.38	II	210	7950.80	I
50	4465.45	II	185	7952.16	I
50 d	4466.28	II	110	7981.94	I
50	4467.83	II	135	7982.40	I
50	4469.41	II	190	7986.98	I
360	4590.97	II	135	7987.33	I
285	4596.17	II	250	7995.07	I
80 d	4609.39	II	400	8221.82	I
160	4638.85	II	265	8227.65	I
360	4641.81	II	265	8230.02	I
450	4649.14	II	325	8233.00	I
160	4650.84	II	120	8235.35	I
360	4661.64	II	120	8426.16	I
285	4676.23	II	810	8446.25	I
220	4699.21	II	1000	8446.36	I
285	4705.36	II	935	8446.76	I
160	4924.60	II	325	8820.43	I
230 w	4930.27	V	160 d	9057.01	I
220	4943.06	II	120	9118.29	I
135	5329.10	I	80	9134.71	I
160	5329.68	I	80	9150.14	I
190	5330.74	I	80	9151.48	I
90	5435.18	I	235	9156.01	I
110	5435.78	I	450	9260.81	I
135	5436.86	I	490	9260.84	I
120	5577.34	I	450	9260.94	I
110	5592.37	III	400	9262.58	I
130	5597.91	V	540	9262.67	I
160	5958.39	I	590	9262.77	I
190	5958.58	I	490	9265.94	I
80	5995.28	I	640	9266.01	I
160	6046.23	I	185	9399.19	I
190	6046.44	I	120	9481.16	I
110	6046.49	I	120 d	9482.88	I
100	6106.27	I	235	9487.43	I
400	6155.98	I	140	9492.71	I
450	6156.77	I	265	9497.97	I
490	6158.18	I	160	9499.30	I
80	6256.83	I	235	9505.59	I
100	6261.55	I	210	9521.96	I
100	6366.34	I	120	9523.36	I
100	6374.32	I	120	9523.96	I
320	6453.60	I	100	9528.28	I
360	6454.44	I			
400	6455.98	I			

Intensity	Wavelength/Å		Intensity	Wavelength/Å	
100	9622.13	I	4000	1914.62	III
120	9625.29	I	1000	1930.33	III
160	9677.38	I	2000	1941.64	III
80	9694.66	I	800	2002.16	III
65	9694.91	I	1000	2004.47	III
235	9741.50	I	500	2055.11	III
235	9760.65	I	500	2149.82	III
120	9909.05	I	500	2177.55	III
140	9936.98	I	500	2177.63	III
120	9940.41	I	100 r	2231.59	II
160	9995.31	I	200 r	2296.53	II
120 d	10421.18	I	100	2426.87	II
590	11286.34	I	100	2430.94	II
640	11286.91	I	100	2433.11	II
490	11287.02	I	100	2435.32	II
490	11287.32	I	150	2446.17	II
490	11295.10	I	1100	2447.91	I
540	11297.68	I	100	2457.29	II
590	11302.38	I	150	2469.29	II
265	11358.69	I	100	2471.18	II
490	12464.02	I	1700	2476.42	I
450	12570.04	I	250	2486.52	II
120	12990.77	I	300	2488.92	II
160	13076.91	I	200	2498.81	II
700	13163.89	I	150	2505.73	II
750	13164.85	I	150	2551.84	II
640	13165.11	I	150	2565.51	II
160	16212.06	I	100	2569.56	II
120	17966.70	I	150	2658.75	II
590	18021.21	I	1900	2763.09	I
120	18041.48	I	150 h	2776.85	II
120	18042.19	I	100 h	2787.92	II
120	18046.23	I	200	2854.59	II
140	18229.23	I	100 h	2871.37	II
540	18243.63	I	100 h	2878.01	II
140	26173.56	I	520	2922.49	I
Palladium			650	3002.65	I
Pd Z = 46			1500	3027.91	I
200	705.49	III	1100	3065.31	I
200	727.72	III	2600	3114.04	I
500	763.06	III	11000	3242.70	I
500	766.42	III	2700	3251.64	I
2000	781.02	III	3500	3258.78	I
500	794.08	III	3600	3302.13	I
500	797.52	III	5000	3373.00	I
500	800.03	III	24000	3404.58	I
500	800.10	III	13000	3421.24	I
500	803.67	III	5000	3433.45	I
500	825.35	III	6400	3441.40	I
500	840.58	III	7700	3460.77	I
500	856.47	III	10000	3481.15	I
500	864.04	III	2000	3489.77	I
500	880.59	III	12000	3516.94	I
500	888.84	III	12000	3553.08	I
1000	889.29	III	4500	3571.16	I
300	1596.89	III	20000	3609.55	I
500	1741.62	III	20000	3634.70	I
4000	1782.55	III	5500	3690.34	I
400	1843.49	III	1400	3718.91	I
1500	1851.59	III	1500	3799.19	I
2000	1852.27	III	1500	3832.29	I
1000	1859.21	III	2200	3894.20	I
1500	1874.63	III	1500	3958.64	I
2000	1885.83	III	290	4087.34	I
1000	1887.40	III	2500	4212.95	I
1500	1891.34	III	180	4473.59	I

Intensity	Wavelength/Å		Intensity	Wavelength/Å		Intensity	Wavelength/Å		Intensity	Wavelength/Å	
160	5163.84	I	15	1373.500	I	520	3204.04	V	150	6992.690	III
120	5295.63	I	10	1374.732	I	300	3219.307	III	100	7102.200	I
55	5542.80	I	15	1377.080	I	400	3233.602	III	100	7158.367	I
75	5670.07	I	15	1377.937	I	650	3347.736	IV	180	7165.465	I
55 h	5695.09	I	25	1379.429	I	570	3364.467	IV	180	7175.102	I
65	6784.52	I	25	1381.469	I	400	3371.122	IV	180	7176.660	I
75	7368.12	I	15	1381.637	I	300	3957.641	III	200	7443.657	IV
120	7764.03	I	500	1484.507	IV	350	3978.307	III	250	7845.63	II
45	7915.80	I	400	1487.788	IV	400	4059.312	III	100	8046.801	I
55	8132.82	I	350	1502.228	III	300	4080.084	III	150	8113.528	III
45	8300.83	I	80	1532.51	II	500	4222.195	III	140	8278.058	I
65	8761.35	I	120	1535.90	II	350	4246.720	III	100	8367.856	I
	Phosphorus		450	1610.50	V	400	4420.71	II	140	8531.475	I
	P Z = 15		150	1618.632	III	250	4479.776	III	140	8613.835	I
250	328.78	V	200	1618.907	III	250	4540.288	IV	180	8637.578	I
150	359.899	IV	140	1671.070	I	250	4541.112	IV	400	8741.529	I
500	388.318	IV	100	1671.510	I	500	4588.04	II	100	8872.174	I
250	389.50	V	180	1671.680	I	500	4589.86	II	180	9175.819	I
300	390.70	V	140	1672.035	I	600	4602.08	II	950	9193.85	I
300	445.158	IV	140	1672.474	I	300	4626.70	II	600	9278.88	I
375	475.60	V	600	1674.591	I	300	4658.31	II	1250	9304.94	I
120	498.180	III	600	1679.695	I	500	4943.53	II	500	9323.50	I
520	542.57	V	140	1685.976	I	300	4954.39	II	950	9435.069	I
600	544.92	V	100	1694.028	I	300	4969.71	II	950	9441.86	I
200	569.853	III	100	1694.486	I	100	5079.381	I	600	9452.83	I
200	581.831	III	100	1706.376	I	100	5098.221	I	1250	9493.56	I
350	629.008	IV	100	1707.553	I	100	5100.974	I	1700	9525.73	I
400	629.914	IV	600	1774.951	I	140	5109.628	I	1500	9545.18	I
500	631.779	IV	500	1782.838	I	140	5154.844	I	280	9556.81	I
450	673.90	V	400	1787.656	I	180	5162.290	I	1700	9563.439	I
10	810.24	II	140	1834.801	I	300	5253.52	II	280	9593.50	I
650	823.179	IV	140	1847.165	I	140	5293.539	I	750	9609.04	I
700	824.730	IV	100	1849.820	I	400	5296.13	II	400	9638.939	I
800	827.932	IV	140	1851.194	I	250	5316.07	II	500	9676.24	I
300	847.669	III	100	1852.069	I	300	5344.75	II	180	9706.533	I
350	855.624	III	500	1858.886	I	180	5345.851	I	1500	9734.750	I
500	859.652	III	400	1859.393	I	100	5364.631	I	280	9736.680	I
10	865.44	II	140	1864.348	I	250	5378.20	II	1500	9750.77	I
450	865.45	V	650	1888.523	IV	300	5386.88	II	600	9790.21	I
600	871.39	V	180	1905.481	I	400	5425.91	II	1700	9796.85	I
700	877.476	IV	140	1906.403	I	100	5428.094	I	280	9834.80	I
300	913.971	III	280	1907.665	I	400	5450.74	II	400	9903.68	I
300	917.120	III	280	2023.489	I	140	5458.305	I	280	9976.67	I
350	918.665	III	180	2024.516	I	180	5477.672	I	229	10084.27	I
1000	950.655	IV	400	2032.432	I	140	5477.860	I	458	10511.58	I
250	1003.598	III	400	2033.477	I	140	5478.267	I	962	10529.52	I
570	1025.563	IV	400	2135.465	I	100	5514.774	I	1235	10581.57	I
500	1028.096	IV	400	2136.182	I	100	5516.997	I	415	10596.90	I
570	1030.517	IV	400	2149.145	I	250	5588.34	II	435	10681.40	I
500	1033.111	IV	280	2152.940	I	500	6024.18	II	265	10813.13	I
500	1035.517	IV	500	2154.080	I	400	6034.04	II	764	11183.23	I
900	1117.98	V	180	2235.732	I	500	6043.12	II	402	11186.75	I
570	1118.551	IV	450	2440.93	V	250	6055.50	II	479	14241.64	I
700	1128.01	V	250	2478.256	IV	150	6083.409	III	256	14307.83	I
20	1249.82	II	750	2533.976	I	350	6087.82	II	714	15711.52	I
20	1301.87	II	950	2535.603	I	180	6097.690	I	228	15962.53	I
20	1304.47	II	750	2553.262	I	350	6165.59	II	296	16254.77	I
15	1304.68	II	500	2554.915	I	500	6199.024	I	203	16292.97	I
35	1305.48	II	250	2605.506	IV	180	6210.499	I	1627	16482.92	I
60	1310.70	II	300	2632.713	III	140	6375.681	I	588	16590.07	I
500	1334.808	III	400	2644.295	IV	100	6388.579	I	225	16613.05	I
650	1344.327	III	400	2728.770	IV	250	6435.32	II	221	16738.68	I
300	1344.845	III	500	2739.309	IV	600	6459.99	II	419	16803.39	I
500	1366.695	IV	250	2739.872	IV	600	6503.46	II	471	17112.48	I
15	1372.033	I	450	2978.55	V	600	6507.97	II	289	17286.91	I
400	1372.674	IV	700	3175.09	V	100	6717.411	I	299	17423.67	I

Intensity	Wavelength/Å		Intensity	Wavelength/Å		Intensity	Wavelength/Å		Intensity	Wavelength/Å	
287	23844.97	I	100	2418.06	I	40 h	2865.05	II	14	4445.55	I
311	29097.16	I	50	2424.87	II	40 h	2875.85	II	25	4498.76	I
Platinum			80	2428.04	I	100 h	2877.52	II	12	4520.90	I
Pt Z = 78			50	2428.20	I	25	2888.20	I	35	4552.42	I
30	1621.66	II	25	2429.10	I	25	2893.22	I	12	4879.53	I
30	1723.13	II	180	2436.69	I	600	2893.86	I	14	5044.04	I
30	1751.70	II	650	2440.06	I	300	2897.87	I	30	5059.48	I
50 r	1777.09	II	60	2450.97	I	60	2905.90	I	35	5227.66	I
30	1781.86	II	440	2467.44	I	120	2912.26	I	40	5301.02	I
30	1879.09	II	35	2471.01	I	120	2913.54	I	12	5368.99	I
40	1883.05	II	1000	2487.17	I	70	2919.34	I	12	5390.79	I
50	1889.52	II	25	2488.74	II	30	2921.38	I	14	5475.77	I
50	1911.70	II	200	2490.12	I	1700	2929.79	I	14	5478.50	I
30	1929.25	II	160	2495.82	I	30	2942.76	I	6	5763.57	I
30	1929.68	II	240	2498.50	I	30	2944.75	I	20	5840.12	I
30	1939.80	II	50	2505.93	I	25	2959.10	I	8	5844.84	I
30	1949.90	II	120	2508.50	I	60	2960.75	I	6	6026.04	I
30	1983.74	II	50	2514.07	I	1800	2997.97	I	7	6318.37	I
40	2014.93	II	60	2515.03	I	35	3001.17	II	8	6326.58	I
3200	2030.63	I	240	2515.58	I	220	3002.27	I	9	6523.45	I
4400	2032.41	I	140	2524.30	I	30	3017.88	I	10	6710.42	I
100	2036.46	II	40	2529.41	I	30 h	3031.22	II	20	6760.02	I
40	2041.57	II	50	2536.49	I	130	3036.45	I	60	6842.60	I
5500	2049.37	I	160	2539.20	I	800	3042.64	I	20	7113.73	I
1500	2067.50	I	18	2549.46	I	3200	3064.71	I	10	8224.74	I
3000	2084.59	I	50	2552.25	I	30	3071.94	I	**Plutonium**		
1000	2103.33	I	50	2596.00	I	130	3100.04	I	**Pu Z = 94**		
30	2115.57	II	70	2603.14	I	320	3139.39	I	10000	2806.11	II
950	2128.61	I	30	2616.76	II	140	3156.56	I	10000	2950.06	II
30	2130.69	II	50	2619.57	I	120	3200.71	I	10000	3000.31	II
1900	2144.23	I	30	2625.34	II	320	3204.04	I	10000	3200.23	II
100	2144.24	II	1100	2628.03	I	30	3230.29	I	10000	3418.88	II
600	2165.17	I	130	2639.35	I	20	3233.42	I	10000	3805.93	I
1500	2174.67	I	1000	2646.89	I	20	3250.36	I	10000	4097.12	I
30	2190.32	II	500	2650.86	I	40	3251.98	I	10000	4170.95	I
400	2202.22	I	20	2658.17	I	160	3255.92	I	10000	4367.41	I
50 h	2202.58	II	2800	2659.45	I	25	3268.42	I	10000	5590.54	I
320	2222.61	I	40	2674.57	I	25	3281.97	I	10000	7068.90	I
50 h	2233.11	II	440	2677.15	I	120	3290.22	I	10000	8691.94	I
30 h	2240.99	II	200	2698.43	I	500	3301.86	I	3000	9533.07	I
100	2245.52	II	2000	2702.40	I	60	3315.05	I	3000	12144.46	I
150	2249.30	I	1600	2705.89	I	35	3323.80	I	3000	16897.38	I
30	2251.52	II	60	2713.13	I	340	3408.13	I	**Polonium**		
30 h	2251.92	II	1300	2719.04	I	35	3427.93	I	**Po Z = 84**		
190	2268.84	I	130	2729.92	I	60	3483.43	I	1500 w	2450.08	I
30 h	2271.72	II	1800	2733.96	I	160	3485.27	I	1500 w	2558.01	I
280	2274.38	I	70	2738.48	I	120	3628.11	I	2500 w	3003.21	I
50 h	2287.50	II	70	2747.61	I	70	3638.79	I	1200	4170.52	I
30	2288.20	II	80	2753.86	I	70	3643.17	I	800	4493.21	I
150	2289.27	I	200	2754.92	I	50	3663.10	I	500	8618.26	I
150	2292.40	I	30	2769.84	I	80	3671.99	I	**Potassium**		
240	2308.04	I	500	2771.67	I	80	3674.04	I	**K Z = 19**		
50	2310.96	II	40	2773.24	I	35	3699.91	I	100	214.35	V
90	2315.50	I	20	2774.00	I	18	3706.53	I	150	271.82	IV
220	2318.29	I	50	2774.77	II	80	3818.69	I	100	273.06	IV
100	2326.10	I	50	2793.27	I	40	3900.73	I	150	282.35	V
170	2340.18	I	100	2794.21	II	110	3922.96	I	150	293.33	V
280	2357.10	I	40 h	2799.98	II	35	3948.40	I	300	294.84	V
180	2368.28	I	140	2803.24	I	100	3966.36	I	200	296.17	V
50	2377.28	II	10	2808.51	I	20	3996.57	I	200	297.06	V
130	2383.64	I	50	2818.25	I	110	4118.69	I	200	300.25	V
40	2386.81	I	30 h	2822.27	II	80	4164.56	I	200	300.50	V
120	2389.53	I	1400	2830.30	I	40	4192.43	I	200	311.24	V
35	2396.17	I	70	2834.71	I	18	4327.06	I	250	312.77	V
70	2401.87	I	16	2853.11	I	18	4391.83	I	200	315.18	V
200	2403.09	I	80 h	2860.68	II	80	4442.55	I	250	327.38	V

Intensity	Wavelength/Å		Intensity	Wavelength/Å		Intensity	Wavelength/Å		Intensity	Wavelength/Å	
25	330.68	III	75	434.72	III	15	874.04	III	11	8503.45	I
300	340.46	IV	50	435.68	III	6	2550.02	III	10	8505.11	I
150	340.74	IV	250	438.02	V	5	2635.11	III	4	8763.96	I
30	341.92	III	25	441.81	II	5	2689.90	III	3	8767.05	I
15	348.00	III	200	442.30	IV	5	2938.45	III	13	8902.19	I
200	349.50	V	300	443.57	IV	5	2986.20	III	12	8904.02	I
300	354.93	IV	75	444.34	III	6	2992.42	III	5	8923.31	I
150	356.26	IV	200	445.61	IV	6	3052.07	III	4	8925.44	I
300	359.73	IV	250	446.83	IV	5	3056.84	III	7	9347.24	I
200	359.91	IV	75	448.60	III	5	3062.18	II	3	9349.25	I
250	362.08	IV	750	448.60	IV	4	3101.79	I	6	9351.59	I
150	362.15	IV	200	449.71	V	3	3102.04	I	15	9595.70	I
150	363.02	IV	200	452.90	V	7	3217.16	I	14	9597.83	I
500	372.15	V	250	455.67	V	6	3217.62	I	6	9949.67	I
200	372.46	V	400	456.33	IV	11	3446.37	I	5	9954.14	I
200	372.77	V	400	456.33	V	10	3447.38	I	9	10479.63	I
300	375.96	IV	75	466.79	III	3	3648.84	I	5	10482.15	I
300	375.96	V	100	470.09	III	4	3648.98	I	8	10487.11	I
250	377.76	V	75	471.57	III	18	4044.14	I	17	11019.87	I
30	379.12	III	45	474.92	III	17	4047.21	I	16	11022.67	I
300	379.12	V	10	476.03	II	10	4641.88	I	17	11690.21	I
300	379.88	IV	40	479.18	III	11	4642.37	I	16	11769.62	I
25	380.48	III	10	482.11	III	4	4740.91	I	17	11772.83	I
250	380.48	IV	10	482.41	III	6	4744.35	I		12522.11	I
200	381.70	IV	200	482.71	V	5	4753.93	I		13377.86	I
30	382.23	III	200	483.75	V	7	4757.39	I		13397.09	I
300	382.23	IV	30	495.14	II	5	4786.49	I		15163.08	I
150	382.49	IV	75	497.10	III	7	4791.05	I		15168.40	I
200	382.65	IV	10	514.94	III	6	4799.75	I		40158.37	I
300	382.91	IV	50	520.61	III	8	4804.35	I	**Praseodymium**		
250	384.10	IV	250	523.00	IV	7	4849.86	I	**Pr Z = 59**		
200	386.61	IV	25	523.79	III	8	4856.09	I	7000	865.90	V
300	387.80	V	200	526.45	IV	8	4863.48	I	5000	869.17	V
250	388.92	IV	150	527.62	IV	9	4869.76	I	2000	1228.59	IV
250	389.07	IV	40	529.80	III	8	4942.02	I	5000	1293.22	IV
250	389.07	V	15	539.71	III	9	4950.82	I	5000	1295.28	IV
250	390.11	V	15	546.12	III	9	4956.15	I	5000	1321.36	IV
250	390.42	IV	750	580.32	V	10	4965.03	I	5000	1333.57	IV
300	390.57	IV	250	585.51	V	10	5084.23	I	5000	1354.66	IV
200	391.46	IV	500	586.32	V	11	5097.17	I	2000	1360.64	IV
200	392.47	IV	30	600.77	II	11	5099.20	I	2000	1365.77	IV
500	393.14	IV	250	602.27	V	12	5112.25	I	5000	1374.41	IV
250	395.40	V	400	603.43	V	12	5323.28	I	5000	1435.56	IV
200	398.36	V	25	607.93	II	13	5339.69	I	2000	1520.98	IV
15	398.63	III	30	612.62	II	12	5342.97	I	5000	1574.55	IV
200	398.88	V	250	638.67	V	14	5359.57	I	5000	1575.10	IV
200	399.75	V	750	646.19	IV	16	5782.38	I	3000	1578.38	IV
400	400.21	IV	300	687.50	V	17	5801.75	I	2000	1622.30	IV
20	402.10	III	20	708.84	III	15	5812.15	I	10000	1884.87	IV
300	402.91	IV	300	720.43	V	17	5831.89	I	2000	2083.23	IV
250	403.97	IV	400	724.42	V	8	6120.27	II	3300	2246.20	V
150	404.41	IV	600	731.86	V	7	6307.29	II	2000 c	2378.98	IV
30	406.48	III	500	737.14	IV	19	6911.08	I	40 h	2598.04	II
250	408.08	IV	500	741.95	IV	12	6936.28	I	100 h	2707.37	II
40	408.96	III	500	745.26	IV	20	6938.77	I	60	2760.35	II
50	413.79	III	400	746.35	IV	7	6964.18	I	270	3168.24	II
30	414.87	III	300	749.99	IV	12	6964.67	I	200 d	3195.99	II
250	415.05	V	150	754.19	IV	25	7664.90	I	190	3219.48	II
200	415.79	V	400	754.67	IV	24	7698.96	I	200	3584.21	II
30	416.00	III	20	765.31	III	5	7955.37	I	250	3645.66	II
150	417.28	IV	30	765.64	III	4	7956.83	I	250	3646.30	II
30	417.54	III	150	770.29	V	7	8078.11	I	370	3668.83	II
30	418.62	III	150	771.46	V	6	8079.62	I	290	3714.05	II
400	422.18	V	35	778.53	III	9	8250.18	I	410	3739.18	II
300	425.16	V	20	872.31	III	8	8251.74	I	680	3761.87	II
500	425.59	V	10	873.86	III	3	8390.22	I	680	3800.30	II

Intensity	Wavelength/Å		Intensity	Wavelength/Å		Intensity	Wavelength/Å		Intensity	Wavelength/Å	
390	3811.84	II	560	4096.82	II	680	5259.73	II	1000 r	3998.96	II
1300 h	3816.02	II	380	4098.40	II	340 c	5292.02	II	1000	4417.96	II
680	3818.28	II	2900 c	4100.72	II	340	5292.62	II	900 r	4728.36	I
310	3821.80	II	1700 c	4118.46	II	430	5322.76	II	900	6100.21	I
960	3830.72	II	340	4130.77	II	65	5509.15	II	1000 d	6520.45	I
480	3840.99	II	1500 c	4141.22	II	150	5535.17	II		**Protactinium**	
580	3846.59	II	2700	4143.11	II	110	5623.05	II		**Pa Z = 91**	
1200	3850.79	II	1700 c	4164.16	II	90	5624.45	II	3000	2599.16	II
720 c	3851.55	II	620	4171.82	II	90	5756.17	II	3000	2699.22	II
960	3852.80	II	730	4172.25	II	90	5779.28	I	3000	2822.79	II
480 c	3865.45	II	5200	4179.39	II	160 d	5815.17	II	3000 h	2871.42	II
480	3876.19	II	2500	4189.48	II	90	5823.72	II	3000 h	2891.14	II
1700 c	3877.18	II	560 c	4191.60	II	90	5859.68	II	3000 l	3011.10	II
680	3880.47	II	2500 c	4206.72	II	160	5939.90	II	3000 s	3033.59	II
440 c	3885.19	II	500	4208.32	II	7000 w	5956.05	III	3000 l	3071.24	II
440 c	3889.34	II	320	4211.86	II	90	5956.60	II	3000 l	3093.23	II
770 c	3908.05	II	320	4217.81	II	110	5967.82	II	3000 l	3126.23	II
630	3912.90	II	3800	4222.93	II	90	6006.33	II	3000 l	3146.28	II
310	3913.55	II	3800	4225.35	II	150	6017.80	II	3000 l	3170.89	II
1300 c	3918.85	II	320	4233.11	II	150	6025.72	II	3000 l	3171.54	II
420	3919.63	II	320 c	4236.15	II	140	6055.13	I	3000 l	3240.58	II
960	3925.47	II	960	4241.01	II	65	6087.52	II	3000	3274.46	II
480	3927.46	II	340	4243.51	II	9000 w	6090.02	III	3000 l	3332.69	II
370	3929.29	II	840 c	4247.63	II	65	6114.38	II	3000 s	3346.66	II
370	3935.82	II	500	4254.40	II	65	6148.23	I	3000 l	3452.82	II
730 c	3947.63	II	320	4269.09	II	5000	6160.24	III	3000	3504.97	I
900 c	3949.43	II	790 c	4272.27	II	190	6161.18	II	3000 s	3530.65	II
900 c	3953.51	II	470 c	4280.07	II	270	6165.94	II	3000	3570.56	I
380	3956.75	II	790 c	4282.42	II	45	6244.35	II	3000	3571.82	I
470	3962.45	II	450 c	4298.98	II	110	6281.28	II	3000	3618.07	I
560	3964.26	II	1500	4305.76	II	55 c	6359.03	I	10000	3636.52	I
1600 c	3964.81	II	1300	4333.97	II	55	6411.23	I	3000	3702.74	I
560 c	3966.57	II	360	4338.70	II	45	6429.63	II	3000	3752.67	I
500	3971.16	II	620 cw	4344.30	II	45	6431.84	II	3000	3873.35	I
320	3971.67	II	470 c	4347.49	II	45	6486.55	I	3000	3931.83	I
620 c	3972.14	II	340	4350.40	II	45	6566.77	II	3000 s	3952.62	II
320	3974.85	II	450	4354.91	II	55	6616.67	I	100001	3957.85	II
1300 c	3989.68	II	410 c	4359.79	II	75	6656.83	II	3000 s	3970.07	II
340	3992.16	II	1200	4368.33	II	55	6673.41	II	3000	3981.82	I
1600	3994.79	II	320	4371.62	II	75	6673.78	II	10000	3982.23	I
560 c	3997.04	II	430	4405.83	II	35 c	6747.09	I	30001	4012.96	II
320	3999.12	II	1700	4408.82	II	55 cw	6798.60	I	3000 s	4018.21	II
620 c	4000.17	II	410	4413.77	II	35 cw	6827.60	II	3000	4030.16	II
730	4004.70	II	1200 c	4429.13	II	7000	6910.14	III	3000 s	4046.93	II
1900	4008.69	II	730	4449.83	II	40	7021.51	II	10000 s	4056.20	II
620	4010.60	II	960	4468.66	II	5000	7030.39	III	10000 s	4070.40	II
730	4015.39	II	1100	4496.46	II	4500	7076.62	III	30001	4176.18	II
620	4020.96	II	790	4510.15	II	20	7114.55	I	100001	4217.23	II
470	4022.71	II	340 c	4534.15	II	24	7227.70	II	10000 s	4248.08	II
360	4025.54	II	340	4535.92	II	16	7407.56	II	3000 s	4291.34	II
360 c	4029.72	II	270 c	4628.74	II	20 c	7451.74	II	3000 s	4601.43	II
730 c	4031.75	II	270 c	4672.09	II	14	7541.02	II	30001	6035.78	I
960	4033.83	II	290	4695.77	I	20	7645.66	II	3000	6162.56	I
730	4038.45	II	250	4736.69	I	16	7721.84	I	30001	6358.61	I
470	4039.34	II	200	4924.60	I	14	7871.67	I	3000	6379.25	I
1300	4044.81	II	320	4939.74	I	14	8067.44	I	30001	6438.97	I
340	4047.08	II	380	4951.37	I	10 cw	8122.78	II	3000 h	6792.75	I
450	4051.13	II	270	5034.41	II	11	8141.10	I	10000	6945.72	I
2200	4054.88	II	320	5045.52	I	5000 w	8602.74	III	3000	6960.09	I
2200	4056.54	II	360	5110.38	II	10	8714.59	II	3000 h	6961.78	I
450	4058.80	II	560	5110.76	II		**Promethium**		3000 s	6992.73	I
3400	4062.81	II	410	5129.52	II		**Pm Z = 61**		3000	7076.27	I
500 c	4079.77	II	620	5173.90	II	1000	3892.15	II	3000 h	7100.94	I
500 c	4080.98	II	360	5206.55	II	1000	3910.26	II	10000 s	7114.89	I
790	4081.85	II	360	5219.05	II	1000	3919.10	II	3000 h	7171.55	I
500	4083.34	II	560	5220.11	II	1000	3957.74	II	3000	7227.13	I

Intensity	Wavelength/Å		Intensity	Wavelength/Å		Intensity	Wavelength/Å		Intensity	Wavelength/Å	
3000	7318.79	I	680	2306.54	I	270	2649.05	I	440	3177.71	I
10000 l	7368.25	I	800	2322.49	I	660	2651.90	I	260	3178.61	I
3000 h	7471.89	I	300	2328.66	I	400	2654.12	I	600	3182.87	I
10000 h	7493.15	I	860	2344.78	I	220	2663.63	I	1100	3184.76	I
3000 h	7558.26	I	230	2349.39	I	940	2674.34	I	1100	3185.57	I
10000 h	7608.20	I	680	2352.07	I	220	2688.53	I	260	3190.78	I
10000	7626.79	I	250	2356.50	I	1300	2715.47	I	260	3192.36	I
10000 s	7635.18	I	1200	2365.90	I	220	2732.21	I	220	3198.58	I
10000	7669.34	I	570	2367.68	I	610	2733.04	II	1100 c	3204.25	I
3000	7679.20	I	520	2369.27	I	220	2758.00	I	380	3235.94	I
10000 h	7749.19	I	220	2370.76	II	210	2763.79	I	600	3258.85	I
3000	7872.95	I	320	2375.07	I	310	2767.74	I	600	3259.55	I
3000 l	7945.56	I	370	2379.77	I	220	2768.85	I	300	3268.89	I
10000	8039.34	I	340	2388.57	I	220	2769.32	I	280	3296.70	I
10000 h	8099.84	I	230	2393.65	I	350	2770.42	I	280	3296.99	I
10000	8199.04	I	320	2394.37	I	550	2783.57	I	280	3301.60	I
10000	8271.87	I	320	2396.79	I	220	2791.29	I	240	3302.23	I
3000 s	8358.98	I	210 d	2400.72	I	220	2814.68	I	320	3303.21	II
3000 s	8369.60	I	210	2401.68	I	880	2819.95	I	280	3303.75	I
3000 h	8441.04	I	1500	2405.06	I	310	2834.08	I	240	3313.95	I
10000 h	8532.66	I	740	2405.60	I	220	2843.00	I	600	3322.48	I
10000 s	8572.96	I	320	2406.70	I	270	2850.98	I	2000	3338.18	I
3000 h	8639.91	I	270	2410.37	I	240	2867.19	I	1600	3342.24	I
3000 h	8653.51	I	1200	2419.81	I	2900	2887.68	I	810	3344.32	I
10000	8735.27	I	300	2421.73	I	490	2896.01	I	320	3346.20	I
3000	10923.32	I	300	2421.88	I	830 c	2902.48	I	240 d	3356.33	I
10000	11791.73	I	2500	2428.58	I	210	2905.58	I	240	3377.74	I
10000	14344.76	I	490	2431.54	I	550	2909.82	I	320	3379.06	II
3000	18478.61	I	420	2432.18	I	830 c	2927.42	I	320	3379.70	I
Radium			340 c	2441.47	I	270	2930.61	I	240	3389.43	I
Ra Z = 88			230	2442.51	I	440	2943.14	I	4000	3399.30	I
100	3649.55	II	250	2444.94	I	270	2962.27	I	650	3404.72	I
200	3814.42	II	610	2446.98	I	720	2965.11	I	650	3405.89	I
100	4340.64	II	610	2449.71	I	1500	2965.76	I	240	3408.67	I
100	4682.28	II	390	2461.20	I	310	2976.29	I	320	3409.83	I
100	4825.91	I	800 c	2461.84	II	210	2978.15	I	320	3417.77	I
50	5660.81	I	1200	2483.92	I	220	2980.82	I	810	3419.41	I
50	7141.21	I	390	2485.81	I	220	2982.19	I	8000	3424.62	I
50	8019.70	II	980	2487.33	I	220	2988.47	I	400	3426.19	I
Radon			370	2496.04	I	1800	2992.36	I	300	3427.61	I
Rn Z = 86			370	2501.72	I	5500	2999.60	I	320	3437.71	I
100	4349.60	I	570	2502.35	II	350	3001.14	I	400	3449.37	I
200	7055.42	I	230	2504.60	II	220	3004.14	I	16000 c	3451.88	I
100	7268.11	I	270	2505.94	I	500	3016.02	I	240	3453.50	I
300	7450.00	I	1800 c	2508.99	I	300	3016.49	I	55000 c	3460.46	I
100	7809.82	I	570	2520.01	I	380	3030.45	I	40000 c	3464.73	I
100	8099.51	I	540	2521.50	I	240	3047.25	I	400	3467.96	I
100	8270.96	I	370	2534.80	I	1600	3067.40	I	240	3476.44	I
100	8600.07	I	570	2540.51	I	320	3069.94	I	400	3480.38	I
Rhenium			740 d	2544.74	I	260	3071.16	I	320	3480.85	I
Re Z = 75			370	2545.48	I	550	3082.43	I	240	3482.23	I
25000	2003.53	I	300	2552.02	I	340	3088.76	I	560	3503.06	I
16000	2017.87	I	370	2554.63	II	700	3100.67	I	320	3516.65	I
27000	2049.08	I	1000	2556.51	I	700	3108.81	I	320	3517.33	I
10000	2085.59	I	250	2559.08	I	340	3110.86	I	320	3537.46	I
9800	2097.12	I	340	2564.19	I	340 c	3118.19	I	240	3549.89	I
3400	2139.04	II	540	2568.64	II	340	3121.36	I	240	3570.26	I
3700	2156.67	I	370	2571.81	II	420	3128.94	I	360	3579.12	I
4900	2167.94	I	380	2586.79	I	260	3134.02	I	810 c	3580.15	II
3400	2176.21	I	290	2599.86	I	250	3141.38	I	650	3580.97	I
4200 c	2214.26	II	290	2603.89	I	440	3151.64	I	810	3583.02	I
5200 c	2275.25	II	660	2608.50	II	330	3153.79	I	320	3617.08	I
2900	2287.51	I	610 d	2611.54	I	360 c	3158.31	I	810	3637.84	I
2700	2294.49	I	310	2635.83	II	220	3164.52	I	440	3651.97	I
390	2298.09	II	550	2636.64	I	700	3168.37	I	320	3670.53	I
610	2302.99	I	270	2642.75	I	220	3174.61	I	860 c	3689.50	I

Intensity	Wavelength/Å		Intensity	Wavelength/Å		Intensity	Wavelength/Å		Intensity	Wavelength/Å	
1500 c	3691.48	I	800	1880.66	III	160	2728.94	I	880 d	3538.14	I
520	3703.24	I	500	1884.91	III	100	2771.51	I	280	3541.91	I
240	3709.93	I	500	1887.36	III	130	2783.03	I	1200	3543.95	I
360 c	3717.28	I	700	1888.62	III	150	2826.43	I	1800	3549.54	I
4000	3725.76	I	800	1901.32	III	180	2826.68	I	1200	3570.18	I
240 c	3735.01	I	500	1910.16	III	280	2862.94	I	4700	3583.10	I
810	3735.31	I	600	1919.37	III	110	2878.66	I	4700	3596.19	I
910	3740.10	I	500	1927.07	III	140	2882.37	I	5900	3597.15	I
300 cw	3745.44	I	700	1931.79	III	160	2907.21	I	3100	3612.47	I
700	3787.52	I	500	1954.25	III	65	2910.17	II	1800	3626.59	I
240	3869.94	I	500	1994.26	III	180	2924.02	I	8200	3657.99	I
240	3875.26	I	800	2013.71	III	130	2929.11	I	1300	3666.22	I
240	3876.86	I	500	2017.47	III	130	2931.94	I	560	3681.04	I
380 c	3917.27	I	500	2028.53	III	230	2968.66	I	1900	3690.70	I
550	3929.85	I	800	2036.72	III	160	2977.68	I	9400	3692.36	I
280	3961.04	I	600	2037.61	III	450	2986.20	I	940	3695.52	I
350 c	3962.48	I	1000	2040.18	III	110	3004.46	I	280	3698.26	I
220	4033.31	I	3000	2048.67	III	50	3006.43	III	380	3698.60	I
240	4081.43	I	2000	2064.11	III	130	3023.91	I	7600	3700.91	I
240 c	4110.89	I	800	2076.84	III	50	3052.44	III	940	3713.02	I
240 cw	4133.42	I	1000	2118.53	III	180	3083.96	I	650	3735.28	I
1800	4136.45	I	1000	2118.63	III	140	3121.76	I	420	3737.27	I
700	4144.36	I	1000	2139.44	III	240	3123.70	I	420	3744.17	I
220	4182.90	I	1000	2152.23	III	130	3155.78	I	1200	3748.22	I
220	4183.06	I	3000	2158.17	III	140	3189.05	I	240	3754.12	I
650	4221.08	I	3000	2163.19	III	470	3191.19	I	380	3754.27	I
3600 c	4227.46	I	3000	2167.33	III	190	3197.13	I	490	3755.58	I
260 c	4257.60	I	150	2276.21	II	520	3263.14	I	1000	3760.40	I
380	4358.69	I	140	2288.57	I	520	3271.61	I	2300	3765.08	I
360 cw	4394.38	I	110	2309.82	I	2300	3280.55	I	490	3769.97	I
2600	4513.31	I	350	2322.58	I	2300	3283.57	I	380	3778.13	I
260	4516.64	I	140	2326.47	I	280	3289.14	I	1000	3788.47	I
500	4522.73	I	190	2334.77	II	210	3294.28	I	1300	3792.18	I
2200 cw	4889.14	I	300	2361.92	I	260	3300.46	I	3800	3793.22	I
220	4923.90	I	110	2368.34	I	50	3310.69	III	4900	3799.31	I
1300	5270.95	I	270	2382.89	I	4200	3323.09	I	760	3805.92	I
1600 cw	5275.56	I	230	2383.40	I	330	3338.54	I	1300	3806.76	I
100	5667.88	I	270	2386.14	II	280	3360.80	I	470	3815.01	I
110 c	5752.93	I	80	2415.84	II	420	3368.38	I	760	3816.47	I
110 cw	5776.83	I	130	2427.68	I	1100	3372.25	I	1300	3818.19	I
550	5834.31	I	230	2429.52	I	110	3377.14	I	3800	3822.26	I
200	6307.70	I	110	2437.90	I	110	3385.78	I	2300	3828.48	I
200	6321.90	I	330	2440.34	I	5600	3396.82	I	2000	3833.89	I
100 cw	6605.19	I	90	2461.04	II	820	3399.70	I	5900	3856.52	I
180 c	6813.41	I	130	2473.09	I	160	3406.55	I	490	3870.01	I
260	6829.90	I	150	2487.47	I	820	3412.27	I	380	3877.34	I
50 cw	7640.94	I	100	2490.77	II	330	3421.22	I	120	3913.51	I
65 cw	7912.94	I	130	2502.46	I	120 d	3424.38	I	240	3922.19	I
Rhodium			300	2504.29	II	8200	3434.89	I	2000	3934.23	I
Rh Z = 45			150	2505.67	I	1400	3440.53	I	590	3942.72	I
50	813.44	III	350	2509.70	I	120	3447.74	I	3800	3958.86	I
80	882.51	III	300	2511.03	II	120	3450.29	I	45	3964.54	II
100	925.75	III	200	2515.75	I	400	3455.22	I	380	3975.31	I
150	937.28	III	130	2520.53	II	180	3457.07	I	240	3984.40	I
500	991.62	III	110	2537.04	II	220	3457.93	I	240	3995.61	I
400	992.48	III	350	2545.70	I	5900	3462.04	I	380	3996.15	I
500 d	1009.60	III	550	2555.36	I	180	3469.62	I	120	4023.14	I
200	1012.22	III	150	2622.58	I	4700	3470.66	I	560	4082.78	I
200	1015.17	III	230	2625.88	I	120	3472.25	I	140	4097.52	I
200	1073.87	III	100	2630.42	I	4700	3474.78	I	120	4119.68	I
150	1784.24	III	110	2647.28	I	2100	3478.91	I	1100	4121.68	I
200	1784.94	III	400	2652.66	I	110	3494.44	I	1500	4128.87	I
150	1796.50	III	100	2680.63	I	1200	3498.73	I	2100	4135.27	I
200	1816.03	III	400	2703.73	I	5900	3502.52	I	240	4154.37	I
1000	1832.05	III	100	2715.31	II	2800	3507.32	I	330	4196.50	I
500	1859.85	III	180	2718.54	I	8800	3528.02	I	3300	4211.14	I

Intensity	Wavelength/Å		Intensity	Wavelength/Å		Intensity	Wavelength/Å		Intensity	Wavelength/Å	
820	4288.71	I	1200	598.49	III	5	6299.224	I	370	2456.57	II
4200	4374.80	I	1500	643.878	II	10000	6458.33	II	280	2478.93	II
130	4569.00	I	25	663.76	IV	5000	6560.81	II	140	2498.42	II
150	4675.03	I	3000	697.049	II	100 1	7279.997	I	140	2498.57	II
70	4745.11	I	6000	711.187	II	150	7408.173	I	260	2507.01	II
70	5090.63	I	25	716.24	IV	200 1	7618.933	I	110	2513.32	II
60	5155.54	I	50	740.85	IV	300	7757.651	I	110	2517.32	II
60	5175.97	I	10000	741.456	II	60	7759.436	I	150	2535.59	II
95	5193.14	I	5000	769.04	III	90000 c	7800.27	I	550	2549.58	I
130	5354.40	I	25	776.89	IV	45000 c	7947.60	I	370	2609.06	I
95	5390.44	I	2500	815.28	III	40 1	8271.41	I	830	2612.07	I
160	5599.42	I	15	850.18	IV	30	8271.71	I	460	2642.96	I
40	5686.38	I	1000	1604.12	II	40 1	8868.512	I	330	2651.84	I
29	5792.66	I	5000	1760.50	II	30	8868.852	I	400	2659.62	I
40	5806.91	I	2000	2068.92	II	30 1	9522.65	I	330	2661.61	II
35	5831.58	I	10000	2075.95	II	20 1	9540.18	I	690	2678.76	II
130	5983.60	I	30000	2143.83	II	2000 c	9689.05	II	330	2692.06	II
35	6102.72	I	10000	2217.08	II	35 1	10075.282	I	200	2712.41	II
14	6199.99	I	5000	2291.71	II	30 1	10075.708	I	690	2719.52	I
16	6253.72	I	50000	2472.20	II	100	13235.17	I	140	2725.47	II
29	6319.53	I	1000	2631.75	III	20	13442.81	I	310	2734.35	II
40	6752.35	I	2000	2956.07	III	30	13443.57	I	1800	2735.72	I
13	6827.33	I	500	3086.84	III	75	13665.01	I	100	2778.38	II
11	6857.68	I	500	3111.36	III	1000	14752.41	I	110	2787.83	II
20	6879.94	I	5000 c	3148.90	II	800	15288.43	I	350	2810.03	I
65	6965.67	I	25	3157.54	I	150	15289.48	I	1700	2810.55	I
16	6979.15	I	50	3227.98	I	20	22529.65	I	350	2818.36	I
16	7001.58	I	500	3286.41	III	4	27314.31	I	400	2829.16	I
18	7101.64	I	60	3348.72	I	**Ruthenium**			640	2854.07	I
15	7104.45	I	75	3350.82	I	**Ru Z = 44**			420	2861.41	I
18	7268.18	I	100	3587.05	I	250	850.09	III	550	2866.64	I
35	7270.82	I	40	3591.57	I	200	850.30	III	1800	2874.98	I
18 h	7442.39	I	5000	3600.60	II	250	919.74	III	740	2886.54	I
12	7475.74	I	10000	3600.64	II	500	940.09	III	370	2908.88	I
12	7495.24	I	25000	3940.51	II	500	966.54	III	1100	2916.26	I
11	7557.67	I	1000	4201.80	I	750	974.14	III	180	2945.67	II
29	7791.61	I	500	4215.53	I	900	979.43	III	370	2949.50	I
55	7824.91	I	90000	4244.40	II	500	981.35	III	550	2965.16	I
21	8029.91	I	15000	4273.14	II	900	986.84	III	170	2965.55	II
29	8045.36	I	20000	4571.77	II	900	994.56	III	140	2976.59	II
15	8136.20	I	10000	4648.57	II	300	1001.65	III	550	2976.92	I
8	8425.59	I	30000	4775.95	II	500	1009.13	III	1400	2988.95	I
Rubidium			2	5087.987	I	900	1009.87	III	460	2994.96	I
Rb Z = 37			2	5132.471	I	500	1014.68	III	440	3006.59	I
30	465.85	III	10	5150.134	I	800	1190.51	III	330	3017.24	I
40	481.118	II	10000	5152.08	II	500	1200.07	III	310	3020.88	I
500	482.83	I	1	5165.023	I	500	1207.17	III	390	3064.84	I
500	489.66	III	2	5165.142	I	500	1209.77	III	330	3096.57	I
600	493.48	III	15	5195.278	I	300	1211.31	III	830	3099.28	I
90	497.430	II	2	5233.968	I	500	1941.35	III	740	3100.84	I
20	508.434	II	20	5260.034	I	500	2009.28	III	490	3294.11	I
150	513.266	II	1	5260.228	I	2400	2076.43	I	370	3301.59	I
300	530.173	II	3	5322.380	I	2600	2083.77	I	930	3339.55	I
75	533.801	II	40	5362.601	I	2400	2090.89	I	3100	3417.35	I
1200	535.86	III	4	5390.568	I	690	2255.52	I	4900	3428.31	I
40	542.887	II	75	5431.532	I	780	2272.09	I	6400	3436.74	I
200	555.036	II	3	5431.830	I	780	2279.57	I	8300	3498.94	I
1200	556.19	III	6	5578.788	I	480	2317.80	I	640	3514.49	I
1500	566.71	III	40	5647.774	I	120	2334.96	II	790	3539.37	I
1000	572.82	III	20	5653.750	I	190 h	2342.85	II	690	3570.59	I
1500	576.65	III	60	5724.121	I	310	2351.33	I	6400	3589.22	I
2500	579.63	III	3	5724.614	I	170	2357.91	II	6900	3593.02	I
1500	581.26	III	75	6070.755	I	780	2402.72	II	6400	3596.18	I
2500	589.419	II	30 c	6159.626	I	150	2407.92	II	1300	3599.76	I
1000	594.94	III	75 c	6206.309	I	180	2455.53	II	3100	3634.93	I
1300	595.88	III	120 c	6298.325	I	150	2456.44	II	6200	3661.35	I

Intensity	Wavelength/Å		Intensity	Wavelength/Å		Intensity	Wavelength/Å		Intensity	Wavelength/Å	
830	3663.37	I	930	4307.60	I	18	7722.87	I	1500	3990.00	II
650	3669.49	I	550	4319.87	I	22	7791.86	I	1400	4064.58	II
550	3726.10	I	550	4342.07	I	30	7847.80	I	1000	4092.27	II
8700	3726.93	I	710	4354.13	I	80	7881.49	I	1900	4118.55	II
11000	3728.03	I	870	4361.21	I	18	8264.96	I	1200	4152.21	II
7100	3730.43	I	2400	4372.21	I	22	8710.84	I	1000	4188.13	II
3500	3742.28	I	870	4385.39	I	**Samarium**			1100	4203.05	II
870	3742.78	I	1300	4385.65	I	**Sm Z = 62**			1000	4225.33	II
2800	3745.59	I	1700	4390.44	I	150	2789.38	II	1200	4236.74	II
760	3753.54	I	1600	4410.03	I	410	3152.52	II	2100	4256.39	II
870	3755.93	I	1100	4460.04	I	720	3183.92	II	1300	4262.68	II
1200	3759.84	I	5400	4554.51	I	600	3211.73	II	1200	4279.68	II
600	3761.51	I	1700	4584.44	I	530	3216.85	II	2200	4280.79	II
600	3767.35	I	720	4647.61	I	600	3218.61	II	710	4282.21	I
1500	3777.59	I	1400	4709.48	I	720	3230.56	II	470	4282.83	I
600	3782.74	I	500	4757.84	I	720	3236.64	II	1600	4296.74	I
3900	3786.06	I	550	4869.15	I	720	3239.66	II	1900	4318.94	II
6000	3790.51	I	160	5011.23	I	720	3250.37	II	470	4319.53	I
760	3798.05	I	450	5057.33	I	850	3254.38	II	1800	4329.02	II
7600	3798.90	I	120	5076.32	I	1700	3306.39	II	440	4330.02	I
7600	3799.35	I	200	5093.83	I	1200	3321.18	II	1300	4334.15	II
600	3812.72	I	530	5136.55	I	1200	3365.86	II	880	4336.14	I
760	3817.27	I	170	5142.76	I	1200	3382.40	II	1100	4347.80	II
760	3819.03	I	250	5147.24	I	4200	3568.27	II	440	4362.91	I
650	3822.09	I	110	5151.07	I	4200	3592.60	II	530	4380.42	I
550	3824.93	I	500	5155.14	I	1700	3604.28	II	1600	4390.86	II
760	3831.80	I	920	5171.03	I	3400	3609.49	II	410	4401.17	I
930	3839.70	I	180	5195.02	I	1700	3621.23	II	470	4419.33	I
760	3850.43	I	130	5284.08	I	3400	3634.29	II	1500	4420.53	II
1300	3857.55	I	260	5309.27	I	2200	3661.36	II	2900	4424.34	II
650	3862.69	I	110	5335.93	I	2200	3670.84	II	470	4429.66	I
1300	3867.84	I	130	5361.77	I	1100	3693.99	II	1600	4433.88	II
650	3892.21	I	110 h	5401.04	I	1600	3728.47	II	1800	4434.32	II
760	3909.08	I	80	5484.32	I	2100	3731.26	II	530	4441.81	I
1500	3923.47	I	130	5510.71	I	1600	3735.98	II	440	4442.28	I
3300	3925.92	I	90	5559.75	I	2900	3739.12	II	710	4445.15	I
600	3931.76	I	290	5636.24	I	1200	3743.87	II	1300	4452.73	II
760	3945.57	I	180	5699.05	I	930	3745.46	I	1200	4454.63	II
600	3978.44	I	65	5814.98	I	800	3756.41	I	1000	4458.52	II
600	3979.42	I	55	5919.34	I	1200	3757.53	II	2200	4467.34	II
870	3984.86	I	80	5921.45	I	1900	3760.69	II	810	4470.89	I
1500	4022.16	I	21 h	5973.38	I	1100	3764.37	II	370	4499.11	I
600	4023.83	I	16	5988.67	I	370 d	3773.33	I	440	4581.73	I
1400	4051.40	I	35	5993.65	I	1100	3778.14	II	380	4649.49	I
710	4054.05	I	18	6116.77	I	1500	3788.12	II	470 d	4670.75	I
760	4068.37	I	26	6199.42	I	1600	3793.97	II	1100	4674.60	II
980	4076.73	I	26	6225.20	I	1600	3797.73	II	370	4688.73	I
6000	4080.60	I	18	6295.22	I	1600	3826.20	II	530	4704.40	II
930	4097.79	I	16	6390.23	I	1100	3831.50	II	730	4716.10	I
1900	4112.74	I	26 h	6444.84	I	560	3834.48	I	770	4728.42	I
2000	4144.16	I	21	6663.14	I	1600	3843.50	II	470	4745.68	II
650	4145.74	I	55	6690.00	I	530	3853.30	I	730	4760.27	I
870	4167.51	I	21	6766.95	I	2700	3854.21	II	580	4783.10	I
550	4197.58	I	30	6775.02	I	480	3854.56	I	350	4785.86	I
550	4198.88	I	21	6824.17	I	400	3858.74	I	970	4841.70	I
7600	4199.90	I	26	6911.48	I	3700	3885.29	II	730	4883.97	I
1500	4206.02	I	110	6923.23	I	1600	3896.98	II	630	4910.40	I
5400	4212.06	I	26	6982.01	I	1300	3903.42	II	350	4913.25	II
760	4214.44	I	26	7027.98	I	2500	3922.40	II	430	4918.99	I
930	4217.27	I	35	7238.92	I	1900	3928.28	II	400	5044.28	I
550	4230.31	I	16	7393.93	I	1300	3941.87	II	540	5071.20	I
760	4241.05	I	18	7468.91	I	470	3951.89	I	510	5117.16	I
760	4243.06	I	26	7485.79	I	1500	3963.00	II	350	5122.14	I
760	4284.33	I	70	7499.75	I	1500	3971.40	II	360	5155.03	II
550	4295.93	I	26	7559.61	I	620	3974.66	II	470	5175.42	I
3700	4297.71	I	18	7621.50	I	1000	3976.43	II	250	5200.59	I

Intensity	Wavelength/Å		Intensity	Wavelength/Å		Intensity	Wavelength/Å		Intensity	Wavelength/Å	
260	5251.92	I	500	252.85	V	130 d	3469.65	I	160 h	4573.99	I
400	5271.40	I	500	253.73	V	110	3471.13	I	350	4670.40	II
250	5282.91	I	900	283.91	V	200	3498.91	I	120	4706.97	I
220	5453.00	I	800	284.45	V	2700	3535.73	II	120	4709.34	I
230	5493.72	I	600	288.29	V	6600	3558.55	II	200	4728.77	I
230	5516.09	I	900	289.59	V	6100	3567.70	II	490	4729.23	I
140	5550.40	I	15	289.85	IV	13000	3572.53	II	590	4734.10	I
140	5659.86	I	1000 d	291.93	V	9900	3576.35	II	690	4737.65	I
120	5696.73	I	800	293.25	V	7700	3580.94	II	790	4741.02	I
85	5706.20	I	15	296.31	IV	4000	3589.64	II	1200	4743.81	I
70	5773.77	I	15	299.04	IV	4000	3590.48	II	200	4753.16	I
60	5778.33	I	700	300.00	V	28000	3613.84	II	220	4779.35	I
70 d	5786.98	II	1000	573.36	V	110	3617.43	I	170	4839.44	I
60	5788.38	I	600	587.94	V	20000	3630.75	II	90	4909.76	I
60	5800.52	I	10	785.12	IV	13000	3642.79	II	90	4922.84	I
65	5802.84	I	25	1168.61	III	6600	3645.31	II	90	4934.25	I
65	5867.79	I	15	1550.80	IV	110	3646.90	I	170	4954.06	I
50	5874.21	I	180	1603.06	III	5300	3651.80	II	120	4973.66	I
50	5898.96	I	150	1610.19	III	110	3664.25	II	150	4980.37	I
65	5965.71	II	160	2010.42	III	290	3666.54	II	140	4991.92	I
50	6045.00	I	12	2118.97	IV	75 h	3717.10	I	530	5031.02	II
50	6070.06	I	11	2185.43	IV	270	3833.07	II	250	5064.32	I
45	6084.12	I	11	2205.46	IV	610	3843.03	II	530	5070.23	I
45 h	6159.56	I	14	2222.22	IV	20000	3907.49	I	250	5075.81	I
45	6256.54	I	11	2271.33	IV	23000	3911.81	I	2100	5081.56	I
100	6267.28	II	110	2438.62	I	4400	3933.38	I	1200	5083.72	I
140	6569.31	II	560	2545.22	II	5500	3996.61	I	1100	5085.55	I
110	6589.72	II	2900	2552.37	II	530	4014.49	II	750	5086.95	I
50	6671.51	I	560	2555.82	II	20000	4020.40	I	390	5087.14	I
120 d	6731.84	II	2300	2560.25	II	20000	4023.69	I	270	5089.89	I
95	6794.20	II	1100	2563.21	II	220	4030.67	I	390	5096.73	I
120	6860.93	I	11	2586.93	IV	140	4031.39	I	620	5099.23	I
120	6955.29	II	120	2692.78	I	220	4043.80	I	370	5101.12	I
90	7020.44	II	350	2699.07	III	200	4046.48	I	180	5109.06	I
90	7039.22	II	360	2706.77	I	2700	4047.79	I	150	5112.86	I
90	7042.24	II	210	2707.95	I	120	4049.95	I	320	5116.69	I
90	7051.52	II	580	2711.35	I	5500	4054.55	I	390	5210.52	I
90	7082.37	II	230	2734.05	III	220	4056.59	I	280	5219.67	I
26	7088.30	I	340	2965.86	I	100	4068.66	III	350	5239.82	II
30	7095.50	I	1200	2974.01	I	160 h	4074.97	I	280	5258.33	I
30	7104.54	I	1400	2980.57	I	160	4078.57	I	210	5285.76	I
26	7115.96	I	340	2988.95	I	6100	4082.40	I	120	5341.05	I
85 d	7149.60	II	2200	3015.36	I	200	4086.67	I	350	5349.30	I
23	7213.82	I	2700	3019.34	I	400	4087.16	I	120	5349.71	I
60	7240.90	II	360	3030.76	I	440 h	4133.00	I	210	5355.75	I
26	7347.30	I	120 h	3056.31	I	530 h	4140.30	I	530	5356.10	I
30	7444.56	I	130	3065.11	II	720	4152.36	I	270	5375.35	I
26	7445.41	I	990	3251.32	II	1100 h	4165.19	I	370	5392.08	I
13	7470.76	I	1500	3255.69	I	110 h	4218.26	I	270	5446.20	I
45	7645.09	II	4400	3269.91	I	110 h	4219.73	I	120	5451.34	I
12	7645.82	I	5500	3273.63	I	180	4231.93	I	750	5481.99	II
40 w	7835.08	II	110 d	3343.28	II	200	4233.61	I	530	5484.62	I
16	7895.96	I	270	3352.05	II	400	4238.05	I	570	5514.22	I
90	7928.14	I	9900	3353.73	II	15000	4246.83	II	660	5520.50	I
40	8048.70	II	2000	3359.68	II	290	4294.77	II	660	5526.82	II
16	8065.16	I	1700	3361.27	II	350	4305.71	II	70	5564.86	I
45	8068.46	II	1700	3361.94	II	4200	4314.09	II	110	5591.33	I
40 w	8305.79	II	4000	3368.95	II	3300	4320.74	II	80	5640.98	II
19	8383.71	I	6600	3372.15	II	2400	4325.01	II	250	5657.88	II
45 w	8485.99	II	130	3418.51	I	180	4354.61	II	1500	5671.81	I
45 w	8708.43	II	200	3429.21	I	110	4358.64	I	1200	5686.84	I
95	8913.66	II	200	3429.48	I	2000	4374.46	II	1100	5700.21	I
Scandium			270	3431.36	I	130	4384.81	II	10	5706.82	IV
Sc Z = 21			530	3435.56	I	1100	4400.37	II	190	5708.61	I
350	180.14	V	270	3457.45	I	880	4415.56	II	880	5711.75	I
500	243.87	V	180	3462.19	I	120 h	4557.24	I	230	5717.28	I

Intensity	Wavelength/Å	
180	5724.08	I
14	5771.63	IV
620	6210.68	I
320	6239.78	I
120	6245.63	II
110	6249.96	I
80	6256.01	III
250	6258.96	I
750	6305.67	I
60	6378.82	I
90	6413.35	I
60	6604.60	II
65	6737.87	I
50	6819.52	I
50	6835.03	I
90	7449.16	III
55 h	7741.17	I
30	7800.44	I
19 h	8761.40	I
50	8829.78	III
30 h	8834.35	I
400	22051.86	I
150	22065.05	I

Selenium
Se Z = 34

Intensity	Wavelength/Å	
360	613.0	V
360	652.7	IV
450	670.1	IV
360	724.3	III
450	746.4	IV
450	759.1	V
360	808.7	V
360	830.3	V
360	832.7	II
450	839.5	V
360	843.0	III
360	845.8	V
360	912.9	II
360	959.6	IV
360	974.8	III
450	996.7	IV
360	1013.4	II
360	1014.0	II
450	1033.6	II
450	1049.6	II
360	1057.4	II
360	1094.7	V
360	1099.1	III
450	1119.2	III
360	1141.9	II
450	1192.3	II
450	1227.6	V
285	1291.0	II
285	1308.9	II
285	1314.4	IV
120	1435.3	I
120	1435.8	I
150	1449.2	I
150	1500.9	I
250	1530.4	I
150	1531.3	I
200	1531.8	I
150	1575.3	I
150	1577.6	I
150	1577.9	I
150	1579.5	I

Intensity	Wavelength/Å	
200	1580.0	I
150	1587.5	I
150	1593.2	I
250	1606.5	I
200	1617.4	I
150	1621.2	I
150	1643.4	I
250	1671.2	I
250	1675.3	I
250	1690.7	I
250	1793.3	I
300	1795.3	I
300	1855.2	I
250	1858.8	I
400	1898.6	I
350	1913.8	I
300	1919.2	I
500	1960.9	I
500	2039.8	I
285	2057.5	III
500	2074.8	I
285	2136.6	IV
500	2164.2	I
600	2413.5	I
300	2548.0	I
360	2665.5	IV
285	2724.3	IV
285	2767.2	III
220	2773.8	III
160	2951.6	IV
450	3387.2	III
450	3413.9	III
450	3457.8	III
450	3637.6	III
450	3738.7	III
450	3800.9	III
450	4169.1	III
360	4175.3	II
450	4180.9	II
285	4382.9	II
285	4446.0	II
220	4449.2	II
285	4467.6	II
500	4730.8	I
400	4739.0	I
300	4742.2	I
285	4840.6	II
360	4845.0	II
450	5227.5	II
360	5305.4	II
100	5365.5	I
120	5369.9	I
110	5374.1	I
285	5522.4	II
285	5566.9	II
285	5866.3	II
450	6056.0	II
285	6303.8	III
200	6325.6	I
360	6444.2	II
285	6490.5	II
285	6535.0	II
150	6831.3	I
120	6990.690	I
100	6991.792	I
200	7010.809	I

Intensity	Wavelength/Å	
150	7013.875	I
300	7062.065	I
200	7575.1	I
250	7583.4	I
150	7592.2	I
300	8001.0	I
200	8036.4	I
150	8093.2	I
150	8094.7	I
180	8149.3	I
150	8152.0	I
200	8157.7	I
180	8163.1	I
150	8182.9	I
150	8440.47	I
150	8450.38	I
150	8742.33	I
300	8918.86	I
200	9001.97	I
200	9038.61	I
100	9432.50	I
200	10217.25	I
377	10307.45	I
900	10327.26	I
640	10386.36	I
275	11946.87	I
170	11952.64	I
205	11972.93	I
315	14817.93	I
410	14917.47	I
500	15151.44	I
320	15471.00	I
265	15520.97	I
395	15618.40	I
360	16659.44	I
505	16813.78	I
205	16866.54	I
235	21374.24	I
680	21442.56	I
415	21473.48	I
270	21716.36	I
240	21730.60	I
150	23388.85	I
265	24148.18	I
375	24385.99	I
255	25017.51	I
510	25127.43	I

Silicon
Si Z = 14

Intensity	Wavelength/Å	
10	85.18	V
15	96.44	V
10	97.14	V
20	117.86	V
20	118.97	V
4	457.82	IV
8	566.61	III
8	653.33	III
7	815.05	IV
8	818.13	IV
9	823.41	III
40 h	845.77	II
100	889.72	II
200	892.00	II
9	967.95	III
100	989.87	II
200	992.68	II

Intensity	Wavelength/Å	
10	993.52	III
13	994.79	III
16	997.39	III
50	1023.69	II
8	1066.63	IV
14	1108.37	III
16	1109.97	III
18	1113.23	III
8	1122.49	IV
10	1128.34	IV
100	1190.42	II
200	1193.28	II
250	1194.50	II
100	1197.39	II
30	1206.51	III
30	1206.53	III
9	1207.52	III
10	1210.46	II
50	1226.81	II
100	1227.60	II
150	1228.75	II
200	1229.39	II
100	1246.74	II
150	1248.43	II
100	1250.09	II
150	1250.43	II
200	1251.16	II
40	1256.49	I
50	1258.80	I
1000	1260.42	I
2000	1264.73	II
200	1265.02	II
17	1294.54	III
14	1296.73	III
15	1298.89	III
18	1298.96	III
14	1301.15	III
16	1303.32	III
100	1304.37	II
50 h	1305.59	II
200	1309.27	II
13	1312.59	III
100	1346.87	II
100	1348.54	II
150	1350.06	II
100	1352.64	II
100	1353.72	II
15	1393.76	IV
12	1402.77	IV
13	1417.24	III
90 h	1485.02	II
100 h	1485.51	II
12	1500.24	III
10	1501.19	III
9	1501.87	III
100 h	1509.10	II
50 h	1512.07	II
60 p	1516.91	II
500	1526.72	II
1000	1533.45	II
150	1594.55	I
100	1622.87	I
300	1629.43	I
200	1629.92	I
100	1667.62	I
100	1668.52	I

Intensity	Wavelength/Å		Intensity	Wavelength/Å		Intensity	Wavelength/Å		Intensity	Wavelength/Å	
100	1672.59	I	300	2904.28	II	16	4819.72	III	1000	6347.10	II
200	1675.20	I	500	2905.69	II	18	4828.97	III	1000	6371.36	II
200	1696.20	I	55	2970.355	I	30	4947.607	I	45	6526.609	I
200	1697.94	I	150	2987.645	I	40	5006.061	I	45	6527.199	I
100 h	1770.92	I	50	3006.739	I	1000	5041.03	I	45	6555.462	I
100 h	1776.83	I	75	3020.004	I	1000	5055.98	II	50 h	6660.52	II
100 h	1799.12	I	100 h	3030.00	II	10 h	5091.42	III	100	6671.88	II
150	1808.00	II	9	3040.93	III	100	5181.90	II	7	6701.21	IV
500 h	1814.07	I	100 h	3043.69	II	100 h	5185.25	II	50 h	6717.04	II
200	1816.92	II	50 h	3048.30	II	200 h	5192.86	II	100	6721.853	I
200	1836.51	I	150 h	3053.18	II	500 h	5202.41	II	50	6829.82	II
200	1841.44	I	25	3086.24	III	100 h	5405.34	II	30	6848.568	I
9	1842.55	III	20	3093.42	III	100 h	5438.62	II	80	6976.523	I
200	1843.77	I	16	3096.83	III	100 h	5456.45	II	180	7003.567	I
300	1845.51	I	9	3165.71	IV	500 h	5466.43	II	180	7005.883	I
400	1847.47	I	16	3185.13	III	500 h	5466.87	II	90	7017.646	I
200	1848.14	I	13	3186.02	III	100 h	5469.21	II	250	7034.903	I
500	1850.67	I	150	3188.97	II	200 h	5496.45	II	6 h	7047.94	IV
200	1852.46	I	150	3193.09	II	35	5517.535	I	200	7165.545	I
500 h	1874.84	I	100	3195.41	II	100 h	5540.74	II	100	7226.206	I
200	1881.85	I	14	3196.50	III	150 h	5576.66	II	100	7235.326	I
200	1887.70	I	200	3199.51	II	30	5622.221	I	180	7250.625	I
200 h	1893.25	I	100 h	3203.87	II	100 h	5632.97	II	160	7275.294	I
1000 h	1901.33	I	200 h	3210.03	II	200 h	5639.48	II	400	7289.173	I
100 h	1902.46	II	15	3210.55	III	90	5645.611	I	375	7405.774	I
50 h	1910.62	II	75	3214.66	II	150 h	5660.66	II	200	7409.082	I
50	1941.67	II	12	3230.50	III	80	5665.554	I	275	7415.946	I
100	1949.56	II	14	3233.95	III	1000 h	5669.56	II	425	7423.497	I
100	1954.97	I	15	3241.62	III	120	5684.484	I	9 h	7466.32	III
50	2058.65	II	12	3258.66	III	300 h	5688.81	II	12 h	7612.36	III
50	2059.01	II	10	3276.26	III	100	5690.425	I	100	7680.267	I
200	2072.02	II	300	3333.14	II	90	5701.105	I	6 h	7723.82	IV
200	2072.70	II	500	3339.82	II	200 h	5701.37	II	30	7800.008	I
100	2124.12	I	15	3486.91	III	100 h	5706.37	II	400	7848.80	II
50 h	2136.56	II	9	3525.94	III	160	5708.397	I	500	7849.72	II
110	2207.98	I	20	3590.47	III	20	5739.73	III	30	7849.967	I
115	2210.89	I	8	3762.44	IV	45	5747.667	I	90	7918.386	I
110	2211.74	I	20 c	3791.41	III	45	5753.625	I	120	7932.349	I
120	2216.67	I	25	3796.11	III	45	5754.220	I	140	7944.001	I
120	2218.06	I	30	3806.54	III	45	5762.977	I	35	7970.306	I
10	2296.87	III	100 h	3853.66	II	70	5772.145	I	35	8035.619	I
10	2308.19	III	500 h	3856.02	II	70	5780.384	I	70	8093.241	I
100 h	2356.30	II	200 h	3862.60	II	90	5793.071	I	9 h	8102.86	III
30 h	2357.18	II	300	3905.523	I	100	5797.859	I	11 h	8103.45	III
50 h	2357.97	II	20	3924.47	III	150 h	5800.47	II	35	8230.642	I
300	2435.15	I	10	4088.85	IV	200	5806.74	II	9 h	8262.57	III
11	2449.48	III	70	4102.936	I	50	5846.13	II	40	8443.982	I
425	2506.90	I	9	4116.10	IV	300 h	5868.40	II	40	8501.547	I
375	2514.32	I	300 h	4128.07	II	40	5873.764	I	60	8502.221	I
500	2516.113	I	500 h	4130.89	II	10 h	5898.79	III	40	8536.165	I
7	2517.51	IV	100 h	4190.72	II	150	5915.22	II	120	8556.780	I
350	2519.202	I	50	4198.13	II	200	5948.545	I	50	8648.462	I
425	2524.108	I	9	4338.50	III	500	5957.56	II	40	8728.011	I
450	2528.509	I	30	4552.62	III	500	5978.93	II	75	8742.451	I
110	2532.381	I	25	4567.82	III	90	6125.021	I	100	8752.009	I
25	2541.82	III	20	4574.76	III	85	6131.574	I	35	8790.389	I
10	2546.09	III	100	4621.42	II	90	6131.850	I	100	9412.72	II
14	2559.21	III	150	4621.72	II	100	6142.487	I	100	9413.506	I
30	2563.679	I	9 h	4631.24	IV	100	6145.015	I	30	10371.269	I
85	2568.641	I	10 h	4654.32	IV	160	6155.134	I	120	10585.141	I
45	2577.151	I	9	4683.02	III	160	6237.320	I	120	10603.431	I
190	2631.282	I	16	4716.65	III	40	6238.287	I	120	10660.975	I
11	2640.79	III	50	4782.991	I	125	6243.813	I	30	10694.251	I
14	2655.51	III	35	4792.212	I	125	6244.468	I	30	10727.408	I
9	2817.11	III	80	4792.324	I	180	6254.188	I	60	10749.384	I
1000	2881.579	I	15	4813.33	III	45	6331.954	I	30	10784.550	I

Intensity	Wavelength/Å		Intensity	Wavelength/Å		Intensity	Wavelength/Å		Intensity	Wavelength/Å	
80	10786.856	I	600	1917.08	III	100	4212.82	I	15	410.372	IV
140	10827.091	I	700	1957.62	III	50	4311.07	I	10	411.334	IV
60	10843.854	I	100	1967.38	II	50 h	4476.04	I	13	412.242	IV
130	10869.541	I	600	1975.92	III	30 h	4615.69	I	11	1582.18	IV
30	10882.802	I	500	1977.03	III	80	4620.04	II	11 d	1583.98	IV
30	10885.336	I	600	2000.24	III	50	4620.46	II	12	1584.14	IV
80	10979.308	I	150	2015.96	II	60 h	4668.48	I	12 d	1587.05	IV
30	10982.061	I	150	2033.98	II	30 h	4677.60	I	11	1615.92	IV
80	11017.965	I	200	2061.17	I	100	4788.40	II	12	1618.57	IV
370	11984.19	I	100	2069.85	I	30 h	4847.82	I	11	1655.47	IV
220	11991.57	I	80 r	2113.82	I	100	4874.10	I	15 c	1701.97	IV
440	12031.51	I	60	2145.60	II	80	5027.35	II	20 d	1887.47	III
190	15888.39	I	600	2161.89	III	1000	5209.08	I	12	1960.76	IV
95	16060.03	I	50	2186.76	II	1000	5465.50	I	11	1965.08	IV
110	19722.50	I	60	2229.53	II	100	5471.55	I	12 d	2106.33	IV
	Silver		100 r	2246.43	II	100	5667.34	I	30	2230.33	III
	Ag Z = 47		500	2246.51	II	10 h	6268.50	I	16	2232.19	III
25	730.83	II	75 r	2248.74	II	320	7687.78	I	20 h	2246.70	III
30	752.80	II	75	2280.03	II	25	8005.4	II	300	2315.65	II
400	799.41	III	30 h	2309.56	I	500	8273.52	I	18	2386.99	III
15	1005.32	II	700	2310.04	III	25	8403.8	II	17	2394.03	III
10	1065.49	II	70 r	2317.05	II	30 h	8645.70	I	300	2420.99	II
12	1072.23	II	80 r	2320.29	II	10 h	8704.85	I	300	2424.73	II
250	1074.22	II	70 r	2324.68	II	12	8747.6	II	25	2459.31	III
150	1107.03	II	80 r	2331.40	II	15	9000.9	II	18	2468.85	III
150	1112.46	II	70	2357.92	II	10	12551.0	I	20	2474.73	III
60	1195.83	II	50 h	2375.02	I	60	16819.5	I	1000	2493.15	II
50	1223.33	II	75	2411.41	II	20	17416.7	I	25	2497.03	III
50	1240.80	II	90 r	2413.23	II	15	18307.9	I	17	2510.26	III
50	1246.87	II	100 r	2437.81	II	15	18382.3	I	20	2543.84	I
55	1256.81	II	80	2447.93	II		**Sodium**		10	2543.87	I
55	1257.55	II	80	2473.84	II		**Na Z = 11**		70	2593.87	I
50	1266.63	II	60	2506.63	II	7	142.232	IV	35	2593.92	I
70	1273.67	II	50 h	2575.63	I	8	146.064	IV	850	2611.81	II
65	1297.51	II	60	2660.49	II	9	150.298	IV	850	2661.00	II
85	1311.20	II	60	2721.77	I	8	150.687	IV	1000	2671.83	II
55	1313.81	II	75	2767.54	II	8	155.510	IV	200	2680.34	I
50	1314.61	II	100 h	2824.39	I	8	156.537	IV	100	2680.43	I
60	1323.84	II	30 h	3130.02	I	12	162.448	IV	1000	2841.72	II
60	1342.09	II	90	3180.70	II	10	163.190	IV	400	2852.81	I
50	1342.57	II	100	3267.35	II	12	168.411	IV	200	2853.01	I
70	1346.62	II	55000 r	3280.68	I	10	168.546	IV	2	2893.62	I
50	1353.54	II	28000 r	3382.89	I	5	183.95	III	1100	2904.92	II
150	1364.50	II	30	3469.16	I	10	190.445	IV	1100	2917.52	II
100	1396.00	II	70	3475.82	II	10	199.772	IV	1100	2919.05	II
100	1410.93	II	80	3495.28	II	8	202.49	III	1200	2919.85	II
90	1419.72	II	50	3542.61	I	8	202.76	III	1300	2920.95	II
95	1432.60	II	50 h	3624.68	I	8 p	203.06	III	1000	2923.49	II
100	1464.72	II	75	3682.46	II	8	203.28	III	1200	2951.24	II
50	1466.23	II	30	3682.50	I	8	203.33	III	1100	2952.40	II
50 r	1507.37	I	80	3683.34	II	15	229.87	III	1000	2977.13	II
100 r	1515.63	I	50 h	3709.20	I	50 c	250.52	III	1100	2979.66	II
50 r	1548.58	I	200	3810.94	I	30	251.37	III	1100	2980.63	II
100	1555.16	II	50	3811.78	II	25	266.90	III	1300	2984.19	II
100	1644.50	II	100 h	3840.74	I	70	267.65	III	1700	3124.42	II
60	1651.52	I	50 h	3907.41	I	50	267.87	III	2500	3135.48	II
50	1652.10	I	50	3909.31	II	50	268.63	III	1700	3137.86	II
700	1656.18	III	50 h	3914.40	I	20 p	272.08	III	2000	3149.28	II
120	1682.82	II	70	3920.10	I	20	272.45	III	2000	3163.74	II
500	1693.51	III	60	3949.43	II	10	319.644	IV	1000	3179.06	II
10	1708.11	I	100 h	3981.58	I	300	372.08	II	1700	3189.79	II
50	1709.27	I	70	3985.19	II	350	376.38	II	1600	3212.19	II
125	1736.44	II	100 h	4055.48	I	100	378.14	III	1500	3257.96	II
750	1751.03	III	80	4085.91	II	70	380.10	III	1700	3285.60	II
10 h	1766.14	I	100	4185.48	II	12	408.684	IV	1700	3301.35	II
75	1790.37	II	90 h	4210.96	I	10	409.614	IV	1200	3302.37	I

Intensity	Wavelength/Å		Intensity	Wavelength/Å		Intensity	Wavelength/Å		Intensity	Wavelength/Å	
600	3302.98	I	400	12679.17	I	340	4305.45	II		**Sulfur**	
1500	3304.96	II	60	14767.48	I	65000	4607.33	I		**S Z = 16**	
1000	3318.04	II	100	14779.73	I	9	4685.08	IV	5	437.4	V
50	3426.86	I	60	16373.85	I	3200	4722.28	I	5	438.2	V
1500	3533.05	II	100	16388.85	I	2200	4741.92	I	5	439.6	V
1200	3631.27	II	400	18465.25	I	1400	4784.32	I	20	519.3	IV
6	4238.99	I	50	22056.44	I	4800	4811.88	I	20	520.1	IV
10	4242.08	I	25	22083.67	I	3600	4832.08	I	40	520.8	IV
1	4249.41	I	60	23348.41	I	3000	4872.49	I	20	522.0	IV
2	4252.52	I	100	23379.13	I	2000	4876.32	I	20	522.5	IV
15	4273.64	I		**Strontium**		1000	4891.98	I	20	551.2	IV
20	4276.79	I		**Sr Z = 38**		8000	4962.26	I	40	652.5	IV
2	4287.84	I	15	298.12	IV	1300	4967.94	I	40	653.0	IV
3	4291.01	I	15	300.12	IV	800 h	5156.07	I	70	653.6	IV
30	4321.40	I	125	330.67	III	1400	5222.20	I	40	654.0	IV
40	4324.62	I	500	351.62	III	2000	5225.11	I	70	655.6	IV
3	4341.49	I	75	358.80	III	2000	5229.27	I	20	655.9	IV
5	4344.74	I	250	363.49	III	2800	5238.55	I	110	657.3	IV
40	4390.03	I	150	371.21	III	4800	5256.90	I	40	658.3	V
60	4393.34	I	20	378.53	IV	40	5257.71	III	70	659.8	V
5	4419.88	I	75	392.44	IV	40	5443.48	III	40	660.9	IV
8	4423.25	I	50	393.00	IV	1500	5450.84	I	160	661.4	IV
60	4494.18	I	50	396.22	IV	7000	5480.84	I	110	663.2	V
100	4497.66	I	1000	437.24	III	3500	5504.17	I	40	663.7	IV
10	4541.63	I	1875	491.79	III	2600	5521.83	I	40	664.8	IV
15	4545.19	I	1250	507.04	III	2000	5534.81	I	70	666.1	IV
120	4664.811	I	3750	514.38	III	2000	5540.05	I	20	678.1	V
200	4668.560	I	10	517.28	V	1000	6380.75	I	40	680.3	V
20	4747.941	I	2500	562.75	III	900 h	6386.50	I	110	680.9	V
30	4751.822	I	25	578.01	V	600 h	6388.24	I	40	681.6	V
200	4978.541	I	30	624.93	V	9000	6408.47	I	20	693.5	V
400	4982.813	I	25	642.23	V	5500	6504.00	I	70	729.5	III
40	5148.838	I	50	649.21	V	1000	6546.79	I	110	732.42	III
80	5153.402	I	25	660.94	V	1700	6550.26	I	70	735.2	III
280	5682.633	I	200	664.43	III	3000	6617.26	I	70	738.5	III
70	5688.193	I	35	686.23	V	1800	6791.05	I	110	744.9	IV
560	5688.205	I	100	710.35	IV	4800	6878.38	I	110	748.4	IV
80000	5889.950	I	12	747.82	V	1200	6892.59	I	110	750.2	IV
40000	5895.924	I	50	1025.23	III	5500	7070.10	I	110	753.8	IV
120	6154.225	I	35	1125.49	III	2500	7309.41	I	285	786.5	V
240	6160.747	I	50	1236.23	III	500	7621.50	I	70	789.0	III
130	6530.70	II	1400	2152.84	II	400 h	7673.06	I	70	796.7	III
130	6544.04	II	1400	2165.96	II	200 h	8422.80	I	70	800.5	IV
130	6545.75	II	100	2273.71	III	120	8505.69	II	70	804.0	IV
20	7373.23	I	100	2340.13	III	200	8688.91	II	70	809.7	IV
10	7373.49	I	50	2346.97	IV	100	9294.10	I	110	816.0	IV
50	7809.78	I	160	2428.10	I	400 h	9448.95	I	70	824.9	III
25	7810.24	I	100	2486.52	III	600	9596.00	I	70	836.3	III
4400	8183.256	I	40	2555.60	IV	300	9624.70	I	160	849.2	V
800	8194.790	I	40	2571.04	IV	100	9638.10	I	110	852.2	V
8800	8194.824	I	100	3002.61	III	100 h	9647.70	II	220	854.8	V
100	8649.92	I	200	3012.32	III	300	10036.66	II	110	857.9	V
60	8650.89	I	10	3019.29	IV	1000	10327.31	II	110	860.5	V
25	8942.96	I	100	3021.73	III	200	10914.88	II	40	906.9	II
40	9153.88	I	50	3061.43	III	700	11241.25	I	40	910.5	II
60	9465.94	I	50	3182.61	III	100	12014.76	II	40	912.7	II
80	9961.28	I	100	3235.39	III	60	12445.90	II	40	937.4	II
20	10566.00	I	400	3351.25	I	40	12495.00	I	40	937.7	II
60	10572.28	I	650	3380.71	II	75	12974.70	II	160	1062.7	IV
200	10746.44	I	50	3430.76	II	100	13123.80	II	160	1073.0	IV
80	10749.29	I	950	3464.46	II	50	17447.40	I	70	1073.5	IV
120	10834.87	I	600	3969.26	I	230	20261.40	I	285	1077.1	III
35	11190.19	I	1300	4030.38	I	120	20700.70	I	40	1102.3	II
50	11197.21	I	46000	4077.71	II	30	26023.60	I	70	1194.0	III
400	11381.45	I	32000	4215.52	II				70	1201.0	III
1000	11403.78	I	9	4298.57	IV				40	1234.1	II

Intensity	Wavelength/Å		Intensity	Wavelength/Å		Intensity	Wavelength/Å		Intensity	Wavelength/Å	
40	1250.5	II	110	3097.5	IV	110	8882.5	I	1500	2199.67	II
110	1253.8	II	110	3497.3	III	220	8884.2	I	90	2207.64	IV
110	1259.5	II	160	3632.0	III	160	9035.9	I	1400 d	2210.03	II
275	1270.782	I	110	3709.4	III	450	9212.9	I	1400	2239.48	II
250	1277.216	I	160	3717.8	III	450	9228.1	I	1200	2250.76	II
280	1295.653	I	160	3838.3	III	450	9237.5	I	840	2261.42	II
275	1302.337	I	285	3867.6	I	285	9413.5	I	990	2262.30	II
235	1302.863	I	285	3902.0	I	285	9421.9	I	990	2272.59	II
235	1303.110	I	160	3928.6	III	285	9437.1	I	790	2285.25	II
245	1303.430	I	360	3933.3	II	650	9649.9	I	600	2286.59	II
260	1305.883	I	450	4120.8	I	450	9672.3	I	990	2289.16	II
265	1310.194	I	280	4142.3	II	450	9680.8	I	440	2302.24	II
355	1316.542	I	360	4145.1	II	450	9693.7	I	440	2302.93	II
290	1316.618	I	450	4153.1	II	285	9697.3	I	440	2312.60	II
375	1323.515	I	450	4162.7	II	285	9739.7	I	420	2315.46	II
355	1326.643	I	360	4253.6	III	285	9932.3	I	690	2331.98	II
775	1381.552	I	450	4694.1	I	285	9949.8	I	550	2332.19	II
710	1385.510	I	285	4695.4	I	285	9958.9	I	250	2357.30	I
960	1388.435	I	160	4696.2	I	285	10455.5	I	260	2361.09	I
640	1389.154	I	280	4716.2	II	285	10459.5	I	600	2364.24	II
775	1392.588	I	450	4815.5	II		Tantalum		320	2371.58	I
1000	1396.112	I	360	4924.1	II		Ta Z = 73		1400	2387.06	II
300	1409.337	I	450	4925.3	II	60	493.07	V	2400	2400.63	II
510	1425.030	I	285	4993.5	I	1000	890.87	V	320	2416.89	II
425	1433.280	I	360	5428.6	II	500	947.30	V	360	2427.64	I
300	1436.968	I	650	5432.8	II	67	999.34	IV	360	2429.71	II
300	1448.229	I	1000	5453.8	II	79	1116.10	IV	480	2432.70	II
425	1472.972	I	1000	5473.6	II	78	1136.17	IV	380	2470.90	II
550	1473.995	I	1000	5509.7	II	85	1175.51	IV	600	2474.62	I
300	1474.380	I	280	5564.9	II	80	1189.28	IV	500	2484.95	I
355	1481.665	I	1000	5606.1	II	80	1192.67	IV	600	2488.70	II
485	1483.039	I	450	5640.0	II	85	1213.09	IV	500	2490.46	I
300	1483.233	I	450	5640.3	II	500	1213.42	V	600	2504.45	I
330	1485.622	I	280	5647.0	II	85	1215.53	IV	600	2507.45	I
390	1487.150	I	650	5659.9	II	90	1223.73	IV	1200 d	2526.35	I
20	1624.0	IV	450	5664.7	II	88	1238.12	IV	600	2532.12	II
20	1629.2	IV	160	5706.1	I	95	1240.06	IV	1200	2559.43	I
680	1666.688	I	450	5819.2	II	87	1258.34	IV	460	2562.10	I
640	1687.530	I	450	6052.7	I	94	1264.91	IV	600	2577.37	II
710	1807.311	I	280	6286.4	II	98	1272.42	IV	600	2603.49	II
680	1820.343	I	450	6287.1	II	94	1275.48	IV	1400	2608.63	I
640	1826.245	I	450	6305.5	II	86	1275.94	IV	1200	2635.58	II
710	1900.286	I	450	6312.7	II	92	1308.51	IV	860	2636.90	I
550	1914.698	I	280	6384.9	II	87	1315.58	IV	2400	2647.47	I
20	2387.0	IV	280	6397.3	II	92	1332.38	IV	2600	2653.27	I
40	2398.9	IV	280	6398.0	II	86	1343.30	IV	1900	2656.61	I
110	2460.5	III	360	6413.7	II	92	1365.88	IV	1500	2661.34	I
110	2489.6	III	160	6743.6	I	5000	1392.56	V	770	2675.90	II
160	2496.2	III	285	6748.8	I	91	1398.78	IV	1500	2685.17	II
160	2499.1	III	450	6757.2	I	93	1413.40	IV	470	2694.52	II
220	2508.2	III	450	7579.0	I	91	1454.32	IV	1000	2698.30	I
70	2636.9	III	450	7629.8	I	92	1464.41	IV	1200	2710.13	I
220	2665.4	III	285	7686.1	I	93	1469.82	IV	2600	2714.67	I
110	2691.8	III	450	7696.7	I	90	1495.25	IV	470	2727.44	II
110	2702.8	III	1000	7924.0	I	95	1514.19	IV	1200	2748.78	I
220	2718.9	III	160	7928.8	I	85	1607.70	IV	860	2749.83	I
110	2721.4	III	285	7930.3	I	7000	1709.10	V	410	2752.49	II
220	2726.8	III	450	7931.7	I	85	1712.16	IV	1000	2758.31	I
220	2731.1	III	450	7967.4	I	85	1716.13	IV	430	2761.68	II
110	2741.0	III	450	7967.4	II	85	2055.75	IV	770	2775.88	I
285	2756.9	III	450	8314.7	I	1100	2140.13	II	680	2796.34	I
110	2775.2	III	450	8314.7	II	1500	2146.87	II	680	2797.76	II
160	2785.5	III	450	8585.6	I	1200	2182.71	II	510	2806.58	I
110	2863.5	III	285	8680.5	I	1100	2193.88	II	640	2844.25	I
160	2904.3	III	450	8694.7	I	1500	2196.03	II	560	2848.52	I
160	2986.0	III	360	8874.5	I	90	2199.58	IV	1500	2850.49	I

Intensity	Wavelength/Å		Intensity	Wavelength/Å		Intensity	Wavelength/Å		Intensity	Wavelength/Å	
1900	2850.98	I	150	5645.91	I	20000	4262.27	I	200	4686.91	II
360	2861.98	I	130	5664.90	I	30000	4297.06	I	100	4696.38	II
470	2871.42	I	130	5776.77	I	20000	4853.59	I	100	4706.53	II
380	2880.02	I	90	5780.71	I		**Tellurium**		100	4766.05	II
770	2891.84	I	130	5811.10	I		**Te Z = 52**		100	4784.87	II
560	2902.05	I	240	5877.36	I	8	802.28	II	100	4827.14	II
310	2915.49	I	130	5882.30	I	8	1059.51	II	150	4831.28	II
410	2925.19	I	90	5901.91	I	8	1077.66	II	150	4842.90	II
310	2932.70	I	90	5918.95	I	10	1161.42	II	130	4865.12	II
1700	2933.55	I	130	5939.76	I	10	1174.34	II	200	4866.24	II
470	2940.06	I	240	5944.02	I	12	1175.79	II	8	5083.0	I
1200	2940.22	I	190 c	5997.23	I	9	1208.54	II	50	5449.84	II
510	2951.92	I	100	6020.72	I	9	1220.98	II	50	5487.95	II
340	2953.56	I	250	6045.39	I	9	1253.62	II	150	5576.35	II
1500	2963.32	I	100	6047.25	I	9	1270.52	II	150	5649.26	II
770	2965.13	II	100	6101.58	I	10	1324.92	II	100	5666.20	II
770	2965.54	I	65	6144.56	I	9	1363.24	II	200	5708.12	II
340	2969.47	I	130	6154.50	I	8	1366.73	II	150	5755.85	II
430	2975.56	I	150	6256.68	I	10	1374.80	II	100	5974.68	II
1800	3012.54	II	150	6268.70	I	10	1608.41	II	50	6367.13	II
290 d	3027.48	I	150	6309.58	I	10	1613.15	II	10 h	6790.0	I
530	3049.56	I	75	6325.08	I	5	1655.4	I	20 h	6837.6	I
530	3069.24	I	65	6341.17	I	5	1688.5	I	20 h	6854.7	I
360	3077.24	I	75	6356.16	I	6	1700.0	I	15 h	7191.1	I
560	3103.25	I	65	6360.84	I	5	1708.0	I	20 h	7263.5	I
380	3124.97	I	90	6389.45	I	10	1822.4	I	12	7460.98	II
380	3130.58	I	65	6428.60	I	26000	2002.02	I	15	7468.75	II
270	3132.64	I	250	6430.79	I	6500	2081.16	I	15	7921.69	II
320	3170.29	I	200	6450.36	I	18000	2142.81	I	15	7943.14	II
270	3173.59	I	380	6485.37	I	3200	2147.25	I	10	7950.34	II
600	3180.95	I	65	6505.52	I	500	2259.02	I	30 h	8061.4	I
300	3223.83	I	100	6514.39	I	1200	2383.26	I	10	8122.44	II
1100	3311.16	I	100	6516.10	I	1500	2385.78	I	20	8186.44	II
680	3318.84	I	100	6574.84	I	50	2438.69	II	15	8273.53	II
330 d	3330.99	II	110	6611.95	I	120	2530.72	I	15	8672.95	II
640	3371.54	I	75	6621.30	I	100	2649.66	II	10	8733.81	II
360	3385.05	I	100	6673.73	I	80	2661.10	II	205	8758.18	I
450	3406.94	I	180	6675.53	I	110	2677.13	I	81	9004.37	I
490	3480.52	I	75 c	6740.73	I	100	2858.29	II	5660	9722.74	I
380	3497.85	I	75	6771.74	I	150	2895.41	II	532	9868.92	I
490	3511.04	I	160 c	6813.25	I	70	2967.29	II	689	9956.30	I
750	3607.41	I	210	6866.23	I	70	3047.00	II	325	9977.13	I
980	3626.62	I	180	6875.27	I	100	3175.14	I	5950	10051.41	I
500	3642.06	I	150	6902.10	I	60	3256.80	II	4097	10091.01	I
210	3918.51	I	140	6927.38	I	60	3329.22	II	381	10118.08	I
210	3970.10	I	140	6928.54	I	150	3406.79	II	397	10300.56	I
210	3996.17	I	65	6951.26	I	50	3442.25	II	745	10493.57	I
410	4061.40	I	180	6966.13	I	50	3521.11	II	1880	10918.34	I
310	4067.91	I	110 d	6995.22	I	50	3552.19	II	10200	11089.56	I
300	4205.88	I	75	7006.96	I	100	3611.78	I	508	11163.74	I
360 c	4510.98	I	150	7148.63	I	50	3617.57	II	6620	11487.23	I
340	4574.31	I	110	7172.90	I	50	4006.52	II	1580	13247.75	I
260	4619.51	I	140	7301.74	I	70	4127.32	II	1050	14513.51	I
450	4681.88	I	160	7346.41	I	100	4169.77	II	1480	15452.45	I
200	5037.37	I	140 c	7352.86	I	80	4225.73	II	2430	15546.23	I
100	5067.87	I	100	7356.96	I	100	4261.11	II	3760	16403.90	I
110	5115.84	I	90 cw	7369.09	I	60	4273.43	II	1960	17303.54	I
100	5141.62	I	160	7407.89	I	80	4285.85	II	2780	18291.59	I
100	5143.69	I	100	7882.37	I	150	4364.00	II	1020	21043.73	I
330	5156.56	I	75	8026.50	I	75	4385.10	II	464	21602.50	I
110	5212.74	I	75	8281.62	I	170	4478.63	II	74	22555.29	I
110 d	5218.45	I		**Technetium**		80	4537.07	II	38	26539.17	I
140	5341.05	I		**Tc Z = 43**		100	4557.78	II		**Terbium**	
200	5402.51	I	10000 c	3636.07	I	70	4630.62	II		**Tb Z = 65**	
130	5419.19	I	20000 c	4031.63	I	100	4641.12	II	1000	1259.40	IV
90	5518.91	I	15000	4095.67	I	180	4654.37	II	1000	1327.67	IV

Intensity	Wavelength/Å		Intensity	Wavelength/Å		Intensity	Wavelength/Å		Intensity	Wavelength/Å	
1000	1373.86	IV	1700	3765.14	I	65	4632.07	I	30	7737.63	I
5000	1595.39	IV	2100	3776.49	II	65 h	4636.59	I	30	7855.79	II
2000	1633.19	IV	600	3783.53	I	85	4641.00	II	27	7927.90	II
2000	2027.79	IV	410	3789.92	I	210	4641.98	II	30	8025.42	II
1000	2089.98	IV	760 d	3806.85	II	260 cw	4645.31	II	30	8085.06	II
1000	2332.54	IV	1500	3830.26	I	80	4647.23	II	65	8194.82	II
110	2584.61	II	540	3833.42	I	80	4662.79	I	95	8212.57	I
110	2608.57	II	920 d	3842.50	II	80	4676.90	I	40	8450.06	II
130	2628.69	II	3700	3848.73	II	70 c	4681.87	I	30 h	8511.80	I
140	2669.29	II	3500 w	3874.17	II	80	4688.63	II	45	8583.45	II
190	2704.07	II	480	3888.22	I	80	4693.11	II	30	8603.40	I
270	2769.53	II	490	3894.64	I	200	4702.41	II	65	8765.74	II
320	2897.44	II	2400	3899.20	II	110	4707.94	II		**Thallium**	
250	2956.21	II	1600	3901.33	I	80	4739.93	I		**Tl Z = 81**	
230	3010.59	II	480	3908.06	I	70	4747.80	I	10	570.49	IV
230	3016.18	II	650	3915.43	I	410 cw	4752.53	II	5 r	670.87	II
460	3053.55	II	760	3925.45	II	180	4786.78	I	15 r	696.30	II
460	3070.05	II	810 d	3939.52	II	100	4813.77	I	5 r	709.23	II
670	3078.86	II	2200 d	3976.84	II	80	4875.57	II	10 r	817.18	II
480	3082.36	II	1800	3981.87	II	80	4881.15	II	5 r	836.34	II
480	3089.58	II	970	4002.59	II	95	4915.90	I	8 r	1018.85	II
480	3102.96	II	1900	4005.47	II	65	4931.79	I	30	1028.69	IV
440	3139.64	II	760	4012.75	II	85	4993.82	II	20	1034.73	IV
480	3187.26	II	870	4032.28	I	110	5078.25	I	20	1036.61	IV
480	3199.56	II	2100	4033.03	II	75	5089.12	II	10 r	1049.73	II
1100	3218.93	II	430	4054.12	I	85	5186.13	I	8 r	1050.30	II
1200	3219.98	II	410	4060.37	I	120	5228.12	I	5 r	1074.97	II
480	3252.32	II	1300	4061.58	I	75	5248.71	I	30	1079.68	IV
760	3280.31	II	650	4105.37	I	75 w	5262.11	II	10 r	1130.17	II
760	3281.40	II	1100	4144.41	II	75	5281.05	I	15 r	1162.55	II
1000	3285.04	II	350	4158.53	I	65	5304.72	I	10 r	1167.43	II
1500	3293.07	II	390	4196.74	I	110	5319.23	I	10 r	1183.41	II
3800	3324.40	II	650	4203.74	I	65 w	5337.90	I	12 r	1194.84	II
760	3349.42	II	600	4206.49	I	75	5354.88	I	5 r	1246.00	II
760	3364.93	II	480	4215.09	I	160	5369.72	I	10	1266.33	III
810	3454.06	II	480	4232.82	I	75	5375.98	I	15 r	1307.50	II
810 d	3472.79	II	650	4266.34	I	75	5424.10	II	8 r	1310.20	II
810	3500.84	II	760 cw	4278.52	II	50	5459.81	I	25 r	1321.71	II
5700	3509.17	II	450	4310.42	I	55	5509.61	I	8 r	1330.40	II
1300	3523.66	II	2200	4318.83	I	55	5514.54	I	10 r	1373.52	II
1100	3540.24	II	600	4322.23	I	50	5524.12	I	10	1477.14	III
810	3543.89	II	600	4325.83	II	65	5747.58	I	8 r	1489.65	I
3200	3561.74	II	3000	4326.43	I	85 c	5795.64	I	10 r	1499.30	II
810	3567.35	II	600	4332.12	I	75	5803.13	II	10 r	1507.82	II
4200	3568.52	II	870	4336.43	I	75	5815.36	I	15 r	1561.58	II
1600	3568.98	II	600	4337.64	I	65	5851.07	I	10 r	1568.57	II
1100	3579.20	II	1700	4338.41	I	65	5870.62	I	7 r	1593.26	II
710	3585.03	II	700	4340.62	I	65	5920.78	I	5 h	1616.	
810	3596.38	II	870	4356.81	I	65 c	5967.34	II	5	1685.40	I
1600	3600.44	II	330	4382.45	I	75	6331.68	II	10 r	1792.76	II
810	3625.54	II	300	4388.23	I	35	6518.68	I	12 r	1814.85	II
2300	3650.40	II	260	4390.91	I	35 cw	6581.82	I	25 r	1908.64	II
810	3654.88	II	350	4423.10	I	35	6677.94	II	100 r	2007.56	I
2000	3658.88	II	240	4436.12	I	90	6702.61	I	100 r	2210.71	I
3800	3676.35	II	240	4448.04	I	40 cw	6794.58	II	30	2298.04	II
810	3682.26	II	430	4493.07	I	130	6896.37	II	140	2315.98	I
450	3693.58	I	75	4514.31	II	55	6899.95	I	900 h	2379.69	I
450	3700.12	I	110	4549.07	I	45 h	6901.98	I	20	2530.86	II
4700	3702.86	II	110	4550.45	I	40	7204.28	I	700	2580.14	I
2400	3703.92	II	110	4556.96	I	65	7257.73	I	420	2709.23	I
1000 d	3711.76	II	110	4563.69	II	40	7348.88	II	4400 d	2767.87	I
650	3745.04	I	210	4578.69	II	45	7496.12	I	10	2849.80	II
870	3747.17	II	65	4584.84	II	45	7582.03	II	2800	2918.32	I
870	3747.34	II	65	4591.56	II	27 h	7590.24	I	20	3091.56	II
1100	3755.24	II	75 d	4626.32	II	45	7596.44	I	15	3185.51	II
650	3759.35	I	95	4626.94	II	65	7627.81	I	15	3186.56	II

Intensity	Wavelength/Å		Intensity	Wavelength/Å		Intensity	Wavelength/Å		Intensity	Wavelength/Å	
15	3187.74	II	510	3119.526	II	310	3747.539	I	30	7045.795	II
1200	3229.75	I	510	3122.963	II	650	3752.569	II	30	7084.171	I
15	3291.01	II	480	3125.507	II	180	3770.056	I	30	7168.896	I
15	3369.15	II	100	3136.216	I	590	3803.075	I	40	7191.132	II
9	3456.34	III	420	3139.306	II	450	3828.384	I	35	7208.006	I
20000	3519.24	I	420	3142.835	II	840	3839.746	I	50	7525.508	II
5000	3529.43	I	420	3175.726	II	450	3863.405	II	30	7647.380	I
8	3540.08	II	1100	3180.193	II	210	3875.374	I	30	8330.451	I
9	3560.68	II	770	3188.233	II	340	3895.419	I	40	8967.641	I
12000 w	3775.72	I	560	3221.292	II	590	3929.669	II	20	9833.42	I
10	3832.30	II	560	3229.009	II	200	3932.911	I	20	10726.93	I
10	3887.15	II	480	3235.84	II	390	3967.392	I	20	10942.24	II
7	4109.85	III	590	3238.116	II	200	3972.155	I	30	11230.259	I
6	4269.81	III	910	3256.274	II	150	3980.089	I	20	11984.67	II
20	4274.98	II	180	3257.366	I	530	3994.549	II	20	17208.22	II
40	4306.80	II	910	3262.668	II	220	4008.210	I	15	18811.88	I
20	4737.05	II	620	3287.789	II	220	4009.056	I	10	22264.35	II
15	4981.35	II	910	3291.739	II	280	4012.495	I		**Thulium**	
25	5078.54	II	620	3292.520	II	4200	4019.129	II		**Tm Z = 69**	
25	5152.14	II	240	3301.650	I	250	4030.842	I	5000	2185.94	III
18000	5350.46	I	480	3304.238	I	250	4036.047	I	360	2284.79	II
15 d	5384.85	II	510	3321.450	II	250	4063.407	I	20000	2296.21	III
10	5410.97	II	840	3325.120	II	700	4086.520	II	5000	2305.03	III
25	5949.48	II	250	3330.476	I	700	4094.747	II	20000	2311.16	III
10	6179.98	II	620	3334.604	II	150	4100.341	I	5000	2312.72	III
10	6378.32	II	620	3337.870	II	840	4108.421	II	5000	2326.19	III
16 h	6549.84	I	310	3348.768	I	240	4112.754	I	6000	2328.50	III
10	6966.5	II	980	3351.228	II	280	4115.758	I	6000	2329.29	III
10	7815.80	I	620	3358.602	II	1100	4116.713	II	3000	2331.80	III
20	8373.6	I	250	3374.974	I	200	4127.411	I	3000	2357.05	III
10	8474.27	I	1300	3392.035	II	200	4134.067	I	4000	2406.63	III
10	8664.1	II	200	3396.727	I	450	4149.986	II	450	2409.02	II
20	9130.	II	250	3398.544	I	620	4178.060	II	450	2426.17	II
20	9130.5	I	200	3405.558	I	620	4208.890	II	770	2480.13	II
40	9509.4	I	250	3413.012	I	110	4253.538	I	30000	2489.44	III
20	9930.4	I	390	3421.210	I	110	4260.333	I	2000	2504.71	II
30	10011.9	I	270	3423.989	I	480	4277.313	II	1300	2509.08	II
40	10488.80	I	980	3433.998	II	700	4282.042	II	3000	2519.78	III
1000	11512.82	I	770	3435.976	II	130	4337.277	I	130	2527.02	I
150	12736.4	I	1300	3469.920	II	1300	4381.860	II	10000	2552.46	III
700	13013.2	I	170	3471.218	I	1100	4391.110	II	360	2552.76	I
	Thorium		200	3486.552	I	110	4498.940	I	540	2561.65	II
	Th Z = 90		670	3539.587	II	280	4510.527	II	430	2588.27	II
150	1707.37	IV	180	3544.018	I	90	4723.438	I	170 h	2596.49	I
200	1959.02	IV	170	3549.595	I	50	4840.843	I	810	2607.06	II
200	2002.34	IV	200	3555.013	I	280	4863.163	II	730	2624.33	II
200	2413.50	III	530	3559.451	II	260	5017.255	II	5000	2682.32	III
200	2427.94	III	200	3576.557	I	110	5067.974	I	2000	2707.03	III
200	2431.68	III	270	3592.780	I	120	5148.211	II	3000	2719.47	III
200	2441.24	III	270	3598.120	I	95	5216.596	II	540	2721.19	II
500	2565.593	II	980	3609.445	II	110	5231.160	I	3000	2724.44	III
480	2692.415	II	200	3612.427	I	95	5247.654	II	4000	2727.56	III
520	2747.156	II	480	3615.133	II	60	5343.581	I	680	2794.60	II
410	2752.166	II	270	3635.943	I	60	5587.026	II	730	2797.27	II
800	2832.315	II	210	3642.248	I	95	5707.103	II	2000	2806.77	III
1200	2837.295	II	170	3649.735	I	70	5760.551	I	580	2827.92	II
100	2848.084	I	220	3663.202	I	85	5989.044	II	200	2854.17	I
550	2870.406	II	280	3669.968	I	60	6169.822	I	1600	2869.23	II
100	2936.086	I	700	3675.567	II	50	6182.622	I	1000	2947.72	III
100	2943.729	I	150	3682.486	I	50	6274.116	I	490	2973.22	I
420	3049.092	II	170	3692.566	I	50	6274.117	II	1000	2998.28	III
450	3067.729	II	180	3698.105	I	50	6355.911	II	1500	3015.30	II
670	3078.828	II	340	3706.767	I	60	6457.283	I	360	3081.12	I
480	3080.217	II	590	3719.435	I	50	6462.614	I	7400	3131.26	II
510	3108.296	II	770	3721.825	II	50 h	6531.342	I	2300	3133.89	II
100	3116.263	I	1300	3741.183	II	55	6989.656	I	1900	3151.04	II

Intensity	Wavelength/Å		Intensity	Wavelength/Å		Intensity	Wavelength/Å		Intensity	Wavelength/Å	
1500	3157.34	II	1500	3958.10	II	40	7856.08	I	70	2058.31	I
450	3172.65	I	1800	3996.52	II	55	7927.51	I	80	2068.58	I
2300	3172.83	II	220	4024.23	I	110	7930.84	I	100	2072.89	I
1200	3236.81	II	380	4044.47	I	95	8017.90	I	100	2073.08	I
1600	3240.23	II	10000	4094.19	I	27	8472.01	II	200	2096.39	I
2300	3241.54	II	9500	4105.84	I		Tin		100	2100.93	I
320	3246.96	I	1100	4138.33	I		Sn Z = 50		100 r	2113.93	I
1900	3258.05	II	8800	4187.62	I	7	169.47	II	50	2121.26	I
1600	3266.64	II	6000	4203.73	I	150	361.01	V	40 r	2148.73	I
1200	3267.40	II	380	4222.67	I	100	753.01	III	20 r	2151.43	I
1100	3276.81	II	3000	4242.15	II	200	910.92	III	30	2151.54	II
1200	3283.40	II	270	4271.71	I	500	956.25	IV	80	2171.32	I
1200	3285.61	II	150	4298.36	I	7	985.13	II	150 r	2194.49	I
2300	3291.00	II	2700	4359.93	I	500	1019.72	IV	300 r	2199.34	I
2000	3302.46	II	1400	4386.43	I	1000	1044.49	IV	400 r	2209.65	I
1200	3309.80	II	200	4394.42	I	1000	1073.41	IV	80 r	2231.72	I
230	3349.99	I	140	4396.50	I	200	1089.35	V	400 r	2246.05	I
4000	3362.61	II	120	4454.03	I	8	1108.19	II	60	2251.17	I
1700	3397.50	II	540	4481.26	II	1000	1119.34	IV	400 r	2268.91	I
850	3410.05	I	150	4519.60	I	1000	1139.29	III	200 r	2286.68	I
340	3412.59	I	260	4522.57	II	1000	1158.33	III	600 r	2317.23	I
340	3416.59	I	110	4548.60	I	200	1160.74	V	300 r	2334.80	I
6400	3425.08	II	270	4599.02	I	10	1161.43	I	1000 r	2354.84	I
340	3429.33	I	300	4615.94	II	1000	1184.25	III	22	2368.33	II
4900	3441.50	II	80	4626.33	II	2000	1210.52	III	100	2408.15	I
4900	3453.66	II	95	4626.56	II	9	1219.07	II	800 r	2421.70	I
8500	3462.20	II	110	4634.26	II	13	1223.70	II	1000 r	2429.49	I
210	3467.51	I	120	4655.09	I	11	1243.00	I	15	2448.98	II
340	3476.69	I	160	4681.92	I	2000	1251.38	V	300	2483.39	I
340	3480.98	I	120	4691.11	I	1000	1259.92	III	13	2483.48	II
420	3487.38	I	110	4724.26	I	20	1290.86	II	10	2486.99	II
340	3499.95	I	680	4733.34	I	200	1294.36	V	200	2495.70	I
250	3517.60	I	70	4759.90	I	1000	1305.97	III	400	2546.55	I
1700	3535.52	II	80	4831.20	II	1000	1314.55	IV	500 r	2571.58	I
420	3537.91	I	140	4957.18	I	20	1316.59	II	200	2594.42	I
210	3555.82	I	160	5009.77	II	1000	1327.34	III	200 r	2661.24	I
340	3560.92	I	160	5034.22	II	1000	1347.65	III	700 r	2706.51	I
420	3563.88	I	150	5060.90	I	1000	1386.74	III	150	2779.81	I
1300	3566.47	II	95	5113.97	I	25	1400.52	II	1400 r	2839.99	I
420	3567.36	I	80	5213.38	I	1000	1437.52	IV	1000 r	2863.32	I
280	3586.07	I	650	5307.12	I	20	1475.15	II	700 r	3009.14	I
2100	3608.77	II	80	5346.49	II	9	1489.22	II	850 r	3034.12	I
1000	3629.09	III	270	5631.41	I	1000	1570.36	III	12	3047.50	II
380	3638.41	I	520	5675.84	I	10 r	1737.21	I	550 r	3175.05	I
1100	3668.09	I	40	5684.76	II	15 r	1751.46	I	550 r	3262.34	I
4800	3700.26	II	35	5709.97	II	20 r	1764.98	I	50	3283.21	II
3800	3701.36	II	190	5764.29	I	30 r	1790.75	I	110	3330.62	I
7700	3717.91	I	35	5838.76	II	80 r	1804.60	I	60	3351.97	II
2400	3734.12	II	240	5895.63	I	15	1811.34	II	10	3472.46	II
5000	3744.06	I	140	5971.26	I	500	1811.71	III	11	3575.45	II
1700	3751.81	I	200	6460.26	I	40 r	1815.74	I	280 r	3801.02	I
6000	3761.33	II	95	6604.96	I	120 r	1823.00	I	10	5332.36	II
4800	3761.91	II	110	6779.77	I	9	1831.89	II	20	5561.95	II
7100	3795.75	II	120	6844.26	I	50 r	1848.75	I	25	5588.92	II
770	3798.54	I	80	6845.76	I	200 r	1860.32	I	500	5631.71	I
600	3807.72	I	10	6937.37	I	80	1886.05	I	15	5799.18	II
290	3826.39	I	10	7017.90	I	100	1891.40	I	50	5925.44	I
1300	3838.20	II	12	7034.34	I	12	1899.91	II	100	5970.30	I
290	3840.87	I	10	7106.14	I	50	1909.30	I	150	6037.70	I
8900	3848.02	II	17	7272.62	I	80	1925.31	I	250	6069.00	I
6800	3883.13	I	14	7310.51	I	500	1941.86	III	100	6073.46	I
1800	3883.44	II	14	7432.18	I	150	1952.15	I	400	6149.71	I
5400	3887.35	I	75	7481.08	I	50 h	1977.6	I	200	6154.60	I
440	3896.62	I	75	7490.20	I	80	1984.20	I	150	6171.50	I
3500	3916.48	I	140	7558.33	I	50	2040.66	I	100	6310.78	I
1500	3949.27	I	80	7731.53	I	50	2054.03	I	70	6453.50	II

Intensity	Wavelength/Å		Intensity	Wavelength/Å		Intensity	Wavelength/Å		Intensity	Wavelength/Å	
25	6844.05	II	24	2563.44	III	600	3461.50	II	6000	4533.24	I
20	7191.40	II	23	2565.42	III	600	3477.18	II	240	4533.97	II
10	7387.79	II	22	2567.56	III	480	3491.05	II	3600	4534.78	I
13	7741.80	II	270	2599.92	I	890	3504.89	II	2400	4535.58	I
100	7754.97	I	340	2605.15	I	600	3510.84	II	1200	4535.92	I
100 h	8030.5	I	510	2611.28	I	17	3576.44	IV	1200	4536.05	I
200	8114.09	I	300	2619.94	I	600	3610.16	I	720	4544.69	I
80	8357.04	I	640	2641.10	I	4800	3635.46	I	950	4548.77	I
300	8422.72	I	800	2644.26	I	6600	3642.68	I	240	4549.63	II
400	8552.60	I	950	2646.64	I	7200	3653.50	I	15	4549.84	III
50 h	8681.7	I	250	2742.32	I	600	3671.67	I	950	4552.46	I
50 h	9410.86	I	250	2802.50	I	3100	3685.20	II	720	4555.49	I
80 h	9415.37	I	190	2841.94	II	600	3689.91	I	240	4571.98	II
150	9616.40	I	180	2877.44	II	2900	3729.82	I	15 d	4572.20	III
50	9741.1	I	280	2884.11	II	3300	3741.06	I	950	4617.27	I
100 h	9742.8	I	450	2912.08	I	330	3741.64	II	480	4623.09	I
300 h	9805.38	I	340	2928.34	I	5200	3752.86	I	720	4656.47	I
500	9850.52	I	1100	2942.00	I	3300	3759.30	II	840	4667.59	I
54	10894.00	I	1300	2948.26	I	2900	3761.32	II	950	4681.92	I
70	11191.85	I	1600	2956.13	I	840	3786.04	I	470	4840.87	I
56	11277.66	I	22	2984.75	III	500	3882.89	I	400	4885.08	I
200	11454.59	I	1300 d	3066.22	II	530	3900.54	II	380	4899.91	I
200	11616.26	I	1100	3072.97	II	2600	3904.78	I	5800	4981.73	I
258	11739.78	I	1600	3075.22	II	500	3913.46	II	4600	4991.07	I
96	11825.18	I	2300	3078.64	II	500	3914.34	I	4000	4999.51	I
106	11835.82	I	3600	3088.02	II	15	3915.47	III	3600	5007.21	I
254	11932.99	I	720	3119.72	I	1100	3924.53	I	3200 d	5014.19	I
48	12009.50	I	500	3161.20	II	890	3929.88	I	840	5020.03	I
111	12313.24	I	780	3161.77	II	1100	3947.78	I	840	5022.87	I
42	12530.87	I	1000	3162.57	II	4500	3948.67	I	1200	5035.91	I
42	12536.5	I	1600	3168.52	II	4500	3956.34	I	840	5036.47	I
89	12888.5	I	2400	3186.45	I	5200	3958.21	I	740	5038.40	I
187	12981.7	I	1000	3190.87	II	950	3962.85	I	1200	5039.95	I
187	13018.5	I	3100	3191.99	I	950	3964.27	I	1400	5064.66	I
68	13081.5	I	3800	3199.92	I	4800	3981.76	I	1100	5173.75	I
378	13460.2	I	780	3202.54	II	570	3982.48	I	1300	5192.98	I
144	13608.2	I	1100	3217.06	II	5700	3989.76	I	1400	5210.39	I
40	20861.7	I	1300	3222.84	II	7800	3998.64	I	17	5278.12	III
4	24738.2	I	6600	3234.52	II	950	4008.93	I	20	5398.93	IV
Titanium			5200	3236.57	II	1200	4024.57	I	340	5512.53	I
Ti Z = 22			4100	3239.04	II	840	4078.47	I	270	5514.35	I
17	252.96	V	2600	3241.99	II	890	4286.01	I	320	5514.54	I
15	498.26	V	1200	3248.60	II	840	4287.40	I	250	5644.14	I
14	502.08	V	1200	3252.91	II	950	4289.07	I	130	5675.44	I
13	526.57	V	1200	3254.25	II	840	4290.94	I	95	5689.47	I
18	779.07	IV	1200	3261.60	II	840	4295.76	I	95	5715.13	I
20	1298.66	III	840	3314.42	I	2000	4298.66	I	85	5739.51	I
20	1298.97	III	2900	3322.94	II	200	4300.05	II	400	5866.46	I
23	1455.19	III	2100	3329.46	II	2900	4300.56	I	230	5899.32	I
20	1467.34	IV	1800	3335.20	II	4100	4301.09	I	120	5918.55	I
11	1717.40	V	1100	3340.34	II	6000	4305.92	I	150	5922.12	I
10	1841.49	V	5700	3341.88	I	1200	4314.80	I	120	5941.76	I
20	2067.56	IV	4300	3349.04	II	330	4395.04	II	300	5953.17	I
18	2103.16	IV	12000	3349.41	II	890	4427.10	I	200	5965.84	I
180	2273.28	I	4100	3354.64	I	230	4443.80	II	270	5978.56	I
190	2279.96	I	7200	3361.21	II	840	4449.15	I	340	5999.04	I
190	2305.67	I	1100	3370.44	I	550	4450.90	I	110	6064.63	I
22	2413.99	III	4300	3371.45	I	840	4453.32	I	120	6085.23	I
25	2516.05	III	5700	3372.80	II	950	4455.33	I	120	6091.17	I
360	2525.60	III	2900 d	3377.48	I	1100	4457.43	I	120	6126.22	I
24	2527.84	III	1400	3380.28	II	240	4468.50	II	17	6246.65	IV
210	2529.85	I	5700	3383.76	II	530	4481.26	I	380	6258.10	I
190	2531.25	II	1400	3385.95	I	780	4512.74	I	380	6258.70	I
190	2534.62	II	1400	3387.84	II	1000	4518.03	I	300	6261.10	I
130	2535.87	II	1100	3394.58	II	1000	4522.80	I	55	6546.28	I
23	2540.06	III	890	3444.31	II	780	4527.31	I	65	6554.23	I

Intensity	Wavelength/Å		Intensity	Wavelength/Å		Intensity	Wavelength/Å		Intensity	Wavelength/Å	
75	6556.07	I	1400	2466.85	I	810	2848.02	I	1000	4102.70	I
18	6621.58	III	480	2472.51	I	1500	2896.44	I	540	4137.46	I
18	6667.99	III	1200	2474.15	I	690	2935.00	I	450	4171.17	I
80	6743.12	I	870	2480.13	I	2400	2944.40	I	220	4207.05	I
20	7072.64	III	1500	2481.44	I	2400	2946.99	I	250	4219.37	I
18	7084.57	III	480 d	2482.10	I	730 d	2979.71	I	540	4244.36	I
260	7209.44	I	580	2484.74	I	360	3013.79	I	1400	4269.38	I
130	7244.86	I	390	2487.50	I	520	3016.47	I	4100	4294.61	I
130	7251.72	I	390	2489.23	II	770	3017.44	I	2200	4302.11	I
120	7344.72	I	630	2495.26	I	210	3024.93	I	200	4378.48	I
90	7357.74	I	680	2504.70	I	310 d	3026.67	I	180	4384.85	I
60	7364.11	I	75	2510.47	II	440 d	3041.73	I	200	4408.28	I
60	7978.88	I	310	2520.46	I	270	3043.80	I	640	4484.19	I
55	8024.84	I	780	2521.32	I	440	3046.44	I	170	4588.73	I
75	8364.24	I	270	2522.04	II	810	3049.69	I	640	4659.87	I
100	8377.85	I	780	2523.41	I	180	3073.28	I	640	4680.51	I
100	8382.54	I	430	2527.76	I	180 d	3084.83	I	790	4843.81	I
75	8396.87	I	780	2533.64	I	370	3093.50	I	380	4886.90	I
120	8412.36	I	1200	2547.14	I	240	3107.23	I	220	4982.59	I
170	8426.52	I	780	2550.38	I	240	3108.02	I	820	5053.28	I
490	8434.94	I	2700	2551.35	I	230	3117.57	I	770	5224.66	I
240	8435.70	I	730	2561.97	I	260	3120.18	I	220	5514.68	I
20	8466.87	III	870	2580.49	I	290	3163.42	I	65	5648.37	I
90	8675.39	I	390	2584.39	I	320	3176.60	I	55	5735.09	I
	Tungsten		390	2589.17	II	190	3181.82	I	45	5804.85	I
	W Z = 74		370	2601.96	I	390	3191.57	I	40	5902.64	I
5800	2001.71	II	680	2606.39	I	390	3198.84	I	55	5947.57	I
13000	2008.07	II	370	2608.32	I	520	3207.25	I	55	5965.86	I
5100	2009.98	II	970	2613.08	I	1000	3215.56	I	55	6012.78	I
4100	2010.23	II	480	2613.82	I	190	3232.49	I	40	6021.52	I
4100	2014.23	II	400	2620.25	I	210	3254.36	I	45	6292.02	I
7300	2026.08	II	400	2622.21	I	210	3259.66	I	35	6404.21	I
15000	2029.98	II	400	2625.22	I	210 d	3266.62	I	40	6445.12	I
5300	2049.63	II	400	2632.48	I	730	3300.82	I	17	6611.62	I
9700	2079.11	II	400	2632.70	I	440	3311.38	I	13	6678.42	I
6100	2094.75	II	810	2633.13	I	440	3326.20	I	15	6693.08	I
2100	2118.87	II	400 d	2638.62	I	440	3331.69	I	13	6984.27	I
2400	2121.59	II	650	2646.18	I	390	3373.75	I	15	7140.52	I
1500	2166.32	II	400	2646.73	I	230	3429.59	I	9	7162.64	I
1300	2204.48	II	1600	2656.54	I	240	3443.00	I	11	7200.16	I
460	2249.80	I	810	2662.84	I	400	3495.24	I	10	7278.24	I
510	2277.58	I	810	2671.47	I	650	3545.22	I	15	7285.81	I
530 d	2294.49	I	650	2677.28	I	240	3570.65	I	15	7296.55	I
340	2309.02	I	2100	2681.42	I	1900	3617.52	I	10	7509.00	I
440	2313.17	I	650	2695.67	I	650	3682.08	I	17	7569.92	I
460	2321.63	I	650	2699.59	I	400	3683.30	I	17	7614.15	I
390 d	2326.56	I	400	2700.01	I	570	3688.06	I	13	7688.97	I
320	2354.61	I	400	2706.58	I	810	3707.92	I	11	7784.15	I
580	2360.44	I	400	2708.59	I	510	3757.92	I	22	8017.19	I
850	2363.07	I	400 d	2708.80	I	680	3760.13	I	22	8055.64	I
510	2374.47	I	400	2715.50	I	1000	3768.45	I	13	8123.82	I
670	2384.82	I	2100	2718.91	I	340	3773.71	I	10	8338.08	I
730	2397.09	II	2600	2724.35	I	1000	3780.77	I	27	8585.11	I
560	2397.73	I	400	2725.03	I	290	3809.22	I	10	8594.42	I
560	2397.98	I	650	2748.84	I	190	3810.38	I	13	8865.53	I
1700 d	2405.58	I	400	2762.34	I	260	3810.79	I		**Uranium**	
610	2415.68	I	400	2764.27	II	1400	3817.48	I		**U Z = 92**	
870	2424.21	I	400	2769.74	I	1100	3835.06	I	440	2565.41	II
1800	2435.96	I	810	2770.88	I	730	3846.22	I	610	2635.53	II
580	2444.06	I	810	2774.00	I	1800	3867.99	I	830	2793.94	II
780	2451.48	II	810	2774.48	I	730	3881.41	I	870	2802.56	II
870	2452.00	I	810	2792.70	I	8600	4008.75	I	630	2807.05	II
630	2454.98	I	400	2799.93	I	540	4015.22	I	630	2817.96	II
780	2455.51	I	810	2818.06	I	910	4045.59	I	870	2821.12	II
780	2456.53	I	1600	2831.38	I	730	4069.95	I	680	2828.90	II
1100	2459.30	I	810	2833.63	I	5000	4074.36	I	920	2832.06	II

Intensity	Wavelength/Å		Intensity	Wavelength/Å		Intensity	Wavelength/Å		Intensity	Wavelength/Å	
970	2865.68	II	1000	3881.45	II	100	1680.20	V	490	3592.02	II
1200	2889.62	II	2200	3890.36	II	1000	1694.78	III	560	3592.53	I
780	2906.80	II	2000	3932.02	II	1000	1760.07	III	100	3679.86	III
780	2908.28	II	1200	3943.82	I	1000	1788.26	III	1300	3688.07	I
580	2931.41	II	1200	3985.79	II	1000	1794.60	III	1000	3690.28	I
530 p	2940.37	II	1000	4042.75	I	1000	1812.19	III	1500	3692.22	I
1300	2941.92	II	1600	4050.04	II	300	1861.56	IV	1000	3695.86	I
830	2943.90	II	880	4062.54	II	500	1939.06	IV	3800	3703.58	I
580	2956.06	II	2200	4090.13	II	400	1951.43	IV	1800	3704.70	I
580	2967.94	II	810	4116.10	II	500	1997.72	IV	320	3715.47	II
580	2971.06	II	880	4153.97	I	2100	2092.44	I	250	3727.34	II
530	2984.61	II	1400	4171.59	II	500	2268.30	IV	280	3732.76	II
630	3022.21	II	1000	4241.67	II	1000	2292.86	III	520	3790.32	I
630	3031.99	II	600	4472.33	II	2500	2330.42	III	1100	3794.96	I
580	3050.20	II	620	4543.63	II	2500	2371.06	III	570	3799.91	I
630	3057.91	II	170	4689.07	II	1000	2382.46	III	570	3803.47	I
630	3062.54	II	150	4756.81	I	240	2507.78	I	1000	3813.49	I
580	3072.78	II	110	5008.21	II	410	2526.22	I	1300	3818.24	I
580	3093.01	II	170	5027.38	I	210	2527.90	II	1700	3828.56	I
580	3102.39	II	80	5160.32	II	80 h	2570.72	IV	2600	3840.75	I
970	3111.62	II	70	5280.38	I	230	2574.02	I	1200	3855.37	I
530	3119.35	II	80	5475.70	II	250	2593.05	III	3000	3855.84	I
680	3124.95	II	70	5480.26	II	250	2595.10	III	1300	3864.86	I
530	3139.61	II	70	5481.20	II	80 h	2645.54	IV	1500	3875.08	I
680	3149.24	II	160	5492.95	I	180	2661.42	I	700	3890.18	I
530	3153.11	II	70	5780.59	I	1100	2687.96	II	2400	3902.25	I
730	3229.50	II	70	5798.53	II	680	2700.94	II	700	3909.89	I
680	3232.16	II	230	5915.39	I	530	2706.17	II	540	3990.57	I
730	3291.33	II	100	5976.32	I	640	2715.69	II	430	3998.73	I
1100	3305.89	II	90	6077.29	I	180	2731.35	I	170	4005.71	II
730	3390.38	I	55	6372.46	I	240	2864.36	I	1100	4090.58	I
580	3424.56	II	90	6395.42	I	900	2891.64	II	1800	4092.69	I
580	3435.49	I	110	6449.16	I	900	2892.66	II	890	4095.49	I
630	3466.30	I	90	6826.92	I	1400	2893.32	II	2800	4099.80	I
680	3482.49	II	45	7533.93	I	900	2906.46	II	590	4102.16	I
1600	3489.37	I	50	7881.94	I	2400	2908.82	II	2800	4105.17	I
530	3496.41	II	35	8445.39	I	710	2923.62	I	2300	4109.79	I
630	3500.08	I	75	8607.95	I	2400	2924.02	II	8900	4111.78	I
780	3507.34	I	30	8757.76	I	1700	2924.64	II	4300	4115.18	I
1600	3514.61	I	100	10554.93	I	900	2941.37	II	1800	4116.47	I
630	3533.57	II	75	11167.84	I	1100	2944.57	II	2000	4123.57	I
530	3540.47	II	100	11384.13	I	410	2962.77	I	3100	4128.07	I
1200	3550.82	II	100	11859.42	I	600	2968.38	II	3100	4132.02	I
680	3555.32	I	100	11908.83	I	1200	3056.33	I	2300	4134.49	I
1200	3561.80	I	100	12250.46	I	1400	3060.46	I	20	4200.32	V
2300	3566.59	I	100	13185.16	I	2400	3066.38	I	360	4232.46	I
530	3569.08	I	75	13306.23	I	3800	3093.11	II	560	4268.64	I
630	3578.72	II	100	13961.58	I	3000	3102.30	II	460	4271.55	I
3200	3584.88	I	75	18634.43	I	2600	3110.71	II	460	4276.96	I
840	3638.20	I	75	21910.22	I	2000	3118.38	II	430	4284.06	I
2800	3670.07	II		**Vanadium**		1500	3125.28	II	460	4330.02	I
1100	3701.52	II		**V Z = 23**		3200	3183.41	I	510	4332.82	I
600	3738.04	II	20	225.46	V	5300	3183.98	I	760	4341.01	I
680	3746.42	II	20	251.66	V	3800	3185.40	I	1000	4352.87	I
950	3748.68	II	20	286.84	V	410	3187.71	II	12000	4379.24	I
600	3751.17	I	35	483.01	V	530	3188.51	II	7000	4384.72	I
1900	3782.84	II	50	633.94	III	750	3190.68	II	4800	4389.97	I
570	3793.10	II	200	677.34	IV	1100	3267.70	II	3600	4395.23	I
1900	3811.99	I	500	684.37	IV	900	3271.12	II	1400	4400.58	I
750	3826.51	II	400	737.85	IV	750	3276.12	II	2300	4406.64	I
2000	3831.46	II	100	864.27	III	80 h	3514.25	IV	2800	4407.64	I
1200	3839.63	I	500	1006.46	III	560	3517.30	II	3600	4408.20	I
2400	3854.64	II	500	1149.94	III	560	3533.68	I	4600	4408.51	I
4900	3859.57	II	100	1426.65	IV	560	3545.20	II	640	4416.47	I
1900	3865.92	II	1000	1643.03	III	560	3556.80	II	640	4421.57	I
1500	3871.03	I	1000	1650.14	III	560	3589.76	II	640	4437.84	I

Intensity	Wavelength/Å		Intensity	Wavelength/Å		Intensity	Wavelength/Å		Intensity	Wavelength/Å	
830	4441.68	I	29 c	8027.39	I	30	2827.45	III	10	3669.91	I
640	4444.21	I	120 w	8116.80	I	40	2847.65	III	50	3676.67	III
610	4452.01	I	70 c	8161.07	I	30	2862.40	III	40	3685.90	I
1000	4459.76	I	60 c	8919.85	I	200	2864.73	II	40	3693.49	I
2000	4460.29	I		**Xenon**		80 w	2871.10	III	40	3776.3	III
610	4462.36	I		**Xe Z = 54**		60 w	2871.24	III	300	3781.02	III
510	4577.17	I	8	657.8	III	30	2871.7	III	100	3841.5	III
640	4580.40	I	8	660.1	III	150 h	2895.22	II	200	3877.8	III
830	4586.36	I	9	673.8	III	30	2896.62	III	60	3880.5	III
1300	4594.11	I	9	674.0	III	50	2906.6	III	100 l	3907.91	II
230	4619.77	I	9	676.6	III	40	2911.89	III	500	3922.55	III
100	4635.18	I	10	694.0	III	80 w	2912.36	III	300	3950.59	III
130	4646.40	I	20	698.5	III	40	2940.2	III	100	4037.59	II
160	4670.49	I	12	705.1	III	60	2945.2	III	200	4050.07	III
130	4776.36	I	10	721.2	III	40	2947.5	III	200 l	4057.46	II
110	4786.51	I	15	731.0	III	40	2948.1	III	60	4060.4	III
130	4796.92	I	10	733.3	III	80 w	2970.47	III	100 h	4098.89	II
130	4807.53	I	350	740.41	II	400	2979.32	II	100	4109.1	III
130	4827.45	I	15	742.6	III	40	2992.87	III	100	4145.7	III
150	4831.64	I	10	756.0	III	30	3004.25	III	200 l	4158.04	II
120	4832.43	I	10	761.5	III	100 h	3017.43	II	1000 h	4180.10	II
320	4851.48	I	10	769.1	III	100	3023.81	III	500 h	4193.15	II
480	4864.74	I	25	779.1	III	40	3083.5	III	300 h	4208.48	II
620	4875.48	I	15	792.9	III	50	3091.1	III	100 h	4209.47	II
740	4881.56	I	12	796.1	III	30	3106.46	III	300 h	4213.72	II
110	5128.53	I	15	802.0	III	300	3128.87	II	100	4215.60	II
110	5138.42	I	350	803.07	II	100 w	3138.3	III	300 h	4223.00	II
110	5192.99	I	25	823.2	III	80 c	3150.82	III	400 h	4238.25	II
110	5194.83	I	30	824.9	III	40	3185.2	III	500 h	4245.38	II
110	5234.07	I	25	853.0	III	100	3242.86	III	100 l	4251.57	II
110	5240.87	I	600	880.80	II	80	3268.98	III	30	4285.9	III
100	5401.93	I	350	885.54	II	30	3287.82	III	500 h	4296.40	II
140	5415.26	I	15	889.3	III	80 w	3301.55	III	500 h	4310.51	II
140	5584.50	I	20	894.0	III	40	3331.6	III	1000 l	4330.52	II
100	5592.42	I	20	896.0	III	30	3358.0	III	200 h	4369.20	II
200	5624.60	I	600	925.87	II	200 h	3366.72	II	100 l	4373.78	II
400	5627.64	I	250	935.40	II	80	3384.12	III	500 h	4393.20	II
110	5657.44	I	10	965.5	III	2	3400.07	I	500 l	4395.77	II
110	5668.36	I	800	972.77	II	2	3418.37	I	200 l	4406.88	II
310	5670.85	I	700	976.68	II	2	3420.00	I	150 l	4416.07	II
1200	5698.52	I	35	1003.4	III	3	3442.66	I	50	4434.2	II
920	5703.56	I	35	1017.7	III	60	3444.2	III	500 h	4448.13	II
570	5706.98	I	500	1032.44	II	70	3454.2	III	100 w	4462.1	III
850	5727.03	I	700	1037.68	II	100 w	3458.7	III	1000 h	4462.19	II
230	5731.25	I	1100	1041.31	II	100 h	3461.26	II	500 l	4480.86	II
230	5737.06	I	10	1047.8	III	40	3468.22	III	100 l	4521.86	II
450	6039.73	I	1000	1048.27	II	4	3469.81	I	100 w	4569.1	III
480	6081.44	I	1200	1051.92	II	4	3472.36	I	100 w	4570.1	III
1300	6090.22	I	12	1066.4	III	5	3506.74	I	100 w	4641.4	III
600	6119.52	I	2000	1074.48	II	80	3522.83	III	30	4673.7	III
450	6199.19	I	600	1083.86	II	50	3542.3	III	60	4683.57	III
450	6216.37	I	1200	1100.43	II	10	3549.86	I	30	4723.60	III
430	6230.74	I	30	1130.3	III	50	3552.1	III	600	4734.152	I
710	6243.10	I	600	1158.47	II	10	3554.04	I	100 w	4757.3	III
280	6251.82	I	250	1169.63	II	40	3561.4	III	150	4792.619	I
130	6268.82	I	800 p	1183.05	II	100	3579.7	III	500	4807.02	I
170	6274.65	I	250	1192.04	I	80	3583.6	III	400	4829.71	I
200	6285.16	I	25	1232.1	III	100 w	3595.4	III	300	4843.29	I
200	6292.83	I	600	1244.76	II	100	3606.06	III	40	4869.5	III
170	6296.49	I	250	1250.20	I	40	3607.0	III	500	4916.51	I
110	6531.43	I	1000	1295.59	I	15	3610.32	I	500	4923.152	I
65 c	6753.00	I	600	1469.61	I	8	3613.06	I	200 l	4971.71	II
50 c	6766.49	I	80	2668.98	III	100 w	3615.9	III	400	4972.71	II
40	6784.98	I	100	2717.33	III	40	3623.1	III	300	4988.77	II
40	7338.92	I	30	2814.45	III	600	3624.08	III	100 l	4991.17	II
35	7356.54	I	40	2815.91	III	6	3633.06	I	200	5028.280	I

Intensity	Wavelength/Å		Intensity	Wavelength/Å		Intensity	Wavelength/Å		Intensity	Wavelength/Å	
200	5044.92	II	100	6198.26	I	100	8101.98	I	175	25145.84	I
1000	5080.62	II	60	6205.97	III	150 h	8151.80	II	2000	26269.08	I
300	5122.42	II	100	6220.02	II	100	8171.02	I	2500	26510.86	I
100	5125.70	II	25	6221.7	III	700	8206.34	I	250	28381.54	I
100	5178.82	II	60	6238.2	III	10000	8231.635	I	750	28582.25	I
300	5188.04	II	60	6259.05	III	500	8266.52	I	300	29384.41	I
400	5191.37	II	500	6270.82	I	7000	8280.116	I	150	29448.06	I
100	5192.10	II	400	6277.54	II	2000	8346.82	I	100	29649.58	I
60	5239.0	III	100	6284.41	II	100	8347.24	II	100	29813.62	I
500	5260.44	II	100	6286.01	I	2000	8409.19	I	600	30253.14	I
500	5261.95	II	250	6300.86	II	50 h	8515.19	II	1500	30475.46	I
2000	5292.22	II	500	6318.06	I	200	8576.01	I	100	30504.12	I
300	5309.27	II	400	6343.96	II	50 h	8604.23	II	500	30794.18	I
1000	5313.87	II	600	6356.35	II	250	8648.54	I	6000	31069.23	I
2000	5339.33	II	200	6375.28	II	100	8692.20	I	125	31336.01	I
200	5363.20	II	100	6397.99	II	200	8696.86	I	550	31607.91	I
30	5367.1	III	300	6469.70	I	50 h	8716.19	II	100	32293.08	I
200	5368.07	II	150	6472.84	I	300	8739.39	I	1800	32739.26	I
500	5372.39	II	120	6487.76	I	100	8758.20	I	3500	33666.69	I
100	5392.80	I	100	6498.72	I	5000	8819.41	I	150	34014.67	I
50	5401.0	III	200 h	6504.18	I	300	8862.32	I	450	34335.27	I
3000	5419.15	II	300	6512.83	II	200	8908.73	I	170	34744.00	I
800	5438.96	II	200	6528.65	II	200	8930.83	I	5000	35070.25	I
300	5445.45	II	100	6533.16	I	1000	8952.25	I	110	35246.92	I
200	5450.45	II	1000	6595.01	II	100	8981.05	I	250	36209.21	I
400	5460.39	II	100	6595.56	I	200	8987.57	I	150	36231.74	I
1000	5472.61	II	400	6597.25	II	400	9045.45	I	450	36508.36	I
100 1	5494.86	II	100	6598.84	II	500	9162.65	I	850	36788.83	I
40	5524.4	III	150	6668.92	I	100	9167.52	I	140	38685.98	I
200	5525.53	II	300	6694.32	II	100	9374.76	I	175	38737.82	I
600	5531.07	II	200	6728.01	I	200	9513.38	I	270	38939.60	I
100	5566.62	I	150	6788.71	II	50 h	9591.35	II	120	39955.14	I
300	5616.67	II	100	6790.37	II	150	9685.32	I	**Ytterbium**		
300	5659.38	II	1000	6805.74	II	50 1	9698.68	II	**Yb Z = 70**		
600	5667.56	II	200	6827.32	I	100	9718.16	I	1000	1050.24	IV
150	5670.91	II	100	6872.11	I	2000	9799.70	I	1000	1054.46	IV
100	5695.75	I	300	6882.16	I	3000	9923.19	I	5000	1134.43	IV
200	5699.61	II	80	6910.22	II	100	10838.37	I	900	1316.04	IV
200	5716.10	II	100	6925.53	I	90	11742.01	I	800	1326.36	IV
500	5726.91	II	800 h	6942.11	II	375	12235.24	I	900	1350.26	IV
500	5751.03	II	100	6976.18	I	100	12257.76	I	80	1561.42	III
300	5758.65	II	2000	6990.88	II	300	12590.20	I	80 h	1765.21	IV
300	5776.39	II	150	7082.15	II	2500	12623.391	I	800	1791.06	IV
100	5815.96	II	500	7119.60	I	250	13544.15	I	100	1863.32	III
300	5823.89	I	50 s	7147.50	II	2000	13657.055	I	800	1873.91	III
150	5824.80	I	200	7149.03	II	1250	14142.444	I	500	1898.25	III
100	5875.02	I	500	7164.83	II	800	14240.96	I	500	1998.82	III
300	5893.29	II	100	7284.34	II	375	14364.99	I	900	2116.65	IV
100	5894.99	I	200	7301.80	I	140	14660.81	I	2500	2116.67	II
200	5905.13	II	200	7339.30	II	3000	14732.806	I	800	2123.32	IV
100	5934.17	I	100	7386.00	I	100	15099.72	I	3000	2126.74	II
500	5945.53	II	150	7393.79	I	2500	15418.394	I	800	2139.99	IV
300	5971.13	II	300	7548.45	II	150	15557.13	I	20000	2144.77	IV
2000	5976.46	II	200	7584.68	I	250	15979.54	I	15000	2154.18	II
200	6008.92	II	80	7618.57	II	100	16039.90	I	370	2161.60	II
1000	6036.20	II	500	7642.02	I	1000	16053.28	I	850	2185.71	II
2000	6051.15	II	100	7643.91	I	125	16554.49	I	640	2224.46	II
600	6093.50	II	200	7670.66	II	1500	16728.15	I	300	2240.11	III
1500	6097.59	II	60	7787.04	II	1500	17325.77	I	300	2305.32	III
400	6101.43	II	100	7802.65	I	350	18788.13	I	140	2320.81	II
100	6115.08	II	100	7881.32	I	150	20187.19	I	170	2390.74	II
100	6146.45	II	300	7887.40	I	3000	20262.242	I	460	2464.50	I
150	6178.30	I	500	7967.34	I	250	21470.09	I	140	2512.06	II
120	6179.66	I	100	8029.67	I	1250	23193.33	I	270	2538.67	II
300	6182.42	I	200	8057.26	I	110	23279.54	I	2000	2567.61	III
500	6194.07	II	150	8061.34	I	1800	24824.71	I	1000	2579.57	III

Intensity	Wavelength/Å		Intensity	Wavelength/Å		Intensity	Wavelength/Å		Intensity	Wavelength/Å	
800	2599.14	III	2000	3384.01	III	40	4837.46	I	300	403.45	V
600	2621.11	III	140	3387.50	I	40 h	4894.60	I	300	420.74	V
1000	2642.56	III	50	3412.45	I	27	4912.36	I	600	425.03	IV
1000	2651.74	III	140	3418.39	I	710	4935.50	I	300	473.10	IV
700	2652.25	III	360	3426.04	I	140	4966.90	I	4000	584.98	V
990	2653.75	II	240	3431.11	I	30	5067.80	I	2000	630.97	V
200	2665.04	II	85	3452.40	I	70	5069.14	I	5000	805.20	III
2000	2666.13	III	500	3454.08	II	220	5074.34	I	7000	809.92	III
2000	2666.99	III	190 d	3458.29	II	50	5076.74	I	15000	989.21	III
390	2671.96	I	360	3460.27	I	60	5196.08	I	25000	996.37	III
390	2672.66	II	2400	3464.37	I	85	5211.60	I	5000	1314.51	III
170	2718.35	II	500	3476.30	II	100	5244.11	I	4000	1334.04	III
230	2748.66	II	500	3478.84	II	150 h	5277.04	I	4000	2068.98	III
1300	2750.48	II	50	3517.00	I	170	5335.15	II	10000	2127.98	III
170	2776.28	II	230	3520.29	II	30 h	5351.29	I	16000	2191.16	III
600	2795.60	III	35	3559.03	I	150	5352.95	II	350	2243.06	II
1000	2803.43	III	200	3560.33	II	30	5363.66	I	10000	2284.34	III
600	2816.92	III	170	3560.70	II	40	5449.27	II	10000	2327.31	III
1000	2818.72	III	360	3585.47	II	60	5481.92	I	50	2354.20	I
140	2821.15	II	200	3619.80	II	40	5505.49	I	50000	2367.23	III
190	2830.99	II	240	3637.76	II	17	5524.54	I	40000	2414.64	III
230 h	2847.18	II	70	3648.15	I	85 h	5539.05	I	560	2422.20	II
360	2851.13	II	90	3655.73	I	2400	5556.47	I	60	2694.21	I
430	2859.80	II	240	3669.69	II	60	5651.98	II	95	2723.00	I
140	2861.21	II	140	3675.08	II	220	5719.99	I	70	2742.53	I
200	2867.06	II	32000	3694.19	II	27	5771.66	II	140	2760.10	I
45	2873.49	I	70	3700.58	I	35	5833.99	II	90000	2817.04	III
200	2888.04	II	400	3711.91	III	35	5837.14	II	45	2822.56	I
3600	2891.38	II	180	3734.69	I	27	5854.51	I	70	2854.43	II
600	2898.30	III	550	3770.10	I	17	5989.33	I	95	2886.48	I
1000	2906.31	III	80	3774.32	I	40	5991.51	II	160	2919.05	I
170	2914.21	II	60 h	3791.74	I	60	6152.57	II	99000	2946.01	III
140	2915.28	II	170	3839.91	I	60	6274.78	II	390	2948.40	I
280	2919.35	II	340	3872.85	I	200	6328.52	III	350	2964.96	I
35	2934.36	I	340	3900.85	I	35 h	6400.35	I	480	2974.59	I
140	2945.91	II	140	3911.27	I	35 h	6417.91	I	750	2984.26	I
2000	2970.56	II	500	3931.23	III	340	6489.06	I	140	2996.94	I
200	2983.99	II	32000	3987.99	I	180	6667.82	I	130	3021.73	I
170	2994.80	II	930	3990.88	I	25	6727.61	II	190	3045.37	I
800	2998.00	III	50	4007.36	I	690	6799.60	I	95	3095.88	II
310	3005.77	II	2000	4028.14	III	9 h	7244.41	I	110	3173.06	II
160	3017.56	II	70	4052.28	I	8 h	7305.22	I	220	3179.41	II
160	3026.67	II	440	4089.68	I	10 h	7313.05	I	70	3191.31	I
2000	3029.49	III	470	4149.07	I	16 h	7350.04	I	2300	3195.62	II
920	3031.11	II	120	4174.56	I	25	7448.28	I	2200	3200.27	II
3000	3092.50	III	340	4180.81	II	30 h	7527.46	I	2200	3203.32	II
28	3100.74	I	300	4213.64	III	750	7699.48	I	3900	3216.69	II
170	3107.90	II	150 d	4218.56	II	100	7971.46	III	6200	3242.28	II
190	3117.81	II	120	4231.97	I	70 h	8922.56	II	4700	3327.89	II
4000	3126.01	III	70	4277.74	I	200	10110.60	III	85	3388.59	I
1000	3138.58	III	120	4305.97	I	100	10830.36	III	85	3412.47	I
230	3140.94	II	60 h	4393.69	I	**Yttrium**			170	3485.73	I
28	3162.29	I	60 h	4430.21	I	**Y Z = 39**			1700	3496.09	II
800	3191.35	III	440	4439.19	I	150	264.64	IV	3900	3549.01	II
390	3192.88	II	85 h	4482.42	I	150	273.03	IV	130	3551.80	I
240	3201.16	II	100	4517.58	III	900	333.09	V	540	3552.69	I
2000	3228.58	III	85 h	4563.95	I	500	333.80	V	170	3558.76	I
35	3239.58	I	640	4576.21	I	400	335.14	V	190	3571.43	I
18000	3289.37	II	200	4582.36	I	500	336.62	V	260	3576.05	I
130	3305.25	I	70	4589.21	I	500	339.02	V	3300	3584.52	II
140	3305.73	II	140	4590.83	I	500	344.59	V	300	3587.75	I
80	3319.41	I	40	4684.27	I	900	355.86	IV	100	3589.69	I
2000	3325.51	III	190	4726.08	II	300	370.42	IV	2800	3592.92	I
240	3337.17	II	170 h	4781.87	I	300	372.05	V	10000	3600.73	II
280 d	3342.93	II	170	4786.61	II	400	379.96	V	6200	3601.92	II
240	3375.48	II	35	4816.43	I	500	386.82	IV	7800	3611.05	II

Intensity	Wavelength/Å	Intensity	Wavelength/Å	Intensity	Wavelength/Å	Intensity	Wavelength/Å
4300	3620.94 I	170	4786.89 I	24 h	6950.31 I	200	2670.53 I
1900	3628.71 II	180	4799.30 II	24	6979.88 I	300	2684.16 I
7800	3633.12 II	140	4819.64 I	29	7052.94 I	300	2712.49 I
3000	3664.61 II	120	4822.13 I	35	7191.66 I	200	2756.45 I
170	3692.53 I	770	4839.87 I	35	7264.17 II	300	2770.86 I
13000	3710.30 II	550	4845.68 I	50	7346.46 I	300	2770.98 I
1200	3747.55 II	410	4852.69 I	29	7450.30 II	400	2800.87 I
10000	3774.33 II	120	4854.25 I	9000	7558.71 III	100	2801.06 I
1400	3776.56 II	890	4854.87 II	35	7563.13 I	200	3035.78 I
7400	3788.70 II	330	4859.84 I	29	7855.52 I	200	3072.06 I
1300	3818.35 II	1900	4883.69 II	110	7881.90 II	300	3196.31 II
4000	3832.88 II	95	4893.44 I	10000	7991.43 III	500 r	3282.33 I
80	3876.82 I	1100	4900.12 II	24	8344.43 I	800	3302.58 I
480	3878.28 II	100	4906.11 I	10000	8796.21 III	700 r	3302.94 I
4400	3950.36 II	150	4921.87 I	95	8800.62 I	800	3345.02 I
3600	3982.60 II	120	4974.30 I	19 h	8835.85 II	500	3345.57 I
940	4039.83 I	100	5006.97 I		**Zinc**	50	3883.34 I
2400	4047.64 I	75	5070.21 I		**Zn Z = 30**	300	4680.14 I
9400	4077.38 I	75	5072.19 I	200	425.90 IV	400	4722.15 I
2000	4083.71 I	1100	5087.42 II	200	428.54 IV	400	4810.53 I
9900	4102.38 I	180	5135.20 I	200	430.59 IV	800	4911.62 II
8900	4128.31 I	960	5200.41 II	1000	677.63 III	500	4924.03 II
7500	4142.85 I	1500	5205.72 II	750	677.96 III	200	5181.98 I
100 h	4157.63 I	10000	5238.10 III	200	713.90 III	500	5894.33 I
2400	4167.52 I	180	5240.81 I	60	1193.23 II	500	6021.18 II
2000	4174.14 I	75	5380.62 I	50	1239.12 IV	500	6102.49 II
8000	4177.54 II	220	5402.78 II	50	1249.69 IV	500	6214.61 II
160	4217.80 I	90	5424.37 I	500	1265.74 IV	1000 h	6362.34 I
280 h	4220.63 I	190	5438.24 I	500	1306.66 IV	300	7588.5 II
600	4235.73 II	710	5466.46 I	200	1456.72 III	300	7732.5 I
2200	4235.94 I	100	5468.47 I	200	1459.98 IV	100	11054.25 I
300	4251.20 I	240	5497.41 II	300	1499.42 III	100	13053.63 I
360 h	4302.30 I	300	5503.45 I	300	1500.42 III	100	13150.59 I
2800	4309.63 II	250	5509.90 II	300	1505.92 III	100	14038.70 I
110	4330.78 I	120	5521.63 I	300	1515.85 III	20	16483.45 I
440 h	4348.79 I	740	5527.54 I	300	1552.30 III	20	16491.98 I
120	4357.73 I	120	5544.50 I	90	1572.99 II	20	16505.23 I
800	4358.73 II	180	5577.42 I	200	1629.19 III	10	24375.02 I
120	4366.03 I	620	5581.87 I	200	1639.33 III		**Zirconium**
12000	4374.94 II	120	5606.33 I	200	1673.05 III		**Zr Z = 40**
150 h	4375.61 I	560	5630.13 I	80 d	1735.61 II	500	304.01 V
100	4387.74 I	120	5644.69 I	100	1767.69 III	60	480.66 IV
1800	4398.02 I	120	5648.47 I	100	1797.64 II	60	497.23 IV
890	4422.59 II	740	5662.94 II	100 d	1811.05 II	60	500.22 IV
100	4443.66 I	90	5675.27 I	100 d	1833.57 II	600	628.66 IV
130	4446.63 I	160	5706.73 I	100	1864.12 II	500	633.56 IV
170	4475.72 I	90	5743.85 I	100	1866.08 II	50	690.39 III
180	4476.96 I	75	5765.64 I	100	1872.13 II	2000	740.61 V
160	4477.45 I	100	5781.69 II	100 d	1918.96 II	10000	800.00 V
110	4487.28 I	120	6009.19 I	100 d	1929.67 II	10000	806.89 V
300	4487.47 I	120	6023.41 I	100	1969.40 II	10000	812.05 V
500	4505.95 I	120	6135.04 I	100	1982.11 II	3000	841.40 V
890	4527.25 I	150	6138.43 I	100	1986.99 II	300	863.65 IV
440	4527.80 I	1200	6191.73 I	500	2025.48 II	500	864.59 IV
100	4544.32 I	300	6222.59 I	500	2062.00 II	9000	1183.97 IV
100	4559.37 I	1000	6435.00 I	200	2064.23 II	9000	1201.77 IV
130	4596.55 I	90	6538.60 I	120	2079.08 I	10000	1219.86 IV
95	4604.80 I	70	6557.39 I	300	2099.94 II	500	1303.93 V
2000	4643.70 I	95	6613.75 II	200	2102.18 II	500 p	1323.81 V
200 h	4658.32 I	40	6650.61 I	800 r	2138.56 I	1000	1469.47 IV
2000	4674.84 I	150	6687.58 I	1000	2501.99 II	10000	1546.17 IV
180	4696.81 I	70	6700.71 I	150	2515.81 I	10000	1598.95 IV
170	4728.53 I	190	6793.71 I	1000	2557.95 II	5000	1607.95 IV
160	4752.79 I	21	6815.16 I	300	2582.49 I	100	1612.38 III
410	4760.98 I	45	6845.24 I	200	2608.56 I	700	1725.02 V
120	4781.04 I	29	6887.22 I	300	2608.64 I	200	1790.19 III

Intensity	Wavelength/Å		Intensity	Wavelength/Å		Intensity	Wavelength/Å		Intensity	Wavelength/Å	
150	1793.56	III	350	3120.74	I	3500	3601.19	I	660	4187.56	I
125	1798.13	III	500	3129.18	II	690	3611.89	II	400	4194.76	I
600	1860.86	V	500	3129.76	II	1100	3613.10	II	610	4199.09	I
200	1940.25	III	350	3132.07	I	1100	3614.77	II	610	4201.46	I
600	2028.54	V	690	3138.68	II	1100	3623.86	I	610	4208.98	II
125	2070.43	III	540	3164.31	II	1100	3663.65	I	400	4213.86	I
200	2086.78	III	880	3165.97	II	390	3671.27	II	2000	4227.76	I
10000	2091.49	IV	880	3182.86	II	800	3674.72	II	2000	4239.31	I
10000	2092.36	IV	540	3191.21	I	390	3697.46	II	770	4240.34	I
600	2132.42	V	540	3212.01	I	960	3698.17	II	770	4241.20	I
10000	2163.68	IV	760	3214.19	II	720	3709.26	II	1200	4241.69	I
100	2175.80	III	630	3231.69	II	560	3745.98	II	550	4282.20	I
100	2191.15	III	630	3234.12	I	880	3751.60	II	550	4294.79	I
10000	2286.67	IV	760	3241.05	II	480	3764.39	I	550	4341.13	I
100	2301.60	III	1000	3273.05	II	480	3766.72	I	1000	4347.89	I
90	2539.65	I	1300	3279.26	II	340	3766.82	II	290	4359.74	II
570	2567.64	II	880	3284.71	II	720	3780.54	I	310	4360.81	I
1600	2568.87	II	540	3305.15	II	560	3791.40	I	350	4366.45	I
2100	2571.39	II	880	3306.28	II	560	3822.41	I	550	4507.12	I
250	2620.56	III	380	3322.99	II	2200	3835.96	I	610	4535.75	I
200	2643.79	III	380	3326.80	II	1300	3836.76	II	490	4542.22	I
150	2664.26	III	380	3334.25	II	550	3843.02	II	490	4575.52	I
1800	2678.63	II	760	3340.56	II	550	3847.01	I	350	4602.57	I
90	2687.75	I	380	3344.79	II	550	3849.25	I	700	4633.98	I
750	2700.13	II	760	3356.09	II	2900	3863.87	I	2300	4687.80	I
1300	2722.61	II	540	3357.26	II	770	3864.34	I	510	4688.45	I
800	2726.49	II	380	3374.73	II	990	3877.60	I	1900	4710.08	I
1400	2734.86	II	570	3387.87	II	1500	3885.42	I	1400	4739.48	I
1100	2742.56	II	760	3388.30	II	2900	3890.32	I	870	4772.31	I
660	2745.86	II	5700	3391.98	II	2000	3891.38	I	700	4815.63	I
660	2752.21	II	570	3393.12	II	610	3921.79	I	250	5046.58	I
530	2758.81	II	570	3404.83	II	1200	3929.53	I	360	5064.91	I
620	2814.90	I	760	3410.25	II	940	3958.22	II	470	5078.25	I
390	2818.74	II	380	3414.66	I	490	3966.66	I	300	5155.45	I
530	2825.56	II	1000	3430.53	II	990	3968.26	I	200	5158.00	I
710	2837.23	I	4700	3438.23	II	660	3973.50	I	100	5191.60	II
660	2844.58	II	600	3447.36	I	770	3991.13	II	270	5385.14	I
350	2848.52	I	410	3457.56	II	770	3998.97	II	160	5664.51	I
350	2851.97	II	820	3463.02	II	400	4023.98	I	160	5797.74	I
340	2869.81	II	600	3471.19	I	770	4024.92	I	340	5879.80	I
490	2875.98	I	1200	3479.39	II	990	4027.20	I	170	6045.85	I
300	2915.99	II	1300	3481.15	II	400	4029.68	II	170	6121.91	I
270	2918.24	II	4100	3496.21	II	490	4030.04	I	680	6127.44	I
320	2926.99	II	820	3505.67	II	400	4035.89	I	340	6134.55	I
320	2948.94	II	1000	3509.32	I	610	4043.58	I	440	6143.20	I
320	2955.78	II	2000	3519.60	I	490	4044.56	I	300	6313.02	I
320	2960.87	I	440	3525.81	I	400	4045.61	II	150	6953.84	I
320	2962.68	II	440	3533.22	I	610	4048.67	II	150	6990.84	I
320	2968.96	II	630	3542.62	II	770	4055.03	I	540	7097.70	I
320	2978.05	II	1800	3547.68	I	600	4055.71	I	280	7102.91	I
820	2985.39	I	630	3550.46	I	1500	4064.16	I	170	7103.72	I
320	3003.74	II	1800	3551.95	II	2000	4072.70	I	590	7169.09	I
820	3011.75	I	2100	3556.60	II	240	4078.31	I	160	7944.61	I
350	3020.47	II	1100	3566.10	I	2000	4081.22	I	160	8005.27	I
500	3028.04	II	2100	3572.47	II	400	4121.46	I	150	8063.09	I
880	3029.52	I	1100	3575.79	I	1200	4149.20	II	790	8070.08	I
350 d	3036.39	II	1300	3576.85	II	400	4161.21	II	390	8132.99	I
690	3054.84	II	880	3586.29	I	400	4166.36	I	280	8212.53	I
690	3106.58	II									

SOURCES OF DATA FOR EACH ELEMENT

Numbers following the element name refer to the references on the following pages.

Actinium: 193
Aluminum: 6,8,81,89,127,144,146,227,228,282
Americium: 92
Antimony: 164,167,194,386,406
Argon: 190,203,204,219,367,368,372,373,374,375,414,421
Arsenic: 163,168,197,244,280
Astatine: 188
Barium: 1,78,111,252,259,277,279
Berkelium: 53,339
Beryllium: 15,44,73,102,115,134,135,171,175,198,335
Bismuth: 1,357,358,359,360,361
Boron: 66,69,74,94,104,171,221,222
Bromine: 42,122,124,139,142,240,243,246,248,249,250,316
Cadmium: 44,285,296,353,399
Calcium: 16,25,70,150,270
Californium: 52,331
Carbon: 22,66,211
Cerium: 1,136,166,261,305
Cesium: 78,82,154,155,200,201,259,263,325
Chlorine: 11,28,30,31,85,233,238,239
Chromium: 1,379,380,412
Cobalt: 1,100,125,159,236,276,291
Copper: 199,273,290,295,324
Curium: 51,332
Dysprosium: 1
Einsteinium: 333
Erbium: 1,301
Europium: 1,312
Fluorine: 68,169,224,225,226
Francium: 408
Gadolinium: 1,46,137,151,152
Gallium: 2,19,62,132,140,141,143,195,281
Germanium: 5,119,293,340,341,342
Gold: 38,72,234,393,395
Hafnium: 1,369,404,410,425
Helium: 16,94,173,183,317
Holmium: 1
Hydrogen: 214
Indium: 1,132,348,349,350,351,352,353,435,436
Iodine: 20,21,58,84,124,153,161,176,184
Iridium: 1
Iron: 56,63,71,101,105,138,174,278,381,382
Krypton: 61,121,123,147,208,232,366,390,409,417,421
Lanthanum: 1,78,79,220,309
Lead: 54,64,106,256,274,297,283,329,330
Lithium: 3,15,17,18,37,44,112,284,321,335
Lutetium: 1,148,310,401
Magnesium: 4,7,49,83,103,128,129,177,217,269,315,335
Manganese: 1,126,385,405,433
Mercury (198): 43,50,69,145,229,242

Mercury (Natural): 34,45,90,117,133,189,235,304,327,328,343
Molybdenum: 1,383,420
Neodymium: 1
Neon: 56,58,69,118,150,230,364,365,371,388,389,400,402,413,430
Neptunium: 93
Nickel: 1,294,415,416,422
Niobium: 1,392,407,431
Nitrogen: 66,107,108,212,213,318
Osmium: 1
Oxygen: 23,24,36,66,69,209,210,215
Palladium: 1,287,424
Phosphorus: 179,180,182,336
Platinum: 1,288
Plutonium: 91
Polonium: 47,48
Potassium: 32,59,60,75,76,86,150,160,172,268,314,322
Praseodymium: 1,149,306,308,337,338
Promethium: 196,260
Protactinium: 96
Radium: 253,254
Radon: 251
Rhenium: 1
Rhodium: 1,396
Rubidium: 12,109,130,241,257,258,262,264
Ruthenium: 1,423
Samarium: 1
Scandium: 1,88,150,298,323
Selenium: 9,80,181,216,245,247,275
Silicon: 87,170,237,292,319,320
Silver: 13,99,255,286,289,363,387,398
Sodium: 178,205,206,207,268,299,334
Strontium: 1,109,110,218,231,265,279,313
Sulfur: 29,144,202,209,210,266
Tantalum: 1,411,426
Technetium: 35
Tellurium: 1,344,345,346,347
Terbium: 1,302
Thallium: 1,195,348,354,355,356
Thorium: 1,97,98,156,157,165,434
Thulium: 1,307
Tin: 187,191,399,423
Titanium: 1,378,427,428
Tungsten: 1
Uranium: 1,303
Vanadium: 1,394,397,432
Xenon: 33,116,118,120,232,384,391,429
Ytterbium: 1,40,192,311
Yttrium: 1,77,265,419
Zinc: 39,55,113,131,185,186,370,376,377
Zirconium: 1,362,403,418

REFERENCES

1. Meggers, W. F., Corliss, C. H., and Scribner, B. F., *Natl. Bur. Stand. (U.S.) Monogr.*, 145, Washington, D.C., 1975.
2. Aksenov, V. P. and Ryabtsev, A. N., *Opt. Spectrosc.*, 37, 860, 1970.
3. Andersen, N., Bickel, W. S., Carriveau, G. W., Jensen, K., and Veje, E., *Phys. Scr.*, 4, 113, 1971.
4. Andersson, E. and Johannesson, G. A., *Phys. Scr.*, 3, 203, 1971.
5. Andrew, K. L. and Meissner, K. W., *J. Opt. Soc. Am.*, 49, 146, 1959.
6. Artru, M. C. and Brillet, W. U. L., *J. Opt. Soc. Am.*, 64, 1063, 1974.
7. Artru, M. C. and Kaufman, V., *J. Opt. Soc. Am.*, 62, 949, 1972.
8. Artru, M. C. and Kaufman, V., *J. Opt. Soc. Am.*, 65, 594, 1975.
9. Badami, J. S. and Rao, K. R., *Proc. R. Soc. London*, 140(A), 387, 1933.
10. Baird, K. M. and Smith, D. S., *J. Opt. Soc. Am.*, 48, 300, 1958.
11. Bashkin, S. and Martinson, I., *J. Opt. Soc. Am.*, 61, 1686, 1971.
12. Beacham, J. R., Ph.D. thesis, Purdue University, 1970.
13. Benschop, H., Joshi, Y. N., and van Kleef, T. A. M., *Can. J. Phys.*, 53, 700, 1975.
14. Berry, H. G., Bromander, J., and Buchta, R., *Phys. Scr.*, 1, 181, 1970.
15. Berry, H. G., Bromander, J., Martinson, I., and Buchta, R., *Phys. Scr.*, 3, 63, 1971.
16. Berry, H. G., Desesquelles, J., and Dufay, M., *Phys. Rev. Sect. A.*, 6, 600, 1972.
17. Berry, H. G., Desesquelles, J., and Dufay, M., *Nucl. Instrum. Methods*, 110, 43, 1973.
18. Berry, H. G., Pinnington, E. H., and Subtil, J. L., *J. Opt. Soc. Am.*, 62, 767, 1972.
19. Bidelman, W. P. and Corliss, C. H., *Astrophys. J.*, 135, 968, 1962.
20. Bloch, L. and Bloch, E., *Ann. Phys.* (Paris), 10(11), 141, 1929.
21. Bloch, L., Bloch, E., and Felici, N., *J. Phys. Radium*, 8, 355, 1937.
22. Bockasten, K., *Ark. Fys.*, 9, 457, 1955.
23. Bockasten, K., Hallin, R., Johansson, K. B., and Tsui, P., *Phys. Lett.* (Netherlands), 8, 181, 1964.
24. Bockasten, K. and Johansson, K. B., *Ark. Fys.*, 38, 563, 1969.
25. Borgstrom, A., *Ark. Fys.*, 38, 243, 1968.
26. Borgstrom, A., *Phys. Scr.*, 3, 157, 1971.
27. Bowen, I. S., *Phys. Rev.*, 29, 231, 1927.
28. Bowen, I. S., *Phys. Rev.*, 31, 34, 1928.
29. Bowen, I. S., *Phys. Rev.*, 39, 8, 1932.
30. Bowen, I. S., *Phys. Rev.*, 45, 401, 1934.
31. Bowen, I. S., *Phys. Rev.*, 46, 377, 1934.
32. Bowen, I. S., *Phys. Rev.*, 46, 791, 1934.
33. Boyce, J. C., *Phys. Rev.*, 49, 730, 1936.
34. Boyce, J. C. and Robinson, H. A., *J. Opt. Soc. Am.*, 26, 133, 1936.
35. Bozman, W. R., Meggers, W. F., and Corliss, C. H., *J. Res. Natl. Bur. Stand. Sect. A*, 71, 547, 1967.
36. Bromander, J., *Ark. Fys.*, 40, 257, 1969.
37. Bromander, J. and Buchta, R., *Phys. Scr.*, 1, 184, 1970.
38. Brown, C. M. and Ginter, M. L., *J. Opt. Soc. Am.*, 68, 243, 1978.
39. Brown, C. M., Tilford, S. G., and Ginter, M. L., *J. Opt. Soc. Am.*, 65, 1404, 1975.
40. Bryant, B. W., *Johns Hopkins Spectroscopic Report* No. 21, 1961.
41. Buchet, J. P., Buchet-Poulizac, M. C., Berry, H. G., and Drake, G. W. F., *Phys. Rev. Sect. A*, 7, 922, 1973.
42. Budhiraja, C. J. and Joshi, Y. N., *Can. J. Phys.*, 49, 391, 1971.
43. Burns, K. and Adams, K. B., *J. Opt. Soc. Am.*, 42, 56, 1952.
44. Burns, K. and Adams, K. B., *J. Opt. Soc. Am.*, 46, 94, 1956.
45. Burns, K., Adams, K. B., and Longwell, J., *J. Opt. Soc. Am.*, 40, 339, 1950.
46. Callahan, W. R., Ph.D. thesis, Johns Hopkins University, 1962.
47. Charles, G. W., *J. Opt. Soc. Am.*, 56, 1292, 1966.
48. Charles, G. W., Hunt, D. J., Pish, G., and Timma, D. L., *J. Opt. Soc. Am.*, 45, 869, 1955.
49. Codling, K., *Proc. Phys. Soc.*, 77, 797, 1961.
50. Comite Consultatif Pour La Definition du Metre, *J. Phys. Chem. Ref. Data*, 3, 852, 1974.
51. Conway, J. G., Blaise, J., and Verges, J., *Spectrochim. Acta Part B*, 31, 31, 1976.
52. Conway, J. G., Worden, E. F., Blaise, J., and Verges, J., *Spectrochim. Acta Part B*, 32, 97, 1977.
53. Conway, J. G., Worden, E. F., Blaise, J., Camus, P., and Verges, J., *Spectrochim. Acta Part B*, 32, 101, 1977.
54. Crooker, A. M., *Can. J. Res. Sect. A*, 14, 115, 1936.
55. Crooker, A. M. and Dick, K. A., *Can. J. Phys.*, 46, 1241, 1968.
56. Crosswhite, H. M., *J. Res. Natl. Bur. Stand. Sect. A*, 79, 17, 1975.
57. Crosswhite, H. M. and Dieke, G. H., *American Institute of Physics Handbook*, Section 7, 1972.
59. de Bruin, T. L., *Z. Phys.*, 38, 94, 1926.
60. de Bruin, T. L., *Z. Phys.*, 53, 658, 1929.
61. de Bruin, T. L., Humphreys, C. J., and Meggers, W. F., *J. Res. Natl. Bur. Stand.*, 11, 409, 1933.
62. Dick, K. A., *J. Opt. Soc. Am.*, 64, 702, 1973.
63. Dobbie, J. C., *Ann. Solar Phys. Observ.* (Cambridge), 5, 1, 1938.
64. Earls, L. T. and Sawyer, R. A., *Phys. Rev.*, 47, 115, 1935.
65. Edlen, B., *Z. Phys.*, 85, 85, 1933.
66. Edlen, B., *Nova Acta Reglae Soc. Sci. Ups.*, (IV) 9, No. 6, 1934.
67. Edlen, B., *Z. Phys.*, 93, 726, 1935.
68. Edlen, B., *Z. Phys.*, 94, 47, 1935.
69. Edlen, B., *Rep. Prog. Phys.*, 26, 181, 1963.
70. Edlen, B. and Risberg, P., *Ark. Fys.*, 10, 553, 1956.
71. Edlen, B. and Swings, P., *Astrophys. J.*, 95, 532, 1942.
72. Ehrhardt, J. C. and Davis, S. P., *J. Opt. Soc. Am.*, 61, 1342, 1971.
73. Eidelsberg, M., *J. Phys. B*, 5, 1031, 1972.
74. Eidelsberg, M., *J. Phys. B*, 7, 1476, 1974.
75. Ekberg, J. O. and Svensson, L. A., *Phys. Scr.*, 2, 283, 1970.
76. Ekefors, E., *Z. Phys.*, 71, 53, 1931.
77. Epstein, G. L. and Reader, J., *J. Opt. Soc. Am.*, 65, 310, 1975.
78. Epstein, G. L. and Reader, J., *J. Opt. Soc. Am.*, 66, 590, 1976.
79. Epstein, G. L. and Reader, J., unpublished.
80. Eriksson, K. B. S., *Phys. Lett. A.*, 41, 97, 1972.
81. Eriksson, K. B. S. and Isberg, H. B. S., *Ark. Fys.*, 23, 527, 1963.
82. Eriksson, K. B. S. and Wenaker, I., *Phys. Scr.*, 1, 21, 1970.
83. Esteva, J. M. and Mehlman, G., *Astrophys. J.*, 193, 747, 1974.
84. Even-Zohar, M. and Fraenkel, B. S., *J. Phys. B*, 5, 1596, 1972.
85. Fawcett, B. C., *J. Phys. B*, 3, 1732, 1970.
86. Fawcett, B. C., Culham Laboratory Report ARU-R4, 1971.
87. Ferner, E., *Ark. Mat. Astron. Fys.*, 28(A), 4, 1941.
88. Fischer, R. A., Knopf, W. C., and Kinney, F. E., *Astrophys. J.*, 130, 683, 1959.

89. Fowler, A., *Report on Series in Line Spectra*, Fleetway Press, London, 1922.

90. Fowles, G. R., *J. Opt. Soc. Am.*, 44, 760, 1954.

91. Fred, M., *Argonne Natl. Lab.*, unpublished, 1977.

92. Fred, M. and Tomkins, F. S., *J. Opt. Soc. Am.*, 47, 1076, 1957.

93. Fred, M., Tomkins, F. S., Blaise, J. E., Camus, P., and Verges, J., Argonne National Laboratory Report No. 76-68, 1976.

94. Garcia, J. D. and Mack, J. E., *J. Opt. Soc. Am.*, 55, 654, 1965.

96. Giacchetti, A., *Argonne Natl. Lab.*, unpublished, 1975.

97. Giacchetti, A., Blaise, J., Corliss, C. H., and Zalubas, R., *J. Res. Natl. Bur. Stand. Sect. A*, 78, 247, 1974.

98. Giacchetti, A., Stanley, R. W., and Zalubas, R., *J. Opt. Soc. Am.*, 69, 474, 1970.

99. Gilbert, W. P., *Phys. Rev.*, 47, 847, 1935.

100. Gilroy, H. T., *Phys. Rev.*, 38, 2217, 1931.

101. Glad, S., *Ark. Fys.*, 10, 291, 1956.

102. Goldsmith, S., *J. Phys. B*, 2, 1075, 1969.

103. Goorvitch, D., Mehlmam-Balloffet, G., and Valero, F. P. J., *J. Opt. Soc. Am.*, 60, 1458, 1970.

104. Goorvitch, D. and Valero, F. P. J., *Astrophys. J.*, 171, 643, 1972.

105. Green, L. C., *Phys. Rev.*, 55, 1209, 1939.

106. Gutman, F., *Diss. Abstr. Int. B*, 31, 363, 1970.

107. Hallin, R., *Ark. Fys.*, 31, 511, 1966.

108. Hallin, R., *Ark. Fys.*, 32, 201, 1966.

109. Hansen, J. E. and Persson, W., *J. Opt. Soc. Am.*, 64, 696, 1974.

110. Hansen, J. E. and Persson, W., *Phys. Scr.*, 13, 166, 1976.

111. Hellintin, P., *Phys. Scr.*, 13, 155, 1976.

112. Herzberg, G. and Moore, H. R., *Can. J. Phys.*, 37, 1293, 1959.

113. Hetzler, C. W., Boreman, R. W., and Burns, K., *Phys. Rev.*, 48, 656, 1935.

114. Holmstrom, J. E. and Johansson, L., *Ark. Fys.*, 40, 133, 1969.

115. Hontzeas, S., Martinson, I., Erman, P., and Buchta, R., *Nucl. Instrum. Methods*, 110, 51, 1973.

116. Humphreys, C. J., *J. Res. Natl. Bur. Stand.*, 22, 19, 1939.

117. Humphreys, C. J., *J. Opt. Soc. Am.*, 43, 1027, 1953.

118. Humphreys, C. J., *J. Phys. Chem. Ref. Data*, 2, 519, 1973.

119. Humphreys, C. J. and Andrew, K. L., *J. Opt. Soc. Am.*, 54, 1134, 1964.

120. Humphreys, C. J. and Meggers, W. F., *J. Res. Natl. Bur. Stand.*, 10, 139, 1933.

121. Humphreys, C. J. and Paul, E., Jr., *J. Opt. Soc. Am.*, 60, 200, 1970.

122. Humphreys, C. J. and Paul, E., Jr., *J. Opt. Soc. Am.*, 62, 432, 1972.

123. Humphreys, C. J., Paul, E., Jr., Cowan, R. D., and Andrew, K. L., *J. Opt. Soc. Am.*, 57, 855, 1967.

124. Humphreys, C. J., Paul, E., Jr., and Minnhagen, L., *J. Opt. Soc. Am.*, 61, 110, 1971.

125. Iglesias, L., Inst. of Optics, Madrid, unpublished, 1977.

126. Iglesias, L. and Velasco, R., *Publ. Inst. Opt. Madrid*, No. 23, 1964.

127. Isberg, B., *Ark. Fys.*, 35, 551, 1967.

128. Johannesson, G. A., Lundstrom, T., and Minnhagen, L., *Phys. Scr.*, 6, 129, 1972.

129. Johannesson, G. A. and Lundstrom, T., *Phys. Scr.*, 8, 53, 1973.

130. Johansson, I. *Ark. Fys.*, 20, 135, 1961.

131. Johansson, I. and Contreras, R., *Ark. Fys.*, 37, 513, 1968.

132. Johansson, I. and Litzen, U., *Ark. Fys.*, 34, 573, 1967.

133. Johansson, I. and Svensson, K. F., *Ark. Fys.*, 16, 353, 1960.

134. Johansson, L., *Ark. Fys.*, 20, 489, 1961.

135. Johansson, L., *Ark. Fys.*, 23, 119, 1963.

136. Johansson, S. and Litzen, U., *Phys. Scr.*, 6, 139, 1972.

137. Johansson, S. and Litzen, U., *Phys. Scr.*, 8, 43, 1973.

138. Johansson, S. and Litzen, U., *Phys. Scr.*, 10, 121, 1974.

139. Joshi, Y. N., St. Francis Xavier Univ., Nova Scotia, unpublished.

140. Joshi, Y. N., Bhatia, K. S., and Jones, W. E., *Sci. Light Tokyo*, 21, 113, 1972.

141. Joshi, Y. N., Bhatia, K. S., and Jones, W. E., *Spectrochim. Acta Part B*, 28, 149, 1973.

142. Joshi, Y. N. and Budhiraja, C. J., *Can. J. Phys.*, 49, 670, 1971.

143. Joshi, Y. N. and van Kleef, T. A. M., *Can. J. Phys.*, 52, 1891, 1974.

144. Kaufman, V., *Natl. Bur. Stand.*, unpublished.

145. Kaufman, V., *J. Opt. Soc. Am.*, 52, 866, 1962.

146. Kaufman, V., Artru, M. C., and Brillet, W. U. L., *J. Opt. Soc. Am.*, 64, 197, 1974.

147. Kaufman, V. and Humphreys, C. J., *J. Opt. Soc. Am.*, 59, 1614, 1969.

148. Kaufman, V. and Sugar, J., *J. Opt. Soc. Am.*, 61, 1693, 1971.

149. Kaufman, V. and Sugar, J., *J. Res. Natl. Bur. Stand. Sect. A*, 71, 583, 1967.

150. Kelly, R. L. and Palumbo, L. J., *Naval Research Laboratory Report 7599*, Washington, DC., 1973.

151. Kielkopf, J. F., *Univ. of Louisville*, unpublished. 1975.

152. Kielkopf, J. F., *Univ. of Louisville*, unpublished, 1976.

153. Kiess, C. C. and Corliss, C. H., *J. Res. Natl. Bur. Stand. Sect. A*, 63, 1, 1959.

154. Kleiman, H., *J. Opt. Soc. Am.*, 52, 441, 1962.

155. Eriksson, K. B., Johansson, I., and Norlen, G., *Ark. Fys.*, 28, 233, 1964.

156. Klinkenberg, P. F. A., *Physica*, 15, 774, 1949.

157. Klinkenberg, P. F. A., *Physica*, 16, 618, 1950.

158. Krishnamurty, S. G., *Proc. Phys. Soc. London*, 48, 277, 1936.

159. Kruger, P. G. and Gilroy, H. T., *Phys. Rev.*, 48, 720, 1935.

160. Kruger, P. G. and Pattin, H. S., *Phys. Rev.*, 52, 621, 1937.

161. Lacroute, P., *Ann. Phys.* (Paris), 3, 5, 1935.

162. Lang, R. J., *Phys. Rev.*, 30, 762, 1927.

163. Lang, R. J., *Phys. Rev.*, 32, 737, 1928.

164. Lang, R. J., *Phys. Rev.*, 35, 445, 1930.

165. Lang, R. J., *Can. J. Res. Sect. A*, 14, 43, 1936.

166. Lang, R. J., *Can. J. Res. Sect. A*, 14, 127, 1936.

167. Lang, R. J. and Vestine, E. H., *Phys. Rev.*, 42, 233, 1932.

168. Li, H. and Andrew, K. L., *J. Opt. Soc. Am.*, 61, 96, 1971.

169. Liden, K., *Ark. Fys.*, 1, 229, 1949.

170. Litzen, U., *Ark. Fys.*, 28, 239, 1965.

171. Litzen, U., *Phys. Scr.*, 1, 251, 1970.

172. Litzen, U., *Phys. Scr.*, 1, 253, 1970.

173. Litzen, U., *Phys. Scr.*, 2, 103, 1970.

174. Litzen, U. and Verges, I., *Phys. Scr.*, 13, 240, 1976.

175. Lofstrand, B., *Phys. Scr.*, 8, 57, 1973.

176. Luc-Koenig, E., Morillon, C., and Verges, J., *Phys. Scr.*, 12, 199, 1975.

177. Lundstrom, T., *Phys. Scr.*, 7, 62, 1973.

178. Lundstrom, T. and Minnhagen, L., *Phys. Scr.*, 5, 243, 1972.

179. Magnusson, C. E. and Zetterberg, P. O., *Phys. Scr.*, 10, 177, 1974.

180. Magnusson, C. E., and Zetterberg, P. O., *Phys. Scr.*, 15, 237, 1977.

181. Martin, D. C., *Phys. Rev.*, 48. 938, 1935.

182. Svendenius, N., *Phys. Scr.*, 22, 240, 1980.

183. Martin, W. C., *J. Res. Natl. Bur. Stand. Sect. A*, 64, 19, 1960.

184. Martin, W. C. and Corliss, C. H., *J. Res. Natl. Bur. Stand. Sect. A*, 64, 443, 1960.

185. Martin, W. C. and Kaufman, V., *J. Res. Natl. Bur. Stand. Sect. A*, 74, 11, 1970.

186. Martin, W. C. and Kaufman, V., *J. Opt. Soc. Am.*, 60, 1096, 1970.

187. McCormick, W. W. and Sawyer, R. A., *Phys. Rev.*, 54, 71, 1938.
188. McLaughlin, R., *J. Opt. Soc. Am.*, 54, 965, 1964.
189. McLennan, J. C., McLay, A. B., and Crawford, M. F., *Proc. R. Soc. London Ser. A*, 134, 41, 1931.
190. Meissner, K. W., *Z. Phys.*, 39, 172, 1926.
191. Meggers, W. F., *J. Res. Natl. Bur. Stand.*, 24, 153, 1940.
192. Meggers, W. F. and Corliss, C. H., *J. Res. Natl. Bur. Stand. Sect. A*, 70, 63, 1966.
193. Meggers, W. F., Fred, M., and Tomkins, F. S., *J. Res. Natl. Bur. Stand.*, 58, 297, 1957.
194. Meggers, W. F. and Humphreys, C. J., *J. Res. Natl. Bur. Stand.*, 28, 463, 1942.
195. Meggers, W. F. and Murphy, R. J., *J. Res. Natl. Bur. Stand.*, 48, 334, 1952.
196. Meggers, W. F., Scribner, B. F., and Bozman, W. R., *J. Res. Natl. Bur. Stand.*, 46, 85, 1951.
197. Meggers, W. F., Shenstone, A. G., and Moore, C. E., *J. Res. Natl. Bur. Stand.*, 45, 346, 1950.
198. Mehlman, G. and Esteva, J. M., *Astrophys. J.*, 188, 191, 1974.
199. Meinders, E., *Physica*, 84(C), 117, 1976.
200. Sansonetti, C. J., Dissertation, Purdue University, 1981.
201. Sansonetti, C. J., *Natl. Bur. Stand. (U.S.)*, unpublished.
202. Millikan, R. A. and Bowen, I. S., *Phys. Rev.*, 25, 600, 1925.
203. Minnhagen, L., *J. Opt. Soc. Am.*, 61, 1257, 1925.
204. Minnhagen, L., *J. Opt. Soc. Am.*, 63, 1185, 1973.
205. Minnhagen, L., *Phys. Scr.*, 11, 38, 1975.
206. Minnhagen, L., *J. Opt. Soc. Am.*, 66, 659, 1976.
207. Minnhagen, L. and Nietsche, H., *Phys. Scr.*, 5, 237, 1972.
208. Minnhagen. L., Strihed, H., and Petersson, B., *Ark. Fys.*, 39, 471, 1969.
209. Moore, C. E., *Natl. Bur. Stand. (U.S.) Circ.*, 488, 1950.
210. Moore, C. E., *Revised Multiplet Table*, Princeton University Observatory No. 20, 1945.
211. Moore, C. E., National Standard Reference Data Series - National Bureau of Standards 3, Sect. 3, 1970.
212. Moore, C. E., National Standard Reference Data Series - National Bureau of Standards 3, Sect. 4, 1971.
213. Moore, C. E., National Standard Reference Data Series - National Bureau of Standards 3, Sect. 5, 1975.
214. Moore, C. E., National Standard Reference Data Series - National Bureau of Standards 3, Sect. 6, 1972.
215. Moore, C. E., National Standard Reference Data Series - National Bureau of Standards 3, Sect. 7, 1975.
216. Morillon, C. and Verges, J., *Phys. Scr.*, 10, 227, 1974.
217. Newsom, G. H., *Astrophys. J.*, 166, 243, 1971.
218. Newsom, G. H., O'Connor, S., and Learner, R. C. M., *J. Phys. B*, 6, 2162, 1973.
219. Norlen, G., *Phys. Scr.*, 8, 249, 1973.
220. Odabasi, H., *J. Opt. Soc. Am.*, 57, 1459, 1967.
221. Olme, A., *Ark. Fys.*, 40, 35, 1969.
222. Olme, A., *Phys. Scr.*, 1, 256, 1970.
223. Johansson, S., and Litzen, U., *J. Opt. Soc. Am.*, 61, 1427, 1971.
224. Palenius, H. P., *Ark. Fys.*, 39, 15, 1969.
225. Palenius, H. P., *Phys. Scr.*, 1, 113, 1970.
226. Palenius, H. P., *Univ. of Lund, Sweden*, unpublished.
227. Paschen, F., *Ann. Phys.*, Series 5, 12, 509, 1932.
228. Paschen, F. and Ritschl, R., *Ann. Phys.*, Series 5, 18, 867, 1933.
229. Peck, E. R., Khanna, B. N., and Anderholm, N. C., *J. Opt. Soc. Am.*, 52, 53, 1962.
230. Persson, W., *Phys. Scr.*, 3, 133, 1971.
231. Persson, W. and Valind, S., *Phys. Scr.*, 5, 187, 1972.
232. Petersson, B., *Ark. Fys.*, 27, 317, 1964.
233. Phillips, L. W. and Parker, W. L., *Phys. Rev.*, 60, 301, 1941.
234. Platt, J. R. and Sawyer, R. A., Phys. Rev., 60, 866, 1941.
235. Plyer, E. K., Blaine, L. R., and Tidwell, E., *J. Res. Natl. Bur. Stand.*, 55, 279, 1955.
236. Poppe, R., van Kleef, T. A. M., and Raassen, A. J. J., *Physica*, 77, 165, 1974.
237. Radziemski, L. J., Jr. and Andrew, K. L., *J. Opt. Soc. Am.*, 55, 474, 1965.
238. Radziemski, L. J., Jr. and Kaufman, V., *J. Opt. Soc. Am.*, 59, 424, 1969.
239. Radziemski, L. J., Jr. and Kaufman, V., *J. Opt. Soc. Am.*, 64, 366, 1974.
240. Ramanadham, R. and Rao, K. R., *Indian J. Phys.*, 18, 317, 1944.
241. Ramb, R., *Ann. Phys.*, 10, 311, 1931.
242. Rank, D. H., Bennett, J. M., and Bennett, H. E., *J. Opt. Soc. Am.*, 40, 477, 1950.
243. Rao, A. S. and Krishnamurty, S. G., *Proc. Phys. Soc. London*, 46, 531, 1943.
244. Rao, K. R., *Proc. R. Soc. London, Ser. A*, 134, 604, 1932.
245. Rao, K. R. and Badami, J. S., *Proc. R. Soc. London Ser. A*, 131, 154, 1931.
246. Rao, K. R. and Krishnamurty, S. G., *Proc. R. Soc. London Ser. A*, 161, 38, 1937.
247. Rao, K. R. and Murti, S. G. K., *Proc. R. Soc. London Ser. A*, 145, 681, 1934.
248. Rao, Y. B., *Indian J. Phys.*, 32, 497, 1958.
249. Rao, Y. B., *Indian J. Phys.*, 33, 546, 1959.
250. Rao, Y. B., *Indian J. Phys.*, 35, 386, 1961.
251. Rasmussen, E., *Z. Phys.*, 80, 726, 1933.
252. Rasmussen, E., *Z. Phys.*, 83, 404, 1933.
253. Rasmussen, E., *Z. Phys.*, 86, 24, 1934.
254. Rasmussen, E., *Z. Phys.*, 87, 607, 1934.
255. Rasmussen, E., *Phys. Rev.*, 57, 840, 1940.
256. Rau, A. S. and Narayan, A. L., *Z. Phys.*, 59, 687, 1930.
257. Reader, J., *J. Opt. Soc. Am.*, 65, 286, 1975.
258. Reader, J., *J. Opt. Soc. Am.*, 65, 988, 1975.
259. Reader, J., *J. Opt Soc. Am.*, 73, 349, 1983.
260. Reader, J. and Davis, S., *J. Res. Natl. Bur. Stand. Sect. A*, 71, 587, 1967, and unpublished.
261. Reader, J. and Ekberg, J. O., *J. Opt. Soc. Am.*, 62, 464, 1972.
262. Reader, J. and Epstein, G. L., *J. Opt. Soc. Am.*, 62, 1467, 1972.
263. Reader, J. and Epstein, G. L., *J. Opt. Soc. Am.*, 65, 638, 1975.
264. Reader, J. and Epstein, G. L., *Natl. Bur. Stand.*, unpublished.
265. Reader, J., Epstein, G. L., and Ekberg, J. O., *J. Opt. Soc. Am.*, 62, 273, 1972.
266. Kaufman, V., *Phys. Scr.*, 26, 439, 1982.
267. Ricard, R., Givord, M., and George, F., *C. R. Acad. Sci. Paris*, 205, 1229, 1937.
268. Risberg, P., *Ark. Fys.*, 10, 583, 1956.
269. Risberg, G., *Ark. Fys.*, 28, 381, 1965.
270. Risberg, G., *Ark. Fys.*, 37, 231, 1968.
271. Robinson, H. A., *Phys. Rev.*, 49, 297, 1936.
272. Robinson, H. A., *Phys. Rev.*, 50, 99, 1936.
273. Ross, C. B., Jr., Doctoral dissertation, Purdue University, 1969.
274. Ross, C. B., Wood, D. R., and Scholl, P. S., *J. Opt. Soc. Am.*, 66, 36, 1976.
275. Ruedy, J. E. and Gibbs, R. C., *Phys. Rev.*, 46, 880, 1934.
276. Russell, H. N., King, R. B., and Moore, C. E., *Phys. Rev.*, 58, 407, 1940.
277. Russell, H. N. and Moore, C. E., *J. Res. Natl. Bur. Stand.*, 55, 299, 1955.
278. Russell, H. N., Moore, C. E., and Weeks, D. W., *Trans. Am. Philos. Soc.*, 34(2), 111, 1944.

279. Saunders, F., Schneider, E., and Buckingham, E., *Proc. Natl. Acad. Sci.*, 20, 291, 1934.
280. Sawyer, R. A. and Humphreys, C. J., *Phys. Rev.*, 32, 583, 1928.
281. Sawyer, R. A. and Lang, R. J., *Phys. Rev.*, 34, 712, 1929.
282. Sawyer, R. A. and Paschen, F., *Ann. Phys.*, 84(4),1, 1927.
283. Scholl, P. S., M.S. thesis, Wright State Univ., 1975.
284. Schurmann, D., *Z. Phys.*, 17, 4, 1975.
285. Seguier, J., *C. R. Acad. Sci. Paris*, 256, 1703, 1963.
286. Shenstone, A. G., *Phys. Rev.*, 31, 317, 1928.
287. Shenstone, A. G., *Phys. Rev.*, 32, 30, 1928.
288. Shenstone, A. G., *Trans. R. Soc. London*, 237(A), 57, 1938.
289. Shenstone, A. G., *Phys. Rev.*, 57, 894, 1940.
290. Shenstone, A. G. *Philos. Trans. R. Soc. London Ser. A*, 241, 297, 1948.
291. Shenstone, A. G., *Can. J. Phys.*, 38, 677, 1960.
292. Shenstone, A. G., *Proc. R. Soc. London*, 261(A), 153, 1961.
293. Shenstone, A. G., *Proc. R. Soc. London*, 276(A), 293, 1963.
294. Shenstone, A. G., *J. Res. Natl. Bur. Stand. Sect. A*, 74, 801, 1970.
295. Shenstone, A. G., *J. Res. Natl. Bur. Stand. Sect. A*, 79, 497, 1975.
296. Shenstone, A. G. and Pittenger, J. T., *J. Opt. Soc. Am.*, 39, 219, 1949.
297. Smith, S., *Phys. Rev.*, 36, 1, 1930.
298. Smitt, R., *Phys. Scr.*, 8, 292, 1973.
299. Soderqvist, J., *Ark. Mat. Astronom. Fys.*, 32(A), 1, 1946.
300. Sommer, L. A., *Ann. Phys.*, 75, 163, 1924.
301. Spector, N., *J. Opt. Soc. Am.*, 63, 358, 1973.
302. Spector, N. and Sugar, J., *J. Opt. Soc. Am.*, 66, 436, 1976.
303. Steinhaus, D. W., Radziemski, L. J., Jr., and Blaise, J., *Los Alamos Sci. Lab.*, unpublished, 1975.
304. Subbaraya, T. S., *Z. Phys.*, 78, 541, 1932.
305. Sugar, J., *J. Opt. Soc. Am.*, 55, 33, 1965.
306. Sugar, J., *J. Res. Natl. Bur. Stand. Sect. A*, 73, 333, 1969.
307. Sugar, J., *J. Opt. Soc. Am.*, 60, 454, 1970.
308. Sugar, J., *J. Res. Natl. Bur. Stand. Sect. A*, 78, 555, 1974.
309. Sugar, J. and Kaufman, V., *J. Opt. Soc. Am.*, 55, 1283, 1965.
310. Sugar, J. and Kaufman, V., *J. Opt. Soc. Am.*, 62, 562, 1972.
311. Sugar, J., Kaufman, V., and Spector, N., *J. Res. Natl. Bur. Stand., Sect. A*, 83, 233, 1978.
312. Sugar, J. and Spector, N., *J. Opt. Soc. Am.*, 64, 1484, 1974.
313. Sullivan, F. J. *Univ. Pittsburgh Bull.*, 35, 1, 1938.
314. Svensson, L. A. and Ekberg, J. O., *Ark. Fys.*, 37, 65, 1968.
315. Swensson, J. W. and Risberg, G., *Ark. Fys.*, 31, 237, 1966.
316. Tech, J. L., *J. Res. Natl. Bur. Stand. Sect. A*, 67, 505, 1963.
317. Tech, J. L. and Ward, J. F., *Phys. Rev. Lett.*, 27, 367, 1971.
318. Tilford, S. G., *J. Opt. Soc. Am.*, 53, 1051, 1963.
319. Toresson, Y. G., *Ark. Fys.*, 17, 179, 1960.
320. Toresson, Y. G., *Ark. Fys.*, 18, 389, 1960.
321. Toresson, Y. G. and Edlen, B., *Ark. Fys.*, 23, 117, 1963.
322. Tsien, W. Z., *Chin. J. Phys.*, Peiping, 3, 117, 1939.
323. van Deurzen, C. H. H., Conway, J., and Davis, S. P., *J. Opt. Soc. Am.*, 63, 158, 1973.
324. van Kleef, T. A. M., Raassen, A. J. J., and Joshi, Y. N., *Physica*, 84(C), 401, 1976.
325. Sansonetti, C. J., Andrew, K. L., and Verges, J., *J. Opt. Soc. Am.*, 71, 423, 1981.
326. Wheatley, M. A. and Sawyer, R. A., *Phys. Rev.*, 61, 591, 1942.
327. Wilkinson, P. G., *J. Opt. Soc. Am.*, 45, 862, 1955.
328. Wilkinson, P. G. and Andrew, K. L., *J. Opt. Soc. Am.*, 53, 710, 1963.
329. Wood, D. and Andrew, K. L., *J. Opt. Soc. Am.*, 58, 818, 1968.
330. Wood, D. R., Ron, C. B., Scholl, P. S., and Hoke, M., *J. Opt. Soc. Am.*, 64, 1159, 1974.

331. Worden, E. F. and Conway, J. G., *Lawrence Livermore Lab.*, unpublished, 1977.
332. Worden, E. F., Hulet, E. K., Gutmacher, R. G., Conway, J. G., *At. Data Nucl. Data Tables*, 18, 459, 1976.
333. Worden, E. F., Lougheed, R. W., Gutmacher, R. G., and Conway, J. G., *J. Opt. Soc. Am.*, 64, 77, 1974.
334. Wu, C. M., Ph.D. thesis, University of British Columbia, 1971.
335. Zaidel, A. N., Prokofev, V. K., Raiskii, S. M., Slavnyi, V. A., and Schreider, E. Y., *Tables of Spectral Lines*, 3rd ed., Plenum, New York, 1970.
336. Zetterberg, P. O. and Magnusson, C. E., *Phys. Scr.*, 15, 189, 1977.
337. Sugar, J., *J. Opt. Soc. Am.*, 55, 1058, 1965.
338. Sugar, J., *J. Opt. Soc. Am.*, 61, 727, 1971.
339. Worden, E. F., and Conway, J. G., *At. Data Nucl. Data Tables*, 22, 329, 1978.
340. Kaufman, V. and Edlen, B., *J. Phys. Chem. Ref. Data*, 3, 825, 1974.
341. Lang, R. J., *Phys. Rev.*, 34, 697, 1929.
342. Ryabtsev, A. N., *Opt. Spectros.*, 39, 455, 1975.
343. Foster, E. W., *Proc. R. Soc. London*, 200(A), 429, 1950.
344. Morillon, C. and Verges, J., *Phys. Scr.*, 12, 129, 1975.
345. Ruedy, J. E., *Phys. Rev.*, 41, 588, 1932.
346. McLennan, J. C., McLay, A. B., and McLeod, J. H., *Philos. Mag.*, 4 ,486, 1927.
347. Handrup, M. B. and Mack, J. E., *Physica*, 30, 1245, 1964.
348. Clearman, H. E., *J. Opt. Soc. Am.*, 42, 373, 1952.
349. Paschen, F., *Ann. Physik*, 424, 148, 1938.
350. Paschen, F. and Campbell, J. S., *Ann. Phys.*, 31(5), 29, 1938.
351. Nodwell, R., *Univ. of British Columbia, Vancouver*, unpublished, 1955.
352. Gibbs, R. C. and White, H. E., *Phys. Rev.*, 31, 776, 1928.
353. Green, M., *Phys. Rev.*, 60, 117, 1941.
354. Ellis, C. B. and Sawyer, R. A., *Phys. Rev.*, 49, 145, 1936.
355. McLennan, J. C., McLay, A. B., and Crawford, M. F., *Proc. R. Soc. London Ser. A*, 125, 50, 1929.
356. Mack, J. E. and Fromer, M., *Phys. Rev.*, 48, 346, 1935.
357. Humphreys, C. J. and Paul, E., U.S. Nav. Ord. Lab., Navord Rep. 4589, 25, 1956.
358. Walters, F. M., *Sci. Pap. Bur. Stand.*, 17, 161, 1921.
359. Crawford, M. F. and McLay, A. B., *Proc. R. Soc. London Ser. A*, 143, 540, 1934.
360. McLay, A. D. and Crawford, M. F., Phys. Rev., 44, 986, 1933.
361. Schoepfle, G. K., *Phys. Rev.*, 47, 232, 1935.
362. Acquista, N., and Reader, J., *J. Opt. Soc. Am.*, 70, 789, 1980.
363. Benschop, H., Joshi, Y. N., and van Kleef, T. A. M., *Can. J. Phys.*, 53, 498, 1975.
364. Bockasten, K., Hallin, R., and Hughes, T. P., *Proc. Phys. Soc.*, 81, 522, 1963.
365. Boyce, J. C., *Phys. Rev.*, 46, 378, 1934.
366. Boyce, J. C., *Phys. Rev.*, 47, 718, 1935.
367. Boyce, J. C., *Phys. Rev.*, 48, 396, 1935.
368. Boyce, J. C., *Phys. Rev.*, 49, 351, 1936.
369. Corliss, C. H. and Meggers, W. F., *J. Res. Natl. Bur. Stand.*, 61, 269, 1958.
370. Crooker, A. M. and Dick, K. A., *Can. J. Phys.*, 42, 766, 1964.
371. De Bruin, T. L., *Z. Physik*, 77, 505, 1932.
372. De Bruin, T. L., *Proc. Roy. Acad. Amsterdam*, 36, 727, 1933.
373. De Bruin, T. L., *Zeeman Verhandelingen*, (The Hague), 1935, p. 415.
374. De Bruin, T. L., *Physica*, 3, 809, 1936.
375. De Bruin, T. L., *Proc. Roy. Acad. Amsterdam*, 40, 339, 1937.
376. Dick, K. A., *Can. J. Phys.*, 46, 1291, 1968.
377. Dick, K. A., unpublished, 1978.

378. Edlen, B. and Swensson, J. W., *Phys. Scr.*, 12, 21, 1975.
379. Ekberg, J. O., *Phys. Scr.*, 7, 55, 1973.
380. Ekberg, J. O., *Phys. Scr.*, 7, 59, 1973.
381. Ekberg, J. O., *Phys. Scr.*, 12, 42, 1975.
382. Ekberg, J. O. and Edlen, B., *Phys. Scr.*, 18, 107, 1978.
383. Eliason, A. Y., *Phys. Rev.*, 43, 745, 1933.
384. Gallardo, M., Massone, C. A., Tagliaferri, A. A., Garavaglia, M., and Persson, W., *Phys. Scr.*, 19, 538, 1979.
385. Garcia-Riquelme, O., *Optica Pura Y Aplicada*, 1, 53, 1968.
386. Gibbs, R. C., Vieweg, A. M., and Gartlein, C. W., *Phys. Rev.*, 34, 406, 1929.
387. Gilbert, W. P., *Phys. Rev.*, 48, 338, 1935.
388. Goldsmith, S. and Kaufman, A. S., *Proc. Phys. Soc.*, 81, 544, 1963.
389. Hermansdorfer, H., *J. Opt. Soc. Am.*, 62, 1149, 1972.
390. Humphreys, C. J., *Phys. Rev.*, 47, 712, 1935.
391. Humphreys, C. J., *J. Res. Natl. Bur. Stand.*, 16, 639, 1936.
392. Iglesias, L., *J. Opt. Soc. Am.*, 45, 856, 1955.
393. Iglesias, L., *J. Res. Natl. Bur. Stand.*, 64A, 481, 1960.
394. Iglesias, L., *Anales Fisica Y Quimica*, 58A, 191, 1962.
395. Iglesias, L., *J. Res. Natl. Bur. Stand.*, 70A, 465, 1966.
396. Iglesias, L., *Can. J. Phys.*, 44, 895, 1966.
397. Iglesias, L., *J. Res. Natl. Bur. Stand.*, 72A, 295, 1968.
398. Joshi, Y. N., *Can. Spectrosc.*, 15, 96, 1970.
399. Joshi, Y. N. and van Kleef, T. A. M., *Can. J. Phys.*, 55, 714, 1977.
400. Kaufman, A. S., Hughes, T. P., and Williams, R. V., *Proc. Phys. Soc.*, 76, 17, 1960.
401. Kaufman, V. and Sugar, J., *J. Opt. Soc. Am.*, 68, 1529, 1978.
402. Keussler, V., *Z. Physik*, 85, 1, 1933.
403. Kiess, C. C., *J. Res. Natl. Bur. Stand.*, 56, 167, 1956.
404. Klinkenberg, P. F. A., van Kleef, T. A. M., and Noorman, P. E., *Physica*, 27, 1177, 1961.
405. Kovalev, V. I., Romanos, A. A., and Ryabtsev, A. N., *Opt. Spectrosc.*, 43, 10, 1977.
406. Lang, R. J., *Proc. Natl. Acad. Sci.*, 13, 341, 1927.
407. Lang, R. J., *Zeeman Verhandelingen*, (The Hague), 44, 1935.
408. Liberman, S., et al., *C. R. Acad. Sci.* (Paris), 286, 253, 1978.
409. Livingston, A. E., *J. Phys.*, B9, L215, 1976.
410. Meijer, F. G., *Physica*, 72, 431, 1974.
411. Meijer, F. G. and Metsch, B. C., *Physica*, 94C, 259, 1978.
412. Moore, F. L., thesis, Princeton, 1949.
413. Paul, F. W. and Polster, H. D., *Phys. Rev.*, 59, 424, 1941.
414. Phillips, L. W. and Parker, W. L., *Phys. Rev.*, 60, 301, 1941.
415. Poppe, R., *Physica*, 81C, 351, 1976.
416. Raassen, A. J. J., van Kleef, T. A. M., and Metsch, B. C., *Physica*, 84C, 133, 1976.
417. Rao, A. B. and Krishnamurty, S. G., *Proc. Phys. Soc. (London)*, 51, 772, 1939.
418. Reader, J. and Acquista, N., *J. Opt. Soc. Am.*, 69, 239, 1979.
419. Reader, J. and Epstein, G. L., *J. Opt. Soc. Am.*, 62, 619, 1972.
420. Rico, F. R., *Anales, Real Soc. Esp. Fis. Quim.*, 61, 103, 1965.
421. Schonheit, E., *Optik*, 23, 409, 1966.
422. Shenstone, A. G., *J. Opt. Soc. Am.*, 44, 749, 1954.
423. Shenstone, A. G., unpublished, 1958.
424. Shenstone, A. G., *J. Res. Natl. Bur. Stand.*, 67A, 87, 1963.
425. Sugar, J. and Kaufman, V., *J. Opt. Soc. Am.*, 64, 1656, 1974.
426. Sugar, J. and Kaufman, V., *Phys. Rev.*, C12, 1336, 1975.
427. Svensson, L. A., *Phys. Scr.*, 13, 235, 1976.
428. Swensson, J. W. and Edlen, B., *Phys. Scr.*, 9, 335, 1974.
429. Tagliaferri, A. A., Gallego Lluesma, E., Garavaglia, M., Gallardo, M., and Massone, C. A., *Optica Pura Y Aplica*, 7, 89.
430. Tilford, S. G. and Giddings, L. E., *Astrophys. J.*, 141, 1222, 1965.
431. Trawick, M. W., *Phys. Rev.*, 46, 63, 1934.
432. Van Deurzen, C. H. H., *J. Opt. Soc. Am.*, 67, 476, 1977.
433. Yarosewick, S. L. and Moore, F. L., *J. Opt. Soc. Am.*, 57, 1381, 1967.
434. Zalubas, R., unpublished, 1979.
435. Bhatia, K. S., Jones, W. E., and Crooker, A. M., *Can. J. Phys.*, 50, 2421, 1972.
436. van Kleef, T. A. M. and Joshi, Y. N., *Phys. Scr.*, 24, 557, 1981.

NIST ATOMIC TRANSITION PROBABILITY TABLES

J.R. Fuhr and W.L. Wiese

These tables substantially update and enlarge our earlier tables in this *Handbook*. The new tables contain critically evaluated atomic transition probabilities for about 9000 selected lines of all elements for which reliable data are available on an absolute scale. The material is largely for neutral and singly ionized spectra, but also includes a number of prominent lines of more highly charged ions of important elements.

Many of the data are obtained from comprehensive compilations of the Data Center on Atomic Transition Probabilities at the National Institute of Standards and Technology (formerly the National Bureau of Standards). Specifically, data have been taken from three recent comprehensive critical compilations on C, N and O,[1] on Sc through Mn,[2] and Fe through Ni.[3] Material from earlier compilations for the elements H through Ne[4] and Na through Ca[5] was supplemented by more recent material taken directly from the original literature. For the highly charged ions, some of the data were derived from studies of the systematic behavior of transition probabilities.[6-8] Most of the original literature is cited in the above tables and in recent bibliographies[9,10]; for lack of space, individual literature references are not cited here.

The wavelength range for the neutral species is normally the visible spectrum or shorter wavelengths; only the very prominent near infrared lines are included. For the higher ions, most of the strong lines are located in the far UV. The tabulation is limited to electric dipole — including intercombination — lines and comprises essentially the fairly strong transitions with estimated uncertainties of 50% or less. With the exception of hydrogen, helium, and the alkalis, most transitions are between states with low principal quantum numbers.

The transition probability, A, is given in units of 10^8 s^{-1} and is listed to as many digits as is consistent with the indicated accuracy. The power of 10 is indicated by the E notation (i.e., E-02 means 10^{-2}). Generally, the estimated uncertainties of the A-values are ± 25 to 50% for two-digit numbers, ± 10 to 25% for three-digit numbers and $\pm 1\%$ or better for four- and five-digit numbers.

Each transition is identified by the wavelength, λ, in angstroms; and the statistical weights, g_i and g_k, of the lower (i) and upper (k) states [the product $g_k A$ (or $g_i f$) is needed for many applications]. Whenever the wavelengths of individual lines within a multiplet are extremely close, only an average wavelength for the multiplet as well as the multiplet A-value are given, and this is indicated by an asterisk (*) to the left of the wavelength. This also has been done when the transition probability for an entire multiplet has been taken from the literature and values for individual lines cannot be determined because of insufficient knowledge of the coupling of electrons. The wavelength data have been taken either from recent compilations or from the original literature cited in bibliographies published by the Atomic Energy Levels Data Center[11,12] at the National Institute of Standards and Technology. Wavelength values are consistent with those given in the table "Line Spectra of the Elements", which appears elsewhere in this *Handbook*.

The transition probabilities for hydrogen and hydrogen-like ions are known precisely. Because of the hydrogen degeneracy, a "transition" is actually the sum of all fine-structure transitions between the principal quantum numbers listed in the transition column; therefore, the special hydrogen table which appears below gives weighted average A-values.

In addition to the transition probability A, the atomic oscillator strength f and the line strength S are often used in the literature. The conversion factors between these quantities are (for electric-dipole transitions):

$$g_i f = 1.499 \times 10^{-8} \lambda^2 g_k A = 303.8 \lambda^{-1} S$$

where λ is in angstroms, A is in 10^8 s^{-1}, and S is in atomic units, which are
$a_0^2 e^2 = 7.188 \times 10^{-59}$ m^2 C^2.

After the special table for hydrogen, the tables for other elements appear in alphabetical sequence by element name (not symbol). Within each element, the tables are ordered by increasing ionization stage (e.g., Al I, Al II, etc.).

REFERENCES

1. Wiese, W.L., Fuhr, J.R., and Deters, T.M., *Atomic Transition Probabilities of Carbon, Nitrogen and Oxygen, J. Phys. Chem. Ref. Data, Monograph 7*, 1996.
2. Martin, G.A., Fuhr, J.R., and Wiese, W.L., *Atomic Transition Probabilities—Scandium through Manganese, J. Phys. Chem. Ref. Data, 17, Suppl.* 3, 1988.
3. Fuhr, J.R., Martin, G.A., and Wiese, W.L., *Atomic Transition Probabilities—Iron through Nickel, J. Phys. Chem. Ref. Data, 17, Suppl.* 4, 1988.
4. Wiese, W.L., Smith, M.W., and Glennon, B.M., *Atomic Transition Probabilities (H through Ne—A Critical Data Compilation)*, National Standard Reference Data Series, National Bureau of Standards 4, Vol. I, U.S. Government Printing Office, Washington, D.C., 1966.
5. Wiese, W.L., Smith, M.W., and Miles, B.M., *Atomic Transition Probabilities (Na through Ca—A Critical Data Compilation)*, National Standard Reference Data Series, National Bureau of Standards 22, Vol. II, U. S. Government Printing Office, Washington, D.C., 1969.
6. Wiese, W.L. and Weiss, A.W., *Phys. Rev.*, 175, 50, 1968.
7. Smith, M.W. and Wiese, M.L., *Astrophys. J., Suppl. Ser.*, 23, No. 196, 103, 1971.
8. Martin, G.A., and Wiese, W.L., *J. Phys. Chem. Ref. Data*, 5, 537, 1976.
9. Fuhr, J.R., Miller, B.J., and Martin, G.A., *Bibliography on Atomic Transition Probabilities (1914 through October 1997)*, National Bureau of Standards Special Publication 505, 1978; Miller, B.J., Fuhr, J.R., and Martin, G.A., *Bibliography on Atomic Transition Probabilities (November 1977 through February 1980)*, National Bureau of Standards Special Publication 505, Supplement 1, 1980.
10. Wiese, W.L., Reports on Astronomy, *Trans. Int. Astron. Union*, 18A, 116—123, 1982; 19A, 122—138, 1985.; 20A, 117—123, 1988, Reidel, D., Ed., Kluwer, Dordrecht, Holland.
11. Moore, C.E., *Bibliography on the Analyses of Optical Atomic Spectra*, National Bureau of Standards Special Publication 306—Section 1, 1968; Sections 2—4, 1969.

12. Hagan, L. and Martin, W.C., *Bibliography on Atomic Energy Levels and Spectra (July 1968 through June 1971)*, National Bureau of Standards Special Publication 363, 1972; Hagan, L., *Bibliography on Atomic Energy Levels and Spectra (July 1971 through June 1975)*, National Bureau of Standards Special Publication 363, Supplement 1, 1977; Zalubas, R. and Albright, A., *Bibliography on Atomic Energy Levels and Spectra (July 1975 through June 1979)*, National Bureau of Standards Special Publication 363, Supplement 2, 1980; Musgrove, A. and Zalubas, R., *Bibliography on Atomic Energy Levels and Spectra (July 1979 through December 1983)*, National Bureau of Standards Special Publication 363, Supplement 3, 1985.

13. Younger, S.M. and Weiss, A., *J. Res. Natl. Bur. Stand.*, 79A, 629, 1975.

Transition Probabilities for Allowed Lines of Hydrogen

λ Å	Weights g_i	g_k	A 10^8 s^{-1}	λ Å	Weights g_i	g_k	A 10^8 s^{-1}	λ Å	Weights g_i	g_k	A 10^8 s^{-1}
Hydrogen				3664.68	8	1568	4.022E-06	8598.40	18	392	9.211E-05
H I				3666.10	8	1458	4.826E-06	8665.02	18	338	1.343E-04
912.768	2	1800	5.167E-06	3667.68	8	1352	5.830E-06	8750.48	18	288	2.021E-04
912.839	2	1682	6.122E-06	3669.46	8	1250	7.096E-06	8862.79	18	242	3.156E-04
912.918	2	1568	7.297E-06	3671.48	8	1152	8.707E-06	9014.91	18	200	5.156E-04
913.006	2	1458	8.753E-06	3673.76	8	1058	1.078E-05	9229.02	18	162	8.905E-04
913.104	2	1352	1.057E-05	3676.36	8	968	1.347E-05	9545.97	18	128	1.651E-03
913.215	2	1250	1.286E-05	3679.35	8	882	1.700E-05	10049.4	18	98	3.358E-03
913.339	2	1152	1.578E-05	3682.81	8	800	2.172E-05	10938.1	18	72	7.783E-03
913.480	2	1058	1.952E-05	3686.83	8	722	2.809E-05	12818.1	18	50	2.201E-02
913.641	2	968	2.438E-05	3691.55	8	648	3.685E-05	16407.2	32	288	1.620E-04
913.826	2	882	3.077E-05	3697.15	8	578	4.910E-05	16806.5	32	242	2.556E-04
914.039	2	800	3.928E-05	3703.85	8	512	6.658E-05	17362.1	32	200	4.235E-04
914.286	2	722	5.077E-05	3711.97	8	450	9.210E-05	18174.1	32	162	7.459E-04
914.576	2	648	6.654E-05	3721.94	8	392	1.303E-04	18751.0	18	32	8.986E-02
914.919	2	578	8.858E-05	3734.37	8	338	1.893E-04	19445.6	32	128	1.424E-03
915.329	2	512	1.200E-04	3750.15	8	288	2.834E-04	21655.3	32	98	3.041E-03
915.824	2	450	1.657E-04	3770.63	8	242	4.397E-04	26251.5	32	72	7.711E-03
916.429	2	392	2.341E-04	3797.90	8	200	7.122E-04	27575	50	288	1.402E-04
917.181	2	338	3.393E-04	3835.38	8	162	1.216E-03	28722	50	242	2.246E-04
918.129	2	288	5.066E-04	3889.05	8	128	2.215E-03	30384	50	200	3.800E-04
919.351	2	242	7.834E-04	3970.07	8	98	4.389E-03	32961	50	162	6.908E-04
920.963	2	200	1.263E-03	4101.73	8	72	9.732E-03	37395	50	128	1.388E-03
923.150	2	162	2.143E-03	4340.46	8	50	2.530E-02	40511.5	32	50	2.699E-02
926.226	2	128	3.869E-03	4861.32	8	32	8.419E-02	43753	72	288	1.288E-04
930.748	2	98	7.568E-03	6562.80	8	18	4.410E-01	46525	50	98	3.253E-03
937.803	2	72	1.644E-02	8392.40	18	800	1.517E-05	46712	72	242	2.110E-04
949.743	2	50	4.125E-02	8413.32	18	722	1.964E-05	51273	72	200	3.688E-04
972.537	2	32	1.278E-01	8437.96	18	648	2.580E-05	59066	72	162	7.065E-04
1025.72	2	18	5.575E-01	8467.26	18	578	3.444E-05	74578	50	72	1.025E-02
1215.67	2	8	4.699E+00	8502.49	18	512	4.680E-05	75004	72	128	1.561E-03
3662.26	8	1800	2.847E-06	8545.39	18	450	6.490E-05	123680	72	98	4.561E-03
3663.40	8	1682	3.374E-06								

For hydrogen-like ions of nuclear charge Z, the following scaling laws hold:

$$A_Z = Z^4 A_{\text{Hydrogen}}; f_Z = f_H; S_Z = Z^{-2}S_H$$
$$(\text{For wavelengths, } \lambda_Z = Z^{-2}\lambda_H)$$

For very highly charged hydrogen-like ions, starting at about Z>25, relativistic corrections[13] must be applied.

Transition Probabilities for Other Elements

λ Å	Weights g_i	g_k	A 10^8 s^{-1}	λ Å	Weights g_i	g_k	A 10^8 s^{-1}	λ Å	Weights g_i	g_k	A 10^8 s^{-1}
Aluminum				1384.1	4	2	9.1E+00	*4761	2	6	2.55E-01
Al I				1605.8	2	4	1.22E+01	5172	2	4	3.95E-02
2263.5	2	4	6.6E-01	1611.8	4	4	2.42E+00	5551	4	6	3.85E-02
2269.1	4	6	7.9E-01	1611.9	4	6	1.45E+01	5687	4	4	6.0E-03
2269.2	4	4	1.3E-01	1854.7	2	4	5.40E+00				
2367.1	2	4	7.2E-01	1862.8	2	2	5.33E+00	**Argon**			
2373.1	4	6	8.6E-01	*1935.9	10	14	1.22E+01	**Ar I**			
2373.4	4	4	1.4E-01	3601.6	6	4	1.34E+00	1048.22	1	3	5.36E+00
2568.0	2	4	2.3E-01	3601.9	4	4	1.49E-01	1066.66	1	3	1.29E+00
2575.1	4	6	2.8E-01	3612.4	4	2	1.5E+00	3406.18	3	1	3.9E-03
2575.4	4	4	4.4E-02					3461.08	3	5	6.7E-04
2652.5	2	2	1.33E-01	**Al X**				3554.30	5	5	2.7E-03
2660.4	4	2	2.64E-01	39.925	1	3	2.22E+03	3563.29	1	3	1.2E-03
3082.2	2	4	6.3E-01	51.979	1	3	4.8E+03	3567.66	5	7	1.1E-03
3092.7	4	6	7.4E-01	55.227	1	3	5.2E+03	3572.30	3	1	5.1E-03
3092.8	4	4	1.2E-01	55.272	3	5	7.2E+03	3606.52	3	1	7.6E-03
3944.0	2	2	4.93E-01	55.376	5	7	9.5E+03	3632.68	3	5	6.6E-04
3961.5	4	2	9.8E-01	59.107	3	5	4.6E+03	3634.46	3	3	1.3E-03
6696.0	2	4	1.69E-02	332.78	1	3	5.6E+01	3643.12	3	5	2.4E-04
6698.7	2	2	1.69E-02	394.83	3	1	8.3E+01	3649.83	3	1	8.0E-03
7835.3	4	6	5.7E-02	395.36	3	5	1.2E+01	3659.53	3	3	4.4E-04
7836.1	6	8	6.2E-02	397.76	1	3	1.7E+01	3670.67	3	5	3.1E-04
				400.43	3	3	1.3E+01	3675.23	3	3	4.9E-04
Al II				401.12	5	5	3.6E+01	3770.37	1	3	7.0E-04
1047.9	1	3	3.6E-01	403.55	3	1	4.9E+01	3834.68	3	1	7.5E-03
1048.6	3	5	4.8E-01	406.31	5	3	1.9E+01	3894.66	3	3	5.7E-04
1539.8	3	5	8.8E+00	670.06	3	5	9.8E+00	3947.50	5	5	5.6E-04
1670.8	1	3	1.46E+01	2535	1	3	3.8E-01	3948.98	5	3	4.55E-03
1719.4	1	3	6.79E+00					4044.42	3	5	3.33E-03
1764.0	5	5	9.8E+00	**Al XI**				4045.96	3	3	4.1E-04
1772.8	1	3	9.5E+00	*36.675	2	6	1.5E+03	4054.53	3	3	2.7E-04
1777.0	5	7	1.7E+01	39.091	2	4	2.6E+03	4158.59	5	5	1.40E-02
*1819.0	15	15	5.6E+00	39.180	4	6	3.1E+03	4164.18	5	3	2.88E-03
1855.9	1	3	8.32E-01	39.530	2	2	1.8E+02	4181.88	1	3	5.61E-03
1858.0	3	3	2.48E+00	39.623	4	2	3.7E+02	4190.71	5	5	2.80E-03
1862.3	5	3	4.12E+00	48.298	2	4	3.09E+03	4191.03	1	3	5.39E-03
1931.0	3	1	1.08E+01	48.338	2	2	3.08E+03	4198.32	3	1	2.57E-02
1990.5	3	5	1.47E+01	52.299	2	4	8.1E+03	4200.67	5	7	9.67E-03
2816.2	3	1	3.83E+00	52.446	4	6	9.6E+03	4251.18	5	3	1.11E-03
4663.1	5	3	5.3E-01	52.458	4	4	1.6E+03	4259.36	3	1	3.98E-02
6226.2	1	3	6.2E-01	54.217	2	2	4.8E+02	4266.29	3	5	3.12E-03
6231.8	3	5	8.4E-01	54.388	4	2	9.6E+02	4272.17	3	3	7.97E-03
6243.4	5	7	1.1E+00	*99.083	2	6	2.2E+02	4300.10	3	5	3.77E-03
6335.7	5	3	1.4E-01	103.6	2	4	4.2E+02	4333.56	3	5	5.68E-03
6823.4	3	3	3.4E-01	103.8	4	6	5.0E+02	4335.34	3	3	3.87E-03
6837.1	5	3	5.7E-01	*141.6	2	6	4.07E+02	4345.17	3	3	2.97E-03
6920.3	3	1	9.6E-01	150.31	2	4	8.5E+02	4363.79	3	3	1.2E-04
7042.1	3	5	5.9E-01	150.61	4	6	9.9E+02	4424.00	1	3	7.3E-05
7056.7	3	3	5.8E-01	157.0	2	2	1.3E+02	4510.73	3	1	1.18E-02
7471.4	5	7	9.4E-01	157.4	4	2	2.6E+02	4522.32	1	3	8.98E-04
				*205.0	2	6	6.3E+01	4544.75	3	3	8.3E-04
Al III				*308.6	2	6	9.9E+01	4554.32	3	5	3.8E-04
*560.36	2	6	4.0E-01	*341.3	6	2	1.3E+02	4584.96	3	5	1.6E-03
695.83	2	4	7.4E-01	550.05	2	4	8.55E+00	4586.61	3	3	2.3E-03
696.22	2	2	7.2E-01	568.12	2	2	7.73E+00	4587.21	3	1	4.9E-03
*1352.8	10	14	4.40E+00	1997	2	4	1.07E+00	4589.29	3	5	6.2E-05
1379.7	2	2	4.59E+00	2069	2	2	9.7E-01	4596.10	3	3	9.47E-04

λ Å	Weights g_i	g_k	A 10^8 s⁻¹	λ Å	Weights g_i	g_k	A 10^8 s⁻¹	λ Å	Weights g_i	g_k	A 10^8 s⁻¹
4628.44	3	5	3.83E-04	5473.46	5	3	2.0E-03	6025.15	5	3	9.0E-03
4642.15	3	5	9.6E-04	5490.12	5	5	8.5E-04	6043.22	5	7	1.47E-02
4647.49	3	3	1.2E-03	5492.09	3	1	5.6E-03	6052.73	3	5	1.9E-03
4702.32	3	3	1.09E-03	5495.87	7	9	1.69E-02	6064.76	5	7	5.8E-04
4746.82	3	1	3.6E-03	5506.11	5	7	3.6E-03	6081.25	3	3	7.5E-04
4752.94	3	3	4.5E-03	5524.96	7	7	1.7E-03	6085.86	3	3	9.0E-05
4768.68	3	5	8.6E-04	5528.97	1	3	1.2E-03	6090.79	1	3	3.0E-03
4798.74	7	9	8.8E-04	5534.49	5	3	2.7E-03	6098.81	3	3	5.2E-03
4835.97	7	9	9.3E-04	5540.87	7	5	4.1E-04	6101.16	3	3	3.3E-03
4836.70	3	5	1.02E-03	5552.77	3	3	7.9E-04	6104.58	3	1	3.4E-03
4876.26	3	5	7.8E-03	5558.70	3	5	1.42E-02	6105.64	3	5	1.21E-02
4886.29	7	9	1.2E-03	5559.66	3	5	2.2E-03	6113.46	3	5	4.7E-04
4887.95	3	3	1.3E-02	5572.54	5	7	6.6E-03	6119.66	3	3	5.1E-03
4894.69	3	1	1.8E-02	5574.22	3	5	4.6E-04	6121.86	3	5	1.3E-04
4921.04	5	7	5.9E-04	5581.87	7	5	5.6E-04	6127.42	5	3	1.1E-03
4937.72	7	5	3.6E-04	5588.72	5	5	1.5E-03	6128.73	3	5	8.6E-04
4956.75	7	9	1.8E-03	5597.48	5	7	4.2E-03	6145.44	5	7	7.6E-03
4989.95	5	7	1.1E-03	5606.73	3	3	2.20E-02	6155.24	5	3	5.1E-03
5032.03	7	5	8.2E-04	5618.01	3	3	2.1E-03	6165.12	5	5	9.89E-04
5048.81	3	5	4.6E-03	5620.92	3	1	3.6E-03	6170.17	5	5	5.0E-03
5054.18	3	3	4.5E-03	5623.78	5	5	1.4E-03	6173.10	3	5	6.7E-03
5056.53	3	1	5.7E-03	5635.58	3	5	9.6E-04	6179.41	5	3	6.6E-04
5060.08	7	9	3.7E-03	5637.33	1	3	9.1E-04	6212.50	5	7	3.9E-03
5070.99	5	3	2.6E-03	5639.12	1	3	2.1E-03	6215.94	5	5	5.7E-03
5073.08	3	5	5.9E-04	5641.39	3	5	8.7E-04	6230.93	5	5	1.2E-04
5078.03	7	7	4.7E-04	5648.69	5	3	1.2E-03	6243.40	3	1	1.3E-03
5087.09	5	7	1.6E-03	5650.70	3	1	3.20E-02	6244.73	3	5	2.0E-04
5104.74	3	5	8.7E-04	5659.13	5	5	2.6E-03	6248.41	3	5	6.8E-04
5118.21	5	7	2.7E-03	5681.90	5	7	2.0E-03	6278.65	5	7	2.0E-04
5127.80	5	5	3.3E-04	5683.73	5	5	2.0E-03	6296.87	3	5	9.0E-03
5151.39	3	1	2.39E-02	5700.87	5	7	5.9E-03	6307.66	5	5	6.0E-03
5152.30	3	5	1.1E-03	5712.51	1	3	8.7E-04	6309.14	3	3	7.6E-04
5162.29	3	3	1.90E-02	5739.52	3	5	8.7E-04	6364.89	3	1	5.6E-03
5177.54	7	5	2.4E-03	5772.11	5	7	2.0E-03	6369.58	5	3	4.2E-03
5192.72	7	7	1.2E-04	5773.99	5	5	1.1E-03	6384.72	3	3	4.21E-03
5194.02	3	1	7.8E-03	5783.54	3	5	8.1E-04	6416.31	3	5	1.16E-02
5210.49	7	7	1.1E-03	5789.48	5	5	4.6E-04	6431.56	5	3	5.1E-04
5214.77	5	3	2.1E-03	5790.40	5	3	3.4E-04	6466.55	1	3	1.5E-03
5216.28	5	3	1.3E-03	5802.08	5	3	4.2E-04	6481.14	1	3	9.4E-04
5221.27	7	9	8.8E-03	5843.77	3	5	3.3E-04	6513.85	3	3	5.4E-04
5241.09	5	5	1.3E-03	5882.62	3	1	1.23E-02	6538.11	7	7	1.1E-03
5246.24	5	7	1.2E-03	5888.58	7	5	1.29E-02	6596.12	7	5	2.3E-04
5249.20	5	5	7.9E-04	5916.58	5	3	5.9E-04	6598.68	5	5	3.6E-04
5252.79	5	7	5.4E-03	5927.11	7	7	3.7E-04	6604.02	7	5	2.8E-03
5254.47	3	5	3.6E-03	5928.81	5	3	1.1E-02	6604.85	5	7	1.3E-04
5286.07	5	7	9.6E-04	5940.86	1	3	1.2E-03	6632.09	3	3	5.3E-04
5290.00	5	3	9.0E-04	5942.67	5	5	1.8E-03	6656.88	3	3	3.1E-04
5309.52	5	5	1.2E-03	5943.89	7	5	3.6E-04	6660.68	3	1	7.8E-03
5317.73	5	7	2.6E-03	5949.26	3	3	1.5E-03	6664.05	5	5	1.5E-03
5373.50	3	5	2.7E-03	5964.48	1	3	7.7E-04	6677.28	3	1	2.36E-03
5393.27	5	5	9.6E-04	5968.32	3	3	1.8E-03	6684.73	3	5	3.9E-04
5410.48	5	7	2.0E-03	5971.60	3	1	1.1E-02	6698.47	3	3	2.5E-04
5421.35	7	5	6.0E-03	5981.90	5	7	1.2E-04	6698.88	5	3	1.6E-03
5439.99	3	3	1.9E-03	5987.30	7	7	1.2E-03	6719.22	1	3	2.4E-03
5442.24	7	7	9.3E-04	5988.13	3	5	6.1E-04	6722.88	5	7	3.2E-04
5451.65	3	5	4.7E-03	5994.66	3	5	2.6E-04	6752.84	3	5	1.93E-02
5457.42	5	3	3.6E-03	5999.00	5	5	1.4E-03	6754.37	3	3	2.1E-03
5459.65	7	7	3.8E-04	6005.73	5	3	1.4E-03	6756.10	5	5	3.6E-03
5467.16	5	5	7.6E-04	6013.68	7	5	1.4E-03	6766.61	5	3	4.0E-03

λ Å	Weights g_i	g_k	A 10^8 s^{-1}	λ Å	Weights g_i	g_k	A 10^8 s^{-1}	λ Å	Weights g_i	g_k	A 10^8 s^{-1}
6779.93	1	3	1.21E-03	8014.79	5	5	9.28E-02	13622.4	3	5	7.3E-02
6818.29	3	1	2.0E-03	8037.23	1	3	3.59E-03	13678.5	3	5	6.2E-02
6827.25	5	3	2.4E-03	8046.13	3	1	1.12E-02	14093.6	1	3	4.3E-02
6851.88	3	5	6.7E-04	8053.31	5	3	8.6E-03	14739.1	5	7	8.8E-04
6871.29	3	3	2.78E-02	8066.60	5	5	1.4E-03	15046.4	1	3	5.2E-02
6879.59	3	5	1.8E-03	8103.69	3	3	2.5E-01	15172.3	1	3	1.3E-02
6887.10	5	7	1.3E-03	8115.31	5	7	3.31E-01	15329.6	5	5	1.2E-03
6888.17	3	5	2.5E-03	8264.52	3	3	1.53E-01	15555.5	5	7	9.8E-05
6925.01	3	3	1.2E-03	8384.73	5	7	2.4E-03	15734.9	5	3	2.9E-04
6937.67	3	1	3.08E-02	8408.21	3	5	2.23E-01	15816.8	5	3	8.7E-04
6951.46	5	5	2.2E-03	8424.65	3	5	2.15E-01	15989.3	1	3	1.9E-02
6960.23	5	5	2.4E-03	8490.30	3	5	9.6E-04	16122.7	5	3	3.9E-04
6965.43	5	3	6.39E-02	8521.44	3	3	1.39E-01	16180.0	5	5	1.2E-03
6992.17	3	1	7.5E-03	8605.78	5	5	1.04E-02	16264.1	3	3	3.0E-04
7030.25	7	5	2.67E-02	8620.46	1	3	9.2E-03	16520.1	3	5	2.6E-03
7067.22	5	5	3.80E-02	8667.94	1	3	2.43E-02	16739.8	3	5	3.1E-03
7068.73	5	3	2.0E-02	8761.69	3	5	9.5E-03	16940.4	5	5	2.5E-02
7086.70	1	3	1.5E-03	8784.61	3	1	2.4E-03	20317.0	1	3	1.6E-03
7107.48	5	5	4.5E-03	8799.08	5	3	4.6E-03	20616.5	5	5	3.9E-03
7125.83	3	3	6.0E-03	8962.19	3	3	1.6E-03	20812.0	5	7	7.6E-04
7147.04	5	3	6.25E-03	9075.42	3	1	1.2E-02	21332.2	3	3	3.2E-04
7158.83	3	1	2.1E-02	9122.97	5	3	1.89E-01	21534.9	3	5	1.1E-03
7162.57	1	3	5.8E-04	9194.64	3	3	1.76E-02	22039.2	3	1	1.2E-03
7206.98	5	3	2.48E-02	9224.50	3	5	5.03E-02	22077.4	5	3	1.4E-03
7229.93	5	5	6.6E-04	9291.53	3	1	3.26E-02	23133.4	3	3	1.7E-03
7265.17	3	3	1.7E-03	9354.22	3	3	1.06E-02	23844.8	9	7	1.1E-02
7270.66	7	7	1.1E-03	9657.78	3	3	5.43E-02	23967.5	3	1	3.6E-03
7272.93	3	3	1.83E-02	9784.50	3	5	1.47E-02				
7285.44	5	3	1.2E-03	10470.05	1	3	9.8E-03	**Ar II**			
7311.72	3	3	1.7E-02	10478.0	3	3	2.44E-02	2317.7	6	4	1.4E-01
7316.01	3	3	9.6E-03	10950.7	5	3	3.96E-03	2891.6	4	2	1.82E-01
7350.78	3	1	1.2E-02	11078.9	5	5	8.3E-03	2942.9	4	4	5.3E-01
7353.32	5	7	9.6E-03	11393.7	3	1	2.22E-02	2979.1	2	2	4.16E-01
7372.12	7	9	1.9E-02	11441.8	5	3	1.39E-02	3033.5	2	4	9.9E-02
7383.98	3	5	8.47E-02	11467.5	3	5	3.69E-03	3139.0	6	6	5.2E-01
7392.97	5	3	7.2E-03	11488.11	3	3	1.9E-03	3169.7	4	6	4.9E-01
7412.33	3	5	3.9E-03	11668.7	5	5	3.76E-02	3181.0	6	4	3.7E-01
7422.26	3	5	6.6E-04	11719.5	5	3	9.52E-03	3212.5	4	4	5.2E-02
7425.29	5	7	3.1E-03	12026.6	1	3	4.2E-03	3221.6	6	6	1.8E-02
7435.33	5	5	9.0E-03	12112.2	7	7	3.1E-03	3226.0	4	4	2.1E-02
7436.25	7	5	2.7E-03	12139.8	3	3	4.5E-02	3243.7	4	2	1.06E+00
7471.17	3	3	2.2E-04	12343.7	5	7	2.0E-02	3249.8	2	4	6.3E-01
7484.24	3	5	3.4E-03	12402.9	3	3	1.1E-01	3263.6	2	4	1.55E-01
7503.84	3	1	4.45E-01	12439.2	3	5	4.9E-02	3281.7	2	2	4.2E-01
7510.42	5	5	4.5E-03	12456.1	5	5	8.9E-02	3430.4	6	8	6.2E-02
7514.65	3	1	4.02E-01	12487.6	7	5	1.1E-01	3454.1	6	4	3.14E-01
7618.33	3	5	2.9E-03	12554.4	7	5	1.2E-03	3466.3	8	6	3.0E-02
7628.86	3	5	2.9E-03	12702.4	3	3	7.1E-02	3476.7	6	6	1.25E+00
7635.11	5	5	2.45E-01	12733.6	5	5	1.1E-01	3491.2	4	4	1.79E+00
7670.04	5	3	2.8E-03	12746.3	3	3	2.0E-02	3491.5	6	8	2.31E+00
7704.81	5	7	6.3E-04	12802.7	5	5	5.7E-02	3509.8	2	2	2.55E+00
7723.76	5	3	5.18E-02	12933.3	3	1	1.0E-01	3514.4	4	6	1.36E+00
7724.21	1	3	1.17E-01	12956.6	3	3	7.4E-02	3520.0	6	6	5.2E-01
7798.55	3	5	8.7E-04	13008.5	5	3	8.9E-02	3521.3	8	8	2.27E-01
7868.20	1	3	3.50E-03	13214.7	3	1	8.1E-02	3535.3	2	4	5.7E-01
7891.08	5	5	9.5E-03	13273.1	5	7	1.5E-01	3548.5	4	4	8.7E-01
7916.45	3	3	1.2E-03	13313.4	3	5	1.3E-01	3550.0	6	6	2.6E-02
7948.18	1	3	1.86E-01	13504.0	5	7	1.1E-01	3556.9	2	2	5.0E-02
8006.16	3	5	4.90E-02	13599.2	5	5	2.2E-02	3559.5	6	8	2.88E+00

λ Å	Weights g_i	g_k	A 10^8 s^{-1}	λ Å	Weights g_i	g_k	A 10^8 s^{-1}	λ Å	Weights g_i	g_k	A 10^8 s^{-1}
3565.0	2	4	5.5E-01	4228.2	4	6	1.31E-01	6483.1	4	2	1.06E-01
3576.6	6	8	2.75E+00	4237.2	4	4	1.12E-01	6638.2	6	4	1.37E-01
3581.6	2	4	1.76E+00	4266.5	6	6	1.64E-01	6639.7	4	2	1.69E-01
3582.4	4	6	2.53E+00	4277.5	6	4	8.0E-01	6643.7	10	8	1.47E-01
3588.4	8	10	3.03E+00	4282.9	4	2	1.32E-01	6666.4	2	2	8.8E-02
3605.9	4	6	4.4E-02	4300.6	6	6	5.7E-02	6684.3	8	6	1.07E-01
3656.0	6	6	7.6E-02	4331.2	4	4	5.74E-01	6756.6	4	4	2.0E-02
3682.5	4	2	1.7E-02	4332.0	4	2	1.92E-01	6863.5	6	6	2.5E-02
3709.9	4	4	4.7E-02	4348.1	6	8	1.17E+00	7233.5	2	4	3.7E-02
3717.2	6	8	5.2E-02	4352.2	2	2	2.12E-01	7380.4	4	4	5.6E-02
3729.3	6	4	4.80E-01	4362.1	4	6	5.5E-02	7589.3	6	4	1.07E-01
3746.9	4	6	2.1E-02	4370.8	4	4	6.6E-01				
3763.5	8	6	1.78E-01	4371.3	6	4	2.21E-01	**Ar III**			
3766.1	4	4	7.4E-02	4376.0	4	2	2.05E-01	769.15	5	3	6.0E+00
3777.5	2	2	1.1E-02	4379.7	2	2	1.00E+00	871.10	5	3	1.59E+00
3780.8	8	8	7.7E-01	4383.8	4	4	1.1E-02	875.53	3	1	3.74E+00
3786.4	8	6	1.5E-02	4400.1	4	4	1.60E-01	878.73	5	5	2.79E+00
3799.4	6	4	1.7E-01	4401.0	8	6	3.04E-01	879.62	3	3	9.2E-01
3808.6	6	6	1.0E-02	4412.9	6	8	6.1E-02	883.18	1	3	1.22E+00
3826.8	6	6	2.81E-01	4420.9	2	4	3.1E-02	887.40	3	5	9.0E-01
3841.5	4	2	2.69E-01	4426.0	4	6	8.17E-01	3024.1	5	7	2.6E+00
3844.7	6	8	4.8E-02	4430.2	2	4	5.69E-01	3027.2	5	5	6.4E-01
3845.4	6	4	1.6E-02	4431.0	6	6	1.09E-01	3054.8	3	5	1.9E+00
3850.6	4	4	3.87E-01	4460.6	4	6	1.5E-02	3064.8	3	3	1.0E+00
3868.5	4	6	1.4E+00	4474.8	4	2	2.90E-01	3078.2	1	3	1.4E+00
3872.1	4	4	1.5E-01	4481.8	6	6	4.55E-01	3285.9	5	7	2.0E+00
3875.3	4	2	8.2E-02	4491.0	6	4	4.6E-01	3301.9	5	5	2.0E+00
3880.3	2	2	2.32E-01	4530.5	6	4	2.1E-02	3311.3	5	3	2.0E+00
3891.4	2	2	4.3E-02	4545.1	4	4	4.71E-01	3336.1	7	9	2.0E+00
3892.0	6	4	6.3E-02	4579.4	2	2	8.0E-01	3344.7	5	7	1.8E+00
3900.6	4	6	7.2E-02	4589.9	4	6	6.64E-01	3352.1	7	7	2.2E-01
3911.6	2	4	7.7E-02	4598.8	4	4	6.7E-02	3358.5	3	5	1.6E+00
3914.8	4	4	3.7E-02	4609.6	6	8	7.89E-01	3361.3	5	5	3.0E-01
3928.6	2	4	2.44E-01	4637.2	6	6	7.1E-02	3472.6	5	7	2.0E-01
3931.2	2	4	2.0E-02	4657.9	4	2	8.92E-01	3480.6	7	7	1.6E+00
3932.5	4	4	9.3E-01	4726.9	4	4	5.88E-01	3499.7	3	3	1.3E+00
3944.3	8	6	4.1E-02	4732.1	6	4	6.7E-02	3500.6	3	5	2.6E-01
3952.7	4	4	2.08E-01	4735.9	6	4	5.80E-01	3502.7	5	3	4.3E-01
3958.4	6	4	3.8E-02	4764.9	2	4	6.4E-01	3503.6	5	5	1.2E+00
3968.4	6	6	4.8E-02	4806.0	6	6	7.80E-01	3511.7	7	5	2.6E-01
3979.4	4	2	9.8E-01	4847.8	4	2	8.49E-01				
3988.2	6	6	4.1E-02	4879.9	4	6	8.23E-01	**Ar IV**			
3992.1	4	6	1.6E-02	4889.0	2	2	1.9E-01	840.03	4	2	2.73E+00
4013.9	8	8	1.05E-01	4904.8	6	8	3.7E-02	843.77	4	4	2.70E+00
4031.4	4	2	7.5E-02	4933.2	4	4	1.44E-01	850.60	4	6	2.63E+00
4035.5	4	6	4.4E-02	4965.1	2	4	3.94E-01				
4038.8	6	8	1.2E-02	4972.2	2	2	9.7E-02	**Ar VI**			
4042.9	4	4	4.06E-01	5009.3	4	6	1.51E-01	292.15	2	2	6.9E+01
4045.7	4	4	1.6E-02	5017.2	4	6	2.07E-01	294.05	4	2	1.36E+02
4052.9	2	4	6.7E-01	5017.6	4	4	1.1E-02				
4065.1	4	4	1.1E-02	5062.0	2	4	2.23E-01	**Ar VII**			
4072.0	6	6	5.8E-01	5141.8	6	8	8.1E-02	*250.41	9	3	2.78E+02
4079.6	6	4	1.19E-01	5145.3	4	6	1.06E-01	*477.54	9	15	9.92E+01
4082.4	6	6	2.9E-02	5176.2	6	6	1.7E-02	585.75	1	3	7.83E+01
4112.8	4	4	1.1E-02	6103.5	2	2	1.7E-02	*637.30	9	9	6.7E+01
4128.6	8	6	1.4E-02	6114.9	10	8	2.00E-01				
4131.7	4	2	8.5E-01	6138.7	6	4	1.2E-02	**Ar VIII**			
4178.4	6	4	1.2E-02	6172.3	8	6	2.00E-01	158.92	2	4	1.1E+02
4202.0	2	4	2.1E-02	6243.1	8	6	3.0E-02	159.18	2	2	1.11E+02

λ Å	g_i	g_k	A 10^8 s⁻¹	λ Å	g_i	g_k	A 10^8 s⁻¹	λ Å	g_i	g_k	A 10^8 s⁻¹
229.44	2	2	1.12E+02	2860.4	2	2	5.5E-01	5777.6	5	7	6.5E-01
230.88	4	2	2.21E+02	2898.7	4	2	9.9E-02	5784.0	3	5	2.1E-01
337.09	4	4	1.2E+01					5800.2	5	5	9.9E-02
337.26	6	4	1.0E+02	**Barium**				5805.7	7	7	1.1E-02
338.22	4	2	1.1E+02	**Ba I**				5826.3	5	3	5.6E-01
519.43	2	4	6.3E+01	2409.2	1	3	8.6E-04	5907.6	3	5	1.5E-02
526.46	4	6	7.2E+01	2414.1	1	3	1.5E-03	5971.7	5	5	1.8E-01
526.87	4	4	1.2E+01	2420.1	1	3	2.3E-03	5997.1	3	3	2.7E-01
700.24	2	4	2.55E+01	2427.4	1	3	5.6E-03	6019.5	3	1	1.4E+00
713.81	2	2	2.4E+01	2432.5	1	3	7.2E-03	6063.1	5	3	5.7E-01
				2438.8	1	3	1.4E-03	6083.4	3	1	1.1E-01
Ar IX				2444.6	1	3	4.5E-03	6110.8	7	5	5.5E-01
48.739	1	3	1.69E+03	2452.4	1	3	8.1E-04	6129.2	3	1	6.0E-02
				2473.2	1	3	4.6E-03	6341.7	5	7	1.9E-01
Ar XIII				2500.2	1	3	1.5E-02	6450.9	3	5	1.1E-01
162.96	5	3	3.4E+02	2543.2	1	3	4.1E-02	6482.9	5	7	4.4E-01
*163.08	9	3	5.3E+02	2596.6	1	3	1.2E-01	6498.8	7	7	8.6E-01
184.90	5	5	1.66E+02	2646.5	1	3	1.1E-02	6527.3	5	5	5.9E-01
186.38	1	3	8.8E+01	2702.6	1	3	2.5E-02	6595.3	3	3	3.9E-01
*207.89	9	9	9.5E+01	2739.2	1	3	9.1E-03	6675.3	5	3	1.9E-01
*245.10	9	15	3.7E+01	2785.3	1	3	2.8E-02	6693.8	7	5	2.8E-01
				3071.6	1	3	4.1E-01	6865.7	5	5	2.3E-02
Ar XIV				3501.1	1	3	1.9E-01	7059.9	7	9	7.1E-01
180.29	2	4	4.5E+01	3889.3	1	3	8.8E-03	7120.3	3	5	2.1E-01
183.41	2	2	1.69E+02	3909.9	3	5	4.9E-01	7195.2	1	3	2.4E-01
187.95	4	2	1.97E+02	3935.7	5	7	4.7E-01	7280.3	5	7	5.3E-01
191.35	4	2	7.5E+01	3937.9	5	5	1.1E-01	7392.4	3	3	5.0E-01
194.39	2	2	4.6E+01	3993.4	7	9	5.5E-01	7417.5	7	5	2.5E-02
203.35	4	2	7.8E+01	3995.7	7	7	8.8E-02	7488.1	7	7	1.0E-01
				4132.4	1	3	7.1E-03	7528.2	5	5	2.7E-02
Ar XV				4239.6	5	3	2.4E-01	7672.1	3	5	3.1E-01
25.05	1	3	1.7E+04	4242.6	3	5	5.6E-02	7780.5	5	5	1.3E-01
221.10	1	3	9.55E+01	4264.4	1	3	1.5E-01	7905.8	5	3	6.3E-01
*265.3	9	9	8.1E+01	4283.1	5	7	6.4E-01	7911.3	1	3	2.98E-03
				4323.0	3	5	1.5E-01	8147.7	5	5	6.3E-01
Ar XVI				4325.2	5	7	7.1E-02	9645.6	7	5	1.1E-01
*23.52	2	6	1.43E+04	4332.9	3	3	1.5E-01	9704.3	3	1	1.6E-01
*24.96	6	10	4.4E+04	4350.3	3	5	6.0E-01	9821.5	3	1	5.5E-02
353.88	2	4	1.5E+01	4402.5	3	5	2.7E-01	10370.3	3	5	1.3E-02
389.11	2	2	1.1E+01	4406.8	5	5	1.0E-01	10649.1	5	5	2.7E-02
1268	2	4	1.9E+00	4431.9	1	3	1.2E+00	11075.7	3	3	3.6E-05
1401	2	2	1.4E+00	4467.1	5	7	6.6E-02	11303.1	5	3	1.2E-03
2975	2	4	9.0E-02	4489.0	5	7	4.2E-01	11373.8	3	1	1.3E-01
3514	4	6	6.5E-02	4493.6	5	5	3.6E-01	14158.4	9	7	2.0E-03
				4505.9	3	3	1.1E+00	14723.2	3	5	8.6E-03
Arsenic				4523.2	5	5	9.6E-01	14999.9	5	3	2.8E-03
As I				4573.9	3	1	1.21E+00	17123.7	7	7	3.3E-03
1890.4	4	6	2.0E+00	4579.6	5	5	7.0E-01	17187.1	3	1	2.7E-02
1937.6	4	4	2.0E+00	4591.8	5	5	1.6E-02	20563.9	5	7	2.6E-03
1972.6	4	2	2.0E+00	4599.7	3	1	4.07E-01				
2288.1	6	4	2.8E+00	4605.0	3	1	7.7E-02	**Ba II**			
2344.0	2	4	3.5E-01	4619.9	1	3	9.3E-02	1413.4	6	8	1.7E-02
2349.8	4	2	3.1E-01	4628.3	5	3	6.0E-02	1417.1	4	6	3.8E-02
2369.7	4	4	6.0E-01	4673.6	7	5	6.5E-02	1444.9	4	6	8.1E-02
2370.8	4	6	4.2E-01	4691.6	5	3	1.6E+00	1461.5	6	8	8.7E-02
2456.5	6	4	7.2E-02	4700.4	3	3	2.4E-01	1487.0	4	6	1.4E-01
2492.9	4	2	1.2E-01	4726.4	5	3	4.6E-01	1503.9	6	8	1.5E-01
2745.0	2	4	2.6E-01	5519.1	3	5	5.0E-01	1554.4	4	6	2.6E-01
2780.2	4	4	7.8E-01	5535.5	1	3	1.19E+00	1572.7	6	8	2.4E-01

λ Å	g_i	g_k	A 10^8 s^{-1}	λ Å	g_i	g_k	A 10^8 s^{-1}	λ Å	g_i	g_k	A 10^8 s^{-1}
1573.9	6	6	1.6E-02	5428.8	6	4	2.3E-02	2061.7	4	6	9.9E-01
1630.4	2	2	1.7E-02	5480.3	8	6	1.8E-02	2110.3	4	2	9.1E-01
1674.5	4	6	2.2E-01	5784.2	2	4	2.0E-01	2177.3	4	2	2.6E-02
1694.4	6	8	2.1E-01	5853.7	4	4	4.8E-02	2228.3	4	4	8.9E-01
1697.2	6	6	1.7E-02	5981.3	4	6	1.6E-01	2230.6	4	6	2.6E+00
1761.8	4	4	3.9E-03	5999.9	4	4	2.6E-02	2276.6	4	4	2.5E-01
1771.0	4	2	3.4E-02	6135.8	2	2	8.5E-02	2515.7	4	6	4.3E-02
1786.9	6	4	4.4E-02	6141.7	6	4	3.7E-01	2627.9	4	4	4.7E-01
1892.7	2	4	9.0E-02	6363.2	6	4	2.9E-03	2696.8	4	6	6.4E-02
1904.2	4	6	1.1E-02	6372.9	4	4	6.7E-04	2780.5	4	2	3.09E-01
1906.8	2	2	5.1E-02	6378.9	4	2	9.9E-02	2798.7	6	6	3.6E-02
1924.7	6	8	3.1E-02	6457.7	6	4	3.0E-03	2898.0	4	2	1.53E+00
1954.2	4	6	1.3E-01	6496.9	4	2	3.32E-01	2938.3	6	4	1.23E+00
1955.1	4	4	1.8E-02	7556.8	6	4	1.6E-03	2989.0	4	4	5.5E-01
1970.2	4	2	6.7E-02	7678.2	8	6	6.6E-04	2993.3	4	6	1.6E-01
1985.6	2	4	2.5E-01	8710.7	6	8	8.0E-01	3024.6	6	6	8.8E-01
1999.5	2	4	1.0E-01	8737.7	4	6	9.3E-01	3067.7	4	2	2.07E+00
2009.2	2	2	8.6E-02					3076.7	4	4	3.5E-02
2052.7	4	6	2.0E-01	**Beryllium**				3397.2	6	4	1.81E-01
2054.6	4	4	2.9E-02	**Be I**				3402.9	6	6	1.6E-02
2080.0	4	2	1.0E-01	1491.8	1	3	1.3E-02	3510.9	6	4	6.8E-02
2153.9	2	4	5.3E-01	1661.5	1	3	2.0E-01	3596.1	2	4	1.98E-01
2200.9	2	2	2.0E-01	2348.6	1	3	5.55E+00	3888.2	2	2	6.9E-02
2232.8	4	6	2.9E-01	*2494.7	9	15	1.6E+00	4121.5	2	2	1.64E-01
2235.4	4	4	4.4E-02	*2650.6	9	9	4.24E+00	4308.5	2	4	1.6E-02
2286.0	4	2	1.3E-01	4572.7	3	5	7.9E-01	4493.0	2	4	1.5E-02
2528.5	2	4	7.1E-01					4722.5	4	2	1.17E-01
2634.8	4	6	7.6E-01	**Be II**				6134.8	4	4	1.8E-02
2641.4	4	4	1.2E-01	1197.1	2	2	4.7E-01				
2647.3	2	2	2.0E-01	1197.2	4	2	9.4E-01	**Boron**			
2771.4	4	2	4.0E-01	1512.3	2	4	9.2E+00	**B I**			
3816.7	4	6	2.3E-03	1512.4	4	6	1.1E+01	1378.6	2	4	3.50E+00
3842.8	6	8	2.2E-03	1776.1	2	2	1.4E+00	1378.9	2	2	1.40E+01
3891.8	2	4	1.67E+00	1776.3	4	2	2.9E+00	1378.9	4	4	1.75E+01
4024.1	6	4	5.3E-03	*2453.8	2	6	1.42E-01	1379.2	4	2	7.0E+00
4057.5	8	6	1.2E-02	3046.5	2	4	4.8E-01	1465.5	2	4	3.34E+00
4130.7	4	6	1.80E+00	3046.7	4	6	5.9E-01	1465.7	4	4	6.7E+00
4166.0	4	4	3.7E-01	3130.4	2	4	1.14E+00	1465.8	6	4	1.00E+01
4216.0	2	4	5.8E-02	3131.1	2	2	1.15E+00	1825.9	2	4	1.76E+00
4287.8	2	2	2.4E-02	3241.6	2	2	1.41E-01	1826.4	4	6	2.11E+00
4325.7	4	6	5.9E-02	3241.8	4	2	2.8E-01	2088.9	2	4	2.8E-01
4329.6	4	4	8.8E-03	3274.6	2	4	1.9E-01	2089.6	4	6	3.3E-01
4405.2	4	2	3.9E-02	3274.7	2	2	1.9E-01	2496.8	2	2	8.64E-01
4470.7	6	4	1.4E-02	4360.7	2	4	9.2E-01	2497.7	4	2	1.73E+00
4509.6	8	6	1.2E-02	4361.0	4	6	1.1E+00				
4524.9	2	2	7.2E-01	*5255.9	2	6	2.56E-02	**Bromine**			
4554.0	2	4	1.17E+00	5270.3	2	2	3.30E-01	**Br I**			
4708.9	2	4	9.7E-02	5270.8	4	2	6.6E-01	1488.5	4	4	1.2E+00
4843.5	4	6	9.3E-02	6279.4	2	4	1.2E-01	1540.7	4	4	1.4E+00
4847.1	2	4	4.1E-02	6279.7	4	6	1.43E-01	1574.8	2	4	2.0E-01
4850.8	4	4	1.4E-01	6756.7	2	2	5.1E-01	1576.4	4	6	2.1E-02
4900.0	4	2	7.75E-01	6757.1	4	2	1.02E-01	1633.4	2	4	8.1E-02
4934.1	2	2	9.55E-01	7401.2	2	4	3.0E-02	4365.1	2	4	7.5E-03
4997.8	4	2	6.1E-02	7401.4	2	2	3.0E-02	4425.1	4	2	4.2E-03
5185.0	2	4	1.8E-02					4441.7	6	4	7.5E-03
5361.4	4	6	4.8E-02	**Bismuth**				4472.6	4	4	9.3E-03
5391.6	6	8	5.2E-02	**Bi I**				4477.7	6	8	1.3E-02
5413.6	6	6	8.4E-04	1954.5	4	6	1.2E+00	4513.4	6	4	2.8E-03
5421.1	6	6	1.9E-03	2021.2	4	4	6.0E-02				

λ Å	Weights g_i	g_k	A 10^8 s^{-1}	λ Å	Weights g_i	g_k	A 10^8 s^{-1}	λ Å	Weights g_i	g_k	A 10^8 s^{-1}
4525.6	6	6	7.2E-03	3361.9	5	7	2.23E-01	6493.8	3	5	4.4E-01
4575.7	4	4	1.6E-02	3624.1	1	3	2.12E-01	6499.7	5	5	8.1E-02
4614.6	4	6	5.4E-03	3630.8	3	5	2.97E-01				
4979.8	4	4	2.6E-03	3631.0	3	3	1.53E-01	**Ca II**			
5245.1	2	4	3.1E-03	3644.4	5	7	3.55E-01	1341.9	2	4	1.5E-02
5345.4	2	4	7.6E-03	3644.8	5	5	9.4E-02	1342.5	2	2	1.5E-02
7348.5	4	6	1.2E-01	3870.5	3	5	7.2E-02	1649.9	2	4	3.2E-03
7513.0	6	4	1.2E-01	3957.1	3	3	9.8E-02	1652.0	2	2	3.1E-03
7803.0	2	4	5.3E-02	3973.7	5	3	1.75E-01	1673.9	2	4	2.24E+00
7938.7	6	6	1.9E-01	4092.6	3	5	1.1E-01	1680.1	4	6	2.65E-01
8131.5	2	4	3.8E-02	4094.9	5	7	1.2E-01	1680.1	4	4	4.41E-02
8343.7	2	2	2.2E-01	4098.5	7	9	1.3E-01	1807.3	2	4	3.54E-01
8446.6	4	4	1.2E-01	4108.5	5	7	9.0E-01	1814.5	4	6	4.2E-01
8638.7	6	4	9.7E-02	4226.7	1	3	2.18E+00	1814.7	4	4	7.0E-02
				4283.0	3	5	4.34E-01	1843.1	2	2	1.6E-01
Br II				4289.4	1	3	6.0E-01	1850.7	4	2	3.08E-01
4704.9	5	7	1.1E+00	4299.0	3	3	4.66E-01	2103.2	2	4	8.2E-01
4785.5	5	5	9.4E-01	4302.5	5	5	1.36E+00	2112.8	4	6	9.7E-01
4816.7	5	3	1.1E+00	4307.7	3	1	1.99E+00	2113.2	4	4	1.6E-01
				4318.7	5	3	7.4E-01	2197.8	2	2	3.1E-01
				4355.1	5	7	1.9E-01	2208.6	4	2	6.2E-01
Cadmium				4425.4	1	3	4.98E-01	3158.9	2	4	3.1E+00
Cd I				4435.0	3	5	6.7E-01	3179.3	4	6	3.6E+00
2288.0	1	3	5.3E+00	4435.7	3	3	3.42E-01	3181.3	4	4	5.8E-01
2836.9	1	3	2.8E-01	4454.8	5	7	8.7E-01	3706.0	2	2	8.8E-01
2880.8	3	5	4.2E-01	4455.9	5	5	2.0E-01	3736.9	4	2	1.7E+00
2881.2	3	3	2.4E-01	4526.9	5	3	4.1E-01	3933.7	2	4	1.47E+00
2980.6	5	7	5.9E-01	4578.6	3	5	1.76E-01	3968.5	2	2	1.4E+00
2981.4	5	5	1.5E-01	4581.4	5	7	2.09E-01				
3261.1	1	3	4.06E-03	4585.9	7	9	2.29E-01	**Ca III**			
3403.7	1	3	7.7E-01	4685.3	3	5	8.0E-02	357.97	1	3	8.8E+02
3466.2	3	5	1.2E+00	4878.1	5	7	1.88E-01	439.69	1	3	1.9E-01
3467.7	3	3	6.7E-01	5041.6	5	3	3.3E-01	490.55	1	3	1.6E-02
3610.5	5	7	1.3E+00	5188.9	3	5	4.0E-01				
3612.9	5	5	3.5E-01	5261.7	3	3	1.5E-01	**Ca V**			
4140.5	3	5	4.7E-02	5262.2	3	1	6.0E-01	558.60	5	3	2.2E+01
4662.4	3	5	5.5E-02	5264.2	5	5	9.1E-02	637.93	5	3	3.9E+00
4678.1	1	3	1.3E+00	5265.6	5	3	4.4E-01	643.12	3	1	9.1E+00
4799.9	3	3	4.1E-01	5270.3	7	5	5.0E-01	646.57	5	5	6.9E+00
5085.8	5	3	5.6E-01	5582.0	5	7	6.0E-02	647.88	3	3	2.3E+00
6438.5	3	5	5.9E-01	5588.8	7	7	4.9E-01	651.55	1	3	2.9E+00
				5590.1	3	5	8.3E-02	656.76	3	5	2.1E+00
Cd II				5594.5	5	5	3.8E-01				
2144.4	2	4	2.8E+00	5598.5	3	3	4.3E-01	**Ca VII**			
2265.0	2	2	3.0E+00	5601.3	7	5	8.6E-02	550.20	5	5	1.8E+01
2572.9	2	2	1.7E+00	5602.9	5	3	1.4E+00	624.39	1	3	3.3E+00
2748.5	4	2	2.8E+00	5857.5	3	5	6.6E-01	630.54	3	5	4.5E+00
4415.6	4	6	1.4E-02	6102.7	1	3	9.6E-02	630.79	3	3	2.2E+00
				6122.2	3	3	2.87E-01	639.15	5	7	5.7E+00
Calcium				6161.3	5	5	3.3E-02	640.41	5	5	1.3E+00
Ca I				6162.2	5	3	3.54E-01				
2275.5	1	3	3.01E-01	6163.8	3	3	5.6E-02	**Ca VIII**			
2995.0	1	3	3.67E-01	6166.4	3	1	2.2E-01	182.71	2	2	1.6E+02
2997.3	3	5	2.41E-01	6169.1	5	3	1.7E-01	184.16	4	2	3.2E+02
2999.6	3	3	2.79E-01	6169.6	7	5	1.9E-01				
3000.9	3	1	1.58E+00	6439.1	7	9	5.3E-01	**Ca IX**			
3006.9	5	5	7.5E-01	6449.8	3	5	9.0E-02	163.23	5	3	3.76E+02
3009.2	5	3	4.30E-01	6462.6	5	7	4.7E-01	371.89	1	3	8.8E+01
3344.5	1	3	1.51E-01	6471.7	7	7	5.9E-02	373.81	3	5	1.16E+02
3350.2	3	5	1.78E-01								

λ (Å)	g_i	g_k	A (10^8 s^{-1})	λ (Å)	g_i	g_k	A (10^8 s^{-1})	λ (Å)	g_i	g_k	A (10^8 s^{-1})
378.08	5	7	1.5E+02	1261.00	3	3	4.42E-01	4817.37	3	3	8.76E-04
395.03	3	5	2.2E+02	1261.12	3	5	3.71E-01	4826.80	5	3	6.28E-04
466.24	1	3	1.12E+02	1261.43	5	3	7.06E-01	4932.05	3	1	6.02E-02
498.01	3	5	2.49E+01	1261.55	5	5	1.27E+00	5023.84	7	9	1.81E-03
506.18	5	5	7.2E+01	1274.11	5	7	1.03E-02	5039.06	7	9	4.73E-03
515.57	5	3	3.75E+01	1277.25	1	3	1.27E+00	5041.48	3	5	5.25E-03
				1277.28	3	5	1.73E+00	5041.79	5	7	3.28E-03
Ca X				1277.51	3	3	9.12E-01	5052.17	3	5	2.60E-02
110.96	2	4	2.9E+02	1277.55	5	7	2.31E+00	5380.34	3	3	1.86E-02
111.20	2	2	2.92E+02	1277.72	5	5	6.35E-01	5545.05	3	3	3.04E-03
151.84	2	2	2.3E+02	1277.95	5	3	5.56E-02	5668.94	3	3	2.35E-02
153.02	4	2	4.5E+02	1279.23	5	7	1.10E-01	5793.12	7	5	3.44E-03
206.57	4	4	2.9E+01	1279.89	3	5	3.08E-01	5794.47	5	5	6.44E-04
206.75	6	4	2.6E+02	1280.14	1	3	3.11E-01	5800.23	3	3	1.04E-03
207.39	4	2	2.8E+02	1280.33	5	5	5.77E-01	5800.60	5	3	3.04E-03
411.70	2	4	8.3E+01	1280.40	3	3	1.73E-01	5805.20	3	1	4.12E-03
419.75	4	6	9.5E+01	1280.60	3	1	8.22E-01	6001.12	5	5	3.22E-03
420.47	4	4	1.6E+01	1280.85	5	3	3.33E-01	6006.02	7	5	1.79E-02
557.76	2	4	3.50E+01	1328.83	1	3	7.95E-01	6007.18	3	3	5.34E-03
574.01	2	2	3.2E+01	1329.09	3	1	2.41E+00	6010.68	3	1	2.13E-02
				1329.58	5	5	1.79E+00	6013.17	7	5	1.79E-02
Ca XI				1329.60	5	3	1.00E+00	6013.21	7	9	4.35E-03
30.448	1	3	6.2E+03	1355.84	5	7	1.04E+00	6014.83	5	3	1.60E-02
30.867	1	3	4.9E+04	1364.16	5	5	1.57E-01	6016.45	5	7	3.86E-03
35.212	1	3	2.0E+03	1431.60	5	7	2.11E+00	6587.61	3	3	5.09E-02
				1432.10	5	5	2.01E+00	6655.52	3	3	5.03E-03
Ca XII				1432.53	5	3	2.11E+00	6828.12	3	5	9.89E-03
140.05	4	2	3.7E+02	1459.03	5	3	4.76E-01	7111.47	3	5	2.17E-02
147.27	2	2	1.6E+02	1463.34	5	7	1.88E+00	7113.18	7	9	2.47E-02
				1467.40	5	3	5.49E-01	7115.17	5	7	2.19E-02
Ca XV				1468.41	5	3	3.90E-02	7115.18	3	1	4.43E-02
141.69	5	3	4.08E+02	1470.09	5	7	1.37E-02	7116.99	7	5	3.26E-02
*142.23	9	3	6.3E+02	1472.23	5	3	8.01E-03	7119.66	5	3	3.12E-02
161.00	5	5	1.9E+02	1481.76	5	5	3.92E-01	7860.88	5	5	1.53E-02
				1560.31	1	3	6.57E-01	8058.62	5	5	1.09E-02
Ca XVII				1561.34	5	5	2.94E-01	8335.15	3	1	3.51E-01
19.558	1	3	3.8E+04	1561.44	5	7	1.18E+00	9061.44	3	5	7.31E-02
21.198	3	5	4.9E+04	1656.27	3	5	8.58E-01	9062.49	1	3	9.48E-02
192.82	1	3	1.21E+02	1656.93	1	3	1.13E+00	9078.29	3	3	7.07E-02
218.82	3	5	2.76E+01	1657.01	5	5	2.52E+00	9088.51	3	1	3.00E-01
223.02	1	3	3.44E+01	1657.38	3	3	8.64E-01	9094.83	5	5	2.28E-01
228.72	3	3	2.37E+01	1657.91	3	1	3.43E+00	9111.81	5	3	1.35E-01
232.83	5	5	6.5E+01	1658.12	5	3	1.44E+00	9405.73	3	5	2.91E-01
244.06	5	3	3.28E+01	1751.83	1	3	9.07E-01	9603.03	1	3	3.06E-02
				1763.91	1	3	3.59E-02	9620.78	3	3	8.62E-02
Ca XVIII				1765.37	1	3	1.04E-02	9658.43	5	3	1.25E-01
*18.71	2	6	2.31E+04	1930.90	5	3	3.51E+00				
*19.74	6	10	7.0E+04	2478.56	1	3	3.40E-01	**C II**			
302.19	2	4	2.0E+01	2902.23	1	3	4.32E-03	687.345	4	6	2.84E+01
344.76	2	2	1.3E+01	2903.27	3	3	1.29E-02	858.092	2	2	1.18E+00
				2905.00	5	3	2.15E-02	858.559	4	2	2.35E+00
Carbon				4371.37	3	3	1.27E-02	903.623	2	4	6.85E+00
C I				4762.31	1	3	3.37E-03	903.962	2	2	2.74E+01
945.191	1	3	3.79E+00	4762.53	3	5	2.72E-03	904.142	4	4	3.42E+01
945.338	3	3	1.14E+01	4766.67	3	3	2.36E-03	904.480	4	2	1.37E+01
945.579	5	3	1.89E+01	4770.03	3	1	1.07E-02	1009.86	2	4	5.71E+00
1193.24	5	7	1.22E+00	4771.74	5	5	7.97E-03	1010.08	4	4	1.14E+01
1260.74	1	3	5.32E-01	4775.90	5	3	4.84E-03	1010.37	6	4	1.71E+01
1260.93	3	1	1.70E+00	4812.92	1	3	4.03E-04	1036.34	2	2	7.61E+00

λ Å	Weights g_i	g_k	A 10^8 s^{-1}	λ Å	Weights g_i	g_k	A 10^8 s^{-1}	λ Å	Weights g_i	g_k	A 10^8 s^{-1}
1037.02	4	2	1.52E+01	7046.25	4	2	3.20E-01	5695.92	3	5	4.27E-01
1323.91	4	4	4.33E+00	7053.09	4	4	3.19E-01	5858.34	3	1	1.34E-01
1323.95	6	6	4.49E+00	7063.68	4	6	3.17E-01	5863.25	3	3	3.35E-02
1334.53	2	4	2.37E+00	7112.48	2	4	2.94E-01	5871.68	5	3	1.00E-01
1335.71	4	6	2.84E+00	7113.04	4	6	3.15E-01	5880.56	5	5	1.99E-02
2091.14	2	4	1.00E-01	7115.63	6	8	3.60E-01	5894.07	7	5	1.11E-01
2091.19	4	6	1.69E-01	7119.76	4	4	1.17E-01	6727.48	1	3	1.12E-01
2091.65	6	8	2.41E-01	7119.91	8	10	4.19E-01	6731.04	3	5	1.50E-01
2093.16	6	6	7.20E-02	7125.72	6	6	1.02E-01	6742.15	3	3	8.32E-02
2173.85	2	4	2.31E-01	7132.47	6	4	8.33E-03	6744.39	5	7	1.99E-01
2174.17	2	2	2.31E-01	7134.10	8	8	5.93E-02	6762.17	5	5	4.95E-02
2509.13	2	4	4.53E-02	7231.33	2	4	3.52E-01	6773.39	5	3	5.47E-03
2511.74	4	4	9.04E-02	7236.42	4	6	4.22E-01	6851.18	3	5	7.60E-03
2512.06	4	6	5.42E-02	7237.17	4	4	7.03E-02	6853.68	5	7	5.64E-03
2727.31	2	4	6.63E-02	8028.85	2	2	1.71E-02	6857.24	3	3	3.79E-02
2728.72	4	4	3.31E-01	8037.73	2	4	4.26E-02	6862.69	5	5	3.51E-02
2729.21	2	2	2.65E-01	8039.40	4	2	8.51E-02	6868.78	5	3	1.26E-02
2730.63	4	2	1.32E-01	8048.31	4	4	1.36E-02	6872.04	7	7	4.46E-02
5132.95	2	4	3.89E-01	8062.10	4	6	3.04E-02	6881.10	7	5	7.80E-03
5133.28	4	6	2.80E-01	8062.80	6	4	4.56E-02	7353.88	5	3	3.09E-02
5137.26	2	2	1.55E-01	8076.64	6	6	7.05E-02	7707.43	3	5	1.30E-01
5139.17	4	4	1.24E-01	9238.30	4	6	3.34E-02	7771.76	3	1	1.77E-01
5143.49	4	2	7.73E-01	9251.01	2	4	2.77E-02	7780.41	3	3	1.76E-01
5145.16	6	6	6.49E-01	9863.06	2	4	5.56E-02	7796.00	3	5	1.75E-01
5151.08	6	4	4.16E-01	9870.78	4	6	9.31E-02	8500.32	1	3	1.01E-01
5640.55	2	4	9.89E-02	9882.68	6	8	1.33E-01	9593.32	3	3	5.32E-03
5648.07	4	4	1.97E-01					9651.47	5	5	1.57E-02
5662.46	6	4	2.93E-01	**C III**				9696.48	5	7	7.53E-03
5818.31	2	2	3.38E-02	310.170	1	3	6.56E+00	9696.54	3	5	7.12E-03
5822.98	2	4	3.38E-03	386.203	1	3	3.46E+01	9699.57	7	9	8.47E-03
5823.18	4	2	3.38E-02	459.466	1	3	5.91E+01	9701.10	1	3	4.40E-02
5827.85	4	4	2.16E-02	459.514	3	5	7.97E+01	9705.41	3	5	5.93E-02
5836.37	6	4	4.22E-02	459.627	5	7	1.06E+02	9706.44	3	3	3.29E-02
5843.62	6	6	1.20E-02	574.281	3	5	6.24E+01	9715.09	5	7	7.88E-02
5856.06	8	6	5.31E-02	977.020	1	3	1.767E+01	9717.75	5	5	1.97E-02
6095.29	2	4	4.20E-01	1174.93	3	5	3.293E+00	9718.79	5	3	2.19E-03
6098.51	4	6	5.03E-01	1175.26	1	3	4.385E+00				
6102.56	4	4	8.37E-02	1175.59	3	3	3.287E+00	**C IV**			
6578.05	2	4	3.63E-01	1175.71	5	5	9.856E+00	*312.43	2	6	4.63E+01
6582.88	2	2	3.62E-01	1175.99	3	1	1.313E+01	*384.13	6	10	1.76E+02
6724.56	2	4	3.17E-02	1176.37	5	3	5.468E+00	1548.19	2	4	2.65E+00
6727.07	2	2	6.34E-02	1247.38	3	1	2.082E+01	1550.77	2	2	2.64E+00
6727.26	4	6	2.96E-02	2296.87	3	5	1.376E+00	5801.31	2	4	3.17E-01
6731.07	4	4	5.06E-02	2849.05	3	1	1.95E-01	5811.97	2	2	3.16E-01
6733.58	4	2	6.32E-02	3703.70	3	3	5.90E-01				
6734.00	6	8	1.80E-02	4325.56	3	5	1.24E-01	**C V**			
6738.61	6	6	7.23E-02	4647.42	3	5	7.26E-01	34.9728	1	3	2.554E+03
6742.43	6	4	4.41E-02	4650.25	3	3	7.25E-01	40.2678	1	3	8.873E+03
6750.54	8	8	1.08E-01	4651.02	3	5	2.28E-01	*227.19	3	9	1.363E+02
6755.16	8	6	2.38E-02	4651.47	3	1	7.24E-01	247.315	1	3	1.278E+02
6779.94	4	6	2.56E-02	4652.05	1	3	3.04E-01	*248.71	9	15	4.247E+02
6780.59	2	4	1.52E-01	4659.06	3	3	2.27E-01	*260.19	9	3	6.680E+01
6783.91	6	8	3.65E-01	4663.64	3	1	9.05E-01	267.267	3	5	3.947E+02
6787.21	2	2	3.04E-01	4665.86	5	5	6.78E-01	*2273.9	3	9	5.646E-01
6791.47	4	4	1.94E-01	4673.95	5	3	3.75E-01	3526.66	1	3	1.663E-01
6798.10	4	2	6.04E-02	5244.66	1	3	5.30E-02	8420.72	3	5	6.898E-02
6800.69	6	6	1.09E-01	5253.58	3	3	1.58E-01	*8433.2	3	9	6.868E-02
6812.28	6	4	1.80E-02	5272.52	5	3	2.61E-01	8448.12	3	1	6.832E-02

λ Å	Weights g_i	Weights g_k	A 10^8 s^{-1}	λ Å	Weights g_i	Weights g_k	A 10^8 s^{-1}	λ Å	Weights g_i	Weights g_k	A 10^8 s^{-1}
8449.19	3	3	6.829E-02	7256.6	6	4	1.5E-01	3530.0	6	8	1.8E+00
				7414.1	6	4	4.7E-02	3560.7	4	6	1.7E+00
Cesium				7547.1	4	4	1.2E-01	3602.1	6	8	1.7E+00
Cs I				7717.6	4	4	3.0E-02	3612.9	4	6	1.2E+00
3203.5	2	4	7.6E-06	7745.0	2	4	6.3E-02	3720.5	4	6	1.7E+00
3205.3	2	4	7.9E-06	7769.2	6	6	6.0E-02				
3207.5	2	4	8.5E-06	7821.4	6	8	9.8E-02	**Chromium**			
3210.0	2	4	9.4E-06	7830.8	4	4	9.7E-02	**Cr I**			
3212.8	2	4	1.19E-05	7878.2	6	6	1.8E-02	1999.95	9	9	1.4E+00
3216.2	2	4	1.49E-05	7899.3	4	6	5.1E-02	2383.30	9	11	4.1E-01
3220.1	2	4	1.7E-05	7924.6	2	4	2.1E-02	2389.21	3	5	2.3E-01
3220.2	2	2	1.07E-07	7935.0	6	8	3.9E-02	2408.60	9	7	6.7E-01
3224.8	2	4	2.0E-05	7997.9	4	4	2.1E-02	2408.72	7	5	2.9E-01
3225.0	2	2	1.43E-07					2492.57	3	5	4.5E-01
3230.5	2	4	2.5E-05	**Cl II**				2495.08	3	3	2.7E-01
3230.7	2	2	1.97E-07	3329.1	5	7	1.5E+00	2496.30	5	7	5.6E-01
3237.4	2	4	2.8E-05	3522.1	7	7	1.4E+00	2502.55	7	9	2.2E-01
3237.6	2	2	2.63E-07	3798.8	5	7	1.6E+00	2504.31	7	9	4.5E-01
3245.9	2	4	3.45E-05	3805.2	7	9	1.8E+00	2508.11	5	5	2.1E-01
3246.2	2	2	3.7E-07	3809.5	3	5	1.5E+00	2508.97	5	3	3.8E-01
3256.7	2	4	4.25E-05	3851.0	5	7	1.8E+00	2527.11	9	9	5.3E-01
3257.1	2	2	7.0E-07	3851.4	5	5	1.6E+00	2549.55	3	3	4.8E-01
3270.5	2	4	5.6E-05	3854.7	3	5	2.2E+00	2560.70	5	5	4.3E-01
3271.0	2	2	9.8E-07	3861.9	5	7	2.4E+00	2571.74	7	5	6.4E-01
3288.6	2	4	1.0E-04	3868.6	7	9	2.7E+00	2577.66	7	7	2.6E-01
3289.3	2	2	2.7E-06	3913.9	9	9	8.2E-01	2591.84	9	7	6.5E-01
3313.1	2	4	1.6E-04	3990.2	5	7	8.4E-01	2620.48	5	3	1.9E-01
3314.0	2	2	5.2E-06	4132.5	5	5	1.6E+00	2673.64	3	3	1.8E-01
3347.5	2	4	2.2E-04	4276.5	9	7	7.6E-01	2701.99	9	11	2.1E-01
3348.8	2	2	1.1E-05	4768.7	3	5	7.7E-01	2726.50	5	7	7.5E-01
3397.9	2	4	4.0E-04	4781.3	5	7	1.0E+00	2731.90	5	5	7.8E-01
3400.0	2	2	2.4E-05	4794.6	5	7	1.04E+00	2736.46	5	3	7.5E-01
3476.8	2	4	6.6E-04	4810.1	5	5	9.9E-01	2752.85	3	3	8.7E-01
3480.0	2	2	6.6E-05	4819.5	5	5	1.00E+00	2757.09	5	5	6.8E-01
3611.4	2	4	1.5E-03	4904.8	5	7	8.1E-01	2761.74	5	3	6.8E-01
3617.3	2	2	2.5E-04	4917.7	3	5	7.5E-01	2764.36	7	7	3.7E-01
3876.1	2	4	3.8E-03	5078.3	7	7	7.7E-01	2769.90	7	5	1.1E+00
3888.6	2	2	9.7E-04	5219.1	3	9	8.6E-01	2780.70	9	7	1.4E+00
4555.3	2	4	1.88E-02	5392.1	5	7	1.0E+00	2879.27	5	7	2.1E-01
4593.2	2	2	8.0E-03					2887.00	3	5	2.7E-01
8521.1	2	4	3.276E-01	**Cl III**				2889.22	9	9	6.6E-01
8943.5	2	2	2.87E-01	2298.5	4	4	4.2E+00	2893.25	7	7	5.2E-01
				2340.6	6	6	4.2E+00	2894.17	1	3	3.3E-01
Chlorine				2370.4	8	6	2.8E+00	2896.76	5	5	3.0E-01
Cl I				2531.8	2	4	4.4E+00	2905.48	3	1	1.3E+00
1188.8	4	6	2.33E+00	2532.5	4	6	5.3E+00	2909.05	5	3	6.8E-01
1188.8	4	4	2.71E-01	2577.1	4	6	4.3E+00	2910.89	7	5	3.4E-01
1201.4	2	4	2.39E+00	2580.7	6	8	4.7E+00	2911.15	9	7	2.6E-01
1335.7	4	2	1.74E+00	2601.2	2	4	4.6E+00	2967.64	7	9	3.9E-01
1347.2	4	4	4.19E+00	2603.6	4	6	5.0E+00	2971.10	5	7	7.1E-01
1351.7	2	2	3.23E+00	2609.5	6	8	5.7E+00	2975.48	3	5	8.9E-01
1363.4	2	4	7.5E-01	2617.0	8	10	6.6E+00	2980.78	1	3	5.10E-01
4323.3	4	4	1.1E-02	2661.6	4	6	3.4E+00	2988.64	5	7	5.2E-01
4363.5	4	6	6.8E-03	2665.5	6	8	4.8E+00	2991.88	3	1	3.0E+00
4379.9	4	4	1.4E-02	2691.5	4	4	3.5E+00	2994.06	5	5	2.5E-01
4389.8	6	8	1.4E-02	2710.4	4	6	3.5E+00	2995.09	5	5	4.3E-01
4526.2	4	4	5.1E-02	3340.4	6	6	1.5E+00	2996.57	5	3	2.0E+00
4601.0	2	2	4.2E-02	3392.9	4	4	1.9E+00	2998.78	5	3	4.07E-01
4661.2	2	4	1.2E-02	3393.5	6	6	1.9E+00	3000.88	7	5	1.6E+00

λ Å	Weights g_i	Weights g_k	A 10^8 s⁻¹	λ Å	Weights g_i	Weights g_k	A 10^8 s⁻¹	λ Å	Weights g_i	Weights g_k	A 10^8 s⁻¹
3005.06	9	7	9.2E-01	4432.77	15	15	4.9E-01	2857.40	6	8	2.8E-01
3013.72	3	5	8.3E-01	4443.72	3	1	4.5E-01	2860.92	2	4	6.9E-01
3015.20	1	3	1.63E+00	4482.88	3	3	3.0E-01	2862.57	8	8	6.3E-01
3020.67	3	3	1.5E+00	4490.55	9	7	3.9E-01	2866.72	4	4	1.2E+00
3021.58	9	11	2.91E+00	4492.31	5	3	4.47E-01	2867.09	4	4	1.1E+00
3024.36	5	5	1.27E+00	4495.28	9	7	2.0E-01	2867.65	2	2	1.1E+00
3029.17	5	3	3.8E-01	4500.29	7	7	2.1E-01	2870.43	6	6	1.3E+00
3030.25	7	7	1.1E+00	4506.84	13	11	2.7E-01	2873.81	4	2	8.8E-01
3031.35	5	3	3.1E-01	4540.72	11	11	3.14E-01	2880.86	6	4	7.9E-01
3034.19	7	7	3.5E-01	4564.17	11	13	5.1E-01	2898.53	10	12	1.2E+00
3037.05	9	9	5.4E-01	4595.60	13	13	4.7E-01	2921.81	8	10	9.0E-01
3040.84	7	5	7.4E-01	4622.47	7	7	4.1E-01	2930.83	2	4	1.1E+00
3053.87	9	7	7.97E-01	4663.33	3	3	2.0E-01	2935.12	6	8	1.8E+00
3148.44	9	11	5.6E-01	4665.90	3	3	3.0E-01	2953.34	2	2	1.8E+00
3155.16	11	13	5.7E-01	4689.38	7	5	2.3E-01	2966.03	10	8	5.4E-01
3163.76	13	15	6.0E-01	4698.46	9	7	2.2E-01	2971.90	14	14	2.0E+00
3237.73	9	9	1.3E+00	4708.02	11	9	4.31E-01	2979.73	12	12	1.8E+00
3238.09	11	11	2.0E-01	4718.43	13	11	3.4E-01	2985.32	10	10	2.2E+00
3578.68	7	9	1.48E+00	4730.69	7	5	3.83E-01	2989.18	8	8	2.2E+00
3593.48	7	7	1.50E+00	4737.33	9	7	3.38E-01	3118.64	2	4	1.7E+00
3605.32	7	5	1.62E+00	4741.09	3	5	2.2E-01	3120.36	4	6	1.5E+00
3639.80	13	11	1.8E+00	4752.07	13	13	6.2E-01	3122.59	12	12	4.4E-01
3743.89	13	13	7.61E-01	4756.09	11	11	4.0E-01	3128.69	4	4	8.1E-01
3757.66	7	7	4.13E-01	4792.49	7	5	2.6E-01	3136.68	6	6	6.4E-01
3768.24	5	5	5.10E-01	4801.02	9	7	3.06E-01	4588.22	8	6	1.2E-01
3804.80	9	9	6.9E-01	4816.13	9	9	1.8E-01				
3963.69	13	15	1.3E+00	4870.79	7	9	3.5E-01	**Cr V**			
3969.75	11	13	1.2E+00	4887.01	9	11	3.2E-01	434.306	9	9	1.5E+01
3983.90	7	9	1.05E+00	4922.28	11	13	4.0E-01	436.351	9	7	2.4E+01
3991.12	5	7	1.07E+00	4966.80	3	1	3.0E-01	436.601	7	5	2.1E+01
4001.44	9	11	6.8E-01	5204.51	5	3	5.09E-01	437.420	7	7	1.4E+01
4039.10	15	15	6.7E-01	5206.02	5	5	5.14E-01	437.655	5	5	1.3E+01
4048.78	13	13	6.4E-01	5208.42	5	7	5.06E-01	441.056	5	3	2.3E+01
4058.78	11	11	6.7E-01	5243.38	5	3	2.19E-01	456.357	1	3	9.5E+00
4065.71	9	11	3.5E-01	5297.37	7	9	3.88E-01	456.637	3	1	3.3E+01
4165.52	11	13	7.5E-01	5297.99	7	7	3.0E-01	456.743	3	3	9.1E+00
4204.48	13	11	3.1E-01	5328.36	9	11	6.2E-01	457.028	5	5	2.7E+01
4254.33	7	9	3.15E-01	5329.17	9	9	2.25E-01	457.504	5	3	1.2E+01
4263.15	15	17	6.4E-01	5783.11	3	3	2.1E-01	464.015	9	7	3.6E+01
4274.81	7	7	3.07E-01	5783.89	5	5	2.02E-01	469.634	5	5	2.3E+01
4275.98	11	11	2.2E-01	5787.97	5	7	2.35E-01	1106.25	7	9	1.2E+01
4280.42	13	15	4.7E-01					1121.07	7	9	2.1E+01
4289.73	7	5	3.16E-01	**Cr II**				1127.63	9	11	3.5E+01
4291.97	7	5	2.4E-01	2653.57	4	6	3.5E-01	1465.86	5	3	1.1E+01
4297.75	11	13	4.9E-01	2658.59	2	4	5.8E-01	1481.65	3	1	1.0E+01
4298.05	9	9	2.6E-01	2666.02	6	8	5.9E-01	1519.03	5	7	9.5E+00
4300.52	9	7	1.9E-01	2668.71	4	2	1.4E+00	1579.70	7	9	8.6E+00
4301.19	11	9	2.6E-01	2671.80	6	4	1.0E+00				
4302.78	11	11	2.5E-01	2672.83	8	6	5.5E-01	**Cr VI**			
4319.66	5	3	1.8E-01	2744.97	4	6	8.5E-01	161.687	6	6	1.7E+02
4337.25	5	7	2.0E-01	2787.61	6	6	1.5E+00	168.088	4	6	2.0E+02
4373.65	9	9	2.8E-01	2822.38	14	16	2.3E+00	201.007	4	4	2.5E+03
4376.80	13	13	3.2E-01	2835.63	10	12	2.0E+00	201.224	4	6	1.8E+02
4413.86	7	5	2.7E-01	2840.01	10	12	2.7E+00	201.388	6	4	2.7E+02
4422.70	5	5	2.7E-01	2843.24	8	10	6.4E-01	201.606	6	6	2.6E+03
4424.29	9	7	2.1E-01	2849.83	6	8	9.2E-01	202.442	6	4	1.0E+03
4429.93	3	3	2.4E-01	2851.35	8	10	2.2E+00	202.739	4	2	1.2E+03
4432.16	1	3	1.8E-01	2856.77	4	6	4.3E-01	226.241	6	8	7.2E+02

λ Å	g_i	g_k	A 10^8 s^{-1}
227.202	4	6	6.6E+02
Cr X			
216.72	6	8	9.0E+02
223.86	4	2	7.7E+02
224.74	4	4	7.6E+02
226.24	4	6	7.3E+02
227.42	4	4	5.2E+02
227.50	4	6	1.8E+01
228.63	6	4	8.1E+01
228.71	6	6	4.5E+02
231.21	2	4	1.2E+02
232.96	4	4	4.4E+02
242.20	2	4	5.0E+01
244.19	4	6	5.8E+01
395.984	4	4	2.4E+01
398.150	6	6	2.1E+01
Cr XI			
214.31	5	7	1.4E+01
226.45	5	7	6.0E+02
232	3	1	4.1E+02
235.53	5	7	5.5E+02
240.76	1	3	4.8E+02
250.28	5	7	1.0E+01
366.491	3	3	1.2E+01
366.942	3	1	3.0E+01
374.927	5	5	2.3E+01
422.083	3	5	1.0E+01
Cr XII			
216	4	6	2.4E+02
218	6	8	2.4E+02
239	2	2	1.6E+02
244.70	2	4	3.0E+02
247	4	2	2.4E+02
247	2	2	3.3E+02
248	6	8	1.4E+02
250	6	8	3.5E+02
250	6	6	2.2E+02
251.52	4	6	3.4E+02
252	4	6	2.0E+02
256	2	2	1.5E+02
259	2	4	3.2E+02
269	2	2	2.1E+02
300.32	2	2	1.4E+02
305.81	4	4	2.76E+02
309	4	2	2.7E+02
309	6	6	1.6E+02
311.55	4	2	1.6E+02
324	4	6	2.2E+02
327	6	8	2.2E+02
332.06	6	4	1.4E+02
Cr XIII			
49.59	1	3	9.9E+02
67.01	1	3	1.67E+03
228	5	7	1.8E+02
267.73	5	7	1.9E+02

λ Å	g_i	g_k	A 10^8 s^{-1}
270	3	1	1.7E+02
276.4	5	7	2.2E+02
277	1	3	2.1E+02
279.32	3	5	3.5E+02
286	3	1	4.6E+02
328.29	1	3	1.86E+02
345	7	9	1.74E+02
Cr XIV			
*38.036	2	6	2.47E+02
39.796	2	4	3.05E+02
40.018	4	6	3.6E+02
40.782	2	4	3.9E+02
40.800	2	2	3.9E+02
41.556	2	4	4.5E+02
41.788	4	6	5.3E+02
44.597	2	4	7.1E+02
44.869	4	6	8.3E+02
46.125	4	2	3.1E+02
46.468	2	4	6.6E+02
46.527	2	2	6.7E+02
48.300	4	6	5.9E+02
48.338	6	8	6.3E+02
50.821	2	4	1.2E+03
51.172	4	6	1.4E+03
51.180	4	4	2.3E+02
52.321	4	6	1.0E+03
52.363	6	8	1.1E+03
53.760	2	2	3.0E+02
54.164	4	2	5.9E+02
60.699	4	6	2.05E+03
60.756	6	8	2.19E+03
63.324	2	4	1.07E+03
63.539	2	2	1.13E+03
68.594	2	4	1.98E+03
69.213	4	6	2.31E+03
69.247	4	4	3.8E+02
86.060	4	6	5.3E+03
86.169	6	8	5.9E+03
86.185	6	6	3.9E+02
101.05	6	6	4.4E+02
101.42	4	2	4.83E+02
104.4	4	6	3.0E+02
104.5	6	8	3.1E+02
109.8	2	4	2.3E+02
110.4	4	6	2.8E+02
118.3	4	2	2.1E+02
125.2	4	6	5.0E+02
125.3	6	8	5.4E+02
148.5	2	4	2.18E+02
149.1	2	2	2.1E+02
157.1	2	4	3.3E+02
158.4	4	6	3.7E+02
187.02	4	6	9.3E+02
187.30	6	8	9.6E+02
189.1	2	2	2.13E+02
191.0	4	2	4.11E+02
222.9	4	2	2.2E+02
346.3	4	6	2.4E+02

λ Å	g_i	g_k	A 10^8 s^{-1}
346.5	6	8	2.5E+02
Cr XV			
18.497	1	3	1.62E+05
18.782	1	3	2.8E+04
19.015	1	3	6.3E+02
20.863	1	3	6.0E+03
21.153	1	3	5.6E+03
102	3	3	1.6E+02
102.18	5	3	7.0E+02
103	3	1	3.8E+02
105	7	5	5.3E+02
111.27	3	3	1.7E+02
Cr XVI			
17.073	4	6	1.2E+04
17.242	2	4	8.6E+04
17.299	4	4	2.5E+04
17.372	4	4	1.4E+05
17.438	4	2	1.1E+05
17.514	2	4	1.1E+05
17.587	2	4	2.0E+04
17.656	2	2	2.0E+04
19.442	4	2	9.9E+03
19.714	2	2	1.1E+04
Cr XVII			
16.31	5	3	9.6E+03
16.32	5	7	3.2E+04
16.37	3	1	9.7E+04
16.44	5	7	1.3E+05
16.59	3	1	5.7E+04
16.65	5	5	1.1E+04
16.66	1	3	1.8E+05
16.68	5	7	6.8E+04
16.80	5	7	4.4E+04
16.97	1	3	2.63E+04
16.97	3	3	1.5E+04
17.968	5	3	8.6E+03
18.336	5	3	1.7E+04
18.336	5	5	1.6E+04
18.389	1	3	9.2E+03
Cr XVIII			
95.77	4	2	3.08E+02
102.32	4	4	1.54E+02
104.98	6	4	8.7E+02
106.84	4	2	3.4E+02
110.41	4	2	7.9E+02
112.27	4	2	4.24E+02
119.62	2	2	3.2E+02
123.87	6	4	3.9E+02
125.51	4	4	3.4E+02
128.10	6	6	2.8E+02
136.52	4	2	1.66E+02
139.87	4	4	1.49E+02
140.82	4	2	2.66E+02
155.46	2	2	2.84E+02
157.40	4	4	2.83E+02

λ Å	Weights g_i	g_k	A 10^8 s⁻¹	λ Å	Weights g_i	g_k	A 10^8 s⁻¹	λ Å	Weights g_i	g_k	A 10^8 s⁻¹
Cr XIX				14.04	3	5	1.2E+05	2467.69	6	8	7.0E-02
14.73	3	3	7.1E+04	14.24	1	3	1.41E+05	2470.27	10	12	1.5E-01
14.80	1	3	1.3E+05					2476.64	10	8	2.2E-01
14.81	5	3	3.4E+04	**Cr XXII**				2504.52	10	8	1.8E-01
14.84	5	7	1.3E+05	2.190	4	2	1.7E+06	2511.02	10	10	9.2E-01
109.64	3	3	2.46E+02	2.191	2	2	2.5E+06	2521.36	10	8	3.0E+00
110.37	5	3	6.0E+02	2.198	4	4	4.5E+06	2528.97	8	6	2.8E+00
113.97	5	3	5.5E+02	2.199	2	4	2.3E+06	2530.13	6	6	7.1E-02
118.31	3	1	3.29E+02	2.202	4	6	1.6E+06	2535.96	6	4	1.9E+00
118.67	5	3	2.1E+02	2.203	4	2	1.3E+06	2536.50	8	8	3.0E+00
118.83	3	3	1.35E+02	13.149	2	4	1.29E+05	2544.25	4	2	3.0E+00
126.30	1	3	1.56E+02	13.292	4	6	1.54E+05	2562.12	4	4	3.9E-01
126.33	5	5	4.35E+02					2567.34	6	6	3.0E-01
130.99	7	5	2.9E+02	**Cr XXIII**				2574.35	8	8	1.7E-01
134.89	3	1	1.98E+02	1.7632	1	3	3.68E+05	2685.34	6	8	5.5E-02
138.15	3	1	1.75E+02	1.8557	1	3	8.97E+05	3017.55	8	6	6.9E-02
138.45	5	5	1.71E+02	2.095	3	1	3.5E+06	3044.00	10	10	1.9E-01
140.92	5	3	1.38E+02	2.101	1	3	2.0E+06	3048.89	6	4	7.5E-02
143.57	3	1	7.2E+02	2.101	5	5	7.9E+05	3061.82	8	8	1.6E-01
163.94	5	5	3.1E+02	2.102	3	5	2.1E+06	3072.34	6	6	1.5E-01
179.18	3	1	1.45E+02	2.103	3	5	1.2E+06	3086.78	4	4	1.9E-01
				2.104	1	3	1.4E+06	3354.37	8	6	1.1E-01
Cr XX				2.105	3	1	9.6E+05	3367.11	10	8	6.0E-02
14.13	2	4	1.1E+05	2.106	3	3	2.0E+06	3385.22	8	6	1.1E-01
14.26	4	6	1.3E+05	2.107	5	5	2.3E+06	3388.16	6	4	2.4E-01
128.42	4	4	3.8E+02	2.107	3	5	3.3E+06	3395.37	6	8	2.9E-01
131.31	6	4	1.27E+02	2.109	5	3	1.7E+06	3405.12	10	10	1.0E+00
133.82	2	4	8.3E+01	2.113	3	5	5.9E+05	3409.17	8	8	4.2E-01
135.26	4	2	2.41E+02	2.119	3	1	2.7E+05	3412.34	8	10	6.1E-01
140.75	4	4	1.35E+02	2.129	3	1	5.1E+05	3412.63	10	8	1.2E-01
148.99	6	4	1.75E+02	2.1818	1	3	3.37E+06	3414.74	4	4	8.8E-02
156.00	2	4	8.4E+01	2.1923	1	3	2.34E+05	3417.15	6	6	3.2E-01
167.97	6	6	1.12E+02					3431.58	8	6	1.1E-01
180.85	4	4	1.6E+02	**Cobalt**				3433.05	4	4	1.0E+00
				Co I				3442.92	6	4	1.2E-01
Cr XXI				2287.80	8	8	8.6E-01	3443.64	8	8	6.9E-01
12.97	3	1	4.8E+04	2295.22	10	8	2.2E-01	3449.17	6	6	7.6E-01
12.98	5	5	3.9E+04	2309.03	10	10	5.6E-01	3449.44	10	10	1.8E-01
13.02	3	5	3.8E+04	2323.13	8	8	5.0E-01	3453.51	10	12	1.1E+00
13.02	5	7	3.9E+04	2325.53	6	8	1.1E-01	3455.24	4	2	1.9E-01
13.08	1	3	5.2E+04	2335.98	6	6	5.1E-01	3462.80	4	6	7.9E-01
13.22	3	1	4.6E+04	2338.66	4	4	7.7E-01	3465.79	10	12	9.2E-02
13.34	3	5	5.2E+04	2353.36	8	10	1.5E-01	3474.02	6	8	5.6E-01
13.49	1	3	9.0E+04	2355.48	6	8	1.3E-01	3483.41	8	10	5.5E-02
13.53	3	3	6.6E+04	2358.18	4	6	1.4E-01	3489.40	8	6	1.3E+00
13.55	3	5	1.2E+05	2365.06	10	10	1.3E-01	3491.32	4	4	5.0E-02
13.65	5	7	1.5E+05	2371.85	6	8	7.3E-02	3495.68	4	6	4.9E-01
13.66	3	1	1.2E+05	2384.86	10	8	2.4E-01	3502.28	10	8	8.0E-01
13.67	5	5	3.9E+04	2392.03	6	6	4.0E-01	3502.63	6	6	5.2E-02
13.68	3	3	8.2E+04	2402.06	8	6	5.1E-01	3506.32	8	6	8.2E-01
13.75	5	3	4.5E+04	2407.25	10	12	3.6E+00	3509.84	6	8	3.2E-01
13.75	5	5	9.5E+04	2412.76	4	6	6.5E-01	3512.64	6	4	1.0E+00
13.76	1	3	1.51E+05	2414.46	6	8	3.4E+00	3513.48	8	10	7.8E-02
13.78	5	7	1.7E+05	2415.29	4	6	3.6E+00	3518.34	6	4	1.6E+00
13.84	5	7	2.59E+05	2424.93	10	10	3.2E+00	3521.58	10	8	1.8E-01
13.87	3	5	8.5E+04	2432.21	8	8	2.6E+00	3523.42	4	2	9.8E-01
13.92	3	5	8.5E+04	2436.66	6	6	2.6E+00	3526.85	10	10	1.3E-01
13.93	5	7	4.2E+04	2439.04	4	4	2.7E+00	3529.03	6	8	8.8E-02
13.95	5	5	3.8E+04	2460.80	4	6	1.2E-01	3529.82	8	10	4.6E-01

λ Å	Weights g_i	g_k	A 10^8 s^{-1}
3533.36	4	6	9.1E-02
3560.89	4	4	2.3E-01
3564.95	6	8	7.0E-02
3569.37	8	8	1.6E+00
3574.97	6	6	1.5E-01
3575.36	8	8	9.6E-02
3585.15	8	8	7.1E-02
3587.19	6	6	1.4E+00
3594.87	6	6	9.2E-02
3602.08	4	4	1.0E-01
3704.06	6	8	1.2E-01
3745.49	8	8	7.5E-02
3842.05	8	6	1.3E-01
3845.47	8	10	4.6E-01
3861.16	6	4	1.4E-01
3873.12	10	8	1.2E-01
3873.95	8	6	1.0E-01
3881.87	6	4	8.2E-02
3894.07	6	8	6.9E-01
3894.98	4	2	8.8E-02
3935.96	8	10	6.2E-02
3995.31	8	10	2.5E-01
3997.90	6	8	7.0E-02
4092.39	8	8	5.7E-02
4110.53	6	6	5.5E-02
4118.77	6	8	1.6E-01
4121.32	8	10	1.9E-01
5146.75	8	8	1.5E-01
5212.70	10	10	1.9E-01
5265.79	6	8	5.0E-02
5280.63	10	8	2.8E-01
5352.05	12	10	2.7E-01
5477.09	6	8	6.8E-02
5483.96	8	10	7.3E-02
6082.43	10	10	5.4E-02
6455.00	8	10	9.0E-02
7838.12	8	10	5.4E-02
8093.93	12	10	2.0E-01
8372.79	10	10	8.7E-02

Co II

λ Å	g_i	g_k	A 10^8 s^{-1}
2286.15	11	13	3.3E+00
2307.85	9	11	2.6E+00
2311.61	7	9	2.8E+00
2314.05	5	7	2.8E+00
2314.97	3	5	2.7E+00
2330.36	5	3	1.32E+00
2344.28	3	3	1.5E+00
2353.41	7	7	1.9E+00
2363.80	9	9	2.1E+00
2378.62	11	9	1.9E+00
2383.45	9	7	1.8E+00
2388.92	11	11	2.8E+00
2389.54	5	3	1.5E+00
2404.17	3	3	1.5E+00
2417.66	9	9	8.5E-01

Copper Cu I

λ Å	Weights g_i	g_k	A 10^8 s^{-1}
*2024.3	2	6	9.8E-02
2165.1	2	4	5.1E-01
2178.9	2	4	9.13E-01
2181.7	2	2	1.0E+00
2225.7	2	2	4.6E-01
2244.3	2	4	1.19E-02
2441.6	2	2	2.0E-02
2492.2	2	4	3.11E-02
2618.4	6	4	3.07E-01
2766.4	4	4	9.6E-02
2824.4	6	6	7.8E-02
2961.2	6	8	3.76E-02
3063.4	4	4	1.55E-02
3194.1	4	4	1.55E-02
3247.5	2	4	1.39E+00
3274.0	2	2	1.37E+00
3337.8	6	8	3.8E-03
4022.6	2	4	1.90E-01
4062.6	4	6	2.10E-01
4249.0	2	2	1.95E-01
4275.1	6	8	3.45E-01
4480.4	2	2	3.0E-02
4509.4	4	2	2.75E-01
4530.8	4	2	8.4E-02
4539.7	6	4	2.12E-01
4587.0	8	6	3.20E-01
4651.1	10	8	3.80E-01
4704.6	8	8	5.5E-02
5105.5	6	4	2.0E-02
5153.2	2	4	6.0E-01
5218.2	4	6	7.5E-01
5220.1	4	4	1.50E-01
5292.5	8	8	1.09E-01
5700.2	4	4	2.4E-03
5782.1	4	2	1.65E-02

Cu II

λ Å	g_i	g_k	A 10^8 s^{-1}
2489.7	5	5	1.5E-02
2544.8	9	7	1.1E+00
2689.3	7	7	4.1E-01
2701.0	5	5	6.7E-01
2703.2	3	3	1.2E+00
2713.5	5	5	6.8E-01
2769.7	7	7	6.1E-01

Dysprosium Dy I

λ Å	g_i	g_k	A 10^8 s^{-1}
2862.7	17	15	6.5E-02
2964.6	17	17	6.5E-02
3147.7	15	17	1.1E-01
3263.2	15	13	1.4E-01
3511.0	15	13	3.1E-01
3571.4	15	13	2.0E-01
3757.1	17	19	3.0E+00

λ Å	Weights g_i	g_k	A 10^8 s^{-1}
3868.8	17	17	3.1E+00
3967.5	17	19	8.7E-01
4046.0	17	15	1.5E+00
4103.9	13	11	1.7E+00
4186.8	17	17	1.32E+00
4194.8	17	17	7.2E-01
4211.7	17	19	2.08E+00
4218.1	15	15	1.85E+00
4221.1	15	17	1.52E+00
4225.2	13	15	4.5E+00
4268.3	15	15	3.6E-02
4276.7	13	13	7.3E-01
4292.0	15	15	5.8E-02
4577.8	17	19	2.2E-02
4589.4	17	15	1.3E-01
4612.3	17	15	8.2E-02
5077.7	17	17	5.7E-03
5301.6	17	15	1.1E-02
5547.3	17	17	2.7E-03
5639.5	17	19	4.7E-03
5974.5	17	17	4.0E-03
5988.6	17	15	5.3E-03
6010.8	15	15	2.6E-02
6088.3	15	13	3.5E-02
6168.4	15	17	2.5E-02
6259.1	17	19	8.5E-03
6579.4	17	15	7.5E-03

Erbium Er I

λ Å	g_i	g_k	A 10^8 s^{-1}
3862.9	13	13	2.5E+00
4008.0	13	15	2.6E+00
4151.1	13	11	1.8E+00

Europium Eu I

λ Å	g_i	g_k	A 10^8 s^{-1}
2372.9	8	6	1.9E-01
2375.3	8	8	2.0E-01
2379.7	8	10	2.0E-01
2619.3	8	10	7.0E-03
2643.8	8	8	6.6E-03
2659.4	8	10	1.2E-02
2682.6	8	6	1.2E-02
2710.0	8	10	1.4E-01
2724.0	8	8	1.2E-01
2731.4	8	8	3.1E-02
2732.6	8	6	3.7E-02
2735.3	8	10	4.7E-02
2738.6	8	10	1.3E-02
2743.3	8	6	1.1E-01
2745.6	8	6	5.0E-02
2747.8	8	8	5.2E-02
2772.9	8	6	1.0E-02
2878.9	8	10	2.8E-02
2892.5	8	8	1.0E-01
2893.0	8	6	1.0E-01

λ Å	Weights g_i	Weights g_k	A 10^8 s^{-1}	λ Å	Weights g_i	Weights g_k	A 10^8 s^{-1}	λ Å	Weights g_i	Weights g_k	A 10^8 s^{-1}
2909.0	8	10	6.9E-02	7425.7	4	2	3.4E-01	2754.6	5	3	1.1E+00
2958.9	8	6	1.6E-02	7482.7	4	4	5.6E-02	3039.1	5	3	2.8E+00
3059.0	8	8	3.8E-02	7489.2	2	2	1.1E-01	3124.8	5	5	3.1E-02
3067.0	8	10	9.1E-03	7514.9	2	2	5.2E-02	3269.5	5	3	2.9E-01
3106.2	8	10	5.5E-02	7552.2	4	6	7.8E-02	4226.6	1	3	2.1E-01
3111.4	8	10	3.0E-01	7573.4	2	4	1.0E-01	4685.8	1	3	9.5E-02
3168.3	8	10	6.9E-02	7607.2	4	4	7.0E-02				
3185.5	8	10	5.8E-03	7754.7	4	6	3.82E-01	**Ge II**			
3210.6	8	8	1.1E-01	7800.2	2	4	2.1E-01	999.10	2	4	1.9E+00
3212.8	8	8	2.9E-01					1016.6	4	6	2.1E+00
3213.8	8	6	1.8E-01					1017.1	4	2	3.5E-01
3235.1	8	10	1.0E-02	**Gallium**				1055.0	2	2	6.9E-01
3241.4	8	8	2.3E-02	**Ga I**				1075.1	4	2	1.3E+00
3246.0	8	6	1.4E-02	2195.4	2	2	1.9E-02	1237.1	2	4	1.9E+01
3247.6	8	8	2.3E-02	2199.7	4	2	3.3E-02	1261.9	4	6	2.2E+01
3322.3	8	6	3.5E-02	2214.4	4	6	1.2E-02	1264.7	4	4	3.5E+00
3334.3	8	6	3.4E-01	2235.9	4	2	4.3E-02	1602.5	2	2	3.4E+00
3350.4	8	10	1.5E-02	2255.0	2	2	3.1E-02	1649.2	4	2	6.5E+00
3353.7	8	8	5.8E-03	2259.2	4	6	3.1E-02	4741.8	2	4	4.6E-01
3457.1	8	8	8.4E-03	2294.2	2	4	7.0E-02	4814.6	4	6	5.1E-01
3467.9	8	8	1.0E-02	2297.9	4	2	5.8E-02	4824.1	4	4	8.6E-02
3589.3	8	6	6.9E-03	2338.2	4	6	9.8E-02	5131.8	4	6	1.9E+00
4594.0	8	10	1.4E+00	2371.3	2	2	5.7E-02	5178.5	6	6	1.3E-01
4627.2	8	8	1.3E+00	2418.7	4	2	1.0E-01	5178.6	6	8	2.0E+00
4661.9	8	6	1.3E+00	2450.1	2	2	2.8E-01	5893.4	2	4	9.2E-01
5645.8	8	6	5.4E-03	2500.2	4	6	3.4E-01	6021.0	2	2	8.4E-01
5765.2	8	8	1.1E-02	2659.9	2	2	1.2E-01	6336.4	2	2	4.4E-01
6018.2	8	10	8.5E-03	2719.7	4	2	2.3E-01	6484.2	4	2	8.5E-01
6291.3	8	6	1.8E-02	2874.2	2	4	1.2E+00				
6864.5	8	10	5.8E-03	2943.6	4	6	1.4E+00	**Gold**			
7106.5	8	8	2.6E-03	2944.2	4	4	2.7E-01	**Au I**			
				4033.0	2	2	4.9E-01	2427.95	2	4	1.99E+00
Fluorine				4172.0	4	2	9.2E-01	2675.95	2	2	1.64E+00
F I								3122.78	6	4	1.90E-01
806.96	4	6	3.3E+00	**Ga II**				6278.30	4	2	3.4E-02
809.60	2	4	2.8E+00	829.60	1	3	2.2E-01				
951.87	4	2	2.6E+00	1414.4	1	3	1.88E+01	**Helium**			
954.83	4	4	5.77E+00					**He I**			
955.55	2	2	5.1E+00					510.00	1	3	4.6224E-01
958.52	2	4	1.3E+00	**Germanium**				512.10	1	3	7.3174E-01
6239.7	6	4	2.5E-01	**Ge I**				515.62	1	3	1.2582E+00
6348.5	4	4	1.8E-01	1944.7	3	1	7.0E-01	522.21	1	3	2.4356E+00
6413.7	2	4	1.1E-01	1955.1	3	3	2.8E-01	537.03	1	3	5.6634E+00
6708.3	6	4	1.4E-02	1988.3	5	3	2.5E-01	584.33	1	3	1.7989E+01
6774.0	6	6	1.0E-01	1998.9	5	5	5.5E-01	*2677.1	3	9	4.4174E-03
6795.5	4	2	5.2E-02	2041.7	1	3	1.1E+00	*2696.1	3	9	6.0234E-03
6834.3	4	4	2.1E-01	2065.2	3	3	8.5E-01	*2723.2	3	9	8.4996E-03
6856.0	6	8	4.94E-01	2068.7	3	5	1.2E+00	*2763.8	3	9	1.2508E-02
6870.2	2	2	3.8E-01	2086.0	3	5	4.0E-01	*2829.1	3	9	1.9389E-02
6902.5	4	6	3.2E-01	2094.3	5	7	9.7E-01	*2945.1	3	9	3.2006E-02
6909.8	2	4	2.2E-01	2105.8	5	5	1.7E-01	*3187.7	3	9	5.6361E-02
6966.4	4	2	1.1E-01	2256.0	5	5	3.2E-02	3231.3	1	3	5.1015E-03
7037.5	4	4	3.0E-01	2417.4	5	5	9.6E-01	3258.3	1	3	6.9627E-03
7127.9	2	2	3.8E-01	2498.0	1	3	1.3E-01	3296.8	1	3	9.8432E-03
7309.0	6	8	4.7E-01	2533.2	3	3	1.0E-01	3354.6	1	3	1.4537E-02
7311.0	4	2	3.9E-01	2589.2	5	3	5.1E-02	3447.6	1	3	2.2691E-02
7314.3	4	6	4.8E-01	2592.5	3	5	7.1E-01	*3554.4	9	15	7.5971E-03
7332.0	6	4	3.1E-01	2651.2	5	5	2.0E+00	*3563.0	9	3	4.8362E-03
7398.7	6	6	2.85E-01	2651.6	1	3	8.5E-01	*3587.3	9	15	1.8107E-02
				2691.3	3	3	6.1E-01				
				2709.6	3	1	2.8E+00				

λ Å	Weights g_i	g_k	A 10^8 s^{-1}	λ Å	Weights g_i	g_k	A 10^8 s^{-1}	λ Å	Weights g_i	g_k	A 10^8 s^{-1}
3613.6	1	3	3.8022E-02	**In II**				2276.03	9	7	1.7E-01
*3634.2	9	15	2.6062E-02	2941.1	3	1	1.4E+00	2277.11	7	5	3.7E+01
*3652.0	9	3	9.7444E-03					2287.25	5	3	3.4E-01
*3705.0	9	15	3.9528E-02	**Iodine**				2292.52	7	9	4.3E-02
*3819.6	9	15	6.4351E-02	**I I**				2294.41	3	1	6.1E-01
3833.6	3	5	9.6470E-03	1782.8	4	4	2.71E+00	2300.14	5	7	8.0E-02
*3867.5	9	3	2.4465E-02	1830.4	4	6	1.6E-01	2301.68	1	3	1.3E-01
3871.8	3	5	1.3386E-02					2303.42	1	3	9.4E-02
*3888.7	3	9	9.4746E-02	**Iridium**				2303.58	3	5	7.6E-02
3926.5	3	5	1.9371E-02	**Ir I**				2309.00	3	5	1.5E-01
3935.9	3	1	7.4475E-03	2475.12	10	10	2.1E-01	2313.10	5	7	1.4E-01
3964.7	1	3	6.9507E-02	2502.98	10	12	3.2E-01	2320.36	7	9	1.2E-01
4009.3	3	5	2.9612E-02	2639.71	10	10	4.7E-01	2371.43	5	5	5.2E-02
4024.0	3	1	1.1281E-02	2661.98	10	10	2.5E-01	2373.62	7	7	6.7E-02
*4026.2	9	15	1.1600E-01	2664.79	10	8	4.0E-01	2374.52	1	3	2.9E-01
*4120.8	9	3	4.4529E-02	2694.23	10	12	4.8E-01	2381.83	3	5	5.4E-02
4143.8	3	5	4.8812E-02	2849.72	10	10	2.2E-01	2389.97	5	7	5.0E-02
4169.0	3	1	1.8298E-02	2853.31	10	10	2.0E-03	2462.18	7	5	1.5E-01
4387.9	3	5	8.9889E-02	2882.64	10	8	7.2E-01	2462.65	9	9	5.8E-01
4437.6	3	1	3.2689E-02	2924.79	10	12	1.42E-01	2479.78	5	5	1.8E+00
*4471.5	9	15	2.4578E-01	2934.64	8	10	2.0E-01	2483.27	9	11	4.9E+00
*4713.2	9	3	9.5209E-02	2951.22	10	8	2.8E-02	2488.14	7	9	4.7E+00
4921.9	3	5	1.9863E-02	3003.63	8	10	5.9E-02	2490.64	5	7	3.8E+00
5015.7	1	3	1.3372E-01	3168.88	8	10	5.47E-02	2491.15	3	5	3.0E+00
5047.7	3	1	6.7712E-02	3220.78	10	8	2.4E-01	2501.13	9	7	6.8E-01
*5875.7	9	15	7.0703E-01	3558.99	6	8	1.5E-02	2510.83	7	5	1.3E+00
6678.2	3	5	6.3705E-01	3573.72	8	10	5.4E-02	2518.10	5	3	1.9E+00
*7065.2	9	3	2.7853E-01	3617.21	6	8	2.0E-02	2522.85	9	9	2.9E+00
7281.4	3	1	1.8299E-01	3628.67	8	8	2.8E-02	2524.29	3	1	3.4E+00
*8361.7	3	9	3.8126E-03	3661.71	8	10	4.0E-02	2527.43	7	7	1.9E+00
*9463.6	3	9	5.6868E-03	3734.77	8	8	2.7E-02	2529.13	5	5	9.8E-01
9603.4	1	3	5.8286E-03	4033.76	8	10	2.7E-02	2535.61	1	3	9.7E-01
*9702.6	9	3	8.6511E-03	4069.92	6	8	3.6E-02	2540.97	3	5	9.2E-01
*10311	9	15	1.9945E-02	4913.35	12	12	3.3E-02	2545.98	5	7	6.7E-01
*10668	9	3	1.4471E-02	4939.24	10	12	2.5E-03	2549.61	7	9	3.6E-01
*10830	3	9	1.0216E-01					2584.54	11	13	4.6E-01
*10913	15	21	1.9801E-02	**Iron**				2606.83	9	11	4.2E-01
10917	5	7	1.6083E-02	**Fe I**				2618.02	7	7	4.0E-01
*10997	15	9	1.4253E-03	1934.54	9	7	2.5E-01	2623.53	7	9	3.3E-01
11013	1	3	9.2496E-03	1937.27	9	7	2.2E-01	2656.15	13	15	2.8E-01
11045	3	5	1.8457E-02	1940.66	7	5	2.6E-01	2669.49	11	13	1.7E-01
11226	3	1	1.1168E-02	2084.12	9	7	3.7E-01	2679.06	11	11	1.9E-01
*11969	9	15	3.4781E-02	2102.35	7	7	8.8E-02	2719.03	9	7	1.4E+00
*12528	3	9	7.0932E-03	2112.97	1	3	1.9E-01	2720.90	7	5	1.1E+00
12756	5	3	1.2754E-03	2132.02	9	9	7.6E-02	2723.58	5	3	6.4E-01
*12785	15	21	4.1339E-02	2145.19	7	7	5.7E-02	2733.58	11	9	8.6E-01
12791	5	7	3.2475E-02	2153.01	5	5	6.9E-02	2735.48	9	7	6.2E-01
*12846	9	3	2.7317E-02	2161.58	3	5	5.0E-02	2737.31	3	3	8.5E-01
12968	3	5	3.3615E-02	2166.77	9	7	2.7E+00	2742.41	5	5	6.3E-01
*12985	15	9	2.7292E-03	2171.30	5	7	5.1E-02	2744.07	1	3	3.5E-01
				2173.21	3	5	8.3E-02	2750.14	7	7	3.9E-01
Indium				2176.84	1	3	1.0E-01	2756.33	3	5	2.0E-01
In I				2191.20	1	3	7.3E-02	2788.10	11	13	6.3E-01
2560.2	2	4	4.0E-01	2191.84	5	5	1.2E+00	2894.50	5	5	6.2E-01
2710.3	4	6	4.0E-01	2196.04	3	3	1.2E+00	2899.42	5	3	5.9E-01
3039.4	2	4	1.3E+00	2200.72	3	5	2.8E-01	2920.69	5	5	5.2E-02
3256.1	4	6	1.3E+00	2259.51	9	11	7.0E-02	2923.29	11	11	1.6E+00
4101.8	2	2	5.6E-01	2267.08	7	5	7.1E-02	2925.36	7	9	1.8E-01
4511.3	4	2	1.02E+00	2272.07	7	9	3.8E-02	2929.01	7	5	7.3E-02

λ Å	g_i	g_k	A 10^8 s^{-1}	λ Å	g_i	g_k	A 10^8 s^{-1}	λ Å	g_i	g_k	A 10^8 s^{-1}
2936.90	9	9	1.3E-01	3215.94	5	5	8.0E-01	3442.36	5	5	4.55E-02
2941.34	5	3	5.6E-02	3217.38	11	9	2.2E-01	3443.88	5	3	6.2E-02
2947.88	7	7	2.0E-01	3219.58	7	9	6.2E-01	3445.15	5	7	2.8E-01
2953.94	5	5	1.89E-01	3222.07	11	11	3.3E-01	3447.28	5	5	9.1E-02
2954.65	5	7	1.0E-01	3225.79	11	13	8.8E-01	3450.33	3	3	2.0E-01
2957.36	3	3	1.77E-01	3227.80	9	7	1.4E+00	3476.70	1	3	5.4E-02
2965.25	1	3	1.16E-01	3228.25	5	3	4.5E-01	3477.85	3	1	4.2E-02
2966.90	9	11	2.72E-01	3229.99	9	11	4.5E-01	3485.34	5	3	1.4E-01
2969.36	3	1	3.66E-02	3230.21	5	5	1.9E-01	3495.29	9	7	9.46E-01
2973.13	5	7	1.35E-01	3230.96	7	5	3.9E-01	3497.10	7	7	1.4E-01
2973.24	7	9	1.83E-01	3233.05	13	15	5.4E-01	3505.07	5	3	9.9E-02
2980.53	7	7	2.2E-01	3233.97	9	9	2.0E-01	3506.50	5	5	7.1E-02
2981.45	7	5	6.54E-02	3246.96	5	3	9.9E-02	3508.49	9	11	5.7E-02
2983.57	9	7	2.80E-01	3248.20	7	7	2.2E-01	3510.44	1	3	4.4E-02
2987.29	9	7	6.6E-02	3253.60	7	9	1.8E-01	3516.56	7	5	3.7E-02
2990.39	9	11	3.9E-01	3254.36	11	13	5.1E-01	3521.84	3	5	9.6E-02
2994.43	7	5	4.4E-01	3257.59	7	5	1.4E-01	3523.31	5	3	7.6E-02
2996.39	3	5	1.6E-01	3265.62	7	5	3.8E-01	3524.08	7	5	7.5E-02
2999.51	11	11	2.3E-01	3268.23	3	3	5.9E-02	3524.24	5	7	4.2E-02
3000.95	5	3	6.42E-01	3271.00	5	3	6.6E-01	3527.79	9	9	2.0E-01
3008.14	3	1	1.07E+00	3280.26	9	11	5.4E-01	3529.82	3	3	7.6E-01
3009.09	13	11	6.7E-02	3282.89	5	3	3.0E-01	3536.56	5	7	7.8E-01
3009.57	9	9	1.7E-01	3284.59	5	5	5.4E-02	3537.73	5	3	1.1E-01
3011.48	7	9	4.7E-01	3290.99	3	5	6.0E-02	3537.90	11	11	8.4E-02
3015.92	11	9	5.9E-02	3292.02	7	9	6.1E-01	3540.12	7	9	1.2E-01
3016.18	5	3	8.5E-02	3292.59	3	3	2.6E-01	3541.08	9	11	6.2E-01
3017.63	3	3	6.82E-02	3298.13	3	5	8.1E-02	3542.08	7	9	7.4E-01
3018.98	7	7	1.3E-01	3305.97	5	7	4.7E-01	3543.67	3	5	1.8E-01
3021.07	7	7	4.56E-01	3306.36	3	5	6.1E-01	3548.02	5	3	9.7E-02
3024.03	3	5	4.88E-02	3307.23	13	13	2.0E-01	3552.11	3	5	4.5E-02
3025.84	1	3	3.48E-01	3314.74	5	7	6.9E-01	3552.83	5	5	1.5E-01
3026.46	5	5	1.1E-01	3322.47	9	11	6.2E-02	3553.74	11	9	8.1E-01
3031.63	3	3	1.5E-01	3323.74	5	5	3.0E-01	3556.88	9	11	4.4E-01
3037.39	3	5	3.2E-01	3328.87	11	11	2.7E-01	3559.50	3	3	1.9E-01
3042.02	3	5	4.9E-02	3337.66	11	9	5.7E-02	3560.70	7	9	6.5E-02
3042.66	5	7	5.7E-02	3347.93	5	5	4.0E-02	3565.38	7	9	3.8E-01
3047.60	5	7	2.84E-01	3354.06	1	3	7.7E-02	3567.03	5	7	6.5E-01
3053.07	3	5	1.5E-01	3355.23	9	9	3.2E-01	3568.42	5	3	5.3E-02
3057.45	11	9	4.4E-01	3369.55	9	9	2.4E-01	3568.82	7	9	5.6E-02
3059.09	7	9	1.7E-01	3370.78	11	11	3.3E-01	3570.10	9	11	6.77E-01
3067.24	9	7	3.4E-01	3380.11	7	7	2.4E-01	3572.00	11	11	2.4E-01
3068.17	5	3	9.8E-02	3383.98	7	7	9.3E-02	3573.39	5	7	7.5E-02
3075.72	7	5	2.9E-01	3392.65	7	7	2.6E-01	3576.76	11	9	9.6E-02
3083.74	5	3	3.0E-01	3394.58	5	3	9.9E-02	3578.38	1	3	6.3E-02
3091.58	3	1	5.4E-01	3399.33	5	5	3.8E-01	3581.19	11	13	1.02E+00
3098.19	11	11	1.1E-01	3402.26	13	13	2.8E-01	3582.20	13	11	2.5E-01
3100.67	7	7	1.4E-01	3406.44	3	5	3.0E-01	3583.33	1	3	2.3E-01
3119.49	11	9	8.2E-02	3407.46	7	9	5.8E-01	3585.32	7	7	1.3E-01
3120.43	9	7	8.9E-02	3410.17	3	5	4.7E-01	3585.71	9	9	3.75E-02
3156.27	7	7	5.4E-01	3411.35	9	9	5.5E-02	3586.98	5	5	1.6E-01
3160.66	9	9	1.9E-01	3413.13	5	5	3.6E-01	3591.48	1	3	6.0E-02
3161.95	11	13	1.2E-01	3417.84	3	3	5.1E-01	3592.67	7	5	4.0E-02
3166.44	9	7	1.14E-01	3418.51	3	1	1.3E+00	3594.63	9	9	2.7E-01
3168.85	5	7	5.7E-02	3424.28	7	7	2.0E-01	3595.30	5	5	5.4E-02
3175.45	11	11	1.3E-01	3425.01	9	7	2.8E-01	3597.02	5	3	1.7E-01
3176.36	5	3	9.2E-02	3427.12	7	9	5.5E-01	3599.62	11	9	1.8E-01
3196.93	9	11	9.0E-01	3428.19	5	5	2.1E-01	3603.20	11	11	2.6E-01
3199.53	9	9	2.6E-01	3428.75	7	5	2.7E-01	3603.82	3	3	1.7E-01
3205.40	3	3	1.2E+00	3440.99	7	5	8.4E-02	3605.45	9	9	6.4E-01

λ Å	Weights g_i	g_k	A 10^8 s^{-1}	λ Å	Weights g_i	g_k	A 10^8 s^{-1}	λ Å	Weights g_i	g_k	A 10^8 s^{-1}
3606.68	11	13	8.2E-01	3724.38	5	7	1.3E-01	3820.43	11	9	6.68E-01
3608.86	3	5	8.14E-01	3726.93	5	5	4.6E-01	3821.18	11	13	7.0E-01
3610.16	13	13	4.8E-01	3727.09	9	7	2.0E-01	3821.83	5	5	7.8E-02
3610.70	5	3	7.1E-02	3727.62	7	5	2.25E-01	3825.88	9	7	5.98E-01
3612.07	11	13	7.5E-02	3730.39	9	11	1.3E-01	3827.82	7	5	1.05E+00
3613.45	7	7	6.7E-02	3730.95	5	7	3.8E-02	3833.31	9	9	4.69E-02
3615.19	3	3	5.8E-02	3732.40	5	5	2.8E-01	3834.22	7	5	4.53E-01
3617.79	5	7	6.5E-01	3733.32	3	3	6.2E-02	3836.33	5	5	3.7E-01
3618.77	5	7	7.3E-01	3734.86	11	11	9.02E-01	3839.26	9	9	2.8E-01
3621.46	9	11	5.1E-01	3735.32	9	9	2.4E-01	3839.61	3	5	3.9E-01
3622.00	7	7	5.1E-01	3737.13	7	9	1.42E-01	3840.44	5	3	4.70E-01
3623.19	13	13	7.4E-02	3738.31	11	13	3.8E-01	3841.05	5	3	1.3E+00
3624.06	5	3	5.4E-02	3740.24	7	9	1.4E-01	3843.26	9	7	4.7E-01
3630.35	9	7	7.6E-02	3742.62	9	9	1.0E-01	3845.17	3	3	6.8E-02
3631.46	7	9	5.17E-01	3743.36	5	3	2.60E-01	3845.69	5	7	4.9E-02
3632.04	3	5	4.8E-01	3744.10	5	3	3.6E-01	3846.00	9	7	4.3E-02
3632.55	11	9	5.2E-02	3745.56	5	7	1.15E-01	3846.41	11	9	1.9E-01
3635.19	5	3	1.4E-01	3745.90	1	3	7.33E-02	3846.80	7	7	6.6E-01
3637.86	9	9	5.5E-02	3746.93	7	7	2.2E-01	3849.96	3	1	6.06E-01
3638.30	7	9	2.6E-01	3748.26	3	5	9.15E-02	3856.37	7	5	4.64E-02
3640.39	9	11	3.8E-01	3749.48	9	9	7.64E-01	3859.21	13	11	8.5E-02
3644.80	7	5	7.8E-02	3753.61	7	5	9.3E-02	3859.91	9	9	9.70E-02
3645.82	1	3	5.7E-01	3756.94	11	11	2.4E-01	3865.52	3	3	1.55E-01
3647.84	9	11	2.92E-01	3757.45	5	3	1.2E-01	3867.22	5	5	3.4E-01
3649.51	11	9	4.2E-01	3758.23	7	7	6.34E-01	3871.75	11	11	6.7E-02
3650.03	7	7	9.9E-02	3760.05	13	15	4.47E-02	3872.50	5	5	1.05E-01
3651.47	7	9	6.2E-01	3760.53	3	5	4.8E-02	3873.76	11	9	8.0E-02
3655.46	5	5	1.0E-01	3763.79	5	5	5.44E-01	3878.02	7	7	7.72E-02
3659.52	9	9	5.8E-02	3765.54	13	15	9.8E-01	3878.57	5	3	6.6E-02
3667.25	9	7	1.4E-01	3766.67	5	3	9.7E-02	3883.28	7	7	1.6E-01
3669.15	9	7	7.4E-02	3767.19	3	3	6.40E-01	3884.36	11	9	3.5E-02
3669.52	9	7	3.0E-01	3768.03	3	1	8.4E-02	3885.51	3	5	5.8E-02
3670.09	11	13	7.6E-02	3774.82	3	3	4.7E-02	3886.28	7	7	5.30E-02
3674.77	5	3	6.7E-02	3778.51	7	5	1.2E-01	3887.05	9	9	3.52E-02
3676.31	9	11	4.63E-02	3781.94	5	7	3.7E-02	3888.51	5	5	2.6E-01
3677.31	5	7	3.1E-01	3785.95	11	13	4.2E-02	3888.82	5	3	2.7E-01
3677.63	7	5	8.0E-01	3786.19	5	5	1.2E-01	3891.93	3	3	4.0E-01
3678.86	3	5	4.1E-02	3787.16	5	5	1.0E-01	3893.39	11	11	1.3E-01
3682.24	5	5	1.7E+00	3787.88	3	5	1.29E-01	3895.66	3	1	9.40E-02
3684.11	9	7	3.4E-01	3789.82	9	7	3.9E-02	3900.52	7	7	7.5E-02
3686.00	9	11	2.6E-01	3791.73	5	3	6.3E-02	3902.95	7	7	2.14E-01
3686.26	3	1	1.2E-01	3793.87	3	3	7.4E-02	3903.90	9	9	9.6E-02
3687.46	11	9	8.01E-02	3794.34	9	11	3.8E-02	3906.75	5	7	6.7E-02
3688.48	7	9	6.9E-02	3795.00	5	7	1.15E-01	3907.93	7	5	6.7E-02
3690.73	11	11	2.7E-01	3799.55	7	9	7.32E-02	3909.66	3	5	5.3E-02
3694.01	5	7	6.8E-01	3801.68	5	7	6.6E-02	3909.83	3	3	6.5E-02
3697.43	7	7	2.1E-01	3802.00	11	13	3.5E-02	3914.27	3	3	5.4E-02
3698.60	5	7	3.8E-02	3802.28	5	5	5.0E-02	3916.73	13	11	1.2E-01
3699.15	5	7	4.5E-02	3804.01	11	9	4.7E-02	3919.07	9	9	3.9E-02
3701.09	7	9	4.8E-01	3805.35	9	11	9.8E-01	3925.20	1	3	5.7E-02
3702.03	3	1	3.5E-01	3806.22	3	3	2.3E-01	3931.12	5	7	4.5E-02
3703.69	9	11	5.3E-02	3806.70	11	11	5.4E-01	3941.28	5	5	8.4E-02
3703.82	1	3	1.2E-01	3807.54	3	5	8.0E-02	3942.44	3	5	9.0E-02
3704.46	11	9	1.3E-01	3808.73	9	9	3.54E-02	3946.99	9	11	4.4E-02
3709.25	9	7	1.56E-01	3810.76	5	3	2.0E-01	3948.77	11	9	2.2E-01
3711.41	3	5	7.3E-02	3813.88	13	11	8.7E-02	3949.14	3	3	3.9E-02
3718.41	7	7	5.3E-02	3815.84	9	7	1.3E+00	3949.95	7	5	5.9E-02
3719.93	9	11	1.62E-01	3817.64	11	11	8.3E-02	3951.16	3	5	3.6E-01
3722.56	5	5	4.97E-02	3819.50	7	5	4.6E-02	3952.60	11	11	4.1E-02

λ Å	Weights g_i	Weights g_k	A 10^8 s^{-1}	λ Å	Weights g_i	Weights g_k	A 10^8 s^{-1}	λ Å	Weights g_i	Weights g_k	A 10^8 s^{-1}
3953.15	7	9	3.7E-02	4109.07	1	3	4.5E-02	4250.79	7	7	1.0E-01
3955.34	3	3	1.4E-01	4109.80	3	3	1.6E-01	4260.47	11	11	3.2E-01
3955.96	3	3	5.7E-02	4112.96	11	13	1.4E-01	4267.83	1	3	9.4E-02
3956.45	13	11	2.1E-01	4114.45	5	5	4.7E-02	4268.75	5	3	4.2E-02
3957.02	5	7	1.6E-01	4118.54	11	13	5.8E-01	4271.15	7	9	1.82E-01
3960.28	5	7	4.2E-02	4126.18	11	11	3.9E-02	4271.76	9	11	2.28E-01
3963.10	3	5	1.7E-01	4127.61	1	3	1.3E-01	4282.40	7	5	1.1E-01
3967.42	9	7	2.3E-01	4132.06	5	7	1.2E-01	4300.83	5	5	4.7E-02
3967.96	7	9	6.3E-02	4132.90	3	5	9.4E-02	4305.45	5	3	6.0E-02
3969.26	9	7	2.3E-01	4134.68	5	7	1.8E-01	4307.90	7	9	3.4E-01
3970.39	3	1	3.5E-01	4137.00	3	5	2.2E-01	4315.08	5	5	7.7E-02
3971.32	11	9	5.7E-02	4137.42	5	7	6.1E-02	4325.76	5	7	5.0E-01
3973.65	5	7	6.6E-02	4142.63	3	5	7.4E-02	4327.09	5	5	7.8E-02
3976.61	3	5	1.8E-01	4143.87	7	9	1.5E-01	4352.73	3	5	3.9E-02
3977.74	5	5	7.0E-02	4149.37	11	13	3.6E-02	4369.77	9	9	7.2E-02
3981.77	9	9	3.9E-02	4150.25	3	3	7.1E-02	4383.54	9	11	5.00E-01
3983.96	9	7	7.6E-02	4153.90	7	9	2.3E-01	4387.89	3	3	3.9E-02
3985.39	5	5	6.7E-02	4154.80	9	11	1.5E-01	4388.41	7	7	1.3E-01
3989.86	5	7	5.0E-02	4156.80	5	5	1.9E-01	4401.29	7	7	5.9E-02
3996.97	9	9	6.7E-02	4158.79	3	5	1.6E-01	4404.75	7	9	2.75E-01
3997.39	9	11	1.5E-01	4170.90	5	5	6.1E-02	4415.12	5	7	1.19E-01
3998.05	11	9	6.6E-02	4172.12	7	5	9.7E-02	4422.57	3	3	8.8E-02
4003.76	3	3	7.1E-02	4175.64	3	5	1.6E-01	4430.61	3	1	7.45E-02
4005.24	7	5	2.04E-01	4181.75	5	7	3.6E-02	4433.22	5	3	2.3E-01
4006.31	11	9	4.7E-02	4182.38	5	5	4.9E-02	4438.34	3	1	7.9E-02
4007.27	7	5	4.2E-02	4184.89	5	5	1.1E-01	4442.34	5	5	3.76E-02
4009.71	3	5	5.2E-02	4187.04	7	5	2.15E-01	4443.19	1	3	1.1E-01
4014.53	11	11	2.4E-01	4187.79	9	7	1.52E-01	4446.83	3	3	5.3E-02
4017.15	9	11	4.5E-02	4191.68	1	3	4.8E-02	4447.72	3	3	5.11E-02
4021.87	7	9	1.0E-01	4196.21	7	7	9.8E-02	4454.38	5	5	3.8E-02
4024.72	7	9	8.9E-02	4198.30	11	9	8.03E-02	4455.03	9	7	3.9E-02
4031.96	3	5	7.1E-02	4198.64	5	5	1.3E-01	4466.55	5	7	1.2E-01
4040.64	5	7	4.4E-02	4199.09	9	11	6.1E-01	4469.37	5	7	2.6E-01
4044.61	5	3	1.1E-01	4200.09	7	7	4.0E-02	4481.61	3	3	4.2E-02
4045.81	9	9	8.63E-01	4200.92	7	9	4.2E-02	4484.22	7	9	7.0E-02
4054.87	5	3	1.6E-01	4202.03	9	9	8.22E-02	4485.67	3	3	1.1E-01
4058.22	9	7	4.9E-02	4203.67	7	9	8.6E-02	4528.61	7	9	5.44E-02
4059.73	5	3	8.1E-02	4203.94	13	13	1.3E-01	4533.13	3	1	3.7E-02
4062.44	3	3	2.2E-01	4205.54	5	5	3.6E-02	4547.85	5	7	7.6E-02
4063.59	7	7	6.8E-01	4207.13	5	3	4.3E-02	4619.29	7	5	4.7E-02
4065.40	3	1	1.9E-01	4210.34	3	3	1.7E-01	4669.17	5	3	4.0E-02
4067.98	9	9	1.7E-01	4213.65	3	1	1.9E-01	4673.16	5	7	4.6E-02
4070.77	7	5	1.3E-01	4217.55	3	5	2.3E-01	4678.85	7	9	7.4E-02
4071.74	5	5	7.65E-01	4219.36	11	13	3.8E-01	4704.95	3	1	8.1E-02
4073.76	5	3	1.6E-01	4220.34	3	1	1.9E-01	4736.77	9	11	4.9E-02
4074.79	9	9	4.8E-02	4222.21	7	7	5.77E-02	4789.65	5	5	7.2E-02
4076.63	9	9	1.9E-01	4224.17	9	11	1.3E-01	4859.74	5	3	1.3E-01
4078.35	5	3	4.2E-02	4224.51	3	5	7.1E-02	4871.32	7	5	2.2E-01
4079.18	5	5	5.1E-02	4225.45	5	7	1.7E-01	4872.14	3	3	2.4E-01
4079.84	1	3	6.3E-02	4226.42	3	3	3.7E-02	4878.21	1	3	9.1E-02
4080.21	3	1	2.4E-01	4233.60	3	5	1.85E-01	4890.75	5	5	2.1E-01
4082.44	3	3	3.8E-02	4235.94	9	9	1.88E-01	4891.49	9	7	2.9E-01
4084.49	11	9	1.1E-01	4238.81	7	9	2.2E-01	4892.87	3	3	4.8E-02
4085.00	3	5	4.2E-02	4240.37	5	3	5.7E-02	4903.31	3	5	4.7E-02
4085.30	7	7	1.1E-01	4245.26	1	3	8.3E-02	4917.23	5	3	6.1E-02
4085.98	7	5	5.0E-02	4246.08	7	5	5.7E-02	4918.01	1	3	4.0E-02
4088.57	5	3	3.9E-02	4247.43	9	11	2.0E-01	4918.99	7	7	1.7E-01
4098.18	7	7	6.8E-02	4248.22	3	5	3.5E-02	4920.50	11	9	3.5E-01
4107.49	5	3	2.5E-01	4250.12	5	7	2.08E-01	4930.31	3	3	4.1E-02

λ Å	Weights g_i	g_k	A 10^8 s^{-1}	λ Å	Weights g_i	g_k	A 10^8 s^{-1}	λ Å	Weights g_i	g_k	A 10^8 s^{-1}
4969.92	3	3	1.8E-01	5679.02	5	7	3.6E-02	2370.50	4	4	1.4E-01
4973.10	3	3	1.0E-01	5686.53	9	11	4.4E-02	2373.74	10	10	3.3E-01
4978.60	5	3	1.1E-01	5691.51	3	1	6.2E-02	2375.19	4	2	9.8E-01
4988.95	7	7	4.9E-02	5705.99	7	9	6.7E-02	2379.27	8	8	1.5E-01
4991.27	5	7	8.2E-02	5717.85	1	3	5.0E-02	2380.76	6	8	3.1E-01
5001.86	9	7	3.9E-01	5753.12	3	5	7.0E-02	2382.04	10	12	3.8E+00
5004.04	5	3	3.5E-02	5762.99	5	7	1.0E-01	2382.90	12	14	2.2E-01
5014.94	7	5	3.0E-01	5816.36	9	11	3.7E-02	2383.25	6	6	3.4E-01
5022.24	5	3	2.6E-01	5905.67	5	3	1.2E-01	2384.39	4	4	2.3E-01
5074.75	9	11	1.5E-01	5927.80	5	3	5.1E-02	2388.37	10	12	2.2E-01
5090.78	7	5	2.0E-01	5930.17	5	7	1.6E-01	2388.63	8	8	1.0E+00
5109.65	3	5	5.4E-02	6020.17	7	9	1.1E-01	2390.10	14	16	5.5E+00
5121.64	5	5	7.9E-02	6024.07	9	11	1.3E-01	2390.77	6	6	9.3E-01
5125.11	9	7	2.6E-01	6055.99	7	9	7.0E-02	2395.42	6	4	3.3E-01
5133.69	11	13	2.7E-01	6170.49	5	5	1.3E-01	2395.62	8	10	2.5E+00
5137.38	11	9	1.1E-01	6336.84	3	3	4.9E-02	2399.24	6	6	1.4E+00
5159.06	5	3	1.3E-01	6338.90	5	3	4.8E-02	2400.06	12	14	5.2E+00
5162.27	11	11	2.4E-01	6400.00	7	9	5.5E-02	2401.29	6	8	2.5E+00
5184.26	5	7	3.5E-02	6411.65	5	7	3.5E-02	2404.43	4	2	7.1E-01
5208.59	7	5	5.2E-02	6419.98	7	7	1.3E-01	2404.89	6	8	1.7E+00
5232.94	9	11	1.4E-01	6469.21	3	3	9.0E-02	2406.66	4	4	1.6E+00
5263.30	5	5	5.2E-02	6495.78	3	3	6.0E-02	2410.52	4	6	1.5E+00
5266.55	7	9	8.6E-02	6496.46	5	5	8.5E-02	2411.07	2	2	2.4E+00
5283.62	7	7	8.0E-02	6569.23	7	9	6.5E-02	2413.31	2	4	1.1E+00
5302.30	3	5	6.3E-02	6633.76	7	7	3.6E-02	2416.45	8	10	1.6E+00
5324.18	9	9	1.5E-01	6733.16	3	1	3.9E-02	2418.44	6	8	1.6E+00
5339.93	5	7	7.0E-02	6841.35	5	7	3.6E-02	2423.21	4	6	1.4E+00
5353.39	9	7	4.8E-02	7130.94	3	5	4.3E-02	2428.36	8	10	2.7E+00
5364.87	5	7	5.5E-01					2432.87	14	14	3.2E+00
5367.47	7	9	5.8E-01	**Fe II**				2434.06	8	6	7.0E-01
5369.96	9	11	4.7E-01	1144.94	10	12	4.8E+00	2434.24	8	10	2.0E+00
5373.71	7	9	3.5E-02	1635.40	8	6	2.4E+00	2434.73	12	12	3.2E+00
5383.37	11	13	5.6E-01	1641.76	6	4	1.8E+00	2439.30	12	14	2.8E+00
5389.48	7	7	1.3E-01	1647.16	6	6	5.2E-01	2445.11	12	12	1.9E+00
5398.29	5	5	9.8E-02	2208.41	10	10	1.8E+00	2445.80	4	6	1.5E+00
5400.50	9	9	1.8E-01	2213.66	14	14	4.4E-01	2446.47	12	14	2.9E-01
5410.91	7	9	4.8E-01	2218.27	8	10	1.9E+00	2447.20	6	6	1.2E+00
5415.20	11	13	5.6E-01	2327.40	6	4	5.9E-01	2453.98	8	10	7.3E-01
5424.07	13	15	5.0E-01	2331.31	10	8	2.9E-01	2455.71	8	8	1.0E+00
5432.95	5	5	4.1E-02	2332.80	8	6	1.5E+00	2458.78	10	12	2.7E+00
5445.04	11	11	2.0E-01	2338.01	4	4	1.1E+00	2458.97	6	4	2.0E+00
5463.27	9	9	3.2E-01	2343.49	10	8	1.7E+00	2460.44	10	12	5.3E+00
5466.39	9	7	7.5E-02	2343.96	8	6	2.9E-01	2461.28	6	8	2.6E+00
5473.90	7	7	5.5E-02	2344.28	2	4	8.2E-01	2461.86	8	10	2.6E+00
5480.87	3	1	1.2E-01	2348.11	10	8	5.1E-01	2466.52	2	4	2.1E-01
5487.74	7	5	8.6E-02	2348.30	6	6	1.2E+00	2469.51	8	6	2.8E+00
5554.89	9	9	8.7E-02	2351.67	6	6	1.7E+00	2472.61	8	10	3.7E+00
5569.62	5	3	2.1E-01	2352.31	2	4	4.2E+00	2475.12	4	6	3.9E+00
5572.84	7	5	2.1E-01	2353.68	8	8	1.3E+00	2475.54	6	8	3.5E+00
5576.09	3	1	2.1E-01	2354.89	6	4	2.4E-01	2481.05	12	12	1.9E-01
5586.76	9	7	1.9E-01	2360.00	10	10	2.4E-01	2484.44	8	8	2.3E+00
5598.30	5	5	1.8E-01	2360.29	8	6	5.9E-01	2492.34	10	12	1.6E-01
5615.64	11	9	1.7E-01	2362.02	8	8	1.3E-01	2493.26	14	16	3.4E+00
5624.54	5	5	5.3E-02	2363.86	8	10	5.1E+00	2501.31	2	2	1.4E+00
5633.97	11	13	8.7E-02	2364.83	8	8	6.1E-01	2503.87	10	10	2.4E+00
5638.27	9	7	4.0E-02	2365.77	6	6	2.1E+00	2508.34	8	10	2.7E+00
5650.01	3	5	5.0E-02	2366.59	6	6	9.9E-02	2533.63	12	12	1.3E+00
5655.18	7	9	5.3E-02	2368.60	6	4	5.9E-01	2534.42	8	8	1.2E+00
5658.82	7	7	3.6E-02	2369.95	10	12	5.7E+00	2535.36	6	4	3.3E+00

λ Å	g_i	g_k	A 10^8 s^{-1}	λ Å	g_i	g_k	A 10^8 s^{-1}	λ Å	g_i	g_k	A 10^8 s^{-1}
2535.49	10	8	5.4E-01	2592.78	14	16	2.1E+00	2712.39	10	12	1.3E-01
2536.67	12	12	4.0E-01	2593.72	2	4	1.3E-01	2714.41	8	6	5.5E-01
2537.14	10	10	1.4E+00	2594.96	8	8	1.0E-01	2716.22	6	6	1.1E+00
2538.20	14	12	1.2E+00	2598.37	8	6	1.3E+00	2716.56	14	12	1.6E+00
2538.50	8	6	3.3E-01	2599.40	10	10	2.2E+00	2717.87	16	14	1.4E+00
2538.80	12	10	8.2E-01	2604.05	8	8	1.1E-01	2718.64	10	8	1.3E+00
2538.91	10	8	7.8E-01	2605.04	6	8	2.1E+00	2719.30	6	8	3.7E-01
2538.99	14	12	1.2E+00	2605.34	4	4	1.6E+00	2722.06	8	8	1.1E-01
2540.52	2	2	1.5E+00	2605.42	6	6	2.6E+00	2722.74	6	8	7.8E-01
2541.10	8	6	7.3E-01	2605.90	4	2	1.2E+00	2724.88	6	6	9.7E-01
2541.84	8	6	7.7E-01	2606.51	6	6	1.8E+00	2727.38	12	10	3.2E-01
2542.73	2	2	1.9E+00	2607.09	6	4	1.7E+00	2727.54	6	4	8.5E-01
2543.38	10	12	4.4E-01	2609.13	8	10	3.0E+00	2728.91	8	10	8.8E-02
2543.43	6	4	7.1E-01	2609.87	8	8	1.8E+00	2730.73	4	4	2.5E-01
2544.97	4	6	4.0E-01	2611.87	8	8	1.1E+00	2732.94	8	6	7.8E-01
2545.22	8	10	3.3E-01	2613.82	4	2	2.0E+00	2739.55	8	8	1.9E+00
2545.44	8	10	1.4E-01	2617.62	6	6	4.4E-01	2741.40	6	6	1.7E-01
2546.67	8	8	6.2E-01	2619.07	10	10	2.7E+00	2743.20	2	4	1.8E+00
2547.34	8	8	2.0E-01	2620.17	6	6	1.3E+00	2746.48	4	6	1.9E+00
2548.33	4	6	2.0E-01	2620.70	8	8	3.3E-01	2746.98	6	6	1.6E+00
2548.59	10	10	1.9E-01	2621.67	2	2	4.9E-01	2749.18	4	4	1.1E+00
2548.74	4	2	1.7E+00	2623.11	14	14	1.1E-01	2749.32	6	8	2.1E+00
2548.92	12	10	4.8E-01	2623.73	6	6	2.2E-01	2749.49	2	2	1.1E+00
2549.08	10	8	1.5E+00	2625.49	12	14	2.2E+00	2753.29	10	12	1.2E+00
2549.40	4	4	1.3E+00	2625.67	8	10	3.4E-01	2754.91	8	6	8.4E-01
2549.46	6	6	8.0E-01	2626.50	4	6	3.4E-01	2755.73	8	10	2.1E+00
2549.77	8	6	2.5E-01	2628.29	2	4	8.6E-01	2761.81	2	4	1.1E-01
2550.03	10	10	1.2E+00	2629.59	6	8	6.2E-01	2762.34	6	6	3.7E-01
2550.15	8	10	4.0E-01	2630.07	4	6	5.7E-01	2763.66	14	12	1.3E+00
2550.68	12	12	8.9E-01	2631.05	4	6	7.7E-01	2765.13	10	8	1.2E+00
2551.21	10	8	3.2E-01	2631.32	6	8	6.0E-01	2767.50	12	14	1.9E+00
2555.07	6	8	1.8E-01	2631.61	10	12	5.3E-01	2769.36	12	14	1.6E-01
2555.45	4	6	2.5E-01	2633.20	6	4	1.7E+00	2774.69	2	4	2.4E-01
2557.51	10	8	1.3E-01	2636.69	4	4	1.2E+00	2776.91	8	8	3.0E-01
2559.77	6	8	2.4E-01	2637.50	6	6	5.2E-01	2779.30	10	8	7.6E-01
2559.92	6	8	2.4E-01	2637.64	2	4	8.3E-01	2779.91	2	4	2.3E-01
2560.28	4	4	1.5E+00	2639.56	2	2	1.1E+00	2780.04	2	2	2.9E-01
2562.09	4	2	1.5E+00	2642.01	6	6	3.6E-01	2783.69	12	10	7.0E-01
2562.54	8	6	1.5E+00	2649.47	6	8	1.8E+00	2785.19	12	10	1.0E+00
2563.48	6	4	1.3E+00	2650.48	6	8	1.6E+00	2787.24	8	6	1.3E-01
2566.22	8	10	2.5E+00	2654.63	4	4	7.7E-01	2793.89	10	12	9.6E-02
2566.40	8	6	2.1E+00	2658.25	8	8	3.2E-01	2796.63	10	10	1.0E-01
2566.91	4	2	1.1E+00	2662.56	2	2	9.6E-01	2799.29	10	8	1.1E-01
2568.41	2	4	4.4E-01	2664.66	8	10	1.5E+00	2809.78	8	8	1.6E-01
2569.78	2	4	1.2E+00	2666.64	6	8	1.7E+00	2817.09	6	4	2.1E-01
2570.53	6	8	1.2E+00	2667.22	4	6	9.2E-01	2831.56	4	6	5.8E-01
2570.85	8	6	1.7E+00	2669.93	2	4	4.7E-01	2833.09	6	6	2.7E-01
2573.21	8	10	1.4E+00	2671.40	2	4	5.6E-01	2835.71	4	6	3.1E-01
2574.36	6	4	1.6E+00	2682.51	8	10	7.0E-01	2838.22	4	2	4.2E-01
2576.86	10	12	1.1E+00	2683.00	4	6	6.4E-01	2839.51	10	8	9.9E-01
2577.92	2	2	1.3E+00	2684.75	8	10	1.4E+00	2839.80	8	10	4.1E-01
2582.41	6	8	2.4E-01	2692.60	10	12	1.2E+00	2840.65	2	4	5.3E-01
2582.58	4	4	7.7E-01	2697.33	4	4	2.7E-01	2840.76	10	12	1.1E-01
2585.63	10	10	3.6E-01	2697.46	4	2	1.8E-01	2844.96	2	2	4.5E-01
2585.88	10	8	8.1E-01	2699.20	4	4	6.6E-01	2847.77	4	4	3.3E-01
2587.95	8	10	1.4E+00	2703.99	8	8	1.2E+00	2848.11	6	6	7.0E-01
2588.18	2	2	1.6E-01	2707.13	4	6	8.5E-01	2848.32	6	4	1.1E+00
2590.55	4	6	9.1E-02	2709.05	4	6	3.7E-01	2855.69	8	10	1.0E-01
2591.54	6	6	5.1E-01	2711.84	12	14	3.8E-01	2856.38	6	8	2.7E-01

λ Å	g_i	g_k	A 10^8 s^{-1}	λ Å	g_i	g_k	A 10^8 s^{-1}	λ Å	g_i	g_k	A 10^8 s^{-1}
2856.91	8	8	8.7E-01	**Fe VII**				175.266	2	4	1.72E+03
2857.17	6	8	9.5E-02	150.807	5	7	1.3E+03				
2872.39	10	8	1.5E-01	150.852	7	9	1.3E+03	**Fe XI**			
2873.40	8	10	3.4E-01	151.023	9	11	1.6E+03	72.166	5	7	2.9E+03
2875.35	8	10	9.5E-02	151.046	7	7	2.2E+02	72.310	5	5	1.5E+03
2883.71	12	14	1.0E-01	151.145	9	9	2.1E+02	72.635	5	7	1.6E+03
2884.77	6	8	1.4E-01	151.432	5	7	2.2E+02	91.394	5	7	2.6E+03
2895.22	8	10	8.0E-02	151.512	5	5	5.3E+02	91.472	7	9	2.5E+03
2897.27	6	4	1.4E-01	151.675	7	7	3.9E+02	91.63	3	5	2.3E+03
2944.40	4	2	4.6E-01	151.782	9	9	2.4E+02	91.63	7	9	3.4E+03
2947.66	6	4	2.0E-01	154.307	3	1	8.9E+02	91.63	5	7	2.8E+03
2949.18	10	8	2.0E-01	154.335	5	7	1.2E+03	91.733	9	11	4.1E+03
2959.84	8	6	1.6E-01	154.363	3	3	4.2E+02	92.81	9	11	3.7E+03
2964.63	2	2	9.3E-02	154.565	5	3	3.5E+02	92.87	11	13	3.9E+03
2969.93	8	6	1.8E-01	154.650	5	5	8.8E+02	93.433	9	11	3.2E+03
2982.06	4	6	2.1E-01	154.848	1	3	7.7E+02	179.762	5	7	1.67E+03
2984.82	6	6	3.6E-01	154.921	3	5	9.7E+02				
2985.55	2	4	1.8E-01	154.941	3	3	2.4E+02	**Fe XII**			
2997.30	6	8	8.3E-02	154.949	5	7	1.0E+03	65.905	4	4	2.0E+03
3002.65	4	6	1.4E-01	155.994	9	11	1.8E+03	66.526	6	8	1.7E+03
3036.96	6	6	1.6E-01	158.481	9	9	2.3E+02	66.960	4	6	1.6E+03
3048.99	4	4	2.8E-01	165.087	1	3	6.9E+02	67.164	4	2	1.1E+03
3062.23	12	10	1.2E-01	165.919	7	5	2.8E+03	67.821	4	6	1.4E+03
3071.12	2	4	1.9E-01	166.365	9	7	2.9E+03	68.382	2	4	1.7E+03
3076.44	4	6	2.8E-01	173.441	9	9	3.6E+03	80.541	6	6	8.7E+02
3077.17	14	12	1.1E-01	176.744	9	9	2.7E+03	81.943	6	4	1.4E+03
3078.68	6	8	4.2E-01	176.928	7	7	2.4E+03	82.226	4	2	1.9E+03
3135.36	6	6	8.4E-02	177.172	5	5	1.5E+03	84.48	4	6	4.5E+03
3154.20	10	10	1.5E-01	235.221	5	3	1.7E+02	84.48	8	10	4.9E+03
3167.86	8	8	1.3E-01	240.053	3	1	1.3E+02	84.52	10	12	5.2E+03
3177.54	8	8	8.1E-02	243.379	9	7	2.1E+02	84.52	6	8	4.0E+03
3179.50	6	8	9.9E-02					84.85	6	8	2.3E+03
5247.95	4	6	1.7E+00	**Fe VIII**				85.14	8	10	3.4E+03
5506.20	12	14	1.4E+00	112.472	4	4	3.6E+02	85.477	10	12	4.6E+03
5961.71	10	12	7.7E-01	112.486	6	6	4.3E+02	186.880	6	8	1.0E+03
				116.196	4	6	4.5E+02	192.394	4	2	9.0E+02
				117.197	6	8	3.8E+02	193.509	4	4	9.1E+02
Fe III				167.486	4	4	3.0E+03	195.119	4	6	8.6E+02
1843.4	9	7	4.8E+00	168.172	6	6	3.1E+03				
1844.3	7	5	4.9E+00	168.545	6	4	2.0E+03	**Fe XIII**			
1846.9	5	3	5.5E+00	168.929	4	2	2.1E+03	62.353	1	3	2.0E+03
1854.38	3	1	5.7E+00	185.213	6	8	1.0E+03	62.46	5	7	1.2E+03
1865.20	7	7	6.1E+00	186.601	4	6	9.4E+02	62.699	3	5	2.3E+03
1893.98	11	9	5.5E+00					63.188	5	7	3.9E+03
1896.80	13	11	5.0E+00	**Fe X**				64.139	1	3	2.1E+03
1904.3	5	5	5.7E+00	76.822	2	2	1.8E+03	74.845	5	5	1.0E+03
1907.58	15	13	5.3E+00	77.865	4	6	1.6E+03	75.892	5	3	7.7E+02
1915.08	13	15	6.0E+00	100.026	8	10	2.6E+03	76.117	5	3	2.1E+03
1922.79	11	13	5.5E+00	101.733	6	8	1.8E+03	78.452	9	11	6.3E+03
1930.39	9	11	5.1E+00	101.846	4	6	1.7E+03	84.270	7	9	5.5E+03
1931.51	9	11	5.3E+00	102.095	10	12	2.9E+03	107.384	7	5	1.8E+03
1937.35	7	9	5.1E+00	102.192	10	12	2.9E+03				
1943.48	5	7	5.0E+00	102.829	4	6	2.1E+03	**Fe XIV**			
1950.33	13	15	5.5E+00	103.319	6	8	2.6E+03	58.963	2	4	2.7E+03
1951.01	11	11	5.3E+00	103.724	6	8	1.7E+03	59.579	4	6	3.1E+03
1952.65	9	9	4.9E+00	104.638	8	10	2.1E+03	69.176	4	6	5.6E+02
1953.32	7	7	5.1E+00	174.534	4	6	1.8E+03	69.386	2	4	7.6E+02
1987.50	13	13	4.9E+00								

λ Å	g_i	g_k	A 10^8 s^{-1}	λ Å	g_i	g_k	A 10^8 s^{-1}	λ Å	g_i	g_k	A 10^8 s^{-1}
69.66	2	2	8.9E+02	248	3	1	5.4E+02	12.264	1	3	5.9E+04
69.66	6	6	1.3E+03	284.160	1	3	2.28E+02	12.526	1	3	3.0E+03
70.251	6	4	8.1E+02					12.681	1	3	3.5E+03
70.613	4	2	1.7E+03	**Fe XVI**				13.823	1	3	3.3E+04
72.80	10	12	7.9E+03	31.041	2	4	5.2E+02	13.891	1	3	3.4E+03
76.022	4	6	6.6E+03	31.242	4	6	6.1E+02	15.015	1	3	2.28E+05
76.152	6	8	7.0E+03	32.166	2	4	6.8E+02	15.262	1	3	6.0E+04
91.009	6	4	5.1E+02	32.192	2	2	6.7E+02	16.777	1	3	8.29E+03
91.273	4	2	5.6E+02	32.433	2	4	7.7E+02	17.054	1	3	9.33E+03
188	4	6	2.7E+02	32.652	4	6	9.1E+02	41.37	9	11	4.8E+03
190	6	8	2.8E+02	34.857	2	4	1.23E+03	49.427	3	3	4.0E+03
207	2	2	2.1E+02	35.106	4	6	1.44E+03	50.26	7	9	6.0E+03
211.316	2	4	3.6E+02	35.333	4	6	6.4E+02	58.76	9	11	1.2E+04
213	4	2	2.8E+02	35.368	6	8	6.8E+02				
214	2	2	4.0E+02	36.01	4	2	5.0E+02	**Fe XIX**			
216	6	8	1.7E+02	36.749	2	4	1.1E+03	13.413	5	3	1.3E+04
217	6	8	4.0E+02	36.803	2	2	1.2E+03	13.426	5	7	4.8E+04
217	6	6	2.6E+02	37.096	4	6	1.0E+03	13.47	3	1	1.5E+05
219	2	4	4.8E+02	37.138	6	8	1.07E+03	13.520	5	7	2.0E+05
219	4	6	2.4E+02	39.827	2	4	2.1E+03	13.56	3	5	1.0E+04
219.123	4	6	3.9E+02	40.153	4	6	2.5E+03	13.68	3	1	8.0E+04
220	4	4	3.2E+02	40.161	4	4	4.1E+03	13.69	5	7	2.3E+04
221	4	6	5.9E+02	40.199	4	6	1.7E+03	13.700	1	3	2.7E+05
226	2	4	3.9E+02	40.245	6	8	1.8E+03	13.71	5	5	2.2E+04
234	2	2	2.8E+02	41.91	2	2	4.72E+02	13.738	5	7	1.0E+04
264.787	4	4	3.38E+02	42.30	4	2	9.2E+02	13.796	5	7	7.0E+04
265	4	4	1.5E+02	46.661	4	6	3.46E+03	13.83	5	5	1.4E+04
266	6	4	1.7E+02	46.718	6	8	3.7E+03	13.934	1	3	4.51E+04
268	6	6	2.1E+02	50.350	2	4	1.86E+03	13.961	3	3	2.0E+04
268	4	2	3.3E+02	50.555	2	2	1.98E+03	14.668	5	7	1.1E+04
270.524	4	2	2.1E+02	54.142	2	4	3.41E+03	14.671	5	3	1.1E+04
274.203	2	2	1.8E+02	54.728	4	6	4.16E+03	14.929	3	3	1.2E+04
280	4	6	2.8E+02	54.769	4	6	6.97E+02	14.966	5	3	2.5E+04
283	6	8	2.7E+02	62.879	2	2	1.05E+03	14.995	5	5	2.2E+04
288.45	6	4	1.6E+02	63.719	4	2	2.18E+03	15.015	1	3	1.4E+04
				66.263	4	6	9.39E+03	16.668	3	1	1.1E+04
Fe XV				66.368	6	8	1.00E+04				
38.95	1	3	1.69E+03	66.392	6	6	6.69E+02	**Fe XX**			
52.911	1	3	2.94E+03	76.502	6	4	6.7E+02	12.67	6	6	1.0E+04
59.404	3	5	3.4E+03	76.796	4	2	7.72E+02	12.69	4	6	1.2E+04
63.959	5	7	1.6E+03	80.192	4	6	5.2E+02	12.73	4	2	4.0E+04
65.370	1	3	3.2E+02	80.270	6	8	5.4E+02	12.77	4	4	2.1E+05
65.612	3	3	9.8E+02	85.587	2	4	4.0E+02	12.78	4	2	6.9E+04
66.238	5	3	1.6E+03	86.133	4	6	4.8E+02	12.78	2	4	1.4E+05
68.860	9	11	9.2E+03	96.256	4	6	8.7E+02	12.79	6	4	1.7E+04
69.7	3	1	1.9E+03	96.348	6	8	9.3E+02	12.82	4	4	1.1E+05
69.942	3	5	7.4E+03	117.2	2	4	3.93E+02	12.88	6	4	2.7E+04
69.989	5	7	7.9E+03	117.7	2	2	3.9E+02	12.89	4	4	4.4E+04
70.052	7	9	8.8E+03	123.4	2	4	5.9E+02	12.90	4	2	6.2E+03
70.224	1	3	4.13E+03	124.5	4	6	7.0E+02	12.90	4	6	1.4E+05
70.53	7	5	2.6E+02	144.06	4	6	1.6E+03	12.92	2	4	1.7E+04
70.59	7	7	1.7E+03	144.25	6	8	1.6E+03	12.93	4	6	1.6E+05
73.199	7	9	8.8E+03	148	4	2	6.5E+02	12.93	2	2	1.2E+04
73.473	5	7	6.2E+03	266.7	4	6	3.9E+02	12.98	2	2	6.7E+04
233.857	5	7	2.2E+02	267.0	6	8	4.3E+02	12.99	6	6	5.1E+04
235	1	3	2.5E+02					13.00	6	4	1.1E+04
243	1	3	2.4E+02	**Fe XVII**				13.01	2	4	3.0E+04
243	5	7	2.3E+02	11.023	1	3	2.1E+04	13.03	4	2	8.6E+04
243.790	3	5	4.2E+02	12.123	1	3	8.0E+04	13.07	6	4	8.2E+03

λ Å	g_i	g_k	A 10^8 s^{-1}	λ Å	g_i	g_k	A 10^8 s^{-1}	λ Å	g_i	g_k	A 10^8 s^{-1}
13.13	2	4	8.9E+04	13.14	3	1	2.0E+04	11.325	3	5	1.7E+05
13.24	4	4	1.2E+04	13.41	1	3	7.3E+03	11.338	3	3	9.3E+04
13.28	4	4	6.1E+03					11.429	3	1	1.7E+05
13.70	4	6	1.1E+04	**Fe XXII**				11.433	3	3	1.2E+05
13.71	2	2	9.9E+03	9.002	4	6	5.5E+04	11.441	5	7	2.2E+05
13.78	4	4	1.0E+04	9.006	6	8	5.7E+04	11.445	5	5	5.6E+04
13.79	6	6	1.2E+04	9.006	6	6	5.3E+04	11.485	3	5	1.40E+05
13.83	4	2	9.8E+03	9.163	4	6	6.9E+04	11.491	5	3	5.9E+04
13.90	4	2	1.2E+04	9.183	6	8	8.3E+04	11.519	5	5	1.16E+05
13.98	6	4	1.6E+04	9.241	4	6	5.1E+04	11.520	1	3	2.16E+05
13.99	4	2	2.2E+04	11.748	4	4	1.2E+05	11.524	5	7	2.3E+05
14.05	4	4	1.7E+04	11.748	4	6	1.6E+05	11.593	5	7	3.58E+05
14.23	2	2	6.3E+03	11.748	4	2	1.8E+05	11.613	3	5	1.0E+05
				11.763	2	4	1.6E+05	11.615	3	3	4.4E+04
Fe XXI				11.789	2	2	2.6E+05	11.691	5	7	7.7E+04
8.53	3	1	1.8E+04	11.789	6	8	1.2E+05	11.698	5	5	7.3E+04
8.53	3	5	6.1E+03	11.797	2	4	1.7E+05	11.737	3	5	1.8E+05
8.53	3	3	1.5E+04	11.823	6	4	7.9E+04	11.898	1	3	2.03E+05
8.56	5	7	2.0E+04	11.837	6	8	2.3E+05				
8.56	1	3	2.1E+04	11.837	6	6	1.7E+05	**Fe XXIV**			
8.56	5	3	6.5E+03	11.886	4	6	1.3E+05	1.8523	2	2	1.0E+05
8.64	5	7	1.5E+04	11.898	2	4	8.2E+04	1.8552	2	4	4.82E+06
8.65	5	7	3.9E+04	11.922	4	6	1.8E+05	1.8563	4	2	2.43E+06
8.66	5	5	4.4E+03	11.976	6	8	5.9E+04	1.8572	2	2	3.06E+06
8.74	1	3	2.5E+04	12.027	2	4	6.9E+04	1.858	2	4	1.2E+05
9.42	3	1	4.3E+04	12.045	6	8	2.4E+05	1.8614	4	4	6.24E+06
9.42	3	3	3.3E+04	12.045	4	4	9.7E+04	1.8626	2	4	3.16E+06
9.44	3	5	1.7E+04	12.053	4	6	6.1E+04	1.8627	2	2	5.47E+06
9.45	1	3	5.2E+04	12.077	2	4	1.0E+05	1.8637	2	2	1.91E+06
9.46	5	3	1.5E+04	12.077	4	6	2.4E+05	1.8655	4	6	2.14E+06
9.47	5	7	4.9E+04	12.095	6	6	7.8E+04	1.8672	4	2	1.63E+06
9.47	5	5	6.1E+03	12.193	2	4	7.2E+04	1.8678	4	4	3.5E+05
9.52	3	3	8.1E+03	12.193	4	6	9.9E+04	1.8721	4	6	3.2E+05
9.58	5	5	5.2E+03	12.325	2	2	1.5E+05	1.8721	2	2	2.0E+05
9.59	5	5	1.0E+04					1.8730	2	4	1.5E+05
9.67	1	3	5.7E+04	**Fe XXIII**				1.8739	4	4	8.3E+04
9.68	5	7	4.0E+03	7.733	5	7	3.0E+04	1.891	2	2	9.7E+04
9.74	5	3	5.3E+03	7.849	5	7	4.9E+04	1.897	4	2	9.8E+04
12.02	1	3	1.3E+04	8.307	1	3	4.8E+04	8.231	2	4	6.10E+04
12.13	3	3	1.8E+04	8.529	1	3	4.3E+04	8.316	4	6	7.07E+04
12.18	5	7	2.2E+04	8.550	3	5	6.0E+04	10.619	2	4	7.28E+04
12.19	5	3	6.4E+03	8.552	3	3	3.2E+04	10.663	2	2	7.51E+04
12.21	3	1	1.5E+05	8.614	5	7	7.7E+04	11.030	2	4	1.84E+05
12.21	3	3	1.2E+05	8.664	3	3	4.4E+04	11.171	4	6	2.18E+05
12.25	1	3	2.1E+05	8.669	5	7	6.1E+04				
12.28	5	3	5.2E+04	8.672	1	3	6.8E+04	**Fe XXV**			
12.30	5	7	2.1E+05	8.752	5	7	1.2E+05	1.4607	1	3	2.54E+05
12.36	3	3	3.6E+04	8.764	5	7	4.6E+04	1.4945	1	3	5.05E+05
12.37	5	7	3.1E+05	8.814	3	5	6.2E+04	1.5730	1	3	1.24E+06
12.38	5	3	6.9E+03	10.902	5	5	5.3E+04	1.5749	1	3	1.5E+05
12.47	5	7	5.8E+04	10.910	3	1	6.7E+04	1.778	3	3	8.7E+04
12.47	5	3	1.3E+04	10.927	5	7	6.0E+04	1.782	3	1	4.69E+06
12.49	5	7	1.3E+04	10.934	3	5	5.4E+04	1.787	1	3	2.57E+06
12.53	5	5	1.5E+04	10.979	1	3	7.9E+04	1.787	5	5	1.19E+06
12.57	1	3	7.2E+04	11.018	1	3	4.9E+04	1.788	3	5	2.68E+06
12.73	5	5	8.2E+03	11.086	3	1	6.5E+04	1.788	3	5	1.63E+06
12.95	3	5	6.2E+03	11.165	3	5	6.7E+04	1.789	1	3	1.78E+06
13.00	1	3	7.2E+03	11.255	3	3	3.7E+04	1.790	3	3	1.23E+06
13.03	5	5	1.3E+04	11.298	1	3	1.3E+05	1.791	3	5	4.10E+06

λ Å	Weights g_i	Weights g_k	A 10^8 s^{-1}	λ Å	Weights g_i	Weights g_k	A 10^8 s^{-1}	λ Å	Weights g_i	Weights g_k	A 10^8 s^{-1}
1.791	3	3	2.59E+06	4619.2	4	6	8.1E-01				
1.792	3	1	4.92E+06	4633.9	4	6	7.1E-01	**Magnesium**			
1.792	5	5	2.81E+06	4658.9	6	4	6.5E-01	**Mg I**			
1.793	3	1	2.67E+06	4739.0	6	6	7.6E-01	2025.8	1	3	8.4E-01
1.794	5	3	2.22E+06	4762.4	2	4	4.2E-01	*2779.8	9	9	5.2E+00
1.797	3	5	8.8E+05	4765.7	4	6	6.7E-01	*2850.0	9	15	2.3E-01
1.798	3	3	1.0E+05	4811.8	2	4	1.7E-01	2852.1	1	3	4.95E+00
1.800	1	3	8.6E+04	4825.2	2	4	1.9E-01	*3094.9	9	15	5.2E-01
1.802	3	1	4.1E+05	4832.1	4	2	7.3E-01	3329.9	1	3	3.3E-02
1.810	3	1	5.9E+05	5208.3	4	4	1.4E-01	3332.2	3	3	9.7E-02
1.8502	1	3	4.57E+06	5308.7	4	6	2.4E-02	3336.7	5	3	1.6E-01
1.8593	1	3	4.42E+05	7407.0	6	6	7.0E-02	*3835.3	9	15	1.68E+00
10.038	3	3	8.08E+04					4703.0	3	5	2.55E-01
				Lead				5167.3	1	3	1.16E-01
Krypton				**Pb I**				5172.7	3	3	3.46E-01
Kr I				2022.0	1	3	5.2E-02	5183.6	5	3	5.75E-01
1164.9	1	3	3.16E+00	2053.3	1	3	1.2E-01	5528.4	3	5	1.99E-01
1235.8	1	3	3.12E+00	2170.0	1	3	1.5E+00				
4274.0	5	5	2.6E-02	2401.9	3	3	1.9E-01	**Mg II**			
4351.4	3	1	3.2E-02	2446.2	3	3	2.5E-01	1239.9	2	4	1.4E-02
4362.6	5	3	8.4E-03	2476.4	3	5	2.8E-01	1240.4	2	2	1.4E-02
4376.1	3	1	5.6E-02	2577.3	5	3	5.0E-01	*2660.8	10	14	3.8E-01
4400.0	3	5	2.3E-02	2613.7	3	3	2.7E-01	2790.8	2	4	4.0E+00
4410.4	3	3	4.4E-03	2614.2	3	5	1.9E+00	2795.5	2	4	2.6E+00
4425.2	3	3	9.7E-03	2628.3	5	3	3.1E-02	2797.9	4	4	7.9E-01
4453.9	3	5	7.8E-03	2657.1	3	5	9.8E-04	2798.1	4	6	4.8E+00
4463.7	3	3	2.3E-02	2663.2	5	5	7.1E-01	2802.7	2	2	2.6E+00
4502.4	3	5	9.2E-03	2802.0	5	7	1.6E+00	2928.8	2	2	1.2E+00
5562.2	5	5	2.8E-03	2823.2	5	5	2.6E-01	2936.5	4	2	2.3E+00
5570.3	5	3	2.1E-02	2833.1	1	3	5.8E-01	*3104.8	10	14	8.1E-01
5649.6	1	3	3.7E-03	2873.3	5	5	3.7E-01	3848.2	6	4	2.8E-02
5870.9	3	5	1.8E-02	3572.7	5	3	9.9E-01	3848.3	4	4	3.0E-03
6904.7	3	5	1.3E-02	3639.6	3	3	3.4E-01	3850.4	4	2	3.0E-02
7224.1	3	5	1.4E-02	3671.5	5	3	4.4E-01	*4481.2	10	14	2.23E+00
7587.4	3	1	5.1E-01	3683.5	3	1	1.5E+00	9218.3	2	4	3.6E-01
7601.5	5	5	3.1E-01	3739.9	5	5	7.3E-01	9244.3	2	2	3.6E-01
7685.2	3	1	4.9E-01	4019.6	5	7	3.5E-02				
7694.5	5	3	5.6E-02	4057.8	5	3	8.9E-01	**Mg IV**			
7854.8	1	3	2.3E-01	4062.1	5	3	9.2E-01	320.99	4	2	1.2E+02
8059.5	1	3	1.9E-01	4168.0	5	5	1.2E-02	323.31	2	2	5.9E+01
8104.4	5	5	1.3E-01	5005.4	1	3	2.7E-01	1219.0	6	6	5.9E+00
8112.9	5	7	3.6E-01	5201.4	1	3	1.9E-01	1375.5	4	4	4.5E+00
8190.1	3	5	1.1E-1	7229.0	5	3	8.9E-03	1459.6	6	4	4.6E+00
8263.2	3	5	3.5E-01					1495.5	4	6	6.4E+00
8281.1	3	3	1.9E-01	**Lithium**				1510.7	4	4	6.7E+00
8298.1	3	3	3.2E-01	**Li I**				1683.0	6	8	5.8E+00
8508.9	3	3	2.4E-01	*2741.2	2	6	1.3E-02	1698.8	4	6	3.9E+00
8776.7	3	3	2.7E-01	*3232.7	2	6	1.17E-02	1893.9	6	6	2.8E+00
8928.7	5	3	3.7E-01	*4602.9	6	10	2.23E-01				
				*6103.6	6	10	6.860E-01	**Mg VI**			
Kr II				*6707.8	2	6	3.691E-01	*269.92	10	6	3.1E+02
4250.6	4	4	1.2E-01					*292.53	6	6	9.0E+01
4292.9	4	4	9.6E-01	**Lutetium**				*314.64	6	2	1.8E+02
4355.5	6	8	1.0E+00	**Lu I**				*349.15	10	10	6.1E+01
4431.7	2	2	1.8E+00	3376.5	4	4	2.23E+00	*387.94	6	10	1.3E+01
4436.8	2	4	6.6E-01	3567.8	4	6	5.9E-01	399.29	4	2	2.8E+01
4577.2	6	8	9.6E-01	3620.3	6	4	1.1E-02	400.68	4	4	2.8E+01
4583.0	6	4	7.6E-01	3841.2	6	6	2.5E-01	403.32	4	6	2.7E+01
4615.3	4	4	5.4E-01	4518.6	4	4	2.1E-01				

λ Å	Weights g_i	Weights g_k	A 10^8 s^{-1}
Mg VII			
277.01	3	3	9.5E+01
278.41	5	3	1.5E+02
280.74	5	3	2.0E+02
319.02	5	5	8.9E+01
*366.42	9	9	4.4E+01
*433.04	9	15	1.6E+01
1334.3	5	5	5.3E+00
1410.0	5	5	2.57E+00
1487.0	3	5	3.02E+00
1487.9	5	7	3.66E+00
Mg VIII			
*74.976	6	10	4.3E+03
315.02	4	4	1.2E+02
*342.29	10	6	6.3E+01
353.86	4	4	3.89E+01
356.00	6	4	5.7E+01
*428.52	10	10	3.24E+01
*434.62	6	10	1.6E+01
*489.33	6	6	3.9E+01
*686.92	6	10	9.4E+00
Mg IX			
62.751	1	3	2.87E+03
*67.189	9	15	6.20E+03
*71.965	9	3	1.22E+03
72.312	3	5	4.43E+03
77.737	3	1	3.92E+02
368.07	1	3	5.27E+01
438.69	3	1	7.9E+01
*443.74	9	9	4.19E+01
749.55	3	5	8.2E+00
1639.8	3	5	2.1E+00
2814.2	1	3	3.35E-01
Mg X			
57.876	2	4	2.09E+03
57.920	2	2	2.09E+03
63.152	2	4	5.6E+03
63.295	4	6	6.7E+03
609.79	2	4	7.53E+00
624.94	2	2	7.01E+00
2212.5	2	4	9.64E-01
2278.7	2	2	8.82E-01
5918.7	2	4	3.20E-02
6229.6	4	6	3.30E-02
Mg XI			
7.310	1	3	1.15E+04
7.473	1	3	2.27E+04
7.850	1	3	5.50E+04
9.169	1	3	1.97E+05
Manganese			
Mn I			
2794.82	6	8	3.7E+00
2798.27	6	6	3.6E+00
2801.08	6	4	3.7E+00

λ Å	Weights g_i	Weights g_k	A 10^8 s^{-1}
3007.65	6	8	1.8E-01
3011.38	8	10	3.1E-01
3016.45	10	12	2.9E-01
3043.36	8	8	5.9E-01
3044.57	10	8	5.7E-01
3045.59	10	10	6.7E-01
3045.80	8	10	1.7E-01
3047.03	12	12	6.1E-01
3054.36	8	6	4.6E-01
3070.27	6	6	1.9E-01
3073.18	4	4	3.7E-01
3082.71	14	14	2.9E-01
3110.68	6	8	2.7E-01
3113.80	12	10	2.6E-01
3118.10	4	6	1.7E-01
3122.88	10	10	1.9E-01
3126.85	8	6	2.3E-01
3132.28	10	10	2.1E-01
3132.79	8	8	2.7E-01
3175.58	8	10	1.8E-01
3201.11	4	6	2.2E-01
3228.09	10	12	6.4E-01
3230.23	10	12	1.9E-01
3230.72	8	8	3.5E-01
3240.88	6	4	2.2E-01
3243.78	6	6	5.3E-01
3251.13	4	2	2.3E-01
3252.95	4	4	1.8E-01
3256.14	4	6	5.0E-01
3258.41	2	2	9.7E-01
3260.24	2	4	3.8E-01
3267.79	14	14	3.5E-01
3268.72	6	8	3.3E-01
3270.35	12	12	2.6E-01
3273.02	10	10	2.7E-01
3298.23	6	4	2.8E-01
3303.28	4	4	1.9E-01
3463.66	8	8	3.2E-01
3470.01	6	8	2.4E-01
3511.83	12	12	2.7E-01
3535.30	10	10	1.7E-01
3559.81	6	6	2.1E-01
3577.87	10	8	9.4E-01
3595.11	6	4	1.8E-01
3601.27	12	10	2.3E-01
3607.53	8	8	2.3E-01
3608.49	6	6	3.6E-01
3610.30	4	4	4.2E-01
3635.70	10	8	2.1E-01
3660.40	12	14	9.1E-01
3675.67	6	8	2.2E-01
3676.96	10	12	7.3E-01
3680.15	12	10	1.9E-01
3682.09	8	10	7.6E-01
3684.87	6	8	2.6E-01
3706.08	12	14	1.4E+00
3718.92	10	12	9.6E-01
3731.94	8	10	1.0E+00
3771.44	14	14	1.9E-01

λ Å	Weights g_i	Weights g_k	A 10^8 s^{-1}
3773.86	12	12	2.5E-01
3800.55	6	8	2.7E-01
3806.72	10	12	5.9E-01
3823.51	8	10	5.21E-01
3823.89	6	6	2.31E-01
3833.87	4	4	3.14E-01
3834.37	6	8	4.29E-01
3839.78	2	2	4.64E-01
3841.07	4	6	3.3E-01
3843.99	2	4	2.11E-01
3889.46	12	14	3.1E-01
3898.37	6	8	1.7E-01
3899.34	4	6	2.4E-01
3924.08	2	4	9.4E-01
3926.48	6	8	5.4E-01
3951.98	2	2	3.1E-01
3952.84	6	6	4.1E-01
3975.88	2	4	1.8E-01
3982.16	4	2	3.5E-01
3982.58	6	4	2.3E-01
3982.90	6	4	5.5E-01
3991.60	2	2	2.1E-01
4011.91	8	8	2.3E-01
4018.11	10	8	2.54E-01
4030.76	6	8	1.7E-01
4033.07	6	6	1.65E-01
4034.49	6	4	1.58E-01
4041.36	10	10	7.87E-01
4048.75	6	4	7.5E-01
4052.48	6	8	3.8E-01
4055.55	8	8	4.31E-01
4058.94	4	2	7.25E-01
4061.74	8	6	1.9E-01
4063.53	6	6	1.69E-01
4065.08	12	14	2.5E-01
4066.24	10	8	2.2E-01
4070.28	2	2	2.3E-01
4079.42	2	4	3.8E-01
4082.95	4	6	2.95E-01
4083.63	6	8	2.8E-01
4089.94	8	10	1.7E-01
4105.37	10	8	1.7E-01
4135.03	12	12	3.0E-01
4141.06	10	10	2.6E-01
4148.80	8	8	2.3E-01
4176.61	14	12	2.4E-01
4189.99	12	10	2.0E-01
4201.78	10	8	2.3E-01
4235.30	8	6	9.17E-01
4239.74	4	2	3.9E-01
4257.67	2	2	3.7E-01
4265.93	4	4	4.92E-01
4281.10	6	6	2.3E-01
4411.87	12	10	2.6E-01
4414.89	8	6	2.93E-01
4419.77	10	8	2.1E-01
4436.36	6	4	4.37E-01
4451.58	8	8	7.98E-01
4453.01	4	2	5.44E-01

λ Å	g_i	g_k	A 10^8 s^{-1}	λ Å	g_i	g_k	A 10^8 s^{-1}	λ Å	g_i	g_k	A 10^8 s^{-1}
4455.82	4	6	1.7E-01	1285.10	5	7	1.1E+01	2751.47	7	9	2.54E-01
4457.04	6	4	2.34E-01	1333.87	7	9	1.0E+01	2756.26	5	3	1.18E-01
4457.55	6	6	4.27E-01					2761.53	9	11	2.06E-01
4458.26	6	8	4.62E-01	**Mercury**				2763.02	3	1	4.44E-01
4461.09	8	8	1.7E-01	**Hg I**				2766.25	3	5	1.17E-01
4462.03	8	10	7.00E-01	2536.52	1	3	8.00E-02	2787.83	9	7	2.85E-01
4464.68	6	6	4.39E-01	2652.04	3	5	3.88E-01	2792.96	5	3	1.53E-01
4470.14	4	4	3.00E-01	2655.13	3	5	1.1E-01	2798.02	7	5	1.22E-01
4472.79	2	2	4.35E-01	2752.78	1	3	6.10E-02	2801.47	5	7	1.24E-01
4479.40	8	10	3.4E-01	2856.94	3	1	1.1E-02	2825.68	5	7	2.53E-01
4490.08	2	4	2.49E-01	2893.60	3	3	1.6E-01	2826.75	7	7	4.23E-01
4498.90	4	6	2.49E-01	2925.4	5	3	7.7E-02	2876.54	9	9	2.84E-01
4502.22	6	8	1.86E-01	2967.3	1	3	4.5E-01	2886.60	11	11	4.74E-01
4605.37	10	12	3.6E-01	3021.50	5	7	5.09E-01	2906.06	3	3	8.04E-01
4626.54	12	14	3.6E-01	3023.48	5	5	9.4E-02	2913.52	5	3	1.38E-01
4709.71	8	8	1.72E-01	3027.49	5	5	2.0E-02	2915.38	5	3	7.31E-01
4727.46	6	6	1.7E-01	3125.66	3	5	6.56E-01	2918.84	5	3	3.79E-01
4739.11	4	4	2.40E-01	3341.48	5	3	1.68E-01	2930.39	1	3	1.91E-01
4754.05	6	8	3.03E-01	3650.15	5	7	1.3E+00	2936.50	11	11	2.33E-01
4761.53	2	4	5.35E-01	3654.83	5	5	1.8E-01	2945.43	7	7	3.66E-01
4762.38	8	10	7.83E-01	4046.56	1	3	2.1E-01	2945.66	3	3	4.08E-01
4765.86	4	6	4.1E-01	4077.81	3	1	4.0E-02	2946.01	5	5	1.68E-01
4766.43	6	8	4.6E-01	4108.1	3	1	3.0E-02	2951.45	9	9	1.43E-01
4783.43	8	8	4.01E-01	4339.22	3	5	2.88E-02	2959.48	9	11	1.75E-01
4823.53	10	8	4.99E-01	4347.50	3	5	8.4E-02	2972.96	5	3	2.69E-01
6013.48	4	6	1.72E-01	4358.34	3	3	5.57E-01	2977.27	9	7	3.28E-01
6021.79	8	6	3.32E-01	4916.07	3	1	5.8E-02	2978.28	7	5	1.50E-01
				5025.64	3	3	2.7E-04	2983.04	1	3	2.82E-01
Mn II				5460.75	5	3	4.87E-01	2987.92	3	5	8.43E-01
2593.72	7	7	2.6E+00	5769.59	3	5	2.36E-01	2988.23	5	7	4.28E-01
2605.68	7	5	2.7E+00	6234.4	1	3	5.3E-03	2988.68	7	9	1.61E-01
2933.05	5	3	2.0E+00	6716.4	1	3	4.3E-03	2989.80	9	7	9.27E-01
2939.31	5	5	1.9E+00	6907.5	3	5	2.8E-02	3000.24	9	9	1.40E-01
2949.20	5	7	1.9E+00	7728.8	1	3	9.7E-03	3000.44	5	5	1.25E-01
3441.99	9	7	4.3E-01	10139.79	3	1	2.71E-01	3000.85	5	7	2.58E-01
3460.32	7	5	3.2E-01					3001.43	5	5	2.31E-01
3474.13	5	3	1.5E-01	**Molybdenum**				3007.71	7	5	1.90E-01
3482.90	5	5	2.0E-01	**Mo I**				3013.39	7	5	6.06E-01
3488.68	3	3	2.5E-01	2616.79	3	5	7.34E-01	3016.78	9	9	2.75E-01
				2621.06	7	7	1.16E-01	3025.00	5	5	8.49E-01
Mn VI				2628.96	3	3	2.81E-01	3036.31	3	5	5.81E-01
307.999	9	9	3.7E+01	2629.85	5	7	7.75E-01	3041.70	13	11	5.94E-01
309.440	9	7	5.7E+01	2631.50	1	3	2.54E-01	3046.80	13	11	1.63E-01
309.579	7	5	4.4E+01	2638.30	5	5	7.57E-01	3047.31	11	9	5.01E-01
310.058	7	7	3.4E+01	2640.98	7	5	1.20E+00	3055.32	9	7	4.29E-01
310.182	5	5	2.8E+01	2644.36	5	7	1.96E-01	3057.56	7	5	2.64E-01
311.748	5	3	5.7E+01	2649.46	7	9	9.84E-01	3061.59	7	5	4.41E-01
320.598	3	5	1.5E+01	2655.02	9	7	4.08E-01	3064.27	13	13	8.46E-01
320.681	1	3	2.2E+01	2658.11	7	7	6.43E-01	3065.04	13	13	3.08E-01
320.874	3	1	7.8E+01	2665.09	7	9	1.32E-01	3069.51	5	5	1.52E-01
320.979	3	3	2.2E+01	2679.85	9	11	1.31E+00	3069.96	11	11	2.72E-01
321.176	5	5	6.0E+01	2684.16	9	9	4.18E-01	3070.89	9	11	1.87E-01
321.541	5	3	2.7E+01	2706.11	3	5	2.03E-01	3074.37	11	11	1.42E+00
325.146	9	7	1.3E+02	2710.74	3	3	1.57E-01	3079.88	9	11	9.55E-01
328.431	5	5	4.4E+01	2725.15	3	5	2.79E-01	3080.40	7	9	3.61E-01
328.558	3	5	1.2E+01	2728.71	3	3	1.26E-01	3081.16	3	5	2.35E-01
329.043	1	3	1.1E+01	2733.39	5	7	2.95E-01	3085.62	9	9	1.63E+00
1236.23	5	3	1.3E+01	2743.71	1	3	2.47E-01	3089.13	11	9	1.53E-01
1255.77	3	1	1.2E+01	2745.38	13	11	1.29E-01	3089.71	5	7	2.34E-01

λ Å	Weights g_i	g_k	A 10^8 s^{-1}	λ Å	Weights g_i	g_k	A 10^8 s^{-1}	λ Å	Weights g_i	g_k	A 10^8 s^{-1}
3094.66	7	7	1.63E+00	3285.35	9	7	4.49E-01	3456.52	3	3	2.96E-01
3099.92	9	7	1.45E-01	3287.38	5	5	1.38E-01	3460.22	5	3	2.77E-01
3100.88	7	9	1.20E+00	3289.01	9	9	5.08E-01	3460.78	9	7	6.03E-01
3101.34	5	5	1.92E+00	3290.82	7	5	5.44E-01	3465.84	3	1	9.99E-01
3106.34	7	5	2.21E-01	3305.56	5	3	1.74E-01	3466.19	9	7	2.11E-01
3117.54	13	13	1.89E-01	3305.91	7	9	3.06E-01	3466.96	7	7	1.52E-01
3123.03	3	3	2.81E-01	3307.13	7	9	1.25E-01	3467.85	5	7	2.63E-01
3125.03	5	3	1.98E-01	3312.33	7	5	1.62E-01	3469.22	5	3	6.96E-01
3132.59	7	9	1.79E+00	3323.95	9	7	2.82E-01	3469.63	13	15	1.51E-01
3135.90	9	11	3.68E-01	3325.13	5	3	2.26E-01	3470.92	3	5	2.91E-01
3136.75	9	11	1.57E-01	3325.67	5	5	1.72E-01	3475.03	3	3	4.68E-01
3142.75	3	5	4.10E-01	3327.30	1	3	2.88E-01	3479.42	7	5	2.26E-01
3147.35	13	11	2.41E-01	3336.56	9	9	1.64E-01	3483.67	7	7	1.13E-01
3155.19	7	7	2.75E-01	3340.16	5	3	1.20E-01	3483.83	7	5	1.41E-01
3158.17	7	7	4.63E-01	3344.73	3	5	6.04E-01	3489.43	7	7	3.27E-01
3170.34	7	7	1.37E+00	3346.83	11	11	1.13E-01	3504.41	7	9	8.06E-01
3171.38	5	7	2.03E-01	3347.00	3	3	2.72E-01	3505.31	7	9	2.25E-01
3175.59	13	11	8.40E-01	3358.12	5	7	7.59E-01	3508.11	9	9	1.59E-01
3179.78	11	13	2.33E-01	3361.37	9	9	1.38E-01	3510.77	13	13	4.75E-01
3183.03	11	9	3.98E-01	3363.78	5	7	2.74E-01	3517.55	11	11	5.41E-01
3184.58	7	5	2.77E-01	3363.87	5	7	1.39E-01	3518.21	3	3	3.64E-01
3185.10	7	7	2.54E-01	3373.81	3	3	2.03E-01	3521.38	9	9	1.39E-01
3185.71	5	3	6.10E-01	3375.22	7	7	1.38E-01	3521.41	9	11	6.06E-01
3188.10	7	9	3.45E-01	3375.65	7	9	1.56E-01	3524.65	5	3	3.10E-01
3188.41	5	7	4.40E-01	3378.19	3	1	1.88E-01	3524.98	7	9	2.25E-01
3192.79	9	11	1.88E-01	3378.46	13	13	3.75E-01	3538.92	11	11	2.24E-01
3193.98	7	5	1.53E+00	3379.96	5	5	4.11E-01	3540.57	5	3	4.46E-01
3194.88	9	11	1.75E-01	3382.48	3	3	2.66E-01	3542.17	7	5	4.93E-01
3195.96	9	7	4.10E-01	3384.61	7	9	7.32E-01	3552.71	9	7	3.64E-01
3197.18	1	3	1.47E-01	3385.87	9	11	3.30E-01	3555.64	3	3	3.46E-01
3198.85	15	13	7.22E-01	3389.79	5	7	1.85E-01	3558.09	5	7	5.43E-01
3200.89	3	5	1.82E-01	3392.17	9	9	1.97E-01	3563.75	1	3	1.53E-01
3205.22	1	3	4.27E-01	3393.65	11	11	2.08E-01	3566.05	9	9	2.67E-01
3205.43	9	11	2.55E-01	3404.33	7	7	2.10E-01	3566.74	7	7	1.43E-01
3205.89	9	9	5.35E-01	3413.37	11	11	1.25E-01	3570.64	15	15	7.18E-01
3208.84	7	5	2.77E-01	3415.27	9	9	1.83E-01	3573.88	3	5	3.58E-01
3210.97	7	5	6.94E-01	3415.61	7	9	1.29E-01	3580.54	13	11	5.49E-01
3214.44	9	7	2.01E-01	3416.14	9	11	2.45E-01	3581.88	11	13	3.81E-01
3215.07	3	5	4.20E-01	3418.52	5	3	1.41E-01	3584.25	3	3	1.73E-01
3216.78	15	13	2.10E-01	3419.69	7	7	1.15E-01	3585.57	7	5	3.95E-01
3221.73	3	1	1.41E+00	3420.04	5	5	3.28E-01	3588.95	7	7	1.18E-01
3228.21	5	7	3.85E-01	3422.31	9	9	2.52E-01	3590.74	7	9	2.23E-01
3229.79	9	11	1.44E-01	3425.13	11	11	2.29E-01	3595.55	5	5	2.32E-01
3233.14	13	13	6.33E-01	3427.90	11	13	4.09E-01	3598.88	13	11	5.67E-01
3237.06	7	9	2.95E-01	3434.79	7	7	1.75E-01	3600.73	9	9	2.07E-01
3244.47	5	3	2.80E-01	3435.45	15	15	1.50E+00	3601.88	7	9	1.15E-01
3247.61	5	5	1.71E-01	3437.21	11	9	8.06E-01	3602.94	5	7	2.96E-01
3249.93	5	3	1.87E-01	3438.87	1	3	2.34E-01	3604.07	9	7	3.25E-01
3251.65	3	5	3.05E-01	3441.87	5	3	1.34E-01	3610.61	5	3	1.78E-01
3256.21	5	3	6.89E-01	3442.66	3	3	2.94E-01	3611.99	7	7	1.16E-01
3256.72	3	3	1.31E-01	3445.03	7	9	1.53E-01	3615.16	7	9	1.96E-01
3259.16	11	13	1.62E-01	3445.26	7	5	2.96E-01	3623.22	11	9	5.58E-01
3262.63	7	9	3.62E-01	3445.80	9	9	1.14E-01	3624.46	9	11	5.27E-01
3264.40	11	9	5.42E-01	3447.12	9	11	8.75E-01	3624.62	5	7	1.37E-01
3265.14	5	7	2.60E-01	3447.29	5	3	1.79E-01	3638.20	5	3	3.51E-01
3266.16	9	11	1.95E-01	3449.07	7	9	1.52E-01	3638.21	5	3	3.33E-01
3270.90	7	7	3.59E-01	3449.85	5	7	1.65E-01	3640.62	7	5	1.94E-01
3276.07	11	9	1.18E-01	3452.60	7	7	2.48E-01	3647.84	7	7	2.11E-01
3285.03	1	3	1.41E-01	3456.15	5	5	3.60E-01	3648.70	7	5	1.15E-01

λ Å	Weights g_i	g_k	A $10^8 s^{-1}$	λ Å	Weights g_i	g_k	A $10^8 s^{-1}$	λ Å	Weights g_i	g_k	A $10^8 s^{-1}$
3654.58	3	3	1.80E-01	3834.64	3	5	1.20E-01	4536.80	13	15	5.03E-01
3657.36	5	7	2.03E-01	3846.18	7	7	1.26E-01	4598.23	1	3	1.47E-01
3658.13	9	9	1.86E-01	3847.25	3	1	2.41E-01	4624.23	9	9	1.32E-01
3659.36	7	9	6.70E-01	3848.30	9	9	1.26E-01	4633.08	3	5	2.35E-01
3660.92	3	5	1.34E-01	3851.99	11	9	1.78E-01	4649.06	3	1	1.25E-01
3662.15	7	9	1.45E-01	3864.10	7	7	6.24E-01	4652.24	5	7	1.55E-01
3662.99	11	11	3.48E-01	3866.69	3	5	1.74E-01	4686.08	3	3	1.72E-01
3663.27	7	5	2.30E-01	3867.67	5	3	2.22E-01	4688.21	13	15	1.54E-01
3664.81	11	13	9.54E-01	3869.08	5	3	1.35E-01	4707.25	7	9	3.63E-01
3664.88	1	3	1.92E-01	3874.15	7	5	1.67E-01	4718.86	5	5	2.17E-01
3669.34	9	7	2.16E-01	3902.95	7	5	6.17E-01	4723.05	9	9	1.23E-01
3672.81	9	11	1.95E-01	3909.54	9	7	1.13E-01	4731.44	9	11	4.49E-01
3672.82	9	9	1.13E-01	3911.94	5	5	1.15E-01	4758.50	11	9	3.01E-01
3676.23	3	1	5.22E-01	3915.43	5	5	1.40E-01	4760.18	11	13	4.67E-01
3680.68	11	11	2.96E-01	3916.43	5	3	1.78E-01	4764.11	9	7	2.16E-01
3681.72	9	7	1.68E-01	3919.55	11	13	2.24E-01	4811.05	13	11	4.36E-01
3683.01	3	5	1.20E-01	3955.48	13	11	1.71E-01	4819.25	11	9	2.71E-01
3687.96	5	7	2.12E-01	3973.76	11	13	4.39E-01	4830.51	9	7	4.07E-01
3688.97	11	9	3.26E-01	3977.90	9	7	1.35E-01	4858.39	13	11	1.24E-01
3690.59	11	9	2.07E-01	3980.20	5	3	2.70E-01	4868.02	7	5	3.11E-01
3694.94	5	7	6.36E-01	3991.85	11	9	1.29E-01	5037.18	9	7	1.14E-01
3696.04	11	11	3.59E-01	4010.13	5	3	4.38E-01	5044.36	7	5	1.31E-01
3698.07	7	5	1.48E-01	4021.01	9	11	2.65E-01	5047.70	3	1	2.61E-01
3708.55	7	9	1.28E-01	4051.18	13	11	1.36E-01	5163.18	9	11	2.03E-01
3715.75	9	7	2.38E-01	4062.08	11	9	1.96E-01	5171.06	5	7	1.84E-01
3718.48	5	7	1.34E-01	4069.88	13	11	3.25E-01	5172.94	5	5	4.11E-01
3720.25	7	9	2.86E-01	4076.19	9	9	1.16E-01	5174.18	5	3	5.83E-01
3725.55	7	7	1.60E-01	4084.37	9	7	1.94E-01	5191.45	7	9	1.62E-01
3727.68	9	11	1.51E-01	4102.15	5	3	1.22E-01	5238.21	7	9	3.74E-01
3728.30	7	5	1.55E-01	4107.46	7	5	2.02E-01	5240.87	7	7	3.89E-01
3728.50	7	9	2.20E-01	4120.09	13	15	6.05E-01	5242.80	7	5	2.01E-01
3733.02	7	7	1.45E-01	4131.92	9	11	1.56E-01	5261.53	5	7	1.13E-01
3733.41	13	13	2.80E-01	4148.98	9	11	1.56E-01	5280.85	5	5	1.28E-01
3735.62	11	11	1.66E-01	4157.40	13	11	2.17E-01	5355.52	9	9	1.21E-01
3742.28	7	7	1.56E-01	4157.90	9	11	1.60E-01	5356.46	11	11	2.11E-01
3747.19	5	7	3.07E-01	4185.82	11	13	3.82E-01	5360.51	9	11	6.19E-01
3748.48	9	11	3.95E-01	4188.32	11	13	3.32E-01	5364.28	9	9	2.26E-01
3755.10	3	5	1.41E-01	4194.56	11	11	2.70E-01	5460.50	5	3	3.46E-01
3755.16	9	9	2.48E-01	4232.59	9	11	3.17E-01	5493.76	7	5	2.13E-01
3758.52	9	9	1.22E-01	4240.83	5	5	1.68E-01	5506.49	5	7	3.61E-01
3759.60	9	7	1.82E-01	4246.02	11	13	2.00E-01	5533.03	5	5	3.72E-01
3760.88	9	9	2.16E-01	4251.88	13	11	1.76E-01	5570.44	5	3	3.30E-01
3768.73	9	9	2.88E-01	4254.95	7	9	2.01E-01	5849.71	3	3	3.02E-01
3769.99	7	9	2.46E-01	4269.28	11	11	1.36E-01	5851.50	3	5	1.55E-01
3777.72	13	11	1.66E-01	4276.91	7	9	2.85E-01	5893.36	5	5	2.60E-01
3788.25	7	9	2.87E-01	4277.24	9	11	1.35E-01	5895.93	5	7	3.12E-01
3794.43	9	9	1.22E-01	4317.92	15	15	1.28E-01	5926.37	7	7	1.63E-01
3797.47	7	5	1.48E-01	4325.80	3	3	1.84E-01	5928.88	7	9	5.32E-01
3798.25	7	9	6.90E-01	4326.14	5	7	2.56E-01	7154.11	9	9	3.45E-01
3801.84	9	7	3.16E-01	4340.74	5	7	1.23E-01				
3805.99	5	5	2.44E-01	4381.63	13	13	2.93E-01	**Neodymium**			
3819.78	9	11	1.47E-01	4382.41	11	13	3.83E-01	**Nd II**			
3824.78	5	7	1.40E-01	4409.94	13	13	1.38E-01	3780.4	16	18	1.4E-01
3827.15	7	7	1.94E-01	4411.69	11	11	2.63E-01	3805.4	14	16	6.9E-01
3828.88	7	7	1.35E-01	4434.95	9	9	2.51E-01	3807.2	10	12	4.9E-02
3830.81	5	5	1.83E-01	4446.42	11	11	1.90E-01	3863.3	8	10	1.5E-01
3831.07	7	9	1.20E-01	4457.35	7	7	1.28E-01	3941.5	10	10	6.1E-01
3832.11	9	9	3.05E-01	4474.57	5	5	2.10E-01	3951.2	12	12	6.0E-01
3833.75	9	9	1.70E-01	4491.65	11	11	2.09E-01	3973.3	18	18	6.3E-01

λ Å	g_i	g_k	A 10^8 s^{-1}	λ Å	g_i	g_k	A 10^8 s^{-1}	λ Å	g_i	g_k	A 10^8 s^{-1}
3979.5	10	12	2.7E-01	3464.3	5	5	6.7E-03	6402.2	5	7	5.14E-01
3990.1	16	16	5.2E-01	3466.6	1	3	1.3E-02	6506.5	3	5	3.00E-01
4012.3	18	20	5.5E-01	3472.6	5	7	1.7E-02	6532.9	1	3	1.08E-01
4061.1	16	18	4.4E-01	3498.1	3	5	5.1E-03	6599.0	3	3	2.32E-01
4106.6	14	16	6.8E-02	3501.2	3	3	1.2E-02	6602.9	3	3	5.9E-03
4109.5	14	16	3.7E-01	3510.7	5	3	2.2E-03	6652.1	3	1	2.9E-03
4133.4	14	12	1.5E-01	3515.2	3	5	6.9E-03	6678.3	3	5	2.33E-01
4156.1	12	14	3.4E-01	3520.5	3	1	9.3E-02	6717.0	3	3	2.17E-01
4205.6	18	16	1.8E-01	3593.5	3	5	9.9E-03	6721.1	3	3	4.9E-04
4284.5	18	18	8.5E-02	3593.6	3	3	6.6E-03	6929.5	3	5	1.74E-01
4303.6	8	10	4.7E-01	3600.2	3	3	4.3E-03	7024.1	3	3	1.89E-02
4325.8	16	16	1.6E-01	3633.7	3	1	1.1E-02	7032.4	5	3	2.53E-01
4358.2	14	14	1.5E-01	3682.2	3	5	1.6E-03	7051.3	3	3	3.0E-02
4382.7	12	10	4.0E-02	3685.7	3	3	3.9E-03	7059.1	3	5	6.8E-02
4400.8	10	10	6.8E-02	3701.2	3	5	2.2E-03	7173.9	3	5	2.87E-02
4451.6	12	14	2.5E-01	4536.3	3	3	5.0E-03	7245.2	3	3	9.35E-02
4456.4	16	18	6.4E-02	4702.5	3	3	2.1E-03	7304.8	1	3	2.55E-03
4463.0	14	16	1.8E-01	4708.9	3	3	4.2E-02	7438.9	1	3	2.31E-02
4958.1	12	10	1.2E-02	4955.4	3	3	3.3E-03	7472.4	3	3	4.0E-02
5130.6	22	20	1.6E-01	5113.7	3	3	1.0E-02	7535.8	3	3	4.3E-01
5192.6	20	18	1.7E-01	5120.5	3	3	5.6E-03	7937.0	5	5	7.8E-03
5249.6	18	16	1.8E-01	5154.4	3	3	1.9E-02	8082.5	3	3	1.2E-03
5276.9	12	10	1.2E-01	5191.3	3	3	1.3E-02	8118.5	3	3	4.9E-02
5293.2	16	14	1.2E-01	5326.4	3	3	6.8E-03	8128.9	3	5	7.2E-03
5302.3	20	18	1.1E-01	5333.3	3	3	5.3E-03	8259.4	5	5	2.03E-02
5311.5	14	12	1.1E-01	5341.1	3	3	1.1E-01	8571.4	3	3	5.5E-02
5319.8	12	10	1.6E-01	5400.6	3	1	9.0E-02	8582.9	3	5	1.00E-01
5357.0	18	16	1.8E-01	5418.6	3	3	5.2E-03	8647.0	5	5	3.91E-02
5371.9	20	20	5.1E-02	5433.7	3	5	2.83E-03	8681.9	3	3	2.1E-01
5485.7	18	18	5.7E-02	5652.6	3	3	8.9E-03	8767.5	3	3	1.1E-03
5594.4	16	16	7.0E-02	5662.5	3	3	6.9E-03	8771.7	3	3	1.6E-01
5620.6	18	18	1.3E-01	5852.5	3	1	6.82E-01	8783.8	3	5	3.13E-01
5688.5	14	14	5.9E-02	5868.4	3	3	1.4E-02	8865.3	3	3	9.4E-03
5718.1	16	16	8.7E-02	5881.9	5	3	1.15E-01	9201.8	3	3	9.1E-02
5726.8	10	10	5.6E-02	5913.6	3	3	4.8E-02	9433.0	3	3	1.1E-03
5740.9	12	12	7.2E-02	5939.3	5	3	2.00E-03	9486.7	3	3	2.5E-02
5804.0	10	10	4.6E-02	5944.8	5	5	1.13E-01	9534.2	3	3	6.3E-02
5865.1	16	18	1.3E-02	5961.6	3	3	3.3E-02	10621	3	3	2.4E-03
6051.9	12	10	1.1E-02	5975.5	5	3	3.51E-02	11409	3	3	4.2E-02
				6030.0	3	3	5.61E-02	11525	3	3	8.4E-02
Neon				6046.1	3	3	2.26E-02	11767	3	3	6.9E-02
Ne I				6074.3	3	1	6.03E-01	12459	3	3	1.5E-02
615.63	1	3	3.8E-01	6096.2	3	5	1.81E-01				
618.67	1	3	9.3E-01	6118.0	5	3	6.09E-03	**Ne II**			
619.10	1	3	3.3E-01	6128.5	3	3	6.7E-03	*357.03	6	10	3.8E+01
626.82	1	3	7.4E-01	6143.1	5	5	2.82E-01	*361.77	6	2	1.6E+01
629.74	1	3	4.8E-01	6150.3	3	3	1.5E-02	*406.28	6	10	1.8E+01
735.90	1	3	6.11E+00	6163.6	1	3	1.46E-01	*446.37	6	6	4.07E+01
743.72	1	3	4.86E-01	6217.3	5	3	6.37E-02	460.73	4	2	4.7E+01
3369.8	5	5	1.0E-03	6266.5	1	3	2.49E-01	462.39	2	2	2.3E+01
3369.9	5	3	7.6E-03	6273.0	3	3	9.7E-03	1907.5	4	2	2.8E-01
3375.6	5	3	2.2E-03	6293.7	3	3	6.39E-03	1916.1	4	4	6.9E-01
3417.9	3	5	9.2E-03	6304.8	3	5	4.16E-02	1930.0	2	2	5.7E-01
3418.0	3	3	2.2E-03	6328.2	5	3	3.39E-02	1938.8	2	4	1.3E-01
3423.9	3	3	1.0E-02	6330.9	3	3	2.3E-02	2858.0	6	6	7.9E-01
3447.7	5	5	2.1E-02	6334.4	5	5	1.61E-01	2870.0	6	6	1.7E-01
3450.8	5	3	4.9E-03	6351.9	1	3	3.45E-03	2873.0	6	4	3.8E-01
3454.2	3	1	3.7E-02	6383.0	3	3	3.21E-01	2876.3	4	6	7.8E-01
3460.5	1	3	7.0E-03	6401.1	3	3	1.39E-02	2876.5	6	4	3.3E-01

λ Å	g_i	g_k	A 10^8 s^{-1}	λ Å	g_i	g_k	A 10^8 s^{-1}	λ Å	g_i	g_k	A 10^8 s^{-1}
2878.1	2	2	6.9E-02	3309.7	4	2	3.1E-01	3568.5	6	8	1.4E+00
2888.4	4	6	7.0E-02	3310.5	4	4	6.9E-02	3571.2	4	4	6.3E-01
2891.5	4	4	6.1E-02	3311.3	4	2	2.6E-01	3574.2	6	6	1.0E+00
2897.0	6	8	5.2E-02	3314.7	6	6	4.4E-02	3574.6	4	6	1.3E+00
2906.8	2	4	5.5E-01	3319.7	4	2	1.6E+00	3590.4	4	6	3.6E-02
2910.1	4	2	1.7E+00	3320.2	8	6	2.1E-01	3594.2	4	2	1.3E+00
2910.4	2	4	5.9E-01	3323.7	4	4	1.6E+00	3612.3	2	4	2.6E-01
2916.2	6	4	9.6E-02	3327.2	4	4	9.1E-01	3628.0	4	4	6.0E-01
2925.6	2	2	5.6E-01	3329.2	8	8	8.8E-01	3632.7	4	4	1.3E-01
2933.7	6	6	6.9E-02	3330.7	6	6	3.9E-02	3643.9	4	4	3.2E-01
2955.7	6	4	1.2E+00	3334.8	6	8	1.8E+00	3644.9	2	4	9.9E-01
3001.7	4	4	8.7E-01	3336.1	4	6	1.1E+00	3659.9	4	6	6.7E-02
3017.3	6	4	3.5E-01	3344.4	2	2	1.5E+00	3664.1	6	4	7.0E-01
3027.0	6	6	1.4E+00	3345.5	6	4	1.4E+00	3679.8	4	2	3.2E-01
3028.7	4	2	8.5E-01	3345.8	4	4	2.2E-01	3694.2	6	6	1.0E+00
3028.9	2	4	4.7E-01	3353.6	4	2	1.2E-01	3697.1	2	2	2.8E-01
3034.5	6	8	3.1E+00	3355.0	4	6	1.3E+00	3701.8	4	6	2.7E-01
3037.7	4	4	2.1E+00	3356.3	6	6	2.0E-01	3709.6	4	2	1.1E+00
3045.6	2	2	2.5E+00	3357.8	6	6	5.0E-01	3713.1	4	6	1.3E+00
3047.6	4	6	1.8E+00	3360.3	2	4	8.6E-01	3721.8	4	6	2.0E-01
3054.7	2	4	9.4E-01	3360.6	2	4	8.2E-01	3726.9	4	4	1.2E-01
3092.9	6	6	1.3E+00	3362.9	4	2	3.5E-01	3727.1	2	4	9.8E-01
3097.1	8	8	1.3E+00	3371.8	4	6	2.2E-01	3734.9	4	4	1.9E-01
3118.0	8	6	4.2E-02	3374.1	4	4	3.0E-01	3744.6	2	4	2.6E-01
3134.1	6	4	2.6E-01	3378.2	2	2	1.7E+00	3751.2	2	2	1.8E-01
3140.4	8	6	2.4E-01	3379.3	2	2	3.0E-01	3753.8	4	6	4.5E-01
3151.1	6	6	4.8E-02	3386.2	4	6	5.5E-02	3766.3	4	6	2.9E-01
3154.8	8	6	1.8E-02	3388.4	4	6	2.2E+00	3777.1	2	4	4.2E-01
3164.4	8	8	1.6E-01	3390.6	2	4	7.7E-02	3800.0	4	4	3.7E-01
3165.7	6	6	1.2E-01	3392.8	2	4	4.4E-01	3818.4	2	4	6.1E-01
3173.6	6	4	4.5E-02	3404.8	4	6	1.9E+00	3829.8	4	6	8.4E-01
3176.1	4	6	6.0E-02	3407.0	6	8	2.3E+00	3942.3	4	6	1.0E-02
3187.6	4	6	1.4E-02	3411.4	4	2	6.1E-01				
3188.7	6	6	3.9E-01	3413.2	4	4	1.8E+00	**Ne V**			
3190.9	4	6	1.5E-01	3414.9	4	6	1.8E-02	*142.61	9	9	6.7E+02
3194.6	4	4	5.2E-01	3416.9	6	6	6.4E-01	*143.32	9	15	1.2E+03
3198.6	6	8	1.7E+00	3417.7	6	8	1.6E+00	147.13	5	7	1.5E+03
3198.9	4	4	2.3E-01	3438.9	2	2	1.4E+00	151.23	5	5	3.38E+02
3209.0	8	8	1.6E-01	3440.7	2	4	3.5E-01	154.50	1	3	7.0E+02
3209.4	2	4	6.0E-01	3453.1	4	4	4.6E-01	*167.69	9	9	1.5E+02
3213.7	2	4	1.7E+00	3454.8	4	4	1.6E+00	*358.93	9	3	2.1E+02
3214.3	4	6	2.2E+00	3456.6	2	4	9.6E-01	365.59	5	3	1.35E+02
3218.2	8	10	3.6E+00	3457.1	4	6	9.9E-02	*482.15	9	9	3.01E+01
3224.8	6	8	3.5E+00	3459.3	6	6	1.6E+00	*571.04	9	15	1.0E+01
3229.5	8	8	1.3E-01	3475.2	4	4	1.2E-02	2259.6	3	5	1.9E+00
3229.6	8	10	3.6E+00	3477.6	4	6	4.3E-01	2265.7	5	7	2.4E+00
3230.1	6	6	1.8E+00	3481.9	4	2	1.4E+00				
3230.4	4	6	1.4E-01	3503.6	2	2	2.0E+00	**Ne VII**			
3232.0	6	4	2.7E-01	3522.7	4	2	2.3E-02	97.502	1	3	1.07E+03
3232.4	4	4	1.6E+00	3538.0	4	2	7.6E-01	*115.46	9	3	4.8E+02
3243.4	6	6	2.3E-01	3539.9	4	4	3.6E-02	116.69	3	5	1.6E+03
3244.1	6	8	1.5E+00	3542.2	6	4	6.0E-01	127.66	3	1	1.9E+02
3248.1	4	4	2.4E-01	3542.9	4	6	1.2E+00	465.22	1	3	4.09E+01
3255.4	6	4	3.8E-02	3546.2	2	4	6.3E-02	558.61	3	5	8.11E+00
3263.4	2	4	3.9E-01	3551.6	2	4	3.7E-02	559.95	1	3	1.07E+01
3269.9	4	6	5.1E-01	3557.8	2	2	1.9E-01	561.38	3	3	7.99E+00
3270.8	6	4	5.7E-02	3561.2	4	6	2.1E-01	561.73	5	5	2.39E+01
3297.7	6	6	4.3E-01	3565.8	4	4	6.2E-01	562.99	3	1	3.17E+01

λ Å	g_i	g_k	A 10^8 s^{-1}	λ Å	g_i	g_k	A 10^8 s^{-1}	λ Å	g_i	g_k	A 10^8 s^{-1}
564.53	5	3	1.31E+01	2313.98	5	5	5.0E+00	4752.43	3	3	2.0E-01
				2317.16	7	5	3.8E+00	4756.52	9	9	1.5E-01
Ne VIII				2320.03	9	11	6.9E+00	4786.54	11	11	1.8E-01
*88.09	2	6	8.4E+02	2321.38	5	7	5.6E+00	4812.00	3	1	9.5E-02
*98.208	6	10	2.77E+03	2324.65	7	9	1.8E-01	4829.03	5	7	1.9E-01
770.41	2	4	5.90E+00	2325.79	7	9	3.5E+00	4831.18	9	7	1.6E-01
780.32	2	2	5.69E+00	2329.96	5	3	5.3E+00	4838.64	9	7	2.2E-01
2820.7	2	4	7.20E-01	2345.54	9	7	2.2E+00	4855.41	5	5	5.7E-01
2860.1	2	2	6.88E-01	2346.63	7	5	5.5E-01	4904.41	5	3	6.2E-0i
				2347.51	9	9	2.2E-01	4912.03	3	3	1.5E-01
Nickel				2348.73	7	7	2.2E-01	4913.97	1	3	2.2E-01
Ni I				2419.31	7	5	2.0E-01	4918.36	9	7	2.3E-01
1963.85	7	7	1.1E-01	2943.91	7	5	1.1E-01	4935.83	7	5	2.4E-01
1976.87	7	9	1.1E+00	2981.65	5	3	2.8E-01	4937.34	9	9	1.2E-01
1981.61	5	5	1.3E-01	3002.48	7	7	8.0E-01	4953.20	5	5	1.2E-01
1990.25	5	7	8.3E-01	3003.62	5	5	6.9E-01	4980.17	9	11	1.9E-01
2007.01	5	5	1.7E-01	3012.00	5	5	1.3E+00	5000.34	7	7	1.4E-01
2007.69	7	7	9.0E-02	3037.93	7	7	2.8E-01	5012.46	7	7	1.1E-01
2014.25	3	5	9.3E-01	3050.82	7	9	6.0E-01	5017.58	11	11	2.0E-01
2025.40	7	5	2.3E-01	3054.31	5	5	4.0E-01	5035.37	7	9	5.7E-01
2026.62	9	7	2.4E-01	3057.64	3	3	1.0E+00	5042.20	3	5	1.4E-01
2047.35	7	5	1.8E-01	3064.62	5	7	1.1E-01	5048.85	7	7	1.6E-01
2052.04	9	9	9.7E-02	3101.56	5	7	6.3E-01	5080.53	9	11	3.2E-01
2055.50	5	3	3.3E-01	3101.88	5	7	4.9E-01	5081.11	7	9	5.7E-01
2059.92	7	5	2.1E-01	3134.11	3	5	7.3E-01	5082.35	3	3	2.5E-01
2060.20	5	3	2.3E-01	3225.02	5	3	9.3E-02	5084.08	7	9	3.1E-01
2064.39	3	1	4.0E-01	3369.56	9	7	1.8E-01	5099.95	7	7	2.9E-01
2069.52	5	5	1.1E-01	3380.57	5	3	1.3E+00	5115.40	11	9	2.2E-01
2085.57	5	5	2.6E+00	3392.98	7	7	2.4E-01	5129.37	7	5	1.2E-01
2089.09	7	5	9.7E-02	3414.76	7	9	5.5E-01	5155.14	5	5	1.1E-01
2095.13	5	7	1.1E-01	3423.71	3	3	3.3E-01	5155.76	5	7	2.9E-01
2114.43	5	5	9.7E-02	3433.56	7	7	1.7E-01	5176.57	5	5	1.8E-01
2121.40	7	5	2.8E-01	3446.26	5	5	4.4E-01	5371.33	7	7	1.6E-01
2124.80	5	3	3.8E-01	3452.88	5	7	9.8E-02	5476.91	1	3	9.5E-02
2147.80	5	3	4.7E-01	3458.46	3	5	6.1E-01	5637.12	3	3	1.1E-01
2157.83	5	3	4.1E-01	3461.66	7	9	2.7E-01	5664.02	5	7	1.1E-01
2158.31	7	5	6.9E-01	3472.55	5	7	1.2E-01	5695.00	3	3	1.7E-01
2161.04	5	5	1.3E-01	3483.77	5	3	1.4E-01	6086.29	3	5	1.1E-01
2173.54	5	3	1.5E-01	3492.96	5	3	9.8E-01	6175.42	3	3	1.7E-01
2174.48	3	1	8.9E-01	3510.33	3	1	1.2E+00	7122.24	5	7	2.1E-01
2182.38	7	5	1.3E-01	3515.05	5	7	4.2E-01	7381.94	9	11	9.7E-02
2183.91	5	5	1.2E-01	3524.54	7	5	1.0E+00	7422.30	7	5	1.8E-01
2190.22	5	5	3.0E-01	3566.37	5	5	5.6E-01	7727.66	7	7	1.1E-01
2197.35	3	3	7.8E-01	3597.70	3	3	1.4E-01				
2201.59	5	3	7.3E-01	3619.39	5	7	6.6E-01	**Ni II**			
2221.94	5	3	2.2E-01	4027.67	5	7	1.3E-01	2165.55	10	10	2.4E+00
2244.46	5	5	3.8E-01	4295.88	9	7	1.7E-01	2169.10	8	8	1.58E+00
2253.57	7	7	1.9E-01	4401.54	9	11	3.8E-01	2174.67	8	10	1.43E+00
2254.81	9	9	9.6E-02	4462.46	3	5	1.7E-01	2175.15	6	6	1.77E+00
2258.15	7	5	1.7E-01	4470.48	5	7	1.9E-01	2184.61	4	4	2.90E+00
2259.56	5	3	2.0E-01	4600.37	5	3	2.6E-01	2201.41	4	6	1.3E+00
2261.42	9	7	9.1E-02	4604.99	9	7	2.3E-01	2206.72	6	8	1.66E+00
2287.32	3	5	1.8E-01	4606.23	5	3	1.0E-01	2216.48	10	12	3.4E+00
2289.98	9	7	2.1E+00	4648.66	11	9	2.4E-01	2220.40	6	8	2.3E-01
2293.11	5	5	3.8E-01	4686.22	5	5	1.4E-01	2222.96	10	10	9.8E-01
2300.77	7	7	7.5E-01	4701.54	9	9	1.4E-01	2224.86	8	8	1.55E+00
2302.97	3	3	4.5E-01	4714.42	13	11	4.6E-01	2226.33	6	6	1.3E+00
2307.35	5	7	1.6E-01	4715.78	7	7	2.0E-01	2253.85	4	6	1.98E+00
2312.34	7	7	5.5E+00	4732.47	7	9	9.3E-02	2264.46	6	8	1.43E+00

λ Å	Weights g_i	g_k	A 10^8 s⁻¹	λ Å	Weights g_i	g_k	A 10^8 s⁻¹	λ Å	Weights g_i	g_k	A 10^8 s⁻¹
2270.21	8	10	1.56E+00	168	6	8	3.2E+02	292	5	7	2.2E+02
2278.77	8	6	2.8E+00	182	2	2	2.5E+02				
2287.09	6	4	2.8E+00	185.23	2	4	4.2E+02	**Ni XVIII**			
2296.55	8	8	1.98E+00	187	4	6	1.2E+02	24.881	2	4	8.6E+02
2297.14	6	4	2.70E+00	187	4	2	3.3E+02	25.070	4	6	9.9E+02
2297.49	4	2	3.0E+00	188	2	2	4.7E+02	26.02	2	4	1.26E+03
2298.27	6	6	2.8E+00	190	6	8	2.0E+02	26.020	2	4	1.1E+03
2303.00	8	6	2.9E+00	192	6	8	4.54E+02	26.046	2	2	1.1E+03
2316.04	10	8	2.88E+00	192	6	6	3.1E+02	26.218	4	6	1.5E+03
2334.58	8	8	8.0E-01	194	4	6	2.8E+02	27.98	4	6	1.0E+03
2375.42	6	8	6.6E-01	194	2	4	5.5E+02	27.982	2	4	2.0E+03
2394.52	8	10	1.70E+00	194	2	2	1.1E+02	28.018	6	8	1.1E+03
2416.13	6	8	2.1E+00	194	4	4	3.5E+02	28.220	4	6	2.33E+03
2437.89	8	10	5.4E-01	194.04	4	6	4.6E+02	29.383	4	6	1.58E+03
2510.87	8	10	5.8E-01	195.27	4	4	9.5E+01	29.422	6	8	1.69E+03
				196	4	6	6.7E+02	29.779	2	4	1.9E+03
Ni III				197	4	6	1.5E+02	29.829	2	2	1.9E+03
1692.51	11	13	7.9E+00	197	4	2	1.2E+02	31.845	4	6	2.7E+03
1709.90	9	11	6.3E+00	199	2	4	4.9E+02	31.890	6	8	3.0E+03
1719.46	5	7	6.0E+00	206	2	2	3.7E+02	32.034	2	4	3.4E+03
1722.28	3	5	5.9E+00	217	4	4	1.1E+02	32.340	4	6	4.0E+03
1724.52	3	1	6.7E+00	218.391	2	4	9.5E+01	36.990	4	6	5.5E+03
1741.96	9	7	5.7E+00	223.119	2	2	1.3E+02	37.049	6	8	5.9E+03
1752.43	7	5	5.5E+00	231	4	4	1.6E+02	41.015	2	4	2.97E+03
1760.56	5	3	6.5E+00	232.475	4	4	4.07E+02	41.218	2	2	3.2E+03
1769.64	11	11	6.2E+00	233	6	4	2.4E+02	43.814	2	4	5.5E+03
1823.06	9	9	5.6E+00	235	4	2	3.8E+02	44.365	4	6	6.8E+03
				235	6	6	2.5E+02	44.405	4	4	1.14E+03
Ni XIV				236	4	4	1.2E+02	52.615	4	6	1.5E+04
164.13	6	8	1.2E+03	237.875	4	2	2.6E+02	52.720	6	8	1.6E+04
168	2	4	2.4E+02	238	6	4	1.3E+02	52.745	6	6	1.06E+03
168.12	4	2	8.5E+02	239.550	2	2	2.6E+02	59.950	6	4	9.6E+02
169.69	4	4	9.8E+02	245	4	4	1.4E+02	60.212	4	2	1.1E+03
170.50	4	4	7.1E+02	245	4	6	3.2E+02	63.512	4	6	7.9E+02
171.37	4	6	9.4E+02	249	6	8	3.3E+02	63.589	6	8	8.5E+02
172.16	6	6	4.7E+02	249	6	4	1.2E+02	69.075	4	6	8.0E+02
172.80	6	4	1.4E+02	250	4	2	1.6E+02	76.254	4	6	1.38E+03
177.28	4	4	5.6E+02	254	6	4	1.8E+02	76.359	6	8	1.47E+03
178	2	4	8.9E+01					99.275	2	4	1.0E+03
181	4	6	7.4E+01	**Ni XVII**				100.4	4	6	1.2E+03
182.14	4	2	1.5E+02	30.919	1	3	2.77E+03	114.46	4	6	2.5E+03
196	4	2	3.8E+01	42.855	1	3	4.75E+03	114.74	6	8	2.7E+03
288.894	4	4	4.6E+01	54.451	9	11	1.5E+04				
292.399	6	6	3.6E+01	55.361	1	3	6.7E+03	**Ni XIX**			
				57.348	7	9	1.4E+04	9.140	1	3	3.1E+04
Ni XV				197.39	1	3	1.6E+02	9.153	1	3	5.2E+03
50.249	5	7	6.8E+03	199.87	3	5	2.1E+02	9.977	1	3	1.1E+05
60.890	9	11	1.0E+04	204	3	3	1.8E+02	10.110	1	3	9.4E+04
64.635	7	9	9.6E+03	205	3	1	2.4E+02	10.283	1	3	4.7E+03
163.64	5	7	5.6E+01	206	1	3	3.0E+02	10.433	1	3	5.1E+03
173.73	5	7	7.6E+02	207.50	5	7	2.5E+02	11.539	1	3	4.8E+04
175	3	1	5.7E+02	215.89	3	5	4.8E+02	11.599	1	3	6.3E+03
179.28	5	7	7.5E+02	216	1	3	2.7E+02	12.435	1	3	3.66E+05
181	1	3	6.8E+02	217	5	7	2.4E+02	12.656	1	3	1.0E+05
269	3	1	5.3E+01	227	5	5	1.6E+02	13.779	1	3	1.23E+04
278.386	5	5	4.3E+01	249.180	1	3	2.75E+02	14.043	1	3	1.31E+04
				281.50	3	1	2.1E+02	40.7	3	3	6.4E+03
Ni XVI				282	3	1	2.4E+02	40.7	3	1	8.4E+03
166	4	6	3.1E+02	284	5	3	1.5E+02	41.132	7	9	9.4E+03

λ Å	Weights g_i	g_k	A 10^8 s^{-1}	λ Å	Weights g_i	g_k	A 10^8 s^{-1}	λ Å	Weights g_i	g_k	A 10^8 s^{-1}
Ni XXI				103.53	4	2	4.17E+02	1.539	1	3	2.6E+06
11.13	3	3	1.7E+04	104.64	2	2	4.7E+02	1.539	3	5	2.6E+06
11.23	5	3	1.7E+04	106.68	4	2	3.67E+02	1.540	3	3	1.7E+06
11.239	5	7	5.7E+04	113.14	4	4	1.65E+02	1.541	3	5	5.5E+06
11.28	3	1	2.2E+05	118.52	2	4	1.5E+02	1.542	3	3	3.6E+06
11.318	5	7	2.8E+05	122.72	6	4	2.17E+02	1.542	5	5	3.5E+06
11.48	3	1	1.1E+05	134.73	6	6	1.44E+02	1.544	5	3	3.2E+06
11.48	1	3	4.0E+05	135.47	4	4	8.0E+01	1.546	3	5	1.6E+06
11.517	5	7	1.4E+05	137.01	4	4	2.6E+02	1.547	3	3	2.1E+05
11.539	5	7	1.2E+05	138.80	4	6	7.2E+01	1.549	1	3	2.0E+05
11.67	1	3	8.0E+04	153.47	2	2	1.27E+02	1.551	3	1	8.2E+05
11.72	3	3	2.3E+04	159.69	2	4	8.9E+01	1.558	3	1	6.5E+05
12.454	5	3	3.3E+04					1.5883	1	3	6.02E+06
12.472	3	3	1.8E+04	**Ni XXV**				1.5963	1	3	7.70E+05
12.502	5	5	2.8E+04	9.30	3	1	9.3E+04				
				9.31	5	7	8.2E+04	**Nitrogen**			
Ni XXII				9.32	3	5	7.8E+04	**N I**			
72.52	4	2	2.84E+02	9.34	1	3	1.1E+05	1163.88	6	6	7.52E-01
84.06	6	4	1.2E+03	9.42	3	1	9.0E+04	1164.00	4	6	1.27E-02
84.24	4	2	5.6E+02	9.49	3	5	8.9E+04	1164.21	6	4	5.17E-02
85.86	4	2	4.9E+02	9.60	1	3	1.8E+05	1164.32	4	4	6.94E-01
88.00	4	2	1.2E+03	9.63	3	5	2.4E+05	1167.45	6	8	1.29E+00
95.95	2	2	4.4E+02	9.64	3	3	1.3E+05	1168.42	6	6	4.24E-02
98.16	4	4	5.2E+02	9.71	3	1	2.3E+05	1168.54	4	6	1.24E+00
98.58	4	4	2.45E+02	9.71	3	3	1.8E+05	1176.51	6	4	9.22E-01
100.60	6	6	3.9E+02	9.74	5	7	3.0E+05	1176.63	4	4	1.02E-01
101.31	6	4	4.83E+02	9.75	3	5	1.3E+05	1177.69	4	2	1.02E+00
103.31	4	2	2.66E+02	9.76	1	3	3.03E+05	1199.55	4	6	4.01E+00
106.04	4	4	2.36E+02	9.76	5	3	7.5E+04	1200.22	4	4	3.99E+00
106.16	4	2	5.1E+02	9.78	5	7	2.9E+05	1200.71	4	2	3.98E+00
124.31	2	2	3.7E+02	9.86	5	7	4.8E+05	1310.54	4	6	8.42E-01
126.32	4	4	3.3E+02	9.87	3	5	2.03E+05	1316.29	4	6	1.42E-02
				9.92	5	5	1.3E+05	1492.63	6	4	3.13E+00
Ni XXIII				9.94	5	7	1.29E+05	1492.82	4	4	3.51E-01
87.66	3	3	2.8E+02	9.97	3	5	2.5E+05	1494.68	4	2	3.72E+00
88.11	5	3	8.3E+02	10.08	1	3	2.80E+05	3822.03	2	2	3.70E-02
90.49	3	3	1.77E+02					3830.43	4	4	4.67E-02
90.96	5	3	2.5E+02	**Ni XXVI**				3834.22	4	2	1.89E-02
91.83	5	3	7.5E+02	1.5930	4	2	3.4E+06	4099.94	2	4	3.48E-02
92.32	3	1	4.39E+02	1.5935	2	2	4.0E+06	4109.95	4	6	3.90E-02
100.42	1	3	2.1E+02	1.5973	4	4	8.1E+06	4113.97	4	4	6.62E-03
102.08	5	5	5.3E+02	1.5977	2	4	4.4E+06	4137.64	2	4	2.80E-03
103.23	3	3	2.4E+02	1.5982	2	2	7.3E+06	4143.43	4	4	6.09E-03
103.67	5	5	1.78E+02	1.5996	2	2	2.7E+06	4151.48	6	4	1.01E-02
104.70	3	1	2.94E+02	1.6005	4	6	2.7E+06	4249.87	4	2	2.59E-02
106.02	5	5	2.87E+02	1.6036	4	2	2.1E+06	4264.00	6	4	2.26E-02
108.27	7	5	3.32E+02	9.390	2	4	2.59E+05	4356.29	6	8	5.10E-02
111.23	3	1	2.26E+02	9.535	4	6	2.96E+05	4385.54	2	2	8.84E-03
111.78	5	3	2.19E+02					4392.41	4	2	1.76E-02
111.86	1	3	1.7E+02	**Ni XXVII**				4435.43	2	4	7.51E-03
112.55	3	1	1.0E+03	1.2534	1	3	3.35E+05	4442.45	4	4	3.81E-02
128.87	5	5	4.02E+02	1.2824	1	3	6.38E+05	4669.89	4	4	7.49E-03
133.54	3	3	1.86E+02	1.3500	1	3	1.63E+06	4914.94	2	2	8.08E-03
137.55	3	1	2.53E+02	1.3516	1	3	2.4E+05	4935.12	4	2	1.76E-02
				1.531	3	3	2.0E+05	5199.84	2	2	1.87E-02
Ni XXIV				1.534	3	1	6.9E+06	5201.61	2	4	1.87E-02
101.13	6	4	1.63E+02	1.537	5	5	2.3E+06	5281.20	6	6	2.45E-03
102.11	4	4	5.4E+02	1.537	1	3	3.7E+06	5344.05	6	6	6.10E-04
103.43	2	4	1.3E+02	1.538	3	5	3.9E+06	5356.62	4	6	1.41E-03

λ Å	Weights g_i	g_k	A 10^8 s^{-1}	λ Å	Weights g_i	g_k	A 10^8 s^{-1}	λ Å	Weights g_i	g_k	A 10^8 s^{-1}
5367.01	4	4	1.07E-03	9810.01	4	2	5.30E-02	1085.55	5	5	9.47E-01
5372.61	2	4	8.34E-04	9814.02	6	8	6.56E-03	1085.70	5	7	3.87E+00
5378.27	2	2	1.66E-03	9822.75	6	6	4.95E-02	3408.13	3	1	2.19E-01
6606.18	4	6	8.87E-04	9834.61	6	4	4.50E-02	3437.14	3	1	2.07E+00
6622.54	6	6	7.93E-03	9863.33	8	8	9.62E-02	3593.60	3	5	1.21E-01
6626.99	2	4	2.20E-03	9872.15	8	6	2.97E-02	3609.10	3	3	1.41E-01
6636.94	4	4	1.40E-02	9883.38	2	2	2.93E-02	3615.86	3	1	1.53E-01
6644.96	8	6	3.49E-02	9905.52	4	2	3.11E-03	3829.80	3	5	2.42E-01
6646.50	2	2	2.18E-02	9909.22	2	4	7.58E-03	3838.37	5	5	6.98E-01
6653.46	6	4	2.74E-02	9931.47	4	4	3.64E-02	3842.19	1	3	3.06E-01
6656.51	6	4	2.17E-02	9947.07	6	8	1.08E-02	3847.40	3	3	2.22E-01
6926.67	4	6	7.75E-03	9965.75	4	6	7.60E-03	3855.10	3	1	8.82E-01
6945.18	6	6	1.83E-02	9968.51	6	4	4.50E-03	3856.06	5	3	3.71E-01
6951.60	2	4	1.03E-02	9980.42	4	6	8.10E-03	3919.00	3	3	6.76E-01
6960.50	4	4	4.67E-03	9997.73	8	8	9.20E-03	3955.85	3	5	1.31E-01
6973.07	2	2	3.83E-03					3995.00	3	5	1.35E+00
6979.18	6	4	9.83E-03	**N II**				4114.33	3	3	1.42E-03
6982.03	4	2	2.04E-02	474.891	5	5	9.66E+00	4124.08	3	5	3.20E-01
7423.64	2	4	5.95E-02	475.647	1	3	1.17E+01	4133.67	5	5	5.30E-01
7442.30	4	4	1.24E-01	475.698	3	5	1.58E+01	4145.77	7	5	7.36E-01
7468.31	6	4	1.93E-01	475.757	3	3	8.75E+00	4374.99	3	5	5.55E-03
7898.98	6	4	2.82E-01	475.803	5	7	2.10E+01	4447.03	3	5	1.14E+00
7899.28	4	4	3.28E-02	475.884	5	5	5.25E+00	4459.94	3	1	1.12E-01
7915.42	4	2	3.13E-01	508.697	5	5	1.91E+00	4465.53	3	3	2.36E-02
8184.86	4	6	8.58E-02	510.758	5	7	1.87E+01	4477.68	5	3	8.85E-02
8188.01	2	4	1.27E-01	513.849	5	5	1.24E+01	4488.09	5	5	1.30E-02
8200.36	2	2	4.95E-02	529.355	1	3	7.23E+00	4507.56	7	5	1.00E-01
8210.72	4	4	4.84E-02	529.413	3	1	2.43E+01	4564.76	3	5	1.41E-02
8216.34	6	6	2.23E-01	529.491	3	3	6.75E+00	4601.48	3	5	2.35E-01
8223.13	4	2	2.64E-01	529.637	3	5	4.92E+00	4607.15	1	3	3.26E-01
8242.39	6	4	1.36E-01	529.722	5	3	1.03E+01	4613.87	3	3	2.26E-01
8567.74	2	4	4.92E-02	529.867	5	5	1.94E+01	4621.39	3	1	9.55E-01
8594.00	2	2	2.09E-01	533.511	1	3	2.39E+01	4630.54	5	5	7.72E-01
8629.24	4	4	2.66E-01	533.581	3	5	3.20E+01	4643.09	5	3	4.51E-01
8655.88	4	2	1.05E-01	533.650	3	3	1.66E+01	4654.53	3	5	2.43E-02
8680.28	6	8	2.46E-01	533.729	5	7	4.13E+01	4667.21	3	3	2.99E-02
8683.40	4	6	1.80E-01	533.815	5	5	9.19E+00	4674.91	3	1	1.05E-01
8686.15	2	4	1.09E-01	547.818	5	3	2.16E+00	4694.27	1	3	1.23E-01
8703.25	2	2	2.10E-01	559.762	1	3	1.14E+01	4695.90	3	5	1.29E-01
8711.70	4	4	1.28E-01	574.650	5	7	3.60E+01	4697.64	3	3	3.06E-02
8718.84	6	6	6.75E-02	582.156	5	5	2.85E+01	4698.55	3	1	3.67E-01
8728.90	4	2	3.76E-02	635.197	1	3	2.33E+01	4700.03	5	7	1.05E-01
8747.37	6	4	1.04E-02	644.634	1	3	1.21E+01	4702.50	5	5	9.15E-02
9028.92	2	2	3.02E-01	644.837	3	3	3.64E+01	4704.25	5	3	2.13E-01
9045.88	6	8	2.80E-01	645.178	5	3	6.07E+01	4706.40	7	9	6.09E-02
9049.49	6	6	1.88E-02	660.286	5	3	3.69E+01	4709.58	7	7	1.82E-01
9049.89	4	6	2.60E-01	671.016	3	5	2.47E+00	4712.07	7	5	1.46E-01
9060.48	2	4	2.95E-01	671.386	5	5	7.40E+00	4718.38	9	9	3.02E-01
9187.45	6	6	2.44E-01	671.411	1	3	3.04E+00	4721.58	9	7	7.75E-02
9187.86	4	6	1.76E-02	671.630	3	3	2.27E+00	4774.24	3	5	3.24E-02
9207.59	6	4	2.70E-02	671.773	3	1	9.85E+00	4779.72	3	3	2.52E-01
9208.00	4	4	2.33E-01	672.001	5	3	3.87E+00	4781.19	5	7	2.05E-02
9386.81	2	4	2.24E-01	745.841	1	3	1.25E+01	4788.14	5	5	2.52E-01
9392.79	4	6	2.63E-01	746.984	5	3	3.85E+01	4793.65	5	3	7.77E-02
9460.68	4	4	3.98E-02	748.369	5	3	3.83E+00	4803.29	7	7	3.18E-01
9776.90	2	4	1.18E-02	775.965	5	5	3.08E+01	4810.30	7	5	4.75E-02
9786.78	4	6	1.13E-02	915.612	1	3	4.38E+00	4860.17	3	5	1.61E-02
9788.29	2	2	2.99E-02	915.962	3	1	1.32E+01	4987.38	3	1	7.48E-01
9798.56	4	4	2.75E-02	1083.99	1	3	2.18E+00	4991.24	3	5	3.54E-01

λ Å	Weights g_i	g_k	A 10^8 s⁻¹	λ Å	Weights g_i	g_k	A 10^8 s⁻¹	λ Å	Weights g_i	g_k	A 10^8 s⁻¹
4994.36	5	7	2.62E-01	5551.92	7	7	2.00E-01	2978.84	2	4	1.66E-01
4994.37	3	3	7.60E-01	5552.68	5	3	1.50E-01	2983.64	4	4	8.24E-01
4997.22	3	3	1.96E-01	5565.26	7	5	3.97E-02	3342.76	2	2	3.80E-01
5001.13	3	5	9.76E-01	5666.63	3	5	3.74E-01	3353.98	2	4	7.66E-01
5001.47	5	7	1.05E+00	5676.02	1	3	2.96E-01	3354.32	4	6	5.51E-01
5002.70	1	3	8.45E-02	5679.56	5	7	5.25E-01	3355.46	4	2	7.51E-01
5005.15	7	9	1.16E+00	5686.21	3	3	1.94E-01	3358.78	2	2	3.05E-01
5005.30	5	5	6.51E-02	5710.77	5	5	1.24E-01	3360.98	4	4	2.44E-01
5007.33	3	5	7.89E-01	5730.66	5	3	1.34E-02	3365.80	4	2	1.52E+00
5010.62	3	3	2.19E-01	5747.30	3	5	3.40E-02	3367.36	6	6	1.27E+00
5011.31	5	3	5.84E-01	5767.45	3	3	2.44E-02	3374.07	6	4	8.13E-01
5012.04	7	7	5.19E-01	5893.15	5	7	2.88E-01	3745.95	2	4	1.90E-01
5016.38	5	5	1.62E-01	5897.25	3	5	2.16E-01	3752.63	2	2	6.67E-02
5023.05	7	5	3.61E-01	5899.83	1	3	1.60E-01	3754.69	4	4	3.78E-01
5025.66	7	7	1.07E-01	5927.81	1	3	3.22E-01	3762.60	4	4	4.24E-02
5040.71	7	5	3.78E-03	5931.78	3	5	4.27E-01	3771.03	6	4	5.59E-01
5045.10	5	3	3.42E-01	5940.24	3	3	2.26E-01	3771.36	6	4	8.28E-02
5073.59	3	3	2.59E-02	5941.65	5	7	5.54E-01	3792.97	8	6	1.03E-01
5168.05	3	5	3.06E-01	5952.39	5	5	1.27E-01	3934.50	2	4	7.49E-01
5170.16	3	3	6.54E-01	5960.91	5	3	1.34E-02	3938.51	4	6	8.96E-01
5171.27	3	1	8.71E-01	6065.00	5	3	2.21E-03	3942.88	4	4	1.49E-01
5171.47	5	7	5.81E-01	6284.32	5	3	7.74E-02	4097.36	2	4	8.70E-01
5172.34	3	5	6.01E-01	6379.62	3	3	6.11E-02	4103.39	2	2	8.67E-01
5172.97	1	3	5.01E-01	6482.05	3	3	3.01E-01	4195.74	2	4	9.37E-01
5173.39	5	7	7.36E-01	6610.56	5	7	6.34E-01	4200.07	4	6	1.12E+00
5174.46	5	5	5.07E-01	6857.03	5	3	2.53E-01	4215.77	4	4	1.85E-01
5175.89	7	9	8.93E-01	6869.58	5	5	2.51E-01	4318.78	2	4	5.40E-02
5176.57	5	3	2.17E-01	6887.83	5	7	2.49E-01	4321.22	2	2	1.08E-01
5177.06	3	3	5.00E-01	7762.24	5	5	8.74E-02	4321.39	4	6	5.03E-02
5179.34	7	9	8.67E-01	8438.74	1	3	2.24E-01	4325.43	4	4	8.60E-02
5179.52	9	11	1.07E+00	8831.75	1	3	8.42E-03	4327.69	6	8	3.06E-02
5180.36	5	5	4.28E-01	8855.30	3	3	2.51E-02	4327.88	4	2	1.07E-01
5183.20	7	7	2.88E-01	8893.29	5	3	4.12E-02	4332.95	6	6	1.23E-01
5184.96	7	7	3.20E-01					4337.01	6	4	7.47E-02
5185.09	5	3	7.11E-02	**N III**				4345.81	8	8	1.82E-01
5186.21	7	5	5.76E-02	374.198	2	4	9.89E+01	4351.11	8	6	4.01E-02
5190.38	9	9	1.77E-01	451.871	2	2	1.03E+01	4510.88	2	4	2.84E-01
5191.96	7	5	4.25E-02	452.227	4	2	2.05E+01	4510.96	4	6	4.77E-01
5199.50	9	7	1.51E-02	684.998	2	4	9.63E+00	4514.85	6	8	6.80E-01
5313.42	3	3	1.41E-01	685.515	2	2	3.83E+01	4518.14	2	2	5.65E-01
5320.20	5	3	4.20E-01	685.817	4	4	4.54E+01	4523.56	4	4	3.61E-01
5320.96	3	5	2.52E-01	686.336	4	2	1.95E+01	4530.86	4	2	1.12E-01
5327.76	5	5	4.65E-02	763.334	2	2	9.58E+00	4534.58	6	6	2.01E-01
5338.73	5	7	1.85E-01	764.351	4	2	1.85E+01	4547.30	6	4	3.33E-02
5340.21	7	5	2.59E-01	771.545	2	4	8.19E+00	4634.13	2	4	6.36E-01
5351.23	7	7	3.67E-01	771.901	4	4	1.64E+01	4640.64	4	6	7.60E-01
5383.72	3	5	3.31E-03	772.384	6	4	2.45E+01	4641.85	4	4	1.26E-01
5452.07	1	3	8.89E-02	772.889	6	4	2.09E+01	4858.70	2	4	4.35E-01
5454.22	3	1	3.34E-01	772.955	4	2	2.34E+01	4858.98	4	6	4.66E-01
5462.58	3	3	1.00E-01	979.832	4	4	8.84E+00	4861.27	6	8	5.32E-01
5478.09	3	5	4.75E-02	979.905	6	6	9.21E+00	4867.12	4	4	1.73E-01
5480.05	5	3	1.30E-01	989.799	2	4	4.18E+00	4867.17	8	10	6.18E-01
5495.65	5	5	2.40E-01	991.511	4	4	8.17E-01	4873.60	6	6	1.50E-01
5526.23	3	5	2.13E-01	991.577	4	6	4.97E+00	4881.78	6	4	1.22E-02
5530.24	5	7	4.04E-01	1747.85	2	4	1.28E+00	4884.14	8	8	8.71E-02
5535.35	7	9	6.04E-01	1751.22	4	4	2.48E-01	4896.58	8	6	5.86E-03
5535.38	3	3	4.53E-01	1751.66	4	6	1.51E+00	5260.86	2	2	2.80E-02
5540.06	3	1	6.03E-01	2972.55	2	2	6.67E-01	5270.57	2	4	6.95E-02
5543.47	5	5	3.51E-01	2977.33	4	2	3.32E-01	5272.68	4	2	1.39E-01

λ (Å)	Weights g_i	Weights g_k	A (10^8 s⁻¹)	λ (Å)	Weights g_i	Weights g_k	A (10^8 s⁻¹)	λ (Å)	Weights g_i	Weights g_k	A (10^8 s⁻¹)
5282.43	4	4	2.21E-02	4769.86	5	3	2.50E-02	935.193	5	5	1.33E+00
5297.75	4	6	4.93E-02	4786.92	7	7	8.79E-02	1028.16	1	3	4.22E-01
5298.95	6	4	7.38E-02	4796.66	7	5	1.53E-02	1152.15	5	5	5.28E+00
5314.36	6	6	1.14E-01	5200.41	3	5	2.67E-01	1217.65	1	3	2.06E+00
5320.87	6	8	5.68E-01	5204.28	5	7	3.55E-01	1302.17	5	3	3.41E+00
5327.19	4	6	5.29E-01	5205.15	1	3	1.97E-01	1304.86	3	3	2.03E+00
5352.46	6	6	3.72E-02	5226.70	3	3	1.46E-01	1306.03	1	3	6.76E-01
6365.84	2	2	2.18E-01	5245.60	5	5	8.66E-02	3823.41	7	7	6.63E-03
6394.75	2	4	2.15E-01	5272.35	5	3	9.48E-03	3823.87	5	3	1.87E-03
6445.34	2	4	8.89E-02	5288.25	5	3	3.22E-02	3824.35	5	5	5.19E-03
6450.79	2	2	1.77E-01	5736.93	3	5	1.84E-01	3825.02	3	3	5.59E-03
6454.08	4	6	1.49E-01	5776.31	1	3	1.85E-02	3825.19	5	7	8.31E-04
6463.09	4	4	1.13E-01	5784.76	3	1	5.51E-02	3855.01	5	5	1.63E-02
6467.02	6	8	2.11E-01	5795.09	3	3	1.37E-02	3947.29	5	7	4.91E-03
6468.57	4	2	3.52E-02	5812.31	3	5	1.36E-02	3947.48	5	5	4.88E-03
6478.76	6	6	6.31E-02	5826.43	5	3	2.25E-02	3947.59	5	3	4.87E-03
6487.84	6	4	1.05E-02	5843.84	5	5	4.01E-02	3951.93	3	1	3.10E-03
7371.51	4	4	3.53E-02	6380.75	1	3	1.42E-01	3952.98	5	3	1.29E-03
7404.54	6	6	3.61E-02	7103.24	1	3	6.28E-02	3953.00	1	3	1.03E-03
8307.51	2	4	1.65E-02	7109.35	3	5	8.46E-02	3954.52	3	5	7.73E-04
8344.95	2	2	6.52E-02	7111.28	3	3	4.70E-02	3954.61	5	5	2.32E-03
8386.39	4	4	8.03E-02	*7116.8	9	15	1.12E-01	3997.95	5	3	2.41E-02
8424.56	4	2	3.17E-02	7122.98	5	7	1.12E-01	4217.09	3	1	5.44E-03
				7127.25	5	5	2.80E-01	4222.77	5	3	2.26E-03
N IV				7127.25	5	5	3.11E-03	4222.82	1	3	1.81E-03
247.205	1	3	1.19E+02	9165.07	3	5	4.23E-02	4233.27	5	5	4.04E-03
*283.52	9	15	3.05E+02	9182.16	5	7	4.45E-02	4368.19	3	1	7.56E-03
*322.64	9	3	8.99E+01	9222.99	7	9	4.95E-02	4368.24	3	5	7.59E-03
335.047	3	5	1.845E+02	9247.04	5	5	7.66E-03	4967.38	3	5	4.43E-03
387.356	3	1	2.55E+01	9311.55	7	7	5.36E-03	4967.88	5	7	8.44E-03
765.147	1	3	2.320E+01					4968.79	7	9	1.27E-02
*923.16	9	9	1.759E+01	**N V**				5019.29	5	5	7.13E-03
955.334	3	1	2.919E+01	*209.29	2	6	1.21E+02	5020.22	7	5	9.98E-03
1718.55	3	5	2.321E+00	*247.66	6	10	4.26E+02	5329.11	3	5	9.48E-03
2649.88	3	3	1.07E+00	1238.82	2	4	3.40E+00	5329.69	5	7	1.81E-02
3052.20	1	3	1.33E-01	1242.80	2	2	3.37E+00	5330.74	7	9	2.71E-02
3059.60	3	3	3.95E-01	4603.74	2	4	4.14E-01	5435.18	3	5	7.74E-03
3075.19	5	3	6.48E-01	4619.97	2	2	4.10E-01	5435.77	5	5	1.29E-02
3443.61	3	5	3.46E-01					5436.86	7	5	1.80E-02
3445.22	1	3	4.60E-01	**N VI**				5512.60	3	5	2.69E-03
3454.65	3	3	3.42E-01	24.8980	1	3	5.158E+03	5512.77	5	7	3.58E-03
3461.36	3	1	1.36E+00	28.7870	1	3	1.809E+04	5554.83	3	3	5.83E-03
3463.36	5	5	1.02E+00	*161.220	3	9	2.859E+02	5555.00	5	3	9.71E-03
3474.53	5	3	5.61E-01	173.275	1	3	2.697E+02	5958.39	3	5	6.80E-03
3478.72	3	5	1.06E+00	*173.93	9	15	8.756E+02	5958.58	5	7	9.06E-03
*3480.8	3	9	1.06E+00	185.192	3	5	8.205E+02	6046.23	3	3	1.05E-02
3483.00	3	3	1.06E+00	*1901	3	9	6.780E-01	6046.44	5	3	1.75E-02
3484.93	3	1	1.06E+00	2896.4	1	3	2.079E-01	6046.49	1	3	3.50E-03
3689.94	3	1	9.10E-02	*6991.1	3	9	8.384E-02	6155.99	3	5	2.67E-02
3694.14	3	3	2.27E-02	9622.0	1	3	3.276E-02	6156.78	5	7	5.08E-02
3707.39	5	5	6.73E-02					6158.19	7	9	7.62E-02
3714.43	5	5	1.34E-02	**Oxygen**				6324.84	7	5	3.76E-05
3735.43	7	5	7.37E-02	**O I**				6453.60	3	5	1.65E-02
3747.54	3	5	9.92E-01	791.973	5	5	4.94E+00	6454.44	5	5	2.75E-02
4057.76	3	5	6.62E-01	792.938	1	3	2.19E+00	6455.98	7	5	3.85E-02
4740.26	3	5	1.53E-02	792.967	3	5	1.64E+00	6726.28	5	5	1.18E-05
4747.96	3	3	7.60E-02	877.798	5	3	2.85E+00	6726.54	5	3	6.44E-06
4752.49	5	7	1.13E-02	877.879	5	5	5.12E+00	7001.92	3	5	2.65E-02
4762.09	5	5	6.99E-02	922.008	5	7	1.23E+00	7002.23	5	7	3.53E-02

λ Å	Weights g_i	Weights g_k	A 10^8 s⁻¹	λ Å	Weights g_i	Weights g_k	A 10^8 s⁻¹	λ Å	Weights g_i	Weights g_k	A 10^8 s⁻¹
7254.15	3	3	2.24E-02	2418.46	6	6	2.30E-01	4104.72	4	6	3.14E-01
7254.45	5	3	3.73E-02	2425.57	6	6	1.77E-01	4104.99	4	4	9.14E-01
7254.53	1	3	7.45E-03	2433.54	2	4	4.21E-01	4106.02	8	6	1.70E-02
7771.94	5	7	3.69E-01	2436.06	4	4	1.69E-01	4109.84	6	6	1.21E-02
7774.17	5	5	3.69E-01	2444.25	4	4	7.56E-02	4110.19	6	4	2.54E-01
7775.39	5	3	3.69E-01	2445.53	4	6	4.98E-01	4110.79	4	2	7.70E-01
7981.94	3	3	2.33E-04	2517.96	4	6	7.72E-02	4112.02	6	6	1.81E-01
7982.40	1	3	3.09E-04	2523.21	2	2	9.63E-02	4113.83	8	6	2.41E-01
7986.98	3	5	4.19E-04	2526.87	4	4	1.20E-01	4119.22	6	8	1.33E+00
7987.33	5	5	1.41E-04	2530.28	6	8	8.16E-02	4120.28	6	6	2.15E-01
7995.07	5	7	5.63E-04	2571.46	2	4	1.15E-01	4120.55	6	4	2.60E-01
8221.82	7	7	2.89E-01	2575.28	4	6	1.37E-01	4121.46	2	2	5.60E-01
8227.65	5	3	8.13E-02	3134.73	8	6	1.23E+00	4129.32	4	2	1.79E-01
8230.00	5	5	2.26E-01	3273.43	8	6	9.99E-01	4132.80	2	4	9.13E-01
8233.00	3	3	2.43E-01	3377.15	2	2	1.27E+00	4140.70	4	4	4.09E-02
8235.35	3	5	4.86E-02	3390.21	2	4	1.22E+00	4153.30	4	6	7.91E-01
8446.25	3	1	3.22E-01	3407.28	6	6	1.02E+00	4156.53	6	4	2.11E-01
8446.36	3	5	3.22E-01	3712.74	2	4	2.84E-01	4169.22	6	6	2.71E-01
8446.76	3	3	3.22E-01	3727.32	4	4	5.81E-01	4185.44	6	8	1.91E+00
8820.42	5	7	2.93E-01	3749.48	6	4	9.31E-01	4189.58	8	8	7.06E-02
9260.81	3	1	4.46E-01	3833.07	6	8	1.02E-02	4189.79	8	10	1.98E+00
9260.85	3	3	3.34E-01	3842.81	2	4	7.45E-02	4192.51	6	4	3.21E-01
9260.94	3	5	1.56E-01	3843.58	4	6	3.55E-02	4196.27	4	4	3.56E-02
9262.58	5	3	1.11E-01	3847.89	2	2	1.95E-01	4196.70	4	2	3.56E-01
9262.67	5	5	2.60E-01	3850.80	4	6	6.00E-03	4317.14	2	4	3.70E-01
9262.78	5	7	2.97E-01	3851.03	4	6	1.59E-01	4319.63	4	6	2.55E-01
9265.83	7	5	2.97E-02	3851.47	8	8	2.72E-02	4319.87	2	2	5.62E-01
9265.93	7	7	1.48E-01	3856.13	4	2	2.28E-01	4325.76	2	2	1.47E-01
9266.01	7	9	4.45E-01	3857.16	6	6	6.59E-02	4327.46	6	6	6.76E-01
9482.89	5	3	2.34E-01	3863.50	6	8	6.49E-02	4327.85	6	4	7.24E-02
9622.11	5	3	5.22E-04	3864.13	2	2	9.12E-02	4328.59	4	2	1.12E+00
9622.16	3	3	1.57E-03	3864.43	6	6	2.15E-01	4331.47	4	6	4.82E-02
9625.26	7	5	3.25E-04	3864.67	6	4	1.80E-01	4331.86	4	4	6.50E-01
9625.30	7	7	1.85E-03	3874.09	2	4	3.26E-02	4336.86	4	4	1.57E-01
9694.66	5	7	4.54E-04	3875.80	8	6	3.38E-02	4345.56	4	2	8.31E-01
9694.91	5	5	4.54E-04	3882.19	8	8	5.50E-01	4347.22	6	4	1.19E-01
9695.06	5	3	4.54E-04	3882.45	4	4	8.94E-02	4347.41	4	4	9.32E-01
				3883.14	8	6	1.13E-01	4349.43	6	6	6.91E-01
O II				3893.52	4	6	1.89E-02	4351.26	6	6	9.89E-01
429.918	4	2	4.25E+01	3907.45	6	6	8.64E-02	4351.46	4	6	5.82E-02
430.041	4	4	4.13E+01	3911.96	6	4	1.09E+00	4359.40	4	6	1.44E-02
430.176	4	6	4.36E+01	3912.12	4	4	1.41E+00	4366.89	6	4	3.98E-01
483.760	4	2	2.05E+01	3919.27	4	2	1.22E+00	4369.27	4	4	3.57E-01
483.980	6	4	1.80E+01	3945.04	2	4	2.05E-01	4395.93	6	6	3.91E-01
484.027	4	4	3.22E+00	3954.36	2	2	8.57E-01	4405.98	6	4	4.30E-02
485.087	6	8	2.60E+01	3973.26	4	4	1.04E+00	4414.90	4	6	8.34E-01
485.470	6	6	1.20E+00	3982.71	4	2	4.27E-01	4416.97	2	4	7.13E-01
485.518	4	6	1.93E+01	4069.62	2	4	1.52E+00	4443.01	6	6	5.05E-01
2290.85	2	4	7.41E-02	4069.88	4	6	1.53E+00	4443.52	6	8	1.89E-02
2293.30	2	2	3.25E-01	4072.15	6	8	1.98E+00	4447.68	8	6	2.52E-02
2300.33	4	4	4.17E-01	4075.86	8	10	2.11E+00	4448.19	8	8	5.10E-01
2302.81	4	2	1.67E-01	4078.84	4	4	5.52E-01	4452.38	4	4	1.37E-01
2365.14	4	2	1.52E-01	4084.65	6	8	7.28E-02	4466.24	2	4	9.00E-01
2375.72	6	4	1.35E-01	4085.11	6	6	4.55E-01	4467.46	2	2	9.00E-01
2406.38	6	4	1.85E-01	4092.93	8	8	2.65E-01	4563.18	4	4	7.18E-03
2407.48	4	4	2.25E-01	4094.14	6	4	4.70E-02	4590.97	6	8	8.85E-01
2411.60	4	2	2.05E-01	4096.53	4	6	1.73E-01	4595.96	6	6	4.87E-02
2411.64	2	2	1.10E-01	4097.22	2	4	3.62E-01	4596.18	4	6	8.34E-01
2415.13	4	2	2.20E-01	4103.00	2	2	5.09E-01	4638.86	2	4	3.71E-01

λ Å	Weights g_i	g_k	A $10^8 s^{-1}$	λ Å	Weights g_i	g_k	A $10^8 s^{-1}$	λ Å	Weights g_i	g_k	A $10^8 s^{-1}$
4641.81	4	6	5.96E-01	263.817	5	7	5.97E+01	3017.62	7	7	5.38E-01
4649.13	6	8	7.81E-01	263.861	5	5	1.49E+01	3023.43	3	5	4.79E-01
4650.84	2	2	6.86E-01	277.386	5	7	9.43E+01	3024.36	7	5	9.39E-02
4661.63	4	4	4.10E-01	279.788	5	5	4.25E+01	3024.54	1	3	6.16E-01
4673.73	4	2	1.35E-01	295.942	1	3	5.56E+01	3035.41	3	3	4.59E-01
4676.23	6	6	2.05E-01	303.413	1	3	4.29E+01	3042.07	3	1	1.94E+00
4690.89	2	4	1.86E-01	303.461	3	1	1.29E+02	3047.10	5	3	1.49E+00
4691.42	2	2	7.43E-01	303.517	3	3	3.21E+01	3059.28	5	3	8.72E-01
4696.35	6	4	3.25E-02	303.622	3	5	3.21E+01	3064.98	1	3	2.17E-01
4698.44	6	6	6.59E-02	303.695	5	3	5.34E+01	3068.13	3	1	6.49E-01
4699.01	6	8	9.88E-01	303.800	5	5	9.61E+01	3068.26	3	3	5.41E-02
4699.22	4	6	9.36E-01	305.596	1	3	1.20E+02	3068.67	3	5	2.27E-01
4701.18	4	4	9.23E-01	305.656	3	5	1.62E+02	3074.14	5	7	1.84E-01
4701.71	4	2	3.69E-01	305.702	3	3	9.01E+01	3074.72	5	3	3.76E-01
4703.16	4	6	9.20E-01	305.767	5	7	2.16E+02	3075.13	5	5	1.61E-01
4705.35	6	8	1.10E+00	305.836	5	5	5.40E+01	3075.95	7	9	1.07E-01
4710.01	4	6	2.98E-01	320.978	5	7	2.17E+02	3083.65	7	7	3.20E-01
4741.70	6	6	4.71E-02	328.448	5	5	1.04E+02	3084.64	7	5	2.55E-01
4751.28	6	8	6.39E-02	345.312	1	3	1.35E+02	3088.04	9	9	5.30E-01
4752.69	6	6	1.45E-02	374.073	5	5	2.85E+01	3095.79	9	7	1.35E-01
4844.92	4	6	1.02E-02	395.557	5	3	2.80E+01	3115.67	3	1	1.39E+00
4856.39	4	6	5.58E-02	507.388	1	3	1.61E+01	3121.63	3	3	1.38E+00
4856.76	4	4	1.00E-01	507.680	3	3	4.82E+01	3132.79	3	5	1.37E+00
4860.97	2	4	4.70E-01	508.178	5	3	8.04E+01	3198.18	3	5	9.57E-02
4864.88	4	2	8.07E-02	525.794	5	3	9.60E+01	3201.14	3	3	4.77E-01
4871.52	4	6	5.60E-01	597.814	1	3	1.49E+01	3202.51	5	7	7.08E-02
4872.02	4	4	9.34E-02	599.590	5	5	5.41E+01	3207.61	5	5	4.40E-01
4890.86	4	2	4.80E-01	702.337	1	3	6.06E+00	3210.58	5	3	1.58E-01
4906.83	4	4	4.54E-01	702.838	3	1	1.83E+01	3216.07	7	7	5.58E-01
4924.53	4	6	5.43E-01	832.929	1	3	3.41E+00	3221.21	7	5	9.75E-02
4941.07	2	4	5.87E-01	835.092	5	5	1.44E+00	3260.86	5	7	1.68E+00
4943.01	4	6	7.78E-01	835.289	5	7	5.99E+00	3265.33	7	9	1.88E+00
4955.71	4	4	1.82E-01	1679.03	3	5	6.57E-01	3267.20	3	5	1.58E+00
5159.94	2	2	3.29E-01	1686.73	3	3	6.48E-01	3281.83	5	5	2.89E-01
5175.90	4	2	1.49E-01	1760.41	3	5	8.38E-01	3284.45	7	7	2.06E-01
5190.50	2	4	1.26E-01	1764.46	5	5	2.50E+00	3299.39	1	3	1.64E-01
5206.65	4	4	3.58E-01	1766.63	1	3	1.11E+00	3312.33	3	3	4.60E-01
5583.22	2	4	2.17E-02	1772.28	3	1	3.29E+00	3326.06	3	3	2.65E-01
5611.07	2	2	2.14E-02	1772.97	5	3	1.37E+00	3330.30	3	5	6.81E-01
6627.37	4	4	1.73E-01	2390.43	3	3	1.62E+00	3330.32	3	5	4.76E-01
6641.03	2	2	9.88E-02	2454.97	3	1	3.43E+00	3332.41	5	3	7.92E-01
6666.66	4	2	6.78E-02	2665.68	3	5	6.75E-01	3332.93	5	7	5.04E-01
6677.87	2	4	3.37E-02	2674.58	5	5	1.11E+00	3336.67	3	3	3.76E-01
6717.75	2	2	1.33E-01	2683.66	3	1	1.85E+00	3336.69	5	5	8.77E-02
6721.39	4	2	1.81E-01	2686.15	7	5	1.54E+00	3340.76	5	3	6.57E-01
6810.48	6	8	1.64E-03	2687.55	3	3	1.84E+00	3344.20	5	5	1.25E-01
6844.10	4	6	2.97E-03	2695.48	3	5	1.82E+00	3344.51	5	7	3.48E-01
6846.80	8	8	3.17E-02	2794.14	3	1	1.82E-01	3347.98	7	5	4.86E-01
6869.48	6	6	5.35E-02	2798.93	3	3	4.52E-02	3350.62	5	3	1.12E+00
6884.88	4	4	6.12E-02	2809.66	5	3	1.34E-01	3350.92	7	7	9.91E-01
6895.10	10	8	2.72E-01	2818.70	5	5	2.66E-02	3355.86	7	7	6.89E-01
6906.44	8	6	2.48E-01	2836.31	7	5	1.46E-01	3362.31	7	5	6.87E-01
6907.87	4	2	3.03E-01	2959.69	3	5	1.83E+00	3376.61	3	1	1.49E+00
6910.56	6	4	2.43E-01	2983.78	3	5	2.15E+00	3376.76	3	3	1.12E+00
				2992.08	3	5	9.32E-02	3377.26	3	5	5.20E-01
O III				2996.48	3	3	4.64E-01	3382.61	5	7	9.86E-01
263.694	1	3	3.32E+01	2997.69	5	7	6.88E-02	3383.31	5	3	3.70E-01
263.727	3	5	4.48E+01	3004.34	5	5	4.27E-01	3383.81	5	5	8.62E-01
263.773	3	3	2.49E+01	3008.78	5	3	1.53E-01	3384.90	7	9	1.48E+00

λ Å	g_i	g_k	A 10^8 s^{-1}	λ Å	g_i	g_k	A 10^8 s^{-1}	λ Å	g_i	g_k	A 10^8 s^{-1}
3394.22	7	7	4.88E-01	4440.09	5	3	4.42E-01	2816.53	4	4	5.74E-01
3395.43	7	5	9.75E-02	4447.69	5	5	4.40E-01	2829.17	8	6	1.56E-01
3406.88	1	3	1.93E-01	4461.61	5	7	4.36E-01	2836.27	6	4	8.43E-01
3408.13	3	1	5.79E-01	4524.22	3	1	3.38E-01	2916.31	2	4	1.06E+00
3415.26	3	3	1.44E-01	4532.78	5	3	1.40E-01	2921.46	4	6	1.27E+00
3428.63	3	5	1.42E-01	4535.29	3	3	8.40E-02	2926.18	4	4	2.11E-01
3430.57	5	3	2.37E-01	4555.39	5	5	2.49E-01	3063.43	2	4	1.30E+00
3444.05	5	5	4.21E-01	4557.91	3	5	8.27E-02	3071.60	2	2	1.29E+00
3446.68	3	5	9.71E-01	5268.30	1	3	3.50E-01	3177.89	2	4	7.59E-02
3447.15	1	3	8.09E-01	5508.24	5	5	1.06E-01	3180.77	2	2	1.51E-01
3447.97	5	7	1.19E+00	5592.25	3	3	3.27E-01	3180.99	4	6	7.06E-02
3450.91	7	9	1.44E+00					3185.74	4	4	1.21E-01
3451.30	3	3	8.06E-01	**O IV**				3188.22	6	8	4.28E-02
3454.84	5	5	6.89E-01	238.360	2	4	2.96E+02	3188.64	4	2	1.50E-01
3454.99	9	11	1.72E+00	238.570	4	6	3.54E+02	3194.78	6	6	1.71E-01
3459.48	5	3	1.14E-01	238.579	4	4	5.90E+01	3199.58	6	4	1.04E-01
3459.94	7	7	5.14E-01	279.631	2	2	2.68E+01	3209.65	8	8	2.53E-01
3466.13	9	9	2.84E-01	279.933	4	2	5.34E+01	3216.31	8	6	5.56E-02
3466.85	7	5	6.82E-02	553.329	2	4	1.22E+01	3348.06	2	4	8.51E-01
3475.24	9	7	2.42E-02	554.076	2	2	4.86E+01	3349.11	4	6	1.02E+00
3520.94	1	3	1.50E-01	554.513	4	4	6.06E+01	3354.27	4	2	7.71E-01
3531.22	3	1	4.45E-01	555.263	4	2	2.41E+01	3362.55	4	4	7.65E-01
3533.38	3	3	1.11E-01	608.397	2	2	1.21E+01	3375.40	4	6	7.56E-01
3534.90	3	5	1.11E-01	609.829	4	2	2.40E+01	3378.02	4	4	1.66E-01
3555.24	5	3	1.82E-01	616.952	6	4	2.60E+01	3381.21	4	6	7.19E-01
3556.78	5	5	3.26E-01	617.005	4	4	2.89E+00	3381.30	2	4	4.28E-01
3695.38	3	5	4.01E-01	617.036	4	2	2.89E+01	3385.52	6	8	1.02E+00
3698.72	5	7	7.62E-01	624.619	2	4	1.07E+01	3390.19	2	2	8.49E-01
3703.36	7	9	1.14E+00	625.127	4	4	2.13E+01	3396.80	4	4	5.40E-01
3704.75	3	3	8.53E-01	625.853	6	4	3.19E+01	3405.77	4	2	1.67E-01
3707.27	3	5	7.34E-01	779.736	6	4	1.46E+00	3409.70	6	6	3.00E-01
3709.54	3	1	1.13E+00	779.820	4	4	1.31E+01	3411.30	4	4	1.69E-01
3712.49	5	5	6.59E-01	779.912	6	6	1.36E+01	3411.69	4	6	1.02E+00
3714.03	3	3	4.06E-01	779.997	4	6	9.70E-01	3425.55	6	4	4.94E-02
3715.09	5	7	9.73E-01	787.710	2	4	5.95E+00	3489.89	4	6	7.29E-01
3720.89	7	7	3.74E-01	790.112	4	4	1.18E+00	3492.21	2	4	6.06E-01
3721.95	5	3	2.80E-01	790.199	4	6	7.08E+00	3493.43	4	4	1.21E-01
3725.31	5	5	2.41E-01	921.296	2	4	2.21E+00	3560.39	4	6	1.03E+00
3728.51	5	7	1.29E+00	921.365	2	2	8.83E+00	3563.33	6	8	1.10E+00
3728.84	7	9	1.45E+00	923.367	4	4	1.10E+01	3593.08	6	6	7.15E-02
3729.80	3	5	1.22E+00	923.436	4	2	4.39E+00	3725.89	2	4	5.61E-01
3732.13	5	3	2.67E-02	1338.61	2	4	2.17E+00	3725.94	4	6	6.01E-01
3734.83	7	5	7.40E-02	1342.99	4	4	4.29E-01	3729.03	6	8	6.86E-01
3742.63	5	5	2.24E-01	1343.51	4	6	2.57E+00	3736.68	4	4	2.23E-01
3746.90	7	7	1.59E-01	2120.58	2	2	1.05E+00	3736.85	8	10	7.95E-01
3754.70	3	5	7.53E-01	2132.64	4	4	1.29E+00	3744.89	6	6	1.92E-01
3757.23	1	3	5.56E-01	2493.39	2	4	1.18E+00	3758.39	8	8	1.11E-01
3759.88	5	7	9.79E-01	2493.75	4	6	8.48E-01	3930.68	2	2	3.80E-02
3774.03	3	3	3.91E-01	2493.99	2	2	6.09E-01	3942.06	2	4	9.42E-02
3791.28	5	5	2.24E-01	2499.27	2	2	4.68E-01	3945.31	4	2	1.88E-01
3810.98	5	3	2.37E-02	2501.81	4	4	3.73E-01	3956.77	4	4	2.98E-02
3816.75	5	3	9.63E-02	2507.73	4	2	2.32E+00	3974.58	4	6	6.62E-02
3961.57	5	7	1.25E+00	2509.22	6	6	1.94E+00	3977.09	6	4	9.91E-02
4072.64	1	3	3.37E-01	2510.58	4	2	1.19E+00	3995.08	6	6	1.52E-01
4073.98	3	5	4.54E-01	2517.37	6	4	1.24E+00	4687.03	2	4	2.79E-01
4081.02	5	7	6.02E-01	2781.22	2	2	1.03E-01	4772.60	2	4	1.23E-01
4089.30	3	3	2.49E-01	2803.57	6	4	1.26E-01	4779.10	2	2	2.45E-01
4103.07	5	5	1.48E-01	2805.87	2	4	2.90E-01	4783.42	4	6	2.06E-01
4118.60	5	3	1.63E-02	2812.50	6	6	3.58E-02	4794.18	4	4	1.56E-01

λ Å	Weights g_i	g_k	A 10^8 s^{-1}	λ Å	Weights g_i	g_k	A 10^8 s^{-1}	λ Å	Weights g_i	g_k	A 10^8 s^{-1}
4798.27	6	8	2.91E-01	4213.35	5	3	1.19E-02	2149.1	4	2	3.18E+00
4813.15	6	6	8.65E-02	4522.66	5	3	1.02E-02	2152.9	2	4	4.85E-01
5305.51	4	4	6.10E-02	4554.53	3	5	2.41E-01	2154.1	4	4	1.73E-01
5362.51	6	6	6.12E-02	5114.06	1	3	1.80E-01	2154.1	4	6	5.8E-01
6876.49	2	4	1.88E-02	5339.94	1	3	1.85E-01	2534.0	2	4	2.00E-01
6931.60	2	2	7.35E-02	5349.74	3	1	7.04E-02	2535.6	4	4	9.5E-01
7004.11	4	4	8.90E-02	5372.71	3	3	1.42E-02	2553.3	2	2	7.1E-01
7061.30	4	2	3.48E-02	5414.59	3	5	9.29E-03	2554.9	4	2	3.00E-01
				5428.38	5	3	2.68E-02				
O V				5471.12	5	5	4.86E-02	**P II**			
172.169	1	3	2.94E+02	5571.81	1	3	8.33E-02	1301.9	1	3	5.0E-01
*192.85	9	15	6.90E+02	5580.12	3	5	1.11E-01	1304.5	3	1	1.5E+00
*215.17	9	3	1.83E+02	5583.23	3	3	6.20E-02	1304.7	3	3	3.7E-01
220.353	3	5	4.292E+02	*5589.9	9	15	1.49E-01	1305.5	3	5	3.8E-01
248.460	3	1	5.59E+01	5597.89	5	7	1.48E-01	1309.9	5	3	6.2E-01
629.732	1	3	2.872E+01	5604.27	5	5	3.68E-02	1310.7	5	5	1.1E+00
758.677	3	5	5.547E+00	5607.41	5	3	4.08E-03	4475.3	5	7	1.3E+00
759.442	1	3	7.373E+00	6330.05	5	7	1.21E-01	4499.2	5	7	1.4E+00
760.227	3	3	5.514E+00	6460.12	3	5	9.37E-02	4530.8	3	5	1.0E+00
760.446	5	5	1.652E+01	6466.14	5	7	1.01E-01	4554.8	3	5	9.6E-01
761.128	3	1	2.197E+01	6500.24	7	9	1.11E-01	4588.0	5	7	1.7E+00
762.004	5	3	9.125E+00	6543.77	5	5	1.64E-02	4589.9	3	5	1.6E+00
774.518	3	1	3.804E+01	6601.28	7	7	1.14E-01	4602.1	7	9	1.9E+00
1371.30	3	5	3.336E+00	6764.72	1	3	4.37E-02	4943.5	7	5	6.3E-01
2729.31	3	5	4.52E-01	6789.62	3	5	5.79E-02	5253.5	3	5	1.0E+00
2731.45	1	3	5.90E-01	6817.40	3	3	3.00E-02	5425.9	5	5	6.9E-01
2743.61	3	3	4.38E-01	6828.95	5	7	7.35E-02	6024.2	3	5	5.1E-01
2752.23	3	1	1.82E+00	6878.76	5	5	1.65E-02	6043.1	5	7	6.8E-01
2755.13	5	5	1.37E+00								
2769.69	5	3	7.88E-01	**O VI**				**P III**			
2781.01	3	5	1.40E+00	*150.10	2	6	2.62E+02	1334.8	2	4	5.5E-01
*2784.0	3	9	1.40E+00	*173.03	6	10	8.78E+02	1344.3	4	6	6.4E-01
2786.99	3	3	1.39E+00	1031.91	2	4	4.16E+00	1344.8	4	4	1.1E-01
2789.85	3	1	1.38E+00	1037.61	2	2	4.09E+00	4057.4	4	4	1.0E-01
3058.68	3	5	1.39E+00	3811.35	2	4	5.14E-01	4059.3	6	4	9.0E-01
3144.66	3	5	8.86E-01	3834.24	2	2	5.05E-01	4080.1	4	2	9.9E-01
3219.24	3	1	1.54E-01								
3222.29	1	3	1.16E-01	**O VII**				**Potassium**			
3227.54	3	3	3.38E-02	18.6270	1	3	9.365E+03	**K I**			
3239.21	3	3	3.28E-01	21.6020	1	3	3.309E+04	4044.1	2	4	1.24E-02
3248.28	5	3	1.18E-01	*120.33	3	9	5.334E+02	4047.2	2	2	1.24E-02
3263.54	5	3	1.86E-02	128.411	1	3	8.982E+02	5084.2	4	2	3.50E-03
3275.64	5	3	4.76E-01	*128.46	9	15	1.615E+03	5099.2	4	2	7.0E-03
3297.62	7	5	1.30E-01	135.820	3	5	1.523E+03	5323.3	2	2	6.3E-03
3690.17	3	5	1.97E-02	*1630.3	3	9	7.935E-01	5339.7	4	2	1.26E-02
3698.36	3	3	1.03E-01	2448.98	1	3	2.514E-01	5343.0	2	4	4.0E-03
3702.72	5	7	1.41E-02	*5933.1	3	9	1.002E-01	5359.6	4	6	4.6E-03
3717.31	5	5	9.63E-02	8241.76	1	3	3.864E-02	5782.4	2	2	1.23E-02
3725.63	5	3	2.91E-02					5801.8	4	2	2.46E-02
3746.64	7	7	1.18E-01	**Phosphorus**				5812.2	2	4	2.8E-03
3761.58	7	5	1.61E-02	**P I**				5831.9	4	6	3.2E-03
4119.37	3	5	3.66E-01	1671.7	4	2	3.9E-01	6911.1	2	2	2.72E-02
4120.49	3	1	3.33E-01	1674.6	4	4	4.0E-01	6938.8	4	2	5.4E-02
4123.96	5	7	4.81E-01	1679.7	4	6	3.9E-01	7664.9	2	4	3.87E-01
4125.49	1	3	2.70E-01	1775.0	4	6	2.17E+00	7699.0	2	2	3.82E-01
4134.11	3	3	3.34E-01	1782.9	4	4	2.14E+00				
4153.27	3	3	1.92E-01	1787.7	4	2	2.13E+00	**K II**			
4158.86	3	5	3.39E-01	2135.5	4	4	2.11E-01	607.93	1	3	1.3E-02
4178.46	5	5	1.12E-01	2136.2	6	4	2.83E+00				

λ Å	Weights g_i	g_k	A 10^8 s^{-1}	λ Å	Weights g_i	g_k	A 10^8 s^{-1}	λ Å	Weights g_i	g_k	A 10^8 s^{-1}
K III				3121.76	6	6	1.1E-01	4056.34	6	4	9.5E-03
2550.0	6	4	2.0E+00	3123.70	10	8	4.6E-02	4082.78	6	4	1.4E-01
2635.1	4	4	1.2E+00	3137.71	4	6	3.3E-02	4097.52	2	4	7.0E-02
2992.4	6	8	2.5E+00	3189.05	6	6	3.03E-01	4121.68	6	6	9.8E-02
3052.1	4	6	1.7E+00	3197.13	6	4	4.35E-02	4128.87	6	8	1.73E-01
3202.0	4	4	1.8E+00	3263.14	6	6	1.3E-01	4135.27	8	8	1.0E-01
3289.1	4	6	2.0E+00	3271.61	6	4	2.0E-01	4196.50	6	8	3.9E-02
3322.4	6	6	1.3E+00	3280.55	8	8	2.36E-01	4211.14	8	10	1.62E-01
3421.8	2	4	1.5E+00	3283.57	6	8	4.4E-01	4244.44	4	4	6.5E-03
				3289.14	4	4	1.0E-01	4278.60	4	6	9.2E-03
K XVI				3323.09	8	10	6.3E-01	4288.71	6	8	6.1E-02
206.27	1	3	9.4E+01	3331.09	4	2	5.40E-02	4373.04	2	4	1.8E-02
				3338.54	8	6	3.5E-02	4374.80	8	10	1.64E-01
K XVII				3360.80	4	4	1.2E-01	4379.92	6	6	2.48E-02
22.020	2	4	4.7E+04	3368.38	6	4	1.1E-01	4492.47	6	6	4.5E-03
22.163	4	6	5.6E+04	3396.82	10	10	6.5E-01	4528.72	6	8	1.35E-02
22.18	4	4	9.3E+03	3399.70	6	8	1.2E-01	4548.73	4	6	5.5E-03
22.60	2	2	2.5E+03	3462.04	6	6	6.2E-01	4551.64	4	4	4.00E-02
22.76	4	2	4.7E+03	3470.66	4	4	8.5E-01	4565.19	4	4	1.1E-02
				3478.91	6	6	3.32E-01	4569.00	6	8	1.0E-02
Praseodymium				3484.04	6	8	9.3E-03	4608.12	2	2	2.1E-02
Pr II				3498.73	4	6	2.12E-01	4675.03	8	8	6.4E-03
3997.0	15	15	1.87E-01	3502.52	10	10	4.3E-01	4721.00	6	4	3.43E-03
4062.8	13	15	1.00E+00	3507.32	6	8	3.4E-01	4745.11	6	6	5.2E-03
4100.7	17	19	8.4E-01	3528.02	8	8	8.5E-01	4755.58	4	4	6.0E-03
4143.1	15	17	5.8E-01	3543.95	4	4	4.65E-01	4842.43	6	8	1.6E-03
4179.4	13	15	5.2E-01	3549.54	6	6	2.22E-01	4963.71	2	2	3.0E-02
4222.9	11	13	3.91E-01	3570.18	4	6	1.82E-01	4977.75	4	4	9.8E-03
4241.0	17	15	2.30E-01	3583.10	8	10	2.6E-01	4979.18	4	6	1.0E-02
4359.8	15	15	1.1E-01	3596.19	6	4	5.5E-01	5090.63	6	6	5.0E-03
4405.8	17	17	9.0E-02	3597.15	6	8	5.9E-01	5120.69	6	8	3.1E-03
4429.3	15	15	2.28E-01	3612.47	4	2	8.90E-01	5130.76	4	4	4.35E-03
4449.8	13	13	1.24E-01	3620.46	6	4	8.5E-02	5155.54	2	4	9.8E-03
4468.7	11	13	1.54E-01	3654.87	8	8	6.0E-02	5184.19	6	8	1.6E-03
4510.2	13	15	1.16E-01	3657.99	8	6	8.8E-01	5212.73	4	2	5.95E-03
4534.2	15	17	4.9E-02	3666.22	6	8	8.4E-02	5292.14	10	10	3.7E-03
4734.2	15	13	2.5E-02	3690.70	6	4	3.23E-01	5390.44	4	6	9.5E-03
4879.1	15	15	1.8E-02	3692.36	10	8	9.1E-01	5424.72	4	4	5.0E-03
4886.0	15	15	1.3E-02	3700.91	8	10	3.9E-01	5599.42	6	8	1.3E-02
4912.6	17	15	5.7E-02	3713.02	4	4	8.3E-02	5983.60	10	10	2.1E-02
5034.4	19	19	1.1E-01	3788.47	4	6	1.4E-01				
5110.8	21	19	2.78E-01	3793.22	8	6	4.2E-01	**Rubidium**			
5135.1	17	17	1.25E-01	3799.31	8	8	5.5E-01	**Rb I**			
5173.9	19	17	3.18E-01	3806.76	6	6	6.2E-02	3022.5	2	4	4.13E-05
5219.1	15	15	9.5E-02	3818.19	6	4	5.8E-01	3032.0	2	4	4.93E-05
5220.1	17	15	2.35E-01	3822.26	6	6	8.5E-01	3044.2	2	4	8.2E-05
5251.7	15	13	1.1E-02	3828.48	6	6	6.2E-01	3060.2	2	4	1.05E-04
5259.7	15	13	2.24E-01	3833.89	6	4	5.8E-01	3082.0	2	4	1.49E-04
5292.6	13	13	9.3E-02	3856.52	8	10	5.9E-01	3112.6	2	4	2.5E-04
5810.6	17	19	2.3E-02	3872.39	4	6	6.7E-03	3113.1	2	2	1.3E-04
5879.3	15	15	7.6E-02	3877.34	8	6	3.7E-02	3157.5	2	4	3.38E-04
6200.8	15	17	1.8E-02	3913.51	8	8	2.5E-03	3158.3	2	2	2.0E-04
6278.7	13	15	2.6E-02	3922.19	4	2	6.25E-02	3228.0	2	4	6.4E-04
6398.0	11	13	1.9E-02	3934.23	8	8	1.58E-01	3229.2	2	2	3.8E-04
				3942.72	4	4	7.15E-01	3348.7	2	4	1.37E-03
Rhodium				3958.86	6	8	5.5E-01	3350.8	2	2	8.9E-04
Rh I				3984.40	4	4	1.1E-01	3587.1	2	4	3.97E-03
3083.96	8	6	4.8E-02	3995.61	4	6	4.7E-02	3591.6	2	2	2.9E-03
3114.91	6	4	4.45E-02	4053.44	2	2	2.8E-02	4201.8	2	4	1.8E-02

λ Å	Weights g_i	Weights g_k	A 10^8 s^{-1}	λ Å	Weights g_i	Weights g_k	A 10^8 s^{-1}	λ Å	Weights g_i	Weights g_k	A 10^8 s^{-1}
4215.5	2	2	1.5E-02	4093.12	4	4	1.23E-01	5339.43	6	6	1.06E-01
7800.3	2	4	3.70E-01	4094.86	6	6	1.44E-01	5341.07	4	2	3.8E-01
7947.6	2	2	3.40E-01	4098.36	8	8	8.7E-02	5349.34	6	4	5.9E-01
				4132.98	4	6	1.19E+00	5350.28	8	8	6.8E-02
Scandium				4140.27	6	8	1.17E+00	5355.79	6	4	3.0E-01
Sc I				4147.38	6	6	1.74E-01	5356.10	8	6	5.7E-01
2116.7	4	4	2.0E-01	4161.85	8	8	1.77E-01	5375.37	8	6	3.4E-01
2120.4	6	6	2.0E-01	4171.53	6	4	1.36E-01	5392.06	10	8	4.2E-01
2262.3	4	4	5.8E-02	4186.42	6	8	8.4E-02	5416.16	4	6	4.4E-02
2266.6	4	2	4.8E-01	4187.61	8	6	1.28E-01	5416.41	6	6	2.0E-02
2270.9	6	4	4.6E-01	4193.53	4	6	6.1E-02	5425.55	6	8	4.5E-02
2280.8	4	6	2.8E-01	4204.52	6	8	3.5E-02	5429.42	2	4	9.0E-02
2289.6	6	6	4.1E-02	4205.20	10	8	1.12E-01	5432.98	4	4	5.4E-02
2311.29	4	6	4.1E-02	4212.32	4	6	1.58E-01	5433.25	6	4	9.7E-02
2315.69	4	4	2.5E-01	4212.48	6	6	8.6E-02	5438.28	4	6	3.4E-02
2320.32	6	6	2.4E-01	4216.08	2	4	2.36E-01	5439.04	2	2	1.74E-01
2324.75	6	4	4.1E-01	4218.23	4	2	2.26E-01	5442.62	4	2	2.15E-01
2328.19	4	6	4.6E-01	4225.54	6	8	9.5E-02	5446.20	8	8	2.8E-01
2334.67	4	2	1.7E-01	4225.69	4	6	7.6E-02	5451.37	6	6	1.50E-01
2346.03	6	4	1.3E-01	4231.64	4	4	1.31E-01	5455.24	4	4	6.6E-02
2429.19	4	4	2.8E-01	4233.59	6	6	4.0E-01	5464.95	4	2	3.2E-02
2438.63	6	6	2.1E-01	4238.05	8	8	7.1E-01	5468.40	6	4	9.7E-02
2468.40	4	2	4.9E-02	4239.55	6	4	2.27E-01	5472.19	8	6	9.7E-02
2692.78	4	2	1.61E-01	4246.14	8	6	1.15E-01	5482.01	8	8	5.2E-01
2699.02	4	6	2.4E-02	4542.55	6	4	1.28E-01	5484.63	6	6	5.2E-01
2706.74	4	4	3.1E-01	4544.67	8	6	1.33E-01	5514.23	6	8	4.1E-01
2707.93	6	4	1.49E-01	4706.94	4	6	2.81E-01	5520.52	8	10	4.3E-01
2711.34	6	6	3.2E-02	4709.31	6	8	4.0E-01	5526.10	4	4	7.1E-02
2965.88	4	6	7.5E-02	4711.72	2	4	1.81E-01	5541.07	6	6	5.5E-02
2974.01	4	4	5.5E-01	4714.30	4	4	2.14E-01	5631.04	2	4	3.0E-02
2980.76	6	6	5.4E-01	4719.31	6	6	1.04E+00	5671.83	10	12	5.4E-01
2988.97	6	4	6.9E-02	4728.77	8	8	1.16E+00	5686.86	8	10	4.9E-01
3015.37	4	6	7.8E-02	4729.20	4	4	2.20E-01	5700.19	6	8	4.6E-01
3019.35	6	8	8.7E-01	4729.24	6	6	1.93E-01	5708.64	10	10	4.7E-02
3030.76	6	6	1.00E-01	4734.11	4	2	1.10E+00	5711.79	4	6	4.5E-01
3255.68	4	4	3.2E-01	4737.65	6	4	8.8E-01	5717.31	8	8	7.5E-02
3269.90	4	2	3.13E+00	4741.02	8	6	9.1E-01	5724.13	6	6	7.4E-02
3273.63	6	4	2.81E+00	4743.82	10	8	9.8E-01	5988.43	6	6	6.6E-02
3907.48	4	6	1.66E+00	4973.67	4	2	8.4E-01	6026.16	4	4	7.2E-02
3911.81	6	8	1.79E+00	4980.36	6	4	5.6E-01	6146.20	6	8	4.2E-02
3933.38	6	6	1.62E-01	4983.43	4	4	2.58E-01	6198.43	4	6	3.5E-02
3996.60	4	6	1.65E-01	4991.91	6	6	3.8E-01	6249.96	6	8	3.2E-02
4020.39	4	4	1.63E+00	4995.00	4	6	5.9E-02	6262.22	4	6	8.4E-01
4023.22	4	4	3.0E-01	5018.41	6	4	2.09E-01	6280.16	2	4	4.0E-02
4023.68	6	6	1.65E+00	5021.52	4	4	2.30E-01	6284.16	6	6	3.9E-02
4031.38	6	6	2.9E-01	5064.31	8	10	7.3E-02	6284.73	4	4	7.1E-02
4036.86	6	4	7.9E-02	5066.38	6	6	3.6E-02	6293.02	2	2	1.04E-01
4043.80	8	8	3.11E-01	5070.17	6	8	1.16E-01	7741.16	10	10	3.8E-02
4047.80	6	4	1.54E-01	5072.71	2	4	2.0E-02	7800.42	8	8	5.1E-02
4051.83	8	6	7.7E-02	5075.82	4	6	1.15E-01				
4054.54	4	2	1.67E-01	5080.22	4	4	4.1E-02	**Sc II**			
4067.00	6	8	1.91E-01	5081.56	10	10	7.6E-01	1880.6	5	3	5.0E+00
4067.63	10	8	4.1E-02	5083.72	8	8	6.2E-01	2064.3	7	5	2.2E+00
4074.96	4	6	3.7E-01	5085.55	6	6	5.7E-01	2068.0	5	3	2.0E+00
4078.56	2	4	4.3E-01	5086.94	4	4	6.6E-01	2273.1	1	3	7.7E+00
4080.57	4	4	6.6E-02	5096.72	6	4	1.69E-01	2545.20	5	5	4.0E-01
4082.39	6	4	2.73E-01	5099.27	4	6	1.50E-01	2552.35	7	5	2.21E+00
4086.66	6	8	3.7E-01	5101.12	10	8	8.8E-02	2555.79	3	3	6.9E-01
4087.47	4	6	1.12E-01	5331.79	4	4	1.11E-01	2560.23	5	3	2.01E+00

λ Å	Weights g_i	Weights g_k	A 10^8 s^{-1}	λ Å	Weights g_i	Weights g_k	A 10^8 s^{-1}	λ Å	Weights g_i	Weights g_k	A 10^8 s^{-1}
2563.19	3	1	2.70E+00	5526.79	9	7	3.3E-01	1260.4	2	4	2.0E+01
2611.19	5	5	2.2E+00	5657.91	5	5	1.04E-01	1264.7	4	6	2.3E+01
2667.70	3	5	1.5E+00	5669.06	3	1	1.31E-01	1304.4	2	2	3.6E+00
2746.36	3	1	3.9E+00					1309.3	4	2	7.0E+00
2782.31	5	5	1.3E+00	**Silicon**				1526.7	2	2	3.73E+00
2789.15	7	7	1.3E+00	**Si I**				1533.5	4	2	7.4E+00
2801.31	9	9	1.3E+00	1977.6	1	3	1.8E-01	1808.0	2	4	3.7E-02
2819.49	3	5	2.3E+00	1979.2	3	1	5.1E-01	2904.3	4	6	6.7E-01
2822.12	5	7	2.5E+00	1980.6	3	3	1.3E-01	2905.7	6	8	7.1E-01
2826.64	7	9	2.8E+00	1983.2	3	5	1.4E-01	3210.0	4	6	4.6E-01
2870.85	5	3	1.1E+00	1986.4	5	3	2.1E-01	4128.1	4	6	1.32E+00
2912.98	5	3	1.1E+00	1989.0	5	5	4.1E-01	4130.9	6	8	1.42E+00
2979.68	3	5	1.2E+00	2208.0	1	3	3.11E-01	5041.0	2	4	9.8E-01
2988.92	5	7	2.9E+00	2210.9	3	5	4.16E-01	5056.0	4	6	1.2E+00
3039.92	7	9	3.5E+00	2211.7	3	3	2.32E-01	5957.6	2	2	4.2E-01
3045.73	5	7	3.68E+00	2216.7	5	7	5.5E-01	5978.9	4	2	8.1E-01
3052.92	7	9	3.92E+00	2218.1	5	5	1.38E-01	6347.1	2	4	7.0E-01
3060.54	7	7	3.0E-01	2506.9	3	5	4.66E-01	6371.4	2	2	6.9E-01
3065.12	9	11	4.00E+00	2514.3	1	3	6.1E-01	7848.8	4	6	3.9E-01
3075.36	9	9	2.5E-01	2516.1	5	5	1.21E+00	7849.7	6	8	4.2E-01
3128.27	3	3	1.9E+00	2519.2	3	3	4.56E-01				
3133.07	5	5	1.8E+00	2524.1	3	1	1.81E+00	**Si III**			
3139.72	7	7	2.1E+00	2528.5	5	3	7.7E-01	883.40	5	7	6.3E+01
3190.98	3	3	1.1E+00	2532.4	1	3	2.6E-01	994.79	3	3	7.89E+00
3199.33	5	3	1.9E+00	2631.3	1	3	9.7E-01	997.39	5	3	1.31E+01
3312.72	5	7	1.2E+00	2881.6	5	3	1.89E+00	1141.6	3	5	3.0E-01
3320.40	5	3	1.2E+00	3905.5	1	3	1.18E+00	1144.3	5	7	3.9E+01
3343.23	9	7	1.1E+00	4738.8	3	3	1.0E-02	1161.6	5	5	1.6E+01
3353.72	5	7	1.51E+00	4783.0	5	5	1.7E-02	1206.5	1	3	2.59E+01
3359.67	5	5	2.16E-01	4792.3	5	5	1.7E-02	1206.5	3	5	4.89E+01
3361.26	3	3	3.4E-01	4818.1	5	7	1.1E-02	1207.5	5	5	1.9E+01
3361.93	3	1	1.17E+00	4821.2	3	5	8.0E-03	1294.5	3	5	5.42E+00
3368.94	5	3	8.3E-01	4947.6	3	1	4.2E-02	1296.7	1	3	7.19E+00
3372.15	7	5	9.9E-01	5006.1	3	5	2.8E-02	1298.9	3	3	5.36E+00
3379.16	3	3	2.5E+00	5622.2	3	3	1.6E-02	1299.0	5	5	1.61E+01
3535.71	5	3	6.1E-01	5690.4	3	3	1.2E-02	1301.2	3	1	2.13E+01
3558.53	5	7	3.0E-01	5708.4	5	3	1.4E-02	1303.3	5	3	8.85E+00
3567.70	3	5	3.5E-01	5754.2	5	3	1.5E-02	1328.8	1	3	2.7E+01
3572.53	7	7	1.38E+00	5772.1	3	1	3.6E-02	1417.2	3	1	2.60E+01
3576.34	5	5	1.06E+00	5948.5	3	5	2.2E-02	1435.8	5	7	2.1E+01
3580.93	3	3	1.23E+00	7226.2	3	5	7.9E-03	1589.0	5	3	1.1E+01
3589.63	5	3	4.6E-01	7405.8	3	5	3.7E-02	1778.7	7	9	4.4E+00
3590.47	7	5	2.9E-01	7409.1	5	7	2.3E-02	1783.1	5	7	3.8E+00
3613.83	7	9	1.48E+00	7680.3	3	5	4.6E-02	3241.6	5	3	2.3E+00
3630.74	5	7	1.20E+00	7918.4	3	5	5.2E-02	*3486.9	15	21	1.8E+00
3642.78	3	5	1.13E+00	7932.3	5	7	5.1E-02	3590.5	3	5	3.9E+00
3645.31	7	7	2.74E-01	7944.0	7	9	5.8E-02	4552.6	3	5	1.26E+00
3651.80	5	5	3.0E-01	7970.3	5	5	7.1E-03	4554.0	5	3	7.6E-01
3859.59	7	5	1.1E+00					4567.8	3	3	1.25E+00
4246.82	5	5	1.29E+00	**Si II**				4683.0	5	5	9.5E-01
4314.08	9	7	4.1E-01	989.87	2	4	6.7E+00	4716.7	5	7	2.8E+00
4320.75	7	5	4.0E-01	992.68	4	6	8.0E+00	5451.5	3	5	6.0E-01
4325.00	5	3	4.3E-01	1020.7	2	2	1.3E+00	5473.1	5	7	7.9E-01
4374.46	9	9	1.48E+00	1190.4	2	4	6.9E+00	5716.3	9	7	1.9E-01
4400.39	7	7	1.43E-01	1193.3	2	2	2.8E+01	5739.7	1	3	4.7E-01
4415.54	5	5	1.47E-01	1194.5	4	4	3.6E+01	7462.6	5	3	4.9E-01
4670.41	5	7	1.16E-01	1197.4	4	2	1.4E+01	7466.3	7	5	5.4E-01
5031.01	5	3	3.5E-01	1248.4	4	4	1.3E+01	7612.4	3	5	1.1E+00
5239.81	1	3	1.39E-01	1251.2	6	4	1.9E+01				

λ Å	Weights g_i	g_k	A 10^8 s^{-1}	λ Å	Weights g_i	g_k	A 10^8 s^{-1}	λ Å	Weights g_i	g_k	A 10^8 s^{-1}
Si IV				258.35	4	4	1.4E+02	4747.9	2	2	6.3E-03
457.82	2	4	3.6E+00	261.05	4	2	5.4E+01	4751.8	4	2	1.27E-02
458.16	2	2	3.6E+00	272.00	2	2	3.0E+01	4978.5	2	4	4.1E-02
515.12	2	2	4.1E+00	277.26	4	2	5.7E+01	4982.8	4	4	8.2E-03
516.35	4	2	8.2E+00	287.08	2	4	2.6E+01	4982.8	4	6	4.89E-02
*560.50	6	10	1.0E+00	289.19	4	4	5.0E+01	5148.8	2	2	1.17E-02
*749.94	10	14	1.45E+01	292.22	6	4	7.3E+01	5153.4	4	2	2.33E-02
815.05	2	2	1.23E+01	*347.73	10	10	4.3E+01	5682.6	2	4	1.03E-01
818.13	4	2	2.44E+01	*353.09	6	10	2.1E+01	5688.2	4	6	1.2E-01
*860.74	10	6	1.8E+00					5688.2	4	4	2.1E-02
*1066.6	10	14	3.91E+01	**Si XI**				5890.0	2	4	6.11E-01
1122.5	2	4	2.05E+01	43.763	1	3	6.11E+03	5895.9	2	2	6.10E-01
1128.3	4	4	4.03E+00	*49.116	9	3	2.45E+03	6154.2	2	2	2.6E-02
1128.3	4	6	2.42E+01	49.222	3	5	8.9E+03	6160.8	4	2	5.2E-02
1393.8	2	4	7.73E+00	52.296	3	1	7.6E+02	8183.3	2	4	4.53E-01
1402.8	2	2	7.58E+00	303.30	1	3	6.42E+02	8194.8	4	6	5.4E-01
*1724.1	10	6	5.5E+00	358.29	3	1	1.03E+02	8194.8	4	4	9.0E-02
				358.63	3	5	1.38E+01	11381	2	2	8.9E-02
Si V				361.41	1	3	1.80E+01	11404	4	2	1.76E-01
96.439	1	3	4.8E+02	364.50	3	3	1.32E+01				
97.143	1	3	2.0E+03	365.42	5	5	3.90E+01	**Na II**			
117.86	1	3	3.0E+02	368.28	3	1	5.1E+01	300.15	1	3	3.0E+01
				371.48	5	3	2.07E+01	301.44	1	3	4.9E+01
Si VI				604.14	3	5	1.12E+01	372.08	1	3	3.4E+01
246.00	4	2	1.7E+02	2300.8	1	3	4.34E-01				
249.12	2	2	8.5E+01					**Na III**			
				Si XII				378.14	4	2	7.7E+01
Si VII				*40.924	2	6	4.42E+03	380.10	2	2	3.7E+01
217.83	5	3	4.3E+02	*44.118	6	10	1.4E+04	1991.0	4	6	8.3E+00
272.64	5	3	5.1E+01	499.43	2	4	9.56E+00	2004.2	2	4	4.6E+00
274.18	3	1	1.2E+02	520.72	2	2	8.47E+00	2011.9	6	8	8.4E+00
275.35	5	5	8.9E+01	1862	2	4	1.15E+00	2151.5	2	4	4.4E+00
275.67	3	3	3.0E+01	1949	2	2	1.0E+00	2174.5	4	6	5.3E+00
276.84	1	3	3.9E+01	4620	2	4	4.6E-02	2230.3	6	8	3.7E+00
278.45	3	5	2.9E+01	4942	4	6	4.5E-02	2232.2	4	4	3.3E+00
								2246.7	4	6	2.4E+00
Si VIII				**Silver**				2459.3	4	6	3.0E+00
214.76	4	2	4.1E+02	**Ag I**				2468.9	2	4	2.4E+00
216.92	6	4	3.6E+02	2061.2	2	4	3.1E-02	2497.0	6	6	1.7E+00
232.86	2	2	8.0E+01	2069.9	2	2	1.5E-02				
235.56	4	4	9.7E+01	3280.7	2	4	1.4E+00	**Na V**			
250.45	2	2	7.7E+01	3382.9	2	2	1.3E+00	*307.89	10	6	2.0E+02
250.79	4	2	1.6E+02	5209.1	2	4	7.5E-01	*333.46	6	6	5.6E+01
314.31	4	2	5.2E+01	5465.5	4	6	8.6E-01	*369.01	10	6	1.2E+02
316.20	4	4	5.0E+01	5471.6	4	4	1.4E-01	*400.72	10	10	5.0E+01
319.83	4	6	4.9E+01					*445.14	6	10	7.1E+00
				Sodium				459.90	4	2	2.3E+01
Si IX				**Na I**				461.05	4	4	2.3E+01
223.73	1	3	4.2E+01	3302.4	2	4	2.81E-02	463.26	4	6	2.2E+01
225.03	3	3	1.2E+02	3303.0	2	2	2.81E-02	510.10	2	2	5.6E+01
227.01	5	3	2.0E+02	4390.0	2	4	7.7E-03	511.19	4	4	6.8E+01
227.30	5	3	2.3E+02	4393.3	4	4	1.6E-03				
258.10	5	5	1.04E+02	4393.3	4	6	9.2E-03	**Na VI**			
*294.37	9	9	5.9E+01	4494.2	2	4	1.2E-02	313.75	5	3	1.3E+02
*347.36	9	15	2.2E+01	4497.7	4	6	1.4E-02	361.25	5	5	7.7E+01
				4497.7	4	4	2.4E-02	*416.53	9	9	3.7E+01
Si X				4664.8	2	4	2.33E-02	*492.80	9	15	1.3E+01
253.77	2	4	2.9E+01	4668.6	4	4	4.1E-03	1550.6	5	5	4.35E+00
256.57	2	2	1.1E+02	4668.6	4	6	2.5E-02	1567.8	5	3	2.68E+00

λ Å	g_i	g_k	A 10^8 s^{-1}	λ Å	g_i	g_k	A 10^8 s^{-1}	λ Å	g_i	g_k	A 10^8 s^{-1}
1608.5	3	1	2.6E+00	4607.3	1	3	2.01E+00	4694.1	5	7	6.7E-03
1649.4	5	5	2.05E+00					4695.4	5	5	6.7E-03
1741.5	3	5	2.59E+00	**Sr II**				4696.2	5	3	6.5E-03
1747.5	5	7	3.1E+00	2018.7	2	2	1.2E-01	6403.6	3	5	5.7E-03
				2051.9	4	2	2.4E-01	6408.1	5	5	9.5E-03
Na VII				2282.0	2	4	8.3E-01	6415.5	7	5	1.3E-02
*94.409	6	10	2.7E+03	2322.4	4	6	9.1E-01	*6751.2	15	25	7.9E-02
*105.27	6	2	4.5E+02	2324.5	4	4	1.5E-01	7679.6	3	5	1.2E-02
353.29	4	4	1.0E+02	2423.5	2	2	2.4E-01	7686.1	5	5	2.0E-02
381.30	4	2	4.0E+01	2471.6	4	2	4.8E-01	7696.7	7	5	2.8E-02
397.49	4	4	3.5E+01	3464.5	4	6	3.1E+00				
399.18	6	4	5.2E+01	3474.9	4	4	5.1E+00	**S II**			
*483.28	10	10	2.9E+01	4077.7	2	4	1.42E+00	1124.4	2	4	1.0E+00
486.74	2	4	1.1E+01	4161.8	2	2	6.5E-01	1125.0	4	4	4.6E+00
491.95	4	6	1.3E+01	4215.5	2	2	1.27E+00	1131.0	2	2	3.5E+00
555.80	4	4	2.3E+01	4305.5	4	2	1.4E+00	1131.6	4	2	1.4E+00
777.83	4	6	6.8E+00	4414.8	4	6	1.1E-01	1250.5	4	2	4.6E-01
				4417.5	4	4	1.8E-02	1253.8	4	4	4.2E-01
Na VIII				4585.9	4	2	7.0E-02	1259.5	4	6	3.4E-01
*83.34	9	15	3.94E+03	5303.1	2	4	1.9E-01	4463.6	8	6	5.3E-01
*89.88	9	3	8.09E+02	5379.1	4	6	2.2E-01	4483.4	6	4	3.1E-01
90.536	3	5	2.86E+03	5385.5	4	4	3.7E-02	4486.7	4	2	6.6E-01
411.15	1	3	4.42E+01	5723.7	2	2	7.1E-02	4524.7	4	4	9.3E-02
1239.4	3	3	3.02E+00	5819.0	4	2	1.4E-01	4525.0	6	4	1.2E+00
1802.7	3	1	2.70E+00	8688.9	4	6	5.5E-01	4552.4	4	2	1.2E+00
1867.7	3	5	2.01E+00	8719.6	4	4	9.7E-02	4656.7	2	4	9.0E-02
2059.1	3	5	1.80E+00					4716.2	4	4	2.9E-01
2558.2	5	3	2.26E-02	**Sulfur**				4815.5	6	4	8.8E-01
2772.0	3	5	4.19E-01	**S I**				4885.6	2	4	1.7E-01
3021.0	5	7	4.90E-01	1295.7	5	5	4.9E+00	4917.2	2	2	6.6E-01
3108.9	1	3	2.58E-01	1296.2	5	3	2.7E+00	4924.1	4	6	2.2E-01
3182.3	1	3	2.92E-01	1302.3	3	5	1.8E+00	4925.3	2	4	2.4E-01
				1302.9	3	3	1.6E+00	4942.5	2	2	1.5E-01
Na IX				1303.1	3	1	6.6E+00	4991.9	4	4	1.5E-01
70.615	2	4	1.35E+03	1303.4	5	3	1.9E+00	5009.5	4	2	7.0E-01
70.653	2	2	1.35E+03	1305.9	1	3	2.4E+00	5014.0	4	4	8.4E-01
77.764	2	4	3.6E+03	1401.5	5	3	9.1E-01	5027.2	4	2	2.6E-01
77.911	4	6	4.3E+03	1409.3	3	3	5.0E-01	5032.4	6	6	8.1E-01
681.72	2	4	6.63E+00	1412.9	1	3	1.6E-01	5047.3	4	2	3.6E-01
694.17	2	2	6.30E+00	1425.0	5	7	4.5E+00	5103.3	6	4	5.0E-01
2487.7	2	4	8.32E-01	1425.2	5	5	1.2E+00	5142.3	2	2	1.9E-01
2535.8	2	2	7.89E-01	1433.3	3	5	3.3E+00	5201.0	4	4	7.5E-01
6841.8	2	4	2.59E-02	1433.3	3	3	1.9E+00	5201.3	6	4	6.5E-02
7103.4	4	6	2.78E-02	1437.0	1	3	2.4E+00	5212.6	4	6	9.8E-02
				1448.2	5	3	7.3E+00	5212.6	6	6	8.5E-01
Strontium				1473.0	5	7	4.2E-01	5320.7	6	8	9.2E-01
Sr I				1474.0	5	7	1.6E+00	5345.7	4	6	8.8E-01
2206.2	1	3	6.6E-03	1474.4	5	5	5.0E-01	5345.7	6	6	1.1E-01
2211.3	1	3	8.5E-03	1474.6	5	3	6.2E-02	5428.6	2	4	4.2E-01
2217.8	1	3	1.2E-02	1481.7	3	5	1.7E+00	5432.8	4	6	6.8E-01
2226.3	1	3	1.6E-02	1483.0	3	5	1.2E+00	5453.8	6	8	8.5E-01
2237.7	1	3	2.3E-02	1483.2	3	3	7.5E-01	5473.6	2	2	7.3E-01
2253.3	1	3	3.7E-02	1487.2	1	3	8.7E-01	5509.7	4	4	4.0E-01
2275.3	1	3	6.7E-02	1666.7	5	5	6.3E+00	5526.2	8	8	8.1E-02
2307.3	1	3	1.2E-01	1687.5	1	3	9.4E-01	5536.8	4	6	6.6E-02
2354.3	1	3	1.8E-01	1782.3	1	3	1.9E+00	5556.0	4	2	1.1E-01
2428.1	1	3	1.7E-01	1807.3	5	3	3.8E+00	5564.9	6	6	1.7E-01
2569.5	1	3	5.3E-02	1820.3	3	3	2.2E+00	5578.8	6	6	1.1E-01
2931.8	1	3	1.9E-02	1826.2	1	3	7.2E-01	5606.1	10	8	5.4E-01

λ Å	Weights g_i	g_k	A 10^8 s^{-1}	λ Å	Weights g_i	g_k	A 10^8 s^{-1}	λ Å	Weights g_i	g_k	A 10^8 s^{-1}
5616.6	4	4	1.2E-01	249.27	2	2	3.1E+01	4153	4	6	5.7E-02
5640.0	4	6	6.6E-01	388.94	2	2	4.5E+01				
5645.6	6	4	1.8E-02	390.86	4	2	8.8E+01	**Tantalum**			
5647.0	2	4	5.7E-01	706.48	2	4	4.17E+01	**Ta I**			
5659.9	6	4	4.6E-01	712.68	4	6	4.85E+01	3127.9	4	6	5.7E-03
5664.7	4	2	5.8E-01	712.84	4	4	8.1E+00	3168.3	4	4	6.0E-03
5819.2	4	4	8.5E-02	933.38	2	4	1.7E+01	3170.3	8	10	8.5E-02
6305.5	8	6	1.8E-01	944.52	2	2	1.6E+01	3205.5	6	8	5.6E-03
6312.7	6	4	3.0E-01					3260.2	4	4	5.8E-03
				S VII				3337.8	6	6	1.3E-02
S III				60.161	1	3	9.46E+03	3383.9	6	4	5.3E-03
2496.2	7	5	2.5E+00	60.804	1	3	5.1E+02	3406.9	4	6	6.8E-03
2508.2	5	3	2.3E+00	72.029	1	3	8.61E+02	3419.7	8	8	1.91E-02
2636.9	3	5	4.5E-01					3463.8	4	6	2.62E-02
2665.4	5	5	1.4E+00	**S VIII**				3484.6	4	4	8.5E-03
2680.5	1	3	6.2E-01	198.55	4	2	2.5E+02	3488.8	6	4	7.3E-03
2691.8	3	3	4.6E-01	202.61	2	2	1.2E+02	3497.9	6	8	4.9E-02
2702.8	3	1	1.9E+00					3505.0	8	6	2.72E-02
2718.9	3	3	1.2E+00	**S XI**				3553.4	4	6	3.3E-03
2721.4	5	3	7.7E-01	*189.90	9	3	4.3E+02	3607.4	6	8	4.6E-02
2726.8	3	5	6.0E-01	190.37	5	3	2.8E+02	3625.2	10	8	1.0E-02
2731.1	5	5	1.1E+00	215.95	5	5	1.4E+02	3626.6	8	10	7.1E-02
2756.9	7	7	1.4E+00	217.63	1	3	7.2E+01	3642.1	10	12	5.5E-02
2785.5	3	3	6.1E-01	239.81	1	3	2.6E+01	3657.5	6	6	4.3E-03
2856.0	5	7	5.1E+00	242.57	3	5	1.9E+01	3731.0	4	6	5.3E-03
2863.5	7	9	5.7E+00	242.82	3	3	1.9E+01	3754.5	8	8	6.5E-03
2872.0	3	5	4.7E+00	246.90	5	5	5.4E+01	3784.3	4	6	4.3E-03
2950.2	3	5	3.0E+00	247.12	5	3	3.0E+01	3792.1	4	4	9.0E-03
2964.8	5	7	4.0E+00	*288.49	9	15	2.9E+01	3826.9	6	6	5.2E-03
3662.0	3	3	6.4E-01					3836.6	8	10	4.0E-03
3717.8	5	3	1.0E+00	**S XII**				3848.1	10	8	1.30E-02
3778.9	3	5	4.4E-01	212.14	2	4	3.7E+01	3858.6	10	10	2.5E-03
3831.8	1	3	5.6E-01	215.18	2	2	1.4E+02	3918.5	4	2	2.5E-02
3837.8	3	3	4.2E-01	218.20	4	4	1.7E+02	3922.8	4	4	3.98E-02
3838.3	5	5	1.3E+00	221.44	4	2	6.4E+01	3996.2	2	4	3.35E-02
3860.6	3	1	1.6E+00	227.50	2	2	3.7E+01	3999.3	4	4	1.8E-02
3899.1	5	3	6.7E-01	234.48	4	2	6.8E+01	4003.7	10	8	3.1E-03
4253.6	5	7	1.2E+00					4006.8	6	8	7.6E-03
4285.0	3	5	9.0E-01	**S XIII**				4026.9	4	4	3.60E-02
				32.236	1	3	1.09E+04	4029.9	10	10	2.8E-02
S IV				37.600	3	1	1.3E+03	4030.7	8	10	2.3E-03
551.17	2	2	2.06E+01	256.66	1	3	8.7E+01	4040.9	10	12	7.3E-03
554.07	4	2	4.08E+01	299.89	3	5	1.78E+01	4061.4	2	4	6.5E-02
3097.5	2	4	2.6E+00	303.37	1	3	2.28E+01	4064.6	4	4	3.83E-02
3117.7	2	2	2.5E+00	307.36	3	3	1.64E+01	4067.2	6	4	6.8E-03
				308.91	5	5	4.82E+01	4067.9	6	8	8.4E-03
S V				312.68	3	1	6.3E+01	4097.2	10	10	2.1E-02
437.37	1	3	1.12E+01	316.84	5	3	2.50E+01	4105.0	6	4	1.1E-02
438.19	3	3	3.33E+01	500.42	3	5	1.43E+01	4136.2	8	6	1.82E-02
439.65	5	3	5.5E+01					4147.9	10	8	1.79E-02
*661.52	9	15	6.44E+01	**S XIV**				4175.2	6	8	2.8E-02
*679.01	9	15	8.6E+01	*30.434	2	6	8.28E+03	4205.9	8	10	8.9E-03
*690.75	9	9	5.0E+01	*32.517	6	10	2.6E+04	4303.0	6	6	2.08E-02
786.48	1	3	5.25E+01	417.67	2	4	1.2E+01	4378.8	8	6	4.8E-03
*854.85	9	9	4.18E+01	445.71	2	2	1.0E+01	4386.1	4	6	1.0E-02
				1550	2	4	1.4E+00	4402.5	6	6	2.28E-02
S VI				1663	2	2	1.2E+00	4415.7	2	4	2.53E-02
248.99	2	4	3.1E+01	3967	2	4	5.4E-02	4441.0	4	6	7.5E-03

λ Å	g_i	g_k	A 10^8 s^{-1}	λ Å	g_i	g_k	A 10^8 s^{-1}	λ Å	g_i	g_k	A 10^8 s^{-1}
4441.7	10	8	9.0E-03	5518.9	8	10	3.8E-02	2921.5	4	4	7.6E-02
4473.5	6	8	1.36E-02	5620.7	8	10	6.0E-03	3229.8	4	2	1.73E-01
4511.0	10	12	1.56E-02	5640.2	6	8	4.9E-03	3519.2	4	6	1.24E+00
4511.5	10	8	3.6E-03	5645.9	6	8	1.43E-02	3529.4	4	4	2.20E-01
4514.2	10	10	3.1E-03	5699.2	6	6	4.2E-03	3775.7	2	2	6.25E-01
4521.1	10	10	2.3E-03	5767.9	6	8	2.6E-03	5350.5	4	2	7.05E-01
4530.9	4	6	2.42E-02	5780.7	4	6	3.3E-03				
4547.2	4	6	5.3E-03	5811.1	8	6	5.7E-03	**Thulium**			
4553.7	6	8	9.5E-03	5849.7	10	8	2.8E-03	**Tm I**			
4565.9	8	8	2.5E-02	5877.4	10	12	2.3E-02	2513.8	8	10	6.9E-02
4574.3	4	4	1.2E-02	5939.8	2	4	1.6E-02	2527.0	8	8	1.7E-01
4580.7	8	10	2.1E-03	5944.0	4	6	2.13E-02	2596.5	8	10	1.6E-01
4619.5	6	4	5.3E-02	5997.2	10	10	2.4E-02	2601.1	8	6	1.7E-01
4633.1	4	4	1.2E-02	6020.7	2	4	1.0E-02	2622.5	8	10	6.1E-02
4669.1	6	4	2.85E-02	6045.4	6	8	2.6E-02	2841.1	6	6	2.0E-01
4681.9	6	6	1.5E-02	6047.3	8	10	9.0E-03	2854.2	8	6	2.7E-01
4684.9	10	8	2.8E-03	6249.8	6	6	3.5E-02	2914.8	8	8	7.7E-02
4685.3	6	8	3.4E-03	6258.7	6	8	3.3E-02	2933.0	8	6	1.0E-01
4691.9	2	4	4.08E-02	6309.6	4	6	1.83E-02	2973.2	8	8	2.3E-01
4706.1	6	6	1.4E-02	6360.8	6	8	4.6E-03	3046.9	8	8	1.8E-01
4740.2	4	4	5.0E-02	6428.6	6	6	6.0E-03	3081.1	8	8	1.9E-01
4758.0	4	6	7.5E-03	6430.8	8	8	2.9E-02	3122.5	6	6	5.2E-01
4769.0	8	8	2.8E-02	6450.4	8	10	2.2E-02	3142.4	6	6	8.8E-02
4780.9	10	8	2.16E-02	6485.4	10	10	5.8E-02	3172.7	8	8	1.8E-01
4812.8	4	4	1.2E-02	6514.4	6	4	2.2E-02	3233.7	8	10	5.1E-02
4825.4	6	6	2.63E-02	6516.1	6	8	1.25E-02	3247.0	6	8	3.0E-01
4832.2	4	4	1.7E-02	6612.0	6	4	1.9E-02	3251.8	6	4	5.2E-01
4852.2	4	4	1.7E-02	6673.7	2	4	9.0E-03	3380.7	6	8	2.0E-01
4884.0	6	8	1.1E-02	6771.7	4	4	5.8E-03	3406.0	6	8	1.5E-01
4904.6	12	10	1.95E-02	6866.2	8	6	2.58E-02	3410.1	8	10	1.0E-01
4920.9	8	10	2.1E-03	6927.4	10	12	1.01E-02	3416.6	8	8	5.7E-02
4921.3	2	4	1.2E-02	6928.5	10	8	1.69E-02	3418.6	6	6	1.1E-01
4926.0	4	4	1.5E-02	6951.3	10	10	3.7E-03	3563.9	8	6	9.8E-02
4936.4	8	6	4.5E-02	6953.9	6	8	8.3E-03	3567.4	8	10	4.2E-02
4969.7	4	4	1.0E-02	6966.1	8	8	1.2E-02	3744.1	8	8	9.5E-01
5012.5	4	4	1.9E-02	6969.5	10	10	2.9E-03	3751.8	8	10	1.9E-01
5037.4	10	8	4.4E-02	7407.9	6	4	2.0E-02	3798.5	6	4	1.2E+00
5043.3	6	4	2.73E-02					3807.7	6	6	3.9E-01
5067.9	8	6	2.92E-02					3883.1	8	6	1.0E+00
5069.9	10	12	1.7E-03	**Thallium**				3887.4	8	8	3.8E-01
5082.3	10	12	1.9E-03	**Tl I**				3916.5	6	8	1.5E+00
5087.4	6	4	1.5E-02	2104.6	2	4	4.0E-02	3949.3	6	6	1.0E+00
5090.7	8	6	9.5E-03	2118.9	2	2	2.0E-02	4022.6	6	8	4.0E-02
5095.3	6	6	5.0E-03	2129.3	2	4	5.8E-02	4044.5	6	4	2.9E-01
5136.5	2	2	4.5E-02	2151.9	2	2	3.1E-02	4094.2	8	6	9.0E-01
5141.6	4	2	1.2E-02	2168.6	2	4	9.8E-02	4105.8	8	10	6.0E-01
5143.7	6	4	1.7E-02	2237.8	2	4	1.9E-01	4138.3	6	4	7.0E-01
5147.6	6	4	9.0E-03	2316.0	2	2	7.8E-02	4158.6	6	8	5.5E-02
5161.8	4	6	6.3E-03	2379.7	2	4	4.4E-01	4187.6	8	8	6.1E-01
5218.7	8	6	8.2E-03	2507.9	4	2	1.1E-02	4203.7	8	10	2.5E-01
5235.4	6	6	4.7E-03	2538.2	4	2	1.6E-02	4222.7	6	8	1.5E-01
5295.0	6	6	7.5E-03	2580.1	2	2	1.8E-01	4271.7	6	6	1.1E-01
5336.1	6	8	5.5E-03	2609.0	4	6	1.0E-01	4359.9	8	6	1.3E-01
5349.6	6	4	2.2E-02	2609.8	4	4	1.9E-02	4386.4	8	8	4.2E-02
5354.7	4	4	6.5E-03	2665.6	4	2	5.7E-02	4394.4	6	4	1.1E-01
5396.0	6	8	2.5E-02	2709.2	4	6	1.7E-01	4643.1	6	6	3.4E-01
5402.5	4	2	1.4E-02	2710.7	4	4	3.7E-02	4681.9	6	8	3.9E-02
5435.3	4	6	1.1E-02	2767.9	2	4	1.26E+00	4691.1	6	6	3.9E-02
5499.4	10	10	6.1E-03	2826.2	4	2	8.0E-02	5307.1	8	10	2.3E-02
				2918.3	4	6	4.2E-01				

λ Å	Weights g_i	g_k	A 10^8 s^{-1}	λ Å	Weights g_i	g_k	A 10^8 s^{-1}	λ Å	Weights g_i	g_k	A 10^8 s^{-1}
5658.3	6	8	1.0E-02	6073.5	3	1	6.3E-02	3642.68	7	9	7.74E-01
5675.8	8	10	1.3E-02	6171.5	3	3	4.9E-02	3653.50	9	11	7.54E-01
5760.2	6	6	1.3E-02					3724.57	9	9	9.1E-01
				Sn II				3725.16	5	3	7.3E-01
Tin				2368.3	4	2	4.4E-03	3729.81	5	5	4.27E-01
Sn I				2449.0	4	6	3.7E-01	3741.06	7	7	4.17E-01
2073.1	1	3	3.6E-02	2487.0	6	8	5.5E-01	3752.86	9	9	5.04E-01
2199.3	3	5	2.9E-01	3283.2	4	6	1.0E+00	3786.04	5	3	1.4E+00
2209.7	5	5	5.6E-01	3352.0	6	8	1.0E+00	3948.67	5	3	4.85E-01
2246.1	1	3	1.6E+00	3472.5	2	4	1.6E-01	3956.34	7	5	3.00E-01
2268.9	5	7	1.2E+00	3575.5	4	6	1.3E-01	3958.21	9	7	4.05E-01
2286.7	5	5	3.1E-01	5332.4	2	4	8.6E-01	3981.76	5	5	3.76E-01
2317.2	5	7	2.0E+00	5562.0	4	6	1.2E+00	3989.76	7	7	3.79E-01
2334.8	3	3	6.6E-01	5588.9	4	6	8.5E-01	3998.64	9	9	4.08E-01
2354.8	3	5	1.7E+00	5596.2	4	4	1.5E-01	4013.24	7	5	2.0E-01
2380.7	3	5	3.1E-02	5797.2	6	6	2.8E-01	4055.01	1	3	2.8E-01
2408.2	5	3	1.8E-01	5799.2	6	8	8.1E-01	4060.26	3	5	2.4E-01
2421.7	5	7	2.5E+00	6453.5	2	4	1.2E+00	4064.20	3	3	2.4E-01
2429.5	5	7	1.5E+00	6761.5	2	2	3.2E-01	4065.09	3	1	7.0E-01
2433.5	5	3	8.0E-03	6844.1	2	2	6.6E-01	4186.12	9	9	2.10E-01
2455.2	5	5	1.1E-02					4266.23	5	5	3.1E-01
2476.4	5	3	1.1E-02	**Titanium**				4284.99	5	5	3.2E-01
2483.4	5	5	2.1E-01	**Ti I**				4289.07	5	5	3.0E-01
2491.8	1	3	1.7E-01	2276.75	7	5	1.3E+00	4290.93	3	3	4.5E-01
2495.7	5	5	6.2E-01	2280.00	9	7	9.4E-01	4295.75	3	1	1.3E+00
2523.9	5	3	7.4E-02	2299.86	5	5	6.9E-01	4393.93	9	11	3.3E-01
2546.6	1	3	2.1E-01	2302.75	7	7	5.7E-01	4417.27	11	9	3.6E-01
2558.0	1	3	3.4E-01	2305.69	9	9	5.2E-01	4449.14	11	11	9.7E-01
2571.6	5	7	4.5E-01	2424.26	9	9	1.7E-01	4450.90	9	9	9.6E-01
2594.4	5	5	3.0E-01	2520.54	5	3	3.8E-01	4453.31	5	5	5.98E-01
2636.9	1	3	1.1E-01	2529.87	7	5	3.8E-01	4453.71	7	7	4.7E-01
2661.2	3	3	1.1E-01	2541.92	9	7	4.3E-01	4455.32	7	7	4.8E-01
2706.5	3	5	6.6E-01	2599.91	5	5	6.7E-01	4457.43	9	9	5.6E-01
2761.8	5	5	3.7E-03	2605.16	7	7	6.4E-01	4465.81	5	7	3.28E-01
2779.8	5	7	1.8E-01	2611.29	9	9	6.4E-01	4481.26	7	7	5.7E-01
2785.0	5	3	1.4E-01	2611.47	7	5	3.3E-01	4496.15	7	5	4.4E-01
2788.0	1	3	1.4E-01	2619.94	9	7	2.1E-01	4518.02	7	9	1.72E-01
2812.6	1	3	2.3E-01	2631.55	7	7	1.7E-01	4522.80	5	7	1.9E-01
2813.6	5	5	1.2E-01	2632.42	5	5	2.7E-01	4527.31	3	5	2.2E-01
2840.0	5	5	1.7E+00	2641.12	5	3	1.8E+00	4533.24	11	11	8.83E-01
2850.6	5	5	3.3E-01	2644.28	7	5	1.4E+00	4534.78	9	9	6.87E-01
2863.3	1	3	5.4E-01	2646.65	9	7	1.5E+00	4544.69	5	3	3.3E-01
2913.5	1	3	8.3E-01	2733.27	5	5	1.9E+00	4548.76	7	5	2.85E-01
3009.1	3	3	3.8E-01	2735.30	3	1	4.1E+00	4552.45	9	7	2.1E-01
3032.8	1	3	6.2E-01	2912.07	5	7	1.3E+00	4563.43	9	11	2.1E-01
3034.1	3	1	2.0E+00	2942.00	5	5	1.0E+00	4617.27	7	9	8.51E-01
3141.8	1	3	1.9E-01	2948.26	7	7	9.3E-01	4623.10	5	7	5.74E-01
3175.1	5	3	1.0E+00	2956.13	9	9	9.7E-01	4639.94	3	3	6.64E-01
3218.7	1	3	4.7E-02	2956.80	7	5	1.8E-01	4640.43	3	1	5.0E-01
3223.6	5	5	1.2E-03	3186.45	5	7	8.0E-01	4645.19	3	1	8.57E-01
3262.3	5	3	2.7E+00	3191.99	7	9	8.5E-01	4650.02	5	3	2.6E-01
3330.6	5	5	2.0E-01	3199.92	9	11	9.4E-01	4742.79	9	9	5.3E-01
3655.8	1	3	4.1E-02	3341.88	5	7	6.5E-01	4758.12	11	11	7.13E-01
3801.0	5	3	2.8E-01	3354.63	7	9	6.9E-01	4759.27	13	13	7.40E-01
4524.7	1	3	2.6E-01	3370.44	5	3	7.6E-01	4778.26	9	9	2.0E-01
5631.7	1	3	2.4E-02	3371.45	9	11	7.2E-01	4805.42	5	7	5.8E-01
5970.3	5	3	9.6E-02	3377.58	7	5	6.9E-01	4840.87	5	5	1.76E-01
6037.7	5	5	5.0E-02	3385.94	9	7	5.0E-01	4856.01	13	15	5.2E-01
6069.0	1	3	4.6E-02	3635.46	5	7	8.04E-01	4885.08	11	13	4.90E-01

λ Å	Weights g_i	g_k	A 10^8 s^{-1}	λ Å	Weights g_i	g_k	A 10^8 s^{-1}	λ Å	Weights g_i	g_k	A 10^8 s^{-1}
4913.62	7	9	4.44E-01	2877.47	8	8	5.7E-01	3278.91	6	4	1.0E+00
4928.34	3	5	6.2E-01	2884.13	10	10	5.2E-01	3282.32	2	2	1.6E+00
4981.73	11	13	6.60E-01	2910.65	8	8	4.6E-01	3287.66	8	10	1.4E+00
4989.14	7	5	3.25E-01	2926.64	10	8	8.9E-01	3315.32	2	4	3.8E-01
4991.07	9	11	5.84E-01	2931.10	6	6	3.2E+00	3321.70	4	4	7.2E-01
4999.50	7	9	5.27E-01	2936.02	4	6	2.7E+00	3322.94	10	10	3.96E-01
5000.99	9	7	3.52E-01	2938.57	6	8	2.4E+00	3329.46	8	8	3.25E-01
5007.21	5	7	4.92E-01	2941.90	8	10	1.8E+00	3332.11	6	4	1.1E+00
5014.28	3	5	6.8E-01	2942.97	8	8	1.1E+00	3340.34	4	4	3.6E-01
5036.47	7	9	3.94E-01	2945.30	10	12	2.7E+00	3361.23	8	10	1.1E+00
5038.40	5	7	3.87E-01	2952.00	8	8	3.0E-01	3372.80	6	8	1.11E+00
5062.11	5	3	2.98E-01	2954.59	10	12	4.0E+00	3383.77	4	6	1.09E+00
5210.39	9	9	3.57E-02	2958.80	8	10	4.0E+00	3452.49	2	2	7.7E-01
5222.69	3	3	1.95E-01	2979.06	4	6	1.2E+00	3456.40	4	4	8.2E-01
5224.30	11	11	3.6E-01	2990.06	6	8	5.6E-01	3465.56	4	2	4.1E-01
5259.98	5	7	2.3E-01	3017.17	12	12	3.6E-01	3483.63	10	8	9.7E-01
5351.07	7	7	3.4E-01	3022.64	10	10	1.2E+00	3492.37	8	6	9.8E-01
5503.90	11	9	2.6E-01	3023.67	8	8	1.0E+00	3504.90	10	10	8.2E-01
5774.04	9	11	5.5E-01	3029.76	10	10	3.5E-01	3510.86	8	8	9.3E-01
5785.98	11	13	6.1E-01	3056.75	2	4	3.2E-01	3520.27	2	4	4.8E-01
5804.27	13	15	6.8E-01	3058.08	6	6	5.0E-01	3535.41	4	6	5.5E-01
6098.66	9	7	2.5E-01	3066.34	4	4	3.3E-01	3641.33	4	2	4.9E-01
6220.46	9	7	1.8E-01	3071.25	6	4	3.6E-01	3706.23	4	4	3.1E-01
				3072.99	4	2	1.6E+00	3741.64	6	6	6.2E-01
Ti II				3075.23	6	4	1.13E+00	3757.70	4	4	4.1E-01
2440.91	4	4	5.1E-01	3078.65	8	6	1.09E+00	3759.30	8	8	9.4E-01
2451.18	6	6	4.5E-01	3081.52	10	8	1.1E+00	3761.33	6	6	9.9E-01
2525.59	10	8	5.6E-01	3088.04	10	8	1.25E+00	4911.18	6	4	3.2E-01
2531.28	8	6	4.9E-01	3089.44	8	6	1.3E+00				
2534.63	6	4	5.4E-01	3097.20	4	6	4.4E-01	**Ti III**			
2535.89	4	2	6.8E-01	3103.81	10	8	1.1E+00	865.79	5	3	6.6E+01
2555.99	6	8	3.2E-01	3105.10	2	4	6.3E-01	1002.37	5	5	7.6E+00
2635.44	4	4	1.9E+00	3106.26	6	6	7.8E-01	1004.67	7	5	4.3E+01
2638.56	6	6	1.7E+00	3117.67	4	2	1.1E+00	1005.80	3	3	1.3E+01
2642.02	8	8	1.9E+00	3119.83	6	4	5.9E-01	1007.16	5	3	3.8E+01
2645.86	10	10	2.7E+00	3127.86	6	6	1.6E+00	1008.12	3	1	5.1E+01
2746.54	6	8	2.6E+00	3128.50	8	8	1.1E+00	1286.37	9	9	2.0E+00
2751.59	8	10	3.7E+00	3161.23	4	2	5.9E-01	1289.30	7	7	2.2E+00
2752.68	8	10	1.1E+00	3161.80	6	4	4.6E-01	1291.62	5	5	2.4E+00
2757.62	6	8	7.2E-01	3162.59	8	6	3.9E-01	1293.23	9	7	1.0E+00
2758.35	4	6	9.9E-01	3168.55	10	8	4.1E-01	1298.97	7	5	4.9E+00
2758.79	2	4	4.4E-01	3181.73	6	8	4.6E-01	1327.59	5	3	3.2E+00
2764.28	4	4	7.4E-01	3182.54	4	6	4.3E-01	1420.44	1	3	1.2E+00
2804.82	6	8	4.6E+00	3189.49	4	4	9.2E-01	1421.63	3	1	4.0E+00
2810.30	8	10	5.1E+00	3190.91	6	8	1.3E+00	1422.41	5	5	3.0E+00
2817.83	10	12	3.8E+00	3202.56	4	6	1.1E+00	1424.14	5	3	1.6E+00
2819.87	8	8	6.5E-01	3224.25	12	10	7.0E-01	1455.19	9	7	6.4E+00
2821.26	6	8	7.9E-01	3228.62	4	2	2.0E+00	1498.70	5	5	2.8E+00
2827.12	8	10	1.0E+00	3232.29	8	6	6.0E-01	2007.36	3	3	3.4E+00
2828.06	12	14	4.4E+00	3234.51	10	10	1.38E+00	2007.60	1	3	1.2E+00
2828.64	6	6	1.2E+00	3236.13	4	4	7.0E-01	2010.80	5	3	5.4E+00
2828.83	10	10	9.1E-01	3236.58	8	8	1.11E+00	2097.30	5	7	3.3E+00
2834.02	10	12	7.9E-01	3239.04	6	6	9.87E-01	2099.86	3	5	2.5E+00
2836.47	8	8	1.2E+00	3239.66	6	4	9.4E-01	2104.86	3	3	1.1E+00
2839.64	12	12	8.3E-01	3241.99	4	4	1.16E+00	2105.09	1	3	1.7E+00
2845.93	10	10	1.2E+00	3251.91	6	4	3.38E-01	2199.22	3	3	5.7E+00
2851.11	2	4	4.1E-01	3252.92	8	6	3.9E-01	2237.77	7	7	2.4E+00
2856.10	12	12	1.5E+00	3272.07	2	4	3.2E+00	2331.35	3	1	4.3E+00
2862.33	4	6	4.0E-01	3278.28	4	4	9.6E-01	2331.66	3	3	1.2E+00

λ Å	Weights g_i	Weights g_k	A 10^8 s^{-1}	λ Å	Weights g_i	Weights g_k	A 10^8 s^{-1}	λ Å	Weights g_i	Weights g_k	A 10^8 s^{-1}
2339.00	5	3	3.0E+00	2862.60	4	2	4.1E+00	308.250	3	5	1.3E+02
2346.79	7	5	3.3E+00	3576.44	4	6	4.6E+00	313.229	5	7	1.6E+02
2374.99	5	3	4.0E+00					318	3	1	1.4E+02
2413.99	5	7	3.8E+00	**Ti VIII**				322.75	5	7	1.99E+02
2516.05	7	9	3.4E+00	249	6	4	1.0E+01	323	1	3	1.8E+02
2567.56	3	3	2.3E+00	258.610	6	8	7.5E+02	327.192	3	5	2.9E+02
2984.75	5	5	1.9E+00	269.533	4	6	6.0E+02	332	3	1	3.25E+02
3066.51	3	3	2.5E+00	272.037	4	4	4.3E+02	386.140	1	3	1.48E+02
3228.89	3	3	1.5E+00	272.843	6	4	6.2E+01	408	7	9	1.37E+02
3278.31	7	9	3.4E+00	276.701	2	4	9.3E+01	425.74	3	1	1.2E+02
3320.94	3	5	2.8E+00	277.813	4	4	3.8E+02	446.69	3	1	1.2E+02
3340.20	7	9	3.7E+00	289.375	2	4	3.6E+01	453	5	7	1.3E+02
3346.18	9	11	3.7E+00	478.971	4	4	1.7E+01				
3354.71	11	13	4.4E+00	480.376	6	6	1.5E+01	**Ti XII**			
3397.24	3	1	1.8E+00					52.896	2	4	1.61E+02
3404.46	3	3	1.8E+00	**Ti IX**				53.140	4	6	1.9E+02
3417.62	3	5	1.9E+00	267.941	5	7	5.1E+02	53.433	2	4	2.1E+02
3915.47	9	11	2.1E+00	278.713	5	7	4.7E+02	53.457	2	2	2.1E+02
4119.14	5	5	9.9E-01	281.446	3	1	3.2E+02	55.181	2	4	2.4E+02
4213.26	9	11	2.2E+00	285.128	1	3	4.1E+02	55.443	4	6	2.81E+02
4215.53	9	11	2.2E+00	433.567	1	3	6.9E+00	59.133	2	4	3.72E+02
4247.62	11	13	1.1E+00	439.513	3	3	7.5E+00	59.435	4	6	4.41E+02
248.54	5	7	2.3E+00	439.745	3	1	2.1E+01	60.701	2	4	3.4E+02
250.09	3	5	9.5E-01	447.484	5	5	1.6E+01	60.762	2	2	3.5E+02
259.01	11	13	9.4E-01	447.701	5	3	6.5E+00	61.286	4	2	1.8E+02
4269.84	9	11	1.7E+00	507.174	3	5	6.5E+00	62.433	4	6	2.08E+02
4285.61	13	15	3.0E+00	516.215	5	7	6.9E+00	62.470	6	8	2.22E+02
4288.66	11	13	1.1E+00					65.540	4	6	3.2E+02
4296.70	11	13	1.6E+00	**Ti X**				65.577	6	8	3.5E+02
4319.56	9	11	1.1E+00	253	4	6	2.1E+02	67.171	2	4	6.2E+02
4343.25	3	1	1.0E+00	254	6	8	2.3E+02	67.555	4	6	7.2E+02
4378.94	3	5	1.6E+00	281	2	2	1.1E+02	70.986	4	6	5.7E+02
4433.91	11	13	1.8E+00	289.579	2	4	2.5E+02	71.031	6	8	6.1E+02
4440.66	1	3	1.2E+00	290.294	4	6	1.1E+02	71.545	2	2	1.8E+02
4533.26	3	5	1.5E+00	291	4	2	1.8E+02	71.987	4	2	3.48E+02
4576.53	9	7	1.3E+00	291	2	2	2.3E+02	82.121	2	4	5.9E+02
4628.07	3	1	1.5E+00	292	6	8	1.1E+02	82.307	4	6	1.13E+03
4652.86	7	9	2.6E+00	293.684	6	8	2.97E+02	82.344	2	2	5.8E+02
4874.00	5	7	1.5E+00	293.798	6	6	1.7E+02	82.368	6	8	1.2E+03
4914.32	3	3	1.1E+00	295.584	4	6	2.9E+02	89.844	2	4	9.9E+02
4971.19	9	11	2.1E+00	296	4	6	1.4E+02	90.512	4	6	1.16E+03
5083.80	5	3	9.7E-01	297	4	6	9.9E+01	90.547	4	6	1.9E+02
5278.33	3	3	9.4E-01	298	4	6	4.3E+02	116.497	4	6	3.0E+03
7506.87	11	13	1.1E+00	302	2	2	1.6E+02	116.597	6	8	3.2E+03
				305	2	4	2.5E+02	116.62	6	6	2.1E+02
Ti IV				317	2	2	1.5E+02	139.884	6	4	2.6E+02
423.49	4	6	4.9E+01	355.815	2	2	1.3E+02	140.361	4	2	2.9E+02
424.16	6	8	5.3E+01	360.133	4	4	2.19E+02	141.6	4	6	1.7E+02
433.63	4	2	5.5E+00	363	4	2	2.1E+02	141.7	6	8	1.7E+02
433.76	6	4	5.0E+00	363	6	6	1.3E+02	169.7	4	6	2.8E+02
729.36	4	2	5.7E+00	365.628	4	2	1.2E+02	169.8	6	8	2.9E+02
1183.64	2	2	6.9E+00	382	4	6	1.8E+02	207.2	2	4	1.5E+02
1195.21	4	2	1.4E+01	385	6	8	1.8E+02	208.5	4	6	1.8E+02
1451.74	2	4	1.8E+01	389.99	6	4	1.1E+02	252.8	4	6	4.8E+02
1467.34	4	6	2.1E+01					253.1	6	8	5.2E+02
2067.56	2	4	5.1E+00	**Ti XI**				257.5	4	2	2.4E+02
2103.16	2	2	5.0E+00	65.403	1	3	5.1E+02				
2541.79	4	6	6.9E+00	87.725	1	3	8.5E+02	**Ti XIII**			
2546.88	6	8	7.4E+00	266	5	7	1.8E+02	23.356	1	3	1.02E+05

λ Å	Weights g_i	g_k	A 10^8 s^{-1}	λ Å	Weights g_i	g_k	A 10^8 s^{-1}	λ Å	Weights g_i	g_k	A 10^8 s^{-1}
23.698	1	3	1.2E+04	145.665	6	6	2.3E+02	16.46	3	3	4.4E+04
23.991	1	3	3.4E+02	157.812	4	2	1.32E+02	16.51	5	7	1.0E+05
26.641	1	3	4.06E+03	161.168	4	4	1.2E+02	16.55	5	5	2.7E+04
26.960	1	3	3.06E+03	163.610	4	2	1.92E+02	16.61	3	1	8.0E+04
117.1	3	3	1.3E+02	169.740	4	6	1.0E+02	16.64	3	3	5.3E+04
117.3	3	1	2.8E+02	176.267	2	2	2.45E+02	16.69	1	3	1.02E+05
120.2	5	3	5.4E+02	178.240	4	4	2.52E+02	16.71	3	5	7.3E+04
120.2	7	5	4.4E+02					16.72	5	3	3.3E+04
128.7	3	3	1.2E+02	**Ti XVII**				16.72	5	5	7.3E+04
				18.05	3	3	4.5E+04	16.74	5	7	1.2E+05
Ti XIV				18.13	5	3	2.4E+04	16.77	3	3	2.6E+04
21.341	4	6	9.8E+03	18.13	1	3	8.1E+04	16.80	5	7	1.81E+05
21.522	2	4	4.5E+04	18.176	5	7	9.2E+04	16.85	3	5	4.4E+04
21.657	4	4	1.3E+04	123.654	3	3	2.3E+02	17.08	3	5	8.3E+04
21.733	4	4	8.8E+04	124.553	5	3	5.2E+02	17.36	1	3	9.5E+04
21.82	4	2	6.4E+04	127.782	5	3	4.6E+02				
21.883	2	4	7.0E+04	135.202	3	1	2.93E+02	**Ti XX**			
21.958	2	4	1.2E+04	136.160	5	3	1.95E+02	2.629	2	4	4.9E+04
22.05	2	2	1.4E+04	136.393	3	3	1.14E+02	2.6295	4	4	3.2E+06
24.592	4	2	6.1E+03	141.948	5	5	3.87E+02	2.631	2	2	6.1E+05
24.891	2	2	7.5E+03	142.589	1	3	1.35E+02	2.6319	2	4	1.5E+06
				144.405	5	5	9.4E+01	2.632	2	2	2.7E+06
Ti XV				146.067	7	5	2.6E+02	2.6355	4	6	1.2E+06
20.19	5	7	6.9E+03	154.133	3	1	1.63E+02	8.621	4	2	1.1E+06
20.234	5	7	1.9E+04	156.54	3	1	1.44E+02	9.788	4	6	5.26E+03
20.234	3	3	4.9E+04	158.469	5	5	1.4E+02	10.046	2	4	7.29E+03
20.246	1	3	4.2E+04	159.62	5	3	1.03E+02	10.109	4	6	8.6E+03
20.250	5	3	6.5E+03	163.049	3	1	6.2E+02	*10.278	2	6	8.4E+03
20.29	3	3	1.1E+04	186.863	5	5	2.66E+02	10.620	2	4	1.34E+04
20.30	1	3	3.4E+04	207.73	3	1	1.07E+02	10.690	4	6	1.58E+04
20.30	1	1	5.8E+04					*11.452	2	6	1.7E+04
20.313	5	3	7.5E+04	**Ti XVIII**				11.872	2	4	2.8E+04
20.418	5	7	8.0E+04	17.22	2	4	7.3E+04	11.958	4	6	3.4E+04
20.538	3	3	3.8E+04	17.365	4	6	8.6E+04	11.958	4	4	5.6E+03
20.54	3	1	4.1E+04	17.39	4	4	1.4E+04	15.211	2	4	3.50E+04
20.551	1	3	1.3E+04	133.852	2	4	5.2E+01	15.253	2	2	3.58E+04
20.689	5	7	4.3E+04	144.759	4	4	3.2E+02	15.907	2	4	8.84E+04
20.698	1	3	1.1E+05	150.15	6	4	1.15E+02	16.049	4	6	1.05E+05
20.771	5	3	1.1E+04	153.15	4	2	1.97E+02	16.067	4	4	1.8E+04
20.897	5	7	2.85E+04	153.23	2	4	6.7E+01	31.586	4	6	5.49E+03
20.928	5	5	8.4E+03	159.00	4	4	1.16E+02	45.650	2	4	9.6E+03
21.065	3	3	1.1E+04	166.225	6	4	1.54E+02	45.996	4	6	1.1E+04
21.079	1	3	1.58E+04	179.902	2	4	6.3E+01				
21.102	3	5	1.3E+04	189.663	6	6	9.6E+01	**Ti XXI**			
22.482	5	3	6.4E+03	191.23	4	4	6.6E+01	2.0633	1	3	1.32E+05
22.936	5	5	1.1E+04	197.838	4	6	4.56E+01	2.1108	1	3	2.60E+05
22.966	5	3	1.1E+04	208.07	4	4	1.2E+02	2.2211	1	3	6.35E+05
23.034	1	3	6.3E+03					2.497	3	1	2.4E+06
				Ti XIX				2.505	5	5	3.5E+06
Ti XVI				15.67	3	1	3.3E+04	2.505	1	3	1.4E+06
110.561	4	2	3.36E+02	15.68	5	5	2.7E+04	2.507	3	5	1.4E+06
116.198	4	4	1.45E+02	15.74	5	7	2.7E+04	2.508	3	5	7.9E+05
118.215	6	4	7.4E+02	15.75	3	5	2.4E+04	2.510	3	3	6.9E+05
121.382	4	2	2.4E+02	15.83	1	3	3.2E+04	2.510	1	3	9.6E+05
124.805	4	2	6.1E+02	15.86	1	3	2.9E+04	2.511	3	3	1.4E+06
129.075	4	2	3.81E+02	16.02	3	1	3.1E+04	2.512	5	5	1.8E+06
134.724	2	2	2.6E+02	16.18	3	5	3.8E+04	2.512	3	1	1.4E+06
138.800	6	4	3.5E+02	16.41	1	3	6.1E+04	2.513	3	1	2.7E+06
143.459	4	4	2.8E+02	16.43	3	5	8.2E+04	2.513	3	5	2.4E+06

λ Å	Weights g_i	g_k	A 10^8 s^{-1}	λ Å	Weights g_i	g_k	A 10^8 s^{-1}	λ Å	Weights g_i	g_k	A 10^8 s^{-1}
2.514	5	3	1.2E+06	3780.8	7	5	4.2E-02	4924.6	13	11	1.75E-03
2.520	3	5	2.6E+05	3809.2	7	5	9.0E-03	4931.6	7	5	1.0E-02
2.527	3	1	1.2E+05	3817.5	7	7	3.1E-02	4948.6	9	11	1.36E-03
2.539	3	1	4.1E+05	3829.1	3	3	3.83E-03	4972.6	9	11	3.9E-03
2.6102	1	3	2.40E+06	3835.1	5	5	5.2E-02	4982.6	1	3	4.17E-03
2.6227	1	3	1.12E+05	3846.3	3	5	2.14E-02	4986.9	11	9	6.3E-03
				3847.5	1	3	8.3E-03	5006.2	9	7	1.2E-02
Tungsten				3864.3	5	5	5.6E-03	5015.3	7	9	5.4E-03
W I				3868.0	7	9	4.6E-02	5040.4	3	5	5.2E-03
2879.4	1	3	2.4E-01	3881.4	7	7	3.6E-02	5053.3	3	3	1.9E-02
2911.0	1	3	7.7E-02	3968.5	1	3	5.07E-03	5071.5	13	11	3.4E-03
2923.5	7	9	1.54E-02	3975.5	9	11	4.1E-03	5117.6	11	11	1.61E-03
2935.0	3	5	1.5E-01	4001.4	9	9	5.6E-03	5124.2	5	5	4.0E-03
3013.8	7	9	6.4E-02	4008.8	7	9	1.63E-01	5141.2	7	9	1.12E-02
3016.5	9	11	9.27E-02	4019.3	5	3	6.7E-03	5224.7	7	5	1.2E-02
3017.4	7	9	1.21E-01	4028.8	1	3	2.0E-02	5243.0	9	7	1.1E-02
3024.9	3	3	1.4E-01	4045.6	7	5	2.88E-02	5254.5	7	5	3.86E-03
3046.4	3	5	5.8E-02	4055.2	7	9	1.79E-03	5268.6	9	9	1.4E-03
3049.7	7	5	1.7E-01	4070.0	7	5	3.60E-02	5500.5	11	9	6.9E-03
3064.9	5	7	1.1E-02	4070.6	3	5	5.6E-03	5514.7	5	3	7.3E-03
3084.9	5	5	1.3E-02	4074.4	7	7	1.0E-01	5537.7	9	11	2.2E-03
3093.5	7	9	4.4E-02	4088.3	5	3	4.13E-03	5617.1	7	7	1.47E-03
3107.2	5	7	2.33E-02	4102.7	9	7	4.9E-03	5631.9	9	7	1.43E-03
3108.0	7	9	1.58E-02	4115.6	11	11	4.8E-03	5660.7	13	11	6.8E-03
3145.5	9	9	4.8E-03	4137.5	5	7	8.4E-03	5675.4	5	5	2.20E-03
3170.2	7	5	6.0E-03	4171.2	7	9	8.6E-03	5796.5	9	7	2.21E-03
3176.6	3	5	2.12E-02	4203.8	9	7	4.9E-03	5891.6	7	7	1.47E-03
3183.5	7	7	2.64E-03	4219.4	9	7	6.1E-03	5947.6	5	7	2.40E-03
3184.4	5	3	2.3E-02	4244.4	9	11	1.38E-02	5965.9	7	5	1.0E-02
3191.6	1	3	3.2E-02	4269.4	7	5	3.04E-02	6021.5	5	3	8.7E-03
3198.8	7	9	4.6E-02	4283.8	9	7	1.69E-03	6081.4	5	3	4.7E-03
3207.3	7	9	3.0E-02	4294.6	7	5	1.2E-01	6203.5	7	7	3.0E-03
3208.3	5	5	4.4E-02	4302.1	7	7	3.6E-02	6285.9	7	5	6.6E-03
3215.6	9	11	2.1E-01	4355.2	9	9	5.1E-03	6292.0	3	5	2.26E-03
3221.9	5	7	1.61E-02	4361.8	9	7	1.64E-03	6303.2	9	9	1.84E-03
3223.1	5	3	3.53E-03	4378.5	7	5	3.48E-03	6404.2	5	7	1.50E-03
3232.5	9	9	2.4E-02	4458.1	3	5	4.2E-02	6439.7	9	9	1.29E-03
3235.1	7	5	2.68E-03	4466.3	7	5	1.5E-02	6445.1	7	5	6.4E-03
3259.7	7	7	1.3E-02	4472.5	13	11	1.55E-03	6532.4	3	5	4.6E-03
3300.8	7	9	8.1E-02	4484.2	3	5	5.6E-03	6538.1	11	9	2.7E-03
3311.4	7	5	5.6E-02	4492.3	9	11	3.6E-03	6563.2	5	5	2.04E-03
3363.3	9	7	6.6E-03	4495.3	11	11	3.3E-03	6814.9	9	9	1.46E-03
3371.0	7	5	1.0E-02	4504.8	9	7	7.0E-03	7285.8	13	11	1.47E-03
3371.4	3	3	6.7E-03	4552.5	9	9	1.42E-03	7569.9	5	3	3.73E-03
3386.1	7	7	2.64E-03	4586.8	1	3	4.20E-03	7664.9	5	3	3.80E-03
3413.0	7	9	9.7E-03	4592.6	7	9	3.4E-03	8017.2	5	7	1.6E-03
3459.5	9	9	2.04E-03	4609.9	7	9	1.42E-02	8358.7	5	7	1.89E-03
3510.0	7	9	5.2E-03	4613.3	9	9	2.9E-03	9381.4	9	7	1.53E-03
3545.2	1	3	3.2E-02	4634.8	9	9	8.8E-03				
3570.6	5	3	6.7E-03	4659.9	1	3	1.0E-02	**Uranium**			
3606.1	3	5	9.6E-03	4680.5	7	7	1.4E-02	**U I**			
3617.5	7	7	1.1E-01	4720.4	3	5	3.22E-03	3553.0	13	13	2.0E-02
3631.9	3	5	1.3E-02	4729.6	7	5	7.8E-03	3553.0	9	7	1.4E-02
3675.6	9	11	1.20E-02	4752.6	3	3	5.20E-03	3553.4	15	13	2.2E-02
3682.1	9	11	2.0E-02	4757.5	7	5	2.72E-03	3554.5	11	9	8.4E-03
3707.9	7	7	2.9E-02	4757.8	11	9	4.1E-03	3554.9	15	17	7.9E-03
3757.9	7	9	1.38E-02	4788.4	9	11	2.6E-03	3555.3	13	15	2.7E-02
3760.1	5	7	1.99E-02	4843.8	5	5	1.9E-02	3555.8	13	11	4.1E-03
3768.5	3	3	3.47E-02	4886.9	9	11	8.1E-03	3556.9	13	11	7.5E-03

λ Å	Weights g_i	g_k	A 10^8 s^{-1}	λ Å	Weights g_i	g_k	A 10^8 s^{-1}	λ Å	Weights g_i	g_k	A 10^8 s^{-1}
3557.8	13	13	2.9E-02	3075.93	4	6	2.8E-01	3818.24	4	2	6.73E-01
3558.0	11	13	1.6E-02	3080.33	2	4	2.7E-01	3828.56	6	4	5.33E-01
3558.6	9	7	3.9E-02	3083.54	6	8	2.5E-01	3840.75	8	6	5.48E-01
3559.4	7	9	1.5E-02	3087.06	2	2	9.2E-01	3855.36	4	4	3.30E-01
3560.3	9	7	6.4E-02	3088.11	4	6	4.9E-01	3855.85	10	8	5.78E-01
3561.4	15	13	5.5E-02	3089.13	4	4	5.3E-01	3863.86	8	6	3.1E-01
3561.5	9	9	2.5E-02	3093.79	6	6	4.1E-01	3864.86	6	6	2.70E-01
3561.8	13	11	5.7E-02	3094.69	2	4	4.3E-01	3871.07	10	8	2.8E-01
3563.7	13	13	2.9E-02	3112.92	4	2	5.0E-01	3875.07	8	8	2.36E-01
3563.8	7	7	1.1E-02	3183.41	6	8	2.4E+00	3902.26	10	10	2.68E-01
3565.0	13	11	2.9E-02	3183.96	8	10	2.5E+00	3921.86	4	2	2.7E-01
3566.0	13	15	1.7E-02	3183.98	4	6	2.4E+00	3922.43	6	6	2.6E-01
3566.6	11	11	2.4E-01	3185.38	10	12	2.7E+00	3930.02	10	10	3.3E-01
3568.8	13	13	3.8E-02	3198.01	6	6	3.9E-01	3934.01	8	8	6.2E-01
3569.1	17	15	1.1E-01	3202.39	8	8	4.0E-01	3992.80	12	10	1.2E+00
3569.4	9	9	1.5E-02	3205.58	8	10	1.3E+00	3998.73	14	12	1.0E+00
3570.1	13	11	1.3E-02	3207.41	10	10	2.6E-01	4050.96	10	10	1.4E+00
3570.2	11	9	5.3E-03	3212.43	10	12	1.4E+00	4051.35	12	12	1.3E+00
3570.6	13	15	2.7E-02	3218.87	8	6	3.5E-01	4090.57	8	10	8.5E-01
3570.7	15	15	1.2E-02	3233.19	10	8	3.2E-01	4092.68	8	10	2.30E-01
3571.2	11	11	6.3E-03	3273.03	8	8	2.7E-01	4095.48	6	8	7.2E-01
3571.6	17	15	1.3E-01	3284.36	10	10	2.8E-01	4099.78	6	8	4.10E-01
3572.9	13	15	1.5E-02	3309.18	4	4	3.2E-01	4102.15	4	6	7.1E-01
3573.9	13	11	4.0E-02	3329.85	6	4	7.7E-01	4104.77	10	8	2.1E+00
3574.1	13	15	3.5E-02	3356.35	4	6	3.1E-01	4105.16	4	6	4.9E-01
3574.8	13	15	1.9E-02	3365.55	2	4	4.8E-01	4109.78	2	4	5.00E-01
3577.1	17	15	4.3E-02	3376.05	4	4	3.2E-01	4111.78	10	10	1.01E+00
3577.5	15	13	7.8E-03	3377.39	4	2	9.0E-01	4115.18	8	8	5.80E-01
3577.8	11	11	8.3E-03	3377.62	6	6	6.0E-01	4116.47	6	6	3.2E-01
3577.9	13	13	2.3E-02	3397.58	6	4	2.3E-01	4116.59	2	2	2.90E-01
3578.3	13	11	2.0E-02	3400.39	8	8	2.5E-01	4123.50	4	2	1.00E+00
3580.0	9	9	1.2E-02	3529.73	4	6	4.1E-01	4128.06	6	4	7.70E-01
3580.2	11	9	2.9E-02	3533.68	6	8	5.2E-01	4131.99	8	6	5.5E-01
3580.4	11	13	7.5E-03	3533.76	2	2	3.7E-01	4134.49	10	8	2.90E-01
3580.9	13	13	2.1E-02	3543.49	2	2	6.7E-01	4232.46	10	10	9.8E-01
3582.6	13	13	2.9E-02	3545.33	4	4	3.7E-01	4232.95	8	8	7.7E-01
3584.6	7	5	2.4E-02	3553.27	6	6	2.2E-01	4268.64	14	14	1.2E+00
3584.9	13	15	1.8E-01	3555.14	4	2	2.6E-01	4271.55	12	12	9.6E-01
3585.4	11	11	1.9E-02	3663.60	4	6	3.1E+00	4276.95	10	10	9.4E-01
3585.8	11	9	2.8E-02	3667.74	6	8	2.7E+00	4284.05	8	8	1.2E+00
3587.8	9	11	1.3E-02	3672.41	12	12	9.2E-01	4291.82	12	14	8.8E-01
3588.3	7	9	1.8E-02	3673.41	8	10	2.7E+00	4296.10	10	12	7.7E-01
3589.7	11	13	2.1E-02	3676.70	14	14	1.3E+00	4297.67	8	10	7.0E-01
3589.8	15	13	5.9E-02	3680.12	10	12	2.2E+00	4298.03	6	8	7.8E-01
3590.7	9	7	2.2E-02	3686.26	10	12	2.3E-01	4379.23	10	12	1.1E+00
3591.7	11	9	5.3E-02	3687.50	12	14	2.9E+00	4384.71	8	10	1.1E+00
3593.0	11	11	1.4E-02	3688.07	8	8	3.5E-01	4389.98	6	8	6.9E-01
3593.2	13	15	4.2E-02	3690.28	2	4	4.5E-01	4395.22	4	6	5.5E-01
3593.7	11	11	7.2E-02	3692.22	6	6	5.4E-01	4400.57	2	4	3.4E-01
				3695.34	14	16	2.8E+00	4406.64	10	10	2.2E-01
Vanadium				3695.86	4	4	6.6E-01	4407.63	8	8	4.4E-01
V I				3703.57	10	8	9.2E-01	4408.20	6	6	6.0E-01
3043.12	6	8	2.3E-01	3704.70	8	6	6.6E-01	4416.47	4	2	2.6E-01
3050.39	10	8	5.3E-01	3705.04	6	4	3.6E-01	4452.01	14	16	9.2E-01
3053.65	4	4	1.3E+00	3706.03	10	10	5.2E-01	4457.75	10	12	2.7E-01
3056.33	6	6	1.3E+00	3708.71	12	12	4.4E-01	4460.33	10	8	3.0E-01
3060.46	8	8	1.4E+00	3790.46	10	8	2.3E-01	4462.36	12	14	7.6E-01
3066.37	10	10	2.1E+00	3794.96	10	10	2.3E-01	4468.00	8	10	2.3E-01
3066.53	6	4	3.2E-01	3806.79	10	10	2.5E-01	4469.71	10	12	6.2E-01

λ Å	g_i	g_k	A 10^8 s⁻¹	λ Å	g_i	g_k	A 10^8 s⁻¹	λ Å	g_i	g_k	A 10^8 s⁻¹
4474.04	10	8	4.7E-01	2893.31	9	7	1.2E+00	2371.06	10	12	5.2E+00
4496.06	8	6	4.0E-01	2903.07	3	5	3.4E-01	2373.06	4	6	2.9E+00
4514.18	6	4	3.3E-01	2906.45	7	7	7.8E-01	2382.46	8	10	5.0E+00
4524.21	12	10	3.0E-01	2908.81	11	9	1.6E+00	2393.58	6	8	4.3E+00
4525.17	4	2	4.1E-01	2910.01	5	5	1.1E+00	2404.18	4	6	2.5E+00
4529.58	10	8	2.4E-01	2910.38	3	3	1.2E+00	2516.14	10	10	3.7E+00
4545.40	10	12	7.6E-01	2911.05	7	9	3.7E-01	2521.55	8	8	3.5E+00
4560.72	8	10	7.0E-01	2912.46	11	9	5.0E-01	2548.21	6	4	2.0E+00
4571.79	6	8	6.0E-01	2915.88	9	7	4.9E-01	2554.22	8	6	1.2E+00
4578.73	4	6	6.8E-01	2924.02	11	11	1.7E+00	2593.05	6	6	2.8E+00
4706.16	6	4	2.4E-01	2924.63	9	9	1.2E+00	2595.10	8	8	2.8E+00
4757.47	4	2	7.6E-01	2930.80	7	7	5.8E-01				
4766.62	6	4	5.6E-01	2941.37	11	9	3.5E-01	**V IV**			
4776.36	8	6	5.1E-01	2944.57	9	7	7.6E-01	677.345	9	9	6.7E+00
4786.50	10	8	4.7E-01	2948.08	9	11	4.0E-01	680.632	9	7	1.2E+00
4796.92	12	10	4.8E-01	2952.07	7	5	7.2E-01	681.145	7	5	1.1E+01
4807.52	14	12	5.8E-01	2955.58	7	9	3.3E-01	682.455	7	7	6.5E+00
5193.00	12	12	4.0E-01	2968.37	7	9	7.0E-01	682.923	5	5	6.9E+00
5195.39	8	8	2.3E-01	2972.26	5	7	5.2E-01	684.450	7	5	7.7E+00
5234.08	10	10	4.9E-01	2973.98	9	11	3.5E-01	691.530	5	3	1.1E+01
5240.87	12	12	4.3E-01	2985.18	7	9	4.4E-01	723.537	3	1	1.5E+01
5415.25	12	14	3.1E-01	3001.20	7	7	7.5E-01	724.068	5	5	1.1E+01
5487.91	12	10	2.9E-01	3014.82	5	3	8.9E-01	724.809	5	3	5.6E+00
5507.75	10	8	3.5E-01	3016.78	7	5	5.0E-01	737.854	9	7	2.4E+01
6090.21	8	6	2.60E-01	3020.21	9	7	5.0E-01	750.110	5	5	1.0E+01
				3048.21	11	13	7.0E-01	884.146	1	3	4.7E+00
V II				3063.25	9	11	1.0E+00	1071.05	5	5	6.1E+00
2527.90	13	13	6.1E-01	3100.94	7	7	5.8E-01	1110.72	3	3	5.0E+00
2528.47	9	9	5.2E-01	3113.56	11	11	5.0E-01	1112.20	7	7	6.3E+00
2528.83	11	11	5.3E-01	3122.89	11	13	7.6E-01	1112.44	5	5	5.0E+00
2554.04	9	9	5.4E-01	3134.93	13	13	5.9E-01	1127.84	7	5	8.9E+00
2589.10	9	9	7.7E-01	3136.50	11	11	5.3E-01	1131.26	9	7	9.4E+00
2640.86	5	7	1.2E+00	3139.73	9	9	5.2E-01	1194.46	7	5	1.0E+01
2677.80	3	5	3.4E-01	3151.32	3	5	4.4E-01	1226.52	5	5	1.5E+01
2679.33	7	7	3.4E-01	3190.69	9	9	3.3E-01	1243.72	3	1	9.4E+00
2683.09	1	3	3.4E-01	3250.78	11	9	5.2E-01	1247.07	5	3	4.7E+00
2687.96	9	9	7.6E-01	3251.87	5	7	3.5E-01	1272.97	3	1	2.7E+01
2689.88	3	1	9.2E-01	3271.12	7	9	6.9E-01	1304.17	3	5	1.5E+01
2690.25	7	5	3.4E-01	3276.12	9	11	5.2E-01	1305.42	5	7	7.0E+00
2690.79	5	3	5.2E-01	3279.84	9	11	5.8E-01	1308.06	7	9	7.9E+00
2700.94	9	11	3.5E-01	3287.71	5	7	7.5E-01	1309.50	5	5	8.7E+00
2706.17	7	9	3.4E-01	3337.85	5	7	5.3E-01	1312.72	7	7	8.6E+00
2734.22	9	7	6.2E-01	3517.30	9	7	3.8E-01	1317.57	5	7	8.7E+00
2753.41	13	11	4.2E-01	3530.77	5	3	4.5E-01	1321.92	7	9	9.9E+00
2784.20	9	9	1.3E+00	3545.19	7	5	4.3E-01	1326.81	3	5	4.0E+00
2787.91	7	9	5.0E-01	3556.80	9	7	5.1E-01	1329.29	5	5	1.5E+01
2825.86	9	7	1.2E+00	3592.01	7	5	4.4E-01	1329.97	3	3	4.8E+00
2843.82	7	5	9.9E-01	3618.92	3	5	3.3E-01	1330.36	1	3	6.0E+00
2847.57	9	7	4.6E-01					1331.67	3	1	1.7E+01
2854.34	11	9	5.0E-01	**V III**				1332.46	5	3	7.5E+00
2862.31	11	11	3.6E-01	2318.06	8	10	4.6E+00	1334.49	9	9	8.3E+00
2868.11	5	3	2.1E+00	2323.82	6	8	3.8E+00	1355.13	7	9	2.5E+01
2869.13	13	11	4.8E-01	2330.42	10	12	3.2E+00	1356.53	5	3	4.9E+00
2882.49	5	5	4.2E-01	2331.75	8	8	2.5E+00	1395.00	5	7	1.4E+01
2884.78	3	3	5.6E-01	2334.21	6	6	2.2E+00	1400.42	5	7	7.5E+00
2889.61	3	1	1.9E+00	2337.13	4	4	2.7E+00	1403.62	7	9	8.4E+00
2891.64	5	3	1.4E+00	2343.10	6	8	3.6E+00	1412.69	3	3	1.1E+01
2892.43	9	9	3.6E-01	2358.73	6	8	4.2E+00	1414.41	5	7	1.2E+01
2892.65	7	5	1.3E+00	2366.31	8	10	4.2E+00	1414.84	5	5	4.6E+00

λ Å	g_i	g_k	A 10^8 s^{-1}	λ Å	g_i	g_k	A 10^8 s^{-1}	λ Å	g_i	g_k	A 10^8 s^{-1}
1418.53	7	7	5.2E+00	**Xe II**				4128.30	6	6	1.6E+00
1419.58	7	9	1.3E+01	4180.1	4	4	2.2E+00	4142.84	4	4	1.6E+00
1423.72	3	5	7.1E+00	4330.5	6	8	1.4E+00	4167.51	6	6	2.38E-01
1426.65	9	11	2.2E+01	4414.8	6	6	1.0E+00	4235.93	6	4	3.0E-01
1429.11	5	5	5.0E+00	4603.0	4	4	8.2E-01	4352.40	4	4	6.7E-03
1434.84	7	7	5.4E+00	4844.3	6	8	1.1E+00	4379.33	6	4	7.83E-01
1451.04	3	3	7.0E+00	4876.5	6	6	6.3E-01	4385.47	4	4	6.9E-02
1454.00	5	3	1.1E+01	5260.4	2	4	2.2E-01	4394.01	8	8	1.9E-02
1520.14	5	7	7.2E+00	5262.0	4	4	8.5E-01	4409.70	4	6	2.7E-03
1522.49	3	5	5.5E+00	5292.2	6	6	8.9E-01	4417.43	10	8	3.2E-02
1601.92	3	3	1.2E+01	5372.4	4	2	7.1E-01	4437.34	6	6	8.64E-02
1611.88	7	7	5.2E+00	5419.2	4	6	6.2E-01	4443.65	10	8	1.1E-01
1806.18	5	3	7.3E+00	5439.0	4	2	7.4E-01	4459.01	4	6	1.8E-02
1809.85	3	1	7.2E+00	5472.6	8	8	9.9E-02	4476.95	8	6	2.8E-01
1817.68	5	3	4.8E+00	5531.1	8	6	8.8E-02	4491.74	10	10	2.3E-02
1825.84	7	5	5.3E+00	5719.6	4	6	6.1E-02	4514.01	4	6	3.34E-01
1861.56	5	7	6.6E+00	5976.5	4	4	2.8E-01	4527.78	8	6	8.33E-01
1939.07	7	9	5.8E+00	6036.2	6	6	7.5E-02	4534.09	6	8	4.4E-02
1951.43	5	7	5.0E+00	6051.2	8	6	1.7E-01	4544.31	6	6	4.10E-01
1963.10	3	5	4.8E+00	6097.6	6	4	2.6E-01	4559.36	2	4	4.0E-01
1997.72	7	7	4.7E+00	6270.8	4	6	1.8E-01	4581.33	6	4	1.5E-01
2084.43	5	5	4.0E+00	6277.5	4	6	3.6E-02	4613.00	6	4	1.8E-01
2120.05	7	9	8.1E+00	6805.7	8	6	6.1E-02	4643.70	4	6	1.8E-01
2141.20	3	5	7.0E+00	6990.9	10	8	2.7E-01	4653.78	4	6	1.6E-01
2146.83	7	9	6.6E+00					4674.85	6	8	1.3E-01
2149.85	5	7	5.1E+00	**Ytterbium**				4725.84	4	4	1.5E-01
2151.09	7	9	4.3E+00	**Yb I**				4762.96	6	4	4.2E-02
2155.34	11	13	1.2E+01	2464.5	1	3	9.1E-01	4780.16	2	4	8.9E-02
2446.80	9	11	5.3E+00	2672.0	1	3	1.18E-01	4781.03	8	10	1.0E-01
2570.72	9	11	7.6E+00	3464.4	1	3	6.2E-01	4799.30	6	8	1.6E-01
3284.56	7	9	5.3E+00	3988.0	1	3	1.76E+00	4804.31	6	4	2.6E-01
3496.42	7	9	4.4E+00	5556.5	1	3	1.14E-02	4804.80	4	4	3.84E-01
3514.25	9	11	4.7E+00					4821.63	6	6	1.0E-01
				Yb II				4845.67	8	8	6.8E-01
				3289.4	2	4	1.8E+00	4852.68	6	6	6.2E-01
Xenon				3694.2	2	2	1.4E+00	4856.71	6	6	2.0E-01
Xe I								4859.84	4	4	7.26E-01
1043.8	1	3	5.9E-01	**Yttrium**				4893.44	6	4	2.2E-01
1047.1	1	3	1.3E+00	**Y I**				4900.08	8	6	2.0E-01
1050.1	1	3	8.5E-02	2948.41	4	4	3.5E-01	4906.11	10	8	1.2E-01
1056.1	1	3	2.45E+00	2974.59	4	6	3.5E-01	4950.01	8	6	2.0E-02
1061.2	1	3	1.9E-01	2984.25	6	8	4.8E-01	4963.49	4	4	1.4E-02
1068.2	1	3	3.99E+00	2995.26	6	4	5.1E-02	4981.97	4	6	4.7E-03
1085.4	1	3	4.10E-01	2996.94	4	6	8.4E-02	5004.44	6	4	1.2E-02
1099.7	1	3	4.34E-01	3005.26	4	4	4.8E-02	5205.01	4	4	8.4E-03
1110.7	1	3	1.5E+00	3022.28	6	6	6.6E-02	5258.47	6	6	2.9E-03
1129.3	1	3	4.4E-02	3045.36	6	6	1.07E-01	5271.82	8	6	1.1E-02
1170.4	1	3	1.6E+00	3053.95	6	4	1.9E-03	5380.63	6	4	3.2E-01
1192.0	1	3	6.2E+00	3155.65	4	6	2.7E-03	5381.24	4	4	9.9E-03
1250.2	1	3	1.4E-01	3172.84	4	4	9.9E-03	5388.39	6	8	1.1E-02
1295.6	1	3	2.46E+00	3185.96	6	8	1.2E-03	5390.81	8	6	2.9E-02
1469.6	1	3	2.81E+00	3209.38	6	6	3.0E-03	5401.88	6	8	6.0E-03
4501.0	5	3	6.2E-03	3227.16	6	4	1.10E-03	5424.36	6	4	3.47E-01
4524.7	5	5	2.1E-03	3484.05	4	6	1.2E-02	5466.24	4	4	1.0E-01
4624.3	5	5	7.2E-03	3549.66	6	6	1.0E-03	5466.47	10	12	6.3E-01
4671.2	5	7	1.0E-02	3552.69	4	4	2.3E-01	5469.10	4	6	3.6E-03
4807.0	3	1	2.4E-02	4077.36	4	6	1.1E+00	5513.65	6	6	2.39E-01
7119.6	7	9	6.6E-02	4083.71	4	4	2.5E-01	5519.88	4	6	1.2E-02
7967.3	1	3	3.0E-03	4102.36	6	8	1.3E+00	5526.43	6	4	3.9E-03
8409.2	5	3	1.0E-02								

λ Å	Weights g_i	Weights g_k	A 10^8 s^{-1}	λ Å	Weights g_i	Weights g_k	A 10^8 s^{-1}	λ Å	Weights g_i	Weights g_k	A 10^8 s^{-1}
5527.56	8	10	5.4E-01	3448.81	5	5	4.1E-02	5119.11	5	7	1.6E-02
5541.63	8	8	5.2E-02	3467.88	5	3	2.7E-02	5200.41	5	5	1.3E-01
5551.00	4	4	6.9E-02	3496.08	1	3	3.49E-01	5205.73	7	7	1.6E-01
5573.03	6	4	1.8E-02	3549.01	5	7	3.97E-01	5289.82	7	5	6.7E-03
5594.12	6	8	5.0E-02	3584.51	3	5	4.02E-01	5320.78	9	7	3.9E-03
5606.34	10	10	5.84E-02	3600.74	7	7	1.4E+00	5473.39	3	5	4.3E-02
5619.96	6	4	2.0E-02	3601.91	3	3	1.13E+00	5480.73	1	3	7.62E-02
5630.14	4	6	4.9E-01	3611.04	5	5	1.04E+00	5497.41	5	5	1.2E-01
5641.78	2	4	1.9E-02	3628.70	5	3	3.3E-01	5509.90	5	5	4.24E-02
5675.27	6	6	9.3E-02	3664.62	7	5	3.7E-01	5544.61	3	1	1.8E-01
5675.64	4	6	4.3E-02	3710.29	7	9	1.5E+00	5546.01	5	3	5.8E-02
5693.63	4	4	1.1E-01	3747.55	3	3	1.9E-01	5728.89	5	5	3.0E-02
5714.94	8	6	2.0E-02	3774.34	5	7	1.1E+00	6613.74	5	7	1.7E-02
5729.25	6	6	2.2E-03	3776.56	5	3	2.42E-01	6832.48	5	5	3.3E-03
5732.09	6	6	7.5E-02	3788.70	3	5	8.1E-01	7264.16	5	3	1.3E-02
5740.22	8	6	4.0E-02	3818.34	5	5	9.70E-02				
5757.59	4	6	7.6E-03	3832.90	7	7	3.0E-01	**Zinc**			
5788.36	4	4	9.4E-03	3878.29	7	5	2.9E-02	**Zn I**			
5844.13	6	4	5.6E-03	3930.66	5	5	2.1E-02	748.29	1	3	6.0E-02
5879.93	4	2	8.5E-02	3950.36	3	5	2.80E-01	765.60	1	3	7.6E-02
5902.91	6	8	4.0E-02	3951.59	5	3	1.5E-02	792.05	1	3	5.7E-02
6087.94	6	4	1.1E-01	3982.60	5	5	2.7E-01	793.85	1	3	1.8E-01
6191.72	4	4	4.7E-02	4124.91	5	7	1.8E-02	809.92	1	3	2.6E-01
6222.58	4	6	5.9E-03	4177.54	5	5	5.27E-01	1109.1	1	3	3.05E-01
6402.01	6	4	2.7E-01	4199.27	3	5	5.36E-03	2138.6	1	3	7.09E+00
6435.02	6	6	4.0E-02	4204.69	1	3	2.20E-02	3075.9	1	3	3.29E-04
6437.17	10	8	4.8E-02	4235.73	5	5	2.3E-02	3282.3	1	3	9.0E-01
6538.57	10	10	1.5E-01	4309.62	7	5	1.29E-01	3302.6	3	5	1.2E+00
6622.48	8	6	4.5E-03	4358.73	3	3	5.55E-02	3302.9	3	3	6.7E-01
6815.15	2	4	7.18E-02	4374.95	5	5	9.97E-01	3345.0	5	7	1.7E+00
7009.89	2	4	4.4E-02	4398.01	5	3	1.16E-01	3345.6	5	5	4.0E-01
7035.15	4	4	6.3E-02	4422.59	3	1	1.83E-01	3345.9	5	3	4.5E-02
				4682.33	5	5	1.9E-02	6362.3	3	5	4.74E-01
Y II				4786.58	7	7	2.1E-02	11054	3	1	2.43E-01
3112.03	1	3	1.3E-02	4823.31	5	5	4.3E-02				
3179.42	3	5	3.8E-02	4854.87	5	3	3.9E-01	**Zn II**			
3195.62	3	3	8.23E-01	4881.44	5	3	1.5E-03	2025.5	2	4	3.3E+00
3200.27	5	5	4.8E-01	4883.69	9	7	4.7E-01	2064.2	2	4	4.6E+00
3203.32	3	1	2.77E+00	4900.11	7	5	4.51E-01	2099.9	4	6	5.6E+00
3216.69	5	3	2.0E+00	4982.13	7	9	1.5E-02	2102.2	4	4	9.3E-01
3242.28	7	5	2.0E+00	5087.42	9	9	2.0E-01	4911.6	4	6	1.6E+00

ELECTRON AFFINITIES
Thomas M. Miller

Electron affinity is defined as the energy difference between the lowest (ground) state of the neutral and the lowest state of the corresponding negative ion. The accuracy of electron affinity measurements has been greatly improved since the advent of laser photodetachment experiments with negative ions. Electron affinities can be determined with optical precision, though a detailed understanding of atomic and molecular states and splittings is required to specify the photodetachment threshold corresponding to the electron affinity.

Atomic and molecular electron affinities are discussed in two excellent articles reviewing photodetachment studies which appear in *Gas Phase Ion Chemistry*, Vol. 3, Bowers, M. T., Ed., Academic Press, Orlando, 1984: Chapter 21 by Drzaic, P. S., Marks, J., and Brauman, J. I., "Electron Photodetachment from Gas Phase Negative Ions", p. 167, and Chapter 22 by Mead, R. D., Stevens, A. E., and Lineberger, W. C., "Photodetachment in Negative Ion Beams", p. 213. Persons interested in photodetachment details should consult these articles and the critical review of Hotop, H., and Lineberger, W. C., *J. Phys. Chem. Ref. Data*, 14, 731, 1985. For simplicity in the tables below, any electron affinity which was discussed in the articles by Drzaic *et al.* or Hotop and Lineberger is referenced to these sources, where original references are given. A great many additional electron affinities have been provided here by G. B. Ellison, W. C. Lineberger, H. Hotop, D. G. Leopold, and K. H. Bowen. Electron affinities for the lanthanides and actinides have not been measured, but theoretical estimates have been made by Bratch, S. G., *Chem. Phys. Lett.*, 98, 113, 1983, and Bratch, S. G., and Lagowski, J. J., *Chem. Phys. Lett.*, 107, 136, 1984. The development of cluster-ion photodetachment apparatuses has brought an explosion of electron affinity estimates for atomic and molecular clusters. [See Arnold, S. T., Eaton, J. G., Patel-Mistra, D., Sarkas, H. W., and Bowen, K. H., in *Ion and Cluster Ion Spectroscopy and Structure*, Maier, J. P., Ed., Elsevier Science, New York, 1989, p. 417.] The policy in this tabulation is to list the electron affinities for the atoms, diatoms, and triatoms, if adiabatic electron affinities have been determined, but to refer the reader to original sources for higher-order clusters. Additional data on molecular electron affinities may be found in Lias, S. G., Bartmess, J. E., Liebman, J. F., Holmes, J. L., Levin, R. D., and Mallard, W. G., *Gas Phase Ion and Neutral Thermochemistry, J. Phys. Chem. Ref. Data*, 17, (Supplement No. 1), 1988.

For the present tabulation the value $e/hc = 8065.5410 \pm 0.0024$ cm^{-1} eV^{-1} (Cohen, E. R., and Taylor, B. N., *Rev. Mod. Phys.*, 59, 1121, 1987) has been used to convert electron affinities from the units used in spectroscopic work, cm^{-1}, into eV for these tables. The uncertainties in the electron affinities given in the tables do not include the 0.3 ppm uncertainty in e/hc; only in a few cases, *e.g.*, atomic oxygen, is this additional uncertainty significant.

Abbreviations used in the tables: calc = calculated value; PT = photodetachment threshold using a lamp as a light source; LPT = laser photodetachment threshold; LPES = laser photoelectron spectroscopy; DA = dissociative attachment; e-scat = electron scattering; CT = charge transfer; and CD = collisional detachment.

Table 1
Atomic Electron Affinities

Atomic number	Atom	Electron affinity in eV	Uncertainty in eV	Method	Ref.
1	D	0.754593	0.000074	LPT	89
	H	0.754195	0.000019	LPT	89
		0.754209	0.000003	calc	1
2	He	not stable	—	calc	1
3	Li	0.6180	0.0005	LPT	1
4	Be	not stable	—	calc	1
5	B	0.277	0.010	LPES	1
6	C	1.2629	0.0003	LPT	1
7	N	not stable	—	DA	1
8	O	1.4611103	0.0000007	LPT	4
9	F	3.401190	0.000004	LPT	74
10	Ne	not stable	—	calc	1
11	Na	0.547926	0.000025	LPT	1
12	Mg	not stable	—	e-scat	1
13	Al	0.441	0.010	LPES	1
14	Si	1.385	0.005	LPES	1
15	P	0.7465	0.0003	LPT	1
16	S	2.077104	0.000001	LPT	1
17	Cl	3.61269	0.00006	LPT	52
18	Ar	not stable	—	calc	1
19	K	0.50147	0.00010	LPT	1
20	Ca	0.02455	0.00010	LPT	44
21	Sc	0.188	0.020	LPES	1
22	Ti	0.079	0.014	LPES	1
23	V	0.525	0.012	LPES	1
24	Cr	0.666	0.012	LPES	1
25	Mn	not stable	—	calc	1

Table 1
Atomic Electron Affinities (continued)

Atomic number	Atom	Electron affinity in eV	Uncertainty in eV	Method	Ref.
26	Fe	0.151	0.003	LPES	27
27	Co	0.662	0.003	LPES	27
28	Ni	1.156	0.010	LPES	1
29	Cu	1.235	0.005	LPES	37
30	Zn	not stable	—	e-scat	1
31	Ga	0.3	0.15	PT	1
32	Ge	1.233	0.003	LPES	28
33	As	0.81	0.03	PT	1
34	Se	2.020670	0.000025	LPT	1
35	Br	3.363590	0.000003	LPT	74
36	Kr	not stable	—	calc	1
37	Rb	0.48592	0.00002	LPT	1
38	Sr	0.048	0.006	LPT	122
39	Y	0.307	0.012	LPES	1
40	Zr	0.426	0.014	LPES	1
41	Nb	0.893	0.025	LPES	1
42	Mo	0.748	0.002	LPES	127
43	Tc	0.55	0.20	calc	1
44	Ru	1.05	0.15	calc	1
45	Rh	1.137	0.008	LPES	1
46	Pd	0.562	0.005	LPES	116
47	Ag	1.302	0.007	LPES	1
48	Cd	not stable	—	e-scat	1
49	In	0.3	0.2	PT	1
50	Sn	1.112	0.004	LPES	28
51	Sb	1.046	0.005	LPES	108
52	Te	1.9708	0.0003	LPT	1
53	I	3.059038	0.000010	LPT	92
54	Xe	not stable	—	calc	1
55	Cs	0.471626	0.000025	LPT	1
56	Ba	0.15	—	calc	57
57	La	0.5	0.3	calc	1
72	Hf	≈0	—	calc	1
73	Ta	0.322	0.012	LPES	1
74	W	0.815	0.002	LPES	37
75	Re	0.15	0.15	calc	1
76	Os	1.1	0.2	calc	1
77	Ir	1.5638	0.0005	LPT	141
78	Pt	2.128	0.002	LPT	1
79	Au	2.30863	0.00003	LPT	1
80	Hg	not stable	—	e-scat	1
81	Tl	0.2	0.2	PT	1
82	Pb	0.364	0.008	LPES	1
83	Bi	0.946	0.010	LPES	1
84	Po	1.9	0.3	calc	1
85	At	2.8	0.2	calc	1
86	Rn	not stable	—	calc	1
87	Fr	0.46	—	calc	82
118	ekaradon	0.056	—	calc	140

Table 2
Electron Affinities for Diatomic Molecules

Molecule	Electron affinity in eV	Uncertainty in eV	Method	Ref.	Molecule	Electron affinity in eV	Uncertainty in eV	Method	Ref.
Ag_2	1.023	0.007	LPES	37	MgH	1.05	0.06	PT	2
Al_2	1.10	0.15	LPES	68	MgI	1.899	0.018	LPES	31
AlO	2.60	0.02	LPES	143	MnD	0.866	0.010	LPES	9
AlS	2.60	0.03	LPES	129	MnH	0.869	0.010	LPES	9
As_2	0	—	PT	2	MoO	1.290	0.006	LPES	127
AsH	1.0	0.1	PT	2	NH	0.370	0.004	LPT	32
Au_2	1.938	0.007	LPES	37	NO	0.026	0.005	LPES	73
BO	3.12	0.09	PT	6	NS	1.194	0.011	LPES	2
BeH	0.7	0.1	PT	2	Na_2	0.430	0.015	LPES	104
Bi_2	1.271	0.008	LPES	119	NaBr	0.788	0.010	LPES	30
Br_2	2.55	0.10	CT	2	NaCl	0.727	0.010	LPES	30
BrO	2.353	0.006	LPES	88	NaF	0.520	0.010	LPES	30
C_2	3.269	0.006	LPES	87	NaI	0.865	0.010	LPES	30
CH	1.238	0.008	LPES	2	NaK	0.465	0.030	LPES	104
CN	3.862	0.004	LPES	111	Ni_2	0.926	0.010	LPES	112
CS	0.205	0.021	LPES	2	NiCu	0.889	0.010	LPES	128
CaH	0.93	0.05	PT	2	NiAg	0.979	0.010	LPES	128
Cl_2	2.38	0.10	CT	2	NiD	0.477	0.007	LPES	29
ClO	2.275	0.006	LPES	88	NiH	0.481	0.007	LPES	29
Co_2	1.110	0.008	LPES	27	NiO	1.470	0.003	LPES	146
CoD	0.680	0.010	LPES	29	O_2	0.451	0.007	LPES	73
CoH	0.671	0.010	LPES	29	OD	1.825534	0.000037	LPT	142
Cr_2	0.505	0.005	LPES	114	OH	1.8276542	0.0000037	LPT	142
CrD	0.568	0.010	LPES	29	P_2	0.589	0.025	LPES	42
CrH	0.563	0.010	LPES	29	PH	1.028	0.010	LPES	2
CrO	1.221	0.006	LPES	5	PO	1.092	0.010	LPES	2
Cs_2	0.469	0.015	LPES	104	Pb_2	1.366	0.010	LPES	117
CsCl	0.455	0.010	LPES	30	PbO	0.722	0.006	LPES	105
CsO	0.273	0.012	LPES	133	Pd_2	1.685	0.008	LPES	112
Cu_2	0.836	0.006	LPES	37	Pt_2	1.898	0.008	LPES	112
CuO	1.777	0.006	LPES	118	PtN	1.240	0.010	LPES	46
F_2	3.08	0.10	CT	2	Rb_2	0.498	0.015	LPES	104
FO	2.272	0.006	LPES	88	RbCl	0.544	0.010	LPES	30
Fe_2	0.902	0.008	LPES	27	RbCs	0.478	0.020	LPES	104
FeD	0.932	0.015	LPES	9	Re_2	1.571	0.008	LPES	33
FeH	0.934	0.011	LPES	9	S_2	1.670	0.015	LPES	53
FeO	1.493	0.005	LPES	45	SD	2.315	0.002	LPES	10
Ge_2	2.035	0.001	LPES	123	SF	2.285	0.006	LPES	93
I_2	2.55	0.05	CT	2	SH	2.314344	0.000004	LPT	47
IBr	2.55	0.10	CT	2	SO	1.125	0.005	LPES	84
IO	2.378	0.006	LPES	88	Sb_2	1.282	0.008	LPES	108
InP	1.95	0.05	LPES	137	Se_2	1.94	0.07	LPES	38
K_2	0.497	0.012	LPES	104	SeH	2.212520	0.000025	LPT	48
KBr	0.642	0.010	LPES	30	SeO	1.456	0.020	LPES	41
KCl	0.582	0.010	LPES	30	Si_2	2.201	0.010	LPES	100
KCs	0.471	0.020	LPES	104	SiH	1.277	0.009	LPES	2
KI	0.728	0.010	LPES	30	Sn_2	1.962	0.010	LPES	117
KRb	0.486	0.020	LPES	104	SnPb	1.569	0.008	LPES	117
LiCl	0.593	0.010	LPES	30	Te_2	1.92	0.07	LPES	38
LiD	0.337	0.012	LPES	102	TeH	2.102	0.015	LPES	39
LiH	0.342	0.012	LPES	102	TeO	1.697	0.022	LPES	40
MgCl	1.589	0.011	LPES	31	ZnH	≤0.95	—	PT	2

Table 3
Electron Affinities for Triatomic Molecules

Molecule	Electron affinity in eV	Uncertainty in eV	Method	Ref.	Molecule	Electron affinity in eV	Uncertainty in eV	Method	Ref.
Ag_3	2.32	0.05	LPES	37	Ge_3	2.23	0.01	LPES	123
Al_3	1.4	0.15	LPES	68	GeH_2	1.097	0.015	LPES	28
AlO_2	4.23	0.02	LPES	143	HCO	0.313	0.005	LPES	35
Al_2S	0.80	0.12	LPES	129	HCl_2	4.896	0.005	LPES	69
AsH_2	1.27	0.03	PT	2	HNO	0.338	0.015	LPES	14
Au_3	3.7	0.3	LPES	37	HO_2	1.078	0.017	LPES	15
BO_2	3.57	0.13	CT	6	InP_2	1.61	0.05	LPES	137
BO_2	4.3	0.2	CT	98	In_2P	2.36	0.05	LPES	137
Bi_3	1.60	0.03	LPES	119	K_3	0.956	0.050	LPES	18
C_3	1.981	0.020	LPES	11	MnD_2	0.465	0.014	LPES	34
CCl_2	1.591	0.010	LPES	95	MnH_2	0.444	0.016	LPES	34
CD_2	0.645	0.006	LPES	12	N_3	2.70	0.12	PT	2
CDF	0.535	0.005	LPES	95	NCN	2.484	0.006	LPES	154
CF_2	0.165	0.010	LPES	95	NCND	2.622	0.005	LPES	154
CH_2	0.652	0.006	LPES	12	NCNH	2.622	0.005	LPES	154
CHBr	1.454	0.005	LPES	95	NCO	3.609	0.005	LPES	111
CHCl	1.210	0.005	LPES	95	NCS	3.537	0.005	LPES	111
CHF	0.542	0.005	LPES	95	NH_2	0.771	0.005	LPES	58
CHI	1.42	0.17	LPES	95	N_2O	0.22	0.10	CT	59
C_2H	2.969	0.006	LPES	87	NO_2	2.273	0.005	LPES	63
C_2O	1.848	0.027	LPES	13	(NO)R	R=Ar,Kr,Xe	—	LPES	90
COS	0.46	0.20	CD	2	Na_3	1.019	0.060	LPES	18
CS_2	0.895	0.020	LPES	11	Ni_3	1.41	0.05	LPES	55
C_2Ti	1.542	0.020	LPES	147	NiCO	0.804	0.012	LPES	2
CoD_2	1.465	0.013	LPES	34	NiD_2	1.926	0.007	LPES	34
CoH_2	1.450	0.014	LPES	34	NiH_2	1.934	0.008	LPES	34
CrH_2	>2.5	—	LPES	34	O_3	2.1028	0.0025	LPT	2
Cr_2D	1.464	0.005	LPES	107	O_2Ar	0.52	0.02	LPES	75
Cr_2H	1.474	0.005	LPES	107	OClO	2.140	0.008	LPES	88
CrO_2	2.413	0.008	LPES	144	OIO	2.577	0.008	LPES	88
Cs_3	0.864	0.030	LPES	18	PH_2	1.271	0.010	LPES	2
Cu_3	2.11	0.05	LPES	37	PO_2	3.42	0.01	LPES	124
DCO	0.301	0.005	LPES	35	Pt_3	1.87	0.02	LPES	55
DNO	0.330	0.015	LPES	14	Pd_3	<1.5	0.1	LPES	55
DO_2	1.089	0.017	LPES	15	Rb_3	0.920	0.030	LPES	18
DS_2	1.912	0.015	LPES	53	S_3	2.093	0.025	LPES	16
HS_2	1.907	0.015	LPES	53	SO_2	1.107	0.008	LPES	16
Fe_3	1.47	0.08	LPES	149	S_2O	1.877	0.008	LPES	16
FeCO	1.157	0.005	LPES	103	Sb_3	1.85	0.03	LPES	108
FeD_2	1.038	0.013	LPES	34	SeO_2	1.823	0.050	LPES	38
FeH_2	1.049	0.014	LPES	34	SiH_2	1.124	0.020	LPES	2
FeO_2	2.358	0.030	LPES	130	VO_2	2.3	0.2	CT	101

Table 4
Electron Affinities for Larger Polyatomic Molecules

Molecule	Electron Affinity in eV	Uncertainty in eV	Method	Ref.	Name
Ag_n	n = 1-10	—	LPES	37	
Al_n	n = 3-32	—	LPES	68	
Al_nO_m	n=1,2	m=1-5	LPES	143	
Al_3O	1.00	0.15	LPES	68	
Al_nS_m	n = 1-5	m = 1-3	LPES	129	
$Ar(H_2O)_n$	n = 2,6,7	—	LPES	77	

Table 4
Electron Affinities for Larger Polyatomic Molecules (continued)

Molecule	Electron Affinity in eV	Uncertainty in eV	Method	Ref.	Name
Au_n	$n = 2\text{-}5$	—	LPES	37	
AuF_6	7.5	estimate	CT	98	
BD_3	0.027	0.014	LPES	62	
BH_3	0.038	0.015	LPES	62	
Bi_4	1.05	0.010	LPES	119	
$Br(CO_2)$	3.582	0.017	LPES	131	
C_n	$n = 2\text{-}84$	—	LPES	70	
$(CO_2)_n$	$n = 1,2$	—	LPES	75	
$(CS)_n$	$n = 2$	—	LPES	75	
$(CS_2)_n$	$n = 1,2$	—	LPES	75	
CF_3Br	0.91	0.2	CD	2	
CF_3I	1.57	0.2	CD	2	
CDO_2	3.510	0.015	LPES	109	
CHO_2	3.498	0.015	LPES	109	
CH_2S	0.465	0.023	LPES	53	
CH_3	0.08	0.03	LPES	2	
CH_3I	0.2	0.1	CT	2	
CH_3NO_2	0.48	0.10	CT	61	
CH_3Si	0.852	0.010	LPES	97	$CH_3\text{-}Si$
CH_3Si	2.010	0.010	LPES	97	$CH_2 = SiH$
CD_3O	1.552	0.022	LPES	2	
CH_3O	1.570	0.022	LPES	2	
CD_3S	1.856	0.006	LPT	2	
CH_3S	1.861	0.004	LPT	2	
CD_3S_2	1.748	0.022	LPES	53	
CH_3S_2	1.757	0.022	LPES	53	
CH_3SiH_2	1.19	0.04	LPT	65	
CFO_2	4.277	0.030	LPES	131	
CO_3	2.69	0.14	LPES	2	
$CO_3(H_2O)$	2.1	0.2	PT	2	
C_2F_2	2.255	0.006	LPES	106	difluorovinylidene
C_2DO	2.350	0.020	LPES	13	
C_2HO	2.350	0.020	LPES	13	
C_2D_2	0.492	0.006	LPES	83	vinylidene-d_2
C_2HD	0.489	0.006	LPES	83	vinylidene-d_1
C_2HF	1.718	0.006	LPES	106	monofluorovinylidene
C_2H_2	0.490	0.006	LPES	83	vinylidene
C_2H_2FO	2.22	0.09	PT	2	acetyl fluoride enolate
C_2D_2N	1.538	0.012	LPES	21	cyanomethyl-d_2 radical
C_2D_2N	1.070	0.024	LPES	21	isocyanomethyl-d_2 radical
C_2H_2N	1.543	0.014	LPES	21	cyanomethyl radical
C_2H_2N	1.059	0.024	LPES	21	isocyanomethyl radical
C_2H_3	0.667	0.024	LPES	90	vinyl
C_2D_3O	1.81897	0.00012	LPT	22	acetaldehyde-d_3 enolate
C_2H_3O	1.82476	0.00012	LPT	22	acetaldehyde enolate
C_2H_5N	0.56	0.01	PT	2	ethyl nitrine
C_2D_5O	1.702	0.033	LPES	23	ethoxide-d_3
C_2H_5O	1.726	0.033	LPES	23	ethoxide
C_2H_5S	1.953	0.006	LPT	2	ethyl sulfide
C_2H_5S	0.868	0.051	LPES	53	CH_3SCH_2
$C_2H_7O_2$	2.26	0.08	PT	50	MeOHOMe
C_3Fe	1.69	0.08	LPES	132	
C_3H	1.858	0.023	LPES	11	
C_3HFe	1.58	0.06	LPES	132	
C_3H_2	1.794	0.008	LPES	153	
$C_3H_2F_3O$	2.625	0.010	LPT	113	1,1,1-trifluoroacetone enolate

Table 4
Electron Affinities for Larger Polyatomic Molecules (continued)

Molecule	Electron Affinity in eV	Uncertainty in eV	Method	Ref.	Name
C_3H_3	0.893	0.025	LPES	24	propargyl radical
C_3H_2D	0.88	0.15	LPES	24	propargyl-d_1 radical
C_3D_2H	0.907	0.023	LPES	24	propargyl-d_2 radical
C_3H_3N	1.247	0.012	LPES	21	$CH_3CH\text{-}CN$
C_3D_5	0.464	0.006	LPES	138	allyl-d_5
C_3H_5	0.481	0.008	LPES	138	allyl
C_3H_4D	0.373	0.019	LPES	25	allyl-d_1
C_3H_5O	1.758	0.019	LPT	113	acetone enolate
C_3H_5O	1.621	0.006	LPT	113	propionaldehyde enolate
$C_3H_5O_2$	1.80	0.06	PT	2	methyl acetate enolate
C_3H_7O	1.789	0.033	LPES	23	n-propyl oxide
C_3H_7O	1.839	0.029	LPES	23	isopropyl oxide
C_3H_7S	2.00	0.02	PT	2	n-propyl sulfide
C_3H_7S	2.02	0.02	PT	2	isopropyl sulfide
C_3O	1.34	0.15	LPES	11	
C_3O_2	0.85	0.15	LPES	11	
C_3Ti	1.561	0.015	LPES	147	
$C_4F_4O_3$	0.5	0.2	CD	2	tetrafluorosuccinic anhydride
C_4Fe	≤2.2	0.2	LPES	132	
C_4HFe	1.67	0.06	LPES	132	
$C_4H_2O_3$	1.44	0.10	CT	61	maleic anhydride
C_4D_4	0.909	0.015	LPES	125	vinylvinylidene-d_4
C_4H_4	0.914	0.015	LPES	125	vinylvinylidene
C_4H_4N	2.39	0.13	PT	2	pyrrolate
C_4H_5O	1.801	0.008	LPT	113	cyclobutanone enolate
C_4H_6	0.431	0.006	LPES	135	trimethylenemethane
$C_4H_6O_2$	0.69	0.10	CT	61	2,3-butanedione
C_4H_6D	0.493	0.008	LPES	138	2-methylallyl-d_7
C_4H_7	0.505	0.006	LPES	138	2-methylallyl
C_4H_7O	1.67	0.05	PT	2	butyraldehyde enolate
C_4H_5DO	1.67	0.05	PT	2	2-butanone-3-d_1 enolate
$C_4H_5D_2O$	1.75	0.06	PT	2	2-butanone-3,3-d_2 enolate
C_4H_9O	1.912	0.054	LPES	23	t-butoxyl
C_4H_9S	2.03	0.02	PT	2	n-butyl sulfide
C_4H_9S	2.07	0.02	PT	2	t-butyl sulfide
C_4O	2.05	0.15	LPES	11	
C_4O_2	2.0	0.2	LPES	11	
C_4Ti	1.494	0.020	LPES	147	
C_5	2.853	0.001	LPT	99	
C_5F_5N	0.68	0.11	CT	67	pentafluoropyridine
$C_5F_6O_3$	1.5	0.2	CD	2	hexafluoroglutaric anhydride
C_5D_5	1.790	0.008	LPES	11	cyclopentadienyl-d_5
C_5H_5	1.804	0.007	LPES	11	cyclopentadienyl
C_5H_7	0.91	0.03	PT	2	pentadienyl
C_5H_7O	1.598	0.007	LPT	113	cyclopentanone enolate
C_5H_9O	1.69	0.05	PT	2	3-penanone enolate
$C_5H_{11}O$	1.93	0.05	LPT	2	neopentoxyl
$C_5H_{11}S$	2.09	0.02	PT	2	n-pentyl sulfide
C_5O_2	1.2	0.2	LPES	11	
C_5Ti	1.748	0.050	LPES	147	
C_6	4.180	0.001	LPT	8	
$C_6Br_4O_2$	2.44	0.20	CT	2	tetrabromo-BQ
$C_6Cl_4O_2$	2.78	0.10	CT	61	tetrachloro-BQ
$C_6F_4O_2$	2.70	0.10	CT	61	tetrafluoro-BQ
C_6F_5Br	1.15	0.11	CT	67	pentafluorobromobenzene

Table 4
Electron Affinities for Larger Polyatomic Molecules (continued)

Molecule	Electron Affinity in eV	Uncertainty in eV	Method	Ref.	Name
C_6F_5Cl	0.82	0.11	CT	67	pentafluorochlorobenzene
C_6F_5I	1.41	0.11	CT	67	pentafluoroiodobenzene
$C_6F_5NO_2$	1.52	0.11	CT	67	pentafluoro-NB
C_6F_6	0.52	0.10	CT	51	hexafluorobenzene
C_6F_{10}	>1.4	0.3	CT	2	perfluorocyclohexene
$C_6H_2Cl_2O_2$	2.48	0.10	CT	61	2,6-dichloro-BQ
$C_6H_3F_2NO_2$	1.17	0.10	CT	61	2,4-difluoro-NB
C_6D_4	0.551	0.010	LPES	36	o-benzyne-d$_4$
C_6H_4	0.560	0.010	LPES	36	o-benzyne
$C_6H_4BrNO_2$	1.16	0.10	CT	61	o-bromo-NB
$C_6H_4BrNO_2$	1.32	0.10	CT	61	m-bromo-NB
$C_6H_4BrNO_2$	1.29	0.10	CT	61	p-bromo-NB
$C_6H_4ClNO_2$	1.14	0.10	CT	61	o-chloro-NB
$C_6H_4ClNO_2$	1.28	0.10	CT	61	m-chloro-NB
$C_6H_4ClNO_2$	1.26	0.10	CT	61	p-chloro-NB
C_6H_4ClO	≤2.58	0.08	PT	2	o-chloroperoxide
$C_6H_4FNO_2$	1.07	0.10	CT	61	o-fluoro-NB
$C_6H_4FNO_2$	1.23	0.10	CT	61	m-fluoro-NB
$C_6H_4FNO_2$	1.12	0.10	CT	61	p-fluoro-NB
$C_6H_4N_2O_4$	1.65	0.10	CT	61	o-diNB
$C_6H_4N_2O_4$	1.65	0.10	CT	61	m-diNB
$C_6H_4N_2O_4$	2.00	0.10	CT	61	p-diNB
$C_6H_4O_2$	1.91	0.10	CT	61	1,4-benzoquinone (BQ)
C_6D_5	1.092	0.020	LPES	26	phenyl-d$_5$
C_6D_5N	1.44	0.02	LPES	96	phenylnitrene-d$_5$
C_6H_5	1.096	0.006	LPES	26	phenyl
C_6H_5N	1.429	0.011	LPT	115	phenylnitrene
$C_6H_5NO_2$	1.01	0.10	CT	61	nitrobenzene (NB)
C_6H_5O	2.253	0.006	LPES	26	phenoxyl
C_6H_5S	≤2.47	0.06	PT	2	thiophenoxide
C_6H_5NH	1.70	0.03	PT	2	anilide
C_6H_7	<1.67	0.04	PT	2	methylchylopentadienyl
C_6H_8Si	1.435	0.004	LPT	65	$C_6H_5SiH_3$
C_6H_9O	1.526	0.010	LPT	113	cyclohexanone enolate
C_6H_{10}	0.645	0.015	LPES	126	t-butyl vinylidene
$C_6H_{11}O$	1.755	+0.05/-0.005	LPT	113	pinacolone enolate
$C_6H_{11}O$	1.82	0.06	PT	2	3,3-dimethylbutananl enolate
C_6N_4	2.3	0.3	PT	2	TCNE
C_7F_5N	1.11	0.11	CT	67	pentafluorobenzonitrile
C_7F_8	0.86	0.11	CT	67	octafluorotoluene
C_7F_{14}	1.08	0.10	CT	61	perfluoromethylcyclohexane
C_7HF_5O	1.10	0.11	CT	67	pentafluorobenzaldehyde
$C_7H_3N_3O_4$	2.16	0.10	CT	61	3,5-$(NO_2)_2$-benzonitrile
$C_7H_4F_3NO_2$	1.41	0.10	CT	61	m-trifluoromethyl-NB
$C_7H_4N_2O_2$	1.61	0.10	CT	61	o-cyano-NB
$C_7H_4N_2O_2$	1.56	0.10	CT	61	m-cyno-NB
$C_7H_4N_2O_2$	1.72	0.10	CT	61	p-cyano-NB
C_7H_6FO	2.218	0.010	LPT	2	m-fluoroacetophenone enolate
C_7H_6FO	2.176	0.010	LPT	2	p-fluoroacetophenone enolate
$C_7H_6FeO_3$	0.990	0.10	CT	120	η^4-1,3-butadiene-Fe(CO)$_3$
$C_7H_6N_2O_4$	1.77	0.05	PT	60	3,4-dintrotoluene
$C_7H_6N_2O_4$	1.77	0.05	PT	60	2,3-dinitrotoluene
$C_7H_6N_2O_4$	1.60	0.05	PT	60	2,4-dinitrotoluene
$C_7H_6N_2O_4$	1.55	0.05	PT	60	2,6-dinitrotoluene
C_7H_7	0.912	0.006	LPES	26	benzyl

Molecule	Electron Affinity in eV	Uncertainty in eV	Method	Ref.	Name
C_7H_7	0.868	0.006	LPES	136	1-quadricyclanide
C_7H_7	0.962	0.006	LPES	136	2-quadricyclanide
C_7H_7	1.286	0.006	LPES	136	norbornadienide
C_7H_7	0.39	0.04	LPES	136	cycloheptatrienide
C_7H_7	3.046	0.006	LPES	136	1-(1,6-heptadiynide)
C_7H_7	≥1.140	0.006	LPES	136	3-(1,6-heptadiynide)
$C_7H_7NO_2$	0.92	0.10	CT	61	o-methyl-NB
$C_7H_7NO_2$	0.99	0.10	CT	61	m-methyl-NB
$C_7H_7NO_2$	0.95	0.10	CT	61	p-methyl-NB
$C_7H_7NO_3$	1.04	0.10	CT	61	m-OCH$_3$-NB
$C_7H_7NO_3$	0.91	0.10	CT	61	p-OCH$_3$-NB
C_7H_7O	≤2.36	0.06	PT	2	o-methyl phenoxide
C_7H_7O	2.14	0.02	PT	50	benzyloxide
$C_7H_7O_2$	1.85	0.10	CT	61	o-CH$_3$-BQ
C_7H_8FO	<3.05	0.06	PT	50	PhCH$_2$OHF
C_7H_9	1.27	0.03	PT	2	heptatrienyl
C_7H_9O	1.61	0.05	PT	2	2-norbornanone enolate
C_7H_9Si	1.33	0.04	LPT	65	C$_6$H$_5$(CH$_3$)SiH
$C_7H_{11}O$	1.598	0.007	LPT	113	cycloheptanone enolate
$C_7H_{11}O$	1.49	0.04	PT	2	2,5-dimethylcyclopentanone enolate
$C_7H_{13}O$	1.72	0.06	PT	2	4-heptanone enolate
$C_7H_{13}O$	1.46	0.04	PT	2	di-isopropyl ketone enolate
$C_8F_{14}N_2$	1.89	0.10	CT	51	1,4-(CN)$_2$C$_6$F$_4$
$C_8H_3F_5O$	0.88	0.11	CT	67	pentafluoroacetophenone
$C_8H_3F_6NO_2$	1.79	0.10	CT	61	3,5-(CF$_3$)$_2$-NB
$C_8H_4O_3$	1.21	0.10	CT	61	phthalic anhydride
C_8H_6	1.044	0.008	LPES	148	
C_8H_7	1.091	0.008	LPES	134	
C_8H_7O	2.057	0.010	PT	2	acetophenone enolate
C_8H_7O	2.10	0.08	LPT	2	phenylacetaldehyde enolate
C_8H_8	0.55	0.02	CT	134	cycooctatetraene
C_8H_8	0.65	0.04	LPES	134	cycooctatetraene
C_8H_8	0.919	0.008	LPES	139	m-xylylene
$C_8H_9NO_2$	1.21	0.05	PT	60	3,5-dimethyl-NB
$C_8H_9NO_2$	2.61	0.05	PT	60	2,6-dimethyl-NB
$C_8H_9NO_2$	0.86	0.10	CT	61	2,3-dimethyl-NB
$C_8H_{13}O$	1.63	0.06	PT	2	cyclooctanone enolate
$C_9H_8FeO_3$	0.76	0.10	CT	120	η^4-1,3-cyclohexadiene-Fe(CO)$_3$
C_9H_9O	2.030	0.010	LPT	2	m-methylacetophenone enolate
C_9H_9SiN	1.43	0.10	PT	2	trimethylsilylnitrene
$C_9H_{11}NO_2$	0.70	0.10	CT	61	2,4,6-trimethyl-NB
$C_9H_{15}O$	1.69	0.06	PT	2	cyclononanone enolate
$C_{10}H_4Cl_2O_2$	2.19	0.10	CT	61	2,3-dichloro-1,4-naphthoquinone
$C_{10}H_6N_2O_4$	1.78	0.10	CT	61	1,3-dinitronaphthalene
$C_{10}H_6N_2O_4$	1.77	0.10	CT	61	1,5-dinitronaphthalene
$C_{10}H_6O_2$	1.81	0.10	CT	61	1,4-naphthoquinone
$C_{10}H_7NO_2$	1.23	0.10	CT	61	1-nitronaphthalene
$C_{10}H_7NO_2$	1.18	0.10	CT	61	2-nitronaphthalene
$C_{10}H_8$	0.69	0.10	CT	61	azulene
$C_{10}H_8CrO_3$	0.93	0.10	CT	120	η^4-1,3,5-cycloheptatriene Cr(CO)$_3$
$C_{10}H_8FeO_3$	0.98	0.10	CT	120	η^4-1,3,5-cycloheptatriene-Fe(CO)$_3$
$C_{10}H_{17}O$	1.83	0.06	PT	2	cyclodecanone enolate
$C_{11}H_8FeO_3$	1.29	0.10	CT	120	η^4-1,3-butadiene-Fe(CO)$_3$
$C_{12}F_{10}$	0.82	0.11	CT	67	decafluorobiphenyl
$C_{12}H_4N_4$	2.8	0.3	CD	2	TCNQ

Molecule	Electron Affinity in eV	Uncertainty in eV	Method	Ref.	Name
$C_{12}H_9$	1.07	0.10	PT	2	perinaphthenyl
$C_{12}H_{15}O$	2.032	0.010	LPT	2	*t*-butylacetophenone enolate
$C_{12}H_{21}O$	1.90	0.07	PT	2	cyclododecanone enolate
$C_{13}F_{10}O$	1.52	0.11	CT	67	decafluorobenzophenone
$C_{13}H_9FO$	0.64	0.10	CT	61	4-fluorobenzophenone
$C_{13}H_{10}O$	0.62	0.10	CT	61	benzophenone
$C_{14}H_9NO_2$	1.43	0.10	CT	61	9-nitroanthracene
$C_{14}H_{10}$	0.57	0.10	CT	66	anthracene
$C_{18}H_{12}$	1.04	0.10	CT	66	tetracene
$C_{20}H_{12}$	0.79	0.10	CT	66	benz[*a*]pyrene
$C_{20}H_{12}$	0.97	0.10	CT	66	perylene
$C_{22}H_{14}$	1.35	0.10	CT	66	pentacene
CeF_4	3.8	0.4	CT	98	
$Cl(CO_2)$	3.907	0.010	LPES	131	
CoF_4	6.4	0.3	CT	98	
$Cr(CO)_3$	1.349	0.006	LPES	94	
CrO_3	3.6	0.2	CT	98	
Cu_n	$n = 1\text{-}41$	—	LPES	37	
Fe_n	$n = 3\text{-}24$	—	LPES	149	
$Fe(CO)_2$	1.22	0.02	LPES	2	
$Fe(CO)_3$	1.8	0.2	LPES	2	
$Fe(CO)_4$	2.4	0.3	LPES	2	
FeF_3	3.6	0.1	CT	98	
FeF_4	6.0	estimate	CT	98	
Fe_nO_m	$n = 1\text{-}4$	$m = 1\text{-}6$	LPES	152	
Ge_n	$n = 3\text{-}15$	—	LPES	71	
Ge_xAs_y	$n = 5\text{-}30$	$n = x + y$	LPES	72	
GeH_3	≤1.74	0.04	PT	2	
$H(NH_3)_n$	$n = 1,2$	—	LPES	76	
HNO_3	0.57	0.15	CD	2	
$(H_2O)_n$	$n = 2\text{-}19$	—	LPES	77	
$I(CO_2)$	3.225	0.001	LPES	131	
In_xP_y	$n = 2\text{-}8$	$n = x + y$	LPES	137	
IrF_4	4.7	0.3	CT	98	
IrF_6	6.5	0.4	CT	98	
K_n	$n = 2\text{-}7$	—	LPES	18	
MnF_4	5.5	0.2	CT	98	
$Mo(CO)_3$	1.337	0.006	LPES	94	
MoF_5	3.5	0.2	CT	98	
MoF_6	3.8	0.2	CT	98	
MoO_3	2.9	0.2	CT	98	
$(NH_3)_n$	$n = 41\text{-}1100$	—	LPES	77	
$NH_2(NH_3)_n$	$n = 1,2$	—	LPES	78	
$NO(H_2O)_n$	$n = 1,2$	—	LPES	75	
NO_3	3.937	0.014	LPES	85	
$NO(N_2O)_n$	$n = 1,2$	—	LPES	79	
$(NO)_2$	≥2.1	—	LPES	75	
$(N_2O)_n$	$n = 1,2$	—	LPES	81	
Na_n	$n = 2\text{-}5$	—	LPES	18	
$(NaF)_n$	$n = 1\text{-}7,12$	—	LPES	64	
$Na(NaF)_n$	$n = 5,7\text{-}12$	—	LPES	64	
$Ni(CO)_2$	0.643	0.014	LPES	2	
$Ni(CO)_3$	1.077	0.013	LPES	2	
NiO_2	3.05	0.01	LPES	145	ONiO
NiO_2	0.82	0.03	LPES	145	Ni(O$_2$)

Molecule	Electron Affinity in eV	Uncertainty in eV	Method	Ref.	Name
$OH(H_2O)$	<2.95	0.15	PT	2	
OsF_4	3.9	0.3	CT	98	
OsF_6	6.0	0.3	CT	98	
PBr_3	1.59	0.15	CD	2	
PBr_2Cl	1.63	0.20	CD	2	
PCl_2Br	1.52	0.20	CD	2	
PCl_3	0.82	0.10	CD	2	
PF_5	0.75	0.15	CT	121	
PO_3	4.5	0.5	CT	98	
$POCl_2$	3.83	0.25	CD	2	
$POCl_3$	1.41	0.20	CD	2	
PtF_4	5.5	0.3	CT	98	
PtF_6	7.0	0.4	CT	98	
ReF_6	4.7	estimate	CT	98	
RhF_4	5.4	0.3	CT	98	
RuF_4	4.8	0.3	CT	98	
RuF_5	5.2	0.4	CT	98	
RuF_6	7.5	0.3	CT	98	
SF_4	1.5	0.2	CT	91	
SF_6	1.05	0.10	CT	56	
SO_3	≥1.70	0.15	CD	2	
$(SO_2)_2$	0.6	0.2	LPES	80	
SeF_6	2.9	0.2	CD	2	
Si_4	2.17	0.01	LPES	110	
Si_n	n = 3-20	—	LPES	71	
SiD_3	1.386	0.022	LPES	43	
SiF_3	≤2.95	0.10	PT	17	
SiH_3	1.406	0.014	LPES	43	
TeF_6	3.34	0.17	CD	2	
Ti_n	n=3-65	—	LPES	151	
UF_5	3.7	0.2	CT	98	
UF_6	5.1	0.2	CT	98	
UO_3	<2.1	—	CT	98	
V_n	n=3-65	—	LPES	150	
VF_4	3.5	0.2	CT	98	
V_4O_{10}	4.2	0.6	CT	101	
$W(CO)_3$	1.859	0.006	LPES	94	
WF_5	1.25	0.3	CD	18	
WF_6	3.36	+0.04/-0.20	CT	19	
WO_3	3.33	+0.04/-0.15	LPT	86	
WO_3	3.9	0.2	CT	98	

REFERENCES

1. Hotop, H., and Lineberger, W. C., *J. Phys. Chem. Ref. Data*, 14, 731, 1985.
2. Drzaic, P. S., Marks, J., and Brauman, J. I., in *Gas Phase Ion Chemistry*, Vol. 3, Bowers, M. T., Ed., Academic Press, Orlando, 1984, p. 167.
3. Schulz, P. A., Mead, R. D., Jones, P. L., and Lineberger, W. C., *J. Chem. Phys.*, 77, 1153, 1982.
4. Neumark, D. M., Lykke, K. R., Anderson, T., and Lineberger, W. C., *Phys. Rev. A*, 32, 1890, 1985.
5. Wenthold, P. G., Gunion, R. F., and Lineberger, W. C., *Chem. Phys. Lett.*, 258, 101, 1996.
6. Srivastava, R. D., Uy, O. M., and Farber, M., *Trans. Faraday Soc.*, 67, 2941, 1971.
7. Klein, R., McGinnis, R. P., and Leone, S. R., *Chem. Phys. Lett.*, 100, 475, 1983.
8. Arnold, C. C., Zhao, Y., Kitsopoulos, T. N., and Neumark, D. M., *J. Chem. Phys.*, 97, 6121, 1992; linear C_6^- to linear C_6.
9. Stevens, A. E., Fiegerle, C. S., and Lineberger, W. C., *J. Chem. Phys.*, 78, 5420, 1983.
10. Breyer, F., Frey, P., and Hotop, H., *Z. Phys.*, A 300, 7, 1981.
11. Oakes, J. M., and Ellison, G. B., *Tetrahedron*, 42, 6263, 1986.
12. Leopold, D. G., Murray, K. K., Miller, A. E. S., and Lineberger, W. C., *J. Chem. Phys.*, 83, 4849, 1985.

ELECTRON AFFINITIES (continued)

13. Oakes, J. M., Jones, M.E., Bierbaum, V. M., and Ellison, G. B., *J. Chem. Phys.*, 87, 4810, 1983.
14. Ellis, H. B., Jr. and Ellison, G. B., *J. Chem. Phys.*, 78, 6541, 1983.
15. Oakes, J. M., Harding, L. B., and Ellison, G. B., *J. Chem. Phys.*, 83, 5400, 1985.
16. Nimlos, M. E., and Ellison, G. B., *J. Chem. Phys.*, 90, 2574, 1986.
17. Richardson, L. M., Stephenson, L. M., and Brauman, J. I., *Chem. Phys. Lett.*, 30, 17, 1975.
18. McHugh, K. M., Eaton, J. G., Lee, G. H., Sarkas, H. W., Kidder, L. H., Snodgrass, J. T., Manaa, M. R., and Bowen, K. H., *J. Chem. Phys.*, 91, 3792, 1989. See also Ref. 104.
19. Viggiano, A. A., Paulson, J. F., Dale, F., Henchman, M., Adams, N. G., and Smith, D., *J. Phys. Chem.*, 89, 2264, 1985.
20. Burnett, S. M., Stevens, A. E., Fiegerle, C. S., and Lineberger, W. C., *Chem. Phys. Lett.*, 100, 124, 1983.
21. Moran, S., Ellis, H. B., DeFrees, D. J., McLean, A. D., and Ellison, G. B., *J. Am. Chem. Soc.*, 109, 5996, 1987; Moran, S., Ellis, H. B., DeFrees, D. J., McLean, A. D., Paulson, S. E., and Ellison, G. B., *J. Am. Chem. Soc.*, 109, 6004, 1987; see also Lykke, K. R., Neumark, D. M., Andersen, T., Trapa, V. J., and Lineberger, W. C., *J. Chem. Phys.*, 87, 6842, 1987.
22. Mead, R. D., Lykke, K. R., Lineberger, W. C., Marks, J., and Brauman, J. I., *J. Chem. Phys.*, 81, 4883, 1984; Lykke, K. R., Mead, R. D., and Lineberger, W. C., *Phys. Rev. Lett.*, 52, 2221, 1984. The EAs are 14717.7 ± 1.0 cm^{-1} for actaldehyde enolate and 14671.0 ± 1.0 cm^{-1} for acetaldehyde-d$_3$ enolate.
23. Ellison, G. B., Engelking, P. C., and Lineberger, W. C., *J. Chem. Phys.*, 86, 4873, 1982.
24. Oakes, J. M., and Ellison, G. B., *J. Am. Chem. Soc.*, 105, 2969, 1983.
25. Ellison, G. B., and Oakes, J. M., *J. Am. Chem. Soc.*, 106, 7734, 1984. EA(allyl) and EA(allyl-d$_5$) are 0.119 and 0.083 eV too low, respectively, in this work, according to Reference 138. Therefore, EA(allyl-d$_1$) is likely too low by a similar amount.
26. Gunion, R. F., Gilles, M. K., Polak, M. L., and Lineberger, W. C., *Int. J. Mass Spectrom. Ion Processes*, 117, 601, 1992; also, Miller, A. E. S., and Lineberger, W. C., unpublished. See also Reference 136.
27. Leopold, D. G., and Lineberger, W. C., *J. Chem. Phys.*, 85, 51, 1986.
28. Miller, T. M., Miller, A. E. S., and Lineberger, W. C., *Phys. Rev. A*, 33, 3558, 1986.
29. Miller, A. E. S., Fiegerle, C. S., and Lineberger, W. C., *J. Chem. Phys.*, 87, 1549, 1987.
30. Miller, T. M., Leopold, D. G., Murray, K. K., and Lineberger, W. C., *J. Chem. Phys.*, 85, 2368, 1986.
31. Miller, T. M., and Lineberger, W. C., *Chem. Phys. Lett.*, 146, 364, 1988.
32. Neumark, D. M., Lykke, K. R., Andersen, T., and Lineberger, W. C., *J. Chem. Phys.*, 83, 4364, 1985.
33. Leopold, D. G., Miller, T. M., and Lineberger, W. C., *J. Am. Chem. Soc.*, 108, 178, 1986.
34. Miller, A. E. S., Fiegerle, C. S., and Lineberger, W. C., *J. Chem. Phys.*, 84, 4127, 1986.
35. Murray, K. K., Miller, T. M., Leopold, D. G., and Lineberger, W. C., *J. Chem. Phys.*, 84, 2520, 1986.
36. Leopold, D. G., Miller, A. E. S., and Lineberger, W. C., *J. Am. Chem. Soc.*, 108, 1379, 1986.
37. Ho, J., Ervin, K. M., and Lineberger, W. C., *J. Chem. Phys.*, 93, 6987, 1990; Leopold, D. G., Ho, J., and Lineberger, W. C., *J. Chem. Phys.*, 86, 1715, 1987; Pettiette, C. L., Yang, S. H., Craycraft, M. J., Conceicao, J., Laaksonen, R. T., Cheshnovsky, O., and Smalley, R. E., *J. Chem. Phys.*, 88, 5377, 1988.
38. Snodgrass, J. T., Coe, J. V., McHugh, K. M., Friedhoff, C. B., and Bowen, K. H., *J. Phys. Chem.*, 93, 1249, 1989.
39. Friedhoff, C. B., Snodgrass, J. T., Coe, J. V., McHugh, K. M., and Bowen, K. H., *J. Chem. Phys.*, 84, 1051, 1986.
40. Friedhoff, C. B., Coe, J. V., Snodgrass, J. T., McHugh, K. M., and Bowen, K. H., *Chem. Phys. Lett.*, 124, 268, 1986.
41. Coe, J. V., Snodgrass, J. T., Friedhoff, C. B., McHugh, K. M., and Bowen, K. H., *J. Chem. Phys.*, 84, 619, 1986.
42. Snodgrass, J. T., Coe, J. V., Friedhoff, C. B., McHugh, K. M., and Bowen, K. H., *Chem. Phys. Lett.*, 122, 352, 1985.
43. Nimlos, M. R., and Ellison, G. B., *J. Am. Chem. Soc.*, 108, 6522, 1986.
44. Petrunin, V., Andersen, H., Balling, P., and Andersen, T., *Phys. Rev. Lett.*, 76, 744, 1996
45. Andersen, T., Lykke, K. R., Neumark, D. M., and Lineberger, W. C., *J. Chem. Phys.*, 86, 1858, 1987.
46. Murray, K. K., Lykke, K. R., and Lineberger, W. C., *Phys. Rev. A*, 36, 699, 1987.
47. Mansour, N. B., and Larson, D. J., *Abstracts of the XV Int. Conf. on the Phys. of Electronic and Atomic Collisions*, p. 70, 1987.
48. Stonemann, R. C., and Larson, D. J., *Phys. Rev. A*, 35, 2928, 1987.
49. Nimlos, M. R., Harding, L. B., and Ellison, G. B., *J. Chem. Phys.*, 87, 5116, 1987.
50. Moylan, C. R., Dodd, J. A., Han, C.-C., and Braumann, J. I., *J. Chem. Phys.*, 86, 5350, 1987.
51. Chowdhury, S., Grimsrud, E. P., Heinis, T., and Kebarle, P., *J. Am. Chem. Soc.*, 108, 3630, 1986.
52. Trainham, R., Fletcher, G. D., and Larson, D. J., *J. Phys. B: At. Mol. Phys.*, 20, L777, 1987.
53. Moran, S., and Ellison, G. B., *J. Phys. Chem.*, 92, 1794, 1988.
54. Murray, K. K., Leopold, D. G., Miller, T. M., and Lineberger, W. C., *J. Chem. Phys.*, 89, 5442, 1988.
55. Ervin, K. M., Ho, J., and Lineberger, W. C., *J. Chem. Phys.*, 89, 4514, 1988.
56. Grimsrud, E. P., Chowdhury, S., and Kebarle, P., *J. Chem. Phys.*, 85, 4989, 1985.
57. Fischer, C. F., *Phys. Rev. A*, 39, 963, 1989.
58. Wickham-Jones, C. T., Ervin, K. M., Ellision, G. B., and Lineberger, W. C., *J. Chem. Phys.*, 91, 2762, 1989.
59. Hopper, D. G., Wahl, A. C., Wu, R. L. C., and Tiernan, T. O., *J. Chem. Phys.*, 65, 5474, 1976.
60. Mock, R. S., and Grimsrud, E. P., *J. Am. Chem. Soc.*, 111, 2861, 1989.
61. Chowdhury, S., Heinis, T., Grimsrud, E. P., and Kebarle, P., *J. Phys. Chem.*, 90, 2747, 1986. The uncertainty and other results are quoted in Ref. 60.
62. Wickham-Jones, C. T., Moran, S., and Ellison, G. B., *J. Chem. Phys.*, 90, 795, 1989.
63. Ervin, K. M., Ho, J., and Lineberger, W. C., *J. Phys. Chem.*, 92, 5405, 1988.
64. Miller, T. M., and Lineberger, W. C., *Int. J. Mass Spectrom. Ion Processes*, 102, 239, 1990.
65. Wetzel, D. M., Salomon, K. E., Berger, S., and Brauman, J. I., *J. Am. Chem. Soc.*, 111, 3835, 1989.
66. Crocker, L., Wang, T., and Kebarle, P., *J. Am. Chem. Soc.*, 115, 7818, 1993.
67. Dillow, G. W., and Kebarle, P., *J. Am. Chem. Soc.*, 111, 5592, 1989.
68. Gantefor, G., Gausa, M., Meiwes-Broer, K. H., and Lutz, H. O., *Z. Phys. D*, 9, 253, 1988; Taylor, K. J., Petteitte, C. L., Craycraft, M. J., Chesnovsky, O., and Smalley, R. E., *Chem. Phys. Lett.*, 152, 347, 1988.
69. Metz, R. B., Kitsopoulos, T., Weaver, A., and Neumark, D. M., *J. Chem. Phys.*, 88, 1463, 1988.
70. Yang, S., Pettiette, C. L., Conceicao, J., Cheshnovsky, O., and Smalley, R. E., *Chem. Phys. Lett.*, 139, 233, 1987; Yang, S., Taylor, K. J., Craycraft, M. J., Conceicao, J., Pettiette, C. L., Cheshnovsky, O., and Smalley, R. E., *Chem. Phys. Lett.*, 144, 431, 1988; Arnold, D. W., Bradforth, S. E., Kitsopoulos, T. N., and Neumark, D. M., *J. Chem. Phys.*, 95, 5479, 1991.
71. Cheshnovsky, O., Yang, S., Pettiette, C. L., Craycraft, M. J., Liu, Y., and Smalley, R. E., *Chem. Phys. Lett.*, 138, 119, 1987.

73. Travers, M. J., Cowles, D. C., and Ellison, G. B., *Chem. Phys. Lett.*, 164, 449, 1989.
74. Blondel, C., Cacciani, P., Delsart, C., and Trainham, R., *Phys. Rev. A*, 40, 3698, 1989.
75. Bowen, K. H., and Eaton, J. G., in *The Structure of Small Molecules and Ions*, Naaman, R., and Vager, Z., Eds., Plenum, New York, 1988, pp. 147-169; Arnold, S. T., Eaton, J. G., Patel-Mistra, D., Sarkas, H. W., and Bowen, K. H., in *Ion and Cluster Ion Spectroscopy and Structure*, Maier, J. P., Ed., Elsevier Science, New York, 1989, p. 417.
76. Snodgrass, J. T., Coe, J. V., Freidhoff, C. B., McHugh, K. M., and Bowen, K. H., *Faraday Disc. Chem. Soc.*, 88, 1988.
77. Lee, G. H., Arnold, S. T., Eaton, J. G., Sarkas, H. W., Bowen, K. H., Ludewigt, C., and Haberland, H., *Z. Phys. D - Atoms, Mol. and Clusters*, 20, 9, 1991; Coe, J. V., Lee, G. H., Eaton, J. G., Arnold, S. T., Sarkas, H. W., Bowen, K. H., Ludewigt, C., Haberland, H., and Worsnop, D. R., *J. Chem. Phys.*, 92, 3980, 1990.
78. Snodgrass, J. T., Coe, J. V., Freidhoff, C. B., McHugh, K. M., and Bowen, K. H., to be published, quoted in Ref. 75.
79. Coe, J. V., Snodgrass, J. T., Freidhoff, C. B., McHugh, K. M., and Bowen, K. H., *J. Chem. Phys.*, 87, 4302, 1987.
80. Friedhoff, C. B., Snodgrass, J. T., and Bowen, K. H., to be published, quoted in Ref. 75.
81. Coe, J. V., Snodgrass, J. T., Friedhoff, C. B., McHugh, K. M., and Bowen, K. H., *Chem. Phys. Lett.*, 124, 274, 1986.
82. Greene, C. H., *Phys. Rev. A*, 42, 1405, 1990.
83. Ervin, K. M., Ho, J., and Lineberger, W. C., *J. Chem. Phys.*, 91, 5974, 1989.
84. Polak, M. L., Fiala, B. L., Ervin, K. M., and Lineberger, W. C., *J. Chem. Phys.*, 94, 6924, 1991.
85. Weaver, A., Arnold, D. W., Bradforth, S. E., Neumark, D. M., *J. Chem. Phys.*, 94, 1740, 1991.
86. Walter, C. W., Devynck, P., Hertzler, C. F., Bae, Y. K., Smith, G. P., and Peterson, J. R., *Bull. Am. Phys. Soc.*, 35, 1163, 1990.
87. Ervin, K. M., and Lineberger, W. C., *J. Phys. Chem.*, 95, 1167, 1991.
88. Gilles, M. K., Polak, M. L., and Lineberger, W. C., *J. Chem. Phys.*, 96, 8012, 1992.
89. Lykke, K. R., Murray, K. K., and Lineberger, W. C., *Phys. Rev. A*, 43, 6104, 1991.
90. Ervin, K. M., Gronert, S., Barlow, S. E., Gilles, M. K., Harrison, A. G., Bierbaum, V. M., DePuy, C. H., Lineberger, W. C., and Ellison, G. B., *J. Am. Chem. Soc.*, 112, 5750, 1990.
91. Viggiano, A. A., Miller, T. M., Miller, A. E. S., Morris, R. A., Van Doren, J. M., and Paulson, J. F., *Int. J. Mass Spectrom. Ion Processes*, 109, 327, 1991.
92. Hanstorp, D., and Gustafsson, M., *J. Phys. B: At. Mol. Opt. Phys.*, 25, 1773, 1992.
93. Polak, M. L., Gilles, M. K., and Lineberger, W. C., *J. Chem. Phys.*, 96, 7191, 1992.
94. Bengali, A. A., Casey, S. M., Cheng, C.-L., Dick, J. P., Fenn, P. T., Villalta, P. W., and Leopold, D. G., *J. Am. Chem. Soc.*, 114, 5257, 1992.
95. Gilles, M. K., Ervin, K. M., Ho, J., and Lineberger, W. C., *J. Phys. Chem.*, 96, 1130, 1992.
96. Travers, M. J., Cowles, D. C., Clifford, E. P., and Ellison, G. B., *J. Am. Chem. Soc.*, 114, 8699, 1992.
97. Bengali, A. A., and Leopold, D. G., *J. Am. Chem. Soc.*, 114, 9192, 1992.
98. Rudnyi, E. B., Kaibicheva, E. A., and Sidorov, L. N., *Rapid Comm. in Mass Spectrom.*, 6, 356, 1992; Sidorov, L. N., *High Temp. Sci.*, 29, 153, 1990.
99. Kitsopoulos, T. N., Chick, C. J., Zhao, Y., and Neumark, D. M., *J. Chem. Phys.*, 95, 5479, 1991.
100. Arnold, C. C., Kitsopoulos, T. N., and Neumark, D. M., *J. Chem. Phys.*, 99, 766, 1993.
101. Rudnyi, E. B., Kaibicheva, E. A., and Sidorov, L. N., *J. Chem. Thermodynamics*, 25, 929, 1993.
102. Sarkas, H. W., Hendricks, J. H., Arnold, S. T., and Bowen, K. H., *J. Chem. Phys.* 100, 1884, 1994.
103. Villalta, P. W., and Leopold, D. G., *J. Chem. Phys.* 98, 7730, 1993.
104. Eaton, J. G., Sarkas, H. W., Arnold, S. T., McHugh, K. M., and Bowen, K. H., *Chem. Phys. Lett.*, 193, 141, 1992. See also Ref. 18.
105. Polak, M. L., Gilles, M. K., Gunion, R. F., and Lineberger, W. C., *Chem. Phys. Lett.*, 210, 55, 1993.
106. Gilles, M. K., Lineberger, W. C., and Ervin, K. M., *J. Am. Chem. Soc.*, 115, 1031, 1993.
107. Casey, S. M., and Leopold, D. G., *Chem. Phys. Lett.*, 201, 205, 1993.
108. Polak, M. L., Gerber, G., Ho, J., and Lineberger, W. C., *J. Chem. Phys.*, 97, 8990, 1992.
109. Kim, E. H., Bradforth, S. E., Arnold, D. W., Metz, R. B., and Neumark, D. M., *J. Chem. Phys.*, 103, 7801, 1995.
110. Arnold, C. C., and Neumark, D. M., *J. Chem. Phys.*, 99, 3353, 1993.
111. Bradforth, S. E., Kim, E. H., Arnold, D. W., and Neumark, D. M., *J. Chem. Phys.*, 98, 800, 1993.
112. Ho, J., Polak, M. L., Ervin, K. M., and Lineberger, W. C., *J. Chem. Phys.*, 99, 8542, 1993.
113. Brinkman, E. A., Berger, S., Marks, J., and Brauman, J. I., *J. Chem. Phys.*, 99, 7586, 1993.
114. Casey, S. M., and Leopold, D. G., *J. Phys. Chem.*, 97, 816, 1993.
115. McDonald, R. N., and Davidson, S. J., *J. Am. Chem. Soc.*, 115, 10857, 1993.
116. Ho, J., Ervin, K. M., Polak, M. L., Gilles, M. K., and Lineberger, W. C., *J. Chem. Phys.*, 95, 4845, 1991.
117. Ho, J., Polak, M. L., and Lineberger, W. C., *J. Chem. Phys.*, 96, 144, 1992.
118. Polak, M. L., Gilles, M. K., Ho, J., and Lineberger, W. C., *J. Phys. Chem.*, 95, 3460, 1991.
119. Polak, M. L., Ho, J., Gerber, G., and Lineberger, W. C., *J. Chem. Phys.*, 95, 3053, 1991.
120. Sharpe, P., and Kebarle, P., *J. Am. Chem. Soc.*, 115, 782, 1993.
121. Miller, T. M., Miller, A. E. S., Viggiano, A. A., Morris, R. A., and Paulson, J. F., *J. Chem. Phys.*, 100, 7200, 1994.
122. Berkovits, D., Boaretto, E., Gehlberg, S., Heber, O., and Paul, M., *Phys. Rev. Lett.*, 75, 414, 1995.
123. Arnold, C. C., Xu, C., Burton, G. R., and Neumark, D. M., *J. Chem. Phys.*, 102, 6982, 1995. Burton, G. R., Xu, C., Arnold, C. C., and Neumark, D. M., *J. Chem. Phys.*, in press.
124. Xu, C., de Beer, E., and Neumark, D. M., *J. Chem. Phys.*, 104, 2749, 1996.
125. Gunion, R. F., Koppel, H., Leach, G. W., and Lineberger, W. C., *J. Chem. Phys.*, 103, 1250, 1995.
126. Gunion, R. F., and Lineberger, W. C., *J. Phys. Chem.*, 100, 4395, 1996.
127. Gunion, R. F., Dixon-Warren, St. J., Lineberger, W. C., and Morse, M. D., *J. Chem. Phys.*, in press.
128. Dixon-Warren, St. J., Gunion, R. F., and Lineberger, W. C., *J. Chem. Phys.*, 104, 4902, 1996.
129. Nakajima, A., Zhang, N., Kawamata, H., Hayase, T., Nakao, K., and Kaya, K., *Chem. Phys. Lett.*, 241, 295, 1995; Nakajima, A., Taguwa, T., Nakao, K., Hoshino, K., Iwata, S., and Kaya, K., *J. Chem. Phys.*, 102, 660, 1995.
130. Fan, J., and Wang, L.-S., *J. Chem. Phys.*, 102, 8714, 1995.
131. Arnold, D. W., Bradforth, S. E., Kim, E. H., and Neumark, D. M., *J. Chem. Phys.*, 102, 3493, 1995; Zhao, Y., Arnold, C. C., and Neumark, D. M., *J. Chem. Soc. Faraday Trans. 2*, 89, 1449, 1992.
132. Fan, J., Lou, L., and Wang, L.-S., *J. Chem. Phys.*, 102, 2701, 1995.

133. Sarkas, H. W., Hendricks, J. H., Arnold, S. T., Slager, V. L., and Bowen, K. H., *J. Chem. Phys.*, 100, 3358, 1994.

134. Kato, S., Lee, H. S., Gareyev, R., Wenthold, P. G., Lineberger, W. C., DePuy, C. H., and Bierbaum, V. M., *J. Am. Chem. Soc.*, 119, 7863, 1997.

135. Wenthold, P. G., Hu, J., Squires, R. R., and Lineberger, W. C., *J. Am. Chem. Soc.*, 118, 475, 1996.

136. Gunion, R. F., Karney, W., Wenthold, P. G., Borden, W. T., and Lineberger, W. C., *J. Am. Chem. Soc.*, 118, 5074, 1996. The numbers in the abstract for 1,6-heptadiyne were misprinted. EA(cycloheptatrienide) quoted here derives from the LPES data combined with other thermochemical data in Reference 136.

137. Xu, C., de Beer, E., Arnold, D. W., Arnold, C. C., and Neumark, D. M., *J. Chem. Phys.*, 101, 5406, 1996.

138. Wenthold, P. G., Polak, M. L., and Lineberger, W. C., *J. Phys. Chem.*, 100, 6920, 1996.

139. Wenthold, P. G., Kim, J. B., and Lineberger, W. C., *J. Am. Chem. Soc.*, 119, 1354, 1997.

140. Eliav, E., Kaldor, U., Ishikawa, Y., and Pyykkoe, *Phys. Rev. Lett.*, 77, 5350, 1996.

141. Davies, B. J., Ingram, C. W., Larson, D. J., and Ljungblad, U., *J. Chem. Phys.* 106, 5783, 1997.

142. Smith, J. R., Kim, J. B., and Lineberger, W. C., *Phys. Rev. A.*, 55, 2036, 1997. Schulz, P.A., Mead, R. D., Jones, P. L., and Lineberger, W. C., *J. Chem. Phys.*, 77, 1153, 1982. See also Rudmin, J. D., Ratliff, L. P., Yukich, J. N., and Larson, D. J., *J. Phys. B: At. Mol. Opt. Phys.*, 29, L881, 1996.

143. Desai, S. R., Wu, H., Rohlfing, C. M., and Wantg, L.-S., *Int. J. Chem. Phys.*, 106, 1309, 1997.

144. Wenthold, P. G., Joans, K.-L., and Lineberger, W. C., *J. Chem. Phys.*, 106, 9961, 1997.

145. Wu, H. and Wang, L.-S., *J. Chem. Phys.*, 107, 16, 1997.

146. Moravec, V. D. and Jarrold, C. C., *J. Chem. Phys.*, in press.

147. Wang, X.-B., Ding, C.-F., and Wang, L.-S., *J. Phys. Chem. A*, 101, 7699, 1997.

148. Wenthold, P. G. and Lineberger, W. C., *J. Am. Chem. Soc.*, 19, 7772, 1997.

149. Wang, L.-S., Cheng, H.-S., and Fan, J., *J. Chem. Phys.*, 102, 9480, 1995.

150. Wu, H., Desai, S. R., and Wang, L.-S., *Phys. Rev. Lett.*, 77, 2436, 1996.

151. Wu, H., Desai, S. R., and Wang, L.-S., *Phys. Rev. Lett.*, 76, 212, 1996.

152. Wang, L.-S., Wu, H., and Desai, S. R., *Phys. Rev. Lett.*, 76, 4853, 1996.

153. Robinson, M. S., Polak, M. L., Bierbaum, V. M., DePuy, C. H., and Lineberger, W. C., *J. Am. Chem. Soc.*, 117, 6766, 1995.

154. Clifford, E. P., Wenthold, P. G., Lineberger, W. C., Petersson, G. A., and Ellison, G. B., *J. Phys. Chem.*, 101, 4338, 1997.

155. Wenthold, P. G., Kim, J. B., Jonas, K., and Lineberger, W. C., *J. Phys. Chem.*, in press, 1997.

ATOMIC AND MOLECULAR POLARIZABILITIES
Thomas M. Miller

The *polarizability* of an atom or molecule describes the response of the electron cloud to an external field. The atomic or molecular energy shift ΔW due to an external electric field E is proportional to E^2 for external fields which are weak compared to the internal electric fields between the nucleus and electron cloud. The *electric dipole polarizability* α is the constant of proportionality defined by $\Delta W = -\alpha E^2/2$. The induced electric dipole moment is αE. *Hyperpolarizabilities*, coefficients of higher powers of E, are less often required. Technically, the polarizability is a tensor quantity but for spherically symmetric charge distributions reduces to a single number. In any case, an *average polarizability* is usually adequate in calculations. Frequency-dependent or *dynamic polarizabilities* are needed for electric fields which vary in time, except for frequencies which are much lower than electron orbital frequencies, where *static polarizabilities* suffice.

Polarizabilities for atoms and molecules in excited states are found to be larger than for ground states and may be positive or negative. Molecular polarizabilities are very slightly temperature dependent since the size of the molecule depends on its rovibrational state. Only in the case of dihydrogen has this effect been studied enough to warrant consideration in Table 3.

Polarizabilities are normally expressed in cgs units of cm^3. Ground state polarizabilities are in the range of 10^{-24} $cm^3 = 1$ $Å^3$ and hence are often given in $Å^3$ units. Theorists tend to use atomic units of a_o^3 where a_o is the Bohr radius. The conversion is $\alpha(cm^3) = 0.148184 \times 10^{-24} \times \alpha$ (a_o^3). Polarizabilities are only recently encountered in SI units, $C \cdot m^2/V = J/(V/m)^2$. The conversion from cgs units to SI units is $\alpha(C \cdot m^2/V) = 4\pi\varepsilon_o \times 10^{-6} \times \alpha(cm^3)$, where ε_o is the permittivity of free space in SI units and the factor 10^{-6} simply converts cm^3 into m^3. Thus, $\alpha(C \cdot m^2/V) = 1.11265 \times 10^{-16} \times \alpha(cm^3)$. Persons measuring excited state polarizabilities by optical methods tend to use units of $MHz/(V/cm)^2$, where the energy shift, ΔW, is expressed in frequency units with a factor of h understood. The polarizability is $-2 \Delta W/E^2$. The conversion into cgs units is $\alpha(cm^3) = 5.95531 \times 10^{-16} \times \alpha[MHz/(V/cm)^2]$.

The polarizability appears in many formulas for low-energy processes involving the valence electrons of atoms or molecules. These formulas are given below in cgs units: the polarizability α is in cm^3; masses m or μ are in grams; energies are in ergs; and electric charges are in esu, where $e = 4.8032 \times 10^{-10}$ esu. The symbol $\alpha(\nu)$ denotes a frequency (ν) dependent polarizability, where $\alpha(\nu)$ reduces to the static polarizability α for $\nu = 0$. For further information and references, see Miller, T. M., and Bederson, B., *Advances in Atomic and Molecular Physics*, 13, 1, 1977. Details on polarizability-related interactions, especially in regard to hyperpolarizabilities and nonlinear optical phenomena, are given by Bogaard, M. P., and Orr, B. J., in *Physical Chemistry*, Series Two, Vol. 2, *Molecular Structure and Properties*, Buckingham, A. D., Ed., Butterworths, London, 1975, pp. 149-194. A tabulation of tensor and hyperpolarizabilities is included. The gas number density, n, in Table 1 is usually taken to be that of 1 atm at 0°C in reporting experimental data.

Table 1
Formulas Involving Polarizability

Description	Formula	Remarks
Lorentz-Lorenz relation	$\alpha(\nu) = \dfrac{3}{4\pi n}\left[\dfrac{\eta^2(\nu)-1}{\eta^2(\nu)+2}\right]$	For a gas of atoms or nonpolar molecules; the index of refraction is $\eta(\nu)$
Refraction by polar molecules	$\alpha(\nu) + \dfrac{d^2}{3kT} = \dfrac{3}{4\pi n}\left[\dfrac{\eta^2(\nu)-1}{\eta^2(\nu)+2}\right]$	The dipole moment is d, in esu·cm ($= 10^{-18}$ D)
Dielectric constant (dimensionless)	$\kappa(\nu) = 1 + 4\pi n \ \alpha(\nu)$	From the Lorentz-Lorenz relation for the usual case of $\kappa(\nu)\approx 1$
Index of refraction (dimensionless)	$\eta(\nu) = 1 + 2\pi n \ \alpha(\nu)$	From $\eta^2(\nu) = \kappa(\nu)$
Diamagnetic susceptibility	$\chi_m = e^2\left(a_o N\alpha\right)^{1/2} / 4m_e c^2$	From the approximation that the static polarizability is given by the variational formula $\alpha = (4/9a_o)\Sigma(N_i r_i^2)^2$; N is the number of electrons, m_e is the electron mass; a crude approximation is $\chi_m=(E_i/4m_e c^2)\alpha$, where E_i is the ionization energy
Long-range electron- or ion-molecule interaction energy	$V(r) = -e^2\alpha / 2r^4$	The target molecule polarizability is α
Ion mobility in a gas	$\kappa = 13.87/(\alpha\mu)^{1/2} \ cm^2 / V \cdot s$	This one formula is not in cgs units. Enter α in $Å^3$ or 10^{-24} cm^3 units and the reduced mass μ of the ion-molecule pair in amu. Classical limit; pure polarization potential

Table 1
Formulas Involving Polarizability (continued)

Description	Formula	Remarks
Langevin capture cross section	$\sigma(v_o) = (2\pi e/v_o)(\alpha/\mu)^{1/2}$	The relative velocity of approach for an ion-molecule pair is v_o; the target molecular polarizability is α and the reduced mass of the ion-molecule pair is μ
Langevin reaction rate coefficient	$k = 2\pi e(\alpha/\mu)^{1/2}$	Collisional rate coefficient for an ion-molecule reaction
Rate coefficient for polar molecules	$k_d = 2\pi e\left[(\alpha/\mu)^{1/2} + cd(2/\mu\pi kT)^{1/2}\right]$	The dipole moment of the neutral is d in esu·cm; the number c is a "locking factor" that depends on α and d, and is between 0 and 1
Modified effective range cross section for electron-neutral scattering	$\sigma(k) = 4\pi A^2$ $+32\pi^4\mu e^2\alpha Ak/3h^2$ $+L$	Here, k is the electron momentum divided by $h/2\pi$, where h is Planck's constant; A is called the "scattering length"; the reduced mass is μ
van der Waals constant between two systems A, B	$C_6 = \dfrac{3}{2}\left[\dfrac{\alpha^A\alpha^B E^A E^B}{E^A + E^B}\right]$	For the interaction potential term $V_6(r) = -C_6 r^{-6}$; $E^{A,B}$ represents average dipole transition energies and $\alpha^{A,B}$ the respective polarizabilities of A, B
Dipole-quadrupole constant between two systems A, B	$C_8 = \dfrac{15}{4}\left[\dfrac{\alpha^A\alpha_q^{\,B} E^A E_q^{\,B}}{E^A + E_q^{\,B}}\right]$ $+ \dfrac{15}{4}\left[\dfrac{\alpha_q^{\,A}\alpha^B E_q^{\,A} E^B}{E_q^{\,A} + E^B}\right]$	For the interaction potential term $V_8(r) = -C_8 r^{-8}$; $E_q^{A,B}$ represents average quadrupole transition energies and $\alpha_q^{A,B}$ are the respective quadrupole polarizabilities of A, B
van der Waals constant between an atom and a surface	$C_3 = \dfrac{\alpha g E^A E^S}{8(E^A + E^S)}$	For an interaction potential $V_3(r) = -C_3 r^{-3}$; $E^{A,S}$ are characteristic energies of the atom and surface; $g = 1$ for a free-electron metal and $g = (\varepsilon_\infty - 1)/(\varepsilon_\infty + 1)$ for an ionic crystal
Relation between $\alpha(\nu)$ and oscillator strengths	$\alpha(\nu) = \dfrac{e^2 h^2}{4\pi^2 m_e}\Sigma\dfrac{f_k}{E_k^2 - (h\nu)^2}$	Here, f_k is the oscillator strength from the ground state to an excited state k, with excitation energy E_k. This formula is often used to estimate static polarizabilities ($\nu = 0$)
Dynamic polarizability	$\alpha(\nu) = \dfrac{\alpha E_r^{\,2}}{E_r^{\,2} - (h\nu)^2}$	Approximate variation of the frequency-dependent polarizability $\alpha(\nu)$ from $\nu = 0$ up to the first dipole-allowed electronic transition, of energy E_r; the static dipole polarizability is $\alpha(0)$; infrared contributions ignored
Rayleigh scattering cross section	$\alpha(\nu) = \dfrac{8\pi}{9c^4}(2\pi\nu)^4$ $\times\left[3\alpha^2(\nu) + 2\gamma^2(\nu)/3\right]$	The photon frequency is ν; the polarizability anisotropy (the difference between polarizabilities parallel and perpendicular to the molecular axis) is $\gamma(\nu)$
Verdet constant	$V(\nu) = \dfrac{\nu n}{2m_e c^2}\left[\dfrac{d\alpha(\nu)}{d\nu}\right]$	Defined from $\theta = V(\nu)B$, where θ is the angle of rotation of linearly polarized light through a medium of number density n, per unit length, for a longitudinal magnetic field strength B (Faraday effect)

Table 2
Static Average Electric Dipole Polarizabiilities for Ground State Atoms (in Units of 10^{-24} cm^3)

Atomic number	Atom	Polarizability	Estimated accuracy (%)	Method	Ref.
1	H	0.666793	"exact"	calc	MB77
2	He	0.204956	"exact"	calc	MB77
		0.2050	0.1	index/diel	NB65/OC67
3	Li	24.3	2	beam	MB77
4	Be	5.60	2	calc	MB77
5	B	3.03	2	calc	MB77
6	C	1.76	2	calc	MB77
7	N	1.10	2	calc/index	MB77
8	O	0.802	2	calc/index	MB77
9	F	0.557	2	calc	MB77
10	Ne	0.3956	0.1	diel	OC67
11	Na	24.08	0.4	interferom	ESCHP94
12	Mg	10.6	2	calc	MB77
13	Al	6.8	4.4	beam	MMD90
14	Si	5.38	2	calc	MB77
15	P	3.63	2	calc	MB77
16	S	2.90	2	calc	MB77
17	Cl	2.18	2	calc	MB77
18	Ar	1.6411	0.05	index/diel	NB65/OC67
19	K	43.4	2	beam	MB77
20	Ca	22.8	2	calc	MB77
		25.0	8	beam	MB77
21	Sc	17.8	25	calc	D84
22	Ti	14.6	25	calc	D84
23	V	12.4	25	calc	D84
24	Cr	11.6	25	calc	D84
25	Mn	9.4	25	calc	D84
26	Fe	8.4	25	calc	D84
27	Co	7.5	25	calc	D84
28	Ni	6.8	25	calc	D84
29	Cu	6.1	25	calc	D84
		7.31	25	calc	G84
30	Zn	7.1	25	calc	MB77
		5.6	25	calc	D84
31	Ga	8.12	2	calc	MB77
32	Ge	6.07	2	calc	MB77
33	As	4.31	2	calc	MB77
34	Se	3.77	2	calc	MB77
35	Br	3.05	2	calc	MB77
36	Kr	2.4844	0.05	diel	OC67
37	Rb	47.3	2	beam	MB77
38	Sr	27.6	8	beam	MB77
39	Y	22.7	25	calc	D84.
40	Zr	17.9	25	calc	D84
41	Nb	15.7	25	calc	D84
42	Mo	12.8	25	calc	D84
43	Tc	11.4	25	calc	D84
44	Ru	9.6	25	calc	D84
45	Rh	8.6	25	calc	D84
46	Pd	4.8	25	calc	D84
47	Ag	7.2	25	calc	D84
		8.56	25	calc	G84
48	Cd	7.2	25	calc	D84
49	In	10.2	12	beam	GMBSJ84
		9.1	25	calc	D84

Table 2
Static Average Electric Dipole Polarizabiilities for Ground State Atoms (in Units of 10^{-24} cm^3)
(continued)

Atomic number	Atom	Polarizability	Estimated accuracy (%)	Method	Ref.
50	Sn	7.7	25	calc	D84
51	Sb	6.6	25	calc	D84
52	Te	5.5	25	calc	D84
53	I	5.35	25	index	A56
		4.7	25	calc	D84
54	Xe	4.044	0.5	diel	MB77
55	Cs	59.6	2	beam	MB77
56	Ba	39.7	8	beam	MB77
57	La	31.1	25	calc	D84
58	Ce	29.6	25	calc	D84
59	Pr	28.2	25	calc	D84
60	Nd	31.4	25	calc	D84
61	Pm	30.1	25	calc	D84
62	Sm	28.8	25	calc	D84
63	Eu	27.7	25	calc	D84
64	Gd	23.5	25	calc	D84
65	Tb	25.5	25	calc	D84
66	Dy	24.5	25	calc	D84
67	Ho	23.6	25	calc	D84
68	Er	22.7	25	calc	D84
69	Tm	21.8	25	calc	D84
70	Yb	21.0	25	calc	D84
71	Lu	21.9	25	calc	D84
72	Hf	16.2	25	calc	D84
73	Ta	13.1	25	calc	D84
74	W	11.1	25	calc	D84
75	Re	9.7	25	calc	D84
76	Os	8.5	25	calc	D84
77	Ir	7.6	25	calc	D84
78	Pt	6.5	25	calc	D84
79	Au	5.8	25	calc	D84
		6.48	25	calc	G84
80	Hg	5.7	25	calc	D84
		5.1	15	diel	MB77
81	Tl	7.6	15	beam	NYU84
		7.5	25	calc	D84
82	Pb	6.8	25	calc	D84
83	Bi	7.4	25	calc	D84
84	Po	6.8	25	calc	D84
85	At	6.0	25	calc	D84
86	Rn	5.3	25	calc	D84
87	Fr	48.7	25	calc	D84
88	Ra	38.3	25	calc	D84
89	Ac	32.1	25	calc	D84
90	Th	32.1	25	calc	D84
91	Pa	25.4	25	calc	D84
92	U	24.9	6	beam	KB94
93	Np	24.8	25	calc	D84
94	Pu	24.5	25	calc	D84
95	Am	23.3	25	calc	D84
96	Cm	23.0	25	calc	D84
97	Bk	22.7	25	calc	D84
98	Cf	20.5	25	calc	D84
99	Es	19.7	25	calc	D84

Table 2
Static Average Electric Dipole Polarizabiilities for Ground State Atoms (in Units of 10^{-24} cm^3)
(continued)

Atomic number	Atom	Polarizability	Estimated accuracy (%)	Method	Ref.
100	Fm	23.8	25	calc	D84
101	Md	18.2	25	calc	D84
102	No	17.5	25	calc	D84

Note: calc = calculated value; beam = atomic beam deflection technique; interferom = atomic beam interference; index = determination based on the measured index of refraction; diel = determination based on the measured dielectric constant.

REFERENCES

A56. Atoji, M., *J. Chem. Phys.*, 25, 174, 1956. Semiempirical method based on molecular polarizabilities and atomic radii.

D84. Doolen, G. D., Los Alamos National Laboratory, unpublished. A relativistic linear response method was used. The method is that described by Zangwill, A., and Soven, P., *Phys. Rev. A*, 21, 1561, 1980. Adjustments of less than 10% across the periodic table have been made to these results to bring them into agreement with accurate experimental values where avaialble, for the purpose of presenting "recommended" polarizabilities in Table 2.

ESCHP94. Ekstrom, C. R., Schmiedmayer, J., Chapman, M. S., Hammond, T. D., and Pritchard, D. E., *Phys. Rev. Lett.*, in press, 1994.

G84. Gollisch, H., *J. Phys. B: Atom. Mol. Phys.*, 17, 1463, 1984. Other results and useful references are contained in this paper.

GMBSJ84. Guella, T. P., Miller, T. M., Bederson, B., Stockdale, J. A. D., and Jaduszliwer, B., *Phys. Rev. A*, 29, 2977, 1984.

KB94. Kadar-Kallen, M. A., and Bonin, K. D., *Phys. Rev. Lett.*, 72, 828, 1994.

MB77. Miller, T. M., and Bederson, B., *Adv. At. Mol. Phys.*, 13, 1, 1977. For simplicity, any value in Table 2 which has not changed since this 1977 review is referenced as MB77. Persons interested in original references and further details should consult MB77.

MMD90. Milani, P., Moullet, I., and de Heer, W. A., *Phys. Rev. A*, 42, 5150, 1990.

NB65. Newell, A. C., and Baird, R. D., *J. Appl. Phys.*, 36, 3751, 1965.

NYU84. Preliminary value from the New York University group. See GMBSJ84.

OC67. Orcutt, R. H., and Cole, R. H., *J. Chem. Phys.*, 46, 697, 1967; see also the later references from this group, given following the tables.

Table 3
Average Electric Dipole Polarizabilities for Ground State Diatomic Molecules (in Units of 10^{-24} cm^3)

Molecule	Polarizability	Ref.	Molecule	Polarizability	Ref.
Al$_2$	19	23	HI	5.44	3
BH	3.32*	1		5.35	2
Br$_2$	7.02	2	HgCl	7.4*	9
CO	1.95	3	ICl	12.3	2
Cl$_2$	4.61	3	K$_2$	77	22
Cs$_2$	104	22		72	21
CsK	89	22	Li$_2$	34	22
D$_2$ ($v = 0, J = 0$)	0.7921*	5	LiCl	3.46*	10
D$_2$ (293 K)	0.7954	6	LiF	10.8*	11
DCl	2.84	2	LiH	3.84*	12
F$_2$	1.38*	7		3.68*	13
H$_2$ ($v = 0, J = 0$)	0.8023*	5		3.88*	14
H$_2$ (293 K)	0.8045*	5	N$_2$	1.7403	6,8
H$_2$ (293 K)	0.8042	6	NO	1.70	2
H$_2$ (322 K)	0.8059	8	Na$_2$	40	22
HBr	3.61	3		38	21
HCl	2.63	3	NaK	51	22
	2.77	2	NaLi	40	4
HD ($v = 0, J = 0$)	0.7976*	5	O$_2$	1.5812	6
HF	0.80	27	Rb$_2$	79	22

Table 4
Average Electric Dipole Polarizabilities for Ground State Triatomic Molecules (in Units of 10^{-24} cm^3)

Molecule	Polarizability	Ref.	Molecule	Polarizability	Ref.	Molecule	Polarizability	Ref.
BeH_2	4.34*	14	HCN	2.59	3	O_3	3.21	2
CO_2	2.911	8		2.46	2	OCS	5.71	2
CS_2	8.74	3	$HgBr_2$	14.5	2		5.2	15
	8.86	2	$HgCl_2$	11.6	2	SO_2	3.72	3
D_2O	1.26	2	HgI_2	19.1	2		4.28	2
H_2O	1.45	2	N_2O	3.03	8			
H_2S	3.78	3	NO_2	3.02	2†			
	3.95	2	Na_3	70	21			

Table 5
Average Electric Dipole Polarizabilities for Ground State Inorganic Polyatomic Molecules (Larger than Triatomic) (in Units of 10^{-24} cm^3)

Molecule	Polarizability	Ref.	Molecule	Polarizability	Ref.
$AsCl_3$	14.9	2	$(NaBr)_2$	26.8	16
AsN_3	5.75	2	$(NaCl)_2$	23.4	16
BCl_3	9.38	20	$(NaF)_2$	20.7	16
BF_3	3.31	2	$(NaI)_2$	26.9	16
$(BN_3)_2$	5.73	2	OsO_4	8.17	2
$(BH_2N)_3$	8.0	2†	PCl_3	12.8	2
ClF_3	6.32	2	PF_5	6.10	2
$(CsBr)_2$	54.5	16	PH_3	4.84	2
$(CsCl)_2$	42.4	16	$(RbBr)_2$	48.2	16
$(CsF)_2$	28.4	16	$(RbCl)_2$	43.2	16
$(CsI)_2$	51.8	16	$(RbF)_2$	40.7	16
$GeCl_4$	15.1	2	$(RbI)_2$	46.3	16
GeH_3Cl	6.7	2†	SF_6	6.54	8
$(HgCl)_2$	14.7	9	$(SF_5)_2$	13.2	2
K_n	$n = 2,5,7-9,11,20$	21	SO_3	4.84	2
$(KBr)_2$	42.0	16	SO_2Cl_2	10.5	2
$(KCl)_2$	32.1	16	SeF_6	7.33	2
$(KF)_2$	21.0	16	SiF_4	5.45	2
$(KI)_2$	36.3	16	SiH_4	5.44	2
$(LiBr)_2$	18.9	16	$(SiH_3)_2$	11.1	2
$(LiCl)_2$	13.1	16	$SiHCl_3$	10.7	2
$(LiF)_2$	6.9	16	SiH_2Cl_2	8.92	2
$(LiI)_2$	23.4	16	SiH_3Cl	7.02	2
ND_3	1.70	2	$SnBr_4$	22.0	2
NF_3	3.62	2	$SnCl_4$	18.0	2
NH_3	2.81	20		13.8	15
	2.10	2	SnI_4	32.3	2
	2.26	3	TeF_6	9.00	2
$(NO_2)_2$	6.69	2	$TiCl_4$	16.4	2
Na_n	$n = 1-40$	21	UF_6	12.5	2

Table 6
Average Electric Dipole Polarizabilities for Ground State Hydrocarbon Molecules (in Units of 10^{-24} cm^3)

Molecule	Name	Polarizability	Ref.	Molecule	Name	Polarizability	Ref.
CH_4	methane	2.593	8	C_8H_{10}	ethylbenzene	14.2	2
C_2H_2	acetylene	3.33	3		o-xylene	14.9	2
		3.93	2			14.1	15
C_2H_4	ethylene	4.252	8		p-xylene	13.7	25
C_2H_6	ethane	4.47	3			14.2	15
		4.43	2			14.9	2
C_3H_4	propyne	6.18	2		m-xylene	14.2	15
C_3H_6	propene	6.26	2	C_8H_{16}	ethylcyclohexane	15.9	2
	cyclopropane	5.66	2	C_8H_{18}	n-octane	15.9	2
C_3H_8	propane	6.29	3		3-methylheptane	15.4	27
		6.37	2		2,2,4-trimethylpentane	15.4	27
C_4H_6	1-butyne	7.41	2[†]	C_9H_{10}	α-methylstyrene	16.05	27
	1,3-butadiene	8.64	2	C_9H_{12}	isopropylbenzene	16.0	2–
C_4H_8	1-butene	7.97	2		mesitylene	15.5	25
		8.52	2			16.1	27
	trans-2-butene	8.49	2	C_9H_{18}	isopropylcyclohexane	17.2	2
	2-methylpropene	8.29	2	C_9H_{20}	nonane	17.4	27
C_4H_{10}	n-butane	8.20	2	$C_{10}H_8$	napthalene	16.5	17
	isobutane	8.14	27			17.5	27
C_5H_6	1,3-cyclopentadiene	8.64	2	$C_{10}H_{14}$	durene	17.3	25
C_5H_8	1-pentyne	9.12	2		t-butylbenzene	17.2	25
	trans-1,3-pentadiene	10.0	2			17.8	2[†]
	isoprene	9.99	2	$C_{10}H_{20}$	t-butylcyclohexane	19.8	2
C_5H_{10}	cyclopentane	9.15	18	$C_{10}H_{22}$	decane	19.1	27
	1-pentene	9.65	27	$C_{11}H_{10}$	α-methylnaphthalene	19.35	27
	2-pentene	9.84	27		β-methylnaphthalene	19.52	27
C_5H_{12}	pentane	9.99	2	$C_{11}H_{14}$	α,β,β-trimethylstyrene	19.64	27
	neopentane	10.20	18	$C_{11}H_{16}$	pentamethylbenzene	19.1	25
C_6H_6	benzene	10.0	25	$C_{11}H_{24}$	undecane	21.0	27
		10.32	3	$C_{12}H_{10}$	acenaphthene	20.6	27
		10.74	2	$C_{12}H_{12}$	α-ethylnaphthalene	21.19	27
C_6H_{10}	1-hexyne	10.9	2[†]		β-ethylnaphthalene	21.36	27
	2-ethyl-1,3-butadiene	11.8	2[†]	$C_{12}H_{18}$	hexamethylbenzene	20.9	25
	3-methyl-1,3-pentadiene	11.8	2[†]	$C_{12}H_{26}$	dodecane	22.8	27
	2-methyl-1,3-pentadiene	12.1	2[†]	$C_{13}H_{10}$	fluorene	21.7	27
	2,3-dimethyl-1,3-butadiene	11.8	2[†]	$C_{14}H_{10}$	anthracene	25.4	17
	cyclohexene	10.7	2[†]			25.9	27
C_6H_{12}	cyclohexane	11.0	18		phenanthrene	36.8*	17
		10.87	15			24.7	27
	1-hexene	11.7	27	$C_{14}H_{22}$	p-di-t-butylbenzene	24.5	25
C_6H_{14}	n-hexane	11.9	2	$C_{16}H_{10}$	pyrene	28.2	27
C_7H_8	toluene	11.8	25	$C_{17}H_{12}$	2,3-benzfluorene	30.2	27
		12.26	15	$C_{18}H_{12}$	napthacene	32.3	27
		12.3	2		1,2-benzanthracene	32.9	27
C_7H_{12}	1-heptyne	12.8	2[†]		chrysene	33.1	27
C_7H_{14}	methylcyclohexane	13.1	2		triphenylene	31.1	27
	1-heptene	13.5	27	$C_{18}H_{30}$	1,3,5-tri-t-butylbenzene	31.8	25
C_7H_{16}	n-heptane	13.7	2	$C_{24}H_{12}$	coronene		
C_8H_8	styrene	15.0	2				
		14.4	27				

Table 7
Average Electric Dipole Polarizabilities for Ground State Organic Halides (in Units of 10^{-24} cm^3)

Molecule	Name	Polarizability	Ref.
CBr_2F_2	dibromodifluoromethane	9.0	2[†]
$CClF_3$	chlorotrifluoromethane	5.72	20
		5.59	2
CCl_2F_2	dichlorodifluoromethane	7.93	20
		7.81	2
CCl_2O	phosgene	7.29	2
CCl_2S	thiophosgene	10.2	2
CCl_3F	trichlorofluoromethane	9.47	2
CCl_3NO_2	trichloronitromethane	10.8	2[†]
CCl_4	carbon tetrachloride	11.2	2
		10.5	3
CF_4	carbon tetrafluoride	3.838	8
CF_2O	carbonylfluoride	1.88*	17
$CHBr_3$	bromoform	11.8	17
$CHBrF_2$	bromodifluoromethane	5.7	2[†]
$CHClF_2$	chlorodifluoromethane	6.38	20
		5.91	2
$CHCl_2F$	dichlorofluoromethane	6.82	2
$CHCl_3$	chloroform	9.5	8
		8.23	27
CHF_3	fluoroform	3.52	20
		3.57	8
$CHFO$	fluoroformaldehyde	1.76*	17
CHI_3	iodoform	18.0	17
CH_2Br_2	dibromomethane	9.32	2
		8.62	27
CH_2ClNO_2	chloronitromethane	6.9	2[†]
CH_2Cl_2	dichloromethane	6.48	3
		7.93	2
CH_2I_2	diiodomethane	12.90	27
CH_3Br	bromomethane	5.87	20
		6.03	2
		5.55	15
CH_3Cl	chloromethane	5.35	20
		4.72	8
CH_3F	fluoromethane	2.97	8
CH_3I	iodomethane	7.97	2
C_2ClF_5	chloropentafluoroethane	6.3	2[†]
$C_2Cl_2F_4$	1,2-dichlorotetrafluoroethane	8.5	2[†]
C_2Cl_3N	trichloroacetonitrile	6.10	18
C_2F_6	hexafluoroethane	6.82	2
C_2HBr	bromoacetylene	7.39	2
C_2HCl	chloroacetylene	6.07	2
C_2HCl_5	pentachloroethane	14.0	2
$C_2H_2Cl_2$	dichloroethylene	10.03	27
	trans-dichloroethylene	9.28	27
	cis-dichloroethylene	9.19	27
$C_2H_2Cl_2F_2$	1,1-dichloro-2,2-difluoroethane	8.4	2[†]
$C_2H_2Cl_2O$	chloroacetyl chloride	8.92	2
$C_2H_2Cl_3F$	1,2,2-trichloro-1-fluoroethane	10.2	2[†]
$C_2H_2Cl_4$	1,1,2,2-tetrachloroethane	12.1	2[†]
C_2H_2ClN	chloroacetonitrile	6.10	18
$C_2H_2F_2$	1,1-difluoroethylene	5.01	20
C_2H_3Br	bromoethylene	7.59	2
C_2H_3Cl	chloroethylene	6.41	2

Table 7
Average Electric Dipole Polarizabilities for Ground State Organic Halides (in Units of 10^{-24} cm^3) (continued)

Molecule	Name	Polarizability	Ref.
$C_2H_3ClF_2$	1-chloro-1,1-difluoroethane	8.05	2
C_2H_3ClO	acetyl chloride	6.62	2
$C_2H_3ClO_2$	ethyl chloroformate	7.1	2[†]
$C_2H_3Cl_3$	1,1,1-trichloroethane	10.7	2
$C_2H_3F_3$	1,1,1-trifluoroethane	4.4	2[†]
C_2H_3I	iodoethylene	9.3	2[†]
C_2H_4BrCl	1-bromo-2-chloroethane	9.5	2[†]
$C_2H_4Br_2$	1,2-dibromomethane	10.7	2[†]
C_2H_4ClF	1-chloro-2-fluoroethane	6.5	2[†]
$C_2H_4ClNO_2$	1-chloro-1-nitroethane	10.9	2
$C_2H_4Cl_2$	1,1-dichloroethane	8.64	2
	1,2-dichloroethane	8.0	2[†]
C_2H_5Br	bromoethane	8.05	2
		7.28	27
C_2H_5Cl	chloroethane	7.27	20
		8.29	2
		6.4	15
C_2H_5ClO	2-chloroethanol	7.1	2[†]
		6.88	27
	chloromethyl methyl ether	7.1	2[†]
C_2H_5F	fluoroethane	4.96	2
C_2H_5I	iodoethane	10.0	2
$C_3H_4Cl_2$	dichloropropene	10.1	2[†]
C_3H_5Cl	chloropropene	8.3	2
C_3H_5ClO	chloroacetone	8.4	2[†]
$C_3H_3ClO_2$	ethyl chloroformate	9.0	2[†]
$C_3H_6ClNO_2$	1-chloro-1-nitropropane	10.4	2[†]
$C_3H_6Cl_2$	dichloropropane	10.9	2[†]
C_3H_7Br	1-bromopropane	9.4	2[†]
		9.07	27
	2-bromopropane	9.6	2[†]
C_3H_7Cl	chloropropane	10.0	2
C_3H_7ClO	β-chloroethyl methyl ether	8.71	27
	2-chloro-1-propanol	8.89	27
	3-chloro-1-propanol	8.84	27
C_3H_7I	1-iodopropane	11.5	2[†]
C_4H_5Cl	4-chloro-1,2-butadiene	10.0	2[†]
C_4H_7Cl	1-chloro-2-methylpropene	10.8	2
$C_4H_7ClO_2$	2-chlorobutyric acid	10.7	27
	3-chlorobutyric acid	10.7	27
	4-chlorobutyric acid	10.6	27
$C_4H_8Cl_2$	1,4-dichlorobutane	12.0	2[†]
C_4H_9Br	bromobutane	13.9	2
		10.86	27
C_4H_9Cl	1-chlorobutane	11.3	2
	1-chloro-2-methylpropane	11.1	2
	2-chloro-2-methylpropane	12.5	2[†]
	2-chlorobutane	12.4	2
C_4H_9ClO	β-chloroethyl ethyl ether	10.56	27
	2-chloro-1-butanol	10.70	27
	3-chloro-1-butanol	10.38	27
C_4H_9I	1-iodobutane	13.3	2[†]
		12.65	27
$C_5H_9ClO_2$	2-chlorobutanoate acid	12.33	27
	3-chlorobutanoate acid	12.31	27

Table 7
Average Electric Dipole Polarizabilities for Ground State Organic Halides (in Units of 10^{-24} cm^3) (continued)

Molecule	Name	Polarizability	Ref.
	4-chlorobutanoate acid	12.27	27
	2-chloropentanoic acid	12.69	27
	3-chloropentanoic acid	12.57	27
	4-chloropentanoic acid	12.53	27
$C_5H_{11}Br$	1-bromopentane	13.1	2[†]
$C_5H_{11}Cl$	1-chloropentane	12.0	2[†]
$C_5H_{11}F$	fluoropentane	9.95	27
C_6F_6	hexafluorobenzene	9.58	27
C_6HF_5	pentafluorobenzene	9.63	27
$C_6H_2Cl_2O_2$	2,5-dichloro-1,4-benzoquinone	18.4	2
$C_6H_2F_4$	1,2,3,4-tetrafluorobenzene	9.69	27
	1,2,4,5-tetrafluorobenzene	9.69	27
$C_6H_3F_3$	1,3,5-trifluorobenzene	9.74	27
C_6H_4BrF	p-bromofluorobenzene	13.4	2[†]
$C_6H_4ClNO_2$	chloronitrobenzene	14.6	2[†]
$C_6H_4Cl_2$	o-dichlorobenzene	14.17	27
	m-dichlorobenzene	14.23	27
	p-dichlorobenzene	14.20	27
C_6H_4FI	p-fluoroiodobenzene	15.5	2[†]
$C_6H_4FNO_2$	p-fluoronitrobenzene	12.8	2[†]
$C_6H_4F_2$	o-difluorobenzene	9.80	27
	m-difluorobenzene	10.3	2[†]
	p-difluorobenzene	9.80	27
C_6H_5Br	bromobenzene	14.7	2
		13.62	27
C_6H_5Cl	chlorobenzene	14.1	2
		12.3	15
C_6H_5ClO	chlorophenol	13.0	2[†]
C_6H_5F	fluorobenzene	10.3	2
C_6H_5I	iodobenzene	15.5	2[†]
$C_6H_{11}ClO_2$	2-chlorobutanoate acid	14.16	27
	3-chlorobutanoate acid	14.13	27
	4-chlorobutanoate acid	14.11	27
$C_6H_{13}Br$	bromohexane	14.44	27
$C_6H_{13}F$	fluorohexane	11.80	27
C_7H_7Br	p-bromotoluene	14.80	27
C_7H_7Cl	p-chlorotoluene	13.70	27
C_7H_7F	p-fluorotoluene	11.70	27
C_7H_7I	p-iodotoluene	17.10	27
$C_7H_{15}Br$	1-bromoheptane	16.8	2[†]
		16.23	27
$C_7H_{15}F$	fluoroheptane	13.66	27
$C_8H_{17}Br$	bromooctane	18.02	27
$C_8H_{17}F$	fluorooctane	15.46	27
$C_9H_{19}Br$	bromononane	19.81	27
$C_9H_{19}F$	fluorononane	17.34	27
$C_{10}F_8$	octafluoronaphthalene	17.64	27
$C_{10}H_7Br$	α-bromonaphthalene	20.34	27
$C_{10}H_7Cl$	α-chloronaphthalene	19.30	27
	β-chloronaphthalene	19.58	27
$C_{10}H_7I$	α-iodonaphthalene	22.41	27
	β-iodonaphthalene	22.95	27
$C_{10}H_{21}Br$	bromodecane	21.60	27
$C_{10}H_{21}F$	fluorodecane	19.18	27

Table 7
Average Electric Dipole Polarizabilities for Ground State Organic Halides (in Units of 10^{-24} cm^3) (continued)

Molecule	Name	Polarizability	Ref.
$C_{11}H_{23}F$	fluoroundecane	21.00	27
$C_{12}H_{25}Br$	bromododecane	25.18	27
$C_{12}H_{25}F$	fluorododecane	22.83	27
$C_{12}H_8Br_2O$	4,4′–dibromodiphenyl ether	27.8	2[†]
$C_{12}H_9BrO$	4-bromodiphenyl ether	24.2	2[†]
$C_{13}H_{11}BrO$	p-bromophenyl-p-tolyl ether	26.6	2[†]
$C_{14}H_9Br$	9-bromoanthracene	28.32	27
$C_{14}H_9Cl$	9-chloroanthracene	27.35	27
$C_{14}H_9F$	fluoranthracene	28.34	27
$C_{14}H_{29}F$	fluorotetradecane	26.57	27
$C_{16}H_{33}Br$	bromohexadecane	32.34	27
$C_{18}H_{37}Br$	bromooctadecane	35.92	27

Table 8
Static Average Electric Dipole Polarizabilities for Other Ground State Organic Molecules (in Units of 10^{-24} cm^3)

Molecule	Name	Polarizability	Ref.
CN_4O_8	tetranitromethane	15.3	2
CH_2O	formaldehyde	2.8	2[†]
		2.45	18
CH_2O_2	formic acid	3.4	2[†]
CH_3NO	formamide	4.2	2[†]
		4.08	18
CH_3NO_2	nitromethane	7.37	2
CH_4O	methanol	3.29	2
		3.23	15
		3.32	18
CH_5N	methyl amine	4.7	2
		4.01	19
C_2N_2	cyanogen	7.99	2
C_2H_2O	ketene	4.4	2[†]
C_2H_3N	acetonitrile	4.40	2[†]
		4.48	18
C_2H_4O	acetaldehyde	4.6	2[†]
		4.59	18
	ethylene oxide	4.43	18
$C_2H_4O_2$	acetic acid	5.1	2[†]
	methyl formate	5.05	27
$C_2H_4O_4$	formic acid dimer	12.7	2
C_2H_5NO	acetamide	5.67	18
	N-methyl formamide	5.91	18
$C_2H_5NO_2$	nitroethane	9.63	2
	ethyl nitrite	7.0	15
C_2H_6O	ethanol	5.41	2
		5.11	18
	methyl ether	5.29	20
		5.84	2
		5.16	15
$C_2H_6O_2$	ethylene glycol	5.7	2[†]
		5.61	27
$C_2H_6O_2S$	dimethyl sulfone	7.3	2[†]

Table 8
Static Average Electric Dipole Polarizabilities for Other Ground State Organic Molecules (in Units of 10^{-24} cm^3) (continued)

Molecule	Name	Polarizability	Ref.
C_2H_6S	ethanethiol	7.41	2
C_2H_7N	ethyl amine	7.10	2
	dimethyl amine	6.37	2
$C_2H_8N_2$	ethylene diamine	7.2	2†
$C_3H_2N_2$	malononitrile	5.79	18
C_3H_3N	acrylonitrile	8.05	2
$C_3H_4N_2$	pyrazole	7.23	27
C_3H_4O	propenal	6.38	2†
C_3H_5N	propionitrile	6.70	2
		6.24	18
C_3H_6O	acetone	6.33	15
		6.4	2†
		6.39	18
	allyl alcohol	7.65	2
	propionaldehyde	6.50	2
$C_3H_6O_2$	propionic acid	6.9	2†
	ethyl formate	8.01	2
		6.88	27
	methyl acetate	6.94	2
		6.81	27
$C_3H_6O_3$	dimethyl carbonate	7.7	2†
C_3H_7NO	N-methyl acetamide	7.82	18
	N,N-dimethyl formamide	7.81	18
$C_3H_7NO_2$	nitropropane	8.5	2†
C_3H_8O	2-propanol	7.61	2
		6.97	18
	1-propanol	6.74	2
	ethyl methyl ether	7.93	2
$C_3H_8O_2$	dimethoxymethane	7.7	2†
	monomethylether ethylene glycol	7.44	27
C_3H_9N	n-propylamine	7.70	27
		9.20	2
	isopropylamine	7.77	27
	trimethylamine	8.15	2
$C_4H_2N_2$	fumaronitrile	11.8	2
$C_4H_4N_2$	succinonitrile	8.1	2†
	pyrimidene	8.53*	17
	pyridazine	9.27*	17
$C_4H_4O_2$	diketene	8.0	2†
C_4H_4S	thiophene	9.67	2
C_4H_5N	methacrylonitrile	8.0	2†
	trans-crotononitrile	8.2	2†
$C_4H_6N_2$	N-methylpyrazole	8.99	27
C_4H_6O	crotonaldehyde	8.5	2†
	methacrylaldehyde	8.3	2†
$C_4H_6O_2$	biacetyl	8.2	2†
$C_4H_6O_3$	acetic anhydride	8.9	2†
C_4H_6S	divinyl sulfide	10.9	2†
C_4H_7N	butyronitrile	8.4	2†
	isobutyronitrile	8.05	18
C_4H_8O	butanal	8.2	2†
	methyl ethyl ketone	8.13	15
	trans-2,3-epoxy butane	8.22*	17
$C_4H_8O_2$	ethyl acetate	9.7	2
		8.62	27

Table 8
Static Average Electric Dipole Polarizabilities for Other Ground State Organic Molecules (in Units of 10^{-24} cm^3) (continued)

Molecule	Name	Polarizability	Ref.
	1,4-dioxane	10.0	2
	p-dioxane	8.60	18
	2-methyl-1,3-dioxolane	9.44	15
	butyric acid	8.38	27
	methyl propionate	8.97	27
$C_4H_9NO_2$	1-nitrobutane	10.4	2[†]
	2-methyl-2-nitropropane	10.3	2[†]
$C_4H_{10}O$	ethyl ether	10.2	2
		8.73	15
	1-butanol	8.88	2
	2-methylpropanol	8.92	2
	n-propyl methyl ether	8.86	27
$C_4H_{10}O_2$	monoethylether ethylene glycol	9.28	27
$C_4H_{10}S$	ethyl sulfide	10.8	2
$C_4H_{11}N$	n-butylamine	13.5	2
	diethylamine	10.2	2
		9.61	27
C_5H_5N	pyridine	9.5	15
		9.18	27
	4-cyano-1,2-butadiene	10.5	2[†]
$C_5H_8N_2$	1,5-dimethylpyrazole	10.72	27
$C_5H_8O_2$	acetyl acetone	10.5	2[†]
C_5H_9N	valeronitrile	10.4	2
	22-DMPN	9.59	18
$C_5H_{10}O$	diethyl ketone	9.93	15
	methyl propyl ketone	9.93	15
$C_5H_{10}O_2$	ethyl propionate	10.41	27
	methyl butanoate	10.41	27
$C_5H_{10}O_3$	diethyl carbonate	11.3	2
$C_5H_{12}O$	ethyl propyl ether	10.68	27
$C_5H_{12}O_4$	tetramethyl orthocarbonate	13.0	2[†]
$C_6H_4N_2O_4$	p-dinitrobenzene	18.4	2
$C_6H_4O_2$	p-benzoquinone	14.5	2
$C_6H_5NO_2$	nitrobenzene	14.7	2
		12.92	15
C_6H_6O	phenol	11.1	2[†]
		9.94*	17
C_6H_7N	aniline	12.1	2[†]
$C_6H_8N_2$	phenylenediamine	13.8	2[†]
	phenylhydrazine	12.91	27
$C_6H_{10}N_2$	1-ethyl-5-methylpyrazole	12.50	27
$C_6H_{10}O_3$	ethyl acetoacetate	12.9	2[†]
$C_6H_{12}N_2$	dimethylketazine	15.6	2
$C_6H_{12}O$	cyclohexanol	11.56	18
$C_6H_{12}O_2$	amyl formate	14.2	2
$C_6H_{12}O_3$	paraldehyde	17.9	2
$C_6H_{14}O$	propyl ether	12.8	2
		12.5	15
$C_6H_{14}O_2$	1,1-diethoxyethane	13.2	2[†]
	1,2-diethoxyethane	11.3	2[†]
$C_6H_{15}N$	triethylamine	13.1	2
		13.38	27
	di-n-propylamine	13.29	27
$C_7H_4N_2O_2$	p-cyanonitrobenzene	19.0	2
C_7H_5N	benzonitrile	12.5	2[†]

Table 8
Static Average Electric Dipole Polarizabilities for Other Ground State Organic Molecules (in Units of 10^{-24} cm^3) (continued)

Molecule	Name	Polarizability	Ref.
$C_7H_7NO_3$	nitroanisole	15.7	2[†]
C_7H_7O	anisole	13.1	2[†]
C_7H_9NO	o-anisidine	14.2	2[†]
$C_7H_{10}N_2$	1,1-methylphenylhydrazine	14.81	27
$C_7H_{14}O$	cyclohexyl methyl ether	13.4	2[†]
	2,4-dimethyl-3-pentanone	13.5	15
$C_7H_{14}O_2$	pentyl acetate	14.9	2
$C_8H_4N_2$	p-dicyanobenzene	19.2	2
$C_8H_6O_2$	quinoxaline	15.13	27
C_8H_8O	acetophenone	15.0	2
$C_8H_8O_2$	2,5-dimethyl-1,4-benzoquinone	18.8	2
$C_8H_{10}O$	phenetole	14.9	2
$C_8H_{11}N$	N-dimethylaniline	16.2	2[†]
$C_8H_{12}N_2$	1,1-ethylphenylhydrazine	16.62	27
$C_8H_{12}O_2$	ethyl sorbate	17.2	2[†]
	tetramethylcyclobutane-1,3-dione	18.6	2
$C_8H_{14}O_4$	diethyl succinate	16.8	2[†]
$C_8H_{18}O$	n-butyl ether	17.2	2
C_9H_7N	quinoline	15.70	27
	isoquinoline	16.43	27
$C_9H_{10}O_2$	ethyl benzoate	16.9	2[†]
$C_9H_{21}N$	tri-n-propylamine	18.87	27
$C_{10}H_9N$	α-naphthylamine	19.50	27
	β-naphthylamine	19.73	27
	1-methylquinoline	18.65	27
	1-methylisoquinoline	18.28	27
$C_{10}H_{10}Fe$	ferrocene	17.1	26
$C_{10}H_{10}N_2$	2,3-dimethylquinoxaline	18.70	27
$C_{10}H_{14}BeO_4$	beryllium acetylacetonate	34.1	2
$C_{11}H_8O$	1-naphthaldehyde	19.75	27
	2-naphthaldehyde	20.06	27
$C_{14}H_8O_2$	anthraquinone	24.46	27
$C_{12}H_8N_2$	phenazine	23.43	27
$C_{12}H_9NO_3$	4-nitrodiphenyl ether	24.7	2[†]
$C_{14}H_{14}O$	di-p-tolyl ether	24.9	2[†]
$C_{15}H_{21}AlO_6$	aluminum acetylacetonate	51.9	2
$C_{15}H_{21}CrO_6$	chromium acetylacetonate	53.7	2
$C_{15}H_{21}FeO_6$	ferric acetylacetonate	58.1	2
$C_{20}H_{28}O_8Th$	thorium acetylacetonate	79.0	2
C_{60}	buckminsterfullerene	~80	24

Note: All polarizabilities in the tables are experimental values except those values marked by an asterisk (*), which indicates a calculated result. The experimental polarizabilities are mostly determined by measurements of a dielectric constant or refractive index which are quite accurate (0.5% or better). However, one should treat many of the results with several percent of caution because of the age of the data and because some of the results refer to optical frequencies rather than static. Comments given with the references are intended to allow one to judge the degree of caution required. Interested persons should consult these references. In many cases, the reference given is to a theoretical paper in which the experimental results are quoted. These papers, noted in the References, contain valuable information on polarizability calculations and experimental data which often includes the tensor components of the polarizability.

REFERENCES

1. McCullough, E. A., Jr., *J. Chem. Phys.*, 63, 5050, 1975. This calculation is for the parallel component, not the average polarizability.

2. Maryott, A. A., and Buckley, F., U. S. National Bureau of Standards Circular No. 537, 1953. A tabulation of dipole moments, dielectric constants, and molar refractions measured between 1910 and 1952, and used here to determine polarizabilities if no more recent result exists. The polarizability is $3/(4\pi N_0)$ times the molar polarization or molar refraction, where N_0 is Avogadro's number. The value $3/(4\pi N_0) = 0.3964308 \times 10^{-24}$ cm^3 was used for this conversion. A dagger (†) following the reference number in the tables indicates that the polarizability was derived from the molar refraction and hence may not include some low-frequency contributions to the static polarizability; these "static" polarizabilities are therefore low by 1 to 30%.

3. Hirschfelder, J. O., Curtis, C. F., and Bird, R. B., *Molecular Theory of Gases and Liquids*, Wiley, New York, 1954, p. 950. Fundamental information on molecular polarizabilities.

4. Miller, T. M., and Bederson, B., *Adv. At. Mol. Phys.*, 13, 1, 1977. Review emphasizing atomic polarizabilities and measurement techniques. The data quoted in Table 3 are accurate to 8 to 12%.

5. Kolos, W., and Wolniewicz, L., *J. Chem. Phys.*, 46, 1426, 1967. Highly accurate molecular hydrogen calculations.

6. Newell, A. C., and Baird, R. C., *J. Appl. Phys.*, 36, 3751, 1965. Highly accurate refractive index measurements at 47.7 GHz (essentially static).

7. Jao, T. C., Beebe, N. H. F., Person, W. B., and Sabin, J. R., *Chem. Phys. Lett.*, 26, 47, 1974. Tensor polarizabilities, derivatives, and other results are reported.

8. Orcutt, R. H., and Cole, R. H., *J. Chem. Phys.*, 46, 697, 1967 (He, Ne, Ar, Kr, H$_2$, N$_2$); Sutter, H., and Cole, R. H., *J. Chem. Phys.*, 52, 132, 1970 (CF$_3$H, CFH$_3$, CClF$_3$, CClH$_3$); Bose, T. K., and Cole, R. H., *J. Chem. Phys.*, 52, 140, 1970 (CO$_2$), and 54, 3829, 1971 (C$_2$H$_2$, C$_2$H$_4$); Nelson, R. D., and Cole, R. H., *J. Chem. Phys.*, 54, 4033, 1971 (SF$_6$, CClF$_3$); Bose, T. K., Sochanski, J. S., and Cole, R. H., *J. Chem. Phys.*, 57, 3592, 1972 (CH$_4$, CF$_4$); Kirouac, S., and Bose, T. K., *J. Chem. Phys.*, 59, 3043, 1973 (N$_2$O), and 64, 1580, 1976 (He). Highly accurate dielectric constant measurements. These modern data give the most accurate polarizabilities available. A criticism of the interpretation of these data in the case of polar molecules is given in Ref. 20, p. 2905.

9. Huestis, D. L., Technical Report #MP 78-25, SRI International (project PYU 6158), Menlo Park, CA 94025. Molar refractions for mercury-chlorine compounds are analyzed.

10. Bounds, D. G., Clarke, J. H. R., and Hinchliffe, A., *Chem. Phys. Lett.*, 45, 367, 1977. Theoretical tensor polarizability for LiCl.

11. Kolker, H. J., and Karplus, M., *J. Chem. Phys.*, 39, 2011, 1963. Theoretical.

12. Cutschick, V. P., and McKoy, V., *J. Chem. Phys.*, 58, 2397, 1973. Theoretical tensor polarizabilities.

13. Gready, J. E., Bacskay, G. B., and Hush, N. S., *Chem. Phys.*, 22, 141, 1977, and 23, 9, 1977. Theoretical.

14. Amos, A. T., and Yoffe, J. A., *J. Chem. Phys.*, 63, 4723, 1975. Theoretical.

15. Stuart, H. A., *Landolt-Börnstein Zahlenwerte and Funktionen*, Vol. 1, Part 3, Eucken, A., and Hellwege, K. H., Eds., Springer-Verlag, Berlin, 1951, p. 511. Tabulation of molecular polarizabilities. Two misprints in the chemical symbols have been corrected.

16. Guella, T., Miller, T. M., Stockdale, J. A. D., Bederson, B., and Vuskovic, L., *J. Chem. Phys.*, 94, 6857, 1991. Beam measurements with accuracies between 12-24%.

17. Marchese, F. T., and Jaffé, *Theoret. Chim. Acta (Berlin)*, 45, 241, 1977. Theoretical and experimental tensor polarizabilities are tabulated in this paper.

18. Applequist, J., Carl, J. R., and Fung, K.-K., *J. Am. Chem. Soc.*, 94, 2952, 1972. Excellent reference on the calculation of molecular polarizabilities, including extensive tables of tensor polarizabilities, both theoretical and experimental, at 589.3 nm wavelength.

19. Bridge, N. J., and Buckingham, A. D., *Proc. Roy. Soc. (London)*, A295, 334, 1966. Measured tensor polarizabilities at 633 nm wavelength.

20. Barnes, A. N. M., Turner, D. J., and Sutton, L. E., *Trans. Faraday Soc.*, 67, 2902, 1971. Dielectric constants yielding polarizabilities accurate from 0.3-8%.

21. Knight, W. D., Clemenger, K., de Heer, W. A., and Saunders, W. A., *Phys. Rev. B*, 31, 2539, 1985. These data probably correspond to a very low internal temperature.

22. Tarnovsky, V., Bunimovicz, M., Vuskovic, L., Stumpf, B., and Bederson, B., *J. Chem. Phys.*, 98, 3894, 1993. These data correspond to internal temperatures 480-948 K.

23. Milani, P., Moullet, I., and de Heer, W. A., *Phys. Rev. A*, 42, 5150, 1990. Beam measurements accurate to 11%.

24. Bonin, K. D., and Kadar-Kallen, M. A., *Int. J. Mod. Phys.*, 8, 3313, 1994. Review article.

25. Aroney, M. J., and Pratten, S. J., *J. Chem. Soc., Faraday Trans. 1*, 80, 1201, 1984. Uncertainties in the range 1-3%.

26. Le Fevre, R. J. W., Murthy, D. S. N., and Saxby, J. D., *Aust. J. Chem.*, 24, 1057, 1971. Kerr effect.

27. No, K. T., Cho, K. H., Jhon, M. S., and Scheraga, H. A., *J. Am. Chem. Soc.*, 115, 2005, 1993. Theoretical; these results are quoted in numerous valuable papers on calculated polarizabilities, e.g., Miller, K. J., and Savchik, J. A., *J. Am. Chem. Soc.*, 101, 7206, 1979.

IONIZATION POTENTIALS OF ATOMS AND ATOMIC IONS

The ionization potentials of neutral and partially ionized atoms are listed in this table. Data were obtained from the compilations cited below, supplemented by results from the recent research literature. All values have been corrected to the currently recommended value of the conversion factor from wave number to energy, namely 1 eV = 8065.541 cm^{-1} (Reference 5). Values are given in eV.

Following the traditional spectroscopic notation, columns are headed I, II, III, etc. up to XXX, where I indicates the neutral atom, II the singly ionized atom, III the doubly ionized atom, etc. The first section of the table includes spectra I to VIII of all the elements; subsequent sections cover higher spectra (ionization stages) for those elements for which data are available.

REFERENCES

1. Moore, C. E., *Ionization Potentials and Ionization Limits Derived from the Analysis of Optical Spectra*, Natl. Stand. Ref. Data Ser. — Natl. Bur. Stand. (U.S.) No. 34, 1970.
2. Martin, W. C., Zalubas, R., and Hagan, L., *Atomic Energy Levels — The Rare Earth Elements*, Natl. Stand. Ref. Data Ser. — Natl. Bur. Stand. (U.S.), No. 60, 1978.
3. Sugar, J. and Corliss, C., *Atomic Energy Levels of the Iron Period Elements: Potassium through Nickel*, J. Phys. Chem. Ref. Data, Vol.14, Suppl. 2, 1985.
4. References to papers in *J. Phys. Chem. Ref. Data*, in the period 1973—91 covering other elements may be found in the cumulative index to that journal.
5. Cohen, E. R. and Taylor, B. N., *J. Phys. Chem. Ref. Data*, 17, 1795, 1988.
6. Martin, W.C., and Wiese, W.L., in *Atomic, Molecular, and Optical Physics Handbook*, Drake, G.W.F., Ed., AIP Press, New York, 1996.

Neutral Atoms to +7 Ions

Z	Element	I	II	III	IV	V	VI	VII	VIII
1	H	13.59844							
2	He	24.58741	54.41778						
3	Li	5.39172	75.64018	122.45429					
4	Be	9.3227	18.21116	153.89661	217.71865				
5	B	8.29803	25.15484	37.93064	259.37521	340.22580			
6	C	11.26030	24.38332	47.8878	64.4939	392.087	489.99334		
7	N	14.53414	29.6013	47.44924	77.4735	97.8902	552.0718	667.046	
8	O	13.61806	35.11730	54.9355	77.41353	113.8990	138.1197	739.29	871.4101
9	F	17.42282	34.97082	62.7084	87.1398	114.2428	157.1651	185.186	953.9112
10	Ne	21.5646	40.96328	63.45	97.12	126.21	157.93	207.2759	239.0989
11	Na	5.13908	47.2864	71.6200	98.91	138.40	172.18	208.50	264.25
12	Mg	7.64624	15.03528	80.1437	109.2655	141.27	186.76	225.02	265.96
13	Al	5.98577	18.82856	28.44765	119.992	153.825	190.49	241.76	284.66
14	Si	8.15169	16.34585	33.49302	45.14181	166.767	205.27	246.5	303.54
15	P	10.48669	19.7694	30.2027	51.4439	65.0251	220.421	263.57	309.60
16	S	10.36001	23.3379	34.79	47.222	72.5945	88.0530	280.948	328.75
17	Cl	12.96764	23.814	39.61	53.4652	67.8	97.03	114.1958	348.28
18	Ar	15.75962	27.62967	40.74	59.81	75.02	91.009	124.323	143.460
19	K	4.34066	31.63	45.806	60.91	82.66	99.4	117.56	154.88
20	Ca	6.11316	11.87172	50.9131	67.27	84.50	108.78	127.2	147.24
21	Sc	6.5615	12.79967	24.75666	73.4894	91.65	110.68	138.0	158.1
22	Ti	6.8281	13.5755	27.4917	43.2672	99.30	119.53	140.8	170.4
23	V	6.7463	14.66	29.311	46.709	65.2817	128.13	150.6	173.4
24	Cr	6.7665	16.4857	30.96	49.16	69.46	90.6349	160.18	184.7
25	Mn	7.43402	15.63999	33.668	51.2	72.4	95.6	119.203	194.5
26	Fe	7.9024	16.1878	30.652	54.8	75.0	99.1	124.98	151.06
27	Co	7.8810	17.083	33.50	51.3	79.5	102.0	128.9	157.8
28	Ni	7.6398	18.16884	35.19	54.9	76.06	108	133	162
29	Cu	7.72638	20.29240	36.841	57.38	79.8	103	139	166
30	Zn	9.3942	17.96440	39.723	59.4	82.6	108	134	174
31	Ga	5.99930	20.5142	30.71	64				
32	Ge	7.8994	15.93462	34.2241	45.7131	93.5			
33	As	9.7886	18.633	28.351	50.13	62.63	127.6		
34	Se	9.75238	21.19	30.8204	42.9450	68.3	81.7	155.4	
35	Br	11.81381	21.8	36	47.3	59.7	88.6	103.0	192.8
36	Kr	13.99961	24.35985	36.950	52.5	64.7	78.5	111.0	125.802
37	Rb	4.17713	27.285	40	52.6	71.0	84.4	99.2	136
38	Sr	5.6949	11.03013	42.89	57	71.6	90.8	106	122.3
39	Y	6.2171	12.24	20.52	60.597	77.0	93.0	116	129
40	Zr	6.63390	13.13	22.99	34.34	80.348			
41	Nb	6.75885	14.32	25.04	38.3	50.55	102.057	125	

Z	Element	I	II	III	IV	V	VI	VII	VIII
42	Mo	7.09243	16.16	27.13	46.4	54.49	68.8276	125.664	143.6
43	Tc	7.28	15.26	29.54					
44	Ru	7.36050	16.76	28.47					
45	Rh	7.45890	18.08	31.06					
46	Pd	8.3369	19.43	32.93					
47	Ag	7.5763	21.49	34.83					
48	Cd	8.9938	16.90832	37.48					
49	In	5.78636	18.8698	28.03	54				
50	Sn	7.3439	14.63225	30.50260	40.73502	72.28			
51	Sb	8.6084	16.53051	25.3	44.2	56	108		
52	Te	9.0096	18.6	27.96	37.41	58.75	70.7	137	
53	I	10.45126	19.1313	33					
54	Xe	12.1298	21.20979	32.1230					
55	Cs	3.89390	23.15745						
56	Ba	5.21170	10.00390						
57	La	5.5770	11.060	19.1773	49.95	61.6			
58	Ce	5.5387	10.85	20.198	36.758	65.55	77.6		
59	Pr	5.464	10.55	21.624	38.98	57.53			
60	Nd	5.5250	10.73	22.1	40.41				
61	Pm	5.58	10.90	22.3	41.1				
62	Sm	5.6436	11.07	23.4	41.4				
63	Eu	5.6704	11.241	24.92	42.7				
64	Gd	6.1501	12.09	20.63	44.0				
65	Tb	5.8638	11.52	21.91	39.79				
66	Dy	5.9389	11.67	22.8	41.47				
67	Ho	6.0215	11.80	22.84	42.5				
68	Er	6.1077	11.93	22.74	42.7				
69	Tm	6.18431	12.05	23.68	42.7				
70	Yb	6.25416	12.1761	25.05	43.56				
71	Lu	5.4259	13.9	20.9594	45.25	66.8			
72	Hf	6.82507	14.9	23.3	33.33				
73	Ta	7.5496							
74	W	7.8640							
75	Re	7.8335							
76	Os	8.4382							
77	Ir	8.9670							
78	Pt	8.9587	18.563						
79	Au	9.2255	20.5						
80	Hg	10.43750	18.756	34.2					
81	Tl	6.1082	20.428	29.83					
82	Pb	7.41666	15.0322	31.9373	42.32	68.8			
83	Bi	7.2856	16.69	25.56	45.3	56.0	88.3		
84	Po	8.41671							
85	At								
86	Rn	10.74850							
87	Fr	4.0727							
88	Ra	5.2784	10.14716						
89	Ac	5.17	12.1						
90	Th	6.3067	11.5	20.0	28.8				
91	Pa	5.89							
92	U	6.19405							
93	Np	6.2657							
94	Pu	6.0262							
95	Am	5.9738							
96	Cm	6.02							
97	Bk	6.23							
98	Cf	6.30							
99	Es	6.42							
100	Fm	6.50							
101	Md	6.58							
102	No	6.65							

+8 Ions to +15 Ions

Z	Element	IX	X	XI	XII	XIII	XIV	XV	XVI
9	F	1103.1176							
10	Ne	1195.8286	1362.1995						
11	Na	299.864	1465.121	1648.702					
12	Mg	328.06	367.50	1761.805	1962.6650				
13	Al	330.13	398.75	442.00	2085.98	2304.1410			
14	Si	351.12	401.37	476.36	523.42	2437.63	2673.182		
15	P	372.13	424.4	479.46	560.8	611.74	2816.91	3069.842	
16	S	379.55	447.5	504.8	564.44	652.2	707.01	3223.78	3494.1892
17	Cl	400.06	455.63	529.28	591.99	656.71	749.76	809.40	3658.521
18	Ar	422.45	478.69	538.96	618.26	686.10	755.74	854.77	918.03
19	K	175.8174	503.8	564.7	629.4	714.6	786.6	861.1	968
20	Ca	188.54	211.275	591.9	657.2	726.6	817.6	894.5	974
21	Sc	180.03	225.18	249.798	687.36	756.7	830.8	927.5	1009
22	Ti	192.1	215.92	265.07	291.500	787.84	863.1	941.9	1044
23	V	205.8	230.5	255.7	308.1	336.277	896.0	976	1060
24	Cr	209.3	244.4	270.8	298.0	354.8	384.168	1010.6	1097
25	Mn	221.8	248.3	286.0	314.4	343.6	403.0	435.163	1134.7
26	Fe	233.6	262.1	290.2	330.8	361.0	392.2	457	489.256
27	Co	186.13	275.4	305	336	379	411	444	511.96
28	Ni	193	224.6	321.0	352	384	430	464	499
29	Cu	199	232	265.3	369	401	435	484	520
30	Zn	203	238	274	310.8	419.7	454	490	542
36	Kr	230.85	268.2	308	350	391	447	492	541
37	Rb	150	277.1						
38	Sr	162	177	324.1					
39	Y	146.2	191	206		374.0			
42	Mo	164.12	186.4	209.3	230.28	279.1	302.60	544.0	570

+16 Ions to +23 Ions

Z	Element	XVII	XVIII	XIX	XX	XXI	XXII	XXIII	XXIV
17	Cl	3946.2960							
18	Ar	4120.8857	4426.2296						
19	K	1033.4	4610.8	4934.046					
20	Ca	1087	1157.8	5128.8	5469.864				
21	Sc	1094	1213	1287.97	5674.8	6033.712			
22	Ti	1131	1221	1346	1425.4	6249.0	6625.82		
23	V	1168	1260	1355	1486	1569.6	6851.3	7246.12	
24	Cr	1185	1299	1396	1496	1634	1721.4	7481.7	7894.81
25	Mn	1224	1317	1437	1539	1644	1788	1879.9	8140.6
26	Fe	1266	1358	1456	1582	1689	1799	1950	2023
27	Co	546.58	1397.2	1504.6	1603	1735	1846	1962	2119
28	Ni	571.08	607.06	1541	1648	1756	1894	2011	2131
29	Cu	557	633	670.588	1697	1804	1916	2060	2182
30	Zn	579	619	698	738	1856			
36	Kr	592	641	786	833	884	937	998	1051
42	Mo	636	702	767	833	902	968	1020	1082

+24 Ions to +29 Ions

Z	Element	XXV	XXVI	XXVII	XXVIII	XXIX	XXX
25	Mn	8571.94					
26	Fe	8828	9277.69				
27	Co	2219.0	9544.1	10012.12			
28	Ni	2295	2399.2	10288.8	10775.40		
29	Cu	2308	2478	2587.5	11062.38	11567.617	
36	Kr	1151	1205.3	2928	3070	3227	3381
42	Mo	1263	1323	1387	1449	1535	1601

IONIZATION ENERGIES OF GAS-PHASE MOLECULES
Sharon G. Lias

This table presents values for the first ionization energies (IP) of approximately 1000 molecules and atoms. Substances are listed by molecular formula in the modified Hill order (see introduction). Values enclosed in parentheses are considered not to be well established. Data appearing in the 1988 reference, were updated in 1996 for inclusion in the database of ionization energies available at the Internet site of the Standard Reference Data program of the National Institute of Standards and Technology (http://webbook.nist.gov). The list appearing here includes these updates.

The list also includes values for enthalpies of formation of the ions at 298 K, $\Delta_f H_{ion}$, given according to the ion convention used by mass spectrometrists; to convert these values to the electron convention used by thermodynamicists, add 6 kJ/mol. Details on the calculation of $\Delta_f H_{ion}$ as well as data for a much larger number of molecules, may be found in the reference and on the Internet site.

REFERENCE

Lias, S.G., Bartmess, J.E., Liebman, J.F., Holmes, J.L., Levin, R.D., and Mallard, W.G., *Gas-Phase Ion and Neutral Thermochemistry, J. Phys. Chem. Ref. Data*, Vol. 17, Suppl. No. 1, 1988.

Mol. Form.	Name	IP/eV	$\Delta_f H_{ion}$ kJ/mol
	Substances not containing carbon		
Ac	Actinium	5.17	905
Ag	Silver	7.57624	1016
AgCl	Silver(I) chloride	(≤ 10.08)	≤ 1065
AgF	Silver(I) fluoride	(11.0 ± 0.3)	1071
Al	Aluminum	5.98577	905
AlBr	Aluminum monobromide	(9.3)	913
AlBr$_3$	Aluminum tribromide	(10.4)	593
AlCl	Aluminum monochloride	9.4	855
AlCl$_3$	Aluminum trichloride	(12.01)	573
AlF	Aluminum monofluoride	9.73 ± 0.01	673
AlF$_3$	Aluminum trifluoride	≤ 15.45	≤ 282
AlI	Aluminum monoiodide	9.3 ± 0.3	965
AlI$_3$	Aluminum triiodide	(9.1)	673
Am	Americium	5.9738 ± 0.0002	860
Ar	Argon	15.75962	1521
As	Arsenic	9.8152	1250
AsCl$_3$	Arsenic(III) chloride	(10.55 ± 0.025)	754
AsF$_3$	Arsenic(III) fluoride	(12.84 ± 0.05)	452
AsH$_3$	Arsine	(9.89)	1021
Au	Gold	9.22567	1254
B	Boron	8.29803	1363
BBr$_3$	Boron tribromide	(10.51)	809
BCl$_3$	Boron trichloride	11.60 ± 0.02	718
BF	Fluoroborane	11.12 ± 0.01	957
	Difluoroborane	(9.4)	317
BF$_3$	Boron trifluoride	15.7 ± 0.3	365
BH	Boron monohydride	(9.77)	1385
BH$_3$	Borane	12.026 ± 0.024	1261
BI$_3$	Boron triiodide	(9.25 ± 0.03)	964
BO$_2$	Boron dioxide	(13.5 ± 0.3)	1001
B$_2$H$_6$	Diborane	11.38 ± 0.05	1134
B$_2$O$_3$	Boron oxide	13.5 ± 0.15	460
B$_4$H$_{10}$	Tetraborane	10.76 ± 0.04	1105
B$_5$H$_9$	Pentaborane(9)	9.90 ± 0.04	1028
B$_6$H$_{10}$	Hexaborane	(9.0)	965
Ba	Barium	5.21170	683
BaO	Barium oxide	6.91 ± 0.06	543
Be	Beryllium	9.32263	1224
BeO	Beryllium oxide	(10.1 ± 0.4)	1111
Bi	Bismuth	7.2855	908
BiCl$_3$	Bismuth trichloride	(10.4)	736
Bk	Berkelium	6.23	911

Mol. Form.	Name	IP/eV	$\Delta_f H_{ion}$ kJ/mol
Br	Bromine (atomic)	11.81381	1252
BrCl	Bromine chloride	11.01	1079
BrF	Bromine fluoride	11.86	1086
BrF$_5$	Bromine pentafluoride	13.172 ± 0.002	840
BrH	Hydrogen bromide	11.66 ± 0.03	1087
BrH$_3$Si	Bromosilane	10.6	943
BrI	Iodine bromide	9.790 ± 0.004	986
BrK	Potassium bromide	7.85 ± 0.1	578
BrLi	Lithium bromide	(8.7)	685
BrNO	Nitrosyl bromide	10.17 ± 0.03	1065
BrNa	Sodium bromide	8.31 ± 0.1	660
BrO	Bromine monoxide	10.46 ± 0.02	1135
BrRb	Rubidium bromide	7.94 ± 0.03	583
BrTl	Thallium(I) bromide	9.14 ± 0.02	844
Br$_2$	Bromine	10.516 ± 0.005	1046
Br$_2$Hg	Mercury(II) bromide	10.560 ± 0.003	935
Br$_2$Sn	Tin(II) bromide	9.0	839
Br$_3$Ga	Gallium(III) bromide	10.40	711
Br$_3$P	Phosphorus(III) bromide	9.7	798
Br$_4$Hf	Hafnium(IV) bromide	(10.9)	366
Br$_4$Sn	Tin(IV) bromide	10.6	709
Br$_4$Ti	Titanium(IV) bromide	10.3	375
Br$_4$Zr	Zirconium(IV) bromide	(10.7)	388
Ca	Calcium	6.11316	768
CaCl	Calcium monochloride	5.86 ± 0.07	462
CaO	Calcium oxide	6.66 ± 0.18	668
Cd	Cadmium	8.99367	980
Ce	Cerium	5.5387	957
Cf	Californium	6.30	805
Cl	Chlorine (atomic)	12.96764	1373
ClCs	Cesium chloride	(7.84 ± 0.05)	510
ClF	Chlorine fluoride	12.66 ± 0.01	1171
ClFO$_3$	Perchloryl fluoride	(12.945 ± 0.005)	1224
ClF$_2$	Chlorine difluoride	(12.77 ± 0.05)	1128
ClF$_3$	Chlorine trifluoride	(12.65 ± 0.05)	1057
ClF$_5$S	Sulfur chloride pentafluoride	(12.335 ± 0.005)	144
ClH	Hydrogen chloride	12.749 ± 0.009	1137
ClHO	Hypochlorous acid	(11.12 ± 0.01)	993
ClH$_3$Si	Chlorosilane	11.4	899
ClI	Iodine chloride	10.088 ± 0.01	991
ClIn	Indium(I) chloride	(9.51)	842
ClK	Potassium chloride	(8.0 ± 0.4)	557
ClLi	Lithium chloride	9.57	727
ClNO	Nitrosyl chloride	10.87 ± 0.01	1099
ClNO$_2$	Nitryl chloride	(11.84)	1155
ClNa	Sodium chloride	8.92 ± 0.06	681
ClO	Chlorine monoxide	10.95	1159
ClO$_2$	Chlorine dioxide	10.33 ± 0.02	1093
ClRb	Rubidium chloride	(8.50 ± 0.03)	590
ClTl	Thallium(I) chloride	9.70 ± 0.03	869
Cl$_2$	Chlorine	11.480 ± 0.005	1108
Cl$_2$CrO$_2$	Chromyl chloride	11.6	580
Cl$_2$Ge	Germanium(II) chloride	(10.20 ± 0.05)	813
Cl$_2$H$_2$Si	Dichlorosilane	11.4	765
Cl$_2$Hg	Mercury(II) chloride	11.380 ± 0.003	952
Cl$_2$O	Chlorine oxide	10.94	1135
Cl$_2$OS	Thionyl chloride	10.96	844
Cl$_2$O$_2$S	Sulfuryl chloride	12.05	807
Cl$_2$Pb	Lead(II) chloride	(10.2)	791
Cl$_2$S	Sulfur dichloride	9.45 ± 0.03	895

Mol. Form.	Name	IP/eV	$\Delta_f H_{ion}$ kJ/mol
Cl_2Si	Dichlorosilylene	(10.93 ± 0.10)	887
Cl_2Sn	Tin(II) chloride	(10.0)	760
Cl_3Ga	Gallium(III) chloride	11.52	664
Cl_3HSi	Trichlorosilane	(11.7)	648
Cl_3N	Nitrogen trichloride	(10.12 ± 0.1)	1244
Cl_3OP	Phosphorus(V) oxychloride	11.36 ± 0.02	540
Cl_3OV	Vanadyl trichloride	(11.6)	425
Cl_3P	Phosphorus(III) chloride	9.91	668
Cl_3PS	Phosphorus(V) sulfide trichloride	9.71 ± 0.03	573
Cl_3Sb	Antimony(III) chloride	(≤ 10.7)	s719
Cl_4Ge	Germanium(IV) chloride	11.68 ± 0.05	629
Cl_4Hf	Hafnium(IV) chloride	(11.7)	246
Cl_4Si	Tetrachlorosilane	11.79 ± 0.01	527
Cl_4Sn	Tin(IV) chloride	11.7 ± 0.2	656
Cl_4Ti	Titanium(IV) chloride	(11.5)	349
Cl_4V	Vanadium(IV) chloride	(9.2)	361
Cl_4Zr	Zirconium(IV) chloride	(11.2)	210
Cl_5Mo	Molybdenum(V) chloride	(8.7)	392
Cl_5Nb	Niobium(V) chloride	(10.97)	356
Cl_5P	Phosphorus(V) chloride	(10.2)	608
Cl_5Ta	Tantalum(V) chloride	(11.08)	303
Cl_6W	Tungsten(VI) chloride	(9.5)	348
Cm	Curium	6.02	966
Co	Cobalt	7.8810	1187
Cr	Chromium	6.76664	1050
Cs	Cesium	3.89390	452
CsF	Cesium fluoride	(8.80 ± 0.10)	489
$CsNa$	Cesium sodium	(4.05 ± 0.04)	535
Cu	Copper	7.72638	1084
CuF	Copper(I) fluoride	10.15 ± 0.02	984
Dy	Dysprosium	5.9389	862
Er	Erbium	6.1078	907
Es	Einsteinium	6.42	753
Eu	Europium	5.6704	723
F	Fluorine (atomic)	17.42282	1761
FGa	Gallium monofluoride	(9.6 ± 0.5)	700
FH	Hydrogen fluoride	16.044 ± 0.003	1276
FHO	Hypofluorous acid	12.71 ± 0.01	1130
FH_3Si	Fluorosilane	11.7	752
FI	Iodine fluoride	10.54 ± 0.01	922
FIn	Indium monofluoride	(9.6 ± 0.5)	740
FNO	Nitrosyl fluoride	12.63 ± 0.03	1152
FNO_2	Nitryl fluoride	(13.09)	1154
FNS	Thionitrosyl fluoride (NSF)	11.51 ± 0.04	1090
FO	Fluorine monoxide	12.78 ± 0.03	1342
FO_2	Fluorine superoxide (FOO)	(12.6 ± 0.2)	1228
FS	Sulfur fluoride	10.09	986
FTl	Thallium(I) fluoride	10.52	835
F_2	Fluorine	15.697 ± 0.003	1515
F_2Ge	Germanium(II) fluoride	(≤ 11.65)	551
F_2HN	Difluoramine	(11.53 ± 0.08)	1046
F_2H_2Si	Difluorosilane	(12.2)	386
F_2Mg	Magnesium fluoride	(13.6 ± 0.3)	588
F_2N	Difluoroamidogen	11.628 ± 0.01	1155
F_2N_2	*trans*-Difluorodiazine	(12.8)	1315
F_2O	Fluorine monoxide	13.11 ± 0.01	1290
F_2OS	Thionyl fluoride	12.25	688
F_2O_2S	Sulfuryl fluoride	13.04 ± 0.01	501
F_2Pb	Lead(II) fluoride	(11.5)	679
F_2S	Sulfur difluoride	(10.08)	676

Mol. Form.	Name	IP/eV	$\Delta_f H_{ion}$ kJ/mol
F_2Si	Difluorosilylene	10.78 ± 0.05	450
F_2Sn	Tin(II) fluoride	(11.1)	586
F_2Xe	Xenon difluoride	12.35 ± 0.01	1083
F_3HSi	Trifluorosilane	(14.0)	150
F_3N	Nitrogen trifluoride	13.00 ± 0.02	1125
F_3NO	Trifluoramine oxide	13.31 ± 0.06	1121
F_3OP	Phosphorus(V) oxyfluoride	12.76 ± 0.01	-24
F_3P	Phosphorus(III) fluoride	11.60 ± 0.05	161
F_3PS	Phosphorus(V) sulfide trifluoride	$\leq 11.05 \pm 0.035$	≤ 58
F_3Si	Trifluorosilyl	(9.99)	- 32
F_4Ge	Germanium(IV) fluoride	(15.5)	307
F_4N_2	Tetrafluorohydrazine	11.94 ± 0.03	1119
F_4S	Sulfur tetrafluoride	12.0 ± 0.3	399
F_4Si	Tetrafluorosilane	15.24 ± 0.14	-144
F_4Xe	Xenon tetrafluoride	12.65 ± 0.1	1016
F_5I	Iodine pentafluoride	12.943 ± 0.005	408
F_5P	Phosphorus(V) fluoride	(15.1)	-137
F_5S	Sulfur pentafluoride	9.60 ± 0.05	10
F_6Mo	Molybdenum(VI) fluoride	(14.5 ± 0.1)	-159
F_6S	Sulfur hexafluoride	15.32 ± 0.02	258
F_6U	Uranium(VI) fluoride	14.00 ± 0.10	-796
Fe	Iron	7.9024	1177
Fm	Fermium	6.50	627
Ga	Gallium	5.99930	851
GaI_3	Gallium(III) iodide	9.40	765
Gd	Gadolinium	6.1500	991
Ge	Germanium	7.900	1139
GeH_4	Germane	≤ 10.53	≤ 1108
GeI_4	Germanium(IV) iodide	(9.42)	850
GeO	Germanium(II) oxide	11.25 ± 0.01	1044
GeS	Germanium(II) sulfide	(9.98)	1055
H	Hydrogen (atomic)	13.59844	1530
HI	Hydrogen iodide	10.386 ± 0.001	1028
HLi	Lithium hydride	7.7	882
HN	Imidogen	$\leq 13.49 \pm 0.01$	1678
HNO	Nitrosyl hydride	(10.1)	1075
HNO_2	Nitrous acid	≤ 11.3	≤ 1011
HNO_3	Nitric acid	11.95 ± 0.01	1019
HN_3	Hydrazoic acid	10.72 ± 0.025	1328
HO	Hydroxyl	13.0170 ± 0.0002	1294
HO_2	Hydroperoxy	11.35 ± 0.01	1106
HS	Mercapto	10.4219 ± 0.0004	1145
H_2	Hydrogen	15.42593 ± 0.00005	1488
H_2N	Amidogen	11.14 .01	1264
H_2O	Water	12.6206 ± 0.0020	976
H_2O_2	Hydrogen peroxide	10.58 ± 0.04	885
H_2S	Hydrogen sulfide	10.457 ± 0.012	989
H_2Se	Hydrogen selenide	9.892 ± 0.005	984
H_2Si	Silylene	8.244 ± 0.025	1084
H_3N	Ammonia	10.070 ± 0.020	925
H_3NO	Hydroxylamine	(10.00)	923
H_3P	Phosphine	9.869 ± 0.002	958
H_3Sb	Stibine	9.54 ± 0.03	1067
H_4N_2	Hydrazine	8.1 ± 0.15	877
H_4Si	Silane	11.00 ± 0.02	1095
H_4Sn	Stannane	(10.75)	1200
H_6Si_2	Disilane	9.74 ± 0.02	1019
H_8Si_3	Trisilane	(9.2)	1009
He	Helium	24.58741	2372
Hf	Hafnium	6.82507 ± 0.00004	1278

Mol. Form.	Name	IP/eV	$\Delta_f H_{ion}$ kJ/mol
Hg	Mercury	10.43750	1069
HgI$_2$	Mercury(II) iodide	9.5088 ± 0.0022	900
Ho	Holmium	6.0216	882
I	Iodine (atomic)	10.45126	1115
IK	Potassium iodide	(7.21 ± 0.3)	570
ILi	Lithium iodide	(7.5)	633
INa	Sodium iodide	7.64 ± 0.02	659
ITl	Thallium(I) iodide	8.47 ± 0.02	826
I$_2$	Iodine	9.3074 ± 0.0002	960
I$_4$Ti	Titanium(IV) iodide	(9.1)	602
I$_4$Zr	Zirconium(IV) iodide	(9.3)	500
In	Indium	5.78636	802
Ir	Iridium	9.1	1543
K	Potassium	4.34066	508
KLi	Lithium potassium	4.57 ± 0.04	512
KNa	Potassium sodium	4.41636 ± 0.00017	561
K$_2$	Dipotassium	4.0637 ± 0.0002	519
Kr	Krypton	13.99961	1351
La	Lanthanum	5.5770	969
Li	Lithium	5.39172	680
LiNa	Lithium sodium	5.05 ± 0.04	571
LiO	Lithium monoxide	(8.44)	894
LiRb	Lithium rubidium	4.3 ± 0.1	486
Li$_2$	Dilithium	5.1127 ± 0.0003	709
Lu	Lutetium	5.42585	950
Md	Mendelevium	6.58	635
Mg	Magnesium	7.64624	885
MgO	Magnesium oxide	(8.76 ± 0.22)	901
Mn	Manganese	7.43402	998
Mo	Molybdenum	7.09243	1343
N	Nitrogen (atomic)	14.53414	1875
NO	Nitric oxide	9.26438 ± 0.00005	985
NO$_2$	Nitrogen dioxide	9.586 ± 0.002	958
NP	Phosphorus nitride	11.84 ± 0.04	1247
NS	Nitrogen sulfide	8.87 ± 0.01	1119
N$_2$	Nitrogen	15.5808	1503
N$_2$O	Nitrous oxide	12.886	1325
N$_2$O$_4$	Nitrogen tetroxide	(10.8)	1050
N$_2$O$_5$	Nitrogen pentoxide	(11.9)	1161
Na	Sodium	5.13908	603
NaRb	Rubidium sodium	4.32 ± 0.04	480
Na$_2$	Disodium	4.894 ± 0.003	614
Nb	Niobium	6.75885	1384
Nd	Neodymium	5.5250	859
Ne	Neon	21.56454	2081
Ni	Nickel	7.6398	1167
No	Nobelium	6.65	642
Np	Neptunium	6.2657 ± 0.0003	1069
O	Oxygen (atomic)	13.61806	1563
OPb	Lead(II) oxide	9.08 ± 0.10	939
OS	Sulfur monoxide	10.294 ± 0.004	998
OS$_2$	Sulfur oxide (SSO)	10.584 ± 0.005	971
OSi	Silicon monoxide	11.49 ± 0.20	1008
OSn	Tin(II) oxide	9.60 ± 0.02	944
OSr	Strontium oxide	6.6 ± 0.2	623
O$_2$	Oxygen	12.0697 ± 0.0002	1165
O$_2$S	Sulfur dioxide	12.349 ± 0.001	894
O$_2$Th	Thorium(IV) oxide	(8.7 ± 0.15)	342
O$_2$Ti	Titanium(IV) oxide	(9.54 ± 0.1)	623
O$_2$U	Uranium(IV) oxide	(5.4 ± 0.1)	57

Mol. Form.	Name	IP/eV	$\Delta_f H_{ion}$ kJ/mol
O_3	Ozone	12.43	1342
O_3S	Sulfur trioxide	12.82 ± 0.03	841
O_3U	Uranium(VI) oxide	(10.5 ± 0.5)	214
O_4Os	Osmium(VIII) oxide	(12.32)	850
O_4Ru	Ruthenium(VIII) oxide	12.15 ± 0.03	988
O_7Re_2	Rhenium(VII) oxide	(12.7 ± 0.2)	125
Os	Osmium	8.7	1630
P	Phosphorus	10.48669	1328
P_2	Diphosphorus	10.53	1160
Pa	Protactinium	5.89	1133
Pb	Lead	7.41666	911
PbS	Lead(II) sulfide	(8.5 ± 0.5)	954
Pd	Palladium	8.3367	1181
Pm	Promethium	5.55	536
Pr	Praseodymium	5.464	883
Pt	Platinum	9.0	1433
Pu	Plutonium	6.025	926
Ra	Radium	5.27892	668
Rb	Rubidium	4.17713	484
Re	Rhenium	7.88	1530
Rh	Rhodium	7.45890	1276
Rn	Radon	10.74850	1037
Ru	Ruthenium	7.36050	1355
S	Sulfur	10.36001	1277
SSn	Tin(II) sulfide	(8.8)	966
S_2	Disulfur	9.356 ± 0.002	1031
Sb	Antimony	8.64	1096
Sc	Scandium	6.56144	1010
Se	Selenium	9.75238	1168
Si	Silicon	8.15169	1238
Sm	Samarium	5.6437	751
Sn	Tin	7.34381	1011
Sr	Strontium	5.69484	713
Ta	Tantalum	7.89	1544
Tb	Terbium	5.8639	955
Tc	Technetium	7.28	1380
Te	Tellurium	9.0096	1066
Th	Thorium	6.308 ± 0.003	1207
Ti	Titanium	6.8282	1127
Tl	Thallium	6.10829	771
Tm	Thulium	6.18431	827
U	Uranium	6.19405	1129
V	Vanadium	6.746 ± 0.002	1166
W	Tungsten	7.98	1621
Xe	Xenon	12.12987	1170
Y	Yttrium	6.217	1022
Yb	Ytterbium	6.25416	754
Zn	Zinc	9.39405	1037
Zr	Zirconium	6.63390	1251

Substances containing carbon

Mol. Form.	Name	IP/eV	$\Delta_f H_{ion}$ kJ/mol
C	Carbon	11.26030	1803
$CBrClF_2$	Bromochlorodifluoromethane	(11.21)	642
$CBrCl_3$	Bromotrichloromethane	(10.6)	980
$CBrF_3$	Bromotrifluoromethane	(11.40)	451
CBr_2F_2	Dibromodifluoromethane	11.03 ± 0.04	683
CBr_4	Tetrabromomethane	(10.31 ± 0.02)	1079
CCl	Chloromethylidene	(8.9 ± 0.2)	1244
$CClF_3$	Chlorotrifluoromethane	12.6 ± 0.2	505

Mol. Form.	Name	IP/eV	$\Delta_f H_{ion}$ kJ/mol
CClN	Cyanogen chloride	12.34 ± 0.01	1329
CCl$_2$	Dichloromethylene	(9.27)	1058
CCl$_2$F$_2$	Dichlorodifluoromethane	12.05 ± 0.24	685
CCl$_2$O	Carbonyl chloride	(11.5)	888
CCl$_3$F	Trichlorofluoromethane	11.77 ± 0.02	868
CCl$_4$	Tetrachloromethane	11.47 ± 0.01	1010
CF	Fluoromethylidene	9.11 ± 0.01	1134
CFN	Cyanogen fluoride	13.34 ± 0.02	1325
CF$_2$	Difluoromethylene	11.44 ± 0.03	899
CF$_2$O	Carbonyl fluoride	13.035 ± 0.030	617
CF$_3$	Trifluoromethyl	8.7 ± 0.2	379
CF$_3$I	Trifluoroiodomethane	10.23	397
CH	Methylidyne	10.64 ± 0.01	1622
CHBrCl$_2$	Bromodichloromethane	10.6	973
CHBr$_2$Cl	Chlorodibromomethane	10.59 ± 0.01	1030
CHBr$_3$	Tribromomethane	10.48 ± 0.02	1035
CHCl	Chloromethylene	9.84	1247
CHClF$_2$	Chlorodifluoromethane	(12.2)	693
CHCl$_2$F	Dichlorofluoromethane	(11.5)	829
CHCl$_3$	Trichloromethane	11.37 ± 0.02	992
CHF	Fluoromethylene	10.06 ± 0.05	1121
CHF$_3$	Trifluoromethane	(13.86)	643
CHI$_3$	Triiodomethane	9.25 ± 0.02	1010
CHN	Hydrogen cyanide	13.60 ± 0.01	1447
CHN	Hydrogen isocyanide	(12.5 ± 0.1)	1407
CHNO	Isocyanic acid	11.595 ± 0.005	1016
CHNO	Fulminic acid	(10.83)	1263
CHO	Oxomethyl (HCO)	(8.55)	826
CH$_2$	Methylene	10.396 ± .003	1392
CH$_2$BrCl	Bromochloromethane	10.77 ± 0.01	1085
CH$_2$Br$_2$	Dibromomethane	(10.50 ± 0.02)	1013
CH$_2$ClF	Chlorofluoromethane	11.71 ± 0.01	870
CH$_2$Cl$_2$	Dichloromethane	11.32 ± .01	996
CH$_2$F$_2$	Difluoromethane	12.71	774
CH$_2$I$_2$	Diiodomethane	9.46 ± 0.02	1030
CH$_2$N$_2$	Diazomethane	8.999 ± 0.001	1098
CH$_2$N$_2$	Cyanamide	(10.4)	1137
CH$_2$O	Formaldehyde	10.88 ± 0.01	941
CH$_2$O$_2$	Formic acid	11.33 ± 0.01	715
CH$_3$	Methyl	9.843 ± 0.002	1095
CH$_3$BO	Borane carbonyl	11.14 ± 0.02	962
CH$_3$Br	Bromomethane	10.541 ± 0.003	979
CH$_3$Cl	Chloromethane	11.22 ± 0.01	1001
CH$_3$Cl$_3$Si	Methyltrichlorosilane	(11.36 ± 0.03)	548
CH$_3$F	Fluoromethane	12.47 ± 0.02	956
CH$_3$I	Iodomethane	9.538	936
CH$_3$NO	Formamide	10.16 ± 0.06	796
CH$_3$NO$_2$	Nitromethane	11.08 ± 0.07	994
CH$_3$N$_3$	Methyl azide	9.81 ± 0.02	1227
CH$_3$O	Methoxy	(10.72)	1050
CH$_4$	Methane	12.61 ± 0.01	1143
CH$_4$N$_2$O	Urea	9.7	690
CH$_4$O	Methanol	10.85 ± 0.01	845
CH$_4$S	Methanethiol	9.44 ± 0.005	888
CH$_5$N	Methylamine	(8.80)	826
CH$_6$N$_2$	Methylhydrazine	7.7 ± 0.15	835
CH$_6$Si	Methylsilane	(10.7)	1003
CN	Cyanide	13.5984	1748
CNO	Cyanate	11.76 ± 0.01	1290
CO	Carbon monoxide	14.014 ± 0.0003	1242

Mol. Form.	Name	IP/eV	$\Delta_f H_{ion}$ kJ/mol
COS	Carbon oxysulfide	11.18 ± 0.01	936
COSe	Carbon oxyselenide	10.36 ± 0.01	929
CO_2	Carbon dioxide	13.773 ± 0.002	935
CS	Carbon sulfide	11.33 ± 0.01	1361
CS_2	Carbon disulfide	10.0685 ± 0.0020	1089
C_2	Dicarbon	(11.4 ± 0.3)	2000
$C_2Br_2F_4$	1,2-Dibromotetrafluoroethane	(11.1)	280
C_2ClF_3	Chlorotrifluoroethylene	9.81 ± 0.03	373
C_2ClF_5	Chloropentafluoroethane	(12.6)	99
C_2Cl_2	Dichloroacetylene	9.9	1165
$C_2Cl_2F_4$	1,2-Dichlorotetrafluoroethane	12.2	252
$C_2Cl_3F_3$	1,1,1-Trichlorotrifluoroethane	11.5	386
$C_2Cl_3F_3$	1,1,2-Trichlorotrifluoroethane	11.99 ± 0.02	429
C_2Cl_4	Tetrachloroethylene	9.326 ± 0.001	887
$C_2Cl_4F_2$	1,1,2,2-Tetrachloro-1,2-difluoroethane	(11.3)	563
C_2Cl_4O	Trichloroacetyl chloride	(11.0)	827
C_2Cl_6	Hexachloroethane	(11.1)	920
C_2F_3N	Trifluoroacetonitrile	13.93 ± 0.07	845
C_2F_4	Tetrafluoroethylene	10.12 ± 0.02	315
C_2F_6	Hexafluoroethane	(13.6)	-30
C_2H	Ethynyl	(11.61 ± 0.07)	1685
C_2HBr	Bromoacetylene	10.31 ± 0.02	1242
$C_2HBrClF_3$	2-Bromo-2-chloro-1,1,1-trifluoroethane	(11.0)	363
C_2HCl	Chloroacetylene	10.58 ± 0.02	1276
C_2HClF_2	1-Chloro-2,2-difluoroethylene	9.80 ± 0.04	628
C_2HCl_3	Trichloroethylene	9.46 ± 0.02	894
C_2HCl_3O	Dichloroacetyl chloride	(10.9)	809
C_2HCl_5	Pentachloroethane	(11.0)	919
C_2HF	Fluoroacetylene	11.26	1195
C_2HF_3	Trifluoroethylene	10.14	489
$C_2HF_3O_2$	Trifluoroacetic acid	11.46	75
C_2H_2	Acetylene	11.400 ± 0.002	1328
$C_2H_2Cl_2$	1,1-Dichloroethylene	9.81 ± 0.04	949
$C_2H_2Cl_2$	cis-1,2-Dichloroethylene	9.66 ± 0.01	936
$C_2H_2Cl_2$	trans-1,2-Dichloroethylene	9.64 ± 0.02	934
$C_2H_2Cl_2O$	Chloroacetyl chloride	(≤ 10.3)	815
$C_2H_2Cl_4$	1,1,1,2-Tetrachloroethane	(11.1)	920
$C_2H_2Cl_4$	1,1,2,2-Tetrachloroethane	(≤ 11.62)	≤ 971
$C_2H_2F_2$	1,1-Difluoroethylene	10.29 ± 0.01	650
$C_2H_2F_2$	cis-1,2-Difluoroethylene	10.23 ± 0.02	690
C_2H_2O	Ketene	9.617 ± 0.003	880
$C_2H_2O_2$	Glyoxal	10.2	773
$C_2H_2S_2$	Thiirene	8.61	892
C_2H_3Br	Bromoethylene	9.83 ± 0.02	1028
C_2H_3Cl	Chloroethylene	9.99 ± 0.02	985
$C_2H_3ClF_2$	1-Chloro-1,1-difluoroethane	(11.98)	626
C_2H_3ClO	Acetyl chloride	10.82 ± 0.04	801
C_2H_3ClO	Chloroacetaldehyde	(10.48)	815
$C_2H_3ClO_2$	Chloroacetic acid	(10.7)	597
$C_2H_3Cl_3$	1,1,1-Trichloroethane	(11.0)	917
$C_2H_3Cl_3$	1,1,2-Trichloroethane	(11.0)	911
C_2H_3F	Fluoroethylene	10.36 ± 0.01	861
C_2H_3FO	Acetyl fluoride	(11.5)	667
$C_2H_3F_3$	1,1,1-Trifluoroethane	13.3 ± 0.5	536
C_2H_3N	Acetonitrile	12.20 ± 0.01	1253
C_2H_3NO	Methylisocyanate	(10.67)	900
C_2H_4	Ethylene	10.5138 ± 0.0006	1067
$C_2H_4Br_2$	1,2-Dibromoethane	10.35 ± 0.04	961
$C_2H_4Cl_2$	1,1-Dichloroethane	11.04 ± 0.02	935
$C_2H_4Cl_2$	1,2-Dichloroethane	11.04 ± 0.02	931

Mol. Form.	Name	IP/eV	$\Delta_f H_{ion}$ kJ/mol
$C_2H_4F_2$	1,1-Difluoroethane	(11.87)	643
C_2H_4O	Acetaldehyde	10.229 ± 0.0007	821
C_2H_4O	Ethylene oxide	10.56 ± 0.01	966
$C_2H_4O_2$	Acetic acid	10.65 ± 0.02	595
$C_2H_4O_2$	Methyl formate	10.835 ± 0.005	690
C_2H_5Br	Bromoethane	10.29 ± 0.01	931
C_2H_5Cl	Chloroethane	10.98 ± 0.02	947
C_2H_5ClO	2-Chloroethanol	(10.5)	756
C_2H_5F	Fluoroethane	(11.78)	873
C_2H_5I	Iodoethane	9.3492 ± 0.0006	893
C_2H_5N	Ethyleneimine	(9.5 ± 0.3)	1044
C_2H_5NO	Acetamide	9.65 ± 0.03	693
C_2H_5NO	N-Methylformamide	9.83 ± 0.04	760
$C_2H_5NO_2$	Nitroethane	10.88 ± 0.05	948
C_2H_6	Ethane	11.56 ± 0.02	1031
$C_2H_6Cl_2Si$	Dichlorodimethylsilane	(10.7)	576
C_2H_6O	Ethanol	10.43 ± 0.05	772
C_2H_6O	Dimethyl ether	10.025 ± 0.025	783
C_2H_6OS	Dimethyl sulfoxide	9.10 ± 0.03	727
$C_2H_6O_2$	Ethylene glycol	10.16	593
C_2H_6S	Ethanethiol	9.31 ± 0.03	851
C_2H_6S	Dimethyl sulfide	8.69 ± 0.02	801
$C_2H_6S_2$	Dimethyl disulfide	(7.4 ± 0.3)	690
C_2H_7N	Ethylamine	8.86 ± 0.02	808
C_2H_7N	Dimethylamine	8.24 ± 0.08	777
C_2H_7NO	Ethanolamine	8.96	664
$C_2H_8N_2$	1,2-Ethanediamine	(8.6)	812
$C_2H_8N_2$	1,1-Dimethylhydrazine	7.29 ± 0.05	787
C_2N_2	Cyanogen	13.37 ± 0.01	1597
C_3F_6	Perfluoropropene	10.60 ± 0.03	-103
C_3F_6O	Perfluoroacetone	(11.57 ± 0.13)	-282
C_3F_8	Perfluoropropane	(13.38)	-491
C_3HN	Cyanoacetylene	11.64 ± 0.01	1475
C_3H_2O	2-Propynal	(10.7 ± 0.1)	1145
$C_3H_3F_3$	3,3,3-Trifluoropropene	(10.9)	437
C_3H_3N	2-Propenenitrile	10.91 ± 0.01	1237
C_3H_3NO	Oxazole	(9.9)	940
C_3H_3NO	Isoxazole	(9.93)	1038
C_3H_4	Allene	9.692 ± 0.004	1126
C_3H_4	Propyne	10.37 ± 0.01	1187
C_3H_4	Cyclopropene	9.67 ± 0.01	1209
$C_3H_4N_2$	Imidazole	(8.81)	997
C_3H_4O	Propargyl alcohol	10.49 ± 0.02	1060
C_3H_4O	Acrolein	10.103 ± 0.006	900
C_3H_4O	Cyclopropanone	(9.1 ± 0.1)	895
$C_3H_4O_2$	Propenoic acid	10.60	701
$C_3H_4O_2$	2-Oxetanone	(9.70 ± 0.01)	653
C_3H_5Br	3-Bromopropene	(9.96)	1008
C_3H_5Cl	3-Chloropropene	10.04 ± 0.01	965
C_3H_5ClO	Epichlorohydrin	(10.64)	919
$C_3H_5ClO_2$	Methyl chloroacetate	(10.3)	575
C_3H_5F	3-Fluoropropene	(10.11)	821
C_3H_5N	Propanenitrile	11.84 ± 0.02	1194
C_3H_5NO	Acrylamide	(9.5)	720
C_3H_6	Propene	9.73 ± 0.02	959
C_3H_6	Cyclopropane	9.86	1005
$C_3H_6Br_2$	1,2-Dibromopropane	(10.1)	903
$C_3H_6Br_2$	1,3-Dibromopropane	(≤ 10.2)	≤ 919
$C_3H_6Cl_2$	1,2-Dichloropropane	10.8 ± 0.1	886
$C_3H_6Cl_2$	1,3-Dichloropropane	10.89 ± 0.04	892

Mol. Form.	Name	IP/eV	$\Delta_f H_{ion}$ kJ/mol
C_3H_6O	Allyl alcohol	9.67 ± 0.05	808
C_3H_6O	Methyl vinyl ether	8.95 ± 0.01	763
C_3H_6O	Propanal	9.96 ± 0.01	772
C_3H_6O	Acetone	9.703 ± 0.006	719
C_3H_6O	Methyloxirane	(10.22)	892
C_3H_6O	Oxetane	9.65 ± 0.01	851
$C_3H_6O_2$	Propanoic acid	10.525 ± 0.003	568
$C_3H_6O_2$	Ethyl formate	10.61 ± 0.01	639
$C_3H_6O_2$	Methyl acetate	10.25 ± 0.02	579
$C_3H_6O_2$	1,3-Dioxolane	(9.9)	658
$C_3H_6O_3$	1,3,5-Trioxane	(10.3)	528
C_3H_7Br	1-Bromopropane	10.18 ± 0.01	898
C_3H_7Br	2-Bromopropane	10.10 ± 0.03	877
C_3H_7Cl	1-Chloropropane	10.81 ± 0.01	911
C_3H_7Cl	2-Chloropropane	10.79 ± 0.02	896
C_3H_7F	1-Fluoropropane	(11.3)	806
C_3H_7F	2-Fluoropropane	(11.08)	776
C_3H_7I	1-Iodopropane	9.25 ± 0.01	860
C_3H_7I	2-Iodopropane	9.19 ± 0.02	845
C_3H_7N	Allylamine	(8.76)	891
C_3H_7N	Cyclopropylamine	(8.8)	926
C_3H_7N	Propyleneimine	(9.0)	960
C_3H_7NO	N,N-Dimethylformamide	(9.12)	688
$C_3H_7NO_2$	1-Nitropropane	(10.81)	919
$C_3H_7NO_2$	2-Nitropropane	(10.71)	894
C_3H_8	Propane	10.95 ± 0.05	952
C_3H_8O	1-Propanol	10.18 ± 0.06	727
C_3H_8O	2-Propanol	10.17 ± 0.02	709
C_3H_8O	Ethyl methyl ether	9.72 ± 0.07	722
$C_3H_8O_2$	Dimethoxymethane	9.7	588
C_3H_8S	1-Propanethiol	9.20 ± 0.01	819
C_3H_8S	2-Propanethiol	9.145 ± 0.005	806
C_3H_8S	Ethyl methyl sulfide	(8.55)	765
$C_3H_9BO_3$	Trimethyl borate	(10.0)	65
C_3H_9ClSi	Trimethylchlorosilane	(10.15)	624
C_3H_9N	Propylamine	(8.78)	777
C_3H_9N	Isopropylamine	(8.72)	758
C_3H_9N	Trimethylamine	7.82 ± 0.06	731
C_3H_9NO	3-Amino-1-propanol	(9.0)	651
$C_4H_2O_3$	Maleic anhydride	(10.8)	645
C_4H_4	1-Buten-3-yne	9.58 ± 0.02	1230
$C_4H_4N_2$	Succinonitrile	(12.1 ± 0.25)	1377
$C_4H_4N_2$	Pyrimidine	9.23	1087
$C_4H_4N_2$	Pyridazine	8.67 ± 0.03	1112
C_4H_4O	Furan	8.883 ± 0.003	822
$C_4H_4O_2$	Diketene	(9.6 ± 0.02)	736
$C_4H_4O_3$	Succinic anhydride	(10.6)	500
$C_4H_4O_4$	Fumaric acid	(10.7)	355
C_4H_4S	Thiophene	8.86 ± 0.02	970
C_4H_5N	Methylacrylonitrile	10.34	1127
C_4H_5N	Pyrrole	8.207 ± 0.005	900
C_4H_5N	Cyclopropanecarbonitrile	(10.25)	1173
C_4H_6	1,2-Butadiene	(9.03)	1034
C_4H_6	1,3-Butadiene	9.082 ± 0.004	986
C_4H_6	1-Butyne	10.19 ± 0.02	1148
C_4H_6	2-Butyne	9.59 ± 0.03	1071
C_4H_6	Cyclobutene	9.43 ± 0.02	1067
C_4H_6O	Divinyl ether	(8.7)	827
C_4H_6O	trans-2-Butenal	9.73 ± 0.01	835
C_4H_6O	2-Methylpropenal	(9.92)	834

Mol. Form.	Name	IP/eV	$\Delta_f H_{ion}$ kJ/mol
C_4H_6O	Cyclobutanone	(9.35	815
$C_4H_6O_2$	cis-Crotonic acid	(10.08)	625
$C_4H_6O_2$	trans-Crotonic acid	(9.9)	604
$C_4H_6O_2$	Methacrylic acid	(10.15)	611
$C_4H_6O_2$	Vinyl acetate	9.19 ± 0.05	572
$C_4H_6O_2$	Methyl acrylate	(9.9)	641
$C_4H_6O_3$	Acetic anhydride	(10.0)	398
$C_4H_6O_4$	Dimethyl oxalate	(10.0)	287
C_4H_6S	2,5-Dihydrothiophene	(8.4)	898
C_4H_7N	Butanenitrile	(11.2)	1110
C_4H_7N	2-Methylpropanenitrile	(11.3)	1115
C_4H_7NO	2-Pyrrolidone	(9.2)	674
C_4H_8	1-Butene	9.55 ± 0.06	921
C_4H_8	cis-2-Butene	9.11 ± 0.01	871
C_4H_8	trans-2-Butene	9.10 ± 0.01	866
C_4H_8	Isobutene	9.239 ± 0.003	875
C_4H_8	Cyclobutane	(9.82 ± 0.05)	976
C_4H_8	Methylcyclopropane	(9.46)	936
$C_4H_8Br_2$	1,4-Dibromobutane	(10.15)	879
C_4H_8O	Ethyl vinyl ether	(8.98)	709
C_4H_8O	1,2-Epoxybutane	(≤ 10.15)	862
C_4H_8O	Butanal	9.84 ± 0.02	742
C_4H_8O	Isobutanal	9.71 ± 0.01	721
C_4H_8O	2-Butanone	9.52 ± 0.04	678
C_4H_8O	Tetrahydrofuran	9.38 ± 0.05	721
$C_4H_8O_2$	Butanoic acid	10.17 ± 0.05	509
$C_4H_8O_2$	2-Methylpropanoic acid	10.33 ± 0.03	516
$C_4H_8O_2$	Propyl formate	10.52 ± 0.02	555
$C_4H_8O_2$	Ethyl acetate	10.01 ± 0.05	522
$C_4H_8O_2$	Methyl propanoate	10.15 ± 0.03	548
$C_4H_8O_2$	1,3-Dioxane	9.8	607
$C_4H_8O_2$	1,4-Dioxane	9.19 ± 0.01	571
$C_4H_8O_2S$	Sulfolane	(9.8)	577
C_4H_8S	Tetrahydrothiophene	8.38	774
C_4H_9Br	1-Bromobutane	(10.12)	869
C_4H_9Br	2-Bromobutane	10.01 ± 0.02	845
C_4H_9Br	1-Bromo-2-methylpropane	10.09 ± 0.02	861
C_4H_9Br	2-Bromo-2-methylpropane	9.92 ± 0.03	823
C_4H_9Cl	1-Chlorobutane	10.67 ± 0.03	875
C_4H_9Cl	2-Chlorobutane	10.53	857
C_4H_9Cl	1-Chloro-2-methylpropane	10.73 ± 0.07	877
C_4H_9Cl	2-Chloro-2-methylpropane	(10.61)	842
C_4H_9I	1-Iodobutane	9.23 ± 0.01	840
C_4H_9I	2-Iodobutane	9.10 ± 0.02	815
C_4H_9I	1-Iodo-2-methylpropane	9.19 ± 0.01	824
C_4H_9I	2-Iodo-2-methylpropane	(9.02)	798
C_4H_9N	Pyrrolidine	(8.0)	769
C_4H_9NO	N,N-Dimethylacetamide	8.81 ± 0.03	616
C_4H_9NO	Morpholine	(8.2)	841
C_4H_{10}	Butane	10.53 ± 0.10	890
C_4H_{10}	Isobutane	(10.57)	886
$C_4H_{10}O$	1-Butanol	9.99 ± 0.05	689
$C_4H_{10}O$	2-Butanol	9.88 ± 0.03	658
$C_4H_{10}O$	2-Methyl-1-propanol	10.02 ± 0.04	683
$C_4H_{10}O$	2-Methyl-2-propanol	9.90 ± 0.02	642
$C_4H_{10}O$	Diethyl ether	9.51 ± 0.03	666
$C_4H_{10}O$	Methyl propyl ether	9.41 ± 0.07	670
$C_4H_{10}O$	Isopropyl methyl ether	9.45 ± 0.04	661
$C_4H_{10}O_2$	Ethylene glycol monoethyl ether	(9.6)	529
$C_4H_{10}O_2$	Ethylene glycol dimethyl ether	(9.3)	558

Mol. Form.	Name	IP/eV	$\Delta_f H_{ion}$ kJ/mol
$C_4H_{10}S$	1-Butanethiol	9.14 ± 0.01	794
$C_4H_{10}S$	2-Butanethiol	(9.10)	781
$C_4H_{10}S$	2-Methyl-1-propanethiol	(9.12)	783
$C_4H_{10}S$	2-Methyl-2-propanethiol	(9.03)	762
$C_4H_{10}S$	Diethyl sulfide	(8.43)	730
$C_4H_{10}S$	Methyl propyl sulfide	(8.8)	767
$C_4H_{10}S$	Isopropyl methyl sulfide	(8.7)	749
$C_4H_{10}S_2$	Diethyl disulfide	(8.27)	724
$C_4H_{11}N$	Butylamine	8.7 ± 0.1	748
$C_4H_{11}N$	sec-Butylamine	8.46 ± 0.1	711
$C_4H_{11}N$	tert-Butylamine	8.46 ± 0.1	695
$C_4H_{11}N$	Isobutylamine	8.50 ± 0.1	721
$C_4H_{11}N$	Diethylamine	7.85 ± 0.1	684
$C_4H_{12}Si$	Tetramethylsilane	9.80 ± 0.04	713
$C_4H_{12}Sn$	Tetramethylstannane	8.89 ± 0.05	837
C_4NiO_4	Nickel carbonyl	8.27 ± 0.04	200
$C_5H_4O_2$	Furfural	9.22 ± 0.01	739
C_5H_5N	Pyridine	9.25	1031
C_5H_6	1-Penten-3-yne	9.00 ± 0.01	1119
C_5H_6	cis-3-Penten-1-yne	9.14 ± 0.04	1137
C_5H_6	trans-3-Penten-1-yne	9.05 ± 0.01	1128
C_5H_6	2-Methyl-1-buten-3-yne	9.25 ± 0.02	1152
C_5H_6	1,3-Cyclopentadiene	8.55 ± 0.02	955
C_5H_6O	2-Methylfuran	8.38 ± 0.02	729
C_5H_6O	3-Methylfuran	(8.64)	763
C_5H_6S	2-Methylthiophene	(8.14)	867
C_5H_6S	3-Methylthiophene	(8.40)	893
C_5H_8	cis-1,3-Pentadiene	8.63 ± 0.03	914
C_5H_8	trans-1,3-Pentadiene	8.59 ± 0.02	905
C_5H_8	1,4-Pentadiene	9.60 ± 0.02	1032
C_5H_8	2-Methyl-1,3-butadiene	8.84 ± 0.01	928
C_5H_8	1-Pentyne	10.10 ± 0.01	1119
C_5H_8	Cyclopentene	9.01 ± 0.01	905
C_5H_8	Spiropentane	(9.26)	1078
C_5H_8O	Cyclopropyl methyl ketone	(≤ 9.46)	796
C_5H_8O	Cyclopentanone	9.26 ± 0.01	701
C_5H_8O	3,4-Dihydro-2H-pyran	8.35 ± 0.01	681
$C_5H_8O_2$	Ethyl acrylate	(≤ 10.3)	617
$C_5H_8O_2$	Methyl methacrylate	(9.7)	589
$C_5H_8O_2$	2,4-Pentanedione	8.85 ± 0.01	469
C_5H_9NO	N-Methyl-2-pyrrolidone	(≤ 9.17)	≤ 676
C_5H_{10}	1-Pentene	9.51 ± 0.01	896
C_5H_{10}	cis-2-Pentene	9.01 ± 0.03	843
C_5H_{10}	trans-2-Pentene	9.04 ± 0.01	841
C_5H_{10}	2-Methyl-1-butene	9.12 ± 0.01	844
C_5H_{10}	3-Methyl-1-butene	9.52 ± 0.01	891
C_5H_{10}	2-Methyl-2-butene	8.69 ± 0.01	796
C_5H_{10}	Cyclopentane	(10.33 ± 0.15)	918
$C_5H_{10}O$	2,2-Dimethylpropanal	9.51 ± 0.01	675
$C_5H_{10}O$	Cyclopentanol	(9.72)	695
$C_5H_{10}O$	Pentanal	9.74 ± 0.04	709
$C_5H_{10}O$	2-Pentanone	9.38 ± 0.01	646
$C_5H_{10}O$	3-Pentanone	9.31 ± 0.01	640
$C_5H_{10}O$	3-Methyl-2-butanone	9.30 ± 0.01	635
$C_5H_{10}O$	Tetrahydropyran	9.25 ± 0.01	670
$C_5H_{10}O_2$	Pentanoic acid	(≤ 10.53)	≤ 527
$C_5H_{10}O_2$	3-Methylbutanoic acid	(≤ 10.51)	≤ 499
$C_5H_{10}O_2$	Butyl formate	10.52 ± 0.02	584
$C_5H_{10}O_2$	Propyl acetate	(≤ 9.92)	501
$C_5H_{10}O_2$	Isopropyl acetate	9.99 ± 0.03	482

Mol. Form.	Name	IP/eV	$\Delta_f H_{ion}$ kJ/mol
$C_5H_{10}O_2$	Ethyl propanoate	(10.00)	500
$C_5H_{10}O_2$	Methyl butanoate	(10.07)	520
$C_5H_{10}S$	Thiacyclohexane	(8.2)	728
$C_5H_{11}Br$	1-Bromopentane	10.10±0.01	846
$C_5H_{11}I$	1-Iodopentane	9.20±0.01	817
$C_5H_{11}N$	Piperidine	8.03±0.11	726
$C_5H_{11}N$	N-Methylpyrrolidine	≤8.41±0.02	≤809
C_5H_{12}	Pentane	10.28±0.10	845
C_5H_{12}	Isopentane	10.32±0.05	843
C_5H_{12}	Neopentane	(≤10.2)	≤818
$C_5H_{12}O$	1-Pentanol	(10.00)	668
$C_5H_{12}O$	2-Pentanol	(9.78)	630
$C_5H_{12}O$	3-Pentanol	9.78	628
$C_5H_{12}O$	2-Methyl-1-butanol	(9.86)	649
$C_5H_{12}O$	2-Methyl-2-butanol	(9.8)	615
$C_5H_{12}O$	3-Methyl-2-butanol	(9.88±0.13)	637
$C_5H_{12}O$	Butyl methyl ether	(9.4±0.1)	648
$C_5H_{12}O$	Methyl tert-butyl ether	(9.24)	608
$C_5H_{12}O$	Ethyl propyl ether	(9.45)	640
$C_5H_{12}S$	tert-Butyl methyl sulfide	(8.38)	687
$C_5H_{12}S$	Ethyl propyl sulfide	(8.50)	716
$C_5H_{12}S$	Ethyl isopropyl sulfide	(8.35)	689
C_6BrF_5	Bromopentafluorobenzene	(9.67)	222
C_6ClF_5	Chloropentafluorobenzene	(9.72)	126
C_6Cl_6	Hexachlorobenzene	(8.98)	822
C_6F_6	Hexafluorobenzene	9.89±0.04	8
C_6F_{12}	Perfluorocyclohexane	(13.2)	-1095
C_6HF_5	Pentafluorobenzene	(9.63)	122
C_6HF_5O	Pentafluorophenol	(9.20)	-71
$C_6H_2F_4$	1,2,3,4-Tetrafluorobenzene	(9.53)	284
$C_6H_2F_4$	1,2,3,5-Tetrafluorobenzene	(9.53)	263
$C_6H_2F_4$	1,2,4,5-Tetrafluorobenzene	(9.35)	254
$C_6H_3Cl_3$	1,2,4-Trichlorobenzene	(9.04)	880
$C_6H_3Cl_3$	1,3,5-Trichlorobenzene	9.32±0.02	899
$C_6H_4ClNO_2$	1-Chloro-3-nitrobenzene	(9.92±0.1)	995
$C_6H_4ClNO_2$	1-Chloro-4-nitrobenzene	(9.96±0.1)	999
$C_6H_4Cl_2$	o-Dichlorobenzene	9.06±0.02	907
$C_6H_4Cl_2$	m-Dichlorobenzene	9.10±0.02	906
$C_6H_4Cl_2$	p-Dichlorobenzene	8.92±0.02	885
$C_6H_4FNO_2$	1-Fluoro-4-nitrobenzene	(9.90)	826
$C_6H_4F_2$	o-Difluorobenzene	9.29±0.01	602
$C_6H_4F_2$	m-Difluorobenzene	9.33±0.01	591
$C_6H_4F_2$	p-Difluorobenzene	9.1589±0.0003	577
$C_6H_4O_2$	p-Benzoquinone	10.01±0.06	844
C_6H_5Br	Bromobenzene	9.00±0.02	971
C_6H_5Cl	Chlorobenzene	9.07±0.02	930
C_6H_5ClO	m-Chlorophenol	8.655±0.001	680
C_6H_5ClO	p-Chlorophenol	(≤8.69)	≤692
C_6H_5F	Fluorobenzene	9.20±0.01	772
C_6H_5I	Iodobenzene	8.685	1003
$C_6H_5NO_2$	Nitrobenzene	9.86±0.02	1019
$C_6H_5NO_3$	o-Nitrophenol	(9.1)	782
$C_6H_5NO_3$	m-Nitrophenol	(9.0)	755
$C_6H_5NO_3$	p-Nitrophenol	(9.1)	761
C_6H_6	Benzene	9.24378±0.00007	975
C_6H_6	Fulvene	(8.36)	1031
C_6H_6ClN	o-Chloroaniline	(8.50)	883
C_6H_6ClN	m-Chloroaniline	(8.09)	835
C_6H_6ClN	p-Chloroaniline	(≤8.18)	≤844
$C_6H_6N_2O_2$	o-Nitroaniline	(8.27)	861

Mol. Form.	Name	IP/eV	$\Delta_f H_{ion}$ kJ/mol
$C_6H_6N_2O_2$	m-Nitroaniline	(8.31)	865
$C_6H_6N_2O_2$	p-Nitroaniline	(8.34)	859
C_6H_6O	Phenol	8.49 ± 0.02	723
$C_6H_6O_2$	p-Hydroquinone	7.94 ± 0.01	503
C_6H_6S	Benzenethiol	(8.32)	915
C_6H_7N	Aniline	7.720 ± 0.002	832
C_6H_7N	2-Methylpyridine	(9.02)	970
C_6H_7N	3-Methylpyridine	(9.04)	979
C_6H_7N	4-Methylpyridine	(9.04)	976
$C_6H_8N_2$	o-Phenylenediamine	(7.2)	787
$C_6H_8N_2$	m-Phenylenediamine	(7.14)	777
$C_6H_8N_2$	p-Phenylenediamine	(6.87 ± 0.05)	759
C_6H_{10}	1,5-Hexadiene	9.27 ± 0.05	978
C_6H_{10}	1-Hexyne	10.03 ± 0.05	1089
C_6H_{10}	3,3-Dimethyl-1-butyne	9.90 ± 0.04	1060
C_6H_{10}	Cyclohexene	8.945 ± 0.01	859
$C_6H_{10}O$	Cyclohexanone	9.14 ± 0.01	656
$C_6H_{10}O$	Mesityl oxide	9.10 ± 0.01	694
$C_6H_{10}O_4$	Diethyl oxalate	(9.8)	205
$C_6H_{11}NO$	Caprolactam	(9.07 ± 0.02)	629
C_6H_{12}	1-Hexene	9.44 ± 0.04	869
C_6H_{12}	cis-2-Hexene	(8.97 ± 0.01)	818
C_6H_{12}	trans-2-Hexene	(8.97 ± 0.01)	814
C_6H_{12}	2-Methyl-1-pentene	(9.08 ± 0.01)	817
C_6H_{12}	4-Methyl-1-pentene	9.45 ± 0.01	862
C_6H_{12}	2-Methyl-2-pentene	(8.58)	761
C_6H_{12}	4-Methyl-cis-2-pentene	8.98 ± 0.01	809
C_6H_{12}	4-Methyl-trans-2-pentene	(8.97 ± 0.01)	804
C_6H_{12}	2-Ethyl-1-butene	(9.06 ± 0.02)	818
C_6H_{12}	2,3-Dimethyl-1-butene	(9.07 ± 0.01)	812
C_6H_{12}	2,3-Dimethyl-2-butene	8.27 ± 0.01	729
C_6H_{12}	Cyclohexane	9.86 ± 0.03	828
C_6H_{12}	Methylcyclopentane	(9.85)	845
$C_6H_{12}O$	Hexanal	9.72 ± 0.05	691
$C_6H_{12}O$	2-Hexanone	9.3 ± 0.1	626
$C_6H_{12}O$	3-Hexanone	9.12 ± 0.02	600
$C_6H_{12}O$	3-Methyl-2-pentanone	9.21 ± 0.01	600
$C_6H_{12}O$	4-Methyl-2-pentanone	9.30 ± 0.01	609
$C_6H_{12}O$	2-Methyl-3-pentanone	9.10 ± 0.01	592
$C_6H_{12}O$	3,3-Dimethyl-2-butanone	9.12 ± 0.02	589
$C_6H_{12}O$	Cyclohexanol	(9.75)	651
$C_6H_{12}O_2$	Hexanoic acid	≤ 10.12	≤ 463
$C_6H_{12}O_2$	Butyl acetate	(9.92 ± .05)	471
$C_6H_{12}O_2$	sec-Butyl acetate	9.90	453
$C_6H_{12}O_2$	Methyl 2,2-dimethylpropanoate	(9.90 ± 0.04)	466
$C_6H_{13}I$	1-Iodohexane	9.179	794
$C_6H_{13}N$	Cyclohexylamine	(8.86)	750
C_6H_{14}	Hexane	10.13	810
C_6H_{14}	2-Methylpentane	(10.12)	802
C_6H_{14}	3-Methylpentane	(10.08)	801
C_6H_{14}	2,2-Dimethylbutane	(10.06)	787
C_6H_{14}	2,3-Dimethylbutane	(10.02)	791
$C_6H_{14}O$	1-Hexanol	(9.89)	639
$C_6H_{14}O$	2-Hexanol	(9.80 ± 0.03)	611
$C_6H_{14}O$	3-Hexanol	(9.63 ± 0.03)	599
$C_6H_{14}O$	Dipropyl ether	(9.27)	602
$C_6H_{14}O$	Diisopropyl ether	9.20 ± 0.05	569
$C_6H_{14}O$	Butyl ethyl ether	(9.36)	610
$C_6H_{14}O$	Methyl pentyl ether	(≤ 9.67)	≤ 657
$C_6H_{14}O_2$	1,1-Diethoxyethane	(9.2)	434

Mol. Form.	Name	IP/eV	$\Delta_f H_{ion}$ kJ/mol
$C_6H_{14}O_3$	Diethylene glycol dimethyl ether	≤ 9.8	≤ 448
$C_6H_{14}S$	Dipropyl sulfide	8.30 ± 0.02	676
$C_6H_{14}S$	Diisopropyl sulfide	(8.2 ± 0.2)	649
$C_6H_{15}N$	Hexylamine	(8.63 ± 0.05)	699
$C_6H_{15}N$	Dipropylamine	(7.84 ± 0.02)	641
$C_6H_{15}N$	Diisopropylamine	(7.73 ± 0.03)	602
$C_6H_{15}N$	Triethylamine	(7.50 ± 0.02)	631
$C_6H_{15}NO_3$	Triethanolamine	(7.9)	206
$C_7H_3F_5$	2,3,4,5,6-Pentafluorotoluene	(9.4)	64
C_7H_5ClO	Benzoyl chloride	(9.53)	815
$C_7H_5Cl_3$	(Trichloromethyl)benzene	(≤ 9.60)	≤ 914
$C_7H_5F_3$	(Trifluoromethyl)benzene	9.685 ± 0.005	335
C_7H_5N	Benzonitrile	9.70 ± 0.01	1154
C_7H_6O	Benzaldehyde	9.49 ± 0.02	878
$C_7H_6O_2$	Benzoic acid	(9.3)	604
C_7H_7Br	p-Bromotoluene	8.67 ± 0.02	908
C_7H_7Cl	o-Chlorotoluene	(8.7 ± 0.1)	856
C_7H_7Cl	m-Chlorotoluene	(8.83)	869
C_7H_7Cl	p-Chlorotoluene	(8.69)	855
C_7H_7Cl	(Chloromethyl)benzene	9.10 ± 0.02	897
C_7H_7F	o-Fluorotoluene	8.91 ± 0.01	709
C_7H_7F	m-Fluorotoluene	8.91 ± 0.01	709
C_7H_7F	p-Fluorotoluene	8.79 ± 0.01	701
C_7H_7NO	Benzamide	(9.25)	792
$C_7H_7NO_2$	o-Nitrotoluene	9.24	946
$C_7H_7NO_2$	m-Nitrotoluene	9.45 ± 0.1	941
$C_7H_7NO_2$	p-Nitrotoluene	9.46 ± 0.05	942
C_7H_8	Toluene	8.8276 ± 0.0006	901
C_7H_8O	o-Cresol	(8.24)	670
C_7H_8O	m-Cresol	8.29 ± 0.07	668
C_7H_8O	p-Cresol	(8.3)	675
C_7H_8O	Benzyl alcohol	(8.3)	701
C_7H_8O	Anisole	8.22 ± 0.03	725
C_7H_9N	Benzylamine	(8.64)	917
C_7H_9N	o-Methylaniline	(7.44 ± 0.02)	772
C_7H_9N	m-Methylaniline	(7.50 ± 0.02)	778
C_7H_9N	p-Methylaniline	(7.24 ± 0.02)	753
C_7H_9N	N-Methylaniline	7.34 ± 0.04	792
C_7H_9N	2,3-Dimethylpyridine	(8.85 ± 0.02)	922
C_7H_9N	2,4-Dimethylpyridine	(8.85 ± 0.03)	918
C_7H_9N	2,5-Dimethylpyridine	(≤ 8.80 ± 0.05)	≤ 916
C_7H_9N	2,6-Dimethylpyridine	8.86 ± 0.03	913
C_7H_9N	3,4-Dimethylpyridine	(≤ 9.15)	≤ 953
C_7H_9N	3,5-Dimethylpyridine	(≤ 9.25)	≤ 965
$C_7H_{10}O$	Dicyclopropyl ketone	(9.1)	1041
C_7H_{14}	1-Heptene	9.34 ± 0.10	839
C_7H_{14}	trans-3-Heptene	(8.92)	790
C_7H_{14}	Cycloheptane	9.97	844
C_7H_{14}	Methylcyclohexane	9.64	775
C_7H_{14}	cis-1,2-Dimethylcyclopentane	(9.92 ± 0.05)	828
C_7H_{14}	trans-1,2-Dimethylcyclopentane	9.7 ± 0.2	799
$C_7H_{14}O$	1-Heptanal	(9.65)	668
$C_7H_{14}O$	2-Heptanone	9.28 ± 0.10	594
$C_7H_{14}O$	3-Heptanone	9.18 ± 0.08	589
$C_7H_{14}O$	4-Heptanone	9.10 ± 0.06	577
$C_7H_{14}O$	5-Methyl-2-hexanone	(9.28)	586
$C_7H_{14}O$	2,4-Dimethyl-3-pentanone	8.95 ± 0.01	552
$C_7H_{14}O$	1-Methylcyclohexanol	(9.8 ± 0.2)	586
C_7H_{16}	Heptane	9.93 ± 0.10	771
$C_7H_{16}O$	1-Heptanol	(9.84)	614

Mol. Form.	Name	IP/eV	$\Delta_f H_{ion}$ kJ/mol
$C_7H_{16}O$	2-Heptanol	(9.70)	580
$C_7H_{16}O$	3-Heptanol	(9.68)	578
$C_7H_{16}O$	4-Heptanol	(9.61)	572
$C_7H_{16}O$	Ethyl pentyl ether	($\leq$ 9.49)	$\leq$ 602
$C_8H_4O_3$	Phthalic anhydride	(10.1)	603
$C_8H_6O_4$	Isophthalic acid	(9.98)	268
$C_8H_6O_4$	Terephthalic acid	(9.86)	232
C_8H_7N	2-Methylbenzonitrile	($\leq$ 9.38)	1085
C_8H_7N	3-Methylbenzonitrile	($\leq$ 9.34)	1085
C_8H_7N	4-Methylbenzonitrile	9.32 ± 0.02	1083
C_8H_7N	Indole	7.7602 ± 0.0006	908
C_8H_8	Styrene	8.464 ± 0.001	964
C_8H_8O	p-Tolualdehyde	(9.33)	825
C_8H_8O	Acetophenone	9.29 ± 0.03	810
$C_8H_8O_2$	o-Toluic acid	(9.1)	558
$C_8H_8O_2$	m-Toluic acid	(9.43)	579
$C_8H_8O_2$	p-Toluic acid	(9.23)	560
$C_8H_8O_2$	Benzeneacetic acid	(8.26)	479
$C_8H_8O_2$	Methyl benzoate	9.32 ± 0.03	611
C_8H_{10}	Ethylbenzene	8.77 ± 0.01	876
C_8H_{10}	o-Xylene	8.56 ± 0.01	844
C_8H_{10}	m-Xylene	8.56 ± 0.01	843
C_8H_{10}	p-Xylene	8.44 ± 0.01	832
$C_8H_{10}O$	p-Ethylphenol	(7.84)	613
$C_8H_{10}O$	2,3-Xylenol	(8.26)	640
$C_8H_{10}O$	2,4-Xylenol	(8.0)	609
$C_8H_{10}O$	2,6-Xylenol	(8.05)	615
$C_8H_{10}O$	3,4-Xylenol	(8.09)	624
$C_8H_{10}O$	Phenetole	(8.13)	683
$C_8H_{11}N$	2,4,6-Trimethylpyridine	($\leq$ 8.9)	$\leq$ 880
$C_8H_{11}N$	N-Ethylaniline	($\leq$ 7.67)	$\leq$ 794
$C_8H_{11}N$	N,N-Dimethylaniline	7.12 ± 0.02	787
C_8H_{14}	1-Octyne	(9.95 ± 0.02)	1040
C_8H_{14}	2-Octyne	9.31 ± 0.01	961
C_8H_{14}	3-Octyne	9.22 ± 0.01	952
C_8H_{14}	4-Octyne	9.20 ± 0.01	946
C_8H_{16}	1-Octene	9.43 ± 0.01	829
C_8H_{16}	Cyclooctane	9.75 ± 0.05	816
C_8H_{16}	Ethylcyclohexane	(9.54)	748
C_8H_{16}	1,1-Dimethylcyclohexane	(9.42)	728
C_8H_{16}	cis-1,2-Dimethylcyclohexane	(<9.78)	772
C_8H_{16}	trans-1,2-Dimethylcyclohexane	9.41	728
C_8H_{16}	cis-1,3-Dimethylcyclohexane	(<9.98)	778
C_8H_{16}	trans-1,3-Dimethylcyclohexane	9.53	743
C_8H_{16}	cis-1,4-Dimethylcyclohexane	(<9.93)	782
C_8H_{16}	trans-1,4-Dimethylcyclohexane	(9.56)	738
C_8H_{16}	Propylcyclopentane	(9.34)	753
$C_8H_{16}O$	2,2,4-Trimethyl-3-pentanone	(8.80)	511
C_8H_{18}	Octane	9.80 ± 0.10	737
C_8H_{18}	2-Methylheptane	(9.84)	734
C_8H_{18}	2,2,4-Trimethylpentane	(9.86)	713
C_8H_{18}	2,2,3,3-Tetramethylbutane	9.8	720
$C_8H_{18}O$	Dibutyl ether	(9.28)	s 560
$C_8H_{18}O$	Di-sec-butyl ether	(9.11)	511
$C_8H_{18}O$	Di-tert-butyl ether	8.88 ± 0.07	493
$C_8H_{18}S$	Dibutyl sulfide	(8.2)	624
$C_8H_{18}S$	Di-tert-butyl sulfide	(8.0)	583
$C_8H_{18}S$	Diisobutyl sulfide	(8.34)	625
$C_8H_{19}N$	Dibutylamine	(7.69)	586
$C_8H_{19}N$	Diisobutylamine	(7.8)	574

Mol. Form.	Name	IP/eV	$\Delta_f H_{ion}$ kJ/mol
$C_8H_{20}Si$	Tetraethylsilane	(8.9)	595
C_9H_7N	Quinoline	8.62 ± 0.01	1041
C_9H_7N	Isoquinoline	8.53 ± 0.03	1032
C_9H_8	Indene	8.14 ± 0.01	949
C_9H_{10}	o-Methylstyrene	(8.20)	908
C_9H_{10}	m-Methylstyrene	(8.15)	899
C_9H_{10}	p-Methylstyrene	(8.1)	895
C_9H_{10}	Cyclopropylbenzene	(8.35)	956
C_9H_{10}	Indan	(8.3)	864
$C_9H_{10}O_2$	Ethyl benzoate	(8.9)	537
C_9H_{12}	Propylbenzene	8.713 ± 0.010	848
C_9H_{12}	Isopropylbenzene	8.73 ± 0.01	847
C_9H_{12}	1,2,3-Trimethylbenzene	8.42 ± 0.02	803
C_9H_{12}	1,2,4-Trimethylbenzene	8.27 ± 0.01	784
C_9H_{12}	1,3,5-Trimethylbenzene	8.41 ± 0.01	796
$C_9H_{13}N$	N,N-Dimethyl-o-toluidine	7.40 ± 0.02	814
$C_9H_{14}O$	Isophorone	(≤ 9.07)	≤ 670
C_9H_{18}	Butylcyclopentane	(9.95)	793
C_9H_{18}	Propylcyclohexane	(9.46)	720
C_9H_{18}	Isopropylcyclohexane	(9.33)	704
$C_9H_{18}O$	2-Nonanone	(9.16)	545
$C_9H_{18}O$	5-Nonanone	(9.07)	530
$C_9H_{18}O$	2,6-Dimethyl-4-heptanone	9.01 ± 0.06	512
C_9H_{20}	Nonane	9.71 ± 0.10	709
$C_{10}F_8$	Perfluoronaphthalene	8.85	-368
$C_{10}H_7Br$	1-Bromonaphthalene	8.08 ± 0.03	955
$C_{10}H_7Cl$	1-Chloronaphthalene	(8.13)	906
$C_{10}H_8$	Naphthalene	8.1442 ± 0.0009	936
$C_{10}H_8$	Azulene	7.38 ± 0.05	1001
$C_{10}H_8O$	1-Naphthol	7.76 ± 0.03	719
$C_{10}H_8O$	2-Naphthol	7.87 ± 0.06	729
$C_{10}H_{10}O_4$	Dimethyl phthalate	(9.64 ± 0.07)	277
$C_{10}H_{12}$	1,2,3,4-Tetrahydronaphthalene	8.46 ± 0.02	841
$C_{10}H_{14}$	Butylbenzene	8.69 ± 0.02	826
$C_{10}H_{14}$	sec-Butylbenzene	8.68 ± 0.02	820
$C_{10}H_{14}$	tert-Butylbenzene	8.68 ± 0.05	816
$C_{10}H_{14}$	Isobutylbenzene	8.69 ± 0.02	817
$C_{10}H_{14}$	p-Cymene	(8.29)	771
$C_{10}H_{14}$	o-Diethylbenzene	(≤ 8.51)	≤ 804
$C_{10}H_{14}$	m-Diethylbenzene	(8.49)	798
$C_{10}H_{14}$	p-Diethylbenzene	(8.40)	790
$C_{10}H_{14}$	1,2,4,5-Tetramethylbenzene	8.04 ± 0.02	730
$C_{10}H_{14}O$	p-tert-Butylphenol	(7.8)	552
$C_{10}H_{16}$	α-Pinene	(8.07)	808
$C_{10}H_{16}O$	Camphor	(8.76)	577
$C_{10}H_{18}$	cis-Decahydronaphthalene	9.36 ± 0.04	734
$C_{10}H_{18}$	trans-Decahydronaphthalene	9.34 ± 0.04	720
$C_{10}H_{20}$	1-Decene	9.42 ± 0.05	786
$C_{10}H_{20}$	Butylcyclohexane	(9.41)	695
$C_{10}H_{22}$	Decane	(9.65)	682
$C_{11}H_{10}$	1-Methylnaphthalene	7.97 ± 0.03	882
$C_{11}H_{10}$	2-Methylnaphthalene	7.91 ± 0.08	877
$C_{11}H_{16}$	p-tert-Butyltoluene	(8.12)	730
$C_{11}H_{24}$	Undecane	(9.56)	650
$C_{11}H_{24}$	2-Methyldecane	(9.7)	658
$C_{12}H_8$	Acenaphthylene	(8.22)	1053
$C_{12}H_9N$	Carbazole	(7.57)	961
$C_{12}H_{10}$	Acenaphthene	7.75 ± 0.07	903
$C_{12}H_{10}$	Biphenyl	8.23 ± 0.10	977
$C_{12}H_{10}N_2O$	trans-Azoxybenzene	(8.1)	1123

Mol. Form.	Name	IP/eV	$\Delta_f H_{ion}$ kJ/mol
$C_{12}H_{10}O$	Diphenyl ether	(8.09)	766
$C_{12}H_{11}N$	Diphenylamine	7.16 ± 0.04	908
$C_{12}H_{18}$	5,7-Dodecadiyne	(8.67)	1079
$C_{12}H_{18}$	Hexamethylbenzene	7.85 ± 0.01	670
$C_{12}H_{22}$	Cyclohexylcyclohexane	(9.41)	690
$C_{12}H_{27}N$	Tributylamine	(7.4)	492
$C_{13}H_{10}$	9H-Fluorene	7.91 ± 0.02	952
$C_{13}H_{10}O$	Benzophenone	9.08 ± 0.05	926
$C_{13}H_{12}$	Diphenylmethane	(8.55)	963
$C_{14}H_{10}$	Anthracene	7.439 ± 0.006	948
$C_{14}H_{10}$	Phenanthrene	7.8914 ± 0.0006	966
$C_{14}H_{10}$	Diphenylacetylene	7.94 ± 0.03	1168
$C_{14}H_{12}$	cis-Stilbene	(7.80)	1005
$C_{14}H_{12}$	trans-Stilbene	7.656 ± 0.001	973
$C_{14}H_{14}$	1,2-Diphenylethane	8.9 ± 0.1	1002
$C_{16}H_{10}$	Fluoranthene	7.9 ± 0.1	1052
$C_{16}H_{10}$	Pyrene	7.4256 ± 0.0006	935
$C_{18}H_{12}$	Chrysene	7.60 ± 0.01	1017
$C_{18}H_{14}$	o-Terphenyl	(7.99)	1056
$C_{18}H_{14}$	m-Terphenyl	(8.01)	1057
$C_{18}H_{14}$	p-Terphenyl	7.80 ± 0.03	1037
$C_{20}H_{12}$	Perylene	6.960 ± 0.001	981
$C_{24}H_{12}$	Coronene	7.29 ± 0.01	1026

The energy levels in this tables are the values recommended by Bearden and Burr on the basis of a thorough review of the literature on x-ray wavelengths and related data. All values are in electron volts (eV). Values in parentheses are interpolated, and an asterisk * indicates a level which is not resolved from the level above it. See Reference 1 for uncertainties in the levels and a complete description of how the recommended values were obtained.

REFERENCES

1. Bearden, J. A., and Burr, A. F., *Rev. Mod. Phys.*, 39, 125, 1967; also published as *X-Ray Wavelengths and X-Ray Atomic Energy Levels*, Natl. Stand. Ref. Data Sys.- Natl. Bur. Standards (U.S.), No. 14, 1967.
2. Gray, D. E., Editor, *American Institute of Physics Handbook, Third Edition*, pp. 7-158 to 7-167, McGraw-Hill, New York, 1972.

Level	^{1}H	^{2}He	^{3}Li	^{4}Be	^{5}B	^{6}C	^{7}N	^{8}O
K	13.59811	24.58678	54.75	111.0	188.0	283.8	401.6	532.0
L_I								23.7
$L_{II,III}$					4.7	6.4	9.2	7.1

Level	^{9}F	^{10}Ne	^{11}Na	^{12}Mg	^{13}Al	^{14}Si	^{15}P	^{16}S
K	685.4	866.9	1072.1	1305.0	1559.6	1838.9	2145.5	2472.0
L_I	(31)	(45)	63.3	89.4	117.7	148.7	189.3	229.2
$L_{II,III}$	8.6	18.3	31.1	51.4	73.1	99.2	132.2	164.8

Level	^{17}Cl	^{18}Ar	^{19}K	^{20}Ca	^{21}Sc	^{22}Ti	^{23}V	^{24}Cr
K	2822.4	3202.9	3607.4	4038.1	4492.8	4966.4	5465.1	5989.2
L_I	270.2	320	377.1	437.8	500.4	563.7	628.2	694.6
L_{II}	201.6	247.3	296.3	350.0	406.7	461.5	520.5	583.7
L_{III}	200.0	245.2	293.6	346.4	402.2	455.5	512.9	574.5
M_I	17.5	25.3	33.9	43.7	53.8	60.3	66.5	74.1
$M_{II,III}$	6.8	12.4	17.8	25.4	32.3	34.6	37.8	42.5
$M_{IV,V}$					6.6	3.7	2.2	2.3

Level	^{25}Mn	^{26}Fe	^{27}Co	^{28}Ni	^{29}Cu	^{30}Zn	^{31}Ga	^{32}Ge
K	6539.0	7112.0	7708.9	8332.8	8978.9	9658.6	10367.1	11103.1
L_I	769.0	846.1	925.6	1008.1	1096.1	1193.6	1297.7	1414.3
L_{II}	651.4	721.1	793.6	871.9	951.0	1042.8	1142.3	1247.8
L_{III}	640.3	708.1	778.6	854.7	931.1	1019.7	1115.4	1216.7
M_I	83.9	92.9	100.7	111.8	119.8	135.9	158.1	180.0
M_{II}	48.6	54.0	59.5	68.1	73.6	86.6	106.8	127.9
M_{III}	48.6*	54.0*	59.5*	68.1*	73.6*	86.6*	102.9	120.8
$M_{IV,V}$	3.3	3.6	2.9	3.6	1.6	8.1	17.4	28.7

Level	^{33}As	^{34}Se	^{35}Br	^{36}Kr	^{37}Rb	^{38}Sr	^{39}Y	^{40}Zr
K	11866.7	12657.8	13473.7	14325.6	15199.7	16104.6	17038.4	17997.6
L_I	1526.5	1653.9	1782.0	1921.0	2065.1	2216.3	2372.5	2531.6
L_{II}	1358.6	1476.2	1596.0	1727.2	1863.9	2006.8	2155.5	2306.7
L_{III}	1323.1	1435.8	1549.9	1674.9	1804.4	1939.6	2080.0	2222.3
M_I	203.5	231.5	256.5		322.1	357.5	393.6	430.3
M_{II}	146.4	168.2	189.3	222.7	247.4	279.8	312.4	344.2
M_{III}	140.5	161.9	181.5	213.8	238.5	269.1	300.3	330.5
M_{IV}	41.2	56.7	70.1	88.9	111.8	135.0	159.6	182.4
M_V	41.2*	56.7*	69.0	88.9*	110.3	133.1	157.4	180.0
N_I		27.3	24.0		29.3	37.7	45.4	51.3
N_{II}	2.5	5.6	5.2	10.6	14.8	19.9	25.6	28.7
N_{III}	2.5*	5.6*	4.6	10.6*	14.0	19.9*	25.6*	28.7*

Level	^{41}Nb	^{42}Mo	^{43}Tc	^{44}Ru	^{45}Rh	^{46}Pd	^{47}Ag	^{48}Cd
K	18985.6	19999.5	21044.0	22117.2	23219.9	24350.3	25514.0	26711.2
L_I	2697.7	2865.5	3042.5	3224.0	3411.9	3604.3	3805.8	4018.0
L_{II}	2464.7	2625.1	2793.2	2966.9	3146.1	3330.3	3523.7	3727.0
L_{III}	2370.5	2520.2	2676.9	2837.9	3003.8	3173.3	3351.1	3537.5
M_I	468.4	504.6		585.0	627.1	669.9	717.5	770.2
M_{II}	378.4	409.7	444.9	482.8	521.0	559.1	602.4	650.7
M_{III}	363.0	392.3	425.0	460.6	496.2	531.5	571.4	616.5
M_{IV}	207.4	230.3	256.4	283.6	311.7	340.0	372.8	410.5
M_V	204.6	227.0	252.9	279.4	307.0	334.7	366.7	403.7
N_I	58.1	61.8		74.9	81.0	86.4	95.2	107.6
N_{II}	33.9	34.8	38.9	43.1	47.9	51.1	62.6	66.9
N_{III}	33.9*	34.8*	38.9*	43.1*	47.9*	51.1*	55.9	66.9*
$N_{IV,V}$	3.2	1.8		2.0	2.5	1.5	3.3	9.3

Level	^{49}In	^{50}Sn	^{51}Sb	^{52}Te	^{53}I	^{54}Xe	^{55}Cs	^{56}Ba
K	27939.9	29200.1	30491.2	31813.8	33169.4	34561.4	35984.6	37440.6
L_I	4237.5	4464.7	4698.3	4939.2	5188.1	5452.8	5714.3	5988.8
L_{II}	3938.0	4156.1	4380.4	4612.0	4852.1	5103.7	5359.4	5623.6
L_{III}	3730.1	3928.8	4132.2	4341.4	4557.1	4782.2	5011.9	5247.0
M_I	825.6	883.8	943.7	1006.0	1072.1		1217.1	1292.8
M_{II}	702.2	756.4	811.9	869.7	930.5	999.0	1065.0	1136.7
M_{III}	664.3	714.4	765.6	818.7	874.6	937.0	997.6	1062.2
M_{IV}	450.8	493.3	536.9	582.5	631.3		739.5	796.1
M_V	443.1	484.8	527.5	572.1	619.4	672.3	725.5	780.7
N_I	121.9	136.5	152.0	168.3	186.4		230.8	253.0
N_{II}	77.4	88.6	98.4	110.2	122.7	146.7	172.3	191.8
N_{III}	77.4*	88.6*	98.4*	110.2*	122.7*	146.7*	161.6	179.7
N_{IV}	16.2	23.9	31.4	39.8	49.6		78.8	92.5
N_V	16.2*	23.9*	31.4*	39.8*	49.6*		76.5	89.9
O_I	0.1	0.9	6.7	11.6	13.6		22.7	39.1
O_{II}	0.8	1.1	2.1	2.3	3.3		13.1	16.6
O_{III}	0.8*	1.1*	2.1*	2.3*	3.3*		11.4	14.6

Level	^{57}La	^{58}Ce	^{59}Pr	^{60}Nd	^{61}Pm	^{62}Sm	^{63}Eu	^{64}Gd
K	38924.6	40443.0	41990.6	43568.9	45184.0	46834.2	48519.0	50239.1
L_I	6266.3	6548.8	6834.8	7126.0	7427.9	7736.8	8052.0	8375.6
L_{II}	5890.6	6164.2	6440.4	6721.5	7012.8	7311.8	7617.1	7930.3
L_{III}	5482.7	5723.4	5964.3	6207.9	6459.3	6716.2	6976.9	7242.8
M_I	1361.3	1434.6	1511.0	1575.3		1722.8	1800.0	1880.8
M_{II}	1204.4	1272.8	1337.4	1402.8	1471.4	1540.7	1613.9	1688.3
M_{III}	1123.4	1185.4	1242.2	1297.4	1356.9	1419.8	1480.6	1544.0
M_{IV}	848.5	901.3	951.1	999.9	1051.5	1106.0	1160.6	1217.2
M_V	831.7	883.3	931.0	977.7	1026.9	1080.2	1130.9	1185.2
N_I	270.4	289.6	304.5	315.2		345.7	360.2	375.8
N_{II}	205.8	223.3	236.3	243.3	242	265.6	283.9	288.5
N_{III}	191.4	207.2	217.6	224.6	242*	247.4	256.6	270.9
$N_{IV,V}$	98.9	110.0	113.2	117.5	120.4	129.0	133.2	140.5
$N_{VI,VII}$		0.1	2.0	1.5		5.5	0.0	0.1
O_I	32.3	37.8	37.4	37.5		37.4	31.8	36.1
$O_{II,III}$	14.4	19.8	22.3	21.1		21.3	22.0	20.3

Level	^{65}Tb	^{66}Dy	^{67}Ho	^{68}Er	^{69}Tm	^{70}Yb	^{71}Lu	^{72}Hf
K	51995.7	53788.5	55617.7	57485.5	59389.6	61332.3	63313.8	65350.8
L_I	8708.0	9045.8	9394.2	9751.3	10115.7	10486.4	10870.4	11270.7
L_{II}	8251.6	8580.6	8917.8	9264.3	9616.9	9978.2	10348.6	10739.4
L_{III}	7514.0	7790.1	8071.1	8357.9	8648.0	8943.6	9244.1	9560.7
M_I	1967.5	2046.8	2128.3	2206.5	2306.8	2398.1	2491.2	2600.9

Level	^{65}Tb	^{66}Dy	^{67}Ho	^{68}Er	^{69}Tm	^{70}Yb	^{71}Lu	^{72}Hf
M_{II}	1767.7	1841.8	1922.8	2005.8	2089.8	2173.0	2263.5	2365.4
M_{III}	1611.3	1675.6	1741.2	1811.8	1884.5	1949.8	2023.6	2107.6
M_{IV}	1275.0	1332.5	1391.5	1453.3	1514.6	1576.3	1639.4	1716.4
M_V	1241.2	1294.9	1351.4	1409.3	1467.7	1527.8	1588.5	1661.7
N_I	397.9	416.3	435.7	449.1	471.7	487.2	506.2	538.1
N_{II}	310.2	331.8	343.5	366.2	385.9	396.7	410.1	437.0
N_{III}	385.0	292.9	306.6	320.0	336.6	343.5	359.3	380.4
N_{IV}	147.0	154.2	161.0	176.7	179.6	198.1	204.8	223.8
N_V	147.0*	154.2*	161.0*	167.6	179.6*	184.9	195.0	213.7
$N_{VI,VII}$	2.6	4.2	3.7	4.3	5.3	6.3	6.9	17.1
O_I	39.0	62.9	51.2	59.8	53.2	54.1	56.8	64.9
O_{II}	25.4	26.3	20.3	29.4	32.3	23.4	28.0	38.1
O_{III}	25.4*	26.3*	20.3*	29.4*	32.3*	23.4*	28.0*	30.6

Level	^{73}Ta	^{74}W	^{75}Re	^{76}Os	^{77}Ir	^{78}Pt	^{79}Au	^{80}Hg
K	67416.4	69525.0	71676.4	73870.8	76111.0	78394.8	80724.9	83102.3
L_I	11681.5	12099.8	12526.7	12968.0	13418.5	13879.9	14352.8	14839.3
L_{II}	11136.1	11544.0	11958.7	12385.0	12824.1	13272.6	13733.6	14208.7
L_{III}	9881.1	10206.8	10535.3	10870.9	11215.2	11563.7	11918.7	12283.9
M_I	2708.0	2819.6	2931.7	3048.5	3173.7	3296.0	3424.9	3561.6
M_{II}	2468.7	2574.9	2681.6	2792.2	2908.7	3026.5	3147.8	3278.5
M_{III}	2194.0	2281.0	2367.3	2457.2	2550.7	2645.4	2743.0	2847.1
M_{IV}	1793.2	1871.6	1948.9	2030.8	2116.1	2201.9	2291.1	2384.9
M_V	1735.1	1809.2	1882.9	1960.1	2040.4	2121.6	2205.7	2294.9
N_I	565.5	595.0	625.0	654.3	690.1	722.0	758.8	800.3
N_{II}	464.8	491.6	517.9	546.5	577.1	609.2	643.7	676.9
N_{III}	404.5	425.3	444.4	468.2	494.3	519.0	545.4	571.0
N_{IV}	241.3	258.8	273.7	289.4	311.4	330.8	352.0	378.3
N_V	229.3	245.4	260.2	272.8	294.9	313.3	333.9	359.8
N_{VI}	25.0	36.5	40.6	46.3	63.4	74.3	86.4	102.2
N_{VII}	25.0*	33.6	40.6*	46.3*	60.5	71.1	82.8	98.5
O_I	71.1	77.1	82.8	83.7	95.2	101.7	107.8	120.3
O_{II}	44.9	46.8	45.6	58.0	63.0	65.3	71.7	80.5
O_{III}	36.4	35.6	34.6	45.4	50.5	51.7	53.7	57.6
$O_{IV,V}$	5.7	6.1	3.5		3.8	2.2	2.5	6.4

Level	^{81}Tl	^{82}Pb	^{83}Bi	^{84}Po	^{85}At	^{86}Rn	^{87}Fr	^{88}Ra
K	85530.4	88004.5	90525.9	93105.0	95729.9	98404	101137	103921.9
L_I	15346.7	15860.8	16387.5	16939.3	17493	18049	18639	19236.7
L_{II}	14697.9	15200.0	15711.1	16244.3	16784.7	17337.1	17906.5	18484.3
L_{III}	12657.5	13035.2	13418.6	13813.8	14213.5	14619.4	15031.2	15444.4
M_I	3704.1	3850.7	3999.1	4149.4	(4317)	(4482)	(4652)	4822.0
M_{II}	3415.7	3554.2	3696.3	3854.1	4008	4159	4327	4489.5
M_{III}	2956.6	3066.4	3176.9	3301.9	3426	3538	3663	3791.8
M_{IV}	2485.1	2585.6	2687.6	2798.0	2908.7	3021.5	3136.2	3248.4
M_V	2389.3	2484.0	2579.6	2683.0	2786.7	2892.4	2999.9	3104.9
N_I	845.5	893.6	938.2	995.3	(1042)	(1097)	(1153)	1208.4
N_{II}	721.3	763.9	805.3	851	886	929	980	1057.6
N_{III}	609.0	644.5	678.9	705	740	768	810	879.1
N_{IV}	406.6	435.2	463.6	500.2	533.2	566.6	603.3	635.9
N_V	386.2	412.9	440.0	473.4			577	602.7
N_{VI}	122.8	142.9	161.9					298.9
N_{VII}	118.5	138.1	157.4					298.9*
O_I	136.3	147.3	159.3					254.4
O_{II}	99.6	104.8	116.8					200.4
O_{III}	75.4	86.0	92.8					152.8

Level	81Tl	82Pb	83Bi	84Po	85At	86Rn	87Fr	88Ra
O_IV	15.3	21.8	26.5	31.4				67.2
O_V	13.1	19.2	24.4	31.4*				67.2*
P_I		3.1						43.5
P_II,III		0.7	2.7					18.8

Level	89Ac	90Th	91Pa	92U	93Np	94Pu	95Am	96Cm
K	106755.3	109650.9	112601.4	115606.1	118678	121818	125027	128220
L_I	19840	20472.1	21104.6	21757.4	22426.8	23097.2	23772.9	24460
L_II	19083.2	19693.2	20313.7	20947.6	21600.5	22266.2	22944.0	23779
L_III	15871.0	16300.3	16733.1	17166.3	17610.0	18056.8	18504.1	18930
M_I	(5002)	5182.3	5366.9	5548.0	5723.2	5932.9	6120.5	6288
M_II	4656	4830.4	5000.9	5182.2	5366.2	5541.2	5710.2	5895
M_III	3909	4046.1	4173.8	4303.4	4434.7	4556.6	4667.0	4797
M_IV	3370.2	3490.8	3611.2	3727.6	3850.3	3972.6	4092.1	4227
M_V	3219.0	3332.0	3441.8	3551.7	3665.8	3778.1	3886.9	3971
N_I	(1269)	1329.5	1387.1	1440.8	1500.7	1558.6	1617.1	1643
N_II	1080	1168.2	1224.3	1272.6	1327.7	1372.1	1411.8	1440
N_III	890	967.3	1006.7	1044.9	1086.8	1114.8	(1135.7)	1154
N_IV	674.9	714.1	743.4	780.4	815.9	848.9	878.7	
N_V		676.4	708.2	737.7	770.3	801.4	827.6	
N_VI		344.4	371.2	391.3	415.0	445.8		
N_VII		335.2	359.5	380.9	404.4	432.4		
O_I		290.2	309.6	323.7		351.9		
O_II		229.4	222.9	259.3	283.4	274.1		385
O_III		181.8	222.9*	195.1	206.1	206.5		
O_IV		94.3	94.1	105.0	109.3	116.0	115.8	
O_V		87.9	94.1*	96.3	101.3	105.4	103.3	
P_I		59.5		70.7				
P_II		49.0		42.3				
P_III		43.0		32.3				

Level	97Bk	98Cf	99Es	100Fm	101Md	102No	103Lr
K	131590	135960	139490	143090	146780	150540	154380
L_I	25275	26110	26900	27700	28530	29380	30240
L_II	24385	25250	26020	26810	27610	28440	29280
L_III	19452	19930	20410	20900	21390	21880	22360
M_I	6556	6754	6977	7205	7441	7675	7900
M_II	6147	6359	6574	6793	7019	7245	7460
M_III	4977	5109	5252	5397	5546	5688	5710
M_IV	4366	4497	4630	4766	4903	5037	5150
M_V	4132	4253	4374	4498	4622	4741	4860
N_I	1755	1799	1868	1937	2010	2078	2140
N_II	1554	1616	1680	1747	1814	1876	1930
N_III	1235	1279	1321	1366	1410	1448	1480
O_I	398	419	435	454	472	484	490

ELECTRON BINDING ENERGIES OF THE ELEMENTS

Gwyn P. Williams

This table gives the binding energies in electron volts (eV) for selected electronic levels of the elements. For metallic elements the binding energy is referred to the Fermi level; for semiconductors, to the valence band maximum; and for gases and insulators, to the vacuum level. The atomic number is listed after the element name.

REFERENCES

1. Fluggle and Martensson, J. *Elect. Spect.*, 21, 275, 1980.
2. Cardona, M. and Ley, L., *Photoemission from Solids*, Springer Verlag, Heidelberg, 1978.
3. Bearden, J. A. and Burr, A. F., *Rev. Mod. Phys.*, 39, 125, 1967.

Actinium (89)

K	1s	106755
L I	2s	19840
L II	$2p_{1/2}$	19083
L III	$2p_{3/2}$	15871
M I	3s	5002
M II	$3p_{1/2}$	4656
M III	$3p_{3/2}$	3909
M IV	$3d_{3/2}$	3370
M V	$3d_{5/2}$	3219
N I	4s	1269[a]
N II	$4p_{1/2}$	1080[a]
N III	$4p_{3/2}$	890[a]
N IV	$4d_{3/2}$	675[a]
N V	$4d_{5/2}$	639[a]
N VI	$4f_{5/2}$	319[a]
N VII	$4f_{7/2}$	319[a]
O I	5s	272[a]
O II	$5p_{1/2}$	215[a]
O III	$5p_{3/2}$	167[a]
O IV	$5d_{3/2}$	80[a]
O V	$5d_{5/2}$	80[a]
P I	6s	—
P II	$6p_{1/2}$	—
P III	$6p_{3/2}$	—

Aluminum (13)

K	1s	1559.0
L I	2s	117.8[a]
L II	$2p_{1/2}$	72.9[a]
L III	$2p_{3/2}$	72.5[a]

Antimony (51)

K	1s	30419
L I	2s	4698
L II	$2p_{1/2}$	4380
L III	$2p_{3/2}$	4132
M I	3s	946[b]
M II	$3p_{1/2}$	812.7[b]
M III	$3p_{3/2}$	766.4[b]
M IV	$3d_{3/2}$	537.5[b]
M V	$3d_{5/2}$	528.2[b]
N I	4s	153.2[b]
N II	$4p_{1/2}$	95.6[b,c]
N III	$4p_{3/2}$	95.6[b]
N IV	$4d_{3/2}$	33.3[b]
N V	$4d_{5/2}$	32.1[b]

Argon (18)

K	1s	3205.9[a]
L I	2s	326.3[a]
L II	$2p_{1/2}$	250.6[a]
L III	$2p_{3/2}$	248.4[a]
M I	3s	29.3[a]
M II	$3p_{1/2}$	15.9[a]
M III	$3p_{3/2}$	15.7[a]

Arsenic (33)

K	1s	11867
L I	2s	1527.0[a,d]
L II	$2p_{1/2}$	1359.1[a,d]
L III	$2p_{3/2}$	1323.6[a,d]
M I	3s	204.7[a]
M II	$3p_{1/2}$	146.2[a]
M III	$3p_{3/2}$	141.2[a]
M IV	$3d_{3/2}$	41.7[a]
M V	$3d_{5/2}$	41.7[a]

Astatine (85)

K	1s	95730
L I	2s	17493
L II	$2p_{1/2}$	16785
L III	$2p_{3/2}$	14214
M I	3s	4317
M II	$3p_{1/2}$	4008
M III	$3p_{3/2}$	3426
M IV	$3d_{3/2}$	2909
M V	$3d_{5/2}$	2787
N I	4s	1042[a]
N II	$4p_{1/2}$	886[a]
N III	$4p_{3/2}$	740[a]
N IV	$4d_{3/2}$	533[a]
N V	$4d_{5/2}$	507[a]
N VI	$4f_{5/2}$	210[a]
N VII	$4f_{7/2}$	210[a]
O I	5s	195[a]
O II	$5p_{1/2}$	148[a]
O III	$5p_{3/2}$	115[a]
O IV	$5d_{3/2}$	40[a]
O V	$5d_{5/2}$	40[a]

Barium (56)

K	1s	37441
L I	2s	5989
L II	$2p_{1/2}$	5624

L III	$2p_{3/2}$	5247
M I	3s	1293[a,d]
M II	$3p_{1/2}$	1137[a,d]
M III	$3p_{3/2}$	1063[a,d]
M IV	$3d_{3/2}$	795.7[a]
M V	$3d_{5/2}$	780.5[a]
N I	4s	253.5[b]
N II	$4p_{1/2}$	192
N III	$4p_{3/2}$	178.6[b]
N IV	$4d_{3/2}$	92.6[b]
N V	$4d_{5/2}$	89.9[b]
N VI	$4f_{5/2}$	—
N VII	$4f_{7/2}$	—
O I	5s	30.3[b]
O II	$5p_{1/2}$	17.0[b]
O III	$5p_{3/2}$	14.8[b]

Beryllium (4)

K	1s	111.5[a]

Bismuth (83)

K	1s	90526
L I	2s	16388
L II	$2p_{1/2}$	15711
L III	$2p_{3/2}$	13419
M I	3s	3999
M II	$3p_{1/2}$	3696
M III	$3p_{3/2}$	3177
M IV	$3d_{3/2}$	2688
M V	$3d_{5/2}$	2580
N I	4s	939[b]
N II	$4p_{1/2}$	805.2[b]
N III	$4p_{3/2}$	678.8[b]
N IV	$4d_{3/2}$	464.0[b]
N V	$4d_{5/2}$	440.1[b]
N VI	$4f_{5/2}$	162.3[b]
N VII	$4f_{7/2}$	157.0[b]
O I	5s	159.3[a,d]
O II	$5p_{1/2}$	119.0[b]
O III	$5p_{3/2}$	92.6[b]
O IV	$5d_{3/2}$	26.9[b]
O V	$5d_{5/2}$	23.8[b]

Boron (5)

K	1s	188[a]

Bromine (35)

K	1s	13474

Cadmium (48)

K	1s	26711
L I	2s	4018
L II	$2p_{1/2}$	3727
L III	$2p_{3/2}$	3538
M I	3s	772.0[b]
M II	$3p_{1/2}$	652.6[b]
M III	$3p_{3/2}$	618.4[b]
M IV	$3d_{3/2}$	411.9[b]
M V	$3d_{5/2}$	405.2[b]
N I	4s	109.8[b]
N II	$4p_{1/2}$	63.9[b,c]
N III	$4p_{3/2}$	63.9[b,c]
N IV	$4d_{3/2}$	11.7[b]
N V	$4d_{5/2}$	10.7[b]

Calcium (20)

K	1s	4038.5[a]
L I	2s	438.4[b]
L II	$2p_{1/2}$	349.7[b]
L III	$2p_{3/2}$	346.2[b]
M I	3s	44.3[b]
M II	$3p_{1/2}$	25.4[b]
M III	$3p_{3/2}$	25.4[b]

Carbon (6)

K	1s	284.2[a]

Cerium (58)

K	1s	40443
L I	2s	6548
L II	$2p_{1/2}$	6164
L III	$2p_{3/2}$	5723
M I	3s	1436[a,d]
M II	$3p_{1/2}$	1274[a,d]
M III	$3p_{3/2}$	1187[a,d]
M IV	$3d_{3/2}$	902.4[a]
M V	$3d_{5/2}$	883.8[a]

N I	4s	291.0[a]
N II	4p$_{1/2}$	223.3
N III	4p$_{3/2}$	206.5[a]
N IV	4d$_{3/2}$	109[a]
N V	4d$_{5/2}$	—
N VI	4f$_{5/2}$	0.1
N VII	4f$_{7/2}$	0.1
O I	5s	37.8
O II	5p$_{1/2}$	19.8[a]
O III	5p$_{3/2}$	17.0[a]

Cesium (55)

K	1s	35985
L I	2s	5714
L II	2p$_{1/2}$	5359
L III	2p$_{3/2}$	5012
M I	3s	1211[a,d]
M II	3p$_{1/2}$	1071[a]
M III	3p$_{3/2}$	1003[a]
M IV	3d$_{3/2}$	740.5[a]
M V	3d$_{5/2}$	726.6[a]
N I	4s	232.3[a]
N II	4p$_{1/2}$	172.4[a]
N III	4p$_{3/2}$	161.3[a]
N IV	4d$_{3/2}$	79.8[a]
N V	4d$_{5/2}$	77.5[a]
N VI	4f$_{5/2}$	—
N VII	4f$_{7/2}$	—
O I	5s	22.7
O II	5p$_{1/2}$	14.2[a]
O III	5p$_{3/2}$	12.1[a]

Chlorine (17)

K	1s	2822.0
L I	2s	270[a]
L II	2p$_{1/2}$	202[a]
L III	2p$_{3/2}$	200[a]

Chromium(24)

K	1s	5989
L I	2s	696.0[b]
L II	2p$_{1/2}$	583.8[b]
L III	2p$_{3/2}$	574.1[b]
M I	3s	74.1[b]
M II	3p$_{1/2}$	42.2[b]
M III	3p$_{3/2}$	42.2[b]

Cobalt (27)

K	1s	7709
L I	2s	925.1[b]
L II	2p$_{1/2}$	793.2[b]
L III	2p$_{3/2}$	778.1[b]
M I	3s	101.0[b]
M II	3p$_{1/2}$	58.9[b]
M III	3p$_{3/2}$	58.9[b]

Copper (29)

K	1s	8979
L I	2s	1096.7[b]

L II	2p$_{1/2}$	952.3[b]
L III	2p$_{3/2}$	932.5[b]
M I	3s	122.5[b]
M II	3p$_{1/2}$	77.3[b]
M III	3p$_{3/2}$	75.1[b]

Dysprosium (66)

K	1s	53789
L I	2s	9046
L II	2p$_{1/2}$	8581
L III	2p$_{3/2}$	7790
M I	3s	2047
M II	3p$_{1/2}$	1842
M III	3p$_{3/2}$	1676
M IV	3d$_{3/2}$	1333
M V	3d$_{5/2}$	1292[a]
N I	4s	414.2[a]
N II	4p$_{1/2}$	333.5[a]
N III	4p$_{3/2}$	293.2[a]
N IV	4d$_{3/2}$	153.6[a]
N V	4d$_{5/2}$	153.6[a]
N VI	4f$_{5/2}$	8.0[a]
N VII	4f$_{7/2}$	4.3[a]
O I	5s	49.9[a]
O II	5p$_{1/2}$	26.3
O III	5p$_{3/2}$	26.3

Erbium (68)

K	1s	57486
L I	2s	9751
L II	2p$_{1/2}$	9264
L III	2p$_{3/2}$	8358
M I	3s	2206
M II	3p$_{1/2}$	2006
M III	3p$_{3/2}$	1812
M IV	3d$_{3/2}$	1453
M V	3d$_{5/2}$	1409
N I	4s	449.8[a]
N II	4p$_{1/2}$	366.2
N III	4p$_{3/2}$	320.2[a]
N IV	4d$_{3/2}$	167.6[a]
N V	4d$_{5/2}$	167.6[a]
N VI	4f$_{5/2}$	—
N VII	4f$_{7/2}$	4.7[a]
O I	5s	50.6[a]
O II	5p$_{1/2}$	31.4[a]
O III	5p$_{3/2}$	24.7[a]

Europium (63)

K	1s	48519
L I	2s	8052
L II	2p$_{1/2}$	7617
L III	2p$_{3/2}$	6977
M I	3s	1800
M II	3p$_{1/2}$	1614
M III	3p$_{3/2}$	1481
M IV	3d$_{3/2}$	1158.6[a]
M V	3d$_{5/2}$	1127.5[a]
N I	4s	360

N II	4p$_{1/2}$	284
N III	4p$_{3/2}$	257
N IV	4d$_{3/2}$	133
N V	4d$_{5/2}$	1227[a]
N VI	4f$_{5/2}$	0
N VII	4f$_{7/2}$	0
O I	5s	32
O II	5p$_{1/2}$	22
O III	5p$_{3/2}$	22

Fluorine (9)

K	1s	696.7[a]

Francium (87)

K	1s	101137
L I	2s	18639
L II	2p$_{1/2}$	17907
L III	2p$_{3/2}$	15031
M I	3s	4652
M II	3p$_{1/2}$	4327
M III	3p$_{3/2}$	3663
M IV	3d$_{3/2}$	3136
M V	3d$_{5/2}$	3000
N I	4s	1153[a]
N II	4p$_{1/2}$	980[a]
N III	4p$_{3/2}$	810[a]
N IV	4d$_{3/2}$	603[a]
N V	4d$_{5/2}$	577[a]
N VI	4f$_{5/2}$	268[a]
N VII	4f$_{7/2}$	268[a]
O I	5s	234[a]
O II	5p$_{1/2}$	182[a]
O III	5p$_{3/2}$	140[a]
O IV	5d$_{3/2}$	58[a]
O V	5d$_{5/2}$	58[a]
P I	6s	34
P II	6p$_{1/2}$	15
P III	6p$_{3/2}$	15

Gadolinium (64)

K	1s	50239
L I	2s	8376
L II	2p$_{1/2}$	7930
L III	2p$_{3/2}$	7243
M I	3s	1881
M II	3p$_{1/2}$	1688
M III	3p$_{3/2}$	1544
M IV	3d$_{3/2}$	1221.9[a]
M V	3d$_{5/2}$	1189.6[a]
N I	4s	378.6[a]
N II	4p$_{1/2}$	286
N III	4p$_{3/2}$	271
N IV	4d$_{3/2}$	—
N V	4d$_{5/2}$	142.6[a]
N VI	4f$_{5/2}$	8.6[a]
N VII	4f$_{7/2}$	8.6[a]
O I	5s	36
O II	5p$_{1/2}$	20
O III	5p$_{3/2}$	20

Gallium (31)

K	1s	10367
L I	2s	1299.0[a,d]
L II	2p$_{1/2}$	1143.2[b]
L III	2p$_{3/2}$	1116.4[b]
M I	3s	159.5[b]
M II	3p$_{1/2}$	103.5[b]
M III	3p$_{3/2}$	100.0[b]
M IV	3d$_{3/2}$	18.7[b]
M V	3d$_{5/2}$	18.7[b]

Germanium (32)

K	1s	11103
L I	2s	1414.6[a,d]
L II	2p$_{1/2}$	1248.1[a,d]
L III	2p$_{3/2}$	1217.0[a,d]
M I	3s	180.1[a]
M II	3p$_{1/2}$	124.9[a]
M III	3p$_{3/2}$	120.8[a]
M IV	3d$_{3/2}$	29.8[a]
M V	3d$_{5/2}$	29.2[a]

Gold (79)

K	1s	80725
L I	2s	14353
L II	2p$_{1/2}$	13734
L III	2p$_{3/2}$	11919
M I	3s	3425
M II	3p$_{1/2}$	3148
M III	3p$_{3/2}$	2743
M IV	3d$_{3/2}$	2291
M V	3d$_{5/2}$	2206
N I	4s	762.1[b]
N II	4p$_{1/2}$	642.7[b]
N III	4p$_{3/2}$	546.3[b]
N IV	4d$_{3/2}$	353.2[b]
N V	4d$_{5/2}$	335.1[b]
N VI	4f$_{5/2}$	87.6[b]
N VII	4f$_{7/2}$	83.9[b]
O I	5s	107.2[a,d]
O II	5p$_{1/2}$	74.2[b]
O III	5p$_{3/2}$	57.2[b]

Hafnium (72)

K	1s	65351
L I	2s	11271
L II	2p$_{1/2}$	10739
L III	2p$_{3/2}$	9561
M I	3s	2601
M II	3p$_{1/2}$	2365
M III	3p$_{3/2}$	2107
M IV	3d$_{3/2}$	1176
M V	3d$_{5/2}$	1662
N I	4s	538[a]
N II	4p$_{1/2}$	438.2[b]
N III	4p$_{3/2}$	380.7[b]
N IV	4d$_{3/2}$	220.0[b]
N V	4d$_{5/2}$	211.5[b]
N VI	4f$_{5/2}$	15.9[b]

N VII	$4f_{7/2}$	14.2[b]
O I	5s	64.2[b]
O II	$5p_{1/2}$	38[a]
O III	$5p_{3/2}$	29.9[b]

Helium (2)

K	1s	24.6[a]

Holmium (67)

K	1s	55618
L I	2s	9394
L II	$2p_{1/2}$	8918
L III	$2p_{3/2}$	8071
M I	3s	2128
M II	$3p_{1/2}$	1923
M III	$3p_{3/2}$	1741
M IV	$3d_{3/2}$	1392
M V	$3d_{5/2}$	1351
N I	4s	432.4[a]
N II	$4p_{1/2}$	343.5
N III	$4p_{3/2}$	308.2[a]
N IV	$4d_{3/2}$	160[a]
N V	$4d_{5/2}$	160[a]
N VI	$4f_{5/2}$	8.6[a]
N VII	$4f_{7/2}$	5.2[a]
O I	5s	49.3[a]
O II	$5p_{1/2}$	30.8[a]
O III	$5p_{3/2}$	24.1[a]

Hydrogen (1)

K	1s	13.6

Indium (49)

K	1s	27940
L I	2s	4238
L II	$2p_{1/2}$	3938
L III	$2p_{3/2}$	3730
M I	3s	827.2[b]
M II	$3p_{1/2}$	703.2[b]
M III	$3p_{3/2}$	665.3[b]
M IV	$3d_{3/2}$	451.4[b]
M V	$3d_{5/2}$	443.9[b]
N I	4s	122.9[b]
N II	$4p_{1/2}$	73.5[b,c]
N III	$4p_{3/2}$	73.5[b,c]
N IV	$4d_{3/2}$	17.7[b]
N V	$4d_{5/2}$	16.9[b]

Iodine (53)

K	1s	33169
L I	2s	5188
L II	$2p_{1/2}$	4852
L III	$2p_{3/2}$	4557
M I	3s	1072[a]
M II	$3p_{1/2}$	931[a]
M III	$3p_{3/2}$	875[a]
M IV	$3d_{3/2}$	631[a]
M V	$3d_{5/2}$	620[a]
N I	4s	186[a]
N II	$4p_{1/2}$	123[a]

N III	$4p_{3/2}$	123[a]
N IV	$4d_{3/2}$	50[a]
N V	$4d_{5/2}$	50[a]

Iridium (77)

K	1s	76111
L I	2s	13419
L II	$2p_{1/2}$	12824
L III	$2p_{3/2}$	11215
M I	3s	3174
M II	$3p_{1/2}$	2909
M III	$3p_{3/2}$	2551
M IV	$3d_{3/2}$	2116
M V	$3d_{5/2}$	2040
N I	4s	691.1[b]
N II	$4p_{1/2}$	577.8[b]
N III	$4p_{3/2}$	495.8[b]
N IV	$4d_{3/2}$	311.9[b]
N V	$4d_{5/2}$	296.3[b]
N VI	$4f_{5/2}$	63.8[b]
N VII	$4f_{7/2}$	60.8[b]
O I	52	95.2[a,d]
O II	$5p_{1/2}$	63.0[a,d]
O III	$5p_{3/2}$	48.0[b]

Iron (26)

K	1s	7112
L I	2s	844.6[b]
L II	$2p_{1/2}$	719.9[b]
L III	$2p_{3/2}$	706.8[b]
M I	3s	91.3[b]
M II	$3p_{1/2}$	52.7[b]
M III	$3p_{3/2}$	52.7[b]

Krypton (36)

K	1s	14326
L I	2s	1921
L II	$2p_{1/2}$	1730.9[a]
L III	$2p_{3/2}$	1678.4[a]
M I	3s	292.8[a]
M II	$3p_{1/2}$	222.2[a]
M III	$3p_{3/2}$	214.4[a]
M IV	$3d_{3/2}$	95.0[a]
M V	$3d_{5/2}$	93.8[a]
N I	4s	27.5[a]
N II	$4p_{1/2}$	14.1[a]
N III	$4p_{3/2}$	14.1[a]

Lanthanum (57)

K	1s	38925
L I	2s	6266
L II	$2p_{1/2}$	5891
L III	$2p_{3/2}$	5483
M I	3s	1362[a,d]
M II	$3p_{1/2}$	1209[a,d]
M III	$3p_{3/2}$	1128[a,d]
M IV	$3d_{3/2}$	853[a]
M V	$3d_{5/2}$	836[a]
N I	4s	247.7[a]
N II	$4p_{1/2}$	205.8

N III	$4p_{3/2}$	196.0[a]
N IV	$4d_{3/2}$	105.3[a]
N V	$4d_{5/2}$	102.5[a]
N VI	$4f_{5/2}$	—
N VII	$4f_{7/2}$	—
O I	5s	34.3[a]
O II	$5p_{1/2}$	19.3[a]
O III	$5p_{3/2}$	16.8[a]

Lead (82)

K	1s	88005
L I	2s	15861
L II	$2p_{1/2}$	15200
L III	$2p_{3/2}$	13055
M I	3s	3851
M II	$3p_{1/2}$	3554
M III	$3p_{3/2}$	3066
M IV	$3d_{3/2}$	2586
M V	$3d_{5/2}$	2484
N I	4s	891.8[b]
N II	$4p_{1/2}$	761.9[b]
N III	$4p_{3/2}$	643.5[b]
N IV	$4d_{3/2}$	434.3[b]
N V	$4d_{5/2}$	412.2[b]
N VI	$4f_{5/2}$	141.7[b]
N VII	$4f_{7/2}$	136.9[b]
O I	5s	147[a,d]
O II	$5p_{1/2}$	106.4[b]
O III	$5p_{3/2}$	83.3[b]
O IV	$5d_{3/2}$	20.7[b]
O V	$5d_{5/2}$	18.1[b]

Lithium (3)

K	1s	54.7[a]

Lutetium

K	1s	63314
L I	2s	10870
L II	$2p_{1/2}$	10349
L III	$2p_{3/2}$	9244
M I	3s	2491
M II	$3p_{1/2}$	2264
M III	$3p_{3/2}$	2024
M IV	$3d_{3/2}$	1639
M V	$3d_{5/2}$	1589
N I	4s	506.8[a]
N II	$4p_{1/2}$	412.4[a]
N III	$4p_{3/2}$	359.2[a]
N IV	$4d_{3/2}$	206.1[a]
N V	$4d_{5/2}$	196.3[a]
N VI	$4f_{5/2}$	8.9[a]
N VII	$4f_{7/2}$	7.5[a]
O I	5s	57.3[a]
O II	$5p_{1/2}$	33.6[a]
O III	$5p_{3/2}$	26.7[a]

Magnesium (12)

K	1s	1303.0[b]
L I	2s	88.6[a]
L II	$2p_{1/2}$	49.6[b]
L III	$2p_{3/2}$	49.2[a]

Manganese (25)

K	1s	6539
L I	2s	769.1[b]
L II	$2p_{1/2}$	649.9[b]
L III	$2p_{3/2}$	638.7[b]
M I	3s	82.3[b]
M II	$3p_{1/2}$	47.2[b]
M III	$3p_{3/2}$	47.2[b]

Mercury (80)

K	1s	83102
L I	2s	14839
L II	$2p_{1/2}$	14209
L III	$2p_{3/2}$	12284
M I	3s	3562
M II	$3p_{1/2}$	3279
M III	$3p_{3/2}$	2847
M IV	$3d_{3/2}$	2385
M V	$3d_{5/2}$	2295
N I	4s	802.2[b]
N II	$4p_{1/2}$	680.2[b]
N III	$4p_{3/2}$	576.6[b]
N IV	$4d_{3/2}$	378.2[b]
N V	$4d_{5/2}$	358.8[b]
N VI	$4f_{5/2}$	104.0[b]
N VII	$4f_{7/2}$	99.9[b]
O I	5s	127[b]
O II	$5p_{3/2}$	83.1[b]
O III	$5p_{3/2}$	64.5[b]
O IV	$5d_{3/2}$	9.6[b]
O V	$5d_{5/2}$	7.8[b]

Molybdenum (42)

K	1s	20000
L I	2s	2866
L II	$2p_{1/2}$	2625
L III	$2p_{3/2}$	2520
M I	3s	506.3[b]
M II	$3p_{1/2}$	411.6[b]
M III	$3p_{3/2}$	394.0[b]
M IV	$3d_{3/2}$	231.1[b]
M V	$3d_{5/2}$	227.9[b]
N I	4s	63.2[b]
N II	$4p_{1/2}$	37.6[b]
N III	$4p_{3/2}$	35.5[b]

Neodymium (60)

K	1s	43569
L I	2s	7126
L II	$2p_{1/2}$	6722
L III	$2p_{3/2}$	6208
M I	3s	1575
M II	$3p_{1/2}$	1403
M III	$3p_{3/2}$	1297
M IV	$3d_{3/2}$	1003.3[a]
M V	$3d_{5/2}$	980.4[a]
N I	4s	319.2[a]
N II	$4p_{1/2}$	243.3
N III	$4p_{3/2}$	224.6
N IV	$4d_{3/2}$	120.5[a]

N V	$4d_{5/2}$	120.5[a]
N VI	$4f_{5/2}$	1.5
N VII	$4f_{7/2}$	1.5
O I	$5s$	37.5
O II	$5p_{1/2}$	21.1
O III	$5p_{3/2}$	21.1

Neon (10)

K	$1s$	870.2[a]
L I	$2s$	48.5[a]
L II	$2p_{1/2}$	21.7[a]
L III	$2p_{3/2}$	21.6[a]

Nickel (28)

K	$1s$	8333
L I	$2s$	1008.6[b]
L II	$2p_{1/2}$	870.0[b]
L III	$2p_{3/2}$	852.7[b]
M I	$3s$	110.8[b]
M II	$3p_{1/2}$	68.0[b]
M III	$3p_{3/2}$	66.2[b]

Niobium (41)

K	$1s$	18986
L I	$2s$	2698
L II	$2p_{1/2}$	2465
L III	$2p_{3/2}$	2371
M I	$3s$	466.6[b]
M II	$3p_{1/2}$	376.1[b]
M III	$3p_{3/2}$	360.6[b]
M IV	$3d_{3/2}$	205.0[b]
M V	$3d_{5/2}$	202.3[b]
N I	$4s$	56.4[b]
N II	$4p_{1/2}$	32.6[b]
N III	$4p_{3/2}$	30.8[b]

Nitrogen (7)

K	$1s$	409.9[a]
L I	$2s$	37.3[a]

Osmium (76)

K	$1s$	73871
L I	$2s$	12968
L II	$2p_{1/2}$	12385
L III	$2p_{3/2}$	10871
M I	$3s$	3049
M II	$3p_{1/2}$	2792
M III	$3p_{3/2}$	2457
M IV	$3d_{3/2}$	2031
M V	$3d_{5/2}$	1960
N I	$4s$	658.2[b]
N II	$4p_{1/2}$	549.1[b]
N III	$4p_{3/2}$	470.7[b]
N IV	$4d_{3/2}$	293.1[b]
N V	$4d_{5/2}$	278.5[b]
N VI	$4f_{5/2}$	53.4[b]
N VII	$4f_{7/2}$	50.7[b]
O I	$5s$	84[a]
O II	$5p_{1/2}$	58[a]

O III	$5p_{3/2}$	44.5[b]

Oxygen (8)

K	$1s$	543.1[a]
L I	$2s$	41.6[a]

Palladium (46)

K	$1s$	24350
L I	$2s$	3604
L II	$2p_{1/2}$	3330
L III	$2p_{3/2}$	3173
M I	$3s$	671.6[b]
M II	$3p_{1/2}$	559.9[b]
M III	$3p_{3/2}$	532.3[b]
M IV	$3d_{3/2}$	340.5[b]
M V	$3d_{5/2}$	335.2[b]
N I	$4s$	87.1[a,d]
N II	$4p_{1/2}$	55.7[b,c]
N III	$4p_{3/2}$	50.9[b,c]

Phosphorus (15)

K	$1s$	2145.5
L I	$2s$	189[a]
L II	$2p_{1/2}$	136[a]
L III	$2p_{3/2}$	135[a]

Platinum (78)

K	$1s$	78395
L I	$2s$	13880
L II	$2p_{1/2}$	13273
L III	$2p_{3/2}$	11564
M I	$3s$	3296
M II	$3p_{1/2}$	3027
M III	$3p_{3/2}$	2645
M IV	$3d_{3/2}$	2202
M V	$3d_{5/2}$	2122
N I	$4s$	725.4[b]
N II	$4p_{1/2}$	609.1[b]
N III	$4p_{3/2}$	519.4[b]
N IV	$4d_{3/2}$	331.6[b]
N V	$4d_{5/2}$	314.6[b]
N VI	$4f_{5/2}$	74.5[b]
N VII	$4f_{7/2}$	71.2[b]
O I	$5s$	101.7[a,d]
O II	$5p_{1/2}$	65.3[a,b]
O III	$5p_{3/2}$	51.7[b]

Polonium (84)

K	$1s$	93105
L I	$2s$	16939
L II	$2p_{1/2}$	16244
L III	$2p_{3/2}$	13814
M I	$3s$	4149
M II	$3p_{1/2}$	3854
M III	$3p_{3/2}$	3302
M IV	$3d_{3/2}$	2798
M V	$3d_{5/2}$	2683
N I	$4s$	995[a]

N II	$4p_{1/2}$	851[a]
N III	$4p_{3/2}$	705[a]
N IV	$4d_{3/2}$	500[a]
N V	$4d_{5/2}$	473[a]
N VI	$4f_{5/2}$	184[a]
N VII	$4f_{7/2}$	184[a]
O I	$5s$	177[a]
O II	$5p_{1/2}$	132[a]
O III	$5p_{3/2}$	104[a]
O IV	$5d_{3/2}$	31[a]
O V	$5d_{5/2}$	31[a]

Potassium (19)

K	$1s$	3608.4[a]
L I	$2s$	378.6[a]
L II	$2p_{1/2}$	297.3[a]
L III	$2p_{3/2}$	294.6[a]
M I	$3s$	34.8[a]
M II	$3p_{1/2}$	18.3[a]
M III	$3p_{3/2}$	18.3[a]

Praseodymium (59)

K	$1s$	41991
L I	$2s$	6835
L II	$2p_{1/2}$	6440
L III	$2p_{3/2}$	5964
M I	$3s$	1511
M II	$3p_{1/2}$	1337
M III	$3p_{3/2}$	1242
M IV	$3d_{3/2}$	948.3[a]
M V	$3d_{5/2}$	928.8[a]
N I	$4s$	304.5
N II	$4p_{1/2}$	236.3
N III	$4p_{3/2}$	217.6
N IV	$4d_{3/2}$	115.1[a]
N V	$4d_{5/2}$	115.1[a]
N VI	$4f_{5/2}$	2.0
N VII	$4f_{7/2}$	2.0
O I	$5s$	37.4
O II	$5p_{1/2}$	22.3
O III	$5p_{3/2}$	22.3

Promethium (61)

K	$1s$	45184
L I	$2s$	7428
L II	$2p_{1/2}$	7013
L III	$2p_{3/2}$	6459
M I	$3s$	—
M II	$3p_{1/2}$	1471.4
M III	$3p_{3/2}$	1357
M IV	$3d_{3/2}$	1052
M V	$3d_{5/2}$	1027
N I	$4s$	—
N II	$4p_{1/2}$	242
N III	$4p_{3/2}$	242
N IV	$4d_{3/2}$	120
N V	$4d_{5/2}$	120

Protactinium (91)

K	$1s$	112601
L I	$2s$	21105
L II	$2p_{1/2}$	20314
L III	$2p_{3/2}$	16733
M I	$3s$	5367
M II	$3p_{1/2}$	5001
M III	$3p_{3/2}$	4174
M IV	$3d_{3/2}$	3611
M V	$3d_{5/2}$	3442
N I	$4s$	1387[a]
N II	$4p_{1/2}$	1224[a]
N III	$4p_{3/2}$	1007[a]
N IV	$4d_{3/2}$	743[a]
N V	$4d_{5/2}$	708[a]
N VI	$4f_{5/2}$	371[a]
N VII	$4f_{7/2}$	360[a]
O I	$5s$	310[a]
O II	$5p_{1/2}$	232[a]
O III	$5p_{3/2}$	232[a]
O IV	$5d_{3/2}$	94[a]
O V	$5d_{5/2}$	94[a]
P I	$6s$	—
P II	$6p_{1/2}$	—
P III	$6p_{3/2}$	—

Radium (88)

K	$1s$	103922
L I	$2s$	19237
L II	$2p_{1/2}$	18484
L III	$2p_{3/2}$	15444
M I	$3s$	4822
M II	$3p_{1/2}$	4490
M III	$3p_{3/2}$	3792
M IV	$3d_{3/2}$	3248
M V	$3d_{5/2}$	3105
N I	$4s$	1208[a]
N II	$4p_{1/2}$	1058
N III	$4p_{3/2}$	879[a]
N IV	$4d_{3/2}$	636[a]
N V	$4d_{5/2}$	603[a]
N VI	$4f_{5/2}$	299[a]
N VII	$4f_{7/2}$	299[a]
O I	$5s$	254[a]
O II	$5p_{1/2}$	200[a]
O III	$5p_{3/2}$	153[a]
O IV	$5d_{3/2}$	68[a]
O V	$5d_{5/2}$	68[a]
P I	$6s$	44
P II	$6p_{1/2}$	19
P III	$6p_{3/2}$	19

Radon (86)

K	$1s$	98404
L I	$2s$	18049
L II	$2p_{1/2}$	17337
L III	$2p_{3/2}$	14619
M I	$3s$	4482
M II	$3p_{1/2}$	4159
M III	$3p_{3/2}$	3538
M IV	$3d_{3/2}$	3022
M V	$3d_{5/2}$	2892

N I	4s	1097[a]
N II	4p$_{1/2}$	929[a]
N III	4p$_{3/2}$	768[a]
N IV	4d$_{3/2}$	567[a]
N V	4d$_{5/2}$	541[a]
N VI	4f$_{5/2}$	238[a]
N VII	4f$_{7/2}$	238[a]
O I	5s	214[a]
O II	5p$_{1/2}$	164[a]
O III	5p$_{3/2}$	127[a]
O IV	5d$_{3/2}$	48[a]
O V	5d$_{5/2}$	48[a]
P I	6s	26

Rhenium (75)

K	1s	71676
L I	2s	12527
L II	2p$_{1/2}$	11959
L III	2p$_{3/2}$	10535
M I	3s	2932
M II	3p$_{1/2}$	2682
M III	3p$_{3/2}$	2367
M IV	3d$_{3/2}$	1949
M V	3d$_{5/2}$	1883
N I	4s	625.4[b]
N II	4p$_{1/2}$	518.7[b]
N III	4p$_{3/2}$	446.8[b]
N IV	4d$_{3/2}$	273.9[b]
N V	4d$_{5/2}$	260.5[b]
N VI	4f$_{5/2}$	42.9[a]
N VII	4f$_{7/2}$	40.5[a]
O I	5s	83[b]
O II	5p$_{1/2}$	45.6[b]
O III	5p$_{3/2}$	34.6[a,d]

Rhodium (45)

K	1s	23220
L I	2s	3412
L II	2p$_{1/2}$	3146
L III	2p$_{3/2}$	3004
M I	3s	628.1[b]
M II	3p$_{1/2}$	521.3[b]
M III	3p$_{3/2}$	496.5[b]
M IV	3d$_{3/2}$	311.9[b]
M V	3d$_{5/2}$	307.2[b]
N I	4s	81.4[a,d]
N II	4p$_{1/2}$	50.5[b]
N III	4p$_{3/2}$	47.3[b]

Rubidium (37)

K	1s	15200
L I	2s	2065
L II	2p$_{1/2}$	1864
L III	2p$_{3/2}$	1804
M I	3s	326.7[a]
M II	3p$_{1/2}$	248.7[a]
M III	3p$_{3/2}$	239.1[a]
M IV	3d$_{3/2}$	113.0[a]
M V	3d$_{5/2}$	112[a]

N I	4s	30.5[a]
N II	4p$_{1/2}$	16.3[a]
N III	4p$_{3/2}$	15.3[a]

Ruthenium (44)

K	1s	22117
L I	2s	3224
L II	2p$_{1/2}$	2967
L III	2p$_{3/2}$	2838
M I	3s	586.2[b]
M II	3p$_{1/2}$	483.3[b]
M III	3p$_{3/2}$	461.5[b]
M IV	3d$_{3/2}$	284.2[b]
M V	3d$_{5/2}$	280.0[b]
N I	4s	75.0[b]
N II	4p$_{1/2}$	46.5[b]
N III	4p$_{3/2}$	43.2[b]

Samarium (62)

K	1s	46834
L I	2s	7737
L II	2p$_{1/2}$	7312
L III	2p$_{3/2}$	6716
M I	3s	1723
M II	3p$_{1/2}$	1541
M III	3p$_{3/2}$	1419.8
M IV	3d$_{3/2}$	1110.9[a]
M V	3d$_{5/2}$	1083.4[a]
N I	4s	347.2[a]
N II	4p$_{1/2}$	265.6
N III	4p$_{3/2}$	247.4
N IV	4d$_{3/2}$	129.0
N V	4d$_{5/2}$	129.0
N VI	4f$_{5/2}$	5.2
N VII	4f$_{7/2}$	5.2
O I	5s	37.4
O II	5p$_{1/2}$	21.3
O III	5p$_{3/2}$	21.3

Scandium (21)

K	1s	4492
L I	2s	498.0[a]
L II	2p$_{1/2}$	403.6[a]
L III	2p$_{3/2}$	389.7[a]
M I	3s	51.1[a]
M II	3p$_{1/2}$	28.3[a]
M III	3p$_{3/2}$	28.3[a]

Selenium (34)

K	1s	12658
L I	2s	1652.0[a,d]
L II	2p$_{1/2}$	1474.3[a,d]
L III	2p$_{3/2}$	1433.9[a,d]
M I	3s	229.6[a]
M II	3p$_{1/2}$	166.5[a]
M III	3p$_{3/2}$	160.7[a]
M IV	3d$_{3/2}$	55.5[a]
M V	3d$_{5/2}$	54.6[a]

Silicon (14)

K	1s	1839
L I	2s	149.7[a,d]
L II	2p$_{1/2}$	99.8[a]
L III	2p$_{3/2}$	99.2[a]

Silver (47)

K	1s	25514
L I	2s	3806
L II	2p$_{1/2}$	3524
L III	2p$_{3/2}$	3351
M I	3s	719.0[b]
M II	3p$_{1/2}$	603.8[b]
M III	3p$_{3/2}$	573.0[b]
M IV	3d$_{3/2}$	374.0[b]
M V	3d$_{5/2}$	368.0[b]
N I	4s	97.0[b]
N II	4p$_{1/2}$	63.7[b]
N III	4p$_{3/2}$	58.3[b]

Sodium (11)

K	1s	1070.8[b]
L I	2s	63.5[b]
L II	2p$_{1/2}$	30.4[b]
L III	2p$_{3/2}$	30.5[a]

Strontium (38)

K	1s	16105
L I	2s	2216
L II	2p$_{1/2}$	2007
L III	2p$_{3/2}$	1940
M I	3s	358.7[b]
M II	3p$_{1/2}$	280.3[b]
M III	3p$_{3/2}$	270.0[b]
M IV	3d$_{3/2}$	136.0[b]
M V	3d$_{5/2}$	134.2[b]
N I	4s	38.9[b]
N II	4p$_{1/2}$	21.6[b]
N III	4p$_{3/2}$	20.1[b]

Sulfur (16)

K	1s	2472
L I	2s	230.9[a,d]
L II	2p$_{1/2}$	163.6[a]
L III	2p$_{3/2}$	162.5[a]

Tantalum (73)

K	1s	67416
L I	2s	11682
L II	2p$_{1/2}$	11136
L III	2p$_{3/2}$	9881
M I	3s	2708
M II	3p$_{1/2}$	2469
M III	3p$_{3/2}$	2194
M IV	3d$_{3/2}$	1793
M V	3d$_{5/2}$	1735
N I	4s	563.4[b]
N II	4p$_{1/2}$	463.4[b]

N III	4p$_{3/2}$	400.9[b]
N IV	4d$_{3/2}$	237.9[b]
N V	4d$_{5/2}$	226.4[b]
N VI	4f$_{5/2}$	23.5[b]
N VII	4f$_{7/2}$	21.6[b]
O I	5s	69.7[b]
O II	5p$_{1/2}$	42.2[a]
O III	5p$_{3/2}$	32.7[b]

Technetium (43)

K	1s	21044
L I	2s	3043
L II	2p$_{1/2}$	2793
L III	2p$_{3/2}$	2677
M I	3s	586.1[a]
M II	3p$_{1/2}$	447.6[a]
M III	3p$_{3/2}$	417.7[a]
M IV	3d$_{3/2}$	257.6[a]
M V	3d$_{5/2}$	253.9[a]
N I	4s	69.5[a]
N II	4p$_{1/2}$	42.3[a]
N III	4p$_{3/2}$	39.9[a]

Tellurium (52)

K	1s	31814
L I	2s	4939
L II	2p$_{1/2}$	4612
L III	2p$_{3/2}$	4341
M I	3s	1006[b]
M II	3p$_{1/2}$	870.8[b]
M III	3p$_{3/2}$	820.0[b]
M IV	3d$_{3/2}$	583.4[b]
M V	3d$_{5/2}$	573.0[b]
N I	4s	169.4[b]
N II	4p$_{1/2}$	103.3[b,c]
N III	4p$_{3/2}$	103.3[b,c]
N IV	4d$_{3/2}$	41.9[b]
N V	4d$_{5/2}$	40.4[b]

Terbium (65)

K	1s	51996
L I	2s	8708
L II	2p$_{1/2}$	8252
L III	2p$_{3/2}$	7514
M I	3s	1968
M II	3p$_{1/2}$	1768
M III	3p$_{3/2}$	1611
M IV	3d$_{3/2}$	1267.9[a]
M V	3d$_{5/2}$	1241.1[a]
N I	4s	396.0[a]
N II	4p$_{1/2}$	322.4[a]
N III	4p$_{3/2}$	284.1[a]
N IV	4d$_{3/2}$	150.5[a]
N V	4d$_{5/2}$	150.5[a]
N VI	4f$_{5/2}$	7.7[a]
N VII	4f$_{7/2}$	2.4[a]
O I	5s	45.6[a]
O II	5p$_{1/2}$	28.7[a]
O III	5p$_{3/2}$	22.6[a]

Thallium (81)

K	1s	85530
L I	2s	15347
L II	2p1/2	14698
L III	2p3/2	12658
M I	3s	3704
M II	3p1/2	3416
M III	3p3/2	2957
M IV	3d3/2	2485
M V	3d5/2	2389
N I	4s	846.2[b]
N II	4p1/2	720.5[b]
N III	4p3/2	609.5[b]
N IV	4d3/2	405.7[b]
N V	4d5/2	385.0[b]
N VI	4f5/2	122.2[b]
N VII	4f7/2	117.8[b]
O I	5s	136[a,d]
O II	5p1/2	94.6[b]
O III	5p3/2	73.5[b]
O IV	5d3/2	14.7[b]
O V	5d5/2	12.5[b]

Thorium (90)

K	1s	109651
L I	2s	20472
L II	2p1/2	19693
L III	2p3/2	16300
M I	3s	5182
M II	3p1/2	4830
M III	3p3/2	4046
M IV	3d3/2	3491
M V	3d5/2	3332
N I	4s	1330[a]
N II	4p1/2	1168[a]
N III	4p3/2	966.4[b]
N IV	4d3/2	712.1[b]
N V	4d5/2	675.2[b]
N VI	4f5/2	342.4[b]
N VII	4f7/2	333.1[b]
O I	5s	290[a,c]
O II	5p1/2	229[a,c]
O III	5p3/2	182[a,c]
O IV	5d3/2	92.5[b]
O V	5d5/2	85.4[b]
P I	6s	41.4[b]
P II	6p1/2	24.5[b]
P III	6p3/2	16.6[b]

Thulium (69)

K	1s	59390
L I	2s	10116
L II	2p1/2	9617
L III	2p3/2	8648
M I	3s	2307
M II	3p1/2	2090
M III	3p3/2	1885
M IV	3d3/2	1515
M V	3d5/2	1468
N I	4s	470.9[a]
N II	4p1/2	385.9[a]
N III	4p3/2	332.6[a]
N IV	4d3/2	175.5[a]
N V	4d5/2	175.5[a]
N VI	4f5/2	—
N VII	4f7/2	4.6
O I	5s	54.7[a]
O II	5p1/2	31.8[a]
O III	5p3/2	25.0[a]

Tin (50)

K	1s	29200
L I	2s	4465
L II	2p1/2	4156
L III	2p3/2	3929
M I	3s	884.7[b]
M II	3p1/2	756.5[b]
M III	3p3/2	714.6[b]
M IV	3d3/2	493.2[b]
M V	3d5/2	484.9[b]
N I	4s	137.1[b]
N II	4p1/2	83.6[b,c]
N III	4p3/2	83.6[b,c]
N IV	4d3/2	24.9[b]
N V	4d5/2	23.9[b]

Titanium (22)

K	1s	4966
L I	2s	560.9[b]
L II	2p1/2	460.2[b]
L III	2p3/2	453.8[b]
M I	3s	58.7[b]
M II	3p1/2	32.6[b]
M III	3p3/2	32.6[b]

Tungsten (74)

K	1s	69525
L I	2s	12100
L II	2p1/2	11544
L III	2p3/2	10207
M I	3s	2820
M II	3p1/2	2575
M III	3p3/2	2281
M IV	3d3/2	1949
M V	3d5/2	1809
N I	4s	594.1[b]
N II	4p1/2	490.4[b]
N III	4p3/2	423.6[b]
N IV	4d3/2	255.9[b]
N V	4d5/2	243.5[b]
N VI	4f5/2	33.6[a]
N VII	4f7/2	31.4[b]
O I	5s	75.6[b]
O II	5p1/2	453[a,d]
O III	5p3/2	36.8[b]

Uranium (92)

K	1s	115606
L I	2s	21757
L II	2p1/2	20948
L III	2p3/2	17166
M I	3s	5548
M II	3p1/2	5182
M III	3p3/2	4303
M IV	3d3/2	3728
M V	3d5/2	3552
N I	4s	1439[a,d]
N II	4p1/2	1271[a,d]
N III	4p3/2	1043[b]
N IV	4d3/2	778.3[b]
N V	4d5/2	736.2[b]
N VI	4f5/2	388.2[a]
N VII	4f7/2	377.4[b]
O I	5s	321[a,c,d]
O II	5p1/2	257[a,c,d]
O III	5p3/2	192[a,c,d]
O IV	5d3/2	102.8[b]
O V	5d5/2	94.2[b]
P I	6s	43.9[b]
P II	6p1/2	26.8[b]
P III	6p3/2	16.8[b]

Vanadium (23)

K	1s	5465
L I	2s	626.7[b]
L II	2p1/2	519.8[b]
L III	2p3/2	521.1[b]
M I	3s	66.3[b]
M II	3p1/2	37.2[b]
M III	3p3/2	37.2[b]

Xenon (54)

K	1s	34561
L I	2s	5453
L II	2p1/2	5107
L III	2p3/2	4786
M I	3s	1148.7[a]
M II	3p1/2	1002.1[a]
M III	3p3/2	940.6[a]
M IV	3d3/2	689.0[a]
M V	3d5/2	676.4[a]
N I	4s	213.2[a]
N II	4p1/2	146.7
N III	4p3/2	145.5[a]
N IV	4d3/2	69.5[a]
N V	4d5/2	67.5[a]
N VI	4f5/2	—
N VII	4f7/2	—
O I	5s	23.3[a]
O II	5p1/2	13.4[a]
O III	5p3/2	12.1[a]

Ytterbium (70)

K	1s	61332
L I	2s	10486
L II	2p1/2	9978
L III	2p3/2	8944
M I	3s	2398
M II	3p1/2	2173
M III	3p3/2	1950
M IV	3d3/2	1576
M V	3d5/2	1528
N I	4s	480.5[a]
N II	4p1/2	388.7[a]
N III	4p3/2	339.7[a]
N IV	4d3/2	191.2[a]
N V	4d5/2	182.4[a]
N VI	4f5/2	2.5[a]
N VII	4f7/2	1.3[a]
O I	5s	52.0[a]
O II	5p1/2	30.3[a]
O III	5p3/2	24.1[a]

Yttrium (39)

K	1s	17038
L I	2s	2373
L II	2p1/2	2156
L III	2p3/2	2080
M I	3s	392.0[a,d]
M II	3p1/2	310.6[a]
M III	3p3/2	298.8[a]
M IV	3d3/2	157.7[b]
M V	3d5/2	155.8[b]
N I	4s	43.8[a]
N II	4p1/2	24.4[a]
N III	4p3/2	23.1[a]

Zinc (30)

K	1s	9659
L I	2s	1196.2[a]
L II	2p1/2	1044.9[a]
L III	2p3/2	1021.8[a]
M I	3s	139.8[a]
M II	3p1/2	91.4[a]
M III	3p3/2	88.6[a]
M IV	3d3/2	10.2[a]
M V	3d5/2	10.1[a]

Zirconium (40)

K	1s	17998
L I	2s	2532
L II	2p1/2	2307
L III	2p3/2	2223
M I	3s	430.3[b]
M II	3p1/2	343.5[b]
M III	3p3/2	329.8[b]
M IV	3d3/2	181.1[b]
M V	3d5/2	178.8[b]
N I	4s	50.6[b]
N II	4p1/2	28.5[b]
N III	4p3/2	27.1[b]

[a] Reference 1.
[b] Reference 2 (remaining values from Reference 3).
[c] One-particle approximation not valid.
[d] Derived using energy differences from Reference 3.

NATURAL WIDTH OF X-RAY LINES

Natural widths of K X-ray lines in eV:

Element	Kα₁	Kα₂	Kβ₁	Kβ₃	Element	Kα₁	Kα₂	Kβ₁	Kβ₃
Ca	1.00	0.98			Ce	18.60	19.50	20.60	18.60
Ti	1.45	2.13			Nd	21.50	21.50	23.25	21.33
Cr	2.05	2.64			Sm	26.00	24.70	25.65	24.65
Fe	2.45	3.20			Gd	29.50	28.00	29.37	28.00
Ni	3.00	3.70			Dy	33.90	32.20	32.73	32.00
Zn	3.40	3.96			Er	35.00	35.50	36.20	35.70
Ge	3.75	4.18			Yb	38.80	40.60	41.43	41.15
Se	4.10	4.43			Hf	42.70	44.30	46.00	46.10
Kr	4.23	4.62			W	46.80	48.00	51.83	51.50
Sr	5.17	4.97			Os	49.00	49.40	55.90	55.95
Zr	5.70	5.25			Pt	54.10	54.30	59.98	62.13
Mo	6.82	6.80			Hg	64.75	68.20	65.75	68.95
Ru	7.41	7.96			Pb	67.10	72.30	72.20	73.80
Pd	8.80	9.20			Po	73.20	75.10	78.60	80.10
Cd	9.80	10.40			Rn	80.00	81.50	85.50	86.50
Sn	11.20	12.40	11.80	11.00	Ra	87.00	88.20	94.20	95.50
Te	12.80	14.20	13.30	13.10	Th	94.70	95.00	99.70	101.00
Xe	14.20	15.10	15.30	14.50	U	103.00	104.30	105.00	107.30
Ba	16.10	16.80	18.15	16.70					

From Salem, S. I. and Lee, P. L., *At. Data Nucl. Data Tables,* 18, 233, 1976.

Natural widths of L X-ray lines in eV:

Element	Lα₁	Lα₂	Lβ₁	Lβ₂	Lβ₃	Lβ₄	Lγ₁
Zr	1.68	1.52	1.87	5.13	5.50	5.60	3.34
Mo	1.86	1.80	2.03	5.30	5.90	5.78	3.76
Ru	2.03	1.98	2.18	5.45	6.35	5.96	4.15
Pd	2.21	2.16	2.36	5.63	6.80	6.18	4.50
Gd	2.43	2.40	2.54	5.82	7.23	6.28	4.83
Sn	2.62	2.62	2.75	6.10	7.70	6.60	5.23
Tc	2.88	2.88	2.96	6.25	8.22	6.82	5.60
Xe	3.15	3.15	3.20	6.43	8.70	7.15	5.95
Ba	3.39	3.45	3.45	6.70	9.20	7.42	6.35
Ce	3.70	3.78	3.73	6.86	9.70	7.82	6.75
Nd	3.93	4.08	4.00	7.18	10.30	8.15	7.16
Sm	4.13	4.50	4.33	7.42	10.80	8.60	7.50
Cd	4.46	4.90	4.63	7.70	11.20	9.08	7.83
Dy	4.81	5.35	5.03	7.90	11.50	9.60	8.30
Er	5.17	5.73	5.45	8.28	11.85	10.03	8.75
Yb	5.40	6.22	5.90	8.58	12.20	11.00	9.20
Hf	5.83	6.70	6.36	8.92	12.40	12.80	9.63
W	6.50	7.20	6.90	9.06	13.10	14.60	10.20
Os	7.04	7.70	7.42	9.60	14.60	16.50	10.65
Pt	7.60	8.28	8.00	9.95	16.10	18.00	11.20
Hg	8.10	8.80	8.70	10.40	17.40	19.70	11.80
Pb	8.82	9.35	9.35	10.75	18.65	21.30	12.30
Po	9.50	9.95	10.10	11.25	19.90	22.70	13.05
Rn	10.03	10.50	10.65	11.65	21.00	24.00	13.55
Ra	11.00	11.20	11.60	12.20	22.00	25.20	14.30
Th	11.90	11.80	12.40	12.80	22.85	26.35	15.00
U	12.40	12.40	13.50	13.30	23.70	27.50	15.70
Pu	13.20	13.00	14.10	13.90	24.10	28.30	16.40
Cm	14.80	13.60	15.70	14.60	25.00	29.40	17.10

PHOTON ATTENUATION COEFFICIENTS

Martin J. Berger and John H. Hubbell

This table gives mass attenuation coefficients for photons for all elements at energies between 1 keV (soft x-rays) and 1 GeV (hard gamma rays). The mass attenuation coefficient μ describes the attenuation of radiation as it passes through matter by the relation

$$I(x)/I_0 = e^{-\mu\rho x}$$

where I_0 is the initial intensity, $I(x)$ the intensity after path length x, and ρ is the mass density of the element in question. To a high approximation the mass attenuation coefficient is additive for the elements present, independent of the way in which they are bound in chemical compounds.

The power of ten is indicated beside each number in the table; i.e., $7.41 + 03$ means 7.41×10^3. A vertical line between two columns indicates that an absorption edge lies between those energy values. The various edges are labeled at the bottom of the table.

The attenuation coefficients were calculated with the computer program XCOM (Reference 1), which uses a cross-section database compiled at the Photon and Charged Particle Data Center at the National Institute of Standards and Technology. Their accuracy has been confirmed at all energies by extensive comparisons with experimental attenuation coefficients. Such comparisons for X-ray energies up to 100 keV can be found in Reference 2.

REFERENCES

1. Berger, M. J. and Hubbell, J. H., National Bureau of Standards Report NBSIR-87-3597, 1987.
2. Saloman, E. B., Hubbell, J. H., and Scofield, J. H., *Atomic Data and Nuclear Data Tables*, 38, 1, 1988.

Mass attenuation coefficient, cm²/g

	Atomic no.	Photon energy, MeV								
		0.001	0.002	0.005	0.01	0.02	0.05	0.1	0.2	0.5
H	1	7.21 + 00	1.06 + 00	4.19-01	3.85-01	3.69-01	3.36-01	2.94-01	2.43-01	1.73-01
He	2	6.08 + 01	6.86 + 00	5.77-01	2.48-01	1.96-01	1.70-01	1.49-01	1.22-01	8.71-02
Li	3	2.34 + 02	2.71 + 01	1.62 + 00	3.40-01	1.86-01	1.49-01	1.29-01	1.06-01	7.53-02
Be	4	6.04 + 02	7.47 + 01	4.37 + 00	6.47-01	2.25-01	1.55-01	1.33-01	1.09-01	7.74-02
B	5	1.23 + 03	1.60 + 02	9.68 + 00	1.25 + 00	3.01-01	1.66-01	1.39-01	1.14-01	8.07-02
C	6	2.21 + 03	3.03 + 02	1.91 + 01	2.37 + 00	4.42-01	1.87-01	1.51-01	1.23-01	8.72-02
N	7	3.31 + 03	4.77 + 02	3.14 + 01	3.88 + 00	6.18-01	1.98-01	1.53-01	1.23-01	8.72-02
O	8	4.59 + 03	6.95 + 02	4.79 + 01	5.95 + 00	8.65-01	2.13-01	1.55-01	1.24-01	8.73-02
F	9	5.65 + 03	9.05 + 02	6.51 + 01	8.21 + 00	1.13 + 00	2.21-01	1.50-01	1.18-01	8.27-02
Ne	10	7.41 + 03	1.24 + 03	9.34 + 01	1.20 + 01	1.61 + 00	2.58-01	1.60-01	1.24-01	8.66-02
Na	11	6.54 + 02	1.52 + 03	1.19 + 02	1.56 + 01	2.06 + 00	2.80-01	1.59-01	1.20-01	8.37-02
Mg	12	9.22 + 02	1.93 + 03	1.58 + 02	2.11 + 01	2.76 + 00	3.29-01	1.69-01	1.24-01	8.65-02
Al	13	1.19 + 03	2.26 + 03	1.93 + 02	2.62 + 01	3.44 + 00	3.68-01	1.70-01	1.22-01	8.44-02
Si	14	1.57 + 03	2.78 + 03	2.45 + 02	3.39 + 01	4.46 + 00	4.38-01	1.84-01	1.28-01	8.75-02
P	15	1.91 + 03	3.02 + 02	2.86 + 02	4.04 + 01	5.35 + 00	4.92-01	1.87-01	1.25-01	8.51-02
S	16	2.43 + 03	3.85 + 02	3.49 + 02	5.01 + 01	6.71 + 00	5.85-01	2.02-01	1.30-01	8.78-02
Cl	17	2.83 + 03	4.52 + 02	3.90 + 02	5.73 + 01	7.74 + 00	6.48-01	2.05-01	1.27-01	8.45-02
Ar	18	3.18 + 03	5.12 + 02	4.23 + 02	6.32 + 01	8.63 + 00	7.01-01	2.04-01	1.20-01	7.96-02
K	19	4.06 + 03	6.59 + 02	5.19 + 02	7.91 + 01	1.09 + 01	8.68-01	2.34-01	1.32-01	8.60-02
Ca	20	4.87 + 03	8.00 + 02	6.03 + 02	9.34 + 01	1.31 + 01	1.02 + 00	2.57-01	1.38-01	8.85-02
Sc	21	5.24 + 03	8.70 + 02	6.31 + 02	9.95 + 01	1.41 + 01	1.09 + 00	2.58-01	1.31-01	8.31-02
Ti	22	5.87 + 03	9.86 + 02	6.84 + 02	1.11 + 02	1.59 + 01	1.21 + 00	2.72-01	1.31-01	8.19-02
V	23	6.50 + 03	1.11 + 03	9.29 + 01	1.22 + 02	1.77 + 01	1.35 + 00	2.88-01	1.32-01	8.07-02
Cr	24	7.40 + 03	1.28 + 03	1.08 + 02	1.39 + 02	2.04 + 01	1.55 + 00	3.17-01	1.38-01	8.28-02
Mn	25	8.09 + 03	1.42 + 03	1.21 + 02	1.51 + 02	2.25 + 01	1.71 + 00	3.37-01	1.39-01	8.19-02
Fe	26	9.09 + 03	1.63 + 03	1.40 + 02	1.71 + 02	2.57 + 01	1.96 + 00	3.72-01	1.46-01	8.41-02
Co	27	9.80 + 03	1.78 + 03	1.54 + 02	1.84 + 02	2.80 + 01	2.14 + 00	3.95-01	1.48-01	8.32-02
Ni	28	9.86 + 03	2.05 + 03	1.79 + 02	2.09 + 02	3.22 + 01	2.47 + 00	4.44-01	1.58-01	8.70-02
Cu	29	1.06 + 04	2.15 + 03	1.90 + 02	2.16 + 02	3.38 + 01	2.61 + 00	4.58-01	1.56-01	8.36-02
Zn	30	1.55 + 03	2.37 + 03	2.12 + 02	2.33 + 02	3.72 + 01	2.89 + 00	4.97-01	1.62-01	8.45-02
Ga	31	1.70 + 03	2.52 + 03	2.27 + 02	3.42 + 01	3.93 + 01	3.08 + 00	5.20-01	1.62-01	8.24-02
Ge	32	1.89 + 03	2.71 + 03	2.47 + 02	3.74 + 01	4.22 + 01	3.34 + 00	5.55-01	1.66-01	8.21-02
As	33	2.12 + 03	2.93 + 03	2.71 + 02	4.12 + 01	4.56 + 01	3.63 + 00	5.97-01	1.72-01	8.26-02
Se	34	2.32 + 03	3.10 + 03	2.90 + 02	4.41 + 01	4.82 + 01	3.86 + 00	6.28-01	1.74-01	8.13-02
Br	35	2.62 + 03	3.41 + 03	3.21 + 02	4.91 + 01	5.27 + 01	4.26 + 00	6.86-01	1.84-01	8.33-02
Kr	36	2.85 + 03	3.60 + 03	3.43 + 02	5.26 + 01	5.55 + 01	4.52 + 00	7.22-01	1.87-01	8.23-02
Rb	37	3.17 + 03	3.41 + 03	3.74 + 02	5.77 + 01	5.98 + 01	4.92 + 00	7.80-01	1.96-01	8.36-02
Sr	38	3.49 + 03	2.59 + 03	4.06 + 02	6.27 + 01	6.39 + 01	5.31 + 00	8.37-01	2.04-01	8.44-02
Y	39	3.86 + 03	7.42 + 02	4.42 + 02	6.87 + 01	6.86 + 01	5.76 + 00	9.05-01	2.15-01	8.61-02
Zr	40	4.21 + 03	8.12 + 02	4.76 + 02	7.42 + 01	7.24 + 01	6.17 + 00	9.66-01	2.24-01	8.69-02
Nb	41	4.60 + 03	8.89 + 02	5.13 + 02	8.04 + 01	7.71 + 01	6.64 + 00	1.04 + 00	2.34-01	8.83-02
Mo	42	4.94 + 03	9.60 + 02	5.45 + 02	8.58 + 01	1.31 + 01	7.04 + 00	1.10 + 00	2.42-01	8.85-02
Tc	43	5.36 + 03	1.04 + 03	5.84 + 02	9.23 + 01	1.41 + 01	7.52 + 00	1.17 + 00	2.53-01	8.97-02
Ru	44	5.72 + 03	1.12 + 03	6.17 + 02	9.80 + 01	1.50 + 01	7.92 + 00	1.23 + 00	2.62-01	8.99-02
Rh	45	6.17 + 03	1.21 + 03	6.59 + 02	1.05 + 02	1.61 + 01	8.45 + 00	1.31 + 00	2.74-01	9.13-02
Pd	46	6.54 + 03	1.29 + 03	6.91 + 02	1.11 + 02	1.70 + 01	8.85 + 00	1.38 + 00	2.83-01	9.13-02
Ag	47	7.04 + 03	1.40 + 03	7.39 + 02	1.19 + 02	1.84 + 01	9.45 + 00	1.47 + 00	2.97-01	9.32-02
Cd	48	7.35 + 03	1.47 + 03	7.69 + 02	1.24 + 02	1.92 + 01	9.78 + 00	1.52 + 00	3.04-01	9.25-02
In	49	7.81 + 03	1.58 + 03	8.13 + 02	1.32 + 02	2.04 + 01	1.03 + 01	1.61 + 00	3.17-01	9.37-02
Sn	50	8.16 + 03	1.66 + 03	8.47 + 02	1.38 + 02	2.15 + 01	1.07 + 01	1.68 + 00	3.26-01	9.37-02

L₃ L₁
L₂   K EDGE

Mass attenuation coefficient, cm²/g

	Atomic no.	Photon energy, MeV								
		1.0	2.0	5.0	10.0	20.0	50.0	100.0	500.0	1000.0
H	1	1.26-01	8.77-02	5.05-02	3.25-02	2.15-02	1.42-02	1.19-02	1.14-02	1.16-02
He	2	6.36-02	4.42-02	2.58-02	1.70-02	1.18-02	8.61-03	7.78-03	7.79-03	7.95-03
Li	3	5.50-02	3.83-02	2.26-02	1.53-02	1.11-02	1.02-02	9.94-03	1.08-02	1.12-02
Be	4	5.65-02	3.94-02	2.35-02	1.63-02	1.23-02	1.19-02	1.19-02	1.32-02	1.37-02
B	5	5.89-02	4.11-02	2.48-02	1.76-02	1.37-02	1.43-02	1.46-02	1.64-02	1.70-02
C	6	6.36-02	4.44-02	2.71-02	1.96-02	1.58-02	1.43-02	1.46-02	1.64-02	1.70-02
N	7	6.36-02	4.45-02	2.74-02	2.02-02	1.67-02	1.57-02	1.63-02	1.85-02	1.92-02
O	8	6.37-02	4.46-02	2.78-02	2.09-02	1.77-02	1.71-02	1.79-02	2.06-02	2.13-02
F	9	6.04-02	4.23-02	2.66-02	2.04-02	1.77-02	1.75-02	1.86-02	2.14-02	2.21-02
Ne	10	6.32-02	4.43-02	2.82-02	2.20-02	1.95-02	1.96-02	2.11-02	2.43-02	2.51-02
Na	11	6.10-02	4.28-02	2.75-02	2.18-02	1.97-02	2.03-02	2.19-02	2.53-02	2.62-02
Mg	12	6.30-02	4.43-02	2.87-02	2.31-02	2.13-02	2.23-02	2.42-02	2.81-02	2.90-02
Al	13	6.15-02	4.32-02	2.84-02	2.32-02	2.17-02	2.31-02	2.52-02	2.93-02	3.03-02
Si	14	6.36-02	4.48-02	2.97-02	2.46-02	2.34-02	2.52-02	2.76-02	3.23-02	3.34-02
P	15	6.18-02	4.36-02	2.91-02	2.45-02	2.36-02	2.58-02	2.84-02	3.33-02	3.45-02
S	16	6.37-02	4.50-02	3.04-02	2.59-02	2.53-02	2.79-02	3.08-02	3.62-02	3.75-02
Cl	17	6.13-02	4.33-02	2.95-02	2.55-02	2.52-02	2.81-02	3.11-02	3.67-02	3.80-02
Ar	18	5.76-02	4.07-02	2.80-02	2.45-02	2.45-02	2.76-02	3.07-02	3.62-02	3.75-02
K	19	6.22-02	4.40-02	3.05-02	2.70-02	2.74-02	3.11-02	3.46-02	4.09-02	4.24-02
Ca	20	6.39-02	4.52-02	3.17-02	2.84-02	2.90-02	3.32-02	3.71-02	4.40-02	4.56-02
Sc	21	5.98-02	4.24-02	3.00-02	2.72-02	2.80-02	3.23-02	3.62-02	4.30-02	4.45-02
Ti	22	5.89-02	4.18-02	2.98-02	2.73-02	2.84-02	3.30-02	3.71-02	4.40-02	4.56-02
V	23	5.79-02	4.11-02	2.96-02	2.74-02	2.88-02	3.36-02	3.78-02	4.49-02	4.65-02
Cr	24	5.93-02	4.21-02	3.06-02	2.86-02	3.03-02	3.56-02	4.01-02	4.76-02	4.93-02
Mn	25	5.85-02	4.16-02	3.04-02	2.87-02	3.07-02	3.63-02	4.09-02	4.86-02	5.04-02
Fe	26	5.99-02	4.26-02	3.15-02	2.99-02	3.22-02	3.83-02	4.33-02	5.15-02	5.33-02
Co	27	5.91-02	4.20-02	3.13-02	3.00-02	3.26-02	3.88-02	4.40-02	5.23-02	5.41-02
Ni	28	6.16-02	4.39-02	3.29-02	3.18-02	3.48-02	4.17-02	4.73-02	5.61-02	5.81-02
Cu	29	5.90-02	4.20-02	3.18-02	3.10-02	3.41-02	4.10-02	4.66-02	5.53-02	5.72-02
Zn	30	5.94-02	4.24-02	3.22-02	3.18-02	3.51-02	4.24-02	4.82-02	5.72-02	5.91-02
Ga	31	5.77-02	4.11-02	3.16-02	3.13-02	3.48-02	4.22-02	4.80-02	5.70-02	5.89-02
Ge	32	5.73-02	4.09-02	3.16-02	3.16-02	3.53-02	4.30-02	4.89-02	5.80-02	6.00-02
As	33	5.73-02	4.09-02	3.19-02	3.21-02	3.60-02	4.40-02	5.01-02	5.95-02	6.15-02
Se	34	5.62-02	4.01-02	3.14-02	3.19-02	3.60-02	4.41-02	5.03-02	5.97-02	6.17-02
Br	35	5.73-02	4.09-02	3.23-02	3.29-02	3.74-02	4.60-02	5.24-02	6.22-02	6.43-02
Kr	36	5.63-02	4.02-02	3.20-02	3.28-02	3.74-02	4.61-02	5.26-02	6.25-02	6.46-02
Rb	37	5.69-02	4.06-02	3.25-02	3.36-02	3.85-02	4.75-02	5.43-02	6.45-02	6.67-02
Sr	38	5.71-02	4.08-02	3.29-02	3.41-02	3.93-02	4.87-02	5.56-02	6.61-02	6.83-02
Y	39	5.80-02	4.14-02	3.35-02	3.50-02	4.05-02	5.03-02	5.75-02	6.83-02	7.06-02
Zr	40	5.81-02	4.15-02	3.38-02	3.55-02	4.12-02	5.13-02	5.87-02	6.98-02	7.22-02
Nb	41	5.87-02	4.18-02	3.44-02	3.63-02	4.22-02	5.27-02	6.03-02	7.17-02	7.42-02
Mo	42	5.84-02	4.16-02	3.44-02	3.65-02	4.26-02	5.33-02	6.10-02	7.26-02	7.51-02
Tc	43	5.88-02	4.19-02	3.48-02	3.71-02	4.35-02	5.45-02	6.24-02	7.43-02	7.68-02
Ru	44	5.85-02	4.16-02	3.48-02	3.73-02	4.39-02	5.50-02	6.30-02	7.51-02	7.77-02
Rh	45	5.89-02	4.20-02	3.53-02	3.80-02	4.48-02	5.63-02	6.45-02	7.69-02	7.94-02
Pd	46	5.85-02	4.16-02	3.52-02	3.80-02	4.50-02	5.66-02	6.49-02	7.73-02	8.00-02
Ag	47	5.92-02	4.21-02	3.58-02	3.88-02	4.61-02	5.81-02	6.67-02	7.93-02	8.20-02
Cd	48	5.83-02	4.14-02	3.54-02	3.85-02	4.59-02	5.79-02	6.64-02	7.91-02	8.18-02
In	49	5.85-02	4.15-02	3.56-02	3.90-02	4.65-02	5.88-02	6.75-02	8.04-02	8.32-02
Sn	50	5.80-02	4.11-02	3.55-02	3.90-02	4.66-02	5.90-02	6.78-02	8.07-02	8.35-02

Mass attenuation coefficients, cm²/g

	Atomic no.	Photon energy, MeV								
		0.001	0.002	0.005	0.01	0.02	0.05	0.1	0.2	0.5
Sb	51	8.58 + 03	1.77 + 03	8.85 + 02	1.46 + 02	2.27 + 01	1.12 + 01	1.76 + 00	3.38-01	9.45-02
Te	52	8.43 + 03	1.83 + 03	9.01 + 02	1.50 + 02	2.34 + 01	1.14 + 01	1.80 + 00	3.43-01	9.33-02
I	53	9.10 + 03	2.00 + 03	8.43 + 02	1.63 + 02	2.54 + 01	1.23 + 01	1.94 + 00	3.66-01	9.70-02
Xe	54	9.41 + 03	2.09 + 03	6.39 + 02	1.69 + 02	2.65 + 01	1.27 + 01	2.01 + 00	3.76-01	9.70-02
Cs	55	9.37 + 03	2.23 + 03	2.30 + 02	1.79 + 02	2.82 + 01	1.34 + 01	2.12 + 00	3.94-01	9.91-02
ZB	56	8.54 + 03	2.32 + 03	2.41 + 02	1.86 + 02	2.94 + 01	1.38 + 01	2.20 + 00	4.05-01	9.92-02
La	57	9.09 + 03	2.46 + 03	2.58 + 02	1.97 + 02	3.12 + 01	1.45 + 01	2.32 + 00	4.24-01	1.01-01
Ce	58	9.71 + 03	2.61 + 03	2.74 + 02	2.08 + 02	3.31 + 01	1.52 + 01	2.45 + 00	4.45-01	1.04-01
Pr	59	1.06 + 04	2.77 + 03	2.92 + 02	2.21 + 02	3.53 + 01	1.60 + 01	2.59 + 00	4.69-01	1.07-01
Nd	60	6.63 + 03	2.88 + 03	3.06 + 02	2.30 + 02	3.68 + 01	1.65 + 01	2.69 + 00	4.84-01	1.08-01
Pm	61	2.06 + 03	3.05 + 03	3.26 + 02	2.44 + 02	3.92 + 01	1.73 + 01	2.84 + 00	5.10-01	1.12-01
Sm	62	2.11 + 03	3.12 + 03	3.36 + 02	2.50 + 02	4.03 + 01	1.77 + 01	2.90 + 00	5.19-01	1.11-01
Eu	63	2.22 + 03	3.28 + 03	3.54 + 02	2.63 + 02	4.24 + 01	1.85 + 01	3.04 + 00	5.43-01	1.14-01
Gd	64	2.29 + 03	3.36 + 03	3.65 + 02	2.69 + 02	4.36 + 01	3.86 + 00	3.11 + 00	5.54-01	1.14-01
Tb	65	2.40 + 03	3.51 + 03	3.84 + 02	2.82 + 02	4.59 + 01	4.06 + 00	3.25 + 00	5.77-01	1.17-01
Dy	66	2.49 + 03	3.47 + 03	3.99 + 02	2.90 + 02	4.76 + 01	4.23 + 00	3.36 + 00	5.95-01	1.18-01
Ho	67	2.62 + 03	3.59 + 03	4.17 + 02	3.01 + 02	4.98 + 01	4.43 + 00	3.49 + 00	6.18-01	1.20-01
Er	68	2.75 + 03	3.52 + 03	4.36 + 02	3.13 + 02	5.20 + 01	4.63 + 00	3.63 + 00	6.41-01	1.23-01
Tm	69	2.90 + 03	3.69 + 03	4.57 + 02	2.83 + 02	5.45 + 01	4.87 + 00	3.78 + 00	6.68-01	1.26-01
Yb	70	3.02 + 03	3.80 + 03	4.72 + 02	2.94 + 02	5.63 + 01	5.04 + 00	3.88 + 00	6.86-01	1.27-01
Lu	71	3.19 + 03	3.45 + 03	4.94 + 02	2.21 + 02	5.88 + 01	5.28 + 00	4.03 + 00	7.13-01	1.30-01
Hf	72	3.34 + 03	3.60 + 03	5.11 + 02	2.30 + 02	6.09 + 01	5.48 + 00	4.15 + 00	7.34-01	1.32-01
Ta	73	3.51 + 03	3.77 + 03	5.33 + 02	2.38 + 02	6.33 + 01	5.72 + 00	4.30 + 00	7.60-01	1.35-01
W	74	3.68 + 03	3.92 + 03	5.53 + 02	9.69 + 01	6.57 + 01	5.95 + 00	4.44 + 00	7.84-01	1.38-01
Re	75	3.87 + 03	3.77 + 03	5.76 + 02	1.01 + 02	6.84 + 01	6.21 + 00	4.59 + 00	8.12-01	1.41-01
Os	76	4.03 + 03	2.22 + 03	5.93 + 02	1.04 + 02	7.04 + 01	6.41 + 00	4.70 + 00	8.33-01	1.43-01
Ir	77	4.24 + 03	1.03 + 03	6.18 + 02	1.09 + 02	7.32 + 01	6.69 + 00	4.86 + 00	8.63-01	1.46-01
Pt	78	4.43 + 03	1.08 + 03	6.40 + 02	1.13 + 02	7.57 + 01	6.95 + 00	4.99 + 00	8.90-01	1.49-01
Au	79	4.65 + 03	1.14 + 03	6.66 + 02	1.18 + 02	7.88 + 01	7.26 + 00	5.16 + 00	9.22-01	1.53-01
Hg	80	4.83 + 03	1.18 + 03	6.87 + 02	1.22 + 02	8.12 + 01	7.50 + 00	5.28 + 00	9.46-01	1.56-01
Tl	81	5.01 + 03	1.23 + 03	7.07 + 02	1.26 + 02	8.36 + 01	7.75 + 00	5.40 + 00	9.69-01	1.58-01
Pb	82	5.21 + 03	1.29 + 03	7.30 + 02	1.31 + 02	8.64 + 01	8.04 + 00	5.55 + 00	9.99-01	1.61-01
Bi	83	5.44 + 03	1.35 + 03	7.58 + 02	1.36 + 02	8.95 + 01	8.38 + 00	5.74 + 00	1.03 + 00	1.66-01
Po	84	5.72 + 03	1.42 + 03	7.93 + 02	1.43 + 02	9.35 + 01	8.80 + 00	5.99 + 00	1.08 + 00	1.71-01
At	85	5.87 + 03	1.49 + 03	8.25 + 02	1.49 + 02	9.70 + 01	9.19 + 00	6.17 + 00	1.12 + 00	1.77-01
Rn	86	5.83 + 03	1.49 + 03	8.16 + 02	1.48 + 02	9.56 + 01	9.12 + 00	6.09 + 00	1.10 + 00	1.73-01
Fr	87	6.08 + 03	1.56 + 03	8.49 + 02	1.54 + 02	9.93 + 01	9.52 + 00	1.66 + 00	1.14 + 00	1.78-01
Ra	88	6.20 + 03	1.62 + 03	8.74 + 02	1.59 + 02	1.02 + 02	9.85 + 00	1.71 + 00	1.17 + 00	1.82-01
Ac	89	6.47 + 03	1.70 + 03	8.69 + 02	1.65 + 02	1.06 + 02	1.03 + 01	1.79 + 00	1.21 + 00	1.87-01
Th	90	6.61 + 03	1.74 + 03	8.88 + 02	1.69 + 02	9.37 + 01	1.05 + 01	1.83 + 00	1.23 + 00	1.90-01
Pa	91	6.53 + 03	1.83 + 03	8.76 + 02	1.77 + 02	7.03 + 01	1.10 + 01	1.92 + 00	1.29 + 00	1.97-01
U	92	6.63 + 03	1.86 + 03	8.89 + 02	1.79 + 02	7.11 + 01	1.12 + 01	1.95 + 00	1.30 + 00	1.98-01
Np	93	6.95 + 03	1.96 + 03	9.32 + 02	1.87 + 02	7.45 + 01	1.18 + 01	2.05 + 00	1.35 + 00	2.05-01
Pu	94	7.19 + 03	2.04 + 03	9.65 + 02	1.94 + 02	7.71 + 01	1.22 + 01	2.13 + 00	1.39 + 00	2.10-01
Am	95	7.37 + 03	2.10 + 03	9.90 + 02	1.98 + 02	7.93 + 01	1.25 + 01	2.19 + 00	1.42 + 00	2.14-01
Cm	96	7.54 + 03	2.15 + 03	1.02 + 03	2.03 + 02	8.14 + 01	1.28 + 01	2.25 + 00	1.44 + 00	2.18-01
Bk	97	7.84 + 03	2.25 + 03	1.06 + 03	2.10 + 02	8.39 + 01	1.34 + 01	2.35 + 00	1.50 + 00	2.25-01
Cf	98	7.89 + 03	2.31 + 03	9.27 + 02	2.15 + 02	8.58 + 01	1.37 + 01	2.41 + 00	1.52 + 00	2.29-01
Es	99	7.79 + 03	2.40 + 03	9.59 + 02	2.22 + 02	4.01 + 01	1.42 + 01	2.51 + 00	1.57 + 00	2.36-01
Fm	100	7.13 + 03	2.46 + 03	9.77 + 02	2.26 + 02	4.09 + 01	1.45 + 01	2.57 + 00	1.59 + 00	2.39-01

N_5 M_5 M_3 L_3

N_4 M_4 M_2 L_2

N_3 M_1 L_1

N_2

N_1

K EDGE

Mass attenuation coefficients, cm^2/g

	Atomic no.	Photon Energy, MeV								
		1.0	2.0	5.0	10.0	20.0	50.0	100.0	500.0	1000.0
Sb	51	5.80-02	4.10-02	3.56-02	3.92-02	4.70-02	5.96-02	6.85-02	8.16-02	8.44-02
Te	52	5.67-02	4.01-02	3.49-02	3.86-02	4.64-02	5.89-02	6.77-02	8.07-02	8.35-02
I	53	5.84-02	4.12-02	3.61-02	4.00-02	4.82-02	6.13-02	7.04-02	8.40-02	8.69-02
Xe	54	5.78-02	4.08-02	3.58-02	3.99-02	4.82-02	6.12-02	7.04-02	8.40-02	8.69-02
Cs	55	5.85-02	4.12-02	3.64-02	4.06-02	4.91-02	6.25-02	7.19-02	8.58-02	8.88-02
ZB	56	5.80-02	4.08-02	3.61-02	4.04-02	4.90-02	6.25-02	7.19-02	8.58-02	8.88-02
La	57	5.88-02	4.12-02	3.66-02	4.11-02	5.00-02	6.37-02	7.34-02	8.76-02	9.06-02
Ce	58	5.96-02	4.18-02	3.73-02	4.19-02	5.10-02	6.52-02	7.50-02	8.96-02	9.27-02
Pr	59	6.07-02	4.24-02	3.80-02	4.29-02	5.23-02	6.68-02	7.69-02	9.19-02	9.50-02
Nd	60	6.07-02	4.24-02	3.81-02	4.30-02	5.26-02	6.72-02	7.74-02	9.25-02	9.56-02
Pm	61	6.19-02	4.31-02	3.88-02	4.40-02	5.38-02	6.89-02	7.94-02	9.48-02	9.81-02
Sm	62	6.11-02	4.24-02	3.83-02	4.35-02	5.34-02	6.84-02	7.88-02	9.41-02	9.73-02
Eu	63	6.19-02	4.28-02	3.88-02	4.42-02	5.42-02	6.96-02	8.02-02	9.57-02	9.90-02
Gd	64	6.12-02	4.23-02	3.84-02	4.38-02	5.38-02	6.91-02	7.97-02	9.51-02	9.83-02
Tb	65	6.20-02	4.27-02	3.89-02	4.45-02	5.47-02	7.03-02	8.11-02	9.67-02	1.00-01
Dy	66	6.20-02	4.26-02	3.90-02	4.46-02	5.49-02	7.06-02	8.15-02	9.72-02	1.00-01
Ho	67	6.26-02	4.29-02	3.93-02	4.50-02	5.55-02	7.14-02	8.24-02	9.83-02	1.02-01
Er	68	6.32-02	4.32-02	3.96-02	4.55-02	5.61-02	7.23-02	8.34-02	9.95-02	1.03-01
Tm	69	6.40-02	4.36-02	4.01-02	4.61-02	5.70-02	7.35-02	8.48-02	1.01-01	1.04-01
Yb	70	6.40-02	4.35-02	4.00-02	4.61-02	5.70-02	7.35-02	8.49-02	1.01-01	1.04-01
Lu	71	6.48-02	4.39-02	4.05-02	4.66-02	5.77-02	7.45-02	8.60-02	1.02-01	1.06-01
Hf	72	6.50-02	4.39-02	4.05-02	4.68-02	5.80-02	7.48-02	8.64-02	1.03-01	1.06-01
Ta	73	6.57-02	4.41-02	4.08-02	4.72-02	5.85-02	7.56-02	8.73-02	1.04-01	1.07-01
W	74	6.62-02	4.43-02	4.10-02	4.75-02	5.89-02	7.62-02	8.80-02	1.05-01	1.08-01
Re	75	6.69-02	4.46-02	4.14-02	4.79-02	5.95-02	7.70-02	8.89-02	1.06-01	1.09-01
Os	76	6.71-02	4.46-02	4.13-02	4.79-02	5.96-02	7.71-02	8.90-02	1.06-01	1.10-01
Ir	77	6.79-02	4.50-02	4.17-02	4.84-02	6.02-02	7.80-02	9.01-02	1.07-01	1.11-01
Pt	78	6.86-02	4.52-02	4.20-02	4.87-02	6.06-02	7.86-02	9.08-02	1.08-01	1.12-01
Au	79	6.95-02	4.57-02	4.24-02	4.93-02	6.14-02	7.95-02	9.19-02	1.09-01	1.13-01
Hg	80	6.99-02	4.57-02	4.25-02	4.94-02	6.15-02	7.98-02	9.22-02	1.10-01	1.13-01
Tl	81	7.03-02	4.58-02	4.25-02	4.94-02	6.16-02	8.00-02	9.24-02	1.10-01	1.14-01
Pb	82	7.10-02	4.61-02	4.27-02	4.97-02	6.21-02	8.06-02	9.31-02	1.11-01	1.15-01
Bi	83	7.21-02	4.66-02	4.32-02	5.03-02	6.28-02	8.15-02	9.42-02	1.12-01	1.16-01
Po	84	7.39-02	4.75-02	4.40-02	5.12-02	6.40-02	8.32-02	9.61-02	1.15-01	1.18-01
At	85	7.54-02	4.82-02	4.46-02	5.20-02	6.49-02	8.44-02	9.76-02	1.16-01	1.20-01
Rn	86	7.30-02	4.65-02	4.30-02	5.01-02	6.26-02	8.14-02	9.42-02	1.12-01	1.16-01
Fr	87	7.45-02	4.72-02	4.36-02	5.08-02	6.35-02	8.26-02	9.56-02	1.14-01	1.18-01
Ra	88	7.53-02	4.75-02	4.38-02	5.10-02	6.38-02	8.31-02	9.61-02	1.15-01	1.19-01
Ac	89	7.69-02	4.82-02	4.44-02	5.17-02	6.47-02	8.43-02	9.75-02	1.16-01	1.20-01
Th	90	7.71-02	4.81-02	4.42-02	5.15-02	6.45-02	8.40-02	9.72-02	1.16-01	1.20-01
Pa	91	7.94-02	4.93-02	4.52-02	5.26-02	6.59-02	8.60-02	9.95-02	1.19-01	1.23-01
U	92	7.90-02	4.88-02	4.46-02	5.19-02	6.51-02	8.49-02	9.83-02	1.17-01	1.21-01
Np	93	8.13-02	4.99-02	4.56-02	5.30-02	6.65-02	8.68-02	1.01-01	1.20-01	1.24-01
Pu	94	8.26-02	5.05-02	4.60-02	5.34-02	6.71-02	8.76-02	1.01-01	1.21-01	1.25-01
Am	95	8.33-02	5.06-02	4.60-02	5.34-02	6.70-02	8.77-02	1.02-01	1.21-01	1.25-01
Cm	96	8.41-02	5.08-02	4.60-02	5.34-02	6.70-02	8.77-02	1.02-01	1.21-01	1.26-01
Bk	97	8.62-02	5.18-02	4.68-02	5.42-02	6.81-02	8.92-02	1.03-01	1.24-01	1.28-01
Cf	98	8.70-02	5.20-02	4.68-02	5.42-02	6.81-02	8.92-02	1.04-01	1.24-01	1.28-01
Es	99	8.89-02	5.28-02	4.74-02	5.48-02	6.89-02	9.04-02	1.05-01	1.25-01	1.29-01
Fm	100	8.94-02	5.28-02	4.72-02	5.45-02	6.86-02	9.00-02	1.05-01	1.25-01	1.29-01

CLASSIFICATION OF ELECTROMAGNETIC RADIATION

Hans Dolezalek

Basic Conversions: $c = \lambda\nu = \nu/k$; $\nu = c/\lambda = ck$; $\lambda = c/\nu = 1/k$; $k = \nu/c = 1/\lambda$
c = speed of light = 2.99792458×10^8 m/s

Frequency (ν)	Wavelength (λ)	Wave number (k)	Names of bands	Approximate photon energies
$3 \times 10^0 - 3 \times 10^1$ Hz 3 — 30 Hz	$10^8 - 10^7$ m 100 — 10 Mm	$10^{-8} - 10^{-7}$ m^{-1} 10 — 100 Gm^{-1}	ELF-(ELF 1), ITU band no. 1	
$3 \times 10^1 - 3 \times 10^2$ Hz 30 — 300 Hz	$10^7 - 10^6$ m 10 — 1 Mm	$10^{-7} - 10^{-6}$ m^{-1} 100 Gm^{-1} — 1 Mm^{-1}	SLF-(ELF 2), ITU band no. 2, mega-meter waves	
$3 \times 10^2 - 3 \times 10^3$ Hz 300 Hz — 3 kHz	$10^6 - 10^5$ m 1 Mm — 100 km	$10^{-6} - 10^{-5}$ m^{-1} 1 — 10 Mm^{-1}	ULF-(ELF 3), ITU band no. 3	
$3 \times 10^3 - 3 \times 10^4$ Hz 3 — 30 kHz	$10^5 - 10^4$ m 100 — 10 km	$10^{-5} - 10^{-4}$ m^{-1} 10 — 100 Mm^{-1}	VLF, ITU band no. 4, myriameter waves	
$3 \times 10^4 - 3 \times 10^5$ Hz 30 — 300 kHz	$10^4 - 10^3$ m 10 — 1 km	$10^{-4} - 10^{-3}$ m^{-1} 100 Mm^{-1} — 1 km^{-1}	LF, ITU band no. 5, kilometer waves	
$3 \times 10^5 - 3 \times 10^6$ Hz 300 kHz — 3 MHz	$10^3 - 10^2$ m 1 km — 100 m	$10^{-3} - 10^{-2}$ m^{-1} 1 — 10 km^{-1}	MF, ITU band no. 6, hectometer waves	
$3 \times 10^6 - 3 \times 10^7$ Hz 3 — 30 MHz	$10^2 - 10^1$ m 100 — 10 m	$10^{-2} - 10^{-1}$ m^{-1} 10 — 100 km^{-1}	HF, ITU band no. 7, decameter waves	
$3 \times 10^7 - 3 \times 10^8$ Hz 30 — 300 MHz	$10^1 - 10^0$ m 10 — 1 m	$10^{-1} - 10^0$ m^{-1} 100 km^{-1} — 1 m^{-1}	VHF, ITU band no. 8, meter waves	
$3 \times 10^8 - 3 \times 10^9$ Hz 300 MHz — 3 GHz	$10^0 - 10^{-1}$ m 1 m — 100 mm	$10^0 - 10^1$ m^{-1} 1 — 10 m^{-1}	UHF, ITU band no. 9, decimeter waves[a]	
$3 \times 10^9 - 3 \times 10^{10}$ Hz 3 — 30 GHz	$10^{-1} - 10^{-2}$ m 100 — 10 mm	$10^1 - 10^2$ m^{-1} 10 — 100 m^{-1}	SHF, ITU band no. 10, centimeter waves[a]	
$3 \times 10^{10} - 3 \times 10^{11}$ Hz 30 — 300 GHz	$10^{-2} - 10^{-3}$ m 10 — 1 mm	$10^2 - 10^3$ m^{-1} 100 m^{-1} — 1 mm^{-1} (1 — 10 cm^{-1})	EHF, ITU band no. 11, millimeter waves	
$3 \times 10^{11} - 3 \times 10^{12}$ Hz 300 GHz — 3 THz	$10^{-3} - 10^{-4}$ m 1 mm — 100 μm	$10^3 - 10^4$ m^{-1} 1 — 10 mm^{-1} (10 — 100 cm^{-1})	Part of micrometer waves, includes part of far or thermal infrared; ITU band no. 12	
$3 \times 10^{12} - 3 \times 10^{13}$ Hz 3 — 30 THz	$10^{-4} - 10^{-5}$ m 100 — 10 μm	$10^4 - 10^5$ m^{-1} 10 — 100 mm^{-1} (100 — 1000 cm^{-1})	Part of micrometer waves includes part of far (thermal) infrared	
$3 \times 10^{13} - 3 \times 10^{14}$ Hz 30 — 300 THz	$10^{-5} - 10^{-6}$ m 10 — 1 μm (100,000 — 10,000 Å)	$10^5 - 10^6$ m^{-1} 100 mm^{-1} — 1 μm^{-1}	Part of μm waves, part of infrared	$(1.6-16) \times 10^{-20}$ joule {0.1 — 1 eV}
$3 \times 10^{14} - 3 \times 10^{15}$ Hz 300 THz — 3 PHz	$10^{-6} - 10^{-7}$ m 1 μm — 100 nm (10,000 — 1000 Å)	$10^6 - 10^7$ m^{-1} 1 — 10 μm^{-1}	Near infrared, visible, near ultraviolet	$(1.6-16) \times 10^{-19}$ joule {1 — 10 eV}
$3 \times 10^{15} - 3 \times 10^{16}$ Hz 3 — 30 PHz	$10^{-7} - 10^{-8}$ m 100 — 10 nm (1000 — 100 Å)	$10^7 - 10^8$ m^{-1} 10 — 100 μm^{-1}	Part of "vacuum" - ultraviolet	$(1.6-16) \times 10^{-18}$ joule {10 — 100 eV}
$3 \times 10^{16} - 3 \times 10^{17}$ Hz 30 — 300 PHz	$10^{-8} - 10^{-9}$ m 10 — 1 nm (100 — 10 Å)	$10^8 - 10^9$ m^{-1} 100 μm^{-1} — 1 nm^{-1}	Part of soft X-rays	$(1.6-16) \times 10^{-17}$ joule {100 — 1000 eV)}
$3 \times 10^{17} - 3 \times 10^{18}$ Hz 300 PHz — 3 EHz	$10^{-9} - 10^{-10}$ m 1 nm — 100 pm (10 — 1 Å)	$10^9 - 10^{10}$ m^{-1} 1 — 10 nm^{-1}	Part of soft X-rays	$(1.6-16) \times 10^{-16}$ joule {1 — 10 keV}
$3 \times 10^{18} - 3 \times 10^{19}$ Hz 3 — 30 EHz	$10^{-10} - 10^{-11}$ m 100 — 10 pm (1 — 0.1 Å)	$10^{10} - 10^{11}$ m^{-1} 10 — 100 nm^{-1}	Hard X-rays and part of soft γ-rays	$(1.6-16) \times 10^{-15}$ joule {10 — 100 keV}
$3 \times 10^{19} - 3 \times 10^{20}$ Hz 30 — 300 EHz	$10^{-11} - 10^{-12}$ m 10 — 1 pm (0.1 — 0.01 Å)	$10^{11} - 10^{12}$ m^{-1} 100 nm^{-1} — 1 pm^{-1}	Part of soft and part of hard γ-rays (limit at 510 keV)	$(1.6-16) \times 10^{-14}$ joule {100 keV — 1 MeV}
$3 \times 10^{20} - 3 \times 10^{21}$ Hz 300 — 3000 EHz	$10^{-12} - 10^{-13}$ m 1 pm — 100 fm (0.01 — 0.001 Å)	$10^{12} - 10^{13}$ m^{-1} 1 — 10 pm^{-1}	Part of hard γ-rays and part of "cosmic" γ-rays	$(1.6-16) \times 10^{-13}$ joule {1 — 10 MeV}
$3 \times 10^{21} - 3 \times 10^{22}$ Hz 3000 — 30,000 EHz	$10^{-13} - 10^{-14}$ m 100 — 10 fm (0.001 — 0.0001 Å)	$10^{13} - 10^{14}$ m^{-1} 10 — 100 pm^{-1}	γ-rays produced by cosmic rays	$(1.6-16) \times 10^{-12}$ joule {10 — 100 MeV}

CLASSIFICATION OF ELECTROMAGNETIC RADIATION (continued)

Note: Abbreviations used in this table: Å—ångstrom (1 Å = 10^{-10} m); EHz—exahertz (10^{18} hertz); EHF—extremely high frequency; ELF—extremely low frequency; eV—electron volt (1 eV = 1.60219×10^{-19} joule); PHz—petahertz (10^{15} hertz); fm—femtometer (10^{-15} m); GHz—gigahertz (10^{9} hertz); Gm—gigameter (10^{9} m); HF—high frequency; Hz—hertz (s^{-1}); ITU—International Telecommunications Union; keV—kiloelectron volt (10^{3} eV); km—kilometer (10^{3} m); LF—low frequency; m—meter; MeV—megaelectron volt (10^{6} eV); MF—medium frequency; MHz—megahertz (10^{6} hertz); Mm—megameter (10^{6} meter); mm—millimeter (10^{-3} meter); μm—micrometer (10^{-6} meter); nm—nanometer (10^{-9} meter); pm—picometer (10^{-12} meter); SHF—super high frequency; SLF—super low frequency; THz—terahertz; UHF—ultra high frequency; ULF—ultra low frequency; VHF—very high frequency; VLF—very low frequency.

[a] Also called "microwaves"; not to be confused with "micrometer waves".

LETTER DESIGNATIONS OF MICROWAVE BANDS

Frequency (GHz)	Wavelength (cm)	Wavenumber (cm^{-1})	Band
1—2	30—15	0.033—0.067	L-Band
1—4	15—7.5	0.067—0.133	S-Band
4—8	7.5—3.7	0.133—0.267	C-Band
8—12	3.7—2.5	0.267—0.4	X-Band
12—18	2.5—1.7	0.4—0.6	Ku-Band
18—27	1.7—1.1	0.6—0.9	K-Band
27—40	1.1—0.75	0.9—1.33	Ka-Band

BLACK BODY RADIATION

The total power radiated from an ideal black body and the wavelength corresponding to maximum power are given here as a function of absolute temperature. Constants used in the calculation are taken from the table "Fundamental Physical Constants" in Section 1. The radiated power in a band $\Delta\lambda$ at λ_{max} may be calculated from:

$$P_{max} = 0.657548 \, (\Delta\lambda/\lambda_{max}) \, P_{tot}$$

T/K	P_{tot}	$\lambda_{max}/\mu m$	T/K	P_{tot}	$\lambda_{max}/\mu m$	T/K	P_{tot}	$\lambda_{max}/\mu m$
50	0.354 W/m²	57.955	740	17.004	3.916	1520	302.689	1.906
100	5.671	28.978	750	17.942	3.864	1540	318.937	1.882
150	28.707	19.318	760	18.918	3.813	1560	335.831	1.858
200	90.728	14.489	770	19.934	3.763	1580	353.387	1.834
250	221.504	11.591	780	20.989	3.715	1600	371.623	1.811
273	314.973	10.614	790	22.087	3.668	1620	390.555	1.789
280	348.541	10.349	800	23.226	3.622	1640	410.202	1.767
290	401.064	9.992	810	24.410	3.577	1660	430.581	1.746
300	459.311	9.659	820	25.638	3.534	1680	451.710	1.725
310	523.684	9.348	830	26.911	3.491	1700	473.607	1.705
320	594.596	9.055	840	28.232	3.450	1720	496.290	1.685
330	672.478	8.781	850	29.600	3.409	1740	519.779	1.665
340	757.771	8.523	860	31.018	3.369	1760	544.093	1.646
350	850.931	8.279	870	32.486	3.331	1780	569.249	1.628
360	952.428	8.049	880	34.006	3.293	1800	595.267	1.610
370	1.063 kW/m²	7.832	890	35.578	3.256	1820	622.168	1.592
380	1.182	7.626	900	37.204	3.220	1840	649.970	1.575
390	1.312	7.430	910	38.886	3.184	1860	678.694	1.558
400	1.452	7.244	920	40.623	3.150	1880	708.359	1.541
410	1.602	7.068	930	42.418	3.116	1900	738.987	1.525
420	1.764	6.899	940	44.272	3.083	1920	770.597	1.509
430	1.939	6.739	950	46.187	3.050	1940	803.210	1.494
440	2.125	6.586	960	48.162	3.018	1960	836.848	1.478
450	2.325	6.439	970	50.201	2.987	1980	871.531	1.464
460	2.539	6.299	980	52.303	2.957	2000	907.282	1.449
470	2.767	6.165	990	54.471	2.927	2020	944.121	1.435
480	3.010	6.037	1000	56.705	2.898	2040	982.071	1.420
490	3.269	5.914	1020	61.379	2.841	2060	1.021 MW/m²	1.407
500	3.544	5.796	1040	66.337	2.786	2080	1.061	1.393
510	3.836	5.682	1060	71.589	2.734	2100	1.103	1.380
520	4.146	5.573	1080	77.147	2.683	2120	1.145	1.367
530	4.474	5.467	1100	83.022	2.634	2140	1.189	1.354
540	4.822	5.366	1120	89.227	2.587	2160	1.234	1.342
550	5.189	5.269	1140	95.773	2.542	2180	1.281	1.329
560	5.577	5.175	1160	102.672	2.498	2200	1.328	1.317
570	5.986	5.084	1180	109.939	2.456	2220	1.377	1.305
580	6.417	4.996	1200	117.584	2.415	2240	1.428	1.294
590	6.871	4.911	1220	125.621	2.375	2260	1.479	1.282
600	7.349	4.830	1240	134.063	2.337	2280	1.532	1.271
610	7.851	4.750	1260	142.924	2.300	2300	1.587	1.260
620	8.379	4.674	1280	152.217	2.264	2320	1.643	1.249
630	8.933	4.600	1300	161.955	2.229	2340	1.700	1.238
640	9.514	4.528	1320	172.154	2.195	2360	1.759	1.228
650	10.122	4.458	1340	182.827	2.163	2380	1.819	1.218
660	10.760	4.391	1360	193.989	2.131	2400	1.881	1.207
670	11.427	4.325	1380	205.655	2.100	2420	1.945	1.197
680	12.124	4.261	1400	217.838	2.070	2440	2.010	1.188
690	12.853	4.200	1420	230.556	2.041	2460	2.077	1.178
700	13.615	4.140	1440	243.822	2.012	2480	2.145	1.168
710	14.410	4.081	1460	257.652	1.985	2500	2.215	1.159
720	15.239	4.025	1480	272.063	1.958	2550	2.398	1.136
730	16.103	3.970	1500	287.070	1.932	2600	2.591	1.115

T/K	P_{tot}	$\lambda_{max}/\mu m$	T/K	P_{tot}	$\lambda_{max}/\mu m$	T/K	P_{tot}	$\lambda_{max}/\mu m$
2650	2.796	1.093	3600	9.524	0.805	5100	38.362	0.568
2700	3.014	1.073	3650	10.065	0.794	5200	41.461	0.557
2750	3.243	1.054	3700	10.627	0.783	5300	44.743	0.547
2800	3.485	1.035	3750	11.214	0.773	5400	48.217	0.537
2850	3.741	1.017	3800	11.824	0.763	5500	51.889	0.527
2900	4.011	0.999	3850	12.458	0.753	5600	55.767	0.517
2950	4.294	0.982	3900	13.118	0.743	5700	59.858	0.508
3000	4.593	0.966	3950	13.804	0.734	5800	64.170	0.500
3050	4.907	0.950	4000	14.517	0.724	5900	68.712	0.491
3100	5.237	0.935	4100	16.024	0.707	6000	73.490	0.483
3150	5.583	0.920	4200	17.645	0.690	6500	101.222	0.446
3200	5.946	0.906	4300	19.386	0.674	7000	136.149	0.414
3250	6.326	0.892	4400	21.254	0.659	7500	179.418	0.386
3300	6.725	0.878	4500	23.253	0.644	8000	232.264	0.362
3350	7.142	0.865	4600	25.389	0.630	8500	296.004	0.341
3400	7.578	0.852	4700	27.670	0.617	9000	372.042	0.322
3450	8.033	0.840	4800	30.101	0.604	9500	461.867	0.305
3500	8.509	0.828	4900	32.689	0.591	10000	567.051	0.290
3550	9.006	0.816	5000	35.441	0.580			

The curves below show, for various temperatures, the fraction of radiant power as a function of wavelength. The function plotted is $P_\lambda/\Delta\lambda\, P_{tot}$, where P_λ is the power at wavelength λ in a small interval $\Delta\lambda$ (in μm), and P_{tot} is the total power.

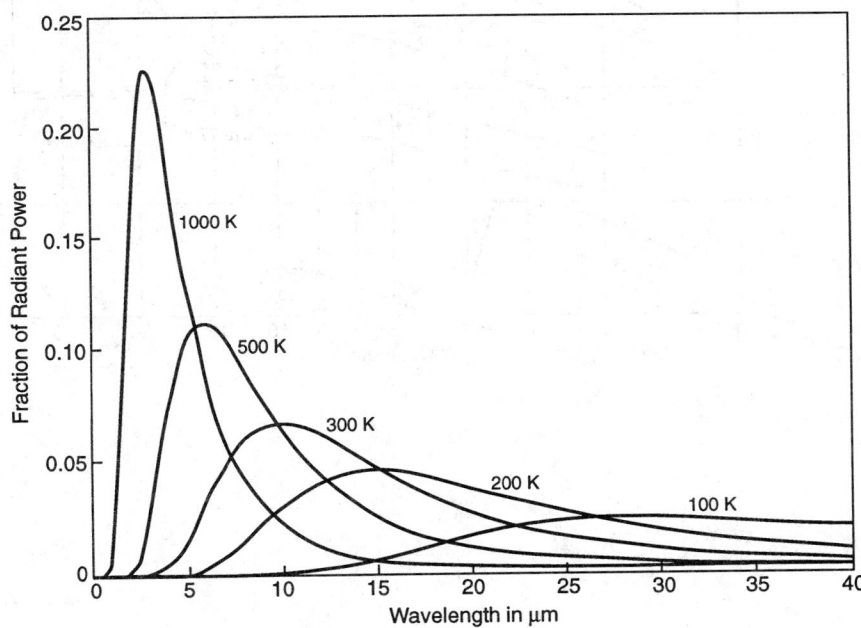

CHARACTERISTICS OF INFRARED DETECTORS

This graph summarizes the wavelength response of some semiconductors used as detectors for infrared radiation. The quantity $D^*(\lambda)$ is the signal to noise ratio for an incident radiant power density of 1 W/cm^2 and a bandwidth of 1 Hz (60° field of view). The Ge, InAs, and InSb detectors are photovoltaics, while the HgCdTe series are photoconductive devices. The cutoff wavelength of the latter can be varied by adjusting the relative amounts of Hg, Cd, and Te (three examples are shown at 77 K). The graph also shows the theoretical background limited sensitivity for ideal detectors which introduce no intrinsic noise.

REFERENCE

Infrared Detectors 1995, EG&G Judson, Montgomeryville, PA.

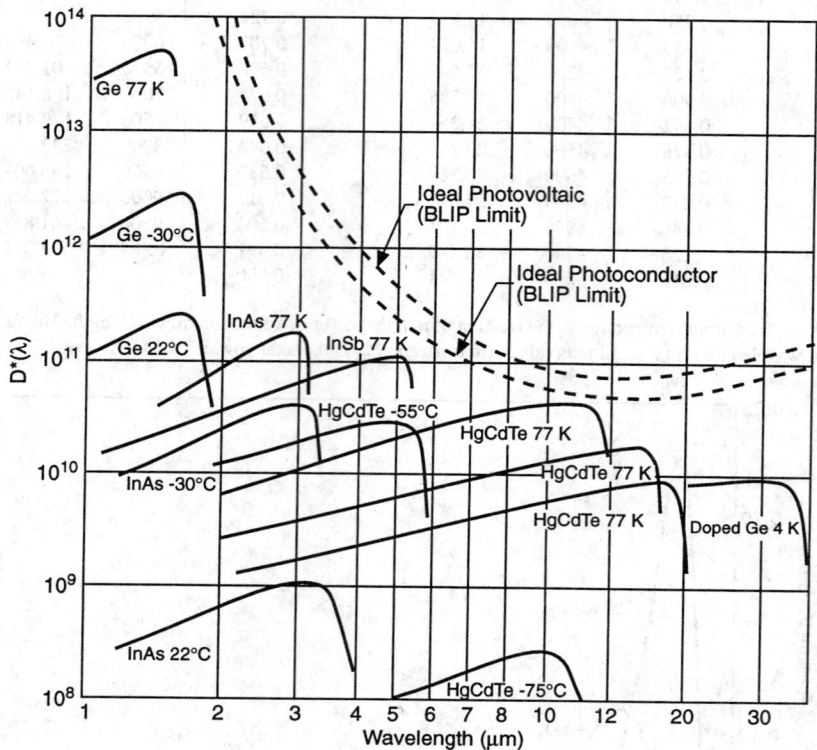

REFRACTIVE INDEX AND TRANSMITTANCE OF REPRESENTATIVE GLASSES

Typical values of the index of refraction and internal transmittance (fraction of light transmitted through a one centimeter thickness) are tabulated here for selected types of glasses, as well as for synthetic fused (vitreous) silica. Nominal compositions are given in the first part of the table. The second part gives the index of refraction, relative to air, and the internal transmittance for representative samples of each glass at wavelengths in the infrared, visible, and near-ultraviolet regions. It should be emphasized that wide variation of these parameters may be found among subtypes of each glass. More detailed data may be found in Reference 3.

Assuming that the Lambert-Beer Law is followed, the transmittance of a glass plate of thickness d (in centimeters) can be obtained by raising the transmittance value in the table to the power d.

REFERENCES

1. Weber, M.J., *CRC Handbook of Laser Science and Technology*, Vol. IV, Part 2, CRC Press, Boca Raton, FL ,1988.
2. Gray, D.E., Ed., *American Institute of Physics Handbook, Third Edition*, McGraw Hill, New York, 1972.
3. *Schott Optical Glass*, Schott Glass Technologies, Inc., 400 York Ave., Duryea, PA 18642.
4. Kaye, G.W.C., and Laby, T.H., *Tables of Physical and Chemical Constants, 15th Edition*, Longman, London, 1986.

Type	Name	Composition in percent by mass									
		SiO_2	B_2O_3	Al_2O_3	Na_2O	K_2O	CaO	BaO	ZnO	PbO	P_2O_5
PK	Phosphate crown		3	10		12	5				70
PSK	Dense phosphate crown		3	5			4	28			60
BK	Borosilicate crown	70	10		8	8	1	3			
K	Crown	74			9	11	6				
ZK	Zinc crown	71			17				12		
BaK	Barium crown	60	3		3	10		19	5		
SK	Dense crown	39	15	5				41			
KF	Crown flint	67		2	16				3	12	
BaLF	Barium light flint	51			6	5		20		14	4
SSK	Extra dense crown	35	10	5				42	8		
LLF	Extra light flint	63			5	8				24	
BaF	Barium flint	46				8		16	8	22	
LF	Light flint	53			5	8				34	
F	Flint	47			2	7				44	
BaSF	Dense barium flint	43			1	7		11	5	33	
SF	Dense flint	33				5				62	
KzFS	Short flint										
SiO_2	Fused silica	100									

Type	Index of refraction				Transmittance of 1 cm plate			
	1.060 μm	546.1 nm	365.0 nm	312.6 nm	1.060 μm	546.1 nm	365.0 nm	310 nm
PK	1.51519	1.52736	1.54503	1.5574	0.997	0.998	0.987	0.46
PSK	1.54154	1.55440	1.57342	1.5868	0.996	0.998	0.984	0.46
BK	1.50669	1.51872	1.53627	1.5486	0.999	0.998	0.987	0.35
K	1.50091	1.51314	1.53189	1.5454	0.998	0.998	0.988	0.40
ZK	1.52220	1.53534	1.55588	1.5708	0.996	0.998	0.976	0.27
BaK	1.55695	1.57124	1.59407	1.6108	0.998	0.997	0.986	0.28
SK	1.59490	1.60994	1.63398		0.998	0.998	0.959	0.28
KF	1.50586	1.51978	1.54251	1.5600	0.998	0.996	0.989	0.49
BaLF	1.57579	1.59166	1.61804		0.996	0.998	0.933	0.010
SSK	1.60402	1.61993	1.64595		0.999	0.998	0.915	0.010
LaK	1.69710	1.71616	1.74573		0.999	0.998	0.882	0.17
LLF	1.52775	1.54344	1.57038		0.998	0.997	0.990	0.32
BaF	1.56873	1.58565	1.61524		0.999	0.997	0.992	0.004
LF	1.56594	1.58482	1.61926		0.999	0.998	0.981	0.008
F	1.58636	1.60718	1.64606		0.997	0.998	0.959	
BaSF	1.60889	1.62987	1.66926		0.999	0.998	0.857	
SF	1.71350	1.74620	1.8145		0.998	0.997	0.650	
KzFS	1.59680	1.61639	1.64849	1.6739		0.998	0.672	0.012
SiO_2	1.44968	1.46008	1.47435[a]	1.53430[b]				

[a] At 366.3 nm.

[b] At 213.9 nm.

INDEX OF REFRACTION OF WATER

This table gives the index of refraction of liquid water at atmospheric pressure, relative to a vacuum, at several temperatures and wavelengths. It is generated from the formulation in Reference 1, which covers a wide range of temperature, pressure, and wavelength. The wavelengths listed here correspond to prominent lines of cadmium (226.50 and 361.05 nm), potassium (404.41 nm), sodium (589.00 nm), Ne (632.80 nm, from a helium - neon laser), and mercury (1.01398 μm).

REFERENCES

1. Schiebener, P., Straub, J., Levelt Sengers, J.M.H., and Gallagher, J.S., *J. Phys. Chem. Ref. Data*, 19, 677 (1990); 19, 1617, 1990.
2. Marsh, K.N., Editor, *Recommended Reference Materials for the Realization of Physicochemical Properties*, Blackwell Scientific Publications, Oxford, 1987.

$T/°C$	226.50 nm	361.05 nm	404.41 nm	589.00 nm	632.80 nm	1.01398 μm
0	1.39450	1.34896	1.34415	1.33432	1.33306	1.32612
10	1.39422	1.34870	1.34389	1.33408	1.33282	1.32591
20	1.39336	1.34795	1.34315	1.33336	1.33211	1.32524
30	1.39208	1.34682	1.34205	1.33230	1.33105	1.32424
40	1.39046	1.34540	1.34065	1.33095	1.32972	1.32296
50	1.38854	1.34373	1.33901	1.32937	1.32814	1.32145
60	1.38636	1.34184	1.33714	1.32757	1.32636	1.31974
70	1.38395	1.33974	1.33508	1.32559	1.32438	1.31784
80	1.38132	1.33746	1.33284	1.32342	1.32223	1.31576
90	1.37849	1.33501	1.33042	1.32109	1.31991	1.31353
100	1.37547	1.33239	1.32784	1.31861	1.31744	1.31114

INDEX OF REFRACTION OF LIQUIDS FOR CALIBRATION PURPOSES

This table gives the index of refraction of six liquids which are available in highly pure form and whose index of refraction has been accurately measured as a function of wavelength and temperature. They are therefore useful for calibration of refractometers. The estimated uncertainty in the values is:

2,2,4-Trimethylpentane	±0.00003
Hexadecane	±0.00008
trans-Bicyclo[4.0.0]decane	±0.00008
1-Methylnaphthalene	±0.00008
Toluene	±0.00003
Methylcyclohexane	±0.00003

Full details are given in the references. This table is reprinted from Reference 1 by permission of the International Union of Pure and Applied Chemistry.

REFERENCES

1. Marsh, K. N., Editor, *Recommended Reference Materials for the Realization of Physicochemical Properties,* Blackwell Scientific Publications, Oxford, 1987.
2. Tilton, L. W., J. *Opt. Soc. Am.*, 32, 71, 1941.

λ	2,2,4-Trimethylpentane			Hexadecane		
nm	20°C	25°C	30°C	20°C	25°C	30°C
667.81	1.38916	1.38670	1.38424	1.43204	1.43001	1.42798
656.28	1.38945	1.38698	1.38452	1.43235	1.43032	1.42829
589.26	1.39145	1.38898	1.38650	1.43453	1.43250	1.43047
546.07	1.39316	1.39068	1.38820	1.43640	1.43436	1.43232
501.57	1.39544	1.39294	1.39044	1.43888	1.43684	1.43480
486.13	1.39639	1.39389	1.39138	1.43993	1.43788	1.43583
435.83	1.40029	1.39776	1.39523	1.44419	1.44213	1.44007

λ	*trans*-Bicyclo[4.4.0]decane			1-Methylnaphthalene		
nm	20°C	25°C	30°C	20°C	25°C	30°C
667.81	1.46654	1.46438	1.46222	1.60828	1.60592	1.60360
656.28	1.46688	1.46472	1.46256	1.60940	1.60703	1.60471
589.26	1.46932	1.46715	1.46498	1.61755	1.61512	1.61278
546.07	1.47141	1.46923	1.46705	1.62488	1.62240	1.62005
501.57	1.47420	1.47200	1.46980	1.63513	1.63259	1.63022
486.13	1.47535	1.47315	1.47095	1.63958	1.63701	1.63463
435.83	1.48011	1.47789	1.47567		1.65627	1.65386

λ	Toluene			Methylcyclohexane		
nm	20°C	25°C	30°C	20°C	25°C	30°C
667.81	1.49180	1.48903	1.48619	1.42064	1.41812	1.41560
656.28	1.49243	1.48966	1.48682	1.42094	1.41842	1.41591
589.26	1.49693	1.49413	1.49126	1.42312	1.42058	1.41806
546.07	1.50086	1.49803	1.49514	1.42497	1.42243	1.41989
501.57	1.50620	1.50334	1.50041	1.42744	1.42488	1.42233
486.13	1.50847	1.50559	1.50265	1.42847	1.42590	1.42334
435.83	1.51800	1.51506	1.51206	1.43269	1.43010	1.42752

INDEX OF REFRACTION OF AIR

This is a table of the index of refraction n of dry air at 15°C and a pressure of 101.325 kPa and containing 0.03% by volume of carbon dioxide ("standard air"). The index of refraction is defined by $n = \lambda_{vac}/\lambda_{air}$, where λ is the wavelength of the radiation. The index is calculated from the expression:

where $\sigma = 1/\lambda_{vac}$ and λ_{vac} has units of μm. The equation is valid for λ_{vac} from 200 nm to 2 μm.
The table also gives the correction $(n-1)\lambda_{air}$ which must be added to the wavelength in air to obtain λ_{vac}.
If the air is at a temperature t in °C and a pressure p in pascals, a value of $(n-1)$ from this table should be multiplied by

$$\frac{p\left[1 + p(61.3 - t)\times 10^{-10}\right]}{96095.4(1 + 0.003661t)}$$

REFERENCE

Edlen, B., *Metrologia*, 2, 71, 1966.

λ_{vac}	$(n-1)\times 10^8$	$\lambda_{vac} - \lambda_{air}$	λ_{vac}	$(n-1)\times 10^8$	$\lambda_{vac} - \lambda_{air}$	λ_{vac}	$(n-1)\times 10^8$	$\lambda_{vac} - \lambda_{air}$
200 nm	32408	0.0648 nm	540	27803	0.1501	880	27461	0.2416
210	31746	0.0666	550	27782	0.1528	890	27457	0.2443
220	31224	0.0687	560	27763	0.1554	900	27452	0.2470
230	30799	0.0708	570	27745	0.1581	910	27448	0.2497
240	30445	0.0730	580	27728	0.1608	920	27444	0.2524
250	30146	0.0753	590	27712	0.1635	930	27440	0.2551
260	29890	0.0777	600	27697	0.1661	940	27436	0.2578
270	29669	0.0801	610	27682	0.1688	950	27432	0.2605
280	29475	0.0825	620	27669	0.1715	960	27429	0.2632
290	29306	0.0850	630	27656	0.1742	970	27425	0.2660
300	29155	0.0874	640	27643	0.1769	980	27422	0.2687
310	29022	0.0899	650	27631	0.1796	990 nm	27419	0.2714 nm
320	28902	0.0925	660	27620	0.1822			
330	28795	0.0950	670	27609	0.1849	1.00 μm	27416	0.000274 μm
340	28698	0.0975	680	27599	0.1876	1.05	27401	0.000288
350	28611	0.1001	690	27589	0.1903	1.10	27389	0.000301
360	28531	0.1027	700	27579	0.1930	1.15	27378	0.000315
370	28458	0.1053	710	27570	0.1957	1.20	27368	0.000328
380	28392	0.1079	720	27562	0.1984	1.25	27360	0.000342
390	28331	0.1105	730	27553	0.2011	1.30	27352	0.000355
400	28275	0.1131	740	27545	0.2038	1.35	27346	0.000369
410	28223	0.1157	750	27538	0.2065	1.40	27340	0.000383
420	28175	0.1183	760	27530	0.2092	1.45	27334	0.000396
430	28131	0.1209	770	27523	0.2119	1.50	27330	0.000410
440	28090	0.1236	780	27516	0.2146	1.55	27325	0.000423
450	28052	0.1262	790	27510	0.2173	1.60	27321	0.000437
460	28016	0.1288	800	27504	0.2200	1.65	27318	0.000451
470	27983	0.1315	810	27498	0.2227	1.70	27314	0.000464
480	27952	0.1341	820	27492	0.2254	1.75	27311	0.000478
490	27923	0.1368	830	27486	0.2281	1.80	27309	0.000491
500	27896	0.1394	840	27481	0.2308	1.85	27306	0.000505
510	27870	0.1421	850	27476	0.2335	1.90	27304	0.000519
520	27846	0.1448	860	27471	0.2362	1.95	27302	0.000532
530	27824	0.1474	870	27466	0.2389	2.00 μm	27300	0.000546 μm

CHARACTERISTICS OF LASER SOURCES
William F. Krupke

Light Amplification by Stimulated Emission of Radiation was first demonstrated by Maiman in 1960, the result of a population inversion produced between energy levels of chromium ions in a ruby crystal when irradiated with a xenon flashlamp. Since then population inversions and coherent emission have been generated in literally thousands of substances (neutral and ionized gases, liquids, and solids) using a variety of incoherent excitation techniques (optical pumping, electrical discharges, gas-dynamic flow, electron-beams, chemical reactions, nuclear decay).

The extrema of laser output parameters which have been demonstrated to date and the laser media used are summarized in Table 1. Note that the extreme power and energy parameters listed in this table were attained with laser *systems* rather than with simple laser oscillators.

Laser sources are commonly classified in terms of the state-of-matter of the active medium: gas, liquid, and solid. Each of these classes is further subdivided into one or more types as shown in Table 2. A well-known representative example of each type of laser is also given in Table 2 together with its nominal operation wavelength and the methods by which it is pumped.

The various lasers together cover a wide spectral range from the far ultraviolet to the far infrared. The particular wavelength of emission (usually a narrow line) is presented for some six dozen lasers in Figures 1A and 1B.

By suitably designing the excitation source and/or by controlling the laser resonator structure, laser systems can provide continuous or pulsed radiation as shown in Table 3.

Besides the method of excitation and the temporal behavior of a laser, there are many other parameters that characterize its operation and efficiency, as shown in Tables 4 and 5.

Although many lasers only emit in one or more narrow spectral "lines", an increasing number of lasers can be tuned by changing the composition or the pressure of the medium, or by varying the wavelength of the pump bands. The spectral regions in which these tunable lasers operate are presented in Figure 2.

REFERENCE

Krupke, W. F., in *Handbook of Laser Science and Technology,* Vol. I, Weber, M. J., Ed., CRC Press, Boca Raton, FL, 1986.

TABLE 1
Extrema of Output Parameters of Laser Devices or Systems

Parameter	Value	Laser medium
Peak power	1×10^{14} W (collimated)	Nd:glass
Peak power density	10^{18} W/cm^2 (focused)	Nd:glass
Pulse energy	$>10^5$ J	CO_2, Nd:glass
Average power	10^5 W	CO_2
Pulse duration	3×10^{-15} s continuous wave (cw)	Rh6G dye; various gases, liquids, solids
Wavelength	60 nm $\leftrightarrow$ 385 μm	Many required
Efficiency (nonlaser pumped)	70%	CO
Beam quality	Diffraction limited	Various gases, liquids, solids
Spectral linewidth	20 Hz (for 10^{-1} s)	Neon-helium
Spatial coherence	10 m	Ruby

TABLE 2
Classes, Types, and Representative Examples of Laser Sources

Class	Type (characteristic)	Representative example	Nominal operating wavelength (nm)	Method(s) of excitation
Gas	Atom, neutral (electronic transition)	Neon-Helium (Ne-He)	633	Glow discharge
	Atom, ionic (electronic transition)	Argon (Ar^+)	488	Arc discharge
	Molecule, neutral (electronic transition)	Krypton fluoride (KrF)	248	Glow discharge; e-beam
	Molecule, neutral (vibrational transition)	Carbon dioxide (CO_2)	10600	Glow discharge; gas-dynamic flow
	Molecule, neutral (rotational transition)	Methyl fluoride (CH_3F)	496000	Laser pumping
	Molecule, ionic (electronic transition)	Nitrogen ion (N_2^+)	420	E-beam
Liquid	Organic solvent (dye-chromophore)	Rhodamine dye (Rh6G)	580–610	Flashlamp; laser pumping
	Organic solvent (rare earth chelate)	Europium:TTF	612	Flashlamp
Solid	Inorganic solvent (trivalent rare earth ion)	Neodymium:$POCl_4$	1060	Flashlamp
	Insulator, crystal (impurity)	Neodymium:YAG	1064	Flashlamp, arc lamp
	Insulator, crystal (stoichiometric)	Neodymium:UP(NdP_5O_{14})	1052	Flashlamp
	Insulator, crystal (color center)	F_2^-:LiF	1120	Laser pumping
	Insulator, amorphous (impurity)	Neodymium:glass	1061	Flashlamp
	Semiconductor (p-n junction)	GaAs	820	Injection current
	Semiconductor (electron-hole plasma)	GaAs	890	E-beam, laser pumping

Table 3
Temporal Characteristics of Lasers and Laser Systems

Form	Technique	Pulse width range (s)
Continuous wave	Excitation is continuous; resonator Q is held constant at some moderate value	∞
Pulsed	Excitation is pulsed; resonator Q is held constant at some moderate value	10^{-8}–10^{-3}
Q-Switched	Excitation is continuous or pulsed; resonator Q is switched from a very low value to a moderate value	10^{-8}–10^{-6}
Cavity dumped	Excitation is continuous or pulsed; resonator Q is switched from a very high value to a low value	10^{-7}–10^{-5}
Mode locked	Excitation is continuous or pulsed; phase or loss of the resonator modes is modulated at a rate related to the resonator transit time	10^{-12}–10^{-9}

Table 4
Properties and Performance of Some Continuous Wave (CW) Lasers

Parameter	Unit	Gas			Liquid	Solid	
		Neon helium	Argon ion	Carbon dioxide	Rhodamine 6G dye	Nd:YAG	GaAs
Excitation method		DC discharge	DC discharge	DC discharge	Ar$^+$ laser pump	Krypton arc lamp	DC injection
Gain medium composition		Neon:helium	Argon	CO_2:N_2:He	Rh 6G:H_2O	Nd:YAG	p:n:GaAs
Gain medium density	Torr; ions/cm^3	0.1:1.0	0.4	0.4:0.8:5.0	2(18):2(22)	1.5(20):2(22)	2(19):3(18):3(22)
Wavelength	nm	633	488	10600	590	1064	810
Laser cross-section	cm^{-2}	3(-13)	1.6(-12)	1.5(-16)	1.8(-16)	7(-19)	~6(-15)
Radiative lifetime (upper level)	s	~1(-7)	7.5(-9)	4(-3)	6.5(-9)	2.6(-4)	~1(-9)
Decay lifetime (upper level)	s	~1(-7)	~5.0(-9)	~4(-3)	6.0(-9)	2.3(-4)	~1(-9)
Gain bandwidth	nm	2(-3)	5(-3)	1.6(-2)	80	0.5	10
Type, gain saturation		Inhomogeneous	Inhomogeneous	Homogeneous	Homogeneous	Homogeneous	Homogeneous
Homogeneous saturation flux	W cm^{-2}	~1(-8)	~4(-10)	~20	3(5)	2.3(3)	~2(4)
Decay lifetime (lower level)	s	~1(9)	2(10)	~5(-6)[b]	<1(-12)	<1(-7)	<1(-12)
Inversion density	cm^{-3}	~1(-3)	~3(-2)	2(15)	2(16)	6(16)	1(16)
Small signal gain coefficient	cm^{-1}			1(-2)	4	5(-2)	40
Pump power density	W cm^{-3}	3	900	0.15	1(6)	150	7(7)
Output power density	W cm^{-3}	2.6(-3)	~1	2(-2)	3(5)	95	5(6)
Laser size (diameter:length)	cm:cm	0.5:100	0.3:100	5.0:600	1(-3):0.3	0.6:10	5(-4):7(-3):2(-2)[a]
Excitation current/voltage	A/V	3(-2):2(3)	30:300	0.1:1.5(4)		90:125	1.0/1.7
Excitation current density	A cm^{-2}	0.15	600	6(-3)		140	4.5(3)
Excitation power	W	60	9(3)	1.5(3)	4	1.1(4)	1.7
Output power	W	0.06	10	240	0.3	300	0.12
Efficiency	%	0.1	0.1	13	7	2.6	7

[a] Junction thickness:width:length.
[b] Pressure dependent.

Table 5
Properties and Performance of Some Pulsed Lasers

Parameter	Unit	Gas				Liquid	Solid	
		Carbon dioxide		Krypton fluoride		Rhodamine 6G	Nd:YAG	Nd:glass
Excitation method		TEA-discharge	E-beam/sust.	Glow discharge	E-beam	Xenon flashlamp	Xenon flashlamp	Xenon flashlamp
Gain medium composition		CO_2:N_2:He	CO_2:N_2:He	He:Kr:F_2	Ar:Kr:F_2	RhG:alcohol	Nd:YAG	Nd:Glass
Gain medium density	torr / ions/cm^3	100:50:600	240:240:320	1070:70:3	1235:52:3	1(18):1.5(22)	1.5(20):1(22)	3(20):2(22)
Wavelength	nm	10600	10600	249	249	590	1064	1061
Laser cross-section	cm^{-2}	2(-18)	2(-18)	2(-16)	2(-16)	1.8(-16)	7(-19)	2.8(-20)
Radiative lifetime (upper level)	s	4(-3)	4(-3)	7(-9)	7(-9)	6.5(-9)	2.6(-4)	4.1(-4)
Decay lifetime (upper level)	s	~1(-4)	5(-5)	3(-9)	3(-9)	6.0(-9)	2.3(-4)	3.7(-4)
Gain bandwidth	nm	1	1	2	2	80	0.5	26
Homogeneous saturation fluence	J/cm^2	0.2	0.2	4(-3)	4(-3)	2(-3)	0.6	~5
Decay lifetime (lower level)	s	5(-8)[a]	1(-8)[a]	<1(-12)	<1(-12)	<1(-12)	<1(-7)	<1(-8)
Inversion density	cm^{-3}	3(17)	6(17)	4(14)	2(14)	2(16)	4(17)	3(18)
Small signal gain coefficient	cm^{-1}	2(-2)	4(-2)	8-92	4(-2)	4	0.3	8(-2)
Medium excitation energy density	J/cm^3	0.1	0.36	0.15	0.13	2.8	0.15	0.6
Output energy density	J/cm^3	1.8(-2)	1.8(-2)	1.5(-3)	1.2(-2)	0.85	5(-2)	2(-2)
Laser dimensions	cm:cm:cm	4.5:4.5:87	10:10:100	1.5:4.5:100	8.5:10:100	1.2Φ25	0.6Φ7.5	0.6Φ8.3
Excitation current/voltage	A/V	6(4)/3.3(3)	2.4(4)/4(4)	2.5(4)/1.5(5)	1.2(4)/2.5(5)	2(5)/2.5(4)		
Excitation current density	A cm^2	8.5	22	170	11.5	2.6(3)		
Excitation peak power	W	2(8)	9(8)	4(9)	3(9)	5.4(9)	4(4)	9(4)
Output pulse energy	J	35	180	1	102	32	0.1	1.0
Output pulse length	s	1(-6)	4(-6)	2.5(-8)	6(-7)	3.2(-6)	2(-8)	1(-4)
Output pulse power	W	3.5(7)	4(7)	4(7)	2(8)	1(7)	5(6)	1(4)
Efficiency	%	17	5	1	10[b]	0.2	1.5	3.7

[a] Pressure dependent.

[b] Intrinsic efficiency ≡ energy output/energy deposited in gas.

FIGURE 1A. Wavelengths of lasers operating in the 120 to 1200 nm spectral region.

FIGURE 1B. Wavelength of lasers operating in the 1300 to 12,000 nm spectral region.

FIGURE 2. Spectral tuning ranges of various types of tunable lasers.

INFRARED LASER FREQUENCIES
Arthur Maki

The CO_2 laser has been the subject of a number of very accurate frequency measurements. Most of the earlier measurements are given by Bradley et al.[1] That analysis was based on a single absolute frequency measurement and many laser frequency differences. New measurements of the methane frequency[2-4] have made it necessary to slightly revise that single absolute frequency measurement. In addition, there have been several other absolute frequency measurements[5-7] that have been used here to improve the accuracy of the present tables. New frequency difference measurements have also been added to the database used for the present tables.[8]

REFERENCES

1. Bradley, L. C., Soohoo, K. L., and Freed, C., *IEEE J. Quantum Electron.*, QE-22, 234-267, 1986.
2. Clairon, A., Dahmani, B., Filimon, A., and Rutman, J., *IEEE Trans. Inst. Meas.*, IM-34, 265-268, 1985.
3. Weiss, C. O., Kramer, G., Lipphardt, B., and Garcia, E., *IEEE J. Quantum Electron.*, QE-24, 1970-1972, 1988.
4. Bagayev, S. N., Baklanov, A. E., Chebotayev, V. P., and Dychkov, A. S., *Appl. Phys.*, B-48, 31-35, 1989.
5. Blaney, T. G., Bradley, C. C., Edwards, G. J., Jolliffe, B. W., Knight, D. J. E., Rowley, W. R. C., Shotten, K. C., and Woods, P. T., *Proc. R. Soc. Lond.*, A-355, 61-88, 1977.
6. Chardonnet, Ch., Van Lerberghe, A., and Bordé, Ch. J., *Opt. Comm.*, 58, 333-337, 1986.
7. Clairon, A., Acef, O., Chardonnet, Ch., and Bordé, Ch. J., *Frequency Standards and Metrology*, De Marchi, A., Ed., Springer-Verlag, Berlin, Heidelberg, 1989, p. 212.
8. Evenson, K., private communication.

Frequencies for the $00°1$-$(10°0,02°0)_I$ and $00°1$-$(10°0,02°0)_{II}$ Bands of $^{12}C^{16}O_2$ with the Estimated 2-σ Uncertainties

| | Band I | | | Band II | |
| | Frequency | Uncertainty | | Frequency | Uncertainty |
Line	(MHz)	(MHz)	Line	(MHz)	(MHz)
P(70)	26721305.4647	0.1680	P(70)	29789856.3783	0.0308
P(68)	26794232.6712	0.1217	P(68)	29861850.7690	0.0192
P(66)	26866318.8073	0.0867	P(66)	29933216.1760	0.0122
P(64)	26937571.7234	0.0606	P(64)	30003944.2861	0.0086
P(62)	27007998.9216	0.0415	P(62)	30074026.9127	0.0072
P(60)	27077607.5643	0.0279	P(60)	30143456.0039	0.0066
P(58)	27146404.4834	0.0185	P(58)	30212223.6504	0.0061
P(56)	27214396.1873	0.0121	P(56)	30280322.0930	0.0055
P(54)	27281588.8696	0.0081	P(54)	30347743.7306	0.0049
P(52)	27347988.4161	0.0057	P(52)	30414481.1273	0.0044
P(50)	27413600.4119	0.0043	P(50)	30480527.0196	0.0041
P(48)	27478430.1487	0.0036	P(48)	30545874.3239	0.0039
P(46)	27542482.6310	0.0032	P(46)	30610516.1429	0.0039
P(44)	27605762.5826	0.0030	P(44)	30674445.7724	0.0039
P(42)	27668274.4525	0.0028	P(42)	30737656.7080	0.0039
P(40)	27730022.4206	0.0027	P(40)	30800142.6511	0.0039
P(38)	27791010.4036	0.0026	P(38)	30861897.5150	0.0038
P(36)	27851242.0594	0.0025	P(36)	30922915.4310	0.0037
P(34)	27910720.7927	0.0024	P(34)	30983190.7534	0.0037
P(32)	27969449.7593	0.0023	P(32)	31042718.0652	0.0037
P(30)	28027431.8708	0.0022	P(30)	31101492.1833	0.0036
P(28)	28084669.7981	0.0021	P(28)	31159508.1631	0.0037
P(26)	28141165.9762	0.0020	P(26)	31216761.3029	0.0037
P(24)	28196922.6067	0.0019	P(24)	31273247.1487	0.0037
P(22)	28251941.6622	0.0017	P(22)	31328961.4978	0.0037
P(20)	28306224.8888	0.0016	P(20)	31383900.4028	0.0037
P(18)	28359773.8090	0.0014	P(18)	31438060.1749	0.0037
P(16)	28412589.7245	0.0012	P(16)	31491437.3872	0.0036
P(14)	28464673.7184	0.0011	P(14)	31544028.8776	0.0036
P(12)	28516026.6574	0.0009	P(12)	31595831.7516	0.0036
P(10)	28566649.1935	0.0008	P(10)	31646843.3843	0.0035
P(8)	28616541.7661	0.0008	P(8)	31697061.4225	0.0035
P(6)	28665704.6027	0.0008	P(6)	31746483.7868	0.0035

Line	Band I Frequency (MHz)	Uncertainty (MHz)	Line	Band II Frequency (MHz)	Uncertainty (MHz)
P(4)	28714137.7205	0.0008	P(4)	31795108.6724	0.0035
P(2)	28761840.9272	0.0008	P(2)	31842934.5511	0.0035
R(0)	28832026.2198	0.0008	R(0)	31913172.5691	0.0035
R(2)	28877902.4382	0.0007	R(2)	31958996.0621	0.0034
R(4)	28923046.4303	0.0006	R(4)	32004017.3822	0.0034
R(6)	28967457.0657	0.0005	R(6)	32048236.2498	0.0034
R(8)	29011133.0054	0.0003	R(8)	32091652.6619	0.0034
R(10)	29054072.7010	0.0001	R(10)	32134266.8917	0.0034
R(12)	29096274.3935	0.0003	R(12)	32176079.4878	0.0034
R(14)	29137736.1129	0.0005	R(14)	32217091.2721	0.0035
R(16)	29178455.6759	0.0007	R(16)	32257303.3386	0.0036
R(18)	29218430.6852	0.0009	R(18)	32296717.0510	0.0037
R(20)	29257658.5269	0.0010	R(20)	32335334.0408	0.0038
R(22)	29296136.3689	0.0011	R(22)	32373156.2044	0.0039
R(24)	29333861.1583	0.0012	R(24)	32410185.7003	0.0041
R(26)	29370829.6191	0.0011	R(26)	32446424.9459	0.0042
R(28)	29407038.2491	0.0011	R(28)	32481876.6140	0.0042
R(30)	29442483.3168	0.0011	R(30)	32516543.6293	0.0042
R(32)	29477160.8582	0.0012	R(32)	32550429.1641	0.0042
R(34)	29511066.6733	0.0013	R(34)	32583536.6340	0.0042
R(36)	29544196.3221	0.0015	R(36)	32615869.6937	0.0041
R(38)	29576545.1205	0.0017	R(38)	32647432.2320	0.0040
R(40)	29608108.1360	0.0019	R(40)	32678228.3665	0.0039
R(42)	29638880.1831	0.0022	R(42)	32708262.4386	0.0038
R(44)	29668855.8183	0.0024	R(44)	32737539.0081	0.0039
R(46)	29698029.3350	0.0027	R(46)	32766062.8469	0.0041
R(48)	29726394.7582	0.0032	R(48)	32793838.9334	0.0045
R(50)	29753945.8385	0.0037	R(50)	32820872.4463	0.0055
R(52)	29780676.0464	0.0042	R(52)	32847168.7576	0.0071
R(54)	29806578.5659	0.0047	R(54)	32872733.4269	0.0099
R(56)	29831646.2878	0.0052	R(56)	32897572.1935	0.0141
R(58)	29855871.8032	0.0058	R(58)	32921690.9701	0.0202
R(60)	29879247.3960	0.0074	R(60)	32945095.8355	0.0288
R(62)	29901765.0357	0.0113	R(62)	32967793.0268	0.0407
R(64)	29923416.3695	0.0186	R(64)	32989788.9322	0.0567
R(66)	29944192.7145	0.0302	R(66)	33011090.0831	0.0780
R(68)	29964085.0488	0.0475	R(68)	33031703.1467	0.1060
R(70)	29983084.0036	0.0720	R(70)	33051634.9172	0.1423

Frequencies for the $00°1$-$(10°0,02°0)_I$ and $00°1$-$(10°0,02°0)_{II}$ Bands of $^{13}C^{16}O_2$ with the Estimated 2-σ Uncertainties

Line	Band I Frequency (MHz)	Uncertainty (MHz)	Line	Band II Frequency (MHz)	Uncertainty (MHz)
P(66)	25523832.1808	0.7836	P(66)	28512082.5283	1.2894
P(64)	25590013.4703	0.5415	P(64)	28585121.9396	0.9194
P(62)	25655543.6502	0.3629	P(62)	28657449.4180	0.6420
P(60)	25720428.2487	0.2339	P(60)	28729056.6374	0.4375
P(58)	25784672.4840	0.1430	P(58)	28799935.4147	0.2897
P(56)	25848281.2771	0.0810	P(56)	28870077.7187	0.1853
P(54)	25911259.2627	0.0405	P(54)	28939475.6771	0.1135
P(52)	25973610.8005	0.0157	P(52)	29008121.5846	0.0659
P(50)	26035339.9857	0.0045	P(50)	29076007.9109	0.0357
P(48)	26096450.6582	0.0079	P(48)	29143127.3077	0.0180
P(46)	26156946.4123	0.0101	P(46)	29209472.6164	0.0090

INFRARED LASER FREQUENCIES (continued)

Line	Band I Frequency (MHz)	Uncertainty (MHz)	Line	Band II Frequency (MHz)	Uncertainty (MHz)
P(44)	26216830.6053	0.0101	P(44)	29275036.8754	0.0058
P(42)	26276106.3655	0.0090	P(42)	29339813.3270	0.0050
P(40)	26334776.6003	0.0077	P(40)	29403795.4243	0.0044
P(38)	26392844.0030	0.0068	P(38)	29466976.8383	0.0037
P(36)	26450311.0599	0.0063	P(36)	29529351.4635	0.0032
P(34)	26507180.0565	0.0061	P(34)	29590913.4252	0.0029
P(32)	26563453.0836	0.0060	P(32)	29651657.0844	0.0028
P(30)	26619132.0428	0.0058	P(30)	29711577.0447	0.0028
P(28)	26674218.6515	0.0055	P(28)	29770668.1566	0.0031
P(26)	26728714.4479	0.0054	P(26)	29828925.5239	0.0035
P(24)	26782620.7952	0.0054	P(24)	29886344.5074	0.0041
P(22)	26835938.8858	0.0054	P(22)	29942920.7308	0.0046
P(20)	26888669.7451	0.0055	P(20)	29998650.0838	0.0051
P(18)	26940814.2347	0.0055	P(18)	30053528.7271	0.0054
P(16)	26992373.0555	0.0055	P(16)	30107553.0955	0.0055
P(14)	27043346.7508	0.0054	P(14)	30160719.9016	0.0055
P(12)	27093735.7083	0.0052	P(12)	30213026.1388	0.0054
P(10)	27143540.1624	0.0051	P(10)	30264469.0839	0.0054
P(8)	27192760.1962	0.0049	P(8)	30315046.2994	0.0054
P(6)	27241395.7431	0.0048	P(6)	30364755.6359	0.0055
P(4)	27289446.5880	0.0047	P(4)	30413595.2335	0.0056
P(2)	27336912.3682	0.0046	P(2)	30461563.5231	0.0057
R(0)	27407012.8882	0.0045	P(0)	30531879.5415	0.0057
R(2)	27453013.4589	0.0043	P(2)	30577664.6138	0.0056
R(4)	27498426.5430	0.0040	P(4)	30622575.1885	0.0054
R(6)	27543251.1200	0.0037	P(6)	30666611.0128	0.0051
R(8)	27587486.0225	0.0034	P(8)	30709772.1257	0.0047
R(10)	27631129.9356	0.0031	P(10)	30752058.8571	0.0045
R(12)	27674181.3963	0.0029	P(12)	30793471.8269	0.0044
R(14)	27716638.7917	0.0029	P(14)	30834011.9425	0.0043
R(16)	27758500.3577	0.0029	P(16)	30873680.3976	0.0044
R(18)	27799764.1770	0.0029	P(18)	30912478.6694	0.0044
R(20)	27840428.1773	0.0030	P(20)	30950408.5159	0.0044
R(22)	27880490.1283	0.0029	P(22)	30987471.9732	0.0043
R(24)	27919947.6395	0.0029	P(24)	31023671.3517	0.0042
R(26)	27958798.1567	0.0028	P(26)	31059009.2327	0.0042
R(28)	27997038.9591	0.0028	P(28)	31093488.4642	0.0042
R(30)	28034667.1551	0.0027	P(30)	31127112.1569	0.0043
R(32)	28071679.6785	0.0027	P(32)	31159883.6793	0.0045
R(34)	28108073.2842	0.0026	P(34)	31191806.6529	0.0046
R(36)	28143844.5432	0.0026	P(36)	31222884.9469	0.0048
R(38)	28178989.8377	0.0026	P(38)	31253122.6730	0.0053
R(40)	28213505.3554	0.0028	P(40)	31282524.1795	0.0061
R(42)	28247387.0838	0.0033	P(42)	31311094.0452	0.0077
R(44)	28280630.8035	0.0046	P(44)	31338837.0736	0.0108
R(46)	28313232.0818	0.0083	P(46)	31365758.2858	0.0173
R(48)	28345186.2652	0.0161	P(48)	31391862.9147	0.0295
R(50)	28376488.4720	0.0301	P(50)	31417156.3972	0.0505
R(52)	28407133.5839	0.0531	P(52)	31441644.3679	0.0845
R(54)	28437116.2372	0.0887	P(54)	31465332.6516	0.1366
R(56)	28466430.8141	0.1419	P(56)	31488227.2557	0.2138
R(58)	28495071.4324	0.2188	P(58)	31510334.3631	0.3247
R(60)	28523031.9357	0.3271	P(60)	31531660.3243	0.4800
R(62)	28550305.8819	0.4763	P(62)	31552211.6497	0.6932
R(64)	28576886.5323	0.6781	P(64)	31571995.0017	0.9805
R(66)	28602766.8393	0.9467	P(66)	31591017.1868	1.3619

Frequencies for the $00^\circ1$-$(10^\circ0,02^\circ0)_I$ and $00^\circ1$-$(10^\circ0,02^\circ0)_{II}$ Bands of $^{12}C^{18}O_2$ with the Estimated 2-σ Uncertainties

Line	Band I Frequency (MHz)	Uncertainty (MHz)	Line	Band II Frequency (MHz)	Uncertainty (MHz)
P(70)	27045326.3119	0.4540	P(70)	30695237.5856	0.0858
P(68)	27114914.0922	0.3324	P(68)	30755520.2231	0.0570
P(66)	27183635.7945	0.2392	P(66)	30815311.4928	0.0364
P(64)	27251496.4118	0.1688	P(64)	30874607.2084	0.0223
P(62)	27318500.7361	0.1165	P(62)	30933403.2309	0.0131
P(60)	27384653.3618	0.0783	P(60)	30991695.4724	0.0075
P(58)	27449958.6881	0.0510	P(58)	31049479.9009	0.0049
P(56)	27514420.9224	0.0319	P(56)	31106752.5446	0.0041
P(54)	27578044.0828	0.0191	P(54)	31163509.4964	0.0040
P(52)	27640832.0010	0.0108	P(52)	31219746.9183	0.0040
P(50)	27702788.3248	0.0059	P(50)	31275461.0455	0.0039
P(48)	27763916.5206	0.0035	P(48)	31330648.1908	0.0039
P(46)	27824219.8762	0.0028	P(46)	31385304.7490	0.0039
P(44)	27883701.5029	0.0026	P(44)	31439427.2006	0.0039
P(42)	27942364.3379	0.0025	P(42)	31493012.1163	0.0038
P(40)	28000211.1464	0.0024	P(40)	31546056.1605	0.0038
P(38)	28057244.5242	0.0022	P(38)	31598556.0954	0.0037
P(36)	28113466.8992	0.0021	P(36)	31650508.7847	0.0037
P(34)	28168880.5335	0.0020	P(34)	31701911.1970	0.0037
P(32)	28223487.5256	0.0019	P(32)	31752760.4093	0.0037
P(30)	28277289.8118	0.0017	P(30)	31803053.6105	0.0037
P(28)	28330289.1679	0.0016	P(28)	31852788.1043	0.0038
P(26)	28382487.2111	0.0015	P(26)	31901961.3125	0.0038
P(24)	28433885.4012	0.0013	P(24)	31950570.7773	0.0038
P(22)	28484485.0420	0.0012	P(22)	31998614.1649	0.0038
P(20)	28534287.2828	0.0011	P(20)	32046089.2669	0.0037
P(18)	28583293.1193	0.0010	P(18)	32092994.0036	0.0037
P(16)	28631503.3952	0.0010	P(16)	32139326.4254	0.0036
P(14)	28678918.8025	0.0009	P(14)	32185084.7154	0.0036
P(12)	28725539.8830	0.0010	P(12)	32230267.1907	0.0036
P(10)	28771367.0288	0.0010	P(10)	32274872.3041	0.0037
P(8)	28816400.4829	0.0010	P(8)	32318898.6455	0.0038
P(6)	28860640.3403	0.0011	P(6)	32362344.9434	0.0039
P(4)	28904086.5477	0.0011	P(4)	32405210.0652	0.0041
P(2)	28946738.9048	0.0011	P(2)	32447493.0185	0.0041
R(0)	29009228.1702	0.0010	P(0)	32509824.0580	0.0042
R(2)	29049894.0586	0.0010	P(2)	32550648.1723	0.0042
R(4)	29089764.2368	0.0009	P(4)	32590887.7542	0.0042
R(6)	29128837.8426	0.0008	P(6)	32630542.4457	0.0041
R(8)	29167113.8668	0.0008	P(8)	32669612.0295	0.0041
R(10)	29204591.1529	0.0009	P(10)	32708096.4282	0.0040
R(12)	29241268.3964	0.0010	P(12)	32745995.7040	0.0040
R(14)	29277144.1444	0.0011	P(14)	32783310.0573	0.0040
R(16)	29312216.7955	0.0012	P(16)	32820039.8258	0.0040
R(18)	29346484.5984	0.0012	P(18)	32856185.4827	0.0040
R(20)	29379945.6517	0.0013	P(20)	32891747.6358	0.0040
R(22)	29412597.9024	0.0013	P(22)	32926727.0254	0.0040
R(24)	29444439.1458	0.0013	P(24)	32961124.5220	0.0040
R(26)	29475467.0236	0.0014	P(26)	32994941.1249	0.0040
R(28)	29505679.0230	0.0015	P(28)	33028177.9594	0.0040
R(30)	29535072.4755	0.0016	P(30)	33060836.2743	0.0040
R(32)	29563644.5557	0.0018	P(32)	33092917.4394	0.0041
R(34)	29591392.2794	0.0020	P(34)	33124422.9429	0.0043
R(36)	29618312.5023	0.0023	P(36)	33155354.3878	0.0046
R(38)	29644401.9182	0.0028	P(38)	33185713.4894	0.0049

Line	Band I Frequency (MHz)	Uncertainty (MHz)	Line	Band II Frequency (MHz)	Uncertainty (MHz)
R(40)	29669657.0575	0.0036	P(40)	33215502.0716	0.0056
R(42)	29694074.2853	0.0053	P(42)	33244722.0637	0.0068
R(44)	29717649.7992	0.0082	P(44)	33273375.4969	0.0092
R(46)	29740379.6276	0.0128	P(46)	33301464.5003	0.0134
R(48)	29762259.6274	0.0200	P(48)	33328991.2976	0.0199
R(50)	29783285.4820	0.0307	P(50)	33355958.2027	0.0294
R(52)	29803452.6988	0.0461	P(52)	33382367.6161	0.0427
R(54)	29822756.6072	0.0681	P(54)	33408222.0209	0.0607
R(56)	29841192.3558	0.0985	P(56)	33433523.9780	0.0848
R(58)	29858754.9100	0.1401	P(58)	33458276.1228	0.1165
R(60)	29875439.0495	0.1960	P(60)	33482481.1601	0.1576
R(62)	29891239.3658	0.2702	P(62)	33506141.8605	0.2104
R(64)	29906150.2589	0.3673	P(64)	33529261.0556	0.2775
R(66)	29920165.9352	0.4930	P(66)	33551841.6335	0.3621
R(68)	29933280.4042	0.6540	P(68)	33573886.5352	0.4679
R(70)	29945487.4756	0.8581	P(70)	33595398.7493	0.5992

Frequencies for the $00^\circ 1$-$(10^\circ 0, 02^\circ 0)_I$ and $00^\circ 1$-$(10^\circ 0, 02^\circ 0)_{II}$ Bands of $^{13}C^{18}O_2$ with the Estimated 2-σ Uncertainties

Line	Band I Frequency (MHz)	Uncertainty (MHz)	Line	Band II Frequency (MHz)	Uncertainty (MHz)
P(70)	25967863.7652	1.1146	P(70)	28960476.2278	0.4069
P(68)	26033448.2798	0.8152	P(68)	29022326.9578	0.2861
P(66)	26098273.9159	0.5860	P(66)	29083661.3546	0.1961
P(64)	26162346.4813	0.4129	P(64)	29144473.5795	0.1303
P(62)	26225671.5466	0.2844	P(62)	29204757.8761	0.0833
P(60)	26288254.4494	0.1906	P(60)	29264508.5768	0.0507
P(58)	26350100.2984	0.1237	P(58)	29323720.1086	0.0290
P(56)	26411213.9778	0.0772	P(56)	29382386.9988	0.0152
P(54)	26471600.1504	0.0459	P(54)	29440503.8809	0.0073
P(52)	26531263.2618	0.0258	P(52)	29498065.4997	0.0038
P(50)	26590207.5442	0.0138	P(50)	29555066.7172	0.0032
P(48)	26648437.0195	0.0077	P(48)	29611502.5178	0.0031
P(46)	26705955.5026	0.0057	P(46)	29667368.0132	0.0031
P(44)	26762766.6051	0.0055	P(44)	29722658.4475	0.0034
P(42)	26818873.7378	0.0055	P(42)	29777369.2022	0.0039
P(40)	26874280.1143	0.0056	P(40)	29831495.8006	0.0044
P(38)	26928988.7531	0.0056	P(38)	29885033.9125	0.0049
P(36)	26983002.4809	0.0056	P(36)	29937979.3584	0.0053
P(34)	27036323.9351	0.0055	P(34)	29990328.1139	0.0054
P(32)	27088955.5657	0.0054	P(32)	30042076.3132	0.0055
P(30)	27140899.6384	0.0051	P(30)	30093220.2534	0.0055
P(28)	27192158.2363	0.0049	P(28)	30143756.3978	0.0054
P(26)	27242733.2620	0.0047	P(26)	30193681.3793	0.0053
P(24)	27292626.4396	0.0044	P(24)	30242992.0038	0.0052
P(22)	27341839.3165	0.0042	P(22)	30291685.2529	0.0051
P(20)	27390373.2651	0.0040	P(20)	30339758.2870	0.0049
P(18)	27438229.4843	0.0037	P(18)	30387208.4477	0.0048
P(16)	27485409.0008	0.0035	P(16)	30434033.2603	0.0046
P(14)	27531912.6704	0.0033	P(14)	30480230.4356	0.0045
P(12)	27577741.1795	0.0031	P(12)	30525797.8725	0.0044
P(10)	27622895.0455	0.0031	P(10)	30570733.6593	0.0043
P(8)	27667374.6182	0.0031	P(8)	30615036.0750	0.0043
P(6)	27711180.0803	0.0033	P(6)	30658703.5912	0.0044
P(4)	27754311.4480	0.0034	P(4)	30701734.8727	0.0045

Line	Band I Frequency (MHz)	Uncertainty (MHz)	Line	Band II Frequency (MHz)	Uncertainty (MHz)
P(2)	27796768.5718	0.0036	P(2)	30744128.7785	0.0045
R(0)	27859189.3155	0.0036	P(0)	30806522.5414	0.0045
R(2)	27899959.0889	0.0035	P(2)	30847319.2956	0.0044
R(4)	27940052.7921	0.0033	P(4)	30887476.2168	0.0043
R(6)	27979469.5315	0.0031	P(6)	30926993.0424	0.0042
R(8)	28018208.2478	0.0028	P(8)	30965869.7046	0.0041
R(10)	28056267.7161	0.0026	P(10)	31004106.3298	0.0040
R(12)	28093646.5448	0.0025	P(12)	31041703.2379	0.0040
R(14)	28130343.1757	0.0025	P(14)	31078660.9408	0.0040
R(16)	28166355.8825	0.0025	P(16)	31114980.1420	0.0040
R(18)	28201682.7706	0.0025	P(18)	31150661.7340	0.0041
R(20)	28236321.7757	0.0025	P(20)	31185706.7976	0.0042
R(22)	28270270.6628	0.0024	P(22)	31220116.5992	0.0043
R(24)	28303527.0249	0.0024	P(24)	31253892.5891	0.0043
R(26)	28336088.2817	0.0023	P(26)	31287036.3991	0.0044
R(28)	28367951.6781	0.0024	P(28)	31319549.8396	0.0043
R(30)	28399114.2823	0.0025	P(30)	31351434.8973	0.0043
R(32)	28429572.9843	0.0026	P(32)	31382693.7318	0.0042
R(34)	28459324.4940	0.0028	P(34)	31413328.6728	0.0042
R(36)	28488365.3390	0.0029	P(36)	31443342.2165	0.0041
R(38)	28516691.8625	0.0029	P(38)	31472737.0219	0.0040
R(40)	28544300.2211	0.0031	P(40)	31501515.9074	0.0039
R(42)	28571186.3823	0.0032	P(42)	31529681.8467	0.0040
R(44)	28597346.1222	0.0032	P(44)	31557237.9646	0.0042
R(46)	28622775.0223	0.0038	P(46)	31584187.5329	0.0046
R(48)	28647468.4672	0.0071	P(48)	31610533.9656	0.0057
R(50)	28671421.6417	0.0148	P(50)	31636280.8146	0.0088
R(52)	28694629.5272	0.0286	P(52)	31661431.7650	0.0151
R(54)	28717086.8993	0.0510	P(54)	31685990.6298	0.0261
R(56)	28738788.3239	0.0852	P(56)	31709961.3449	0.0434
R(58)	28759728.1540	0.1355	P(58)	31733347.9642	0.0693
R(60)	28779900.5263	0.2075	P(60)	31756154.6537	0.1068
R(62)	28799299.3572	0.3078	P(62)	31778385.6867	0.1594
R(64)	28817918.3393	0.4447	P(64)	31800045.4375	0.2317
R(66)	28835750.9374	0.6283	P(66)	31821138.3761	0.3291
R(68)	28852790.3843	0.8707	P(68)	31841669.0622	0.4581
R(70)	28869029.6768	1.1863	P(70)	31861642.1394	0.6268

Frequencies for the $01^{1e}1$-$(11^{1e}0,03^{1e}0)_I$ and $01^{1e}1$-$(11^{1e}0,03^{1e}0)_{II}$ Bands of $^{12}C^{16}O_2$ with the Estimated 2-σ Uncertainties

Line	Band I Frequency (MHz)	Uncertainty (MHz)	Line	Band II Frequency (MHz)	Uncertainty (MHz)
P(59)	26125213.2723	1.6633	P(59)	30427055.2899	0.1962
P(57)	26191576.6703	1.0880	P(57)	30494640.3229	0.1332
P(55)	26257240.7898	0.6844	P(55)	30561557.5929	0.0865
P(53)	26322208.2302	0.4094	P(53)	30627802.0344	0.0530
P(51)	26386481.4313	0.2286	P(51)	30693368.7014	0.0306
P(49)	26450062.6783	0.1155	P(49)	30758252.7710	0.0175
P(47)	26512954.1076	0.0498	P(47)	30822449.5469	0.0123
P(45)	26575157.7109	0.0191	P(45)	30885954.4624	0.0114
P(43)	26636675.3402	0.0160	P(43)	30948763.0834	0.0109
P(41)	26697508.7115	0.0182	P(41)	31010871.1119	0.0100
P(39)	26757659.4084	0.0177	P(39)	31072274.3882	0.0091
P(37)	26817128.8857	0.0160	P(37)	31132968.8940	0.0091
P(35)	26875918.4726	0.0144	P(35)	31192950.7549	0.0102

Line	Band I Frequency (MHz)	Uncertainty (MHz)	Line	Band II Frequency (MHz)	Uncertainty (MHz)
P(33)	26934029.3751	0.0131	P(33)	31252216.2430	0.0118
P(31)	26991462.6787	0.0119	P(31)	31310761.7788	0.0134
P(29)	27048219.3509	0.0106	P(29)	31368583.9339	0.0147
P(27)	27104300.2431	0.0096	P(27)	31425679.4328	0.0155
P(25)	27159706.0925	0.0093	P(25)	31482045.1550	0.0157
P(23)	27214437.5237	0.0097	P(23)	31537678.1367	0.0154
P(21)	27268495.0505	0.0104	P(21)	31592575.5725	0.0147
P(19)	27321879.0769	0.0108	P(19)	31646734.8172	0.0137
P(17)	27374589.8987	0.0108	P(17)	31700153.3868	0.0127
P(15)	27426627.7040	0.0104	P(15)	31752828.9602	0.0119
P(13)	27477992.5747	0.0098	P(13)	31804759.3803	0.0113
P(11)	27528684.4867	0.0096	P(11)	31855942.6551	0.0113
P(9)	27578703.3113	0.0101	P(9)	31906376.9582	0.0116
P(7)	27628048.8151	0.0113	P(7)	31956060.6304	0.0122
P(5)	27676720.6609	0.0127	P(5)	32004992.1796	0.0129
P(3)	27724718.4080	0.0141	P(3)	32053170.2819	0.0136
R(1)	27841759.7696	0.0152	P(1)	32170312.0391	0.0149
R(3)	27887393.2105	0.0146	P(3)	32215845.0845	0.0151
R(5)	27932349.2934	0.0135	P(5)	32260620.8121	0.0152
R(7)	27976627.0108	0.0124	P(7)	32304638.8261	0.0152
R(9)	28020225.2521	0.0115	P(9)	32347898.8990	0.0150
R(11)	28063142.8031	0.0110	P(11)	32390400.9714	0.0148
R(13)	28105378.3457	0.0109	P(13)	32432145.1513	0.0145
R(15)	28146930.4576	0.0109	P(15)	32473131.7137	0.0142
R(17)	28187797.6116	0.0107	P(17)	32513361.0997	0.0140
R(19)	28227978.1750	0.0103	P(19)	32552833.9153	0.0140
R(21)	28267470.4088	0.0099	P(21)	32591550.9309	0.0141
R(23)	28306272.4666	0.0099	P(23)	32629513.0796	0.0143
R(25)	28344382.3939	0.0107	P(25)	32666721.4564	0.0144
R(27)	28381798.1267	0.0122	P(27)	32703177.3164	0.0142
R(29)	28418517.4902	0.0141	P(29)	32738882.0732	0.0136
R(31)	28454538.1976	0.0165	P(31)	32773837.2976	0.0136
R(33)	28489857.8477	0.0213	P(33)	32808044.7156	0.0174
R(35)	28524473.9240	0.0312	P(35)	32841506.2063	0.0279
R(37)	28558383.7917	0.0486	P(37)	32874223.8000	0.0462
R(39)	28591584.6963	0.0754	P(39)	32906199.6761	0.0735
R(41)	28624073.7602	0.1131	P(41)	32937436.1606	0.1114
R(43)	28655847.9806	0.1644	P(43)	32967935.7238	0.1624
R(45)	28686904.2261	0.2328	P(45)	32997700.9775	0.2292
R(47)	28717239.2334	0.3239	P(47)	33026734.6728	0.3151
R(49)	28746849.6038	0.4465	P(49)	33055039.6965	0.4238
R(51)	28775731.7988	0.6142	P(51)	33082619.0689	0.5595
R(53)	28803882.1361	0.8465	P(53)	33109475.9403	0.7272

Frequencies for the $01^{1f}1$-$(11^{1f}0,03^{1f}0)_I$ and $01^{1f}1$-$(11^{1f}0,03^{1f}0)_{II}$ Bands of $^{12}C^{16}O_2$ with the Estimated 2-σ Uncertainties

Line	Band I Frequency (MHz)	Uncertainty (MHz)	Line	Band II Frequency (MHz)	Uncertainty (MHz)
P(60)	26051570.0104	4.4521	P(60)	30355115.0204	0.2752
P(58)	26120964.4932	3.0629	P(58)	30425283.5969	0.1926
P(56)	26189552.8496	2.0516	P(56)	30494732.8293	0.1301
P(54)	26257339.6006	1.3305	P(54)	30563455.6325	0.0840
P(52)	26324329.0344	0.8289	P(52)	30631445.1076	0.0512
P(50)	26390525.2136	0.4901	P(50)	30698694.5456	0.0292
P(48)	26455931.9824	0.2698	P(48)	30765197.4310	0.0163

INFRARED LASER FREQUENCIES (continued)

Line	Band I Frequency (MHz)	Uncertainty (MHz)	Line	Band II Frequency (MHz)	Uncertainty (MHz)
P(46)	26520552.9722	0.1334	P(46)	30830947.4444	0.0111
P(44)	26584391.6075	0.0551	P(44)	30895938.4662	0.0104
P(42)	26647451.1105	0.0181	P(42)	30960164.5794	0.0105
P(40)	26709734.5057	0.0151	P(40)	31023620.0723	0.0105
P(38)	26771244.6242	0.0174	P(38)	31086299.4415	0.0107
P(36)	26831984.1067	0.0157	P(36)	31148197.3941	0.0114
P(34)	26891955.4069	0.0126	P(34)	31209308.8510	0.0126
P(32)	26951160.7945	0.0105	P(32)	31269628.9481	0.0138
P(30)	27009602.3576	0.0096	P(30)	31329153.0395	0.0147
P(28)	27067282.0045	0.0092	P(28)	31387876.6994	0.0151
P(26)	27124201.4662	0.0090	P(26)	31445795.7236	0.0149
P(24)	27180362.2977	0.0089	P(24)	31502906.1318	0.0141
P(22)	27235765.8792	0.0090	P(22)	31559204.1695	0.0128
P(20)	27290413.4182	0.0093	P(20)	31614686.3091	0.0113
P(18)	27344305.9494	0.0096	P(18)	31669349.2515	0.0098
P(16)	27397444.3368	0.0097	P(16)	31723189.9280	0.0086
P(14)	27449829.2733	0.0096	P(14)	31776205.5007	0.0081
P(12)	27501461.2824	0.0096	P(12)	31828393.3642	0.0085
P(10)	27552340.7179	0.0101	P(10)	31879751.1463	0.0095
P(8)	27602467.7649	0.0111	P(8)	31930276.7092	0.0107
P(6)	27651842.4399	0.0125	P(6)	31979968.1497	0.0120
P(4)	27700464.5912	0.0139	P(4)	32028823.8002	0.0131
P(2)	27748333.8988	0.0148	P(2)	32076842.2290	0.0139
R(2)	27864709.8633	0.0146	P(2)	32193218.1935	0.0150
R(4)	27909939.2762	0.0135	P(4)	32238298.4853	0.0151
R(6)	27954412.3294	0.0122	P(6)	32282538.0393	0.0153
R(8)	27998127.7801	0.0112	P(8)	32325936.7244	0.0153
R(10)	28041084.2173	0.0107	P(10)	32368494.6458	0.0154
R(12)	28083280.0620	0.0108	P(12)	32410212.1438	0.0155
R(14)	28124713.5668	0.0110	P(14)	32451089.7941	0.0155
R(16)	28165382.8151	0.0112	P(16)	32491128.4063	0.0154
R(18)	28205285.7213	0.0111	P(18)	32530329.0234	0.0152
R(20)	28244420.0302	0.0110	P(20)	32568692.9211	0.0149
R(22)	28282783.3158	0.0114	P(22)	32606221.6061	0.0147
R(24)	28320372.9812	0.0129	P(24)	32642916.8154	0.0144
R(26)	28357186.2574	0.0149	P(26)	32678780.5147	0.0140
R(28)	28393220.2023	0.0168	P(28)	32713814.8971	0.0133
R(30)	28428471.6994	0.0175	P(30)	32748022.3813	0.0120
R(32)	28462937.4565	0.0165	P(32)	32781405.6101	0.0106
R(34)	28496614.0042	0.0142	P(34)	32813967.4482	0.0122
R(36)	28529497.6934	0.0163	P(36)	32845710.9809	0.0212
R(38)	28561584.6939	0.0309	P(38)	32876639.5111	0.0385
R(40)	28592870.9914	0.0593	P(40)	32906756.5580	0.0646
R(42)	28623352.3850	0.1054	P(42)	32936065.8540	0.1015
R(44)	28653024.4839	0.1779	P(44)	32964571.3426	0.1518
R(46)	28681882.7038	0.2907	P(46)	32992277.1760	0.2184
R(48)	28709922.2632	0.4635	P(48)	33019187.7118	0.3048
R(50)	28737138.1785	0.7231	P(50)	33045307.5105	0.4152

INFRARED AND FAR-INFRARED ABSORPTION FREQUENCY STANDARDS
Arthur Maki

Aside from the CO_2 laser transitions, the absorption spectrum of CO has been more accurately and thoroughly measured than any other spectrum. A bibliography of earlier measurements on CO is given by Maki and Wells,[1] and the present tables were calculated from the measurements referred to in that work. In addition, some new and very accurate frequency measurements[2,3] have been made and were incorporated in the present tables. The frequencies of the rotational transitions of HF and HCl were calculated from constants obtained from fitting the measurements of Evenson et al.[4,5] and Jennings and Wells.[6]

A new report on infrared wavenumber standards from the International Union of Pure and Applied Chemistry, Commission on Molecular Structure and Spectroscopy, may be found in Reference 7.

REFERENCES

1. Maki, A. G. and Wells, J. S., *Wavenumber Calibration Tables from Heterodyne Frequency Measurements*, NIST Special Publication 821, U.S. Dept. of Commerce, Washington, D.C., 1991.
2. Evenson, K. and Stroh, F., private communication.
3. George, T., Urban, W., and co-workers, private communication.
4. Jennings, D. A., Evenson, K. M., Zink, L. R., Demuynck, C., Destombes, J. L., Lemoine, B., and Johns, J. W. C., *J. Mol. Spectrosc.*, 122, 477-480, 1987.
5. Nolt, I. G., Radostitz, J. V., DiLonardo, G., Evenson, K. M., Jennings, D. A., Leopold, K. R., Vanek, M. D., Zink, L. R., Hinz, A., and Chance, K. V., *J. Mol. Spectrosc.*, 125, 274-287, 1987.
6. Jennings, D. A. and Wells, J. S., *J. Mol. Spectrosc.*, 130, 267-268, 1988.
7. *High Resolution Wavenumber Standards for the Infrared, Pure Appl. Chemistry,* 68, 193, 1996.

Wavenumbers for the v = 1-0 Band of CO

Wavenumber (unc)* cm^{-1}	Transition	Wavenumber (unc) cm^{-1}	Transition
		2147.081132(01)	R(0)
2139.426071(01)	P(1)	2150.856006(01)	R(1)
2135.546178(01)	P(2)	2154.595581(01)	R(2)
2131.631574(01)	P(3)	2158.299710(01)	R(3)
2127.682404(01)	P(4)	2161.968245(01)	R(4)
2123.698816(01)	P(5)	2165.601041(01)	R(5)
2119.680957(01)	P(6)	2169.197949(01)	R(6)
2115.628973(01)	P(7)	2172.758824(01)	R(7)
2111.543012(01)	P(8)	2176.283519(01)	R(8)
2107.423221(01)	P(9)	2179.771887(01)	R(9)
2103.269746(01)	P(10)	2183.223782(01)	R(10)
2099.082734(01)	P(11)	2186.639057(01)	R(11)
2094.862333(01)	P(12)	2190.017565(01)	R(12)
2090.608688(01)	P(13)	2193.359161(01)	R(13)
2086.321947(01)	P(14)	2196.663698(01)	R(14)
2082.002256(01)	P(15)	2199.931030(01)	R(15)
2077.649762(01)	P(16)	2203.161010(01)	R(16)
2073.264612(01)	P(17)	2206.353492(01)	R(17)
2068.846952(01)	P(18)	2209.508331(02)	R(18)
2064.396929(01)	P(19)	2212.625379(02)	R(19)
2059.914688(02)	P(20)	2215.704492(02)	R(20)
2055.400377(02)	P(21)	2218.745522(02)	R(21)
2050.854140(02)	P(22)	2221.748326(03)	R(22)
2046.276126(03)	P(23)	2224.712755(03)	R(23)
2041.666479(03)	P(24)	2227.638666(03)	R(24)
2037.025345(03)	P(25)	2230.525912(04)	R(25)
2032.352870(04)	P(26)	2233.374349(04)	R(26)
2027.649200(04)	P(27)	2236.183829(04)	R(27)
2022.914480(04)	P(28)	2238.954210(05)	R(28)
2018.148857(05)	P(29)	2241.685344(05)	R(29)
2013.352474(05)	P(30)	2244.377088(06)	R(30)
2008.525477(06)	P(31)	2247.029296(07)	R(31)
2003.668012(06)	P(32)	2249.641824(08)	R(32)
1998.780224(07)	P(33)	2252.214527(10)	R(33)
1993.862257(09)	P(34)	2254.747262(14)	R(34)

Wavenumber (unc)* cm^{-1}	Transition	Wavenumber (unc) cm^{-1}	Transition
1988.914257(11)	P(35)	2257.239883(18)	R(35)
1983.936367(14)	P(36)	2259.692248(24)	R(36)
1978.928733(18)	P(37)	2262.104213(33)	R(37)
1973.891500(25)	P(38)	2264.475634(45)	R(38)
1968.824811(34)	P(39)	2266.806368(61)	R(39)
1963.728813(46)	P(40)	2269.096273(81)	R(40)
1958.603648(61)	P(41)	2271.345206(106)	R(41)
1953.449462(82)	P(42)	2273.553027(139)	R(42)

* The uncertainty in the last digits (twice the standard error) is given in parentheses.

Wavenumbers for the v = 2-0 Band of CO

Wavenumber (unc)* cm^{-1}	Transition	Wavenumber (unc) cm^{-1}	Transition
		4263.837198(02)	R(0)
4256.217140(02)	P(1)	4267.542066(02)	R(1)
4252.302244(02)	P(2)	4271.176630(02)	R(2)
4248.317633(02)	P(3)	4274.740746(02)	R(3)
4244.263453(02)	P(4)	4278.234264(02)	R(4)
4240.139852(02)	P(5)	4281.657039(02)	R(5)
4235.946975(02)	P(6)	4285.008924(02)	R(6)
4231.684972(02)	P(7)	4288.289772(02)	R(7)
4227.353987(02)	P(8)	4291.499437(02)	R(8)
4222.954169(02)	P(9)	4294.637773(02)	R(9)
4218.485665(02)	P(10)	4297.704631(02)	R(10)
4213.948620(02)	P(11)	4300.699868(02)	R(11)
4209.343182(02)	P(12)	4303.623334(02)	R(12)
4204.669499(02)	P(13)	4306.474886(02)	R(13)
4199.927716(02)	P(14)	4309.254375(02)	R(14)
4195.117980(02)	P(15)	4311.961657(02)	R(15)
4190.240439(02)	P(16)	4314.596584(02)	R(16)
4185.295239(02)	P(17)	4317.159011(02)	R(17)
4180.282526(02)	P(18)	4319.648791(02)	R(18)
4175.202447(02)	P(19)	4322.065779(03)	R(19)
4170.055149(03)	P(20)	4324.409829(03)	R(20)
4164.840777(03)	P(21)	4326.680794(03)	R(21)
4159.559478(03)	P(22)	4328.878530(03)	R(22)
4154.211398(03)	P(23)	4331.002889(04)	R(23)
4148.796683(04)	P(24)	4333.053728(04)	R(24)
4143.315479(04)	P(25)	4335.030899(05)	R(25)
4137.767932(04)	P(26)	4336.934259(06)	R(26)
4132.154187(05)	P(27)	4338.763661(07)	R(27)
4126.474391(06)	P(28)	4340.518961(09)	R(28)
4120.728689(07)	P(29)	4342.200014(11)	R(29)
4114.917226(09)	P(30)	4343.806675(16)	R(30)
4109.040148(12)	P(31)	4345.338799(21)	R(31)
4103.097600(16)	P(32)	4346.796243(29)	R(32)
4097.089728(21)	P(33)	4348.178862(40)	R(33)
4091.016676(29)	P(34)	4349.486513(54)	R(34)
4084.878591(40)	P(35)	4350.719052(73)	R(35)
4078.675618(54)	P(36)	4351.876336(96)	R(36)
4072.407901(73)	P(37)	4352.958224(127)	R(37)
4066.075588(97)	P(38)	4353.964572(166)	R(38)
4059.678822(127)	P(39)	4354.895240(214)	R(39)

* The uncertainty in the last digits (twice the standard error) is given in parentheses.

Wavenumbers for the $v = 3$-0 Band of CO

Wavenumber (unc)* cm^{-1}	Transition	Wavenumber (unc) cm^{-1}	Transition
		6354.179057(13)	R(0)
6346.594000(13)	P(1)	6357.813923(13)	R(1)
6342.644103(13)	P(2)	6361.343487(13)	R(2)
6338.589491(13)	P(3)	6364.767599(13)	R(3)
6334.430309(13)	P(4)	6368.086115(13)	R(4)
6330.166705(13)	P(5)	6371.298887(13)	R(5)
6325.798826(13)	P(6)	6374.405768(12)	R(6)
6321.326819(13)	P(7)	6377.406611(12)	R(7)
6316.750831(12)	P(8)	6380.301271(12)	R(8)
6312.071008(12)	P(9)	6383.089600(12)	R(9)
6307.287498(12)	P(10)	6385.771452(12)	R(10)
6302.400447(12)	P(11)	6388.346680(13)	R(11)
6297.410003(12)	P(12)	6390.815139(13)	R(12)
6292.316311(13)	P(13)	6393.176681(13)	R(13)
6287.119520(13)	P(14)	6395.431160(13)	R(14)
6281.819775(13)	P(15)	6397.578430(13)	R(15)
6276.417224(13)	P(16)	6399.618344(13)	R(16)
6270.912012(13)	P(17)	6401.550757(13)	R(17)
6265.304287(13)	P(18)	6403.375523(13)	R(18)
6259.594194(13)	P(19)	6405.092495(14)	R(19)
6253.781880(13)	P(20)	6406.701527(14)	R(20)
6247.867492(14)	P(21)	6408.202474(14)	R(21)
6241.851176(14)	P(22)	6409.595189(15)	R(22)
6235.733077(14)	P(23)	6410.879527(15)	R(23)
6229.513342(15)	P(24)	6412.055343(16)	R(24)
6223.192117(15)	P(25)	6413.122491(17)	R(25)
6216.769547(16)	P(26)	6414.080825(19)	R(26)
6210.245778(17)	P(27)	6414.930201(23)	R(27)
6203.620957(19)	P(28)	6415.670474(28)	R(28)
6196.895229(23)	P(29)	6416.301500(37)	R(29)
6190.068739(28)	P(30)	6416.823133(50)	R(30)
6183.141633(37)	P(31)	6417.235231(67)	R(31)
6176.114058(50)	P(32)	6417.537649(90)	R(32)
6168.986159(67)	P(33)		
6161.758082(90)	P(34)		

* The uncertainty in the last digits (twice the standard error) is given in parentheses.

Frequencies and Wavenumbers for the Rotational Lines of CO

Frequency MHz	Uncertainty* MHz	J'	J''	Wavenumber cm^{-1}	Uncertainty* cm^{-1}
115271.2029	0.0004	1	0	3.84503345	0.00000001
230538.0016	0.0008	2	1	7.68991999	0.00000003
345795.9923	0.0012	3	2	11.53451273	0.00000004
461040.7712	0.0016	4	3	15.37866477	0.00000005
576267.9350	0.0019	5	4	19.22222923	0.00000006
691473.0809	0.0021	6	5	23.06505926	0.00000007
806651.8065	0.0023	7	6	26.90700800	0.00000008
921799.7104	0.0025	8	7	30.74792863	0.00000008
1036912.3919	0.0027	9	8	34.58767438	0.00000009
1151985.4515	0.0029	10	9	38.42609848	0.00000010
1267014.4906	0.0031	11	10	42.26305422	0.00000010
1381995.1119	0.0034	12	11	46.09839491	0.00000011
1496922.9195	0.0038	13	12	49.93197392	0.00000013

Frequency MHz	Uncertainty* MHz	J'	J''	Wavenumber cm^{-1}	Uncertainty* cm^{-1}
1611793.5189	0.0042	14	13	53.76364468	0.00000014
1726602.5173	0.0047	15	14	57.59326065	0.00000016
1841345.5237	0.0052	16	15	61.42067535	0.00000017
1956018.1486	0.0057	17	16	65.24574239	0.00000019
2070616.0050	0.0061	18	17	69.06831542	0.00000020
2185134.7075	0.0065	19	18	72.88824816	0.00000022
2299569.8733	0.0069	20	19	76.70539441	0.00000023
2413917.1217	0.0071	21	20	80.51960806	0.00000024
2528172.0747	0.0073	22	21	84.33074306	0.00000024
2642330.3567	0.0074	23	22	88.13865346	0.00000025
2756387.5949	0.0075	24	23	91.94319341	0.00000025
2870339.4194	0.0077	25	24	95.74421713	0.00000026
2984181.4631	0.0080	26	25	99.54157896	0.00000027
3097909.3621	0.0085	27	26	103.33513334	0.00000028
3211518.7558	0.0090	28	27	107.12473480	0.00000030
3325005.2869	0.0096	29	28	110.91023800	0.00000032
3438364.6013	0.0102	30	29	114.69149772	0.00000034
3551592.3489	0.0107	31	30	118.46836884	0.00000036
3664684.1829	0.0111	32	31	122.24070637	0.00000037
3777635.7608	0.0118	33	32	126.00836545	0.00000039
3890442.7435	0.0137	34	33	129.77120137	0.00000046
4003100.7965	0.0179	35	34	133.52906952	0.00000060
4115605.5892	0.0254	36	35	137.28182546	0.00000085
4227952.7954	0.0370	37	36	141.02932487	0.00000123
4340138.0932	0.0531	38	37	144.77142361	0.00000177
4452157.1657	0.0746	39	38	148.50797766	0.00000249
4564005.7001	0.1025	40	39	152.23884318	0.00000342

* The uncertainty given is twice the standard error.

Frequencies and Wavenumbers for the Rotational Lines of HF

Frequency MHz	Uncertainty* MHz	J'	J''	Wavenumber cm^{-1}	Uncertainty* cm^{-1}
1232476.21	0.12	1	0	41.110981	0.000004
2463428.09	0.19	2	1	82.171116	0.000006
3691334.81	0.25	3	2	123.129676	0.000008
4914682.58	0.51	4	3	163.936165	0.000017
6131968.11	1.10	5	4	204.540439	0.000037
7341702.00	2.00	6	5	244.892818	0.000067
8542412.1	3.21	7	6	284.944197	0.000107
9732646.8	4.72	8	7	324.646153	0.000157
10910978.2	6.51	9	8	363.951056	0.000217
12076004.8	8.55	10	9	402.81216	0.000285
13226355.2	10.81	11	10	441.18372	0.000361
14360689.8	13.25	12	11	479.02105	0.00044
15477704.4	15.86	13	12	516.28065	0.00053
16576131.8	18.61	14	13	552.92024	0.00062
17654744.4	21.48	15	14	588.89888	0.00072
18712356.5	24.44	16	15	624.17703	0.00082
19747825.6	27.43	17	16	658.71656	0.00092
20760054.3	30.32	18	17	692.4809	0.00101
21747991.7	32.91	19	18	725.4349	0.00110
22710634.7	34.94	20	19	757.5452	0.00117
23647028.7	36.08	21	20	788.7800	0.00120
24556268.8	35.93	22	21	819.1090	0.00120
25437499.9	34.12	23	22	848.5037	0.00114

Frequency MHz	Uncertainty* MHz	J'	J''	Wavenumber cm⁻¹	Uncertainty* cm⁻¹
26289917.4	30.32	24	23	876.9373	0.00101
27112767.2	24.41	25	24	904.38457	0.00081
27905345.6	16.88	26	25	930.82214	0.00056
28666999.3	10.80	27	26	956.22817	0.00036
29397124.8	14.65	28	27	980.58253	0.00049
30095168.2	24.62	29	28	1003.86676	0.00082
30760624.2	33.36	30	29	1026.0640	0.00111
31393035.7	36.17	31	30	1047.1590	0.00121

* The uncertainty given is twice the standard error.

Frequencies and Wavenumbers for the Rotational Lines of H³⁵Cl

Frequency MHz	Uncertainty* MHz	J'	J''	Wavemumber cm⁻¹	Uncertainty* cm⁻¹
1876226.517	0.065	3	2	62.584180	0.000002
2499864.439	0.066	4	3	83.386502	0.000002
3121986.563	0.064	5	4	104.138262	0.000002
3742216.601	0.076	6	5	124.826909	0.000003
4360180.042	0.098	7	6	145.439951	0.000003
4975504.51	0.11	8	7	165.964966	0.000004
5587820.10	0.12	9	8	186.389615	0.000004
6196759.76	0.22	10	9	206.701656	0.000007
6801959.63	0.50	11	10	226.888951	0.000017
7403059.41	1.02	12	11	246.939481	0.000034
7999702.7	1.8	13	12	266.841359	0.000062
8591537.3	3.1	14	13	286.582837	0.000103
9178215.8	4.8	15	14	306.152324	0.000161

* The uncertainty given is twice the standard error.

Frequencies and Wavenumbers for the Rotational Lines of H³⁷Cl

Frequency MHz	Uncertainty* MHz	J'	J''	Wavenumber cm⁻¹	Uncertainty* cm⁻¹
1873410.72	0.05	3	2	62.490255	0.000002
2496115.33	0.05	4	3	83.261445	0.000002
3117308.69	0.05	5	4	103.982225	0.000002
3736615.64	0.06	6	5	124.640082	0.000002
4353662.84	0.08	7	6	145.222561	0.000003
4968079.04	0.09	8	7	165.717279	0.000003
5579495.53	0.10	9	8	186.111938	0.000003
6187546.42	0.19	10	9	206.394332	0.000006
6791869.04	0.45	11	10	226.552365	0.000015
7392104.3	0.9	12	11	246.574057	0.000030
7987896.9	1.6	13	12	266.447561	0.000054
8578896.1	2.7	14	13	286.161170	0.000089

* The uncertainty given is twice the standard error.

Section 11
Nuclear and Particle Physics

SUMMARY TABLES OF PARTICLE PROPERTIES

July 1996

Particle Data Group

R.M. Barnett, C.D. Carone, D.E. Groom, T.G. Trippe, C.G. Wohl,
B. Armstrong*, P.S. Gee*, G.S. Wagman*†, F. James, M. Mangano,
K. Mönig, L. Montanet, J.L. Feng, H. Murayama, J.J. Hernández,
A. Manohar, M. Aguilar-Benitez, C. Caso, R.L. Crawford, M. Roos,
N.A. Törnqvist, K.G. Hayes, K. Hagiwara, K. Nakamura, M. Tanabashi,
K. Olive, K. Honscheid, P.R. Burchat, R.E. Shrock, S. Eidelman,
R.H. Schindler, A. Gurtu, K. Hikasa, G. Conforto, R.L. Workman,
C. Grab, and C. Amsler
*Technical Associate
†Deceased

Reprinted from Physical Review D54 1 (1996)

(Approximate closing date for data: January 1, 1996)

GAUGE AND HIGGS BOSONS

γ

$$I(J^{PC}) = 0,1(1^{--})$$

Mass $m < 6 \times 10^{-16}$ eV, CL = 99.7%
Charge $q < 5 \times 10^{-30}$ e
Mean life τ = Stable

g or gluon

$$I(J^P) = 0(1^-)$$

Mass $m = 0$ [a]
SU(3) color octet

W

$$J = 1$$

Charge $= \pm 1$ e
Mass $m = 80.33 \pm 0.15$ GeV
$m_Z - m_W = 10.85 \pm 0.15$ GeV
$m_{W^+} - m_{W^-} = -0.2 \pm 0.6$ GeV
Full width $\Gamma = 2.07 \pm 0.06$ GeV

W^- modes are charge conjugates of the modes below.

W^+ DECAY MODES		Fraction (Γ_i/Γ)	Confidence level	p (MeV/c)
$\ell^+\nu$	[b]	(10.8 ± 0.4) %		40110
$e^+\nu$		(10.8 ± 0.4) %		40110
$\mu^+\nu$		(10.4 ± 0.6) %		40110
$\tau^+\nu$		(10.9 ± 1.0) %		40110
hadrons		(67.9 ± 1.5) %		–
$\pi^+\gamma$		$< 5 \quad \times 10^{-4}$	95%	40110

Z

$$J = 1$$

Charge = 0
Mass $m = 91.187 \pm 0.007$ GeV [c]
Full width $\Gamma = 2.490 \pm 0.007$ GeV
$\Gamma(\ell^+\ell^-) = 83.83 \pm 0.27$ MeV [b]
$\Gamma(\text{invisible}) = 498.3 \pm 4.2$ MeV [d]
$\Gamma(\text{hadrons}) = 1740.7 \pm 5.9$ MeV
$\Gamma(\mu^+\mu^-)/\Gamma(e^+e^-) = 1.000 \pm 0.005$
$\Gamma(\tau^+\tau^-)/\Gamma(e^+e^-) = 0.998 \pm 0.005$ [e]

Average charged multiplicity

$\langle N_{charged} \rangle = 20.99 \pm 0.14$

Couplings to leptons

$g_V^\ell = -0.0376 \pm 0.0012$
$g_A^\ell = -0.5008 \pm 0.0008$
$g^{\nu e} = 0.53 \pm 0.09$
$g^{\nu\mu} = 0.502 \pm 0.017$

Asymmetry parameters [f]

$A_e = 0.156 \pm 0.008$ (S = 1.2)
$A_\tau = 0.145 \pm 0.009$
$A_c = 0.59 \pm 0.19$
$A_b = 0.89 \pm 0.11$

Charge asymmetry (%) at Z pole

$A_{FB}^{(0\ell)} = 1.59 \pm 0.18$
$A_{FB}^{(0s)} = 13 \pm 4$
$A_{FB}^{(0c)} = 7.22 \pm 0.67$
$A_{FB}^{(0b)} = 9.92 \pm 0.35$

Z DECAY MODES		Fraction (Γ_i/Γ)	Confidence level	p (MeV/c)
e^+e^-		(3.366 ± 0.008) %		45600
$\mu^+\mu^-$		(3.367 ± 0.013) %		45600
$\tau^+\tau^-$		(3.360 ± 0.015) %		45600
$\ell^+\ell^-$	[b]	(3.366 ± 0.006) %		45600
invisible		(20.01 ± 0.16) %		–
hadrons		(69.90 ± 0.15) %		–
$(u\bar{u}+c\bar{c})/2$		(9.6 ± 1.3) %		–
$(d\bar{d}+s\bar{s}+b\bar{b})/3$		(16.9 ± 0.9) %		–
$c\bar{c}$		(11.0 ± 0.7) %		–
$b\bar{b}$		(15.46 ± 0.14) %		–
$\pi^0\gamma$		$< 5.2 \quad \times 10^{-5}$	95%	45600
$\eta\gamma$		$< 5.1 \quad \times 10^{-5}$	95%	45600
$\omega\gamma$		$< 6.5 \quad \times 10^{-4}$	95%	45600
$\eta'(958)\gamma$		$< 4.2 \quad \times 10^{-5}$	95%	45600
$\gamma\gamma$		$< 5.2 \quad \times 10^{-5}$	95%	45600
$\gamma\gamma\gamma$		$< 1.0 \quad \times 10^{-5}$	95%	45600
$\pi^\pm W^\mp$	[g]	$< 7 \quad \times 10^{-5}$	95%	10300
$\rho^\pm W^\mp$	[g]	$< 8.3 \quad \times 10^{-5}$	95%	10300
$J/\psi(1S)X$		$(3.80 \pm 0.27) \times 10^{-3}$		–
$\psi(2S)X$		$(1.60 \pm 0.33) \times 10^{-3}$		–
$\chi_{c1}(1P)X$		$(6.0 \pm 1.9) \times 10^{-3}$		–
ΥX		$(1.0 \pm 0.5) \times 10^{-4}$		–
$(D^0/\overline{D}^0)X$		(20.7 ± 2.0) %		–
$D^\pm X$		(12.2 ± 1.7) %		–
$D^*(2010)^\pm X$	[g]	(11.4 ± 1.3) %		–
$B_s^0 X$		seen		–
anomalous γ + hadrons	[h]	$< 3.2 \quad \times 10^{-3}$	95%	–
$e^+e^-\gamma$	[h]	$< 5.2 \quad \times 10^{-4}$	95%	45600
$\mu^+\mu^-\gamma$	[h]	$< 5.6 \quad \times 10^{-4}$	95%	45600
$\tau^+\tau^-\gamma$	[h]	$< 7.3 \quad \times 10^{-4}$	95%	45600
$\ell^+\ell^-\gamma\gamma$	[i]	$< 6.8 \quad \times 10^{-6}$	95%	45600
$q\bar{q}\gamma\gamma$	[i]	$< 5.5 \quad \times 10^{-6}$	95%	–
$\nu\bar{\nu}\gamma\gamma$	[i]	$< 3.1 \quad \times 10^{-6}$	95%	45600
$e^\pm\mu^\mp$	LF [g]	$< 1.7 \quad \times 10^{-6}$	95%	45600
$e^\pm\tau^\mp$	LF [g]	$< 9.8 \quad \times 10^{-6}$	95%	45600
$\mu^\pm\tau^\mp$	LF [g]	$< 1.7 \quad \times 10^{-5}$	95%	45600

Gauge & Higgs Boson Summary Table

Higgs Bosons — H^0 and $H^\pm$, Searches for

H^0 Mass $m > 58.4$ GeV, CL = 95%

H_1^0 in Supersymmetric Models $(m_{H_1^0} < m_{H_2^0})$ [j]

Mass $m > 44$ GeV, CL = 95%

A^0 Pseudoscalar Higgs Boson in Supersymmetric Models [j]

Mass $m > 24.3$ GeV, CL = 95% $\tan\beta > 1$, $m_t < 200$ GeV

$H^\pm$ Mass $m > 43.5$ GeV, CL = 95%

See the Particle Listings for a Note giving details of Higgs Bosons.

Heavy Bosons Other Than Higgs Bosons, Searches for

Additional W Bosons

W_R — right-handed W

Mass $m > 406$ GeV, CL = 90%
(assuming light right-handed neutrino)

W' with standard couplings decaying to $e\nu$, $\mu\nu$

Mass $m > 652$ GeV, CL = 95%

Additional Z Bosons

Z'_{SM} with standard couplings

Mass $m > 505$ GeV, CL = 95%   ($p\bar{p}$ direct search)

Mass $m > 779$ GeV, CL = 95%   (electroweak fit)

Z_{LR} of $SU(2)_L \times SU(2)_R \times U(1)$

(with $g_L = g_R$)

Mass $m > 445$ GeV, CL = 95%   ($p\bar{p}$ direct search)

Mass $m > 389$ GeV, CL = 95%   (electroweak fit)

Z_χ of $SO(10) \rightarrow SU(5) \times U(1)_\chi$

(coupling constant derived from G.U.T.)

Mass $m > 425$ GeV, CL = 95%   ($p\bar{p}$ direct search)

Mass $m > 321$ GeV, CL = 95%   (electroweak fit)

Z_ψ of $E_6 \rightarrow SO(10) \times U(1)_\psi$

(coupling constant derived from G.U.T.)

Mass $m > 415$ GeV, CL = 95%   ($p\bar{p}$ direct search)

Mass $m > 160$ GeV, CL = 95%   (electroweak fit)

Z_η of $E_6 \rightarrow SU(3) \times SU(2) \times U(1) \times U(1)_\eta$

(coupling constant derived from G.U.T.);

charges are $Q_\eta = \sqrt{3/8}Q_\chi - \sqrt{5/8}Q_\psi$)

Mass $m > 440$ GeV, CL = 95%   ($p\bar{p}$ direct search)

Mass $m > 182$ GeV, CL = 95%   (electroweak fit)

Scalar Leptoquarks

Mass $m > 116$ GeV, CL = 95%   (1st generation, pair prod.)

Mass $m > 230$ GeV, CL = 95%   (1st gener., single prod.)

Mass $m > 97$ GeV, CL = 95%   (2nd gener., pair prod.)

Mass $m > 73$ GeV, CL = 95%   (2nd gener., single prod.)

Mass $m > 45$ GeV, CL = 95%   (3rd gener., pair prod.)

(The second, fourth, and fifth limits above are for charge $-1/3$, weak isoscalar.)

Axions (A^0) and Other Very Light Bosons, Searches for

The standard Peccei-Quinn axion is ruled out. Variants with reduced couplings or much smaller masses are constrained by various data. The Particle Listings in the full *Review* contain a Note discussing axion searches.

The best limit for the half-life of neutrinoless double beta decay with Majoron emission is $> 7.2 \times 10^{24}$ years (CL = 90%).

NOTES

In this Summary Table:

When a quantity has "(S = ...)" to its right, the error on the quantity has been enlarged by the "scale factor" S, defined as $S = \sqrt{\chi^2/(N-1)}$, where N is the number of measurements used in calculating the quantity. We do this when $S > 1$, which often indicates that the measurements are inconsistent. When $S > 1.25$, we also show in the Particle Listings an ideogram of the measurements. For more about S, see the Introduction.

A decay momentum p is given for each decay mode. For a 2-body decay, p is the momentum of each decay product in the rest frame of the decaying particle. For a 3-or-more-body decay, p is the largest momentum any of the products can have in this frame.

[a] Theoretical value. A mass as large as a few MeV may not be precluded.

[b] ℓ indicates each type of lepton (e, μ, and τ), not sum over them.

[c] The Z-boson mass listed here corresponds to a Breit-Wigner resonance parameter. It lies approximately 34 MeV above the real part of the position of the pole (in the energy-squared plane) in the Z-boson propagator.

[d] This partial width takes into account Z decays into $\nu\bar{\nu}$ and any other possible undetected modes.

[e] This ratio has not been corrected for the τ mass.

[f] Here $A \equiv 2g_V g_A/(g_V^2 + g_A^2)$.

[g] The value is for the sum of the charge states of particle/antiparticle states indicated.

[h] See the Z Particle Listings for the γ energy range used in this measurement.

[i] For $m_{\gamma\gamma} = (60 \pm 5)$ GeV.

[j] The limits assume no invisible decays.

LEPTONS

e

$$J = \tfrac{1}{2}$$

Mass $m = 0.51099907 \pm 0.00000015$ MeV [a]
$= (5.48579903 \pm 0.00000013) \times 10^{-4}$ u
$(m_{e^+} - m_{e^-})/m < 4 \times 10^{-8}$, CL = 90%
$|q_{e^+} + q_{e^-}|/e < 4 \times 10^{-8}$
Magnetic moment $\mu = 1.001159652193 \pm 0.000000000010 \; \mu_B$
$(g_{e^+} - g_{e^-}) \, / \, g_{\text{average}} = (-0.5 \pm 2.1) \times 10^{-12}$
Electric dipole moment $d = (-0.3 \pm 0.8) \times 10^{-26} \, e\,\text{cm}$
Mean life $\tau > 4.3 \times 10^{23}$ yr, CL = 68% [b]

μ

$$J = \tfrac{1}{2}$$

Mass $m = 105.658389 \pm 0.000034$ MeV [c]
$= 0.113428913 \pm 0.000000017$ u
Mean life $\tau = (2.19703 \pm 0.00004) \times 10^{-6}$ s
$\tau_{\mu^+}/\tau_{\mu^-} = 1.00002 \pm 0.00008$
$c\tau = 658.654$ m
Magnetic moment $\mu = 1.001165923 \pm 0.000000008 \; e\hbar/2m_\mu$
$(g_{\mu^+} - g_{\mu^-}) \, / \, g_{\text{average}} = (-2.6 \pm 1.6) \times 10^{-8}$
Electric dipole moment $d = (3.7 \pm 3.4) \times 10^{-19} \, e\,\text{cm}$

Decay parameters [d]

$\rho = 0.7518 \pm 0.0026$
$\eta = -0.007 \pm 0.013$
$\delta = 0.749 \pm 0.004$
$\xi P_\mu = 1.003 \pm 0.008$ [e]
$\xi P_\mu \delta/\rho > 0.99682$, CL = 90% [e]
$\xi' = 1.00 \pm 0.04$
$\xi'' = 0.7 \pm 0.4$
$\alpha/A = (0 \pm 4) \times 10^{-3}$
$\alpha'/A = (0 \pm 4) \times 10^{-3}$
$\beta/A = (4 \pm 6) \times 10^{-3}$
$\beta'/A = (2 \pm 6) \times 10^{-3}$
$\bar{\eta} = 0.02 \pm 0.08$

μ^+ modes are charge conjugates of the modes below.

μ^- DECAY MODES	Fraction (Γ_i/Γ)		Confidence level	p (MeV/c)
$e^- \bar{\nu}_e \nu_\mu$	$\approx 100\%$			53
$e^- \bar{\nu}_e \nu_\mu \gamma$	$(1.4 \pm 0.4)\%$	[f]		53
$e^- \bar{\nu}_e \nu_\mu e^+ e^-$	$(3.4 \pm 0.4) \times 10^{-5}$	[g]		53

Lepton Family number (LF) violating modes

$e^- \nu_e \bar{\nu}_\mu$	LF	< 1.2 %	[h]	90%	53
$e^- \gamma$	LF	$< 4.9 \times 10^{-11}$		90%	53
$e^- e^+ e^-$	LF	$< 1.0 \times 10^{-12}$		90%	53
$e^- 2\gamma$	LF	$< 7.2 \times 10^{-11}$		90%	53

τ

$$J = \tfrac{1}{2}$$

Mass $m = 1777.00^{+0.30}_{-0.27}$ MeV
Mean life $\tau = (291.0 \pm 1.5) \times 10^{-15}$ s
$c\tau = 87.2 \; \mu$m
Electric dipole moment $d < 5 \times 10^{-17} \, e\,\text{cm}$, CL = 95%

Weak dipole moment

$\text{Re}(d_\tau^w) < 7.8 \times 10^{-18} \, e\,\text{cm}$, CL = 95%
$\text{Im}(d_\tau^w) < 4.5 \times 10^{-17} \, e\,\text{cm}$, CL = 95%

Decay parameters

See the τ Particle Listings for a note concerning τ-decay parameters.

$\rho^\tau(e \text{ or } \mu) = 0.742 \pm 0.027$
$\rho^\tau(e) = 0.736 \pm 0.028$
$\rho^\tau(\mu) = 0.74 \pm 0.04$
$\xi^\tau(e \text{ or } \mu) = 1.03 \pm 0.12$
$\xi^\tau(e) = 1.03 \pm 0.25$
$\xi^\tau(\mu) = 1.23 \pm 0.24$
$\eta^\tau(e \text{ or } \mu) = -0.01 \pm 0.14$
$\eta^\tau(\mu) = -0.24 \pm 0.29$
$(\delta\xi)^\tau(e \text{ or } \mu) = 0.76 \pm 0.11$ (S = 1.3)
$(\delta\xi)^\tau(e) = 1.11 \pm 0.18$
$(\delta\xi)^\tau(\mu) = 0.71 \pm 0.15$
$\xi^\tau(\pi) = 0.99 \pm 0.06$
$\xi^\tau(\rho) = 1.04 \pm 0.07$
$\xi^\tau(a_1) = 1.01 \pm 0.04$
$\xi^\tau(\text{all hadronic modes}) = 1.011 \pm 0.027$

τ^+ modes are charge conjugates of the modes below. "$h^\pm$" stands for $\pi^\pm$ or $K^\pm$. "ℓ" stands for e or μ. "Neutral" means neutral hadron whose decay products include γ's and/or π^0's.

τ^- DECAY MODES	Fraction (Γ_i/Γ)		Scale factor/ Confidence level	p (MeV/c)
Modes with one charged particle				
particle$^- \geq 0$ neutrals $\geq 0 K^0_L \, \nu_\tau$ ("1-prong")	$(84.96 \pm 0.14)\%$		S=1.3	–
particle$^- \geq 0$ neutrals $\geq 0 K^0 \, \nu_\tau$	$(85.53 \pm 0.14)\%$		S=1.3	–
$\mu^- \bar{\nu}_\mu \nu_\tau$	$(17.35 \pm 0.10)\%$	[l]		885
$\mu^- \bar{\nu}_\mu \nu_\tau \gamma$ ($E_\gamma > 37$ MeV)	$(2.3 \pm 1.0) \times 10^{-3}$			–
$e^- \bar{\nu}_e \nu_\tau$	$(17.83 \pm 0.08)\%$	[l]		888
$h^- \geq 0$ neutrals $\geq 0 K^0_L \, \nu_\tau$	$(49.78 \pm 0.17)\%$		S=1.2	–
$h^- \geq 0 K^0_L \, \nu_\tau$	$(12.51 \pm 0.13)\%$		S=1.1	–
$h^- \nu_\tau$	$(12.03 \pm 0.14)\%$		S=1.1	–
$\pi^- \nu_\tau$	$(11.31 \pm 0.15)\%$	[l]	S=1.1	883
$K^- \nu_\tau$	$(7.1 \pm 0.5) \times 10^{-3}$			820
$h^- \geq 1\pi^0 \, \nu_\tau$	$(36.97 \pm 0.18)\%$		S=1.1	–
$h^- \pi^0 \nu_\tau$	$(25.76 \pm 0.15)\%$		S=1.1	–
$\pi^- \pi^0 \nu_\tau$	$(25.24 \pm 0.16)\%$	[l]	S=1.1	878
$\pi^- \pi^0 \text{non-}\rho(770) \, \nu_\tau$	$(3.0 \pm 3.2) \times 10^{-3}$			878
$K^- \pi^0 \nu_\tau$	$(5.2 \pm 0.5) \times 10^{-3}$	[l]		814
$h^- \geq 2\pi^0 \nu_\tau$	$(10.95 \pm 0.16)\%$		S=1.1	–
$h^- 2\pi^0 \nu_\tau$	$(9.50 \pm 0.14)\%$		S=1.1	–
$h^- 2\pi^0 \nu_\tau \,(\text{ex.}K^0)$	$(9.35 \pm 0.14)\%$		S=1.1	–
$\pi^- 2\pi^0 \nu_\tau \,(\text{ex.}K^0)$	$(9.27 \pm 0.14)\%$	[l]	S=1.1	862
$K^- 2\pi^0 \nu_\tau \,(\text{ex.}K^0)$	$(8.1 \pm 2.7) \times 10^{-4}$	[l]		796
$h^- \geq 3\pi^0 \nu_\tau$	$(1.46 \pm 0.11)\%$		S=1.1	–
$h^- 3\pi^0 \nu_\tau$	$(1.28 \pm 0.10)\%$			–
$\pi^- 3\pi^0 \nu_\tau \,(\text{ex.}K^0)$	$(1.14 \pm 0.14)\%$			836
$K^- 3\pi^0 \nu_\tau \,(\text{ex.}K^0)$	$(5.0^{+10.0}_{-3.3}) \times 10^{-4}$	[l]		766
$h^- 4\pi^0 \nu_\tau \,(\text{ex.}K^0)$	$(1.8 \pm 0.6) \times 10^{-3}$			–
$h^- 4\pi^0 \nu_\tau \,(\text{ex.}K^0,\eta)$	$(1.2 \pm 0.6) \times 10^{-3}$	[l]		–
$K^- \geq 1 \, (\pi^0 \text{ or } K^0) \, \nu_\tau$	$(9.4 \pm 1.0) \times 10^{-3}$			–
Modes with K^0's				
$h^- \overline{K}^0 \geq 0$ neutrals $\geq 0 K^0_L \nu_\tau$	$(1.54 \pm 0.10)\%$		S=1.3	–
$h^- \overline{K}^0 \nu_\tau$	$(9.2 \pm 0.8) \times 10^{-3}$		S=1.3	–
$\pi^- \overline{K}^0 \nu_\tau$	$(7.7 \pm 0.8) \times 10^{-3}$	[l]	S=1.3	812
$\pi^- \overline{K}^0 \,(\text{non-}K^*(892)^-) \, \nu_\tau$	$< 1.7 \times 10^{-3}$		CL=95%	812
$K^- K^0 \nu_\tau$	$(1.55 \pm 0.28) \times 10^{-3}$	[l]		737
$h^- \overline{K}^0 \pi^0 \nu_\tau$	$(5.5 \pm 0.5) \times 10^{-3}$			–
$\pi^- \overline{K}^0 \pi^0 \nu_\tau$	$(4.1 \pm 0.6) \times 10^{-3}$	[l]		794
$K^- K^0 \pi^0 \nu_\tau$	$(1.38 \pm 0.32) \times 10^{-3}$	[l]		685
$h^- K^0_S K^0_S \nu_\tau$	$(2.5 \pm 0.6) \times 10^{-4}$			–
$\pi^- K^0 \overline{K}^0 \nu_\tau$	$(1.01 \pm 0.23) \times 10^{-3}$	[l]		682
$K^- K^0 \geq 0$ neutrals ν_τ	$(2.9 \pm 0.4) \times 10^{-3}$			–
$K^- \geq 0\pi^0 \geq 0 K^0 \, \nu_\tau$	$(1.65 \pm 0.10)\%$			–
$K^0 (\text{particles})^- \nu_\tau$	$(1.58 \pm 0.10)\%$		S=1.2	–
$K^0 h^+ h^- h^- \geq 0$ neut. ν_τ	$< 1.7 \times 10^{-3}$		CL=95%	–

Modes with three charged particles

Mode	Value		
$h^- h^- h^+ \geq 0$ neut. ν_τ ("3-prong")	(14.91 ± 0.14) %	S=1.3	–
$h^- h^- h^+ \geq 0$ neutrals ν_τ (ex. $K^0_S \rightarrow \pi^+\pi^-$)	(14.36 ± 0.14) %	S=1.3	
$\pi^- \pi^+ \pi^- \geq 0$ neutrals ν_τ	(14.09 ± 0.31) %		
$h^- h^- h^+ \nu_\tau$	(9.80 ± 0.10) %	S=1.1	
$h^- h^- h^+ \nu_\tau$ (ex. K^0)	(9.48 ± 0.10) %	S=1.1	
$h^- h^- h^+ \nu_\tau$ (ex. K^0, ω) [l]	(9.44 ± 0.10) %	S=1.1	
$h^- h^- h^+ \geq 1$ neutrals ν_τ	(5.08 ± 0.11) %	S=1.2	
$h^- h^- h^+ \geq 1$ neutrals ν_τ (ex. $K^0_S \rightarrow \pi^+\pi^-$)	(4.88 ± 0.11) %	S=1.2	
$h^- h^- h^+ \pi^0 \nu_\tau$	(4.44 ± 0.09) %	S=1.1	
$h^- h^- h^+ \pi^0 \nu_\tau$ (ex. K^0)	(4.25 ± 0.09) %	S=1.1	
$h^- h^- h^+ \pi^0 \nu_\tau$ (ex. K^0, ω) [l]	(2.55 ± 0.09) %		
$h^- (\rho\pi)^0 \nu_\tau$	(2.84 ± 0.34) %		
$(a_1(1260) h)^- \nu_\tau$	< 2.0 %	CL=95%	
$h^- \rho\pi^0 \nu_\tau$	(1.33 ± 0.20) %		
$h^- \rho^+ h^- \nu_\tau$	$(4.4 \pm 2.2) \times 10^{-3}$		
$h^- \rho^- h^+ \nu_\tau$	(1.15 ± 0.23) %		
$h^- h^- h^+ 2\pi^0 \nu_\tau$	$(5.2 \pm 0.5) \times 10^{-3}$		
$h^- h^- h^+ 2\pi^0 \nu_\tau$ (ex. K^0)	$(5.1 \pm 0.5) \times 10^{-3}$		
$h^- h^- h^+ 2\pi^0 \nu_\tau$ (ex. K^0, ω, η) [l]	$(1.0 \pm 0.4) \times 10^{-3}$		
$h^- h^- h^+ \geq 3\pi^0 \nu_\tau$ [l]	$(1.1 \pm 0.6) \times 10^{-3}$		
$K^- h^+ h^- \geq 0$ neutrals ν_τ	$< 6 \times 10^{-3}$	CL=90%	
$K^- \pi^+ \pi^- \geq 0$ neut. ν_τ	$(3.9 ^{+1.9}_{-1.6}) \times 10^{-3}$	S=1.5	
$K^- \pi^+ K^- \geq 0$ neut. ν_τ	$< 9 \times 10^{-4}$	CL=95%	
$K^- K^+ \pi^- \geq 0$ neut. ν_τ	$(1.5 ^{+0.9}_{-0.8}) \times 10^{-3}$		
$K^- K^+ \pi^- \nu_\tau$	$(2.2 ^{+1.8}_{-1.2}) \times 10^{-3}$		685
$\phi\pi^- \nu_\tau$	$< 3.5 \times 10^{-4}$	CL=90%	585
$K^- K^+ K^- \geq 0$ neut. ν_τ	$< 2.1 \times 10^{-3}$	CL=95%	–
$\pi^- K^+ \pi^- \geq 0$ neut. ν_τ	$< 2.5 \times 10^{-3}$	CL=95%	
$e^- e^- e^+ \overline{\nu}_e \nu_\tau$	$(2.8 \pm 1.5) \times 10^{-5}$		888
$\mu^- e^- e^+ \overline{\nu}_\mu \nu_\tau$	$< 3.6 \times 10^{-5}$	CL=90%	885

Modes with five charged particles

Mode	Value		
$3h^- 2h^+ \geq 0$ neutrals ν_τ (ex. $K^0_S \rightarrow \pi^-\pi^+$) ("5-prong")	$(9.7 \pm 0.7) \times 10^{-4}$		–
$3h^- 2h^+ \nu_\tau$ (ex. K^0) [l]	$(7.5 \pm 0.7) \times 10^{-4}$		–
$3h^- 2h^+ \pi^0 \nu_\tau$ (ex. K^0) [l]	$(2.2 \pm 0.5) \times 10^{-4}$		–
$3h^- 2h^+ 2\pi^0 \nu_\tau$	$< 1.1 \times 10^{-4}$	CL=90%	

Miscellaneous other allowed modes

Mode	Value		
$(5\pi)^- \nu_\tau$	$(3.3 \pm 0.7) \times 10^{-3}$		–
$4h^- 3h^+ \geq 0$ neutrals ν_τ ("7-prong")	$< 1.9 \times 10^{-4}$	CL=90%	–
$K^*(892)^- \geq 0 (h^0 \neq K^0_S) \nu_\tau$	(1.94 ± 0.31) %		–
$K^*(892)^- \geq 0$ neutrals ν_τ	(1.33 ± 0.13) %		–
$K^*(892)^- \nu_\tau$	(1.28 ± 0.08) %		665
$K^*(892)^0 K^- \geq 0$ neutrals ν_τ	$(3.2 \pm 1.4) \times 10^{-3}$		–
$K^*(892)^0 K^- \nu_\tau$	$(2.0 \pm 0.6) \times 10^{-3}$		539
$\overline{K}^*(892)^0 \pi^- \geq 0$ neutrals ν_τ	$(3.8 \pm 1.7) \times 10^{-3}$		–
$\overline{K}^*(892)^0 \pi^- \nu_\tau$	$(2.5 \pm 1.1) \times 10^{-3}$		653
$K_1(1270)^- \nu_\tau$	$(4 \pm 4) \times 10^{-3}$		433
$K_1(1400)^- \nu_\tau$	$(8 \pm 4) \times 10^{-3}$		335
$K^*_2(1430)^- \nu_\tau$	$< 3 \times 10^{-3}$	CL=95%	317
$\eta\pi^- \nu_\tau$	$< 1.4 \times 10^{-4}$	CL=95%	798
$\eta\pi^- \pi^0 \nu_\tau$ [l]	$(1.71 \pm 0.28) \times 10^{-4}$		778
$\eta\pi^- \pi^0 \pi^0 \nu_\tau$	$< 4.3 \times 10^{-4}$	CL=95%	746
$\eta K^- \nu_\tau$	$(2.6 \pm 0.7) \times 10^{-4}$		720
$\eta\pi^+ \pi^- \pi^- \geq 0$ neutrals ν_τ	$< 3 \times 10^{-4}$	CL=90%	–
$\eta\eta\pi^- \nu_\tau$	$< 1.1 \times 10^{-4}$	CL=95%	637
$\eta\eta\pi^- \pi^0 \nu_\tau$	$< 2.0 \times 10^{-4}$	CL=95%	559
$h^- \omega \geq 0$ neutrals ν_τ	(2.32 ± 0.11) %		–
$h^- \omega\nu_\tau$	(1.91 ± 0.09) %	[l]	–
$h^- \omega\pi^0 \nu_\tau$	$(4.1 \pm 0.6) \times 10^{-3}$	[l]	–

Lepton Family number (LF), Lepton number (L), or Baryon number (B) violating modes
(In the modes below, ℓ means a sum over e and μ modes)

L means lepton number violation (e.g. $\tau^- \rightarrow e^+\pi^-\pi^-$). Following common usage, LF means lepton family violation *and not* lepton number violation (e.g. $\tau^- \rightarrow e^-\pi^+\pi^-$).

Mode	Type	Value		CL	
$e^- \gamma$	LF	$< 1.1 \times 10^{-4}$		CL=90%	888
$\mu^- \gamma$	LF	$< 4.2 \times 10^{-6}$		CL=90%	885
$e^- \pi^0$	LF	$< 1.4 \times 10^{-4}$		CL=90%	883
$\mu^- \pi^0$	LF	$< 4.4 \times 10^{-5}$		CL=90%	880
$e^- K^0$	LF	$< 1.3 \times 10^{-3}$		CL=90%	819
$\mu^- K^0$	LF	$< 1.0 \times 10^{-3}$		CL=90%	815
$e^- \eta$	LF	$< 6.3 \times 10^{-5}$		CL=90%	804
$\mu^- \eta$	LF	$< 7.3 \times 10^{-5}$		CL=90%	800
$e^- \rho^0$	LF	$< 4.2 \times 10^{-6}$		CL=90%	722
$\mu^- \rho^0$	LF	$< 5.7 \times 10^{-6}$		CL=90%	718
$e^- K^*(892)^0$	LF	$< 6.3 \times 10^{-6}$		CL=90%	663
$\mu^- K^*(892)^0$	LF	$< 9.4 \times 10^{-6}$		CL=90%	657
$\pi^- \gamma$	L	$< 2.8 \times 10^{-4}$		CL=90%	883
$\pi^- \pi^0$	L	$< 3.7 \times 10^{-4}$		CL=90%	878
$e^- e^+ e^-$	LF	$< 3.3 \times 10^{-6}$		CL=90%	888
$e^- \mu^+ \mu^-$	LF	$< 3.6 \times 10^{-6}$		CL=90%	882
$e^+ \mu^- \mu^-$	LF	$< 3.5 \times 10^{-6}$		CL=90%	882
$\mu^- e^+ e^-$	LF	$< 3.4 \times 10^{-6}$		CL=90%	885
$\mu^+ e^- e^-$	L	$< 3.4 \times 10^{-6}$		CL=90%	885
$\mu^- \mu^+ \mu^-$	LF	$< 1.9 \times 10^{-6}$		CL=90%	873
$e^- \pi^+ \pi^-$	LF	$< 4.4 \times 10^{-6}$		CL=90%	877
$e^+ \pi^- \pi^-$	L	$< 4.4 \times 10^{-6}$		CL=90%	877
$\mu^- \pi^+ \pi^-$	LF	$< 7.4 \times 10^{-6}$		CL=90%	866
$\mu^+ \pi^- \pi^-$	L	$< 6.9 \times 10^{-6}$		CL=90%	866
$e^- \pi^+ K^-$	LF	$< 7.7 \times 10^{-6}$		CL=90%	813
$e^- \pi^- K^+$	LF	$< 4.6 \times 10^{-6}$		CL=90%	813
$e^+ \pi^- K^-$	L	$< 4.5 \times 10^{-6}$		CL=90%	813
$\mu^- \pi^+ K^-$	LF	$< 8.7 \times 10^{-6}$		CL=90%	800
$\mu^- \pi^- K^+$	LF	$< 1.5 \times 10^{-5}$		CL=90%	800
$\mu^+ \pi^- K^-$	L	$< 2.0 \times 10^{-5}$		CL=90%	800
$\overline{p}\gamma$	L,B	$< 2.9 \times 10^{-4}$		CL=90%	641
$\overline{p}\pi^0$	L,B	$< 6.6 \times 10^{-4}$		CL=90%	632
$\overline{p}\eta$	L,B	$< 1.30 \times 10^{-3}$		CL=90%	475
$e^- \overline{K}^*(892)^0$	LF	$< 1.1 \times 10^{-5}$		CL=90%	663
$\mu^- \overline{K}^*(892)^0$	LF	$< 8.7 \times 10^{-6}$		CL=90%	657
e^- light boson	LF	$< 2.7 \times 10^{-3}$		CL=95%	–
μ^- light boson	LF	$< 5 \times 10^{-3}$		CL=95%	–

Heavy Charged Lepton Searches

$L^\pm$ – charged lepton
Mass $m > 42.7$ GeV, CL = 95% $m_\nu \approx 0$

$L^\pm$ – stable charged heavy lepton
Mass $m > 42.8$ GeV, CL = 95%

Neutrinos

See the Particle Listings for a Note giving details of neutrinos, masses, mixing, and the status of experimental searches.

ν_e $J = \frac{1}{2}$

Mass m: Unexplained effects have resulted in significantly negative m^2 in the new, precise tritium beta decay experiments. It is felt that a real neutrino mass as large as 10–15 eV would cause observable spectral distortions even in the presence of the end-point count excesses.

Mean life/mass, $\tau/m_{\nu_e} > 300$ s/eV, CL = 90%

Magnetic moment $\mu < 1.8 \times 10^{-10}\ \mu_B$, CL = 90%

ν_μ $J = \frac{1}{2}$

Mass $m < 0.17$ MeV, CL = 90%

Mean life/mass, $\tau/m_{\nu_\mu} > 15.4$ s/eV, CL = 90%

Magnetic moment $\mu < 7.4 \times 10^{-10}\ \mu_B$, CL = 90%

ν_τ		$J = \frac{1}{2}$

Mass $m < 24$ MeV, CL = 95%
Magnetic moment $\mu < 5.4 \times 10^{-7}\ \mu_B$, CL = 90%

Number of Light Neutrino Types

(including ν_e, ν_μ, and ν_τ)

Number $N = 2.991 \pm 0.016$ (Standard Model fits to LEP data)
Number $N = 3.09 \pm 0.13$ (Direct measurement of invisible Z width)

Massive Neutrinos and Lepton Mixing, Searches for

For excited leptons, see Compositeness Limits below.

See the Particle Listings for a Note giving details of neutrinos, masses, mixing, and the status of experimental searches.

No direct, uncontested evidence for massive neutrinos or lepton mixing has been obtained. Sample limits are:

Mass $m > 45.0$, CL = 95%   (Dirac)
Mass $m > 39.5$, CL = 95%   (Majorana)

ν oscillation: $\nu_\mu \to \nu_e$ ($\theta = $ mixing angle)

Mass $m > 19.6$ GeV, CL = 95%   (all $|U_{\ell J}|^2$) (Dirac)
Mass $m > 45.7$ GeV   or $m < 25$, CL = 95% ($|U_{\ell J}|^2 > 10^{-13}$) (Dirac)

ν oscillation: $\overline{\nu}_e \not\to \overline{\nu}_e$

$\Delta(m^2) < 0.0075$ eV^2, CL = 90%   (if $\sin^2 2\theta = 1$)
$\sin^2 2\theta < 0.02$, CL = 90%   (if $\Delta(m^2)$ is large)

ν oscillation: $\nu_\mu \to \nu_e$ ($\theta = $ mixing angle)

$\Delta(m^2) < 0.09$ eV^2, CL = 90%   (if $\sin^2 2\theta = 1$)
$\sin^2 2\theta < 2.5 \times 10^{-3}$, CL = 90%   (if $\Delta(m^2)$ is large)

NOTES

In this Summary Table:

When a quantity has "(S = ...)" to its right, the error on the quantity has been enlarged by the "scale factor" S, defined as $S = \sqrt{\chi^2/(N-1)}$, where N is the number of measurements used in calculating the quantity. We do this when $S > 1$, which often indicates that the measurements are inconsistent. When $S > 1.25$, we also show in the Particle Listings an ideogram of the measurements. For more about S, see the Introduction.

A decay momentum p is given for each decay mode. For a 2-body decay, p is the momentum of each decay product in the rest frame of the decaying particle. For a 3-or-more-body decay, p is the largest momentum any of the products can have in this frame.

[a] The uncertainty in the electron mass in unified atomic mass units (u) is ten times smaller than that given by the 1986 CODATA adjustment, quoted in the Table of Physical Contants (Section 1). The conversion to MeV via the factor 931.49432(28) MeV/u is more uncertain because of the electron charge uncertainty. Our value in MeV differs slightly from the 1986 CODATA result.

[b] This is the best "electron disappearance" limit. The best limit for the mode $e^- \to \nu\gamma$ is $> 2.35 \times 10^{25}$ yr (CL=68%).

[c] The muon mass is most precisely known in u (unified atomic mass units). The conversion factor to MeV via the factor 931.49432(28) MeV/u is more uncertain because of the electron charge uncertainty.

[d] See the "Note on Muon Decay Parameters" in the μ Particle Listings for definitions and details.

[e] P_μ is the longitudinal polarization of the muon from pion decay. In standard $V-A$ theory, $P_\mu = 1$ and $\rho = \delta = 3/4$.

[f] This only includes events with the γ energy > 10 MeV. Since the $e^- \overline{\nu}_e \nu_\mu$ and $e^- \overline{\nu}_e \nu_\mu \gamma$ modes cannot be clearly separated, we regard the latter mode as a subset of the former.

[g] See the μ Particle Listings for the energy limits used in this measurement.

[h] A test of additive vs. multiplicative lepton family number conservation.

[i] Basis mode for the τ.

QUARKS

The u-, d-, and s-quark masses are estimates of so-called "current-quark masses," in a mass-independent subtraction scheme such as $\overline{MS}$ at a scale $\mu \approx 1$ GeV. The c- and b-quark masses are estimated from charmonium, bottomonium, D, and B masses. They are the "running" masses in the $\overline{MS}$ scheme. These can be different from the heavy quark masses obtained in potential models.

u

$$I(J^P) = \tfrac{1}{2}(\tfrac{1}{2}^+)$$

Mass $m = 2$ to 8 MeV [a]   Charge $= \tfrac{2}{3} e$ $I_z = +\tfrac{1}{2}$
$m_u/m_d = 0.25$ to 0.70

d

$$I(J^P) = \tfrac{1}{2}(\tfrac{1}{2}^+)$$

Mass $m = 5$ to 15 MeV [a]   Charge $= -\tfrac{1}{3} e$ $I_z = -\tfrac{1}{2}$
$m_s/m_d = 17$ to 25

s

$$I(J^P) = 0(\tfrac{1}{2}^+)$$

Mass $m = 100$ to 300 MeV [a]   Charge $= -\tfrac{1}{3} e$ Strangeness $= -1$
$(m_s - (m_u + m_d)/2)/(m_d - m_u) = 34$ to 51

c

$$I(J^P) = 0(\tfrac{1}{2}^+)$$

Mass $m = 1.0$ to 1.6 GeV   Charge $= \tfrac{2}{3} e$ Charm $= +1$

b

$$I(J^P) = 0(\tfrac{1}{2}^+)$$

Mass $m = 4.1$ to 4.5 GeV   Charge $= -\tfrac{1}{3} e$ Bottom $= -1$

t

$$I(J^P) = 0(\tfrac{1}{2}^+)$$

Charge $= \tfrac{2}{3} e$ Top $= +1$

Mass $m = 180 \pm 12$ GeV   (direct observation of top events)
Mass $m = 179 \pm 8^{+17}_{-20}$ GeV   (Standard Model electroweak fit)

b' (4th Generation) Quark, Searches for

Mass $m > 85$ GeV, CL $= 95\%$ ($p\bar{p}$, charged current decays)
Mass $m > 46.0$ GeV, CL $= 95\%$ (e^+e^-, all decays)

Free Quark Searches

All searches since 1977 have had negative results.

NOTES

[a] The ratios m_u/m_d and m_s/m_d are extracted from pion and kaon masses using chiral symmetry. The estimates of u and d masses are not without controversy and remain under active investigation. Within the literature there are even suggestions that the u quark could be essentially massless. The s-quark mass is estimated from SU(3) splittings in hadron masses.

LIGHT UNFLAVORED MESONS
$(S = C = B = 0)$

For $I = 1$ (π, b, ρ, a): $u\overline{d}$, $(u\overline{u}-d\overline{d})/\sqrt{2}$, $d\overline{u}$;
for $I = 0$ $(\eta, \eta', h, h', \omega, \phi, f, f')$: $c_1(u\overline{u} + d\overline{d}) + c_2(s\overline{s})$

$I^G(J^P) = 1^-(0^-)$

Mass $m = 139.56995 \pm 0.00035$ MeV
Mean life $\tau = (2.6033 \pm 0.0005) \times 10^{-8}$ s $(S = 1.2)$
$c\tau = 7.8045$ m

$\pi^\pm \to \ell^\pm \nu \gamma$ form factors [a]
$F_V = 0.017 \pm 0.008$
$F_A = 0.0116 \pm 0.0016$ $(S = 1.3)$
$R = 0.059^{+0.009}_{-0.008}$

π^- modes are charge conjugates of the modes below.

π^+ DECAY MODES		Fraction (Γ_i/Γ)	Confidence level	p (MeV/c)
$\mu^+ \nu_\mu$	[b]	(99.98770 ± 0.00004) %		30
$\mu^+ \nu_\mu \gamma$	[c]	$(1.24 \pm 0.25) \times 10^{-4}$		30
$e^+ \nu_e$	[b]	$(1.230 \pm 0.004) \times 10^{-4}$		70
$e^+ \nu_e \gamma$	[c]	$(1.61 \pm 0.23) \times 10^{-7}$		70
$e^+ \nu_e \pi^0$		$(1.025 \pm 0.034) \times 10^{-8}$		4
$e^+ \nu_e e^+ e^-$		$(3.2 \pm 0.5) \times 10^{-9}$		70
$e^+ \nu_e \nu \overline{\nu}$		$< 5 \times 10^{-6}$	90%	70

Lepton Family number (LF) or Lepton number (L) violating modes

$\mu^+ \overline{\nu}_e$	L	[d]	$< 1.5 \times 10^{-3}$	90%	30
$\mu^+ \nu_e$	LF	[d]	$< 8.0 \times 10^{-3}$	90%	30
$\mu^- e^+ e^+ \nu$	LF		$< 1.6 \times 10^{-6}$	90%	30

π^0

$I^G(J^{PC}) = 1^-(0^{-+})$

Mass $m = 134.9764 \pm 0.0006$ MeV
$m_{\pi^\pm} - m_{\pi^0} = 4.5936 \pm 0.0005$ MeV
Mean life $\tau = (8.4 \pm 0.6) \times 10^{-17}$ s $(S = 3.0)$
$c\tau = 25.1$ nm

π^0 DECAY MODES		Fraction (Γ_i/Γ)	Scale factor/ Confidence level	p (MeV/c)
2γ		(98.798 ± 0.032) %	S=1.1	67
$e^+ e^- \gamma$		(1.198 ± 0.032) %	S=1.1	67
γ positronium		$(1.82 \pm 0.29) \times 10^{-9}$		67
$e^+ e^+ e^- e^-$		$(3.14 \pm 0.30) \times 10^{-5}$		67
$e^+ e^-$		$(7.5 \pm 2.0) \times 10^{-8}$		67
4γ		$< 2 \times 10^{-8}$	CL=90%	67
$\nu \overline{\nu}$	[e]	$< 8.3 \times 10^{-7}$	CL=90%	67
$\nu_e \overline{\nu}_e$		$< 1.7 \times 10^{-6}$	CL=90%	67
$\nu_\mu \overline{\nu}_\mu$		$< 3.1 \times 10^{-6}$	CL=90%	67
$\nu_\tau \overline{\nu}_\tau$		$< 2.1 \times 10^{-6}$	CL=90%	67

Charge conjugation (C) or Lepton Family number (LF) violating modes

3γ	C	$< 3.1 \times 10^{-8}$	CL=90%	67
$\mu^+ e^- + e^- \mu^+$	LF	$< 1.72 \times 10^{-8}$	CL=90%	26

η

$I^G(J^{PC}) = 0^+(0^{-+})$

Mass $m = 547.45 \pm 0.19$ MeV $(S = 1.6)$
Full width $\Gamma = 1.18 \pm 0.11$ keV [f] $(S = 1.8)$

C-nonconserving decay parameters [g]

$\pi^+ \pi^- \pi^0$	Left-right asymmetry =	$(0.09 \pm 0.17) \times 10^{-2}$
$\pi^+ \pi^- \pi^0$	Sextant asymmetry =	$(0.18 \pm 0.16) \times 10^{-2}$
$\pi^+ \pi^- \pi^0$	Quadrant asymmetry =	$(-0.17 \pm 0.17) \times 10^{-2}$
$\pi^+ \pi^- \gamma$	Left-right asymmetry =	$(0.9 \pm 0.4) \times 10^{-2}$
$\pi^+ \pi^- \gamma$	β (D-wave) = 0.05 ± 0.06 $(S = 1.5)$	

η DECAY MODES		Fraction (Γ_i/Γ)	Scale factor/ Confidence level	p (MeV/c)
neutral modes		(71.4 ± 0.6) %	S=1.3	–
2γ	[f]	(39.25 ± 0.31) %	S=1.3	274
$3\pi^0$		(32.1 ± 0.4) %	S=1.2	180
$\pi^0 2\gamma$		$(7.1 \pm 1.4) \times 10^{-4}$		258
other neutral modes		< 2.8 %	CL=90%	–

charged modes	Fraction	Scale factor	p (MeV/c)
$\pi^+ \pi^- \pi^0$	(28.6 ± 0.6) %	S=1.3	–
$\pi^+ \pi^- \gamma$	(23.2 ± 0.5) %	S=1.3	175
$e^+ e^- \gamma$	(4.78 ± 0.12) %	S=1.2	236
$\mu^+ \mu^- \gamma$	$(4.9 \pm 1.1) \times 10^{-3}$		274
$e^+ e^-$	$(3.1 \pm 0.4) \times 10^{-4}$		253
$\mu^+ \mu^-$	$< 3 \times 10^{-4}$ CL=90%		274
$\pi^+ \pi^- e^+ e^-$	$(5.8 \pm 0.8) \times 10^{-6}$		253
$\pi^+ \pi^- 2\gamma$	$(1.3 ^{+1.2}_{-0.8}) \times 10^{-3}$		236
$\pi^+ \pi^- \pi^0 \gamma$	$< 2.1 \times 10^{-3}$		236
$\pi^0 \mu^+ \mu^- \gamma$	$< 6 \times 10^{-4}$ CL=90%		175
	$< 3 \times 10^{-6}$ CL=90%		211

Charge conjugation (C), Parity (P), or Charge conjugation × Parity (CP) violating modes

$\pi^+ \pi^-$	P,CP		$< 1.5 \times 10^{-3}$		236
3γ	C		$< 5 \times 10^{-4}$	CL=95%	274
$\pi^0 e^+ e^-$	C	[h]	$< 4 \times 10^{-5}$	CL=90%	258
$\pi^0 \mu^+ \mu^-$	C	[h]	$< 5 \times 10^{-6}$	CL=90%	211

$f_0(400\text{–}1200)$ [i]
or σ

$I^G(J^{PC}) = 0^+(0^{++})$

The interpretation of this entry as a particle is controversial. See the "Note on scalar mesons" in the Particle Listings under the $f_0(1370)$.
Mass $m = (400\text{–}1200)$ MeV
Full width $\Gamma = (600\text{–}1000)$ MeV

$f_0(400\text{–}1200)$ DECAY MODES	Fraction (Γ_i/Γ)	p (MeV/c)
$\pi \pi$	dominant	–
$\gamma \gamma$	seen	–

$\rho(770)$ [j]

$I^G(J^{PC}) = 1^+(1^{--})$

Mass $m = 768.5 \pm 0.6$ MeV $(S = 1.2)$
Full width $\Gamma = 150.7 \pm 1.2$ MeV
$\Gamma_{ee} = 6.77 \pm 0.32$ keV

$\rho(770)$ DECAY MODES		Fraction (Γ_i/Γ)	Scale factor/ Confidence level	p (MeV/c)
$\pi \pi$		~ 100 %		358

$\rho(770)^\pm$ decays

$\pi^\pm \gamma$		$(4.5 \pm 0.5) \times 10^{-4}$	S=2.2	372
$\pi^\pm \eta$		$< 6 \times 10^{-3}$	CL=84%	146
$\pi^\pm \pi^+ \pi^- \pi^0$		$< 2.0 \times 10^{-3}$	CL=84%	249

$\rho(770)^0$ decays

$\pi^+ \pi^- \gamma$		$(9.9 \pm 1.6) \times 10^{-3}$		358
$\pi^0 \gamma$		$(7.9 \pm 2.0) \times 10^{-4}$		372
$\eta \gamma$		$(3.8 \pm 0.7) \times 10^{-4}$		189
$\mu^+ \mu^-$	[k]	$(4.60 \pm 0.28) \times 10^{-5}$		369
$e^+ e^-$	[k]	$(4.48 \pm 0.22) \times 10^{-5}$		384
$\pi^+ \pi^- \pi^0$		$< 1.2 \times 10^{-4}$	CL=90%	319
$\pi^+ \pi^- \pi^+ \pi^-$		$< 2 \times 10^{-4}$	CL=90%	246
$\pi^+ \pi^- \pi^0 \pi^0$		$< 4 \times 10^{-5}$	CL=90%	252

$\omega(782)$

$I^G(J^{PC}) = 0^-(1^{--})$

Mass $m = 781.94 \pm 0.12$ MeV $(S = 1.5)$
Full width $\Gamma = 8.43 \pm 0.10$ MeV
$\Gamma_{ee} = 0.60 \pm 0.02$ keV

$\omega(782)$ DECAY MODES	Fraction (Γ_i/Γ)	Confidence level	p (MeV/c)
$\pi^+ \pi^- \pi^0$	(88.8 ± 0.7) %		327
$\pi^0 \gamma$	(8.5 ± 0.5) %		379
$\pi^+ \pi^-$	(2.21 ± 0.30) %		365
neutrals (excluding $\pi^0 \gamma$)	$(5.3 ^{+8.7}_{-3.5}) \times 10^{-3}$		–
$\eta \gamma$	$(8.3 \pm 2.1) \times 10^{-4}$		199
$\pi^0 e^+ e^-$	$(5.9 \pm 1.9) \times 10^{-4}$		379
$\pi^0 \mu^+ \mu^-$	$(9.6 \pm 2.3) \times 10^{-5}$		349

e^+e^-	$(7.15 \pm 0.19) \times 10^{-5}$		391
$\pi^+\pi^-\pi^0\pi^0$	< 2 %	90%	261
$\pi^+\pi^-\gamma$	$< 3.6 \times 10^{-3}$	95%	365
$\pi^+\pi^-\pi^+\pi^-$	$< 1 \times 10^{-3}$	90%	256
$\pi^0\pi^0\gamma$	$(7.2 \pm 2.5) \times 10^{-5}$		367
$\mu^+\mu^-$	$< 1.8 \times 10^{-4}$	90%	376
3γ	$< 2 \times 10^{-4}$	90%	391

Charge conjugation (C)

$\eta\pi^0$	C	$< 1 \times 10^{-3}$	90%	162
$3\pi^0$	C	$< 3 \times 10^{-4}$	90%	329

$\eta'(958)$ $I^G(J^{PC}) = 0^+(0^{-+})$

Mass $m = 957.77 \pm 0.14$ MeV
Full width $\Gamma = 0.201 \pm 0.016$ MeV   (S = 1.3)

$\eta'(958)$ DECAY MODES	Fraction (Γ_i/Γ)	Scale factor/ Confidence level	p (MeV/c)
$\pi^+\pi^-\eta$	(43.7 ± 1.5) %	S=1.1	232
$\rho^0\gamma$	(30.2 ± 1.3) %	S=1.1	169
$\pi^0\pi^0\eta$	(20.8 ± 1.3) %	S=1.2	239
$\omega\gamma$	(3.02 ± 0.30) %		160
$\gamma\gamma$	(2.12 ± 0.13) %	S=1.2	479
$3\pi^0$	$(1.55 \pm 0.26) \times 10^{-3}$		430
$\mu^+\mu^-\gamma$	$(1.04 \pm 0.26) \times 10^{-4}$		467
$\pi^+\pi^-\pi^0$	< 5 %	CL=90%	427
$\pi^0\rho^0$	< 4 %	CL=90%	118
$\pi^+\pi^-$	< 2 %	CL=90%	458
$\pi^0 e^+ e^-$	< 1.3 %	CL=90%	469
$\eta e^+ e^-$	< 1.1 %	CL=90%	322
$\pi^+\pi^+\pi^-\pi^-$	< 1 %	CL=90%	372
$\pi^+\pi^+\pi^-\pi^-$ neutrals	< 1 %	CL=95%	
$\pi^+\pi^+\pi^-\pi^-\pi^0$	< 1 %	CL=90%	298
6π	< 1 %	CL=90%	189
$\pi^+\pi^- e^+ e^-$	$< 6 \times 10^{-3}$	CL=90%	458
$\pi^0\pi^0$	$< 9 \times 10^{-4}$	CL=90%	459
$\pi^0\gamma\gamma$	$< 8 \times 10^{-4}$	CL=90%	469
$4\pi^0$	$< 5 \times 10^{-4}$	CL=90%	379
3γ	$< 1.0 \times 10^{-4}$	CL=90%	479
$\mu^+\mu^-\pi^0$	$< 6.0 \times 10^{-5}$	CL=90%	445
$\mu^+\mu^-\eta$	$< 1.5 \times 10^{-5}$	CL=90%	274
e^+e^-	$< 2.1 \times 10^{-7}$	CL=90%	479

$f_0(980)$ [l] $I^G(J^{PC}) = 0^+(0^{++})$

Mass $m = 980 \pm 10$ MeV
Full width $\Gamma = 40$ to 100 MeV

$f_0(980)$ DECAY MODES	Fraction (Γ_i/Γ)	Confidence level	p (MeV/c)
$\pi\pi$	(78.1 ± 2.4) %		470
$K\overline{K}$	(21.9 ± 2.4) %		–
$\gamma\gamma$	$(1.19 \pm 0.33) \times 10^{-5}$		490
e^+e^-	$< 3 \times 10^{-7}$	90%	490

$a_0(980)$ [l] $I^G(J^{PC}) = 1^-(0^{++})$

Mass $m = 983.5 \pm 0.9$ MeV
Full width $\Gamma = 50$ to 100 MeV

$a_0(980)$ DECAY MODES	Fraction (Γ_i/Γ)	p (MeV/c)
$\eta\pi$	dominant	321
$K\overline{K}$	seen	–
$\gamma\gamma$	seen	492

$\phi(1020)$ $I^G(J^{PC}) = 0^-(1^{--})$

Mass $m = 1019.413 \pm 0.008$ MeV
Full width $\Gamma = 4.43 \pm 0.05$ MeV
$\Gamma_{ee} = 1.37 \pm 0.05$ keV

$\phi(1020)$ DECAY MODES	Fraction (Γ_i/Γ)	Scale factor/ Confidence level	p (MeV/c)
K^+K^-	(49.1 ± 0.6) %	S=1.2	127
$K_L^0 K_S^0$	(34.1 ± 0.5) %	S=1.1	110
$\rho\pi$	(12.9 ± 0.7) %		181

$\pi^+\pi^-\pi^0$	(2.7 ± 0.9) %	S=1.1	462
$\eta\gamma$	(1.26 ± 0.06) %	S=1.1	363
$\pi^0\gamma$	$(1.31 \pm 0.13) \times 10^{-3}$		501
e^+e^-	$(3.00 \pm 0.06) \times 10^{-4}$	S=1.1	510
$\mu^+\mu^-$	$(2.48 \pm 0.34) \times 10^{-4}$		499
$\eta e^+ e^-$	$(1.3 ^{+0.8}_{-0.6}) \times 10^{-4}$		363
$\pi^+\pi^-$	$(8 ^{+5}_{-4}) \times 10^{-5}$	S=1.5	490
$\omega\gamma$	< 5 %	CL=84%	210
$\rho\gamma$	< 2 %	CL=84%	219
$\pi^+\pi^-\gamma$	$< 7 \times 10^{-3}$	CL=90%	490
$\pi^0\pi^0\gamma$	$< 1 \times 10^{-3}$	CL=90%	492
$\pi^+\pi^-\pi^+\pi^-$	$< 8.7 \times 10^{-4}$	CL=90%	410
$\eta'(958)\gamma$	$< 4.1 \times 10^{-4}$	CL=90%	60
$\pi^+\pi^+\pi^-\pi^-\pi^0$	$< 1.5 \times 10^{-4}$	CL=95%	341
$\pi^0 e^+ e^-$	$< 1.2 \times 10^{-4}$	CL=90%	501
$a_0(980)\gamma$	$< 5 \times 10^{-3}$	CL=90%	36

$h_1(1170)$ $I^G(J^{PC}) = 0^-(1^{+-})$

Mass $m = 1170 \pm 20$ MeV
Full width $\Gamma = 360 \pm 40$ MeV

$h_1(1170)$ DECAY MODES	Fraction (Γ_i/Γ)	p (MeV/c)
$\rho\pi$	seen	310

$b_1(1235)$ $I^G(J^{PC}) = 1^+(1^{+-})$

Mass $m = 1231 \pm 10$ MeV [l]
Full width $\Gamma = 142 \pm 8$ MeV   (S = 1.1)

$b_1(1235)$ DECAY MODES	Fraction (Γ_i/Γ)	Confidence level	p (MeV/c)
$\omega\pi$	dominant		348
[D/S amplitude ratio = 0.26 ± 0.04]			
$\pi^\pm\gamma$	$(1.6 \pm 0.4) \times 10^{-3}$		608
$\eta\rho$	seen		–
$\pi^+\pi^+\pi^-\pi^0$	< 50 %	84%	536
$(K\overline{K})^\pm\pi^0$	< 8 %	90%	248
$K_S^0 K_S^0 \pi^\pm$	< 6 %	90%	238
$K_S^0 K_S^0 \pi^\pm$	< 2 %	90%	238
$\pi\phi$	< 1.5 %	84%	146

$a_1(1260)$ [m] $I^G(J^{PC}) = 1^-(1^{++})$

Mass $m = 1230 \pm 40$ MeV [l]
Full width $\Gamma \sim 400$ MeV

$a_1(1260)$ DECAY MODES	Fraction (Γ_i/Γ)	p (MeV/c)
$\rho\pi$	dominant	356
$\pi\gamma$	seen	607
$K\overline{K}^*(892)$	possibly seen	–

$f_2(1270)$ $I^G(J^{PC}) = 0^+(2^{++})$

Mass $m = 1275 \pm 5$ MeV [l]
Full width $\Gamma = 185 \pm 20$ MeV [l]

$f_2(1270)$ DECAY MODES	Fraction (Γ_i/Γ)	Scale factor/ Confidence level	p (MeV/c)
$\pi\pi$	$(84.7 ^{+2.6}_{-1.2})$ %	S=1.3	622
$\pi^+\pi^-2\pi^0$	$(7.2 ^{+1.4}_{-2.9})$ %	S=1.3	562
$K\overline{K}$	(4.6 ± 0.5) %	S=2.8	403
$2\pi^+2\pi^-$	(2.8 ± 0.4) %	S=1.2	559
$\eta\eta$	$(4.5 \pm 1.0) \times 10^{-3}$	S=2.4	327
$4\pi^0$	$(3.0 \pm 1.0) \times 10^{-3}$		564
$\gamma\gamma$	$(1.32 ^{+0.18}_{-0.16}) \times 10^{-5}$		637
$\eta\pi\pi$	$< 8 \times 10^{-3}$	CL=95%	475
$K^0 K^-\pi^+ + $ c.c.	$< 3.4 \times 10^{-3}$	CL=95%	293
e^+e^-	$< 9 \times 10^{-9}$	CL=90%	637

$f_1(1285)$ $I^G(J^{PC}) = 0^+(1^{++})$

Mass $m = 1282.2 \pm 0.7$ MeV [i]   (S = 1.7)
Full width $\Gamma = 24.8 \pm 1.3$ MeV [i]   (S = 1.3)

$f_1(1285)$ DECAY MODES	Fraction (Γ_i/Γ)	Scale factor/ Confidence level	p (MeV/c)
4π	(29 ± 6) %		563
$\pi^0\pi^0\pi^+\pi^-$	$(15 \, {}^{+\,9}_{-\,8})$ %	S=1.1	566
$2\pi^+2\pi^-$	(15 ± 6) %		563
$\rho^0\pi^+\pi^-$	dominates $2\pi^+2\pi^-$		340
$4\pi^0$	$< 7 \quad \times 10^{-4}$	CL=90%	568
$\eta\pi\pi$	(54 ± 15) %		479
$a_0(980)\pi$ [ignoring $a_0(980) \to K\overline{K}$]	(44 ± 7) %	S=1.1	234
$\eta\pi\pi$ [excluding $a_0(980)\pi$]	$(10 \, {}^{+\,7}_{-\,6})$ %	S=1.1	–
$K\overline{K}\pi$	(9.7 ± 1.6) %	S=1.2	308
$K\overline{K}^*(892)$	not seen		–
$\gamma\rho^0$	(6.6 ± 1.3) %	S=1.5	410
$\phi\gamma$	$(8.0 \pm 3.1) \times 10^{-4}$		236

$\eta(1295)$ $I^G(J^{PC}) = 0^+(0^{-+})$

Mass $m = 1295 \pm 4$ MeV
Full width $\Gamma = 53 \pm 6$ MeV

$\eta(1295)$ DECAY MODES	Fraction (Γ_i/Γ)	p (MeV/c)
$\eta\pi^+\pi^-$	seen	488
$a_0(980)\pi$	seen	245

$\pi(1300)$ $I^G(J^{PC}) = 1^-(0^{-+})$

Mass $m = 1300 \pm 100$ MeV [i]
Full width $\Gamma = 200$ to 600 MeV

$\pi(1300)$ DECAY MODES	Fraction (Γ_i/Γ)	p (MeV/c)
$\rho\pi$	seen	406
$\pi(\pi\pi)_{S\text{-wave}}$	seen	612

$a_2(1320)$ $I^G(J^{PC}) = 1^-(2^{++})$

Mass $m = 1318.1 \pm 0.7$ MeV   (S = 1.2)
Full width $\Gamma = 107 \pm 5$ MeV [i]   ($K^\pm K^0_S$ and $\eta\pi$ modes)

$a_2(1320)$ DECAY MODES	Fraction (Γ_i/Γ)	Scale factor/ Confidence level	p (MeV/c)
$\rho\pi$	(70.1 ± 2.7) %	S=1.2	419
$\eta\pi$	(14.5 ± 1.2) %		535
$\omega\pi\pi$	(10.6 ± 3.2) %	S=1.3	362
$K\overline{K}$	(4.9 ± 0.8) %		437
$\eta'(958)\pi$	$(5.7 \pm 1.1) \times 10^{-3}$		287
$\pi^\pm\gamma$	$(2.8 \pm 0.6) \times 10^{-3}$		652
$\gamma\gamma$	$(9.7 \pm 1.0) \times 10^{-6}$		659
$\pi^+\pi^-\pi^-$	$< 8 \quad$ %	CL=90%	621
e^+e^-	$< 2.3 \quad \times 10^{-7}$	CL=90%	659

$f_0(1370)$ [i]   was $f_0(1300)$ $I^G(J^{PC}) = 0^+(0^{++})$

Mass $m = 1200$ to 1500 MeV
Full width $\Gamma = 300$ to 500 MeV

In two-particle decay modes the $\pi\pi$ decay is dominant. We include here the resonance observed in 4π under the same entry as the one decaying to 2 pseudoscalars. See also the minireview under non-$q\overline{q}$ candidates.

$f_0(1370)$ DECAY MODES	Fraction (Γ_i/Γ)	p (MeV/c)
$\pi\pi$	seen	–
4π	seen	–
$2\pi^+2\pi^-$	seen	–
$\pi^+\pi^-2\pi^0$	seen	–

$\eta\eta$	seen	–
$K\overline{K}$	seen	–
$\gamma\gamma$	seen	–
e^+e^-	not seen	–

$f_1(1420)$ [n] $I^G(J^{PC}) = 0^+(1^{++})$

Mass $m = 1426.8 \pm 2.3$ MeV   (S = 1.3)
Full width $\Gamma = 53 \pm 5$ MeV

$f_1(1420)$ DECAY MODES	Fraction (Γ_i/Γ)	p (MeV/c)
$K\overline{K}\pi$	dominant	439
$\eta\pi\pi$	possibly seen	571

$\omega(1420)$ [o] $I^G(J^{PC}) = 0^-(1^{--})$

Mass $m = 1419 \pm 31$ MeV
Full width $\Gamma = 174 \pm 60$ MeV

$\omega(1420)$ DECAY MODES	Fraction (Γ_i/Γ)	p (MeV/c)
$\rho\pi$	dominant	488

$\eta(1440)$ [p] $I^G(J^{PC}) = 0^+(0^{-+})$

Mass $m = 1415 \pm 10$ MeV [i]
Full width $\Gamma = 60 \pm 20$ MeV [i]

$\eta(1440)$ DECAY MODES	Fraction (Γ_i/Γ)	p (MeV/c)
$K\overline{K}\pi$	seen	429
$\eta\pi\pi$	seen	564
$a_0(980)\pi$	seen	347
4π	seen	637

$\rho(1450)$ [q] $I^G(J^{PC}) = 1^+(1^{--})$

Mass $m = 1465 \pm 25$ MeV [i]
Full width $\Gamma = 310 \pm 60$ MeV [i]

$\rho(1450)$ DECAY MODES	Fraction (Γ_i/Γ)	Confidence level	p (MeV/c)
$\pi\pi$	seen		719
4π	seen		665
e^+e^-	seen		732
$\eta\rho$	<4 %		317
$\omega\pi$	<2.0 %	95%	512
$\phi\pi$	<1 %		358
$K\overline{K}$	$<1.6 \times 10^{-3}$	95%	541

$f_0(1500)$ [r]   was $f_0(1525)$ and $f_0(1590)$ $I^G(J^{PC}) = 0^+(0^{++})$

Mass $m = 1503 \pm 11$ MeV
Full width $\Gamma = 120 \pm 19$ MeV

$f_0(1500)$ DECAY MODES	Fraction (Γ_i/Γ)	p (MeV/c)
$\eta\eta'(958)$	seen	–
$\eta\eta$	seen	515
$4\pi^0$	seen	690
$\pi^0\pi^0$	seen	739
$2\pi^+2\pi^-$	seen	686

$f_1(1510)$ $I^G(J^{PC}) = 0^+(1^{++})$

Mass $m = 1512 \pm 4$ MeV
Full width $\Gamma = 35 \pm 15$ MeV

$f_1(1510)$ DECAY MODES	Fraction (Γ_i/Γ)	p (MeV/c)
$K\overline{K}^*(892) + $ c.c.	seen	292

$f_2'(1525)$ $I^G(J^{PC}) = 0^+(2^{++})$

Mass $m = 1525 \pm 5$ MeV [l]
Full width $\Gamma = 76 \pm 10$ MeV [l]

$f_2'(1525)$ DECAY MODES	Fraction (Γ_i/Γ)	p (MeV/c)
$K\overline{K}$	(88.8 ± 3.1) %	581
$\eta\eta$	(10.3 ± 3.1) %	531
$\pi\pi$	$(8.2 \pm 1.5) \times 10^{-3}$	750
$\gamma\gamma$	$(1.32 \pm 0.21) \times 10^{-6}$	763

$\omega(1600)$ [s] $I^G(J^{PC}) = 0^-(1^{--})$

Mass $m = 1649 \pm 24$ MeV   (S = 2.3)
Full width $\Gamma = 220 \pm 35$ MeV   (S = 1.6)

$\omega(1600)$ DECAY MODES	Fraction (Γ_i/Γ)	p (MeV/c)
$\rho\pi$	seen	637
$\omega\pi\pi$	seen	601
$e^+ e^-$	seen	824

$\omega_3(1670)$ $I^G(J^{PC}) = 0^-(3^{--})$

Mass $m = 1667 \pm 4$ MeV
Full width $\Gamma = 168 \pm 10$ MeV [l]

$\omega_3(1670)$ DECAY MODES	Fraction (Γ_i/Γ)	p (MeV/c)
$\rho\pi$	seen	647
$\omega\pi\pi$	seen	614
$b_1(1235)\pi$	possibly seen	359

$\pi_2(1670)$ $I^G(J^{PC}) = 1^-(2^{-+})$

Mass $m = 1670 \pm 20$ MeV [l]
Full width $\Gamma = 258 \pm 18$ MeV [l]   (S = 1.7)
$\Gamma_{ee} = 1.35 \pm 0.26$ keV

$\pi_2(1670)$ DECAY MODES	Fraction (Γ_i/Γ)	p (MeV/c)
3π	(95.8 ± 1.4) %	806
$f_2(1270)\pi$	(56.2 ± 3.2) %	325
$\rho\pi$	(31 ± 4) %	649
$f_0(1370)\pi$	(8.7 ± 3.4) %	–
$K\overline{K}^*(892)+$ c.c.	(4.2 ± 1.4) %	453
$\gamma\gamma$	$(5.2 \pm 1.1) \times 10^{-6}$	835

$\phi(1680)$ $I^G(J^{PC}) = 0^-(1^{--})$

Mass $m = 1680 \pm 20$ MeV [l]
Full width $\Gamma = 150 \pm 50$ MeV [l]

$\phi(1680)$ DECAY MODES	Fraction (Γ_i/Γ)	p (MeV/c)
$K\overline{K}^*(892)+$ c.c.	dominant	463
$K_S^0 K\pi$	seen	620
$K\overline{K}$	seen	681
$e^+ e^-$	seen	840
$\omega\pi\pi$	not seen	622

$\rho_3(1690)$ $I^G(J^{PC}) = 1^+(3^{--})$

J^P from the 2π and $K\overline{K}$ modes.
Mass $m = 1691 \pm 5$ MeV [l]
Full width $\Gamma = 160 \pm 10$ MeV [l]   (S = 1.5)

$\rho_3(1690)$ DECAY MODES	Fraction (Γ_i/Γ)	Scale factor	p (MeV/c)
4π	(71.1 ± 1.9) %		788
$\pi^\pm \pi^+ \pi^- \pi^0$	(67 ± 22) %		788
$\pi\pi$	(23.6 ± 1.3) %		834
$\omega\pi$	(16 ± 6) %		656
$K\overline{K}\pi$	(3.8 ± 1.2) %		628
$K\overline{K}$	(1.58 ± 0.26) %	1.2	686
$\eta\pi^+\pi^-$	seen		728

$\rho(1700)$ [q] $I^G(J^{PC}) = 1^+(1^{--})$

Mass $m = 1700 \pm 20$ MeV [l]   ($\eta\rho^0$ and $\pi^+\pi^-$ modes)
Full width $\Gamma = 235 \pm 50$ MeV [l]   ($\eta\rho^0$ and $\pi^+\pi^-$ modes)

$\rho(1700)$ DECAY MODES	Fraction (Γ_i/Γ)	p (MeV/c)
$\rho\pi\pi$	dominant	640
$\rho^0\pi^+\pi^-$	large	640
$\rho^\pm\pi^\mp\pi^0$	large	642
$2(\pi^+\pi^-)$	large	792
$\pi^+\pi^-$	seen	838
$K\overline{K}^*(892)+$ c.c.	seen	479
$\eta\rho$	seen	533
$K\overline{K}$	seen	692
$e^+ e^-$	seen	850

$f_J(1710)$ [t] $I^G(J^{PC}) = 0^+(\text{even}^{++})$

Mass $m = 1697 \pm 4$ MeV   (S = 1.4)
Full width $\Gamma = 175 \pm 9$ MeV   (S = 1.7)

$f_J(1710)$ DECAY MODES	Fraction (Γ_i/Γ)	p (MeV/c)
$K\overline{K}$	seen	690
$\eta\eta$	seen	648
$\pi\pi$	seen	837

$\phi_3(1850)$ $I^G(J^{PC}) = 0^-(3^{--})$

Mass $m = 1854 \pm 7$ MeV
Full width $\Gamma = 87^{+28}_{-23}$ MeV   (S = 1.2)

$\phi_3(1850)$ DECAY MODES	Fraction (Γ_i/Γ)	p (MeV/c)
$K\overline{K}$	seen	785
$K\overline{K}^*(892)+$ c.c.	seen	602

$f_2(2010)$ $I^G(J^{PC}) = 0^+(2^{++})$

Seen by one group only.
Mass $m = 2011^{+60}_{-80}$ MeV
Full width $\Gamma = 202 \pm 60$ MeV

$f_2(2010)$ DECAY MODES	Fraction (Γ_i/Γ)	p (MeV/c)
$\phi\phi$	seen	–

$f_4(2050)$ $I^G(J^{PC}) = 0^+(4^{++})$

Mass $m = 2044 \pm 11$ MeV   (S = 1.4)
Full width $\Gamma = 208 \pm 13$ MeV   (S = 1.2)

$f_4(2050)$ DECAY MODES	Fraction (Γ_i/Γ)	p (MeV/c)
$\omega\omega$	(26 ± 6) %	658
$\pi\pi$	(17.0 ± 1.5) %	1012
$K\overline{K}$	$(6.8^{+3.4}_{-1.8}) \times 10^{-3}$	895
$\eta\eta$	$(2.1 \pm 0.8) \times 10^{-3}$	863
$4\pi^0$	< 1.2 %	977

$f_2(2300)$ $I^G(J^{PC}) = 0^+(2^{++})$

Mass $m = 2297 \pm 28$ MeV
Full width $\Gamma = 149 \pm 40$ MeV

$f_2(2300)$ DECAY MODES	Fraction (Γ_i/Γ)	p (MeV/c)
$\phi\phi$	seen	529

$f_2(2340)$	$I^G(J^{PC}) = 0^+(2^{++})$

Mass $m = 2339 \pm 60$ MeV
Full width $\Gamma = 319^{+80}_{-70}$ MeV

$f_2(2340)$ DECAY MODES	Fraction (Γ_i/Γ)	p (MeV/c)
$\phi\phi$	seen	573

STRANGE MESONS
$(S = \pm 1, C = B = 0)$

$K^+ = u\bar{s}$, $K^0 = d\bar{s}$, $\overline{K^0} = \bar{d}s$, $K^- = \bar{u}s$, similarly for K^*'s

---

$K^\pm$	$I(J^P) = \frac{1}{2}(0^-)$

Mass $m = 493.677 \pm 0.016$ MeV [u] (S = 2.8)
Mean life $\tau = (1.2386 \pm 0.0024) \times 10^{-8}$ s (S = 2.0)
$c\tau = 3.713$ m

Slope parameter g [v]

(See Particle Listings for quadratic coefficients)

$K^+ \to \pi^+\pi^+\pi^- = -0.2154 \pm 0.0035$ (S = 1.4)
$K^- \to \pi^-\pi^-\pi^+ = -0.217 \pm 0.007$ (S = 2.5)
$K^\pm \to \pi^\pm\pi^0\pi^0 = 0.594 \pm 0.019$ (S = 1.3)

$K^\pm$ decay form factors [a,w]

$K^+_{e3} \quad \lambda_+ = 0.0286 \pm 0.0022$
$K^+_{\mu3} \quad \lambda_+ = 0.033 \pm 0.008$ (S = 1.6)
$K^+_{\mu3} \quad \lambda_0 = 0.004 \pm 0.007$ (S = 1.6)
$K^+_{e3} \quad |f_S/f_+| = 0.084 \pm 0.023$ (S = 1.2)
$K^+_{e3} \quad |f_T/f_+| = 0.38 \pm 0.11$ (S = 1.1)
$K^+_{\mu3} \quad |f_T/f_+| = 0.02 \pm 0.12$
$K^+ \to e^+\nu_e\gamma \quad |F_A + F_V| = 0.148 \pm 0.010$
$K^+ \to \mu^+\nu_\mu\gamma \quad |F_A + F_V| < 0.23$, CL = 90%
$K^+ \to e^+\nu_e\gamma \quad |F_A - F_V| < 0.49$
$K^+ \to \mu^+\nu_\mu\gamma \quad |F_A - F_V| = -2.2$ to 0.3

K^- modes are charge conjugates of the modes below.

K^+ DECAY MODES		Fraction (Γ_i/Γ)	Scale factor/ Confidence level	p (MeV/c)
$\mu^+\nu_\mu$		(63.51 ± 0.18) %	S=1.3	236
$e^+\nu_e$		$(1.55\pm0.07)\times 10^{-5}$		247
$\pi^+\pi^0$		(21.16 ± 0.14) %	S=1.1	205
$\pi^+\pi^+\pi^-$		(5.59 ± 0.05) %	S=1.8	125
$\pi^+\pi^0\pi^0$		(1.73 ± 0.04) %	S=1.2	133
$\pi^0\mu^+\nu_\mu$ Called $K^+_{\mu3}$.		(3.18 ± 0.08) %	S=1.5	215
$\pi^0 e^+\nu_e$ Called K^+_{e3}.		(4.82 ± 0.06) %	S=1.3	228
$\pi^0\pi^0 e^+\nu_e$		$(2.1\pm0.4)\times 10^{-5}$		206
$\pi^+\pi^- e^+\nu_e$		$(3.91\pm0.17)\times 10^{-5}$		203
$\pi^+\pi^-\mu^+\nu_\mu$		$(1.4\pm0.9)\times 10^{-5}$		151
$\pi^0\pi^0\pi^0 e^+\nu_e$		< 3.5 $\times 10^{-6}$	CL=90%	135
$\pi^+\gamma\gamma$	[x]	< 1 $\times 10^{-6}$	CL=90%	227
$\pi^+ 3\gamma$	[x]	< 1.0 $\times 10^{-4}$	CL=90%	227
$\mu^+\nu_\mu\nu\bar{\nu}$		< 6.0 $\times 10^{-6}$	CL=90%	236
$e^+\nu_e\nu\bar{\nu}$		< 6 $\times 10^{-6}$	CL=90%	247
$\mu^+\nu_\mu e^+e^-$		$(1.06\pm0.32)\times 10^{-6}$		236
$e^+\nu_e e^+e^-$		$(2.1^{+2.1}_{-1.1})\times 10^{-7}$		247
$\mu^+\nu_\mu\mu^+\mu^-$		< 4.1 $\times 10^{-7}$	CL=90%	185
$\mu^+\nu_\mu\gamma$	[x,y]	$(5.50\pm0.28)\times 10^{-3}$		236
$\pi^+\pi^0\gamma$	[x,y]	$(2.75\pm0.15)\times 10^{-4}$		205
$\pi^+\pi^0\gamma$(DE)	[x,z]	$(1.8\pm0.4)\times 10^{-5}$		205
$\pi^+\pi^+\pi^-\gamma$	[x,y]	$(1.04\pm0.31)\times 10^{-4}$		125
$\pi^+\pi^0\pi^0\gamma$	[x,y]	$(7.5^{+5.5}_{-3.0})\times 10^{-6}$		133
$\pi^0\mu^+\nu_\mu\gamma$	[x,y]	< 6.1 $\times 10^{-5}$	CL=90%	215
$\pi^0 e^+\nu_e\gamma$	[x,y]	$(2.62\pm0.20)\times 10^{-4}$		228
$\pi^0 e^+\nu_e\gamma$(SD)	[aa]	< 5.3 $\times 10^{-5}$	CL=90%	228
$\pi^0\pi^0 e^+\nu_e\gamma$		< 5 $\times 10^{-6}$	CL=90%	206

Lepton Family number (LF), Lepton number (L), $\Delta S = \Delta Q$ (SQ) violating modes, or $\Delta S = 1$ weak neutral current (S1) modes

$\pi^+\pi^+ e^-\bar{\nu}_e$	SQ	< 1.2	$\times 10^{-8}$	CL=90%	203
$\pi^+\pi^+\mu^-\bar{\nu}_\mu$	SQ	< 3.0	$\times 10^{-6}$	CL=95%	151
$\pi^+ e^+ e^-$	S1	(2.74 ± 0.23)	$\times 10^{-7}$		227
$\pi^+\mu^+\mu^-$	S1	< 2.3	$\times 10^{-7}$	CL=90%	172
$\pi^+\nu\bar{\nu}$	S1	< 2.4	$\times 10^{-9}$	CL=90%	227
$\mu^- \nu e^+ e^+$	LF	< 2.0	$\times 10^{-8}$	CL=90%	236
$\mu^+\nu_e$	LF	[d] < 4	$\times 10^{-3}$	CL=90%	236
$\pi^+\mu^+ e^-$	LF	< 2.1	$\times 10^{-10}$	CL=90%	214
$\pi^+\mu^- e^+$	LF	< 7	$\times 10^{-9}$	CL=90%	214
$\pi^-\mu^+ e^+$	L	< 7	$\times 10^{-9}$	CL=90%	214
$\pi^- e^+ e^+$	L	< 1.0	$\times 10^{-8}$	CL=90%	227
$\pi^-\mu^+\mu^+$	L	< 1.5	$\times 10^{-4}$	CL=90%	172
$\mu^+\bar{\nu}_e$	L	[d] < 3.3	$\times 10^{-3}$	CL=90%	236
$\pi^0 e^+\bar{\nu}_e$	L	[d] < 3	$\times 10^{-3}$	CL=90%	228

K^0	$I(J^P) = \frac{1}{2}(0^-)$

50% K_S, 50% K_L

Mass $m = 497.672 \pm 0.031$ MeV
$m_{K^0} - m_{K^\pm} = 3.995 \pm 0.034$ MeV (S = 1.1)
$|m_{K^0} - m_{\overline{K^0}}| / m_{average} < 9 \times 10^{-19}$ [bb]

K^0_S	$I(J^P) = \frac{1}{2}(0^-)$

Mean life $\tau = (0.8927 \pm 0.0009) \times 10^{-10}$ s
$c\tau = 2.6762$ cm

CP-violation parameters [cc]

$\text{Im}(\eta_{+-0}) = -0.015 \pm 0.030$
$\text{Im}(\eta_{000})^2 < 0.1$, CL = 90%

K^0_S DECAY MODES		Fraction (Γ_i/Γ)	Scale factor/ Confidence level	p (MeV/c)
$\pi^+\pi^-$		(68.61 ± 0.28) %	S=1.2	206
$\pi^0\pi^0$		(31.39 ± 0.28) %	S=1.2	209
$\pi^+\pi^-\gamma$	[y,dd]	$(1.78\pm0.05)\times 10^{-3}$		206
$\gamma\gamma$		$(2.4\pm0.9)\times 10^{-6}$		249
$\pi^+\pi^-\pi^0$		$(3.9^{+5.5}_{-1.9})\times 10^{-7}$		133
$3\pi^0$		< 3.7 $\times 10^{-5}$	CL=90%	139
$\pi^\pm e^\mp\nu$	[ee]	$(6.70\pm0.07)\times 10^{-4}$	S=1.3	229
$\pi^\pm\mu^\mp\nu$	[ee]	$(4.69\pm0.06)\times 10^{-4}$	S=1.2	216

$\Delta S = 1$ weak neutral current (S1) modes

$\mu^+\mu^-$	S1	< 3.2	$\times 10^{-7}$	CL=90%	225
$e^+ e^-$	S1	< 2.8	$\times 10^{-6}$	CL=90%	249
$\pi^0 e^+ e^-$	S1	< 1.1	$\times 10^{-6}$	CL=90%	231

K^0_L	$I(J^P) = \frac{1}{2}(0^-)$

$m_{K_L} - m_{K_S} = (0.5304 \pm 0.0014) \times 10^{10} \hbar$ s^{-1}
$= (3.491 \pm 0.009) \times 10^{-12}$ MeV
Mean life $\tau = (5.17 \pm 0.04) \times 10^{-8}$ s (S = 1.1)
$c\tau = 15.51$ m

Slope parameter g [v]

(See Particle Listings for quadratic coefficients)

$K^0_L \to \pi^+\pi^-\pi^0 = 0.670 \pm 0.014$ (S = 1.6)

K_L decay form factors [w]

$K^0_{e3} \quad \lambda_+ = 0.0300 \pm 0.0016$ (S = 1.2)
$K^0_{\mu3} \quad \lambda_+ = 0.034 \pm 0.005$ (S = 2.3)
$K^0_{\mu3} \quad \lambda_0 = 0.025 \pm 0.006$ (S = 2.3)
$K^0_{e3} \quad |f_S/f_+| < 0.04$, CL = 68%
$K^0_{e3} \quad |f_T/f_+| < 0.23$, CL = 68%
$K^0_{\mu3} \quad |f_T/f_+| = 0.12 \pm 0.12$
$K_L \to e^+ e^-\gamma: \quad \alpha_{K^*} = -0.28 \pm 0.08$

CP-violation parameters [cc]

$\delta = (0.327 \pm 0.012)\%$

$|\eta_{00}| = (2.275 \pm 0.019) \times 10^{-3}$ (S = 1.1)

$|\eta_{+-}| = (2.285 \pm 0.019) \times 10^{-3}$

$|\eta_{00}/\eta_{+-}| = 0.9956 \pm 0.0023$ [ff] (S = 1.8)

$\epsilon'/\epsilon = (1.5 \pm 0.8) \times 10^{-3}$ [ff] (S = 1.8)

$\phi_{+-} = (43.7 \pm 0.6)°$

$\phi_{00} = (43.5 \pm 1.0)°$

$\phi_{00} - \phi_{+-} = (-0.2 \pm 0.8)°$

j for $K_L^0 \rightarrow \pi^+\pi^-\pi^0 = 0.0011 \pm 0.0008$

$|\eta_{+-\gamma}| = (2.35 \pm 0.07) \times 10^{-3}$

$\phi_{+-\gamma} = (44 \pm 4)°$

$|\epsilon'_{+-\gamma}|/\epsilon < 0.3$, CL = 90%

$\Delta S = -\Delta Q$ in $K_{\ell3}^0$ decay

Re $x = 0.006 \pm 0.018$ (S = 1.3)

Im $x = -0.003 \pm 0.026$ (S = 1.2)

CPT-violation parameters

Re $\Delta = 0.018 \pm 0.020$

Im $\Delta = 0.02 \pm 0.04$

K_L^0 DECAY MODES		Fraction (Γ_i/Γ)	Scale factor/ Confidence level	p (MeV/c)
$3\pi^0$		$(21.12 \pm 0.27)\%$	S=1.1	139
$\pi^+\pi^-\pi^0$		$(12.56 \pm 0.20)\%$	S=1.7	133
$\pi^\pm\mu^\mp\nu$	[gg]	$(27.17 \pm 0.25)\%$	S=1.1	216
Called $K_{\mu3}^0$.				
$\pi^\pm e^\mp\nu_e$	[gg]	$(38.78 \pm 0.27)\%$	S=1.1	229
Called K_{e3}^0.				
2γ		$(5.92 \pm 0.15) \times 10^{-4}$		249
3γ		$< 2.4 \times 10^{-7}$	CL=90%	249
$\pi^0 2\gamma$	[hh]	$(1.70 \pm 0.28) \times 10^{-6}$		231
$\pi^0\pi^\pm e^\mp\nu$	[gg]	$(5.18 \pm 0.29) \times 10^{-5}$		207
$(\pi\mu\text{atom})\nu$		$(1.06 \pm 0.11) \times 10^{-7}$		–
$\pi^\pm e^\mp\nu_e\gamma$	[y,gg,hh]	$(1.3 \pm 0.8)\%$		229
$\pi^+\pi^-\gamma$	[y,hh]	$(4.61 \pm 0.14) \times 10^{-5}$		206
$\pi^0\pi^0\gamma$		$< 5.6 \times 10^{-6}$		209

Charge conjugation × Parity (CP, CPV) or Lepton Family number (LF) violating modes, or $\Delta S = 1$ weak neutral current (S1) modes

		Fraction (Γ_i/Γ)			p (MeV/c)
$\pi^+\pi^-$	CPV	$(2.067 \pm 0.035) \times 10^{-3}$	S=1.1		206
$\pi^0\pi^0$	CPV	$(9.36 \pm 0.20) \times 10^{-4}$			209
$\mu^+\mu^-$	S1	$(7.2 \pm 0.5) \times 10^{-9}$	S=1.4		225
$\mu^+\mu^-\gamma$	S1	$(3.23 \pm 0.30) \times 10^{-7}$			225
e^+e^-	S1	$< 4.1 \times 10^{-11}$	CL=90%		249
$e^+e^-\gamma$	S1	$(9.1 \pm 0.5) \times 10^{-6}$			249
$e^+e^-\gamma\gamma$	S1 [hh]	$(6.5 \pm 1.2) \times 10^{-7}$			249
$\pi^+\pi^-e^+e^-$	S1	$< 2.5 \times 10^{-6}$	CL=90%		206
$\mu^+\mu^-e^+e^-$	S1	$< 4.9 \times 10^{-6}$	CL=90%		225
$e^+e^-e^+e^-$	S1 [ii]	$(4.1 \pm 0.8) \times 10^{-8}$	S=1.2		249
$\pi^0\mu^+\mu^-$	CP,S1 [jj]	$< 5.1 \times 10^{-9}$	CL=90%		177
$\pi^0 e^+e^-$	CP,S1 [jj]	$< 4.3 \times 10^{-9}$	CL=90%		231
$\pi^0\nu\bar{\nu}$	CP,S1 [kk]	$< 5.8 \times 10^{-5}$	CL=90%		231
$e^\pm\mu^\mp$	LF [gg]	$< 3.3 \times 10^{-11}$	CL=90%		238

$\boxed{K^*(892)}$ $\qquad I(J^P) = \frac{1}{2}(1^-)$

$K^*(892)^\pm$ mass $m = 891.59 \pm 0.24$ MeV (S = 1.1)

$K^*(892)^0$ mass $m = 896.10 \pm 0.28$ MeV (S = 1.4)

$K^*(892)^\pm$ full width $\Gamma = 49.8 \pm 0.8$ MeV

$K^*(892)^0$ full width $\Gamma = 50.5 \pm 0.6$ MeV (S = 1.1)

$K^*(892)$ DECAY MODES	Fraction (Γ_i/Γ)	Confidence level	p (MeV/c)
$K\pi$	$\sim 100\%$		291
$K^0\gamma$	$(2.30 \pm 0.20) \times 10^{-3}$		310
$K^\pm\gamma$	$(1.01 \pm 0.09) \times 10^{-3}$		309
$K\pi\pi$	$< 7 \times 10^{-4}$	95%	224

$\boxed{K_1(1270)}$ $\qquad I(J^P) = \frac{1}{2}(1^+)$

Mass $m = 1273 \pm 7$ MeV [l]

Full width $\Gamma = 90 \pm 20$ MeV [l]

$K_1(1270)$ DECAY MODES	Fraction (Γ_i/Γ)	p (MeV/c)
$K\rho$	$(42 \pm 6)\%$	76
$K_0^*(1430)\pi$	$(28 \pm 4)\%$	–
$K^*(892)\pi$	$(16 \pm 5)\%$	301
$K\omega$	$(11.0 \pm 2.0)\%$	–
$K f_0(1370)$	$(3.0 \pm 2.0)\%$	–

$\boxed{K_1(1400)}$ $\qquad I(J^P) = \frac{1}{2}(1^+)$

Mass $m = 1402 \pm 7$ MeV

Full width $\Gamma = 174 \pm 13$ MeV (S = 1.6)

$K_1(1400)$ DECAY MODES	Fraction (Γ_i/Γ)	p (MeV/c)
$K^*(892)\pi$	$(94 \pm 6)\%$	401
$K\rho$	$(3.0 \pm 3.0)\%$	298
$K f_0(1370)$	$(2.0 \pm 2.0)\%$	–
$K\omega$	$(1.0 \pm 1.0)\%$	285

$\boxed{K^*(1410)}$ $\qquad I(J^P) = \frac{1}{2}(1^-)$

Mass $m = 1412 \pm 12$ MeV (S = 1.1)

Full width $\Gamma = 227 \pm 22$ MeV (S = 1.1)

$K^*(1410)$ DECAY MODES	Fraction (Γ_i/Γ)	Confidence level	p (MeV/c)
$K^*(892)\pi$	$> 40\%$	95%	408
$K\pi$	$(6.6 \pm 1.3)\%$		611
$K\rho$	$< 7\%$	95%	309

$\boxed{K_0^*(1430)}$ [ll] $\qquad I(J^P) = \frac{1}{2}(0^+)$

Mass $m = 1429 \pm 6$ MeV

Full width $\Gamma = 287 \pm 23$ MeV

$K_0^*(1430)$ DECAY MODES	Fraction (Γ_i/Γ)	p (MeV/c)
$K\pi$	$(93 \pm 10)\%$	621

$\boxed{K_2^*(1430)}$ $\qquad I(J^P) = \frac{1}{2}(2^+)$

$K_2^*(1430)^\pm$ mass $m = 1425.4 \pm 1.3$ MeV (S = 1.1)

$K_2^*(1430)^0$ mass $m = 1432.4 \pm 1.3$ MeV

$K_2^*(1430)^\pm$ full width $\Gamma = 98.4 \pm 2.3$ MeV

$K_2^*(1430)^0$ full width $\Gamma = 109 \pm 5$ MeV (S = 1.9)

$K_2^*(1430)$ DECAY MODES	Fraction (Γ_i/Γ)	Scale factor/ Confidence level	p (MeV/c)
$K\pi$	$(49.7 \pm 1.2)\%$		622
$K^*(892)\pi$	$(25.2 \pm 1.7)\%$		423
$K^*(892)\pi\pi$	$(13.0 \pm 2.3)\%$		375
$K\rho$	$(8.8 \pm 0.8)\%$	S=1.2	331
$K\omega$	$(2.9 \pm 0.8)\%$		319
$K^+\gamma$	$(2.4 \pm 0.5) \times 10^{-3}$		627
$K\eta$	$(1.4^{+2.8}_{-0.9}) \times 10^{-3}$	S=1.1	492
$K\omega\pi$	$< 7.2 \times 10^{-4}$	CL=95%	110
$K^0\gamma$	$< 9 \times 10^{-4}$	CL=90%	631

$K^*(1680)$ $I(J^P) = \frac{1}{2}(1^-)$

Mass $m = 1714 \pm 20$ MeV   (S = 1.1)
Full width $\Gamma = 323 \pm 110$ MeV   (S = 4.2)

$K^*(1680)$ DECAY MODES	Fraction (Γ_i/Γ)	p (MeV/c)
$K\pi$	(38.7 ± 2.5) %	779
$K\rho$	$(31.4^{+4.7}_{-2.1})$ %	571
$K^*(892)\pi$	$(29.9^{+2.2}_{-4.7})$ %	615

$K_2(1770)$ [mm] $I(J^P) = \frac{1}{2}(2^-)$

Mass $m = 1773 \pm 8$ MeV
Full width $\Gamma = 186 \pm 14$ MeV

$K_2(1770)$ DECAY MODES	Fraction (Γ_i/Γ)	p (MeV/c)
$K\pi\pi$		–
$K_2^*(1430)\pi$	dominant	287
$K^*(892)\pi$	seen	653
$K f_2(1270)$	seen	–
$K\phi$	seen	441
$K\omega$	seen	608

$K_3^*(1780)$ $I(J^P) = \frac{1}{2}(3^-)$

Mass $m = 1770 \pm 10$ MeV   (S = 1.7)
Full width $\Gamma = 164 \pm 17$ MeV   (S = 1.1)

$K_3^*(1780)$ DECAY MODES	Fraction (Γ_i/Γ)	Scale factor/ Confidence level	p (MeV/c)
$K\rho$	(45 ± 4) %	S=1.4	612
$K^*(892)\pi$	(27.3 ± 3.2) %	S=1.5	651
$K\pi$	(19.3 ± 1.0) %		810
$K\eta$	(8.0 ± 1.5) %	S=1.4	715
$K_2^*(1430)\pi$	< 21 %	CL=95%	284

$K_2(1820)$ [nn] $I(J^P) = \frac{1}{2}(2^-)$

Mass $m = 1816 \pm 13$ MeV
Full width $\Gamma = 276 \pm 35$ MeV

$K_2(1820)$ DECAY MODES	Fraction (Γ_i/Γ)	p (MeV/c)
$K\phi$	possibly seen	481
$K_2^*(1430)\pi$	seen	325
$K^*(892)\pi$	seen	680
$K f_2(1270)$	seen	186
$K\omega$	seen	638

$K_4^*(2045)$ $I(J^P) = \frac{1}{2}(4^+)$

Mass $m = 2045 \pm 9$ MeV   (S = 1.1)
Full width $\Gamma = 198 \pm 30$ MeV

$K_4^*(2045)$ DECAY MODES	Fraction (Γ_i/Γ)	p (MeV/c)
$K\pi$	(9.9 ± 1.2) %	958
$K^*(892)\pi\pi$	(9 ± 5) %	800
$K^*(892)\pi\pi\pi$	(7 ± 5) %	764
$\rho K\pi$	(5.7 ± 3.2) %	742
$\omega K\pi$	(5.0 ± 3.0) %	736
$\phi K\pi$	(2.8 ± 1.4) %	591
$\phi K^*(892)$	(1.4 ± 0.7) %	363

CHARMED MESONS ($C = \pm 1$)

$D^+ = c\bar{d}$, $D^0 = c\bar{u}$, $\overline{D}^0 = \bar{c}u$, $D^- = \bar{c}d$,   similarly for D^*'s

$D^\pm$ $I(J^P) = \frac{1}{2}(0^-)$

Mass $m = 1869.3 \pm 0.5$ MeV   (S = 1.1)
Mean life $\tau = (1.057 \pm 0.015) \times 10^{-12}$ s
$c\tau = 317$ μm

CP-violation decay-rate asymmetries
$A_{CP}(K^+ K^- \pi^\pm) = -0.03 \pm 0.07$
$A_{CP}(K^\pm K^0) = -0.12 \pm 0.13$
$A_{CP}(\phi\pi^\pm) = 0.07 \pm 0.09$

$D^+ \rightarrow \overline{K}^*(892)^0 \ell^+ \nu_\ell$ form factors
$r_2 = 0.73 \pm 0.15$
$r_V = 1.90 \pm 0.25$
$\Gamma_L/\Gamma_T = 1.23 \pm 0.13$
$\Gamma_+/\Gamma_- = 0.16 \pm 0.04$

D^- modes are charge conjugates of the modes below.

D^+ DECAY MODES	Fraction (Γ_i/Γ)	Scale factor/ Confidence level	p (MeV/c)
Inclusive modes			
e^+ anything	(17.2 ± 1.9) %		–
K^- anything	(24.2 ± 2.8) %	S=1.4	–
$\overline{K}^0$ anything + K^0 anything	(59 ± 7) %		–
K^+ anything	(5.8 ± 1.4) %		–
η anything	[oo] < 13 %	CL=90%	–
Leptonic and semileptonic modes			
$\mu^+ \nu_\mu$	< 7.2 $\times 10^{-4}$	CL=90%	932
$\overline{K}^0 \ell^+ \nu_\ell$	[pp] (6.7 ± 0.8) %		868
$\overline{K}^0 e^+ \nu_e$	(6.6 ± 0.9) %		868
$\overline{K}^0 \mu^+ \nu_\mu$	$(7.0 ^{+3.0}_{-2.0})$ %		865
$K^- \pi^+ e^+ \nu_e$	$(4.2 ^{+0.9}_{-0.7})$ %		863
$\overline{K}^*(892)^0 e^+ \nu_e$	(3.2 ± 0.33) %		720
$\times$ B$(\overline{K}^{*0} \rightarrow K^- \pi^+)$			
$K^- \pi^+ e^+ \nu_e$ nonresonant	< 7 $\times 10^{-3}$	CL=90%	863
$K^- \pi^+ \mu^+ \nu_\mu$	(3.2 ± 0.4) %	S=1.1	851
$\overline{K}^*(892)^0 \mu^+ \nu_\mu$	(3.0 ± 0.4) %		715
$\times$ B$(\overline{K}^{*0} \rightarrow K^- \pi^+)$			
$K^- \pi^+ \mu^+ \nu_\mu$ nonresonant	$(2.7 \pm 1.1) \times 10^{-3}$		851
$(\overline{K}^*(892)\pi)^0 e^+ \nu_e$	< 1.2 %	CL=90%	714
$(\overline{K}\pi\pi)^0 e^+ \nu_e$ non-$\overline{K}^*(892)$	< 9 $\times 10^{-3}$	CL=90%	846
$K^- \pi^+ \pi^0 \mu^+ \nu_\mu$	< 1.4 $\times 10^{-3}$	CL=90%	825
$\pi^0 \ell^+ \nu_\ell$	[qq] $(5.7 \pm 2.2) \times 10^{-3}$		930
Fractions of some of the following modes with resonances have already appeared above as submodes of particular charged-particle modes.			
$\overline{K}^*(892)^0 \ell^+ \nu_\ell$	[pp] (4.8 ± 0.4) %		720
$\overline{K}^*(892)^0 e^+ \nu_e$	(4.8 ± 0.5) %		720
$\overline{K}^*(892)^0 \mu^+ \nu_\mu$	(4.5 ± 0.6) %	S=1.1	715
$\rho^0 e^+ \nu_e$	< 3.7 $\times 10^{-3}$	CL=90%	776
$\rho^0 \mu^+ \nu_\mu$	$(2.0 ^{+1.5}_{-1.3}) \times 10^{-3}$		772
$\phi e^+ \nu_e$	< 2.09 %	CL=90%	657
$\phi \mu^+ \nu_\mu$	< 3.72 %	CL=90%	651
$\eta'(958) \mu^+ \nu_\mu$	< 9 $\times 10^{-3}$	CL=90%	684
Hadronic modes with a $\overline{K}$ or $\overline{K}K\overline{K}$			
$\overline{K}^0 \pi^+$	(2.74 ± 0.29) %		862
$K^- \pi^+ \pi^+$	[rr] (9.1 ± 0.6) %		845
$\overline{K}^*(892)^0 \pi^+$	(1.28 ± 0.13) %		712
$\times$ B$(\overline{K}^{*0} \rightarrow K^- \pi^+)$			
$\overline{K}_0^*(1430)^0 \pi^+$	(2.3 ± 0.3) %		368
$\times$ B$(\overline{K}_0^*(1430)^0 \rightarrow K^- \pi^+)$			
$\overline{K}^*(1680)^0 \pi^+$	$(3.7 \pm 0.8) \times 10^{-3}$		65
$\times$ B$(\overline{K}^*(1680)^0 \rightarrow K^- \pi^+)$			
$K^- \pi^+ \pi^+$ nonresonant	(8.6 ± 0.9) %		845

Mode		Fraction		Scale	p
$\overline{K}^0 \pi^+ \pi^0$	[rr]	(9.7 ±3.0) %		S=1.1	845
$\overline{K}^0 \rho^+$		(6.6 ±2.5) %			680
$\overline{K}^*(892)^0 \pi^+$		(6.4 ±0.6) × 10^{-3}			712
$\times\ B(\overline{K}^{*0} \to\ \overline{K}^0 \pi^0)$					
$\overline{K}^0 \pi^+ \pi^0$ nonresonant		(1.3 ±1.1) %			845
$K^- \pi^+ \pi^+ \pi^0$	[rr]	(6.4 ±1.1) %			816
$\overline{K}^*(892)^0 \rho^+$ total		(1.4 ±0.9) %			423
$\times\ B(\overline{K}^{*0} \to\ K^- \pi^+)$					
$\overline{K}_1(1400)^0 \pi^+$		(2.2 ±0.6) %			390
$\times\ B(\overline{K}_1(1400)^0 \to\ \cdot K^- \pi^+ \pi^0)$					
$K^- \rho^+ \pi^+$ total		(3.1 ±1.1) %			616
$K^- \rho^+ \pi^+$ 3-body		(1.1 ±0.4) %			616
$\overline{K}^*(892)^0 \pi^+ \pi^0$ total		(4.5 ±0.9) %			687
$\times\ B(\overline{K}^{*0} \to\ K^- \pi^+)$					
$\overline{K}^*(892)^0 \pi^+ \pi^0$ 3-body		(2.8 ±0.9) %			687
$\times\ B(\overline{K}^{*0} \to\ K^- \pi^+)$					
$K^*(892)^- \pi^+ \pi^+$ 3-body		(7 ±3) × 10^{-3}			688
$\times\ B(K^{*-} \to\ \overline{K}^0 \pi^-)$					
$K^- \pi^+ \pi^+ \pi^0$ nonresonant	[ss]	(1.2 ±0.6) %			816
$\overline{K}^0 \pi^+ \pi^+ \pi^-$	[rr]	(7.0 ±1.0) %			814
$\overline{K}^0 a_1(1260)^+$		(4.0 ±0.9) %			328
$\times\ B(a_1(1260)^+ \to\ \pi^+ \pi^+ \pi^-)$					
$\overline{K}_1(1400)^0 \pi^+$		(2.2 ±0.6) %			390
$\times\ B(\overline{K}_1(1400)^0 \to\ \overline{K}^0 \pi^+ \pi^-)$					
$K^*(892)^- \pi^+ \pi^+$ 3-body		(1.4 ±0.6) %			688
$\times\ B(K^{*-} \to\ \overline{K}^0 \pi^-)$					
$\overline{K}^0 \rho^0 \pi^+$ total		(4.2 ±0.9) %			614
$\overline{K}^0 \rho^0 \pi^+$ 3-body		(5 ±5) × 10^{-3}			614
$\overline{K}^0 \pi^+ \pi^+ \pi^-$ nonresonant		(8 ±4) × 10^{-3}			814
$K^- \pi^+ \pi^+ \pi^+ \pi^-$		(8.2 ±1.4) × 10^{-3}			772
$\overline{K}^*(892)^0 \pi^+ \pi^+ \pi^-$		(6.8 ±1.8) × 10^{-3}			642
$\times\ B(\overline{K}^{*0} \to\ K^- \pi^+)$					
$\overline{K}^*(892)^0 \rho^0 \pi^+$		(5.1 ±2.2) × 10^{-3}			242
$\times\ B(\overline{K}^{*0} \to\ K^- \pi^+)$					
$K^- \pi^+ \pi^+ \pi^0 \pi^0$		(2.2 $^{+5.0}_{-0.9}$) %			775
$\overline{K}^0 \pi^+ \pi^+ \pi^- \pi^0$		(5.4 $^{+3.0}_{-1.4}$) %			773
$\overline{K}^0 \pi^+ \pi^+ \pi^- \pi^- \pi^-$		(8 ±7) × 10^{-4}			714
$K^- \pi^+ \pi^+ \pi^+ \pi^- \pi^0$		(2.0 ±1.8) × 10^{-3}			718
$\overline{K}^0 \overline{K}^0 K^+$		(1.8 ±0.8) %			545

Fractions of some of the following modes with resonances have already appeared above as submodes of particular charged-particle modes.

Mode		Fraction		Scale	p
$\overline{K}^0 \rho^+$		(6.6 ±2.5) %			680
$\overline{K}^0 a_1(1260)^+$		(8.1 ±1.7) %			328
$\overline{K}^0 a_2(1320)^+$		< 3 × 10^{-3}	CL=90%		199
$\overline{K}^*(892)^0 \pi^+$		(1.92±0.19) %			712
$\overline{K}^*(892)^0 \rho^+$ total		(2.1 ±1.4) %			423
$\overline{K}^*(892)^0 \rho^+$ S-wave	[ss]	(1.7 ±1.6) %			423
$\overline{K}^*(892)^0 \rho^+$ P-wave		< 1 × 10^{-3}	CL=90%		423
$\overline{K}^*(892)^0 \rho^+$ D-wave		(10 ±7) × 10^{-3}			423
$\overline{K}^*(892)^0 \rho^+$ D-wave longitudinal		< 7 × 10^{-3}	CL=90%		423
$\overline{K}_1(1270)^0 \pi^+$		< 7 × 10^{-3}	CL=90%		487
$\overline{K}_1(1400)^0 \pi^+$		(5.0 ±1.3) %			390
$\overline{K}^*(1410)^0 \pi^+$		< 7 × 10^{-3}	CL=90%		382
$\overline{K}^*_0(1430)^0 \pi^+$		(3.7 ±0.4) %			368
$\overline{K}^*(1680)^0 \pi^+$		(1.45±0.31) %			65
$\overline{K}^*(892)^0 \pi^+ \pi^0$ total		(6.7 ±1.4) %			687
$\overline{K}^*(892)^0 \pi^+ \pi^0$ 3-body		(4.2 ±1.4) %			687
$K^*(892)^- \pi^+ \pi^+$ 3-body		(2.1 ±0.9) %			688
$K^- \rho^+ \pi^+$ total		(3.1 ±1.1) %			616
$K^- \rho^+ \pi^+$ 3-body		(1.1 ±0.4) %			616
$\overline{K}^0 \rho^0 \pi^+$ total		(4.2 ±0.9) %	CL=90%		614
$\overline{K}^0 \rho^0 \pi^+$ 3-body		(5 ±5) × 10^{-3}			614
$\overline{K}^0 f_0(980) \pi^+$		< 5 × 10^{-3}	CL=90%		461
$\overline{K}^*(892)^0 \pi^+ \pi^+ \pi^-$		(1.02±0.27) %			642
$\overline{K}^*(892)^0 \rho^0 \pi^+$		(7.7 ±3.3) × 10^{-3}			242

Pionic modes

Mode	Fraction		p
$\pi^+ \pi^0$	(2.5 ±0.7) × 10^{-3}		925
$\pi^+ \pi^+ \pi^-$	(3.2 ±0.6) × 10^{-3}		908
$\rho^0 \pi^+$	< 1.4 × 10^{-3}	CL=90%	769
$\pi^+ \pi^+ \pi^-$ nonresonant	(2.5 ±0.7) × 10^{-3}		908

Mode	Fraction		p
$\pi^+ \pi^+ \pi^- \pi^0$	(1.9 $^{+1.5}_{-1.2}$) %		882
$\eta \pi^+ \times B(\eta \to\ \pi^+ \pi^- \pi^0)$	(1.8 ±0.6) × 10^{-3}		848
$\omega \pi^+ \times B(\omega \to\ \pi^+ \pi^- \pi^0)$	< 6 × 10^{-3}	CL=90%	764
$\pi^+ \pi^+ \pi^+ \pi^- \pi^-$	(1.0 $^{+0.8}_{-0.7}$) × 10^{-3}		845
$\pi^+ \pi^+ \pi^+ \pi^- \pi^- \pi^0$	(2.9 $^{+2.9}_{-2.0}$) × 10^{-3}		799

Fractions of some of the following modes with resonances have already appeared above as submodes of particular charged-particle modes.

Mode	Fraction		p
$\eta \pi^+$	(7.5 ±2.5) × 10^{-3}		848
$\rho^0 \pi^+$	< 1.4 × 10^{-3}	CL=90%	769
$\omega \pi^+$	< 7 × 10^{-3}	CL=90%	764
$\eta \rho^+$	< 1.2 %	CL=90%	658
$\eta'(958) \pi^+$	< 9 × 10^{-3}	CL=90%	680
$\eta'(958) \rho^+$	< 1.5 %	CL=90%	355

Hadronic modes with a $K\overline{K}$ pair

Mode		Fraction		p
$K^+ \overline{K}^0$		(7.2 ±1.2) × 10^{-3}		792
$K^+ K^- \pi^+$	[rr]	(8.9 ±0.8) × 10^{-3}		744
$\phi \pi^+ \times B(\phi \to\ K^+ K^-)$		(3.0 ±0.3) × 10^{-3}		647
$K^+ \overline{K}^*(892)^0$		(2.8 ±0.4) × 10^{-3}		610
$\times\ B(\overline{K}^{*0} \to\ K^- \pi^+)$				
$K^+ K^- \pi^+$ nonresonant		(4.6 ±0.9) × 10^{-3}		744
$K^0 \overline{K}^0 \pi^+$				741
$K^*(892)^+ \overline{K}^0$		(2.0 ±0.9) %		611
$\times\ B(K^{*+} \to\ K^0 \pi^+)$				
$K^+ K^- \pi^+ \pi^0$				682
$\phi \pi^+ \pi^0 \times B(\phi \to\ K^+ K^-)$		(1.1 ±0.5) %		619
$\phi \rho^+ \times B(\phi \to\ K^+ K^-)$		< 7 × 10^{-3}	CL=90%	268
$K^+ K^- \pi^+ \pi^0$ non-ϕ		(1.5 $^{+0.7}_{-0.6}$) %		682
$K^+ \overline{K}^0 \pi^+ \pi^-$		< 2 %	CL=90%	678
$K^0 K^- \pi^+ \pi^+$		(1.0 ±0.5) %		678
$K^*(892)^+ \overline{K}^*(892)^0$		(1.2 ±0.5) %		273
$\times\ B^2(K^* \to\ K\pi^+)$				
$K^0 K^- \pi^+ \pi^+$ non-$K^{*+} \overline{K}^{*0}$		< 7.9 × 10^{-3}	CL=90%	678
$K^+ K^- \pi^+ \pi^+ \pi^-$				600
$\phi \pi^+ \pi^+ \pi^-$		< 1 × 10^{-3}	CL=90%	565
$\times\ B(\phi \to\ K^+ K^-)$				
$K^+ K^- \pi^+ \pi^+ \pi^-$ nonresonant		< 3 %	CL=90%	600

Fractions of the following modes with resonances have already appeared above as submodes of particular charged-particle modes.

Mode	Fraction		p
$\phi \pi^+$	(6.1 ±0.6) × 10^{-3}		647
$\phi \pi^+ \pi^0$	(2.3 ±1.0) × 10^{-3}		619
$\phi \rho^+$	< 1.5 %	CL=90%	268
$\phi \pi^+ \pi^+ \pi^-$	< 2 × 10^{-3}	CL=90%	565
$K^+ \overline{K}^*(892)^0$	(4.2 ±0.5) × 10^{-3}		610
$K^*(892)^+ \overline{K}^0$	(3.0 ±1.4) %		611
$K^*(892)^+ \overline{K}^*(892)^0$	(2.6 ±1.1) %		273

Doubly Cabibbo suppressed (DC) modes, $\Delta C = 1$ weak neutral current (C1) modes, or Lepton Family number (LF) or Lepton number (L) violating modes

Mode			Fraction		p
$K^+ \pi^+ \pi^-$	DC		(6.5 ±2.6) × 10^{-4}		845
$K^+ \rho^0$	DC		< 6 × 10^{-4}	CL=90%	681
$K^*(892)^0 \pi^+$	DC		< 1.9 × 10^{-4}	CL=90%	712
$K^+ K^+ K^-$	DC		< 1.5 × 10^{-4}	CL=90%	550
ϕK^+	DC		< 1.3 × 10^{-4}	CL=90%	527
$\pi^+ e^+ e^-$	C1		< 6.6 × 10^{-5}	CL=90%	929
$\pi^+ \mu^+ \mu^-$	C1		< 1.8 × 10^{-5}	CL=90%	917
$\rho^+ \mu^+ \mu^-$	C1		< 5.6 × 10^{-4}	CL=90%	759
$K^+ e^+ e^-$		[tt]	< 4.8 × 10^{-3}	CL=90%	869
$K^+ \mu^+ \mu^-$		[tt]	< 3.2 × 10^{-4}	CL=90%	856
$\pi^+ e^+ \mu^\mp$	LF	[gg]	< 3.8 × 10^{-3}	CL=90%	926
$\pi^+ e^+ \mu^-$	LF		< 3.3 × 10^{-3}	CL=90%	926
$\pi^+ e^- \mu^+$	LF		< 3.3 × 10^{-3}	CL=90%	926
$K^+ e^+ \mu^-$	LF		< 3.4 × 10^{-3}	CL=90%	866
$K^+ e^- \mu^+$	LF		< 3.4 × 10^{-3}	CL=90%	866
$\pi^- e^+ e^+$	L		< 4.8 × 10^{-3}	CL=90%	929
$\pi^- \mu^+ \mu^+$	L		< 2.2 × 10^{-4}	CL=90%	917
$\pi^- e^+ \mu^+$	L		< 3.7 × 10^{-3}	CL=90%	926
$\rho^- \mu^+ \mu^+$	L		< 5.6 × 10^{-4}	CL=90%	759
$K^- e^+ e^+$	L		< 9.1 × 10^{-3}	CL=90%	869
$K^- \mu^+ \mu^+$	L		< 3.2 × 10^{-4}	CL=90%	856
$K^- e^+ \mu^+$	L		< 4.0 × 10^{-3}	CL=90%	866
$K^*(892)^- \mu^+ \mu^+$	L		< 8.5 × 10^{-4}	CL=90%	703

D⁰

$$I(J^P) = \tfrac{1}{2}(0^-)$$

Mass $m = 1864.5 \pm 0.5$ MeV (S = 1.1)

$|m_{D_1^0} - m_{D_2^0}| < 21 \times 10^{10}\,\hbar$ s⁻¹, CL = 90% [uu]

$m_{D^\pm} - m_{D^0} = 4.78 \pm 0.10$ MeV

Mean life $\tau = (0.415 \pm 0.004) \times 10^{-12}$ s

$c\tau = 124.4$ μm

$|\Gamma_{D_1^0} - \Gamma_{D_2^0}|/\Gamma_{D^0} < 0.17$, CL = 90% [uu]

$\dfrac{\Gamma(K^-\pi^- \text{ or } K^+\pi^-\pi^+\pi^-(\text{via}\,\overline{D}^0))}{\Gamma(K^-\pi^+ \text{ or } K^-\pi^+\pi^+\pi^-)} < 0.0037$, CL = 90%

$\Gamma(\mu^- X \text{ (via } \overline{D}^0))/\Gamma(\mu^+ X) < 0.0056$, CL = 90%

CP-violation decay-rate asymmetries

$A_{CP}(K^+K^-) = 0.06 \pm 0.05$

$A_{CP}(K^0_S\phi) = -0.03 \pm 0.09$

$A_{CP}(K^0_S\pi^0) = -0.018 \pm 0.030$

$\overline{D}^0$ modes are charge conjugates of the modes below.

D⁰ DECAY MODES		Fraction (Γ_i/Γ)	Scale factor/ Confidence level	p (MeV/c)
Inclusive modes				
e^+ anything		(7.7 ±1.2) %	S=1.1	–
μ^+ anything	[vv]	(6.8 ±1.0) %		–
K^- anything		(53 ±4) %	S=1.3	–
$\overline{K}^0$ anything + K^0 anything		(42 ±5) %		–
K^+ anything		(3.4 $^{+0.6}_{-0.4}$) %		–
η anything	[oo]	< 13 %	CL=90%	–
Semileptonic modes				
$K^-\ell^+\nu_\ell$	[pp]	(3.48±0.16) %	S=1.1	867
$K^-e^+\nu_e$		(3.64±0.20) %	S=1.1	867
$K^-\mu^+\nu_\mu$		(3.23±0.19) %		863
$K^-\pi^0 e^+\nu_e$		(1.6 $^{+1.3}_{-0.5}$) %		861
$\overline{K}^0\pi^- e^+\nu_e$		(2.8 $^{+1.7}_{-0.9}$) %		860
$\overline{K}^*(892)^- e^+\nu_e$		(1.34±0.22) %		719
$\quad \times$ B($K^{*-} \to \overline{K}^0\pi^-$)				
$K^-\pi^+\pi^- \mu^+\nu_\mu$		< 1.2 ×10⁻³	CL=90%	821
$(\overline{K}^*(892)\pi)^-\mu^+\nu_\mu$		< 1.4 ×10⁻³	CL=90%	693
$\pi^- e^+\nu_e$		(3.8 $^{+1.2}_{-1.0}$) × 10⁻³		927

A fraction of the following resonance mode has already appeared above as a submode of a charged-particle mode.

$\overline{K}^*(892)^- e^+\nu_e$		(2.01±0.33) %		719

Hadronic modes with a $\overline{K}$ or $\overline{K}K\overline{K}$				
$K^-\pi^+$		(3.83±0.12) %		861
$\overline{K}^0\pi^0$		(2.11±0.21) %	S=1.1	860
$\overline{K}^0\pi^+\pi^-$	[rr]	(5.4 ±0.4) %	S=1.2	842
$\overline{K}^0\rho^0$		(1.20±0.17) %		676
$\overline{K}^0 f_0(980)$		(3.0 ±0.8) × 10⁻³		549
$\quad \times$ B($f_0 \to \pi^+\pi^-$)				
$\overline{K}^0 f_2(1270)$		(2.3 ±0.9) × 10⁻³		263
$\quad \times$ B($f_2 \to \pi^+\pi^-$)				
$\overline{K}^0 f_0(1370)$		(4.3 ±1.3) × 10⁻³		–
$\quad \times$ B($f_0 \to \pi^+\pi^-$)				
$K^*(892)^-\pi^+$		(3.3 ±0.3) %		711
$\quad \times$ B($K^{*-} \to \overline{K}^0\pi^-$)				
$K^*_0(1430)^-\pi^+$		(6.4 ±1.6) × 10⁻³		364
$\quad \times$ B($K^*_0(1430)^- \to \overline{K}^0\pi^-$)				
$\overline{K}^0\pi^+\pi^-$ nonresonant		(1.46±0.24) %		842
$K^-\pi^+\pi^0$	[rr]	(13.9 ±0.9) %	S=1.3	844
$K^-\rho^+$		(10.8 ±1.0) %		678
$K^*(892)^-\pi^+$		(1.7 ±0.2) %		711
$\quad \times$ B($K^{*-} \to K^-\pi^0$)				
$\overline{K}^*(892)^0\pi^0$		(2.1 ±0.3) %		709
$\quad \times$ B($\overline{K}^{*0} \to K^-\pi^+$)				
$K^-\pi^+\pi^0$ nonresonant		(6.9 ±2.5) × 10⁻³		844
$\overline{K}^0\pi^0\pi^0$				843
$\overline{K}^*(892)^0\pi^0$		(1.0 ±0.2) %		709
$\quad \times$ B($\overline{K}^{*0} \to \overline{K}^0\pi^0$)				
$\overline{K}^0\pi^0\pi^0$ nonresonant		(7.8 ±2.0) × 10⁻³		843

$K^-\pi^+\pi^+\pi^-$	[rr]	(7.5 ±0.4) %	S=1.1	812
$K^-\pi^+\rho^0$ total		(6.3 ±0.4) %		612
$K^-\pi^+\rho^0$ 3-body		(4.7 ±2.1) × 10⁻³		612
$\overline{K}^*(892)^0\rho^0$		(9.8 ±2.2) × 10⁻³		418
$\quad \times$ B($\overline{K}^{*0} \to K^-\pi^+$)				
$K^-a_1(1260)^+$		(3.6 ±0.6) %		327
$\quad \times$ B($a_1(1260)^+ \to \pi^+\pi^+\pi^-$)				
$\overline{K}^*(892)^0\pi^+\pi^-$ total		(1.5 ±0.4) %		683
$\quad \times$ B($\overline{K}^{*0} \to K^-\pi^+$)				
$\overline{K}^*(892)^0\pi^+\pi^-$ 3-body		(9.5 ±2.1) × 10⁻³		683
$\quad \times$ B($\overline{K}^{*0} \to K^-\pi^+$)				
$K_1(1270)^-\pi^+$	[ss]	(3.6 ±1.0) × 10⁻³		483
$\quad \times$ B($K_1(1270)^- \to K^-\pi^+\pi^-$)				
$K^-\pi^+\pi^+\pi^-$ nonresonant		(1.75±0.25) %		812
$\overline{K}^0\pi^+\pi^-\pi^0$	[rr]	(10.0 ±1.2) %		812
$\overline{K}^0\eta \times$ B($\eta \to \pi^+\pi^-\pi^0$)		(1.6 ±0.3) × 10⁻³		772
$\overline{K}^0\omega \times$ B($\omega \to \pi^+\pi^-\pi^0$)		(1.9 ±0.4) %		670
$K^*(892)^-\rho^+$		(4.0 ±1.6) %		422
$\quad \times$ B($K^{*-} \to \overline{K}^0\pi^-$)				
$\overline{K}^*(892)^0\rho^0$		(4.9 ±1.1) × 10⁻³		418
$\quad \times$ B($\overline{K}^{*0} \to \overline{K}^0\pi^0$)				
$K_1(1270)^-\pi^+$	[ss]	(5.1 ±1.4) × 10⁻³		483
$\quad \times$ B($K_1(1270)^- \to \overline{K}^0\pi^-\pi^0$)				
$\overline{K}^*(892)^0\pi^+\pi^-$ 3-body		(4.7 ±1.1) × 10⁻³		683
$\quad \times$ B($\overline{K}^{*0} \to \overline{K}^0\pi^0$)				
$\overline{K}^0\pi^+\pi^-\pi^0$ nonresonant		(2.1 ±2.1) %		812
$K^-\pi^+\pi^0\pi^0$		(15 ±5) %		815
$K^-\pi^+\pi^+\pi^-\pi^0$		(4.0 ±0.4) %		771
$\overline{K}^*(892)^0\pi^+\pi^-\pi^0$		(1.2 ±0.6) %		641
$\quad \times$ B($\overline{K}^{*0} \to K^-\pi^+$)				
$\overline{K}^*(892)^0\eta$		(3.0 ±0.8) × 10⁻³		580
$\quad \times$ B($\overline{K}^{*0} \to K^-\pi^+$)				
$\quad \times$ B($\eta \to \pi^+\pi^-\pi^0$)				
$K^-\pi^+\omega \times$ B($\omega \to \pi^+\pi^-\pi^0$)		(2.7 ±0.5) %		605
$\overline{K}^*(892)^0\omega$		(7 ±3) × 10⁻³		406
$\quad \times$ B($\overline{K}^{*0} \to K^-\pi^+$)				
$\quad \times$ B($\omega \to \pi^+\pi^-\pi^0$)				
$\overline{K}^0\pi^+\pi^+\pi^-\pi^-$		(5.8 ±1.6) × 10⁻³		768
$\overline{K}^0\pi^+\pi^-\pi^0\pi^0(\pi^0)$		(10.6 $^{+7.3}_{-3.0}$) %		771
$\overline{K}^0 K^+K^-$		(9.3 ±1.0) × 10⁻³		544
$\overline{K}^0\phi \times$ B($\phi \to K^+K^-$)		(4.2 ±0.5) × 10⁻³		520
$\overline{K}^0 K^+K^-$ non-ϕ		(5.0 ±0.8) × 10⁻³		544
$K^0_S K^0_S K^0_S$		(9.7 ±2.3) × 10⁻⁴		538
$K^+K^-K^-\pi^+$		(2.1 ±0.5) × 10⁻⁴		434
$K^+K^-\overline{K}^0\pi^0$		(7.2 $^{+4.8}_{-3.5}$) × 10⁻³		435

Fractions of many of the following modes with resonances have already appeared above as submodes of particular charged-particle modes. (Modes for which there are only upper limits and $\overline{K}^*(892)\rho$ submodes only appear below.)

$\overline{K}^0\eta$		(7.0 ±1.0) × 10⁻³		772
$\overline{K}^0\rho^0$		(1.20±0.17) %		676
$K^-\rho^+$		(10.8 ±1.0) %	S=1.2	678
$\overline{K}^0\omega$		(2.1 ±0.4) %		670
$\overline{K}^0\eta'(958)$		(1.70±0.26) %		565
$\overline{K}^0 f_0(980)$		(5.7 ±1.6) × 10⁻³		549
$\overline{K}^0\phi$		(8.5 ±1.0) × 10⁻³		520
$K^-a_1(1260)^+$		(7.3 ±1.1) %		327
$\overline{K}^0 a_1(1260)^0$		< 1.9 %	CL=90%	322
$\overline{K}^0 f_2(1270)$		(4.1 ±1.5) × 10⁻³		263
$\overline{K}^0 f_0(1370)$		(6.9 ±2.1) × 10⁻³		–
$K^-a_2(1320)^+$		< 2 × 10⁻³	CL=90%	197
$K^*(892)^-\pi^+$		(5.0 ±0.4) %	S=1.2	711
$\overline{K}^*(892)^0\pi^0$		(3.1 ±0.4) %		709
$\overline{K}^*(892)^0\pi^+\pi^-$ total		(2.3 ±0.5) %		683
$\overline{K}^*(892)^0\pi^+\pi^-$ 3-body		(1.42±0.32) %		683
$K^-\pi^+\rho^0$ total		(6.3 ±0.4) %		612
$K^-\pi^+\rho^0$ 3-body		(4.7 ±2.1) × 10⁻³		612
$\overline{K}^*(892)^0\rho^0$		(1.47±0.33) %		418
$\overline{K}^*(892)^0\rho^0$ transverse		(1.5 ±0.5) %		418
$\overline{K}^*(892)^0\rho^0$ S-wave		(2.8 ±0.6) %		418
$\overline{K}^*(892)^0\rho^0$ S-wave long.		< 3 × 10⁻³	CL=90%	418
$\overline{K}^*(892)^0\rho^0$ P-wave		< 3 × 10⁻³	CL=90%	418
$\overline{K}^*(892)^0\rho^0$ D-wave		(1.9 ±0.6) %		418

$K^*(892)^- \rho^+$	(6.0 ±2.4) %		422
$K^*(892)^- \rho^+$ longitudinal	(2.9 ±1.2) %		422
$K^*(892)^- \rho^+$ transverse	(3.2 ±1.8) %		422
$K^*(892)^- \rho^+$ P-wave	< 1.5 %	CL=90%	422
$K^- \pi^+ f_0(980)$	< 1.1 %	CL=90%	459
$\overline{K}^*(892)^0 f_0(980)$	< 7 × 10^{-3}	CL=90%	–
$K_1(1270)^- \pi^+$	[ss] (1.06±0.29) %		483
$K_1(1400)^- \pi^+$	< 1.2 %	CL=90%	386
$\overline{K}_1(1400)^0 \pi^0$	< 3.7 %	CL=90%	387
$K^*(1410)^- \pi^+$	< 1.2 %	CL=90%	378
$K_0^*(1430)^- \pi^+$	(1.04±0.26) %		364
$K_2^*(1430)^- \pi^+$	< 8 × 10^{-3}	CL=90%	367
$\overline{K}_2^*(1430)^0 \pi^0$	< 4 × 10^{-3}	CL=90%	363
$\overline{K}^*(892)^0 \pi^+ \pi^- \pi^0$	(1.8 ±0.9) %		641
$\overline{K}^*(892)^0 \eta$	(1.9 ±0.5) %		580
$K^- \pi^+ \omega$	(3.0 ±0.6) %		605
$\overline{K}^*(892)^0 \omega$	(1.1 ±0.4) %		406
$K^- \pi^+ \eta'(958)$	(7.0 ±1.8) × 10^{-3}		479
$\overline{K}^*(892)^0 \eta'(958)$	< 1.1 × 10^{-3}	CL=90%	99

Pionic modes

$\pi^+ \pi^-$	(1.52±0.11) × 10^{-3}		922
$\pi^0 \pi^0$	(8.4 ±2.2) × 10^{-4}		922
$\pi^+ \pi^- \pi^0$	(1.6 ±1.1) %	S=2.7	907
$\pi^+ \pi^+ \pi^- \pi^-$	(7.4 ±0.6) × 10^{-3}		879
$\pi^+ \pi^+ \pi^- \pi^- \pi^0$	(1.9 ±0.4) %		844
$\pi^+ \pi^+ \pi^+ \pi^- \pi^- \pi^-$	(4.0 ±3.0) × 10^{-4}		795

Hadronic modes with a $K\overline{K}$ pair

$K^+ K^-$	(4.33±0.27) × 10^{-3}		791
$K^0 \overline{K}^0$	(1.3 ±0.4) × 10^{-3}		788
$K^0 K^- \pi^+$	(6.4 ±1.0) × 10^{-3}	S=1.1	739
$K^*(892)^0 K^0$	< 1.1 × 10^{-3}	CL=90%	605
× B($\overline{K}^{*0} \to K^- \pi^+$)			
$K^*(892)^+ K^-$	(2.3 ±0.5) × 10^{-3}		610
× B($K^{*+} \to K^0 \pi^+$)			
$K^0 K^- \pi^+$ nonresonant	(2.3 ±2.3) × 10^{-3}		739
$\overline{K}^0 K^+ \pi^-$	(4.9 ±1.0) × 10^{-3}		739
$K^*(892)^0 \overline{K}^0$	< 5 × 10^{-4}	CL=90%	605
× B($K^{*0} \to K^+ \pi^-$)			
$K^*(892)^- K^+$	(1.2 ±0.7) × 10^{-3}		610
× B($K^{*-} \to \overline{K}^0 \pi^-$)			
$\overline{K}^0 K^+ \pi^-$ nonresonant	(3.8 $^{+2.3}_{-1.9}$) × 10^{-3}		739
$K^+ K^- \pi^+ \pi^-$	[ww] (2.58±0.28) × 10^{-3}		676
$\phi \pi^+ \pi^- \times$ B($\phi \to K^+ K^-$)	(5.3 ±1.4) × 10^{-4}		614
$\phi \rho^0 \times$ B($\phi \to K^+ K^-$)	(5.3 ±1.4) × 10^{-4}		260
$K^+ K^- \rho^0$ 3-body	(9.0 ±2.3) × 10^{-3}		309
$K^*(892)^0 K^- \pi^+$	(2.1 ±0.9) × 10^{-3}		528
× B($K^{*0} \to K^+ \pi^-$)			
$\overline{K}^*(892)^0 K^+ \pi^-$	(1.1 ±0.8) × 10^{-3}		528
× B($\overline{K}^{*0} \to K^- \pi^+$)			
$K^*(892)^0 \overline{K}^*(892)^0$	(6 ±2) × 10^{-4}		257
× B^2($K^{*0} \to K^+ \pi^-$)			
$K^+ K^- \pi^+ \pi^-$ non-ϕ	(1.7 ±0.5) × 10^{-3}		676
$K^+ K^- \pi^+ \pi^-$ nonresonant	< 8 × 10^{-4}	CL=90%	676
$K^0 \overline{K}^0 \pi^+ \pi^-$	(6.8 ±2.7) × 10^{-3}		673
$K^+ K^- \pi^+ \pi^- \pi^0$	(3.1 ±2.0) × 10^{-3}		600

Fractions of most of the following modes with resonances have already
appeared above as submodes of particular charged-particle modes.

$\overline{K}^*(892)^0 K^0$	< 1.6 × 10^{-3}	CL=90%	605
$K^*(892)^+ K^-$	(3.5 ±0.8) × 10^{-3}		610
$K^*(892)^0 \overline{K}^0$	< 8 × 10^{-4}	CL=90%	605
$K^*(892)^- K^+$	(1.8 ±1.0) × 10^{-3}		610
$\phi \pi^0$	< 1.4 × 10^{-3}	CL=90%	644
$\phi \eta$	< 2.8 × 10^{-3}	CL=90%	489
$\phi \omega$	< 2.1 × 10^{-3}	CL=90%	239
$\phi \pi^+ \pi^-$	(1.07±0.29) × 10^{-3}		614
$\phi \rho^0$	(1.07±0.29) × 10^{-3}		260
$\phi \pi^+ \pi^-$ 3-body	< 5 × 10^{-4}	CL=90%	614
$K^*(892)^0 K^- \pi^+$	(3.2 ±1.3) × 10^{-3}		528
$\overline{K}^*(892)^0 K^+ \pi^-$	(1.7 ±1.2) × 10^{-3}		528
$K^*(892)^0 \overline{K}^*(892)^0$	(1.4 ±0.5) × 10^{-3}		257

Doubly Cabibbo suppressed (DC) modes,
$\Delta C = 2$ forbidden via mixing (C2M) modes,
$\Delta C = 1$ weak neutral current (C1) modes, or
Lepton Family number (LF) violating modes

$K^+ \pi^-$	DC	(2.9 ±1.4) × 10^{-4}		861
$K^+ \pi^-$ (via $\overline{D}^0$)	C2M	< 1.9 × 10^{-4}	CL=90%	861
$K^+ \pi^- \pi^+ \pi^-$	DC	< 1.4 × 10^{-3}	CL=90%	812
$K^+ \pi^- \pi^+ \pi^-$ (via $\overline{D}^0$)	C2M	< 4 × 10^{-4}	CL=90%	812
μ^- anything (via $\overline{D}^0$)	C2M	< 4 × 10^{-4}	CL=90%	–
$e^+ e^-$	C1	< 1.3 × 10^{-5}	CL=90%	932
$\mu^+ \mu^-$	C1	< 7.6 × 10^{-6}	CL=90%	926
$\pi^0 e^+ e^-$	C1	< 4.5 × 10^{-5}	CL=90%	927
$\pi^0 \mu^+ \mu^-$	C1	< 1.8 × 10^{-4}	CL=90%	915
$\eta e^+ e^-$	C1	< 1.1 × 10^{-4}	CL=90%	852
$\eta \mu^+ \mu^-$	C1	< 5.3 × 10^{-4}	CL=90%	838
$\rho^0 e^+ e^-$	C1	< 1.0 × 10^{-4}	CL=90%	773
$\rho^0 \mu^+ \mu^-$	C1	< 2.3 × 10^{-4}	CL=90%	756
$\omega e^+ e^-$	C1	< 1.8 × 10^{-4}	CL=90%	768
$\omega \mu^+ \mu^-$	C1	< 8.3 × 10^{-4}	CL=90%	751
$\phi e^+ e^-$	C1	< 5.2 × 10^{-5}	CL=90%	654
$\phi \mu^+ \mu^-$	C1	< 4.1 × 10^{-4}	CL=90%	631
$\overline{K}^0 e^+ e^-$	[tt] < 1.1 × 10^{-4}	CL=90%	866
$\overline{K}^0 \mu^+ \mu^-$	[tt] < 2.6 × 10^{-4}	CL=90%	852
$\overline{K}^*(892)^0 e^+ e^-$	[tt] < 1.4 × 10^{-4}	CL=90%	717
$\overline{K}^*(892)^0 \mu^+ \mu^-$	[tt] < 1.18 × 10^{-3}	CL=90%	698
$\pi^+ \pi^- \pi^0 \mu^+ \mu^-$	C1	< 8.1 × 10^{-3}	CL=90%	863
$\mu^\pm e^\mp$	LF [gg] < 1.9 × 10^{-5}	CL=90%	929
$\pi^0 e^\pm \mu^\mp$	LF [gg] < 8.6 × 10^{-5}	CL=90%	924
$\eta e^\pm \mu^\mp$	LF [gg] < 1.0 × 10^{-4}	CL=90%	848
$\rho^0 e^\pm \mu^\mp$	LF [gg] < 4.9 × 10^{-5}	CL=90%	769
$\omega e^\pm \mu^\mp$	LF [gg] < 1.2 × 10^{-4}	CL=90%	764
$\phi e^\pm \mu^\mp$	LF [gg] < 3.4 × 10^{-5}	CL=90%	648
$\overline{K}^0 e^\pm \mu^\mp$	LF [gg] < 1.0 × 10^{-4}	CL=90%	862
$\overline{K}^*(892)^0 e^\pm \mu^\mp$	LF [gg] < 1.0 × 10^{-4}	CL=90%	712

$D^*(2007)^0$	$I(J^P) = \frac{1}{2}(1^-)$
	I, J, P need confirmation.

Mass $m = 2006.7 \pm 0.5$ MeV   (S = 1.1)
$m_{D^{*0}} - m_{D^0} = 142.12 \pm 0.07$ MeV
Full width Γ < 2.1 MeV, CL = 90%

$\overline{D}^*(2007)^0$ modes are charge conjugates of modes below.

$D^*(2007)^0$ DECAY MODES	Fraction (Γ_i/Γ)	p (MeV/c)
$D^0 \pi^0$	(61.9±2.9) %	43
$D^0 \gamma$	(38.1±2.9) %	137

$D^*(2010)^\pm$	$I(J^P) = \frac{1}{2}(1^-)$
	I, J, P need confirmation.

Mass $m = 2010.0 \pm 0.5$ MeV   (S = 1.1)
$m_{D^*(2010)^+} - m_{D^+} = 140.64 \pm 0.09$ MeV
$m_{D^*(2010)^+} - m_{D^0} = 145.42 \pm 0.05$ MeV
Full width Γ < 0.131 MeV, CL = 90%

$D^*(2010)^-$ modes are charge conjugates of the modes below.

$D^*(2010)^\pm$ DECAY MODES	Fraction (Γ_i/Γ)	p (MeV/c)
$D^0 \pi^+$	(68.3±1.4) %	39
$D^+ \pi^0$	(30.6±2.5) %	38
$D^+ \gamma$	(1.1 $^{+2.1}_{-0.7}$) %	136

$D_1(2420)^0$	$I(J^P) = \frac{1}{2}(1^+)$
	I, J, P need confirmation.

Mass $m = 2422.2 \pm 1.8$ MeV   (S = 1.2)
Full width $\Gamma = 18.9 ^{+4.6}_{-3.5}$ MeV

$\overline{D}_1(2420)^0$ modes are charge conjugates of modes below.

$D_1(2420)^0$ DECAY MODES	Fraction (Γ_i/Γ)	p (MeV/c)
$D^*(2010)^+ \pi^-$	seen	355
$D^+ \pi^-$	not seen	474

$D_2^*(2460)^0$ $I(J^P) = \frac{1}{2}(2^+)$

$J^P = 2^+$ assignment strongly favored (ALBRECHT 89B).

Mass $m = 2458.9 \pm 2.0$ MeV   (S = 1.2)
Full width $\Gamma = 23 \pm 5$ MeV

$\overline{D}_2^*(2460)^0$ modes are charge conjugates of modes below.

$D_2^*(2460)^0$ DECAY MODES	Fraction (Γ_i/Γ)	p (MeV/c)
$D^+\pi^-$	seen	503
$D^*(2010)^+\pi^-$	seen	387

$D_2^*(2460)^\pm$ $I(J^P) = \frac{1}{2}(2^+)$

$J^P = 2^+$ assignment strongly favored (ALBRECHT 89B).
Mass $m = 2459 \pm 4$ MeV   (S = 1.7)
$m_{D_2^*(2460)^\pm} - m_{D_2^*(2460)^0} = 0.9 \pm 3.3$ MeV   (S = 1.1)
Full width $\Gamma = 25^{+8}_{-7}$ MeV

$D_2^*(2460)^-$ modes are charge conjugates of modes below.

$D_2^*(2460)^\pm$ DECAY MODES	Fraction (Γ_i/Γ)	p (MeV/c)
$D^0\pi^+$	seen	508
$D^{*0}\pi^+$	seen	390

CHARMED, STRANGE MESONS ($C = S = \pm1$)

$D_s^+ = c\overline{s}$, $D_s^- = \overline{c}s$,   similarly for D_s^*'s

$D_s^\pm$ was $F^\pm$ $I(J^P) = 0(0^-)$

Mass $m = 1968.5 \pm 0.6$ MeV   (S = 1.1)
$m_{D_s^\pm} - m_{D^\pm} = 99.2 \pm 0.5$ MeV   (S = 1.1)
Mean life $\tau = (0.467 \pm 0.017) \times 10^{-12}$ s
$c\tau = 140 \ \mu m$

D_s^+ form factors
$r_2 = 1.6 \pm 0.4$
$r_v = 1.5 \pm 0.5$
$\Gamma_L/\Gamma_T = 0.72 \pm 0.18$

Branching fractions for modes with a resonance in the final state include all the decay modes of the resonance. D_s^- modes are charge conjugates of the modes below.

D_s^+ DECAY MODES	Fraction (Γ_i/Γ)	Confidence level	p (MeV/c)
Inclusive modes			
K^- anything	(13^{+14}_{-12}) %		–
$\overline{K}^0$ anything + K^0 anything	(39 ± 28) %		–
K^+ anything	(20^{+18}_{-14}) %		–
non-$K\overline{K}$ anything	(64 ± 17) %		–
e^+ anything	< 20 %	90%	–
Leptonic and semileptonic modes			
$\mu^+\nu_\mu$	$(9 \pm 4) \times 10^{-3}$		981
$\phi\ell^+\nu_\ell$	[xx] (1.9 ± 0.5) %		–
$\eta\ell^+\nu_\ell + \eta'(958)\ell^+\nu_\ell$	(3.3 ± 1.0) %		–
$\eta\ell^+\nu_\ell$	(2.5 ± 0.7) %		–
$\eta'(958)\ell^+\nu_\ell$	$(8.7 \pm 3.4) \times 10^{-3}$		–

Hadronic modes with a $K\overline{K}$ pair (including from a ϕ)			
$K^+\overline{K}^0$	(3.6 ± 1.1) %		850
$K^+K^-\pi^+$	[rr,yy] (4.6 ± 1.2) %		805
$\phi\pi^+$	(3.6 ± 0.9) %		712
$K^+\overline{K}^*(892)^0$	(3.4 ± 0.9) %		682
$f_0(980)\pi^+$	(1.1 ± 0.4) %		732
$K^+\overline{K}_0^*(1430)^0$	$(7 \pm 4) \times 10^{-3}$		186
$f_J(1710)\pi^+ \to K^+K^-\pi^+$	[zz] $(1.5 \pm 2.0) \times 10^{-3}$		204
$K^+K^-\pi^+$ nonresonant	$(9 \pm 4) \times 10^{-3}$		805
$K^0\overline{K}^0\pi^+$			802
$K^*(892)^+\overline{K}^0$	(4.3 ± 1.4) %		683
$K^+K^-\pi^+\pi^0$			748
$\phi\pi^+\pi^0$	(9 ± 5) %		687
$\phi\pi^+\pi^0$ 3-body	(6.7 ± 2.3) %		407
$K^+K^-\pi^+\pi^0$ non-ϕ	< 2.6 %	90%	687
$K^+\overline{K}^0\pi^+\pi^-$	< 9 %	90%	748
$K^0K^-\pi^+\pi^+$	< 2.8 %	90%	744
$K^*(892)^+\overline{K}^*(892)^0$	(4.3 ± 1.5) %		744
$K^0K^-\pi^+\pi^+$ non-$K^{*+}\overline{K}^{*0}$	(5.8 ± 2.5) %		412
$K^+K^-\pi^+\pi^+\pi^-$	< 2.9 %	90%	744
$\phi\pi^+\pi^+\pi^-$	(1.8 ± 0.6) %		673
$K^+K^-\pi^+\pi^+\pi^-$ non-ϕ	$(3.0^{+3.0}_{-2.0}) \times 10^{-3}$		640
			673

Other hadronic modes (0, 1, or 3 K's)			
$\pi^+\pi^+\pi^-$	(1.4 ± 0.4) %		959
$\rho^0\pi^+$	< 2.9 $\times 10^{-3}$	90%	827
$f_0(980)\pi^+$	(1.2 ± 0.5) %		732
$\pi^+\pi^+\pi^-$ nonresonant	(1.0 ± 0.4) %		959
$\pi^+\pi^+\pi^-\pi^0$	< 12 %	90%	935
$\eta\pi^+$	(2.0 ± 0.6) %		902
$\omega\pi^+$	< 1.8 %	90%	822
$\pi^+\pi^+\pi^+\pi^-\pi^-$	$(3.0^{+4.0}_{-3.0}) \times 10^{-3}$		899
$\pi^+\pi^+\pi^-\pi^0\pi^0$			902
$\eta\rho^+$	(10.3 ± 3.2) %		727
$\eta\pi^+\pi^0$ 3-body	< 3.0 %	90%	886
$\pi^+\pi^+\pi^+\pi^-\pi^-\pi^0$	(4.9 ± 3.2) %		856
$\eta'(958)\pi^+$	(4.9 ± 1.8) %		743
$\pi^+\pi^+\pi^+\pi^-\pi^-\pi^0\pi^0$			803
$\eta'(958)\rho^+$	(12 ± 4) %		470
$\eta'(958)\pi^+\pi^0$ 3-body	< 3.1 %	90%	720
$K^0\pi^+$	< 8 $\times 10^{-3}$	90%	916
$K^+\pi^+\pi^-$	(1.0 ± 0.4) %		900
$K^+\rho^0$	< 2.9 $\times 10^{-3}$	90%	747
$K^*(892)^0\pi^+$	$(6.5 \pm 2.8) \times 10^{-3}$		773
$K^+K^+K^-$	< 6 $\times 10^{-4}$	90%	628
ϕK^+	< 5 $\times 10^{-4}$	90%	607

$\Delta C = 1$ weak neutral current ($C1$) modes, or Lepton number (L) violating modes			
$\pi^+\mu^+\mu^-$	[aaa] < 4.3 $\times 10^{-4}$	90%	968
$K^+\mu^+\mu^-$	$C1$ < 5.9 $\times 10^{-4}$	90%	909
$K^*(892)^+\mu^+\mu^-$	$C1$ < 1.4 $\times 10^{-3}$	90%	765
$\pi^-\mu^+\mu^+$	L < 4.3 $\times 10^{-4}$	90%	968
$K^-\mu^+\mu^+$	L < 5.9 $\times 10^{-4}$	90%	909
$K^*(892)^-\mu^+\mu^+$	L < 1.4 $\times 10^{-3}$	90%	765

Meson Summary Table

$D_s^{*\pm}$

$I(J^P) = ?(?^?)$

Mass $m = 2112.4 \pm 0.7$ MeV (S = 1.1)

$m_{D_s^{*\pm}} - m_{D_s^\pm} = 143.8 \pm 0.4$ MeV

Full width $\Gamma < 1.9$ MeV, CL = 90%

D_s^{*-} modes are charge conjugates of the modes below.

D_s^{*+} DECAY MODES	Fraction (Γ_i/Γ)	p (MeV/c)
$D_s^+ \gamma$	seen	139
$D_s^+ \pi^0$	seen	48

$D_{s1}(2536)^\pm$

$I(J^P) = 0(1^+)$

I, J, P need confirmation.

Mass $m = 2535.35 \pm 0.34$ MeV

Full width $\Gamma < 2.3$ MeV, CL = 90%

$D_{s1}(2536)^-$ modes are charge conjugates of the modes below.

$D_{s1}(2536)^+$ DECAY MODES	Fraction (Γ_i/Γ)	p (MeV/c)
$D^*(2010)^+ K^0$	seen	150
$D^*(2007)^0 K^+$	seen	169
$D^+ K^0$	not seen	382
$D^0 K^+$	not seen	392
$D_s^{*+} \gamma$	possibly seen	389

$D_{sJ}(2573)^\pm$

$I(J^P) = ?(?^?)$

J^P is natural, width and decay modes consistent with 2^{+2}.

Mass $m = 2573.5 \pm 1.7$ MeV

Full width $\Gamma = 15^{+5}_{-4}$ MeV

$D_{sJ}(2573)^-$ modes are charge conjugates of the modes below.

$D_{sJ}(2573)^+$ DECAY MODES	Fraction (Γ_i/Γ)	p (MeV/c)
$D^0 K^+$	seen	436
$D^*(2007)^0 K^+$	seen	245

BOTTOM MESONS
($B = \pm 1$)

$B^+ = u\overline{b}$, $B^0 = d\overline{b}$, $\overline{B}^0 = \overline{d}b$, $B^- = \overline{u}b$, similarly for B^*'s

B-particle organization

Many measurements of B decays involve admixtures of B hadrons. Previously we arbitrarily included such admixtures in the $B^\pm$ section, but because of their importance we have created two new sections: "$B^\pm/B^0$ Admixture" for $\Upsilon(4S)$ results and "$B^\pm/B^0/B_s^0/b$-baryon Admixture" for results at higher energies. Most inclusive decay branching fractions are found in the Admixture sections. B^0-$\overline{B}^0$ mixing data are found in the B^0 section, while B_s^0-$\overline{B}_s^0$ mixing data and B-$\overline{B}$ mixing data for a B^0/B_s^0 admixture are found in the B_s^0 section. CP-violation data are found in the B^0 section. b-baryons are found near the end of the Baryon section.

The organization of the B sections is now as follows, where bullets indicate particle sections and brackets indicate reviews.

[Production and Decay of b-flavored Hadrons]

[Semileptonic Decays of B Mesons]

- $B^\pm$
  - mass
  - mean life
  - branching fractions
- B^0
  - mass
  - mean life
  - branching fractions
  - polarization in B^0 decay
  - B^0-$\overline{B}^0$ mixing
  - [B^0-$\overline{B}^0$ Mixing and CP Violation in B Decay]
  - CP violation
- $B^\pm B^0$ Admixtures
  - branching fractions
- $B^\pm/B^0/B_s^0/b$-baryon Admixtures
  - mean life
  - production fractions
  - branching fractions
- B^*
  - mass
- B_s^0
  - mass
  - mean life
  - branching fractions
  - polarization in B_s^0 decay
  - B_s^0-$\overline{B}_s^0$ mixing
  - B-$\overline{B}$ mixing (admixture of B^0, B_s^0)

At end of Baryon Listings:

- Λ_b
  - mass
  - mean life
  - branching fractions

$B^\pm$		$I(J^P) = \frac{1}{2}(0^-)$

I, J, P need confirmation. Quantum numbers shown are quark-model predictions.

Mass $m_{B^\pm} = 5278.9 \pm 1.8$ MeV
Mean life $\tau_{B^\pm} = (1.62 \pm 0.06) \times 10^{-12}$ s
$c\tau = 462$ μm

B^- modes are charge conjugates of the modes below. Modes which do not identify the charge state of the B are listed in the $B^\pm/B^0$ ADMIXTURE section.

The branching fractions listed below assume 50% $B^0 \overline{B}^0$ and 50% $B^+ B^-$ production at the $\Upsilon(4S)$. We have attempted to bring older measurements up to date by rescaling their assumed $\Upsilon(4S)$ production ratio to 50:50 and their assumed D, D_s, D^*, and ψ branching ratios to current values whenever this would affect our averages and best limits significantly.

Indentation is used to indicate a subchannel of a previous reaction. All resonant subchannels have been corrected for resonance branching fractions to the final state so the sum of the subchannel branching fractions can exceed that of the final state.

B^+ DECAY MODES	Fraction (Γ_i/Γ)	Scale factor/ Confidence level	p (MeV/c)
Semileptonic and leptonic modes			
$\ell^+ \nu_\ell$ anything	[qq] (10.1 $\pm$2.3) %		–
$\overline{D}^0 \ell^+ \nu_\ell$	[qq] (1.6 $\pm$0.7) %		–
$\overline{D}^*(2007)^0 \ell^+ \nu_\ell$	[qq] (5.3 $\pm$0.8) %		–
$\pi^0 e^+ \nu_e$	< 2.2 $\times 10^{-3}$	CL=90%	2638
$\omega \ell^+ \nu_\ell$	[qq] < 2.1 $\times 10^{-4}$	CL=90%	–
$\rho^0 \ell^+ \nu_\ell$	[qq] < 2.1 $\times 10^{-4}$	CL=90%	–
$e^+ \nu_e$	< 1.5 $\times 10^{-5}$	CL=90%	2639
$\mu^+ \nu_\mu$	< 2.1 $\times 10^{-5}$	CL=90%	2638
$\tau^+ \nu_\tau$	< 1.8 $\times 10^{-3}$	CL=90%	2340
D, D^*, or D_s modes			
$\overline{D}^0 \pi^+$	(5.3 $\pm$0.5) $\times 10^{-3}$		2308
$\overline{D}^0 \rho^+$	(1.34$\pm$0.18) %		2238
$\overline{D}^0 \pi^+ \pi^+ \pi^-$	(1.1 $\pm$0.4) %		2289
$\overline{D}^0 \pi^+ \pi^+ \pi^-$ nonresonant	(5 $\pm$4) $\times 10^{-3}$		2289
$\overline{D}^0 \pi^+ \rho^0$	(4.2 $\pm$3.0) $\times 10^{-3}$		2209
$\overline{D}^0 a_1(1260)^+$	(5 $\pm$4) $\times 10^{-3}$		2123
$D^*(2010)^- \pi^+ \pi^+$	(2.1 $\pm$0.6) $\times 10^{-3}$		2247
$D^- \pi^+ \pi^+$	< 1.4 $\times 10^{-3}$	CL=90%	2299
$\overline{D}^*(2007)^0 \pi^+$	(5.2 $\pm$0.8) $\times 10^{-3}$		2256
$\overline{D}^*(2007)^0 \rho^+$	(1.55$\pm$0.31) %		2183
$\overline{D}^*(2007)^0 \pi^+ \pi^+ \pi^-$	(9.4 $\pm$2.6) $\times 10^{-3}$		2236
$\overline{D}^*(2007)^0 a_1(1260)^+$	(1.9 $\pm$0.5) %		2062
$D^*(2010)^- \pi^+ \pi^+ \pi^0$	(1.5 $\pm$0.7) %		2235
$D^*(2010)^- \pi^+ \pi^+ \pi^+ \pi^-$	< 1 %	CL=90%	2217
$\overline{D}_1(2420)^0 \pi^+$	(1.5 $\pm$0.6) $\times 10^{-3}$	S=1.3	2081
$\overline{D}_1(2420)^0 \rho^+$	< 1.4 $\times 10^{-3}$	CL=90%	1997
$\overline{D}_2^*(2460)^0 \pi^+$	< 1.3 $\times 10^{-3}$	CL=90%	2064
$\overline{D}_2^*(2460)^0 \rho^+$	< 4.7 $\times 10^{-3}$	CL=90%	1979
$\overline{D}^0 D_s^+$	(1.7 $\pm$0.6) %		1815
$\overline{D}^0 D_s^{*+}$	(1.2 $\pm$1.0) %		1734
$\overline{D}^*(2007)^0 D_s^+$	(10 $\pm$7) $\times 10^{-3}$		1737
$\overline{D}^*(2007)^0 D_s^{*+}$	(2.3 $\pm$1.4) %		1650
$D_s^+ \pi^0$	< 2.0 $\times 10^{-4}$	CL=90%	2270
$D_s^{*+} \pi^0$	< 3.3 $\times 10^{-4}$	CL=90%	2214
$D_s^+ \eta$	< 5 $\times 10^{-4}$	CL=90%	2235
$D_s^{*+} \eta$	< 8 $\times 10^{-4}$	CL=90%	2177
$D_s^+ \rho^0$	< 4 $\times 10^{-4}$	CL=90%	2198
$D_s^{*+} \rho^0$	< 5 $\times 10^{-4}$	CL=90%	2139
$D_s^+ \omega$	< 5 $\times 10^{-4}$	CL=90%	2195
$D_s^{*+} \omega$	< 7 $\times 10^{-4}$	CL=90%	2136
$D_s^+ a_1(1260)^0$	< 2.2 $\times 10^{-3}$	CL=90%	2079
$D_s^{*+} a_1(1260)^0$	< 1.6 $\times 10^{-3}$	CL=90%	2014
$D_s^+ \phi$	< 3.2 $\times 10^{-4}$	CL=90%	2141
$D_s^{*+} \phi$	< 4 $\times 10^{-4}$	CL=90%	2079
$D_s^+ \overline{K}^0$	< 1.1 $\times 10^{-3}$	CL=90%	2241
$D_s^{*+} \overline{K}^0$	< 1.1 $\times 10^{-3}$	CL=90%	2184
$D_s^+ \overline{K}^*(892)^0$	< 5 $\times 10^{-4}$	CL=90%	2171
$D_s^{*+} \overline{K}^*(892)^0$	< 4 $\times 10^{-4}$	CL=90%	2110
$D_s^- \pi^+ K^+$	< 8 $\times 10^{-4}$	CL=90%	2222
$D_s^{*-} \pi^+ K^+$	< 1.2 $\times 10^{-3}$	CL=90%	2164
$D_s^- \pi^+ K^*(892)^+$	< 6 $\times 10^{-3}$	CL=90%	2137
$D_s^{*-} \pi^+ K^*(892)^+$	< 8 $\times 10^{-3}$	CL=90%	2075
Charmonium modes			
$J/\psi(1S) K^+$	(1.01$\pm$0.14) $\times 10^{-3}$		1683
$J/\psi(1S) K^+ \pi^+ \pi^-$	(1.4 $\pm$0.6) $\times 10^{-3}$		1612
$J/\psi(1S) K^*(892)^+$	(1.7 $\pm$0.5) $\times 10^{-3}$		1571
$J/\psi(1S) \pi^+$	(4.4 $\pm$2.4) $\times 10^{-5}$		1727
$\psi(2S) K^+$	(6.9 $\pm$3.1) $\times 10^{-4}$	S=1.3	1284
$\psi(2S) K^*(892)^+$	< 3.0 $\times 10^{-3}$	CL=90%	1115
$\psi(2S) K^*(892)^+ \pi^+ \pi^-$	(1.9 $\pm$1.2) $\times 10^{-3}$		909
$\chi_{c1}(1P) K^+$	(1.0 $\pm$0.4) $\times 10^{-3}$		1411
$\chi_{c1}(1P) K^*(892)^+$	< 2.1 $\times 10^{-3}$	CL=90%	1265
K or K^* modes			
$K^0 \pi^+$	< 4.8 $\times 10^{-5}$	CL=90%	2614
$K^+ \pi^0$	< 1.4 $\times 10^{-5}$	CL=90%	2615
$K^*(892)^0 \pi^+$	< 4.1 $\times 10^{-5}$	CL=90%	2561
$K^*(892)^+ \pi^0$	< 9.9 $\times 10^{-5}$	CL=90%	2562
$K^+ \pi^- \pi^+$ (no charm)	< 1.9 $\times 10^{-4}$	CL=90%	2609
$K_1(1400)^0 \pi^+$	< 2.6 $\times 10^{-3}$	CL=90%	2451
$K_2^*(1430)^0 \pi^+$	< 6.8 $\times 10^{-4}$	CL=90%	2443
$K^+ \rho^0$	< 1.9 $\times 10^{-5}$	CL=90%	2559
$K^0 \rho^+$	< 4.8 $\times 10^{-5}$	CL=90%	2559
$K^*(892)^+ \pi^+ \pi^-$	< 1.1 $\times 10^{-3}$	CL=90%	2556
$K^*(892)^+ \rho^0$	< 9.0 $\times 10^{-4}$	CL=90%	2505
$K_1(1400)^+ \rho^0$	< 7.8 $\times 10^{-3}$	CL=90%	2389
$K_2^*(1430)^+ \rho^0$	< 1.5 $\times 10^{-3}$	CL=90%	2382
$K^+ K^- K^+$	< 3.1 $\times 10^{-4}$	CL=90%	2522
$K^+ \phi$	< 1.2 $\times 10^{-5}$	CL=90%	2516
$K^*(892)^+ K^+ K^-$	< 1.6 $\times 10^{-3}$	CL=90%	2466
$K^*(892)^+ \phi$	< 7.0 $\times 10^{-5}$	CL=90%	2460
$K_1(1400)^+ \phi$	< 1.1 $\times 10^{-3}$	CL=90%	2339
$K_2^*(1430)^+ \phi$	< 3.4 $\times 10^{-3}$	CL=90%	2332
$K^+ f_0(980)$	< 8 $\times 10^{-5}$	CL=90%	2524
$K^*(892)^+ \gamma$	(5.7 $\pm$3.3) $\times 10^{-5}$		2564
$K_1(1270)^+ \gamma$	< 7.3 $\times 10^{-3}$	CL=90%	2486
$K_1(1400)^+ \gamma$	< 2.2 $\times 10^{-3}$	CL=90%	2453
$K_2^*(1430)^+ \gamma$	< 1.4 $\times 10^{-3}$	CL=90%	2447
$K^*(1680)^+ \gamma$	< 1.9 $\times 10^{-3}$	CL=90%	2361
$K_3^*(1780)^+ \gamma$	< 5.5 $\times 10^{-3}$	CL=90%	2343
$K_4^*(2045)^+ \gamma$	< 9.9 $\times 10^{-3}$	CL=90%	2243
Light unflavored meson modes			
$\pi^+ \pi^0$	< 1.7 $\times 10^{-5}$	CL=90%	2636
$\pi^+ \pi^+ \pi^-$	< 1.9 $\times 10^{-4}$	CL=90%	2630
$\rho^0 \pi^+$	< 4.3 $\times 10^{-5}$	CL=90%	2582
$\pi^+ f_0(980)$	< 1.4 $\times 10^{-4}$	CL=90%	2547
$\pi^+ f_2(1270)$	< 2.4 $\times 10^{-4}$	CL=90%	2483
$\pi^+ \pi^0 \pi^0$	< 8.9 $\times 10^{-4}$	CL=90%	2631
$\rho^+ \pi^0$	< 7.7 $\times 10^{-5}$	CL=90%	2582
$\pi^+ \pi^- \pi^+ \pi^0$	< 4.0 $\times 10^{-3}$	CL=90%	2621
$\rho^+ \rho^0$	< 1.0 $\times 10^{-3}$	CL=90%	2525
$a_1(1260)^+ \pi^0$	< 1.7 $\times 10^{-3}$	CL=90%	2494
$a_1(1260)^0 \pi^+$	< 9.0 $\times 10^{-4}$	CL=90%	2494
$\omega \pi^+$	< 4.0 $\times 10^{-4}$	CL=90%	2580
$\eta \pi^+$	< 7.0 $\times 10^{-5}$	CL=90%	2609
$\pi^+ \pi^+ \pi^+ \pi^- \pi^-$	< 8.6 $\times 10^{-4}$	CL=90%	2608
$\rho^0 a_1(1260)^+$	< 6.2 $\times 10^{-4}$	CL=90%	2434
$\rho^0 a_2(1320)^+$	< 7.2 $\times 10^{-4}$	CL=90%	2411
$\pi^+ \pi^+ \pi^+ \pi^- \pi^- \pi^0$	< 6.3 $\times 10^{-3}$	CL=90%	2592
$a_1(1260)^+ a_1(1260)^0$	< 1.3 %	CL=90%	2335
Baryon modes			
$p \overline{p} \pi^+$	< 1.6 $\times 10^{-4}$	CL=90%	2439
$p \overline{p} \pi^+ \pi^+ \pi^-$	< 5.2 $\times 10^{-4}$	CL=90%	2369
$p \overline{\Lambda}$	< 6 $\times 10^{-5}$	CL=90%	2430
$p \overline{\Lambda} \pi^+ \pi^-$	< 2.0 $\times 10^{-4}$	CL=90%	2367
$\overline{\Delta}^0 p$	< 3.8 $\times 10^{-4}$	CL=90%	2402
$\Delta^{++} \overline{p}$	< 1.5 $\times 10^{-4}$	CL=90%	2402

Lepton Family number (LF) or Lepton number (L) violating modes, or $\Delta B = 1$ weak neutral current (B1) modes

Mode					p
$\pi^+ e^+ e^-$	B1	< 3.9	$\times 10^{-3}$	CL=90%	2638
$\pi^+ \mu^+ \mu^-$	B1	< 9.1	$\times 10^{-3}$	CL=90%	2633
$K^+ \mu^+ \mu^-$	B1	< 1.7	$\times 10^{-4}$	CL=90%	2612
$K^*(892)^+ e^+ e^-$	B1	< 6.9	$\times 10^{-4}$	CL=90%	2564
$K^*(892)^+ \mu^+ \mu^-$	B1	< 1.2	$\times 10^{-3}$	CL=90%	2560
$\pi^+ e^+ \mu^-$	LF	< 6.4	$\times 10^{-3}$	CL=90%	2637
$\pi^+ e^- \mu^+$	LF	< 6.4	$\times 10^{-3}$	CL=90%	2637
$K^+ e^+ \mu^-$	LF	< 6.4	$\times 10^{-3}$	CL=90%	2615
$K^+ e^- \mu^+$	LF	< 6.4	$\times 10^{-3}$	CL=90%	2615
$\pi^- e^+ e^+$	L	< 3.9	$\times 10^{-3}$	CL=90%	2638
$\pi^- \mu^+ \mu^+$	L	< 9.1	$\times 10^{-3}$	CL=90%	2633
$\pi^- e^+ \mu^+$	L	< 6.4	$\times 10^{-3}$	CL=90%	2637
$K^- e^+ e^+$	L	< 3.9	$\times 10^{-3}$	CL=90%	2616
$K^- \mu^+ \mu^+$	L	< 9.1	$\times 10^{-3}$	CL=90%	2612
$K^- e^+ \mu^+$	L	< 6.4	$\times 10^{-3}$	CL=90%	2615

B^0 $I(J^P) = \frac{1}{2}(0^-)$

I, J, P need confirmation. Quantum numbers shown are quark-model predictions.

Mass $m_{B^0} = 5279.2 \pm 1.8$ MeV
$m_{B^0} - m_{B^\pm} = 0.35 \pm 0.29$ MeV   (S = 1.1)
Mean life $\tau_{B^0} = (1.56 \pm 0.06) \times 10^{-12}$ s
    $c\tau = 468$ μm
$\tau_{B^+}/\tau_{B^0} = 1.02 \pm 0.05$ (average of direct and inferred)
$\tau_{B^+}/\tau_{B^0} = 1.03 \pm 0.06$ (direct measurements)
$\tau_{B^+}/\tau_{B^0} = 0.93 \pm 0.22$ (inferred from branching fractions)

B^0-$\overline{B}^0$ mixing parameters

$\chi_d = 0.175 \pm 0.016$
$\Delta m_{B^0} = m_{B^0_H} - m_{B^0_L} = (0.474 \pm 0.031) \times 10^{12}\, \hbar\, s^{-1}$
$x_d = \Delta m_{B^0}/\Gamma_{B^0} = 0.73 \pm 0.05$

CP violation parameters

$|Re(\epsilon_{B^0})| < 0.045$

$\overline{B}^0$ modes are charge conjugates of the modes below. Reactions indicate the weak decay vertex and do not include mixing. Modes which do not identify the charge state of the B are listed in the $B^\pm/B^0$ ADMIXTURE section.

The branching fractions listed below assume 50% $B^0\overline{B}^0$ and 50% B^+B^- production at the $\Upsilon(4S)$. We have attempted to bring older measurements up to date by rescaling their assumed $\Upsilon(4S)$ production ratio to 50:50 and their assumed D, D_s, D^*, and ψ branching ratios to current values whenever this would affect our averages and best limits significantly.

Indentation is used to indicate a subchannel of a previous reaction. All resonant subchannels have been corrected for resonance branching fractions to the final state so the sum of the subchannel branching fractions can exceed that of the final state.

B^0 DECAY MODES	Fraction (Γ_i/Γ)	Scale factor/ Confidence level	p (MeV/c)
Semileptonic and leptonic modes			
$\ell^+ \nu_\ell$ anything	[qq] (10.3 ±1.0) %		–
$D^- \ell^+ \nu_\ell$	[qq] (1.9 ±0.5) %		–
$D^*(2010)^- \ell^+ \nu_\ell$	[qq] (4.56±0.27) %		–
$\rho^- \ell^+ \nu_\ell$	[qq] < 4.1 $\times 10^{-4}$	CL=90%	–
D, D*, or D_s modes			
$D^- \pi^+$	(3.0 ±0.4) $\times 10^{-3}$		2306
$D^- \rho^+$	(7.8 ±1.4) $\times 10^{-3}$		2236
$\overline{D}^0 \pi^+ \pi^-$	< 1.6 $\times 10^{-3}$	CL=90%	2301
$D^*(2010)^- \pi^+$	(2.6 ±0.4) $\times 10^{-3}$		2254
$D^- \pi^+ \pi^+ \pi^-$	(8.0 ±2.5) $\times 10^{-3}$		2287
$\quad(D^- \pi^+ \pi^+ \pi^-)$ nonresonant	(3.9 ±1.9) $\times 10^{-3}$		2287
$D^- \pi^+ \pi^0$	(1.1 ±1.0) $\times 10^{-3}$		2207
$\quad D^- a_1(1260)^+$	(6.0 ±3.3) $\times 10^{-3}$		2121
$D^*(2010)^- \pi^+ \pi^0$	(1.5 ±0.5) %		2247
$D^*(2010)^- \rho^+$	(7.3 ±1.5) $\times 10^{-3}$		2181
$D^*(2010)^- \pi^+ \pi^+ \pi^-$	(7.6 ±1.7) $\times 10^{-3}$	S=1.3	2235
$\quad(D^*(2010)^- \pi^+ \pi^+ \pi^-)$ non-resonant	(0.0 ±2.5) $\times 10^{-3}$		2235
$\quad D^*(2010)^- \pi^+ \rho^0$	(5.7 ±3.1) $\times 10^{-3}$		2151
$\quad D^*(2010)^- a_1(1260)^+$	(1.30±0.27) %		2061
$D^*(2010)^- \pi^+ \pi^+ \pi^- \pi^0$	(3.4 ±1.8) %		2218
$\overline{D}_2^*(2460)^- \pi^+$	< 2.2 $\times 10^{-3}$	CL=90%	2064
$\overline{D}_2^*(2460)^- \rho^+$	< 4.9 $\times 10^{-3}$	CL=90%	1979
$D^- D_s^+$	(7 ±4) $\times 10^{-3}$		1812
$D^*(2010)^- D_s^+$	(1.2 ±0.6) %		1735
$D^- D_s^{*+}$	(2.0 ±1.5) %		1731
$D^*(2010)^- D_s^{*+}$	(1.9 ±1.2) %		1649
$D_s^+ \pi^-$	< 2.8 $\times 10^{-4}$	CL=90%	2270
$D_s^{*+} \pi^-$	< 5 $\times 10^{-4}$	CL=90%	2214
$D_s^+ \rho^-$	< 7 $\times 10^{-4}$	CL=90%	2198
$D_s^{*+} \rho^-$	< 8 $\times 10^{-4}$	CL=90%	2139
$D_s^+ a_1(1260)^-$	< 2.6 $\times 10^{-3}$	CL=90%	2079
$D_s^{*+} a_1(1260)^-$	< 2.2 $\times 10^{-3}$	CL=90%	2014
$D_s^- K^+$	< 2.4 $\times 10^{-4}$	CL=90%	2242
$D_s^{*-} K^+$	< 1.7 $\times 10^{-4}$	CL=90%	2185
$D_s^- K^*(892)^+$	< 9.9 $\times 10^{-4}$	CL=90%	2172
$D_s^{*-} K^*(892)^+$	< 1.1 $\times 10^{-3}$	CL=90%	2112
$D_s^- \pi^+ K^0$	< 5 $\times 10^{-3}$	CL=90%	2221
$D_s^{*-} \pi^+ K^0$	< 3.1 $\times 10^{-3}$	CL=90%	2164
$D_s^- \pi^+ K^*(892)^0$	< 4 $\times 10^{-3}$	CL=90%	2136
$D_s^{*-} \pi^+ K^*(892)^0$	< 2.0 $\times 10^{-3}$	CL=90%	2074
$\overline{D}^0 \pi^0$	< 4.8 $\times 10^{-4}$	CL=90%	2308
$\overline{D}^0 \rho^0$	< 5.5 $\times 10^{-4}$	CL=90%	2238
$\overline{D}^0 \eta$	< 6.8 $\times 10^{-4}$	CL=90%	2274
$\overline{D}^0 \eta'$	< 8.6 $\times 10^{-4}$	CL=90%	2198
$\overline{D}^0 \omega$	< 6.3 $\times 10^{-4}$	CL=90%	2235
$\overline{D}^*(2007)^0 \pi^0$	< 9.7 $\times 10^{-4}$	CL=90%	2256
$\overline{D}^*(2007)^0 \rho^0$	< 1.17 $\times 10^{-3}$	CL=90%	2183
$\overline{D}^*(2007)^0 \eta$	< 6.9 $\times 10^{-4}$	CL=90%	2220
$\overline{D}^*(2007)^0 \eta'$	< 2.7 $\times 10^{-3}$	CL=90%	2141
$\overline{D}^*(2007)^0 \omega$	< 2.1 $\times 10^{-3}$	CL=90%	2180
Charmonium modes			
$J/\psi(1S) K^0$	(7.5 ±2.1) $\times 10^{-4}$		1683
$J/\psi(1S) K^+ \pi^-$	(1.1 ±0.6) $\times 10^{-3}$		1652
$J/\psi(1S) K^*(892)^0$	(1.58±0.27) $\times 10^{-3}$		1570
$J/\psi(1S) \pi^0$	< 6.9 $\times 10^{-4}$	CL=90%	1728
$\psi(2S) K^0$	< 8 $\times 10^{-4}$	CL=90%	1283
$\psi(2S) K^+ \pi^-$	< 1 $\times 10^{-3}$	CL=90%	1238
$\psi(2S) K^*(892)^0$	(1.4 ±0.9) $\times 10^{-3}$		1113
$\chi_{c1}(1P) K^0$	< 2.7 $\times 10^{-3}$	CL=90%	1411
$\chi_{c1}(1P) K^*(892)^0$	< 2.1 $\times 10^{-3}$	CL=90%	1263
K or K* modes			
$K^+ \pi^-$	< 1.7 $\times 10^{-5}$	CL=90%	2615
$K^0 \pi^0$	< 4.0 $\times 10^{-5}$	CL=90%	2614
$K^+ K^-$	< 4 $\times 10^{-6}$	CL=90%	2593
$K^+ \rho^-$	< 3.5 $\times 10^{-5}$	CL=90%	2559
$K^0 \rho^0$	< 3.9 $\times 10^{-5}$	CL=90%	2559
$K^0 f_0(980)$	< 3.6 $\times 10^{-4}$	CL=90%	2523
$K^*(892)^+ \pi^-$	< 7.2 $\times 10^{-5}$	CL=90%	2562
$K^*(892)^0 \pi^0$	< 2.8 $\times 10^{-5}$	CL=90%	2562
$K_2^*(1430)^+ \pi^-$	< 2.6 $\times 10^{-3}$	CL=90%	2445
$K^0 K^+ K^-$	< 1.3 $\times 10^{-3}$	CL=90%	2522
$\quad K^0 \phi$	< 8.8 $\times 10^{-5}$	CL=90%	2516
$K^- \pi^+ \pi^+ \pi^-$	[bbb] < 2.1 $\times 10^{-4}$	CL=90%	2600
$K^*(892)^0 \pi^+ \pi^-$	< 1.4 $\times 10^{-3}$	CL=90%	2556
$\quad K^*(892)^0 \rho^0$	< 4.6 $\times 10^{-4}$	CL=90%	2504
$\quad K^*(892)^0 f_0(980)$	< 1.7 $\times 10^{-4}$	CL=90%	2467
$K_1(1400)^+ \pi^-$	< 1.1 $\times 10^{-3}$	CL=90%	2451
$K^- a_1(1260)^+$	[bbb] < 3.9 $\times 10^{-4}$	CL=90%	2471
$K^*(892)^0 K^+ K^-$	< 6.1 $\times 10^{-4}$	CL=90%	2466
$\quad K^*(892)^0 \phi$	< 4.3 $\times 10^{-5}$	CL=90%	2459
$K_1(1400)^0 \rho^0$	< 3.0 $\times 10^{-3}$	CL=90%	2389
$K_1(1400)^0 \phi$	< 5.0 $\times 10^{-3}$	CL=90%	2339
$K_2^*(1430)^0 \rho^0$	< 1.1 $\times 10^{-3}$	CL=90%	2380
$K_2^*(1430)^0 \phi$	< 1.4 $\times 10^{-3}$	CL=90%	2330
$K^*(892)^0 \gamma$	(4.0 ±1.9) $\times 10^{-5}$		2564
$K_1(1270)^0 \gamma$	< 7.0 $\times 10^{-3}$	CL=90%	2486
$K_1(1400)^0 \gamma$	< 4.3 $\times 10^{-3}$	CL=90%	2453
$K_2^*(1430)^0 \gamma$	< 4.0 $\times 10^{-4}$	CL=90%	2445
$K^*(1680)^0 \gamma$	< 2.0 $\times 10^{-3}$	CL=90%	2361
$K_3^*(1780)^0 \gamma$	< 1.0 %	CL=90%	2343
$K_4^*(2045)^0 \gamma$	< 4.3 $\times 10^{-3}$	CL=90%	2244
$\phi\phi$	< 3.9 $\times 10^{-5}$	CL=90%	2435

Light unflavored meson modes

					p (MeV/c)
$\pi^+\pi^-$		$<$ 2.0	$\times 10^{-5}$	CL=90%	2636
$\pi^0\pi^0$		$<$ 9.1	$\times 10^{-6}$	CL=90%	2636
$\eta\pi^0$		$<$ 2.5	$\times 10^{-4}$	CL=90%	2609
$\eta\eta$		$<$ 4.1	$\times 10^{-4}$	CL=90%	2582
$\pi^+\pi^-\pi^0$		$<$ 7.2	$\times 10^{-4}$	CL=90%	2631
$\rho^0\pi^0$		$<$ 2.4	$\times 10^{-5}$	CL=90%	2582
$\rho^\mp\pi^\pm$	[gg]	$<$ 8.8	$\times 10^{-5}$	CL=90%	2582
$\pi^+\pi^-\pi^+\pi^-$		$<$ 2.8	$\times 10^{-4}$	CL=90%	2621
$\rho^0\rho^0$		$<$ 2.8	$\times 10^{-4}$	CL=90%	2525
$a_1(1260)^\mp\pi^\pm$	[gg]	$<$ 4.9	$\times 10^{-4}$	CL=90%	2494
$a_2(1320)^\mp\pi^\pm$	[gg]	$<$ 3.0	$\times 10^{-4}$	CL=90%	2473
$\pi^+\pi^-\pi^0\pi^0$		$<$ 3.1	$\times 10^{-3}$	CL=90%	2622
$\rho^+\rho^-$		$<$ 2.2	$\times 10^{-3}$	CL=90%	2525
$a_1(1260)^0\pi^0$		$<$ 1.1	$\times 10^{-3}$	CL=90%	2494
$\omega\pi^0$		$<$ 4.6	$\times 10^{-4}$	CL=90%	2580
$\pi^+\pi^+\pi^-\pi^-\pi^0$		$<$ 9.0	$\times 10^{-3}$	CL=90%	2609
$a_1(1260)^+\rho^-$		$<$ 3.4	$\times 10^{-3}$	CL=90%	2434
$a_1(1260)^0\rho^0$		$<$ 2.4	$\times 10^{-3}$	CL=90%	2434
$\pi^+\pi^+\pi^+\pi^-\pi^-\pi^-$		$<$ 3.0	$\times 10^{-3}$	CL=90%	2592
$a_1(1260)^+a_1(1260)^-$		$<$ 2.8	$\times 10^{-3}$	CL=90%	2336
$\pi^+\pi^+\pi^+\pi^-\pi^-\pi^-\pi^0$		$<$ 1.1	%	CL=90%	2572

Baryon modes

$p\bar{p}$		$<$ 3.4	$\times 10^{-5}$	CL=90%	2467
$p\bar{p}\pi^+\pi^-$		$<$ 2.5	$\times 10^{-4}$	CL=90%	2406
$p\bar{\Lambda}\pi^-$		$<$ 1.8	$\times 10^{-4}$	CL=90%	2401
$\Delta^0\bar{\Delta}^0$		$<$ 1.5	$\times 10^{-3}$	CL=90%	2334
$\Delta^{++}\Delta^{--}$		$<$ 1.1	$\times 10^{-4}$	CL=90%	2334
$\overline{\Sigma}_c\Delta^{++}$		$<$ 1.2	$\times 10^{-4}$	CL=90%	1839

Lepton Family number (LF) violating modes, or $\Delta B = 1$ weak neutral current (B1) modes

$\gamma\gamma$	B1		$<$ 3.9	$\times 10^{-5}$	CL=90%	2640
e^+e^-	B1		$<$ 5.9	$\times 10^{-6}$	CL=90%	2640
$\mu^+\mu^-$	B1		$<$ 5.9	$\times 10^{-6}$	CL=90%	2637
$K^0 e^+e^-$	B1		$<$ 3.0	$\times 10^{-4}$	CL=90%	2616
$K^0\mu^+\mu^-$	B1		$<$ 3.6	$\times 10^{-4}$	CL=90%	2612
$K^*(892)^0 e^+e^-$	B1		$<$ 2.9	$\times 10^{-4}$	CL=90%	2564
$K^*(892)^0\mu^+\mu^-$	B1		$<$ 2.3	$\times 10^{-5}$	CL=90%	2559
$e^\pm\mu^\mp$	LF	[gg]	$<$ 5.9	$\times 10^{-6}$	CL=90%	2639
$e^\pm\tau^\mp$	LF	[gg]	$<$ 5.3	$\times 10^{-4}$	CL=90%	2341
$\mu^\pm\tau^\mp$	LF	[gg]	$<$ 8.3	$\times 10^{-4}$	CL=90%	2339

$B^\pm/B^0$ ADMIXTURE

The branching fraction measurements are for an admixture of B mesons at the $\Upsilon(4S)$. The values quoted assume that $B(\Upsilon(4S) \to B\bar{B}) = 100\%$.

For inclusive branching fractions, e.g., $B \to D^\pm$anything, the treatment of multiple D's in the final state must be defined. One possibility would be to count the number of events with one-or-more D's and divide by the total number of B's. Another possibility would be to count the total number of D's and divide by the total number of B's, which is the definition of average multiplicity. The two definitions are identical when only one of the specified particles is allowed in the final state. Even though the "one-or-more" definition seems sensible, for practical reasons inclusive branching fractions are almost always measured using the multiplicity definition. For heavy final state particles, authors call their results inclusive branching fractions while for light particles some authors call their results multiplicities. In the B sections, we list all results as inclusive branching fractions, adopting a multiplicity definition. This means that inclusive branching fractions can exceed 100% and that inclusive partial widths can exceed total widths, just as inclusive cross sections can exceed total cross sections.

$\bar{B}$ modes are charge conjugates of the modes below. Reactions indicate the weak decay vertex and do not include mixing.

B DECAY MODES	Fraction (Γ_i/Γ)		Scale factor/ Confidence level	p (MeV/c)
Semileptonic and leptonic modes				
$e^+\nu_e$ anything	[ccc]	(10.4 $\pm$0.4) %	S=1.3	–
$\bar{p}e^+\nu_e$ anything		$<$ 1.6 $\times 10^{-3}$	CL=90%	–
$\mu^+\nu_\mu$ anything	[ccc]	(10.3 $\pm$0.5) %		–
$\ell^+\nu_\ell$ anything	[qq,ccc]	(10.43$\pm$0.24) %		–
$D^-\ell^+\nu_\ell$ anything	[qq]	(2.7 $\pm$0.8) %		–
$\overline{D}^0\ell^+\nu_\ell$ anything	[qq]	(7.0 $\pm$1.4) %		–
$\overline{D}^{**}\ell^+\nu_\ell$	[qq,ddd]	(2.7 $\pm$0.7) %		–
$\overline{D}(1)(2420)^0\ell^+\nu_\ell$ anything		seen		–
$\overline{D}(2)^*(2460)^0\ell^+\nu_\ell$ anything		not seen		–
$D^{*-}\pi^+\ell^+\nu_\ell$ anything		(1.00$\pm$0.34) %		–
$D_s^-\ell^+\nu_\ell$ anything	[qq]	$<$ 9 $\times 10^{-3}$	CL=90%	–
$D_s^-\ell^+\nu_\ell K^+$ anything	[qq]	$<$ 6 $\times 10^{-3}$	CL=90%	–
$D_s^-\ell^+\nu_\ell K^0$ anything	[qq]	$<$ 9 $\times 10^{-3}$	CL=90%	–
$K^+\ell^+\nu_\ell$ anything	[qq]	(6.0 $\pm$0.5) %		–
$K^-\ell^+\nu_\ell$ anything	[qq]	(10 $\pm$4) $\times 10^{-3}$		–
$K^0/\overline{K}^0\ell^+\nu_\ell$ anything	[qq]	(4.4 $\pm$0.5) %		–
D, D^*, or D_s modes				
$D^\pm$ anything		(24.2 $\pm$3.3) %		–
$D^0/\overline{D}^0$ anything		(58 $\pm$5) %	S=1.1	–
$D^*(2010)^\pm$ anything		(23.1 $\pm$3.3) %	S=1.1	–
$D_s^\pm$ anything	[gg]	(8.6 $\pm$1.6) %		–
$D_s D$, $D_s^* D$, $D_s D^*$, or $D_s^* D^*$	[gg]	(4.9 $\pm$1.1) %		–
$D^*(2010)\gamma$		$<$ 1.1 $\times 10^{-3}$	CL=90%	–
$D_s^+\pi^-$, $D_s^{*+}\pi^-$, $D_s^+\rho^-$, $D_s^{*+}\rho^-$, $D_s^+\pi^0$, $D_s^{*+}\pi^0$, $D_s^+\eta$, $D_s^{*+}\eta$, $D_s^+\rho^0$, $D_s^{*+}\rho^0$, $D_s^+\omega$, $D_s^{*+}\omega$	[gg]	$<$ 5 $\times 10^{-4}$	CL=90%	–
Charmonium modes				
$J/\psi(1S)$ anything		(1.14$\pm$0.06) %		–
$J/\psi(1S)$ (direct) anything		(8.0 $\pm$0.8) $\times 10^{-3}$		–
$\psi(2S)$ anything		(3.5 $\pm$0.5) $\times 10^{-3}$		–
$\chi_{c1}(1P)$ anything		(4.2 $\pm$0.7) $\times 10^{-3}$		–
$\chi_{c1}(1P)$ (direct) anything		(3.7 $\pm$0.7) $\times 10^{-3}$		–
$\chi_{c2}(1P)$ anything		$<$ 3.8 $\times 10^{-3}$	CL=90%	–
$\eta_c(1S)$ anything		$<$ 9 $\times 10^{-3}$	CL=90%	–
K or K^* modes				
$K^\pm$ anything	[gg]	(78.9 $\pm$2.5) %		–
K^+ anything		(66 $\pm$5) %		–
K^- anything		(13 $\pm$4) %		–
$K^0/\overline{K}^0$ anything	[gg]	(64 $\pm$4) %		–
$K^*(892)^\pm$ anything		(18 $\pm$6) %		–
$K^*(892)^0/\overline{K}^*(892)^0$ anything	[gg]	(14.6 $\pm$2.6) %		–
$K_1(1400)\gamma$		$<$ 4.1 $\times 10^{-4}$	CL=90%	–
$K_2^*(1430)\gamma$		$<$ 8.3 $\times 10^{-4}$	CL=90%	–
$K_2(1770)\gamma$		$<$ 1.2 $\times 10^{-3}$	CL=90%	–
$K_3^*(1780)\gamma$		$<$ 3.0 $\times 10^{-3}$	CL=90%	–
$K_4^*(2045)\gamma$		$<$ 1.0 $\times 10^{-3}$	CL=90%	–
$b \to s\gamma$		(2.3 $\pm$0.7) $\times 10^{-4}$		–
Light unflavored meson modes				
$\pi^\pm$ anything	[gg,eee]	(359 $\pm$7) %		–
ρ^0 anything		(21 $\pm$5) %		–
ω anything		$<$ 81 %	CL=90%	–
ϕ anything		(3.5 $\pm$0.7) %	S=1.8	–
Baryon modes				
charmed-baryon anything		(6.4 $\pm$1.1) %		–
$\overline{\Sigma}_c^{--}$ anything		(4.8 $\pm$2.5) $\times 10^{-3}$		–
$\overline{\Sigma}_c^-$ anything		$<$ 1.1 %	CL=90%	–
$\overline{\Sigma}_c^0$ anything		(5.2 $\pm$2.5) $\times 10^{-3}$		–
$\overline{\Sigma}_c^0 N$ ($N = p$ or n)		$<$ 1.7 $\times 10^{-3}$	CL=90%	–
$p/\bar{p}$ anything	[gg]	(8.0 $\pm$0.4) %		–
$p/\bar{p}$ (direct) anything	[gg]	(5.5 $\pm$0.5) %		–
$\Lambda/\overline{\Lambda}$ anything	[gg]	(4.0 $\pm$0.5) %		–
$\Xi^-/\overline{\Xi}^+$ anything	[gg]	(2.7 $\pm$0.6) $\times 10^{-3}$		–
baryons anything		(6.8 $\pm$0.6) %		–
$p\bar{p}$ anything		(2.47$\pm$0.23) %		–
$\Lambda\bar{p}/\overline{\Lambda}p$ anything	[gg]	(2.5 $\pm$0.4) %		–
$\Lambda\overline{\Lambda}$ anything		$<$ 5 $\times 10^{-3}$	CL=90%	–

$\Delta B = 1$ weak neutral current ($B1$) modes

$e^+ e^-$ anything	$B1$	<	2.4	$\times 10^{-3}$ CL=90%	—
$\mu^+ \mu^-$ anything	$B1$	<	2.4	$\times 10^{-3}$ CL=90%	—

$B^\pm / B^0 / B_s^0 / b$-baryon ADMIXTURE

These measurements are for an admixture of bottom particles at high energy (LEP, Tevatron, $Sp\bar{p}S$).

Mean life $\tau := (1.549 \pm 0.020) \times 10^{-12}$ s

Mean life $\tau = (1.72 \pm 0.10) \times 10^{-12}$ s   Charged b-hadron admixture

Mean life $\tau = (1.58 \pm 0.14) \times 10^{-12}$ s   Neutral b-hadron admixture

$\tau_{\text{charged } b-\text{hadron}} / \tau_{\text{neutral } b-\text{hadron}} = 1.09 \pm 0.13$

The branching fraction measurements are for an admixture of B mesons and baryons at energies above the $\Upsilon(4S)$. Only the highest energy results (LEP, Tevatron, $Sp\bar{p}S$) are used in the branching fraction averages. The production fractions give our best current estimate of the admixture at LEP.

For inclusive branching fractions, e.g., $B \rightarrow D^\pm$ anything, the treatment of multiple D's in the final state must be defined. One possibility would be to count the number of events with one-or-more D's and divide by the total number of B's. Another possibility would be to count the total number of D's and divide by the total number of B's, which is the definition of average multiplicity. The two definitions are identical when only one of the specified particles is allowed in the final state. Even though the "one-or-more" definition seems sensible, for practical reasons inclusive branching fractions are almost always measured using the multiplicity definition. For heavy final state particles, authors call their results inclusive branching fractions while for light particles some authors call their results multiplicities. In the B sections, we list all results as inclusive branching fractions, adopting a multiplicity definition. This means that inclusive branching fractions can exceed 100% and that inclusive partial widths can exceed total widths, just as inclusive cross sections can exceed total cross sections.

The modes below are listed for a $\bar{b}$ initial state. b modes are their charge conjugates. Reactions indicate the weak decay vertex and do not include mixing.

$\bar{b}$ DECAY MODES		Fraction (Γ_i/Γ)	Confidence level	p (MeV/c)

PRODUCTION FRACTIONS

The production fractions for weakly decaying b-hadrons at the Z have been calculated from the best values of mean lives, mixing parameters, and branching fractions in this edition by O. Hayes (CERN) and M. Jimack (U. Birmingham) as described in the note "Production and Decay of b-Flavored Hadrons" in the $B^\pm$ Particle Listings. Values assume

$B(\bar{b} \rightarrow B^+) = B(\bar{b} \rightarrow B^0)$
$B(\bar{b} \rightarrow B^+) + B(\bar{b} \rightarrow B^0) + B(\bar{b} \rightarrow B_s^0) + B(b \rightarrow \Lambda_b) = 100\%$.

The notation for production fractions varies in the literature (f_{B^0}, $f(b \rightarrow \bar{B}^0)$, $Br(b \rightarrow \bar{B}^0)$). We use our own branching fraction notation here, $B(\bar{b} \rightarrow B^0)$.

$\bar{b} \rightarrow B^+$		(37.8 $\pm$ 2.2) %	—
$\bar{b} \rightarrow B^0$		(37.8 $\pm$ 2.2) %	—
$\bar{b} \rightarrow B_s^0$		(11.2 $^{+1.8}_{-1.9}$) %	—
$b \rightarrow \Lambda_b$		(13.2 $\pm$ 4.1) %	—

DECAY MODES

Semileptonic and leptonic modes

$\bar{b} \rightarrow e^+ \nu_e$ anything	[ccc]	(11.1 $\pm$ 1.0) %	—
$\bar{b} \rightarrow \mu^+ \nu_\mu$ anything	[ccc]	(10.7 $\pm$ 0.7) %	—
$\bar{b} \rightarrow \ell^+ \nu_\ell$ anything	[qq,ccc]	(11.13 $\pm$ 0.29) %	—
$\bar{b} \rightarrow D^- \ell^+ \nu_\ell$ anything	[qq]	(2.01 $\pm$ 0.29) %	—
$\bar{b} \rightarrow \overline{D}^0 \ell^+ \nu_\ell$ anything	[qq]	(6.6 $\pm$ 0.6) %	—
$\bar{b} \rightarrow D^{*-} \ell^+ \nu_\ell$ anything	[qq]	(2.76 $\pm$ 0.29) %	—
$\bar{b} \rightarrow \overline{D}_j^0 \ell^+ \nu_\ell$ anything	[qq,fff]	seen	—
$\bar{b} \rightarrow D_j^- \ell^+ \nu_\ell$ anything	[qq,fff]	seen	—
$\bar{b} \rightarrow \overline{D}_2^*(2460)^0 \ell^+ \nu_\ell$ anything		seen	—
$\bar{b} \rightarrow D_2^*(2460)^- \ell^+ \nu_\ell$ anything		seen	—
$\bar{b} \rightarrow \tau^+ \nu_\tau$ anything		(2.7 $\pm$ 0.4) %	—
$\bar{b} \rightarrow \bar{b} \rightarrow \bar{c} \rightarrow \ell^- \bar{\nu}_\ell$ anything	[qq]	(7.9 $\pm$ 0.8) %	—

Charmonium modes

$\bar{b} \rightarrow J/\psi(1S)$ anything		(1.16 $\pm$ 0.10) %	—
$\bar{b} \rightarrow \psi(2S)$ anything		(4.8 $\pm$ 2.4) $\times 10^{-3}$	—
$\bar{b} \rightarrow \chi_{c1}(1P)$ anything		(1.8 $\pm$ 0.5) %	—

K or K^* modes

$\bar{b} \rightarrow \bar{s}\gamma$		< 1.2 $\times 10^{-3}$ 90%	—
$\bar{b} \rightarrow K^\pm$ anything		(88 $\pm$19) %	—
$\bar{b} \rightarrow K_S^0$ anything		(29.0 $\pm$ 2.9) %	—

Baryon modes

$\bar{b} \rightarrow p/\bar{p}$ anything		(14 $\pm$ 6) %	—
$\bar{b} \rightarrow \Lambda/\overline{\Lambda}$ anything		(5.9 $\pm$ 1.1) %	—

Other modes

$\bar{b} \rightarrow$ charged anything	[eee]	(584 $\pm$40) %	—

$\Delta B = 1$ weak neutral current ($B1$) modes

$\bar{b} \rightarrow \mu^+ \mu^-$ anything	$B1$	< 5.0 $\times 10^{-5}$ 90%	—
$\bar{b} \rightarrow \nu\bar{\nu}$ anything	$B1$	< 3.9 $\times 10^{-4}$	—

B^*		$I(J^P) = \frac{1}{2}(1^-)$

I, J, P need confirmation. Quantum numbers shown are quark-model predictions.

Mass $m_{B^*} = 5324.8 \pm 1.8$ MeV
$m_{B^*} - m_B = 45.7 \pm 0.4$ MeV

B^* DECAY MODES	Fraction (Γ_i/Γ)	p (MeV/c)
$B\gamma$	dominant	46

BOTTOM, STRANGE MESONS
($B = \pm 1$, $S = \mp 1$)

$B_s^0 = s\bar{b}$, $\overline{B}_s^0 = \bar{s}b$,   similarly for B_s^*'s

B_s^0		$I(J^P) = 0(0^-)$

I, J, P need confirmation. Quantum numbers shown are quark-model predictions.

Mass $m_{B_s^0} = 5369.3 \pm 2.0$ MeV
Mean life $\tau = (1.61 ^{+0.10}_{-0.09}) \times 10^{-12}$ s
$c\tau = 483$ μm

B_s^0-$\overline{B}_s^0$ mixing parameters

$\chi_s > 0.49$, CL = 95%
χ_B at high energy $= f_d \chi_d + f_s \chi_s = 0.126 \pm 0.008$
$\Delta m_{B_s^0} = m_{B_{sH}^0} - m_{B_{sL}^0} > 5.9 \times 10^{12}$ $\hbar$ s^{-1}, CL = 95%
$x_s = \Delta m_{B_s^0} / \Gamma_{B_s^0} > 9.5$, CL = 95%

These branching fractions all scale with $B(\bar{b} \rightarrow B_s^0)$, the LEP B_s^0 production fraction. The first four were evaluated using $B(\bar{b} \rightarrow B_s^0) = (11.2^{+1.8}_{-1.9})\%$ and the rest assume $B(\bar{b} \rightarrow B_s^0) = 12\%$.

The branching fraction $B(B_s^0 \rightarrow D_s^- \ell^+ \nu_\ell$ anything) is not a pure measurement since the measured product branching fraction $B(\bar{b} \rightarrow B_s^0) \times B(B_s^0 \rightarrow D_s^- \ell^+ \nu_\ell$ anything) was used to determine $B(\bar{b} \rightarrow B_s^0)$, as described in the note on "Production and Decay of b-Flavored Hadrons."

B_s^0 DECAY MODES		Fraction (Γ_i/Γ)	Confidence level	p (MeV/c)
D_s^- anything		(87 $\pm$31) %		—
$D_s^- \ell^+ \nu_\ell$ anything	[ggg]	(7.6 $\pm$ 2.4) %		—
$D_s^- \pi^+$		< 12 %		2321
$J/\psi(1S)\phi$		< 6 $\times 10^{-3}$		1590
$\psi(2S)\phi$		seen		1122
$\pi^0 \pi^0$		< 2.1 $\times 10^{-4}$	90%	2861
$\eta \pi^0$		< 1.0 $\times 10^{-3}$	90%	2655
$\eta \eta$		< 1.5 $\times 10^{-3}$	90%	2628
$\pi^+ K^-$		< 2.6 $\times 10^{-4}$	90%	2660
$K^+ K^-$		< 1.4 $\times 10^{-4}$	90%	2639

$\Delta B = 1$ weak neutral current ($B1$) modes

$\gamma\gamma$	$B1$	< 1.48 $\times 10^{-4}$	90%	2685

$c\bar{c}$ MESONS

$\eta_c(1S)$

$I^G(J^{PC}) = 0^+(0^{-+})$

Mass $m = 2979.8 \pm 2.1$ MeV (S = 2.1)
Full width $\Gamma = 13.2^{+3.8}_{-3.2}$ MeV

$\eta_c(1S)$ DECAY MODES	Fraction (Γ_i/Γ)	Confidence level	p (MeV/c)
Decays involving hadronic resonances			
$\eta'(958)\pi\pi$	(4.1 ± 1.7) %		1319
$\rho\rho$	(2.6 ± 0.9) %		1275
$K^*(892)^0 K^-\pi^+ +$ c.c.	(2.0 ± 0.7) %		1273
$K^*(892)\overline{K}^*(892)$	$(8.5 \pm 3.1) \times 10^{-3}$		1193
$\phi\phi$	$(7.1 \pm 2.8) \times 10^{-3}$		1086
$a_0(980)\pi$	< 2 %	90%	1323
$a_2(1320)\pi$	< 2 %	90%	1193
$K^*(892)\overline{K} +$ c.c.	< 1.28 %	90%	1307
$f_2(1270)\eta$	< 1.1 %	90%	1142
$\omega\omega$	$< 3.1 \times 10^{-3}$	90%	1268
Decays into stable hadrons			
$K\overline{K}\pi$	(5.5 ± 1.7) %		1378
$\eta\pi\pi$	(4.9 ± 1.8) %		1425
$\pi^+\pi^- K^+ K^-$	$(2.0^{+0.7}_{-0.6})$ %		1342
$2(K^+ K^-)$	(2.1 ± 1.2) %		1053
$2(\pi^+\pi^-)$	(1.2 ± 0.4) %		1457
$p\overline{p}$	$(1.2 \pm 0.4) \times 10^{-3}$		1157
$K\overline{K}\eta$	< 3.1 %	90%	1262
$\pi^+\pi^- p\overline{p}$	< 1.2 %	90%	1023
$\Lambda\overline{\Lambda}$	$< 2 \times 10^{-3}$	90%	987
Radiative decays			
$\gamma\gamma$	$(3.0 \pm 1.2) \times 10^{-4}$		1489

$J/\psi(1S)$

$I^G(J^{PC}) = 0^-(1^{--})$

Mass $m = 3096.88 \pm 0.04$ MeV
Full width $\Gamma = 87 \pm 5$ keV
$\Gamma_{ee} = 5.26 \pm 0.37$ keV (Assuming $\Gamma_{ee} = \Gamma_{\mu\mu}$)

$J/\psi(1S)$ DECAY MODES	Fraction (Γ_i/Γ)	Scale factor/ Confidence level	p (MeV/c)
hadrons	(87.7 ± 0.5) %		—
virtual $\gamma \rightarrow$ hadrons	(17.0 ± 2.0) %		—
$e^+ e^-$	(6.02 ± 0.19) %		1548
$\mu^+ \mu^-$	(6.01 ± 0.19) %		1545
Decays involving hadronic resonances			
$\rho\pi$	(1.28 ± 0.10) %		1449
$\rho^0\pi^0$	$(4.2 \pm 0.5) \times 10^{-3}$		1449
$a_2(1320)\rho$	(1.09 ± 0.22) %		1125
$\omega\pi^+\pi^+\pi^-\pi^-$	$(8.5 \pm 3.4) \times 10^{-3}$		1392
$\omega\pi^+\pi^-$	$(7.2 \pm 1.0) \times 10^{-3}$		1435
$K^*(892)^0\overline{K}_2^*(1430)^0 +$ c.c.	$(6.7 \pm 2.6) \times 10^{-3}$		1005
$\omega K^*(892)\overline{K} +$ c.c.	$(5.3 \pm 2.0) \times 10^{-3}$		1098
$\omega f_2(1270)$	$(4.3 \pm 0.6) \times 10^{-3}$		1143
$K^+\overline{K}^*(892)^- +$ c.c.	$(5.0 \pm 0.4) \times 10^{-3}$		1373
$K^0\overline{K}^*(892)^0 +$ c.c.	$(4.2 \pm 0.4) \times 10^{-3}$		1371
$\omega\pi^0\pi^0$	$(3.4 \pm 0.8) \times 10^{-3}$		1436
$b_1(1235)^\pm\pi^\mp$	$(3.0 \pm 0.5) \times 10^{-3}$ [gg]		1299
$\omega K^\pm K_S^0\pi^\mp$	$(3.0 \pm 0.7) \times 10^{-3}$ [gg]		1210
$b_1(1235)^0\pi^0$	$(2.3 \pm 0.6) \times 10^{-3}$		1299
$\phi K^*(892)\overline{K} +$ c.c.	$(2.04 \pm 0.28) \times 10^{-3}$		969
$\omega K\overline{K}$	$(1.9 \pm 0.4) \times 10^{-3}$		1268
$\omega f_J(1710) \rightarrow \omega K\overline{K}$	$(4.8 \pm 1.1) \times 10^{-4}$		878
$\phi 2(\pi^+\pi^-)$	$(1.60 \pm 0.32) \times 10^{-3}$		1318
$\Delta(1232)^{++}\overline{p}\pi^-$	$(1.6 \pm 0.5) \times 10^{-3}$		1030
$\omega\eta$	$(1.58 \pm 0.16) \times 10^{-3}$		1394
$\phi K\overline{K}$	$(1.48 \pm 0.22) \times 10^{-3}$		1179
$\phi f_J(1710) \rightarrow \phi K\overline{K}$	$(3.6 \pm 0.6) \times 10^{-4}$		875
$p\overline{p}\omega$	$(1.30 \pm 0.25) \times 10^{-3}$	S=1.3	769
$\Delta(1232)^{++}\overline{\Delta}(1232)^{--}$	$(1.10 \pm 0.29) \times 10^{-3}$		938
$\Sigma(1385)^-\overline{\Sigma}(1385)^+$ (or c.c.)	$(1.03 \pm 0.13) \times 10^{-3}$ [gg]		692
$p\overline{p}\eta'(958)$	$(9 \pm 4) \times 10^{-4}$	S=1.7	596
$\phi f_2'(1525)$	$(8 \pm 4) \times 10^{-4}$	S=2.7	871
$\phi\pi^+\pi^-$	$(8.0 \pm 1.2) \times 10^{-4}$		1365
$\phi K^\pm K_S^0\pi^\mp$	$(7.2 \pm 0.9) \times 10^{-4}$ [gg]		1114
$\omega f_1(1420)$	$(6.8 \pm 2.4) \times 10^{-4}$		1062
$\phi\eta$	$(6.5 \pm 0.7) \times 10^{-4}$		1320
$\Xi(1530)^-\overline{\Xi}^+$	$(5.9 \pm 1.5) \times 10^{-4}$		597
$pK^-\overline{\Sigma}(1385)^0$	$(5.1 \pm 3.2) \times 10^{-4}$		645
$\omega\pi^0$	$(4.2 \pm 0.6) \times 10^{-4}$	S=1.4	1447
$\phi\eta'(958)$	$(3.3 \pm 0.4) \times 10^{-4}$		1192
$\phi f_0(980)$	$(3.2 \pm 0.9) \times 10^{-4}$	S=1.9	1182
$\Xi(1530)^0\overline{\Xi}^0$	$(3.2 \pm 1.4) \times 10^{-4}$		608
$\Sigma(1385)^-\overline{\Sigma}^+$ (or c.c.)	$(3.1 \pm 0.5) \times 10^{-4}$ [gg]		857
$\phi f_1(1285)$	$(2.6 \pm 0.5) \times 10^{-4}$	S=1.1	1032
$\rho\eta$	$(1.93 \pm 0.23) \times 10^{-4}$		1398
$\omega\eta'(958)$	$(1.67 \pm 0.25) \times 10^{-4}$		1279
$\omega f_0(980)$	$(1.4 \pm 0.5) \times 10^{-4}$		1271
$\rho\eta'(958)$	$(1.05 \pm 0.18) \times 10^{-4}$		1283
$p\overline{p}\phi$	$(4.5 \pm 1.5) \times 10^{-5}$		527
$a_2(1320)^\pm\pi^\mp$	$< 4.3 \times 10^{-3}$ [gg]	CL=90%	1263
$K\overline{K}_2^*(1430) +$ c.c.	$< 4.0 \times 10^{-3}$	CL=90%	1159
$K_2^*(1430)^0\overline{K}_2^*(1430)^0$	$< 2.9 \times 10^{-3}$	CL=90%	588
$K^*(892)^0\overline{K}^*(892)^0$	$< 5 \times 10^{-4}$	CL=90%	1263
$\phi f_2(1270)$	$< 3.7 \times 10^{-4}$	CL=90%	1036
$p\overline{p}\rho$	$< 3.1 \times 10^{-4}$	CL=90%	779
$\phi\eta(1440) \rightarrow \phi\eta\pi\pi$	$< 2.5 \times 10^{-4}$	CL=90%	946
$\omega f_2'(1525)$	$< 2.2 \times 10^{-4}$	CL=90%	1003
$\Sigma(1385)^0\overline{\Lambda}$	$< 2 \times 10^{-4}$	CL=90%	911
$\Delta(1232)^+\overline{p}$	$< 1 \times 10^{-4}$	CL=90%	1100
$\Sigma^0\overline{\Lambda}$	$< 9 \times 10^{-5}$	CL=90%	1032
$\phi\pi^0$	$< 6.8 \times 10^{-6}$	CL=90%	1377
Decays into stable hadrons			
$2(\pi^+\pi^-)\pi^0$	(3.37 ± 0.26) %		1496
$3(\pi^+\pi^-)\pi^0$	(2.9 ± 0.6) %		1433
$\pi^+\pi^-\pi^0$	(1.50 ± 0.20) %		1533
$\pi^+\pi^-\pi^0 K^+ K^-$	(1.20 ± 0.30) %		1368
$4(\pi^+\pi^-)\pi^0$	$(9.0 \pm 3.0) \times 10^{-3}$		1345
$\pi^+\pi^- K^+ K^-$	$(7.2 \pm 2.3) \times 10^{-3}$		1407
$K\overline{K}\pi$	$(6.1 \pm 1.0) \times 10^{-3}$		1440
$p\overline{p}\pi^+\pi^-$	$(6.0 \pm 0.5) \times 10^{-3}$	S=1.3	1107
$2(\pi^+\pi^-)$	$(4.0 \pm 1.0) \times 10^{-3}$		1517
$3(\pi^+\pi^-)$	$(4.0 \pm 2.0) \times 10^{-3}$		1466
$n\overline{n}\pi^+\pi^-$	$(4 \pm 4) \times 10^{-3}$		1106
$\Sigma\overline{\Sigma}$	$(3.8 \pm 0.5) \times 10^{-3}$		992
$2(\pi^+\pi^-)K^+ K^-$	$(3.1 \pm 1.3) \times 10^{-3}$		1320
$p\overline{p}\pi^+\pi^-\pi^0$	$(2.3 \pm 0.9) \times 10^{-3}$ [hhh]	S=1.9	1033
$p\overline{p}$	$(2.14 \pm 0.10) \times 10^{-3}$		1232
$p\overline{p}\eta$	$(2.09 \pm 0.18) \times 10^{-3}$		948
$p\overline{n}\pi^-$	$(2.00 \pm 0.10) \times 10^{-3}$		1174
$n\overline{n}$	$(1.9 \pm 0.5) \times 10^{-3}$		1231
$\Xi\overline{\Xi}$	$(1.8 \pm 0.4) \times 10^{-3}$	S=1.8	818
$\Lambda\overline{\Lambda}$	$(1.35 \pm 0.14) \times 10^{-3}$	S=1.2	1074
$p\overline{p}\pi^0$	$(1.09 \pm 0.09) \times 10^{-3}$		1176
$\Lambda\overline{\Sigma}^-\pi^+$ (or c.c.)	$(1.06 \pm 0.12) \times 10^{-3}$ [gg]		945
$pK^-\overline{\Lambda}$	$(8.9 \pm 1.6) \times 10^{-4}$		876
$2(K^+ K^-)$	$(7.0 \pm 3.0) \times 10^{-4}$		1131
$pK^-\overline{\Sigma}^0$	$(2.9 \pm 0.8) \times 10^{-4}$		820
$K^+ K^-$	$(2.37 \pm 0.31) \times 10^{-4}$		1468
$\Lambda\overline{\Lambda}\pi^0$	$(2.2 \pm 0.7) \times 10^{-4}$		998
$\pi^+\pi^-$	$(1.47 \pm 0.23) \times 10^{-4}$		1542
$K_S^0 K_L^0$	$(1.08 \pm 0.14) \times 10^{-4}$		1466
$\Lambda\overline{\Sigma} +$ c.c.	$< 1.5 \times 10^{-4}$	CL=90%	1032
$K_S^0 K_S^0$	$< 5.2 \times 10^{-6}$	CL=90%	1466
Radiative decays			
$\gamma\eta_c(1S)$	(1.3 ± 0.4) %		116
$\gamma\pi^+\pi^- 2\pi^0$	$(8.3 \pm 3.1) \times 10^{-3}$		1518
$\gamma\eta\pi\pi$	$(6.1 \pm 1.0) \times 10^{-3}$		1487
$\gamma\eta(1440) \rightarrow \gamma K\overline{K}\pi$	$(9.1 \pm 1.8) \times 10^{-4}$ [p]		1223
$\gamma\eta(1440) \rightarrow \gamma\gamma\rho^0$	$(6.4 \pm 1.4) \times 10^{-5}$		1223
$\gamma\rho\rho$	$(4.5 \pm 0.8) \times 10^{-3}$		1343
$\gamma\eta'(958)$	$(4.31 \pm 0.30) \times 10^{-3}$		1400
$\gamma 2\pi^+ 2\pi^-$	$(2.8 \pm 0.5) \times 10^{-3}$	S=1.9	1517
$\gamma f_4(2050)$	$(2.7 \pm 0.7) \times 10^{-3}$		874
$\gamma\omega\omega$	$(1.59 \pm 0.33) \times 10^{-3}$		1337
$\gamma\eta(1440) \rightarrow \gamma\rho^0\rho^0$	$(1.7 \pm 0.4) \times 10^{-3}$	S=1.3	1223
$\gamma f_2(1270)$	$(1.38 \pm 0.14) \times 10^{-3}$		1286
$\gamma f_J(1710) \rightarrow \gamma K\overline{K}$	$(9.7 \pm 1.2) \times 10^{-4}$		1075
$\gamma\eta$	$(8.6 \pm 0.8) \times 10^{-4}$		1500

$\gamma f_1(1420) \to \gamma K\overline{K}\pi$	$(8.3 \pm 1.5) \times 10^{-4}$		1220
$\gamma f_1(1285)$	$(6.5 \pm 1.0) \times 10^{-4}$		1283
$\gamma f_2'(1525)$	$(6.3 \pm 1.0) \times 10^{-4}$		1173
$\gamma \phi\phi$	$(4.0 \pm 1.2) \times 10^{-4}$	S=2.1	1166
$\gamma p\overline{p}$	$(3.8 \pm 1.0) \times 10^{-4}$		1232
$\gamma\eta(2225)$	$(2.9 \pm 0.6) \times 10^{-4}$		834
$\gamma\eta(1760) \to \gamma\rho^0\rho^0$	$(1.3 \pm 0.9) \times 10^{-4}$		1048
$\gamma\pi^0$	$(3.9 \pm 1.3) \times 10^{-5}$		1546
$\gamma p\overline{p}\pi^+\pi^-$	$< 7.9 \times 10^{-4}$	CL=90%	1107
$\gamma\gamma$	$< 5 \times 10^{-4}$	CL=90%	1548
$\gamma\Lambda\overline{\Lambda}$	$< 1.3 \times 10^{-4}$	CL=90%	1074
3γ	$< 5.5 \times 10^{-5}$	CL=90%	1548
$\gamma f_0(1370)$	$(3.4 \pm 0.7) \times 10^{-4}$		–
$\gamma f_0(1500)$	$(8.2 \pm 1.5) \times 10^{-4}$		1184

$\chi_{c0}(1P)$ $I^G(J^{PC}) = 0^+(0^{++})$

Mass $m = 3415.1 \pm 1.0$ MeV
Full width $\Gamma = 14 \pm 5$ MeV

$\chi_{c0}(1P)$ DECAY MODES	Fraction (Γ_i/Γ)	Confidence level	p (MeV/c)
Hadronic decays			
$2(\pi^+\pi^-)$	(3.7 ± 0.7) %		1679
$\pi^+\pi^- K^+ K^-$	(3.0 ± 0.7) %		1580
$\rho^0\pi^+\pi^-$	(1.6 ± 0.5) %		1608
$3(\pi^+\pi^-)$	(1.5 ± 0.5) %		1633
$K^+\overline{K}^*(892)^0\pi^- + $ c.c.	(1.2 ± 0.4) %		1522
$\pi^+\pi^-$	$(7.5 \pm 2.1) \times 10^{-3}$		1702
K^+K^-	$(7.1 \pm 2.4) \times 10^{-3}$		1635
$\pi^+\pi^- p\overline{p}$	$(5.0 \pm 2.0) \times 10^{-3}$		1320
$\pi^0\pi^0$	$(3.1 \pm 0.6) \times 10^{-3}$		1702
$\eta\eta$	$(2.5 \pm 1.1) \times 10^{-3}$		1617
$p\overline{p}$	$< 9.0 \times 10^{-4}$	90%	1427
Radiative decays			
$\gamma J/\psi(1S)$	$(6.6 \pm 1.8) \times 10^{-3}$		303
$\gamma\gamma$	$(4.0 \pm 2.3) \times 10^{-4}$		1708

$\chi_{c1}(1P)$ $I^G(J^{PC}) = 0^+(1^{++})$

Mass $m = 3510.53 \pm 0.12$ MeV
Full width $\Gamma = 0.88 \pm 0.14$ MeV

$\chi_{c1}(1P)$ DECAY MODES	Fraction (Γ_i/Γ)		p (MeV/c)
Hadronic decays			
$3(\pi^+\pi^-)$	(2.2 ± 0.8) %		1683
$2(\pi^+\pi^-)$	(1.6 ± 0.5) %		1727
$\pi^+\pi^- K^+ K^-$	$(9 \pm 4) \times 10^{-3}$		1632
$\rho^0\pi^+\pi^-$	$(3.9 \pm 3.5) \times 10^{-3}$		1659
$K^+\overline{K}^*(892)^0\pi^- + $ c.c.	$(3.2 \pm 2.1) \times 10^{-3}$		1576
$\pi^+\pi^- p\overline{p}$	$(1.4 \pm 0.9) \times 10^{-3}$		1381
$p\overline{p}$	$(8.6 \pm 1.2) \times 10^{-5}$		1483
$\pi^+\pi^- + K^+ K^-$	$< 2.1 \times 10^{-3}$		–
Radiative decays			
$\gamma J/\psi(1S)$	(27.3 ± 1.6) %		389

$\chi_{c2}(1P)$ $I^G(J^{PC}) = 0^+(2^{++})$

Mass $m = 3556.17 \pm 0.13$ MeV
Full width $\Gamma = 2.00 \pm 0.18$ MeV

$\chi_{c2}(1P)$ DECAY MODES	Fraction (Γ_i/Γ)	Confidence level	p (MeV/c)
Hadronic decays			
$2(\pi^+\pi^-)$	(2.2 ± 0.5) %		1751
$\pi^+\pi^- K^+ K^-$	(1.9 ± 0.5) %		1656
$3(\pi^+\pi^-)$	(1.2 ± 0.8) %		1707
$\rho^0\pi^+\pi^-$	$(7 \pm 4) \times 10^{-3}$		1683
$K^+\overline{K}^*(892)^0\pi^- + $ c.c.	$(4.8 \pm 2.8) \times 10^{-3}$		1601
$\pi^+\pi^- p\overline{p}$	$(3.3 \pm 1.3) \times 10^{-3}$		1410
$\pi^+\pi^-$	$(1.9 \pm 1.0) \times 10^{-3}$		1773
K^+K^-	$(1.5 \pm 1.1) \times 10^{-3}$		1708
$p\overline{p}$	$(10.0 \pm 1.0) \times 10^{-5}$		1510
$\pi^0\pi^0$	$(1.10 \pm 0.28) \times 10^{-3}$		1773
$\eta\eta$	$(8 \pm 5) \times 10^{-4}$		1692
$J/\psi(1S)\pi^+\pi^-\pi^0$	< 1.5 %	90%	185

Radiative decays			
$\gamma J/\psi(1S)$	(13.5 ± 1.1) %		430
$\gamma\gamma$	$(1.6 \pm 0.5) \times 10^{-4}$		1778

$\psi(2S)$ $I^G(J^{PC}) = 0^-(1^{--})$

Mass $m = 3686.00 \pm 0.09$ MeV
Full width $\Gamma = 277 \pm 31$ keV   (S = 1.1)
$\Gamma_{ee} = 2.14 \pm 0.21$ keV   (Assuming $\Gamma_{ee} = \Gamma_{\mu\mu}$)

$\psi(2S)$ DECAY MODES	Fraction (Γ_i/Γ)	Scale factor/ Confidence level	p (MeV/c)
hadrons	(98.10 ± 0.30) %		–
virtual-$\gamma \to$ hadrons	(2.9 ± 0.4) %		–
$e^+ e^-$	$(8.8 \pm 1.3) \times 10^{-3}$		1843
$\mu^+\mu^-$	$(7.7 \pm 1.7) \times 10^{-3}$		1840
Decays into $J/\psi(1S)$ and anything			
$J/\psi(1S)$ anything	(57 ± 4) %		–
$J/\psi(1S)$ neutrals	(23.2 ± 2.6) %		–
$J/\psi(1S)\pi^+\pi^-$	(32.4 ± 2.6) %		477
$J/\psi(1S)\pi^0\pi^0$	(18.4 ± 2.7) %		481
$J/\psi(1S)\eta$	(2.7 ± 0.4) %	S=1.7	200
$J/\psi(1S)\pi^0$	$(9.7 \pm 2.1) \times 10^{-4}$		527
Hadronic decays			
$3(\pi^+\pi^-)\pi^0$	$(3.5 \pm 1.6) \times 10^{-3}$		1746
$2(\pi^+\pi^-)\pi^0$	$(3.1 \pm 0.7) \times 10^{-3}$		1799
$\pi^+\pi^- K^+ K^-$	$(1.6 \pm 0.4) \times 10^{-3}$		1726
$\pi^+\pi^- p\overline{p}$	$(8.0 \pm 2.0) \times 10^{-4}$		1491
$K^+\overline{K}^*(892)^0\pi^- + $ c.c.	$(6.7 \pm 2.5) \times 10^{-4}$		1673
$2(\pi^+\pi^-)$	$(4.5 \pm 1.0) \times 10^{-4}$		1817
$\rho^0\pi^+\pi^-$	$(4.2 \pm 1.5) \times 10^{-4}$		1751
$\overline{p}p$	$(1.9 \pm 0.5) \times 10^{-4}$		1586
$3(\pi^+\pi^-)$	$(1.5 \pm 1.0) \times 10^{-4}$		1774
$\overline{p}p\pi^0$	$(1.4 \pm 0.5) \times 10^{-4}$		1543
K^+K^-	$(1.0 \pm 0.7) \times 10^{-4}$		1776
$\pi^+\pi^-\pi^0$	$(9 \pm 5) \times 10^{-5}$		1830
$\pi^+\pi^-$	$(8 \pm 5) \times 10^{-5}$		1838
$\Lambda\overline{\Lambda}$	$< 4 \times 10^{-4}$	CL=90%	1467
$\Xi^- \overline{\Xi}{}^+$	$< 2 \times 10^{-4}$	CL=90%	1285
$\rho\pi$	$< 8.3 \times 10^{-5}$	CL=90%	1760
$K^+K^-\pi^0$	$< 2.96 \times 10^{-5}$	CL=90%	1754
$K^+\overline{K}^*(892)^- + $ c.c.	$< 5.4 \times 10^{-5}$	CL=90%	1698
Radiative decays			
$\gamma\chi_{c0}(1P)$	(9.3 ± 0.8) %		261
$\gamma\chi_{c1}(1P)$	(8.7 ± 0.8) %		171
$\gamma\chi_{c2}(1P)$	(7.8 ± 0.8) %		127
$\gamma\eta_c(1S)$	$(2.8 \pm 0.6) \times 10^{-3}$		639
$\gamma\pi^0$	$< 5.4 \times 10^{-3}$	CL=95%	1841
$\gamma\eta'(958)$	$< 1.1 \times 10^{-4}$	CL=90%	1719
$\gamma\gamma$	$< 1.6 \times 10^{-4}$	CL=90%	1843
$\gamma\eta(1440) \to \gamma K\overline{K}\pi$	$< 1.2 \times 10^{-4}$	CL=90%	1569

$\psi(3770)$ $I^G(J^{PC}) = ?^?(1^{--})$

Mass $m = 3769.9 \pm 2.5$ MeV   (S = 1.8)
Full width $\Gamma = 23.6 \pm 2.7$ MeV   (S = 1.1)
$\Gamma_{ee} = 0.26 \pm 0.04$ keV   (S = 1.2)

$\psi(3770)$ DECAY MODES	Fraction (Γ_i/Γ)	Scale factor	p (MeV/c)
$D\overline{D}$	dominant		242
$e^+ e^-$	$(1.12 \pm 0.17) \times 10^{-5}$	1.2	1885

$\psi(4040)$ [lll] $I^G(J^{PC}) = ?^?(1^{--})$

Mass $m = 4040 \pm 10$ MeV
Full width $\Gamma = 52 \pm 10$ MeV
$\Gamma_{ee} = 0.75 \pm 0.15$ keV

$\psi(4040)$ DECAY MODES	Fraction (Γ_i/Γ)		p (MeV/c)
$e^+ e^-$	$(1.4 \pm 0.4) \times 10^{-5}$		2020
$D^0\overline{D}{}^0$	seen		777
$D^*(2007)^0\overline{D}{}^0 + $ c.c.	seen		578
$D^*(2007)^0\overline{D}{}^*(2007)^0$	seen		232

$\psi(4160)$ [IlI]

$I^G(J^{PC}) = ?^?(1^{--})$

Mass $m = 4159 \pm 20$ MeV
Full width $\Gamma = 78 \pm 20$ MeV
$\Gamma_{ee} = 0.77 \pm 0.23$ keV

$\psi(4160)$ DECAY MODES	Fraction (Γ_i/Γ)	p (MeV/c)
e^+e^-	$(10\pm4) \times 10^{-6}$	2079

$\psi(4415)$ [IlI]

$I^G(J^{PC}) = ?^?(1^{--})$

Mass $m = 4415 \pm 6$ MeV
Full width $\Gamma = 43 \pm 15$ MeV (S = 1.8)
$\Gamma_{ee} = 0.47 \pm 0.10$ keV

$\psi(4415)$ DECAY MODES	Fraction (Γ_i/Γ)	p (MeV/c)
hadrons	dominant	–
e^+e^-	$(1.1\pm0.4) \times 10^{-5}$	2207

$b\bar{b}$ MESONS

$\Upsilon(1S)$

$I^G(J^{PC}) = 0^-(1^{--})$

Mass $m = 9460.37 \pm 0.21$ MeV (S = 2.7)
Full width $\Gamma = 52.5 \pm 1.8$ keV
$\Gamma_{ee} = 1.32 \pm 0.05$ keV

$\Upsilon(1S)$ DECAY MODES	Fraction (Γ_i/Γ)	Scale factor/ Confidence level	p (MeV/c)
$\tau^+\tau^-$	$(2.67^{+0.14}_{-0.16})$ %		4384
e^+e^-	(2.52 ± 0.17) %		4730
$\mu^+\mu^-$	(2.48 ± 0.07) %	S=1.1	4729
Hadronic decays			
$J/\psi(1S)$ anything	$(1.1\pm0.4) \times 10^{-3}$		4223
$\rho\pi$	< 2 $\times 10^{-4}$	CL=90%	4698
$\pi^+\pi^-$	< 5 $\times 10^{-4}$	CL=90%	4728
K^+K^-	< 5 $\times 10^{-4}$	CL=90%	4704
$p\bar{p}$	< 5 $\times 10^{-4}$	CL=90%	4636
Radiative decays			
$\gamma 2h^+ 2h^-$	$(7.0\pm1.5) \times 10^{-4}$		4720
$\gamma 3h^+ 3h^-$	$(5.4\pm2.0) \times 10^{-4}$		4703
$\gamma 4h^+ 4h^-$	$(7.4\pm3.5) \times 10^{-4}$		4679
$\gamma \pi^+\pi^- K^+K^-$	$(2.9\pm0.9) \times 10^{-4}$		4686
$\gamma 2\pi^+ 2\pi^-$	$(2.5\pm0.9) \times 10^{-4}$		4720
$\gamma 3\pi^+ 3\pi^-$	$(2.5\pm1.2) \times 10^{-4}$		4703
$\gamma 2\pi^+ 2\pi^- K^+K^-$	$(2.4\pm1.2) \times 10^{-4}$		4658
$\gamma \pi^+\pi^- p\bar{p}$	$(1.5\pm0.6) \times 10^{-4}$		4604
$\gamma 2\pi^+ 2\pi^- p\bar{p}$	$(4\pm6) \times 10^{-5}$		4563
$\gamma 2K^+ 2K^-$	$(2.0\pm2.0) \times 10^{-5}$		4601
$\gamma \eta'(958)$	< 1.3 $\times 10^{-3}$	CL=90%	4682
$\gamma \eta$	< 3.5 $\times 10^{-4}$	CL=90%	4714
$\gamma f_2'(1525)$	< 1.4 $\times 10^{-4}$	CL=90%	4607
$\gamma f_2(1270)$	< 1.3 $\times 10^{-4}$	CL=90%	4644
$\gamma \eta(1440)$	< 8.2 $\times 10^{-5}$	CL=90%	4624
$\gamma f_J(1710) \to \gamma K\bar{K}$	< 2.6 $\times 10^{-4}$	CL=90%	4576
$\gamma f_0(2200) \to \gamma K^+K^-$	< 2 $\times 10^{-4}$	CL=90%	4475
$\gamma f_J(2220) \to \gamma K^+K^-$	< 1.5 $\times 10^{-5}$	CL=90%	4469
$\gamma \eta(2225) \to \gamma \phi\phi$	< 3 $\times 10^{-3}$	CL=90%	4469
γX	< 3 $\times 10^{-5}$	CL=90%	–
X = pseudoscalar with $m< 7.2$ GeV)			
$\gamma X\bar{X}$	< 1 $\times 10^{-3}$	CL=90%	–
$X\bar{X}$ = vectors with $m< 3.1$ GeV)			

$\chi_{b0}(1P)$ [IlI]

$I^G(J^{PC}) = 0^+(0^{++})$
J needs confirmation.

Mass $m = 9859.8 \pm 1.3$ MeV

$\chi_{b0}(1P)$ DECAY MODES	Fraction (Γ_i/Γ)	Confidence level	p (MeV/c)
$\gamma \Upsilon(1S)$	<6 %	90%	391

$\chi_{b1}(1P)$ [IlI]

$I^G(J^{PC}) = 0^+(1^{++})$
J needs confirmation.

Mass $m = 9891.9 \pm 0.7$ MeV

$\chi_{b1}(1P)$ DECAY MODES	Fraction (Γ_i/Γ)	p (MeV/c)
$\gamma \Upsilon(1S)$	(35 ± 8) %	422

$\chi_{b2}(1P)$ [IlI]

$I^G(J^{PC}) = 0^+(2^{++})$
J needs confirmation.

Mass $m = 9913.2 \pm 0.6$ MeV

$\chi_{b2}(1P)$ DECAY MODES	Fraction (Γ_i/Γ)	p (MeV/c)
$\gamma \Upsilon(1S)$	(22 ± 4) %	443

$\Upsilon(2S)$

$I^G(J^{PC}) = 0^-(1^{--})$

Mass $m = 10.02330 \pm 0.00031$ GeV
Full width $\Gamma = 44 \pm 7$ keV
$\Gamma_{ee} = 0.52 \pm 0.03$ keV

$\Upsilon(2S)$ DECAY MODES	Fraction (Γ_i/Γ)	Confidence level	p (MeV/c)
$\Upsilon(1S)\pi^+\pi^-$	(18.5 ± 0.8) %		475
$\Upsilon(1S)\pi^0\pi^0$	(8.8 ± 1.1) %		480
$\tau^+\tau^-$	(1.7 ± 1.6) %		4686
$\mu^+\mu^-$	(1.31 ± 0.21) %		5011
e^+e^-	seen		5012
$\Upsilon(1S)\pi^0$	< 8 $\times 10^{-3}$	90%	531
$\Upsilon(1S)\eta$	< 2 $\times 10^{-3}$	90%	127
$J/\psi(1S)$ anything	< 6 $\times 10^{-3}$	90%	4533
Radiative decays			
$\gamma \chi_{b1}(1P)$	(6.7 ± 0.9) %		131
$\gamma \chi_{b2}(1P)$	(6.6 ± 0.9) %		110
$\gamma \chi_{b0}(1P)$	(4.3 ± 1.0) %		162
$\gamma f_J(1710)$	< 5.9 $\times 10^{-4}$	90%	4866
$\gamma f_2'(1525)$	< 5.3 $\times 10^{-4}$	90%	4896
$\gamma f_2(1270)$	< 2.41 $\times 10^{-4}$	90%	4931

$\chi_{b0}(2P)$ [IlI]

$I^G(J^{PC}) = 0^+(0^{++})$
J needs confirmation.

Mass $m = 10.2321 \pm 0.0006$ GeV

$\chi_{b0}(2P)$ DECAY MODES	Fraction (Γ_i/Γ)	p (MeV/c)
$\gamma \Upsilon(2S)$	(4.6 ± 2.1) %	210
$\gamma \Upsilon(1S)$	$(9\pm6) \times 10^{-3}$	746

$\chi_{b1}(2P)$ [IlI]

$I^G(J^{PC}) = 0^+(1^{++})$
J needs confirmation.

Mass $m = 10.2552 \pm 0.0005$ GeV
$m_{\chi_{b1}(2P)} - m_{\chi_{b0}(2P)} = 23.5 \pm 1.0$ MeV

$\chi_{b1}(2P)$ DECAY MODES	Fraction (Γ_i/Γ)	Scale factor	p (MeV/c)
$\gamma \Upsilon(2S)$	(21 ± 4) %	1.5	229
$\gamma \Upsilon(1S)$	(8.5 ± 1.3) %	1.3	764

$\chi_{b2}(2P)$ [IlI]

$I^G(J^{PC}) = 0^+(2^{++})$
J needs confirmation.

Mass $m = 10.2685 \pm 0.0004$ GeV
$m_{\chi_{b2}(2P)} - m_{\chi_{b1}(2P)} = 13.5 \pm 0.6$ MeV

$\chi_{b2}(2P)$ DECAY MODES	Fraction (Γ_i/Γ)	p (MeV/c)
$\gamma \Upsilon(2S)$	(16.2 ± 2.4) %	242
$\gamma \Upsilon(1S)$	(7.1 ± 1.0) %	776

Meson Summary Table

$\Upsilon(3S)$ $I^G(J^{PC}) = 0^-(1^{--})$

Mass $m = 10.3553 \pm 0.0005$ GeV
Full width $\Gamma = 26.3 \pm 3.5$ keV

$\Upsilon(3S)$ DECAY MODES	Fraction (Γ_i/Γ)	Scale factor	p (MeV/c)
$\Upsilon(2S)$ anything	(10.6 ± 0.8) %		296
$\Upsilon(2S)\pi^+\pi^-$	(2.8 ± 0.6) %	2.2	177
$\Upsilon(2S)\pi^0\pi^0$	(2.00 ± 0.32) %		190
$\Upsilon(2S)\gamma\gamma$	(5.0 ± 0.7) %		327
$\Upsilon(1S)\pi^+\pi^-$	(4.48 ± 0.21) %		814
$\Upsilon(1S)\pi^0\pi^0$	(2.06 ± 0.28) %		816
$\mu^+\mu^-$	(1.81 ± 0.17) %		5177
e^+e^-	seen		5177

Radiative decays

$\gamma\chi_{b2}(2P)$	(11.4 ± 0.8) %	1.3	87
$\gamma\chi_{b1}(2P)$	(11.3 ± 0.6) %		100
$\gamma\chi_{b0}(2P)$	(5.4 ± 0.6) %	1.1	123

$\Upsilon(4S)$ or $\Upsilon(10580)$ $I^G(J^{PC}) = ?^?(1^{--})$

Mass $m = 10.5800 \pm 0.0035$ GeV
Full width $\Gamma = 21 \pm 4$ MeV $(S = 2.3)$
$\Gamma_{ee} = 0.248 \pm 0.031$ keV $(S = 1.3)$

$\Upsilon(4S)$ DECAY MODES	Fraction (Γ_i/Γ)	Confidence level	p (MeV/c)
$B\overline{B}$	dominant		–
e^+e^-	$(2.8 \pm 0.7) \times 10^{-5}$		5290
$J/\psi(3097)$ anything	$(2.2 \pm 0.7) \times 10^{-3}$		
D^{*+} anything + c.c.	< 7.4 %	90%	5099
ϕ anything	$< 2.3 \times 10^{-3}$	90%	5240
$\Upsilon(1S)$ anything	$< 4 \times 10^{-3}$	90%	1053
non-$B\overline{B}$	< 4 %	95%	–

$\Upsilon(10860)$ $I^G(J^{PC}) = ?^?(1^{--})$

Mass $m = 10.865 \pm 0.008$ GeV $(S = 1.1)$
Full width $\Gamma = 110 \pm 13$ MeV
$\Gamma_{ee} = 0.31 \pm 0.07$ keV $(S = 1.3)$

$\Upsilon(10860)$ DECAY MODES	Fraction (Γ_i/Γ)	p (MeV/c)
e^+e^-	$(2.8 \pm 0.7) \times 10^{-6}$	5432

$\Upsilon(11020)$ $I^G(J^{PC}) = ?^?(1^{--})$

Mass $m = 11.019 \pm 0.008$ GeV
Full width $\Gamma = 79 \pm 16$ MeV
$\Gamma_{ee} = 0.130 \pm 0.030$ keV

$\Upsilon(11020)$ DECAY MODES	Fraction (Γ_i/Γ)	p (MeV/c)
e^+e^-	$(1.6 \pm 0.5) \times 10^{-6}$	5509

NOTES

In this Summary Table:

When a quantity has "(S = ...)" to its right, the error on the quantity has been enlarged by the "scale factor" S, defined as $S = \sqrt{\chi^2/(N-1)}$, where N is the number of measurements used in calculating the quantity. We do this when $S > 1$, which often indicates that the measurements are inconsistent. When $S > 1.25$, we also show in the Particle Listings an ideogram of the measurements. For more about S, see the Introduction.

A decay momentum p is given for each decay mode. For a 2-body decay, p is the momentum of each decay product in the rest frame of the decaying particle. For a 3-or-more-body decay, p is the largest momentum any of the products can have in this frame.

[a] See the "Note on $\pi^\pm \rightarrow \ell^\pm \nu \gamma$ and $K^\pm \rightarrow \ell^\pm \nu \gamma$ Form Factors" in the $\pi^\pm$ Particle Listings for definitions and details.

[b] Measurements of $\Gamma(e^+\nu_e)/\Gamma(\mu^+\nu_\mu)$ always include decays with γ's, and measurements of $\Gamma(e^+\nu_e\gamma)$ and $\Gamma(\mu^+\nu_\mu\gamma)$ never include low-energy γ's. Therefore, since no clean separation is possible, we consider the modes with γ's to be subreactions of the modes without them, and let $[\Gamma(e^+\nu_e) + \Gamma(\mu^+\nu_\mu)]/\Gamma_{total} = 100\%$.

[c] See the $\pi^\pm$ Particle Listings for the energy limits used in this measurement; low-energy γ's are not included.

[d] Derived from an analysis of neutrino-oscillation experiments.

[e] Astrophysical and cosmological arguments give limits of order 10^{-13}; see the π^0 Particle Listings.

[f] See the "Note on the Decay Width $\Gamma(\eta \rightarrow \gamma\gamma)$" in our 1994 edition, Phys. Rev. **D50**, 1 August 1994, Part I, p. 1451.

[g] See the "Note on η Decay Parameters" in the η Particle Listings.

[h] C parity forbids this to occur as a single-photon process.

[i] See the "Note on scalar mesons" in the $f_0(1370)$ Particle Listings.

[j] See the "Note on $\rho(770)$" in the $\rho(770)$ Particle Listings.

[k] The e^+e^- branching fraction is from $e^+e^- \rightarrow \pi^+\pi^-$ experiments only. The $\omega\rho$ interference is then due to $\omega\rho$ mixing only, and is expected to be small. If $e\mu$ universality holds, $\Gamma(\rho^0 \rightarrow \mu^+\mu^-) = \Gamma(\rho^0 \rightarrow e^+e^-) \times 0.99785$.

[l] This is only an educated guess; the error given is larger than the error on the average of the published values. See the Particle Listings for details.

[m] See the "Note on $a_1(1260)$" in the $a_1(1260)$ Particle Listings.

[n] See the "Note on the $f_1(1420)$" in the $f_1(1420)$ Particle Listings.

[o] See also the $\omega(1600)$ Particle Listings.

[p] See the "Note on the $\eta(1440)$" in the $\eta(1440)$ Particle Listings.

[q] See the "Note on the $\rho(1450)$ and the $\rho(1700)$" in the $\rho(1700)$ Particle Listings.

[r] See the "Note on non-$q\overline{q}$ mesons" in the Particle Listings (see the index for the page number).

[s] See also the $\omega(1420)$ Particle Listings.

[t] See the "Note on $f_J(1710)$" in the $f_J(1710)$ Particle Listings.

[u] See the note in the $K^\pm$ Particle Listings.

[v] The definition of the slope parameter g of the $K \rightarrow 3\pi$ Dalitz plot is as follows (see also "Note on Dalitz Plot Parameters for $K \rightarrow 3\pi$ Decays" in the $K^\pm$ Particle Listings):
$$|M|^2 = 1 + g(s_3 - s_0)/m_{\pi^+}^2 + \cdots.$$

[w] For more details and definitions of parameters see the Particle Listings.

[x] See the $K^\pm$ Particle Listings for the energy limits used in this measurement.

[y] Most of this radiative mode, the low-momentum γ part, is also included in the parent mode listed without γ's.

[z] Direct-emission branching fraction.

[aa] Structure-dependent part.

[bb] Derived from measured values of ϕ_{+-}, ϕ_{00}, $|\eta|$, $\tau_{K_S^0}$, and $|m_{K_L^0} - m_{K_S^0}|$, as described in the introduction to "Tests of Conservation Laws."

[cc] The *CP*-violation parameters are defined as follows (see also "Note on *CP* Violation in $K_S \to 3\pi$" and "Note on *CP* Violation in K_L^0 Decay" in the Particle Listings):

$$\eta_{+-} = |\eta_{+-}|e^{i\phi_{+-}} = \frac{A(K_L^0 \to \pi^+\pi^-)}{A(K_S^0 \to \pi^+\pi^-)} = \epsilon + \epsilon'$$

$$\eta_{00} = |\eta_{00}|e^{i\phi_{00}} = \frac{A(K_L^0 \to \pi^0\pi^0)}{A(K_S^0 \to \pi^0\pi^0)} = \epsilon - 2\epsilon'$$

$$\delta = \frac{\Gamma(K_L^0 \to \pi^-\ell^+\nu) - \Gamma(K_L^0 \to \pi^+\ell^-\nu)}{\Gamma(K_L^0 \to \pi^-\ell^+\nu) + \Gamma(K_L^0 \to \pi^+\ell^-\nu)},$$

$$\mathrm{Im}(\eta_{+-0})^2 = \frac{\Gamma(K_S^0 \to \pi^+\pi^-\pi^0)^{CP\ \mathrm{viol.}}}{\Gamma(K_L^0 \to \pi^+\pi^-\pi^0)},$$

$$\mathrm{Im}(\eta_{000})^2 = \frac{\Gamma(K_S^0 \to \pi^0\pi^0\pi^0)}{\Gamma(K_L^0 \to \pi^0\pi^0\pi^0)}.$$

where for the last two relations *CPT* is assumed valid, *i.e.*, $\mathrm{Re}(\eta_{+-0}) \simeq 0$ and $\mathrm{Re}(\eta_{000}) \simeq 0$.

[dd] See the K_S^0 Particle Listings for the energy limits used in this measurement.

[ee] Calculated from K_L^0 semileptonic rates and the K_S^0 lifetime assuming $\Delta S = \Delta Q$.

[ff] ϵ'/ϵ is derived from $|\eta_{00}/\eta_{+-}|$ measurements using theoretical input on phases.

[gg] The value is for the sum of the charge states of particle/antiparticle states indicated.

[hh] See the K_L^0 Particle Listings for the energy limits used in this measurement.

[ii] $m_{e^+e^-} > 470$ MeV.

[jj] Allowed by higher-order electroweak interactions.

[kk] Violates *CP* in leading order. Test of direct *CP* violation since the indirect *CP*-violating and *CP*-conserving contributions are expected to be suppressed.

[ll] See the "Note on $f_0(1370)$" in the $f_0(1370)$ Particle Listings and in the 1994 edition.

[mm] See the note in the $L(1770)$ Particle Listings in Reviews of Modern Physics **56** No. 2 Pt. II (1984), p. S200. See also the "Note on $K_2(1770)$ and the $K_2(1820)$" in the $K_2(1770)$ Particle Listings.

[nn] See the "Note on $K_2(1770)$ and the $K_2(1820)$" in the $K_2(1770)$ Particle Listings.

[oo] This is a weighted average of $D^\pm$ (44%) and D^0 (56%) branching fractions. See "D^+ and $D^0 \to (\eta\,\mathrm{anything}) / (\mathrm{total}\ D^+\ \mathrm{and}\ D^0)$" under "$D^+$ Branching Ratios" in the Particle Listings.

[pp] This value averages the e^+ and μ^+ branching fractions, after making a small phase-space adjustment to the μ^+ fraction to be able to use it as an e^+ fraction; hence our ℓ^+ is really an e^+.

[qq] ℓ indicates e or μ mode, not sum over modes.

[rr] The branching fractions for this mode may differ from the sum of the submodes that contribute to it, due to interference effects. See the relevant papers in the Particle Listings.

[ss] The two experiments determining this ratio are in serious disagreement. See the Particle Listings.

[tt] This mode is not a useful test for a $\Delta C=1$ weak neutral current because both quarks must change flavor in this decay.

[uu] The D_1^0-D_2^0 limits are inferred from the D^0-$\overline{D}^0$ mixing ratio $\Gamma(K^+\pi^-$ or $K^+\pi^-\pi^+\pi^-$ via $\overline{D}^0) / \Gamma(K^-\pi^+$ or $K^-\pi^+\pi^+\pi^-)$.

[vv] This value is calculated from the ratio $\Gamma(K^-\mu^+\nu_\mu)/\Gamma(\mu^+\,\mathrm{anything})$ in the D^0 Particle Listings.

[ww] The experiments on the division of this charge mode amongst its submodes disagree, and the submode branching fractions here add up to considerably more than the charged-mode fraction.

[xx] For now, we average together measurements of the $\phi e^+\nu_e$ and $\phi\mu^+\nu_\mu$ branching fractions. This is the *average*, not the *sum*.

[yy] This branching fraction is calculated from appropriate fractions of the next three branching fractions.

[zz] This value includes only K^+K^- decays of the $f_J(1710)$, because branching fractions of this resonance are not known.

[aaa] This mode is not a useful test for a $\Delta C=1$ weak neutral current because both quarks must change flavor in this decay.

[bbb] B^0 and B_s^0 contributions not separated. Limit is on weighted average of the two decay rates.

[ccc] These values are model dependent. See 'Note on Semileptonic Decays' in the B^+ Particle Listings.

[ddd] D^{**} stands for the sum of the $D(1\,^1P_1)$, $D(1\,^3P_0)$, $D(1\,^3P_1)$, $D(1\,^3P_2)$, $D(2\,^1S_0)$, and $D(2\,^1S_1)$ resonances.

[eee] Inclusive branching fractions have a multiplicity definition and can be greater than 100%.

[fff] D_J represents an unresolved mixture of pseudoscalar and tensor D^{**} (P-wave) states.

[ggg] Not a pure measurement. See note at head of B_s^0 Decay Modes.

[hhh] Includes $p\overline{p}\pi^+\pi^-\gamma$ and excludes $p\overline{p}\eta$, $p\overline{p}\omega$, $p\overline{p}\eta'$.

[iii] J^{PC} known by production in e^+e^- via single photon annihilation. I^G is not known; interpretation of this state as a single resonance is unclear because of the expectation of substantial threshold effects in this energy region.

[jjj] Spectroscopic labeling for these states is theoretical, pending experimental information.

Meson Summary Table

See also the table of suggested $q\bar{q}$ quark-model assignments in the Quark Model section.

• Indicates particles that appear in the preceding Meson Summary Table. We do not regard the other entries as being established.

† Indicates that the value of J given is preferred, but needs confirmation.

LIGHT UNFLAVORED ($S = C = B = 0$)

Particle	$I^G(J^{PC})$	Particle	$I^G(J^{PC})$
• $\pi^\pm$	$1^-(0^-)$	$f_2(1640)$	$0^+(2^{++})$
• π^0	$1^-(0^{-+})$	• $\omega_3(1670)$	$0^-(3^{--})$
• η	$0^+(0^{-+})$	• $\pi_2(1670)$	$1^-(2^{-+})$
• $f_0(400–1200)$	$0^+(0^{++})$	• $\phi(1680)$	$0^-(1^{--})$
• $\rho(770)$	$1^+(1^{--})$	• $\rho_3(1690)$	$1^+(3^{--})$
• $\omega(782)$	$0^-(1^{--})$	• $\rho(1700)$	$1^+(1^{--})$
• $\eta'(958)$	$0^+(0^{-+})$	• $f_J(1710)$	$0^+(\text{even}^{++})$
• $f_0(980)$	$0^+(0^{++})$	$X(1740)$	$0^+(\text{even}^{++})$
• $a_0(980)$	$1^-(0^{++})$	$\eta(1760)$	$0^+(0^{-+})$
• $\phi(1020)$	$0^-(1^{--})$	$\pi(1800)$	$1^-(0^{-+})$
• $h_1(1170)$	$0^-(1^{+-})$	$X(1775)$	$1^-(?^{-+})$
• $b_1(1235)$	$1^+(1^{+-})$	$f_2(1810)$	$0^+(2^{++})$
• $a_1(1260)$	$1^-(1^{++})$	• $\phi_3(1850)$	$0^-(3^{--})$
• $f_2(1270)$	$0^+(2^{++})$	$\eta_2(1870)$	$0^+(2^{-+})$
• $f_1(1285)$	$0^+(1^{++})$	$X(1910)$	$0^+(?^+)$
• $\eta(1295)$	$0^+(0^{-+})$	$f_2(1950)$	$0^+(2^{++})$
• $\pi(1300)$	$1^-(0^{-+})$	$X(2000)$	$1^-(?^+)$
• $a_2(1320)$	$1^-(2^{++})$	• $f_2(2010)$	$0^+(2^{++})$
• $f_0(1370)$	$0^+(0^{++})$	$a_4(2040)$	$1^-(4^{++})$
$h_1(1380)$	$?^-(1^{+?})$	• $f_4(2050)$	$0^+(4^{++})$
$\hat{\rho}(1405)$	$1^-(1^{-+})$	$\pi_2(2100)$	$1^-(2^{-+})$
• $f_1(1420)$	$0^+(1^{++})$	$f_2(2150)$	$0^+(2^{++})$
• $\omega(1420)$	$0^-(1^{--})$	$\rho(2150)$	$1^+(1^{--})$
$f_2(1430)$	$0^+(2^{++})$	$f_0(2200)$	$0^+(0^{++})$
• $\eta(1440)$	$0^+(0^{-+})$	$f_J(2220)$	$0^+(2^{++}$ or $4^{++})$
$a_0(1450)$	$1^-(0^{++})$	$\eta(2225)$	$0^+(0^{-+})$
• $\rho(1450)$	$1^+(1^{--})$	$\rho_3(2250)$	$1^+(3^{--})$
• $f_0(1500)$	$0^+(0^{++})$	• $f_2(2300)$	$0^+(2^{++})$
• $f_1(1510)$	$0^+(1^{++})$	$f_4(2300)$	$0^+(4^{++})$
• $f'_2(1525)$	$0^+(2^{++})$	• $f_2(2340)$	$0^+(2^{++})$
$f_2(1565)$	$0^+(2^{++})$	$\rho_5(2350)$	$1^+(5^{--})$
• $\omega(1600)$	$0^-(1^{--})$	$a_6(2450)$	$1^-(6^{++})$
$X(1600)$	$2^+(2^{++})$	$f_6(2510)$	$0^+(6^{++})$
		$X(3250)$	$?^?(?^{??})$

OTHER LIGHT UNFLAVORED ($S = C = B = 0$)

Particle	$I^G(J^{PC})$
$e^+e^-(1100–2200)$	$?^?(1^{--})$
$\overline{N}N(1100–3600)$	
$X(1900–3600)$	

STRANGE ($S = \pm1, C = B = 0$)

Particle	$I(J^P)$
• $K^\pm$	$1/2(0^-)$
• K^0	$1/2(0^-)$
• K^0_S	$1/2(0^-)$
• K^0_L	$1/2(0^-)$
• $K^*(892)$	$1/2(1^-)$
• $K_1(1270)$	$1/2(1^+)$
• $K_1(1400)$	$1/2(1^+)$
• $K^*(1410)$	$1/2(1^-)$
• $K^*_0(1430)$	$1/2(0^+)$
• $K^*_2(1430)$	$1/2(2^+)$
$K(1460)$	$1/2(0^-)$
$K_2(1580)$	$1/2(2^-)$
$K_1(1650)$	$1/2(1^+)$
• $K^*(1680)$	$1/2(1^-)$
• $K_2(1770)$	$1/2(2^-)$
• $K^*_3(1780)$	$1/2(3^-)$
• $K_2(1820)$	$1/2(2^-)$
$K(1830)$	$1/2(0^-)$
$K^*_0(1950)$	$1/2(0^+)$
$K^*_2(1980)$	$1/2(2^+)$
• $K^*_4(2045)$	$1/2(4^+)$
$K_2(2250)$	$1/2(2^-)$
$K_3(2320)$	$1/2(3^+)$
$K^*_5(2380)$	$1/2(5^-)$
$K_4(2500)$	$1/2(4^-)$
$K(3100)$	$?^?(?^{??})$

CHARMED ($C = \pm1$)

Particle	$I(J^P)$
• $D^\pm$	$1/2(0^-)$
• D^0	$1/2(0^-)$
• $D^*(2007)^0$	$1/2(1^-)$
• $D^*(2010)^\pm$	$1/2(1^-)$
• $D_1(2420)^0$	$1/2(1^+)$
$D_1(2420)^\pm$	$1/2(?)$
• $D^*_2(2460)^0$	$1/2(2^+)$
• $D^*_2(2460)^+$	$1/2(2^+)$

CHARMED, STRANGE ($C = S = \pm1$)

Particle	$I(J^P)$
• $D^\pm_s$	$0(0^-)$
• $D^{*\pm}_s$	$?(?)$
• $D_{s1}(2536)^\pm$	$0(1^+)$
$D_{sJ}(2573)^\pm$	$?(?)$

BOTTOM ($B = \pm1$)

Particle	$I(J^P)$
• $B^\pm$	$1/2(0^-)$
• B^0	$1/2(0^-)$
• B^*	$1/2(1^-)$
$B^*_J(5732)$	$?(?^?)$

BOTTOM, STRANGE ($B = \pm1, S = \mp1$)

Particle	$I^G(J^{PC})$
• B^0_s	$0(0^-)$
B^*_s	$?(?^?)$
$B^*_{sJ}(5850)$	$?(?^?)$

$c\bar{c}$

Particle	$I^G(J^{PC})$
• $\eta_c(1S)$	$0^+(0^{-+})$
• $J/\psi(1S)$	$0^-(1^{--})$
• $\chi_{c0}(1P)$	$0^+(0^{++})$
• $\chi_{c1}(1P)$	$0^+(1^{++})$
$h_c(1P)$	$?^?(?^{??})$
• $\chi_{c2}(1P)$	$0^+(2^{++})$
$\eta_c(2S)$	$?^?(?^{?+})$
• $\psi(2S)$	$0^-(1^{--})$
• $\psi(3770)$	$?^?(1^{--})$
• $\psi(4040)$	$?^?(1^{--})$
• $\psi(4160)$	$?^?(1^{--})$
• $\psi(4415)$	$?^?(1^{--})$

$b\bar{b}$

Particle	$I^G(J^{PC})$
• $\Upsilon(1S)$	$0^-(1^{--})$
• $\chi_{b0}(1P)$	$0^+(0^{++})$
• $\chi_{b1}(1P)$	$0^+(1^{++})$
• $\chi_{b2}(1P)$	$0^+(2^{++})$
• $\Upsilon(2S)$	$0^-(1^{--})$
• $\chi_{b0}(2P)$	$0^+(0^{++})$
• $\chi_{b1}(2P)$	$0^+(1^{++})$
• $\chi_{b2}(2P)$	$0^+(2^{++})$
• $\Upsilon(3S)$	$0^-(1^{--})$
• $\Upsilon(4S)$	$?^?(1^{--})$
• $\Upsilon(10860)$	$?^?(1^{--})$
• $\Upsilon(11020)$	$?^?(1^{--})$

NON-$q\bar{q}$ CANDIDATES

Non-$q\bar{q}$ Candidates

This short table gives the name, the quantum numbers (where known), and the status of baryons in the Review. Only the baryons with 3- or 4-star status are included in the main Baryon Summary Table. Due to insufficient data or uncertain interpretation, the other entries in the short table are not established as baryons. The names with masses are of baryons that decay strongly. See our 1986 edition (Physics Letters **170B**) for listings of evidence for Z baryons (KN resonances).

p	P_{11}	****	$\Delta(1232)$	P_{33}	****	Λ	P_{01}	****	Σ^+	P_{11}	****	Ξ^0	P_{11}	****
n	P_{11}	****	$\Delta(1600)$	P_{33}	***	$\Lambda(1405)$	S_{01}	****	Σ^0	P_{11}	****	Ξ^-	P_{11}	****
$N(1440)$	P_{11}	****	$\Delta(1620)$	S_{31}	****	$\Lambda(1520)$	D_{03}	****	Σ^-	P_{11}	****	$\Xi(1530)$	P_{13}	****
$N(1520)$	D_{13}	****	$\Delta(1700)$	D_{33}	****	$\Lambda(1600)$	P_{01}	***	$\Sigma(1385)$	P_{13}	****	$\Xi(1620)$		*
$N(1535)$	S_{11}	****	$\Delta(1750)$	P_{31}	*	$\Lambda(1670)$	S_{01}	****	$\Sigma(1480)$		*	$\Xi(1690)$		***
$N(1650)$	S_{11}	****	$\Delta(1900)$	S_{31}	***	$\Lambda(1690)$	D_{03}	****	$\Sigma(1560)$		**	$\Xi(1820)$	D_{13}	***
$N(1675)$	D_{15}	****	$\Delta(1905)$	F_{35}	****	$\Lambda(1800)$	S_{01}	***	$\Sigma(1580)$	D_{13}	**	$\Xi(1950)$		***
$N(1680)$	F_{15}	****	$\Delta(1910)$	P_{31}	****	$\Lambda(1810)$	P_{01}	***	$\Sigma(1620)$	S_{11}	**	$\Xi(2030)$		***
$N(1700)$	D_{13}	***	$\Delta(1920)$	P_{33}	***	$\Lambda(1820)$	F_{05}	****	$\Sigma(1660)$	P_{11}	***	$\Xi(2120)$		*
$N(1710)$	P_{11}	***	$\Delta(1930)$	D_{35}	***	$\Lambda(1830)$	D_{05}	****	$\Sigma(1670)$	D_{13}	****	$\Xi(2250)$		**
$N(1720)$	P_{13}	****	$\Delta(1940)$	D_{33}	*	$\Lambda(1890)$	P_{03}	****	$\Sigma(1690)$		**	$\Xi(2370)$		**
$N(1900)$	P_{13}	**	$\Delta(1950)$	F_{37}	****	$\Lambda(2000)$		*	$\Sigma(1750)$	S_{11}	***	$\Xi(2500)$		*
$N(1990)$	F_{17}	**	$\Delta(2000)$	F_{35}	**	$\Lambda(2020)$	F_{07}	*	$\Sigma(1770)$	P_{11}	*			
$N(2000)$	F_{15}	**	$\Delta(2150)$	S_{31}	*	$\Lambda(2100)$	G_{07}	****	$\Sigma(1775)$	D_{15}	****	Ω^-		****
$N(2080)$	D_{13}	**	$\Delta(2200)$	G_{37}	*	$\Lambda(2110)$	F_{05}	***	$\Sigma(1840)$	P_{13}	*	$\Omega(2250)^-$		***
$N(2090)$	S_{11}	*	$\Delta(2300)$	H_{39}	**	$\Lambda(2325)$	D_{03}	*	$\Sigma(1880)$	P_{11}	**	$\Omega(2380)^-$		**
$N(2100)$	P_{11}	*	$\Delta(2350)$	D_{35}	*	$\Lambda(2350)$	H_{09}	***	$\Sigma(1915)$	F_{15}	****	$\Omega(2470)^-$		**
$N(2190)$	G_{17}	****	$\Delta(2390)$	F_{37}	*	$\Lambda(2585)$		**	$\Sigma(1940)$	D_{13}	***			
$N(2200)$	D_{15}	**	$\Delta(2400)$	G_{39}	**				$\Sigma(2000)$	S_{11}	*	Λ_c^+		****
$N(2220)$	H_{19}	****	$\Delta(2420)$	$H_{3,11}$	****				$\Sigma(2030)$	F_{17}	****	$\Lambda_c(2593)^+$		***
$N(2250)$	G_{19}	****	$\Delta(2750)$	$I_{3,13}$	**				$\Sigma(2070)$	F_{15}	*	$\Lambda_c(2625)^+$		***
$N(2600)$	$I_{1,11}$	***	$\Delta(2950)$	$K_{3,15}$	**				$\Sigma(2080)$	P_{13}	**	$\Sigma_c(2455)$		****
$N(2700)$	$K_{1,13}$	**							$\Sigma(2100)$	G_{17}	*	$\Sigma_c(2530)$		*
									$\Sigma(2250)$		***	Ξ_c^+		***
									$\Sigma(2455)$		**	Ξ_c^0		***
									$\Sigma(2620)$		**	$\Xi_c(2645)$		***
									$\Sigma(3000)$		*	Ω_c^0		***
									$\Sigma(3170)$		*			
												Λ_b^0		***
												Ξ_b^0, Ξ_b^-		*

****   Existence is certain, and properties are at least fairly well explored.

***   Existence ranges from very likely to certain, but further confirmation is desirable and/or quantum numbers, branching fractions, etc. are not well determined.

**   Evidence of existence is only fair.

*   Evidence of existence is poor.

N BARYONS
$(S = 0, I = 1/2)$

$p, N^+ = uud;\quad n, N^0 = udd$

$\boxed{p}$ $I(J^P) = \frac{1}{2}(\frac{1}{2}^+)$

Mass $m = 938.27231 \pm 0.00028$ MeV [a]
    $= 1.007276470 \pm 0.000000012$ u
$\left|\frac{q_p}{m_p}\right|/\left(\frac{q_p}{m_p}\right) = 1.0000000015 \pm 0.0000000011$
$|q_p + q_{\bar{p}}|/e < 2 \times 10^{-5}$
$|q_p + q_e|/e < 1.0 \times 10^{-21}$ [b]
Magnetic moment $\mu = 2.79284739 \pm 0.00000006\ \mu_N$
Electric dipole moment $d = (-4 \pm 6) \times 10^{-23}$ ecm
Electric polarizability $\bar{\alpha} = (12.1 \pm 0.9) \times 10^{-4}$ fm^3
Magnetic polarizability $\bar{\beta} = (2.1 \pm 0.9) \times 10^{-4}$ fm^3
Mean life $\tau > 1.6 \times 10^{25}$ years (independent of mode)
    $=> 10^{31} - 5 \times 10^{32}$ years [c] (mode dependent)

Below, for N decays, p and n distinguish proton and neutron partial lifetimes. See also the "Note on Nucleon Decay" in our 1994 edition (Phys. Rev. D50, 1673) for a short review.

The "partial mean life" limits tabulated here are the limits on τ/B_i, where τ is the total mean life and B_i is the branching fraction for the mode in question.

p DECAY MODES	Partial mean life (10^{30} years)	Confidence level	p (MeV/c)
Antilepton + meson			
$N \to e^+ \pi$	> 130 (n), > 550 (p)	90%	459
$N \to \mu^+ \pi$	> 100 (n), > 270 (p)	90%	453
$N \to \nu \pi$	> 100 (n), > 25 (p)	90%	459
$p \to e^+ \eta$	> 140	90%	309
$p \to \mu^+ \eta$	> 69	90%	296
$n \to \nu \eta$	> 54	90%	310
$N \to e^+ \rho$	> 58 (n), > 75 (p)	90%	153
$N \to \mu^+ \rho$	> 23 (n), > 110 (p)	90%	119
$N \to \nu \rho$	> 19 (n), > 27 (p)	90%	153
$p \to e^+ \omega$	> 45	90%	142
$p \to \mu^+ \omega$	> 57	90%	104
$n \to \nu \omega$	> 43	90%	144
$N \to e^+ K$	> 1.3 (n), > 150 (p)	90%	337
$p \to e^+ K^0_S$	> 76	90%	337
$p \to e^+ K^0_L$	> 44	90%	337
$N \to \mu^+ K$	> 1.1 (n), > 120 (p)	90%	326
$p \to \mu^+ K^0_S$	> 64	90%	326
$p \to \mu^+ K^0_L$	> 44	90%	326
$N \to \nu K$	> 86 (n), > 100 (p)	90%	339
$p \to e^+ K^*(892)^0$	> 52	90%	45
$N \to \nu K^*(892)$	> 22 (n), > 20 (p)	90%	45
Antilepton + mesons			
$p \to e^+ \pi^+ \pi^-$	> 21	90%	448
$p \to e^+ \pi^0 \pi^0$	> 38	90%	449
$n \to e^+ \pi^- \pi^0$	> 32	90%	449
$p \to \mu^+ \pi^+ \pi^-$	> 17	90%	425
$p \to \mu^+ \pi^0 \pi^0$	> 33	90%	427
$n \to \mu^+ \pi^- \pi^0$	> 33	90%	427
$n \to e^+ K^0 \pi^-$	> 18	90%	319
Lepton + meson			
$n \to e^- \pi^+$	> 65	90%	459
$n \to \mu^- \pi^+$	> 49	90%	453
$n \to e^- \rho^+$	> 62	90%	154
$n \to \mu^- \rho^+$	> 7	90%	120
$n \to e^- K^+$	> 32	90%	340
$n \to \mu^- K^+$	> 57	90%	330
Lepton + mesons			
$p \to e^- \pi^+ \pi^+$	> 30	90%	448
$n \to e^- \pi^+ \pi^0$	> 29	90%	449
$p \to \mu^- \pi^+ \pi^+$	> 17	90%	425
$n \to \mu^- \pi^+ \pi^0$	> 34	90%	427
$p \to e^- \pi^+ K^+$	> 20	90%	320
$p \to \mu^- \pi^+ K^+$	> 5	90%	279

	Partial mean life	Confidence	p
Antilepton + photon(s)			
$p \to e^+ \gamma$	> 460	90%	469
$p \to \mu^+ \gamma$	> 380	90%	463
$n \to \nu \gamma$	> 24	90%	470
$p \to e^+ \gamma \gamma$	> 100	90%	469
Three leptons			
$p \to e^+ e^+ e^-$	> 510	90%	469
$p \to e^+ \mu^+ \mu^-$	> 81	90%	457
$p \to e^+ \nu \nu$	> 11	90%	469
$n \to e^+ e^- \nu$	> 74	90%	470
$n \to \mu^+ e^- \nu$	> 47	90%	464
$n \to \mu^+ \mu^- \nu$	> 42	90%	458
$p \to \mu^+ e^+ e^-$	> 91	90%	464
$p \to \mu^+ \mu^+ \mu^-$	> 190	90%	439
$p \to \mu^+ \nu \nu$	> 21	90%	463
$p \to e^- \mu^+ \mu^+$	> 6	90%	457
$n \to 3\nu$	> 0.0005	90%	470
Inclusive modes			
$N \to e^+$ anything	> 0.6 (n, p)	90%	–
$N \to \mu^+$ anything	> 12 (n, p)	90%	–
$N \to e^+ \pi^0$ anything	> 0.6 (n, p)	90%	–

$\Delta B = 2$ dinucleon modes

The following are lifetime limits per iron nucleus.

	Partial mean life	Confidence	p
$pp \to \pi^+ \pi^+$	> 0.7	90%	–
$pn \to \pi^+ \pi^0$	> 2	90%	–
$nn \to \pi^+ \pi^-$	> 0.7	90%	–
$nn \to \pi^0 \pi^0$	> 3.4	90%	–
$pp \to e^+ e^+$	> 5.8	90%	–
$pp \to e^+ \mu^+$	> 3.6	90%	–
$pp \to \mu^+ \mu^+$	> 1.7	90%	–
$pn \to e^+ \bar{\nu}$	> 2.8	90%	–
$pn \to \mu^+ \bar{\nu}$	> 1.6	90%	–
$nn \to \nu_e \bar{\nu}_e$	> 0.000012	90%	–
$nn \to \nu_\mu \bar{\nu}_\mu$	> 0.000006	90%	–

$\bar{p}$ DECAY MODES

$\bar{p}$ DECAY MODES	Partial mean life (years)	Confidence level	p (MeV/c)
$\bar{p} \to e^- \gamma$	> 1848	95%	469
$\bar{p} \to e^- \pi^0$	> 554	95%	459
$\bar{p} \to e^- \eta$	> 171	95%	309
$\bar{p} \to e^- K^0_S$	> 29	95%	337
$\bar{p} \to e^- K^0_L$	> 9	95%	337

$\boxed{n}$ $I(J^P) = \frac{1}{2}(\frac{1}{2}^+)$

Mass $m = 939.56563 \pm 0.00028$ MeV [a]
    $= 1.008664904 \pm 0.000000014$ u
$m_n - m_p = 1.293318 \pm 0.000009$ MeV
    $= 0.001388434 \pm 0.000000009$ u
Mean life $\tau = 887.0 \pm 2.0$ s (S = 1.3)
    $c\tau = 2.659 \times 10^8$ km
Magnetic moment $\mu = -1.9130428 \pm 0.0000005\ \mu_N$
Electric dipole moment $d < 1.1 \times 10^{-25}$ ecm, CL = 95%
Electric polarizability $\alpha = (0.98^{+0.19}_{-0.23}) \times 10^{-3}$ fm^3 (S = 1.1)
Charge $q = (-0.4 \pm 1.1) \times 10^{-21}\ e$
Mean $n\bar{n}$-oscillation time $> 1.2 \times 10^8$ s, CL = 90% [d] (bound n)
    $> 0.86 \times 10^8$ s, CL = 90% (free n)

Decay parameters [e]

$pe^-\bar{\nu}_e$	$g_A/g_V = -1.2601 \pm 0.0025$ (S = 1.1)
"	$A = -0.1139 \pm 0.0011$ (S = 1.3)
"	$B = 0.990 \pm 0.008$
"	$a = -0.102 \pm 0.005$
"	$\phi_{AV} = (180.07 \pm 0.18)°$ [f]
"	$D = (-0.5 \pm 1.4) \times 10^{-3}$

n DECAY MODES	Fraction (Γ_i/Γ)	Confidence level	p (MeV/c)
$pe^-\bar{\nu}_e$	100 %		1.19
Charge conservation (Q) violating mode			
$p\nu_e\bar{\nu}_e$	Q $< 9 \times 10^{-24}$	90%	1.29

$N(1440)\ P_{11}$ $\qquad I(J^P) = \frac{1}{2}(\frac{1}{2}^+)$

Mass m = 1430 to 1470 ($\approx$ 1440) MeV
Full width Γ = 250 to 450 ($\approx$ 350) MeV
p_{beam} = 0.61 GeV/c $\qquad 4\pi\lambda^2$ = 31.0 mb

$N(1440)$ DECAY MODES	Fraction (Γ_I/Γ)	p (MeV/c)
$N\pi$	60–70 %	397
$N\pi\pi$	30–40 %	342
$\Delta\pi$	20–30 %	143
$N\rho$	<8 %	†
$N(\pi\pi)_{S\text{-wave}}^{I=0}$	5–10 %	–
$p\gamma$	0.035–0.048 %	414
$p\gamma$, helicity=1/2	0.035–0.048 %	414
$n\gamma$	0.009–0.032 %	413
$n\gamma$, helicity=1/2	0.009–0.032 %	413

$N(1520)\ D_{13}$ $\qquad I(J^P) = \frac{1}{2}(\frac{3}{2}^-)$

Mass m = 1515 to 1530 ($\approx$ 1520) MeV
Full width Γ = 110 to 135 ($\approx$ 120) MeV
p_{beam} = 0.74 GeV/c $\qquad 4\pi\lambda^2$ = 23.5 mb

$N(1520)$ DECAY MODES	Fraction (Γ_I/Γ)	p (MeV/c)
$N\pi$	50–60 %	456
$N\pi\pi$	40–50 %	410
$\Delta\pi$	15–25 %	228
$N\rho$	15–25 %	†
$N(\pi\pi)_{S\text{-wave}}^{I=0}$	<8 %	–
$p\gamma$	0.46–0.56 %	470
$p\gamma$, helicity=1/2	0.001–0.034 %	470
$p\gamma$, helicity=3/2	0.44–0.53 %	470
$n\gamma$	0.30–0.53 %	470
$n\gamma$, helicity=1/2	0.04–0.10 %	470
$n\gamma$, helicity=3/2	0.25–0.45 %	470

$N(1535)\ S_{11}$ $\qquad I(J^P) = \frac{1}{2}(\frac{1}{2}^-)$

Mass m = 1520 to 1555 ($\approx$ 1535) MeV
Full width Γ = 100 to 250 ($\approx$ 150) MeV
p_{beam} = 0.76 GeV/c $\qquad 4\pi\lambda^2$ = 22.5 mb

$N(1535)$ DECAY MODES	Fraction (Γ_I/Γ)	p (MeV/c)
$N\pi$	35–55 %	467
$N\eta$	30–55 %	182
$N\pi\pi$	1–10 %	422
$\Delta\pi$	<1 %	242
$N\rho$	<4 %	†
$N(\pi\pi)_{S\text{-wave}}^{I=0}$	<3 %	–
$N(1440)\pi$	<7 %	†
$p\gamma$	0.08–0.27 %	481
$p\gamma$, helicity=1/2	0.08–0.27 %	481
$n\gamma$	0.004–0.29 %	480
$n\gamma$, helicity=1/2	0.004–0.29 %	480

$N(1650)\ S_{11}$ $\qquad I(J^P) = \frac{1}{2}(\frac{1}{2}^-)$

Mass m = 1640 to 1680 ($\approx$ 1650) MeV
Full width Γ = 145 to 190 ($\approx$ 150) MeV
p_{beam} = 0.96 GeV/c $\qquad 4\pi\lambda^2$ = 16.4 mb

$N(1650)$ DECAY MODES	Fraction (Γ_I/Γ)	p (MeV/c)
$N\pi$	55–90 %	547
$N\eta$	3–10 %	346
ΛK	3–11 %	161
$N\pi\pi$	10–20 %	511
$\Delta\pi$	1–7 %	344
$N\rho$	4–12 %	†

$N(\pi\pi)_{S\text{-wave}}^{I=0}$	<4 %	–
$N(1440)\pi$	<5 %	147
$p\gamma$	0.04–0.18 %	558
$p\gamma$, helicity=1/2	0.04–0.18 %	558
$n\gamma$	0.003–0.17 %	557
$n\gamma$, helicity=1/2	0.003–0.17 %	557

$N(1675)\ D_{15}$ $\qquad I(J^P) = \frac{1}{2}(\frac{5}{2}^-)$

Mass m = 1670 to 1685 ($\approx$ 1675) MeV
Full width Γ = 140 to 180 ($\approx$ 150) MeV
p_{beam} = 1.01 GeV/c $\qquad 4\pi\lambda^2$ = 15.4 mb

$N(1675)$ DECAY MODES	Fraction (Γ_I/Γ)	p (MeV/c)
$N\pi$	40–50 %	563
ΛK	<1 %	209
$N\pi\pi$	50–60 %	529
$\Delta\pi$	50–60 %	364
$N\rho$	< 1–3 %	†
$p\gamma$	0.004–0.023 %	575
$p\gamma$, helicity=1/2	0.0–0.015 %	575
$p\gamma$, helicity=3/2	0.0–0.011 %	575
$n\gamma$	0.02–0.12 %	574
$n\gamma$, helicity=1/2	0.006–0.046 %	574
$n\gamma$, helicity=3/2	0.01–0.08 %	574

$N(1680)\ F_{15}$ $\qquad I(J^P) = \frac{1}{2}(\frac{5}{2}^+)$

Mass m = 1675 to 1690 ($\approx$ 1680) MeV
Full width Γ = 120 to 140 ($\approx$ 130) MeV
p_{beam} = 1.01 GeV/c $\qquad 4\pi\lambda^2$ = 15.2 mb

$N(1680)$ DECAY MODES	Fraction (Γ_I/Γ)	p (MeV/c)
$N\pi$	60–70 %	567
$N\pi\pi$	30–40 %	532
$\Delta\pi$	5–15 %	369
$N\rho$	3–15 %	†
$N(\pi\pi)_{S\text{-wave}}^{I=0}$	5–20 %	–
$p\gamma$	0.21–0.32 %	578
$p\gamma$, helicity=1/2	0.001–0.011 %	578
$p\gamma$, helicity=3/2	0.20–0.32 %	578
$n\gamma$	0.021–0.046 %	577
$n\gamma$, helicity=1/2	0.004–0.029 %	577
$n\gamma$, helicity=3/2	0.01–0.024 %	577

$N(1700)\ D_{13}$ $\qquad I(J^P) = \frac{1}{2}(\frac{3}{2}^-)$

Mass m = 1650 to 1750 ($\approx$ 1700) MeV
Full width Γ = 50 to 150 ($\approx$ 100) MeV
p_{beam} = 1.05 GeV/c $\qquad 4\pi\lambda^2$ = 14.5 mb

$N(1700)$ DECAY MODES	Fraction (Γ_I/Γ)	p (MeV/c)
$N\pi$	5–15 %	580
ΛK	<3 %	250
$N\pi\pi$	85–95 %	547
$N\rho$	<35 %	†
$p\gamma$	0.01–0.05 %	591
$p\gamma$, helicity=1/2	0.0–0.024 %	591
$p\gamma$, helicity=3/2	0.002–0.026 %	591
$n\gamma$	0.01–0.13 %	590
$n\gamma$, helicity=1/2	0.0–0.09 %	590
$n\gamma$, helicity=3/2	0.01–0.05 %	590

$N(1710)$ P_{11} $I(J^P) = \frac{1}{2}(\frac{1}{2}^+)$

Mass $m = 1680$ to 1740 (≈ 1710) MeV
Full width $\Gamma = 50$ to 250 (≈ 100) MeV
$p_{beam} = 1.07$ GeV/c $4\pi\lambdabar^2 = 14.2$ mb

$N(1710)$ DECAY MODES	Fraction (Γ_i/Γ)	p (MeV/c)
$N\pi$	10–20 %	587
ΛK	5–25 %	264
$N\pi\pi$	40–90 %	554
$\Delta\pi$	15–40 %	393
$N\rho$	5–25 %	48
$N(\pi\pi)^{I=0}_{S\text{-wave}}$	10–40 %	–
$p\gamma$	0.002–0.05%	598
$p\gamma$, helicity=1/2	0.002–0.05%	598
$n\gamma$	0.0–0.02%	597
$n\gamma$, helicity=1/2	0.0–0.02%	597

$N(1720)$ P_{13} $I(J^P) = \frac{1}{2}(\frac{3}{2}^+)$

Mass $m = 1650$ to 1750 (≈ 1720) MeV
Full width $\Gamma = 100$ to 200 (≈ 150) MeV
$p_{beam} = 1.09$ GeV/c $4\pi\lambdabar^2 = 13.9$ mb

$N(1720)$ DECAY MODES	Fraction (Γ_i/Γ)	p (MeV/c)
$N\pi$	10–20 %	594
ΛK	1–15 %	278
$N\pi\pi$	>70 %	561
$N\rho$	70–85 %	104
$p\gamma$	0.003–0.10 %	604
$p\gamma$, helicity=1/2	0.003–0.08 %	604
$p\gamma$, helicity=3/2	0.001–0.03 %	604
$n\gamma$	0.002–0.39 %	603
$n\gamma$, helicity=1/2	0.0–0.002 %	603
$n\gamma$, helicity=3/2	0.001–0.39 %	603

$N(2190)$ G_{17} $I(J^P) = \frac{1}{2}(\frac{7}{2}^-)$

Mass $m = 2100$ to 2200 (≈ 2190) MeV
Full width $\Gamma = 350$ to 550 (≈ 450) MeV
$p_{beam} = 2.07$ GeV/c $4\pi\lambdabar^2 = 6.21$ mb

$N(2190)$ DECAY MODES	Fraction (Γ_i/Γ)	p (MeV/c)
$N\pi$	10–20 %	888

$N(2220)$ H_{19} $I(J^P) = \frac{1}{2}(\frac{9}{2}^+)$

Mass $m = 2180$ to 2310 (≈ 2220) MeV
Full width $\Gamma = 320$ to 550 (≈ 400) MeV
$p_{beam} = 2.14$ GeV/c $4\pi\lambdabar^2 = 5.97$ mb

$N(2220)$ DECAY MODES	Fraction (Γ_i/Γ)	p (MeV/c)
$N\pi$	10–20 %	905

$N(2250)$ G_{19} $I(J^P) = \frac{1}{2}(\frac{9}{2}^-)$

Mass $m = 2170$ to 2310 (≈ 2250) MeV
Full width $\Gamma = 290$ to 470 (≈ 400) MeV
$p_{beam} = 2.21$ GeV/c $4\pi\lambdabar^2 = 5.74$ mb

$N(2250)$ DECAY MODES	Fraction (Γ_i/Γ)	p (MeV/c)
$N\pi$	5–15 %	923

$N(2600)$ $I_{1,11}$ $I(J^P) = \frac{1}{2}(\frac{11}{2}^-)$

Mass $m = 2550$ to 2750 (≈ 2600) MeV
Full width $\Gamma = 500$ to 800 (≈ 650) MeV
$p_{beam} = 3.12$ GeV/c $4\pi\lambdabar^2 = 3.86$ mb

$N(2600)$ DECAY MODES	Fraction (Γ_i/Γ)	p (MeV/c)
$N\pi$	5–10 %	1126

Δ BARYONS
$(S = 0, I = 3/2)$

$\Delta^{++} = uuu, \quad \Delta^+ = uud, \quad \Delta^0 = udd, \quad \Delta^- = ddd$

$\Delta(1232)$ P_{33} $I(J^P) = \frac{3}{2}(\frac{3}{2}^+)$

Mass $m = 1230$ to 1234 (≈ 1232) MeV
Full width $\Gamma = 115$ to 125 (≈ 120) MeV
$p_{beam} = 0.30$ GeV/c $4\pi\lambdabar^2 = 94.8$ mb

$\Delta(1232)$ DECAY MODES	Fraction (Γ_i/Γ)	p (MeV/c)
$N\pi$	>99 %	227
$N\gamma$	0.54–0.61 %	259
$N\gamma$, helicity=1/2	0.12–0.14 %	259
$N\gamma$, helicity=3/2	0.41–0.47 %	259

$\Delta(1600)$ P_{33} $I(J^P) = \frac{3}{2}(\frac{3}{2}^+)$

Mass $m = 1550$ to 1700 (≈ 1600) MeV
Full width $\Gamma = 250$ to 450 (≈ 350) MeV
$p_{beam} = 0.87$ GeV/c $4\pi\lambdabar^2 = 18.6$ mb

$\Delta(1600)$ DECAY MODES	Fraction (Γ_i/Γ)	p (MeV/c)
$N\pi$	10–25 %	512
$N\pi\pi$	75–90 %	473
$\Delta\pi$	40–70 %	301
$N\rho$	<25 %	†
$N(1440)\pi$	10–35 %	74
$N\gamma$	0.001–0.02 %	525
$N\gamma$, helicity=1/2	0.0–0.02 %	525
$N\gamma$, helicity=3/2	0.001–0.005 %	525

$\Delta(1620)$ S_{31} $I(J^P) = \frac{3}{2}(\frac{1}{2}^-)$

Mass $m = 1615$ to 1675 (≈ 1620) MeV
Full width $\Gamma = 120$ to 180 (≈ 150) MeV
$p_{beam} = 0.91$ GeV/c $4\pi\lambdabar^2 = 17.7$ mb

$\Delta(1620)$ DECAY MODES	Fraction (Γ_i/Γ)	p (MeV/c)
$N\pi$	20–30 %	526
$N\pi\pi$	70–80 %	488
$\Delta\pi$	30–60 %	318
$N\rho$	7–25 %	†
$N\gamma$	0.004–0.044 %	538
$N\gamma$, helicity=1/2	0.004–0.044 %	538

$\Delta(1700)$ D_{33} $I(J^P) = \frac{3}{2}(\frac{3}{2}^-)$

Mass $m = 1670$ to 1770 (≈ 1700) MeV
Full width $\Gamma = 200$ to 400 (≈ 300) MeV
$p_{beam} = 1.05$ GeV/c $4\pi\lambdabar^2 = 14.5$ mb

$\Delta(1700)$ DECAY MODES	Fraction (Γ_i/Γ)	p (MeV/c)
$N\pi$	10–20 %	580
$N\pi\pi$	80–90 %	547
$\Delta\pi$	30–60 %	385
$N\rho$	30–55 %	†
$N\gamma$	0.12–0.26 %	591
$N\gamma$, helicity=1/2	0.08–0.16 %	591
$N\gamma$, helicity=3/2	0.025–0.12 %	591

$\Delta(1900)$ S_{31} $I(J^P) = \frac{3}{2}(\frac{1}{2}^-)$

Mass $m = 1850$ to 1950 (≈ 1900) MeV
Full width $\Gamma = 140$ to 240 (≈ 200) MeV
$p_{beam} = 1.44$ GeV/c $4\pi\lambdabar^2 = 9.71$ mb

$\Delta(1900)$ DECAY MODES	Fraction (Γ_i/Γ)	p (MeV/c)
$N\pi$	10–30 %	710

| $\Delta(1905)$ F_{35} | | $I(J^P) = \frac{3}{2}(\frac{5}{2}^+)$ |

Mass m = 1870 to 1920 ($\approx$ 1905) MeV
Full width Γ = 280 to 440 ($\approx$ 350) MeV
p_{beam} = 1.45 GeV/c $4\pi\lambda^2$ = 9.62 mb

$\Delta(1905)$ DECAY MODES	Fraction (Γ_i/Γ)	p (MeV/c)
$N\pi$	5–15 %	713
$N\pi\pi$	85–95 %	687
$\Delta\pi$	<25 %	542
$N\rho$	>60 %	421
$N\gamma$	0.01–0.03 %	721
$N\gamma$, helicity=1/2	0.0–0.1 %	721
$N\gamma$, helicity=3/2	0.004–0.03 %	721

| $\Delta(1910)$ P_{31} | | $I(J^P) = \frac{3}{2}(\frac{1}{2}^+)$ |

Mass m = 1870 to 1920 ($\approx$ 1910) MeV
Full width Γ = 190 to 270 ($\approx$ 250) MeV
p_{beam} = 1.46 GeV/c $4\pi\lambda^2$ = 9.54 mb

$\Delta(1910)$ DECAY MODES	Fraction (Γ_i/Γ)	p (MeV/c)
$N\pi$	15–30 %	716
$N\gamma$	0.0–0.2 %	725
$N\gamma$, helicity=1/2	0.0–0.2 %	725

| $\Delta(1920)$ P_{33} | | $I(J^P) = \frac{3}{2}(\frac{3}{2}^+)$ |

Mass m = 1900 to 1970 ($\approx$ 1920) MeV
Full width Γ = 150 to 300 ($\approx$ 200) MeV
p_{beam} = 1.48 GeV/c $4\pi\lambda^2$ = 9.37 mb

$\Delta(1920)$ DECAY MODES	Fraction (Γ_i/Γ)	p (MeV/c)
$N\pi$	5–20 %	722

| $\Delta(1930)$ D_{35} | | $I(J^P) = \frac{3}{2}(\frac{5}{2}^-)$ |

Mass m = 1920 to 1970 ($\approx$ 1930) MeV
Full width Γ = 250 to 450 ($\approx$ 350) MeV
p_{beam} = 1.50 GeV/c $4\pi\lambda^2$ = 9.21 mb

$\Delta(1930)$ DECAY MODES	Fraction (Γ_i/Γ)	p (MeV/c)
$N\pi$	10–20 %	729
$N\gamma$	0.0–0.02 %	737
$N\gamma$, helicity=1/2	0.0–0.01 %	737
$N\gamma$, helicity=3/2	0.0–0.01 %	737

| $\Delta(1950)$ F_{37} | | $I(J^P) = \frac{3}{2}(\frac{7}{2}^+)$ |

Mass m = 1940 to 1960 ($\approx$ 1950) MeV
Full width Γ = 290 to 350 ($\approx$ 300) MeV
p_{beam} = 1.54 GeV/c $4\pi\lambda^2$ = 8.91 mb

$\Delta(1950)$ DECAY MODES	Fraction (Γ_i/Γ)	p (MeV/c)
$N\pi$	35–40 %	741
$N\pi\pi$		716
$\Delta\pi$	20–30 %	574
$N\rho$	<10 %	469
$N\gamma$	0.08–0.13 %	749
$N\gamma$, helicity=1/2	0.03–0.055 %	749
$N\gamma$, helicity=3/2	0.05–0.075 %	749

| $\Delta(2420)$ $H_{3,11}$ | | $I(J^P) = \frac{3}{2}(\frac{11}{2}^+)$ |

Mass m = 2300 to 2500 ($\approx$ 2420) MeV
Full width Γ = 300 to 500 ($\approx$ 400) MeV
p_{beam} = 2.64 GeV/c $4\pi\lambda^2$ = 4.68 mb

$\Delta(2420)$ DECAY MODES	Fraction (Γ_i/Γ)	p (MeV/c)
$N\pi$	5–15 %	1023

| Λ BARYONS
($S = -1$, $I = 0$)
$\Lambda^0 = uds$ |

| Λ | | $I(J^P) = 0(\frac{1}{2}^+)$ |

Mass m = 1115.684 ± 0.006 MeV
Mean life τ = (2.632 ± 0.020) × 10^{-10} s   (S = 1.6)
    $c\tau$ = 7.89 cm
Magnetic moment μ = −0.613 ± 0.004 μ_N
Electric dipole moment d < 1.5×10^{-16} e cm, CL = 95%

Decay parameters

$p\pi^-$	α_- = 0.642 ± 0.013
"	ϕ_- = (−6.5 ± 3.5)°
"	γ_- = 0.76 [g]
"	Δ_- = (8 ± 4)° [g]
$n\pi^0$	α_0 = +0.65 ± 0.05
$pe^-\bar{\nu}_e$	g_A/g_V = −0.718 ± 0.015 [e]

Λ DECAY MODES	Fraction (Γ_i/Γ)	p (MeV/c)
$p\pi^-$	(63.9 ±0.5) %	101
$n\pi^0$	(35.8 ±0.5) %	104
$n\gamma$	(1.75±0.15) × 10^{-3}	162
$p\pi^-\gamma$	[h](8.4 ±1.4) × 10^{-4}	101
$pe^-\bar{\nu}_e$	(8.32±0.14) × 10^{-4}	163
$p\mu^-\bar{\nu}_\mu$	(1.57±0.35) × 10^{-4}	131

| $\Lambda(1405)$ S_{01} | | $I(J^P) = 0(\frac{1}{2}^-)$ |

Mass m = 1407 ± 4 MeV
Full width Γ = 50.0 ± 2.0 MeV
    Below $\overline{K}N$ threshold

$\Lambda(1405)$ DECAY MODES	Fraction (Γ_i/Γ)	p (MeV/c)
$\Sigma\pi$	100 %	152

| $\Lambda(1520)$ D_{03} | | $I(J^P) = 0(\frac{3}{2}^-)$ |

Mass m = 1519.5 ± 1.0 MeV [l]
Full width Γ = 15.6 ± 1.0 MeV [l]
    p_{beam} = 0.39 GeV/c $4\pi\lambda^2$ = 82.8 mb

$\Lambda(1520)$ DECAY MODES	Fraction (Γ_i/Γ)	p (MeV/c)
$N\overline{K}$	45 ± 1%	244
$\Sigma\pi$	42 ± 1%	267
$\Lambda\pi\pi$	10 ± 1%	252
$\Sigma\pi\pi$	0.9 ± 0.1%	152
$\Lambda\gamma$	0.8 ± 0.2%	351

| $\Lambda(1600)$ P_{01} | | $I(J^P) = 0(\frac{1}{2}^+)$ |

Mass m = 1560 to 1700 ($\approx$ 1600) MeV
Full width Γ = 50 to 250 ($\approx$ 150) MeV
    p_{beam} = 0.58 GeV/c $4\pi\lambda^2$ = 41.6 mb

$\Lambda(1600)$ DECAY MODES	Fraction (Γ_i/Γ)	p (MeV/c)
$N\overline{K}$	15–30 %	343
$\Sigma\pi$	10–60 %	336

$\Lambda(1670)$ S_{01} $\qquad I(J^P) = 0(\frac{1}{2}^-)$

Mass m = 1660 to 1680 ($\approx$ 1670) MeV
Full width Γ = 25 to 50 ($\approx$ 35) MeV
p_{beam} = 0.74 GeV/c $\qquad 4\pi\lambda^2$ = 28.5 mb

$\Lambda(1670)$ DECAY MODES	Fraction (Γ_i/Γ)	p (MeV/c)
$N\overline{K}$	15–25 %	414
$\Sigma\pi$	20–60 %	393
$\Lambda\eta$	15–35 %	64

$\Lambda(1690)$ D_{03} $\qquad I(J^P) = 0(\frac{3}{2}^-)$

Mass m = 1685 to 1695 ($\approx$ 1690) MeV
Full width Γ = 50 to 70 ($\approx$ 60) MeV
p_{beam} = 0.78 GeV/c $\qquad 4\pi\lambda^2$ = 26.1 mb

$\Lambda(1690)$ DECAY MODES	Fraction (Γ_i/Γ)	p (MeV/c)
$N\overline{K}$	20–30 %	433
$\Sigma\pi$	20–40 %	409
$\Lambda\pi\pi$	$\sim$ 25 %	415
$\Sigma\pi\pi$	$\sim$ 20 %	350

$\Lambda(1800)$ S_{01} $\qquad I(J^P) = 0(\frac{1}{2}^-)$

Mass m = 1720 to 1850 ($\approx$ 1800) MeV
Full width Γ = 200 to 400 ($\approx$ 300) MeV
p_{beam} = 1.01 GeV/c $\qquad 4\pi\lambda^2$ = 17.5 mb

$\Lambda(1800)$ DECAY MODES	Fraction (Γ_i/Γ)	p (MeV/c)
$N\overline{K}$	25–40 %	528
$\Sigma\pi$	seen	493
$\Sigma(1385)\pi$	seen	345
$N\overline{K}^*(892)$	seen	†

$\Lambda(1810)$ P_{01} $\qquad I(J^P) = 0(\frac{1}{2}^+)$

Mass m = 1750 to 1850 ($\approx$ 1810) MeV
Full width Γ = 50 to 250 ($\approx$ 150) MeV
p_{beam} = 1.04 GeV/c $\qquad 4\pi\lambda^2$ = 17.0 mb

$\Lambda(1810)$ DECAY MODES	Fraction (Γ_i/Γ)	p (MeV/c)
$N\overline{K}$	20–50 %	537
$\Sigma\pi$	10–40 %	501
$\Sigma(1385)\pi$	seen	356
$N\overline{K}^*(892)$	30–60 %	†

$\Lambda(1820)$ F_{05} $\qquad I(J^P) = 0(\frac{5}{2}^+)$

Mass m = 1815 to 1825 ($\approx$ 1820) MeV
Full width Γ = 70 to 90 ($\approx$ 80) MeV
p_{beam} = 1.06 GeV/c $\qquad 4\pi\lambda^2$ = 16.5 mb

$\Lambda(1820)$ DECAY MODES	Fraction (Γ_i/Γ)	p (MeV/c)
$N\overline{K}$	55–65 %	545
$\Sigma\pi$	8–14 %	508
$\Sigma(1385)\pi$	5–10 %	362

$\Lambda(1830)$ D_{05} $\qquad I(J^P) = 0(\frac{5}{2}^-)$

Mass m = 1810 to 1830 ($\approx$ 1830) MeV
Full width Γ = 60 to 110 ($\approx$ 95) MeV
p_{beam} = 1.08 GeV/c $\qquad 4\pi\lambda^2$ = 16.0 mb

$\Lambda(1830)$ DECAY MODES	Fraction (Γ_i/Γ)	p (MeV/c)
$N\overline{K}$	3–10 %	553
$\Sigma\pi$	35–75 %	515
$\Sigma(1385)\pi$	>15 %	371

$\Lambda(1890)$ P_{03} $\qquad I(J^P) = 0(\frac{3}{2}^+)$

Mass m = 1850 to 1910 ($\approx$ 1890) MeV
Full width Γ = 60 to 200 ($\approx$ 100) MeV
p_{beam} = 1.21 GeV/c $\qquad 4\pi\lambda^2$ = 13.6 mb

$\Lambda(1890)$ DECAY MODES	Fraction (Γ_i/Γ)	p (MeV/c)
$N\overline{K}$	20–35 %	599
$\Sigma\pi$	3–10 %	559
$\Sigma(1385)\pi$	seen	420
$N\overline{K}^*(892)$	seen	233

$\Lambda(2100)$ G_{07} $\qquad I(J^P) = 0(\frac{7}{2}^-)$

Mass m = 2090 to 2110 ($\approx$ 2100) MeV
Full width Γ = 100 to 250 ($\approx$ 200) MeV
p_{beam} = 1.68 GeV/c $\qquad 4\pi\lambda^2$ = 8.68 mb

$\Lambda(2100)$ DECAY MODES	Fraction (Γ_i/Γ)	p (MeV/c)
$N\overline{K}$	25–35 %	751
$\Sigma\pi$	$\sim$ 5 %	704
$\Lambda\eta$	<3 %	617
ΞK	<3 %	483
$\Lambda\omega$	<8 %	443
$N\overline{K}^*(892)$	10–20 %	514

$\Lambda(2110)$ F_{05} $\qquad I(J^P) = 0(\frac{5}{2}^+)$

Mass m = 2090 to 2140 ($\approx$ 2110) MeV
Full width Γ = 150 to 250 ($\approx$ 200) MeV
p_{beam} = 1.70 GeV/c $\qquad 4\pi\lambda^2$ = 8.53 mb

$\Lambda(2110)$ DECAY MODES	Fraction (Γ_i/Γ)	p (MeV/c)
$N\overline{K}$	5–25 %	757
$\Sigma\pi$	10–40 %	711
$\Lambda\omega$	seen	455
$\Sigma(1385)\pi$	seen	589
$N\overline{K}^*(892)$	10–60 %	524

$\Lambda(2350)$ H_{09} $\qquad I(J^P) = 0(\frac{9}{2}^+)$

Mass m = 2340 to 2370 ($\approx$ 2350) MeV
Full width Γ = 100 to 250 ($\approx$ 150) MeV
p_{beam} = 2.29 GeV/c $\qquad 4\pi\lambda^2$ = 5.85 mb

$\Lambda(2350)$ DECAY MODES	Fraction (Γ_i/Γ)	p (MeV/c)
$N\overline{K}$	$\sim$ 12 %	915
$\Sigma\pi$	$\sim$ 10 %	867

Σ BARYONS
$(S = -1, I = 1)$
$\Sigma^+ = uus, \quad \Sigma^0 = uds, \quad \Sigma^- = dds$

Σ^+ $I(J^P) = 1(\frac{1}{2}^+)$

Mass $m = 1189.37 \pm 0.07$ MeV   (S = 2.2)
Mean life $\tau = (0.799 \pm 0.004) \times 10^{-10}$ s
   $c\tau = 2.396$ cm
Magnetic moment $\mu = 2.458 \pm 0.010$ μ_N (S = 2.1)
$\Gamma(\Sigma^+ \to n\ell^+\nu)/\Gamma(\Sigma^- \to n\ell^-\bar\nu) < 0.043$

Decay parameters

$p\pi^0$	$\alpha_0 = -0.980^{+0.017}_{-0.015}$
"	$\phi_0 = (36 \pm 34)^\circ$
"	$\gamma_0 = 0.16$ [g]
"	$\Delta_0 = (187 \pm 6)^\circ$ [g]
$n\pi^+$	$\alpha_+ = 0.068 \pm 0.013$
"	$\phi_+ = (167 \pm 20)^\circ$ (S = 1.1)
"	$\gamma_+ = -0.97$ [g]
"	$\Delta_+ = (-73^{+133}_{-10})^\circ$ [g]
$p\gamma$	$\alpha_\gamma = -0.76 \pm 0.08$

Σ^+ DECAY MODES	Fraction (Γ_i/Γ)	Confidence level	p (MeV/c)
$p\pi^0$	(51.57 ± 0.30) %		189
$n\pi^+$	(48.31 ± 0.30) %		185
$p\gamma$	$(1.23 \pm 0.05) \times 10^{-3}$		225
$n\pi^+\gamma$	[h] $(4.5 \pm 0.5) \times 10^{-4}$		185
$\Lambda e^+\nu_e$	$(2.0 \pm 0.5) \times 10^{-5}$		71

$\Delta S = \Delta Q$ (SQ) violating modes or
$\Delta S = 1$ weak neutral current (S1) modes

$ne^+\nu_e$	SQ	< 5	$\times 10^{-6}$	90%	224
$n\mu^+\nu_\mu$	SQ	< 3.0	$\times 10^{-5}$	90%	202
pe^+e^-	S1	< 7	$\times 10^{-6}$		225

Σ^0 $I(J^P) = 1(\frac{1}{2}^+)$

J^P not measured; assumed to be the same as for the Σ^+ and Σ^-.
Mass $m = 1192.55 \pm 0.08$ MeV   (S = 1.2)
$m_{\Sigma^-} - m_{\Sigma^0} = 4.88 \pm 0.08$ MeV   (S = 1.2)
$m_{\Sigma^0} - m_\Lambda = 76.87 \pm 0.08$ MeV   (S = 1.2)
Mean life $\tau = (7.4 \pm 0.7) \times 10^{-20}$ s
   $c\tau = 2.22 \times 10^{-11}$ m
Transition magnetic moment $|\mu_{\Sigma\Lambda}| = 1.61 \pm 0.08$ μ_N

Σ^0 DECAY MODES	Fraction (Γ_i/Γ)	Confidence level	p (MeV/c)
$\Lambda\gamma$	100 %		74
$\Lambda\gamma\gamma$	< 3 %	90%	74
Λe^+e^-	[j] 5×10^{-3}		74

Σ^- $I(J^P) = 1(\frac{1}{2}^+)$

Mass $m = 1197.436 \pm 0.033$ MeV   (S = 1.2)
$m_{\Sigma^-} - m_{\Sigma^+} = 8.07 \pm 0.08$ MeV   (S = 1.9)
$m_{\Sigma^-} - m_\Lambda = 81.752 \pm 0.034$ MeV   (S = 1.2)
Mean life $\tau = (1.479 \pm 0.011) \times 10^{-10}$ s   (S = 1.3)
   $c\tau = 4.434$ cm
Magnetic moment $\mu = -1.160 \pm 0.025$ μ_N (S = 1.7)

Decay parameters

$n\pi^-$	$\alpha_- = -0.068 \pm 0.008$
"	$\phi_- = (10 \pm 15)^\circ$
"	$\gamma_- = 0.98$ [g]
"	$\Delta_- = (249^{+12}_{-120})^\circ$ [g]
$ne^-\bar\nu_e$	$g_A/g_V = 0.340 \pm 0.017$ [e]
"	$f_2(0)/f_1(0) = 0.97 \pm 0.14$
"	$D = 0.11 \pm 0.10$
$\Lambda e^-\bar\nu_e$	$g_V/g_A = 0.01 \pm 0.10$ [e] (S = 1.5)
"	$g_{WM}/g_A = 2.4 \pm 1.7$ [e]

Σ^- DECAY MODES	Fraction (Γ_i/Γ)	p (MeV/c)
$n\pi^-$	(99.848 ± 0.005) %	193
$n\pi^-\gamma$	[h] $(4.6 \pm 0.6) \times 10^{-4}$	193
$ne^-\bar\nu_e$	$(1.017 \pm 0.034) \times 10^{-3}$	230
$n\mu^-\bar\nu_\mu$	$(4.5 \pm 0.4) \times 10^{-4}$	210
$\Lambda e^-\bar\nu_e$	$(5.73 \pm 0.27) \times 10^{-5}$	79

$\Sigma(1385)$ P_{13} $I(J^P) = 1(\frac{3}{2}^+)$

$\Sigma(1385)^+$ mass $m = 1382.8 \pm 0.4$ MeV   (S = 2.0)
$\Sigma(1385)^0$ mass $m = 1383.7 \pm 1.0$ MeV   (S = 1.4)
$\Sigma(1385)^-$ mass $m = 1387.2 \pm 0.5$ MeV   (S = 2.2)
$\Sigma(1385)^+$ full width $\Gamma = 35.8 \pm 0.8$ MeV
$\Sigma(1385)^0$ full width $\Gamma = 36 \pm 5$ MeV
$\Sigma(1385)^-$ full width $\Gamma = 39.4 \pm 2.1$ MeV   (S = 1.7)
   Below $\overline{K}N$ threshold

$\Sigma(1385)$ DECAY MODES	Fraction (Γ_i/Γ)	p (MeV/c)
$\Lambda\pi$	88 ± 2 %	208
$\Sigma\pi$	12 ± 2 %	127

$\Sigma(1660)$ P_{11} $I(J^P) = 1(\frac{1}{2}^+)$

Mass $m = 1630$ to 1690 (≈ 1660) MeV
Full width $\Gamma = 40$ to 200 (≈ 100) MeV
$p_{beam} = 0.72$ GeV/c $4\pi\lambda^2 = 29.9$ mb

$\Sigma(1660)$ DECAY MODES	Fraction (Γ_i/Γ)	p (MeV/c)
$N\overline{K}$	10–30 %	405
$\Lambda\pi$	seen	439
$\Sigma\pi$	seen	385

$\Sigma(1670)$ D_{13} $I(J^P) = 1(\frac{3}{2}^-)$

Mass $m = 1665$ to 1685 (≈ 1670) MeV
Full width $\Gamma = 40$ to 80 (≈ 60) MeV
$p_{beam} = 0.74$ GeV/c $4\pi\lambda^2 = 28.5$ mb

$\Sigma(1670)$ DECAY MODES	Fraction (Γ_i/Γ)	p (MeV/c)
$N\overline{K}$	7–13 %	414
$\Lambda\pi$	5–15 %	447
$\Sigma\pi$	30–60 %	393

$\Sigma(1750)$ S_{11} $I(J^P) = 1(\frac{1}{2}^-)$

Mass $m = 1730$ to 1800 (≈ 1750) MeV
Full width $\Gamma = 60$ to 160 (≈ 90) MeV
$p_{beam} = 0.91$ GeV/c $4\pi\lambda^2 = 20.7$ mb

$\Sigma(1750)$ DECAY MODES	Fraction (Γ_i/Γ)	p (MeV/c)
$N\overline{K}$	10–40 %	486
$\Lambda\pi$	seen	507
$\Sigma\pi$	<8 %	455
$\Sigma\eta$	15–55 %	81

$\Sigma(1775) \, D_{15}$ $I(J^P) = 1(\frac{5}{2}^-)$

Mass $m = 1770$ to 1780 (≈ 1775) MeV
Full width $\Gamma = 105$ to 135 (≈ 120) MeV
$p_{beam} = 0.96$ GeV/c $4\pi \lambdabar^2 = 19.0$ mb

$\Sigma(1775)$ DECAY MODES	Fraction (Γ_i/Γ)	p (MeV/c)
$N\overline{K}$	37–43%	508
$\Lambda\pi$	14–20%	525
$\Sigma\pi$	2–5%	474
$\Sigma(1385)\pi$	8–12%	324
$\Lambda(1520)\pi$	17–23%	198

$\Sigma(1915) \, F_{15}$ $I(J^P) = 1(\frac{5}{2}^+)$

Mass $m = 1900$ to 1935 (≈ 1915) MeV
Full width $\Gamma = 80$ to 160 (≈ 120) MeV
$p_{beam} = 1.26$ GeV/c $4\pi \lambdabar^2 = 12.8$ mb

$\Sigma(1915)$ DECAY MODES	Fraction (Γ_i/Γ)	p (MeV/c)
$N\overline{K}$	5–15 %	618
$\Lambda\pi$	seen	622
$\Sigma\pi$	seen	577
$\Sigma(1385)\pi$	<5 %	440

$\Sigma(1940) \, D_{13}$ $I(J^P) = 1(\frac{3}{2}^-)$

Mass $m = 1900$ to 1950 (≈ 1940) MeV
Full width $\Gamma = 150$ to 300 (≈ 220) MeV
$p_{beam} = 1.32$ GeV/c $4\pi \lambdabar^2 = 12.1$ mb

$\Sigma(1940)$ DECAY MODES	Fraction (Γ_i/Γ)	p (MeV/c)
$N\overline{K}$	<20 %	637
$\Lambda\pi$	seen	639
$\Sigma\pi$	seen	594
$\Sigma(1385)\pi$	seen	460
$\Lambda(1520)\pi$	seen	354
$\Delta(1232)\overline{K}$	seen	410
$N\overline{K}^*(892)$	seen	320

$\Sigma(2030) \, F_{17}$ $I(J^P) = 1(\frac{7}{2}^+)$

Mass $m = 2025$ to 2040 (≈ 2030) MeV
Full width $\Gamma = 150$ to 200 (≈ 180) MeV
$p_{beam} = 1.52$ GeV/c $4\pi \lambdabar^2 = 9.93$ mb

$\Sigma(2030)$ DECAY MODES	Fraction (Γ_i/Γ)	p (MeV/c)
$N\overline{K}$	17–23 %	702
$\Lambda\pi$	17–23 %	700
$\Sigma\pi$	5–10 %	657
ΞK	<2 %	412
$\Sigma(1385)\pi$	5–15 %	529
$\Lambda(1520)\pi$	10–20 %	430
$\Delta(1232)\overline{K}$	10–20 %	498
$N\overline{K}^*(892)$	<5 %	438

$\Sigma(2250)$ $I(J^P) = 1(?^?)$

Mass $m = 2210$ to 2280 (≈ 2250) MeV
Full width $\Gamma = 60$ to 150 (≈ 100) MeV
$p_{beam} = 2.04$ GeV/c $4\pi \lambdabar^2 = 6.76$ mb

$\Sigma(2250)$ DECAY MODES	Fraction (Γ_i/Γ)	p (MeV/c)
$N\overline{K}$	<10 %	851
$\Lambda\pi$	seen	842
$\Sigma\pi$	seen	803

Ξ BARYONS
$(S = -2, I = 1/2)$
$\Xi^0 = uss, \quad \Xi^- = dss$

Ξ^0 $I(J^P) = \frac{1}{2}(\frac{1}{2}^+)$

P is not yet measured; $+$ is the quark model prediction.
Mass $m = 1314.9 \pm 0.6$ MeV
$m_{\Xi^-} - m_{\Xi^0} = 6.4 \pm 0.6$ MeV
Mean life $\tau = (2.90 \pm 0.09) \times 10^{-10}$ s
$c\tau = 8.71$ cm
Magnetic moment $\mu = -1.250 \pm 0.014 \, \mu_N$

Decay parameters

$\Lambda\pi^0$	$\alpha = -0.411 \pm 0.022$	$(S = 2.1)$
"	$\phi = (21 \pm 12)^\circ$	
"	$\gamma = 0.85$ [g]	
"	$\Delta = (218^{+12}_{-19})^\circ$ [g]	
$\Lambda\gamma$	$\alpha = 0.4 \pm 0.4$	
$\Sigma^0\gamma$	$\alpha = 0.20 \pm 0.32$	

Ξ^0 DECAY MODES	Fraction (Γ_i/Γ)	Confidence level	p (MeV/c)
$\Lambda\pi^0$	(99.54 ± 0.05) %		135
$\Lambda\gamma$	$(1.06 \pm 0.16) \times 10^{-3}$		184
$\Sigma^0\gamma$	$(3.5 \pm 0.4) \times 10^{-3}$		117
$\Sigma^+ e^- \overline{\nu}_e$	$< 1.1 \times 10^{-3}$	90%	120
$\Sigma^+ \mu^- \overline{\nu}_\mu$	$< 1.1 \times 10^{-3}$	90%	64

$\Delta S = \Delta Q$ (SQ) violating modes or
$\Delta S = 2$ forbidden ($S2$) modes

$\Sigma^- e^+ \nu_e$	SQ	$< 9 \times 10^{-4}$	90%	112
$\Sigma^- \mu^+ \nu_\mu$	SQ	$< 9 \times 10^{-4}$	90%	49
$p\pi^-$	$S2$	$< 4 \times 10^{-5}$	90%	299
$p e^- \overline{\nu}_e$	$S2$	$< 1.3 \times 10^{-3}$		323
$p \mu^- \overline{\nu}_\mu$	$S2$	$< 1.3 \times 10^{-3}$		309

Ξ^- $I(J^P) = \frac{1}{2}(\frac{1}{2}^+)$

P is not yet measured; $+$ is the quark model prediction.
Mass $m = 1321.32 \pm 0.13$ MeV
Mean life $\tau = (1.639 \pm 0.015) \times 10^{-10}$ s
$c\tau = 4.91$ cm
Magnetic moment $\mu = -0.6507 \pm 0.0025 \, \mu_N$

Decay parameters

$\Lambda\pi^-$	$\alpha = -0.456 \pm 0.014$	$(S = 1.8)$
"	$\phi = (4 \pm 4)^\circ$	
"	$\gamma = 0.89$ [g]	
"	$\Delta = (188 \pm 8)^\circ$ [g]	
$\Lambda e^- \overline{\nu}_e$	$g_A/g_V = -0.25 \pm 0.05$ [e]	

Ξ^- DECAY MODES	Fraction (Γ_i/Γ)	Confidence level	p (MeV/c)
$\Lambda\pi^-$	(99.887 ± 0.035) %		139
$\Sigma^-\gamma$	$(1.27 \pm 0.23) \times 10^{-4}$		118
$\Lambda e^- \overline{\nu}_e$	$(5.63 \pm 0.31) \times 10^{-4}$		190
$\Lambda \mu^- \overline{\nu}_\mu$	$(3.5 ^{+3.5}_{-2.2}) \times 10^{-4}$		163
$\Sigma^0 e^- \overline{\nu}_e$	$(8.7 \pm 1.7) \times 10^{-5}$		122
$\Sigma^0 \mu^- \overline{\nu}_\mu$	$< 8 \times 10^{-4}$	90%	70
$\Xi^0 e^- \overline{\nu}_e$	$< 2.3 \times 10^{-3}$	90%	6

$\Delta S = 2$ forbidden ($S2$) modes

$n\pi^-$	$S2$	$< 1.9 \times 10^{-5}$	90%	303
$n e^- \overline{\nu}_e$	$S2$	$< 3.2 \times 10^{-3}$	90%	327
$n \mu^- \overline{\nu}_\mu$	$S2$	< 1.5 %	90%	314
$p\pi^-\pi^-$	$S2$	$< 4 \times 10^{-4}$	90%	223
$p\pi^- e^- \overline{\nu}_e$	$S2$	$< 4 \times 10^{-4}$	90%	304
$p\pi^- \mu^- \overline{\nu}_\mu$	$S2$	$< 4 \times 10^{-4}$	90%	250
$p\mu^-\mu^-$	L	$< 4 \times 10^{-4}$	90%	272

$\Xi(1530)\ P_{13}$ $I(J^P) = \frac{1}{2}(\frac{3}{2}^+)$

$\Xi(1530)^0$ mass $m = 1531.80 \pm 0.32$ MeV   (S = 1.3)
$\Xi(1530)^-$ mass $m = 1535.0 \pm 0.6$ MeV
$\Xi(1530)^0$ full width $\Gamma = 9.1 \pm 0.5$ MeV
$\Xi(1530)^-$ full width $\Gamma = 9.9^{+1.7}_{-1.9}$ MeV

$\Xi(1530)$ DECAY MODES	Fraction (Γ_i/Γ)	Confidence level	p (MeV/c)
$\Xi\pi$	100 %		152
$\Xi\gamma$	<4 %	90%	200

$\Xi(1690)$ $I(J^P) = \frac{1}{2}(?^?)$

Mass $m = 1690 \pm 10$ MeV [l]
Full width $\Gamma < 50$ MeV

$\Xi(1690)$ DECAY MODES	Fraction (Γ_i/Γ)	p (MeV/c)
$\Lambda\overline{K}$	seen	240
$\Sigma\overline{K}$	seen	51
$\Xi^-\pi^+\pi^-$	possibly seen	214

$\Xi(1820)\ D_{13}$ $I(J^P) = \frac{1}{2}(\frac{3}{2}^-)$

Mass $m = 1823 \pm 5$ MeV [l]
Full width $\Gamma = 24^{+15}_{-10}$ MeV [l]

$\Xi(1820)$ DECAY MODES	Fraction (Γ_i/Γ)	p (MeV/c)
$\Lambda\overline{K}$	large	400
$\Sigma\overline{K}$	small	320
$\Xi\pi$	small	413
$\Xi(1530)\pi$	small	234

$\Xi(1950)$ $I(J^P) = \frac{1}{2}(?^?)$

Mass $m = 1950 \pm 15$ MeV [l]
Full width $\Gamma = 60 \pm 20$ MeV [l]

$\Xi(1950)$ DECAY MODES	Fraction (Γ_i/Γ)	p (MeV/c)
$\Lambda\overline{K}$	seen	522
$\Sigma\overline{K}$	possibly seen	460
$\Xi\pi$	seen	518

$\Xi(2030)$ $I(J^P) = \frac{1}{2}(\geq \frac{5}{2}^?)$

Mass $m = 2025 \pm 5$ MeV [l]
Full width $\Gamma = 20^{+15}_{-5}$ MeV [l]

$\Xi(2030)$ DECAY MODES	Fraction (Γ_i/Γ)	p (MeV/c)
$\Lambda\overline{K}$	~ 20 %	589
$\Sigma\overline{K}$	~ 80 %	533
$\Xi\pi$	small	573
$\Xi(1530)\pi$	small	421
$\Lambda\overline{K}\pi$	small	501
$\Sigma\overline{K}\pi$	small	430

Ω BARYONS
$(S = -3, I = 0)$
$\Omega^- = sss$

Ω^- $I(J^P) = 0(\frac{3}{2}^+)$

J^P is not yet measured; $\frac{3}{2}^+$ is the quark model prediction.

Mass $m = 1672.45 \pm 0.29$ MeV
Mean life $\tau = (0.822 \pm 0.012) \times 10^{-10}$ s
   $c\tau = 2.46$ cm
Magnetic moment $\mu = -2.02 \pm 0.05\ \mu_N$

Decay parameters

ΛK^-	$\alpha = -0.026 \pm 0.026$
$\Xi^0\pi^-$	$\alpha = 0.09 \pm 0.14$
$\Xi^-\pi^0$	$\alpha = 0.05 \pm 0.21$

Ω^- DECAY MODES		Fraction (Γ_i/Γ)	Confidence level	p (MeV/c)
ΛK^-		(67.8 ± 0.7) %		211
$\Xi^0\pi^-$		(23.6 ± 0.7) %		294
$\Xi^-\pi^0$		(8.6 ± 0.4) %		290
$\Xi^-\pi^+\pi^-$		$(4.3^{+3.4}_{-1.3}) \times 10^{-4}$		190
$\Xi(1530)^0\pi^-$		$(6.4^{+5.1}_{-2.0}) \times 10^{-4}$		17
$\Xi^0 e^-\overline{\nu}_e$		$(5.6 \pm 2.8) \times 10^{-3}$		319
$\Xi^-\gamma$		$< 4.6 \quad \times 10^{-4}$	90%	314

$\Delta S = 2$ forbidden ($S2$) modes

$\Lambda\pi^-$	$S2$	$< 1.9 \quad \times 10^{-4}$	90%	449

$\Omega(2250)^-$ $I(J^P) = 0(?^?)$

Mass $m = 2252 \pm 9$ MeV
Full width $\Gamma = 55 \pm 18$ MeV

$\Omega(2250)^-$ DECAY MODES	Fraction (Γ_i/Γ)	p (MeV/c)
$\Xi^-\pi^+ K^-$	seen	531
$\Xi(1530)^0 K^-$	seen	437

CHARMED BARYONS
$(C = +1)$

$\Lambda_c^+ = udc,\ \ \Sigma_c^{++} = uuc,\ \ \Sigma_c^+ = udc,\ \ \Sigma_c^0 = ddc,$
$\Xi_c^+ = usc,\ \ \Xi_c^0 = dsc,\ \ \Omega_c^0 = ssc$

Λ_c^+ $I(J^P) = 0(\frac{1}{2}^+)$

J not confirmed; $\frac{1}{2}$ is the quark model prediction.

Mass $m = 2284.9 \pm 0.6$ MeV
Mean life $\tau = (0.206 \pm 0.012) \times 10^{-12}$ s
   $c\tau = 61.8\ \mu$m

Decay asymmetry parameters

$\Lambda\pi^+$	$\alpha = -0.98 \pm 0.19$
$\Sigma^+\pi^0$	$\alpha = -0.45 \pm 0.32$
$\Lambda\ell^+\nu_\ell$	$\alpha = -0.82^{+0.11}_{-0.07}$

Λ_c^+ DECAY MODES		Fraction (Γ_i/Γ)	Scale factor/ Confidence level	p (MeV/c)
Hadronic modes with a p and one $\overline{K}$				
$p\overline{K}^0$		(2.2 ± 0.4) %		872
$p K^-\pi^+$		(4.4 ± 0.6) %		822
$p\overline{K}^*(892)^0$	[k]	(1.6 ± 0.4) %		681
$\Delta(1232)^{++}K^-$		$(7 \pm 4) \times 10^{-3}$		709
$\Lambda(1520)\pi^+$	[k]	$(4.0^{+2.0}_{-1.7}) \times 10^{-3}$		626
$p K^-\pi^+$ nonresonant		$(2.5^{+0.5}_{-0.6})$ %		822
$p\overline{K}^0\eta$		(1.10 ± 0.29) %		567

$p \overline{K}^0 \pi^+ \pi^-$		$(2.1 \pm 0.8)\%$	753
$p K^- \pi^+ \pi^0$		seen	758
$p K^*(892)^- \pi^+$	[k]	$(9 \pm 5) \times 10^{-3}$	579
$p (K^- \pi^+)_{nonresonant} \pi^0$		$(3.2 \pm 0.7)\%$	758
$\Delta(1232) \overline{K}^*(892)$		seen	416
$p K^- \pi^+ \pi^+ \pi^-$		$(10 \pm 7) \times 10^{-4}$	670
$p K^- \pi^+ \pi^0 \pi^0$		$(7.0 \pm 3.5) \times 10^{-3}$	676
$p K^- \pi^+ \pi^0 \pi^0 \pi^0$		$(4.4 \pm 2.8) \times 10^{-3}$	573

Hadronic modes with a p and zero or two K's

$p \pi^+ \pi^-$		$(3.0 \pm 1.6) \times 10^{-3}$	926
$p f_0(980)$	[k]	$(2.4 \pm 1.6) \times 10^{-3}$	621
$p \pi^+ \pi^+ \pi^- \pi^-$		$(1.6 \pm 1.0) \times 10^{-3}$	851
$p K^+ K^-$		$(2.0 \pm 0.6) \times 10^{-3}$	615
$p \phi$	[k]	$(1.06 \pm 0.33) \times 10^{-3}$	589

Hadronic modes with a hyperon

$\Lambda \pi^+$		$(7.9 \pm 1.8) \times 10^{-3}$		863
$\Lambda \pi^+ \pi^0$		$(3.2 \pm 0.9)\%$		843
$\Lambda \rho^0$		$< 4 \%$	CL=95%	638
$\Lambda \pi^+ \pi^+ \pi^-$		$(2.9 \pm 0.6)\%$		806
$\Lambda \pi^+ \eta$		$(1.5 \pm 0.4)\%$		690
$\Sigma(1385)^+ \eta$	[k]	$(7.5 \pm 2.4) \times 10^{-3}$		569
$\Lambda K^+ \overline{K}^0$		$(5.3 \pm 1.4) \times 10^{-3}$		441
$\Sigma^0 \pi^+$		$(8.8 \pm 2.0) \times 10^{-3}$		824
$\Sigma^+ \pi^0$		$(8.8 \pm 2.2) \times 10^{-3}$		826
$\Sigma^+ \eta$		$(4.8 \pm 1.7) \times 10^{-3}$		712
$\Sigma^+ \pi^+ \pi^-$		$(3.0 \pm 0.6)\%$		803
$\Sigma^+ \rho^0$		$< 1.2 \%$	CL=95%	578
$\Sigma^- \pi^+ \pi^+$		$(1.6 \pm 0.6)\%$		798
$\Sigma^0 \pi^+ \pi^0$		$(1.6 \pm 0.6)\%$		802
$\Sigma^0 \pi^+ \pi^+ \pi^-$		$(9.2 \pm 3.4) \times 10^{-3}$		762
$\Sigma^+ \pi^+ \pi^- \pi^0$				766
$\Sigma^+ \omega$	[k]	$(2.4 \pm 0.7)\%$		568
$\Sigma^+ \pi^+ \pi^+ \pi^- \pi^-$		$(2.6 \, {}^{+3.5}_{-1.8}) \times 10^{-3}$		707
$\Sigma^+ K^+ K^-$		$(3.1 \pm 0.8) \times 10^{-3}$		346
$\Sigma^+ \phi$	[k]	$(3.0 \pm 1.3) \times 10^{-3}$		292
$\Sigma^+ K^+ \pi^-$		$(5.7 \, {}^{+5.3}_{-3.2}) \times 10^{-3}$		668
$\Xi^0 K^+$		$(3.4 \pm 0.9) \times 10^{-3}$		652
$\Xi^- K^+ \pi^+$		$(4.3 \pm 1.1) \times 10^{-3}$		564
$\Xi(1530)^0 K^+$	[k]	$(2.3 \pm 0.7) \times 10^{-3}$		471

Semileptonic modes

$\Lambda \ell^+ \nu_\ell$	[l]	$(2.3 \pm 0.5)\%$	–
e^+ anything		$(4.5 \pm 1.7)\%$	–
$p e^+$ anything		$(1.8 \pm 0.9)\%$	–
Λe^+ anything		$(1.6 \pm 0.6)\%$	–
$\Lambda \mu^+$ anything		$(1.5 \pm 0.9)\%$	–

Inclusive modes

p anything		$(50 \pm 16)\%$	–
p anything (no Λ)		$(12 \pm 19)\%$	–
n anything		$(50 \pm 16)\%$	–
n anything (no Λ)		$(29 \pm 17)\%$	–
Λ anything		$(35 \pm 11)\%$	S=1.4 –
$\Sigma^\pm$ anything	[m]	$(10 \pm 5)\%$	–

$\Delta C = 1$ weak neutral current ($C1$) modes, or Lepton number (L) violating modes

$p \mu^+ \mu^-$	$C1$	< 3.4	$\times 10^{-4}$ CL=90%	936
$\Sigma^- \mu^+ \mu^+$	L	< 7.0	$\times 10^{-4}$ CL=90%	811

$\Lambda_c(2593)^+$ $\qquad I(J^P) = 0(\tfrac{1}{2}^-)$

The spin-parity follows from the fact that $\Sigma_c(2455)\pi$ decays, with little available phase space, dominate.

Mass $m = 2593.6 \pm 1.0$ MeV $(S = 1.2)$
$m - m_{\Lambda_c^+} = 308.6 \pm 0.8$ MeV $(S = 1.3)$
Full width $\Gamma = 3.9 \, {}^{+2.4}_{-1.6}$ MeV

$\Lambda_c^+ \pi\pi$ and $\Sigma_c(2455)\pi$ — the latter just barely — are the only strong decays allowed to an excited Λ_c^+ having this mass; and the $\Lambda_c^+ \pi^+ \pi^-$ mode seems to be largely via $\Sigma_c^{++}\pi^-$ or $\Sigma_c^0 \pi^+$.

$\Lambda_c(2593)^+$ DECAY MODES	Fraction (Γ_i/Γ)	p (MeV/c)
$\Lambda_c^+ \pi^+ \pi^-$	seen	124
$\Sigma_c(2455)^{++} \pi^-$	large	17
$\Sigma_c(2455)^0 \pi^+$	large	23
$\Lambda_c^+ \pi^+ \pi^-$ 3-body	small	124
$\Lambda_c^+ \pi^0$	not seen	261
$\Lambda_c^+ \gamma$	not seen	290

$\Lambda_c(2625)^+$ $\qquad I(J^P) = 0(?^?)$

J^P is expected to be $3/2^-$.

Mass $m = 2626.4 \pm 0.9$ MeV $(S = 1.3)$
$m - m_{\Lambda_c^+} = 341.5 \pm 0.8$ MeV $(S = 1.9)$
Full width $\Gamma < 1.9$ MeV, CL = 90%

$\Lambda_c^+ \pi\pi$ and $\Sigma(2455)\pi$ are the only strong decays allowed to an excited Λ_c^+ having this mass.

$\Lambda_c(2625)^+$ DECAY MODES	Fraction (Γ_i/Γ)	p (MeV/c)
$\Lambda_c^+ \pi^+ \pi^-$	seen	184
$\Sigma_c(2455)^{++} \pi^-$	small	100
$\Sigma_c(2455)^0 \pi^+$	small	101
$\Lambda_c^+ \pi^+ \pi^-$ 3-body	large	184
$\Lambda_c^+ \pi^0$	not seen	293
$\Lambda_c^+ \gamma$	not seen	319

$\Sigma_c(2455)$ $\qquad I(J^P) = 1(\tfrac{1}{2}^+)$

J^P not confirmed; $\tfrac{1}{2}^+$ is the quark model prediction.

$\Sigma_c(2455)^{++}$ mass $m = 2452.9 \pm 0.6$ MeV
$\Sigma_c(2455)^+$ mass $m = 2453.5 \pm 0.9$ MeV
$\Sigma_c(2455)^0$ mass $m = 2452.1 \pm 0.7$ MeV
$m_{\Sigma_c^{++}} - m_{\Lambda_c^+} = 167.95 \pm 0.25$ MeV
$m_{\Sigma_c^+} - m_{\Lambda_c^+} = 168.5 \pm 0.7$ MeV $(S = 1.1)$
$m_{\Sigma_c^0} - m_{\Lambda_c^+} = 167.2 \pm 0.4$ MeV $(S = 1.1)$
$m_{\Sigma_c^{++}} - m_{\Sigma_c^0} = 0.79 \pm 0.33$ MeV $(S = 1.2)$
$m_{\Sigma_c^+} - m_{\Sigma_c^0} = 1.4 \pm 0.6$ MeV

$\Lambda_c^+ \pi$ is the only strong decay allowed to a Σ_c having this mass.

$\Sigma_c(2455)$ DECAY MODES	Fraction (Γ_i/Γ)	p (MeV/c)
$\Lambda_c^+ \pi$	$\approx 100\%$	90

Ξ_c^+ $\qquad I(J^P) = \tfrac{1}{2}(\tfrac{1}{2}^+)$

$I(J^P)$ not confirmed; $\tfrac{1}{2}(\tfrac{1}{2}^+)$ is the quark model prediction.

Mass $m = 2465.6 \pm 1.4$ MeV
Mean life $\tau = (0.35 \, {}^{+0.07}_{-0.04}) \times 10^{-12}$ s
$c\tau = 106 \; \mu m$

Ξ_c^+ DECAY MODES	Fraction (Γ_i/Γ)	p (MeV/c)
$\Lambda K^- \pi^+ \pi^+$	seen	784
$\Lambda \overline{K}^*(892)^0 \pi^+$	not seen	601
$\Sigma(1385)^+ K^- \pi^+$	not seen	676
$\Sigma^+ K^- \pi^+$	seen	808
$\Sigma^+ \overline{K}^*(892)^0$	seen	653
$\Sigma^0 K^- \pi^+ \pi^+$	seen	733
$\Xi^0 \pi^+$	seen	875
$\Xi^- \pi^+ \pi^+$	seen	850
$\Xi(1530)^0 \pi^+$	not seen	748
$\Xi^0 \pi^+ \pi^0$	seen	854
$\Xi^0 \pi^+ \pi^+ \pi^-$	seen	817
$\Xi^0 e^+ \nu_e$	seen	882

Ξ_c^0 $I(J^P) = \frac{1}{2}(\frac{1}{2}^+)$

$I(J^P)$ not confirmed; $\frac{1}{2}(\frac{1}{2}^+)$ is the quark model prediction.

Mass $m = 2470.3 \pm 1.8$ MeV   (S = 1.3)

$m_{\Xi_c^0} - m_{\Xi_c^+} = 4.7 \pm 2.1$ MeV   (S = 1.2)

Mean life $\tau = (0.098^{+0.023}_{-0.015}) \times 10^{-12}$ s

$c\tau = 29$ μm

Ξ_c^0 DECAY MODES	Fraction (Γ_i/Γ)	p (MeV/c)
$\Lambda \overline{K}^0$	seen	864
$\Xi^- \pi^+$	seen	875
$\Xi^- \pi^+ \pi^+ \pi^-$	seen	816
$p K^- \overline{K}^*(892)^0$	seen	406
$\Omega^- K^+$	seen	522
$\Xi^- e^+ \nu_e$	seen	882
$\Xi^- \ell^+$ anything	seen	–

$\Xi_c(2645)$ $I(J^P) = ?(?^?)$

Mass $m = 2643.8 \pm 1.8$ MeV

$m_{\Xi_c(2645)^0} - m_{\Xi_c^+} = 178.2 \pm 1.1$ MeV

Full width $\Gamma < 5.5$ MeV, CL = 90%

$\Xi_c \pi$ is the only strong decay allowed to a Ξ_c resonance having this mass.

$\Xi_c(2645)$ DECAY MODES	Fraction (Γ_i/Γ)	p (MeV/c)
$\Xi_c^+ \pi^-$	seen	107

Ω_c^0 $I(J^P) = 0(\frac{1}{2}^+)$

$I(J^P)$ not confirmed; $0(\frac{1}{2}^+)$ is the quark model prediction.

Mass $m = 2704 \pm 4$ MeV   (S = 1.8)

Mean life $\tau = (0.064 \pm 0.020) \times 10^{-12}$ s

$c\tau = 19$ μm

Ω_c^0 DECAY MODES	Fraction (Γ_i/Γ)	p (MeV/c)
$\Sigma^+ K^- K^- \pi^+$	seen	697
$\Xi^- K^- \pi^+ \pi^+$	seen	838
$\Omega^- \pi^+$	seen	827
$\Omega^- \pi^- \pi^+ \pi^+$	seen	759

BOTTOM (BEAUTY) BARYONS
(B = −1)

$\Lambda_b^0 = udb, \; \Xi_b^0 = usb, \; \Xi_b^- = dsb$

Λ_b^0 $I(J^P) = 0(\frac{1}{2}^+)$

$I(J^P)$ not yet measured; $0(\frac{1}{2}^+)$ is the quark model prediction.

Mass $m = 5641 \pm 50$ MeV

Mean life $\tau = (1.14 \pm 0.08) \times 10^{-12}$ s

$c\tau = 342$ μm

These branching fractions are actually an average over weakly decaying b-baryons weighted by their production rates in Z decay (or high-energy $p\overline{p}$), branching ratios, and detection efficiencies. They scale with the LEP Λ_b production fraction B($b \to \Lambda_b$) and are evaluated for our value B($b \to \Lambda_b$) = (13.2 ± 4.1)%.

The branching fractions B($\Lambda_b^0 \to \Lambda \ell^- \overline{\nu}_\ell$anything) and B($\Lambda_b^0 \to \Lambda_c^+ \ell^- \overline{\nu}_\ell$anything) are not pure measurements because the underlying measured products of these with B($b \to \Lambda_b$) were used to determine B($b \to \Lambda_b$), as described in the note "Production and Decay of b-Flavored Hadrons."

Λ_b^0 DECAY MODES	Fraction (Γ_i/Γ)	p (MeV/c)
$J/\psi(1S) \Lambda$	(1.4 ± 0.9) %	1756
$p D^0 \pi^-$	seen	2383
$\Lambda_c^+ \pi^+ \pi^- \pi^-$	seen	2336
$p \mu^- \overline{\nu}$anything	(3.7 ± 1.7) %	
$\Lambda \ell^- \overline{\nu}_\ell$anything	[n](2.5 ± 0.5) %	
$\Lambda_c^+ \ell^- \overline{\nu}_\ell$anything	[n](10.0 ± 3.0) %	
$\Lambda / \overline{\Lambda}$anything	(17 $^{+11}_{-8}$) %	

NOTES

This Summary Table only includes established baryons. The Particle Listings include evidence for other baryons. The masses, widths, and branching fractions for the resonances in this Table are Breit-Wigner parameters. The Particle Listings also give, where available, pole parameters. See, in particular, the *Note on N and Δ Resonances*.

For most of the resonances, the parameters come from various partial-wave analyses of more or less the same sets of data, and it is not appropriate to treat the results of the analyses as independent or to average them together. Furthermore, the systematic errors on the results are not well understood. Thus, we usually only give ranges for the parameters. We then also give a best guess for the mass (as part of the name of the resonance) and for the width. The *Note on N and Δ Resonances* and the *Note on Λ and Σ Resonances* in the Particle Listings review the partial-wave analyses.

When a quantity has "(S = ...)" to its right, the error on the quantity has been enlarged by the "scale factor" S, defined as $S = \sqrt{\chi^2/(N-1)}$, where N is the number of measurements used in calculating the quantity. We do this when S > 1, which often indicates that the measurements are inconsistent. When S > 1.25, we also show in the Particle Listings an ideogram of the measurements. For more about S, see the Introduction.

A decay momentum p is given for each decay mode. For a 2-body decay, p is the momentum of each decay product in the rest frame of the decaying particle. For a 3-or-more-body decay, p is the largest momentum any of the products can have in this frame. For any resonance, the *nominal* mass is used in calculating p. A dagger ("†") in this column indicates that the mode is forbidden when the nominal masses of resonances are used, but is in fact allowed due to the nonzero widths of the resonances.

[a] The masses of the p and n are most precisely known in u (unified atomic mass units). The conversion factor to MeV, 1 u = 931.49432 ± 0.00028 MeV, is less well known in u.

[b] The limit is from neutrality-of-matter experiments; it assumes $q_n = q_p + q_e$. See also the charge of the neutron.

[c] The first limit is geochemical and independent of decay mode. The second entry, a range of limits, assumes the dominant decay modes are among those investigated. For antiprotons the best limit, inferred from the observation of cosmic ray $\overline{p}$'s is $\tau_{\overline{p}} > 10^7$ yr, the cosmic-ray storage time, but this limit depends on a number of assumptions. The best direct observation of stored antiprotons gives $\tau_{\overline{p}}/B(\overline{p} \to e^- \gamma) > 1848$ yr.

[d] There is some controversy about whether nuclear physics and model dependence complicate the analysis for bound neutrons (from which the best limit comes). The second limit here is from reactor experiments with free neutrons.

[e] The parameters g_A, g_V, and g_{WM} for semileptonic modes are defined by $\overline{B}_f[\gamma_\lambda(g_V + g_A\gamma_5) + i(g_{WM}/m_{B_i}) \sigma_{\lambda\nu} q^\nu]B_i$, and ϕ_{AV} is defined by $g_A/g_V = |g_A/g_V|e^{i\phi_{AV}}$. See the "Note on Baryon Decay Parameters" in the neutron Particle Listings.

[f] Time-reversal invariance requires this to be 0° or 180°.

[g] The decay parameters γ and Δ are calculated from α and ϕ using

$$\gamma = \sqrt{1-\alpha^2}\,\cos\phi, \qquad \tan\Delta = -\frac{1}{\alpha}\,\sqrt{1-\alpha^2}\,\sin\phi.$$

See the "Note on Baryon Decay Parameters" in the neutron Particle Listings.

[h] See the Particle Listings for the pion momentum range used in this measurement.

[i] The error given here is only an educated guess. It is larger than the error on the weighted average of the published values.

[j] A theoretical value using QED.

[k] This branching fraction includes all the decay modes of the final-state resonance.

[l] ℓ indicates e or μ mode, not sum over modes.

[m] The value is for the sum of the charge states of particle/antiparticle states indicated.

[n] Not a pure measurement. See note at head of Λ_b^0 Decay Modes.

MONOPOLES, SUPERSYMMETRY, COMPOSITENESS, etc., SEARCHES FOR

Magnetic Monopole Searches

Isolated candidate events have not been confirmed. Most experiments obtain negative results.

Supersymmetric Particle Searches

Limits are based on the Minimal Supersymmetric Standard Model. Assumptions include: 1) $\tilde{\chi}_1^0$ (or $\tilde{\gamma}$) is lightest supersymmetric particle; 2) R-parity is conserved; 3) $m_{\tilde{t}_L} = m_{\tilde{t}_R}$, and all scalar quarks (except $\tilde{t}_L$ and $\tilde{t}_R$) are degenerate in mass.

See the Particle Listings for a Note giving details of supersymmetry.

$\tilde{\chi}_i^0$ — neutralinos (mixtures of $\tilde{\gamma}$, $\tilde{Z}^0$, and $\tilde{H}_i^0$)

Mass $m_{\tilde{\gamma}} > 15$ GeV, CL = 90%   [if $m_{\tilde{q}} = 100$ GeV (from cosmology)]

Mass $m_{\tilde{\chi}_1^0} > 23$ GeV, CL = 95%   [$\tan\beta > 3$]

Mass $m_{\tilde{\chi}_2^0} > 52$ GeV, CL = 95%   [$\tan\beta > 3$]

Mass $m_{\tilde{\chi}_3^0} > 84$ GeV, CL = 95%   [$\tan\beta > 3$]

Mass $m_{\tilde{\chi}_4^0} > 127$ GeV, CL = 95%   [$\tan\beta > 3$]

$\tilde{\chi}_i^\pm$ — charginos (mixtures of $\tilde{W}^\pm$ and $\tilde{H}_i^\pm$)

Mass $m_{\tilde{\chi}_1^\pm} > 45$ GeV, CL = 95%   [all $m_{\tilde{\chi}_1^0}$]

Mass $m_{\tilde{\chi}_2^\pm} > 99$ GeV, CL = 95%   [GUT relations assumed]

$\tilde{\nu}$ — scalar neutrino (sneutrino)

Mass $m > 37.1$ GeV, CL = 95%   [one flavor]

Mass $m > 41.8$ GeV, CL = 95%   [three degenerate flavors]

$\tilde{e}$ — scalar electron (selectron)

Mass $m > 65$ GeV, CL = 95%   [if $m_{\tilde{\gamma}} = 0$]

Mass $m > 50$ GeV, CL = 95%   [if $m_{\tilde{\gamma}} < 5$ GeV]

Mass $m > 45$ GeV, CL = 95%   [if $m_{\tilde{\chi}_1^0} < 41$ GeV]

$\tilde{\mu}$ — scalar muon (smuon)

Mass $m > 45$ GeV, CL = 95%   [if $m_{\tilde{\chi}_1^0} < 41$ GeV]

$\tilde{\tau}$ — scalar tau (stau)

Mass $m > 45$ GeV, CL = 95%   [if $m_{\tilde{\chi}_1^0} < 38$ GeV]

$\tilde{q}$ — scalar quark (squark)

These limits include the effects of cascade decays, evaluated assuming a fixed value of the parameters μ and $\tan\beta$. The limits are weakly sensitive to these parameters over much of parameter space. Limits assume GUT relations between gaugino masses and the gauge coupling; in particular that for $|\mu|$ not small, $m_{\tilde{\chi}_1^0} \approx m_{\tilde{g}}/6$.

Mass $m > 176$ GeV, CL = 95%   [any $m_{\tilde{g}} < 300$ GeV, $\mu = -250$ GeV, $\tan\beta = 2$]

Mass $m > 224$ GeV, CL = 95%   [$m_{\tilde{g}} \le m_{\tilde{q}}$, $\mu = -400$ GeV, $\tan\beta = 4$]

$\tilde{g}$ — gluino

There is some controversy about a low-mass window ($1 \lesssim m_{\tilde{g}} \lesssim 4$ GeV). Several experiments cast doubt on the existence of this window.

These limits include the effects of cascade decays, evaluated assuming a fixed value of the parameters μ and $\tan\beta$. The limits are weakly sensitive to these parameters over much of parameter space. Limits assume GUT relations between gaugino masses and the gauge coupling; in particular that for $|\mu|$ not small, $m_{\tilde{\chi}_1^0} \approx m_{\tilde{g}}/6$.

Mass $m > 154$ GeV, CL = 95%   [$m_{\tilde{g}} \le m_{\tilde{q}}$, $\mu = -400$ GeV, $\tan\beta = 4$]

Mass $m > 212$ GeV, CL = 95%   [$m_{\tilde{g}} \ge m_{\tilde{q}}$, $\mu = -250$ GeV, $\tan\beta = 2$]

Quark and Lepton Compositeness, Searches for

Scale Limits Λ for Contact Interactions (the lowest dimensional interactions with four fermions)

If the Lagrangian has the form

$$\pm \frac{g^2}{2\Lambda^2} \overline{\psi}_L \gamma_\mu \psi_L \overline{\psi}_L \gamma^\mu \psi_L$$

(with $g^2/4\pi$ set equal to 1), then we define $\Lambda \equiv \Lambda_{LL}^\pm$. For the full definitions and for other forms, see the Note in the Listings on Searches for Quark and Lepton Compositeness in the full *Review* and the original literature.

$\Lambda_{LL}^+(eeee)$ > 1.6 TeV, CL = 95%

$\Lambda_{LL}^-(eeee)$ > 3.6 TeV, CL = 95%

$\Lambda_{LL}^+(ee\mu\mu)$ > 2.6 TeV, CL = 95%

$\Lambda_{LL}^-(ee\mu\mu)$ > 1.9 TeV, CL = 95%

$\Lambda_{LL}^+(ee\tau\tau)$ > 1.9 TeV, CL = 95%

$\Lambda_{LL}^-(ee\tau\tau)$ > 2.9 TeV, CL = 95%

$\Lambda_{LL}^+(\ell\ell\ell\ell)$ > 3.5 TeV, CL = 95%

$\Lambda_{LL}^-(\ell\ell\ell\ell)$ > 2.8 TeV, CL = 95%

$\Lambda_{LL}^+(eeqq)$ > 2.3 TeV, CL = 95%

$\Lambda_{LL}^-(eeqq)$ > 2.2 TeV, CL = 95%

$\Lambda_{LL}^+(\mu\mu qq)$ > 1.4 TeV, CL = 95%

$\Lambda_{LL}^-(\mu\mu qq)$ > 1.6 TeV, CL = 95%

$\Lambda_{LR}^\pm(\nu_\mu \nu_e \mu e)$ > 3.1 TeV, CL = 90%

$\Lambda_{LL}^\pm(qqqq)$ > 1.4 TeV, CL = 95%

Recent CDF measurements of the inclusive jet cross section in $p\overline{p}$ collisions could be interpreted as tentative evidence for a four-quark contact interaction with $\Lambda_{LL}^\pm(qqqq) \sim 1.6$ TeV. However, CDF notes that uncertainty in the parton distribution functions, higher-order QCD corrections, and detector calibration may possibly account for the effect.

Excited Leptons

The limits from $\ell^{*+}\ell^{*-}$ do not depend on λ (where λ is the $\ell\ell^*$ transition coupling). The λ-dependent limits assume chiral coupling, except for the third limit for e^* which is for nonchiral coupling. For chiral coupling, this limit corresponds to $\lambda_\gamma = \sqrt{2}$.

$e^{*\pm}$ — excited electron

Mass $m > 46.1$ GeV, CL = 95%   (from $e^{*+}e^{*-}$)

Mass $m > 91$ GeV, CL = 95%   (if $\lambda_Z > 1$)

Mass $m > 146$ GeV, CL = 95%   (if $\lambda_\gamma = 1$)

$\mu^{*\pm}$ — excited muon

Mass $m > 46.1$ GeV, CL = 95%   (from $\mu^{*+}\mu^{*-}$)

Mass $m > 91$ GeV, CL = 95%   (if $\lambda_Z > 1$)

$\tau^{*\pm}$ — excited tau

Mass $m > 46.0$ GeV, CL = 95%   (from $\tau^{*+}\tau^{*-}$)

Mass $m > 90$ GeV, CL = 95%   (if $\lambda_Z > 0.18$)

ν^* — excited neutrino

Mass $m > 47$ GeV, CL = 95%   (from $\nu^*\overline{\nu}^*$)

Mass $m > 91$ GeV, CL = 95%   (if $\lambda_Z > 1$)

q^* — excited quark

Mass $m > 45.6$ GeV, CL = 95%   (from $q^*\overline{q}^*$)

Mass $m > 88$ GeV, CL = 95%   (if $\lambda_Z > 1$)

Mass $m > 570$ GeV, CL = 95%   ($p\overline{p} \rightarrow q^*X$)

Color Sextet and Octet Particles

Color Sextet Quarks (q_6)

Mass $m > 84$ GeV, CL = 95%   (Stable q_6)

Color Octet Charged Leptons (ℓ_8)

Mass $m > 86$ GeV, CL = 95%   (Stable ℓ_8)

Color Octet Neutrinos (ν_8)

Mass $m > 110$ GeV, CL = 90%   ($\nu_8 \rightarrow \nu g$)

Norman E. Holden

Reactor Division

Brookhaven National Laboratory

Upton, New York 11973

This table presents an evaluated set of values for the experimental quantities which characterize the decay of radioactive nuclides. A list of the major references used in this evaluation is given below. When uncertainties are not listed, they are assumed to be five or less in the last digit quoted. If they exceed five in the last digit, the value is prefaced by an approximate sign. The effective literature cutoff date for data in this edition of the Table is January, 1998. The assistance of Felicia Kramer in assembling this file is gratefully acknowledged.

TABLE LAYOUT

Column No.	Column Title	Description
1	Isotope or Element	For elements, the atomic number and chemical symbol are listed. For nuclides, the mass number and chemical symbol are listed. Isomers are indicated by the addition of m, m1, or m2.
2	Isotopic Abundance	in atom percent.
3	Atomic Mass or Atomic Weight	Atomic mass relative to $^{12}C = 12$. Atomic weight is given on the same scale.
4	Half-life	Half-life in decimal notation. μs = microseconds; ms = milliseconds; s = seconds; m = minutes; h = hours; d = days; and y = years.
5	Decay Mode/Energy	Decay modes are α = alpha particle emission; β^- = negative beta emission; β^+ = positron emission; EC = orbital electron capture; IT = isomeric transition from upper to lower isomeric state; n = neutron emission; SF = spontaneous fission. Total disintegration energy in MeV units.
6	Particle Energy/Intensity	End point energies of beta transitions and discrete energies of alpha particles in MeV and their intensities in percent.
7	Spin and Parity	Nuclear spin or angular momentum of the nuclides in units of $h/2\pi$; parity is positive or negative.
8	Magnetic Dipole Moment	Magnetic dipole moments in nuclear magneton units.
9	Electric Quadrupole Moment	Electric quadrupole moments in barn units $(10^{-24}\ cm^2)$.
10	Gamma Ray Energy/Intensity	Gamma ray energies in MeV and intensities in percent. Ann. rad. refers to the 511.006 keV photons emitted in the annihilation of positrons in matter.

GENERAL NUCLEAR DATA REFERENCES

The following references represent the major sources of the nuclear data presented, along with subsequent published journals and reports:

1. G. Audi, O. Bersillon, J. Blachot, A.H. Wapstra, *The Nubase Evaluation of Nuclear and Decay Properties*, Nuclear Physics A624, 1 (1997).
2. International Commission on Atomic Weights, *Atomic Weights of the Elements - 1995*, Pure & Applied Chemistry 68, 2339 (1996).
3. J.R. Parrington, H.D. Knox, S. Breneman, E.M. Baum, F. Feiner, *Chart of the Nuclides, 15th Edition*, Knolls Atomic Power Lab. (1996)
4. N.E. Holden, *Total and Spontaneous Fission Half-lives for Uranium, Plutonium, Americium and Curium Nuclides*, Pure & Applied Chemistry 61, 1483 (1989).
5. N.E. Holden, *Half-lives of Selected Nuclides*, Pure & Applied Chemistry 62, 941 (1990).
6. N.E. Holden, *Review of Thermal Neutron Cross Sections and Isotopic Composition of the Elements*, BNL-NCS-42224 (March 1989).
7. P. Raghavan, *Table of Nuclear Moments*, Atomic Data Nuclear Data Tables 42, 189 (1989).
8. E. Brown, R. Firestone, *Radioactivity Handbook*, Wiley Interscience Press (1986).
9. J.K. Tuli, *Nuclear Wallet Cards*, Brookhaven National Laboratory (July 1995).

*This research was carried out under the auspices of the US Department of Engery Contract No. DE-AC02-76CH00016.

Elem. or Isot.	Natural Abundance (%)	Atomic Mass or Weight	Half-Life	Decay Mode/Energy (/MeV)	Particle Energy /Intensity (MeV / %)	Spin (h/2π)	Nuclear Magnetic Mom.(nm)	Elect. Quadr. Mom. (b)	γ-ray Energy /Intensity (MeV / %)
₀n		1.008664924	10.3 m	β⁻/0.78235	0.782/100.	1/2+	-1.913043		
₁H	1.00794(7)								
¹H	99.985(1)	1.007825032				1/2+	+2.79285		
²H	0.015(1)	2.014101778				1+	+0.85744	+2.86 mb	
³H		3.016049268	12.32 y	β⁻/0.01859	0.01860/100.	1/2+	+2.97896		
₂He		4.002602(2)							
³He	1.37x10⁻⁴	3.016029309				1/2+	-2.12762		
⁴He	≈100.	4.002603250				0+			
⁵He		5.01222	7.6x10⁻²² s	n,α		3/2-			
⁶He		6.018888	0.807 s	β⁻/3.508,d	3.510/100.	0+			
⁷He		7.02803	3.x10⁻²¹ s	n		(3/2)-			
⁸He		8.03392	0.119 s	β⁻/10.65, t n/	13/88. /12.	0+			0.9807/84. 0.4776/5.
⁹He		9.0438		n					
¹⁰He		10.0524		2n					
₃Li		6.941(2)							
⁵Li		5.01254	≈3.x10⁻²² s			3/2-			
⁶Li	7.5(2)	6.0151223				1+	+0.82205	-0.8 mb	
⁷Li	92.5(2)	7.0160041				3/2-	+3.25644	-0.041	
⁸Li		8.022486	0.84 s	β⁻/16.004 α/	12.5/100. α(1.6)	2+	+1.6536	+0.032	
⁹Li		9.026789	0.178 s	β⁻/13.606 β⁻/	13.5/75. 11/25.	3/2-	3.439	0.028	
¹⁰Li		10.03590	4. x 10⁻²² s	β⁻/20.84					
¹¹Li		11.04380	8.4 ms	β⁻/20.6 n,2n,3n,α		3/2(-)	3.668	0.031	3.367/35. (0.22-2.81)
₄Be		9.012182(3)							
⁶Be		6.01973	5.0x10⁻²¹ s	2p,α		0+			
⁷Be		7.0169293	53.28 d	EC/0.8618		3/2-			0.4776/10.4
⁸Be		8.00530509	≈ 7.x10⁻¹⁷ s	2α/0.046		0+			
⁹Be	100.	9.0121822				3/2-	-1.1776	+0.053	
¹⁰Be		10.0135338	1.52x10⁶ y	β⁻/0.5559	0.555/100.	0+			
¹¹Be		11.02166	13.8 s	β⁻,β⁻α/11.51	11.48/61.	1/2+			2.125/35.5 (0.478-7.97)
¹²Be		12.02692	24. ms	β⁻,(n)/11.71		0+			(0.95 - 4.4)
¹³Be		13.0361							
¹⁴Be		14.0428	4.3 ms	β⁻,(n)/16.2		0+			
₅B		10.811(7)							
⁷B		7.0299	4.x10⁻²² s	p		2+	1.0355	0.068	ann.rad.
⁸B		8.024607	0.770 s	β⁺, 2α/17.979	13.7(β⁺)/93.	3/2-			
⁹B		9.013329	8x10⁻¹⁹ s	p2α/		3+	+1.8006	+0.085	
¹⁰B	19.9(2)	10.0129371							
¹¹B	80.1(2)	11.0093055				3/2-	+2.6886	+0.0406	
¹²B		12.014352	0.0202 s	β⁻/13.369 β⁻α/1.6/		1+	+1.0031	0.0132	4.438/1.3 3.215/0.00065
¹³B		13.017780	0.0174 s	β⁻/13.437 β⁻n/0.25/	13.4 2.43(n)/0.09 3.55(n)/0.16	3/2-	+3.17778	0.037	3.68/7.6
¹⁴B		14.02540	14. ms	β⁻/20.64		2-	1.185	0.0298	6.094/90.
¹⁵B		15.03110	10.4 ms	β⁻,(n)/19.09		(3/2-)	2.66	0.038	
¹⁶B		16.0398	<1.9x10⁻¹⁰ s						
¹⁷B		17.0469	5.1 ms	β⁻,(n)/22.7			2.55		
¹⁹B		19.0637		β⁻/26.5					
₆C		12.0107(8)							
⁸C		8.03768	2.0x10⁻²¹ s	p					
⁹C		9.031040	127. ms	β⁺,p, 2α/16.498		(3/2-)	-1.391		ann.rad.
¹⁰C		10.0168532	19.3 s	β⁺/3.648	1.865	0+			ann.rad. 0.71829/100.
¹¹C		11.011433	20.3 m	β⁺,EC/1.982	0.9608/99.	3/2-	-0.964	0.0333	ann.rad.
¹²C	98.93(8)	12.000000000				0+			
¹³C	1.07(8)	13.003354838				1/2-	+0.70241		
¹⁴C		14.003241991	5715. y	β⁻/0.15648	0.1565/100.	0+			
¹⁵C		15.010599	2.45 s	β⁻/9.772	4.51/68. 9.82/32.	1/2+	1.32		5.298/68. (7.30-9.05)

Elem. or Isot.	Natural Abundance (%)	Atomic Mass or Weight	Half-Life	Decay Mode/Energy (/MeV)	Particle Energy /Intensity (MeV / %)	Spin (h/2π)	Nuclear Magnetic Mom.(nm)	Elect. Quadr. Mom. (b)	γ-ray Energy /Intensity (MeV / %)
^{16}C		16.014701	0.75 s	β^- ,n/8.012					
^{17}C		17.02258	0.19 s	β^- ,n/13.17					1.375
									1.849
^{18}C		18.02676	0.09 s	β^- ,n/11.81					1.906
^{19}C		19.0353	0.05 s	n					
^{20}C		20.0403	0.01 s						
^{21}C		21.0493							
^{22}C		22.056							
,N		**14.00674(7)**							
^{12}N		12.018613	11.00 ms	β^+ ,$\beta^+ \alpha$/17.338	16.38/95.	1+	+0.457	+0.026	ann.rad.
									4.438/2.
^{13}N		13.0057386	9.97 m	β^+ /2.2204	1.190/100.	1/2-	0.3222		
^{14}N	99.632(7)	14.003074007				1+	+0.40376	+0.020	
^{15}N	0.368(7)	15.00010897				1/2-	-0.28319		
^{16}N		16.006100	7.13 s	β^- /10.419	4.27/68.	2-			6.129/68.8
					10.44/26.				7.115/4.7
				β^- , α	1.85/.0012				(0.99-8.87)
^{17}N		17.00845	4.17 s	β^- ,β^- n/8.68	3.7/100.	1/2-	0.352		0.871/3.
				0.4-1.7n/95.					2.1842/0.3
				β^- α/	8.0, 8.2				
^{18}N		18.01408	0.62 s	β^- /13.90	9.4/100.	1-			0.822/61.
									1.65/60.5
									1.982/98.
									(0.535-7.13)
^{19}N		19.01703	0.32 s	β^- /12.53					(0.096-3.14)
^{20}N		20.02337	0.14 s	β^- /17.97					
^{21}N		21.0271	0.08 s						
^{22}N		22.0344	0.02s						
^{23}N		23.0405							
,O		**15.9994(3)**							
^{12}O		12.03440	$\approx 1.\times10^{21}$ s	2p					
^{13}O		13.02481	8.9 ms	β^+ ,p/17.77	1.56 (p)/	(3/2-)	1.389		ann.rad.
									4.438/0.56
^{14}O		14.0085953	70.60 s	β^+ /5.1430	1.81/99.	0+			ann.rad.
									2.312/99.4
^{15}O		15.0030655	122.2 s	β^+ /2.754	1.723/100.	1/2-	0.719		ann.rad.
^{16}O	99.757(16)	15.994914622				0+			
^{17}O	0.038(1)	16.9991315				5/2+	-1.8938	-0.026	
^{18}O	0.205(14)	17.999160				0+			
^{19}O		19.003579	26.9 s	β^- /4.820	3.25/60.	5/2+			0.197/95.9
					4.60/40.				1.3569/50.4
									(0.11-4.18)
^{20}O		20.004076	13.5 s	β^- /3.814		0+			1.057/100.
^{21}O		21.00866	3.4 s	β^- /8.11					(0.28-4.6)
^{22}O		22.00997	2.2 s	β^- /6.5					(0.64-1.86)
^{23}O		23.0157	0.08 s						
^{24}O		24.0204	0.06 s						
,F		**18.9984032(5)**							
^{15}F		15.0180	$5.\times10^{22}$ s	p		(1/2+)			
^{16}F		16.01147	$\approx 1.\times10^{20}$ s	p		0-			
^{17}F		17.0020952	64.5 s	β^+ /2.761	1.75/	5/2+	+4.722	0.058	ann.rad.
^{18}F		18.000938	1.830 h	β^+ ,EC/1.656	0.635/97.	1+			
^{19}F	100.	18.9984032				1/2+	+2.62887	0.072	
^{20}F		19.9999813	11.00 s	β^- /7.0245	5.398/100.	2+	+2.094	0.042	1.634/100.
									3.33/0.009
^{21}F		20.999949	4.16 s	β^- /5.684	3.7/8.	5/2+			0.3507/90.
					5.0/63.				1.395/15.
					5.4/29.				(1.746-4.684)
^{22}F		22.00300	4.23 s	β^- /10.82	3.48/15.	4+			1.2746/100.
					4.67/7.				2.0826/82.
					5.50/62.				(0.82-4.37)
^{23}F		23.00357	2.2 s	β^- /8.5		5/2+			1.701/48.
									2.129/34.
									(0.493-3.83)
^{24}F		24.0081	0.3 s	β^- /13.5					1.9816/
^{25}F		25.0121	0.06 s	β^- ,(n)					
^{26}F		26.0196	0.03 s						
^{27}F		27.0269	5. ms						
^{29}F		29.043	2. ms						

Elem. or Isot.	Natural Abundance (%)	Atomic Mass or Weight	Half-Life	Decay Mode/Energy (/MeV)	Particle Energy /Intensity (MeV / %)	Spin (h/2π)	Nuclear Magnetic Mom.(nm)	Elect. Quadr. Mom. (b)	γ-ray Energy /Intensity (MeV / %)
$_{10}$Ne		20.1797(6)							
^{16}Ne		16.02575	$4.\times10^{21}$ s	2p					ann.rad./
^{17}Ne		17.01770	109. ms	β^+ .p/14.53	1.4-10.6/	1/2-			0.495
^{18}Ne		18.005697	1.67 s	β^+ /4.446	3.416/92.	0+			ann.rad./ 1.0413/7.8 (0.658-1.70)
^{19}Ne		19.001880	17.22 s	β^+ /3.238	2.24/99.	1/2+	-1.885		ann.rad./ (0.11-1.55)
^{20}Ne	90.48(3)	19.992440176				0+			
^{21}Ne	0.27(1)	20.99384674				3/2+	-0.66180	+0.103	
^{22}Ne	9.25(3)	21.9913855				0+		-0.19	
^{23}Ne		22.9944673	37.2 s	β^- /4.376	3.95/32. 4.39/67.	5/2+	-1.08		0.440/33. (1.64-2.98)
^{24}Ne		23.99362	3.38 m	β^- /2.47	1.10/8. 1.98/92.	0+			0.4723/100. 0.874/7.9
^{25}Ne		24.99779	0.61 s	β^- /7.30	6.3/ 7.3/	1/2+			0.0895/96. (0.98-3.69) 0.233/
^{26}Ne		26.00046	197 ms	β^- /7.3					
^{27}Ne		27.0076	32 ms	β^-, n/12.7					
^{28}Ne		28.0121	18. ms	β^-, n/12.3					
^{29}Ne		29.0194	15. ms	β^- /15.4					
^{30}Ne		30.024	7. ms						
^{31}Ne		31.033							
^{32}Ne		32.040							
$_{11}$Na		22.989770(2)							
^{19}Na		19.01388	0.03 s	β^+ ,p/11.18					
^{20}Na		20.00735	0.446 s	β^+ /13.89 α	2.15/	2+	+0.3694		ann.rad./ 1.634/79.
^{21}Na		20.997655	22.48 s	β^+ /3.547	2.50/95.	3/2+	+2.3863	+0.05	ann.rad./ 0.351/5.
^{22}Na		21.9944366	2.605 y	β^+ /90/2.842 EC/10/	0.545/90.	3+	+1.746		ann.rad./ 1.2745/99.9
^{23}Na	100.	22.9897697				3/2+	+2.21752	+0.101	0.4723/100.
^{24m}Na			20.2 ms	I.T.,β^-		1+			1.3686/100.
^{24}Na		23.9909633	14.96 h	β^- /5.5158	1.389/ >99.	4+	+1.690		2.754/100. (0.997-4.238)
^{25}Na		24.989954	59.3 s	β^- /3.835	2.6/7. 3.15/25. 4.0/65.	5/2+	+3.683	-0.10	0.3897/12.7 0.5850/13. 0.9747/14.9 (0.836-2.80) 1.809/98.9
^{26}Na		25.99259	1.07 s	β^- /9.31		3+	+2.851	-0.08	0.9847/87.4 1.698/11.9
^{27}Na		26.99401	0.290 s	β^- /9.01 β^- ,n/	7.95/	5/2+	+3.90	-0.06	1.473/37. 2.389/18.6
^{28}Na		27.9989	31. ms	β^- /14.0 β^- ,n/	12.3/	1+	+2.43	-0.02	2.560/36. (1.04-3.99)
^{29}Na		29.0028	44. ms	β^- ,n/13.3	11.5/	3/2+	+2.45	-0.03	1.483/46.
^{30}Na		30.0092	50. ms	β^- /17.5 /15.9		2	+2.08		1.483/14. (0.05-3.54)
^{31}Na		31.0136	17.2 ms	β^- ,n		(3/2-)	+2.31		0.886/60.
^{32}Na		32.0197	13.5 ms	β^- /19.1					0.886/60.
^{33}Na		33.027	8.1 ms	β^- /20.					0.886/16.
^{34}Na		34.035	5. ms	β^- /24.					
^{35}Na		35.044	1.5 ms	β^- /24					
$_{12}$Mg		24.3050(6)							
^{20}Mg		20.01886	96. ms	β^+ ,p/10.73					0.332/51.
^{21}Mg		21.01171	122. ms	β^+ ,p/13.10		5/2+			0.0729/60.
^{22}Mg		21.999574	3.86 s	β^+ /4.786	3.05/	0+			0.5820/100. (1.28-1.93)
^{23}Mg		22.994125	11.32 s	β^+ /4.057	3.09/92.	3/2+	0.536		0.440/8.2
^{24}Mg	78.99(4)	23.9850419				0+			
^{25}Mg	10.00(1)	24.9858370				5/2+	-0.85545	+0.199	
^{26}Mg	11.01(3)	25.9825930				0+			
^{27}Mg		26.9843407	9.45 m	β^- /2.6103	1.59/41. 1.75/58. 2.65/0.3	1/2+			0.17068/0.8 0.84376/73. 1.01443/29.
^{28}Mg		27.983877	20.9 h	β^- /1.832	0.459/95.	0+			0.0306/95. 0.4006/36. 0.9418/36. 1.342/54.

Elem. or Isot.	Natural Abundance (%)	Atomic Mass or Weight	Half-Life	Decay Mode/Energy (/MeV)	Particle Energy /Intensity (MeV / %)	Spin (h/2π)	Nuclear Magnetic Mom.(nm)	Elect. Quadr. Mom. (b)	γ-ray Energy /Intensity (MeV / %)
^{29}Mg		28.98855	1.3 s	β^- /7.55	5.4/	3/2+			0.960/15. 1.398/16.
^{30}Mg		29.9905	0.32 s	β^- /7.0		0+			2.224/36.
^{31}Mg		30.9966	0.24 s	β^- /11.7		(3/2+)			0.224/85.
^{32}Mg		31.9992	0.12 s	β^- /10.3					1.61/26.
^{33}Mg		33.0056	0.09 s	β^- /13.7					2.765/25.
^{34}Mg		34.0091	0.02 s	β^- /11.3					1.848/
^{35}Mg		35.0175							
^{36}Mg		36.022							
^{37}Mg		37.031							
^{38}Mg									
$_{13}$Al		26.981538(2)							
^{22}Al		22.0195	59. ms	β^+ /18.6 β^+,p,2p,α/		4+			ann.rad./
^{23}Al		23.00727	0.47 s	β^+ /12.24 β^+,p/					ann.rad./
^{24m}Al			0.129 s	I.T./0.4259 β^+	13.3	1+			
^{24}Al		23.999941	2.07 s	β^+ /13.878,p	3.40/48. 4.42/41. 6.80/3. 8.74/8.	4+			1.3686/5.3 1.078(2)/16. 1.368(2)/96. 2.753(2)/43. 4.315(3)/15. 5.392(3)/20. 7.0662(2)/41.
^{25}Al		24.990429	7.17 s	β^+ /4.277	3.27/	5/2+	3.646		ann.rad./ 1.6115(2)/100.
^{26m}Al			6.345 s	β^+ /	3.2/	0+			0.975(2)/5. ann.rad./
^{26}Al		25.9868917	7.1x10^5 y	β^+ /82/4.0042 EC/18/	1.16/	5+	+2.804		ann.rad./ 1.80865(7)/99.8
^{27}Al	100.	26.9815384				5/2+	+3.64151	+0.15	
^{28}Al		27.9819102	2.25 m	β^- /4.6422	2.865/100.	3+	3.24	0.18	1.7778(6)/100.
^{29}Al		28.980445	6.5 m	β^- /3.680	1.4/30. 2.5/70.	5/2+			1.2732(8)/89. 2.0282(8)/4. 2.4262(8)/7.
^{30}Al		29.98296	3.68 s	β^- /8.56	5.05/	3+			1.26313(3)/35. 2.23525(5)/65.
^{31}Al		30.98395	0.64 s	β^- /8.00	6.25/				0.75223(3)/18. 1.69473(3)/59. 2.31664(4)/73.
^{32}Al		31.9881	33. ms	β^- /13.0		1+			
^{33}Al		32.9909	41. ms						
^{34}Al		33.9969	≈42. ms	β^- /17.1					
^{35}Al		34.9999	30 ms						
^{36}Al		36.0064	0.09 s						
^{37}Al		37.010							
^{38}Al		38.0169							
^{39}Al		39.022							
^{40}Al									
^{41}Al									
$_{14}$Si		28.0855(3)							
^{22}Si		22.0345	29. ms	β^+,p	1.99/20				
^{23}Si		23.0255	40.7 ms	β^+,p/5.9	1.32,2.40,2.83				
^{24}Si		24.01155	0.14 s	β^+ ,p/10.81					ann.rad./
^{25}Si		25.00411	221 ms	β^+ ,p/12.74		5/2+			ann.rad./
^{26}Si		25.992330	2.23 s	β^+ /5.066	3.282/	0+			ann.rad./ 0.8294(8)/22.
^{27}Si		26.9867048	4.14 s	β^+ /4.8118	3.85/100.	5/2+	-0.8554		ann.rad./ 2.211(5)/0.2
^{28}Si	92.22(2)	27.97692653				0+			
^{29}Si	4.69(1)	28.97649472				1/2+	-0.5553		
^{30}Si	3.09(1)	29.97377022				0+			
^{31}Si		30.9753633	2.62 h	β^- /1.4920	1.471/99.9	3/2+			1.2662(5)/0.05
^{32}Si		31.974148	1.6x10^2 y	β^- /0.224	0.213/100.	0+			
^{33}Si		32.97800	6.1 s	β^- /5.85	3.92	(3/2+)			1.4313(5)/13. 1.84769(5)/100. 2.538(2)/10.
^{34}Si		33.97858	2.8 s	β^- /4.60	3.09/	0+			0.42907(5)/60. 1.17852(2)/64. 1.60756(5)/36.
^{35}Si		34.98458	0.9 s	β^- /10.50					
^{36}Si		35.9867	0.5 s	β^- /7.9					
^{37}Si		36.9930	≈0.09 s						
^{38}Si		37.9960							

Elem. or Isot.	Natural Abundance (%)	Atomic Mass or Weight	Half-Life	Decay Mode/Energy (/MeV)	Particle Energy /Intensity (MeV / %)	Spin (h/2π)	Nuclear Magnetic Mom.(nm)	Elect. Quadr. Mom. (b)	γ-ray Energy /Intensity (MeV / %)
³⁹Si		39.0023							
⁴⁰Si		40.0058							
⁴¹Si		41.013							
⁴²Si		42.016							
₁₅P		30.973761(2)							
²⁶P		26.0118	≈20. ms	β⁺,p/18.1		3+			
²⁷P		26.99919	0.3 s	β⁺,p /11.63		1/2+			
²⁸P		27.992312	270. ms	β⁺ /14.332	3.94/13. 5.25/13. 6.96/16. 8.8/7. 11.49/52.	3+			ann.rad./ 1.779(2)/98. 2.839(2)/2.8 3.040(2)/3.2 4.498(2)/12. 7.537(2)/9.
²⁹P		28.981801	4.14 s	β⁺ /4.9431	3.945/98.	1/2+	1.2349		ann.rad./ 1.273/1.32 2.426/0.39
³⁰P		29.9783138	2.50 m	β⁺ /4.2323	3.245/99.9	1+			ann.rad./ 2.230(3)/0.07
³¹P	100.	30.9737615				1/2+	+1.13160		
³²P		31.9739071	14.28 d	β⁻ /1.7106	1.710/100.	1+	-0.2524		
³³P		32.971725	25.3 d	β⁻ /0.249	0.249/100.	1/2+			
³⁴P		33.973636	12.4 s	β⁻ /5.374	3.2/15. 5.1/85.	1+			1.78-4.1/ 2.127(5)/15.
³⁵P		34.973314	47. s	β⁻ /3.989	2.34/100.	1/2+			1.572(1)/100.
³⁶P		35.97826	5.7 s	β⁻ /10.41					0.902/77. 3.291/100.
³⁷P		36.97961	2.3 s	β⁻ /7.90					0.6462/ 1.5829/
³⁸P		37.9845	0.6 s	β⁻ /12.4					1.2923/ 2.224/
³⁹P		38.9864	≈ 0.16 s						
⁴⁰P		39.9911							
⁴¹P		40.9948	0.12 s						
⁴²P		42.0001	0.11 s						
⁴³P		43.0033	33. ms	β⁻,(n)/					
⁴⁴P		44.010							
⁴⁵P		45.015							
⁴⁶P		46.024							
₁₆S		32.066(6)							
²⁷S		27.0188	21. ms	β⁺, 2p /18.3					
²⁹S		28.99661	0.188 s	β⁺ /13.79 β⁺,p/		5/2+			ann.rad./
³⁰S		29.984903	1.18 s	β⁺ /6.138	4.42/78. 5.08/20.	0+			ann.rad./ 0.678/79.
³¹S		30.979555	2.56 s	β⁺ /5.396	4.39/99.	1/2+	0.48793		ann.rad./ 1.2662(5)/1.2
³²S	94.93(31)	31.9720707				0+			
³³S	0.76(2)	32.9714585				3/2+	+0.64382	-0.07	
³⁴S	4.29(28)	33.9678668				0+			
³⁵S		34.9690321	87.2 d	β⁻ /0.1672	0.1674/100.	3/2+	+1.00	+0.047	
³⁶S	0.02(1)	35.9670809				0+			
³⁷S		36.9711257	5.05 m	β⁻ /4.8653	1.64/94. 4.75/5.6				0.9083(4)/0.06 3.1033(2)/94.2
³⁸S		37.97116	2.84 h	β⁻ /2.94	1.00/	0+			0.1962(4)/0.2 1.9421(3)/84.
³⁹S		38.97514	11.5 s	β⁻ /6.64					1.301/52. 1.697/44.
⁴⁰S		39.9755	9. s	β⁻ /4.7					
⁴¹S		40.9800							
⁴²S		41.9815	≈0.56 s	β⁻,(n)/					
⁴³S		42.987							
⁴⁴S		43.9883	0.12 s	β⁻, n/9.					
⁴⁵S		44.9948	0.08 s	β⁻,n/					
⁴⁶S		45.9996							
⁴⁷S		47.008							
⁴⁸S		48.013							
₁₇Cl		35.4527(9)							
³¹Cl		30.99242	0.15 s	β⁺,p/11.98	1.52	3/2+			ann.rad./

Elem. or Isot.	Natural Abundance (%)	Atomic Mass or Weight	Half-Life	Decay Mode/Energy (/MeV)	Particle Energy /Intensity (MeV / %)	Spin (h/2π)	Nuclear Magnetic Mom.(nm)	Elect. Quadr. Mom. (b)	γ-ray Energy /Intensity (MeV / %)
^{32}Cl		31.98569	297. ms	β^+ /12.69	4.75/25. 6.18/10. 7.48/14. 9.47/50. 11.6/1.	1+			ann.rad./ 1.548(2)/3.5 2.2305(1)/92. 2.4638(1)/4. 2.885(1)/1. 4.770(1)/20.
^{33}Cl		32.977452	2.511 s	β^+ /5.583	4.51/98.	3/2+	+0.752		ann.rad./ 0.8409/0.52 1.966/0.45 2.866/0.44
^{34m}Cl			32.2 m	β^+ / I.T./	1.35/24. 2.47/28.	3+			ann.rad./ 0.1457(8)/42. 2.1276(5)/42.
^{34}Cl		33.9737620	1.528 s	β^+ /5.4922	4.50/100.	0+			ann.rad./
^{35}Cl	75.78(4)	34.96885271				3/2+	+0.82187	-0.08	
^{36}Cl		35.9683069	3.01x10⁵ y	β^- /0.7086 β^+,EC/1.1421	0.7093/98. 0.115/0.002	0+	+1.28547	-0.018	
^{37}Cl	24.22(4)	36.96590260				3/2+	+0.68412	-0.065	ann.rad./
^{38m}Cl			0.715 s	I.T./		5-			0.67138(2)/100.
^{38}Cl		37.9680106	37.2 m	β^- /4.9168	1.11/31. 2.77/11. 4.91/58.	2-	2.05		1.64216(1)/31. 2.16760(2)/42.
^{39}Cl		38.968008	55.6 m	β^- /3.442	1.91/85. 2.18/8. 3.45/7.	3/2+			0.25026(1)/47. 1.26720(5)/54. 0.986-1.517
^{40}Cl		39.97042	1.38 m	β^- /7.48		2-			0.6431(3)/6. 1.4608(1)/77. 2.8402(2)/17. (0.167-1.359)
^{41}Cl		40.9707	34. s	β^- /5.7	3.8/				
^{42}Cl		41.9732	6.8 s	β^- /9.4					
^{43}Cl		42.9742	3.3 s	β^- /8.0					
^{44}Cl		43.9785	≈0.43 s	β^-,n/12.3					
^{45}Cl		44.980	0.40 s	β^-, n/11.					
^{46}Cl		45.984	0.22 s	β^-, n/14.9					
^{47}Cl		46.988		β^-, n/15.					
^{48}Cl		47.995							
^{49}Cl		48.9999							
^{51}Cl		51.014							
$_{18}$Ar		39.948(1)							
^{31}Ar		31.0121	15. ms	β^+ /18.4 β^+, 2p β^+, 3p					
^{32}Ar		31.99766	98. ms	β^+,p/11.2					ann.rad./
^{33}Ar		32.98993	174. ms	β^+ /11.62 β^+,p/	3.12/	1/2+	-0.72		ann.rad./ 0.810(2)/48.
^{34}Ar		33.980270	0.844 s	β^+/6.061	5.0/95.	0+			ann.rad./ 0.6658(1)/2.5 3.1290(1)/1.3
^{35}Ar		34.975257	1.77 s	β^+/5.965	4.94/93.	3/2+	+0.633	-0.08	ann.rad./ 1.2185(5)/1.22 1.763(1)/0.25 2.964(1)/0.2
^{36}Ar	0.3365(30)	35.9675463				0+			
^{37}Ar		36.9667759	35.0 d	EC/.813		3/2+	+1.15	+0.076	
^{38}Ar	0.0632(5)	37.9627322				0+			
^{39}Ar		38.964313	268. y	β^-/0.565	0.565/100.	7/2-	-1.3	-0.12	
^{40}Ar	99.6003(30)	39.962383123				0+			
^{41}Ar		40.964501	1.82 h	β^-/2.492	1.198/	7/2-			1.29364(5)/99. 1.6770(3)/0.05
^{42}Ar		41.96305	33. y	β^-/0.60	0.60/100.	0+			
^{43}Ar		42.9657	5.4 m	β^-/4.6					0.4791(2)/10. 0.7380(1)/43. 0.9752(1)/100. 1.4400(3)/39.
^{44}Ar		43.96537	11.87 m	β^-/3.55		0+			0.182-1.866
^{45}Ar		44.96809	21.5 s	β^-/6.9		7/2-			0.0610/25. 1.020/35. 3.707/34.
^{46}Ar		45.96809	8.4 s	β^-/5.70		0+			1.944/
^{47}Ar		46.9722	≈0.7 s	β^-					
^{49}Ar		48.9822							
^{50}Ar		49.986							
^{51}Ar		50.993							

Elem. or Isot.	Natural Abundance (%)	Atomic Mass or Weight	Half-Life	Decay Mode/Energy (/MeV)	Particle Energy /Intensity (MeV / %)	Spin (h/2π)	Nuclear Magnetic Mom.(nm)	Elect. Quadr. Mom. (b)	γ-ray Energy /Intensity (MeV / %)
₁₉K		39.0983(1)							
³⁵K		34.98801	0.19 s	β⁺ /11.88 β⁺ ,p/		3/2+			ann.rad./ 1.751/14. 2.5698/26. 2.9827/51.
³⁶K		35.98129	0.342 s	β⁺ /12.81	5.3/42. 9.9/44.	2+	+0.548		ann.rad./ 1.97044(5)/82. 2.20783(5)/30. 2.43343(2)/32.
³⁷K		36.9733769	1.23 s	β⁺ /6.149	5.13/	3/2+	+0.2032		ann.rad./ 2.7944(8)/2. 3.602(2)/0.05
³⁸ᵐK			0.924 s	β⁺ /6.742	5.02/100.	0+			ann.rad./
³⁸K		37.969080	7.63 m	β⁺ /5.913	2.60/99.8	3+	+1.37		ann.rad./ 2.1675(3)/99.8 3.9356(5)/0.2
³⁹K	93.2581(44)	38.9637069				3/2+	+0.39146	+0.049	
⁴⁰K	0.0117(1)	39.9639987	1.26x10⁹ y	β⁻ /1.3111 β⁺ ,EC/1.505	1.312/89. 1.50/10.7	4-	-1.298	-0.061	ann.rad./ 1.46081(5)/10.5
⁴¹K	6.7302(44)	40.9618260				3/2+	+0.21487	+0.060	
⁴²K		41.9624031	12.36 h	β⁻ /3.525	1.97/19. 3.523/81.	2-	-1.1425		0.31260(2)/0.3 1.5246(3)/18.
⁴³K		42.96072	22.3 h	β⁻ /1.82	0.465/8. 0.825/87. 1.24/3.5 1.814/1.3	3/2+	+0.163		0.2211(2)/4. 0.3729(2)/88. 0.3971(2)/11. 0.6178(2)/81.
⁴⁴K		43.96156	22.1 m	β⁻ /5.66	5.66/34.	2-	-0.856		0.368207(1)/2.2 1.15700(1)/58. 2.15079(2)/22.
⁴⁵K		44.96070	17.8 m	β⁻ /4.20	1.1/23. 2.1/69. 4.0/8.	3/2+	+0.1734		0.1743(5)/80. 1.2607(8)/7. 1.7056(6)/69. 2.3542(5)/14.
⁴⁶K		45.96198	1.8 m	β⁻ /7.72	6.3/	2-	-1.05		1.347(1)/91. 3.700(5)/28.
⁴⁷K		46.96168	17.5 s	β⁻ /6.64	4.1/99. 6.0/1.	1/2+	+1.93		0.56474(3)/15. 0.58575(3)/85. 2.01313(3)/100.
⁴⁸K		47.96551	6.8 s	β⁻ /12.09	5.0/	(2-)			0.67122(1)/4. 0.6723(5)/20. 0.78016(1)/32. 3.83153(7)/80. 2.025/ 2.252/
⁴⁹K		48.9675	1.26 s	β⁻ /11.0					
⁵⁰K		49.9728	0.472 s	β⁻ /14.2					
⁵¹K		50.9764	0.365 s	β⁻ /					
⁵²K		51.983	0.105 s	β⁻					
⁵³K		52.987	30. ms	β⁻		3/2+			
⁵⁴K		53.994	10. ms	β⁻					
₂₀Ca		40.078(4)							
³⁵Ca		35.0048	0.05 s	β⁺ /15.6					
³⁶Ca		35.99309	0.10 s	β⁺,(p)/10.99 β⁺ ,n/	2.52				ann.rad./
³⁷Ca		36.98587	0.18 s	β⁺ /11.64 β⁺ ,n/	3.103	3/2+			ann.rad./ 1.369
³⁸Ca		37.976319	0.44 s	β⁺ /6.74		0+			ann.rad./ 1.5677(5)/25. 3.210(2)/1.
³⁹Ca		38.970718	0.861 s	β⁺ /6.531	5.49/100.	3/2+	1.02168		ann.rad./
⁴⁰Ca	96.941(156)	39.9625912				0+			
⁴¹Ca		40.9622783	1.02x10⁵ y	EC/0.4214		7/2-	-1.595	-0.08	
⁴²Ca	0.647(23)	41.9586183				0+			
⁴³Ca	0.135(10)	42.9587668				7/2-	-1.3173	-0.05	
⁴⁴Ca	2.086(110)	43.955481				0+			
⁴⁵Ca		44.956186	162.7 d	β⁻ /0.257	0.257/100.	7/2-	-1.327	+0.05	
⁴⁶Ca	0.004(3)	45.953693				0+			
⁴⁷Ca		46.954546	4.536 d	β⁻ /1.992	0.684/84. 1.98/16.	7/2-	-1.38	+0.02	1.297/75 (0.041-1.88)
⁴⁸Ca	0.187(21)	47.952533				0+			
⁴⁹Ca		48.955673	8.72 m	β⁻ /5.262	0.89/7. 1.95/92.	3/2-			3.0844(1)/92. 4.0719(1)/7.
⁵⁰Ca		49.95752	14. s	β⁻ /4.97	3.12/	0+			0.2569/98. (0.0715 - 1.59)
⁵¹Ca		50.9615	10. s	β⁻ /7.3		(3/2-)			
⁵²Ca		51.9651	4.6 s	β⁻ /8.0					
⁵³Ca		52.9701	0.09 s	β⁻ /10.9					

Elem. or Isot.	Natural Abundance (%)	Atomic Mass or Weight	Half-Life	Decay Mode/Energy (/MeV)	Particle Energy /Intensity (MeV / %)	Spin (h/2π)	Nuclear Magnetic Mom.(nm)	Elect. Quadr. Mom. (b)	γ-ray Energy /Intensity (MeV / %)
⁵⁴Ca		53.975							
⁵⁵Ca		54.981							
⁵⁶Ca		55.986							
₂₁Sc		**44.955910(8)**							
³⁹Sc		38.98479		p					
⁴⁰Sc		39.977964	0.182 s	β⁺ /14.320	5.73/50. 7.53/15. 8.76/15. 9.58/20.	4-			ann.rad./ 0.752/41. 3.732/99.5 (1.12-3.92)
⁴¹Sc		40.9692513	0.596 s	β⁺ /6.4953	5.61/100.	7/2-	5.431	0.166	ann.rad./
⁴²ᵐSc			61.6 s	β⁺ /	2.82/	7+			ann.rad./ 0.4375(5)/100. 1.2270(5)/100. 1.5245(5)/100.
⁴²Sc		41.9655168	0.682 s	β⁺ /6.4259	5.32/100.	0+			ann.rad./
⁴³Sc		42.961151	3.89 h	β⁺ ,EC/2.221	0.82/22. 1.22/78.	7/2-	+4.62	-0.26	ann.rad./ 0.3729(1)/22.
⁴⁴ᵐSc			2.44 d	I.T./0.27 EC/3.926		6+	+3.88		0.27124(1)/87. (1.00-1.16)
⁴⁴Sc		43.959403	3.93 h	β⁺, EC/3.653	1.47/	2+	+2.56	+0.10	ann.rad./ 1.15700(1)/100.
⁴⁵Sc	100.	44.955910				7/2-	+4.75649	-0.22	
⁴⁶ᵐSc			18.7 s	I.T./0.14253		1-			0.14253(2)/62.
⁴⁶Sc		45.955170	83.81 d	β⁻ /2.367	0.357/100.	4+	+3.03	+0.12	0.88925(3)/100. 1.12051(1)/100.
⁴⁷Sc		46.952408	3.349 d	β⁻ /0.600	0.439/69. 0.601/31.	7/2-	+5.34	-0.22	0.15938(1)/68.
⁴⁸Sc		47.95224	43.7 h	β⁻ /3.99	0.655/	6+			0.98350(1)/100. 1.03750(1)/97. 1.31209(3)/100.
⁴⁹Sc		48.950024	57.3 m	β⁻ /2.006	2.00/99.9.	7/2-			1.7619(3)/0.05
⁵⁰Sc		49.95219	1.71 m	β⁻ /6.89	3.05/76. 3.60/24.	(5+)			0.5235(1)/88. 1.1210(1)/100. 1.5537(2)/100.
⁵¹Sc		50.95360	12.4 s	β⁻ /6.51	4.4/ 5.0/	7/2-			1.4373(4)/52. 0.718-2.144
⁵²Sc		51.9566	8.2 s	β⁻ /9.0		(3+)			
⁵³Sc		52.9592	< 1.6 s	β⁻ /8.1					
⁵⁴Sc		53.9630	≈0.23 s	β⁻ /11.6					
⁵⁵Sc		54.967	0.13 s	β⁻ /13					
⁵⁶Sc		55.973							
⁵⁷Sc		56.977							
⁵⁸Sc		57.983							
₂₂Ti		**47.867(1)**							
³⁹Ti		39.0013	28. ms	β⁺ /15.4					
⁴⁰Ti		39.9905	52. ms	β⁺ /11.7					
⁴¹Ti		40.98313	80. ms	β⁺ ,p/12.93		3/2 +			ann.rad./
⁴²Ti		41.97303	0.20 s	β⁺ /7.000	6.0/				ann.rad./ 0.6107(5)/56.
⁴³Ti		42.96852	0.50 s	β⁺ /6.87	5.80/	7/2-			ann.rad./
⁴⁴Ti		43.959690	66. y	EC/0.268		0+	0.85		0.067868(1)/91. 0.078322(1)/97.
⁴⁵Ti		44.958124	3.078 h	β⁺/86/2.062 EC/14/	1.04	7/2-	0.095	0.015	ann.rad./ (0.36-1.66)/weak
⁴⁶Ti	8.25(3)	45.952630				0+			
⁴⁷Ti	7.44(2)	46.951764				5/2-	-0.78848	+0.30	
⁴⁸Ti	73.72(3)	47.947947				0+			
⁴⁹Ti	5.41(2)	48.947871				7/2-	-1.10417	+0.24	
⁵⁰Ti	5.18(2)	49.944792				0+			
⁵¹Ti		50.946616	5.76 m	β⁻ /2.471	1.50/92. 2.13/	3/2-			0.3197(2)/93. 0.6094-0.9291
⁵²Ti		51.94690	1.7 m	β⁻ /1.97	1.8/100.	0+			0.0170(5)/100. 0.12445(5)/100.
⁵³Ti		52.9497	33. s	β⁻ /5.0	(2.2-3)/	3/2-			0.1008(1)/20. 0.1276(1)/45. 0.2284(1)/39. 1.6755(5)/45. (1.72-2.8)/
⁵⁴Ti		53.9509	1.5 s	β⁻ /4.3					
⁵⁵Ti		54.9551	0.60 s	β⁻ /7.4					
⁵⁶Ti		55.9580	0.15 s	β⁻ /7.0					
⁵⁷Ti		56.963	≈ 56. ms	β⁻ /11.					
⁵⁸Ti		57.966							
⁵⁹Ti		58.972							
⁶⁰Ti		59.976							

Elem. or Isot.	Natural Abundance (%)	Atomic Mass or Weight	Half-Life	Decay Mode/Energy (/MeV)	Particle Energy /Intensity (MeV / %)	Spin (h/2π)	Nuclear Magnetic Mom.(nm)	Elect. Quadr. Mom. (b)	γ-ray Energy /Intensity (MeV / %)
^{61}Ti		60.982							
$_{23}$V		50.9415(1)							
^{43}V		42.9807	>0.8 s	β^+ /11.3					
^{44}V		43.9744	0.09 s	β^+ ,α/13.7					ann.rad./
^{45}V		44.96578	0.54 s	β^+ /7.13		7/2-			
^{46}V		45.960200	0.4223 s	β^+ /7.051	6.03/100.	0+			ann.rad./
^{47}V		46.954907	32.6 m	β^+ ,EC/2.928	1.90/99.+	3/2-			ann.rad./ 1.7949(8)/0.19 (0.2-2.16)
^{48}V		47.952254	15.98 d	β^+ /4.012	0.698/50.	4+	2.01		ann.rad./ 0.98350(2)/100. (1.3-2.4)
^{49}V		48.948517	337. d	EC/0.602		7/2-	4.47		
^{50}V	0.250(4)	49.947163	>1.4x10^{17}y	EC, β^-		6+	+3.34569	+0.21	
^{51}V	99.750(4)	50.943964				7/2-	+5.148706	-0.04	
^{52}V		51.944780	3.76 m	β^- /3.976	2.47/	3+			1.4341(1)/100.
^{53}V		52.944342	1.61 m	β^- /3.436	2.52/	7/2-			1.0060(5)/90. 1.2891(3)/10.
^{54}V		53.94644	49.8 s	β^- /7.04	1.00/5. 2.00/12. 2.95/45. 5.20/11.	3+			0.8348/97. 0.9887/80. 2.259/46. (0.56-3.38)
^{55}V		54.9472	6.5 s	β^- /6.0	6.0/	(7/2-)			0.5177/73. (0.224-1.21)
^{56}V		55.9504	0.23 s	β^- /9.1					
^{57}V		56.9524	0.33 s	β^- /8.1					
^{58}V		57.9567	0.20 s	β^- /11.6					
^{59}V		58.9593	0.07 s	β^- /9.9					
^{60}V		59.965		β^- /14.					
^{61}V		60.967							
^{62}V		61.973							
^{63}V		62.977							
^{64}V									
$_{24}$Cr		51.9961(6)							
^{43}Cr		42.9977	21. ms						
^{44}Cr		43.9855	53. ms	β^+ ,(p) /10.3	p/0.95-3.1				
^{45}Cr		44.9792	0.05 s	β^+ ,p/12.5		7/2-			ann.rad./
^{46}Cr		45.96836	0.3 s	β^+ /7.60					ann.rad./
^{47}Cr		46.96291	0.51 s	β^+ /7.45		3/2-			ann.rad./
^{48}Cr		47.95404	21.6 h	EC/1.66					ann.rad./ 0.116(2)/95. 0.305(10)/100.
^{49}Cr		48.951341	42.3 m	β^+,EC/2.631	1.39/ 1.45/ 1.54/	5/2-	0.476		ann.rad./ 0.09064(1)/51. 0.15293(1)/27. (0.062-1.6)
^{50}Cr	4.345(13)	49.946050				0+			
^{51}Cr		50.944772	27.70 d	EC/0.7527		7/2-	-0.934		0.320076(1)/10.2
^{52}Cr	83.789(18)	51.940512				0+			
^{53}Cr	9.501(17)	52.940653				3/2-	-0.47454	-0.15	
^{54}Cr	2.365(7)	53.938885				0+		-0.2	
^{55}Cr		54.940844	3.497 m	β^- /2.603	2.5/	3/2-			1.5282(2)/0.04 (0.13-2.37)
^{56}Cr		55.94065	5.9 m	β^- /1.62	1.50/100.	0+			0.026(2)/100. 0.083(3)/100.
^{57}Cr		56.9438	21. s	β^- /5.1	3.3/ 3.5/	3/2-	0.0834		0.850/8. (0.083-2.62)
^{58}Cr		57.9443	7.0 s	β^- /4.0					(0.131-0.683)
^{59}Cr		58.9487	1.0 s	β^- /7.7					1.236
^{60}Cr		59.9497	0.6 s	β^- /6.0					
^{61}Cr		60.9541		β^- /8.8					
^{62}Cr		61.9558	0.20 s	β^- /7.3					
^{63}Cr		62.962							
^{65}Cr		64.970							
^{66}Cr									
^{67}Cr									
$_{25}$Mn		54.938049(9)							
^{46}Mn		45.9867	≈41. ms	β^+ /17.1					
^{47}Mn		46.9761	≈0.1 s	β^+ /12.3					
^{48}Mn		47.9689	0.15 s	β^+ /13.5	5.79/58. 4.43/10.	4+			
^{49}Mn		48.95962	0.38 s	β^+ /7.72	6.69/	5/2-			ann.rad./

Elem. or Isot.	Natural Abundance (%)	Atomic Mass or Weight	Half-Life	Decay Mode/Energy (/MeV)	Particle Energy /Intensity (MeV / %)	Spin (h/2π)	Nuclear Magnetic Mom.(nm)	Elect. Quadr. Mom. (b)	γ-ray Energy /Intensity (MeV / %)
^{50m}Mn			1.74 m	β^+ /7.887	3.54/	5+			ann.rad./ 1.0980/94. 0.783/91.
^{50}Mn		49.954244	0.283 s	β^+ /7.6330	6.61/	0+			(0.66-3.11)
^{51}Mn		50.948215	46.2 m	β^+ ,EC/3.208	2.2/	5/2-	3.568	0.4	ann.rad./ 0.7491(1)/0.26
^{52m}Mn			21.1 m	β^+ /98/5.09 I.T./2/0.378	2.631/	2+	0.0076		(1.148-1.164) ann.rad./ 0.3778 (I.T.) 1.43406(1)/98.
^{52}Mn		51.945570	5.591 d	β^+ /4.712 EC/	0.575/	6+	+3.063	+0.5	(0.7-4.8) ann.rad./ 0.74421(1)/90. 1.43406(1)/100.
^{53}Mn		52.941294	3.7×10^6 y	EC/0.5970		7/2-	5.024		
^{54}Mn		53.940363	312.1 d	EC/1.377		3+	+3.282	+0.33	0.83403(5)/100.
^{55}Mn	100.	54.938049				5/2-	+3.4687	+0.32	
^{56}Mn		55.938909	2.579 h	β^- /3.6954	0.718/18. 1.028/34.	3+	+3.2266		0.84675(2)/98.9 1.81072(4)/27. 2.11305(5)/14.5
^{57}Mn		56.938287	1.45 m	β^- /2.691		5/2-			
^{58}Mn		57.93999	65 s	β^- /6.25	3.8/ 5.1/	3+			0.45916(2)/20. 0.81076(1)/82. 1.32309(5)/53.
^{59}Mn		58.94045	4.6 s	β^- /5.19	4.5/				0.471/ 0.531-0.726
^{60m}Mn			1.77 s	β^- /IT	5.7/	3+			0.824/
^{60}Mn		59.9433	50. s	β^- /8.6		0+			1.969/
^{61}Mn		60.9446	0.71 s	β^- /7.4		(5/2)-			
^{62}Mn		61.9480	0.9 s	β^- /10.4		(3+)			0.877/
^{63}Mn		62.9498	0.25 s	β^- /8.8					0.942-1.299
^{64}Mn		63.9537	0.17 s	β^- /11.8					0.356
^{65}Mn		64.9561		β^- /11.8					
^{66}Mn		65.961		β^- /10.					
^{67}Mn		66.964							
^{68}Mn									
^{69}Mn									
$_{26}$Fe		55.845(2)							
^{45}Fe		45.0146							
^{46}Fe		46.0008	$\approx$0.02 s	β^+ /13.1					
^{47}Fe		46.9929	$\approx$0.03 s	β^+ /15.6					
^{48}Fe		47.9806	$\approx$ 44. ms	β^+ /11.2					
^{49}Fe		48.9763	70. ms	β^+ /13.0		(7/2-)			ann.rad./
^{50}Fe		49.9630	0.15 s	β^+ /8.2					0.651
^{51}Fe		50.95683	0.31 s	β^+ /8.02		(5/2-)			ann.rad./
^{52m}Fe			46. s	β^+ /4.4		(12+)			ann.rad./ (0.622-2.286)/
^{52}Fe		51.94812	8.28 h	β^+ /57/2.37 EC/43/ I.T./	0.804/	0+			ann.rad./ 0.16868(1)/99. 0.377 (I.T.)/
^{53m}Fe			2.6 m	I.T./3.0407		19/2-			0.7011(1)/99. 1.0115(1)/87. 1.3281(1)/87. 2.3396(1)/13.
^{53}Fe		52.945312	8.51 m	β^+ /3.743	2.40/42. 2.80/57.	7/2-			ann.rad./ 0.3779(1)/42.
^{54}Fe	5.845(35)	53.939615				0+			(1.2 - 3.2)
^{55}Fe		54.938298	2.73 y	EC/0.2314		3/2-			
^{56}Fe	91.754(36)	55.934942				0+			
^{57}Fe	2.119(10)	56.935398				1/2-	+0.0906	0.16	
^{58}Fe	0.282(4)	57.933280				0+			
^{59}Fe		58.934880	44.51 d	β^- /1.565	0.273/48. 0.475/51.	3/2-	- 0.336		1.099/57 1.292/43.
^{60}Fe		59.934077	1.5×10^6 y	β^- /0.237	0.184/100.	0+			(0.14-1.48) 0.0586/100.(IT)
^{61}Fe		60.93675	6.0 m	β^- /3.98	2.5/13. 2.63/54. 2.80/31.				1.205/44. 1.028/43. (0.12-3.37)
^{62}Fe		61.93677	68. s	β^- /2.53	2.5/100.	0+			0.5061(1)/100.
^{63}Fe		62.9404	6. s	β^- /6.3		5/2-			0.995/ (1.365-1.427)
^{64}Fe		63.9411	2.0 s	β^- /4.9					
^{65}Fe		64.9449	$\approx$0.45 s	β^- /7.9					
^{66}Fe		65.9460		β^- /5.7					

Elem. or Isot.	Natural Abundance (%)	Atomic Mass or Weight	Half-Life	Decay Mode/Energy (/MeV)	Particle Energy /Intensity (MeV / %)	Spin (h/2π)	Nuclear Magnetic Mom.(nm)	Elect. Quadr. Mom. (b)	γ-ray Energy /Intensity (MeV / %)
^{67}Fe		66.9500		β^- /8.8					
^{68}Fe		67.953	≈0.10 s	β^- / ≈7.6					
^{69}Fe		68.958							
^{70}Fe									
^{71}Fe									
^{72}Fe									
$_{27}$Co		58.933200(9)							
^{50}Co		49.9812	44. ms	β^+ /17.0	2.03-2.79				
^{51}Co		50.9705		β^+ /12.8					
^{52}Co		51.9632	0.12 s	β^+ /14.0					0.849-1.942 ann.rad./
^{53}Co			0.25 s	β^+ ,p/		19/2-			ann.rad./
^{53m}Co			0.26 s	β^+ /8.30		7/2-			ann.rad./
^{55}Co		52.95423	1.46 m	β^+ /8.44	4.25/100.	7+			0.411(1)/99. 1.130(1)/100. 1.408(1)/100. ann.rad./
^{54}Co		53.948464	0.1932 s	β^+ /8.2430	7.34/100.	0+			ann.rad./
^{55}Co		54.942003	17.53 h	β^+ /3.4513 EC/	0.53/ 1.03/ 1.50/	7/2-	+4.822		0.9312/75. 0.4772/20. (0.092-3.11)
^{56}Co		55.939844	77.3 d	β^+ /4.566 EC/	1.459/18.	4 +	3.85	+0.25	ann.rad./ 0.84678(6)/99.9 1.2383/68. (0.26-3.61)
^{57}Co		56.936296	271.8 d	EC/0.8361		7/2-	+4.72	+0.5	0.12206(2)/85.6 (0.014-0.706) 0.02489/0.035
^{58m}Co			9.1 h	I.T./		5 +			
^{58}Co		57.935757	70.88 d	β^+ /2.307 EC/		2 +	+4.04	+0.22	ann.rad./ 0.810755(3)/99.
^{59}Co	100.	58.933200				7/2-	+4.63	+0.41	
^{60m}Co			10.47 m	I.T./99.8/0.059 β^- /0.2/1.56		2 +	+4.40	+0.3	0.058603(1)/2.0
^{60}Co		59.933822	5.271 y	β^- /2.824	0.315/99.7	5 +	+3.799	+0.44	1.173210(2)/100. 1.332470(2)/100.
^{61}Co		60.932479	1.650 h	β^- /1.322	1.22/95.	7/2-			0.067415(1)/86. (0.8417-0.9092)
^{62m}Co			13.9 m	β^- /	0.88/25. 2.88/75.	5 +			1.1635(3)/70. 1.1730(3)/98. 2.0039(3)/19.
^{62}Co		61.93405	1.50 m	β^- /5.32	1.03/10. 1.76/5. 2.9/20. 4.05/60.	2 +			1.1292(3)/13. 1.1730(3)/83. 1.9851(1)/3. 2.3020(1)/19.
^{63}Co		62.93362	27.5 s	β^- /3.67	3.6/	7/2-			0.08713(1)/49. 0.9817(3)/2.6 0.156-2.17
^{64}Co		63.93581	0.30 s	β^- /7.31	7.0/	1 +			
^{65}Co		64.93648	1.14 s	β^- /5.96		(7/2)-			
^{66}Co		65.9398	0.25 s	β^- /10.0					(1.245-1.425)
^{67}Co		66.9406	0.32 s	β^- /8.4					0.694
^{68}Co		67.9444	≈ 0.18 s	β^- /11.7					
^{69}Co		68.9452	0.27 s	β^- /9.3					
^{70}Co		69.950	0.17 s	β^- 13.					
^{71}Co		70.952	0.20 s	β					
^{72}Co		71.956	0.12 s	β					
^{73}Co									
^{74}Co									
^{75}Co									
$_{28}$Ni		58.6934(2)							
^{49}Ni		49.9959							
^{50}Ni		50.9877							
^{51}Ni		51.9757		β^+ /16.0					
^{52}Ni		51.9757	38. ms	β^+ /11.7					
^{53}Ni		52.9685	0.05 s	β^+ ,p/13.3		7/2-			ann.rad./ 0.937
^{54}Ni		53.95791	0.14 s	β^+ /8.80					
^{55}Ni		54.95134	0.20 s	β^+ /8.70	7.66/	7/2-			ann.rad./ 0.15838(3)/98.8
^{56}Ni		55.94214	6.08 d	EC/2.14		0+			0.81185(3)/87. 0.2695-0.7500
^{57}Ni		56.939800	35.6 h	β^+ /3.264 EC/	0.712/10. 0.849/76.	3/2-	- 0.798		ann.rad./ 1.3776/78. (0.127-3.177)
^{58}Ni	68.0769(89)	57.935348				0+			
^{59}Ni		58.934351	≈7.6x10^4 y	EC/		3/2-			

Elem. or Isot.	Natural Abundance (%)	Atomic Mass or Weight	Half-Life	Decay Mode/Energy (/MeV)	Particle Energy /Intensity (MeV / %)	Spin (h/2π)	Nuclear Magnetic Mom.(nm)	Elect. Quadr. Mom. (b)	γ-ray Energy /Intensity (MeV / %)
⁶⁰Ni	26.2231(77)	59.930790				0+			
⁶¹Ni	1.1399(6)	60.931060				3/2-	-0.75002	+0.16	
⁶²Ni	3.6345(17)	61.928348				0+			
⁶³Ni		62.929673	100. y	β⁻ /0.066945	0.065/	1/2-			
⁶⁴Ni	0.9256(9)	63.927969				0+			
⁶⁵Ni		64.930088	2.517 h	β⁻ /2.137	0.65/30. 1.020/11. 2.140/58.	5/2-	0.69		0.36627(3)/5. 1.11553(4)/16. 1.48184(5)/23.
⁶⁶Ni		65.92912	54.6 h	β⁻ /0.23		0+			
⁶⁷Ni		66.93157	21. s	β⁻ /3.56	3.8/	1/2-			1.0722/100. 1.6539/100. (0.10-1.98)
⁶⁸Ni		67.93185	19. s	β⁻ /2.06					
⁶⁹Ni		68.9352	10. s	β⁻ /5.4					0.6807(3)/100. (0.207-1.213)
⁷⁰Ni		69.9361		β⁻ /3.5					
⁷¹Ni		70.9400	1.9 s	β⁻ /6.9					
⁷²Ni		71.9413	2.1 s	β⁻ /5.2					
⁷³Ni		72.946	0.9 s	β⁻ /9.					
⁷⁴Ni		73.948	1.1 s	β⁻ /7.					
⁷⁵Ni		74.953	≈ 0.47 s						
⁷⁶Ni		75.955	≈ 0.24 s						
⁷⁷Ni		76.961							
⁷⁸Ni		77.964							
₂₉Cu		63.546(3)							
⁵⁵Cu		54.9655		β⁺ /13.2					
⁵⁶Cu		55.9586		β⁺ /15.3					
⁵⁷Cu		56.94922	196. ms	β⁺ /8.77		3/2-			
⁵⁸Cu		57.944541	3.21 s	β⁺ /8.563 EC/	4.5/15. 7.439/83.	1+			0.77-3.01 ann.rad./ 0.0403(4)/5. 1.4483(2)/11. 1.4546(2)/16. ann.rad./
⁵⁹Cu		58.939504	1.36 m	β⁺ /4.800	1.9/ 3.75/	3/2-			0.3393(1)/8. 0.8780(1)/12. 1.3015(1)/15. (0.4 - 2.6) ann.rad./
⁶⁰Cu		59.937368	23.7 m	β⁺ /6.127 EC/	2.00/69. 3.00/18. 3.92/6.	2+	+1.219		1.3325/88. 1.7915/45. (0.12-5.048) ann.rad./
⁶¹Cu		60.933462	3.35 h	β⁺ /2.237	0.56/3. 0.94/5. 1.15/2. 1.220/51.	3/2-	+2.14		0.2830/13. 0.6560/11. (0.067-2.123) ann.rad./
⁶²Cu		61.932587	9.67 m	β⁺ /98/3.948 EC/	2.93/98.	1+	-0.380		1.17302(1)/0.6 (0.87-3.37)
⁶³Cu	69.17(3)	62.929601				3/2-	+2.2233	-0.211	
⁶⁴Cu		63.929768	12.701 h	β⁻ /39/0.579 β⁺ /19/1.6751 EC/41/	0.578/ 0.65/	1+	-0.217		ann.rad./ 1.3459(3)/0.6
⁶⁵Cu	30.83(3)	64.927794				3/2-	+2.3817	-0.195	
⁶⁶Cu		65.928873	5.09 m	β⁻ /2.642	1.65/6. 2.7/94.	1+	-0.282		0.8330(1)/0.22 1.0392(2)/9.2
⁶⁷Cu		66.92775	2.580 d	β⁻ /0.58	0.395/56. 0.484/23. 0.577/20.	3/2-			0.09125(1)/7. 0.09325(1)/17. 0.18453(1)/47.
⁶⁸ᵐCu			3.79 m	I.T./86/ β⁻ /14/1.8		6-			0.0843(5)/70. 0.1112(5)/18. 0.5259(5)/74. (0.64-1.34)
⁶⁸Cu		67.92964	31. s	β⁻ /4.46	3.5/40. 4.6/31.	1+			1.0774(5)/58. 1.2613(5)/17. (0.15-2.34)
⁶⁹Cu		68.92943	2.8 m	β⁻ /2.68	2.48/80.	3/2-			0.5307(3)/3. 0.8340(5)/6. 1.0065(8)/10.
⁷⁰ᵐCu			47. s	β⁻ /	2.52/10.	5-			0.8848(2)/100. 0.9017(2)/90. 1.2517(5)/60. (0.39-3.06)
⁷⁰Cu		69.93241	5. s	β⁻ /6.60	5.42/54. 6.09/46.	1+			0.8848(2)/54.
⁷¹Cu		70.93262	20. s	β⁻ /4.56		3/2-			0.490/
⁷²Cu		71.9357	6.6 s	β⁻ /8.2		(1+)			0.652/
⁷³Cu		72.9365	3.9 s	β⁻ /6.3					0.450/

Elem. or Isot.	Natural Abundance (%)	Atomic Mass or Weight	Half-Life	Decay Mode/Energy (/MeV)	Particle Energy /Intensity (MeV / %)	Spin (h/2π)	Nuclear Magnetic Mom.(nm)	Elect. Quadr. Mom. (b)	γ-ray Energy /Intensity (MeV / %)
⁷⁴Cu		73.9401	1.6 s	β⁻ /9.9					
⁷⁵Cu		74.9414	1.2 s	β⁻ /7.9					
⁷⁶Cu		75.9455	0.64 s	β⁻ /11.					
⁷⁷Cu		76.947	0.47 s	β⁻ /≈10.					
⁷⁸Cu		77.952	0.34 s	β⁻ /12.					
⁷⁹Cu		78.954	0.19 s	β⁻ /11.					
⁸⁰Cu		79.962							
₃₀Zn		65.39(2)							
⁵⁷Zn		56.9649	0.04 s	β⁺,p/14.6		(7/2-)			ann.rad./
⁵⁹Zn		58.94927	183. ms	β⁺,p/9.09	8.1/	3/2-			ann.rad./ (0.491-0.914)
⁶⁰Zn		59.94183	2.40 m	β⁺ /97/4.16 EC/3/		0+			ann.rad./ 0.669/47. (0.062-0.947)
⁶¹Zn		60.93951	1.485 m	β⁺ /5.64	4.38/68.	3/2-			ann.rad./ 0.4748/17. (0.15-3.52)
⁶²Zn		61.93433	9.22 h	β⁺ /3/1.63 EC/93/	0.66/7.	0+			ann.rad./ 0.0408/25 0.5967/26. (0.20-1.526)/
⁶³Zn		62.933215	38.5 m	β⁺ /93/3.367 EC/7/	1.02/ 1.40/ 1.71/ 2.36/84.	3/2-	-0.28164	+0.29	ann.rad./ 0.66962(5)/8.4 0.96206(5)/6.6 (0.24-3.1)
⁶⁴Zn	48.63(60)	63.929146				0+			
⁶⁵Zn		64.929245	243.8 d	β⁺ /98/1.3514 EC/1.5/	0.325/	5/2-	+0.7690	-0.023	ann.rad./ 1.11552(2)/50.8
⁶⁶Zn	27.90(27)	65.926036				0+			
⁶⁷Zn	4.10(13)	66.927131				5/2-	+0.8755	+0.15	
⁶⁸Zn	18.75(51)	67.924847				0+			
⁶⁹ᵐZn			13.76 h	I.T./99+/0.439		9/2+			0.4390(2)/95.
⁶⁹Zn		68.926553	56. m	β⁻ /0.906	0.905/99.9	1/2-			0.318/
⁷⁰Zn	0.62(3)	69.925325				0+			
⁷¹ᵐZn			3.97 h	β⁻ /	1.45/	9/2+			0.3864/93. 0.4874/62. 0.6203/57. (0.099-2.489)
⁷¹Zn		70.92773	2.4 m	β⁻ /2.81		1/2-			0.5116(1)/30. 0.9103(1)/7.5 (0.12-2.29)
⁷²Zn		71.92686	46.5 h	β⁻ /0.46	0.25/14. 0.30/86.	0+			0.0164(3)/8. 0.1447(1)/83. 0.1915(2)/9.4
⁷³ᵐZn			6. s	I.T./0.196		(7/2+)			0.042 0.216(1)/100. 0.496-0.911
⁷³Zn		72.92978	24. s	β⁻ /4.29	4.7/	(1/2-)			0.0565/ 0.1401/ (0.05-0.35)
⁷⁴Zn		73.92946	1.60 m	β⁻ /2.3	2.1/				0.229/
⁷⁵Zn		74.9329	10.2 s	β⁻ /6.0					0.119/
⁷⁶Zn		75.9334	5.7 s	β⁻ /4.2	3.6/				0.772
⁷⁷ᵐZn			1.0 s	β⁻ /		(1/2-)			0.189/
⁷⁷Zn		76.9371	2.1 s	β⁻ /7.3	4.8/				0.225/
⁷⁸Zn		77.9386	1.5 s	β⁻ /6.4					0.702/
⁷⁹Zn		78.9421	1.0 s	β⁻ /8.6					0.713/
⁸⁰Zn		79.9444	0.54 s	β⁻ /7.3					0.2248/
⁸¹Zn		80.9505	0.29 s	β⁻ /11.9					
⁸²Zn		81.9548							
⁸³Zn									
₃₁Ga		69.723(1)							
⁶⁰Ga		59.9571							
⁶¹Ga		60.9492	0.15 s	β⁺ /9.0					
⁶²Ga		61.94418	0.116 s	β⁺ /9.17 EC/	8.3/	0+			ann.rad./
⁶³Ga		62.9391	32. s	β⁺ /5.5 EC/	4.5/				ann.rad./ 0.6271(2)/10. 0.6370(2)/11. 1.0652(4)/45.
⁶⁴Ga		63.936838	2.63 m	β⁺ /7.165	2.79/ 6.05/	0+			ann.rad./ 0.80785(1)/14. 0.99152(1)/43. 1.38727(1)/12. 3.3659(1)/13.

Elem. or Isot.	Natural Abundance (%)	Atomic Mass or Weight	Half-Life	Decay Mode/Energy (/MeV)	Particle Energy /Intensity (MeV / %)	Spin (h/2π)	Nuclear Magnetic Mom.(nm)	Elect. Quadr. Mom. (b)	γ-ray Energy /Intensity (MeV / %)
⁶⁵Ga		64.9394	15.2 m	β⁺ /86/3.255 EC/	0.82/10. 1.39/19. 2.113/56. 2.237/15.	3/2-			ann. rad./ 0.1151(2)/55. 0.1530(2)/96. 0.2069(2)/39. (0.06-2.4)
⁶⁶Ga		65.931592	9.5 h	β⁺ /56/5.175 EC/43/	0.74/1. 1.84/54. 4.153/51.	0+			ann.rad./ 1.03935(8)/38. 2.7523(1)/23. (0.28-5.01)
⁶⁷Ga		66.928205	3.260 d	EC/1.001		3/2-	+1.8507	0.20	0.09332/37. 0.18459/20. 0.30024/17. (0.091-0.89)
⁶⁸Ga		67.927983	1.130 h	β⁺ /90/2.921 EC/10/	1.83/	1+	0.01175	0.028	ann.rad./ 1.0774(1)/3. (0.57-2.33)/
⁶⁹Ga	60.108(9)	68.925581				3/2-	+2.01659	+0.17	
⁷⁰Ga		69.926027	21.1 m	EC/0.2/0.655 β⁻ /99.8/1.656	1.65/99.	1+			0.1755(5)/0.15 1.042(5)/0.48
⁷¹Ga	39.892(9)	70.924707				3/2-	+2.56227	+0.11	
⁷²Ga		71.926372	14.10 h	β⁻ /4.001	0.64/40. 1.51/9. 2.52/8. 3.15/11.	3-	-0.13224	+0.5	0.62986(5)/24. 2.2016(2)/26. 2.5077(2)/12.8 (0.11-3.3)/
⁷³Ga		72.92517	74.87 h	β⁻ /1.59		3/2-			0.05344(5)/10. 0.29732(5)/47. (0.01-1.00)/
⁷⁴ᵐGa			10. s	I.T./		1+			0.0565(1)/75.
⁷⁴Ga		73.92694	8.1 m	β⁻ /5.4	2.6/	3-			0.5959/92. 2.354/45. (0.23-3.99)
⁷⁵Ga		74.92650	2.10 m	β⁻ /3.39	3.3/	3/2-			0.2529/ 0.5746/ (0.12-2.10)
⁷⁶Ga		75.9289	29. s	β⁻ /7.0		3-			0.5629/66. 0.5455/26. (0.34-4.25)
⁷⁷Ga		76.9293	13.0 s	β⁻ /5.3	5.2/				0.469/ 0.459/
⁷⁸Ga		77.9317	5.09 s	β⁻ /8.2		3+			0.619/77. 1.187/20.
⁷⁹Ga		78.9329	2.85 s	β⁻ /7.0	4.6/				0.465/
⁸⁰Ga		79.9366	1.68 s	β⁻ /10.4	10./				0.659/
⁸¹Ga		80.9377	1.22 s	β⁻ /8.3	5.1/				0.217/
⁸²Ga		81.9432	0.599 s	β⁻ /12.6					1.348/
⁸³Ga		82.9469	0.308 s	β⁻ / ≈ 11.5					
⁸⁴Ga		83.952	≈0.085 s	β⁻ /14					
⁸⁵Ga									
⁸⁶Ga									
₃₂Ge		72.61(2)							
⁶¹Ge		60.9638	0.04 s	β⁺ /13.6					
⁶³Ge		62.9496	0.10 s	β⁻ /9.8					
⁶⁴Ge		63.9416	1.06 m	β⁺ /4.4 EC/	3.0/	0+			ann.rad./ 0.1282(2)/11. 0.4270(3)/37. 0.6671(3)/17.
⁶⁵Ge		64.9394	31. s	β⁺ /6.2 EC/	0.82/10. 1.39/19. 2.113/56. 2.237/15.				ann.rad./ 0.0620/27. 0.6497/33. 0.8091/21. (0.19-3.28)
⁶⁶Ge		65.93385	2.26 h	β⁺ /27/2.10 EC/73/		0+			ann.rad./ 0.0438/29. 0.3819/28. (0.022-1.77)
⁶⁷Ge		66.932738	19.0 m	β⁺ /96/4.225 EC/4/	1.6/ 2.3/ 3.15/	1/2-			ann.rad./ 0.1670/84. (0.25-3.73)
⁶⁸Ge		67.92810	270.8 d	EC/0.11		0+			Ga k x-ray/39.
⁶⁹Ge		68.927973	1.63 d	β⁺ /36/2.2273 EC/64/	0.70/ 1.2/	5/2-	0.735	0.02	ann.rad./ 0.574/13. 1.1068/36. (0.2-2.04)
⁷⁰Ge	20.84(87)	69.924250				0+			
⁷¹ᵐGe			20.4 ms		I.T./0.0234	9/2+			0.1749
⁷¹Ge		70.924954	11.2 d	EC/0.229		1/2-	+0.547		
⁷²Ge	27.54(34)	71.922076				0+			

Elem. or Isot.	Natural Abundance (%)	Atomic Mass or Weight	Half-Life	Decay Mode/Energy (/MeV)	Particle Energy /Intensity (MeV / %)	Spin (h/2π)	Nuclear Magnetic Mom.(nm)	Elect. Quadr. Mom. (b)	γ-ray Energy /Intensity (MeV / %)
⁷³Ge	7.73(5)	72.923460				9/2+	-0.879467	-0.17	
⁷⁴Ge	36.28(73)	73.921178				0+			
⁷⁵ᵐGe			48. s	I.T./		7/2+			0.13968(3)/39.
⁷⁵Ge		74.922860	1.380 h	β⁻/1.177	1.19/	1/2-	+0.510		0.26461(5)/11. 0.41931(5)/0.2
⁷⁶Ge	7.61(38)	75.921403				0+			
⁷⁷ᵐGe			53. s	I.T./20/ β⁻/80/2.861	2.9/	1/2-			0.1597(1)/11. 0.21551/21. (0.19-0.61)
⁷⁷Ge		76.923549	11.25 h	β⁻/2.702	0.71/23. 1.38/35. 2.19/42.	7/2+			0.2110/29. 0.2155/27. 0.2644/51. (0.15-2.35)
⁷⁸Ge		77.922853	1.45 h	β⁻/0.95	0.70/	0+			0.2773(5)/96. 0.2939(5)/4.
⁷⁹ᵐGe			39. s	β⁻/IT		7/2+			
⁷⁹Ge		78.9254	19.1 s	β⁻/4.2	4.0/20. 4.3/80.	1/2-			0.1096/21. (0.10-2.59) 0.5427(4)/15.
⁸⁰Ge		79.92545	29.5 s	β⁻/2.67	2.4/	0+			0.1104(4)/6. 0.2656(4)/25.
⁸¹ᵐGe			≈7.6 s	β⁻/	3.75/	1/2+			0.3362(4)/ 0.7935(4)/
⁸¹Ge		80.9288	≈7.6 s	β⁻/6.2	3.44/	9/2+			0.1976(4)/21. 0.3362(4)/100.
⁸²Ge		81.9296	4.6 s	β⁻/4.7		0+			1.093/
⁸³Ge		82.9345	1.9 s	β⁻/8.9					
⁸⁴Ge		83.9373	0.98 s	β⁻/7.7					
⁸⁵Ge		84.943	0.54 s	β⁻/10.					
⁸⁶Ge		85.946							
⁸⁷Ge									
⁸⁸Ge									
⁸⁹Ge									
₃₃As		74.92160(2)							
⁶⁴As		63.9576							
⁶⁵As		64.9495	0.19 s	β⁺/9.4					
⁶⁶As		65.94410	95.8 ms	β⁺/9.55					
⁶⁷As		66.9392	42. s	β⁺/6.0 EC/	5.0/	5/2-			0.121/ 0.123/ 0.244/
⁶⁸As		67.9368	2.53 m	β⁺/8.1		3+			ann.rad./ 0.652/32. 0.762/33. 1.016/77. (0.61-3.55)
⁶⁹As		68.93228	15.2 m	β⁺/98/4.01 EC/2/	2.95/	5/2-	1.2		ann.rad./ 0.0868(5)/1.5 0.1458(3)/2.4
⁷⁰As		69.93093	52.6 m	β⁺/84/6.22 EC/16/2.14 /2.89	1.44/	4+	+2.1061	+0.09	ann.rad./ 1.0395(7)/82. (0.17-4.4)/
⁷¹As		70.927114	2.72 d	β⁺/32/2.013 EC/68/		5/2-	+1.6735	-0.02	ann.rad./ 0.1749(2)/84. 1.0957(2)/4.2
⁷²As		71.926753	26.0 h	β⁺/77/4.356	0.669/5. 1.884/12. 2.498/62. 3.339/19.	2-	-2.1566	-0.08	ann.rad./ 0.83395(5)/80. 1.0507(1)/9.6 (0.1-4.0)
⁷³As		72.923825	80.3 d	EC/0.341		3/2-			0.013263(1)/0.1 0.053437(1)/10.5 Se k x-ray/90.
⁷⁴As		73.923829	17.78 d	β⁺/31/2.562 EC/37/ β⁻/1.353	0.94/26. 1.53/3. 0.71/16. 1.35/16.	2-	-1.597		ann.rad./ 0.59588(1)/60. 0.6084(1)/0.6 0.6348(1)/15.
⁷⁵As	100.	74.921597				3/2-	+1.43947	+0.31	
⁷⁶As		75.922394	26.3 h	β⁻/2.962	0.54/3. 1.785/8. 2.410/36. 2.97/51.	2-	-0.906		0.5591(1)/45. 0.65703(5)/6.2 1.21602(1)/3.4 (0.3-2.67)
⁷⁷As		76.920648	38.8 h	β⁻/0.683	0.70/98.	3/2-			0.2391(2)/1.6 0.2500(3)/0.4 0.52078(1)/0.43
⁷⁸As		77.92183	1.512 h	β⁻/4.21	3.00/12. 3.70/17. 4.42/37.	2-			0.6136(3)/54. 0.6954(3)/18. 1.3088(3)/10.

Elem. or Isot.	Natural Abundance (%)	Atomic Mass or Weight	Half-Life	Decay Mode/Energy (/MeV)	Particle Energy /Intensity (MeV / %)	Spin (h/2π)	Nuclear Magnetic Mom.(nm)	Elect. Quadr. Mom. (b)	γ-ray Energy /Intensity (MeV / %)
^{79}As		78.92095	9.0 m	β^- /2.28	1.80/95.	3/2-			0.0955(5)/16.
^{80}As		79.92258	16. s	β^- /5.64	3.38/	1+			0.3645(5)/1.9
^{81}As		80.92213	33. s	β^- /3.856		3/2-			0.6662(2)/42.
									(2.5-3.0)
									0.4676(2)/20.
^{82m}As			13.7 s	β^- /	3.6/	5-			0.4911(2)/8.
									0.6544(1)/72.
									0.8186(4)/27.
									1.7313(2)/27.
									1.8954(2)/38.
^{82}As		81.9246	19. s	β^- /7.4	7.2/80.	1+			0.6544(1)/15.
^{83}As		82.9250	13.4 s	β^- /5.5					0.7345/100.
									1.1131/34.
									2.0767/28.
^{84m}As			0.6 s	β^-					
^{84}As		83.9291	4. s	β^-, n/7.2		1-			0.6671(2)/21.
									1.4439(5)/49.
									(0.325-5.150)
^{85}As		84.9318	2.03 s	β^-, n/8.9		3/2-			0.667(1)/42.
									1.4551(2)/100.
^{86}As		85.9362	0.95 s	β^-, n/11.4					0.704/
^{87}As		86.9396	0.49 s	β^-, n/10.					0.704/
^{88}As		87.945							
^{89}As		88.949							
^{90}As									
^{91}As									
^{92}As									
$_{34}$Se		78.96(3)							
^{65}Se		64.965	0.011 s	β^+ /60/14.	3.55/				
				β^+, p					
^{66}Se		65.9552							
^{67}Se		66.9501	0.06 s	β^+ /10.2					
				β^+ ,(p)/					ann.rad./
^{68}Se		67.9419	36. s	β^+ /4.7					0.352
									ann.rad./
^{69}Se		68.93956	27.4 s	β^+ /6.78	5.006/				(0.050-0.426)
				EC/					ann.rad./
									0.0664(4)/27.
^{70}Se		69.9335	41.1 m	β^+ /2.4		0+			0.0982(4)/63.
									ann.rad
									0.04951(5)/35.
^{71}Se		70.9319	4.7 m	β^+ /4.4	3.4/36.	5/2-			0.4262(2)/29.
				EC/					ann.rad
									0.1472(3)/47.
									0.8309(3)/13.
^{72}Se		71.92711	8.5 d	EC/0.34		0+			1.0960(3)/10.
^{73m}Se			40. m	I.T./73/0.0257	0.85	3/2-			0.0460(2)/57.
				β^+ /27/2.77	1.45/				ann.rad.
					1.70/				0.0257(2)/27.
^{73}Se		72.92678	7.1 h	β^+ /65/2.74	0.80/	9/2+			0.2538(1)/2.5
				EC/35/	1.32/95.				ann.rad
					1.68/1.				0.0670(1)/72.
									0.3609(1)/97.
									(0.6-1.5)
^{74}Se	0.89(4)	73.922477				0+			
^{75}Se		74.922524	119.78 d	EC/0.864		5/2+	0.67	1.0	0.136000(1)/55.
									0.264651(1)/58.
									(0.024-0.821)
^{76}Se	9.37(29)	75.919214				0+			
^{77m}Se			17.4 s	I.T./		7/2+			0.1619(2)/52.
^{77}Se	7.63(16)	76.919915				1/2-	+0.53506		
^{78}Se	23.77(28)	77.917310				0+			
^{79m}Se			3.92 m	I.T./		7/2+			0.09573(3)/9.5
^{79}Se		78.918500	1.1x10^6 y	β^- /0.151		7/2+	-1.018	+0.8	
^{80}Se	49.61(41)	79.916522				0+			
^{81m}Se			57.3 m	I.T./99/0.1031		7/2+			0.1031(3)/9.7
									0.2602(2)/0.06
^{81}Se		80.917993	18.5 m	β^- /1.585	1.6/98.	1/2-			0.27599(1)/0.06
									0.27594(5)/0.85
									0.29008(5)/0.75
									0.82827(5)/0.32
^{82}Se	8.73(22)	81.916700				0+			
^{83m}Se			1.17 m	β^- /3.96	2.88/	1/2-			0.35666(6)/17.
					3.92/				0.9879(1)/15.
									1.0305(1)/21.
									2.0514(2)/11.
									(0.19-3.1)

Elem. or Isot.	Natural Abundance (%)	Atomic Mass or Weight	Half-Life	Decay Mode/Energy (/MeV)	Particle Energy /Intensity (MeV / %)	Spin (h/2π)	Nuclear Magnetic Mom.(nm)	Elect. Quadr. Mom. (b)	γ-ray Energy /Intensity (MeV / %)
^{83}Sc		82.919119	22.3 m	β^- /3.668	0.93/ 1.51/	9/2+			0.22516(6)/33. 0.35666(6)/69. 0.51004(8)/45. (0.21-2.42)
^{84}Sc		83.91847	3.3 m	β^- /1.83	1.41/100.	0+			0.4088(5)/100.
^{85}Sc		84.92225	32. s	β^- /6.18	5.9/	5/2+			0.3450(1)/22. 0.6094(1)/41.
^{86}Sc		85.92428	15. s	β^- /5.10		5/2+			2.0124(1)/24. 2.4433(8)/100. 2.6619(1)/49.
^{87}Sc		86.92853	5.4 s	β^- /7.28 n/					0.468(1)/100. 1.4979(1)/23.
^{88}Sc		87.93143	1.5 s	β^- ,n/6.85					0.5346/
^{89}Sc		88.9360	0.41s	β^- ,n/9.0					
^{90}Sc		89.9394							
^{91}Sc		90.945	0.27 s	β^- ,n/8.					
^{92}Sc		91.949							
^{93}Sc									
^{94}Sc									
$_{35}$Br		79.904(1)							
^{69}Br		68.9502		β^+ /9.6					
^{70}Br		69.9446	79. ms	β^+ /10.0					
^{71}Br		70.9392	21. s	β^+ /6.9					
^{72}Br		71.9365	1.31 m	β^+ /8.7		3	≈0.55		0.4547-1.3167
^{73}Br		72.9318	3.4 m	β^+ /4.7	3.7/	3/2-			ann. rad (0.0649-0.6995)
^{74m}Br			46. m	β^+ /	4.5/	4-	1.8		ann.rad 0.6348 0.7285 (0.2 - 4.38)
^{74}Br		73.92989	25.4 m	β^+ /6.91					ann.rad 0.6341 0.6348 (0.2-4.7)
^{75}Br		74.92578	1.62	β^+ /76/3.03		3/2-	0.75(11)		ann.rad 0.28650 (0.1-1.56)
^{76m}Br			1.4 s	I.T./5.05		4+			0.104548 0.05711
^{76}Br		75.92454	16.0 h	β^+ /57 /4.96	1.9/ 3.68/	1-	0.54821	0.270	ann.rad 0.55911 1.85368 (0.4-4.6)
^{77m}Br			4.3 m	I.T./0.1059		9/2+	0.97		0.1059
^{77}Br		76.921380	2.376 d	EC/99 /1.365		3/2-			ann. rad. 0.23898 0.52069 (0.08-1.2)
^{78}Br		77.921146	6.45 m	β^+ /92/3.574 EC/8 /	1.2/ 2.5/	1+	0.1		ann. rad. 0.61363 (0.7-3.0)
^{79m}Br			4.86 s	I.T./0.207		9/2+	+2.106		0.2072
^{79}Br	50.69(7)	78.918338				3/2-	+2.106400	+0.331	
^{80m}Br			4.42 h	I.T./0.04885		5-	+1.3177		Br k x-ray 0.03705/39.1 0.04885/0.3
^{80}Br		79.918530	17.66 m	β^- /92 /2.004 EC/5.7/1.8706 β^+ /2.6/	1.38 β^- /7.6 1.99 β^- /82 0.85 β^+ /2.8	1+	0.5140	0.196	ann.rad. 0.6169/6.7 (0.64-1.45)
^{81}Br	49.31(7)	80.916291				3/2-	+2.270562	+0.276	
^{82m}Br			6.1 m	I.T./98/0.046 β^- /2 /3.139	0.444/	2-			0.046/0.24 (0.62-2.66)
^{82}Br		81.916805	1.471 d	β^- /3.093		5-	+1.6270	0.751	0.5544/71 0.61905/43 0.77649/84 (0.013-1.96)
^{83}Br		82.915181	2.40 h	β^- /0.972	0.395/1 0.925/99	3/2-			0.52964 (0.12-0.68)
^{84m}Br			6.0 m	β^- /4.97	2.2/100	(6-)			0.4240/100 0.8817/98 1.4637/101
^{84}Br		83.91651	31.8 m	β^- /4.65	2.70/11 3.81/20 4.63/34	2-	2.		0.8816/41 1.8976/13 (0.23-4.12)
^{85}Br		84.91561	2.87 m	β^- /2.87	2.57	3/2-			0.80241/2.56 0.92463/1.6 (0.09-2.4)

Elem. or Isot.	Natural Abundance (%)	Atomic Mass or Weight	Half-Life	Decay Mode/Energy (/MeV)	Particle Energy /Intensity (MeV / %)	Spin (h/2π)	Nuclear Magnetic Mom.(nm)	Elect. Quadr. Mom. (b)	γ-ray Energy /Intensity (MeV / %)
⁸⁶Br		85.91880	55.5 s	β⁻ /7.63	3.3 7.4	(2-)			1.56460/64 2.75106/21
⁸⁷Br		86.92072	55.6 s	β⁻ /6.85 n/	6.1/	3/2-			(0.5-6.8) 1.41983 1.4762
⁸⁸Br		87.92407	16.3 s	β⁻ /8.96 n/		1-			(0.2-6.1) 0.7649 0.7753 0.8021
⁸⁹Br		88.92640	4.35 s	β⁻ /8.16 n/		3/2-			(0.1-6.99) 0.7753 1.0978
⁹⁰Br		89.9306	1.91 s	β⁻ /10.4 n/	8.3/ 9.8/	2-			0.6555 0.7071 1.3626
⁹¹Br		90.9339	0.54 s	β⁻ /90 /9.80 β⁻ n/10 /					0.263 0.803
⁹²Br		91.9392	0.31 s	β⁻ /12.20 β⁻ n/					0.740
⁹³Br		92.9431	0.10 s	β⁻ n/11.1					
⁹⁴Br		93.9487	0.07 s	β⁻ n/					
⁹⁵Br									
⁹⁶Br									
⁹⁷Br									
₃₆Kr		83.80(1)							
⁶⁹Kr		68.9653	0.03 s	β⁺,(p)	4.07/				
⁷⁰Kr		69.9560							
⁷¹Kr		70.9505	100. ms	β⁺,EC/10.1					
⁷²Kr		71.9419	17. s	β⁺ /5.0 EC/		0+			(0.198-0.207) ann.rad 0.3100/29 0.4150/36
⁷³Kr		72.9389	27. s	β⁺ /6.7 EC/ β⁺,p/		5/2-			(0.12-0.58) ann.rad. 0.1781/66
⁷⁴Kr		73.9333	11.5 m	β⁺ /3.1 EC/		0+			(0.06-0.86) ann.rad. 0.08970/31 0.2030/20
⁷⁵Kr		74.93104	4.3 m	β⁺ /4.90 EC/	3.2/		- 0.531	+ 1.1	(0.010-1.06) ann.rad. 0.1325/68 0.1547/21
⁷⁶Kr		75.92595	14.8 h	EC/1.31		0+			(0.02-1.7) Br k x-ray 0.270/21 0.3158/39
⁷⁷Kr		76.92467	1.24 h	β⁺ /80 /3.06 EC/20 /	1.55/ 1.70/ 1.87/	5/2+	- 0.583	+ 0.9	(0.03-1.07) ann.rad. 0.1297/80 0.1465/38
⁷⁸Kr	0.35(1)	77.92039				0+			(0.02-2.3)
⁷⁹ᵐKr			53. s	I.T./0.1299		7/2+	- 0.786	+ 0.40	Kr x-ray
⁷⁹Kr		78.920083	1.455 d	β⁺ /7 /1.626 EC/93 /		1/2-	+ 0.536		ann.rad. 0.2613/13 0.39756/19 0.6061/8
⁸⁰Kr	2.28(6)	79.916379				0+			(0.04-1.3)
⁸¹ᵐKr			13.1 s	I.T./0.1904		1/2-	+ 0.586		0.1904
⁸¹Kr		80.916593	2.1x10⁵ y	EC/0.2807		7/2+	- 0.908	+ 0.64	Br k x-ray 0.2760
⁸²Kr	11.58(14)	81.913485				0+			
⁸³ᵐKr			1.86 h	I.T./0.0416		1/2-	+ 0.591		Kr k x-ray 0.00940 0.03216
⁸³Kr	11.49(6)	82.914137				9/2+	-0.970699	+0.253	
⁸⁴Kr	57.00(4)	83.911508				0+			
⁸⁵ᵐKr			4.48 h	β⁻ /79 / I.T./21 /0.305	0.83/79	1/2-	+ 0.633		0.30487 0.15118
⁸⁵Kr		84.912530	10.73 y	β⁻ /0.687	0.15/0.4	9/2+	1.005	+0.43	0.51399
⁸⁶Kr	17.30(22)	85.910615				0+			
⁸⁷Kr		86.913359	1.27 h	β⁻ /3.887	1.33/8 3.49/43 3.89/30	5/2+	-1.023	- 0.30	0.40258/49.6 2.5548/9.2 (0.13-3.31)

Elem. or Isot.	Natural Abundance (%)	Atomic Mass or Weight	Half-Life	Decay Mode/Energy (/MeV)	Particle Energy /Intensity (MeV / %)	Spin (h/2π)	Nuclear Magnetic Mom.(nm)	Elect. Quadr. Mom. (b)	γ-ray Energy /Intensity (MeV / %)
^{88}Kr		87.91445	2.84 h	β^- /2.91		0+			0.19632/26. 2.392/34.6 (0.03-2.8)
^{89}Kr		88.91764	3.15 m	β^- /4.99	3.8/ 4.6/ 4.9/	5/2+	- 0.330	+ 0.16	0.19746 0.2209/19.9 0.5858/16.4 1.4728/6.8 (0.2-4.7)
^{90}Kr		89.91953	32.3 s	β^- /4.39	2.6/77 2.8/6	0+			0.12182/32.9 0.5395/28.6 1.1187/36.2 (0.1 - 4.2)
^{91}Kr		90.9234	8.6 s	β^- /6.4	4.33/ 4.59/	5/2+	- 0.583	+ 0.30	0.10878/43.5 0.50658/19. (0.2-4.4)
^{92}Kr		91.92611	1.84 s	β^- /5.99 n/					0.1424/66. (0.14 - 3.7)
^{93}Kr		92.9312	1.29 s	β^- /8.6 n/	7.1/	1/2+	- 0.413		0.1820 0.2534/42. 0.32309/24.6 (0.057-4.03)
^{94}Kr		93.9343	0.21 s	β^- /7.3					0.2196/67 0.6293/100. (0.098-0.985)
^{95}Kr		94.9397	0.78 s	β^- /9.7			- 0.410		
^{96}Kr		95.9431							
^{97}Kr		96.9486	< 0.1 s	β^-					
^{98}Kr									
^{99}Kr									
^{100}Kr									
$_{37}$Rb		85.4678(3)							
^{74}Rb		73.9445	65. ms	β^+ /10.4					
^{75}Rb		74.93857	19. s	β^+ /7.02	2.31/				ann. rad. 0.179
^{76}Rb		75.93508	39. s	β^+ /8.50	4.7/	1-	-0.372623	+0.4	ann.rad. 0.4240/92. (0.064-1.68)
^{77}Rb		76.93041	3.8 m	β^+ /5.34	3.86/	3/2-	+0.654468	+0.70	ann.rad. 0.0665/59 (0.04 - 2.82)
^{78m}Rb			5.7 m	I.T./0.1034 β^+ / EC/	3.4	4-			ann.rad. 0.4553/81. (0.103-4.01)
^{78}Rb		77.92814	17.7 m	β^+ /7.22 EC/		0+			ann.rad. 0.4553/63. (0.42-5.57)
^{79}Rb		78.92400	23. m	β^+ /84 /3.65 EC/16 /		5/2+	+0.3358	-0.10	ann.rad. 0.68812/23. (0.017-3.02)
^{80}Rb		79.92252	34. s	β^+ /5.72	4.1/22 4.7/74	1+	-0.0836	+0.35	ann.rad. 0.6167/25. (0.085-1.9)
^{81m}Rb			30.5 m	I.T./0.85 β^+,EC/	1.4	9/2+			ann.rad./ (0.05 - 1.9)
^{81}Rb		80.91900	4.57 h	β^+/27 /2.24 EC/73	1.05/	3/2-	+2.060	+0.40	ann.rad./ 0.19030/64. (0.05 - 1.9)
^{82m}Rb			6.47 h	β^+/26 / EC/74 /	0.80/	5-	+1.5		ann.rad./ 0.5544/63. 0.7765/85. (0.092 - 2.3)
^{82}Rb		81.91821	1.258 m	β^+/96 /4.40 EC/4 /	3.3/	1+	+0.554508	+0.19	ann.rad./ 0.7665/13. (0.47 - 3.96)
^{83}Rb		82.91511	86.2 d	EC/0.91		5/2-	+1.425	+0.20	Kr x-ray 0.5205/46. (0.03-0.80)
^{84m}Rb			20.3 m	I.T./0.216		6-			0.2163/34. 0.2482/63. 0.4645/32.
^{84}Rb		83.914387	32.9 d	β^+/22 /2.681 EC/75 / β^-/3 /0.894	0.780/11 1.658/11 0.893/	2-	-1.32412	-0.015	ann.rad./ 0.8817/68. (1.02-1.9)
^{85}Rb	72.17(2)	84.911792				5/2-	+1.353	+0.23	
^{86m}Rb			1.018 m	I.T./0.5560		6-			0.556/98.
^{86}Rb		85.911170	18.65 d	β^-/1.775	1.774/8.8	2-	-1.6920		1.0768/8.8
^{87}Rb	27.83(2)	86.909186	4.88x10^{10}y	β^-/0.283	0.273/100	3/2-	+2.7512		

Elem. or Isot.	Natural Abundance (%)	Atomic Mass or Weight	Half-Life	Decay Mode/Energy (/MeV)	Particle Energy /Intensity (MeV / %)	Spin (h/2π)	Nuclear Magnetic Mom.(nm)	Elect. Quadr. Mom. (b)	γ-ray Energy /Intensity (MeV / %)
^{88}Rb		87.911323	17.7 m	β⁻/5.316	5.31	2-	0.508		0.8980/14. 1.8360/21. (0.34-4.85)
^{89}Rb		88.91229	15.4 m	β⁻/4.50	1.26/38 1.9/5 2.2/34 4.49/18	3/2-	+2.304	+0.14	1.032/58. 1.248/42. 2.1960/13 (0.12-4.09)
^{90m}Rb			4.3 m	β⁻/4.50	1.7/ 6.5/	4-			0.1069(IT) 0.8317/94 (0.20-5.00)
^{90}Rb		89.91481	2.6 m	β⁻/6.59	6.6	1-			0.8317/28. (0.31-5.60)
^{91}Rb		90.91649	58.0 s	β⁻/5.861	5.9	3/2-	+2.182	+0.15	0.0936/34. (0.35-4.70)
^{92}Rb		91.91968	4.48 s	β⁻/8.11	8.1/94	1-			0.8148/8. (0.1-6.1)
^{93}Rb		92.92195	5.85 s	β⁻/7.46 n/1	7.4/	5/2	+1.410	+0.18	0.2134/4.8 0.4326/12.5 0.9861/4.9 (0.16-5.41)
^{94}Rb		93.92643	2.71 s	β⁻/10.31 n/10	9.5/	3	+1.498	+0.16	0.8369/87. 1.5775/32. (0.12-6.35)
^{95}Rb		94.92929	0.377 s	β⁻/9.30 n/8	8.6/	5/2	+1.334	+0.21	0.352/65. 0.680/22. (0.20-2.27)
^{96}Rb		95.93427	0.199 s	β⁻/11.76 n/13/	10.8/	2+	+1.466	+0.25	0.815/76. (0.20-5.42)
^{97}Rb		96.93733	0.169 s	β⁻/10.42 n/27/	10.0	3/2	+1.841	+0.58	0.167/100. 0.585/79. 0.599/56. 1.258/52. (0.14-2.08)
^{98}Rb		97.94174	0.107 s	β⁻/12.34 n/13	0.144/				(0.07-3.68)
^{99}Rb		98.9453	59. ms	β⁻/11.3					
^{100}Rb		99.9499	53. ms	β⁻/13.5					
^{101}Rb		100.9532	0.03 s	β⁻/11.8					
^{102}Rb		101.9592	0.09 s	β⁻					
$_{38}$Sr		87.62(1)							
^{73}Sr		72.966							
^{74}Sr		73.9563							
^{75}Sr		74.9499	≈ 0.07 s						
^{76}Sr		75.9416	8.9 s	β⁺/6.1					
^{77}Sr		76.9378	9.0 s	β⁺/6.9	5.6		-0.35	1.4	0.147 (0.047-0.793)
^{78}Sr		77.93218	2.7 m	β⁺/3.76					
^{79}Sr		78.92971	2.1 m	β⁺/5.32	4.1	3/2-	-0.474	+0.74	ann.rad./ 0.039/28. 0.105/22. (0.135-0.612)
^{80}Sr		79.92453	1.77 h	β⁺/1.87		0+			ann.rad./ 0.174/10. 0.589/39. (0.24-0.55)
^{81}Sr		80.92322	22.3 m	β⁺/87 /3.93 EC/13 /	2.43/ 2.68/	1/2-	+0.544		ann.rad./ 0.148/31. 0.1534/35 (0.06-1.7)
^{82}Sr		81.91840	25.36 d	EC/0.18					Rb x-ray
^{83m}Sr			5.0 s	I.T./0.2591		1/2-			0.2591/87.5
^{83}Sr		82.91756	1.350 d	β⁺/24 /2.28 EC/76/	0.465/ 0.803/ 1.227/	7/2+	-0.898	+0.8	ann.rad./ 0.3816/12. 0.3816 0.7627/30. (0.094-2.15)
^{84}Sr	0.56(1)	83.913426				0+			
^{85m}Sr			1.127 h	I.T./87 /0.2387 EC/13		1/2-			0.2318/84. (0.15-0.24)
^{85}Sr		84.912936	64.85 d	EC/1.065		9/2+	-1.001	+0.32	0.51399/99.3
^{86}Sr	9.86(1)	85.909265				0+			
^{87m}Sr			2.81 h	I.T./0.3884		1/2-			0.3884(IT)
^{87}Sr	7.00(1)	86.908882				9/2+	-1.09360	+0.34	
^{88}Sr	82.58(1)	87.905617				0+			
^{89}Sr		88.907455	50.52 d	β⁻/1.497	1.492/100	5/2+	-1.149	-0.3	0.9092
^{90}Sr		89.907738	29.1 y	β⁻/0.546	0.546/100	0+			

Elem. or Isot.	Natural Abundance (%)	Atomic Mass or Weight	Half-Life	Decay Mode/Energy (/MeV)	Particle Energy /Intensity (MeV / %)	Spin (h/2π)	Nuclear Magnetic Mom.(nm)	Elect. Quadr. Mom. (b)	γ-ray Energy /Intensity (MeV / %)
⁹¹Sr		90.91020	9.5 h	β⁻/2.70	0.61/7 1.09/33 1.36/29 2.66/26	5/2+	-0.887	+0.044	0.5556/61. 0.7498/24. 1.0243/33. (0.12-2.4)
⁹²Sr		91.91098	2.71 h	β⁻/1.91	0.55/96 1.5/3	0+			1.3831/90. (0.24-1.1)
⁹³Sr		92.91394	7.4 m	β⁻/4.08	2.2/10 2.6/25 3.2/65	5/2+	-0.794	+0.27	0.5903/ 0.7104 0.87573 0.8883/ (0.17-3.97)
⁹⁴Sr		93.91537	1.25 m	β⁻/3.511	2.1/ 3.3/	0+			0.6219 0.7043 0.7241 0.8064 1.4283
⁹⁵Sr		94.91931	25.1 s	β⁻/6.08	6.1/50	1/2+	-0.5379		0.6859 0.8269 2.7173 2.9332
⁹⁶Sr		95.92165	1.06 s	β⁻/5.37	4.2/	0+			0.1222 0.5305 0.8094 0.9318
⁹⁷Sr		96.92615	0.42 s	β⁻/7.47	5.3	(1/2+)	-0.500		0.2164 0.3071 0.6522 0.9538 1.2580 1.9050
⁹⁸Sr		97.92845	0.65 s	β⁻/5.83	5.1				0.0365 0.1190 0.4286 0.4447 0.5636
⁹⁹Sr		98.9333	0.27 s	β⁻/8.0					
¹⁰⁰Sr		99.9354	0.201 s	β⁻/7.1					
¹⁰¹Sr		100.9405	0.115 s	β⁻/9.5					
¹⁰²Sr		101.9430	68. ms	β⁻/8.8					
¹⁰³Sr		102.9490							
¹⁰⁴Sr		103.952							
¹⁰⁵Sr									
₃₉Y		88.90585(2)							
⁷⁸Y		77.9435		β⁺/10.5					
⁷⁹Y		78.9374	15. s	β⁺/7.1					(0.152-1.106)
⁸⁰Y		79.9320	36. s	β⁺/7.0	5.5 5.0/	(4)			ann.rad./ 0.3858 0.5951 1.1852
⁸¹Y		80.9291	1.21 m	β⁺/5.5	3.7/ 4.2/				ann.rad./ 0.428 0.469
⁸²Y		81.9268	9.5 s	β⁺/7.8	6.3/	1+			ann.rad./ 0.5736 0.6017 0.7375
⁸³ᵐY			2.85 m	β⁺/95/4.6 EC/5 /	2.9	1/2-			ann.rad./ 0.2591 0.4218 0.4945
⁸³Y		82.92235	7.1 m	β⁺/4.47 EC/	3.3	9/2+			ann.rad./ 0.0355 0.4899 0.8821 (0.03 - 3.4)
⁸⁴ᵐY			4.6 s	β⁺/ EC/		1+			ann.rad./ 0.7930
⁸⁴Y		83.9203	40. m	β⁺/6.4 EC/	1.64/47 2.24/25 2.64/21 3.15/7	5-			ann.rad./ 0.4628 0.6606 0.7931 0.9744 1.0398 (0.2 - 3.3)

Elem. or Isot.	Natural Abundance (%)	Atomic Mass or Weight	Half-Life	Decay Mode/Energy (/MeV)	Particle Energy /Intensity (MeV / %)	Spin (h/2π)	Nuclear Magnetic Mom.(nm)	Elect. Quadr. Mom. (b)	γ-ray Energy /Intensity (MeV / %)
^{83m}Y			4.9 h	β⁺/70 / EC/30 /		9/2+			ann.rad./ 0.2317 0.5356 0.7673 2.1238 (0.1 - 3.1)
^{85}Y		84.91643	2.6 h	β⁺/55 /3.26 EC/45 /	1.54/	1/2-			ann.rad./ 0.2317 0.5045 0.9140 (0.07 - 1.4)
^{86m}Y			48. m	I.T./99 / β⁺/ EC/		8+			ann.rad./ 0.0102(IT) 0.2080 (0.09 - 1.1)
^{86}Y		85.91489	14.74 h	β⁺/5.24 EC/		4-	<0.6		ann.rad./ 0.3070 0.6277 1.0766 1.1531 1.9207 (0.1 - 3.8)
^{87m}Y			13. h	I.T./98 / β⁺/0.7 / EC/	1.15/0.7	9/2+			0.3807
^{87}Y		86.910880	3.35 d	EC/99+/1.862	0.78/	1/2-			0.3880 0.4870
^{88}Y		87.909506	106.6 d	EC/99+ /3.623 β⁺/0.2 /	0.76/	4-			ann.rad./ 0.89802 1.83601 2.73404 3.2190
^{89m}Y			15.7 s	I.T./0.909		9/2+			0.9092(IT)
^{89}Y	100.	88.905849				1/2-	-0.13742		
^{90m}Y			3.24 h	I.T./99+ /0.68204 β⁻/0.002/		7+			0.2025 0.4794 0.6820
^{90}Y		89.907152	2.67 d	β⁻/2.282	2.28/	2-	-1.630	-0.155	
^{91m}Y			49.7 m	I.T./0.555		9/2+			0.5556(IT)
^{91}Y		90.907301	58.5 d	β⁻/1.544	1.545/	1/2-	6.0		1.208
^{92}Y		91.90893	3.54 h	β⁻/3.63	3.64/	2-	0.1641		0.4485 0.5611 0.9345 1.4054 (0.4 - 3.3)
^{93m}Y			0.82 s	I.T./0.759		9/2+			0.1686(IT)
^{93}Y		92.90956	10.2 h	β⁻/2.87	2.88/90	1/2-			0.5902 0.2669 0.9471 1.9178
^{94}Y		93.91160	18.7 m	β⁻/4.919	4.92/	2-			0.3816 0.9188 1.1389 (0.3 - 4.1)
^{95}Y		94.91279	10.3 m	β⁻/4.42		1/2-			0.4324 0.9542 2.1760 3.5770
^{96m}Y			9.6 s	β⁻/		(3+)			0.1467 0.6174 0.9150 1.1071 1.7507
^{96}Y		95.91588	6.2 s	β⁻ /7.09	7.12/	0-			1.594
^{97m}Y			1.21 s	β⁻ /7.4	4.8/ 6.0/	9/2+			0.1614 0.9700 1.1030
^{97}Y		96.91813	3.76 s	β⁻ /6.69	6.7	1/2-			0.2969 1.9960 3.2876 3.4013
^{98m}Y			2.1 s	β⁻ /9.8	5.5/	(4-)			0.2415 0.6205 0.6473 1.2228 1.8016

Elem. or Isot.	Natural Abundance (%)	Atomic Mass or Weight	Half-Life	Decay Mode/Energy (/MeV)	Particle Energy /Intensity (MeV / %)	Spin (h/2π)	Nuclear Magnetic Mom.(nm)	Elect. Quadr. Mom. (b)	γ-ray Energy /Intensity (MeV / %)
⁹⁸Y		97.92224	0.59 s	β⁻ /8.83	8.7/	1+			0.2131 1.2228 1.5907 2.9413 4.4501
⁹⁹Y		98.92463	1.47 s	β⁻ /7.57 n	/2.5/	1/2-			0.1218/43.8 0.5362 0.7242 1.0130
¹⁰⁰ᵐY			0.94 s	β⁻,n /		3+			
¹⁰⁰Y		99.9278	0.73 s	β⁻,n /9.3	n/1.8/	1+			
¹⁰¹Y		100.9303	0.43 s	β⁻,n /8.6	n/1.5/	(5/2)			
¹⁰²Y		101.9336	0.36 s	β⁻,n /9.9	n/4.0/				
¹⁰³Y		102.9369	0.23 s	β⁻,n	n/8.3/				
¹⁰⁴Y		103.9414							
¹⁰⁵Y		104.9451							
¹⁰⁶Y		105.950							
¹⁰⁷Y									
¹⁰⁸Y									
₄₀Zr		91.224(2)							
⁸⁰Zr		79.9406		β⁺ /8.0					0.290 0.538
⁸¹Zr		80.9368	15. s	β⁺ /7.2	6.1				ann.rad./
⁸²Zr		81.9311	32. s	β⁺ /4.0	3.				ann.rad./
⁸³ᵐZr			7. s	β⁺ /7.0		(7/2+)			ann.rad./
⁸³Zr		82.9287	44. s	β⁺ /5.9 EC	4.8	(1/2-)			ann.rad./ 0.0556 0.1050 0.2560 0.474 1.525
⁸⁴Zr		83.9233	26. m	β⁺ /2.7 EC/		0+			ann.rad./ 0.0449 0.1125 0.3729 0.667
⁸⁵ᵐZr			10.9 s	I.T./0.2922 β⁺ ,EC/		1/2-			ann.rad./ 0.2922(IT) 0.4165
⁸⁵Zr		84.9215	7.9 m	β⁺ /4.7 EC/	3.1	7/2+			ann.rad./ 0.2663 0.4163 0.4543
⁸⁶Zr		85.91647	16.5 h	EC/1.47		0+			0.0280 0.243 0.612
⁸⁷ᵐZr			14.0 s	I.T./0.3362		1/2-			0.1352(IT) 0.2010
⁸⁷Zr		86.91482	1.73 h	β⁺ /3.67 EC/	2.26	9/2+			ann.rad./ 0.3811 1.228
⁸⁸Zr		87.91023	83.4 d	EC/0.67		0+			0.3929
⁸⁹ᵐZr			4.18 m	I.T./94 /0.5877 β⁺ /1.5 / EC/4.7 /		1/2-			ann.rad./ 0.5877(IT) 1.507
⁸⁹Zr		88.908889	3.27 d	β⁺ /23 /2.832 EC/77 /	0.9/	9/2+	-1.07		ann.rad./ 0.9092
⁹⁰ᵐZr			0.809 s	I.T./		5-			0.1326 2.1862 2.3189(IT)
⁹⁰Zr	51.45(40)	89.904702				0+			
⁹¹Zr	11.22(5)	90.905643				5/2+	-1.30362	-0.21	
⁹²Zr	17.15(8)	91.905039				0+			
⁹³Zr		92.906474	1.5x10⁶y	β⁻ /0.091		5/2+			0.0304
⁹⁴Zr	17.38(28)	93.906314				0+			
⁹⁵Zr		94.908041	64.02 d	β⁻ /1.125	0.366/55 0.400/44	5/2+	1.10	+0.29	0.7242 0.7567
⁹⁶Zr	2.80(9)	95.908275				0+			
⁹⁷Zr		96.910950	16.8 h	β⁻ /2.658	1.91/	1/2-			0.7434
⁹⁸Zr		97.91276	30.7 s	β⁻ /2.26	2.2/100	0+			
⁹⁹Zr		98.91651	2.2 s	β⁻ /4.56	3.9/ 3.5/	1/2+			0.4692/55.2 0.5459 0.5940
¹⁰⁰Zr		99.91776	7.1 s	β⁻ /3.34		0+			0.4006 0.5043

Elem. or Isot.	Natural Abundance (%)	Atomic Mass or Weight	Half-Life	Decay Mode/Energy (/MeV)	Particle Energy /Intensity (MeV / %)	Spin (h/2π)	Nuclear Magnetic Mom.(nm)	Elect. Quadr. Mom. (b)	γ-ray Energy /Intensity (MeV / %)
^{101}Zr		100.92114	2.1 s	β^- /5.49	6.2/	3/2-			0.1194
									0.2057
									0.2089
^{102}Zr		101.92298	2.9 s	β^- /4.61					
^{103}Zr		102.9266	1.3 s	β^- /7.0					
^{104}Zr		103.9288	1.2 s	β^- /5.9					
^{105}Zr		104.9331	≈1. s	β^- /8.5					
^{106}Zr		105.9359							
^{107}Zr		106.941							
^{108}Zr		107.944							
^{109}Zr									
^{110}Zr									
$_{41}$Nb		92.90638(2)							
^{82}Nb		81.9431		β^+ /11.					
^{83}Nb		82.9367		β^+ /7.5					
^{84}Nb		83.9336	12. s	β^+ ,EC/9.6		(3+)			
^{85}Nb		84.9279	2.3 m	β^+ /6.0					
^{86m}Nb			56. s	β^+					
^{86}Nb		85.9250	1.46 m	β^+ /8.0					ann.rad./
									0.751
									1.003
^{87m}Nb			3.7 m	β^+ / EC/		1/2-			ann.rad./
									0.1352
									0.2010
^{87}Nb		86.92036	2.6 m	β^+ 5.2/ EC/		(9/2+)			ann.rad./
									0.2010
									0.4706
									0.6165
									1.0665
									1.8842
^{88m}Nb			7.7 m	β^+ / EC/		4-			ann.rad./
									0.2625
									0.3996
									1.0569
^{88}Nb		87.9183	14.3 m	β^+ /7.6 EC/	3.2/	8+			1.0825
									ann.rad./
									1.0570
									1.0828
^{89m}Nb			2.0 h	β^+ / EC/	3.3/	9/2+			(0.07 - 2.5)
									0.5880/10(D)
^{89}Nb		88.91349	1.10 h	β^+ /74 /4.29 EC/26 /	2.8/	1/2-	+6.216		(0.17 - 4.0)
									ann.rad./
									0.5074
									0.5880
									0.7696
									1.2775
^{90m}Nb			18.8 s	I.T./0.1246		4-			0.002
									0.1225
^{90}Nb		89.911263	14.6 h	β^+ /53 /6.111 EC/47 /	0.86/5 1.5/92	8+	4.961	+0.005	ann.rad./
									0.1412
									1.1292
									2.1862
									2.3189
^{91m}Nb			62. d	I.T./97 / EC/3 /		1/2-			(0.1 - 3.3)
									0.1045(TT)
									1.2050
^{91}Nb		90.906989	7x10^{2}y	EC/1.253		9/2+			Mo k x-ray
^{92m}Nb			10.13 d	EC/99+ /		2+	6.114		0.9126
									0.9345
									1.8475
^{92}Nb		91.907192	3.7x10^{7}y	EC/2.006		7+			0.5611
									0.9345
^{93m}Nb			16.1 y	I.T./0.0304		1/2-			Nb x-ray
									0.0304
^{93}Nb	100.	92.906376				9/2+	+6.1705	-0.32	
^{94m}Nb			6.26 m	I.T./99+ /2.086 β^- /0.5 /		3+			Nb k x-ray
									0.0409
									0.87109
^{94}Nb		93.907282	2.4x10^{4}y	β^- /2.045	0.47/	6+			0.70263
									0.87109
^{95m}Nb			3.61 d	I.T./97.5 /0.2357 β^- /2.5 /		1/2-			0.2040
									0.2356
^{95}Nb		94.906834	34.97 d	β^- /0.926	0.160/	9/2+	6.141		0.76578
^{96}Nb		95.908099	23.4 h	β^- /3.187	0.5/10 0.75/90	6+	4.976		0.7782
									0.2191-1.498
^{97m}Nb			58.1 s	I.T./0.7434	0.734/98	1/2-			0.7434

Elem. or Isot.	Natural Abundance (%)	Atomic Mass or Weight	Half-Life	Decay Mode/Energy (/MeV)	Particle Energy /Intensity (MeV / %)	Spin (h/2π)	Nuclear Magnetic Mom.(nm)	Elect. Quadr. Mom. (b)	γ-ray Energy /Intensity (MeV / %)
⁹⁷Nb		96.908096	1.23 h	β⁻ /1.934	1.27/98	9/2+	6.15		0.4809
									0.6579
⁹⁸ᵐNb			51. m	β⁻ /4.67		5+			0.7874
									0.1726-1.89
⁹⁸Nb		97.91033	2.9 s	β⁻ /4.59	4.6/	1+			0.6451
									0.7874
									1.0243
⁹⁹ᵐNb			2.6 m	β⁻ /	3.2/	1/2-			0.0978/100
									(0.138-3.010)
⁹⁹Nb		98.91162	15.0 s	β⁻ /3.64	3.5/100	9/2+			0.0977
									0.1378/3.1
¹⁰⁰ᵐNb			3.0 s	β⁻ /6.74	5.8				Nb k x-ray
									0.159
									0.6364
									1.0637
¹⁰⁰Nb		99.91418	1.5 s	β⁻ /6.25	6.2/				0.5354
					5.3/				0.6001-1.566
¹⁰¹Nb		100.91525	7.1 s	β⁻ /4.57	4.3/				0.1105-0.810
¹⁰²ᵐNb			4.3 s	β⁻ /					
¹⁰²Nb		101.91804	1.3 s	β⁻ /7.21	7.2/				0.2960-2.184
¹⁰³Nb		102.91914	1.5 s	β⁻ /5.53	5.3/	5/2+			
¹⁰⁴ᵐNb			0.9 s	β⁻,n/	n/0.06				
¹⁰⁴Nb		103.9225	4.8 s	β⁻,n/8.1	n/0.05				
¹⁰⁵Nb		104.9239	3.0 s	β⁻,n/6.5	n/1.7				
¹⁰⁶Nb		105.9282	1.0 s	β⁻,n/9.3	n/4.5				
¹⁰⁷Nb		106.9303	0.30 s	β⁻,n/7.9	n/6.0				
¹⁰⁸Nb		107.9350	0.19 s	β ,n/	n/6.2				(0.193-0.590)
¹⁰⁹Nb		108.9376	0.19 s	β ,n/	n/31				
¹¹⁰Nb		109.943	0.17 s	β ,n/	n/40				
¹¹¹Nb									
¹¹²Nb									
¹¹³Nb									
₄₂Mo		95.94(1)							
⁸⁴Mo		83.9401		β⁺ /6.					
⁸⁵Mo		84.9366	3.2 s	β⁺/8.1					
⁸⁶Mo		85.9302	20. s	β⁺ /4.8					
⁸⁷Mo		86.9273	14. s	EC, β⁺/6.5					(0.752-1.004)
⁸⁸Mo		87.92195	8.0 m	β⁺ /3.4		0+			ann.rad./
				EC					0.0800
									0.1399
									0.1707
⁸⁹ᵐMo			0.19 s	I.T./0.118		1/2-			0.118(IT)
									0.268
⁸⁹Mo		88.91948	2.2 m	β⁺ /5.58		9/2+			ann.rad./
				EC/					0.659
									0.803
									1.155
									1.272
⁹⁰Mo		89.91394	5.7 h	β⁺ /25 /2.489	1.085/	0+			ann.rad./
				EC/75 /					0.04274
									0.12237
									0.25734
⁹¹ᵐMo			1.08 m	I.T./50 /0.653		1/2-			ann.rad./
				β⁺ ,EC/50 /	2.5/				0.6529
					2.8/				1.2081
					4.0/				1.5080
									2.2407
⁹¹Mo		90.91175	15.5 m	β⁺ /94 /4.43	3.44/94	9/2-			ann.rad./
				EC/6 /					1.6373
									2.6321
									3.0286
⁹²Mo	14.84(35)	91.906810				0+			(0.1 - 4.2)
⁹³ᵐMo			6.9 h	I.T./99+ /2.425	21/2+		+9.21		0.26306(IT)
									0.68461
									1.47711
⁹³Mo		92.906811	3.5x10³y	EC/0.405		5/2+			0.0304
⁹⁴Mo	9.25(12)	93.905087				0+			
⁹⁵Mo	15.92(13)	94.905841				5/2+	-0.9142	-0.02	
⁹⁶Mo	16.68(2)	95.904678				0+			
⁹⁷Mo	9.55(8)	96.906020				5/2+	-0.9335	+0.26	
⁹⁸Mo	24.13(31)	97.905407				0+			
⁹⁹Mo		98.907711	2.7476 d	β⁻ /1.357	0.45/14	1/2+	0.375		0.144048
					0.84/2				0.18109
					1.21/84				0.36644
									0.73947
¹⁰⁰Mo	9.63(23)	99.90748				0+			

11-66

Elem. or Isot.	Natural Abundance (%)	Atomic Mass or Weight	Half-Life	Decay Mode/Energy (/MeV)	Particle Energy /Intensity (MeV / %)	Spin (h/2π)	Nuclear Magnetic Mom.(nm)	Elect. Quadr. Mom. (b)	γ-ray Energy /Intensity (MeV / %)
101Mo		100.91035	14.6 m	β⁻ /2.82	2.23/ 0.7/	1/2+			0.0063 0.19193 0.5909 (0.0809-2.405)
102Mo		101.91030	11.3 m	β⁻ /1.01	1.2/	0+			0.1493/89. 0.2116/100. 0.2243/32.
103Mo		102.91320	1.13 m	β⁻ /3.8		3/2+			0.1028(2)/ 0.1440(2)
104Mo		103.91376	1.00 m	β⁻ /2.16		0+			0.2511(2) 0.0686(1)/100.
105Mo		104.9170	36. s	β⁻ /4.95		3/2+			0.4239(4)/21. 0.0642/ 0.0856/ 0.2495/
106Mo		105.91814	8.4 s	β⁻ /3.52		0+			0.1894(2)/22. 0.3644(2)/6. 0.3723(2)/12.
107Mo		106.9217	3.5 s	β⁻ /6.2					
108Mo		107.9236	1.1 s	β⁻ /5.1					(0.028-0.636)
109Mo		108.9278	0.5 s	β⁻ /7.2					
110Mo		109.9297	0.30 s	β⁻ /5.7					Tc k x-ray 0.142 (0.039-0.599)
111Mo		110.9345							
112Mo		111.937							
113Mo		112.942							
114Mo									
115Mo									
116Mo									
117Mo									
43Tc									
86Tc		85.9430		β⁺ /11.9					
87Tc		86.9365		β⁺ /8.6					
88Tc		87.9328		β⁺ /10.1					
89mTc			13. s						
89Tc		88.9275	13. s	β⁺ /7.5					
90mTc			49.2 s	β⁺	5.3/	6+			ann.rad./ 0.9479/ 1.0542/
90Tc		89.9235	8.3 s	β⁺ /8.9	7.0/15 7.9/95.	1+			ann.rad./ 0.9479/
91mTc			3.3 m	β⁺ EC		1/2+			ann.rad./170. 0.3375(2)/1.1 0.8110(5)/5. 1.6052(1)/7.8 1.6339(1)/9.1 1.9023(1)/6. 2.4509(1)/13.5
91Tc		90.9184	3.14 m	β⁺ /6.2	5.2	9/2+			ann.rad./200.
92Tc		91.91526	4.4 m	β⁺ /7.87 EC	4.1	8+			ann.rad./200. 0.0850/ 0.1475 0.3293 0.7731 1.5096
93mTc			43. m	I.T./13 EC/20		1/2-			0.3924(TT) 0.9437 2.6445
93Tc		92.910248	2.73 h	β⁺ /13/3.201 EC/87/	0.81	9/2+	6.26		ann.rad./ 1.3629 1.4771 1.5203 (0.1 - 3.0)
94mTc			52. m	β⁺ /72/4.33 EC/28/		2+			ann.rad./ 0.8710 1.8686
94Tc		93.909655	4.88 h	β⁺ /11/4.256 EC/89/		7+	5.08		ann.rad./ 0.4491 0.7026 0.8496 0.8710

Elem. or Isot.	Natural Abundance (%)	Atomic Mass or Weight	Half-Life	Decay Mode/Energy (/MeV)	Particle Energy /Intensity (MeV / %)	Spin (h/2π)	Nuclear Magnetic Mom.(nm)	Elect. Quadr. Mom. (b)	γ-ray Energy /Intensity (MeV / %)
^{95m}Tc			61. d	I.T./4/ β^+ /0.3 EC/96	0.5/ 0.7/	1/2-			ann.rad./ 0.0389(IT) 0.2041 0.5821 0.5821 0.8351
^{95}Tc		94.90766	20.0 h	EC/100/1.691		9/2+	5.89		0.7657 1.0738
^{96m}Tc			52. m	I.T./90/ β^+ ,EC/2/		4+			0.0342(IT) 0.7782 1.2002
^{96}Tc		95.90787	4.3 d	EC/2.973		7+	+5.04		Mo k x-ray 0.7782 0.8125 0.8498 1.12168
^{97m}Tc			90. d	I.T./0.0965		1/2-			Tc k x-ray 0.0965
^{97}Tc		96.906364	2.6×10^6 y	EC/100/0.320		9/2+			Mo k x-ray
^{98}Tc		97.907215	4.2×10^6 y	β^- /1.80	0.40/100	6+			0.65241 0.74535
^{99m}Tc			6.01 h	I.T./100/0.142		1/2-			Tc k x-ray 0.14049 0.14261
^{99}Tc		98.906254	2.13×10^5 y	β^- /0.294	0.293/100	9/2+	+5.6847	-0.129	
^{100}Tc		99.907657	15.8 s	β^- /3.202 EC /1.8(10)$^{-3}$/0.17	2.2/ 2.9/ 3.3	1+			0.5396 0.5908 1.5122 (0.3 - 2.6)
^{101}Tc		100.90731	14.2 m	β^- /1.61	1.32/	9/2+			0.1272 0.1841 0.3068 0.5451 (0.073-0.969)
^{102m}Tc			4.4 m	I.T./2/4.8 β^- /98/	1.8/				0.4184 0.4752 0.6281 0.6302 1.0464 1.1033 1.6163 2.2447
^{102}Tc		101.90921	5.3 s	β^- /4.53	3.4/ 4.2 2.2/	1+			0.4686 0.4751 1.1055
^{103}Tc		102.90918	54. s	β^- /2.66	2.0/ 2.2/	5/2+			0.1361 0.1743 0.2104 0.3464 0.5629 (0.13 - 1.0)
^{104}Tc		103.91144	18.2 m	β^- /5.60	5.3/	(3+)			0.3483 0.3580 0.5305 0.5351 0.8844 0.8931 1.6768 (0.3 - 3.7)
^{105}Tc		104.91166	7.6 m	β^- /3.6	3.4/	5/2+			0.1079 0.1432 0.3215
^{106}Tc		105.91436	36. s	β^- /6.55		2+			0.2703 0.5222 1.9694 2.2393 2.7893
^{107}Tc		106.9151	21.2 s	β^- /4.8					0.1027 0.1063 0.1770 0.4587
^{108}Tc		107.9185	5.1 s	β^- /7.72		(3)			0.2422 0.4656 0.7078 0.7326 1.5835
^{109}Tc		108.9200	1.4 s	β^- /6.3	p/0.08				
^{110}Tc		109.9234	0.83 s	β^- /8.8	p/0.04				0.2407
^{111}Tc		110.9250	0.30 s	β^-.n /7.0	n/0.85				

Elem. or Isot.	Natural Abundance (%)	Atomic Mass or Weight	Half-Life	Decay Mode/Energy (/MeV)	Particle Energy /Intensity (MeV / %)	Spin (h/2π)	Nuclear Magnetic Mom.(nm)	Elect. Quadr. Mom. (b)	γ-ray Energy /Intensity (MeV / %)
^{112}Tc		111.9292	0.21 s	β,n	n/2.6				
^{113}Tc		112.931	0.13 s	β^- /8.					
^{114}Tc		113.936							
^{115}Tc		114.938							
^{116}Tc									
^{117}Tc									
^{118}Tc									
$_{44}$Ru		101.07(2)							
^{87}Ru		86.949							
^{88}Ru		87.9404							
^{89}Ru		88.936		β^+ /8.					
^{90}Ru		89.9298	11. s	β^+ /5.9					ann.rad./ 0.155 - 1.551
^{91}Ru		90.9264	9. s	β^+ ,EC/7.4		9/2+			ann.rad./
^{92}Ru		91.9201	3.7 m	β^+ /53/4.5 EC/47/		0+			ann.rad./ 0.1346 0.2138 0.2593
^{93m}Ru			10.8 s	I.T./21/ β^+ ,EC/79/	5.3/	1/2-			ann.rad./ 0.7344 1.1112 1.3962 2.0931
^{93}Ru		92.9171	1.0 m	β^+ /6.3 EC/		9/2+			ann.rad./ 0.6807 1.4349
^{94}Ru		93.91137	52. m	EC/100/1.59					(0.5- 4.2)weak 0.3672 0.5247 0.8922
^{95}Ru		94.91042	1.64 h	EC/85/2.57 β^+ /15/	1.20/ 0.91/	5/2+	0.86		ann.rad./ 0.2904 0.3364 0.6268
^{96}Ru	5.52(20)	95.90760				0+			
^{97}Ru		96.90756	2.89 d	EC/1.12		5/2+	-0.78		Tc k x-ray 0.2157 0.3245 0.4606
^{98}Ru	1.88(9)	97.90529				0+			
^{99}Ru	12.74(26)	98.905939				5/2+	-0.6413	+0.079	
^{100}Ru	12.60(19)	99.904219				0+			
^{101}Ru	17.05(7)	100.905582				5/2+	-0.7188	+0.46	
^{102}Ru	31.57(31)	101.904349				0+			
^{103}Ru		102.906323	39.27 d	β^- /0.763	0.223	3/2+	0.20	+0.62	0.05329 0.29498 0.4438 0.49708 0.55704 0.61033 (0.04 - 1.6)
^{104}Ru	18.66(44)	103.905430				0+			
^{105}Ru		104.907750	4.44 h	β^- /1.917	1.11/22 1.134/13 1.187/49	3/2+	-0.3		0.12968 / 0.0 0.1491 0.2629 0.31664 0.46943 0.67634 0.72420 (0.1 - 1.8)
^{106}Ru		105.90733	1.020 y	β^- /0.0394	0.0394/100	0+			
^{107}Ru		106.9099	3.8 m	β^- /2.9	2.1/ 3.2/				0.1939 0.3741 0.4625 0.8488
^{108}Ru		107.9102	4.5 m	β^- /1.4	1.2/				0.0923 0.1651 0.4339 0.4975 0.6189
^{109}Ru		108.91320	34.5 s	β^- /4.2					0.1164 0.3584
^{110}Ru		109.9140	15. s	β^- /2.81					0.1121 0.3737 0.4397 0.7967
^{111}Ru		110.9176	1.5 s	β^- /5.5					

Elem. or Isot.	Natural Abundance (%)	Atomic Mass or Weight	Half-Life	Decay Mode/Energy (/MeV)	Particle Energy /Intensity (MeV / %)	Spin (h/2π)	Nuclear Magnetic Mom.(nm)	Elect. Quadr. Mom. (b)	γ-ray Energy /Intensity (MeV / %)
^{112}Ru		111.9188	4.5 s	β^- /4.5					
^{113}Ru		112.9225	2.7 s	β^- /7.					
^{114}Ru		113.9239	0.57 s	β^- /6.1					0.127/24 (0.053-0.180)
^{115}Ru		114.928	≈0.74 s	β^- /8.					
^{116}Ru		115.930							
^{117}Ru		116.935							
^{118}Ru		117.937							
^{119}Ru									
^{120}Ru									
$_{45}$Rh		102.90550(2)							
^{89}Rh		88.9494							
^{90}Rh		89.9429							
^{91}Rh		90.9366							
^{92}Rh		91.9320		β^+ /11.1					
^{93}Rh		92.9257		β^+ /8.1					
^{94m}Rh			25.8 s	β^+ /		8+			(0.138-1.493) ann.rad./ 0.1264 0.3117 0.7562 1.0752 1.4307
^{94}Rh		93.9217	1.18 m	β^+ /9.6	6.4/	3+			ann.rad./ 0.1461 0.3117 0.7562 1.4307
^{95m}Rh			1.96 m	I.T./88/ β^+ ,EC/12/		1/2+			ann.rad./ 0.5433(IT) 0.7837
^{95}Rh		94.9159	5.0 m	β^+ /5.1	3.2	9/2+			ann.rad./ 0.2293 0.4103 0.6610 0.9416 1.3520
^{96m}Rh			1.51 m	I.T./60 /0.052 β^+ ,EC/40/	4.70/	2+			(0.2 - 3.8) ann.rad./ Tc,Ru x-rays 0.8326 1.0985 1.6921
^{96}Rh		95.91452	9.6 m	β^+/6.45 EC/	3.3/	5+			(0.4 - 3.3) ann.rad./ 0.4299 0.6315 0.6853 0.7418 0.8326
^{97m}Rh			46.m	I.T./5 / β^+,EC/95 /	2.6/	1/2-			(0.2 - 3.4) ann.rad./ 0.1886 0.4215 2.2452
^{97}Rh		96.91134	31.0m	β^+ /3.52	2.1/	9/2+			ann.rad./ 0.1886 0.3892 0.4515 0.8398 0.8788
^{98m}Rh			3.5 m	β^+ /		5+			(0.2 - 3.5) ann.rad./ 0.6154 0.6524 0.7452
^{98}Rh		97.91072	8.7 m	β^+ /90 /5.06	3.4/	2+			ann.rad./ 0.6524 0.7623
^{99m}Rh			4.7 h	β^+ /8 / EC/92 /	.74/	9/2+			ann.rad./ 0.2766/ 0.3408 0.6178 1.2612

Elem. or Isot.	Natural Abundance (%)	Atomic Mass or Weight	Half-Life	Decay Mode/Energy (/MeV)	Particle Energy /Intensity (MeV / %)	Spin (h/2π)	Nuclear Magnetic Mom.(nm)	Elect. Quadr. Mom. (b)	γ-ray Energy /Intensity (MeV / %)
⁹⁹Rh		98.90820	16. d	β⁺/4 /2.10 EC/97 /	0.54/ 0.68/	1/2-			ann.rad./ 0.0894/ 0.3530 0.5277 (0.1 - 2.0)
¹⁰⁰ᵐRh			4.7 m	I.T./99 / β⁺ /0.4 /		5+			ann.rad./ 0.0748/ 0.2647(IT)
¹⁰⁰Rh		99.90812	20.8 h	β⁺ /3.63 EC/	2.62/ 2.07/	1-			0.4462 0.5396 0.5882 0.8225 1.5534 2.3761
¹⁰¹ᵐRh			4.35 d	EC/92 / I.T./8 /0.1573		9/2+	+5.51		Rh k x-ray 0.1272/ 0.3069 0.5451
¹⁰¹Rh		100.90616	3.3 y	EC/0.54		1/2-			Ru k x-ray 0.1272 0.1980 0.3252
¹⁰²ᵐRh			207. d	I.T./5 / β⁻ /19/ β⁺ /14/ EC/62/					ann.rad./ 0.4686 0.4751 0.5566 0.6280 1.1032 (0.4 - 1.6)
¹⁰²Rh		101.906842	2.9 y	EC/2.323		6+	4.04		Ru k x-ray 0.4751 0.6313 0.6975 0.7668 1.0466 1.1032
¹⁰³ᵐRh			56.12 m	IT		7/2+			
¹⁰³Rh	100.	102.905504				1/2-	-0.0884		
¹⁰⁴ᵐRh			4.36 m	I.T./99+ / β⁻	1.3/	5+			Rh k x-ray 0.0514 0.0971 0.5558
¹⁰⁴Rh		103.906655	42.3 s	β⁻ /99+ /2.441 EC/0.4/1.141	1.88/2 2.44/98	1+			0.3581 0.5558 1.2370 (0.35 - 1.8)
¹⁰⁵ᵐRh			40. s	I.T./1.296		1/2-			Rh k x-ray 0.1296
¹⁰⁵Rh		104.905692	35.4 h	β⁻ /0.567	0.247/30 0.567/70	7/2+	+4.45		0.2801 0.3061 0.3189
¹⁰⁶ᵐRh			2.18 h	β⁻ /	0.92/	6+			0.2217 0.4510 0.5119 0.6162 0.7173 0.7484 1.0458 1.5277
¹⁰⁶Rh		105.90729	29.9 s	β⁻ /3.54	2.4/2 3.0/12 3.54/79	1+	2.58		0.51186/ 0.61612 0.62187 (0.05 - 3.04)
¹⁰⁷Rh		106.90675	21.7 m	β⁻ /1.51	1.20/65 1.5/17	7/2+			0.2776 0.3028 0.3925 0.4339 0.4973 0.6189
¹⁰⁸ᵐRh			6.0 m	β⁻ /	1.57/				0.4046 0.4339 0.4973 0.5811 0.6146 0.9014 0.9471
¹⁰⁸Rh		107.9087	17. s	β⁻ /4.5		1+			

Elem. or Isot.	Natural Abundance (%)	Atomic Mass or Weight	Half-Life	Decay Mode/Energy (/MeV)	Particle Energy /Intensity (MeV / %)	Spin (h/2π)	Nuclear Magnetic Mom.(nm)	Elect. Quadr. Mom. (b)	γ-ray Energy /Intensity (MeV / %)
[109]Rh		108.90874	1.34 m	β⁻ /2.59	2.25/	7/2+			0.1134 0.1780 0.2914 0.3254 0.3268 0.4261 (0.1 - 1.6)
[110m]Rh			29. s	β⁻ /	[.6/				0.3737 0.4397 0.7967
[110]Rh		109.9110	3.1 s	β⁻ /5.4	5.5/	1+			0.3737 0.4400 0.5463 0.6877 0.8381 0.9045
[111]Rh		110.9117	11. s	β⁻ /3.7					0.275
[112m]Rh			6.8 s	β⁻ /					
[112]Rh		111.9140	4. s	β⁻ /6.2		1+			0.3489
[113]Rh		112.9154	0.9 s	β⁻ /4.9					0.1285
[114m]Rh			1.8 s	β⁻ /					
[114]Rh		113.9173	1.8 s	β⁻ /6.5		1+			
[115]Rh		114.9201	0.99 s	β⁻ /6.0					
[116m]Rh			0.9 s	β⁻ /					0.3405
[116]Rh		115.9228	0.7 s	β⁻ /8.0		1+			
[117]Rh		116.925	0.44 s	β⁻ /7.					0.0346 0.1317
[118]Rh		117.929							
[119]Rh		118.931							
[120]Rh		119.936							
[121]Rh		120.938							
[122]Rh									
₄₆Pd		106.42(1)							
[91]Pd		90.949							
[92]Pd		91.9404							
[93]Pd		92.9359							
[94]Pd		93.9288	9. s	EC,β⁺ /≈ 6.6					0.5582 (0.0546-0.798)
[95m]Pd		94.92684	13.4 s	EC,β⁺ /10.2		21/2+			
[95]Pd									
[96]Pd		95.9182	2.03 m	EC,β⁺ /3.5	1.15/				0.1248 0.4995
[97]Pd		96.9165	3.1 m	β⁺ ,EC/4.8	3.5/	5/2+			ann.rad./ 0.2653 0.4752 0.7927 (0.2 - 3.4)
[98]Pd		97.91273	17.7 m	β⁺ /1.87 EC/		0+			ann.rad./ 0.0677 0.1125 0.6630 0.8379
[99]Pd		98.91181	21.4 m	β⁺ /49 /3.37 EC/51 /	2.18/	5/2+			ann.rad./ 0.1360 0.2636 0.6734 (0.2 - 2.85)
[100]Pd		99.90851	3.7 d	EC/0.36		0+			0.03271 0.0748 0.0840
[101]Pd		100.90829	8.4 h	β⁺ /5 /1.980 EC/95 /	0.776/	5/2+	-0.66		ann.rad./ 0.0244 0.2963 0.5904
[102]Pd	1.02(1)	101.905607				0+			
[103]Pd		102.906087	16.99 d	EC/0.543		5/2+			Rh k x-ray 0.03975 0.3575 0.4971
[104]Pd	11.14(8)	103.904034				0+			
[105]Pd	22.33(8)	104.905083				5/2+	-0.642	+0.66	
[106]Pd	27.33(3)	105.903484				0+			
[107m]Pd			20.9 s	I.T./0.2149		11/2-			Pd k x-ray 0.2149(IT)
[107]Pd		106.90513	6.5x10⁶y	β⁻ /0.033	0.03/	5/2+			
[108]Pd	26.46(9)	107.903895				0+			

Elem. or Isot.	Natural Abundance (%)	Atomic Mass or Weight	Half-Life	Decay Mode/Energy (/MeV)	Particle Energy /Intensity (MeV / %)	Spin (h/2π)	Nuclear Magnetic Mom.(nm)	Elect. Quadr. Mom. (b)	γ-ray Energy /Intensity (MeV / %)
^{109m}Pd			4.75 m	I.T./0.1889		11/2-			Pd x-ray 0.1889(IT)
^{109}Pd		108.905954	13.5 h	β⁻ /1.116	1.028	5/2+			0.0880
^{110}Pd	11.72(9)	109.905153				0+			(0.08 - 1.0)
^{111m}Pd			5.5 h	I.T./73 /0.172 β⁻ /27 /	0.35 0.77	11/2-			0.0704 0.1722 0.3912
^{111}Pd		110.90764	23.4 m	β⁻ /2.19	2.2/95	5/2+			(0.1 - 1.97) 0.0598 0.2454 0.5800 0.6504 1.3885 1.4590
^{112}Pd		111.90731	21.04 h	β⁻ /0.29	0.28/	0+			0.018
^{113m}Pd			1.48 m	β⁻ /		5/2+			0.0959
^{113}Pd		112.91015	1.64 m	β⁻ /3.34					0.0958 0.4824 0.6436 0.7394
^{114}Pd		113.91037	2.48 m	β⁻ /1.45		0+			0.1266 0.2320 0.5582 0.5760
^{115}Pd		114.9137	47. s	β⁻ /4.58					0.1255 0.2554 0.3428
^{116}Pd		115.9142	12.7 s	β⁻ /2.61					0.1015 0.1147 0.1778
^{117}Pd		116.9178	4.4 s	β⁻ /5.7					0.2473 (0.0766-0.4025)
^{118}Pd		117.9189	2.4 s	β⁻ /4.1					0.1254 (0.0284-0.5957)
^{119}Pd		118.9227	0.9 s	β⁻ /6.5					0.2566 (0.0699-0.3261)
^{120}Pd		119.9240	0.5 s	β⁻ /5.0					0.1581 (0.0525-0.5952)
^{121}Pd		120.9282							
^{122}Pd		121.9298							
^{123}Pd									
$_{47}$Ag		107.8682(2)							
^{94}Ag		93.9428	0.42 s	β⁺, p/					
^{95}Ag		94.9355	2.0 s	β⁺, p/					(0.539-2.025)
^{96}Ag		95.9307	5.1 s	β⁺ /11.6 EC/					ann.rad./ 0.1248 0.4995
^{97}Ag		96.9240	19. s	β⁺ /7.0 EC/					(0.1066-1.416) ann.rad./ 0.6862 1.2941
^{98}Ag		97.9218	47. s	β⁺ /8.4 EC/		5+			(0.352-3.294) ann.rad./ 0.5711 0.6786 0.8631
^{99m}Ag			11. s	I.T./100/		1/2-			(0.153-1.185) Ag k x-ray 0.1636(IT)
^{99}Ag		98.9176	2.07 m	β⁺ /87 /5.4 EC/13 /		9/2+			0.3426 ann.rad./ 0.2199 0.2645 0.8056 0.8323
^{100m}Ag			2.3 m	β⁺ / EC/		2+			(0.2 - 3.5) ann.rad./ 0.6657 1.6941
^{100}Ag		99.9161	2.0m	β⁺/7.1 EC/	4.7/	5+			ann.rad./ 0.2807 0.4503 0.6657 0.7508 0.7732

Elem. or Isot.	Natural Abundance (%)	Atomic Mass or Weight	Half-Life	Decay Mode/Energy (/MeV)	Particle Energy /Intensity (MeV / %)	Spin (h/2π)	Nuclear Magnetic Mom.(nm)	Elect. Quadr. Mom. (b)	γ-ray Energy /Intensity (MeV / %)
[101m]Ag			3.1 s	I.T./0.23		1/2-			Ag k x-ray 0.0981 0.176(IT)
[101]Ag		100.9128	11.1 m	β⁺ /69 /4.2 EC/31 /	2.7/ 2.18/ 2.73/ 3.38/	9/2+	5.7		ann.rad./ 0.2610 0.2747 0.3269 0.4392 0.6673 1.1739 (0.2 - 3.1)
[102m]Ag			7.8 m	β⁺ /38 / EC/13 / I.T./49 /	3.4	2+	+4.14		ann.rad./ 0.5567 0.9777 1.8347 2.0545 2.1594 3.2386
[102]Ag		101.91197	13.0 m	β⁺ /78 /5.92 EC/22 /	2.26/	5+	4.6		ann.rad./ 0.5567 0.7194 0.8354 1.2571 1.5816 1.7446
[103m]Ag			5.7 s	I.T./0.134		1/2-			Ag k x-ray 0.1344
[103]Ag		102.90897	1.10 h	β⁺ /28 /2.69 EC/72 /	1.7 1.3	7/2+	+4.47		ann.rad./ 0.1187 0.1482
[104m]Ag			33. m	β⁺ /64/ EC/36/ I.T./0.07/	2.71/	2+	+3.7		ann.rad./ 0.5558 0.7657 (0.5 - 3.4)
[104]Ag		103.90863	69. m	β⁺ /16 /4.28 EC/84 /	0.99/	5+	3.92		ann.rad./ 0.5558 0.9259 0.9416 (0.18 - 2.27)
[105m]Ag			7.2 m	I.T./98 /0.0255 EC/2 /		7/2+			Ag x-ray 0.3063 0.3192 (0.1 - 1.0)
[105]Ag		104.90653	41.3 d	EC/1.35		1/2-	0.1014		0.0640 0.2804 0.3445 0.4434
[106m]Ag			8.4 d	EC/		6+	3.71		Pd k x-ray 0.4510 0.5118 0.7173 1.0458
[106]Ag		105.90667	24.0 m	β⁺ /59 /2.965 EC/41 /	/1.96	1+	+2.85		ann.rad./ 0.5119
[107m]Ag			44.2 s	I.T./0.093		7/2+			Ag x-ray 0.0931
[107]Ag	51.839(8)	106.905093				1/2-	-0.11357		
[108m]Ag			1.3x10²y	EC/92 / I.T./8 /0.079		6+	3.580		Ag k x-ray Pd k x-ray 0.43392 0.61427 0.72290
[108]Ag		107.905954	2.39 m	β⁻ /97/1.65 EC/2/ β⁻ /1/1.92	1.02/1.7 1.65/96 0.88/0.3	1+	+2.6884		ann.rad./ 0.43392 0.61885 0.63298
[109m]Ag			39.8 s	I.T./0.088		7/2+			Ag k x-ray 0.0880
[109]Ag	48.161(8)	108.904756				1/2-	-0.13056		
[110m]Ag			249.8 d	β⁻ /99 / I.T./1 /0.1164	0.087 0.530	6+	+3.607		0.65774 0.76393 0.88467 0.93748 1.38427 (0.447-1.56)
[110]Ag		109.906111	24.6 s	β⁻ /2.892	2.22/5 2.89/95	1+	+2.7271		0.65774 0.8154 1.1257

Elem. or Isot.	Natural Abundance (%)	Atomic Mass or Weight	Half-Life	Decay Mode/Energy (/MeV)	Particle Energy /Intensity (MeV / %)	Spin (h/2π)	Nuclear Magnetic Mom.(nm)	Elect. Quadr. Mom. (b)	γ-ray Energy /Intensity (MeV / %)
^{111m}Ag			1.08 m	I.T./99 /0.0598 β⁻ /1/		7/2+			Ag k x-ray 0.0598 0.2454
^{111}Ag		110.905295	7.47 d	β⁻ /1.037	1.035/	1/2-	-0.146		0.2454 0.3421
^{112}Ag		111.90701	3.13 h	β⁻ /3.96	3.94/ 3.4	2-	0.0547		0.6067 0.6174 1.3877 (0.4 - 2.9)
^{113m}Ag			1.14 m	I.T./80 /0.043 β⁻ /20 /	1.5	7/2+			0.1422 0.2983 0.3161 0.3923
^{113}Ag		112.90657	5.3 h	β⁻ /2.02	2.01/	1/2-	0.159		0.2588 0.2986
^{114}Ag		113.90881	4.6 s	β⁻ /5.08	4.9/	1+			0.5582 0.5760 1.9946
^{115m}Ag			18.7 s	β⁻ /		7/2+			0.1134 0.1315 0.2288 0.3887
^{115}Ag		114.90876	20. m	β⁻ /3.10		1/2-			0.1316 0.2128 0.2291 0.4727 (0.13 - 2.49)
^{116m}Ag			10.5 s	I.T./2 / β⁻ /98 /	3.2/ 2.9	5+			0.1027 0.2549 0.5134 0.7055 1.0289
^{116}Ag		115.91137	2.68 m	β⁻ /6.16	5.3	2-			0.5134 0.6993 2.4779
^{117m}Ag			5.3 s	β⁻ /	3.2/	7/2+			0.1354 0.2981 0.3868
^{117}Ag		116.91171	1.22 m	β⁻ /4.18	2.3	1/2-			0.1354 0.1571 0.3377
^{118m}Ag			2.8 s	β⁻ /59 / I.T./41 /0.1277					0.1277 0.4878 0.6771 0.7709 1.0586
^{118}Ag		117.9145	4.0 s	β⁻ /7.1					0.4878 0.6771 3.2259
^{119}Ag		118.9157	2.1 s	β⁻ /5.35		7/2+			0.0674 0.3662 0.3991 0.6264
^{120m}Ag			0.32 s	β⁻ / I.T./					0.2030 0.5059 0.6978 0.8300 0.9258
^{120}Ag		119.9188	1.23 s	β⁻ /8.2					0.5059 0.6978 0.8171 1.3231
^{121}Ag		120.9198	0.78 s	β⁻ /6.4					0.1150 0.3148 0.3537 0.3696 0.5007 1.5105 (0.11 - 2.5)
^{122m}Ag			1. s	β⁻ /					
^{122}Ag		121.9233	0.44 s	β⁻ /9.2					
^{123}Ag		122.9249	0.31 s	β⁻ /7.4					
^{124}Ag		123.9285	0.22 s	β⁻ /10.1					
^{125}Ag		124.9305	0.17 s						
^{126}Ag		125.9345	0.11 s						
^{127}Ag		126.9369	0.11 s						

Elem. or Isot.	Natural Abundance (%)	Atomic Mass or Weight	Half-Life	Decay Mode/Energy (/MeV)	Particle Energy /Intensity (MeV / %)	Spin (h/2π)	Nuclear Magnetic Mom.(nm)	Elect. Quadr. Mom. (b)	γ-ray Energy /Intensity (MeV / %)
₄₈Cd		112.411(8)							
⁹⁷Cd		96.9349	3. s	β⁺,(p)					
⁹⁸Cd		97.9276	9.2 s	β⁺ /5.4 (p)	/0.025				
⁹⁹Cd		98.9250	16. s	β⁺ ,EC/6.9					ann.rad./
¹⁰⁰Cd		99.9203	1.1 m	β⁺ ,EC/3.9					ann.rad./ (0.090-1.043)
¹⁰¹Cd		100.9187	1.2 m	β⁺ /83 /5.5 EC/17 /	4.5	5/2+			In k x-ray 0.0985 1.7225 0.31 - 2.84)
¹⁰²Cd		101.91474	5.8 m	β⁺ /27 /2.59 EC/73		0+			ann.rad./ 0.0974 0.4810 1.0366 1.3598
¹⁰³Cd		102.91342	7.5 m	β⁺ /33 /4.14 EC/67 /		5/2+	-0.81	-0.8	ann.rad./ Ag k x-ray 1.0799 1.4487 1.4618 (0.1 - 2.8)
¹⁰⁴Cd		103.90985	58. m	EC/1.14		0+			Ag k x-ray 0.0835 0.7093
¹⁰⁵Cd		104.90947	55.5 m	β⁺ /26 /2.739 EC/74 /	1.69/	5/2+	-0.7393	+0.43	Ag k x-ray 0.3469 0.6072 0.9618 1.3025 (0.25 - 2.4)
¹⁰⁶Cd	1.25(6)	105.90646	>2.6x10¹⁷ y	β⁺,EC		0+			
¹⁰⁷Cd		106.90661	6.52 h	EC/99+ /1.417 β⁺ /		5/2+	-0.615055	+0.68	Ag k x-ray 0.0931 0.8289
¹⁰⁸Cd	0.89(3)	107.90418				0+			
¹⁰⁹Cd		108.904985	462.0 d	EC/0.214		5/2+	-0.827846	+0.69	Ag k x-ray 0.08804
¹¹⁰Cd	12.49(18)	109.903006				0+			
¹¹¹ᵐCd			48.5 m	I.T./		11/2-			Cd k x-ray 0.1508(IT) 0.2454
¹¹¹Cd	12.80(12)	110.904182				1/2+	-0.594886		
¹¹²Cd	24.13(21)	111.902758				0+			
¹¹⁵ᵐCd			14.1 y	β⁻ /99.9 /0.59	0.59/99.9	11/2-	-1.087		0.2637
¹¹³Cd	12.22(12)	112.904401	7.7x10¹⁵y	β⁻		1/2+	-0.622301		
¹¹⁴Cd	28.73(42)	113.903359				0+			
¹¹⁵ᵐCd			44.6 d	β⁻ /1.629	0.68/1.6 1.62/97	11/2-	-1.042		0.48450 0.93381 1.29064
¹¹⁵Cd		114.905431	2.228 d	β⁻ /1.446	0.593/42 1.11/58	1/2+	-0.648426		0.23141 0.26085 0.33624 0.49227 0.52780
¹¹⁶Cd	7.49(18)	115.904756				0+			
¹¹⁷ᵐCd			3.4 h	β⁻ /2.66	0.72/	11/2-			0.1586 0.5529 0.37 - 2.42
¹¹⁷Cd		116.907219	2.49 h	β⁻ /2.52	0.67/51 2.2/10	1/2+			0.2209 0.2733 0.3445 1.3033
¹¹⁸Cd		117.90692	50.3 m	β⁻ /0.52		0+			
¹¹⁹ᵐCd			2.20 m	β⁻ /		11/2-			0.1056 0.7208 1.0250 2.0213
¹¹⁹Cd		118.90992	2.69 m	β⁻ /3.8	≈ 3.5/	1/2+			0.1340 0.2929 0.3429
¹²⁰Cd		119.90985	50.8 s	β⁻ /1.76	1.5/	0+			
¹²¹ᵐCd			8. s	β⁻ /		11/2-			0.1008 0.9878 1.0209 1.1815 2.0594

Elem. or Isot.	Natural Abundance (%)	Atomic Mass or Weight	Half-Life	Decay Mode/Energy (/MeV)	Particle Energy /Intensity (MeV / %)	Spin (h/2π)	Nuclear Magnetic Mom.(nm)	Elect. Quadr. Mom. (b)	γ-ray Energy /Intensity (MeV / %)
¹²¹Cd		120.9131	13.5 s	β⁻ /4.9		(3/2+)			0.2102 0.3242 0.3492 1.0403
¹²²Cd		121.9135	5.3 s	β⁻ /3.0		0+			
¹²³ᵐCd			1.9 s	β⁻ /					
¹²³Cd		122.91770	2.09 s	β⁻ /6.12		3+			
¹²⁴Cd		123.9177	1.24 s	β⁻ /4.17		0+			0.0365 0.0628 0.1799
¹²⁵ᵐCd			0.66 s	β⁻ /					
¹²⁵Cd		124.92129	0.68 s	β⁻ /7.16		3/+			
¹²⁶Cd		125.9224	0.52 s	β⁻ /5.49		0+			0.2601
¹²⁷Cd		126.9264	0.4 s	β⁻ /8.5		3/+			
¹²⁸Cd		127.9278	0.28 s	β⁻ /7.1		0+			0.247
¹²⁹Cd		128.9323	0.27 s	β⁻ /5.9					0.281
¹³⁰Cd		129.9340	0.20 s	β⁻ /		0+			
₄₉In		114.818(3)							
⁹⁸In		97.9422							
⁹⁹In		98.9346		β⁺ /8.9					
¹⁰⁰In		99.9316	6. s	β⁺,(p)/10.5					
¹⁰¹In		100.9266	15. s	β⁺ /7.3					
¹⁰²In		101.9243	22. s	EC/8.9		(5)			0.1566 0.7767 (0.397-0.923)
¹⁰³ᵐIn			34. s						
¹⁰³In		102.91991	1.1 m	β⁺,EC/6.05	4.2	9/2+			ann.rad./ 0.1879 (0.157-3.98)
¹⁰⁴ᵐIn			16. s	IT/0.0935					
¹⁰⁴In		103.9183	1.84 m	β⁺,EC/7.9	4.8	5+	+4.44	+0.7	ann.rad./ 0.6580 0.8341 0.8781
¹⁰⁵ᵐIn			43. s	I.T.		1/2-			In k x-ray 0.6740
¹⁰⁵In		104.91467	5.1 m	β⁺,EC/4.85	3.7	9/2+	+5.675	+0.83	0.1310 0.2600 0.6038
¹⁰⁶ᵐIn			5.3 m	β⁺ /85 / EC/15 /	4.90	3+			ann.rad./ 0.6326 0.8611 1.7164
¹⁰⁶In		105.91346	6.2 m	β⁺ /65 /6.52 EC/35 /	2.6	7+	+4.92	+0.97	ann.rad./ 0.2259 0.6327 0.8611 0.9978 1.0091
¹⁰⁷ᵐIn			51. s	I.T./0.6786		1/2-			In k x-ray 0.6785
¹⁰⁷In		106.91029	32.4 m	β⁺ /35 /3.43 E.C/65 /	2.20/	9/2+	+5.59	+0.81	ann.rad./ Cd k x-ray 0.2050 0.3209 0.5055 (0.2 - 2.99)
¹⁰⁸ᵐIn			57. m	β⁺ /53 / EC/47 /	1.3	6+			ann.rad./ Cd k x-ray 0.6329 1.9863 3.4522
¹⁰⁸In		107.90971	40. m	β⁺ /33 /5.15 EC/67 /	3.49/	3+	+4.56	+1.01	ann.rad./ Cd k x-ray 0.2429 0.6331 0.8756
¹⁰⁹ᵐIn			1.3 m	I.T./0.650		1/2-			In k x-ray 0.6498
¹⁰⁹In		108.90715	4.2 h	β⁺ /8 /2.02 EC/92 /	0.79/	9/2+	+5.54	+0.84	ann.rad./ Cd k x-ray 0.2035 0.6235

Elem. or Isot.	Natural Abundance (%)	Atomic Mass or Weight	Half-Life	Decay Mode/Energy (/MeV)	Particle Energy /Intensity (MeV / %)	Spin (h/2π)	Nuclear Magnetic Mom.(nm)	Elect. Quadr. Mom. (b)	γ-ray Energy /Intensity (MeV / %)
¹¹⁰ᵐIn			4.9 h	EC/		7+	+4.72	+1.00	Cd k x-ray 0.6577 0.8847 0.9375 (0.1 - 1.98)
¹¹⁰In		109.90717	1.15 h	β⁺ /62 /3.88 EC/38 /	2.22/	2+	+4.37	+0.35	ann.rad./ Cd k x-ray 0.6577 (0.6 - 3.6)
¹¹¹ᵐIn			7.7 m	I.T./0.537		1/2-	+5.53		In k x-ray 0.537
¹¹¹In		110.90511	2.8049 d	EC/0.866		9/2+	+5.50	+0.80	Cd k x-ray 0.1712 0.2453
¹¹²ᵐIn			20.8 m	I.T./0.155		4+			In k x-ray 0.1555
¹¹²In		111.90553	14.4 m	β⁺ /22 /2.586 EC/34 / β⁻ /0.663		1+	+2.82	+0.09	ann.rad./ Cd k x-ray 0.6171
¹¹³ᵐIn			1.658 h	I.T./0.3917		1/2-	-0.210		In k x-ray 0.3917
¹¹³In	4.29(5)	112.904062				9/2+	+5.529	+0.80	
¹¹⁴ᵐIn			49.51 d	I.T./97 /0.190 EC/3 /		5+	+4.7		In k x-ray 0.19027
¹¹⁴In		113.904918	1.198 m	β⁻ /97 /1.989 EC/3 /1.453	1.984/	1+	+2.82		Cd k x-ray 0.5584 0.5727 1.2998
¹¹⁵ᵐIn			4.486 h	I.T./95 /0.336 β⁻ /5 /0.83		1/2-	-0.255		In k x-ray 0.3362 0.4974
¹¹⁵In	95.71(5)	114.903879	4.4x10¹⁴ y	β⁻ /0.495		9/2+	+5.541	+0.81	
¹¹⁶ᵐ²In			2.16 s	I.T./0.162		8-			In k x-ray 0.1624
¹¹⁶ᵐ¹In			54.1 m	β⁻ /	1.0	5+	+4.3		0.13792 0.41688/27 1.09723/58.5 1.29349/85
¹¹⁶In		115.905261	14.1 s	β⁻ /3.274	3.3/99	1+	2.788	0.11	0.46313 1.2526 1.29349
¹¹⁷ᵐIn			1.94 h	β⁻ /53 /1.769 I.T./47 /	1.77/	1/2-	0.25		In k x-ray 0.15855 0.31531 0.55294
¹¹⁷In		116.90452	44. m	β⁻ /1.455	0.74/	9/2+	+5.52	+0.83	0.15855 0.3966 0.55294
¹¹⁸ᵐ²In			8.5 s	I.T./98 / β⁻ /2 /		(8-)			In k x-ray 0.1382
¹¹⁸ᵐ¹In			4.40 m	β⁻ /	1.3 2.0	5+			0.2086 0.6833 1.2295
¹¹⁸In		117.90636	5.0 s	β⁻ /4.42	4.2/	1+			0.5282 1.1734 1.2295 2.0432
¹¹⁹ᵐIn			17.9 m	β⁻ /97 / I.T./3 /0.311	2.7/	1/2-			0.3114 0.7631
¹¹⁹In		118.90585	2.3 m	β⁻ /2.36	1.6/	9/2+	+5.52	+0.85	0.0239 0.6495 0.7631 1.2149
¹²⁰ᵐIn			47. s	β⁻ /	2.2/	8-			0.7042 1.1725 2.0398 2.3902
¹²⁰In		119.90796	3.1 s	β⁻ /5.37	5.6/ 3.1/	(1+)	+4.30	+0.81	0.4146 0.5924 0.8637 1.0232 1.1714 (0.4 - 2.7)
¹²¹ᵐIn			3.8 m	β⁻ /99 / I.T./1 /0.313	3.7/	1/2-			0.0601 0.3136 0.9256 1.0412 1.1022 1.1204

Elem. or Isot.	Natural Abundance (%)	Atomic Mass or Weight	Half-Life	Decay Mode/Energy (/MeV)	Particle Energy /Intensity (MeV / %)	Spin (h/2π)	Nuclear Magnetic Mom.(nm)	Elect. Quadr. Mom. (b)	γ-ray Energy /Intensity (MeV / %)
^{121}In		120.90785	23. s	β⁻ /3.36	2.5	9/2+	+5.50	+0.81	0.2620 / 0.6573 / 0.9256
^{122m}In			10. s	β⁻ /	4.4/	8-			1.0014 / 1.1403
^{122}In		121.91028	1.5 s	β⁻ /6.37	5.3/	(1+)			0.2391 / 1.0014 / 1.1403 / 1.164 / 1.1903
^{123m}In			47. s	β⁻ /	4.6/	(1/2-)	-0.40		0.1258 / 1.170 / 3.234
^{123}In		122.91044	6.0 s	β⁻ /4.39	3.3/	(9/2+)	+5.49	+0.76	0.6188 / 1.0197 / 1.1305
^{124m}In			3.4 s	β⁻		8-	+3.89	+0.66	0.1029 / 0.9699 / 1.0729 / 1.1316
^{124}In		123.91318	3.18 s	β⁻ /7.36	5/	3+	+4.04	+0.61	0.7070 / 0.9978 / 1.1316 / 3.2142 / (0.3 - 4.6)
^{125m}In			12.2 s	β⁻ /	5.5/	1/2-	-0.43		0.1876
^{125}In		124.91360	2.33 s	β⁻ /5.42	4.1/	9/2+	+5.50	+0.71	0.4260 / 1.0318 / 1.3350
^{126m}In			1.53 s		4.9/	3+	+4.03	+0.49	0.9086 / 0.9696 / 1.1411
^{126}In		125.91646	1.63 s	β⁻ /8.21	4.2/	8-	+4.06		0.1118 / 0.9086 / 1.1411
^{127m}In			3.73 s	β⁻ /	6.4/	(1/2-)			0.2523 / 3.074
^{127}In		126.91734	1.14 s	β⁻ /6.51	4.9/	(9/2+)	+5.52	+0.59	0.4680 / 0.6461 / 0.8051 / 1.5977
^{128m}In			0.7 s	β⁻ /	5.4/	(8-)			1.8670 / 1.9739 / (0.1205-2.12)
^{128}In		127.92017	0.80 s	β⁻ /8.98	5.0/	3+			0.9352 / 1.1688 / 3.5198 / 4.2970
^{129m}In			1.23 s	β⁻ /98 / n/2 /	≈ 7.5/	1/2-			0.3153 / 0.9067 / 1.2220
^{129}In		128.9217	0.63 s	β⁻ /7.66	5.5/	9/2+			0.2853 / 0.7693 / 1.8650 / 2.1180
^{130m2}In			0.53 s	β⁻ /	8.8/	5+			0.0892 / 0.7744 / 1.2212
^{130m1}In			0.51 s	β⁻ /	6.1/	10-			0.0892 / 0.1298 / 0.7744 / 1.2212 / 1.9052
^{130}In		129.92486	0.29 s	β⁻ /10.25	10.0/	1-			
^{131m2}In			0.3 s	β⁻ /		(21/2+)			
^{131m1}In			0.35 s	β⁻ /		(1/2-)			
^{131}In		130.9268	0.28 s	β⁻ /9.18	6.4/	(9/2+)			0.3328 / 2.433
^{132}In		131.9323	0.20 s	β⁻ /13.6	6.0/ 8.8/	(7-)			0.1320 / 0.2992 / 0.3747 / 4.0406
^{133}In		132.9383	0.18 s	β⁻ ,(n)					
^{134}In		133.9447	0.14 s						(0.354-2.005)
$_{50}$Sn		118.710(7)							
^{100}Sn		99.9394	≈ 0.9 s	β⁺ /7.3	3.4/				
^{101}Sn		100.9361	3. s	β⁺ /9.					

Elem. or Isot.	Natural Abundance (%)	Atomic Mass or Weight	Half-Life	Decay Mode/Energy (/MeV)	Particle Energy /Intensity (MeV / %)	Spin (h/2π)	Nuclear Magnetic Mom.(nm)	Elect. Quadr. Mom. (b)	γ-ray Energy /Intensity (MeV / %)
¹⁰²Sn		101.9243	≈ 5 s	β⁺ /5.8					
¹⁰³Sn		102.9281	7. s	β⁺ /7.7					
¹⁰⁴Sn		103.9232	21. s	β⁺, EC/4.5					
¹⁰⁵Sn		104.9214	28. s	β⁺ /6.3					In-x-ray (0.2879-3.819)
¹⁰⁶Sn		105.91688	2.0 m	β⁺ /20 /3.18 EC/80 /					ann.rad./ In k x-ray 0.3865 0.4772 0.4218 0.6105 0.6785 1.0013 1.1290 1.542
¹⁰⁷Sn		106.9157	2.92 m	EC/5.0 β⁺ /	1.2/				
¹⁰⁸Sn		107.91196	10.3 m	β⁺ /1 /2.09 EC/99 /	0.36/	0+			In k x-ray 0.2724 0.3965 (0.105-1.68)
¹⁰⁹Sn		108.91129	18.0 m	β⁺ /9 /3.85 EC/91 /	1.52/	7/2+	-1.08	+0.3	ann.rad./ In k x-ray 0.6498 1.0992
¹¹⁰Sn		109.90785	4.1 h	EC/0.64		0+			In k x-ray 0.283
¹¹¹Sn		110.90774	35. m	β⁺ /31 /2.45 EC/69 /	1.5/	7/2+	+0.61	+0.2	In k x-ray 0.7620 1.1530 1.9147
¹¹²Sn	0.97(1)	111.904822				0+			
¹¹³ᵐSn			21.4 m	I.T./92 /0.077 EC/8 /		7/2+			Sn k x-ray In x-ray 0.0774
¹¹³Sn		112.905174	115.1 d	EC/1.036		1/2+	-0.879		In k x-ray 0.25511 0.39169
¹¹⁴Sn	0.65(1)	113.902783				0+			
¹¹⁵Sn	0.34(1)	114.903347				1/2+	-0.9188		
¹¹⁶Sn	14.54(9)	115.901745				0+			
¹¹⁷ᵐSn			13.60 d	I.T./0.3146		11/2-	-1.40	-0.4	Sn k x-ray 0.15856
¹¹⁷Sn	7.68(7)	116.902955				1/2+	-1.0010		
¹¹⁸Sn	24.22(9)	117.901608				0+			
¹¹⁹ᵐSn			293. d	I.T./0.0896		11/2-	-1.4	0.21	Sn k x-ray 0.02387
¹¹⁹Sn	8.59(4)	118.903311				1/2+	-1.0473		
¹²⁰Sn	32.59(9)	119.902199				0+			
¹²¹ᵐSn			≈ 55. y	I.T./78 /0.006 β⁻ /22 /	0.354/	11/2-			Sn k x-ray 0.03715
¹²¹Sn		120.904239	1.128 d	β⁻ /0.388	0.383/100	3/2+	0.698	-0.02	
¹²²Sn	4.63(3)	121.903441				0+			
¹²³ᵐSn			40.1 m	β⁻ /1.428	1.26/99	3/2+			0.1603 0.3814
¹²³Sn		122.905723	129.2 d	β⁻ /1.404	1.42/99.4	11/2-	-1.370	+0.03	0.1603 1.0302 1.0886
¹²⁴Sn	5.79(5)	123.905275				0+			
¹²⁵ᵐSn			9.51 m	β⁻ /2.387	2.03/98	3/2+			0.3321 1.4040
¹²⁵Sn		124.907785	9.63 d	β⁻ /2.364	2.35/82	11/2-	-1.35	+0.1	1.0671 (0.2-2.3)
¹²⁶Sn		125.90765	2.1x10⁵y	β⁻ /0.38	0.25/100	0+			0.0643 0.0876 0.4148 0.6663 0.6950
¹²⁷ᵐSn			4.15 m	β⁻ /3.21	2.72/	3/2+			0.4909 1.3480 1.5640
¹²⁷Sn		126.91035	2.12 h	β⁻ /3.20	2.42/ 3.2/	11/2-			0.8231 1.0956 (0.120-2.84)
¹²⁸ᵐSn			6.5 s	IT/0.091		(7-)			
¹²⁸Sn		127.91054	59.1 m	β⁻ /1.27	0.48/ 0.63/	0+			0.4823 0.5573 0.6805 1.1611
¹²⁹ᵐSn			6.9 m	β⁻ /		11/2-			0.6456
¹²⁹Sn		128.9134	2.4 m	β⁻ /4.0		3/2+			

Elem. or Isot.	Natural Abundance (%)	Atomic Mass or Weight	Half-Life	Decay Mode/Energy (/MeV)	Particle Energy /Intensity (MeV / %)	Spin (h/2π)	Nuclear Magnetic Mom.(nm)	Elect. Quadr. Mom. (b)	γ-ray Energy /Intensity (MeV / %)
^{130m}Sn			1.7 m	β⁻ /		(7-)			0.1449 0.8992
^{130}Sn		129.91386	3.7 m	β⁻ /2.15	1.10/	0+			0.0700 0.1925 0.7798
^{131m}Sn			1.02m	β⁻/	3.4/	11/2-			0.3043 0.4500 0.7985 1.2260 (0.08 - 3.21) see ^{131m} Sn
^{131}Sn		130.9169	39. s	β⁻ /4.64	3.8/	3/2+			
^{132}Sn		131.91775	40. s	β⁻ /3.3	1.8/				0.0855 0.2467 0.3402 0.8985
^{133}Sn		132.9236	1.44 s	β⁻ /7.8	7.5/	7/2-			
^{134}Sn		133.9278	1.04 s	β⁻ /6.8					
$_{51}$Sb		121.760(1)							
^{103}Sb		102.9401							
^{104}Sb		103.9363	0.5 s						
^{105}Sb		104.9315	1.1 s						
^{106}Sb		105.9288	0.6 s	β⁺ /10.5					
^{107}Sb		106.9242	≈ 4.6 s	β⁺ /7.9					
^{108}Sb		107.9222	7.0 s	β⁺ /9.5					(0.253-2.154)
^{109}Sb		108.91814	17. s	β⁺ /6.38 EC/	4.42/ 4.67/ 4.33/	5/2+			(0.151-1.280) ann.rad./ 0.2467 0.6645 0.9254 1.0617 1.4958
^{110}Sb		109.9175	24. s	β⁺ /9.0 EC/	6.8/	3+			ann.rad./ 0.6365 0.9847 1.2117 1.2433
^{111}Sb		110.91254	1.25 m	β⁺ /87 /4.47 EC/13 /	3.3/	5/2+			ann.rad./ 0.1002 0.1545 0.4891 1.0326
^{112}Sb		111.91240	51.4 s	β⁺ /90 /7.06 EC/10 /	4.75/	3+			ann.rad./ 0.6700 0.9909 1.2571 (0.3 - 3.6)
^{113}Sb		112.90937	6.7 m	β⁺ /65 /3.91 EC/35 /	2.42/	5/2+			ann.rad./ Sn k x-ray 0.3324 0.4980
^{114}Sb		113.9091	3.49 m	β⁺ /78 /5.9 EC/22 /	3.4/	3+	1.7		ann.rad./ Sn k x-ray 0.8876 1.2999
^{115}Sb		114.90660	32.1 m	β⁺ /67 /3.03 EC/33 /	1.51/	5/2+	+3.46	-0.4	ann.rad./ Sn k x-ray 0.4973
^{116m}Sb			1.00 h	β⁺ /78 / EC/22 /	1.16/	8-			ann.rad./ Sn k x-ray 0.4073 0.5429 0.9725 1.2935 (0.0998-1.501)
^{116}Sb		115.90680	16. m	β⁺ /50 /4.707 EC/50 /	1.3/ 2.3/	3+	2.72		ann.rad./ Sn k x-ray 0.93180 1.29354 (0.138-3.903)
^{117}Sb		116.90484	2.80 h	β⁺ /2 /1.76 EC/98 /	0.57/	5/2+	+3.4		Sn k x-ray 0.1586
^{118m}Sb			5.00 h	EC/99 /		8-	2.3		0.25368 1.05069 1.22964
^{118}Sb		117.905533	3.6 m	β⁺ /74 /3.657 EC/26 /	2.65/	1+	2.5		ann.rad./ Sn k x-ray 1.22964

Elem. or Isot.	Natural Abundance (%)	Atomic Mass or Weight	Half-Life	Decay Mode/Energy (/MeV)	Particle Energy /Intensity (MeV / %)	Spin (h/2π)	Nuclear Magnetic Mom.(nm)	Elect. Quadr. Mom. (b)	γ-ray Energy /Intensity (MeV / %)
^{119}Sb		118.90395	38.1 h	EC/0.59		5/2+	+3.45	-0.4	Sn k x-ray 0.0239
^{120m}Sb			5.76 d	EC/		8-	2.34		Sn k x-ray 0.0898 0.19730 1.02301 1.17121
^{120}Sb		119.90508	15.89 m	β$^+$ /41 /2.68 EC/59 /	1.72/	1+	+2.3		ann.rad./ Sn k x-ray 0.7038 1.17121
^{121}Sb	57.21(5)	120.903822				5/2+	+3.363	-0.4	
^{122m}Sb			4.19 m	I.T./0.162		8-			Sb x-ray 0.0614 0.0761
^{122}Sb		121.90518	2.72 d	β$^-$ /98 /1.979 β$^+$ /2 /1.620	1.414/65 1.980/26	2-	-1.90	+0.9	0.56409 0.69277 1.14050 1.2569
^{123}Sb	42.79(5)	122.904216				7/2+	+2.550	-0.5	
^{124m2}Sb			20.3 m	I.T./0.035		8-			
^{124m1}Sb			1.6 m	I.T./80 / β$^-$ /20 /	1.2/ 1.7/	5+			0.4984 0.6027 0.6458 1.1010
^{124}Sb		123.905938	60.20 d	β$^-$ /2.905	0.61/52 2.301/23	3-	1.2	+1.9	0.60271/97.8 0.64583/7.4 0.72277/10.5 1.69094/48.2
^{125}Sb		124.905247	2.758 y	β$^-$ /0.767	0.13/30 0.302/45 0.62/13	7/2+	+2.63		(0.0274-2.808) 0.0355 0.17632 0.38044 0.42786 0.46336 0.60060 0.63595
^{126m2}Sb			11. s	I.T./		3-			L x-ray 0.0227 0.4148 0.6663 0.6950
^{126m1}Sb			19.0 m	β$^-$ /86 / I.T./14 /	1.9	5+			0.2786 0.4148 0.6663 0.6950 0.7205
^{126}Sb		125.90725	12.4 d	β$^-$ /3.67	1.9	8-	1.3		0.2524 0.2908 0.4121 0.4370 0.6857 0.7837
^{127}Sb		126.906914	3.84 d	β$^-$ /1.581	0.89/ 1.10/ 1.50/	7/2+	2.70		0.3140 0.5941 0.7432 0.7539
^{128m}Sb			10.1 m	β$^-$ /96 / I.T./4 /	2.6/	5+			0.2148 0.3141 0.5265 0.7433 0.7540
^{128}Sb		127.90917	9.1 h	β$^-$ /4.38	2.3/	8-	1.3		0.4338 0.6578 0.7598
^{129m}Sb			17.7 m	β$^-$ /					0.0278 0.1808 0.3594 0.4596 0.5447 0.8128 0.9146 1.0301
^{129}Sb		128.90915	4.40 h	β$^-$ /2.38	0.65/	7/2-	2.82		0.1023 0.7934 / 0.8394
^{130m}Sb			6.5 m	β$^-$ /2.6	2.12/				

Elem. or Isot.	Natural Abundance (%)	Atomic Mass or Weight	Half-Life	Decay Mode/Energy (/MeV)	Particle Energy /Intensity (MeV / %)	Spin (h/2π)	Nuclear Magnetic Mom.(nm)	Elect. Quadr. Mom. (b)	γ-ray Energy /Intensity (MeV / %)
[130]Sb		129.91155 ·	38.4 m	β⁻ /4.96	2.9/	8-			0.1823 0.3309 0.4680 0.7394 0.8394
[131]Sb		130.9120	23.0 m	β⁻ /3.2	1.31/ 3.0/	7/2+			0.6423 0.6579 0.9331 0.9434
[132m]Sb			2.8 m	β⁻ /	3.9/	4+			0.1034 0.3538 0.6968 0.9739 0.9896
[132]Sb		131.91420	4.2 m	β⁻ /5.29		8-			0.1034 0.1506 0.6968 0.9739
[133]Sb		132.9152	2.5 m	β⁻ /4.00	1.20/	7/2+	3.00		0.4235 0.6318 0.8165 1.0764
[134m]Sb			10.4 s	β⁻ /	6.1	7-			
[134]Sb		133.9206	0.8 s	β⁻ /8.4	8.4	0-			0.1152 0.2970 0.7063 1.2791
[135]Sb		134.9252	1.71 s	β⁻ /8.12		7/2+			1.127 1.279
[136]Sb		135.9301	0.82 s	β⁻ /9.3					
₅₂Te		127.60(3)							
[106]Te		105.9377	0.06 ms	α/4.32					
[107]Te		106.9350	3.1 ms	α/ 70/ β⁺ ,EC/10.1	3.86(1)/				
[108]Te		107.9295	2.1 s	α/68 / β⁺ ,EC/32 /6.8	3.314(4)/	0+			
[109]Te		108.9275	4.6 s	β⁺ EC/96 /8.7 α/4 /	3.107(4)/				
[110]Te		109.9224	19. s	β⁺ ,EC/4.5		0+			ann.rad./ 0.2191 0.6059
[111]Te		110.9211	19.3 s	β⁺ ,EC/8.0		(7/2+)			ann.rad./ 0.267 0.322 0.341
[112]Te		111.9171	2.0 m	β⁺ ,EC/4.3		0+			ann.rad./ 0.2962 0.3727 0.4187
[113]Te		112.9154	1.7 s	β⁺ /85 /5.7 EC/15 /	4.5/	(7/2+)			ann.rad./ Sb k x-ray 0.8144 1.0181 1.1812
[114]Te		113.9125	15. m	β⁺ /40 /3.2 EC/60 /		0+			ann.rad./ Sb k x-ray 0.0838 0.0903
[115m]Te			6.7 m	β⁺ /45 / EC/55 /		(1/2+)			ann.rad./ Sb k x-ray 0.7236 0.7704
[115]Te		114.9116	5.8 m	β⁺ /45 /4.6 EC/55 /	2.7/	7/2+			ann.rad./ Sb k x-ray 0.7236 1.3268 1.3806 (0.22 - 2.7)
[116]Te		115.9084	2.49 h	EC/1.5		0+			Sb k x-ray 0.0937
[117]Te		116.90864	1.03 h	EC/75 /3.54 β⁺ /25 /	1.78/	1/2+			ann.rad./ Sb k x-ray 0.9197 1.7164 2.3000
[118]Te		117.90583	6.00 d	EC/0.28		0+			Sb k x-ray

Elem. or Isot.	Natural Abundance (%)	Atomic Mass or Weight	Half-Life	Decay Mode/Energy (/MeV)	Particle Energy/Intensity (MeV / %)	Spin (h/2π)	Nuclear Magnetic Mom.(nm)	Elect. Quadr. Mom. (b)	γ-ray Energy/Intensity (MeV / %)
^{119m}Te			4.69 d	EC/		11/2-	0.89		Sb k x-ray 0.15360 0.2705 1.21271
^{119}Te		118.90641	16.0 h	β+ /2 /2.293 EC/98 /	0.627/	1/2+	0.25		ann.rad./ Sb k x-ray 0.6440 0.6998
^{120}Te	0.09(1)	119.90403				0+			
^{121m}Te			≈ 154. d	I.T.(89%) EC(11%)		11/2-			Te k x-ray 0.2122
^{121}Te		120.90494	16.8 d	EC/1.04		1/2+			Sb k x-ray 0.5076 0.5731
^{122}Te	2.55(12)	121.903056				0+			
^{123m}Te			119.7 d	I.T./0.247		11/2-	-0.93		Te k x-ray 0.1590/84.1
^{123}Te	0.89(3)	122.904271	2.4x10^{19} y	EC/0.051		1/2+	-0.73695		
^{124}Te	4.74(14)	123.902819				0+			
^{125m}Te			58. d	I.T./0.145		11/2-	-0.99	-0.06	Te k x-ray 0.0355
^{125}Te	7.07(15)	124.904424				1/2+	-0.8885		
^{126}Te	18.84(25)	125.903305				0+			
^{127m}Te			109. d	I.T./98 /0.088 β⁻ /2 /0.77		11/2-	-1.04		Te k x-ray 0.0883 0.3603
^{127}Te		126.905217	9.4 h	β⁻ /0.698	0.696/	3/2+	0.64		
^{128}Te	31.74(8)	127.904462				0+			
^{129m}Te			33.6 d	I.T./63 /0.105 β⁻ /37 /	1.60/	11/2-			Te k x-ray 0.45984 0.6959
^{129}Te		128.906596	1.16 h	β⁻ /1.498	0.99/9 1.45/89	3/2+	0.70	0.06	0.0278 0.45984 0.48728
^{130}Te	34.08(62)	129.906223	2.5x10^{21} y			0+			
^{131m}Te			1.35 d	β⁻ /78 /2.4 I.T./22 /0.18	0.42/	11/2-	-1.04		0.0811 0.1021 0.14973 0.77369 0.79375 0.85225
^{131}Te		130.908522	25.0 m	β⁻ /2.233	1.35/12 1.69/22 2.14/60	3/2+	0.70		0.14973 0.45327 0.49269
^{132}Te		131.90852	3.26 d	β⁻ /0.493	0.215	0+			0.049725 0.11198 0.22830
^{133m}Te			55.4 m	β⁻ /82 / I.T./18 /0.334	2.4/30	11/2-			Te k x-ray 0.0949 0.1689 0.3121 0.3341
^{133}Te		132.9109	12.4 m	β⁻ /2.9	2.25/25 2.65	3/2+			0.3121 0.4079 1.3334
^{134}Te		133.9116	42. m	β⁻ /1.6	0.6/ 0.7/	0+			0.7672/29 0.0794-0.9255
^{135}Te		134.9165	19.0 s	β⁻ /6.0	5.4/ 6.0				0.267 0.603 0.870
^{136}Te		135.92010	17.5 s	β⁻ /5.1	2.5/	0+			2.0779/25 0.0873-3.235
^{137}Te		136.9253	2.5 s	β⁻ /98 /6.9 n/2 /	6.8	7/2-			0.2436
^{138}Te		137.9292	1.4 s	β⁻ /6.4					
$_{53}$I		126.90447(3)							
^{108}I		107.9436	0.04 s	α/91 /4. p	3.95				
^{109}I		108.9382	0.11 ms	p					
^{110}I		109.9346	0.65 s	β+ ,EC/83 /11.4 α/17 / ≈3.6 p /11 /	3.457(10)/				ann.rad./
^{111}I		110.9303	2.5 s	β+ ,E../8.5					ann.rad./ 0.2665 0.3215 0.3412
^{112}I		111.9280	3.4 s	β+ ,EC/10.2					ann.rad./ 0.6889 0.7869

Elem. or Isot.	Natural Abundance (%)	Atomic Mass or Weight	Half-Life	Decay Mode/Energy (/MeV)	Particle Energy /Intensity (MeV / %)	Spin (h/2π)	Nuclear Magnetic Mom.(nm)	Elect. Quadr. Mom. (b)	γ-ray Energy /Intensity (MeV / %)
113I		112.9237	5.9 s	β+ ,EC/7.6					ann.rad./0.4625/100 0.6224/74
114I		113.9219	2.1 s	β+ ,EC/8.7					0.0550-1.422 ann.rad./0.6826 0.7088
115I		114.9188	1.3 m	β+ ,EC/6.7		5/2+			ann.rad./0.275 0.284 0.460 0.709
116I		115.9167	2.9 s	β+ /97 /7.8 EC/3 /	6.7/	1+			ann.rad./0.5402 0.6789
117I		116.9136	2.22 m	β+ ,EC/4.7	3.2/	(5/2+)	3.1		ann.rad./0.2744 0.3259
118mI			8.5 m	β+ ,EC/ I.T.	4.9/	7-	4.2		ann.rad./0.104 0.5998 0.6052 0.6138
118I		117.9134	14. m	β+ ,EC/7.0		2-	2.0		ann.rad./0.5448 0.6052 1.3384
119I		118.9102	19. m	β+ /54 /3.5 EC/46 /	2.4/	(5/2+)	+2.9		ann.rad./Te k x-ray 0.2575
120mI			53. m	β+ /80 / EC/20 /	3.8		4.2		ann.rad. Te k x-ray 0.4257 0.5604 0.6147 1.3459
120I		119.91005	1.35 h	β+ /81 /5.62 EC/19 /	4.03 4.60	2-	1.23		ann.rad./Tek x-ray 0.5604 0.6411 1.5230 (0.43 - 3.1)
121I		120.90737	2.12 h	β+ /13 /2.27 EC/87 /	1.2/	5/2+	2.3		ann.rad./Te k x-ray 0.2122 (0.14 - 1.1)
122I		121.90760	3.6 m	β+ /4.234 EC/	3.1/	1+	0.94		ann.rad./Te k x-ray 0.5641
123I		122.905605	13.2 h	EC/1.242		5/2+	2.82		Te k x-ray 0.1590
124I		123.906211	4.18 d	β+ /23 /3.160 EC/77 /	1.54/ 2.14/ 0.75/	2-	1.44		ann.rad./Te k x-ray 0.6027/62.9 0.7228/10.3 1.6910/11.2 (0.31-1.73)
125I		124.904624	59.4 d	EC/0.1861		5/2+	2.82	-0.89	Te k x-ray 0.0355
126I		125.905619	13.0 d	EC/ β+ /2.155 β- /1.258	1.13/ 0.87/ 1.25/	2-	1.44		ann.rad./Te k x-ray 0.3887 0.6622
127I	100.	126.904468				5/2+	+2.8133	-0.79	
128I		127.905805	25.00 m	β- /2.118 EC/1.251	2.13/	1+			Te k x-ray 0.44287 0.52658
129I		128.904988	1.7x10^7 y	β- /0.194	0.15/	7/2+	+2.621	-0.55	Xe k x-ray 0.0396
130mI			9.0 m	I.T./83 /0.048 β- /17 /		2+			I k x-ray 0.5361
130I		129.906674	12.36 h	β- /2.949	1.04/ 0.62	5+	3.35		0.4180 0.5361 0.6685 0.7395
131I		130.906125	8.040 d	β- /0.971	0.606/	7/2+	+2.742	-0.40	0.08017 0.28431 0.36446 0.63699

Elem. or Isot.	Natural Abundance (%)	Atomic Mass or Weight	Half-Life	Decay Mode/Energy (/MeV)	Particle Energy /Intensity (MeV / %)	Spin (h/2π)	Nuclear Magnetic Mom.(nm)	Elect. Quadr. Mom. (b)	γ-ray Energy /Intensity (MeV / %)
¹³²ᵐI			1.39 h	IT		8-			
¹³²I		131.90800	2.28 h	β⁻ /14 /3.58 I.T./86 /	0.80/ 1.03/ 1.2/ 1.6/ 2.16/	4+	3.09	0.09	I k x-ray 0.0980 0.5059 0.52264 0.63019 0.6506 0.66768 0.77260 0.95457
¹³³ᵐI			9. s	I.T./1.63		19/2-			I kx-ray 0.0730 0.6474 0.9126
¹³³I		132.90781	20.8 h	β⁻ /1.77	1.24/85	7/2+	+2.86	-0.27	0.51056 0.52989 0.87537
¹³⁴ᵐI			3.7 m	I.T./98 /0.316 β⁻ /2 /		8-			I k x-ray 0.0444 0.2719
¹³⁴I		133.9099	52.6 m	β⁻ /4.2	1.2/	4+			0.1354 0.84702 0.88409
¹³⁵I		134.91005	6.57 h	β⁻ /2.65	0.9/ 1.3/	7/2+			0.2884 0.41768 0.52658 1.13156 1.26046
¹³⁶ᵐI			47. s	β⁻ /	4.7/ 5.2/	6-			0.1973 0.3468 0.3701 0.3814 1.3130 (0.16 - 2.36)
¹³⁶I		135.91466	1.39 m	β⁻ /6.93	4.3/ 5.6/	2-			0.3447 1.3130 1.3211 2.2896 (0.3 - 6.1)
¹³⁷I		136.91787	24.5 s	β⁻ /5.88	5.0/	(7/2+)			0.6010 1.2180 1.2201 1.3026 1.5343 (0.25 - 4.4)
¹³⁸I		137.9224	6.5 s	β⁻ /7.8	6.9/ 7.4/	2-			0.4836 0.5888 0.8752 (0.4 - 5.3)
¹³⁹I		138.92609	2.30 s	β⁻ /6.81 n/					0.192 0.198 0.273 0.382 0.386 0.468 0.683 1.313
¹⁴⁰I		139.9310	0.86 s	β⁻ /8.8 n/		(3)			0.372 0.377 0.457
¹⁴¹I		140.9351	0.45 s	β⁻ /7.8					
¹⁴²I		141.9402	≈ 0.2 s	β⁻					
₅₄Xe		131.29(2)							
¹¹⁰Xe		109.9445	0.2 s	β⁺ /9.2					
¹¹¹ᵐXe			0.9 s	EC,β⁺					
¹¹¹Xe		110.9416	0.7 s	EC,β⁺ /10.6 α/	3.58(1)/				
¹¹²Xe		111.9357	3. s	EC,β⁺ /7.2 α/0.8 /					
¹¹³Xe		112.9334	2.8 s	EC,β⁺ /9.1					
¹¹⁴Xe		113.9281	10.0 s	β⁺ ,EC/5.9		0+			ann.rad./ 0.1031 0.1616 0.3085 0.6826 0.7088
¹¹⁵Xe		114.9270	18. s	β⁺ ,EC/7.6		(5/2+)			ann.rad./

Elem. or Isot.	Natural Abundance (%)	Atomic Mass or Weight	Half-Life	Decay Mode/Energy (/MeV)	Particle Energy /Intensity (MeV / %)	Spin (h/2π)	Nuclear Magnetic Mom.(nm)	Elect. Quadr. Mom. (b)	γ-ray Energy /Intensity (MeV / %)
^{116}Xe		115.9214	56. s	β$^+$,EC/4.3	3.3/	0+			ann.rad./ 0.1042 0.1916 0.2477 0.3107 0.4127
^{117}Xe		116.9206	1.02 m	β$^+$,EC/6.5		(5/2+)			ann.rad./ 0.2214 0.5190 0.6389 0.6613
^{118}Xe		117.917	≈ 4. m	β$^+$,EC/3.	2.7/	0+			ann.rad./ 0.0535 0.0600 0.1199
^{119}Xe		118.9156	5.8 m	β$^+$,EC/5.0	3.5/	7/2+	0.6		0.0873 0.1000 0.2318 0.4615
^{120}Xe		119.91216	40. m	β$^+$,EC/97 /1.96 β$^+$ /3 /		0+			I k x-ray 0.0251 0.0726 0.1781 (0.1 - 1.03)
^{121}Xe		120.91138	39. m	β$^+$ /44 /3.73 EC/56 /	2.8/	5/2+	0.65		ann.rad./ I k x-ray 0.1328 0.2527 0.4452 (0.1 - 3.1)
^{122}Xe		121.9086	20.1 h	EC/0.9		0+			I k x-ray 0.3501
^{123}Xe		122.90848	2.00 h	β$^+$ /23 /2.68 EC/77 /	1.51/	1/2+			ann.rad./ I k x-ray 0.1489 0.1781 (0.1 - 2.1)
^{124}Xe	0.09(1)	123.905895							
^{125m}Xe			57. s	I.T./0.252		(9/2-)			Xe k x-ray 0.1111 0.141
^{125}Xe		124.906398	17.1 h	EC/1.653	0.47/	1/2+			I k x-ray 0.1884 0.2434
^{126}Xe	0.09(1)	125.90427				0+			
^{127m}Xe			1.15 m	I.T./0.297		(9/2-)			Xe k x-ray 0.1246 0.1725
^{127}Xe		126.905179	36.4 d	EC/0.662		1/2+	-0.504		I k x-ray 0.1721 0.2029 0.3750
^{128}Xe	1.92(3)	127.903531				0+			
^{129m}Xe			8.89 d	I.T./0.236		11/2-			Xe k x-ray 0.0396 0.1966
^{129}Xe	26.44(24)	128.904780				1/2+	-0.7780		
^{130}Xe	4.08(2)	129.903509				0+			
^{131m}Xe			11.9 d	I.T./0.164		11/2-	+0.6908		Xe k x-ray 0.16398
^{131}Xe	21.18(3)	130.905083				3/2+	+0.69186	-0.12	
^{132}Xe	26.89(6)	131.904155				0+			
^{133m}Xe			2.19 d	I.T./0.233		11/2-	-1.082		Xe k x-ray 0.23325
^{133}Xe		132.905906	5.243 d	β$^-$ /0.427	0.346/99	3/2+	+0.813	+0.14	Cs k x-ray 0.080998 0.1606
^{134}Xe	10.44(10)	133.905395				0+			
^{135m}Xe			15.3 m	I.T./		11/2-	1.103		Xe k x-ray 0.52658
^{135}Xe		134.90721	9.10 h	β$^-$ /1.15	0.91/	3/2+	0.903		0.24975 0.60807
^{136}Xe	8.87(16)	135.90722				0+			
^{137}Xe		136.91156	3.82 m	β$^-$ /4.17	4.1/ 3.6/	7/2-			0.45549 0.8489 0.9822 1.2732 1.7834 2.8498

Elem. or Isot.	Natural Abundance (%)	Atomic Mass or Weight	Half-Life	Decay Mode/Energy (/MeV)	Particle Energy /Intensity (MeV / %)	Spin (h/2π)	Nuclear Magnetic Mom.(nm)	Elect. Quadr. Mom. (b)	γ-ray Energy /Intensity (MeV / %)
¹³⁸Xe		137.91399	14.1 m	β⁻ /2.77	0.8/ 2.4/	0+			0.1538 0.2426 0.2583 0.4345 1.76826 2.0158
¹³⁹Xe		138.91879	39.7 s	β⁻ /5.06	4.5/ 5.0/				0.1750 0.2186 0.2965 (0.1 - 3.37)
¹⁴⁰Xe		139.9216	13.6 s	β⁻ /4.1	2.6	0+			0.0801 0.6220 0.8055 1.4137 (0.04 - 2.3)
¹⁴¹Xe		140.9267	1.72 s	β⁻ /6.2	6.2/	5/2+			0.1187 0.9095 (0.05 - 2.55)
¹⁴²Xe		141.9297	1.22 s	β⁻ /5.0	3.7/ 4.2/	0+			0.0338 0.0729 0.2038 0.3091 0.4145 0.5382 0.5718 0.6181 0.6448
¹⁴³ᵐXe			0.96 s	β⁻					
¹⁴³Xe		142.9352	0.30 s	β⁻ /7.3					
¹⁴⁴Xe		143.9385	1.2 s	β⁻ /6.1					
¹⁴⁵Xe		144.9437	0.9 s	β⁻ ,(n)					
₅₅Cs		132.90545(2)							
¹¹²Cs		111.9503	0.5 ms	p	0.81				
¹¹³Cs		112.9445	17. μs	p	0.96				
¹¹⁴Cs		113.9408	0.58 s	β⁺ ,EC/11.8		1+			ann.rad./ 0.6826 0.7088
¹¹⁵Cs		114.9359	≈ 1.4 s	β⁺ ,EC/8.4					ann.rad./
¹¹⁶ᵐCs			0.7 s	β ,EC/					ann.rad./ 0.3935
¹¹⁶Cs		115.9330	3.8 s	β⁺ ,EC/10.8					ann.rad./ 0.3935 0.5243 0.6151 0.6223
¹¹⁷ᵐCs			6.5 s	β⁺ ,EC/					
¹¹⁷Cs		116.9286	≈ 8.4 s	β⁺ ,EC/7.5					ann.rad./
¹¹⁸ᵐCs			17. s	β⁺ ,EC/					
¹¹⁸Cs		117.92654	14. s	β⁺ ,EC/9.		2			ann.rad./ 0.3372 0.4727 0.5865 0.5906
¹¹⁹ᵐCs			28. s			3/2			
¹¹⁹Cs		118.92234	38. s	β⁺ ,EC/6.3		9/2+	+5.5	+2.8	ann.rad./ 0.169 0.176 0.224 0.257
¹²⁰ᵐCs			60. s	β⁺ ,EC/					
¹²⁰Cs		119.92066	64. s	β⁺ ,EC/7.92		2+	+3.87	+1.45	ann.rad./ 0.3224 0.4735 0.5534 (0.3 - 3.28)
¹²¹ᵐCs			2.0 m	I.T./60 / β⁺ /40 /	4.4	(9/2+)	+5.41	+2.7	ann.rad./ 0.1794 0.1961
¹²¹Cs		120.91718	2.3 m	β⁺ ,EC/5.40	4.38/	3/2+	+0.77	+0.84	ann.rad./ 0.1537 (0.08 - 0.56)
¹²²ᵐ²Cs			4.4 m	β⁺ ,EC		8-	+4.77	+3.3	ann.rad./
¹²²ᵐ¹Cs			0.36 s	IT					0.3311 0.4971 0.6385 (0.27 - 2.22)

Elem. or Isot.	Natural Abundance (%)	Atomic Mass or Weight	Half-Life	Decay Mode/Energy (/MeV)	Particle Energy /Intensity (MeV / %)	Spin (h/2π)	Nuclear Magnetic Mom.(nm)	Elect. Quadr. Mom. (b)	γ-ray Energy /Intensity (MeV / %)
¹²²Cs		121.91614	21. s	β⁺ ,EC/7.1	5.8/	(1+)	-0.133	-0.19	ann.rad./ 0.3311 0.5120 0.8179
¹²³ᵐCs			1.6 s	I.T./		11/2-			Cs k x-ray 0.0946
¹²³Cs		122.91299	5.87 m	β⁺ /75 /4.20 EC/25 /	3.0/	1/2+	+1.38		ann.rad./ Xe k x-ray 0.0974 0.5964
¹²⁴ᵐCs			6.3 s	IT		7+			
¹²⁴Cs		123.91225	30. s	β⁺ /92 /5.92 EC/8 /	≈ 5.	1+	+0.673	-0.74	ann.rad./ Xe k x-ray 0.3539 0.4925 0.9418
¹²⁵Cs		124.90972	45. m	β⁺ /40 /3.09 EC/60 /	2.06/	1/2+	+1.41		ann.rad./ Xe k x-ray 0.112 0.526
¹²⁶Cs		125.90945	1.64 m.	β⁺ /81 /4.83 EC/19 /	3.4 3.7/	1+	+0.78	-0.68	ann.rad./ Xe k x-ray 0.3886 0.4912 0.9252
¹²⁷Cs		126.90741	6.2 h	β⁺ /96 /2.08 EC/4 /	0.65/ 1.06	1/2+	+1.46		Xe k x-ray 0.1247 0.4119
¹²⁸Cs		127.90775	3.62 m	β⁺ /68 /3.930 EC/32 /	2.44/ 2.88/	1+	+0.97	-0.57	ann.rad./ Xe k x-ray 0.4429
¹²⁹Cs		128.90606	1.336 d	EC/1.195		1/2+	+1.49		Xe k x-ray 0.3719 0.4115
¹³⁰ᵐCs			3.5 m	IT,β⁺ ,EC		5-			
¹³⁰Cs		129.90671	29.21 m	β⁺ /55 /2.98 EC/43 / β⁻ /1.6 /0.37	1.98/ 0.44/1.6	1+	+1.46	-0.06	ann.rad./ Xe k x-ray 0.5361
¹³¹Cs		130.90546	9.69 d	EC/0.352		5/2+	+3.54	-0.58	Xe k x-ray
¹³²Cs		131.906430	6.48 d	EC/98 / β⁺ /0.3 /2.120 β⁻ / /1.280		2-	+2.22	+0.51	Xe k x-ray 0.4646 0.6302 0.66769
¹³³Cs	100.	132.905447				7/2+	+2.582	-0.0037	
¹³⁴ᵐCs			2.91 h	I.T./0.139		8-	+1.098	+1.0	Cs k x-ray 0.12749
¹³⁴Cs		133.906714	2.065 y	β⁻ /2.059 EC/1.22	0.089/27 0.658/70	4+	+2.994	+0.39	0.56327 0.56935 0.60473 0.79584
¹³⁵ᵐCs			53. m	I.T./1.627		19/2-	+2.18	+0.9	0.7869 0.8402
¹³⁵Cs		134.905972	2.3x10⁶ y	β⁻ /0.269	0.205/100	7/2+	+2.732	+0.05	
¹³⁶ᵐCs			19. s	I.T./		8			
¹³⁶Cs		135.907307	13.16 d	β⁻ /2.548	0.341/	5+	+3.71	+0.2	0.06691 0.34057 0.81850 1.04807
¹³⁷Cs		136.907085	30.2 y	β⁻ /1.176	0.514/95	7/2+	+2.84	+0.05	Ba k x-ray 0.66164
¹³⁸ᵐCs			2.9 m	I.T./75 /0.080 β⁻ /25 /	3.3	6-	+1.71	-0.40	Cs k x-ray 0.0799 0.1917 0.4628 1.43579
¹³⁸Cs		137.91101	32.2 m	β⁻ /5.37	2.9/	3-	+0.700	+0.12	0.1381 0.46269 1.00969 1.43579 2.21788
¹³⁹Cs		138.913359	9.3 m	β⁻ /4.213	4.21	7/2+	+2.70	-0.07	0.6272 1.2832 (0.4 - 3.66)
¹⁴⁰Cs		139.91727	1.06 m	β⁻ /6.22	5.7/ 6.21/	1-	+0.13390	-0.11	0.5283 0.6023 0.9084 (0.41 - 3.94)

Elem. or Isot.	Natural Abundance (%)	Atomic Mass or Weight	Half-Life	Decay Mode/Energy (/MeV)	Particle Energy /Intensity (MeV / %)	Spin (h/2π)	Nuclear Magnetic Mom.(nm)	Elect. Quadr. Mom. (b)	γ-ray Energy /Intensity (MeV / %)
¹⁴¹Cs		140.92005	24.9 s	β⁻ /5.26	5.20/	7/2+	+2.44	-0.4	Ba k x-ray 0.0485 0.5616 0.5887 1.1940 (0.05 - 3.33)
¹⁴²Cs		141.92430	1.8 s	β⁻ /7.31	6.9/ 7.28				0.3596 0.9668 1.1759 1.3265
¹⁴³Cs		142.92732	1.78 s	β⁻ /6.24	6.1	(3/2+)	+0.87	+0.47	0.1955 0.2324 0.3064 (0.17 - 1.98)
¹⁴⁴Cs		143.93203	1.01 s	β⁻ /8.47	8.46/ 7.9/	1	-0.546	+0.30	0.1993 0.5598 0.6392 0.7587
¹⁴⁵Cs		144.93541	0.59 s	β⁻ /7.89	7.4/ 7.9/	3/2+	+0.784	+0.6	0.1126 0.1755 0.1990
¹⁴⁶Cs		145.94024	0.322 s	β⁻ ,(n)/9.38	≈ 9.0	2-			
¹⁴⁷Cs		146.9439	0.227 s	β⁻ ,(n)/9.3					
¹⁴⁸Cs		147.9490	0.15 s	β⁻ ,(n)/10.5					
₅₆Ba		137.327(7)							
¹¹⁴Ba		113.9509	0.43 s	β⁺,(p)	p/20				
¹¹⁵Ba		114.948	0.45 s	β⁺,(p)	p/ < 15				
¹¹⁶Ba		115.9417	1.3 s	β⁺,(p)	p/3				
¹¹⁷Ba		116.9377	1.8 s	β⁺,(p),EC/8.4	p/13	(3/2-)			(0.0457-0.364)
¹¹⁸Ba		117.9466	5.2 s	β⁺,					(0.040-0.156)
¹¹⁹Ba		118.931	5.4 s	β⁺ ,EC/8.					
¹²⁰Ba		119.9260	24. s	β⁺ ,EC/5.0		0+			ann.rad./ 0.140 (0.075-0.146)
¹²¹Ba		120.9245	30. s	β⁺ ,EC/6.8		5/2	+0.660	+1.8	ann.rad./
¹²²Ba		121.9203	2.0 m	β⁺ ,EC/3.8		0+			ann.rad./
¹²³Ba		122.9189	2.7 m	β⁺ ,EC/5.5			-0.68	+1.5	ann.rad./ 0.0306 0.0927 0.1161 0.1235
¹²⁴Ba		123.91509	12. m	β⁺ ,EC/2.65					ann.rad./ 0.1695 0.1888 1.2160
¹²⁵ᵐBa			8. m	β⁺ ,EC/	4.5				
¹²⁵Ba		124.9146	3.5 m	β⁺ ,EC/4.6	3.4	1/2+	+0.18		ann.rad./ 0.0550 0.0776 0.0854 0.1409
¹²⁶Ba		125.91124	1.65 h	β⁺ /2 /1.67 EC/98 /		0+			Cs k x-ray 0.2179 0.2336 0.2576
¹²⁷Ba		126.9111	12.9 m	β⁺ /54 /3.5 EC/46 /		1/2+	+0.09		ann.rad./ Cs k x-ray 0.1148 0.1808 (0.07 - 2.5)
¹²⁸Ba		127.90831	2.43 d	EC/0.52		0+			Cs k x-ray 0.27344
¹²⁹ᵐBa			2.17 h	EC/98 / β⁺ /2 /		7/2+			Cs k x-ray 0.1769 0.1823 0.2023 1.4593
¹²⁹Ba		128.90868	2.2 h	β⁺ /20 /2.43 EC/80 /	1.42/	1/2+	-0.40		ann.rad./ Cs k x-ray 0.1291 0.2143 0.2208
¹³⁰Ba	0.106(1)	129.90631				0+			
¹³¹ᵐBa			14.6 m	I.T./0.187		9/2-	-0.87	+1.5	Ba k x-ray 0.1085

Elem. or Isot.	Natural Abundance (%)	Atomic Mass or Weight	Half-Life	Decay Mode/Energy (/MeV)	Particle Energy /Intensity (MeV / %)	Spin (h/2π)	Nuclear Magnetic Mom.(nm)	Elect. Quadr. Mom. (b)	γ-ray Energy /Intensity (MeV / %)
^{131}Ba		130.90693	11.7 d	EC/1.37		1/2+	0.7081		Cs k x-ray 0.12381/28.4 0.21608/21.3 0.49636/42.9 (0.0549-1.171)
^{132}Ba	0.101(1)	131.905056				0+			
^{133m}Ba			1.621 d	I.T./0.288		11/2-	-0.91	+0.9	Ba k x-ray 0.2761
^{133}Ba		132.906003	10.53 y	EC/0.517		1/2+	0.7717		Cs k x-ray 0.08099 0.35600
^{134}Ba	2.417(18)	133.904504				0+			
^{135m}Ba			1.20 d	I.T./0.2682		11/2-	-1.00	+1.0	Ba k x-ray 0.2682
^{135}Ba	6.592(12)	134.905684				3/2+	+0.838	+0.16	
^{136m}Ba			0.308 s	I.T./2.0305		7-			Ba k x-ray 0.8185 1.0481
^{136}Ba	7.854(24)	135.904571				0+			
^{137m}Ba			2.552 m	I.T./0.6617		11/2-	-0.99	+0.8	Ba k x-ray 0.66164
^{137}Ba	11.232(24)	136.905822				3/2+	+0.9374	+0.245	
^{138}Ba	71.698(42)	137.905242				0+			
^{139}Ba		138.908836	1.396 h	β⁻/2.317	2.14/27 2.27/72	7/2-	-0.97	-0.57	0.16585 1.2544 1.42033
^{140}Ba		139.91060	12.75 d	β⁻/1.05	0.48 1.0/ 1.02/	0+			0.16268 0.30485 0.53727
^{141}Ba		140.91441	18.3 m	β⁻/3.22	2.59/ 2.73/	3/2-	-0.34	+0.45	0.1903 0.2770 0.3042 (0.1 - 2.5)
^{142}Ba		141.91645	10.7 m	β⁻/2.212	1.0/ 1.10/	0+			0.23152 0.25512 0.3090 1.2040
^{143}Ba		142.92061	14.3 s	β⁻/4.24	4.2/	5/2+	+0.44	-0.88	0.1786 0.21148 0.7988 (0.17 - 2.4)
^{144}Ba		143.92294	11.4 s	β⁻/3.1	2.4/ 2.9/	0+			La k x-ray 0.10386 0.1566 0.1728 0.3882 0.43048
^{145}Ba		144.9269	4.0 s	β⁻/4.9	4.9/	(5/2-)	-0.28	+1.22	La k x-ray 0.0918 0.09709
^{146}Ba		145.9302	2.20 s	β⁻/4.12	3.9/	0+			0.0644 0.2513 0.3270 0.3329 0.3622
^{147}Ba		146.9340	0.892 s	β⁻/5.75	5.5/				
^{148}Ba		147.9377	0.64 s	β⁻,n/5.11					
^{149}Ba		148.9421	0.36 s	β⁻,(n)/7.3					
$_{57}$La		138.9055(2)							
^{120}La		119.938	2.8 s	EC,β⁺/11.					
^{121}La		120.9330	5.3 s						
^{122}La		121.9307	9. s	EC,β⁺/ ≈ 9.7					
^{123}La		122.9262	17. s	EC /7.					
^{124}La		123.9245	30. s	EC/ ≈ 8.8		(7+)			
^{125}La		124.9207	1.2 m	β⁺,EC/5.6		11/2-			ann.rad./ 0.0436 0.0676
^{126}La		125.9194	1.0 m	β⁺,EC/7.6					ann.rad./ 0.2561 0.340 0.4555 0.6214
^{127}La		126.9162	3.8 m	β⁺,EC/4.7		3/2+			ann.rad./ 0.025 0.0562

Elem. or Isot.	Natural Abundance (%)	Atomic Mass or Weight	Half-Life	Decay Mode/Energy (/MeV)	Particle Energy /Intensity (MeV / %)	Spin (h/2π)	Nuclear Magnetic Mom.(nm)	Elect. Quadr. Mom. (b)	γ-ray Energy /Intensity (MeV / %)
^{128}La		127.9155	5.0 m	β^+ /80 /6.7 EC/20 /		(5-)			ann.rad./ Ba k x-ray 0.2841/87 0.4793/54 (0.315-2.212)
^{129m}La			0.56 s	IT		(11/2-)			
^{129}La		128.91267	11.6 m	β^+ /58 /3.72 EC/42 /	2.42/	3/2+			ann.rad./ Ba k x-ray 0.1105 0.2786 (0.1 - 1.8)
^{130}La		129.9123	8.7 m	β^+ /78 /5.6 EC/22 /		3+			ann.rad./ Ba k x-ray 0.3573/81 0.5506/27 (0.1965-1.989)
^{131}La		130.9101	59. m	β^+ /76 /3.0 EC/24 /	1.42/ 1.94/	3/2+			ann.rad./ Ba k x-ray 0.1085 0.3658 0.5263
^{132m}La			24. m	I.T./76 / β^+ ,EC/24 /		6-			La k x-ray 0.1352 0.4645
^{132}La		131.91011	4.8 h	β^+ /40 /4.71 EC/60 /	2.6/ 3.2 3.7/	2-			ann.rad./ Ba k x-ray 0.4645 0.5671
^{133}La		132.9084	3.91 h	β^+ /4 /2.2 EC/96 /	1.2/	5/2+			Ba k x-ray 0.2788 0.2901 0.3024
^{134}La		133.90849	6.5 m	β^+ /63 /3.71 EC/37/	2.67/	1+			ann.rad./ Ba k x-ray 0.6047 (0.5 - 1.9)
^{135}La		134.90697	19.5 h	EC/1.20		5/2+			Ba k x-ray 0.4805
^{136}La		135.9077	9.87 m	β^+ /36 /2.9 EC/64 /	1.8/	1+			ann.rad./ Ba k x-ray 0.8185
^{137}La		136.90647	6x10^4 y	EC/0.60		7/2+			0.2836
^{138}La	0.090(1)	137.907107	1.06x10^{11} y			5+	+2.70	+0.2	1.4358/65 0.7887/35
^{139}La	99.910(1)	138.906349				7/2+			
^{140}La		139.909473	1.678 d	β^- /3.762	1.35 1.24/ 1.67/	3-			
^{141}La		140.910958	3.90 h	β^- /2.502	2.43/	7/2+			
^{142}La		141.91408	1.54 h	β^- /4.505	2.11/ 2.98/ 4.52/	2-			
^{143}La		142.91606	14.1 m	β^- /3.43	3.3/	7/2-			
^{144}La		143.9196	40.7 s	β^- /5.5	4.1/				
^{145}La		144.9217	24. s	β^- /4.1	4.1/	3/2+			
^{146m}La			10.0 s	β^- /6.7	5.5/	(6)			
^{146}La		145.9258	6.3 s	β^- /6.6	6.2/	(2-)			
^{147}La		146.9278	4.02 s	β^- /5.0	4.6/				
^{148}La		147.9322	1.1 s	β^- /7.26		2-			
^{149}La		148.9342	1.10 m	β^- /5.5					
^{150}La		149.9386	0.51 s						x-ray (0.097-0.209)
$_{58}$Ce		140.116(1)							
^{121}Ce		121.944	1.1 s	β^+,p					
^{123}Ce		122.936	3.8 s	β^+ ,EC/ ≈ 8.6					
^{124}Ce		123.931	6. s	EC / ≈5.6					ann.rad./
^{125}Ce		124.929	10. s	β^+ ,EC/7.		(5/2+)			
^{126}Ce		125.9241	50. s	EC/4.					ann.rad./
^{127}Ce		126.9228	32. s	β^+ ,EC/6.1					ann.rad./ (0.058-1.148)
^{128}Ce		127.9189	4.1 m	β^+ ,EC/3.2					ann.rad./ (0.023-0.880)
^{129}Ce		128.9187	3.5 m	β^+ ,EC/5.6					ann.rad./ (0.0675-1.015)
^{130}Ce		129.9147	26. m	β^+ ,EC/2.2		0+			ann.rad./ La k x-ray (0.047-1.431)

Elem. or Isot.	Natural Abundance (%)	Atomic Mass or Weight	Half-Life	Decay Mode/Energy (/MeV)	Particle Energy /Intensity (MeV / %)	Spin (h/2π)	Nuclear Magnetic Mom.(nm)	Elect. Quadr. Mom. (b)	γ-ray Energy /Intensity (MeV / %)
^{131m}Ce			5. m	β^+ ,EC/					ann.rad./ 0.2304 0.3955 0.4213
^{131}Ce		130.9144	10. m	β^+ ,EC/4.0	2.8/				ann.rad./ 0.119 0.169 0.414
^{132}Ce		131.9115	3.5 h	EC/1.3		0+			La k x-ray 0.1554 0.1821
^{133m}Ce			1.6 h	β^+ ,EC/		1/2+			ann.rad./ 0.0769 0.0973 0.5577
^{133}Ce		132.9116	5.4 h	β^+ /8 /2.9 EC/92 /	1.3/	9/2-			ann.rad./ La k x-ray 0.0584 0.1308 0.4722 0.5104
^{134}Ce		133.9090	3.16 d	EC/0.5		0+			La k x-ray 0.1304 0.1623 0.6047
^{135m}Ce			20. s	I.T./0.446		11/2-			Ce k x-ray 0.0826 0.1497 0.2134
^{135}Ce		134.90915	17.7 h	β^+ /1 /2.026 EC/99 /	0.8/	1/2+			La k x-ray 0.0345 0.2656 0.3001 0.6068
^{136}Ce	0.19(1)	135.90714				0+			
^{137m}Ce			1.43 d	I.T./99 /0.254 EC/0.8 /		11/2-	0.70		Ce k x-ray 0.1693 0.2543
^{137}Ce		136.90778	9.0 h	β^+ /1.222		3/2+	0.9		La k x-ray 0.4472
^{138}Ce	0.25(1)	137.90599				0+			
^{139m}Ce			56.4 s	I.T./0.7542		11/2-			Ce k x-ray 0.7542
^{139}Ce		138.90665	137.6 d	EC/0.28		3/2+	0.9		La k x-ray 0.16585
^{140}Ce	88.48(10)	139.905435				0+			
^{141}Ce		140.908272	32.50 d	β^- /0.581	0.436/69 0.581/31	7/2-	1.1		Pr k x-ray 0.14544/48.0
^{142}Ce	11.08(10)	141.909241				0+			
^{143}Ce		142.912382	1.38 d	β^- /1.462	1.404/ 1.110/47	3/2-	≈ 1.		Pr k x-ray 0.0574 0.2933
^{144}Ce		143.913643	284.6 d	β^- /0.319	0.185/20 0.318/	0+			Pr k x-ray 0.0801 0.1335
^{145}Ce		144.91723	3.00 m	β^- /2.54	1.7/24 1.3	3/2-			Pr k x-ray 0.0627 0.7245
^{146}Ce		145.9187	13.5 m	β^- /1.04	0.7/90	0+			Pr k x-ray 0.0986 0.2182 0.3167
^{147}Ce		146.9225	56. s	β^- /3.29	3.3/				0.0930 0.2687
^{148}Ce		147.9244	56. s	β^- /2.1	1.66/	0+			0.0904 0.0985 0.1212 0.2918
^{149}Ce		148.9283	5.2 s	β^- /4.2					0.0577 0.0864 0.3800
^{150}Ce		149.9302	4.4 s	β^- /3.0					0.1099
^{151}Ce		150.9340	1.0 s	β^- /5.3					0.0526
^{152}Ce		151.9366	1.4 s	β^- /4.4					Pr k x-ray 0.098 0.115

Elem. or Isot.	Natural Abundance (%)	Atomic Mass or Weight	Half-Life	Decay Mode/Energy (/MeV)	Particle Energy /Intensity (MeV / %)	Spin (h/2π)	Nuclear Magnetic Mom.(nm)	Elect. Quadr. Mom. (b)	γ-ray Energy /Intensity (MeV / %)
₅₉Pr		140.90765(2)							
¹²⁴Pr		123.943	1.2 s	β⁺ ,EC/12.					ann.rad./
¹²⁵Pr		124.9378	≈ 3.3 s	β⁺					ann.rad./ 0.1358
¹²⁶Pr		125.9353	3.1 s	β⁺ ,EC/ ≈10.4					ann.rad./ (0.170-0.985)
¹²⁷Pr		126.9308	4.2 s	β⁺ / ≈7.5					ann.rad./ (0.0288-0.8949)
¹²⁸Pr		127.9288	3.2 s	β⁺ ,EC/ ≈9.3					ann.rad./
¹²⁹Pr		128.9249	32 s	β⁺ ,EC/5.8					ann.rad./ (0.0395-1.865)
¹³⁰Pr		129.9234	40. s	β⁺ ,EC/8.1					ann.rad./
¹³¹ᵐPr			5.7 s						(0.06 - 0.16)
¹³¹Pr		130.9201	1.7 m	β⁺ ,EC/5.3				≈5.5	ann.rad./ (0.059-0.980)
¹³²Pr		131.9191	1.6 m	β⁺ ,EC/7.1					ann.rad./ 0.325 0.496 0.533
¹³³Pr		132.9162	6.5 m	β⁺ ,EC/4.3		5/2+			ann.rad./ 0.074 0.1343 0.2419 0.3156 0.3308 0.4650
¹³⁴ᵐPr			≈ 11. m	β⁺ ,EC/					ann.rad./ 0.294 0.460 0.495 0.632
¹³⁴Pr		133.9157	17. m	β⁺ ,EC/6.2		2+			ann.rad./ 0.294 0.495
¹³⁵Pr		134.9131	24. m	β⁺ ,EC/3.7	2.5/	3/2+			ann.rad./ 0.0826 0.2135 0.2961 0.5832
¹³⁶Pr		135.91265	13.1 m	β⁺ /57 /5.13 EC/43	2.98/	2+			ann.rad./ Ce k x-ray 0.5398 0.5522
¹³⁷Pr		136.91068	1.28 h	β⁺ /26 /2.70 EC/74 /	1.68/	5/2+			ann.rad./ Ce k x-ray 0.4339 0.5140 0.8367 (0.16 - 1.8)
¹³⁸ᵐPr			2.1 h	β⁺ /24 / EC/76 /	1.65/	7-			ann.rad./ Ce k x-ray 0.3027 0.7887 1.0378 (0.07 - 2.0)
¹³⁸Pr		137.91075	1.45 m	β⁺ /75 /4.44 EC/25 /	3.42/	1+			ann.rad./ Ce k x-ray 0.7887
¹³⁹Pr		138.90893	4.41 h	β⁺ /8 /2.129 EC/92 /	1.09/	5/2+			ann.rad./ Ce k x-ray 0.2551 1.3473 1.6307
¹⁴⁰Pr		139.90907	3.39 m	β⁺ /51 /3.39 EC/49 /	2.37/	1+			ann.rad./ Ce k x-ray 0.3069 1.5965
¹⁴¹Pr	100.	140.907648				5/2+	+4.275	-0.059	
¹⁴²ᵐPr			14.6 m	I.T./0.004	c.e./	5-	2.2		
¹⁴²Pr		141.910041	19.12 h	β⁻ /2.162 EC/0.744	0.58/4 2.16/96	2-	+0.234	+0.030	0.5088 1.57580
¹⁴³Pr		142.910813	13.57 d	β⁻ /0.934	0.933/	7/2+	+2.7	+0.8	0.7420
¹⁴⁴Pr			7.2 m	I.T./99+ /0.059 β⁻ /		3-			Pr k x-ray 0.0590 0.6965 0.8142

Elem. or Isot.	Natural Abundance (%)	Atomic Mass or Weight	Half-Life	Decay Mode/Energy (/MeV)	Particle Energy /Intensity (MeV / %)	Spin (h/2π)	Nuclear Magnetic Mom.(nm)	Elect. Quadr. Mom. (b)	γ-ray Energy /Intensity (MeV / %)
^{144}Pr		143.913301	17.28 m	β⁻ /2.998	0.807/1 2.30/ 2.996/98	0-			0.69649 1.48912 2.18562
^{145}Pr		144.91451	5.98 h	β⁻ /1.81	1.80/97	7/2+			0.0725 0.6758 0.7483
^{146}Pr		145.9176	24.2 m	β⁻ /4.2	2.2/30 3.7/10 4.2/40	2-			0.4539/48 1.5247
^{147}Pr		146.91898	13.4 m	β⁻ /2.69	1.5/ 2.1/	3/2+			0.3146/24. 0.5779/16 0.6413/19.
^{148m}Pr			2.0 m	β⁻ /	4.0/ 3.8/	(4)			0.3016 0.4506 0.6975
^{148}Pr		147.9222	2.27 m	β⁻ /4.9	4.8/ 4.5/	1-			0.3017
^{149}Pr		148.92379	2.3 m	β⁻ /3.40	3.0	(5/2+)			0.1085 0.1385 0.1651
^{150}Pr		149.9270	6.2 s	β⁻ /5.7	≈ 5.5	1-			0.1302 0.8044 0.8527
^{151}Pr		150.9283	22.4 s	β⁻ /4.2					0.0726
^{152}Pr		151.9319	3.2 s	β⁻ /6.7		(3)			0.164 0.285
^{153}Pr		152.9339	4.3 s	β⁻ /5.5					
^{154}Pr		153.9381	2.3 s	β⁻ /7.9					
$_{60}$Nd		144.24(3)							
^{127}Nd		126.941	1.8 s	β⁺ ,EC/9.		(5/2)			ann.rad./
^{128}Nd		127.935	4. s	β⁺ ,EC/6.					ann.rad./
^{129}Nd		128.933	4.9 s	β⁺ ,EC/8.		5/2(-)			ann.rad./ (0.091-0.875)
^{130}Nd		129.929	28. s	β⁺ ,EC/5.					ann.rad./
^{131}Nd		130.9271	0.5 m	β⁺ ,EC/6.6					ann.rad./ (0.09-0.36)
^{132}Nd		131.9231	1.5 m	β⁺ ,EC/3.7					ann.rad./ (0.099-0.567)
^{133}Nd		132.9222	1.2 m	β⁺ ,EC/5.6					ann.rad./ (0.06-0.37)
^{134}Nd		133.9187	≈ 8.5 m	β⁺ /17 /2.8 EC/83 /		0+			ann.rad./ Pr k x-ray 0.1631/58 (0.09-1.00)
^{135m}Nd			5.5 m	β⁺ /					
^{135}Nd		134.9182	12. m	β⁺ /65 /4.8 EC/35 /		9/2-	0.78	+2.1	ann.rad./ Pr k x-ray 0.0415/23. 0.204/51. (0.11-1.8)
^{136}Nd		135.9150	50.6 m	EC/94 /2.21 β⁺ /6 /	1.04/	0+			Pr kx-ray 0.0401/21. 0.1091/35. (0.10-0.97)
^{137m}Nd			1.6 s	I.T./0.5196		11/2-			Nd k x-ray 0.1084 0.1775 0.2337
^{137}Nd		136.9146	38. m	β⁺ /40 /3.69 EC/60 /	1.7/20 2.40/20	1/2+	-0.63		ann.rad./ Pr k x-ray 0.0755 0.5806
^{138}Nd		137.9119	5.1 h	EC/1.1		0+			Pr k x-ray 0.1995 0.3258
^{139m}Nd			5.5 h	I.T./12 /0.231 β⁺ /88 /	1.17/	11/2-			Nd k x-ray Pr k x-ray 0.1139/34. 0.7382/30.
^{139}Nd		138.91192	30. m	β⁺ /25 /2.79 EC/75 /	1.77/	3/2+	0.91	+0.3	ann.rad./ Pr k x-ray 0.4050
^{140}Nd		139.90931	3.37 d	EC /0.22		0+			Pr k x-ray Nd k x-ray
^{141m}Nd			1.04 m	I.T./99+ /0.756		11/2-			0.7565

Elem. or Isot.	Natural Abundance (%)	Atomic Mass or Weight	Half-Life	Decay Mode/Energy (/MeV)	Particle Energy /Intensity (MeV / %)	Spin (h/2π)	Nuclear Magnetic Mom.(nm)	Elect. Quadr. Mom. (b)	γ-ray Energy /Intensity (MeV / %)
141Nd		140.909605	2.49 h	EC/98 /1.823 β+ /2 /	0.802/	3/2+	1.01	+0.3	Pr k x-ray (0.15-1.7)
142Nd	27.13(12)	141.907719				0+			
143Nd	12.18(6)	142.909810				7/2-	-1.07	-0.6	
144Nd	23.80(12)	143.910083	2.1x10^15 y			0+			
145Nd	8.30(6)	144.912569				7/2-	-0.66	-0.33	
146Nd	17.19(9)	145.913113				0+			
147Nd		146.916096	10.98 d	β- /0.896	0.805/	5/2-	0.58	0.9	Pr k x-ray 0.09111 0.53102
148Nd	5.76(3)	147.916889				0+			
149Nd		148.920145	1.73 h	β- /1.691	1.03/25 1.13/26 1.42/	5/2-	0.35	1.3	Pr k x-ray 0.1143/19. 0.2113/27. (0.06 - 1.6)
150Nd	5.64(3)	149.920887				0+			
151Nd		150.923825	12.4 m	β- /2.442	1.2/	(3/2+)			Pm k x-ray 0.1168 0.2557 1.1806 (0.10 - 1.9)m
152Nd		151.92468	11.4 m	β- /1.1		0+			0.2785/29. 0.2501/18. (0.016 - 0.66)
153Nd		152.9280	28.9 s	β- /3.6					0.418
154Nd		153.9296	25.9 s	β- /2.8					0.1519 0.7998
155Nd		154.9334	8.9 s	β- /5.0					0.1807
156Nd		155.9355	5.5 s	β- /4.1					0.0848
61 Pm									
130Pm		129.940	2.2 s	β+ ,EC/11.					ann.rad./
132Pm		131.934	6. s	β+ ,EC/10.					ann.rad./
133Pm		132.930	12. s	β+ ,EC/ ≈ 7.0					ann.rad./
134Pm		133.9282	24. s	β+ ,EC/ ≈ 8.9		(5+)			ann.rad./ 0.294 0.495 (0.13-0.47)
135Pm		134.9247	0.8 m	β+ ,EC/6.0		11/2-			ann.rad./ Nd k x-ray 0.3735 0.6027
136Pm		135.9235	1.8 m	β+ /89 /7.9 EC/11 /		(3+)			ann.rad./ 0.1086 0.1775
137Pm		136.9206	2.4 m	β+ ,EC/5.6		(11/2-)			ann.rad./ Nd k x-ray 0.5209 0.7290
138mPm			3.2 m	β+ /50 / ≈ 7.0 EC/50 /	3.9/	3+			ann.rad./ 0.1887
138Pm		137.9193	10. s	β+ /6.9	6.1/	1+			ann.rad./ Nd k x-ray 0.4028 (0.27 - 2.4)
139mPm			0.18 s	IT/		(11/2-)			
139Pm		138.91678	4.14 m	β+ /68 /4.52 EC/32 /	3.52/	(5/2+)			ann.rad./ Nd k x-ray 0.4199 0.7738 1.0283
140mPm			5.87 m	β+ /70 / EC/30 /	3.2	7/2-			ann.rad./ Nd k x-ray 0.7738 1.4898
140Pm		139.91585	9.2 s	β+ /89 /6.09 EC/11 /	5.07/74	1+			ann.rad./ Nd k x-ray 0.8862 1.2233
141Pm		140.91359	20.9 m	β+ /52 /3.72 EC/48 /	2.71	5/2+			ann.rad./ Nd k x-ray 0.6414 1.5758
142Pm		141.91295	40.5 s	β+ /86 /4.87 EC/20 /	3.8/	1+			Nd k x-ray 0.7420
143Pm		142.910928	265. d	EC/1.041 β+ / <6 x 10^4/		5/2+	3.8		Nd k x-ray 0.6180 0.6965
144Pm		143.912586	360. d	EC/2.332 β+ / <8 x 10^5/		5-	1.7		

Elem. or Isot.	Natural Abundance (%)	Atomic Mass or Weight	Half-Life	Decay Mode/Energy (/MeV)	Particle Energy /Intensity (MeV / %)	Spin (h/2π)	Nuclear Magnetic Mom.(nm)	Elect. Quadr. Mom. (b)	γ-ray Energy /Intensity (MeV / %)
¹⁴⁵Pm		144.912745	17.7 y	EC/0.163		5/2+			Nd k x-ray 0.0723
¹⁴⁶Pm		145.914693	5.53 y	EC/63 /1.472 β⁻ /37 /1.542	0.795/	3-			Nd k x-ray 0.4538 0.7362 0.7474
¹⁴⁷Pm		146.915134	2.6234 y	β⁻ /0.224	0.224/	7/2+	+2.6	+0.7	0.1213 0.1974
¹⁴⁸ᵐPm			41.3 d	β⁻ /95 /2.6 I.T./5 /0.137	0.4/60 0.5/17 0.7/21	6-	1.8		0.5503/94. 0.6300/89. 0.7257/33
¹⁴⁸Pm		147.91747	5.37 d	β⁻ /2.47	1.02/ 2.47/	1-	+2.0	+0.2	0.5503 0.9149 1.4651
¹⁴⁹Pm		148.918330	2.212 d	β⁻ /1.071	0.78/9 1.072/90	7/2+	3.3		0.2859 0.5909 0.8594
¹⁵⁰Pm		149.92098	2.68 h	β⁻ /3.45	1.6/ 2.3/ 1.8/	(1-)			0.3339/69. 1.1658/16. 1.3245/17. (0.25 - 2.9)
¹⁵¹Pm		150.92120	1.183 d	β⁻ /1.187	0.84/	5/2+	+1.8	1.9	0.1677/8 0.2751/7 0.3401/22 (0.14-1.4)
¹⁵²ᵐ²Pm			15. m	β⁻ ,I.T./		(>6)			
¹⁵²ᵐ¹Pm			7.5 m	β⁻ /		(4-)			0.1218 0.2447 0.3404 1.0971 1.4375
¹⁵²Pm		151.9235	4.1 m	β⁻ /3.5	3.5/20 3.50/60	1+			0.1218 (0.12 - 2.1)
¹⁵³Pm		152.92414	5.4 m	β⁻ /1.90	1.7/	(5/2-)			0.0910 0.1198 0.1273
¹⁵⁴ᵐPm			2.7 m	β⁻ /	2.0/				0.0820 0.1848 1.4403
¹⁵⁴Pm		153.9266	1.7 m	β⁻ /4.1	1.9/				0.0820 0.8396 1.3940 2.0589 (0.08 - 2.8)
¹⁵⁵Pm		154.9280	48. s	β⁻ /3.2		(5/2-)			(0.05-0.78)
¹⁵⁶Pm		155.93106	26.7 s	β⁻ /5.16					
¹⁵⁷Pm		156.9332	10.9 s	β⁻ /4.6					
¹⁵⁸Pm		157.9367	5. s	β⁻ /6.3					
₆₂Sm		150.36(3)							
¹³¹Sm		130.946	1.2 s	β⁺ ,EC/					ann.rad./
¹³²Sm		131.941	4.0 s	β⁺					
¹³³Sm		132.939	2.9 s	β⁺ ,EC/ ≈ 8.4		5/2+			ann.rad./
¹³⁴Sm		133.934	11. s	β⁺ ,EC/5.		0+			ann.rad./
¹³⁵Sm		134.932	10. s	β⁺ ,EC/7.		7/2+			ann.rad./
¹³⁶Sm		135.9283	42. s	β⁺ ,EC/4.5		0+			ann.rad./
¹³⁷Sm		136.9271	45. s	β⁺ ,EC/6.1					ann.rad./
¹³⁸Sm		137.9235	3.0 m	β⁺ ,EC/3.9		0+			ann.rad./ 0.0536 0.0747
¹³⁹ᵐSm			10. s	I.T./94 /0.457 β⁺ /6 /	4.7	(11/2-)			Sm k x-ray 0.1118 0.1553 0.1901 0.2673
¹³⁹Sm		138.9226	2.6 m	β⁺ /75 /5.5 EC/25 /	4.1/	1/2+	-0.53		Pm k x-ray 0.3678 0.4028 (0.27 - 2.4)
¹⁴⁰Sm		139.9195	14.8 m	β⁺ ,EC/3.4	1.9/	0+			ann.rad./ Pm k x-ray 0.1396 0.2255 (0.07 - 1.7)
¹⁴¹ᵐSm			22.6 m	β⁺ /32 / EC/68 / I.T./0.3 /0.1758	1.6/ 2.19/	11/2-	-0.83	+1.6	ann.rad./ Pm k x-ray 0.1966 0.4318 0.7774

Elem. or Isot.	Natural Abundance (%)	Atomic Mass or Weight	Half-Life	Decay Mode/Energy (/MeV)	Particle Energy /Intensity (MeV / %)	Spin (h/2π)	Nuclear Magnetic Mom.(nm)	Elect. Quadr. Mom. (b)	γ-ray Energy /Intensity (MeV / %)
¹⁴¹Sm		140.91847	10.2 m	β⁺ /52 /4.54 EC/48 /	3.2/	1/2+	-0.74		ann.rad./ Pm k x-ray 0.4382
¹⁴²Sm		141.91520	1.208 h	β⁺ /6 /2.10 EC/94 /	1.0/	0+			ann.rad./ Pm k x-ray 0.7540
¹⁴³ᵐSm			1.10 m	I.T./99 /0.7540		11/2-			Sm k x-ray 0.7540
¹⁴³Sm		142.914624	8.83 m	β⁺ /46 /3.443 EC/54 /	2.47/	3/2+	+1.01	+0.4	ann.rad./ Pm k x-ray 1.0565
¹⁴⁴Sm	3.1(1)	143.911996				0+			
¹⁴⁵Sm		144.913407	340. d	EC/0.617		7/2-	-1.1	-0.6	Pm k x-ray 0.0613 0.4924
¹⁴⁶Sm		145.913038	1.03x10⁸y	α/	2.50/	0+			
¹⁴⁷Sm	15.0(2)	146.914894	1.06x10¹¹y	α/	2.23/	7/2-	-0.815	-0.26	
¹⁴⁸Sm	11.3(1)	147.914818	7x10¹⁵y	α/	1.96/	0+			
¹⁴⁹Sm	13.8(1)	148.917180	10¹⁶y	α/		7/2-	-0.672	+0.075	
¹⁵⁰Sm	7.4(1)	149.917272				0+			
¹⁵¹Sm		150.919929	90. y	β⁻ /0.0768	0.076/	5/2-	-0.363	+0.7	0.02154
¹⁵²Sm	26.7(2)	151.919729				0+			
¹⁵³Sm		152.922094	1.929 d	β⁻ /0.808	0.64/ 0.69/	3/2+	-0.0216	+1.3	Eu k x-ray 0.069676 0.10318
¹⁵⁴Sm	22.7(2)	153.922206				0+			
¹⁵⁵Sm		154.924636	22.2 m	β⁻ /1.627	1.52	3/2-		1.1	Eu k x-ray 0.1043/75.
¹⁵⁶Sm		155.92553	9.4 h	β⁻ /0.72	0.43/ 0.71/	0+			0.0872 0.1657 0.2038
¹⁵⁷Sm		156.9283	8.0 m	β⁻ /2.7	2.4/	3/2-			Eu k x-ray 0.1964 0.1978 0.3942
¹⁵⁸Sm		157.9299	5.5 m	β⁻ /2.0		0+			0.1894/100. 0.3636/82.
¹⁵⁹Sm		158.9332	11.3 s	β⁻ /3.8					0.1898
¹⁶⁰Sm		159.9353	9.6 s	β⁻ /3.6		0+			0.110
₆₃Eu		151.964(1)							
¹³⁴Eu		133.946	0.5 s	EC,β⁺					ann.rad./
¹³⁵Eu		134.942	1.5 s	EC,β⁺ / ≈8.7					ann.rad./
¹³⁶ᵐEu			≈ 3.2 s			7+			0.255
¹³⁶Eu		135.940	≈ 3.9 s	EC,β⁺ /10.		1+			ann.rad./
¹³⁷Eu		136.935	11. s	EC/ ≈7.5		11/2-			ann.rad./
¹³⁸Eu		137.9335	12. s	EC,β⁺ / ≈ 9.2		7+			ann.rad./
¹³⁹Eu		138.9298	18. s	EC,β⁺ /6.7					ann.rad./
¹⁴⁰ᵐEu			0.125 s	EC,β⁺					ann.rad./
¹⁴⁰Eu		139.9285	1.51 s	EC,β⁺ /8.4		1-			ann.rad./
¹⁴¹ᵐEu			3.0 s	β⁺ /58 / EC/9 / I.T./33 /0.0964		11/2-			ann.rad./ Eu k x-ray (0.09 - 1.6)
¹⁴¹Eu		140.9244	40. s	β⁺/81 /5.6 EC/15 /		5/2+	+3.49	+0.85	ann.rad./ Sm k x-ray 0.3845 0.3940
¹⁴²ᵐEu			1.22 m	β⁺/83 / EC/17 /	4.8/	8-	+2.98	+1.4	ann.rad./ Sm k x-ray 0.5566 0.7680 1.0233
¹⁴²Eu		141.9231	2.4 s	β⁻ /94/7.4 EC/6 /	7.0/	1+	+1.54	+0.12	ann.rad./ 0.7680
¹⁴³Eu		142.92017	2.62 m	β⁺ /72/5.17 EC/28/	4.1/ 5.1/	5/2+	+3.67	+0.51	ann.rad./ Sm k x-ray 0.1107/7 1.5368/3. 1.9127/2.
¹⁴⁴Eu		143.91879	10.2 s	β⁺ /86 /6.33 EC/13 /	5.31/	1+	+1.89	+0.10	ann.rad./ Sm k x-ray 1.6601
¹⁴⁵Eu		144.916263	5.93 d	β⁺ /2 /2.660 EC/98 /1.71	0.79/	5/2+	+3.99	+0.29	ann.rad./ Sm k x-ray 0.6535 0.8937 1.6587

Elem. or Isot.	Natural Abundance (%)	Atomic Mass or Weight	Half-Life	Decay Mode/Energy (/MeV)	Particle Energy /Intensity (MeV / %)	Spin (h/2π)	Nuclear Magnetic Mom.(nm)	Elect. Quadr. Mom. (b)	γ-ray Energy /Intensity (MeV / %)
¹⁴⁶Eu		145.91720	4.57 d	β⁺ /5 /3.88 EC/95 /	1.47/	4-	+1.43	-0.18	ann.rad./ Sm k x-ray 0.6336 0.6341 0.7470 (0.27 - 2.64)
¹⁴⁷Eu		146.916742	24.4 d	EC/99. /1.722 β⁺ /0.4 /		5/2+	+3.72	+0.55	Sm k x-ray 0.12113 0.19725 0.6776
¹⁴⁸Eu		147.91815	54.5 d	EC/3.11	0.92	5-	+2.34	+0.35	Sm k x-ray 0.5503/99. 0.6299/71. (0.067-2.17)
¹⁴⁹Eu		148.91792	93.1 d	EC/0.692		5/2+	+3.57	+0.75	Sm k x-ray 0.2770 0.3275
¹⁵⁰Eu		149.91970	36. y	EC/2.26		5-	+2.71	+1.13	Sm k x-ray 0.3340 0.4394 0.5843 (0.25 - 1.8)
¹⁵⁰ᵐEu			12.8 h	β⁻ /92 / β⁺ /0.4 / EC/8 /	1.013/ 1.24/	0-			Sm k x-ray 0.3339 0.4065
¹⁵¹Eu	47.8(1.5)	150.919846				5/2+	+3.472	+0.90	
¹⁵²ᵐ²Eu			1.60 h	I.T./0.1478		8-			Eu k x-ray 0.0898
¹⁵²ᵐ¹Eu			9.30 h	β⁻ /72 / EC/28 /	1.85/ 0.89/	0-			Sm k x-ray 0.12178 0.84153 0.96334
¹⁵²Eu		151.921741	13.5 y	EC/72 /1.874 β⁻ /28 /1.818	0.69/ 1.47/	3-	-1.941	+2.71	Sm k x-ray Gd k x-ray 0.12178 0.34427 1.40802 (0.252-1.528)
¹⁵³Eu	52.2(15)	152.921227				5/2+	+1.533	+2.41	
¹⁵⁴ᵐEu			46.1 m	I.T./≈ 0.16		8-			Eu k x-ray 0.0682 0.1009
¹⁵⁴Eu		153.922976	8.59 y	β⁻ /99.9 /1.969 EC./0.02 /0.717	0.27/29 0.58/38 0.84/17 0.98/4 1.87/11	3-	-2.01	+2.8	Gd k x-ray 0.12299/40. 0.72331/20. 1.2745/36 (0.059-1.90)
¹⁵⁵Eu		154.922890	4.76 y	β⁻ /0.252	0.15/	5/2+	+1.6	+2.3	Gd k x-ray 0.0865/30 0.1053/20
¹⁵⁶Eu		155.92475	15.2 d	β⁻ /2.451	0.30/11 0.49/30 1.2/12 2.45/31	1+	≈ 1.1		0.08899/9. 0.64623/7. 0.723441/6. 0.8118/10.
¹⁵⁷Eu		156.92542	15.13 h	β⁻ /1.36	0.98/ 1.30/41	(5/2+)			Gd k x-ray 0.0639/100. 0.3705/48. 0.4107/76.
¹⁵⁸Eu		157.9278	45.9 m	β⁻ /3.5	2.5/	(1-)			0.0795 0.8976 0.9442 0.9771
¹⁵⁹Eu		158.92909	18.1 m	β⁻ /2.51	2.4/ 2.57/	(5/2+)			0.0678 0.0786 0.0957
¹⁶⁰Eu		159.9315	38. s	β⁻ /4.1	2.7/ 4.1/	(0-)			0.0753 0.1735 0.4131 0.5155 0.8217 0.9110 0.9246
¹⁶¹Eu		160.9337	27. s	β⁻ /3.7					0.0719
¹⁶²Eu		161.9370	11. s	β⁻ /5.6					
₆₄Gd		157.25(3)							
¹³⁵Gd			1.1 s	β⁺					(0.163-0.360) ann.rad./
¹³⁷Gd		136.945	7. s	EC,β⁺ / ≈ 8.8					

Elem. or Isot.	Natural Abundance (%)	Atomic Mass or Weight	Half-Life	Decay Mode/Energy (/MeV)	Particle Energy /Intensity (MeV / %)	Spin (h/2π)	Nuclear Magnetic Mom.(nm)	Elect. Quadr. Mom. (b)	γ-ray Energy /Intensity (MeV / %)
^{139}Gd		138.9381	5. s	EC,β^+ / ≈7.7					
^{140}Gd		139.934	16. s	EC/4.8		0+		ann.rad./	
^{141m}Gd			25. s	EC,β^+ /		11/2-			0.1748
^{141}Gd		140.9322	21. s	β^+ /7.3		0+			ann.rad./
^{142}Gd		141.9276	1.17 m	EC,β^+ /4.2		1/2+			ann.rad./
^{143m}Gd			1.84 m	β^+ /67 / EC/33 / I.T./		11/2-			ann.rad./ Eu k x-ray 0.1176 0.2719 0.5880 0.6681 0.7999
^{143}Gd		142.9266	39. s	β^+ /82 /6.0 EC/18 /		1/2+			ann.rad./ Eu k x-ray 0.2048 0.2588
^{144}Gd		143.9234	4.5 m	β^+ /45 /4.3 EC/55 /	3.3/	0+			ann.rad./ Eu k x-ray 0.3332
^{145m}Gd			1.44 m	I.T./95 /0.749 β^+ /4 /5.7		11/2-			0.0273 0.3295 0.3866 0.7214
^{145}Gd		144.92169	23.4 m	β^+ /33 /5.05 EC/67 /	2.5/	1/2+			ann.rad./ Eu k x-ray 1.7579 1.8806 (0.32 - 3.69)
^{146}Gd		145.91831	48.3 d	EC/99.9 /1.03 β^+ /0.2	0.35/	0+			Eu k x-ray 0.1147 0.1155 0.1546
^{147}Gd		146.919090	1.588 d	EC/99.8 /2.188 EC/0.2 /	0.93/	7/2-	1.0		Eu k x-ray 0.2293 0.3699 0.3960 0.9289 (0.1 - 1.8)
^{148}Gd		147.918111	75. y	α/3.27	3.1828/	0+			
^{149}Gd		148.919339	9.3 d	EC/1.32		7/2-	0.9		Eu k x-ray 0.1496 0.2985 0.3465
^{150}Gd		149.91866	1.8x10⁶ y	α/2.80	2.73/	0+			
^{151}Gd		150.920345	124. d	EC/0.464		7/2-	0.8		Eu k x-ray 0.1536 0.2432
^{152}Gd	0.20(1)	151.919789				0+			
^{153}Gd		152.921747	241.6 d	EC/0.485		3/2-	0.4		Eu k x-ray 0.09743 0.10318
^{154}Gd	2.18(3)	153.920862				0+			
^{155}Gd	14.80(5)	154.922619				3/2-	-2.59	+1.30	
^{156}Gd	20.47(4)	155.922120				0+			
^{157}Gd	15.65(3)	156.923957				3/2-	-3.40	+1.36	
^{158}Gd	24.84(12)	157.924101				0+			
^{159}Gd		158.926385	18.6 h	β^- 0.971	0.60/11 0.89/26 0.96/63	3/2-	-0.44		Tb k x-ray 0.36351 (0.05845-0.8550)
^{160}Gd	21.86(4)	159.927051				0+			
^{161}Gd		160.929666	3.66 m	β^- /1.956	1.56/85	5/2-			Tb k x-ray 0.1023 0.3149 0.3609 0.4030 0.4421
^{162}Gd		161.930981	8.4 m	β^- /1.39	1.0/	0+			
^{163}Gd		162.9340	1.13 m	β^- /3.1					0.2868 0.214 1.685
^{164}Gd		163.9359	45. s	β^- /2.3					
$_{65}$Tb		158.92534(2)							
^{140}Tb		139.946	2.4 s	β^+ ,EC/11					
^{141}Tb		140.941	3.5 s	β^+ ,EC/ ≈ 8.3					
^{142m}Tb			0.30 s	β^+ ,EC/		4-			
^{142}Tb		141.939	0.60 s	β^+ ,EC/10.		0+			
^{143}Tb		142.9346	12. s	β^+ ,EC/7.4		11/2-			
^{144m}Tb			4.1 s	IT		5-			

Elem. or Isot.	Natural Abundance (%)	Atomic Mass or Weight	Half-Life	Decay Mode/Energy (/MeV)	Particle Energy /Intensity (MeV / %)	Spin (h/2π)	Nuclear Magnetic Mom.(nm)	Elect. Quadr. Mom. (b)	γ-ray Energy /Intensity (MeV / %)
^{144}Tb		143.9324	< 1.5 s	β^+ ,EC/8.4		1+			
^{145m}Tb			30. s	β^+ ,EC/ ≈ 6.6		11/2-			ann.rad./ 0.2577 0.5370 0.9876
^{145}Tb		144.9287		β^+ ,EC/6.5		1/2+			
^{146m}Tb			23. s	β^+ /76 / EC/24 /		(5-)			ann.rad./ Gd k x-ray 1.0789 1.5795
^{146}Tb		145.9270	≈ 8. s	β^+ /8.1		1+			
^{147m}Tb			1.8 m	β^+ /35 / EC/65 /		11/2-			ann.rad./ Gd k x-ray 1.3977 1.7978
^{147}Tb		146.92404	1.6 h	β^+ /42 /4.61 EC/58 /		5/2+			ann.rad./ Gd k x-ray 0.6944 1.1522 (0.120-3.318)
^{148m}Tb			2.3 m	β^+ /25 / EC/75 /		9+			ann.rad./ Gd k x-ray 0.3945 0.6319 0.7845 0.8824
^{148}Tb		147.92422	1.00 h	β^+ ,EC/5.69		2-			ann.rad./ Gd k x-ray 0.4888 0.7845 (0.14 - 3.8)
^{149m}Tb			4.16 m	EC/88 / β^+ /12 /		11/2-			ann.rad./ Gd k x-ray 0.1650 0.7960
^{149}Tb		148.923243	4.13 h	β^+ /4 /3.636 α/16/	1.8/ 3.97/	1/2+		<0.9	Gd k x-ray 0.1650 0.3522 0.3886 (0.1 - 3.2)
^{150m}Tb			6.0 m	β^+ /17 / EC/83 /					ann.rad./ Gd k x-ray 0.4384 0.6380 0.6504 0.8275
^{150}Tb		149.92366	3.3 h	β^+ ,EC/4.66		2-			ann.rad./ 0.4963 0.6380 (0.3 - 4.29)
^{151m}Tb			25. s	I.T./95 / β^+ ,EC/7 /		11/2-			0.0229 0.0495 0.3797 0.8305
^{151}Tb		150.923099	17.61 h	β^+ /1 /2.565 EC/99 /	0.70/	1/2+			Gd k x-ray 0.1083 0.2517 0.2870 (0.1 - 1.8)
^{152m}Tb			4.3 m	I.T./79 /0.5018 EC/21 /4.35		(8+)			Tb k x-ray Gd k x-ray 0.2833 0.3443 0.4111
^{152}Tb		151.92407	17.5 h	β^+ /20 /3.99 EC/80 /	2.5/ 2.8/	2-	0.9		ann.rad./ Gd k x-ray 0.3443 (0.2 - 2.88)
^{153}Tb		152.923433	2.34 d	EC/1.570		5/2+	3.5		Gd k x-ray 0.2119 (0.05 - 1.1)
^{154m2}Tb			23.1 h	EC/98 / I.T./2 /		(7-)			Gd k x-ray 0.1231 0.2479 0.3467 1.4199

Elem. or Isot.	Natural Abundance (%)	Atomic Mass or Weight	Half-Life	Decay Mode/Energy (/MeV)	Particle Energy /Intensity (MeV / %)	Spin (h/2π)	Nuclear Magnetic Mom.(nm)	Elect. Quadr. Mom. (b)	γ-ray Energy /Intensity (MeV / %)
^{154m1}Tb			9. h	β^+ /78 / I.T./22 /		(3-)	1.8	+3.	Gd k x-ray 0.1231 0.2479 0.5401 (0.12 - 2.57)
^{154}Tb		153.92469	21.5 h	EC/99 /3.56 β^+ /1 /	1.86/ 2.45	0-	0.9		Gd k x-ray 0.1231 1.2744 2.1872 (0.12 - 3.14)
^{155}Tb		154.92350	5.3 d	EC/0.82		3/2+	2.0		Gd k x-ray 0.08654 0.10530
^{156m2}Tb			1.02 d	I.T./		(7-)			Tb k x-ray 0.0496
^{156m1}Tb			5.3 h	I.T./0.0884		(0+)			Tb k x-ray 0.0884
^{156}Tb		155.924744	5.3 d	EC/2.444		3-	1.4	+2.3	Gd k x-ray 0.08896 0.19921 0.53435 1.22245
^{157}Tb		156.924021	1.1×10^2 y	EC/0.0601		3/2+	2.0		Gd k x-ray 0.0545
^{158m}Tb			10.5 s	I.T./0.11		0-			Gd k x-ray 0.0110
^{158}Tb		157.925410	1.8×10^2 y	EC/80 /1.220 β^- /20 /0.937		3-	+1.76	+2.7	Gd k x-ray 0.0795 0.9442 0.9621
^{159}Tb	100.	158.925343				3/2+	+2.014	+1.43	
^{160}Tb		159.927164	72.3 d	β^- /1.835	0.57/47 0.86/27	3-	+1.79	3.8	Dy k x-ray 0.08678 0.29857 0.87936 0.96615
^{161}Tb		160.927566	6.91 d	β^- /0.593	0.46/23 0.52/66 0.6/10	3/2+	2.2	+1.2	Dy k x-ray 0.02565 0.04892 0.07458
^{162}Tb		161.92948	7.6 m	β^- /2.51	1.4	(1/2-)			Dy k x-ray 0.2600 0.8075 0.8882
^{163}Tb		162.930644	19.5 m	β^- /1.785	0.80/	3/2+			Dy k x-ray 0.3511 0.3897 0.4945
^{164}Tb		163.9334	3.0 m	β^- /3.9	1.7/	(5+)			Dy k x-ray 0.1689 0.2157 0.6110 0.6885 0.7548
^{165}Tb		164.9349	2.1 m	β^- /3.0		3/2+			0.5389 1.1785 1.2920 1.6648
^{166}Tb		165.9380	≈ 21 s	β^-/					
$_{66}$Dy		162.50(3)							
^{141}Dy		140.951	0.9 s	EC,β^+ /9.					
^{142}Dy		141.946	2.3 s	EC,β^+ /7.1					
^{143}Dy		142.9440	3.9 s	EC,β^+ / ≈ 8.8					
^{144}Dy		143.9391	9.1 s	EC,β^+ / ≈ 6.2					
^{145m}Dy		144.9365	14. s	EC,β^+		11/2-			
^{146m}Dy			0.15 s	I.T.		10+			
^{146}Dy		145.9325	30. s	EC,β^+ /5.2					
^{147m}Dy			56. s	I.T./40 / β^+ ,EC/60 /		(11/2-)	-0.66	+0.7	Dy k x-ray 0.072 0.6787 ann.rad./
^{147}Dy		146.9309	75. s	EC,β^+ /6.37		1/2+	-0.92		0.2534 0.3653
								0.1007	ann.rad./
^{148}Dy		147.92710	3.1 m	β^+ /4 /2.68 EC/96 /	1.2/	0+			Tb k x-ray 0.6202

Elem. or Isot.	Natural Abundance (%)	Atomic Mass or Weight	Half-Life	Decay Mode/Energy (/MeV)	Particle Energy /Intensity (MeV / %)	Spin (h/2π)	Nuclear Magnetic Mom.(nm)	Elect. Quadr. Mom. (b)	γ-ray Energy /Intensity (MeV / %)
^{149}Dy		148.92734	4.2 m	β^+ ,EC/3.81		(7/2-)	-0.12	-0.62	ann.rad./ 0.1008 0.1063 0.2534 0.6536 0.7894 1.7765 1.8062 Tb k x-ray
^{150}Dy		149.92558	7.18 m	β^+ ,EC/67 /1.79 α /33 /	4.233/	0+			0.3967 Tb k x-ray
^{151}Dy		150.926181	17. m	β^+ /5 /2.871 EC/89 / α /6 /	4.067/	7/2-	-0.95	-0.30	0.1764 0.3030 0.3861 0.5463 (0.16 - 2.09) Tb k x-ray
^{152}Dy		151.92472	2.37 h	EC/0.60 α /	3.63/	0+			0.2569 Tb k x-ray
^{153}Dy		152.925763	6.3 h	β^+ /1 /2.171 EC/99 / α /0.01 /	0.89/ 3.46/	(7/2-)	-0.78	-0.02	0.0807 0.0997 0.2137 (0.08 - 1.66) Tb k x-ray
^{154}Dy		153.92442	3.x10^6 y	α/2.95	2.87/	0+			
^{155}Dy		154.92575	9.9 h	β^+ /2 /2.095 EC/98 /	0.845/	3/2-	-0.385	+1.04	Tb k x-ray 0.0655 0.2269
^{156}Dy	0.06(1)	155.92428				0+			
^{157}Dy		156.92546	8.1 h	EC/1.34		3/2-	-0.301	+1.30	Tb k x-ray (0.0609-1.319)
^{158}Dy	0.10(1)	157.924405				0+			
^{159}Dy		158.925736	144. d	EC/0.366		3/2-	-0.354	+1.37	Tb k x-ray 0.3262
^{160}Dy	2.34(6)	159.925194				0+			
^{161}Dy	18.9(2)	160.926930				5/2+	-0.480	+2.51	
^{162}Dy	25.5(2)	161.926795				0+			
^{163}Dy	24.9(2)	162.928728				5/2-	+0.673	+2.65	
^{164}Dy	28.2(2)	163.929171				0+			
^{165m}Dy			1.26 m	I.T./98 /0.108 β^- /2 /		1/2-			Dy k x-ray 0.1082 0.5155
^{165}Dy		164.931700	2.33 h	β^- /1.286	1.29/	7/2+	-0.52	-3.5	Ho k x-ray 0.09468
^{166}Dy		165.932803	3.400 d	β^- /0.486	0.40/	0+			Ho k x-ray 0.0282 0.0825
^{167}Dy		166.9357	6.2 m	β^- / $\approx$ 2.35	1.78	(1/2-)			Ho k x-ray 0.2593 0.3103 0.5697 (0.06 - 1.4)
^{168}Dy		167.9372	8.5 m	β^- /1.6		0+			Ho k x-ray 0.1925 0.4867
^{169}Dy		168.9403	$\approx$ 39. s	β^- /3.2					
$_{67}$Ho		164.93032(2)							
^{144}Ho		143.952	0.7 s	β^+ ,EC/12					
^{145}Ho		144.947	2.4 s	β^+					
^{146}Ho		145.9440	3.3 s	β^+ ,EC/10.7		(10+)			ann.rad./
^{147}Ho		146.9396	5.8 s	β^+ ,EC/8.2		11/2-			ann.rad./
^{148m}Ho			9. s	β^+ ,EC/		4-			ann.rad./
^{148}Ho		147.9372	2. s	β^+ ,EC/9.4		1+			ann.rad./ 0.6615 1.6883
^{149m}Ho			21. s	β^+ ,EC/		11/2-			ann.rad./ 1.0733 1.0911
^{149}Ho		148.93379	> 30. s	β^+ ,EC/6.01		1/2+			
^{150m}Ho			25. s	β^+ ,EC/		(9+)			ann.rad./ 0.3939 0.5511 0.6534 0.8034
^{150}Ho		149.9326	1.3 m	β^+ ,EC/6.6					ann.rad./ 0.5913 0.6534 0.8034

Elem. or Isot.	Natural Abundance (%)	Atomic Mass or Weight	Half-Life	Decay Mode/Energy (/MeV)	Particle Energy /Intensity (MeV / %)	Spin (h/2π)	Nuclear Magnetic Mom.(nm)	Elect. Quadr. Mom. (b)	γ-ray Energy /Intensity (MeV / %)
^{151m}Ho			47. s	β^+ ,EC/87 / α/13	4.605/				ann.rad./ 0.2102 0.4889 0.6948 0.7762
^{151}Ho		150.93169	35.2 s	β^+ ,EC/80/5.13 α/20 /	4.519/				ann.rad./ 0.3522 0.5274 0.9676 1.0471
^{152m}Ho			50. s	β^+ ,EC/90 / α/10 /	4.453/	(9+)	+5.9	-1.	ann.rad./ 0.4929 0.6138 0.6474 0.6835
^{152}Ho		151.93166	2.4 m	β^+ ,EC/88 /6.47 α/12 /	4.387/	(3+)	-1.02	+0.1	ann.rad./ 0.6140 0.6476
^{153m}Ho			9.3 m	β^+ ,EC/99+ /4.12 α/	4.01/	5/2	+1.19		ann.rad./ 0.0905 0.1089 0.1618 0.2302 0.2707 0.3659 0.4565
^{153}Ho		152.93020	2.0 m	β^+ ,EC/99+ /4.13 α/	3.91/	11/2-	+6.8	-1.	ann.rad./ 0.2958 0.3346 0.4381 0.6383
^{154m}Ho			3.3 m	β^+ ,EC/		(8+)	5.7	-.1.	ann.rad./ 0.3346 0.4124 0.4771
^{154}Ho		153.93060	12. m	β^+ ,EC/5.75		1-	-0.64	+0.2	ann.rad./ Dy k x-ray 0.3346 0.5700 0.8734
^{155}Ho		154.92908	48. m	β^+ /6 /3.10 EC/94 /		(5/2+)	+3.51	+1.5	ann.rad./ Dy k x-ray 0.0474 0.1363 0.3254
^{156m}Ho			5.8 m	I.T./0.0352 β^+ /25 / EC/75 /	1.8/ 2.9/		+2.99	+2.3	(0.06 - 2.24) ann.rad./ Dy k x-ray 0.1378 0.2666
^{156}Ho		155.9290	56. m	β^+ ,EC/4.4		(5+)			(0.28 - 2.9) ann.rad./ 0.1378 0.2665
^{157}Ho		156.92819	12.6 m	β^+ /5 /2.54 EC/95 /	1.18/	7/2-	+4.35	+3.0	ann.rad./ Dy k x-ray 0.2800 0.3411
^{158m2}Ho			28. m	I.T./44 / EC/56 /		2-	+2.44	+1.6	ann.rad./ Dy k x-ray 0.0989 0.2182
^{158m1}Ho			21. m	β^+ ,EC/		(9+)			ann.rad./ 0.0981 0.1664 0.2182 0.3205 0.4062 0.9774 1.0532 0.4846
^{158}Ho		157.92895	11.3 m	β^+ /8 /4.24 EC/92 /	1.30/	5+	+3.77	+4.1	ann.rad./ Dy k x-ray 0.0989 0.2182 0.9488
^{159m}Ho			8.3 s	I.T./0.206		1/2+			Ho k x-ray 0.1660 0.2059

Elem. or Isot.	Natural Abundance (%)	Atomic Mass or Weight	Half-Life	Decay Mode/Energy (/MeV)	Particle Energy /Intensity (MeV / %)	Spin (h/2π)	Nuclear Magnetic Mom.(nm)	Elect. Quadr. Mom. (b)	γ-ray Energy /Intensity (MeV / %)
¹⁵⁹Ho		158.927708	33.0 m	EC/1.838		7/2-	+4.28	+3.2	Dy k x-ray 0.1210 0.1320 0.2529 0.3096 (0.06 - 1.2)
^{160m2}Ho			3. s			1+			
^{160m}Ho			5.0 h	I.T./67 /0.060 EC/33 /3.35		2-	+2.52	+1.8	0.0868 0.1970 0.6464 0.7281 0.8791 0.9619 0.9658
¹⁶⁰Ho		159.92873	25.6 m	β⁺ ,EC/3.29	0.57/	5+	+3.71	+4.0	See Ho[166m] 0.7282 0.8794
^{161m}Ho			6.8 s	I.T./0.211					Ho k x-ray 0.2112
¹⁶¹Ho		160.927852	2.48 h	EC/0.859		7/2-	+4.25	+3.2	Dy k x-ray 0.0256 0.0592 0.0774 0.1031
^{162m}Ho			1.12 h	I.T./61 / EC/39 /		6-	+3.60	+4.	Dy k x-ray Ho k x-ray 0.0807 0.1850 0.2828 0.9372 1.2200
¹⁶²Ho		161.929092	15. m	EC/96 /0.295 β⁺ /4 /		1+			Dy k x-ray 0.0807 1.3196 1.3728
^{163m}Ho			1.09 s	I.T./0.298		(1/2+)			Ho k x-ray 0.2798
¹⁶³Ho		162.928730	4.57x10³ y	EC/0.00258		7/2-	+4.23	+3.6	Dy M x-rays
^{164m}Ho			38. m	I.T./0.140		(6-)			Ho k x-ray 0.0373 0.0566 0.0940
¹⁶⁴Ho		163.930231	29. m	EC/58 /0.987 β⁻ /42 /0.963		1+			Dy k x-ray 0.0734 0.0914
¹⁶⁵Ho	100.	164.930319				7/2-	+4.127	+3.49	
^{166m}Ho			1.2x10³ y	β⁻ /		7-	3.6	-3.	Er k x-ray 0.18407 0.71169 0.81031
¹⁶⁶Ho		165.932281	1.117 d	β⁻ /1.855	1.776/48 1.855/51	0-			Er k x-ray 0.08057 1.37943
¹⁶⁷Ho		166.933127	3.1 h	β⁻ /1.007	0.31/43 0.61/21 0.96/15 0.97/15	(7/2-)			Er k x-ray 0.0793 0.0835 0.2379 0.3213 0.3465
^{168m}Ho			2.2 m	I.T./					Er k x-ray
¹⁶⁸Ho		167.93550	3.0 m	β⁻ /2.91	2.0/	3+			0.7413 0.8159 0.8211 (0.08 - 2.34)
¹⁶⁹Ho		168.93687	4.7 m	β⁻ /2.12	1.2/ 2.0/	(7/2-)			0.1496 0.7610 0.7784 0.7884 0.8529
^{170m}Ho			43. s	β⁻ /		1+			0.0787 0.8123 1.8940 1.9726

Elem. or Isot.	Natural Abundance (%)	Atomic Mass or Weight	Half-Life	Decay Mode/Energy (/MeV)	Particle Energy /Intensity (MeV / %)	Spin (h/2π)	Nuclear Magnetic Mom.(nm)	Elect. Quadr. Mom. (b)	γ-ray Energy /Intensity (MeV / %)
^{170}Ho		169.93962	2.8 m	β^- /3.87		6+			Er k x-ray 0.1816 0.2582 0.8902 0.9321 0.9414 1.1387
^{171}Ho		170.941	53 s	β/					
^{172}Ho		171.9448	25. s	β^- /					Er k x-ray (0.077-1.186)
$_{68}$Er		167.26(3)							
^{145}Er		144.957	0.9 s	β^+					
^{146}Er		145.952	≈ 1.7 s	β^+					
^{147}Er		146.9494	2.5 s	E.C,β^+ / ≈ 9.1					
^{148}Er		147.9444	4.5 s	β^+ ,EC/6.8					
^{149m}Er			10. s	IT		11/2-			
^{149}Er		148.9425	10.7 s	ECβ^+ /8.1		1/2+			
^{150}Er		149.9370	18. s	β^+ /36 /4.11 EC/64 /		0+			ann.rad./ Ho k x-ray 0.4758
^{151}Er		150.9373	23. s	β^+ ,EC/5.2		7/2-			ann.rad./
^{152}Er		151.93500	10.2 s	β^+ ,EC/10 /3.11 α/90 /	4.804/	0+			ann.rad./
^{153}Er		152.93509	37.1 s	α/ β^+ ,EC/47 /4.56	4.674 4.35/				0.351 (0.0945-1.700)
^{154}Er		153.93278	3.7 m	β^+ ,EC/99+ /2.03 α/0.5 /	4.166/	0+			ann.rad./
^{155}Er		154.93321	5.3 m	β^+ ,EC/47 /3.84 EC/53 /		(7/2-)	-0.669	-0.27	ann.rad./ Ho k x-ray 0.1101 0.2415
^{156}Er		155.9308	20. m	β^+ ,EC/1.7		0+			ann.rad./ 0.0298 0.0352 0.0522 0.1336
^{157}Er		156.9319	25. m	β^+ ,EC/3.5		3/2-	-0.412	+0.92	ann.rad./ 0.117 0.385 1.320 1.660 1.820 2.000
^{158}Er		157.93087	2.2 h	EC/99.5 /1.78 β^+ /0.5 /	0.74/	0+			Ho k x-ray 0.0719 0.2486 0.3868
^{159}Er		158.930681	36. m	β^+ /7 /2.769 EC/93 /		3/2-	-0.304	+1.17	ann.rad./ Ho k x-ray 0.6245 0.6493 (0.07 - 2.5)
^{160}Er		159.92908	1.191 d	EC/0.33		0+			Ho k x-ray (0.05 - 0.96)
^{161}Er		160.93000	3.21 h	EC/2.00		3/2-	-0.37	+1.36	Ho k x-ray 0.8265 (0.07 - 1.74)
^{162}Er	0.14(1)	161.928775				0+			
^{163}Er		162.93003	1.25 h	EC/1.210		5/2-	+0.557	+2.55	Ho k x-ray 0.4361 0.4399 1.1135
^{164}Er	1.61(2)	163.929197				0+			
^{165}Er		164.930723	10.36 h	EC/0.376		5/2-	+0.643	+2.71	Ho k x-ray
^{166}Er	33.6(2)	165.930290				0+			
^{167m}Er			2.27 s	I.T./0.208		1/2-			Er k x-ray 0.2078
^{167}Er	22.95(15)	166.932046				7/2+	-0.5639	+3.57	
^{168}Er	26.8(2)	167.932368				0+			
^{169}Er		168.934588	9.40 d	β^- /0.351	0.35/ ≈ 100	1/2-	+0.515		Tm k x-ray 0.1098 0.1182
^{170}Er	14.9(2)	169.935461				0+			

Elem. or Isot.	Natural Abundance (%)	Atomic Mass or Weight	Half-Life	Decay Mode/Energy (/MeV)	Particle Energy /Intensity (MeV / %)	Spin (h/2π)	Nuclear Magnetic Mom.(nm)	Elect. Quadr. Mom. (b)	γ-ray Energy /Intensity (MeV / %)
^{171}Er		170.938026	7.52 h	β⁻ /1.491		5/2-	0.66	2.9	Tm k x-ray 0.11160 0.29591 0.30832 (0.08 - 1.4)
^{172}Er		171.939352	2.05 d	β⁻ /0.891	0.28/48 0.36/46				Tm k x-ray 0.0597 0.4073 0.6101
^{173}Er		172.9424	1.4 m	β⁻ /2.6		(7/2-)			Tm k x-ray 0.1928 0.1992 0.8952
^{174}Er		173.9441	3.1 m	β⁻ /1.8					Tm k x-ray (0.100-0.152)
^{175}Er		174.9479	1.2 m	β⁻					(0.0765-1.168)
$_{69}$Tm		168.93421(2)							
^{146m}Tm			0.21 s	β⁺,p	p/1.118				
^{146}Tm		145.967	0.06 s	β⁺ /14. p	1.119/ 1.189/				
^{147m}Tm			0.4 ms	β⁺,p	p/1.115				
^{147}Tm		146.961	0.56 s	EC, β⁺ /85/ ≈10.7 p/15/	1.052/				
^{148m}Tm		147.9573	0.7 s	β⁺ ,EC/12.					ann.rad./
^{148}Tm									
^{149}Tm		148.9524	0.9 s	β⁺ ,EC/ ≈9.2		11/2-			
^{150}Tm		149.9494	2.3 s	β⁺ ,EC/ ≈11.5		6-			(0.1007-2.177)
^{151}Tm		150.9454	4. s	β⁺ ,EC/7.5					ann.rad./
^{152m}Tm			8. s	β⁺ ,EC/		9+			
^{152}Tm		151.9443	5. s	β⁺ ,EC/8.8					ann.rad./
^{153}Tm		152.94203	1.6 s	β⁺ ,EC/10 /6.46 α/90 /	5.109/				ann.rad./
^{154m}Tm			3.3 s	β⁺ ,EC/15 / α /	α/5.031/100 4.84/0.24				ann.rad./ (0.4605-0.7960)
^{154}Tm		153.9407	8.1 s	β⁺ ,EC/56 /7.4 α /44 /	α/4.956/100 4.83/0.45				ann.rad./
^{155}Tm		154.93919	30. s	β⁺ ,EC/5.58 α/	4.46/				0.0315 0.0638 0.0881 0.2268 0.5320 0.6067
^{156m}Tm			19. s	α/	4.46/				ann.rad./
^{156}Tm		155.9389	1.40 m	β⁺ ,EC/7.6 α/	4.23/	2-	+0.40	-0.5	0.3446 0.4529 0.5860
^{157}Tm		156.9367	3.6 m	β⁺ ,EC/4.5 α/	2.6 3.97/	1/2	+0.48		ann.rad./ 0.1104 0.3484 0.3855 0.4550 (0.1 - 1.58)
^{158}Tm		157.9379	4.0 m	β⁺ ,EC/74 /6.5 EC/26 /		(2-)	+0.04	+0.7	ann.rad./ Er k x-ray 0.1921 0.3351 0.6280 1.1498 (0.18 - 2.81)
^{159}Tm		158.9348	9.1 m	β⁺ /23 /3.9 EC/77 /		5/2+	+3.42	+1.9	ann.rad./ Er k x-ray 0.0591 0.0848 0.2713 (0.05 - 1.27)
^{160m}Tm			1.24 m	IT		(5)			
^{160}Tm		159.9354	9.4 m	β⁺ /15 /5.9 EC/85 /		1-	+0.16	+0.58	ann.rad./ Er k x-ray 0.1264 0.2642 0.7285 0.8544 0.8614 1.3685

Elem. or Isot.	Natural Abundance (%)	Atomic Mass or Weight	Half-Life	Decay Mode/Energy (/MeV)	Particle Energy /Intensity (MeV / %)	Spin (h/2π)	Nuclear Magnetic Mom.(nm)	Elect. Quadr. Mom. (b)	γ-ray Energy /Intensity (MeV / %)
¹⁶¹Tm		160.9334	31. m	β⁺ ,EC/3.2		7/2+	+2.40	+2.9	ann.rad./ Er k x-ray 0.0595 0.0844 1.6481 (0.04 - 2.15)
¹⁶²ᵐTm			24. s	I.T./90 / β⁺ ,EC/10 /		5+			Tm k x-ray Er k x-ray 0.0669 0.8115 0.9003
¹⁶²Tm		161.93394	21.7 m	β⁺ /8 /4.81 EC/92 /		1-	+0.07	+0.69	ann.rad./ Er k x-ray 0.1020 0.7987 (0.1 - 3.75)m
¹⁶³Tm		162.93265	1.81 h	EC/98 /2.439 β⁺ /1 /		1/2+	0.082		Er k x-ray 0.0692 0.1043 0.2414 0.0914 0.1394 0.2081 0.2405 0.3149
¹⁶⁴ᵐTm			5.1 m	I.T./80 / β⁺ ,EC/20 /		6-			
¹⁶⁴Tm		163.93345	2.0 m	β⁺ /36 /3.96 EC/64 /	2.94/	1+	+2.38	+0.71	ann.rad./ Er k x-ray 0.0914
¹⁶⁵Tm		164.932433	1.253 d	EC/1.593		1/2+	0.139		Er k x-ray 0.0472 0.0544 0.29728 0.80636
¹⁶⁶Tm		165.93355	7.70 h	EC/98 /3.04 β⁺ /2 /		2+	+0.092	+2.14	Er k x-ray 0.0806 0.1844 0.7789 1.2734 2.0524
¹⁶⁷Tm		166.932849	9.24 d	EC/0.748		1/2+	-0.197		Er k x-ray 0.0571 0.20778
¹⁶⁸Tm		167.934171	93.1 d	EC/1.679		3+	0.23	+3.2	Er k x-ray 0.19825 0.4475 0.81595
¹⁶⁹Tm	100	168.934211				1/2+	-0.2316	-1.2	
¹⁷⁰Tm		169.935798	128.6 d	β⁻ /99.8 /0.968 EC/0.2 /0.314	0.883/24 0.968/76	1-	0.2476	+0.74	Yb k x-ray 0.08425 0.06674
¹⁷¹Tm		170.936426	1.92 y	β⁻ /0.096	0.03/2 0.096/98	1/2+	-0.2303		
¹⁷²Tm		171.93840	2.65 d	β⁻ /1.88	1.79/36 1.88/29	2-			Yb k x-ray 0.07879 1.38722 1.46601 1.52982 1.60861
¹⁷³Tm		172.93960	8.2 h	β⁻ /1.298	0.80/21 0.86/71	1/2+			Yb k x-ray 0.3988 0.4613
¹⁷⁴Tm		173.94216	5.4 m	β⁻ /3.08	0.70/14 1.20/83	(4-)			Yb k x-ray 0.07664 0.17669 0.27332 0.3666 0.99205 (0.08 - 1.6)
¹⁷⁵Tm		174.94383	15.2 m	β⁻ /2.39	0.9/36 1.9/23	(1/2+)			Yb k x-ray 0.36396 0.51487 0.94125 0.98247
¹⁷⁶Tm		175.9471	1.9 m	β⁻ /4.2	2.0/ 1.2/	(4+)			Yb k x-ray 0.1898 0.3819 1.0691
¹⁷⁷Tm		176.9490	1.4 m	β⁻					

T A B L E   O F   T H E   I S O T O P E S

Elem. or Isot.	Natural Abundance (%)	Atomic Mass or Weight	Half-Life	Decay Mode/Energy (/MeV)	Particle Energy /Intensity (MeV / %)	Spin (h/2π)	Nuclear Magnetic Mom.(nm)	Elect. Quadr. Mom. (b)	γ-ray Energy /Intensity (MeV / %)
₇₀Yb		173.04(3)							
¹⁵¹Yb		150.9545	1.6 s	β⁺ /8.5					
¹⁵²Yb		151.9502	3.2 s	β⁺ EC/5.5					
¹⁵³Yb		152.9492	4. s	β⁺ EC/6.7					
¹⁵⁴Yb		153.9455	0.40 s	β⁺ EC/7 /4.49 α/93 /	5.32/				ann.rad./
¹⁵⁵Yb		154.9456	1.7 s	β⁺ ,EC/16 /6.0 α/84 /	5.19/		-0.8		ann.rad./
¹⁵⁶Yb		155.94277	26. s	β⁺ ,EC/21 /3.57 α/79 /	4.69/	0+			ann.rad./
¹⁵⁷Yb		156.9427	39. s	β⁺ ,EC/99+ /5.5 α/0.5 /	4.69/				ann.rad./ 0.231 (0.035-0.670)
¹⁵⁸Yb		157.93986	1.5 m	β⁺ ,EC/1.9		0+			ann.rad./ 0.0741 0.2526
¹⁵⁹Yb		158.9402	1.4 m	EC,β⁺ /5.1					Tm k x-ray 0.1661 0.1772 0.3297 0.3903
¹⁶⁰Yb		159.9376	4.8 m	β⁺ ,EC/2.0		0+			ann.rad./ 0.1404 0.1737 0.2158
¹⁶¹Yb		160.9375	4.2 m	β⁺ ,EC/3.9		3/2-	-0.33	+1.03	ann.rad./ Tm k x-ray 0.0782 0.5999 0.6315
¹⁶²Yb		161.9358	18.9 m	β⁺ ,EC/1.7		0+			ann.rad./ Tm k x-ray 0.1188 0.1635
¹⁶³Yb		162.9363	11.1 m	β⁺ /26 /3.4	1.4/	3/2-	-0.37	+1.2	ann.rad./ Tm k x-ray 0.0636 0.8603 (0.06 - 1.9)
¹⁶⁴Yb		163.9345	1.26 h	EC/1.0		0+			Tm k x-ray 0.0914 0.6752
¹⁶⁵Yb		164.93540	9.9 m	β⁺ /10 /2.76 EC/90 /	1.58/	(5/2-)	+0.48	+2.48	ann.rad./ Tm k x-ray 0.0801 1.0903
¹⁶⁶Yb		165.93388	2.363 d	EC/0.30		0+			Tm k x-ray 0.0828 0.1844 0.7789 1.2734 2.0524
¹⁶⁷Yb		166.934947	17.5 m	β⁺ /0.5 /1.954 EC/99.5 /	0.639/	5/2-	+0.62	+2.70	Tm k x-ray 0.06296 0.10616 0.11337 0.17633
¹⁶⁸Yb	0.13(1)	167.933895				0+			
¹⁶⁹ᵐYb			46. s	I.T./0.0242		1/2-			Yb L x-ray 0.0242
¹⁶⁹Yb		168.935187	32.03 d	EC/0.909		7/2+	-0.63	+3.5	Tm k x-ray (0.0498-0.3078)
¹⁷⁰Yb	3.05(6)	169.934759				0+			
¹⁷¹Yb	14.3(2)	170.936323				1/2-	+0.49367		
¹⁷²Yb	21.9(3)	171.936378				0+			
¹⁷³Yb	16.12(21)	172.938207				5/2-	-0.67989	+2.80	
¹⁷⁴Yb	31.8(4)	173.938858				0+			
¹⁷⁵Yb		174.941273	4.19 d	β⁻ /0.470	0.466/73 0.071/21 0.353/6.2	7/2-	0.6		Lu k x-ray 0.3963/13 (0.114 - 0.28)
¹⁷⁶ᵐYb			11.4 s	I.T./1.051		(8-)			Yb k x-ray 0.0961 0.1901 0.2929 0.3897
¹⁷⁶Yb	12.7(2)	175.942569				0+			

Elem. or Isot.	Natural Abundance (%)	Atomic Mass or Weight	Half-Life	Decay Mode/Energy (/MeV)	Particle Energy /Intensity (MeV / %)	Spin (h/2π)	Nuclear Magnetic Mom.(nm)	Elect. Quadr. Mom. (b)	γ-ray Energy /Intensity (MeV / %)
^{177m}Yb			6.41 s	I.T./0.3315		1/2-			Yb k x-ray 0.1131 0.2084
^{177}Yb		176.945257	1.9 h	β⁻ /1.399	1.40	9/2+			Lu k x-ray 0.1504
^{178}Yb		177.94664	1.23 h	β⁻ /0.65	0.25/	0+			0.1415 0.3246 0.3516 0.3815 0.6125
^{179}Yb		178.9499	8. m	β⁻ /2.4					
^{180}Yb		179.9523	2. m	β⁻					(0.1028-0.4423)
$_{71}$Lu		174.967(1)							
^{150}Lu		149.973	≈ 35. ms	p					
^{151}Lu		150.967	0.08 s	p/1.231					
^{152}Lu		151.963	0.7 s						
^{153}Lu		152.959							
^{154}Lu		153.9571	1.0 s	β⁺ ,EC/10.8					
^{155m}Lu			2.6 ms	α/7.41					
^{155}Lu		154.9542	0.07 s	EC/8.0 α/	5.66/90 5.57/				
^{156m}Lu			0.20 s	α/	5.45/				
^{156}Lu		155.9529	≈ 0.5 s	β⁺ , EC/9.5 α/					ann.rad./
^{157m}Lu			≈9.6 s	α	4.925/				
^{157}Lu		156.95010	4.8 s	β⁺ , EC/94 /6.93 α/	5.00/				ann.rad./
^{158}Lu		157.94984	10.4 s	β⁺ , EC/99 /8.0 α/	4.67/				ann.rad./ 0.3682 0.4770
^{159}Lu		158.9467	12.3 s	β⁺ , EC/6.0					ann.rad./ 0.1505 0.1875 0.3693
^{160}Lu		159.94654	36.1 s	β⁺ , EC/7.3					ann.rad./ 0.2434 0.3957 0.5773
^{161}Lu		160.9432	1.2 m	β⁺ , EC/5.3					ann.rad./ 0.0437 0.0671 0.1003 0.1108 0.1562 0.2562
^{162m}Lu			≈ 1.5 m	EC/		4-			
^{162}Lu		161.9432	1.37 m	β⁺ , EC/6.9		1-			ann.rad./ 0.1666 0.6314
^{163}Lu		162.9412	4.1 m	β⁺ , EC/4.6					ann.rad./ 0.0539 0.0581 0.1504 0.1631 0.3717
^{164}Lu		163.9412	3.14 m	β⁺ , EC/6.3	1.6/ 3.8/				0.1238 0.2621 0.7404 0.8639 0.8804
^{165}Lu		164.9396	10.7 m	β⁺ , EC/3.9	2.06/	1/2+			ann.rad./ 0.1206 0.1324 0.1742 0.2036 (0.04 - 2.0)
^{166m2}Lu			2.1 m	β⁺ /35 / EC/65 /		(0-)			ann.rad./ Yb k x-ray 1.0673 1.2566 2.0986
^{166m1}Lu			1.4 m	β⁺ , EC/58 / I.T./42 /0.0344		(3-)			ann.rad./ 0.1024 0.2281 0.2861 0.8119 0.8301

Elem. or Isot.	Natural Abundance (%)	Atomic Mass or Weight	Half-Life	Decay Mode/Energy (/MeV)	Particle Energy /Intensity (MeV / %)	Spin (h/2π)	Nuclear Magnetic Mom.(nm)	Elect. Quadr. Mom. (b)	γ-ray Energy /Intensity (MeV / %)
^{166}Lu		165.9398	2.8 m	β^+ /25 /5.5 EC/75 /		(6-)			ann.rad./ Yb k x-ray 0.1024 0.2281 0.3375 0.3679
^{167}Lu		166.9383	52. m	β^+ /2 /3.1 EC/98 /	2.1/	7/2+			Yb k x-ray 0.0297 0.2392 (0.03 - 2.0)
^{168m}Lu			6.7 m	β^+ /12 / EC/88 /		3+			ann.rad./ Yb k x-ray 0.1988 0.8960 0.9792
^{168}Lu		167.9387	5.5 m	β^+ /6 /4.5 EC/94 /	1.2/	(6-)			ann.rad./ Yb k x-ray 0.1114 0.1124 0.2286 0.3483 1.4836
^{169m}Lu			2.7 m	I.T./0.0290		1/2-			Lu L x-ray 0.0290
^{169}Lu		168.93765	1.419 d	EC/2.293	1.271/	7/2+	2.30	3.4	Yb k x-ray 0.19121 0.9606 (0.08 - 2.1)
^{170m}Lu			0.7 s	I.T./0.0929		4-			Lu L x-ray 0.04449 0.0484
^{170}Lu		169.93847	2.01 d	EC/3.46	2.44/	0+			Yb k x-ray 0.58711 0.5908 1.28029 (0.1 - 3.38)
^{171m}Lu			1.31 m	I.T./0.0711		1/2-			Lu k x-ray 0.07119
^{171}Lu		170.937910	8.24 d	EC/1.479	0.362/	7/2+	2.31	3.38	Yb k x-ray 0.01939 0.66744 (0.02 - 1.3)
^{172m}Lu			3.7 m	I.T./0.0419		1-			Lu L x-rays 0.04186
^{172}Lu		171.939082	6.70 d	EC/2.519		4-	2.89	3.79	Yb k x-ray 0.18156 1.09367 (0.07 - 2.2)
^{173}Lu		172.938927	1.37 y	EC/0.671		7/2+	2.28	3.56	Yb k x-ray 0.07860 0.27198
^{174m}Lu			142. d	I.T./99.3 /0.17086 EC/0.7 /		6-	1.5		Lu k x-ray 0.067055
^{174}Lu		173.940334	3.3 y	EC/1.374		1-	1.9		Yb k x-ray 0.07664 1.2419
^{175}Lu	97.41(2)	174.940768				7/2+	+2.2327	+3.49	
^{176m}Lu			3.66 h	β^- /1.315	1.229/ 1.317/	1-	+0.318	-1.47	Hf k x-ray 0.088372
^{176}Lu	2.59(2)	175.942683	3.8×10^{10} y	β^- /1.192		7-	+3.169	+4.92	Hf k x-ray 0.20187 0.30691
^{177m}Lu			160.7 d	I.T./22 /0.9702 β^- /78		23/2-	2.34	5.2	Lu k x-ray Hf k x-ray 0.11295 0.20836 0.37850 0.41853
^{177}Lu		176.943755	6.75 d	β^- /0.498	0.497/	7/2+	+2.239	+3.39	0.11295 0.20836 0.2166 0.3317
^{178m}Lu			23.1 m	β^- /		(9-)			Hf k x-ray 0.0932 1.3099 1.3408
^{178}Lu		177.945952	28.5 m	β^- /2.099	2.03/	1+			(0.09 - 1.7) 0.2143 0.3377
^{179}Lu		178.94732	4.6 h	β^- /1.405	1.35/	7/2+			

Elem. or Isot.	Natural Abundance (%)	Atomic Mass or Weight	Half-Life	Decay Mode/Energy (/MeV)	Particle Energy /Intensity (MeV / %)	Spin (h/2π)	Nuclear Magnetic Mom.(nm)	Elect. Quadr. Mom. (b)	γ-ray Energy /Intensity (MeV / %)
^{180}Lu		179.9499	5.7 m	$β^-$ /3.1	1.49/				0.40795/50. (0.07-1.9)
^{181}Lu		180.9518	3.5 m	$β^-$ /2.5		(7/2+)			0.0458 0.2059 0.5749
^{182}Lu		181.9552	2.0 m	$β^-$ / ≈ 4.1					0.0978 0.7208 0.8182
^{183}Lu		182.9576	58. s	$β^-$ /		7/2+			
^{184}Lu		183.9612	20 s	$β^-$					
$_{72}$Hf		178.49(2)							
^{154}Hf		153.964	2. s	EC,$β^+$ / ≈ 6.7					
^{155}Hf		154.963	0.9 s	EC,$β^+$ /8.					
^{156}Hf		155.9593	25. ms	α/					
^{157}Hf		156.9581	0.11 s	α/					
^{158}Hf		157.9539	2.9 s	EC/54 /5.1 α/46 /	5.27/	0+			
^{159}Hf		158.9538	5.6 s	$β^+$,EC/88 /6.9 α/12 /	5.09/				ann.rad./
^{160}Hf		159.95063	≈ 12. s	$β^+$,EC/97 /4.9 α /4.78		0+			ann.rad./
^{161}Hf		160.9503	17. s	α/	4.60/				
^{162}Hf		161.94720	38. s	$β^+$,EC/3.7		0+			ann.rad./ 0.1739 0.1963 0.4101
^{163}Hf		162.9471	40. s	$β^+$,EC/5.5					ann.rad./ 0.0454 0.0621 0.0710 0.6882
^{164}Hf		163.9536	2.8 m	EC,$β^+$ /3.0					
^{165}Hf		164.9445	1.32 m	EC/4.6		11/2-			
^{166}Hf		165.9423	6.8 m	EC/93 /2.3 $β^+$ /7 /					ann.rad./ Lu k x-ray 0.0788
^{167}Hf		166.9426	2.0 m	$β^+$ /40 /4.0 EC/60 /		(5/2-)			ann.rad./ Lu k x-ray 0.1754 0.3152
^{168}Hf		167.9406	25.9 m	$β^+$,EC/1.8		0+			ann.rad./ (0.0144-1.311)
^{169}Hf		168.9412	3.25 m	EC/85 /3.3 $β^+$ /15 /		(5/2-)			ann.rad./ Lu k x-ray 0.3695 0.4929
^{170}Hf		169.9397	16.0 h	EC/1.1		0+			Lu k x-ray 0.0985 0.1202 0.1647 0.5729 0.6207
^{171m}Hf			30. s						
^{171}Hf		170.9405	12.2 h	EC,$β^+$ /2.4		7/2+			ann.rad./ Lu k x-ray 0.1221 0.6620 1.0714
^{172}Hf		171.93946	1.87 y	EC/0.35		0+			Lu k x-ray 0.02399 0.12582 (0.0818-0.123)
^{173}Hf		172.9407	23.6 h	EC/1.6		1/2-			Lu k x-ray 0.12367 0.13963 0.29697 0.31124 (0.1 - 2.1)
^{174}Hf	0.162(3)	173.940042	2.0x10^{15} y			0+			
^{175}Hf		174.941504	70. d	EC/0.686		5/2-	-0.62	+2.6	Lu k x-ray 0.08936 0.34340
^{176}Hf	5.206(5)	175.941403				0+			

Elem. or Isot.	Natural Abundance (%)	Atomic Mass or Weight	Half-Life	Decay Mode/Energy (/MeV)	Particle Energy /Intensity (MeV / %)	Spin (h/2π)	Nuclear Magnetic Mom.(nm)	Elect. Quadr. Mom. (b)	γ-ray Energy /Intensity (MeV / %)
^{177m2}Hf			51.4 m	I.T./2.740		37/2-			Hf k x-ray 0.2140 0.2951 0.3115 0.3267
^{177m1}Hf			1.1 s	I.T./		23/2+			Hf k x-ray 0.20836 0.22847 0.37851
^{177}Hf	18.606(4)	176.943220				7/2-	+0.7935	+0.337	
^{178m2}Hf			31. y	I.T./		16+	+8.2	+6.0	Hf k x-ray 0.32555 0.42635 (0.0889-0.5742)
^{178m1}Hf			4.0 s	I.T./		8-			Hf k x-ray 0.21342 0.32555 0.42635
^{178}Hf	27.297(4)	177.943698				0+			
^{179m2}Hf			25.1 d	I.T./1.1057		25/2-	7.4		Hf k x-ray 0.1227 0.1461 0.3626 0.4537
^{179m1}Hf			18.7 s	I.T./0.375		1/2-			Hf k x-ray 0.1607 0.2141
^{179}Hf	13.629(6)	178.945815				9/2+	-0.6409	+3.79	
^{180m}Hf			5.52 h	I.T./1.1416		8-	+8.7	+4.6	Hf k x-ray 0.2152 0.3323 0.4432
^{180}Hf	35.100(7)	179.946549				0+			
^{181}Hf		180.949099	42.4 d	β⁻ /1.027	0.408/	1/2-			Ta k x-ray 0.13294 0.48200
^{182m}Hf			62. m	β⁻ /54 /1.60 I.T./46 /1.1729	0.49/43 0.95/10	8-			Hf k x-ray 0.0509 0.2244 0.3441 0.4558 0.5066 0.9428
^{182}Hf		181.95055	9.x10⁶ y	β⁻ /0.37		0+			Ta k x-ray 0.2704
^{183}Hf		182.95353	1.07 h	β⁻ /2.01	1.18/68 1.54/25	3/2-			Ta k x-ray 0.0732 0.4591 0.7837
^{184}Hf		183.95545	4.1 h	β⁻ /1.34	0.74/38 0.85/16 1.10/46	0+			Ta k x-ray 0.0414 0.1391 0.3449
^{185}Hf		184.9588	≈3.5m	β⁻ /					0.165
$_{73}$Ta		180.9479(1)							
^{156}Ta		155.972	0.11 s	β⁺ / ≈11.6 p/	1.02/ ≈100				
^{157}Ta		156.968	10 ms	α/ p/	6.117 0.927/3.4				
^{158}Ta		157.9664	37. ms	α/	6.05/100 5.97/100				
^{159}Ta		158.9629	0.6 s	β⁺ ,EC/20 /8.5 α/80 /	α/5.52/34 5.60/55				ann.rad./
^{160}Ta		159.9615	1.4 s	β⁺ ,EC/10.1 α	5.41/				ann.rad./
^{161}Ta		160.9584	2.9 s	β⁺ ,EC/7.5 α/	5.15				ann.rad./
^{162}Ta		161.9564	4. s	EC/8.6					
^{163}Ta		162.9544	10.6 s	EC/6.8					
^{164}Ta		163.9536	14.2 s	β⁺ /8.5 α/	4.62/	3+			ann.rad./ 0.2110 0.3768
^{165}Ta		164.9508	31. s	ECβ⁺ /5.9					

Elem. or Isot.	Natural Abundance (%)	Atomic Mass or Weight	Half-Life	Decay Mode/Energy (/MeV)	Particle Energy /Intensity (MeV / %)	Spin (h/2π)	Nuclear Magnetic Mom.(nm)	Elect. Quadr. Mom. (b)	γ-ray Energy /Intensity (MeV / %)
^{166}Ta		165.9505	34. s	β$^+$ /82 /7.7 EC/18 /					ann.rad./ Hf k x-ray 0.1587 0.3117 0.8101
^{167}Ta		166.9486	1.4 m	β$^+$,EC/5.6					ann.rad./
^{168}Ta		167.9478	2.4 m	β$^+$ /77 /6.7 EC/23 /		3+			ann.rad./ Hf k x-ray 0.1239 0.2615 0.7502
^{169}Ta		168.9459	4.9 m	β$^+$,EC/4.4					ann.rad./ 0.0288 0.1535 0.1924
^{170}Ta		169.9461	6.8 m	β$^+$ /70 /6.0 EC/35 /		(3+)			ann.rad./ Hf k x-ray 0.1008 0.2212
^{171}Ta		170.9445	23.3 m	β$^+$,EC/3.7		(5/2-)			0.0496 0.5018 0.5064 (0.05 - 1.02)
^{172}Ta		171.9447	36.8 m	β$^+$ /25 /4.9 EC/75 /		(3-)			ann.rad./ Hf k x-ray 0.21396 1.10923 (0.09 - 3.8)
^{173}Ta		172.9446	3.6 h	β$^+$ /24 /3.7 EC/76 /		(5/2-)	1.7	-1.9	ann.rad./ Hf k x-ray 0.06972 0.17219 (0.06 - 2.7)
^{174}Ta		173.9442	1.12 h	β$^+$ /27 /3.8 EC/73 /		(3+)			ann.rad./ Hf k x-ray 0.09089 0.20638 (0.09 - 3.64)
^{175}Ta		174.9437	10.5 h	EC/2.0		7/2+	2.27	+3.7	Hf k x-ray 0.2077 0.2671 0.3487
^{176}Ta		175.9447	8.1 h	EC/3.1		1-			Hf k x-ray 0.08837 1.15735
^{177}Ta		176.944472	2.356 d	EC/1.166		7/2+	2.25		Hf k x-ray 0.11295 (0.07 - 1.06)
^{178m}Ta			2.4 h	EC/		(7-)			Hf k x-ray 0.08886 0.21342 0.32555 0.42635
^{178}Ta		177.9458	9.29 m	EC/99 /1.9 β$^+$ /1 /		1+	+2.74	+0.65	ann.rad./ Hf k x-ray 0.09316
^{179}Ta		178.94593	1.8 y	EC/0.110		7/2+	2.29	3.37	Hf k x-ray
^{180m}Ta	0.012(2)		> 1.2x10^{15} y			(9-)	4.77		
^{180}Ta		179.947466	8.15 h	EC/87 /0.854 β$^-$ /13 /0.708	0.61/3 0.71/10	1+			Hf k x-ray W k x-ray 0.09333 0.10340
^{181}Ta	99.988(2)	180.947996				7/2+	+2.370	+3.3	
^{182m}Ta			15.8 m	I.T./0.5198		10-			Ta k x-ray 0.14678 0.17157
^{182}Ta		181.950152	114.43 d	β$^-$ /1.814	0.25/30 0.44/20 0.52/40	3-	+3.02		W k x-ray 0.06775 0.10010 1.12127 1.22138
^{183}Ta		182.951373	5.1 d	β$^-$ /1.070	0.45/5 0.62/91	7/2+	+2.36		W k x-ray 0.0847 0.0991 0.1079 0.2461 0.3540

Elem. or Isot.	Natural Abundance (%)	Atomic Mass or Weight	Half-Life	Decay Mode/Energy (/MeV)	Particle Energy /Intensity (MeV / %)	Spin (h/2π)	Nuclear Magnetic Mom.(nm)	Elect. Quadr. Mom. (b)	γ-ray Energy /Intensity (MeV / %)
¹⁸⁴Ta		183.95401	8.7 h	β⁻ /2.87	1.11/15 1.17/81	(5-)			W k x-ray 0.2528/44. 0.4140/74. (0.09-1.4)
¹⁸⁵Ta		184.95556	49. m	β⁻ /1.99	1.21/5 1.77/81	(7/2+)			W k x-ray 0.0697 0.1739 0.1776
¹⁸⁶Ta		185.9586	10.5 m	β⁻ /3.9	2.2/	(3-)			W k x-ray 0.1979 0.2149 0.5106 (0.09 - 1.5)
₇₄W		183.84(1)							
¹⁵⁸ᵐW			<1. ms	α	8.28(3)/				
¹⁵⁸W		157.974	0.9 ms	α/	6.433				
¹⁵⁹W		158.972	7. ms	α/					
¹⁶⁰W		159.9684	0.08 s	α/	5.92/	0+			
¹⁶¹W		160.9671	0.41 s	β⁺ ,EC/18 /8.1 α/82 /	5.78/				
¹⁶²W		161.9626	1.39 s	β⁺ ,EC/54 /5.8 α/46 /	5.54/	0+			
¹⁶³W		162.9624	2.8 s	β⁺ ,EC/59 /7.5 α/41 /	5.38/				
¹⁶⁴W		163.95890	6. s	β⁺ ,EC/97 /5.0 α/3 /	5.15/	0+			ann.rad./
¹⁶⁵W		164.9583	5.1 s	β⁺ ,EC/99 /7.0 α/1 /	4.91/				ann.rad./
¹⁶⁶W		165.95502	16. s	β⁺ ,EC/99 /4.2 α/1 /	4.74/	0+			ann.rad./
¹⁶⁷W		166.9547	20. s	EC/5.6					ann.rad./
¹⁶⁸W		167.9519	53. s	EC/3.8 α/10⁻³/	4.40(1)				Ta k x-ray 0.1755 (0.037-0.573)
¹⁶⁹W		168.9518	1.3 m	EC/5.4					ann.rad./ Ta k x-ray 0.123 (0.097-0.699)
¹⁷⁰W		169.9485	2.4 m	EC/2.2					ann.rad./ Ta k x-ray 0.3162 (0.060-0.144)
¹⁷¹W		170.9494	2.4 m	EC/4.6					ann.rad./ Ta k x-ray 0.1842 (0.052-0.479)
¹⁷²W		171.9474	6.6 m	β⁺ ,EC/2.5					ann.rad./ Ta k x-ray 0.0389 (0.034-0.674)
¹⁷³W		172.9489	6.3 m	EC/4.0					ann.rad./ Ta k x-ray 0.4576 (0.035-0.623)
¹⁷⁴W		173.9462	35. m	EC/1.9		0+			ann.rad./ Ta k x-ray 0.3287 0.4288 (0.056-0.429)
¹⁷⁵W		174.9468	35. m	EC/2.9		1/2-			(0.015-0.27) 0.03358 0.06129 0.09487 0.10020
¹⁷⁶W		175.9456	2.5 h	β⁺ ,EC/0.8		0+			
¹⁷⁷W		176.9466	2.21 h	EC/2.0		(1/2-)			Ta k x-ray 0.15505 0.18569 0.42694
¹⁷⁸W		177.9459	21.6 d	EC/0.091		0+			Ta k x-ray
¹⁷⁹ᵐW			6.4 m	I.T./99.7 /0.222 EC/0.3 /		(1/2-)			W k x-ray 0.2220
¹⁷⁹W		178.94707	38. m	EC/1.06		(7/2-)			Ta k x-ray 0.0307
¹⁸⁰W	0.120(1)	179.946706	7.4x10¹⁶ y	α/		0+			

Elem. or Isot.	Natural Abundance (%)	Atomic Mass or Weight	Half-Life	Decay Mode/Energy (/MeV)	Particle Energy /Intensity (MeV / %)	Spin (h/2π)	Nuclear Magnetic Mom.(nm)	Elect. Quadr. Mom. (b)	γ-ray Energy /Intensity (MeV / %)
^{181}W		180.94820	121.2 d	EC/0.188		9/2+			Ta k x-ray 0.13617 0.15221
^{182}W	26.498(29)	181.948205	8.3x10^{18} y	α/		0+			
^{183m}W			5.15 s	I.T./		(11/2+)			W k x-ray 0.0465 0.0526 0.0991 0.1605
^{183}W	14.314(4)	182.950224	1.9x10^{18} y	α/		1/2-	+0.1177848		
^{184}W	30.642(8)	183.950932	4.0x10^{18} y	α/		0+			
^{185m}W			1.6 m	I.T./0.1974		11/2+			W k x-ray 0.0659 0.1315 0.1737 0.12536
^{185}W		184.953420	74.8 d	β$^-$ /0.433	0.433/99.9	3/2-			
^{186}W	28.426(37)	185.954362	6.5x10^{18} y	α/		0+			
^{187}W		186.957158	23.9 h	β$^-$ /1.311	0.624/ 1.315/	3/2-	0.62		Re k x-ray 0.0725 0.47951 0.68572
^{188}W		187.958487	69.4 d	β$^-$ /0.349	0.349/99	0+			0.0636 0.2271
^{189}W		188.9619	11.5 m	β$^-$ /2.5	1.4/ 2.5/	(3/2-)			0.2907 (0.1262-1.466)
^{190}W		189.9632	30. m	β$^-$ /1.3	0.95/	0+			Re k x-ray 0.1576 0.1621
$_{75}$Re		186.207(1)							
^{160}Re		159.981	0.7 ms	p/ α/	1.261(6)/91 6.54/				
^{161}Re		160.978	14 ms	α/ p	6.24 1.35				
^{162}Re		161.9757	0.10 s	α/	6.12/94 6.09/94				
^{163}Re		162.9721	0.26 s	β$^+$,EC/9.0 α/	α/5.87/32 5.92/66				
^{164}Re		163.9704	0.9 s	β$^+$,EC/10.7 α/	5.78/				
^{165}Re		164.9671	2. s	β$^+$,EC/87 /8.1 α/	5.51/				
^{166}Re		165.9651	2.5 s	β$^+$,EC/9.4 α/	5.50/				
^{167m}Re			6.2 s	α, EC/					
^{167}Re		166.9626	3.4 s	β$^+$,EC/7.4 α/	5.015/				
^{168}Re		167.9616	4.4 s	β$^+$,EC/9.1 α/	4.833/				
^{169m}Re			8.1 s	α	4.70/ 4.87/				0.1117
^{169}Re		168.9588	16. s						
^{170}Re		169.9582	9.2 s	β$^+$, EC/9.0					0.1560 0.3055 0.4125
^{171}Re		170.9555	15.2 s	EC/ ≈ 5.7					ann.rad./ 0.1234 0.2537 0.3504
^{172m}Re			55. s	β$^+$,EC/		(2)			ann.rad./ 0.1234 0.2537
^{172}Re		171.9553	15. s	β$^+$,EC/7.3					ann.rad./ 0.1234
^{173}Re		172.9531	2.0 m	EC/ ≈3.9					0.2537 ann.rad./
^{174}Re		173.9521	2.4 m	β$^+$,EC/5.6					0.1119 0.2430
^{175}Re		174.9514	5.8 m	β$^+$,EC/4.3					ann.rad./ 0.1089
^{176}Re		175.9516	5.3 m	β$^+$,EC/5.6		(3+)			0.2406 ann.rad./
^{177}Re		176.9503	14. m	EC/78 /3.4 β$^+$ /22 /		(5/2-)			W k x-ray 0.0797 0.0843 0.1968

Elem. or Isot.	Natural Abundance (%)	Atomic Mass or Weight	Half-Life	Decay Mode/Energy (/MeV)	Particle Energy /Intensity (MeV / %)	Spin (h/2π)	Nuclear Magnetic Mom.(nm)	Elect. Quadr. Mom. (b)	γ-ray Energy /Intensity (MeV / %)
^{178}Re		177.9509	13.2 m	β⁺ /11 /4.7 EC/89 /	3.3/	(3)			ann.rad./ W k x-ray 0.1059 0.2373 0.9391
^{179}Re		178.9500	19.7 m	EC/99 /2.71 β⁺ /1 /	0.95/	(5/2+)			W k x-ray 0.1199 0.2900 0.4154 0.4302 1.6803
^{180}Re		179.95079	2.45 m	EC/92 /3.80 β⁺ /8 /	1.76/	1-			ann.rad./ W k x-ray 0.1036 0.9028 (0.07 - 2.2)
^{181}Re		180.95006	20. h	EC /1.74		5/2+	3.19		W k x-ray 0.3607 0.3655 0.6390
^{182m}Re			12.7 h	EC/	0.55/ 1.74/	2+	3.3	+1.8	W k x-ray 0.0677 1.1214 1.2215 (0.06 - 2.2)
^{182}Re		181.9512	2.67 d	EC/2.8		(7+)	2.8	+4.1	W k x-ray 0.0678 0.2293 1.1213 1.2214
^{183}Re		182.95082	70. d	EC/0.56		(5/2+)	+3.17	+2.3	W k x-ray 0.16232
^{184m}Re			165. d	I.T./75 /0.188 EC/25 /		8+	+2.9		Re k x-ray 0.1047 0.2165 0.92093 (0.10 - 1.1)
^{184}Re		183.95252	38. d	EC/1.48		3-	+2.53	+2.8	W k x-ray 0.79207 0.90328 (0.1 - 1.4)
^{185}Re	37.40(2)	184.952955				5/2+	+3.1871	+2.18	
^{186m}Re			2.0x10⁵ y	I.T./0.150		8+			Re k x-ray 0.0590
^{186}Re		185.954986	3.718 d	β⁻ /92 /1.070 EC/8 /0.582	0.973/21 1.07/71	1-	+1.739	+0.62	W k x-ray 0.1227/0.6 0.1372/9.5 (0.63-0.77)
^{187}Re	62.60(2)	186.955751	4.4x10¹⁰ y	β⁻ /0.00266	0.0025/	5/2+	+3.2197	+2.07	
^{187m}Re			18.6 m	I.T./0.172		(6-)			Re k x-ray 0.0925 0.1059
^{188}Re		187.958112	16.94 h	β⁻ /2.120	1.962/20 2.118/79	1-	+1.788	+0.57	Os k x-ray 0.15502
^{189}Re		188.959228	24. h	β⁻ /1.01	1.01/	(5/2+)			0.1471 0.2167 0.2194 0.2451
^{190m}Re			3.0 h	β⁻ /51 / I.T./49 /		(6-)			Re k x-ray 0.1191 0.2238 0.6731 (0.1 - 1.79)
^{190}Re		189.9618	3.0 m	β⁻ /3.2	1.8/	(2-)			Os k x-ray 0.1867 0.5580 0.6051
^{191}Re		190.96312	9.7 m	β⁻ /2.05	1.8/				
^{192}Re		191.9660	16. s	β⁻ /4.2	≈ 2.5/				(0.2-0.75)
$_{76}$Os		190.23(3)							
^{162}Os		161.984	1.5 ms	α/	6.62				
^{163}Os		162.982	5.5 ms	α/	6.51				
^{164}Os		163.9779	0.04 s	α					
^{165}Os		164.9765	0.07 s	α					
^{166}Os		165.9718	0.18 s	β⁺ ,EC/28 /6.3 α/72 /	6.27/ 5.98/	0+			ann. rad./

Elem. or Isot.	Natural Abundance (%)	Atomic Mass or Weight	Half-Life	Decay Mode/Energy (/MeV)	Particle Energy /Intensity (MeV / %)	Spin (h/2π)	Nuclear Magnetic Mom.(nm)	Elect. Quadr. Mom. (b)	γ-ray Energy /Intensity (MeV / %)
¹⁶⁷Os		166.9714	0.7 s	β⁺ ,EC/76 /8.2 α/24 /	5.84/				ann.rad./
¹⁶⁸Os		167.96775	2.2 s	β⁺ ,EC/51 /5.7 α/49 /		0+			ann. rad./
¹⁶⁹Os		168.9671	3.3 s	β⁺ ,EC/89 /7.7 α/13 /	5.57/80 5.51/12 5.54/8				ann.rad./
¹⁷⁰Os		169.96357	7.1 s	β⁺ ,EC/5.0 α/	5.40/	0+			ann.rad./ (0.162-0.216)
¹⁷¹Os		170.9630	8.4 s	β⁺ ,EC/98 /7.1 α/19 /	α/5.24/93.5 5.17/6.5				ann.rad./ (0.1898-0.7050)
¹⁷²Os		171.9601	19. s	β⁺ ,EC/99 /4.5 α/1.1/	5.10/	0+			ann.rad./ (0.063-1.120)
¹⁷³Os		172.9598	16. s	β⁺ ,EC/6.3 α/0.4 /	4.94/				ann.rad./ (0.1416-0.2993)
¹⁷⁴Os		173.9563	44. s	β⁺ ,EC/3.9 α/0.02 /	4.76/	0+			0.118 0.138 / 0.001 0.158 0.325
¹⁷⁵Os		174.9570	1.4 m	β⁺ ,EC/5.3					0.125 0.181 0.248
¹⁷⁶Os		175.9550	3.6 m	β⁺ ,EC/3.2		0+			0.8155 0.7758 0.8573 1.2093 1.2909
¹⁷⁷Os		176.9551	2.8 m	β⁺ ,EC/4.5		(1/2-)			0.0848 0.1958 0.3002 1.2686
¹⁷⁸Os		177.9534	5.0 m	β⁺ ,EC/2.3		0+			ann.rad./ 0.5946 0.6850 0.9687 1.3311
¹⁷⁹Os		178.9539	7. m	β⁺ ,EC/3.7					ann.rad./ 0.0654 0.2186 0.5938
¹⁸⁰Os		179.9524	21.5 m	β⁺ ,EC/1.5		0+			Re k x-ray 0.0202-0.7174
¹⁸¹ᵐOs			1.75 h	EC/		(1/2-)			ann.rad./ 0.0489
¹⁸¹Os		180.9532	2.7 m	EC/2.9		(7/2-)			ann.rad./ 0.11794 0.23868 0.8267 (0.07 - 2.64)
¹⁸²Os		181.95219	21.5 h	EC/0.9		0+			Re k x-ray 0.1802 0.5100
¹⁸³ᵐOs			9.9 h	EC/84 / I.T./16 /		1/2-			Os k x-ray Re k x-ray 1.1020 1.1080
¹⁸³Os		182.9531	13. h	EC/2.1		9/2+	-0.79	+3.1	Re k x-ray 0.1144 0.3818
¹⁸⁴Os	0.020(3)	183.952491				0+			
¹⁸⁵Os		184.954043	93.6 d	EC/1.013		1/2-			Re k x-ray 0.6461 0.8748 0.8805
¹⁸⁶Os	1.58(10)	185.953838	2.x10¹⁵ y	α/	≈ 2.75/	0+			
¹⁸⁷Os	1.6(1)	186.955748				1/2-	+0.0646519		
¹⁸⁸Os	13.3(2)	187.955836				0+			
¹⁸⁹ᵐOs			5.8 h	I.T./0.0308		9/2-			Os L x-ray 0.0308
¹⁸⁹Os	16.1(3)	188.958145				3/2+	+0.65993	+0.86	
¹⁹⁰ᵐOs			9.9 m	I.T./1.705		10-	-0.6		Os k x-ray 0.1867 0.3611 0.5026 0.6161
¹⁹⁰Os	26.4(4)	189.958445				0+			
¹⁹¹ᵐOs			13.1 h	I.T./0.0744		3/2-			Os k x-ray 0.0744

Elem. or Isot.	Natural Abundance (%)	Atomic Mass or Weight	Half-Life	Decay Mode/Energy (/MeV)	Particle Energy /Intensity (MeV / %)	Spin (h/2π)	Nuclear Magnetic Mom.(nm)	Elect. Quadr. Mom. (b)	γ-ray Energy /Intensity (MeV / %)
¹⁹¹Os		190.960928	15.4 d	β⁻ /0.314	0.140/100	9/2-		+2.5	Ir k x-ray 0.1294
¹⁹²ᵐOs			6.0 s	I.T./2.0154		(10-)			Os k x-ray 0.2058/65.9 0.5692/70 (0.201-1.000)
¹⁹²Os	41.0(3)	191.961479				0+			
¹⁹³Os		192.964148	30.5 h	β⁻ /1.141	1.04/20	3/2-	+0.730	+0.47	Ir k x-ray 0.1389 0.4605
¹⁹⁴Os		193.965179	6.0 y	β⁻ /0.097	0.054/33 0.096/67	0+			Ir L x-ray 0.0429
¹⁹⁵Os		194.9681	6.5 m	β⁻ /2.0	2.0/				
¹⁹⁶Os		195.96962	34.9 m	β⁻ /1.16	0.84/	0+			0.1262/5 0.4079/5.9
₇₇Ir		192.217(3)							
¹⁶³Ir		164.9876	0.3 ms	p/87 α/13	1.71 6.72				
¹⁶⁶ᵐIr			15 ms	α/98.2 p/1.8	6.56 1.32				
¹⁶⁶Ir		165.9855	≈ 11 ms	α/93 p/6.9	6.56 1.15				
¹⁶⁷ᵐIr			30 ms	α/48,β⁺ p/32	6.35 1.06				
¹⁶⁷Ir		166.9817	35. ms	α/80,β⁺ p/0.4	6.41 1.24				
¹⁶⁸Ir		167.9799		α/					
¹⁶⁹Ir		168.9764	0.4 s	α/					
¹⁷⁰Ir		169.9743	1.0 s	α/	6.03/				
¹⁷¹Ir		170.9718	1.5 s	α/	5.91/				
¹⁷²Ir		171.9706	2.1 s	α/	5.811/				0.228 (0.379-0.475)
¹⁷³Ir		172.9677	3.0 s	α/	5.665/				0.0493 (0.092-0.296)
¹⁷⁴Ir		173.9668	4. s	α/	5.478/				0.1587 (0.276-1.33)
¹⁷⁵Ir		174.9641	≈ 4.5 s	α/	5.393/				0.1056
¹⁷⁶Ir		175.9635	8. s	EC, β⁺/80 α/3.2/	5.118/				0.260 (0.135-0.415)
¹⁷⁷Ir		176.9612	30. s	EC, β⁺/5.7 α/0.06/	5.011/				0.184 (0.062-0.194)
¹⁷⁸Ir		177.9601	12. s	β⁺ ,EC/6.3					0.1320 0.2667 0.3633
¹⁷⁹Ir		178.9592	4. m	EC/4.9					0.0975 (0.045-0.220)
¹⁸⁰Ir		179.9593	1.5 m	EC/6.4					0.2765 ((0.132-1.106)
¹⁸¹Ir		180.9576	4.9 m	β⁺ ,EC/4.1		(7/2+)			ann.rad./ 0.1076 (0.0196-1.715)
¹⁸²Ir		181.9582	15. m	β⁺ /44 /5.6 EC/56 /					ann.rad./ Os k x-ray 0.1273 0.2370
¹⁸³Ir		182.9568	57. m	β⁺ ,EC/3.5					ann.rad./ 0.0877 0.2285 0.2824
¹⁸⁴Ir		183.9574	3.0 h	β⁺ /12 /4.6 EC/88 /	2.3/ 2.9/	5-	0.70	+2.0	ann.rad./ Os k x-ray 0.11968 0.2640 0.3904
¹⁸⁵Ir		184.9566	14. h	β⁺ /3 /2.4 EC/97 /		(5/2-)	2.60	-2.1	ann.rad./ Os k x-ray 0.2543 1.8288
¹⁸⁶ᵐIr			1.7 h	EC /		(2-)			Os k x-ray 0.1371 0.7675
¹⁸⁶Ir		185.95795	15.7 h	EC/98 /3.83 β⁺ /2 /		(5+)	3.9	-2.5	Os k x-ray 0.1372 0.2968 0.4348 (0.13 - 3.0)

Elem. or Isot.	Natural Abundance (%)	Atomic Mass or Weight	Half-Life	Decay Mode/Energy (/MeV)	Particle Energy /Intensity (MeV / %)	Spin (h/2π)	Nuclear Magnetic Mom.(nm)	Elect. Quadr. Mom. (b)	γ-ray Energy /Intensity (MeV / %)
¹⁸⁷Ir		186.95736	10.5 h	EC/1.50		3/2+			Os k x-ray 0.0743 0.4009 0.4271 0.6109 0.9128
¹⁸⁸Ir		187.95885	1.72 d	β⁺ /2.81 EC/99+ /	1.13/ 1.64/	(2-)	0.30	+5.4	Os k x-ray 0.1550 0.4780 0.6330 2.2146
¹⁸⁹Ir		188.95872	13.2 d	EC/0.53		3/2+	0.13	+1.0	Os k x-ray 0.2449
¹⁹⁰ᵐ²Ir			3.09 h	β⁺ ,EC/95 / I.T./5 /		(11-)			0.376
¹⁹⁰ᵐ¹Ir			1.12 h	I.T. /0.0263		7+			Ir L x-ray
¹⁹⁰Ir		189.9606	11.8 d	EC/2.0		(4+)	0.04	+2.8	Os k x-ray 0.1867 0.4072 0.5186 0.5580 0.6051 (0.2 - 1.4)
¹⁹¹ᵐIr			4.93 s	I.T./0.1714		11/2-	0.603		Ir k x-ray 0.1294
¹⁹¹Ir	37.3(5)	190.960591				3/2+	+0.151	+0.82	
¹⁹²ᵐ²Ir			241. y	I.T./0.161		(9+)			Ir k x-ray
¹⁹²ᵐ¹Ir			1.44 m	I.T./0.0580		(1+)			Ir L x-ray 0.0580 0.3165
¹⁹²Ir		191.962602	73.83 d	β⁻ /1.460		(4-)	+1.92	+2.3	Pt k x-ray 0.31649/83. 0.46806/48.
¹⁹³ᵐIr			10.53 d	I.T./0.0802		11/2-			Ir L x-ray 0.0803
¹⁹³Ir	62.7(5)	192.962923				3/2+	+0.164	+0.75	
¹⁹⁴ᵐIr			170. d	β⁻ /		11			Pt k x-ray 0.3284 0.4829 0.5624
¹⁹⁴Ir		193.965075	19.3 h	β⁻ /2.247	1.92/9 2.25/86	1-	0.39	+0.34	0.2935 0.3284 0.6451 (0.1 - 2.2)
¹⁹⁵ᵐIr			3.9 h	β⁻ /	0.41/ 0.97/	(11/2-)			Pt k x-ray 0.3199/9.6 0.3649/9.5 0.4329/9.6 0.6849/9.6
¹⁹⁵Ir		194.965976	2.8 h	β⁻ /1.120	1.0/80 1.11/13	(3/2+)			Pt k x-ray 0.0989/9.7
¹⁹⁶ᵐIr			1.40 h	β⁻ /	1.16/				Pt k x-ray 0.3557 0.3935 0.4471 0.5214 0.6473
¹⁹⁶Ir		195.96838	52.s	β⁻ /3.21	2.1/15 3.2/80	0-			0.3329 0.3557 0.7796
¹⁹⁷ᵐIr			8.9 m	β⁻ / I.T./		(11/2-)			0.3465 See Ir[197]
¹⁹⁷Ir		196.96964	5.8 m	β⁻ /2.16	1.5/ 2.0/	(3/2+)			0.0531 0.1351 0.4306 0.4697
¹⁹⁸Ir		197.9723	8. s	β⁻ /4.1					0.4074 0.5070
₇₈Pt		195.078(2)							
¹⁶⁶Pt			0.3 ms	α/	7.11/				
¹⁶⁷Pt			0.7 ms	α/	6.99/				
¹⁶⁸Pt		167.9880	2.0 ms	α	6.83				
¹⁶⁹Pt		168.9864	3. ms	α					
¹⁷⁰Pt		169.9816	15 ms	α	6.55				
¹⁷¹Pt		170.9811	0.03 s	α					
¹⁷²Pt		171.97730	0.10 s	α /	6.31/94	0+			

Elem. or Isot.	Natural Abundance (%)	Atomic Mass or Weight	Half-Life	Decay Mode/Energy (/MeV)	Particle Energy /Intensity (MeV / %)	Spin (h/2π)	Nuclear Magnetic Mom.(nm)	Elect. Quadr. Mom. (b)	γ-ray Energy /Intensity (MeV / %)
173Pt		172.9765	0.34 s	β+,EC/8.2 α/	6.20/				
174Pt		173.97281	0.89 s	β+,EC/17 /5.6 α/83 /	6.040/	0+			
175Pt		174.9723	2.5 s	β+,EC/65 /7.6 α/35 /	5.831/5 5.96/54 6.038/				0.0774 0.1354 0.2128
176Pt		175.9690	6.3 s	β+,EC/60 /5.1 α/40 /	5.528/0.6 5.750/41	0+			ann.rad./ 0.2277
177Pt		176.9685	11. s	EC/91 /6.8 α/9 /	5.53/ 5.485/3 5.525/6				0.0908
178Pt		177.9649	21. s	EC/93 /4.5 α/7 /	5.286/0.2 5.442/7	0+			
179Pt		178.9653	33. s	β+,EC/5.7 α /	5.16/				
180Pt		179.9632	52. s	β+,EC/99.7 /3.7 α/0.3 /	5.140/	0+			
181Pt		180.9632	51. s	β+,EC/5.2					
182Pt		181.9613	2.7 m	β+,EC/2.9		0+			ann.rad./ 0.1360 0.1460 0.2100
183mPt			43. s	β+,EC/ I.T./		(7/2-)			ann.rad./ 0.3132 0.3164 0.6296
183Pt		182.9617	7. m	β+,EC/4.6					ann.rad./
184Pt		183.9599	17.3 m	β+,EC/2.3					ann.rad./ 0.1549 0.1919 0.5484
185mPt			33. m	β+,EC/		1/2-			
185Pt		184.9607	1.18 h	β+,EC/3.8		(9/2+)			ann.rad./ 0.1353 0.1974 0.2296 0.2551
186Pt		185.95943	2.0 h	β+,EC/1.38		0+			ann.rad./ 0.6115 0.6892
187Pt		186.9607	2.35 h	β+,EC/3.1		3/2			ann.rad./ Ir k x-ray 0.1064 0.1100 0.2015 0.2849 0.7092
188Pt		187.95940	10.2 d	EC/0.51		0+			Ir k x-ray 0.1876 0.1951
189Pt		188.96083	10.9 h	β+,EC/1.97		3/2-	0.43	-0.6	Ir k x-ray 0.0943 0.6076 0.7214 (0.09 - 1.47)
190Pt	0.01(1)	189.95993	4.5x10^11 y			0+			
191Pt		190.961684	2.96 d	EC/1.02		(3/2-)	0.50	-0.6	Ir k x-ray 0.3599 0.4094 0.5389
192Pt	0.79(6)	191.961035				0+			
193mPt			4.33 d	I.T./0.1498		13/2+	-0.75		Pt k x-ray 0.1355
193Pt		192.962984	60. y	EC/0.0566		(1/2-)			Ir k x-rays
194Pt	32.9(6)	193.962663				0+			
195mPt			4.02 d	I.T./0.2952		13/2+	0.61	+1.4	Pt k x-ray 0.0989
195Pt	33.8(6)	194.964774				1/2-	+0.6095		
196Pt	25.3(6)	195.964934				0+			
197mPt			1.590 h	I.T./97 / β-/3 /		13/2+			Pt k x-ray 0.0530 0.3465
197Pt		196.967323	18.3 h	β-/0.719		1/2-	0.51		Au k x-ray 0.1914 0.2688

Elem. or Isot.	Natural Abundance (%)	Atomic Mass or Weight	Half-Life	Decay Mode/Energy (/MeV)	Particle Energy /Intensity (MeV / %)	Spin (h/2π)	Nuclear Magnetic Mom.(nm)	Elect. Quadr. Mom. (b)	γ-ray Energy /Intensity (MeV / %)
¹⁹⁸Pt	7.2(2)	197.967875				0+			
¹⁹⁹ᵐPt			13.6 s	I.T./0.424		13/2+			Pt k x-ray 0.3919
¹⁹⁹Pt		198.970576	30.8 m	β⁻ /1.70	0.90/18 1.14/14	(5/2-)			0.3170/4.9 0.49375/5.7 0.5430/14.8 (0.055-1.293)
²⁰⁰Pt		199.97142	12.5 h	β⁻ /≈0.66		0+			Au k x-ray 0.13590 0.22747 0.24371
²⁰¹Pt		200.9745	2.5 m	β⁻ /2.66		(5/2-)			0.070 0.152 0.222 1.760
²⁰²Pt		201.9757	1.8 d						0.440
₇₉Au		196.96655(2)							
¹⁷¹Au		170.9918	1.0 ms	p/46 α/54	1.69 7.00				
¹⁷²Au		171.9901	4 ms	α/7.02	6.86				
¹⁷³Au		172.9864	0.06 s	α					
¹⁷⁴Au		173.9842	0.12 s	α					
¹⁷⁵Au		174.9817	0.20 s	α					
¹⁷⁶Au		175.9803	1.2 s	β⁺ ,EC/10.5 α /	6.260/80 6.290/20				
¹⁷⁷Au		176.9772	1.2 s	α /	6.115/ 6.150/				
¹⁷⁸Au		177.9760	2.6 s	α /	5.920/				
¹⁷⁹Au		178.9732	7.5 s	α /	5.85/				
¹⁸⁰Au		179.9724	8.1 s	EC/8.6 α/	5.65 5.61 5.50				0.1522 0.2564 0.5242 0.6765 0.8084 0.8597
¹⁸¹Au		180.9700	11.4 s	EC/97.5/6.3 α/2.7/	5.482/				
¹⁸²Au		181.9686	21. s	β⁺ ,EC/6.9 α/0.13/					ann.rad./ 0.1549 0.2649 (0.13 - 1.4)
¹⁸³Au		182.9676	42. s	EC/5.5 α/0.8/			+1.97		0.1630 0.2730 0.3625
¹⁸⁴ᵐAu			48 s			(2+)	+1.44	+1.9	0.069(IT)
¹⁸⁴Au		183.9675	21. s	EC,β⁺ /7.1 α/0.013/		(5+)	+2.07	+4.7	
¹⁸⁵ᵐAu			6.8 m	β⁺ ,EC/ I.T./0.145					
¹⁸⁵Au		184.9657	4.3 m	β⁺ ,EC/4.71 α/0.26/		(5/2-)	+2.17	-1.1	ann.rad./ 0.1915
¹⁸⁶ᵐAu			< 2. m	β⁺ ,EC/					0.1915
¹⁸⁶Au		185.9659	10.7 m	β⁺ ,EC/6.0 α/8(10)⁻⁴/		3-	-1.26	+3.1	ann.rad./ 0.1915 0.2988
¹⁸⁷ᵐAu			2.3 s	IT		9/2-			
¹⁸⁷Au		186.9646	8.3 m	β⁺ ,EC/3.60		1/2+	+0.54		ann.rad./ 0.9152 1.2668 1.3321 1.4081
¹⁸⁸Au		187.9651	8.8 m	β⁺ ,EC/5.3		(1-)	-0.07		ann.rad./ 0.2660 0.3404 0.6061
¹⁸⁹ᵐAu			4.6 m	β⁺ ,EC/		11/2-	+6.19		0.1667
¹⁸⁹Au		188.9642	28.7 m	EC/96 /3.2 β⁺ /4 /		1/2+	+0.49		ann.rad./ Pt k x-ray 0.4478 0.7133 0.8128
¹⁹⁰Au		189.96470	43. m	β⁺ /2 /4.44 EC/98 /		1-	-0.07		ann.rad./ Pt k x-ray 0.2958 0.3018 0.5977

Elem. or Isot.	Natural Abundance (%)	Atomic Mass or Weight	Half-Life	Decay Mode/Energy (/MeV)	Particle Energy /Intensity (MeV / %)	Spin (h/2π)	Nuclear Magnetic Mom.(nm)	Elect. Quadr. Mom. (b)	γ-ray Energy /Intensity (MeV / %)
¹⁹¹ᵐAu			0.9 s	I.T./0.2663		(11/2-	6.6		Au k x-ray 0.2414 0.2526
¹⁹¹Au		190.96365	3.2 h	EC/1.83		3/2+	+0.137	+0.72	Pt k x-ray 0.5864/16 (0.088-1.30)
¹⁹²Au		191.96481	4.9 h	β⁺ /5 /3.52 EC/95 /	2.19/ 2.49/	1-	-0.011	-0.23	ann.rad./ Pt k x-ray 0.2959 0.3165
¹⁹³ᵐAu			3.9 s	I.T./0.2901		11/2-	6.2	+1.98	Au k x-ray 0.2580
¹⁹³Au		192.96413	17.6 h	EC/1.07		3/2+	+0.140	+0.66	Pt k x-ray 0.1862 0.2556
¹⁹⁴Au		193.96534	1.64 d	β⁺ /3 /2.49 EC/97 /	1.49/	1-	+0.075	-0.24	ann.rad./ Pt k x-ray 0.2935 0.3284/61
¹⁹⁵ᵐAu			30.5 s	I.T./0.3186		11/2-	6.2	+1.9	Au k x-ray 0.2617
¹⁹⁵Au		194.965017	186.12 d	EC/0.227		3/2+	+0.149	+0.61	Pt k x-ray
¹⁹⁶²Au			9.7 h	I.T./0.5954		12-	5.7		Au k x-ray 0.1478 0.1883
¹⁹⁶ᵐ¹Au			8.1 s	I.T./0.0846		8+			0.0847
¹⁹⁶Au		195.966551	6.18 d	EC/92 /1.506		2-	+0.591	0.81	Pt k x-ray
¹⁹⁷ᵐAu			7.8 s	I.T./0.4094 β⁻ /8 /0.686		11/2-	+6.0	+1.7	Au k x-ray 0.1302 0.2790
¹⁹⁷Au	100.	196.966551				3/2+	+0.14575	+0.55	
¹⁹⁸ᵐAu			2.30 d	I.T./0.812		(12-)			Au k x-ray 0.0972 0.1803 0.2419
¹⁹⁸Au		197.968225	2.694 d	β⁻ /1.372	0.290/1 0.961/99	2-	+0.5934	+0.68	Hg k x-ray 0.411794
¹⁹⁹Au		198.968748	3.14 d	β⁻ /0.453	0.25/22 0.292/72 0.462/6	3/2+	+0.2715	0.55	Hg k x-ray 0.15837 0.20820
²⁰⁰ᵐAu			18.7 h	β⁻ /84 /1.0 I.T./16 /	0.56/	12-	5.9		Au k x-ray 0.2559/71 0.3680/77 0.4978/73 0.5793/72 0.084-0.904)
²⁰⁰Au		199.97072	48.4 m	β⁻ /2.24	0.7/15 2.2/77	1-			0.3679/19 1.2254/10.6 (0.077-1.570)
²⁰¹Au		200.97165	26. m	β⁻ /1.28	1.27/82	3/2+			(0.027-0.732)
²⁰²Au		201.9738	29. s	β⁻ /3.0		(1-)			0.4396
²⁰³Au		202.97515	1.0 m	β⁻ /2.14	≈ 1.9/	3/2+			(0.04-0.37)
²⁰⁴Au		203.9783	40. s	β⁻ /4.5		(2-)			0.4366 1.5113
²⁰⁵Au		204.9796	31. s	β⁻ /					(0.38 - 1.33)
₈₀Hg		200.59(2)							
¹⁷⁴Hg			≈ 2.7 ms	α	7.07				
¹⁷⁵Hg		174.9912	0.02 s	α					
¹⁷⁶Hg		175.98733	0.03 s	α	6.77				
¹⁷⁷Hg		176.9863	0.13 s	α					
¹⁷⁸Hg		177.98248	0.26 s	EC/50 /6.1 α/50 /	6.43/	0+			
¹⁷⁹Hg		178.9818	1.09 s	EC/8.0 α /	6.29/				
¹⁸⁰Hg		179.9783	2.6 s	EC/5.5 α /	6.12/33 5.69/.03	0+			0.1250 0.3005 0.3812
¹⁸¹Hg		180.9778	3.6 s	β⁺ EC/74 / ≈7.3 α/26 /		(1/2-)	+0.5071		0.0663 0.0811 0.0924 0.1474 0.1587 0.2142 0.2398

Elem. or Isot.	Natural Abundance (%)	Atomic Mass or Weight	Half-Life	Decay Mode/Energy (/MeV)	Particle Energy /Intensity (MeV / %)	Spin (h/2π)	Nuclear Magnetic Mom.(nm)	Elect. Quadr. Mom. (b)	γ-ray Energy /Intensity (MeV / %)
[182]Hg		181.9739	10.8 s	β⁺ ,EC/85 /5.0 α/15 /	5.87/8.6 5.45/0.03	0+			0.1289 0.2168 0.4126
[183]Hg		182.9744	9. s	β⁺ ,EC/77 /6.3 α /	5.83/ 5.91/	1/2-	+0.524		0.0714 0.0874 0.1538
[184]Hg		183.9719	30.9 s	β⁺ ,EC/99 /4.1 α/1 /	5.54/1.3 5.07/2 x 10⁻³	0+			0.0915 0.1265 0.1560 0.2362
[185m]Hg			21. s	β⁺ ,EC,IT,α /	5.37/	13/2+	-1.02	+0.2	0.211 0.292 (0.02 - 0.55)
[185]Hg		184.9720	51. s	β⁺ ,EC/95 /5.8		1/2-	+0.509		0.1119
[186]Hg		185.9695	1.4 m	β⁺ ,EC/3.3 α	5.09/0.02	0+			0.2518
[187m]Hg			1.7 m	β⁺ ,EC/		13/2+	-1.04	+0.5	See Hg[187]
[187]Hg		186.9698	2.4 m	β⁺ ,EC/4.9		3/2-	-0.594		0.1034/32. 0.2334/100. 0.2403/33. 0.27151/31. 0.3763/38. 0.5254/30. (0.10-2.18)
[188]Hg		187.9676	3.2 m	β⁺ ,EC/2.3 α	4.61	0+			0.0988 0.1148 0.1424 0.1900
[189m]Hg			8.6 m	EC/		13/2+	-1.06	+0.7	0.0780 0.3210 0.4345 0.5655 (0.08 - 2.170)
[189]Hg		188.9687	7.6 m	EC/4.2		3/2-	-0.6086	-0.8	0.2005 0.2038 0.2386 0.2485
[190]Hg		189.9663	20.0 m	EC/1.5		0+			0.1296 0.1426
[191m]Hg			51. m	β⁺ /6 / EC/94 /		13/2+	-1.07	+0.6	ann.rad./ Au k x-ray 0.2741 0.4203 0.5787 (0.07 - 1.9)
[191]Hg		190.9671	50. m	β⁺ ,EC/3.2		(3/2-)	-0.62	-0.8	0.1963 0.2247 0.2524 Au k x-ray
[192]Hg		191.9653	5.0 h	EC/≈0.5		0+			0.1572 0.2748 0.3065 Hg k x-ray
[193m]Hg			11.8 h	β⁺ ,EC/91 / I.T./9 /0.2901		13/2+	-1.05843	+0.92	0.1866 0.2580 0.4076 0.5733 0.9324 (0.1 - 1.96)
[193]Hg		192.96664	3.8 h	EC,B⁺/2.34		3/2-	-0.6276		0.1866 0.2580 0.8611 Au L x-rays
[194]Hg		193.96538	520. y	EC/0.04		0+			Hg k x-ray
[195m]Hg			1.67 d	I.T./(54)/0.3186 EC/(46)/		13/2+	-1.04465	+1.1	Au k x-ray 0.2617 0.5603 0.7798 Au k x-ray
[195]Hg		194.96664	9.5 h	EC/1.51		1/2-	+0.541475		0.0614 0.7798 Au k x-ray
[196]Hg	0.15(1)	195.965814				0+			
[197m]Hg			23.8 h	I.T./(93)/0.2989		13/2+	-1.02768	+1.2	Hg k x-ray Au k x-ray 0.13398 Au k x-ray
[197]Hg		196.967195	2.672 d	EC/0.600		1/2-	+0.527374		0.07735
[198]Hg	9.97(8)	197.966752				0+			
[199m]Hg			42.6 m	I.T./0.532		13/2+	-1.014703	+1.2	Hg k x-ray 0.15841

Elem. or Isot.	Natural Abundance (%)	Atomic Mass or Weight	Half-Life	Decay Mode/Energy (/MeV)	Particle Energy /Intensity (MeV / %)	Spin (h/2π)	Nuclear Magnetic Mom.(nm)	Elect. Quadr. Mom. (b)	γ-ray Energy /Intensity (MeV / %)
^{199}Hg	16.87(10)	198.968262				1/2-	+0.505885		
^{200}Hg	23.10(16)	199.968309				0+			
^{201}Hg	13.18(8)	200.970285				3/2-	-0.560226	+0.39	
^{202}Hg	29.86(20)	201.970625				0+			
^{203}Hg		202.972857	46.61 d	β^- /0.492	0.213/100	5/2-	+0.8489	+0.34	Tl k x-ray 0.279188
^{204}Hg	6.87(4)	203.973475				0+			
^{205}Hg		204.976056	5.2 m	β^- /1.531	1.33/4	1/2-	+0.6010		0.20378 (0.2 - 1.4)
^{206}Hg		205.97750	8.2 m	β^- /1.31	0.935/34 1.3/63	0+			Tl k x-ray 0.3052 0.6502
^{207}Hg		206.9825	2.9 m	β^- /4.8		(9/2+)			
^{208}Hg		207.9859	0.7 h	β^-					0.474
$_{81}$Tl		204.3833(2)							
^{179}Tl		178.9917	0.2 s	α					
^{182}Tl		181.9856	3. s	β^+, EC/10.9					0.351 (0.26 - 0.41)
^{183}Tl			0.06 s	α		9/2-			
^{183}Tl		182.9826	5. s	β^+, EC/7.7		1/2+			0.208
^{184}Tl		183.9818	11. s	β^+, EC/(98)/9.2 α/(2)/	6.16/				0.2868 0.3399 0.3667
^{185m}Tl			1.8 s	I.T./0.453 α/5.97	6.01/	(9/2-)			0.1688 0.2840
^{185}Tl		184.9791	20. s	EC/β^+/6.6					
^{186m}Tl			4. s	I.T./0.374					
^{186}Tl		185.9776	28. s	β^+, EC/7.5					0.3738 0.3567 0.4026 0.4053
^{187m}Tl			15.6 s	I.T./$\approx$ 0.33		(9/2+)	3.8	-2.4	0.2995
^{187}Tl		186.9762	50. s	β^+, EC/6.0		1/2+	1.6		
^{188m}Tl			1.18 m	β^+, EC/		(7+)			Hg k x-ray 0.4129 0.5043 0.5921
^{188}Tl		187.9759	1.2 m	β^+, EC/7.8		(2-)	0.48	0.13	See Tl[188m] 0.4129
^{189m}Tl			1.4 m	β^+, EC/		(9/2-)	+3.878	-2.29	0.2156 0.2284 0.3175 0.4452
^{189}Tl		188.9743	2.3 m	β^+, EC/5.2		(1/2+)			0.3337 0.4510 0.5223 0.9422
^{190m}Tl			3.7 m	β^+, EC/	4.2/	(7+)	+0.495	0.29	0.1968 0.4164 0.7311
^{190}Tl		189.9738	2.6 m	β^+, EC/7.0	5.7/	(2-)	+0.25	-0.33	0.4164 0.6254 0.6838 1.0999
^{191m}Tl			5.2 m	β^+, EC/(98)/		(9/2+)	+3.903	-2.3	0.2157 0.2647 0.3256 0.3359
^{191}Tl		190.9723				(1/2)	1.59		
^{192m}Tl			10.8 m	β^+, EC/		(7+)	+0.518	0.46	0.1740 0.4228 0.6348 0.7863 0.7455
^{192}Tl		191.972	9.6 m	β^+, EC/6.4		(2-)	+0.20	-0.33	0.3975 0.4228 0.6908
^{193m}Tl			2.1 m	I.T./(75)/		(9/2-)	+3.948	-2.2	0.3650
^{193}Tl		192.9706	22. m	β^+, EC/3.6		(1/2+)	+1.591		0.2077 0.3244 0.3440 0.6761 1.0447 1.5793

Elem. or Isot.	Natural Abundance (%)	Atomic Mass or Weight	Half-Life	Decay Mode/Energy (/MeV)	Particle Energy /Intensity (MeV / %)	Spin (h/2π)	Nuclear Magnetic Mom.(nm)	Elect. Quadr. Mom. (b)	γ-ray Energy /Intensity (MeV / %)
¹⁹⁴ᵐTl			32.8 m	β⁺ /(20)/ ≈ 0.30 EC/(80)/		(7+)	+0.540	0.62	ann.rad./ Hg k x-ray 0.4282 0.6363 0.7490
¹⁹⁴Tl		193.9711	34. m	β⁺ ,EC/5.3		2-	0.14	-0.28	0.3955 0.4282 0.6363
¹⁹⁵ᵐTl			3.6 s	I.T./0.483		9/2-			Tl k x-ray 0.0990 0.3836
¹⁹⁵Tl		194.9697	1.16 h	EC/(97)/2.8 β⁺ /(3)/		1/2+	+1.58		ann.rad./ Hg k x-ray 0.2422 0.5635 0.8845 1.3639 (0.13 - 2.5)
¹⁹⁶ᵐTl			1.41 h	β⁺ ,EC/(95)/4.9		(7+)	0.55	+0.76	0.0840 0.4261 0.6353 0.6954 (0.08 - 1.0)
¹⁹⁶Tl		195.9705	1.84 h	β⁺ /(15)/4.4 EC/(85)/		2-	0.07	-0.18	ann.rad./ Hg k x-ray 0.4257 0.6105 (0.03 - 2.4)
¹⁹⁷ᵐTl			0.54 s	I.T./(53)/0.608 β⁺ ,EC/(47)/		9/2-			Tl k x-ray 0.2262 0.4118 0.5872 0.6367
¹⁹⁷Tl		196.96954	2.83 h	β⁺ /(1)/2.18 EC/(99)/		1/2+	+1.58		Hg k x-ray 0.1522/8.2 0.4258
¹⁹⁸ᵐTl			1.87 h	β⁺ ,EC/(53)/ I.T./(47)/0.5347		7+	+0.64		Hg k x-ray Tl k x-ray 0.4118 0.5872 0.6367
¹⁹⁸Tl		197.9405	5.3 h	EC,β⁺ /(1)/3.5	1.4/ 2.1/ 2.4/	2-	0.00(1)		Hg k x-ray 0.4118 0.6367 0.6759 (0.23 - 2.8)
¹⁹⁹Tl		198.9698	7.4 h	EC/1.4		1/2-	+1.60		Hg k x-ray 0.2082 0.2473 0.4555
²⁰⁰Tl		199.97095	1.087 d	EC/2.46	1.07/ 1.44/	2-	0.04		Hg k x-ray 0.36799 1.2057 (0.11 - 2.3)
²⁰¹Tl		200.97080	3.040 d	EC/0.48		1/2+	+1.605		Hg k x-ray 0.13528 0.16740/10.0
²⁰²Tl		201.97209	12.23 d	EC/1.36		2-	0.06		Hg k x-ray 0.43957
²⁰³Tl	29.524(14)	202.972329				1/2+	+1.622258		
²⁰⁴Tl		203.973848	3.78 y	β⁻ /(97)/0.7637 EC/(3)/0.347	0.763/97	2-	0.09		Hg k x-ray
²⁰⁵Tl	70.476(14)	204.974412				1/2+	+1.638215		
²⁰⁶ᵐTl			3.76 m	I.T./2.644		12-			Tl k x-ray 0.2166 0.2661 0.4534 0.6866 1.0219
²⁰⁶Tl		205.976095	4.20 m	β⁻ /1.533	1.53/99.9	0-			Pb k x-ray 0.80313
²⁰⁷ᵐTl			1.3 s	I.T./1.350		11/2-			Tl k x-ray 0.3501 1.0000
²⁰⁷Tl		206.97741	4.77 m	β⁻ /1.423	1.43/99.8	1/2+	+1.88		0.89723

Elem. or Isot.	Natural Abundance (%)	Atomic Mass or Weight	Half-Life	Decay Mode/Energy (/MeV)	Particle Energy /Intensity (MeV / %)	Spin (h/2π)	Nuclear Magnetic Mom.(nm)	Elect. Quadr. Mom. (b)	γ-ray Energy /Intensity (MeV / %)
²⁰⁸Tl		207.982004	3.053 m	β⁻ /5.001	1.28/23 1.52/22 1.796/51	(5+)	+0.29		Pb k x-ray 0.27728 0.51061 0.58302 2.61448
²⁰⁹Tl		208.98535	2.16 m	β⁻ /3.98	1.8 /100	(1/2+)			Pb k x-ray 1.5670/100 0.4651/95 (0.12 - 1.33)
²¹⁰Tl		209.99006	1.30 m	β⁻ /5.48	1.3/25 1.9/56	(5+)			Pb k x-ray 0.081 0.2981 0.79788
₈₂Pb		207.2(1)							
¹⁸⁰Pb			5 ms	α/	7.23				
¹⁸¹Pb		180.9967	0.05 s	α/	7.07				
¹⁸²Pb		181.99268	0.06 s	α	6.92				
¹⁸³Pb		182.9919	0.3 s	α/		1/2+			
¹⁸⁴Pb		183.9882	0.6 s	α/	6.63/	0+			
¹⁸⁵Pb		184.9876	4.1 s	α/	6.34/ 6.40/ 6.48/				
¹⁸⁶Pb		185.9835	5. s	β⁺ ,EC/(95)/5.5 α/(5)/	6.32/ 6.34/ < 100 6.01/ < 0.2	0+			
¹⁸⁷ᵐPb			15.2 s	β⁺ ,EC/	5.99/ 6.19/	(1/2-)			0.0674 0.2080 0.2755 0.2995 0.4487 0.7477
¹⁸⁷Pb		186.9839	18.3 s	EC/7.2 α/	6.08/	13/2+			0.1930 0.3314 0.3435 0.3934
¹⁸⁸Pb		187.9811	23. s	EC/(78)/4.8 α/(22)/	5.98/ < 10 5.61/ < 0.1	0+			0.1850 0.7582
¹⁸⁹Pb		188.9809	51. s	EC/6.1 α/	5.58/				
¹⁹⁰Pb		189.9782	1.2 m	β⁺ (13)/4.1 EC/(86)/ α/(0.9)/	5.58/	0+			ann.rad./ Tl k x-ray 0.1415 0.1512 0.9422
¹⁹¹ᵐPb			2.2 m	β⁺ ,EC/		13/2+			ann.rad./ 0.3871 0.6135 0.7122
¹⁹¹Pb		190.9782	1.3 m	β⁺ ,EC/5.5					ann.rad./ 0.9368
¹⁹²Pb		191.9758	3.5 m	β⁺ ,EC/ ≈3.4 α/.006/	5.11	0+			ann.rad./ 0.1675 0.6082 1.1954
¹⁹³ᵐPb			5.8 m	β⁺ ,EC/		13/2+	-1.16		ann.rad./ 0.3650 0.3922
¹⁹³Pb		192.9761	≈ 2. m	EC/5.2		3/2			
¹⁹⁴Pb		193.9740	11. m	β⁺ ,EC/2.7 α	4.64	0+			ann.rad./ 0.2036
¹⁹⁵ᵐPb			15. m	β⁺ /(8)/ EC/(92)/		13/2+	-1.132	+0.29	ann.rad./ Tl k x-ray 0.3836 0.3942 0.8784
¹⁹⁵Pb		194.976	≈ 15. m	β⁺,EC/5.8					ann.rad./ 0.3836 0.3937 0.7776
¹⁹⁶Pb		195.9727	37. m	β⁺ ,EC/2.1		0+			Tl k x-ray 0.2531 0.5021

Elem. or Isot.	Natural Abundance (%)	Atomic Mass or Weight	Half-Life	Decay Mode/Energy (/MeV)	Particle Energy /Intensity (MeV / %)	Spin (h/2π)	Nuclear Magnetic Mom.(nm)	Elect. Quadr. Mom. (b)	γ-ray Energy /Intensity (MeV / %)
¹⁹⁷ᵐPb			43. m	EC/(79)/ β⁺/(2)/ I.T./(19)/0.3193		13/2+	-1.105	+0.5	Tl k x-ray 0.3079 0.3877 0.7743 (0.2 - 2.2)
¹⁹⁷Pb		196.9734	≈ 8. m	EC/(97)/3.6 β⁺/(3)/		(3/2-)	-1.075	-0.08	Tl k x-ray 0.3755 0.3858 0.7611
¹⁹⁸Pb		197.9720	2.4 h	EC/1.4		0+			Tl k x-ray 0.1734 0.2903 0.3654
¹⁹⁹ᵐPb			12.2 m	I.T./(93)/0.4248 β⁺,EC/(7)/		13/2+	-1.074	+0.08	Pb k x-ray 0.4255
¹⁹⁹Pb		198.9729	1.5 h	EC/(99)/2.9 β⁺/(1)/		5/2-			Tl k x-ray 0.3534 0.7202 1.1350 (0.22 - 2.4)
²⁰⁰Pb		199.97182	21.5 h	EC/0.81		0+			Tl k x-ray 0.14763
²⁰¹ᵐPb			1.02 m	I.T./0.6291		13/2+			Pb k x-ray 0.6288
²⁰¹Pb		200.97285	9.33 h	EC/1.90		5/2-	+0.675	-0.009	Tl k x-ray 0.33120 0.36131 (0.11 - 1.8)
²⁰²ᵐPb			3.53 h	I.T./(90)/2.170 β⁺/(10)/		9-	-0.228	+0.58	Pb k x-ray Tl k x-ray 0.42219 0.78700 0.96271
²⁰²Pb		201.97214	5.3x10⁴ y	EC/0.05		0+			Tl L x-ray
²⁰³ᵐPb			6.2 s	I.T./0.8252		13/2+			Pb k x-ray 0.8203 0.8252
²⁰³Pb		202.97338	2.1615 d	EC/0.98		5/2-	+0.686	+0.10	Tl k x-ray 0.279188
²⁰⁴ᵐPb			1.12 h	I.T./2.185		9-			Pb k x-ray 0.37481 0.89922 0.91175
²⁰⁴Pb	1.4(1)	203.973028				0+			
²⁰⁵Pb		204.974467	1.51x10⁷ y	EC/0.0512		5/2-	+0.712	+0.23	Tl L x-ray
²⁰⁶Pb	24.1(1)	205.974449				0+			
²⁰⁷ᵐPb			0.80 s	I.T./1.632		13/2+			Pb k x-ray 0.56915 1.06310
²⁰⁷Pb	22.1(1)	206.975880				1/2-	+0.59258		
²⁰⁸Pb	52.4(1)	207.976636	>2x10¹⁹ y	SF		0+			
²⁰⁹Pb		208.981075	3.25 h	β⁻/0.644	0.645/100	9/2+	-1.474	-0.3	
²¹⁰Pb		209.984174	22.6 y	β⁻/0.0635 α	0.017/81 0.061/19 3.72	0+			
²¹¹Pb		210.988732	36.1 m	β⁻/1.37	0.57/5 1.36/92	(9/2+)	-1.404	+0.09	0.40486 0.42700 0.83186 (0.09 - 1.27)
²¹²Pb		211.991887	10.64 h	β⁻/0.574	0.28/83 0.57/12	0+			Bi k x-ray 0.23858
²¹³Pb		212.9966	10.2 m	β⁻/2.1					
²¹⁴Pb		213.999797	26.9 m	β⁻/1.0	0.67/48 0.73/42	0+			Bi k x-ray 0.24192 0.29509 0.35187
²¹⁵Pb			36 s						
₈₃Bi		208.98038(2)							
¹⁸⁵Bi		184.9977	0.04 ms	p/86 α/14	1.59 7.16				
¹⁸⁶Bi		185.9965	10 ms	α	7.16 7.26				
¹⁸⁷ᵐBi			≈ 8. ms	α					
¹⁸⁷Bi		186.9935	35. ms	α					
¹⁸⁸Bi		187.9922		α					
¹⁸⁹ᵐBi			7.0 ms	α	7.30				
¹⁸⁹Bi		188.9895	0.68 s	α					

Elem. or Isot.	Natural Abundance (%)	Atomic Mass or Weight	Half-Life	Decay Mode/Energy (/MeV)	Particle Energy /Intensity (MeV / %)	Spin (h/2π)	Nuclear Magnetic Mom.(nm)	Elect. Quadr. Mom. (b)	γ-ray Energy /Intensity (MeV / %)
¹⁹⁰Bi		189.9875	5. s	β⁺ ,EC/(10)/8.7 α/(90)/	6.45/				
¹⁹¹Bi		190.9861	12. s	β⁺ ,EC/(60)/7.3 α/(40)/	6.32/				
¹⁹²Bi		191.9854	40. s	β⁺ ,EC/(80)/9.0 α/(20)/	6.06/				
¹⁹¹ᵐBi			3.2 s	β⁺ ,EC/ α/	6.48/	1/2+			
¹⁹³Bi		192.9837	1.11 m	β⁺ ,EC/(40)/7.1 α/(60)/	5.91/	9/2+			
¹⁹⁴Bi		193.9828	1.8 m	β⁺ ,EC/(99.9)/8.2 α/0.1/		(10-)			0.1661 0.1740 0.2802 0.421 0.5754 0.9650
¹⁹⁵ᵐBi			1.45 m	β⁺ ,EC/(94)/ α/(6)/	6.11/				
¹⁹⁵Bi		194.9811	2.9 m	β⁺ ,EC/(99.8)/5.8 α/(0.2)	5.45/	3/2-			
¹⁹⁶Bi		195.9806	5. m	EC/≈7.4					0.1376 0.3720 0.6880 1.0486
¹⁹⁷Bi		196.9789	5. m	β⁺ ,EC/5.2		1/2+			
¹⁹⁸ᵐBi			7.7 s	I.T./0.2485		(10-)			0.2485
¹⁹⁸Bi		197.9790	11.8 m	β⁺ ,EC/6.6		(7+)			0.0900 0.1976 0.5624 1.0635
¹⁹⁹ᵐBi			24.7 m	β⁺ ,EC/					ann.rad./
¹⁹⁹Bi		198.9776	27. m	β⁺ ,EC/4.3		9/2-			0.7203 0.8374 0.8417 0.9460 1.0528 1.3056 (0.12 - 3.2)
²⁰⁰ᵐBi			31. m	β⁺ ,EC/		(2+)			0.2453 0.4198 0.4624 1.0265 ann.rad./
²⁰⁰Bi		199.9781	36. m	EC/(90)/5.9 β⁺ /(10)/		7+			Pb k x-ray 0.4198 0.4623 1.0265 Bi k x-ray
²⁰¹ᵐBi			59.1 m	I.T./0.846 β⁺ ,EC/		(1/2+)			0.8464
²⁰¹Bi		200.97697	1.8 h	EC/3.84		9/2-			Pb k x-ray 0.6288 0.9357 1.0138 (0.13 - 2.4)
²⁰²Bi		201.97768	1.72 h	β⁺ /(3)/5.16 EC/(97)/		5+	+4.26	-0.72	ann.rad./ Pb k x-ray 0.57860 0.92734 (0.08 - 3.5)
²⁰³Bi		202.97687	11.8 h	EC/(99.8)/3.25 β⁺ /(0.2)/	1.35/	9/2-	+4.02	-0.69	Pb k x-ray 0.1865 0.8203 0.8969 1.8475 (0.1 - 2.9)
²⁰⁴Bi		203.97779	11.2 h	EC/4.44		6+	+4.32	-0.43	Pb k x-ray 0.37481 0.89922 0.98409
²⁰⁵Bi		204.97737	15.31 d	EC/2.71		9/2-	+4.07	-0.59	Pb k x-ray 0.70347 1.76435
²⁰⁶Bi		205.97848	6.243 d	EC/3.76		6+	+4.36	-0.39	Pb k x-ray 0.51619 0.80313 0.88100
²⁰⁷Bi		206.978456	35. y	EC/2.399		9/2-	4.08	-0.6	Pb k x-ray 0.56915 1.06310

Elem. or Isot.	Natural Abundance (%)	Atomic Mass or Weight	Half-Life	Decay Mode/Energy (/MeV)	Particle Energy /Intensity (MeV / %)	Spin (h/2π)	Nuclear Magnetic Mom.(nm)	Elect. Quadr. Mom. (b)	γ-ray Energy /Intensity (MeV / %)
²⁰⁸Bi		207.979727	3.68x10⁵ y	EC/2.880		5+			Pb k x-ray 2.61435
²⁰⁹Bi	100.	208.980384				9/2-	+4.111	-0.37	
²¹⁰ᵐBi			3.0x10⁶ y	α/	4.420(3)/0.29 4.569(3)/3.9 4.584(3)/1.4 4.908(4)/39 4.946(3)/55	9-	+2.73	-0.47	Tl k x-ray 0.2661 0.3052 0.6502
²¹⁰Bi		209.984105	5.01 d	β⁻ /1.163	1.16/99	1-	-0.0445	+0.136	0.2661 0.3.52
²¹¹Bi		210.98726	2.14 m	α/(99.7)/ β⁻ /(0.3)/0.58	6.279/16 6.623/84	9/2-			Tl k x-ray 0.3501
²¹²ᵐ²Bi			7. m	β⁻ /		(15-)			
²¹²ᵐ¹Bi			25.0 m	α/(93)/ β⁻ /(7)/	6.300/40 6.340/53	(9-)			0.120 0.233 0.275 0.404 0.727
²¹²Bi		211.991271	1.009 h	β⁻ /(64)/2.254 α/(36)/	6.051/25 6.090/9.6	(1-)	+0.32	+0.1	Tl k x-ray Po k x-ray 0.2881 0.72725 0.78551 1.62066
²¹³Bi		212.99437	45.6 m	β⁻ /(98)/1.43 α/(2)/	1.02/31 1.42/66 5.549/0.16 5.869/2.0	9/2-	+3.72	-0.60	Po k x-ray 0.4404 (0.15 - 1.328)
²¹⁴Bi		213.99870	19.7 m	β⁻ /3.27					1.10006 0.60931 1.12027 1.76449
²¹⁵Bi		215.0018	7.7 m	β⁻ /2.3					(0.19 - 3.2) 0.2937
²¹⁶Bi		216.0062	3.6 m	β⁻/4.0					(0.27 - 0.835) 0.5498
²¹⁷Bi			97 s	β/					0.4192
₈₄Po									
¹⁹⁰Po		189.9951	2.4 ms	α/	7.53				
¹⁹¹Po		190.9947	17 ms	α/	7.33				
¹⁹²Po		191.9915	34. ms	α/8.5	7.17				
¹⁹³ᵐPo			0.24 s	α/	7.00				
¹⁹³Po		192.9911	0.45 s	α/	6.95				
¹⁹⁴Po		193.9883	0.39 s	α/	6.84/93 6.19/0.22	0+			
¹⁹⁵ᵐPo			1.9 s	α/	6.70/				
¹⁹⁵Po		194.9881	4.6 s	α/	6.61/				
¹⁹⁶Po		195.9855	5.8 s	α/(95)/ β⁺ ,EC/(5)/ ≈4.6	6.52/94 5.77/0.02	0+			
¹⁹⁷ᵐPo			25.8 s	α/(84)/ β⁺ ,EC/(16)/	6.385(3)/55	13/2+			
¹⁹⁷Po		196.9856	53. s	α/(44)/ β⁺ ,EC/(56)/6.2	6.282(4)/76	(3/2-)			
¹⁹⁸Po		197.9834	1.76 m	α/(70)/ β⁺ ,EC/(30)/4.0	6.18/57 5.27/7.6 x 10⁻⁴	0+			
¹⁹⁹ᵐPo			4.2 m	β⁺ ,EC/(51)/ α/(39)/	6.059/24	13/2+	0.99		ann.rad./ 0.2745 0.4998 1.0020 Bi k x-ray 0.1877 0.3616 1.0214
¹⁹⁹Po		198.985	5.2 m	β⁺ ,EC/(88)/7. α/(12)/	5.952/7.5	(3/2-)			1.0344 0.14748 0.32792 0.6176
²⁰⁰Po		199.9817	11.5 m	β⁺ ,EC/(85)/3.4 α/(15)/	5.863/11.1	0+			0.6709 Bi k x-ray Po k x-ray 0.2726
²⁰¹ᵐPo			8.9 m	β⁺ ,EC/(57)/ I.T./(40)/0.418 α/(3)/	5.786/ ≈3.	13/2+	1.0		0.4123 0.4179 0.9670

Elem. or Isot.	Natural Abundance (%)	Atomic Mass or Weight	Half-Life	Decay Mode/Energy (/MeV)	Particle Energy /Intensity (MeV / %)	Spin (h/2π)	Nuclear Magnetic Mom.(nm)	Elect. Quadr. Mom. (b)	γ-ray Energy /Intensity (MeV / %)
²⁰¹Po		200.9822	15.3 m	β⁺,EC/(98)/4.9 α/(2)/	5.683(3)/1.1	3/2-	0.94		Bi k x-ray 0.2056 0.2250 0.8483 0.9048
²⁰²Po		201.9807	45. m	β⁺,EC/(98)/2.8 α/(2)/	5.588/1.9	0+			0.0410 0.1656 0.3158 0.6884
²⁰³ᵐPo			1.2 m	I.T./(96)/0.6414 β⁻ EC/(4)/		13/2+			Bi k x-ray Po k x-ray 0.6414
²⁰³Po		202.9814	35. m	β⁺,EC/4.2		5/2-	+0.74		0.17516 0.21477 0.89350 0.90863 1.09095
²⁰⁴Po		203.98031	3.53 h	EC/2.34 α	5.377/0.66	0+			Bi k x-ray 0.2702 0.8844 1.0162 (0.11 - 1.9)
²⁰⁵Po		204.98117	1.7 h	β⁺,EC/3.53		5/2-	+0.76	+0.17	Bi k x-ray 0.83681 0.84983 0.87241 1.00124 (0.12 - 2.7)
²⁰⁶Po		205.98047	8.8 d	EC/(95)/1.85 α/(5)/	5.223/5.5	0+			Bi k x-ray 0.28644 0.31156 0.51134 0.80737 1.03228 (0.11 - 1.5)
²⁰⁷ᵐPo			2.8 s	I.T./1.383		19/2-			Po k x-ray 0.2682 0.30074 0.81448
²⁰⁷Po		206.98158	5.80 h	EC,β⁺ /2.91		5/2-	+0.79	+0.28	Bi k x-ray 0.74263 0.91176 0.99225
²⁰⁸Po		207.981231	2.898 y	α/5.213	4.233/0.0002 5.1158/100	0+			
²⁰⁹Po		208.982415	102. y	α/4.976	4.624/0.56 4.879/99.2	1/2-	+0.77		0.26049 0.8964
²¹⁰Po		209.982857	138.38 d	α/5.407	4.516/0.001 5.304/100	0+			0.80313
²¹¹ᵐPo			25.2 s	α/	7.273/91 7.994/1.7 8.316/0.25 8.875/7.0	25/2+			Pb k x-ray 0.32808 0.56915 0.89723 1.06310
²¹¹Po		210.986637	0.516 s	α/7.594	6.570/0.54 6.892/0.55 7.450/98.9	9/2+			0.56915 0.89723
²¹²ᵐPo			45. s	α/	8.514/2.0 9.086/1.0 11.650/97	16+			
²¹²Po		211.988852	0.298 μs	α/8.953	8.784/100	0+			
²¹³Po		212.992843	4. μs	α/8.537	7.614/0.003 8.375/100	9/2+			
²¹⁴Po		213.995186	163.7 μs	α/7.833	6.904/0.01 7.686/99.99	0+			0.7995 0.298
²¹⁵Po		214.999415	1.780 ms	α/7.526	6.950/0.02 6.957/0.03 7.386/100	(9/2+)			
²¹⁶Po		216.001905	0.145 s	α/6.906	5.895/0.002 6.778/99.99	0+			
²¹⁷Po		217.0064	< 10. s	α/6.662	6.539/				
²¹⁸Po		218.008965	3.04 m	α/6.114	5.181/1.00	0+			
₈₅At									
¹⁹³At		192.9998	40 ms	α/					
¹⁹⁴At		193.9990	40 ms	α/					
¹⁹⁵At		194.9965	140 ms	α/					
¹⁹⁶At		195.9957	0.3 s	α/	7.05/				

Elem. or Isot.	Natural Abundance (%)	Atomic Mass or Weight	Half-Life	Decay Mode/Energy (/MeV)	Particle Energy /Intensity (MeV / %)	Spin (h/2π)	Nuclear Magnetic Mom.(nm)	Elect. Quadr. Mom. (b)	γ-ray Energy /Intensity (MeV / %)
¹⁹⁷ᵐAt			4. s	α		(1/2+)			
¹⁹⁷At		196.9939	0.35 s	β⁺ ,EC/7.8 α/	6.96/	(9/2-)			
¹⁹⁸ᵐAt			1.5 s	β⁺ ,EC/(75)/ α/(25)/	6.85/86				
¹⁹⁸At		197.9928	5. s	α/	6.75/94				
¹⁹⁹At		198.9910	7.1 s	β⁺ ,EC/(8)/5.6 α/(92)/	6.64/	9/2-			
²⁰⁰ᵐAt			4.3 s	β⁺ ,EC/(80) α/(20)/	6.536/12	10-			
²⁰⁰At		199.990	43. s	β⁺ ,EC/(65)/ ≈ 8.0 α/(35)/	6.412/44 6.465/57	5+			
²⁰¹At		200.9885	1.48 s	β⁺ ,EC/(29)/5.9 α/(71)/6.474	6.344/	9/2-			
²⁰²ᵐAt			≤ 1.5 s	I.T./0.391					
²⁰²At		201.9885	3.02 m	β⁺ ,EC/(88)/7.2 α/(12)/	6.135/7.7 6.225/4.3	5+			ann.rad./ 0.4413 0.5697 0.6753
²⁰³At		202.9868	7.4 m	β⁺ ,EC/(69)/5.1 α/(31)/6.210	6.088/	9/2-			0.1458 0.2459 0.6414 1.0020 1.0340
²⁰⁴At		203.9873	9.1 m	β⁺ ,EC/(95)/6.5 α/(5)/	5.951/	(5+)			Po k x-ray 0.3271 0.4254 0.5156 0.6837
²⁰⁵At		204.98604	26. m	β⁺ ,EC/(90)/4.54 α/(10)/6.020	5.902/	(9/2-)			Po k x-ray 0.1543 0.6696 0.7194
²⁰⁶At		205.98660	29.4 m	β⁺ ,EC/(99)/5.72 α/(1)/5.881	5.703/	5+			Po k x-ray 0.20186 0.39561 0.47716 0.70071
²⁰⁷At		206.98578	1.81 h	β⁺ ,EC/(90)/3.91 α/(10)/5.873	5.758/	9/2-			Po k x-ray 0.16801 0.58842 0.81448
²⁰⁸At		207.98657	1.63 h	β⁺ ,EC/(99)/4.97 α/(1)/5.752	5.626/0.01 5.641/0.53	(6+)			Po k x-ray 0.1770 0.2060 0.6601 0.6852 0.8450 1.0281
²⁰⁹At		208.98616	5.4 h	β⁺ ,EC/(96)/3.49 α/(4)/5.757	5.647/4.1	(6+)			Po k x-ray 0.10422 0.54503 0.78189 0.79020 (0.1 - 2.6)
²¹⁰At		209.98713	8.1 h	EC/99.8/3.98 α/(0.2)/5.632	5.361/0.05 5.442/0.05	5+			Po k x-ray 0.24535 0.52758 1.18143 1.43678 1.48335 (0.04 - 2.4)
²¹¹At		210.987481	7.21 h	EC/(58)/0.787 α/(42)/5.980	5.211/0.004 5.868/42	9/2-			Po k x-ray 0.66956 0.6870 0.74263
²¹²ᵐAt			0.119 s	α/	7.837/65 7.897/33	(9-)			
²¹²At		211.990735	0.314 s	α/7.828	7.058/0.4 7.088/0.6 7.618/15 7.681/84	(1-)			
²¹³At		212.992922	0.11 μs	α/9.254	9.080/	9/2-			
²¹⁴ᵐAt			0.76 μs	α/8.762		(9-)			
²¹⁴At		213.996357	0.56 μs	α/8.987	8.819/100	(1-)			
²¹⁵At		214.99864	0.10 ms	α/8.178	7.626/0.045 8.023/99.9	(9/2-)			0.40486

Elem. or Isot.	Natural Abundance (%)	Atomic Mass or Weight	Half-Life	Decay Mode/Energy (/MeV)	Particle Energy /Intensity (MeV / %)	Spin (h/2π)	Nuclear Magnetic Mom.(nm)	Elect. Quadr. Mom. (b)	γ-ray Energy /Intensity (MeV / %)
²¹⁶At		216.002408	0.30 ms	α/7.947	7.595/0.2 7.697/2.1 7.800/97	(1-)			
²¹⁷At		217.00471	32. ms	α/7.202	6.812/0.06 7.067/99.9	(9/2-)			0.2595 0.3345 0.5940
²¹⁸At		218.00868	1.6 s	α/6.883	6.654/6 6.695/90 6.748/4				
²¹⁹At		219.0113	50. s	α/6.390	6.275/				
²²⁰At		220.0153	3.71 m	β⁻ /3.7					(0.24-0.70)
²²¹At		221.0181	2.3 m	β					
²²²At		222.0223	0.9 m	β					
²²³At		223.0253	50. s	β					
₈₆Rn									
¹⁹⁶Rn		195.9977	≈ 3 ms	α/	7.49				
^{197m}Rn			0.02 s	α	7.36				
¹⁹⁷Rn		196.9983	0.07 s	α/	7.26				
¹⁹⁸Rn		197.9988	0.05 s	α					
^{199m}Rn			0.3 s	α		(13/2+)			
¹⁹⁹Rn		198.9983	0.62 s	α/		3/2-			
²⁰⁰Rn		199.9957	1.06 s	α/(98)/ EC/(2)/5.	6.901/	0+			0.4329 0.5043
^{201m}Rn			3.8 s	EC/(10)/ α/(90)/	6.773/	13/2+			
²⁰¹Rn		200.9955	7.0 s	α/(80)/ EC/(20)/	6.725/ α/6.778	(3/2-)			
²⁰²Rn		201.9932	9.9 s	α/(12)/ EC/(88)/	6.641/	0+			0.5695 (0.2876-0.6255)
^{203m}Rn			28. s	α/	6.551	13/2+	-0.96	+1.3	
²⁰³Rn		202.9948	45. s	α/(66)/6.629 EC/(34)/≈7.4	6.499/	0			
²⁰⁴Rn		203.9914	1.24 m	α/(68)/ EC/(32)/3.8	6.420/	0+			
²⁰⁵Rn		204.9917	2.8 m	α/(23)/6.390 EC/(77)/5.2	6.123(3)/0.02 6.262(3)/23	(5/2-)	+0.80	+0.06	0.2652 0.3553 0.4648 0.6205 0.6753 0.7300
²⁰⁶Rn		205.9902	5.7 m	α/(68)/6.384 EC/(32)/3.3	6.258(3)/	0+			0.06170 0.0968 0.3245 0.3862 0.4822 0.4973 0.7728
²⁰⁷Rn		206.9907	9.3 m	β⁺ ,EC/(77)/4.6 α/(23)/6.252	5.995(4)/0.02 6.068(3)/0.15 6.126(3)/22.8	5/2-	+0.82	+0.22	At k x-ray 0.32947 0.34455 0.36767 0.40267 0.74723 (0.18 - 1.4)
²⁰⁸Rn		207.98963	24.3 m	α/(60)/6.260 EC/(40)/2.85	5.469(2)/0.003 6.140(2)/60	0+			
²⁰⁹Rn		208.99038	29. m	β⁺ /(83)/3.93 α/(17)/	2.16/2.3 5.887(3)/0.04 5.898(3)/0.02 6.039(2)/16.9	5/2-	+0.8388	+0.31	At k x-ray 0.27933 0.33753 0.40841 0.68942 0.74594 (0.18 - 3.2)
²¹⁰Rn		209.98968	2.4 h	α/(96)/6.157 EC/(4)/2.37	5.351(2)/0.005 6.039(2)/96	0+			At k x-ray 0.19625 0.45824 0.57104 0.64868 (0.14 - 1.7)
²¹¹Rn		210.99059	14.6 h	β⁺ ,EC/(74)/2.89 α/(26)/5.964	5.619(1)/0.7 5.784(1)/16.4 5.851(1)/8.8	1/2-	+0.60		At k x-ray 0.16877 0.25022 0.37049 0.67412 0.67839 1.36298 (0.11 - 2.7)

Elem. or Isot.	Natural Abundance (%)	Atomic Mass or Weight	Half-Life	Decay Mode/Energy (/MeV)	Particle Energy /Intensity (MeV / %)	Spin (h/2π)	Nuclear Magnetic Mom.(nm)	Elect. Quadr. Mom. (b)	γ-ray Energy /Intensity (MeV / %)
²¹²Rn		211.990689	24. m	α/6.385	5.587(4)/0.05 6.260(4)/99.95	0+			
²¹³Rn		212.99387	25.0 ms	α/8.243	7.552(8)/1.0 8.087(8)/99	9/2+			
²¹⁴Rn		213.99535	0.27 μs	α/9.209	9.037(9)/	0+			
²¹⁵Rn		214.99873	2.3 μs	α/8.840	8.674(8)/	(9/2+)			
²¹⁶Rn		216.00026	45. μs	α		0+			
²¹⁷Rn		217.003915	0.6 ms	α/7.885	7.500/0.1 7.742(4)/100	9/2+			
²¹⁸Rn		218.005586	35. ms	α/7.267	6.534(1)/0.16 7.133(1)/99.8	0+			0.6093 0.6653
²¹⁹Rn		219.009475	3.96 s	α/6.946(1)	6.3130(5)/0.05 6.425(3)/7.5 6.5309(4)/0.12 6.5531(3)/12.2 6.8193(3)/81	(5/2+)	-0.44	+0.93	Po K x-ray 0.13057 0.27113 0.40170 (0.1 - 1.05)
²²⁰Rn		220.011384	55.6 s	α/6.404	5.7486(5)/0.07 6.2883(1)/99.9	0+			
²²¹Rn		221.0156	25. m	α/(22)/6.148 β⁻/(78)/1.2	5.778(3)/1.8 5.788(3)/2.2 6.037(3)/18	7/2+	-0.020	-0.38	Fr L x-ray 0.07384 0.08323 0.0610 0.18639 0.510
²²²Rn		222.017570	3.8235 d	α/5.590	4.987(1)/0.08 5.4897(3)/99.9	0+			
²²³Rn		223.0218	23. m	β⁻/					
²²⁴Rn		224.0241	1.8 h	β⁻/		0+			0.1085 0.2601 0.2655
²²⁵Rn		225.0284	4.5 m	β⁻/		7/2	-0.70	+0.84	
²²⁶Rn		226.0309	7.4 m	β⁻/					
²²⁷Rn		227.0354	2. s	β⁻/					
²²⁸Rn		228.0381	65. s	β⁻/					
₈₇Fr									
²⁰⁰Fr		200.0065	≈ 20 ms	α	7.47				
²⁰¹Fr		201.0046	0.05 s	α/	7.36/	(9/2-)			
²⁰²Fr		202.0033	0.34 s	α/7.590	7.237(8)/100				
²⁰³Fr		203.0014	0.55 s	α/7.280	7.132(5)/	(9/2-)			
²⁰⁴Fr		204.001	2.1 s	α/	7.03/96 6.97/90 7.01/74				
²⁰⁵Fr		204.9987	3.9 s	α/7.050	6.914(5)/	(9/2-)			
²⁰⁶ᵐFr			0.7 s	α/	6.93				0.531(IT)
²⁰⁶Fr		205.9985	16.0 s	α/7.416	6.792(5)/84				
²⁰⁷Fr		206.9969	14.8 s	α/6.900	6.766(5)/	9/2-	+3.9	-0.16	
²⁰⁸Fr		207.99713	59.1 s	α/(77)/6.770 EC/(23)/6.99	6.636(5)/	7+	-4.8	+0.004	
²⁰⁹Fr		208.99592	50.0 s	α/(89)/5.1 EC/(11)/5.16	6.646(3)/	9/2-	+3.9	-0.24	0.7978 (0.1103-1.384)
²¹⁰Fr		209.99640	3.2 m	α/6.670 EC/6.26	6.543(5)/	6+	+4.4	+0.19	0.2030 0.6438 0.8175 0.9008
²¹¹Fr		210.99553	3.10 m	α/6.660 EC/4.61	6.534(5)/	9/2-	+4.0	-0.19	0.220 0.2799 0.5389 0.9169
²¹²Fr		211.99618	20. m	EC/(57)/5.12 α/(43)/6.529	6.261(1)/16 6.335(1)/4 6.335(1)/4 6.343(1)/1.3 6.383(1)/10 6.406(1)/9.5 6.08-6.18	(5+)	+4.6	-0.10	Rn x-ray 0.08107 0.08378 0.2277 1.1856 1.2748 0.014-1.178
²¹³Fr		212.99617	34.6 s	α/6.905	8.476(4)/51 8.547(4)/46 6.775-8.046	9/2-	+4.0	-0.14	
²¹⁴ᵐFr			3.4 ms	α/		9-			
²¹⁴Fr		213.99895	5.1 ms	α/8.587	7.409(3)/0.3 7.605(8)/1.0 7.940(3)/1.0 8.355(3)/4.7 8.427(3)/93	(1-)			
²¹⁵Fr		215.00033	0.12 μs	α/9.537	9.360(8)/	(9/2-)			
²¹⁶Fr		216.00319	0.70 μs	α/9.175	9.005(10)/95				
²¹⁷Fr		217.00462	0.016 ms	α/8.471	8.315(8)/	(9/2-)			(0.045-0.160)
²¹⁸ᵐFr			22. ms	α					

Elem. or Isot.	Natural Abundance (%)	Atomic Mass or Weight	Half-Life	Decay Mode/Energy (/MeV)	Particle Energy /Intensity (MeV / %)	Spin (h/2π)	Nuclear Magnetic Mom.(nm)	Elect. Quadr. Mom. (b)	γ-ray Energy /Intensity (MeV / %)
^{218}Fr		218.00756	1. ms	α/8.014	7.384(10)/0.5 7.542(15)/1.0 7.572(10)/5 7.732(10)/0.5 7.867(2)/93	(1-)			
^{219}Fr		219.00924	21. ms	α/8.132	6.802(2)/0.25 6.967(2)/0.6 7.146(2)/0.25 7.313(2)/99	(9/2-)			
^{220}Fr		220.012313	27.4 s	α/6.800	6.582(1)/10 6.630(2)/6 6.641(1)/12 6.686(1)/61 6.39-6.58	1+	-0.67	+0.47	0.0450 0.061 0.1060 0.1539 0.1617 At k x-ray
^{221}Fr		221.01425	4.8 m	α/6.457	5.9393(7)/0.17 5.9797(7)/0.49 6.0751(7)/0.15 6.1270(7)/ 6.2433(3)/1.3 6.3410(7)/83.4	(5/2-)	+1.58	-1.0	0.0995 0.21798 0.4091
^{222}Fr		222.01754	14.3 m	β⁻/2.03 α/5.850	1.78/	2-	+0.63	+0.51	
^{223}Fr		223.019731	22.0 m	β⁻/1.149	1.17/65	(3/2+)	+1.17	+1.17	0.05014 0.07972 (0.13 - 0.9)
^{224}Fr		224.02323	3.0 m	β⁻/2.82		1-	+0.40	+0.517	0.13150 0.21575 0.8367 (0.1 - 2.21)
^{225}Fr		225.02561	3.9 m	β⁻/1.87		3/2	+1.07	+1.3	
^{226}Fr		226.0293	49. s	β⁻/3.6		1	+0.071	-1.35	0.18606 0.25373
^{227}Fr		227.0318	2.48 m	β⁻/2.5		1/2	+1.50		
^{228}Fr		228.0357	39. s	β⁻/≈3.5		2-	-0.76	+2.4	
^{229}Fr		229.0384	50. s	β⁻/					
^{230}Fr		230.0425	19. s	β⁻/		(3)			
^{231}Fr		231.0454	17. s	β⁻/					
^{232}Fr		232.0500	5. s	β⁻/					
$_{88}$Ra									
^{202}Ra			≈ 3 ms	α	7.86				
^{203m}Ra			0.03 s	α	7.62				
^{203}Ra		203.0092	≈ 4 ms	α	7.58				
^{204}Ra		204.0065	0.06 s	α	7.48				
^{205m}Ra			≈ 0.17 s						
^{205}Ra		205.0062	0.22 s	α	7.34				
^{206}Ra		206.0038	0.4 s	α/7.416	7.272(5)/	0+			
^{207}Ra		207.0037	1.3 s	α/7.270	7.133(5)/				
^{208}Ra		208.0018	1.4 s	α/7.273	7.133(5)/	0+			
^{209}Ra		209.0019	4.6 s	α/7.150	7.008(5)/	5/2	+0.87	+0.38	
^{210}Ra		210.0005	3.7 s	α/7.610	7.020(5)/	0+			
^{211}Ra		211.0009	13. s	α/7.046 EC/5.0	6.912(5)/	(5/2-)	+0.878	+0.46	
^{212}Ra		211.99978	13.0 s	α/7.033	6.901(2)/	0+			
^{213m}Ra			2.1 ms	IT					
^{213}Ra		213.00034	2.7 m	EC/(20)/3.88 α/(80)/6.860	6.521(3)/4.8 6.622(3)/39 6.730(3)/36	(1/2-)	+0.613		0.1024 0.11010 0.2125
^{214}Ra		214.00009	2.46 s	α/7.272	7.136(4)/	0+			
^{215}Ra		215.00270	1.6 ms	α/8.864	7.883(6)/2.8 8.171(3)/1.4 8.700(3)/95.9	(9/2+)			
^{216}Ra		216.00352	0.18 μs	α/9.526	9.349(8)/	0+			
^{217}Ra		217.00631	1.6 μs	α/9.161	8.992(8)/	9/2-			
^{218}Ra		218.00712	26. μs	α/8.547	8.390(8)/	0+			
^{219}Ra		219.01006	0.010 s	α/8.132	7.680(10)/65 7.982(9)/35				
^{220}Ra		220.01101	25. ms	α/7.593	6.998(7)/1.0 7.455(7)/99	0+			0.465
^{221}Ra		221.01391	29. s	α/6.879	6.254(10)/0.7 6.578(5)/3 6.585(3)/8 6.608(3)/35 6.669(3)/21 6.758(3)/31	5/2	-0.180	+1.9	
^{222}Ra		222.015361	36.2 s	α/5.590	6.237(2)/3.0 6.556(2)/97	0+			0.324 (0.1448-0.8402)

Elem. or Isot.	Natural Abundance (%)	Atomic Mass or Weight	Half-Life	Decay Mode/Energy (/MeV)	Particle Energy /Intensity (MeV / %)	Spin (h/2π)	Nuclear Magnetic Mom.(nm)	Elect. Quadr. Mom. (b)	γ-ray Energy /Intensity (MeV / %)
^{223}Ra		223.018497	11.435 d	α/5.979	5.287(1)/0.15 5.338(1)/0.13 5.365(1)/0.13 5.433(5)/2.3 5.502(1)/1.0 5.540(1)/9.2 5.607(3)/24 5.716(3)/52 5.747(1)/9 5.857(1)/0.32 5.872(1)/0.85	(3/2+)	+0.271	+1.2	Rn k x-ray 0.12231 0.14418 0.15418 0.15859 0.26939 0.32388 0.33328 0.44494 (0.10 - 0.7)
^{224}Ra		224.020202	3.66 d	α/5.789	5.034(10)/0.003 5.047(1)/0.007 5.164(5)/0.007 5.449(2)/4.9 5.685(2)/95	0+			Rn k x-ray 0.2407 0.4093 0.6501
^{225}Ra		225.023603	14.9 d	β⁻/0.36	0.32/100	(3/2+)	-0.734		Ac k x-ray 0.0434
^{226}Ra		226.025402	1599. y	α/4.870	4.194(1)/0.001 4.343(1)/0.006 4.601(1)/5.5 4.784(1)/94	0+			Rn k x-ray 0.1861 0.2624
^{227}Ra		227.029170	42. m	β⁻/1.325	1.03/ 1.30/	(3/2+)	-0.404	+1.5	Ac L x-ray Ac k x-ray 0.02739
^{228}Ra		228.031063	5.76 y	β⁻/0.046	0.039/50 0.014/30 0.026/20	0+			0.0135 (0.006-0.0306)
^{229}Ra		229.0348	4.0 m	β⁻/1.76	1.76/	(3/2+)	+0.503	+3.0	(0.0145-0.1715)
^{230}Ra		230.03708	1.5 h	β⁻/1.0	0.7/	0+			0.0631 0.0720 0.2028 0.4698 0.4787
^{231}Ra		231.0412	1.7 m	β⁻					
^{232}Ra		232.0437	4. m	β⁻					
^{233}Ra		233.0480	30. s	β⁻					
^{234}Ra		234.051	≈ 30. s	β⁻/					
$_{89}$Ac									
^{206m}Ac			0.04 s	α	7.79				
^{206}Ac			≈ 26 ms	α	7.75				
^{207}Ac		207.0121	27 ms	α/	7.69				
^{208m}Ac			≈25. ms	α/	7.72				
^{208}Ac		208.0115	≈0.1 s	α/	7.62				
^{209}Ac		209.0096	≈0.09 s	α/	7.58				
^{210}Ac		210.0093	0.4 s	α/7.610	7.462(8)/				
^{211}Ac		211.0076	0.3 s	α/7.620	7.480(8)/				
^{212}Ac		212.0078	0.9 s	α/7.520	7.379(8)/				
^{213}Ac		213.0066	0.8 s	α/7.500	7.364(8)/	(9/2-)			
^{214}Ac		214.0069	8.2 s	α/(86)/7.350 EC/(14)/6.34	7.007(8)/3 7.082(5)/38 7.214(5)/45	(5+)			
^{215}Ac		215.0065	0.17 s	α/7.750	7.604(5)/	(9/2-)			
^{216m}Ac			0.33 ms	α/	8.198(8)/1.7 8.283(8)/2.5 9.028(5)/49 9.106(5)/46	(9-)			
^{216}Ac		216.00871	≈ 0.3 ms	α/9.241	8.990(2)/10 9.070(8)/90	(1)			
^{217m}Ac			0.7 μs	α/	10.540/100				
^{217}Ac		217.00933	0.07 μs	α/9.832	9.650(10)/100	9/2-			
^{218}Ac		218.01162	1.1 μs	α/9.380	9.205(15)/				
^{219}Ac		219.01241	0.012 ms	α/8.830	8.664(10)/	(9/2-)			
^{220}Ac		220.0148	26. ms	α/8.350	7.610(20)/23 4.680(20)/21 7.790(10)/13 7.850(10)/24 7.985(10)/4 8.005(10)/5 8.060(10)/6 8.195(10)/3				
^{221}Ac		221.01558	52. ms	α/7.790	7.170(10)/2 7.375(10)/10 7.440(15)/20 7.645(10)/70				

Elem. or Isot.	Natural Abundance (%)	Atomic Mass or Weight	Half-Life	Decay Mode/Energy (/MeV)	Particle Energy /Intensity (MeV / %)	Spin (h/2π)	Nuclear Magnetic Mom.(nm)	Elect. Quadr. Mom. (b)	γ-ray Energy /Intensity (MeV / %)
222m Ac			63. s	α/(>89)/ EC/(1)/ I.T./(<10)/	6.710(20)/7 6.750(20)/13 6.810(20)/24 6.840(20)/9 6.890(20)/13 6.970(20)/7 7.000(20)/13				
222 Ac		222.01782	5. s	α/7.141	6.967(10)/6 7.013(2)/94	1-			
223 Ac		223.01913	2.1 m	α/(>½)/6.783 EC/(1)/0.59	6.131(2)/0.12 6.177(2)/0.94 6.293(1)/0.47 6.326(1)/0.3 6.332(2)/0.14 6.360(1)/0.22 6.397(1)/0.13 6.448(1)/0.2 6.473(1)/3.1 6.523(2)/0.6 6.528(1)/3.1 6.563(1)/13.6 6.582(3)/0.3 6.646(1)/44 6.661(1)/31	(5/2-)			0.0725 0.0839 0.0927 0.0990 0.1917 0.2158 0.3588 0.4768
224 Ac		224.021708	2.7 h	EC/(90)/1.403 α/(10)/6.323	5.841(1)/0.5 5.860(1)/0.75 5.875(1)/1.7 5.941(1)/4.4 6.000(1)/6.7 6.013(1)/1.4 6.056(1)/22 6.138(1)/26 6.154(1)/1.0 6.204(1)/12 6.210(1)/20	0-			Ra L kx-ray Ra k x-ray 0.08426 0.13150 0.1571 0.21575 0.2619 (0.03 - 0.3)
225 Ac		225.02322	10.0 d	α/5.935	5.286(1)/0.2 5.444(3)/0.1 5.554(1)/0.1 5.608(1)/1.1 5.636(1)/4.5 5.681(1)/1.4 5.722(1)/2.9 5.731(1)/10 5.791(1)/9 5.793(1)/18	3/2			Fr k x-ray 0.9958 0.9982 0.1084 0.1116 0.1451 0.1539 0.15724 0.18799 0.19575 0.2162 0.21686 (0.025 - 0.52)
226 Ac		226.026089	1.224 d	EC/(17)/0.640 β⁻/(83)/1.116 α/(0.006)/5.510	5.399(5)/0.006	(1-)			Ra k x-ray Th k x-ray 0.07218 0.15816 0.23034
227 Ac		227.027747	21.77 y	β⁻/(98.6)/0.045 α/(1.4)/5.043	0.045/54 4.869(1)/0.09 4.938(1)/0.52 4.951(1)/0.65	(3/2-)	+1.1	+1.7	0.0838/23. 0.0811/14. 0.2696/13. (0.044 - 1.27)
228 Ac		228.031014	6.15 h	β⁻/2.127	1.11/32 1.85/12 2.18/11	(3+)			Th L x-ray Th k x-ray 0.12903 0.33842 0.91116 0.96897 (0.2 - 1.96)
229 Ac		229.03293	1.04 h	β⁻/1.10	1.1/	(3/2+)			0.07450 0.16451 0.26188 0.5085 0.56916
230 Ac		230.0360	2.03 m	β⁻/2.7	1.4/	1+			Th k x-ray 0.45497 0.50820 (0.12 - 2.5)
231 Ac		231.0386	7.5 m	β⁻/2.1	2.1/100	(1/2+)			0.14379 0.18574 0.22140 0.28250 0.3070
232 Ac		232.0420	2.0 m	β⁻/3.7		(2-)			
233 Ac		233.0446	2.4 m	β⁻/		(1/2+)			

Elem. or Isot.	Natural Abundance (%)	Atomic Mass or Weight	Half-Life	Decay Mode/Energy (/MeV)	Particle Energy /Intensity (MeV / %)	Spin (h/2π)	Nuclear Magnetic Mom.(nm)	Elect. Quadr. Mom. (b)	γ-ray Energy /Intensity (MeV / %)
^{234}Ac		234.0484	40. s	β^- /		(1+)			
$_{90}$Th		232.0381(1)							
^{209}Th			≈ 0.01 s	α	8.08				
^{210}Th		210.0150	≈ 12 ms	α	7.90				
^{211}Th		211.0149	0.04 s	α	7.79				
^{212}Th		212.0129	≈ 30. ms	α/	7.80/	0+			
^{213}Th		213.0130	0.14 s	α/7.840	7.692(10)/				
^{214}Th		214.0115	0.09 s	α/7.825	7.677(10)/	0+			
^{215}Th		215.0117	1.2 s	α/7.660	7.33(10)/8	(1/2-)			
					7.395(8)/52				7.524(8)/40
^{216m}Th			0.18 ms	α					
^{216}Th		216.01105	28. ms	α/8.071	7.921(8)/	0+			
^{217}Th		217.01306	0.25 ms	α/9.424	9.250(10)/				
^{218}Th		218.01327	0.11 μs	α/9.847	9.665(10)/	0+			
^{219}Th		219.01552	1.05 μs	α/9.510	9.340(20)/				
^{220}Th		220.01573	10. μs	α/8.953	8.790(20)/	0+			
^{221}Th		221.01817	1.7 ms	α/8.628	7.743(8)/6				
					8.146(5)/56				
					8.4272(5)/39				
^{222}Th		222.01845	2.8 ms	α/8.129	7.982(8)/9.7	0+			
					7.600(15)/3				
^{223}Th		223.02079	0.65 s	α/7.454	7.29(1)/41(5)				
					7.32(1)/29(5)				
					7.350(15)/20(5)				
					7.390(15)/10(4)				
^{224}Th		224.02146	1.05 s	α/7.305	6.768(5)/1.2				
					6.997(5)/19				
					7.170(5)/79				
^{225}Th		225.02394	8.72 m	EC/(10)/0.68 α/(90)/6.920	6.441(2)/15	(3/2+)			
					6.479(2)/43				
					6.501(3)/14				
					6.627(3)/3				
					6.650(5)/3				
					6.700(5)/2				
					6.743(3)/7				
					6.796(2)/9				
^{226}Th		226.024891	30.83 m	α/6.454	6.026(1)/0.2	0+			Ra k x-ray 0.1112
					6.041(1)/0.19				0.2421
					6.098(1)/1.3				0.1310
					6.2283(4)/23				(0.1733-0.9295)
					6.3375(4)/75				
^{227}Th		227.027699	18.72 d	α/6.146		(3/2+)			Ra L x-ray Ra k x-ray 0.05014 0.23597 0.25624 (0.02 - 1.0)
^{228}Th		228.028731	1.913 y	α/5.520	5.1770(2)/0.18	0+			
					5.2114(1)/0.4				
					5.3405(1)/26.7				
					5.4233(1)/73				
^{229}Th		229.031754	7.9x10³ y	α/5.168	4.814/9.3	5/2+	+0.46	+4.	
					4.845(5)/56				
					4.9008(5)/10.2				
					4.689-5.077				
^{230}Th		230.033126	7.54x10⁴ y	α/4.771	4.4383(6)/0.03	0+			0.0677/0.46 0.1439/0.078
					4.4798(6)/0.12				
					4.6211(6)/23.4				
					4.6876(6)/76.3				
^{231}Th		231.036296	1.063 d	β^-/0.390	0.138/22	5/2+			Pa L x-ray Pa k x-ray 0.02564 0.084203/ (0.02 - 0.3)
					0.218/20				
					0.305/52				
^{232}Th	100.	232.038050	1.4x10¹⁰ y	α/4.081 SF/9x10⁻⁸	3.830(10)/0.2	0+			0.0590 0.124
					3.952(5)/23				
					4.010(5)/77				
^{233}Th		233.041576	22.3 m	β^-/1.245	1.245/	1/2+			Pa L x-ray Pa k x-ray 0.02938 0.08653 0.45930 (0.02 - 1.2)
^{234}Th		234.043596	24.10 d	β^-/0.273	0.102/20	0+			Pa L x-ray 0.06329/4.1 0.09235/2.4
					0.198/72				

Elem. or Isot.	Natural Abundance (%)	Atomic Mass or Weight	Half-Life	Decay Mode/Energy (/MeV)	Particle Energy /Intensity (MeV / %)	Spin (h/2π)	Nuclear Magnetic Mom.(nm)	Elect. Quadr. Mom. (b)	γ-ray Energy /Intensity (MeV / %)
²³⁵Th		235.04751	7.2 m	β⁻ /1.9					0.09278/2.4 0.4162 0.6594 0.7272 0.747 0.9318
²³⁶Th		236.0497	37.5 m	β⁻ / ≈ 1.0					Pa k x-ray 0.1107
²³⁷Th		237.0539	5.0 m	β⁻					
₉₁Pa		231.03588(2)							
²¹²Pa			≈ 5 ms	α	8.27				
²¹³Pa		213.0212	7 ms	α	8.24				
²¹⁴Pa		214.0207	17 ms	α	8.12				
²¹⁵Pa		215.0190	15. ms	α	8.08/100				
²¹⁶Pa		216.0190	0.19 s	α/	7.95/57 7.86/18 7.84/14				
²¹⁷ᵐPa			1.5 ms	α/	10.16/100				
²¹⁷Pa		217.0183	3.4 ms	α/8.490	8.340(10)/100				
²¹⁸Pa		218.0200	0.12 ms	α/	9.54/31 9.61/69				
²²¹Pa		221.0219	6. μs	α	9.08(3)				
²²²Pa		222.0237	≈ 4.3 ms	α/8.700	8.180/50 8.330/20 8.540/30				
²²³Pa		223.0240	≈ 6.5 ms	α/8.340	8.006(10)/55 8.196(10)/45				
²²⁴Pa		224.0256	0.84 s	α/7.630	7.555(10)/75(3) 7.46(1)/25(3)				0.1945 (0.028-0.412)
²²⁵Pa		225.0261	1.8 s	α/7.380	7.195(10)/30 7.245(10)/70				
²²⁶Pa		226.02792	1.8 m	α/(74)/6.987 EC/(26)/2.83	6.728(10)/0.7 6.823(10)/35 6.863(10)/35				
²²⁷Pa		227.02879	38.3 m	α/(85)/6.582 EC/(15)/1.02	6.357(4)/7 6.376(10)/2.2 6.401(4)/8 6.416(4)/13 6.423(10)/10 6.465(4)/43	(5/2-)			0.0649 0.0669 0.1100
²²⁸Pa		228.03100	22. h	EC/(98)/2.111 α/(2)	5.779/0.23 5.805/0.15 6.078/0.4 6.105/0.25 6.118/0.22	(3+)	+3.5		Th k x-ray 0.20939 0.27026 0.28202 0.32767 0.32807 0.46310 0.91116 0.96464 0.96897 (0.1 - 1.96)
²²⁹Pa		229.03209	1.5 d	EC/(99.8)/0.32 α/(0.2)/5.836	5.536(2)/0.02 5.579(2)/0.09 5.668(2)/0.05	(5/2)			0.04244 (0.024 - 0.18)
²³⁰Pa		230.034532	17.4 d	EC/(90)/1.310 β⁻ /(10)/0.563	0.51/	(2-)	2.0		Th L x-ray Th k x-ray 0.4437 0.45477 0.89876 0.91856 0.95199 (0.053-1.07)
²³¹Pa		231.035878	3.25x10⁴ y	α/5.148	4.6781(5)/1.5 4.7102(5)/1.0 4.7343(5)/8.4 4.8513(5)/1.4 4.9339(5)/3 4.9505(5)/22.8 4.9858(5)/1.4 5.0131(5)/25.4 5.0292(5)/20 5.0318(5)/2.5 5.0587(5)/11	3/2-	2.01	-1.	Ac L x-ray Ac k x-ray 0.01899 0.027396 0.03823 0.04639 0.25586 0.26029 0.28367 0.30007 0.30264 0.33007 (0.02 - 0.61)

Elem. or Isot.	Natural Abundance (%)	Atomic Mass or Weight	Half-Life	Decay Mode/Energy (/MeV)	Particle Energy /Intensity (MeV / %)	Spin (h/2π)	Nuclear Magnetic Mom.(nm)	Elect. Quadr. Mom. (b)	γ-ray Energy /Intensity (MeV / %)
²³²Pa		232.03858	1.31 d	β⁻ /1.34		(2-)			U k x-ray 0.10900 0.15009 0.89439 0.96934 (0.10 - 1.17)
²³³Pa		233.040239	27.0 d	β⁻ /0.571	0.15/40 0.256/60	3/2-	+4.0	-3.0	U L x-ray U k x-ray 0.30017 0.31201 0.34059
²³⁴ᵐPa			1.17 m	β⁻ /(99.9)/2.29 IT/(0.13)/		(0-)			U k x-ray 0.25818/0.07 0.76641/0.32 1.0009/0.85 (0.06 - 1.96)
²³⁴Pa		234.043303	6.69 h	β⁻ /2.197	0.51/	(4+)			U L x-ray U k x-ray 0.1312/0.03 0.5695/0.02 0.9256/0.02 (0.02 - 1.99)
²³⁵Pa		235.04544	24.4 m	β⁻ /1.41	1.4/97	(3/2-)			0.0308-0.65893
²³⁶Pa		236.0487	9.1 m	β⁻ /2.9	1.1/40 2.0/50 3.1/10	(1-)			U k x-ray 0.64235 0.68759 1.7630 (0.04 - 2.18)
²³⁷Pa		237.0511	8.7 m	β⁻ /2.3	1.1/60 1.6/30 2.3/10	(1/2+)			0.4986 0.5293 0.5407 0.8536 0.8650 (0.04 - 1.4)
²³⁸Pa		238.0545	2.3 m	β⁻ /3.5	1.2/ 1.7/	(3-)			0.10350 0.1785 0.4484 0.6350 0.6800 1.01446 (0.04 - 2.5)
²³⁹Pa		239.0571	1.8 h						
₉₂U		238.0289(1)							
²¹⁸U		218.0235	≈0.002 s	α	8.63(3)/				
²¹⁹U		219.0249	0.04 ms	α	9.68(4)/				
²²²U		222.0261	≈ 1.μs	α					
²²³U		223.0277	0.02 s	α/	8.78(4)/				
²²⁴U		224.02759	≈ 1. ms	α/	8.46/100				
²²⁵U		225.02938	0.09 s	α/	7.88/85 7.82/15				
²²⁶U		226.02933	0.5 s	α/7.560	7.430/	0+			
²²⁷U		227.03113	1.1 m	α/7.200	6.870/				
²²⁸U		228.03137	9.1 m	α/6.803	6.404(6)/0.6 6.440(5)/0.7 6.589(5)/29 6.681(6)/70	0+			0.095 0.152 0.187 0.246
²²⁹U		229.03350	58. m	EC/(80)/1.31 α/(20)/6.473	6.223/3 6.297(3)/11 6.332(3)/20 6.360(3)/64	(3/2+)			
²³⁰U		230.033927	20.8 d	α/5.992	5.5866(3)/0.01 5.6624(3)/0.26 5.6663(3)/0.38 5.8178(3)/32 5.8887(3)/67	0+			Th L x-ray 0.07218 0.15421 0.23034 (0.081-0.8565)
²³¹U		231.03626	4.2 d	EC/0.36 α/(10⁻³)	5.46/1.6 x 10⁻³ 5.47/1.4 x 10⁻³ 5.40/1. x 10⁻³	(5/2-)			Pa L x-ray Pa k x-ray 0.02564 0.08420
²³²U		232.037146	68.9 y	α/5.414	4.9979(1)/0.003 5.1367(1)/0.3 5.2635(1)/31 5.3203(1)/69	0+			
²³³U		233.039627	1.59x10⁵ y	α/4.909	4.7830(8)/13.2 4.8247(8)/84.4 4.510-4.804	5/2+	+0.59	3.66	Th L x-ray 0.04244 0.09714 (0.0252-1.119)

Elem. or Isot.	Natural Abundance (%)	Atomic Mass or Weight	Half-Life	Decay Mode/Energy (/MeV)	Particle Energy /Intensity (MeV / %)	Spin (h/2π)	Nuclear Magnetic Mom.(nm)	Elect. Quadr. Mom. (b)	γ-ray Energy /Intensity (MeV / %)
²³⁴U	0.0055(5)	234.040945	2.45x10⁵ y	α/4.856	4.604(1)/0.24 4.7231(1)/27.5 4.776(1)/72.5	0+			0.05323/0.156 0.12091
²³⁵ᵐU			26. m	IT/0.0007		1/2+			
²³⁵U	0.720(1)	235.043922	7.04x10⁸ y	α/4.6793	4.1525(9)/0.9 4.2157(9)/5.7 4.3237(9)/4.6 4.3641(9)/11 4.370(4)/6 4.3952(9)/55 4.4144(9)/2.1 4.5025(9)/1.7 4.5558(9)/4.2 4.5970(9)/5.0	7/2-	-0.38	4.6	Th L x-ray Th k x-ray 0.10917 0.14378 0.16338 0.18574 0.20213 0.20533 0.22140 (0.03 - 0.79)
²³⁶U		236.045561	2.34x10⁷ y	α/4.569	4.332(8)/0.26 4.445(5)/26 4.494(3)/74	0+			Th L x-ray 0.04937 0.11275
²³⁷U		237.048723	6.75 d	β⁻/0.519	0.24/ 0.25/	1/2+			Np L x-ray Np k x-ray 0.05953 0.20801
²³⁸U	99.2745(15)	238.050784	4.46x10⁹ y	α	4.039(5)/0.23 4.147(5)/23 4.196(5)/77	0+			Th L x-ray 0.04955/.06 0.1135/.01 (0.522-0.681)
²³⁹U		239.054289	23.5 m	β⁻/1.265	1.2/ 1.3/	5/2+			
²⁴⁰U		240.056585	14.1 h	β⁻/0.39	0.36/	0+			Np L x-ray 0.04410 0.05558 0.06760
²⁴²U		242.0629	16.8 m	β⁻/≈ 1.2					
₉₃Np									
²²⁵Np		225.0339							
²²⁶Np		226.0351	0.03 s	α/	8.04(2)/				
²²⁷Np		227.0350	0.51 s	α/	7.65(2)/ 7.68(1)/				
²²⁸Np		228.0362	61. s	EC/60(7)/ α/40(7)/,SF					
²²⁹Np		229.0363	4.0 m	α/7.010	6.890(20)				
²³⁰Np		230.0378	4.6 m	EC/97 /3.6 α/3	6.660(20)				
²³¹Np		231.03823	48.8 m	EC/98 /1.8 α/2 /6.368	6.280/2	5/2			0.2629 0.3475 0.3703
²³²Np		232.0400	14.7 m	EC/99 /2.7		(4-)			U L x-ray U k x-ray 0.3268 0.81925 0.86683
²³³Np		233.0410	36.2 m	EC/1.2		(5/2+)			U L x-ray U k x-ray 0.29887 0.31201
²³⁴Np		234.04289	4.4 d	β⁺,EC/1.81	0.79/	(0+)			U L x-ray U k x-ray 1.5272 1.5587 1.6022
²³⁵Np		235.044055	1.085 y	EC/99.9 /0.124 α/0.001/5.191		5/2+			U k x-ray
²³⁶ᵐNp			22.5 h	EC/52 / β⁻/48 /		(1-)			U L x-ray Pu L x-ray U k x-ray 0.64235 0.68759
²³⁶Np		236.04657	1.55x10⁵ y	EC/91 /0.94 β⁻/9 /0.49		(6-)			U L x-ray U k x-ray 0.10423 0.16031
²³⁷Np		237.048166	2.14x10⁶ y	α/4.957	4.6395(5)/6.5 4.766(5)/9.7 4.7715(5)/22.7 4.7884(5)/47.8 4.558-4.873	5/2+	+3.14	+3.89	Pa L x-ray Pa k x-ray 0.029378/15 0.08653/12 (0.03-0.28)

Elem. or Isot.	Natural Abundance (%)	Atomic Mass or Weight	Half-Life	Decay Mode/Energy (/MeV)	Particle Energy /Intensity (MeV / %)	Spin (h/2π)	Nuclear Magnetic Mom.(nm)	Elect. Quadr. Mom. (b)	γ-ray Energy /Intensity (MeV / %)
^{238}Np		238.050940	2.117 d	β^- /1.292	1.2/	2+			Pu L x-ray Pu k x-ray 0.98447/25.2 1.02855/18.3 (.044-1.026)
^{239}Np		239.052931	2.355 d	β^- /0.722	0.341/30 0.438/48	5/2+			Pu L x-ray Pu k x-ray 0.10613 0.228186/11 0.27760/15 (0.04-0.50)
^{240m}Np			7.22 m	β^- /99.9 / IT/0.1 /	2.18/	(1+)			0.25143 0.26333 0.55454 0.59735
^{240}Np		240.05617	1.032 h	β^- /2.20	0.89/	5+			0.1471/ 0.5664 0.6008
^{241}Np		241.0583	13.9 m	β^- /1.3	1.3/	5/2+			0.1330/ 0.1740 0.280
^{242m}Np			2.2 m	β^- /		(1+)			0.15910 0.2651/ 0.78570 0.9448/
^{242}Np		242.0616	5.5 m	β^- /2.7	2.7/	6+			0.6209 0.73620 0.78074 1.47340 (0.04-2.37)
^{243}Np		243.0643	1.9 m						
^{244}Np		244.0678	2.3 m						
$_{94}$Pu									
^{228}Pu		228.0387		α/	7.81(2)/				
^{229}Pu		229.0362	4.0 m	α/	7.46(3)/				
^{230}Pu		230.03964	4.6 m	α/	7.05/				
^{232}Pu		232.04118	34. m	EC/ >80/1.1 α/ <20/6.716	6.542(10)/38 6.600(10)/62	0+			
^{233}Pu		233.04299	20.9 m	EC(99.9)/1.9 α/0.1 /6.416	6.300(20)/0.1				0.1503 0.1804 0.2353 0.5002 0.5346/ 1.0352/
^{234}Pu		234.04331	8.8 h	EC/94 /0.39 α/6 /6.310	6.035(3)/0.024 6.149(3)/1.9 6.200(3)/4.0	0+			
^{235}Pu		235.0453	25.3 m	EC/99+ /1.2 α/0.003/5.957	5.850(20)/0.003	(5/2+)			
^{236}Pu		236.046048	2.87 y	α/5.867	5.611/0.21 5.7210/30.5 5.7677(1)/69.3	0+			0.0476/0.07 0.109/0.02 (0.17 - 0.97)
^{237}Pu		237.048403	45.7 d	EC/99.9 /0.220 α/0.003 /5.747	5.334(4)/0.0015 5.356(4)/0.0006 5.650(4)/0.0007	7/2-			Np L x-ray Np k x-ray 0.026344 0.03319 0.05954 (0.03-0.5)
^{238}Pu		238.049553	87.74 y	α/5.593	5.3583(1)/0.10 5.465(1)/28.3 5.4992(1)/71.6	0+			U k x-ray 0.04347 (0.04-1.1)
^{239}Pu		239.052156	2.411x10^4 y	α/5.244	5.055/0.047 5.076/0.078 5.106/11.9 5.144/17.1 5.157/70.8 (4.74 -5.03)	1/2+	+0.203		U k x-ray 0.05162 0.05682 0.12928 0.37502 0.41369
^{240}Pu		240.053807	6537. y	α/5.255	5.0212(1)/0.07 5.1237(1)/26.4 5.1681(1)/73.5	0+			U L x-ray 0.04524 0.10423 (0.04-0.97)
^{241}Pu		241.056844	14.4 y	β^- /99+ /0.0208 α/0.002 /5.139	4.853(7)/3x10^{-4} 4.8966(7)/0.002	5/2+	-0.683	+6.	0.14854 0.1600

Elem. or Isot.	Natural Abundance (%)	Atomic Mass or Weight	Half-Life	Decay Mode/Energy (/MeV)	Particle Energy /Intensity (MeV / %)	Spin (h/2π)	Nuclear Magnetic Mom.(nm)	Elect. Quadr. Mom. (b)	γ-ray Energy /Intensity (MeV / %)
²⁴²Pu		242.058736	3.76x10⁵ y	α/4.983	4.7546(7)/0.098 4.8564(7)/22.4 4.9006(7)/78	0+			U L x-ray 0.04491 0.10350
²⁴³Pu		243.061996	4.956 h	β⁻ /0.582	0.49/21 0.58/60	7/2+			Am L x-ray 0.0417 0.0839
²⁴⁴Pu		244.064197	8.2x10⁷ y	α/99.9/4.665 SF/0.1 /	4.546(1)/19.4 4.589(1)/80.5	0+		U L x-ray	0.0439
²⁴⁵Pu		245.06774	10.5 h	β⁻ /1.21	0.93/57 1.21/11	(9/2-)			Am L x-ray Am k x-ray 0.2804 / 0.30832 0.32752 0.56014 (0.03-1.2)
²⁴⁶Pu		246.07020	10.85 d	β⁻ /0.40	0.150/85 0.35/10	0+			Am L x-ray Am k x-ray 0.04379 0.22371
²⁴⁷Pu		247.0741	2.3 d						
₉₅Am									
²³²Am		232.0466	0.9 m	EC/ ≈ 5.0					
²³⁴Am		234.0478	2.3 m	EC/4.2					
²³⁵Am		235.0480	≈ 15 m	EC					Pu K x-ray
²³⁷Am		237.0503	1.22 h	EC/99.98 /1.7 α/0.02 /6.20	6.042(5)/0.02	(5/2-)			Pu k x-ray 0.14559 0.28026 0.43845
²³⁸Am		238.05198	1.63 h	EC/2.26 α/0.0001 /6.04	5.940/0.0001	1+			Pu L x-ray Pu k x-ray 0.91870 0.96278
²³⁹Am		239.053018	11.9 h	EC/99.99 /0.803 a/0.01 /5.924	5.734(2)/0.001 5.776(2)/0.008	5/2-			Pu L x-ray Pu k x-ray 0.18172 0.22818 0.27760
²⁴⁰Am		240.05529	2.12 d	EC/1.38 α/5.592	5.378(1)/16x10⁻⁴	(3-)			Pu L x-ray Pu k x-ray 0.88878 0.98764 (0.1-1.3)
²⁴¹Am		241.056822	432.2 y	α/5.637	5.2443(1)/0.002 5.3221(1)/0.015 5.3884(1)/1.4 5.4431(1)/12.8 5.4857(1)/85.2 5.5116(1)/0.20 5.5442(1)/0.34	5/2-	+1.61	+4.	Np L x-ray 0.02634 0.033192 0.059536 (0.03-1.128)
²⁴²ᵐAm			141. y	IT/99.5 /0.048 α/0.5 /5.62	5.141(4)/0.026 5.2070(2)/0.4	5-	+1.0	+6.5	Am L x-ray 0.04863 0.08648 0.10944 0.16304
²⁴²Am		242.059542	16.02 h	β⁻ /83 /0.665 EC/17 /0.750	0.63/46 0.67/37	1-	+0.388	-2.4	Pu L x-ray Cm L x-ray Pu k x-ray 0.0422 0.04453
²⁴³Am		243.061372	7.37x10³ y	α/5.438	5.1798(5)/1.1 5.2343(5)/11 5.2766(5)/88 5.394(5)/0.12 5.3500(5)/0.16	5/2-	+1.61	+4.	0.04354 0.07467 0.08657 0.11770 0.14197 0.0429
²⁴⁴ᵐAm			≈ 26. m	β⁻ /1.498		(1-)			
²⁴⁴Am		244.064279	10.1 h	β⁻ /1.428					Am L x-ray Cm k x-ray 0.7460 0.9000
²⁴⁵Am		245.066444	2.05 h	β⁻ /0.894	0.65/19 0.90/77	(5/2+)			Cm L x-ray Cm k x-ray 0.25299

Elem. or Isot.	Natural Abundance (%)	Atomic Mass or Weight	Half-Life	Decay Mode/Energy (/MeV)	Particle Energy /Intensity (MeV / %)	Spin (h/2π)	Nuclear Magnetic Mom.(nm)	Elect. Quadr. Mom. (b)	γ-ray Energy /Intensity (MeV / %)
^{244m}Am			25.0 m	β^- /	1.3/79. 1.60/14 2.1/7	2-			Cm L x-ray Cm k x-ray 0.27002 0.79881 1.06201 1.07885 (0.04-2.29)
^{246}Am		246.06977	39. m	β^- /2.38	1.2/	(7-)			Cm L x-ray Cm k x-ray 0.1529 0.2046 0.6786
^{247}Am		247.0722	22. m	β^- /1.7					Cm L x-ray Cm k x-ray 0.2267 / 0.2853 /
$_{96}$Cm									
^{236}Cm		236.0514		EC/1.7					
^{237}Cm		237.0529		EC/2.5					
^{238}Cm		238.05302	2.4 h	EC/ > 90 /0.97 α/ < 10 /6.632	6.520(50)/ < 10	0+			
^{239}Cm		239.0548	≈ 3. h	EC/1.7					0.0407 0.1466 0.1874
^{240}Cm		240.055519	27. d	α/6.397	5.989/0.014 6.147/0.05 6.2478(6) /28.8 6.2906(6) /70.6	0+			
^{241}Cm		241.057646	32.8 d	EC/99 /0.768 α/1 /6.184	5.8842(4)/0.12 5.9291(4)/0.18 5.9389(4)/0.69	1/2+			Am k x-ray 0.13241 0.16505 0.18028 0.43063 0.47181
^{242}Cm		242.058828	162.8 d	α/6.216	5.9694(1)/0.035 6.069(1)/25 6.1129(1)/74	0+			Pu L x-ray 0.04408 0.10189 (0.04-1.2)
^{243}Cm		243.061381	28.5 y	α/6.167	5.6815(5) /0.2 5.6856(5)/1.6 5.7420(5)/10.6 5.7859(5)/73.3 5.9922(5)/6.5 6.0103(5)/1.0 6.0589(5)/5 6.0666(5)/1.5	5/2+	0.41		Pu L x-ray Pu k x-ray 0.10612 0.20975 0.22819 0.27760 0.28546 0.33431 (0.04-0.7)
^{244}Cm		244.062745	18.11 y	α/5.902	5.6656(1)/0.02 5.7528(1)/24 5.8050(1)/76	0+			Pu L x-ray 0.04282 0.09885 0.15262
^{245}Cm		245.065485	8.5x10^3 y	α /5.623	5.235(10)/0.3 5.3038(10)/5.0 5.3620(7)/93 5.4927(11)/0.8 5.5331(11)/0.6	7/2+	0.5		Pu L x-ray Pu k x-ray 0.04195 0.13299 0.13606 0.17494
^{246}Cm		246.067217	4.78x10^3 y	α/5.476	5.343(3)/21 5.386(3)/79	0+			Pu L x-ray 0.04453
^{247}Cm		247.070346	1.56x10^7 y	α/5.352	4.818(4)/4.7 4.8690(20)/71 4.941(4)/1.6 4.9820(20)/2.0 5.1436(20)/1.2 5.2104(20)/5.7 5.2659(20)/13.8	9/2- 9/2-	0.37		Pu k x-ray 0.2792 0.2886 0.3471 0.4035
^{248}Cm		248.072341	3.4x10^5 y	α/92 /5.162 SF/8/	4.931(5)/0.07 5.0349(2)/16.5 5.0784(2)/(75)/1	0+			
^{249}Cm		249.075946	64.15 m	β^- /0.900	0.9/	1/2+			Bk k x-ray 0.56039 0.63431
^{250}Cm		250.07835	≈ 9.7x10^3 y	SF/ α/5.27		0+			

Elem. or Isot.	Natural Abundance (%)	Atomic Mass or Weight	Half-Life	Decay Mode/Energy (/MeV)	Particle Energy /Intensity (MeV / %)	Spin (h/2π)	Nuclear Magnetic Mom.(nm)	Elect. Quadr. Mom. (b)	γ-ray Energy /Intensity (MeV / %)
^{251}Cm		251.08228	16.8 m	β⁻ /1.42	0.90/16	(1/2+)			0.3896 / 0.5299 0.5425
^{252}Cm		252.0849	< 2 d						
$_{97}$Bk									
^{238}Bk		238.0583	2.4 m	EC/5.0					
^{240}Bk		240.0598	≈ 4.8 m						
^{242}Bk		242.0621	7.0 m	EC/3.0					
^{243}Bk		243.063001	4.5 h	EC/99.8 /1.508 α/0.15 /6.871	6.542(4)/0.03 6.5738(2)/0.04 6.7180(22)/0.02 6.7581(20)/0.02	(3/2-)			0.1466 0.1874 0.755 0.840 0.946
^{244}Bk		244.0652	4.4 h	EC/99.99 /2.26 α/0.01 /6.778	6.625(4)/0.003 6.667(4)/0.003	(4-)			0.1445 0.1876 0.2176 0.9815 0.9215/
^{245}Bk		245.066355	4.94 d	EC/99.9 /0.810 α/0.1 /6.453	5.8851(5)/0.03 6.1176(9)/0.01 6.1467(5)/0.02 6.3087(5)/0.014 6.3492(5)/0.018	3/2-			Cm L x-ray Cm k x-ray 0.25299 0.3809 0.3851
^{246}Bk		246.0687	1.80 d	EC/1.35		(2-)			Cm L x-ray Cm k x-ray 0.79881 1.08142
^{247}Bk		247.07030	1.4x10³ y	α/5.889	5.465(5)/1.5 5.501(5)/7 5.532(5)/45 5.6535(20)/5.5 5.678(2)/13 5.712(2)/17 5.753(2)/4.3 5.794(2)/5.5	(3/2-)			0.04175 0.0839 0.268
^{248}Bk		248.07311	23.7 h	β⁻ /70 /0.87 EC/30 /0.72	0.86/	(1-)			Cm L x-ray Cf L x-ray Cm k x-ray Cf k x-ray 0.5507
^{249}Bk		249.074980	320. d	β⁻ /0.125 α/0.001 /5.525	0.125/100 5.390(1)/0.0002 5.4174(6)/0.001	7/2+	2.0		0.327/10⁻⁵ 0.308/10⁻⁶
^{250}Bk		250.078309	3.217 h	β⁻ /1.780	0.74/	2-			Cf L x-ray Cf k x-ray 0.98912 1.03184 (0.04-1.6)
^{251}Bk		251.08075	56. m	β⁻ /1.09		(3/2-)			0.02481 0.1528 0.1776
^{252}Bk		252.0843	1.8 m						
$_{98}$Cf									
^{237}Cf		237.0621	2.1 s	α,SF/10					
^{238}Cf		238.0614	21 ms	SF					
^{239}Cf		239.0626	≈ 0.7 m	α					
^{240}Cf		240.0623	1.1 m	α/7.719,SF/2	7.590(10)/	0+			
^{241}Cf		241.0637	4. m	EC/3.3 α/7.60	7.335(5)/				
^{242}Cf		242.06369	3.5 m	α/7.509 SF/0.014	7.351(6)/20 7.385(4)/80	0+			
^{243}Cf		243.0654	11. m	EC/86 /2.2 α/14 /7.40	7.060(6)/20 7.170/4	(1/2+)			
^{244}Cf		244.065990	20. m	α/7.328	7.168(5)/25 7.210(5)/75	0+			
^{245}Cf		245.068038	44. m	α/30 /7.255 EC/70 /1.569	7.15/91.7 6.983/0.31 7.09/7 7.065/0.68				Cm K x-ray 0.5709 0.6014 0.6163 Cm L x-ray
^{246}Cf		246.068798	1.49 h	α/6.869	6.6156(10)/0.18 6.7086(7)/21.8 6.7501(7)/78.0	0+			0.04221 0.0945 0.147

Elem. or Isot.	Natural Abundance (%)	Atomic Mass or Weight	Half-Life	Decay Mode/Energy (/MeV)	Particle Energy /Intensity (MeV / %)	Spin (h/2π)	Nuclear Magnetic Mom.(nm)	Elect. Quadr. Mom. (b)	γ-ray Energy /Intensity (MeV / %)
^{247}Cf		247.07099	3.11 h	EC/99.96 /0.65 α/0.04 /6.55	6.301(5)/	7/2+			Bk k x-ray 0.2941 0.4778
^{248}Cf		248.07218	334. d	α/6.369 SF/0.0029	6.220(5)/17 6.262(5)/83	0+			
^{249}Cf		249.074846	351. y	α/6.295	5.7582(2)/3.7 5.8119(2)/84 5.8488(2)/1.0 5.9029(2)/2.8 5.9451(2)/4.0 6.1401(2)/1.1 6.1940(2)/2.2	9/2-			Cm L x-ray Cm k x-ray 0.25299 0.33351 0.38832 (0.0376-1.103)
^{250}Cf		250.076399	13.1 y	α/6.129 SF/0.077	5.8913(4)/0.3 5.9889(4)/15 6.0310(4)/84.5	0+			Cm L x-ray 0.04285
^{251}Cf		251.079579	9.0×10^2 y	α/6.172	5.56448(7)/1.5 5.632(1)/4.5 5.648(1)/3.5 5.6773(6)/35 5.762(3)/3.8 5.7937(7)/2.0 5.8124(8)/4.2 5.8514(6)/27 6.0140(7)/11.6 6.0744(7)/2.7	1/2+			
^{252}Cf		252.081619	2.65 y	α/96.9 /6.217 SF/3.1/	5.7977(1)/0.23 6.0756(4)/15.2 6.1184(4)/81.6	0+			Cm L x-ray 0.04339 0.1002
^{253}Cf		253.08512	17.8 d	β⁻ /99.7 /0.29 α/0.3 /6.126	0.27/100 5.921(5)/0.02	(7/2+)			
^{254}Cf		254.08732	60.5 d	SF/99.7/ α/0.3/5.930	5.792(5)/0.05 5.834(5)/0.26	0+			
^{255}Cf		255.0910	1.4 h	β⁻ /0.7					
^{256}Cf		256.0934	12. m	SF					
$_{99}$Es									
^{241}Es		241.0687	≈ 8 s	α	8.11				
^{242}Es		242.0697	16 s	α	7.92				
^{243}Es		243.0696	21. s	α/ > 30 / EC/ < 70 /4.0	7.89/ > 30				
^{244}Es		244.0709	37. s	EC/76 /4.6 α/4 /	7.57/4				
^{245}Es		245.0713	1.3 m	α/40 /7.858 EC/60 /3.1	7.74				
^{246}Es		246.0730	7.7 m	EC/90 /3.9 α/10 /	7.35				
^{247}Es		247.07365	4.8 m	EC/93 /2.48 α/7 /	7.32				
^{248}Es		248.0755	26. m	EC/99.7 /3.1 α/0.3 /	6.87				
^{249}Es		249.07640	1.70 h	EC/99.4 /1.45 α/0.6 /	6.77	(7/2+)			0.3795 0.8132
^{250m}Es			2.2 h	EC/ β⁺		(1-)			Cf L x-ray Cf k x-ray 0.9891 1.0319
^{250}Es		250.0787	8.6 h	EC/2.1		(6+)			Cf L x-ray Cf k x-ray 0.30339 0.34948 0.82883
^{251}Es		251.07998	1.38 d	EC/99.5 /0.38 α/0.5 /	6.462/0.05 6.492/0.4	(3/2-)			
^{252}Es		252.08297	1.29 y	α/76 / EC/24 /1.26	6.632/61.0 6.562/10.3	(5-)			
^{253}Es		253.084818	20.47 d	α/	6.633/89.8 6.5916/6.6	7/2+	+4.10	7.	0.04180 0.3892
^{254m}Es			1.64 d	β⁻ /99.6 /0.475 α/0.3 /6.67	0.475 6.382	2+ 2+	2.9	3.7	Fm L x-ray Fm k x-ray 0.6488 0.6938
^{254}Es		254.088017	276. d	α/	6.429	(7+)			0.064
^{255}Es		255.09027	40. d	β⁻ /92 /0.29 α/8 / SF/0.004/	6.26 6.300	(7/2+)			

Elem. or Isot.	Natural Abundance (%)	Atomic Mass or Weight	Half-Life	Decay Mode/Energy (/MeV)	Particle Energy /Intensity (MeV / %)	Spin (h/2π)	Nuclear Magnetic Mom.(nm)	Elect. Quadr. Mom. (b)	γ-ray Energy /Intensity (MeV / %)
^{256m}Es			7.6 h	β⁻ /		(8+)			0.218 0.232 0.862
^{256}Es		256.0936	25. m	β⁻ /1.7		(1+)			
$_{100}$Fm									
^{242}Fm		242.0734	0.8 ms	SF					
^{243}Fm		243.0745	0.2 s	α/	8.55				
^{244}Fm		244.0741	3.7 ms	SF/		0+			
^{245}Fm		245.0754	4. s	α/	8.15/				
^{246}Fm		246.07528	1.2 s	α/92 / SF/8/	8.24/	0+			
^{247m}Fm			9. s	α/	8.18/				
^{247}Fm		247.0768	35. s	α/8.20 EC/2.9	7.87/70 7.93/30				
^{248}Fm		248.07718	36. s	α/99.9 /8.001 SF/0.1/	7.83/20 7.87/80	0+			
^{249}Fm		249.0790	3. m	EC/2.4 α/	7.53	(7/2+)			
^{250m}Fm			1.8 s	IT/					
^{250}Fm		250.07951	30. m	α/ EC/0.8	7.43/	0+			
^{251}Fm		251.08157	5.3 h	EC/98 /1.47 α/2 /	6.833	(9/2-)			
^{252}Fm		252.08246	1.058 d	α/7.154 SF	6.998/15 7.039/85	0+			
^{253}Fm		253.085175	3.0 d	EC(88%)/0.333 α/12 /	6.676/ 6.943/	1/2+			Es k x-ray 0.2719
^{254}Fm		254.086847	3.240 h	α/ SF/0.06/	7.150 7.192	0+			
^{255}Fm		255.089955	20.1 h	α/	6.9635(5)/5.0 7.0225(5)/93.4	7/2+			
^{256}Fm		256.09177	2.63 h	SF/92/ α/18 /	6.92/	0+			
^{257}Fm		257.09510	100.5 d	α/99.8 / SF/0.2	6.519	(9/2+)			0.1794 0.2410
^{258}Fm		258.0971	0.37 ms	SF/					
^{259}Fm		259.1006	1.5 s	SF/					
$_{101}$Md									
^{245m}Md			≈ 0.4 s	α	8.64,8.68				
^{245}Md		245.0810	0.9 ms	SF					
^{246}Md		246.0819	1.0 s	α	8.74 8.50-8.56				
^{247}Md		247.0818	3. s	α	8.43				
^{248}Md		248.0828	7. s	EC/80 /5.3 α/20 /	8.32/15 8.36/5				
^{249}Md		249.0830	24. s	EC >/< 80 /3.7 α/ > 20 /8.46	8.030(20)/				
^{250}Md		250.0845	50. s	EC/94 /4.6 α/6 /8.25	7.75/4 7.83/2				
^{251}Md		251.0849	4.0 m	EC/ > 94 /3.1 α/ < 6 /	7.55/				
^{252}Md		252.0866	2. m	EC/ > 50 /3.9 α/ < 50 /	7.73/				
^{253}Md		253.0873	≈6 m	EC/2.0					
^{254m}Md			30. m	EC/					
^{254}Md		254.0897	10. m	EC/2.7					
^{255}Md		255.09108	27. m	EC/92 /1.04 α/8 /	7.33	(7/2-)			0.4531 0.4056
^{256}Md		256.0941	1.30 h	EC/89 /2.13 α/11 /	7.22/47 7.155/21				Fm k x-ray (0.112-0.834)
^{257}Md		257.095535	5.5 h	EC/85 /0.41 α/15 /,SF	7.074 7.014	(7/2-)			Fm k x-ray (0.181-0.389)
^{258m}Md			57. m	EC/		(1-)			Fm k x-ray
^{258}Md		258.098427	51.5 d	α/7.40	6.718(2)/ 6.763(4)/	(8-)			0.3678 (0.057 - 0.448)
^{259}Md		259.1005	1.64 h	SF/ α/ < 1.3		7/2+			
^{260}Md		260.104	≈ 27.8 d	SF/					
$_{102}$No									
^{250}No		250.0875	0.25 ms	SF/		0+			
^{251}No		251.0889	0.8 s	α/	8.60/80 8.68/20				

Elem. or Isot.	Natural Abundance (%)	Atomic Mass or Weight	Half-Life	Decay Mode/Energy (/MeV)	Particle Energy /Intensity (MeV / %)	Spin (h/2π)	Nuclear Magnetic Mom.(nm)	Elect. Quadr. Mom. (b)	γ-ray Energy /Intensity (MeV / %)
^{252}No		252.08897	2.3 s	α/73 /8.551	8.42	0+			
				SF/27/	8.37				
^{253}No		253.0907	1.7 m	α/	8.010(20)	(9/2−)			
				EC/3.2					
^{254m}No			0.28 s	I.T./					
^{254}No		254.09095	55. s	α/	8.09	0+			
				EC/1.1					
^{255}No		255.09323	3.1 m	α/62 /	8.12/	1/2+			0.187
				EC/38/2.01	7.93				
					8.08				
^{256}No		256.09428	2.9 s	α/	8.43	0+			
^{257}No		257.09685	25. s	α/	8.22	(7/2+)			
					8.27				
					8.32				
^{258}No		258.0983	≈ 1.2 ms	SF/		0+			
^{259}No		259.1011	58. m	α/78 /7.794	7.52	(9/2+)			
				EC/22/0.5	7.55				
^{262}No		262.108	≈ 8. ms	SF/					
$_{103}$Lr									
^{253}Lr		253.0953	1.3 s	α/	8.80				
					8.72				
^{254}Lr		254.0965	13. s	α/	8.45				
				EC/5.2					
^{255}Lr		255.0967	22. s	α/	8.37/60				
				EC/3.2	8.43/40				
^{256}Lr		256.0988	28. s	α/99.7 /8.554	8.43/				
				EC/4.2	8.39				
^{257}Lr		257.0996	0.65 s	α/	8.80	7/2+			
				EC/2.5	8.80				
^{258}Lr		258.1019	3.9 s	α/	8.60				
				EC/3.4	8.62				
					8.56				
^{259}Lr		259.1030	6.1 s	α/80	8.44(1)				
				SF/20					
^{260}Lr		260.1056	3. m	α/	8.03				
^{261}Lr		261.1069	40. m	SF					
^{262}Lr		262.110	3.6 h	EC/2.					
$_{104}$Rf									
^{253}Rf		253.1007	≈ 48. μs	SF					
				α/ < 10					
^{254}Rf		254.1002	23. μs	SF					
				α/ < 1.5					
^{255}Rf		255.1015	1.7 s	α	8.72				
				SF/45	8.77				
^{256}Rf		256.10118	6.2 ms	SF					
				α/0.32	8.81				
^{257}Rf		257.1031	4.7 s	α/9.22	8.77				0.117
				EC/11	9.01				
				SF/ < 1.4	8.95				
					8.62				
^{258}Rf		258.1036	12. ms	SF,α					
^{259}Rf		259.1056	3.4 s.	α/9.09/93	8.77(2)/				
				SF/7	8.86/				
^{260}Rf		260.1064	20. ms	SF/					
^{261}Rf		261.1088	1.1 m	α/8.78,SF/ < 10	8.28/				
^{262}Rf		262.1101	2.1 s	SF/					
^{263}Rf		263.1125	< 20. m	SF					
$_{105}$Db									
^{255}Db		255.1074	≈ 1.5 s	α,SF					
^{256}Db		256.1081	≈ 2.6 s	α,SF,E.C.					
^{257}Db		257.1079	1.5 s	α/	8.97/				
				SF	9.07/				
					9.16/				
^{258}Db		258.1094	4.2 s	α/	9.17/				
				E.C/5.3	9.08/				
^{259}Db		259.1097	≈ 1.2 s	SF/					
^{260}Db		260.1114	1.5 s	α/	9.05/				
				SF/	9.08/				
					9.13/				
^{261}Db		261.1121	1.8 s	α/	8.93/				
				SF/					

Elem. or Isot.	Natural Abundance (%)	Atomic Mass or Weight	Half-Life	Decay Mode/Energy (/MeV)	Particle Energy /Intensity (MeV / %)	Spin (h/2π)	Nuclear Magnetic Mom.(nm)	Elect. Quadr. Mom. (b)	γ-ray Energy /Intensity (MeV / %)
^{262}Db		262.1141	34. s	SF/ α/	8.45/ 8.53/ 8.67/				
^{263}Db		263.1151	≈ 0.45 m	SF/57/, α/43/	8.35/43				
$_{106}$Sg									
^{258}Sg		258.1132	≈ 2.9 ms	SF α/ < 20					
^{259}Sg		259.1147	0.5 s	α,SF/	9.62				
^{260}Sg		260.11444	4. ms	α,SF/	9.77 9.72				
^{261}Sg		261.1162	0.3 s	α,SF/	9.56				
^{263}Sg		263.1183	0.8 s	SF,α	9.06 9.25/				
^{265}Sg		265.1211	< 30. s	α/ > 50 SF/ < 50					
^{266}Sg		266.1219	< 30. s	α/ > 50 SF/ < 50	8.63(5)				
$_{107}$Bh									
^{260}Bh		260.122		α					
^{261}Bh		261.1218	12. ms	α/,SF	10.40 10.10 10.03				
^{262m}Bh			8. ms	α/	10.37 10.24				
^{262}Bh		262.1230	0.10 s	α/	10.06 9.91 9.74				
^{264}Bh		264.1247	0.44 s	α	9.47 9.62				
$_{108}$Hs									
^{263}Hs		263.1287		α/					
^{264}Hs		264.1284	≈ 0.08 ms	α/,SF/ ≈ 50	11.0				
^{265m}Hs			0.9 ms	α	10.57				
^{265}Hs		265.1300	≈ 1.6 ms	α/	10.31				
^{267}Hs		267.1371	60 ms	α/ > 88	9.74-9.87				
^{269}Hs		269.1341	19.7 s	α	9.23				
$_{109}$Mt									
^{266m}Mt			≈ 1.2 ms	α	10.46-10.81				
^{266}Mt		266.1379	≈ 0.7 ms	α	10.48-11.31				
^{267}Mt		267.138	19 ms	α					
^{268}Mt		268.1388	0.07 s	α/ > 68	10.10,10.24				
$_{110}$110									
^{267}110		267.1440	≈ 3 μs	α/ > 32	11.6				
^{269}110		269.1451	0.17 ms	α/ > 75	11.11				
^{271m}110			≈ 56 ms	α	10.71				
^{271}110		271.1461	≈ 1.1 ms	α	10.68,10.74				
^{273}110		273.1492	0.3 ms	α/ > 32	11.35				
$_{111}$111									
^{272}111		272.1535	≈ 1.5 ms	α/ > 68	10.8				
$_{112}$112									
^{277}112			≈ 0.24 ms	α	11.45				

NEUTRON SCATTERING AND ABSORPTION PROPERTIES*
(Revised 1996)

Norman E. Holden
Reactor Division
Brookhaven National Laboratory
Upton, New York 11973

This Table presents an evaluated set of values for the experimental quantities, which characterize the properties for scattering and absorption of neutrons. The neutron cross section is given for room temperature neutrons, 20.43° C, corresponds to a thermal neutron energy of 0.0253 electron volts (eV) or a neutron velocity of 2200 meters/second. The neutron resonance integral is defined over the energy range from 0.5 eV to 0.1×10^6 eV, or 0.1 MeV. Bound neutron scattering lengths and neutron cross sections averaged over a maxwellian spectrum at 30 keV for astrophysical applications are also presented. A list of the major references used is given below. The literature cutoff date is December 1996. Uncertainties are given in parentheses. Parentheses with two or more numbers indicate values to the excited state(s) and to the ground state of the product nucleus. Discussions with Frank Feiner and Harold Knox of KAPL on thermal cross sections and resonance integrals and the assistance of Felicia Kramer in assembling this file are gratefully acknowledged.

TABLE LAYOUT

Column No.	Column Title	Description
1	Isotope or Element	For elements, the atomic number and chemical symbol are listed. For nuclides, the mass number and chemical symbol are listed. Isomers are indicated by the addition of m, m1, or m2.
2	Isotopic Abundance	in atom percent.
3	Half-life	Half-life in decimal notation. μs = microseconds; ms = milliseconds; s = seconds; m = minute; h = hours; d = days; and y = years.
4	Thermal Neutron Cross Section	Cross sections for neutron capture reactions in barns (10^{24} cm^2) or in millibarns (mb). Proton, alpha production and fission reactions are designated by σ_p, σ_α, σ_f, respectively. Separate values are listed for isomeric production.
5	Neutron Resonance Integral	Resonance integrals for neutron capture reactions in barns (10^{24} cm^2) or millibarns (mb). Proton, alpha production and fission reactions are designated by RI_p, RI_α, RI_f, respectively. Separate values are listed for isomeric production.
6	Neutron Scattering Length	Bound Coherent scattering length for neutron scattering reactions in units of femtometer (fm), which is equal to fermis (10^{-13} cm).
7	Maxwellian Averaged Cross Section	Astrophysical Cross Sections, averaged over a stellar neutron maxwellian spectrum characterized by a thermal energy of 30 keV, expressed in barns (10^{24} cm^2), millibarns (mb) or microbarns (μb).

GENERAL NUCLEAR DATA REFERENCES

The following references represent the major sources of the neutron data presented:

1. Mughabghab, S.F.; Divadeenam, M.; Holden, N.E.; Neutron Cross Sections, Vol. I *Neutron Resonance Parameters and Thermal Cross Sections* Part A Z=1-60. Academic Press Inc., New York, New York (1981); Mughabghab, S.F.; Part B, Z-61-100. Academic Press Inc., Orlando, Florida (1984).

2. Parrington, J.R.; Knox, H.D.; Breneman, S.L.; Baum, E.M.; Feiner, F.: *Chart of the Nuclides, 15th Edition*, Knolls Atomic Power Laboratory (Oct. 1996).

3. Holden, N.E.; *Fifty Years with Nuclear Fission* Conference, Wash., D.C., Gaithersburg, Md. April 26-29, 1989, p. 946. American Nuclear Society, LaGrange Park, Illinois (1989).

4. Holden, N.E.; *Review of Thermal Neutron Cross Sections and Isotopic Composition of the Elements*, BNL-NCS-42224 (March 1989) and Holden, N.E.; *Thermal Neutron Cross Sections for the 1991 Table of the Isotopes*, BNL-45255 (Feb. 1991).

5. Tuli, J.K.; *Nuclear Wallet Cards*, Brookhaven National Laboratory (July 1995).

6. Holden, N.E.; *Half-lives of Selected Nuclides*, Pure & Applied Chemistry <u>62</u>, 941 (1990).

7. Koester, L; Rauch, H.; Seymann, E.; *Neutron Scattering Lengths: A Survey of Experimental Data and Methods*, Atomic Data Nuclear Data Tables <u>49</u>, 65 (1991).

8. Sears, V.F.; *Neutron Scattering Lengths and Cross Sections*, Neutron News <u>3</u>, (3), 26 (1992).

*Research carried out under auspices of the US Department of Energy, Contract No. DE-AC02-76CH00636

Elem. or Isot.	Natural Abundance (%)	Half-life	Thermal Neut. Cross Section (barns)	Resonance Integral (barns)	Coh. Scat. Length (fm)	σ(30KEV) Maxw. Avg. (barns)
$_1$H			0.332(2)	0.149(1)	- 3.739(1)	
^{1}H	99.985(1)		0.332(2)	0.149(1)	- 3.741(1)	
^{2}H	0.015(1)		0.51(1)mb	0.23(2) mb	6.671(4)	0.25(2) mb*
^{3}H		12.32 y	< 6. μb		4.79(3)	
$_2$He			< 0.05		3.26(3)	
^{3}He	1.37x10^{-6}		σ_p=5.33(1)x10^3	RI$_p$=2.39(1)x10^3	5.74(7)	
			0.05(1) mb			8.(1) μb*
^{4}He	≈ 100.				3.26(3)	
$_3$Li			71.(2)	32.(1)	- 1.90(2)	
^{6}Li	7.5(2)		σ_α=9.4(1)x10^2	RI$_\alpha$=422.(4)	2.0(1)	
			39.(5) mb	17.(2) mb		
^{7}Li	92.5(2)		45.(5) mb	20.(2) mb	- 2.22(2)	39(6) μb
$_4$Be			8.8(4) mb	3.9(2) mb	7.79(1)	
^{7}Be *		53.28 d	σ_p=3.9(1)x10^4	RI$_p$=1.75(5)x10^4		σ_p 16(4)
^{9}Be	100.		σ_α ≈ 0.1 8.8(4) mb	3.9(2) mb	7.79(1)	
^{10}Be		1.52x10^6 y	<1. mb			
$_5$B			7.6(1)x10^2	343.(5)	5.30(4)	
^{10}B	19.9(2)		σ_α=38.4(1)x10^2 0.3(1) σ_p=7.(1) mb σ_t=8.(2) mb	RI$_\alpha$=17.3(1)x10^2 0.13(4)	- 0.1(3)	
^{11}B	80.1(2)		5.(3) mb	2(1) mb	6.65(4)	
$_6$C			3.5(1) mb	1.6(1) mb	6.646(1)	
^{12}C	98.89(1)		3.5(1) mb	1.6(1) mb	6.651(2)	16(1) μb*
^{13}C	1.11(1)		1.4(1) mb	1.7(2) mb	6.19(9)	0.2(1) mb
^{14}C		5715. y	<1. μb			1.5(4) μb*
$_7$N			1.9(1)	0.85(5)	9.36(2)	
^{14}N	99.634(9)		σ_p=1.83(7) 0.080(1)	RI$_p$=0.82(3) 0.034(1)	9.37(2)	σ_p 1.8(2) mb* 0.04(7) mb
^{15}N	0.366(9)		0.04(1) mb	0.11(3) mb	6.44(3)	4.9(5) μb*
$_8$O			0.29(1) mb	0.40(4) mb	5.805(4)	
^{16}O	99.76(1)		0.19(1) mb	0.36(4) mb	5.805(5)	34(4) μb
^{17}O	0.04		σ_α=0.24(1) 0.54(7) mb	0.11(1) 0.39(5) mb	5.8(2)	σ_α 3.9(5) mb*
^{18}O	0.20(1)		0.16(1) mb	0.81(4) mb	5.84(7)	7.2(6) μb*

Elem. or Isot.	Natural Abundance (%)	Half-life	Thermal Neut. Cross Section (barns)	Resonance Integral (barns)	Coh. Scat. Length (fm)	σ(30KEV) Maxw. Avg. (barns)
$_9$F			9.5(7) mb	21.(3) mb	5.65(1)	
^{19}F	100.		9.5(7) mb	21.(3) mb	5.65(1)	6.(1) mb
$_{10}$Ne			42(5) mb	19(3) mb	4.566(6)	
^{20}Ne	90.48(3)		39(5) mb	18(3) mb	4.631(6)	0.12(1) mb
^{21}Ne	0.27(1)		0.7(1), σ_α < 1.5	0.31(5)	6.7(2)	$\approx$ 1.6 mb
^{22}Ne	9.25(3)		51(5) mb	23(3) mb	3.87(1)	60.(5) μb*
$_{11}$Na			0.53(2)	0.32(2)	3.63(2)	2.3(2) mb
^{22}Na		2.605 y	σ_p=2.8(3)x10^4 σ_α=2.6(4)x10^2	< 2x10^5 1.2(2) x 10^2		
^{23}Na	100.		σ_m=0.43(3)	RI$_m$=0.30(6)	3.63(2)	
$_{12}$Mg			66(6) mb	38.(5) mb	5.375(4)	
^{24}Mg	78.99(3)		0.053(6)	32.(4) mb	5.7(2)	4.2(2) mb
^{25}Mg	10.00(1)		.20(1)	98.(15) mb	3.6(2)	6.5(3) mb
^{26}Mg	11.01(2)		.038(1)	25.(2) mb	4.9(2)	66.(3) μb
^{27}Mg		9.45 m	0.07(2)	0.03(1)		
$_{13}$Al			0.230(2)	0.17(1)	3.45(1)	
^{26}Al			σ_p $\approx$ 2.1			0.14(2)
^{27}Al	100.		0.230(2)	0.17(1)	3.45(1)	3.8(3) mb
$_{14}$Si			0.166(9)	0.12(2)	4.15(1)	
^{28}Si	92.21(2)		0.17(1)	0.11(2)	4.11(1)	1.8(2) mb
^{29}Si	4.69(2)		0.12(1)	0.08(2)	4.7(1)	7.9(9) mb
^{30}Si	3.10(1)		0.107(4)	0.62(6)	4.61(1)	6.4(6) mb
^{31}Si		2.62 h	73.(6) mb	33.(3) mb		
$_{15}$P			0.17(1)	0.08(1)	5.13(1)	
^{31}P	100.		0.17(1)	0.08(1)	5.13(1)	1.7(1) mb
$_{16}$S			0.54(2)	0.24(2)	2.847(1)	
^{32}S	95.02(9)		0.55(5) σ_α < 0.5 mb	0.25(2)	2.804(2)	4.7(2) mb
^{33}S	0.75(4)		0.46(3) σ_α = 0.12(1) σ_p = 2 mb	0.21(2) RI$_\alpha$ = 0.05(1)	4.7(2)	7.4(15) mb σ_α = 0.18(1)
^{34}S	4.21(8)		0.30(2)	0.13	3.48(3)	2.9(9) mb
^{36}S	0.02(1)		0.23(2)	0.26(3)		0.17(1) mb
$_{17}$Cl			33.6(3)	15.(2)	9.58(1)	
^{35}Cl	75.77(7)		43.7(4); σ_p=0.44(1) σ_α $\approx$ 0.08 mb	20.(2); RI$_p$=0.2	11.7(1)	10.(3) mb σ_p 1.7(2) mb

Elem. or Isot.	Natural Abundance (%)	Half-life	Thermal Neut. Cross Section (barns)	Resonance Integral (barns)	Coh. Scat. Length (fm)	σ(30KEV) Maxw. Avg. (barns)
^{36}Cl		3.01×10^5 y	$\sigma_p = 46.(2)$ mb $<10.$ $\sigma_\alpha = 0.59(7)$ mb	$RI_p = 0.02$		σ_p 91.(8) mb σ_α 0.9(2) mb
^{37}Cl	24.23(7)		$(0.05 + 0.38)$	$(0.04 + 0.26)$	3.1(1)	2.15(8) mb
$_{18}$Ar			0.66(3)	0.42(5)	1.91(1)	
^{36}Ar	0.337(3)		$5(1); \sigma_p < 1.5$ mb σ_α 6(1) mb	2.(1)	24.9(1)	
^{37}Ar		35.0 d	$\sigma_\alpha = 2.0(3) \times 10^3$ $\sigma_p = 69(14)$	$RI_\alpha = 900.$ $RI_p = 31.$		
^{38}Ar	0.063(1)		0.8(2)	0.4(1)		
^{39}Ar		268. y	$6.(2) \times 10^2$		3.5(35)	
^{40}Ar	99.600(3)		0.64(3)	0.41(5)	1.83(1)	2.3(2)mb
^{41}Ar		1.82 h	0.5(1)	0.2(1)		
$_{19}$K			2.1(1)	1.0(1)	3.67(2)	
^{39}K	93.2581(44)		2.1(2) $\sigma_\alpha = 4.3(5)$ mb $\sigma_p < 0.05$ mb	0.9(1)	3.74(2)	11.8(4)mb
^{40}K	0.0117(1)	1.26×10^9 y	30.(8) $\sigma_p = 4.4(4)$ $\sigma_\alpha = 0.42(8)$	13.(4) 2.0(2)		σ_p .7(1)mb σ_α 40(6)mb
^{41}K	6.7302(44)		1.46(3)	1.4(2)	2.69(8)	22.0(7)mb
$_{20}$Ca			0.43(2)	0.23(2)	4.70(2)	
^{40}Ca	96.941(18)		0.41(3) σ_α 0.13(4) mb	0.22(4)	4.80(2)	6.7(7)mb
^{41}Ca		1.02×10^5 y	$\approx 4.$ $\sigma_\alpha = 0.19$			
^{42}Ca	0.647(9)		0.65(10)	0.39(4)	3.4(1)	16.(2)mb
^{43}Ca	0.135(6)		6.(1)	3.9(2)	- 1.56(9)	51.(6)mb
^{44}Ca	2.086(12)		0.8(2)	0.56(1)	1.42(6)	9.(1)mb
^{45}Ca		162.7 d	$\approx 15.$			
^{46}Ca	0.004(3)		0.70(3)	0.9(1)	3.6(2)	5.7(5)mb
^{48}Ca	0.187(4)		1.1(1)	0.5(1)	0.39(9)	1.0(2)mb
$_{21}$Sc			27.2(2)	12.(1)	12.3(1)	
^{45}Sc	100.		$(10. + 17.)$	$(5.6 + 6.4)$	12.3(1)	90.(6)mb
^{46}Sc		83.81 d	8.(1)	3.6(5)		
$_{22}$Ti			6.1(1)	2.8(2)	- 3.438(2)	
^{46}Ti	8.25(3)		0.6(2)	0.4(1)	4.93(6)	27.(3)mb
^{47}Ti	7.44(2)		1.6(2)	1.6(2)	3.63(1)	64.(8)mb
^{48}Ti	73.72(3)		7.9(9)	3.6(2)	- 6.09(2)	31.(5)mb
^{49}Ti	5.41(2)		1.9(5)	1.2(2)	1.04(5)	22.(2)mb
^{50}Ti	5.18(2)		0.179(3)	0.12(2)	6.18(8)	4.0(5)mb

Elem. or Isot.	Natural Abundance (%)	Half-life	Thermal Neut. Cross Section (barns)	Resonance Integral (barns)	Coh. Scat. Length (fm)	σ(30KEV) Maxw. Avg. (barns)
$_{23}$V			5.0(2)	2.8(1)	- 0.382(1)	
^{50}V	0.250(4)	>1.4x10^{17} y	21(4) $\sigma_p = 0.7(4)$ mb	50.(20)	7.6(6)	
^{51}V	99.750(4)		4.9(1)	2.7(2)	- 0.402(2)	42.(3)mb
$_{24}$Cr			3.0(2)	1.7(1)	3.635(7)	
^{50}Cr	4.345(13)		15.(1)	8.(1)	- 4.5(1)	52.(1)mb
^{51}Cr		27.70 d	< 10.			
^{52}Cr	83.79(2)		0.8(1)	0.6(2)	4.91(2)	8.8(4)mb
^{53}Cr	9.50(2)		18.(2)	9(1)	- 4.2(1)	0.06(1)
^{54}Cr	2.365(7)		0.36(4)	0.25(5)	4.6(1)	7.(2)mb
$_{25}$Mn			13.3(1)	14.0(3)	- 3.75(2)	
^{53}Mn		3.7x10^6 y	70.(10)	32.(5)		
^{54}Mn		312.2 d	< 10.			
^{55}Mn	100.		13.3(1)	14.0(3)	- 3.75(2)	40.(2)mb
$_{26}$Fe			2.6(1)	1.4(2)	9.45(2)	
^{54}Fe	5.85(4)		2.7(5) $\sigma_\alpha = 10$ μb	1.3(2) $RI_\alpha = 1.1(1)$ mb	4.2(1)	29.(2)mb
^{55}Fe		2.73 y	13.(2) $\sigma_\alpha = 0.03$	6.(1)		
^{56}Fe	91.75(4)		2.6(2)	1.4(2)	9.93(3)	13.5(5)mb
^{57}Fe	2.12(1)		2.5(5)	1.6(2)	2.3(1)	35.(4)mb
^{58}Fe	0.28(1)		1.3(1)	1.3(2)	15.(7)	13.(2)mb
^{59}Fe		44.51 d	13.(3)	6.(1)		
$_{27}$Co			37.19(8)	74.(2)	2.49(2)	
^{58m}Co		9.1 h	1.4(1)x10^5	2.5(10)x10^5		
^{58}Co		70.88 d	1.9(2)x10^3	7.(1)x10^3		
^{59}Co	100.		(20.7+16.5)	(39.+35.)	2.49(2)	38.(4)mb
^{60m}Co		10.47 m	58.(3)	230.(50)		
^{60}Co		5.271 y	2.0(2)	4.3(10)		
$_{28}$Ni			4.5(2)	2.3(2)	10.3(1)	
^{58}Ni	68.077(9)		4.6(4) $\sigma_\alpha < 0.03$ mb	2.3(2)	14.4(1)	42.(3)mb
^{59}Ni		≈7.6x10^4 y	σabs=92.(4) $\sigma_\alpha = 14(2)$ $\sigma_p = 2(1)$	RIabs=1.4(1)x10^2		
^{60}Ni	26.223(8)		2.9(3)	1.5(2)	2.8(1)	27.(2)mb
^{61}Ni	1.140(1)		2.5(5) $\sigma_\alpha = 0.03$ mb	1.5(4)	7.60(6)	82.(1)mb
^{62}Ni	3.634(2)		15.(1)	6.8(3)	- 8.7(2)	36.(4)mb
^{63}Ni		100. y	20(5)	9.(2)	26.(8)mb	
^{64}Ni	0.926(1)		1.6(1)	1.(2)	- 0.37(7)	10.(1)mb
^{65}Ni		2.517 h	22.(2)	10.(1)		

Elem. or Isot.	Natural Abundance (%)	Half-life	Thermal Neut. Cross Section (barns)	Resonance Integral (barns)	Coh. Scat. Length (fm)	σ(30KEV) Maxw. Avg. (barns)
$_{29}$Cu			3.8(1)	4.1(4)	7.718(4)	
^{63}Cu	69.17(3)		4.5(2)	5.(1)	6.43(15)	0.09(1)
^{64}Cu		12.701 h	$\approx$ 270.			
^{65}Cu	30.83(3)		2.17(3)	2.2(1)	10.61(19)	53(5)mb
^{66}Cu		5.09 m	$1.4(1) \times 10^2$	60.(20)		
$_{30}$Zn			1.1(2)	2.8(4)	5.680(5)	
^{64}Zn	48.6(6)		0.74(5) $\sigma_p < 12\ \mu b$ $\sigma_\alpha = 11(3)\ \mu b$	1.4(3)	5.23(4)	59(5)mb
^{65}Zn		243.8 d	66.(8) $\sigma_\alpha = 2.0(2)$	30.(4)		
^{66}Zn	27.9(3)		0.9(3) $\sigma_\alpha < 0.02$ mb	1.8(2)	5.98(5)	35(3)mb
^{67}Zn	4.1(1)		6.9(1.4) $\sigma_\alpha = 0.4$ mb	25.(5)	7.58(8)	0.15(2)
^{68}Zn	18.8(5)		(0.072 + 0.8) $\sigma_\alpha < 0.02$ mb	(0.2 + 2.9)	6.04(3)	19(2)mb
^{70}Zn	0.6(3)		(8.1+83.) mb	0.9(2)		
$_{31}$Ga			2.9(1)	22.(3)	7.288(2)	
^{69}Ga	60.11(3)		1.68(7)	16.(2)	7.88(4)	0.15(1)
^{71}Ga	39.89(3)		4.7(2);σ_m=0.15(5)	31.(3)	6.40(3)	0.12(1)
$_{32}$Ge			2.2(1)	6.(2)	8.19(2)	
^{70}Ge	21.23(4)		(0.3 + 2.7)	2.3(1)	10.0(1)	86(5)mb
^{72}Ge	27.66(3)		0.9(2)	0.8(3)	8.5(1)	0.06(2)
^{73}Ge	7.73(1)		15.(1)	66.(20)	5.02(4)	0.28(8)
^{74}Ge	35.94(2)		(0.14 + 0.28)	(0.4+0.5)	7.6(1)	54(6)mb
^{76}Ge	7.44(2)		(0.09 + 0.06)	(1.3+0.6)	8.2(15)	16(1)mb
$_{33}$As			4.0(4)	61.(5)	6.58(1)	
^{75}As	100.		4.0(4)	61.(5)	6.58(1)	0.46(2)
$_{34}$Se			12.(1)	14.(3)	7.970(9)	
^{74}Se	0.89(4)		50.(2)	520(50)	0.8(3)	
^{75}Se		119.78 d	$3.3(10) \times 10^2$			
^{76}Se	9.37(29)		(22. + 63.)	(9.+31.)	12.2(1)	164(8)mb
^{77}Se	7.63(16)		42.(4),σ_α=0.97(3)μb	30.(5)	8.25(8)	
^{78}Se	23.77(28)		σ_m=0.38(2)	RI$_m$=4.3(4)	8.24(9)	0.08
^{80}Se	49.61(41)		(0.06+0.35)	(0.36+1.0)	7.48(3)	44.(3)mb
^{82}Se	8.73(22)		(39.+ 5.2) mb	39.(4)mb	6.34(8)	
$_{35}$Br			6.8(2)	92.(8)	6.79(2)	
^{76}Br		16.0 h	224.(42)			
^{79}Br	50.69(7)		(2.5+8.3)	(36.+96)	6.79(7)	0.74(3)
^{81}Br	49.31(7)		(2.4+0.24)	51.(5)	6.78(7)	0.24(1)

Elem. or Isot.	Natural Abundance (%)	Half-life	Thermal Neut. Cross Section (barns)	Resonance Integral (barns)	Coh. Scat. Length (fm)	σ(30KEV) Maxw. Avg. (barns)
$_{36}$Kr			24.(1)	39.(6)	7.81(2)	
^{78}Kr	0.35(1)		(0.17+6.)	20.(1)		0.34(4)
^{80}Kr	2.28(6)		(4.6+7.)	57.(6)		0.24(1)
^{82}Kr	11.58(14)		(14.+7.)	130.(13)		79(6)mb
^{83}Kr	11.50(6)		183.(30)	183.(20)	0.24(2)	
^{84}Kr	56.99(4)		σ_m=0.09;σ_m+g=0.11	2.4(3)		(17+17)mb
^{85}Kr		10.73 y	1.7(2)	1.8(10)		0.07(2)
^{86}Kr	17.30(22)		3.(2)mb	$\approx$1. mb	8.1(3)	3.5(3)mb
$_{37}$Rb			0.39(4)	6.(3)	7.08(2)	
^{84}Rb		32.9 d	σ_p=12.(2)			
^{85}Rb	72.17(2)		(0.06+0.38)	(0.7+7.)	7.0(1)	240(9)mb
^{86}Rb		18.65 d	<20.			
^{87}Rb	27.83(2)	4.88x10^{10} y	0.10(1)	2.3(4)	7.3(1)	18.0(5)mb
^{88}Rb		17.7 m	1.2(3)	0.5(1)		
$_{38}$Sr			1.2(1)	10.(1)	7.02(2)	
^{84}Sr	0.56(1)		(0.6+0.2)	(9.+1.)		(48+22)mb
^{86}Sr	9.86(1)		σ_m=0.81(4)	RI$_m$=4.(1)	5.68(5)	97.(5)mb
^{87}Sr	7.00(1)		16.(3)	118.(30)	7.41(7)	6.1(2)mb
^{88}Sr	82.58(1)		5.8(4)mb	0.07(3)	7.16(6)	
^{89}Sr		50.52 d	0.42(4)	0.2		
^{90}Sr		29.1 y	9.7(7)mb	4. mb		
$_{39}$Y			1.25(5)	1.0(1)	7.75(2)	
^{89}Y	100.		(0.001+1.25)	(0.006+1.0)	7.75(2)	19.(1)mb
^{90}Y		2.67 d	<6.5			
^{91}Y		58.5 d	1.4(3)	0.6(1)		
$_{40}$Zr			0.19(1);σ_α<0.1mb	0.95(9)	7.16(3)	
^{90}Zr	51.45(40)		$\approx$0.014	0.2(1)	6.4(1)	21.(2)mb
^{91}Zr	11.22(5)		1.2(3)	5.(2)	8.8(1)	60.(8)mb
^{92}Zr	17.15(8)		0.2(1)	0.6(2)	7.5(2)	33.(4)mb
^{93}Zr		1.5x10^6 y	<4.	16.(5)		0.10(1)
^{94}Zr	17.38(28)		0.049(6)	0.25(3)	8.3(2)	26.(1)mb
^{96}Zr	2.80(9)		0.020(3)	5.0(5)	5.5(1)	10.0(3)mb
$_{41}$Nb			1.11(1);σ_α<0.1mb	8.5(6)	7.14(3)	
^{93}Nb	100.		1.1;σ_m=0.86	(6.3+2.2)	7.14(3)	266.(5)mb
^{94}Nb		2.4x10^4 y	σ_m+g=15.(1)	126.(13)		
			σ_m=0.6(1)			
^{95}Nb		34.97 d	<7.	<200.		
$_{42}$Mo			2.5(1),σ_α<0.1mb	26.(5)	6.72(2)	
^{92}Mo	14.84(35)		0.06,σ_m=0.2μb	$\approx$0.8	6.93(8)	
^{94}Mo	9.25(12)		0.02	$\approx$0.8	6.82(7)	
^{95}Mo	15.92(13)		13.4(3),σ_α=30.(4)μb	109.(5)	6.93(6)	0.29(1)

Elem. or Isot.	Natural Abundance (%)	Half-life	Thermal Neut. Cross Section (barns)	Resonance Integral (barns)	Coh. Scat. Length (fm)	σ(30KEV) Maxw. Avg. (barns)
^{96}Mo	16.68(2)		0.5	17.(3)	6.22(6)	112.(5)mb
^{97}Mo	9.55(8)		2.5(2),σ_α=0.4(2)μb	14.(3)	7.26(8)	0.34(1)
^{98}Mo	24.13(31)		0.14(1)	7.2(7)	6.60(7)	99.(5)mb
^{100}Mo	9.63(23)		0.19(1)	3.6(3)	6.75(7)	
$_{43}$Tc						
^{98}Tc		4.2x10^6 y	σ_m = 0.9(2)			
^{99}Tc		2.13x10^5 y	23.(2)	4.0(4)x10^2	6.8(3)	0.78(4)
$_{44}$Ru			2.6 (1)	48.(5)	7.03(3)	
^{96}Ru	5.52(20)		0.23(4)	7.(2)		
^{98}Ru	1.86(9)		< 8.			
^{99}Ru	12.74(26)		4.(1)	195.(20)		
^{100}Ru	12.60(19)		5.8(6)	11.(2)		0.21(1)
^{101}Ru	17.05(7)		5.(1),σ_α<0.15μb	1.1(3)x10^2		1.00(4)
^{102}Ru	31.57(31)		1.2(1)	4.3(5)		0.19(1)
^{103}Ru		39.27 d	<20.	≈30.		
^{104}Ru	18.66(44)		0.49(2)	6.(2)		0.16(1)
^{105}Ru		4.44 h	0.29(3)	0.13(1)		
^{106}Ru		1.020 y	0.15(4)	2.0(6)		
$_{45}$Rh			145.(2)	1.2(1)x10^3	5.88(4)	
^{103}Rh	100.		(11.+ 134.)	(0.08+1.1)x10^3	5.88(4)	0.81(1)
^{104m}Rh		4.36 m	800.(100)			
^{104}Rh		42.3 s	40.(30)			
^{105}Rh		35.4 h	1.1(3)x10^4	1.7(4)x10^4		
$_{46}$Pd			7.(1)	82.(8)	5.91(6)	
^{102}Pd	1.02(1)		3.2(10)	10.(2)		
^{104}Pd	11.14(11)			16.(2)		
^{105}Pd	22.33(13)		22.(2),σ_α=0.5(2)μb	60.(20)	5.5(3)	1.20(6)
^{106}Pd	27.33(6)		(0.013+0.28)	(0.2+5.5)	6.4(4)	0.25(3)
^{107}Pd		6.5x10^6 y	1.8(2)	108.(4)		1.34(6)
^{108}Pd	26.46(15)		(0.19+8.5)	(2.+240.)	4.1(3)	0.20(2)
^{110}Pd	11.72(16)		(0.033+0.7)	(0.7+8.)		0.15(2)
$_{47}$Ag			62.(1)	767.(60)	5.922(7)	
^{107}Ag	51.839(8)		(0.37+35.)	(1.2+105.)	7.56(1)	0.80(2)
^{109}Ag	48.161(8)		(4.2 + 87.)	(0.7+14.1)x10^2	4.17(1)	0.78(2)
^{110m}Ag		249.8 d	82.(11)	20.(4))		
^{111}Ag		7.47 d	3.(2)	105.(20)		
$_{48}$Cd			2.52(5)x10^3	73.(8)	4.87(5)	
^{106}Cd	1.25(6)		0.20(3)	4.(1)		0.56(6)
^{108}Cd	0.89(3)		1.	14.(3)	5.4(1)	0.41(7)
^{109}Cd		462.0 d	≈180.,σ_α<0.05	6.7(12)x10^3		
^{110}Cd	12.49(18)		(0.06+11.)	(6.+34.)	5.9(1)	0.25(3)
^{111}Cd	12.80(12)		3.5(20)	51.(6)	6.5(1)	1.06(13)

Elem. or Isot.	Natural Abundance (%)	Half-life	Thermal Neut. Cross Section (barns)	Resonance Integral (barns)	Coh. Scat. Length (fm)	σ(30KEV) Maxw. Avg. (barns)
^{112}Cd	24.13(21)		(0.012+2.2)	15.	6.4(1)	0.23(3)
^{113}Cd	12.22(12)	9x10^{15} y	2.06(4)x10^4,σ_α<1.μb	390.(40)	- 8.0(2)	0.73(8)
^{114}Cd	28.73(42)		(0.04+0.29)	16.(7)	7.5(1)	0.16(3)
^{116}Cd	7.49(18)		(26.+52.)mb	1.2	6.3(1)	0.09(1)
$_{49}$In			197.(4)	3.3(2)x10^3	4.07(2)	
^{113}In	4.29(5)		(3.1+5.0+3.9)	(220.+90)	5.39(6)	0.79(7)
^{115}In	95.71(5)	4.4x10^{14} y	(88.+73.+44.)	(1.5+1.2+0.7)x10^3	4.01(2)	0.74(7)
$_{50}$Sn			0.61(3)	8.(2)	6.225(2)	
^{112}Sn	0.97(1)		(0.15+0.40)	(8.+19.)		0.20(5)
^{113}Sn		115.1 d	≈ 9.	210.(50)		
^{114}Sn	0.66(1)		≈ 0.12	5.(1)	6.2(3)	134.(3) mb
^{115}Sn	0.34(1)		σ_α=0.06 mb	29.(6)		0.34(1)
^{116}Sn	14.54(9)		(0.006+0.14)	(0.5+11.)	5.93(5)	91.(2)mb
^{117}Sn	7.68(7)		1.1(1)	16.(5)	6.48(5)	319.(7)mb
^{118}Sn	24.22(9)		σ_m=4. mb	4.7(5)	6.07(5)	62.(1)mb
^{119}Sn	8.58(4)		2.(1)	2.9(5)	6.12(5)	0.25(5)
^{120}Sn	32.59(9)		(0.001+0.13)	1.2(3)	6.49(5)	(0.5+36)mb
^{122}Sn	4.63(3)		(0.15+0.001)	0.81(4)	5.74(5)	
^{124}Sn	5.79(5)		(0.13+0.005)	RI$_m$=8.0(2)	5.97(5)	
$_{51}$Sb			5.2(2)	169.(20)	5.57(3)	
^{121}Sb	57.21(5)		(0.4+5.6)	(13.+192.)	5.71(6)	0.53(2)
^{123}Sb	42.79(5)		(0.02+0.04+4.0)	(1.+119.)	5.38(7)	0.30(1)
^{124}Sb		60.20 d	17.(3)	≈8.		
$_{52}$Te			4.2(1)	47.(3)	5.80(3)	
^{120}Te	0.096(2)		(0.25+2.0)	≈1.	5.3(5)	
^{122}Te	2.603(4)		2.4(5)	80.(20)	3.8(2)	295.(3)mb
^{123}Te	0.908(2)	1.3x10^{13} y	370.(40),σ_α=0.05mb	4.5(3)x10^3	- 0.05	0.83(1)
^{124}Te	4.816(6)		(0.05+7.)	5.2(7)	8.0(1)	155.(2)mb
^{125}Te	7.139(6)		1.6(2)	21.(4)	5.02(8)	431.(4)mb
^{126}Te	18.952(11)		(0.12+0.8)	8.2(6)	5.56(7)	81.(1)mb
^{128}Te	31.687(11)		(0.016+0.20)	(0.08+1.5)	5.89(7)	
^{130}Te	33.799(10)	2.5x10^{21} y	(0.03+0.20)	(0.08+0.34)	6.02(7)	
$_{53}$I			6.15(10)	148.(7)	5.28(2)	
^{125}I		59.4 d	900.(100)	1.4(2)x10^4		
^{127}I	100.		6.15(10)	148.(7)	5.28(2)	0.64(3)
^{128}I		25.00 m	22.(4)	≈10.		
^{129}I		1.7x10^7 y	(20.7+10.3)	36.(4)		0.44(2)
^{130}I		12.36 h	18.(3)	≈8.		
^{131}I		8.040 d	≈ 0.7	8.(4)		
$_{54}$Xe			25.(1)	263.(50)	4.92(3)	
^{124}Xe	0.09(1)		(28.+137.)	(0.6+3.0)x10^3		
^{125}Xe		17.1 h	σ_α≤ <0.03			

Elem. or Isot.	Natural Abundance (%)	Half-life	Thermal Neut. Cross Section (barns)	Resonance Integral (barns)	Coh. Scat. Length (fm)	σ(30KEV) Maxw. Avg. (barns)
^{126}Xe	0.09(1)		(0.45+3.)	(8.+52.)		
^{127}Xe		36.4 d	$\sigma_\alpha \leq 0.01$			
^{128}Xe	1.92(3)		σ_m=0.48	RI_m=38.(10)		0.35(11)
^{129}Xe	26.44(24)		22.(5)	250.(50)		0.76(7)
^{130}Xe	4.08(2)		σ_m=0.45	RI_m=16.(4)		0.18(5)
^{131}Xe	21.18(3)		90.(10)	9.(1)x10^2		0.45(8)
^{132}Xe	26.89(6)		(0.05+0.4)	(0.9+3.7)		65.(5)mb
^{133}Xe		5.243 d	190.(90)			
^{134}Xe	10.44(10)		(0.003 + 0.26)	0.40(4)		20.(2)mb
^{135}Xe		9.10 h	2.65(11)x10^6	7.6(5)x10^3		
^{136}Xe	8.87(16)		0.26(2)	0.7(2)		0.9(1)mb
$_{55}$Cs			30.4(8)	422.(50)	5.42(2)	
^{132}Cs		6.48 d	$\sigma_\alpha < 0.15$			
^{133}Cs	100.		(2.7+27.7)	(32.+390.)	5.42(2)	0.51(2)
^{134}Cs		2.065 y	140.(10)	54.(9)		
^{135}Cs		2.3x10^6 y	8.9(5)	90.(20)		
^{137}Cs		30.2 y	0.25(1)	0.36(7)		
$_{56}$Ba			1.3(2)	10.(2)	5.07(3)	
^{130}Ba	0.106(1)		(1.+8.)	(25.+200.)	- 3.6(6)	
^{132}Ba	0.101(1)		(0.84+9.7)	(4.7+24.)	7.8(3)	
^{133}Ba		10.53 y	4.(1)	85.(30)		
^{134}Ba	2.417(18)		(0.1+1.3)	(5.6+18.)	5.7(1)	0.18(6)
^{135}Ba	6.592(12)		(0.014+5.8)	(0.47+131.)	4.7(1)	0.30(6)
^{136}Ba	7.854(24)		(0.010+0.44)	(0.1+1.5)	4.91(8)	62.(2)mb
^{137}Ba	11.232(24)		5.(1)	4.(1)	6.8(1)	0.05(1)
^{138}Ba	71.698(42)		0.41(2)	0.4(1)	4.84(8)	4.1(2)mb
^{139}Ba		1.396 h	5.(1)	2.2(5)		
^{140}Ba		12.75 d	1.6(3)	14.(1)		
$_{57}$La			9.2(2)	12.(1)	8.24(4)	
^{138}La	0.090(1)	1.06x10^{11} y	57.(6)	4.1(9)x10^2		
^{139}La	99.910(1)		9.2(2)	12.(1)	8.24(4)	38.(3)mb
^{140}La		1.678 d	2.7(3)	69.(4)		
$_{58}$Ce			0.64(4)	0.71(6)	4.84(2)	
^{136}Ce	0.185(2)		(1.0+6.5)	58.(12)	5.80(9)	(28+272)mb
^{138}Ce	0.251(2)		(0.018+1.)	(1.5+5.2)	6.70(9)	179.(5)mb
^{140}Ce	88.450(18)		0.58(4)	0.50(5)	4.84(9)	11.0(4)mb
^{141}Ce		32.50 d	29.(3)	13.(2)		
^{142}Ce	11.114(17)		0.97(3)	1.3(3)	4.75(9)	28.(1)mb
^{143}Ce		1.38 d	6.1(7)	2.7(3)		
^{144}Ce		284.6 d	1.0(1)	2.6(3)		
$_{59}$Pr			11.5(4)	14.(3)	4.58(5)	
^{141}Pr	100.		(4.+7.5)	14.(3)	4.58(5)	0.12(2)
^{142}Pr		19.12 h	20.(3)	9.(1)		
^{143}Pr		13.57 d	90.(10)	190.(25)		

Elem. or Isot.	Natural Abundance (%)	Half-life	Thermal Neut. Cross Section (barns)	Resonance Integral (barns)	Coh. Scat. Length (fm)	σ(30KEV) Maxw. Avg. (barns)
$_{60}$Nd			51.(2)	49.(5)	7.69(5)	
^{142}Nd	27.2(5)		19.(1)	34.(11)	7.7(3)	46.(4)mb
^{143}Nd	12.2(2)		330.(10),σ_α=17.mb	128.(30)		0.24(1)
^{144}Nd	23.8(3)	2.1x10^{15} y	3.6(3)	3.9(5)	2.8(3)	0.11(1)
^{145}Nd	8.30(1)		47.(6),σ_α=12.μb	260.(40)		0.49(10)
^{146}Nd	17.2(3)		1.5(2)	3.0(4)	8.7(2)	0.16(4)
^{147}Nd		10.98 d	440.(150)	200.		
^{148}Nd	5.7(1)		2.4(1)	13.(2)	5.7(3)	0.19(4)
^{150}Nd	5.6(2)		1.0(1)	14.(2)	5.3(2)	
$_{61}$Pm						
^{146}Pm		5.53 y	8.4(1.7)x10^3			
^{147}Pm		2.6234 y	(84.+96.)	(1000.+1280.)	12.6(4)	
^{148m}Pm		41.3 d	10600.(800)			
^{148}Pm		5.37 d	$\approx 10^3$	2.6(2.4)x10^3		
^{149}Pm		2.212 d	1400.(200)			
^{151}Pm		1.183 d	$\approx$ 150.			
$_{62}$Sm			5.6(1)x10^3	1.4(2)x10^3		
^{144}Sm	3.08(7)		1.6(1)	2.4(3)		
^{145}Sm		340. d	280.(20)	600.(90)		
^{147}Sm	14.99(17)	1.06x10^{11} y	56.(4),σ_α=0.6mb	710.(50)	14.(3)	0.97(1)
^{148}Sm	11.24(10)	7x10^{15} y	2.4(6)	27.(14)		241.(2)mb
^{149}Sm	13.82(7)	10^{16} y	40100.(600),σ_α=31.mb	3100.(500)		1.82(2)
^{150}Sm	7.38(1)		102.(5)	290.(30)	14.(3)	422.(4)mb
^{151}Sm		90. y	15200.(300)	3520.(60)		
^{152}Sm	26.74(14)		206.(15)	3000.(300)	- 5.0(6)	473.(4)mb
^{153}Sm		1.929 d	420.(180)			
^{154}Sm	22.75(27)		7.5(3)	32.(6)	9.(1)	
$_{63}$Eu			4570.(100)	3.8(5)x10^3	5.3(3)	
^{151}Eu	47.81(6)		(4.+3150.+6000.) σ_α=8.7(3)μb	(2000.+4000.)		4.4(2)
^{152m1}Eu		9.30 h	68000.(15000)	< 100000.		
^{152}Eu		13.5 y	11000.(2000)	1600.(200)		
^{153}Eu	52.19(6)		300.(20),σ_α<1.μb	1800.(400)	8.2(1)	3.2(3)
^{154}Eu		8.59 y	1500.(300)	1600.(200)		4.4(7)
^{155}Eu		4.76 y	3900.(200)	16000.(2000)		
$_{64}$Gd			48.8(6)x10^3	400.(10)	9.5(2)	
^{152}Gd	0.20(1)		700.(200),σ_α<7.mb	700.(200)		1.05(7)
^{153}Gd		241.6 d	20000.(10000),σ_α=0.03			
^{154}Gd	2.18(3)		(0.06+60.)	230.(50)		0.88(3)
^{155}Gd	14.80(12)		61.(1)x10^3,σ_α=.08mb	1540.(100)		2.72(9)
^{156}Gd	20.47(9)		$\approx$2.0	104.(15)	6.3(4)	0.64(6)
^{157}Gd	15.65(2)		2.54(3)x10^5,σ_α<0.05	800.(100)		1.36(4)
^{158}Gd	24.84(7)		2.3(3)	73.(7)	9.(2)	0.22(2)
^{160}Gd	21.86(19)		1.5(7)	6.(1)	9.15(5)	
^{161}Gd		3.66 m	2.0(6)x10^4			

Elem. or Isot.	Natural Abundance (%)	Half-life	Thermal Neut. Cross Section (barns)	Resonance Integral (barns)	Coh. Scat. Length (fm)	σ(30KEV) Maxw. Avg. (barns)
$_{65}$Tb			23.2(5)	420.(50)	7.38(3)	
^{159}Tb	100.		23.2(5)	420.(50)	7.38(3)	1.75(10)
^{160}Tb		72.3 d	570.(110)			
$_{66}$Dy			9.5(2)x10^2	1.5(2)x10^3	16.9(3)	
^{156}Dy	0.06(1)		33.(3),σ_α<9mb	1000.(100)		
^{158}Dy	0.10(1)		43.(6),σ_α<6mb	120.(10)	6.1(5)	
^{159}Dy		144. d	8.(2)x10^3			
^{160}Dy	2.34(8)		60.(10),σ_α<0.3mb	1100.(200)	6.7(4)	0.77(4)
^{161}Dy	18.91(24)		600.(50),σ_α<1μb	1100.(100)	10.3(4)	2.01(6)
^{162}Dy	25.51(26)		170.(20)	2755.(300)	- 1.4(5)	0.47(5)
^{163}Dy	24.90(16)		120.(10),σ_α<20μb	1600.(400)	5.0(4)	1.14(4)
^{164}Dy	28.18(36)		(1.7+1.0)x10^3	(420.+100.)	49.4(2)	0.27(3)
^{165m}Dy		1.26 m	2.0(6)x10^3			
^{165}Dy		2.33 h	3.5(3)x10^3	2.2(3)x10^4		
$_{67}$Ho			61.(2)	670.(40)	8.01(8)	
^{165}Ho	100.		(3.1+58.),σ_α<20μb	(?+670.)	8.01(8)	1.32(7)
^{166}Ho		1.2x10^3 y	9140.(650)	1140.(90)		
$_{68}$Er			169.(20)	730.(10)	7.79(2)	
^{162}Er	0.14(1)		19.(3),σ_α<11mb	480.(50)	8.8(2)	
^{164}Er	1.61(3)		13.(3),σ_α<1.2mb	105.(10)	8.2(2)	0.71(6)
^{166}Er	33.61(35)		(15.+5.),σ_α<70μb	96.(12)	10.6(2)	0.61(6)
^{167}Er	22.93(17)		700.(200),σ_α=3μb	2970.(70)	3.0(3)	1.5(2)
^{168}Er	26.78(26)		2.0(6),σ_α=0.09mb	37.(5)	7.4(4)	0.37(4)
^{170}Er	14.93(27)		6.(1)	26.(4)	9.6(5)	0.22(3)
^{171}Er		7.52 h	370.(40)	170.(20)		
$_{69}$Tm			106.(5)	1.5(2)x10^3	7.07(3)	
^{169}Tm	100		(8.+98.)	1.5(2)x10^3	7.07(3)	1.13(6)
^{170}Tm		128.6 d	100.(20)	460.(50)		
^{171}Tm		1.92 y	$\approx$ 160.	118.(6)		
$_{70}$Yb				52.(10)	1.7(2)x10^2	12.43(3)
^{168}Yb	0.13(1)		2.4(2)x10^3,σ_α<0.1mb	2.0(5)x10^4	-4.07(2)	
^{169}Yb		32.03 d	3.6(3)x10^3	5200.(500)		
^{170}Yb	3.04(15)		12.(2),σ_α<10μb	320.(30)	6.8(1)	0.74(3)
^{171}Yb	14.3(6)		53.(5),σ_α<1.5μb	315.(30)	9.7(1)	1.41(5)
^{172}Yb	21.8(7)		$\approx$ 1.3,σ_α<1μb	25.(3)	9.4(1)	0.41(3)
^{173}Yb	16.13(3)		16.(2),σ_α<1μb	380.(30)	9.56(7)	0.87(7)
^{174}Yb	31.8(9)		(46.+74.),σ_α<0.02mb	(13.+47.)	19.3(1)	0.18(2)
^{176}Yb	12.8(4)		3.1(2),σ_α<1μb	8.(2)	8.7(1)	
$_{71}$Lu			78.(7)	8.3(7)x10^2	7.21(3)	
^{175}Lu	97.41(2)		(16.+8.)	(550.+270.)	7.24(3)	1.18(4)
^{176}Lu	2.59(2)	3.8x10^{10} y	(2.+2100.)	(3.+1100.)	6.1(2)	1.54(6)

Elem. or Isot.	Natural Abundance (%)	Half-life	Thermal Neut. Cross Section (barns)	Resonance Integral (barns)	Coh. Scat. Length (fm)	σ(30KEV) Maxw. Avg. (barns)
^{177m}Lu		160.7 d	3.2(3)	1.4(2)		
^{177}Lu		6.75 d	1000.(300)			
$_{72}$Hf			106.(3)	19.7(5)x10^2	7.8(1)	
^{174}Hf	0.162(3)	2.0x10^{15} y	600.(50)	400.(50)	11.(1)	
^{176}Hf	5.206(5)		23.(4)	700.(100)	6.6(2)	0.46(2)
^{177}Hf	18.606(13)		(1.+375.), σ_α<20μb	7170.(200)		1.37(6)
^{178}Hf	27.297(4)		(53.+32.)	(340.+1500.)	5.9(2)	0.31(1)
^{179}Hf	13.629(6)		(0.43+46.)	(6.8+620.)	7.5(2)	0.99(3)
^{180}Hf	35.100(7)		13.(1), σ_α<13μb	32.(2)	13.2(3)	175.(5)mb
^{181}Hf		42.4 d	30.(25)			
$_{73}$Ta			20.(1)	650(20.)	6.91(7)	
^{180m}Ta	0.012(2)	> 1.2x10^{15} y	$\approx$ 560.	1350.(100)		
^{181}Ta	99.998(2)		(0.012 + 20), σ_α<1μb	(0.4+650.)	6.91(7)	0.77(2)
^{182}Ta		114.43 d	8200.(600)	900.(90)		
$_{74}$W			18.(1)	3.6(3)x10^2	4.86(2)	
^{180}W	0.12(1)		$\approx$ 4.	210.(30)		
^{182}W	26.50(16)		20.(1)	600.(90)	6.97(4)	274.(8)mb
^{183}W	14.31(4)		10.5(3)	340.(50)	6.53(4)	0.52(2)
^{184}W	30.64(2)		(0.002 + 2.0)	15.(2)	7.48(6)	0.23(1)
^{185}W		74.8 d	$\approx$ 3.3	300.(50)		
^{186}W	28.43(19)		37.(2)	510.(50)	- 0.72(4)	176.(5)mb
^{187}W		23.9 h	70.(10)	2760.(550)		
$_{75}$Re			90.(4)	8.4(2)x10^2	9.2(3)	
^{185}Re	37.40(2)		(0.33+110.)	1700.(50)	9.0(3)	1.54(6)
^{187}Re	62.60(2)	4.4x10^{10} y	(2.+72.)	(9.+310.)	9.3(3)	1.16(6)
$_{76}$Os			17.(1)	1.5(1)x10^2	10.7(2)	
^{184}Os	0.020(3)		3.3(3)x10^3, σ_α<10mb	1.4(1)x10^3		
^{186}Os	1.58(2)	2.x10^{15} y	$\approx$ 80., σ_α<0.1mb	280.(40)	12(2)	0.42(2)
^{187}Os	1.6(4)		2.(1)x10^2, σ_α<0.1mb	500.(70)		0.87(3)
^{188}Os	13.3(1)		$\approx$ 5., σ_α<30μb	150.(20)	7.6(3)	0.40(2)
^{189}Os	16.1(1)		(0.00026+40), σ_α<10μb	(0.013+670.)	10.7(3)	1.17(5)
^{190}Os	26.4(2)		(9.+4.), σ_α<20μb	(22.+8.)	11.0(3)	0.30(5)
^{191}Os		15.4 d	3.8(6)x10^2			
^{192}Os	41.0(3)		3.(1), σ_α<10μb	6.(1)	11.5(4)	0.31(5)
^{193}Os		30.5 h	40.(10)			
$_{77}$Ir			4.2(1)x10^2	2.8(4)x10^3	10.6(3)	
^{191}Ir	37.3(2)		(660.+260.)	(1000.+4200.)		1.35(7)
^{192}Ir		73.83 d	1450.(250)	4800.(700)		
^{193}Ir	62.7(2)		(6.+110.)	1400.(200)		0.86(4)
^{194}Ir		19.3 h	1.5(3)x10^3			
$_{78}$Pt			10.(1)	1.3(1)x10^2	9.60(1)	

Elem. or Isot.	Natural Abundance (%)	Half-life	Thermal Neut. Cross Section (barns)	Resonance Integral (barns)	Coh. Scat. Length (fm)	σ(30KEV) Maxw. Avg. (barns)
^{190}Pt	0.014(1)	6.5×10^{11} y	$1.5(1) \times 10^2, \sigma_\alpha < 8$mb	70.(10)	9(1)	
^{192}Pt	0.782(7)		$(2.0+6.), \sigma_\alpha < 0.2$mb	115.(20)	9.9(5)	
^{194}Pt	32.97(10)		$(0.1+1.1), \sigma_\alpha < 5\mu$b	(4. + ?)	10.55(8)	0.20(6)
^{195}Pt	33.832(10)		$28.(1), \sigma_\alpha < 5\mu$b	365.(50)	8.8(1)	0.37(8)
^{196}Pt	25.242(41)		$(0.045+0.55)$	7.(2)	9.89(8)	0.9(2)
^{198}Pt	7.16(6)		$(0.027+3.6)$	$(0.5+56.)$	7.8(1)	0.20(2)
^{199}Pt		30.8 m	$\approx 16.$			
$_{79}$Au			98.7(1)	$1.55(3) \times 10^3$	7.63(6)	
^{197}Au	100.		98.7(1)	$1.55(3) \times 10^3$	7.63(6)	582.(9)mb
^{198}Au		2.694 d	$26.5(15) \times 10^3$	$\approx 4. \times 10^4$		
^{199}Au		3.14 d	$\approx 30.$			
$_{80}$Hg			$3.7(1) \times 10^2$	87.(5)	12.69(2)	
^{196}Hg	0.15(1)		$(105.+3000.)$	$(53.+410.)$	30.(1)	
^{198}Hg	9.97(20)		$(0.017+2.)$	$(1.7+70.)$		0.17(2)
^{199}Hg	16.87(22)		$2.1(2) \times 10^3$	435(20)	16.9(4)	0.37(2)
^{200}Hg	23.10(19)		$< 60.$	2.1(5)		0.12(1)
^{201}Hg	13.18(9)		$< 60.$	30.(3)		0.26(1)
^{202}Hg	29.86(26)		4.9(5)	4.5(2)		74.(6)mb
^{204}Hg	6.87(15)		0.4(1)	0.8(2)		42.(4)mb
$_{81}$Tl			3.3(1)	12.5(8)	8.776(5)	
^{203}Tl	29.52(1)		$11.(1), \sigma_\alpha < 0.3$mb	41.(2)	7.0(2)	124.(8)mb
^{204}Tl		3.78 y	22.(2)	90.(20)		
^{205}Tl	70.48(1)		0.11(2)	0.6(2)	9.52(7)	54.(4)mb
$_{82}$Pb			0.172(2)	0.14(4)	9.402(2)	
^{204}Pb	1.4(1)		0.68(7)	2.0(2)	10.6(20)	90.(6)mb
^{205}Pb		1.51×10^7 y	$\approx 5.$			
^{206}Pb	24.1(1)		0.030(1)	0.10(1)	9.23(5)	14.(1)mb
^{207}Pb	22.1(1)		0.70(1)	0.38(1)	9.28(4)	8.4(4)mb
^{208}Pb	52.4(1)		$0.49(3)$mb$, \sigma_\alpha < 8\mu$b	2.0(2) mb	9.50(6)	0.36(2)mb
^{210}Pb		22.6 y	< 0.5			
$_{83}$Bi			0.034(1)	0.19(2)	8.532(2)	
^{209}Bi	100.		$(11+23)$mb$, \sigma_\alpha < 0.3\mu$b	0.19(2)	8.532(2)	3.1(3)mb
^{210m}Bi		3.0×10^6 y	54(4)mb	0.20(3)		
$_{84}$Po						
^{210}Po		138.38 d	$\sigma_m < 0.5$mb$, \sigma_\alpha < 2.$mb $\sigma_g < 30.$mb$, \sigma_f < 0.1$			
$_{85}$At						
$_{86}$Rn						
^{220}Rn		55.6 s	< 0.2			

Elem. or Isot.	Natural Abundance (%)	Half-life	Thermal Neut. Cross Section (barns)	Resonance Integral (barns)	Coh. Scat. Length (fm)	σ(30KEV) Maxw. Avg. (barns)
^{222}Rn		3.8235 d	0.74(5)			
$_{88}$Ra						
^{223}Ra		11.435d	$1.3(2) \times 10^2, \sigma_f < 0.7$			
^{224}Ra		3.66 d	12.0(5)			
^{226}Ra		1599. y	$\approx 13., \sigma_f < 7.\mu b$	280.(50)	10.(1)	
^{228}Ra		5.76 y	$36.(5), \sigma_f < 2.$			
$_{89}$Ac						
^{227}Ac		21.77 y	$8.8(7) \times 10^2, \sigma_f < .35$ mb	$1.5(4) \times 10^3$		
$_{90}$Th			7.4	85.(3)	10.31(3)	
^{227}Th		18.72 d	$\sigma_f = 2.0(2) \times 10^2$			
^{228}Th		1.913 y	$1.2(2) \times 10^2, \sigma_f < 0.3$	1014.(400)		
^{229}Th		7.9×10^3 y	$\approx 60.$ $\sigma_f = 30(3)$	$1.0(2) \times 10^3$ $RI_f = 466.(75)$		
^{230}Th		7.54×10^4 y	23.4(5) $\sigma_f < 0.5$mb	$1.0(1) \times 10^3$		
^{232}Th	100.	1.4×10^{10}y	7.37(4) $\sigma_f = 3.(1) \mu b$ $\sigma_\alpha < 1. \mu b$	85.(3)	10.31(3)	
^{233}Th		22.3 m	$1.5(1) \times 10^3$ $\sigma_f = 15.(2)$	$4.(1) \times 10^2$		
^{234}Th		24.10 d	1.8(5) $\sigma_f < 0.01$			
$_{91}$Pa						
^{230}Pa		17.4 d	$1.5(3) \times 10^3$			
^{231}Pa		3.25×10^4 y	$2.0(1) \times 10^2$ $\sigma_f = 20.(1)$ mb	750.(80) $RI_f = 0.05(1)$	9.1(3)	
^{232}Pa		1.31 d	$4.6(10) \times 10^2$ $\sigma_f = 7.(1) \times 10^2$	300.(70)		
^{233}Pa		27.0 d	39.(2) $\sigma_m = 20.(4)$ $\sigma_g = 19.(3)$ $\sigma_f < 0.1$	$(460. + 440)$		
$_{92}$U			$3.4(3); \sigma_f = 4.2(1)$	280.(20), $RI_f = 2.0$	8.417(5)	
^{230}U		20.8 d	$\sigma_f \approx 25.$			
^{231}U		4.2 d	$\sigma_f \approx 250.$			
^{232}U		68.9 y	73.(2) $\sigma_f = 74.(8)$	280.(15) $RI_f = 350.(30)$		
^{233}U		1.59×10^5 y	47.(2) $\sigma_f = 5.3(1) \times 10^2$ $\sigma_\alpha < 0.2$ mb	137.(6) $RI_f = 760.(17)$	10.1(2)	
^{234}U	0.0055(5)	2.45×10^5 y	96.(2) $\sigma_f < 5.$ mb	660.(70) $RI_f = 6.5$	12.(4)	

Elem. or Isot.	Natural Abundance (%)	Half-life	Thermal Neut. Cross Section (barns)	Resonance Integral (barns)	Coh. Scat. Length (fm)	σ(30KEV) Maxw. Avg. (barns)
^{235}U	0.720(1)	7.04×10^8 y	95.(5) σ_f=586.(2) σ_α< 0.1 mb	144.(6) RI$_f$=275(5)	10.47(4)	
^{236}U		2.34×10^7 y	5.1(3) σ_f=0.04	360.(15) RI$_f$=4.38(50)		
^{237}U		6.75 d	$\approx 10^2$ σ_f< 0.35	1200.(200)		
^{238}U	99.2745(15)	4.46×10^9 y	2.7(1) $\sigma_f \approx 3. \mu$b σ_α=1.4(5) μb	277.(3) 1.54(15) mb	8.402(5)	
^{239}U		23.5 m	22.(2) σ_f=15.(3)			
$_{93}$Np						
^{234}Np		4.4 d	σ_f=9.(3)$\times 10^2$			
^{235}Np		1.085 y	1.6(1)$\times 10^2$			
^{236m}Np		22.5 h	σ_f=2.7(2)$\times 10^3$	700.(400)		
^{236}Np		1.55×10^5 y	σ_f=2.6(5)$\times 10^3$	1030.(100)		
^{237}Np		2.14×10^6 y	160.(10) σ_f=20.(1)mb	8.(2)$\times 10^2$ RI$_f$=0.32	10.6(1)	
^{238}Np		2.117 d	σ_f=2.1(1)$\times 10^3$	9.(1)$\times 10^2$		
^{239}Np		2.355 d	(32.+19.) σ_f<1.			
$_{94}$Pu						
^{236}Pu		2.87 y	σ_f=1.6(3)$\times 10^2$	1000.(60)		
^{237}Pu		45.7 d	σ_f=2.3(3)$\times 10^3$			
^{238}Pu		87.74 y	5.1(2)$\times 10^2$ σ_f=17.(1)	1.6(2)$\times 10^2$ RI$_f$=33.(5)	14.1(5)	
^{239}Pu		2.411×10^4y	2.7(1)$\times 10^2$ σ_f=752.(3) $\sigma_\alpha \leq 0.3$ mb	2.0(2)$\times 10^2$ 3.0(1)$\times 10^2$	7.7(1)	
^{240}Pu		6537. y	2.9(1)$\times 10^2$ $\sigma_f \approx 44$.mb	8.4(3)$\times 10^3$ RI$_f$=2.4	3.5(1)	
^{241}Pu		14.4 y	3.7(1)$\times 10^2$,σ_α<0.2mb σ_f=1.01(1)$\times 10^3$	1.6(1)$\times 10^2$ 5.7(4)$\times 10^2$		
^{242}Pu		3.76×10^5 y	19.(1) σ_f<0.2	1.1(1)$\times 10^3$ RI$_f$=0.23	8.1(1)	
^{243}Pu		4.956 h	<100. σ_f=2.0(2)$\times 10^2$			
^{244}Pu		8.2×10^7 y	1.7(1)	41.(3)		
^{245}Pu		10.5 h	1.5(3)$\times 10^2$	220(40)		
$_{95}$Am						
^{241}Am		432.2 y	(0.5+5.7)$\times 10^2$ σ_f=3.1(2)	(2.0+12.3)$\times 10^2$ 14.(1)		
^{242m}Am		141. y	1.7(4)$\times 10^3$ σ_f=7.0(3)$\times 10^3$	$\approx$ 200. RI$_f$=1.8(1)$\times 10^3$		
^{242}Am		16.02 h	σ_f=2.1(2)$\times 10^3$	RI$_f$= < 300.		
^{243}Am		7.37×10^3y	(75.+5.) σ_f=74.(4) mb	(17.1+1.0)$\times 10^2$ RI$_f$=0.056	8.3(2)	

Elem. or Isot.	Natural Abundance (%)	Half-life	Thermal Neut. Cross Section (barns)	Resonance Integral (barns)	Coh. Scat. Length (fm)	σ(30KEV) Maxw. Avg. (barns)
^{244m}Am		$\approx$ 26. m	σ_f=1.6(3)x10^3			
^{244}Am		10.1 h	σ_f=2.2(3)x10^3			
$_{96}$Cm						
^{242}Cm		162.8 d	$\approx$ 20. $\sigma_f \approx$ 5.	120.(50)		
^{243}Cm		28.5 y	1.3(1)x10^2 σ_f=6.2(2)x10^2	214.(20) RI$_f$=1.6(1)x10^3		
^{244}Cm		18.11 y	15.(1) σ_f=1.1(2)	640.(50) RI$_f$=10.8(8)	9.5(3)	
^{245}Cm		8.5x10^3 y	3.5(2)x10^2 σ_f=2.1(1)x10^3	110(10) RI$_f$=8.(1)x10^2		
^{246}Cm		4.78x10^{3}y	1.2(2) σ_f=0.16(7)	120.(10) 13.(2)	9.3(2)	
^{247}Cm		1.56x10^{7}y	60.(30) σ_f=82.(5)	5.(1)x10^2 7.3(7)x10^2		
^{248}Cm		3.4x10^5 y	2.6(3) σ_f=0.36(7)	270.(30) 13.(2)	7.7(2)	
^{249}Cm		64.15 m	$\approx$1.6			
^{250}Cm		$\approx$ 9.7x10^3 y	$\approx$80.			
$_{97}$Bk						
^{249}Bk		320. d	7.(1)x10^2 $\sigma_f \approx$0.1	9.(1)x10^2		
^{250}Bk		3.217 h	σ_f=1.0(2)x10^3			
$_{98}$Cf						
^{249}Cf		351. y	5.0(3)x10^2 σ_f=1.7(1)x10^3	7.7(4)x10^2 RI$_f$=2.1(3)x10^3		
^{250}Cf		13.1 y	2.0(2)x10^3 σ_f=110.(90)	12.(2)x10^3 RI$_f$=160.(40)		
^{251}Cf		9.0x10^2 y	2.9(2)x10^3 σ_f=4.5(5)x10^3	1.6(1)x10^3 RI$_f$=5.5(3)x10^3		
^{252}Cf		2.65 y	20.(2) σ_f=32.(4)	43.(3) RI$_f$=1.1(3)x10^2		
^{253}Cf		17.8 d	18.(2) σ_f=1.3(2)x10^3	8.(1)		
^{254}Cf		60.5 d	4.5(10)	2.		
$_{99}$Es						
^{253}Es		20.47 d	(180.+5.8)	(37.5+1.1)x10^2		
^{254m}Es		1.64 d	σ_f=1.8(1)x10^3			
^{254}Es		276. d	28.(3) σ_f=1.8(2)x10^3	18.(2) RI$_f$=1.2(3)x10^3		
^{255}Es		40. d	$\approx$ 55.			
$_{100}$Fm						
^{255}Fm		20.1 h	26.(3) σ_f=3.3(2)x10^3	14.(2)		
^{257}Fm		100.5 d	σ_f=3.0(2)x10^3			

COSMIC RADIATION

A.G. Gregory and R.W. Clay

The Nature of Cosmic Rays

Primary cosmic radiation, in the form of high energy nuclear particles, electrons and photons from outside the solar system and from the Sun, continually bombards our atmosphere. Secondary radiation, resulting from the interaction of the primary cosmic rays with atmospheric gas, is present at sea-level and throughout the atmosphere.

The secondary radiation is collimated by absorption and scattering in the atmosphere and consists of a number of components associated with different particle species. High energy primary particles can produce large numbers of secondary particles forming an extensive air shower. Thus, a number of particles may then be detected simultaneously at sea-level.

Primary particle energies accessible in the vicinity of the earth range from $\sim 10^8$ eV to $\sim 10^{20}$ eV. At the lower energies, the limit is determined by the inability of charged particles to traverse the heliosphere to us through the outward-moving solar wind. The upper energy limit is set by the practicality of building detectors to record particles with the extremely low fluxes found at those energies (J.G. Wilson, 1976; O.C. Allkofer, 1975a).

Primary Cosmic Rays

Primary Particle Energy Spectrum

Figure 1 shows the spectrum of primary particle energies. This includes all particle species. In differential form it is roughly a power law of intensity versus energy with an index of ~ -3. There appears to be a knee (a steepening) at a little above 10^{15} eV and an ankle (a flattening) above $\sim 10^{18}$ eV. Figure 2 emphasizes the features in the spectrum at the highest energies through multiplying the flux with a strongly rising power law of energy. This figure should be used with caution as errors for the two axes are not now independent.

Data on the high energy cosmic ray spectrum are uncertain largely because of limited event statistics due to the very low flux which might best be measured in particles per square kilometer per century. The highest energy event recorded to 1995 had an energy of 3×10^{20} eV (D.J. Bird et al., 1993).

It is expected that the highest energy cosmic rays will interact with the 2.7 K cosmic microwave background through photoproduction or photodisintegration. These interactions will appreciably reduce the observed flux of cosmic rays with energies above 5×10^{19} eV if they travel further than ~ 150 million light years. This process is known as the Greisen-Zatsepin-Kuz'min (GZK) cut off (P. Sokolsky, 1989).

At energies below $\sim 10^{13}$ eV, solar system magnetic fields and plasma can modulate the primary component and Figure 3 shows the extent of this modulation between solar maximum and minimum (E. Juliusson, 1975; J. Linsley, 1981).

Primary Particle Energy Density

If the above spectrum is corrected for solar effects, the energy density above a particle energy of 10^9 eV outside the solar system is found to be $\sim 5 \times 10^5$ eV m^{-3}. As the threshold energy is increased, the energy density decreases rapidly, being 2×10^4 eV m^{-3} above 10^{12} eV and 10^2 eV m^{-3} above 10^{15} eV. The energy density at lower energies outside the heliosphere is unknown but may be substantially greater if the particle rest mass energy is included together with the kinetic energy (A. W. Wolfendale, 1979).

Primary Particle Isotropy

This is measured as an anisotropy $(I_{max} - I_{min})/(I_{max} + I_{min}) \times 100\%$, where I, the intensity (m^{-2}s^{-1}sr^{-1}), is usually measured with an angular resolution of a few degrees.

The measured anisotropy is small and energy dependent. It is roughly constant in amplitude at between 0.05 and 0.1% (with a phase of 0 to 6 hours in right ascension) for energies between 10^{11} eV and 10^{14} eV and appears to increase at higher energies roughly as $0.4 \times (Energy(eV)/10^{16})^{0.5}\%$ up to $\sim 10^{18}$ eV. The latter rise may well be an artifact of the progressively more limited statistics as the flux drops rapidly with energy. It appears possible that a real anisotropy has been observed at the highest energies (above a few times 10^{19} eV) with a directional preference for the supergalactic plane (this plane reflects the directions of galaxies within about 100 million light years) (A.W. Wolfendale, 1979; R.W. Clay, 1987; T. Stanev et al., 1995).

Primary Particle Composition

The composition of low energy cosmic rays is close to universal nuclear abundances except where propagation effects are present. For example, Li, Be, and B which are spallation products, are over-abundant by about six orders of magnitude.

Composition at 10^{11} eV per nucleus

Charge	1	2	(3–5)	(6–8)	(10–14)	(16–24)	(26–28)	≥30
% Composition	50	25	1	12	7	4	4	0.1
(10% uncertainty)								

Measurements at higher energies indicate that there is an increase in the relative abundances of nuclei with charge greater than 6 at energies above 50 TeV/nucleus (K. Asakimori et al., 1993) (1 TeV = 10^{12} eV).

Cosmic ray composition at low energies is often quoted at a fixed energy per nucleon. When presented in this way, protons constitute roughly 90% of the flux, helium nuclei about 10% and the remainder sum to a total of about 1%.

Certain radioactive isotopic ratios show lifetime effects. The ratio of Be^{10}/B^9 abundances is used to measure an "age" of cosmic rays since Be^{10} is unstable with a half life of about 1.6×10^6 years. A ratio of 0.6 is expected in the absence of Be^{10} decay and a ratio of about 0.2 is found experimentally (E. Juliusson, 1975; P. Meyer, 1981).

FIGURE 1. The energy spectrum of cosmic ray particles. This spectrum is of a differential form and can be converted to an integral spectrum by integration over all energies above a required threshold (E). Insofar as the spectrum approximates a power law of index –3, a simple conversion to the integral at an energy $E/1.8$ is obtained by multiplying the differential flux by the energy and dividing by 0.62.

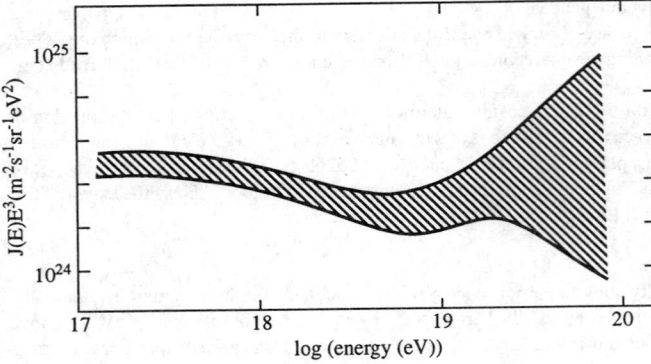

FIGURE 2. Energy spectrum at the highest energies. This spectrum (after Yoshida et al., 1995) has the differential spectrum multiplied by energy cubed. It is from a compilation of a number of measurements and indicates the good general agreement at the lower energies and a spread due to inadequate statistics at the highest energies.

FIGURE 3. Energy spectrum of particles at lower energies. (a) Solar minimum proton energy spectrum. (b) Solar maximum proton energy spectrum. (c) Gamma-ray energy spectrum. (d) Local interstellar electron spectrum.

At higher energies, composition determinations are indirect and are rather contradictory and controversial. Experiments aim to differentiate between broad composition models. The measurement technique is based on studies of cosmic ray shower development. A rather direct technique for such studies is to use fluorescence observations of the shower development to determine the atmospheric depth of maximum development of the shower. Such observations suggest a heavy composition (large atomic number) at energies $\sim 10^{17}$ eV which changes with increasing energy to a light composition (perhaps protonic) above $\sim 10^{19}$ eV (T. K. Gaisser et al., 1993).

Primary Electrons

Primary electrons constitute about 1% of the cosmic ray beam. The positron to negative electron ratio is about 10% (J. M. Clem et al., 1995).

Antimatter in the Primary eam

The ratio of antiprotons to protons in the primary cosmic ray beam (at about 400 MeV) is about 10^{-5}. At about 10 GeV the ratio is about 10^{-3}. At the highest measured energies (10 TeV), the upper limit to the ratio is about 20% (S. Orito et al., 1995; M. Amenomori et al., 1995).

Primary Gamma-Rays

The flux of primary gamma-rays is low at high energies. At 1 GeV the ratio of gamma-rays to protons is about 10^{-6}. The arrival directions of these gamma-rays are strongly concentrated in the plane of the Milky Way although there is a diffuse, near isotropic background flux and some point sources have been detected.

Since the absorption cross section for gamma-rays above 100 MeV is approximately 20 mbarn/electron, less than 10% of gamma-rays reach mountain altitudes (A. W. Wolfendale, 1979; P. F. Michelson, 1994).

Sea Level Cosmic Radiation

The sea level cosmic ray dose is 300 millirad·yr^{-1} and the sea level ionization is 2.2×10^6 ion pairs m^{-3}s^{-1}. The sea level flux has a soft component, which can be absorbed in about 100 mm of lead (about 100 g·cm^{-2} of absorber) and a more penetrating (largely muon) hard component. The sea level radiation is largely produced in the atmosphere and is a secondary component from interactions of the primary particles. The steep primary energy spectrum means that most secondaries at sea level are from rather low energy primaries. Thus the secondary flux is dependent on the solar cycle and the geomagnetic latitude of the observer.

Absolute Flux of the Hard Component

Vertical Integral Intensity $I(0) \sim 100$ m^{-2}s^{-1}sr^{-1}
Angular dependence $I(\theta) \sim I(0) \cos^2(\theta)$
Integrated Intensity ~ 200 m^{-2}s^{-1}
(O.C. Allkofer, 1975b).

Flux of the Soft Component

In free air, the soft component comprises about one third of the total cosmic ray flux.

Latitude Effect

The geomagnetic field influences the trajectories of lower energy cosmic rays approaching the Earth. As a result, the background flux is reduced by about 7% at the geomagnetic equator. The effect decreases towards the poles and is negligible at latitudes above about 40°.

Flux of Protons

The proton component is strongly attenuated by the atmosphere with an attenuation length (reduction by a factor of e) of about 120 g·cm^{-2}. It constitutes about 1% of the total vertical sea level flux.

Absorption

The soft component is absorbed in about 100 g·cm^{-2} of matter. The hard component is absorbed much more slowly:

   Absorption in lead, 6% per 100 g·cm^{-2}
   Absorption in rock, 8.5% per 100 g·cm^{-2}
   Absorption in water, 10% per 100 g·cm^{-2}
   (Absorption for depths less than 100 g·pd cm^{-2} is given by K. Greisen, 1943.)

Altitude Dependence

The cosmic ray background in the atmosphere has a maximum intensity of about 15 times that at sea level at a depth of about 150 g·cm^{-2} (15 km altitude). At maximum intensity, the soft and hard components contribute roughly equally but the hard component is then attenuated more slowly (S. Hayakawa, 1969).

Cosmic Ray Showers

High energy cosmic rays produce particle cascades in the atmosphere which can be detected at sea level provided that their energy exceeds about 100 GeV (such low energy cascades may be detected by using the most sensitive atmospheric Cerenkov detectors). The primary particle progressively loses energy which is transferred through the production of successive generations of secondary particles to a cascade of hadrons, an electromagnetic shower component (both positively and negatively charged electrons and gamma-rays) and muons. The secondary particles are relativistic and all travel effectively at the speed of light. As a result, they reach sea level at approximately the same time but, due to Coulomb scattering (for the electrons) and production angles (for the pions producing the muons), are spread laterally into a disk-like shower front with a characteristic lateral width of several tens of meters and thickness (near the central shower core) of 2 to 3 m. The number of particles at sea level is roughly proportional to the primary particle energy:

Number of particles at sea level $\sim 10^{-10} \times$ energy (eV).

At altitudes below a few kilometers, the number of particles in a shower attenuates with an *attenuation length* of about 200 g·cm^{-2}.

i.e., particle number = original number $\times \exp(-(\text{depth increase})/200)$

The above applies to an individual shower. The rate of observation of showers of a given size (particle number at the detector) at different depths of absorber attenuates with an *absorption length* of about 100 g·cm^{-2} (J.G. Wilson, 1976).

Atmospheric Background Light from Cosmic Rays

Cosmic ray particles produce Cerenkov light in the atmosphere and produce fluorescent light through the excitation of atmospheric molecules.

Cerenkov Light

High energy charged particles will cause the emission of Cerenkov light in air if their energies are above about 30 MeV (electrons). This threshold is pressure (and hence altitude) dependent. A typical Cerenkov light pulse (at sea level, 100 m from the central shower core) has a time spread of a few nanoseconds. Over this time, the photon flux between 430 and 530 nm would be $\sim 10^{14}$ m^{-2}s^{-1} for a primary particle energy of 10^{16} eV. For comparison, the night sky background flux is $\sim 6 \times 10^{11}$ photons m^{-2}s^{-1}sr^{-1} in the same wavelength band (J.V. Jelley, 1967).

Fluorescence Light

Cosmic ray particles in the atmosphere excite atmospheric molecules which then emit fluorescence light. This is weak compared to the highly collimated Cerenkov component when viewed in the direction of the incident cosmic ray particle but is emitted isotropically. Typical pulse widths are longer than 50 ns and may be up to several microseconds for the total pulse from distant large showers (R.M. Baltrusaitis et al., 1985).

Effects of Cosmic Rays

Cerenkov Effects in Transparent Media

Background cosmic ray particles will produce Cerenkov light in transparent material with a photon yield between wavelengths λ_1 and λ_2

$$\sim (2\pi/137)\sin^2\left(\theta_c\right)\int_{\lambda_1}^{\lambda_2} d\lambda/\lambda^2 \text{ photons (unit length)}^{-1}$$

where θ_c (the Cerenkov angle) = cos^{-1} (1/refractive index).

This background light is known to affect light detectors, e.g., photomultipliers, and can be a major source of background noise (R.W. Clay and A.G. Gregory, 1977).

Effects on Electronic Components

If background cosmic ray particles pass through electronic components, they may deposit sufficient energy to affect the state of, e.g., a transistor flip-flop. This effect may be significant where reliability is of great importance or the background flux is high. For instance, it has been estimated that, in communication satellite operation, an error rate of about 2×10^{-3} per transistor per year may be found. Permanent damage may also result. A significant error rate may be found even at sea level in large electronic memories. This error rate is dependent on the sensitivity of the component devices to the deposition of electrons in their sensitive volumes (J.F. Ziegler, 1981).

Biophysical Significance

When cosmic rays interact with living tissue, they produce radiation damage. The amount of the damage depends on the total dose of radiation. At sea level, this dose is small compared with doses from other sources but both the quantity and quality of the radiation change rapidly with altitude. Approximate dose rates under various conditions are:

> Dose rates (mrem·yr^{-1})
> Sea level cosmic rays, 30
> Cosmic rays at 10 km (subsonic jets), 2000
> Cosmic rays at 18 km (supersonic transports), 10,000
> (c.f., mean total sea level dose, 300)

Astronauts would be subject to radiation from galactic (0.05 rads per day) and solar (a few hundred rads per solar flare) cosmic rays as well as large fluxes of low energy radiation when passing through the Van Allen belts (about 0.3 rads per traverse).

Both astronauts and SST travellers would be subject to a small flux of low energy heavy nuclei stopping in the body. Such particles are capable of destroying cell nuclei and could be particularly harmful in the early stages of the development of an embryo. The rates of heavy nuclei stopping in tissue in supersonic transports and spacecraft are approximately as follows:

Stopping nuclei ((cm^3 tissue)$^{-1}$ hr^{-1})
Supersonic transport (16 km), 0.0005
Supersonic transport (20 km), 0.005
Spacecraft, 0.15
(O. C. Allkofer, 1975a; O. C. Allkofer et al., 1974).

Carbon Dating

Radiocarbon is produced in the atmosphere due to the action of cosmic ray slow neutrons. Solar cycle modulation of the very low energy cosmic rays causes an anticorrelation of the atmospheric ^{14}C activity with sunspot number with a mean amplitude of about 0.5%. In the long term, modulation of cosmic rays by a varying magnetic field may be important (A.A. Burchuladze et al., 1979).

Practical Uses of Cosmic Rays

There are few direct practical uses of cosmic rays. Their attenuation in water and snow have, however, enabled automatic monitors of water and snow depth to be constructed. A search for hidden cavities in pyramids has been carried out using a muon "telescope".

Other Effects

Stellar X-rays have been observed to affect the transmission times of radio signals between distant stations by altering the depth of the ionospheric reflecting layer. It has also been suggested that variations in ionization of the atmosphere due to solar modulation may have observable effects on climatic conditions.

REFERENCES

O.C. Allkofer, (1975a) *Introduction to Cosmic Radiation,* Verlag Karl Thiemig, Munchen, Germany.
O.O. Allkofer, (1975b) *J. Phys. G: Nucl. Phys.,* 1, L51.
O.C. Allkofer and W. Heinrich, (1974) *Health Phys.,* 27, 543.
M. Amenomori et al., (1995) Proc. 24th Int. Cosmic Ray Conf. Rome, 3, 85. Universita La Sapienza, Roma.
K. Asakimori et al., (1993) Proc. 23rd Int. Cosmic Ray Conf. Calgary, 2, 25, University of Calgary, Canada.
R.M. Baltrusaitis et al., (1985) *Nucl. Inst. Meth.,* A420, 410.
D.J. Bird et al., (1993) *Phys. Rev. Lett.,* 71, 3401.
A.A. Burchuladze, S.V. Pagava, P. Povinec, G. I. Togonidze, S. Usacev, (1979) Proc. 16th Int. Cosmic Ray Conf. Kyoto, 3, 201, Univ. of Tokyo, Japan.
R.W. Clay, (1987) *Aust. J. Phys.,* 40, 423.
R.W. Clay and A.G. Gregory, (1977) *J. Phys. A: Math. Gen.,* 10, 135.
J.M. Clem et al., (1995) Proc. 24th Int. Cosmic Ray Conf. Rome, 3, 5, Universita La Sapienza, Roma.
T.K. Gaisser et al., (1993) *Phys. Rev. D,* 47, 1919.
K. Greisen, (1943) *Phys. Rev.,* 63, 323.
S. Hayakawa, (1969) *Cosmic Ray Physics,* Wiley-Interscience, New York.
J.V. Jelley, (1967) *Prog. in Elementary Particle and Cosmic Ray Physics,* 9, 41.
E. Juliusson, (1975) Proc. 14th Int. Cosmic Ray Conf. Munich, 8, 2689, Max Planck Institute fur Extraterrestriche Physik, Munchen, Germany.
J. Linsley, (1981) *Origin of Cosmic Rays,* I.A.U. Symposium 94, 53, D. Reidel Publishing Co Dordrecht, Holland.
P. Meyer, (1981) *Origin of Cosmic Rays,* I.A.U. Symposium 94, 7, D. Reidel Publishing Co. Dordrecht, Holland.
P.F. Michelson (1994) in *Towards a Major Atmospheric Cerenkov Detector III,* 257, Ed. T. Kifune, Universal Academy Press Inc., Tokyo, Japan.
P. Sokolsky, (1989) *Introduction to Ultrahigh Energy Cosmic Ray Physics,* Addison Wesley Publishing Company.
T. Stanev et al., (1995) *Phys. Rev. Lett.,* 75, 3056.
S. Orito et al., (1995) Proc. 24th Int. Cosmic Ray Conf. Rome, 3, 76. Universita La Sapienza, Roma.
J.G. Wilson, (1976) *Cosmic Rays,* Wykeham Pub. (London) Lt., U.K.
A.W. Wolfendale, (1979) *Pramana,* 12, 631.
S.Yoshida et al., (1995) *Astroparticle Phys.,* 3, 105.
J.F. Ziegler, (1981) IEEE Trans. Electron Devices, ED-28, 560.

Section 12
Properties of Solids

TECHNIQUES FOR MATERIALS CHARACTERIZATION

EXPERIMENTAL TECHNIQUES USED TO DETERMINE THE COMPOSITION, STRUCTURE, AND ENERGY STATES OF SOLIDS AND LIQUIDS

H.P.R.Frederikse

The many experimental methods, originally designed to study the chemical and physical behavior of solids and liquids, have grown into a new field known as Materials Characterization (or Materials Analysis). During the past 30 years a host of techniques aimed at the study of surfaces and thin films has been added to the many tools for the analysis of bulk samples. The field has benefitted particularly from the development of computers and microprocessors, which have vastly increased the speed and accuracy of the measuring devices and the recording of their output. Materials characterization was and is a very important tool in the search for new physical and chemical phenomena. It plays an essential role in new applications of solids and liquids in industry, communications, and medicine. Many of its techniques are used in quality control, in safety regulations, and in the fight against pollution.

In most Materials Characterization experiments the sample is subjected to some kind of radiation: electromagnetic, acoustic, thermal, or particles (electrons, ions, neutrons, etc.). The surface analysis techniques usually require a high vacuum. As a result of interactions between the solid (or liquid) and the incoming radiation a beam of a similar (or a different) nature will emerge from the sample. Measurement of the physical and/or chemical attributes of this emerging radiation will yield qualitative, and often quantitative, information about the composition and the properties of the material being probed.

The modern tendency of describing practically everything in this world by a combination of a few letters (acronyms) has also penetrated the field of Materials Characterization. The table below gives the meaning of the acronym for every technique listed, the form and size of the required sample (bulk, surface, film, liquid, powder, etc.), the nature of the incoming radiation, the depth and the lateral spatial resolution that can be probed, and the information obtained from the experiment. The last column lists one or two major references to the technique described.

OPTICAL AND MASS SPECTROSCOPIES FOR CHEMICAL ANALYSIS

	Technique	Sample	In	Out	Depth	Lateral resolution	Information obtained	Ref.
1.	AAS Atomic Absorption Spectroscopy	Atomize (flame, electro, thermal, etc.)	Light e.g. glow discharge	Absorption spectrum	—	—	Concentration of atomic species (quantitative, using standards)	1,2
2.	ICP-AES Induct. Coupled Plasma — Atomic Emission Spectroscopy	Atomize (flame, electro, thermal, ICP, etc.)	—	Emission spectrum	·	—	Concentration of atomic species (quantitative, using standards)	3
3.	Dynamic SIMS Dynamic Secondary Ion Mass Spectroscopy	Surface	Ion beam (1–20 keV)	Secondary ions; analysis with mass spectrometer	2 nm–1 µm (or deeper: ion milling)	0.50 nm	Elemental and isotopic analysis; depth profile (all elements); detection limits: ppb-ppm	4
4.	Static SIMS Static Secondary Ion Mass Spectroscopy	Surface	Ion beam (0.5–20 keV)	Secondary ions, analysis with mass spectrometer	0.1–0.5 nm	10 µm	Elemental analysis of surface layers; molecular analysis; detection limits: ppb-ppm	4
5.	SNMS Sputtered Neutral Mass Spectroscopy	Surface, bulk	Plasma discharge; noble gases: 0.5–20 keV	Sputtered atoms ionized by atoms or electrons; then mass analyzed	0.1–0.5 nm (or deeper: ion milling)	1 cm	Elemental analysis Z ≥ 3; depth profile; detection limit: ppm	4,6
6.	SALI Surface Analysis by Laser Ionization	Surface	e-beam, ion-beam, or laser for sputtering	Sputtered atoms ionized by laser; then mass analyzed	0.1–0.5 nm up to 3 µm in milling mode	60 nm	Surface analysis; depth profiling	7
7.	LIMS Laser Ionization Mass Spectroscopy	Surface, bulk	u.v. laser (ns pulses)	Ionized species; analyzed with mass spectrometer	50–150 nm	5 µm–1 mm	Elemental (micro)analysis; detection limits: 1–100 ppm	8
8.	SSMS Spark Source Mass Spectroscopy	Sample in the form of two electrodes	High voltage R.F. spark produces ions	Ions — analyzed in mass spectrometer	1–5 µm	—	Survey of trace elements; detection limit: 0.01–0.05 ppm	9
9.	GDMS Glow Discharge Mass Spectroscopy	Sample forms the cathode for a D.C. glow discharge	Sputtered atoms ionized in plasma	Ions — analyzed in mass spectrometer	0.1–100 µm	3–4 mm	(Bulk) trace element analysis; detection limit: sub-ppb	9,10
10.	ICPMS Induct. Coupled Plasma Mass Spectroscopy	Liquid-dissolved sample carried by gas stream into R.F. induction coil	Ions produced in argon plasma	Ions — analyzed in quadrupole mass spectrometer	—	—	High sensitivity analysis of trace elements	11

TECHNIQUES FOR MATERIALS CHARACTERIZATION (continued)

PHOTONS — ABSORPTION, REFLECTION AND ELECTRON EMISSION

Technique	Sample	In	Out	Depth	Lateral resolution	Information obtained	Ref.
11. IRS Infrared Spectroscopy	Thin crystal, glass, liquid	I.R. light (W-filament, globar, Hg-arc)	I.R. spectrum	—	—	Electronic transitions (mainly in semiconductors and superconductors); vibrational modes (in crystals and molecules)	12,13, 14
12. FTIR Fourier Transform I.R. Spectroscopy	Solid, liquid; transmission or reflection	White light (all frequencies)	Fourier Transform of spectrum (interferometer)	—	—	Spectra obtained at higher speed and resolution	15
13. ATR Attenuated Total Reflection	Surface or thin crystal	—	—	μm's	—	Atomic or molecular spectra of surfaces and films	16
14. (μ-)RS (Micro-) Raman Spectroscopy	Solid, liquid (1 μm–1 cm)	Laser beam, e.g. Ar-line, YAG-line	Raman spectra	0.5 μm	0.5 μm	Molecular and crystal vibrations	12,14, 17
15. CARS Coherent Anti-Stokes Raman Spectroscopy	Solid, liquid (50 μm–3 cm)	Pump beam (ω_0)+ probe beam (ω_a)	Anti-Stokes spectrum	—	—	High resolution Raman spectra	14
16. Ellipsometry	Transparent films, crystals, adsorbed layers	Polarized light	Change in polarization	0.05 nm–5 μm	—	Refractive index and absorption	18,19
17. UPS Ultraviolet Photo-electron Spectroscopy	Surfaces, adsorbed layers	u.v. light, 10–100 eV; 200 eV (synchrotron)	Electrons	0.2–10 nm	25 μm (or sample thickness)	Energies of electronic states of surfaces and free molecules	20,21
18. PSD Photon Stimulated Desorption	Surfaces with adsorbed species	Far u.v. light E > 10 eV	Ions — analyzed with mass spectrometer	0.1–2 nm	0.1–10 nm	Structure and desorption kinetics of adsorbed atoms and molecules	22

X-RAYS

Technique	Sample	In	Out	Depth	Lateral resolution	Information obtained	Ref.
19. XRD X-Ray Diffraction	Single crystals, powders films	X-rays: λ = 0.05–0.2 nm (6–17 keV)	Diffracted X-ray beam	1–1000 μm	0.1–10 mm	Identification of crystallographic structures; all elements (low Z difficult)	23,24
20. XRF/EDS X-Ray Fluorescence/Energy Dispersive Spectroscopy	Thin films, single layer	Prim. X-ray beam λ = 0.02–0.1 nm 12–80 keV	Fluorescent X-rays	1–100 μm	10 mm	Elemental analysis; all elements except H, He, Li — (EDS also used in XRD, SEM, TEM and EPMA)	25,26
21. EXAFS Extended X-Ray Absorption Fine Structure	Films, foils	High intensity X-rays (synchrotron)	Spectrum near absorption edge	nm–μm	—	Local atomic structure: order/disorder in vicinity of absorbing atom	27
22. XPS/ESCA X-Ray Photo-electron Spectroscopy/Electron Spect. for Chemical Analysis	Surfaces, thin films (≈20 atomic layers)	Soft X-rays (1–20 keV)	Core electrons; valence electrons	0.5–10 nm	5 nm–50 μm	(Quantitative) identification of all elements in surface layer or film	28,29

ELECTRONS

Technique	Sample	In	Out	Depth	Lateral resolution	Information obtained	Ref.
23. CL Cathode Luminescence	Insulators, semiconductors	Electrons 5–50 keV	Photons	1 nm–2 μn	1 or 2 μm	Energy levels of impurities and point defects	30
24. APS Appearance Potential Spectroscopy	Surface (≈20 atomic layers)	Electrons (energy scan) 50–2000 eV	X-rays to pinpoint electron energy threshold	—	—	Identification of surface defects	21, see also C

TECHNIQUES FOR MATERIALS CHARACTERIZATION (continued)

	Technique	Sample	In	Out	Depth	Lateral resolution	Information obtained	Ref.
25.	AES Auger Electron Spectroscopy	Thin films, surfaces	Electrons 3-10 keV	Auger electrons 20-2000 eV	0.3-3 nm	≈30 nm	Elemental composition of surface (except H, He); detection limit 0.1-1%	28,29
26.	EELS Electron Energy Loss Spectroscopy	Very thin samples (<200 nm)	Electrons (100-400 keV)	(Retarded) electrons; minus 1-1000 eV	<200 nm	1-100 nm	Local elemental concentration; electronic structure, chem. bonding; interatomic distances	31
27.	EXELFS Extended Electron Energy Loss Fine Structure	Thin films	Electrons (100-400 keV)	Electrons energies 0-30 eV above edge	<200 nm	1-100 nm	Density of states of valence electrons (above Fermi level)	27,32
28.	ESD Electron Stimulated Desorption	Adsorbed species	Electrons E > 10 eV	Ions — analyzed with mass spectrometer	—	—	Structure and desorption properties of adsorbed atoms and molecules	22
29.	ESDIAD ESD-Ion Angular Distribution	(See ESD)	(See ESD)	Directional dependence of emitted ions	—	—	Geometries of adsorbed species (atoms or molecules)	22
30.	EPMA Electron Probe (X-Ray) Micro Analysis	Solid conductors and insulators <1 cm thick	Electrons 5-30 keV	Characteristic X-ray 0.1-15 keV	100 nm-5 μm	1 μm	Elemental analysis, Z ≤ 4, major, minor and trace amounts	33,34
31.	LEED Low Energy Electron Diffraction	Surface	Mono-energetic electron beam 10-1000 eV	Diffracted electrons	0.4-2 nm	<5 μm	Crystallographic structure of surface; resolution: 0.01 nm	35
32.	RHEED Reflection High Energy Electron Diffraction	Surface	Electron beam at grazing angle 5-50 keV	Reflected electrons	0.2-10 nm	<5 μm	Surface symmetry	36,37
33.	SEM Scanning Electron Microscopy	Bulk, films (conducting)	High energy electrons usually ~30 keV	Secondary and backscattered electrons	1 nm-5 μm	1-20 nm	Surface image, defect structure; resolution 5-15 nm; magnification 300,000×	33,34
34.	(S)TEM (Scanning) Transmission Electron Microscopy	Thin specimen — <200 nm	High energy electrons typically 300 keV	Transmitted and diffracted electrons	(Sample thickness)	2-20 nm	(Defect) structure of cryst. solids; microchemistry; high resol.: 0.2 nm	33
35.	FEM Field Emission Microscopy	Metals, alloys (sharp point)	—	Electron emission (with appl. electric field — 50 kV)	≈0.5 nm	10-100 nm	Surface image, crystallographic structure	34
36.	STM Scanning Tunneling Microscopy	Polished or cleaved surface (conducting)	Tunneling current controls distance between sample and very sharp tip		1-5 nm	2-10 nm	Atomic-scale relief map of surface; resolution: vert. 0.002 nm, hor. 0.2 nm	39
37.	SPM Scanned Probe Microscopy	Very flat surface	Any field: e.g. mechan. vibration recorded with laser probe; same with magnetic, electric or thermal field		1-100 nm	1-100 nm	Surface-magnetic field, surface-thermal conductivity, etc.	39a
38.	AFM Atomic Force Microscopy	Very flat surface	Similar to STM; force measured with cantilever spring		0.5-5 nm	0.2-130 nm	Surface topography with atomic resolution; interatomic force	40
	IONS AND NEUTRONS							
39.	ISS (or LEIS) Ion Scattering Spectroscopy (Low Energy Ion Scattering)	Surface	Ion beam He+ or Ne+ <3 keV	Sputtered ions (energy analysis)	0.1-0.5 nm	1-100 μm	Elemental analysis (better for low Z) detection limits: 0.01-1%	41
40.	FIM Field Ion Microscopy	Surface: metals, alloys; very sharp tip	(He gas above sample)	He ions + high electric field produce image	≈0.1 nm	0.1-2 nm	Atomic structure of surface	34,42
41.	RBS Rutherford Back Scattering	Solids, thin films	Mono-energetic ions (H+ or He++) 0.5-3 MeV	Backscattered ions	10 nm-1 μm	1 mm	Element identification (Li to U) detection limit: 0.01-1%	46

TECHNIQUES FOR MATERIALS CHARACTERIZATION (continued)

Technique	Sample	In	Out	Depth	Lateral resolution	Information obtained	Ref.
42. NRA Nuclear Reaction Analysis	Solids, thin films	Mono-energetic ions (Li, Be, B, etc.) 200 keV–6 MeV	Protons, deuterons ³He, α-particles, γ-rays	0.1–5 μm	10 μm–10 mm	Element identification (all) detection limit: 10^{-12}–10^{-2}	47
43. PIXE Particle Induced X-ray Emission	Thin films, surface layers	High energy ions (H⁺ or He⁺⁺)	Characteristic X-rays	<10 μm	1 μm–2 mm	Trace impurities: Z >3 detection limit: 0.1–100 ppm (depending on sample thickness)	48
44. INS Ion Neutralization Spectroscopy	Surface	He-ions (≈5 eV)	Electrons	—	—	Energies of valence electrons	49
45. NAA Neutron Activation Analysis	Bulk, >0.5 g	Thermal neutrons	Characteristic γ-rays, (≈1 MeV)	Bulk	—	Trace concentrations (of isotopes) of elements; trans. metals, Pt-group; detection limit: 10^8–10^{14} atoms/cm³	43
46. N(P)D Neutron (Powder) Diffraction	Crystalline solids	Thermal neutrons E =0.0025 eV	Diffracted neutrons	Bulk	—	Crystallographic structure; porosity, particle size	44
47. SANS Small Angle Neutron Scattering	Inhomogeneous solids; powders; porous samples	Thermal neutrons $2\theta = 10^{-2}$–10^{-4}	Scattered neutrons	1–25 mm	—	Average size of inhomogeneities; range: 1 nm–1 mm	45
ACOUSTIC							
48. SLAM Scanning Laser Acoustic Microscopy	Bulk, film	Acoustic wave produced by laser 1 MHz–1 GHz	Reflected acoustic wave	μm–cm	0.1–20 mm	Defect structure; thickness measurement	50
THERMAL							
49. DTA Differential Thermal Analysis	Specimen and reference sample	Uniform heating	Temperature difference	Bulk	—	Phase transitions, crystallization	51
50. DSC Differential Scanning Calorimetry	Specimen and ref. sample	Controlled heating	Measure heat required for equal temperature	Bulk	—	Phase transitions, crystallization; activation energies	51
51. TGA Thermo Gravimetric Analysis	Bulk, 1–100 g	Controlled heating	Weight as function of temperature (and time)	Bulk	—	Decomposition, non-stoichiometry, kinetics of reaction	52
RESONANCE							
52. EPR (ESR) Electron Paramagnetic (Spin) Resonance	Paramagnetic solids or liquids	Microwave radiation in magnetic field 3–300 GHz; 1–100 kG	Microwave absorption (at resonance)	Bulk	—	Local environment of paramagnetic ion; concentration of paramagnetic species; detection limit: 10^{11} spins/cm³	53,54
53. ECR Electron Cyclotron Resonance	Semiconductors, metals; free electrons (low temperature)	Microwave radiation in magnetic field 10–30 GHz; 5–10 kG	Microwave absorption (at resonance)	Bulk	—	Electronic energy bands, effective masses	55
54. Mössbauer Effect	Source and absorber	Mono-energetic γ-rays: 5–100 keV	Mössbauer spectrum (Doppler shifted lines)	50 m	1 cm	Interaction between nucleus and its environment (local electric, magnetic fields; bonds; valency; diffusion, etc.)	56
55. NMR (MRI) Nuclear Magnetic Resonance (Magnetic Resonance Imaging)	Solids, liquids	R.F. radiation + magnetic field; e.g. for protons: 60 MHz, 14 kG	R.F. absorption	<1 cm	1 cm	Quant. analysis; local magnetic environment; diffusion; imaging	58
56. ENDOR Electron Nuclear Double Resonance	Solids, liquids	R.F. + microwave radiation in magn. field.	Microwave absorption	—	—	Hyperfine interaction → local atomic structure	54

Technique	Sample	In	Out	Depth	Lateral resolution	Information obtained	Ref.
57. NQR Nuclear Quadrupole Resonance	Solids	R.F. radiation 0.5–1000 MHz	R.F. absorption	—	—	Asymmetry of the charge distribution at the nucleus	55,59
OTHER							
58. BET Brunauer-Emmett-Teller	(Large) surface area 1–20 m²/g	Adsorbed gas (e.g., N_2 at low temp.) as function of pressure (monolayer coverage)	—	—	Surface area measurement	60

REFERENCES

General References

A. Wachtman, J. B., *Characterization of Materials*, Butterworth-Heinemann, Boston, 1993.
B. Brundle, C. R., Evans, C. A., and Wilson, S., Eds., *Encyclopedia of Materials*, Butterworth-Heinemann, Boston, 1992.
C. Woodruff, D. P. and Delchar, T. A., *Modern Techniques of Surface Science*, Cambridge University Press, Cambridge, 1986.
D. *Metals Handbook*, 9th Edition, Vol. 10, Materials Characterization, Whan, R. E., Coordinator, American Society for Metals, Metals Park, OH, 1986.

Specific References

1. Slavin, M., *Atomic Absorption Spectroscopy*, 2nd Edition, John Wiley & Sons, New York, 1978.
2. Schrenk, W. G., *Analytical Atomic Spectroscopy*, Plenum Press, New York, 1975.
3. Dean, J. A. and Rains, T. E., *Flame Emission and Atomic Absorption Spectroscopy*, Vols. 1—3, Marcel Dekker, New York, 1969.
4. Benninghoven, A., Rudenauer, F. G., and Werner, H. W., *Secondary Ion Mass Spectroscopy*, John Wiley & Sons, New York, 1987.
5. Bird, J. R. and Williams, J. S., Eds., in *Ion Beams for Materials Analysis*, Academic Press, New York, 1989, pp. 515—537.
6. Smith, G. C., *Quantitative Surface Analysis for Materials Science*, The Institute of Metals, London, 1991.
7. Becker, E. H., in *Ion Spectroscopies for Surface Analysis*, Czanderna, A. W. and Hercules, D. M., Eds., Plenum Press, New York, 1991, p. 273.
8. Simons, D. S., *Int. J. Mass Spectrometry and Ion Processes*, 55, 15, 1983.
9. White, F. A. and Wood, G. M., *Mass Spectrometry: Applications in Science and Engineering*, John Wiley & Sons, New York, 1986.
10. Harrison, W. W. and Bentz, B. L., *Prog. Anal. Spectrometry*, 11, 53, 1988.
11. Bowmans, P. W. J. M., *Inductively Coupled Plasma Emission Spectroscopy*, Parts I and II, John Wiley & Sons, New York, 1987.
12. Brame, Jr., E. G. and Grasselli, J., *Infrared and Raman Spectroscopy*, Practical Spectroscopy Series, Vol. I, Marcel Dekker, New York, 1976.
13. Hollas, J. M., *Modern Spectroscopy*, John Wiley & Sons, New York, 1987.
14. Turrell, G., *Infrared and Raman Spectroscopy of Crystals*, Academic Press, New York and London, 1972.
15. Griffith, P. R. and Haseth, J. A., *Fourier Transform Infrared Spectroscopy*, John Wiley & Sons, New York, 1986.
16. Barnowski, M. K., *Fundamentals of Optical Fiber Communications*, Academic Press, New York, 1976.
17. Long, D. A., *Raman Spectroscopy*, McGraw-Hill, New York, 1977.
18. Azzam, R. M. A., *Ellipsometry and Polarized Light*, Elsevier-North Holland, Amsterdam, 1977.
19. Hecht, E., *Optics*, 2nd Edition, Addison-Wesley, Reading MA, 1987.
20. Brundle, C. R., in *Molecular Spectroscopy*, West, A. R., Ed., Heyden, London, 1976.
21. Park, R. L., in *Experimental Methods in Catalytic Research*, Vol. III, Anderson, R. B. and Dawson, P. T., Academic Press, New York, 1976, pp. 1—39.
22. Madey, T. E. and Stockbauer, R., in *Solid State Physics: Surfaces*, Vol. 22 of Methods of Experimental Physics, Park, R.L. and Lagally, M. G., Eds., Academic Press, New York, 1985.
23. Cullity, B. D., *Elements of X-Ray Diffraction*, 2nd Edition, Addison-Wesley, Reading, MA, 1978.
24. Schwartz, L. H. and Cohen, J. B., *Diffraction from Materials*, Springer Verlag, Berlin, 1987.

TECHNIQUES FOR MATERIALS CHARACTERIZATION (continued)

25. deBoer, D. K. G., in *Advances in X-Ray Analysis*, Vol. 34, Barrett, C. S. et. al., Eds., Plenum Press, New York, 1991.

26. Birks, L. S., *X-Ray Spectrochemical Analysis*, 2nd Edition, John Wiley & Sons, New York, 1969.

27. Bonnelle, C. and Mande, C., *Advances in X-Ray Spectroscopy*, Pergamon Press, Oxford, 1982.

28. *Practical Surface Analysis by Auger and X-Ray Photo-Electric Spectroscopy*, Briggs, D. and Seah, M. P., Eds., John Wiley & Sons, New York, 1983.

29. Powell, C. J. and Seah, M. P., *J. Vac. Sci. Technol. A*, Vol. 8, 735, 1990.

30. Yacobi, G. G. and Holt, D. B., *Cathodeluminescence Microscopy of Inorganic Solids*, Plenum Press, New York, 1990.

31. Egerton, R. F., *Electron Energy Loss Spectroscopy in the Electron Microscope*, Plenum Press, New York, 1986.

32. Disko, M. M., Krivanek, O. L., and Rez, P., *Phys. Rev.*, B25, 4252, 1982.

33. Goldstein, J. I., et. al., *Scanning Electron Microscopy and X-Ray Microanalysis*, 2nd Edition, Plenum Press, New York, 1986.

34. Murr, L. E., *Electron and Ion Microscopy and Microanalysis*, Marcel Dekker, New York, 1982.

35. Armstrong, R. A., in *Experimental Methods in Catalytic Research*, Vol. III, Anderson, R. B., and Dawson, P. T.,Eds., Academic Press, New York, 1976.

36. Dobson, P. J. et. al., *Vacuum*, 33, 593, 1983.

37. Rymer, T. B., *Electron Diffraction*, Methuen, London, 1970.

38. Reimer, L., *Transmission Electron Microscopy*, Springer-Verlag, Berlin, 1984.

39. *Scanning Tunneling Microscopy and Related Methods*, Behm, R. J., Garcia, N., and Rohrer, H., Kluwer, Eds., Academic Publishers, Norwell, MA, 1990.

39a. Wikramasinghe, H.K., *Scientific American*, Vol. 261, No. 4, pp. 98—105, Oct. 1989.

40. Rugar, D. and Hansma, P., *Physics Today*, 43(10), pp. 23—30, 1990.

41. Feldman, C. C. and Mayer, J. W., *Fundamentals of Surface and Thin Film Analysis*, North-Holland, Amsterdam, 1986.

42. Muller, E. W. and Tsong, T. T., *Field Ion Microscopy*, Elsevier, Amsterdam, 1969.

43. Amiel, S., *Nondestructive Activation Analysis*, Elsevier, Amsterdam, 1981.

44. Bacon, G. E., *Neutron Diffraction*, 3rd Edition, Clarendon Press, Oxford, 1975.

45. Neutron Scattering, Part A., in *Methods of Experimental Physics*, Vol. 23, Skold, K. and Price, D. L., Eds., Academic Press, New York, 1986.

46. Chu, W. K., Mayer, J. W., and Nicolet, M. A., *Backscattering Spectroscopy*, Academic Press, New York, 1987.

47. Rickey, F. A., in *High Energy and Heavy Ion Beams in Materials Analysis*, Tesmer, J. R., et. al., Eds., MRS, 1990, pp. 3—26.

48. Johansson, S. A. E. and Campbell, J. L., *PIXE: A Novel Technique for Elemental Analysis*, John Wiley & Sons, New York, 1988.

49. Hagstrum, H. D., in *Inelastic Ion-Surface Collisions*, Tolk, N. H. et. al., Eds., Academic Press, New York, 1977, pp. 1—46.

50. Nikoonahad, M., in *Research Techniques in Nondestructive Testing*, Vol. VI, Sharpe, R.S., Ed., Academic Press, New York, 1984, pp. 217—257.

51. Gallagher, P. K., *Characterization of Materials by Thermoanalytical Techniques*, MRS - Bulletin, Vol. 13, No. 7, pp. 23—27, 1988.

52. Earnest, C. M., *Compositional Analysis of Materials by Thermogravimetry*, ASTM Special Technical Publication 997, 1988.

53. Poole, C. P., *Electron Spin Resonance — A Comprehensive Treatise on Experimental Techniques*, 2nd Edition, John Wiley & Sons, New York, 1983.

54. Atherton, N. M., *Principles of Electron Spin Resonance*, Ellis Horwood Ltd., Chichester, U.K., 1993.

55. Kittel, C., *Introduction to Solid State Physics*, 6th Edition, John Wiley & Sons, New York, 1986, p. 196.

56. Gibb, T. C., *Principles of Mössbauer Spectroscopy*, Chapman & Hall, London, 1976.

57. Slichter, C. P., *Principles of Magnetic Resonance*, 3rd Edition, Springer-Verlag, Berlin, 1990.

58. *NMR Spectroscopy Techniques*, Dybrowski, C. and Lichter, R. L., Eds., Marcel Dekker, New York, 1987.

59. Das, T. P. and Hahn, E. L., *Nuclear Quadrupole Resonance Spectroscopy*, Academic Press, New York, 1958.

60. Somorjai, G. A., *Principles of Surface Chemistry*, Prentice-Hall, Englewood Cliffs, NJ, 1972, p. 216.

SYMMETRY OF CRYSTALS

L. I. Berger

The ability of a body to coincide with itself in its different positions regarding a coordinate system is called its symmetry. This property reveals itself in iteration of the parts of the body in space. The iteration may be done by reflection in mirror planes, rotation about certain axes, inversions and translations. These actions are called the symmetry operations. The planes, axes, points, etc., are known as symmetry elements. Essentially, mirror reflection is the only truly primitive symmetry operation. All other operations may be done by a sequence of reflections in certain mirror planes. Hence, the mirror plane is the only true basic symmetry element. But for clarity, it is convenient to use the other symmetry operations, and accordingly, the other aforementioned symmetry elements. The symmetry elements and operations are presented in Table 1.

The entire set of symmetry elements of a body is called its symmetry class. There are thirty-two symmetry classes that describe all crystals which have ever been noted in mineralogy or been synthesized (more than 150,000). The denominations and symbols of the symmetry classes are presented in Table 2.

There are several known approaches to classification of individual crystals in accordance with their symmetry and crystallochemistry. The particles which form a crystal are distributed in certain points in space. These points are separated by certain distances (translations) equal to each other in any chosen direction in the crystal. Crystal lattice is a diagram that describes the location of particles (individual or groups) in a crystal. The lattice parameters are three non-coplanar translations that form the crystal lattice. Three basic translations form the unit cell of a crystal. August Bravais (1848) has shown that all possible crystal lattice structures belong to one or another of fourteen lattice types (Bravais lattices). The Bravais lattices, both primitive and non-primitive, are the contents of Table 3.

Among the three-dimensional figures, there is a group of polyhedrons that are called regular, which have all faces of the same shape and all edges of the same size (regular polygons). It has been shown that there are only five regular polyhedrons. Because of their importance in crystallography and solid state physics, a brief description of these polyhedrons is included in Table 4.

The systematic description of crystal structures is presented primarily in the well known *Structurbericht*. The classification of crystals by the Structurbericht does not reflect their crystal class, the Bravais lattice, but is based on the crystallochemical type. This makes it inconvenient to use the Structurbericht categories for comparison of some individual crystals. Thus, there have been several attempts to provide a more convenient classification of crystals. Table 5 presents a compilation of different classifications which allows the reader to correlate the Structurbericht type with the international and Schoenflies point and space groups and with Pearson's symbols, based on the Bravais lattice and chemical composition of the class prototype. The information included in Table 5 has been chosen as an introduction to a more detailed crystallophysical and crystallochemical description of solids.

TABLE 1
Symmetry Operations and Elements

Symmetry operation	Name	Symbol International (Hermann-Mauguin)	Schoenflies	Presentation on the stereographic projection Parallel	Perpendicular
Reflection in a plane	Plane	m	C_s		
Rotation by angle $\alpha = 360°/n$ about an axis	Axis	n = 1, 2, 3, 4 or 6	C_n		
		n = 2	C_2		
		n = 3	C_3		
		n = 4	C_4		
		n = 6	C_6		
Rotation about an axis and inversion in a symmetry center lying on the axis	Inversion (improper) axis	$\bar{n} = \bar{3}, \bar{4}, \bar{6}$	C_{ni}		
		$\bar{n} = \bar{3}$	C_{3i}		
		$\bar{n} = \bar{4}$	C_{4i}		

TABLE 1
Symmetry Operations and Elements (continued)

Symmetry operation	Name	Symbol — International (Hermann-Mauguin)	Schoenflies	Presentation on the stereographic projection Parallel	Perpendicular
		$\bar{n} = \bar{6}$	C_{6i}		
Inversion in a point	Center	$\bar{1}$	C_i	● ○	✕
Parallel translation	Translation vector a, b, c				
Reflection in a plane and translation parallel to the plane	Glide–plane	a, b, c, n, d			
Rotation about an axis and translation parallel to the axis	Screw axis	n_m (m = 1, 2, .., n − 1)			
Rotation about an axis and reflection in a plane perpendicular to the axis	Rotatory-reflection axis	ñ $\tilde{n} = \bar{1}, \bar{2}, \bar{3}, \bar{4}, \bar{6}$	S_n		

TABLE 2
The Thirty-Two Symmetry Classes

Crystal symbol	Primitive Int	Sch	Central Int	Sch	Planal Int	Sch	Axial Int	Sch	Plane-axial Int	Sch	Inversion primitive Int	Sch	Inversion-planal Int	Sch
Triclinic	1	C_1	$\bar{1}$	C_i										
Monoclinic					m	C_s	2	C_2	2/m	C_{2h}				
Orthorhombic					mm2	C_{2v}	222	D_2	mmm	D_{2h}				
Trigonal	3	C_3	$\bar{3}$	C_{3i}	3m	C_{3v}	32	D_3	$\bar{3}m$	C_{3d}				
Tetragonal	4	C_4	4/m	C_{4h}	4mm	C_{4v}	422	D_4	4/mmm	D_{4h}	$\bar{4}$	S_4	$\bar{4}2m$	D_{2d}
Hexagonal	6	C_6	6/m	C_{6h}	6mm	C_{6v}	622	D_6	6/mmm	D_{6h}	$\bar{6}$	C_{3h}	$\bar{6}m2$	D_{3h}
Cubic	23	T	m3	T_h	$\bar{4}3m$	T_d	432	O	m3m	O_h				

Class name[a] and its symbol — International (Int) and Schoenflies (Sch)

[a] Per Fedorov Institute of Crystallography, USSR Academy of Sciences, nomenclature.

SYMMETRY OF CRYSTALS (continued)

TABLE 3

The Fourteen Possible Space Lattices (Bravais Lattices)

Crystal system	Metric category of the system	No. of different lattices in the system	Lattice type[a] (marked by +)					No. of identi-points per unit cell	Characteristic parameters (marked by +)						Description of characteristic parameters $a \subset X$, $b \subset Y$, $c \subset Z$; $\alpha \equiv (b,c)$, $\beta \equiv (a,c)$, $\gamma \equiv (a,b)$	Symmetry of the lattice	
			P	C	I	F	R		a	b	c	α	β	γ		Int	Sch
Triclinic	Trimetric	1	+					1	+	+	+	+	+	+	$a \neq b \neq c$, $\alpha \neq \beta \neq \gamma$	1	C
Monoclinic	Trimetric	2	+	+				1 or 2	+	+	+		+		$a \neq b \neq c$, $\alpha = \gamma = 90° \neq \beta$	2/m	C_{2h}
Orthorhombic	Trimetric	4	+	+	+	+		1, 2 or 4	+	+	+				$a \neq b \neq c$, $\alpha = \beta = \gamma = 90°$	mmm	D_{2h}
Trigonal (rhombohedral)	Dimetric	1					+	1	+			+			$a = b = c$, $120° > \alpha = \beta = \gamma \neq 90°$	3m	D_{3d}
Tetragonal	Dimetric	2	+		+			1 or 2	+		+				$a = b \neq c$, $\alpha = \beta = \gamma = 90°$	4/mmm	D_{4h}
Hexagonal	Dimetric	1	+					1	+		+				$a \neq b \neq c$, $\alpha = \beta = 90°$, $\gamma = 120°$	6/mmm	D_{6h}
Isometric (cubic)	Monometric	3	+		+	+		1, 2 or 4	+						$a = b = c$, $\alpha = \beta = \gamma = 90°$	m3m	O_h

[a] Designations of the space-lattice types: P — primitive, C — side-centered (base-centered), I — body-centered, F — face-centered, R — rhombohedral.

TABLE 4
The Five Possible Regular Polyhedrons

	Symmetry (Schoenflies)			Number of[a]		
Polyhedron	Class	Elements	Form of faces	Faces (F)	Edges (E)	Vertices (V)
Tetrahedron	T	$4C_3 3C_2$	Equilateral triangle	4	6	4
Cube (hexahedron)	O	$3C_4 4C_3 6C_2$	Square	6	12	8
Octahedron	O	$3C_4 4C_3 6C_2$	Equilateral triangle	8	12	6
Pentagonal dodecahedron	J	$6C_5 10C_3 15C_2$	Regular pentagon	12	30	20
Icosahedron	J	$6C_5 10C_3 15C_2$	Equilateral triangle	20	30	12

[a] Per formula by Leonhard Euler: $F + V - E = 2$

TABLE 5
Classification of Crystals

Strukturbericht symbol 1	Structure name 2	Symmetry group		Pearson symbol[a] 5	Standard ASTM E157-82a symbol[b] 6
		International 3	Schoenflies 4		
A1	Cu	Fm3m	O^4_h	cF4	F
A2	W	Im3m	O^9_h	cI2	B
A3	Mg	P6$_3$/mmc	D^4_{6h}	hP2	H
A4	C	Fd3m	O^7_h	cF8	F
A5	Sn	If$_1$/amd	D^{19}_{4h}	tI4	U
A6	In	I4/mmm	D^{17}_{4h}	tI2	U
A7	As	R$\bar{3}$m	D^5_{3d}	hR2	R
A8	Se	P3$_1$21 or P3$_2$21	D^4_3 (D^6_3)	hP3	H
A10	Hg	R$\bar{3}$m	D^5_{3d}	hR1	R
A11	Ga	Cmca	D^{18}_{2h}	oC8	Q
A12	α-Mn	I$\bar{4}$3m	T^3_d	cI58	B
A13	β-Mn	P4$_1$32	O^7	cP20	C
A15	OW$_3$	Pm3n	O^3_h	cP8	C
A20	α-U	Cmcm	D^{17}_{2h}	oC4	Q
B1	ClNa	Fm3m	O^5_h	cF8	F
B2	ClCs	Pm3m	O^1_h	cP2	C
B3	SZn	F$\bar{4}$3m	T^2_d	cF8	F
B4	SZn	P6$_3$mc	C^4_{6v}	hP4	H
B8$_1$	AsNi	P6$_3$/mmc	D^4_{6h}	hP4	H
B8$_2$	InNi$_2$	P6$_3$/mmc	D^4_{6h}	hP6	H
B9	HgS	P3$_1$21 or P3$_2$21	D^4_3 or D^6_3	hP6	H
B10	OPb	P4/nmm	D^7_{4h}	tP4	T
B11	γ-CuTi	P4/nmm	D^7_{4h}	tP4	T
B13	NiS	R$\bar{3}$m	D^5_{3d}	hR6	R
B16	GeS	Pnma	D^{16}_{2h}	oP8	O
B17	PtS	P4$_2$/mmc	D^9_{4h}	tP4	T
B18	CuS	P6$_3$/mmc	D^4_{6h}	hP12	H
B19	AuCd	Pmma	D^5_{2h}	oP4	O
B20	FeSi	P2$_1$3	T^4	cP8	C
B27	BFe	Pnma	D^{16}_{2h}	oP8	O
B31	MnP	Pnma	D^{16}_{2h}	oP8	O
B32	NaTl	Fd3m	O^7_h	cF16	F
B34	Pds	P4$_2$/m	C^2_{4h}	tP16	T

TABLE 5
Classification of Crystals (continued)

Strukturbericht symbol	Structure name	Symmetry group		Pearson symbol[a]	Standard ASTM E157-82a symbol[b]
		International	Schoenflies		
1	2	3	4	5	6
B35	CoSn	P6/mmm	D^1_{6h}	hP6	H
B37	SeTl	I4/mcm	D^{18}_{4h}	tI16	U
B_e	CdSb	Pbca	D^{15}_{2h}	oP16	O
B_f (B33)	ξ-BCr	Cmcm	D^{17}_{2h}	oC8	Q
B_g	BMo	$I4_1/amd$	D^{19}_{4h}	tI4	U
B_h	CW	P6m2	D^1_{3h}	hP2	H
B_i	γ'-CMo (AsTi)	$P6_3/mmc$	D^4_{6h}	hP8	H
C1	CaF_2	Fm3m	O^5_h	cF12	F
$C1_b$	AgAsMg	F43m	T^2_d	cF12	F
C2	FeS_2	Pa3	T^6_h	cP12	C
C3	Cu_2O	Pn3m	O^4_h	cP6	C
C4	O_2Ti	$P4_2/mnm$	D^{14}_{4h}	tP6	T
C6	CdI_2	P3m1	D^3_{3d}	hP3	H
C7	MoS_2	$P6_3/mmc$	D^4_{6h}	hP6	H
$C11_a$	C_2Ca	I4/mmm	D^{17}_{4h}	tI6	U
$C11_b$	$MoSi_2$	I4/mmm	D^{17}_{4h}	tI6	U
C12	$CaSi_2$	R3m	D^5_{3d}	hR6	R
C14	$MgZn_2$	$P6_3/mmc$	D^4_{6h}	hP12	H
C15	Cu_2Mg	Fd3m	O^7_h	cF24	F
$C15_b$	$AuBe_5$	F43m or F23	T^2_d or T^2	cF24	F
C16	Al_2Cu	I4/mcm	D^{18}_{4h}	tI12	U
C18	FeS_2	Pnnm	D^{12}_{2h}	oP6	O
C19	$CdCl_2$	R3m	D^5_{3d}	hR3	R
C22	Fe_2P	P26m	D^1_{3h}	hP9	H
C23	Cl_2Pb	Pnma	D^{16}_{2h}	oP12	O
C32	AlB_2	P6/mmm	D^1_{6h}	hP3	H
C33	Bi_2STe_2	R3m	D^5_{3d}	hR5	R
C34	$AuTe_2$	C2/m (P2/m)	C^3_{2h} (C^1_{2h})	mC6	N
C36	$MgNi_2$	$P6_3/mmc$	D^4_{6h}	hP24	H
C38	Cu_2Sb	P4/nmm	D^7_{4h}	tP6	T
C40	$CrSi_2$	$P6_222$	D^4_6	hP9	H
C42	SiS_2	Ibam	D^{26}_{2h}	oI12	P
C44	GeS_2	Fdd2	C^{19}_{2v}	oF72	S
C46	$AuTe_2$	Pma2	C^4_{2v}	oP24	O
C49	Si_2Zr	Cmcm	D^{17}_{2h}	oC12	Q
C54	Si_2Ti	Fddd	D^{24}_{2h}	oF24	S
C_c	Si_2Th	$I4_1/amd$	D^{19}_{4h}	tI12	U
C_e	$CoGe_2$	Aba2	C^{17}_{2v}	oC23	Q
DO_2	As_3Co	Im3	T^5_h	cI32	B
DO_3	BiF_3	Fm3m	O^5_h	cF16	F
DO_9	O_3Re	Pm3m	O^1_h	cP4	C
DO_{11}	CFe_3	Pnma	D^{16}_{2h}	oP16	O
DO_{18}	$AsNa_3$	$P6_3/mmc$	D^4_{6h}	hP8	H
DO_{19}	Ni_3Sn	$P6_3/mmc$	D^4_{6h}	hP8	H
DO_{20}	Al_3Ni	Pnma	D^{16}_{2h}	oP16	O
DO_{21}	Cu_3P	P3c1	D^4_{3d}	hP24	H
DO_{22}	Cu_3P	I4/mmm	D^{17}_{4h}	tI8	U
DO_{23}	Al_3Zr	I4/mmm	D^{17}_{4h}	tI16	U
DO_{24}	Ni_3Ti	$P6_3/mmc$	D^4_{6h}	hP16	H
DO_c	SiU_3	I4/mcm	D^{18}_{4h}	tI16	U
DO_e	Ni_3P	I4	S^2_4	tI32	U
$D1_3$	Al_4Ba	I4/mmm	D^{17}_{4h}	tI10	U
$D1_a$	$MoNi_4$	I4/m	C^5_{4h}	tI10	U

TABLE 5
Classification of Crystals (continued)

Strukturbericht symbol	Structure name	Symmetry group International	Symmetry group Schoenflies	Pearson symbol[a]	Standard ASTM E157-82a symbol[b]
1	2	3	4	5	6
$D1_b$	Al_4U	Imma	D^{28}_{2h}	oI20	P
$D1_c$	$PtSn_4$	Aba2	C^{17}_{2v}	oC20	Q
$D1_e$	B_4Th	P4/mbm	D^5_{4h}	tP20	T
$D1_f$	BMn_4	Fddd	D^{24}_{2h}	oF40	S
$D2_1$	B_6Ca	Pm3m	O^1_h	cP7	C
$D2_3$	$NaZn_{13}$	Fm3m	O^5_h	cF112	F
$D2_b$	$Mn_{12}Th$	I4/mmm	D^{17}_{4h}	tI26	U
$D2_c$	MnU_6	I4/mcm	D^{18}_{4h}	tI28	U
$D2_d$	$CaCu_5$	P6/mmm	D^1_{6h}	hP6	H
$D2_f$	$B_{12}U$	Fm3m	O^5_h	cF52	F
$D2_h$	Al_6Mn	Cmcm	D^{17}_{2h}	oC28	Q
$D5_1$	$\alpha\text{-}Al_2O_3$	R$\bar3$c	D^6_{3d}	hR10	R
$D5_2$	La_2O_3	P$\bar3$m1	D^3_{3d}	hP5	H
$D5_3$	Mn_2O_3	Ia3	T^7_h	cI80	B
$D5_8$	S_3Sb_2	Pnma	D^{16}_{2h}	oP20	O
$D5_9$	P_2Zn_3	P4$_2$/mmc	D^9_{4h}	tP40	T
$D5_{10}$	C_2C_3	Pnma	D^{16}_{2h}	oP20	O
$D5_{13}$	Al_3Ni_2	P$\bar3$m1	D^3_{3d}	hP5	H
$D5_a$	Si_2U_3	P4/mbm	D^5_{4h}	tP10	T
$D5_c$	C_3Pu_2	I$\bar4$3d	T^6_d	cI40	B
$D7_1$	Al_4C_3	R$\bar3$m	D^5_{3d}	hR7	R
$D7_3$	P_4Th_3	I$\bar4$3d	T^6_d	cI28	B
$D7_b$	B_4Ta_3	Immm	D^{25}_{2h}	oI14	P
$D8_1$	Fe_3Zn_{10}	Im3m	O^9_h	cI52	B
$D8_2$	Cu_5Zn_8	I$\bar4$3m	T^3_d	cI52	B
$D8_3$	Al_4Cu_9	P$\bar4$3m	T^1_d	cP52	C
$D8_4$	C_6Cr23	Fm3m	O^5_h	cF116	F
$D8_5$	Fe_7W_6	R$\bar3$m	D^5_{3d}	hR13	R
$D8_6$	$Cu_{15}Si_4$	I$\bar4$3m	T^3_d	cI76	B
$D8_8$	Mn_5Si_3	P6$_3$/mcm	D^3_{6h}	hP16	H
$D8_9$	Co_9S_8	Fm3m	O^5_h	cF68	F
$D8_{10}$	Al_8Cr_5	R3m	C^5_{3v}	hR26	R
$D8_{11}$	Al_5Co_2	P6$_3$/mcm	D^3_{6h}	hP28	H
$D8_a$	$Mn_{23}Th_6$	Fm3m	O^5_h	cF116	F
$D8_b$	σ-phase of Cr-Fe	p4$_2$/mnm	D^{14}_{4h}	tP30	T
$D8_e$	$(Al,Zn)_{49}Mg_{32}$	Im3	T^5_h	cI162	B
$D8_f$	Ge_7Ir_3	Im3m	O^9_h	cI40	B
$D8_h$	B_5W_2	P6$_3$/mmc	D^4_{6h}	hP14	H
$D8_i$	B_5Mo_2	R$\bar3$m	D^5_{3d}	hR7	R
$D8_1$	B_3Cr_5	I4/mcm	D^{18}_{4h}	tI32	U
$D8_m$	Si_3W_5	I4/mcm	D^{18}_{4h}	tI32	U
$D10_1$	C_3Cr_7	P31c	C^4_{3v}	hP80	H
$D10_2$	Fe_3Th_7	P6$_3$mc	C^4_{6v}	hP20	H
$E0_1$	ClFPb	P4/nmm	D^7_{4h}	tP6	T
$E1_1$	$CuFeS_2$	I$\bar4$2d	D^{12}_{2d}	tI16	U
$E2_1$	CaO_3Ti	Pm3m	O^1_h	cP5	C
$E2_4$	S_3Sn_2	Pnma	D^{16}_{2h}	oP20	O
$E3$	Al_2CdS_4	I$\bar4$	S^2_4	tI14	U
$E9_3$	$SiFe_3W_3$	Fd3m	O^7_h	cF112	F
$E9_a$	Al_7Cu_2Fe	P4/mnc	D^6_{4h}	tP40	T
$E9_b$	$AlLi_3N_2$	Ia3	T^7_h	cI96	B
$F0_1$	NiSSb	P2$_1$3	T^4	cP12	C
$F5_1$	$CrNaS_2$	R3m or R32	D^5_{3d} or D^7_3	hR4	R
$F5_6$	CuS_2Sb	Pnma	D^{16}_{2h}	oP16	O

TABLE 5
Classification of Crystals (continued)

Strukturbericht symbol	Structure name	Symmetry group		Pearson symbol[a]	Standard ASTM E157-82a symbol[b]
1	2	International 3	Schoenflies 4	5	6
$H1_1$	Al_2MgO_4	$Fd3m$	O^7_h	cF56	F
$H2_4$	Cu_3S_4V	$P\bar{4}3m$	T^1_d	cP8	C
$H2_5$	$AsCu_3S_4$	$Pmn2_1$	C^7_{2v}	oP16	O
$L1_0$	$AuCu$	$P4/mmm$	D^1_{4h}	tP4	T
$L1_2$	$AlCu_3$	$Pm3m$	O^1_h	cP4	C
$L2_1$	$AlCu_2Mn$	$Fm3m$	O^5_h	cF16	F
$L2_2$	Sb_2Tl_7	$Im3m$	O^9_h	cI54	B
$L'2_b$	H_2Th	$I4/mmm$	D^{17}_{4h}	tI6	U
$L'3$	Fe_2N	$P6_3/mmc$	D^4_{6h}	hP3	H
$L6_0$	$CuTi_3$	$P4/mmm$	D^1_{4h}	tP4	T

[a] The first letter denotes the crystal system: triclinic (a), monoclinic (m), orthorhombic (o), tetragonal (t), hexagonal (h) and cubic (c). Trigonal (rhombohedral) system is presented by combination hR. The second letter of Pearson's symbol denotes lattice type: primitive (P), edge- (base-) centered (C), body-centered (I) or face-centered (F). The following number denotes amount of atoms in the crystal unit cell.

[b] Standard ASTM E157-82a has the Bravais lattices designations as following: C — primitive cubic; B — body-centered cubic; F — face-centered cubic; T — primitive tetragonal; U — body-centered tetragonal; R — rhombohedral; H — hexagonal; O — primitive orthorhombic; P — body-centered orthorhombic; Q — base-centered orthorhombic; S — face-centered orthorhombic; M — primitive monoclinic; N — centered monoclinic; A — triclinic.

REFERENCES

1. A. Schoenflies, *Kristallsysteme und Kristallstructur*, Leipzig, 1891.
2. E. S. Fedorow, Zusammenstellung der kristallographischen Resultate, *Zs. Krist.*, 20, 1892.
3. P. Groth, *Elemente der physikalischen und chemischen Krystallographie*, R. Oldenbourg, München/Berlin, 1921.
4. N. V. Belov, *Class Method of Deriving Space Groups of Symmetry*, Trudy Instituta Kristallodraffi imeni Fedorova (Transactions of the Fedorov Inst. of Crystallography), 5, 25, 1951, in Russian.
5. W. B. Pearson, *Handbook of Lattice Spacings and Structures of Metals and Alloys*, Vol. 1, Pergamon Press, 1958; Vol. 2, 1967.
6. Ch. Kittel, *Introduction to Solid State Physics*, John Wiley & Sons, 1956.
7. G. S. Zhdanov, *Fizika Tverdogo Tela (Solid State Physics)*, Moscow University Press, 1962, in Russian.
8. M. J. Buerger, *Elementary Crystallography*, John Wiley & Sons, 1963.
9. F. D. Bloss, *Crystallography & Crystal Chemistry*, Holt, Rinehart & Winston, 1971.
10. T. Janssen, *Crystallographic Groups*, North-Holland/American Elsevier, 1973.
11. M. P. Shaskolskaya, *Kristallografiya (Crystallography)*, Vysshaya Shkola, Moscow, 1976, in Russian.
12. T. Hahn, Ed., Internat. *Tables for Crystallography*, Vol. A, D. Reidel Publishing, Boston, 1983.
13. Crystal Data. Determinative Tables, Volumes 1—6, 1966—1983, JCPDS-Intern Centre for Diffraction Data and U.S. Dept. of Commerce.
14. R. W. G. Wyckoff, *Crystal Structures*, 2nd ed., Volumes 1–6, Interscience, New York, 1963.
15. C.J. Bradley and A.P. Cracknell, *The Mathematical Theory of Symmetry in Solids*, Clarendon Press, Oxford, 1972.
16. International Tables for Crystallography. Volume A, *Space–Group Symmetry*, T. Hahn, Ed., 1989; Volume B, *Reciprocal Space*, U. Schmueli, Ed.; Volume C, *Mathematical, Physical and Chemical Tables*, A. J. C. Wilson, Ed., Kluwer Academic Publishers, Dordrecht, 1989.
17. G. R. Desiraju, *Crystal Engineering: The Design of Organic Solids*, Elsevier, Amsterdam, 1989.
18. M. Senechal, *Crystalline Symmetries: An Informal Mathematical Introduction*, Adam Hilger Publ., Bristol, 1990.
19. C. Hammond, *Introduction to Crystallography*, Oxford University Press, 1990.
20. N.W. Alcock, *Bonding and Structure: Structural Principles in Inorganic and Organic Chemistry*, Ellis Norwood Publ., 1990.
21. T. C. W. Mak and G. D. Zhou. *Crystallography in Modern Chemistry: A Resource Book of Crystal Structures*, Wiley–Interscience, New York, 1992.
22. S. C. Abrahams, K. Mirsky, and R. M. Nielson, *Acta Cryst*, B52, 806 (1996); B52, 1057 (1996).
23. C. Marcos, A. Panalague, D. B. Morciras, S. Garcia–Granda and M. R. Dias. *Acta Cryst*, B52, 899 (1996).

Crystallographic Computing

24. A. C. Larson, *Crystallographic Computing*, Manksgaard, Copenhagen, 1970.
25. G. M. Sheldrick, SHELXS86. Crystallographic Computing 3, Clarendon Press, Oxford, 1986; SHELXL93. Program for the Refinement of Crystal Structures, University of Göttingen Press, 1993.
26. Inorganic Crystal Structure Database, CD–ROM. Sci. Inf. Service. E-mail: SISI@Delphi.com.

IONIC RADII IN CRYSTALS

Ionic radii are a useful tool for predicting and visualizing crystal structures. This table lists a set of ionic radii R_i in Å units for the most common coordination numbers CN of positive and negative ions. The values are based on experimental crystal structure determinations, supplemented by empirical relationships, and theoretical calculations. The notation sq after the coordination number indicates a square configuration, while py indicates pyramidal.

The advice of Howard T. Evans and Marvin J. Weber in preparing this table is appreciated.

REFERENCES

1. Shannon, R. D., *Acta Crystallogr.* A32, 751, 1976.
2. Jia, Y.Q., *J. Solid State Chem.* 95, 184, 1991.

Ion	CN	R_i/Å	Ion	CN	R_i/Å	Ion	CN	R_i/Å
Anions			C^{+4}	4	0.15	Er^{+3}	6	0.89
F^{-1}	6	1.33		6	0.16		8	1.00
Cl^{-1}	6	1.81	Ca^{+2}	6	1.00	Eu^{+2}	6	1.17
Br^{-1}	6	1.96		8	1.12		8	1.25
I^{-1}	6	2.20		10	1.23		10	1.35
OH^{-1}	4	1.35		12	1.34	Eu^{+3}	6	0.95
	6	1.37	Cd^{+2}	4	0.78		8	1.07
O^{-2}	2	1.21		6	0.95	F^{+7}	6	0.08
	6	1.40		8	1.10	Fe^{+2}	4	0.63
	8	1.42		12	1.31		6	0.61
S^{-2}	6	1.84	Ce^{+3}	6	1.01		8	0.92
Se^{-2}	6	1.98		8	1.14	Fe^{+3}	4	0.49
Te^{-2}	6	2.21		10	1.25		6	0.55
				12	1.34		8	0.78
Cations			Ce^{+4}	6	0.87	Fr^{+1}	6	1.80
Ac^{+3}	6	1.12		8	0.97	Ga^{+3}	4	0.47
Ag^{+1}	4	1.00		10	1.07		6	0.62
	6	1.15		12	1.14	Gd^{+3}	6	0.94
	8	1.28	Cf^{+3}	6	0.95		8	1.05
Ag^{+2}	4sq	0.79	Cf^{+4}	6	0.82	Ge^{+2}	6	0.73
	6	0.94		8	0.92	Ge^{+4}	4	0.39
Al^{+3}	4	0.39	Cl^{+5}	3py	0.12		6	0.53
	5	0.48	Cl^{+7}	4	0.08	Hf^{+4}	4	0.58
	6	0.54	Cm^{+3}	6	0.97		6	0.71
Am^{+3}	6	0.98	Cm^{+4}	6	0.85		8	0.83
	8	1.09		8	0.95	Hg^{+1}	6	1.19
Am^{+4}	6	0.85	Co^{+2}	4	0.56	Hg^{+2}	2	0.69
	8	0.95		6	0.65		4	0.96
As^{+3}	6	0.58		8	0.90		6	1.02
As^{+5}	4	0.34	Co^{+3}	6	0.55		8	1.14
	6	0.46	Cr^{+2}	6	0.73	I^{+5}	3py	0.44
Au^{+1}	6	1.37	Cr^{+3}	6	0.62		6	0.95
Au^{+3}	4sq	0.64	Cr^{+4}	4	0.41	I^{+7}	4	0.42
	6	0.85		6	0.55		6	0.53
Ba^{+2}	6	1.35	Cr^{+6}	4	0.26	In^{+3}	4	0.62
	8	1.42		6	0.44		6	0.80
	12	1.61	Cs^{+1}	6	1.67	Ir^{+3}	6	0.68
Be^{+2}	4	0.27		8	1.74	Ir^{+4}	6	0.63
	6	0.45		10	1.81	Ir^{+5}	6	0.57
Bi^{+3}	5	0.96		12	1.88	K^{+1}	4	1.37
	6	1.03	Cu^{+1}	2	0.46		6	1.38
	8	1.17		4	0.60		8	1.51
Bi^{+5}	6	0.76		6	0.77		12	1.64
Bk^{+3}	6	0.96	Cu^{+2}	4sq	0.57	La^{+3}	6	1.03
Bk^{+4}	6	0.83		6	0.73		8	1.16
	8	0.93	Dy^{+2}	6	1.07		10	1.27
Br^{+5}	3py	0.31		8	1.19		12	1.36
Br^{+7}	4	0.25	Dy^{+3}	6	0.91	Li^{+1}	4	0.59
	6	0.39		8	1.03		6	0.76

Ion	CN	R_i/Å	Ion	CN	R_i/Å	Ion	CN	R_i/Å
	8	0.92		6	0.78	Sr^{+2}	6	1.18
Lu^{+3}	6	0.86		8	0.94		8	1.26
	8	0.97	Pd^{+2}	4sq	0.64		10	1.36
Mg^{+2}	4	0.57		6	0.86		12	1.44
	6	0.72	Pd^{+3}	6	0.76	Ta^{+3}	6	0.72
	8	0.89	Pd^{+4}	6	0.62	Ta^{+4}	6	0.68
Mn^{+2}	4	0.66	Pm^{+3}	6	0.97	Ta^{+5}	6	0.64
	6	0.83		8	1.09	Tb^{+3}	6	0.92
	8	0.96	Po^{+4}	6	0.97		8	1.04
Mn^{+3}	6	0.58	Pr^{+3}	6	0.99	Tb^{+4}	6	0.76
Mn^{+4}	4	0.39		8	1.13		8	0.88
	6	0.53	Pr^{+4}	6	0.85	Tc^{+4}	6	0.65
Mn^{+5}	4	0.33		8	0.96	Te^{+4}	4	0.66
Mn^{+6}	4	0.26	Pt^{+2}	4sq	0.60		6	0.97
Mn^{+7}	4	0.25		6	0.80	Te^{+6}	4	0.43
Mo^{+3}	6	0.69	Pt^{+4}	6	0.63		6	0.56
Mo^{+4}	6	0.65	Pu^{+3}	6	1.00	Th^{+4}	6	0.94
Mo^{+5}	4	0.46	Pu^{+4}	6	0.86		8	1.05
	6	0.61	Pu^{+5}	6	0.74		10	1.13
Mo^{+6}	4	0.41	Pu^{+6}	6	0.71		12	1.21
	6	0.59	Ra^{+2}	8	1.48	Ti^{+2}	6	0.86
	7	0.73		12	1.70	Ti^{+3}	6	0.67
N^{+3}	6	0.16	Rb^{+1}	6	1.52	Ti^{+4}	4	0.42
N^{+5}	6	0.13		8	1.61		6	0.61
Na^{+1}	4	0.99		10	1.66		8	0.74
	6	1.02		12	1.72	Tl^{+1}	6	1.50
	8	1.18	Re^{+4}	6	0.63		8	1.59
	9	1.24	Re^{+5}	6	0.58		12	1.70
	12	1.39	Re^{+6}	6	0.55	Tl^{+3}	4	0.75
Nb^{+3}	6	0.72	Re^{+7}	4	0.38		6	0.89
	8	0.79		6	0.53		8	0.98
Nb^{+4}	6	0.68	Rh^{+3}	6	0.67	Tm^{+2}	6	1.01
Nb^{+5}	4	0.48	Rh^{+4}	6	0.60		7	1.09
	6	0.64	Rh^{+5}	6	0.55	Tm^{+3}	6	0.88
	8	0.74	Ru^{+3}	6	0.68		8	0.99
Nd^{+3}	6	0.98	Ru^{+4}	6	0.62	U^{+3}	6	1.03
	8	1.12	Ru^{+5}	6	0.57	U^{+4}	6	0.89
	9	1.16	Ru^{+7}	4	0.38		8	1.00
	12	1.27	Ru^{+8}	4	0.36		12	1.17
Ni^{+2}	4sq	0.49	S^{+4}	6	0.37	U^{+5}	6	0.76
	6	0.69	S^{+6}	4	0.12	U^{+6}	2	0.45
Ni^{+3}	6	0.56		6	0.29		4	0.52
Np^{+3}	6	1.01	Sb^{+3}	4py	0.76		6	0.73
Np^{+4}	6	0.87		6	0.76		8	0.86
Np^{+5}	6	0.75	Sb^{+5}	6	0.60	V^{+2}	6	0.79
Np^{+6}	6	0.72	Sc^{+3}	6	0.75	V^{+3}	6	0.64
Os^{+4}	6	0.63		8	0.87	V^{+4}	5	0.53
Os^{+5}	6	0.58	Se^{+4}	6	0.50		6	0.58
Os^{+6}	6	0.55	Se^{+6}	4	0.28		8	0.72
Os^{+8}	4	0.39		6	0.42	V^{+5}	4	0.36
P^{+5}	4	0.17	Si^{+4}	4	0.26		5	0.46
	6	0.38		6	0.40		6	0.54
Pa^{+3}	6	1.04	Sm^{+2}	6	1.19	W^{+4}	6	0.66
Pa^{+4}	6	0.90		8	1.27	W^{+5}	6	0.62
Pa^{+5}	6	0.78	Sm^{+3}	6	0.96	W^{+6}	4	0.42
Pb^{+2}	6	1.19		8	1.08		5	0.51
	8	1.29		12	1.24		6	0.60
	10	1.40	Sn^{+4}	4	0.55	Y^{+3}	6	0.90
	12	1.49		6	0.69		8	1.02
Pb^{+4}	4	0.65		8	0.81		9	1.08

Ion	CN	R_i/Å	Ion	CN	R_i/Å	Ion	CN	R_i/Å
Yb^{+2}	6	1.02	Zn^{+2}	4	0.60		6	0.72
	8	1.14		6	0.74		8	0.84
Yb^{+3}	8	0.99		8	0.90		9	0.89
	9	1.04	Zr^{+4}	4	0.59			

POLARIZABILITIES OF ATOMS AND IONS IN SOLIDS
H. P. R. Frederikse

The polarization of a solid dielectric medium, P, is defined as the dipole moment per unit volume averaged over the volume of a crystal cell. A component of P can be expanded as a function of the electric field E:

$$P_i = \sum_j a_j E_j + \sum_{jk} b_{jk} E_j E_k$$

For relatively small electric fields in isotropic substances $P = \chi_e E$, where χ_e is the electric susceptibility. If the medium is made up of N atoms (or ions) per unit volume, the polarization is $P = N p_m$ where p_m is the average dipole moment per atom. The polarizability α can be defined as $p_m = \alpha E_0$, where E_0 is the local field at the position of the atom. Using the Lorentz method to calculate the local field one finds:

$$P = N\alpha(E + 4\pi P) = \chi_e E$$

Together with the definition of the dielectric constant (relative permittivity) , $\varepsilon = 1 + 4\pi\chi_e$, this leads to:

$$\alpha = \frac{3}{4\pi N}\left(\frac{\varepsilon - 1}{\varepsilon + 2}\right)$$

This expression is known as the Clausius-Mossotti equation.

The total polarization associated with atoms, ions, or molecules is due to three different sources:

1. Electronic polarization arises because the center of the local electronic charge cloud around the nucleus is displaced under the action of the field: $P_e = N\alpha_e E_0$ where α_e is the *electronic polarizability*.
2. Ionic polarization occurs in ionic materials because the electric field displaces cations and anions in opposite directions: $P_i = N\alpha_i E_0$, where α_i is the *ionic polarizability*.
3. Orientational polarization can occur in substances composed of molecules that have permanent electric dipoles. The alignment of these dipoles depends on temperature and leads to an *orientational polarizability* per molecule: $\alpha_{or} = p^2/3kT$, where p is the permanent dipole moment per molecule, k is the Boltzmann constant, and T is the temperature.

Because of the different nature of these three polarization processes the response of a dielectric solid to an applied electric field will strongly depend on the frequency of the field. The resonance of the electronic excitation in insulators (dielectrics) takes place in the ultraviolet part of the spectrum; the characteristic frequency of the lattice vibrations is located in the infrared, while the orientation of dipoles requires fields of much lower frequencies (below 10^{10} Hz). This response to electric fields of different frequencies is shown in Figure 1. Values of the electronic polarizabilities for selected atoms and ions are given in Table 1.

REFERENCES

1. Kittel, C., *Introduction to Solid State Physics*, Fourth Edition, John Wiley & Sons, New York, 1971.
2. Lerner, R.G., and Trigg, G.L., Editors, *Encyclopedia of Physics, Second Edition*, VCH Publishers, New York, 1990.
3. Ralls, K.M., Courtney, T.H., and Wulff, J., *An Introduction to Materials Science and Engineering*, John Wiley & Sons, New York, 1976.

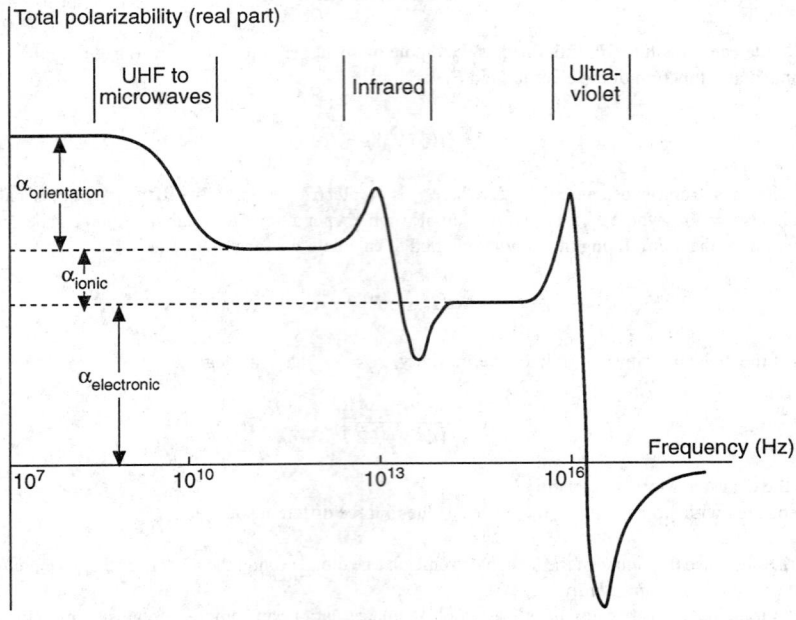

Figure 1. Frequency dependence of the three contributions to the polarizability. (Reprinted with permission from Kittel, C., *Introduction to Solid State Physics*, Fourth Edition, John Wiley & Sons, New York, 1971.)

TABLE 1
Electronic Polarizabilities of Ions, in 10^{-24} cm^3

			He	Li$^+$	Be^{2+}	B^{3+}	C^{4+}
Pauling			0.201	0.029	0.008	0.003	0.0013
JS				0.029			
	O^{2-}	F$^-$	Ne	Na$^+$	Mg^{2+}	Al^{3+}	Si^{4+}
Pauling	3.88	1.04	0.390	0.179	0.094	0.052	0.0165
JS-(TKS)	(2.4)	0.858		0.290			
	S^{2-}	Cl$^-$	Ar	K$^+$	Ca^{2+}	Sc^{3+}	Ti^{4+}
Pauling	10.2	3.66	1.62	0.83	0.47	0.286	0.185
JS-(TKS)	(5.5)	2.947		1.133	(1.1)		(0.19)
	Se^{2-}	Br$^-$	Kr	Rb$^+$	Sr^{2+}	Y^{3+}	Zr^{4+}
Pauling	10.5	4.77	2.46	1.40	0.86	0.55	0.37
JS-(TKS)	(7)	4.091		1.679	(1.6)		
	Te^{2-}	I$^-$	Xe	Cs$^+$	Ba^{2+}	La^{3+}	Ce^{4+}
Pauling	14.0	7.10	3.99	2.42	1.55	1.04	0.73
JS-(TKS)	(9)	6.116		2.743	(2.5)		

Values from L. Pauling, *Proc. R. Soc. London*, A114, 181, 1927; S.S. Jaswal and T.P. Sharma, *J. Phys. Chem. Solids*, 34, 509, 1973; and J. Tessman, A. Kahn, and W. Shockley, *Phys. Rev.*, 92, 890, 1953. The TKS polarizabilities are at the frequency of the D lines of sodium. The values are in CGS; to convert to SI, multiply by $1/9 \times 10^{-15}$. (Reprinted with permission from Kittel, C., *Introduction to Solid State Physics*, Fourth Edition, John Wiley & Sons, New York, 1971.

CRYSTAL STRUCTURES AND LATTICE PARAMETERS OF ALLOTROPES OF THE ELEMENTS

H. W. King

The crystal structures of the allotropic forms of the elements are presented in terms of the Pearson symbol, the Strukturbericht designation, and the prototype of the structure. The temperatures of the phase transformations are listed in degrees Celsius and the pressures are in GPa. A consistent nomenclature is used, whereby all allotropes are labeled by Greek letters. The lattice parameters of the unit cells are given in nanometers (nm) and are considered to be accurate to ±2 in the last reported digit.

This compilation is restricted to changes in crystal structure that occur as a result of a change in temperature or pressure. Low-temperature structures are included for the diatomic and rare gases, which show many similarities with respect to the metallic elements. The elements identified with an asterisk (*) have polymorphic structures based on different molecular configurations. The crystal data given for these elements refer to the most stable structure at room temperature.

Reprinted with the permission of ASM International from T. B. Massalski, Ed., Binary Alloy Phase Diagrams, ASM International, Metals Park, Ohio, 1986; certain data on rare earth elements were provided by K. A. Gschneidner.

Element	Temperature, °C	Pressure, GPa	Pearson symbol	Space group	Strukturbericht designation	Prototype	Lattice parameters, nm a	b	c	Comment, c/a or α or β
Ac	25	atm	cF4	Fm3m	A1	Cu	0.5311	...	...	...
Ag	25	atm	cF4	Fm3m	A1	Cu	0.40857	...	...	...
αAl	25	atm	cF4	Fm3m	A1	Cu	0.40496	...	...	...
βAl	25	>20.5	hP2	P6₃/mmc	A3	Mg	0.2693	...	0.4398	1.6331
α'Am	25	atm	hP4	P6₃/mmc	A3'	αLa	0.34681	...	1.1241	2*1.621
αAm	>769	atm	cF4	Fm3m	A1	Cu	0.4894	...	...	...
βAm	>1074	atm	cI2	Im3m	A2	W	?	...	...	...
γAm	25	>15	oC4	Cmcm	A20	αU	0.3063	0.5968	0.5169	...
αAr	<−189.35	atm	cF4	Fm3m	A1	Cu	0.5316	...	...	...
(βAr)	<−189.40	atm	hP2	P6₃/mmc	A3	Mg	0.3760	...	0.6141	1.633
αAs	25	atm	hR2	R3m	A7	αAs	0.41319	...	...	α = 54.12°
εAs	>448	atm	oC8	Cmca	...	P (black)	0.362	1.085	0.448	...
Au	25	atm	cF4	Fm3m	A1	Cu	0.40782	...	...	...
βB	25	atm	hR105	R3m	...	βB	1.017	...	...	α = 65.12°
αBa	25	atm	cI2	Im3m	A2	W	0.50227	...	...	...
βBa	25	>5.33	hP2	P6₃/mmc	A3	Mg	0.3901	...	0.6154	1.5775
γBa	25	>23	?	?	...	...	...	...	...	...
αBe	25	atm	hP2	P6₃/mmc	A3	Mg	0.22859	...	0.35845	1.5681
βBe	>1270	atm	cI2	Im3m	A2	W	0.25515	...	...	...
γBe	25	>9.3	?	...	...	...	...	...	...	...
αBi	25	atm	hR2	R3m	A7	αAs	0.47460	...	...	α = 57.23°
βBi	25	>2.6	mC4	C2/m	...	βBi	0.6674	0.6117	0.3304	β = 110.33°
γBi	25	>3.0	mP3	?	...	...	0.605	0.42	0.465	β = 85.33°
σBi	25	>4.3	?	?	...	...	...	...	...	...
εBi	25	>6.5	?	?	...	...	...	...	...	...
ζBi	25	>9.0	cI2	Im3m	A2	W	0.3800	...	...	...
αBk	25	atm	hP4	P6₃/mmc	A3'	αLa	0.3416	...	1.1069	2*1.620
βBk	>977	atm	cF4	Fm3m	A1	Cu	0.4997	...	...	...
Br	<−7.25	atm	oC8	Cmca	...	Cl	0.668	0.449	0.874	...
C (graphite)	25	atm	hP4	P6₃/mmc	A9	C (graphite)	0.24612	...	0.6709	2.7258
C (diamond)	25	>60	cF8	Fd3m	A4	C (diamond)	0.35669	...	...	...
C (hd)	25	HP	hP4	P6₃/mmc	...	C (hd)	0.2522	...	0.4119	1.633
αCa	25	atm	cF4	Fm3m	A1	Cu	0.55884	...	...	...
βCa	>443	atm	cI2	Im3m	A2	W	0.4480	...	...	...
γCa	25	>1.5	?	...	...	...	...	...	...	...
Cd	25	atm	hP2	P6₃/mmc	A3	Mg	0.29793	...	0.56196	1.8862
αCe	<−177	atm	cF4	Fm3m	A1	Cu	0.485	...	...	...
βCe	25	atm	hP4	P6₃/mmc	A3'	αLa	0.36810	...	1.1857	2*1.611
γCe	25	atm	cF4	Fm3m	A1	Cu	0.51610	...	...	...
δ-Ce	>726	atm	cI2	Im3m	A2	W	0.412	...	...	...
α'Ce	25	>5.4	oC4	Cmcm	A20	αU	0.3049	0.5998	0.5215	...
αCf	25	atm	hP4	P6₃/mmc	A3'	αLa	0.339	...	1.1015	2*1.625
βCf	>590	atm	cF4	Fm3m	A1	Cu	?	...	...	...
Cl	25	atm	oC8	Cmca	...	Cl	0.624	0.448	0.826	...
αCm	25	atm	hP4	P6₃/mmc	A3'	αLa	0.3496	...	1.1331	2*1.621
βCm	>1277	atm	cF4	Fm3m	A1	Cu	0.4382	...	...	...
εCo	25	atm	hP2	P6₃/mmc	A3	Mg	0.25071	...	0.40686	1.6228
αCo	>422	atm	cF4	Fm3m	A1	Cu	0.35447	...	...	...
αCr	25	atm	cI2	Im3m	A2	W	0.38848	...	...	...
α'Cr	25	HP	tI2	I4/mmm	...	α'Cr	0.2882	...	0.2887	1.002
αCs	25	atm	cI2	Im3m	A2	W	0.6141	...	...	...
βCs	25	>2.37	cF4	Fm3m	A1	Cu	0.6465	...	...	...
β'Cs	25	>4.22	cF4	Fm3m	A1	Cu	0.5800	...	...	...
γCs	25	>4.27	?	...	...	...	...	...	...	...
Cu	25	atm	cF4	Fm3m	A1	Cu	0.36146	...	...	...
α'Dy	<−187	atm	oC4	Cmcm	...	α'Dy	0.3595	0.6184	0.5678	...
αDy	25	atm	hP2	P6₃/mmc	A3	Mg	0.35915	...	0.56501	1.5732

Element	Temperature, °C	Pressure, GPa	Pearson symbol	Space group	Strukturbericht designation	Prototype	Lattice parameters, nm a	b	c	Comment, c/a or α or β
βDy	>1381	atm	cI2	Im3m	A2	W	0.403	...	...	...
γDy	25	>7.5	hR3	R3̄m	...	αSm	0.3436	...	2.483	4.5*1.606
Er	25	atm	hP2	P6₃/mmc	A3	Mg	0.35592	...	0.55850	1.5692
αEs	25	atm	hP4	P6₃/mmc	A3'	αLa	?	...	...	...
βEs	?	atm	cF4	Fm3m	A1	Cu	?	...	...	...
Eu	25	atm	cI2	Im3m	A2	W	0.45827	...	...	...
αF	<−227.6	atm	mC8	C2/c	...	αF	0.550	0.338	0.728	β = 102.17°
βF	<−219.67	atm	cP16	Pm3n	...	γO	0.667	...	...	...
αFe	25	atm	cI2	Im3m	A2	W	0.28665	...	...	...
γFe	>912	atm	cF4	Fm3m	A1	Cu	0.36467	...	...	...
σFe	>1394	atm	cI2	Im3m	A2	W	0.29315	...	...	...
εFe	25	>13	hP2	P6₃/mmc	A3	Mg	0.2468	...	0.396	1.603
αGa	25	atm	oC8	Cmca	A11	αGa	0.45186	0.76570	0.45258	...
βGa	25	>1.2	tI2	I4/mmm	A6	In	0.2808	...	0.4458	1.588
γGa	−53	>3.0	oC40	Cmcm	...	γGa	1.0593	1.3523	0.5203	...
αGd	25	atm	hP2	P6₃/mmc	A3	Mg	0.36336	...	0.57810	1.5910
βGd	>1235	atm	cI2	Im3m	A2	W	0.406	...	...	...
γGd	25	>3.0	hR3	R3m	...	αSm	0.361	...	2.603	4*1.60
αGe	25	atm	cF8	Fd3m	A4	C (diamond)	0.56574	...	...	...
βGe	25	>12	tI4	I4₁/amd	A5	βSn	0.4884	...	0.2692	0.551
γGe	25	>12 → atm	tP12	P4₃2₁2	...	σGe	0.593	...	0.698	1.18
σGe	LT	>12	cI16	Im3m	...	γSi	0.692	...	...	...
αH	<−271.9	atm	cF4	Fm3m	A1	Cu	0.5338	...	...	...
βH	<−259.34	atm	hP2	P6₃/mmc	A3	Mg	0.3776	...	0.6162	1.632
αHe	<−268.94	atm	hP2	P6₃/mmc	A3	Mg	0.3555	...	0.5798	1.631
βHe	>−258	0.125	cF4	Fm3m	A1	Cu	0.4240	...	...	...
γHe	<−271.47	0.03	cI2	Im3m	A2	W	0.4110	...	...	...
αHf	25	atm	hP2	P6₃/mmc	A3	Mg	0.31946	...	0.50510	1.5811
βHf	>1995	atm	cI2	Im3m	A2	W	0.3610	...	...	...
αHg	<−38.84	atm	hR1	R3̄m	A10	αHg	0.3005	...	...	α = 70.53°
βHg	<−194	HP	tI2	I4/mmm	...	βHg	0.3995	...	0.2825	0.707
γHg	<−194	c.w.	hR1	?	...	...	...	...	...	...
αHo	25	atm	hP2	P6₃/mmc	A3	Mg	0.35778	...	0.56178	1.5702
βHo	25	>7.5	hR3	R3̄m	...	αSm	0.334	...	2.45	4.5*1.63
I	25	atm	oC8	Cmca	...	Cl	0.72697	0.47903	0.97942	...
In	25	atm	tI2	I4/mmm	A6	In	0.3253	...	0.49470	1.5210
Ir	25	atm	cF4	Fm3m	A1	Cu	0.38392	...	...	...
K	25	atm	cI2	Im3m	A2	W	0.5321	...	...	...
Kr	<−157.39	atm	cF4	Fm3m	A1	Cu	0.5810	...	...	...
αLa	25	atm	hP4	P6₃/mmc	A3'	αLa	0.37740	...	1.2171	2*1.6125
βLa	>310	atm	cF4	Fm3m	A1	Cu	0.5303	...	...	...
γLa	>865	atm	cI2	Im3m	A2	W	0.426	...	...	...
β'La	25	>2.0	cF4	Fm3m	A1	Cu	0.517	...	...	...
αLi	<−193	atm	hP2	P6₃/mmc	A3	Mg	0.3111	...	0.5093	1.637
βLi	25	atm	cI2	Im3m	A2	W	0.35093	...	...	...
γLi	<−201	c.w.	cF4	Fm3m	A1	Cu	0.4388	...	...	...
Lu	25	atm	hP2	P6₃/mmc	A3	Mg	0.35052	...	0.55494	1.5832
Mg	25	atm	hP2	P6₃/mmc	A3	Mg	0.32094	...	0.52107	1.6236
αMn	25	atm	cI58	I4̄3m	A12	αMn	0.89126	...	...	...
βMn	>710	atm	cP20	P4₁32	A13	βMn	0.63152	...	...	...
γMn	>1079	atm	cF4	Fm3m	A1	Cu	0.3860	...	...	...
σMn	>1143	atm	cI2	Im3m	A2	W	0.3080	...	...	...
Mo	25	atm	cI2	Im3m	A2	W	0.31470	...	...	...
αN	<−237.6	atm	cP8	Pa3	...	αN	0.5661	...	...	...
βN	<−210.00	atm	hP4	P6₃/mmc	...	βN	0.4050	...	0.6604	1.631
γN	<−253	>3.3	tP4	P4₂/mnm	...	γN	0.3957	...	0.5109	1.291
αNa	<−233	atm	hP2	P6₃/mmc	A3	Mg	0.3767	...	0.6154	1.634
βNa	25	atm	cI2	Im3m	A2	W	0.42906	...	...	...
Nb	25	atm	cI2	Im3m	A2	W	0.33004	...	...	...
αNd	25	atm	hP4	P6₃/mmc	A3'	αLa	0.36582	...	1.17966	2*1.6124
βNd	>863	atm	cI2	Im3m	A2	W	0.413	...	...	...
γNd	25	>5.0	cF4	Fm3m	A1	Cu	0.480	...	...	...
Ne	<−243.59	atm	cF4	Fm3m	A1	Cu	0.4462	...	...	...
Ni	25	atm	cF4	Fm3m	A1	Cu	0.35240	...	...	...
αNp	25	atm	oP8	Pnma	A_c	αNp	0.6663	0.4723	0.4887	...
βNp	>280	atm	tP4	P42₁2	A_d	βNp	0.4883	...	0.3389	0.694
γNp	>576	atm	cI2	Im3m	A2	W	0.352	...	...	...
αO	<−243.3	atm	mC4	C2m	...	αO	0.5403	0.3429	0.5086	β = 132.53°
βO	<−229.6	atm	hR2	R3̄m	...	βO	0.4210	...	...	α = 46.27°
γO	<−218.79	atm	cP16	Pm3n	...	γO	0.683	...	...	...
Os	25	atm	hP2	P6₃/mmc	A3	Mg	0.27341	...	0.43918	1.6063
P (black)	25	atm	oC8	Cmca	...	P (black)	0.33136	1.0478	0.43763	...
αPa	25	atm	tI2	I4/mmm	A_a	αPa	0.3921	...	0.3235	0.825
βPa	>1170	atm	cI2	Im3m	A2	W	0.381	...	...	...
αPb	25	atm	cF4	Fm3m	A1	Cu	0.49502	...	...	...
βPb	25	>10.3	hP2	P6₃/mmc	A3	Mg	0.3265	...	0.5387	1.650

Element	Temperature, °C	Pressure, GPa	Pearson symbol	Space group	Strukturbericht designation	Prototype	Lattice parameters, nm — a	b	c	Comment, c/a or α or β
Pd	25	atm	cF4	Fm3m	A1	Cu	0.38903	...	...	...
αPm	25	atm	hP4	P6₃/mmc	A3'	αLa	0.365	...	1.165	2*1.60
βPm	>890	atm	cI2	Im3m	A2	W	(0.410)	...	...	...
αPo	25	atm	cP1	Pm3m	A_h	αPo	0.3366	...	...	...
βPo	>54	atm	hR1	R3m	...	βPo	0.3373	...	...	α = 98.08°
αPr	25	atm	hP4	P6₃/mmc	A3'	αLa	0.36721	...	1.18326	2*1.6111
βPr	>795	atm	cI2	Im3m	A2	W	0.413	...	...	...
γPr	25	>4.0	cF4	Fm3m	A1	Cu	0.488	...	...	...
Pt	25	atm	cF4	Fm3m	A1	Cu	0.39236	...	...	...
αPu	25	atm	mP16	P2₁/m	...	αPu	0.6183	0.4822	1.0963	β = 101.97°
βPu	>125	atm	mI34	I2/m	...	βPu	0.9284	1.0463	0.7859	β = 92.13°
γPu	>215	atm	oF8	Fddd	...	γPu	0.31587	0.57682	1.0162	...
σPu	>320	atm	cF4	Fm3m	A1	Cu	0.46371	...	...	...
σ'Pu	>463	atm	tI2	I4/mmm	A6	In	0.33261	...	0.44630	1.3418
εPu	>483	atm	cI2	Im3m	A2	W	0.36343	...	...	...
Ra	25	atm	cI2	Im3m	A2	W	0.5148	...	...	...
αRb	25	atm	cI2	Im3m	A2	W	0.5705	...	...	...
βRb	25	>1.08	?	...	...	...	...	...	...	...
γRb	25	>2.05	?	...	...	...	...	...	...	...
Re	25	atm	hP2	P6₃/mmc	A3	Mg	0.27609	...	0.4458	1.6145
Rh	25	atm	cF4	Fm3m	A1	Cu	0.38032	...	...	...
Ru	25	atm	hP2	P6₃/mmc	A3	Mg	0.27058	...	0.42816	1.5824
αS	25	atm	oF128	Fddd	A16	αS	1.0464	1.28660	2.44860	...
αSb	25	atm	hR2	R3m	A7	αAs	0.45067	...	...	α = 57.11°
βSb	25	>5.0	cP1	Pm3m	A_h	αPo	0.2992	...	...	...
γSb	25	>7.5	hP2	P6₃/mmc	A3	Mg	0.3376	...	0.5341	1.582
σSb	25	>14.0	mP3	?	...	...	0.556	0.404	0.422	β = 86.0°
αSc	25	atm	hP2	P6₃/mmc	A3	Mg	0.33088	...	0.52680	1.5921
βSc	>1337	atm	cI2	Im3m	A2	W	(0.373)	...	...	...
γSe	25	atm	hP3	P3₁21	A8	γSe	0.43659	...	0.49537	1.1346
αSi	25	atm	cF8	Fd3m	A4	C (diamond)	0.54306	...	...	...
βSi	25	>9.5	tI4	I4₁/amd	A5	βSn	0.4686	...	0.2585	0.552
γSi	25	>16.0	cI16	Im3m	...	γSi	0.6636	...	...	...
σSi	25	>16 → atm	hP4	P6₃/mmc	A3'	αLa	0.380	...	0.628	1.653
αSm	25	atm	hR3	R3m	...	αSm	0.36290	...	2.6207	4*1.6048
βSm	>734	atm	hP2	P6₃/mmc	A3	Mg	0.36630	...	0.58448	1.5956
γ'Sm	>922	atm	cI2	Im3m	A2	W	(0.410)	...	...	...
σSm	25	>4.0	hP4	P6₃/mmc	A3'	αLa	0.3618	...	1.166	2*1.611
αSn	<13	atm	cF8	Fd3m	A4	C (diamond)	0.64892	...	...	...
βSn	25	atm	tI4	I4₁/amd	A5	βSn	0.58318	...	0.31818	0.5456
γSn	25	>9.0	tI2	?	...	γSn	0.370	...	0.337	0.91
αSr	25	atm	cF4	Fm3m	A1	Cu	0.6084	...	...	...
βSr	>547	atm	cI2	Im3m	A2	W	0.487	...	...	...
β'Sr	25	>3.5	cI2	Im3m	A2	W	0.4437	...	...	...
Ta	25	atm	cI2	Im3m	A2	W	0.33030	...	...	...
α"Tb	<-53	atm	oC4	Cmcm	...	α'Dy	0.3605	0.6244	0.5706	...
αTb	25	atm	hP2	P6₃/mmc	A3	Mg	0.36055	...	0.56966	1.5800
βTb	>1289	atm	cI2	Im3m	A2	W	(0.407)	...	...	...
γTb	25	>6.0	hR3	R3m	...	αSm	0.341	...	2.45	4*1.60
Tc	25	atm	hP2	P6₃/mmc	A3	Mg	0.2738	...	0.4393	1.604
αTe	25	atm	hP3	P3₁21	A8	γSe	0.44566	...	0.59264	1.3298
βTe	25	>2.0	hR2	R3m	A7	αAs	0.469	...	...	α = 53.30°
γTe	25	>7.0	hR1	R3m	...	βPo	0.3002	...	...	α = 103.3°
αTh	25	atm	cF4	Fm3m	A1	Cu	0.50842	...	...	...
βTh	>1360	atm	cI2	Im3m	A2	W	0.411	...	...	...
αTi	25	atm	hP2	P6₃/mmc	A3	Mg	0.29506	...	0.46835	1.5873
βTi	>882	atm	cI2	Im3m	A2	W	0.33065	...	...	...
ωTi	25	HP → atm	hP3	P6/mmm	...	ωTi	0.4625	...	0.2813	0.6082
αTl	25	atm	hP2	P6₃/mmc	A3	Mg	0.34566	...	0.55248	1.5983
βTl	>230	atm	cI2	Im3m	A2	W	0.3879	...	...	...
γTl	25	HP	cF4	Fm3m	A1	Cu	?	...	...	...
Tm	25	atm	hP2	P6₃/mmc	A3	Mg	0.35375	...	0.55540	1.5700
αU	25	atm	oC4	Cmcm	A20	αU	0.28537	0.58695	0.49548	...
βU	>668	atm	tP30	P4₂/mnm	A_b	βU	1.0759	...	0.5656	0.526
γU	>776	atm	cI2	Im3m	A2	W	0.3524	...	...	...
V	25	atm	cI2	Im3m	A2	W	0.30240	...	...	...
W	25	atm	cI2	Im3m	A2	W	0.31652	...	...	...
Xe	<-111.76	atm	cF4	Fm3m	A1	Cu	0.6350	...	...	...
αY	25	atm	hP2	P6₃/mmc	A3	Mg	0.36482	...	0.57318	1.5711
βY	>1478	atm	cI2	Im3m	A2	W	(0.410)	...	...	...
αYb	<-3	atm	hP2	P6₃/mmc	A3	Mg	0.38799	...	0.63859	1.6459
βYb	25	atm	cF4	Fm3m	A1	Cu	0.54848	...	...	...
γYb	>795	atm	cI2	Im3m	A2	W	0.444	...	...	...
Zn	25	atm	hP2	P6₃/mmc	A3	Mg	0.26650	...	0.49470	1.8563
αZr	25	atm	hP2	P6₃/mmc	A3	Mg	0.32316	...	0.51475	1.5929
βZr	>863	atm	cI2	Im3m	A2	W	0.36090	...	...	...
ωZr	25	HP → atm	hP2	P6/mmm	...	ωTi	0.5036	...	0.3109	0.617

LATTICE ENERGIES

H. D. B. Jenkins

Table 1 contains calculated values of the lattice energies (total lattice potential energies), U_{POT}, of crystalline salts, M_aX_b.

U_{POT} is expressed in units of kilojoules per mole, kJ mol^{-1}. M and X can be either simple or complex ions.

Also listed in the table is the lattice energy, $U_{POT}{}^{BFHC}$, obtained from the application of the Born-Fajans-Haber cycle (BFHC) described below, using thermochemical data from Reference 1 plus certain other data which are given in Table 2 of this section (Reference 2).

The lattice enthalpy, ΔH_L, is given by the cycle:

where (ss) is the standard state of the element concerned.

The lattice enthalpy, ΔH_L, is obtained using the equation:

$$\Delta H_L = a\Delta H_f{}^{\circ}\left(M^{Za+}\right)(g) + b\Delta H_f{}^{\circ}\left(X^{Zb-}\right)(g) - \Delta H_f{}^{\circ}\left(M_aX_b\right)(c)$$

and is further related to the total lattice potential energy, $U_{POT}{}^{BFHC}$, by the relationship:

$$U_{POT}{}^{BFHC}\left(M_aX_b\right) = \Delta H_L - \left[a\left(\frac{n_{M^{Za+}}}{2} - 2\right) + b\left(\frac{n_{X^{Zb-}}}{2} - 2\right)\right]RT$$

where $n_{M^{Za+}}$ and $n_{X^{Zb-}}$ is equal to 3 for monatomic ions, 5 for linear polyatomic ions and 6 for polyatomic non-linear ions.

Substances are arranged by chemical class.

REFERENCES

1. Wagman, D. D., et al., *The NBS Tables of Chemical Thermodynamic Properties*, J. Phys. Chem. Ref. Data, 11, Suppl. 2, 1982.
2. *Adv. Inorg. Chem. Radiochem.*, 22, 1, 1978.

Table 1
LATTICE ENERGIES

Substance	Calc. U_{POT} (kJ mol^{-1})	$U_{POT}{}^{BFHC}$ (kJ mol^{-1})	Substance	Calc. U_{POT} (kJ mol^{-1})	$U_{POT}{}^{BFHC}$ (kJ mol^{-1})
Acetates			Mn(CH$_3$COO)$_2$	2548	2616
Li(CH$_3$COO)	—	881	Cu(CH$_3$COO)$_2$	—	2835
Na(CH$_3$COO)	761	763	Zn(CH$_3$COO)$_2$	2615	2750
K(CH$_3$COO)	686	682	Hg(CH$_3$COO)$_2$	2368	2595
Rb(CH$_3$COO)	715	656	Pb(CH$_3$COO)$_2$	2247	2225
Cs(CH$_3$COO)	682	—	Acetylides		
NH$_4$(CH$_3$COO)	725	695	CaC$_2$	2911	2904
Ag(CH$_3$COO)	—	863	SrC$_2$	2788	2784
Tl(CH$_3$COO)	—	750	BaC$_2$	2647	2654
Ca(CH$_3$COO)$_2$	2431	2294	Ammonium salts		
Sr(CH$_3$COO)$_2$	2280	2166	NH$_4$HF$_2$	705	705
Ba(CH$_3$COO)$_2$	2180	2033	NH$_4$HCO$_3$	—	741

Substance	Calc. U_{POT} (kJ mol^{-1})	U_{POT}^{BFHC} (kJ mol^{-1})	Substance	Calc. U_{POT} (kJ mol^{-1})	U_{POT}^{BFHC} (kJ mol^{-1})
NH_4HSO_4	640	(645)	$Cd(N_3)_2$	2446	2454
NH_4BF_4	582	—	$Pb(N_3)_2$	—	2173
NH_4NCO	724	—	Bihalide salts		
NH_4CN	617	670	$LiHF_2$	893	866
NH_4HCO_2	715	—	$NaHF_2$	788	788
$NH_4(NH_2CH_2CO_2)$	650	—	KHF_2	703	698
NH_4IO_3	—	808	$RbHF_2$	674	676
$NH_4IO_2F_2$	678	685	$CsHF_2$	646	628
NH_4HS	661	666	NH_4HF_2	705	705
NH_4NO_3	661	676	$CsHCl_2$	601	—
NH_4ClO_4	583	580	Me_4NHCl_2	427	—
$(NH_4)_2S$	2008	(2026)	Et_4NHCl_2	346	—
$(NH_4)_2SO_4$	1766	1777	Bu_4NHCl_2	290	—
$(NH_4)_2GeF_6$	1657	—	Bicarbonates		
$(NH_4)_2IrCl_6$	1442	1440	$NaHCO_3$	820	818
$(NH_4)_2OsCl_6$	1433	—	$KHCO_3$	741	736
$(NH_4)_2PdCl_6$	1481	—	$RbHCO_3$	707	714
$(NH_4)_2PbCl_6$	1355	—	$CsHCO_3$	678	709
$(NH_4)_2PbCl_6$	1468	—	NH_4HCO_3	—	741
$(NH_4)_2ReCl_6$	1402	1390	$Ca(HCO_3)_2$	2402	(2403)
$(NH_4)_2SiF_6$	1657	1727	$Sr(HCO_3)_2$	2255	(2272)
$(NH_4)_2SnCl_6$	1370	1334	$Ba(HCO_3)_2$	2155	(2159)
$(NH_4)_2SnBr_6$	1319	—	Bisulphates		
$(NH_4)_2TeCl_6$	1318	—	NH_4HSO_4	640	(645)
$(NH_4)_2TeBr_6$	1294	—	Borides		
$(NH_4)_2TiCl_6$	1413	—	CaB_6	5146	—
NH_4CNS	605	611	SrB_6	5104	—
$(NH_4)_2SeCl_6$	1420	—	BaB_6	5021	—
$(NH_4)_2SeBr_6$	1380	—	YB_6	7447	—
Arsenates			LaB_6	7406	—
$Mg_3(AsO_4)_2$	10669	10716	CeB_6	10083	—
$Ca_3(AsO_4)_2$	9749	9653	PrB_6	7447	—
$Sr_3(AsO_4)_2$	9330	9266	NdB_6	7447	—
$Ba_3(AsO_4)_2$	8870	8985	PmB_6	7406	—
$AlAsO_4$	7255	—	SmB_6	7447	—
$GaAsO_4$	7243	—	EuB_6	5104	—
Astatides			GdB_6	7489	—
$LiAt$	720	—	TbB_6	7489	—
$NaAt$	657	—	DyB_6	7489	—
KAt	615	—	HoB_6	7489	—
$RbAt$	594	—	ErB_6	7489	—
$CsAt$	586	—	TmB_6	7489	—
$FrAt$	573	—	YbB_6	5146	—
Azides			LuB_6	7489	—
LiN_3	812	818	ThB_6	10167	—
NaN_3	732	731	Borohydrides		
KN_3	659	658	$LiBH_4$	778	—
RbN_3	637	632	$NaBH_4$	703	—
CsN_3	612	604	KBH_4	665	—
AgN_3	854	—	$RbBH_4$	648	—
TlN_3	689	686	$CsBH_4$	628	—
$Ca(N_3)_2$	2186	2162	Borohalides		
$Sr(N_3)_2$	2056	2066	$LiBF_4$	699	—
$Ba(N_3)_2$	2021	1965	$NaBF_4$	657	619
$Mn(N_3)_2$	2408	2416	KBF_4	611	631
$Cu(N_3)_2$	2730	2738	$RbBF_4$	577	605
$Zn(N_3)_2$	2840	2848	$CsBF_4$	556	—

Table 1 (continued)

Substance	Calc. U_{POT} (kJ mol^{-1})	U_{POT}^{BFHC} (kJ mol^{-1})	Substance	Calc. U_{POT} (kJ mol^{-1})	U_{POT}^{BFHC} (kJ mol^{-1})
NH$_4$BF$_4$	582	—	Sr(HCO$_2$)$_2$	2221	2261
Co(BF$_4$)$_2$	2127	—	Ba(HCO$_2$)$_2$	2092	2134
Ni(BF$_4$)$_2$	2136	—	Mn(HCO$_2$)$_2$	2598	2701
Zn(BF$_4$)$_2$	2063	—	Co(HCO$_2$)$_2$	—	2792
Cd(BF$_4$)$_2$	1937	—	Ni(HCO$_2$)$_2$	—	2880
KBCl$_4$	506	—	Cu(HCO$_2$)$_2$	2870	2913
RbBCl$_4$	489	—	Zn(HCO$_2$)$_2$	2791	2847
CsBCl$_4$	473	—	Cd(HCO$_2$)$_2$	2556	—
Carbonates			Pb(HCO$_2$)$_2$	2276	2330
Li$_2$CO$_3$	2523	2269	Germanates		
Na$_2$CO$_3$	2301	2030	Mg$_2$GeO$_4$	7991	—
K$_2$CO$_3$	2084	1858	Ca$_2$GeO$_4$	7301	7306
Rb$_2$CO$_3$	2000	1795	Sr$_2$GeO$_4$	6987	—
Cs$_2$CO$_3$	1920	1702	Ba$_2$GeO$_4$	6653	6643
MgCO$_3$	3180	3122	Glycinates		
CaCO$_3$	2804	2810	Na(NH$_2$CH$_2$COO$_2$)	739	—
SrCO$_3$	2720	2688	K(NH$_2$CH$_2$COO$_2$)	668	—
BaCO$_3$	2615	2554	Rb(NH$_2$CH$_2$COO$_2$)	648	—
MnCO$_3$	3046	3151	NH$_4$(NH$_2$CH$_2$COO$_2$)	650	—
FeCO$_3$	3121	3171	Cu(NH$_2$CH$_2$CO$_2$)$_2$	2694	—
CoCO$_3$	3443	3232	Halates		
NiCO$_3$	—	3297	LiBrO$_3$	883	897
CuCO$_3$	3494	3327	NaBrO$_3$	803	814
ZnCO$_3$	3121	3273	KBrO$_3$	740	745
CdCO$_3$	2929	3052	RbBrO$_3$	720	742
SnCO$_3$	2904	(2853)	CsBrO$_3$	694	681
PbCO$_3$	2728	2750	NaClO$_3$	770	770
Cyanates			KClO$_3$	711	706
LiNCO	770	—	RbClO$_3$	690	687
NaNCO	736	—	CsClO$_3$	—	647
KNCO	653	—	LiIO$_3$	975	—
RbNCO	615	—	NaIO$_3$	883	882
CsNCO	586	—	KIO$_3$	820	806
NH$_4$NCO	724	—	RbIO$_3$	791	—
Cyanides			CsIO$_3$	761	769
LiCN	—	849	NH$_4$IO$_3$	—	808
NaCN	738	739	Mg(ClO$_3$)$_2$	2535	(2475)
KCN	674	669	Ca(ClO$_3$)$_2$	2259	2286
RbCN	646	(640)	Sr(ClO$_3$)$_2$	2138	2155
CsCN	612	602	Ba(ClO$_3$)$_2$	2021	2027
NH$_4$CN	617	670	Halides		
Ca(CN)$_2$	2268	2191	LiF	1030	1036
Sr(CN)$_2$	2138	(2076)	LiCl	834	853
Ba(CN)$_2$	2046	1960	LiBr	788	807
CuCN	—	1035	LiI	730	757
AgCN	741	914	NaF	910	923
Zn(CN)$_2$	2431	2768	NaCl	769	786
Cd(CN)$_2$	2284	2542	NaBr	732	747
Formates			NaI	682	704
Li(HCO$_2$)	865	872	KF	808	821
Na(HCO$_2$)	791	811	KCl	701	715
K(HCO$_2$)	713	729	KBr	671	682
Rb(HCO$_2$)	685	682	KI	632	649
Cs(HCO$_2$)	651	644	RbF	774	785
NH$_4$(HCO$_2$)	715	—	RbCl	680	689
Mg(HCO$_2$)$_2$	2674	—	RbBr	651	660
Ca(HCO$_2$)$_2$	2360	2390	RbI	617	630

Substance	Calc. U_{POT} (kJ mol^{-1})	U_{POT}^{BFHC} (kJ mol^{-1})	Substance	Calc. U_{POT} (kJ mol^{-1})	U_{POT}^{BFHC} (kJ mol^{-1})
CsF	744	740	VBr$_2$	—	2523
CsCl	657	659	VI$_2$	—	2456
CsBr	632	631	CrF$_2$	2778	2917
CsI	600	604	CrCl$_2$	2455	2586
FrF	715	—	CrBr$_2$	2377	2523
FrCl	632	—	CrI$_2$	2269	2425
FrBr	611	—	MoCl$_2$	2485	2733
FrI	582	—	MoBr$_2$	2448	2742
CuCl	921	996	MoI$_2$	2422	2630
CuBr	879	979	MnF$_2$	2644	—
CuI	835	966	MnCl$_2$	2368	2537
AgF	953	967	MnBr$_2$	2304	2471
AgCl	864	915	MnI$_2$	2212	—
AgBr	830	904	FeF$_2$	2769	—
AgI	808	889	FeCl$_2$	2525	2631
AuCl	1013	1066	FeBr$_2$	2464	2569
AuBr	1015	1061	FeI$_2$	2382	2480
AuI	1015	1070	CoF$_2$	2878	3018
InCl	—	763	CoCl$_2$	2709	2691
InBr	—	767	CoBr$_2$	2648	2629
InI	—	732	CoI$_2$	2569	2545
TlF	—	845	NiF$_2$	2845	3066
TlCl	782	751	NiCl$_2$	2753	2772
TlBr	713	735	NiBr$_2$	2699	2709
TlI	687	709	NiI$_2$	2607	2623
BeF$_2$	3150	3505	PdCl$_2$	2766	2778
BeCl$_2$	3004	3020	PdBr$_2$	—	2741
BeBr$_2$	2950	2914	PdI$_2$	—	2748
BeI$_2$	2653	2800	CuF$_2$	3046	3082
MgF$_2$	2913	2957	CuCl$_2$	2774	2811
MgCl$_2$	2326	2526	CuBr$_2$	2711	2763
MgBr$_2$	2097	2440	CuI$_2$	2640	—
MgI$_2$	1944	2327	AgF$_2$	2919	2942
CaF$_2$	2609	2630	ZnF$_2$	2930	3032
CaCl$_2$	2223	2258	ZnCl$_2$	2690	2734
CaBr$_2$	2132	2176	ZnBr$_2$	2632	2678
CaI$_2$	1905	2074	ZnI$_2$	2549	2605
SrF$_2$	2476	2492	CdF$_2$	2740	2809
SrCl$_2$	2127	2156	CdCl$_2$	2226	2552
SrI$_2$	2008	2075	CdBr$_2$	2468	2507
SrI$_2$	1937	1963	CdI$_2$	2406	2441
BaF$_2$	2341	2352	HgF$_2$	2757	(2798)
BaCl$_2$	2033	2056	HgCl$_2$	2569	2651
BaBr$_2$	1950	1985	HgBr$_2$	2598	2628
BaI$_2$	1831	1877	HgI$_2$	2569	2610
RaF$_2$	2284	—	SnF$_2$	2551	—
RaCl$_2$	2004	—	SnCl$_2$	2276	2297
RaBr$_2$	1929	—	SnBr$_2$	2211	2245
RaI$_2$	1803	—	SnI$_2$	2123	2193
ScCl$_2$	2380	—	PbF$_2$	2460	2522
ScBr$_2$	2291	—	PbCl$_2$	2229	2269
ScI$_2$	2201	—	PbBr$_2$	2169	2219
TiF$_2$	2724	—	PbI$_2$	2086	2163
TiCl$_2$	2431	2501	ScF$_3$	5096	5492
TiBr$_2$	2360	2419	ScCl$_3$	4874	4866
TiI$_2$	2259	2329	ScBr$_3$	4711	4729
VCl$_2$	2607	2579	ScI$_3$	4640	—

Substance	Calc. U_{POT} (kJ mol^{-1})	U_{POT}^{BFHC} (kJ mol^{-1})	Substance	Calc. U_{POT} (kJ mol^{-1})	U_{POT}^{BFHC} (kJ mol^{-1})
YF$_3$	4983	—	GaBr$_3$	4966	5552
YCl$_3$	4447	4506	GaI$_3$	4611	5476
YBr$_3$	4410	—	InCl$_3$	4736	5187
YI$_3$	4125	4240	InBr$_3$	4535	5124
TiF$_3$	5644	—	InI$_3$	4234	5005
TiCl$_3$	5134	5134	TlF$_3$	5493	—
TiBr$_3$	5012	5007	TlCl$_3$	5252	5258
TiI$_3$	4845	—	TlBr$_3$	5171	—
ZrF$_3$	—	(5400)	TlI$_3$	5088	—
ZrCl$_3$	—	4791	AsBr$_3$	—	5497
ZrBr$_3$	—	4758	AsI$_3$	(3758)	4824
ZrI$_3$	—	(4591)	SbF$_3$	—	5295
VF$_3$	5895	—	SbCl$_3$	—	5032
VCl$_3$	5322	5315	SbBr$_3$	—	4954
BVr$_3$	5192	5214	SbI$_3$	—	4867
VI$_3$	5058	5121	BiCl$_3$	—	4689
NbCl$_3$	5062	—	BiBr$_3$	—	—
NbBr$_3$	4980	—	BiI$_3$	(3774)	—
NbI$_3$	4860	—	LaF$_3$	4682	—
CrF$_3$	5958	6033	LaCl$_3$	4343	4242
CrCl$_3$	5473	5509	LaBr$_3$	4209	—
CrBr$_3$	5355	—	LaI$_3$	3916	3986
CrI$_3$	5201	5274	CeCl$_3$	4297	4284
MoF$_3$	(6459)	5230	CeI$_3$	—	4029
MoCl$_3$	—	5230	PrCl$_3$	4322	4326
MoBr$_3$	—	5156	PrI$_3$	—	4071
MoI$_3$	—	5073	NdCl$_3$	4343	—
MnF$_3$	6017	—	SmCl$_3$	4376	—
MnCl$_3$	5544	—	EuCl$_3$	4393	—
MnBr$_3$	5448	—	GdCl$_3$	4406	—
MnI$_3$	5330	—	DyCl$_3$	4481	—
TcCl$_3$	5270	—	HoCl$_3$	4501	—
TcBr$_3$	5215	—	ErCl$_3$	4527	—
TcI$_3$	5188	—	TmCl$_3$	4548	4550
FeF$_3$	5870	—	TmI$_3$	—	4314
FeCl$_3$	5364	5359	YbCl$_3$	—	4546
FeBr$_3$	5268	5333	AcCl$_3$	4096	—
FeI$_3$	5117	—	UCl$_3$	4243	—
RuCl$_3$	—	5245	NpCl$_3$	4268	—
RuBr$_3$	—	5223	PuCl$_3$	4289	—
RuI$_3$	—	5222	PuBr$_3$	(3959)	—
CoF$_3$	5991	6118	AmCl$_3$	4293	—
RhCl$_3$	—	5641	TiF$_4$	10012	9908
IrF$_3$	(6112)	—	TiCl$_4$	9431	—
IrBr$_3$	(4794)	—	TiBr$_4$	9288	9039
NiF3	(6111)	—	TiI$_4$	9108	8893
AuF$_3$	(5777)	—	ZrF$_4$	8853	8971
AuCl$_3$	(4605)	—	ZrCl$_4$	8096	8144
ZnCl$_3$	5832	—	ZrBr$_4$	7916	7984
ZnBr$_3$	5732	—	ZrI$_4$	7661	7801
ZnI$_3$	5636	—	MoF$_4$	8795	—
AlF$_3$	5924	5215	MoCl$_4$	8556	9573
AlCl$_3$	5376	5492	MoBr$_4$	8510	9475
AlBr$_3$	5247	5361	MoI$_4$	8427	—
AlI$_3$	5070	5218	SnCl$_4$	8355	—
GaF$_3$	—	6205	SnBr$_4$	7970	8833
GaCl$_3$	5217	5645	PbF$_4$	—	9461

Table 1 (continued)

Substance	Calc. U_{POT} (kJ mol^{-1})	U_{POT}^{BFHC} (kJ mol^{-1})	Substance	Calc. U_{POT} (kJ mol^{-1})	U_{POT}^{BFHC} (kJ mol^{-1})
CrF_2Cl	5795	—	SrH_2	2250	2253
CrF_2Br	5753	—	BaH_2	2121	2121
CrF_2I	5669	—	ScH_2	2711	2659
$CrCl_2Br$	5448	—	YH_2	(2598)	2670
$CrCl_2I$	5381	—	LaH_2	2380	2522
$CrBr_2I$	5330	—	CeH_2	2414	2484
$CuFCl$	2891	—	PrH_2	2448	2398
$CuFBr$	2853	—	NdH_2	2464	2367
$CuFI$	2803	—	PmH_2	2519	—
$CuClBr$	2753	—	SmH_2	2510	2389
$CuClI$	2694	—	GdH_2	2494	2706
$CuBrI$	2669	—	AcH_2	2372	—
FeF_2Cl	5711	—	ThH_2	2711	—
FeF_2Br	5653	—	PuH_2	2519	—
FeF_2I	5569	—	AmH_2	2544	—
$FeCl_2Br$	5339	—	TiH_2	2866	2845
$FeCl_2I$	5272	—	ZrH_2	2711	2999
FBr_2I	5209	—	CuH_2	2941	—
$LiIO_2F_2$	845	—	ZnH_2	2870	—
$NaIO_2F_2$	766	764	HgH_2	2707	—
KIO_2F_2	699	697	AlH_3	5924	—
$RbIO_2F_2$	674	671	FeH_3	5724	—
$CsIO_2F_2$	636	—	ScH_3	5439	—
$NH_4IO_2F_2$	678	—	YH_3	5063	—
$AgIO_2F_2$	736	685	LaH_3	4895	4493
Hydrides			FeH_3	5724	—
LiH	858	920	GaH_3	5690	—
NaH	782	808	InH_3	5092	—
KH	699	714	TlH_3	5092	—
RbH	674	685	**Hydroselenides**		
CsH	648	644	NaHse	703	—
TiH	—	1407	KHSe	644	—
ZrH	—	1590	RbHSe	623	—
VH	—	(1344)	CsHSe	598	—
NbH	—	(1633)	**Hydrosulphides**		
PdH	—	(1368)	LiHS	759	821
CuH	—	1254	NaHS	704	747
TiH	996	1407	KHS	650	659
ZrH	916	1590	RbHS	623	637
HfH	904	—	CsHS	582	595
LaH	828	—	NH_4HS	661	666
VH	1184	(1344)	$Ca(HS)_2$	2184	(2171)
NbH	1163	(1633)	$Sr(HS)_2$	2063	—
TaH	1021	—	$Ba(HS)_2$	1979	(1956)
CrH	1050	—	**Hydroxides**		
NiH	929	—	LiOH	1021	1039
PdH	979	1368	NaOH	887	900
PtH	937	—	KOH	789	804
CuH	828	1254	RbOH	766	773
AgH	941	—	CsOH	721	724
AuH	1033	—	$Be(OH)_2$	3477	3629
TlH	745	—	$Mg(OH)_2$	2870	3006
GeH	950	—	$Ca(OH)_2$	2506	2645
PbH	778	—	$Sr(OH)_2$	2330	2483
BeH_2	3205	3295	$Ba(OH)_2$	2141	2339
MgH_2	2791	2706	$Ti(OH)_2$	—	2962
CaH_2	2410	2394	$Mn(OH)_2$	2909	3008

Table 1 (continued)

Substance	Calc. U_{POT} (kJ mol^{-1})	U_{POT}^{BFHC} (kJ mol^{-1})	Substance	Calc. U_{POT} (kJ mol^{-1})	U_{POT}^{BFHC} (kJ mol^{-1})
$Fe(OH)_2$	2653	3055	$Ba(NO_3)_2$	1975	2016
$Co(OH)_2$	2786	3115	$Mn(NO_3)_2$	2318	2519
$Ni(OH)_2$	2832	3193	$Fe(NO_3)_2$	—	(2563)
$Pd(OH)_2$	—	3175	$CO(NO_3)_2$	2560	2626
$Cu(OH)_2$	2870	3237	$Ni(NO_3)_2$	—	2709
CuOH	1006	—	$Cu(NO_3)_2$	—	2720
AgOH	918	—	$Zn(NO_3)_2$	2376	2628
AuOH	1033	—	$Cd(NO_3)_2$	2238	2443
TlOH	705	—	$Ha(NO_3)_2$	2255	—
$Zn(OH)_2$	2795	3158	$Sn(NO_3)_2$	2155	2254
$Cd(OH)_2$	2607	2918	$Pb(NO_3)_2$	2067	2189
$Hg(OH)_2$	2669	—	**Nitrides**		
$Sn(OH)_2$	2489	2729	ScN	7547	7506
$Pb(OH)_2$	2376	2623	LaN	6876	6793
$Sc(OH)_3$	5063	—	TiN	8130	8033
$Y(OH)_3$	4707	—	ZrN	7633	7723
$La(OH)_3$	4443	—	VN	8283	8233
$Cr(OH)_3$	5556	—	NbN	7939	8022
$Mn(OH)_3$	6213	—	CrN	8269	8358
$Al(OH)_3$	5627	—	**Nitrites**		
$Ga(OH)_3$	5732	—	$NaNO_2$	774	748
$In(OH)_3$	5280	—	KNO_2	660	664
$Tl(OH)_3$	5314	—	$RbNO_2$	638	765
$Ti(OH)_4$	9456	—	$CsNO_2$	598	—
$Zr(OH)_4$	8619	—	$Ca(NO_2)_2$	2460	2225
$Mn(OH)_4$	10933	—	$Sr(NO_2)_2$	2305	2111
$Sn(OH)_4$	9188	—	$Ba(NO_2)_2$	2205	1987
Imides			**Oxides**		
CaNH	3293	—	Li_2O	2799	—
SrNH	3146	—	Na_2O	2481	—
BaNH	2975	—	K_2O	2238	—
Metaniobates			Rb_2O	2163	—
$NaNbO_3$	789	—	Cu_2O	3273	—
$Ca(NbO_3)_2$	2315	—	Ag_2O	3002	—
$Fe(NBO_3)_2$	2502	—	TI_2O	2659	—
Metatantalates			LiO_2	(878)	(872)
$NaTaO_3$	789	—	NaO_2	799	796
$Ca(TaO_3)_2$	2315	—	KO_2	741	725
$Fe(TaO_3)_2$	2502	—	RbO_2	706	695
Metavanadates			CsO_2	679	668
Li_3VO_4	3945	—	Li_2O_2	2592	256
Na_3VO_4	3766	—	Na_2O_2	2309	2305
K_3VO_4	3376	—	K_2O_2	2114	2078
Rb_3VO_4	3243	—	Rb_2O_2	2025	2006
Cs_3VO_4	3137	—	Cs_2O_2	1948	1861
Nitrates			MgO_2	3356	3526
$LiNO_3$	848	848	CaO_2	3144	3133
$NaNO_3$	755	756	SrO_2	3037	2849
KNO_3	685	687	KO_3	697	—
$RbNO_3$	662	658	BeO	4293	4443
$CsNO_3$	648	625	MgO	3795	3791
NH_4NO_3	661	676	CaO	3414	3401
$AgNO_3$	820	822	SrO	3217	3223
$TlNO_3$	690	700	BaO	3029	3054
$Mg(NO_3)_2$	2468	2503	TiO	3832	3811
$Ca(NO_3)_2$	2209	2228	VO	3932	3863
$Sr(NO_3)_2$	2092	2132	MnO	3724	3745

Table 1 (continued)

Substance	Calc. U_{POT} (kJ mol^{-1})	U_{POT}^{BFHC} (kJ mol^{-1})	Substance	Calc. U_{POT} (kJ mol^{-1})	U_{POT}^{BFHC} (kJ mol^{-1})
FeO	3795	3865	$NaClO_4$	643	648
CoO	3837	3910	$KClO_4$	599	602
NiO	3908	4010	$RbClO_4$	564	582
PdO	3736	—	$CsClO_4$	636	(542)
CuO	4135	4050	NH_4ClO_4	583	580
ZnO	4142	3971	$Ca(ClO_4)_2$	1958	1971
CdO	3806	—	$Sr(ClO_4)_2$	1862	1862
HgO	3907	—	$Ba(ClO_4)_2$	1795	1769
GeO	3919	—	$NaMnO_4$	661	—
SnO	3652	—	$KMnO_4$	607	—
PbO	3520	—	$RbMnO_4$	586	—
Sc_2O_3	13557	13708	$CsMnO_4$	565	—
Y_2O_3	12705	—	$Ca(MnO_4)_2$	1937	—
La_2O_3	12452	—	$Sr(MnO_4)_2$	1845	—
Ce_2O_3	12661	—	$Ba(MnO_4)_2$	1778	—
Pr_2O_3	12703	—	Phosphates		
Nd_2O_3	12736	—	$Mg_4(PO_4)_2$	11632	11407
Pm_2O_3	12811	—	$Ca_3(PO_4)_2$	10602	10479
Sm_2O_3	12878	—	$Sr_3(PO_4)_2$	10125	10075
Eu_2O_3	12945	—	$Ba_3(PO_4)_3$	9652	9654
Gd_2O_3	12996	—	$MnPO_4$	7397	—
Tb_2O_3	13071	—	$FePO_4$	7251	7303
Dy_2O_3	13138	—	BPO_4	8201	—
Ho_2O_3	13180	—	$AlPO_4$	7427	7509
Er_2O_3	13263	—	$GaPO_4$	7381	—
Tm_2O_3	13322	—	Phosphonium salts		
Yb_2O_3	13380	—	PH_4Br	616	—
Lu_2O_3	13665	—	PH_4I	590	—
Ac_2O_3	12573	—	Selenides		
Ti_2O_3	—	14149	Li_2Se	2364	—
V_2O_3	15096	14520	Na_2Se	2130	—
Cr_2O_3	15276	14957	K_2Se	1933	—
Mn_2O_3	15146	15035	Rb_2Se	1837	—
Fe_2O_3	14309	14774	CS_2Se	1745	—
Al_2O_3	15916	—	Ag_2Se	2686	—
Ga_2O_3	15590	15220	Tl_2Se	2209	—
In_2O_3	13928	—	BeSe	3431	—
Ti_2O_3	14702	—	MgSe	3071	—
Pb_2O_3	(14841)	—	CaSe	2858	2862
CeO_2	9627	—	SrSe	2736	—
ThO_2	10397	—	BaSe	2611	—
PaO_2	10573	—	MnSe	3176	3194
VO_2	10644	—	FeSe	3499	3396
NpO_2	10707	—	CoSe	3554	3471
PuO_2	10786	—	NiSe	3658	3558
AmO_2	10799	—	CuSe	3736	3662
CmO_2	10832	—	ZnSe	3502	3514
TiO_2	12150	—	CdSe	3330	—
ZrO_2	11188	—	HgSe	3501	—
MoO_2	11648	—	SnSe	3058	—
MnO_2	12970	—	PbSe	3050	—
SiO_2	13125	—	Selenites		
GeO_2	12828	—	Li_2SeO_3	2171	—
SnO_2	11807	—	Na_2SeO_3	1950	1931
PbO_2	11217	—	K_2SeO_3	1774	—
Perchlorates			Rb_2SeO_3	1715	1675
$LiClO_4$	709	723	Cs_2SeO_3	1640	—

Table 1 (continued)

Substance	Calc. U_{POT} (kJ mol^{-1})	U_{POT}^{BFHC} (kJ mol^{-1})	Substance	Calc. U_{POT} (kJ mol^{-1})	U_{POT}^{BFHC} (kJ mol^{-1})
Tl_2SeO_3	1879	—	Rb_2SO_4	1636	1748
Ag_2SeO_3	2113	—	Cs_2SO_4	1596	1658
$BeSeO_3$	3322	—	$(NH_4)_2SO_4$	1766	1777
$MgSeO_3$	3012	2996	Cu_2SO_4	2276	2166
$CaSeO_3$	2732	—	Ag_2SO_4	2104	1989
$SrSeO_3$	2586	2586	Tl_2SO_4	1828	1722
$BaSeO_3$	2460	2448	Hg_2SO_4	—	2127
$RaSeO_3$	2456	—	$CaSO_4$	2489	2480
$MnSeO_3$	2975	—	$SrSO_4$	2577	2484
$FeSeO_3$	2895	—	$BaSO_4$	2469	2374
$CoSeO_3$	3155	—	$MnSO_4$	2920	2825
$NiSeO_3$	2945	—	$FeSO_4$	2983	2921
$CuSeO_3$	3209	—	$CoSO_4$	3088	2917
$ZnSeO_3$	3167	—	$NiSO_4$	3167	3044
$CdSeO_3$	2962	—	$CuSO_4$	3167	3066
$PbSeO_3$	2669	—	$ZnSO_4$	3100	3006
Selenates			$CdSO_4$	2891	—
Li_2SeO_4	2054	—	$PbSO_4$	2635	2534
Na_2SeO_4	1879	—	$Sn(SO_4)_2$	—	9616
K_2SeO_4	1732	—	Ternary salts		
Rb_2SeO_4	1686	—	Cs_2CuCl_4	1393	—
Cs_2SeO_4	1615	—	Rb_2ZnCl_4	1529	—
Cu_2SeO_4	2201	—	Cs_2ZnCl_4	1492	—
Ag_2SeO_4	2033	—	Rb_2ZnBr_4	1498	—
Tl_2SeO_4	1766	—	Cs_2ZnBr_4	1454	—
Hg_2SeO_4	2163	—	$(Me_4N)_2ZnBr_4$	1364	—
$BeSeO_4$	3448	—	Cs_2ZnI_4	1386	—
$MgSeO_4$	2895	—	$CsGaCl_4$	494	—
$CaSeO_4$	2632	—	$NaAlCl_4$	556	—
$SrSeO_4$	2489	—	$CsAlCl_4$	486	—
$BaSeO_4$	2385	—	$NaFeCl_4$	492	—
$RaSeO_4$	2364	—	Rb_2CoCl_4	1447	—
$GdSeO_4$	2753	—	Cs_2CoCl_4	1391	—
$MnSeO_4$	2837	—	K_2PtCl_4	1574	1550
$FeSeO_4$	3008	—	Cs_2GeF_6	1573	—
$CoSeO_4$	2912	—	$(NH_4)_2GeF_6$	1657	—
$NiSeO_4$	3079	—	Cs_2GeCl_6	1375	1375
$CuSeO_4$	3104	—	K_2HfCl_6	1345	1461
$ZnSeO_4$	3021	—	K_2IrCl_6	1442	1440
$CdSeO_4$	2833	—	$(NH_4)_2IrCl_6$	1442	—
$HgSeO_4$	2845	—	Na_2MoCl_6	1526	—
$PbSeO_4$	2561	—	K_2MoCl_6	1418	1433
Sulphides			Rb_2MoCl_6	1399	1399
Li_2S	2464	2472	Cs_2MoCl_6	1347	1347
Na_2S	2192	2203	K_2NbCl_6	1375	1398
K_2S	1979	(2052)	Rb_2NbCl_6	1371	1385
Rb_2S	1929	1949	Cs_2NbCl_6	1381	1344
Cs_2S	1892	1850	K_2OsCl_6	1447	1447
$(NH_4)_2S$	2008	(2026)	Cs_2OsCl_6	1409	—
Cu_2S	2786	2865	$(NH_4)_2OsCl_6$	1433	—
Ag_2S	2606	2677	K_2OsBr_6	1396	—
Au_2S	2908	—	K_2PdCl_6	1450	1466
Tl_2S	2298	2258	Rb_2PdCl_6	1449	—
Sulphates			Cs_2PdCl_6	1426	—
Li_2SO_4	2229	2142	$(NH_4)_2PdCl_6$	1481	—
Na_2SO_4	1827	1938	Rb_2PbCl_6	1343	1343
K_2SO_4	1700	1796	Cs_2PbCl_6	1344	—

Substance	Calc. U_{POT} (kJ mol^{-1})	U_{POT}^{BFHC} (kJ mol^{-1})	Substance	Calc. U_{POT} (kJ mol^{-1})	U_{POT}^{BFHC} (kJ mol^{-1})
$(NH_4)_2PbCl_6$	1355	—	Cs_2TeI_6	1246	—
K_2PtCl_6	1468	—	K_2TiCl_6	1412	1443
Rb_2PtCl_6	1464	—	Rb_2TiCl_6	1415	1425
Cs_2PtCl_6	1444	—	Cs_2TiCl_6	1402	1370
$(NH_4)_2PtCl_6$	1468	—	Tl_2TiCl_6	1560	1553
Tl_2PtCl_6	1546	—	$(NH_4)_2TiCl_6$	1413	—
Ag_2PtCl_6	1773	—	K_2TiBr_6	1379	1379
$BaPtCl_6$	2047	—	Rb_2TiBr_6	1341	1365
K_2PtBr_6	1423	—	Cs_2TiBr_6	1339	1316
Ag_2PtBr_6	1791	—	Na_2UBr_6	1504	—
K_2PtI_6	1421	—	K_2UBr_6	1484	—
K_2ReCl_6	1416	1442	Rb_2UBr_6	1473	—
Rb_2ReCl_6	1414	—	Cs_2UBr_6	1459	—
Cs_2ReCl_6	1398	—	K_2WCl_6	1398	1423
$(NH_4)_2ReCl_6$	1402	1390	Rb_2WCl_6	1397	1434
K_2ReBr_6	1375	1375	Cs_2WCl_6	1392	1366
K_2SiF_6	1670	1628	K_2WBr_6	1408	1408
Rb_2SiF_6	1639	1621	Rb_2WBr_6	1361	1391
Cs_2SiF_6	1604	1498	Cs_2WBr_6	1362	1332
Tl_2SiF_6	1675	—	K_2ZrCl_6	1339	1371
$(NH_4)_2SiF_6$	1657	1727	Rb_2ZrCl_6	1341	—
K_2SnCl_6	1363	1390	Cs_2ZrCl_6	1339	1308
Rb_2SnCl_6	1361	1363	Tetraalkyl ammonium salts		
Cs_2SnCl_6	1358	—	Me_4NCl	566	—
Tl_2SnCl_6	1437	—	Me_4NBr	553	—
$(NH_4)_2SnCl_6$	1370	1334	Me_4NI	544	—
Rb_2SnBr_6	1309	—	Tellurides		
Cl_2SnBr_6	1306	—	Li_2Te	2212	—
$(NH_4)_2SnBr_6$	1319	—	Na_2Te	1997	2095
Rb_2SnI_6	1226	—	K_2Te	1830	—
Cs_2SnI_6	1243	—	Rb_2Te	1837	—
K_2TeCl_6	1318	1320	Cs_2Te	1745	—
Rb_2TeCl_6	1321	—	Cu_2Te	2706	2683
Cs_2TeCl_6	1323	—	Ag_2Te	2607	2600
Tl_2TeCl_6	1392	—	Tl_2Te	2084	2172
$(NH_4)_2TeCl_6$	1318	—	$BeTe$	3319	—
K_2RuCl_6	1451	—	$MgTe$	2878	3081
Rb_2CoF_6	1688	—	$CaTe$	2721	—
Cs_2CoF_6	1632	—	$SrTe$	2599	—
K_2NiF_6	1721	—	$BaTe$	2473	—
Rb_2NiF_6	1688	—	$RaTe$	2481	—
Rb_2SbCl_6	1357	—	$MnTe$	3041	—
Rb_2SeCl_6	1409	—	$FeTe$	3399	—
Cs_2SeCl_6	1397	—	$CoTe$	3429	—
$(NH_4)_2SeCl_6$	1420	—	$NiTe$	3534	—
K_2SeBr_6	1379	—	$CuTe$	3639	—
$(NH_4)_2SeBr_6$	1380	—	$ZnTe$	3416	—
$(NH_4)_2PoCl_6$	1338	—	$CdTe$	3212	—
Cs_2PoBr_6	1286	—	$PbTe$	2930	—
$(NH_4)_2PoBr_6$	1292	—	Thiocyanates		
Cs_2CrF_6	1603	—	$LiCNS$	764	(765)
Rb_2MnF_6	1688	—	$NaCNS$	682	682
Cs_2MnF_6	1620	—	$KCNS$	623	616
K_2MnCl_6	1462	—	$RbCNS$	623	619
Rb_2MnCl_6	1451	—	$CsCNS$	623	568
$(NH_4)_2MnCl_6$	1464	—	NH_4CNS	605	611
Cs_2TeBr_6	1306	—	$Ca(CNS)_2$	2184	2118
$(NH_4)_2TeBr_6$	1294	—	$Sr(CNS)_2$	2063	1957

Table 1 (continued)

Substance	Calc. U_{POT} (kJ mol^{-1})	U_{POT}^{BFHC} (kJ mol^{-1})	Substance	Calc. U_{POT} (kJ mol^{-1})	U_{POT}^{BFHC} (kJ mol^{-1})
Ba(CNS)$_2$	1979	1852	Vanadates		
Mn(CNS)$_2$	2280	2351	LiVO$_3$	810	—
Zn(CNS)$_2$	2335	2560	NaVO$_3$	761	—
Cd(CNS)$_2$	2201	2374	KVO$_3$	686	—
Hg(CNS)$_2$	2146	2492	RbVO$_3$	657	—
Sn(CNS)$_2$	2117	2142	CsVO$_3$	628	—
Pb(CNS)$_2$	2058	—			

Table 2
ANCILLARY THERMODYNAMIC DATA

Substance	State	$\Delta H_f°$ (kJ mol^{-1})	Substance	State	$\Delta H_f°$ (kJ mol^{-1})
Li$^+$	g	687.16	OsCl$_6^{-2}$	g	−752
Na$^+$	g	609.84	IrCl$_6^{-2}$	g	−785
K$^+$	g	514.19	PdCl$_6^{-2}$	g	−749
Cs$^+$	g	452.3	PtCl$_6^{-2}$	g	−774
O$_2^-$	g	−74	PtBr$_6^{-2}$	g	−645
CN$^-$	g	41	SiF$_6^{-2}$	g	−2207
NH$_4^+$	g	630.2	GeCl$_6^{-2}$	g	−981
C$_2^{-2}$	g	918	SuCl$_6^{-2}$	g	−1156
O$_2^{-2}$	g	553	PbCl$_6^{-2}$	g	−940
NO$_3^-$	g	−320.1	TeCl$_6^{-2}$	g	−902
IO$_2$F$_2^-$	g	−693	NO$_2^-$	g	−202
IO$_3^-$	g	−208	LiO$_2$	c	−259.4
ClO$_3^-$	g	−200	NaO$_2$	c	−260.7
BrO$_3^-$	g	−145	KO$_2$	c	−284.5
HSO$_4^-$	g	(−1012)	CsO$_2$	c	−289.5
HCO$_3^-$	g	−738	Li$_2$O$_2$	c	−632.6
NH$_2$CH$_2$CO$_2^-$	g	−564	Na$_2$O$_2$	c	−531.2
N^{-3}	g	(2588)	K$_2$O$_2$	c	−495.8
CO$_3^{-2}$	g	−321	Cs$_2$O$_2$	c	−402.5
PO$_4^{-3}$	g	(291)	ThO$_2$	c	−1230.5
AsO$_4^{-3}$	g	(289)	MgO$_2$	c	−623.0
GeO$_4^{-4}$	g	(1460)	NaIO$_2$F$_2$	c	−848.7
CH$_3$COO$^-$	g	−554	KIO$_2$F$_2$	c	−876.6
ClO$_4^-$	g	−344	RbIO$_2$F$_2$	c	−875.3
NO$_2^-$	g	−219	NH$_4$IO$_2$F$_2$	c	−747.7
SeO$_3^{-2}$	g	(−249)	LiBrO$_3$	c	−356.4
HCO$_2^-$	g	−463	CsBrO$_3$	c	−374.5
BH$_4^-$	g	−96	CsHCO$_3$	c	−995.6
HS$^-$	g	−120	Ca(HCO$_3$)$_2$	c	(−1950)
TiCl$_6^{-2}$	g	−1330	Sr(HCO$_3$)$_2$	c	(−1954)
TiBr$_6^{-2}$	g	−1142	Ba(HCO$_3$)$_2$	c	(−1971)
ZrCl$_6^{-2}$	g	−1526	CH$_3$COONH$_4$	c	−618.4
HfCl$_6^{-2}$	g	−1640	CH$_3$COOLi	c	−748.9
NbCl$_6^{-2}$	g	−1224	CH$_3$COORb	c	−720.9
TaCl$_6^{-2}$	g	−1275	LiClO$_4$	c	−380.7
MoCl$_6^{-2}$	g	−1070	CsClO$_4$	c	(−435.1)
WCl$_6^{-2}$	g	−985	LiH	c	−90.62
WBr$_6^{-2}$	g	−705	NaH	c	−56.4
ReCl$_6^{-2}$	g	−919	KH	c	−57.8
ReBr$_6^{-2}$	g	−689	CsH	c	−49.9

Table 2
ANCILLARY THERMODYNAMIC DATA (continued)

Substance	State	ΔH_f° (kJ mol^{-1})	Substance	State	ΔH_f° (kJ mol^{-1})
LiN$_3$	c	10.8	Ca(ClO$_3$)$_2$	c	−757
NaN$_3$	c	21.25	Sr(ClO$_3$)$_2$	c	−766
KN$_3$	c	−1.38	ScN	c	−284.5
RbN$_3$	c	−0.29	Li$_2$CO$_3$	c	−1216.0
CsN$_3$	c	−9.92	Cs$_2$CO$_3$	c	−1118.8
TlN$_3$	c	233.4	NiCO$_3$	c	−689.1
AgN$_3$	c	310.3	CuCO$_3$	c	−595.0
CsHS	c	−263.2	SnCO$_3$	c	(−740.6)
Ca(HS)$_2$	c	(−481)	HCO$_2$Li	c	−649.3
Ba(HS)$_2$	c	(−531.4)	HCO$_2$Rb	c	−656.0
LiHS	c	−254.4	TiH	c	−130.5
K$_2$TiCl$_6$	c	−1747	ZrH	c	−173.5
Rb$_2$TiCl$_6$	c	−1767	VH	c	(−32)
Cs$_2$TiCl$_6$	c	−1797	NbH	c	(−100)
Tl$_2$TiCl$_6$	c	−1330	PdH	c	(−37)
K$_2$TiBr$_6$	c	−1493	CuH	c	(−21.4)
Rb$_2$TiBr$_6$	c	−1517	LaH$_2$	c	−230.1
Cs$_2$TiBr$_6$	c	−1553	CeH$_2$	c	−141.8
K$_2$ZrCl$_6$	c	−1932	PrH$_2$	c	−200.0
Cs$_2$ZrCl$_6$	c	−1992	NdH$_2$	c	−187.4
K$_2$HfCl$_6$	c	−1957	SmH$_2$	c	−187.4
K$_2$NbCl$_6$	c	−1594	GdH$_2$	c	−196.2
Rb$_2$NbCl$_6$	c	−1619	CsOH	c	−406.7
Cs$_2$NbCl$_6$	c	−1663	Ti(OH)$_2$	c	(−778.0)
K$_2$ReCl$_6$	c	−1333	LiOH	c	−487.2
(NH$_4$)$_2$ReCl$_6$	c	−1056	K$_2$TaCl$_6$	c	−1648
K$_2$ReBr$_6$	c	−1036	Rb$_2$TaCl$_6$	c	−1669
K$_2$OsCl$_6$	c	−1171	Cs$_2$TaCl$_6$	c	−1711
K$_2$IrCl$_6$	c	−1197	Na$_2$MoCl$_6$	c	−1376
K$_2$SiF$_6$	c	−2807	K$_2$MoCl$_6$	c	−1475
Rb$_2$SiF$_6$	c	−2838	Rb$_2$MoCl$_6$	c	−1479
Cs$_2$SiF$_6$	c	−2801	Cs$_2$MoCl$_6$	c	−1512
(NH$_4$)$_2$SiF$_6$	c	−2681	K$_2$WCl$_6$	c	−1380
Rb$_2$PbCl$_6$	c	−1293	Rb$_2$WCl$_6$	c	−1429
Mn(CN)$_2$	c	60.7	Cs$_2$WCl$_6$	c	−1446
RbCN	c	(−108.8)	K$_2$WBr$_6$	c	−1085
CsCN	c	(−108.7)	Rb$_2$WBr$_6$	c	−1106
LiCN	c	−121	Cs$_2$WBr$_6$	c	−1132
Sr(CN)$_2$	c	(−205)	K$_2$PdCl$_6$	c	−1187
LiNO$_3$	c	−482.3	Ag$_2$PtCl$_6$	c	−527
RbNO$_3$	c	−489.7	BaPtCl$_6$	c	−1180
CsNO$_3$	c	−494.2	K$_2$PtBr$_6$	c	−1040
Fe(NO$_3$)$_2$	c	(−447.7)	Ag$_2$PtBr$_6$	c	−398
Sn(NO$_3$)$_2$	c	−456.1	Cs$_2$GeCl$_6$	c	−1451
NaBF$_4$	c	−1774	K$_2$SnCl$_6$	c	−1518
CsIO$_3$	c	−526.3	Rb$_2$SnCl$_6$	c	−1529
CsClO$_3$	c	−395.8	(NH$_4$)$_2$SnCl$_6$	c	−1237
Mg(ClO$_3$)$_2$	c	(−523)	Rb$_2$TeCl$_6$	c	−1237

THE MADELUNG CONSTANT AND CRYSTAL LATTICE ENERGY

If U is the crystal lattice energy and M is the Madelung constant, then[a]

$$U = \frac{N M z_i z_j\; e^2}{r}(1 - 1/n)$$

Substance	Ion type	Crystal form[b]	M
Sodium chloride, NaCl	M^+, X^-	FCC	1.74756
Cesium chloride, CsCl	M^+, X^-	BCC	1.76267
Calcium chloride, $CaCl_2$	M^{++}, $2X^-$	Cubic	2.365
Calcium fluoride (fluorite), CaF_2	M^{++}, $2X^-$	Cubic	2.51939
Cadmium chloride, $CdCl_2$	M^{++}, $2X^-$	Hexagonal	2.244[c]
Cadmium iodide (α), CdI_2	M^{++}, $2X^-$	Hexagonal	2.355[c]
Magnesium fluoride, MgF_2	M^{++}, $2X^-$	Tetragonal	2.381[c]
Cuprous oxide (cuprite), Cu_2O	$2M^+$, X^{--}	Cubic	2.22124
Zinc oxide, ZnO	M^{++}, X^{--}	Hexagonal	1.4985[c]
Sphalerite (zinc blende), ZnS	M^{++}, X^{--}	FCC	1.63806
Wurtzite, ZnS	M^{++}, X^{--}	Hexagonal	1.64132[c]
Titanium dioxide (anatase), TiO_2	M^{4+}, $2X^{--}$	Tetragonal	2.400[c]
Titanium dioxide (rutile), TiO_2	M^{4+}, $2X^{--}$	Tetragonal	2.408[c]
β-Quartz, SiO_2	M^{4+}, $2X^{--}$	Hexagonal	2.2197[c]
Corundum, Al_2O_3	$2M^{3+}$, $3X^{--}$	Rhombohedral	4.1719

[a] N is Avogadro's number, z_i and z_j are the integral charges on the ions(in units of e), and e is the charge on the electron in electrostatic units ($e = 4.803 \times 10^{-10}$ esu). r is the shortest distance between cation-anion pairs in centimeters. Then U is in ergs (1 erg = 10^{-7} J).

[b] FCC = face centered cubic; BCC = body centered cubic.

[c] For tetragonal and hexagonal crystals the value of M depends on the details of the lattice parameters.

The Born Exponent, n is:

Ion type	n
He, Li^+	5
Ne, Na^+, F^-	7
Ar, K^+, Cu^+, Cl^-	9
Kr, Rb^+, Ag^+, Br^-	10
Xe, Cs^+, Au^+, I^-	12

For a crystal with a mixed-ion type, an average of the values of n in this table is to be used (6 for LiF, for example).

ELASTIC CONSTANTS OF SINGLE CRYSTALS
H. P. R. Frederikse

This table gives selected values of elastic constants for single crystals. The values believed most reliable were selected from the original literature. The substances are arranged by crystal system and, within each system, alphabetically by name. A reference to the original literature is given for each value; a useful compilation of published values from many sources may be found in Reference 1 below.

Data are given for the single-crystal density and for the elastic constants c_{ij}, in units of 10^{11} N/m^2, which is equivalent to 10^{12} dyn/cm^2.

GENERAL REFERENCES

1. Simmons, G., and Wang, H., *Single Crystal Elastic Constants and Calculated Aggregate Properties: A Handbook, Second Edition,* The MIT Press, Cambridge, MA, 1971.
2. Gray, D.E., Ed., *American Institute of Physics Handbook, Third Edition*, McGraw-Hill, New York, 1972.

CUBIC CRYSTALS

Name	Formula	ρ/g cm^{-3}	T/K	Ref.	C_{11}	C_{12}	C_{44}
Aluminum	Al	2.6970	298	1	1.0675	0.6041	0.2834
Aluminum antimonide	AlSb	4.3600	300	2	0.8939	0.4427	0.4155
Ammonium bromide	NH$_4$Br	2.4314	300	3	0.3414	0.0782	0.0722
Ammonium chloride	NH$_4$Cl	1.5279	290	4	0.3814	0.0866	0.0903
Argon	Ar	1.7710	4.2	5	0.0529	0.0135	0.0159
Barium fluoride	BaF$_2$	4.8860	298	6	0.9199	0.4157	0.2568
Barium nitrate	Ba(NO$_3$)$_2$	3.2560	293	7	0.2925	0.2065	0.1277
Calcium fluoride	CaF$_2$	3.810	298	8	1.6420	0.4398	0.8406
Calcium telluride	CaTe	5.8544	298	9	0.5351	0.3681	0.1994
Cesium	Cs	1.9800	78	10	0.0247	0.0206	0.0148
Cesium bromide	CsBr	4.4560	298	11	0.3063	0.0807	0.0750
Cesium chloride	CsCl	3.9880	298	11	0.3644	0.0882	0.0804
Cesium iodide	CsI	4.5250	298	11	0.2446	0.0661	0.0629
Chromite	FeCr$_2$O$_4$	4.4500	RT	12	3.2250	1.4370	1.1670
Chromium	Cr	7.20	298	13	3.398	0.586	0.990
Cobalt oxide	CoO	6.44	298	14	2.6123	1.4699	0.8300
Cobalt zinc ferrite	CoZnFeO$_2$	5.43	303	12	2.660	1.530	0.780
Copper	Cu	8.932	298	15	1.683	1.221	0.757
Gallium antimonide	GaSb	5.6137	298	16	0.8839	0.4033	0.4316
Gallium arsenide	GaAs	5.3169	298	17	1.1877	0.5372	0.5944
Gallium phosphide	GaP	4.1297	300	18	1.4120	0.6253	0.7047
Garnet (yttrium-iron)	Y$_3$Fe$_2$(FeO$_4$)$_3$	5.17	298	19	2.680	1.106	0.766
Germanium	Ge	5.313	298	20	1.2835	0.4823	0.6666
Gold	Au	19.283	296.5	21	1.9244	1.6298	0.4200
Indium antimonide	InSb	5.7890	298	22	0.6720	0.3670	0.3020
Indium arsenide	InAs	5.6720	293	23	0.8329	0.4526	0.3959
Indium phosphide	InP	4.78	RT	24	1.0220	0.5760	0.4600
Iridium	Ir	22.52	300	25	5.80	2.42	2.56
Iron	Fe	7.8672	298	26	2.26	1.40	1.16
Lead	Pb	11.34	296	27	0.4966	0.4231	0.1498
Lead fluoride	PbF$_2$	7.79	300	28	0.8880	0.4720	0.2454
Lead nitrate	Pb(NO$_3$)$_2$	4.547	293	29	0.3729	0.2765	0.1347
Lead telluride	PbTe	8.2379	303.2	30	1.0795	0.0764	0.1343
Lithium	Li	0.5326	298	31	0.1350	0.1144	0.0878
Lithium bromide	LiBr	3.47	RT	32	0.3940	0.1880	0.1910
Lithium chloride	LiCl	2.068	295	33	0.4927	0.2310	0.2495
Lithium fluoride	LiF	2.638	RT	34	1.1397	0.4767	0.6364
Lithium iodide	LiI	4.061	RT	32	0.2850	0.1400	0.1350
Magnesium oxide	MgO	3.579	298	20	2.9708	0.9536	1.5613
Magnetite	Fe$_3$O$_4$	5.18	RT	32	2.730	1.060	0.971
Manganese oxide	MnO	5.39	298	35	2.23	1.20	0.79
Mercury telluride	HgTe	8.079	290	36	0.548	0.381	0.204
Molybdenum	Mo	10.2284	273	37	4.637	1.578	1.092

Name	Formula	ρ/g cm^{-3}	T/K	Ref.	C_{11}	C_{12}	C_{44}
Nickel	Ni	8.91	298	15	2.481	1.549	1.242
Niobium	Nb	8.578	300	38	2.4650	1.3450	0.2873
Palladium	Pd	12.038	300	39	2.2710	1.7604	0.7173
Platinum	Pt	21.50	300	40	3.4670	2.5070	0.7650
Potassium	K	0.851	295	41	0.0370	0.0314	0.0188
Potassium bromide	KBr	2.740	298	11	0.3468	0.0580	0.0507
Potassium chloride	KCl	1.984	298	11	0.4069	0.0711	0.0631
Potassium cyanide	KCN	1.553	RT	32	0.1940	0.1180	0.0150
Potassium fluoride	KF	2.480	295	33	0.6490	0.1520	0.1232
Potassium iodide	KI	3.128	300	42	0.2710	0.0450	0.0364
Pyrite	FeS$_2$	5.016	RT	43	3.818	0.310	1.094
Rubidium	Rb	1.58	170	44	0.0296	0.0250	0.0171
Rubidium bromide	RbBr	3.350	300	45	0.3152	0.0500	0.0380
Rubidium chloride	RbCl	2.797	300	45	0.3624	0.0612	0.0468
Rubidium iodide	RbI	3.551	300	45	0.2556	0.0382	0.0278
Silicon	Si	2.331	298	46	1.6578	0.6394	0.7962
Silver	Ag	10.50	300	47	1.2399	0.9367	0.4612
Silver bromide	AgBr	5.585	300	48	0.5920	0.3640	0.0616
Sodium	Na	0.971	299	49	0.0739	0.0622	0.0419
Sodium bromate	NaBrO$_3$	3.339	RT	32	0.5450	0.1910	0.1500
Sodium bromide	NaBr	3.202	300	33	0.3970	0.1001	0.0998
Sodium chlorate	NaClO$_3$	2.485	RT	50	0.4920	0.1420	0.1160
Sodium chloride	NaCl	2.163	298	11	0.4947	0.1288	0.1287
Sodium fluoride	NaF	2.804	300	51	0.9700	0.2380	0.2822
Sodium iodide	NaI	3.6689	300	52	0.3007	0.0912	0.0733
Spinel	MgAl$_2$O$_4$	3.6193	298	53	2.9857	1.5372	1.5758
Strontium fluoride	SrF$_2$	4.277	300	54	1.2350	0.4305	0.3128
Strontium nitrate	Sr(NO$_3$)$_2$	2.989	293	29	0.4255	0.2921	0.1590
Strontium oxide	SrO	4.99	300	55	1.601	0.435	0.590
Strontium titanate	SrTiO$_3$	5.123	RT	56	3.4817	1.0064	4.5455
Tantalum	Ta	16.626	298	57	2.6023	1.5446	0.8255
Tantalum carbide	TaC	14.65	RT	58	5.05	0.73	0.79
Thallium bromide	TlBr	7.4529	298	59	0.3760	0.1458	0.0757
Thorium	Th	11.694	300	60	0.7530	0.4890	0.4780
Thorium oxide	ThO$_2$	9.991	298	61	3.670	1.060	0.797
Tin telluride	SnTe	6.445	300	62	1.1250	0.0750	0.1172
Titanium carbide	TiC	4.940	RT	107	5.00	1.13	1.75
Tungsten	W	19.257	297	64	5.2239	2.0437	1.6083
Uranium carbide	UC	13.63	300	65	3.200	0.850	0.647
Uranium dioxide	UO$_2$	10.97	298	66	3.960	1.210	0.641
Vanadium	V	6.022	300	67	2.287	1.190	0.432
Zinc selenide	ZnSe	5.262	298	68	0.8096	0.4881	0.4405
Zinc sulfide	ZnS	4.088	298	68	1.0462	0.6534	0.4613
Zinc telluride	ZnTe	5.636	298	68	0.7134	0.4078	0.3115
Zirconium carbide	ZrC	6.606	298	63	4.720	0.987	1.593

TETRAGONAL CRYSTALS

Name	Formula	ρ/g cm^{-3}	T/K	Ref.	C_{11}	C_{12}	C_{13}	C_{16}	C_{33}	C_{44}	C_{66}
Ammonium dihydrogen arsenate (ADA)	$NH_4H_2AsO_4$	2.3110	298	69	0.6747	-0.106	0.1652		0.3022	0.0685	0.0639
Ammonium dihydrogen phosphate (ADP)	$NH_4H_2PO_4$	1.8030	293	69	0.6200	-0.050	0.1400		0.3000	0.0910	0.0610
Barium titanate	$BaTiO_3$	5.9988	298	70	2.7512	1.7897	1.5156		1.6486	0.5435	1.1312
Calcium molybdate	$CaMoO_4$	4.255	298	79	1.447	0.664	0.466	0.134	1.265	0.369	0.451
Indium	In	7.300	RT	71	0.4450	0.3950	0.4050		0.4440	0.0655	0.1220
Magnesium fluoride	MgF_2	3.177	RT	72	1.237	0.732	0.536		1.770	0.552	0.978
Nickel sulfate hexahydrate	$NiSO_4 \cdot 6H_2O$	2.070	RT	73	0.3209	0.2315	0.0209		0.2931	0.1156	0.1779
Potassium dihydrogen arsenate (KDA)	KH_2AsO_4	2.867	RT	12	0.530	-0.060	-0.020		0.370	0.120	0.070
Potassium dihydrogen phosphate (KDP)	KH_2PO_4	2.388	RT	71	0.7140	-0.049	0.1290		0.5620	0.1270	0.0628
Rubidium dihydrogen phosphate (RDP)	RbH_2PO_4	2.800	298	74	0.5562	-0.064	0.0279		0.4398	0.1142	0.0350
Rutile	TiO_2	4.260	298	75	2.7143	1.7796	1.4957		4.8395	1.2443	1.9477
Tellurium oxide	TeO_2	5.99	RT	76	0.5320	0.4860	0.2120		1.0850	0.2440	0.5520
Tin (white)	Sn	7.29	288	77	0.7529	0.6156	0.4400		0.9552	0.2193	0.2336
Zircon	$ZrSiO_4$	4.70	RT	78	2.585	1.791	1.542		3.805	0.733	1.113

ORTHORHOMBIC CRYSTALS

Name	Formula	ρ/g cm^{-3}	T/K	Ref.	C_{11}	C_{12}	C_{13}	C_{22}	C_{23}	C_{33}	C_{44}	C_{55}	C_{66}
Acenaphthene	$C_{12}H_{10}$	1.220	293	80	0.1380	0.0210	0.0410	0.1262	0.0460	0.1117	0.0265	0.0290	0.0185
Ammonium sulfate	$(NH_4)_2SO_4$	1.774	293	81	0.3607	0.1651	0.1580	0.2981	0.1456	0.3534	0.1025	0.0717	0.0974
Aragonite	$CaCO_3$	2.93	RT	82	1.5958	0.3663	0.0197	0.8697	0.1597	0.8503	0.4132	0.2564	0.4274
Barite	$BaSO_4$	4.40	RT	82	0.8941	0.4614	0.2691	0.7842	0.2676	1.0548	0.1190	0.2874	0.2778
Benzene	C_6H_6	1.061	250	83	0.0614	0.0352	0.0401	0.0656	0.0390	0.0583	0.0197	0.0378	0.0153
Benzophenone	$(C_6H_5)_2CO$	1.219	RT	32	0.1070	0.0550	0.0169	0.1000	0.0321	0.0710	0.0203	0.0155	0.0353
Bronzite	$(MgFe)SiO_3$	3.38	RT	78	1.876	0.686	0.605	1.578	0.561	2.085	0.700	0.592	0.544
Calcium sulfate	$CaSO_4$	2.962	RT	84	0.9382	0.1650	0.1520	1.845	0.3173	1.1180	0.3247	0.2653	0.0926
Celestite	$SrSO_3$	3.96	RT	12	1.044	0.773	0.605	1.061	0.619	1.286	0.135	0.279	0.266
Cesium sulfate	Cs_2SO_4	4.243	293	81	0.4490	0.1958	0.1815	0.4283	0.1800	0.3785	0.1326	0.1319	0.1323
Fosterite	Mg_2SiO_4	3.224	298	85	3.2848	0.6390	0.6880	1.9980	0.7380	2.3530	0.6515	0.8120	0.8088
Iodic acid	HIO_3	4.630	RT	73	0.3030	0.1194	0.1169	0.5448	0.0548	0.4359	0.1835	0.2193	0.1736
Lithium ammonium tartrate	$LiNH_4C_4H_4O_6 \cdot 4H_2O$	1.71	RT	12	0.3864	0.1655	0.0875	0.5393	0.2007	0.3624	0.1190	0.0667	0.2326
Magnesium sulfate heptahydrate	$MgSO_4 \cdot 7H_2O$	1.68	RT	86	0.325	0.174	0.182	0.288	0.182	0.315	0.078	0.156	0.090
Natrolite	$(Na,Al)SiO_3$	2.25	RT	78	0.716	0.261	0.297	0.632	0.297	1.378	0.196	0.248	0.423
Nickel sulfate heptahydrate	$NiSO_4 \cdot 7H_2O$	1.948	RT	86	0.353	0.198	0.201	0.311	0.201	0.335	0.091	0.172	0.099
Olivine	$(MgFe)SiO_4$	3.324	RT	87	3.240	0.590	0.790	1.980	0.780	2.490	0.667	0.810	0.793
Potassium pentaborate	$KB_5O_8 \cdot 4H_2O$	1.74	RT	71	0.582	0.229	0.174	0.359	0.231	0.255	0.164	0.046	0.057
Potassium sulfate	K_2SO_4	2.665	293	81	0.5357	0.1999	0.2095	0.5653	0.1990	0.5523	0.195	0.1879	0.1424
Rochelle salt	$NaK(C_4H_4O_6) \cdot 4H_2O$	1.79	RT	71	0.255	0.141	0.116	0.381	0.146	0.371	0.134	0.032	0.098
Rubidium sulfate	Rb_2SO_4	3.621	293	81	0.5029	0.1965	0.1999	0.5098	0.1925	0.4761	0.1626	0.1589	0.1407
Sodium ammonium tartrate	$NaNH_4C_4H_4O_6 \cdot 4H_2O$	1.587	RT	12	0.3685	0.2725	0.3083	0.5092	0.3472	0.5541	0.1058	0.0303	0.0870
Sodium tartrate	$Na_2C_4H_4O_6 \cdot 2H_2O$	1.794	RT	12	0.461	0.286	0.320	0.547	0.352	0.665	0.124	0.031	0.098
Strontium formate dihydrate	$Sr(CHO_2)_2 \cdot 2H_2O$	2.25	RT	12	0.4391	0.1037	-0.149	0.3484	-0.014	0.3746	0.1538	0.1075	0.1724
Sulfur	S	2.07	RT	12	0.240	0.133	0.171	0.205	0.159	0.483	0.043	0.087	0.076
Thallium sulfate	$TlSO_4$	6.776	293	81	0.4106	0.2573	0.2288	0.3885	0.2174	0.4268	0.1125	0.1068	0.0751
Topaz	$Al_2SiO_3(OH,F)_2$	3.52	RT	82	2.8136	1.2582	0.8464	3.8495	0.8815	2.9452	1.0811	1.3298	1.3089
Uranium (alpha)	U	19.0453	293	88	2.1486	0.4622	0.2176	1.9983	1.0764	2.6763	1.2479	0.7379	0.7454
Zinc sulfate heptahydrate	$ZnSO_4 \cdot 7H_2O$	1.970	RT	86	0.3320	0.1720	0.2000	0.2930	0.1980	0.3200	0.0780	0.1530	0.0830

MONOCLINIC CRYSTALS

Name	Formula	ρ/g cm^{-3}	T/K	Ref.	C_{11}	C_{12}	C_{13}	C_{15}	C_{22}
Aegirine	$(NaFe)Si_2O_6$	3.50	RT	89	1.858	0.685	0.707	0.098	1.813
Anthracene	$C_{14}H_{10}$	1.258	RT	90	0.0852	0.0672	0.0590	-0.0192	0.1170
Cobalt sulfate heptahydrate	$CoSO_4\cdot 7H_2O$	1.948	RT	86	0.335	0.205	0.158	0.016	0.378
Diopside	$(CaMg)Si_2O_6$	3.31	RT	91	2.040	0.884	0.0883	-0.193	1.750
Dipotassium tartrate	$KHC_4H_4O_6$	1.97	RT	12	0.4294	0.1399	0.3129	-0.0105	0.3460
Feldspar (microceine)	$KAlSi_3O_8$	2.56	RT	92	0.664	0.438	0.259	-0.033	1.710
Ferrous sulfate heptahydrate	$FeSO_4\cdot 7H_2O$	1.898	RT	86	0.349	0.208	0.174	-0.020	0.376
Lithium sulfate monohydrate	$Li_2SO_4\cdot H_2O$	2.221	RT	32	0.5250	0.1715	0.1730	-0.0196	0.5060
Naphthalene	$C_{10}H_8$	1.127	RT	93	0.0780	0.0445	0.0340	-0.006	0.0990
Potassium tartrate	$K_2C_4H_4O_6$	1.987	RT	32	0.3110	0.1720	0.1690	0.0287	0.3900
Sodium thiosulfate	$Na_2S_2O_3$	1.7499	RT	12	0.3323	0.1814	0.1875	0.0225	0.2953
Stilbene	$(C_6H_5CH)_2$	1.60	RT	94	0.0930	0.0570	0.0670	-0.003	0.0920
Triglycine sulfate (TGS)	$(NH_2CH_2COOH)_3\cdot$ H_2SO_4	1.68	RT	32	0.4550	0.1720	0.1980	-0.030	0.3210

Name	C_{23}	C_{25}	C_{33}	C_{35}	C_{44}	C_{46}	C_{55}	C_{66}
Aegirine	0.626	0.094	2.344	0.214	0.692	0.077	0.510	0.474
Anthracene	0.0375	-0.0170	0.1522	-0.0187	0.0272	0.0138	0.0242	0.0399
Cobalt sulfate heptahydrate	0.158	-0.018	0.371	-0.047	0.060	0.016	0.058	0.101
Diopside	0.482	-0.196	2.380	-0.336	0.675	-0.113	0.588	0.705
Dipotassium tartrate	0.1173	0.0176	0.6816	0.0294	0.0961	-0.0044	0.1270	0.0841
Feldspar (microceine)	0.192	-0.148	1.215	-0.131	0.143	-0.015	0.238	0.361
Ferrous sulfate heptahydrate	0.172	-0.019	0.360	-0.014	0.064	0.001	0.056	0.096
Lithium sulfate monohydrate	0.0368	0.0571	0.5400	-0.0254	0.1400	-0.0054	0.1565	0.2770
Naphthalene	0.0230	-0.0270	0.1190	0.0290	0.0330	-0.0050	0.0210	0.0415
Potassium tartrate	0.1330	0.0182	0.5540	0.0710	0.0870	0.0072	0.1040	0.0826
Sodium thiosulfate	0.1713	0.0983	0.4590	-0.0678	0.0569	-0.0268	0.1070	0.0598
Stilbene	0.0485	-0.005	0.0790	-0.005	0.0325	0.0050	0.0640	0.0245
Triglycine sulfate (TGS)	0.2080	-0.0036	0.2630	-0.0500	0.0950	-0.0026	0.1110	0.0620

HEXAGONAL CRYSTALS

Name	Formula	ρ/g cm^{-3}	T/K	Ref.	C_{11}	C_{12}	C_{13}	C_{33}	C_{55}
Apatite	$Ca_5(PO_4)_3(OH,F,Cl)$	3.218	RT	12	1.667	0.131	0.655	1.396	0.663
Beryl	$Be_3Al_2Si_6O_{18}$	2.68	RT	12	2.800	0.990	0.670	2.480	0.658
Beryllium	Be	1.8477	300	95	2.923	0.267	0.140	3.364	1.625
Beryllium oxide	BeO	3.01	RT	96	4.70	1.68	1.19	4.94	1.53
Cadmium	Cd	8.652	300	97	1.1450	0.3950	0.3990	0.5085	0.1985
Cadmium selenide	CdSe	5.655	298	68	0.7046	0.4516	0.3930	0.8355	0.1317
Cadmium sulfide	CdS	4.824	298	98	0.8431	0.5208	0.4567	0.9183	0.1458
Cobalt	Co	8.836	298	99	3.071	1.650	1.027	3.581	0.755
Dysprosium	Dy	8.560	298	100	0.7466	0.2616	0.2233	0.7871	0.2427
Erbium	Er	9.064	298	100	0.8634	0.3050	0.2270	0.8554	0.2809
Gadolinium	Gd	7.888	298	101	0.6667	0.2499	0.2132	0.7191	0.2089
Hafnium	Hf	12.727	298	102	1.881	0.772	0.661	1.969	0.557
Ice	H_2O(solid)	0.920	250	103	0.1410	0.0660	0.0624	0.1515	0.0288
Indium	In	7.2788	300	104	0.4535	0.4006	0.4151	0.4515	0.0651
Magnesium	Mg	1.7364	298	105	0.5950	0.2612	0.2180	0.6155	0.1635
Rhenium	Re	21.024	298	100	6.1820	2.7530	2.0780	6.8350	1.6060
Ruthenium	Ru	12.3615	298	100	5.6260	1.8780	1.6820	6.2420	1.8060
Thallium	Tl	11.560	300	106	0.4080	0.3540	0.2900	0.5280	0.0726
Titanium	Ti	4.5063	298	102	1.6240	0.9200	0.6900	1.8070	0.4670
Titanium diboride	TiB_2	4.95	RT	107	6.90	4.10	3.20	4.40	2.50
Yttrium	Y	4.472	300	108	0.7790	0.2850	0.2100	0.7690	0.2431
Zinc	Zn	7.134	295	109	1.6368	0.3640	0.5300	0.6347	0.3879
Zinc oxide	ZnO	5.6760	298	110	2.0970	1.2110	1.0510	2.1090	0.4247
Zinc sulfide	ZnS	4.089	298	96	1.2420	0.6015	0.4554	1.4000	0.2864
Zirconium	Zr	6.505	298	102	1.434	0.728	0.653	1.648	0.320

TRIGONAL CRYSTALS

Name	Formula	ρ/g cm^{-3}	T/K	Ref.	C_{11}	C_{12}	C_{13}	C_{14}	C_{33}	C_{44}
Aluminum oxide	Al_2O_3	3.986	300	111	4.9735	1.6397	1.1220	-0.2358	4.9911	1.4739
Aluminum phosphate	$AlPO_4$	2.556	RT	73	1.0503	0.2934	0.6927	-0.1271	1.3353	0.2314
Antimony	Sb	6.70	295	112	1.0130	0.3450	0.2920	0.2090	0.4500	0.3930
Bismuth	Bi	9.80	295	112	0.6370	0.2490	0.2470	0.0717	0.3820	0.1123
Calcite	$CaCO_3$	2.712	300	113	1.4806	0.5578	0.5464	-0.2058	0.8557	0.3269
Hematite	Fe_2O_3	5.240	RT	82	2.4243	0.5464	0.1542	-0.1247	2.2734	0.8569
Lithium niobate	$LiNbO_3$	4.70	RT	114	2.030	0.530	0.750	0.090	2.450	0.600
Lithium tantalate	$LiTaO_3$	7.45	RT	114	2.330	0.470	0.800	-0.110	2.750	0.940
Quartz	SiO_2	2.6485	298	115	0.8680	0.0704	0.1191	-0.1804	1.0575	0.5820
Selenium	Se	4.838	300	116	0.1870	0.0710	0.2620	0.0620	0.7410	0.1490
Sodium nitrate	$NaNO_3$	2.27	RT	12	0.8670	0.1630	0.1600	0.0820	0.3740	0.2130
Tourmaline		3.05	RT	82	2.7066	0.6927	0.0872	-0.0774	1.6070	0.6682

REFERENCES

1. Thomas, J. F., *Phys. Rev.*, 175, 955-962, 1968.
2. Bolef, D. I. and M. Menes, *J. Appl. Phys.*, 31, 1426-1427, 1960.
3. Garland, C. W. and C. F. Yarnell, *J. Chem. Phys.*, 44, 1112-1120, 1966.
4. Garland, C. W. and R. Renard, *J. Chem. Phys.*, 44, 1130-1139, 1966.

5. Gsänger, M., H. Egger and E. Lüscher, *Phys. Letters*, 27A, 695-696, 1968.
6. Wong, C. and D. E. Schuele, *J. Phys. Chem. Solids*, 29, 1309-1330, 1968.
7. Haussühl, S., *Phys. Stat. Sol.*, 3, 1072-1076, 1963.
8. Wong, C. and D. E. Schuele, *J. Phys. Chem. Solids*, 28, 1225-1231, 1967.
9. McSkimin, H. J. and D. G. Thomas, *J. Appl. Phys.*, 33, 56-59, 1962.
10. Kollarits, F. J. and J. Trivisonno, *J. Phys. Chem. Solids*, 29, 2133-2139, 1968.
11. Slagle, D. D. and H. A. McKinstry, *J. Appl. Phys.*, 38, 446-458, 1967.
12. Hearmon, R. F. S., *Adv. Phys.*, 5, 323-382, 1956.
13. Sumer, A. and J. F. Smith, *J. Appl. Phys.*, 34, 2691-2694, 1963.
14. Alexandrov, K. S. et. al., *Sov. Phys. Sol. State*, 10, 1316-1321, 1968.
15. Epstein, S. G. and O. N. Carlson, *Acta Metal.*, 13, 487-491, 1965.
16. McSkimin, H. J., et. al., *J. Appl. Phys.*, 39, 4127-4128, 1968.
17. McSkimin, H. J., et. al., *J. Appl. Phys.*, 38, 2362-2364, 1967.
18. Weil, R. and W. O. Groves, *J. Appl. Phys.*, 39, 4049-4051, 1968.
19. Bateman, T. B., *J. Appl. Phys.*, 37, 2194-2195, 1966.
20. Bogardus, E. H., *J. Appl. Phys.*, 36, 2504-2513, 1965.
21. Golding, B., S. C. Moss and B. L. Averbach, *Phys. Rev.*, 158, 637-645, 1967.
22. Bateman, T. B., H. J. McSkimin and J. M. Whelan, *J. Appl. Phys.*, 30, 544-545, 1959.
23. Gerlich, D., *J. Appl. Phys.*, 35, 3062, 1964.
24. Hickernell, F. S. and W. R. Gayton, *J. Appl. Phys.*, 37, 462, 1966.
25. MacFarlane, R. E., et. al., *Phys. Letters*, 20, 234-235, 1966.
26. Leese, J. and A. E. Lord Jr., *J. Appl. Phys.*, 39, 3986-3988, 1968.
27. Miller, R. A. and D. E. Schuele, *J. Phys. Chem. Solids*, 30, 589-600, 1969.
28. Wasilik, J. H. and M. L. Wheat, *J. Appl. Phys.*, 36, 791-793, 1965.
29. Haussühl, S., *Phys. Stat. Sol.*, 3, 1072-1076, 1963.
30. Houston, B., et. al., *J. Appl. Phys.*, 39, 3913-3916, 1968.
31. Trivisonno, J. and C. S. Smith, *Acta Metal.*, 9, 1064-1071, 1961.
32. Alexandrov, K. S. and T. V. Ryzhova, *Sov. Phys. Cryst.*, 6, 228-252, 1961.
33. Lewis, J. T., A. Lehoczky and C. V. Briscoe, *Phys. Rev.*, 161, 877-887, 1967.
34. Drabble, J. R. and R. E. B. Strathen, *Proc. Phys. Soc.*, 92, 1090-1995, 1967.
35. Oliver, D. W., *J. Appl. Phys.*, 40, 893, 1969.
36. Alper, T., and G. A. Saunders, *J. Phys. Chem. Solids*, 28, 1637-1642, 1967.
37. Dickinson, J. M. and P. E. Armstrong, *J. Appl. Phys.*, 38, 602-606, 1967.
38. Bolef, D. I., *J. Appl. Phys.*, 32, 100-105, 1961.
39. Rayne, J. A., *Phys. Rev.*, 112, 1125-1130, 1958.
40. MacFarlane, R. E., et. al., *Phys. Letters*, 18, 91-92, 1965.
41. Smith, P. A. and C. S. Smith, *J. Phys. Chem. Solids*, 26, 279-289, 1965.
42. Norwood, M. H. and C. V. Briscoe, *Phys. Rev.*, 112, 45-48, 1958.
43. Simmons, G. and F. Birch, *J. Appl. Phys.*, 34, 2736-2738, 1963.
44. Gutman, E. J. and J. Trivisonno, *J. Phys. Chem. Sol.*, 28, 805-809, 1967.
45. Ghafelehbashi, M., et. al., *J. Appl. Phys.*, 41, 652-666, 1970.
46. McSkimin, H. J. and P. Andreatch Jr., *J. Appl. Phys.*, 35, 2161-2165, 1964.
47. Neighbours, J. R. and G. A. Alers, *Phys. Rev.*, 111, 707-712, 1958.
48. Hidshaw, W., J. T. Lewis, and C. V. Briscoe, *Phys. Rev.*, 163, 876-881, 1967.
49. Daniels, W. B., *Phys. Rev.*, 119, 1246-1252, 1960.
50. Viswanathan, R., *J. Appl. Phys.*, 37, 884-886, 1966.
51. Miller, R. A. and C. S. Smith, *J. Phys. Chem. Sol.*, 25, 1279-1292, 1964.
52. Claytor, R. N. and B. J. Marshall, *Phys. Rev.*, 120, 332-334, 1960.
53. Schreiber, E., *J. Appl. Phys.*, 38, 2508-2511, 1967.
54. Gerlich, D., *Phys. Rev.*, 136, A1366-A1368, 1964.
55. Johnston, D. L., P. H. Thrasher and R. J. Kearney, *J. Appl. Phys.*, 41, 427-428, 1970.
56. Poindexter, E. and A. A. Giardini, *Phys. Rev.*, 110, 1069, 1958.
57. Soga, N., *J. Appl. Phys.*, 37, 3416-3420, 1966.
58. Bartlett, R. W. and C. W. Smith, *J. Appl. Phys.*, 38, 5428-5429, 1967.
59. Morse, G. E. and A. W. Lawson, *J. Phys. Chem. Sol.*, 28, 939-950, 1967.
60. Armstrong, P. E., O. N. Carlson and J. F. Smith, *J. Appl. Phys.*, 30, 36-41, 1959.
61. Macedo, P. M., W. Capps and J. B. Wachtman, *J. Am. Cer. Soc.*, 47, 651, 1964.
62. Beattie, A. G., *J. Appl. Phys.*, 40, 4818-4821, 1969.
63. Chang, R. and L. J. Graham, *J. Appl Phys.*, 37, 3778-3783, 1966.
64. Lowrie, R. and A. M. Gonas, *J. Appl. Phys.*, 38, 4505-4509. 1967.
65. Graham, L. J., H. Nadler and R. Chang, *J. Appl. Phys.*, 34, 1572-1573, 1963.

66. Wachtman, J. B. Jr., et. al., *J. Nucl. Mat.*, 16, 39-41, 1965.
67. Bolef, D. I., *J. Appl. Phys.*, 32, 100-105, 1961.
68. Berlincourt, D., H. Jaffe and L. R. Shiozawa, *Phys. Rev.*, 129, 1009-1017, 1963.
69. Adhav. R. S. J. *Acoust. Soc. Am.*, 43, 835-838, 1968.
70. Berlincourt, D. and H. Jaffe, *Phys. Rev.*, 111, 143-148, 1958.
71. Huntington, H. B., in *Solid State Pysics, Vol. 7*, Seitz, F., and Turnbull, D., Ed., pp. 213-285; Academic Press, New York 1958.
72. Cutler, H. R., J. J. Gibson and K. A. McCarthy, *Sol. State Comm.*, 6, 431-433, 1968.
73. Mason, W. P., *Piezoelectric Crystals and Their Application to Ultrasonics*, D. Van Nostrand Co., Inc., New York, 1950.
74. Adhav, R. S., *J. Appl. Phys.*, 40, 2725-2727, 1969.
75. Manghnani, M. H., *J. Geophys. Res.*, 74, 4317-4328, 1969.
76. Uchida, N. and Y. Ohmachi, *J. Appl Phys.*, 40, 4692-4695, 1969.
77. House, D. G. and E. Y. Vernon, *Br. J. Appl. Phys.*, 11, 254-259, 1960.
78. Ryzhova, T. V., et. al., *Bull. Acad. Sci. USSR, Earth Phys. Ser.*, English Transl., no. 2, 111-113, 1966.
79. Alton, W. J. and A. J. Barlow, *J. Appl. Phys.*, 38, 3817-3820, 1967.
80. Michard, F., et. al., *C. R. Acad. Sci., Paris*, 265, 565-567, 1967.
81. Haussühl, S., *Acta Cryst.*, 18, 839-842, 1965.
82. Hearmon, R. F. S., *Rev. Mod. Phys.*, 18, 409-440, 1946.
83. Heseltine, J. C. W., D. W. Elliott and O. B. Wilson, *J. Chem. Phys.*, 40, 2584-2587, 1964.
84. Schwerdtner, W. M., et. al., *Canad. J. Earth Sci.*, 2, 673-683, 1965.
85. Kumazawa, M. and O. L. Anderson, *J. Geophys. Res.*, 74, 5961-5972, 1969 .
86. Alexandrov, K. S., et. al., *Sov. Phys. Cryst.*, 7, 753-755, 1963.
87. Verma, R. K., *J. Geophys. Soc.*, 65, 757-766, 1960.
88. McSkimin, H. J. and E. S. Fisher, *J. Appl. Phys.*, 31, 1627-1639, 1960.
89. Alexandrov, K. S. and T.V. Ryzhova, *Bull. Acad. Sci. USSR, Geophys. Ser.*, English Transl., no.8, 871-875, 1961.
90. Afanaseva, G. K., et. al., *Phys. Stat. Sol.*, 24, K61-K63, 1967.
91. Alexandrov, K. S., et. al., *Sov. Phys. Cryst.*, 8, 589-591, 1964.
92. Alexandrov, K. S. and T. V Ryzhova, *Bull Acad. Sci. USSR, Geophys. Ser.*, English Transl., no.2, 129-131, 1962.
93. Alexandrov, K. S., et. al., *Sov. Phys. Cryst.*, 8, 164-166, 1963.
94. Teslenko, V. F., et. al., *Sov. Phys. Cryst.*, 10, 744-747, 1966.
95. Smith, J. F. and C. L. Arbogast, *J. Appl. Phys.*, 31, 99-102, 1960.
96. Cline, C. F., H. L. Dunegan and G. M. Henderson, *J. Appl. Phys.*, 38, 1944-1948, 1967.
97. Chang, Y. A. and L. Himmel, *J. Appl. Phys.*, 37, 3787-3790, 1966.
98. Gerlich, D., *J. Phys. Chem. Solids*, 28, 2575-2579, 1967.
99. McSkimin, H. J., *J. Appl. Phys.*, 26, 406-409, 1955.
100. Fisher, E. S. and D. Dever, *Trans. Met. Soc. AIME*, 239, 48-57, 1967.
101. Fisher, E. S. and D. Dever, *Proc. Conf. Rare Earth Res.*, 6th, Gatlinburg, Tenn., 522-533, 1967.
102. Fisher, E. S. and C. J. Renken, Phys. Rev., 135, A482-A494, 1964.
103. Proctor, T. M. Jr., *J. Acoust. Soc. Am.*, 39, 972-977, 1966.
104. Chandrasekhar, B. S. and J. A. Rayne, *Phys. Rev.*, 124, 1011-1041, 1961.
105. Wazzan, A. R. and L. B. Robinson, *Phys. Rev.*, 155, 586-594, 1967.
106. Ferris, R. W., et. al., *J. Appl. Phys.*, 34, 768-770, 1963.
107. Gilman, J. J. and B. W. Roberts, *J. Appl. Phys.*, 32, 1405, 1961.
108. Smith, J. F. and J. A. Gjevre, *J. Appl. Phys.*, 31, 645-647, 1960.
109. Alers, G. A. and J. R. Neighbours, *J. Phys. Chem. Solids*, 7, 58-64, 1908.
110. Bateman, T. B., *J. Appl. Phys.*, 33, 3309-3312, 1962.
111. Tefft, W. E., *J. Res. Natl. Bur. Stand.*, 70A, 277-280, 1966.
112. DeBretteville, Jr., A. et. al., *Phys. Rev.*, 148, 575-579, 1966.
113. Dandekar, D. P. and A. L. Ruoff, *J. Appl. Phys.*, 39, 6004-6009, 1968.
114. Warner, A. W., M. Onoe and G. A. Coquin, *J. Acoust. Soc. Am.*, 42, 1223-1231, 1967.
115. McSkimin, H. J., P. Andreatch and R. N. Thurston, *J. Appl. Phys.*, 36, 1624-1632, 1965.
116. Mort, J., *J. Appl. Phys.*, 38, 3414-3415, 1967.

ELECTRICAL RESISTIVITY OF PURE METALS

The first part of this table gives the electrical resistivity, in units of 10^{-8} Ω m, for 28 common metallic elements as a function of temperature. The data refer to polycrystalline samples. The number of significant figures indicates the accuracy of the values. However, at low temperatures (especially below 50 K) the electrical resistivity is extremely sensitive to sample purity. Thus the low-temperature values refer to samples of specified purity and treatment. The references should be consulted for further information on this point, as well as for values at additional temperatures.

The second part of the table gives resistivity values in the neighborhood of room temperature for other metallic elements that have not been studied over an extended temperature range.

REFERENCES

1. C. Y. Ho, et al., *J. Phys. Chem. Ref. Data*, 12, 183—322, 1983; 13, 1069—1096, 1984; 13, 1097—1130, 1984, 13, 1131—1172, 1984.
2. R. A. Matula, *J. Phys Chem. Ref. Data*, 8, 1147—1298, 1979.
3. T. C. Chi, *J. Phys. Chem. Ref. Data*, 8, 339—438, 1979; 8, 439—498, 1979.
4. K. H. Hellwege, Ed., *Landolt-Börnstein Numerical Data and Functional Relationships in Science and Technology*, Group III, Vol. 15, Subvolume a, Springer-Verlag, Heidelberg, 1982.
5. L. A. Hall, *Survey of Electrical Resistivity Measurements on 16 Pure Metals in the Temperature Range 0 to 273 K*, NBS Technical Note 365, U.S. Superintendent of Documents, 1968.

ELECTRICAL RESISTIVITY IN 10^{-8} Ω m

T/K	Aluminum	Barium	Beryllium	Calcium	Cesium	Chromium	Copper
1	0.000100	0.081	0.0332	0.045	0.0026		0.00200
10	0.000193	0.189	0.0332	0.047	0.243		0.00202
20	0.000755	0.94	0.0336	0.060	0.86		0.00280
40	0.0181	2.91	0.0367	0.175	1.99		0.0239
60	0.0959	4.86	0.067	0.40	3.07		0.0971
80	0.245	6.83	0.075	0.65	4.16		0.215
100	0.442	8.85	0.133	0.91	5.28	1.6	0.348
150	1.006	14.3	0.510	1.56	8.43	4.5	0.699
200	1.587	20.2	1.29	2.19	12.2	7.7	1.046
273	2.417	30.2	3.02	3.11	18.7	11.8	1.543
293	2.650	33.2	3.56	3.36	20.5	12.5	1.678
298	2.709	34.0	3.70	3.42	20.8	12.6	1.712
300	2.733	34.3	3.76	3.45	21.0	12.7	1.725
400	3.87	51.4	6.76	4.7		15.8	2.402
500	4.99	72.4	9.9	6.0		20.1	3.090
600	6.13	98.2	13.2	7.3		24.7	3.792
700	7.35	130	16.5	8.7		29.5	4.514
800	8.70	168	20.0	10.0		34.6	5.262
900	10.18	216	23.7	11.4		39.9	6.041

T/K	Gold	Hafnium	Iron	Lead	Lithium	Magnesium	Manganese
1	0.0220	1.00	0.0225		0.007	0.0062	7.02
10	0.0226	1.00	0.0238		0.008	0.0069	18.9
20	0.035	1.11	0.0287		0.012	0.0123	54
40	0.141	2.52	0.0758		0.074	0.074	116
60	0.308	4.53	0.271		0.345	0.261	131
80	0.481	6.75	0.693	4.9	1.00	0.557	132
100	0.650	9.12	1.28	6.4	1.73	0.91	132
150	1.061	15.0	3.15	9.9	3.72	1.84	136
200	1.462	21.0	5.20	13.6	5.71	2.75	139
273	2.051	30.4	8.57	19.2	8.53	4.05	143
293	2.214	33.1	9.61	20.8	9.28	4.39	144
298	2.255	33.7	9.87	21.1	9.47	4.48	144
300	2.271	34.0	9.98	21.3	9.55	4.51	144
400	3.107	48.1	16.1	29.6	13.4	6.19	147
500	3.97	63.1	23.7	38.3		7.86	149

T/K	Gold	Hafnium	Iron	Lead	Lithium	Magnesium	Manganese
600	4.87	78.5	32.9			9.52	151
700	5.82		44.0			11.2	152
800	6.81		57.1			12.8	
900	7.86					14.4	

T/K	Molybdenum	Nickel	Palladium	Platinum	Potassium	Rubidium	Silver
1	0.00070	0.0032	0.0200	0.002	0.0008	0.0131	0.00100
10	0.00089	0.0057	0.0242	0.0154	0.0160	0.109	0.00115
20	0.00261	0.0140	0.0563	0.0484	0.117	0.444	0.0042
40	0.0457	0.068	0.334	0.409	0.480	1.21	0.0539
60	0.206	0.242	0.938	1.107	0.90	1.94	0.162
80	0.482	0.545	1.75	1.922	1.34	2.65	0.289
100	0.858	0.96	2.62	2.755	1.79	3.36	0.418
150	1.99	2.21	4.80	4.76	2.99	5.27	0.726
200	3.13	3.67	6.88	6.77	4.26	7.49	1.029
273	4.85	6.16	9.78	9.6	6.49	11.5	1.467
293	5.34	6.93	10.54	10.5	7.20	12.8	1.587
298	5.47	7.12	10.73	10.7	7.39	13.1	1.617
300	5.52	7.20	10.80	10.8	7.47	13.3	1.629
400	8.02	11.8	14.48	14.6			2.241
500	10.6	17.7	17.94	18.3			2.87
600	13.1	25.5	21.2	21.9			3.53
700	15.8	32.1	24.2	25.4			4.21
800	18.4	35.5	27.1	28.7			4.91
900	21.2	38.6	29.4	32.0			5.64

T/K	Sodium	Strontium	Tantalum	Tungsten	Vanadium	Zinc	Zirconium
1	0.0009	0.80	0.10	0.000016		0.0100	0.250
10	0.0015	0.80	0.102	0.000137	0.0145	0.0112	0.253
20	0.016	0.92	0.146	0.00196	0.039	0.0387	0.357
40	0.172	1.70	0.751	0.0544	0.304	0.306	1.44
60	0.447	2.68	1.65	0.266	1.11	0.715	3.75
80	0.80	3.64	2.62	0.606	2.41	1.15	6.64
100	1.16	4.58	3.64	1.02	4.01	1.60	9.79
150	2.03	6.84	6.19	2.09	8.2	2.71	17.8
200	2.89	9.04	8.66	3.18	12.4	3.83	26.3
273	4.33	12.3	12.2	4.82	18.1	5.46	38.8
293	4.77	13.2	13.1	5.28	19.7	5.90	42.1
298	4.88	13.4	13.4	5.39	20.1	6.01	42.9
300	4.93	13.5	13.5	5.44	20.2	6.06	43.3
400		17.8	18.2	7.83	28.0	8.37	60.3
500		22.2	22.9	10.3	34.8	10.82	76.5
600		26.7	27.4	13.0	41.1	13.49	91.5
700		31.2	31.8	15.7	47.2		104.2
800		35.6	35.9	18.6	53.1		114.9
900			40.1	21.5	58.7		123.1

Element	T/K	Electrical resistivity $10^{-8}\ \Omega\,m$
Antimony	273	39
Bismuth	273	107
Cadmium	273	6.8
Cerium (β, hex)	290—300	82.8
Cerium (γ, cub)	298	74.4
Cobalt	273	5.6
Dysprosium	290—300	92.6
Erbium	290—300	86.0
Europium	290—300	90.0
Gadolinium	290—300	131
Gallium	273	13.6
Holmium	290—300	81.4
Indium	273	8.0
Iridium	273	4.7
Lanthanum	290—300	61.5
Lutetium	290—300	58.2
Mercury	273	94.1
Neodymium	290—300	64.3
Niobium	273	15.2
Osmium	273	8.1
Polonium	273	40
Praseodymium	290—300	70.0
Promethium	290—300	75 est.
Protactinium	273	17.7
Rhenium	273	17.2
Rhodium	273	4.3
Ruthenium	273	7.1
Samarium	290—300	94.0
Scandium	290—300	56.2
Terbium	290—300	115
Thallium	273	15
Thorium	273	14.7
Thulium	290—300	67.6
Tin	273	11.5
Titanium	273	39
Uranium	273	28
Ytterbium	290—300	25.0
Yttrium	290—300	59.6

ELECTRICAL RESISTIVITY OF SELECTED ALLOYS

Values of the resistivity are given in units of 10^{-8} Ω m. General comments in the preceeding table for pure metals also apply here.

REFERENCE

C. Y. Ho, et al., *J. Phys. Chem. Ref. Data*, 12, 183—322, 1983.

Alloy Aluminum-Copper

Wt % Al	273 K	293 K	300 K	350 K	400 K
99[a]	2.51	2.74	2.82	3.38	3.95
95[a]	2.88	3.10	3.18	3.75	4.33
90[b]	3.36	3.59	3.67	4.25	4.86
85[b]	3.87	4.10	4.19	4.79	5.42
80[b]	4.33	4.58	4.67	5.31	5.99
70[b]	5.03	5.31	5.41	6.16	6.94
60[b]	5.56	5.88	5.99	6.77	7.63
50[b]	6.22	6.55	6.67	7.55	8.52
40[c]	7.57	7.96	8.10	9.12	10.2
30[c]	11.2	11.8	12.0	13.5	15.2
25[f]	16.3[aa]	17.2	17.6	19.8	22.2
15[h]	—	12.3	—	—	—
19[g]	10.8[aa]	11.0	11.1	11.7	12.3
5[e]	9.43	9.61	9.68	10.2	10.7
1[b]	4.46	4.60	4.65	5.00	5.37

Alloy—Aluminum-Magnesium

Wt % Al	273 K	293 K	300 K	350 K	400 K
99[c]	2.96	3.18	3.26	3.82	4.39
95[c]	5.05	5.28	5.36	5.93	6.51
90[c]	7.52	7.76	7.85	8.43	9.02
85	—	—	—	—	—
80	—	—	—	—	—
70	—	—	—	—	—
60	—	—	—	—	—
50	—	—	—	—	—
40	—	—	—	—	—
30	—	—	—	—	—
25	—	—	—	—	—
15	—	—	—	—	—
10[b]	17.1	17.4	17.6	18.4	19.2
5[b]	13.1	13.4	13.5	14.3	15.2
1[a]	5.92	6.25	6.37	7.20	8.03

Alloy—Copper-Gold

Wt % Cu	273 K	293 K	300 K	350 K	400 K
99[c]	1.73	1.86[aa]	1.91[aa]	2.24[aa]	2.58[aa]
95[c]	2.41	2.54[aa]	2.59[aa]	2.92[aa]	3.26[aa]
90[c]	3.29	4.42[aa]	3.46[aa]	3.79[aa]	4.12[aa]
85[c]	4.20	4.33	4.38[aa]	4.71[aa]	5.05[aa]
80[c]	5.15	5.28	5.32	5.65	5.99
70[c]	7.12	7.25	7.30	7.64	7.99
60[c]	9.18	9.13	9.36	9.70	10.05
50[c]	11.07	11.20	11.25	11.60	11.94
40[c]	12.70	12.85	12.90[aa]	13.27[aa]	13.65[aa]
30[c]	13.77	13.93	13.99[aa]	14.38[aa]	14.78[aa]
25[c]	13.93	14.09	14.14	14.54	14.94
15[c]	12.75	12.91	12.96[aa]	13.36[aa]	13.77
10[c]	10.70	10.86	10.91	11.31	11.72
5[c]	7.25	7.41[aa]	7.46	7.87	8.28
1[c]	3.40	3.57	3.62	4.03	4.45

Alloy—Copper-Nickel

Wt % Cu	273 K	293 K	300 K	350 K	400 K
99[c]	2.71	2.85	2.91	3.27	3.62
95[c]	7.60	7.71	7.82	8.22	8.62
90[c]	13.69	13.89	13.96	14.40	14.81
85[c]	19.63	19.83	19.90	2032	20.70
80[c]	25.46	25.66	25.72	26.12[aa]	26.44[aa]
70[i]	36.67	36.72	36.76	36.85	36.89
60[i]	45.43	45.38	45.35	45.20	45.01
50[i]	50.19	50.05	50.01	49.73	49.50
40[i]	47.42	47.73	47.82	48.28	48.49
30[i]	40.19	41.79	42.34	44.51	45.40
25[c]	33.46	35.11	35.69	39.67[aa]	42.81[aa]
15[c]	22.00	23.35	23.85	27.60	31.38
10[c]	16.65	17.82	18.26	21.51	25.19
5[c]	11.49	12.50	12.90	15.69	18.78
1[c]	7.23	8.08	8.37	10.63[aa]	13.18[aa]

Alloy—Copper-Palladium

Wt % Cu	273 K	293 K	300 K	350 K	400 K
99[c]	2.10	2.23	2.27	2.59	2.92
95[c]	4.21	4.35	4.40	4.74	5.08
90[c]	6.89	7.03	7.08	7.41	7.74
85[c]	9.48	9.61	9.66	10.01	10.36
80[c]	11.99	12.12	12.16	12.51[aa]	12.87
70[c]	16.87	17.01	17.06	17.41	17.78
60[c]	21.73	21.87	21.92	22.30	22.69
50[c]	27.62	27.79	27.86	28.25	28.64
40[c]	35.31	35.51	35.57	36.03	36.47
30[c]	46.50	46.66	46.71	47.11	47.47
25[c]	46.25	46.45	46.52	46.99[aa]	47.43[aa]
15[c]	36.52	36.99	37.16	38.28	39.35
10[c]	28.90	29.51	29.73	31.19[aa]	32.56[aa]
5[c]	20.00	20.75	21.02	22.84[aa]	24.54[aa]
1[c]	11.90	12.67	12.93[aa]	14.82[aa]	16.68[aa]

Alloy—Copper-Zinc

Wt % Cu	273 K	293 K	300 K	350 K	400 K
99[b]	1.84	1.97	2.02	2.36	2.71
95[b]	2.78	2.92	2.97	3.33	3.69
90[b]	3.66	3.81	3.86	4.25	4.63
85[b]	4.37	4.54	4.60	5.02	5.44
80[b]	5.01	5.19	5.26	5.71	6.17
70[b]	5.87	6.08	6.15	6.67	7.19
60	—	—	—	—	—
50	—	—	—	—	—
40	—	—	—	—	—
30	—	—	—	—	—
25	—	—	—	—	—
15	—	—	—	—	—
10	—	—	—	—	—
5	—	—	—	—	—
1	—	—	—	—	—

Alloy—Gold-Palladium

Wt % Au	273 K	293 K	300 K	350 K	400 K
99[c]	2.69	2.86	2.91	3.32	3.73
95[c]	5.21	5.35	5.41	5.79	6.17
90[i]	8.01	8.17	8.22	8.56	8.93
85[b]	10.50[aa]	10.66	10.72[aa]	11.100[aa]	11.48[aa]
80[b]	12.75	12.93	12.99	13.45	13.93
70[c]	18.23	18.46	18.54	19.10	19.67
60[b]	26.70	26.94	27.02	27.63[aa]	28.23[aa]
50[a]	27.23	27.63	27.76	28.64[aa]	29.42[aa]
40[c]	24.65	25.23	25.42	26.74	27.95
30[b]	20.82	21.49	21.72	23.35	24.92
25[b]	18.86	19.53	19.77	21.51	23.19
15[a]	15.08	15.77	16.01	17.80	19.61
10[a]	13.25	13.95	14.20[aa]	16.00[aa]	17.81[aa]
5[a]	11.49	12.21	12.46[aa]	14.26[aa]	16.07[aa]
1[a]	10.07	10.85[aa]	11.12[aa]	12.99[aa]	14.80[aa]

Alloy—Gold-Silver

Wt % Au	273 K	293 K	300 K	350 K	400 K
99[b]	2.58	2.75	2.80[aa]	3.22[aa]	3.63[aa]
95[a]	4.58	4.74	4.79	5.19	5.59
90[j]	6.57	6.73	6.78	7.19	7.58
85[j]	8.14	8.30	8.36[aa]	8.75	9.15
80[j]	9.34	9.50	9.55	9.94	10.33
70[j]	10.70	10.86	10.91	11.29	11.68[aa]
60[j]	10.92	11.07	11.12	11.50	11.87
50[j]	10.23	10.37	10.42	10.78	11.14
40[j]	8.92	9.06	9.11	9.46[aa]	9.81
30[a]	7.34	7.47	7.52	7.85	8.19
25[a]	6.46	6.59	6.63	6.96	7.30[aa]
15[a]	4.55	4.67	4.72	5.03	5.34
10[a]	3.54	3.66	3.71	4.00	4.31
5[j]	2.52	2.64[aa]	2.68[aa]	2.96[aa]	3.25[aa]
1[b]	1.69	1.80	1.84[aa]	2.12[aa]	2.42[aa]

Alloy—Iron-Nickel

Wt % Fe	273 K	293 K	300 K	350 K	400 K
99[a]	10.9	12.0	12.4	—	18.7
95[c]	18.7	19.9	20.2	—	26.8
90[c]	24.2	25.5	25.9	—	33.2
85[c]	27.8	29.2	29.7	—	37.3
80[c]	30.1	31.6	32.2	—	40.0
70[b]	32.3	33.9	34.4	—	42.4
60[c]	53.8	57.1	58.2	—	73.9
50[d]	28.4	30.6	31.4	—	43.7
40[d]	19.6	21.6	22.5	—	34.0
30[c]	15.3	17.1	17.7	—	27.4
25[b]	14.3	15.9	16.4	—	25.1
15[c]	12.6	13.8	14.2	—	21.1
10[c]	11.4	12.5	12.9	—	18.9
5[c]	9.66	10.6	10.9	—	16.1[aa]
1[b]	7.17	7.94	8.12	—	12.8

Alloy—Silver-Palladium

Wt % Ag	273 K	293 K	300 K	350 K	400 K
99[b]	1.891	2.007	2.049	2.35	2.66
95[b]	3.58	3.70	3.74	4.04	4.34
90[b]	5.82	5.94	5.98	6.28	6.59
85[k]	7.92[aa]	8.04[aa]	8.08	8.38[aa]	8.68[aa]
80[k]	10.01	10.13	10.17	10.47	10.78
70[i]	14.53	14.65	14.69	14.99	15.30
60[i]	20.9	21.1	21.2	21.6	22.0
50[k]	31.2	31.4	31.5	32.0	32.4
40[m]	42.2	42.2	42.2	42.3	42.3
30[b]	40.4	40.6	40.7	41.3	41.7
25[k]	36.67[aa]	37.06	37.19	38.1[aa]	38.8[aa]
15[i]	27.08[aa]	26.68[aa]	27.89[aa]	29.3[aa]	30.6[aa]
10[i]	21.69	22.39	22.63	24.3	25.9
5[b]	15.98	16.72	16.98	18.8[aa]	20.5[aa]
1[a]	11.06	11.82	12.08[aa]	13.92[aa]	15.70[aa]

[a] Uncertainty in resistivity is ± 2%.
[b] Uncertainty in resistivity is ± 3%.
[c] Uncertainty in resistivity is ± 5%.
[d] Uncertainty in resistivity is ± 7% below 300 K and ± 5% at 300 and 400 K.
[e] Uncertainty in resistivity is ± 7%.
[f] Uncertainty in resistivity is ± 8%.
[g] Uncertainty in resistivity is ± 10%.
[h] Uncertainty in resistivity is ± 12%.
[i] Uncertainty in resistivity is ± 4%.
[j] Uncertainty in resistivity is ± 1%.
[k] Uncertainty in resistivity is ± 3% up to 300 K and ± 4% above 300 K.
[m] Uncertainty in resistivity is ± 2% up to 300 K and ± 4% above 300 K.
[n] Crystal usually a mixture of α-hcp and fcc lattice.
[aa] In temperature range where no experimental data are available.

PERMITTIVITY (DIELECTRIC CONSTANT) OF INORGANIC SOLIDS

H. P. R. Frederikse

This table lists the permittivity ϵ, frequently called the dielectric constant, of a number of inorganic solids. When the material is not isotropic, the individual components of the permittivity are given. A superscript S indicates a measurement made under constant strain ("clamped" dielectric constant). If the constraint is removed, the measurement yields ϵ^T, the "unclamped" or free dielectric constant.

The temperature of the measurement is given when available; the symbol r.t. indicates a value at nominal room temperature. The frequency of the measurement is given in the last column (i.r. indicates a measurement in the infrared).

Substances are listed in alphabetical order by chemical formula.

REFERENCE

Young, K. F. and Frederikse, H. P. R., *J. Phys. Chem. Ref. Data*, 2, 313, 1973.

Formula	Name	ϵ_{ijk}	T/K	ν/Hz
Ag_3AsS_3	Silver thioarsenate (Proustite)	$\epsilon_{11}^T = 16.5$, $\epsilon_{11}^S = 14.5$	r.t.	2×10^7
		$\epsilon_{33}^T = 20.0$, $\epsilon_{33}^S = 18.0$	r.t.	2×10^7
$AgBr$	Silver bromide	12.50	r.t.	
$AgCN$	Silver cyanide	5.6	r.t.	10^6
$AgCl$	Silver chloride	11.15	r.t.	
$AgNO_3$	Silver nitrate	9.0	293	5×10^5
$AgNa(NO_2)_2$	Silver sodium nitrite	4.5 ± 0.5	r.t.	9.4×10^9
Ag_2O	Silver oxide	8.8	r.t.	
$(AlF)_2SiO_4$	Aluminum fluosilicate (topaz)	$\epsilon_{11} = 6.62$	297	7×10^3
		$\epsilon_{22} = 6.58$	297	7×10^3
		$\epsilon_{33} = 6.95$	297	7×10^3
Al_2O_3	Aluminum oxide (alumina)	$\epsilon_{11} = \epsilon_{22} = 9.34$	298	10^2—8×10^9
		$\epsilon_{33} = 11.54$	298	10^2—8×10^9
$AlPO_4$	Aluminum phosphate	$\epsilon_{11}^T = 6.05$	r.t.	10^3
$AlSb$	Aluminum antimonide	11.21	300	i.r.
AsF_3	Arsenic trifluoride	5.7	r.t.	
BN	Boron nitride	7.1	r.t.	i.r.
$BaCO_3$	Barium carbonate	8.53	291	2×10^5
$Ba(COOH)_2$	Barium formate	$\epsilon_{11} = 7.9$	r.t.	10^3
		$\epsilon_{22} = 5.9$	r.t.	10^3
		$\epsilon_{33} = 7.5$	r.t.	10^3
$BaCl_2$	Barium chloride	9.81	r.t.	
$BaCl_2 \cdot 2H_2O$	Barium chloride dihydrate	9.00	r.t.	10^3
BaF_2	Barium fluoride	7.32	292	5×10^2—10^{11}
$Ba(NO_3)_2$	Barium nitrate	4.95	292	2×10^5
$Ba_2NaNb_5O_{15}$	Barium sodium niobate ("Bananas")	$\epsilon_{11}^S = 222$, $\epsilon_{11}^T = 235$	296	10^4
		$\epsilon_{22}^S = 227$, $\epsilon_{22}^T = 247$	296	
		$\epsilon_{33}^S = 32$, $\epsilon_{33}^T = 51$	296	
BaO	Barium oxide (baria)	34 ± 1	248, 333	60×10^7
BaO_2	Barium peroxide	10.7	r.t.	2×10^6
BaS	Barium sulfide	19.23	r.t.	7.25×10^6
$BaSO_4$	Barium sulfate	11.4	288	10^8
$BaSnO_3$	Barium stannate	18	298	25×10^5
$BaTiO_3$	Barium titanate	$\epsilon_{11}^T = 3600$	298	10^5
		$\epsilon_{11}^S = 2300$	298	2.5×10^8
		$\epsilon_{33}^T = 150$	298	10^5
		$\epsilon_{33}^S = 80$	298	2.5×10^8
$Ba_6Ti_2Nb_8O_{30}$	Barium titanium niobate	$\epsilon_{11} = \epsilon_{22} \approx 190$	298	
		$\epsilon_{33} \approx 220$	298	
$BaWO_4$	Barium tungstate	$\epsilon_{11} = \epsilon_{22} = 35.5 \pm 0.2$	297.5	1.6×10^3
		$\epsilon_{33} = 37.2 \pm 0.2$	297.5	1.6×10^3
$BaZrO_3$	Barium zirconate	43	r.t.	

Formula	Name	ϵ_{ijk}	T/K	ν/Hz
$Be_3Al_2Si_6O_{18}$	Beryllium aluminum silicate (Beryl)	$\epsilon_{33} = 5.95$	297	7×10^3
		$\epsilon_{11} = \epsilon_{22} = 6.86$	297	7×10^3
$BeCO_3$	Beryllium carbonate	9.7	291	2×10^5
BeO	Beryllium oxide (beryllia)	7.35 ± 0.2	293	2×10^6
$BiFeO_3$	Bismuth iron oxide	40 ± 3	300	9.4×10^9
$Bi_{12}GeO_{20}$	Bismuth germanite	$\epsilon_{11}^S = 38$	r.t.	
$Bi(GeO_4)_3$	Bismuth germanate	16	293	
Bi_2O_3	Bismuth sesquioxide	18.2	r.t.	2×10^6
$Bi_4Ti_3O_{12}$	Bismuth titanate	112	r.t.	10^3
C	Diamond			
	Type I	5.87 ± 0.19	300	10^3
	Type IIa	5.66 ± 0.04	300	10^3
$C_4H_4O_6$	Tartaric acid	$\epsilon_{11} = \epsilon_{22} = 4.3$	298	
		$\epsilon_{33} = 4.5$	298	
		$\epsilon_{13} = 0.55$	298	
$C_6H_{14}N_2O_6$	Ethylene diamine tartrate (EDT)	$\epsilon_{11}^T = 5.0$	293	
		$\epsilon_{22}^T = 8.3$	293	
		$\epsilon_{33}^T = 6.0$	293	
		$\epsilon_{13}^T = 0.7$	293	
$C_6H_{12}O_6NaBr$	Dextrose sodium bromide	$\epsilon_{11}^T = 4.0$	r.t.	10^3
$(CH_3NH_3)Al(SO_4)_2 \cdot 2H_2O$	Methyl ammonium alum (MASD)	19	197	
$Ca_2B_6O_{11} \cdot 5H_2O$	Colemanite	$\epsilon_{11} = 20$	293	10^3
		$\epsilon_{33} = 25$	293	10^3
$CaCO_3$	Calcium carbonate	$\epsilon_{11} = 8.67$	r.t.	9.4×10^{10}
		$\epsilon_{22} = 8.69$	r.t.	9.4×10^{10}
		$\epsilon_{33} = 8.31$	r.t.	9.4×10^{10}
$CaCeO_3$	Calcium cerate	21	r.t.	
CaF_2	Calcium fluoride	6.81	300	$5 \times 10^2 - 10^{11}$
$CaMoO_4$	Calcium molybdate	$\epsilon_{11} = \epsilon_{22} = 24.0 \pm 0.2$	297.5	<10
		$\epsilon_{33} = 20.0 \pm 0.2$	297.5	<10
$Ca(NO_3)_2$	Calcium nitrate	6.54	292	2×10^5
$CaNb_2O_6$	Calcium niobate	$\epsilon_{11} = 22.8 \pm 1.9$	r.t.	$(5-500) \times 10^3$
$Ca_2Nb_2O_7$	Calcium pyroniobate	~ 45	r.t.	5×10^7
CaO	Calcium oxide	11.8 ± 0.3	283	2×10^6
CaS	Calcium sulfide	6.699	r.t.	7.25×10^6
$CaSO_4 \cdot 2H_2O$	Calcium sulfate dihydrate	$\epsilon_{11} = 5.10$	r.t.	
		$\epsilon_{22} = 5.24$	r.t.	
		$\epsilon_{33} = 10.30$	r.t.	
$CaTiO_3$	Calcium titanate	165	r.t.	
$CaWO_4$	Calcium tungstate	$\epsilon_{11} = \epsilon_{22} = 11.7 \pm 0.1$	297.5	1.59×10^3
		$\epsilon_{33} = 9.5 \pm 0.2$	297.5	1.59×10^3
Cd_3As_2	Cadmium arsenide	$\epsilon_{33} = 18.5$	4	
$CdBr_2$	Cadmium bromide	8.6	293	5×10^5
CdF_2	Cadmium fluoride	8.33 ± 0.08	300	$10^5 - 10^7$
CdS	Cadmium sulfide	$\epsilon_{11} = \epsilon_{22} = 8.7$	300	i.r.
		$\epsilon_{33} = 9.25$	300	i.r.
		$\epsilon_{11} = \epsilon_{22} = 8.37$	8	i.r.
		$\epsilon_{33} = 9.00$	8	i.r.
		$\epsilon_{11}^T = 8.48$	77	10^4
		$\epsilon_{33}^T = 9.48$	77	10^4
		$\epsilon_{11}^S = 9.02, \epsilon_{11}^T = 9.35$	298	10^4
		$\epsilon_{33}^S = 9.53, \epsilon_{33}^T = 10.33$	298	10^4
$CdSe$	Cadmium selenide	$\epsilon_{11}^S = 9.53, \epsilon_{11}^T = 9.70$	298	10^4
		$\epsilon_{33}^S = 10.2, \epsilon_{33}^T = 10.65$	298	10^4
$CdTe$	Cadmium telluride	$\epsilon_{11} = \epsilon_{22} = 10.60 \pm 0.15$	297	i.r.
		$\epsilon_{33} = 7.05 \pm 0.05$	297	i.r.
$Cd_2Nb_2O_7$	Cadmium pyroniobate	500—580	293	10^3
CeO_2	Cerium oxide	7.0	r.t.	2×10^6

Formula	Name	ϵ_{ijk}	T/K	ν/Hz
$CoNb_2O_6$	Cobalt niobate	$\epsilon_{11} = 18.4 \pm 0.6$	r.t.	$(5—500) \times 10^3$
		$\epsilon_{22} = 21.4 \pm 1.1$	r.t.	$(5—500) \times 10^3$
		$\epsilon_{33} = 33.0 \pm 0.7$	r.t.	$(5—500) \times 10^3$
CoO	Cobalt oxide	12.9	298	$10^2—10^{10}$
Cr_2O_3	Chromic sesquioxide	$\epsilon_{11} = \epsilon_{22} = 13.3$	298.5	10^3
		$\epsilon_{33} = 11.9$	298.5	10^3
		8	315 (T_N)	6×10^{10}
$CsAl(SO_4)_2 \cdot 12H_2O$	Cesium alum	5.0	r.t.	$20—20 \times 10^3$
$CsBr$	Cesium bromide	6.38	298	1.6×10^3
Cs_2CO_3	Cesium carbonate	6.53	291	2×10^5
$CsCl$	Cesium chloride	7.2	298	
$Cs_2H_2AsO_4$	Cesium dihydrogen arsenate (CDA)	4.8	273	9.5×10^9
$Cs_2H_2PO_4$	Cesium dihydrogen phosphate (CDP)	6.15	285	9.5×10^9
$CsH_3(SeO_3)_2$	Cesium trihydrogen selenite	$\epsilon_{11} = 80$	273	10^5
		$\epsilon_{22} = 63$	273	10^5
		$\epsilon_{33} = 12$	273	10^5
CsI	Cesium iodide	6.31	298	1.6×10^3
$CsNO_3$	Cesium nitrate	$\epsilon_{11} = \epsilon_{22} = 9.4$	r.t.	5×10^5
		$\epsilon_{33} = 8.3$	r.t.	5×10^5
$CsPbCl_3$	Cesium lead chloride	14.37	300	$10^5—10^6$
$CuBr$	Cuprous bromide	8.0	293	5×10^5
$CuCl$	Cuprous chloride	9.8 ± 0.5	r.t.	10^3
CuO	Cupric oxide	18.1	r.t.	2×10^6
Cu_2O	Cuprous oxide (Cuprite)	7.60 ± 0.06	r.t.	10^5
$CuSO_4 \cdot 5H_2O$	Cupric sulfate pentahydrate	6.60	r.t.	
EuF_2	Europium fluoride	7.7 ± 0.2	298	$(1—300) \times 10^3$
$Eu_2(MoO_4)_3$	Europium molybdate	9.5	298	
EuS	Europium sulfide	13.10 ± 0.04	80	$5 \times 10^2—10^5$
FeO	Ferrous oxide	14.2	r.t.	2×10^6
Fe_2O_3	Ferric sesquioxide	4.5	r.t.	$10^5—10^7$
$Fe_2O_3\text{-}\alpha$	Ferric sesquioxide (hematite)	12		6×10^{10}
Fe_3O_4	Ferrosoferric oxide (magnetite)	20	r.t.	$10^5—10^7$
$GaAs$	Gallium arsenide	13.13	300	
		12.90	4	i.r.
GaP	Gallium phosphide	11.1	r.t.	
		10.75 ± 0.1	1.6	i.r.
$GaSb$	Gallium antimonide	15.69	r.t.	
		15.7	4	i.r.
$Gd_2(MoO_4)_3$	Gadolinium molybdate	$\epsilon^T = 10$	298	
		$\epsilon^S = 9.5$	298	10^3
Ge	Germanium	16.0 ± 0.3	4	9.2×10^9
		15.8 ± 0.2	r.t.	$500—3 \times 10^{10}$
GeO_2	Germanium dioxide	$\epsilon_{11} = \epsilon_{22} = 7.44$	r.t.	i.r.
HIO_3	Iodic acid	$\epsilon_{11} = 7.5$	r.t.	10^3
		$\epsilon_{22} = 12.4$	r.t.	10^3
		$\epsilon_{33} = 8.1$	r.t.	10^3
$HNH_4(ClCH_2COO)_2$	Hydrogen ammonium dichloroacetate	$\epsilon_{[102]} = 5.9$	r.t.	10^5
H_2O	Ice I (P = 0 kbar)	99	243	
	Ice III (P = 3 kbar)	117	243	
	Ice V(P = 5 kbar)	114	243	
	Ice VI (P = 8 kbar)	193	243	
$HgCl$	Mercurous chloride (Calumel)	$\epsilon_{11} = \epsilon_{22} = 14.0$	r.t.	10^{12}
$HgCl_2$	Mercuric chloride	6.5	r.t.	10^{12}
HgS	Mercurous sulfide (Cinnabar)	$\epsilon_{11} = \epsilon_{22} = 18.0$	r.t.	i.r.
		$\epsilon_{33} = 32.5$	r.t.	i.r.
$HgSe$	Mercurous selenide	25.6	r.t.	$10^4—10^6$
I_2	Iodine	$\epsilon_{11} = 6$	r.t.	$5 \times 10^4—10^7$
		$\epsilon_{22} = 3$	r.t.	$5 \times 10^4—10^7$
		$\epsilon_{33} = 40$	r.t.	$5 \times 10^4—10^7$

Formula	Name	ϵ_{ijk}	T/K	ν/Hz
InAs	Indium arsenide	14.55 ± 0.3	r.t.	i.r.
		15.15	4	i.r.
InP	Indium phosphide	12.61	r.t.	i.r.
InSb	Indium antimonide	17.88	4	i.r.
$KAl(SO_4)_2 \cdot 12H_2O$	Potassium alum	6.5	r.t.	$20—20 \times 10^3$
KBr	Potassium bromide	4.88	300	
		4.53	4.2	
$KBrO_3$	Potassium bromate	7.3	r.t.	2×10^6
KCN	Potassium cyanide	6.15	r.t.	2×10^6
K_2CO_3	Potassium carbonate	4.96	291	2×10^5
$K_2C_4H_4O_6 \cdot \frac{1}{2} H_2O$	Dipotassium tartrate (DKT)	$\epsilon_{11} = 6.44$	r.t.	
		$\epsilon_{22} = 5.80$	r.t.	
		$\epsilon_{33} = 6.49$	r.t.	
		$\epsilon_{13} = 0.005$	r.t.	
KCl	Potassium chloride	4.86 ± 0.02	r.t.	5×10^3
		4.50	4.2	
$KClO_3$	Potassium chlorate	5.1	r.t.	2×10^6
$KClO_4$	Potassium perchlorate	5.9	r.t.	2×10^6
K_2CrO_4	Potassium chromate	7.3	r.t.	6×10^7
$KCr(SO_4)_2 \cdot 12H_2O$	Potassium chrome alum	6.5	100—240	175×10^3
KD_2AsO_4	Potassium dideuterium arsenate (KDDA)	$\epsilon_{11} = 70$	298	
		$\epsilon_{33} = 31$	298	
KD_2PO_4	Potassium dideuterium phosphate (KDDP)	50 ± 2	297	10^3
KF	Potassium fluoride	6.05		2×10^6
KH_2AsO_4	Potassium dihydrogen arsenate (KDA)	$\epsilon_{11} = 60$	298	
		$\epsilon_{33} = 24$	298	
KH_2PO_4	Potassium dihydrogen phosphate (KDP)	46	298	10^3
		$\epsilon_{11} = 42$	r.t.	
		$\epsilon_{33} = 21$	r.t.	
K_2HPO_4	Dipotassium monohydrogen orthophosphate	9.05	r.t.	2×10^6
KI	Potassium iodide	5.00	r.t.	9.4×10^{10}
KIO_3	Potassium iodate	170	255	10^5
		10	293	10^5
		$\epsilon_{[101]} \approx 40,70$	r.t.	10^5
		16.85	r.t.	2×10^6
$(K,H)Al_3(SiO_4)_3$	Mica (muscovite)	5.4	299	$10^2—3 \times 10^9$
$(K,H)Mg_3Al(SiO_4)_3$	Mica (Canadian)	$\epsilon_{11} = \epsilon_{22} = 6.9$	298	$10^2—10^4$
		$\epsilon_{33} = 7.3$	298	10^4
KNO_2	Potassium nitrite	25	305	
KNO_3	Potassium nitrate	4.37	293	2×10^5
$KNbO_3$	Potassium niobate	700	r.t.	
K_3PO_4	Potassium orthophosphate	7.75	r.t.	2×10^6
KSCN	Potassium thiocyanate	7.9	r.t.	2×10^6
K_2SO_4	Potassium sulfate	6.4	r.t.	2×10^6
$K_2S_3O_6$	Potassium trithionate	5.7	293	1.8×10^6
$K_2S_4O_6$	Potassium tetrathionate	5.5	293	1.8×10^6
$K_2S_5O_6 \cdot H_2O$	Potassium pentathionate	7.8	293	1.8×10^6
$K_2S_6O_6$	Potassium hexathionate	7.8	293	1.8×10^6
K_2SeO_4	Potassium selenate	$\epsilon_{11} = 5.9$	r.t.	10^3
		$\epsilon_{22} = 7.7$	r.t.	10^3
$KSr_2Nb_5O_{15}$	Potassium strontium niobate	$\epsilon_{11} = \epsilon_{11} \approx 1200$	298	
		$\epsilon_{33} \approx 800$	298	
$KTaNbO_3$	Potassium tantalate niobate (KTN)	34,000	273	10^4
		6,000	293	10^4
$KTaO_3$	Potassium tantalate	242	298	2×10^5
$LaScO_3$	Lanthanum scandate	30	r.t.	
LiBr	Lithium bromide	12.1	r.t.	2×10^6
Li_2CO_3	Lithium carbonate	4.9	291	2×10^5

Formula	Name	ϵ_{ijk}	T/K	ν/Hz
LiCl	Lithium chloride	11.05	r.t.	2×10^6
LiD	Lithium deuteride	14.0 ± 0.5	r.t.	i.r.
LiF	Lithium fluoride	9.00	298	10^2—10^7
		9.11	353	10^2—10^7
LiGaO$_2$	Lithium metagallate	$\epsilon_{11}^T = 7.0$, $\epsilon_{22}^T = 6.0$	r.t.	
		$\epsilon_{33}^T = 9.5$	r.t.	
		$\epsilon_{11}^S = 6.8$, $\epsilon_{22}^S = 5.8$	r.t.	
Li^{6}H	Lithium-6 hydride	13.2 ± 0.5	r.t.	
Li^{7}H	Lithium-7 hydride	12.9 ± 0.5	r.t.	
LiH$_3$(SeO$_3$)$_2$	Lithium trihydrogen selenite	29	298	10^4
		$\epsilon_{11} = 13.0$	r.t.	
		$\epsilon_{22} = 12.9$	r.t.	
		$\epsilon_{33} = 46$	r.t.	
LiI	Lithium iodide	11.03	r.t.	2×10^6
LiIO$_3$	Lithium iodate	$\epsilon_{11} = \epsilon_{22} = 65$	294.5	10^3
		$\epsilon_{33} = 554$	298	
LiNH$_4$C$_4$H$_4$O$_6 \cdot$ H$_2$O	Lithium ammonium tartrate (LAT)	$\epsilon_{11}^T = 7.2$	298	
		$\epsilon_{22}^T = 8.0$	298	
		$\epsilon_{33}^T = 6.9$	298	
LiNa$_3$CrO$_4 \cdot$ 6H$_2$O	Lithium trisodium chromate	8.0	r.t.	10^3
LiNa$_3$MoO$_4 \cdot$ 6H$_2$O	Lithium trisodium molybdate	$\epsilon_{11} = 6.7$	r.t.	10^3
		$\epsilon_{33} = 5.3$	r.t.	10^3
LiNbO$_3$	Lithium niobate	$\epsilon_{11} = \epsilon_{22} = 82$	298	10^5
		$\epsilon_{33} = 30$	298	10^5
Li$_2$SO$_4 \cdot$ H$_2$O	Lithium sulfate monohydrate	$\epsilon_{11} = 5.6$	298	
		$\epsilon_{22} = 10.3$	298	
		$\epsilon_{33} = 6.5$	298	
		$\epsilon_{13} = 0.07$	298	
LiTaO$_3$	Lithium tantalate	$\epsilon_{11} = \epsilon_{22} = 53$	r.t.	10^5
		$\epsilon_{33} = 46$	r.t.	10^5
		$\epsilon_{11}^S = \epsilon_{22}^S = 41$	r.t.	
		$\epsilon_{33}^S = 43$	r.t.	
		$\epsilon_{11}^T = \epsilon_{22}^T = 51$	r.t.	
		$\epsilon_{33}^T = 45$	r.t.	
LiTlC$_4$H$_4$O$_6 \cdot$ H$_2$O	Lithium thallium tartrate (LTT)	$\epsilon_{11} \approx 20$	80	
Mg$_3$B$_7$O$_{13}$Cl	Magnesium borate monochloride (boracite)	$\epsilon_{11} = 14.1$	r.t.	5×10^5
MgCO$_3$	Magnesium carbonate	8.1	291	2×10^5
MgNb$_2$O$_6$	Magnesium niobate	$\epsilon_{11} = 16.4 \pm 0.5$	r.t.	$(5$—$500) \times 10^3$
		$\epsilon_{22} = 20.9 \pm 0.5$	r.t.	$(5$—$500) \times 10^3$
		$\epsilon_{33} = 32.4 \pm 0.5$	r.t.	$(5$—$500) \times 10^3$
MgO	Magnesium oxide (Periclase)	9.65	298	10^2—10^8
(MgO)$_x$Al$_2$O$_3$	Spinel	8.6	r.t.	
MgSO$_4$	Magnesium sulfate	8.2	r.t.	
MgSO$_4 \cdot$ 7H$_2$O	Magnesium sulfate septa hydrate	5.46	r.t.	
MgTiO$_3$	Magnesium titanate	13.5	r.t.	
MgWO$_4$	Magnesium tungstate	$\epsilon_{11} = 18.0 \pm 1$	r.t.	$(5$—$500) \times 10^3$
		$\epsilon_{22} = 18.0 \pm 1$	r.t.	$(5$—$500) \times 10^3$
MnNb$_2$O$_6$	Manganese niobate	$\epsilon_{11} = 17.4 \pm 2$	r.t.	$(5$—$500) \times 10^3$
		$\epsilon_{22} = 16.1 \pm 0.5$	r.t.	$(5$—$500) \times 10^3$
		$\epsilon_{33} = 30.7 \pm 1$	r.t.	$(5$—$500) \times 10^3$
MnO	Manganese oxide (Pyrolusite)	12.8	r.t.	6×10^{10}
MnO$_2$	Manganese dioxide	$\sim 10^4$	298	10^4
Mn$_2$O$_3$	Manganese sesquioxide	8	r.t.	6×10^{10}
MnWO$_4$	Manganese tungstate	$\epsilon_{11} = 19.3 \pm 1.3$	r.t.	$(5$—$500) \times 10^3$
		$\epsilon_{22} = 14.3 \pm 0.5$	r.t.	$(5$—$500) \times 10^3$
		$\epsilon_{33} = 16.5 \pm 1.1$	r.t.	$(5$—$500) \times 10^3$
N(CH$_3$)$_4$HgBr$_3$	Tetramethylammonium tribromo mercurate (TTM)	~ 10	233—373	

Formula	Name	ϵ_{ijk}	T/K	ν/Hz
$N(CH_3)_4HgI_3$	Tetramethylammonium triiodo mercurate (TTM)	~10	233—373	
$N_4(CH_2)_6$	Hexamethylene tetramine (HMTA)	2.6 ± 0.2	r.t.	10^9—10^{10}
$(ND_4)_2BeF_4$	Deuteroammonium fluoberyllate	$\epsilon_{11} = 10$	r.t.	
		$\epsilon_{22} = 9$	r.t.	
		$\epsilon_{33} = 9$	r.t.	
$(ND_4)_2SO_4$	Deuteroammonium sulfate	$\epsilon_{11} = 9$	r.t.	
		$\epsilon_{22} = 10$	r.t.	
		$\epsilon_{33} = 9$	r.t.	
$(NH_2 \cdot CH_2COOH)_3 \cdot H_2SO_4$	Triglycine sulfate (TGS)	$\epsilon_{11} = 9$	273	10^4
		$\epsilon_{22} = 30$	273	10^4
		$\epsilon_{33} = 6.5$	273	10^4
$(NH_2 \cdot CH_2COOH)_3 \cdot H_2SeO_4$	Triglycine selenate (TGSe)	200	293	1.6×10^3
$(NH_2 \cdot CH_2 COOH)_3 \cdot H_2BeF_4$	Triglycine fluorberyllate (TGFB)	$\epsilon_{22} = 12$	273	10^4
$NH_4Al(SO_4)_2 \cdot 12H_2O$	Ammonium alum	6	r.t.	10^{12}
$(NH_4)_2BeF_4$	Ammonium fluorberyllate	$\epsilon_{11} = \epsilon_{22} = 7.8$	123	10^5
		$\epsilon_{33} = 7.1$	123	10^5
		$\epsilon_{11} = \epsilon_{22} = 8.8$	293	10^5
		$\epsilon_{33} = 9.2$	293	10^5
NH_4Br	Ammonium bromide	7.1	r.t.	7×10^5
NH_4I	Ammonium iodide	9.8	r.t.	
$(NH_4)_2C_2H_6O_6$	Ammonium tartrate	$\epsilon_{11} = 6.45$	r.t.	10^3
		$\epsilon_{22} = 6.8$	r.t.	10^3
		$\epsilon_{33} = 6.0$	r.t.	10^3
$(NH_4)_2Cd_2(SO_4)_3$	Ammonium cadmium sulfate	10.0	r.t.	10^4
NH_4Cl	Ammonium chloride	6.9	r.t.	7×10^5
$NH_4(ClCH_2COO)$	Ammonium monochloroacetate	5	r.t.	2×10^6
$NH_4Cr(SO_4)_2 \cdot 12H_2O$	Ammonium chrome alum	6.5	r.t.	175×10^3
NH_4HSO_4	Ammonium bisulfate	165	273	5×10^4
$NH_4H_2AsO_4$	Ammonium dihydrogen arsenate (ADA)	5.1	265	9.5×10^9
		$\epsilon_{11} = \epsilon_{22} = 85$	298	10^3
		$\epsilon_{33} = 22$	298	
$NH_4H_2PO_4$	Ammonium dihydrogen phosphate (ADP)	$\epsilon_{11} = \epsilon_{22} = 57.1 \pm 0.6$	294.5	10^5—35×10^9
		$\epsilon_{33} = 14.0 \pm 0.3$	294	10^5—36×10^9
$ND_4D_2PO_4$	Ammonium dideuterium phosphate (ADDP)	$\epsilon_{11} = \epsilon_{22} = 74, \epsilon_{33} = 24$	300	
NH_4NO_3	Ammonium nitrate	10.7	322	$(5—50) \times 10^3$
$(NH_4)_2SO_4$	Ammonium sulfate	$\epsilon_{11} = \epsilon_{22} = 8.0$	123	10^5
		$\epsilon_{33} = 6.3$	123	10^5
		$\epsilon_{11} = \epsilon_{22} = 10.0$	293	10^5
		$\epsilon_{33} = 9.3$	293	10^5
$(NH_4)_2UO_2(C_2O_4)_2$	Ammonium uranyl oxalate	8.03	r.t.	10^4—3.3×10^9
$(NH_4)_2UO_2(C_2O_4)_2 \cdot 3H_2O$	Ammonium uranyl oxalate trihydrate	6.06	r.t.	10^4—3.3×10^9
$NaBr$	Sodium bromide	6.44	298	1.6×10^3
$NaBrO_3$	Sodium bromate	$\epsilon_{11}^T = 5.70$	298	10^3
$NaCN$	Sodium cyanide	7.55	293	10^5
$NaCO_3$	Sodium carbonate	8.75	291	2×10^5
$NaCO_3 \cdot 10H_2O$	Sodium carbonate decahydrate	5.3	r.t.	6×10^7
$NaCl$	Sodium chloride	5.9	298	10^2—10^7
		5.45	4.2	
$NaClO_3$	Sodium chlorate	$\epsilon_{11}^T = 5.76$	301	10^3
		5.28	r.t.	10^3
$NaClO_4$	Sodium perchlorate	5.76	r.t.	10^3
NaF	Sodium fluoride	5.08 ± 0.02	r.t.	5×10^3
$NaH_3(SeO_3)_2$	Sodium trihydrogen selenite	$\epsilon_{11} \approx 75$	273	2×10^5
$NaD_3(SeO_3)_2$	Sodium trideuterium selenite	$\epsilon_{11} \approx 220$	273	2×10^5
NaI	Sodium iodide	7.28 ± 0.03	r.t.	
$NaK(C_4H_2D_2O_6) \cdot 4D_2O$	Sodium potassium tartrate tetradeutrate (double deuterated Rochelle salt)	$\epsilon_{11} = 70$	273	10^3
		$\epsilon_{22} = 8.9$	273	10^3

Formula	Name	ϵ_{ijk}	T/K	ν/Hz
$NaK(C_4H_4O_6) \cdot 4H_2O$	Sodium potassium tartrate tetrahydrate (Rochelle salt)	$\epsilon_{11} = 170$	273	10^3
		$\epsilon_{22} = 9.1$	273	10^3
$NaNH_4(C_4H_4O_6) \cdot 4H_2O$	Sodium ammonium tartrate (Ammonium Rochelle salt)	$\epsilon_{11} = 8.4$	298	
		$\epsilon_{22} = 9.2$	298	
		$\epsilon_{33} = 9.5$	298	
$NaNbO_3$	Sodium niobate	$\epsilon_{33} = 670 \pm 13$	r.t.	
		$\epsilon_{11} = \epsilon_{22} = 76 \pm 2$	r.t.	
$NaNO_2$	Sodium nitrite	$\epsilon_{11} = 7.4$	r.t.	5×10^5
		$\epsilon_{22} = 5.5$	r.t.	5×10^5
		$\epsilon_{33} = 5.0$	r.t.	5×10^5
$NaNO_3$	Sodium nitrate	6.85	292	2×10^5
$NaSO_4$	Sodium sulfate	7.90	r.t.	
$NaSO_4 \cdot 10H_2O$	Sodium sulfate decahydrate	5.0	r.t.	
$Na_2S_2O_3 \cdot 5H_2O$	Sodium sulfate pentahydrate	7	250—290	$300—10^4$
$Na_2UO_2(C_2O_4)_2$	Sodium uranyl oxalate	5.18	r.t.	
$NdAlO_3$	Neodymium aluminate	17.5	r.t.	
$NdScO_3$	Neodymium scandate	27	r.t.	
$Ni_3B_7O_{13}I$	Nickel iodine boracite	$\epsilon_{11} = 14$	260	
$NiNb_2O_6$	Nickel niobate	$\epsilon_{11} = 16.0 \pm 0.5$	r.t.	$(5—500) \times 10^3$
		$\epsilon_{22} = 23.8 \pm 1.8$	r.t.	$(5—500) \times 10^3$
		$\epsilon_{33} = 31.3 \pm 2.5$	r.t.	$(5—500) \times 10^3$
NiO	Nickel oxide	11.9	298	10^5
$NiSO_4 \cdot 6H_2O$	Nickel sulfate hexahydrate	$\epsilon_{11} = 6.2$	r.t.	
		$\epsilon_{33} = 6.8$	r.t.	
$NiWO_4$	Nickel tungstate	$\epsilon_{11} = 17.4 \pm 2.4$	r.t.	$(5—500) \times 10^3$
		$\epsilon_{22} = 13.6 \pm 1.0$	r.t.	$(5—500) \times 10^3$
		$\epsilon_{33} = 19.7 \pm 0.6$	r.t.	$(5—500) \times 10^3$
P	Phosphorous (red)	4.1	r.t.	10^8
	Phosphorous (yellow)	3.6	r.t.	10^8
$[P(CH_3)_4]HgBr_3$	Tetramethylphosphonium tribromo mercurate (TTM)	~10	233—373	
$PbBr_2$	Lead bromide	>30	293	$(0.5—3) \times 10^6$
$PbCO_3$	Lead carbonate	18.6	288	10.8
$Pb(C_2H_3O_2)_2$	Lead acetate	2.6	290—295	10^6
$PbCl_2$	Lead chloride	33.5	273	$(0.5—3) \times 10^6$
Pb_2CoWO_6	Lead cobalt tungstate	~250	r.t.	
PbF_2	Lead fluoride	26.3	r.t.	
$PbHfO_3$	Lead hafnate	390	300	10^5
		185	400	
PbI_2	Lead iodide	20.8	293	$(0.5—3) \times 10^6$
$Pb_3MgNb_2O_9$	Lead magnesium niobate	10,000	297	
$PbMoO_4$	Lead molybdate	$\epsilon_{11} = 34.0 \pm 0.4$	297.5	1.6×10^3
		$\epsilon_{33} = 40.6 \pm 0.2$	297.5	1.6×10^3
$Pb(NO_3)_2$	Lead nitrate	16.8	r.t.	$(0.5—3) \times 10^6$
$PbNb_2O_6$	Lead niobate	$\epsilon_{33}^T = 180$	298	
PbO	Lead oxide	25.9	r.t.	2×10^6
PbS	Lead sulfide (Galena)	190	77	i.r.
		200 ± 35	r.t.	i.r.
$PbSO_4$	Lead sulfate	14.3	290—295	10^6
$PbSe$	Lead selenide	280	r.t.	i.r.
$PbTa_2O_6$	Lead metatantalate	$\epsilon_{11} = \epsilon_{22} \approx 300$	r.t.	10^4
		$\epsilon_{33} = 150$	r.t.	10^4
$PbTe$	Lead telluride	450	r.t.	i.r.
		40	77	$10^4—15 \times 10^4$
		430	4.2	$10^4—15 \times 10^4$
$PbTiO_3$	Lead titanate	~200	r.t.	10^3
$PbWO_4$	Lead tungstate	$\epsilon_{11} = \epsilon_{22} = 23.6 \pm 0.3$	297.5	1.59×10^3
		$\epsilon_{33} = 31.0 \pm 0.4$	297.5	1.59×10^3
$Pb(Zn_{1/3}Nb_{2/3})O_3$	Lead zinc niobate	7	300	$10^3, 300 \times 10^3$
$PbZrO_3$	Lead zirconate	200	400	

Formula	Name	ϵ_{ijk}	T/K	ν/Hz
$RbAl(SO_4)_2 \cdot 12H_2O$	Rubidium alum	5.1	r.t.	10^{12}
$RbBr$	Rubidium bromide	4.83	300	
Rb_2CO_3	Rubidium carbonate	4.87 ± 0.02	r.t.	5×10^3
$RbCl$	Rubidium chloride	4.91 ± 0.02	r.t.	5×10^3
$RbCr(SO_4)_2 \cdot 12H_2O$	Rubidium chrome alum	5.0	r.t.	10^{12}
RbF	Rubidium fluoride	5.91	r.t.	2×10^6
$RbHSO_4$	Rubidium bisulfate	$\epsilon_{11} = 7$	r.t.	10^5
		$\epsilon_{22} = 8$	r.t.	10^5
		$\epsilon_{33} = 10$	r.t.	10^5
RbH_2AsO_4	Rubidium dihydrogen arsenate (RDA)	3.90	273	9.5×10^9
RbH_2PO_4	Rubidium dihydrogen phosphate (RDP)	6.15	285	9.5×10^9
RbI	Rubidium iodide	4.94 ± 0.02	r.t.	5×10^3
$RbInSO_4$	Rubidium indium sulfate	6.85	r.t.	
$RbNO_3$	Rubidium nitrate	20—380	433—488	10^6
		30	488—538	10^6
S	Sulfur	$\epsilon_{11} = 3.75$	298	10^2—10^3
		$\epsilon_{22} = 3.95$	298	10^2—10^3
		$\epsilon_{33} = 4.44$	298	10^2—10^3
	sublimed	3.69	298	10^2—10^3
$SC(NH_2)_2$	Thiourea	$\epsilon_{11} = \epsilon_{33} \approx 3$	77—300	10^3
		$\epsilon_{22} = 35$	300	10^3
Sb_2O_3	Antimonous sesquioxide	12.8	r.t.	$(1.5—2) \times 10^3$
Sb_2S_3	Antimonous sulfide (stibnite)	$\epsilon_{11} = \epsilon_{22} = 15$	r.t.	10^3
		$\epsilon_{33} = 180$	r.t.	10^3
Sb_2Se_3	Antimonous selenide	~110	r.t.	$(10—16.5) \times 10^9$
$SbSI$	Antimonous sulfide iodide	2000	273	10^5
		$\epsilon_{11} = \epsilon_{22} \approx 25$	r.t.	10^3—10^5
		$\epsilon_{33} \approx 5 \times 10^4$	295	10^3—10^5
Se	Selenium	$\epsilon_{11} = \epsilon_{22} = 11$	300	24×10^9
	(monocrystal)			
		$\epsilon_{33} = 21$	300	24×10^9
	(amorphous)	6.0	298	10^2—10^{10}
Si	Silicon	12.1	4.2	10^7—10^9
SiC	Silicon carbide			
	cubic	9.72	r.t.	i.r.
	6H	$\epsilon_{11} = \epsilon_{22} = 9.66$	r.t.	i.r.
		$\epsilon_{33} = 10.03$	r.t.	i.r.
		9.7 ± 0.1	1.8	i.r.
Si_3N_4	Silicon nitride	4.2 (film)	r.t.	10^3
SiO	Silicon monoxide	5.8	r.t.	10^3
SiO_2	Silicon dioxide	$\epsilon_{11} = 4.42$	r.t.	9.4×10^{10}
		$\epsilon_{22} = 4.41$	r.t.	9.4×10^{10}
		$\epsilon_{33} = 4.60$	r.t.	9.4×10^{10}
$Sm_2(MoO_4)_3$	Samarium molybdate	12	298	
SnO_2	Stannic dioxide	$\epsilon_{11} = \epsilon_{22} = 14 \pm 2$	r.t.	10^4—10^{10}
		$\epsilon_{33} = 9.0 \pm 0.5$	r.t.	10^4—10^{10}
$SnSb$	Tin antimonide	147	r.t.	10^4—10^6
$SnTe$	Tin telluride	1770 ± 300	r.t.	i.r.
$Sr(COOH)_2 \cdot 2H_2O$	Strontium formate dihydrate	6.1	r.t.	10^3
$SrCO_3$	Strontium carbonate	8.85	298	2×10^5
$SrCl_2$	Strontium chloride	9.19	r.t.	
$Sr_4Cl_2 \cdot 6H_2O$	Strontium chloride hexahydrate	8.52	r.t.	
SrF_2	Strontium fluoride	6.50	300	5×10^2—10^{11}
$SrMoO_4$	Strontium molybdate	$\epsilon_{11} = \epsilon_{22} = 31.7 \pm 0.2$	297.5	1.59×10^3
		$\epsilon_{33} = 41.7 \pm 0.2$	297.5	1.59×10^3
$Sr(NO_3)_2$	Strontium nitrate	5.33	292	2×10^5
$Sr_2Nb_2O_7$	Strontium niobate	$\epsilon_{11} = 75$	r.t.	10^3
		$\epsilon_{22} = 46$	r.t.	10^3
		$\epsilon_{33} = 43$	r.t.	10^3

Formula	Name	ϵ_{ijk}	T/K	ν/Hz
SrO	Strontium oxide	13.3 ± 0.3	273	2×10^6
SrS	Strontium sulfide	11.3	r.t.	7.25×10^6
$SrSO_4$	Strontium sulfate	11.5	r.t.	
$SrTiO_3$	Strontium titanate	332	298	10^3
		2080	78	10^3
$SrWO_4$	Strontium tungstate	$\epsilon_{11} = \epsilon_{22} = 25.7 \pm 0.2$	297.5	1.6×10^3
		$\epsilon_{33} = 34.1 \pm 0.2$	297.5	1.6×10^3
Ta_2O_5	Tantalum pentoxide (tantala)			
	α phase	$\epsilon_{11} = \epsilon_{22} = 30$	77	10^3
		$\epsilon_{33} = 65$	77	10^3
	β phase	24	292	10^3
$Tb(MoO_4)_3$	Terbium molybdate	11	298	
		$\epsilon_{11} = \epsilon_{22} = 33$	100—200	9.4×10^9
		$\epsilon_{33} = 53$	100—200	9.4×10^9
Te	Tellurium	$\epsilon_{11} = \epsilon_{22} = 33$	r.t.	
		$\epsilon_{33} = 54$	r.t.	
	polycrystalline	27.5	r.t.	i.r.
	monocrystalline	28.0	r.t.	i.r.
ThO_2	Thorium dioxide	18.9 ± 0.4	r.t.	3×10^5
TiO_2	Titanium dioxide (rutile)	$\epsilon_{11} = \epsilon_{22} = 86$	300	10^4—10^6
		$\epsilon_{33} = 170$	300	10^4—10^6
Ti_2O_3	Titanium sesquioxide	30	77	6×10^{10}
TlBr	Thallium bromide	30	293	10^3—10^7
TlCl	Thallous chloride	32.2 ± 0.2	293	10^3—10^5
TlI	Thallous iodide (orthorhombic)	20.7 ± 0.2	293	10^4
		37.3	193	10^7
$TlNO_3$	Thallous nitrate	16.5	293	5×10^5
$TlSO_4$	Thallous sulfate	25.5	293	5×10^5
UO_2	Uranium dioxide	24	r.t.	3×10^5
WO_3	Tungsten trioxide	300		
$YMnO_3$	Yttrium manganate	20	r.t.	2×10^7
Y_2O_3	Yttrium sesquioxide	10	r.t.	10^6
$YbMnO_3$	Ytterbium manganate	20	r.t.	2×10^7
Yb_2O_3	Ytterbium sesquioxide	5.0 (film)	r.t.	10^3
ZnO	Zinc monoxide	$\epsilon_{11}^S = 8.33$	r.t.	
		$\epsilon_{33}^S = 8.84$	r.t.	
		$\epsilon_{11}^T = 9.26$	r.t.	
		$\epsilon_{33}^T = 11.0$	r.t.	
		$\epsilon_{11} = 9.26$	r.t.	
		$\epsilon_{33} = 8.2$	r.t.	
		8.15	r.t.	i.r.
ZnS	Zinc sulfide	$\epsilon_{11}^S = 8.08 \pm 2\%$	77	10^4
		$\epsilon_{11}^S = 8.32 \pm 2\%$	298	10^4
		$\epsilon_{11}^T = 8.14 \pm 2\%$	77	10^4
		$\epsilon_{11}^T = 8.37 \pm 2\%$	298	10^4
ZnSe	Zinc selenide	$\epsilon_{11}^T = \epsilon_{11}^S = 9.12 \pm 2\%$	298	10^4
ZnTe	Zinc telluride	$\epsilon_{11}^T = \epsilon_{11}^S = 10.10 \pm 2\%$	r.t.	
$ZnWO_4$	Zinc tungstate	$\epsilon_{22} = 16.1 \pm 0.5$	r.t.	$(5—500) \times 10^3$
ZrO_2	Zirconium dioxide (zirconia)	12.5	r.t.	2×10^6

CURIE TEMPERATURE OF SELECTED FERROELECTRIC CRYSTALS

H. P. R. Frederikse

The following table lists the major ferroelectric crystals and their Curie temperatures, T_C.

REFERENCE

Young, K. F. and Frederikse, H. P. R., *J. Phys. Chem. Ref. Data*, 2, 313, 1973.

Name or acronym	Formula	T_C/K
Potassium dihydrogen phosphate group		
KDP	KH_2PO_4	123
KDA	KH_2AsO_4	97
KDDP	KD_2PO_4	213
KDDA	KD_2AsO_4	162
RDP	RbH_2PO_4	146
RDA	RbH_2AsO_4	111
RDDP	RbD_2PO_4	218
RDDA	RbD_2AsO_4	178
CDP	CsH_2PO_4	159
CDA	CsH_2AsO_4	143
CDDA	CsD_2AsO_4	212
Rochelle salt group		
Rochelle salt	$NaKC_4H_4O_6 \cdot 4H_2O$	255—297
Deuterated Rochelle salt	$NaKC_4H_2D_2O_6 \cdot 4H_2O$	251—308
Ammonium Rochelle salt	$NaNH_4C_4H_4O_6 \cdot 4H_2O$	109
LAT	$LiNH_4C_4H_4O_6 \cdot H_2O$	106
Triglycine sulfate group		
TGS	$(NH_2CH_2COOH)_3 \cdot H_2SO_4$	322
TGSe	$(NH_2CH_2COOH)_3 \cdot H_2SeO_4$	295
TGFB	$(NH_2CH_2COOH)_3 \cdot H_2BeF_4$	346
AFB	$(NH_4)_2BeF_4$	176
HADA	$HNH_4(ClCH_2COO)_2$	128
Perovskites and related compounds		
Barium titanate	$BaTiO_3$	406, 278, 193
Lead titanate	$PbTiO_3$	765
Potassium niobate	$KNbO_3$	712
Potassium tantalate niobate	$KTa_{2/3}Nb_{1/3}O_3$	271, 220, 170
Lithium niobate	$LiNbO_3$	1483
Lithium tantalate	$LiTaO_3$	891
Barium titanium niobate	$Ba_6Ti_2Nb_8O_{30}$	521
Ba-Na niobate ("Bananas")	$Ba_2NaNb_5O_{15}$	833
Potassium iodate	KIO_3	485, 343, 257—263, 83
Lithium iodate	$LiIO_3$	529
Potassium nitrate	KNO_3	397
Sodium nitrate	$NaNO_3$	548
Rubidium nitrate	$RbNO_3$	437—487
Miscellaneous compounds		
Cesium trihydrogen selenite	$CsH_3(SeO_3)_2$	143
Lithium trihydrogen selenite	$LiH_3(SeO_3)_2$	$T_C > T_{mp}$

Name or acronym	Formula	T_C/K
Potassium selenate	K_2SeO4	93
Methyl ammonium alum (MASD)	$CH_3NH_3Al(SO_4)_2 \cdot 12H_2O$	177
Ammonium cadmium sulfate	$(NH_4)_2Cd_2(SO_4)_3$	95
Ammonium bisulfate	$(NH_4)HSO_4$	271
Ammonium sulfate	$(NH_4)_2SO_4$	224
Ammonium nitrate	NH_4NO_3	398, 357, 305, 255
Colemanite	$CaB_3O_4(OH)_3 \cdot H_2O$	266
Cadmium pyroniobite	$Cd_2Nb_2O_7$	185
Gadolinium molybdate	$Gd_2(MoO_4)_3$	432

PROPERTIES OF ANTIFERROELECTRIC CRYSTALS

H. P. R. Frederikse

Some important antiferroelectric crystals are listed here with their Curie Temperatures T_C. The last column gives the constant T_0 which appears in the Curie-Weiss law describing the dielectric constant of these materials above the Curie Temperature:

$$\varepsilon = \text{const.}/(T - T_0)$$

Name or acronym	Formula	T_C/K	T_0/K
ADP	$NH_4H_2PO_4$	148	
ADA	$NH_4H_2AsO_4$	216	
ADDP	$NH_4D_2PO_4$	242, 245	
ADDA	$NH_4D_2AsO_4$	299	
A_dDDP	$ND_4D_2PO_4$	243	
A_dDDA	$ND_4D_2AsO_4$	304	
Sodium niobate	$NaNbO_3$	911, 793	
Lead hafnate	$PbHfO_3$	476	378
Lead zirconate	$PbZrO_3$	503	475
Lead metaniobate	$PbNb_2O_6$	843	530
Lead metatantalate	$PbTa_2O_6$	543	533
Tungsten trioxide	WO_3	1010	
Potassium strontium niobate	$KSr_2Nb_5O_{15}$	427	413
Sodium nitrite	$NaNO_2$	437	437
Sodium trihydrogen selenite	$NaH_3(SeO_3)_2$	193	192
Sodium trideuterium selenite	$NaD_3(SeO_3)_2$	271	245
Ammonium trihydrogen iodate	$(NH_4)_2H_3IO_3$	245	

DIELECTRIC CONSTANTS OF GLASSES

Type	Dielectric constant at 100 MHz (20°C)	Volume resistivity (350°C megohm-cm)	Loss factor[a]
Corning 0010	6.32	10	0.015
Corning 0080	6.75	0.13	0.058
Corning 0120	6.65	100	0.012
Pyrex 1710	6.00	2,500	0.025
Pyrex 3320	4.71	—	0.019
Pyrex 7040	4.65	80	0.013
Pyrex 7050	4.77	16	0.017
Pyrex 7052	5.07	25	0.019
Pyrex 7060	4.70	13	0.018
Pyrex 7070	4.00	1,300	0.0048
Vycor 7230	3.83	—	0.0061
Pyrex 7720	4.50	16	0.014
Pyrex 7740	5.00	4	0.040
Pyrex 7750	4.28	50	0.011
Pyrex 7760	4.50	50	0.0081
Vycor 7900	3.9	130	0.0023
Vycor 7910	3.8	1,600	0.00091
Vycor 7911	3.8	4,000	0.00072
Corning 8870	9.5	5,000	0.0085
G. E. Clear (silica glass)	3.81	4,000—30,000	0.00038
Quartz (fused)	3.75 4.1 (1 MHz)	—	0.0002 (1 MHz)

[a]   Power factor × dielectric constant equals loss factor.

PROPERTIES OF SUPERCONDUCTORS

L. I. Berger and B. W. Roberts

The following tables include superconductive properties of selected elements, compounds, and alloys. Individual tables are given for thin films, elements at high pressures, superconductors with high critical magnetic fields, and high critical temperature superconductors.

The historically first observed and most distinctive property of a superconductive body is the near total loss of resistance at a critical temperature (T_c) that is characteristic of each material. Figure 1(a) below illustrates schematically two types of possible transitions. The sharp vertical discontinuity in resistance is indicative of that found for a single crystal of a very pure element or one of a few well annealed alloy compositions. The broad transition, illustrated by broken lines, suggests the transition shape seen for materials that are not homogeneous and contain unusual strain distributions. Careful testing of the resistivity limit for superconductors shows that it is less than 4×10^{-23} ohm cm, while the lowest resistivity observed in metals is of the order of 10^{-13} ohm cm. If one compares the resistivity of a superconductive body to that of copper at room temperature, the superconductive body is at least 10^{17} times less resistive.

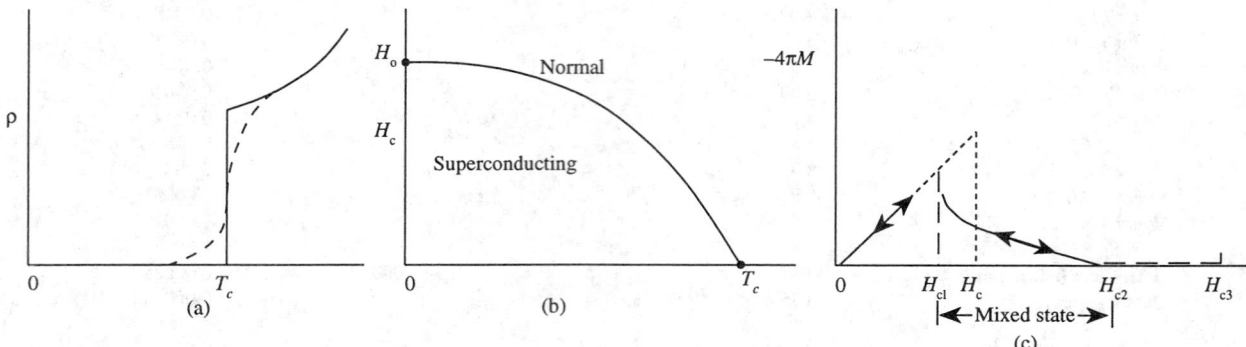

FIGURE 1. Physical properties of superconductors. (a) Resistivity vs. temperature for a pure and perfect lattice (solid line); impure and/or imperfect lattice (broken line). (b) Magnetic-field temperature dependence for Type-I or "soft" superconductors. (c) Schematic magnetization curve for "hard" or Type-II superconductors.

The temperature interval ΔT_c, over which the transition between the normal and superconductive states takes place, may be of the order of as little as 2×10^{-5} K or several K in width, depending on the material state. The narrow transition width was attained in 99.9999% pure gallium single crystals.

A Type-I superconductor below T_c, as exemplified by a pure metal, exhibits perfect diamagnetism and excludes a magnetic field up to some critical field H_c, whereupon it reverts to the normal state as shown in the H-T diagram of Figure 1(b).

The magnetization of a typical high-field superconductor is shown in Figure 1(c). The discovery of the large current-carrying capability of Nb_3Sn and other similar alloys has led to an extensive study of the physical properties of these alloys. In brief, a high-field superconductor, or Type-II superconductor, passes from the perfect diamagnetic state at low magnetic fields to a mixed state and finally to a sheathed state before attaining the normal resistive state of the metal. The magnetic field values separating the four stages are given as H_{c1}, H_{c2}, and H_{c3}. The superconductive state below H_{c1} is perfectly diamagnetic, identical to the state of most pure metals of the "soft" or Type-I superconductor. Between H_{c1} and H_{c2} a "mixed superconductive state" is found in which fluxons (a minimal unit of magnetic flux) create lines of normal flux in a superconductive matrix. The volume of the normal state is proportional to $-4\pi M$ in the "mixed state" region. Thus at H_{c2} the fluxon density has become so great as to drive the interior volume of the superconductive body completely normal. Between H_{c2} and H_{c3} the superconductor has a sheath of current-carrying superconductive material at the body surface, and above H_{c3} the normal state exists. With several types of careful measurement, it is possible to determine H_{c1}, H_{c2}, and H_{c3}. Table 6 contains some of the available data on high-field superconductive materials.

High-field superconductive phenomena are also related to specimen dimension and configuration. For example, the Type-I superconductor, Hg, has entirely different magnetization behavior in high magnetic fields when contained in the very fine sets of filamentary tunnels found in an unprocessed Vycor glass. The great majority of superconductive materials are Type-II. The elements in very pure form and a very few precisely stoichiometric and well annealed compounds are Type I with the possible exceptions of vanadium and niobium.

Metallurgical Aspects. The sensitivity of superconductive properties to the material state is most pronounced and has been used in a reverse sense to study and specify the detailed state of alloys. The mechanical state, the homogeneity, and the presence of impurity atoms and other electron-scattering centers are all capable of controlling the critical temperature and the current-carrying capabilities in high-magnetic fields. Well annealed specimens tend to show sharper transitions than those that are strained or inhomogeneous. This sensitivity to mechanical state underlines a general problem in the tabulation of properties for superconductive materials. The occasional divergent values of the critical temperature and of the critical fields quoted for a Type-II superconductor may lie in the variation in sample preparation. Critical temperatures of materials studied early in the history of superconductivity must be evaluated in light of the probable metallurgical state of the material, as well as the availability of less pure starting elements. It has been noted that recent work has given extended consideration to the metallurgical aspects of sample preparation.

Symbols in tables: T_c: Critical temperature; H_o: Critical magnetic field in the $T = 0$ limit; θ_D: Debye temperature; and γ: Electronic specific heat.

TABLE 1
Selective Properties of Superconductive Elements

Element	T_c(K)	H_o(oersted)	θ_D(K)	γ(mJ mol^{-1}K^{-1})
Al	1.175 ± 0.002	104.9 ± 0.3	420	1.35
Am* (α,?)	0.6			
Am* (β,?)	1.0			
Be	0.026			0.21
Cd	0.517 ± 0.002	28 ± 1	209	0.69
Gd	1.083 ± 0.001	58.3 ± 0.2	325	0.60
Gd (β)	5.9, 6.2	560		
Gd (γ)	7	950, HF[a]		
Gd (Δ)	7.85	815, HF		
Hf	0.128	12.7		2.21
Hg (α)	4.154 ± 0.001	411 ± 2	87, 71.9	1.81
Hg (β)	3.949	339	93	1.37
In	3.408 ± 0.001	281.5 ± 2	109	1.672
Ir	0.1125 ± 0.001	16 ± 0.05	425	3.19
La (α)	4.88 ± 0.02	800 ± 10	151	9.8
La (β)	6.00 ± 0.1	1096, 1600	139	11.3
Lu	0.1 ± 0.03	350 ± 50		
Mo	0.915 ± 0.005	96 ± 3	460	1.83
Nb	9.25 ± 0.02	2060 ± 50, HF	276	7.80
Os	0.66 ± 0.03	70	500	2.35
Pa	1.4			
Pb	7.196 ± 0.006	803 ± 1	96	3.1
Re	1.697 ± 0.006	200 ± 5	4.5	2.35
Ru	0.49 ± 0.015	69 ± 2	580	2.8
Sn	3.722 ± 0.001	305 ± 2	195	1.78
Ta	4.47 ± 0.04	829 ± 6	258	6.15
Tc	7.8 ± 0.1	1410, HF	411	6.28
Th	1.38 ± 0.02	1.60 ± 3	165	4.32
Ti	0.40 ± 0.04	56	415	3.3
Tl	2.38 ± 0.02	178 ± 2	78.5	1.47
U	0.2			
V	5.40 ± 0.05	1408	383	9.82
W	0.0154 ± 0.0005	1.15 ± 0.03	383	0.90
Zn	0.85 ± 0.01	54 ± 0.3	310	0.66
Zr	0.61 ± 0.15	47	290	2.77
Zr (ω)	0.65, 0.95			

TABLE 2
Range of Critical Temperatures Observed for Superconductive Elements in Thin Films Condensed Usually at Low Temperatures

Element	T_c Range (K)	Comments	Element	T_c Range (K)	Comments
Al	1.15—5.7	HF[a]	Nb	2.0—10.1	
Be	5—9.75	HF	Pb	1.8—7.5	
Bi	6.17—6.6		Re	1.7—7	
Cd			Sn	3.5—6	
(Disordered)	0.79—0.91		Ta	<1.7—4.51	HF[a]
(Ordered)	0.53—0.59		Tc	4.6—7.7	
Ga	2.5—8.5	HF	Ti	1.3 Max	
Hg	3.87—4.5		Tl	2.33—2.96	
In	2.2—5.6	HF	V	1.8—6.02	
La	3.55—6.74		W	<1.0—4.1	
Mo	3.3—8.0		Zn	0.77—1.9	

[a]   HF denotes high magnetic field superconductive properties.

TABLE 3
Elements Exhibiting Superconductivity Under or After Application of High Pressure

Element	T_c Range (K)	Pressure (kbar)	Element	T_c Range (K)	Pressure (kbar)
Al	1.98—0.075	0—62	Pb II	3.55	160
As	0.31—0.5	220—140	Re II	2.3 Max.	"Plastic"
	0.2—0.25	140—100			compression
Ba II	1—1.8	55—85	Sb (prepared 120	2.6—2.7	
III	1.8—5	85—144	kbar, held below		
IV	4.5—5.4	144—190	77K)		
Bi II	3.9	25—27	Sb II	3.55—3.40	85—150
III	6.55—7.25	28—38	Se II	6.75, 6.95	130
IV	7.0, 8.7—6.0	43, 43—62	Si	6.7—7.1	120—130
V	6.7, 8.3	48—80	Sn II	5.2—4.85	125—160
VI	8.55	90, 92—101	III	5.30	113
VII(?)	8.2	30	Te II	2.4—5.1	38—55
Ce (α)	0.020—0.045	20—35		4.1—4.2	53—62
Ce (α')	1.9—1.3	45—125	IV	4.72—4	63—80
Cs V	1.5	>125	( )	3.3—2.8	100—260
Ga II	6.38	≥35	Tl (cubic form)	1.45	35
II'	7.5	≥35 then P	(hexagonal form)	1.95	35
		removed	U	2.4—0.4	10—85
Ge	5.35	115	Y	1.7—2.5	110—160
La	5.5—12.9	0—210	Zr (omega form,	1—1.7	60—130
Lu	0.022—1.0	45—190	metastable)		
P	5.8	170			

TABLE 4
Superconductive Compounds and Alloys

All compositions are denoted on an atomic basis, i.e., AB, AB$_2$, or AB$_3$ for compounds, unless noted. Solid solutions or odd compositions may be denoted as A$_z$B$_{1-z}$ or A$_z$B. A series of three or more alloys is indicated as A$_x$B$_{1-x}$ or by actual indication of the atomic fraction range, such as A$_{0—0.6}$B$_{1—0.4}$. The critical temperature of such a series of alloys is denoted by a range of values or possibly the maximum value.

The selection of the critical temperature from a transition in the effective permeability, or the change in resistance, or possibly the incremental changes in frequency observed by certain techniques is not often obvious from the literature. Most authors choose the mid-point of such curves as the probable critical temperature of the idealized material, while others will choose the highest temperature at which a deviation from the normal state property is observed. In view of the previous discussion concerning the variability of the superconductive properties as a function of purity and other metallurgical aspects, it is recommended that appropriate literature be checked to determine the most probable critical temperature or critical field of a given alloy.

A very limited amount of data on critical fields, H_o, is available for these compounds and alloys; these values are given at the end of the table.

A. SUPERCONDUCTORS WITH T_c < 10 K

Substance	T_c, K	Crystal structure type
Ag$_{3.3}$Al	0.34	A12-cI58 (Mn)
Ag$_x$Al$_y$Zn$_{1-x-y}$	0.15	Cubic
AgBi$_2$	2.87—3.0	
Ag$_7$F$_{0.25}$N$_{0.75}$O$_{10.25}$	0.85—0.90	
Ag$_2$F	0.0.066	
Ag$_7$FO$_8$	0.3	Cubic
Ag$_{0.8—0.3}$Ga$_{0.2—0.7}$	6.5—8	
Ag$_4$Ge	0.85	Hex., c.p.
Ag$_{0.438}$Hg$_{0.562}$	0.64	D8$_2$
AgIn$_2$	~2.4	C16

TABLE 4
Superconductive Compounds and Alloys (continued)

Substance	T_c, K	Crystal structure type
$Ag_{0.1}In_{0.9}Te$ ($n = 1.4 \times 10^{22}$)*	1.2—1.89	B1
$Ag_{0.2}In_{0.8}Te$ ($n = 1.07 \times 10^{22}$)	0.77—1.00	B1
AgLa	0.94	B2-cP2 (CsCl)
AgLa (9.5 kbar)	1.2	B2
AgLu	0.33	B2-cP2
$AgMo_4S_5$	9.1	hR15 (Mo_6PbS_8)
$Ag_{1.2}Mo_6Se_8$	5.9	Same
Ag_7NO_{11}	1.04	Cubic
Ag_xPb_{1-x}	7.2 max.	
Ag_4Sn	0.1	h**
Ag_xSn_{1-x}	1.5—3.7	
Ag_xSn_{1-x} (film)	2.0—3.8	
$AgTe_3$	2.6	Cubic
AgTh	2.2	C16-tI12 (Al_2Cu)
$AgTh_2$	2.26	C16
$Ag_{0.03}Tl_{0.97}$	2.67	
$Ag_{0.94}Tl_{0.06}$	2.32	
AgY	0.33	B2-cP2 (CsCl)
Ag_xZn_{1-x}	0.5—0.845	
$AlAu_4$	0.4—0.7	Like A13
Al_2Au	0.1	C1-cF12 (CaF_2)
Al_2CMo_3	9.8—10.2	A13+trace 2nd. phase)
Al_2CaSi	5.8	
$Al_{0.131}Cr_{0.088}V_{0.781}$	1.46	Cubic
$AlGe_2$	1.75	
Al_2Ge_2U	1.6	LI_2-cP4 (Cu_3Au)
$AlLa_3$	5.57	DO_{19}
Al_2La	3.23	C15
Al_2Lu	1.02	C15-cF24 (Cu_2Mg)
Al_3Mg_2	0.84	F.C.C.
$AlMo_3$	0.58	A15
$AlMo_6Pd$	2.1	
AlN	1.55	B4
Al_2NNb_3	1.3	A13
Al_3Nb	0.64	tI8 (Al_3Ti)
AlOs	0.39	B2
Al_3Os	5.90	
AlPb (film)	1.2—7	
Al_2Pt	0.48—0.55	C1
Al_5Re_{24}	3.35	A12
AlSb	2.8	B4-tI4 (Sn)
Al_2Sc	1.02	C15-cF24 (Cu_2Mg)
Al_2Si_2U	1.34	LI_2-cP4 (Cu_3Au)
$AlTh_2$	0.1	C16-tI12 (Al_2Cu)
Al_3Th	0.75	DO_{19}
$Al_xTi_yV_{1-x-y}$	2.05—3.62	Cubic
$Al_{0.108}V_{0.892}$	1.82	Cubic
Al_2Y	0.35	C15-cF24 (Cu_2Mg)
Al_3Yb	0.94	LI_2-cP4 (Cu_3Au)
Al_xZn_{1-x}	0.5—0.845	
$AlZr_3$	0.73	LI_2
AsBiPb	9.0	
AsBiPbSb	9.0	
AsHfOs	3.2	C22-hP9 (Fe_2P)
AsHfRu	4.9	same

TABLE 4
Superconductive Compounds and Alloys (continued)

Substance	T_c, K	Crystal structure type
$As_{0.33}InTe_{0.67}$ ($n = 1.24 \times 10^{22}$)	0.85—1.15	B1
$As_{0.5}InTe_{0.5}$ ($n = 0.97 \times 10^{22}$)	0.44—0.62	B1
As_4La_3	0.6	cI28 (Th_3P_4)
$AsNb_3$	0.3	$L1_2$-tP32
$As_{0.50}Ni_{0.06}Pd_{0.44}$	1.39	C2
$AsNi_{0.25}Pd_{0.75}$	1.6	$B8_1$-hP4 (NiAs)
$AsOsZr$	8.0	C22-hP9 (Fe_2P)
$AsPb$	8.4	
$AsPd_2$ (low-temp. phase)	0.60	Hexagonal
$AsPd_2$ (high-temp. phase)	1.70	C22
$AsPd_5$	0.46	Complex
As_3Pd_5	1.9	
$AsRh$	0.58	B31
$AsRh_{1.4—1.6}$	< 0.03—0.56	Hexagonal
$AsSn$	4.10	
$AsSn$ ($n = 2.14 \times 10^{22}$)	3.41—3.65	B1
$As_{-2}Sn_{-3}$	3.5—3.6; 1.21—1.17	
As_3Sn_4 ($n = 0.56 \times 10^{22}$)	1.16—1.19	Rhombohedral
AsV_3	0.20	A15-cP8 (Cr_3Si)
Au_5Ba	0.4—0.7	$D2_d$
$AuBe$	2.64	B20
Au_2Bi	1.80	C15
Au_5Ca	0.34—0.38	$C15_b$
$AuGa_2$	1.6	C1-cF12 (CaF_2)
$AuGa$	1.2	B31
$Au_{0.40—0.92}Ge_{0.60—0.08}$	<0.32—1.63	Complex
$AuIn_2$	0.2	C1-cF12
$AuIn$	0.4—0.6	Complex
$AuLu$	<0.35	B2
$AuNb_3$	1.2	A2
$AuPb_2$	3.15	
$AuPb_2$ (film)	4.3	
$AuPb_3$	4.40	
$AuPb_3$ (film)	4.25	
Au_2Pb	1.18; 6—7	C15
$AuSb_2$	0.58	C2
$AuSn$	1.25	$B8_1$
Au_xSn_{1-x} (film)	2.0—3.8	
Au_5Sn	0.7—1.1	A3
$AuTa_{4.3}$	0.55	A15-cP8 (Cr_3Si)
Au_3Te_5	1.62	Cubic
$AuTh_2$	3.08	C16
$AuTl$	1.92	
AuV_3	0.74	A15
Au_xZn_{1-x}	0.50—0.845	
$AuZn_3$	1.21	Cubic
Au_xZr_y	1.7—2.8	A3
$AuZr_3$	0.92	A15
$B_2Ba_{0.67}Pt_3$	5.60	hP12 (B_2BaPt_3)
$BCMo_2$	5.4	Orthorhombic
$BCMo_2$	5.3—7.0	Same
$B_2Ca_{0.67}Pt_3$	1.57	hP12
B_4ErIr_4	2.1	tP18 (B_4CeCo_4)
B_4ErRh_4	4.3	oC108 (B_4LuRh_4)
B_4ErRh_4	8.7	tP18 (B_4CeCo_4)

TABLE 4
Superconductive Compounds and Alloys (continued)

Substance	T_c, K	Crystal structure type
BHf	3.1	Cubic
B_4HoIr_4	2.0	tP18
B_4HoRh_4	1.4	oC108
B_2Ir_3La	1.65	hP6 ($CaCu_5$)
B_2Ir_3Th	2.09	Same
B_4Ir_4Tm	1.6	tP18
B_6La	5.7	
B_2LaRh_3	2.82	hP6
$B_{12}Lu$	0.48	
B_2LuOs	2.66	oP16 (B_2LuRu)
B_2LuOs_3	4.62	hP6
B_4LuRh_4	6.2	oC108
B_2LuRu	9.86	oP16
B_4LuRu_4	2.0	tI72 (B_4LuRu_4)
BMo	0.5 (extrapol.)	
BMo_2	4.74	C16
BNb	8.25	B_f
B_4NdRh_4	5.3	tP18
B_2OsSc	1.34	oP16
B_2OsY	2.22	oP16
$B_2Pt_3Sr_{0.67}$	2.78	hP12 (B_2BaPt_3)
BRe_2	2.80; 4.6	
$B_4Rh_{3.4}Ru_{0.6}$	8.38	tI72
B_4Rh_4Sm	2.7	tP18
B_4Rh_4Th	4.3	Same
B_4Rh_4Tm	9.8	Same
B_4Rh_4Tm	5.4	oC108
$B_{0.3}Ru_{0.7}$	2.58	$D10_2$
B_4Ru_4Sc	7.2	tI72
B_2Ru_3Th	1.79	hP6
B_2Ru_3Y	2.85	Same
$B_2Ru\ Y$	7.80	oP16
B_4Ru_4Y	1.4	tI72
$B_{12}Sc$	0.39	
BTa	4.0	B_f
BTa_2	3.12	C16-tI12 (Al_2Cu)
B_6Th	0.74	
BW_2	3.1	C16
B_6Y	6.5—7.1	
$B_{12}Y$	4.7	
BZr	3.4	Cubic
$B_{12}Zr$	5.82	
$BaBi_3$	5.69	Tetragonal
$Ba_2Mo_{15}Se_{19}$	2.75	hP15 (Mo_6PbS_8)
$Ba_xO_3Sr_{1-x}Ti$ ($n = 4.2 \times 10^{19}$)	<0.1—0.55	
$Ba_{0.13}O_3W$	1.9	Tetragonal
$Ba_{0.14}O_3W$	<1.25—2.2	Hexagonal
$BaRh_2$	6.0	C15
$Be_{22}Mo$	2.51	Cubic ($Be_{22}Re$)
$Be_8Nb_5Zr_2$	5.2	
$Be_{0.98—0.92}Re_{0.02—0.08}$ (quenched)	9.5—9.75	Cubic
$Be_{0.957}Re_{0.043}$	9.62	Cubic ($Be_{22}Re$)
BeTc	5.21	Cubic
$Be_{22}W$	4.12	Cubic ($Be_{22}Re$)

TABLE 4
Superconductive Compounds and Alloys (continued)

Substance	T_c, K	Crystal structure type
$Be_{13}W$	4.1	Tetragonal
Bi_3Ca	2.0	
$Bi_{0.5}Cd_{0.13}Pb_{0.25}Sn_{0.12}$ (weight fractions)	8.2	
BiCo	0.42—0.49	
Bi_2Cs	4.75	C15
Bi_xCu_{1-x} (electrodeposited)	2.2	
BiCu	1.33—1.40	
Bi_3Fe	1.0	m**
$Bi_{0.019}In_{0.981}$	3.86	
$Bi_{0.05}In_{0.95}$	4.65	α-phase
$Bi_{0.10}In_{0.90}$	5.05	Same
$Bi_{0.15—0.30}In_{0.85—0.70}$	5.3—5.4	α- and β-phases
$Bi_{0.34—0.48}In_{0.66—0.52}$	4.0—4.1	
Bi_3In_5	4.1	
$BiIn_2$	5.65	β-phase
Bi_2Ir	1.7—2.3	
Bi_2Ir (quenched)	3.0—3.96	
BiK	3.6	
Bi_2K	3.58	C15
BiLi	2.47	$L1_o$, α-phase
$Bi_{4—9}Mg$	0.7—~1.0	
Bi_3Mo	3—3.7	
BiNa	2.25	$L1_o$
$BiNb_3$	4.5	A15-cP8 (Cr_3Si)
$BiNb_3$ (high pressure and temperature)	3.05	A15
BiNi	4.25	$B8_1$
Bi_3Ni	4.06	Orthorhombic
$BiNi_{0.5}Rh_{0.5}$	3.0	$B8_1$-hP4 (AsNi)
$Bi_{0.5}NiSb_{0.5}$	2.0	Same
$Bi_{1-0}Pb_{0-1}$	7.26—9.14	
$Bi_{1-0}Pb_{0-1}$ (film)	7.25—8.67	
$Bi_{0.05—0.40}Pb_{0.95—0.60}$	7.35—8.4	H.C.P. to ε-phase
Bi_2Pb	4.25	t**
BiPbSb	8.9	
$Bi_{0.5}Pb_{0.31}Sn_{0.19}$ (weight fractions)	8.5	
$Bi_{0.5}Pb_{0.25}Sn_{0.25}$	8.5	
$BiPd_2$	4.0	
$Bi_{0.4}Pd_{0.6}$	3.7—4	Hexagonal, ordered
BiPd	3.7	Orthorhombic
Bi_2Pd	1.70	Monoclinic, α-phase
Bi_2Pd	4.25	Tetragonal, β-phase
$BiPd_{0.45}Pt_{0.55}$	3.7	$B8_1$-hP4 (NiAs)
BiPdSe	1.0	C2
BiPdTe	1.2	C2
BiPt	1.21	$B8_1$
$Bi_{0.1}PtSb_{0.9}$	2.05; 1.5	$B8_1$-hP4 (NiAs)
BiPtSe	1.45	C2
BiPtTe	1.15	C2
Bi_2Pt	0.155	Hexagonal
Bi_2Rb	4.25	C15
$BiRe_2$	1.9—2.2	
BiRh	2.06	$B8_1$

TABLE 4
Superconductive Compounds and Alloys (continued)

Substance	T_c, K	Crystal structure type
Bi_3Rh	3.2	Orthorhombic (NiB_3)
Bi_4Rh	2.7	Hexagonal
$BiRu$	5.7	m**
Bi_3Sn	3.6—3.8	
$BiSn$	3.8	
Bi_xSn_y	3.85—4.18	
Bi_3Sr	5.62	$L1_2$
Bi_3Te	0.75—1.0	
Bi_5Tl_3	6.4	
$Bi_{0.26}Tl_{0.74}$	4.4	Cubic, disordered
$Bi_{0.26}Tl_{0.74}$	4.15	$L1_2$, ordered (?)
Bi_2Y_3	2.25	
Bi_3Zn	0.8—0.9	
$Bi_{0.3}Zr_{0.7}$	1.51	
$BiZr_3$	2.4—2.8	
$BrMo_6Se_7$	7.1	hP15 (Mo_6PbS_8)
$Br_3Mo_6Se_5$	7.1	Same
CCs_x	0.020—0.135	Hexagonal
CFe_3	1.30	DO_{11}-oP16 (Fe_3C)
$CGaMo_2$	3.7—4.1	Hexagonal
$CHf_{0.5}Mo_{0.5}$	3.4	B1
$CHf_{0.3}Mo_{0.7}$	5.5	B1
$CHf_{0.25}Mo_{0.75}$	6.6	B1
$CHf_{0.7}Nb_{0.3}$	6.1	B1
$CHf_{0.6}Nb_{0.4}$	4.5	B1
$CHf_{0.5}Nb_{0.5}$	4.8	B1
$CHf_{0.4}Nb_{0.6}$	5.6	B1
$CHf_{0.25}Nb_{0.75}$	7.0	B1
$CHf_{0.2}Nb_{0.8}$	7.8	B1
$CHf_{0.9—0.1}Ta_{0.1—0.9}$	5.0—9.0	B1
CK (excess K)	0.55	Hexagonal
C_8K	0.39	Hexagonal
C_2La	1.66	tI6 (CaC_2)
C_2Lu	3.33	Same
$C_{0.40—0.44}Mo_{0.60—0.56}$	9—13	
C_3MoRe	3.8	B1-cF8
$C_{0.6}Mo_{4.8}Si_3$	7.6	$D8_8$
$CMo_{0.2}Ta_{0.8}$	7.5	B1
$CMo_{0.5}Ta_{0.5}$	7.7	B1
$CMo_{0.75}Ta_{0.25}$	8.5	B1
$CMo_{0.8}Ta_{0.2}$	8.7	B1
$CMo_{0.85}Ta_{0.15}$	8.9	B1
CMo_xV_{1-x}	2.9—9.3	B1
CMo_xZr_{1-x}	9.8	B1
$C_{0.984}Nb$	9.8	B1
CNb_2	9.1	
CNb_xTi_{1-x}	<4.2—8.8	B1
$CNb_{0.1—0.9}Zr_{0.9—0.1}$	4.2—8.4	B1
CRb_x (Au)	0.023—0.151	Hexagonal
$CRe_{0.06}W$	5.0	
CRu	2.00	hP2 (CW)
$C_{0.98}7Ta$	9.7	
$C_{0.848—0.987}$	2.04—9.7	
CTa (film)	5.09	B1
CTa_2	3.26	L'_3

TABLE 4
Superconductive Compounds and Alloys (continued)

Substance	T_c, K	Crystal structure type
$CTa_{0.4}Ti_{0.6}$	4.8	B1
$Cta_{1-0.4}W_{0-0.6}$	8.5—10.5	B1
$CTa_{0.2-0.9}Zr_{0.8-0.1}$	4.6—8.3	B1
CTc (excess C)	3.85	Cubic
$CTi_{0.5-0.7}W_{0.5-0.3}$	6.7—2.1	B1
CW	1.0	
CW_2	2.74	L'_3
CW_2	5.2	F.C.C.
C_2Y	3.88	tI6 (CaC_2)
$Ca_3Co_4Sn_{13}$	5.9	cP40 ($Pr_3Rh_2Sn_{13}$)
$Ca_3Ge_{13}Rh_4$	2.1	Same
CaHg	1.6	B2-cP2 (CsCl)
$CaHg_3$	1.6	hP8 (Ni_3Sn)
$CaIr_2$	6.15	C15
$Ca_3Ir_4Sn_{13}$	7.1	cP40
$Ca_xO_3Sr_{1-x}Ti$ ($n = 3.7-11 \times 10^{19}$)	<0.1—0.55	
$Ca_{0.1}O_3W$	1.4—3.4	Hexagonal
CaPb	7.0	
$CaRh_2$	6.40	C15
$CaRh_{1.2}Sn_{4.5}$	8.7	cP40
$CaTl_3$	2.0	B2-cP2
$Cd_{0.3-0.5}Hg_{0.7-0.5}$	1.70—1.92	
CdHg	1.77; 2.15	Tetragonal
$Cd_{0.0075-0.05}In_{0.9925-0.95}$	3.24—3.36	Tetragonal
$Cd_{0.97}Pb_{0.03}$	4.2	
CdSn	3.65	
$Cd_{0.17}Tl_{0.83}$	2.3	
$Cd_{0.18}Tl_{0.82}$	2.54	
$CeCo_2$	0.84	C15
$CeCo_{1.67}Ni_{0.33}$	0.46	C15
$CeCo_{1.67}Rh_{0.33}$	0.47	C15
$Ce_xGd_{1-x}Ru_2$	3.2—5.2	C15
$CeIr_3$	3.34	
$CeIr_5$	1.82	
$Ce_{0.005}La_{0.995}$	4.6	
Ce_xLa_{1-x}	1.3—6.3	
$Ce_xPr_{1-x}Ru_2$	1.4—5.3	C15
Ce_xPt_{1-x}	0.7—1.55	
$CeRu_2$	6.0	C15
$Ce_3Mo_6Se_5$	5.7	hR15 (Mo_6PbS_8)
$Ce_2Mo_6Te_6$	1.7	Same
$Co_xFe_{1-x}Si_2$	1.4 (max.)	C1
$CoHf_2$	0.56	$E9_3$
$CoLa_3$	4.28	
$Co_4La_3Sn_{13}$	2.8	cP40
$CoLu_3$	~0.35	
Co_xLuSn_y	1.5	cP40
$Co_{0-0.01}Mo_{0.8}Re_{0.2}$	2—10	
$Co_{0.02-0.10}Nb_3Rh_{0.98-0.90}$	2.28—1.90	A15
$Co_xNi_{1-x}Si_2$	1.4 (max.)	C1
$Co_{0.5}Rh_{0.5}Si_2$	2.5	
$Co_xRh_{1-x}Si_2$	3.65 (max.)	
$Co_{-0.3}So_{-0.7}$	~0.35	
$Co_4Sc_5Si_{10}$	5.0	tP38 ($Co_4Sc_5Si_{10}$)
$CoSi_2$	1.40; 1.22	C1

TABLE 4
Superconductive Compounds and Alloys (continued)

Substance	T_c, K	Crystal structure type
Co_xSn_yYb	2.5	cP40
Co_3Th_7	1.83	$D10_2$
Co_xTi_{1-x}	2.8 (max.)	Co in α-Ti
Co_xTi_{1-x}	3.8 (max.)	Co in β-Ti
$CoTi_2$	3.44	$E9_3$
$CoTi$	0.71	A2
CoU	1.7	B2, distorted
CoU_6	2.29	$D2_c$
$Co_{0.28}Y_{0.72}$	0.34	
CoY_3	<0.34	
$CoZr_2$	6.3	C16
$Co_{0.1}Zr_{0.9}$	3.9	A3
$Cr_{0.6}Ir_{0.4}$	0.4	H.C.P.
$Cr_{0.65}Ir_{0.35}$	0.59	H.C.P.
$Cr_{0.7}Ir_{0.3}$	0.76	H.C.P.
$Cr_{0.72}Ir_{0.28}$	0.83	
Cr_3Ir	0.45	A15
$Cr_{0—0.1}Nb_{1—0.9}$	4.6—9.2	A2
$Cr_{0.80}Os_{0.20}$	2.5	Cubic
Cr_3Os	4.68	A15-cP8 (Cr_3Si)
Cr_xRe_{1-x}	1.2—5.2	
$Cr_{0.4}Re_{0.6}$	2.15	$D8_b$
$Cr_{0.8—0.6}Rh_{0.2—0:4}$	0.5—1.10	A3
Cr_3Rh	0.3	A15-cP8
Cr_3Ru (annealed)	3.3	A15
Cr_2Ru	2.02	$D8_b$
Cr_3Ru_2	2.10	$D8_b$-tP30 (CrFe)
$Cr_{0.1—0.5}Ru_{0.9—0.5}$	0.34—1.65	A3
Cr_xTi_{1-x}	3.6 (max.)	Cr in α-Ti
Cr_xTi_{1-x}	4.2 (max.)	Cr in β-Ti
$Cr_{0.1}Ti_{0.3}V_{0.6}$	5.6	
$Cr_{0.0175}U_{0.9825}$	0.75	β-phase
$Cs_{0.32}O_3W$	1.12	Hexagonal
$Cu_{0.15}In_{0.85}$ (film)	3.75	
$Cu_{0.04—0.08}In_{0.94—0.92}$	4.4	
$CuLa$	5.85	
$Cu_2Mo_6O_2S_6$	9	hR15 (Mo_6PbS_8)
$Cu_2Mo_6Se_8$	5.9	Same
Cu_xPb_{1-x}	5.7—7.7	
CuS	1.62	B18
CuS_2	1.48—1.53	C18
$CuSSe$	1.5—2.0	C18
$CuSe_2$	2.3—2.43	C18
$CuSeTe$	1.6—2.0	C18
Cu_xSn_{1-x}	3.2—3.7	
Cu_xSn_{1-x} (film, made at 10K)	3.6—7	
Cu_xSn_{1-x} (film, made at 300K)	2.8—3.7	
$CuTe_2$	<1.25—1.3	C18
$CuTh_2$	3.49	C16
$Cu_{0—0.027}V$	3.9—5.3	A2
CuY	0.33	B2-cP2 (CsCl)
Cu_xZn_{1-x}	0.5—0.845	
$DyMo_6S_8$	2.1	hR15
Er_xLa_{1-x}	1.4—6.3	
$ErMo_6S_8$	2.2	hR15

TABLE 4
Superconductive Compounds and Alloys (continued)

Substance	T_c, K	Crystal structure type
$ErMo_6Se_8$	6.2	hR15
$Fe_3Lu_2Si_5$	6.1	tP40 ($Fe_3Sc_2Si_5$)
$Fe_{0-0.04}Mo_{0.8}Re_{0.2}$	1—10	
$Fe_{0.05}Ni_{0.05}Zr_{0.90}$	~3.9	
Fe_3Re_2	6.55	$D8_b$-tP30 (FeCr)
$Fe_3Sc_2Si_5$	4.52	tP40
Fe_3Si_5Tm	1.3	Same
$Fe_3Si_5Y_2$	2.4	Same
Fe_3Th_7	1.86	D10
Fe_xTi_{1-x}	3.2 (max.)	Fe in α-Ti
Fe_xTi_{1-x}	3.7 (max.)	Fe in β-Ti
$Fe_xTi_{0.6}V_{1-x}$	6.8 (max.)	
FeU_6	3.86	$D2_c$
$Fe_{0.1}Zr_{0.9}$	1.0	A3
$Ga_{0.5}Ge_{0.5}Nb_3$	7.3	A15
Ga_2Ge_2U	0.87	B2-cP2
$GaHf_2$	0.21	C16-tI12 (Al_2Cu)
$GaLa_3$	5.84	
Ga_3Lu	2.3	B2-cP2
Ga_2Mo	9.5	
$GaMo_3$	0.76	A15
GaN (black)	5.85	B4
$Ga_{0.7}Pt_{0.3}$	2.9	C1
GaPt	1.74	B20
GaSb (120kbar, 77K, annealed)	4.24	A5
GaSb (unannealed)	~5.9	
$Ga_{0-1}Sn_{1-0}$ (quenched)	3.47—4.18	
$Ga_{0-1}Sn_{1-0}$ (annealed)	2.6—3.85	
GaTe	0.17	mC24 (GaTe)
Ga_5V_2	3.55	Tetragonal (Mn_2Hg_5)
$GaV_{4.5}$	9.15	
Ga_3Zr	1.38	
Ga_3Zr_5	3.8	$D8_b$-hP16 (Mn_5Si_3)
Gd_xLa_{1-x}	< 1.0—5.5	
$GdMo_6S_8$	3.5	hR15
$GdMo_6Se_8$	5.6	hR15
$Gd_xOs_2Y_{1-x}$	1.4—4.7	
$Gd_xRu_2Th_{1-x}$	3.6 (max.)	C15
$Ge_{10}As_4Y_5$	9.06	tP38 ($C0_4Sc_5Si_{10}$)
GeIr	4.7	B31
GeIrLa	1.64	tI12 (LaPtSi)
$Ge_{10}Ir_4Lu_5$	2.60	tP38
$Ge_{10}Ir_4Y_5$	2.62	tP38
Ge_2La	1.49; 2.2	Orthorhombic, distorted (Mn_2Hg_5)
GeLaPt	3.53	tI12
$Ge_{13}Lu_3Os_4$	3.6	cP40 ($Pr_3Rh_2Sn_{13}$)
$Ge_{10}Lu_5Rh_4$	2.79	tP38
$Ge_{13}Lu_3Ru_4$	2.3	cP40
$GeMo_3$	1.43	A15
$GeNb_2$	1.9	
$Ge_{0.29}Nb_{0.71}$	6	A15
GePt	0.40	B31
Ge_3Rh_5	2.12	Orthorhombic, related to $InNi_2$

TABLE 4
Superconductive Compounds and Alloys (continued)

Substance	T_c, K	Crystal structure type
GeRh	0.96	B31-oP8 (MnP)
$Ge_{13}Rh_4Sc_3$	1.9	c P40
$Ge_{10}Rh_4Y_5$	1.35	tP38
$Ge_{13}Ru_4Y_3$	1.7	cP40
Ge_2So	1.3	
$GeTa_3$	8.0	A15-cP8 (Cr_3Si)
Ge_3Te_4 ($n = 1.06 \times 10^{22}$)	1.55—1.80	Rhombohedral
Ge_xTe_{1-x} ($n = 8.5—64 \times 10^{20}$)	0.07—0.41	R1
GeV_3	6.01	A15
Ge_2Y	3.80	C_c
$Ge_{1.62}Y$	2.4	
Ge_2Zr	0.30	oC12 ($ZrSi_2$)
$GeZr_3$	0.4	$L1_2$-tP32 (Ti_3P)
$H_{0.33}Nb_{0.67}$	7.28	B.C.C.
$H_{0.1}Nb_{0.9}$	7.38	Same
$H_{0.05}Nb_{0.95}$	7.83	Same
$H_{0.12}Ta_{0.88}$	2.81	B.C.C.
$H_{0.08}Ta_{0.92}$	3.26	Same
$H_{0.04}Ta_{0.96}$	3.62	Same
HfIrSi	3.50	C37-cP12 (Co_2Si)
$HfMo_2$	0.05	hP24 (Ni_2Mn)
$HfN_{0.989}$	6.6	B1
$Hf_{0—0.5}Nb_{1—0.5}$	8.3—9.5	A2
$Hf_{0.75}Nb_{0.25}$	>4.2	
$HfOs_2$	2.69	C14
HfOsP	6.1	C22-hP9 (Fe_2P)
HfPRu	9.9	Same
$HfRe_2$	4.80	C14
$Hf_{0.14}Re_{0.86}$	5.86	A12
$Hf_{0.99—0.96}Rh_{0.01—0.04}$	0.85—1.51	
$Hf_{0—0.55}Ta_{1—0.45}$	4.4—6.5	A2
HfV_2	8.9—9.6	C15
Hg_xIn_{1-x}	3.14—4.55	
HgIn	3.81	
Hg_2K	1.20	Orthorhombic
Hg_3K	3.18	
Hg_4K	3.27	
Hg_8K	3.42	
Hg_3Li	1.7	Hexagonal
$HgMg_3$	0.17	hP8 (Na_3As)
Hg_2Mg	4.0	tI6 ($MoSi_2$)
Hg_3Mg_5	0.48	$D8_b$-hP16 (Mn_5Si_3)
Hg_2Na	1.62	Hexagonal
Hg_4Na	3.05	
Hg_xPb_{1-x}	4.14—7.26	
HgSn	4.2	
Hg_xTl_{1-x}	2.30—4.19	
Hg_5Tl_2	3.86	
Ho_xLa_{1-x}	1.3—6.3	
$Ho_{1.2}Mo_6Se_8$	6.1	$D10_2$-hR12 (Be_3Nb)
$In_{1—0.86}Mg_{0—0.14}$	3.395—3.363	
$In_2Mo_6Te_6$	2.6	hR15 (Mo_6PbS_8)
$InNb_3$ (high pressure and temp.)	4—8; 9.2	A15
$In_{0.5}Nb_3Zr_{0.5}$	6.4	
$In_{0.11}O_3W$	< 1.25—2.8	Hexagonal

TABLE 4
Superconductive Compounds and Alloys (continued)

Substance	T_c, K	Crystal structure type
$In_{0.95-0.85}Pb_{0.05-0.15}$	3.6—5.05	
$In_{0.98-0.91}Pb_{0.02-0.09}$	3.45—4.2	
InPb	6.65	
InPd	0.7	B2
InSb (quenched from 170 kbar into liquid N_2)	4.8	Like A5
InSb	2.1	B4
$(InSb)_{0.95-0.10}Sn_{0.05-0.90}$ (various heat treatments)	3.8—5.1	
$(InSb)_{0-0.07}Sn_{1-0.93}$	3.67—3.74	
In_3Sn	~5.5	
In_xSn_{1-x}	3.4—7.3	
$In_{0.82-1}Te$ ($n = 0.83-1.71 \times 10^{22}$)	1.02—3.45	B1
$In_{1.000}Te_{1.002}$	3.5—3.7	B1
In_3Te_4 ($n = 4.7 \times 10^{21}$)	1.15—1.25	Rhombohedral
In_xTl_{1-x}	2.7—3.374	
$In_{0.8}Tl_{0.2}$	3.223	
$In_{0.62}Tl_{0.38}$	2.760	
$In_{0.78-0.69}Tl_{0.22-0.31}$	3.18—3.32	Tetragonal
$In_{0.69-0.62}Tl_{0.31-0.38}$	2.98—3.3	F.C.C.
Ir_2La	0.48	C15
Ir_3La	2.32	$D10_2$
Ir_3La_7	2.24	$D10_2$
Ir_5La	2.13	
$IrLaSi_2$	2.03	oC16 ($CeNiSi_2$)
$IrLaSi_3$	2.7	tI10 ($BaNiSn_3$)
Ir_2Lu	2.47	C15
Ir_3Lu	2.89	C15
$Ir_4Lu_5Si_{10}$	3.9	tP38 ($Co_4Sc_5Si_{10}$)
IrMo	< 1.0	A3
$IrMo_3$	9.6	A15
$IrMo_3$	6.8	$D8_b$
$IrNb_3$	1.9	A15
$Ir_{0.4}Nb_{0.6}$	9.8	$D8_b$
$Ir_{0.37}Nb_{0.63}$	2.32	$D8_b$
IrNb	7.9	$D8_b$
$Ir_{1.15}Nb_{0.85}$	4.6	oP12 (IrTa)
$Ir_{0.02}Nb_3Rh_{0.98}$	2.43	A15
$Ir_{0.05}Nb_3Rh_{0.95}$	2.38	A15
$Ir_{0.287}O_{0.14}Ti_{0.573}$	5.5	$E9_3$
$Ir_{0.265}O_{0.035}Ti_{0.65}$	2.30	$E9_3$
Ir_xOs_{1-x}	0.3—0.98	
$Ir_{1.5}Os_{0.5}$	2.4	C14
IrOsY	2.6	C15
IrSiY	2.70	C37-oP12 (Co_2Si)
IrSiZr	2.04	Same
Ir_2Sc	2.07	C15
$Ir_{2.5}Sc$	2.46	C15
$Ir_4Sc_5Si_{10}$	8.46	tP38
Ir_2Si_2Th	2.14	tI10
$IrSi_3Th$	1.75	tI10
IrSiTh	6.50	tI12 (LaPtSi)
Ir_2Si_2Y	2.60	tI10 (Al4Ba)
$Ir_4Si_{10}Y_5$	3.10	tP38
$Ir_3Si_5Y_2$	2.83	oI40

TABLE 4
Superconductive Compounds and Alloys (continued)

Substance	T_c, K	Crystal structure type
$IrSn_2$	0.65—0.78	C1
Ir_2Sr	5.70	C15
Ir_7Ta_{13}	1.2	$D8_b$-tP30 (FeCr)
$Ir_{0.5}Te_{0.5}$	~3	
$IrTe_3$	1.18	C2
IrTh	< 0.37	B_f
Ir_2Th	6.50	C15
Ir_3Th	4.71	
Ir_3Th_7	1.52	$D10_2$
Ir_5Th	3.93	$D2_d$
$IrTi_3$	5.40	A15
IrV_2	1.39	A15
IrW_3	3.82	
$Ir_{0.28}W_{0.72}$	4.49	
Ir_2Y	2.18; 1.38	C15
$Ir_{0.69}Y_{0.31}$	1.98; 1.44	C15
$Ir_{0.70}Y_{0.30}$	2.16	C15
Ir_2Y_3	1.61	
Ir_3Y	3.50	$D10_2$-hR13 (Be_3Nb)
Ir_xY_{1-x}	0.3—3.7	
Ir_2Zr	4.10	C15
$Ir_{0.1}Zr_{0.9}$	5.5	A3
$K_2Mo_{15}S_{19}$	3.32	hR15
$K_{0.27—0.31}O_3W$	0.50	Hexagonal
$K_{0.40—0.57}O_3W$	1.5	Tetragonal
$La_{0.55}Lu_{0.45}$	2.2	Hexagonal, La type
$La_{0.8}Lu_{0.2}$	3.4	Same
$LaMg_2$	1.05	C15
$LaMo_6S_8$	7.1	hR15
LaN	1.35	
$LaOs_2$	6.5	C15
$LaPt_2$	0.46	C15
$La_{0.28}Pt_{0.72}$	0.54	C15
LaPtSi	3.48	tI12
$LaRh_3$	2.60	
$LaRh_5$	1.62	
La_7Rh_3	2.58	$D10_2$
$LaRhSi_2$	3.42	oC16 ($CeNiSi_2$)
$La_2Rh_3Si_5$	4.45	oI40 ($Co_3Si_5U_2$)
$LaRhSi_3$	2.7	tI10 ($BaNiSn_3$)
$LaRh_2Si_2$	3.90	tI10 (Al_4Ba)
$LaRu_2$	1.63	C15
La_3S_4	6.5	$D7_3$
La_3Se_4	8.6	$D7_3$
$LaSi_2$	2.3	C_c
La_xY_{1-x}	1.7—5.4	
LaZn	1.04	B2
$Li_2Mo_6S_8$	4.2	hR15
LiPb	7.2	
$LuOs_2$	3.49	C14
$Lu_{0.275}Rh_{0.725}$	1.27	C15
$LuRh_5$	0.49	
$Lu_5Rh_4Si_{10}$	3.95	tP38 ($Co_4So_5Si_{10}$)
$LuRu_2$	0.86	C14
$Mg_{1.14}Mo_{6.6}S_8$	3.5	hR15

TABLE 4
Superconductive Compounds and Alloys (continued)

Substance	T_c, K	Crystal structure type
Mg_2Nb	5.6	
$Mg_{-0.47}Tl_{-0.53}$	2.75	B2
MgZn	0.9	A3-oP4 (AuCd)
Mn_xTi_{1-x}	2.3 (max.)	Mn in -Ti
Mn_xTi_{1-x}	1.1—3.0	Mn in -Ti
MnU_6	2.32	$D2_c$
Mo_2N	5.0	F.C.C.
$Mo_6Na_2S_8$	8.6	hR15
Mo_xNb_{1-x}	0.016—9.2	
$Mo_{5.25}Nb_{0.75}Se_8$	6.2	hR15
Mo_6NdSa_8	8.2	hR15
Mo_3Os	7.2	A15
$Mo_{0.62}Cs_{0.38}$	5.65	$D8_b$
Mo_3P	5.31	DO_e
$Mo_6Pb_{1.2}Se_8$	6.75	hR15
$Mo_{0.5}Pd_{0.5}$	3.52	A3
Mo_6PrSe_8	9.2	hR15
MoRe	7.8	$D8_b$-tP30
$MoRe_3$	9.25; 9.89	A12
Mo_xRe_{1-x}	1.2—12.2	
$Mo_{0.42}Re_{0.58}$	6.35	$D8_b$
MoRh	1.97	A3
Mo_xRh_{1-x}	1.5—8.2	B.C.C.
MoRu	9.5—10.5	A3
$Mo_{0.61}Ru_{0.39}$	7.18	$D8_b$
$Mo_{0.2}Ru_{0.8}$	1.66	A3
Mo_3Ru_2	7.0	$D8_b$-tP30
$Mo_4Ru_2Te_8$	1.7	hR15
Mo_6S_8	1.85	hR15
Mo_6S_8Sc	3.6	hR15
$Mo_6S_8Sm_{1.2}$	2.9	hR15
Mo_6S_8Tb	2.0	hR15
Mo_6S_8Tl	8.7	hR15
$Mo_6S_8Tm_{1.2}$	2.1	hR15
$Mo_6S_8Y_{1.2}$	3.0	hR15
Mo_6S_8Yb	9.2	hR15
$Mo_{6.6}S_8Zn_{11}$	3.6	hR15
Mo_3Sb_4	2.1	
Mo_6Se_8	6.3	hR15
$Mo_6Se_8Sm_{1.2}$	6.8	hR15
$Mo_6Se_8Sn_{1.2}$	6.8	hR15
Mo_6Se_8Tb	5.7	hR15
Mo_3Se_3Tl	4.0	hP14
$Mo_6Se_8Tm_{1.2}$	6.3	hR15
Mo_6Se_8Yb	6.2	hR15
Mo_3Si	1.30	A15
$MoSi_{0.7}$	1.34	
Mo_xSiV_{3-x}	4.54—16.0	A15
$Mo_{5.25}Ta_{0.75}Te_8$	1.7	hR15
Mo_6Te_8	1.7	hR15
$Mo_{0.16}Ti_{0.84}$	4.18; 4.25	
$Mo_{0.913}Ti_{0.087}$	2.95	
$Mo_{0.04}Ti_{0.96}$	2.0	Cubic
$Mo_{0.025}Ti_{0.975}$	1.8	
Mo_xU_{1-x}	0.7—2.1	

TABLE 4
Superconductive Compounds and Alloys (continued)

Substance	T_c, K	Crystal structure type
Mo_xV_{1-x}	$0-\sim5.3$	
Mo_2Zr	$4.25-4.75$	C15
NNb (film)	$6-9$	B1
$N_xO_yTi_z$	$2.9-5.6$	Cubic
$N_xO_yV_z$	$5.8-8.2$	Cubic
$N_{0.34}Re$	$4-5$	F.C.C.
NTa (film)	4.84	B1
$N_{0.6-0.987}Ti$	$<1.17-5.8$	B1
$N_{0.82-0.99}V$	$2.9-7.9$	B1
NZr	9.8	B1
$N_{0.906-0.984}Zr$	$3.0-9.5$	B1
$Na_{0.28-0.35}O_3W$	0.56	Tetragonal
$Na_{0.28}Pb_{0.72}$	7.2	
NbO	1.25	
$NbOs_2$	2.52	A12
Nb_3Os	1.05	A15
$Nb_{0.6}Os_{0.4}$	1.89; 1.78	$D8_b$
$Nb_3Os_{0.02-0.10}Rh_{0.98-0.90}$	$2.42-2.30$	A15
Nb_3P	1.8	$L1_2$-tP32 (Ti_3P)
NbPRh	4.08	C37-oP12 (Co_2Si)
$Nb_{0.6}Pd_{0.4}$	1.60	$D8_f$ plus cubic
$Nb_3Pd_{0.02-0.10}Rh_{0.92-0.90}$	$2.49-2.55$	A15
$Nb_{0.62}Pt_{0.38}$	4.21	$D8_b$
Nb_5Pt_3	3.73	$D8_b$
$Nb_3Pt_{0.02-0.98}Rh_{0.98-0.02}$	$2.52-9.6$	A15
$NbRe_3$	5.27	$D8_b$-tP30 (FeCr)
$Nb_{0.38-0.18}Re_{0.62-0.82}$	$2.43-9.70$	A15
NbRe	3.8	$D8_b$-tP30
NbReSi	5.1	oI36 (FeTiSi)
Nb_3Rh	2.64	A15
$Nb_{0.6}Rh_{0.40}$	4.21	$D8_b$ plus other
$Nb_{0.9}Rh_{1.1}$	3.07	A3-oP4 (AuCd)
$Nb_3Rh_{0.98-0.90}Ru_{0.02-0.10}$	$2.42-2.44$	A15
Nb_xRu_{1-x}	$1.2-4.8$	
NbRuSi	2.65	oI36
NbS_2	$6.1-6.3$	Hexagonal, $NbSe_2$ type
NbS_2	$5.0-5.5$	Hexagonal, three-layer type
Nb_3Sb	0.2	$L1_2$-tP32 (Ti_3P)
$Nb_3Sb_{0-0.7}Sn_{1-0.3}$	$6.8-18$	A15
$NbSe_2$	$5.15-5.62$	Hexagonal
$Nb_{1-1.05}Se_2$	$2.2-7.0$	Same
Nb_3Se_4	2.0	hP14
Nb_3Si	1.5	$L1_2$
Nb_3SiSnV_3	4.0	
$NbSn_2$	2.60	Orthorhombic
Nb_6Sn_5	2.8	oI44 (Sn_5Ti_6)
NbSnTaV	6.2	A15
$NbSnV_2$	5.5	A15
Nb_2SnV	9.8	A15
Nb_xTa_{1-x}	$4.4-9.2$	A2
Nb_3Te_4	1.8	hP14
Nb_xTi_{1-x}	$0.6-9.8$	
$Nb_{0.6}Ti_{0.4}$	9.8	
Nb_xU_{1-x}	1.95 (max.)	
$Nb_{0.88}V_{0.12}$	5.7	A2

TABLE 4
Superconductive Compounds and Alloys (continued)

Substance	T_c, K	Crystal structure type
$Nb_{0.5}V_{1.5}Zr$	4.3	C15-hP12 ($MgZn_2$)
$Ni_{0.3}Th_{0.7}$	1.98	$D10_2$
$NiZr_2$	1.52	
$Ni_{0.1}Zr_{0.9}$	1.5	A3
$O_3Rb_{0.27—0.29}W$	1.98	Hexagonal
OSn	3.81	tP4 (PbO)
O_3SrTi (n = 1.7—12.0 × 10^{19})	0.12—0.37	
O_3SrTi (n = 10^{18}—10^{21})	0.05—0.47	
O_3SrTi (n = 10^{20})	0.47	
$O_3Sr_{0.08}W$	2—4	Hexagonal
OTi	0.58	
$O_3Tl_{0.30}W$	2.0—2.14	Hexagonal
OV_3Zr_3	7.5	$E9_3$
OW_3 (film)	3.35; 1.1	A15
OsPti	1.2	C22-hP9 (Fe_2P)
OsPZr	7.4	Same
OsReY	2.0	C14
Os_2Sc	4.6	C14
OsTa	1.95	A12
Os_3Th_7	1.51	$D10_2$
Os_xW_{1-x}	0.9—4.1	
OsW_3	~3	
Os_2Y	4.7	C14
Os_2Zr	3.0	C14
Os_xZr_{1-x}	1.5—5.6	
PPb	7.8	
OsW_2	3.81	$D8_b$-tP30 (FeCr)
$PPd_{3.0—3.2}$	<0.35—0.7	DO_{11}
P_3Pd_7 (high temperature)	1.0	Rhombohedral
P_3Pd_7 (low temperature)	0.70	Complex
PRh	1.22	
PRh_2	1.3	C1
P_4Rh_5	1.22	oP28 ($CaFe_2O_4$)
PRhTa	4.41	C37-oP12 (Co_2Si)
PRhZr	1.55	Same
PRuTi	1.3	C22-hP9 (Fe_2P)
PRuZr	3.46	C37-oP12
PW_3	2.26	DO_e
Pb_2Pd	2.95	C16
Pb_4Pt	2.80	Related to C16
Pb_2Rh	2.66	C16
PbSb	6.6	
PbTe (plus 0.1 w/o Pb)	5.19	
PbTe (plus 0.1 w/o Te)	5.24—5.27	
$PbTl_{0.27}$	6.43	
$PbTl_{0.17}$	6.73	
$PbTl_{0.12}$	6.88	
$PbTl_{0.075}$	6.98	
$PbTl_{0.04}$	7.06	
$Pb_{1—0.26}Tl_{0—0.74}$	7.20—3.68	
$PbTl_2$	3.75—4.1	
Pb_3Zr_5	4.60	$D8_8$
$PbZr_3$	0.76	A15
$Pd_{0.9}Pt_{0.1}Te_2$	1.65	C6
$Pd_{0.05}Ru_{0.05}Zr_{0.9}$	~9	

TABLE 4
Superconductive Compounds and Alloys (continued)

Substance	T_c, K	Crystal structure type
$Pd_{2.2}S$ (quenched)	1.63	Cubic
$PdSb_2$	1.25	C2
PdSb	1.5	$B8_1$
PdSbSe	1.0	C2
PdSbTe	1.2	C2
Pd_4Se	0.42	Tetragonal
$Pd_{6-7}Se$	0.66	Like Pd_4Te
$Pd_{2.8}Se$	2.3	
Pd_xSe_{1-x}	2.5 (max.)	
PdSi	0.93	B31
PdSn	0.41	B31
$PdSn_2$	3.34	
Pd_2Sn	0.41	C37
Pd_3Sn	0.47—0.64	$B8_2$
Pd_2SnTm	1.77	DO_3-cF16 (BiF_3)
Pd_2SnY	4.92	Same
Pd_2SnYb	1.79	Same
PdTe	2.3; 3.85	$B8_1$
$PdTe_{1.02—1.08}$	2.56—1.88	$B8_1$
$PdTe_2$	1.69	C6
$PdTe_{2.1}$	1.89	C6
$PdTe_{2.3}$	1.85	C6
$Pd_{1.1}Te$	4.07	$B8_1$
Pd_3Te	0.76	cI2 (W)
$PdTh_2$	0.85	C16
$Pd_{0.1}Zr_{0.9}$	7.5	A3
PtSb	2.1	$B8_1$
PtSi	0.88	B31
PtSn	0.37	$B8_1$
$PtSn_4$	2.38	C16-oC20 ($PdSn_4$)
Pt_3Ta_7	1.5	$D8_b$-tP30
$PtTa_3$	0.4	A15-cP8 (Cr_3Si)
PtTe	0.59	Orthorhombic
PtTh	0.44	B_f
Pt_3Th_7	0.98	$D10_2$
Pt_5Th	3.13	
$PtTi_3$	0.58	A15
$Pt_{0.02}U_{0.98}$	0.87	β-phase
$PtV_{2.5}$	1.36	A15
PtV_3	2.87—3.20	A15
$PtV_{3.5}$	1.26	A15
$Pt_{0.5}W_{0.5}$	1.45	A1
Pt_xW_{1-x}	0.4—2.7	
Pt_2Y_3	0.90	
Pt_2Y	1.57; 1.70	C15
Pt_3Y_7	0.82	$D10_2$
PtZr	3.0	A3
Re_2Sc	4.2	C15-hP12 ($MgZn_2$)
$Re_{24}Sc_5$	2.2	A12-cI58 (Mg)
ReSiTa	4.4	oI36 (FeTiSi)
$Re_3Si_5Y_2$	1.76	tP40 ($Fe_3Sc_2Si_5$)
Re_3Ta_2	1.4	$D8_b$-tP30 (FeCr)
$Re_{0.64}Ta_{0.36}$	1.46	A12
Re_3Ta	6.78	A12-cI58 (Mn)
$Re_{24}Ti_5$	6.60	A12

TABLE 4
Superconductive Compounds and Alloys (continued)

Substance	T_c, K	Crystal structue type
Re_xTi_{1-x}	6.6 (max.)	
$Re_{0.76}V_{0.24}$	4.52	$D8_b$
Re_3V	6.26	$D8_b$-tP30
$Re_{0.92}V_{0.08}$	6.8	A3
$Re_{0.6}W_{0.4}$	6.0	
$Re_{0.5}W_{0.5}$	5.12	$D8_b$
$Re_{13}W_{12}$	5.2	$D8_b$-tP30
Re_3W	9.0	A12-cI58
Re_2Y	1.83	C14
Re_2Zr	5.9	C14
Re_3Zr	7.40	A12-cI58
Re_6Zr	7.40	Same
$Rh_{17}S_{15}$	5.8	Cubic
$Rh_{-0.24}Sc_{-0.76}$	0.88; 0.92	
$Rh_4Sc_5Si_{10}$	8.54	tP38
$Rh_4Sc_3Sn_{13}$	4.5	cP40
Rh_xSe_{1-x}	6.0 (max.)	
$RhSi_3Th$	1.76	tI10
$Rh_{0.86}Sc_{1.04}Th$	6.45	tI12
Rh_2Si_2Y	3.11	tI10
$Rh_3Si_5Y_2$	2.70	oI40
$Rh_4Sn_{13}Sr_3$	4.3	cP40
Rh_xSn_yTh	1.9	cI2 (W)
Rh_xSn_yTm	2.3	cP40
$Rh_4Sn_{13}Y_3$	3.2	cP40
Rh_2Sr	6.2	C15
$Rh_{0.4}Ta_{0.6}$	2.35	$D8_b$
$RhTe_2$	1.51	C2
$Rh_{0.67}Te_{0.33}$	0.49	
Rh_xTe_{1-x}	1.51 (max.)	
$RhTh$	0.36	B_f
Rh_3Th_7	2.15	$D10_2$
Rh_5Th	1.07	
Rh_xTi_{1-x}	2.25—3.95	
$Rh_{0.02}U_{0.98}$	0.96	
RhV_3	0.38	A15
RhW	~3.4	A3
RhY_3	0.65	
Rh_2Y_3	1.48	
Rh_3Y	1.07	C15
Rh_5Y	0.56	
Rh_3Y_7	0.32	hP20 (Fe_3Th_7)
$Rh_{0.005}Zr$ (annealed)	5.8	
$Rh_{0—0.45}Zr_{1—0.55}$	2.1—10.8	
$Rh_{0.1}Zr_{0.9}$	9.0	H.C.P.
Ru_2Sc	1.67	C14
$RuSiTa$	3.15	oI36
Ru_3Si_2Th	3.98	hP12
Ru_3Si_2Y	3.51	hP12
$Ru_{1.1}Sn_{3.1}Y$	1.3	cP40
Ru_2Th	3.56	C15
$RuTi$	1.07	B2
$Ru_{0.05}Ti_{0.95}$	2.5	
$Ru_{0.1}Ti_{0.9}$	3.5	
$Ru_xTi_{0.6}V_y$	6.6 (max.)	

TABLE 4
Superconductive Compounds and Alloys (continued)

Substance	T_c, K	Crystal structue type
Ru_3U	0.15	$L1_2$-cP4
$Ru_{0.45}V_{0.55}$	4.0	B2
RuW	7.5	A3
Ru_2Y	1.52	C14
Ru_2Zr	1.84	C14
$Ru_{0.1}Zr_{0.9}$	5.7	A3
STh	0.5	B1-cF8 (NaCl)
SbSn	1.30—1.42	B1 or distorted
$SbTa_3$	0.72	A15-cP8 (Cr_3Si)
$SbTi_3$	5.8	Same
Sb_2Ti_7	5.2	
$Sb_{0.01—0.03}V_{0.99—0.97}$	3.76—2.63	A2
SbV_3	0.80	A15
SeTh	1.7	B1-cF8
$SiMo_3$	1.4	A15-cP8
Si_2Th	3.2	C_c, α-phase
Si_2Th	2.4	C32, β-phase
$SiV_{2.7}Ru_{0.3}$	2.9	A15
Si_2W_3	2.8; 2.84	
$SiZr_3$	0.5	$L1_2$-tP32 (Ti_3P)
$Sn_{0.174—0.104}Ta_{0.826—0.896}$	6.5—< 4.2	A15
$SnTa_3$	8.35	A15, highly ordered
$SnTa_3$	6.2	A15, partially ordered
$SnTaV_2$	2.8	A15
$SnTa_2V$	3.7	A15
Sn_xTe_{1-x} (n = 10.5—20 × 10^{20})	0.07—0.22	B1
Sn_3Th	3.33	$L1_2$-cP4
$SnTi_3$	5.80	A15-cP8
Sn_xTl_{1-x}	2.37—5.2	
SnV_3	3.8	A15
$Sn_{0.02—0.057}V_{0.98—0.943}$	2.87—~1.6	A2
$SnZr_3$	0.92	A15-cP8
$Ta_{0.025}Ti_{0.975}$	1.3	Hexagonal
$Ta_{0.05}Ti_{0.95}$	2.9	Hexagonal
$Ta_{0.05—0.75}V_{0.95—0.25}$	4.30—2.65	A2
$Ta_{0.8—1}W_{0.2—0}$	1.2—4.4	A2
$Tc_{0.1—0.4}W_{0.9—0.6}$	1.25—7.18	Cubic
$Tc_{0.50}W_{0.50}$	7.52	α plus
$Tc_{0.60}W_{0.40}$	7.88	plus α
Tc_6Zr	9.7	A12
TeY	1.02	B1-cF8
$ThTl_3$	0.87	$L1_2$-cP4
$Th_{0—0.55}Y_{1—0.45}$	1.2—1.8	
$Ti_{0.70}V_{0.30}$	6.14	Cubic
Ti_xV_{1-x}	0.2—7.5	
$Ti_{0.5}Zr_{0.5}$ (annealed)	1.23	
$Ti_{0.5}Zr_{0.5}$ (quenched)	2.0	
Tl_3Y	1.52	$L1_2$-cP4
V_2Zr	8.80	C15
$V_{0.26}Zr_{0.74}$	5.9	
W_2Zr	2.16	C15
YZn	0.33	B2-cP2 (CsCl)

* n denotes current carriers concentration in cm^{-3}.

TABLE 4
Superconductive Compounds and Alloys (continued)

B. SUPERCONDUCTORS WITH $T_c > 10K$

Substance	T_c,K	Crystal structure type	
Al_2CMo_3	10.0	A13	
$Al_{0.5}Ge_{0.5}Nb$	12.6	A15	
$Al_{-0.8}Ge_{-0.2}Nb_3$	20.7	A15	
$AlNb_3$	18.0	A15	(Cr_3Si)
$AlNb_3$	12.0		$(FeCr)$
Al_xNb_{1-x}	<4.2—13.5	$D8_b$	
Al_xNb_{1-x}	12—17.5	A15	
$Al_{0.27}Nb_{0.73—0.48}V_{0—0.25}$	14.5—17.5	A15	
$Al\ Nb_xV_{1-x}$	4.4—13.5		
$Al_{0.1}Si_{0.9}V_3$	14.05		
AlV_3	11.8	A15	(Cr_3Si)
$AuNb_3$	11.5	A15	
$Au_{0—0.3}Nb_{1—0.7}$	1.1—11.0		
$Au_{0.02—0.98}Nb_3Rh_{0.98—0.02}$	2.53—10.9	A15	
$AuNb_{3(1-x)}V_{3x}$	1.5—11.0	A15	
$B_{0.03}C_{0.51}Mo_{0.47}$	12.5		
B_4LuRh_4	11.7		(B_4CeCo_4)
B_2LuRu	10		
B_4Rh_4Y	11.3		(B_4CeCo_4)
$B_{0.1}Si_{0.9}V_3$	15.8	A15	
$BaBi_{0.2}O_3Pb_{0.8}$	13.2		
$Ba_2CaCu_2O_8Tl_2$	120		
$Ba_2Cu_3LaO_6$	80		
$Ba_2Cu_3O_7Tm$	101		
$Ba_2Cu_3O_7Y$	90		
$(Ba,La)_2CuO_4$	36	A15	(K_2NiF_4)
$Bi_2CaCu_2O_8Sr_2$	110		
$Br_2Mo_6S_6$	13.8		(Mo_6PbS_8)
C_3La	11.0		(C_3Pu_2)
CMo	14.3	B1	$(NaCl)$
CMo_2	12.2	$o**$	
$C_{0.5}Mo_xNb_{1-x}$	10.8—12.5	B1	
CMo_xTi_{1-x}	10.2(max)	B1	
$CMo_{0.83}Ti_{0.17}$	10.2	B1	
$C_{0—0.38}N_{1—0.62}Ta$	10.0—11.3		
CNb (whiskers)	7.5—10.5		
CNb	11.5	B1	
$C_{0.7—1.0}Nb_{0.3—0}$	6—11	B1	
CNb_xTa_{1-x}	8.2—13.9		
$CNb_{0.6—0.9}W_{0.4—0.1}$	12.5—11.6	B1	
$C_{0.1}Si_{0.9}V_3$	16.4	A15	
CTa	10.3	B1	
$CTa_{1—0.4}W_{0—0.6}$	8.5—10.5	B1	
$C_{0.66}Th_{0.13}Y_{0.21}$	17		(C_3Pu_2)
C_3Y_2	11.5		(C_3Pu_2)
CW	10	B1	
$(Ca,La)_2CuO_4$	18		(K_2NiF_4)
$Cu(La,Sr)_2O_4$	39		
$Cu_{1.8}Mo_6S_8$	10.8		(Mo_6PbS_8)
$Cr_{0.3}SiV_{2.7}$	11.3	A15	
$GaNb_3$	14.5	A15	(Cr_3Si)
$Ga_xNb_3Sn_{1-x}$	14—18.37	A15	

TABLE 4
Superconductive Compounds and Alloys (continued)

Substance	T_c, K	Crystal structure type	
GaV_3	16.8	A15	
$GaV_{2.1—3.5}$	6.3—14.45	A15	
$GeNb_3$	23.2	A15	
$GeNb_3$ (quenched)	6—17	A15	
$Ge_xNb_3Sn_{1-x}$	17.6—18.0	A15	
$Ge_{0.5}Nb_3Sn_{0.5}$	11.3		
$Ge_{0.1}Si_{0.9}V_3$	14.0	A15	
GeV_3	11	A15	
$InLa_3$	9.83; 10.4	LI$_2$	(AuCu$_3$)
$InLa_3$ (0—35 kbar)	9.75—10.55		
$In_{0—0.3}Nb_3Sn_{1—0.7}$	18.0—18.19	A15	
InV_3	13.9	A15	
$Ir_{0.4}Nb_{0.6}$	10		(FeCr)
$LaMo_6Se_8$	11.4		(Mo$_6$PbS$_8$)
LiO_4Ti_2	13.7		(Al$_2$MgO$_4$)
MoN	12; 14.8	h*	
Mo_3Os	12.7	A15	
$Mo_6Pb_{0.9}S_{7.5}$	15.2		(Mo$_6$PbS$_8$)
Mo_3Re	10.0; 15	A15	
Mo_xRe_{1-x}	1.2—12.2		
$Mo_{0.52}Re_{0.48}$	11.1		
$Mo_{0.57}Re_{0.43}$	14.0		
$Mo_{-0.60}Re_{0.395}$	10.6		
MoRu	9.5—10.5	A3	
Mo_3Ru	10.6	A15	
Mo_6Se_8Tl	12.2		(Mo$_6$PbS$_8$)
$Mo_{0.3}SiV_{2.7}$	11.7	A15	
Mn_3Si	12.5	A15	
Mo_3Tc	15	A15	
$Mo_{0.3}Tc_{0.7}$	12.0	A15	
Mo_xTc_{1-x}	10.8—15.8		
$MoTc_3$	15.8		
NNb (whiskers)	10—14.5		
NNb (diffusion wires)	16.10		
$N_{0.988}Nb$	14.9; 17.3	B1	
$N_{0.824—0.988}Nb$	14.4—15.3	B1	
$N_{0.7—0.795}Nb$	11.3—12.9		
NNb_xO_y	13.5—17.0	B1	
NNb_xO_y	6.0—11		
$N_{100—42w/o}Nb_{0—58w/o}Ti$	15—16.8		
$N_{100—75w/o}Nb_{0—25w/o}Zr$	12.5—16.35		
NNb_xZr_{1-x}	9.8—13.8	B1	
$N_{0.93}Nb_{0.85}Zr_{0.15}$	13.8	B1	
NTa	12—14	B1	
NZr	10.7	B1	
Nb_3Pt	10.9	A15	
$Nb_{0.18}Re_{0.82}$	10		(Mn)
Nb_3Si	19	A15	
$Nb_{0.3}SiV_{2.7}$	12.8	A15	
Nb_3Sn	18.05	A15	
$Nb_{0.8}Sn_{0.2}$	18.18; 18.5	A15	
Nb_xSn_{1-x} (film)	2.6—18.5	o*	
Nb_3Sn_2	16.6	t*	
$NbSnTa_2$	10.8	A15	
Nb_2SnTa	16.4	A15	

TABLE 4
Superconductive Compounds and Alloys (continued)

Substance	T_c, K	Crystal structure type
$Nb_{2.5}SnTa_{0.5}$	17.6	A15
$Nb_{2.75}SnTa_{0.25}$	17.8	A15
$Nb_{3x}SnTa_{3(1-x)}$	6.0—18.0	
$Nb_2SnTa_{0.5}V_{0.5}$	12.2	A15
$NbTc_3$	10.5	A12
$Nb_{0.75}Zr_{0.25}$	10.8	
$Nb_{0.66}Zr_{0.33}$	10.8	
$PbTa_3$	17	A15
$RhTa_3$	10	A15
$RhZr_2$	10.8; 11.3	C16 (A1$_2$Cu)
$Rh_{0—0.45}Zr_{1—0.55}$	2.1—10.8	
$SiTi_{0.3}V_{2.7}$	10.9	A15
SiV_3	17.1	A15
$SiV_{2.7}Zr_{0.3}$	13.2	A15

TABLE 5
Critical Field Data

Substance	H_o oersteds	Substance	H_o oersteds
Ag_2F	2.5	InSb	1100
Ag_7NO_{11}	57	In_xTl_{1-x}	252—284
Al_2CMo_3	1700	$In_{0.8}Tl_{0.2}$	252
$BaBi_3$	740	$Mg_{0.47}Tl_{0.53}$	220
Bi_2Pt	10	$Mo_{0.16}Ti_{0.84}$	<985
Bi_3Sr	530	$NbSn_2$	620
Bi_5Tl_3	>400	$PbTl_{0.27}$	756
CdSn	>266	$PbTl_{0.17}$	796
$CoSi_2$	105	$PbTl_{0.12}$	849
$Cr_{0.1}Ti_{0.3}V_{0.6}$	1360	$PbTl_{0.075}$	880
$In_{1-0.86}Mg_{0-0.14}$	272.4—259.2	$PbTl_{0.04}$	864

TABLE 6
High Critical Magnetic-Field Superconductive Compounds and Alloys

Substance	T_c, K	H_{c1}, kOe	H_{c2}, kOe	H_{c3}, kOe	T_{obs}, K[a]
Al_2CMo_3	9.8—10.2	0.091	156		1.2
$AlNb_3$		0.375			
$Ba_xO_3Sr_{1-x}Ti$	<0.1—0.55	0.0039 max.			
$Bi_{0.5}Cd_{0.1}Pb_{0.27}Sn_{0.13}$			>24		3.06
Bi_xPb_{1-x}	7.35—8.4	0.122 max.	30 max.		4.2
$Bi_{0.56}Pb_{0.44}$	8.8		15		4.2
$Bi_{7.5w/o}Pb_{92.5w/o}$[b]			2.32		
$Bi_{0.099}Pb_{0.901}$		0.29	2.8		
$Bi_{0.02}Pb_{0.98}$		0.46	0.73		
$Bi_{0.53}Pb_{0.32}Sn_{0.16}$			>25		3.06
$Bi_{1-0.93}Sn_{0-0.07}$			0—0.032		3.7
Bi_5Tl_3	6.4		>5.6		3.35
C_8K (excess K)	0.55		0.160 (H$\perp$c)		0.32
			0.730 (H$\parallel$c)		0.32

TABLE 6
High Critical Magnetic-Field Superconductive Compounds and Alloys (continued)

Substance	T_c, K	H_{c1}, kOe	H_{c2}, kOe	H_{c3}, kOe	T_{obs}, K[a]
C_8K	0.39		0.025 (H⊥c)		0.32
			0.250 (H∥c)		0.32
$C_{0.44}Mo_{0.56}$	12.5—13.5	0.087	98.5		1.2
CNb	8—10	0.12	16.9		4.2
$CNb_{0.4}Ta_{0.6}$	10—13.6	0.19	14.1		1.2
CTa	9—11.4	0.22	4.6		1.2
$Ca_xO_3Sr_{1-x}Ti$	<0.1—0.55	0.002—0.004			
$Cd_{0.1}Hg_{0.9}$ (by weight)		0.23	0.34		2.04
$Cd_{0.05}Hg_{0.95}$		0.28	0.31		2.16
$Cr_{0.10}Ti_{0.30}V_{0.60}$	5.6	0.071	84.4		0
GaN	5.85	0.725			4.2
Ga_xNb_{1-x}			>28		4.2
GaSb (annealed)	4.24		2.64		3.5
$GaV_{1.95}$	5.3		73[e]		
$GaV_{2.1-3.5}$	6.3—14.45		230—300[d]		0
GaV_3		0.4	350[e]		0
			500[d]		
$GaV_{4.5}$	9.15		121[c]		0
Hf_xNb_y			>52—>102		1.2
Hf_xTa_y			>28—>86		1.2
$Hg_{0.05}Pb_{0.95}$		0.235	2.3		
$Hg_{0.101}Pb_{0.899}$		0.23	4.3		4.2
$Hg_{0.15}Pb_{0.85}$	6.75		>13		2.93
$In_{0.98}Pb_{0.02}$	3.45	0.1		0.12	2.76
$In_{0.96}Pb_{0.04}$	3.68	0.1	0.12	0.25	2.94
$In_{0.94}Pb_{0.06}$	3.90	0.095	0.18	0.35	3.12
$In_{0.913}Pb_{0.087}$	4.2	~10.17	0.55	2.65	
$In_{0.316}Pb_{0.684}$		0.155	3.7		4.2
$In_{0.17}Pb_{0.83}$			2.8	5.5	4.2
$In_{1.000}Te_{1.002}$	3.5—3.7		1.2[c]		0
$In_{0.95}Tl_{0.05}$		0.263	0.263		3.3
$In_{0.90}Tl_{0.10}$		0.257	0.257		3.25
$In_{0.83}Tl_{0.17}$		0.242	0.39		3.21
$In_{0.75}Tl_{0.25}$		0.216	0.50		3.16
LaN	1.35	0.45			0.76
La_3S_4	6.5	≈0.15	>25		1.3
La_3Se_4	8.6	≈0.2	>25		1.25
$Mo_{0.52}Re_{0.48}$	11.1		14—21	22—33	4.2
			18—28	37—43	1.3
$Mo_{0.6}Re_{0.395}$	10.6		14—20	20—37	4.2
			19—26	26—37	1.3
$Mo_{0.5}Ti_{0.5}$			75[c]		0
$Mo_{0.16}Ti_{0.84}$	4.18	0.028	98.7[c]		0
			36—38		3.0
$Mo_{0.913}Ti_{0.087}$	2.95	0.060	15		4.2
$Mo_{0.1-0.3}U_{0.9-0.7}$	1.85—2.06		>25		
$Mo_{0.17}Zr_{0.83}$			30		
$N_{(12.8\ w/o)}Nb$	15.2		>9.5		13.2
NNb (wires)	16.1		153[c]		0
			132		4.2
			95		8
			53		12
NNb_xO_{1-x}	13.5—17.0		38		

TABLE 6
High Critical Magnetic-Field Superconductive Compounds and Alloys (continued)

Substance	T_c, K	H_{c1}, kOe	H_{c2}, kOe	H_{c3}, kOe	T_{obs}, K[a]
NNb$_x$Zr$_{1-x}$	9.8–13.8		4- >130		4.2
N$_{0.93}$Nb$_{0.85}$Zr$_{0.15}$	13.8		>130		4.2
Na$_{0.086}$Pb$_{0.914}$		0.19	6.0		
Na$_{0.016}$Pb$_{0.984}$		0.28	2.05		
Nb	9.15		2.020		1.4
			1.710		4.2
Nb		0.4—1.1	3—5.5		4.2
Nb (unstrained)		1.1—1.8	3.40	6—9.1	4.2
Nb (strained)		1.25—1.92	3.44	6.0—8.7	4.2
Nb (cold-drawn wire)		2.48	4.10	≈10	4.2
Nb (film)			>25		4.2
NbSc			>30		
Nb$_3$Sn		0.170	221		4.2
			70		14.15
			54		15
			34		16
			17		17
Nb$_{0.1}$Ta$_{0.9}$		0.084	0.154		4.195
Nb$_{0.2}$Ta$_{0.8}$			10		4.2
Nb$_{0.65-0.73}$Ta$_{0.02-0.10}$Zr$_{0.25}$			>70—>90		4.2
Nb$_x$Ti$_{1-x}$			148 max.		1.2
			120 max.		4.2
Nb$_{0.222}$U$_{0.778}$		1.98	23		1.2
Nb$_x$Zr$_{1-x}$			127 max.		1.2
			94 max.		4.2
O$_3$SrTi	0.43	0.0049c	0.504c		0
O$_3$SrTi	0.33	0.00195c	0.420c		0
PbSb$_{1 \text{ w/o}}$(quenched)			>1.5		4.2
PbSb$_{1 \text{ w/o}}$(annealed)			>0.7		4.2
PbSb$_{2.8 \text{ w/o}}$(quenched)			>2.3		4.2
PbSb$_{2.8 \text{ w/o}}$(annealed)			>0.7		4.2
Pb$_{0.871}$Sn$_{0.129}$		0.45	1.1		
Pb$_{0.965}$Sn$_{0.035}$		0.53	0.56		
Pb$_{1-0.26}$Tl$_{0-0.74}$	7.20—3.68		2—6.9c		0
PbTl$_{0.17}$	6.73		4.5c		0
Re$_{0.26}$W$_{0.74}$			>30		
Sb$_{0.93}$Sn$_{0.07}$			0.12		3.7
SiV$_3$	17.0	0.55	156e		
Sn$_x$Te$_{1-x}$		0.00043—0.00236	0.005—0.0775		0.012—0.079
Ta (99.95%)		0.425	1.850		1.3
		0.325	1.425		2.27
		0.275	1.175		2.66
		0.090	0.375		3.72
Ta$_{0.5}$Nb$_{0.5}$			3.55		4.2
Ta$_{0.65-0}$Ti$_{0.35-1}$	4.4—7.8		>14—138		1.2
Ta$_{0.5}$Ti$_{0.5}$			138		1.2
Te	3.3	0.25c			0
Tc$_x$W$_{1-x}$	5.75—7.88		8—44		4.2
Ti				2.7	4.2
Ti$_{0.75}$V$_{0.25}$	5.3	0.029c	199c		0
Ti$_{0.775}$V$_{0.225}$	4.7	0.024c	172c		0
Ti$_{0.615}$V$_{0.385}$	7.07	0.050	34		4.2
Ti$_{0.516}$V$_{0.484}$	7.20	0.062	28		4.2
Ti$_{0.415}$V$_{0.585}$	7.49	0.078	25		4.2

TABLE 6
High Critical Magnetic-Field Superconductive Compounds and Alloys (continued)

Substance	T_c, K	H_{c1}, kOe	H_{c2}, kOe	H_{c3}, kOe	T_{obs}, K[a]
$Ti_{0.12}V_{0.88}$			17.3	28.1	4.2
$Ti_{0.09}V_{0.91}$			14.3	16.4	4.2
$Ti_{0.06}V_{0.94}$			8.2	12.7	4.2
$Ti_{0.03}V_{0.97}$			3.8	6.8	4.2
Ti_xV_{1-x}			108 max.		1.2
V	5.31	0.8	3.4		1.79
		0.75	3.15		2
		0.45	2.2		3
		0.30	1.2		4
$V_{0.26}Zr_{0.74}$	≈5.9	0.238			1.05
		0.227			1.78
		0.185			3.04
		0.165			3.5
W (film)	1.7—4.1		>34		1

[a] Temperature of critical field measurement.
[b] w/o denotes weight percent.
[c] Extrapolated.
[d] Linear extrapolation.
[e] Parabolic extrapolation.

REFERENCES

1. B. W. Roberts, in *Superconductive Materials and Some of Their Properties. Progress in Cryogenics*, Vol. IV, 1964, pp. 160—231.
2. B. W. Roberts, Superconductive Materials and Some of Their Properties, NBS Technical Notes 408 and 482, U.S. Government Printing Office, 1966 and 1969; B. W. Roberts, *J. Phys. Chem. Ref. Data*, 5, 581, 1976.
3. B. W. Roberts, Properties of Selected Superconductive Materials, 1978 Supplement, NBS Technical Note 983, 1978.
4. T. Claeson, *Phys. Rev.*, 147, 340, 1966.
5. C. J. Raub, W. H. Zachariasen, T. H. Geballe, and B. T. Matthias, *J. Phys. Chem. Solids*, 24, 1093, 1963.
6. T. H. Geballe, B. T. Matthias, V. B. Compton, E. Corenzwit, G. W. Hull, Jr., and L. D. Longinotti, *Phys. Rev.*, 1A, 119, 1965.
7. C. J. Raub, V. B. Compton, T. H. Geballe, B. T. Matthias, J. P. Maita, and G. W. Hull, Jr., *J. Phys. Chem. Solids*, 26, 2051, 1965.
8. R. D. Blaugher, J. K. Hulm, and P. N. Yocom, *J. Phys. Chem. Solids*, 26, 2037, 1965.
9. T. Claeson and H. L. Luo, *J. Phys. Chem. Solids*, 27, 1081, 1966.
10. S. C. Ng and B. N. Brockhouse, *Solid State Comm.*, 5, 79, 1967.
11. O. I. Shulishova and I. A. Shcherbak, *Izv. AN SSSR, Neorg. Materials*, 3, 1495, 1967.
12. T. F. Smith and H. L. Luo, *J. Phys. Chem. Solids*, 28, 569, 1967.
13. A. C. Lawson, *J. Less–Common Metals*, 23, 103, 1971.
14. R. Chevrel, M. Sergent, and J. Prigent, *J. Solid State Chem.*, 3, 515, 1971.
15. M. Marezio, P. D. Dernier, J. P. Remeika, and B. T. Matthias, *Mat. Res. Bull.*, 8, 657, 1973.
16. J. K. Hulm and R. D., *Blaugher in Superconductivity in d– and f–Band Metals*, D. H. Douglass, Ed., American Institute of Physics, 4, 1, 1972.
17. R. N. Shelton, A. C. Lawson, and D. C. Johnston, *Mat. Res. Bull.*, 10, 297, 1975.
18. H. D. Wiesinger, *Phys. Status Sol.*, 41A, 465, 1977.
19. O. Fisher, *Applied Phys.*, 16, 1, 1978.
20. D. C. Johnston, *Solid State Comm.*, 24, 699, 1977.
21. H. C. Ku and R. H. Shelton, *Mat. Res. Bull.*, 15, 1441, 1980.
22. H. Barz, *Mat. Res. Bull.*, 15, 1489, 1980.
23. G. P. Espinosa, A. S. Cooper, H. Barz, and J. P. Remeika, *Mat. Res. Bull.*, 15, 1635, 1980.
24. E. M. Savitskii, V. V. Baron, Yu. V. Efimov, M. I. Bychkova, and L. F. Myzenkova, in *Superconducting Materials*, Plenum Press, 1981, p. 107.
25. R. Fluckiger and R. Baillif, in Topics in *Current Physics*, O. Fischer and M. B. Maple, Eds., Springer Verlag, 34, 113, 1982.
26. R. N. Shelton, in *Superconductivity in d– and f–Band Metals*, W. Buckel and W. Weber, Eds., Kernforschungszentrum, Karlsruhe, 1982, p. 123.
27. D. C. Johnston and H. F. Braun, *Topics in Current Phys.*, 32, 11, 1982.
28. R. Chevrel and M. Sergent, *Topics in Current Phys.*, 32, 25, 1982.

29. G. P. Espinosa, A. S. Cooper, and H. Barz, *Mat. Res. Bull.*, 17, 963, 1982.
30. R. Muller, R. N. Shelton, J. W. Richardson, Jr., and R. A. Jacobson, *J. Less–Comm. Met.*, 92, 177, 1983.
31. You–Xian Zhao and Shou–An He, in *High Pressure in Science and Technology*, North Holland, 22, 51, 1983.
32. You–Xian Zhao and Shou–An He, *Solid State Comm.*, 24, 699, 1983.
33. G. P. Meisner and H. C. Ku, *Appl. Phys.*, A31, 201, 1983.
34. R. J. Cava, D. W. Murphy, and S. M. Zahurak, *J. Electrochem. Soc.*, 130, 2345, 1983.
35. R. N. Shelton, *J. Less–Comm. Met.*, 94, 69, 1983.
36. B. Chevalier, P. Lejay, B. Lloret, Wang Xian–Zhong, J. Etourneau, and P. Hagenmuller, *Annales de Chemie*, 9, 191, 1984.
37. G. Venturini, M. Meot–Meyer, E. McRae, J. F. Mareche, and B. Rogues, *Mat. Res. Bull.*, 19, 1647, 1984.
38. J. M. Tarascon, F. G. DiSalvo, D. W. Murphy, G. Hull, and J. V. Waszczak, *Phys. Rev.*, 29B, 172, 1984.
39. G. V. Subba and G. Balakrishnan, *Bull. Mat. Sci.*, 6, 283, 1984.
40. B. Batlog, *Physica*, 126B, 275, 1984.
41. M. J. Johnson, Ames Lab (USA) Report IS-T-1140, 1984.
42. I. M. Chapnik, *J. Mat. Sci. Lett.*, 4, 370, 1985.
43. W. Rong–Yao, L. Qi–Guang, and Z. Xiao, *Phys. Status Sol.*, 90A, 763, 1985.
44. W. Xian–Zhong, B. Chevalier, J. Etourneau, and P. Hagenmuller, *Mat. Res. Bull.*, 20, 517, 1985.
45. H. R. Ott, F. Hulliger, H. Rudigier, and Z. Fisk, *Phys. Rev.*, 31B, 1329, 1985.
46. P. Villars and L. D. Calver, *Pearson's Handbook of Crystallographic Data for Intermetallic Phases*, Vol. 1—3, ASM, 1985.
47. G. V. Subba Rao, K. Wagner, G. Balakrishnan, J. Jakani, W. Paulus, and R. Scollhorn, *Bull. Mat. Sci.*, 7, 215, 1985.
48. J. G. Bednorz and K. A. Muller, *Zs. Physik*, B64, 189, 1986.
49. W. Rong–Yao, *Phys. Status Sol.*, 94A, 445, 1986.
50. H. D. Yang, R. N. Shelton, and H. F. Braun, *Phys. Rev.*, 33B, 5062, 1986.
51. G. Venturini, M. Kanta, E. McRae, J. F. Mareche, B. Malaman, and B. Roques, *Mat. Res. Bull.*, 21, 1203, 1986.
52. W. Rong–Yao, *J. Mat. Sci. Lett.*, 5, 87, 1986.
53. M. K. Wu, J. R. Ashburn, C. J. Torng, P. H. Hor, R. L. Meng, L. Gao, Z. J. Huang, Y. Q. Wang, and C. W. Chu, *Phys. Rev. Lett.*, 58, 908, 1987.
54. R. J. Cava, R. B. Van Dover, B. Batlog, and E. A. Rietman, *Phys. Rev. Lett.*, 58, 408, 1987.
55. L. C. Porter, T. J. Thorn, U. Geiser, A. Umezawa, H. H. Wang, W. K. Kwok, H–C. I. Kao, M. R. Monaghan, G. W. Crabtree, K. D. Carlson, and J. M. Williams, *Inorg. Chem.*, 26, 1645, 1987.
56. A. M. Kini, U. Geiser, H–C. I. Kao, K. D. Carlson, H. H. Wang, M. R. Monaghan, and K. M. Williams, *Inorg. Chem.*, 26, 1834, 1987.
57. T. Penney, S. von Molnar, D. Kaiser, F. Holtzberg, and A. W. Kleinsasser, *Phys. Rev.*, B38, 2918, 1988.
58. Y. K. Tao, J. S. Swinnea, A. Manthiram, J. S. Kim, J. B. Goodenoug, and H. Steinfink, *J. Mat. Res.*, 3, 248, 1988.
59. G. G. Peterson, B. R. Weinberger, L. Lynds, and H. A. Krasinski, *J. Mat. Res.*, 3, 605, 1988.
60. J. B. Torrance, Y. Tokura, A. Nazzai, and S. S. P. Parkin, *Phys. Rev. Lett.*, 60, 542, 1988.
61. K. Kourtakis, M. Robbins, P. K. Gallagher, and T. Teifel, *J. Mat. Res.*, 4, 1289, 1989.
62. J. C. Phillips, *Physics of High-T_c Superconductors*, Academic Press, 1989, p. 336.
63. Shui Wai Lin and L. I. Berger, *Rev. Sci. Instrum.*, 60, 507, 1989.
64. M. Tinkham, *Introduction to Superconductivity*, McGraw–Hill, New York, 1975.
65. O. Fischer and M.B. Maple, Eds., *Topics in Current Physics*, Volume 32: Superconductivity in Ternary Compounds I; Volume 34: Superconductivity in Ternary Compounds II, Springer–Verlag, Berlin, 1982.
66. K. J. Dunn and F. P. Bundy, *Phys. Rev.*, B25, 194, 1982.
67. A. Barone and G. Paterno, *Physics and Applications of the Josephson Effect*, Wiley, New York, 1982.
68. D. H. Douglass, Ed., *Superconductivity in d- and f-band Metals*, Plenum Press, New York, 1976.
69. D. M. Ginsberg, Ed., *Physical Properties of High Temperature Superconductors*, (Volume II, 1990; Volume III, 1992; Volume V, 1996), World Scientific, Singapore.
70. T. Ishiguro and K. Yamji, *Organic Superconductors*, Springer Verlag, Berlin, 1990.
71. Sh. Okada, K. Shimizu, T. C. Kobayashi, K. Amaya, and Sh. Endo., *J. Phys. Soc. Jpn.*, 65, 1924, 1996.
72. A. Bourdillon and N. X. Tan Bourdillon, *High Temperature Superconductors: Processing and Science*, Academic Press, 1994.
73. J. M. Williams, J. R. Ferraro, R. J. Thorn, K. Carlson, U. Geiser, H. H. Wang, A. M. Kini, and M.-H. Whangbo, *Organic Superconductors (Including Fullerenes): Synthesis, structure, Properties, and Theory*, Prentice–Hall, 1992.

HIGH TEMPERATURE SUPERCONDUCTORS
C. N. R. Rao and A. K. Raychaudhuri

The following tables give properties of a number of high temperature superconductors. Table 1 lists the crystal structure (space group and lattice constants) and the critical transition temperature T_c for the more important high temperature superconductors so far studied. Table 2 gives energy gap, critical current density, and penetration depth in the superconducting state. Table 3 gives electrical and thermal properties of some of these materials in the normal state. The tables were prepared in November 1992 and updated in November 1994.

REFERENCES

1. Ginsburg, D.M., Ed., *Physical Properties of High-Temperature Superconductors*, Vols. I—III, World Scientific, Singapore, 1989—1992.
2. Rao, C.N.R., Ed., *Chemistry of High-Temperature Superconductors*, World Scientific, Singapore, 1991.
3. Shackelford, J.F., *The CRC Materials Science and Engineering Handbook*, CRC Press, Boca Raton, 1992, 98—99 and 122—123.
4. Kaldis, E., Ed., *Materials and Crystallographic Aspects of HT$_c$-Superconductivity*, Kluwer Academic Publ., Dordrecht, The Netherlands, 1992.
5. Malik, S.K. and Shah, S.S., Ed., *Physical and Material Properties of High Temperature Superconductors*, Nova Science Publ., Commack, N.Y., 1994.
6. Chmaissem, O. et. al., *Physica*, C230, 231—238, 1994.
7. Antipov, E.V. et. al., *Physica*, C215, 1—10, 1993.

Table 1
Structural Parameters and Approximate T_c Values of High-Temperature Superconductors

Material	Structure	T_c/K (maximum value)
$La_2CuO_{4+\delta}$	Bmab; $a = 5.355$, $b = 5.401$, $c = 13.15$ Å	39
$La_{2-x}Sr_x(Ba_x)CuO_4$	I4/mmm; $a = 3.779$, $c = 13.23$ Å	35
$La_2Ca_{1-x}Sr_xCu_2O_6$	I4/mmm; $a = 3.825$, $c = 19.42$ Å	60
$YBa_2Cu_3O_7$	Pmmm; $a = 3.821$, $b = 3.885$, $c = 11.676$ Å	93
$YBa_2Cu_4O_8$	Ammm; $a = 3.84$, $b = 3.87$, $c = 27.24$ Å	80
$Y_2Ba_4Cu_7O_{15}$	Ammm; $a = 3.851$, $b = 3.869$, $c = 50.29$ Å	93
$Bi_2Sr_2CuO_6$	Amaa; $a = 5.362$, $b = 5.374$, $c = 24.622$ Å	10
$Bi_2CaSr_2Cu_2O_8$	A$_2$aa; $a = 5.409$, $b = 5.420$, $c = 30.93$ Å	92
$Bi_2Ca_2Sr_2Cu_3O_{10}$	A$_2$aa; $a = 5.39$, $b = 5.40$, $c = 37$ Å	110
$Bi_2Sr_2(Ln_{1-x}Ce_x)_2Cu_2O_{10}$	P4/mmm; $a = 3.888$, $c = 17.28$ Å	25
$Tl_2Ba_2CuO_6$	A$_2$aa; $a = 5.468$, $b = 5.472$, $c = 23.238$ Å;	
	I4/mmm; $a = 3.866$, $c = 23.239$ Å	92
$Tl_2CaBa_2Cu_2O_8$	I4/mmm; $a = 3.855$, $c = 29.318$ Å	119
$Tl_2Ca_2Ba_2Cu_3O_{10}$	I4/mmm; $a = 3.85$, $c = 35.9$ Å	128
$Tl(BaLa)CuO_5$	P4/mmm; $a = 3.83$, $c = 9.55$ Å	40
$Tl(SrLa)CuO_5$	P4/mmm; $a = 3.7$, $c = 9$ Å	40
$(Tl_{0.5}Pb_{0.5})Sr_2CuO_5$	P4/mmm; $a = 3.738$, $c = 9.01$ Å	40
$TlCaBa_2Cu_2O_7$	P4/mmm; $a = 3.856$, $c = 12.754$ Å	103
$(Tl_{0.5}Pb_{0.5})CaSr_2Cu_2O_7$	P4/mmm; $a = 3.80$, $c = 12.05$ Å	90
$TlSr_2Y_{0.5}Ca_{0.5}Cu_2O_7$	P4/mmm; $a = 3.80$, $c = 12.10$ Å	90
$TlCa_2Ba_2Cu_3O_8$	P4/mmm; $a = 3.853$, $c = 15.913$ Å	110
$(Tl_{0.5}Pb_{0.5})Sr_2Ca_2Cu_3O_9$	P4/mmm; $a = 3.81$, $c = 15.23$ Å	120
$TlBa_2(La_{1-x}Ce_x)_2Cu_2O_9$	I4/mmm; $a = 3.8$, $c = 29.5$ Å	40
$Pb_2Sr_2La_{0.5}Ca_{0.5}Cu_3O_8$	Cmmm; $a = 5.435$, $b = 5.463$, $c = 15.817$ Å	70
$Pb_2(Sr,La)_2Cu_2O_6$	P22$_1$2; $a = 5.333$, $b = 5.421$, $c = 12.609$ Å	32
$(Pb,Cu)Sr_2(La,Ca)Cu_2O_7$	P4/mmm; $a = 3.820$, $c = 11.826$ Å	50
$(Pb,Cu)(Sr,Eu)(Eu,Ce)Cu_2O_x$	I4/mmm; $a = 3.837$, $c = 29.01$ Å	25
$Nd_{2-x}Ce_xCuO_4$	I4/mmm; $a = 3.95$, $c = 12.07$ Å	30
$Ca_{1-x}Sr_xCuO_2$	P4/mmm; $a = 3.902$, $c = 3.35$ Å	110
$Sr_{1-x}Nd_xCuO_2$	P4/mmm; $a = 3.942$, $c = 3.393$ Å	40
$Ba_{0.6}K_{0.4}BiO_3$	Pm3m; $a = 4.287$ Å	31
Rb_2CsC_{60}	$a = 14.493$ Å	31
$NdBa_2Cu_3O_7$	Pmmm; $a = 3.878$, $b = 3.913$, $c = 11.753$	58

Table 1
Structural Parameters and Approximate T_c Values of High-Temperature Superconductors
(continued)

Material	Structure	T_c/K (maximum value)
$SmBaSrCu_3O_7$	I4/mmm; $a = 3.854$, $c = 11.62$	84
$EuBaSrCu_3O_7$	I4/mmm; $a = 3.845$, $c = 11.59$	88
$GdBaSrCu_3O_7$	I4/mmm; $a = 3.849$, $c = 11.53$	86
$DyBaSrCu_3O_7$	Pmmm; $a = 3.802$, $b = 3.850$, $c = 11.56$	90
$HoBaSrCu_3O_7$	Pmmm; $a = 3.794$, $b = 3.849$, $c = 11.55$	87
$ErBaSrCu_3O_7$ (multiphase)	Pmmm; $a = 3.787$, $b = 3.846$, $c = 11.54$	82
$TmBaSrCu_3O_7$ (multiphase)	Pmmm; $a = 3.784$, $b = 3.849$, $c = 11.55$	88
$YBaSrCu_3O_7$	Pmmm; $a = 3.803$, $b = 3.842$, $c = 11.54$	84
$HgBa_2CuO_4$	I4/mmm; $a = 3.878$, $c = 9.507$	94
$HgBa_2CaCu_2O_6$ (annealed in O_2)	I4/mmm; $a = 3.862$, $c = 12.705$	127
$HgBa_2Ca_2Cu_3O_8$	Pmmm; $a = 3.85$, $c = 15.85$	133
$HgBa_2Ca_3Cu_4O_{10}$	Pmmm; $a = 3.854$, $c = 19.008$	126

Table 2
Superconducting Properties

J_c (0): Critical current density extrapolated to 0 K
λ_{ab}: Penetration depth in a-b plane
k_B: Boltzmann constant

Material	Form	Energy gap (Δ)		$10^{-6} \times J_c$ (0)/A cm^{-2}	λ_{ab}/Å
		$2\Delta_{pp}/k_B T_c$*	$2\Delta_{fit}/k_B T_c$†		
$Y Ba_2Cu_3O_7$	Single Crystal	5–6	4–5	30 (film)	1400
$Bi_2Sr_2CaCu_2O_8$	Single Crystal	8–9	5.5–6.5	2	2700
$Tl_2Ba_3CaCu_2O_8$	Ceramic	6–7	4–6	10 (film, 80 K)	2000
$La_{2-x}Sr_xCuO_4$, $x = 0.15$	Ceramic	7–9	4–6		
$Nd_{2-x}Ce_xCuO_4$	Ceramic	8	4–5	0.2 (film)	

* Obtained from peak to peak value.
† Obtained from fit to BCS-type relation.

Table 3
Normal State Properties

ρ_{ab}: Resistivity in the a-b plane
ρ_c: Resistivity along the c axis
+ve: ρ_c has positive temperature coefficient of resistivity
−ve: ρ_c has negative temperature coefficient of resistivity
n_H: Hall density
k: Thermal conductivity
in plane: Along a-b plane
out of plane: Perpendicular to a-b plane

Material	Form	$\rho_{ab}/\mu\Omega\,cm$ 300 K	100 K	$\rho_c/m\Omega\,cm$ 300 K	$d\rho_c/dT$	$10^{-21} \times n_H/cm^{-3}$ 300 K	100 K	$k/(mW/cm\,K)$ at 300 K in plane	out of plane
$YBa_2Cu_3O_7$	Single Crystal	110	35	5	+ve	11–16	4–6	120	3
	film	200–300	60–100			5–9	2–3		
$YBa_2Cu_4O_8$	Single Crystal	75	20	10	−ve	14	17		
	film	100–200	20–50						
$Bi_2Sr_2CuO_6$	Single Crystal	300	150	5000	−ve	6	5		
$Bi_2Sr_2CaCu_2O_8$	Single Crystal	150	50	>1000	−ve	4	3	60	8
$Tl_2Ba_2CuO_6$	Single Crystal	300–400	50–75	200–300	+ve	3.1	2.5		
$Tl_2Ba_2Ca_2Cu_3O_{10}$	Ceramic	***	**				≈ 2*		
$La_{2-x}Sr_xCuO_4$, $x = 0.12$	Single Crystal	900	350	200	+ve for T >225 K	2.5		50 (for x = 0.04)	20
$La_{2-x}Sr_xCuO_4$, $x = 0.20$	Single Crystal	400	200	80	+ve for T >150 K	10			
	film	400	160			8.4	6.3		
$Nd_{2-x}Ce_xCuO_4$, $x = 0.17$	Single Crystal	500	275			53	17	250 (for x = 0.15)	
$x = 0.15$	film	140–180	35			32	11		

*   At 200 K
** ρ ~0.4 mΩ cm at 120 K
*** ρ ~1.5 mΩ cm at 300 K

ORGANIC SUPERCONDUCTORS
H.P.R. Frederikse

Although the vast majority of organic compounds are insulating gases, liquids or solids, a small number of organic solids show a considerable amount of electrical conductivity. Some of these appear to be superconductors. The superconducting organics fall primarily into two groups: those containing fulvalenes (pentagonal rings containing sulfur or selenium) and fullerenes based on the nearly spherical cluster C_{60}.

The transition temperatures T_c of the fulvalene derivatives are shown in Table 1. The abbreviations of the various molecular groups are listed in Table 2 and their chemical structures are depicted in Figure 1. Most of the T_c's are between 1 and 12 K. Several of the compounds only show superconductivity under pressure.

The fullerenes are A_3C_{60} compounds, where A represents a single or a combination of alkali atoms. The C_{60} cluster is shown in Figure 2a, while Figure 2b illustrates how the alkali atoms fit into the A_3C_{60} molecule to form the A15 crystallographic structure. Their superconducting transition temperatures range from 8 to 31.3 K (see Table 3).

REFERENCES

1. Ishigura, T. and Yamaji, K., *Organic Superconductors*, Springer-Verlag, Berlin, 1990.
2. Williams, Jack M. et al., *Organic Superconductors (Including Fullerenes)*, Prentice Hall, Englewood Cliffs, N.J., 1992.
3. *The Fullerenes*, Ed.: Krato, H.W., Fisher, J.E., and Cox, D.E., Pergamon Press, Oxford, 1993.
4. Schluter, M. et al., in *The Fullerenes* (Ref. 3), p. 303.

Table 1
Critical Pressure and Maximum Critical Temperature of Organic Superconductors

Material	P_c/kbar	T_c/K	Material	P_c/kbar	T_c/K
$(TMTSF)_2PF_6$	6.5	1.2	β-$(ET)_2IBr_2$	0	2.8
$(TMTSF)_2AsF_6$	9	1.3	β-$(ET)_2AuI_2$	0	4.8
$(TMTSF)_2SbF_6$	11	0.4	$(ET)_4Hg_{2.89}Cl_8$	0	4.2
$(TMTSF)_2TaF_6$	12	1.4	$(ET)_4Hg_{2.89}Br_8$	12	1.8
$(TMTSF)_2ClO_4$	0	1.4	$(ET)_3Cl_2(H_2O)_2$	16	2
$(TMTSF)_2ReO_4$	9.5	1.3	κ-$(ET)_2Cu(NCS)_2$	0	10.4
$(TMTSF)_2FSO_3$	5	3	κ-$(d$-$ET)_2Cu(NCS)_2$	0	11.4
$(ET)_4(ReO_4)_2$	4.5	2	$(DMET)_2Au(CN)_2$	1.5	0.9
β_L-$(ET)_2I_3$	0	1.4	$(DMET)_2AuI_2$	5	0.6
β_H-$(ET)_2I_3$	0	8.1	$(DMET)_2AuBr_2$	0	1.9
γ-$(ET)_3I_{2.5}$	0	2.5	$(DMET)_2AuCl_2$	0	0.9
ε-$(ET)_2I_3(I_8)_{0.5}$	0	2.5	$(DMET)_2I_3$	0	0.6
α-$(ET)_2I_3I_2$-doped	0	3.3	$(DMET)_2IBr_2$	0	0.7
α_t-$(ET)_2I_3$	0	8	$(MDT$-$TTF)_2AuI_2$	0	3.5
$\varepsilon \rightarrow \beta$-$(ET)_2I_3$[a]	0	6	$TTF[Ni(dmit)_2]_2$	2	1.6[b]
θ-$(ET)_2I_3$	0	3.6	$TTF[Pd(dmit)_2]_2$	20	6.5
κ-$(ET)_2I_3$	0	3.6	$(CH_3)_4N[Ni(dmit)_2]_2$	7	5

[a] Converted form ε-type to β-type by thermal treatment.
[b] For 7 kbar.

From Ishigura, T. and Yamaji, K., *Organic Superconductors*, Springer-Verlag, Berlin, 1990. With permission.

Table 2
List of Symbols and Abbreviations

TTF	tetrathiafulvalene
TMTSF	tetramethyltetraselenafulvalene
BEDT-TTF or "ET"	bis(ethylenedithio)tetrathiafulvalene
MDT-TTF	methylenedithiotetrathiafulvalene
DMET	[dimethyl(ethylenedithio)diselenadithiafulvalene]
dmit	4,5-dimercapto-1,3-dithiole-2-thione
T_c	transition temperature to superconducting state
P_c	minimum pressure required for superconducting transition

TMTSF

Tetramethyltetraselenafulvalene

TTF

Tetrathiafulvalene

BEDT – TTF or ET

Bis(ethylenedithio)tetrathiafulvalene

DMET

Dimethyl(ethylenedithio)diselenadithiafulvalene

MDT – TTF

Methylenedithiotetrathiafulvalene

M=Ni, Pd, Pt

$M(dmit)_2^{2-}$

Ligand is 4,5-dimercapto-1.3-dithiole-2-thione

FIGURE 1. Structures of various donor molecules and acceptor species.

a

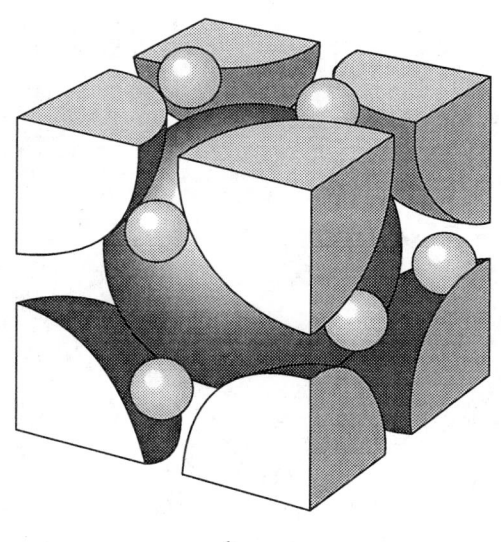

b

FIGURE 2. (a) C_{60} cluster placed in a fcc lattice. Each crystal axis crosses a double bond shared by two hexagons. (b) A hypothetical A_3C_{60} with the A15 structure. The structure can be seen to be an ordered defect structure of A_6C_{60}.

Table 3
Unit Cell and T_c for FCC-A$_3$C$_{60}$

	Lattice parameter(s) (Å)	T_c/K
Na$_2$Rb$_{0.5}$Cs$_{0.5}$C$_{60}$	14.148(3)	8.0
Na$_2$CsC$_{60}$ No. 1[a]	14.132(2)	10.5
Na$_2$CsC$_{60}$ No. 2[a]	14.176(9)	14.0
K$_3$C$_{60}$	14.253(3)	19.3
K$_2$RbC$_{60}$	14.299(2)	21.8
Rb$_2$KC$_{60}$ No. 1[a]	14.336(1)	24.4
Rb$_2$KC$_{60}$ No. 2[a]	14.364(5)	26.4
Rb$_3$C$_{60}$	14.436(2)	29.4
Rb$_2$CsC$_{60}$	14.493(2)	31.3

[a] Samples labeled No. 1 and No. 2 have the same nominal composition.

From Schluter, M et. al., *The Fullerenes*, Ed.: Krato, H.W., Fisher, J.E., and Cox, D.E., Pergamon Press, Oxford, 1993. With permission.

PROPERTIES OF SEMICONDUCTORS

L. I. Berger and B. R. Pamplin

The term "semiconductor" is applied to a material in which electric current is carried by electrons or holes and whose electrical conductivity, when extremely pure, rises exponentially with temperature and may be increased from its low "intrinsic" value by many orders of magnitude by "doping" with electrically active impurities.

Semiconductors are characterized by an energy gap in the allowed energies of electrons in the material which separates the normally filled energy levels of the *valence band* (where "missing" electrons behave like positively charged current carriers "holes") and the *conduction band* (where electrons behave rather like a gas of free negatively charged carriers with an effective mass dependent on the material and the direction of the electrons' motion). This energy gap depends on the nature of the material and varies with direction in anisotropic crystals. It is slightly dependent on temperature and pressure, and this dependence is usually almost linear at normal temperatures and pressures.

Data are presented in three tables. Table I "General Properties of Semiconductors" lists the main crystallographic and semiconducting properties of a large number of semiconducting materials in three main categories: "Tetrahedral Semiconductors" in which every atom is tetrahedrally co-ordinated to four nearest neighbor atoms (or atomic sites) as for example in the diamond structure; "Octahedral Semiconductors" in which every atom is octahedrally co-ordinated to six nearest neighbor atoms — as for examples the halite structure; and "Other Semiconductors."

Table II gives more detailed information about some better known semiconductors, while Table III gives some information about the electronic energy band structure parameters of the best known materials.

Table I
PHYSICO-CHEMICAL PROPERTIES OF SEMICONDUCTORS
(LISTED BY CRYSTAL STRUCTURE)

Substance	Molecular mass	Average atomic mass	Lattice parameters (Å, room temp.)	Density (g/cm³)	Melting point (K)	Microhardness, N/mm² (M-Mohs Scale)	Specific heat, J/kg·K (300 K)	Debye temp. (K)	Coefficient of thermal linear expansion [10^{-6} K^{-1} (300K)]	Thermal conductivity [mW/cm·K (300K)]

PART A. ADAMANTINE SEMICONDUCTORS

§A1. Diamond Structure Elements (Strukturbericht symbol A4, Space Group Fd3m-O_h^7)

Substance	Molecular mass	Average atomic mass	Lattice parameters (Å, room temp.)	Density (g/cm³)	Melting point (K)	Microhardness, N/mm² (M-Mohs Scale)	Specific heat, J/kg·K (300 K)	Debye temp. (K)	Coeff. thermal expansion [10^{-6} K^{-1} (300K)]	Thermal conductivity [mW/cm·K (300K)]
C	12.01		3.56683	3.51	≈3850 Transition to graphite > 980	10(M)	471.5	2340	1.18	9900(I) 23200(IIA) 13600(IIB)
Si	28.09		5.43072	2.3283	1685 ± 2	11270	702	645	2.49	1240
Ge	72.59		5.65754	5.3234	1231	7644	321.9	374	6.1	640
α-Sn	118.69		6.4912	5.765 (281 K)	505.2 (Tr. 286.4)		213	230	5.4 (220 K)	

§A2. Sphalerite (Zinc Blende) Structure Compounds (Strukturbericht symbol B3 Space Group F $\overline{4}$ 3m-T_d^2)

Substance	Molecular mass	Average atomic mass	Lattice parameters (Å, room temp.)	Density (g/cm³)	Melting point (K)	Microhardness, N/mm² (M-Mohs Scale)	Specific heat, J/kg·K (300 K)	Debye temp. (K)	Coeff. thermal expansion [10^{-6} K^{-1} (300K)]	Thermal conductivity [mW/cm·K (300K)]
I VII Compounds										
CuF	82.54	41.27	4.255		1181					
CuCl	98.99	49.49	5.4057	3.53	695	2.3 (M)	490	240	12.1	8.4
CuBr	143.36	71.73	5.6905	4.98	770	2.5(M)	381	207	15.4	12.5
CuI	190.46	95.23	6.60427	5.63	878	192	276	181	19.2	16.8
AgBr	187.78	93.89		6.473	>1570 (Tr. 410)	2.5(M)		270		
AgI	234.77	117.39	6.502	5.67	831	2.5(M)	232	134	−2.5	4.2
II VI Compounds										
BeS	41.08	20.54	4.865	2.36						
BeSe	87.97	43.99	5.139	4.315						
BeTe	136.61	68.31	5.626	5.090						
BePo	(218)	(109)	5.838	7.3						
ZnO	81.37	40.69	4.63	5.675	2248	5.0 (M)	494	416	2.9	234
ZnS	97.43	48.72	5.4093	4.079	2100 (Tr. 1295)	1780	472	530	6.36	251
ZnSe	144.34	72.17	5.6676	5.42	1790	1350	339	400	7.2	140
ZnTe	192.99	96.5	6.101	6.34	1568	900	264	223	8.19	108
ZnPo	(274)	(137)	6.309							
CdS	144.46	72.23	5.832	4.826	1750	1250	330	219	4.7	200
CdSe	191.36	95.68	6.05	5.674	1512	1300	255	181	3.8	90
CdTe	240.00	120.00	6.477	5.86	1365	600	205	200	4.9	58.5
CdPo	(321)	(161)	6.665							
HgS	232.65	116.33	5.8517	7.73	1820	3(M)	210			
HgSe	279.55	139.78	6.084	8.25	1070	2.5(M)	178	151	5.46	10
HgTe	328.19	164.10	6.4623	8.17	943	300	164	242	4.6	20
III V Compounds										
BN	24.82	12.41	3.615	3.49	≈3300	10(M)	793	≈1900		200
BP(L.T.)	41.78	20.87	4.538	2.9	≈2800	37000		≈980		
BAs	85.73	42.87	4.777		≈2300	19000		≈625		

Table I
PHYSICO-CHEMICAL PROPERTIES OF SEMICONDUCTORS
(LISTED BY CRYSTAL STRUCTURE) (continued)

Substance	Molecular mass	Average atomic mass	Lattice parameters (Å, room temp.)	Density (g/cm³)	Melting point (K)	Microhardness, N/mm² (M-Mohs Scale)	Specific heat, J/kg·K (300 K)	Debye temp. (K)	Coefficient of thermal linear expansion [10^{-6} K^{-1} (300K)]	Thermal conductivity [mW/cm·K (300K)]
AlP	57.95	28.98	5.451	2.42	≈2100	5.5(M)		588		920
AlAs	101.90	50.95	5.6622	3.81	2013	5000		417	3.5	840
AlSb	148.73	74.37	6.1355	4.218	1330	4000		292	4.2	600
GaP	100.69	50.35	5.4505	4.13	1750	9450		446	5.3	752
GaAs	144.64	72.32	5.65315	5.316	1510	7500		344	5.4	560
GaSb	191.47	95.74	6.0954	5.619	980	4480	320	265	6.1	270
InP	145.79	72.90	5.86875	4.787	1330	4100		321	4.6	800
InAs	189.74	94.87	6.05838	5.66	1215	3300	268	249	4.7	290
InSb	236.57	118.29	6.47877	5.775	798	2200	144	202	4.7	160
Other sphalerite structure compounds										
MnS	87.0	43.5	5.011							
MnSe	133.9	66.95	5.82							
β-SiC	40.1	20.1	4.348	3.21	3070					
Ga$_2$Se$_3$	376.32	75.26	5.429	4.92	1020	3160			8.9	50
Ga$_2$Te$_3$	522.24	104.45	5.899	5.75	1063	2370				47
In$_2$Te$_3$(H.T.)	608.44	121.7	6.150	5.8	940	1660				69
MgGeP$_2$	158.84	39.71	5.652							
ZnSnP$_2$	246.00	61.5	5.65		1200					
ZnSnAs$_2$(H.T.)	333.90	82.38	5.851	5.53	1050					76
ZnSnSb$_2$	427.56	106.89	6.281	5.67	870	2500				76

§A3. Wurtzite (Zincite) Structure Compounds (Strukturbericht symbol B4, Space Group P 6_3mc-C_{6v}^4)

Substance	Molecular mass	Average atomic mass	Lattice parameters (Å, room temp.)	Density (g/cm³)	Melting point (K)	Microhardness, N/mm² (M-Mohs Scale)	Specific heat, J/kg·K (300 K)	Debye temp. (K)	Coefficient of thermal linear expansion [10^{-6} K^{-1} (300K)]	Thermal conductivity [mW/cm·K (300K)]
I VII Compounds										
CuCl	99.0	49.5	3.91 6.42		T$_c$680K					
CuBr	143.46	71.73	4.06 6.66		T$_c$658K					
CuI	190.46	95.23	4.31 7.09							
AgI	234.80	117.40	4.580 7.494							
II VI Compounds										
BeO	25.01	12.51	2.698 4.380		2800					
MgTe	151.9	76.0	4.54 7.39	3.85	≈2800					
ZnO	81.37	40.69	3.24950 5.2069	5.66	2250					600
ZnS	97.43	48.72	3.8140 6.2576	4.1	2100					460
ZnTe	192.99	46.50	4.27 6.99		1568					
CdS	144.46	72.23	4.1348 6.7490	4.82	1748					401
CdSe	191.36	95.68	4.299 7.010	5.66	1512					316
CdTe	240.00	120.00	4.57 7.47							
III V Compounds										
BP(H.T.)	41.79	20.90	3.562 5.900							
AlN	40.99	20.50	3.111 4.978	3.26	≈2500					823
GaN	83.73	41.87	3.190 5.189	6.10	1500					656
InN	128.83	64.42	3.533 5.693	6.88	1200					556
Other wurtzite structure compounds										
MnS	87.0	43.5	3.985 6.45	3.248						
MnSe	133.9	66.95	4.12 6.72							
SiC	40.1	20.1	3.076 5.048							
MnTe	182.54	91.27	4.078 6.701							
Al$_2$S$_3$	150.14	30.03	3.579 5.829	2.55	1400					
Al$_2$Se$_3$	290.84	58.17	3.890 6.30	3.91	1250					

§A4. Chalcopyrite Structure Compounds (Strukturbericht symbol E1$_1$, Space Group I $\bar{4}$2d-D_{2d}^{12})

Substance	Molecular mass	Average atomic mass	Lattice parameters (Å, room temp.)	Density (g/cm³)	Melting point (K)	Microhardness, N/mm² (M-Mohs Scale)	Specific heat, J/kg·K (300 K)	Debye temp. (K)	Coefficient of thermal linear expansion [10^{-6} K^{-1} (300K)]	Thermal conductivity [mW/cm·K (300K)]
I III VI$_2$ Compounds										
CuAlS$_2$	154.65	38.66	5.323 10.44	3.47	2500					
CuAlSe$_2$	248.45	62.11	5.617 10.92	4.70	2260					
CuAlTe$_2$	345.73	86.43	5.976 11.80	5.50	2550					
CuGaS$_2$	197.39	49.35	5.360 10.49	4.35	2300					
CuGaSe$_2$	291.19	72.80	5.618 11.01	5.56	1970	4200		275	5.4	42
CuGaTe$_2$	388.47	97.12	6.013 11.93	5.99	2400	3500			6.9	27
CuInS$_2$	242.49	60.62	5.528 11.08	4.75	1400	2550				
CuInSe$_2$	336.29	84.07	5.785 11.56	5.77	1600	2050			6.6	37
CuInTe$_2$	433.57	108.39	6.179 12.365	6.10	1660	400		195	7.1	49
CuTlS$_2$	332.05	83.01	5.580 11.17	6.32						
CuTlSe$_2$(L.T.)	425.85	106.46	5.844 11.65	7.11	900					
CuFeS$_2$	183.51	45.88	5.25 10.32	4.088						
CuFeSe$_2$	277.31	69.33			850					

Substance	Molecular mass	Average atomic mass	Lattice parameters (Å, room temp.)		Density (g/cm³)	Melting point (K)	Microhardness, N/mm² (M-Mohs Scale)	Specific heat, J/kg·K (300 K)	Debye temp. (K)	Coefficient of thermal linear expansion [10⁻⁶ K⁻¹] (300K)	Thermal conductivity [mW/cm·K] (300K)
CuLaS₂	266.58	66.65	5.65	10.86							
AgAlS₂	198.97	49.74	5.707	10.28	3.94						
AgAlSe₂	292.77	73.19	5.968	10.77	5.07	1220					
AgAlTe₂	390.05	97.51	6.309	11.85	6.18	1000					
AgGaS₂	241.71	60.43	5.755	10.28	4.72						
AgGaSe₂	335.51	83.88	5.985	10.90	5.84	1120	4400				
AgGaTe₂	432.79	108.2	6.301	11.96	6.05	990	1800		212		10
AgInS₂(L.T.)	286.87	71.70	5.828	11.19	5.00		2250				
AgInSe₂	380.61	95.15	6.102	11.69	5.81	1053	1850				30
AgInTe₂	477.89	119.47	6.42	12.59	6.12	965				9.49, 0.69	
AgFeS₂	227.83	56.96	5.66	10.30	4.53						
II IV V₂ Compounds											
ZnSiP₂	155.40	38.85	5.400	10.441	3.39	1640	11000				
ZnGeP₂	199.90	49.98	5.465	10.771	4.17	1295	8100				180
ZnSnP₂	246.00	61.5					6500				
CdSiP₂	202.43	50.61	5.678	10.431	4.00	≈1470	10500		282		
CdGeP₂	246.94	61.74	5.741	10.775	4.48	1049	5650				110
CdSnP₂	243.03	73.26	5.900	11.518			5000		195		140
ZnSiAs₂	242.20	60.55	5.61	10.88	4.70	1311	9200				
ZnGeAs₂	287.80	71.95	5.672	11.153	5.32	1150	6800		263		110
ZnSnAs₂	333.90	83.48	5.8515	11.704	5.53	1048	4550		271		150
CdSiAs₂	290.34	72.58	5.884	10.882			6850				
CdGeAs₂	334.83	83.71	5.9427	11.2172	5.60	938	4700				48
CdSnAs₂	380.93	95.23	6.0944	11.9182	5.72	880	3450				40

§A5. Other Ternary Semiconductors with Tetrahedral Coordination

Substance	Molecular mass	Average atomic mass	Lattice parameters (Å, room temp.)		Density (g/cm³)	Melting point (K)	Microhardness, N/mm² (M-Mohs Scale)	Specific heat, J/kg·K (300 K)	Debye temp. (K)	Coefficient of thermal linear expansion [10⁻⁶ K⁻¹] (300K)	Thermal conductivity [mW/cm·K] (300K)
I₂ IV VI₃ Compounds											
Cu₂SiS₃(H.T.)	251.36	41.89	3.684	6.004	3.81	1200					23
Cu₂SiS₃(L.T.)			5.290	10.156	3.63						
Cu₂SiTe₃	537.98	89.66	5.93		5.47						
Cu₂GeS₃(H.T.)	295.88	49.31	5.317		4.45	1210	4550	510	254	7.2	12
Cu₂GeS₃(L.T.)			5.327	5.215	4.46						
Cu₂GeSe₃	436.56	72.76	5.589	5.485	5.57	1030	3840	340	168	8.4	24
Cu₂GeTe₃	582.51	97.09	5.958	5.935	5.92		2890				130
Cu₂SnS₃	341.94	57.00	5.436		5.02	1110	2770	440	214	7.8	28
Cu₂SnSe₃	482.66	80.44	5.687		5.94	960	2510	310	148	8.9	35
Cu₂SnTe₃	628.61	104.77	6.048		6.51	680	1970				144
Ag₂GeSe₃	525.21	87.54									
Ag₂SnSe₃	571.31	95.22									
Ag₂GeTe₃	671.13	111.86									
Ag₂SnTe₃	717.23	119.54									
I₃ V VI₄ Compounds											
Cu₃PS₄	349.85	40.73	7.44		6.19						
Cu₃AsS₄	393.79	49.22	6.43		6.14	4.37				3.2	30.2
Cu₃AsSe₄	581.37	72.67	5.570	10.957	5.61				169	9.5	19
Cu₃SbS₄	440.64	55.08	5.38	16.76	4.90						
Cu₃SbSe₄	628.22	78.53	5.654	11.256	6.0				131	12.4	14.6
I IV₂ V₃ Compounds											
CuSi₂P₃	212.64	35.44	5.25								
CuGe₂P₃	301.65	50.28	5.375		4.318	1113	8500	429		8.21	37.6
AgGe₂P₃	345.97	57.66				1015	6150				

§A6. "Defect Chalcopyrite" Structure Compounds (Strukturbericht symbol E3, Space Group I $\overline{4}$-S₄²)

Substance	Molecular mass	Average atomic mass	Lattice parameters (Å, room temp.)		Density (g/cm³)	Melting point (K)	Microhardness	Specific heat	Debye temp.	Coeff. thermal expansion	Thermal conductivity
ZnAl₂Se₄	435.18	62.17	5.503	10.90	4.37						
ZnAl₂Te₄(?)	629.74	84.96	5.904	12.05	4.95						
ZnGa₂S₄(?)	333.06	47.58	5.274	10.44	3.80						
ZnGa₂Se₄(?)	520.66	74.38	5.496	10.99	5.21						
ZnGa₂Te₄(?)	715.22	102.17	5.937	11.87	5.67						
ZnIn₂Se₄	610.86	87.27	5.711	11.42	5.44	1250					
ZnIn₂Te₄	805.42	115.06	6.122	12.24	5.83	1075					
CdAl₂S₄	294.61	42.09	5.564	10.32	3.06						
CdAl₂Se₄	482.21	68.89	5.747	10.68	4.54						
CdAl₂Te₄(?)	676.77	97.68	6.011	12.21	5.10						

Table I
PHYSICO-CHEMICAL PROPERTIES OF SEMICONDUCTORS
(LISTED BY CRYSTAL STRUCTURE) (continued)

Substance	Molecular mass	Average atomic mass	Lattice parameters (Å, room temp.)		Density (g/cm³)	Melting point (K)	Microhardness, N/mm² (M-Mohs Scale)	Specific heat, J/kg·K (300 K)	Debye temp. (K)	Coefficient of thermal linear expansion [10⁻⁶ K⁻¹ (300K)]	Thermal conductivity [mW/cm·K (300K)]
$CdGa_2S_4$	380.09	54.30	5.577	10.08	4.03						
$CdGa_2Se_4$	567.69	81.10	5.743	10.73	5.32						
$CdGa_2Te_4$	762.25	108.89	6.093	11.81	5.77						
$CdIn_2Te_4$	852.45	121.78	6.205	12.41	5.9	1060					
$HgAl_2S_4$	382.79	54.68	5.488	10.26	4.11						
$HgAl_2Se_4$	570.39	82.48	5.708	10.74	5.05						
$HgAl_2Te_4(?)$	764.48	109.28	6.004	12.11	5.81						
$HgGa_2S_4$	468.27	66.90	5.507	10.23	5.00						
$HgGa_2Se_4$	655.87	93.70	5.715	10.78	6.18						
$HgIn_2Se_4$	746.07	106.58	5.764	11.80	6.3	1100					
$HgIn_2Te_4(?)$	940.63	134.38	6.186	12.37	6.3	980					

§A7. Other Adamantine Compounds

Substance	Molecular mass	Average atomic mass	Lattice parameters (Å, room temp.)		Density (g/cm³)	Melting point (K)					
αSiC	40.1	20.1	3.0817 15.1183		3.21	3070					
$Hg_5Ga_2Te_8$	2163.19	144.21	6.235								
$Hg_5In_2Te_8$	2253.39	150.23	6.328								
$CdIn_2Se_4$	657.89	93.98	$a = c = 5.823$								

PART B. OCTAHEDRAL SEMICONDUCTORS
§B1. HALITE STRUCTURE SEMICONDUCTORS (Strukturbericht symbol B1, Space Group Fm3m-O_h^5)

Substance	Molecular mass	Average atomic mass	Lattice parameters (Å, room temp.)		Density (g/cm³)	Melting point (K)					Thermal conductivity [mW/cm·K (300K)]
GeTe	200.19	100.1	5.98		6.14						
SnSe	197.65	98.83	6.020			1133					
SnTe	246.29	123.15	6.313		6.45	1080 (max)					91
PbS	239.26	119.63	5.9362		7.61	1390					23
PbSe	286.16	143.08	6.1243		8.15	1340					17
PbTe	334.8	167.4	6.454		8.16	1180					23

Selected other binary halites

Substance	Molecular mass	Average atomic mass	Lattice parameters (Å, room temp.)		Density (g/cm³)	Melting point (K)					Thermal conductivity [mW/cm·K (300K)]
BiSe	287.94	143.97	5.99		7.98	880					
BiTe	336.58	168.29	6.47								
EuSe	230.92	115.46	6.191			2300					2.4
GdSe	236.21	118.11	5.771			2400					
NiD	60.71	30.35	4.1684		6.6	2260					
CdO	128.41	64.21	4.6953			1700					7
SrS	119.68	59.84	6.0199		3.643	3000					

PART C. OTHER SEMICONDUCTORS
§C1. Antifluorite Structure Compounds (Fm3m - O_h^5)

Substance	Molecular mass	Average atomic mass	Lattice parameters (Å, room temp.)		Density (g/cm³)	Melting point (K)				Coefficient of thermal linear expansion [10⁻⁶ K⁻¹ (300K)]	Thermal conductivity [mW/cm·K (300K)]
Mg_2Si	76.70	25.57	6.338		1.88	1375				11.5	
Mg_2Ge	121.20	40.4	6.380		3.08	1388				15.0	
Mg_2Sn	167.3	55.77	6.765		3.53	1051				9.9	92
Mg_2Pb	225.81	85.27	6.836		5.1	823				10.0	

§C2. Tetradymite Structure Compounds ($\bar{R}3m - D_{3d}^5$)

Substance	Molecular mass	Average atomic mass	Lattice parameters (Å, room temp.)		Density (g/cm³)	Melting point (K)	Microhardness, N/mm² (M-Mohs Scale)	Specific heat, J/kg·K (300 K)	Debye temp. (K)		Thermal conductivity [mW/cm·K (300K)]
Sb_2Te_3	626.3	125.26	4.25	30.3	6.44	895					
Bi_2Se_3	654.84	130.97	4.14	28.7	7.51	979	167				24
Bi_2Te_3	800.76	160.15	4.38	30.45	7.73	858	155	16			30

§C3. Skutterudite Structure Compounds (Im3 - T_h^5)

Substance	Molecular mass	Average atomic mass	Lattice parameters (Å, room temp.)		Density (g/cm³)	Melting point (K)			Debye temp. (K)		Thermal conductivity [mW/cm·K (300K)]
CoP_3	151.85	37.96	7.7073			>1270					
$CoAs_3$	286.70	71.65	8.2060		6.73	1230					
$CoSb_3$	424.18	106.05	9.0385			1123			307		50
$NiAs_3$	283.45	70.86	8.330		6.43						
RhP_3	195.83	48.96	7.9951			>1470					
$RhAs_3$	327.67	81.92	8.4427			>1270					
$RhSb_3$	468.16	117.04	9.2322			1170					100
IrP_3	285.14	71.29	8.0151		7.36	>1470					
$IrAs_3$	416.98	104.25	8.4673		9.12	>1470					
$IrSb_3$	557.47	139.37	9.2533		9.35	1170			303		90

§C4. Selected Multinary Compounds

Substance	Molecular mass	Average atomic mass	Lattice parameters (Å, room temp.)		Density (g/cm³)	Melting point (K)					Thermal conductivity [mW/cm·K (300K)]
$AgSbSe_2$	387.54	96.88	5.786		6.60	910					10.5
$AgSbTe_2$ (or $Ag_{19}Sb_{29}Te_{52}$)	484.82	121.2	6.078		7.12	830					86, 0.3
$AgBiS_2$(H.T.)	380.97	95.24	5.648								
$AgBiSe_2$(H.T.)	474.77	118.69	5.82								
$AgBiTe_2$(H.T.)	572.05	143.01	6.155								
Cu_2CdSnS_4	486.43	60.80	5.586		10.83						

Substance	Molecular mass	Average atomic mass	Lattice parameters (Å, room temp.)	Density (g/cm³)	Melting point (K)	Microhardness, N/mm² (M-Mohs Scale)	Specific heat, J/kg·K (300 K)	Debye temp. (K)	Coefficient of thermal linear expansion [10^{-6} K^{-1} (300K)]	Thermal conductivity [mW/cm·K (300K)]
			§C5. Some Elemental Semiconductors							
B	10.81	4.91	12.6	2.34	2348	9.5(M)	1277	1370	8.3	600
Se(gray)	78.96	4.36	4.95	4.81	493	350	292.6		(∥C) 17.89	(∥C) 45.2
									(⊥C) 74.09	(⊥C) 13.1
Te	127.6	4.45	5.91	6.23	723		196.5		16.8	(∥C) 33.8
										(⊥C) 19.7

Table II
BASIC THERMODYNAMIC, ELECTRICAL, AND MAGNETIC PROPERTIES OF
SEMICONDUCTORS (LISTED BY CRYSTAL STRUCTURE)

Substance	Heat of formation [kJ/mole (300K)]	Volume compressibility (10^{-10} m²/N)	Static dielectric constant	Atomic magnetic susceptibility (10^{-6} CGS)	Index of refraction	Minimum room temperature energy gap (eV)	Mobility (Room temp.) (cm²/V·s) Electrons	Holes	Optical transition	Remarks
PART A. ADAMANTINE SEMICONDUCTORS										
§A1. Diamond Structure Elements (Strukturbericht symbol A4, Space Group Fd3m-O$_h^7$)										
C	714.4	18	5.7	−5.88	2.419 (589 nm)	5.4	1800	1400	i*	
Si	324	0.306	11.8	−3.9	3.49 (589 nm)	1.107	1900	500	i	
Ge	291	0.768	16	−0.12	3.99 (589 nm)	0.67	3800	1820	i	
α-Sn	267.5		24		2.75 (589 nm)	0.0; 0.8	2500	2400		
§A2. Sphalerite (Zinc Blende) Structure Compounds (Strukturbericht symbol B3 Space Group F $\bar{4}$ 3m-T$_d^2$)										
I VII Compounds										
CuF										
CuCl	481	0.26	7.9		1.93	3.17			d	Nantokite
CuBr	481	0.26	7.9		2.12	2.91			d	
CuI	439	0.27	6.5		2.346	2.95			d	Marshite
AgBr	486		12.4		2.253	2.50	4000		i	Bromirite
AgI	389	0.41	10		2.22	2.22	30		d	Miersite
II VI Compounds										
BeS						4.17			i	
BeSe						3.61			i	
BeTe						1.45			d	
BePo							20			
ZnO										See A3
ZnS	477		8.9	−9.9	2.356	3.54	180	5(400°C)	d	See also A3
ZnSe	422		9.2		2.89	2.58	540	28	d	
ZnTe	376		10.4		3.56	2.26	340	100	d	
ZnPo										
CdS										See A3
CdSe										See A3
CdTe	339		7.2		2.50	1.44	1200	50	d	
CdPo										
HgS						2.85	250		d	Metacinnabarite
HgSe	247					2.10 (α)	20000	≈1.5	s	Tiemannite
HgTe	242					−0.06	25000	350	s	Coloradoite
III V Compounds										
BN	815					4.6				Borazone
BP(L.T.)						≈2.1	500	70		Ignites 470K
BAs						≈1.5				
AlP						2.45	80		i	
AlAs	627		10.9			2.16	1200	420	i	
AlSb	585	0.571	11		3.2	1.60	200—400	550	i	
GaP	635	0.110	11.1	−13.8	3.2	2.24	300	150	i	
GaAs	535	0.771	13.2	−16.2	3.30	1.35	8800	400	d	
GaSb	493	0.457	15.7	−14.2	3.8	0.67	4000	1400	d	
InP	560	0.735	12.4	−22.8	3.1	1.27	4600	150	d	
InAs	477	0.549	14.6	−27.7	3.5	0.36	33000	460	d	
InSb	447	0.442	17.7	−32.9	3.96	0.163	78000	750	d	

* i = indirect, d = direct, s = semimetal.

Substance	Heat of formation [kJ/mole] (300K)	Volume compressibility (10^{-10} m²/N)	Static dielectric constant	Atomic magnetic susceptibility (10^{-6} CGS)	Index of refraction	Minimum room temperature energy gap (eV)	Mobility (Room temp.) (cm²/V·s)		Optical transition	Remarks
							Electrons	Holes		
Other sphalerite structure compounds										
MnS										See also §A3
MnSe										See also §A3
β-SiC					2.697	2.3	4000			
Ga_2Te_3	271			−13.5		1.35	50			
In_2Te_3(H.T.)	198			−13.6		1.04	50			
$MgGeP_2$										E1—T^{d12}
$ZnSnP_2$						2.1				Same
$ZnSnAs_2$(H.T.)						≈0.7				Same
$ZnSnSb_2$						0.4				Same

§A3. Wurtzite (Zincite) Structure Compounds (Strukturbericht symbol B4, Space Group P 6_3mc-C_{6v}^4)

Substance	Heat of formation [kJ/mole] (300K)	Volume compressibility (10^{-10} m²/N)	Static dielectric constant	Atomic magnetic susceptibility (10^{-6} CGS)	Index of refraction	Minimum room temperature energy gap (eV)	Electrons	Holes	Optical transition	Remarks
I VII Compounds										
CuCl										
CuBr										
CuI										
AgI						2.63				Iodargirite
II VI Compounds										
BeO										
MgTe										
ZnO	−350					3.2	180			
ZnS	−206					3.67				
ZnTe	−163									
CdS			8.45; 9.12		2.32	2.42	350	40	d	Greenockide
CdSe						1.74	900	50	d	Cadmoselite
CdTe						1.50	650			
III V Compounds										
BP(H.T.)										
AlN						6.02				
GaN						3.34				
InN						2.0				
Other wurtzite structure compounds										
MnS										
MnSe										
SiC					2.654					
MnTe						≈1.0				
Al_2S_3	426					4.1				
Al_2Se_3	367					3.1				

§A4. Chalcopyrite Structure Compounds (Strukturbericht symbol E1₁, Space Group I $\bar{4}$ 2d-D_{2d}^{12})

Substance	Heat of formation [kJ/mole] (300K)	Volume compressibility (10^{-10} m²/N)	Static dielectric constant	Atomic magnetic susceptibility (10^{-6} CGS)	Index of refraction	Minimum room temperature energy gap (eV)	Electrons	Holes	Optical transition	Remarks
I III VI₂ Compounds										
$CuAlS_2$		0.106				2.5				
$CuAlSe_2$						1.1				
$CuAlTe_2$						0.88				
$CuGaS_2$		0.106				2.38				
$CuGaSe_2$		0.141				0.96, 1.63				
$CuGaTe_2$		0.227				0.82, 1.0				
$CuInS_2$		0.141				1.2				
$CuInSe_2$		0.187				0.86, 0.92				
$CuInTe_2$		0.278				0.95				
$CuTlS_2$										
$CuTlSe_2$(L.T.)						1.07				
$CuFeS_2$						0.53				Chalcopyrite
$CuFeSe_2$						0.16				
$CuLaS_2$										
$AgAlS_2$										
$AgAlSe_2$						0.7				
$AgAlTe_2$						0.56				
$AgGaS_2$		0.150				1.66				
$AgGaSe_2$		0.182				1.1				
$AgGaTe_2$		0.280				1.9				
$AgInS_2$(L.T.)		0.185				1.18				
$AgInSe_2$		0.238				0.96, 0.52				
$AgInTe_2$		0.338								
$AgFeS_2$										

Substance	Heat of formation [kJ/mole] (300K)	Volume compressibility (10^{-10} m²/N)	Static dielectric constant	Atomic magnetic susceptibility (10^{-6} CGS)	Index of refraction	Minimum room temperature energy gap (eV)	Mobility (Room temp.) (cm²/V·s)		Optical transition	Remarks
							Electrons	Holes		
II IV V_2 Compounds										
$ZnSiP_2$	312					2.3	1000			
$ZnGeP_2$	293					2.2				
$ZnSnP_2$	275					1.45				
$CdSiP_2$		0.103				2.2	1000			
$CdGeP_2$	289					1.8				
$CdSnP_2$	270					1.5				
$ZnSiAs_2$	290					1.7		50		
$ZnGeAs_2$	271			-14.4		0.85				
$ZnSnAs_2$	252			-18.4		0.65		300		Disorders at 910K
$CdSiAs_2$		0.143				1.6				
$CdGeAs_2$	266			-23.4		0.53	70	25		Disorders at 903
$CdSnAs_2$	247		13.7	-21.5		0.26	22000	250		

§A5. Other Ternary Semiconductors with Tetrahedral Coordination

Substance	Heat of formation [kJ/mole] (300K)	Volume compressibility (10^{-10} m²/N)	Static dielectric constant	Atomic magnetic susceptibility (10^{-6} CGS)	Index of refraction	Minimum room temperature energy gap (eV)	Mobility Electrons	Mobility Holes	Optical transition	Remarks
I_2 IV VI_3 Compounds										
Cu_2SiS_3(H.T.)										Wurtzite
Cu_2SiS_3(L.T.)										Tetragonal
Cu_2SiTe_3										Cubic
Cu_2GeS_3(H.T.)				-18.7						Cubic
Cu_2GeS_3(L.T.)							360			Tetragonal
Cu_2GeSe_3	211.5			-21.3		0.94	238			Same
Cu_2GeTe_3	190.2			-23.4						Same
Cu_2SnS_3				-18.2		0.91	405			Cubic
$CuSnSe_3$				-21.0		0.66	870			Cubic
Cu_2SnTe_3				-28.4						Cubic
Ag_2GeSe_3				-29.6		0.91 (77K)				
Ag_2SnSe_3				-29.5		0.81				
Ag_2GeTe_3				-31.4		0.25				
Ag_2SnTe_3				-31.0		0.08				
I_3 V VI_4 Compounds										
Cu_3PS_4						1.24				Enargite
Cu_3AsS_4	269.6			-15.8		0.88				Famatinite
Cu_3AsSe_4	161.3			-13.1		0.74				Famatinite
Cu_3SbS_4				-8.3		0.31				
Cu_3SbSe_4	127.1			-20.5						
I IV_2 V_3 Compounds										
$CuSi_2P_3$										E1
$CuGe_2P_3$		0.12				0.9				E1
$AgGe_2P_3$										

§A6. "Defect Chalcopyrite" Structure Compounds (Strukturbericht symbol E3, Space Group I $\overline{4}$-S$_2^4$)

Substance	Heat of formation [kJ/mole] (300K)	Volume compressibility (10^{-10} m²/N)	Static dielectric constant	Atomic magnetic susceptibility (10^{-6} CGS)	Index of refraction	Minimum room temperature energy gap (eV)	Mobility Electrons	Mobility Holes	Optical transition	Remarks
$ZnAl_2Se_4$										
$ZnAl_2Te_4$(?)										
$ZnGa_2S_4$(?)						≈ 3.4				
$ZnGa_2Se_4$(?)						≈ 2.2				
$ZnGa_2Te_4$(?)						1.35				
$ZnIn_2Se_4$	206					1.82	35			
$ZnIn_2Te_4$	198					1.2				
$CdAl_2S_4$										
$CdAl_2Se_4$										
$CdAl_2Te_4$(?)										
$CdGa_2S_4$	256					3.44	60			
$CdGa_2Se_4$	216					2.43	33			
$CdGa_2Te_4$										
$CdIn_2Te_4$	195					(1.26 or 0.9)	4000			
$HgAl_2S_4$										
$HgAl_2Se_4$										
$HgAl_2Te_4$(?)										
$HgGa_2S_4$	249					2.84				
$HgGa_2Se_4$	204					1.95	400			
$HgIn_2Se_4$	196					0.6	290			
$HgIn_2Te_4$(?)	188					0.86	200			

Substance	Heat of formation [kJ/mole (300K)]	Volume compressibility (10^{-10} m²/N)	Static dielectric constant	Atomic magnetic susceptibility (10^{-6} CGS)	Index of refraction	Minimum room temperature energy gap (eV)	Mobility (Room temp.) (cm²/V·s) Electrons	Holes	Optical transition	Remarks

§A7. Other Adamantine Compounds

Substance	Heat	Vol	Static	Mag	Index	Gap	Electrons	Holes	Optical	Remarks
αSiC			10.2	−6.4	2.67	2.86	400			6H structure
$Hg_5Ga_2Te_8$										B3 with superlattice
$Hg_5In_2Te_8$						0.7	2000			B3 with superlattice
$CdIn_2Se_4$						1.55				

PART B. OCTAHEDRAL SEMICONDUCTORS

§B1. HALITE STRUCTURE SEMICONDUCTORS (Strukturbericht symbol B1, Space Group Fm3m-O_h^5)

Substance	Heat	Vol	Static	Mag	Index	Gap	Electrons	Holes	Optical	Remarks
GeTe										
SnSe										
SnTe										
PbS	435					0.5	600	600		
PbSe	393		161			0.37	1000	900		
PbTe	393		280			0.26	1600	600		Altaite
			360			0.25				

Selected other binary halites

Substance	Heat	Vol	Static	Mag	Index	Gap	Electrons	Holes	Optical	Remarks
BiSe										
BiTe						0.4				
EuSe										
GdSe						1.8	4			
NiD						2.0 or 3.7	100			
CdO	531					2.5				
SrSW						4.1				

PART C. OTHER SEMICONDUCTORS

§C1. Antifluorite Structure Compounds (Fm3m - O_h^5)

Substance	Heat	Vol	Static	Mag	Index	Gap	Electrons	Holes	Optical	Remarks
Mg_2Si	79.08					0.77	405	70		
Mg_2Ge						0.74	520	110		
Mg_2Sn	76.57					0.36	320	260		
Mg_2Pb	52.72					0.1				

§C2. Tetradymite Structure Compounds (R$\bar{3}$m − D_{3d}^5)

Substance	Heat	Vol	Static	Mag	Index	Gap	Electrons	Holes	Optical	Remarks
Sb_2Te_3						0.3		360		
Bi_2Se_3						0.35	600			
Bi_2Te_3						0.21	1140	680		R3m (166)

§C3. Skutterudite Structure Compounds (Im3-T_h^5)

Substance	Heat	Vol	Static	Mag	Index	Gap	Electrons	Holes	Optical	Remarks
CoP_3						0.43				
$CoAs_3$						0.69		~4000		
$CoSb_3$						0.63	70	~3000		
RhP_3								700		
$RhAs_3$						0.85		~3000		
$RhSb_3$						0.80		~7000		
$IrSb_3$						1.18		1500		

§C4. Selected Multinary Compounds

Substance	Heat	Vol	Static	Mag	Index	Gap	Electrons	Holes	Optical	Remarks
$AgSbSe_2$						0.58				
$AgSbTe_2$ (or $Ag_{19}Sb_{29}Te_{52}$)						0.7, 0.27				
$AgBiS_2$(H.T.)										
$AgBiSe_2$(H.T.)										
$AgBiTe_2$(H.T.)										
Cu_2CdSnS_4						1.16	<2			

§C5. Some Elemental Semiconductors

Substance	Heat	Vol	Static	Mag	Index	Gap	Electrons	Holes	Optical	Remarks
B	397.1			−6.7	3.4	1.55	10			
Se(gray)		6.6 (0.1 GHz)		−22.1	2.5	1.5		5		P3₁21(152)
Te				−39.5	3.3	0.33	1700	1200		Same

Table III
SEMICONDUCTING PROPERTIES OF SELECTED MATERIALS

Substance	Minimum Energy Gap (eV) R.T.	0 K	$\frac{dE_g}{dT}$ x 10⁴ eV/°C	$\frac{dE_g}{dP}$ x 10⁶ eV·cm²/kg	Density of States Electron Effective Mass m_{d_n} (m_o)	Electron Mobility and Temperature Dependence μ_n cm²/V·s	−x	Density of States Hole Effective Mass m_{d_p} (m_o)	Hole Mobility and Temperature Dependence μ_p cm²/V·s	−x
Si	1.107	1.153	−2.3	−2.0	1.1	1,900	2.6	0.56	500	2.3
Ge	0.67	0.744	−3.7	+7.3	0.55	3,800	1.66	0.3	1,820	2.33
αSn	0.08	0.094	−0.5		0.02	2,500	1.65	0.3	2,400	2.0
Te	0.33				0.68	1,100		0.19	560	
III–V Compounds										
AlAs	2.2	2.3				1,200			420	
AlSb	1.6	1.7	−3.5	−1.6	0.09	200	1.5	0.4	500	1.8
GaP	2.24	2.40	−5.4	−1.7	0.35	300	1.5	0.5	150	1.5
GaAs	1.35	1.53	−5.0	+9.4	0.068	9,000	1.0	0.5	500	2.1
GaSb	0.67	0.78	−3.5	+12	0.050	5,000	2.0	0.23	1,400	0.9
InP	1.27	1.41	−4.6	+4.6	0.067	5,000	2.0		200	2.4
InAs	0.36	0.43	−2.8	+8	0.022	33,000	1.2	0.41	460	2.3
InSb	0.165	0.23	−2.8	+15	0.014	78,000	1.6	0.4	750	2.1
II–VI Compounds										
ZnO	3.2		−9.5	+0.6	0.38	180	1.5			
ZnS	3.54		−5.3	+5.7		180			5(400°C)	
ZnSe	2.58	2.80	−7.2	+6		540			28	
ZnTe	2.26			+6		340			100	
CdO	2.5 ± .1		−6		0.1	120				
CdS	2.42		−5	+3.3	0.165	400		0.8		
CdSe	1.74	1.85	−4.6		0.13	650	1.0	0.6		
CdTe	1.44	1.56	−4.1	+8	0.14	1,200		0.35	50	
HgSe	0.30				0.030	20,000	2.0			
HgTe	0.15		−1		0.017	25,000		0.5	350	
Halite Structure Compounds										
PbS	0.37	0.28	+4		0.16	800		0.1	1,000	2.2
PbSe	0.26	0.16	+4		0.3	1,500		0.34	1,500	2.2
PbTe	0.25	0.19	+4	−7	0.21	1,600		0.14	750	2.2
Others										
ZnSb	0.50	0.56			0.15	10				1.5
CdSb	0.45	0.57	−5.4		0.15	300			2,000	1.5
Bi₂S₃	1.3					200			1,100	
Bi₂Se₃	0.27					600			675	
Bi₂Te₃	0.13		−0.95		0.58	1,200	1.68	1.07	510	1.95
Mg₂Si		0.77	−6.4		0.46	400	2.5		70	
Mg₂Ge		0.74	−9			280	2		110	
Mg₂Sn	0.21	0.33	−3.5		0.37	320			260	
Mg₃Sb₂		0.32				20			82	
Zn₃As₂	0.93					10	1.1		10	
Cd₃As₂	0.55				0.046	100,000	0.88			
GaSe	2.05		3.8						20	
GaTe	1.66	1.80	−3.6			14	−5			
InSe	1.8					900				
TlSe	0.57		−3.9		0.3	30		0.6	20	1.5
CdSnAs₂	0.23				0.05	25,000	1.7			
Ga₂Te₃	1.1	1.55	−4.8							
α-In₂Te₃	1.1	1.2			0.7				50	1.1
β-In₂Te₃	1.0								5	
Hg₃In₂Te₆	0.5								11,000	
SnO₂									78	

Table IV
BAND PROPERTIES OF SEMICONDUCTORS

PART A. DATA ON VALENCE BANDS OF SEMICONDUCTORS (ROOM TEMPERATURES)

Substance	Band Curvature Effective Mass Heavy Holes	Light Holes	"Split-off" Band Holes	Energy Separation of "Split-off" Band (eV)	Measured (Light) Hole Mobility cm²/V·s
	(Expressed as fraction of free electron mass)				
Semiconductors with Valence Band Maximum at the Center of the Brillouin Zone ('Γ'')					
Si	0.52	0.16	0.25	0.044	500
Ge	0.34	0.043	0.08	0.3	1,820
Sn	0.3				2,400
AlAs					
AlSb	0.4			0.7	550
GaP				0.13	100
GaAs	0.8	0.12	0.20	0.34	400
GaSb	0.23	0.06		0.7	1,400
InP				0.21	150
InAs	0.41	0.025	0.083	0.43	460
InSb	0.4	0.015		0.85	750
CdTe	0.35				50
HgTe	0.5				350

Table IV
BAND PROPERTIES OF SEMICONDUCTORS (continued)

Semiconductors with Multiple Valence Band Maxima

Substance	Number of Equivalent Valleys and Direction	Band Curvature Effective Masses		Anisotropy $K = m_L/m_T$	Measured (Light) Hole Mobility cm^2/V·s
		Longitudinal m_L	Transverse m_T		
PbSe	4 "L" [111]	0.095	0.047	2.0	1,500
PbTe	4 "L" [111]	0.27	0.02	10	750
Bi$_2$Te$_3$	6	0.207	~0.045	4.5	515

PART B. DATA ON CONDUCTION BANDS OF SEMICONDUCTORS (Room Temperature Data)

Single Valley Semiconductors

Substance	Energy Gap (eV)	Effective Mass (m$_0$)	Mobility (cm^2/V.s)	Comments
GaAs	1.35	0.067	8,500	3(or 6?) equivalent [100] valleys 0.36 eV above this maximum with a mobility of ~50
InP	1.27	0.067	5,000	3(or 6?) equivalent [100] valleys 0.4 eV above this minimum.
InAs	0.36	0.022	33,000	equivalent valleys ~1.0 eV above this minimum.
InSb	0.165	0.014	78,000	
CdTe	1.44	0.11	1,000	4(or 8?) equivalent [111] valleys 0.51 eV above this minimum.

Multivalley Semiconductors

Substance	Energy Gap	Number of Equivalent Valleys and Direction	Band Curvature Effective Mass		Anisotropy $K = m_L/m_T$	Comments
			Longitudinal m_L	Transverse m_T		
Si	1.107	6 in [100] "△"	0.90	0.192	4.7	
Ge	0.67	4 in [111] at "L"	1.588	0.0815	19.5	
GaSb	0.67	as Ge (?)	~1.0	~0.2	~5	
PbSe	0.26	4 in [111] at "L"	0.085	0.05	1.7	
PbTe	0.25	4 in [111] at "L"	0.21	0.029	5.5	
Bi$_2$Te$_3$	0.13	6			~0.05	

Table V
RESISTIVITY OF SEMICONDUCTING MINERALS

Mineral	ρ (ohm · m)	Mineral	ρ (ohm · m)
Diamond (C)	2.7	Gersdorffite, NiAsS	1 to 160 × 10^{-6}
Sulfides		Glaucodote, (Co, Fe)AsS	5 to 100 × 10^{-6}
Argentite, Ag$_2$S	1.5 to 2.0 × 10^{-3}	Antimonide	
Bismuthinite, Bi$_2$S$_3$	3 to 570	Dyscrasite, Ag$_3$Sb	0.12 to 1.2 × 10^{-6}
Bornite, Fe$_2$S$_3$ · nCu$_2$S	1.6 to 6000 × 10^{-6}	Arsenides	
Chalcocite, Cu$_2$S	80 to 100 × 10^{-6}	Allemonite, SbAs$_2$	70 to 60,000
Chalcopyrite, Fe$_2$S$_3$ · Cu$_2$S	150 to 9000 × 10^{-6}	Lollingite, FeAs$_2$	2 to 270 × 10^{-6}
Covellite, CuS	0.30 to 83 × 10^{-6}	Nicollite, NiAs	0.1 to 2 × 10^{-6}
Galena, PbS	6.8 × 10^{-6} to 9.0 × 10^{-2}	Skutterudite, CoAs$_3$	1 to 400 × 10^{-6}
Haverite, MnS$_2$	10 to 20	Smaltite, CoAs$_2$	1 to 12 × 10^{-6}
Marcasite, FeS$_2$	1 to 150 × 10^{-3}	Tellurides	
Metacinnabarite, 4HgS	2 × 10^{-6} to 1 × 10^{-3}	Altaite, PbTe	20 to 200 × 10^{-6}
Millerite, NiS	2 to 4 × 10^{-7}	Calavarite, AuTe$_2$	6 to 12 × 10^{-6}
Molybdenite, MoS$_2$	0.12 to 7.5	Coloradoite, HgTe	4 to 100 × 10^{-6}
Pentlandite, (Fe, Ni)$_9$S$_8$	1 to 11 × 10^{-6}	Hessite, Ag$_2$Te	4 to 100 × 10^{-6}
Pyrrhotite, Fe$_7$S$_8$	2 to 160 × 10^{-6}	Nagyagite, Pb$_6$Au(S, Te)$_{14}$	20 to 80 × 10^{-6}
Pyrite, FeS$_2$	1.2 to 600 × 10^{-3}	Sylvanite, AgAuTe$_4$	4 to 20 × 10^{-6}
Sphalerite, ZnS	2.7 × 10^{-3} to 1.2 × 10^4	Oxides	
Antimony-sulfur compounds		Braunite, Mn$_2$O$_3$	0.16 to 1.0
Berthierite, FeSb$_2$S$_4$	0.0083 to 2.0	Cassiterite, SnO$_2$	4.5 × 10^{-4} to 10,000
Boulangerite, Pb$_5$Sb$_4$S$_{11}$	2 × 10^3 to 4 × 10^6	Cuprite, Cu$_2$O	10 to 50
Cylindrite, Pb$_3$Sn$_4$Sb$_2$S$_{14}$	2.5 to 60	Hollandite, (Ba, Na, K)Mn$_8$O$_{16}$	2 to 100 × 10^{-3}
Franckeite, Pb$_5$Sn$_3$Sb$_2$S$_{14}$	1.2 to 4	Ilmenite, FeTiO$_3$	0.001 to 4
Hauchecornite, Ni$_9$(Bi, Sb)$_2$S$_8$	1 to 83 × 10^{-6}	Magnetite, Fe$_3$O$_4$	52 × 10^{-6}
Jamesonite, Pb$_4$FeSb$_6$S$_{14}$	0.020 to 0.15	Manganite, MnO · OH	0.018 to 0.5
Tetrahedrite, Cu$_3$SbS$_3$	0.30 to 30,000	Melaconite, CuO	6000
Arsenic-sulfur compounds		Psilomelane, KMnO · MnO$_2$ · nH$_2$O	0.04 to 6000
Arsenopyrite, FeAsS	20 to 300 × 10^{-6}	Pyrolusite, MnO$_2$	0.007 to 30
Cobaltite, CoAsS	6.5 to 130 × 10^{-3}	Rutile, TiO$_2$	29 to 910
Enargite, Cu$_3$AsS$_4$	0.2 to 40 × 10^{-3}	Uraninite, UO	1.5 to 200

From Carmichael, R. S., ed., *Handbook of Physical Properties of Rocks*, Vol. I, CRC Press, 1982.

REFERENCES

1. Beer, A. C., *Galvanomagnetic Effects in Semiconductors*, Academic Press, 1963.
2. Goryunova, N. A., *The Chemistry of Diamond-Like Semiconductors*, The MIT Press, 1965.
3. Abrikosov, N. Kh., Bankina, V. F., Poretskaya, L. E., Shelimova, L. E., and Skudnova, E. V., *Semiconducting II-VI, IV-VI, and V-VI Compounds*, Plenum Press, 1969.
4. Berger, L. I. and Prochukhan, V. D., *Ternary Diamond-Like Semiconductors*, Cons. Bureau/Plenum Press, 1969.
5. Shay, J. L. and Wernick, J. H., *Ternary Chalcopyrite Semiconductors: Growth, Electronic Properties, and Applications*, Pergamon Press, 1975.
6. Bergman, R., *Thermal Conductivity in Solids*, Clarendon, Oxford, 1976.
7. Handbook of Semiconductors, Vol. 1, Moss, T. S. and Paul, W., Eds., *Band Theory and Transport Properties*; Vol. 2, Moss, T. S. and Balkanski, M., Eds., *Optical Properties of Solids*; Vol. 3, Moss, T. S. and Keller, S. P., Eds., *Materials Properties and Preparation*, North Holland Publ. Co., 1980.
8. Böer, K. W., *Survey of Semiconductor Physics*, Van Nostrand Reinhold, 1990.
9. Rowe, D. M., Ed., *CRC Handbook of Thermoelectrics*, CRC Press, Boca Raton, FL, 1995.
10. Berger, L. I., *Semiconductor Materials*, CRC Press, Boca Raton, FL, 1997.
11. Glazov, V. M., Chizhevskaya, S. N., and Glagoleva, N. N., *Liquid Semiconductors*, Plenum Press, New York, 1969.
12. Phillips, J. C., *Bonds and Bands in Semiconductors*, Academic Press, New York, 1973.
13. Harrison, W. A., *Electronic Structure and the Properties of Solids*, Freeman Publ. House, San Francisco, 1980.
14. Balkanski, M., Ed., *Optical Properties of Solids*, North-Holland, Amsterdam, 1980.
15. *Landolt-Börnstein. Numerical Data and Functional Relationships in Science and Technology, New Series, Group III: Crystal and Solid State Physics*, Hellwege, K.-H., and Madelung, O., Eds., Volumes 17 and 22, Springer Verlag, Berlin, 1984 (and further).
16. Shklovskii, B. L., and Efros, A. L., *Electronic Processes in Doped Semiconductors*, Springer Verlag, Berlin, 1984.
17. Cohen, M. L., and Chelikowsky, J. R., *Electronic Structure and Optical Properties of Semiconductors*, Springer Verlag, New York, 1988.
18. Glass, J. T., Messier, R. F., and Fujimori, N., Eds., *Diamond, Silicon Carbide, and Related Wide Bandgap Semiconductors*, MRS Symposia Proc. 1652, Mater. Res. Soc., Pittsburgh, 1990.
19. Palik, E., Ed., *Handbook of Optical Constants of Solids II*, Academic Press, New York, 1991.
20. Reed, M., Ed., *Semiconductors and Semimetals*, Volume 35, Academic Press, Boston, 1992.
21. Haug, H., and Koch, S. W., *Quantum Theory of the Optical and Electronic Properties of Semiconductors*, 2nd Edition, World Scientific, Singapore, 1993.
22. Lockwood, D. J., Ed., *Proc. 22nd Intl. Conf. on the Physics of Semiconductors, Vancouver, 1994*, World Scientific, Singapore, 1994.
23. Morelli, D. T., Caillat, T., Fleurial, J.-P., Borshchevsky, A., Vandersande, J., Chen, B., and Uher, C., *Phys. Rev.*, B51, 9622, 1995.
24. Caillat, T., Borshchevsky, A., and Fleurial, J.-P., *J. Appl. Phys.*, 80, 4442, 1996.
25. Fleurial, J.-P., Caillat, T., and Borshchevsky, A., *Proc. XVI Intl. Conf. Thermoelectrics*, Dresden, Germany, August 26–29, 1997 (in print).

DIFFUSION DATA FOR SEMICONDUCTORS
B. L. Sharma

The diffusion coefficient D in many semiconductors may be expressed by an Arrhenius-type relation

$$D = D_o \exp(-Q/kT)$$

where D_o is a frequency factor, Q is the activation energy for diffusion, k is the Boltzmann constant, and T is the absolute temperature. This table lists D_o and Q for various diffusants in common semiconductors.

Abbreviations used in the table are

AES — Auger Electron Spectroscopy
DLTS — Deep Level Transient Spectroscopy
SEM — Scanning Electron Microscopy
SIMS — Secondary Ion Mass Spectrometry
D(c) — Concentration Dependent Diffusion Coefficient
D_{max} — Maximum Diffusion Coefficient

(f) — Fast Diffusion Component
(i) — Interstitial Diffusion Component
(s) — Slow Diffusion Component
($\parallel$) — Parallel to c Direction
($\perp$) — Perpendicular to c Direction

Semiconductor	Diffusant	Frequency factor, D_o (cm^2/s)	Activation energy, Q (eV)	Temperature range (°C)	Method of measurement	Ref.
Si	Li	2.5×10^{-3}	0.65	25—1350	Electrical	1
	Na	1.65×10^{-3}	0.77	800—1100	Electrical and flame photometry	2
	K	1.1×10^{-3}	0.76	800—1100	Electrical and flame photometry	2
	Cu	4×10^{-2}	1.0	800—1100	Radioactive	3
		4.7×10^{-3}	0.43 (i)	300—700	Radioactive	4
	Ag	2×10^{-3}	1.6	1100—1350	Radioactive	5
	Au	2.4×10^{-4}	0.39 (i)	700—1300	Radioactive	6
		2.75×10^{-3}	2.05 (s)			
	Be	$(D \sim 10^{-7})$	—	1050	Electrical	7
	Zn	1×10^{-1}	1.4	980—1270	Electrical	8
	B	2.46	3.59	1100—1250	Electrical	9
		2.4×10^{1}	3.87	840—1250	Electrical	10
	Al	1.38	3.41	1119—1390	Electrical	11
		1.8	3.2	1025—1175	Electrical	12
	Ga	3.74×10^{-1}	3.39	1143—1393	Electrical	11
		6×10^{1}	3.89	900—1050	Radioactive	13
	In	7.85×10^{-1}	3.63	1180—1389	Electrical	11
		1.94×10^{1}	3.86	1150—1242	Radioactive	14
	Tl	1.37	3.7	1244—1338	Electrical	11
		1.65×10^{1}	3.9	1105—1360	Electrical	15
	Ti	1.45×10^{-2}	1.79	950—1200	DLTS	16
	C	3.3×10^{-1}	2.92	1070—1400	Radioactive	17
	Si (self)	1.54×10^{2}	4.65	855—1175	SIMS	18
		1.6×10^{3}	4.77	1200—1400	Radioactive	19
	Ge	3.5×10^{-1}	3.92	855—1000	Radioactive	20
		2.5×10^{3}	4.97	1030—1302	Radioactive	20
		7.55×10^{3}	5.08	1100—1300	SIMS	21
	Sn	3.2×10^{1}	4.25	1050—1200	Neutron activation	22
	P	2.02×10^{1}	3.87	1100—1250	Electrical	9
		1.1	3.4	900—1200	Radioactive	23
		7.4×10^{-2}	3.3	1130—1405	Electrical	24
	As	6.0×10^{1}	4.2	950—1350	Radioactive	25
		6.55×10^{-2}	3.44	1167—1394	Electrical	26
		2.29×10^{1}	4.1	900—1250	Electrical	27
	Sb	1.29×10^{1}	3.98	1190—1398	Radioactive	28
		2.14×10^{-1}	3.65	1190—1405	Electrical	26
	Bi	1.03×10^{3}	4.64	1220—1380	Electrical	15
		1.08	3.85	1190—1394	Electrical	26

Semiconductor	Diffusant	Frequency factor, D_0 (cm²/s)	Activation energy, Q (eV)	Temperature range (°C)	Method of measurement	Ref.
	Cr	1×10^{-2}	1	1100—1250	Radioactive	29
	O	7×10^{-2}	2.44	700—1250	SIMS	30
		1.4×10^{-1}	2.53	700—1160	SIMS	31
	S	5.95×10^{-3}	1.83	975—1200	Radioactive	32
	Se	9.5×10^{-1}	2.6	1050—1250	Electrical	33
	Te	$(D < 10^{-11})$	—	1050—1250	Electrical	33
	Mn	6.9×10^{-4}	0.63	900—1200	Radioactive	34
	Fe	1.3×10^{-3}	0.68	30—1250	Radioactive	35
	Co	2×10^{-3}	0.69	700—1300	Radioactive	36
	Ni	2×10^{-3}	0.47	800—1300	Radioactive	37
	Rh	$(D \sim 10^{-6} - 10^{-4})$	—	1000—1200	Electrical	38
	Os	$(D \sim 2 \times 10^{-6})$	—	1280	Electrical	39
	Ir	4.2×10^{-2}	1.3	950—1250	Electrical	40
	Pt	$(D \sim 9.8 \times 10^{-16} - 3.6 \times 10^{-14})$	—	700—800	DLTS	41
Ga	Li	1.3×10^{-3}	0.46	350—800	Electrical	42
		9.1×10^{-3}	0.57	800—500	Electrical	43
	Na	3.95×10^{-1}	2.03	700—850	Radioactive	44
	Cu	1.9×10^{-4}	0.18 (i)	750—900	Radioactive	45
		4×10^{-2}	0.99 (s)	600—700		
		4×10^{-3}	0.33 (i)	350—750	Radioactive	4
	Ag	4.4×10^{-2}	1.0 (i)	700—900	Radioactive	46, 47
		4×10^{-2}	2.23 (s)	800—900	Radioactive	48
	Au	2.25×10^{2}	2.5	600—900	Radioactive	49
	Be	5×10^{-1}	2.5	720—900	Electrical	50
	Zn	5	2.7	600—900	Radioactive and electrical	51
	Cd	1.75×10^{9}	4.4	760—915	Radioactive	52
	B	1.8×10^{9}	4.55	600—900	Electrical	51
	Al	1.0×10^{3}	3.45	554—905	SIMS	53
		$\sim 1.6 \times 10^{2}$	~ 3.24	750—850	Electrical	54
	Ga	1.4×10^{2}	3.35	554—916	SIMS	55
		3.4×10^{1}	3.1	600—900	Electrical	51
	In	1.8×10^{4}	3.67	554—919	SIMS	56
		3.3×10^{1}	3.02	700—855	Radioactive	57
	Tl	1.7×10^{3}	3.4	800—930	Radioactive	58
	Si	2.4×10^{-1}	2.9	650—900	(γ) resonance	59
	Ge (self)	2.48×10^{1}	3.14	549—891	Radioactive	60
		7.8	2.95	766—928	Radioactive	61
	Sn	1.7×10^{-2}	1.9	—	Radioactive	45
	P	3.3	2.5	600—900	Electrical	51
	As	2.1	2.39	700—900	Electrical	62
	Sb	3.2	2.41	700—855	Radioactive	57
		1.0×10^{1}	2.5	600—900	Radioactive and electrical	51
	Bi	3.3	2.57	650—850	—	63
	O	4×10^{-1}	2.08	—	Optical	64
	S	$(D \sim 10^{-9})$	—	920	—	65
	Se	$(D \sim 10^{-10})$	—	920	—	65
	Te	5.6	2.43	750—900	Radioactive	66
	Fe	1.3×10^{-1}	1.08	750—900	Radioactive	67
	Co	1.6×10^{-1}	1.12	750—850	Radioactive	47
	Ni	8×10^{-1}	0.9	670—900	Electrical	68
GaAs	Li	5.3×10^{-1}	1.0	250—500	Electrical and chemical	69
	Cu	3×10^{-2}	0.53	100—500	Radioactive	70

Semiconductor	Diffusant	Frequency factor, D_0 (cm^2/s)	Activation energy, Q (eV)	Temperature range (°C)	Method of measurement	Ref.
		6×10^{-2}	0.98	450—750	Ultrasonic	71
		1.5×10^{-3}	0.6	800—1000	Radioactive	72
	Ag	4×10^{-4}	0.8	500—1150	Radioactive	73
	Au	1×10^{-3}	1.0	740—1025	Radioactive	74
	Be	7.3×10^{-6}	1.2	800—990	Electrical	75
	Mg	4×10^{-5}	1.22	800—1200	Electrical	76
	Zn	1.5×10^{1}	2.49	600—980	Radioactive	77
		2.5×10^{-1}	3.0	750—1000	Radioactive	78
	Cd	1.3×10^{-3}	2.2	800—1100	Radioactive	79
		5×10^{-2}	2.43	868—1149	Radioactive	77
	Hg	$(D \sim 5 \times 10^{-14})$	—	1100	Radioactive	80
	Al	$(D \sim 4 \times 10^{-18}$— $10^{-14})$	4.3	850—1100	AES	81
	Ga (self)	4×10^{-5}	2.6	1025—1100	Radioactive	82
		1×10^{7}	5.6	1125—1230	Radioactive	83
	Si	1.1×10^{-1}	2.5	850—1050	SIMS	84
	Ge	1.6×10^{-5}	2.06	650—850	SIMS	85
	Sn	6×10^{-4}	2.5	1060—1200	Radioactive	86
		1×10^{-5}	2	800—1000	Radioactive	87
	P	$(D \sim 10^{-12}$—$10^{-10})$	2.9	800—1150	Reflectance measurements	88
	As (self)	7×10^{-1}	3.2	—	Radioactive	78
	Cr	2.04×10^{-6}	0.83 (f)	750—1000	SIMS	89
			1.7 (s)	700—900		
		7.9×10^{-3}	2.2	800—1100	Chemical analysis	90
	O	2×10^{-3}	1.1	700—900	Mass spectroscopy	91
	S	1.85×10^{-2}	2.6	1000—1300	Radioactive	92
		1.1×10^{1}	2.95	750—900	Electrical	93
	Se	3×10^{3}	4.16	1025—1200	Radioactive	83
	Te	1.5×10^{-1}	3.5	1000—1150	Radioactive	94
	Mn	6.5×10^{-1}	2.49	850—1100	Radioactive	95
	Fe	4.2×10^{-2}	1.8	850—1150	Radioactive	96
		2.2×10^{-3}	2.32	750—1050	Radioactive	97
	Co	5×10^{2}	2.5	800—1000	Radioactive	98
		1.2×10^{-1}	2.64	750—1050	Radioactive	97
	Tm	2.3×10^{-16}	1.0	800—1000	Radioactive	99
GaSb	Li	2.3×10^{-4}	1.9 (s)	527—657	Electrical and flame photometry	100
		1.2×10^{-1}	0.7 (f)	277—657		
	Cu	4.7×10^{-3}	0.9	470—650	Radioactive	101
	Zn	$(D \sim 2 \times 10^{-13}$— $1 \times 10^{-11})$	2	510—600	Radioactive	102
	Cd	1.5×10^{-6}	0.72	640—800	Electrical	103
	Ga (self)	3.2×10^{3}	3.15	658—700	Radioactive	104
	In	1.2×10^{-7}	0.53	320—650	Radioactive	105
	Sn	2.4×10^{-5}	0.8	320—650	Radioactive	106
		1.3×10^{-5}	1.1	500—650	Radioactive	107
	Sb (self)	3.4×10^{4}	3.45	658—700	Radioactive	104
	Se	$(D \sim 2.4 \times 10^{-13}$— $1.37 \times 10^{-11})$	—	400—500	Radioactive	106
	Te	3.8×10^{-4}	1.20	320—650	Radioactive	106
	Fe	5×10^{-2}	1.9 (I)	500—650	Radioactive	108
		5×10^{2}	2.3 (II)	500—650		
GaP	Ag	—	—	1000—1300	Radioactive	109
	Au	8	2.5 (I)	1050—1250	Radioactive	110
		20	2.4 (II)	1100—1250	Diffusion (I) A face and (II) B face	

Semiconductor	Diffusant	Frequency factor, D_o (cm²/s)	Activation energy, Q (eV)	Temperature range (°C)	Method of measurement	Ref.
	Be	$(D_{max} \sim 2.4 \times 10^{-9}$— $8.5 \times 10^{-8})$	—	900—1000	Atomic absorption analysis	111
	Mg	5×10^{-5}	1.4	700—1050	Electrical	112
	Zn	1.0	2.1	700—1300	Radioactive	113
	Ge	—	—	900—1000	Radioactive	114
	Cr	6.2×10^{-4}	1.2	900—1130	Radioactive; ESR	115
	S	3.2×10^3	4.7	1120—1305	Radioactive	92
	Mn	2.1×10^9	4.7	$T < 950$	Radioactive; ESR	116
		1.1×10^{-6}	0.9	950—1130		
	Fe	1.6×10^{-1}	2.3	980—1180	Radioactive	117
	Co	2.8×10^{-3}	2.9	850—1100	Radioactive	118
InP	Cu	3.8×10^{-3}	0.69	600—900	Radioactive	119
	Ag	3.6×10^{-4}	0.59	500—900	Radioactive	120
	Au	1.32×10^{-5}	0.48	600—820	Radioactive	121
		1.37×10^{-4}	0.73	600—900	Radioactive	122
	Zn	1.6×10^{-8}	0.3	750—900	Electrical	123
		$(D \sim 2 \times 10^{-9}$— $4 \times 10^{-8})$	—	700—900	Radioactive	124
	Cd	1.8	1.9	700—900	Radioactive	125
		1.1×10^{-7}	0.72	700—900	Electrical	123
		$(D \sim 7 \times 10^{-13}$— $2 \times 10^{-10})$	—	450—650	Electrical	126
	In (self)	1×10^5	3.85	830—990	Radioactive	83
	Sn	$(D \sim 3 \times 10^{-8})$	—	550	Etching and cathodo-luminescence	127
	P (self)	7×10^{10}	5.65	900—1000	Radioactive	83
	Cr	—	—	600—900	Radioactive	128
	S	—	—	607—710	Electrical	129
	Mn	—	2.9	650—750	SIMS	130
	Fe	3	2	600—950	Radioactive	131
		6.8×10^5	3.4	600—700	SIMS	132
	Co	9×10^{-1}	1.8	600—950	Radioactive	131
InAs	Cu	3.6×10^{-3}	0.52	342—875	Radioactive	133
		2.2×10^{-2}	0.54	525—890	Radioactive	134
	Ag	7.3×10^{-4}	0.26	450—900	Radioactive	135
	Au	5.8×10^{-3}	0.65	600—900	Radioactive	136
	Mg	1.98×10^{-6}	1.17	600—900	Electrical	137
	Zn	4.2×10^{-3}	0.96	600—900	Radioactive	138
		3.11×10^{-3}	1.17	600—900	Electrical	137
	Cd	7.4×10^{-4}	1.15	650—900	Radioactive	139
	Hg	1.45×10^{-5}	1.32	650—850	Radioactive	140
	In (self)	6×10^5	4.0	740—900	Radioactive	141
	Ge	3.74×10^{-6}	1.17	600—900	Electrical	137
	Sn	1.49×10^{-6}	1.17	600—900	Electrical	137
	As (self)	3×10^7	4.45	740—900	Radioactive	141
	S	6.78	2.2	600—900	Electrical	137
	Se	12.6	2.2	600—900	Electrical	137
	Te	3.43×10^{-5}	1.28	600—900	Electrical	137
InSb	Li	7×10^{-4}	0.28	0—210	Electrical	142
	Cu	9×10^{-4}	1.08	200—500	Radioactive	143
		3×10^{-5}	0.37	230—490	Radioactive	144
	Ag	1×10^{-7}	0.25	440—510	Radioactive	145
	Au	7×10^{-4}	0.32	140—510	Radioactive	144
	Zn	5×10^{-1}	1.35	362—508	Radioactive	146
		—	1.5	355—455	SIMS	147
	Cd	1×10^{-5}	1.1	250—500	Radioactive	148

Semiconductor	Diffusant	Frequency factor, D_0 (cm²/s)	Activation energy, Q (eV)	Temperature range (°C)	Method of measurement	Ref.
		1.3×10^{-4}	1.2	360—500	Electrical	149
	Hg	4×10^{-6}	1.17	425—500	Radioactive	150
	In (self)	5×10^{-2}	1.82	478—520	Radioactive	104
		1.8×10^{13}	4.3	475—517	Radioactive	151
	Sn	5.5×10^{-8}	0.75	390—512	Radioactive	152
	Pb	$(D \sim 2.7 \times 10^{-15})$	—	500	Radioactive	153
	Sb (self)	5×10^{-2}	1.94	478—520	Radioactive	104
		3.1×10^{13}	4.3	475—517	Radioactive	151
	S	9×10^{-2}	1.4	360—500	Electrical	154
	Se	1.6	1.87	380—500	Electrical	155
	Te	1.7×10^{-7}	0.57	300—500	Radioactive	156
	Fe	1×10^{-7}	0.25	440—510	Radioactive	145
	Co	2.7×10^{-11}	0.39	420—500	Radioactive	157
AlAs	Ga	$(D \sim 2 \times 10^{-18}$— $10^{-15})$	3.6	850—1100	AES	81
	Zn	$(D \sim 9 \times 10^{-11})$	—	557	SEM	158
AlSb	Cu	3.5×10^{-3}	0.36	150—500	Radioactive	159
	Zn	3.3×10^{-1}	1.93	660—860	Radioactive	160
	Cd	$D(c) \sim 4 \times 10^{-12}$— 3×10^{-10}	—	900	Radioactive	161
	Al (self)	2	1.88	570—620	X-ray	162
	Sb (self)	1	1.7	570—620	X-ray	162
ZnS	Cu	2.6×10^{-3}	0.79	470—750	Radioactive	163
		4.3×10^{-4}	0.64	250—1200	Electroluminescence	164
		9.75×10^{-3}	1.04	400—800	Luminescence	165
	Au	1.75×10^{-4}	1.16	500—800	Radioactive	166
	Zn (self)	3×10^{-4}	1.5	925<T<940	Radioactive	167
		1.5×10^{4}	3.26	940<T<1030		
		1×10^{16}	6.5	1030<T<1075		
	Cd	$(D \sim 10^{-10})$	—	1100	Luminescence	168
	Al	5.69×10^{-4}	1.28	800—1000	Luminescence	167
	In	3×10^{1}	2.2	750—1000	Radioactive	163
	S (self)	2.16×10^{4}	3.15	600—800	Radioactive	166
		8×10^{-5}	2.2	740—1100	Radioactive	169
	Se	$(D \sim 5 \times 10^{-13})$	—	1070	X-ray microprobe	170
	Mn	2.3×10^{3}	2.46	500—800	Radioactive	166
ZnSe	Cu	1×10^{-4}	0.66	400—800	Luminescence	171
		1.7×10^{-5}	0.56	200—570	Radioactive	172
	Ag	2.2×10^{-2}	1.18	400—800	Luminescence	171
	Zn (self)	9.8	3.0	760—1150	Radioactive	173
	Cd	6.39×10^{-4}	1.87	700—950	Photoluminescence	174
	Al	2.3×10^{-2}	1.8	800—1100	Luminescence	171
	Ga	1.81×10^{2}	3.0	900—1100	Luminescence	165
		—	1.3	700—850	Electron probe	175
	In	$(D \sim 2 \times 10^{-12})$	—	940	—	176
	S	$(D \sim 8 \times 10^{-12})$	—	1060	X-ray microprobe	170
	Se (self)	1.3×10^{1}	2.5	860—1020	Radioactive	173
		2.3×10^{-1}	2.7	1000—1050	Radioactive	177
	Ni	$(D \sim 1.5 \times 10^{-8}$— $1.7 \times 10^{-7})$	—	740—910	Luminescence	165
ZnTe	Li	2.9×10^{-2}	1.22 (s)	400—700	Nuclear and chemical analysis	178
		1.7×10^{-4}	0.78 (f)			
	Zn (self)	2.34	2.56	760—860	Radioactive	179
		1.4×10^{1}	2.69	667—1077	Radioactive	180
	Al	—	2.0	700—1000	Electrical and optical	181

Semiconductor	Diffusant	Frequency factor, D_0 (cm²/s)	Activation energy, Q (eV)	Temperature range (°C)	Method of measurement	Ref.
CdS	In	4	1.96	1100—1300	Radioactive	182
	Te (self)	2×10^4	3.8	727—977	Radioactive	180
	Li	3×10^{-6}	0.68	610—960	Microhardness	183
	Na	$(D \sim 3 \times 10^{-7})$	—	800	Radioactive	184
	Cu	1.5×10^{-3}	0.76	400—700	Radioactive	185
		1.2×10^{-2}	1.05	300—700	Ultrasonic	186
		8×10^{-5}	0.72	20—200	Electrical	187
	Ag	2.5×10^1	1.2 (s)	300—500	Radioactive	188
		2.4×10^{-1}	0.8 (f)			
	Au	2×10^2	1.8	500—800	Radioactive	189
	Zn	1.27×10^{-9}	0.86 (s)	720—1000	Radioactive	190
		1.22×10^{-8}	0.66 (f)			
	Cd (self)	3.4	2.0	700—1100	Radioactive	191
	Ga	—	—	667—967	Optical and microprobe	192
	In	6×10^1	2.3 (∥)	650—930	Radioactive, optical and microprobe	193
		1×10^1	2.03 (⊥)			
	P	6.5×10^{-4}	1.6	800—1100	Radioactive	194
	S (self)	1.6×10^{-2}	2.05	800—900	Radioactive	195
		—	2.4	750—1050	Radioactive	196
	Se	$(D \sim 1.2 \times 10^{-9})$	—	900	Radioactive	177
	Te	1.3×10^{-7}	10.4	700—1000	Radioactive	197
	Cl	$(D \sim 3 \times 10^{-10})$	—	800	Electrical	184
	I	$(D \sim 5 \times 10^{-12})$	—	1000	Radioactive	184
	Ni	6.75×10^{-3}	10.9	570—900	Luminescence	165
	Yb	$(D \sim 1.3 \times 10^{-9})$	—	960	Photoluminescence	198
CdSe	Ag	2×10^{-4}	0.53	22—400	Ultrasonic	199
	Cd (self)	1.6×10^{-3}	1.5	700—1000	Radioactive	200
		6.3×10^{-2}	1.25 (I)	600—900	Radioactive;	201
		4.12×10^{-2}	2.18 (II)	600—900	(I) saturated Cd and (II) saturated Se pressure	
	P	$(D \sim 5.3 \times 10^{-12}-$ $6 \times 10^{-11})$	—	900—1000	Radioactive	202
	Se (self)	2.6×10^3	1.55	700—1000	Radioactive; saturated Se pressure	177
CdTe	Li	$(D \sim 1.5 \times 10^{-10})$	—	300	Ion microprobe	203
	Cu	3.7×10^{-4}	0.67	97—300	Radioactive	204
		8.2×10^{-8}	0.64	290—350	Ion backscattering	205
	Ag	—	—	700—800	Electrical and photo-luminescence	206
	Au	6.7×10^1	2.0	600—1000	Radioactive	207
	Cd (self)	1.26	2.07	700—1000	Radioactive	208
		3.26×10^2	2.67 (I)	650—900	Radioactive;	209
		1.58×10^1	2.44 (II)		(I) saturated Cd and (II) saturated Te pressure	
	In	8×10^{-2}	1.61	650—1000	Radioactive	210
		1.17×2	2.21 (I)	500—850	Radioactive; (I) saturated Cd and (II) saturated Te pressure	211
		6.48×10^{-4}	1.15 (II)			
	Sn	8.3×10^{-2}	2.2	700—925	Radioactive	212
	P	$(D \sim 1.2 \times 10^{-10})$	—	900	Radioactive	202
	As	—	—	850	—	213
	O	5.6×10^{-9}	1.22	200—650	Mass spectrometry	214
		6.0×10^{-10}	0.29	650—900		
	Se	1.7×10^{-4}	1.35	700—1000	Radioactive	215

DIFFUSION DATA FOR SEMICONDUCTORS (continued)

Semiconductor	Diffusant	Frequency factor, D_0 (cm^2/s)	Activation energy, Q (eV)	Temperature range (°C)	Method of measurement	Ref.
	Te (self)	8.54×10^{-7}	1.42 (I)	600—900	Radioactive; (I) saturated Cd and (II) saturated Te pressure	209
		1.66×10^{-4}	1.38 (II)	500—800		
	Cl	7.1×10^{-2}	1.6	520—800	Radioactive	216
	Fe	$(D \sim 4 \times 10^{-8})$	0.77	900	Radioactive	217
HgSe	Sb	6.3×10^{-5}	0.85	540—630	Radioactive	218
	Se (self)	—	—	200—400	Radioactive	219
HgTe	Ag	6×10^{-4}	0.8	250—350	Radioactive	220
	Zn	5×10^{-8}	0.6	250—350	Radioactive	220
	Cd	3.1×10^{-4}	0.66	250—350	Radioactive	221
	Hg (self)	2×10^{-8}	0.6	200—350	Radioactive	220
	In	6×10^{-6}	0.9	200—300	Radioactive	220
	Sn	1.72×10^{-6}	0.66 (s)	200—300	Radioactive	222
		1.8×10^{-3}	0.80 (f)			
	Te (self)	10^{-6}	1.4	200—400	Radioactive	220
	Mn	1.5×10^{-4}	1.3	250—350	Radioactive	220
PbS	Cu	4.6×10^{-4}	0.36	150—450	Electrical	223
		5×10^{-3}	0.31	100—400	Electrical	224
	Pb (self)	8.6×10^{-5}	1.52	500—800	Radioactive	225
	S (self)	6.8×10^{-5}	1.38	500—750	Radioactive	226
	Ni	1.78×10^{1}	0.95	200—500	Electrical	227
PbSe	Na	1.5×10^{1}	1.74 (s)	400—850	Radioactive	228
		5.6×10^{-6}	0.4 (f)			
	Cu	2×10^{-5}	0.31	93—520	Radioactive	229
	Ag	7.4×10^{-4}	0.35	400—850	Radioactive	228
	Pb (self)	4.98×10^{-6}	0.83	400—800	Radioactive	230
	Sb	3.4×10^{-1}	2.0	650—850	Radioactive	231
	Se (self)	2.1×10^{-5}	1.2	650—850	Radioactive	231
	Cl	1.6×10^{-8}	0.45	400—850	Radioactive	232
	Ni	$(D \sim 1 \times 10^{-10})$	—	700	Radioactive	233
PbTe	Na	1.7×10^{-1}	1.91	600—850	Radioactive	234
	Sn	3.1×10^{-2}	1.56	500—800	Radioactive	231
	Pb (self)	2.9×10^{-5}	0.6	250—500	Radioactive	231
	Sb	4.9×10^{-2}	1.54	500—800	Radioactive	231
	Te	2.7×10^{-6}	0.75	500—800	Radioactive	231
	Cl	$(D > 2.3 \times 10^{-10})$	—	700	Radioactive	235
	Ni	$(D > 1 \times 10^{-6})$	—	700	Radioactive	235

REFERENCES

1. E. M. Pell, *Phys. Rev.,* 119, 1960; 119, 1014, 1960.
2. L. Svob, *Solid State Electron,* 10, 991, 1967.
3. B. I. Boltaks and I. I. Sosinov, *Zh. Tekh. Fiz.,* 28, 3, 1958.
4. R. N. Hall and J. N. Racette, *J. Appl. Phys.,* 35, 379, 1964.
5. B. I. Boltaks and Hsueh Shih-Yin, *Sov. Phys. Solid State,* 2, 2383, 1961.
6. W. R. Wilcox and T. J. LaChapelle, *J. Appl. Phys.,* 35, 240, 1964.
7. E. A. Taft and R. O. Carlson, *J. Electrochem. Soc.,* 117, 711, 1970.
8. R. Sh. Malkovich and N. A. Alimbarashvili, *Sov. Phys. Solid State,* 4, 1725, 1963.
9. R. N. Ghoshtagore, *Solid State Electron,* 15, 1113, 1972.
10. C. Hill, *Semiconductor Silicon 1981,* H. R. Huff, R. J. Kreiger, and Y. Takeishi, Eds., p. 988, *Electrochem. Soc.,* 1981.
11. R. N. Ghoshtagore, *Phys. Rev. B,* 3, 2507, 1971.
12. W. Rosnowski, *J. Electrochem. Soc.,* 125, 957, 1978.

13. J. S. Makris and B. J. Masters, *J. Appl. Phys.,* 42, 3750, 1971.
14. M. F. Millea, *J. Phys. Chem. Solids,* 27, 315, 1965 (refer Reference 2).
15. C. S. Fuller and J. A. Ditzenberger, *J. Appl. Phys.,* 27, 544, 1956.
16. S. Hocine and D. Mathiot, *Appl. Phys. Lett.,* 53, 1269, 1988.
17. R. C. Newman and J. Wakefield, *J. Phys. Chem. Solids,* 19, 230, 1961.
18. L. Kalinowski and R. Seguin, *Appl. Phys. Lett.,* 35, 211, 1979; *Appl. Phys. Lett.,* 36, 171, 1980.
19. R. F. Peart, *Phys. Stat. Sol.,* 15, K 119, 1966.
20. G. Hettich, H. Mehrer and K. Maler, *Inst. Phys. Conf. Ser.,* 46, 500, 1979.
21. M. Ogina, Y. Oana and M. Watanabe, *Phys. Stat. Sol. (a),* 72, 535, 1982.
22. T. H. Yeh, S. M. Hu, and R. H. Kastl, *Appl. Phys.,* 39, 4266, 1968.
23. I. Franz and W. Langheinrich, *Solid State Electron,* 14, 835, 1971.
24. R. N. Ghoshtagore, *Hys. Rev. B,* 3, 389, 1971.
25. B. J. Masters and J. M. Fairfield, *J. Appl. Phys.,* 40, 2390, 1969.
26. R. N. Goshtagore, *Phys. Rev. B,* 3, 397, 1971.
27. R. S. Fair and J. C. C. Tsai, *J. Electrochem. Soc.,* 122, 1689, 1975.
28. J. J. Rohan, N. E. Pickering and J. Kennedy, *J. Electrochem. Soc.,* 106, 705, 1969.
29. W. Wuerker, K. Roy, and J. Hesse, *Matsr. Res. Bull.,* 9, 971, 1974.
30. J. C. Mikkelsen, Jr., *Appl. Phys. Lett.,* 40, 336, 1982.
31. S. Tang Lee and D. Nicols, *Appl. Phys. Lett.,* 47, 1001, 1985.
32. P. L. Gruzin, S. V. Zemskii, A. D. Bullkin, and N. M. Makarov, *Sov. Phys. Sem.,* 7, 1241, 1974.
33. N. S. Zhdanovich and Yu. I. Kozlov, *Svoistva Legir, Poluprovodn.,* V. S. Zemskov, Ed., Nauka, Moscow, 1977, 115-120; *Fiz Tekh. Poluprovod.,* 9, 1594, 1975.
34. D. Gilles, W. Bergholze, and W. Schroeter, *J. Appl. Phys.,* 59, 3590, 1986.
35. E. R. Weber, *Appl. Phys. A,* 30, 1, 1983.
36. E. R. Weber, Properties of Silicon, EMIS Datareviews Ser. No. 4, INSPEC Publications, 1988, 409-451.
37. M. K. Bakhadyrkhanov, S. Zainabidinov, and A. Khamidov, *Sov. Phys. Sem.,* 14, 243, 1980.
38. S. A. Azimov, M. S. Yunosov, F. K. Khatamkulov, and G. Nasyrov, *Poluprovod.,* N. Kh. Abrikosov and V. S. Zemskov, Eds., Nauka, Moscow, 1975, 21-23.
39. S. A. Azimov, M. S. Yunosov, G. Nurkuziev, and F. R. Karimov, *Sov. Phys. Sem.,* 12, 981, 1978.
40. S. A. Azimov, B. V. Umarov, and M. S. Yunusov, *Sov. Phys. Sem.,* 10, 842, 1976.
41. S. Mantovani, F. Nava, C. Nobili, and G. Ottaviani, *Phys. Rev. B,* 33, 5536, 1986.
42. C. S. Fuller and J. A. Ditzenberger, *Phys. Rev.,* 91, 193, 1953.
43. B. Pratt and F. Friedman, *J. Appl. Phys.,* 37, 1893, 1966.
44. M. Stojic, V. Spiric, and D. Kostoski, *Inst. Phys. Conf. Ser.,* 31, 304, 1976.
45. B. I. Boltaks, *Diffusion in Semiconductors,* Inforsearch, London, 1963, 162.
46. A. A. Bugai, V. E. Kosenko, and E. G. Miselyuk, *Zh. Tekh. Fiz.,* 27, 67, 1957.
47. L. Y. Wei, *J. Phys. Chem. Solids,* 18, 162, 1961.
48. V. E. Kosenko, *Sov. Phys. Solid State,* 4, 42, 1962.
49. W. C. Dunlap, Jr., *Phys. Rev.,* 97, 614, 1955
50. Yu. I. Belyaev and V. A. Zhidkov, *Sov. Phys. Solid State,* 3, 133, 1961.
51. W. C. Dunlap, Jr. *Phys. Rev.,* 94, 1531, 1954.
52. V. E. Kosenko, *Sov. Phys. Solid State,* 1, 1481, 1960.
53. P. Dorner, W. Gust, A. Lodding, H. Odelius, B. Predel, and U. Roll, *Acta Metall.,* 30, 941, 1982.
54. W. Meer and D. Pommerrening, *Z. Agnew. Phys.,* 23, 369, 1967.
55. U. Sodervall, H. Odelius, A. Lodding, U. Roll, B. Predel, W. Gust, and P. Dorner, *Phil. Mag. A,* 54, 539, 1986.
56. P. Dorner, W. Gust, A. Lodding, H. Odelius, B. Predel, and U. Roll, *Z. Metalkd.,* 73, 325, 1982.
57. P. V. Pavlov, *Sov. Phys. Solid State,* 8, 2377, 1967.
58. V. I. Tagirov and A. A. Kuliev, *Sov. Phys. Solid State,* 4, 196, 1962.
59. J. Raisanen, J. Hirvonen, and A. Anttila, *Solid State Electron.,* 24, 333, 1981.
60. C. Vogel, G. Hettich, and H. Mehrer, *J. Phys. C.,* 16, 6197, 1983.
61. H. Letaw, Jr., W. M. Portnoy, and L. Slifkin, *Phys. Rev.,* 102, 363, 1956.
62. W. Bosenberg, *Z. Naturforsch.,* 10a, 285, 1955.
63. V. M. Glazov and V. S. Zemskov, Physicochemical Principles of Semiconductor Doping, Israel Program for Scientific Translation, Jerusalem, 1968.
64. J. W. Corbett, R. S. McDonald, and G. D. Watkins, *J. Phys. Chem. Solids,* 25, 873, 1964.
65. W. W. Tyler, *J. Phys. Chem. Solids,* 8, 59, 1959.
66. V. D. Ignatkov and V. E. Kosenko, *Sov. Phys. Solid State,* 4, 1193, 1962.
67. A. A. Bugal, V. E. Kosenko, and E. G. Miseluk, *Zh. Tekh. Fiz.,* 27, 210, 1957.
68. F. van der Maesen and J. A. Brenkman, *Phillips Res. Rep.,* 9, 255, 1954.
69. C. S. Fuller and K. B. Wolfstrin, *J. Appl. Phys.,* 33, 2507, 1962.

70. R. N. Hall and J. H. Racette, *J. Appl. Phys.*, 25, 379, 1964.
72. V. S. Vasilev, I. N. Kamevoskii, and V. B. Osvenskii, *Sov. Phys. Sem.*, 2, 1495, 1969.
73. B. I. Boltaks and F. S. Shishiyanu, *Sov. Phys. Solid State*, 5, 1680, 1964.
74. V. I. Sokolov and F. S. Shishiyanu, *Sov. Phys. Solid State*, 6, 265, 1964.
75. E. A. Poltoratskii and V. M. Stuchebnikov, *Sov. Phys. Solid State*, 8, 770, 1966.
76. E. I. Andreivckii, S. V. Mashkin, and S. S. Zludkov, *Diffusion in Semiconductor*, (Russian), Gorkii, 1969.
77. B. Goldstein, *Phys. Rev.*, 118, 1024, 1960.
78. D. L. Kendall, *Semiconductors and Semimetals*, Vol. 4, R. K. Willardson and A. C. Beer, Eds., Academic, 163-259.
79. M. Fujimoto, K. Kudo, and N. Hishinuma, *Jpn. J. Appl. Phys.*, 8, 725, 1969.
80. J. A. Kanz, (p. 227 Reference 78).
81. L. L. Chang and A. Koma, *Appl. Physics Lett.*, 29, 138, 1976.
82. H. D. Palfrey, M. Brown, and A. F. W. Willoughby, *J. Electrochem. Soc.*, 128, 2224, 1981.
83. B. Goldstein, *Phys. Rev.*, 121, 1305, 1961.
84. M. E. Greiner and J. B. Gibbons, *Appl. Phys. Lett.*, 44, 750, 1984.
85. K. Sarma, R. Dalby, K. Rose, O. Aina, W. Katz, and N. Lewis, *J. Appl. Phys.*, 56, 2703, 1984.
86. S. Goldstein and H. Keller, *J. Appl. Phys.*, 32, 1180, 1961.
87. H. Yamazaki, Y. Kawaski, M. Futjimoto, and K. Kudo, *Jpn. J. Appl. Phys.*, 14, 717, 1975.
88. G. C. Jain, D. K. Sadana, and B. K. Das, *Solid State Electron.*, 19, 731, 1976.
89. M. D. Deal and D. A. Stevenson, *J. Appl. Phys.*, 59, 2398, 1986.
90. S. S. Khludkov, G. L., Prikhodko, and T. A. Karchina, *Izv. Akad. Nauk SSR, Heorg. Mater.*, 8, 1044, 1972.
91. J. Rachmann and R. Biermann, *Solid State Commun.*, 7, 1771, 1969.
92. A. S. Young and G. L. Pearson, *J. Phys. Chem. Solids*, 31, 517, 1970.
93. H. Matino, *Solid State Electron.*, 17, 35, 1974.
94. T. A. Karelina, T. T. Lavrischev, G. L. Prokhodko, and S. S. Khludkov, *Izv. Akad. Nauk SSSR, Neorg. Mater.*, 10, 228, 1974.
95. M. S. Seltzer, *J. Phys. Chem. Solids*, 26, 243, 1965.
96. B. I. Boltaks, G. S. Kulikov, I. N. Nikulista, and F. S. Shishiyanu, *Inorg. Mater.*, 11, 292, 1975.
97. V. A. Uskov and V. P. Sorvina, *Izv. Akad. Nauk SSSR, Neorg. Mater.*, 8, 758, 1972.
98. T. D. Dzhafarov, E. A. Skoryatina, E. S. Guds, and I. E. Moronchuk, *Phys. Stat. Sol. (a)*, 51, K221, 1979.
99. H. C. Casey, Jr. and G. L. Pearson, *J. Appl. Phys.*, 35, 3401, 1965.
100. M. H. van Maaren, *Phys. Stat. Sol.*, 24, K125, 1967.
101. A. I. Blashku and T. D. Dzhafarov, *Sov. Phys. Solid State*, 15, 536, 1973.
102. A. I. Blashku, B. I. Boltaks, I. I. Burdiyan, T. D. Dzhafarov, and M. A. Rzaev, *Sov. Phys. Sem.*, 6, 402, 1972.
103. J. Bougnot, L. Szepessy, and S. F. DaCunka, *Phys. Stat. Sol.*, 26, K127, 1968.
104. F. H. Eissen and C. E. Birchenall, *Acta Met.*, 5, 265, 1957.
105. R. K. Willardson, *Ultrapurification of Materials*, (Conf. Report, Boston), Macmillan, 1962.
106. B. I. Boltaks and Yu. A. Gutorov, *Sov. Phys. Solid State*, 1, 930, 1960.
107. V. A. Uskov, *Fiz. i Tekhn. Poluprov.*, 8, 2414, 1974.
108. F. S. Shishiyanu, *Diffusion and Degradation in Semiconductor Materials*, (Russian), Kishinev, 1978, 55.
109. S. V. Shudyakov, *Sov. Phys. Sem.*, 15, 4, 1981.
110. T. D. Dzhafarov, A. A. Litvin, and S. V. Khudyakov, *Sov. Phys. Solid State*, 20, 152, 1978.
111. M. Ilegems and W. C. O'Mara, *J. Appl. Phys.*, 43, 1190, 1972.
112. F. S. Shishiyanu, *Diffusion and Degradation in Semiconductor Materials*, (Russian), Kishinev, 1978, 78.
113. H. W. Allison, *J. Appl. Phys.*, 34, 231, 1963.
114. M. Schneider and E. Nebaur, *Phys. Stat. Sol. (a)*, 32, 333, 1975.
115. V. I. Kirillov, N. N. Pribylov, S. I. Rembeza, and A. I. Spirin, *Sov. Phys. Sem.*, 12, 1342, 1978.
116. V. I. Kirillov, N. N. Pribylov, S. I. Rembeza, and A. I. Spirin, *Sov. Phys. Solid State*, 22, 1945, 1980.
117. F. S. Shishiyanu and V. G. Georgiu, *Sov. Phys. Sem.*, 10, 1301, 1976.
118. T. D. Dzhafarov, Yu. P. Demakov, and N. N. Pribylov, *Fiz. Tverd. Tela*, 17, 3110, 1975.
119. K. A. Arseni, *Sov. Phys. Solids State*, 10, 2263, 1969.
120. K. A. Arseni and B. I. Boltaks, *Sov. Phys. Solid State*, 10, 2190, 1969.
121. S. I. Rembeza, *Sov. Phys. Sem.*, 3, 519, 1969.
122. K. A. Arseni, *Sov. Phys. Sem.*, 2, 1454, 1969.
123. R. M. Kundukhov, S. G. Metreveli, and N. V. Siukaev, *Sov. Phys. Sem.*, 1, 765, 1967.
124. L. L. Chang and H. Y. Casey, Jr., *Solid State Electron*, 7, 481, 1964.
125. K. A. Arseni, B. I. Boltaks, V. I. Gordin, and Ya. A. Ugai, *Izv. Akad. Nauk SSSR, Neorg. Mater.*, 3, 1679, 1967.
126. K. Ohtsuka, T. Nishino, and Y. Hamakawa, *Jpn. J. Appl. Phys.*, 21, 1170, 1982.
127. A. K. Chin, I. Camlibel, B. V. Dutt, V. Swaminathan, W. A. Bonner, and A. A. Ballman, *Appl. Phys. Lett.*, 42, 901, 1983.
128. M. R. Brozel, E. J. Foulkes, and B. Tuck, *Phys. Stat. Sol. (a)*, 72, K159, 1982; *Electron. Lett.*, 17, 532, 1981.
129. T. Kawakami and M. Okamura, *Electron. Lett.*, 15, 503, 1979.
130. R. Chaplain, M. Gaunear, H. H. L'Haridon, and R. Rupert, *J. Appl. Phys.*, 58, 1803, 1985.

131. F. S. Shishiyanu, V. Gh. Gheorgiu, and S. K. Palazov, *Phys. Stat. Sol. (a)*, 40, 29, 1977.
132. R. E. Holmes and R. G. Wilson, *J. Appl. Phys.*, 52, 3396, 1981.
133. C. S. Fuller and K. B. Wolfstirn, *J. Electrochem. Soc.*, 114, 856, 1967.
134. B. I. Boltaks, S. I. Rembeza, and M. K. Bakhadyrkhanov, *Sov. Phys. Solid State*, 10, 432, 1968.
135. B. I. Boltaks, S. I. Rembeza, and B. L. Sharma, *Sov. Phys. Sem.*, 1, 196, 1967.
136. S. I. Rembeza, *Sov. Phys. Sem.*, 1, 516, 1967.
137. E. Schillman, *Z. Naturforsch.*, 11a, 472, 1956.
138. B. I. Boltaks and S. I. Rembeza, *Sov. Phys. Solid State*, 8, 2177, 1967.
139. K. A. Arseni, B. I. Boltaks and S. I. Rembeza, *Sov. Phys. Solid State*, 8, 2248, 1967.
140. B. L. Sharma, R. K. Purohit, and S. N. Mukerjee, *J. Phys. Chem. Solids*, 32, 1397, 1971.
141. H. Kato, M. Yokozawa, R. Kohara, Y. Okabayashi, and S. Takayanagi, *Solid State Electron.*, 12, 137, 1969.
142. T. T. Takabatake, H. Ikari, and Y. Uyeda, *Jpn. J. Appl. Phys.*, 5, 839, 966.
143. H. J. Stocker, *Phys. Rev.*, 130, 2160, 1963.
144. B. I. Boltaks and V. I. Sokolov, *Sov. Phys. Solid State*, 6, 600, 1964.
145. L. A. K. Watt and W. S. Chen, *Bull. Am. Phys. Soc.*, 7, 89, 1962.
146. B. Goldstein, *Properties of Elemental and Compound Semiconductors*, Interscience, 1960, 155.
147. K. Nishitani, K. Hagahama, and T. Murotani, *Jpn. J. Appl. Phys.*, 22, 836, 1983.
148. B. I. Boltaks and V. I. Sokolov, *Sov. Phys. Solid State*, 5, 785, 1963.
149. R. B. Wilson and E. L. Heasell, *Proc. Soc. (London)*, 79, 403, 1962.
150. I. A. Gusev and A. N. Murin, *Sov. Phys. Solid State*, 6, 1229, 1964.
151. D. L. Kendall and R. A. Huggins, *J. Appl. Phys.*, 40, 2750, 1969.
152. S. M. Sze and L. Y. Wei, *Phys. Rev.*, 124, 84, 1961.
153. D. L. Kendall, *Semiconductors and Semimetals*, Vol. 4, R. K. Willardson and A. C. Beer, Eds., Academic, 1968, 255.
154. G. I. Rekalova, U. Kebe, and L. A. Mezrinea, *Sov. Phys. Sem.*, 5, 685, 1971.
155. G. I. Rekalova, A. A. Shakov, and V. V. Gaurushko, *Sov. Phys. Sem.*, 2, 1452, 1969.
156. B. I. Boltaks and G. S. Kulikov, *Sov. Phys. Tech. Phys.*, 2, 67, 1957.
157. I. A. Gusev and A. N. Nurin, *Sov. Phys. Solid State*, 6, 2274, 1965.
158. S. W. Kirchoefer, N. Holonyak, J. J. Coleman, and P. D. Dapkus, *J. Appl. Phys.*, 53, 766, 1982.
159. R. H. Wieber, H. C. Gorton, and C. S. Peet, *J. Appl. Phys.*, 31, 608, 1960.
160. D. Shaw, P. Joans, and D. Hazelby, *Proc. Phys. Soc. (London)*, 80, 167, 1962.
161. D. Shaw and S. R. Showan, *Phys. Stat. Sol.*, 34, 475, 1969.
162. B. Y. Pines and E. F. Shaikovskii, *Sov. Phys. Solid State*, 1, 864, 1959.
163. H. Nelkowski and G. Bollman, *Z. Naturforsch.*, A24, 1302, 1969.
164. V. M. Korsun and A. M. Nemchenko, *Sov. Phys. Solid State*, 8, 2988, 1967.
165. T. Lukaszewicz, *Phys. Stat. Sol. (a)*, 73, 611, 1982.
166. V. A. Williams, *J. Mat. Sci.*, 7, 807, 1972.
167. E. A. Secco, *J. Chem. Phys.*, 29, 406, 1958.
168. H. J. Biter and F. Williams, *J. Luminescence*, 3, 395, 1971.
169. H. Gobrecht, H. Nelkowski, J. W. Bears, and M. Weight, *Solid State Comm.*, 5, 777, 1967.
170. S. Asami, A. Ebina, and T. Takahashi, *Jpn. J. Appl. Phys.*, 17, 779, 1978.
171. T. Lukaszewicz and J. Zmija, *Phys. Stat. Sol. (A)*, 62, 695, 1980.
172. M. Aven and R. E. Halsted, *Phys. Rev.*, 137, A228, 1965.
173. M. M. Henneberg and D. A. Stevenson, *Phys. Stat. Sol. (b)*, 48, 255, 1971.
174. W. E. Martin, *J. Appl. Phys.*, 44, 5639, 1973.
175. T. Muranoi and M. Furukoshi, *Thin Solid Films*, 86, 307, 1981.
176. A. K. Kun and R. J. Robinson, *J. Electron. Mat.*, 5, 23, 1976.
177. H. H. Woodbury and R. B. Hall, *Phys. Rev.*, 157, 641, 1967.
178. P. Martin and A. Bontemps, *J. Phys. Chem. Solids*, 41, 1171, 1980.
179. E. A. Secco and S. C. Yeo, *Can. J. Chem.*, 49, 1953, 1971.
180. R. A. Reynold and D. A. Stevenson, *J. Phys. Chem. Solids*, 30, 139, 1969.
181. M. Aven and E. L. Kreiger, *J. Appl. Phys.*, 41, 1930, 1970.
182. M. Yokozawa, H. Kato, and S. Takayanagi, *Denki Kagaku*, 36, 282, 1968.
183. E. R. Dobrovinska, N. I. Krainyukov, Ya. A. Obukhouski, and L. A. Sysoev, *Ukr, Fiz. Zh.*, 13, 861, 1968.
184. H. H. Woodbury, *II-VI Semiconducting Compounds*, D. G. Thomas, Ed., Benjamin, NY, 1967, 244-276.
185. R. L. Clarke, *J. Appl. Phys.*, 30, 957, 1959.
186. J. L. Sullivan, *J. Phys. D: Appl. Phys.*, 6, 552, 1973.
187. B. Lepley, P. H. Nguyen, C. Boutrit, and S. Revelet, *J. Phys. D: Appl. Phys.*, 12, 145, 1979.
188. H. H. Woodbury, *J. Appl. Phys.*, 36, 2287, 1965.
189. E. Nebauer, *Phys. Stat. Sol.*, 29, 269, 1968.
190. J. Zimija and L. Sados, *Biul. Wojek, Akad. Tech.*, 20, No. 4, 105, 1971.

191. H. H. Woodbury, *Phys. Rev.,* 134, A492, 1964.
192. E. D. Jones and H. Mykura, *J. Phys. Chem. Solids,* 41, 1261, 1980.
193. E. D. Jones and H. Mykura, *J. Phys. Chem. Solids,* 39, 11, 1978.
194. E. Nebauer, *Phys. Stat. Sol. (b),* 60, K57, 1973.
195. V. Kumar and F. A. Kroger, *J. Solid State Chem.,* 3, 387, 1971.
196. L. A. Sysoev, A. Ya, Gelfman, A. D. Kovaleva, and N. G. Kravchencko, *Izv. Akad. Nauk. SSSR., Neorg. Mater.,* 5, 2208, 1969.
197. E. Nebauer and J. Lautenbach, *Phys. Stat. Sol. (b),* 48, 657, 1971.
198. D. G. Girton and W. E. Anderson, *Trans. Met. Soc. AIME,* 245, 465, 1969.
199. J. L. Sullivan, *Thin Solid Films,* 25, 245, 1975.
200. N. V. Dmitrieva, A. V. Vanyukov, and S. G. Yakovlev, *Elektron. Tekh. Nauk Tekh. Sb. Mater.,* No. 5, 150, 1970.
201. J. Zmija, *Acta Phys. Pol.,* A43, 345, 1973.
202. R. B. Hall and H. H. Woodbury, *J. Appl. Phys.,* 39, 5361, 1968.
203. L. Svob and C. Grattepain, *Calloq. Metall.,* 19, 725, 1976.
204. H. H. Woodbury and M. Aven, *J. Appl. Phys.,* 39, 5485, 1968.
205. H. Mann, G. Linker, and O. Meyer, *Solid State Comm.,* 11, 475, 1972.
206. J. P. Chamonal, E. Molva, J. L. Pautrat, and L. Revoil, *J. Cryst. Growth,* 59, 297, 1982.
207. I. Teramoto and S. Takayanagi, *J. Phys. Soc. Jpn.,* 17, 1137, 1982.
208. R. C. Whelan and D. Shaw, *II-IV Semiconducting Compounds,* D. C. Thomas, Ed., Benjamin, NY, 1967, 451.
209. P. M. Borsenberger and D. A. Stevenson, *J. Phys. Chem. Solids,* 29, 1277, 1968.
210. M. Yokozawa, H. Kato, and S. Takayanagi, *Denki Kagaku,* 34, 828, 1966.
211. E. Watson and D. Shaw, *J. Phys. C: Solid State Phys.,* 16, 515, 1983.
212. O. E. Panchuk, L. P. Shcherbak, P. I. Felchuk, and A. V. Savitskii, *Izv. Akad. Nauk SSSR, Neorg. Mater.,* 14, 54, 1978.
213. F. F. Morehead and G. Mandel, *Phys. Lett.,* 10, 5, 1964.
214. F. F. Vodovatov, G. V. Indenbaum, and A. V. Vanyukov, *Sov. Phys. Solid State,* 12, 17, 1970.
215. H. Kato, M. Yokozawa, and S. Takayanagi, *Jpn. J. Appl. Phys.,* 4, 1019, 1965.
216. D. Shaw and E. Watson, *J. Phys. C: Solid State Phys.,* 17, 4945, 1985.
217. O. E. Panchuk, R. N. Fesh, A. V. Savitskii, and L. P. Shcherbak, *Izv. Akad. Nauk SSSR, Neorg. Mater.,* 17, 1354, 1981.
218. B. I. Boltaks, *Diffusion in Semiconductors,* Inforsearch, 1963, 300.
219. F. F. Kharakhorin, *Izv. Akad. Nauk SSSR, Neorg. Mater.,* 5, 2212, 1969.
220. F. A. Zaitov, V. I. Stafeev, G. S., Khodakov, *Sov. Phys. Solid State,* 14, 2628, 1973.
221. V. V. Belov, F. A. Zaitov, and G. E. Popovyan, *Sov. Phys. Solid State,* 11, 1627, 1970.
222. F. A. Zaitov, *Sov. Phys. Solid State,* 13, 219, 1971.
223. V. S. Andramonov, N. S. Baryshev, and I. S. Aver'yanov, *Soviet Phys. Solid State,* 4, 1626, 1963.
224. J. Bloem and F. A. Kroger, *Philips Res. Rep.,* 12, 281, 1957.
225. G. Simkovich and J. B. Wagner, Jr., *J. Chem. Phys.,* 38, 1368, 1963.
226. M. S. Seltzer and J. B. Wagner, Jr., *J. Phys. Chem. Solids,* 26, 233, 1965.
227. J. Bloem and F. A. Kroger, Philips Res. Rep., 12, 303, 1957.
228. N. A. Fedorovich, *Soviet Phys. Solid State,* 7, 1289, 1965.
229. F. F. Kharakhorin, D. A. Gambarova, and V. V. Aksenov, *Soviet Phys. Solid State,* 7, 2813, 1966.
230. M. S. Seltzer and J. B. Wagner, Jr., *J. Chem. Phys.,* 36, 130, 1962.
231. B. I. Boltaks and Yu. N. Mokhov, *Zh. Tekh. Fiz.,* 28, 1046, 1958; *Zh. Tekh. Fiz.,* 26, 2448, 1956.
232. N. A. Fedorovich, *Soviet Phys. Solid State,* 7, 1291, 1965.
233. B. Swaroop and J. B. Wagner, Jr., *Appl. Phys. Lett.,* 12, 267, 1968.
234. A. J. Crocker and B. F. Dorning, *J. Phys. Chem. Solids,* 29, 155, 1968.
235. T. D. George and J. B. Wagner, Jr., *J. Electrochem. Soc.,* 115, 956, 1968; 116, 848, 1969.

PROPERTIES OF MAGNETIC MATERIALS

H. P. R. Frederikse

Glossary of Symbols

Quantity	Symbol	Units SI	emu
Magnetic field	H	A m^{-1}	Oe (oersted)
Magnetic induction	B	T (tesla)	G (gauss)
Magnetization	M	A m^{-1}	emu cm^{-3}
Spontaneous magnetization	M_s	A m^{-1}	emu cm^{-3}
Saturation magnetization	M_0	A m^{-1}	emu cm^{-3}
Magnetic flux	Φ	Wb (weber)	maxwell
Magnetic moment	m, μ	A m^2	erg/G
Coercive field	H_c	A m^{-1}	Oe
Remanence	B_r	T	G
Saturation magnetic polarization	J_s	T	G
Magnetic susceptibility	χ		
Magnetic permeability	μ	H m^{-1} (henry/meter)	
Magnetic permeability of free space	μ_0	H m^{-1}	
Saturation magnetostriction	λ ($\Delta l/l$)		
Curie temperature	T_C	K	K
Néel temperature	T_N	K	K

Magnetic moment $\mu = \gamma \hbar J = g \mu_B J$
where

γ = gyromagnetic ratio; J = angular momentum; g = spectroscopic splitting factor (~2)
μ_B = bohr magneton = $9.2741 \cdot 10^{-24}$ J/T = $9.2741 \cdot 10^{-21}$ erg/G

Earth's magnetic field $H = 56$ A m^{-1} = 0.7 Oe
For iron: $M_0 = 1.7 \cdot 10^6$ A m^{-1}; $B_r = 0.8 \cdot 10^6$ A m^{-1}
1 Oe = $(1000/4\pi)$ A m^{-1}; 1 G = 10^{-4} T; 1 emu cm^{-3} = 10^3 A m^{-1}
1 maxwell = 10^{-8} Wb
$\mu_0 = 4\pi \cdot 10^{-7}$ H m^{-1}

Relation Between Magnetic Induction and Magnetic Field

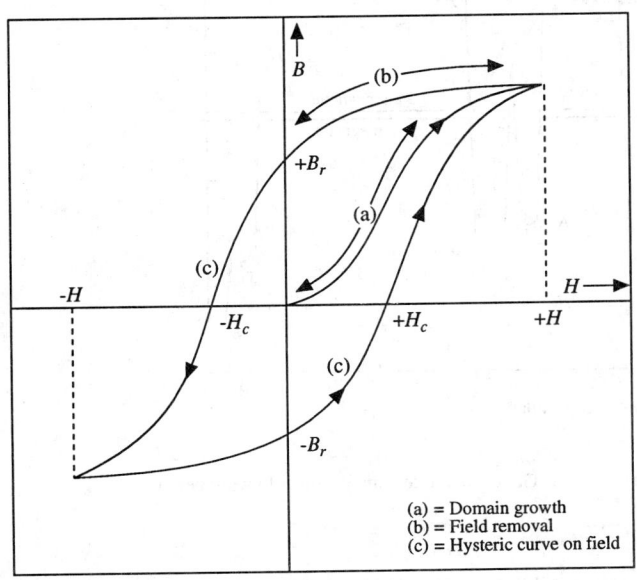

(a) = Domain growth
(b) = Field removal
(c) = Hysteric curve on field

Figure 1. Magnetic induction B vs. magnetic field H relationship for a ferromagnetic material. Along (a), starting from the demagnetized state, the more favorably oriented domains grow at the expense of less favored domains. At high fields, nearing saturation, the induction increases only by domain rotation. Upon removal of the applied field [curve (b)], the material maintains a remnant induction, B, and can only be demagnetized by imposing a negative field equal in magnitude to the coercivity, H_c. Cyclic applications of fields [such as in curve (c)] lead to a hysteresis loop for ferromagnetic materials. The area within the B-H loop represents an energy loss per unit volume of material per cycle.

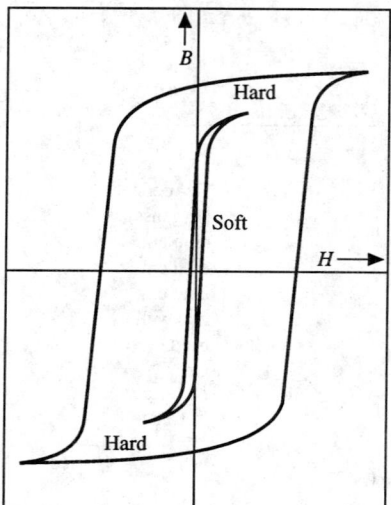

Figure 2. Schematic induction-field curves for a hard magnetic material and for a soft material. Hard magnetic materials are characterized by high remnant inductions, high coercivities, and large hysteresis losses per cycle.

REFERENCE

Ralls, K.M., Courtney, T.H., and Wulff, J., *Introduction to Materials Science and Engineering*, J. Wiley & Sons, New York, 1976, p. 577, 582. With permission.

Magnetic Susceptibility of the Elements

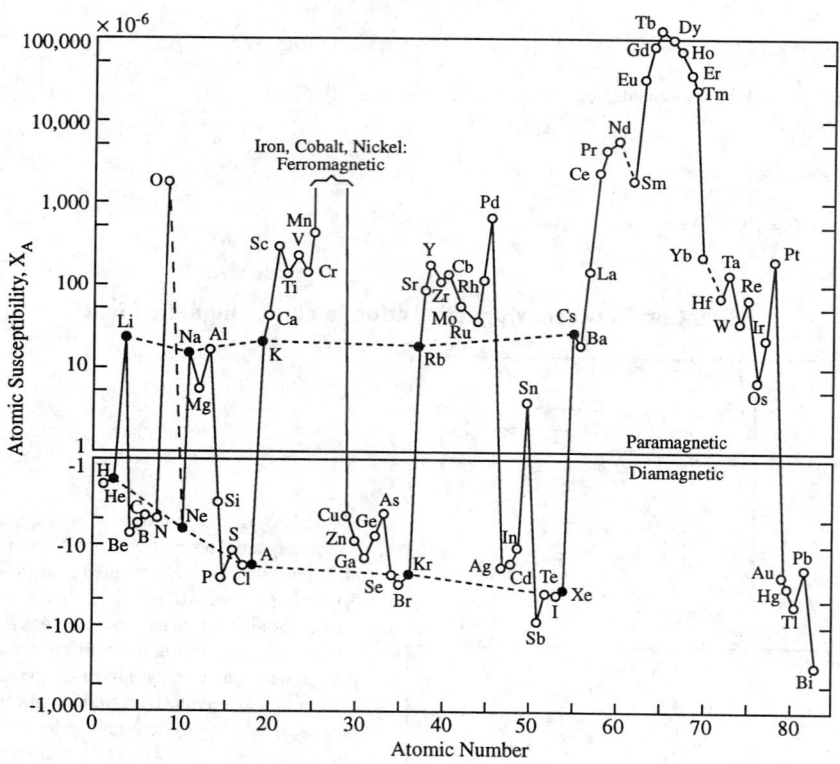

Figure 3. Atomic susceptibility of the elements at room temperature. The elements Gd to Tm are ferromagnetic at low temperatures.

REFERENCE

Gray, D.E., Ed., *American Institute of Physics Handbook, Third Edition*, McGraw Hill, New York, 1972, p. 5-224. With permission.

Ground State of Ions with Partly Filled d or f Shells

Z	Element	n	S	L	J	Gr. state	p_{calc}[a]	p_{calc}[b]	p_{meas}
22	Ti^{3+}	1	1/2	2	3/2	$^2D_{3/2}$	1.73	1.55	1.8
23	V^{4+}	1	1/2	2	3/2	$^2D_{3/2}$	1.73	1.55	1.8
23	V^{3+}	2	1	3	2	$^3F_2$	2.83	1.63	2.8
23	V^{2+}	3	3/2	3	3/2	$^4F_{3/2}$	3.87	0.77	3.8
24	Cr^{3+}	3	3/2	3	3/2	$^4F_{3/2}$	3.87	0.77	3.7
25	Mn^{4+}	3	3/2	3	3/2	$^4F_{3/2}$	3.87	0.77	4.0
24	Cr^{2+}	4	2	2	0	$^5D_0$	4.90	0	4.9
25	Mn^{3+}	4	2	2	0	$^5D_0$	4.90	0	5.0
25	Mn^{2+}	5	5/2	0	5/2	$^6S_{5/2}$	5.92	5.92	5.9
26	Fe^{3+}	5	5/2	0	5/2	$^6S_{5/2}$	5.92	5.92	5.9
26	Fe^{2+}	6	2	2	4	$^5D_4$	4.90	6.70	5.4
27	Co^{2+}	7	3/2	3	9/2	$^4F_{9/2}$	3.87	6.54	4.8
28	Ni^{2+}	8	1	3	4	$^3F_4$	2.83	5.59	3.2
29	Cu^{2+}	9	1/2	2	5/2	$^2D_{5/2}$	1.73	3.55	1.9

Z	Element	n	S	L	J	Gr. state	p_{calc}[c]		p_{meas}
58	Ce^{3+}	1	1/2	3	5/2	$^2F_{5/2}$	2.54		2.4
59	Pr^{3+}	2	1	5	4	$^3H_4$	3.58		3.5
60	Nd^{3+}	3	3/2	6	9/2	$^4I_{9/2}$	3.62		3.5
61	Pm^{3+}	4	2	6	4	$^5I_4$	2.68		
62	Sm^{3+}	5	5/2	5	5/2	$^6H_{5/2}$	0.84		1.5
63	Eu^{3+}	6	3	3	0	$^7F_0$	0.0		3.4
64	Gd^{3+}	7	7/2	0	7/2	$^8S_{7/2}$	7.94		8.0
65	Tb^{3+}	8	3	3	6	$^7F_6$	9.72		9.5
66	Dy^{3+}	9	5/2	5	15/2	$^6H_{15/2}$	10.63		10.6
67	Ho^{3+}	10	2	6	8	$^5I_8$	10.60		10.4
68	Er^{3+}	11	3/2	6	15/2	$^4I_{15/2}$	9.59		9.5
69	Tm^{3+}	12	1	5	6	$^3H_6$	7.57		7.3
70	Yb^{3+}	13	1/2	3	7/2	$^2F_{7/2}$	4.54		4.5

[a] $p_{calc} = 2[S(S + 1)]^{1/2}$
[b] $p_{calc} = 2[J(J + 1)]^{1/2}$
[c] $p_{calc} = g[J(J + 1)]^{1/2}$

REFERENCES

1. Jiles, D., *Magnetism and Magnetic Materials*, Chapman & Hall, London, 1991, p. 243.
2. Kittel, C., *Introduction to Solid State Physics, 6th Edition*, J. Wiley & Sons, New York, 1986, pp. 405—406.
3. Ashcroft, N.W. and Mermin, N.D., *Solid State Physics*, Holt, Rinehart, and Winston, New York, 1976, p. 652.

Ferro- and Antiferromagnetic Elements

M_0 is the saturation magnetization at $T = 0$ K
n_B is the number of Bohr magnetons per atom
T_C is the Curie temperature
T_N is the Néel temperature

	M_0/gauss	n_B	T_C/K	T_N/K	Comments
Fe	1752	2.22	1043		
Co	1446	1.72	1388		
Ni	510	0.62	627		
Cr				311	
Mn				100	
Ce				12.5	c-Axis antiferromagnetic
Nd				19.2	Basal plane modulation on hexagonal sites

Ferro- and Antiferromagnetic Elements (continued)

	M_0/gauss	n_B	T_C/K	T_N/K	Comments
				7.8	Cubic sites order (periodicity different from high-T phase)
Sm				106	Ordering on hexagonal sites
				13.8	Cubic site order
Eu				90.5	Spiral along cube axis
Gd	1980	7	293		
Tb		9	220		Basal plane ferromagnet
				230.2	Basal plane spiral
Dy		10	87		Basal plane ferromagnet
				176	Basal plane spiral
Ho		10	20		Bunched cone structure
				133	Basal plane spiral
Er		9	32		c-Axis ferrimagnetic cone structure
				80	c-Axis modulated structure
Tm		7	32		c-Axis ferrimagnetic cone structure
				56	c-Axis modulated structure

REFERENCES

1. Ashcroft, N.W., and Mermin, N.D., *Solid State Physics,* Holt, Rinehart, and Winston, New York, 1976, p.652.
2. Gschneidner, K.A., and Eyring, L., *Handbook on the Physics and Chemistry of Rare Earths*, North Holland Publishing Co., Amsterdam, 1978.

Selected Ferromagnetic Compounds

M_0 is the saturation magnetization at $T = 293$ K
T_C is the Curie temperature

Compound	M_0/gauss	T_C/K	Crystal system
MnB	152	578	orthorh(FeB)
MnAs	670	318	hex(FeB)
MnBi	620	630	hex(FeB)
MnSb	710	587	hex(FeB)
M_4N	183	743	
MnSi		34	cub(FeSi)
CrTe	247	339	hex(NiAs)
$CrBr_3$	270	37	hex(BiI_3)
CrI_3		68	hex(BiI_3)
CrO_2	515	386	tetr(TiO_2)
EuO	1910*	77	cub
EuS	1184*	16.5	cub
$GdCl_3$	550*	2.2	orthorh
FeB		598	orthorh
Fe_2B		1043	tetr ($CuAl_2$)
$FeBe_5$		75	cub($MgCu_2$)
Fe_3C		483	orthorh
FeP		215	orthorh (MnP)

* At $T = 0$ K

REFERENCES

1. Kittel, C., *Introduction to Solid State Physics, 6th Edition*, J. Wiley & Sons, New York, 1986.
2. Ashcroft, N.W., and Mermin, N.D., *Solid State Physics*, Holt, Rinehart, and Winston, New York, 1976.

Magnetic Properties of High-Permeability Metals and Alloys (Soft)

μ_i is the initial permeability
μ_m is the maximum permeability
H_c is the coercive force
J_s is the saturation polarization
W_H is the hysteresis loss per cycle
T_C is the Curie temperature

Material	Composition (mass %)	μ_i/μ_0	μ_m/μ_0	$H_c/A\ m^{-1}$	J_s/T	$W_H/J\ m^{-3}$	T_C/K
Iron	Commercial 99Fe	200	6000	70	2.16	500	1043
Iron	Pure 99.9Fe	25000	350000	0.8	2.16	60	1043
Silicon-iron	96Fe-4Si	500	7000	40	1.95	50-150	1008
Silicon-iron (110) [001]	97Fe-3Si	9000	40000	12	2.01	35-140	1015
Silicon-iron {100} <100>	97Fe-3Si		100000	6	2.01		1015
Mild steel	Fe-0.1C-0.1Si-0.4Mn	800	1100	200			
Hypernik	50Fe-50Ni	4000	70000	4	1.60	22	753
Deltamax {100} <100>	50Fe-50Ni	500	200000	16	1.55		773
Isoperm {100} <100>	50Fe-50Ni	90	100	480	1.60		
78 Permalloy	78Ni-22Fe	4000	100000	4	1.05	50	651
Supermalloy	79Ni-16Fe-5Mo	100000	1000000	0.15	0.79	2	673
Mumetal	77Ni-16Fe-5Cu-2Cr	20000	100000	4	0.75	20	673
Hyperco	64Fe-35Co-0.5Cr	650	10000	80	2.42	300	1243
Permendur	50Fe-50Co	500	6000	160	2.46	1200	1253
2V-Permendur	49Fe-49Co-2V	800	4000	160	2.45	600	1253
Supermendur	49Fe-49Co-2V		60000	16	2.40	1150	1253
25Perminvar	45Ni-30Fe-25Co	400	2000	100	1.55		
7Perminvar	70Ni-23Fe-7Co	850	4000	50	1.25		
Perminvar (magnet. annealed)	43Ni-34Fe-23Co		400000	2.4	1.50		
Alfenol (or Alperm)	84Fe-16Al	3000	55000	3.2	0.8		723
Alfer	87Fe-13Al	700	3700	53	1.20		673
Aluminum-Iron	96.5Fe-3.5Al	500	19000	24	1.90		
Sendust	85Fe-10Si-5Al	36000	120000	1.6	0.89		753

REFERENCES

1. McCurrie, R.A., *Structure and Properties of Ferromagnetic Materials*, Academic Press, London, 1994, p. 42.
2. Gray, D.E., Ed., *American Institute of Physics Handbook, Third Edition*, McGraw Hill, New York, 1972, p. 5-224.

Applications of High-Permeability Materials

Applications	Requirements
POWER APPLICATIONS	
Distribution and power transformers	Low core losses, high permeability, high saturation magnetic polarization
High-quality motors and generators, stators and armatures, switched-mode power supplies	
INSTRUMENT TRANSFORMERS	
Audiofrequency transformers	Low core losses, high permeability, high magnetic polarization
Pulse transformers	High permeability

Applications	Requirements
CORES FOR INDUCTOR COILS	
Audiofrequency	Low hysteresis, high permeability
Carrier frequency	Very low hysteresis and eddy current loss
Radiofrequency	High permeability at low fields
MISCELLANEOUS	
Relays, switches ⎫ Earth leakage circuit ⎭	High permeability, low remanence, low coercivity
Magnetic shielding	Low core loss for AC applications
Magnetic recording heads	High initial permeability, low or zero remanence
Magnetic amplifiers ⎫ Saturable reactors ⎪ Saturable transformers ⎬ Transformer cores ⎭	Rectangular hysteresis loops, low hysteresis loss
Magnetic shunts for temperature compensation in magnetic circuits	Low Curie temperature, appropriate decrease in permeability with increase in temperature
Electromagnets in indicating instruments, fire detection, quartz watches, electromechanical devices	High permeability, high saturation magnetic polarization
Magnetic yokes in permanent magnet devices, such as lifting and holding magnets, loudspeakers	High permeability, high saturation magnetic polarization

REFERENCE

McCurrie, R.A., *Structure and Properties of Ferromagnetic Materials*, Academic Press, London, 1994. With permission.

Saturation Magnetostriction of Selected Materials

The tabulated parameter λ_s is related to the fractional change in length $\Delta l/l$ by $\Delta l/l = (3/2)\lambda_s(\cos^2\theta - 1/3)$, where θ is the angle of rotation.

Material	$\lambda_s \times 10^6$
Iron	-7
Fe - 3.2% Si	+9
Nickel	-33
Cobalt	-62
45 Permalloy, 45% Ni - 55% Fe	+27
Permalloy, 82% Ni - 18% Fe	0
Permendur, 49% Co - 49% Fe - 2% V	+70
Alfer, 87% Fe - 13% Al	+30
Magnetite, Fe_3O_4	+40
Cobalt ferrite, $CoFe_2O_4$	-110
$SmFe_2$	-1560
$TbFe_2$	+1753
$Tb_{0.3}Dy_{0.7}Fe_{1.93}$ (Terfenol D)	+2000
$Fe_{66}Co_{18}B_{15}Si$ (amorphous)	+35
$Co_{72}Fe_3B_6A_{13}$ (amorphous)	0

REFERENCE

McCurrie, R.A., *Structure and Properties of Ferromagnetic Materials*, Academic Press, London, 1994, p. 91; additional data provided by A.E. Clark, Adelphi, MD.

Properties of Various Permanent Magnetic Materials (Hard)

B_r is the remanence
$_BH_c$ is the flux coercivity
$_iH_c$ is the intrinsic coercivity
$(BH)_{max}$ is the maximum energy product
T_C is the Curie temperature
T_{max} is the maximum operating temperature

Composition	B_r/T	$_BH_c$/10³ A m⁻¹	$_iH_c$/10³ A m⁻¹	$(BH)_{max}$/ kJ m⁻³	T_C/°C	T_{max}/°C
Alnico1 20Ni;12Al;5Co	0.72		35	25		
Alnico2 17Ni;10Al;12.5Co;6Cu	0.72		40-50	13-14		
Alnico3 24-30Ni;12-14Al;0-3Cu	0.5-0.6		40-54	10		
Alnico4 21-28Ni;11-13Al;3-5Co;2-4Cu	0.55-0.75		36-56	11-12		
Alnico5 14Ni;8Al;24Co;3Cu	1.25	53	54	40	850	520
Alnico6 16Ni;8Al;24Co;3Cu;2Ti	1.05		75	52		
Alnico8 15Ni;7Al;35Co;4Cu;5Ti	0.83	1.6	160	45		
Alnico9 15Ni;7Al;35Co;4Cu;5Ti	1.10	1.45	1.45	75	850	520
Alnico12 13.5Ni;8Al;24.5Co;2Nb	1.20		64	76.8		
BaFe₁₂O₁₉ (Ferroxdur)	0.4	1.6	192	29	450	400
SrFe₁₂O₁₉	0.4	2.95	3.3	30	450	400
LaCo₅	0.91			164	567	
CeCo₅	0.77			117	380	
PrCo₅	1.20			286	620	
NdCo₅	1.22			295	637	
SmCo₅	1.00	7.9	696	196	700	250
Sm(Co₀.₇₆Fe₀.₁₀Cu₀.₁₄)₆.₈	1.04	4.8	5	212	800	300
Sm(Co₀.₆₅Fe₀.₂₈Cu₀.₀₅Zr₀.₀₂)₇.₇	1.2	10	16	264	800	300
Nd₂Fe₁₄B sintered	1.22	8.4	1120	280	300	100
Fe;52Co;14V (Vicalloy II)	1.0	42		28	700	500
Fe;24Cr;15Co;3Mo (anisotropic)	1.54	67		76	630	500
Fe;28Cr;10.5Co (Chromindur II)	0.98	32		16	630	500
Fe;23Cr;15Co;3V;2Ti	1.35	4		44	630	500
Cu;20Ni;20Fe (Cunife)	0.55	4		12	410	350
Cu;21Ni;29Fe (Cunico)	0.34	0.5		8		
Pt;23Co	0.64	4		76	480	350
Mn;29.5Al;0.5C (anisotropic)	0.61	2.16	2.4	56	300	120

REFERENCES

1. McCurrie, R.A., *Structure and Properties of Ferromagnetic Materials*, Academic Press, London, 1994, p. 204.
2. Gray, D.E., Ed., *American Institute of Physics Handbook, Third Edition*, McGraw Hill, New York, 1972, p. 5-165.
3. Jiles, D., *Magnetism and Magnetic Materials,* Chapman & Hall, London, 1991.

Selected Ferrites

J_s is the saturation magnetic polarization
T_C is the Curie temperature
ΔH is the line width

Material	J_s/T	T_C/°C	ΔH/ kA m⁻¹	Applications
Spinels				
γ-Fe₂O₃	0.52	575		
Fe₃O₄	0.60	585		
NiFe₂O₄	0.34	575	350	Microwave devices
MgFe₂O₄	0.14	440	70	
NiZnFe₂O₄	0.50	375	120	Transformer cores
MnFe₂O₄	0.50	300	50	Microwave devices

Material	J_s/T	T_C/°C	ΔH/ kA m^{-1}	Applications
NiCoFe$_2$O$_4$	0.31	590	140	Microwave devices
NiCoAlFe$_2$O$_4$	0.15	450	330	Microwave devices
NiAl$_{0.35}$Fe$_{1.65}$O$_4$	0.12	430	67	Microwave devices
NiAlFe$_2$O$_4$	0.05	1860	32	Microwave devices
Mg$_{0.9}$Mn$_{0.1}$Fe$_2$O$_4$	0.25	290	56	Microwave devices
Ni$_{0.5}$Zn$_{0.5}$Al$_{0.8}$Fe$_{1.2}$O$_4$	0.14		17	Microwave devices
CuFe$_2$O$_4$	0.17	455		Electromechanical transducers
CoFe$_2$O$_4$	0.53	520		
LiFe$_5$O$_8$	0.39	670		Microwave devices
Garnets				
Y$_3$Fe$_5$O$_{12}$	0.178	280	55	Microwave devices
Y$_3$Fe$_5$O$_{12}$ (single crys.)	0.178	292	0.5	Microwave devices
(Y,Al)$_3$Fe$_5$O$_{12}$	0.12	250	80	Microwave devices
(Y,Gd)$_3$Fe$_5$O$_{12}$	0.06	250	150	Microwave devices
Sm$_3$Fe$_5$O$_{12}$	0.170	305		Microwave devices
Eu$_3$Fe$_5$O$_{12}$	0.116	293		Microwave devices
GdFe$_5$O$_{12}$	0.017	291		Microwave devices
Hexagonal crystals				
BaFe$_{12}$O$_{19}$	0.45	430	1.5	Permanent magnets
Ba$_3$Co$_2$Fe$_{24}$O$_{41}$	0.34	470	12	Microwave devices
Ba$_2$Zn$_2$Fe$_{12}$O$_{22}$	0.28	130	25	Microwave devices
Ba$_3$Co$_{1.35}$Zn$_{0.65}$Fe$_{24}$O$_{41}$		390	16	Microwave devices
Ba$_2$Ni$_2$Fe$_{12}$O$_{22}$	0.16	500	8	Microwave devices
SrFe$_{12}$O$_{19}$	0.4	450		Permanent magnets

REFERENCE

McCurrie, R.A., *Structure and Properties of Ferromagnetic Materials*, Academic Press, London, 1994.

Spinel Structure (AB$_2$O$_4$)

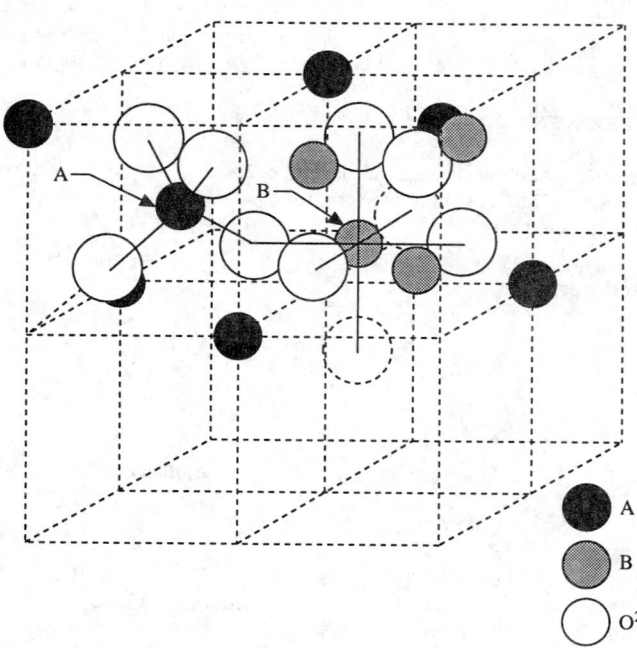

Figure 4. Arrangement of metal ions in the two octants A and B, showing tetrahedrally (A) and octahedrally (B) coordinated sites. (Reprinted from McCurrie, R.A., *Ferromagnetic Materials*, Academic Press, London, 1994. With permission.)

Selected Antiferromagnetic Solids

T_N is the Néel temperature

Material	Structure	T_N/K	Material	Structure	T_N/K
Binary oxides			NiAs and related structures		
MnO	cub(fcc)	122	CrAs	orth	300
FeO	cub(fcc)	198	CrSb	hex	705-723
CoO	cub(fcc)	291	CrSe	hex	300
NiO	cub(fcc)	525	MnTe	hex	320-323
α-Mn_2O_3	cub	90	NiS	hex	263
CuO	monocl	230	CrS	monocl	460
UO_2	cub	30.8	Rutile and related structures		
Er_2O_3	cub	3.4	CoF_2	tetr	38
Gd_2O_3	cub	1.6	CrF_2	monocl	53
Perovskites			FeF_2	tetr	79
$LaCrO_3$	orth	282	MnF_2	tetr	67
$LaMnO_3$	orth	100	NiF_2	tetr	83
$LaFeO_3$	orth	750	$CrCl_2$	orth	20
$NdCrO_3$	orth	224	MnO_2	tetr	84
$NdFeO_3$	orth	760	FeOF	tetr	315
$YbCrO_3$	orth	118	Corundum and related structures		
$CaMnO_3$	cub	110	Cr_2O_3	rhomb	318
$EuTiO_3$	cub	5.3	α-Fe_2O_3	rhomb	948
$YCrO_3$	orth	141	$FeTiO_3$	rhomb	68
$BiFeO_3$	cub*	673	$MnTiO_3$	rhomb	41
$KCoF_3$	cub	125	$CoTiO_3$	rhomb	38
$KMnF_3$	cub*	88.3	VF_3 and related structures		
$KFeF_3$	cub	115	CoF_3	rhomb	460
$KNiF_3$	cub	275	CrF_3	rhomb	80
$NaMnF_3$	cub*	60	FeF_3	rhomb	394
$NaNiF_3$	orth	149	MnF_3	monocl	43
$RbMnF_3$	cub	82	MoF_3	rhomb	185
Spinels			Miscellaneous		
Co_3O_4	cub	40	K_2NiF_4	tetr	97
$NiCr_2O_4$	tetr	65	MnI_2	hex	3.4
$ZnCr_2O_4$	cub	15	$CoUO_4$	orth	12
$ZnFe_2O_4$	cub	9	$CaMn_2O_4$	orth	225
$GeFe_2O_4$	cub	10	CrN	cub*	273
MgV_2O_4	cub	45	CeC_2	tetr	33
$MnGa_2O_4$	cub	33	FeSn	hex	373
			Mn_2P	hex	103

* Distorted.

REFERENCES

1. Gray, D.E., Ed., *American Institute of Physics Handbook, Third Edition*, McGraw Hill, New York, 1972, p. 5-168 to 183.
2. Kittel, C., *Introduction to Solid State Physics, 6th Edition*, J. Wiley & Sons, New York, 1986.
3. Ashcroft, N.W., and Mermin, N.D., *Solid State Physics*, Holt, Rinehart, and Winston, New York, 1976, p. 697.

ELECTRON WORK FUNCTION OF THE ELEMENTS

The electron work function Φ is a measure of the minimum energy required to extract an electron from the surface of a solid. It is defined more precisely as the energy difference between the state in which an electron has been removed to a distance from the surface of a single crystal face that is large enough that the image force is negligible but small compared to the distance to any other face (typically about 10^{-4} cm) and the state in which the electron is in the bulk solid. In general, Φ differs for each face of a monocrystalline sample.

Since Φ is dependent on the cleanliness of the surface, measurements reported in the literature often cover a considerable range. This table contains selected values for the electron work function of the elements which may be regarded as typical values for a reasonably clean surface. The method of measurement is indicated for each value. The following abbreviations appear:

TE — Thermionic emission
PE — Photoelectric effect
FE — Field emission
CPD — Contact potential difference
polycr — Polycrystalline sample
amorp — Amorphous sample

Values in parentheses are only approximate.

REFERENCES

1. Hölzl, J., and Schulte, F. K., Work Functions of Metals, in *Solid Surface Physics*, Höhler, G., Editor, Springer-Verlag, Berlin, 1979.
2. Riviere, J. C., Work Function: Measurements and Results, in *Solid State Surface Science, Vol.1*, Green, M., Editor, Decker, New York, 1969.
3. Michaelson, H. B., *J. Appl. Phys.*, 48, 4729, 1977.

Element	Plane	Φ/eV	Method	Element	Plane	Φ/eV	Method	Element	Plane	Φ/eV	Method
Ag	100	4.64	PE		210	5.00	PE	Ru	polycr	4.71	PE
	110	4.52	PE	K	polycr	2.29	PE	Sb	amorp	4.55	
	111	4.74	PE	La	polycr	3.5	PE		100	4.7	
Al	100	4.20	PE	Li	polycr	2.93	FE	Sc	polycr	3.5	PE
	110	4.06	PE	Lu	polycr	(3.3)	CPD	Se	polycr	5.9	PE
	111	4.26	PE	Mg	polycr	3.66	PE	Si	n	4.85	CPD
As	polycr	(3.75)	PE	Mn	polycr	4.1	PE		p 100	(4.91)	CPD
Au	100	5.47	PE	Mo	100	4.53	PE		p 111	4.60	PE
	110	5.37	PE		110	4.95	PE	Sm	polycr	2.7	PE
	111	5.31	PE		111	4.55	PE	Sn	polycr	4.42	CPD
B	polycr	(4.45)	TH		112	4.36	PE	Sr	polycr	(2.59)	TH
Ba	polycr	2.52	TH		114	4.50	PE	Ta	polycr	4.25	TH
Be	polycr	4.98	PE		332	4.55	PE		100	4.15	TH
Bi	polycr	4.34	PE	Na	polycr	2.36	PE		110	4.80	TH
C	polycr	(5.0)	CPD	Nb	001	4.02	TH		111	4.00	TH
Ca	polycr	2.87	PE		110	4.87	TH	Tb	polycr	3.0	PE
Cd	polycr	4.08	CPD		111	4.36	TH	Te	polycr	4.95	PE
Ce	polycr	2.9	PE		112	4.63	TH	Th	polycr	3.4	TH
Co	polycr	5.0	PE		113	4.29	TH	Ti	polycr	4.33	PE
Cr	polycr	4.5	PE		116	3.95	TH	Tl	polycr	(3.84)	CPD
Cs	polycr	1.95	PE		310	4.18	TH	U	polycr	3.63	PE
Cu	100	5.10	FE	Nd	polycr	3.2	PE		100	3.73	PE
	110	4.48	PE	Ni	100	5.22	PE		110	3.90	PE
	111	4.94	PE		110	5.04	PE		113	3.67	PE
	112	4.53	PE		111	5.35	PE	V	polycr	4.3	PE
Eu	polycr	2.5	PE	Os	polycr	5.93	PE	W	polycr	4.55	CPD
Fe	100	4.67	PE	Pb	polycr	4.25	PE		100	4.63	FE
	111	4.81	PE	Pd	polycr	5.22	PE		110	5.22	FE
Ga	polycr	4.32	PE		111	5.6	PE		111	4.45	FE
Gd	polycr	2.90	CPD	Pt	polycr	5.64	PE		113	4.46	FE
Ge	polycr	5.0	CPD		110	5.84	FE		116	4.32	TH
Hf	polycr	3.9	PE		111	5.93	FE	Y	polycr	3.1	PE
Hg	liquid	4.475	PE		320	5.22	FE	Zn	polycr	3.63	PE
In	polycr	4.09	PE		331	5.12	FE		polycr	(4.9)	CPD
Ir	100	5.67	PE	Rb	polycr	2.261	PE	Zr	polycr	4.05	PE
	110	5.42	PE	Re	polycr	4.72	TE				
	111	5.76	PE	Rh	polycr	4.98	PE				

SECONDARY ELECTRON EMISSION

The secondary emission yield, or secondary emission ratio, δ, is the average number of secondary electrons emitted from a bombarded material for every incident primary electron. It is a function of the primary electron energy E_p. The maximum yield δ_{max} corresponds to a primary electron energy E_{pmax} (see figure). The two primary electron energies corresponding to a yield of unity are denoted the first and second crossovers (E_I and E_{II}). An insulating target, or a conducting target that is electrically floating, will charge positively or negatively depending on the primary electron energy. For $E_I < E_p < E_{II}$, $\delta > 1$ and the surface charges positively provided there is a collector present that is positive with respect to the target. For $E_p < E_I$ or $E_p > E_{II}$, $\delta < 1$, and the surface charges negatively with respect to the potential of the source of primary electrons.

Element	δ_{max}	E_{pmax} (eV)	E_I (eV)	E_{II} (eV)	Element	δ_{max}	E_{pmax} (eV)	E_I (eV)	E_{II} (eV)
Ag	1.5	800	200	>2000	Li	0.5	85	None	None
Al	1.0	300	300	300	Mg	0.95	300	None	None
Au	1.4	800	150	>2000	Mo	1.25	375	150	1200
B	1.2	150	50	600	Na	0.82	300	None	None
Ba	0.8	400	None	None	Nb	1.2	375	150	1050
Bi	1.2	550	None	None	Ni	1.3	550	150	>1500
Be	0.5	200	None	None	Pb	1.1	500	250	1000
C (diamond)	2.8	750	None	>5000	Pd	>1.3	>250	120	None
C (graphite)	1.0	300	300	300	Pt	1.8	700	350	3000
C (soot)	0.45	500	None	None	Rb	0.9	350	None	None
Cd	1.1	450	300	700	Sb	1.3	600	250	2000
Co	1.2	600	200	None	Si	1.1	250	125	500
Cs	0.7	400	None	None	Sn	1.35	500	None	None
Cu	1.3	600	200	1500	Ta	1.3	600	250	>2000
Fe	1.3	400	120	1400	Th	1.1	800	None	None
Ga	1.55	500	75	None	Ti	0.9	280	None	None
Ge	1.15	500	150	900	Tl	1.7	650	70	>1500
Hg	1.3	600	350	>1200	W	1.4	650	250	>1500
K	0.7	200	None	None	Zr	1.1	350	None	None

Compound	δ_{max}	E_{pmax} (eV)	Compound	δ_{max}	E_{pmax} (eV)
Alkali halides			NaCl (layer)	6.8	600
CsCl	6.5		NaF (crystal)	14	1200
KBr (crystal)	14	1800	NaF (layer)	5.7	
KCl (crystal)	12	1600	NaI (crystal)	19	1300
KCl (layer)	7.5	1200	NaI (layer)	5.5	
KI (crystal)	10	1600	RbCl (layer)	5.8	
KI (layer)	5.6		Oxides		
LiF (crystal)	8.5		Ag_2O	1.0	
LiF (layer)	5.6	700	Al_2O_3 (layer)	2—9	
NaBr (crystal)	24	1800	BaO (layer)	2.3—4.8	400
NaBr (layer)	6.3		BeO	3.4	2000
NaCl (crystal)	14	1200	CaO	2.2	500

Compound	δ_{max}	E_{pmax} (eV)	Compound	δ_{max}	E_{pmax} (eV)
Cu_2O	1.2	400	Others		
MgO (crystal)	20—25	1500	BaF_2 (layer)	4.5	
MgO (layer)	3—15	400—1500	CaF_2 (layer)	3.2	
MoO_2	1.2		$BiCs_3$	6	1000
SiO_2 (quartz)	2.1—4	400	BiCs	1.9	1000
SnO_2	3.2	640	GeCs	7	700
Sulfides			Rb_3Sb	7.1	450
MoS_2	1.1		$SbCs_3$	6	700
PbS	1.2	500	Mica	2.4	350
WS_2	1.0		Glasses	2—3	300—450
ZnS	1.8	350			

OPTICAL PROPERTIES OF METALS AND SEMICONDUCTORS
J. H. Weaver and H. P. R. Frederikse

These tables list the index of refraction n, the extinction coefficient k, and the normal incidence reflection R ($\phi = 0$) as a function of photon energy E, which is expressed in electron volts (eV). To convert the energy in eV to wavelength in μm, use $\lambda = 1.2398/E$. To compute the dielectric function $\tilde{\varepsilon} = \varepsilon_1 + i\varepsilon_2$ from the complex index of refraction $\tilde{N} = n + ik$, use $\varepsilon_1 = n^2 - k^2$ and $\varepsilon_2 = 2nk$.

The optical constants in these tables are abridged from three more extensive tabulations:

- *Optical Properties of Metals* (OPM), Volumes I and II, *Physics Data, Nr.* 18-1 and 18-2, J. H. Weaver, C. Krafka, D. W. Lynch, and E. E. Koch, Fachinformationzentrum, Karlsruhe, Germany.
- *Handbook of Optical Constants* (HOC), Vol. I, 1985, and Vol. II, 1991. Edited by E. D. Palik, published by Academic Press, Inc.
- *American Institute of Physics Handbook* (AIPH), 3rd Edition, Coord. Editor D. E. Gray, published by McGraw-Hill Book Co., New York, 1972.

The first two of these major sources provide detailed comparisons of all optical data available in the literature at the time of the compilation. For critical applications the reader should refer to the original work. References for individual metals and semiconductors are listed at the end of the tables. Generally, tabulated values for the optical properties are accurate to better than 10%. Data in parentheses are extrapolated or interpolated values. For most elements the spectral range covered is from the far infrared (0.010 or 0.10 eV) to the far ultraviolet (10, 30 or 300 eV). The intervals between successive energies in the tables are chosen in such a way that the major spectral features are preserved.

Very small values of k are expressed in exponential notation, e.g., 1.23E-5 means 1.23×10^{-5}.

The following table is convenient for identifying the energy entries in these tables with the corresponding wavelengths:

λ	E/eV	λ	E/eV
1 mm	0.00124	6000 Å	2.066
500 µm	0.00248	5000 Å	2.480
100 µm	0.01240	4000 Å	3.100
50 µm	0.02480	3000 Å	4.133
10 µm	0.12398	2000 Å	6.199
5 µm	0.24797	1000 Å	12.398
1 µm	1.240	400 Å	30.996

Energy (eV)	n	k	$R(\phi=0)$	Energy (eV)	n	k	$R(\phi=0)$	Energy (eV)	n	k	$R(\phi=0)$
Aluminium[1]				2.200	1.018	6.846	0.9200	14.200	0.053	0.373	0.8312
0.040	98.595	203.701	0.9923	2.400	0.826	6.283	0.9228	14.400	0.058	0.327	0.8102
0.050	74.997	172.199	0.9915	2.600	0.695	5.800	0.9238	14.600	0.067	0.273	0.7802
0.060	62.852	150.799	0.9906	2.800	0.598	5.385	0.9242	14.800	0.086	0.211	0.7202
0.070	53.790	135.500	0.9899	3.000	0.523	5.024	0.9241	15.000	0.125	0.153	0.6119
0.080	45.784	123.734	0.9895	3.200	0.460	4.708	0.9243	15.200	0.178	0.108	0.4903
0.090	39.651	114.102	0.9892	3.400	0.407	4.426	0.9245	15.400	0.234	0.184	0.3881
0.100	34.464	105.600	0.9889	3.600	0.363	4.174	0.9246	15.600	0.280	0.073	0.3182
0.125	24.965	89.250	0.9884	3.800	0.326	3.946	0.9247	15.800	0.318	0.065	0.2694
0.150	18.572	76.960	0.9882	4.000	0.294	3.740	0.9248	16.000	0.351	0.060	0.2326
0.175	14.274	66.930	0.9879	4.200	0.267	3.552	0.9248	16.200	0.380	0.055	0.2031
0.200	11.733	59.370	0.9873	4.400	0.244	3.380	0.9249	16.400	0.407	0.050	0.1789
0.250	8.586	48.235	0.9858	4.600	0.223	3.222	0.9249	16.750	0.448	0.045	0.1460
0.300	6.759	40.960	0.9844	4.800	0.205	3.076	0.9249	17.000	0.474	0.042	0.1278
0.350	5.438	35.599	0.9834	5.000	0.190	2.942	0.9244	17.250	0.498	0.040	0.1129
0.400	4.454	31.485	0.9826	6.000	0.130	2.391	0.9257	17.500	0.520	0.038	0.1005
0.500	3.072	25.581	0.9817	6.500	0.110	2.173	0.9260	17.750	0.540	0.036	0.0899
0.600	2.273	21.403	0.9806	7.000	0.095	1.983	0.9262	18.000	0.558	0.035	0.0809
0.700	1.770	18.328	0.9794	7.500	0.082	1.814	0.9265	18.500	0.591	0.032	0.0664
0.800	1.444	15.955	0.9778	8.000	0.072	1.663	0.9269	19.000	0.620	0.030	0.0554
0.900	1.264	14.021	0.9749	8.500	0.063	1.527	0.9272	19.500	0.646	0.028	0.0467
1.000	1.212	12.464	0.9697	9.000	0.056	1.402	0.9277	20.000	0.668	0.027	0.0398
1.100	1.201	11.181	0.9630	9.500	0.049	1.286	0.9282	20.500	0.689	0.025	0.0342
1.200	1.260	10.010	0.9521	10.000	0.044	1.178	0.9286	21.000	0.707	0.024	0.0296
1.300	1.468	8.949	0.9318	10.500	0.040	1.076	0.9293	21.500	0.724	0.023	0.0258
1.400	2.237	8.212	0.8852	11.000	0.036	0.979	0.9298	22.000	0.739	0.022	0.0226
1.500	2.745	8.309	0.8678	11.500	0.033	0.883	0.9283	22.500	0.753	0.021	0.0199
1.600	2.625	8.597	0.8794	12.000	0.033	0.791	0.9224	23.000	0.766	0.021	0.0177
1.700	2.143	8.573	0.8972	12.500	0.034	0.700	0.9118	23.500	0.778	0.020	0.0157
1.800	1.741	8.205	0.9069	13.000	0.038	0.609	0.8960	24.000	0.789	0.019	0.0140
1.900	1.488	7.821	0.9116	13.500	0.041	0.517	0.8789	24.500	0.799	0.018	0.0126
2.000	1.304	7.479	0.9148	14.000	0.048	0.417	0.8486	25.000	0.809	0.018	0.0113

Energy (eV)	n	k	R(φ = 0)
25.500	0.817	0.017	0.0102
26.000			
	0.826	0.016	0.0092
27.000	0.840	0.015	0.0076
28.000	0.854	0.014	0.0063
29.000	0.865	0.014	0.0053
30.000	0.876	0.013	0.0044
35.000	0.915	0.010	0.0020
40.000	0.940	0.008	0.0010
45.000	0.957	0.007	0.0005
50.000	0.969	0.006	0.0003
55.000	0.979	0.005	0.0001
60.000	0.987	0.004	0.0000
65.000	0.995	0.004	0.0000
70.000	1.006	0.004	0.0000
72.500	1.025	0.004	0.0002
75.000	1.011	0.024	0.0002
77.500	1.008	0.025	0.0002
80.000	1.007	0.024	0.0002
85.000	1.007	0.028	0.0002
90.000	1.005	0.031	0.0002
95.000	0.999	0.036	0.0003
100.000	0.991	0.030	0.0002
110.000	0.994	0.025	0.0002
120.000	0.991	0.024	0.0002
130.000	0.987	0.021	0.0001
140.000	0.989	0.016	0.0001
150.000	0.990	0.015	0.0001
160.000	0.989	0.014	0.0001
170.000	0.989	0.011	0.0001
180.000	0.990	0.010	0.0000
190.000	0.990	0.009	0.0000
200.000	0.991	0.007	0.0000
220.000	0.992	0.006	0.0000
240.000	0.993	0.005	0.0000
260.000	0.993	0.004	0.0000
280.000	0.994	0.003	0.0000
300.000	0.995	0.002	0.0000

Carbon (diamond)[2]

Energy (eV)	n	k	R(φ = 0)
0.06199	2.3741		0.166
0.06888	2.3741		0.166
0.07749	2.3745		0.166
0.08856	2.3750		0.166
0.1033	2.3757		0.166
0.1240	2.3765		0.166
0.1550	2.3772		0.166
0.1907		3.1 E-05	
0.2066	2.3779	5.7 E-05	0.166
0.22		1.21E-04	
0.23		2.36E-04	
0.24		3.82E-04	
0.25		5.21E-04	
0.26		2.96E-04	
0.27		4.39E-04	
0.28		2.75E-04	
0.29		7.82E-05	
0.30		1.32E-04	
0.31	2.3787	1.30E-04	0.167
0.32		1.11E-04	
0.33		2.99E-05	
0.34		1.89E-05	
0.35		2.11E-05	
0.36		2.47E-05	
0.37		2.80E-05	
0.38		3.11E-05	

Energy (eV)	n	k	R(φ = 0)
0.39		3.67E-05	
0.40		3.58E-05	
0.41		3.25E-05	
0.4133	2.3795		0.167
0.42		2.94E-05	
0.43		2.87E-05	
0.44		3.14E-05	
0.45		3.62E-05	
0.46		3.22E-05	
0.47		1.57E-05	
0.48		6.17E-06	
0.4959	2.3801		0.167
0.6199	2.3813		0.167
0.8266	2.3837		0.167
1.240	2.3905		0.168
1.378	2.3934		0.169
1.459	2.3953		0.169
1.550	2.3975		0.169
1.653	2.4003		0.170
1.771	2.4036		0.170
1.889	2.4073		0.171
1.926	2.4084		0.171
2.066	2.4133		0.171
2.105	2.4147		0.172
2.271	2.4210		0.173
2.480	2.4299		0.174
2.650	2.4380		0.175
2.845		3.82E-07	
3.100	2.4627		0.178
3.434	2.4849		0.182
3.576	2.4955		0.183
3.961		8.97E-07	
4.160	2.5465		0.190
4.511		1.29E-06	
4.8187	2.6205	1.47E-06	0.200
5.00			
	2.6383		
			0.203
5.30		2.98E-06	
5.35		6.45E-06	
5.40		1.04E-05	
5.50		3.41E-05	
5.55		5.48E-04	
5.60	2.740	1.48E-03	0.216
5.80	2.780	5.02E-03	0.222
6.00	2.826	7.99E-03	0.228
6.10	2.852	8.62E-03	0.231
6.20	2.879	9.30E-03	0.235
6.30	2.910	9.74E-03	0.239
6.40	2.944	9.87E-03	0.243
6.50	2.985	1.10E-02	0.248
6.60	3.031	1.47E-02	0.254
6.70	3.085	2.20E-02	0.261
6.80	3.146	3.44E-02	0.268
6.90	3.220	5.24E-02	0.277
7.00	3.322	9.35E-02	0.289
7.10	3.444	0.210	0.304
7.15	3.464	0.307	0.308
7.20	3.437	0.388	0.307
7.30	3.376	0.473	0.303
7.40	3.335	0.515	0.300
7.50	3.321	0.533	0.299
7.60	3.306	0.592	0.300
7.80	3.276	0.659	0.300
8.00	3.251	0.712	0.300
8.25	3.232	0.765	0.301

Energy (eV)	n	k	R(φ = 0)
8.75	3.247	0.855	0.308
9.00	3.272	0.910	0.314
9.25	3.308	0.978	0.322
9.50	3.348	1.055	0.331
9.75	3.398	1.147	0.342
10.00	3.453	1.258	0.355
10.25	3.514	1.403	0.371
10.50	3.565	1.581	0.389
10.75	3.600	1.813	0.411
11.00	3.582	2.078	0.434
11.25	3.507	2.380	0.460
11.50	3.346	2.693	0.488
11.75	3.090	2.986	0.518
12.00	2.736	3.228	0.551
12.20	2.383	3.354	0.580
12.40	1.983	3.382	0.610
12.60	1.532	3.265	0.641
12.80	1.312	2.953	0.627
13.00	1.223	2.722	0.604
13.50	1.129	2.379	0.557
14.00	1.070	2.178	0.526
14.50	1.018	2.034	0.504
15.00	0.972	1.929	0.489
15.50	0.917	1.845	0.482
16.00	0.861	1.767	0.477
16.50	0.805	1.692	0.474
17.00	0.753	1.619	0.471
17.50	0.707	1.546	0.467
18.00	0.665	1.476	0.463
18.50	0.626	1.408	0.459
19.00	0.589	1.341	0.455
19.50	0.557	1.273	0.449
20.00	0.527	1.203	0.442
21.00	0.487	1.052	0.413
22.00	0.518	0.888	0.330
23.00	0.597	0.850	0.270
24.00	0.586	0.829	0.268
25.00	0.562	0.787	0.265
26.00	0.538	0.736	0.260
27.00	0.516	0.679	0.252
28.00	0.501	0.616	0.239
29.00	0.494	0.552	0.221
30.00	0.493	0.490	0.201

Cesium (evaporated)[3]

Energy (eV)	n	k	R(φ = 0)
2.145	0.264	1.123	0.631
2.271	0.278	0.950	0.561
2.845	0.425	0.438	0.235
3.064	0.540	0.320	0.127
3.397	0.671	0.233	0.057
3.966	0.827	0.174	0.018
4.889	0.916	0.143	0.007

Chromium[4]

Energy (eV)	n	k	R(φ = 0)
0.06	21.19	42.00	0.962
0.10	11.81	29.76	0.955
0.14	15.31	26.36	0.936
0.18	8.73	25.37	0.53
0.22	5.30	20.62	0.954
0.26	3.91	17.12	0.951
0.30	3.15	14.28	0.943
0.42	3.47	8.97	0.862
0.54	3.92	7.06	0.788
0.66	3.96	5.95	0.736
0.78	4.13	5.03	0.680

Energy (eV)	n	k	$R(\phi = 0)$
0.90	4.43	4.60	0.650
1.00	4.47	4.43	0.639
1.12	4.53	4.31	0.631
1.24	4.50	4.28	0.629
1.36	4.42	4.30	0.631
1.46	4.31	4.32	0.632
1.77	3.84	4.37	0.639
2.00	3.48	4.36	0.644
2.20	3.18	4.41	0.656
2.40	2.75	4.46	0.677
2.60	2.22	4.36	0.698
2.80	1.80	4.06	0.703
3.00	1.54	3.71	0.695
3.20	1.44	3.40	0.670
3.40	1.39	3.24	0.657
3.60	1.26	3.12	0.661
3.80	1.12	2.95	0.660
4.00	1.02	2.76	0.651
4.20	0.94	2.58	0.639
4.40	0.90	2.42	0.620
4.50	0.89	2.35	0.607
4.60	0.88	2.28	0.598
4.70	0.86	2.21	0.586
4.80	0.86	2.13	0.572
4.90	0.86	2.07	0.557
5.00	0.85	2.01	0.542
5.10	0.86	1.94	0.523
5.20	0.87	1.87	0.503
5.40	0.93	1.80	0.466
5.60	0.95	1.74	0.443
5.80	0.97	1.74	0.437
6.00	0.94	1.73	0.444
6.20	0.89	1.69	0.446
6.40	0.85	1.66	0.447
6.60	0.80	1.59	0.444
6.80	0.75	1.51	0.439
7.00	0.74	1.45	0.425
7.20	0.71	1.39	0.414
7.40	0.69	1.33	0.404
7.60	0.66	1.23	0.378
7.80	0.67	1.15	0.347
8.00	0.68	1.07	0.315
8.20	0.71	1.00	0.278
8.50	0.74	0.92	0.235
9.0	0.83	0.81	0.170
9.50	0.92	0.74	0.132
10.00	0.98	0.73	0.120
10.50	1.01	0.72	0.112
11.00	1.05	0.69	0.103
11.50	1.09	0.69	0.100
12.00	1.13	0.70	0.101
12.50	1.15	0.73	0.108
13.00	1.15	0.77	0.119
13.50	1.12	0.80	0.128
14.00	1.09	0.82	0.135
14.50	1.03	0.82	0.142
15.00	1.00	0.82	0.143
15.50	0.96	0.80	0.141
16.00	0.92	0.77	0.139
16.50	0.31	0.75	0.134
17.00	0.90	0.73	0.132
17.50	0.88	0.72	0.130
18.00	0.87	0.70	0.129
18.50	0.84	0.69	0.130
19.00	0.82	0.68	0.131
20.00	0.77	0.64	0.130
20.5	0.76	0.63	0.129
21.0	0.74	0.58	0.121
21.5	0.72	0.55	0.116
22.0	0.71	0.52	0.112
22.5	0.70	0.50	0.109
23.0	0.69	0.48	0.105
23.5	0.68	0.45	0.101
24.0	0.68	0.43	0.096
24.5	0.67	0.39	0.089
25.0	0.68	0.36	0.080
25.5	0.68	0.33	0.072
26.0	0.70	0.31	0.063
26.5	0.71	0.28	0.055
27.0	0.72	0.26	0.048
27.5	0.73	0.25	0.043
28.0	0.75	0.23	0.037
29.0	0.77	0.22	0.032
30.0	0.78	0.21	0.030

Cobalt, single crystal, $\overset{\perp}{E} \parallel \hat{c}$ [5]

Energy (eV)	n	k	$R(\phi = 0)$
0.10	6.71	37.87	0.982
0.15	4.66	25.47	0.973
0.20	3.55	18.78	0.962
0.25	3.98	14.59	0.933
0.30	4.04	12.16	0.907
0.40	4.24	9.13	0.847
0.50	4.41	7.19	0.782
0.60	4.91	6.13	0.729
0.70	5.24	5.85	0.713
0.80	5.17	5.89	0.716
0.90	4.94	5.95	0.720
1.00	4.46	5.86	0.722
1.10	4.07	5.61	0.715
1.20	3.81	5.36	0.706
1.30	3.60	5.20	0.701
1.40	3.37	5.09	0.701
1.50	3.10	4.96	0.701
1.60	2.84	4.77	0.697
1.70	2.66	4.57	0.690
1.80			
	2.45	4.41	0.687
1.90	2.31	4.18	0.675
2.00	2.21	4.00	0.664
2.10	2.13	3.85	0.654
2.20	2.07	3.70	0.642
2.30	2.01	3.59	0.634
2.40	1.95	3.49	0.627
2.50	1.88	3.40	0.622
2.60	1.81	3.32	0.618
2.70	1.73	3.24	0.615
2.80	1.66	3.13	0.607
2.90	1.61	3.05	0.600
3.00	1.55	2.96	0.594
3.20	1.46	2.80	0.579
3.40	1.38	2.64	0.563
3.60	1.31	2.48	0.544
3.80	1.28	2.33	0.519
4.00	1.26	2.20	0.495
4.20	1.25	2.10	0.471
4.40	1.24	2.01	0.452
4.60	1.24	1.94	0.435
4.80	1.23	1.88	0.423
5.00	1.22	1.83	0.411
5.20	1.21	1.79	0.403
5.40	1.19	1.77	0.399
5.60	1.16	1.75	0.400
5.80	1.10	1.73	0.406
6.00	1.03	1.68	0.407
6.20	0.97	1.62	0.401
6.40	0.94	1.53	0.386
6.60	0.91	1.46	0.368
6.80	0.91	1.38	0.345
7.00	0.91	1.32	0.326
7.00	0.91	1.26	0.305
7.40	0.92	1.21	0.286
7.60	0.93	1.17	0.269
7.80	0.94	1.13	0.253
8.00	0.95	1.09	0.239

Cobalt, single crystal, $\overset{\perp}{E} \perp \hat{c}$ [5]

Energy (eV)	n	k	$R(\phi = 0)$
0.10	5.83	32.36	0.979
0.15	4.24	21.37	0.965
0.20	3.87	15.53	0.042
0.30	4.34	10.01	0.865
0.40	4.66	7.39	0.785
0.50	5.17	5.75	0.709
0.60	5.77	5.17	0.682
0.70	6.15	5.20	0.685
0.80	6.08	5.61	0.702
0.90	5.57	5.93	0.715
1.00	4.83	5.94	0.721
1.10	4.31	5.60	0.711
1.20	4.02	5.34	0.701
1.30	3.78	5.16	0.694
1.40	3.55	5.05	0.692
1.50	3.26	4.93	0.692
1.60	3.03	4.74	0.687
1.70	2.83	4.60	0.684
1.80	2.61	4.45	0.683
1.90	2.41	4.27	0.677
2.00	2.25	4.09	0.670
2.10	2.13	3.89	0.659
2.20	2.04	3.72	0.646
2.30	1.99	3.56	0.632
2.40	1.95	3.44	0.620
2.50	1.90	3.34	0.611
2.60	1.86	3.26	0.605
2.70	1.79	3.19	0.602
2.80	1.72	3.11	0.596
2.90	1.66	3.03	0.591
3.00	1.60	2.94	0.586
3.20	1.50	2.78	0.571
3.40	1.42	2.62	0.553
3.60	1.36	2.47	0.533
3.80	1.33	2.33	0.511
4.00	1.31	2.21	0.488
4.20	1.28	2.12	0.471
4.40	1.27	2.03	0.452
4.60	1.26	1.95	0.435
4.80	1.25	1.90	0.423
5.00	1.24	1.84	0.411
5.20	1.22	1.80	0.403
5.40	1.21	1.78	0.399
5.60	1.17	1.76	0.400
5.80	1.11	1.74	0.406
6.00	1.04	1.69	0.407
6.20	0.98	1.62	0.401
6.40	0.94	1.54	0.386
6.60	0.92	1.46	0.368
6.80	0.91	1.38	0.345

Energy (eV)	n	k	R(φ = 0)
7.00	0.91	1.32	0.326
7.20	0.91	1.26	0.305
7.40	0.92	1.21	0.285
7.60	0.93	1.17	0.269
7.80	0.94	1.13	0.253

Copper[6]

Energy (eV)	n	k	R(φ = 0)
0.10	29.69	71.57	0.980
0.50	1.71	17.63	0.979
1.00	0.44	8.48	0.976
1.50			
	0.26		
		5.26	
			0.965
1.70	0.22	4.43	0.958
1.75	0.21	4.25	0.956
1.80	0.21	4.04	0.952
1.85	0.22	3.85	0.947
1.90	0.21	3.67	0.943
2.00	0.27	3.24	0.910
2.10	0.47	2.81	0.814
2.20	0.83	2.60	0.673
2.30	1.04	2.59	0.618
2.40	1.12	2.60	0.602
2.60	1.15	2.50	0.577
2.80	1.17	2.36	0.545
3.00	1.18	2.21	0.509
3.20	1.23	2.07	0.468
3.40	1.27	1.95	0.434
3.60	1.31	1.87	0.407
3.80	1.34	1.81	0.387
4.00	1.34	1.72	0.364
4.20	1.42	1.64	0.336
4.40	1.49	1.64	0.329
4.60	1.52	1.67	0.334
4.80	1.53	1.71	0.345
5.00	1.47	1.78	0.366
5.20	1.38	1.80	0.380
5.40	1.28	1.78	0.389
5.60	1.18	1.74	0.391
5.80	1.10	1.67	0.389
6.00	1.04	1.59	0.380
6.50	0.96	1.37	0.329
7.00	0.97	1.20	0.271
7.50	1.00	1.09	0.230
8.00	1.03	1.03	0.206
8.50	1.03	0.98	0.189
9.00	1.03	0.92	0.171
9.50	1.03	0.87	0.154
10.00	1.04	0.82	0.139
11.00	1.07	0.75	0.118
12.00	1.09	0.73	0.111
13.00	1.08	0.72	0.109
14.00	1.06	0.72	0.111
14.50	1.03	0.72	0.111
15.00	1.01	0.71	0.111
15.50	0.98	0.69	0.109
16.00	0.95	0.67	0.106
17.00	0.91	0.62	0.097
18.00	0.89	0.56	0.084
19.00	0.88	0.51	0.071
20.00	0.88	0.45	0.059
21.00	0.90	0.41	0.048
22.00	0.92	0.38	0.040
23.00	0.94	0.37	0.035
24.00	0.96	0.37	0.035

Energy (eV)	n	k	R(φ = 0)
27.00	0.88	0.38	0.043
28.00	0.86	0.35	0.039
29.00	0.85	0.30	0.032
30.00	0.86	0.26	0.025
31.00	0.88	0.24	0.020
32.00	0.89	0.22	0.017
33.00	0.90	0.21	0.015
34.00	0.91	0.20	0.014
35.00	0.92	0.20	0.013
36.00	0.92	0.19	0.012
37.00	0.92	0.19	0.011
38.00	0.93	0.18	0.010
39.00	0.93	0.17	0.009
40.00	0.93	0.17	0.009
41.00	0.94	0.16	0.008
42.00	0.94	0.16	0.007
43.00	0.94	0.15	0.007
44.00	0.95	0.15	0.007
45.00	0.95	0.15	0.006
46.00	0.95	0.15	0.006
47.00	0.95	0.14	0.006
48.00	0.95	0.14	0.006
49.00	0.95	0.14	0.005
50.00	0.95	0.13	0.005
51.00	0.95	0.13	0.005
52.00	0.95	0.13	0.005
53.00	0.96	0.12	0.004
54.00	0.96	0.12	0.004
55.00	0.96	0.12	0.004
56.00	0.96	0.11	0.004
57.00	0.96	0.11	0.004
58.00	0.96	0.11	0.004
59.00	0.97	0.11	0.003
60.00	0.97	0.11	0.003
61.00	0.97	0.11	0.003
62.00	0.97	0.11	0.003
63.00	0.96	0.10	0.003
64.00	0.96	0.10	0.003
65.00	0.97	0.10	0.003
66.00	0.97	0.10	0.003
67.00	0.97	0.09	0.003
68.00	0.97	0.09	0.002
69.00	0.97	0.09	0.002
70.00			
	0.97		
		0.09	0.002
75.00	0.98	0.09	0.002
80.00	0.98	0.09	0.002
85.00	0.97	0.09	0.002
90.00	0.96	0.08	0.002

Gallium (liquid)[7]

Energy (eV)	n	k	R(φ = 0)
1.425	2.40	9.20	0.900
1.550	2.09	8.50	0.898
1.771	1.65	7.60	0.898
2.066	1.25	6.60	0.897
2.480	0.89	5.60	0.898
3.100	0.59	4.50	0.896

Germanium, single crystal[8]

Energy (eV)	n	k	R(φ = 0)
0.01240	(4.0065)	3.00E-03	0.361
0.01364	4.0063	2.40E-03	0.361
0.01488	(4.0060)	1.70E-03	0.361
0.01612	(4.0060)	1.55E-03	0.361
0.01736	(4.0060)	1.50E-03	0.361
0.01860		1.50E-03	

Energy (eV)	n	k	R(φ = 0)
0.02108		1.60E-03	
0.02232		1.55E-03	
0.02356		1.53E-03	
0.02480		1.50E-03	
0.02604		1.25E-03	
0.02728		8.50E-04	
0.02852		6.50E-04	
0.02976		7.00E-04	
0.03100	3.9827	8.50E-04	0.358
0.03224		1.55E-03	
0.03348		2.75E-03	
0.03472		3.55E-03	
0.03596	(3.9900)	3.05E-03	0.359
0.03720		2.75E-03	
0.03844		2.70E-03	
0.03968	(3.9930)	2.90E-03	0.359
0.04092		2.95E-03	
0.04215		3.20E-03	
0.04339		6.30E-03	
0.04463		3.40E-03	
0.04587	(3.9955)	2.50E-03	0.360
0.04711		2.10E-03	
0.04835		2.00E-03	
0.04959		8.00E-04	
0.05083		1.40E-03	
0.05207		1.35E-03	
0.05331		1.10E-03	
0.05455		8.00E-04	
0.05579		6.00E-04	
0.05703		9.0 E-04	
0.05827		6.5 E-04	
0.05951		4.6 E-04	
0.06075		4.0 E-04	
0.06199	3.9992	3.98E-04	0.360
0.06323		4.0 E-04	
0.06447		4.3 E-04	
0.06571		4.4 E-04	
0.06695	(4.0000)	4.3 E-04	0.360
0.06819		3.1 E-04	
0.06943		3.3 E-04	
0.07067		3.8 E-04	
0.07191		3.3 E-04	
0.07315		2.5 E-04	
0.07439		1.9 E-04	
0.07514		1.58E-04	
0.07749	4.0009	9.55E-05	0.360
0.07999	4.0011	1.71E-04	0.360
0.08266	4.0013	9.78E-05	0.360
0.08551	4.0015	5.77E-05	0.360
0.08920		3.98E-05	
0.09460		4.59E-05	
0.09840		3.51E-05	
0.1	4.0063	3.70E-05	0.361
0.2	4.0108		0.361
0.3	4.0246		0.362
0.4	4.0429		0.364
0.5	(4.074)		0.367
0.6	(4.104)	6.58E-07	0.370
0.7	4.180	1.27E-04	0.377
0.8	4.275	5.67E-03	0.385
0.9	4.285	7.45E-02	0.386
1.0	4.325	8.09E-02	0.390
1.1	4.385	0.103	0.395
1.2	4.420	0.123	0.398
1.3	4.495	0.167	0.405

Energy (eV)	n	k	R(φ = 0)	Energy (eV)	n	k	R(φ = 0)	Energy (eV)	n	k	R(φ = 0)
1.4	4.560	0.190	0.411	40.0		0.0604		8.80	1.31	0.86	0.140
1.5	4.635	0.298	0.418					9.00	1.30	0.83	0.133
1.6	4.763	0.345	0.428	**Gold, electropolished, Au (110)[9]**				9.20	1.31	0.81	0.126
1.7	4.897	0.401	0.439	0.10	8.17	82.83	0.995	9.40	1.33	0.78	0.122
1.8	5.067	0.500	0.453	0.20	2.13	41.73	0.995	9.60	1.36	0.78	0.121
1.9	5.380	0.540	0.475	0.30	0.99	27.82	0.995	9.80	1.37	0.79	0.124
2.0	5.588	0.933	0.495	0.40	0.59	20.83	0.995	10.00	1.37	0.80	0.126
2.1	5.748	1.634	0.523	0.50	0.39	16.61	0.994	10.20	1.36	0.80	0.127
2.2	5.283	2.049	0.516	0.60	0.28	13.78	0.994	10.40	1.35	0.80	0.125
2.3	5.062	2.318	0.519	0.70	0.22	11.75	0.994	10.60	1.34	0.79	0.123
2.4	4.610	2.455	0.508	0.80	0.18	10.21	0.993	10.80	1.34	0.77	0.120
2.5	4.340	2.384	0.492	0.90	0.15	9.01	0.993	11.00	1.34	0.76	0.116
2.6	4.180	2.309	0.480	1.00	0.13	8.03	0.992	11.20	1.34	0.74	0.113
2.7	4.082	2.240	0.471	1.20	0.10	6.54	0.991	11.40	1.35	0.73	0.111
2.8	4.035	2.181	0.464	1.40	0.08	5.44	0.989	11.60	1.36	0.72	0.109
2.9	4.037	2.140	0.461	1.60	0.08	4.56	0.986	11.80	1.38	0.71	0.108
3.0	4.082	2.145	0.463	1.80	0.09	3.82	0.979	12.00	1.39	0.71	0.109
3.1	4.141	2.215	0.471	2.00	0.13	3.16	0.953	12.40	1.44	0.73	0.115
3.2	4.157	2.340	0.482	2.10	0.18	2.84	0.925	12.80	1.45	0.79	0.127
3.3	4.128	2.469	0.490	2.20	0.24	2.54	0.880	13.20	1.42	0.84	0.137
3.4	4.070	2.579	0.497	2.40	0.50	1.86	0.647	13.60	1.37	0.86	0.140
3.5	4.020	2.667	0.502	2.50	0.82	1.59	0.438	14.00	1.33	0.86	0.140
3.6	3.985	2.759	0.509	2.60	1.24	1.54	0.331	14.40	1.29	0.86	0.139
3.7	3.958	2.863	0.517	2.70	1.43	1.72	0.356	14.80	1.26	0.84	0.135
3.8	3.936	2.986	0.527	2.80	1.46	1.77	0.368	15.20	1.24	0.83	0.132
3.9	3.920	3.137	0.539	2.90	1.50	1.79	0.368	15.60	1.22	0.81	0.127
4.0	3.905	3.336	0.556	3.00	1.54	1.80	0.369	16.00	1.21	0.79	0.123
4.1	3.869	3.614	0.579	3.10	1.54	1.81	0.371	16.40	1.20	0.78	0.119
4.2	3.745	4.009	0.612	3.20	1.54	1.80	0.368	16.80	1.19	0.76	0.116
4.3	3.338	4.507	0.659	3.30	1.55	1.78	0.362	17.20	1.19	0.75	0.114
4.4	2.516	4.669	0.705	3.40	1.56	1.76	0.356	17.60	1.19	0.74	0.111
4.5	1.953	4.297	0.713	3.50	1.58	1.73	0.349	18.00	1.19	0.74	0.109
4.6	1.720	3.960	0.702	3.60	1.62	1.73	0.346	18.40	1.19	0.73	0.109
4.7	1.586	3.709	0.690	3.70	1.64	1.75	0.351	18.80	1.20	0.74	0.110
4.8	1.498	3.509	0.677	3.80	1.63	1.79	0.360	19.20	1.21	0.76	0.116
4.9	1.435	3.342	0.664	3.90	1.59	1.81	0.366	19.60	1.21	0.80	0.125
5.0	1.394	3.197	0.650	4.00	1.55	1.81	0.369	20.00	1.18	0.83	0.133
5.1	1.370	3.073	0.636	4.10	1.51	1.79	0.368	20.40	1.14	0.85	0.141
5.2	1.364	2.973	0.622	4.20	1.48	1.78	0.367	20.80	1.10	0.87	0.149
5.3	1.371	2.897	0.609	4.30	1.45	1.77	0.368	21.20	1.05	0.88	0.156
5.4	1.383	2.854	0.600	4.40	1.41	1.76	0.370	21.60	1.00	0.88	0.162
5.5	1.380	2.842	0.598	4.50	1.35	1.74	0.370	22.00	0.94	0.86	0.164
5.6	1.360	2.846	0.602	4.60	1.30	1.69	0.364	22.40	0.89	0.83	0.163
5.7	1.293	2.163	0.479	4.70	1.27	1.64	0.354	22.80	0.85	0.79	0.157
5.8	1.209	2.873	0.632	4.80	1.25	1.59	0.344	23.20	0.82	0.75	0.149
5.9	1.108	2.813	0.641	4.90	1.23	1.54	0.332	23.60	0.80	0.70	0.138
6.0	1.30	2.34	0.517	5.00	1.22	1.49	0.319	24.00	0.80	0.66	0.125
6.5	1.10	2.05	0.489	5.20	1.21	1.40	0.295	24.40	0.80	0.62	0.113
7.0	1.00	1.80	0.448	5.40	1.21	1.33	0.275	24.80	0.80	0.58	0.101
7.5		1.60		5.60	1.21	1.27	0.256	25.20	0.82	0.56	0.090
8.0	0.92	1.40	0.348	5.80				25.60	0.83	0.54	0.084
8.5	0.92	1.20	0.282			1.21		26.00	0.84	0.52	0.079
9.0	0.92	1.14	0.262					26.40	0.85	0.51	0.074
9.5		1.00				1.20		26.80	0.85	0.50	0.071
10.0	0.93	0.86	0.167	6.00	1.22	1.14	0.236	27.20	0.86	0.49	0.068
20.0		0.237		6.20	1.24	1.09	0.218	27.60	0.86	0.49	0.065
22.0		0.179		6.40	1.25	1.05	0.203	28.00	0.87	0.48	0.063
24.0		0.144		6.60	1.27	1.01	0.190	28.40	0.88	0.48	0.062
26.0		0.110		6.80	1.30	0.97	0.177	28.80	0.88	0.48	0.062
28.0		0.0747		7.00	1.34	0.95	0.167	29.20	0.88	0.48	0.062
30.0		0.1020		7.20	1.36	0.95	0.162	29.60	0.87	0.48	0.064
32.0		0.0999		7.40	1.38	0.96	0.161	30.00	0.86	0.48	0.064
34.0		0.0856		7.60	1.38	0.98	0.164				
36.0		0.0740		7.80	1.35	0.99	0.169	**Hafnium, single crystal, $\overset{\perp}{E} \parallel \hat{c}$ [10]**			
38.0		0.0651		8.00	1.31	0.96	0.171	0.52	1.48	4.11	0.747
				8.20	1.30	0.92	0.165				

Energy (eV)	n	k	$R(\phi=0)$	Energy (eV)	n	k	$R(\phi=0)$	Energy (eV)	n	k	$R(\phi=0)$
0.56	1.84	3.29	0.615	8.20	1.43	1.01	0.178	1.40	3.61	3.36	0.561
0.60	2.34	2.62	0.486	8.40	1.45	1.01	0.180	1.50	3.55	3.13	0.540
0.66	3.21	2.13	0.428	8.60	1.47	1.02	0.183	1.60	3.58	3.01	0.529
0.70	3.70	2.03	0.441	8.80	1.48	1.04	0.186	1.70	3.63	2.98	0.526
0.76	4.31	2.10	0.476	9.00	1.49	1.07	0.193	1.80	3.66	3.02	0.530
0.80	4.61	2.31	0.504	9.20	1.50	1.10	0.201	1.90	3.63	3.14	0.541
0.86	4.71	2.70	0.533	9.40	1.48	1.14	0.211	2.00	3.51	3.26	0.551
0.90	4.64	2.85	0.541	9.60	1.46	1.18	0.222	2.10	3.35	3.33	0.558
0.95	4.54	2.96	0.545	9.80	1.41	1.21	0.230	2.20	3.18	3.36	0.563
1.00	4.45	3.00	0.545	10.00	1.36	1.22	0.235	2.30	2.99	3.39	0.568
1.10	4.28	3.08	0.547	10.20	1.32	1.22	0.238	2.40	2.78	3.35	0.569
1.20	4.08	3.10	0.544	10.40	1.28	1.22	0.240	2.50	2.65	3.26	0.562
1.30	3.87	3.04	0.536	10.60	1.24	1.21	0.241	2.60	2.54	3.22	0.560
1.40	3.72	2.95	0.525	10.80	1.20	1.20	0.242	2.70	2.42	3.17	0.559
1.50	3.60	2.85	0.514	11.00	1.16	1.19	0.242	2.80	2.31	3.13	0.558
1.60	3.52	2.73	0.500	11.20	1.13	1.17	0.241	2.90	2.20	3.08	0.558
1.70	3.52	2.61	0.488	11.40	1.10	1.16	0.241	3.00	2.08	3.05	0.561
1.80	3.57	2.56	0.485	11.60	1.07	1.14	0.239	3.10	1.94	2.98	0.560
1.90	3.63	2.59	0.489	11.80	1.04	1.12	0.238	3.20	1.83	2.88	0.555
2.00				12.00	1.02	1.10	0.236	3.30	1.74	2.78	0.547
	3.65			12.40	0.96	1.06	0.232	3.40	1.68	2.69	0.538
		2.67		12.80	0.92	1.01	0.225	3.50	1.62	2.61	0.529
			0.498	13.20	0.88	0.96	0.218	3.60	1.57	2.52	0.519
2.10	3.64	2.81	0.511	13.60	0.84	0.90	0.205	3.70	1.53	2.45	0.510
2.20	3.53	2.99	0.526	14.00	0.83	0.83	0.186	3.80	1.49	2.38	0.501
2.30	3.34	3.09	0.534	14.40	0.83	0.80	0.172	3.90	1.45	2.32	0.493
2.40	3.15	3.11	0.537	14.80	0.81	0.76	0.167	4.00	1.41	2.25	0.484
2.50	2.99	3.13	0.540	15.20	0.79	0.70	0.153	4.10	1.38	2.18	0.474
2.60	2.83	3.12	0.542	15.60	0.79	0.64	0.132	4.20	1.35	2.11	0.462
2.70	2.68	3.10	0.542	16.00	0.83	0.60	0.111	4.30	1.33	2.05	0.451
2.80	2.54	3.08	0.543	16.40	0.81	0.60	0.114	4.40	1.31	1.99	0.438
2.90	2.40	3.04	0.544	16.80	0.79	0.55	0.105	4.50	1.30	1.93	0.427
3.00	2.27	3.00	0.544	17.20	0.79	0.50	0.089	4.60	1.29	1.88	0.415
3.10	2.14	2.95	0.544	17.60	0.80	0.46	0.077	4.70	1.28	1.82	0.402
3.20	2.00	2.89	0.544	18.00	0.81	0.42	0.064	4.80	1.28	1.77	0.389
3.30	1.87	2.79	0.538	18.40	0.84	0.38	0.051	4.90	1.27	1.73	0.379
3.40	1.78	2.68	0.528	18.80	0.87	0.34	0.040	5.00	1.27	1.69	0.367
3.50	1.71	2.58	0.517	19.00	0.89	0.33	0.036	5.20	1.27	1.62	0.349
3.60	1.66	2.48	0.503	19.60	0.93	0.32	0.030	5.40	1.27	1.57	0.335
3.70	1.63	2.40	0.491	20.00	0.94	0.31	0.027	5.60	1.26	1.52	0.322
3.80	1.60	2.33	0.481	20.60	0.97	0.30	0.023	5.80	1.24	1.48	0.313
3.90	1.56	2.27	0.473	21.00	0.99	0.29	0.022	6.00	1.21	1.42	0.302
4.00	1.52	2.21	0.466	21.60	1.01	0.28	0.020	6.20	1.19	1.36	0.285
4.10	1.48	2.14	0.455	22.00	1.03	0.28	0.020	6.40	1.18	1.29	0.265
4.20	1.45	2.07	0.442	22.60	1.06	0.28	0.020	6.60	1.19	1.22	0.244
4.30	1.43	2.01	0.431	23.00	1.07	0.28	0.021	6.80	1.21	1.18	0.230
4.40	1.41	1.95	0.420	23.60	1.09	0.29	0.022	7.00	1.22	1.14	0.217
4.50	1.39	1.89	0.407	24.00	1.09	0.30	0.023	7.20	1.23	1.10	0.206
4.60	1.39	1.83	0.394	24.60	1.10	0.31	0.024	7.40	1.26	1.06	0.194
4.70	1.39	1.79	0.382					7.60	1.28	1.04	0.187
4.80	1.38	1.75	0.373	**Hafnium, single crystal,** $\mathbf{E} \perp \hat{c}$ [10]				7.80	1.30	1.02	0.180
4.90	1.38	1.71	0.364	0.52	2.25	4.65	0.723	8.00	1.33	1.00	0.174
5.00	1.37	1.68	0.356	0.56	2.34	3.66	0.623	8.20	1.35	0.99	0.173
5.20	1.36	1.61	0.341	0.60				8.40	1.38	0.99	0.173
5.40	1.35	1.55	0.324		2.84			8.60	1.40	1.00	0.174
5.60	1.35	1.51	0.314			2.89		8.80	1.42	1.02	0.178
5.80	1.32	1.48	0.308				0.512	9.00	1.43	1.04	0.184
6.00	1.28	1.41	0.295	0.66	3.71	2.35	0.469	9.20	1.45	1.08	0.193
6.20	1.26	1.35	0.278	0.70	4.26	2.21	0.482	9.40	1.43	1.12	0.204
6.40	1.26	1.28	0.258	0.76	4.97	2.33	0.521	9.60	1.40	1.16	0.214
6.60	1.27	1.22	0.240	0.80	5.41	2.62	0.554	9.80	1.37	1.19	0.223
6.80	1.28	1.16	0.224	0.86	5.46	3.36	0.593	10.00	1.32	1.21	0.230
7.00	1.31	1.13	0.212	0.90	5.22	3.62	0.601	10.20	1.27	1.21	0.234
7.20	1.33	1.10	0.204	0.95	4.95	3.72	0.602	10.40	1.23	1.20	0.235
7.40	1.34	1.07	0.197	1.00	4.76	3.76	0.602	10.60	1.19	1.20	0.237
7.60	1.36	1.05	0.191	1.10	4.43	3.80	0.601				

Energy (eV)	n	k	$R(\phi=0)$	Energy (eV)	n	k	$R(\phi=0)$	Energy (eV)	n	k	$R(\phi=0)$
10.80	1.15	1.19	0.237	2.40	2.07	4.14	0.689	17.60	1.30	0.87	0.140
11.00	1.12	1.17	0.237	2.50	1.98	4.00	0.682	18.00	1.30	0.93	0.154
11.20	1.08	1.16	0.237	2.60	1.91	3.86	0.673	18.40	1.27	0.97	0.166
11.40	1.05	1.14	0.236	2.70	1.85	3.73	0.665	18.80	1.24	1.00	0.176
11.60	1.03	1.12	0.235	2.80	1.81	3.61	0.655	19.20	1.20	1.03	0.187
11.80	1.00	1.10	0.233	2.90	1.77	3.51	0.646	19.60	1.15	1.05	0.197
12.00	0.97	1.08	0.231	3.00	1.73	3.43	0.640	20.00	1.10	1.06	0.205
12.40	0.92	1.04	0.226	3.20	1.62	3.26	0.629	20.50	1.04	1.05	0.210
12.80	0.88	0.99	0.219	3.40	1.53	3.05	0.610	21.00	0.99	1.04	0.215
13.20	0.83	0.94	0.211	3.60	1.52	2.81	0.573	21.50	0.94	1.02	0.220
13.60	0.80	0.88	0.196	3.80	1.61	2.69	0.541	22.00	0.89	1.00	0.222
14.00	0.79	0.81	0.177	4.00	1.64	2.68	0.535	22.50	0.84	0.99	0.228
14.40	0.80	0.77	0.160	4.20	1.58	2.71	0.549	23.00	0.79	0.96	0.232
14.80	0.77	0.73	0.154	4.40	1.45	2.68	0.561	23.50	0.76	0.92	0.228
15.20	0.76	0.68	0.140	4.60	1.31	2.60	0.567	24.00	0.73	0.87	0.223
15.60	0.76	0.61	0.119	4.80	1.18	2.49	0.570	24.50	0.70	0.83	0.218
16.00	0.81	0.58	0.099	5.00	1.10	2.35	0.559	25.00	0.69	0.79	0.209
16.40	0.78	0.57	0.102	5.20	1.04	2.22	0.543	25.50	0.68	0.76	0.200
16.80	0.77	0.53	0.092	5.40	1.00	2.09	0.522	26.00	0.67	0.72	0.192
17.20	0.77	0.48	0.077	5.60	0.98	1.98	0.499	26.50	0.67	0.69	0.181
17.60	0.79	0.44	0.065	5.80	0.96	1.86	0.474	27.00	0.66	0.66	0.174
18.00				6.00	0.95	1.78	0.454	27.50	0.66	0.63	0.166
	0.80	0.39	0.053	6.20	0.94	1.68	0.427	28.00	0.66	0.61	0.158
18.40	0.82	0.36	0.041	6.40	0.94	1.59	0.401	28.50	0.66	0.59	0.151
18.80	0.86	0.33	0.032	6.60	0.94	1.50	0.375	29.00	0.65	0.57	0.148
19.00	0.88	0.32	0.030	6.80	0.95	1.42	0.345	29.50	0.64	0.55	0.145
19.60	0.91	0.31	0.025	7.00	0.97	1.34	0.318	30.00	0.64	0.53	0.140
20.00	0.93	0.30	0.023	7.20	0.99	1.27	0.290	32.00	0.62	0.44	0.119
20.60	0.96	0.29	0.021	7.40	1.02	1.20	0.262	34.00	0.64	0.35	0.091
21.00	0.97	0.29	0.020	7.60	1.03	1.14	0.241	36.00	0.69	0.27	0.059
21.60	1.00	0.28	0.019	7.80	1.08	1.06	0.208	38.00	0.73	0.24	0.044
22.00	1.01	0.28	0.019	8.00	1.13	1.03	0.191	40.00	0.76	0.22	0.034
22.60	1.03	0.27	0.018	8.20	1.18	1.00	0.179				
23.00	1.05	0.28	0.019	8.40	1.22	0.98	0.171	**Iron[5]**			
23.60	1.06	0.28	0.020	8.60	1.26	0.96	0.164	0.10	6.41	33.07	0.978
24.00	1.07	0.29	0.021	8.80	1.29	0.95	0.160	0.15	6.26	22.82	0.956
24.60	1.09	0.30	0.022	9.00	1.33	0.94	0.157	0.20	3.68	18.23	0.958
				9.20	1.36	0.95	0.159	0.26	4.98	13.68	0.911
Iridium[11]				9.40	1.39	0.95	0.161	0.30	4.87	12.05	0.892
0.10	28.49	60.62	0.975	9.60	1.42	0.97	0.163	0.36	4.68	10.44	0.867
0.15	15.32	45.15	0.973	9.80	1.44	0.99	0.169	0.40	4.42	9.75	0.858
0.20	9.69	35.34	0.972	10.00	1.45	1.01	0.175	0.50	4.14	8.02	0.817
0.25	6.86	28.84	0.969	10.20	1.45	1.04	0.182	0.60	3.93	6.95	0.783
0.30	5.16	24.25	0.967	10.40	1.44	1.07	0.187	0.70	3.78	6.17	0.752
0.35	4.11	20.79	0.964	10.60	1.43	1.09	0.193	0.80	3.65	5.60	0.725
0.40	3.42	18.06	0.960	10.80	1.41	1.12	0.200	0.90	3.52	5.16	0.700
0.45	3.05	15.82	0.954	11.00	1.38	1.13	0.206	1.00	3.43	4.79	0.678
0.50	2.98	14.06	0.944	11.20	1.34	1.14	0.208	1.10	3.33	4.52	0.660
0.60	2.79	11.58	0.925	11.40	1.31	1.13	0.208	1.20	3.24	4.26	0.641
0.70	2.93	9.78	0.895	11.60	1.28	1.12	0.206	1.30	3.16	4.07	0.626
0.80	3.14	8.61	0.862	11.80	1.25	1.10	0.203	1.40	3.12	3.87	0.609
0.90	3.19	7.88	0.840	12.00	1.24	1.08	0.199	1.50	3.05	3.77	0.601
1.00	3.15	7.31	0.822	12.40	1.21	1.05	0.191	1.60	3.00	3.60	0.585
1.10	3.04	6.84	0.808	12.80	1.19	1.01	0.181	1.70	2.98	3.52	0.577
1.20	2.96	6.41	0.791	13.20	1.18	0.98	0.173	1.80	2.92	3.46	0.573
1.30	2.85	6.07	0.779	13.60				1.90	2.89	3.37	0.563
1.40	2.72	5.74	0.767		1.17			2.00	2.85	3.36	0.563
1.50	2.65	5.39	0.750			0.95	0.165	2.10	2.80	3.34	0.562
1.60	2.68	5.08	0.728	14.00	1.16	0.91	0.155	2.20	2.74	3.33	0.563
1.70	2.69	4.92	0.716	14.40	1.17	0.88	0.147	2.30	2.65	3.34	0.567
1.80	2.64	4.81	0.710	14.80	1.18	0.87	0.142	2.40	2.56	3.31	0.567
1.90	2.57	4.68	0.704	15.20	1.19	0.84	0.136	2.50	2.46	3.31	0.570
2.00	2.50	4.57	0.699	15.60	1.20	0.83	0.133	2.60	2.34	3.30	0.576
2.10	2.40	4.48	0.697	16.00	1.21	0.83	0.131	2.70	2.23	3.25	0.575
2.20	2.29	4.38	0.695	16.40	1.23	0.82	0.129	2.80	2.12	3.23	0.580
2.30	2.18	4.26	0.692	16.80	1.25	0.82	0.127				

Energy (eV)	n	k	R(φ = 0)	Energy (eV)	n	k	R(φ = 0)	Energy (eV)	n	k	R(φ = 0)
2.90	2.01	3.17	0.580	13.17	0.84	0.79	0.161	24.00	0.74	0.27	0.045
3.00	1.88	3.12	0.583	13.33	0.84	0.78	0.160	24.17	0.74	0.26	0.044
3.10	1.78	3.04	0.580	13.50	0.83	0.77	0.159	24.33	0.74	0.26	0.043
3.20	1.70	2.96	0.576	13.67	0.82	0.76	0.157	24.50	0.74	0.25	0.042
3.30	1.62	2.87	0.572	13.83	0.81	0.75	0.154	24.67	0.75	0.25	0.040
3.40	1.55	2.79	0.565	14.00	0.81	0.73	0.151	24.83	0.75	0.24	0.039
3.50	1.50	2.70	0.556	14.17	0.80	0.72	0.149	25.00	0.75	0.24	0.038
3.60	1.47	2.63	0.548	14.33	0.80	0.71	0.146	26.00	0.76	0.21	0.031
3.70	1.43	2.56	0.542	14.50	0.79	0.79	0.144	27.00	0.78	0.18	0.026
3.83	1.38	2.49	0.534	14.67	0.79	0.69	0.141	28.00	0.79	0.16	0.021
4.00	1.30	2.39	0.527	14.83	0.78	0.67	0.138	29.00	0.81	0.14	0.017
4.17	1.26	2.27	0.510	15.00	0.78	0.66	0.135	30.00	0.82	0.13	0.014
4.33	1.23	2.18	0.494	15.17	0.78	0.65	0.131				
4.50	1.20	2.10	0.482	15.33	0.78	0.64	0.238	**Lithium[12]**			
4.67	1.16	2.02	0.470	15.50	0.77	0.63	0.126	0.14	0.659	38.0	0.998
4.83	1.14	1.93	0.451	15.67	0.77	0.62	0.123	0.54	0.661	12.6	0.984
5.00	1.14	1.87	0.435	15.83	0.77	0.61	0.119	0.75	0.561	7.68	0.963
5.17	1.12	1.81	0.425	16.00	0.77	0.60	0.116	1.05	0.448	5.58	0.946
5.33	1.11	1.75	0.408	16.17	0.78	0.58	0.112	1.35	0.338	4.36	0.935
5.50	1.09	1.17	0.401	16.33	0.78	0.58	0.110	1.65	0.265	3.55	0.925
5.67	1.09	1.65	0.383	16.50	0.78	0.57	0.107	1.95	0.221	2.94	0.913
5.83	1.10	1.61	0.373	16.67	0.77	0.56	0.106	2.25	0.206	2.48	0.892
6.00	1.09	1.59	0.366	16.83	0.78	0.55	0.103	2.55	0.217	2.11	0.854
6.17	1.08	1.57	0.365	17.00	0.78	0.55	0.102	2.85	0.247	1.82	0.797
6.33				17.17	0.78	0.54	0.100	3.15	0.304	1.60	0.715
	1.04			17.33	0.78	0.54	0.098	3.45	0.334	1.45	0.656
		1.55	0.365	17.50	0.77	0.53	0.097	3.75	0.345	1.32	0.611
6.50	1.02	1.51	0.358	17.67	0.77	0.52	0.095	4.05	0.346	1.21	0.578
6.67	1.00	1.47	0.351	17.83	0.78	0.51	0.092	4.35	0.333	1.11	0.557
6.83	0.97	1.43	0.346	18.00	0.78	0.51	0.091	4.65	0.317	1.01	0.540
7.00	0.96	1.39	0.333	18.17	0.78	0.51	0.090	4.95	0.302	0.906	0.520
7.17	0.94	1.35	0.327	18.33	0.78	0.50	0.089	5.25	0.299	0.795	0.484
7.33	0.94	1.30	0.311	18.50	0.77	0.50	0.089	5.55	0.310	0.688	0.434
7.50	0.94	1.26	0.298	18.67	0.77	0.50	0.088	5.85	0.342	0.594	0.365
7.67	0.94	1.23	0.288	18.83	0.77	0.49	0.087	6.15	0.376	0.522	0.306
7.83	0.94	1.21	0.279	19.00	0.77	0.49	0.087	6.45	0.408	0.460	0.256
8.00	0.94	1.18	0.272	19.17	0.76	0.49	0.088	6.75	0.440	0.407	0.214
8.17	0.94	1.16	0.265	19.33	0.76	0.48	0.087	7.05	0.466	0.364	0.183
8.33	0.94	1.14	0.258	19.50	0.75	0.47	0.086	7.35	0.492	0.320	0.155
8.50	0.94	1.12	0.251	19.67	0.75	0.47	0.085	7.65	0.517	0.282	0.131
8.67	0.94	1.10	0.246	19.83	0.75	0.46	0.084	7.95	0.545	0.246	0.109
8.83	0.92	1.08	0.240	20.00	0.74	0.45	0.083	8.25	0.572	0.214	0.091
9.00	0.93	1.07	0.236	20.17	0.74	0.44	0.081	8.55	0.601	0.189	0.075
9.17	0.92	1.06	0.233	20.33	0.74	0.44	0.081	8.85	0.624	0.163	0.063
9.33	0.91	1.04	0.231	20.50	0.74	0.42	0.080	9.15	0.657	0.144	0.050
9.50	0.90	1.02	0.226	20.67	0.73	0.43	0.079	9.45	0.680	0.130	0.042
9.67	0.90	1.00	0.221	20.83	0.73	0.42	0.078	9.75	0.708	0.119	0.034
9.83	0.89	0.99	0.218	21.00	0.73	0.41	0.077	10.1	0.726	0.108	0.029
10.00	0.88	0.97	0.213	21.17	0.72	0.40	0.076	10.4	0.743	0.102	0.025
10.17	0.87	0.94	0.203	21.33	0.72	0.39	0.074	10.6	0.753	0.080	0.022
10.33	0.87	0.91	0.196	21.50	0.72	0.38	0.073				
10.50	0.87	0.89	0.189	21.67	0.72	0.38	0.071	**Magnesium (evaporated)[13]**			
10.67	0.88	0.87	0.179	21.83	0.72	0.37	0.070	2.145	0.48	3.71	0.880
10.83	0.89	0.85	0.170	22.00	0.72	0.36	0.068	2.270	0.57	3.47	0.843
11.00	0.91	0.83	0.162	22.17	0.71	0.35	0.067	2.522	0.53	2.92	0.805
11.17	0.92	0.83	0.159	22.33	0.72	0.34	0.064	2.845	0.52	2.65	0.777
11.33	0.93	0.84	0.159	22.50	0.72	0.34	0.063	3.064	0.52	2.05	0.681
11.50	0.93	0.84	0.160	22.67	0.72	0.33	0.062	5.167	0.10	1.60	0.894
11.67	0.93	0.84	0.162	22.83	0.72	0.32	0.059	5.636	0.15	1.50	0.832
11.83	0.92	0.84	0.163	23.00	0.72	0.31	0.058	6.200	0.20	1.40	0.765
12.00	0.91	0.84	0.163	23.17	0.72	0.30	0.056	6.889	0.25	1.30	0.693
12.17	0.90	0.84	0.165	23.33	0.72	0.29	0.054	7.750	0.20	1.20	0.722
12.33	0.89	0.83	0.164	23.50	0.73	0.28	0.050	8.857	0.15	0.95	0.730
12.50	0.98	0.83	0.165	23.67	0.73	0.28	0.049	10.335	0.25	0.40	0.419
12.67	0.87	0.82	0.166	23.83	0.74	0.27	0.047				
12.83	0.86	0.81	0.166								

Energy (eV)	n	k	R(φ = 0)	Energy (eV)	n	k	R(φ = 0)	Energy (eV)	n	k	R(φ = 0)
Manganese[14]				2.4	1.384	4.407	0.779	17.0	1.177	0.367	0.034
0.64	3.89	5.95	0.738	2.6	1.186	4.090	0.779	17.5	1.184	0.366	0.034
0.77	3.78	5.41	0.710	2.8	1.027	3.802	0.779	18.0	1.191	0.367	0.035
0.89	3.65	5.02	0.688	3.0	0.898	3.538	0.777	18.5	1.195	0.367	0.035
1.02	3.48	4.74	0.673	3.2	0.798	3.294	0.773	19.0	1.200	0.366	0.035
1.14	3.30	4.53	0.662	3.4	0.713	3.074	0.770	19.5	1.208	0.364	0.035
1.26	3.10	4.35	0.653	3.6	0.644	2.860	0.763				
1.39	2.97	4.18	0.643	3.8	0.589	2.665	0.755	**Molybdenum**[16]			
1.51	2.83	4.03	0.634	4.0	0.542	2.502	0.749	0.10	18.53	68.51	0.985
1.64	2.70	3.91	0.627	4.2	0.507	2.341	0.738	0.15	8.78	47.54	0.985
1.76	2.62	3.78	0.617	4.4	0.477	2.195	0.727	0.20	5.10	35.99	0.985
1.88	2.56	3.65	0.606	4.6	0.452	2.058	0.715	0.25	3.36	28.75	0.984
2.01	2.51	3.54	0.596	4.8	0.431	1.929	0.701	0.30	2.44	23.80	0.983
2.13	2.47	3.43	0.585	5.0	0.414	1.806	0.685	0.34	2.00	20.84	0.982
2.26	2.39	3.33	0.577	5.2	0.401	1.687	0.666	0.38	1.70	18.44	0.980
2.38	2.32	3.23	0.567	5.4	0.394	1.569	0.642	0.42	1.57	16.50	0.978
2.50	2.25	3.14	0.559	5.6	0.386	1.454	0.617	0.46	1.46	14.91	0.975
2.63	2.19	3.06	0.552	5.7	0.386	1.396	0.601	0.50	1.37	13.55	0.971
2.75	2.11	2.98	0.545	5.8	0.386	1.341	0.585	0.54	1.35	12.36	0.966
2.88	2.06	2.90	0.536	5.9	0.385	1.287	0.569	0.58	1.34	11.34	0.960
3.00	2.00	2.82	0.528	6.0	0.386	1.232	0.551	0.62	1.38	10.44	0.952
3.12	1.96	2.74	0.518	6.1	0.388	1.176	0.531	0.66	1.43	9.67	0.942
3.25	1.92	2.67	0.509	6.2	0.390	1.118	0.510	0.70	1.48	8.99	0.932
3.37	1.89	2.59	0.498	6.3	0.399	1.058	0.481	0.74	1.51	8.38	0.921
3.50	1.89	2.51	0.484	6.4	0.412	1.002	0.450	0.78	1.60	7.83	0.906
3.62	1.87	2.45	0.475	6.5	0.428	0.949	0.418	0.82	1.64	7.35	0.892
3.74	1.86	2.38	0.463	6.6	0.436	0.898	0.392	0.86	1.70	6.89	0.876
3.87	1.86	2.32	0.451	6.7	0.438	0.836	0.367	9.90	1.74	6.48	0.859
3.99	1.86	2.25	0.438	6.8	0.459	0.756	0.320	1.00	1.94	5.58	0.805
4.12	1.86	2.19	0.427	6.9	0.510	0.676	0.255	1.10	2.15	4.85	0.743
4.24	1.85	2.14	0.417	7.0	0.585	0.617	0.191	1.20	2.44	4.22	0.671
4.36	1.85	2.08	0.406	7.1	0.663	0.589	0.148	1.30	2.77	3.74	0.608
4.49	1.86	2.03	0.395	7.2	0.717	0.584	0.128	1.40	3.15	3.40	0.562
4.61	1.85	1.99	0.388	7.3	0.769	0.575	0.111	1.50	3.53	3.30	0.550
4.74	1.84	1.94	0.378	7.4	0.817	0.574	0.100	1.60	3.77	3.41	0.562
4.86	1.83	1.91	0.372	7.5	0.860	0.580	0.094	1.70	3.84	3.51	0.570
4.98	1.82	1.86	0.362	7.6	0.893	0.597	0.093	1.80	3.81	3.58	0.576
5.11	1.82	1.82	0.354	7.8	0.929	0.623	0.096	1.90	3.74	3.58	0.576
5.23	1.81	1.79	0.348	8.0	0.946	0.639	0.098	2.00	3.68	3.52	0.571
5.36	1.78	1.76	0.342	8.2	0.952	0.645	0.099	2.10	3.68	3.45	0.565
5.48	1.74	1.73	0.337	8.4	0.953	0.638	0.097	2.20	3.76	3.41	0.562
5.60	1.73	1.70	0.331	8.6	0.956	0.624	0.093	2.30	3.79	3.61	0.578
5.73	1.72	1.67	0.325	8.8	0.965	0.607	0.087	2.40	3.59	3.78	0.594
5.85	1.70	1.64	0.319	9.0	0.975	0.588	0.082	2.50	3.36	3.73	0.591
5.98	1.67	1.61	0.313	9.2	0.988	0.568	0.076	2.60	3.22	3.61	0.582
6.10	1.63	1.58	0.307	9.4	1.009	0.548	0.069	2.70	3.13	3.51	0.573
6.22	1.62	1.55	0.301	9.6	1.044	0.541	0.066	2.80	3.08	3.42	0.565
6.35	1.59	1.52	0.295	9.8	1.061	0.557	0.069	2.90	3.05	3.33	0.566
6.47	1.55	1.50	0.292	10.0	1.062	0.567	0.071	3.00	3.04	3.27	0.550
6.60	1.48	1.47	0.288	10.2	1.054	0.569	0.072	3.10	3.03	3.21	0.544
				10.4	1.045	0.561	0.070	3.20	3.05	3.18	0.540
Mercury (liquid)[15]				10.6	1.041	0.550	0.068	3.30	3.06	3.18	0.540
0.2	13.99	14.27	0.869	10.8	1.039	0.537	0.065	3.40	3.06	3.19	0.541
0.3	11.37	11.95	0.846	11.0	1.039	0.523	0.062	3.50	3.06	3.21	0.543
0.4	9.741	10.65	0.830	11.5	1.050	0.491	0.055	3.60	3.05	3.23	0.546
0.5	8.528	9.805	0.818	12.0	1.064	0.467	0.050	3.70	3.04	3.27	0.550
0.6	7.574	9.195	0.808	12.5	1.078	0.445	0.045	3.80	3.04	3.31	0.554
0.8	6.086	8.312	0.796	13.0	1.092	0.430	0.042	3.90	3.04	3.40	0.564
1.0	4.962	7.643	0.789	13.5	1.104	0.416	0.040	4.00	3.01	3.51	0.576
1.2	4.050	7.082	0.786	14.0	1.115	0.404	0.038	4.20	2.77	3.77	0.610
1.4	3.324	6.558	0.785	14.5	1.125	0.394	0.037	4.40	2.39	3.88	0.640
1.6	2.746	6.054	0.783	15.0	1.135	0.383	0.035	4.60	2.06	3.84	0.658
1.8	2.284	5.582	0.782	15.5	1.146	0.374	0.034	4.80	1.75	3.76	0.678
2.0	1.910	5.150	0.782	16.0	1.159	0.368	0.034	5.00	1.46	3.62	0.695
2.2	1.620	4.751	0.780	16.5				5.20	1.22	3.42	0.706
					1.170	0.367	0.034				

Energy (eV)	n	k	$R(\phi = 0)$	Energy (eV)	n	k	$R(\phi = 0)$	Energy (eV)	n	k	$R(\phi = 0)$
5.40	1.07	3.20	0.706	27.00	0.73	0.29	0.050	4.80	1.53	2.11	0.435
5.60	0.96	2.99	0.700	27.50	0.76	0.28	0.041	5.00	1.40	2.10	0.449
5.80	0.89	2.80	0.688	28.00	0.79	0.27	0.036	5.20			
6.00	0.85	2.64	0.674	28.50	0.81	0.26	0.031		1.27	2.04	0.454
6.20	0.81	2.50	0.660	29.00	0.83	0.26	0.028	5.40	1.16	1.94	0.449
6.40	0.79	2.36	0.641	29.50	0.86	0.26	0.025	5.60	1.09	1.83	0.435
6.60	0.78	2.24	0.619	30.00	0.88	0.26	0.023	5.80	1.04	1.73	0.417
6.80	0.78	2.13	0.592	31.00	0.92	0.29	0.024	6.20	1.00	1.54	0.371
7.00	0.80	2.04	0.568	32.00	0.92	0.32	0.030	6.40	1.01	1.46	0.345
7.20	0.81	1.98	0.548	33.00	0.90	0.33	0.032	6.60	1.01	1.40	0.325
7.40	0.81	1.95	0.542	34.00	0.91	0.34	0.034	6.80	1.02	1.35	0.308
7.60	0.75	1.90	0.552	35.00	0.87	0.37	0.043	7.00	1.03	1.30	0.291
7.80	0.71	1.81	0.542	36.00	0.82	0.34	0.043	7.20	1.03	1.27	0.282
8.00	0.69	1.73	0.530	37.00	0.81	0.30	0.038	7.40	1.03	1.24	0.273
8.20	0.67	1.65	0.512	38.00	0.81	0.27	0.033	7.60	1.02	1.22	0.265
8.40	0.66	1.57	0.495	39.00	0.82	0.25	0.029	7.80	1.01	1.18	0.256
8.60	0.65	1.49	0.475	40.00	0.83	0.23	0.025	8.00	1.01	1.15	0.248
8.80	0.65	1.41	0.450					8.20	1.00	1.13	0.242
9.00	0.65	1.33	0.420	**Nickel[17]**				8.40	0.99	1.11	0.235
9.20	0.67	1.25	0.385	0.10	9.54	45.82	0.983	8.60	0.98	1.08	0.228
9.40	0.69	1.19	0.355	0.15	5.45	30.56	0.978	8.80	0.97	1.05	0.220
9.60	0.71	1.12	0.320	0.20	4.12	22.48	0.969	9.00	0.97	1.01	0.211
9.80	0.74	1.05	0.285	0.25	4.25	17.68	0.950	9.20	0.96	0.99	0.203
10.00	0.77	0.99	0.250	0.30	4.19	15.05	0.934	9.40	0.95	0.96	0.194
10.20	0.81	0.93	0.217	0.35	4.03	13.05	0.918	9.60	0.95	0.93	0.185
10.40	0.86	0.88	0.188	0.40	3.84	11.43	0.900	9.80	0.95	0.89	0.175
10.60	0.91	0.83	0.162	0.50	4.03	9.64	0.864	10.00	0.95	0.87	0.166
10.80	0.98	0.79	0.138	0.60	3.84	8.35	0.835	10.20	0.95	0.83	0.155
11.00	1.05	0.77	0.125	0.70	3.59	7.48	0.813	10.40	0.95	0.80	0.145
11.20	1.12	0.78	0.123	0.80	3.38	6.82	0.794	10.60	0.97	0.76	0.129
11.40	1.18	0.80	0.125	0.90	3.18	6.23	0.774	10.80	0.99	0.75	0.123
11.60	1.23	0.85	0.135	1.00	3.06	5.74	0.753	11.00	1.01	0.73	0.115
11.80	1.25	0.89	0.145	1.10	2.97	5.38	0.734	11.25	1.04	0.72	0.111
12.00				1.20	2.85	5.10	0.721	11.50	1.05	0.71	0.109
	1.26	0.92	0.154	1.30	2.74	4.85	0.708	11.75	1.07	0.71	0.108
12.40	1.25	0.98	0.168	1.40	2.65	4.63	0.695	12.00	1.07	0.71	0.108
12.80	1.23	1.00	0.178	1.50	2.53	4.47	0.688	12.25	1.07	0.71	0.107
13.20	1.20	1.02	0.185	1.60	2.43	4.31	0.679	12.50	1.08	0.71	0.106
13.60	1.17	1.02	0.187	1.70	2.28	4.18	0.677	12.75	1.08	0.71	0.106
14.00	1.15	1.01	0.185	1.80	2.14	4.01	0.670	13.00	1.08	0.71	0.105
14.40	1.13	1.00	0.182	1.90	2.02	3.82	0.659	13.25	1.08	0.71	0.105
14.80	1.13	0.99	0.179	2.00	1.92	3.65	0.649	13.50	1.07	0.70	0.105
15.00	1.14	0.99	0.179	2.10	1.85	3.48	0.634	13.75	1.07	0.70	0.105
15.60	1.15	1.01	0.184	2.20	1.80	3.33	0.620	14.00	1.07	0.71	0.106
16.00	1.14	1.04	0.194	2.30	1.75	3.19	0.605	14.25	1.06	0.70	0.106
16.60	1.10	1.10	0.216	2.40	1.71	3.06	0.590	14.50	1.05	0.70	0.106
17.00	1.04	1.12	0.233	2.50	1.67	2.93	0.575	14.75	1.04	0.70	0.107
17.60	0.94	1.14	0.257	2.60	1.65	2.81	0.557	15.00	1.03	0.70	0.107
18.00	0.87	1.12	0.270	2.70	1.64	2.71	0.542	15.25	1.02	0.69	0.106
18.60	0.77	1.08	0.283	2.80	1.63	2.61	0.525	15.50	1.01	0.69	0.105
19.00	0.71	1.02	0.284	2.90	1.62	2.52	0.509	15.75	1.00	0.68	0.104
19.60	0.66	0.94	0.275	3.00	1.61	2.44	0.495	16.00	0.99	0.67	0.103
20.00	0.64	0.89	0.264	3.10	1.61	2.36	0.480	16.50	0.98	0.66	0.101
20.60	0.62	0.81	0.245	3.20	1.61	2.30	0.467	17.00	0.96	0.64	0.098
21.00	0.61	0.77	0.234	3.30	1.61	2.23	0.454	17.50	0.94	0.63	0.096
21.60	0.61	0.71	0.215	3.40	1.62	2.17	0.441	18.00	0.92	0.61	0.092
22.00	0.60	0.69	0.207	3.50	1.63	2.11	0.428	18.50	0.91	0.58	0.087
22.60	0.59	0.63	0.195	3.60	1.64	2.07	0.416	19.00	0.90	0.56	0.082
23.00	0.58	0.60	0.185	3.70	1.66	2.02	0.405	19.50	0.90	0.54	0.077
23.60	0.58	0.53	0.166	3.80	1.69	1.99	0.397	20.00	0.89	0.51	0.071
24.00	0.58	0.49	0.151	3.90	1.72	1.98	0.393	20.50	0.89	0.49	0.066
24.60	0.60	0.43	0.124	4.00	1.73	1.98	0.392	21.00	0.90	0.47	0.061
25.00	0.62	0.39	0.106	4.20	1.74	2.01	0.396	21.50	0.91	0.46	0.057
25.60	0.66	0.35	0.085	4.40	1.71	2.06	0.409	22.00	0.91	0.45	0.055
26.00	0.68	0.33	0.072	4.60	1.63	2.09	0.421	22.50	0.91	0.44	0.053
26.50	0.71	0.31	0.060					23.00	0.92	0.44	0.051

Energy (eV)	n	k	R($\phi = 0$)	Energy (eV)	n	k	R($\phi = 0$)	Energy (eV)	n	k	R($\phi = 0$)
23.50	0.91	0.44	0.052	4.60	2.39	2.56	0.470	23.60	0.88	0.30	0.029
24.00	0.90	0.43	0.051	4.80	2.32	2.52	0.465	24.00	0.91	0.29	0.025
24.50	0.90	0.43	0.051	5.00	2.26	2.57	0.475	24.60	0.94	0.28	0.022
25.00	0.89	0.42	0.050	5.20	2.16	2.62	0.487	25.00			
26.00	0.88	0.39	0.046	5.40	2.00	2.68	0.505		0.96		
27.00	0.87	0.37	0.042	5.60	1.81	2.67	0.518			0.27	
28.00	0.87	0.35	0.040	5.80	1.63	2.60	0.522				0.020
29.00	0.86	0.34	0.037	6.00	1.49	2.49	0.520	25.60	0.99	0.26	0.018
30.00	0.86	0.32	0.034	6.20	1.38	2.38	0.512	26.00	1.00	0.26	0.017
35.00	0.86	0.24	0.022	6.40	1.31	2.25	0.496	26.60	1.03	0.25	0.016
40.00	0.87	0.18	0.014	6.60	1.26	2.14	0.480	27.00	1.04	0.25	0.015
45.00	0.88	0.13	0.008	6.80	1.24	2.04	0.460	27.60	1.06	0.25	0.015
50.00	0.92	0.10	0.004	7.00	1.23	1.96	0.441	28.00	1.08	0.24	0.015
60.00	0.96	0.08	0.002	7.20	1.22	1.91	0.430	28.60	1.11	0.24	0.016
65.00	0.98	0.09	0.002	7.40	1.20	1.88	0.427	29.00	1.13	0.25	0.017
68.00	0.96	0.12	0.004	7.60	1.14	1.85	0.430	29.60	1.16	0.26	0.020
70.00	0.94	0.11	0.004	7.80	1.07	1.78	0.428	30.00	1.18	0.28	0.023
75.00	0.94	0.09	0.003	8.00	1.02	1.69	0.412	31.00	1.18	0.31	0.026
80.00	0.94	0.07	0.002	8.20	1.00	1.60	0.390	32.00	1.20	0.34	0.031
90.00	0.94	0.06	0.002	8.40	0.99	1.51	0.365	33.00	1.21	0.38	0.038
				8.60	0.99	1.43	0.340	34.00	1.20	0.42	0.044
Niobium[18]				8.70	0.99	1.39	0.328	35.20	1.17	0.47	0.051
0.12	15.99	53.20	0.979	8.80	1.00	1.36	0.315	36.00	1.15	0.50	0.056
0.20	7.25	34.14	0.976	9.00	1.01	1.29	0.290	37.50	1.07	0.53	0.064
0.24	5.47	28.88	0.975	9.20	1.04	1.22	0.265	39.50	0.95	0.50	0.063
0.28	4.26	24.95	0.974	9.40	1.07	1.18	0.245	40.50	0.92	0.47	0.059
0.35	3.11	20.03	0.970	9.60	1.10	1.13	0.227				
0.45	2.28	15.58	0.964	9.80	1.13	1.09	0.209	**Osmium (Polycrystalline)**[9]			
0.55	1.83	12.67	0.956	10.00	1.18	1.05	0.194	0.10	4.08	50.23	0.994
0.65	1.57	10.59	0.947	10.20	1.23	1.04	0.187	0.15	2.90	33.60	0.990
0.75	1.41	9.00	0.935	10.40	1.27	1.04	0.185	0.20	2.44	25.11	0.985
0.85	1.35	7.74	0.918	10.60	1.30	1.06	0.190	0.25	2.35	19.99	0.977
0.95	1.35	6.70	0.893	10.80	1.32	1.08	0.195	0.30	2.23	16.54	0.969
1.05	1.44	5.86	0.857	11.00	1.32	1.10	0.200	0.35	2.33	14.06	0.955
1.15	1.55	5.18	0.814	11.20	1.31	1.12	0.204	0.40	2.45	12.32	0.940
1.25	1.65	4.63	0.768	11.40	1.30	1.13	0.207	0.45	2.43	11.02	0.927
1.35				11.60	1.28	1.13	0.209	0.50	2.41	9.97	0.913
	1.76			11.80	1.27	1.13	0.210	0.55	2.33	9.12	0.901
		4.13	0.715	12.00	1.25	1.12	0.209	0.60	2.21	8.37	0.890
1.45	1.95	3.68	0.650	12.40	1.24	1.10	0.204	0.65	2.11	7.68	0.877
1.55	2.15	3.37	0.595	12.80	1.24	1.09	0.200	0.70	2.02	7.04	0.862
1.65	2.36	3.13	0.552	13.20	1.24	1.09	0.201	0.75	2.00	6.46	0.842
1.75	2.54	2.99	0.527	13.60	1.23	1.12	0.208	0.80	2.00	5.95	0.820
1.85	2.69	2.89	0.510	14.00	1.20	1.13	0.216	0.85	2.01	5.51	0.796
1.95	2.82	2.86	0.505	14.40	1.16	1.15	0.225	0.90	2.03	5.10	0.769
2.05	2.89	2.87	0.505	14.80	1.11	1.16	0.234	0.95	2.05	4.74	0.742
2.15	2.92	2.87	0.505	15.00	1.08	1.16	0.238	1.00	2.09	4.41	0.712
2.25	2.93	2.87	0.505	15.60	0.99	1.14	0.247	1.10	2.15	3.84	0.651
2.35	2.92	2.88	0.506	16.00	0.92	1.11	0.250	1.20	2.16	3.35	0.592
2.45	2.89	2.90	0.509	16.60	0.85	1.04	0.245	1.30	2.25	2.77	0.506
2.55	2.83	2.92	0.512	17.00	0.80	0.99	0.240	1.40	2.49	2.23	0.419
2.65	2.74	2.90	0.511	17.20	0.79	0.96	0.236	1.50	2.84	1.80	0.369
2.75	2.66	2.86	0.507	17.40	0.77	0.93	0.230	1.60	3.36	1.62	0.379
2.85	2.58	2.80	0.500	17.80	0.75	0.87	0.217	1.70	3.70	1.75	0.411
3.00	2.51	2.68	0.485	18.00	0.74	0.85	0.209	1.80	3.78	1.83	0.423
3.10	2.48	2.60	0.475	18.60	0.73	0.77	0.185	1.90	3.81	1.75	0.418
3.20	2.45	2.53	0.465	19.00	0.72	0.72	0.170	2.00	3.98	1.60	0.418
3.30	2.44	2.45	0.453	19.60	0.72	0.66	0.150	2.10	4.26	1.54	0.432
3.40	2.46	2.38	0.442	20.00	0.72	0.62	0.137	2.20	4.58	1.62	0.457
3.50	2.48	2.33	0.435	20.60	0.71	0.55	0.119	2.30	4.84	1.76	0.479
3.60	2.52	2.29	0.428	21.00	0.72	0.50	0.100	2.40	5.10	2.01	0.506
3.70	2.56	2.27	0.426	21.60	0.75	0.43	0.075	2.50	5.28	2.38	0.532
3.80	2.59	2.28	0.427	22.00	0.78	0.40	0.063	2.60	5.36	2.82	0.557
3.90	2.62	2.29	0.429	22.60	0.82	0.35	0.045	2.70	5.30	3.29	0.580
4.00	2.64	2.33	0.434	23.00	0.85	0.33	0.038	2.80	5.07	3.78	0.603
4.20	2.64	2.42	0.447					2.90	4.65	4.18	0.624

Energy (eV)	n	k	R(φ = 0)	Energy (eV)	n	k	R(φ = 0)	Energy (eV)	n	k	R(φ = 0)
3.40	2.93	3.79	0.607	20.00	0.96	1.10	0.239	2.50	1.41	3.48	0.685
3.60	2.75	3.45	0.577	20.40	0.93	1.09	0.240	2.60	1.37	3.36	0.676
3.80	2.73	3.32	0.562	20.80	0.89	1.05	0.240	2.70	1.32	3.25	0.668
4.00	2.71	3.34	0.565	21.20	0.86	1.02	0.237	2.80	1.29	3.13	0.658
4.20	2.53	3.44	0.584	21.60	0.83	0.99	0.235	2.90	1.26	3.03	0.648
4.40	2.24	3.44	0.599	22.00	0.80	0.96	0.230	3.00	1.23	2.94	0.639
4.60	2.01	3.31	0.598	22.40	0.78	0.93	0.226	3.10	1.20	2.85	0.630
4.80	1.88	3.19	0.592	22.80	0.77	0.90	0.220	3.20	1.17	2.77	0.622
5.00	1.74	3.12	0.596	23.20	0.75	0.88	0.217	3.30	1.14	2.68	0.613
5.20	1.58	3.00	0.597	23.60	0.75	0.86	0.211	3.40	1.12	2.60	0.602
5.40	1.46	2.88	0.593	24.00	0.73	0.84	0.209	3.50	1.10	2.52	0.591
5.60	1.36	2.77	0.589	24.40	0.72	0.82	0.207	3.60	1.08	2.45	0.581
5.80	1.27	2.65	0.582	24.80	0.70	0.80	0.205	3.70	1.07	2.38	0.570
6.00	1.20	2.54	0.575	25.20	0.69	0.77	0.202	3.80	1.06	2.31	0.558
6.20	1.13	2.44	0.571	25.60	0.67	0.75	0.199	3.90	1.05	2.25	0.547
6.40	1.06	2.33	0.562	26.00	0.66	0.72	0.195	4.00	1.03	2.19	0.537
6.60	1.01	2.21	0.548	26.40	0.65	0.69	0.189	4.20	1.04	2.09	0.510
6.80	0.97	2.11	0.532	26.80	0.63	0.66	0.183	4.40	1.03	2.01	0.493
7.00	0.95	2.00	0.514	27.20	0.65	0.62	0.165	4.60	1.03	1.94	0.476
7.20	0.92	1.91	0.497	28.00	0.64	0.59	0.156	4.80	1.01	1.90	0.470
7.40	0.91	1.81	0.476	28.40	0.64	0.57	0.148	5.00	0.96	1.86	0.472
7.60	0.90	1.72	0.451	28.80	0.65	0.55	0.140	5.20	0.90	1.79	0.474
7.80	0.90	1.63	0.426	29.20	0.65	0.53	0.134	5.40	0.85	1.70	0.463
8.00	0.91	1.55	0.400	29.60	0.65	0.51	0.128	5.60	0.81	1.62	0.449
8.20	0.91	1.48	0.375	30.00	0.65	0.49	0.121	5.80	0.78	1.54	0.437
8.40	0.94	1.40	0.344	31.00	0.65	0.45	0.111	6.00	0.76	1.45	0.418
8.60	0.96	1.34	0.319	32.00	0.66	0.41	0.095	6.20	0.74	1.37	0.397
8.80	0.98	1.29	0.296	33.00	0.68	0.37	0.079	6.40	0.73	1.29	0.375
9.00	1.01	1.24	0.274	34.00	0.70	0.34	0.068	6.60	0.72	1.21	0.350
9.20	1.04	1.19	0.255	35.00	0.72	0.31	0.057	6.80	0.73	1.13	0.316
9.40	1.08	1.16	0.238	36.00	0.74	0.29	0.048	7.00	0.73	1.05	0.287
9.60	1.10	1.14	0.229	37.00	0.77	0.27	0.040	7.20	0.75	0.98	0.255
9.80	1.13	1.11	0.217	38.00	0.79	0.26	0.035	7.40	0.77	0.91	0.223
10.00	1.16	1.10	0.209	39.00	0.81	0.26	0.031	7.60	0.79	0.85	0.195
10.20	1.19	1.08	0.203	40.00	0.84	0.26	0.026	7.80	0.83	0.78	0.163
10.30	1.20	1.08	0.201					8.00	0.88	0.73	0.133
10.40	1.22	1.08	0.200	**Palladium**[19]				8.20	0.94	0.70	0.117
10.50	1.23	1.09	0.201	0.10	4.13	54.15	0.994	8.40	0.96	0.70	0.114
10.60	1.24	1.10	0.203	0.15	3.13	35.82	0.990	8.60	1.00	0.65	0.097
10.80	1.25	1.11	0.206	0.20	3.07	26.59	0.983	8.80	1.04	0.65	0.094
11.00	1.24	1.13	0.213	0.26	3.11	20.15	0.971	9.00	1.07	0.64	0.090
11.20	1.23	1.14	0.217	0.30	3.56	17.27	0.955	9.50	1.12	0.65	0.089
11.40	1.19	1.15	0.223	0.36	3.98	14.41	0.932	10.00	1.14	0.65	0.088
11.60	1.17	1.12	0.216	0.40	4.27	13.27	0.916	10.50	1.16	0.65	0.087
11.80	1.16	1.10	0.211	0.46	4.27	12.11	0.902	11.00	1.18	0.64	0.086
12.00	1.15	1.08	0.205	0.50	4.10	11.44	0.896	11.50	1.19	0.65	0.087
12.40	1.14	1.03	0.191	0.56	3.92	10.49	0.883	12.00	1.20	0.66	0.089
12.80	1.15	1.01	0.183	0.60	3.80	9.96	0.876	12.50	1.19	0.67	0.091
13.20	1.16	0.98	0.174	0.72	3.51	8.70	0.854	13.00	1.18	0.67	0.091
13.60	1.17	0.97	0.170	0.80	3.35	8.06	0.840	13.50	1.18	0.67	0.092
14.00	1.17	0.96	0.169	1.00	2.99	6.89	0.811	14.00	1.17	0.67	0.093
14.40	1.16	0.94	0.165	1.10	2.81	6.46	0.800	14.50	1.15	0.68	0.095
14.80	1.16	0.91	0.156	1.20	2.65	6.10	0.790	15.00	1.13	0.69	0.098
15.20	1.17	0.89	0.148	1.30	2.50	5.78	0.781	15.50	1.10	0.68	0.096
15.60	1.20	0.86	0.140	1.40	2.34	5.50	0.774	16.00	1.08	0.66	0.092
16.00	1.25	0.87	0.140	1.50	2.17	5.22	0.767	16.50	1.06	0.63	0.086
16.40	1.28	0.90	0.147	1.60	2.08	4.95	0.755	17.00	1.07	0.61	0.081
16.80	1.28	0.94	0.157	1.70	2.00	4.72	0.745	17.50	1.06	0.61	0.080
17.20	1.27	0.97	0.167	1.80	1.92	4.54	0.737	18.00	1.07	0.59	0.077
17.60	1.26	1.01	0.178	1.90	1.82	4.35	0.729	18.50	1.07	0.59	0.077
18.00	1.23	1.04	0.189	2.00	1.75	4.18	0.721	19.00	1.08	0.59	0.077
18.40	1.19	1.08	0.200	2.10	1.67	4.03	0.714	19.50	1.08	0.61	0.080
18.80	1.14	1.10	0.210	2.20	1.60	3.88	0.707	20.00	1.07	0.65	0.090
19.20	1.10	1.10	0.219	2.30	1.53	3.75	0.700	20.50	1.03	0.67	0.098
19.60	1.05	1.11	0.227	2.40	1.47	3.61	0.693	21.00	0.99	0.67	0.103

Energy (eV)	n	k	$R(\phi = 0)$	Energy (eV)	n	k	$R(\phi = 0)$	Energy (eV)	n	k	$R(\phi = 0)$
21.50	0.95	0.66	0.103	6.00	1.38	1.40	0.276	28.50	0.75	0.59	0.121
22.00	0.91	0.64	0.103	6.20	1.39	1.35	0.261	29.00	0.75	0.58	0.118
22.50	0.88	0.62	0.101	6.40	1.42	1.29	0.246	29.50	0.74	0.58	0.120
23.00	0.86	0.59	0.097	6.60	1.45	1.26	0.236	30.00	0.73	0.58	0.124
23.50	0.85	0.56	0.091	6.80	1.48	1.24	0.231				
24.00	0.84	0.54	0.086	7.00	1.50	1.24	0.230	**Potassium[21]**			
25.00	0.81	0.51	0.084	7.20	1.50	1.25	0.231	0.55	0.139	7.10	0.989
26.40	0.80	0.43	0.066	7.40	1.49	1.23	0.228	0.58	0.119	6.72	0.990
27.80	0.81	0.38	0.052	7.60	1.48	1.22	0.225	0.63	0.106	6.32	0.990
29.20	0.82	0.35	0.046	7.80	1.48	1.20	0.221	0.67	0.091	5.79	0.990
				8.00	1.47	1.18	0.216	0.73			
Platinum[20]				8.20	1.47	1.17	0.212		0.079	5.30	0.989
0.10	13.21	44.72	0.976	8.40	1.47	1.15	0.209	0.81	0.066	4.75	0.989
0.15	8.18	31.16	0.969	8.60	1.47	1.14	0.205	0.92	0.056	4.19	0.988
0.20	5.90	23.95	0.962	8.80	1.47	1.13	0.202	1.05	0.044	3.58	0.987
0.25	4.70	19.40	0.954	9.00	1.48	1.12	0.200	1.23	0.040	3.04	0.985
0.30	3.92	16.16	0.945	9.20	1.49	1.11	0.198	1.44	0.040	2.56	0.979
0.35	3.28	13.66	0.936	9.40	1.49	1.12	0.200	1.65	0.044	2.19	0.970
0.40	2.81	11.38	0.922	9.60	1.49	1.13	0.203	1.87	0.050	1.84	0.955
0.45	3.03	9.31	0.882	9.80	1.48	1.15	0.207	2.07	0.053	1.62	0.943
0.50	3.91	7.71	0.813	10.00	1.46	1.15	0.209	2.27	0.049	1.43	0.938
0.55	4.58	7.14	0.777	10.20	1.43	1.16	0.211	2.45	0.046	1.28	0.933
0.60	5.13	6.75	0.753	10.40	1.40	1.15	0.210	2.64	0.043	1.14	0.928
0.65	5.52	6.66	0.746	10.60	1.37	1.14	0.207	2.82	0.043	1.02	0.919
0.70	5.71	6.83	0.751	10.80	1.35	1.12	0.203	2.95	0.041	0.898	0.913
0.75	5.57	7.02	0.759	11.00	1.33	1.10	0.199	3.06	0.041	0.799	0.905
0.80	5.31	7.04	0.762	11.20	1.31	1.08	0.194	3.40	0.052	0.549	0.852
0.85	5.05	6.98	0.763	11.40	1.30	1.06	0.188	3.71	0.089	0.288	0.719
0.90	4.77	6.91	0.765	11.60	1.29	1.04	0.183	3.97	0.287	0.091	0.310
0.95	4.50	6.77	0.763	11.80	1.29	1.01	0.177	4.00	0.34	0.08	0.245
1.00	4.25	6.62	0.762	12.00	1.29	1.00	0.173	4.065	0.38	0.07	0.204
1.10	3.86	6.24	0.753	12.40	1.29	0.97	0.165	4.133	0.41	0.07	0.177
1.20	3.55	5.92	0.746	12.80	1.29	0.94	0.158	4.203	0.45	0.06	0.145
1.30	3.29	5.61	0.736	13.20	1.31	0.93	0.155	4.275	0.48	0.06	0.125
1.40	3.10	5.32	0.725	13.60	1.31	0.93	0.155	4.350	0.52	0.05	0.101
1.50	2.92	5.07	0.716	14.00	1.31	0.93	0.155	4.428	0.55	0.05	0.085
1.60	2.76	4.84	0.706	14.40	1.30	0.93	0.156	4.509	0.58	0.05	0.072
1.70	2.63	4.64	0.697	14.80	1.27	0.93	0.157	4.592	0.61	0.05	0.060
1.80	2.51	4.43	0.686	15.20	1.27	0.93	0.155	4.679	0.64	0.04	0.049
1.90	2.38	4.26	0.678	15.60	1.25	0.92	0.151	4.769	0.66	0.04	0.043
2.00	2.30	4.07	0.664	16.00	1.24	0.89	0.146	4.862	0.68	0.04	0.037
2.10				16.50	1.24	0.87	0.142	4.959	0.70	0.04	0.032
	2.23	3.92	0.654	17.00	1.25	0.86	0.138	5.061	0.72	0.04	0.027
2.20	2.17	3.77	0.642	17.50	1.27	0.85	0.135	5.166	0.74	0.04	0.023
2.30	2.10	3.67	0.636	18.00	1.31	0.88	0.142	5.276	0.76	0.04	0.019
2.40	2.03	3.54	0.626	18.50	1.30	0.94	0.157	5.391	0.78	0.04	0.016
2.50	1.96	3.42	0.616	19.00	1.28	0.99	0.171	5.510	0.79	0.05	0.015
2.60	1.91	3.30	0.605	19.50	1.23	1.03	0.184	5.637	0.81	0.05	0.012
2.70	1.87	3.20	0.595	20.00	1.18	1.06	0.197	5.767	0.83	0.05	0.009
2.80	1.83	3.10	0.585	20.50	1.11	1.09	0.212	6.048	0.85	0.05	0.007
2.90	1.79	3.01	0.575	21.00	1.03	1.10	0.226	6.199	0.87	0.05	0.006
3.00	1.75	2.92	0.565	21.50	0.94	1.08	0.238	6.358	0.88	0.05	0.005
3.20	1.68	2.76	0.546	22.00	0.87	1.04	0.240	6.526	0.90	0.06	0.004
3.40	1.63	2.62	0.527	22.50	0.81	0.98	0.235	6.702	0.91	0.06	0.003
3.60	1.58	2.48	0.507	23.00	0.77	0.92	0.226	6.888	0.92	0.06	0.003
3.80	1.53	2.37	0.491	23.50	0.75	0.87	0.213	7.085	0.92	0.06	0.003
4.00	1.49	2.25	0.472	24.00	0.74	0.82	0.201	7.293	0.93	0.06	0.002
4.20	1.45	2.14	0.452	24.50	0.73	0.77	0.187	7.514	0.93	0.06	0.002
4.40	1.43	2.04	0.432	25.00	0.73	0.73	0.174	7.749	0.94	0.06	0.002
4.60	1.39	1.95	0.415	25.50	0.73	0.70	0.162	7.999	0.94	0.06	0.002
4.80	1.38	1.85	0.392	26.00	0.74	0.67	0.150	8.260	0.94	0.06	0.002
5.00	1.36	1.76	0.372	26.50	0.74	0.65	0.142	8.551	0.94	0.06	0.002
5.20	1.36	1.67	0.350	27.00	0.74	0.63	0.136	8.856	0.94	0.05	0.002
5.40	1.36	1.61	0.332	27.50	0.74	0.62	0.130	9.184	0.94	0.05	0.002
5.60	1.36	1.54	0.315	28.00	0.75	0.60	0.125	9.537	0.94	0.04	0.001
5.80	1.36	1.47	0.295					9.919	0.94	0.04	0.001

Energy (eV)	n	k	$R(\phi = 0)$	Energy (eV)	n	k	$R(\phi = 0)$	Energy (eV)	n	k	$R(\phi = 0)$
10.33	0.94	0.03	0.001	7.60	1.12	1.99	0.470	31.00	0.62	0.29	0.086
11.0		0.03		7.80	1.08	1.89	0.454	32.00	0.66	0.26	0.065
12.0		0.028		8.00	1.05	1.80	0.435	33.00	0.68	0.24	0.054
				8.20	1.05	1.71	0.411	34.00	0.72	0.21	0.041
Rhenium, single crystal, $\perp$ **E** $\parallel \hat{c}$ [9]				8.40	1.05	1.62	0.386	35.00	0.76	0.20	0.031
0.10	6.06	51.03	0.991	8.60	1.06	1.55	0.360	36.00	0.79	0.20	0.025
0.15	4.66	33.96	0.984	8.80	1.09	1.48	0.336	37.00	0.82	0.19	0.021
0.20	4.16	25.36	0.975	9.00	1.11	1.43	0.317	38.00	0.85	0.20	0.018
0.25	4.03	20.10	0.962	9.20	1.13	1.39	0.301	39.00	0.89	0.21	0.016
0.30	4.37	16.69	0.943	9.40	1.16	1.34	0.281	40.00	0.88	0.26	0.022
0.35	4.50	14.53	0.925	9.60	1.18	1.32	0.274	42.00	0.88	0.26	0.022
0.40	4.53	12.96	0.909	9.80	1.20	1.29	0.264	44.00			
0.45	4.53	11.78	0.893	10.00	1.23	1.26	0.252				
0.50	4.53	10.88	0.878	10.20	1.25	1.25	0.246		0.89		
0.55	4.50	10.26	0.867	10.40	1.28	1.25	0.242			0.29	
0.60	4.29	9.75	0.861	10.60	1.29	1.25	0.242				0.026
0.65	4.07	9.35	0.856	10.80	1.30	1.26	0.244	46.00	0.85	0.32	0.035
0.70	3.80	8.94	0.853	11.00	1.30	1.27	0.247	48.00	0.82	0.30	0.036
0.75	3.48	8.55	0.850	11.20	1.29	1.28	0.249	50.00	0.80	0.30	0.038
0.80	3.21	8.10	0.846	11.40	1.28	1.28	0.252	52.00	0.78	0.30	0.044
0.85	2.96	7.68	0.841	11.60	1.26	1.28	0.252	54.00	0.72	0.30	0.055
0.90	2.73	7.24	0.835	11.80	1.24	1.26	0.249	56.00	0.66	0.24	0.061
0.95	2.56	6.79	0.826	12.00	1.23	1.24	0.244	58.00	0.65	0.16	0.055
1.00	2.45	6.36	0.813	12.40	1.22	1.21	0.237				
1.10	2.38	5.61	0.778	12.80	1.21	1.18	0.230	**Rhenium, single crystal,** $\perp$ **E** $\perp \hat{c}$ [9]			
1.20	2.35	5.02	0.742	13.20	1.22	1.16	0.222	0.10	4.25	42.83	0.991
1.30	2.39	4.54	0.702	13.60	1.22	1.13	0.215	0.15	3.28	28.08	0.984
1.40	2.44	4.13	0.662	14.00	1.24	1.12	0.209	0.20	3.28	20.66	0.971
1.50	2.50	3.79	0.624	14.40	1.27	1.11	0.204	0.25	3.47	16.27	0.951
1.60	2.59	3.49	0.587	14.80	1.29	1.15	0.213	0.30	3.73	13.44	0.926
1.70	2.70	3.27	0.557	15.20	1.29	1.19	0.225	0.35	3.93	11.54	0.900
1.80	2.82	3.10	0.535	15.60	1.26	1.22	0.236	0.40	3.99	10.15	0.875
1.90	2.90	3.00	0.520	16.00	1.23	1.25	0.248	0.45	4.17	9.03	0.846
2.00	2.97	2.91	0.510	16.40	1.19	1.27	0.259	0.50	4.34	8.26	0.821
2.10	3.03	2.86	0.504	16.80	1.14	1.29	0.269	0.55	4.45	7.73	0.801
2.20	3.06	2.84	0.501	17.00	1.12	1.30	0.275	0.60	4.53	7.40	0.788
2.30	3.07	2.82	0.499	17.40	1.07	1.30	0.286	0.65	4.44	7.26	0.784
2.40	3.06	2.81	0.498	18.00	0.99	1.30	0.300	0.70	4.13	7.09	0.784
2.50	3.02	2.80	0.497	18.40	0.93	1.29	0.311	0.75	3.77	6.75	0.779
2.60	2.96	2.77	0.493	18.80	0.87	1.28	0.321	0.80	3.55	6.32	0.766
2.70	2.89	2.68	0.482	19.20	0.81	1.25	0.330	0.85	3.39	5.95	0.752
2.80	2.89	2.57	0.468	19.60	0.77	1.21	0.332	0.90	3.26	5.61	0.737
2.90	2.99	2.47	0.457	20.00	0.73	1.18	0.333	0.95	3.17	5.27	0.719
3.00	3.11	2.57	0.470	20.40	0.70	1.14	0.332	1.00	3.09	4.96	0.701
3.20				20.80	0.67	1.11	0.332	1.10	3.05	4.39	0.658
	2.90			21.20	0.64	1.08	0.334	1.20	3.08	3.89	0.613
		2.68	0.482	21.60	0.61	1.04	0.335	1.30	3.20	3.56	0.578
3.40	2.83	2.50	0.459	22.00	0.58	1.01	0.340	1.40	3.23	3.38	0.559
3.60	2.93	2.48	0.457	22.40	0.55	0.97	0.341	1.50	3.23	3.12	0.532
3.80	2.86	2.56	0.467	22.80	0.53	0.93	0.338	1.60	3.29	2.88	0.507
4.00	2.81	2.51	0.460	23.20	0.51	0.89	0.334	1.70	3.38	2.72	0.491
4.20	2.86	2.55	0.466	23.60	0.50	0.85	0.329	1.80	3.47	2.59	0.480
4.40	2.81	2.74	0.489	24.00	0.48	0.80	0.319	1.90	3.54	2.50	0.473
4.60	2.56	2.83	0.504	24.40	0.48	0.76	0.207	2.00	3.63	2.43	0.469
4.80	2.41	2.71	0.493	24.80	0.47	0.72	0.296	2.10	3.74	2.40	0.470
5.00	2.39	2.68	0.488	25.20	0.47	0.68	0.282	2.20	3.83	2.38	0.472
5.20	2.34	2.75	0.500	25.60	0.47	0.65	0.270	2.30	3.93	2.44	0.481
5.40	2.20	2.81	0.515	26.00	0.47	0.61	0.255	2.40	4.00	2.55	0.492
5.60	2.02	2.84	0.530	26.40	0.48	0.57	0.240	2.50	4.01	2.70	0.505
5.80	1.83	2.80	0.538	26.80	0.48	0.54	0.225	2.60	3.90	2.84	0.514
6.00	1.65	2.71	0.541	27.20	0.49	0.51	0.208	2.70	3.74	2.92	0.517
6.20	1.54	2.59	0.532	27.60	0.50	0.48	0.193	2.80	3.57	2.88	0.511
6.40	1.45	2.50	0.526	28.00	0.51	0.45	0.176	2.90	3.49	2.75	0.497
6.80	1.32	2.31	0.508	29.00	0.54	0.39	0.145	3.00	3.53	2.71	0.493
7.00	1.26	2.23	0.500	30.00	0.57	0.33	0.114	3.20	3.55	2.84	0.506
7.20	1.20	2.15	0.493					3.40	3.34	2.88	0.508
								3.60	3.25	2.83	0.501

Energy (eV)	n	k	$R(\phi = 0)$	Energy (eV)	n	k	$R(\phi = 0)$	Energy (eV)	n	k	$R(\phi = 0)$
4.20	3.05	3.06	0.526	22.80	0.55	0.92	0.325	3.00	1.53	4.29	0.753
4.40	2.88	3.15	0.539	23.20	0.53	0.89	0.322	3.10	1.41	4.20	0.760
4.60	2.67	3.18	0.548	23.60	0.52	0.85	0.317	3.20	1.30	4.09	0.764
4.80	2.44	3.17	0.554	24.00	0.50	0.82	0.314	3.30	1.20	3.97	0.767
5.00	2.25	3.12	0.556	24.40	0.49	0.79	0.309	3.40	1.11	3.84	0.769
5.20	2.10	3.04	0.555	24.80	0.48	0.75	0.303	3.50	1.04	3.71	0.768
5.40	1.96	2.96	0.553	25.20	0.47	0.72	0.295	3.60	0.99	3.58	0.764
5.60	1.84	2.88	0.551	25.60	0.47	0.68	0.286	3.70	0.95	3.45	0.759
5.80	1.73	2.81	0.549	26.00	0.46	0.64	0.276	3.80	0.91	3.34	0.753
6.00	1.61	2.74	0.549	26.40	0.46	0.61	0.263	3.90	0.88	3.23	0.747
6.20	1.51	2.64	0.545	26.80	0.46	0.57	0.249	4.00	0.86	3.12	0.739
6.40	1.42	2.56	0.541	27.20	0.47	0.53	0.231	4.20	0.83	2.94	0.722
6.80	1.28	2.37	0.526	27.60	0.48	0.50	0.216	4.40	0.80	2.76	0.706
7.00	1.22	2.28	0.517	28.00	0.49	0.47	0.198	4.60	0.78	2.60	0.684
7.20	1.16	2.19	0.508	29.00	0.51	0.41	0.164	4.80	0.79	2.46	0.659
7.40	1.12	2.08	0.493	30.00	0.55	0.34	0.129	5.00	0.79	2.34	0.635
7.60	1.12	1.98	0.468	31.00	0.59	0.29	0.097	5.20	0.79	2.23	0.613
7.80	1.08	1.93	0.463	32.00	0.64	0.26	0.072	5.40	0.80	2.14	0.591
8.00	1.05	1.83	0.443	33.00	0.67	0.24	0.060	5.60	0.80	2.06	0.573
8.20	1.05	1.74	0.418	34.00	0.70	0.22	0.047	5.80	0.79	2.00	0.561
8.40	1.05	1.66	0.397	35.00	0.74	0.20	0.036	6.00	0.76	1.93	0.556
8.60	1.06	1.58	0.372	36.00	0.77	0.19	0.029	6.20	0.73	1.85	0.544
8.80	1.07	1.52	0.351	37.00	0.80	0.19	0.023	6.40	0.70	1.77	0.534
9.00	1.09	1.46	0.327	38.00	0.84	0.19	0.018	6.60	0.68	1.69	0.518
9.20	1.11	1.41	0.309	39.00	0.88	0.21	0.016	6.80	0.67	1.60	0.498
9.40	1.14	1.36	0.290	40.00	0.87	0.25	0.023	7.00	0.66	1.52	0.476
9.60	1.17	1.31	0.273	42.00	0.87	0.25	0.023	7.20	0.66	1.43	0.452
9.80	1.20	1.27	0.258	44.00	0.88	0.28	0.026	7.40	0.66	1.35	0.423
10.00	1.24	1.24	0.244	46.00	0.84	0.31	0.035	7.60	0.67	1.27	0.394
10.20	1.29	1.22	0.234	48.00	0.82	0.30	0.036	7.80	0.68	1.20	0.363
10.40	1.33	1.23	0.233	50.00	0.80	0.30	0.039	8.00	0.69	1.12	0.329
10.60	1.36	1.25	0.238	52.00	0.77	0.30	0.044	8.20	0.71	1.04	0.288
10.80	1.38	1.28	0.245	54.00	0.71	0.29	0.055	8.40	0.74	0.97	0.252
11.00	1.37	1.31	0.253	56.00	0.66	0.23	0.061	8.60	0.78	0.89	0.212
11.20	1.36	1.33	0.259	58.00	0.64	0.16	0.055	8.80	0.83	0.83	0.179
11.40	1.33	1.34	0.264					9.00	0.88	0.77	0.148
11.60	1.31	1.34	0.266	**Rhodium**[11]				9.20	0.95	0.73	0.125
11.80	1.28	1.33	0.266	0.10	18.48	69.43	0.986	9.40	1.01	0.71	0.110
12.00	1.26	1.32	0.264	0.20	8.66	37.46	0.977	9.60	1.07	0.69	0.102
12.40	1.23	1.29	0.257	0.30	5.85	25.94	0.967	9.80	1.12	0.69	0.098
12.80	1.22	1.26	0.251	0.40	4.74	19.80	0.955	10.00	1.17	0.69	0.098
13.20	1.20	1.23	0.245	0.50	4.20	16.07	0.941	10.60	1.26	0.73	0.106
13.60	1.19	1.20	0.236	0.60	3.87	13.51	0.925	11.00	1.29	0.76	0.113
14.00	1.20	1.16	0.225	0.70	3.67	11.72	0.908	11.60	1.32	0.80	0.124
14.40	1.22	1.13	0.214	0.80	3.63	10.34	0.887	12.00	1.32	0.82	0.127
14.80	1.27	1.12	0.207	0.90	3.62	9.36	0.867	12.60	1.32	0.82	0.129
15.20	1.31	1.17	0.218	1.00	3.71	8.67	0.848	13.00	1.32	0.83	0.131
15.60	1.31	1.23	0.234	1.10	3.67	8.26	0.837	13.60	1.32	0.85	0.134
16.00	1.28	1.28	0.251	1.20	3.51	7.94	0.832	14.00	1.32	0.86	0.138
16.40	1.24	1.33	0.270	1.30	3.26	7.63	0.829	14.60	1.30	0.89	0.144
16.80	1.17	1.37	0.288	1.40	3.01	7.31	0.827	15.00	1.28	0.90	0.147
17.00	1.14	1.38	0.297	1.50	2.78	6.97	0.823	15.60	1.25	0.90	0.147
17.40	1.06	1.39	0.314	1.60	2.60	6.64	0.818	16.00	1.24	0.89	0.147
18.00	0.95	1.38	0.334	1.70	2.42	6.33	0.813	16.50	1.23	0.88	0.145
18.40	0.88	1.36	0.346	1.80	2.30	6.02	0.805	17.00	1.22	0.88	0.144
18.80	0.82	1.33	0.355	1.90	2.20	5.76	0.798	17.50	1.22	0.87	0.143
19.20	0.76	1.29	0.360	2.00	2.12	5.51	0.789	18.00	1.23	0.88	0.145
19.60	0.72	1.25	0.363	2.10	2.05	5.30	0.780	18.50	1.25	0.92	0.155
20.00	0.67	1.21	0.369	2.20	2.00	5.11	0.772	19.00	1.24	0.98	0.172
20.40	0.64	1.15	0.364	2.30	1.94	4.94	0.765	19.50	1.18	1.05	0.193
20.80	0.61	1.10	0.357	2.40	1.90	4.78	0.756	20.00	1.10	1.09	0.213
21.20	0.60	1.06	0.349	2.50	1.88	4.65	0.748	20.50	1.00	1.09	0.230
21.60	0.58	1.02	0.342	2.60	1.85	4.55	0.743	21.00	0.91	1.05	0.234
22.00	0.57	0.98	0.336	2.70	1.80	4.49	0.742	21.50	0.86	1.00	0.228
22.40	0.56	0.95	0.328	2.90	1.63	4.36	0.748	22.00	0.83	0.95	0.219

Energy (eV)	n	k	R(φ = 0)	Energy (eV)	n	k	R(φ = 0)	Energy (eV)	n	k	R(φ = 0)
22.50	0.81	0.92	0.214	4.20	1.29	3.08	0.649	26.00	0.68	0.63	0.154
23.00	0.79	0.90	0.213	4.40	1.22	2.93	0.639	27.00	0.67	0.57	0.140
23.50	0.75	0.87	0.214	4.60	1.16	2.79	0.628	28.00	0.66	0.51	0.124
24.00	0.73	0.84	0.210	4.80	1.11	2.67	0.617	29.00	0.67	0.46	0.107
24.50	0.70	0.81	0.208	5.00	1.06	2.56	0.607	30.00	0.67	0.43	0.097
25.00	0.69	0.77	0.202	5.20	1.01	2.46	0.600	31.00	0.67	0.37	0.084
25.50	0.67	0.74	0.195	5.40	0.95	2.35	0.593	32.00	0.69	0.33	0.070
26.00	0.66	0.70	0.188	5.60	0.92	2.23	0.576	33.00	0.71	0.30	0.058
26.50	0.65	0.66	0.176	5.80	0.90	2.14	0.559	34.00	0.73	0.27	0.048
27.00	0.65	0.64	0.168	6.00	0.88	2.05	0.545	35.00	0.75	0.25	0.039
27.50	0.65	0.61	0.159	6.20	0.87	1.98	0.531	36.00	0.77	0.24	0.035
28.00	0.65	0.59	0.152	6.40	0.84	1.91	0.521	37.00	0.79	0.23	0.039
29.00	0.65	0.54	0.137	6.60	0.82	1.84	0.510	38.00	0.80	0.22	0.027
30.00	0.66	0.51	0.127	6.80	0.79	1.77	0.500	39.00	0.82	0.22	0.024
31.00	0.64	0.49	0.127	7.00	0.76	1.69	0.489	40.00	0.83	0.22	0.022
32.00	0.61	0.44	0.126	7.20	0.75	1.61	0.472				
33.00	0.60	0.37	0.110	7.40	0.73	1.54	0.455	Ruthenium, single crystal, E ⊥ ĉ [5]			
34.00	0.65	0.30	0.074	7.60	0.73	1.46	0.433	0.10	11.85	50.81	0.983
35.00	0.69	0.28	0.058	7.80	0.73	1.39	0.411	0.20			
36.00	0.73	0.27	0.049	8.00	0.72	1.33	0.391		6.68	27.18	0.966
37.00	0.74	0.28	0.047	8.20	0.72	1.26	0.366	0.30	4.94	18.92	0.950
38.00	0.74	0.27	0.045	8.40	0.73	1.20	0.342	0.40	3.90	14.51	0.933
39.00	0.75	0.25	0.041	8.60	0.74	1.14	0.318	0.50	3.27	11.63	0.915
				8.80	0.74	1.08	0.295	0.60	2.98	9.54	0.888
Ruthenium, single crystal, E ∥ ĉ [9]				9.00	0.75	1.02	0.267	0.70	2.82	7.99	0.856
0.10	11.50	51.38	0.984	9.20	0.77	0.97	0.243	0.80	2.73	6.71	0.815
0.20	5.93	27.14	0.970	9.40	0.79	0.91	0.217	0.90	2.82	5.54	0.751
0.30	4.33	18.50	0.953	9.60	0.82	0.86	0.190	1.00	3.17	4.59	0.670
0.40	3.60	13.97	0.933	9.80	0.85	0.81	0.167	1.10	3.69	3.91	0.604
0.50	3.18	11.04	0.909	10.00	0.88	0.76	0.144	1.20	4.28	3.66	0.585
0.60	3.28	8.89	0.865	10.20	0.92	0.72	0.125	1.30	4.66	3.72	0.593
0.70	3.62	7.73	0.822	10.40	0.96	0.69	0.110	1.40	4.86	3.79	0.601
0.80	3.42	7.02	0.801	10.60	1.01	0.67	0.100	1.50	4.99	3.89	0.609
0.90	3.25	6.12	0.766	10.80	1.05	0.66	0.094	1.60	5.08	4.03	0.618
1.00	3.39	5.33	0.715	11.00	1.09	0.65	0.090	1.70	5.12	4.22	0.629
1.10	3.66	4.83	0.675	11.20	1.12	0.65	0.088	1.80	5.10	4.45	0.642
1.20	3.84	4.57	0.654	11.40	1.15	0.65	0.087	1.90	4.96	4.78	0.660
1.30	3.94	4.38	0.638	11.60	1.18	0.65	0.088	2.00	4.61	5.06	0.677
1.40	4.02	4.19	0.624	11.80	1.21	0.66	0.090	2.10	4.21	5.09	0.682
1.50	4.16	4.07	0.614	12.00	1.23	0.67	0.092	2.20	3.94	5.00	0.681
1.60	4.33	4.08	0.615	12.40	1.26	0.69	0.098	2.30	3.69	4.97	0.684
1.70	4.42	4.21	0.624	12.80	1.27	0.72	0.104	2.40	3.44	4.88	0.684
1.80	4.40	4.38	0.636	13.20	1.28	0.74	0.108	2.50	3.27	4.77	0.681
1.90	4.29	4.61	0.651	13.60	1.28	0.75	0.111	2.60	3.14	4.66	0.677
2.00	4.04	4.81	0.667	14.00	1.28	0.76	0.114	2.70	3.06	4.59	0.674
2.10	3.69	4.90	0.679	14.40	1.27	0.76	0.114	2.80	2.99	4.59	0.676
2.20	3.35	4.82	0.683	14.80	1.27	0.76	0.114	2.90	2.87	4.64	0.686
2.30	3.09	4.70	0.681	15.00	1.27	0.76	0.114	3.00	2.64	4.69	0.701
2.40	2.89	4.55	0.677	15.60	1.28	0.77	0.115	3.10	2.40	4.64	0.710
2.50				16.00	1.30	0.78	0.118	3.20	2.18	4.55	0.717
	2.74	4.40	0.671	16.50	1.32	0.80	0.123	3.30	2.00	4.43	0.721
2.60	2.64	4.25	0.663	17.00	1.34	0.85	0.136	3.40	1.84	4.30	0.723
2.70	2.58	4.14	0.656	17.50	1.32	0.93	0.155	3.50	1.71	4.16	0.723
2.80	2.54	4.05	0.650	18.00	1.26	0.99	0.173	3.60	1.60	4.03	0.722
2.90	2.48	4.03	0.650	18.50	1.18	1.02	0.185	3.70	1.50	3.90	0.721
3.00	2.38	4.03	0.656	19.00	1.11	1.02	0.192	3.80	1.41	3.77	0.718
3.10	2.26	4.00	0.661	19.50	1.05	1.02	0.199	3.90	1.35	3.64	0.713
3.20	2.13	3.96	0.666	20.00	0.99	1.02	0.208	4.00	1.29	3.53	0.707
3.30	2.00	3.91	0.671	20.50	0.92	0.99	0.212	4.20	1.21	3.31	0.694
3.40	1.87	3.83	0.673	21.00	0.86	0.94	0.209	4.40	1.16	3.13	0.679
3.50	1.76	3.74	0.674	21.50	0.83	0.90	0.203	4.60	1.13	2.97	0.662
3.60	1.66	3.65	0.675	22.00	0.81	0.86	0.193	4.80	1.09	2.86	0.652
3.70	1.57	3.55	0.673	23.00	0.77	0.79	0.182	5.00	1.03	2.75	0.648
3.80	1.49	3.45	0.672	24.00	0.74	0.74	0.171	5.20	0.97	2.64	0.643
3.90	1.42	3.35	0.668	25.00	0.71	0.69	0.163	5.40	0.91	2.52	0.635
4.00	1.37	3.24	0.661					5.60	0.88	2.40	0.622

Energy (eV)	n	k	$R(\phi = 0)$
5.80	0.86	2.29	0.605
6.00	0.84	2.20	0.591
6.20	0.82	2.11	0.576
6.40	0.81	2.04	0.564
6.60	0.78	1.97	0.556
6.80	0.76	1.89	0.545
7.00	0.73	1.82	0.538
7.20	0.70	1.75	0.527
7.40	0.68	1.67	0.513
7.60	0.67	1.59	0.496
7.80	0.66	1.51	0.476
8.00	0.66	1.44	0.454
8.20	0.65	1.36	0.430
8.40	0.66	1.29	0.403
8.60	0.66	1.22	0.378
8.80	0.68	1.15	0.346
9.00	0.69	1.09	0.317
9.20	0.70	1.02	0.286
9.40	0.73	0.95	0.251
9.60	0.77	0.89	0.216
9.80	0.82	0.84	0.185
10.00	0.86	0.81	0.163
10.20	0.90	0.77	0.143
10.40	0.94	0.74	0.127
10.60	0.99	0.72	0.115
10.80	1.04	0.71	0.108
11.00	1.08	0.70	0.104
11.20	1.11	0.70	0.102
11.40	1.14	0.70	0.101
11.60	1.17	0.71	0.102
11.80	1.20	0.72	0.104
12.00	1.22	0.73	0.107
12.40	1.25	0.76	0.113
12.80	1.26	0.78	0.118
13.20	1.27	0.81	0.124
13.60	1.27	0.83	0.129
14.00	1.26	0.84	0.132
14.40	1.25	0.84	0.132
14.80	1.25	0.84	0.133
15.00	1.25	0.84	0.133
15.60	1.25	0.85	0.134
16.00	1.27	0.85	0.134
16.50	1.28	0.89	0.145
17.00	1.28	0.94	0.158
17.50	1.25	1.00	0.175
18.00	1.19	1.04	0.190
18.50	1.12	1.05	0.200
19.00	1.07	1.05	0.205
19.50	1.02	1.04	0.212
20.00	0.97	1.04	0.219
20.50			
	0.91		
		1.03	
			0.228
21.00	0.85	1.01	0.234
21.50	0.80	0.97	0.234
22.00	0.77	0.94	0.233
23.00	0.71	0.87	0.229
24.00	0.67	0.79	0.218
25.00	0.64	0.73	0.205
26.00	0.61	0.66	0.194
27.00	0.60	0.59	0.177
28.00	0.60	0.53	0.155
29.00	0.61	0.48	0.134
30.00	0.62	0.45	0.123
31.00	0.61	0.40	0.114

Energy (eV)	n	k	$R(\phi = 0)$
34.00	0.67	0.28	0.065
35.00	0.70	0.26	0.054
36.00	0.72	0.25	0.047
37.00	0.73	0.23	0.041
38.00	0.75	0.22	0.035
39.00	0.77	0.22	0.031
40.00	0.79	0.22	0.028

Selenium, single crystal, $\overset{\perp}{E} \parallel \hat{c}$ [22]

Energy (eV)	n	k	$R(\phi = 0)$
0.01364	2.914	0.248	0.242
0.01488	3.175	9.95E-02	0.272
0.01612	3.263	2.13E-03	0.282
0.01736	3.306	3.81E-02	0.287
0.01860	3.330	7.04E-03	0.290
0.01984	3.346	4.23E-02	0.291
0.02108	3.358	3.40E-03	0.293
0.02232	3.366	5.31E-02	0.294
0.02356	3.372	1.96E-03	0.294
0.02480	3.377	2.39E-02	0.295
0.02604	3.380		0.295
0.02728		1.16E-02	
0.02976		7.96E-03	
0.03224		8.57E-03	
0.03472		2.70E-02	
0.03720	3.397	1.72E-02	0.297
0.04463		1.13E-02	
0.04959	3.403	2.79E-03	0.298
0.05703		1.56E-03	
0.06199	3.405	1.35E-03	0.298
0.06819		5.79E-04	
0.07439	3.407	4.44E-04	0.298
0.08059		4.41E-04	
0.08679	3.408	4.32E-04	0.298
0.09299		2.44E-04	
0.09919	3.409	3.23E-04	0.299
0.1116	3.409	2.87E-04	0.299
0.1240	3.410	2.71E-04	0.299
0.2480	3.417	2.67E-04	0.299
0.3720	3.427	1.90E-04	0.301
0.4959	3.442	1.41E-04	0.302
0.6199	3.462	1.12E-04	0.304
0.7439	3.486	9.42E-05	0.307
0.8679	3.516	8.07E-05	0.310
0.9919	3.551	7.11E-05	0.314
1.116	3.592	6.37E-05	0.319
1.240	3.640	5.81E-05	0.324
1.50		1.33E-04	
1.60		1.59E-04	
1.70		6.27E-04	
1.80	4.46	2.20E-02	0.402
2.0	4.79	0.76	0.438
2.2	4.49	1.19	0.431
2.4	4.28	1.21	0.417
2.6	4.40	1.32	0.430
2.8	4.59	1.70	0.462
3.0	4.44	2.29	0.490
3.2	3.92	2.59	0.493
3.4	3.69	2.76	0.502
3.6	3.39	3.01	0.521
3.8	(3.00)		
4.0	(2.65)		
4.2	(2.30)		
4.5	1.92	2.78	0.528
5.0	1.50	2.31	0.482
6.0	1.57	1.49	0.288

Energy (eV)	n	k	$R(\phi = 0)$
7.0	1.84	1.45	0.276
8.0	1.35	1.68	0.353
9.0	1.35	1.64	0.342
10.0	0.92	1.07	0.238
12.0	1.00	1.10	0.232
14.0	0.81	0.91	0.211
16.0	0.65	0.61	0.160
18.0	0.65	0.48	0.120
20.0	0.69	0.36	0.076
22.0	0.81	0.25	0.030
24.0	0.91	0.18	0.011
26.0	0.86	0.15	0.012
28.0	0.85	0.13	0.011
30.0	0.87	0.11	0.008

Selenium, single crystal, $\overset{\perp}{E} \perp \hat{c}$ [22]

Energy (eV)	n	k	$R(\phi = 0)$
0.01364	2.854	0.0239	0.231
0.01488	2.932	0.0325	0.241
0.01612	3.140	0.1750	0.269
0.01736	2.959	1.3300	0.321
0.01860	2.111	0.2550	0.133
0.01984			
	2.356		
		0.0746	0.164
0.02108	2.462	0.0276	0.178
0.02232	2.502	0.0442	0.184
0.02356	2.543	0.0097	0.190
0.02480	2.550	0.0239	0.191
0.02604	2.582		0.195
0.02728	2.600	0.0101	0.198
0.02976	2.576	9.95E-03	0.194
0.03224	2.598	1.16E-02	0.197
0.03472	2.607	1.68E-02	0.199
0.03720	2.613	1.54E-02	0.199
0.04463		1.17E-02	
0.04959	2.627	3.58E-03	0.201
0.05703		8.65E-04	
0.06199	2.632	2.07E-03	0.202
0.06819		2.89E-04	
0.07439	2.635	1.59E-04	0.202
0.08059		1.35E-04	
0.08679	2.636	1.42E-04	0.202
0.09299		1.04E-04	
0.09919	2.637	8.95E-05	0.203
0.1116	2.638	8.84E-05	0.203
0.1240	2.639	8.51E-05	0.203
0.2480	2.645	5.97E-05	0.204
0.3720	2.652	5.44E-05	0.205
0.4959	2.654	4.58E-05	0.205
0.6199	2.675	3.82E-05	0.208
0.7439	2.692	3.32E-05	0.210
0.8679	2.713	2.96E-05	0.213
0.9919	2.739	2.69E-05	0.216
1.116	2.772	2.48E-05	0.221
1.240	2.816	2.31E-05	0.226
1.50		7.37E-05	
1.60		8.63E-05	
1.70		3.60E-04	
1.80	3.32	0.11	0.289
2.00	3.38	0.65	0.310
2.20	3.07	0.73	0.282
2.40	2.93	0.61	0.259
2.60	3.00	0.53	0.263
2.80	3.12	0.58	0.279
3.00	3.30	0.70	0.305
3.20	3.35	1.01	0.328

Energy (eV)	n	k	R(φ = 0)
3.60	3.06	1.47	0.344
3.80	2.84	1.66	0.351
4.00	2.51	1.81	0.356
4.20	2.18	1.83	0.352
4.50	1.75	1.94	0.382
5.00	1.25	1.50	0.316
6.00	1.32	0.73	0.107
7.00	1.62	0.61	0.105
8.00	1.81	0.69	0.135
9.00	1.66	1.02	0.182
10.00	1.72	0.95	0.171
12.00	1.25	1.02	0.181
14.00	0.98	0.92	0.178
16.00	0.68	0.96	0.274
18.00	0.61	0.65	0.191
20.00	0.73	0.48	0.094
22.00	0.78	0.39	0.060
24.00	0.78	0.32	0.046
26.00	0.78	0.26	0.036
28.00	0.80	0.19	0.023
30.00	0.79	0.14	0.020

Silicon, single crystal[23]

Energy (eV)	n	k	R(φ = 0)
0.01240	3.4185	2.90E-04	0.300
0.01488	3.4190	2.30E-04	0.300
0.01736	3.4192	1.90E-04	0.300
0.01984	3.4195	1.70E-04	0.300
0.02480	3.4197		0.300
0.03100	3.4199		0.300
0.04092	3.4200		0.300
0.04463		1.08E-04	
0.04959	3.4201	9.15E-05	0.300
0.05703		1.56E-04	
0.06199	3.4204	2.86E-04	0.300
0.06943		3.84E-04	
0.07439		7.16E-04	
0.08059	(3.4207)	1.52E-04	0.300
0.08679		1.02E-04	
0.09299		2.59E-04	
0.09919		1.77E-04	
0.1054		1.53E-04	
0.1116		2.02E-04	
0.1178		1.22E-04	
0.1240	3.4215	6.76E-05	0.300
0.1364		5.49E-05	
0.1488		2.41E-05	
0.1612		2.49E-05	
0.1736	(3.4230)	1.68E-05	0.300
0.1798		2.45E-05	
0.1860		2.66E-06	
0.1922		1.74E-06	
0.1984		8.46E-07	
0.2046		5.64E-07	
0.2108	(3.4244)	4.17E-07	0.300
0.2170		4.05E-07	
0.2232			
		3.94E-07	
0.2294		3.26E-07	
0.2356		2.97E-07	
0.2418		2.82E-07	
0.2480	3.4261	1.99E-07	0.300
0.3100	3.4294		0.301
0.3626	3.4327		0.301
0.4568	3.4393	2.50E-09	0.302

Energy (eV)	n	k	R(φ = 0)
1.033	3.5193		0.311
1.1	(3.5341)	1.30E-05	0.312
1.2		1.80E-04	
1.3		2.26E-03	
1.4		7.75E-03	
1.5	3.673	5.00E-03	0.327
1.6	3.714	8.00E-03	0.331
1.7	3.752	1.00E-02	0.335
1.8	3.796	0.013	0.340
1.9	3.847	0.016	0.345
2.0	3.906	0.022	0.351
2.1	3.969	0.030	0.357
2.2	4.042	0.032	0.364
2.3	4.123	0.048	0.372
2.4	4.215	0.060	0.380
2.5	4.320	0.073	0.390
2.6	4.442	0.090	0.400
2.7	4.583	0.130	0.412
2.8	4.753	0.163	0.426
2.9	4.961	0.203	0.442
3.0	5.222	0.269	0.461
3.1	5.570	0.387	0.486
3.2	6.062	0.630	0.518
3.3	6.709	1.321	0.561
3.4	6.522	2.705	0.592
3.5	5.610	3.014	0.575
3.6	5.296	2.987	0.564
3.7	5.156	3.058	0.563
3.8	5.065	3.182	0.568
3.9	5.016	3.346	0.577
4.0	5.010	3.587	0.591
4.1	5.020	3.979	0.614
4.2	4.888	4.639	0.652
4.3	4.086	5.395	0.703
4.4	3.120	5.344	0.726
4.5	2.451	5.082	0.740
4.6	1.988	4.678	0.742
4.7	1.764	4.278	0.728
4.8	1.658	3.979	0.710
4.9	1.597	3.749	0.693
5.0	1.570	3.565	0.675
5.1	1.571	3.429	0.658
5.2	1.589	3.354	0.646
5.3	1.579	3.353	0.647
5.4	1.471	3.366	0.663
5.5	1.340	3.302	0.673
5.6	1.247	3.206	0.675
5.7	1.180	3.112	0.673
5.8	1.133	3.045	0.672
5.9	1.083	2.982	0.673
6.0	1.010	2.909	0.677
6.5	0.847	2.73	0.688
7.0	0.682	2.45	0.691
7.5	0.563	2.21	0.693
8.0	0.478	2.00	0.691
8.5	0.414	1.82	0.688
9.0	0.367	1.66	0.683
9.5	0.332	1.51	0.672
10.0	0.306	1.38	0.661
12.0	0.257	0.963	0.590
14.0	0.275	0.641	0.460
16.0	0.345	0.394	0.297
18.0	0.455	0.219	0.159
20.0	0.567	0.0835	0.079
22.14	0.675	0.0405	0.038

Energy (eV)	n	k	R(φ = 0)
24.31	0.752	0.0243	0.020
26.38	0.803	0.0178	0.012
28.18	0.834	0.0152	0.008
30.24	0.860	0.0138	0.006
31.79	0.877	0.0132	0.004
34.44	0.899	0.0121	0.003
36.47	0.913	0.0113	0.002
38.75	0.925	0.0104	0.002
40.00	0.930	0.0100	0.001

Silver[6]

Energy (eV)	n	k	R(φ = 0)
0.10	9.91	90.27	0.995
0.20	2.84	45.70	0.995
0.30	1.41	30.51	0.994
0.40	0.91	22.89	0.993
0.50	0.67	18.32	0.992
1.00	0.28	9.03	0.987
1.50	0.27	5.79	0.969
2.00	0.27	4.18	0.944
2.50	0.24	3.09	0.914
3.00	0.23	2.27	0.864
3.25	0.23	1.86	0.816
3.50	0.21	1.42	0.756
3.60	0.23	1.13	0.671
3.70			
	0.30		
		0.77	
			0.475
3.77	0.53	0.40	0.154
3.80	0.73	0.30	0.053
3.90	1.30	0.36	0.040
4.00	1.61	0.60	0.103
4.10	1.73	0.85	0.153
4.20	1.75	1.06	0.194
4.30	1.73	1.13	0.208
4.50	1.69	1.28	0.238
4.75	1.61	1.34	0.252
5.00	1.55	1.36	0.257
5.50	1.45	1.34	0.257
6.00	1.34	1.28	0.246
6.50	1.25	1.18	0.225
7.00	1.18	1.06	0.196
7.50	1.14	0.91	0.157
8.00	1.16	0.75	0.114
9.00	1.33	0.56	0.074
10.00	1.46	0.56	0.082
11.00	1.52	0.56	0.088
12.00	1.61	0.59	0.100
13.00	1.66	0.64	0.112
14.00	1.72	0.78	0.141
14.50	1.64	0.88	0.152
15.00	1.56	0.92	0.156
16.00	1.42	0.91	0.151
17.00	1.33	0.86	0.139
18.00	1.28	0.80	0.124
19.00	1.27	0.75	0.111
20.00	1.29	0.71	0.103
21.00	1.35	0.75	0.112
21.50	1.37	0.80	0.124
22.00	1.34	0.87	0.141
22.50	1.26	0.93	0.157
23.00	1.17	0.94	0.163
23.50	1.10	0.93	0.165
24.00	1.04	0.90	0.165
24.50	0.99	0.87	0.160
25.00	0.95	0.83	0.154

Energy (eV)	n	k	R(φ = 0)	Energy (eV)	n	k	R(φ = 0)	Energy (eV)	n	k	R(φ = 0)
26.50	0.89	0.69	0.121	6.358	0.454		0.141	3.20	2.73	2.31	0.432
27.00	0.89	0.65	0.109	6.526	0.485		0.120	3.40	2.61	2.33	0.435
27.50	0.89	0.62	0.099	6.702	0.533		0.093	3.60	2.49	2.30	0.430
28.00	0.90	0.59	0.090	6.888	0.574		0.073	3.80	2.40	2.22	0.418
28.50	0.91	0.57	0.084	7.130	0.616		0.056	4.00	2.36	2.14	0.406
29.00	0.92	0.56	0.079	7.328	0.641		0.048	4.20	2.35	2.06	0.392
30.00	0.93	0.54	0.074	7.583	0.674		0.038	4.40	2.39	2.01	0.384
31.00	0.93	0.53	0.072	7.847	0.700		0.031	4.60	2.45	2.00	0.384
32.00	0.92	0.53	0.072	8.015	0.710		0.029	4.80	2.53	2.06	0.394
33.00	0.90	0.51	0.071	8.634	0.762		0.018	5.00	2.58	2.20	0.416
34.00	0.88	0.49	0.067	9.143	0.800		0.012	5.20	2.52	2.44	0.450
35.00	0.86	0.45	0.061	9.709	0.819		0.010	5.40	2.31	2.61	0.480
36.00	0.89	0.44	0.055	10.20	0.843		0.007	5.60	2.06	2.67	0.501
38.00	0.89	0.39	0.043	11.08	0.870		0.005	5.80	1.83	2.63	0.510
40.00	0.90	0.37	0.039	11.83	0.887		0.004	6.00	1.63	2.56	0.515
42.00	0.90	0.35	0.036	12.73	0.907		0.002	6.20	1.48	2.45	0.512
44.00	0.90	0.33	0.033	13.05	0.913		0.002	6.40	1.37	2.33	0.504
46.00	0.90	0.32	0.031	13.42	0.914		0.002	6.60	1.29	2.22	0.492
48.00	0.89	0.31	0.030	13.73	0.917		0.002	6.80	1.23	2.11	0.478
50.00	0.88	0.29	0.027	14.07	0.922		0.002	7.00	1.18	2.01	0.462
52.00	0.89	0.28	0.024	14.83	0.934		0.001	7.20	1.15	1.91	0.445
54.00	0.88	0.17	0.024	15.05	0.936		0.001	7.40	1.13	1.82	0.425
56.00	0.87	0.26	0.024	15.46	0.942		0.001	7.60	1.12	1.75	0.406
58.00	0.87	0.24	0.021	16.21	0.948		0.001	7.80	1.11	1.68	0.390
60.00	0.87	0.22	0.018	18.10	0.964		0.000	8.00	1.11	1.61	0.370
62.00	0.88	0.21	0.016	21.12	0.979		0.000	8.20	1.12	1.55	0.350
64.00	0.88	0.21	0.016	25.51	0.993		0.000	8.40	1.13	1.50	0.332
66.00	0.88	0.21	0.016	26.95	1.00		0.000	8.60	1.14	1.45	0.317
68.00	0.87	0.21	0.017	27.68	1.01		0.000				
70.00	0.83	0.20	0.021	28.37	1.01		0.000				
72.00	0.85	0.18	0.016	29.52	1.02		0.000				
74.00	0.85	0.17	0.014					8.80	1.17	1.41	0.301
76.00	0.85	0.16	0.013	**Tantalum[16]**				9.00	1.19	1.40	0.294
78.00	0.85	0.15	0.013	0.10	10.14	66.39	0.984	9.20	1.21	1.38	0.289
80.00	0.85	0.14	0.012	0.15	9.45	46.41	0.9834	9.40	1.21	1.38	0.287
85.00	0.85	0.11	0.011	0.20	5.77	35.46	0.982	9.60	1.21	1.38	0.285
90.00	0.85	0.08	0.009	0.26	3.67	27.53	0.981	9.80	1.21	1.37	0.285
95.00	0.86	0.06	0.007	0.30	2.87	23.90	0.980	10.00	1.20	1.37	0.286
100.00	0.87	0.04	0.005	0.38	2.03	18.87	0.978	10.20	1.19	1.37	0.286
				0.50	1.37	14.26	0.974	10.40	1.18	1.37	0.287
Sodium[24]				0.58	1.15	12.19	0.970	10.60	1.16	1.36	0.288
0.55	0.262	9.97	0.990	0.70	0.96	9.92	0.962	10.80	1.15	1.36	0.289
0.58	0.241	9.45	0.989	0.78	0.89	8.77	0.956	11.00	1.13	1.35	0.290
0.63	0.207	8.80	0.990	0.90	0.84	7.38	0.942	11.20	1.11	1.35	0.292
0.67	0.175	8.09	0.990	1.00	0.89	6.47	0.992	11.40	1.09	1.34	0.293
0.73	0.147	7.42	0.990	1.10	0.93	5.75	0.899	11.60	1.07	1.33	0.294
0.81	0.123	6.67	0.989	1.20	0.98	5.14	0.872	11.80	1.05	1.32	0.295
0.92	0.099	5.82	0.989	1.30	1.00	4.62	0.842	12.00	1.02	1.31	0.296
1.05	0.078	5.11	0.989	1.40	1.04	4.15	0.805	12.20	1.00	1.29	0.295
1.23	0.064	4.35	0.987	1.50	1.09	3.73	0.762	12.40	0.98	1.28	0.294
1.44	0.053	3.72	0.986	1.60	1.15	3.33	0.707	12.60	0.96	1.26	0.292
1.65	0.050	3.22	0.983	1.70	1.24	2.95	0.640	12.80	0.94	1.24	0.289
1.87	0.049	2.76	0.978	1.80	1.35	2.60	0.560	13.00	0.93	1.22	0.286
2.07	0.053	2.48	0.971	1.90	1.57	2.24	0.460	13.60	0.91	1.16	0.272
2.27	0.059	2.23	0.961	2.00	1.83	1.99	0.388	14.00	0.90	1.15	0.272
2.45	0.063	2.07	0.953	2.10	2.10	1.84	0.354	14.60	0.85	1.15	0.285
2.64	0.066	1.88	0.943	2.20	2.36	1.81	0.351	15.00	0.80	1.13	0.293
2.82	0.068	1.76	0.936	2.30	2.56	1.86	0.365	15.60	0.72	1.08	0.301
2.95	0.068	1.63	0.928	2.40	2.68	1.92	0.378	16.00	0.68	1.04	0.304
				2.50	2.75	1.98	0.388	16.60	0.63	0.97	0.301
				2.60	2.80	2.02	0.395	17.00	0.60	0.92	0.296
				2.70	2.84	2.08	0.405	17.60	0.60	0.92	0.296
3.06	0.069	1.54	0.921	2.80	2.85	2.14	0.412	18.00	0.55	0.79	0.274
3.20	0.065	1.47	0.921	2.90	2.84	2.20	0.420	18.60	0.53	0.71	0.254
3.40	0.061	1.33	0.916	3.00	2.81	2.24	0.425	19.00	0.53	0.65	0.236
3.71	0.055	1.13	0.908					19.60	0.53	0.57	0.207

Energy (eV)	n	k	$R(\phi = 0)$
21.00	0.57	0.39	0.127
21.60	0.64	0.34	0.089
22.00	0.64	0.32	0.081
22.60	0.69	0.27	0.058
23.00	0.73	0.24	0.043
23.60	0.80	0.26	0.033
24.00	0.80	0.26	0.034
24.60	0.82	0.25	0.029
25.00	0.83	0.25	0.026
25.60	0.86	0.24	0.022
26.00	0.88	0.25	0.022
26.60	0.87	0.26	0.023
27.00	0.87	0.25	0.022
27.60	0.89	0.23	0.019
28.00	0.90	0.23	0.017
28.60	0.91	0.22	0.015
29.00	0.92	0.22	0.014
29.60	0.94	0.22	0.014
30.00	0.95	0.22	0.014
31.00	0.97	0.23	0.014
32.00	0.98	0.24	0.015
33.00	0.98	0.25	0.015
34.00	0.99	0.25	0.016
35.00	0.99	0.26	0.017
36.00	0.99	0.27	0.018
37.00	0.99	0.28	0.019
38.00	0.98	0.28	0.021
39.00	0.97	0.29	0.022
40.00	0.95	0.29	0.023

Tellurium, $\overset{\perp}{E} \parallel \hat{c}$ [25]

Energy (eV)	n	k	$R(\phi = 0)$
0.01364	4.82	0.118	0.431
0.01488	5.26	0.0505	0.463
0.01612	5.47	0.0278	0.477
0.01736	5.59	0.0174	0.485
0.01860		0.0796	
0.01984		0.0696	
0.02108		0.0749	
0.02232		0.1900	
0.02356		0.2220	
0.02480		0.0716	
0.02604		0.0682	
0.02728		0.0832	
0.02976		0.0149	
0.03224		2.14E-03	
0.03472		1.71E-03	
0.03720	5.94	3.71E-03	0.507
0.03968		2.44E-03	
0.04339	5.96	1.59E-03	0.508
0.04711		7.85E-04	
0.05083		7.38E-04	
0.05579		3.89E-04	
0.06199	5.98	3.09E-04	0.509
0.07439		2.52E-04	
0.08679		2.96E-04	
0.09919		3.68E-04	
0.12400	6.246	3.34E-04	0.524
0.15500	6.253		0.525
0.20660	6.286		0.526
0.24800	6.316	7.48E-05	0.528
0.31	6.372	1.18E-05	0.531
0.35		4.93E-04	
0.41		6.74E-03	
0.5	6.53	2.30E-02	0.539
0.6	6.71	7.50E-02	0.549

Energy (eV)	n	k	$R(\phi = 0)$
0.7	7.00	0.24	0.563
0.8	7.23	0.48	0.574
0.9	7.48	0.94	0.589
1.0	7.70	1.56	0.606
1.2	6.99	2.22	0.593
1.4	7.11	2.46	0.604
1.6	6.75	2.91	0.606
1.8	6.89	3.70	0.637
2.0	4.67	4.67	0.654
2.2	4.94	5.16	0.681
2.4	3.94	5.08	0.686
2.6	3.25	4.77	0.681
2.8	2.73	4.42	0.674
3.0	2.30	4.16	0.674
3.5	1.69	3.44	0.646
4.0	1.33	2.64	0.571
4.5	1.32	1.96	0.428
5.0	1.63	1.60	0.312
5.5	1.72	1.57	0.302
6.0	1.73	1.45	0.276
6.5	1.78	1.36	0.257
7.0	1.83	1.36	0.257
7.5	1.72	1.51	0.289
8.0	1.54	1.37	0.260
8.5	1.55	1.23	0.226
9.0	0.99	0.93	0.179
9.5	1.47	1.25	0.233
10.0	0.86	0.86	0.181
11.0	0.80	0.77	0.165
12.0	0.79	0.76	0.164
14.0	0.67	0.59	0.146
16.0	0.59	0.49	0.147
18.0	0.48	0.31	0.160
20.0	0.74	0.20	0.035
22.0	0.83	0.18	0.018
24.0	0.85	0.15	0.013
26.0	0.87	0.12	0.009
28.0	0.89	0.090	0.006
30.0	0.90	0.045	0.003

Tellurium, $\overset{\perp}{E} \perp \hat{c}$ [25]

Energy (eV)	n	k	$R(\phi = 0)$
0.01364	2.61	0.2980	0.204
0.01488	3.65	0.0894	0.325
0.01612	4.10	0.0535	0.370
0.01736	4.63	0.4990	0.420
0.01860		0.1170	
0.01984		0.0343	
0.02108	(4.42)	0.0421	0.398
0.02232		0.1060	
0.02356		0.0880	
0.02480		0.0458	
0.02604		0.0928	
0.02728		0.0886	
0.02976		0.0232	
0.03224		3.06E-03	
0.03472		1.25E-02	
0.03720	4.71	2.65E-03	0.422
0.03968		1.89E-03	
0.04339	4.74	1.41E-03	0.425
0.04711		8.38E-04	
0.05083		6.79E-04	
0.05579		1.59E-04	
0.06199	4.77	1.16E-04	0.427
0.07439		7.23E-05	
0.08679		5.34E-05	

Energy (eV)	n	k	$R(\phi = 0)$
0.09919		4.28E-05	
0.1240	4.796	3.18E-05	0.429
0.1550	4.809		0.430
0.2066	4.838		0.432
0.2480	4.864	2.19E-05	0.434
0.31	4.929	3.18E-05	0.439
0.35		7.89E-02	
0.41		0.149	
0.5	4.90		0.437
0.6	4.93		0.439
0.7	4.95	0.11	0.441
0.8	5.10	0.13	0.452
0.9	5.22	0.22	0.461
1.0	5.35	0.45	0.472
1.2	5.17	0.63	0.462
1.4	5.56	0.63	0.488
1.6	5.88	1.15	0.517
1.8	6.10	1.80	0.545
2.0	5.94	2.69	0.571
2.2	5.10	3.61	0.594
2.4	4.24	3.77	0.593
2.6	3.57	3.75	0.591
2.8	3.03	3.63	0.588
3.0	2.51	3.39	0.578
3.5	1.72	2.70	0.532
4.0	1.32	2.01	0.440
4.5	1.28	1.28	0.251
5.0	1.47	0.82	0.132
5.5	1.74	0.51	0.104
6.0	1.94	0.39	0.118
6.5	2.19	0.32	0.148
7.0	2.48	0.40	0.192
7.5	2.60	0.69	0.226
8.0	2.59	0.91	0.245
8.5	2.39	1.00	0.235
9.0	1.11	1.24	0.259
9.5	2.08	1.11	0.224
10.0	0.99	1.04	0.215
11.0	0.84	1.01	0.237
12.0	0.87	0.87	0.182
14.0	0.59	0.87	0.282
16.0	0.64	0.55	0.144
18.0	0.52	0.41	0.161
20.0	0.50	0.38	0.165
22.0	0.56	0.29	0.110
24.0	0.54	0.25	0.113
26.0	0.50	0.20	0.127
28.0	0.48	0.17	0.135
30.0	0.46	0.088	0.140

Titanium (Polycrystalline)[14]

Energy (eV)	n	k	$R(\phi = 0)$
0.10	5.03	23.38	0.965
0.15	3.00	15.72	0.954
0.20	2.12	11.34	0.939
0.25	2.05	8.10	0.890
0.30	6.39	9.94	0.833
0.35	2.74	6.21	0.792
0.40	2.49	4.68	0.708
0.45	3.35	3.25	0.545
0.50	4.43	3.22	0.555
0.60	4.71	3.77	0.597
0.70	4.38	3.89	0.603
0.80	4.04	3.82	0.596
0.90	3.80	3.65	0.582
1.00	3.62	3.52	0.570

OPTICAL PROPERTIES OF METALS AND SEMICONDUCTORS (continued)

Energy (eV)	n	k	R(φ = 0)	Energy (eV)	n	k	R(φ = 0)	Energy (eV)	n	k	R(φ = 0)
1.10	3.47	3.40	0.560	11.00	0.79	0.72	0.152	0.86	2.92	4.37	0.661
1.20	3.35	3.30	0.550	11.20	0.81	0.69	0.139	0.90	3.11	4.44	0.660
1.30	3.28	3.25	0.546	11.40	0.81	0.69	0.139	0.94	3.15	4.43	0.658
1.40	3.17	3.28	0.549	11.60	0.79	0.68	0.139	0.98	3.15	4.36	0.653
1.50	2.98	3.32	0.557	11.80	0.78	0.67	0.137	1.00	3.14	4.32	0.649
1.60	2.74	3.30	0.559	12.00	0.77	0.65	0.132	1.10	3.05	4.04	0.627
1.70	2.54	3.23	0.557	12.80	0.76	0.55	0.106	1.20	3.00	3.64	0.590
1.80	2.36	3.11	0.550	13.20	0.76	0.52	0.097	1.30	3.12	3.24	0.545
1.90	2.22	2.99	0.540	13.60	0.76	0.48	0.087	1.40	3.29	2.96	0.515
2.00	2.11	2.88	0.530	14.00	0.77	0.45	0.077	1.50	3.48	2.79	0.500
2.10	2.01	2.77	0.520	14.40	0.77	0.42	0.069	1.60	3.67	2.68	0.494
2.20	1.92	2.67	0.509	14.80	0.79	0.38	0.058	1.70	3.84	2.79	0.507
2.30	1.86	2.56	0.495	15.20	0.79	0.36	0.052	1.80	3.82	2.91	0.518
2.40	1.81	2.47	0.483	15.60	0.79	0.32	0.045	1.90	3.70	2.94	0.518
2.50	1.78	2.39	0.471	16.00	0.83	0.31	0.037	2.00	3.60	2.89	0.512
2.60	1.75	2.34	0.462	16.40	0.84	0.28	0.030	2.10	3.54	2.84	0.506
2.70	1.71	2.29	0.456	16.80	0.87	0.27	0.025	2.20	3.49	2.76	0.497
2.80	1.68	2.25	0.451	17.20	0.90	0.25	0.020	2.30	3.49	2.72	0.494
2.90	1.63	2.21	0.447	17.60	0.93	0.25	0.017	2.40	3.45	2.72	0.493
3.00	1.59	2.17	0.444	18.00	0.94	0.24	0.165	2.50	3.38	2.68	0.487
3.10	1.55	2.15	0.442	18.40	0.94	0.23	0.017	2.60	3.34	2.62	0.480
3.20	1.50	2.12	0.442	18.80	0.95	0.24	0.016	2.70	3.31	2.55	0.472
3.30	1.44	2.09	0.442	19.20	0.96	0.25	0.016	2.80	3.31	2.49	0.466
3.40	1.37	2.06	0.443	19.60	0.97	0.25	0.017	2.90	3.32	2.45	0.461
3.50	1.30	2.01	0.443	20.00	0.98	0.27	0.018	3.00	3.35	2.42	0.459
3.60	1.24	1.96	0.441	20.40	0.98	0.27	0.019	3.10	3.39	2.41	0.460
3.70	1.17	1.90	0.436	20.60	1.00	0.29	0.020	3.20	3.43	2.45	0.465
3.80	1.11	1.83	0.430	21.20	0.99	0.31	0.023	3.30	3.45	2.55	0.476
3.85	1.08	1.78	0.423	21.60	0.99	0.31	0.024	3.40	3.39	2.66	0.485
3.90	1.06	1.73	0.413	22.00	0.98	0.32	0.025	3.50	3.24	2.70	0.488
4.00	1.04	1.62	0.389	22.40	0.98	0.33	0.027	3.60	3.13	2.67	0.482
4.20	1.05	1.45	0.333	22.80	0.97	0.33	0.028	3.70	3.05	2.62	0.476
4.40	1.13	1.33	0.284	23.20	0.96	0.34	0.030	3.80	2.99	2.56	0.468
4.60	1.17	1.29	0.265	23.60	0.95	0.35	0.031	3.90	2.96	2.50	0.460
4.80	1.21	1.23	0.244	24.00	0.92	0.35	0.033	4.00			
5.00	1.24	1.21	0.236	24.5	0.91	0.34	0.032		2.95	2.43	0.451
5.20	1.27	1.20	0.228	25.0	0.91	0.33	0.032	4.20	3.02	2.33	0.440
5.40	1.17	1.16	0.228	25.5	0.89	0.33	0.032	4.40	3.13	2.32	0.442
5.60	1.24	1.21	0.234	26.0	0.89	0.33	0.032	4.60	3.24	2.41	0.455
5.80	1.21	1.22	0.241	26.5	0.88	0.32	0.032	4.80	3.33	2.57	0.475
6.00	1.15	1.21	0.244	27.0	0.86	0.31	0.032	5.00	3.40	2.85	0.505
6.20	1.11	1.18	0.240	27.5	0.85	0.30	0.033	5.20	3.27	3.27	0.548
6.40	1.08	1.14	0.232	28.0	0.84	0.29	0.033	5.40	2.92	3.58	0.586
6.60	1.04	1.06	0.212	28.5	0.82	0.26	0.029	5.60	2.43	3.70	0.618
6.80	1.05	1.02	0.198	29.0	0.83	0.25	0.027	5.80	2.00	3.61	0.637
7.00	1.06	0.97	0.182	30.0	0.84	0.22	0.022	6.00	1.70	3.42	0.643
7.20	1.07	0.95	0.175					6.20	1.47	3.24	0.646
7.40	1.11	0.94	0.167	**Tungsten[27]**				6.40	1.32	3.04	0.640
7.60	1.09	0.92	0.165	0.10	14.06	54.71	0.983	6.60	1.21	2.87	0.631
7.80	1.11	0.93	0.165	0.20	3.87	28.30	0.981	6.80	1.12	2.70	0.619
8.00	1.10	0.94	0.169	0.25	2.56	22.44	0.980	7.00	1.06	2.56	0.607
8.20	1.10	0.95	0.171	0.30	1.83	18.32	0.979	7.20	1.01	2.43	0.593
8.40	1.08	0.95	0.175	0.34	1.71	15.71	0.973	7.40	0.98	2.30	0.573
8.60	1.04	0.96	0.181	0.38	1.86	13.88	0.963	7.60	0.95	2.18	0.556
8.80	1.02	0.95	0.181	0.42	1.92	12.63	0.954	7.80	0.93	2.06	0.533
9.00	1.00	0.94	0.182	0.46	1.69	11.59	0.952	8.00	0.94	1.95	0.505
9.20	0.97	0.93	0.182	0.50	1.40	10.52	0.952	8.20	0.94	1.86	0.481
9.40	0.95	0.91	0.181	0.54	1.23	9.45	0.948	8.40	0.96	1.76	0.449
9.60	0.94	0.90	0.179	0.58	1.17	8.44	0.938	8.60	0.99	1.70	0.422
9.80	0.91	0.88	0.179	0.62	1.28	7.52	0.917	8.80	1.01	1.65	0.401
10.00	0.89	0.88	0.180	0.66	1.45	6.78	0.888	9.00	1.01	1.60	0.388
10.20	0.86	0.85	0.178	0.70	1.59	6.13	0.856	9.20	1.02	1.55	0.369
10.40	0.85	0.83	0.175	0.74	1.83	5.52	0.810	9.40	1.03	1.50	0.352
10.60	0.81	0.79	0.167	0.78	2.12	5.00	0.759	9.60	1.05	1.44	0.329
10.80	0.80	0.76	0.162	0.82	2.36	4.61	0.710	9.80	1.09	1.38	0.307
								10.00	1.13	1.34	0.287

Energy (eV)	n	k	$R(\phi = 0)$
10.20	1.19	1.33	0.274
10.40	1.24	1.34	0.270
10.60	1.27	1.36	0.274
10.80	1.29	1.39	0.282
11.00	1.28	1.42	0.290
11.20	1.27	1.44	0.297
11.40	1.25	1.46	0.305
11.60	1.22	1.48	0.313
11.80	1.20	1.48	0.318
12.00	1.16	1.48	0.323
12.40	1.10	1.47	0.329
12.80	1.04	1.44	0.333
13.20	0.98	1.40	0.332
13.60	0.94	1.35	0.325
14.00	0.91	1.28	0.312
14.40	0.90	1.23	0.296
14.80	0.90	1.17	0.276
15.20	0.93	1.13	0.255
15.60	0.97	1.12	0.246
16.00	0.98	1.14	0.249
16.40	0.97	1.17	0.260
16.80	0.94	1.19	0.273
17.20	0.90	1.21	0.289
17.60	0.85	1.21	0.304
18.00	0.80	1.20	0.317
18.40	0.74	1.18	0.330
18.80	0.69	1.15	0.340
19.20	0.64	1.11	0.347
19.60	0.60	1.07	0.353
20.00	0.56	1.02	0.354
20.40	0.54	0.97	0.350
20.80	0.52	0.92	0.342
21.20	0.50	0.87	0.331
21.60	0.50	0.82	0.318
22.00	0.49	0.77	0.303
22.40	0.49	0.73	0.287
22.80	0.49	0.69	0.272
23.20	0.49	0.66	0.263
23.60	0.48	0.62	0.252
24.00	0.49	0.57	0.234
24.40	0.50	0.53	0.213
24.80	0.51	0.49	0.191
25.20	0.53	0.46	0.171
25.60	0.55	0.43	0.150
26.00	0.57	0.40	0.132
26.40	0.59	0.38	0.117
26.80	0.61	0.37	0.105
27.00	0.62	0.36	0.099
27.50	0.64	0.34	0.085
28.00	0.67	0.32	0.073
28.50	0.69	0.31	0.065
29.00	0.71	0.30	0.057
29.50	0.73	0.30	0.052
30.00	0.75	0.29	0.047
31.00	0.78	0.29	0.042
32.00	0.79	0.29	0.040
33.00	0.82	0.28	0.033
34.00	0.84	0.29	0.032
35.00	0.85	0.31	0.033
36.00	0.85	0.32	0.036
37.00	0.84	0.33	0.039
38.00	0.83	0.33	0.040
39.00	0.81	0.33	0.042
40.00	0.80	0.33	0.045

Vanadium[9]

Energy (eV)	n	k	$R(\phi = 0)$
0.10	12.83	45.89	0.978
0.20	3.90	24.30	0.975
0.28			
	2.13	17.35	0.973
0.36	1.54	13.32	0.966
0.44	1.28	10.74	0.957
0.52	1.16	8.93	0.945
0.60	1.10	7.59	0.929
0.68	1.07	6.54	0.909
0.76	1.08	5.67	0.882
0.80	1.10	5.30	0.864
0.90	1.18	4.50	0.811
1.00	1.34	3.80	0.730
1.10	1.60	3.26	0.632
1.20	1.93	2.88	0.543
1.30	2.25	2.71	0.498
1.40	2.48	2.72	0.491
1.50	2.57	2.79	0.499
1.60	2.57	2.84	0.507
1.70	2.52	2.88	0.512
1.80	2.45	2.88	0.515
1.90	2.36	2.85	0.514
2.00	2.34	2.81	0.509
2.10	2.31	2.78	0.506
2.20	2.28	2.80	0.510
2.30	2.23	2.83	0.516
2.40	2.15	2.88	0.528
2.50	2.02	2.91	0.540
2.60	1.89	2.92	0.552
2.70	1.74	2.89	0.561
2.80	1.61	2.85	0.569
2.90	1.48	2.80	0.577
3.00	1.36	2.73	0.582
3.20	1.16	2.55	0.585
3.40	0.99	2.37	0.586
3.60	0.87	2.17	0.575
3.80	0.80	1.96	0.547
4.00	0.78	1.76	0.503
4.20	0.80	1.60	0.449
4.40	0.83	1.47	0.400
4.60	0.87	1.38	0.355
4.80	0.90	1.31	0.326
5.00	0.91	1.26	0.304
5.25	0.93	1.18	0.271
5.50	0.94	1.14	0.258
5.75	0.96	1.09	0.235
6.00	0.98	1.06	0.223
6.25	0.97	1.02	0.212
6.50	0.97	0.98	0.199
6.75	0.97	0.94	0.185
7.00	0.98	0.91	0.175
7.33	0.97	0.89	0.170
7.66	0.98	0.87	0.162
8.00	0.98	0.85	0.155
8.33	0.98	0.81	0.146
8.66	0.98	0.81	0.145
9.00	0.96	0.79	0.142
9.50	0.94	0.77	0.136
10.00	0.91	0.74	0.133
10.50	0.89	0.71	0.126
11.00	0.87	0.65	0.112
11.50	0.88	0.58	0.091
12.00	0.90	0.58	0.089
12.50	0.89	0.57	0.086
13.00	0.88	0.55	0.082

Energy (eV)	n	k	$R(\phi = 0)$
13.50	0.87	0.53	0.079
14.00	0.86	0.51	0.075
14.50	0.86	0.49	0.070
15.00	0.86	0.47	0.065
15.50	0.86	0.46	0.062
16.00	0.85	0.45	0.061
16.50	0.84	0.43	0.059
17.00	0.84	0.41	0.056
17.50	0.83	0.40	0.054
18.00	0.82	0.38	0.051
18.50	0.82	0.37	0.048
19.00	0.82	0.35	0.045
19.50	0.82	0.34	0.043
20.00	0.81	0.32	0.041
20.50	0.81	0.31	0.038
21.00	0.81	0.29	0.036
21.50	0.81	0.28	0.033
22.00	0.81	0.27	0.032
22.50	0.81	0.25	0.029
23.00	0.82	0.24	0.027
23.50	0.82	0.23	0.025
24.00	0.82	0.22	0.024
24.50	0.83	0.21	0.022
25.00	0.83	0.20	0.020
25.50	0.83	0.19	0.019
26.00	0.83	0.18	0.018
26.50	0.84	0.17	0.016
27.00	0.84	0.16	0.015
27.50	0.85	0.16	0.014
28.00	0.85	0.15	0.013
28.50	0.86	0.14	0.012
29.00	0.86	0.14	0.011
29.50	0.86	0.13	0.010
30.00	0.87	0.13	0.009
31.00	0.88	0.12	0.008
32.00			
	0.90		
		0.11	
			0.007
33.00	0.90	0.10	0.005
34.00	0.91	0.10	0.005
35.00	0.92	0.09	0.004
36.00	0.94	0.10	0.004
37.00	0.94	0.10	0.004
38.00	0.95	0.11	0.004
39.00	0.95	0.12	0.004
40.00	0.95	0.13	0.005

Zinc, $\overset{\perp}{\mathbf{E}} \parallel \hat{c}$ [28]

Energy (eV)	n	k	$R(\phi = 0)$
0.7514	1.9241	7.5619	0.883
0.827	1.7921	6.9973	0.874
0.866	1.5571	6.7753	0.881
0.952	1.4824	6.2296	0.868
0.992	1.5762	5.8843	0.847
1.033	1.5407	5.3192	0.823
1.078	1.5853	4.9013	0.793
1.127	1.7768	4.5307	0.748
1.181	1.9808	4.2004	0.701
1.240	2.8821	3.4766	0.575
1.305	3.2042	3.0042	0.520
1.377	2.9459	3.5761	0.584
1.459	3.2523	4.2447	0.640
1.550	3.8086	4.6212	0.657
1.653	3.7577	4.6239	0.659
1.722	3.5908	4.4614	0.650
1.823	3.4234	4.3232	0.642

OPTICAL PROPERTIES OF METALS AND SEMICONDUCTORS (continued)

Energy (eV)	n	k	$R(\phi = 0)$
2.066	1.4856	4.0555	0.737
2.094	1.2525	3.9961	0.762
2.119	1.0017	3.8683	0.789
2.275	0.7737	3.9129	0.832
2.445	0.6395	3.4013	0.821
2.666	0.4430	3.1379	0.851
2.917	0.3589	2.8140	0.853
3.220	0.3069	2.5088	0.847
3.594	0.2737	2.1737	0.828
4.065	0.2510	1.8528	0.799
4.678	0.2354	1.6357	0.776

Zinc, $\overset{\perp}{E} \perp \hat{c}$ [28]

Energy (eV)	n	k	$R(\phi = 0)$
0.751	1.4469	7.4158	0.905
0.827	1.4744	6.9688	0.892
0.866	1.3628	6.6886	0.892
0.952	1.3165	6.2212	0.881
0.992	1.3835	5.8910	0.863
1.033	1.2889	5.4001	0.850
1.078	1.3095	4.9025	0.822
1.127	1.6897	4.4062	0.746
1.181	1.9701	4.0176	0.684
1.240	2.8717	3.2873	0.555
1.305	3.3991	2.7684	0.497
1.377	3.1807	3.4709	0.569
1.459	3.5064	4.1994	0.630
1.550	4.1241	4.7768	0.664
1.653	4.0269	4.8027	0.667
1.722	3.9369	4.6356	0.657
1.823	3.7549	4.3042	0.635
1.937	3.4512	4.1942	0.631
1.984	3.2515	4.2980	0.644
2.066	2.0802	4.7231	0.738
2.094	1.7084	4.7923	0.774
2.119	1.3329	4.4751	0.791
2.275	0.9725	4.2879	0.825
2.455	0.7568	3.7627	0.824
2.666	0.5470	3.4277	0.845
2.917	0.4774	3.0476	0.834
3.220	0.3911	2.7463	0.835
3.594	0.3147	2.3041	0.821
4.065	0.3013	2.0077	0.789
4.678	0.2806	1.7997	0.770

Zirconium (Polycrystalline)[28]

Energy (eV)	n	k	$R(\phi = 0)$
0.10	6.18	1.76	0.300
0.15	3.37	1.30	0.123
0.20	2.34	1.08	0.058
0.26	2.24	1.06	0.052
0.30	2.59	1.14	0.073
0.36	3.17	1.26	0.110
0.40	3.09	1.24	0.105
0.46	3.36	1.30	0.123
0.50	4.13	1.44	0.175
0.56	5.01	1.58	0.231
0.60	5.18	1.61	0.242
0.70	4.54	1.51	0.202

Energy (eV)	n	k	$R(\phi = 0)$
0.80	4.03	1.42	0.168
0.90	3.74	1.37	0.149
0.96	3.69	1.36	0.145
1.00	3.66	1.35	0.143
1.10	3.65	1.35	0.142
1.20	3.53	1.33	0.134
1.30	3.25	1.27	0.116
1.40	3.10	1.25	0.106
1.50	3.02	1.23	0.100
1.60	2.88	1.20	0.091
1.70	2.68	1.16	0.078
1.80	2.49	1.12	0.067
2.00	2.14	1.03	0.047
2.10	1.99	1.00	0.040
2.20	1.87	0.97	0.034
2.30	1.78	0.94	0.030
2.40			
	1.71	0.92	0.027
2.50	1.62	0.90	0.024
2.60	1.54	0.88	0.022
2.70	1.46	0.86	0.019
2.80	1.40	0.84	0.018
2.90	1.34	0.82	0.016
3.00	0.30	0.81	0.016
3.10	1.26	0.80	0.015
3.30	1.19	0.77	0.014
3.40	1.16	0.76	0.013
3.50	1.13	0.75	0.013
3.60	1.10	0.74	0.013
3.70	1.07	0.73	0.013
3.80	1.04	0.72	0.012
3.90	1.01	0.71	0.012
4.00	0.98	0.70	0.012
4.20	0.94	0.68	0.013
4.40	0.89	0.67	0.013
4.60	0.85	0.65	0.014
4.80	0.81	0.64	0.014
5.00	0.78	0.63	0.015
5.20	0.77	0.62	0.016
5.40	0.77	0.62	0.016
5.60	0.80	0.63	0.014
5.80	0.87	0.66	0.013
6.00	1.00	0.71	0.012
6.20	1.11	0.75	0.013
6.40	1.23	0.78	0.014
6.60	1.33	0.81	0.016
6.80	1.42	0.84	0.018
7.00	1.49	0.86	0.020
7.20	1.54	0.88	0.022
7.40	1.58	0.89	0.023
7.60	1.61	0.90	0.024
7.80	1.63	0.90	0.025
8.00	1.66	0.91	0.026
8.20	0.67	0.91	0.026
8.40	1.68	0.92	0.026
8.60	1.68	0.92	0.026
8.80	1.66	0.91	0.026
9.00	1.65	0.91	0.025

Energy (eV)	n	k	$R(\phi = 0)$
9.20	1.63	0.90	0.025
9.40	1.60	0.89	0.024
9.60	1.57	0.89	0.023
9.80	1.52	0.87	0.021
10.00	1.47	0.86	0.020
10.20	1.42	0.84	0.018
10.40	1.35	0.82	0.016
10.50	1.32	0.81	0.016
10.60	1.28	0.80	0.015
10.80	1.23	0.78	0.014
11.00	1.19	0.77	0.014
11.20	1.16	0.76	0.013
11.40	1.13	0.75	0.013
11.60	1.11	0.74	0.013
11.80	1.09	0.74	0.013
12.00	1.08	0.73	0.013
12.40	1.05	0.72	0.012
12.80	1.01	0.71	0.012
13.20	0.98	0.70	0.012
13.60	0.95	0.69	0.013
14.00	0.92	0.68	0.013
14.40	0.89	0.67	0.013
14.80	0.90	0.67	0.013
15.20	0.92	0.68	0.013
15.60	0.95	0.69	0.013
16.00	0.98	0.70	0.012
16.40	1.01	0.71	0.012
16.80	1.04	0.72	0.012
17.20	1.09	0.74	0.013
17.60	1.13	0.75	0.013
18.00	1.17	0.76	0.014
18.40	1.21	0.78	0.014
18.80	1.24	0.79	0.014
19.20	1.27	0.80	0.015
19.60	1.29	0.80	0.015
20.00	1.30	0.81	0.015
20.60	1.29	0.80	0.015
21.00	1.27	0.80	0.015
21.60	1.23	0.78	0.014
22.00	1.20	0.77	0.014
22.60	1.15	0.76	0.013
23.00	1.12	0.75	0.013
23.60	1.08	0.73	0.013
24.00	1.05	0.73	0.013
24.60	1.02	0.71	0.012
25.00	1.00	0.71	0.012
25.60	0.97	0.69	0.012
26.00	0.95	0.69	0.013
26.60	0.91	0.67	0.013
27.00	0.88	0.66	0.013
27.60	0.84	0.65	0.014
28.00	0.83	0.64	0.014
28.60	0.82	0.64	0.014
29.00	0.81	0.64	0.014
29.60	0.82	0.64	0.014
30.00	0.82	0.64	0.014

REFERENCES

1. Shiles, E., Sasaki, T., Inokuti, M., and Smith, D. Y., *Phys. Rev. Sect. B*, 22, 1612, 1980.
2. Edwards, D. F., and Philipp, H. R., in *HOC-I*, p.665.
3. Ives, H. E., and Briggs, N. B., *J. Opt. Soc. Am.*, 27, 395, 1937.
4. Bos, L. W., and Lynch, D. W., *Phys. Rev. Sect. B*, 2, 4567, 1970.
5. Weaver, J. H., Colavita, E., Lynch, D. W., and Rosei, R., *Phys. Rev. Sect. B*, 19, 3850, 1979.
6. Hagemann, H. J., Gudat, W., and Kunz, C., *J. Opt. Soc. Am.*, 65, 742, 1975.
7. Schulz, L. G., *J. Opt. Soc. Am.*, 47, 64, 1957.
8. Potter, R. F., in *HOC-I*, p.465.
9. Olson, C. G., Lynch, D. W., and Weaver, J. H., unpublished.
10. Lynch, D. W., Olson, C. G., and Weaver, J. H., unpublished.
11. Weaver, J. H., Olson, C. G., and Lynch, D. W., *Phys. Rev. Sect. B*, 15, 4115, 1977.
12. Lynch, D. W., and Hunter, W. R., in *HOC-II*, p.345.
13. Priol, M. A., Daudé, A., and Robin, S., *Compt. Rend.*, 264, 935, 1967.
14. Johnson, P. B., and Christy, R. W., *Phys. Rev. Sect. B*, 9, 5056, 1974.
15. Arakawn, E. T., and Inagaki, T., in *HOC-II*, p.461.
16. Weaver, J. H., Lynch, D. W., and Olson, D. G., *Phys. Rev. Sect. B*, 10, 501, 1973.
17. Lynch, D. W., Rosei, R., and Weaver, J. H., *Solid State Commun.*, 9, 2195, 1971.
18. Weaver, J. H., Lynch, D. W., and Olson, C. G., *Phys. Rev. Sect. B*, 7, 4311, 1973.
19. Weaver, J. H., and Benbow, R. L., *Phys. Rev. Sect. B*, 12, 3509, 1975.
20. Weaver, J. H., *Phys. Rev., Sect. B*, 11, 1416, 1975.
21. Lynch, D. W., and Hunter, W. R., in *HOC-II*, p.364.
22. Palik, E. D., in *HOC-II*, p. 691.
23. Edwards, D. F., in *HOC-I*, p. 547.
24. Lynch, D. W., and Hunter, W. R., in *HOC-II*, p.354.
25. Palik, E. D., in *HOC-II*, p. 709.
26. Lynch, D. W., Olson, C. G., and Weaver, J. H., *Phys. Rev. Sect. B*, 11, 3671, 1975.
27. Weaver, J. H., Lynch, D. W., and Olson, C. G., *Phys. Rev. Sect. B*, 12, 1293, 1975.
28. Lanham, A. P., and Terherne, D. M., *Proc. Phys. Soc.*, 83, 1059, 1964.

ELASTO-OPTIC, ELECTRO-OPTIC, AND MAGNETO-OPTIC CONSTANTS

When a crystal is subjected to a stress field, an electric field, or a magnetic field, the resulting optical effects are in general dependent on the orientation of these fields with respect to the crystal axes. It is useful, therefore, to express the optical properties in terms of the refractive index ellipsoid (or indicatrix):

$$\frac{x^2}{n_x^2} + \frac{y^2}{n_y^2} + \frac{z^2}{n_z^2} = 1$$

or

$$\sum_{ij} B_{ij} x_i y_j = 1 \quad (i, j = 1, 2, 3)$$

where

$$B_{ij} = \left[\frac{1}{\varepsilon}\right]_{ij} = \left[\frac{1}{n^2}\right]_{ij}$$

ε is the dielectric constant or permeability; the quantity B_{ij} has the name impermeability.

A crystal exposed to a *stress* S will show a change of its impermeability. The photo-elastic (or elasto-optic) constants, P_{ijkl}, are defined by

$$\Delta\left[\frac{1}{\varepsilon}\right]_{ij} = \Delta\left[\frac{1}{n^2}\right]_{ij} = \sum_{kl} P_{ijkl} S_{kl}$$

where n is the refractive index and S_{kl} are the strain tensor elements; the P_{ijkl} are the elements of a 4th rank tensor.

When a crystal is subjected to an *electric field E* two possible changes of the refractive index may occur depending on the symmetry of the crystal.

1. All materials, including isotropic solids and polar liquids, show an electro-optic birefringence (Kerr effect) which is proportional to the square of the electric field, E:

$$\left[\frac{1}{n^2}\right]_{ij} = \sum_{k,l=1,2,3} K_{ijkl} E_k E_l = \sum_{k,l=1,2,3} g_{ijkl} P_k P_l$$

where E_k and E_l are the components of the electric field and P_k and P_l the electric polarizations. The coefficients, K_{ijkl}, are the quadratic electro-optic coefficients, while the constants g_{ijkl} are known as the Kerr constants.

2. The other electro-optic effect only occurs in the 20 piezo-electric crystal classes (no center of symmetry). This effect is known as the Pockels effect. The optical impermeability changes linearly with the static field

$$\Delta\left[\frac{1}{n^2}\right]_{ij} = \sum_k r_{ij,k} E_k$$

The coefficients $r_{ij,k}$ have the name (linear) electro-optic coefficients.

The values of the electro-optic coefficients depend on the boundary conditions. If the superscripts T and S denote respectively the conditions of zero stress (free) and zero strain (clamped) one finds:

$$r_{ij}^T = r_{ij}^S + q_{ik}^E e_{jk} = r_{ij}^S + P_{ik}^E d_{jk}$$

where $e_{jk} = (\partial T_k/\partial E_j)_S$ and $d_{jk} = (\partial S_k/\partial E_j)_T$ are the appropriate piezo-electric coefficients.

The interaction between a *magnetic field* and a light wave propagating in a solid or in a liquid gives rise to a rotation of the plane of polarization. This effect is known as *Faraday rotation*. It results from a difference in propagation velocity for left and right circular polarized light.

The Faraday rotation, θ_F, is linearly proportional to the magnetic field H:

$$\theta_F = VlH$$

where l is the light path length and V is the *Verdet* constant (minutes/oersted·cm).

For ferromagnetic, ferrimagnetic, and antiferromagnetic materials the magnetic field in the above expression is replaced by the magnetization M and the magneto-optic coefficient in this case is known as the Kund constant K:

$$\text{Specific Faraday rotation } F = KM$$

In the tables below the *Faraday rotation* is listed at the saturation magnetization per unit length, together with the absorption coefficient α, the temperature T, the critical temperature T_C (or T_N), and the wavelength of the measurement.

ELASTO-OPTIC, ELECTRO-OPTIC, AND MAGNETO-OPTIC CONSTANTS (continued)

In the tables which follow, the properties are presented in groups:
- Elasto-optic coefficients (photoelastic constants)
- Linear electro-optic coefficients (Pockels constants)
- Quadratic electro-optic coefficients (Kerr constants)
- Magneto-optic coefficients:
  - Verdet constants
  - Faraday rotation parameters

Within each group, materials are classified by crystal system or physical state. References are given at the end of each group of tables.

ELASTO-OPTIC COEFFICIENTS (PHOTOELASTIC CONSTANTS)

Name

Cubic (43m, 432, m3m)	Formula	$\lambda/\mu m$	p_{11}	p_{12}	p_{44}	$p_{11}\text{-}p_{12}$	Ref.
Sodium fluoride	NaF	0.633	0.08	0.20	-0.03	-0.12	1
Sodium chloride	NaCl	0.589	0.115	0.159	-0.011	-0.042	2
Sodium bromide	NaBr	0.589	0.148	0.184	-0.0036	-0.035	1
Sodium iodide	NaI	0.589	—	—	0.0048	-0.0141	3
Potassium fluoride	KF	0.546	0.26	0.20	-0.029	0.06	1
Potassium chloride	KCl	0.633	0.22	0.16	-0.025	0.06	4
Potassium bromide	KBr	0.589	0.212	0.165	-0.022	0.047	5
Potassium iodide	KI	0.590	0.212	0.171	—	0.041	6
Rubidium chloride	RbCl	0.589	0.288	0.172	-0.041	0.116	7,8
Rubidium bromide	RbBr	0.589	0.293	0.185	-0.034	0.108	7,8
Rubidium iodide	RbI	0.589	0.262	0.167	-0.023	0.095	7,8
Lithium fluoride	LiF	0.589	0.02	0.13	-0.045	-0.11	5
Lithium chloride	LiCl	0.589	—	—	-0.0177	-0.0407	3
Ammonium chloride	NH_4Cl	0.589	0.142	0.245	0.042	-0.103	9
Cadmium telluride	CdTe	1.06	-0.152	-0.017	-0.057	-0.135	10
Calcium fluoride	CaF_2	0.55-0.65	0.038	0.226	0.0254	-0.183	11
Copper chloride	CuCl	0.633	0.120	0.250	-0.082	-0.130	12
Copper bromide	CuBr	0.633	0.072	0.195	-0.083	-0.123	12
Copper iodide	CuI	0.633	0.032	0.151	-0.068	-0.119	12
Diamond	C	0.540-0.589	-0.278	0.123	-0.161	-0.385	13
Germanium	Ge	3.39	-0.151	-0.128	-0.072	-0.023	14
Gallium arsenide	GaAs	1.15	-0.165	-0.140	-0.072	-0.025	15
Gallium phosphide	GaP	0.633	-0.151	-0.082	-0.074	-0.069	15
Strontium fluoride	SrF_2	0.633	0.080	0.269	0.0185	-0.189	16
Strontium titanite	$SrTiO_3$	0.633	0.15	0.095	0.072	—	17
KRS-5	Tl(Br,I)	0.633	-0.140	0.149	-0.0725	-0.289	18,20
KRS-6	Tl(Br,Cl)	0.633	-0.451	-0.337	-0.164	-0.114	19,20
Zinc sulfide	Zn	0.633	0.091	-0.01	0.075	0.101	15

Rare Gases	Formula	$\lambda/\mu m$	p_{11}	p_{12}	p_{44}	$p_{11}\text{-}p_{12}$	Ref.
Neon (T = 24.3 K)	Ne	0.488	0.157	0.168	0.004	-0.011	21
Argon (T = 82.3 K)	Ar	0.488	0.256	0.302	0.015	-0.046	22
Krypton (T = 115.6 K)	Kr	0.488	0.34	0.34	0.037	0	21
Xenon (T = 160.5 K)	Xe	0.488	0.284	0.370	0.029	-0.086	22

Garnets	Formula	$\lambda/\mu m$	p_{11}	p_{12}	p_{44}	$p_{11}\text{-}p_{12}$	Ref.
GGG	$Gd_3Ga_5O_{12}$	0.514	-0.086	-0.027	-0.078	-0.059	23
YIG	$Y_3Fe_5O_{12}$	1.15	0.025	0.073	0.041	—	15
YGG	$Y_3Ga_5O_{12}$	0.633	0.091	0.019	0.079	—	17
YAG	$Y_3Al_5O_{12}$	0.633	-0.029	0.0091	-0.0615	-0.038	15

Name Cubic (23, m3)	Formula	$\lambda/\mu m$	p_{11}	p_{12}	p_{44}	p_{13}	Ref.
Barium nitrate	Ba(NO3)2	0.589	—	$p_{11}\text{-}p_{22} = 0.992$	-0.0205	$p_{11}\text{-}p_{13} = 0.713$	13
Lead nitrate	$Pb(NO_3)_2$	0.589	0.162	0.24	-0.0198	0.20	24,25
Sodium bromate	$NaBrO_3$	0.589	0.185	0.218	-0.0139	0.213	26
Sodium chlorate	$NaClO_3$	0.589	0.162	0.24	-0.0198	0.20	26
Strontium nitrate	$Sr(NO_3)_2$	0.41	0.178	0.362	-0.014	0.316	27

Hexagonal (mmc,6mm)	Formula	$\lambda/\mu m$	p_{11}	p_{12}	p_{13}	p_{31}	p_{33}	p_{44}	Ref.
Beryl	$Be_3Al_2Si_6O_{18}$	0.589	0.0099	0.175	0.191	0.313	0.023	-0.152	28
Cadmium sulfide	CdS	0.633	-0.142	-0.066	-0.057	-0.041	-0.20	-0.099	15,2
Zinc oxide	ZnO	0.633	±0.222	±0.099	-0.111	±0.088	-0.235	0.0585	30
Zinc sulfide	ZnS	0.633	-0.115	0.017	0.025	0.0271	-0.13	-0.0627	31

Trigonal (3m,32,$\overline{3}$m)	Formula	$\lambda/\mu m$	p_{11}	p_{12}	p_{13}	p_{14}	p_{31}
Sapphire	Al_2O_3	0.644	-0.23	-0.03	0.02	0.00	-0.04
Calcite	$CaCO_3$	0.514	0.062	0.147	0.186	-0.011	0.241
Lithium niobate	$LiNbO_3$	0.633	±0.034	±0.072	±0.139	±0.066	±0.178
Lithium tantalate	$LiTaO_3$	0.633	-0.081	0.081	0.093	-0.026	0.089
Cinnabar	HgS	0.633			±0.445		
Quartz	SiO_2	0.589	0.16	0.27	0.27	-0.030	0.29
Proustite	Ag_3AsS_3	0.633	±0.10	±0.19	±0.22		±0.24
Sodium nitrite	$NaNO_3$	0.633		±0.21	±0.215	±0.027	±0.25
Tellurium	Te	10.6	0.155	0.130	—	—	

Trigonal (3m,32,$\overline{3}$m) (continued)	p_{33}	p_{41}	p_{44}	Ref.
Sapphire	-0.20	0.01	-0.10	15,32
Calcite	0.139	-0.036	-0.058	33
Lithium niobate	+-0.060	±0.154	±0.300	15,34
Lithium tantalate	-0.044	-0.085	0.028	15,35
Cinnabar	+-0.115	—	—	36
Quartz	0.10	-0.047	-0.079	37
Proustite	+-0.20	—	—	38
Sodium nitrite		0.055	-0.06	39
Tellurium		—	—	15

Tetragonal (4/mmm,$\overline{4}$2m,422)	Formula	$\lambda/\mu m$	p_{11}	p_{12}	p_{13}	p_{31}
Ammonium dihydrogen phosphate	ADP	0.589	0.319	0.277	0.169	0.197
Barium titanate	$BaTiO_3$	0.633	0.425	—	—	—
Cesium dihydrogen arsenate	CDA	0.633	0.267	0.225	0.200	0.195
Magnesium fluoride	MgF_2	0.546	—	—	—	—
Calomel	Hg_2Cl_2	0.633	±0.551	±0.440	±0.256	±0.137
Potassium dihydrogen phosphate	KDP	0.589	0.287	0.282	0.174	0.241
Rubidium dihydrogen arsenate	RDA	0.633	0.227	0.239	0.200	0.205
Rubidium dihydrogen phosphate	RDP	0.633	0.273	0.240	0.218	0.210
Strontium barium niobate	$Sr_{0.75}Ba_{0.25}Nb_2O_6$	0.633	0.16	0.10	0.08	0.11
Strontium barium niobate	$Sr_{0.5}Ba_{0.5}Nb_2O_6$	0.633	0.06	0.08	0.17	0.09
Tellurium oxide	TeO_2	0.633	0.0074	0.187	0.340	0.090
Rutile	TiO_2	0.633	0.017	0.143	-0.139	-0.080

Name Tetragonal (4/mmm,$\bar{4}$2m,422) (continued)	p_{33}	p_{44}	p_{66}	Ref.
Ammonium dihydrogen phosphate	0.167	-0.058	-0.091	40
Barium titanate	—	—	—	41
Cesium dihydrogen arsenate	0.227	—	—	42
Magnesium fluoride	—	±0.0776	±0.0488	43
Calomel	±0.010	—	±0.047	44
Potassium dihydrogen phosphate	0.122	-0.019	-0.064	45
Rubidium dihydrogen arsenate	0.182	—	—	41
Rubidium dihydrogen phosphate	0.208	—	—	41
Strontium barium niobate	0.47	—	—	46
Strontium barium niobate	0.23	—	—	46
Tellurium oxide	0.240	-0.17	-0.046	47
Rutile	-0.057	-0.009	-0.060	48

Tetragonal (4,$\bar{4}$,4/m)	Formula	$\lambda/\mu m$	p_{11}	p_{12}	p_{13}	p_{16}	p_{31}
Cadmium molybdate	$CdMoO_4$	0.633	0.12	0.10	0.13	—	0.11
Lead molybdate	$PbMoO_4$	0.633	0.24	0.24	0.255	0.017	0.175
Sodium bismuth molybdate	$NaBi(MoO_4)_2$	0.633	0.243	0.205	0.25	—	0.21

Tetragonal (4,$\bar{4}$,4/m) (continued)	p_{33}	p_{44}	p_{45}	p_{61}	p_{66}	Ref.
Cadmium molybdate	0.18	—	—	—	—	49
Lead molybdate	0.300	0.067	-0.01	0.013	0.05	52
Sodium bismuth molybdate	0.29	—	—	—	—	—

Orthorhombic (222,m22,mmm)	Formula	$\lambda/\mu m$	p_{11}	p_{12}	p_{13}	p_{21}	p_{22}	p_{23}
Ammonium chlorate	NH_4ClO_4	0.633	—	0.24	0.18	0.23	—	0.20
Ammonium sulfate	$(NH_4)_2SO_4$	0.633	0.26	0.19	±0.260	±0.230	±0.27	±0.254
Rochelle salt	$NaKC_4H_4O_6$	0.589	0.35	0.41	0.42	0.37	0.28	0.34
Iodic acid (α)	HIO_3	0.633	0.302	0.496	0.339	0.263	0.412	0.304
Sulfur (α)	S	0.633	0.324	0.307	0.268	0.272	0.301	0.310
Barite	$BaSO_4$	0.589	0.21	0.25	0.16	0.34	0.24	0.19
Topaz	$Al_2SiO_4(OH,F)_2$	—	-0.085	0.069	0.052	0.095	-0.120	0.065

Orthorhombic (222,m22,mmm) (continued)	p_{31}	p_{32}	p_{33}	p_{44}	p_{55}	p_{66}	Ref.
Ammonium chlorate	0.19	0.18	±0.02	<±0.02	—	±0.04	51
Ammonium sulfate	0.20	±0.26	0.26	0.015	±0.0015	0.012	52
Rochelle salt	0.36	0.35	0.36	-0.030	0.0046	-0.025	53
Iodic Acid (α)	0.251	0.345	0.336	0.084	-0.030	0.098	54
Sulfur (α)	0.203	0.232	0.270	0.143	0.019	0.118	54
Barite	0.28	0.22	0.31	0.002	-0.012	0.037	55
Topaz	0.095	0.085	-0.083	-0.095	-0.031	0.098	28

Monoclinic (2,m,2/m)	Formula	$\lambda/\mu m$
Taurine	$C_2H_7NO_3S$	0.589

$p_{11} = 0.313$ $p_{25} = -0.0025$ $p_{51} = -0.014$
$p_{12} = 0.251$ $p_{31} = 0.362$ $p_{52} = 0.006$
$p_{13} = 0.270$ $p_{32} = 0.275$ $p_{53} = 0.0048$
$p_{15} = -0.10$ $p_{33} = 0.308$ $p_{55} = 0.047$

Name

Monoclinic (2,m,2/m)	Formula	$\lambda/\mu m$			
Taurine (continued)			$p_{21} = 0.281$	$p_{35} = -0.003$	$p_{64} = 0.0024$
			$p_{22} = 0.252$	$p_{44} = 0.0025$	$p_{66} = 0.0028$
			$p_{23} = 0.272$	$p_{46} = -0.0056$	

Isotropic	Formula	$\lambda/\mu m$	p_{11}	p_{12}	p_{44}	Ref.
Fused silica	SiO_2	0.633	0.121	0.270	-0.075	15
Water	H_2O	0.633	±0.31	±0.31		15
Polystyrene		0.633	±0.30	±0.31		25
Lucite		0.633	±0.30	0.28		25
Orpiment	As_2S_3-glass	1.15	0.308	0.299	0.0045	15
Tellurium oxide	TeO_2-glass	0.633	0.257	0.241	0.0079	56
Laser glasses	LGS-247-2	0.488	±0.168	±0.230		57
	LGS-250-3		±0.135	±0.198		
	LGS-1		±0.214	±0.250		
	KGSS-1621		±0.205	±0.239		
Dense flint glasses	LaSF	0.633	0.088	0.147	-0.030	58
(examples)	SF_4		0.215	0.243	-0.014	
	U10502		0.172	0.179	-0.004	
	$TaFd_7$		0.099	0.138	-0.020	

REFERENCES

A. Narasimhamurty, T. S., *Photoelastic and Electro-Optic Properties of Crystals*, Plenum Press, New York, 1981; pp. 290-293.

B. Weber, M. J., Ed., *CRC Handbook of Laser Science and Technology*, Volume IV, Part 2, CRC Press, Boca Raton, FL, 1986; pp. 324-331.

1. Petterson, H. E., *J. Opt. Soc. Am.*, 63, 1243, 1973.
2. Burstein, E. and Smith, P. L., *Phys. Rev.*, 74, 229, 1948.
3. Pakhnev, A. V., et al., *Sov. Phys. J. (transl.)*, 18, 1662, 1975.
4. Feldman, A., Horovitz, D., and Waxler, R. M., *Appl. Opt.*, 16, 2925, 1977.
5. Iyengar, K. S., *Nature (London)*, 176, 1119, 1955.
6. Bansigir, K. G. and Iyengar, K. S., *Acta Crystallogr.*, 14, 727, 1961.
7. Pakhev, A. V., et al., *Sov. Phys. J. (transl.)*, 20, 648, 1975.
8. Bansigir, K. G., *Acta Crystallogr.*, 23, 505, 1967.
9. Krishna Rao, K. V. and Krishna Murty, V. G., *Ind. J. Phys.*, 41, 150, 1967.
10. Weil, R. and Sun, M. J., *Proc. Int. Symp. CdTe (Detectors)*, Strasbourg Centre de Rech. Nucl., 1971, XIX-1 to 6, 1972.
11. Schmidt, E. D. D. and Vedam, K., *J. Phys. Chem. Solids*, 27, 1563, 1966.
12. Biegelsen, D. K., et al., *Phys. Rev. B*, 14, 3578, 1976.
13. Helwege, K. H., *Landolt-Börnstein, New Series Group III*, Vol. II, Springer-Verlag Berlin, 1979.
14. Feldman, A., Waxler, R. M., and Horovitz, D., *J. Appl. Phys.*, 49, 2589, 1978.
15. Dixon, R. W., *J. Appl. Phys.*, 38, 5149, 1967.
16. Shabin, O. V., et al., *Sov. Phys. Sol. State (transl.)*, 13, 3141, 1972.
17. Reintjes, J. and Schultz, M. B., *J. Appl. Phys.*, 39, 5254, 1968.
18. Rivoallan, L. and Favre, F., *Opt. Commun.*, 8, 404, 1973.
19. Rivoallan, L. and Favre, F., *Opt. Commun.*, 11, 296, 1974.
20. Afanasev, I. I., et al., *Sov. J. Opt. Technol.*, 46, 663, 1979.
21. Rand, S. C., et al., *Phys. Rev. B*, 19, 4205, 1979.
22. Sipe, J. E., *Can J. Phys.*, 56, 199, 1978.
23. Christyi, I. L., et al., *Sov. Phys. Sol. State (transl.)*, 17, 922, 1975.
24. Narasimhamurty, T. S., *Curr. Sci. (India)*, 23, 149, 1954.
25. Smith, T. M. and Korpel, A., *IEEE J. Quant. Electron.*, QE-1, 283, 1965.
26. Narasimhamurty, T. S., *Proc. Ind. Acad. Sci.*, A40, 164, 1954.
27. Rabman, A., *Bhagarantam Commem. Vol.*, Bangalore Print. and Publ., 173, 1969.
28. Eppendahl, R., *Ann. Phys. (IV)*, 61, 591, 1920.
29. Laurenti, J. P. and Rouzeyre, M., *J. Appl. Phys.*, 52, 6484, 1981.
30. Sasaki, H., et al., *J. Appl. Phys.*, 47, 2046, 1976.
31. Uchida, N. and Saito, S., *J. Appl. Phys.*, 43, 971, 1972.

ELASTO-OPTIC COEFFICIENTS (PHOTOELASTIC CONSTANTS) (continued)

32. Waxler, R. M. and Farabaugh, E. M., *J. Res. Natl. Bur. Stand.*, A74, 215, 1970.
33. Nelson, D. F., Lazay, P. D., and Lax, M., *Phys. Rev.*, B6, 3109, 1972.
34. O'Brien, R. J., Rosasco, G. J., and Weber, A., *J. Opt. Soc. Am.*, 60, 716, 1970.
35. Avakyants, L. P., et al., *Sov. Phys.*, 18, 1242, 1976.
36. Sapriel, J., *Appl. Phys. Litt.*, 19, 533, 1971.
37. Narasimhamurty, T. S., *J. Opt. Soc. Am.*, 59, 682, 1969.
38. Zubrinov, I. I., et al., *Sov. Phys. Sol. State (transl.)*, 15, 1921, 1974.
39. Kachalov, O. V. and Shpilko, I. O., *Sov. Phys. JETP (transl.)*, 35, 957, 1972.
40. Narasimhamurty, T. S., et al., *J. Mater. Sci.*, 8, 577, 1973.
41. Tada, K. and Kikuchi, K., *Jpn. J. Appl. Phys.*, 19, 1311, 1980.
42. Aleksandrov, K. S., et al., *Sov. Phys. Sol. State (transl.)*, 19, 1090, 1977.
43. Afanasev, I. I., et al., *Sov. Phys. Sol. State (transl.)*, 17, 2006, 1975.
44. Silvestrova, I. M., et al., *Sov. Phys. Cryst. (transl.)*, 20, 649, 1975.
45. Veerabhadra Rao, K. and Narasimhamurty, T. S., *J. Mater. Sci.*, 10, 1019, 1975.
46. Venturini, E. L., et al., *J. Appl. Phys.*, 40, 1622, 1969.
47. Vehida, N. and Ohmachi, Y., *J. Appl. Phys.*, 40, 4692, 1969.
48. Grimsditch, M. H. and Ramdus, A. K., *Phys. Rev. B*, 22, 4094, 1980.
49. Schinke, D. P. and Viehman, W., unpublished Data.
50. Coquin, G. A., et al., *J. Appl. Phys.*, 42, 2162, 1971.
51. Vasquez, F., et al., *J. Phys. Chem. Solids*, 37, 451, 1976.
52. Luspin, Y. and Hauret, G., *C.R.Ac. Sci. Paris*, B274, 995 1972.
53. Narasimhamurty, T. S., *Phys. Rev.*, 186, 945, 1969.
54. Haussühl, S. and Weber, H. J., *Z. Kristall.*, 132, 266, 1970.
55. Vedam, K., *Proc. Ind. Ac. Sci.*, A34, 161, 1951.
56. Yano, T., Fukumoto, A., and Watanabe, A., *J. Appl. Phys.*, 42, 3674, 1971.
57. Manenkov, A. A. and Ritus, A. I., *Sov. J. Quant. Electr.*, 8, 78, 1978.
58. Eschler, H. and Weidinger, F., *J. Appl. Phys.*, 46, 65, 1975.

LINEAR ELECTRO-OPTIC COEFFICIENTS

Name

Cubic ($\bar{4}3m$)	Formula	$\lambda/\mu m$	r_{41} pm/V
Cuprous bromide	CuBr	0.525	0.85
Cuprous chloride	CuCl	0.633	3.6
Cuprous iodide	CuI	0.55	-5.0
Eulytite (BSO)	$Bi_4Si_3O_{12}$	0.63	0.54
Germanium eulytite (BGO)	$Bi_4Ge_3O_{12}$	0.63	1.0
Gallium arsenide	GaAs	10.6	1.6
Gallium phosphide	GaP	0.56	-1.07
Hexamethylenetetramine	$C_6H_{12}N_4$	0.633	0.78
Sphalerite	ZnS	0.65	2.1
Zinc selenide	ZnSe	0.546	2.0
Zinc telluride	ZnTe	3.41	4.2
Cadmium telluride	CdTe	3.39	6.8

Cubic (23)	Formula	$\lambda/\mu m$	r_{41} pm/V
Ammonium chloride (77 K)	NH_4Cl	—	1.5
Ammonium cadmium langbeinite	$(NH_4)_2Cd_2(SO_4)_3$	0.546	0.70
Ammonium manganese langbeinite	$(NH_4)_2Mn_2(SO_4)_3$	0.546	0.53
Thallium cadmium langbeinite	$Tl_2Cd_2(SO_4)_3$	0.546	0.37
Potassium magnesium langbeinite	$K_2Mg_2(SO_4)_3$	0.546	0.40
Bismuth monogermanate	$Bi_{12}GeO_{20}$	—	3.3
Bismuth monosilicate	$Bi_{12}SiO_{20}$	—	3.3
Sodium chlorate	$NaClO_3$	0.589	0.4
Sodium uranyl acetate	$NaUO_2(CH_3COO)_3$	0.546	0.87

Name

Cubic (23)	Formula	$\lambda/\mu m$	r_{41} pm/V
Trenhydrobromide	$N(CH_2CH_2NH_2)_3$ 3HBr	—	1.5
Trenhydrochloride	$N(CH_2CH_2NH_2)_3$ 3HCl	—	1.7

Tetragonal ($\overline{4}2m$)	Formula	T_{tran} K	r_{41} pm/V	r_{63} pm/V
Ammonium dihydrogen phosphate (ADP)	$NH_4H_2PO_4$	148	24.5	-8.5
Ammonium dideuterium phosphate (AD*P)	$NH_4D_2PO_4$	242	—	11.9
Ammonium dihydrogen arsenate (ADA)	$NH_4H_2AsO_4$	—	—	9.2
Cesium dihydrogen arsenate (CsDA)	CsH_2AsO_4	143	—	18.6
Cesium dideuterium arsenate (CsD*A)	CsD_2AsO_4	212	—	36.6
Potassium dihydrogen phosphate (KDP)	KH_2PO_4	123	8.6	-10.5
Potassium dideuterium phosphate (KD*P)	KD_2PO_4	222	8.8	23.8
Potassium dihydrogen arsenate (KDA)	KH_2AsO_4	97	12.5	10.9
Potassium dideuterium arsenate (KD*A)	KD_2AsO_4	162	—	18.2
Rubidium dihydrogen phosphate (RDP)	RbH_2PO_4	147	—	15.5
Rubidium dihydrogen arsenate (RDA)	RbH_2AsO_4	110	—	13.0
Rubidium dideuterium arsenate (RD*A)	RbD_2AsO_4	178	—	21.4

Tetragonal (4mm)	Formula	T_{tran} K	r_{13} pm/V	r_{33} pm/V	r_{51} pm/V
Barium titanate	$BaTiO_3$	406	8	28	—
Potassium lithium niobate	$K_3Li_2Nb_5O_{15}$	693	8.9	5.9	—
Lead titanate	$PbTiO_3$	765	13.8	5.9	—
Strontium barium niobate (SBN75)	$Sr_{0.75}Ba_{0.25}Nb_2O_6$	330	6.7	1340	42
Strontium barium niobate (SBN46)	$Sr_{0.46}Ba_{0.54}Nb_2O_6$	602	~180	35	—

Hexagonal (6mm)	Formula	r_{13} pm/V	r_{33} pm/V	r_{42} pm/V	r_{51} pm/V
Greenockite	CdS	3.1	2.9	2.0	3.7
Greenockite (const. strain)	CdS	1.1	2.4	—	—
Wurzite	ZnS	0.9	1.8	—	—
Zincite	ZnO	-1.4	+2.6	—	—

Hexagonal (6)	Formula	r_{13} pm/V	r_{33} pm/V	r_{42} pm/V	r_{51} pm/V
Lithium iodate	$LiIO_3$	4.1	6.4	1.4	3.3
Lithium potassium sulfate	$LiKSO_4$	$r_{13}-r_{33} = 1.6$	—	—	—

Trigonal (3m)	Formula	T_{tran} K	r_{13} pm/V	r_{22} pm/V	r_{33} pm/V	r_{42} pm/V
Cesium nitrate	$CsNO_3$	425	—	0.43	—	—
Lithium niobate	$LiNbO_3$	1483	8.6	7.0	30.8	28
Lithium tantalate	$LiTaO_3$	890	8.4	—	30.5	—
Lithium sodium sulfate	$LiNaSO_4$	—	—	<0.02	—	—
Tourmaline	—	—	—	0.3	—	—

Trigonal (32)	Formula	T_{tran} K	r_{11} pm/V	r_{41} pm/V
Cesium tartrate	$Cs_2C_4H_4O_6$	—	1.0	—
Cinnabar	HgS	659	3.1	1.5

Name Trigonal (32)	Formula	T_{tran} K	r_{11} pm/V	r_{41} pm/V		
Potassium dithionate	$K_2S_2O_6$	—	0.26	—		
Strontium dithionate	$SrS_2O_6·4H_2O$	—	0.1	—		
Quartz	SiO_2	1140	-0.47	0.2		
Selenium	Se	398	2.5			

Orthorhombic (222)	Formula	T_{tran} K	r_{41} pm/V	r_{52} pm/V	r_{63} pm/V
Ammonium oxalate	$(NH_4)_2C_2O_4·4H_2O$	—	230	330	250
Rochelle salt	$KNaC_4H_4O_6·4H_2O$	$T_u = 297$ $T_1 = 255$	-2.0	-1.7	+0.32

Orthorhombic (mm2)	Formula	T_{trans} K	r_{13} pm/V	r_{23} pm/V	r_{33} pm/V	r_{42} pm/V	r_{51} pm/V
Barium sodium niobate (BSN)	Ba_2NaNbO_{15}	833	15	13	48	92	90
Potassium niobate	$KNbO_3$	476	28	1.3	64	380	105

Monoclinic (2)	Formula	T_{trans} K	r_{22} pm/V	r_{32} pm/V
Calcium pyroniobate	$Ca_2Nb_2O_7$	—	0.33	13.7
Triglycine sulfate (TGS)	$(NH_2CH_2COOH)_3·H_2SO_4$	322	7.2	13.6

REFERENCES

1. Narasimhamurty, T. S., *Photoelastic and Electro-Optic Properties of Crystals*, Plenum Press, New York, 1981, pp. 405-407.
2. Weber, M. J., Ed., *CRC Handbook of Laser Science and Technology*, Vol. IV, CRC Press, Boca Raton, FL, 1986, pp. 258-278.

QUADRATIC ELECTRO-OPTIC COEFFICIENTS
Kerr Constants of Ferroelectric Crystals[1,2]

Name	Formula	T_{tran} K	λ μm	g_{11} 10^{10} esu	g_{12} 10^{10} esu	$g_{11}-g_{12}$ 10^{10} esu	g_{44} 10^{10} esu
Barium titanate	$BaTiO_3$	406	0.633	1.33	-0.11	1.44	
Strontium titanate	$SrTiO_3$	—	0.633	—	—	1.56	—
Potassium tantalate niobate	$KTa_{0.65}Nb_{0.35}O_3$	330	0.633	1.50	-0.42	1.92	1.63
Potassium tantalate	$KTaO_3$	13	0.633	—	—	1.77	1.33
Lithium niobate	$LiNbO_3$	1483	—	0.94	0.25	0.7	0.6
Lithium tantalate	$LiTaO_3$	938	—	1.0	0.17	0.8	0.7
Barium sodium niobate (BSN)	$Ba_{0.8}Na_{0.4}Nb_2O_6$	833	—	1.55	0.44	1.11	

Kerr Constants of Selected Liquids[2]

K is the Kerr constant at a wavelength of 589 nm and at room temperature; ε is the static dielectric constant; T_m is the melting point; and T_b is the normal boiling point

Name	Molecular formula	K 10^{-7} esu	ε	T_m °C	T_b °C
Carbon disulfide	CS_2	+3.23	2.63	-111.5	+46.3
Acetone	C_3H_6O	+16.3	21.0	-94.8	+56.1
Methyl ethyl ketone	C_4H_8O	+13.6	18.56	-86.67	+79.6

QUADRATIC ELECTRO-OPTIC COEFFICIENTS (continued)
Kerr Constants of Selected Liquids (continued)[2]

Name	Molecular formula	K 10^{-7} esu	ε	T_m °C	T_b °C
Pyridine	C_5H_5N	+20.4	13.26	-42	+115.23
Ethyl cyanoacetate	$C_5H_7NO_2$	+38.8	31.6	-22.5	205
o-Dichlorobenzene	$C_6H_4Cl_2$	+42.6	10.12	-16.7	180
Benzenesulfonyl chloride	$C_6H_5ClO_2S$	+89.9	28.90	+14.5	247
Nitrobenzene	$C_6H_5NO_2$	+326	35.6	+5.7	210.8
Ethyl 3-aminocrotonate	$C_6H_{11}NO_2$	+31.0	—	+33.9	210
Paraldehyde	$C_6H_{12}O_3$	-23.0	14.7 12.0[a]	+12.6	124
Benzaldehyde	C_7H_6O	+80.8	17.85 14.1[a]	-26	179.05
p-Chlorotoluene	C_7H_7Cl	+23.0	6.25	+7.5	162.4
o-Nitrotoluene	$C_7H_7NO_2$	+174	26.26	-10	222.3
m-Nitrotoluene	$C_7H_7NO_2$	+177	24.95	+15.5	232
p-Nitrotoluene	$C_7H_7NO_2$	+222	22.2	+51.6	238.3
Benzyl alcohol	C_7H_8O	-15.4	11.92 10.8[a]	-15.3	205.8
m-Cresol	C_7H_8O	+21.2	12.44 5.0[a]	+11.8	202.27
m-Chloroacetophenone	C_8H_7ClO	+69.1			
Acetophenone	C_8H_8O	+66.6	17.44 15.8[a]	+19.7	202.3
Quinoline	C_9H_7N	+15.0	9.16	-14.78	237.16
Ethyl salicylate	$C_9H_{10}O_3$	+19.6	8.48	+1.3	231.5
Carvone	$C_{10}H_{14}O$	+23.6	11.2	<0	230
Ethyl benzoylacetate	$C_{11}H_{12}O_3$	+16.0	13.50	<0	270
Water	H_2O	+4.0	80.10	0.00	100.0

[a] Dielectric constant at radiofrequencies (10^8-10^9 Hz).

REFERENCES

1. Narasimhamurty, T. S., *Photoelastic and Electro-Optic Properties of Crystals*, Plenum Press, New York, 1981, p. 408.
2. Gray, D. E., Ed., *AIP Handbook of Physics*, McGraw Hill, New York, 1972, p. 6-241.

MAGNETO-OPTIC CONSTANTS
Verdet Constants of Non-Magnetic Crystals[1]
V is the Verdet constant; n is the refractive index; and λ is the wavelength

Material	T K	λ nm	n	V min/Oe cm
Al_2O_3	300	546.1	1.771	0.0240
	300	589.3	1.768	0.0210
$BaTaO_3$	403	427		0.95
	403	496		0.38
	403	620		0.18
	403	826		0.072
$Bi_4Ge_3O_{12}$	300	442	2.077	0.289
	300	632.8	2.048	0.099
	300	1064	2.031	0.026
C (diamond)	300	589.3	2.417	0.0233
$CaCO_3$	300	589.3	1.658	0.019
CaF_2	300	589.3	1.434	0.0088
$Cd_{0.55}Mn_{0.45}Te$	300	632.8		6.87

MAGNETO-OPTIC CONSTANTS (continued)
Verdet Constants of Non-Magnetic Crystals (continued)[1]

Material	T K	λ nm	n	V min/Oe cm
CuCl	300	546.1	1.93	0.20
GaSe	298	632.8		0.80
$KAl(SO_4)_2 \cdot 12H_2O$	300	589.3	1.456	0.0124
KBr	300	546.1	1.564	0.0500
	300	589.3	1.560	0.0425
KCl	300	589.3	1.490	0.0275
KI	300	546.1	1.673	0.083
	300	589.3	1.666	0.070
$KTaO_3$	296	352		0.44
	296	413		0.19
	296	496		0.096
	296	620		0.051
	296	826		0.022
LaF_3	300	325	1.639	0.054
(H∥c)	300	442	1.615	0.028
	300	632.8	1.601	0.012
	300	1064	1.592	0.006
$MgAl_2O_4$	300	589.3	1.718	0.021
$NH_4AlSO_4 \cdot 12H_2O$	300	589.3	1.459	0.0128
NH_4Br	300	589.3	1.711	0.0504
NH_4Cl	300	546.1		0.0410
	300	589.3	1.643	0.0362
NaBr	300	546.1		0.0621
NaCl	300	546.1		0.0410
	300	589.3	1.544	0.0345
$NaClO_3$	300	546.1		0.0105
	300	589.3	1.515	0.0081
$NiSO_4 \cdot 6H_2O$	297	546.1		0.0256
	297	589.3	1.511	0.0221
SiO_2	300	546.1	1.546	0.0195
	300	589.3	1.544	0.0166
$SrTiO_3$	298	413	2.627	0.78
	298	496		0.31
	298	620		0.14
	298	826		0.066
ZnS	300	546.1		0.287
	300	589.3	2.368	0.226
ZnSe	300	476	2.826	1.50
	300	496	2.759	1.04
	300	514	2.721	0.839
	300	587	2.627	0.529
	300	632.8	2.592	0.406

Verdet Constants of Rare-Earth Aluminum Garnets at Various Wavelengths[1]
The absorption coefficient α for these materials ranges from 0.2 to 0.6 cm^{-1} at 300 K

Material	T/K	λ = 405 nm	450 nm	480 nm	520 nm	546 nm	578 nm	635 nm	670 nm
					V in min/Oe cm				
$Tb_2Al_5O_{12}$	300	-2.266	-1.565	-1.290	-1.039	-0.912	-0.787	-0.620	-0.542
	77		-102.16	-83.45	-3.425	-3.051	-2.603	-2.008	-1.815
	4.2				-64.80	-58.35	-53.77	48.39	-45.15
	1.45		-200.95	-172.52	-139.28	-125.07	-111.27	97.47	-93.42
$Dy_3Al_5O_{12}$	300	-1.241	-0.942	-0.803	-0.667	-0.592	-0.518	-0.411	-0.359
$Ho_3Al_5O_{12}$	300	-0.709	-0.320	-0.260	-0.335	-0.304	-0.299		-0.206

MAGNETO-OPTIC CONSTANTS (continued)
Verdet Constants of Rare-Earth Aluminum Garnets at Various Wavelengths (continued)[1]

Material	T/K	V in min/Oe cm							
		λ = 405 nm	450 nm	480 nm	520 nm	546 nm	578 nm	635 nm	670 nm
$Er_3Al_5O_{12}$	300	-0.189	-0.240	-0.154	-0.162	-0.157	-0.145	-0.105	-0.089
$Tm_3Al_5O_{12}$	300	+0.151	+0.103	+0.093	0.076	0.069	+0.059	+0.048	
$Yb_3Al_5O_{12}$	298	0.287	0.215	0.186	0.140	0.133	0.116	0.094	
	77	0.718	0.540	0.481	0.393	0.342	0.302	0.239	

Verdet Constants for KDP-Type Crystals[1]
Measurements refer to T = 298 K and
λ = 632.8 nm, with k ∥ [001]

Material	V min/Oe cm
KH_2PO_4 (KDP)	0.0124
$KH_{0.3}D_{1.7}PO_4$ (KD*P)	0.145
$NH_4H_2PO_4$ (ADP)	0.138
KH_2AsO_4 (KDA)	0.238
$KH_{0.1}D_{1.9}AsO_4$ (KD*A)	0.245
$NH_4H_2AsO_4$ (ADH)	0.244

Verdet Constants of Gases[2]
Values refer to T = 0°C and P = 101.325 kPa (760 mmHg); n_D is the refractive index at a wavelength of 589 nm

Gas	$(n_D - 1) \times 10^3$	$10^6 \times V$ min/Oe cm
He	0.036	+0.40
Ar	2.81	+9.36
H_2		+6.29
N_2	0.297	+6.46
O_2	0.272	+5.69
Air	0.293	+6.27
Cl_2	0.773	+31.9
HCl	0.447	+21.5
H_2S	0.63	+41.5
NH_3	0.376	+19.0
CO	0.34	+11.0
CO_2	0.45	+9.39
NO	0.297	-58
CH4	0.444	+17.4
n-C4H10		+44.0

Verdet Constants of Liquids[2]
n_D is the refractive index at a wavelength of 589 nm and a temperature of 20°C, unless otherwise indicated. V is the Verdet constant

Liquid	λ/nm	T/°C	$10^2 \times V$ min/Oe cm	n_D
P	589	33	+13.3	
S	589	114	+8.1	1.929 (110°C)
H_2O	589	20	+1.309	1.3328
D_2O	589	19.7	+1.257	1.3384
H_3PO_4	578	97.4	+1.35	
CS_2	589	20	+4.255	1.6255
CCl_4	578-589	25.1	+1.60	1.463 (15°C)
$SbCl_5$	578	18	+7.45	1.601 (14°C)
$TiCl_4$	578	17	-1.65	1.61
$TiBr_4$	578	46	-5.3	
Methanol	589	18.7	+0.958	1.3289
Acetone	578-589	20.0	+1.116	1.3585
Toluene	578-589	15.0	+2.71	1.4950
Benzene	578-589	15.0	+3.00	1.5005
Chlorobenzene	589	15	+2.92	1.5246
Nitrobenzene	589	15	+2.17	1.5523
Bromoform	589	17.9	+3.13	1.5960

MAGNETO-OPTIC CONSTANTS (continued)

Verdet Constants of Rare Earth Paramagnetic Crystals[1]

n is the refractive index, and V is the Verdet constant at the wavelength and temperature indicated

Rare Earth	Host	T/K	λ/nm	n	V min/Oe cm
Ce^{3+}(30%)	CaF_2	300	325	1.516	-0.956
		300	442	1.502	-0.297
		300	633	1.494	-0.111
		300	1064	1.489	-0.035
Ce^{3+}	CeF_3	300	442	1.613	-1.05
		300	633	1.598	-0.406
		77	633		-1.418
		300	1064	1.589	-0.113
Pr^{3+}(5%)	CaF_2	300	266	1.471	-0.172
		300	325	1.461	-0.0818
		300	442	1.451	-0.0089
		300	633	1.445	-0.0168
		300	1064	1.441	-0.0045
Nd^{3+}(2.9%)	CaF_2	4.2	426		-0.19
Nd^{3+}	NdF_3	300	442	1.60	-0.553
		290	633	1.59	-0.209
		77	633		-0.755
		300	1064	1.58	-0.097
Eu^{3+}(3%)	CaF_2	4.2	430		29
		4.2	440		22
Eu^{2+}	EuF_2	300	450		-4.5
		300	500		-2.6
		300	550		-1.6
		300	600		-1.1
		300	650		-0.8
		300	1064		-0.19
Tb^{3+}	KTb_3F_{10}	300	325	1.531	-2.174
		300	442	1.518	-0.933
		300	633	1.510	-0.386
		77	633		-1.94
		300	1064	1.505	-0.114
Tb^{3+}	$LiTbF_4$	300	325	1.493	-1.9
		300	442	1.481	-0.98
		300	633	1.473	-0.44
		300	1064	1.469	-0.13
Tb^{3+}	$Tb_3Ga_5O_{12}$	300	500	1.989	-0.749
		300	570	1.981	-0.581
		300	633	1.976	-0.461
		300	830	1.967	-0.21
		300	1060	1.954	-0.12

MAGNETO-OPTIC CONSTANTS (continued)
Verdet Constants of Paramagnetic Glasses[1]

The Verdet constant V is given at room temperature for the wavelengths indicated

Rare earth phosphate glasses of composition $R_2O_3 \cdot xP_2O_5$, where x is given in the second column

		Verdet constant V in min/Oe cm									
R	x	λ=405 nm	λ=436 nm	λ=480 nm	λ=500 nm	λ=520 nm	λ=546 nm	λ=578 nm	λ=600 nm	λ=635 nm	λ=670 nm
La		0.037	0.030	0.024	0.022	0.020	0.018	0.015	-0.014	0.013	—
Ce	2.67	-0.672	0.510	-0.366	-0.326	-0.287	-0.253	-0.217	-0.197	-0.173	-0.150
Pr	3.09	-0.447	-0.332	-0.283	-0.261	-0.236	-0.208	-0.182	-0.170	-0.150	-0.132
Nd	2.92	-0.250	-0.209	-0.167	-0.155	-0.136	-0.134	-0.094	-0.080	-0.080	-0.071
Sm	2.87	0.026	0.024	0.020	0.020	0.017	0.015	0.014	0.012	0.011	0.010
Eu	2.93	-0.025	-0.017	-0.010	-0.006	-0.006	-0.005	-0.004	-0.003	-0.002	-0.002
Gd	3.01	0.018	0.015	0.014	0.012	0.012	0.011	0.011	0.010	0.009	0.009
Tb	2.94	-0.560	-0.458	-0.357	-0.323	-0.295	-0.261	-0.226	-0.206	-0.190	-0.164
Dy	2.51	-0.540	-0.453	-0.359	-0.331	-0.301	0.268	-0.237	-0.217	-0.197	-0.173
Ho	2.94	-0.299	-0.313	-0.156	-0.153	-0.138	-0.138	-0.119	-0.110	-0.098	-0.084
Er	3.01	-0.139	-0.121	-0.100	-0.111	-0.095	-0.062	-0.060	-0.057	-0.051	-0.044
Tm	2.79	0.019	0.013	0.012	0.009	0.008	0.006	0.005	0.004	0.004	0.007
Yb	3.01	0.087	0.072	0.056	0.050	0.045	0.041	0.036	0.032	0.029	0.024

The following are rare earth borate glasses with composition:
for La and Pr: $R_2O_3 \cdot xP_2O_5$; for Tb-Pr and Dy-Pr: $R_2O_3 \cdot xB_2O_3$; and for other elements: $R_2O_3 \cdot 0.85La_2O_3 \cdot xB_2O_3$.

R	x	λ=405 nm	λ=436 nm	λ=480 nm	λ=500 nm	λ=520 nm	λ=546 nm	λ=578 nm	λ=600 nm	λ=635 nm	λ=670 nm
La	3.04	0.043	0.036	0.029	0.026	0.023	0.022	0.019	0.018	0.016	0.014
Pr-La	5.44	-0.380	-0.307	-0.230	-0.220	-0.201	-0.178	-0.153	-0.146	-0.128	-0.110
Nd-La	5.41	-0.180	-0.147	-0.120	-0.111	-0.096	-0.094	-0.100	-0.059	-0.056	-0.046
Sm-La	4.97	0.032	0.030	0.025	0.024	0.022	0.019	0.017	0.016	0.014	0.012
Eu-La	4.69	-0.081	-0.060	-0.038	-0.033	-0.029	-0.024	0.019	-0.016	0.014	-0.012
Gd-La	4.71	0.032	0.026	0.024	0.022	0.021	0.020	0.018	0.017	0.015	0.013
Tb-La	4.73	-0.512	-0.419	-0.319	-0.288	-0.262	-0.234	-0.205	-0.186	-0.167	-0.142
Dy-La	4.88	-0.436	-0.361	-0.299	-0.273	-0.246	-0.220	-0.193	-0.177	-0.159	-0.138
Ho-La	4.36	-0.269	-0.252	-0.123	-0.131	-0.112	-0.128	-0.104	-0.096	—	-0.074
Er-La	4.50	-0.093	-0.078	-0.068	-0.082	—	-0.045	-0.042	-0.040	-0.035	-0.034
Tm-La	4.75	0.060	0.046	0.039	0.034	0.031	0.026	0.023	0.021	0.018	0.016
Yb-La	8.58	0.115	0.094	0.073	0.066	0.060	0.054	0.046	0.043	0.037	0.033
Tb-Pr	4.99	-0.940	-0.786	-0.560	-0.536	-0.489	-0.436	-0.380	-0.348	-0.306	-0.265
Dy-Pr	4.63	-0.850	—	—	-0.497	-0.465	-0.413	-0.358	-0.332	-0.290	-0.252
Pr	2.56	-0.843	-0.646	-0.471	-0.480	-0.432	-0.390	-0.334	-0.317	-0.271	-0.243

MAGNETO-OPTIC CONSTANTS (continued)

Verdet Constants of Diamagnetic Glasses[1]

The Verdet constant V is given at room temperature for the wavelengths indicated

Glass type	Composition (wt. %)	Verdet constant V in min/Oe cm			
		$\lambda = 325$ nm	$\lambda = 442$ nm	$\lambda = 633$ nm	$\lambda = 1064$ nm
SiO_2	100% SiO_2			0.013	
B_2O_3	100% B_2O_3			0.010	
CdO	47.5% CdO, 52.5% P_2O_5	0.079	0.033	0.022	
ZnO	36.4% ZnO, 63.6% P_2O_5	0.072	0.044	0.020	
TeO_2	88.9% TeO_2, 11.1% P_2O_5		0.196	0.076	0.022
ZrF_4	63.1% ZrF_4, 14.9% BaF_2, 7.2% LaF_3, 1.9% AlF_3, 9.1% PbF_2, 3.8% LiF			0.011	

Glass type	Composition (wt. %)	$\lambda = 700$ nm	$\lambda = 853$ nm	$\lambda = 1060$ nm
Bi_2O_3	95% Bi_2O_3, 5% B_2O_3	0.086	0.051	0.033
PbO	95% PbO, 5% B_2O_3	0.093	0.061	0.031
	82% PbO, 18% SiO_2	0.077	0.045	0.027
	50% PbO, 15% K_2O, 35% SiO_2	0.032	0.020	0.011
Tl_2O	95% Tl_2O, 5% B_2O_3	0.092	0.061	0.032
	82% Tl_2O, 18% SiO_2	0.100	0.067	0.043
	50% Tl_2O, 15% K_2O, 35% SiO_2	0.036	0.022	0.012
SnO	76% SnO, 13% B_2O_3, 11% SiO_2	0.071	0.046	0.026
TeO_3	75% TeO_2, 25% Sb_2O_3	0.076	0.052	0.032
	80% TeO_2, 20% $ZnCl_2$	0.073	0.046	0.025
	84% TeO_2, 16% BaO	0.056	0.041	0.029
	70% TeO_2, 30% WO_3	0.052	0.035	0.022
	20% TeO_2, 80% PbO	0.128	0.075	0.048
Sb_2O_3	25% Sb_2O_3, 75% TeO_2	0.076	0.050	0.032
	75% Sb_2O_3, 75% Cs_2O, 5% Al_2O_3	0.074	0.044	0.025
	75% Sb_2O_3, 10% Cs_2O, 10% Rb_2O, 5% Al_2O_3	0.078	0.052	0.030

MAGNETO-OPTIC CONSTANTS (continued)
Verdet Constants of Commercial Glasses[1]

This table gives the density, ρ, refractive index at 589 nm, n_D, and Verdet constant, V, for the wavelengths indicated; the data refer to room temperature

Glass type	ρ g/cm^3	n_D	V in min/Oe cm				
			$\lambda = 365.0$ nm	$\lambda = 404.7$ nm	$\lambda = 435.8$ nm	$\lambda = 546.1$ nm	$\lambda = 578.0$ nm
BSC	2.49	1.5096	0.0499	0.0392	0.0333	0.02034	0.01798
HC	2.53	1.5189	0.0561	0.0440	0.0372	0.0225	0.01995
LBC	2.87	1.5406	0.0609	0.0477	0.0403	0.0245	0.0216
LF	3.23	1.5785	0.1143	0.0850	0.0693	0.0394	0.0344
BLF	3.48	1.6047	0.1112	0.0832	0.0685	0.0393	0.0344
DBC	3.56	1.6122	0.0662	0.0517	0.0435	0.0261	0.0231
DF	3.63	1.6203	0.1473	0.1076	0.0872	0.0485	0.0423
EDF	3.9	1.6533	0.1725	0.1248	0.1007	0.0556	0.0483

The composition of the glasses in weight percent is:

Glass type	SiO_2	B_2O_3	K_2O	CaO	Al_2O_3	As_2O_3	Na_2O	BaO	ZnO	PbO
BSC	69.6	6.7	20.5	2.9	0.3	0.1	—	—	—	—
HC	72.0	—	10.1	11.4	0.3	0.2	6.1	—	—	—
LBC	57.1	1.8	13.7	0.3	0.2	0.1	—	26.9	—	—
LF	52.5	—	9.5	0.3	0.2	0.1	—	—	—	37.6
BLF	45.2	—	7.8	—	—	0.4	—	16.0	8.3	22.2
DBC	36.2	7.7	0.2	0.2	3.5	0.7	—	44.6	6.7	—
DF	46.3	—	1.1	0.3	0.2	0.1	5.0	—	—	47.0
EDF	40.6	—	7.5	0.2	0.2	0.2	0.1	—	—	51.5

REFERENCES

1. Weber, M. J., *CRC Handbook of Laser Science and Technology*, Vol. IV, Part 2, CRC Press, Boca Raton, FL, 1988, p. 299-310.
2. Gray, D. E., Ed., *American Institute of Physics Handbook*, Third Edition, McGraw Hill, New York, 1972, p. 6-230.

FARADAY ROTATION
Ferro-, Ferri-, and Antiferromagnetic Solids

Material	T_c K	$4\pi M_s$ gauss	F deg/cm	α cm^{-1}	$2F/\alpha$	T K	λ nm
Fe	1043	21,800	4.4×10^5	6.5×10^5	1.4	300	500
			6.5×10^5	5.0×10^5	2.6	300	1000
			7×10^5	4.2×10^5	3.3	300	1500
			7×10^5	3.5×10^5	4.0	300	2000
Co	1390	18,200	2.9×10^5	—	—	300	500
			5.5×10^5	6.1×10^5	1.8	300	1000
			5.5×10^5	4.5×10^5	2.4	300	1500
			5.5×10^5	3.6×10^5	2.7	300	2000
Ni	633	6,400	0.8×10^5	—	—	300	500
			2.6×10^5	5.8×10^5	0.9	300	1000
			1.5×10^5	4.8×10^5	0.6	300	1500
			1×10^5	4.1×10^5	0.25	300	2000
Permalloy (Ni/Fe = 82/18)	803	10,700	1.2×10^5	6×10^5	0.4	300	500

Material	T_c K	$4\pi M_s$ gauss	F deg/cm	α cm^{-1}	$2F/\alpha$	T K	λ nm
Ni/Fe = 100/0		6,000	1.2×10^5	7.05×10^5	0.34	300	632.8
Ni/Fe = 80/20		10,800	2.2×10^5	7.10×10^5	0.62	300	632.8
Ni/Fe = 60/40		14,900	2.9×10^5	7.54×10^5	0.77	300	632.8
Ni/Fe = 40/60		14,400	2.2×10^5	8.17×10^5	0.54	300	632.8
Ni/Fe = 20/80		19,400	3.3×10^5	8.10×10^5	0.81	300	632.8
Ni/Fe = 0/100	639	21,600	3.5×10^5	8.13×10^5	0.86	300	632.8
MnBi		7,700	4.2×10^5	6.1×10^5	1.4	300	450
			7.5×10^5	4.2×10^5	3.6	300	900
MnAs	313	—	0.44×10^5	5.0×10^5	0.174	300	500
			0.62×10^5	4.4×10^5	0.28	300	900
CrTe	334	1015	0.5×10^5	2.0×10^5	0.5	300	550
			0.4×10^5	1.2×10^5	0.7	300	900
FeRh	333	—	0.9×10^5	3.3×10^5	0.56	348	700
$Y_3Fe_5O_{12}$ (YIG)	560	2500	2400	1500	3.2	300	555
			1250	1400	1.8	300	625
			750	450	3.3	300	770
			175	<0.06	$>3 \times 10^3$	300	5000 to 1500
$Gd_3Fe_5O_{12}$ (GdIG)	Tn = 564	7300	-2000	6000	0.6	300	500
	T = 286		-1050	900	2.3	300	600
			-300	100	6.0	300	800
			-80	70	2.3	300	1000
$NiFe_2O_4$	858	3350	2.0×10^4	5.9×10^4	0.7	300	286
			-1.0×10^4	10×10^4	0.2	300	500
			-120	38	6	300	1500
			+75	15	10	300	3000
			+110	32	7	300	5000
$CoFe_2O_4$	793	4930	2.75×10^4	12×10^4	0.5	300	286
			3.6×10^4	17×10^4	0.4	300	400
			-2.5×10^4	6×10^4	0.8	300	660
$MgFe_2O_4$	593-713[e]	1450[e]	-60	100	1	300	2500
			0	12	0	300	4000
			+35	6	11	300	6000
$Li_{0.5}Fe_{2.5}O_4$	863-953[e]	3240[e] to 3900	-440	150	6	300	1500
			+10	85	0.2	300	3000
			+110	44	5	300	5000
			+135	80	3	300	7000
$BaFe_{12}O_{19}$	723	—	-50	-38	3	300	2000
			+75	20	7.5	300	3000
			+150	20	15	300	5000
			+165	22	15	300	7000
$Ba_2Zn_2Fe_{12}O_{19}$	—	—	90	120	1.5	300	5000
			75	65	2.0	300	7000
$RbNiF_3$	220	1250	360	35	20	77	450[a]
			70	10	14	77	600[a]
			310	70	9	77	800[a]
			75	25	6	77	1000[a]
$RbNi_{0.75}Co_{0.25}F_3$	109	—	180	9	40	77	600[b]
$RbFeF_3$	102	—	3400	7	900	82	300[c]
			1600	3	1100	82	400[c]
			620	1.5	830	82	600[c]
			300	2.5	240	82	800[c]
FeF_3	365	40 at 300 K	670	14	95	300	349[d]
			180	4.4	82	300	522.5[d]
$CrCl_3$	16.8	3880	2000	200	20	1.5	410

FARADAY ROTATION (continued)
Ferro-, Ferri-, and Antiferromagnetic Solids (continued)

Material	T_c K	$4\pi M_s$ gauss	F deg/cm	α cm^{-1}	$2 F/\alpha$	T K	λ nm
			-500	300	3	1.5	450
			-1000	70	30	1.5	590
$CrBr_3$	32.5	3390	3×10^5	3×10^3	200	1.5	478
			1.6×10^5	1.4×10^4	23	1.5	500
CrI_3	68	2690	1.1×10^5	6.3×10^3	35	1.5	970
			0.8×10^5	3×10^3	53	1.5	1000
$FeBO_3$	348	115	3200	140	45	300	500
		at 300 K	450	38	24	300	700
EuO	69	23700	-1.0×10^5	0.5×10^4	40	5	1100
			5×10^5	9.7×10^4	10	5	700
			0.5×10^5	7.8×10^4	1.3	5	500
			3×10^4	>0.5	~105	20	2500
			660	>1.0	1300	20	10600
EuS	16.3	—	-1.6×10^5	0	—	6	825
			-9.6×10^5	3.3×10^4	58	6	690
			$+5.5 \times 10^5$	1.2×10^5	9.2	6	563
EuSe	7.0	13,200	1.45×10^5	80	3600	4.2	750
			0.95×10^5	60	3170	4.2	800

[a] Measured along the C-axis (magnetic hard axis).
[b] Measured along the C-axis (magnetic easy axis).
[c] Measured along the C-axis ([100]-direction at room temperature).
[d] Strong natural birefringence interferes with the Faraday effect.
[e] Depends on heat treatment.

REFERENCE

1. Weber, M. J., Ed., *CRC Handbook of Laser Science and Technology,* Vol. IV, Part 2, CRC Press, Boca Raton, FL, 1988, pp. 288-296.

NONLINEAR OPTICAL CONSTANTS

H. P. R. Frederikse

The relation between the polarization density P of a dielectric medium and the electric field E is linear when E is small, but becomes nonlinear as E acquires values comparable with interatomic electric fields (10^5 to 10^8 V/cm). Under these conditions the relation between P and E can be expanded in a Taylor's series

$$P = \varepsilon_0 \chi^{(1)} E + 2 \chi^{(2)} E^2 + 4 \chi^{(3)} E^3 + \cdots \tag{1}$$

where ε_o is the permittivity of free space, while $\chi^{(1)}$ is the linear and $\chi^{(2)}$, $\chi^{(3)}$ etc. the nonlinear optical susceptibilities.

If we consider two optical fields, the first $E_j^{\omega_1}$ (along the j-direction at frequency ω_1) and the second $E_k^{\omega_2}$ (along the k-direction at frequency ω_2) one can write the second term of the Taylor's series as follows

$$P_i(\omega_1 \omega_2) = 2 \chi_{ijk}^{\omega_3 = \omega_1 \pm \omega_2} E_j^{\omega_1} E_k^{\omega_2}$$

When $\omega_1 \neq \omega_2$ the (parametric) mixing of the two fields gives rise to two new polarizations at the frequencies $\omega_3 = \omega_1 + \omega_2$ and $\omega_3' = \omega_1 - \omega_2$. When the two frequencies are equal, $\omega_1 = \omega_2 = \omega$, the result is Second Harmonic Generation (SHG) $\chi_{ijk}(2\omega, \omega, \omega)$, while equal and opposite frequencies, $\omega_1 = \omega$ and $\omega_2 = -\omega$ leads to Optical Rectification (OR): $\chi_{ijk}(0, \omega, -\omega)$. In the SHG case the following convention is adopted: the second order nonlinear coefficient d is equal to one half of the second order nonlinear susceptibility

$$d_{ijk} = 1/2 \chi^{(2)}$$

Because of the symmetry of the indices j and k one can replace these two by a single index (subscript) m. Consequently the notation for the SHG nonlinear coefficient in reduced form is d_{im} where m takes the values 1 to 6. Only noncentrosymmetric crystals can possess a nonvanishing d_{ijk} tensor (third rank). The unit of the SHG coefficients is m/V (in the MKSQ/SI system).

In centrosymmetric media the dominant nonlinearity is of the third order. This effect is represented by the third term in the Taylor's series (Equation 1); it is the result of the interaction of a number of optical fields (one to three) producing a new frequency $\omega_4 = \omega_1 + \omega_2 + \omega_3$. The third order polarization is given by

$$P_j(\omega_1 \omega_2 \omega_3) = g_4 \chi_{jklm} E_k^{\omega_1} E_l^{\omega_2} E_m^{\omega_3}$$

Third Harmonic Generation (THG) is achieved when $\omega_1 = \omega_2 = \omega_3 = \omega$. In this case the constant $g_4 = 1/4$. The third order nonlinear coefficient C is related to the third order susceptibility as follows

$$C_{jklm} = 1/4 \chi_{jklm}$$

This coefficient is a fourth rank tensor. In the THG case the matrices must be invariant under permutation of the indices k, l, and m; as a result the notation for the third order nonlinear coefficient can be simplified to C_{jn}. The unit of C_{jn} is $m^2 \cdot V^{-2}$ (in the MKSQ/SI system).

Applications of second order nonlinear optical materials include the generation of higher (up to sixth) optical harmonics, the mixing of monochromatic waves to generate sum or difference frequencies (frequency conversion), the use of two monochromatic waves to amplify a third wave (parametric amplification) and the addition of feedback to such an amplifier to create an oscillation (parametric oscillation).

Third order nonlinear optical materials are used for THG, self-focusing, four wave mixing, optical amplification, and optical conjugation. Many of these effects — as well as the variation and modulation of optical propagation caused by mechanical, electric, and magnetic fields (see the preceeding table on "Elasto-Optic, Electro-Optic, and Magneto-Optic Constants") are used in the areas of optical communication, optical computing, and optical imaging.

REFERENCES (NONLINEAR OPTICS)

1. *Handbook of Laser Science and Technology*, Vol. 111, Part 1; Ed.: Marvin J. Weber, Publ.: CRC Press, Inc., Boca Raton, FL, 1986.
2. Dmitriev, V.G., Gurzadyan, G.G., and Nikogosyan, D., *Handbook of Nonlinear Optical Crystals*, Springer-Verlag, Berlin, 1991.
3. Shen, Y.R., *The Principles of Nonlinear Optics*, John Wiley, New York, 1984.
4. Yariv, A., *Quantum Electronics*, 3rd edition, John Wiley, New York, 1988.
5. Bloembergen, N., *Nonlinear Optics*, W.A. Benjamin, New York, 1965.
6. Zernike F. and Midwinter, J.E., *Applied Nonlinear Optics*, John Wiley, New York, 1973.
7. Hopf, F.A. and Stegeman, G.I., *Applied Classical Electrodynamics*, Volume 2: Nonlinear Optics, John Wiley, New York, 1986.
8. *Nonlinear Optical Properties of Organic Molecules and Crystals*, Eds.: D.S. Chemla and J. Zyss, Publ.: Academic Press, Orlando, FL, 1987.
9. *Optical Phase Conjugation*, Ed.: R.A. Fisher, Publ.: Academic Press, New York, 1983.
10. Zyss, J., *Molecular Nonlinear Optics: Materials, Devices and Physics*, Academic Press, Boston, 1994.
11. Nonlinear Optics, 5 articles in *Physics Today, (Am. Inst. of Phys.)*, Vol. 47, No. 5, May, 1994.

Selected SHG Coefficients of NLO Crystals*

Material	Symmetry class	$d_{im} \times 10^{12}$ m/V	λ μm
GaAs	$\overline{4}3$ m	$d_{14} = 134.1 \pm 42$	10.6
GaP	$\overline{4}3$ m	$d_{14} = 71.8 \pm 12.3$	1.058
InAs	$\overline{4}3$ m	$d_{14} = 364 \pm 47$	1.058
		$d_{14} = 210$	10.6
ZnSe	$\overline{4}3$ m	$d_{14} = 78.4 \pm 29.3$	10.6
		$d_{36} = 26.6 \pm 1.7$	1.058
β-ZnS	$\overline{4}3$ m	$d_{14} = 30.6 \pm 8.4$	10.6
		$d_{36} = 20.7 \pm 1.3$	1.058
ZnTe	$\overline{4}3$ m	$d_{14} = 92.2 \pm 33.5$	10.6
		$d_{14} = 83.2 \pm 8.4$	1.058
		$d_{36} = 89.6 \pm 5.7$	1.058
CdTe	$\overline{4}3$ m	$d_{14} = 167.6 \pm 63$	10.6
Bi_4GeO_{12}	$\overline{4}3$ m	$d_{14} = 1.28$	1.064
$N_4(CH_2)_6$ (hexamine)	$\overline{4}3$ m	$d_{14} = 4.1$	1.06
$LiIO_3$	6	$d_{33} = -7.02$	1.06
		$d_{31} = -5.53 \pm 0.3$	1.064
ZnO	6 mm	$d_{33} = -5.86 \pm 0.16$	1.058
		$d_{31} = 1.76 \pm 0.16$	1.058
		$d_{15} = 1.93 \pm 0.16$	1.058
α-ZnS	6 mm	$d_{33} = 11.37 \pm 0.07$	1.058
		$d_{33} = 37.3 \pm 12.6$	10.6
		$d_{31} = -18.9 \pm 6.3$	10.6
		$d_{15} = 21.37 \pm 8.4$	10.6
CdS	6 mm	$d_{33} = 25.8 \pm 1.6$	1.058
		$d_{31} = -13.1 \pm 0.8$	1.058
		$d_{15} = 14.4 \pm 0.8$	1.058
CdSe	6 mm	$d_{33} = 54.5 \pm 12.6$	10.6
		$d_{31} = -26.8 \pm 2.7$	10.6
$BaTiO_3$	4 mm	$d_{33} = 6.8 \pm 1.0$	1.064
		$d_{31} = 15.7 \pm 1.8$	1.064
		$d_{15} = 17.0 \pm 1.8$	1.064
$PbTiO_3$	4 mm	$d_{33} = 7.5 \pm 1.2$	1.064
		$d_{31} = 37.6 \pm 5.6$	1.064
		$d_{15} = 33.3 \pm 5$	1.064
$K_3Li_2Nb_5O_{15}$	4 mm	$d_{33} = 11.2 \pm 1.6$	1.064
		$d_{31} = 6.18 \pm 1.28$	1.064
		$d_{15} = 5.45 \pm 0.54$	1.064
$K_{0.8}Na_{0.2}Ba_2Nb_5O_{15}$	4 mm	$d_{31} = 13.6 \pm 1.6$	1.064
$SrBaNb_5O_{15}$	4 mm	$d_{33} = 11.3 \pm 3.3$	1.064
		$d_{31} = 4.31 \pm 1.32$	1.064
		$d_{15} = 5.98 \pm 2$	1.064
$NH_4H_2PO_4$ (ADP)	$\overline{4}2$ m	$d_{36} = 0.53$	1.064
		$d_{36} = 0.85$	0.694
KH_2PO_4 (KDP)	$\overline{4}2$ m	$d_{36} = 0.44$	1.064
		$d_{36} = 0.47 \pm 0.07$	0.694
KD_2PO_4 (KD*P)	$\overline{4}2$ m	$d_{36} = 0.38 \pm 0.016$	1.058
		$d_{36} = 0.34 \pm 0.06$	0.694
		$d_{14} = 0.37$	1.058
KH_2AsO_4 (KDA)	$\overline{4}2$ m	$d_{36} = 0.43 \pm 0.025$	1.06
		$d_{36} = 0.39 \pm 0.4$	0.694
$CdGeAs_2$	$\overline{4}2$ m	$d_{36} = 351 \pm 105$	10.6
$AgGaS_2$	$\overline{4}2$ m	$d_{36} = 18 \pm 2.7$	10.6
$AgGaSe_2$	$\overline{4}2$ m	$d_{36} = 37.4 \pm 6.0$	10.6
$(NH_2)_2CO$ (urea)	$\overline{4}2$ m	$d_{36} = 1.3$	1.06
$AlPO_4$	32	$d_{11} = 0.35 \pm 0.03$	1.058
Se	32	$d_{11} = 97 \pm 25$	10.6

Material	Symmetry class	$d_{im} \times 10^{12}$ m/V	λ μm
Te	32	$d_{11} = 650 \pm 30$	10.6
SiO$_2$ (quartz)	32	$d_{11} = 0.335$	1.064
HgS	32	$d_{11} = 50.3 \pm 17$	10.6
(C$_6$H$_5$CO)$_2$ [benzil]	32	$d_{11} = 3.6 \pm 0.5$	1.064
β-BaB$_2$O$_4$ [BBO]	3 m	$d_{22} = 2.22 \pm 0.09$	1.06
		$d_{31} = 0.16 \pm 0.08$	1.06
LiNbO$_3$	3 m	$d_{33} = 34.4$	1.06
		$d_{31} = -5.95$	1.06
		$d_{22} = 2.76$	1.06
LiTaO$_3$	3 m	$d_{33} = -16.4 \pm 2$	1.058
		$d_{31} = -1.07 \pm 0.2$	1.058
		$d_{22} = +1.76 \pm 0.2$	1.058
Ag$_3$AsS$_3$ [proustite]	3 m	$d_{31} = 11.3 \pm 2.5$	10.6
		$d_{22} = 18.0 \pm 2.5$	10.6
Ag$_3$SbS$_3$ [pyrargerite]	3m	$d_{31} = 12.6 \pm 4$	10.6
		$d_{22} = 13.4 \pm 4$	10.6
α-HIO$_3$	222	$d_{36} = 5.15 \pm 0.16$	1.064
NO$_2 \cdot$ CH$_3$NOC$_5$H$_4 \cdot$ (POM)	222	$d_{36} = 6.4 \pm 1.0$	1.064
Ba$_2$NaNb$_5$O$_{15}$ [Banana]	mm 2	$d_{33} = -17.6 \pm 1.28$	1.064
		$d_{31} = -12.8 \pm 1.28$	1.064
C$_6$H$_4$(NO$_2$)$_2$ [MDB]	mm 2	$d_{33} = 0.74$	1.064
		$d_{32} = 2.7$	1.064
		$d_{31} = 1.78$	1.064
Gd$_2$(MoO$_4$)$_3$	mm 2	$d_{33} = -0.044 \pm 0.008$	1.064
		$d_{32} = +2.42 \pm 0.36$	1.064
		$d_{31} = -2.49 \pm 0.37$	1.064
KNbO$_3$	mm 2	$d_{33} = -19.58 \pm 1.03$	1.064
		$d_{32} = +11.34 \pm 1.03$	1.064
		$d_{31} = -12.88 \pm 1.03$	1.064
KTiOPO$_4$ [KTP]	mm 2	$d_{33} = 13.7$	1.06
		$d_{32} = \pm 5.0$	1.06
		$d_{31} = \pm 6.5$	1.06
NO$_2$C$_6$H$_4 \cdot$ NH$_2$ [mNA]	mm 2	$d_{33} = 13.12 \pm 1.28$	1.064
		$d_{32} = 1.02 \pm 0.22$	1.064
		$d_{31} = 12.48 \pm 1.28$	1.064
C$_{10}$H$_{12}$N$_3$O$_6$ [MAP]	2	$d_{23} = 10.67 \pm 1.3$	1.064
		$d_{22} = 11.7 \pm 1.3$	1.064
		$d_{21} = 2.35 \pm 0.5$	1.064
		$d_{25} = -0.35 \pm 0.3$	1.064
(NH$_2$CH$_2$COOH)$_3$H$_2$SO$_4$ [TGS]	2	$d_{23} = 0.32$	0.694

* These data are taken from References 1 and 2.

NONLINEAR OPTICAL CONSTANTS (continued)

Selected THG Coefficients of Some NLO Materials*

Material	NLO process	$C_{jn} \times 10^{20}$ m^2/V^{-2}	λ μm
$NH_4H_2PO_4$ [ADP]	$(-3\omega,\omega,\omega,\omega)$	$C_{11} = 0.0104$	1.06
		$C_{18} = 0.0098$	1.06
C_6H_6 [benzene]	$(-3\omega,\omega,\omega,\omega)$	$C_{11} = 0.0184 \pm 0.0042$	1.89
$CdGeAs_2$	$(-3\omega,\omega,\omega,\omega)$	$C_{11} = 182 \pm 84$	10.6
p-type: 5×10^{16} cm^{-3}		$C_{16} = 175$	10.6
		$C_{18} = -35$	10.6
$C_{40}H_{56}$ [β-carotene]	$(-3\omega,\omega,\omega,\omega)$	C_{11} 0.263 ± 0.08	1.89
GaAs	$(-3\omega,\omega,\omega,-\omega)$	$C_{11} = 62 \pm 31$	1.06
high-resistivity			
Ge	$(-3\omega,\omega,\omega,-\omega)$	$C_{11} = 23.5 \pm 12$	1.06
$LiIO_3$	$(-3\omega,\omega,\omega,-\omega)$	$C_{12} = 0.2285$	1.06
		$C_{35} = 6.66 \pm 1$	1.06
KBr	$(-3\omega,\omega,\omega,-\omega)$	$C_{11} = 0.0392$	1.06
		$C_{18}/C_{11} = 0.3667$	1.06
KCl	$(-3\omega,\omega,\omega,-\omega)$	$C_{11} = 0.0168$	1.06
		$C_{18}/C_{11} = 0.28$	1.06
KH_2PO_4 [KDP]	$(-3\omega,\omega,\omega,-\omega)$	$C_{11}-3C_{18} = 0.04$	1.06
Si	$(-3\omega,\omega,\omega,-\omega)$	$C_{11} = 82.8 \pm 25$	1.06
p-type: 10^{14} cm^{-3}			
NaCl	$(-3,\omega,\omega,\omega,-\omega)$	$C_{11} = 0.0168$	1.06
		$C_{18}/C_{11} = 0.4133$	1.06
NaF	$(-3\omega,\omega,\omega,-\omega)$	$C_{11} = 0.0035$	1.06

* These data are taken from Reference 1.

PHASE DIAGRAMS
H. P. R. Frederikse

A phase is a structurally homogeneous portion of matter. Regardless of the number of chemical constituents of a gas, there is only one vapor phase. This is true also for the liquid form of a pure substance, although a mixture of several liquid substances may exist as one or several phases, depending on the interactions among the substances. On the other hand a pure solid may exist in several phases at different temperatures and pressures because of differences in crystal structure (Reference 1). At the phase transition temperature, T_{tr}, the chemical composition of the solid remains the same, but often a change in the physical properties will take place. Such changes are found in ferroelectric crystals (example $BaTiO_3$) which develop a spontaneous polarization below T_{tr}, in superconductors (example Pb) which loose all electrical resistance below the transition point, and in many other classes of solids.

In quite a few cases it is difficult to bring about the phase transition, and the high- (or low-) temperature phase persists in its metastable form. Many liquids remain in the liquid state for shorter or longer periods of time when cooled below the melting point (supercooling). However, often the slightest disturbance will cause solidification. Persistence of the high temperature phase in solid-solid transitions is usually of much longer duration. An example of this behavior is found in white tin; although gray tin is the thermodynamically stable form below T_{tr} (286.4 K), the metal remains in its undercooled, white tin state all the way to $T = 0$ K, and crystals of gray tin are very difficult to produce.

A *phase diagram* is a map which indicates the areas of stability of the various phases as a function of external conditions (temperature and pressure). Pure materials, such as mercury, helium, water, and methyl alcohol are considered one-component systems and they have *unary* phase diagrams. The equilibrium phases in two-component systems are presented in *binary* phase diagrams. Because many important materials consist of three, four, and more components, many attempts have been made to deduce their multicomponent phase diagrams. However, the vast majority of systems with three or more components are very complex, and no overall maps of the phase relationships have been worked out.

It has been shown during the last 20 to 25 years that very useful partial phase diagrams of complex systems can be obtained by means of thermodynamic modeling (References 2, 3). Especially for complicated, multicomponent alloy systems the CALPHAD method has proved to be a successful approach for producing valuable portions of very intricate phase diagrams (Reference 4). With this method thermodynamic descriptions of the free energy functions of various phases are obtained which are consistent with existing (binary) phase diagram information and other thermodynamic data. Extrapolation methods are then used to extend the thermodynamic functions into a ternary system. Comparison of the results of this procedure with available experimental data is then used to fine-tune the phase diagram and add ternary interaction functions if necessary. In principle this approximation strategy can be extended to four, five, and more component systems.

The nearly two dozen phase diagrams shown below present the reader with examples of some important types of single and multicomponent systems, especially for ceramics and metal alloys. This makes it possible to draw attention to certain features like the kinetic aspects of phase transitions (see Figure 22, which presents a time-temperature-transformation, or TTT, diagram for the precipitation of α-phase particles from the β-phase in a Ti-Mo alloy; Reference 1, pp.358-360). The general references listed below and the references to individual figures contain phase diagrams for many additional systems.

GENERAL REFERENCES

1. Ralls, K.M., Courtney, T.H., and Wulff, J., *Introduction to Materials Science and Engineering*, Chapters 16 and 17, John Wiley & Sons, New York, 1976.
2. Kaufman, L., and Bernstein, H., *Computer Calculation of Phase Diagrams*, Academic Press, New York, 1970.
3. Kattner, U.R., Boettinger, W.J.B., and Coriell, S.R., *Z. Metallkd.*, 87, 9, 1996.
4. Dinsdale, A.T., Editor, *CALPHAD*, Vol. 1–20, Pergamon Press, Oxford, 1977–1996 and continuing.
5. Baker, H., Editor, *ASM Handbook, Volume 3: Alloy Phase Diagrams*, ASM International, Materials Park, OH, 1992.
6. Massalski, T.B., Editor, *Binary Alloy Phase Diagrams, Second Edition*, ASM International, Materials Park, OH, 1990.
7. Roth. R.S., Editor, *Phase Diagrams for Ceramists*, Vol. I (1964) to Volume XI (1995), American Ceramic Society, Waterville, OH.

REFERENCES TO INDIVIDUAL PHASE DIAGRAMS

Figure 1. Carbon: Reference 7, Vol. X (1994), Figure 8930. Reprinted with permission.
Figure 2. Si-Ge : Ref.5, p. 2.231. Reprinted with permission.
Figure 3. H_2O (ice): Lerner, R.G., and Trigg, G.L., Editors, *Encyclopedia of Physics*, VCH Publishers, New York, 1990. Reprinted with permission.
Figure 4. SiO_2: Reference 7, Vol. XI (1995), Figure 9174. Reprinted with permission.
Figure 5. Fe-O: Darken, L.S., and Gurry, R.W., *J. Am. Chem. Soc.*, 68, 798, 1946. Reprinted with permission.
Figure 6. Ti-O: Reference 5, p. 2.324. Reprinted with permission.
Figure 7. $BaO-TiO_2$: Reference 7, Vol. III (1975), Figure 4302. Reprinted with permission.
Figure 8. $MgO-Al_2O_3$: Reference 7, Vol. XI (1995), Figure 9239. Reprinted with permission.
Figure 9. $Y_2O_3-ZrO_2$: Reference 7, Vol. XI (1995), Figure 9348. Reprinted with permission.
Figure 10. Si-N-Al-O (Sialon): Reference 7, Vol. X (1994), Figure 8759. Reprinted with permission.
Figure 11. $PbO-ZrO_2-TiO_2$ (PZT): Reference 7, Vol. III (1975), Figure 4587. Reprinted with permission.
Figure 12. Al-Si-Ca-O: Reference 7 (1964), Vol. I, Figure 630. Reprinted with permission.
Figure 13. Y-Ba-Cu-O: Whitler, J.D., and Roth, R.S., *Phase Diagrams for High T_c Superconductors*, Figure S-082, American Ceramic Society, Waterville, OH, 1990. Reprinted with permission.
Figure 14. Al-Cu: Reference 5, p. 2.44. Reprinted with permission.
Figure 15. Fe-C: Ralls, K.M., Courtney, T.H., and Wulff, J., *Introduction to Materials Science and Engineering*, Figure 16.13 , John Wiley & Sons, New York, 1976. Reprinted with permission.
Figure 16. Fe-Cr: Reference 5, p. 2.152. Reprinted with permission.

PHASE DIAGRAMS (continued)

Figure 17. Cu-Sn: Reference 5, p. 2.178. Reprinted with permission.

Figure 18. Cu-Ni: Reference 5, p. 2.173. Reprinted with permission.

Figure 19. Pb-Sn (solder): Reference 5, p. 2.335. Reprinted with permission.

Figure 20. Cu-Zn (brass): Subramanian, P.R., Chakrabarti, D.J., and Laughlin, D.E., Editors, *Phase Diagrams of Binary Copper Alloys*, p. 487, ASM International, Materials Park, OH, 1994. Reprinted with permission.

Figure 21. Co-Sm: Reference 5, p. 2.148. Reprinted with permission.

Figure 22. Ti-Mo: Reference 5, p. 2.296; Reference 1, p. 359. Reprinted with permission.

Figure 23: Fe-Cr-Ni: Reference 5, Figure 48. Reprinted with permission.

Figure 1

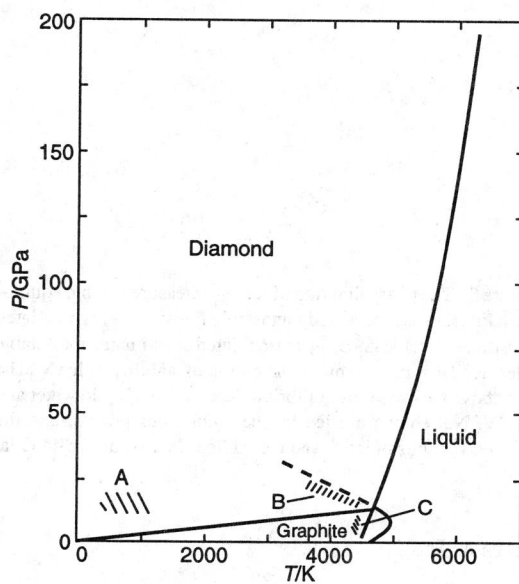

Figure 1. Phase diagram of carbon. (A) Martensitic transition: hex graphite → hex diamond. (B) Fast graphite-to-diamond transition. (C) Fast diamond-to-graphite transition.

Figure 2

Figure 2. Si-Ge system.

Phase	Composition, mass % Si	Pearson symbol	Space group
(Ge,Si)	0 to 100	cF8	$Fd\bar{3}m$
High-pressure phases			
GeII	—	tI4	$I4_1/amd$
SiII	—	tI4	$I4_1/amd$

FIGURE 3

Figure 3. The phase diagram of ice: —, measured stable equilibrium lines; ------, measured metastable lines; - - - -, extrapolated or estimated stable lines; ····, extrapolated or estimated metastable lines. Ice IV is metastable in the region of stability of ice V. The line above ice IV is the equilibrium line between liquid water and ice IV. Not shown are ice Ic, the amorphous phases, and the ordered versions of ice V and ice VI. Ice X exists above 44 GPa.

Structural Data on the Phases of Ice

Ice phase	Crystal system	Space group	Cell dimensions[a] (pm, deg)	No. of molecules in unit cell	No. of nearest neighbors	Density[a] (Mg m^{-3})
Ih	Hexagonal	$P6_3/mmc$	$a = 450; c = 732$	4	4	0.93
		$Fd3m$				0.93
Ic	Cubic	—	$a = 635$	8	4	0.94
a$_I$	Amorphous	—	—	—	4	1.17
a$_{II}$	Amorphous	—	—	—	4	
II	Rhombohedral	$R\bar{3}$	$a = 778; \alpha = 113.1$	12	4	1.18
III	Tetragonal	$P4_12_12$	$a = 673; c = 683$	12	4	1.15
IV	Rhombohedral	$R\bar{3}c$	$a = 760; \alpha = 70.1$	16	4	1.27
V	Monoclinic	$A2/a$	$a = 922; b = 754$	28	4	1.24
			$c = 1035; \beta = 109.2$			
VI	Tetragonal	$P4_2/nmc$	$a = 627; c = 579$	10	4	1.33
VII	Cubic	$Pn3m$	$a = 341$	2	8	1.56
VIII	Tetragonal	$I4_1/amd$	$a = 480; c = 699$	8	8	1.56
IX	Tetragonal	$P4_12_12_1$	$a = 673; c = 683$	12	4	1.16
X	Cubic	$Pn3m$	$a = 283$	2	8	2.51

[a] Cell dimensions and density are at 110 K and atmospheric pressure. Ice X at 44 GPa and 25°C.

Figure 4

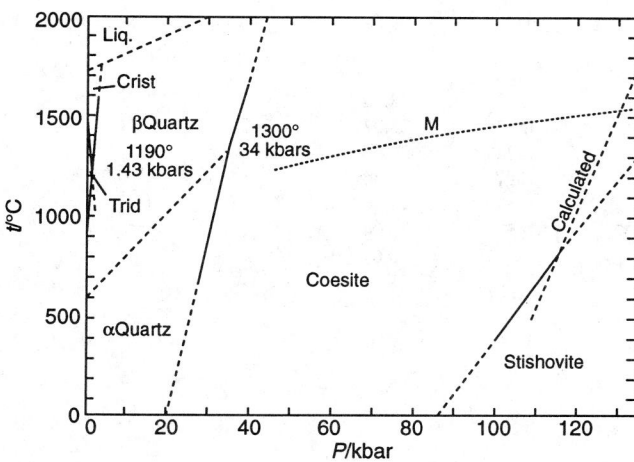

Figure 4. SiO$_2$ system. Crist = cristobalite; Trid = tridymite.

Figure 5

Figure 5. Fe-O system.

Figure 5 (continued)

Point	$t/°C$	% O	p_{CO_2}/p_{CO}	Point	$t/°C$	% O	p_{CO_2}/p_{CO}	p_{O_2}/atm
A	1539			Q	560	23.26	1.05	
B	1528	0.16	0.209	R	1583	28.30		1
C	1528	22.60	0.209	R'	1583	28.07		1
G	1400[a]	22.84	0.263	S	1424	27.64	16.2	
H	1424	25.60	16.2	V	1597	27.64		0.0575
I	1424	25.31	16.2	Y	1457	28.36		1
J	1371	23.16	0.282	Z	1457	30.04		1
L	911[a]	23.10	0.447	Z'		30.6		
N	1371	22.91	0.282					

[a] Values for pure iron.

Figure 6

Figure 6. Ti-O system.

Phase	Composition, mass % O	Pearson symbol	Space group
(βTi)	0 to 3	$cI2$	$Im\bar{3}m$
(αTi)	0 to 13.5	$hP2$	$P6_3/mmc$
Ti_3O	~8 to ~13	$hP\sim16$	$P\bar{3}c$
Ti_2O	~10 to 14.4	$hP3$	$P\bar{3}m1$
γTiO	15.2 to 29.4	$cF8$	$Fm\bar{3}m$
Ti_3O_2	~18	$hP\sim5$	$P6/mmm$
βTiO	~24 to ~29.4	$c**$	—
αTiO	~25.0	$mC16$	$A2/m$ or $B*/*$
$\beta Ti_{1-x}O$	~29.5	$oI12$	$I222$
$\alpha Ti_{1-x}O$	~29.5	$tI18$	$I4/m$
βTi_2O_3	33.2 to 33.6	$hR30$	$R\bar{3}c$
αTi_2O_3	33.2 to 33.6	$hR30$	$R\bar{3}c$
βTi_3O_5	35.8	$m**$	—
αTi_3O_5	35.8	$mC32$	$C2/m$
$\alpha' Ti_3O_5$	35.8	$mC32$	Cc

Figure 6 (continued)

Phase	Composition, mass % O	Pearson symbol	Space group
γTi_4O_7	36.9	aP44	$P\bar{1}$
βTi_4O_7	36.9	aP44	$P\bar{1}$
$\alpha Ti4O_7$	36.9	aP44	$P\bar{1}$
γTi_5O_9	37.6	aP28	$P\bar{1}$
βTi_6O_{11}	38.0	aC68	$A\bar{1}$
Ti_7O_{13}	38.3	aP40	$P\bar{1}$
Ti_8O_{15}	38.5	aC92	$A\bar{1}$
Ti_9O_{17}	38.7	aP52	$P\bar{1}$
Rutile TiO_2	40.1	tP6	$P4_2/mnm$
Metastable phases			
Anatase	—	tI12	$I4_1/amd$
Brookite	—	oP24	Pbca
High-pressure phases			
TiO_2-II	—	oP12	Pbcn
TiO_2-III	—	hP~48	—

FIGURE 7

Figure 7. BaO-TiO_2 system.

Figure 8

Figure 8. MgO-Al₂O₃ system.

Figure 9

Figure 9. Y₂O₃-ZrO₂ system. C$_{ss}$ = cubic ZrO₂ ss (fluorite-type ss); Y$_{ss}$ = cubic Y₂O₃ ss; Tet$_{ss}$ = tetragonal ZrO₂ ss; Mon$_{ss}$ = monoclinic ZrO₂ ss; H$_{ss}$ = hexagonal Y₂O₃ ss; 3:4 = Zr₃Y₄O₁₂; 1:6 = ZrY₆O₁₁ ss.

Figure 10

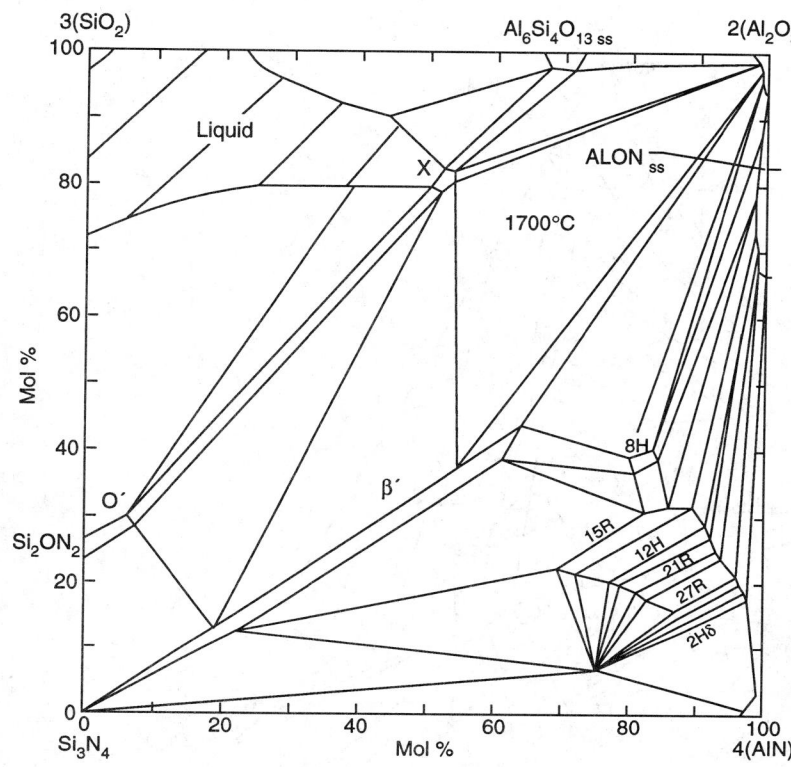

Figure 10. 3(SiO₂)-Si₃N₄-4(AlN)-2(Al₂O₃) system. "Behavior" diagram at 1700°C. The labels 8H, 15R, 12H, 21R, 27R, 2H$^\delta$ indicate defect AlN polytypes. β' = 3-sialon (Si$_{6-x}$Al$_x$O$_x$N$_{8-x}$); O' = sialon of Si₂ON₂ type; X = SiAlO₂N ("nitrogen mullite"). ALON ss = aluminum oxynitride ss extending from approximately Al₇O₉N to Al₃O₃N.

Figure 11

Figure 11. PbO-ZrO₂-TiO₂ (PZT) system, subsolidus at 1100°C. P = PbO; T = TiO₂; Z = ZrO₂.

Figure 12

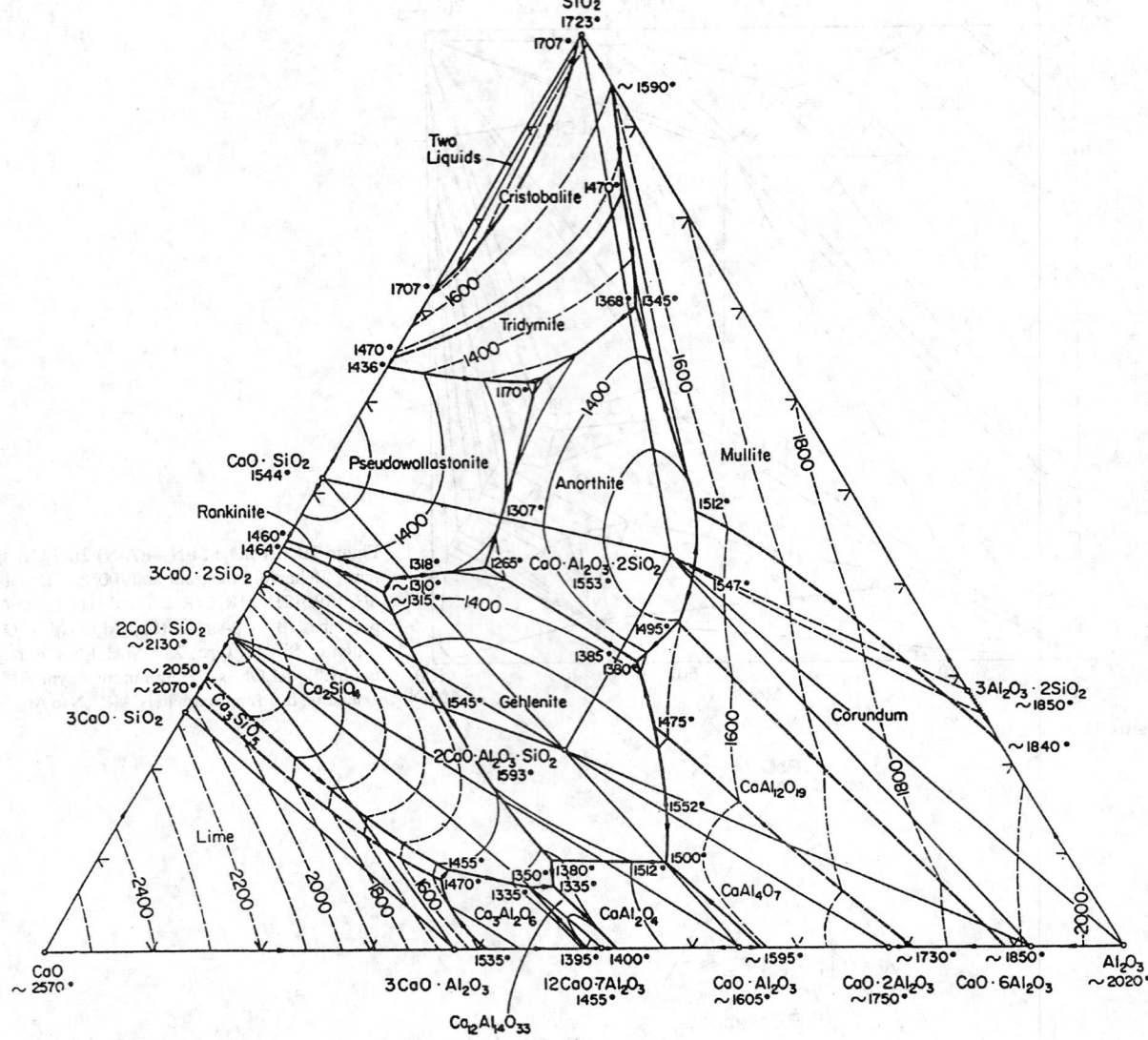

Figure 12. CaO-Al₂O₃-SiO₂ system (temperatures in °C).

Crystalline Phases

Notation	Oxide formula
Cristobalite } Tridymite	SiO₂
Pseudowollastonite	CaO·SiO₂
Rankinite	3CaO·2SiO₂
Lime	CaO
Corundum	Al₂O₃
Mullite	3Al₂O₃·2SiO₂
Anorthite	CaO·Al₂O₃·2SiO₂
Gehlenite	2CaO·Al₂O₃·SiO₂

Temperatures up to approximately 1550°C are on the Geophysical Laboratory Scale; those above 1550°C are on the 1948 International Scale.

Figure 13

Figure 13. BaO-Y$_2$O$_3$-CuO system. 2:1:3 = Ba$_2$YCu$_3$O$_{7-x}$; 1:2:1 = BaY$_2$CuO$_5$; 4:1:2 = Ba$_4$YCu$_2$O$_{7.5+x}$; and 5:1:3 = Ba$_5$YCu$_3$O$_{9.5+x}$. The superconducting 2:1:3 phase was prepared using barium peroxide.

Figure 14

Figure 14. Al-Cu system.

Figure 14 (continued)

Phase	Composition, wt % Cu	Pearson symbol	Space group
(Al)	0 to 5.65	$cF4$	$Fm\bar{3}m$
θ	52.5 to 53.7	$tI12$	$I4/mcm$
η_1	70.0 to 72.2	$oP16$ or $oC16$	$Pban$ or $Cmmm$
η_2	70.0 to 72.1	$mC20$	$C2/m$
ζ_1	74.4 to 77.8	$hP42$	$P6/mmm$
ζ_2	74.4 to 75.2	(a)	—
ε_1	77.5 to 79.4	(b)	—
ε_2	72.2 to 78.7	$hP4$	$P63/mmc$
δ	77.4 to 78.3	(c)	$R\bar{3}m$
γ_0	77.8 to 84	(d)	—
γ_1	79.7 to 84	$cP52$	$P\bar{4}3m$
β_0	83.1 to 84.7	(d)	—
β	85.0 to 91.5	$cI2$	$Im\bar{3}m$
α_2	88.5 to 89	(e)	—
(Cu)	90.6 to 100	$cF4$	$Fm\bar{3}m$
Metastable phases			
θ'	—	$tP6$	—
β'	—	$cF16$	$Fm\bar{3}m$
Al_3Cu_2	61 to 70	$hp5$	$P\bar{3}m1$

(a) Monoclinic? (b) Cubic? (c) Rhombohedral. (d) Unknown. (e) $D0_{22}$-type long-period superlattice.

Figure 15

Figure 15. Fe-C system.

Phase	Composition, mass % C	Pearson symbol	Space group
(δFe)	0 to 0.09	$cI2$	$Im\bar{3}m$
(γFe)	0 to 2.1	$cF4$	$Fm\bar{3}m$
(αFe)	0 to 0.021	$cI2$	$Im\bar{3}m$
(C)	100	$hP4$	$P6_3/mmc$
Metastable/high-pressure phases			
(εFe)	0	$hP2$	$P6_3/mmc$
Martensite	< 2.1	$tI4$	$I4/mmm$

Figure 15 (continued)

Phase	Composition, mass % C	Pearson symbol	Space group
Fe₄C	5.1	*cP*5	*P4̄3m*
Fe₃C (θ)	6.7	*oP*16	*Pnma*
Fe₅C₂ (χ)	7.9	*mC*28	*C2/c*
Fe₇C₃	8.4	*hP*20	*P6₃mc*
Fe₇C₃	8.4	*oP*40	*Pnma*
Fe₂C (η)	9.7	*oP*6	*Pnnm*
Fe₂C (ε)	9.7	*hP**	*P6₃22*
Fe₂C	9.7	*hP**	*P3̄m1*
(C)	100	*cF*8	*Fd3̄m*

Figure 16

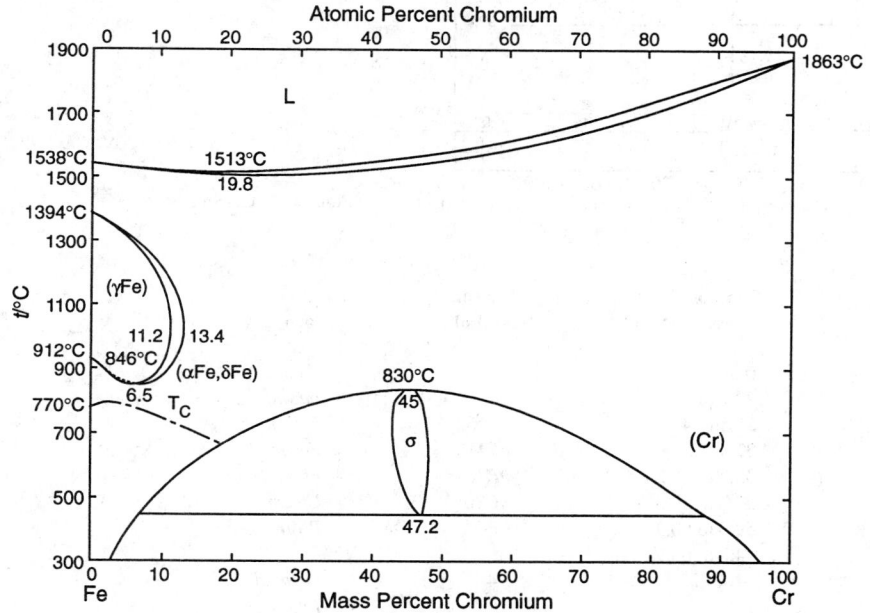

Figure 16. Fe-Cr system.

Phase	Composition, mass % Cr	Pearson symbol	Space group
(aFe, Cr)	0 to 100	*cI*2	*Im3̄m*
(γFe)	0 to 11.2	*cF*4	*Fm3̄m*
σ	42.7 to 48.2	*tP*30	*P4₂/mnm*

Figure 17

Figure 17. Cu-Sn system.

Phase	Composition, mass % Sn	Pearson symbol	Space group
α	0 to 15.8	$cF4$	$Fm\bar{3}m$
β	22.0 to 27.0	$cI2$	$Im\bar{3}m$
γ	25.5 to 41.5	$cF16$	$Fm\bar{3}m$
δ	32 to 33	$cF416$	$F\bar{4}3m$
ζ	32.2 to 35.2	$hP26$	$P6_3$
ε	27.7 to 39.5	$oC80$	$Cmcm$
η	59.0 to 60.9	$hP4$	$P6_3/mmc$
η′	44.8 to 60.9	(a)	—
(βSn)	~100	$tI4$	$I4_1/amd$
(αSn)	100	$cF8$	$Fd\bar{3}m$

(a) Hexagonal; superlattice based on NiAs-type structure.

Figure 18

Figure 18. Cu-Ni system.

Phase	Composition, mass % Ni	Pearson symbol	Space group
(Cu, Ni) (above 354.5°C)	0 to 100	$cF4$	$Fm\bar{3}m$

Figure 19

Figure 19. Pb-Sn system.

Figure 19 (continued)

Phase	Composition, mass % Sn	Pearson symbol	Space group
(Pb)	0 to 18.3	cF4	$Fm\bar{3}m$
(βSn)	97.8 to 100	tI4	$I4_1/amd$
(αSn)	100	cF8	$Fd\bar{3}m$
High-pressure phases			
ε(a)	52 to 74	hP1	$P6/mmm$
ε'(b)	52	hP2	$P6_3/mmc$

(a) From phase diagram calculated at 2500 MPa. (b) This phase was claimed for alloys at 350°C and 5500 MPa.

Figure 20

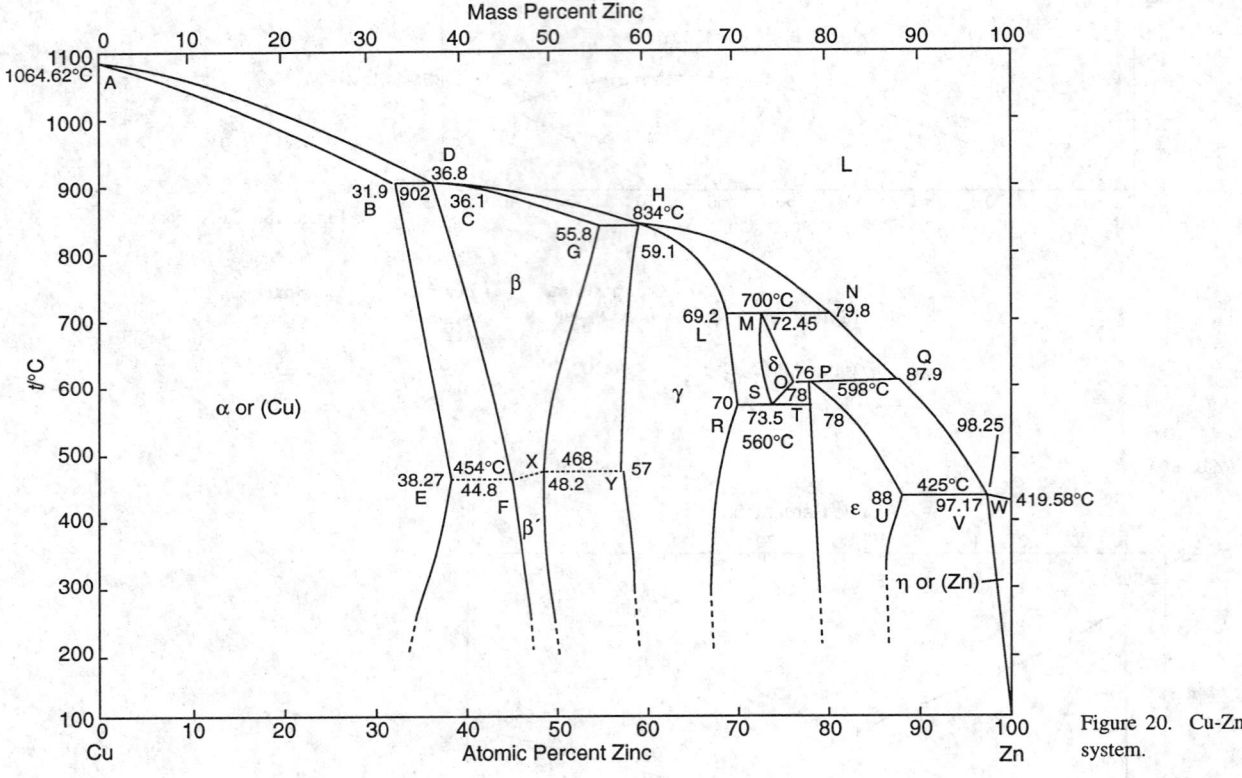

Figure 20. Cu-Zn system.

Phase	Composition, mass % Zn	Pearson symbol	Space group
α or (Cu)	0 to 38.95	cF4	$Fm\bar{3}m$
β	36.8 to 56.5	cI2	$Im\bar{3}m$
β'	45.5 to 50.7	cP2	$Pm\bar{3}m$
γ	57.7 to 70.6	cI52	$I\bar{4}3m$
δ	73.02 to 76.5	hP3	$P\bar{6}$
ε	78.5 to 88.3	hP2	$P6_3/mmc$
η or (Zn)	97.25 to 100	hP2	$P6_3/mmc$

Figure 21

Figure 21. Co-Sm system.

Phase	Composition, mass % Sm	Pearson symbol	Space group
(αCo)	0 to ~3.7	cF4	$Fm\overline{3}m$
(εCo)	~0	hP2	$P6_3/mmc$
$\beta Co_{17}Sm_2$	~23.0	hP38	$P6_3/mmc$
$\alpha Co_{17}Sm_2$	~23.0	hR19	$R\overline{3}m$
		hP8	$P6/mmm$
$Co_{5+x}Sm$	~33 to 34	—	—
$Co_{5-x}Sm$	~34 to 35	—	—
$Co_{19}Sm_5$	~40.1	hR24	$R\overline{3}m$
		hP48	$P6_3/mmc$
αCo_7Sm_2	~42.1	hR18	$R\overline{3}m$
βCo_7Sm_2	~42.1	hP36	$P6_3/mmc$
Co_3Sm	46	hR12	$R\overline{3}m$
Co_2Sm	56.0	hR4	$R\overline{3}m$
		cF24	$Fd\overline{3}m$
Co_4Sm_9	~85.1	o**	—
$CoSm_3$	88	oP16	$Pnma$
(γSm)	~100	cI2	$Im\overline{3}m$
(βSm)	~100	hP2	$P6_3/mmc$
(αSm)	~100	hR3	$R\overline{3}m$
Other reported phases			
Co_5Sm	~33.8	hP6	$P6/mmm$
Co_2Sm_5	~86.4	mC28	$C2/c$

Figure 22

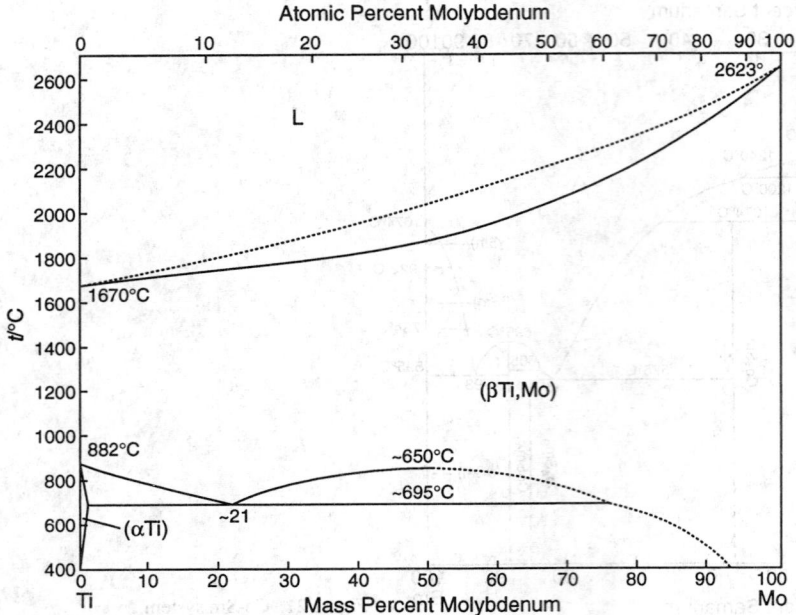

Figure 22. Ti-Mo system.

Phase	Composition, mass % Mo	Pearson symbol	Space group
(βTi, Mo)	0 to 100	$cI2$	$Im\bar{3}m$
(αTi)	0 to 0.8	$hP2$	$P6_3/mmc$
α′	(a)	$hP2$	$P6_3/mmc$
α″	(a)	$oC4$	$Cmcm$
ω	(a)	$hP3$	$P6/mmm$

(a) Metastable.

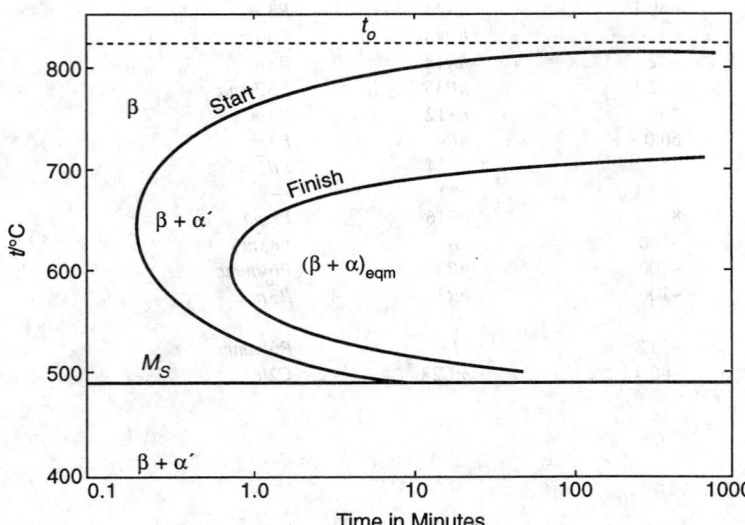

Experimental time-temperature-transformation (TTT) diagram for Ti-Mo. The start and finish times of the isothermal precipitation reaction vary with temperature as a result of the temperature dependence of the nucleation and growth processes. Precipitation is complete, at any temperature, when the equilibrium fraction of α is established in accordance with the lever rule. The solid horizontal line represents the athermal (or nonthermally activated) martensitic transformation that occurs when the β phase is quenched.

Figure 23

Figure 23. The isothermal section at 900°C (1652°F) of the iron-chromium-nickel ternary phase diagram, showing the nominal composition of 18-8 stainless steel.

HEAT CAPACITY OF SELECTED SOLIDS

This table gives the molar heat capacity at constant pressure of representative metals, semiconductors, and other crystalline solids as a function of temperature in the range 200 to 600 K.

REFERENCES

1. Chase, M. W., et al., *JANAF Thermochemical Tables, 3rd ed., J. Phys. Chem. Ref. Data,* 14, Suppl. 1, 1985.
2. Garvin, D., Parker, V. B., and White, H. J., *CODATA Thermodynamic Tables,* Hemisphere Press, New York, 1987.
3. DIPPR Database of Pure Compound Properties, Design Institute for Physical Properties Data, American Institute of Chemical Engineers, New York, 1987.

| Name | C_p in J/mol K | | | | | | |
	200 K	250 K	300 K	350 K	400 K	500 K	600 K
Aluminum	21.33	23.08	24.25	25.11	25.78	26.84	27.89
Aluminum oxide	51.12	67.05	79.45	88.91	96.14	106.17	112.55
Anthracene	138.6	173.9	210.7	248.8	288.4		
Benzoic acid	102.7	123.5	147.4	172.0			
Beryllium	9.98	13.58	16.46	18.53	19.95	21.94	23.34
Biphenyl	131.0	162.5	197.2				
Boron	5.99	8.82	11.40	13.65	15.69	18.72	20.78
Calcium	24.54	25.41	25.94	26.32	26.87	28.49	30.38
Calcium carbonate	66.50	75.66	83.82	91.51	96.97	104.52	109.86
Calcium oxide	33.64	38.59	42.18	45.07	46.98	49.33	50.72
Cesium chloride	50.13	51.34	52.48	53.58	54.68	56.90	59.10
Chromium	19.86	22.30	23.47	24.39	25.23	26.63	27.72
Cobalt	22.23	23.98	24.83	25.68	26.53	28.20	29.66
Copper	22.63	23.77	24.48	24.95	25.33	25.91	26.48
Copper oxide	34.80		42.41	44.95	46.78	49.19	50.83
Copper sulfate	77.01	89.25	99.25	107.65	114.93	127.19	136.31
Germanium			23.25	23.85	24.31	24.96	25.45
Gold			25.41	25.37	25.51	26.06	26.65
Graphite	5.01	6.82	8.58	10.24	11.81	14.62	16.84
Hexachlorobenzene	162.7	183.6	202.4				
Iodine	51.57	53.24	54.51	58.60			
Iron	21.59	23.74	25.15	26.28	27.39	29.70	32.05
Lead	25.87	26.36	26.85	27.30	27.72	28.55	29.40
Lithium	21.57	23.42	24.64	25.96	27.60	29.28	
Lithium chloride	43.35	46.08	48.10	49.66	50.97	53.34	55.59
Magnesium	22.72	24.02	24.90	25.57	26.14	27.17	28.18
Magnesium oxide			37.38	40.59	42.77	45.56	47.30
Manganese	23.05	24.95	26.35	27.52	28.53	30.29	31.90
Naphthalene	105.8	134.1	167.8	204.1			
Potassium	27.00	28.01	29.60				
Potassium chloride	48.44	50.10	51.37	52.31	53.08	54.71	56.35
Silicon	15.64	18.22	20.04	21.28	22.14	23.33	24.15
Silicon dioxide	32.64	39.21	44.77	49.47	53.43	59.64	64.42
Silver			25.36	25.55	25.79	26.36	26.99
Sodium	22.45	27.01	28.20	30.14			
Sodium chloride	46.89	48.85	50.21	51.25	52.14	53.96	55.81
Tantalum	24.08	24.86	25.31	25.60	25.84	26.35	26.84
Titanium	22.37	24.07	25.28	26.17	26.86	27.88	28.60
Tungsten	22.49	23.69	24.30	24.65	24.92	25.36	25.79
Vanadium	21.88	23.70	24.93	25.68	26.23	26.94	27.49
Zinc	24.05	25.02	25.45	25.88	26.35	27.39	28.59
Zirconium	23.87	24.69	25.22	25.61	25.93	26.56	27.28

THERMAL AND PHYSICAL PROPERTIES OF PURE METALS

This table gives the following properties for the metallic elements:

t_m: Melting point in °C
t_b: Normal boiling point in °C, at a pressure of 101.325 kPa (760 Torr)
$\Delta_{fus} H$: Enthalpy of fusion at the melting point in J/g
ρ_{25}: Density at 25°C in g/cm³
α: Coefficient of linear expansion at 25°C in K⁻¹ (the quantity listed is $10^6 \times \alpha$)
c_p: Specific heat capacity at constant pressure at 25°C in J/g K
λ: Thermal conductivity at 27°C in W/cm K

REFERENCES

1. Dinsdale, A. T., *CALPHAD*, 15, 317, 1991 (melting points, enthalpy of fusion).
2. Touloukian, Y. S., *Thermophysical Properties of Matter*, Vol. 12, Thermal Expansion, IFI/Plenum, New York, 1975 (coefficient of expansion, density).
3. Ho, C. Y., Powell, R. W., and Liley, P. E., *J. Phys. Chem. Ref. Data*, 3, Suppl. 1, 1974 (thermal conductivity).
4. Cox, J. D., Wagman, D. D., and Medvedev, V. A., *CODATA Key Values for Thermodynamics*, Hemisphere Publishing Corp., New York, 1989 (heat capacity).
5. Glushko, V. P., Ed., *Thermal Constants of Substances*, VINITI, Moscow, (enthalpy of fusion, heat capacity).
6. Wagman, D. D., et. al., *The NBS Tables of Chemical Thermodynamic Properties, J. Phys. Chem. Ref. Data*, 11, Suppl. 2, 1982 (heat capacity).
7. Chase, M. W., et. al., *JANAF Thermochemical Tables*, 3rd ed., J. Phys. Chem. Ref. Data, 14, Suppl. 1, 1985 (heat capacity, enthalpy of fusion).
8. Gschneidner, K. A., *Bull. Alloy Phase Diagrams*, 11, 216—224, 1990 (various properties of the rare earth metals).
9. Hellwege, K. H., Ed., *Landolt Börnstein, Numerical Values and Functions in Physics, Chemistry, Astronomy, Geophysics, and Technology*, Vol. 2, Part 1, Mechanical-Thermal Properties of State, 1971 (density).

Metal (symbol)	Atomic weight	t_m °C	t_b °C	$\Delta_{fus} H$ J/g	ρ_{25} g/cm³	$\alpha \times 10^6$ K⁻¹	c_p J/g K	λ W/cm K
Actinium (Ac)		1051	3198		10			
Aluminum (Al)	26.98	660.32	2519	397	2.70	23.1	0.897	2.37
Antimony (Sb)	121.76	630.63	1587	163.2	6.68	11.0	0.207	0.243
Barium (Ba)	137.33	727	1897	52	3.62	20.6	0.204	0.184
Beryllium (Be)	9.01	1287	2471	877	1.85	11.3	1.825	2.00
Bismuth (Bi)	208.98	271.40	1564	54.1	9.79	13.4	0.122	0.0787
Cadmium (Cd)	112.41	321.07	767	55.1	8.69	30.8	0.232	0.968
Calcium (Ca)	40.08	842	1484	213	1.54	22.3	0.647	2.00
Cerium (Ce)	140.11	798	3443	39.0	6.77	6.3	0.192	0.113
Cesium (Cs)	132.91	28.44	671	15.8	1.93		0.242	0.359
Chromium (Cr)	52.00	1907	2671	404	7.15	4.9	0.449	0.937
Cobalt (Co)	58.93	1495	2927	275	8.86	13.0	0.421	1.00
Copper (Cu)	63.55	1084.62	2562	208.7	8.96	16.5	0.385	4.01
Dysprosium (Dy)	162.50	1412	2567	68.1	8.55	9.9	0.170	0.107
Erbium (Er)	167.26	1529	2868	119	9.07	12.2	0.168	0.145
Europium (Eu)	151.96	822	1529	60.6	5.24	35.0	0.182	0.139[a]
Gadolinium (Gd)	157.25	1313	3273	63.6	7.90	9.4[b]	0.236	0.105
Gallium (Ga)	69.72	29.76	2204	80.2	5.91		0.371	0.406
Gold (Au)	196.97	1064.18	2856	63.7	19.3	14.2	0.129	3.17
Hafnium (Hf)	178.49	2233	4603	152.4	13.3	5.9	0.144	0.230
Holmium (Ho)	164.93	1474	2700	103[a]	8.80	11.2	0.165	0.162
Indium (In)	114.82	156.60	2072	28.6	7.31	32.1	0.233	0.816
Iridium (Ir)	192.22	2446	4428	213.9	22.5	6.4	0.131	1.47
Iron (Fe)	55.85	1538	2861	247.3	7.87	11.8	0.449	0.802
Lanthanum (La)	138.91	918	3464	44.6	6.15	12.1	0.195	0.134
Lead (Pb)	207.20	327.46	1749	23.0	11.3	28.9	0.129	0.353
Lithium (Li)	6.94	180.5	1342	432	0.534	46	3.582	0.847
Lutetium (Lu)	174.97	1663	3402	126[a]	9.84	9.9	0.154	0.164
Magnesium (Mg)	24.30	650	1090	349	1.74	24.8	1.023	1.56

Metal (symbol)	Atomic weight	t_m °C	t_b °C	$\Delta_{fus}H$ J/g	ρ_{25} g/cm^3	$\alpha \times 10^6$ K^{-1}	c_p J/g K	λ W/cm K
Manganese (Mn)	54.94	1246	2061	235.0	7.3	21.7	0.479	0.0782
Mercury (Hg)	200.59	-38.83	356.73	11.4	13.5336		0.140	0.0834
Molybdenum (Mo)	95.94	2623	4639	390.7	10.2	4.8	0.251	1.38
Neodymium (Nd)	144.24	1021	3074	49.5	7.01	9.6	0.190	0.165
Neptunium (Np)		644			20.2			0.063
Nickel (Ni)	58.69	1455	2913	298	8.90	13.4	0.444	0.907
Niobium (Nb)	92.91	2477	4744	323	8.57	7.3	0.265	0.537
Osmium (Os)	190.23	3033	5012	304.2	22.59	5.1	0.130	0.876
Palladium (Pd)	106.42	1554.9	2963	157.3	12.0	11.8	0.246	0.718
Platinum (Pt)	195.08	1768.4	3825	113.6	21.5	8.8	0.133	0.716
Plutonium (Pu)		640	3228		19.7	46.7		0.0674
Polonium (Po)		254	962		9.20			0.20
Potassium (K)	39.10	63.38	759	59.3	0.89		0.757	1.024
Praseodymium (Pr)	140.91	931	3520	48.9	6.77	6.7	0.193	0.125
Promethium (Pm)		1042	3000[a]	53[a]	7.26	11[a]	0.19[a]	0.15[a]
Protactinium (Pa)	231.04	1572		53.4	15.4			
Radium (Ra)		700			5			
Rhenium (Re)	186.21	3186	5596	324.5	20.8	6.2	0.137	0.479
Rhodium (Rh)	102.91	1964	3695	258.4	12.4	8.2	0.243	1.50
Rubidium (Rb)	85.47	39.31	688	25.6	1.53		0.363	0.582
Ruthenium (Ru)	101.07	2334	4150	381.8	12.1	6.4	0.238	1.17
Samarium (Sm)	150.36	1074	1794	57.3	7.52	12.7	0.197	0.133
Scandium (Sc)	44.96	1541	2836	314	2.99	10.2	0.568	0.158
Silver (Ag)	107.87	961.78	2162	104.8	10.5	18.9	0.235	4.29
Sodium (Na)	22.99	97.72	883	113	0.97	71	1.228	1.41
Strontium (Sr)	87.62	777	1382	84.8	2.64	22.5	0.301	0.353
Tantalum (Ta)	180.95	3017	5458	202.1	16.4	6.3	0.140	0.575
Technetium (Tc)		2157	4265		11			0.506
Terbium (Tb)	158.93	1356	3230	67.9	8.23	10.3	0.182	0.111
Thallium (Tl)	204.38	304	1473	20.3	11.8	29.9	0.129	0.461
Thorium (Th)	232.04	1750	4788	59.5	11.7	11.0	0.113	0.540
Thulium (Tm)	168.93	1545	1950	99.4	9.32	13.3	0.160	0.169
Tin (Sn)	118.71	231.93	2602	59.2	7.26	22.0	0.228	0.666
Titanium (Ti)	47.88	1668	3287	295	4.51	8.6	0.523	0.219
Tungsten (W)	183.84	3422	5555	284.5	19.3	4.5	0.132	1.74
Uranium (U)	238.03	1135	4131	38.4	19.1	13.9	0.116	0.276
Vanadium (V)	50.94	1910	3407	422	6.0	8.4	0.489	0.307
Ytterbium (Yb)	173.04	819	1196	44.3	6.90	26.3	0.155	0.385
Yttrium (Y)	88.91	1522	3345	128	4.47	10.6	0.298	0.172
Zinc (Zn)	65.39	419.53	907	112	7.14	30.2	0.388	1.16
Zirconium (Zr)	91.22	1855	4409	230	6.52	5.7	0.278	0.227

[a] Estimated.
[b] At 100°C.

THERMAL CONDUCTIVITY OF METALS AND SEMICONDUCTORS AS A FUNCTION OF TEMPERATURE

This table gives the temperature dependence of the thermal conductivity of several metals and of carbon, germanium, and silicon. For graphite, separate entries are given for the thermal conductivity parallel ($\parallel$) and perpendicular ($\perp$) to the layer planes. The thermal conductivity of all these materials is very sensitive to impurities at low temperatures, especially below 100 K. Therefore, the values given here should be regarded as typical values for a highly purified specimen; the thermal conductivity of different specimens can vary by more than an order of magnitude in the low-temperature range. See Reference 2 for details.

REFERENCES

1. Ho, C. Y., Powell, R. W., and Liley, P. E., *J. Phys. Chem. Ref. Data*, 1, 279, 1972.
2. White, G. K., and Minges, M. L., *Thermophysical Properties of Some Key Solids*, CODATA Bulletin No. 59, 1985.

Thermal Conductivity in W/cm K

				Carbon (C)						
				Diamond (type)			Pyrolytic graphite			
T/K	Ag	Al	Au	I	IIa	IIb	$\parallel$	$\perp$	Cr	Cu
1	39.4	41.1	5.46						0.402*	42.2
2	78.3	81.8	10.9	0.0138*	0.033*	0.0200*			0.803	84.0
3	115	121	16.1	0.0461	0.111	0.0676			1.20	125
4	147	157	20.9	0.108	0.261	0.160			1.60	162
5	172	188	25.2	0.206	0.494	0.307			2.00	195
6	187	213	28.5	0.344	0.820	0.510			2.39	222
7	193	229	30.9	0.523	1.24	0.778			2.27	239
8	190	237	32.3	0.762	1.77	1.12			3.14	248
9	181	239	32.7	1.05	2.41	1.53			3.50	249
10	168	235	32.4	1.40	3.17	2.03	0.811	0.0116	3.85	243
15	96.0	176	24.6	3.96	8.65	5.66			5.24	171
20	51.0	117	15.8	7.87	16.8	11.2	4.20	0.0397	5.93	108
30	19.3	49.5	7.55	18.8	38.9	26.5	9.86	0.0786	5.49	44.5
40	10.5	24.0	5.15	29.4	65.9	44.0	16.4	0.120	4.25	21.7
50	7.0	13.5	4.21	35.3	92.1	59.1	23.1	0.152	3.17	12.5
60	5.5	8.5	3.74	37.4	112	67.5	29.8	0.173	2.48	8.29
70	4.97	5.85	3.48	36.9	119	69.1	36.6	0.181	2.07	6.47
80	4.71	4.32	3.32	35.1	117	65.7	42.8	0.181	1.84	5.57
90	4.60	3.42	3.28	32.7	109	60.0	47.5	0.176	1.69	5.08
100	4.50	3.02	3.27	30.0	100	54.2	49.7	0.168	1.59	4.82
150	4.32	2.48	3.25	19.5	60.2	32.5	45.1	0.125	1.29	4.29
200	4.30	2.37	3.23	14.1	40.3	22.6	32.3	0.0923	1.11	4.13
250	4.29	2.35	3.21	11.0	29.7	17.0	24.4	0.0711	1.00	4.06
300	4.29	2.37	3.17	8.95	23.0	13.5	19.5	0.0570	0.937	4.01
350	4.27	2.40	3.14	7.55*	18.5*	11.1*	16.2	0.0477	0.929	3.96
400	4.25	2.40	3.11	6.5*	15.4*	9.32*	13.9	0.0409	0.909	3.93
500	4.19	2.36	3.04				10.8	0.0322	0.860	3.86
600	4.12	2.31	2.98				8.92	0.0268	0.807	3.79
800	3.96	2.18	2.84				6.67	0.0201	0.713	3.66
1000	3.79		2.70				5.34	0.0160	0.654	3.52
1200	3.61*		2.55				4.48	0.0134	0.619	3.39
1400							3.84	0.0116	0.588	
1600							3.33	0.0100	0.556	
1800							2.93	0.00895	0.526*	
2000							2.62	0.00807	0.494*	

THERMAL CONDUCTIVITY OF METALS AND SEMICONDUCTORS AS A FUNCTION OF TEMPERATURE (continued)

T/K	Fe	Ge[a]	Mg	Ni	Pb	Pt	Si[a]	Sn	Ti	W
1	1.71	0.274	9.86	2.17	27.9	2.31	0.0693*	183	0.0144*	14.4
2	3.42	2.06	19.6	4.34	44.6	4.60	0.454	323	0.0288*	28.7
3	5.11	5.35	29.0	6.49	35.8	6.79	1.38	297	0.0432	42.8
4	6.77	8.77	37.6	8.59	22.2	8.8	2.97	181	0.0575	56.3
5	8.39	11.6	45.0	10.6	13.8	10.5	5.27	117	0.0719	68.7
6	9.93	13.9	50.8	12.5	8.10	11.8	8.23	76	0.0863	79.5
7	11.4	15.5	54.7	14.2	4.86	12.6	11.7	52	0.101	88.0
8	12.7	16.6	56.7	15.8	3.20	12.9	15.5	36	0.115	93.8
9	13.9	17.3	57.0	17.1	2.30	12.8	19.5	26	0.129	96.8
10	14.8	17.7	55.8	18.1	1.78	12.3	23.3	19.3	0.143	97.1
15	17.0	17.3	41.1	19.5	0.845	8.41	41.6	6.3	0.212	72.0
20	15.4	14.9	27.2	16.5	0.591	4.95	49.8	3.2	0.275	40.5
30	10.0	10.8	12.9	9.56	0.477	2.15	48.1	1.79	0.365	14.4
40	6.23	7.98	7.19	5.82	0.451	1.39	35.3	1.33	0.390	6.92
50	4.05	6.15	4.65	4.00	0.436	1.09	26.8	1.15	0.374	4.27
60	2.85	4.87	3.27	3.08	0.425	0.947	21.1	1.04	0.355	3.14
70	2.16	3.93	2.49	2.50	0.416	0.862	16.8	0.96	0.340	2.58
80	1.75	3.25	2.02	2.10	0.409	0.815	13.4	0.915	0.326	2.29
90	1.50	2.70	1.78	1.83	0.403	0.789	10.8	0.880	0.315	2.17
100	1.34	2.32	1.69	1.64	0.397	0.775	8.84	0.853	0.305	2.08
150	1.04	1.32	1.61	1.22	0.379	0.740	4.09	0.779	0.270	1.92
200	0.94	0.968	1.59	1.07	0.367	0.726	2.64	0.733	0.245	1.85
250	0.865	0.749	1.57	0.975	0.360	0.718	1.91	0.696	0.229	1.80
300	0.802	0.599	1.56	0.907	0.353	0.716	1.48	0.666	0.219	1.74
350	0.744	0.495	1.55	0.850	0.347	0.717	1.19	0.642	0.210	1.67
400	0.695	0.432	1.53	0.802	0.340	0.718	0.989	0.622	0.204	1.59
500	0.613	0.338	1.51	0.722	0.328	0.723	0.762	0.596	0.197	1.46
600	0.547	0.273	1.49	0.656	0.314	0.732	0.619		0.194	1.37
800	0.433	0.198	1.46*	0.676		0.756	0.422		0.197	1.25
1000	0.323	0.174		0.718		0.787	0.312		0.207	1.18
1200	0.283	0.174		0.762		0.826	0.257		0.220	1.12
1400	0.312			0.804		0.871	0.235		0.236	1.08
1600	0.330					0.919	0.221		0.253	1.04
1800	0.345*					0.961			0.270*	1.01
2000						0.994*				0.98

[a] Values below 300 K are typical values.

* Extrapolated.

THERMAL CONDUCTIVITY OF ALLOYS AS A FUNCTION OF TEMPERATURE

This table lists the thermal conductivity of selected alloys at various temperatures. The indicated compositions refer to weight percent. Since the thermal conductivity is sensitive to exact composition and processing history, especially at low temperatures, these values should be considered approximate.

REFERENCES

1. Powell, R. L., and Childs, G. E., in *American Institute of Physics Handbook, 3rd Edition*, Gray, D. E., Ed., McGraw-Hill, New York, 1972.
2. Ho, C. Y., et al., *J. Phys. Chem. Ref. Data*, 7, 959, 1978.

Thermal conductivity in W/m K

	Alloy	4 K	20 K	77 K	194 K	273 K	373 K	573 K	973 K
Aluminum:	1100	50	240	270	220	220			
	2024	3.2	17	56	95	130			
	3003	11	58	140	150	160			
	5052	4.8	25	77	120	140			
	5083, 5086	3	17	55	95	120			
	Duralumin	5.5	30	91	140	160	180		
Bismuth:	Rose metal		5.5	8.3	14	16			
	Wood's metal	4	17	23					
Copper:	electrolytic tough pitch	330	1300	550	400	390	380	370	350
	free cutting, leaded	200	800	460	380	380			
	phosphorus, deoxidized	7.5	42	120	190	220			
	brass, leaded	2.3	12	39	70	120			
	bronze, 68% Cu; 32% Zn	2.3	16	48	92	110			
	beryllium	2	17	36	70	90	113	172	
	german silver	0.75	7.5	17	20	23	25	30	40
	silicon bronze A		3.4	11	23	30			
	manganin	0.48	3.2	14	17	22			
	constantan	0.9	8.6	17	19	22			
Ferrous:	commercial pure iron	15	72	106	82	76	66	54	34
	plain carbon steel(AISI 1020)	13	20	58	65	65			
	plain carbon steel(AISI 1095)		8.5	31	41	45			
	3% Ni; 0.7% Cr; 0.6% Mo		6	22		33	35	36	30
	4% Si					20	24	28	26
	stainless steel	0.3	2	8	13	14	16	19	25
	27% Ni; 15% Cr		1.7	55		11	12	16	21
Gold:	colbalt thermocouple	1.2	8.6	20					
	65% Au; 35% Ag		12	24		61	89		
Indium:	85.5% In; 14.5% Pb	1.9	7.8	24	41				
Lead:	60% Pb; 40% Sn (soft solder)		28	44					
	64.35% Pb; 35.65% In	0.8	3.26	9.1		20.2			
Nickel:	80% Ni; 20% Cr					12	14	17	23
	contracid	0.2	2	7.3	9.5	13			
	inconel	0.5	4.2	12.5	13	15	16	19	26
	monel	0.9	7.1	15	20	21	24	30	43
Platinum:	90% Pt; 10% Ir					31	31.4		
	90% Pt; 10% Rh					30.1	30.5		
Silver:	silver solder		12	34	58				
	normal Ag thermocouple	48	230	310					
Tin:	60% Sn; 40% Pb	16	55	51					
Titanium:	5.5% Al; 2.5% Sn;0.2% Fe		1.8	4.3	6.4	7.8	8.4	10.8	
	4.7% Mn; 3.99% Al; 0.14% C		1.7	4.5	6.5	8.5			

THERMAL CONDUCTIVITY OF CRYSTALLINE DIELECTRICS

This table lists the thermal conductivity of a number of crystalline dielectrics, including some which find use as optical materials. Values are given at temperatures for which data are available.

REFERENCE

Powell, R. L., and Childs, G. E., in *American Institute of Physics Handbook, 3rd Edition*, Gray, D. E., Ed., McGraw-Hill, New York, 1972.

Material	T/K	Ther. cond. W/m K	Material	T/K	Ther. cond. W/m K
AgCl	223	1.3	BeO	4.2	0.3
	273	1.2		20	16
	323	1.1		77	270
	373	1.1		373	210
Al,B silicate (tourmaline)	398	2.9		573	120
‖ to c axis	540	3.2		1273	29
	723	3.5	Bi_2Te_3	80	6.4
Al,Be silicate (beryl)	315	6.4		204	2.8
Al,F silicate (topaz)	315	17.7		303	3.6
‖ to c axis	358	15.6		370	4.6
	417	13.3	C (diamond)	4.2	13
Al,Fe silicate (garnet)	315	35.8	type I	20	800
	358	35.4		77	3550
	377	35.6		194	1450
Al_2O_3 (sapphire):				273	1000
36° to c axis	4.2	110	$CaCO_3$		
	20	3500	‖ to c axis	83	25
	35	6000		273	5.5
	77	1100	⊥ to c axis	83	17
⊥ to c axis	373	2.6		194	6.5
	523	3.9		273	4.6
	773	5.8		373	3.6
Al_2O_3 (sintered)	4.2	0.5	CaF_2	83	39
	20	23		223	18
	77	150		273	10
	194	48		323	9.2
	273	35		373	9
	373	26	$CaWO_4$ (scheelite)	422	11.3
	973	8	CdTe	160	7.0
Ar	8	6.0		297	3.6
	10	3.7		422	2.9
	20	1.4	CsBr	223	1.2
	77	0.31		273	0.94
As_2S_3 (glass)	283	0.16		323	0.81
	323	0.21		373	0.77
	373	0.27	CsI	223	1.4
BN	1047	36.2		273	1.2
	1475	22.7		323	1
	1928	21.9		373	0.95
	2111	18.5	Cu_2O (cuprite)	102	3.74
BaF_2	225	20		163	7.76
	260	13.4		299	5.58
	305	10.9		360	4.86
	370	10.5	Fe_3O_4 (magnetite)	4.5	27.4
$BaTiO_3$	5	4.2		20.5	293.0
	30	24.0		126.5	7.4
	40	25.0		304	7.0
	100	12.0	Glass:		
	250	4.8	phoenix	4.2	0.095
	300	6.2		20	0.13
				77	0.37

Material	T/K	Ther. cond. W/m K	Material	T/K	Ther. cond. W/m K
plastic perspex	4.2	0.058	NaCl	4.2	440
	20	0.074		20	300
pyrex	77	0.44		77	30
	194	0.88		273	6.4
	273	1		323	5.6
H_2 (para + 0.5% ortho)	2.5	100		373	5.4
	3	150	NaF	5	1100
	4	200		50	250
	6	30		100	90
	10	3	Ne	2	3.0
H_2O (ice)	173	3.5		3	4.6
	223	2.8		4.2	4.2
	273	2.2		10	0.8
He^3 (high pressure)	0.6	25		20	0.3
	1	2	NH_4Cl	77	17
	1.5	0.57		194	23
	2	0.21		230	38
He^4 (high pressure)	0.5	42		273	27
	0.8	120	$NH_4H_2PO_4$		
	1	24	‖ to optic axis	315	0.71
	2	0.18		339	0.71
I_2	300	0.45	⊥ to optic axis	313	1.26
	325	0.42		342	1.34
	350	0.4	NiO	4.2	5.9
KBr	2	150		40	400
	4.2	360		194	82
	100	12	SiO_2 (quartz)		
	273	5	‖ to c axis	20	720
	323	4.8		194	20
	373	4.8		273	12
KCl	4.2	500	⊥ to c axis	20	370
	25	140		194	10
	80	35		273	6.8
	194	10	SiO_2 (fused silica)	4.2	0.25
	273	7.0		20	0.7
	323	6.5		77	0.8
	373	6.3		194	1.2
KI	4.2	700		273	1.4
	80	13		373	1.6
	194	4.6		673	1.8
	273	3.1	$SrTiO_3$	5	2.4
Kr	4.2	0.48		30	21.0
	10	1.7		40	19.2
	20	1.2		100	18.5
	77	0.36		250	12.5
LaF_3	78	7.8		300	11.2
	197	5.0	TlBr	316	0.59
	274	5.4	TlCl	311	0.75
LiF	4.2	620	TiO_2 (rutile)		
	20	1800	‖ to optic axis	4.2	200
	77	150		20	1000
$MgO\cdot Al_2O_3$ (spinel)	373	13		273	13
	773	8.5	⊥ to optic axis	4.2	160
MnO	4.2	0.25		20	690
	40	55		273	9
	120	8			
	573	3.5			

THERMAL CONDUCTIVITY OF CERAMICS AND OTHER INSULATING MATERIALS

Thermal conductivity values for ceramics, refractory oxides, and miscellaneous insulating materials are given here. The thermal conductivity refers to samples with density indicated in the second column. Since most of these materials are highly variable, the values should only be considered as a rough guide.

REFERENCES

1. Powell, R. L., and Childs, G. E., in *American Institute of Physics Handbook, 3rd Edition*, Gray, D. E., Ed., McGraw-Hill, New York, 1972.
2. Perry, R. H., and Green, D., *Perry's Chemical Engineers' Handbook, Sixth Edition*, McGraw-Hill, New York, 1984.

Material	Dens. g/cm^3	t °C	Ther. cond. W/m K	Material	Dens. g/cm^3	t °C	Ther. cond. W/m K
Alumina (Al$_2$O$_3$)	3.8	100	30	Diatomite	0.2	0	0.05
		400	13			400	0.09
		1300	6		0.5	0	0.09
		1800	7.4			400	0.16
	3.5	100	17	Ebonite	1.2	0	0.16
		800	7.6	Felt, flax	0.2	30	0.05
Al$_2$O$_3$ + MgO		100	15		0.3	30	0.04
		400	10	Fuller's earth	0.53	30	0.1
		1000	5.6	Glass wool	0.2	-200 to 20	0.005
Asbestos	0.4	-100	0.07			50	0.04
		100	1			100	0.05
		0	0.09			300	0.08
Asbestos + 85% MgO	0.3	30	0.08	Graphite			
Asphalt	2.1	20	0.06	100 mesh	0.48	40	0.18
Beryllia (BeO)	2.8	100	210	20-40 mesh	0.7	40	1.29
		400	90	Linoleum cork	0.54	20	0.08
		1000	20	Magnesia (MgO)		100	36
		1800	15			400	18
	1.85	50	64			1200	5.8
		200	40			1700	9.2
		600	23	MgO + SiO$_2$		100	5.3
Brick, dry	1.54	0	0.04			400	3.5
Brick, refractory:						1500	2.3
alosite		1000	1.3	Mica:			
aluminous	1.99	400	1.2	muscovite		100	0.72
		1000	1.3			300	0.65
diatomaceous	0.77	100	0.2			600	0.69
		500	0.24	phlogopite		100	0.66
	0.4	100	0.08	Canadian		300	0.19
		500	0.1			600	0.2
fireclay	2	400	1	Micanite		30	0.3
		1000	1.2	Mineral wool	0.15	30	0.04
silicon carbide	2	200	2	Perlite, expanded	0.1	-200 to 20	0.002
		600	2.4	Plastics:			
vermiculite	0.77	200	0.26	bakelite	1.3	20	1.4
		600	0.31	celluloid	1.4	30	0.02
Calcium oxide		100	16	polystyrene foam	0.05	-200 to 20	0.033
		400	9	mylar foil	0.05	-200 to 20	0.0001
		1000	7.5	nylon		-253	0.10
Cement mortar	2	90	0.55			-193	0.23
Charcoal	0.2	20	0.055			25	0.30
Coal	1.35	20	0.26	polytetrafluoroethylene		-253	0.13
Concrete	1.6	0	0.8			-193	0.16
Cork	0.05	0	0.03			25	0.26
		100	0.04			230	2.5
	0.35	0	0.06	urethane foam	0.07	20	0.06
		100	0.08	Porcelain		90	1
Cotton wool	0.08	30	0.04				

Material	Dens. g/cm³	t °C	Ther. cond. W/m K	Material	Dens. g/cm³	t °C	Ther. cond. W/m K
Rock:				Uranium dioxide		100	9.8
basalt		20	2			400	5.5
chalk		20	0.92			1000	3.4
granite	2.8	20	2.2	Wood:			
limestone	2	20	1	balsa, ⊥	0.11	30	0.04
sandstone	2.2	20	1.3	fir, ⊥	0.54	20	0.14
slate, ⊥		95	1.4	fir, ∥	0.54	20	0.35
slate, ∥		95	2.5	oak		20	0.16
Rubber:				plywood		20	0.11
sponge	0.2	20	0.05	pine, ⊥	0.45	60	0.11
92 percent		25	0.16	pine, ∥	0.45	60	0.26
Sand, dry	1.5	20	0.33	walnut, ⊥	0.65	20	0.14
Sawdust	0.2	30	0.06	Wool	0.09	30	0.04
Shellac		20	0.23	Zinc oxide		200	17
Silica aerogel	0.1	-200 to 20	0.003			800	5.3
Snow	0.25	0	0.16	Zirconia (ZrO₂)		100	2
Steel wool	0.1	55	0.09			400	2
Thoria (ThO₂)		100	10			1500	2.5
		400	5.8	Zirconia + silica		200	5.6
		1500	2.4			600	4.6
Titanium dioxide		100	6.5			1500	3.7
		400	3.8				
		1200	3.3				

THERMAL CONDUCTIVITY OF GLASSES

This table gives the composition of various types of glasses and the thermal conductivity k as a function of temperature. Because of the variability of glasses, the data should be regarded as only approximate.

Type of glass	Composition			t	k
	SiO_2 (wt%)	Other oxides	(wt%)	°C	W/m K
Vitreous silica	100			−150	0.85
				−100	1.05
				−50	1.20
				0	1.30
				50	1.40
				100	1.50
Vycor glass	96	B_2O_3	3	−100	1.00
				0	1.25
				100	1.40
Pyrex type chemically-resistant borosilicate glasses	80–81	B_2O_3	12–13	−100	0.90
		Na_2O	4	0	1.10
		Al	2	100	1.25
Borosilicate crown glasses	60–65	B_2O_3	15–20	−100	0.65–0.75
				0	0.90–0.95
				100	1.00–1.05
	65–70	B_2O_3	10–15	−100	0.75–0.80
				0	0.95–1.00
				100	1.05–1.15
	70–75	B_3O_3	5–10	−100	0.80–0.85
				0	1.05–1.10
				100	1.15–1.20
Zinc crown glasses (i)	55–65	ZnO Remainder: B_2O_3, Al_2O_3	5–15	−100 0 100	0.88–0.92 1.10–1.15 1.15–1.25
		ZnO Remainder: Na_2O, K_2O	5–15	−100 0 100	0.60–0.70 0.70–0.90 0.85–0.95
		ZnO Remainder: B_2O_3, Al_2O_3	15–25	−100 0 100	0.88–0.92 1.10–1.15 1.15–1.20
		ZnO Remainder: Na_2O, K_2O	15–25	−100 0 100	0.65–0.80 0.85–0.95 0.90–1.05
Zinc crown glasses (ii)	65–75	ZnO Remainder: B_2O_3, Al_2O_3	5–15	−100 0 100	0.88–0.92 1.15–1.15 1.20–1.30
		ZnO Remainder: Na_2O, K_2O	5–15	−100 0 100	0.70–0.85 0.90–1.05 1.00–1.15

Type of glass	Composition SiO$_2$ (wt%)	Other oxides (wt%)		t °C	k W/m K
		ZnO	15–25	−100	0.90–0.95
		Remainder:		0	1.15–1.15
		B$_2$O$_3$, Al$_2$O$_3$		100	1.20–1.25
		ZnO	15–25	−100	0.65–0.85
		Remainder:		0	0.85–1.00
		Na$_2$O, K$_2$O		100	1.05–1.20
Barium crown glasses	31	B$_2$O$_3$	12	−100	0.55
		Al$_2$O$_3$	8	0	0.70
		BaO	48	100	0.80
	41	B$_2$O$_3$	6	−100	0.60
		Al$_2$O$_3$	2	0	0.75
		ZnO	8	100	0.85
		BaO	43		
	47	B$_2$O$_3$	4	−100	0.65
		Na$_2$O	1	0	0.75
		K$_2$O	7	100	0.90
		ZnO	8		
		BaO	32		
	65	B$_2$O$_3$	2	−100	0.70
		Na$_2$O	5	0	0.90
		K$_2$O	15	100	1.00
		ZnO	2		
		BaO	10		
Borate glasses					
Borate flint glass	9	B$_2$O$_3$	36	−100	0.55
		Na$_2$O	1	0	0.65
		K$_2$O	2	100	0.80
		PbO	36		
		Al$_2$O$_3$	10		
		ZnO	6		
Borate flint glass	0	B$_2$O$_3$	56	−100	0.50
		Al$_2$O$_3$	12	0	0.65
		PbO	32	100	0.85
Borate flint glass	0	B$_2$O$_3$	43	−100	0.40
		Al$_2$O$_3$	5	0	0.55
		PbO	52	100	0.70
Borate glass	4	B$_2$O$_3$	55	−100	0.65
		Al$_2$O$_3$	14	0	0.80
		PbO	11	100	0.90
		K$_2$O	4		
		ZnO	12		
Borate crown glass	0	B$_2$O$_3$	64	−100	0.50
		Na$_2$O	8	0	0.65
		K$_2$O	3	100	0.85
		BaO	4		
		PbO	3		
		Al$_2$O$_3$	18		

Type of glass	SiO$_2$ (wt%)	Other oxides (wt%)		t °C	k W/m K
Light borate crown glass	0	B$_2$O$_3$	69	−100	0.55
		Na$_2$O	8	0	0.70
		BaO	5	100	0.90
		Al$_2$O$_3$	18		
Zinc borate glass	0	B$_2$O$_3$	40	−100	0.65
		ZnO	60	0	0.75
				100	0.85
Phosphate crown glasses					
Potash phosphate glass	0	P$_2$O$_5$	70	0	0.75
		B$_2$O$_3$	3	100	0.85
		K$_2$O	12		
		Al$_2$O$_3$	10		
		MgO	4		
Baryta phosphate glass	0	P$_2$O$_5$	60	45	0.75
		B$_2$O$_3$	3		
		Al$_2$O$_3$	8		
		BaO	28		
Soda-lime glasses	75	Na$_2$O	17	−100	0.75
		CaO	8	0	0.95
				100	1.10
	75	Na$_2$O	12	−100	0.90
		CaO	13	0	1.10
				100	1.15
	72	Na$_2$O	15	−100	0.80
		CaO	11	0	1.00
		Al$_2$O$_3$	2	100	1.15
	65	Na$_2$O	25	−100	0.65
		CaO	10	0	0.85
				100	0.95
	65	Na$_2$O	15	−100	0.85
		CaO	20	0	1.00
				100	1.10
	60	Na$_2$O	20	−100	0.75
		CaO	20	0	0.90
				100	1.00
Other crown glasses					
Crown glass	75	Na$_2$O	9	−100	0.80
		K$_2$O	11	0	1.00
		CaO	5	100	1.10
High dispersion crown glass	68	Na$_2$O	16	−100	0.65
		ZnO	3	0	0.85
		PbO	13	100	1.00

Type of glass	SiO$_2$ (wt%)	Other oxides (wt%)		t °C	k W/m K
Miscellaneous flint glasses					
(i) Silicate flint glasses					
Light flint glasses	65	PbO	25	−100	0.65–0.70
		Others	10	0	0.88–0.92
				100	1.00–1.05
	55	PbO	35	−100	0.60–0.65
		Others	10	0	0.75–0.85
				100	0.88–0.92
Ordinary flint glass	45	PbO	45	−100	0.50–0.60
		Others	10	0	0.65–0.75
				100	0.80–0.85
Heavy flint glass	35	PbO	60	−100	0.45–0.50
		Others	5	0	0.60–0.65
				100	0.70–0.75
Very heavy flint glasses	25	PbO	73	−100	0.40–0.45
		Others	2	0	0.55–0.60
				100	0.63–0.67
	20	PbO	80	−100	0.40
				0	0.50
				100	0.60
(ii) Borosilicate flint glass	33	B$_2$O$_3$	31	−100	0.65
		PbO	25	0	0.85
		Al$_2$O$_3$	7	100	0.95
		K$_2$O	3		
		Na$_2$O	1		
(iii) Barium flint glass	50	BaO	24	−100	0.60
		PbO	6	0	0.70
		K$_2$O	8	100	0.85
		Na$_2$O	3		
		ZnO	8		
		Sb$_2$O$_3$	1		
Other glasses					
Potassium glass	59	K$_2$O	33	50	0.88–0.92
		CaO	8		
Iron glasses	63	Fe$_2$O$_3$	10	−100	0.80
		Na$_2$O	17	0	0.95
		MgO	4	100	1.05
		CaO	3		
		Al$_2$O$_3$	2		
	67	Fe$_2$O$_3$	15	0	0.88—0.92
		Na$_2$O$_3$	18	100	1.00—1.05
	62	Fe$_2$O$_3$	20	0	0.85—0.90
		Na$_2$O	18	100	0.95—1.00
Rock glasses					
Obsidian				0	1.35
Artificial diabase				100	1.25

COMMERCIAL METALS AND ALLOYS

This table gives typical values of mechanical, thermal, and electrical properties of several common commercial metals and alloys. Values refer to ambient temperature (0 to 25°C). All values should be regarded as typical, since these properties are dependent on the particular type of alloy, heat treatment, and other factors. Values for individual specimens can vary widely.

REFERENCES

1. *ASM Metals Reference Book, Second Edition*, American Society for Metals, Metals Park, OH, 1983.
2. Lynch, C. T., *CRC Practical Handbook of Materials Science*, CRC Press, Boca Raton, FL, 1989.
3. Shackelford, J. F., and Alexander, W., *CRC Materials Science and Engineering Handbook*, CRC Press, Boca Raton, FL, 1991.

Common name	Thermal conductivity W/cm K	Density g/cm^3	Coeff. of linear expansion 10^{-6}/°C	Electrical resistivity μΩ cm	Modulus of elasticity GPa	Tensile strength MPa	Approx. melting point °C
Ingot iron	0.7	7.86	11.7	9.7	205	-	1540
Plain carbon steel AISI-SAE 1020	0.52	7.86	11.7	18	205	450	1515
Stainless steel type 304	0.15	7.9	17.3	72	195	550	1425
Cast gray iron	0.47	7.2	10.5	67	90	180	1175
Malleable iron		7.3	12	30	170	345	1230
Hastelloy C	0.12	8.94	11.3	125	200	780	1350
Inconel	0.15	8.25	11.5	103	200	800	1370
Aluminum alloy 3003, rolled	1.9	2.73	23.2	3.7	70	110	650
Aluminum alloy 2014, annealed	1.9	2.8	23.0	3.4	70	185	650
Aluminum alloy 360	1.5	2.64	21.0	7.5	70	325	565
Copper, electrolytic (ETP)	3.9	8.94	16.5	1.7	120	300	1080
Yellow brass (high brass)	1.2	8.47	20.3	6.4	100	300-800	930
Aluminum bronze	0.7	7.8	16.4	12	120	400-600	1050
Beryllium copper 25	0.8	8.23	17.8	7	130	500-1400	925
Cupronickel 30%	0.3	8.94	16.2		150	400-600	1200
Red brass, 85%	1.6	8.75	18.7	11	90	300-700	1000
Chemical lead	0.35	11.34	29.3	21	13	17	327
Antimonial lead (hard lead)	0.3	10.9	26.5	23	20	47	290
Solder 50-50	0.5	8.89	23.4	15	-	42	215
Magnesium alloy AZ31B	1.0	1.77	26	9	45	260	620
Monel	0.3	8.84	14.0	58	180	545	1330
Nickel (commercial)	0.9	8.89	13.3	10	200	460	1440
Cupronickel 55-45 (constantan)	0.2	8.9	18.8	49	160	-	1260
Titanium (commercial)	1.8	4.5	8.5	43	110	330-500	1670
Zinc (commercial)	1.1	7.14	32.5	6	-	130	419
Zirconium (commercial)	0.2	6.5	5.85	41	95	450	1855

HARDNESS OF MINERALS AND CERAMICS

There are several hardness scales for describing the resistance of a material to indentation or scratching. This table lists a number of common materials in order of increasing hardness. Values are given, when available, on three different hardness scales: the original Mohs Scale (range 1 to 10); the modified Mohs Scale (range 1 to 15), and the Knoop Hardness Scale. In the last case, a load of 100 g is assumed.

REFERENCE

Shackelford, J. F. and Alexander, W., *CRC Materials Science and Engineering Handbook*, CRC Press, Boca Raton, FL, 1991.

Material	Formula	Mohs	Modified mohs	Knoop
Graphite	C	0.5		
Talc	$3MgO \cdot 4SiO_2 \cdot H_2O$	1	1	
Alabaster	$CaSO_4 \cdot 2H_2O$	1.7		
Gypsum	$CaSO_4 \cdot 2H_2O$	2	2	32
Halite (rock salt)	NaCl	2		
Stibnite (antimonite)	Sb_2S_3	2.0		
Galena	PbS	2.5		
Mica		2.8		
Calcite	$CaCO_3$	3	3	135
Barite	$BaSO_4$	3.3		
Marble		3.5		
Aragonite	$CaCO_3$	3.5		
Dolomite	$CaMg(CO_3)_2$	3.5		
Fluorite	CaF_2	4	4	163
Magnesia	MgO	5		370
Apatite	$CaF_2 \cdot 3Ca_3(PO_4)_2$	5	5	430
Opal		5		
Feldspar (orthoclase)	$K_2O \cdot Al_2O \cdot 6SiO_2$	6	6	560
Augite		6		
Hematite	Fe_2O_3	6		750
Magnetite	Fe_3O_4	6		
Rutile	TiO_2	6.2		
Pyrite	FeS_2	6.3		
Agate	SiO_2	6.5		
Uranium dioxide	UO_2	6.7		600
Silica (fused)	SiO_2		7	
Quartz	SiO_2	7	8	820
Flint		7		
Silicon	Si	7		
Andalusite	Al_2OSiO_4	7.5		
Zircon	$ZrSiO_4$	7.5		
Zirconia	ZrO_2			1200
Aluminum nitride	AlN			1225
Beryl	$Be_3Al_2Si_6O_{18}$	7.8		
Beryllia	BeO			1300
Topaz	$Al_2SiO_4(OH,F)_2$	8	9	1340
Garnet	$Al_2O_3 \cdot 3FeO \cdot 3SiO_2$		10	1360
Emery	Al_2O_3 (impure)	8		
Zirconium nitride	ZrN	8+		1510
Zirconium boride	ZrB_2			1560
Titanium nitride	ZrN	9		1770
Zirconia (fused)	ZrO_2		11	
Tantalum carbide	TaC			1800
Tungsten carbide	WC			1880
Corundum (alumina)	Al_2O_3	9		2025
Zirconium carbide	ZrC			2150
Alumina (fused)	Al_2O_3		12	
Beryllium carbide	Be_2C			2400

Material	Formula	Mohs	Modified mohs	Knoop
Titanium carbide	TiC			2470
Carborundum (silicon carbide)	SiC	9.3	13	2500
Aluminum boride	AlB			2500
Tantalum boride	TaB_2			2600
Boron carbide	B_4C		14	2800
Boron	B	9.5		
Titanium boride	TiB_2			2850
Diamond	C	10	15	7000

Section 13
Polymer Properties

NAMING ORGANIC POLYMERS

Robert B. Fox

Organic polymers have been traditionally named on the basis of the monomer used, a hypothetical monomer, or a semisystematic structure. Alternatively, they may be named in the same way as organic compounds, i.e., on the basis of a structure as written. While the former methods, often called "source-based nomenclature", frequently result in ambiguity and multiple names for a single material, the latter method, termed "structure-based nomenclature", generates a sometimes cumbersome unique name for a given polymer, independent of its source. Within their limitations, both types of names are acceptable and well-documented.[1] The use of stereochemical descriptors with both types of polymer nomenclature has been published.[1]

TRADITIONAL POLYMER NAMES

Monomer-Based Names

"Polystyrene" is the name of a polymer made from the monomer styrene. Generally, the name of the monomer should be enclosed in parentheses, as in "poly(methyl methacrylate)" or "poly(4-bromostyrene)" to more clearly identify the monomer. Many names are often possible for a given polymer with this method: "poly(ethylene glycol)", "poly(ethylene oxide)", and "poly(oxirane)" describe the same polymer. Sometimes, the name of a hypothetical monomer is used, as in "poly(vinyl alcohol)". Even though a name like "polyethylene" covers a multitude of materials, the system does provide understandable names when a single monomer is involved in the synthesis of a single polymer. When one monomer can yield more than one polymer, as with 1,3-butadiene, some sort of structural notation must be used to identify the product, and one is not far from a formal structure-based name.

Copolymers, Block Polymers, and Graft Polymers. Monomer-based names are more complex when more than one monomer is involved. Some common polymers have been given names based on an apparent structure, as with "poly(ethylene terephthalate)". A better system has been approved by the IUPAC.[1] With this method, the arrangement of the monomeric units is introduced through use of an italicized connective placed between the names of the monomers. For monomer names represented by A, B, and C, the various types of arrangements are

Type	Connective	Example
unspecified	*-co-*	poly(A-*co*-B)
statistical	*-stat-*	poly(A-*stat*-B)
random	*-ran-*	poly(A-*ran*-B)
alternating	*-alt-*	poly(A-*alt*-B)
periodic	*-per-*	poly(A-*per*-B-*per*-C)
block	*-block-*	polyA-*block*-polyB
graft	*-graft-*	polyA-*graft*-polyB

A long dash may be used in place of the word "block". These connectives can be used in combination and with small joining units. Names such as "poly[ethylene-*ran*-(vinyl chloride)]" and "polystyrene-*block*-dimethylsilylene-*block*-polybutadiene" result.

STRUCTURE-BASED POLYMER NOMENCLATURE

Regular Single-Strand Polymers

Structure-based nomenclature has been approved by the IUPAC[4] and is currently being updated; it is used by *Chemical Abstracts*.[5] Monomer names are not involved. To the extent that a polymer chain can be described by a repeating segment of the chain, it can be named "poly(repeating unit)", where "repeating unit" is, in most cases, a bivalent group.

Since there are usually many possible repeating units in a given chain, it is necessary to select systematically one, called the "constitutional repeating unit" (CRU), to provide a unique and unambiguous name, "poly(CRU)", where "CRU" is a recitation of the names of successive units as one proceeds through the CRU. For this purpose, a portion of the main chain structure that includes at least two repeating sequences is written out. These sequences will be composed of bivalent subunits such as -CH$_2$-, -O-, and groups from ring systems, each of which can be named by the usual rules of nomenclature.[6,7]

Where a chain is simply one long sequence of a single subunit, that subunit is itself the CRU, as in "poly(methylene)" or "poly(1,4-phenylene)". In chains having more than one kind of subunit, a seniority system is used to determine the beginning of the CRU and the direction to move along the main chain atoms (following the shortest path in rings) to complete the CRU. For citation of the first, most senior, subunit, the descending order of seniority is (1) heterocyclic rings, (2) hetero atoms, (3) carbocyclic rings, and, lowest, (4) acyclic groups containing only carbon atoms.

Within each of these classes, there is a further order of seniority that follows the usual rules of nomenclature. Thus, among hetero atoms, the descending order of seniority is O, S, Se, N, P, As, Si, Ge, B, with other atoms fitted in according to their place in the Periodic Table. The descending order of seniority among heterocylic rings is (a) a ring system containing N; (b) a ring system containing a hetero atom (other than N) as high as possible in the order of hetero atoms given above; (c) a ring system having the greatest number of rings; (d) a ring system with the largest individual ring; (e) a ring system with the largest number of hetero atoms. Among equivalent ring systems, preference is given to the one having lowest locants for the free valences in the subunit, and among otherwise identical ring systems, the one having least hydrogenation is senior. Lowest locants in unsaturated chains are also given preference. Lowest locants for substituents, independent of kind except for alphabetical order, are the final determinant of seniority.

Direction within the repeating unit depends upon the shortest path, in main chain atoms, from the most senior subunit to another subunit of the same kind or to a subunit next lower in seniority. Once identification of the CRU is established, the name of the CRU is formed from the names of largest possible subunits within the CRU. For example, the main chain of the polymer traditionally named "poly(ethylene terephthalate)" has the structure

$$...CH_2CH_2OC(O)—\underset{}{\bigcirc}—C(O)\ [OCH_2CH_2OC(O)—\underset{}{\bigcirc}—C(O)\ ]OCH_2CH_2OC(O)—\underset{}{\bigcirc}—C(O)\ O...$$

The CRU is enclosed in brackets and read from left to right. It is selected because -O- is the most senior subunit, and two carbon atoms is the shortest path to the next -O-. The structure-based name of this polymer is then "poly(oxyethyleneoxyterephthaloyl)", not much longer than the traditional name and much more adaptable to the complexities of substitution. As organic nomenclature evolves, more systematic names can be applied to the subunits, such as "ethane-1,2-diyl" instead of "ethylene".

Structure-based nomenclature can also be used when the CRU has no carbon atoms. An example is the polymer traditionally named "poly(dimethylsiloxane)", which on the basis of structure would be named "poly(oxydimethylsilylene)" or "poly(oxydimethylsilanediyl)". This nomenclature method has also been applied to inorganic and coordination polymers[8] and to double-strand (ladder and spiro) organic polymers.[9]

Additional examples of both structure-based and traditional names will be found in the tables of glass transition temperatures (**13**-4 to 11) in this *Handbook* as well as in the *Chemical Abstracts* indexes and in the IUPAC documents.

Irregular Single-Strand Polymers

Polymers that cannot be described by the repetition of a single CRU or comprise units not all connected identically in a directional sense can also be named on a structure basis.[10] These include copolymers, block and graft polymers, star polymers, and chemically modified polymers. They are given names of the type "poly(A/B/C...)", where A, B, C, etc. are the names of the component constitutional units, the number of which are minimized. The constitutional units may include regular or irregular blocks as well as atoms or atomic groupings, and each is named by the method described above or by the rules of organic nomenclature.

The oblique stroke denotes an unspecified arrangement of the units within the main chain. For example, a statistical copolymer derived from styrene and vinyl chloride with the monomeric units joined head-to-tail is named "poly(1-chloroethylene/1-phenylethylene)". A polymer obtained by 1,4 polymerization and both head-to-head and head-to-tail 1,2 polymerization of 1,3-butadiene would be named "poly(but-2-ene-1,4-diyl/1-vinylethylene/2-vinyl-ethylene)". A long dash is used separate components of block polymers, as in "poly(A)—poly(B)—poly(C)" or "poly(A)—X—poly(B)". Graft polymers are named in the same way as a substituted polymer but without the ending "yl" for the grafted chain: "poly[Z/poly(A)Z]" is the name of a regular polymer consisting of Z units in which some have grafts of "poly(A)". Star polymers are treated as a central unit with substituent blocks, as in "tetrakis(polymethylene)silane".

REFERENCES

1. International Union of Pure and Applied Chemistry, *Compendium of Macromolecular Nomenclature*, Blackwell Scientific Publications, Oxford, 1991.
2. International Union of Pure and Applied Chemistry, Stereochemical Definitions and Notations Relating to Polymers (Recommendations 1980), *Pure & Appl. Chem.*, 53, 733—752, 1981.
3. International Union of Pure and Applied Chemistry, Source-Based Nomenclature for Copolymers (Recommendations 1985), *Pure & Appl. Chem.*, 57, 1427—1440, 1985)
4. International Union of Pure and Applied Chemistry, Nomenclature of Regular Single-Strand Organic Polymers (1975), *Pure & Appl. Chem.*, 48, 373—385, 1976.
5. Chemical Abstracts Service, Naming and Indexing of Chemical Substances for Chemical Abstracts, Appendix IV, *Chemical Abstracts 1994 Index Guide.*
6. International Union of Pure and Applied Chemistry, *A Guide to IUPAC Nomenclature of Organic Compounds (1993)*, Blackwell Scientific Publications, Oxford, 1993.
7. International Union of Pure and Applied Chemistry, *Nomenclature of Organic Chemistry, Sections A, B, C, D, E, F, and H*, Pergamon Press, Oxford, 1979.
8. International Union of Pure and Applied Chemistry, Nomenclature of Regular Double-Strand and Quasi-Single-Strand Inorganic and Coordination Polymers (Recommendations 1984), *Pure & Appl. Chem.*, 57, 149—168, 1985.
9. International Union of Pure and Applied Chemistry, Nomenclature of Regular Double-Strand (Ladder and Spiro) Organic Polymers (Recommendations 1993), *Pure & Appl. Chem.*, 65, 1561—1580, 1993.
10. International Union of Pure and Applied Chemistry, Structure-Based Nomenclature for Irregular Single-Strand Organic Polymers (Recommendations 1994), *Pure & Appl. Chem.*, 66, 873—889, 1994.

SOLVENTS FOR COMMON POLYMERS

Abbreviations: HC: hydrocarbons; MEK: methyl ethyl ketone; THF: tetrahydrofuran; DMF: dimethylformamide; DMSO: dimethylsulfoxide

Polyethylene (HDPE)	HC and halogenated HC
Polypropylene (atactic)	HC and halogenated HC
Polybutadiene	HC, THF, ketones
Polystyrene	ethylbenzene, $CHCl_3$, CCl_4, THF, MEK
Polyacrylates	aromatic HC, chlorinated HC, THF, esters, ketones
Polymethacrylates	aromatic HC, chlorinated HC, THF, esters, MEK
Polyacrylamide	water
Poly(vinyl ethers)	halogenated HC, MEK, butanol
Poly(vinyl alcohol)	glycols (hot), DMF
Poly(vinyl acetate)	aromatic HC, chlorinated HC, THF, esters, DMF
Poly(vinyl chloride)	THF, DMF, DMSO
Poly(vinylidene chloride)	THF (hot), dioxane, DMF
Poly(vinyl fluoride)	DMF, DMSO (hot)
Polyacrylonitrile	DMF, DMSO
Poly(oxyethylene)	aromatic HC, $CHCl_3$, alcohols, esters, DMF
Poly(2,6-dimethylphenylene oxide)	aromatic HC, halogenated HC
Poly(ethylene terephthalate)	phenol, DMSO (hot)
Polyurethanes (linear)	aromatic HC, THF, DMF
Polyureas	phenol, formic acid
Polysiloxanes	HC, THF, DMF
Poly[bis(2,2,2-trifluoroethoxy)-phosphazene]	THF, ketones, ethyl acetate

GLASS TRANSITION TEMPERATURE FOR SELECTED POLYMERS

Robert B. Fox

Polymer names are based on the IUPAC structure-based nomenclature system described in the table "Naming Organic Polymers". Within each category, names are listed in alphabetical order. Source-based and trivial names are also given (in italics) for the most common polymers. The table does not include polymers for which T_g is not clearly defined because of variability of structure or because of reactions taking place near the glass transition.

All values of T_g cited in this table have been determined by differential scanning calorimetry (DSC) except those values indicated by:

(D)	dynamic method
(Dil)	dilatometry
(M)	mechanical method

Polymer name	Glass transition temperature (T_g/K)
ACYCLIC CARBON CHAINS	
Polyalkadienes	
Poly(alkenylene) *Polyalkadiene* $-[CH=CHCH_2CH_2]-$	
Poly(*cis*-1-butenylene)	171
cis-1,3-polybutadiene [PBD]	
Poly(*trans*-1-butenylene)	215
trans-1,3-polybutadiene [PBD]	
Poly(1-chloro-*cis*-1-butenylene)	253
cis-1,3-polychloroprene	
Poly(1-chloro-*trans*-1-butenylene)	233
trans-1,3-polychloroprene	
Poly(1-methyl-*cis*-1-butenylene)	200
cis-1,3-polyisoprene	
Poly(1-methyl-*trans*-1-butenylene)	207
trans-1,3-polyisoprene	
Poly(1,4,4-trifluoro-1-butenylene)	238
Polyalkenes	
Poly(alkylethylene) *Poly(alkylethylene)* $-[RCHCH_2]-$	
Poly(1-benzylethylene)	333
Poly(1-butylethylene)	223
Poly(1-cyclohexylethylene) (atactic)	393
Poly(1-cyclohexylethylene) (isotactic)	406 (D)
Poly(1,1-dimethylethylene)	200
Polyisobutylene [PIB]	
Poly(ethylene)	148
Poly(methylene)	155
Poly(1-phenethylethylene)	283
Poly(propylene) (isotactic)	272
Poly(propylene) (syndiotactic)	ca. 265
Poly[1-(2-pyridyl)ethylene]	377
Poly[1-(4-pyridyl)ethylene]	415
Poly(1-vinylethylene)	273
Polyacrylics	
Poly[1-(alkoxycarbonyl)ethylene] *Poly(alkyl acrylate)* $-[(ROCO)CHCH_2]-$	
Poly[1-(benzyloxycarbonyl)ethylene]	279
Poly[1-(butoxycarbonyl)ethylene]	219 (M)
Poly(butyl acrylate) [PBA]	

Polymer name	Glass transition temperature (T_g/K)
Poly[1-(sec-butoxycarbonyl)ethylene]	251
Poly[1-(butoxycarbonyl)-1-cyanoethylene]	358
Poly[1-(butylcarbamoyl)ethylene]	319 (M)
Poly(1-carbamoylethylene)	438
Polyacrylamide [PAM]	
Poly(1-carboxyethylene)	379
Poly(acrylic acid) [PAA]	
Poly[1-(2-chlorophenoxycarbonyl)ethylene]	326
Poly[1-(4-chlorophenoxycarbonyl)ethylene]	331
Poly[1-(4-cyanobenzyloxycarbonyl)ethylene]	317
Poly[1-(2-cyanoethoxycarbonyl)ethylene]	277
Poly[1-(cyanomethoxycarbonyl)ethylene)]	433 Dil
Poly[1-(4-cyanophenoxycarbonyl)ethylene]	363
Poly[1-(cyclohexyloxycarbonyl)ethylene]	292
Poly[1-(2,4-dichlorophenoxycarbonyl)ethylene]	333
Poly[1-(dimethylcarbamoyl)ethylene]	362
Poly[1-(ethoxycarbonyl)ethylene]	249
Poly(ethyl acrylate) [PEA]	
Poly[1-(ethoxycarbonyl)-1-fluoroethylene]	316
Poly[1-(2-ethoxycarbonylphenoxycarbonyl)ethylene]	303
Poly[1-(3-ethoxycarbonylphenoxycarbonyl)ethylene]	297
Poly[1-(4-ethoxycarbonylphenoxycarbonyl)ethylene]	310
Poly[1-(2-ethoxyethoxycarbonyl)ethylene]	223
Poly[1-(3-ethoxypropoxycarbonyl)ethylene]	218
Poly[1-(isopropoxycarbonyl)ethylene]	267-270
Poly[1-(methoxycarbonyl)ethylene]	283
Poly(methyl acrylate) [PMA]	
Poly[1-(2-methoxycarbonylphenoxycarbonyl)ethylene]	319
Poly[1-(3-methoxycarbonylphenoxycarbonyl)ethylene]	311
Poly[1-(4-methoxycarbonylphenoxycarbonyl)ethylene]	340
Poly[1-(2-methoxyethoxycarbonyl)ethylene]	223
Poly[1-(4-methoxyphenoxycarbonyl)ethylene]	324
Poly[1-(3-methoxypropoxycarbonyl)ethylene]	198
Poly[1-(2-naphthyloxycarbonyl)ethylene]	358
Poly[1-(pentachlorophenoxycarbonyl)ethylene]	420
Poly[1-(phenethoxycarbonyl)ethylene]	270
Poly[1-(phenoxycarbonyl)ethylene]	330
Poly[1-(m-tolyloxycarbonyl)ethylene]	298
Poly[1-(o-tolyloxycarbonyl)ethylene]	325
Poly[1-(p-tolyloxycarbonyl)ethylene]	316
Poly[1-(2,2,2-trifluoroethoxycarbonyl)ethylene]	263

Polymethacrylics

Poly[1-(alkoxycarbonyl)-1-methylethylene] *Poly(alkyl methacrylate)* $-[(ROCO)(Me)CCH_2]-$

Poly[1-(benzyloxycarbonyl)-1-methylethylene]	327
Poly[1-(2-bromoethoxycarbonyl)-1-methylethylene]	325
Poly[(1-(butoxycarbonyl)-1-methylethylene]	293
Poly(butyl methacrylate) [PBMA]	
Poly[1-(sec-butoxycarbonyl)-1-methylethylene]	333
Poly[1-(tert-butoxycarbonyl)-1-methylethylene)]	391
Poly[1-(2-chloroethoxycarbonyl)-1-methylethylene]	ca 315
Poly[1-(2-cyanoethoxycarbonyl)-1-methylethylene]	364
Poly[1-(4-cyanophenoxycarbonyl)-1-methylethylene]	428
Poly[1-(cyclohexyloxycarbonyl)-1-methylethylene] (atactic)	356
Poly[1-(cyclohexyloxycarbonyl)-1-methylethylene)] (isotactic)	324

Polymer name	Glass transition temperature (T_g/K)
Poly[1-(dimethylaminoethoxycarbonyl)-1-methylethylene]	292
Poly[1-(ethoxycarbonyl)-1-ethylethylene]	300
Poly[1-(ethoxycarbonyl)-1-methylethylene] (atactic) *Poly(ethyl methacrylate)* [PEMA]	338
Poly[1-(ethoxycarbonyl)-1-methylethylene] (isotactic)	285
Poly[1-(ethoxycarbonyl)-1-methylethylene]] (syndiotactic)	339
Poly[1-(hexyloxycarbonyl)-1-methylethylene]	268
Poly[1-(isobutoxycarbonyl)-1-methylethylene]	326
Poly[1-(isopropoxycarbonyl)-1-methylethylene]	354
Poly[1-(methoxycarbonyl)-1-methylethylene] (atactic) *Poly(methyl methacrylate)* [PMMA]	378
Poly[1-(methoxycarbonyl)-1-methylethylene]] (isotactic)	311
Poly[1-(methoxycarbonyl)-1-methylethylene]] (syndiotactic)	378
Poly[1-(4-methoxycarbonylphenoxy)-1-methylethylene]	379
Poly[1-(methoxycarbonyl)-1-phenylethylene]] (atactic)	391
Poly[1-(methoxycarbonyl)-1-phenylethylene]] (isotactic)	397
Poly[1-methyl-1-(phenethoxycarbonyl)ethylene]	299
Poly[1-methyl-1-(phenoxycarbonyl)ethylene]	383

Polyvinyl ethers, alcohols, and ketones
    Poly(1-alkoxyethylene) *Poly(alkyl vinyl ether)* –[ROCHCH$_2$]–
    Poly(1-hydroxyethylene) *Poly(vinyl alcohol)* –[HOCHCH$_2$]–
    Poly(1-alkanoylethylene) *Poly(alkyl vinyl ketone)* –[RCOCHCH$_2$]–

Polymer name	Glass transition temperature (T_g/K)
Poly(1-butoxyethylene)	218
Poly(1-*sec*-butoxyethylene)	253
Poly(1-*tert*-butoxyethylene)	361
Poly[1-(butylthio)ethylene]	253
Poly(1-ethoxyethylene)	230
Poly[1-(4-ethylbenzoyl)ethylene]	325
Poly(1-hydroxyethylene) *Poly(vinyl alcohol)* [PVA]	358 (D)
Poly(hydroxymethylene)	407
Poly(1-isopropoxyethylene)	270
Poly[1-(4-methoxybenzoyl)ethylene]	319 (M)
Poly(1-methoxyethylene) *Poly(methyl vinyl ether)* [PMVE]	242
Poly[1-(methylthio)ethylene]	272
Poly(1-propoxyethylene)	224
Poly[1-(trifluoromethoxy)trifluoroethylene]	268

Polyvinyl halides and nitriles
    Poly(1-haloethylene) *Poly(vinyl halide)* –[XCHCH$_2$]–
    Poly(1-cyanoethylene) *Poly(acrylonitrile)* –[NCCHCH$_2$]–

Polymer name	Glass transition temperature (T_g/K)
Poly(1-chloroethylene) *Poly(vinyl chloride)* [PVC]	354
Poly(chlorotrifluoroethylene)	373
Poly(1-cyanoethylene) *Polyacrylonitrile* [PAN]	370
Poly(1-cyano-1-methylethylene) *Polymethacrylonitrile*	393
Poly(1,1-dichloroethylene) *Poly(vinylidene chloride)*	255
Poly(1,1-difluoroethylene) *Poly(vinylidene fluoride)*	ca 233
Poly(1-fluoroethylene) *Poly(vinyl fluoride)*	314 (M)

Polymer name	Glass transition temperature (T_g/K)
Poly(1-hexafluoropropylene)	425
Poly[1-(2-iodoethyl)ethylene]	343
Poly(tetrafluoroethylene)	(160)
Poly[1-(trifluoromethyl)ethylene]	300

Polyvinyl esters
      Poly[1-(alkanoyloxy)ethylene] *Poly(vinyl alkanoate)* –[RCOOCHCH$_2$]–

Polymer name	Glass transition temperature (T_g/K)
Poly(1-acetoxyethylene)	305
Poly(vinyl acetate) [PVAc	
Poly[1-(benzoyloxy)ethylene]	344
Poly[1-(4-bromobenzoyloxy)ethylene]	365
Poly[1-(2-chlorobenzoyloxy)ethylene]	335
Poly[1-(3-chlorobenzoyloxy)ethylene]	338
Poly[1-(4-chlorobenzoyloxy)ethylene]	357
Poly[1-(cyclohexanoyloxy)ethylene]	349 (M)
Poly[1-(4-ethoxybenzoyloxy)ethylene]	343
Poly[1-(4-ethylbenzoyloxy)ethylene]	326
Poly[1-(4-isopropylbenzoyloxy)ethylene]	342
Poly[1-(2-methoxybenzoyloxy)ethylene]	338
Poly[1-(3-methoxybenzoyloxy)ethylene]	ca 317
Poly[1-(4-methoxybenzoyloxy)ethylene]	360
Poly[1-(4-methylbenzoyloxy)ethylene]	343
Poly[1-(4-nitrobenzoyloxy)ethylene]	395
Poly[1-(propionoyloxy)ethylene]	283 (M)

Polystyrenes
      Poly(1-phenylethylene) *Polystyrene* –[C$_6$H$_5$CHCH$_2$]–

Polymer name	Glass transition temperature (T_g/K)
Poly[1-(4-acetylphenyl)ethylene]	389 (M)
Poly[1-(4-benzoylphenyl)ethylene]	371 (M)
Poly[1-(4-bromophenyl)ethylene]	391
Poly[1-(4-butoxyphenyl)ethylene]	ca 320 (M)
Poly[1-(4-butoxycarbonylphenyl)ethylene]	349 (M)
Pol[(1-(4-butylphenyl)ethylene]	279
Poly[1-(4-carboxyphenyl)ethylene]	386 (M)
Poly[1-(2-chlorophenyl)ethylene]	392
Poly[1-(3-chlorophenyl)ethylene]	363
Poly[1-(4-chlorophenyl)ethylene]	383
Poly[1-(2,4-dichlorophenyl)ethylene]	406
Poly[1-(2,5-dichlorophenyl)ethylene]	379
Poly[1-(2,6-dichlorophenyl)ethylene]	440
Poly[1-(3,4-dichlorophenyl)ethylene]	401
Poly[1-(2,4-dimethylphenyl)ethylene]	385
Poly[1-(4-(dimethylamino)phenyl)ethylene]	398 (M)
Poly[1-(4-ethoxyphenyl)ethylene]	ca 359 (M)
Poly[1-(4-ethoxycarbonylphenyl)ethylene]	367 (M)
Poly[1-(4-fluorophenyl)ethylene]	368
Poly[1-(4-iodophenyl)ethylene]	429
Poly[1-(4-methoxyphenyl)ethylene]	386
Poly[1-(4-methoxycarbonylphenyl)ethylene]	386 (M)
Poly(1-methyl-1-phenylethylene)	373
Poly(α-methylstyrene)	
Poly[1-(2-(methylamino)phenyl)ethylene]	462 (M)
Poly(1-phenylethylene)	373
Polystyrene [PS]	

Polymer name	Glass transition temperature (T_g/K)
Poly[1-(4-propoxyphenyl)ethylene]	343 (M)
Poly[1-(4-propoxycarbonylphenyl)ethylene]	365 (M)
Poly(1-o-tolylethylene)	409

CHAINS WITH CARBOCYCLIC UNITS

Poly(arylenealkylene) –[–Ar–(CH₂)ₙ]–

Poly[1-(2-bromo-1,4-phenylene)ethylene]	353 (M)
Poly[1-(2-chloro-1,4-phenylene)ethylene]	343 (M)
Poly[1-(2-cyano-1,4-phenylene)ethylene]	363 (M)
Poly[1-(2,5-dimethyl-1,4-phenylene)ethylene]	373 (M)
Poly[1-(2-ethyl-1,4-phenylene)ethylene]	298 (M)
Poly[1-(1,4-naphthylene)ethylene]	433 (M)
Poly[1-(1,4-phenylene)ethylene]	ca 353 (M)

CHAINS WITH HETEROATOM UNITS

Main chain oxide units

Poly(oxyalkylene) *Poly(alkylene oxide)* –[O(CH₂)ₙ]–

Poly[oxy(1,1-bis(chloromethyl)trimethylene)]	265
Poly[oxy(1-(bromomethyl)ethylene)]	259
Poly[oxy(1-(butoxymethyl)ethylene)]	194
Poly[oxy(1-butylethylene)]	203
Poly[oxy(1-tert-butylethylene)]	308
Poly[oxy(1-(chloromethyl)ethylene)]	251
Poly(epichlorohydrin)	
Poly[oxy(2,6-dimethoxy-1,4-phenylene)]	440
Poly[oxy(1,1-dimethylethylene)]	264
Poly[oxy(2,6-dimethyl-1,4-phenylene)]	482
Poly[oxy(2,6-diphenyl-1,4-phenylene)]	493
Poly[oxy(1-ethylethylene)]	203
Poly(oxyethylidene)	243
Polyacetaldehyde	
Poly[oxy(1-(methoxymethyl)ethylene)]	211
Poly[oxy(2-methyl-6-phenyl-1,4-phenylene)]	428
Poly[oxy(1-methyltrimethylene)]	223 (D)
Poly[oxy(2-methyltrimethylene)]	218
Poly(oxy-1,4-phenylene)	358
Poly(phenylene oxide) [PPO]	
Poly[oxy(1-phenylethylene)]	313
Poly(oxytetramethylene)	189
Poly(tetrahydrofuran) [PTMO]	
Poly(oxytrimethylene)	195

Main-chain ester or anhydride units

Poly(oxyalkyleneoxyalkanedioyl) *Poly(alkylene alkanedioate)*–[O(CH₂)ₘOCO(CH₂)ₙCO]–

Poly(oxyadipoyloxydecamethylene)	217
Poly(oxyadipoyloxy-1,4-phenyleneisopropylidene-1,4-phenylene)	341
Poly(oxycarbonyloxy-1,4-phenylene-isopropylidene-1,4-phenylene)	422
Bisphenol A polycarbonate	
Poly(oxycarbonylpentamethylene)	213
Poly(oxycarbonyl-1,4-phenylenemethylene-1,4-phenylene)	395
Poly(oxycarbonyl-1,4-phenyleneisopropylidene-1,4-phenylene)	333

Polymer name	Glass transition temperature (T_g/K)
Poly[oxy(2,6-dimethyl-1,4-phenyleneisopropylidene-3,5-dimethyl-1,4-phenylene)oxysebacoyl]	318
Poly(oxyethylenecarbonyl-1,4-cyclohexylenecarbonyl) (trans)	291
Poly(oxyethyleneoxycarbonyl-1,4-naphthylenecarbonyl)	337
Poly(oxyethyleneoxycarbonyl-1,5-naphthylenecarbonyl)	344
Poly(oxyethyleneoxycarbonyl-2,6-naphthylenecarbonyl)	386
Poly(oxyethyleneoxycarbonyl-2,7-naphthylenecarbonyl)	392
Poly(oxyethyleneoxyterephthaloyl)	342
Poly(ethylene terephthalate) [PET]	
Poly(oxyisophthaloyl)	403 (D)
Poly(oxy(1-oxo-2,2-dimethyltrimethylene))	263
Poly(pivalolactone)	
Poly(oxy-1,4-phenyleneisopropylidene-1,4-phenyleneoxysebacoyl)	280
Poly(oxy-1,4-phenyleneoxy-1,4-phenyleneoxy-carbonyl-1,4-phenylene) [PEEK]	416
Poly(oxypropyleneoxyterephthaloyl)	341
Poly[oxyterephthaloyloxy(2,6-dimethyl-1,4-phenyleneisopropylidene-3,5-dimethyl-1,4-(D)phenylene)]	498
Poly(oxyterephthaloyloxyoctamethylene)	318 (D)
Poly(oxyterephthaloyloxy-1,4-phenyleneisopropylidene-1,4-phenylene)	478
Poly(bisphenol A terephthalate)	
Poly(oxytetramethyleneoxyterephthaloyl)	323
Poly(butylene terephthalate) [PBT]	

Main-chain amide units
    Poly(iminoalkyleneiminoalkanedioyl) *Poly(alkylene alkanediamide)*–[NH(CH$_2$)$_m$NHCO(CH$_2$)$_n$CO]–

Polymer name	Glass transition temperature (T_g/K)
Poly(iminoadipoyliminodecamethylene)	313
Nylon 10,6	
Poly(iminoadipoyliminohexamethylene)	ca 323
Nylon 6,6	
Poly(iminoadipoyliminooctamethylene)	318
Nylon 8,6	
Poly[iminoadipoyliminotrimethylene(methylimino)trimethylene]	278
Poly(iminocarbonyl-1,4-cyclohexylenemethylene)	466
Poly[iminocarbonyl-1,4-phenylene(2-oxoethylene)iminohexamethylene]	377
Poly(iminoethylene-1,4-phenyleneethyleneiminosebacoyl)	378 (D)
Poly(iminohexamethyleneiminoazelaoyl)	331
Nylon 6,9	
Poly(iminohexamethyleneiminododecanedioyl)	319
Nylon 6, 12	
Poly(iminohexamethyleneiminopimeloyl)	331
Nylon 6,7	
Poly(iminohexamethyleneiminosebacoyl)	323
Nylon 6,10	
Poly(iminohexamethyleneiminosuberoyl)	330
Nylon 6,8	
Poly(iminoisophthaloylimino-4,4'-biphenylylene)	558
Poly(iminoisophthaloyliminohexamethylene)	390
Poly(iminoisophthaloyliminomethylene-1,4-cyclohexylenemethylene)	481
Poly(iminoisophthaloyliminomethylene-1,3-phenylenemethylene)	438 (M)
Poly[iminomethylene(2,5-dimethyl-1,4-phenylene)methyleneiminosuberoyl]	351
Poly(imino-1,5-naphthyleneiminoisophthaloyl)	598
Poly(imino-1,5-naphthyleneiminoterephthaloyl)	578
Poly(iminooctamethyleneiminodecanedioyl)	333
Nylon 8,10	
Poly(iminooxalyliminohexamethylene)	430
Nylon 6,2	
Poly[imino(1-oxohexamethylene)]	326
Nylon 6	

Polymer name	Glass transition temperature (T_g/K)
Poly[imino(1-oxodecamethylene)]	315
Nylon 10	
Poly[imino(1-oxoheptamethylene)]	325
Nylon 7	
Poly[imino(1-oxo-3-methyltrimethylene]	369
Poly[imino(1-oxononamethylene)]	319
Nylon 9	
Poly[imino(1-oxooctamethylene)]	323
Nylon 8	
Poly[imino(1-oxotrimethylene)]	384
Nylon 3	
Poly(iminopentamethyleneiminoadipoyl)	318
Nylon 5,6	
Poly[iminopentamethyleneiminocarbonyl-1,4-phenylene(2-oxoethylene)]	376
Poly(imino-1,3-phenyleneiminoisophthaloyl)	553 (M)
Poly(imino-1,4-phenyleneiminoterephthaloyl)	618
Poly(iminopimeloyliminoheptamethylene)	328
Nylon 7,7	
Poly(iminoterephthaloylimino-4,4′-biphenylylene)	613
Poly(iminotetramethyleneiminoadipoyl)	316
Nylon 4,6	
Poly[iminotetramethyleneiminocarbonyl-1,4-phenylene(2-oxoethylene)]	357
Poly(iminotrimethyleneiminoadipoyliminotrimethylene)	307
Poly[iminotrimethyleneiminocarbonyl-1,4-phenylene(2-oxoethylene)]	382
Poly(oxy-1,4-phenyleneiminoterephthaloyl-imino-1,4-phenylene)	613
Poly(sulfonylimino-1,4-phenyleneiminoadipoylimino-1,4-phenylene)	467

Main-chain urethane units
    Poly(oxyalkyleneoxycarbonyliminoalkyleneiminocarbonyl)–[O(CH$_2$)$_m$OCONH(CH$_2$)$_n$NHCO]–

Polymer name	Glass transition temperature (T_g/K)
Poly(oxyethyleneoxycarbonyliminohexamethyleneiminocarbonyl)	329
Poly[oxyethyleneoxycarbonylimino(6-methyl-1,3-phenylene)iminocarbonyl]	325
Poly(oxyethyleneoxycarbonylimino-1,4-phenylenemethylene-1,4-phenyleneiminocarbonyl)	412
Poly(oxyhexamethyleneoxycarbonyliminohexamethyleneiminocarbonyl)	332
Poly[oxyhexamethyleneoxycarbonylimino(6-methyl-1,3-phenylene)iminocarbonyl]	305
Poly(oxyhexamethyleneoxycarbonylimino-1,4-phenylenemethylene-1,4-phenyleneiminocarbonyl)	364
Poly(oxyoctamethyleneoxycarbonyliminohexamethyleneiminocarbonyl)	331
Poly[oxyoctamethyleneoxycarbonylimino(6-methyl-1,3-phenylene)iminocarbonyl]	337
Poly(oxyoctamethyleneoxycarbonylimino-1,4-phenylenemethylene-1,4-phenyleneiminocarbonyl)	352
Poly(oxytetramethyleneoxycarbonyliminohexamethyleneiminocarbonyl)	332
Poly[oxytetramethyleneoxycarbonylimino(6-methyl-1,3-phenylene)iminocarbonyl]	315
Poly(oxytetramethyleneoxycarbonylimino-1,4-phenylenemethylene-1,4-phenyleneiminocarbonyl)	382

Main-chain siloxanes
    Poly[oxy(dialkylsilylene)] *Poly(dialkylsiloxane)* –[O(R$_2$Si)]–

Polymer name	Glass transition temperature (T_g/K)
Poly[oxy(dimethylsilylene)]	148
Poly(dimethylsiloxane) [PDMS]	
Poly[oxy(dimethylsilylene)oxy-1,4-phenylene]	363 (M)
Poly[oxy(dimethylsilylene)oxy-1,4-phenyleneisopropylidene-1,4-phenylene]	318 (M)
Poly[oxy(diphenylsilylene)]	238
Poly(diphenylsiloxane)	
Poly[oxy(diphenylsilylene)-1,3-phenylene]	ca 331
Poly[oxy((methyl)phenylsilylene)]	187
Poly[oxy((methyl)-3,3,3-trifluoropropylsilylene]	<193

Polymer name	Glass transition temperature (T_g/K)
Main-chain sulfur-containing units	
Poly(dithioethylene)	223
Poly(dithiomethylene-1,4-phenylenemethylene)	296
Poly(oxy-4,4′-biphenylylene-1,4-phenylenesulfonyl-1,4-phenylene)	503 (M)
Poly(oxycarbonyloxy-1,4-phenylenethio-1,4-phenylene)	ca 383
Poly(oxyethylenedithioethylene)	220 (M)
Poly[oxy(2-hydroxytrimethylene)oxy-1,4-phenylenesulfonyl-1,4-phenylene]	428
Poly(oxymethyleneoxyethylenedithioethylene)	214
Poly(oxy-1,4-phenylenesulfinyl-1,4-phenyleneoxy-1,4-phenylenecarbonyl-1,4-phenylene)	478 (M)
Poly(oxy-1,4-phenylenesulfinyl-1,4-phenyleneoxy-1,4-phenyleneisopropylidene-1,4-phenylene)	438 (M)
Poly(oxy-1,4-phenylenesulfonyl-1,4-phenylene)	487
Poly(oxy-1,4-phenylenesulfonyl-4,4′-biphenylylenesulfonyl-1,4-phenylene)	533
Poly[oxy-1,4-phenylenesulfonyl-1,4-phenyleneoxy(2,6-dimethyl-1,4-phenylene)isopropylidene (3,5-dimethyl-1,4-phenylene)]	508 (M)
Poly(oxy-1,4-phenylenesulfonyl-1,4-phenyleneoxy-1,4-phenylenecarbonyl-1,4-phenylene)	478 (M)
Poly[oxy-1,4-phenylenesulfonyl-1,4-phenyleneoxy-1,4-phenylene(hexafluoroisopropylidene)1,4-phenylene]	478 (M)
Poly(oxy-1,4-phenylenesulfonyl-1,4-phenyleneoxy-1,4-phenyleneisopropylidene-1,4-phenylene)	449
Poly(oxy-1,4-phenylenesulfonyl-1,4-phenyleneoxy-1.4-phenylenemethylene-1,4-phenylene)	453 (M)
Poly(oxy-1,4-phenylenesulfonyl-1,4-phenyleneoxy-1.4-phenylenethio-1,4-phenylene)	448 (M)
Poly(oxy-1,4-phenylenesulfonyl-1,4-phenyleneoxyterephthaloyl)	522
Poly(oxytetramethylenedithiotetramethylene)	197
Poly(sulfonyl-1,2-cyclohexylene)	401
Poly(sulfonyl-1,3-cyclohexylene)	381
Poly(sulfonyl-1,4-phenylenemethylene-1,4-phenylene)	497
Poly(thio-1,3-cyclohexylene)	221
Poly[thio(difluoromethylene)]	155
Poly(thioethylene)	223
Poly[thio(1-ethylethylene]	218
Poly[thio(1-methyl-3-oxotrimethylene)]	285
Poly[thio(1-methyltrimethylene)]	214
Pol[(thio(1-oxohexamethylene)]	292
Poly(thio-1,4-phenylene)	370
Poly(thiopropylene)	226
Main-chain heterocyclic units	
Poly(1,3-dioxa-4,6-cyclohexylenemethylene)	378
Poly(vinyl formal)	
Poly[(2,6-dioxopiperidine-1,4-diyl)trimethylene]	363
Poly[(2-methyl-1,3-dioxa-4,6-cyclohexylene)methylene]	355
Poly(vinyl acetal)	
Poly(1,4-piperazinediylcarbonyloxyethyleneoxycarbonyl)	333
Poly(1,4-piperazinediylisophthaloyl)	465 (M)
Poly[(2-propyl-1,3-dioxa-4,6-cyclohexylene)methylene]	322
Poly(vinyl butyral)	
Poly(3,6-pyridazinediyloxy-1,4-phenyleneisopropylidene-1,4-phenyleneoxy)	453 (M)
Poly(2,5-pyridinediylcarbonyliminohexamethyleneiminocarbonyl)	322

DIELECTRIC CONSTANT OF SELECTED POLYMERS

This table lists typical values of the dielectric constant (more properly called relative permittivity) of some important polymers. Values are given for frequencies of 1 kHz, 1 MHz, and 1 GHz; in most cases the dielectric constant at frequencies below 1 kHz does not differ significantly from the value at 1 kHz. Since the dielectric constant of a polymeric material can vary with density, degree of crystallinity, and other details of a particular sample, the values given here should be regarded only as typical or average values.

REFERENCES

1. Gray, D.E., Ed., *American Institute of Physics Handbook, Third Edition*, p. 5-132, McGraw Hill, New York, 1972.
2. Anderson, H.L., Editor, *A Physicist's Desk Reference*, American Institute of Physics, New York, 1989.
3. Brandrup, J., and Immergut, E.H., *Polymer Handbook, Third Edition*, John Wiley & Sons, New York, 1989.

Name	$t/°C$	1 kHz	1 MHz	1 GHz
Polyacrylonitrile	25	5.5	4.2	
Polyamides (nylons)	25	3.50	3.14	2.8
	84	11	4.4	2.8
Polybutadiene	25	2.5		
Polycarbonate	23	2.92	2.8	
Polychloroprene (neoprene)	25	6.6	6.3	4.2
Polychlorotrifluoroethylene	23	2.65	2.46	2.39
Polyethylene	23	2.3		
Poly(ethylene terephthalate) (Mylar)	23	3.25	3.0	2.8
Polyisoprene (natural rubber)	27	2.6	2.5	2.4
Poly(methyl methacrylate)	27	3.12	2.76	2.6
	80	3.80	2.7	2.6
Polyoxymethylene (polyformaldehyde)	25	3.8		
Poly(phenylene oxide)	23	2.59	2.59	
Polypropylene	25	2.3	2.3	2.3
Polystyrene	25	2.6	2.6	2.6
Polysulfones	25	3.13	2.10	
Polytetrafluoroethylene (teflon)	25	2.1	2.1	2.1
Poly(vinyl acetate)	50		3.5	
	150		8.3	
Poly(vinyl chloride)	25	3.39	2.9	2.8
	100	7.8	3.3	2.7
Poly(vinylidene chloride)	23	4.6	3.2	2.7
Poly(vinylidene fluoride)	23	12.2	8.9	4.7

Section 14
Geophysics, Astronomy, and Acoustics

ASTRONOMICAL CONSTANTS
Victor Abalakin

The constants in this table are based primarily on the set of constants adopted by the International Astronomical Union (IAU) in 1976. Updates have been made when new data were available. All values are given in SI Units; thus masses are expressed in kilograms and distances in meters. The astronomical unit of time is a time interval of one day (1 d) equal to 86400 s. An interval of 36525 d is one Julian century (1 cy).

REFERENCES

1. Seidelmann, P. K., *Explanatory Supplement to the Astronomical Almanac*, University Science Books, Mill Valley, CA, 1990.
2. Lang, K. R., *Astrophysical Data: Planets and Stars*, Springer-Verlag, New York, 1992.

Defining constants

Gaussian gravitational constant	$k = 0.0120209895$ m^3 kg^{-1} s^{-2}
Speed of light	$c = 299792458$ m s^{-1}

Primary constants

Light-time for unit distance (1 AU)	$\tau_A = 499.004782$ s
Equatorial radius of earth	$a_e = 6378140$ m
Equatorial radius of earth (IUGG value)	$a_e = 6378136$ m
Dynamical form-factor for earth	$J_2 = 0.001082626$
Geocentric gravitational constant	$GE = 3.986005 \times 10^{14}$ m^{3}s^{-2}
Constant of gravitation	$G = 6.672 \times 10^{-11}$ m^{3}kg^{-1}s^{-2}
Ratio of mass of moon to that of earth	$\mu = 0.01230002$
	$1/\mu = 81.300587$
General precession in longitude, per Julian century, at standard epoch J2000	$\rho = 5029''.0966$
Obliquity of the ecliptic at standard epoch J2000	$\varepsilon = 23°26'21''.448$

Derived constants

Constant of nutation at standard epoch J2000	$N = 9''.2025$
Unit distance (AU $= c\tau_A$)	AU $= 1.49597870 \times 10^{11}$ m
Solar parallax ($\pi_0 = \arcsin(a_e/\text{AU})$)	$\pi_0 = 8''.794148$
Constant of aberration for standard epoch J2000	$\kappa = 20''.49552$
Flattening factor for the earth	$f = 1/298.257 = 0.00335281$
Heliocentric gravitational constant ($GS = A^3k^2/D^2$)	$GS = 1.32712438 \times 10^{20}$ m^3 s^{-2}
Ratio of mass of sun to that of the earth (S/E) $= (GS)/(GE)$)	$S/E = 332946.0$
Ratio of mass of sun to that of earth + moon	$(S/E)/(1 + \mu) = 328900.5$
Mass of the sun ($S = (GS)/G$)	$S = 1.9891 \times 10^{30}$ kg

Ratios of mass of sun to masses of the planets

Mercury	6023600
Venus	408523.5
Earth + moon	328900.5
Mars	3098710
Jupiter	1047.355
Saturn	3498.5
Uranus	22869
Neptune	19314
Pluto	3000000

PROPERTIES OF THE SOLAR SYSTEM

The following tables give various properties of the planets and characteristics of their orbits in the solar system. Certain properties of the sun and of the earth's moon are also included.

Explanations of the column headings:

- *Den.*: mean density in g/cm^3
- *Radius*: radius at the equator in km
- *Flattening*: degree of oblateness, defined as $(r_e - r_p)/r_e$, where r_e and r_p are the equatorial and polar radii, respectively
- *Potential coefficients*: coefficients in the spherical harmonic representation of the gravitational potential U by the equation

$$U(r,\phi) = (GM/r) \left[1 - \sum J_n(a/r)^n P_n(\sin\phi) \right]$$

where G is the gravitational constant, r the distance from the center of the planet, a the radius of the planet, M the mass, ϕ the latitude, and P_n the Legendre polynomial of degree n.

- *Gravity*: acceleration due to gravity at the surface
- *Escape velocity*: velocity needed at the surface of the planet to escape the gravitational pull
- *Dist. to sun*: semi-major axis of the elliptical orbit (1 AU = 1.496×10^8 km)
- *ε*: eccentricity of the orbit
- *Ecliptic angle*: angle between the planetary orbit and the plane of the earth's orbit around the sun
- *Inclin.*: angle between the equatorial plane and the plane of the planetary orbit
- *Rot. period*: period of rotation of the planet measured in earth days
- *Albedo*: ratio of the light reflected from the planet to the light incident on it
- *T_{sur}*: mean temperature at the surface
- *P_{sur}*: pressure of the atmosphere at the surface

The following general information on the solar system is of interest:

Mass of the earth = M_e = 5.9742×10^{24} kg
Total mass of planetary system = 2.669×10^{27} kg = 447 M_e
Total angular momentum of planetary system = 3.148×10^{43} kg m^2/s
Total kinetic energy of the planets = 1.99×10^{35} J
Total rotational energy of planets = 0.7×10^{35} J

Properties of the sun:

Mass = 1.9891×10^{30} kg = 332946.0 M_e
Radius = 6.9599×10^8 m
Surface area = 6.087×10^{18} m^2
Volume = 1.412×10^{27} m^3
Mean density = 1.409 g/cm^3
Gravity at surface = 27398 cm/s^2
Escape velocity at surface = 6.177×10^5 m/s
Effective temperature = 5780 K
Total radiant power emitted (luminosity) = 3.86×10^{26} W
Surface flux of radiant energy = 6.340×10^7 W/m^2
Flux of radiant energy at the earth (Solar Constant) = 1373 W/m^2

REFERENCES

1. Seidelmann, P. K., Editor, *Explanatory Supplement to the Astronomical Almanac*, University Science Books, Mill Valley, CA, 1992.
2. Lang, K. R., *Astrophysical Data: Planets and Stars*, Springer-Verlag, New York, 1992.
3. Allen, C. W., *Astrophysical Quantities, Third Edition*, Athlone Press, London, 1977.

Planet	Mass 10^{24} kg	Den. g/cm^3	Radius km	Flattening	Potential coeffients $10^3 J_2$	$10^6 J_3$	$10^6 J_4$	Gravity cm/s^2	Escape vel. km/s
Mercury	0.33022	5.43	2439.7	0				370	4.25
Venus	4.8690	5.24	6051.9	0	0.027			887	10.4
Earth	5.9742	5.515	6378.140	0.00335364	1.08263	-2.54	-1.61	980	11.2
(Moon)	0.073483	3.34	1738	0	0.2027			162	2.37
Mars	0.64191	3.94	3397	0.00647630	1.964	36		371	5.02
Jupiter	1898.8	1.33	71492	0.0648744	14.75	-580		2312	59.6
Saturn	568.50	0.70	60268	0.0979624	16.45	-1000		896	35.5
Uranus	86.625	1.30	25559	0.0229273	12			777	21.3
Neptune	102.78	1.76	24764	0.0171	4			1100	23.3
Pluto	0.015	1.1	1151	0				72	1.1

Planet	Dist. to sun AU	ε	Ecliptic angle	Inclin.	Rot. period d	Albedo	No. of satellites
Mercury	0.38710	0.2056	7.00°	0°	58.6462	0.106	0
Venus	0.72333	0.0068	3.39°	177.3°	-243.01	0.65	0
Earth	1.00000	0.0167		23.45°	0.99726968	0.367	1
(Moon)				6.68°	27.321661	0.12	
Mars	1.52369	0.0933	1.85°	25.19°	1.02595675	0.150	2
Jupiter	5.20283	0.048	1.31°	3.12°	0.41354	0.52	16
Saturn	9.53876	0.056	2.49°	26.73°	0.4375	0.47	18
Uranus	19.19139	0.046	0.77°	97.86°	-0.65	0.51	15
Neptune	30.06107	0.010	1.77°	29.56°	0.768	0.41	8
Pluto	39.52940	0.248	17.15°	118°	-6.3867	0.3	1

Planet	T_{sur} K	P_{sur} bar	Atmospheric composition CO_2	N_2	O_2	H_2O	H_2	He	Ar	Ne	CO
Mercury	440	2×10^{-15}					2%	98%			
Venus	730	90	96.4%	3.4%	69 ppm	0.1%			4 ppm		20 ppm
Earth	288	1	0.03%	78.08%	20.95%	0 to 3%			0.93%	18 ppm	1 ppm
Mars	218	0.007	95.32%	2.7%	0.13%	0.03%			1.6%	3 ppm	0.07%
Jupiter	129						86.1%	13.8%			
Saturn	97						92.4%	7.4%			
Uranus	58						89%	11%			
Neptune	56						89%	11%			
Pluto	50	1×10^{-5}									

SATELLITES OF THE PLANETS

This table gives characteristics of the known satellites of the planets. The parameters covered are:

- Orbital period in units of earth days. An R following the value indicates a retrograde motion.
- Distance from the planet, as measured by the semi-major axis of the orbit.
- Eccentricity of the orbit.
- Inclination of the satellite orbit with respect to the equator of the planet.
- Mass of the satellite relative to the planet.
- Radius of the satellite in km.
- Mean density of the satellite.
- Geometric albedo, which is a measure of the fraction of incident sunlight reflected by the satellite.

REFERENCES

1. Seidelmann, P. K., Editor, *Explanatory Supplement to the Astronomical Almanac*, University Science Books, Mill Valley, CA, 1992.
2. Lang, K. R., *Astrophysical Data: Planets and Stars*, Springer-Verlag, New York, 1992.
3. Burns, J. A., and Matthews, M. S., Eds., *Satellites*, University of Arizona Press, Tucson, 1986.

Planet		Satellite	Orb. Period d	Distance 10³ km	Eccentricity	Inclination	Rel. mass	Radius km	Den. g/cm³	Albedo
Earth		Moon	27.321661	384.400	0.054900489	18.28–28.58°	0.0123002	1738	3.34	0.12
Mars	I	Phobos	0.31891023	9.378	0.015	1.0°	1.5×10^{-8}	$13.5 \times 10.8 \times 9.4$	<2	0.06
	II	Deimos	1.2624407	23.459	0.0005	0.9–2.7°	3×10^{-9}	$7.5 \times 6.1 \times 5.5$	<2	0.07
Jupiter	I	Io	1.769137786	422	0.004	0.04°	4.68×10^{-5}	1815	3.55	0.61
	II	Europa	3.551181041	671	0.009	0.47°	2.52×10^{-5}	1569	3.04	0.64
	III	Ganymede	7.15455296	1070	0.002	0.21°	7.80×10^{-5}	2631	1.93	0.42
	IV	Callisto	16.6890184	1883	0.007	0.51°	5.66×10^{-5}	2400	1.83	0.20
	V	Amalthea	0.49817905	181	0.003	0.40°	3.8×10^{-9}	$135 \times 83 \times 75$		0.05
	VI	Himalia	250.5662	11480	0.15798	27.63°	5.0×10^{-9}	93		0.03
	VII	Elara	259.6528	11737	0.20719	24.77°	4×10^{-10}	38		0.03
	VIII	Pasiphae	735 R	23500	0.378	145°	1×10^{-10}	25		
	IX	Sinope	758 R	23700	0.275	153°	0.4×10^{-10}	18		
	X	Lysithea	259.22	11720	0.107	29.02°	0.4×10^{-10}	18		
	XI	Carme	692 R	22600	0.20678	164°	0.5×10^{-10}	20		
	XII	Ananke	631 R	21200	0.16870	147°	0.2×10^{-10}	15		
	XIII	Leda	238.72	11094	0.14762	26.07°	0.03×10^{-10}	8		
	XIV	Thebe	0.6745	222	0.015	0.8°	4×10^{-10}	55×45		0.05
	XV	Adrastea	0.29826	129			0.1×10^{-10}	$12.5 \times 10 \times 7.5$		0.05
	XVI	Metis	0.294780	128			0.5×10^{-10}	20		0.05
Saturn	I	Mimas	0.942421813	185.52	0.0202	1.53°	8.0×10^{-8}	196	1.44	0.5
	II	Enceladus	1.370217855	238.02	0.00452	1.86°	1.3×10^{-7}	250	1.13	1.0
	III	Tethys	1.887802160	294.66	0.00000	1.86°	1.3×10^{-6}	530	1.20	0.9
	IV	Dione	2.736914742	377.40	0.002230	0.02°	1.85×10^{-6}	560	1.41	0.7
	V	Rhea	4.51750436	527.04	0.00100	0.35°	4.4×10^{-6}	765	1.33	0.7
	VI	Titan	15.94542068	1221.83	0.029192	0.33°	2.38×10^{-4}	2575	1.88	0.21
	VII	Hyperion	21.2766088	1481.1	0.104	0.43°	3×10^{-8}	$205 \times 130 \times 110$		0.3
	VIII	Iapetus	79.3301825	3561.3	0.02828	14.72°	3.3×10^{-6}	730	1.15	0.2
	IX	Phoebe	550.48 R	12952	0.16326	177°	7×10^{-10}	110		0.06

SATELLITES OF THE PLANETS (continued)

Planet	Satellite		Orb. Period d	Distance 10³ km	Eccentricity	Inclination	Rel. mass	Radius km	Den. g/cm³	Albedo
	X	Janus	0.6945	151.472	0.007	0.14°		110 × 100 × 80		0.8
	XI	Epimetheus	0.6942	151.422	0.009	0.34°		70 × 60 × 50		0.8
	XII	Helene	2.7369	377.40	0.005	0.0°		18 × 16 × 15		0.7
	XIII	Telesto	1.8878	294.66				17 × 14 × 13		0.5
	XIV	Calypso	1.8878	294.66				17 × 11 × 11		0.6
	XV	Atlas	0.6019	137.670	0.000	0.3°		20 × 10		0.9
	XVI	Prometheus	0.6130	139.353	0.003	0.0°		70 × 50 × 40		0.6
	XVII	Pandora	0.6285	141.700	0.004	0.0°		55 × 45 × 35		0.9
	XVIII	Pan	0.5750	133.583				10		0.5
Uranus	I	Ariel	2.52037935	191.02	0.0034	0.3°	1.56×10^{-5}	579	1.55	0.34
	II	Umbriel	4.1441772	266.30	0.0050	0.36°	1.35×10^{-5}	586	1.58	0.18
	III	Titania	8.7058717	435.91	0.0022	0.14°	4.06×10^{-5}	790	1.69	0.27
	IV	Oberon	13.4632389	583.52	0.0008	0.10°	3.47×10^{-5}	762	1.64	0.24
	V	Miranda	1.41347925	129.39	0.0027	4.2°	0.08×10^{-5}	240	1.25	0.27
	VI	Cordelia	0.335033	49.77	<0.001	0.1°		13		0.07
	VII	Ophelia	0.376409	53.79	0.010	0.1°		15		0.07
	VIII	Bianca	0.434577	59.17	<0.001	0.2°		21		0.07
	IX	Cressida	0.463570	61.78	<0.001	0.0°		31		0.07
	X	Desdemona	0.473651	62.68	<0.001	0.2°		27		0.07
	XI	Juliet	0.493066	64.35	<0.001	0.1°		42		0.07
	XII	Portia	0.513196	66.09	<0.001	0.1°		54		0.07
	XIII	Rosalind	0.558459	69.94	<0.001	0.3°		27		0.07
	XIV	Belinda	0.623525	75.26	<0.001	0.0°		33		0.07
	XV	Puck	0.761832	86.01	<0.001	0.31°		77		0.07
Neptune	I	Triton	5.8768541 R	354.76	0.000016	157.345°	2.09×10^{-4}	1353	2.05	0.7
	II	Nereid	360.13619	5513.4	0.7512	27.6°	2×10^{-7}	170		0.4
	III	Naiad	0.294396	117.6	<0.001	4.74°		29		0.06
	IV	Thalassa	0.311485	73.6	<0.001	0.21°		40		0.06
	V	Despina	0.334655	52.6	<0.001	0.07°		74		0.06
	VI	Galatea	0.428745	62.0	<0.001	0.05°		79		0.06
	VII	Larissa	0.554654	50.0	<0.0014	0.20°		104 × 89		0.06
	VIII	Proteus	1.122315	48.2	<0.001	0.55°		218 × 208 × 201		0.06
Pluto	I	Charon	6.38725	19.6	<0.001	99°	0.22	593		0.5

MASS, DIMENSIONS, AND OTHER PARAMETERS OF THE EARTH

This table is a collection of data on various properties of the Earth. Most of the values are given in SI units. Note that 1 AU (astronomical unit) = 149,597,870 km.

REFERENCES

1. Seidelmann, P. K., Editor, *Explanatory Supplement to the Astronomical Almanac*, University Science Books, Mill Valley, CA, 1992.
2. Lang, K. R., *Astrophysical Data: Planets and Stars*, Springer-Verlag, New York, 1992.

Quantity	Symbol	Value	Unit
Mass	M	$5.9742 \cdot 10^{27}$	g
Major orbital semi-axis	a_{orb}	1.000000	AU
		$1.4959787 \cdot 10^8$	km
Distance from sun at perihelion	r_π	0.9833	AU
Distance from sun at aphelion	r_α	1.0167	AU
Moment of perihelion passage	T_π	Jan. 2, 4 h 52 min	
Moment of aphelion passage	T_α	July 4, 5 h 05 min	
Siderial rotation period around sun	P_{orb}	$31.5581 \cdot 10^6$	s
		365.25636	d
Mean rotational velocity	U_{orb}	29.78	km/s
Mean equatorial radius	a	6378.140	km
Mean polar compression (flattening factor)	α	1/298.257	
Difference in equatorial and polar semi-axes	$a - c$	21.385	km
Compression of meridian of major equatorial axis	α_a	1/295.2	
Compression of meridian of minor equatorial axis	α_b	1/298.0	
Equatorial compression	ε	1/30 000	
Difference in equatorial semi-axes	$a - b$	213	m
Difference in polar semi-axes	$c_N - c_S$	~70	m
Polar asymmetry	η	$\sim 1 \cdot 10^{-5}$	
Mean acceleration of gravity at equator	g_e	9.78036	m/s^2
Mean acceleration of gravity at poles	g_p	9.83208	m/s^2
Difference in acceleration of gravity at pole and at equator	$g_p - g_e$	5.172	cm/s^2
Mean acceleration of gravity for entire surface of terrestrial ellipsoid	g	9.7978	m/s^2
Mean radius	R	6371.0	km
Area of surface	S	$5.10 \cdot 10^8$	km^2
Volume	V	$1.0832 \cdot 10^{12}$	km^3
Mean density	ρ	5.515	g/cm^3
Siderial rotational period	P	86,164.09	s
Rotational angular velocity	ω	$7.292116 \cdot 10^{-5}$	rad/s
Mean equatorial rotational velocity	v	0.46512	km/s
Rotational angular momentum	L	$5.861 \cdot 10^{33}$	J s
Rotational energy	E	$2.137 \cdot 10^{29}$	J
Ratio of centrifugal force to force of gravity at equator	q_c	0.0034677 = 1/288	
Moment of inertia	I	$8.070 \cdot 10^{37}$	kg m^2
Relative braking of earth's rotation due to tidal friction	$\Delta\omega_e/\omega$	$-4.2 \cdot 10^{-8}$	century^{-1}
Relative secular acceleration of earth's rotation	$\Delta\omega_i/\omega$	$+1.4 \cdot 10^{-8}$	century^{-1}
Not secular braking of earth's rotation	$\Delta\omega/\omega$	$-2.8 \cdot 10^{-8}$	century^{-1}
Probable value of total energy of tectonic deformation of earth	E_t	$\sim 1 \cdot 10^{23}$	J/century
Secular loss of heat of earth through radiation into space	$\Delta'E_k$	$1 \cdot 10^{23}$	J/century
Portion of earth's kinetic energy transformed into heat as a result of lunar and solar tides in the hydrosphere	$\Delta''E_k$	$1.3 \cdot 10^{23}$	J/century

Quantity	Symbol	Value	Unit
Differences in duration of days in March and August	ΔP	0.0025 (March-August)	s
Corresponding relative annual variation in earth's rotational velocity	$\Delta^*\omega/\omega$	$2.9 \cdot 10^{-8}$ (Aug.-March)	
Presumed variation in earth's radius between August and March	Δ^*R	-9.2 (Aug.-March)	cm
Annual variation in level of world ocean	Δh_o	~10 (Sept.-March)	cm
Area of continents	S_C	$1.49 \cdot 10^8$	km^2
		29.2	% of surface
Area of world ocean	S_o	$3.61 \cdot 10^8$	km^2
		70.8	% of surface
Mean height of continents above sea level	h_C	875	m
Mean depth of world ocean	h_o	3794	m
Mean thickness of lithosphere within the limits of the continents	$h_{c.l.}$	35	km
Mean thickness of lithosphere within the limits of the ocean	$h_{o.l.}$	4.7	km
Mean rate of thickening of continental lithosphere	$\Delta h/\Delta t$	10 - 40	m/10^6 y
Mean rate of horizontal extension of continental lithosphere	$\Delta l/\Delta t$	0.75 - 20	km/10^6 y
Mass of crust	m_1	$2.36 \cdot 10^{22}$	kg
Mass of mantle		$4.05 \cdot 10^{24}$	kg
Amount of water released from the mantle and core in the course of geological time		$3.40 \cdot 10^{21}$	kg
Total reserve of water in the mantle		$2 \cdot 10^{23}$	kg
Present content of free and bound water in the earth's lithosphere		$2.4 \cdot 10^{21}$	kg
Mass of hydrosphere	m_h	$1.664 \cdot 10^{21}$	kg
Amount of oxygen bound in the earth's crust		$1.300 \cdot 10^{21}$	kg
Amount of free oxygen		$1.5 \cdot 10^{18}$	kg
Mass of atmosphere	m_a	$5.136 \cdot 10^{18}$	kg
Mass of biosphere	m_b	$1.148 \cdot 10^{16}$	kg
Mass of living matter in the biosphere		$3.6 \cdot 10^{14}$	kg
Density of living matter on dry land		0.1	g/cm^2
Density of living matter in ocean		$15 \cdot 10^{-8}$	g/cm^3
Age of the earth		$4.55 \cdot 10^9$	y
Age of oldest rocks		$4.0 \cdot 10^9$	y
Age of most ancient fossils		$3.4 \cdot 10^9$	y

GEOLOGICAL TIME SCALE

Period or Epoch	Beginning and end, in 10^6 years	Key events
Cenozoic era		
Quaternian		
Contemporary	0–10,000 y ± 2,000 y	
Pleistocene	10,000–1,000,000 y ± 50,000 y	Homo Erectus breakout
Tertiary		
Pliocene	1.8–5.3	Ape man fossils
Miocene	5–25	Origin of grass
Oligocene	25–37	Rise of cats, dogs, pigs
Eocene	37–55	Debut of hoofed mammals
Paleocene	55–67	Earliest primates
Mesozoic era		
Cretaceous	67–138	Demise of dinosaurs
Jurassic	138–208	First birds
Triassic	208–245	Appearance of dinosaurs
Paleozoic era		
Permian	245–290	Flowers, insect pollination
Carboniferous	290–360	First conifers
Devonian	360–410	First vertebrates ashore
Silurian	410–435	Spore-bearing plants
Ordovician	435–520	First animals ashore
Cambrian	520–570	Vertebrates appear
Pre-Cambrian		
Pre-Cambrian III (Proterozoic)	570–2500	First plants, jellyfish
Pre-Cambrian II (Archean)	2500–3800	Photosynthetic bacteria
Pre-Cambrian I (Hadean)	3800–4450	Earth formed 4600 million years ago

Reference: Calder, N., *Timescale - An Atlas of the Fourth Dimension,* Viking Press, New York, 1983.

ACCELERATION DUE TO GRAVITY

The acceleration due to gravity is tabulated here as a function of latitude and height above the earth's surface. Values were calculated from the expression

$$g/(\text{m/s}^2) = 9.780356 \, (1 + 0.0052885 \sin^2 \phi - 0.0000059 \sin^2 2\phi) - 0.003086 \, H$$

where ϕ is the latitude and H is the height in kilometers.

REFERENCE

Jursa, A. S., Ed., *Handbook of Geophysics and the Space Environment,* 4th ed., Air Force Geophysics Laboratory, 1985, p. 14-17.

ϕ	$H = 0$	$H = 1$ km	$H = 5$ km	$H = 10$ km
0	9.78036	9.77727	9.76493	9.74950
5	9.78075	9.77766	9.76532	9.74989
10	9.78191	9.77882	9.76648	9.75105
15	9.78381	9.78072	9.76838	9.75295
20	9.78638	9.78330	9.77095	9.75552
25	9.78956	9.78647	9.77413	9.75870
30	9.79324	9.79016	9.77781	9.76238
35	9.79732	9.79424	9.78189	9.76646
40	9.80167	9.79858	9.78624	9.77081
45	9.80616	9.80307	9.79073	9.77530
50	9.81065	9.80757	9.79522	9.77979
55	9.81501	9.81193	9.79958	9.78415
60	9.81911	9.81602	9.80368	9.78825
65	9.82281	9.81972	9.80738	9.79195
70	9.82601	9.82292	9.81058	9.79515
75	9.82860	9.82551	9.81317	9.79774
80	9.83051	9.82743	9.81508	9.79965
85	9.83168	9.82860	9.81625	9.80082
90	9.83208	9.82899	9.81665	9.80122

DENSITY, PRESSURE, AND GRAVITY AS A FUNCTION OF DEPTH
WITHIN THE EARTH

This table gives the density ρ, pressure p, and acceleration due to gravity g as a function of depth below the earth's surface, as calculated from the model of the structure of the earth in Reference 1. The model assumes a radius of 6371 km for the earth. The boundary between the crust and mantle (the Mohorovicic discontinuity) is taken as 21 km, while in reality it varies considerable with location.

REFERENCES

1. Anderson, D. L., and Hart, R. S., *J. Geophys. Res.*, 81, 1461, 1976.
2. Carmichael, R. S., *CRC Practical Handbook of Physical Properties of Rocks and Minerals*, p.467, CRC Press, Boca Raton, FL, 1989.

Depth km	ρ g/cm³	p kbar	g cm/s²	Depth km	ρ g/cm³	p kbar	g cm/s²
Crust				1771	4.96	752	994
				2071	5.12	903	1002
0	1.02	0	981	2371	5.31	1061	1017
3	1.02	3	982	2671	5.45	1227	1042
3	2.80	3	982	2886	5.53	1352	1069
21	2.80	5	983				
Mantle (solid)				**Outer core (liquid)**			
21	3.49	5	983	2886	9.96	1352	1069
41	3.51	12	983	2971	10.09	1442	1050
61	3.52	19	984	3371	10.63	1858	953
81	3.48	26	984	3671	11.00	2154	874
101	3.44	33	984	4071	11.36	2520	760
121	3.40	39	985	4471	11.69	2844	641
171	3.37	56	987	4871	11.99	3116	517
221	3.34	73	989	5156	12.12	3281	427
271	3.37	89	991				
321	3.47	106	993	**Inner core (solid)**			
371	3.59	124	994	5156	12.30	3281	427
571	3.95	199	999	5371	12.48	3385	355
871	4.54	328	997	5771	12.52	3529	218
1171	4.67	466	992	6071	12.53	3592	122
1471	4.81	607	991	6371	12.58	3617	0

OCEAN PRESSURE AS A FUNCTION OF DEPTH AND LATITUDE

The following table is based upon an ocean model which takes into account the equation of state of standard seawater and the dependence on latitude of the acceleration of gravity. The tabulated pressure value is the excess pressure over the ambient atmospheric pressure at the surface.

REFERENCES

1. *International Oceanographic Tables, Volume 4*, Unesco Technical Papers in Marine Science No. 40, Unesco, Paris, 1987.
2. Saunders, P.M., and Fofonoff, N.P., *Deep-Sea Res*. 23, 109-111, 1976.

Pressure in MPa at the Specified Latitude

Depth (meters)	0°	15°	30°	45°	60°	75°	90°
0	0.0000	0.0000	0.0000	0.0000	0.0000	0.0000	0.0000
500	5.0338	5.0355	5.0404	5.0471	5.0537	5.0586	5.0605
1000	10.0796	10.0832	10.0930	10.1064	10.1198	10.1296	10.1333
1500	15.1376	15.1431	15.1577	15.1778	15.1980	15.2127	15.2182
2000	20.2076	20.2148	20.2344	20.2613	20.2882	20.3080	20.3153
2500	25.2895	25.2985	25.3231	25.3568	25.3905	25.4153	25.4244
3000	30.3831	30.3940	30.4236	30.4641	30.5047	30.5345	30.5453
3500	35.4886	35.5012	35.5358	35.5832	35.6307	35.6654	35.6782
4000	40.6056	40.6201	40.6598	40.7140	40.7683	40.8082	40.8229
4500	45.7342	45.7505	45.7952	45.8564	45.9176	45.9626	45.9791
5000	50.8742	50.8924	50.9421	51.0102	51.0785	51.1285	51.1469
5500	56.0255	56.0456	56.1004	56.1755	56.2508	56.3059	56.3262
6000	61.1882	61.2100	61.2700	61.3521	61.4344	61.4947	61.5168
6500	66.3619	66.3857	66.4508	66.5399	66.6292	66.6947	66.7187
7000	71.5467	71.5724	71.6427	71.7388	71.8352	71.9059	71.9318
7500	76.7426	76.7701	76.8456	76.9488	77.0523	77.1282	77.1560
8000	81.9493	81.9788	82.0594	82.1697	82.2804	82.3614	82.3911
8500	87.1669	87.1983	87.2841	87.4016	87.5193	87.6057	87.6373
9000	92.3950	92.4284	92.5194	92.6440	92.7689	92.8606	92.8941
9500	97.6346	97.6698	97.7661	97.8978	98.0300	98.1269	98.1624
10000	102.8800	102.9170	103.0185	103.1572	103.2961	103.3981	103.4355

PROPERTIES OF SEAWATER

In addition to the dependence on temperature and pressure, the physical properties of seawater vary with the concentration of the dissolved constituents. A convenient parameter for describing the composition is the salinity, S, which is defined in terms of the electrical conductivity of the seawater sample. The defining equation for the practical salinity is:

$$S = a_0 + a_1 K^{1/2} + a_2 K + a_3 K^{3/2} + a_4 K^2 + a_5 K^{5/2},$$

where K is the ratio of the conductivity of the seawater sample at 15°C and atmospheric pressure to the conductivity of a potassium chloride solution in which the mass fraction of KCl is 0.0324356, at the same temperature and pressure. The values of the coefficients are:

$$a_0 = 0.080 \qquad a_3 = 14.0941$$
$$a_1 = -0.1692 \qquad a_4 = -7.0261$$
$$a_2 = 25.3851 \qquad a_5 = 2.7081$$
$$\Sigma\, a_i = 35.0000$$

Thus when $K = 1$, $S = 35$ exactly (S is normally quoted in units of ‰, i.e., parts per thousand). The value of S can be roughly equated with the mass of dissolved material in grams per kilogram of seawater. Salinity values in the oceans at mid latitudes typically fall between 34 and 36.

The freezing point of seawater at normal atmospheric pressure varies with salinity as follows:

S	0	5	10	15	20	25	30	35	40
t_f/°C	0.000	-0.274	-0.542	-0.812	-1.083	-1.358	-1.636	-1.922	-2.212

The first table below gives several properties of seawater as a function of temperature for a salinity of 35. The second gives electrical conductivity as a function of salinity at several temperatures, and the third lists typical concentrations of the main constituents of seawater as a function of salinity.

REFERENCES

1. *The Practical Salinity Scale 1978 and the International Equation of State of Seawater 1980*, Unesco Technical Papers in Marine Science No. 36, Unesco, Paris, 1981; sections No. 37, 38, 39, and 40 in this series give background papers and detailed tables.
2. Kennish, M.J., *CRC Practical Handbook of Marine Science*, CRC Press, Boca Raton, FL, 1989.
3. Poisson, A. *IEEE J. Ocean. Eng*. OE-5, 50, 1981.

Properties of Seawater as a Function of Temperature at Salinity S=35 and Normal Atmospheric Pressure
(100 kPa)

ρ = density in g/cm^3
$\beta = (1/\rho)\,(d\rho/dS)$ = fractional change in density per unit change in salinity
$\alpha = (1/\rho)\,(d\rho/dt)$ = fractional change in density per unit change in temperature (°C^{-1})
κ = electrical conductivity in S/cm
η = viscosity in mPa s (equal to cP)
c_p = specific heat in J/kg
v = speed of sound in m/s

t/°C	ρ/g cm^{-3}	$10^7 \cdot \beta$	$10^7 \cdot \alpha$/°C^{-1}	κ/S cm^{-1}	η/mPa s	c_p/J kg^{-1} °C^{-1}	v/m s^{-1}
0	1.028106	7854	526	0.029048	1.892	3986.5	1449.1
5	1.027675	7717	1136	0.033468	1.610		
10	1.026952	7606	1668	0.038103	1.388	3986.3	1489.8
15	1.025973	7516	2141	0.042933	1.221		
20	1.024763	7444	2572	0.047934	1.085	3993.9	1521.5
25	1.023343	7385	2970	0.053088	0.966		
30	1.021729	7338	3341	0.058373	0.871	4000.7	1545.6
35	1.019934	7300	3687				
40		7270	4004			4003.5	1563.2

Electrical Conductivity of Seawater in S/cm as a Function of Temperature and Salinity

$t/°C$	$S = 5$	$S = 10$	$S = 15$	$S = 20$	$S = 25$	$S = 30$	$S = 35$	$S = 40$
0	0.004808	0.009171	0.013357	0.017421	0.021385	0.025257	0.029048	0.032775
5	0.005570	0.010616	0.015441	0.020118	0.024674	0.029120	0.033468	0.037734
10	0.006370	0.012131	0.017627	0.022947	0.028123	0.033171	0.038103	0.042935
15	0.007204	0.013709	0.019905	0.025894	0.031716	0.037391	0.042933	0.048355
20	0.008068	0.015346	0.022267	0.028948	0.035438	0.041762	0.047934	0.053968
25	0.008960	0.017035	0.024703	0.032097	0.039276	0.046267	0.053088	0.059751
30	0.009877	0.018771	0.027204	0.035330	0.043213	0.050888	0.058373	0.065683

Composition of Seawater and Ionic Strength at Various Salinities
Concentration of major constituents expressed as molality (moles per kilogram of H_2O)

Constituent	$S = 30$	$S = 35$	$S = 40$
Cl^-	0.482	0.562	0.650
Br^-	0.00074	0.00087	0.00100
F^-		0.00007	
SO_4^{2-}	0.0104	0.0114	0.0122
HCO_3^-	0.00131	0.00143	0.00100
$NaSO_4^-$	0.0085	0.0108	0.0139
KSO_4^-	0.00010	0.00012	0.00015
Na^+	0.405	0.472	0.544
K^+	0.00892	0.01039	0.01200
Mg^{2+}	0.0413	0.0483	0.0561
Ca^{2+}	0.00131	0.00143	0.00154
Sr^+	0.00008	0.00009	0.00011
$MgHCO_3^+$	0.00028	0.00036	0.00045
$MgSO_4$	0.00498	0.00561	0.00614
$CaSO_4$	0.00102	0.00115	0.00126
$NaHCO_3$	0.00015	0.00020	0.00024
H_3BO_3	0.00032	0.00037	0.00042
Ionic strength (mol/kg)	0.5736	0.6675	0.7701

ABUNDANCE OF ELEMENTS IN THE EARTH'S CRUST AND IN THE SEA

This table gives the estimated abundance of the elements in the continental crust (in mg/kg, equivalent to parts per million by mass) and in seawater near the surface (in mg/L). Values represent the median of reported measurements. The concentrations of the less abundant elements may vary with location by several orders of magnitude.

REFERENCES

1. Carmichael, R. S., Ed., *CRC Practical Handbook of Physical Properties of Rocks and Minerals,* CRC Press, Boca Raton, FL, 1989.
2. Bodek, I., et al., *Environmental Inorganic Chemistry,* Pergamon Press, New York, 1988.
3. Ronov, A. B., and Yaroshevsky, A. A., "Earth's Crust Geochemistry", in *Encyclopedia of Geochemistry and Environmental Sciences,* Fairbridge, R. W., Ed., Van Nostrand, New York, 1969.

Element	Abundance Crust mg/kg	Abundance Sea mg/L	Element	Abundance Crust mg/kg	Abundance Sea mg/L
Ac	5.5×10^{-10}		N	1.9×10^1	5×10^{-1}
Ag	7.5×10^{-2}	4×10^{-5}	Na	2.36×10^4	1.08×10^4
Al	8.23×10^4	2×10^{-3}	Nb	2.0×10^1	1×10^{-5}
Ar	3.5	4.5×10^{-1}	Nd	4.15×10^1	2.8×10^{-6}
As	1.8	3.7×10^{-3}	Ne	5×10^{-3}	1.2×10^{-4}
Au	4×10^{-3}	4×10^{-6}	Ni	8.4×10^1	5.6×10^{-4}
B	1.0×10^1	4.44	O	4.61×10^5	8.57×10^5
Ba	4.25×10^2	1.3×10^{-2}	Os	1.5×10^{-3}	
Be	2.8	5.6×10^{-6}	P	1.05×10^3	6×10^{-2}
Bi	8.5×10^{-3}	2×10^{-5}	Pa	1.4×10^{-6}	5×10^{-11}
Br	2.4	6.73×10^1	Pb	1.4×10^1	3×10^{-5}
C	2.00×10^2	2.8×10^1	Pd	1.5×10^{-2}	
Ca	4.15×10^4	4.12×10^2	Po	2×10^{-10}	1.5×10^{-14}
Cd	1.5×10^{-1}	1.1×10^{-4}	Pr	9.2	6.4×10^{-7}
Ce	6.65×10^1	1.2×10^{-6}	Pt	5×10^{-3}	
Cl	1.45×10^2	1.94×10^4	Ra	9×10^{-7}	8.9×10^{-11}
Co	2.5×10^1	2×10^{-5}	Rb	9.0×10^1	1.2×10^{-1}
Cr	1.02×10^2	3×10^{-4}	Re	7×10^{-4}	4×10^{-6}
Cs	3	3×10^{-4}	Rh	1×10^{-3}	
Cu	6.0×10^1	2.5×10^{-4}	Rn	4×10^{-13}	6×10^{-16}
Dy	5.2	9.1×10^{-7}	Ru	1×10^{-3}	7×10^{-7}
Er	3.5	8.7×10^{-7}	S	3.50×10^2	9.05×10^2
Eu	2.0	1.3×10^{-7}	Sb	2×10^{-1}	2.4×10^{-4}
F	5.85×10^2	1.3	Sc	2.2×10^1	6×10^{-7}
Fe	5.63×10^4	2×10^{-3}	Se	5×10^{-2}	2×10^{-4}
Ga	1.9×10^1	3×10^{-5}	Si	2.82×10^5	2.2
Gd	6.2	7×10^{-7}	Sm	7.05	4.5×10^{-7}
Ge	1.5	5×10^{-5}	Sn	2.3	4×10^{-6}
H	1.40×10^3	1.08×10^5	Sr	3.70×10^2	7.9
He	8×10^{-3}	7×10^{-6}	Ta	2.0	2×10^{-6}
Hf	3.0	7×10^{-6}	Tb	1.2	1.4×10^{-7}
Hg	8.5×10^{-2}	3×10^{-5}	Te	1×10^{-3}	
Ho	1.3	2.2×10^{-7}	Th	9.6	1×10^{-6}
I	4.5×10^{-1}	6×10^{-2}	Ti	5.65×10^3	1×10^{-3}
In	2.5×10^{-1}	2×10^{-2}	Tl	8.5×10^{-1}	1.9×10^{-5}
Ir	1×10^{-3}		Tm	5.2×10^{-1}	1.7×10^{-7}
K	2.09×10^4	3.99×10^2	U	2.7	3.2×10^{-3}
Kr	1×10^{-4}	2.1×10^{-4}	V	1.20×10^2	2.5×10^{-3}
La	3.9×10^1	3.4×10^{-6}	W	1.25	1×10^{-4}
Li	2.0×10^1	1.8×10^{-1}	Xe	3×10^{-5}	5×10^{-5}
Lu	8×10^{-1}	1.5×10^{-7}	Y	3.3×10^1	1.3×10^{-5}
Mg	2.33×10^4	1.29×10^3	Yb	3.2	8.2×10^{-7}
Mn	9.50×10^2	2×10^{-4}	Zn	7.0×10^1	4.9×10^{-3}
Mo	1.2	1×10^{-2}	Zr	1.65×10^2	3×10^{-5}

SOLAR SPECTRAL IRRADIANCE

The solar luminosity (total radiant power emitted) is $3.86 \cdot 10^{26}$ W, of which 1373 W/m^2 reaches the top of the earth's atmosphere. To a zeroth approximation the sun can be considered a black body with an effective temperature of 5780 K, which implies a peak in the radiation at around 0.520 µm (5200 Å). The actual solar spectral emission is more complex, especially at ultraviolet and shorter wavelengths. The graph below, which was taken from Reference 1, summarizes the solar irradiance at the top of the atmosphere in the range 0.3 to 10 µm.

REFERENCES

1. Jursa, A.S., ed., *Handbook of Geophysics and the Space Environment*, Air Force Geophysics Laboratory, 1985.
2. Pierce, A.K., and Allen, R.G., "The Solar Spectrum between 0.3 and 10 µm", in *The Solar Output and its Variation*, White, O.R., Ed., Colorado Associated University Press, Boulder, CO, 1977.
3. Lang, K.R., *Astrophysical Data. Planets and Stars*, Springer-Verlag, New York, 1992.

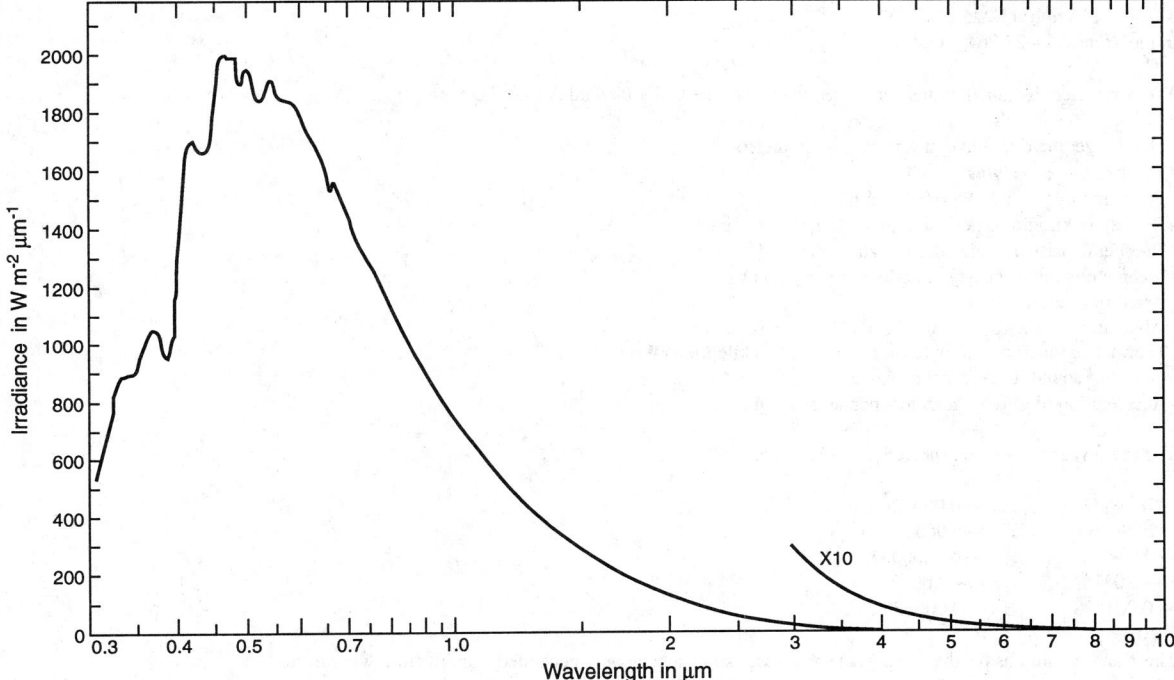

U.S. STANDARD ATMOSPHERE (1976)

A Standard Atmosphere is a hypothetical vertical distribution of atmospheric temperature, pressure, and density which is roughly representative of year-round, midlatitude conditions. Typical uses are to serve as a basis for pressure altimeter calibrations, aircraft performance calculations, aircraft and rocket design, ballistic tables, meteorological diagrams, and various types of atmospheric modeling. The air is assumed to be dry and to obey the perfect gas law and the hydrostatic equation which, taken together, relate temperature, pressure, and density with vertical position. The atmosphere is considered to rotate with the earth and to be an average over the diurnal cycle, the semiannual variation, and the range from active to quiet geomagnetic and sunspot conditions.

The U.S. Standard Atmosphere, 1976 is an idealized, steady-state representation of mean annual conditions of the earth's atmosphere from the surface to 1000 km at latitude 45°N, as it is assumed to exist during a period with moderate solar activity. The defining meteorological elements are sea-level temperature and pressure and a temperature-height profile to 1000 km. The 1976 Standard Atmosphere uses the following sea-level values which have been standard for many decades:

Temperature — 288.15 K (15°C)
Pressure — 101325 Pa (1013.25 mbar, 760 mm of Hg, or 29.92 in. of Hg)
Density — 1225 g/m^3 (1.225 g/L)
Mean molar mass — 28.964 g/mol

The parameters included in this condensed version of the U.S. Standard Atmosphere are:

Z — Height (geometric) above mean sea level in meters
T — Temperature in kelvins
P — Pressure in pascals (1 Pa = 0.01 millibars)
ρ — Density in kilograms per cubic meter (1 kg/m^3 = 1 g/L)
n — Number density in molecules per cubic meter
ν — Mean collision frequency in collisions per second
l — Mean free path in meters
η — Absolute viscosity in pascal seconds (1 Pa s = 1000 cP)
k — Thermal conductivity in joules per meter second kilogram (W/m K)
v_s — Speed of sound in meters per second
g — Acceleration of gravity in meters per second square

The sea-level composition (percent by volume) is taken to be:

N_2 — 78.084%	He — 0.000524
O_2 — 20.9476	Kr — 0.000114
Ar — 0.934	Xe — 0.0000087
CO_2 — 0.0314	CH_4 — 0.0002
Ne — 0.001818	H_2 — 0.00005

The T and P columns for the troposphere and lower stratosphere were generated from the following formulas:

	T/K	P/Pa
$H \leq 1100$ m	288.15 - 0.0065 H	$101325(288.15/T)^{-5.25577}$
11000 m $< H \leq 20000$ m	216.65	$22632\,e^{-0.00015768832(H-11000)}$
20000 m $< H \leq 32000$ m	216.65 + 0.0010(H-20000)	$5474.87(216.65/T)^{34.16319}$

where $H = rZ/(r + Z)$ is the geopotential height in meters and r is the mean earth radius at 45° N latitude, taken as 6356766 m. For altitudes up to 32 km, $\rho = 0.003483677(P/T)$ in the units used here. Formulas for the other quantities may be found in the references.

REFERENCES

1. COESA, U.S. Standard Atmosphere, 1976, U.S. Government Printing Office, Washington, D.C., 1976.
2. Jursa, A.S., ed., *Handbook of Geophysics and the Space Environment*, Air Force Geophysics Laboratory, 1985.

Z/m	T/K	P/Pa	ρ/kg m^{-3}	n/m^{-3}	v/s^{-1}	l/m	η/Pa s	k/J m^{-1}s^{-1}K^{-1}	v_s/m s^{-1}	g/m s^{-2}
-5000	320.68	1.778E+05	1.931	4.015E+25	1.151E+10	4.208E-08	1.942E-05	0.02788	359.0	9.822
-4500	317.42	1.685E+05	1.849	3.845E+25	1.096E+10	4.395E-08	1.927E-05	0.02763	357.2	9.830
-4000	314.17	1.596E+05	1.770	3.680E+25	1.044E+10	4.592E-08	1.912E-05	0.02738	355.3	9.819
-3500	310.91	1.511E+05	1.693	3.520E+25	9.933E+09	4.800E-08	1.897E-05	0.02713	353.5	9.818
-3000	307.66	1.430E+05	1.619	3.366E+25	9.448E+09	5.019E-08	1.882E-05	0.02688	351.6	9.816
-2500	304.41	1.352E+05	1.547	3.217E+25	8.982E+09	5.252E-08	1.867E-05	0.02663	349.8	9.814
-2000	301.15	1.278E+05	1.478	3.102E+25	8.623E+09	5.447E-08	1.852E-05	0.02638	347.9	9.813
-1500	297.90	1.207E+05	1.411	2.935E+25	8.106E+09	5.757E-08	1.836E-05	0.02613	346.0	9.811
-1000	294.65	1.139E+05	1.347	2.801E+25	7.693E+09	6.032E-08	1.821E-05	0.02587	344.1	9.810
-500	291.40	1.075E+05	1.285	2.672E+25	7.298E+09	6.324E-08	1.805E-05	0.02562	342.2	9.808
0	288.15	1.013E+05	1.225	2.547E+25	6.919E+09	6.633E-08	1.789E-05	0.02533	340.3	9.807
500	284.90	9.546E+04	1.167	2.427E+25	6.556E+09	6.961E-08	1.774E-05	0.02511	338.4	9.805
1000	281.65	8.988E+04	1.112	2.311E+25	6.208E+09	7.310E-08	1.758E-05	0.02485	336.4	9.804
1500	278.40	8.456E+04	1.058	2.200E+25	5.874E+09	7.680E-08	1.742E-05	0.02459	334.5	9.802
2000	275.15	7.950E+04	1.007	2.093E+25	5.555E+09	8.073E-08	1.726E-05	0.02433	332.5	9.801
2500	271.91	7.469E+04	0.957	1.990E+25	5.250E+09	8.491E-08	1.710E-05	0.02407	330.6	9.799
3000	268.66	7.012E+04	0.909	1.891E+25	4.959E+09	8.937E-08	1.694E-05	0.02381	328.6	9.797
3500	265.41	6.579E+04	0.863	1.795E+25	4.680E+09	9.411E-08	1.678E-05	0.02355	326.6	9.796
4000	262.17	6.166E+04	0.819	1.704E+25	4.414E+09	9.917E-08	1.661E-05	0.02329	324.6	9.794
4500	258.92	5.775E+04	0.777	1.616E+25	4.160E+09	1.046E-07	1.645E-05	0.02303	322.6	9.793
5000	255.68	5.405E+04	0.736	1.531E+25	3.918E+09	1.103E-07	1.628E-05	0.02277	320.6	9.791
5500	252.43	5.054E+04	0.697	1.450E+25	3.687E+09	1.165E-07	1.612E-05	0.02250	318.5	9.790
6000	249.19	4.722E+04	0.660	1.373E+25	3.467E+09	1.231E-07	1.595E-05	0.02224	316.5	9.788
6500	245.94	4.408E+04	0.664	1.299E+25	3.258E+09	1.302E-07	1.578E-05	0.02197	314.4	9.787
7000	242.70	4.111E+04	0.590	1.227E+25	3.058E+09	1.377E-07	1.561E-05	0.02170	312.3	9.785
7500	239.46	3.830E+04	0.557	1.159E+25	2.869E+09	1.458E-07	1.544E-05	0.02144	310.2	9.784
8000	236.22	3.565E+04	0.526	1.093E+25	2.689E+09	1.545E-07	1.527E-05	0.02117	308.1	9.782
8500	232.97	3.315E+04	0.496	1.031E+25	2.518E+09	1.639E-07	1.510E-05	0.02090	306.0	9.781
9000	229.73	3.080E+04	0.467	9.711E+24	2.356E+09	1.740E-07	1.493E-05	0.02063	303.9	9.779
9500	226.49	2.858E+04	0.440	9.141E+24	2.202E+09	1.848E-07	1.475E-05	0.02036	301.7	9.777
10000	223.25	2.650E+04	0.414	8.598E+24	2.056E+09	1.965E-07	1.458E-05	0.02009	299.5	9.776
10500	220.01	2.454E+04	0.389	8.079E+24	1.918E+09	2.091E-07	1.440E-05	0.01982	297.4	9.774
11000	216.77	2.270E+04	0.365	7.585E+24	1.787E+09	2.227E-07	1.422E-05	0.01954	295.2	9.773
11500	216.65	2.098E+04	0.337	7.016E+24	1.653E+09	2.408E-07	1.422E-05	0.01953	295.1	9.771
12000	216.65	1.940E+04	0.312	6.486E+24	1.528E+09	2.605E-07	1.422E-05	0.01953	295.1	9.770
12500	216.65	1.793E+04	0.288	5.996E+24	1.412E+09	2.818E-07	1.422E-05	0.01953	295.1	9.768
13000	216.65	1.658E+04	0.267	5.543E+24	1.306E+09	3.048E-07	1.422E-05	0.01953	295.1	9.767
13500	216.65	1.533E+04	0.246	5.124E+24	1.207E+09	3.297E-07	1.422E-05	0.01953	295.1	9.765
14000	216.65	1.417E+04	0.228	4.738E+24	1.116E+09	3.566E-07	1.422E-05	0.01953	295.1	9.764
14500	216.65	1.310E+04	0.211	4.380E+24	1.032E+09	3.857E-07	1.422E-05	0.01953	295.1	9.762
15000	216.65	1.211E+04	0.195	4.049E+24	9.538E+08	4.172E-07	1.422E-05	0.01953	295.1	9.761
16000	216.65	1.035E+04	0.166	3.461E+24	8.153E+08	4.881E-07	1.422E-05	0.01953	295.1	9.758
17000	216.65	8.850E+03	0.142	2.959E+24	6.969E+08	5.710E-07	1.422E-05	0.01953	295.1	9.754
18000	216.65	7.565E+03	0.122	2.529E+24	5.958E+08	6.680E-07	1.422E-05	0.01953	295.1	9.751
19000	216.65	6.467E+03	0.104	2.162E+24	5.093E+08	7.814E-07	1.422E-05	0.01953	295.1	9.748

Z/m	T/K	P/Pa	ρ/kg m⁻³	n/m⁻³	v/s⁻¹	l/m	η/Pa s	κ/J m⁻¹ s⁻¹ K⁻¹	v_s/m s⁻¹	g/m s⁻²
20000	216.65	5.529E+03	8.891E-02	1.849E+24	4.354E+08	9.139E-07	1.422E-05	0.01953	295.1	9.745
21000	217.58	4.729E+03	7.572E-02	1.574E+24	3.716E+08	1.073E-06	1.427E-05	0.01961	295.1	9.742
22000	218.57	4.048E+03	6.451E-02	1.341E+24	3.173E+08	1.260E-06	1.432E-05	0.01970	296.4	9.739
23000	219.57	3.467E+03	5.501E-02	1.144E+24	2.712E+08	1.477E-06	1.438E-05	0.01978	297.1	9.736
24000	220.56	2.972E+03	4.694E-02	9.759E+23	2.319E+08	1.731E-06	1.443E-05	0.01986	297.7	9.733
25000	221.55	2.549E+03	4.008E-02	8.334E+23	1.985E+08	2.027E-06	1.448E-05	0.01995	298.4	9.730
26000	222.54	2.188E+03	3.426E-02	7.123E+23	1.700E+08	2.372E-06	1.454E-05	0.02003	299.1	9.727
27000	223.54	1.880E+03	2.930E-02	6.092E+23	1.458E+08	2.773E-06	1.459E-05	0.02011	299.7	9.724
28000	224.53	1.610E+03	2.508E-02	5.214E+23	1.250E+08	3.240E-06	1.465E-05	0.02020	300.4	9.721
29000	225.52	1.390E+03	2.148E-02	4.466E+23	1.073E+08	3.783E-06	1.470E-05	0.02028	301.1	9.718
30000	226.51	1.197E+03	1.841E-02	3.828E+23	9.219E+07	4.414E-06	1.475E-05	0.02036	301.7	9.715
31000	227.50	1.031E+03	1.579E-02	3.283E+23	7.925E+07	5.146E-06	1.481E-05	0.02044	302.4	9.712
32000	228.49	8.891E+02	1.356E-02	2.813E+23	6.818E+07	5.995E-06	1.486E-05	0.02053	303.0	9.709
33000	230.97	7.673E+02	1.157E-02	2.406E+23	5.852E+07	7.021E-06	1.499E-05	0.02073	304.7	9.706
34000	233.74	6.634E+02	9.887E-03	2.056E+23	5.030E+07	8.218E-06	1.514E-05	0.02096	306.5	9.703
35000	236.51	5.746E+02	8.463E-03	1.760E+23	4.331E+07	9.601E-06	1.529E-05	0.02119	308.3	9.700
36000	239.28	4.985E+02	7.258E-03	1.509E+23	3.736E+07	1.120E-05	1.543E-05	0.02142	310.1	9.697
38000	244.82	3.771E+02	5.367E-03	1.116E+23	2.794E+07	1.514E-05	1.572E-05	0.02188	313.7	9.690
40000	250.35	2.871E+02	3.996E-03	8.308E+22	2.104E+07	2.034E-05	1.601E-05	0.02233	317.2	9.684
42000	255.88	2.200E+02	2.995E-03	6.227E+22	1.594E+07	2.713E-05	1.629E-05	0.02278	320.7	9.678
44000	261.40	1.695E+02	2.259E-03	4.697E+22	1.215E+07	3.597E-05	1.657E-05	0.02323	324.1	9.672
46000	266.93	1.313E+02	1.714E-03	3.564E+22	9.318E+06	4.740E-05	1.685E-05	0.02376	327.5	9.666
48000	270.65	1.023E+02	1.317E-03	2.738E+22	7.208E+06	6.171E-05	1.704E-05	0.02397	329.8	9.660
50000	270.65	7.978E+01	1.027E-03	2.135E+22	5.620E+06	7.913E-05	1.704E-05	0.02397	329.8	9.654
52000	269.03	6.221E+01	8.056E-04	1.675E+22	4.397E+06	1.009E-04	1.696E-05	0.02384	328.8	9.648
54000	263.52	4.834E+01	6.390E-04	1.329E+22	3.452E+06	1.272E-04	1.660E-05	0.02340	325.4	9.642
56000	258.02	3.736E+01	5.045E-04	1.049E+22	2.696E+06	1.611E-04	1.640E-05	0.02296	322.0	9.636
58000	252.52	2.872E+01	3.963E-04	8.239E+21	2.095E+06	2.051E-04	1.612E-05	0.02251	318.6	9.632
60000	247.02	2.196E+01	3.097E-04	6.439E+21	1.620E+06	2.624E-04	1.584E-05	0.02206	315.1	9.624
65000	233.29	1.093E+01	1.632E-04	3.393E+21	8.294E+05	4.979E-04	1.512E-05	0.02093	306.2	9.609
70000	219.59	5.221	8.283E-05	1.722E+21	4.084E+05	9.810E-04	1.438E-05	0.01978	297.1	9.594
75000	208.40	2.388	3.992E-05	8.300E+20	1.918E+05	2.035E-03	1.376E-05	0.01883	289.4	9.579
80000	198.64	1.052	1.846E-05	3.838E+20	8.656E+04	4.402E-03	1.321E-05	0.01800	282.5	9.564
85000	188.89	4.457E-01	8.220E-06	1.709E+20	3.766E+04	9.886E-03	1.265E-05	0.01716	275.5	9.550
90000	186.87	1.836E-01	3.416E-06	7.116E+19	1.560E+04	2.370E-02				9.535
95000	188.42	7.597E-02	1.393E-06	2.920E+19	6.440E+03	5.790E-02				9.520
100000	195.08	3.201E-02	5.604E-07	1.189E+19	2.680E+03	1.420E-01				9.505
110000	240.00	7.104E-03	9.708E-08	2.144E+18	5.480E+02	7.880E-01				9.476
120000	360.00	2.538E-03	2.222E-08	5.107E+17	1.630E+02	3.310				9.447
130000	469.27	1.251E-03	8.152E-09	1.930E+17	7.100E+01	8.800				9.418
140000	559.63	7.203E-04	3.831E-09	9.322E+16	3.800E+01	1.800E+01				9.389
150000	634.39	4.542E-04	2.076E-09	5.186E+16	2.300E+01	3.300E+01				9.360
160000	696.29	3.040E-04	1.233E-09	3.162E+16	1.500E+01	5.300E+01				9.331
170000	747.57	2.121E-04	7.815E-10	2.055E+16	1.000E+01	8.200E+01				9.302
180000	790.07	1.527E-04	5.194E-10	1.400E+16	7.200	1.200E+02				9.274

Z/m	T/K	P/Pa	$\rho/kg\ m^{-3}$	n/m^{-3}	v/s^{-1}	l/m	$\eta/Pa\ s$	$k/J\ m^{-1}s^{-1}K^{-1}$	$v_s/m\ s^{-1}$	$g/m\ s^{-2}$
190000	825.16	1.127E−04	3.581E−10	9.887E+15	5.200	1.700E+02				9.246
200000	854.56	8.474E−05	2.541E−10	7.182E+15	3.900	2.400E+02				9.218
220000	899.01	5.015E−05	1.367E−10	4.040E+15	2.300	4.200E+02				9.162
240000	929.73	3.106E−05	7.858E−11	2.420E+15	1.400	7.000E+02				9.106
260000	950.99	1.989E−05	4.742E−11	1.515E+15	9.300E−01	1.100E+03				9.051
280000	965.75	1.308E−05	2.971E−11	9.807E+14	6.100E−01	1.700E+03				8.997
300000	976.01	8.770E−06	1.916E−11	6.509E+14	4.200E−01	2.600E+03				8.943
320000	983.16	5.980E−06	1.264E−11	4.405E+14	2.900E−01	3.800E+03				8.889
340000	988.15	4.132E−06	8.503E−12	3.029E+14	2.000E−01	5.600E+03				8.836
360000	991.65	2.888E−06	5.805E−12	2.109E+14	1.400E−01	8.000E+03				8.784
380000	994.10	2.038E−06	4.013E−12	1.485E+14	1.000E−01	1.100E+04				8.732
400000	995.83	1.452E−06	2.803E−12	1.056E+14	7.200E−02	1.600E+04				8.680
450000	998.22	6.447E−07	1.184E−12	4.678E+13	3.300E−02	3.600E+04				8.553
500000	999.24	3.024E−07	5.215E−13	2.192E+13	1.600E−02	7.700E+04				8.429
550000	999.67	1.514E−07	2.384E−13	1.097E+13	8.400E−03	1.500E+05				8.307
600000	999.85	8.213E−08	1.137E−13	5.950E+12	4.800E−03	2.800E+05				8.188
650000	999.93	4.887E−08	5.712E−14	3.540E+12	3.100E−03	4.800E+05				8.072
700000	999.97	3.191E−08	3.070E−14	2.311E+12	2.200E−03	7.300E+05				7.958
750000	999.98	2.260E−08	1.788E−14	1.637E+12	1.700E−03	1.000E+06				7.846
800000	999.99	1.704E−08	1.136E−14	1.234E+12	1.400E−03	1.400E+06				7.737
850000	1000.00	1.342E−08	7.824E−15	9.717E+11	1.200E−03	1.700E+06				7.630
900000	1000.00	1.087E−08	5.759E−15	7.876E+11	1.000E−03	2.100E+06				7.525
950000	1000.00	8.982E−09	4.453E−15	6.505E+11	8.700E−04	2.600E+06				7.422
1000000	1000.00	7.514E−09	3.561E−15	5.442E+11	7.500E−04	3.100E+06				7.322

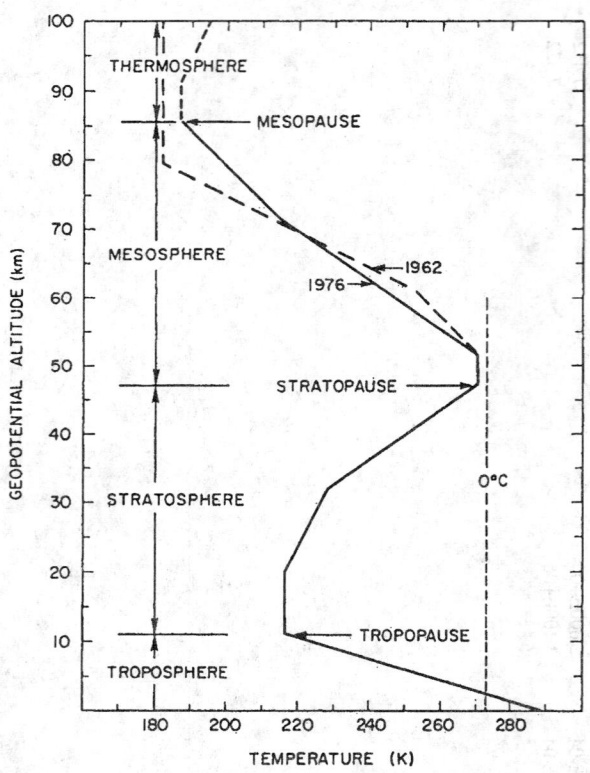

FIGURE 1. Temperature-height profile for U.S. Standard Atmosphere.

FIGURE 2. Total pressure and mass density as a function of geometric altitude.

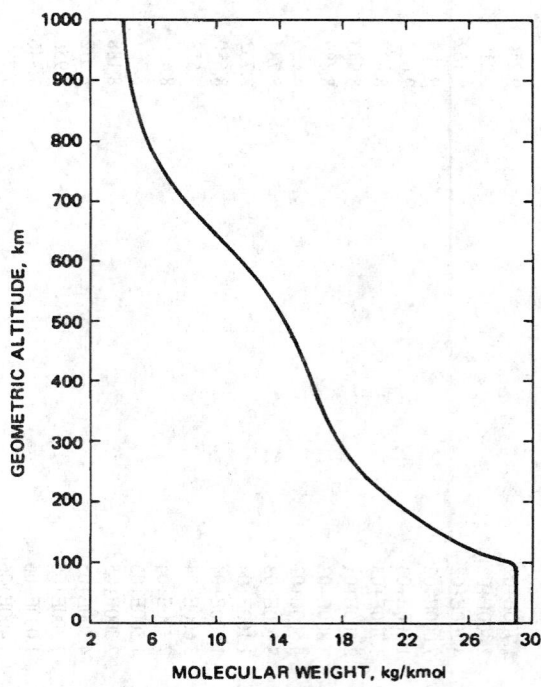

FIGURE 3. Mean molecular weight as a function of geometric altitude.

FIGURE 4. Number density of individual species and total number density as a function of geometric altitude.

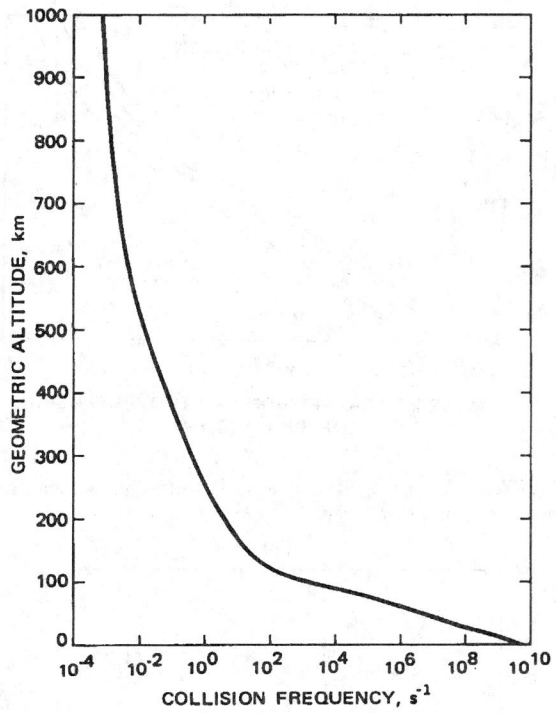

FIGURE 5. Collision frequency as a function of geometric altitude.

FIGURE 7. Mean air-particle speed as a function of geometric altitude.

FIGURE 6. Mean free path as a function of geometric altitude.

FIGURE 8. Dynamic viscosity as a function of geometric altitude.

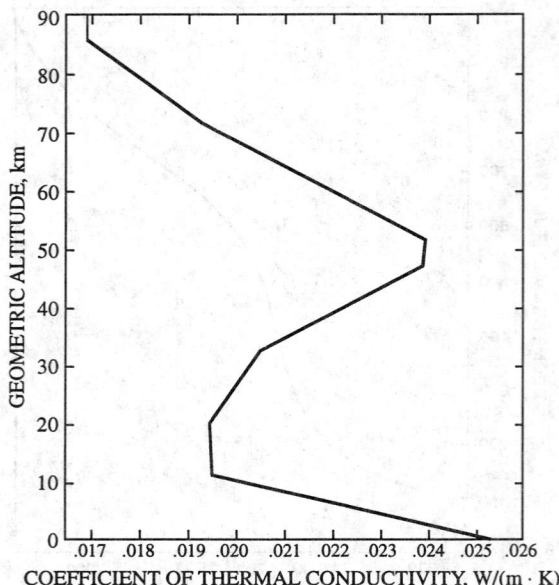

FIGURE 9. Coefficient of thermal conductivity as a function of geometric altitude.

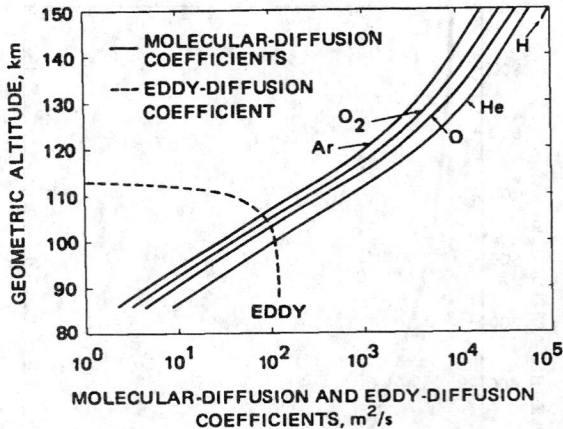

FIGURE 11. Molecular-diffusion and eddy-diffusion coefficients as a function of geometric altitude.

FIGURE 10. Speed of sound as a function of geometric altitude.

FIGURE 12. Acceleration of gravity as a function of geometric altitude.

GEOGRAPHICAL AND SEASONAL VARIATION IN SOLAR RADIATION

This table gives the amount of solar radiation reaching a unit area at the top of the earth's atmosphere per day as a function of latitude and approximate date. It is based upon a solar constant (total energy per unit area at the earth's average orbital distance) of 1373 W/m². Absorption of radiation by the atmosphere is not taken into consideration.

REFERENCE

List, R.J., *Smithsonian Meteorological Tables, Seventh Edition*, Smithsonian Institution Press, Washington, D.C., 1962.

Daily Solar Radiation in MJ/m²

Lat.	Mar. 21	Apr. 13	May 6	May 29	Jun. 2	Jul. 15	Aug. 8	Aug. 31
90°		18.0	32.8	42.4	45.7	42.2	32.5	17.7
80	6.6	18.0	32.3	41.8	45.0	41.6	32.0	17.7
70	13.0	22.3	31.8	39.9	43.0	39.7	31.5	22.0
60	19.0	27.0	34.4	39.7	41.6	39.4	34.0	26.7
50	24.4	31.1	36.8	40.7	42.0	40.5	36.5	30.8
40	29.1	34.3	38.6	41.3	42.1	41.1	38.3	33.9
30	32.9	36.7	39.4	41.1	41.4	40.8	39.1	36.3
20	35.7	38.0	39.2	39.7	39.7	39.5	38.9	37.5
10	37.4	38.1	37.9	37.4	37.1	37.2	37.6	37.7
0	38.0	37.1	35.5	34.1	33.5	34.0	35.2	36.6
−10	37.4	35.0	32.3	30.0	29.2	29.9	32.0	34.6
−20	35.7	31.8	28.0	25.2	24.1	25.1	27.8	31.5
−30	32.9	27.8	23.1	19.7	18.5	19.7	22.8	27.4
−40	29.1	22.8	17.5	14.0	12.6	13.9	17.4	22.6
−50	24.4	17.3	11.7	8.2	7.0	8.2	11.6	17.2
−60	19.0	11.4	5.9	2.9	2.0	2.9	5.9	11.3
−70	13.0	5.4	1.0				1.0	5.3
−80	6.6	0.3						0.3
−90								

Lat.	Sep. 23	Oct. 16	Nov. 8	Nov. 30	Dec. 22	Jan. 13	Feb. 4	Feb. 26
90°								
80	6.5	0.3						0.3
70	12.9	5.5	1.0				1.0	5.6
60	18.8	11.6	6.2	3.1	2.1	3.1	6.2	11.7
50	24.1	17.6	12.1	8.7	7.5	8.7	12.3	17.8
40	28.7	23.1	18.2	14.8	13.5	14.9	18.4	23.5
30	32.5	28.2	23.9	20.9	19.8	21.0	24.1	28.4
20	35.3	32.3	29.1	26.6	25.7	26.7	29.3	32.7
10	37.0	35.5	33.5	31.8	31.1	31.9	33.8	35.9
0	37.6	37.6	36.9	36.1	35.8	36.3	37.3	38.0
−10	37.0	38.6	39.4	39.5	39.6	39.7	39.7	39.1
−20	35.3	38.5	40.7	42.0	42.4	42.2	41.1	39.0
−30	32.5	37.2	40.9	43.3	44.2	43.5	41.3	37.7
−40	28.7	34.8	40.1	43.6	45.0	43.8	40.5	35.2
−50	24.1	31.5	38.3	43.1	44.8	43.2	38.6	31.9
−60	18.8	27.3	35.7	41.9	44.4	42.1	36.0	27.7
−70	12.9	22.6	33.0	42.2	45.9	42.4	33.3	22.9
−80	6.5	18.2	33.5	44.2	48.1	44.4	33.8	18.4
−90		18.2	34.0	44.8	48.8	45.1	34.4	18.4

INFRARED ABSORPTION BY THE EARTH'S ATMOSPHERE

This graph summarizes the absorption by various atmospheric constituents in the wavelength range from 1 to 13 μm (wavenumber range 10,000 to 770 cm⁻¹). The vertical scale is in arbitrary units and does not take into account the wavelength variation of either the solar background radiation or the infrared detector response. Thus the intensities of the absorption bands have only qualitative significance.

WAVELENGTH IN MICROMETERS

ATMOSPHERIC CONCENTRATION OF CARBON DIOXIDE, 1958-1996

The data in this table were taken at Mauna Loa Observatory in Hawaii and represent averages adjusted to the 15th of each month. The concentration of CO_2 is given in parts per million by volume. Data from other measurement sites may be found in the references.

REFERENCES

1. Keeling, C. D., and Whorf, T. P., Carbon Dioxide Information Analysis Center WWW site <http://cdiac.esd.ornl.gov/trends_html/trends/co2>, August 1997.
2. Keeling, C. D., and Whorf, T. P., in *Trends '93: A Compendium of Data on Global Change*, p.16, Boden, T. A., Kaiser, D. P., Sepanski, R. J., and Stoss, F. W., Editors, ORNL/CDIAC-65, Oak Ridge National Laboratory, Oak Ridge, TN, 1994.

Year	Jan.	Feb.	Mar.	Apr.	May	June	July	Aug.	Sept.	Oct.	Nov.	Dec.	Annual
1958	N/A	N/A	315.56	317.29	317.34	N/A	315.69	314.78	313.05	N/A	313.18	314.50	N/A
1959	315.42	316.31	316.50	317.56	318.13	318.00	316.39	314.65	313.68	313.18	314.66	315.43	315.83
1960	316.27	316.81	317.42	318.87	319.87	319.43	318.01	315.74	314.00	313.68	314.84	316.03	316.75
1961	316.73	317.54	318.38	319.31	320.42	319.61	318.42	316.63	314.84	315.16	315.94	316.85	317.49
1962	317.78	318.40	319.53	320.42	320.85	320.45	319.45	317.25	316.11	315.27	316.53	317.53	318.30
1963	318.58	318.93	319.70	321.22	322.08	321.31	319.58	317.61	316.05	315.83	316.91	318.20	318.83
1964	319.41	N/A	N/A	N/A	322.05	321.73	320.27	318.53	316.54	316.72	317.53	318.55	N/A
1965	319.27	320.28	320.73	321.97	322.00	321.71	321.05	318.71	317.66	317.14	318.70	319.25	319.87
1966	320.46	321.43	322.23	323.54	323.91	323.59	322.24	320.20	318.48	317.94	319.63	320.87	321.21
1967	322.17	322.34	322.88	324.25	324.83	323.93	322.38	320.76	319.10	319.24	320.56	321.80	322.02
1968	322.40	322.99	323.73	324.86	325.40	325.20	323.98	321.95	320.18	320.09	321.16	322.74	322.89
1969	323.83	324.27	325.47	326.50	327.21	326.54	325.72	323.50	322.22	321.62	322.69	323.95	324.46
1970	324.89	325.82	326.77	327.97	327.90	327.50	326.18	324.53	322.93	322.90	323.85	324.96	325.52
1971	326.01	326.51	327.02	327.62	328.76	328.40	327.20	325.27	323.20	323.40	324.63	325.85	326.16
1972	326.60	327.47	327.58	329.56	329.90	328.92	327.88	326.16	324.68	325.04	326.34	327.39	327.29
1973	328.37	329.40	330.14	331.33	332.31	331.90	330.70	329.15	327.35	327.02	327.99	328.48	329.51
1974	329.18	330.55	331.32	332.48	332.92	332.08	331.01	329.23	327.27	327.21	328.29	329.41	330.08
1975	330.23	331.25	331.87	333.14	333.80	333.43	331.73	329.90	328.40	328.17	329.32	330.59	330.99
1976	331.58	332.39	333.33	334.41	334.71	334.17	332.89	330.77	329.14	328.78	330.14	331.52	331.98
1977	332.75	333.24	334.53	335.90	336.57	336.10	334.76	332.59	331.42	330.98	332.24	333.68	333.73
1978	334.80	335.22	336.47	337.59	337.84	337.72	336.37	334.51	332.60	332.38	333.75	334.78	335.34
1979	336.05	336.59	337.79	338.71	339.30	339.12	337.56	335.92	333.75	333.70	335.12	336.56	336.68
1980	337.84	338.19	339.91	340.60	341.29	341.00	339.39	337.43	335.72	335.84	336.93	338.04	338.52
1981	339.06	340.30	341.21	342.33	342.74	342.08	340.32	338.26	336.52	336.68	338.19	339.44	339.76
1982	340.57	341.44	342.53	343.39	343.96	343.18	341.88	339.65	337.81	337.69	339.09	340.32	340.96
1983	341.20	342.35	342.93	344.77	345.58	345.14	343.81	342.21	339.69	339.82	340.98	342.82	342.61
1984	343.52	344.33	345.11	346.89	347.25	346.62	345.22	343.11	340.90	341.18	342.80	344.04	344.25
1985	344.79	345.82	347.25	348.17	348.74	348.07	346.38	344.52	342.92	342.62	344.06	345.38	345.73
1986	346.11	346.78	347.68	349.37	350.03	349.37	347.76	345.73	344.68	343.99	345.48	346.72	346.97
1987	347.84	348.29	349.24	350.80	351.66	351.08	349.33	347.92	346.27	346.18	347.64	348.78	348.75
1988	350.25	351.54	352.05	353.41	354.04	353.63	352.22	350.27	348.55	348.72	349.91	351.18	351.31
1989	352.60	352.92	353.53	355.26	355.52	354.97	353.75	351.52	349.64	349.83	351.14	352.37	352.75
1990	353.50	354.55	355.23	356.04	357.00	356.07	354.67	352.76	350.82	351.04	352.70	354.07	354.04
1991	354.59	355.63	357.03	358.48	359.22	358.12	356.06	353.92	352.05	352.11	353.64	354.89	355.48
1992	355.88	356.63	357.72	359.07	359.58	359.17	356.94	354.92	352.94	353.23	354.09	355.33	356.29
1993	356.63	357.10	358.32	359.41	360.23	359.55	357.53	355.48	353.67	353.95	355.30	356.78	356.99
1994	358.34	358.89	359.95	361.25	361.67	360.94	359.55	357.49	355.84	356.00	357.59	359.05	358.88
1995	359.98	361.03	361.66	363.48	363.82	363.30	361.93	359.49	358.08	357.77	359.57	360.69	360.90
1996	362.03	363.22	363.99	364.66	365.36	364.91	363.58	361.41	359.37	359.50	360.65	362.21	362.57

CO$_2$ Concentration in Parts per Million (Annual Average)

MEAN TEMPERATURES IN THE UNITED STATES, 1900-1992

Historical records of atmospheric temperatures have been analyzed to obtain mean temperatures in °C for 23 climatically distinct regions of the United States. The table below gives the average over these 23 regions, which cover completely the contiguous 48 states. Data for the individual regions and for other parts of the world may be found in the references.

The data are presented as temperature anomalies, i.e., as deviations (in °C) from the average temperature at each individual recording station over a 1961-1990 reference period. The trend in the temperature anomaly thus gives an indication of the long-term variation in average temperatures.

CY Mean: Calendar year mean (January-December)
Winter: December-February
Spring: March-May
Summer: June-August
Fall: September-November

REFERENCES

1. Karl, T. R., Easterling, D. R., Knight, R. W., and Hughes, P. Y., in *Trends '93: A Compendium of Data on Global Change*, p. 686, Boden, T. A., Kaiser, D. P., Sepanski, R. J., and Stoss, F. W., Editors, ORNL/CDIAC-65, Oak Ridge National Laboratory, Oak Ridge, TN, 1994.
2. Carbon Dioxide Information Analysis Center, WWW site <http://cdiac.esd.ornl.gov/ftp/trends93>.

Year	CY Mean	Winter	Spring	Summer	Fall
1900	0.46		0.21	0.27	0.85
1901	-0.21	0.07	-0.46	0.48	-0.28
1902	-0.14	-0.69	0.48	-0.58	0.12
1903	-0.77	-0.84	0.10	-0.83	-1.02
1904	-0.72	-1.86	-0.39	-0.92	-0.21
1905	-0.45	-1.86	0.67	-0.32	-0.25
1906	-0.04	0.23	-0.69	-0.42	-0.10
1907	-0.23	1.09	-0.61	-0.85	-0.39
1908	-0.11	0.73	0.36	-0.64	-0.59
1909	-0.36	0.82	-1.06	0.06	0.01
1910	-0.14	-2.08	0.95	-0.55	0.13
1911	0.02	0.52	0.20	-0.19	-0.62
1912	-0.88	-1.50	-0.75	-0.76	-0.51
1913	-0.23	-0.74	-0.65	-0.07	0.22
1914	-0.05	0.48	0.04	0.14	0.32
1915	-0.11	-0.37	-0.18	-1.16	0.31
1916	-0.77	-0.29	-0.36	-0.30	-1.23
1917	-1.34	-1.93	-1.75	-0.73	-1.04
1918	-0.14	-2.02	0.30	0.03	-0.09
1919	-0.16	0.69	0.00	0.12	-0.13
1920	-0.37	-0.83	-0.96	-0.67	0.02
1921	0.87	1.56	0.83	0.52	0.32
1922	0.01	-0.44	0.15	0.16	0.41
1923	-0.10	0.23	-1.02	-0.07	-0.09
1924	-1.01	0.13	-1.27	-0.64	-0.53
1925	0.20	-0.44	0.57	0.05	-0.41
1926	-0.01	0.97	-0.58	-0.27	0.02
1927	0.20	1.11	0.41	-0.83	0.83
1928	-0.08	-0.40	-0.28	-0.43	-0.08
1929	-0.68	-1.94	0.30	-0.39	-0.78
1930	-0.12	0.07	0.09	-0.07	-0.25
1931	0.81	1.16	-0.71	0.51	1.41
1932	-0.14	1.75	-0.58	0.26	-0.78
1933	0.35	-0.60	-0.08	0.45	0.54
1934	0.86	1.45	0.77	0.86	0.82
1935	0.12	0.84	0.12	0.31	-0.47
1936	-0.10	-2.23	0.48	1.00	-0.32
1937	-0.13	-0.65	-0.24	0.66	-0.03
1938	0.71	1.31	0.98	0.39	0.08
1939	0.38	0.36	0.34	0.18	0.15
1940	0.06	0.03	-0.10	0.18	0.12

MEAN TEMPERATURES IN THE UNITED STATES, 1901-1992 (continued)

Year	CY Mean	Winter	Spring	Summer	Fall
1941	0.79	1.56	0.44	0.28	0.85
1942	0.01	0.35	0.22	0.08	0.06
1943	-0.17	0.20	-0.46	0.54	-0.79
1944	0.04	0.61	-0.38	-0.20	0.43
1945	-0.02	0.30	0.25	-0.47	0.08
1946	0.53	-0.26	1.21	-0.21	0.18
1947	0.10	0.47	-0.42	0.03	0.85
1948	-0.22	-0.67	-0.01	0.02	-0.15
1949	0.05	-0.77	0.39	0.33	0.23
1950	-0.37	0.39	-1.02	-0.86	0.13
1951	-0.45	0.05	-0.69	-0.30	-0.70
1952	0.03	0.68	-0.40	0.49	-1.22
1953	0.61	1.91	0.03	0.31	0.31
1954	0.57	1.47	-0.22	0.36	0.70
1955	-0.25	-0.33	0.12	0.19	-0.60
1956	-0.05	-0.16	-0.42	0.01	-0.37
1957	0.45	1.02	0.44	0.25	-0.13
1958	0.14	0.93	0.06	0.11	0.45
1959	0.12	-0.60	0.16	0.50	-0.52
1960	-0.27	0.43	-0.98	0.01	0.56
1961	-0.01	-0.02	-0.17	0.18	-0.29
1962	-0.04	-0.52	-0.09	-0.41	0.65
1963	-0.06	-1.35	0.61	0.12	1.38
1964	-0.27	-1.30	-0.24	-0.08	-0.40
1965	-0.03	-0.07	-0.56	-0.42	0.28
1966	-0.34	-0.31	-0.25	-0.07	-0.13
1967	-0.13	0.23	-0.05	-0.37	-0.32
1968	-0.28	-0.31	-0.16	-0.12	0.01
1969	-0.06	-0.36	-0.54	0.12	-0.16
1970	-0.12	-0.17	-0.44	0.32	-0.07
1971	-0.04	-0.08	-0.99	0.01	0.56
1972	-0.13	0.20	0.26	-0.19	0.05
1973	0.48	-0.23	0.47	0.22	0.90
1974	0.16	0.52	0.74	-0.32	-0.39
1975	-0.09	0.63	-0.79	0.03	-0.20
1976	-0.62	0.88	-0.09	-0.54	-1.72
1977	0.32	-1.95	1.07	0.60	0.68
1978	-0.37	-1.31	0.23	0.09	0.21
1979	-0.53	-2.92	0.09	-0.21	-0.07
1980	0.18	0.72	-0.25	0.43	-0.04
1981	0.64	0.90	0.80	0.57	0.36
1982	-0.08	-0.86	0.03	-0.11	0.06
1983	0.40	2.33	-0.36	0.58	0.94
1984	0.21	-0.78	-0.30	0.31	0.07
1985	-0.26	-0.78	1.24	-0.23	0.05
1986	0.93	0.22	1.22	0.45	0.49
1987	0.67	1.52	0.97	0.33	-0.17
1988	-0.07	-0.26	-0.06	0.57	0.04
1989	-0.30	-0.28	0.36	0.12	-0.27
1990	0.72	0.41	0.72	0.41	0.66
1991	0.77	0.32	1.36	0.56	-0.31
1992		2.48	0.82	-0.70	

GLOBAL TEMPERATURE TREND, 1866-1996

This table and graph summarize the trend in annual mean global temperature from 1886 to 1996. The temperatures were calculated from mean temperature anomalies by assuming an absolute global mean of 14.00 °C (which is the best estimate for the 1951-1980 period). The estimated uncertainty in the year-to-year change in the recent years is about 0.07 °C.

REFERENCES

1. NASA Goddard Institute for Space Studies, World Wide Web site, www.giss.nasa.gov/data/gistemp/
2. Hansen, J., and Wilson, H., *Climatic Change*, 25, 285, 1993.

Year	$t/°C$	Year	$t/°C$	Year	$t/°C$	Year	$t/°C$	Year	$t/°C$
1866	13.52	1897	13.85	1928	14.03	1959	14.05	1990	14.47
1867	13.42	1898	13.63	1929	13.82	1960	14.00	1991	14.42
1868	13.52	1899	13.78	1930	13.93	1961	14.08	1992	14.13
1869	13.64	1900	14.00	1931	14.03	1962	14.03	1993	14.21
1870	13.57	1901	13.92	1932	14.02	1963	14.02	1994	14.31
1871	13.60	1902	13.76	1933	13.86	1964	13.74	1995	14.47
1872	13.76	1903	13.64	1934	14.02	1965	13.85	1996	14.36
1873	13.81	1904	13.51	1935	13.93	1966	13.91		
1874	13.51	1905	13.73	1936	14.02	1967	13.98		
1875	13.41	1906	13.81	1937	14.12	1968	13.88		
1876	13.58	1907	13.57	1938	14.13	1969	14.03		
1877	13.89	1908	13.70	1939	13.97	1970	14.04		
1878	13.98	1909	13.70	1940	14.14	1971	13.89		
1879	13.50	1910	13.71	1941	14.12	1972	13.93		
1880	13.68	1911	13.71	1942	14.09	1973	14.19		
1881	13.67	1912	13.72	1943	14.03	1974	13.93		
1882	13.61	1913	13.77	1944	14.11	1975	13.95		
1883	13.61	1914	13.96	1945	13.96	1976	13.78		
1884	13.35	1915	13.98	1946	14.01	1977	14.16		
1885	13.58	1916	13.76	1947	14.12	1978	14.09		
1886	13.60	1917	13.58	1948	13.99	1979	14.14		
1887	13.48	1918	13.65	1949	13.93	1980	14.28		
1888	13.70	1919	13.83	1950	13.85	1981	14.38		
1889	13.88	1920	13.83	1951	13.98	1982	14.07		
1890	13.60	1921	13.90	1952	14.04	1983	14.29		
1891	13.58	1922	13.87	1953	14.13	1984	14.11		
1892	13.61	1923	13.84	1954	13.93	1985	14.11		
1893	13.55	1924	13.85	1955	13.94	1986	14.16		
1894	13.65	1925	13.85	1956	13.83	1987	14.32		
1895	13.64	1926	14.10	1957	14.10	1988	14.35		
1896	13.81	1927	13.96	1958	14.10	1989	14.25		

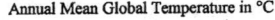

Annual Mean Global Temperature in °C

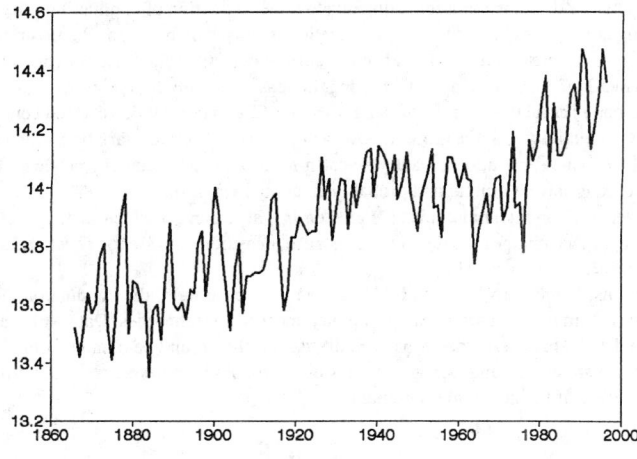

ATMOSPHERIC ELECTRICITY
Hans Dolezalek, Hannes Tammet, John Latham, and Martin A. Uman

I. SURVEY AND GLOBAL CIRCUIT
Hans Dolezalek

The science of atmospheric electricity originated in 1752 by an experimental proof of a related earlier hypothesis (that lightning is an electrical event). In spite of a large effort, in part by such eminent physicists as Coulomb, Lord Kelvin, and many others, an overall, proven theory able to generate models with sufficient resolution is not yet available. Generally accepted and encompassing text books are now more than 20 years old. The voluminous proceedings of the, so far, nine international atmospheric electricity conferences (1954 to 1992) give much valuable detail and demonstrate impressive progress, as do a number of less comprehensive textbooks published in the last 20 years, but a general theory as indicated above is not yet created. Only now, certain related measuring techniques and mathematical possibilities are emerging.

Applications to practical purposes do exist in the field of lightning research (including the electromagnetic radiation emanating from lightning) by the establishment of lightning-location networks and by the now developing possibility to detect electrified clouds which pose hazards to aircraft. Application of atmospheric electricity to other parts of meteorology seems to be promising but so far has seldom been instituted. Because some atmospheric electric signals propagate around the earth and because of the existence of a global circuit, applications for the monitoring of global change processes and conditions are now being proposed. Significant secular changes in the global circuit would indicate a change in the global climate; the availability of many old data (about a span of 100 years) could help detect a long term trend.

The concept of the "global circuit" is based on the theory of the global spherical capacitor: both, the solid (and liquid) earth as one electrode, and the high atmospheric layers (about the ionosphere) as the other, are by orders of magnitude more electrically conductive than the atmosphere between them. According to the "classical picture of atmospheric electricity", this capacitor is continuously charged by the common action of all thunderstorms to a d.c. voltage difference of several hundred kilovolts, the earth being negative. The much smaller but still existing conductivity of the atmosphere allows a current flowing from the ionosphere to the ground, integrated for all sink areas of the whole earth, of the order of 1.5 kA. In this way, a global circuit is created with many generators and sink-areas both interspaced and distributed over the whole globe, all connected to two nodes: ionosphere and ground. Within the scope of the global circuit, for each location, the current density (order of several pA/m^2) is determined by the voltage difference between ionosphere and ground (which is the same for all locations but varying in time) and the columnar resistance reaching from the ground up to the ionosphere (in the order of $10^{17}\ \Omega\ m^2$).

Natural processes, especially meteorological processes and some human activity, which produce or move electric charges ("space charges") or affect the ion distribution, constitute local generators and thereby "local circuits", horziontally and/or in parallel or antiparallel to the local part of the global circuit. In many cases, the local currents are much stronger than the global ones, making the measurement of the global current at a given location and/or during a period of time very difficult or, often, impossible. The strongest local circuits usually occur with certain weather conditions (precipitation, fog, high wind, blown-up dust or snow, heavy cloudiness) which make measurement of the global circuit impossible everywhere; but even in their absence local generators exist in varying magnitudes and of different characters. The separation of the local and global shares in the measured values of current density is a central problem of the science of atmospheric electricity. Aerological measurements are of high value in this regard.

The above description is within the "classical picture" of atmospheric electricity, a group of hypotheses to explain the electricifation of the atmosphere. It is probably fundamentally correct but certainly not complete; it has not yet been confirmed by systems of measurements resulting in no inner contradictions. In particular, extraterrestrial influences must be permitted; their general significance is still under debate.

Within this "classical picture" a kind of electric standard atmosphere may be constructed as shown in Table 1.

Values with a star, *, are rough average values from measurement. A star in parentheses, (*), points to a typical value from one or a few measurements. All other values have been calculated from starred values, under the assumption that at 2 km 50% and at 12 km 90% of the columnar resistance is reached. Voltage drop along one of the partial columns can be calculated by subtracting the value for the lower column from that of the upper one. Columnar resistances, conductances, and capacitances are valid for that particular part of the column which is indicated at the left. Capacitances are calculated with the formula for plate capacitors, and this fact must be considered also for the time constants for columns.

According to measurements, U, the potential difference between 0 m and 65 km may vary by a factor of approximately 2. The total columnar resistance, R_c, is estimated to vary up to a factor of 3, the variation being due to either reduction of conductivity in the exchange layer (about lowest 2 km of this table) or to the presence of high mountains; in both cases the variation is caused in the troposphere. Smaller variations in the stratosphere and mesosphere are being discussed because of aerosols there. The air-earth current density in fair weather varies by a factor of 3 to 6 accordingly. Conductivity near the ground varies by a factor of about 3 but only decreasing; increase of conductivity due to extraordinary radioactivity is a singular event. The field strength near the ground varies as a consequence of variations of air-earth current density and conductivity from about 1/3 to about 10 times of the value quoted in the table. Conductivity near the ground shows a diurnal and an annual variation which depends strongly on the locality: air-earth current density shows a diurnal and annual variation because the earth-ionosphere potential difference undergoes such variations, and also because the columnar resistance is supposed to have a diurnal and probably an annual variation.

Conductivities and air-earth current densities on high mountains are greater than at sea level by factors of up to 10. Conductivity decreases when atmospheric humidity increases. Values for space charges are not quoted because measurements are too few to allow calculation of average values. Values of parameters over the oceans are still rather uncertain.

Theoretically, in fair-weather conditions, Ohm's law must be fulfilled for the electric field, the conduction current density, and the electrical conductivity of the atmosphere. Deviations point to shortcomings in the applied measuring techniques. Data which are representative for a large area (in the extreme, "globally representative data", i.e. data on the global circuit), can on the ground be obtained only by stations on an open plane and only if local generators are either small or constant or are independently measured. Certain measurements with instrumented aircraft provide globally representative information valid for the period of the actual measurement.

TABLE 1

Electrical Parameters of the Clear (Fair-Weather) Atmosphere, Pertinent to the Classical Picture of Atm. Electricity (Electric Standard Atmosphere)

Part of atmosphere for which the values are calculated (elements are in free, cloudless atmosphere)	Currents, I, in A; and current densities, i, in A/m²	Potential differences, U, in V; field strength E in V/m; $U = 0$ at sea level	Resistances, R, in Ω; columnar resistances, R_c in Ω m² and resistivities, ρ, in Ω m	Conductances, G, in Ω^{-1}; columnar conductances G_c in Ω^{-1} m⁻²; total conductivities, γ, in Ω^{-1} m⁻¹	Capacitances, C, in F; columnar capacitances, C_c, in F m⁻² and capacitivities, ε, in F m⁻¹	Time constants τ, in seconds
Volume element at about sea level, 1 m³	$i = 3 \times 10^{-12}$*	$E_0 = 1.2 \times 10^2$*	$\rho_0 = 4 \times 10^{13}$	$\gamma_0 = 2.5 \times 10^{-14}$	$\varepsilon_0 = 8.9 \times 10^{-12}$*	$\tau_0 = 3.6 \times 10^2$
Lower column of 1 m² cross section from sea level to 2 km height	$i = 3 \times 10^{-12}$	At upper end: $U_l = 1.8 \times 10^5$	$R_{cl} = 6 \times 10^{16}$	$G_{cl} = 1.7 \times 10^{-17}$	$C_{cl} = 4.4 \times 10^{-15}$	$\tau_{cl} = 2.6 \times 10^2$
Volume element at about 2 km height, 1 m³	$i = 3 \times 10^{-12}$	$E_2 = 6.6 \times 10^1$	$\rho_2 = 2.2 \times 10^{13(*)}$	$\gamma_2 = 4.5 \times 10^{-14}$	$\varepsilon_2 = 8.9 \times 10^{-12}$	$\tau_2 = 2 \times 10^2$
Center column of 1 m² cross section from 2 to 12 km	$i = 3 \times 10^{-12}$	At upper end: $U_m = 3.15 \times 10^5$	$R_{cm} = 4.5 \times 10^{16}$	$G_{cm} = 5 \times 10^{-17}$	$C_{cm} = 8.8 \times 10^{-16}$	$\tau_{cm} = 1.8 \times 10^1$
Volume element at about 12 km height, 1 m³	$i = 3 \times 10^{-12}$	$E_{12} = 3.9 \times 10^0$	$\rho_{12} = 1.3 \times 10^{12(*)}$	$\gamma_{12} = 7.7 \times 10^{-13}$	$\varepsilon_{12} = 8.9 \times 10^{-12}$	$\tau_{12} = 1.2 \times 10^1$
Upper column of 1 m² cross section from 12 to 65 km height	$i = 3 \times 10^{-12}$	At upper end: $U_u = 3.5 \times 10^5$	$R_{cu} = 1.5 \times 10^{16}$	$G_{cu} = 2.5 \times 10^{-17}$	$C_{cu} = 1.67 \times 10^{-16}$	$\tau_{cm} = 6.7 \times 10^0$
Whole column of 1 m² cross section from 0 to 65 km height	$i = 3 \times 10^{-12}$	At upper end: $U = 3.5 \times 10^5$	$R_c = 1.2 \times 10^{17}$	$G_c = 8.3 \times 10^{-18}$	$C_c = 1.36 \times 10^{-16}$	$\tau_c = 1.64 \times 10^1$
Total spherical capacitor area: 5×10^{14} m²	$1 = 1.5 \times 10^3$	$U = 3.5 \times 10^5$*	$R = 2.4 \times 10^2$	$G = 4.2 \times 10^{-3}$	$C = 6.8 \times 10^{-2}$	$\tau = 1.64 \times 10^1$

Note: All currents and fields listed are part of the global circuit, i.e., circuits of local generators are not included. Values are subject to variations due to latitude and altitude of the point of observation above sea level, locality with respect to sources of disturbances, meteorological and climatological factors, and man-made changes. For more explanations, see text.

II. AIR IONS
Hannes Tammet

The term "air ions" signifies all airborne particles which are the carriers of the electrical current in the air and have drift velocities determined by the electric field.

The probability of electrical dissociation of molecules in the atmospheric air under thermodynamic equilibrium is near to zero. The average ionization at the ground level over the ocean is $2 \cdot 10^6$ ion pairs $m^{-3}s^{-1}$. This ionization is produced mainly by cosmic rays. Over the continents the ionizing radiation from soil and from radioactive substances in the air each add about $4 \cdot 10^6$ $m^{-3}s^{-1}$. The total average ionization rate of 10^7 $m^{-3}s^{-1}$ is equivalent to 17 µR/h which is a customary expression of the background level of the ionizing radiations. The ionization rate over the ground varies in space due to the radioactivity of soil, and in time depending on the exchange of air between the atmosphere and radon-containing soil. Radioactive pollution increases the ionization rate. A temporary increase of about 10 times was registered in Sweden after the Chernobyl accident in 1986. The emission of Kr^{85} from nuclear power plants can noticeably increase the global ionization rate in the next century. The ionization rate decreases with altitude near the ground and increases at higher altitudes up to 15 km, where it has a maximum about $5 \cdot 10^7$ $m^{-3}s^{-1}$. Solar X-ray and extreme UV radiation cause a new increase at altitudes over 60 km.

Local sources of air ions are point discharges in strong electric fields, fluidization of charged drops from waves, etc.

The enhanced chemical activity of an ion results in a chain of ion-molecule reactions with the colliding neutrals, and, in the first microsecond of the life of an air ion, a charged molecular cluster called the *cluster ion* is formed. According to theoretical calculations in the air free from exotic trace gases the following cluster ions should be dominant:

$$NO_3^- \cdot (HNO_3) \cdot H_2O, \ NO_2^- \cdot (H_2O)_2, \ NO_3^- \cdot H_2O, \ O_2^- \cdot (H_2O)_4, \ O_2^- \cdot (H_2O)_5,$$
$$H_3O^+ \cdot (H_2O)_6, \ NH_4^+ \cdot (H_2O)_2, \ NH_4^+ \cdot (H_2O), \ H_3O^+ \cdot (H_2O)_5, \ NH_4^+ \cdot NH_3$$

A measurable parameter of air ions is the electrical mobility k, characterizing the drift velocity in the unit electric field. The mobility is inversely proportional to the density of air, and the results of measurements are as a rule reduced to normal conditions. According to mobility the air ions are called: fast or small or light ions with mobility $k > 5 \cdot 10^{-5}$ $m^2V^{-1}s^{-1}$, intermediate ions, and slow or large or heavy ions with mobility $k < 10^{-6}$ $m^2V^{-1}s^{-1}$. The boundary between intermediate and slow ions is conventional.

Cluster ions are fast ions. The masses of cluster ions may be measured with mass spectrometers, but the possible ion-molecule reactions during the passage of the air through nozzles to the vacuum chamber complicate the measurement. Mass and mobility of cluster ions are highly correlated. The experimental results[5] can be expressed by the empirical formula

$$m \approx \frac{850 \ u}{\left[0.3 + k / \left(10^{-4} \, m^2 V^{-1} s^{-1}\right)\right]^3}$$

where u is the unified atomic mass unit.

The value of the transport cross-section of a cluster ion is needed to calculate its mobility according to the kinetic theory of Chapman and Enskog. The theoretical estimation of transport cross-sections is rough and cannot be used to identify the chemical structures of cluster ions. Mass spectrometry is the main technique of identification of cluster ions.[2]

Märk and Castleman (1986) presented an overview of over 1000 publications on the experimental studies of cluster ions. Most of them present information about ions of millisecond age range. The low concentration makes it difficult to get detailed information about masses and mobilities of the natural atmospheric ions at ground level. The results of a 1-year continuous measurement[6] are as follows:

	+ ions	- ions	unit
Average mobility	1.36	1.56	10^{-4} $m^2V^{-1}s^{-1}$
The corresponding mass	190	130	u
The corresponding diameter	0.69	0.61	nm
The average concentration	400	360	10^6 m^{-3}
The corresponding conductivity	8.7	9.0	fS

The distribution of tropospheric cluster ions according to the mobility and estimated mass is depicted in Figure 1.

The problems and results of direct mass spectrometry of natural cluster ions are analysed by Eisele (1986) for ground level and by Meyerott, Reagan and Joiner (1980) for stratospheric measurements. Air ions in the high atmosphere are a subject of ionospheric physics.

During its lifetime (about 1 min), a cluster ion at ground level collides with nearly 10^{12} molecules. Thus the cluster ions are able to concentrate trace gases of very low concentration if they have an extra high electron or proton affinity. For example, Eisele (1986) demonstrated that a considerable fraction of positive atmospheric cluster ions in the unpolluted atmosphere at ground level probably consist of a molecule derived from pyridine. The concentration of these constituents is estimated to be about 10^{-12}. Therefore, air-ion mass and mobility spectrometry is considered as a promising technique for trace analysis in the air. Mass and mobility spectrometry of millisecond-age air ions has been developed as a technique of chemical analysis known as "plasma chromatography" (Carr, 1984). The sensitivity of the detection grows with the age of the cluster ions measured.

The mechanisms of annihilation of cluster ions are ion-ion recombination (on the average 3%) and sedimentation on aerosol particles (on the average 97% of cluster ions at ground level). The result of the combination of a cluster ion and neutral particle is a charged particle called an *aerosol ion*. In conditions of detailed thermodynamic equilibrium the probability that a spherical particle of diameter d carries q elementary charges is calculated from the Boltzmann distribution:

$$p_q(d) = (2 \ \pi d/d_0)^{1/2} \exp(-q^2 \ d_0/2d)$$

Figure 1. Average mobility and mass spectra of natural tropospheric cluster ions. Concentrations of the mobility fractions were measured in a rural site every 5 min over 1 year.[6] Ion mass is estimated according to the above empirical formula.

where $d_0 = 115$ nm (at 18°C). The supposition about the detailed equilibrium is an approximation and the formula is not valid for particles less than d_0. On the basis of numerical calculations by Hoppel and Frick (1990) the following charge probabilities can be derived:

d	3	10	30	100	300	1000	3000	nm
p_0	98	90	70	42	24	14	8	%
$p_{-1} + p_1$	2	10	30	48	41	25	15	%
$p_{-2} + p_2$	0	0	0	10	23	21	14	%
$p_{q>2}$	0	0	0	0	12	40	63	%
k_1	15000	1900	250	28	5.1	1.11	0.33	$10^{-9} m^2 V^{-1} s^{-1}$

The last line of the table presents the mobility of a particle carrying one elementary charge. The distribution of the atmospheric aerosol ions over mobility is demonstrated in Figure 2.

Although the concentration of aerosol in continental air at ground level is an order of magnitude higher than the concentration of cluster ions, the mobilities of aerosol ions are so small that their percentage in air conductivity is less than 1%.

A specific class of aerosol ions are condensed aerosol ions produced as a result of the condensation of gaseous matter on the cluster ions. In aerosol physics the process is called ion-induced nucleation; it is considered as one among the processes of gas-to-particle conversion. The condensed aerosol ions have an inherent charge. Their sizes and mobilities are between the sizes and mobilities of cluster ions and of ordinary aerosol ions. Water and standard constituents of atmospheric air are not able to condense on the cluster ions in the real atmosphere. Thus the concentration of condensed aerosol ions depends on the trace constituents in the air and is very low in unpolluted air. Knowledge about condensed aerosol ions is poor because of measurement difficulties.

Figure 2. Mobility and size spectra of tropospheric aerosol ions.[6] The wide bars mark the fraction concentrations theoretically estimated on the basis of the standard size distribution of tropospheric aerosol. The pin bars with head + and - mark average values of positive and negative aerosol ion fraction concentrations measured in a rural site every 5 min during 4 months.

REFERENCES

1. Carr, T. W., Ed., *Plasma Chromatography*, Plenum Press, New York and London, XII + 259 pp., 1984.
2. Eisele, F. L., Identification of Tropospheric Ions, *J. Geophys. Res.*, vol. 91, no. D7, pp. 7897-7906, 1986.

3. Hoppel, W. A., and Frick, G. M., The Nonequilibrium Character of the Aerosol Charge Distributions Produced by Neutralizers, *Aerosol Sci. Technol.*, vol. 12, no. 3, pp. 471-496, 1990.

4. Mark, T. D., and Castleman, A. W., Experimental Studies on Cluster Ions, in *Advances in Atomic and Molecular Physics*, vol. 20, pp. 65.-172, Academic Press, 1985.

5. Meyerott, R. E., Reagan, J. B., and Joiner, R. G., The Mobility and Concentration of Ions and the Ionic Conductivity in the Lower Stratosphere, *J. Geophys. Res.*, vol. 85, no. A3, pp. 1273-1278, 1980.

6. Salm, J., Tammet, H., Iher, H., and Hörrak, U., Atmospheric Electrical Measurements in Tahkuse, Estonia (in Russian), in *Voprosy Atmosfernogo Elektrichestva*, pp. 168-175, Gidrometeoizdat, Leningrad, 1990.

III. THUNDERSTORM ELECTRICITY
John Latham

The development of improved radar techniques and instruments for in-cloud electrical and physical measurements, coupled with a much clearer recognition by the research community that establishment of the mechanism or mechanisms responsible for electric field development in thunderclouds, culminating in lightning, is inextricably linked to the concomitant dynamical and microphysical evolution of the clouds, has led to significant progress over the past decade.

Field studies indicate that in most thunderclouds the electrical development is associated with the process of glaciation, which can occur in a variety of incompletely understood ways. In the absence of ice, field growth is slow, individual hydrometeor charges are low, and lightning is produced only rarely. Precipitation — in the solid form, as graupel — also appears to be a necessary ingredient for significant electrification, as does significant convective activity and mixing between the clouds and their environments, via entrainment.

Increasingly, the view is being accepted that charge transfer leading to field-growth is largely a consequence of rebounding collisions between graupel pellets and smaller vapor-grown ice crystals, followed by the separation under gravity of these two types of hydrometeor. These collisions occur predominantly within the temperature range -15 to -30°C, and for significant charge transfer need to occur in the presence of supercooled cloud droplets.

The field evidence is inconsistent with an inductive mechanism, and extensive laboratory studies indicate that the principal charging mechanism is non-inductive and associated — in ways yet to be identified — with differences in surface characteristics of the interacting hydrometeors.

Laboratory studies indicate that the two most favored sites for corona emission leading to the lightning discharge are the tips of ephemeral liquid filaments, produced during the glancing collisions of supercooled raindrops, and protuberances on large ice crystals or graupel pellets. The relative importance of these alternatives will depend on the hydrometeor characteristics and the temperature in the regions of strongest fields; these features are themselves dependent on air-mass characteristics and climatological considerations.

A recently identified but unresolved question is why, in continental Northern Hemisphere thunderclouds at least, the sign of the charge brought to ground by lightning is predominantly negative in summer but more evenly balanced in winter.

IV. LIGHTNING
Martin A. Uman

From both ground-based weather-station data and satellite measurements, it has been estimated that there are about 100 lightning discharges, both cloud and ground flashes, over the whole earth each second; representing an average global lightning flash density of about 6 $km^{-2}yr^{-1}$. Most of this lightning occurs over the earth's land masses. For example, in central Florida, where thunderstorms occur about 90 days/yr, the flash density for discharges to earth is about 15 $km^{-2}yr^{-1}$. Some tropical areas of the earth have thunderstorms up to 300 days/yr.

Lightning can be defined as a transient, high-current electric discharge whose path length is measured in kilometers and whose most common source is the electric charge separated in the ordinary thunderstorm or cumulonimbus cloud. Well over half of all lightning discharges occur totally within individual thunderstorm clouds and are referred to as intracloud discharges. Cloud-to-ground lightning, however, has been studied more extensively than any other lightning form because of its visibility and its more practical interest. Cloud-to-cloud and cloud-to-air discharges are less common than intracloud or cloud-to-ground lightning.

Lightning between the cloud and earth can be categorized in terms of the direction of motion, upward or downward, and the sign of the charge, positive or negative, of the developing discharge (called a *leader*) which initiates the overall event. Over 90% of the worldwide cloud-to-ground discharges is initiated in the thundercloud by downward-moving negatively-charged leaders and subsequently results in the lowering of negative charge to earth. Cloud-to-ground lightning can also be initiated by downward-moving positive leaders, less than 10% of the worldwide cloud-to-ground lightning being of this type although the exact percentage is a function of season and latitude. Lightning between cloud and ground can also be initiated by leaders which develop upward from the earth. These upward-initiated discharges are relatively rare, may be of either polarity, and generally occur from mountaintops and tall man-made structures.

We discuss next the most common type of cloud-to-ground lightning. A negative cloud-to-ground discharge or *flash* has an overall duration of some tenths of a second and is made up of various components, among which are typically three or four high-current pulses called *strokes*. Each stroke lasts about a millisecond, the separation time between strokes being typically several tens of milliseconds. Such lightning often appears to "flicker" because the human eye can just resolve the individual light pulse associated with each stroke. A drawing of the components of a negative cloud-to-ground flash is found in Figure 3. Some values for salient parameters are found in Table 1. The negatively-charged *stepped leader* initiates the first stroke in a flash by propagating from cloud to ground through virgin air in a series of discrete steps. Photographically observed leader steps in clear

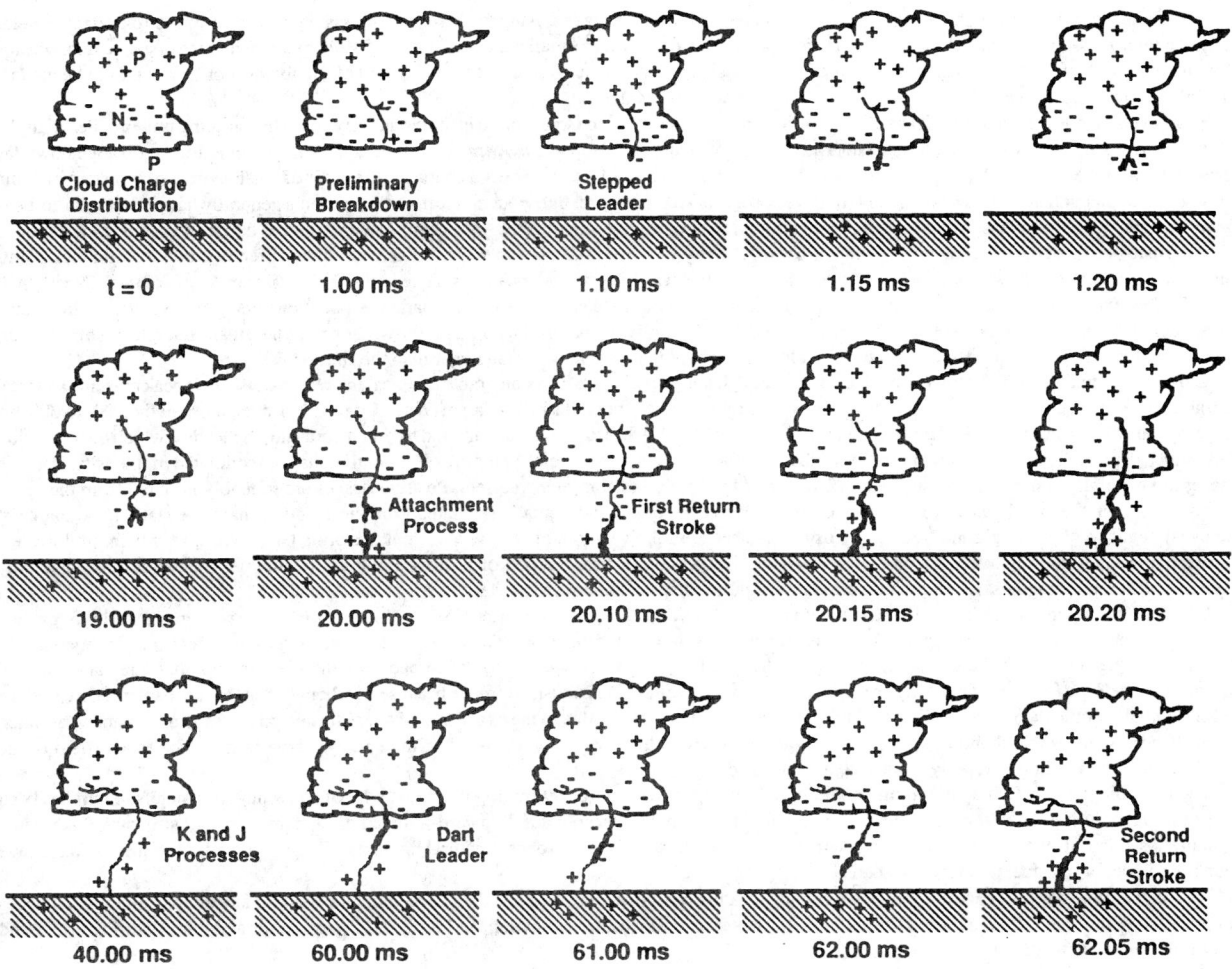

Figure 3. Sequence of steps in cloud-to-ground lightning.

air are typically 1 μs in duration and tens of meters in length, with a pause time between steps of about 50 μs. A fully developed stepped leader lowers up to 10 or more coulombs of negative cloud charge toward ground in tens of milliseconds with an average downward speed of about 2×10^5 m/s. The average leader current is in the 100 to 1000 A range. The steps have pulse currents of at least 1 kA. Associated with these currents are electric- and magnetic-field pulses with widths of about 1 μs or less and risetimes of about 0.1 μs or less. The stepped leader, during its trip toward ground, branches in a downward direction, resulting in the characteristic downward-branched geometrical structure commonly observed. The electric potential of the bottom of the negatively-charged leader channel with respect to ground has a magnitude in excess of 10^7 V. As the leader tip nears ground, the electric field at sharp objects on the ground or at irregularities of the ground itself exceeds the breakdown value of air, and one or more upward-moving discharges (often called upward leaders) are initiated from those points, thus beginning the *attachment process*. An understanding of the physics of the attachment process is central to an understanding of the operation of lightning protection of ground-based objects and the effects of lightning on humans and animals, since it is the attachment process that determines where the lightning connects to objects on the ground and the value of the early currents which flow. When one of the upward-moving discharges from the ground (or from a lightning rod or an individual) contacts the tip of the downward-moving stepped leader, typically some tens of meters above the ground, the leader tip is effectively connected to ground potential. The negatively-charged leader channel is then discharged to earth when a ground potential wave, referred to as the first *return stroke*, propagates continuously up the leader path. The upward speed of a return stroke near the ground is typically near one third the speed of light, and the speed decreases with height. The first return stroke produces a peak current near ground of typically 30 kA, with a time from zero to peak of a few microseconds. Currents measured at the ground fall to half of the peak value in about 50 μs, and currents of the order of hundreds of amperes may flow for times of a few milliseconds up to several hundred milliseconds. The longer-lasting currents are known as *continuing currents*. The rapid release of return stroke energy heats the leader channel to a temperature near 30,000 K and creates a high-pressure channel which expands and generates the shock waves that eventually become thunder, as further discussed later. The return stroke effectively lowers to ground the charge originally deposited onto the stepped-leader channel and additionally initiates the lowering of other charge which may be available to the top of its channel. First return-stroke electric fields exhibit a microsecond scale rise to peak with a typical peak value of 5 V/m, normalized to a distance of 100 km by an inverse distance relationship. Roughly half of the field rise to peak, the so-called "fast transition", takes place in tenths of a microsecond, an observation that can only be made if the field propagation is over a highly conducting surface such as salt water.

After the first return-stroke current has ceased to flow, the flash, including charge motion in the cloud, may end. The lightning is then called a single-stroke flash. On the other hand, if additional charge is made available to the top of the channel, a continuous or *dart leader* may propagate down the residual first-stroke channel at a typical speed of about 1×10^7 m/s. The dart leader lowers a charge of the order of 1 C by virtue of a current of about 1 kA. The dart leader then initiates the second (or any subsequent) return stroke. Subsequent return-stroke currents generally have faster zero-to-peak rise times than do first-stroke currents, but similar maximum rates of change, about 100 kA/μs. Some leaders begin as dart leaders, but toward the end of their trip toward ground become stepped leaders. These leaders are known as *dart-stepped leaders* and may have different ground termination points (and separate upward leaders) from the first stroke. Most often the dart-stepped leaders are associated with the second stroke of the flash. Nearly half of all flashes exhibit more than one termination point on ground with the distance between separate terminations being up to several kilometers. Subsequent return-stroke radiated electric and magnetic fields are similar to, but usually a factor of two or so smaller, than first return-stroke fields. About one third of all multiple-stroke flashes has at least one subsequent stroke which is larger than the first stroke.

Cloud-to-ground flashes that lower positive charge, though not common, are of considerable practical interest because their peak currents and total charge transfer can be much larger than for the more common negative ground flash. The largest recorded peak currents, those in the 200- to 300-kA range, are due to the return strokes of positive lightning. Such positive flashes to ground are initiated by downward-moving leaders which do not exhibit the distinct steps of their negative counterparts. Rather, they show a luminosity which is more or less continuous but modulated in intensity. Positive flashes are generally composed of a single stroke followed by a period of continuing current. Positive flashes are probably initiated from the upper positive charge in the thundercloud charge dipole when that cloud charge is horizontally separated from the negative charge beneath it, the source of the usual negative cloud-to-ground lightning. Positive flashes are relatively common in winter thunderstorms (snow storms), which produce few flashes overall, and are relatively uncommon in summer thunderstorms. The fraction of positive lightning in summer thunderstorms apparently increases with increasing latitude and with increasing height of the ground above sea level.

Distant lightning return stroke fields are often referred to as sferics (called "atmospherics" in the older literature). The peak in the sferics frequency spectrum is near 5 kHz due to the bipolar or ringing nature of the distant return-stroke electromagnetic signal and to the effects of propagation.

Thunder, the acoustic radiation associated with lightning, is sometimes divided into the categories "audible", sounds that one can hear, and "infrasonic", below a few tens of hertz, a frequency range that is inaudible. This division is made because it is thought that the mechanisms that produce audible and infrasonic thunder are different. Audible thunder is thought to be due to the expansion of a rapidly heated return stroke channel, as noted earlier, whereas infrasonic thunder is thought to be associated with the conversion to sound of the energy stored in the electrostatic field of the thundercloud when lightning rapidly reduces that cloud field.

The technology of artificially initiating lightning by firing upward small rocket trailing grounded wire of a few hundred meters length has been well-developed during the past decade. Such "triggered" flashes are similar to natural upward-initiated discharges from tall structure. They often contain subsequent strokes which, when they occur, are similar to the subsequent strokes in natural lightning. These triggered subsequent strokes have been the subject of considerable recent research.

Also in the past 10 years or so sophisticated lightning locating equipment has been installed throughout the world. For example, all ground flashes in the U.S. are now centrally monitored for research, for better overall weather prediction, and for hazard warning for aviation, electric utilities and other lightning-sensitive facilities.

Information on lightning physics can be found in M. A. Uman, *The Lightning Discharge*, Academic Press, San Diego, 1987; on lightning death and injury in *Medical Aspects of Lightning Injury*, editors C. Andrews, M. A. Cooper, M. Darveniza, and D. Mackerras, CRC Press, 1992. Ground flash location information for the U.S., in real time or archived, is available from Geomet Data Service of Tucson, AZ, which is also a source of the names of providers of those data in other countries.

Table 2 has data for cloud-to-ground lightning discharges bringing negative charge to earth. The values listed are intended to convey a rough feeling for the various physical parameters of lightning. No great accuracy is claimed since the results of different investigators are often not in good agreement. These values may, in fact, depend on the particular environment in which the lightning discharge is generated. The choice of some of the entries in the table is arbitrary.

TABLE 2
Data for Cloud-to-Ground Lightning Discharges

	Minimum[a]	Representative values	Maximum[a]
Stepped leader			
Length of step, m	3	50	200
Time interval between steps, μs	30	50	125
Average speed of propagation of stepped leader, m/s[b]	1.0×10^5	2.0×10^5	3.0×10^6
Charged deposited on stepped-leader channel, coulombs	3	5	20
Dart leader			
Speed of propagation, m/s[b]	1.0×10^6	1.0×10^7	2.4×10^7
Charged deposited on dart-leader channel, coulombs	0.2	1	6
Return stroke[c]			
Speed of propagation, m/s[b]	2.0×10^7	1.0×10^8	2.0×10^8
Maximum current rate of increase, kA/μs	<1	100	400
Time to peak current, μs	<1	2	30
Peak current, kA	2	30	200
Time to half of peak current, μs	10	50	250
Charge transferred excluding continuing current, coulombs	0.02	3	20
Channel length, km	2	5	15
Lightning flash			
Number of strokes per flash	1	4	26
Time interval between strokes in absence of continuing current, ms	3	60	100
Time duration of flash, s	10^{-2}	0.5	2
Charge transferred including continuing current, coulombs	3	30	200

[a] The words maximum and minimum are used in the sense that most measured values fall between these limits.

[b] Speeds of propagation are generally determined from photographic data and are "two-dimensional". Since many lightning flashes are not vertical, values stated are probably slight underestimates of actual values.

[c] First return strokes have longer times to current peak and generally larger charge transfer than do subsequent return strokes.

Adapted from Uman, M. A., *Lightning*, Dover Paperbook, New York, 1986, and Uman, M. A., *The Lightning Discharge*, Academic Press, San Diego, 1987.

SPEED OF SOUND IN VARIOUS MEDIA

The speed of sound in various solids, liquids, and gases is given in these tables. While only a single parameter v is needed for liquids and gases, sound propagation in isotropic solids is characterized by three velocity parameters. For a solid of infinite extent (or of finite extent if all dimensions are much larger than a wavelength), there are two relevant quantities,

v_1: velocity of longitudinal waves
v_s: velocity of shear waves.

For a cylindrical rod with diameter much smaller than a wavelength,

v_{ext}: velocity of extensional waves along the rod. (Torsional waves in the rod are propagated at the same speed as sheer waves in an infinite solid.)

Table 1 lists values for a variety of solid materials. Table 2 covers gases liquids and gases; values for cryogenic liquids are given at the normal boiling point. Table 3 gives the speed of sound in pure water and in seawater of salinity $S = 3.5\%$ as a function of temperature. All values are in meters per second and are given for normal atmospheric pressure.

REFERENCES

1. Gray, D.E., Ed., *American Institute of Physics Handbook, Third Edition*, McGraw Hill, New York, 1972.
2. Anderson, H.L., Ed., *A Physicist's Desk Reference*, American Institute of Physics, New York, 1989.
3. Younglove, B.A., *Thermophysical Proeprties of Fluids. Part I, J. Phys. Chem. Ref. Data*, 11, Suppl. 1, 1982.
4. Younglove, B.A., and Ely, J.F., *Thermophysical Properties of Fluids. Part II, J. Phys. Chem. Ref. Data*, 16, 577, 1987.
5. Haar, L., Gallagher, J.S., and Kell, G.S., *NBS/NRC Steam Tables*, Hemisphere Publishing Corp., New York, 1984.
6. Mason, W.P., *Physical Acoustics and the Properties of Solids*, D. Van Nostrand Co., Princeton, N.J., 1958.
7. *Landolt-Börnstein, Numerical Data and Functional Relationships in Science and Technology, New Series, II/5, Molecular Acoustics*, Springer-Verlag, Heidelberg, 1967.

TABLE 1
Speed of Sound in Solids at Room Temperature

Name	v_1/m s^{-1}	v_s/m s^{-1}	v_{ext}/m s^{-1}	Name	v_1/m s^{-1}	v_s/m s^{-1}	v_{ext}/m s^{-1}
Metals				Steel, 347 Stainless	5790	3100	5000
Aluminum, rolled	6420	3040	5000	Steel, K9	5940	3250	5250
Berylium	12890	8880	12870	Tin, rolled	3320	1670	2730
Brass (70 Cu, 30 Zn)	4700	2110	3480	Titanium	6070	3125	5090
Constantan	5177	2625	4270	Tungsten, annealed	5220	2890	4620
Copper, annealed	4760	2325	3810	Tungsten, drawn	5410	2640	4320
Copper, rolled	5010	2270	3750	Zinc, rolled	4210	2440	3850
Duralumin 17S	6320	3130	5150				
Gold, hard-drawn	3240	1200	2030	**Other materials**			
Iron, cast	4994	2809	4480	Fused silica	5968	3764	5760
Iron, electrolytic	5950	3240	5120	Glass, heavy silicate flint	3980	2380	3720
Iron, Armco	5960	3240	5200	Glass, light borate crown	5100	2840	4540
Lead, annealed	2160	700	1190	Glass, pyrex	5640	3280	5170
Lead, rolled	1960	690	1210	Lucite	2680	1100	1840
Magnesium, annealed	5770	3050	4940	Nylon 6-6	2620	1070	1800
Molybdenum	6250	3350	5400	Polyethylene	1950	540	920
Monel metal	5350	2720	4400	Polystyrene	2350	1120	1840
Nickel	6040	3000	4900	Rubber, butyl	1830		
Platinum	3260	1730	2800	Rubber, gum	1550		
Silver	3650	1610	2680	Rubber, neoprene	1600		
Steel (1% C)	5940	3220	5180	Tungsten carbide	6655	3980	6220

TABLE 2
Speed of Sound in Liquids and Gases

Name	$t/°C$	$v/\text{m s}^{-1}$	Name	$t/°C$	$v/\text{m s}^{-1}$
Liquids			Pentane	20	1008
Acetone	20	1203	Propane	-42.1	1158
Argon	-185.9	813	1-Propanol	20	1223
Benzene	25	1310	Tetrachloromethane	25	930
Bromobenzene	20	1169	Trichloromethane	25	987
Butane	-0.5	1034	1-Undecene	20	1275
1-Butanol	20	1258	Water	25	1497
Carbon disulphide	25	1140	Water (sea, $S = 3.5\%$)	25	1535
Chlorobenzene	20	1311			
Cyclohexane	19	1280	**Gases at 1 atm**		
1-Decene	20	1250	Air, dry	25	346
Diethyl ether	25	976	Ammonia	0	415
Ethane	-88.6	1326	Argon	27	323
Ethanol	20	1162	Carbon monoxide	0	338
Ethylene	-103.8	1309	Carbon dioxide	0	259
Ethylene glycol	25	1658	Chlorine	0	206
Fluorobenzene	20	1183	Deuterium	0	890
Glycerol	25	1904	Ethane	27	312
Helium	-268.9	180	Ethylene	27	331
Heptane	20	1162	Helium	0	965
1-Heptene	20	1128	Hydrogen	27	1310
Hexane	20	1083	Hydrogen bromide	0	200
Hydrogen	-252.9	1101	Hydrogen chloride	0	296
Iodobenzene	20	1114	Hydrogen iodide	0	157
Mercury	25	1450	Hydrogen sulfide	0	289
Methane	-161.5	1337	Methane	27	450
Methanol	20	1121	Neon	0	435
Nitrobenzene	25	1463	Nitric oxide	10	325
Nitrogen	-195.8	939	Nitrogen	27	353
1-Nonene	20	1218	Nitrous oxide	0	263
Octane	20	1197	Oxygen	27	330
1-Octene	20	1184	Sulfur dioxide	0	213
Oxygen	-183.0	906	Water (steam)	100	473
1-Pentadecene	20	1351			

TABLE 3
Speed of Sound in Water and Seawater (S = 3.5%) at Different Temperatures

$t/°C$	$v/\text{m s}^{-1}$ Water	$v/\text{m s}^{-1}$ Seawater
0	1401.0	1449.4
10	1447.8	1490.4
20	1483.2	1522.2
25	1497.4	1535.1
30	1509.5	1546.2
40	1528.4	
50	1541.4	
60	1549.5	
70	1553.2	
80	1552.8	

ATTENUATION AND SPEED OF SOUND IN AIR AS A FUNCTION OF HUMIDITY AND FREQUENCY

This table gives the attenuation and speed of sound as a function of frequency at various values of relative humidity. All values refer to still air at 20°C.

REFERENCES

1. Tables of Absorption and Velocity of Sound in Still Air at 68°F (20°C), AD-738576, National Technical Information Service, Springfield, VA.
2. Evans, L. B., Bass, H. E., and Sutherland, L. C., *J. Acoust. Soc. Am.,* 51, 1565, 1972.

Frequency (Hz)	Attenuation (dB/km)	Speed (m/s)	Frequency (Hz)	Attenuation (dB/km)	Speed (m/s)
Relative humidity 0%			**Relative humidity 60%**		
20	0.51	343.477	20	0.02	344.182
40	1.07	343.514	40	0.06	344.183
50	1.26	343.525	50	0.09	344.183
63	1.43	343.536	63	0.15	344.184
100	1.67	343.550	100	0.34	344.185
200	1.84	343.559	200	0.99	344.190
400	1.96	343.561	400	1.94	344.197
630	2.11	343.562	630	2.57	344.200
800	2.27	343.562	800	2.94	344.201
1250	2.82	343.562	1250	4.01	344.202
2000	4.14	343.562	2000	6.55	344.203
4000	8.84	343.564	4000	18.73	344.204
6300	14.89	343.565	6300	42.51	344.204
10000	26.28	343.566	10000	101.84	344.206
12500	35.81	343.566	12500	155.67	344.208
16000	52.15	343.567	16000	247.78	344.211
20000	75.37	343.567	20000	373.78	344.215
40000	267.01	343.567	40000	1195.37	344.238
63000	644.66	343.567	63000	2220.64	344.262
80000	1032.14	343.567	80000	2951.71	344.274
Relative humidity 30%			**Relative humidity 100%**		
20	0.03	343.807	20	0.01	344.685
40	0.11	343.808	40	0.04	344.685
50	0.17	343.810	50	0.06	344.685
63	0.25	343.810	63	0.09	344.685
100	0.50	343.814	100	0.22	344.686
200	1.01	343.821	200	0.77	344.689
400	1.59	343.826	400	2.02	344.695
630	2.24	343.827	630	3.05	344.699
800	2.85	343.828	800	3.57	344.701
1250	5.09	343.828	1250	4.59	344.704
2000	10.93	343.829	2000	6.29	344.705
4000	38.89	343.831	4000	13.58	344.706
6300	90.61	343.836	6300	27.72	344.706
10000	204.98	343.846	10000	63.49	344.706
12500	294.08	343.854	12500	96.63	344.707
16000	422.51	343.865	16000	154.90	344.708
20000	563.66	343.877	20000	237.93	344.709
40000	1110.97	343.911	40000	884.28	344.718
63000	1639.47	343.924	63000	1973.62	344.731
80000	2083.08	343.929	80000	2913.01	344.742

SPEED OF SOUND IN DRY AIR

The values in this table were calculated from the equation of state for dry air (average molecular weight 28.96) treated as a real gas. Values refer to standard atmospheric pressure. The speed of sound varies only slightly with pressure; at two atmospheres and -100°C the value decreases by 0.13%, while at two atmospheres and 80°C the speed increases by 0.04%.

REFERENCE

Sytchev, V.V., Vasserman, A.A., Kozlov, A.D., Spiridonov, G.A., and Tsymarny, V.A., *Thermodynamic Properties of Air*, Hemisphere Publishing Corp., New York, 1987.

$t/°C$	$v_s/m\ s^{-1}$	$t/°C$	$v_s/m\ s^{-1}$	$t/°C$	$v_s/m\ s^{-1}$
-100	263.5	-35	309.5	30	349.1
-95	267.3	-30	312.7	35	352.0
-90	271.1	-25	315.9	40	354.8
-85	274.8	-20	319.1	45	357.6
-80	278.5	-15	322.3	50	360.4
-75	282.1	-10	325.4	55	363.2
-70	285.7	-5	328.4	60	365.9
-65	289.2	0	331.5	65	368.6
-60	292.7	5	334.5	70	371.3
-55	296.1	10	337.5	75	374.0
-50	299.5	15	340.4	80	376.7
-45	302.9	20	343.4		
-40	306.2	25	346.3		

MUSICAL SCALES

EQUAL TEMPERED CHROMATIC SCALE
$A_4 = 440$

American Standard pitch. Adopted by the American Standards Association in 1936

Note	Frequency	Note	Frequency	Note	Frequency	Note	Frequency
C_0	16.35	C_2	65.41	C_4	261.63	C_6	1046.50
$C\#_0$	17.32	$C\#_2$	69.30	$C\#_4$	277.18	$C\#_6$	1108.73
D_0	18.35	D_2	73.42	D_4	293.66	D_6	1174.66
$D\#_0$	19.45	$D\#_2$	77.78	$D\#_4$	311.13	$D\#_6$	1244.51
E_0	20.60	E_2	82.41	E_4	329.63	E_6	1318.51
F_0	21.83	F_2	87.31	F_4	349.23	F_6	1396.91
$F\#_0$	23.12	$F\#_2$	92.50	$F\#_4$	369.99	$F\#_6$	1479.98
G_0	24.50	G_2	98.00	G_4	392.00	G_6	1567.98
$G\#_0$	25.96	$G\#_2$	103.83	$G\#_4$	415.30	$G\#_6$	1661.22
A_0	27.50	A_2	110.00	A_4	440.00	A_6	1760.00
$A\#_0$	29.14	$A\#_2$	116.54	$A\#_4$	466.16	$A\#_6$	1864.66
B_0	30.87	B_2	123.47	B_4	493.88	B_6	1975.53
C_1	32.70	C_3	130.81	C_5	523.25	C_7	2093.00
$C\#_1$	34.65	$C\#_3$	138.59	$C\#_5$	554.37	$C\#_7$	2217.46
D_1	36.71	D_3	146.83	D_5	587.33	D_7	2349.32
$D\#_1$	38.89	$D\#_3$	155.56	$D\#_5$	622.25	$D\#_7$	2489.02
E_1	41.20	E_3	164.81	E_5	659.26	E_7	2637.02
F_1	43.65	F_3	174.61	F_5	698.46	F_7	2793.83
$F\#1$	46.25	$F\#_3$	185.00	$F\#_5$	739.99	$F\#_7$	2959.96
G_1	49.00	G_3	196.00	G_5	783.99	G_7	3135.96
$G\#_1$	51.91	$G\#_3$	207.65	$G\#_5$	830.61	$G\#_7$	3322.44
A_1	55.00	A_3	220.00	A_5	880.00	A_7	3520.00
$A\#_1$	58.27	$A\#_3$	233.08	$A\#_5$	932.33	$A\#_7$	3729.31
B_1	61.74	B_3	246.94	B_5	987.77	B_7	3951.07
						C_8	4186.01

EQUAL TEMPERED CHROMATIC SCALE
$A_4 = 435$

International Pitch, adopted 1891

Note	Frequency	Note	Frequency	Note	Frequency	Note	Frequency
C_0	16.17	C_2	64.66	C_4	258.65	C_6	1034.61
$C\#_0$	17.13	$C\#_2$	68.51	$C\#_4$	274.03	$C\#_6$	1096.13
D_0	18.15	D_2	72.58	D_4	290.33	D_6	1161.31
$D\#_0$	19.22	$D\#_2$	76.90	$D\#_4$	307.59	$D\#_6$	1230.37
E_0	20.37	E_2	81.47	E_4	325.88	E_6	1303.53
F_0	21.58	F_2	86.31	F_4	345.26	F_6	1381.04
$F\#_0$	22.86	$F\#_2$	91.45	$F\#_4$	365.79	$F\#_6$	1463.16
G_0	24.22	G_2	96.89	G_4	387.54	G_6	1550.16
$G\#_0$	25.66	$G\#_2$	102.65	$G\#_4$	410.59	$G\#_6$	1642.34
A_0	27.19	A_2	108.75	A_4	435.00	A_6	1740.00
$A\#_0$	28.80	$A\#_2$	115.22	$A\#_4$	460.87	$A\#_6$	1843.47
B_0	30.52	B_2	122.07	B_4	488.27	B_6	1953.08
C_1	32.33	C_3	129.33	C_5	517.31	C_7	2069.22
$C\#_1$	34.25	$C\#_3$	137.02	$C\#_5$	548.07	$C\#_7$	2192.26
D_1	36.29	D_3	145.16	D_5	580.66	D_7	2322.62
$D\#_1$	38.45	$D\#_3$	153.80	$D\#_5$	615.18	$D\#_7$	2460.73
E_1	40.74	E_3	162.94	E_5	651.76	E_7	2607.05
F_1	43.16	F_3	172.63	F_5	690.52	F_7	2762.08
$F\#_1$	45.72	$F\#_3$	182.89	$F\#_5$	731.58	$F\#_7$	2926.32
G_1	48.44	G_3	193.77	G_5	775.08	G_7	3100.33
$G\#_1$	51.32	$G\#_3$	205.29	$G\#_5$	821.17	$G\#_7$	3284.68
A_1	54.38	A_3	217.50	A_5	870.00	A_7	3480.00
$A\#_1$	57.61	$A\#_3$	230.43	$A\#_5$	921.73	$A\#_7$	3686.93
B_1	61.03	B_3	244.14	B_5	976.54	B_7	3906.17
						C_8	4138.44

SCIENTIFIC OR JUST SCALE
$C_4 = 256$

Note	Frequency	Note	Frequency	Note	Frequency	Note	Frequency
C_0	16	C_2	64	C_4	256	C_6	1024
D_0	18	D_2	72	D_4	288	D_6	1152
E_0	20	E_2	80	E_4	320	E_6	1280
F_0	21.33	F_2	85.33	F_4	341.33	F_6	1365.33
G_0	24	G_2	96	G_4	384	G_6	1536
A_0	26.67	A_2	106.67	A_4	426.67	A_6	1706.67
B_0	30	B_2	120	B_4	480	B_6	1920
C_1	32	C_3	128	C_5	512	C_7	2048
D_1	36	D_3	144	D_5	576	D_7	2304
E_1	40	E_3	160	E_5	640	E_7	2560
F_1	42.67	F_3	170.67	F_5	682.67	F_7	2730.67
G_1	48	G_3	192	G_5	768	G_7	3072
A_1	53.33	A_3	213.33	A_5	853.33	A_7	3413.33
B_1	60	B_3	240	B_5	960	B_7	3840
						C_8	4096

Section 15
Practical Laboratory Data

STANDARD ITS-90 THERMOCOUPLE TABLES

The Instrument Society of America (ISA) has assigned standard letter designations to a number of thermocouple types having specified emf-temperature relations. These designations and the approximate metal compositions which meet the required relations, as well as the useful temperature ranges, are given below:

Type B	(Pt + 30% Rh) vs. (Pt + 6% Rh)	0 to 1820°C
Type E	(Ni + 10% Cr) vs. (Cu + 43% Ni)	-270 to 1000°C
Type J	Fe vs. (Cu + 43%Ni)	-210 to 1200°C
Type K	(Ni + 10% Cr) vs. (Ni + 2% Al + 2% Mn + 1% Si)	-270 to 1372°C
Type N	(Ni + 14% Cr + 1.5% Si) vs. (Ni + 4.5% Si + 0.1% Mg)	-270 to 1300°C
Type R	(Pt + 13% Rh) vs. Pt	-50 to 1768°C
Type S	(Pt + 10% Rh) vs. Pt	-50 to 1768°C
Type T	Cu vs. (Cu + 43% Ni)	-270 to 400°C

The compositions are given in weight percent, and the positive leg is listed first. It should be emphasized that the standard letter designations do not imply a precise composition but rather that the specified emf-temperature relation is satisfied.

The first set of tables below lists, for each thermocouple type, the emf as a function of temperature on the International Temperature Scale of 1990 (ITS-90). The coefficients in the equation used to generate the table are also given. The second set of tables gives the inverse relationships, i.e., the coefficients in the polynomial equation which expresses the temperature as a function of thermocouple emf. The accuracy of these equations is also stated.

Further details and tables at closer intervals may be found in Reference 1.

REFERENCES

1. Burns, G. W., Seroger, M. G., Strouse, G. F., Croarkin, M. C., and Guthrie, W. F., *Temperature-Electromotive Force Reference Functions and Tables for the Letter-Designated Thermocouple Types Based on the ITS-90*, Natl. Inst. Stand. Tech. (U.S.) Monogr. 175, 1993.
2. Schooley, J. F., *Thermometry*, CRC Press, Boca Raton, FL, 1986.

Type B thermocouples: emf-temperature (°C) reference table and equations.

Thermocouple emf as a Function of Temperature in Degrees Celsius (ITS-90)

emf in Millivolts										Reference Junctions at 0 °C	
°C	0	10	20	30	40	50	60	70	80	90	100
0	0.000	−0.002	−0.003	−0.002	−0.000	0.002	0.006	0.011	0.017	0.025	0.033
100	0.033	0.043	0.053	0.065	0.078	0.092	0.107	0.123	0.141	0.159	0.178
200	0.178	0.199	0.220	0.243	0.267	0.291	0.317	0.344	0.372	0.401	0.431
300	0.431	0.462	0.494	0.527	0.561	0.596	0.632	0.669	0.707	0.746	0.787
400	0.787	0.828	0.870	0.913	0.957	1.002	1.048	1.095	1.143	1.192	1.242
500	1.242	1.293	1.344	1.397	1.451	1.505	1.561	1.617	1.675	1.733	1.792
600	1.792	1.852	1.913	1.975	2.037	2.101	2.165	2.230	2.296	2.363	2.431
700	2.431	2.499	2.569	2.639	2.710	2.782	2.854	2.928	3.002	3.078	3.154
800	3.154	3.230	3.308	3.386	3.466	3.546	3.626	3.708	3.790	3.873	3.957
900	3.957	4.041	4.127	4.213	4.299	4.387	4.475	4.564	4.653	4.743	4.834
1000	4.834	4.926	5.018	5.111	5.205	5.299	5.394	5.489	5.585	5.682	5.780
1100	5.780	5.878	5.976	6.075	6.175	6.276	6.377	6.478	6.580	6.683	6.786
1200	6.786	6.890	6.995	7.100	7.205	7.311	7.417	7.524	7.632	7.740	7.848
1300	7.848	7.957	8.066	8.176	8.286	8.397	8.508	8.620	8.731	8.844	8.956
1400	8.956	9.069	9.182	9.296	9.410	9.524	9.639	9.753	9.868	9.984	10.099
1500	10.099	10.215	10.331	10.447	10.563	10.679	10.796	10.913	11.029	11.146	11.263
1600	11.263	11.380	11.497	11.614	11.731	11.848	11.965	12.082	12.199	12.316	12.433
1700	12.433	12.549	12.666	12.782	12.898	13.014	13.130	13.246	13.361	13.476	13.591
1800	13.591	13.706	13.820								

Temperature Ranges and Coefficients of Equations Used to Compute the Above Table
The equations are of the form: $E = c_0 + c_1 t + c_2 t^2 + c_3 t^3 + \ldots c_n t^n$, where E is the emf in millivolts, t is the temperature in degrees Celsius (ITS-90), and c_0, c_1, c_2, c_3, *etc.* are the coefficients. These coefficients are extracted from NIST Monograph 175.

		0 °C to 630.615 °C	630.615 °C to 1820 °C
c_0	=	0.000 000 000 0 . . .	−3.893 816 862 1 . . .
c_1	=	−2.465 081 834 6 X 10^{-4}	2.857 174 747 0 X 10^{-2}
c_2	=	5.904 042 117 1 X 10^{-6}	−8.488 510 478 5 X 10^{-5}
c_3	=	−1.325 793 163 6 X 10^{-9}	1.578 528 016 4 X 10^{-7}
c_4	=	1.566 829 190 1 X 10^{-12}	−1.683 534 486 4 X 10^{-10}
c_5	=	−1.694 452 924 0 X 10^{-15}	1.110 979 401 3 X 10^{-13}
c_6	=	6.299 034 709 4 X 10^{-19}	−4.451 543 103 3 X 10^{-17}
c_7	=	. . . .	9.897 564 082 1 X 10^{-21}
c_8	=	. . . .	−9.379 133 028 9 X 10^{-25}

Type E thermocouples: emf-temperature (°C) reference table and equations.

Thermocouple emf as a Function of Temperature in Degrees Celsius (ITS-90)

emf in Millivolts   Reference Junctions at 0 °C

°C	0	−10	−20	−30	−40	−50	−60	−70	−80	−90	−100
−200	−8.825	−9.063	−9.274	−9.455	−9.604	−9.718	−9.797	−9.835			
−100	−5.237	−5.681	−6.107	−6.516	−6.907	−7.279	−7.632	−7.963	−8.273	−8.561	−8.825
0	0.000	−0.582	−1.152	−1.709	−2.255	−2.787	−3.306	−3.811	−4.302	−4.777	−5.237

°C	0	10	20	30	40	50	60	70	80	90	100
0	0.000	0.591	1.192	1.801	2.420	3.048	3.685	4.330	4.985	5.648	6.319
100	6.319	6.998	7.685	8.379	9.081	9.789	10.503	11.224	11.951	12.684	13.421
200	13.421	14.164	14.912	15.664	16.420	17.181	17.945	18.713	19.484	20.259	21.036
300	21.036	21.817	22.600	23.386	24.174	24.964	25.757	26.552	27.348	28.146	28.946
400	28.946	29.747	30.550	31.354	32.159	32.965	33.772	34.579	35.387	36.196	37.005
500	37.005	37.815	38.624	39.434	40.243	41.053	41.862	42.671	43.479	44.286	45.093
600	45.093	45.900	46.705	47.509	48.313	49.116	49.917	50.718	51.517	52.315	53.112
700	53.112	53.908	54.703	55.497	56.289	57.080	57.870	58.659	59.446	60.232	61.017
800	61.017	61.801	62.583	63.364	64.144	64.922	65.698	66.473	67.246	68.017	68.787
900	68.787	69.554	70.319	71.082	71.844	72.603	73.360	74.115	74.869	75.621	76.373
1000	76.373										

Temperature Ranges and Coefficients of Equations Used to Compute the Above Table
The equations are of the form: $E = c_0 + c_1 t + c_2 t^2 + c_3 t^3 + \ldots c_n t^n$, where E is the emf in millivolts, t is the temperature in degrees Celsius (ITS-90), and c_0, c_1, c_2, c_3, *etc.* are the coefficients. These coefficients are extracted from NIST Monograph 175.

		−270 °C to 0 °C	0 °C to 1000 °C
c_0	=	0.000 000 000 0 . . .	0.000 000 000 0 . . .
c_1	=	5.866 550 870 8 X 10^{-2}	5.866 550 871 0 X 10^{-2}
c_2	=	4.541 097 712 4 X 10^{-5}	4.503 227 558 2 X 10^{-5}
c_3	=	−7.799 804 868 6 X 10^{-7}	2.890 840 721 2 X 10^{-8}
c_4	=	−2.580 016 084 3 X 10^{-8}	−3.305 689 665 2 X 10^{-10}
c_5	=	−5.945 258 305 7 X 10^{-10}	6.502 440 327 0 X 10^{-13}
c_6	=	−9.321 405 866 7 X 10^{-12}	−1.919 749 550 4 X 10^{-16}
c_7	=	−1.028 760 553 4 X 10^{-13}	−1.253 660 049 7 X 10^{-18}
c_8	=	−8.037 012 362 1 X 10^{-16}	2.148 921 756 9 X 10^{-21}
c_9	=	−4.397 949 739 1 X 10^{-18}	−1.438 804 178 2 X 10^{-24}
c_{10}	=	−1.641 477 635 5 X 10^{-20}	3.596 089 948 1 X 10^{-28}
c_{11}	=	−3.967 361 951 6 X 10^{-23}	. . . .
c_{12}	=	−5.582 732 872 1 X 10^{-26}	. . . .
c_{13}	=	−3.465 784 201 3 X 10^{-29}	. . . .

Type J thermocouples: emf-temperature (°C) reference table and equations.

Thermocouple emf as a Function of Temperature in Degrees Celsius (ITS-90)

emf in Millivolts   Reference Junctions at 0 °C

°C	0	−10	−20	−30	−40	−50	−60	−70	−80	−90	−100
−200	−7.890	−8.095									
−100	−4.633	−5.037	−5.426	−5.801	−6.159	−6.500	−6.821	−7.123	−7.403	−7.659	−7.890
0	0.000	−0.501	−0.995	−1.482	−1.961	−2.431	−2.893	−3.344	−3.786	−4.215	−4.633

°C	0	10	20	30	40	50	60	70	80	90	100
0	0.000	0.507	1.019	1.537	2.059	2.585	3.116	3.650	4.187	4.726	5.269
100	5.269	5.814	6.360	6.909	7.459	8.010	8.562	9.115	9.669	10.224	10.779
200	10.779	11.334	11.889	12.445	13.000	13.555	14.110	14.665	15.219	15.773	16.327
300	16.327	16.881	17.434	17.986	18.538	19.090	19.642	20.194	20.745	21.297	21.848
400	21.848	22.400	22.952	23.504	24.057	24.610	25.164	25.720	26.276	26.834	27.393
500	27.393	27.953	28.516	29.080	29.647	30.216	30.788	31.362	31.939	32.519	33.102
600	33.102	33.689	34.279	34.873	35.470	36.071	36.675	37.284	37.896	38.512	39.132
700	39.132	39.755	40.382	41.012	41.645	42.281	42.919	43.559	44.203	44.848	45.494
800	45.494	46.141	46.786	47.431	48.074	48.715	49.353	49.989	50.622	51.251	51.877
900	51.877	52.500	53.119	53.735	54.347	54.956	55.561	56.164	56.763	57.360	57.953
1000	57.953	58.545	59.134	59.721	60.307	60.890	61.473	62.054	62.634	63.214	63.792
1100	63.792	64.370	64.948	65.525	66.102	66.679	67.255	67.831	68.406	68.980	69.553
1200	69.553										

Temperature Ranges and Coefficients of Equations Used to Compute the Above Table

The equations are of the form: $E = c_0 + c_1 t + c_2 t^2 + c_3 t^3 + \ldots c_n t^n$, where E is the emf in millivolts, t is the temperature in degrees Celsius (ITS-90), and c_0, c_1, c_2, c_3, *etc.* are the coefficients. These coefficients are extracted from NIST Monograph 175.

	−210 °C to 760 °C	760 °C to 1200 °C
c_0 =	0.000 000 000 0 . . .	$2.964\ 562\ 568\ 1 \times 10^2$
c_1 =	$5.038\ 118\ 781\ 5 \times 10^{-2}$	$-1.497\ 612\ 778\ 6 \ldots$
c_2 =	$3.047\ 583\ 693\ 0 \times 10^{-5}$	$3.178\ 710\ 392\ 4 \times 10^{-3}$
c_3 =	$-8.568\ 106\ 572\ 0 \times 10^{-8}$	$-3.184\ 768\ 670\ 1 \times 10^{-6}$
c_4 =	$1.322\ 819\ 529\ 5 \times 10^{-10}$	$1.572\ 081\ 900\ 4 \times 10^{-9}$
c_5 =	$-1.705\ 295\ 833\ 7 \times 10^{-13}$	$-3.069\ 136\ 905\ 6 \times 10^{-13}$
c_6 =	$2.094\ 809\ 069\ 7 \times 10^{-16}$	. . . .
c_7 =	$-1.253\ 839\ 533\ 6 \times 10^{-19}$	. . . .
c_8 =	$1.563\ 172\ 569\ 7 \times 10^{-23}$	. . . .

Type K thermocouples: emf-temperature (°C) reference table and equations.

Thermocouple emf as a Function of Temperature in Degrees Celsius (ITS-90)

emf in Millivolts   Reference Junctions at 0 °C

°C	0	−10	−20	−30	−40	−50	−60	−70	−80	−90	−100
−200	−5.891	−6.035	−6.158	−6.262	−6.344	−6.404	−6.441	−6.458			
−100	−3.554	−3.852	−4.138	−4.411	−4.669	−4.913	−5.141	−5.354	−5.550	−5.730	−5.891
0	0.000	−0.392	−0.778	−1.156	−1.527	−1.889	−2.243	−2.587	−2.920	−3.243	−3.554

°C	0	10	20	30	40	50	60	70	80	90	100
0	0.000	0.397	0.798	1.203	1.612	2.023	2.436	2.851	3.267	3.682	4.096
100	4.096	4.509	4.920	5.328	5.735	6.138	6.540	6.941	7.340	7.739	8.138
200	8.138	8.539	8.940	9.343	9.747	10.153	10.561	10.971	11.382	11.795	12.209
300	12.209	12.624	13.040	13.457	13.874	14.293	14.713	15.133	15.554	15.975	16.397
400	16.397	16.820	17.243	17.667	18.091	18.516	18.941	19.366	19.792	20.218	20.644
500	20.644	21.071	21.497	21.924	22.350	22.776	23.203	23.629	24.055	24.480	24.905
600	24.905	25.330	25.755	26.179	26.602	27.025	27.447	27.869	28.289	28.710	29.129
700	29.129	29.548	29.965	30.382	30.798	31.213	31.628	32.041	32.453	32.865	33.275
800	33.275	33.685	34.093	34.501	34.908	35.313	35.718	36.121	36.524	36.925	37.326
900	37.326	37.725	38.124	38.522	38.918	39.314	39.708	40.101	40.494	40.885	41.276
1000	41.276	41.665	42.053	42.440	42.826	43.211	43.595	43.978	44.359	44.740	45.119
1100	45.119	45.497	45.873	46.249	46.623	46.995	47.367	47.737	48.105	48.473	48.838
1200	48.838	49.202	49.565	49.926	50.286	50.644	51.000	51.355	51.708	52.060	52.410
1300	52.410	52.759	53.106	53.451	53.795	54.138	54.479	54.819			

Temperature Ranges and Coefficients of Equations Used to Compute the Above Table

The equations are of the form: $E = c_0 + c_1 t + c_2 t^2 + c_3 t^3 + \ldots c_n t^n$, where E is the emf in millivolts, t is the temperature in degrees Celsius (ITS-90), and c_0, c_1, c_2, c_3, *etc.* are the coefficients. In the 0 °C to 1372 °C range there is also an exponential term that must be evaluated and added to the equation. The exponential term is of the form: $c_0 e^{c_1 (t - 126.9686)^2}$, where t is the temperature in °C, e is the natural logarithm base, and c_0 and c_i are the coefficients. These coefficients are extracted from NIST Monograph 175.

		−270 °C to 0 °C	0 °C to 1372 °C	0 °C to 1372 °C (exponential term)
c_0	=	0.000 000 000 0 . . .	−1.760 041 368 6 X 10^{-2}	1.185 976 X 10^{-1}
c_1	=	3.945 012 802 5 X 10^{-2}	3.892 120 497 5 X 10^{-2}	−1.183 432 X 10^{-4}
c_2	=	2.362 237 359 8 X 10^{-5}	1.855 877 003 2 X 10^{-5}	. . . .
c_3	=	−3.285 890 678 4 X 10^{-7}	−9.945 759 287 4 X 10^{-6}	. . . .
c_4	=	−4.990 482 877 7 X 10^{-9}	3.184 094 571 9 X 10^{-10}	. . . .
c_5	=	−6.750 905 917 3 X 10^{-11}	−5.607 284 488 9 X 10^{-13}	. . . .
c_6	=	−5.741 032 742 8 X 10^{-13}	5.607 505 905 9 X 10^{-16}	. . . .
c_7	=	−3.108 887 289 4 X 10^{-15}	−3.202 072 000 3 X 10^{-19}	. . . .
c_8	=	−1.045 160 936 5 X 10^{-17}	9.715 114 715 2 X 10^{-23}	. . . .
c_9	=	−1.988 926 687 8 X 10^{-20}	−1.210 472 127 5 X 10^{-26}	. . . .
c_{10}	=	−1.632 269 748 6 X 10^{-23}	. . . .	. . . .

Type N thermocouples: emf-temperature (°C) reference table and equations.

Thermocouple emf as a Function of Temperature in Degrees Celsius (ITS-90)

emf in Millivolts   Reference Junctions at 0 °C

°C	0	−10	−20	−30	−40	−50	−60	−70	−80	−90	−100
−200	−3.990	−4.083	−4.162	−4.226	−4.277	−4.313	−4.336	−4.345			
−100	−2.407	−2.612	−2.808	−2.994	−3.171	−3.336	−3.491	−3.634	−3.766	−3.884	−3.990
0	0.000	−0.260	−0.518	−0.772	−1.023	−1.269	−1.509	−1.744	−1.972	−2.193	−2.407

°C	0	10	20	30	40	50	60	70	80	90	100
0	0.000	0.261	0.525	0.793	1.065	1.340	1.619	1.902	2.189	2.480	2.774
100	2.774	3.072	3.374	3.680	3.989	4.302	4.618	4.937	5.259	5.585	5.913
200	5.913	6.245	6.579	6.916	7.255	7.597	7.941	8.288	8.637	8.988	9.341
300	9.341	9.696	10.054	10.413	10.774	11.136	11.501	11.867	12.234	12.603	12.974
400	12.974	13.346	13.719	14.094	14.469	14.846	15.225	15.604	15.984	16.366	16.748
500	16.748	17.131	17.515	17.900	18.286	18.672	19.059	19.447	19.835	20.224	20.613
600	20.613	21.003	21.393	21.784	22.175	22.566	22.958	23.350	23.742	24.134	24.527
700	24.527	24.919	25.312	25.705	26.098	26.491	26.883	27.276	27.669	28.062	28.455
800	28.455	28.847	29.239	29.632	30.024	30.416	30.807	31.199	31.590	31.981	32.371
900	32.371	32.761	33.151	33.541	33.930	34.319	34.707	35.095	35.482	35.869	36.256
1000	36.256	36.641	37.027	37.411	37.795	38.179	38.562	38.944	39.326	39.706	40.087
1100	40.087	40.466	40.845	41.223	41.600	41.976	42.352	42.727	43.101	43.474	43.846
1200	43.846	44.218	44.588	44.958	45.326	45.694	46.060	46.425	46.789	47.152	47.513
1300	47.513										

Temperature Ranges and Coefficients of Equations Used to Compute the Above Table
The equations are of the form: $E = c_0 + c_1t + c_2t^2 + c_3t^3 + \ldots c_nt^n$, where E is the emf in millivolts, t is the temperature in degrees Celsius (ITS-90), and c_0, c_1, c_2, c_3, etc. are the coefficients. These coefficients are extracted from NIST Monograph 175.

	−270 °C to 0 °C	0 °C to 1300 °C
c_0 =	0.000 000 000 0 . . .	0.000 000 000 0 . . .
c_1 =	$2.615\,910\,596\,2 \times 10^{-2}$	$2.592\,939\,460\,1 \times 10^{-2}$
c_2 =	$1.095\,748\,422\,8 \times 10^{-5}$	$1.571\,014\,188\,0 \times 10^{-5}$
c_3 =	$-9.384\,111\,155\,4 \times 10^{-8}$	$4.382\,562\,723\,7 \times 10^{-6}$
c_4 =	$-4.641\,203\,975\,9 \times 10^{-11}$	$-2.526\,116\,979\,4 \times 10^{-10}$
c_5 =	$-2.630\,335\,771\,6 \times 10^{-12}$	$6.431\,181\,933\,9 \times 10^{-13}$
c_6 =	$-2.265\,343\,800\,3 \times 10^{-14}$	$-1.006\,347\,151\,9 \times 10^{-15}$
c_7 =	$-7.608\,930\,079\,1 \times 10^{-17}$	$9.974\,533\,899\,2 \times 10^{-19}$
c_8 =	$-9.341\,966\,783\,5 \times 10^{-20}$	$-6.086\,324\,560\,7 \times 10^{-22}$
c_9 =	. . . .	$2.084\,922\,933\,9 \times 10^{-25}$
c_{10} =	. . . .	$-3.068\,219\,615\,1 \times 10^{-29}$

Type R thermocouples: emf-temperature (°C) reference table and equations.

Thermocouple emf as a Function of Temperature in Degrees Celsius (ITS-90)

emf in Millivolts   Reference Junctions at 0 °C

°C	0	−10	−20	−30	−40	−50	−60	−70	−80	−90	−100
0	0.000	−0.051	−0.100	−0.145	−0.188	−0.226					

°C	0	10	20	30	40	50	60	70	80	90	100
0	0.000	0.054	0.111	0.171	0.232	0.296	0.363	0.431	0.501	0.573	0.647
100	0.647	0.723	0.800	0.879	0.959	1.041	1.124	1.208	1.294	1.381	1.469
200	1.469	1.558	1.648	1.739	1.831	1.923	2.017	2.112	2.207	2.304	2.401
300	2.401	2.498	2.597	2.696	2.796	2.896	2.997	3.099	3.201	3.304	3.408
400	3.408	3.512	3.616	3.721	3.827	3.933	4.040	4.147	4.255	4.363	4.471
500	4.471	4.580	4.690	4.800	4.910	5.021	5.133	5.245	5.357	5.470	5.583
600	5.583	5.697	5.812	5.926	6.041	6.157	6.273	6.390	6.507	6.625	6.743
700	6.743	6.861	6.980	7.100	7.220	7.340	7.461	7.583	7.705	7.827	7.950
800	7.950	8.073	8.197	8.321	8.446	8.571	8.697	8.823	8.950	9.077	9.205
900	9.205	9.333	9.461	9.590	9.720	9.850	9.980	10.111	10.242	10.374	10.506
1000	10.506	10.638	10.771	10.905	11.039	11.173	11.307	11.442	11.578	11.714	11.850
1100	11.850	11.986	12.123	12.260	12.397	12.535	12.673	12.812	12.950	13.089	13.228
1200	13.228	13.367	13.507	13.646	13.786	13.926	14.066	14.207	14.347	14.488	14.629
1300	14.629	14.770	14.911	15.052	15.193	15.334	15.475	15.616	15.758	15.899	16.040
1400	16.040	16.181	16.323	16.464	16.605	16.746	16.887	17.028	17.169	17.310	17.451
1500	17.451	17.591	17.732	17.872	18.012	18.152	18.292	18.431	18.571	18.710	18.849
1600	18.849	18.988	19.126	19.264	19.402	19.540	19.677	19.814	19.951	20.087	20.222
1700	20.222	20.356	20.488	20.620	20.749	20.877	21.003				

Temperature Ranges and Coefficients of Equations Used to Compute the Above Table

The equations are of the form: $E = c_0 + c_1 t + c_2 t^2 + c_3 t^3 + \ldots c_n t^n$, where E is the emf in millivolts, t is the temperature in degrees Celsius (ITS-90), and c_0, c_1, c_2, c_3, etc. are the coefficients. These coefficients are extracted from NIST Monograph 175.

	−50 °C to 1064.18 °C	1064.18 °C to 1664.5 °C	1664.5 °C to 1768.1 °C
c_0 =	0.000 000 000 00 ...	2.951 579 253 16 ...	$1.522\ 321\ 182\ 09 \times 10^2$
c_1 =	$5.289\ 617\ 297\ 65 \times 10^{-3}$	$-2.520\ 612\ 513\ 32 \times 10^{-3}$	$-2.688\ 198\ 885\ 45 \times 10^{-1}$
c_2 =	$1.391\ 665\ 897\ 82 \times 10^{-5}$	$1.595\ 645\ 018\ 65 \times 10^{-5}$	$1.712\ 802\ 804\ 71 \times 10^{-4}$
c_3 =	$-2.388\ 556\ 930\ 17 \times 10^{-8}$	$-7.640\ 859\ 475\ 76 \times 10^{-9}$	$-3.458\ 957\ 064\ 53 \times 10^{-8}$
c_4 =	$3.569\ 160\ 010\ 63 \times 10^{-11}$	$2.053\ 052\ 910\ 24 \times 10^{-12}$	$-9.346\ 339\ 710\ 46 \times 10^{-15}$
c_5 =	$-4.623\ 476\ 662\ 98 \times 10^{-14}$	$-2.933\ 596\ 681\ 73 \times 10^{-16}$	. . . .
c_6 =	$5.007\ 774\ 410\ 34 \times 10^{-17}$	. . . .	. . . .
c_7 =	$-3.731\ 058\ 861\ 91 \times 10^{-20}$	. . . .	. . . .
c_8 =	$1.577\ 164\ 823\ 67 \times 10^{-23}$	. . . .	. . . .
c_9 =	$-2.810\ 386\ 252\ 51 \times 10^{-27}$	. . . .	. . . .

Type S thermocouples: emf-temperature (°C) reference table and equations.

Thermocouple emf as a Function of Temperature in Degrees Celsius (ITS-90)

emf in Millivolts   Reference Junctions at 0 °C

°C	0	−10	−20	−30	−40	−50	−60	−70	−80	−90	−100
0	0.000	−0.053	−0.103	−0.150	−0.194	−0.236					

°C	0	10	20	30	40	50	60	70	80	90	100
0	0.000	0.055	0.113	0.173	0.235	0.299	0.365	0.433	0.502	0.573	0.646
100	0.646	0.720	0.795	0.872	0.950	1.029	1.110	1.191	1.273	1.357	1.441
200	1.441	1.526	1.612	1.698	1.786	1.874	1.962	2.052	2.141	2.232	2.323
300	2.323	2.415	2.507	2.599	2.692	2.786	2.880	2.974	3.069	3.164	3.259
400	3.259	3.355	3.451	3.548	3.645	3.742	3.840	3.938	4.036	4.134	4.233
500	4.233	4.332	4.432	4.532	4.632	4.732	4.833	4.934	5.035	5.137	5.239
600	5.239	5.341	5.443	5.546	5.649	5.753	5.857	5.961	6.065	6.170	6.275
700	6.275	6.381	6.486	6.593	6.699	6.806	6.913	7.020	7.128	7.236	7.345
800	7.345	7.454	7.563	7.673	7.783	7.893	8.003	8.114	8.226	8.337	8.449
900	8.449	8.562	8.674	8.787	8.900	9.014	9.128	9.242	9.357	9.472	9.587
1000	9.587	9.703	9.819	9.935	10.051	10.168	10.285	10.403	10.520	10.638	10.757
1100	10.757	10.875	10.994	11.113	11.232	11.351	11.471	11.590	11.710	11.830	11.951
1200	11.951	12.071	12.191	12.312	12.433	12.554	12.675	12.796	12.917	13.038	13.159
1300	13.159	13.280	13.402	13.523	13.644	13.766	13.887	14.009	14.130	14.251	14.373
1400	14.373	14.494	14.615	14.736	14.857	14.978	15.099	15.220	15.341	15.461	15.582
1500	15.582	15.702	15.822	15.942	16.062	16.182	16.301	16.420	16.539	16.658	16.777
1600	16.777	16.895	17.013	17.131	17.249	17.366	17.483	17.600	17.717	17.832	17.947
1700	17.947	18.061	18.174	18.285	18.395	18.503	18.609				

Temperature Ranges and Coefficients of Equations Used to Compute the Above Table
The equations are of the form: $E = c_0 + c_1 t + c_2 t^2 + c_3 t^3 + \ldots c_n t^n$, where E is the emf in millivolts, t is the temperature in degrees Celsius (ITS-90), and c_0, c_1, c_2, c_3, *etc.* are the coefficients. These coefficients are extracted from NIST Monograph 175.

		−50 °C to 1064.18 °C	1064.18 °C to 1664.5 °C	1664.5 °C to 1768.1 °C
c_0	=	0.000 000 000 00 . . .	1.329 004 440 85 . . .	$1.466\,282\,326\,36 \times 10^2$
c_1	=	$5.403\,133\,086\,31 \times 10^{-3}$	$3.345\,093\,113\,44 \times 10^{-3}$	$-2.584\,305\,167\,52 \times 10^{-1}$
c_2	=	$1.259\,342\,897\,40 \times 10^{-5}$	$6.548\,051\,928\,18 \times 10^{-6}$	$1.636\,935\,746\,41 \times 10^{-4}$
c_3	=	$-2.324\,779\,686\,89 \times 10^{-8}$	$-1.648\,562\,592\,09 \times 10^{-9}$	$-3.304\,390\,469\,87 \times 10^{-8}$
c_4	=	$3.220\,288\,230\,36 \times 10^{-11}$	$1.299\,896\,051\,74 \times 10^{-14}$	$-9.432\,236\,906\,12 \times 10^{-15}$
c_5	=	$-3.314\,651\,963\,89 \times 10^{-14}$	. . . .	. . . .
c_6	=	$2.557\,442\,517\,86 \times 10^{-17}$	. . . .	. . . .
c_7	=	$-1.250\,688\,713\,93 \times 10^{-20}$	. . . .	. . . .
c_8	=	$2.714\,431\,761\,45 \times 10^{-24}$	. . . .	. . . .

Type T thermocouples: emf-temperature (°C) reference table and equations.

Thermocouple emf as a Function of Temperature in Degrees Celsius (ITS-90)

emf in Millivolts — Reference Junctions at 0 °C

°C	0	−10	−20	−30	−40	−50	−60	−70	−80	−90	−100
−200	−5.603	−5.753	−5.888	−6.007	−6.105	−6.180	−6.232	−6.258			
−100	−3.379	−3.657	−3.923	−4.177	−4.419	−4.648	−4.865	−5.070	−5.261	−5.439	−5.603
0	0.000	−0.383	−0.757	−1.121	−1.475	−1.819	−2.153	−2.476	−2.788	−3.089	−3.379

°C	0	10	20	30	40	50	60	70	80	90	100
0	0.000	0.391	0.790	1.196	1.612	2.036	2.468	2.909	3.358	3.814	4.279
100	4.279	4.750	5.228	5.714	6.206	6.704	7.209	7.720	8.237	8.759	9.288
200	9.288	9.822	10.362	10.907	11.458	12.013	12.574	13.139	13.709	14.283	14.862
300	14.862	15.445	16.032	16.624	17.219	17.819	18.422	19.030	19.641	20.255	20.872
400	20.872										

Temperature Ranges and Coefficients of Equations Used to Compute the Above Table

The equations are of the form: $E = c_0 + c_1 t + c_2 t^2 + c_3 t^3 + \ldots c_n t^n$, where E is the emf in millivolts, t is the temperature in degrees Celsius (ITS-90), and c_0, c_1, c_2, c_3, *etc.* are the coefficients. These coefficients are extracted from NIST Monograph 175.

		−270 °C to 0 °C	0 °C to 400 °C
c_0	=	0.000 000 000 0 . . .	0.000 000 000 0 . . .
c_1	=	$3.874\ 810\ 636\ 4 \times 10^{-2}$	$3.874\ 810\ 636\ 4 \times 10^{-2}$
c_2	=	$4.419\ 443\ 434\ 7 \times 10^{-5}$	$3.329\ 222\ 788\ 0 \times 10^{-5}$
c_3	=	$1.184\ 432\ 310\ 5 \times 10^{-7}$	$2.061\ 824\ 340\ 4 \times 10^{-7}$
c_4	=	$2.003\ 297\ 355\ 4 \times 10^{-8}$	$-2.188\ 225\ 684\ 6 \times 10^{-9}$
c_5	=	$9.013\ 801\ 955\ 9 \times 10^{-10}$	$1.099\ 688\ 092\ 8 \times 10^{-11}$
c_6	=	$2.265\ 115\ 659\ 3 \times 10^{-11}$	$-3.081\ 575\ 877\ 2 \times 10^{-14}$
c_7	=	$3.607\ 115\ 420\ 5 \times 10^{-13}$	$4.547\ 913\ 529\ 0 \times 10^{-17}$
c_8	=	$3.849\ 393\ 988\ 3 \times 10^{-15}$	$-2.751\ 290\ 167\ 3 \times 10^{-20}$
c_9	=	$2.821\ 352\ 192\ 5 \times 10^{-17}$	. . . .
c_{10}	=	$1.425\ 159\ 477\ 9 \times 10^{-19}$	. . . .
c_{11}	=	$4.876\ 866\ 228\ 6 \times 10^{-22}$	. . . .
c_{12}	=	$1.079\ 553\ 927\ 0 \times 10^{-24}$	. . . .
c_{13}	=	$1.394\ 502\ 706\ 2 \times 10^{-27}$	. . . .
c_{14}	=	$7.979\ 515\ 392\ 7 \times 10^{-31}$	. . . .

Type B thermocouples: coefficients (c_i) of polynomials for the computation of temperatures in °C as a function of the thermocouple emf in various temperature and emf ranges.

Temperature Range:	250 °C to 700 °C	700 °C to 1820 °C
emf Range:	0.291 mV to 2.431 mV	2.431 mV to 13.820 mV
$c_0 =$	$9.842\ 332\ 1 \times 10^1$	$2.131\ 507\ 1 \times 10^2$
$c_1 =$	$6.997\ 150\ 0 \times 10^2$	$2.851\ 050\ 4 \times 10^2$
$c_2 =$	$-8.476\ 530\ 4 \times 10^2$	$-5.274\ 288\ 7 \times 10^1$
$c_3 =$	$1.005\ 264\ 4 \times 10^3$.	$9.916\ 080\ 4\ \ldots$
$c_4 =$	$-8.334\ 595\ 2 \times 10^2$	$-1.296\ 530\ 3\ \ldots$
$c_5 =$	$4.550\ 854\ 2 \times 10^2$	$1.119\ 587\ 0 \times 10^{-1}$
$c_6 =$	$-1.552\ 303\ 7 \times 10^2$	$-6.062\ 519\ 9 \times 10^{-3}$
$c_7 =$	$2.988\ 675\ 0 \times 10^1$	$1.866\ 169\ 6 \times 10^{-4}$
$c_8 =$	$-2.474\ 286\ 0\ \ldots$	$-2.487\ 858\ 5 \times 10^{-6}$

NOTE—The above coefficients are extracted from NIST Monograph 175 and are for an expression of the form shown in Section 10.3.2. They yield approximate values of temperature that agree within ± 0.03 °C with the values given in Table 10.2.

Type E thermocouples: coefficients (c_i) of polynomials for the computation of temperatures in °C as a function of the thermocouple emf in various temperature and emf ranges.

Temperature Range:	−200 °C to 0 °C	0 °C to 1000 °C
emf Range:	−8.825 mV to 0.0 mV	0.0 mV to 76.373 mV
$c_0 =$	$0.000\ 000\ 0\ \ldots$	$0.000\ 000\ 0\ \ldots$
$c_1 =$	$1.697\ 728\ 8 \times 10^1$	$1.705\ 703\ 5 \times 10^1$
$c_2 =$	$-4.351\ 497\ 0 \times 10^{-1}$	$-2.330\ 175\ 9 \times 10^{-1}$
$c_3 =$	$-1.585\ 969\ 7 \times 10^{-1}$	$6.543\ 558\ 5 \times 10^{-3}$
$c_4 =$	$-9.250\ 287\ 1 \times 10^{-2}$	$-7.356\ 274\ 9 \times 10^{-5}$
$c_5 =$	$-2.608\ 431\ 4 \times 10^{-2}$	$-1.789\ 600\ 1 \times 10^{-6}$
$c_6 =$	$-4.136\ 019\ 9 \times 10^{-3}$	$8.403\ 616\ 5 \times 10^{-8}$
$c_7 =$	$-3.403\ 403\ 0 \times 10^{-4}$	$-1.373\ 587\ 9 \times 10^{-9}$
$c_8 =$	$-1.156\ 489\ 0 \times 10^{-5}$	$1.062\ 982\ 3 \times 10^{-11}$
$c_9 =$	$\ldots\ldots\ldots$	$-3.244\ 708\ 7 \times 10^{-14}$

NOTE—The above coefficients are extracted from NIST Monograph 175 and are for an expression of the form shown in Section 10.3.2. They yield approximate values of temperature that agree within ± 0.02 °C with the values given in Table 10.4.

Type J thermocouples: coefficients (c_i) of polynomials for the computation of temperatures in °C as a function of the thermocouple emf in various temperature and emf ranges.

	-210 °C to 0 °C	0 °C to 760 °C	760 °C to 1200 °C
Temperature Range:			

	-8.095 mV to 0.0 mV	0.0 mV to 42.919 mV	42.919 mV to 69.553 mV
emf Range:			

$c_0 =$	$0.000\ 000\ 0\ \ldots$	$0.000\ 000\ \ldots$	$-3.113\ 581\ 87 \times 10^3$
$c_1 =$	$1.952\ 826\ 8 \times 10^1$	$1.978\ 425 \times 10^1$	$3.005\ 436\ 84 \times 10^2$
$c_2 =$	$-1.228\ 618\ 5\ \ldots$	$-2.001\ 204 \times 10^{-1}$	$-9.947\ 732\ 30\ \ldots$
$c_3 =$	$-1.075\ 217\ 8\ \ldots$	$1.036\ 969 \times 10^{-2}$	$1.702\ 766\ 30 \times 10^{-1}$
$c_4 =$	$-5.908\ 693\ 3 \times 10^{-1}$	$-2.549\ 687 \times 10^{-4}$	$-1.430\ 334\ 68 \times 10^{-3}$
$c_5 =$	$-1.725\ 671\ 3 \times 10^{-1}$	$3.585\ 153 \times 10^{-6}$	$4.738\ 860\ 84 \times 10^{-6}$
$c_6 =$	$-2.813\ 151\ 3 \times 10^{-2}$	$-5.344\ 285 \times 10^{-8}$	$\ldots\ldots\ldots$
$c_7 =$	$-2.396\ 337\ 0 \times 10^{-3}$	$5.099\ 890 \times 10^{-10}$	$\ldots\ldots\ldots$
$c_8 =$	$-8.382\ 332\ 1 \times 10^{-5}$	$\ldots\ldots\ldots$	$\ldots\ldots\ldots$

NOTE—The above coefficients are extracted from NIST Monograph 175 and are for an expression of the form shown in Section 10.3.2. They yield approximate values of temperature that agree within ± 0.05 °C with the values given in Table 10.6.

Type K thermocouples: coefficients (c_i) of polynomials for the computation of temperatures in °C as a function of the thermocouple emf in various temperature and emf ranges.

	-200 °C to 0 °C	0 °C to 500 °C	500 °C to 1372 °C
Temperature Range:			

	-5.891 mV to 0.0 mV	0.0 mV to 20.644 mV	20.644 mV to 54.886 mV
emf Range:			

$c_0 =$	$0.000\ 000\ 0\ \ldots$	$0.000\ 000\ 0\ \ldots$	$-1.318\ 058 \times 10^2$
$c_1 =$	$2.517\ 346\ 2 \times 10^1$	$2.508\ 355 \times 10^1$	$4.830\ 222 \times 10^1$
$c_2 =$	$-1.166\ 287\ 8\ \ldots$	$7.860\ 106 \times 10^{-2}$	$-1.646\ 031\ \ldots$
$c_3 =$	$-1.083\ 363\ 8\ \ldots$	$-2.503\ 131 \times 10^{-1}$	$5.464\ 731 \times 10^{-2}$
$c_4 =$	$-8.977\ 354\ 0 \times 10^{-1}$	$8.315\ 270 \times 10^{-2}$	$-9.650\ 715 \times 10^{-4}$
$c_5 =$	$-3.734\ 237\ 7 \times 10^{-1}$	$-1.228\ 034 \times 10^{-2}$	$8.802\ 193 \times 10^{-6}$
$c_6 =$	$-8.663\ 264\ 3 \times 10^{-2}$	$9.804\ 036 \times 10^{-4}$	$-3.110\ 810 \times 10^{-8}$
$c_7 =$	$-1.045\ 059\ 8 \times 10^{-2}$	$-4.413\ 030 \times 10^{-5}$	$\ldots\ldots\ldots$
$c_8 =$	$-5.192\ 057\ 7 \times 10^{-4}$	$1.057\ 734 \times 10^{-6}$	$\ldots\ldots\ldots$
$c_9 =$	$\ldots\ldots\ldots$	$-1.052\ 755 \times 10^{-8}$	$\ldots\ldots\ldots$

NOTE—The above coefficients are extracted from NIST Monograph 175 and are for an expression of the form shown in Section 10.3.2. They yield approximate values of temperature that agree within ± 0.05 °C with the values given in Table 10.8.

Type N thermocouples: coefficients (c_i) of polynomials for the computation of temperatures in °C as a function of the thermocouple emf in various temperature and emf ranges.

	-200 °C to 0 °C	0 °C to 600 °C	600 °C to 1300 °C
Temperature Range:			
emf Range:	-3.990 mV to 0.0 mV	0.0 mV to 20.613 mV	20.613 mV to 47.513 mV
$c_0 =$	$0.000\,000\,0\ldots$	$0.000\,00\ldots$	$1.972\,485 \times 10^1$
$c_1 =$	$3.843\,684\,7 \times 10^1$	$3.868\,96 \times 10^1$	$3.300\,943 \times 10^1$
$c_2 =$	$1.101\,048\,5\ldots$	$-1.082\,67\ldots$	$-3.915\,159 \times 10^{-1}$
$c_3 =$	$5.222\,931\,2\ldots$	$4.702\,05 \times 10^{-2}$	$9.855\,391 \times 10^{-3}$
$c_4 =$	$7.206\,052\,5\ldots$	$-2.121\,69 \times 10^{-6}$	$-1.274\,371 \times 10^{-4}$
$c_5 =$	$5.848\,858\,6\ldots$	$-1.172\,72 \times 10^{-4}$	$7.767\,022 \times 10^{-7}$
$c_6 =$	$2.775\,491\,6\ldots$	$5.392\,80 \times 10^{-6}$	$\ldots\ldots\ldots$
$c_7 =$	$7.707\,516\,6 \times 10^{-1}$	$-7.981\,56 \times 10^{-8}$	$\ldots\ldots\ldots$
$c_8 =$	$1.158\,266\,5 \times 10^{-1}$	$\ldots\ldots\ldots$	$\ldots\ldots\ldots$
$c_9 =$	$7.313\,886\,8 \times 10^{-3}$	$\ldots\ldots\ldots$	$\ldots\ldots\ldots$

NOTE—The above coefficients are extracted from NIST Monograph 175 and are for an expression of the form shown in Section 10.3.2. They yield approximate values of temperature that agree within ± 0.04 °C with the values given in Table 10.10.

Type R thermocouples: coefficients (c_i) of polynomials for the computation of temperatures in °C as a function of the thermocouple emf in various temperature and emf ranges.

	-50 °C to 250 °C	250 °C to 1200 °C	1064 °C to 1664.5 °C	1664.5 °C to 1768.1 °C
Temperature Range:				
emf Range:	-0.226 mV to 1.923 mV	1.923 mV to 13.228 mV	11.361 mV to 19.739 mV	19.739 mV to 21.103 mV
$c_0 =$	$0.000\,000\,0\ldots$	$1.334\,584\,505 \times 10^1$	$-8.199\,599\,416 \times 10^1$	$3.406\,177\,836 \times 10^4$
$c_1 =$	$1.889\,138\,0 \times 10^2$	$1.472\,644\,573 \times 10^2$	$1.553\,962\,042 \times 10^2$	$-7.023\,729\,171 \times 10^3$
$c_2 =$	$-9.383\,529\,0 \times 10^1$	$-1.844\,024\,844 \times 10^1$	$-8.342\,197\,663\ldots$	$5.582\,903\,813 \times 10^2$
$c_3 =$	$1.306\,861\,9 \times 10^2$	$4.031\,129\,726\ldots$	$4.279\,433\,549 \times 10^{-1}$	$-1.952\,394\,635 \times 10^1$
$c_4 =$	$-2.270\,358\,0 \times 10^2$	$-6.249\,428\,360 \times 10^{-1}$	$-1.191\,577\,910 \times 10^{-2}$	$2.560\,740\,231 \times 10^{-1}$
$c_5 =$	$3.514\,565\,9 \times 10^2$	$6.468\,412\,046 \times 10^{-2}$	$1.492\,290\,091 \times 10^{-4}$	$\ldots\ldots\ldots$
$c_6 =$	$-3.895\,390\,0 \times 10^2$	$-4.458\,750\,426 \times 10^{-3}$	$\ldots\ldots\ldots$	$\ldots\ldots\ldots$
$c_7 =$	$2.823\,947\,1 \times 10^2$	$1.994\,710\,149 \times 10^{-4}$	$\ldots\ldots\ldots$	$\ldots\ldots\ldots$
$c_8 =$	$-1.260\,728\,1 \times 10^2$	$-5.313\,401\,790 \times 10^{-6}$	$\ldots\ldots\ldots$	$\ldots\ldots\ldots$
$c_9 =$	$3.135\,361\,1 \times 10^1$	$6.481\,976\,217 \times 10^{-8}$	$\ldots\ldots\ldots$	$\ldots\ldots\ldots$
$c_{10} =$	$-3.318\,776\,9\ldots$	$\ldots\ldots\ldots$	$\ldots\ldots\ldots$	$\ldots\ldots\ldots$

NOTE—The above coefficients are extracted from NIST Monograph 175 and are for an expression of the form shown in Section 10.3.2. They yield approximate values of temperature that agree within ± 0.02 °C with the values given in Table 10.12.

Type S thermocouples: coefficients (c_i) of polynomials for the computation of temperatures in °C as a function of the thermocouple emf in various temperature and emf ranges.

Temperature Range:	−50 °C to 250 °C	250 °C to 1200 °C	1064 °C to 1664.5 °C	1664.5 °C to 1768.1 °C
emf Range:	−0.235 mV to 1.874 mV	1.874 mV to 11.950 mV	10.332 mV to 17.536 mV	17.536 mV to 18.693 mV
$c_0 =$	0.000 000 00 . . .	$1.291\ 507\ 177 \times 10^1$	$-8.087\ 801\ 117 \times 10^1$	$5.333\ 875\ 126 \times 10^4$
$c_1 =$	$1.849\ 494\ 60 \times 10^2$	$1.466\ 298\ 863 \times 10^2$	$1.621\ 573\ 104 \times 10^2$	$-1.235\ 892\ 298 \times 10^4$
$c_2 =$	$-8.005\ 040\ 62 \times 10^1$	$-1.534\ 713\ 402 \times 10^1$	$-8.536\ 869\ 453$ . . .	$1.092\ 657\ 613 \times 10^3$
$c_3 =$	$1.022\ 374\ 30 \times 10^2$	$3.145\ 945\ 973$ . . .	$4.719\ 686\ 976 \times 10^{-1}$	$-4.265\ 693\ 686 \times 10^1$
$c_4 =$	$-1.522\ 485\ 92 \times 10^2$	$-4.163\ 257\ 839 \times 10^{-1}$	$-1.441\ 693\ 666 \times 10^{-2}$	$6.247\ 205\ 420 \times 10^{-1}$
$c_5 =$	$1.888\ 213\ 43 \times 10^2$	$3.187\ 963\ 771 \times 10^{-2}$	$2.081\ 618\ 890 \times 10^{-4}$	. . . .
$c_6 =$	$-1.590\ 859\ 41 \times 10^2$	$-1.291\ 637\ 500 \times 10^{-3}$	. . . .	. . . .
$c_7 =$	$8.230\ 278\ 80 \times 10^1$	$2.183\ 475\ 087 \times 10^{-5}$	. . . .	. . . .
$c_8 =$	$-2.341\ 819\ 44 \times 10^1$	$-1.447\ 379\ 511 \times 10^{-7}$	. . . .	. . . .
$c_9 =$	2.797 862 60 . . .	$8.211\ 272\ 125 \times 10^{-9}$	. . . .	. . . .

NOTE—The above coefficients are extracted from NIST Monograph 175 and are for an expression of the form shown in Section 10.3.2. They yield approximate values of temperature that agree within ± 0.02 °C with the values given in Table 10.14.

Type T thermocouples: coefficients (c_i) of polynomials for the computation of temperatures in °C as a function of the thermocouple emf in various temperature and emf ranges.

Temperature Range:	−200 °C to 0 °C	0 °C to 400 °C
emf Range:	−5.603 mV to 0.0 mV	0.0 mV to 20.872 mV
$c_0 =$	0.000 000 0 . . .	0.000 000 . . .
$c_1 =$	$2.594\ 919\ 2 \times 10^1$	$2.592\ 800 \times 10^1$
$c_2 =$	$-2.131\ 696\ 7 \times 10^{-1}$	$-7.602\ 961 \times 10^{-1}$
$c_3 =$	$7.901\ 869\ 2 \times 10^{-1}$	$4.637\ 791 \times 10^{-2}$
$c_4 =$	$4.252\ 777\ 7 \times 10^{-1}$	$-2.165\ 394 \times 10^{-3}$
$c_5 =$	$1.330\ 447\ 3 \times 10^{-1}$	$6.048\ 144 \times 10^{-5}$
$c_6 =$	$2.024\ 144\ 6 \times 10^{-2}$	$-7.293\ 422 \times 10^{-7}$
$c_7 =$	$1.266\ 817\ 1 \times 10^{-3}$	. . . .

NOTE—The above coefficients are extracted from NIST Monograph 175 and are for an expression of the form shown in Section 10.3.2. They yield approximate values of temperature that agree within ± 0.04 °C with the values given in Table 10.16.

PROPERTIES OF COMMON LABORATORY SOLVENTS

This table give properties of 200 organic solvents which are frequently used in laboratory and industrial applications. Compounds are listed in alphabetical order by the most common name; synonyms are given in some cases. The properties tabulated are:

MF: Molecular formula
CAS RN: Chemical Abstracts Service Registry Number
M_r: Molecular weight
t_m: Melting point in °C
t_b: Normal boiling point in °C
ρ: Density in g/cm^3 at the temperature in °C indicated by the superscript
c_p: Specific heat capacity of the liquid at constant pressure at 25°C in J/g K
vp: Vapor pressure at 25°C in kPa (1 kPa = 7.50 mmHg)
μ: Electric dipole moment in debye units. Values in parentheses are measurements on the pure liquid or in solution; these are less reliable than the other values, which were obtained in the gas phase.
FP: Flash point temperature in °C. The fact that no flash point is listed does not necessarily mean that the substance is nonflammable, because some liquids will burn if the quantity is large or impurities are present.
Fl. Lim.: Flammable (explosive) range in air in percent by volume
Ign. Temp.: Autoignition temperature in °C
TLV: Threshold limit for allowable airborne concentration, given in parts per million by volume at 25°C and atmospheric pressure (see table "Threshold Limit Values for Airborne Contaminants" in Section 16)

REFERENCES

1. Lide, D.R., *Handbook of Organic Solvents*, CRC Press, Boca Raton, FL, 1994.
2. Lide, D.R., and Kehiaian, H.V., *CRC Handbook of Thermophysical and Thermochemical Data*, CRC Press, Boca Raton, FL, 1994.
3. Riddick, J.A., Bunger, W.B., and Sakano, T.K., *Organic Solvents, Fourth Edition*, John Wiley & Sons, New York, 1986.
4. *Fire Protection Guide to Hazardous Materials*, 10th Edition, National Fire Protection Association, Quincy, MA, 1991.
5. Urben, P.G., Ed., *Bretherick's Handbook of Reactive Chemical Hazards, 5th Edition*, Butterworth-Heinemann, Oxford, 1995.

Name	MF	CAS RN	M_r	t_m/°C	t_b/°C	ρ/g cm^{-3}	c_p/J g^{-1}K^{-1}	vp/kPa	μ/D	FP/°C	Fl. Lim.	Ign. Temp./°C	TLV
Acetal (1,1-Diethoxyethane)	$C_6H_{14}O_2$	105-57-7	118.18	-100	102.2	0.8254[20]	2.01	3.68	(1.4)	-21	2-10%	230	10
Acetic acid	$C_2H_4O_2$	64-19-7	60.05	17	118	1.0492[20]	2.06	2.07	1.70	39	4-20%	463	10
Acetone	C_3H_6O	67-64-1	58.08	-95	56	0.7899[20]	2.18	30.8	2.88	-20	3-13%	465	750
Acetonitrile	C_2H_3N	75-05-8	41.05	-44	82	0.7857[20]	2.23	11.8	3.92	6	3-16%	524	40
Acetylacetone	$C_5H_8O_2$	123-54-6	100.12	-23	138	0.9721[25]	2.08	1.02	(2.8)	34		340	
Acrylonitrile	C_3H_3N	107-13-1	53.06	-83.5	77.3	0.8060[20]	2.05	14.1	3.87	0	3-17%	481	2
Adiponitrile	$C_6H_8N_2$	111-69-3	108.14	1	295	0.9676[20]	1.19	<0.01		93	2-5%	550	2
Allyl alcohol	C_3H_6O	107-18-6	58.08	-129	97.0	0.8540[20]	2.39	3.14	1.60	21	3-18%	378	2
Allylamine	C_3H_7N	107-11-9	57.10	-88.2	53.3	0.758[20]		33.1	1.2	-29	2-22%	374	
2-Aminoisobutanol	$C_4H_{11}NO$	124-68-5	89.14	25.5	165.5	0.934[20]				67			
Benzal chloride	$C_7H_6Cl_2$	98-87-3	161.03	-17	205	1.26[25]		0.06	(2.1)				
Benzaldehyde	C_7H_6O	100-52-7	106.12	-26	179.0	1.0415[10]	1.62	0.17	(3.0)	63		192	
Benzene	C_6H_6	71-43-2	78.11	6	80	0.8765[20]	1.74	12.7	0	-11	1-8%	498	10
Benzonitrile	C_7H_5N	100-47-0	103.12	-12.7	191.1	1.0093[15]	1.60	0.11	4.18				
Benzyl chloride	C_7H_7Cl	100-44-7	126.59	-45	179	1.1004[20]	1.44	0.16	(1.8)	67	1%-	585	1
Bromochloromethane	CH_2BrCl	74-97-5	129.38	-87.9	68.0	1.9344[20]	0.41	19.5	(1.7)				200
Bromoform (Tribromomethane)	$CHBr_3$	75-25-2	252.73	8.0	149	2.899[15]	0.52	0.73	0.99	83			0.5

Name	MF	CAS RN	M_r	t_m/°C	t_b/°C	ρ/g cm^{-3}	c_p/J g^{-1}K^{-1}	vp/kPa	μ/D	FP/°C	Fl. Lim.	Ign. Temp./°C	TLV
Butyl acetate	$C_6H_{12}O_2$	123-86-4	116.16	-78	126	0.8825^{20}	1.96	1.66	(1.9)	22	2-8%	425	150
Butyl alcohol	$C_4H_{10}O$	71-36-3	74.12	-90	118	0.8098^{20}	2.39	0.86	1.66	37	1-11%	343	50
sec-Butyl alcohol	$C_4H_{10}O$	78-92-2	74.12	-114.7	99.5	0.8063^{20}	2.66	2.32	(1.8)	24	2-10%	405	100
tert-Butyl alcohol	$C_4H_{10}O$	75-65-0	74.12	26	82	0.7887^{20}	2.97	5.52	(1.7)	11	2-8%	478	100
Butylamine	$C_4H_{11}N$	109-73-9	73.14	-49	77	0.7414^{20}	2.45	12.2	1.0	-12	2-10%	312	5
tert-Butylamine	$C_4H_{11}N$	75-64-9	73.14	-67	44	0.6958^{20}	2.63	48.4	(1.3)	-9	2-9%	380	
Butyl methyl ketone	$C_6H_{12}O$	591-78-6	100.16	-56	128	0.8113^{20}	2.13	1.54	(2.7)	-25	1-8%	423	5
p-tert-Butyltoluene	$C_{11}H_{16}$	98-51-1	148.25	-52	190	0.8612^{20}		0.09	=0	68			10
γ-Butyrolactone	$C_4H_6O_2$	96-48-0	86.09	-43.3	204	1.1284^{16}	1.64	0.43	4.27	98			
Caprolactam	$C_6H_{11}NO$	105-60-2	113.16	69	270		1.38	<0.01	(3.9)	125			5
Carbon disulfide	CS_2	75-15-0	76.14	-112	46	1.2632^{20}	1.00	48.2	0	-30	1-50%	90	10
Carbon tetrachloride	CCl_4	56-23-5	153.82	-23	77	1.5940^{20}	0.85	15.2	0				5
1-Chloro-1,1-difluoroethane	$C_2H_3ClF_2$	75-68-3	100.50	-131	-10	1.107^{25}	1.30	351	2.14			593	1000
Chlorobenzene	C_6H_5Cl	108-90-7	112.56	-45	132	1.1058^{20}	1.33	1.6	1.69	28	1-10%		10
Chloroform	$CHCl_3$	67-66-3	119.38	-64	61	1.4832^{20}	0.96	26.2	1.04				10
Chloropentafluoroethane	C_2ClF_5	76-15-3	154.47	-99	-38	1.5678^{-42}	1.19	912	0.52				1000
Cumene (Isopropylbenzene)	C_9H_{12}	98-82-8	120.19	-96.0	152	0.8618^{20}	1.75	0.61	0.79	36	1-7%	424	50
Cyclohexane	C_6H_{12}	110-82-7	84.16	7	81	0.7785^{20}	1.84	13.0	=0	-20	1-8%	245	300
Cyclohexanol	$C_6H_{12}O$	108-93-0	100.16	25	161	0.9624^{20}	2.08	0.10		68	1-9%	300	50
Cyclohexanone	$C_6H_{10}O$	108-94-1	98.14	-31	155	0.9478^{20}	1.86	0.53	2.87	44	1-9%	420	25
Cyclohexylamine	$C_6H_{13}N$	108-91-8	99.18	-18	134	0.8191^{20}		1.20	(1.3)	31	1-9%	293	10
Cyclopentane	C_5H_{10}	287-92-3	70.13	-93.8	49.3	0.7457^{20}	1.84	42.3	=0	<-7	2%-	361	600
Cyclopentanone	C_5H_8O	120-92-3	84.12	-51.3	130.5	0.9487^{20}	1.84	1.55	3.3	26			
p-Cymene	$C_{10}H_{14}$	99-87-6	134.22	-69	177	0.8573^{20}	1.76	0.19	=0	47	1-6%	436	
cis-Decalin	$C_{10}H_{18}$	493-01-6	138.25	-42.9	195.8	0.8965^{20}	1.68	0.10	=0	54	1-5%	255	
trans-Decalin	$C_{10}H_{18}$	493-02-7	138.25	-30.3	187.3	0.8699^{20}	1.65	0.16	=0	58	2-7%		
Diacetone alcohol	$C_6H_{12}O_2$	123-42-2	116.16	-44	168	0.9387^{20}	1.91	0.22	(3.2)			643	50
1,2-Dibromoethane	$C_2H_4Br_2$	106-93-4	187.86	9.9	131.6	2.1791^{20}	0.72	1.55	(1.2)				
Dibromofluoromethane	$CHBr_2F$	1868-53-7	191.83	-78	64.9	2.421^{20}			1.43				
Dibromomethane	CH_2Br_2	74-95-3	173.83	-52.5	97	2.4969^{20}	0.61	6.12					
1,2-Dibromotetrafluoroethane	$C_2Br_2F_4$	124-73-2	259.82	-110.4	47.3	2.149^{25}	0.69	43.4					
Dibutylamine	$C_8H_{19}N$	111-92-2	129.25	-62	160	0.7670^{20}	2.27	0.34	(1.0)	47	1-6%		
o-Dichlorobenzene	$C_6H_4Cl_2$	95-50-1	147.00	-17	180	1.3059^{20}	1.10	0.18	2.50	66	2-9%	648	25
1,1-Dichloroethane	$C_2H_4Cl_2$	75-34-3	98.96	-97	57	1.1757^{20}	1.28	30.5	2.06	-17	5-11%	458	100
1,2-Dichloroethane	$C_2H_4Cl_2$	107-06-2	98.96	-36	84	1.2351^{20}	1.30	10.6	(1.8)	13	6-16%	413	10
1,1-Dichloroethylene	$C_2H_2Cl_2$	75-35-4	96.94	-122.5	31.6	1.213^{20}	1.15	80.0	1.34	-15	7-16%	570	5
cis-1,2-Dichloroethylene	$C_2H_2Cl_2$	156-59-2	96.94	-80	60	1.2837^{20}	1.20	26.8	1.90	6	3-15%	460	200
trans-1,2-Dichloroethylene	$C_2H_2Cl_2$	156-60-5	96.94	-50	49	1.2565^{20}	1.20	44.2	0	2	6-13%	460	200
Dichloroethyl ether	$C_4H_8Cl_2O$	111-44-4	143.01	-52	179	1.22^{20}	1.54	0.14	(2.6)	55	3%-	369	5
Dichloromethane	CH_2Cl_2	75-09-2	84.93	-95	40	1.3266^{20}	1.19	58.2	1.60		13-23%	556	50
1,2-Dichloropropane	$C_3H_6Cl_2$	78-87-5	112.99	-100	96	1.1560^{20}	1.32	6.62	(1.8)	16	3-15%	557	75
1,2-Dichlorotetrafluoroethane	$C_2Cl_2F_4$	76-14-2	170.92	-94	4	1.518^{4}	0.96	215	0.5				1000
Diethanolamine	$C_4H_{11}NO_2$	111-42-2	105.14	28	269	1.0966^{20}	2.22	<0.01	(2.8)	172	2-13%	662	0.46
Diethylamine	$C_4H_{11}N$	109-89-7	73.14	-50	55	0.7056^{20}	2.31	30.1	0.92	-23	2-10%	312	5
Diethyl carbonate	$C_5H_{10}O_3$	105-58-8	118.13	-43	126	0.9752^{20}	1.80	1.63	1.10	25			
Diethylene glycol	$C_4H_{10}O_3$	111-46-6	106.12	-10	246	1.1197^{15}	2.31	<0.01	(2.3)	124	2-17%	224	
Diethylene glycol dimethyl ether	$C_6H_{14}O_3$	111-96-6	134.18	-68	162	0.9434^{20}	2.04	0.31	(2.0)	67			
Diethylene glycol monoethyl ether	$C_6H_{14}O_3$	111-90-0	134.18		196	0.9885^{20}	2.24	0.02	(1.6)	96			

PROPERTIES OF COMMON LABORATORY SOLVENTS (continued)

Name	MF	CAS RN	M_r	t_m/°C	t_b/°C	ρ/g cm^{-3}	c_p/J g^{-1}K^{-1}	vp/kPa	μ/D	FP/°C	Fl. Lim.	Ign. Temp./°C	TLV
Diethylene glycol monoethyl ether acetate	$C_8H_{16}O_4$	112-15-2	176.21	-25	218.5	1.0096^{20}		0.03	(1.8)	110		425	
Diethylene glycol monomethyl ether	$C_5H_{12}O_3$	111-77-3	120.15		193	1.035^{20}	2.26	0.02	(1.6)	96	1-23%	240	
Diethylenetriamine	$C_4H_{13}N_3$	111-40-0	103.17	-39	207	0.9569^{20}	2.46	0.03	(1.9)	98	2-7%	358	1
Diethyl ether	$C_4H_{10}O$	60-29-7	74.12	-116	34	0.7138^{20}	2.33	71.7	1.15	-45	2-36%	180	400
Diisobutyl ketone	$C_9H_{18}O$	108-83-8	142.24	-42	169	0.8062^{20}	2.09	0.23	(2.7)	49	1-7%	396	25
Diisopropyl ether	$C_6H_{14}O$	108-20-3	102.18	-87	69	0.7241^{20}	2.12	19.9	1.13	-28	1-8%	443	250
N,N-Dimethylacetamide	C_4H_9NO	127-19-5	87.12	-20	165	0.9366^{25}	2.02	0.07	(3.7)	70	2-12%	490	10
Dimethylamine	C_2H_7N	124-40-3	45.08	-92	7	0.6804^{0}	3.05	203	1.01	20	3-14%	400	5
Dimethyl disulfide	$C_2H_6S_2$	624-92-0	94.20	-85	109.8	1.0625^{20}	1.55	3.82	(1.8)	24		445	
N,N-Dimethylformamide	C_3H_7NO	68-12-2	73.09	-60	153	0.944^{25}	2.06	0.44	3.82	58	2-15%	445	10
Dimethyl sulfoxide	C_2H_6OS	67-68-5	78.14	19	189	1.1014^{20}	1.96	0.08	3.96	95	3-42%	215	
1,4-Dioxane	$C_4H_8O_2$	123-91-1	88.11	12	101	1.0337^{20}	1.74	4.95	0	12	2-22%	180	25
1,3-Dioxolane	$C_3H_6O_2$	646-06-0	74.08	-95	78	1.060^{20}	1.59	14.6	1.19	2			
Dipentene	$C_{10}H_{16}$	7705-14-8	136.24	-95.5	178	0.8402^{21}	1.83	0.26		45		237	
Epichlorohydrin	C_3H_5ClO	106-89-8	92.52	-26	116	1.1812^{20}	1.42	2.2	(1.8)	31	4-21%	411	2
Ethanolamine (Glycinol)	C_2H_7NO	141-43-5	61.08	11	171	1.0180^{20}	3.20	0.05	(2.3)	86	3-24%	410	3
Ethyl acetate	$C_4H_8O_2$	141-78-6	88.11	-84	77	0.9003^{20}	1.94	12.6	1.78	-4	2-12%	426	400
Ethyl acetoacetate	$C_6H_{10}O_3$	141-97-9	130.14	-45	180.8	1.0368^{10}	1.91	0.09		57	1-10%	295	
Ethyl alcohol	C_2H_6O	64-17-5	46.07	-114	78	0.7893^{20}	2.44	7.87	1.69	13	3-19%	363	1000
Ethylamine	C_2H_7N	75-04-7	45.08	-81	17	0.686^{17}	2.88	142	1.22	<-18	4-14%	385	5
Ethylbenzene	C_8H_{10}	100-41-4	106.17	-95	136	0.8670^{20}	1.73	1.28	0.59	21	1-7%	432	100
Ethyl bromide	C_2H_5Br	74-96-4	108.97	-118.6	38.5	1.4604^{20}	0.93	62.5	2.03		7-8%	511	5
Ethyl chloride	C_2H_5Cl	75-00-3	64.51	-139	12	0.909^{12}	1.62	160	2.05	-50	4-15%	519	1000
Ethylene carbonate	$C_3H_4O_3$	96-49-1	88.06	36.4	248	1.3214^{39}	1.52	<0.01	(4.9)	143			
Ethylenediamine	$C_2H_8N_2$	107-15-3	60.10	11	117	0.8979^{20}	2.87	1.62	1.99	40	3-12%	385	10
Ethylene glycol	$C_2H_6O_2$	107-21-1	62.07	-13	197	1.1088^{20}	2.41	0.01	2.28	111	3-22%	398	50
Ethylene glycol diethyl ether	$C_6H_{14}O_2$	629-14-1	118.18	-74	119.4	0.8484^{20}	2.19	4.33		35			
Ethylene glycol dimethyl ether	$C_4H_{10}O_2$	110-71-4	90.12	-58	85	0.8691^{20}	2.14	9.93		-2		202	
Ethylene glycol monobutyl ether	$C_6H_{14}O_2$	111-76-2	118.18	-75	168	0.9015^{20}	2.38	0.15	(2.1)	69	4-13%	238	25
Ethylene glycol monoethyl ether	$C_4H_{10}O_2$	110-80-5	90.12	-70	135	0.9297^{20}	2.34	0.71	(2.1)	43	3-18%	235	5
Ethylene glycol ethyl ether acetate	$C_6H_{12}O_3$	111-15-9	132.16	-62	156	0.9740^{20}	2.85	0.24	(2.2)	56	2-8%	379	5
Ethylene glycol monomethyl ether	$C_3H_8O_2$	109-86-4	76.10	-85	124	0.9647^{20}	2.25	1.31	2.36	39	2-14%	285	5
Ethylene glycol monomethyl ether acetate	$C_5H_{10}O_3$	110-49-6	118.13	-70	143	1.0074^{19}	2.62	0.67	(2.1)	49	2-12%	392	5
Ethyl formate	$C_3H_6O_2$	109-94-4	74.08	-80	54	0.9168^{20}	2.02	32.3	1.9	-20	3-16%	455	100
Furan	C_4H_4O	110-00-9	68.08	-86	31	0.9514^{20}	1.69	80.0	0.66	<0	2-14%		
Furfural	$C_5H_4O_2$	98-01-1	96.09	-37	162	1.1594^{20}	1.70	0.29	(3.5)	60	2-19%	316	2
Furfuryl alcohol	$C_5H_6O_2$	98-00-0	98.10	-31	171	1.1296^{20}	2.08	0.10	(1.9)	75	2-16%	491	10
Glycerol	$C_3H_8O_3$	56-81-5	92.09	18	290	1.2613^{20}	2.38	<0.01	(2.6)	199	3-19%	370	
Heptane	C_7H_{16}	142-82-5	100.20	-91	98	0.6837^{20}	2.24	6.09	=0	-4	1-7%	204	400
1-Heptanol	$C_7H_{16}O$	111-70-6	116.20	-34	176.4	0.8219^{20}	2.34						
Hexane	C_6H_{14}	110-54-3	86.18	-95	69	0.6548^{25}	2.27	20.2	=0	-22	1-8%	225	50
1-Hexanol (Caproyl alcohol)	$C_6H_{14}O$	111-27-3	102.18	-44.6	157.6	0.8136^{20}	2.35	0.11		63			
Hexylene glycol	$C_6H_{14}O_2$	107-41-5	118.18	-50	197	0.923^{15}	2.84	<0.01	(2.9)	102	1-9%	306	25
Hexyl methyl ketone	$C_8H_{16}O$	111-13-7	128.21	-16	172.5	0.820^{20}	2.13		(2.7)	52			
Isobutyl acetate	$C_6H_{12}O_2$	110-19-0	116.16	-99	117	0.8712^{20}	2.01	2.39	(1.9)	18	1-11%	421	150
Isobutyl alcohol	$C_4H_{10}O$	78-83-1	74.12	-108	108	0.8018^{20}	2.44	1.39	1.64	28	2-11%	415	50

PROPERTIES OF COMMON LABORATORY SOLVENTS (continued)

Name	MF	CAS RN	M_r	t_m/°C	t_b/°C	ρ/g cm^{-3}	c_p/J g^{-1} K^{-1}	vp/kPa	μ/D	FP/°C	Fl. Lim.	Ign. Temp./°C	TLV
Isobutylamine	$C_4H_{11}N$	78-81-9	73.14	-87	68	0.724^{25}	2.50	19.0	(1.3)	-9	2-12%	378	
Isopentyl acetate	$C_7H_{14}O_2$	123-92-2	130.19	-79	143	0.876^{15}	1.91	0.73	(1.9)	25	1-8%	360	100
Isophorone	$C_9H_{14}O$	78-59-1	138.21	-8	215	0.9255^{20}	1.83	0.06		84	1-4%	460	5
Isopropyl acetate	$C_5H_{10}O_2$	108-21-4	102.13	-73	89	0.8718^{20}	1.95	8.1		2	2-8%	460	250
Isopropyl alcohol	C_3H_8O	67-63-0	60.10	-90	82	0.7855^{20}	2.58	6.02	1.56	12	2-13%	399	400
Isoquinoline	C_9H_7N	119-65-3	129.16	26.47	243.2	1.0910^{30}	1.52		2.73	49			
d-Limonene (Citrene)	$C_{10}H_{16}$	5989-27-5	136.24	-97	178	0.8411^{20}	1.83	0.28					
2,6-Lutidine	C_7H_9N	108-48-5	107.16	-6.1	144.1	0.9226^{20}	1.73	0.75	(1.7)				
Mesitylene	C_9H_{12}	108-67-8	120.19	-45	165	0.8652^{20}	1.74	0.33	0	50	1-5%	559	25
Mesityl oxide	$C_6H_{10}O$	141-79-7	98.14	-59	130	0.8653^{20}	2.17	1.47	(2.8)	31	1-7%	344	15
Methyl acetate	$C_3H_6O_2$	79-20-9	74.08	-98	57	0.9342^{20}	1.92	28.8	1.72	-10	3-16%	454	200
Methylal	$C_3H_8O_2$	109-87-5	76.10	-105	42	0.8593^{20}	2.12	53.1	(0.7)	-32	2-14%	237	1000
Methyl alcohol	CH_4O	67-56-1	32.04	-98	65	0.7914^{20}	2.53	16.9	1.70	11	6-36%	464	200
Methylamine	CH_5N	74-89-5	31.06	-93	-6	0.656^{25}	3.29	353	1.31	0	5-21%	430	5
Methyl benzoate	$C_8H_8O_2$	93-58-3	136.15	-15	199	1.0933^{15}	1.63	0.05	(1.9)	83			
Methylcyclohexane	C_7H_{14}	108-87-2	98.19	-127	101	0.7694^{20}	1.88	6.18	=0	-4	1-7%	250	400
Methyl ethyl ketone	C_4H_8O	78-93-3	72.11	-87	80	0.8054^{20}	2.20	12.6	2.78	-9	1-11%	404	200
N-Methylformamide	C_2H_5NO	123-39-7	59.07	-3.8	199.5	1.011^{19}	2.10		3.83				
Methyl formate	$C_2H_4O_2$	107-31-3	60.05	-99	32	0.9742^{20}	1.98	78.1	1.77	-19	5-23%	449	100
Methyl iodide	CH_3I	74-88-4	141.94	-66.4	42.5	2.279^{20}	0.89	53.9	1.62				2
Methyl isobutyl ketone	$C_6H_{12}O$	108-10-1	100.16	-84	116	0.7978^{20}	2.13	2.64		18	1-8%	448	50
Methyl isopentyl ketone	$C_7H_{14}O$	110-12-3	114.19		144	0.888^{20}		0.69		36	1-8%	191	50
2-Methylpentane	C_6H_{14}	107-83-5	86.18	-153.7	60.2	0.650^{25}	2.25	28.2	=0	<-29	1-7%	264	
4-Methyl-2-pentanol	$C_6H_{14}O$	108-11-2	102.18	-90	132	0.8075^{20}	2.67	0.70	(2.6)	41	1-6%		25
Methyl pentyl ketone	$C_7H_{14}O$	110-43-0	114.19	-35	151	0.8111^{20}	2.04	0.49	(2.7)	39	1-8%	393	50
Methyl propyl ketone	$C_5H_{10}O$	107-87-9	86.13	-77	102	0.809^{20}	2.14	4.97		7	2-8%	452	200
N-Methyl-2-pyrrolidone	C_5H_9NO	872-50-4	99.13	-24	202	1.0230^{25}	3.11	0.04	(4.1)	96	1-10%	346	
Morpholine	C_4H_9NO	110-91-8	87.12	-5	128	1.0005^{20}	1.89	1.34	1.55	37	1-11%	290	20
Nitrobenzene	$C_6H_5NO_2$	98-95-3	123.11	6	211	1.2037^{20}	1.51	0.03	4.22	88	2-9%	482	1
Nitroethane	$C_2H_5NO_2$	79-24-3	75.07	-90	114	1.0448^{25}	1.79	2.79	3.23	28	3-17%	414	100
Nitromethane	CH_3NO_2	75-52-5	61.04	-29	101	1.1371^{20}	1.75	4.79	3.46	35	7-22%	418	20
1-Nitropropane	$C_3H_7NO_2$	108-03-2	89.09	-108	131.1	0.9961^{25}	1.97	1.36	3.66	36	2%-	421	25
2-Nitropropane	$C_3H_7NO_2$	79-46-9	89.09	-91	120	0.9821^{25}	1.91	2.3	3.73	24	3-11%	428	10
Octane	C_8H_{18}	111-65-9	114.23	-57	126	0.6986^{25}	2.23	1.86	=0	13	1-7%	206	300
1-Octanol	$C_8H_{18}O$	111-87-5	130.23	-15.5	195.1	0.8262^{25}	2.34	0.01	(1.8)	81			
Pentachloroethane	C_2HCl_5	76-01-7	202.29	-29	160	1.6796^{20}	0.86	0.48	0.92				
Pentamethylene glycol	$C_5H_{12}O_2$	111-29-5	104.15	-18	239	0.9914^{20}	3.08		(2.5)	129		335	
Pentane	C_5H_{12}	109-66-0	72.15	-130	36	0.6262^{20}	2.32	68.3	=0	<-40	2-8%	260	600
1-Pentanol	$C_5H_{12}O$	71-41-0	88.15	-79	138	0.8144^{20}	2.36	0.26	(1.7)	33	1-10%	300	
Pentyl acetate	$C_7H_{14}O_2$	628-63-7	130.19	-71	149	0.8756^{20}	2.00	0.60	1.75	16	1-8%	360	100
2-Picoline	C_6H_7N	109-06-8	93.13	-67	129	0.9443^{20}	1.70	1.5	1.85	39		538	
α-Pinene	$C_{10}H_{16}$	80-56-8	136.24	-64	156	0.8539^{25}		0.64		35		275	
β-Pinene	$C_{10}H_{16}$	127-91-3	136.24	-61.5	166	0.860^{25}		0.61		38		275	
Piperidine	$C_5H_{11}N$	110-89-4	85.15	-11	106	0.8606^{20}	2.11	4.28	(1.2)	16	1-10%		
Propanenitrile	C_3H_5N	107-12-0	55.08	-93	97	0.7818^{20}	2.17	6.14	4.05	2	3-14%	512	
Propyl acetate	$C_5H_{10}O_2$	109-60-4	102.13	-93	102	0.8878^{20}	1.92	4.49	(1.8)	13	2-8%	450	200
Propyl alcohol	C_3H_8O	71-23-8	60.10	-126	97	0.8035^{20}	2.39	2.76	1.55	23	2-14%	412	200
Propylamine	C_3H_9N	107-10-8	59.11	-83	47	0.7173^{20}	2.75	42.1	1.17	-37	2-10%	318	
Propylbenzene	C_9H_{12}	103-65-1	120.19	-99.5	159.2	0.8620^{20}	1.79		=0	30	1-6%	450	

PROPERTIES OF COMMON LABORATORY SOLVENTS (continued)

Name	MF	CAS RN	M_r	t_m/°C	t_b/°C	ρ/g cm⁻³	c_p/J g⁻¹K⁻¹	vp/kPa	μ/D	FP/°C	Fl. Lim.	Ign. Temp./°C	TLV
Propylene glycol	C₃H₈O₂	57-55-6	76.10	-60	188	1.0361[20]	2.51	0.02	(2.2)	99	3-13%	371	
Pseudocumene	C₉H₁₂	95-63-6	120.19	-44	169	0.8758[20]	1.79	0.30	=0	44	1-6%	500	25
Pyridine	C₅H₅N	110-86-1	79.10	-42	115	0.9819[20]	1.68	2.76	2.21	20	2-12%	482	5
Pyrrole	C₄H₅N	109-97-7	67.09	-23.4	129.7	0.9698[20]	1.90	1.10	1.74	39			
Pyrrolidine	C₄H₉N	123-75-1	71.12	-57.8	86.5	0.8586[20]	2.20	8.40	(1.6)	3			
2-Pyrrolidone	C₄H₇NO	616-45-5	85.11	25	251	1.120[20]	1.99		(3.5)	129			
Quinoline	C₉H₇N	91-22-5	129.16	-14.78	237.1	1.0977[15]	1.51		2.29			480	
Styrene	C₈H₈	100-42-5	104.15	-31	145	0.9060[20]	1.75	0.81		31	1-7%	490	50
Sulfolane	C₄H₈O₂S	126-33-0	120.17	28	287	1.2723[18]	1.50	<0.01	(4.8)	177			
α-Terpinene	C₁₀H₁₆	99-86-5	136.24		174	0.8375[19]							
1,1,1,2-Tetrachloro-2,2-difluoroethane	C₂Cl₄F₂	76-11-9	203.83	40.6	91.5	1.649[25]		7.36					500
1,1,2,2-Tetrachloro-1,2-difluoroethane	C₂Cl₄F₂	76-12-0	203.83	26	93	1.6447[25]	0.85	7.51			5-12%		500
1,1,1,2-Tetrachloroethane	C₂H₂Cl₄	630-20-6	167.85	-70	131	1.5406[20]	0.92	1.6					
1,1,2,2-Tetrachloroethane	C₂H₂Cl₄	79-34-5	167.85	-44	146	1.5953[20]	0.97	0.62	1.32		20-54%		1
Tetrachloroethylene	C₂Cl₄	127-18-4	165.83	-22	121	1.6227[20]	0.86	2.42	0				50
Tetraethylene glycol	C₈H₁₈O₅	112-60-7	194.23	-6.2	328	1.1285[15]	2.21			182			
Tetrahydrofuran	C₄H₈O	109-99-9	72.11	-108	65	0.8892[20]	1.72	21.6	1.75	-14	2-12%	321	200
1,2,3,4-Tetrahydronaphthalene	C₁₀H₁₂	119-64-2	132.21	-36	208	0.9660[25]	1.65	0.05	=0	71	1-5%	385	
Tetrahydropyran	C₅H₁₀O	142-68-7	86.13	-45	88	0.8814[20]	1.82	9.54	1.74	-20			
Tetramethylsilane	C₄H₁₂Si	75-76-3	88.22	-99.0	26.6	0.648[19]	2.31	94.2	0				
Toluene	C₇H₈	108-88-3	92.14	-95	111	0.8669[20]	1.70	3.79	0.37	4	1-7%	480	50
o-Toluidine	C₇H₉N	95-53-4	107.16	-16.3	200.3	0.9984[20]	1.96	0.04	(1.6)	85		482	2
Triacetin	C₉H₁₄O₆	102-76-1	218.21	-78	259	1.1583[20]	1.76	<0.01		138	1%-	433	
Tributylamine	C₁₂H₂₇N	102-82-9	185.35	-70	217	0.7770[20]		0.01	(0.8)	86	1-5%		
1,1,1-Trichloroethane	C₂H₃Cl₃	71-55-6	133.40	-30	74	1.3390[20]	1.08	16.5	1.76		8-13%	537	350
1,1,2-Trichloroethane	C₂H₃Cl₃	79-00-5	133.40	-37	114	1.4397[20]	1.13	3.1	(1.4)	32	6-28%	460	10
Trichloroethylene	C₂HCl₃	79-01-6	131.39	-85	87	1.4642[20]	0.95	9.91	(0.8)	32	8-11%	420	50
Trichlorofluoromethane	CCl₃F	75-69-4	137.37	-111	24	1.478[24]	0.89	106	0.46				1000
1,1,2-Trichlorotrifluoroethane	C₂Cl₃F₃	76-13-1	187.38	-35	48	1.5635[25]	0.91	44.8					1000
Triethanolamine	C₆H₁₅NO₃	102-71-6	149.19	21	335	1.1242[20]	2.61	<0.01	(3.6)	179	1-10%		
Triethylamine	C₆H₁₅N	121-44-8	101.19	-115	89	0.7275[20]	2.17	7.70	0.66	-7	1-8%	249	0.5
Triethylene glycol	C₆H₁₄O₄	112-27-6	150.17	-7	285	1.1274[15]	2.18			177	1-9%	371	1
Triethyl phosphate	C₆H₁₅O₄P	78-40-0	182.16	-56.4	215.5	1.0695[20]			(3.1)	115		454	
Trimethylamine	C₃H₉N	75-50-3	59.11	-117	3	0.627[25]	2.33	215	0.61	-7	2-12%	190	5
Trimethylene glycol	C₃H₈O₂	504-63-2	76.10	-26.7	214.4	1.0538[20]			(2.5)			400	
Trimethyl phosphate	C₃H₉O₄P	512-56-1	140.08	-46	197.2	1.2144[20]		0.11	(3.2)	107			
Veratrole	C₈H₁₀O₂	91-16-7	138.17	22.5	206	1.0810[25]			(1.3)				
o-Xylene	C₈H₁₀	95-47-6	106.17	-25	144	0.8802[10]	1.75	0.88	0.64	32	1-7%	463	100
m-Xylene	C₈H₁₀	108-38-3	106.17	-48	139	0.8642[20]	1.72	1.13	=0	27	1-7%	527	100
p-Xylene	C₈H₁₀	106-42-3	106.17	13	138	0.8611[20]	1.71	1.19	0	27	1-7%	528	100

DEPENDENCE OF BOILING POINT ON PRESSURE

The normal boiling point of a liquid is defined as the temperature at which the vapor pressure reaches standard atmospheric pressure, 101.325 kPa. The change in boiling point with pressure may be calculated from the representation of the vapor pressure by the Antoine Equation,

$$\ln p = A_1 - A_2/(T + A_3)$$

where p is the vapor pressure, T the absolute temperature, and A_1, A_2, and A_3 are constants. This table, which has been calculated using the Antoine constants in Reference 1, gives values of $\Delta t/\Delta p$ for a number of liquids, in units of both °C/kPa and °C/mmHg. The correction to the boiling point is generally accurate to 0.1 to 0.2 °C as long as the pressure is within 10% of standard atmospheric pressure.

A slightly less accurate estimate of $\Delta t/\Delta p$ may be obtained from the Claussius-Clapeyron equation, with the assumption that the change in volume upon vaporization equals the ideal-gas volume of the vapor. This leads to the equation

$$\Delta t/\Delta p = RT_b^2/p_0 \Delta_{vap}H(T_b)$$

where R is the molar gas constant, p_0 is 101.325 kPa, T_b is the normal boiling point temperature (absolute), and $\Delta_{vap}H(T_b)$ is the molar enthalpy of vaporization at the normal boiling point. Values of the last quantity may be obtained from the table "Enthalpy of Vaporization" in Section 6.

REFERENCE

1. Lide, D.R., and Kehiaian, H.V., *CRC Handbook of Thermophysical and Thermochemical Data*, CRC Press, Boca Raton, FL, 1994, pp. 49-59.

Compound	t_b °C	$\Delta t/\Delta p$ °C/kPa	$\Delta t/\Delta p$ °C/mmHg	Compound	t_b °C	$\Delta t/\Delta p$ °C/kPa	$\Delta t/\Delta p$ °C/mmHg
Acetaldehyde	20.1	0.261	0.0348	1-Hexanol	157.6	0.318	0.0424
Acetic acid	117.9	0.324	0.0432	Hydrogen fluoride	20.1	0.276	0.0368
Acetone	56.0	0.289	0.0385	Iodomethane	42.5	0.291	0.0388
Acetonitrile	81.6	0.316	0.0421	Isobutane	-11.7	0.254	0.0339
Ammonia	-33.33	0.198	0.0264	Methanol	64.6	0.251	0.0335
Aniline	184.1	0.378	0.0504	Methyl acetate	56.8	0.282	0.0376
Anisole	153.7	0.367	0.0489	Methyl formate	31.7	0.582	0.0776
Benzaldehyde	179.0	0.392	0.0523	N-Methylaniline	196.2	0.396	0.0528
Benzene	80.0	0.321	0.0428	N-Methylformamide	199.5	0.371	0.0495
Bromine	58.8	0.300	0.0400	Nitrobenzene	210.8	0.418	0.0557
Butane	-0.5	0.267	0.0356	Nitromethane	101.1	0.320	0.0427
1-Butanol	117.7	0.278	0.0371	1-Octanol	195.1	0.360	0.0480
Carbon disulfide	46.2	0.304	0.0405	Pentane	36.0	0.289	0.0385
Chlorine	-34.04	0.224	0.0299	1-Pentanol	137.9	0.296	0.0395
Chlorobenzene	131.7	0.365	0.0487	Phenol	181.8	0.349	0.0465
1-Chlorobutane	78.6	0.321	0.0428	Propane	-42.1	0.224	0.0299
Chloroethane	12.3	0.262	0.0349	1-Propanol	97.2	0.261	0.0348
Chloroethylene	-13.3	0.241	0.0321	2-Propanol	82.3	0.247	0.0329
Cyclohexane	80.7	0.328	0.0437	Pyridine	115.2	0.340	0.0453
Cyclohexanol	160.8	0.344	0.0459	Pyrrole	129.7	0.330	0.0440
Cyclohexanone	155.4	0.382	0.0509	Pyrrolidine	86.5	0.309	0.0412
Decane	174.1	0.388	0.0517	Styrene	145.1	0.369	0.0492
Dibutyl ether	140.2	0.363	0.0484	Sulfur dioxide	-10.05	0.221	0.0295
Dichloromethane	39.6	0.276	0.0368	Tetrachloroethylene	121.3	0.354	0.0472
Diethyl ether	34.5	0.278	0.0371	Tetrachloromethane	76.8	0.325	0.0433
Dimethyl sulfoxide	189.0	0.379	0.0505	Toluene	110.6	0.353	0.0471
1,4-Dioxane	101.5	0.321	0.0428	Trichloroethylene	87.2	0.330	0.0440
Dipropyl ether	90.0	0.326	0.0435	Trichloromethane	61.1	0.302	0.0403
Ethanol	78.2	0.249	0.0332	Trimethylamine	2.8	0.248	0.0331
Ethyl acetate	77.1	0.300	0.0400	Water	100.0	0.276	0.0368
Ethylene glycol	197.3	0.331	0.0441	o-Xylene	144.5	0.373	0.0497
Heptane	98.5	0.336	0.0448	m-Xylene	139.1	0.368	0.0491
Hexafluorobenzene	80.2	0.305	0.0407	p-Xylene	138.3	0.369	0.0492
Hexane	68.7	0.314	0.0419				

EBULLIOSCOPIC CONSTANTS FOR CALCULATION OF BOILING POINT ELEVATION

The boiling point T_b of a dilute solution of a non-volatile, non-dissociating solute is elevated relative to that of the pure solvent. If the solution is ideal (i.e., follows Raoult's Law), the amount of elevation depends only on the number of particles of solute present. Hence the change in boiling point ΔT_b can be expressed as

$$\Delta T_b = E_b\, m_2$$

where m_2 is the molality (moles of solute per kilogram of solvent) and E_b is the Ebullioscopic Constant, a characteristic property of the solvent. The Ebullioscopic Constant may be calculated from the relation

$$E_b = R\, T_b^2\, M/\Delta_{vap}H$$

where R is the molar gas constant, T_b is the normal boiling point temperature (absolute) of the solvent, M the molar mass of the solvent, and $\Delta_{vap}H$ the molar enthalpy (heat) of vaporization of the solvent at its normal boiling point.

This table lists E_b values for some common solvents, as calculated from data in the table "Enthalpy of Vaporization" in Section 6.

Compound	E_b/K kg mol^{-1}	Compound	E_b/K kg mol^{-1}
Acetic acid	3.22	Hexane	2.90
Acetone	1.80	Iodomethane	4.31
Acetonitrile	1.44	Methanol	0.86
Aniline	3.82	Methyl acetate	2.21
Anisole	4.20	N-Methylaniline	4.3
Benzaldehyde	4.24	N-Methylformamide	2.2
Benzene	2.64	Nitrobenzene	5.2
1-Butanol	2.17	Nitromethane	2.09
Carbon disulfide	2.42	1-Octanol	5.06
Chlorobenzene	4.36	Phenol	3.54
1-Chlorobutane	3.13	1-Propanol	1.66
Cyclohexane	2.92	2-Propanol	1.58
Cyclohexanol	3.5	Pyridine	2.83
Decane	6.10	Pyrrole	2.33
Dichloromethane	2.42	Pyrrolidine	2.32
Diethyl ether	2.20	Tetrachloroethylene	6.18
Dimethyl sulfoxide	3.22	Tetrachloromethane	5.26
1,4-Dioxane	3.01	Toluene	3.40
Ethanol	1.23	Trichloroethylene	4.52
Ethyl acetate	2.82	Trichloromethane	3.80
Ethylene glycol	2.26	Water	0.513
Heptane	3.62	o-Xylene	4.25

CRYOSCOPIC CONSTANTS FOR CALCULATION OF FREEZING POINT DEPRESSION

The freezing point T_f of a dilute solution of a non-volatile, non-dissociating solute is depressed relative to that of the pure solvent. If the solution is ideal (i.e., follows Raoult's Law), this lowering is a function only of the number of particles of solute present. Thus the absolute value of the lowering of freezing point ΔT_f can be expressed as

$$\Delta T_f = E_f m_2$$

where m_2 is the molality (moles of solute per kilogram of solvent) and E_f is the Cryoscopic Constant, a characteristic property of the solvent. The Cryoscopic Constant may be calculated from the relation

$$E_f = R\, T_f^2\, M/\Delta_{fus}H$$

where R is the molar gas constant, T_b is the freezing point temperature (absolute) of the solvent, M the molar mass of the solvent, and $\Delta_{fus}H$ the molar enthalpy (heat) of fusion of the solvent.

This table lists cryscopic constants for selected substances, as calculated from data in the table "Enthalpy of Fusion" in Section 6.

Compound	E_f/K kg mol^{-1}	Compound	E_f/K kg mol^{-1}
Acetamide	3.92	1,4-Dioxane	4.63
Acetic acid	3.63	Diphenylamine	8.38
Acetophenone	5.16	Ethylene glycol	3.11
Aniline	5.23	Formamide	4.25
Benzene	5.07	Formic acid	2.38
Benzonitrile	5.35	Glycerol	3.56
Benzophenone	8.58	Methylcyclohexane	2.60
(+)-Camphor	37.8	Naphthalene	7.45
1-Chloronaphthalene	7.68	Nitrobenzene	6.87
o-Cresol	5.92	Phenol	6.84
m-Cresol	7.76	Pyridine	4.26
p-Cresol	7.20	Quinoline	6.73
Cyclohexane	20.8	Succinonitrile	19.3
Cyclohexanol	42.2	1,1,2,2-Tetrabromoethane	21.4
cis-Decahydronaphthalene	6.42	1,1,2,2-Tetrachloro-1,2-difluoroethane	41.0
trans-Decahydronaphthalene	4.70	Toluene	3.55
Dibenzyl ether	6.17	*p*-Toluidine	4.91
p-Dichlorobenzene	7.57	Tribromomethane	15.0
Diethanolamine	3.16	Water	1.86
Dimethyl sulfoxide	3.85	*p*-Xylene	4.31

FREEZING POINT LOWERING BY ELECTROLYTES IN AQUEOUS SOLUTION

REFERENCE

Forsythe, W. E., *Smithsonian Physical Tables, Ninth Edition*, Smithsonian Institution, Washington, 1956.

Compound	Lowering of freezing point of water (in °C) as function of molality (mol/kg)									
	0.05	**0.10**	**0.25**	**0.50**	**0.75**	**1.00**	**1.50**	**2.00**	**2.50**	**3.00**
$CaCl_2$	0.25	0.49	1.27	2.66	4.28	6.35	10.78	15.27	20.42	28.08
$CuSO_4$	0.13	0.23	0.47	0.96						
HCl	0.18	0.36	0.90	1.86	2.90	4.02	6.63	9.94		
HNO_3	0.18	0.35	0.88	1.80	2.78	3.80	5.98	8.34	10.95	13.92
H_2SO_4	0.20	0.39	0.96	1.95	3.04	4.28	7.35	11.35	16.32	
KBr	0.18	0.36	0.92	1.78						
KCl	0.17	0.35	0.86	1.68	2.49	3.29	4.88	6.50	8.14	9.77
KNO_3	0.17	0.33	0.78	1.47	2.11	2.66				
K_2SO_4	0.23	0.43	1.01	1.87						
LiCl	0.18	0.35	0.88	1.80	2.78					
$MgSO_4$	0.13	0.24	0.55	1.01	1.50	2.08	3.41			
NH_4Cl	0.17	0.34	0.85	1.70	2.55					
NaCl	0.18	0.35	0.85	1.68	2.60					
$NaNO_3$	0.18	0.36	0.80	1.62	2.63	3.10				

DETERMINATION OF RELATIVE HUMIDITY FROM DEW POINT

The relative humidity of a water vapor-air mixture is defined as 100 times the partial pressure of water divided by the saturation vapor pressure of water at the same temperature. The relative humidity may be determined from the dew point t_{dew}, which is the temperature at which liquid water first condenses when the mixture is cooled from an initial temperature t. This table gives relative humidity as a function of the dew point depression $t - t_{dew}$ for several values of the dew point. Values are calculated from the vapor pressure table in Section 6.

$t - t_{dew}$	$t_{dew}/°C$					$t - t_{dew}$	$t_{dew}/°C$				
	-10	**0**	**10**	**20**	**30**		**-10**	**0**	**10**	**20**	**30**
0.0	100	100	100	100	100	8.2	54	56	59	61	63
0.2	99	99	99	99	99	8.4	53	56	58	60	63
0.4	97	97	97	98	98	8.6	52	55	57	60	62
0.6	95	96	96	96	97	8.8	51	54	57	59	61
0.9	94	94	95	95	96	9.0	51	53	56	58	61
1.0	92	93	94	94	94	9.2	50	53	55	58	60
1.2	91	92	92	93	93	9.4	49	52	55	57	59
1.4	90	90	91	92	92	9.6	48	51	54	56	59
1.6	88	89	90	91	91	9.8	48	51	53	56	58
1.8	87	88	89	90	90	10.0	47	50	53	55	57
2.0	86	87	88	88	89	10.5	45	48	51	54	56
2.2	84	85	86	87	89	11.0	44	47	49	52	55
2.4	83	84	85	86	87	11.5	42	45	48	51	53
2.6	82	83	84	85	86	12.0	41	44	47	49	52
2.8	80	82	83	84	85	12.5	39	42	45	48	50
3.0	79	81	82	83	84	13.0	38	41	44	46	49
3.2	78	80	81	82	83	13.5	37	40	43	45	48
3.4	77	79	80	81	82	14.0	35	38	41	44	47
3.6	76	77	79	80	82	14.5	34	37	40	43	45
3.8	75	76	78	79	81	15.0	33	36	39	42	44
4.0	73	75	77	78	80	15.5	32	35	38	40	
4.2	72	74	76	77	79	16.0	31	34	37	39	
4.4	71	73	75	77	78	16.5	30	33	36	38	
4.6	70	72	74	76	77	17.0	29	32	35	37	
4.8	69	71	73	75	76	17.5	28	31	34	36	
5.0	69	70	72	74	75	18.0	27	30	33	35	
5.2	67	69	71	73	75	18.5	26	29	32	34	
5.4	66	68	70	72	74	19.0	25	28	31	33	
5.6	65	67	69	71	73	19.5	24	27	30	33	
5.9	64	66	69	70	72	20.0	24	26	29	32	
6.0	63	66	68	70	71	21.0	22	25	27	30	
6.2	62	65	67	69	71	22.0	21	23	26	29	
6.4	61	64	66	68	70	23.0	19	22	24	27	
6.6	60	63	65	67	69	24.0	18	21	23	26	
6.8	60	62	64	66	68	25.0	17	19	22	24	
7.0	59	61	63	66	68	26.0	16	18	21	23	
7.2	58	60	63	65	67	27.0	15	17	20	22	
7.4	57	60	62	64	66	28.0	14	16	19	21	
7.6	56	59	61	63	65	29.0	13	15	18	20	
7.8	55	58	60	63	65	30.0	12	14	17	19	
8.0	54	57	60	62	64						

DETERMINATION OF RELATIVE HUMIDITY FROM WET AND DRY BULB TEMPERATURES

Relative humidity may be determined by comparing temperature readings of wet and dry bulb thermometers. The following table, extracted from more extensive U.S. National Weather Service tables, gives the relative humidity as a function of air temperature t_d (dry bulb) and the difference $t_d - t_w$ between dry and wet bulb temperatures. The data assume a pressure near normal atmospheric pressure and an instrumental configuration with forced ventilation.

	$(t_d - t_w)/°C$											
$t_d/°C$	0.5	1.0	1.5	2.0	2.5	3.0	3.5	4.0	4.5	5.0	5.5	6.0
-10	83	67	51	35	19							
-8	86	71	57	43	29	15						
-6	88	74	61	49	37	25	8					
-4	89	77	66	55	44	33	23	12				
-2	90	79	69	60	50	40	31	22	12			
0	91	81	72	64	55	46	38	29	21	13	5	
2	91	84	76	68	60	52	44	37	29	22	14	7
4	92	85	78	71	63	57	49	43	36	29	22	16
6	93	86	79	73	66	60	54	48	41	35	29	24
8	93	87	81	75	69	63	57	51	46	40	35	29
10	94	88	82	77	71	66	60	55	50	44	39	34
12	94	89	83	78	73	68	63	58	53	48	43	39
14	95	90	85	79	75	70	65	60	56	51	47	42
16	95	90	85	81	76	71	67	63	58	54	50	46
18	95	91	86	82	77	73	69	65	61	57	53	49
20	96	91	87	83	78	74	70	66	63	59	55	51
22	96	92	87	83	80	76	72	68	64	61	57	54
24	96	92	88	84	80	77	73	69	66	62	59	56
26	96	92	88	85	81	78	74	71	67	64	61	58
28	96	93	89	85	82	78	75	72	69	65	62	59
30	96	93	89	86	83	79	76	73	70	67	64	61
35	97	94	90	87	84	81	78	75	72	69	67	64
40	97	94	91	88	85	82	80	77	74	72	69	67

	$(t_d - t_w)/°C$											
$t_d/°C$	6.5	7.0	7.5	8.0	8.5	9.0	10.0	11.0	12.0	13.0	14.0	15.0
4	9											
6	17	11	5									
8	24	19	14	8								
10	29	24	20	15	10	6						
12	34	29	25	21	16	12	5					
14	38	34	30	26	22	18	10					
16	42	38	34	30	26	23	15	8				
18	45	41	38	34	30	27	20	14	7			
20	48	44	41	37	34	31	24	18	12	6		
22	50	47	44	40	37	34	28	22	17	11	6	
24	53	49	46	43	40	37	31	26	20	15	10	5
26	54	51	49	46	43	40	34	29	24	19	14	10
28	56	53	51	48	45	42	37	32	27	22	18	13
30	58	55	52	50	47	44	39	35	30	25	21	17
32	60	57	54	51	49	46	41	37	32	28	24	20
34	61	58	56	53	51	48	43	39	35	30	26	23
36	62	59	57	54	52	50	45	41	37	33	29	25
38	63	61	58	56	54	51	47	43	39	35	31	27
40	64	62	59	57	54	53	48	44	40	36	33	29

CONSTANT HUMIDITY SOLUTIONS

Anthony Wexler

An excess of a water soluble salt in contact with its saturated solution and contained within an enclosed space produces a constant relative humidity and water vapor pressure according to

$$RH = A \exp(B/T)$$

where RH is the percent relative humidity (generally accurate to $\pm 2\%$), T is the temperature in kelvin, and the constants A and B and the range of valid temperatures are given in the table below. The vapor pressure, p, can be calculated from

$$p = (RH/100) \times p_0$$

where p_0 is the vapor pressure of pure water at temperature T as given in the table in Section 6 titled "Vapor Pressure of Water from 0 to 370°C".

REFERENCES

1. Wexler, A. S. and Seinfeld, J. H., *Atmospheric Environment*, 25A, 2731, 1991.
2. Greenspan, L., *J. Res. National Bureau of Standards*, 81A, 89, 1977.
3. Broul, et al., *Solubility of Inorganic Two-Component Systems*, Elsevier, New York, 1981.
4. Wagman, D. D. et al., *J. Phys. Chem. Ref. Data*, Vol. 11, Suppl. 2, 1982.

Compound	Temperature range (°C)	RH 25°C	A	B
NaOH · H_2O	15—60	6	5.48	27
LiBr · $2H_2O$	10—30	6	0.23	996
$ZnBr_2$ · $2H_2O$	5—30	8	1.69	455
KOH · $2H_2O$	5—30	9	0.014	1924
LiCl · H_2O	20—65	11	14.53	−75
$CaBr_2$ · $6H_2O$	11—22	16	0.17	1360
LiI · $3H_2O$	15—65	18	0.15	1424
$CaCl_2$· $6H_2O$	15—25	29	0.11	1653
$MgCl_2$· $6H_2O$	5—45	33	29.26	34
NaI· $2H_2O$	5—45	38	3.62	702
$Ca(NO_3)_2$ · $4H_2O$	10—30	51	1.89	981
$Mg(NO_3)_2$ · $6H_2O$	5—35	53	25.28	220
NaBr · $2H_2O$	0—35	58	20.49	308
NH_4NO_3	10—40	62	3.54	853
KI	5—30	69	29.35	254
$SrCl_2$ · $6H_2O$	5—30	71	31.58	241
$NaNO_3$	10—40	74	26.94	302
NaCl	10—40	75	69.20	25
NH_4Cl	10—40	79	35.67	235
KBr	5—25	81	40.98	203
$(NH_4)_2SO_4$	10—40	81	62.06	79
KCl	5—25	84	49.38	159
$Sr(NO_3)_2$ · $4H_2O$	5—25	85	28.34	328
$BaCl_2$ · $2H_2O$	5—25	90	69.99	75
CsI	5—25	91	70.77	75
KNO_3	0—50	92	43.22	225
K_2SO_4	10—50	97	86.75	34

STANDARD SALT SOLUTIONS FOR HUMIDITY CALIBRATION

Saturated aqueous solutions of inorganic salts are convenient secondary standards for calibration of instruments for measurement of relative humidity. The International Union of Pure and Applied Chemistry has recommended the salt solutions in this table for calibrations in the range of 10 to 90% relative humidity. Details on the preparation and use of these standards may be found in Reference 1. Data for other salts are given in Reference 2.

REFERENCES

1. Marsh, K.N., Editor, *Recommended Reference Materials for the Realization of Physicochemical Properties*, Blackwell Scientific Publications, Oxford, 1987, pp.157-162.
2. Greenspan, L., *J. Res. Natl. Bur. Stand.*, 81A, 89, 1977.

$t/°C$	LiCl	MgCl$_2$	Mg(NO$_3$)$_2$	NaCl	KCl
0		33.66 ± 0.33	60.35 ± 0.55	75.51 ± 0.34	88.61 ± 0.53
5		33.60 ± 0.28	58.86 ± 0.43	75.65 ± 0.27	87.67 ± 0.45
10		33.47 ± 0.24	57.36 ± 0.33	75.67 ± 0.22	86.77 ± 0.39
15		33.30 ± 0.21	55.87 ± 0.27	75.61 ± 0.18	85.92 ± 0.33
20	11.31 ± 0.31	33.07 ± 0.18	54.38 ± 0.23	75.47 ± 0.14	85.11 ± 0.29
25	11.30 ± 0.27	32.78 ± 0.16	52.89 ± 0.22	75.29 ± 0.12	84.34 ± 0.26
30	11.28 ± 0.24	32.44 ± 0.14	51.40 ± 0.24	75.09 ± 0.11	83.62 ± 0.25
35	11.25 ± 0.22	32.05 ± 0.13	49.91 ± 0.29	74.87 ± 0.12	82.95 ± 0.25
40	11.21 ± 0.21	31.60 ± 0.13	48.42 ± 0.37		82.32 ± 0.25
45	11.16 ± 0.21	31.10 ± 0.13	46.93 ± 0.47		81.74 ± 0.28
50	11.10 ± 0.22	30.54 ± 0.14	45.44 ± 0.60		81.20 ± 0.31
55	11.03 ± 0.23	29.93 ± 0.16			80.70 ± 0.35
60	10.95 ± 0.26	29.26 ± 0.18			80.25 ± 0.41
65	10.86 ± 0.29	28.54 ± 0.21			79.85 ± 0.48
70	10.75 ± 0.33	27.77 ± 0.25			79.49 ± 0.57
75	10.64 ± 0.38	26.94 ± 0.29			79.17 ± 0.66
80	10.51 ± 0.44	26.05 ± 0.34			78.90 ± 0.77

LOW TEMPERATURE BATHS FOR MAINTAINING CONSTANT TEMPERATURE

A liquid-solid slurry is a convenient means of maintaining a constant temperature environment below room temperature. The following is a list of readily available organic liquids suitable for this purpose, arranged in order of their melting (freezing) points t_m. The normal boiling points t_b are also given.

Compound	$t_m/°C$	$t_b/°C$
Isopentane (2-Methylbutane)	-159.9	27.8
Methylcyclopentane	-142.5	71.8
3-Chloropropene (Allyl chloride)	-134.5	45.1
Pentane	-129.7	36.0
Allyl alcohol	-129	97.0
Ethanol	-114.1	78.2
Carbon disulfide	-111.5	46
Isobutyl alcohol	-108	107.8
Toluene	-94.9	110.6
Acetone	-94.8	56.0
Ethyl acetate	-83.6	77.1
Dry ice + acetone	-78	
p-Cymene	-68.9	177.1
Trichloromethane (Chloroform)	-63.6	61.1
N-Methylaniline	-57	196.2
Chlorobenzene	-45.2	131.7
Anisole	-37.5	153.7
Bromobenzene	-30.6	156.0
Tetrachloromethane (Carbon tetrachloride)	-23	76.8
Benzonitrile	-12.7	191.1

WIRE TABLES

The resistance per unit length of wires of various metals is tabulated here. Values were calculated from resistivity values in the tables "Electrical Resistivity of Pure Metals" and "Electrical Resistivity of Selected Alloys", which appear in Section 12. In practice, resistance may vary because of differing heat treatments and metal composition. The values in the table refer to 20°C, but values at other temperatures may be calculated from the following resistivity data:

Metal	Resistivity in 10^{-8} Ω m at temperature			
	0°C	20°C	25°C	100°C
Aluminum	2.417	2.650	2.709	3.56
Brass (70% Cu, 30% Zn)	5.87	6.08	6.13	6.91
Constantan (60% Cu, 40% Ni)	45.43	45.38	45.35	45.11
Copper	1.543	1.678	1.712	2.22
Nichrome (79% Ni, 21% Cr)	107.3	107.5	107.6	108.3
Platinum	9.6	10.5	10.7	13.6
Silver	1.467	1.587	1.617	2.07
Tungsten	4.82	5.28	5.39	7.18

Resistance per unit length at 20°C in Ω/m

B & S Gauge	Diameter (mm)	Aluminum	Brass	Constantan	Copper	Nichrome	Platinum	Silver	Tungsten
0	8.252	0.000495	0.00114	0.00848	0.000314	0.0201	0.00196	0.000297	0.00099
2	6.543	0.000788	0.00181	0.0135	0.000499	0.0320	0.00312	0.000472	0.00157
4	5.189	0.00125	0.00287	0.0214	0.000793	0.0508	0.00496	0.000750	0.00250
6	4.115	0.00199	0.00457	0.0341	0.00126	0.0808	0.00789	0.00119	0.00397
8	3.264	0.00317	0.00727	0.0542	0.00200	0.128	0.0125	0.00190	0.00631
10	2.588	0.00504	0.0115	0.0863	0.00319	0.204	0.0200	0.00302	0.0100
12	2.053	0.00800	0.0184	0.137	0.00507	0.325	0.0317	0.00479	0.0159
14	1.628	0.0127	0.0292	0.218	0.00806	0.516	0.0504	0.00762	0.0254
16	1.291	0.0202	0.0464	0.347	0.0128	0.821	0.0802	0.0121	0.0403
18	1.024	0.0322	0.0738	0.551	0.0204	1.30	0.127	0.0193	0.0641
20	0.8118	0.0512	0.117	0.877	0.0324	2.08	0.203	0.0307	0.102
22	0.6439	0.0814	0.187	1.39	0.0515	3.30	0.322	0.0487	0.162
24	0.5105	0.129	0.297	2.22	0.0820	5.25	0.513	0.0775	0.258
26	0.4049	0.206	0.472	3.52	0.130	8.35	0.815	0.123	0.410
28	0.3211	0.327	0.751	5.60	0.207	13.3	1.30	0.196	0.652
30	0.2548	0.520	1.19	8.90	0.329	21.1	2.06	0.311	1.03
32	0.2019	0.828	1.90	14.2	0.524	33.6	3.28	0.496	1.65
34	0.1601	1.32	3.02	22.5	0.833	53.4	5.22	0.788	2.62
36	0.1270	2.09	4.80	35.8	1.32	84.9	8.29	1.25	4.17
38	0.1007	3.33	7.63	57.0	2.11	135	13.2	1.99	6.63
40	0.07988	5.29	12.1	90.5	3.35	214	20.9	3.17	10.5

CHARACTERISTICS OF PARTICLES AND PARTICLE DISPERSOIDS

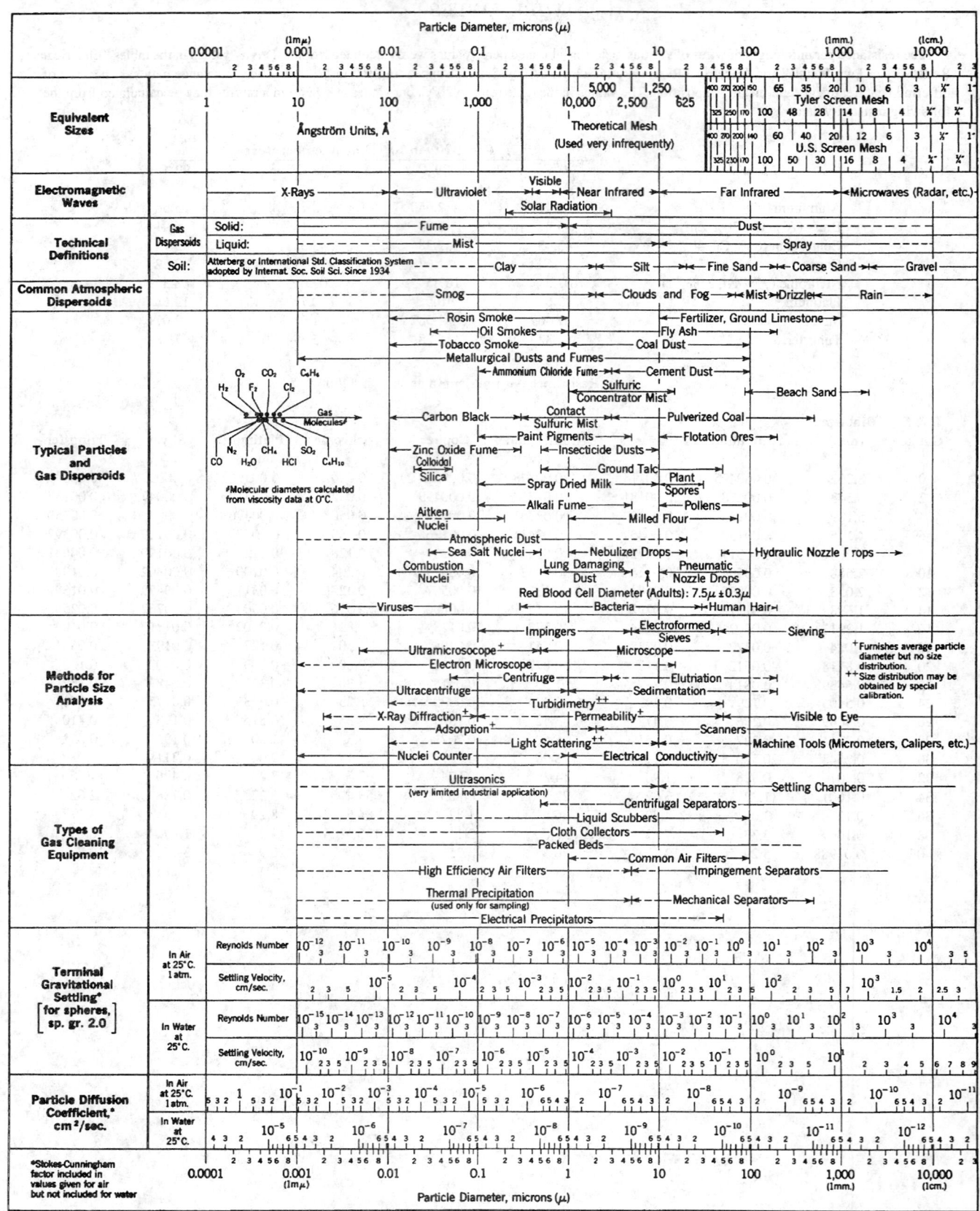

C. E. Lapple, Stanford Research
Institute Journal, Vol. 5, p.95
(Third Quarter, 1961)

DENSITY OF VARIOUS SOLIDS

This table gives the range of density for miscellaneous solid materials whose characteristics depend on the source or method of preparation.

REFERENCES

1. Forsythe, W. E., *Smithsonian Physical Tables, Ninth Edition*, Smithsonian Institution, Washington, 1956.
2. Kaye, G. W. C., and Laby, T. H., *Tables of Physical and Chemical Constants, 16th Edition*, Longman, London, 1995.
3. Brandrup, J., and Immergut, E. H., *Polymer Handbook, Third Edition*, John Wiley & Sons, New York, 1989.

Material	ρ/ g cm^{-3}	Material	ρ/ g cm^{-3}	Material	ρ/ g cm^{-3}
Agate	2.5-2.7	Pyrex	2.23	Soapstone	2.6-2.8
Alabaster,		Granite	2.64-2.76	Solder	8.7-9.4
carbonate	2.69-2.78	Graphite	2.30-2.72	Starch	1.53
sulfate	2.26-2.32	Gum arabic	1.3-1.4	Steel, stainless	7.8
Albite	2.62-2.65	Gypsum	2.31-2.33	Sugar	1.59
Amber	1.06-1.11	Hematite	4.9-5.3	Talc	2.7-2.8
Amphiboles	2.9-3.2	Hornblende	3.0	Tallow, beef	0.94
Anorthite	2.74-2.76	Ice	0.917	Tar	1.02
Asbestos	2.0-2.8	Iron, cast	7.0-7.4	Topaz	3.5-3.6
Asbestos slate	1.8	Ivory	1.83-1.92	Tourmaline	3.0-3.2
Asphalt	1.1-1.5	Kaolin	2.6	Tungsten carbide	14.0-15.0
Basalt	2.4-3.1	Leather, dry	0.86	Wax, sealing	1.8
Beeswax	0.96-0.97	Lime, slaked	1.3-1.4	Wood (seasoned)	
Beryl	2.69-2.70	Limestone	2.68-2.76	alder	0.42-0.68
Biotite	2.7-3.1	Linoleum	1.18	apple	0.66-0.84
Bone	1.7-2.0	Magnetite	4.9-5.2	ash	0.65-0.85
Brasses	8.44-8.75	Malachite	3.7-4.1	balsa	0.11-0.14
Brick	1.4-2.2	Marble	2.6-2.84	bamboo	0.31-0.40
Bronzes	8.74-8.89	Meerschaum	0.99-1.28	basswood	0.32-0.59
Butter	0.86-0.87	Mica	2.6-3.2	beech	0.70-0.90
Calamine	4.1-4.5	Muscovite	2.76-3.00	birch	0.51-0.77
Calcspar	2.6-2.8	Ochre	3.5	blue gum	1.00
Camphor	0.99	Opal	2.2	box	0.95-1.16
Cardboard	0.69	Paper	0.7-1.15	butternut	0.38
Celluloid	1.4	Paraffin	0.87-0.91	cedar	0.49-0.57
Cement, set	2.7-3.0	Peat blocks	0.84	cherry	0.70-0.90
Chalk	1.9-2.8	Pitch	1.07	dogwood	0.76
Charcoal,		Polyamides	1.15-1.25	ebony	1.11-1.33
oak	0.57	Polyethylene	0.92-0.97	elm	0.54-0.60
pine	0.28-0.44	Poly(methyl methacrylate)	1.19	hickory	0.60-0.93
Cinnabar	8.12	Polypropylene	0.91-0.94	holly	0.76
Clay	1.8-2.6	Polystyrene	1.06-1.12	juniper	0.56
Coal,		Polytetrafluoroethylene	2.28-2.30	larch	0.50-0.56
anthracite	1.4-1.8	Poly(vinyl acetate)	1.19	locust	0.67-0.71
bituminous	1.2-1.5	Poly(vinyl chloride)	1.39-1.42	logwood	0.91
Coke	1.0-1.7	Porcelain	2.3-2.5	mahogany	0.66-0.85
Copal	1.04-1.14	Porphyry	2.6-2.9	maple	0.62-0.75
Cork	0.22-0.26	Pyrite	4.95-5.10	oak	0.60-0.90
Corundum	3.9-4.0	Quartz	2.65	pear	0.61-0.73
Diamond	3.51	Resin	1.07	pine, pitch	0.83-0.85
Dolomite	2.84	Rock salt	2.18	white	0.35-0.50
Ebonite	1.15	Rubber,		yellow	0.37-0.60
Emery	4.0	hard	1.19	plum	0.66-0.78
Epidote	3.25-3.50	soft	1.1	poplar	0.35-0.50
Feldspar	2.55-2.75	pure gum	0.91-0.93	satinwood	0.95
Flint	2.63	Neoprene	1.23-1.25	spruce	0.48-0.70
Fluorite	3.18	Sandstone	2.14-2.36	sycamore	0.40-0.60
Galena	7.3-7.6	Serpentine	2.50-2.65	teak, Indian	0.66-0.98
Garnet	3.15-4.3	Silica, fused,	2.21	walnut	0.64-0.70
Gelatin	1.27	Silicon carbide	3.16	water gum	1.00
Glass,		Slag	2.0-3.9	willow	0.40-0.60
common	2.4-2.8	Slate	2.6-3.3	Wood's metal	9.70
lead	3-4				

DIELECTRIC STRENGTH OF INSULATING MATERIALS

L. I. Berger

The loss of the dielectric properties by a sample of a gaseous, liquid, or solid insulator as a result of application to the sample of an electric field* greater than a certain critical magnitude is called *dielectric breakdown*. The critical magnitude of electric field at which the breakdown of a material takes place is called the *dielectric strength* of the material (or *breakdown voltage*). The dielectric strength of a material depends on the specimen thickness (as a rule, thin films have greater dielectric strength than that of thicker samples of a material), the electrode shape**, the rate of the applied voltage increase, the shape of the voltage vs. time curve, and the medium surrounding the sample, e.g., air or other gas (or a liquid — for solid materials only).

Breakdown in Gases

The current carriers in gases are free electrons and ions generated by external radiation. The equilibrium concentration of these particles at normal pressure is about 10^3 cm^{-3}, and hence the electrical conductivity is very small, of the order of 10^{-16} - 10^{-15} S/cm. But in a strong electric field, these particles acquire kinetic energy along their free pass, large enough to ionize the gas molecules. The new charged particles ionize more molecules; this avalanche-like process leads to formation between the electrodes of channels of conducting plasma (streamers), and the electrical resistance of the space between the electrodes decreases virtually to zero.

Because the dielectric strength (breakdown voltage) of gases strongly depends on the electrode geometry and surface condition and the gas pressure, it is generally accepted to present the data for a particular gas as a fraction of the dielectric strength of either nitrogen or sulfur hexafluoride measured at the same conditions. In Table 1, the data are presented in comparison with the dielectric strength of nitrogen, which is considered equal to 1.00. For convenience to the reader, a few average magnitudes of the dielectric strength of some gases are expressed in kilovolts per millimeter. The data in the table relate to the standard conditions, unless indicated otherwise.

Breakdown in Liquids

If a liquid is pure, the breakdown mechanism in it is similar to that in gases. If a liquid contains liquid impurities in the form of small drops with greater dielectric constant than that of the main liquid, the breakdown is the result of formation of ellipsoids from these drops by the electric field. In a strong enough electric field, these ellipsoids merge and form a high-conductivity channel between the electrodes. The current increases the temperature in the channel, liquid boils, and the current along the steam canal leads to breakdown. Formation of a conductive channel (bridge) between the electrodes is observed also in liquids with solid impurities. If a liquid contains gas impurities in the form of small bubbles, breakdown is the result of heating of the liquid in strong electric fields. In the locations with the highest current density, the liquid boils, the size of the gas bubbles increases, they merge and form gaseous channels between the electrodes, and the breakdown medium is again the gas plasma.

Breakdown in Solids

It is known that the current in solid insulators does not obey Ohm's law in strong electric fields. The current density increases almost exponentially with the electric field, and at a certain field magnitude it jumps to very high magnitudes at which a specimen of a material is destroyed. The two known kinds of electric breakdown are thermal and electrical breakdowns. The former is the result of material heating by the electric current. Destruction of a sample of a material happens when, at a certain voltage, the amount of heat produced by the current exceeds the heat release through the sample surface; the breakdown voltage in this case is proportional to the square root of the ratio of the thermal conductivity and electrical conductivity of the material. The electrical breakdown results from the tunneling of the charge carriers from electrodes or from the valence band or from the impurity levels into the conduction band, or by the impact ionization. The tunnel effect breakdown happens mainly in thin layers, e.g., in thin p-n junctions. Otherwise, the impact ionization mechanism dominates. For this mechanism, the dielectric strength of an insulator can be estimated using Boltzmann's kinetic equation for electrons in a crystal.

In the following tables, the dielectric strength values are for room temperature and normal atmospheric pressure, unless indicated otherwise.

---

* The unit of electric field in the SI system is newton per coulomb or volt per meter.

** For example, the U.S. standard ASTM D149 is based on use of symmetrical electrodes, while per U.K. standard BS2918 one electrode is a plane and the other is a rod with the axis normal to the plane.

Table 1
Dielectric Strength of Gases

Material	Dielectric* Strength	Ref.	Material	Dielectric* Strength	Ref.
Nitrogen, N_2	1.00		Trichlorofluoromethane, CCl_3F	3.50	1
Hydrogen, H_2	0.50	1,2		4.53	2
Helium, He	0.15	1	Trichloromethane, $CHCl_3$	4.2	1
Oxygen, O_2	0.92	2		4.39	2
Air	0.97	6	Methylamine, CH_3NH_2	0.81	1
Air (flat electrodes), kV/mm	3.0	3	Difluoromethane, CH_2F_2	0.79	2
Air, kV/mm	0.4-0.7	4	Trifluoromethane, CHF_3	0.71	2
Air, kV/mm	1.40	5	Bromochlorodifluoromethane, CF_2ClBr	3.84	2
Neon, Ne	0.25	1	Chlorodifluoromethane, $CHClF_2$	1.40	1
	0.16	2		1.11	2
Argon, Ar	0.18	2	Dichlorofluoromethane, $CHCl_2F$	1.33	1
Chlorine, Cl_2	1.55	1		2.61	2
Carbon monoxide, CO	1.02	1	Chlorofluoromethane, CH_2ClF	1.03	1
	1.05	2	Hexafluoroethane, C_2F_6	1.82	1
Carbon dioxide, CO_2	0.88	1		2.55	2
	0.82	2	Ethyne (Acetylene), C_2H_2	1.10	1
	0.84	6		1.11	2
Nitrous oxide, N_2O	1.24	2	Chloropentafluoroethane, C_2ClF_5	2.3	1
Sulfur dioxide, SO_2	2.63	2		3.0	6
	2.68	6	Dichlorotetrafluoroethane, $C_2Cl_2F_4$	2.52	1
Sulfur monochloride, S_2Cl_2	1.02	1	Chlorotrifluoroethylene, C_2ClF_3	1.82	2
(at 12.5 Torr)			1,1,1-Trichloro-2,2,2-trifluoroethane	6.55	2
Thionyl fluoride, SOF_2	2.50	1	1,1,2-Trichloro-1,2,2-trifluoroethane	6.05	2
Sulfur hexafluoride, SF_6	2.50	1	Chloroethane, C_2H_5Cl	1.00	1
	2.63	2	1,1-Dichloroethane	2.66	2
Sulfur hexafluoride, SF_6, kV/mm	8.50	7	Trifluoroacetonitrile, CF_3CN	3.5	1
	9.8	8	Acetonitrile, CH_3CN	2.11	2
Perchloryl fluoride, ClO_3F	2.73	1	Dimethylamine, $(CH_3)_2NH$	1.04	1
Tetrachloromethane, CCl_4	6.33	1	Ethylamine, $C_2H_5NH_2$	1.01	1
	6.21	2	Ethylene oxide (oxirane), CH_3CHO	1.01	1
Tetrafluoromethane, CF_4	1.01	1	Perfluoropropene, C_3F_6	2.55	2
Methane, CH_4	1.00	1	Octafluoropropane, C_3F_8	2.19	1
	1.13	2		2.47	2
Bromotrifluoromethane, CF_3Br	1.35	1	3,3,3-Trifluoro-1-propene, CH_2CHCF_3	2.11	2
	1.97	2	Pentafluoroisocyanoethane, C_2F_5NC	4.5	1
Bromomethane, CH_3Br	0.71	2	1,1,1,4,4,4-Hexafluoro-2-butyne, CF_3CCCF_3	5.84	2
Chloromethane, CH_3Cl	1.29	2	Octafluorocyclobutane, C_4F_8	3.34	2
Iodomethane, CH_3I	3.02	2	1,1,1,2,3,4,4,4-Octafluoro-2-butene	2.8	1
Iodomethane, CH_3I, at 370 Torr	2.20	7	Decafluorobutane, C_4F_{10}	3.08	1
Dichloromethane, CH_2Cl_2	1.92	2	Perfluorobutanenitrile, C_3F_7CN	5.5	1
Dichlorodifluoromethane, CCl_2F_2	2.42	1	Perfluoro-2-methyl-1,3-butadiene, C_5F_8	5.5	1
	2.63	2,6	Hexafluorobenzene, C_6F_6	2.11	2
Chlorotrifluoromethane, $CClF_3$	1.43	1	Perfluorocyclohexane, C_6F_{12}, (saturated vapor)	6.18	2
	1.53	2			

*Relative to nitrogen, unless units of kV/mm are indicated.

Table 2
Dielectric Strength of Liquids

Material	Dielectric strength kV/mm	Ref.	Material	Dielectric strength kV/mm	Ref.
Helium, He, liquid, 4.2 K	10	9	Octane, C_8H_{18}	16.6	14
Static	10	11		20.4	15
Dynamic	5	11		179	17,18
	23	12	Ethylbenzene, C_8H_{10}	226	17,18
Nitrogen, N_2, liquid, 77K			Propylbenzene, C_9H_{12}	250	17,18
Coaxial cylinder electrodes	20	10	Isopropylbenzene, C_9H_{12}	238	17,18
Sphere to plane electrodes	60	10	Decane, $C_{10}H_{22}$	192	17,18
Water, H_2O, distilled	65-70	13	Butylbenzene, $C_{10}H_{14}$	275	17,18
Carbon tetrachloride, CCl_4	5.5	14	Isobutylbenzene, $C_{10}H_{14}$	222	17,18
	16.0	15	Silicone oils—polydimethylsiloxanes,		
Hexane, C_6H_{14}	42.0	16	$(CH_3)_3Si-O-[Si(CH_3)_2]_x-O-Si(CH_3)_3$		
Two 2.54 cm diameter spherical			Polydimethylsiloxane silicone fluid	15.4	20
electrodes, 50.8 µm space	156	17,18	Dimethyl silicone	24.0	21,22
Cyclohexane, C_6H_{12}	42-48	16	Phenylmethyl silicone	23.2	22
2-Methylpentane, C_6H_{14}	149	17,18	Silicone oil, Basilone M50	10-15	23
2,2-Dimethylbutane, C_6H_{14}	133	17,18	Mineral insulating oils	11.8	6
2,3-Dimethylbutane, C_6H_{14}	138	17,18	Polybutene oil for capacitors	13.8	6
Benzene, C_6H_6	163	17,18	Transformer dielectric liquid	28-30	6
Chlorobenzene, C_6H_5Cl	7.1	14	Isopropylbiphenyl capacitor oil	23.6	6
	18.8	15	Transformer oil	110.7	24
2,2,4-Trimethylpentane, C_8H_{18}	140	17,18	Transformer oil Agip ITE 360	9-12.6	23
Phenylxylylethane	23.6	19	Perfluorinated hydrocarbons		
Heptane, C_7H_{16}	166	17,18	Fluorinert FC 6001	8.0	23
2,4-Dimethylpentane, C_7H_{16}	133	17,18	Fluorinert FC 77	10.7	23
Toluene, $C_6H_5CH_3$	199	17,18	Perfluorinated polyethers		
	46	16	Galden XAD (Mol. wt. 800)	10.5	23
	12.0	14	Galden D40 (Mol. wt. 2000)	10.2	23
	20.4	15	Castor oil	65	25

Table 3
Dielectric Strength of Solids

Material	Dielectric strength kV/mm	Ref	Material	Dielectric strength kV/mm	Ref
Sodium chloride, NaCl, crystalline	150	26	Phlogopite, amber, natural	118	6
Potassium bromide, KBr, crystalline	80	26	Fluorophlogopite, synthetic	118	6
Ceramics			Glass-bonded mica	14.0-15.7	6
Alumina (99.9% Al_2O_3)	13.4	6,27a	Thermoplastic Polymers		
Aluminum silicate, Al_2SiO_5	5.9	6	Polypropylene	23.6	6
Berillia (99% BeO)	13.8	6,27b	Amide polymer nylon 6/6, dry	23.6	6
Boron nitride, BN	37.4	6	Polyamide-imide copolymer	22.8	6
Cordierite, $Mg_2Al_4Si_5O_{18}$	7.9	6,27c	Modified polyphenylene oxide	21.7	6
Forsterite, Mg_2SiO_4	9.8	28	Polystyrene	19.7	6
Porcelain	35-160	26	Polymethyl methacrylate	19.7	6
Steatite, $Mg_3Si_4O_{11} \cdot H_2O$	9.1-15.4	6	Polyetherimide	18.9	6
Titanates of Mg, Ca, Sr, Ba, and Pb	20-120	3	Amide polymer nylon 11(dry)	16.7	6
Barium titanate, glass bonded	>30	36	Polysulfone	16.7	6
Zirconia, ZrO_2	11.4	29	Styrene-acrylonitrile copolymer	16.7	6
Glasses			Acrylonitrile-butadiene-styrene	16.7	6
Fused silica, SiO_2	470-670	26	Polyethersulfone	15.7	6
Alkali-silicate glass	200	26	Polybutylene terephthalate	15.7	6
Standard window glass	9.8-13.8	28	Polystyrene-butadiene copolymer	15.7	6
Micas			Acetal homopolymer	15.0	6
Muscovite, ruby, natural	118	6	Acetal copolymer	15.0	6

Table 3
Dielectric Strength of Solids (continued)

Material	Dielectric strength kV/mm	Ref.	Material	Dielectric strength kV/mm	Ref.
Polyphenylene sulfide	15.0	6	Varnishes		
Polycarbonate	15.0	6	Vacuum-pressure impregnated baking type solventless polyester varnish		
Acetal homopolymer resin (molding resin)	15.0	6	Rigid, two-part	70.9	6
Acetal copolymer resin	15.0	6	Semiflexible high-bond thixotropic	78.7	6
Thermosetting Molding Compounds			Rigid high-bond high-flash freon-resistant	68.9	6
Glass-filled allyl (Type GDI-30 per MIL-M-14G)	15.7	6	Baking type epoxy varnish		
Glass-filled epoxy, electrical grade	15.4	6	Solventless, rigid, low viscosity, one-part	90.6	6
Glass-filled phenolic (Type GPI-100 per MIL-M-14G)	15.0	6	Solventless, semiflexible, one-part	82.7	6
Glass-filled alkyd/polyester (Type MAI-60 per MIL-M-14G)	14.8	6	Solventless, semirigid, chemical resistant, low dielectric constant	106.3	6
Glass-filled melamine (Type MMI-30 per MIL-M-14G)	13.4	6	Solvable, for hermetic electric motors	181.1	6
Extrusion Compounds for High-Temperature Insulation			Polyurethane coating		
Polytetrafluoroethylene	19.7	6	Clear conformal, fast cure		
Perfluoroalkoxy polymer	21.7	6	Standard conditions	78.7	6
Fluorinated ethylene-propylene copolymer	19.7	6	Immersion conditions	47.2	6
Ethylene-tetrafluoroethylene copolymer	15.7	6	Insulating Films and Tapes		
Polyvinylidene fluoride	10.2	6	Low-density polyethylene film (40 μm thick)	300	31
Ethylene-chlorotrifluoroethylene copolymer	19.3	6	Poly-p-xylylene film	410-590	32
Polychlorotrifluoroethylene	19.7	6	Aromatic polymer films		
Extrusion Compounds for Low-Temperature Insulation			Kapton H (Du Pont)	389-430	33
Polyvinyl chloride			Ultem (GE Plastic and Roem AG)	437-565	33
Flexible	11.8-15.7	30	Hostaphan (Hoechst AG)	338-447	33
Rigid	13.8-19.7	30	Amorphous Stabar K2000 (ICI film)	404-422	33
Polyethylene	18.9	28	Stabar S100 (ICI film)	353-452	33
Polyethylene, low-density	21.7	6	Polyetherimide film (26 μm)	486	34
	300	31	Parylene N/D (poly-p-xylylene/poly-dichloro-p-xylylene) 25 μm film	275	6
Polyethylene, high-density	19.7	6	Cellulose acetate film	157	6
Polypropylene/polyethylene copolymer	23.6	6	Cellulose triacetate film	157	6
Embedding Compounds			Polytetrafluoroethylene film	87-173	6
Basic epoxy resin: bisphenol-A/epichlorohydrin polycondensate	19.7	6	Perfluoroalkoxy film	157-197	6
Cycloaliphatic epoxy: alicyclic diepoxy carboxylate	19.7	6	Fluorinated ethylene-propylene copolymer film	197	6
Polyetherketone	18.9	30	Ethylene-tetrafluoroethylene film	197	6
Polyurethanes			Ethylene-chlorotrifluoroethylene copolymer film	197	6
Two-component, polyol-cured	25.4	6	Polychlorotrifluoroethylene film	118-153.5	6
Two-part solventless, polybutylene-based	24.0	6	High-voltage rubber insulating tape	28	6
Silicones			Various Insulators		
Clear two-part heat curing eletrical grade silicone embedding resin	21.7	6	Rubber, natural	100-215	26
Red insulating enamel (MIL-E-22118)			Butyl rubber	23.6	6
Dry	47.2	6	Neoprene	15.7-27.6	6
Wet	11.8	6	Silicone rubber	26-36	6
Enamels			Room-temperature vulcanized silicone rubber	9.2-10.9	35
Red enamel, fast cure			Ureas (from carbamide to tetraphenylurea)	11.8-15.7	28
Standard conditions	78.7	6	Dielectric papers		
Immersion conditions	47.2	6	Aramid paper, calendered	28.7	6
Black enamel			Aramid paper, uncalendered	12.2	6
Standard conditions	70.9	6	Aramid with Mica	39.4	6
Immersion conditions	47.2	6			

REFERENCES

1. Vijh, A. K. *IEEE Trans.,* EI-12, 313, 1997.
2. Brand, K. P., *IEEE Trans.,* EI-17, 451, 1982.
3. *Encyclopedic Dictionary in Physics,* Vedensky, B. A. and Vul, B. M., Eds., Vol. 4, Soviet Encyclopedia Publishing House, Moscow, 1965.
4. Kubuki, M., Yoshimoto, R., Yoshizumi, K., Tsuru, S., and Hara, M., *IEEE Trans.,* DEI-4, 92, 1997.
5. Al-Arainy, A. A. Malik, N. H., and Cureshi, M. I., *IEEE Trans.,* DEI-1, 305, 1994.
6. Shugg, W. T., *Handbook of Electrical and Electronic Insulating Materials,* Van Nostrand Reinhold, New York, 1986.
7. Devins, J. C., *IEEE Trans.,* EI-15, 81, 1980.
8. Xu, X., Jayaram, S., and Boggs, S. A., *IEEE Trans.,* DEI-3, 836, 1996.
9. Okubo, H., Wakita, M., Chigusa, S., Nayakawa, N., and Hikita, M., *IEEE Trans.,* DEI-4, 120, 1997.
10. Hayakawa, H., Sakakibara, H., Goshima, H., Hikita, M., and Okubo, H., *IEEE Trans.,* DEI-4, 127, 1997.
11. Okubo, H., Wakita, M., Chigusa, S., Hayakawa, N., and Hikita, M., *IEEE Trans.,* DEI-4, 220, 1997.
12. Von Hippel, A. R., *Dielectric Materials and Applications,* MIT Press, Cambridge, MA, 1954.
13. Jones, H. M. and Kunhards, E. E., *IEEE Trans.,* DEI-1, 1016, 1994.
14. Nitta, Y. and Ayhara, Y., *IEEE Trans.,* EI-11, 91, 1976.
15. Gallagher, T. J., *IEEE Trans.,* EI-12, 249, 1977.
16. Wong, P. P. and Forster, E. O., in *Dielectric Materials. Measurements and Applications,* IEE Conf. Publ. 177, 1, 1979.
17. Kao, K. C. *IEEE Trans.,* EI-11, 121, 1976.
18. Sharbaugh, A. H., Crowe, R. W., and Cox, E. B., *J. Appl. Phys.,* 27, 806, 1956.
19. Miller, R. L., Mandelcorn, L., and Mercier, G. E., in *Proc. Intl. Conf. on Properties and Applications of Dielectric Materials,* Xian, China, June 24-28, 1985; cited in Ref. 6, p. 492.
20. Hakim, R. M., Oliver, R. G., and St-Onge, H., *IEEE Trans.,* EI-12, 360, 1977.
21. Hosticka, C., *IEEE Trans.,* 389, 1977.
22. Yasufuku, S., Umemura, T., and Ishioka, Y., *IEEE Trans.,* EI-12, 402, 1977.
23. Forster, E. O., Yamashita, H., Mazzetti, C., Pompini, M., Caroli, L., and Patrissi, S., *IEEE Trans.,* DEI-1, 440, 1994.
24. Bell, W. R., *IEEE Trans.,* 281, 1977.
25. Ramu, T. C. and Narayana Rao, Y., in *Dielectric Materials. Measurements and Applications,* IEE Conf. Publ. 177, 37.
26. Skanavi, G. I., *Fizika Dielektrikov; Oblast Silnykh Polei* (Physics of Dielectrics; Strong Fields). Gos. Izd. Fiz. Mat. Nauk (State Publ. House for Phys. and Math. Scis.), Moscow, 1958.
27. Kleiner, R. N., in *Practical Handbook of Materials Science,* Lynch, C. T., Ed., CRC Press, 1989; 27a: p. 304; 27b: p.300; 27c: p. 316.
28. *Materials Selector Guide. Materials and Methods,* Reinhold Publ., New York, 1973.
29. Flinn, R. A. and Trojan, P. K., *Engineering Materials and Their Applications,* 2nd ed., Houghton Mifflin, 1981, p. 614.
30. Lynch, C. T., Ed., *Practical Handbook of Materials Science,* CRC Press, Boca Raton, FL, 1989.
31. Suzuki, H., Mukai, S., Ohki, Y., Nakamichi, Y., and Ajiki, K., *IEEE Trans.,* DEI-4, 238, 1997.
32. Mori, T., Matsuoka, T., and Muzitani, T., *IEEE Trans.,* DEI-1, 71, 1994.
33. Bjellheim, P. and Helgee, B., *IEEE Trans.,* DEI-1, 89, 1994.
34. Zheng, J. P., Cygan, P. J., and Jow, T. R., *IEEE Trans.,* DEI-3, 144, 1996.
35. Danukas, M. G., *IEEE Trans.,* DEI-1, 1196, 1994.
36. Burn, I. and Smithe, D. H., *J. Mater. Sci.,* 7, 339, 1972.

Review Literature on the Subject

R1. Kuffel, E. and Zaengl, W. S., *HV Engineering Fundamentals,* Pergamon, 1989.
R2. Kok, J. A., *Electrical Breakdown of Insulating Liquids,* Phillips Tech. Library, Cleaver-Hum, Longon, 1961.
R3. Gallagher, T. J., *Simple Dielectric Liquids,* Clarendon, Oxford, 1975.
R4. Meek, J. M. and Craggs, J. D., Eds., *Electric Breakdown in Gases,* John Wiley & Sons, 1976.
R5. Von Hippel, A. R., *Dielectric Materials and Applications,* MIT Press, Cambridge, MA, 1954.

Section 16
Health and Safety Information

HANDLING AND DISPOSAL OF CHEMICALS IN LABORATORIES

Robert Joyce and Blaine C. McKusick

The following material has been extracted from two books prepared under the auspices of the Committee on Hazardous Substances in the Laboratory of the National Academy of Sciences — National Research Council. Readers are referred to these books for full details:

Prudent Practices for Handling Hazardous Chemicals in Laboratories, National Academy Press, Washington, 1981.
Prudent Practices for Disposal of Chemicals from Laboratories, National Academy Press, Washington, 1983.

The permission of the National Academy Press to use these extracts is gratefully acknowledged.

INCOMPATIBLE CHEMICALS

The term "incompatible chemicals" refers to chemicals that can react with each other

- Violently
- With evolution of substantial heat
- To produce flammable products
- To produce toxic products

Good laboratory safety practice requires that incompatible chemicals be stored, transported, and disposed of in ways that will prevent their coming together in the event of an accident. Tables 1 and 2 give some basic guidelines for the safe handling of acids, bases, reactive metals, and other chemicals. Neither of these tables is exhaustive, and additional information on incompatible chemicals can be found in the following references.

1. L. Bretherick, *Handbook of Reactive Chemical Hazards*, 3rd ed., Butterworths, London–Boston, 1985.
2. L. Bretherick, Ed., *Hazards in the Chemical Laboratory*, 3rd ed., Royal Society of Chemistry, London, 1981.
3. *Manual of Hazardous Chemical Reactions, A Compilation of Chemical Reactions Reported to be Potentially Hazardous,* National Fire Protection Association, NFPA 491M, 1975, NFPA, 470 Atlantic Avenue, Boston, MA 02210.

TABLE 1
General Classes of Incompatible Chemicals

A	B
Acids	Bases, reactive metals
Oxidizing agents[a]	Reducing agents[a]
Chlorates	Ammonia, anhydrous and aqueous
Chromates	Carbon
Chromium trioxide	Metals
Dichromates	Metal hydrides
Halogens	Nitrites
Halogenating agents	Organic compounds
Hydrogen peroxide	Phosphorus
Nitric acid	Silicon
Nitrates	Sulfur
Perchlorates	
Peroxides	
Permanganates	
Persulfates	

[a] The examples of oxidizing and reducing agents are illustrative of common laboratory chemicals; they are not intended to be exhaustive.

TABLE 2
Examples of Incompatible Chemicals

Chemical	Is incompatible with
Acetic acid	Chromic acid, nitric acid, hydroxyl compounds, ethylene glycol, perchloric acid, peroxides, permanaganates
Acetylene	Chlorine, bromine, copper, fluorine, silver, mercury
Acetone	Concentrated nitric and sulfuric acid mixtures
Alkali and alkaline earth metals (such as powdered aluminum or magnesium, calcium, lithium, sodium, potassium)	Water, carbon tetrachloride or other chlorinated hydrocarbons, carbon dioxide, halogens
Ammonia (anhydrous)	Mercury (in manometers, for example), chlorine, calcium hypochlorite, iodine, bromine, hydrofluoric acid (anhydrous)
Ammonium nitrate	Acids, powdered metals, flammable liquids, chlorates, nitrites, sulfur, finely divided organic or combustible materials
Aniline	Nitric acid, hydrogen peroxide
Arsenical materials	Any reducing agent
Azides	Acids
Bromine	See Chlorine
Calcium oxide	Water
Carbon (activated)	Calcium hypochlorite, all oxidizing agents
Carbon tetrachloride	Sodium
Chlorates	Ammonium salts, acids, powdered metals, sulfur, finely divided organic or combustible materials
Chromic acid and chromium troixide	Acetic acid, naphthalene, camphor, glycerol, alcohol, flammable liquids in general
Chlorine	Ammonia, acetylene, butadiene, butane, methane, propane (or other petroleum gases), hydrogen, sodium carbide, benzene, finely divided metals, turpentine
Chlorine dioxide	Ammonia, methane, phosphine, hydrogen sulfide
Copper	Acetylene, hydrogen peroxide
Cumene hydroperoxide	Acids (organic or inorganic)
Cyanides	Acids
Flammable liquids	Ammonium nitrate, chromic acid, hydrogen peroxide, nitric acid, sodium peroxide, halogens
Fluorine	Everything
Hydrocarbons (such as butane, propane, benzene)	Fluorine, chlorine, bromine, chromic acid, sodium peroxide
Hydrocyanic acid	Nitric acid, alkali
Hydrofluoric acid (anhydrous)	Ammonia (aqueous or anhydrous)
Hydrogen peroxide	Copper, chromium, iron, most metals or their salts, alcohols, acetone, organic materials, aniline, nitromethane, combustible matierals
Hydrogen sulfide	Fuming nitric acid, oxidizing gases
Hypochlorites	Acids, activated carbon
Iodine	Acetylene, ammonia (aqueous or anhydrous), hydrogen
Mercury	Acetylene, fulminic acid, ammonia
Nitrates	Sulfuric acid
Nitric acid (concentrated)	Acetic acid, aniline, chromic acid, hydrocyanic acid, hydrogen sulfide, flammable liquids, flammable gases, copper, brass, any heavy metals
Nitrites	Acids
Nitroparaffins	Inorganic bases, amines
Oxalic acid	Silver, mercury

TABLE 2
Examples of Incompatible Chemicals (continued)

Chemical	Is incompatible with
Oxygen	Oils, grease, hydrogen, flammable liquids, solids, or gases
Perchloric acid	Acetic anhydride, bismuth and its alloys, alcohol, paper, wood, grease, oils
Peroxides, organic	Acids (organic or mineral), avoid friction, store cold
Phosphorus (white)	Air, oxygen, alkalis, reducing agents
Potassium	Carbon tetrachloride, carbon dioxide, water
Potassium chlorate	Sulfuric and other acids
Potassium perchlorate (see also chlorates)	Sulfuric and other acids
Potassium permanganate	Glycerol, ethylene glycol, benzaldehyde, surfuric acid
Selenides	Reducing agents
Silver	Acetylene, oxalic acid, tartartic acid, ammonium com- pounds, fulminic acid
Sodium	Carbon tetrachloride, carbon dioxide, water
Sodium nitrite	Ammonium nitrate and other ammonium salts
Sodium peroxide	Ethyl or methyl alcohol, glacial acetic acid, acetic anhydride, benzaldehyde, carbon disulfide, glycerin, ethylene glycol, ethyl acetate, methyl acetate, furfural
Sulfides	Acids
Sulfuric acid	Potassium chlorate, potassium perchlorate, potassium permanganate (similar compounds of light metals, such as sodium, lithium)
Tellurides	Reducing agents

EXPLOSION HAZARDS

Table 3 lists some common classes of laboratory chemicals that have potential for producing a violent explosion when subjected to shock or friction. These chemicals should never be disposed of as such, but should be handled by procedures given in *Prudent Practices for Disposal of Chemicals from Laboratories*, National Academy Press, 1983, chapters 6 and 7. Additional information on these, as well as on some less common classes of explosives, can be found in L. Bretherick, *Handbook of Reactive Chemical Hazards,* 3rd ed., Butterworths, London–Boston, 1985.

Table 4 lists some illustrative combinations of common laboratory reagents that can produce explosions when they are brought together or that form reaction products that can explode without any apparent external initiating action. This list is not exhaustive, and additional information on potentially explosive reagent combinations can be found in Manual of Hazardous Chemical Reactions, A Compilation of Chemical Reactions Reported to be Potentially Hazardous, National Fire Protection Association, NFPA 491M, 1975, NFPA, 470 Atlantic Avenue, Boston, MA 02210.

WATER-REACTIVE CHEMICALS

Table 5 lists some common laboratory chemicals that react violently with water and that should always be stored and handled so that they do not come into contact with liquid water or water vapor. Procedures for decomposing laboratory quantities are given in Prudent Practices for Disposal of Chemicals from Laboratories, chapter 6; the pertinent section of that chapter is given in parentheses.

PYROPHORIC CHEMICALS

Many members of the classes of readily oxidized, common laboratory chemicals listed in Table 6 ignite spontaneously in air. A more extensive list can be found in L. Bretherick, *Handbook of Reactive Chemical Hazards*, 3rd ed., Butterworths, London-Boston, 1985. Pyrophoric chemicals should be stored in tightly closed containers under an inert atmosphere (or, for some, an inert liquid), and all transfers and manipulations of them must be carried out under an inert atmosphere or liquid. Suggested procedures for decomposing them are given in Prudent Practices for Disposal of Chemicals from Laboratories, chapter 6; the pertinent section of that chapter is given in parentheses.

TABLE 3
Shock–Sensitive Compounds

Acetylenic compounds, especially polyacetylenes, haloacetylenes, and heavy metal salts of acetylenes (copper, silver, and mercury salts are particularly sensitive)

Acyl nitrates

Alkyl nitrates, particularly polyol nitrates such as nitrocellulose and nitroglycerine

Alkyl and acyl nitrites

Alkyl perchlorates

Amminemetal oxosalts: metal compounds with coordinated ammonia, hydrazine, or similar nitrogenous donors and ionic perchlorate, nitrate, permanganate, or other oxidizing group

Azides, including metal, nonmetal, and organic azides

Chlorite salts of metals, such as $AgClO_2$ and $Hg(ClO_2)_2$

Diazo compounds such as CH_2N_2

Diazonium slats, when dry

Fulminates (silver fulminate, AgCNO, can form in the reaction mixture from the Tollens' test for aldehydes if it is allowed to stand for some time; this can be prevented by adding dilute nitric acid to the test mixture as soon as the test has been completed)

Hydrogen peroxide becomes increasingly treacherous as the concentration rises above 30%, forming explosive mixtures with organic materials and decomposing violently in the presence of traces of transition metals

N–Halogen compounds such as difluoroamino compounds and halogen azides

N–Nitro compounds such as N–nitromethylamine, nitrourea, nitroguanidine, and nitric amide

Oxo salts of nitrogenous bases: perchlorates, dichromates, nitrates, iodates, chlorites, chlorates, and permanganates of ammonia, amines, hydroxylamine, guanidine, etc.

Perchlorate salts. Most metal, nonmetal, and amine perchlorates can be detonated and may undergo violent reaction in contact with combustible materials

Peroxides and hydroperoxides, organic (see Chapter 6, Section II.P)

Peroxides (solid) that crystallize from or are left from evaporation of peroxidizable solvents (see Chapter 6 and Appendix I)

Peroxides, transition–metal salts

Picrates, especially salts of transition and heavy metals, such as Ni, Pb, Hg, Cu, and Zn; picric acid is explosive but is less sensitive to shock or friction than its metal salts and is relatively safe as a water–wet paste (see Chapter 7)

Polynitroalkyl compounds such as tetranitromethane and dinitroacetonitrile

Polynitroaromatic compounds, especially polynitro hydrocarbons, phenols, and amines

TABLE 4
Potentially Explosive Combinations of Some Common Reagents

Acetone + chloroform in the presence of base
Acetylene + copper, silver, mercury, or their salts
Ammonia (including aqueous solutions) + Cl_2, Br_2, or I_2
Carbon disulfide + sodium azide
Chlorine + an alcohol
Chloroform or carbon tetrachloride + powdered Al or Mg
Decolorizing carbon + an oxidizing agent
Diethyl ether + chlorine (including a chlorine atmosphere)
Dimethyl sulfoxide + an acyl halide, $SOCl_2$ or $POCl_3$
Dimethyl sulfoxide + CrO_3
Ethanol + calcium hypochlorite
Ethanol + silver nitrate
Nitric acid + acetic anhydride or acetic acid
Picric acid + a heavy–metal salt, such as of Pb, Hg, or Ag
Silver oxide + ammonia + ethanol
Sodium + a cholrinated hydrocarbon
Sodium hypochlorite + an amine

TABLE 5
Water–Reactive Chemicals

Alkali metals (III.D)
Alkali metal hydrides (III.C.2)
Alkali metal amides (III.C.7)
Metal alkyls, such as lithium alkyls and aluminum alkyls (IV.A)
Grignard reagents (IV.A)
Halides of nonmetals, such as BCl_3, BF_3, PCl_3, PCl_5, $SiCl_4$, S_2Cl_2 (III.F)
Inorganic acid halides, such as $POCl_3$, $SOCl_2$, SO_2Cl_2 (III.F)
Anhydrous metal halides, such as $AlCl_3$, $TiCl_4$, $ZrCl_4$, $SnCl_4$ (III.E)
Phosphorus pentoxide (III.I)
Calcium carbide (IV.E)
Organic acid halides and anhydrides of low molecular weight (II.J)

TABLE 6
Classes of Pyrophoric Chemicals

Grignard reagents, RMgX (IV.A)
Metal alkyls and aryls, such as RLi, RNa, R_3Al, R_2Zn (IV.A)
Metal carbonyls, such as Ni$(CO)_4$, Fe$(CO)_5$, $Co_2(CO)_8$ (IV.B)
Alkali metals such as Na, K (III.D.l)
Metal powders, such as Al, Co, Fe, Mg, Mn, Pd, Pt, Ti, Sn, Zn, Zr (III.D.2)
Metal hydrides, such as NaH, $LiAlH_4$ (IV.C.2)
Nonmetal hydrides, such as B_2H_6 and other boranes, PH_3, AsH_3 (III.G)
Nonmetal alkyls, such as R_3B, R_3P, R_3As (IV.C)
Phosphorus (white) (III.H)

HAZARDS FROM PEROXIDE FORMATION

Many common laboratory chemicals can form peroxides when allowed access to air over a period of time. A single opening of a container to remove some of the contents can introduce enough air for peroxide formation to occur. Some types of compounds form peroxides that are treacherously and violently explosive in concentrated solution or as solids. Accordingly, peroxide–containing liquids should never be evaporated near to or to dryness. Peroxide formation can also occur in many polymerizable unsaturated compounds, and these peroxides can initiate a runaway, sometimes explosive, polymerization reaction. Procedures for testing for peroxides and for removing small amounts from laboratory chemicals are given in Prudent Practices for Disposal of Chemicals from Laboratories, chapter 6, Section II.P.

Table 7 provides a list of structural characteristics in organic compounds that can peroxidize. These structures are listed in approximate order of decreasing hazard. Reports of serious incidents involving the last five structural types are extremely rare, but these structures are listed because laboratory workers should be aware that they can form peroxides that can influence the course of experiments in which they are used.

Table 8 gives examples of common laboratory chemicals that are prone to form peroxides on exposure to air. The lists are not exhaustive, and analogous organic compounds that have any of the structural features given in Table 7 should be tested for peroxides before being used as solvents or reagents, or before being distilled. The recommended retention times begin with the date of synthesis or of opening the original container.

DISPOSAL OF TOXIC CHEMICALS

It is often desirable to precipitate toxic cations or hazardous anions from solution to facilitate recovery or disposal. Table 9 lists precipitants for many common cations, and Table 10 gives precipitants for some hazardous anions. Many cations can be precipitated as sulfides by adding sodium sulfide solution (preferable to the highly toxic hydrogen sulfide) to a neutral solution of the cation (Table 11). Control of pH is important because some sulfides will redissolve in excess sulfide ion. After precipitation, excess sulfide can be destroyed by addition of hypochlorite.

Most metal cations are precipitated as hydroxides or oxides at high pH. Since many of these precipitates will redissolve in excess base, it is often necessary to control pH. Table 12 shows the recommended pH range for precipitating many cations in their most common oxidation state. The notation "1 N" in the right–hand column indicates that the precipitate will not dissolve in 1 N sodium hydroxide (pH 14).

The distinctions between high and low toxicity or hazard are based on toxicological and other data, and are relative. There is no implication of a sharp distinction between high and low, or that any cations or anions are totally without hazard.

TABLE 7
Types of Chemicals That Are Prone to Form Peroxides

A. Organic structures (in approximate order of decreasing hazard)

No.	Structure	Description
1.	$\overset{H}{\underset{}{C}}-O-$	Ethers and acetals with a hydrogen atoms
2.	$C=C-\overset{H}{C}$	Olefins with allylic hydrogen atoms
3.	$\overset{X}{C}=C-$	Chloroolefins and fluoroolefins
4.	$CH_2=C\big\langle$	Vinyl halides, esters, and ethers
5.	$C=C-C=C$	Dienes
6.	$C=\overset{H}{C}-C\equiv CH$	Vinylacetylenes with α hydrogen atoms
7.	$\overset{H}{C}-C\equiv CH$	Alkylacetylenes with α hydrogen atoms
8.	$\overset{H}{C}-Ar$	Alkylarenes that contain tertiary hydrogen atoms
9.	$-\overset{|}{\underset{|}{C}}-H$	Alkanes and cycloalkanes that contain tertiary hydrogen atoms
10.	$C=C-CO_2R$	Acrylates and methacrylates
11.	$\overset{H}{C}-OH$	Secondary alcohols
12.	$-\overset{O}{\underset{}{C}}-\overset{H}{C}$	Ketones that contain a hydrogen atoms
13.	$-\overset{H}{C}=O$	Aldehydes
14.	$-\overset{O}{\underset{}{C}}-\overset{H}{\underset{}{N}}-\overset{H}{C}$	Ureas, amides, and lactams that have a hydrogen atom on a carbon atom attached to nitrogen

TABLE 7
Types of Chemicals That Are Prone to Form Peroxides (continued)

B. Inorganic substances

1. Alkali metals, especially potassium, rubidium, and cesium (see Chapter 6, Section III.D)
2. Metal amides (see Chapter 6, Section III.C.7)
3. Organometallic compounds with a metal atom bonded to carbon (see Chapter 6, Section IV)
4. Metal alkoxides

TABLE 8
Common Peroxide-Forming Chemicals

LIST A

Severe Peroxide Hazard on Storage with Exposure to Air
Discard within 3 months

- Diisopropyl ether (isopropyl ether)
- Divinylacetylene (DVA)[a]
- Potassium metal
- Potassium amide
- Sodium amide (sodamide)
- Vinylidene chloride (1,1–dichloro-ethylene)[a]

LIST B

Peroxide Hazard on Concentration; Do Not Distill or Evaporate Without First Testing for the Presence of Peroxides
Discard or test for peroxides after 6 months

- Acetaldehyde diethyl acetal (acetal)
- Cumene (isopropylbenzene)
- Cyclohexene
- Cyclopentene
- Decalin (decahydronaphthalene)
- Diacetylene
- Dicyclopentadiene
- Diethyl ether (ether)
- Diethylene glycol dimethyl ether (diglyme)
- Dioxane
- Ethylene glycol dimethyl ether (glyme)
- Ethylene glycol ether acetates
- Ethylene glycol monoethers (cello-solves)
- Furan
- Methylacetylene
- Methylcyclopentane
- Methyl isobutyl ketone
- Tetrahydrofuran (THF)
- Tetralin (tetrahydronaphthalene)
- Vinyl ethers[a]

LIST C

Hazard of Rapid Polymerization Initiated by Internally Formed Peroxides[a]
a. Normal Liquids; Discard or test for peroxides after 6 months[b]

- Chloroprene (2–chloro–1,3–buta-diene)[c]
- Styrene
- Vinyl acetate
- Vinylpyridine

b. Normal Gases; Discard after 12 months[d]

- Butadiene[c]
- Tetrafluoroethylene (TFE)[c]
- Vinylacetylene (MVA)[c]
- Vinyl chloride

[a]   Polymerizable monomers should be stored with a polymerization inhibitor from which the monomer can be separated by distillation just before use.

[b]   Although common acrylic monomers such as acrylonitrile, acrylic acid, ethyl acrylate, and methyl methacrylate can form peroxides, they have not been reported to develop hazardous levels in normal use and storage.

[c]   The hazard from peroxides in these compounds is substantially greater when they are stored in the liquid phase, and if so stored without an inhibitor they should be considered as in LIST A.

TABLE 8
Common Peroxide-Forming Chemicals (continued)

[d] Although air will not enter a gas cylinder in which gases are stored under pressure, these gases are sometimes transferred from the original cylinder to another in the laboratory, and it is difficult to be sure that there is no residual air in the receiving cylinder. An inhibitor should be put into any such secondary cylinder before one of these gases is transferred into it; the supplier can suggest inhibitors to be used. The hazard posed by these gases is much greater if there is a liquid phase in such a secondary container, and even inhibited gases that have been put into a secondary container under conditions that create a liquid phase should be discarded within 12 months.

Note: Laboratory workers should label all containers of peroxidizable solvents or reagents with one of the following:

[LIST A]

Peroxidizable compound
Received   Opened
Date   _____   _____
Discard 3 months after opening

[LISTS B AND C]

Peroxidizable compound
Received   Opened
Date   _____   _____
Discard or test for peroxides
6 months after opening

TABLE 9
Relative Toxicity of Cations

High toxic hazard	Precipitant[a]	Low toxic hazard	Precipitant[a]
Antimony	OH^-, S^{2-}	Aluminum	OH^-
Arsenic	S^{2-}	Bismuth	OH^-, S^{2-}
Barium	SO_4^{2-}, CO_3^{2-}	Calcium	SO_4^{2-}, CO_3^{2-}
Beryllium	OH^-	Cerium	OH^-
Cadmium	OH^-, S^{2-}	Cesium	
Chromium (III)[b]	OH^-	Copper[c]	OH^-, S^{2-}
Cobalt (II)[b]	OH^-, S^{2-}	Gold	OH^-, S^{2-}
Gallium	OH^-	Iron[c]	OH^-, S^{2-}
Germanium	OH^-, S^{2-}	Lanthanides	OH^-
Hafnium	OH^-	Lithium	
Indium	OH^-, S^{2-}	Magnesium	OH^-
Iridium	OH^-, S^{2-}	Molybdenum (VI)[b,d]	
Lead	OH^-, S^{2-}	Niobium (V)	OH^-
Manganese (II)[b]	OH^-, S^{2-}	Palladium	OH^-, S^{2-}
Mercury	OH^-, S^{2-}	Potassium	
Nickel	OH^-, S^{2-}	Rubidium	
Osmium (IV)[b,e]	OH^-, S^{2-}	Scandium	OH^-
Platinum (II)[b]	OH^-, S^{2-}	Sodium	
Rhenium (VII)[b]	S^{2-}	Strontium	SO_4^{2-}, CO_3^{2-}
Rhodium (III)[b]	OH^-, S^{2-}	Tantalum	OH^-
Ruthenium (III)[b]	OH^-, S^{2-}	Tin	OH^-, S^{2-}
Selenium	S^{2-}	Titanium	OH^-
Silver	Cl^-, OH^-, S^{2-}	Yttrium	OH^-
Tellurium	S^{2-}	Zinc[c]	OH^-, S^2
Thallium	OH^-, S^{2-}	Zirconium	OH^-
Tungsten (VI)[b,d]			
Vanadium	OH^-, S^{2-}		

TABLE 9
Relative Toxicity of Cations (continued)

[a] Precipitants are listed in order of preference:

OH^- = base (sodium hydroxide or sodium carbonate)

S^{2-} = sulfide

Cl^- = chloride

SO_4^{2-} = sulfate

CO_3^{2-} = carbonate

[b] The precipitant is for the indicated valence state.

[c] Maximum tolerance levels have been set for these low–toxicity ions by the U.S. Public Health Service, and large amounts should not be put into public sewer systems. The small amounts typically used in laboratories will not normally affect water supplies.

[d] These ions are best precipitated as calcium molybdate or calcium tungstate.

[e] CAUTION: OsO_4, a volatile, extremely poisonous substance, is formed from almost any osmium compound under acid conditions in the presence of air.

TABLE 10
Relative Hazard of Anions

High–hazard anions			
Ion	Hazard type[a]	Precipitant	Low-hazard anions
Aluminum hydride, AlH_4	F	—	Bisulfite, HSO_3^-
Amide, NH_2^-	F, E[b]	—	Borate, BO_3^{3-}, $B_4O_7^{2-}$
Arsenate, AsO_3^-, AsO_4^{3-}	T	Cu^{2+}, Fe^{2+}	Bromide, Br^-
Arsenite, AsO_2^-, AsO_3^{3-}	T	Pb^{2+}	Carbonate, CO_3^{2-}
Azide, N_3^-	E, T	—	Chloride, Cl^-
Borohydride, BH_4^-	F	—	Cyanate, OCN^-
Bromate, BrO_3^-	O, E	—	Hydroxide, OH^-
Chlorate, ClO_3^-	O, E	—	Iodide, I^-
Chromate, CrO_4^{2-}, $Cr_2O_7^{2-}$	T, O	[c]	Oxide, O^{2-}
Cyanide, CN^-	T	—	Phosphate, PO_4^{3-}
Ferricyanide, $Fe(CN)_6^{3-}$	T	Fe^{2+}	Sulfate, SO_4^{2-}
Ferrocyanide, $Fe(CN)_6^{4-}$	T	Fe^{3+}	Sulfite, SO_3^{2-}
Fluoride, F^-	T	Ca^{2+}	Thiocyanate, SCN^-
Hydride, H^-	F	—	
Hydroperoxide, O_2H^-	O, E	—	
Hydrosulfide, SH^-	T	—	
Hypochlorite, OCl^-	O	—	
Iodate, IO_3^-	O, E	—	
Nitrate, NO_3^-	O	—	
Nitrite, NO_2^-	T, O	—	
Perchlorate, ClO_4^-	O, E	—	
Permanganate, MnO_4^-	T, O	[d]	
Peroxide, O_2^{2-}	O, E	—	
Persulfate, $S_2O_8^{2-}$	O	—	
Selenate, SeO_4^{2-}	T	Pb^{2+}	
Selenide, Se^{2-}	T	Cu^{2+}	
Sulfide, S^{2-}	T	[e]	

[a] Toxic, T: oxidant, O; flammable, F; explosive, E.

[b] Metal amides readily form explosive peroxides on exposure to air.

[c] Reduce and precipitate as Cr(III); see Table 9.

[d] Reduce and precipitate as Mn(II); see Table 9.

[e] See Table 11.

TABLE 11
Precipitation of Sulfides

Precipitated at pH 7	Not precipitated at low pH	Forms a soluble complex at high pH
Ag^+		
As^{3+a}		X
Au^{+a}		X
Bi^{3+}		
Cd^{2+}		
Co^{2+}	X	
Cr^{3+a}		
Cu^{2+}		
Fe^{2+a}	X	
Ge^{2+}		X
Hg^{2+}		X
In^{3+}	X	
Ir^{4+}		X
Mn^{2+a}	X	
Mo^{3+}		X
Ni^{2+}	X	
Os^{4+}		
Pb^{2+}		
Pd^{2+a}		
Pt^{2+a}		X
Re^{4+}		
Rh^{2+a}		
Ru^{4+}		
Sb^{3+a}		X
Se^{2+}		X
Sn^{2+}		X
Te^{4+}		X
Tl^{+a}	X	
V^{4+a}		
Zn^{2+}	X	

a Higher oxidation states of this ion are reduced by sulfide ion and precipitated as this sulfide.

TABLE 12
pH Range for Precipitation of Metal
Hydroxides and Oxides

	1	2	3	4	5	6	7	8	9	10	
Ag^{1+}									⊢⟶		1 N
Al^{3+}							⊢———⊣				
As^{3+}	Not precipitated (precipitate as sulfide)										
As^{5+}	Not precipitated (precipitate as sulfide)										
Au^{3+}							⊢——⊣				
Be^{2+}							⊢———⊣				
Bi^{3+}							⊢————⟶				1 N
Cd^{2+}							⊢————⟶				1 N
Co^{2+}								⊢———⟶			1 N
Cr^{3+}							⊢————⟶				1 N
Cu^{1+}								⊢———⟶			1 N

TABLE 12
pH Range for Precipitation of Metal
Hydroxides and Oxides (continued)

REFERENCES

L. Erdey, *Gravimetric Analysis*, Part II, Pergamon Press, New York, 1965.

D. T. Burns, A. Towsend, and A. H. Carter, *Inorganic Reaction Chemistry*, Vol. 2, Ellis Horwood, New York, 1981.

FIRE HAZARDS

Flammable solvents are a common source of laboratory fires. The relative ease with which some common laboratory solvents can be ignited is indicated by the following properties.

Flash Point — The lowest temperature, as determined by standard tests, at which a liquid emits vapor in sufficient concentration to form an ignitable mixture with air near the surface of the liquid in a test vessel. Note that many of these common chemicals have flash points below room temperature.

Ignition Temperature — The minimum temperature required to initiate self–sustained combustion, regardless of the heat source.

Flammable Limits — The lower flammable limit is the minimum concentration (percent by volume) of a vapor in air below which a flame is not propagated when an ignition source is present. Below this concentration the mixture is too lean to burn. The upper flammable limit is the maximum concentration (percent by volume) of the vapor in air above which a flame is not propagated. Above this concentration the mixture is too rich to burn. The flammable range comprises all concentrations between these two limits. This range becomes wider with increasing temperature and in oxygen–rich atmospheres. Table 13 lists these properties for a few common laboratory chemicals.

GLOVE MATERIALS

It is good safety practice (and mandated in some laboratories) to wear rubber gloves while handling chemicals that can cause injury when in contact with, or absorbed through, the skin. The various common rubbers are not equally resistant to all chemicals. Table 14 provides guidelines for selecting the best, and avoiding the poorest, glove material for handling a given chemical.

RESPIRATORS

In the event of a laboratory accident or spill, it will be necessary for someone to enter the contaminated area for cleanup. If significant quantities of a chemical are spilled, or even minor quantities of a known toxic material, it is essential to wear the correct kind of respirator equipment when entering the area. If it is not known whether the contamination is of a chemical "Immediately dangerous to life or health", the prudent course is to assume that it is, and to use the corresponding type of respirator. Guidelines are presented in Table 15.

TABLE 13
Flash Points, Boiling Points, Ignition Temperatures, and Flammable Limits of Some Common Laboratory Chemicals

Chemical	Flash point (°C)	Boiling point (°C)	Ignition temp. (°C)	Flammable limit (percent by volume in air)	
				Lower	Upper
Acetaldehyde	−37.8	21.1	175.0	4.0	60.0
Acetone	−19.0	56.0	538.0	2.6	12.8
Benzene	−11.1	80.1	560.0	1.4	8.0
Carbon disulfide	−30.0	45.8	90.0	1.0	44.0
Cyclohexane	−18.0	80.7	260.0	1.3	8.0
Diethyl ether	−45.0	34.4	160.0	1.8	48.0
Ethanol	12.0	78.3	363.0	3.3	19.0
n–Heptane	−3.9	98.4	204.0	1.0	6.7
n–Hexane	−21.7	68.7	223.0	1.2	7.5
Isopropyl alcohol	11.7	82.2	398.9	2.0	12.0
Methanol	11.1	64.5	385.0	6.0	36.5
Methyl ethyl ketone	−6.1	79.6	515.6	1.9	11.0
Pentane	−40.0	36.1	260.0	1.4	7.8
Styrene	31.0	145.0	490.0	1.1	6.1
Toluene	4.4	110.6	530.0	1.3	7.0
p–Xylene	25.0	132.4	529.0	1.1	7.0

Note: For a more extensive listing, see the table "Properties of Common Solvents" in Section 15.

TABLE 14
Resistance to Chemicals of Common Glove Materials
(E = Excellent, G = Good, F = Fair, P = Poor)

Chemical	Natural rubber	Neoprene	Nitrile	Vinyl
Acetaldehyde	G	G	E	G
Acetic acid	E	E	E	E
Acetone	G	G	G	F
Acrylonitrile	P	G	—	F
Ammonium hydroxide (sat)	G	E	E	E
Aniline	F	G	E	G
Benzaldehyde	F	F	E	G
Benzene[a]	P	F	G	F
Benzyl chloride[a]	F	P	G	P
Bromine	G	G	—	G
Butane	P	E	—	P
Butyraldehyde	P	G	—	G
Calcium hypochlorite	P	G	G	G
Carbon disulfide	P	P	G	F
Carbon tetrachloride[a]	P	F	G	F
Chlorine	G	G	—	G
Chloroacetone	F	E	—	P
Chloroform[a]	P	F	G	P
Chromic acid	P	F	F	E
Cyclohexane	F	E	—	P
Dibenzyl ether	F	G	—	P
Dibutyl phthalate	F	G	—	P
Diethanolamine	F	E	—	E
Diethyl ether	F	G	E	P
Dimethyl sulfoxide[b]	—	—	—	—
Ethyl acetate	F	G	G	F
Ethylene dichloride[a]	P	F	G	
Ethylene glycol	G	G	E	E
Ethylene trichloride[a]	P	P	—	P
Fluorine	G	G	—	G
Formaldehyde	G	E	E	E
Formic acid	G	E	E	E
Glycerol	G	G	E	E
Hexane	P	E	—	P
Hydrobromic acid (40%)	G	E	—	E
Hydrochloric acid (conc)	G	G	G	E
Hydrofluoric acid (30%)	G	G	G	E
Hydrogen peroxide	G	G	G	E
Iodine	G	G	—	G
Methylamine	G	G	E	E
Methyl cellosolve	F	E	—	P
Methyl chloride[a]	P	E	—	P
Methyl ethyl ketone	F	G	G	P
Methylene chloride[a]	F	F	G	F
Monoethanolamine	F	E	—	E
Morpholine	F	E	—	E
Naphthalene[a]	G	G	E	G
Nitric acid (conc)	P	P	P	G
Perchloric acid	F	G	F	E
Phenol	G	E	—	E
Phosphoric acid	G	E	—	E
Potassium hydroxide (sat)	G	G	G	E

TABLE 14
Resistance to Chemicals of Common Glove Materials
(E = Excellent, G = Good, F = Fair, P = Poor) (continued)

Chemical	Natural rubber	Neoprene	Nitrile	Vinyl
Propylene dichloride[a]	P	F	—	P
Sodium hydroxide	G	G	G	E
Sodium hypochlorite	G	P	F	G
Sulfuric acid (conc)	G	G	F	G
Toluene[a]	P	F	G	F
Trichloroethylene[a]	P	F	G	F
Tricresyl phosphate	P	F	—	F
Triethanolamine	F	E	E	E
Trinitrotoluene	P	E	—	P

[a] Aromatic and halogenated hydrocarbons will attack all types of natural and synthetic glove materials. Should swelling occur, the user should change to fresh gloves and allow the swollen gloves to dry and return to normal.

[b] No data on the resistance to dimethyl sulfoxide of natural rubber, neoprene, nitrile rubber, or vinyl materials are available; the manufacturer of the substance recommends the use of butyl rubber gloves.

TABLE 15
Guide for Selection of Respirators

Type of hazard	Type of respirator
Oxygen deficiency	Self–contained breathing apparatus Hose mask with blower Combination of air–line respirator and auxiliary self–contained air supply or air–storage receiver with alarm
Gas and vapor contaminants 　Immediately dangerous to life or health	Self–contained breathing apparatus Hose mask with blower Air–purifying full–facepiece respirator with chemical canister (gas mask) Self–rescue mouthpiece respirator (for escape only) Combination of air–line respirator and auxiliary self–contained air supply or air–storage receiver with alarm
Not immediately dangerous to life or health	Air–line respirator Hose mask with blower Air–purifying half–mask or mouthpiece respirator with chemical cartridge
Particulate Contaminants 　Immediately dangerous to life or health	Self–contained breathing apparatus Hose mask with blower Air–purifying full–facepiece respirator with appropriate filter Self–rescue mouthpiece respirator (for escape only) Combination of air–line respirator and auxiliary self–contained air supply or air–storage receiver with alarm

TABLE 15
Guide for Selection of Respirators (continued)

Type of hazard	Type of respirator
Not immediately dangerous to life or health	Air–purifying half–mask or mouthpiece respirator with filter pad or cartridge Air–line respirator Air–line abrasive–blasting respirator Hose mask with blower
Combination of gas, vapor, and particulate contaminants Immediately dangerous to life or health	Self–contained breathing apparatus Hose mask with blower Air–purifying full–facepiece respirator with chemical canister and appropriate filter (gas mask with filter) Self–rescue mouthpiece respirator (for escape only) Combination of air–line respirator and auxiliary self–contained air supply or air–storage receiver with alarm
Not immediately dangerous to life or health	Air–line respirator Hose mask without blower Air–purifying half–mask or mouthpiece respirator with chemical cartridge and appropriate filter

Source: ANSI Standard Z88.2 (1969).

FLAMMABILITY OF CHEMICAL SUBSTANCES

This table gives properties related to the flammability of about 900 chemical substances. The properties listed are:

t_B : Normal boiling point in °C (at 101.325 kPa pressure).

FP: Flash point, which is the minimum temperature at which the vapor pressure of a liquid is sufficient to form an ignitable mixture with air near the surface of the liquid. Flash point is not an intrinsic physical property but depends on the conditions of measurement (see Reference 1).

Fl. Limits: Flammable limits (often called explosive limits), which specify the range of concentration of the vapor in air (in percent by volume) for which a flame can propagate. Below the lower flammable limit, the gas mixture is too lean to burn; above the upper flammable limit, the mixture is too rich. Values refer to ambient temperature and pressure and are dependent on the precise test conditions. A ? indicates that one of the limits is not known.

IT: Ignition temperature (sometimes called autoignition temperature), which is the minimum temperature required for self-sustained combustion in the absence of an external ignition source. As in the case of flash point, the value depends on specified test conditions.

Even in cases where very careful measurements of flash point have been replicated in several laboratories, observed values can differ by 3 to 6°C (Reference 4). For more typical measurements, larger uncertainties should be assumed in both flash points and autoignition temperatures. The absence of a flash point entry in this table does not mean that the substance is nonflammable, but only that no reliable value is available.

Compounds are listed by molecular formula following the Hill convention. Substances not containing carbon are listed first, followed by those that contain carbon. To locate an organic compound by name or CAS Registry Number when the molecular formula is not known, use the table "Physical Constants of Organic Compounds" in Section 3 and its indexes to determine the molecular formula.

REFERENCES

1. *Fire Protection Guide to Hazardous Materials, 11th Edition*, National Fire Protection Association, Quincy, MA, 1994.
2. Urben, P.G., Editor, *Bretherick's Handbook of Reactive Chemical Hazards, 5th Edition*, Butterworth-Heinemann, Oxford, 1995.
3. Daubert, T.E., Danner, R.P., Sibul, H.M., and Stebbins, C.C., *Physical and Thermodynamic Properties of Pure Compounds: Data Compilation*, extant 1994 (core with 4 supplements), Taylor & Francis, Bristol, PA.
4. *Report of Investigation: Flash Point Reference Materials*, National Institute of Standards and Technology, Standard Reference Materials Program, Gaithersburg, MD, 1995.

Mol. Form.	Name	t_B/°C	FP/°C	Fl. Limits	IT/°C
		Compounds not containing carbon			
B_2H_6	Diborane	-92.4	-90	1-98%	≈40
B_5H_9	Pentaborane(9)	60	30	0.4-?	35
BrH_3Si	Bromosilane	1.9	<0		≈20
Br_3HSi	Tribromosilane	109			≈20
Cl_2H_2Si	Dichlorosilane	8.3		4.1-99%	36
Cl_3HSi	Trichlorosilane	33	-50		104
GeH_4	Germane	-88.1			≈20
Ge_2H_6	Digermane	30.8			≈50
H_2	Hydrogen	-252.8		4-74%	
H_2S	Hydrogen sulfide	-59.55		4-44%	260
H_2S_2	Hydrogen disulfide	70.7	<22		
H_2Te	Hydrogen telluride	-2			-50
H_3N	Ammonia	-33.33		16-25%	
H_3P	Phosphine	-87.75		1.8-?	
H_4N_2	Hydrazine	113.55	38	5-100%	
H_4P_2	Diphosphine	63.5			≈20
H_4Si	Silane	-111.9	-112	1.4-?	≈20
H_6Si_2	Disilane	-14.3	-14		≈20
H_8Si_3	Trisilane	52.9	<0		≈20
P	Phosphorus (white)	280.5			38
		Compounds containing carbon			
CHN	Hydrogen cyanide	26	-18	6-40%	538
CH_2Cl_2	Dichloromethane	40		13-23%	556
CH_2N_2	Cyanamide		141		

Mol. Form	Name	t_B/°C	FP/°C	Fl. Limits	IT/°C
CH$_2$O	Formaldehyde	-19.1	85	7.0-73%	424
(CH$_2$O)$_x$	Paraformaldehyde		70	7.0-73%	300
CH$_2$O$_2$	Formic acid	101	50	18-57%	434
CH$_3$Br	Bromomethane	3.5		10-16%	537
CH$_3$Cl	Chloromethane	-24.0		8.1-17.4%	632
CH$_3$Cl$_3$Si	Methyltrichlorosilane	65.6	-9	7.6->20%	>404
CH$_3$NO	Formamide	220	154		
CH$_3$NO$_2$	Nitromethane	101.1	35	7.3-?	418
CH$_4$	Methane	-161.5		5.0-15.0%	537
CH$_4$Cl$_2$Si	Dichloromethylsilane	41	-9	6.0-55%	316
CH$_4$O	Methanol	64.6	11	6.0-36%	464
CH$_4$S	Methanethiol	5.9	-18	3.9-21.8%	
CH$_5$N	Methylamine	-6.3	0	4.9-20.7%	430
CH$_6$N$_2$	Methylhydrazine	87.5	-8	2.5-92%	194
CO	Carbon monoxide	-191.5		12.5-74%	609
COS	Carbon oxysulfide	-50		12-29%	
CS$_2$	Carbon disulfide	46	-30	1.3-50.0%	90
C$_2$ClF$_3$	Chlorotrifluoroethylene	-27.8		8.4-16.0%	
C$_2$F$_4$	Tetrafluoroethylene	-75.9		10.0-50.0%	200
C$_2$HCl$_3$	Trichloroethylene	87.2		8-10.5%	420
C$_2$HCl$_3$O	Dichloroacetyl chloride	108	66		
C$_2$H$_2$	Acetylene	-84.7		2.5-100%	305
C$_2$H$_2$Cl$_2$	1,1-Dichloroethylene	31.6	-28	6.5-15.5%	570
C$_2$H$_2$Cl$_2$	cis-1,2-Dichloroethylene	60.1	6	3-15%	460
C$_2$H$_2$Cl$_2$	trans-1,2-Dichloroethylene	48.7	2	6-13%	460
C$_2$H$_2$F$_2$	1,1-Difluoroethylene	-85.7		5.5-21.3%	
C$_2$H$_3$Br	Bromoethylene	15.8		9-15%	530
C$_2$H$_3$Cl	Chloroethylene	-13.3	-78	3.6-33.0%	472
C$_2$H$_3$ClF$_2$	1-Chloro-1,1-difluoroethane	-9.7		6-18%	632
C$_2$H$_3$ClO	Acetyl chloride	50.7	4		390
C$_2$H$_3$Cl$_2$NO$_2$	1,1-Dichloro-1-nitroethane	123.5	76		
C$_2$H$_3$Cl$_3$	1,1,1-Trichloroethane	74.0		8-10.5%	500
C$_2$H$_3$Cl$_3$	1,1,2-Trichloroethane	113.8	32	6-28%	460
C$_2$H$_3$Cl$_3$Si	Trichlorovinylsilane	91.5	21		
C$_2$H$_3$F	Fluoroethylene	-72		2.6-21.7%	
C$_2$H$_3$N	Acetonitrile	81.6	6	3.0-16.0%	524
C$_2$H$_3$NO	Methyl isocyanate	39.5	-7	5.3-26%	534
C$_2$H$_4$	Ethylene	-103.7		2.7-36%	450
C$_2$H$_4$ClNO$_2$	1-Chloro-1-nitroethane	124.5	56		
C$_2$H$_4$Cl$_2$	1,1-Dichloroethane	57.4	-17	5.4-11.4%	458
C$_2$H$_4$Cl$_2$	1,2-Dichloroethane	83.5	13	6.2-16%	413
C$_2$H$_4$O	Acetaldehyde	20.1	-39	4.0-60%	175
C$_2$H$_4$O	Ethylene oxide	10.6	-20	3.0-100%	429
C$_2$H$_4$O$_2$	Acetic acid	117.9	39	4.0-19.9%	463
C$_2$H$_4$O$_2$	Methyl formate	31.7	-19	4.5-23%	449
C$_2$H$_4$O$_3$	Ethaneperoxoic acid	110	41		
C$_2$H$_5$Br	Bromoethane	38.5		6.8-8.0%	511
C$_2$H$_5$Cl	Chloroethane	12.3	-50	3.8-15.4%	519
C$_2$H$_5$ClO	Ethylene chlorohydrin	128.6	60	4.9-15.9%	425
C$_2$H$_5$Cl$_3$Si	Trichloroethylsilane	100.5	22		
C$_2$H$_5$N	Ethyleneimine	56	-11	3.3-54.8%	320
C$_2$H$_5$NO$_2$	Nitroethane	114.0	28	3.4-17%	414
C$_2$H$_5$NO$_2$	Ethyl nitrite	18	-35	4.0-50%	90
C$_2$H$_5$NO$_3$	Ethyl nitrate	87.2	10	4-?	
C$_2$H$_6$	Ethane	-88.6		3.0-12.5%	472
C$_2$H$_6$Cl$_2$Si	Dichlorodimethylsilane	70.3	<21	3.4-9.5%	
C$_2$H$_6$O	Ethanol	78.2	13	3.3-19%	363
C$_2$H$_6$O	Dimethyl ether	-24.8	-41	3.4-27.0%	350
C$_2$H$_6$OS	2-Mercaptoethanol	158	74		
C$_2$H$_6$OS	Dimethyl sulfoxide	189	95	2.6-42%	215
C$_2$H$_6$O$_2$	Ethylene glycol	197.3	111	3.2-22%	398

Mol. Form.	Name	$t_B/°C$	FP/°C	Fl. Limits	IT/°C
$C_2H_6O_4S$	Dimethyl sulfate		83		188
C_2H_6S	Ethanethiol	35.1	-17	2.8-18.0%	300
C_2H_6S	Dimethyl sulfide	37.3	-37	2.2-19.7%	206
$C_2H_6S_2$	Dimethyl disulfide	109.8	24		
C_2H_7N	Ethylamine	16.5	-16	3.5-14%	385
C_2H_7N	Dimethylamine	6.8	20	2.8-14.4%	400
C_2H_7NO	Ethanolamine	171	86	3.0-23.5%	410
$C_2H_8N_2$	1,2-Ethanediamine	117	40	2.5-12.0%	385
$C_2H_8N_2$	1,1-Dimethylhydrazine	63.9	-15	2-95%	249
C_2N_2	Cyanogen	-21.1		6.6-32%	
C_3H_3Br	3-Bromo-1-propyne	89	10	3.0-?	324
C_3H_3N	2-Propenenitrile	77.3	0	3.0-17.0%	481
C_3H_4	Propyne	-23.2		2.1-12.5%	
C_3H_4ClN	3-Chloropropanenitrile	175.5	76		
$C_3H_4Cl_2$	2,3-Dichloropropene	94	15	2.6-7.8%	
C_3H_4O	Propargyl alcohol	113.6	36		
C_3H_4O	Acrolein	52.6	-26	2.8-31%	220
$C_3H_4O_2$	Propenoic acid	141	50	2.4-8.0%	438
$C_3H_4O_2$	2-Oxetanone	162	74	2.9-?	
$C_3H_4O_3$	Ethylene carbonate	248	143		
C_3H_5Br	3-Bromopropene	70.1	-1	4.4-7.3%	295
C_3H_5Cl	2-Chloropropene	22.6	-37	4.5-16%	
C_3H_5Cl	3-Chloropropene	45.1	-32	2.9-11.1%	485
C_3H_5ClO	Epichlorohydrin	118	31	3.8-21.0%	411
C_3H_5ClO	Propanoyl chloride	80	12		
$C_3H_5ClO_2$	2-Chloropropanoic acid	185	107		500
$C_3H_5ClO_2$	Ethyl chloroformate	95	16		500
$C_3H_5ClO_2$	Methyl chloroacetate	129.5	57	7.5-18.5%	
$C_3H_5Cl_2NO_2$	1,1-Dichloro-1-nitropropane	145	66		
$C_3H_5Cl_3$	1,2,3-Trichloropropane	157	71	3.2-12.6%	
$C_3H_5Cl_3Si$	Trichloro-2-propenylsilane	117.5	35		
C_3H_5N	Propanenitrile	97.1	2	3.1-14%	512
C_3H_5NO	3-Hydroxypropanenitrile	221	129		
$C_3H_5N_3O_9$	Trinitroglycerol				270
C_3H_6	Propene	-47.6		2.0-11.1%	455
C_3H_6	Cyclopropane	-32.8		2.4-10.4%	498
$C_3H_6ClNO_2$	1-Chloro-1-nitropropane	142	62		
$C_3H_6ClNO_2$	2-Chloro-2-nitropropane		57		
$C_3H_6Cl_2$	1,2-Dichloropropane	96.4	21	3.4-14.5%	557
$C_3H_6Cl_2O$	1,3-Dichloro-2-propanol	176	74		
$C_3H_6N_2$	Dimethylcyanamide	163.5	71		
C_3H_6O	Allyl alcohol	97.0	21	2.5-18.0%	378
C_3H_6O	Methyl vinyl ether	5.5			287
C_3H_6O	Propanal	48	-30	2.6-17%	207
C_3H_6O	Acetone	56.0	-20	2.5-12.8%	465
C_3H_6O	Methyloxirane	35	-37	3.1-27.5%	449
$C_3H_6O_2$	Propanoic acid	141.1	52	2.9-12.1%	465
$C_3H_6O_2$	Ethyl formate	54.4	-20	2.8-16.0%	455
$C_3H_6O_2$	Methyl acetate	56.8	-10	3.1-16%	454
$C_3H_6O_2$	1,3-Dioxolane	78	2		
$C_3H_6O_3$	Dimethyl carbonate	90.5	19		
$C_3H_6O_3$	1,3,5-Trioxane	114.5	45	3.6-29%	414
C_3H_7Br	1-Bromopropane	71.1			490
C_3H_7Cl	1-Chloropropane	46.5	<-18	2.6-11.1%	520
C_3H_7Cl	2-Chloropropane	35.7	-32	2.8-10.7%	593
C_3H_7ClO	2-Chloro-1-propanol	133.5	52		
C_3H_7ClO	1-Chloro-2-propanol	127	52		
$C_3H_7Cl_3Si$	Trichloropropylsilane	123.5	37		
C_3H_7N	Allylamine	53.3	-29	2.2-22%	374
C_3H_7NO	*N,N*-Dimethylformamide	153	58	2.2-15.2%	445
$C_3H_7NO_2$	1-Nitropropane	131.1	36	2.2-?	421

Mol. Form.	Name	t_B/°C	FP/°C	Fl. Limits	IT/°C
$C_3H_7NO_2$	2-Nitropropane	120.2	24	2.6-11.0%	428
$C_3H_7NO_3$	Propyl nitrate	110	20	2-100%	175
C_3H_8	Propane	-42.1	-104	2.1-9.5%	450
C_3H_8O	1-Propanol	97.2	23	2.2-13.7%	412
C_3H_8O	2-Propanol	82.3	12	2.0-12.7%	399
C_3H_8O	Ethyl methyl ether	7.4	-37	2.0-10.1%	190
$C_3H_8O_2$	1,2-Propylene glycol	187.6	99	2.6-12.5%	371
$C_3H_8O_2$	1,3-Propylene glycol	214.4			400
$C_3H_8O_2$	Ethylene glycol monomethyl ether	124.1	39	1.8-14%	285
$C_3H_8O_2$	Dimethoxymethane	42	-32	2.2-13.8%	237
$C_3H_8O_3$	Glycerol	290	199	3-19%	370
$C_3H_9BO_3$	Trimethyl borate	67.5	-8		
C_3H_9ClSi	Trimethylchlorosilane	60	-28		395
C_3H_9N	Propylamine	47.2	-37	2.0-10.4%	318
C_3H_9N	Isopropylamine	31.7	-37		402
C_3H_9N	Trimethylamine	2.8	-5	2.0-11.6%	190
C_3H_9NO	3-Amino-1-propanol	187.5	80		
C_3H_9NO	1-Amino-2-propanol	159.4	77		374
C_3H_9NO	N-Methyl-2-ethanolamine	158	74		
$C_3H_9O_3P$	Trimethyl phosphite	111.5	54		
$C_3H_9O_4P$	Trimethyl phosphate	197.2	107		
$C_3H_{10}N_2$	1,3-Propanediamine	139.8	24		
C_4Cl_6	Hexachloro-1,3-butadiene	215			610
$C_4H_2O_3$	Maleic anhydride	202	102	1.4-7.1%	477
C_4H_4	1-Buten-3-yne	5.1		21-100%	
$C_4H_4N_2$	Succinonitrile	266	132		
C_4H_4O	Furan	31.5	-36	2.3-14.3%	
$C_4H_4O_2$	Diketene	126.1	34		
C_4H_4S	Thiophene	84.0	-1		
C_4H_5Cl	2-Chloro-1,3-butadiene	59.4	-20	4.0-20.0%	
C_4H_5N	2-Butenenitrile	120.5	16		
C_4H_5N	Methylacrylonitrile	90.3	1	2-6.8%	
C_4H_5N	Pyrrole	129.7	39		
C_4H_6	1,3-Butadiene	-4.4		2.0-12.0%	420
C_4H_6	2-Butyne	26.9	-31	1.4-?	
C_4H_6O	Divinyl ether	28.3	<-30	1.7-27%	360
C_4H_6O	Ethoxyacetylene	50	<-7		
C_4H_6O	trans-2-Butenal	102.2	13	2.1-15.5%	232
C_4H_6O	3-Buten-2-one	81.4	-7	2.1-15.6%	491
C_4H_6O	Vinyloxirane	68	<-50		
$C_4H_6O_2$	Methacrylic acid	162.5	77	1.6-8.8%	68
$C_4H_6O_2$	Vinyl acetate	72.5	-8	2.6-13.4%	402
$C_4H_6O_2$	Methyl acrylate	80.7	-3	2.8-25%	468
$C_4H_6O_2$	2,3-Butanedione	88	27		
$C_4H_6O_2$	gamma-Butyrolactone	204	98		
$C_4H_6O_3$	Acetic anhydride	139.5	49	2.7-10.3%	316
$C_4H_6O_3$	Propylene carbonate	242	135		
$C_4H_6O_6$	L-Tartaric acid		210		425
C_4H_7Br	1-Bromo-2-butene	104.5		4.6-12.0%	
$C_4H_7BrO_2$	Ethyl bromoacetate	168.5	48		
C_4H_7Cl	2-Chloro-1-butene	58.5	-19	2.3-9.3%	
C_4H_7Cl	3-Chloro-2-methylpropene	71.5	-12	3.2-8.1%	
C_4H_7ClO	2-Chloroethyl vinyl ether	108	27		
$C_4H_7ClO_2$	Ethyl chloroacetate	144.3	64		
C_4H_7N	Butanenitrile	117.6	24	1.6-?	501
C_4H_7N	2-Methylpropanenitrile	103.9	8		482
C_4H_7NO	Acetone cyanohydrin		74	2.2-12.0%	688
C_4H_7NO	2-Pyrrolidone	251	129		
C_4H_8	1-Butene	-6.2		1.6-10.0%	385
C_4H_8	cis-2-Butene	3.7		1.7-9.0%	325
C_4H_8	trans-2-Butene	0.8		1.8-9.7%	324

Mol. Form.	Name	$t_B/°C$	FP/°C	Fl. Limits	IT/°C
C_4H_8	Isobutene	-6.9		1.8-9.6%	465
C_4H_8	Cyclobutane	12.6	<10	1.8-?	
$C_4H_8Cl_2$	1,2-Dichlorobutane	124.1			275
$C_4H_8Cl_2$	1,4-Dichlorobutane	161	52		
$C_4H_8Cl_2O$	Bis(2-chloroethyl) ether	178.5	55	2.7-?	369
C_4H_8O	2-Buten-1-ol	121.5	27	4.2-35.3%	349
C_4H_8O	2-Methyl-2-propenol	114.5	33		
C_4H_8O	Ethyl vinyl ether	35.5	<-46	1.7-28%	202
C_4H_8O	1,2-Epoxybutane	63.4	-22	1.7-19%	439
C_4H_8O	Butanal	74.8	-22	1.9-12.5%	218
C_4H_8O	Isobutanal	64.5	-18	1.6-10.6%	196
C_4H_8O	2-Butanone	79.5	-9	1.4-11.4%	404
C_4H_8O	Tetrahydrofuran	65	-14	2-11.8%	321
C_4H_8OS	1,4-Oxathiane	147	42		
$C_4H_8O_2$	Butanoic acid	163.7	72	2.0-10.0%	443
$C_4H_8O_2$	2-Methylpropanoic acid	154.4	56	2.0-9.2%	481
$C_4H_8O_2$	Propyl formate	80.9	-3		455
$C_4H_8O_2$	Isopropyl formate	68.2	-6		485
$C_4H_8O_2$	Ethyl acetate	77.1	-4	2.0-11.5%	426
$C_4H_8O_2$	Methyl propanoate	79.8	-2	2.5-13%	469
$C_4H_8O_2$	3-Hydroxybutanal		66		250
$C_4H_8O_2$	1,4-Dioxane	101.5	12	2.0-22%	180
$C_4H_8O_2S$	Sulfolane	287.3	177		
$C_4H_8O_3$	Methyl lactate	144.8	49	2.2-?	385
$C_4H_8O_3$	Ethylene glycol monoacetate	188	102		
C_4H_9Br	1-Bromobutane	101.6	18	2.6-6.6%	265
C_4H_9Br	2-Bromobutane	91.2	21		
C_4H_9Cl	1-Chlorobutane	78.6	-12	1.9-10.1%	240
C_4H_9Cl	2-Chlorobutane	68.2	-10		
C_4H_9Cl	1-Chloro-2-methylpropane	68.5	-6	2.0-8.7%	
C_4H_9Cl	2-Chloro-2-methylpropane	50.9	0		
$C_4H_9Cl_3Si$	Butyltrichlorosilane	148.5	54		
C_4H_9N	Pyrrolidine	86.5	3		
C_4H_9NO	N-Ethylacetamide	205	110		
C_4H_9NO	N,N-Dimethylacetamide	165	70	1.8-11.5%	490
C_4H_9NO	Butanal oxime	154	58		
C_4H_9NO	2-Butanone oxime	152.5	≈70		
C_4H_9NO	Morpholine	128	37	1.4-11.2%	290
$C_4H_9NO_2$	N-Acetylethanolamine		179		460
$C_4H_9NO_3$	Butyl nitrate	133	36		
C_4H_{10}	Butane	-0.5	-60	1.9-8.5%	287
C_4H_{10}	Isobutane	-11.7	-87	1.8-8.4%	460
$C_4H_{10}N_2$	Piperazine	146	81		
$C_4H_{10}O$	1-Butanol	117.7	37	1.4-11.2%	343
$C_4H_{10}O$	2-Butanol	99.5	24	1.7-9.8%	405
$C_4H_{10}O$	2-Methyl-1-propanol	107.8	28	1.7-10.6%	415
$C_4H_{10}O$	2-Methyl-2-propanol	82.4	11	2.4-8.0%	478
$C_4H_{10}O$	Diethyl ether	34.5	-45	1.9-36.0%	180
$C_4H_{10}O$	Methyl propyl ether	39.1	-20	2.0-14.8%	
$C_4H_{10}O_2$	1,2-Butanediol	190.5	40		
$C_4H_{10}O_2$	1,3-Butanediol	207.5	121		395
$C_4H_{10}O_2$	1,4-Butanediol	235	121		
$C_4H_{10}O_2$	2,3-Butanediol	182.5			402
$C_4H_{10}O_2$	Ethylene glycol monoethyl ether	135	43	3-18%	235
$C_4H_{10}O_2$	Ethylene glycol dimethyl ether	85	-2		202
$C_4H_{10}O_2$	tert-Butyl hydroperoxide		27		
$C_4H_{10}O_2S$	2,2'-Thiodiethanol	282	160		298
$C_4H_{10}O_3$	Diethylene glycol	245.8	124	2-17%	224
$C_4H_{10}O_4S$	Diethyl sulfate	208	104		436
$C_4H_{10}S$	1-Butanethiol	98.5	2		
$C_4H_{10}S$	2-Butanethiol	85	-23		

Mol. Form.	Name	$t_B/°C$	FP/°C	Fl. Limits	IT/°C
$C_4H_{10}S$	2-Methyl-1-propanethiol	88.5	2		
$C_4H_{10}S$	2-Methyl-2-propanethiol	64.3	<-29		
$C_4H_{10}Se$	Diethyl selenide	108		2.5-?	
$C_4H_{11}N$	Butylamine	77.0	-12	1.7-9.8%	312
$C_4H_{11}N$	sec-Butylamine	63.5	-9		
$C_4H_{11}N$	tert-Butylamine	44.0	-9	1.7-8.9%	380
$C_4H_{11}N$	Isobutylamine	67.7	-9	2-12%	378
$C_4H_{11}N$	Diethylamine	55.5	-23	1.8-10.1%	312
$C_4H_{11}NO$	2-Amino-1-butanol	178	74		
$C_4H_{11}NO$	2-Amino-2-methyl-1-propanol	165.5	67		
$C_4H_{11}NO_2$	Diethanolamine	268.8	172	2-13%	662
$C_4H_{12}Sn$	Tetramethylstannane	78	-12	1.9-?	
$C_4H_{13}N_3$	Diethylenetriamine	207	98	2-6.7%	358
$C_5H_4O_2$	Furfural	161.7	60	2.1-19.3%	316
C_5H_5N	Pyridine	115.2	20	1.8-12.4%	482
C_5H_6	2-Methyl-1-buten-3-yne	32	<-7		
$C_5H_6N_2$	2-Methylpyrazine	137	50		
C_5H_6O	3-Methylfuran	66	-30		
$C_5H_6O_2$	Furfuryl alcohol	171	75	1.8-16.3%	491
C_5H_7N	1-Methylpyrrole	115	16		
C_5H_7NO	2-Furanmethanamine	145.5	37		
$C_5H_7NO_2$	Ethyl cyanoacetate	205	110		
C_5H_8	2-Methyl-1,3-butadiene	34.0	-54	1.5-8.9%	395
C_5H_8	1-Pentyne	40.1	<-20		
C_5H_8	Cyclopentene	44.2	-29		395
C_5H_8O	3-Methyl-3-buten-2-one	98		1.8-9.0%	
C_5H_8O	Cyclopentanone	130.5	26		
C_5H_8O	3,4-Dihydro-2H-pyran	86	-18		
$C_5H_8O_2$	Allyl acetate	103.5	22		374
$C_5H_8O_2$	Isopropenyl acetate	94	26		432
$C_5H_8O_2$	Vinyl propanoate	91.2	1		
$C_5H_8O_2$	Ethyl acrylate	99.4	10	1.4-14%	372
$C_5H_8O_2$	Methyl methacrylate	100.5	10	1.7-8.2%	
$C_5H_8O_2$	2,4-Pentanedione	138	34		340
$C_5H_8O_3$	Methyl acetoacetate	171.7	77		280
C_5H_9NO	N-Methyl-2-pyrrolidone	202	96	1-10%	346
C_5H_{10}	1-Pentene	29.9	-18	1.5-8.7%	275
C_5H_{10}	cis-2-Pentene	36.9	<-20		
C_5H_{10}	trans-2-Pentene	36.3	<-20		
C_5H_{10}	2-Methyl-1-butene	31.2	-20		
C_5H_{10}	3-Methyl-1-butene	20.1	-7	1.5-9.1%	365
C_5H_{10}	2-Methyl-2-butene	38.5	-20		
C_5H_{10}	Cyclopentane	49.3	-25	1.5-?	361
$C_5H_{10}Cl_2$	1,5-Dichloropentane	179	>27		
$C_5H_{10}N_2$	3-(Dimethylamino)propanenitrile	173	65		
$C_5H_{10}O$	Cyclopentanol	140.4	51		
$C_5H_{10}O$	Pentanal	103	12		222
$C_5H_{10}O$	2-Pentanone	102.2	7	1.5-8.2%	452
$C_5H_{10}O$	3-Pentanone	101.9	13	1.6-?	450
$C_5H_{10}O$	Tetrahydropyran	88	-20		
$C_5H_{10}O$	2-Methyltetrahydrofuran	78	-11		
$C_5H_{10}O_2$	Pentanoic acid	186.1	96		400
$C_5H_{10}O_2$	3-Methylbutanoic acid	176.5			416
$C_5H_{10}O_2$	Butyl formate	106.1	18	1.7-8.2%	322
$C_5H_{10}O_2$	Isobutyl formate	98.2	5	2-9%	320
$C_5H_{10}O_2$	Propyl acetate	101.5	13	1.7-8%	450
$C_5H_{10}O_2$	Isopropyl acetate	88.6	2	1.8-8%	460
$C_5H_{10}O_2$	Ethyl propanoate	99.1	12	1.9-11%	440
$C_5H_{10}O_2$	Methyl butanoate	102.8	14		
$C_5H_{10}O_2$	3-Ethoxypropanal	135.2	38		
$C_5H_{10}O_2$	Tetrahydrofurfuryl alcohol	178	75	1.5-9.7%	282

Mol. Form.	Name	$t_B/°C$	FP/°C	Fl. Limits	IT/°C
$C_5H_{10}O_3$	Diethyl carbonate	126	25		
$C_5H_{10}O_3$	Ethylene glycol monomethyl ether acetate	143	49	1.5-12.3%	392
$C_5H_{10}O_3$	Ethyl lactate	154.5	46	1.5-?	400
$C_5H_{11}Br$	1-Bromopentane	129.8	32		
$C_5H_{11}Cl$	1-Chloropentane	107.8	13	1.6-8.6%	260
$C_5H_{11}Cl$	2-Chloro-2-methylbutane	85.6		1.5-7.4%	345
$C_5H_{11}Cl$	1-Chloro-3-methylbutane	98.9	<21	1.5-7.4%	
$C_5H_{11}Cl_3Si$	Trichloropentylsilane	172	63		
$C_5H_{11}N$	Piperidine	106.2	16		
$C_5H_{11}N$	N-Methylpyrrolidine	81	-14		
$C_5H_{11}NO$	4-Methylmorpholine	116	24		
$C_5H_{11}NO_2$	Isopentyl nitrite	99.2			210
C_5H_{12}	Pentane	36.0	-40	1.4-8.0%	260
C_5H_{12}	Isopentane	27.8	-51	1.4-7.6%	420
C_5H_{12}	Neopentane	9.4	-65	1.4-7.5%	450
$C_5H_{12}N_2$	1-Methylpiperazine	138	42		
$C_5H_{12}N_2O$	Tetramethylurea	176.5	77		
$C_5H_{12}O$	1-Pentanol	137.9	33	1.2-10.0%	300
$C_5H_{12}O$	2-Pentanol	119.3	34	1.2-9.0%	343
$C_5H_{12}O$	3-Pentanol	116.2	41	1.2-9.0%	435
$C_5H_{12}O$	2-Methyl-1-butanol	128	50		385
$C_5H_{12}O$	3-Methyl-1-butanol	131.1	43	1.2-9.0%	350
$C_5H_{12}O$	2-Methyl-2-butanol	102.4	19	1.2-9.0%	437
$C_5H_{12}O$	3-Methyl-2-butanol	112.9	38		
$C_5H_{12}O$	2,2-Dimethyl-1-propanol	113.5	37		
$C_5H_{12}O$	Ethyl propyl ether	63.2	<-20	1.7-9.0%	
$C_5H_{12}O_2$	1,5-Pentanediol	239	129		335
$C_5H_{12}O_2$	2-Isopropoxyethanol	145	33		
$C_5H_{12}O_2$	2,2-Dimethyl-1,3-propanediol	208	129		399
$C_5H_{12}O_3$	Diethylene glycol monomethyl ether	193	96	1.38-22.7%	240
$C_5H_{12}S$	1-Pentanethiol	126.6	18		
$C_5H_{12}S$	3-Methyl-2-butanethiol		3		
$C_5H_{13}N$	Pentylamine	104.3	-1	2.2-22%	
$C_5H_{13}N$	Butylmethylamine	91	13		
$C_6H_2Cl_4$	1,2,4,5-Tetrachlorobenzene	244.5	155		
$C_6H_3ClN_2O_4$	1-Chloro-2,4-dinitrobenzene	315	194	2.0-22%	
$C_6H_3Cl_3$	1,2,4-Trichlorobenzene	213.5	105	2.5-6.6%	571
$C_6H_4ClNO_2$	1-Chloro-4-nitrobenzene	242	127		
$C_6H_4Cl_2$	o-Dichlorobenzene	180	66	2.2-9.2%	648
$C_6H_4Cl_2$	m-Dichlorobenzene	173	72		
$C_6H_4Cl_2$	p-Dichlorobenzene	174	66		
$C_6H_4Cl_2O$	2,4-Dichlorophenol	210	114		
C_6H_5Br	Bromobenzene	156.0	51		565
C_6H_5Cl	Chlorobenzene	131.7	28	1.3-9.6%	593
C_6H_5ClO	o-Chlorophenol	174.9	64		
C_6H_5ClO	p-Chlorophenol	220	121		
$C_6H_5Cl_2N$	3,4-Dichloroaniline	272	166		
$C_6H_5Cl_3Si$	Trichlorophenylsilane	201	91		
C_6H_5F	Fluorobenzene	84.7	-15		
$C_6H_5NO_2$	Nitrobenzene	210.8	88	1.8-?	482
$C_6H_5N_3O_4$	2,4-Dinitroaniline		224		
C_6H_6	1,5-Hexadien-3-yne	85	<-20	1.5-?	
C_6H_6	Benzene	80.0	-11	1.2-7.8%	498
$C_6H_6N_2O_2$	p-Nitroaniline	332	199		
C_6H_6O	Phenol	181.8	79	1.8-8.6%	715
$C_6H_6O_2$	1,2-Benzenediol	245	127		
$C_6H_6O_2$	Resorcinol		127	1.4-?	608
$C_6H_6O_2$	p-Hydroquinone	287	165		516
C_6H_7N	Aniline	184.1	70	1.3-11%	615
C_6H_7N	2-Methylpyridine	129.3	39		538
C_6H_7N	4-Methylpyridine	145.3	57		

Mol. Form.	Name	$t_B/°C$	FP/°C	Fl. Limits	IT/°C
C_6H_8ClN	Aniline, hydrochloride		193		
$C_6H_8Cl_2O_2$	Hexanedioyl dichloride		72		
$C_6H_8N_2$	Adiponitrile	295	93	1.0-?	550
$C_6H_8N_2$	o-Phenylenediamine	257	156	1.5-?	
$C_6H_8N_2$	Phenylhydrazine	243.5	88		
$C_6H_8N_2$	2,5-Dimethylpyrazine	155	64		
C_6H_8O	2,5-Dimethylfuran	93.5	7		
$C_6H_8O_4$	Dimethyl maleate	202	113		
C_6H_{10}	1,4-Hexadiene	65	-21	2.0-6.1%	
C_6H_{10}	2-Methyl-1,3-pentadiene	75.8	-12		
C_6H_{10}	4-Methyl-1,3-pentadiene	76.5	-34		
C_6H_{10}	2-Hexyne	84.5	-10		
C_6H_{10}	Cyclohexene	82.9	-12	1.2-?	310
$C_6H_{10}O$	Diallyl ether	94	-7		
$C_6H_{10}O$	Cyclohexanone	155.4	44	1.1-9.4%	420
$C_6H_{10}O$	Mesityl oxide	130	31	1.4-7.2%	344
$C_6H_{10}O_2$	Vinyl butanoate	116.7	20	1.4-8.8%	
$C_6H_{10}O_2$	Ethyl 2-butenoate	136.5	2		
$C_6H_{10}O_2$	Ethyl methacrylate	117	20		
$C_6H_{10}O_2$	2,5-Hexanedione	194	79		499
$C_6H_{10}O_3$	Ethyl acetoacetate	180.8	57	1.4-9.5%	295
$C_6H_{10}O_3$	Propanoic anhydride	170	63	1.3-9.5%	285
$C_6H_{10}O_4$	Adipic acid	337.5	196		420
$C_6H_{10}O_4$	Diethyl oxalate	185.7	76		
$C_6H_{10}O_4$	Ethylene glycol diacetate	190	88	1.6-8.4%	482
$C_6H_{11}Cl$	Chlorocyclohexane	142	32		
$C_6H_{11}NO$	Caprolactam	270	125		
$C_6H_{11}NO_2$	Nitrocyclohexane	205	88		
$C_6H_{11}NO_2$	4-Acetylmorpholine		113		
C_6H_{12}	1-Hexene	63.4	-26	1.2-6.9%	253
C_6H_{12}	cis-2-Hexene	68.8	-21		
C_6H_{12}	2-Methyl-1-pentene	62.1	-28		300
C_6H_{12}	4-Methyl-1-pentene	53.9	-7		300
C_6H_{12}	4-Methyl-cis-2-pentene	56.3	-32		
C_6H_{12}	4-Methyl-trans-2-pentene	58.6	-29		
C_6H_{12}	2-Ethyl-1-butene	64.7	<-20		315
C_6H_{12}	2,3-Dimethyl-1-butene	55.6	<-20		360
C_6H_{12}	2,3-Dimethyl-2-butene	73.3	<-20		401
C_6H_{12}	Cyclohexane	80.7	-20	1.3-8%	245
C_6H_{12}	Methylcyclopentane	71.8	-29	1.0-8.35%	258
C_6H_{12}	Ethylcyclobutane	70.8	-15	1.2-7.7%	210
C_6H_{12}	2-Methyl-2-pentene	67.3	<-7		
$C_6H_{12}Cl_2O_2$	1,2-Bis(2-chloroethoxy)ethane	232	121		
$C_6H_{12}O$	cis-3-Hexen-1-ol	156.5	54		
$C_6H_{12}O$	Butyl vinyl ether	94	-9		255
$C_6H_{12}O$	Isobutyl vinyl ether	83	-9		
$C_6H_{12}O$	Hexanal	131	32		
$C_6H_{12}O$	2-Ethylbutanal		21	1.2-7.7%	
$C_6H_{12}O$	2-Methylpentanal	117	17		199
$C_6H_{12}O$	2-Hexanone	127.6	25	1-8%	423
$C_6H_{12}O$	3-Hexanone	123.5	35	1-8%	
$C_6H_{12}O$	4-Methyl-2-pentanone	116.5	18	1.2-8.0%	448
$C_6H_{12}O$	Cyclohexanol	160.8	68	1-9%	300
$C_6H_{12}O_2$	Hexanoic acid	205.2	102		380
$C_6H_{12}O_2$	2-Methylpentanoic acid	195.6	107		378
$C_6H_{12}O_2$	Diethylacetic acid	194	99		400
$C_6H_{12}O_2$	Pentyl formate	130.4	26		
$C_6H_{12}O_2$	Butyl acetate	126.1	22	1.7-7.6%	425
$C_6H_{12}O_2$	sec-Butyl acetate	112	31	1.7-9.8%	
$C_6H_{12}O_2$	Isobutyl acetate	116.5	18	1.3-10.5%	421
$C_6H_{12}O_2$	Propyl propanoate	122.5	79		

Mol. Form.	Name	t_B/°C	FP/°C	Fl. Limits	IT/°C
$C_6H_{12}O_2$	Ethyl butanoate	121.5	24		463
$C_6H_{12}O_2$	Ethyl 2-methylpropanoate	110.1	13		
$C_6H_{12}O_2$	Diacetone alcohol	167.9	58	1.8-6.9%	643
$C_6H_{12}O_3$	Ethylene glycol monoethyl ether acetate	156.4	56	2-8%	379
$C_6H_{12}O_3$	Paraldehyde	124.3	36	1.3-?	238
$C_6H_{12}S$	Cyclohexanethiol	158.9	43		
$C_6H_{13}Cl$	1-Chlorohexane	135	35		
$C_6H_{13}N$	Cyclohexylamine	134	31	1.9-9.4%	293
$C_6H_{13}NO$	N-Butylacetamide	229	116		
$C_6H_{13}NO$	2,6-Dimethylmorpholine	146.6	44		
$C_6H_{13}NO$	N-Ethylmorpholine	138.5	32		
$C_6H_{13}NO_2$	4-Morpholineethanol	227	99		
C_6H_{14}	Hexane	68.7	-22	1.1-7.5%	225
C_6H_{14}	2-Methylpentane	60.2	<-29	1.0-7.0%	264
C_6H_{14}	3-Methylpentane	63.2	-7	1.2-7.0%	278
C_6H_{14}	2,2-Dimethylbutane	49.7	-48	1.2-7.0%	405
C_6H_{14}	2,3-Dimethylbutane	57.9	-29	1.2-7.0%	405
$C_6H_{14}N_2O$	1-Piperazineethanol	246	124		
$C_6H_{14}O$	1-Hexanol	157.6	63		
$C_6H_{14}O$	2-Methyl-1-pentanol	149	54	1.1-9.65%	310
$C_6H_{14}O$	4-Methyl-2-pentanol	131.6	41	1.0-5.5%	
$C_6H_{14}O$	2-Ethyl-1-butanol	147	57		
$C_6H_{14}O$	Dipropyl ether	90.0	21	1.3-7.0%	188
$C_6H_{14}O$	Diisopropyl ether	68.5	-28	1.4-7.9%	443
$C_6H_{14}O$	Butyl ethyl ether	92.3	4		
$C_6H_{14}O_2$	2,5-Hexanediol	218	110		
$C_6H_{14}O_2$	2-Methyl-2,4-pentanediol	197.1	102	1-9%	306
$C_6H_{14}O_2$	Ethylene glycol monobutyl ether	168.4	69	4-13%	238
$C_6H_{14}O_2$	1,1-Diethoxyethane	102.2	-21	1.6-10.4%	230
$C_6H_{14}O_2$	Ethylene glycol diethyl ether	119.4	27		205
$C_6H_{14}O_3$	1,2,6-Hexanetriol		191		
$C_6H_{14}O_3$	Diethylene glycol monoethyl ether	196	96		
$C_6H_{14}O_3$	Diethylene glycol dimethyl ether	162	67		
$C_6H_{14}O_3$	Trimethylolpropane		149		
$C_6H_{14}O_4$	Triethylene glycol	285	177	0.9-9.2%	371
$C_6H_{15}N$	Hexylamine	132.8	29		
$C_6H_{15}N$	Butylethylamine	107.5	18		
$C_6H_{15}N$	Dipropylamine	109.3	17		299
$C_6H_{15}N$	Diisopropylamine	83.9	-1	1.1-7.1%	316
$C_6H_{15}N$	Triethylamine	89	-7	1.2-8.0%	249
$C_6H_{15}NO_2$	Diisopropanolamine	250	127		374
$C_6H_{15}NO_3$	Triethanolamine	335.4	179	1-10%	
$C_6H_{15}N_3$	1-Piperazineethanamine	220	93		
$C_6H_{15}O_4P$	Triethyl phosphate	215.5	115		454
$C_6H_{16}N_2$	N,N-Diethylethylenediamine	144	46		
$C_7H_3ClF_3NO_2$	1-Chloro-4-nitro-2-(trifluoromethyl)benzene	232	135		
$C_7H_4ClF_3$	1-Chloro-2-(trifluoromethyl)benzene	152.2	59		
$C_7H_4F_3NO_2$	1-Nitro-3-(trifluoromethyl)benzene	202.8	103		
C_7H_5ClO	Benzoyl chloride	197.2	72		
C_7H_5ClO	4-Chlorobenzaldehyde	213.5	88		
$C_7H_5Cl_3$	(Trichloromethyl)benzene	221	127		211
$C_7H_5F_3$	(Trifluoromethyl)benzene	102.1	12		
$C_7H_6N_2O_4$	1-Methyl-2,4-dinitrobenzene		207		
C_7H_6O	Benzaldehyde	179.0	63		192
$C_7H_6O_2$	Benzoic acid	249.2	121		570
$C_7H_6O_2$	Salicylaldehyde	197	78		
$C_7H_6O_3$	Salicylic acid		157	1.1-?	540
C_7H_7Br	o-Bromotoluene	181.7	79		
C_7H_7Br	p-Bromotoluene	184.3	85		
C_7H_7Cl	(Chloromethyl)benzene	179	67	1.1-?	585
$C_7H_7NO_2$	o-Nitrotoluene	222	106		

Mol. Form.	Name	t_B/°C	FP/°C	Fl. Limits	IT/°C
$C_7H_7NO_2$	m-Nitrotoluene	232	106		
$C_7H_7NO_2$	p-Nitrotoluene	238.3	106		
C_7H_8	Toluene	110.6	4	1.1-7.1%	480
C_7H_8	Bicyclo[2.2.1]hepta-2,5-diene	89.5	-21		
C_7H_8O	o-Cresol	191.0	81	1.4-?	599
C_7H_8O	m-Cresol	202.2	86	1.1-?	558
C_7H_8O	p-Cresol	201.9	86	1.1-?	558
C_7H_8O	Benzyl alcohol	205.3	93		436
C_7H_8O	Anisole	153.7	52		475
$C_7H_8O_2$	4-Methoxyphenol	243	132		421
$C_7H_8O_3S$	p-Toluenesulfonic acid		184		
C_7H_9N	o-Methylaniline	200.3	85		482
C_7H_9N	p-Methylaniline	200.4	87		482
C_7H_9NO	o-Anisidine	224	118		
$C_7H_{10}O$	3-Cyclohexene-1-carboxaldehyde	105	57		
$C_7H_{10}O_4$	3,3-Diacetoxy-1-propene	180	82		
C_7H_{12}	4-Methylcyclohexene	102.7	-1		
$C_7H_{12}O_2$	Butyl acrylate	145	29	1.7-9.9%	292
$C_7H_{12}O_2$	Isobutyl acrylate	132	30		427
$C_7H_{12}O_2$	Cyclohexyl formate	162	51		
$C_7H_{12}O_4$	Diethyl malonate	200	93		
C_7H_{14}	1-Heptene	93.6	-1		260
C_7H_{14}	trans-2-Heptene	98	<0		
C_7H_{14}	Cycloheptane	118.4	<21	1.1-6.7%	
C_7H_{14}	Methylcyclohexane	100.9	-4	1.2-6.7%	250
C_7H_{14}	Ethylcyclopentane	103.5	<21	1.1-6.7%	260
$C_7H_{14}O$	2-Heptanone	151.0	39	1.1-7.9%	393
$C_7H_{14}O$	3-Heptanone	147	46		
$C_7H_{14}O$	4-Heptanone	144	49		
$C_7H_{14}O$	5-Methyl-2-hexanone	144	36	1.0-8.2%	191
$C_7H_{14}O$	cis-2-Methylcyclohexanol	165	65		296
$C_7H_{14}O$	trans-2-Methylcyclohexanol	167.5	65		296
$C_7H_{14}O$	cis-3-Methylcyclohexanol	174.5	70		295
$C_7H_{14}O$	trans-3-Methylcyclohexanol	174.5	70		295
$C_7H_{14}O$	cis-4-Methylcyclohexanol	173	70		295
$C_7H_{14}O$	trans-4-Methylcyclohexanol	174	70		295
$C_7H_{14}O_2$	Pentyl acetate	149.2	16	1.1-7.5%	360
$C_7H_{14}O_2$	Isopentyl acetate	142.5	25	1.0-7.5%	360
$C_7H_{14}O_2$	sec-Pentyl acetate	130.5	32		
$C_7H_{14}O_2$	Butyl propanoate	146.8	32		426
$C_7H_{14}O_2$	Propyl butanoate	143.0	37		
$C_7H_{15}NO_2$	Ethyl N-butylcarbamate	202	92		
C_7H_{16}	Heptane	98.5	-4	1.05-6.7%	204
C_7H_{16}	2-Methylhexane	90.0	-1	1.0-6.0%	280
C_7H_{16}	3-Methylhexane	92	-4		280
C_7H_{16}	2,3-Dimethylpentane	89.7	-56	1.1-6.7%	335
C_7H_{16}	2,4-Dimethylpentane	80.4	-12		
C_7H_{16}	2,2,3-Trimethylbutane	80.8	<0		412
$C_7H_{16}N_2O$	4-Morpholinepropanamine	220	104		
$C_7H_{16}O$	2-Heptanol	159	71		
$C_7H_{16}O$	3-Heptanol	157	60		
$C_7H_{16}O$	2,4-Dimethyl-3-pentanol	138.7	49		
$C_7H_{16}O$	2,3,3-Trimethyl-2-butanol	131	<0		375
$C_7H_{17}N$	Heptylamine	156	54		
$C_7H_{18}N_2$	N,N-Diethyl-1,3-propanediamine	168.5	59		
$C_8H_4O_3$	Phthalic anhydride	295	152	1.7-10.5%	570
$C_8H_6O_4$	Phthalic acid		168		
$C_8H_6O_4$	Terephthalic acid		260		496
C_8H_7ClO	α-Chloroacetophenone	247	118		
C_8H_7N	Benzeneacetonitrile	233.5	113		
C_8H_8	Styrene	145	31	0.9-6.8%	490

Mol. Form.	Name	t_B/°C	FP/°C	Fl. Limits	IT/°C
C$_8$H$_8$O	Phenyloxirane	194.1	74		498
C$_8$H$_8$O	Benzeneacetaldehyde	195	71		
C$_8$H$_8$O	Acetophenone	202	77		570
C$_8$H$_8$O$_2$	Benzeneacetic acid	265.5	>100		
C$_8$H$_8$O$_2$	Phenyl acetate	196	80		
C$_8$H$_8$O$_2$	Methyl benzoate	199	83		
C$_8$H$_8$O$_2$	2-Methoxybenzaldehyde	243.5	118		
C$_8$H$_8$O$_3$	Methyl salicylate	222.9	96		454
C$_8$H$_9$Cl	1-Chloro-4-ethylbenzene	184.4	64		
C$_8$H$_9$NO	Acetanilide	304	169		530
C$_8$H$_9$NO$_2$	Methyl 2-aminobenzoate	256	>100		
C$_8$H$_{10}$	Ethylbenzene	136.1	21	0.8-6.7%	432
C$_8$H$_{10}$	o-Xylene	144.5	32	0.9-6.7%	463
C$_8$H$_{10}$	m-Xylene	139.1	27	1.1-7.0%	527
C$_8$H$_{10}$	p-Xylene	138.3	27	1.1-7.0%	528
C$_8$H$_{10}$O	p-Ethylphenol	217.9	104		
C$_8$H$_{10}$O	Benzeneethanol	218.2	96		
C$_8$H$_{10}$O	α-Methylbenzyl alcohol	205	93		
C$_8$H$_{10}$O	Phenetole	169.8	63		
C$_8$H$_{10}$O	Benzyl methyl ether	170	135		
C$_8$H$_{10}$O	4-Methylanisole	175.5	60		
C$_8$H$_{10}$O$_2$	2-Phenoxyethanol	245	121		
C$_8$H$_{11}$N	N-Ethylaniline	203.0	85		
C$_8$H$_{11}$N	N,N-Dimethylaniline	194.1	63		371
C$_8$H$_{11}$N	2,3-Xylidine	221.5	97	1.0-?	
C$_8$H$_{11}$N	2,6-Xylidine	215	96		
C$_8$H$_{11}$N	α-Methylbenzylamine	187	79		
C$_8$H$_{11}$N	5-Ethyl-2-picoline	178.3	68	1.1-6.6%	
C$_8$H$_{11}$NO	N-Phenylethanolamine	279.5	152		
C$_8$H$_{11}$NO	o-Phenetidine	232.5	115		
C$_8$H$_{11}$NO	p-Phenetidine	254	116		
C$_8$H$_{12}$	1,5-Cyclooctadiene	150.8	35		
C$_8$H$_{12}$	4-Vinylcyclohexene	128	16		269
C$_8$H$_{12}$O$_4$	Diethyl maleate	223	121		350
C$_8$H$_{12}$O$_4$	Diethyl fumarate	214	104		
C$_8$H$_{14}$O$_2$	Cyclohexyl acetate	173	58		335
C$_8$H$_{14}$O$_2$	Butyl methacrylate	160	52		
C$_8$H$_{14}$O$_3$	Butanoic anhydride	200	54	0.9-5.8%	279
C$_8$H$_{14}$O$_3$	2-Methylpropanoic anhydride	183	59	1.0-6.2%	329
C$_8$H$_{14}$O$_3$	Butyl acetoacetate		85		
C$_8$H$_{14}$O$_4$	Ethyl succinate	217.7	90		
C$_8$H$_{14}$O$_5$	Diethylene glycol diacetate	200	135		
C$_8$H$_{14}$O$_6$	Diethyl tartrate	281	93		
C$_8$H$_{15}$ClO	Octanoyl chloride	195.6	82		
C$_8$H$_{16}$	1-Octene	121.2	21		230
C$_8$H$_{16}$	2,4,4-Trimethyl-1-pentene	101.4	-5	0.8-4.8%	391
C$_8$H$_{16}$	2,4,4-Trimethylpentene	101	-5	0.8-4.8%	391
C$_8$H$_{16}$	2,4,4-Trimethyl-2-pentene	104.9	2		305
C$_8$H$_{16}$	Ethylcyclohexane	131.9	35	0.9-6.6%	238
C$_8$H$_{16}$	cis-1,2-Dimethylcyclohexane	129.8	16		304
C$_8$H$_{16}$	trans-1,2-Dimethylcyclohexane	123.5	11		304
C$_8$H$_{16}$	cis-1,4-Dimethylcyclohexane	124.4	16		
C$_8$H$_{16}$	Propylcyclopentane	131			269
C$_8$H$_{16}$O	Octanal	171	52		
C$_8$H$_{16}$O	2-Ethylhexanal	163	44	0.85-7.2%	190
C$_8$H$_{16}$O	2-Octanone	172.5	52		
C$_8$H$_{16}$O$_2$	Hexyl acetate	171.5	45		
C$_8$H$_{16}$O$_2$	sec-Hexyl acetate	147.5	45		
C$_8$H$_{16}$O$_2$	2-Ethylbutyl acetate	162.5	54		
C$_8$H$_{16}$O$_2$	Pentyl propanoate	168.6	41		378
C$_8$H$_{16}$O$_2$	Butyl butanoate	166	53		

Mol. Form.	Name	t_B/°C	FP/°C	Fl. Limits	IT/°C
$C_8H_{16}O_2$	Isobutyl butanoate	156.9	50		
$C_8H_{16}O_2$	Isobutyl isobutanoate	148.6	38	0.96-7.59%	432
$C_8H_{16}O_2$	Ethyl hexanoate	167	49		
$C_8H_{16}O_2$	1,4-Cyclohexanedimethanol	283	167		316
$C_8H_{16}O_3$	Pentyl lactate		79		
$C_8H_{16}O_4$	Diethylene glycol monoethyl ether acetate	218.5	110		425
$C_8H_{17}Cl$	1-Chlorooctane	181.5	70		
$C_8H_{17}Cl$	3-(Chloromethyl)heptane	172	60		
C_8H_{18}	Octane	125.6	13	1.0-6.5%	206
C_8H_{18}	2,3-Dimethylhexane	115.6	7		438
C_8H_{18}	2,4-Dimethylhexane	109.5	10		
C_8H_{18}	3-Ethyl-2-methylpentane	115.6	<21		460
C_8H_{18}	2,2,3-Trimethylpentane	110	<21		346
C_8H_{18}	2,2,4-Trimethylpentane	99.2	-12		418
C_8H_{18}	2,3,3-Trimethylpentane	114.8	<21		425
$C_8H_{18}O$	1-Octanol	195.1	81		
$C_8H_{18}O$	2-Octanol	180	88		
$C_8H_{18}O$	2-Ethyl-1-hexanol	184.6	73	0.88-9.7%	231
$C_8H_{18}O$	Dibutyl ether	140.2	25	1.5-7.6%	194
$C_8H_{18}O_2$	2-Ethyl-1,3-hexanediol	244	127		360
$C_8H_{18}O_2$	2,2,4-Trimethyl-1,3-pentanediol	235	113		346
$C_8H_{18}O_2$	Di-tert-butyl peroxide	111	18		
$C_8H_{18}O_3$	Diethylene glycol diethyl ether	188	82		
$C_8H_{18}O_4$	2,5,8,11-Tetraoxadodecane	216	111		
$C_8H_{18}O_5$	Tetraethylene glycol	328	182		
$C_8H_{18}S$	1-Octanethiol	199.1	69		
$C_8H_{18}S$	Dibutyl sulfide	185	76		
$C_8H_{19}N$	Octylamine	179.6	60		
$C_8H_{19}N$	Dibutylamine	159.6	47	1.1-6%	
$C_8H_{19}N$	Diisobutylamine	139.6	29		
$C_8H_{19}N$	2-Ethylhexylamine	169.2	60		
$C_8H_{20}O_4Si$	Ethyl silicate	168.8	52		
$C_8H_{23}N_5$	Tetraethylenepentamine	341.5	163		321
$C_9H_6N_2O_2$	Toluene-2,4-diisocyanate	251	127	0.9-9.5%	
C_9H_7N	Quinoline	237.1			480
C_9H_{10}	o-Methylstyrene	169.8	53	0.8-11.0%	538
C_9H_{10}	m-Methylstyrene	164	53	0.8-11.0%	538
C_9H_{10}	p-Methylstyrene	172.8	53	0.8-11.0%	538
C_9H_{10}	Isopropenylbenzene	165.4	54	1.9-6.1%	574
$C_9H_{10}O$	1-Phenyl-1-propanone	217.5	99		
$C_9H_{10}O$	4-Methylacetophenone	226	96		
$C_9H_{10}O_2$	Ethyl benzoate	212	88		490
$C_9H_{10}O_2$	Benzyl acetate	213	90		460
$C_9H_{10}O_2$	Methyl 2-phenylacetate	216.5	91		
$C_9H_{11}NO$	4-Methylacetanilide	307	168		
C_9H_{12}	Propylbenzene	159.2	30	0.8-6.0%	450
C_9H_{12}	Isopropylbenzene	152.4	36	0.9-6.5%	424
C_9H_{12}	o-Ethyltoluene	165.2			440
C_9H_{12}	m-Ethyltoluene	161.3			480
C_9H_{12}	p-Ethyltoluene	162			475
C_9H_{12}	1,2,3-Trimethylbenzene	176.1	44	0.8-6.6%	470
C_9H_{12}	1,2,4-Trimethylbenzene	169.3	44	0.9-6.4%	500
C_9H_{12}	1,3,5-Trimethylbenzene	164.7	50	1-5%	559
$C_9H_{12}O$	α–Ethylbenzyl alcohol	219	100		
$C_9H_{12}O_2$	Ethylene glycol monobenzyl ether	256	129		352
$C_9H_{12}O_3S$	Ethyl p-toluenesulfonate		158		
$C_9H_{13}N$	Amphetamine	203	<100		
$C_9H_{14}O$	Phorone	197.5	85		
$C_9H_{14}O$	Isophorone	215.2	84	0.8-3.8%	460
$C_9H_{14}O_6$	Triacetin	259	138	1.0-?	433
C_9H_{16}	Octahydroindene	167			296

Mol. Form.	Name	t_B/°C	FP/°C	Fl. Limits	IT/°C
$C_9H_{16}O_2$	Allyl hexanoate	186	66		
C_9H_{18}	1-Nonene	146.9	26		
C_9H_{18}	Propylcyclohexane	156.7			248
C_9H_{18}	Isopropylcyclohexane	154.8			283
C_9H_{18}	Butylcyclopentane	156.6			250
$C_9H_{18}O$	2-Nonanone	195.3	60	0.9-5.9%	360
$C_9H_{18}O$	Diisobutyl ketone	169.4	49	0.8-7.1%	396
$C_9H_{18}O_2$	Pentyl butanoate	186.4	57		
$C_9H_{18}O_2$	Isopentyl butanoate	179	59		
$C_9H_{18}O_2$	Butyl 3-methylbutanoate		53		
C_9H_{20}	Nonane	150.8	31	0.8-2.9%	205
C_9H_{20}	3-Ethyl-4-methylhexane	140	24		
C_9H_{20}	4-Ethyl-2-methylhexane	133.8	<21	0.7-?	280
C_9H_{20}	2,2,5-Trimethylhexane	124.0	13		
C_9H_{20}	3,3-Diethylpentane	146.3		0.7-5.7%	290
C_9H_{20}	3-Ethyl-2,4-dimethylpentane	136.7	390		
C_9H_{20}	2,2,3,3-Tetramethylpentane	140.2	<21	0.8-4.9%	430
C_9H_{20}	2,2,3,4-Tetramethylpentane	133.0	<21		
$C_9H_{21}BO_3$	Triisopropyl borate	140	28		
$C_9H_{21}N$	Tripropylamine	156	41		
$C_9H_{21}NO_3$	Triisopropanolamine		160		320
$C_{10}H_7Cl$	1-Chloronaphthalene	259	121		>558
$C_{10}H_8$	Naphthalene	217.9	79	0.9-5.9%	526
$C_{10}H_8O$	2-Naphthol	285	153		
$C_{10}H_9N$	1-Naphthalenamine	300.8	157		
$C_{10}H_{10}O_2$	Safrole	234.5	100		
$C_{10}H_{10}O_4$	Dimethyl phthalate	283.7	146	0.9-?	490
$C_{10}H_{10}O_4$	Dimethyl isophthalate	282	138		
$C_{10}H_{10}O_4$	Dimethyl terephthalate	288	153		518
$C_{10}H_{11}NO_2$	Acetoacetanilide		185		
$C_{10}H_{12}$	1,2,3,4-Tetrahydronaphthalene	207.6	71	0.8-5.0%	385
$C_{10}H_{12}O_2$	Isopropyl benzoate	216	99		
$C_{10}H_{12}O_2$	Ethyl phenylacetate	227	99		
$C_{10}H_{14}$	Butylbenzene	183.3	71	0.8-5.8%	410
$C_{10}H_{14}$	sec-Butylbenzene	173.3	52	0.8-6.9%	418
$C_{10}H_{14}$	tert-Butylbenzene	169.1	60	0.7-5.7%	450
$C_{10}H_{14}$	Isobutylbenzene	172.7	55	0.8-6.0%	427
$C_{10}H_{14}$	p-Cymene	177.1	47	0.7-5.6%	436
$C_{10}H_{14}$	1,2,3,4-Tetramethylbenzene	205	74		427
$C_{10}H_{14}$	1,2,3,5-Tetramethylbenzene	198	71		427
$C_{10}H_{14}$	1,2,4,5-Tetramethylbenzene	196.8	54		
$C_{10}H_{14}$	o-Diethylbenzene	184	57		395
$C_{10}H_{14}$	m-Diethylbenzene	181.1	56		450
$C_{10}H_{14}$	p-Diethylbenzene	183.7	55	0.7-6.0%	430
$C_{10}H_{14}O$	Butyl phenyl ether	210	82		
$C_{10}H_{14}O_2$	4-tert-Butyl-1,2-benzenediol	285	130		
$C_{10}H_{15}N$	N-Butylaniline	243.5	107		
$C_{10}H_{15}N$	N,N-Diethylaniline	216.3	85		630
$C_{10}H_{15}NO_2$	N-Phenyl-N,N-diethanolamine		196	0.7-?	387
$C_{10}H_{16}$	Dipentene	178	45		237
$C_{10}H_{16}$	d-Limonene	178	45	0.7-6.1%	237
$C_{10}H_{16}$	α-Pinene	156.2	33		255
$C_{10}H_{16}$	β-Pinene	166	38		275
$C_{10}H_{16}$	β-Phellandrene	171.5	49		
$C_{10}H_{16}O$	Camphor	207.4	66	0.6-3.5%	466
$C_{10}H_{18}$	trans-Decahydronaphthalene	187.3	54	0.7-5.4%	255
$C_{10}H_{18}$	Decahydronaphthalene	155.5	58	0.7-4.9%	250
$C_{10}H_{18}O$	Borneol		66		
$C_{10}H_{18}O$	Linalol	198	71		
$C_{10}H_{18}O$	α-Terpineol	220	90		
$C_{10}H_{18}O$	Cineole	176.4	48		

Mol. Form.	Name	t_B/°C	FP/°C	Fl. Limits	IT/°C
$C_{10}H_{18}O$	*trans*-Geraniol	230	>100		
$C_{10}H_{18}O_4$	Dibutyl oxalate	241	104		
$C_{10}H_{19}NO_2$	N-*tert*-Butylaminoethyl methacrylate		96		
$C_{10}H_{20}$	1-Decene	170.5	<55		235
$C_{10}H_{20}$	Butylcyclohexane	180.9			246
$C_{10}H_{20}$	Isobutylcyclohexane	171.3			274
$C_{10}H_{20}$	*tert*-Butylcyclohexane	171.5			342
$C_{10}H_{20}O$	Citronellol	224	96		
$C_{10}H_{20}O_2$	2-Ethylhexyl acetate	199	71	0.76-8.14%	268
$C_{10}H_{20}O_2$	Ethyl octanoate	208.5	79		
$C_{10}H_{21}N$	N-Butylcyclohexanamine		93		
$C_{10}H_{22}$	Decane	174.1	51	0.8-5.4%	210
$C_{10}H_{22}$	2-Methylnonane	167.1			210
$C_{10}H_{22}$	3-Ethyloctane	166.5			230
$C_{10}H_{22}$	4-Ethyloctane	163.7			229
$C_{10}H_{22}O$	1-Decanol	231.1	82		288
$C_{10}H_{22}O$	Dipentyl ether	190	57		170
$C_{10}H_{22}O_2$	Ethylene glycol dibutyl ether	203.3	85		
$C_{10}H_{22}O_5$	Tetraethylene glycol dimethyl ether	275.3	141		
$C_{10}H_{22}S$	Dipentyl sulfide		85		
$C_{10}H_{23}N$	Decylamine	220.5	99		
$C_{10}H_{23}N$	Dipentylamine	202.5	51		
$C_{11}H_{10}$	1-Methylnaphthalene	244.7			529
$C_{11}H_{12}O_3$	Ethyl benzoylacetate		141		
$C_{11}H_{14}O_2$	Butyl benzoate	250.3	107		
$C_{11}H_{16}$	*p-tert*-Butyltoluene	190	68		
$C_{11}H_{16}$	Pentylbenzene	205.4	66		
$C_{11}H_{16}$	1,3-Diethyl-5-methylbenzene	205			455
$C_{11}H_{16}$	Pentamethylbenzene	232	93		427
$C_{11}H_{16}O$	4-*tert*-Butyl-2-methylphenol	237	118		
$C_{11}H_{17}N$	*p-tert*-Pentylaniline	260.5	102		
$C_{11}H_{20}O_2$	2-Ethylhexyl acrylate		82		252
$C_{11}H_{22}$	Pentylcyclohexane	203.7			239
$C_{11}H_{22}O$	2-Undecanone	231.5	89		
$C_{11}H_{22}O_2$	Nonyl acetate	210	68		
$C_{11}H_{24}$	Undecane	195.9	69		
$C_{11}H_{24}$	2-Methyldecane	189.3			225
$C_{11}H_{24}O$	2-Undecanol	228	113		
$C_{12}H_9Br$	4-Bromo-1,1'-Biphenyl	310	144		
$C_{12}H_{10}$	Biphenyl	256.1	113	0.6-5.8%	540
$C_{12}H_{10}Cl_2Si$	Dichlorodiphenylsilane	305	142		
$C_{12}H_{10}O$	*o*-Phenylphenol	286	124		530
$C_{12}H_{10}O$	Diphenyl ether	258.0	112	0.8-1.5%	618
$C_{12}H_{11}N$	2-Aminobiphenyl	299			450
$C_{12}H_{11}N$	Diphenylamine	302	153		634
$C_{12}H_{12}$	1-Ethylnaphthalene	258.6			480
$C_{12}H_{14}O_4$	Diethyl phthalate	295	161	0.7-?	457
$C_{12}H_{14}O_4$	Diethyl terephthalate	302	117		
$C_{12}H_{16}$	Cyclohexylbenzene	240.1	99		
$C_{12}H_{16}O_3$	Pentyl salicylate	270	132		
$C_{12}H_{17}NO$	N-Butyl-N-phenylacetamide	281	141		
$C_{12}H_{18}$	1,5,9-Cyclododecatriene	240	71		
$C_{12}H_{20}O_4$	Dibutyl maleate	280	141		
$C_{12}H_{22}O_4$	Dimethyl sebacate		145		
$C_{12}H_{22}O_6$	Dibutyl tartrate	320	91		284
$C_{12}H_{23}N$	Dicyclohexylamine		>99		
$C_{12}H_{24}$	1-Dodecene	213.8	79		
$C_{12}H_{24}O_2$	Ethyl decanoate	241.5	>100		
$C_{12}H_{25}Br$	1-Bromododecane	276	144		
$C_{12}H_{26}$	Dodecane	216.3	74	0.6-?	203
$C_{12}H_{26}O$	1-Dodecanol	259	127		275

Mol. Form.	Name	$t_B/°C$	FP/°C	Fl. Limits	IT/°C
$C_{12}H_{26}O$	2-Butyl-1-octanol	246.5	110		
$C_{12}H_{26}O_3$	Diethylene glycol dibutyl ether	256	118		310
$C_{12}H_{26}S$	1-Dodecanethiol	277	128		
$C_{12}H_{27}BO_3$	Tributyl borate	234	93		
$C_{12}H_{27}N$	Tributylamine	216.5	63		
$C_{12}H_{27}O_4P$	Tributyl phosphate	289	146		
$C_{13}H_{12}$	2-Methylbiphenyl	255.5	137		502
$C_{13}H_{12}$	Diphenylmethane	265.0	130		485
$C_{13}H_{14}N_2$	p,p′-Diaminodiphenylmethane	398	220		
$C_{13}H_{26}$	1-Tridecene	232.8	79		
$C_{13}H_{26}O$	2-Tridecanone	263	107		
$C_{13}H_{28}$	Tridecane	235.4	79		
$C_{13}H_{28}O$	1-Tridecanol		121		
$C_{14}H_8O_2$	9,10-Anthracenedione	377	185		
$C_{14}H_{10}$	Anthracene	339.9	121	0.6-?	540
$C_{14}H_{10}$	Phenanthrene	340	171		
$C_{14}H_{12}O_2$	Benzyl benzoate	323.5	148		480
$C_{14}H_{12}O_3$	Benzyl salicylate	320	>100		
$C_{14}H_{14}$	1,1-Diphenylethane	272.6	>100		440
$C_{14}H_{14}O$	Dibenzyl ether	298	135		
$C_{14}H_{16}$	1-Butylnaphthalene	289.3	360		
$C_{14}H_{16}N_2O_2$	o-Dianisidine		206		
$C_{14}H_{23}N$	N,N-Dibutylaniline	274.8	110		
$C_{14}H_{28}$	1-Tetradecene	233	110		235
$C_{14}H_{30}$	Tetradecane	253.5	112	0.5-?	200
$C_{14}H_{30}O$	1-Tetradecanol	289	141		
$C_{15}H_{18}$	1-Pentylnaphthalene	307	124		
$C_{15}H_{24}$	Nonylbenzene	280.5	99		
$C_{15}H_{24}O$	2,6-Di-tert-butyl-4-methylphenol	265	127		
$C_{15}H_{26}O_6$	Tributyrin	307.5	180	0.5-?	407
$C_{15}H_{33}N$	Tripentylamine	242.5	102		
$C_{16}H_{14}O$	1,3-Diphenyl-2-buten-1-one	342.5	177		
$C_{16}H_{18}$	2-Butyl-1,1'-biphenyl		>100		430
$C_{16}H_{22}O_4$	Dibutyl phthalate	340	157	0.5-?	402
$C_{16}H_{26}$	Decylbenzene	298	107		
$C_{16}H_{34}$	Hexadecane	286.8	136		202
$C_{16}H_{34}O$	Dioctyl ether	283	>100		205
$C_{16}H_{35}N$	Bis(2-ethylhexyl)amine		132		
$C_{17}H_{20}N_2O$	N,N′-Diethylcarbanilide		150		
$C_{17}H_{34}O$	2-Heptadecanone	320	120		
$C_{17}H_{36}O$	1-Heptadecanol	333	154		
$C_{18}H_{14}$	o-Terphenyl	332	163		
$C_{18}H_{14}$	m-Terphenyl	363	191		
$C_{18}H_{15}O_3P$	Triphenyl phosphite	360	218		
$C_{18}H_{15}O_4P$	Triphenyl phosphate		220		
$C_{18}H_{15}P$	Triphenylphosphine		180		
$C_{18}H_{30}$	Dodecylbenzene	328	140		
$C_{18}H_{32}O_7$	Butyl citrate		157		368
$C_{18}H_{34}O_2$	Oleic acid	360	189		363
$C_{18}H_{34}O_4$	Dibutyl sebacate	344.5	178	0.4-?	365
$C_{18}H_{36}O_2$	Stearic acid		196		395
$C_{18}H_{37}Cl_3Si$	Trichlorooctadecylsilane		89		
$C_{18}H_{38}$	Octadecane	316.3	>100		227
$C_{18}H_{38}O$	1-Octadecanol				450
$C_{19}H_{16}$	Triphenylmethane	359	>100		
$C_{19}H_{38}O$	2-Nonadecanone		124		
$C_{19}H_{38}O_2$	Methyl stearate	443	153		
$C_{19}H_{40}$	Nonadecane	329.9	>100		230
$C_{20}H_{14}O_4$	Diphenyl phthalate		224		
$C_{20}H_{28}$	1-Decylnaphthalene	379	177		
$C_{20}H_{42}$	Eicosane	343	>100		232

Mol. Form.	Name	t_B/°C	FP/°C	Fl. Limits	IT/°C
$C_{21}H_{21}O_4P$	Tri-o-cresyl phosphate	410	225		385
$C_{21}H_{26}O_3$	4-Octylphenyl salicylate		216		416
$C_{21}H_{32}O_2$	Methyl abietate		180		
$C_{22}H_{42}O_2$	Butyl oleate		180		
$C_{22}H_{42}O_4$	Bis(2-ethylhexyl) adipate		206	0.4-?	377
$C_{22}H_{44}O_2$	Butyl stearate	343	160		355
$C_{23}H_{46}O_2$	Pentyl stearate		185		
$C_{24}H_{20}Sn$	Tetraphenylstannane	420	232		
$C_{24}H_{38}O_4$	Bis(2-ethylhexyl) phthalate	384	218		
$C_{25}H_{48}O_4$	Bis(2-ethylhexyl) azelate		227	0.3-?	374

THRESHOLD LIMITS FOR AIRBORNE CONTAMINANTS

Several organizations recommend limits of exposure to airborne contaminants in the workplace. These include the Occupational Safety and Health Administration (OSHA), the National Institute for Occupational Safety and Health (NIOSH), and the non-governmental organization, American Conference of Governmental Industrial Hygienists (ACGIH). The threshold limit value (TLV) for a substance is defined as the concentration level under which the majority of workers may be repeatedly exposed, day after day, without adverse effects. The TLV recommendations are given in two forms:

Time-weighted average (TWA) concentration for a normal 8-hr workday and 40-hr workweek.
Short-term exposure limit (STEL), which should not be exceeded for more than 15 min.

Both kinds of limits are specified for some substances.

The following table gives threshold limit values for a number of substances that may be encountered in the atmosphere of a typical chemical laboratory. All values refer to the concentration in air at 25°C and normal atmospheric pressure. Data for gases are given both in parts per million by volume (ppm) and in mass concentration (mg/m³). Values for liquids refer to mists or aerosols, and those for solids to dusts or fumes. A "C" following a value indicates a ceiling limit which should not be exceeded even for very brief periods because of acute toxic effects of the substance.

Substances are listed by systematic name, which is followed by molecular formula in the Hill format and Chemical Abstracts Service Registry Number. A synonym index which includes common names follows the table.

REFERENCES

1. *Threshold Limit Values for Chemical Substances and Physical Agents; Biological Exposure Indices,* 1997 Edition, American Conference of Governmental Industrial Hygienists, 1330 Kemper Meadow Drive, Cincinnati, OH 45240-1634.
2. *NIOSH Pocket Guide to Chemical Hazards,* U.S. Department of Health and Human Services, National Institute for Occupational Health and Safety, U.S. Government Printing Office, Washington, DC, 1994.
3. *Chemical Information Manual,* U.S. Department of Labor, Occupational Safety and Health Administration, Washington, DC, 1991.

Substance	Molecular Formula	CAS Reg. No.	TWA ppm	TWA mg/m³	STEL ppm	STEL mg/m³
Acetaldehyde	C_2H_4O	75-07-0			25 C	45 C
Acetic acid	$C_2H_4O_2$	64-19-7	10	25	15	37
Acetic anhydride	$C_4H_6O_3$	108-24-7	5	21		
Acetone	C_3H_6O	67-64-1	500	1188	750	1782
Acetonitrile	C_2H_3N	75-05-8	40	67	60	101
2-(Acetyloxy)benzoic acid	$C_9H_8O_4$	50-78-2		5		
Acrolein	C_3H_4O	107-02-8	0.1	0.23	0.3	0.69
Acrylamide	C_3H_5NO	79-06-1		0.03		
Acrylic acid	$C_3H_4O_2$	79-10-7	2	5.9		
Acrylonitrile	C_3H_3N	107-13-1	2	4.3		
Allyl alcohol	C_3H_6O	107-18-6	2	4.8	4	9.5
Allyl glycidyl ether	$C_6H_{10}O_2$	106-92-3	5	23	10	47
Aluminum oxide	Al_2O_3	1344-28-1		10		
Ammonia	H_3N	7664-41-7	25	17	35	24
Ammonium chloride	ClH_4N	12125-02-9		10		20
Aniline	C_6H_7N	62-53-3	2	7.6		
Arsenic	As	7440-38-2		0.01		
Arsine	AsH_3	7784-42-1	0.05	0.16		
Barium sulfate	BaO_4S	7727-43-7		10		
Benzene	C_6H_6	71-43-2	0.5	1.6	2.5	8
1,3-Benzenedimethanamine	$C_8H_{12}N_2$	1477-55-0				0.1 C
Benzenethiol	C_6H_6S	108-98-5	0.5	2.3		
p-Benzoquinone	$C_6H_4O_2$	106-51-4	0.1	0.44		
Benzoyl chloride	C_7H_5ClO	98-88-4			0.5 C	2.8 C
Benzoyl peroxide	$C_{14}H_{10}O_4$	94-36-0		5		
Benzyl acetate	$C_9H_{10}O_2$	140-11-4	10	61		
Beryllium	Be	7440-41-7		0.002		0.01
Biphenyl	$C_{12}H_{10}$	92-52-4	0.2	1.3		
Bis(4-amino-3-chlorophenyl)methane	$C_{13}H_{12}Cl_2N_2$	101-14-4	0.01	0.11		
Bis(2-chloroethyl) ether	$C_4H_8Cl_2O$	111-44-4	5	29	10	58
Bis(chloromethyl) ether	$C_2H_4Cl_2O$	542-88-1	0.001	0.0047		
Bis(2-ethylhexyl) phthalate	$C_{24}H_{38}O_4$	117-81-7		5		10
Boron oxide	B_2O_3	1303-86-2		10		

Substance	Molecular Formula	CAS Reg. No.	TWA		STEL	
			ppm	mg/m^3	ppm	mg/m^3
Boron tribromide	BBr$_3$	10294-33-4			1 C	10 C
Boron trifluoride	BF$_3$	7637-07-2			1 C	2.8 C
Bromine	Br$_2$	7726-95-6	0.1	0.66	0.2	1.3
Bromine pentafluoride	BrF$_5$	7789-30-2	0.1	0.72		
Bromochloromethane	CH$_2$BrCl	74-97-5	200	1060		
2-Bromo-2-chloro-1,1,1-trifluoroethane	C$_2$HBrClF$_3$	151-67-7	50	404		
Bromoethane	C$_2$H$_5$Br	74-96-4	5	22		
Bromoethylene	C$_2$H$_3$Br	593-60-2	5	22		
Bromomethane	CH$_3$Br	74-83-9	1	3.9		
Bromotrifluoromethane	CBrF$_3$	75-63-8	1000	6090		
1,3-Butadiene	C$_4$H$_6$	106-99-0	2	4.4		
Butane	C$_4$H$_{10}$	106-97-8	800	1900		
1-Butanethiol	C$_4$H$_{10}$S	109-79-5	0.5	1.8		
1-Butanol	C$_4$H$_{10}$O	71-36-3			50 C	152 C
sec-Butyl alcohol	C$_4$H$_{10}$O	78-92-2	100	303		
tert-Butyl alcohol	C$_4$H$_{10}$O	75-65-0	100	303		
2-Butanone	C$_4$H$_8$O	78-93-3	200	590	300	885
trans-2-Butenal	C$_4$H$_6$O	4170-30-3	2	5.7		
Butyl acetate	C$_6$H$_{12}$O$_2$	123-86-4	150	713	200	950
sec-Butyl acetate	C$_6$H$_{12}$O$_2$	105-46-4	200	950		
tert-Butyl acetate	C$_6$H$_{12}$O$_2$	540-88-5	200	950		
Butyl acrylate	C$_7$H$_{12}$O$_2$	141-32-2	10	52		
Butylamine	C$_4$H$_{11}$N	109-73-9			5 C	15 C
Butyl glycidyl ether	C$_7$H$_{14}$O$_2$	2426-08-6	25	133		
o-sec-Butylphenol	C$_{10}$H$_{14}$O	89-72-5	5	31		
p-tert-Butyltoluene	C$_{11}$H$_{16}$	98-51-1	1	6.1		
Cadmium	Cd	7440-43-9		0.01		
Calcium metasilicate	CaO$_3$Si	1344-95-2		10		
Calcium oxide	CaO	1305-78-8		2		
Calcium sulfate	CaO$_4$S	7778-18-9		10		
Camphor	C$_{10}$H$_{16}$O	76-22-2	2	12	3	19
Caprolactam	C$_6$H$_{11}$NO	105-60-2	5	23	10	46
Carbon dioxide	CO$_2$	124-38-9	5000	9000	30,000	54,000
Carbon disulfide	CS$_2$	75-15-0	10	31		
Carbon monoxide	CO	630-08-0	25	29		
Carbonyl chloride	CCl$_2$O	75-44-5	0.1	0.40		
Carbonyl fluoride	CF$_2$O	353-50-4	2	5.4	5	13
Chlorine	Cl$_2$	7782-50-5	0.5	1.5	1	2.9
Chlorine dioxide	ClO$_2$	10049-04-4	0.1	0.28	0.3	0.83
Chlorine trifluoride	ClF$_3$	7790-91-2			0.1 C	0.38 C
Chloroacetaldehyde	C$_2$H$_3$ClO	107-20-0			1 C	3.2 C
Chloroacetone	C$_3$H$_5$ClO	78-95-5			1 C	3.8 C
α-Chloroacetophenone	C$_8$H$_7$ClO	532-27-4	0.05	0.32		
Chloroacetyl chloride	C$_2$H$_2$Cl$_2$O	79-04-9	0.05	0.23	0.15	0.69
Chlorobenzene	C$_6$H$_5$Cl	108-90-7	10	46		
2-Chloro-1,3-butadiene	C$_4$H$_5$Cl	126-99-8	10	36		
Chlorodifluoromethane	CHClF$_2$	75-45-6	1000	3540		
Chloroethane	C$_2$H$_5$Cl	75-00-3	100	264		
2-Chloroethanol	C$_2$H$_5$ClO	107-07-3			1 C	3.3 C
Chloroethylene	C$_2$H$_3$Cl	75-01-4	5	13		
Chloromethane	CH$_3$Cl	74-87-3	50	103	100	207
(Chloromethyl)benzene	C$_7$H$_7$Cl	100-44-7	1	5.2		
1-Chloro-4-nitrobenzene	C$_6$H$_4$ClNO$_2$	100-00-5	0.1	0.64		
1-Chloro-1-nitropropane	C$_3$H$_6$ClNO$_2$	600-25-9	2	10		
Chloropentafluoroethane	C$_2$ClF$_5$	76-15-3	1000	6320		
2-Chloropropanoic acid	C$_3$H$_5$ClO$_2$	598-78-7	0.1	0.44		
3-Chloropropene	C$_3$H$_5$Cl	107-05-1	1	3	2	6
2-Chlorostyrene	C$_8$H$_7$Cl	2039-87-4	50	283	75	425

Substance	Molecular Formula	CAS Reg. No.	TWA		STEL	
			ppm	mg/m^3	ppm	mg/m^3
o-Chlorotoluene	C_7H_7Cl	95-49-8	50	259		
Chromyl chloride	Cl_2CrO_2	14977-61-8	0.025	0.16		
Cobalt	Co	7440-48-4		0.02		
Cobalt carbonyl	$C_8Co_2O_8$	10210-68-1		0.1		
Cresol (all isomers)	C_7H_8O	1319-77-3	5	22		
Cyanamide	CH_2N_2	420-04-2		2		
Cyanogen	C_2N_2	460-19-5	10	21		
Cyanogen chloride	CClN	506-77-4			0.3 C	0.75 C
Cyclohexane	C_6H_{12}	110-82-7	300	1030		
Cyclohexanol	$C_6H_{12}O$	108-93-0	50	206		
Cyclohexanone	$C_6H_{10}O$	108-94-1	25	100		
Cyclohexene	C_6H_{10}	110-83-8	300	1010		
Cyclohexylamine	$C_6H_{13}N$	108-91-8	10	41		
1,3-Cyclopentadiene	C_5H_6	542-92-7	75	203		
Cyclopentane	C_5H_{10}	287-92-3	600	1720		
Decaborane(14)	$B_{10}H_{14}$	17702-41-9	0.05	0.25	0.15	0.75
Diacetone alcohol	$C_6H_{12}O_2$	123-42-2	50	238		
4,4'-Diaminodiphenylmethane	$C_{13}H_{14}N_2$	101-77-9	0.1	0.81		
Diazomethane	CH_2N_2	334-88-3	0.2	0.34		
Diborane	B_2H_6	19287-45-7	0.1	0.11		
Dibromodifluoromethane	CBr_2F_2	75-61-6	100	858		
2-Dibutylaminoethanol	$C_{10}H_{23}NO$	102-81-8	0.5	3.5		
2,6-Di-tert-butyl-4-methylphenol	$C_{15}H_{24}O$	128-37-0		10		
Dibutyl phthalate	$C_{16}H_{22}O_4$	84-74-2		5		
Dichloroacetylene	C_2Cl_2	7572-29-4			0.1 C	0.39 C
o-Dichlorobenzene	$C_6H_4Cl_2$	95-50-1	25	150	50	301
p-Dichlorobenzene	$C_6H_4Cl_2$	106-46-7	10	60		
Dichlorodifluoromethane	CCl_2F_2	75-71-8	1000	4950		
1,1-Dichloroethane	$C_2H_4Cl_2$	75-34-3	100	405		
1,2-Dichloroethane	$C_2H_4Cl_2$	107-06-2	10	40		
1,1-Dichloroethylene	$C_2H_2Cl_2$	75-35-4	5	20	20	79
1,2-Dichloroethylene (both isomers)	$C_2H_2Cl_2$	540-59-0	200	793		
Dichlorofluoromethane	$CHCl_2F$	75-43-4	10	42		
Dichloromethane	CH_2Cl_2	75-09-2	50	174		
1,1-Dichloro-1-nitroethane	$C_2H_3Cl_2NO_2$	594-72-9	2	12		
1,2-Dichloropropane	$C_3H_6Cl_2$	78-87-5	75	347	110	508
2,2-Dichloropropanoic acid	$C_3H_4Cl_2O_2$	75-99-0	1	5.8		
1,3-Dichloropropene (both isomers)	$C_3H_4Cl_2$	542-75-6	1	4.5		
1,2-Dichlorotetrafluoroethane	$C_2Cl_2F_4$	76-14-2	1000	6990		
m-Dicyanobenzene	$C_8H_4N_2$	626-17-5		5		
Dicyclopentadiene	$C_{10}H_{12}$	77-73-6	5	27		
Diethanolamine	$C_4H_{11}NO_2$	111-42-2	0.46	2		
Diethylamine	$C_4H_{11}N$	109-89-7	5	15	15	45
2-Diethylaminoethanol	$C_6H_{15}NO$	100-37-8	2	9.6		
Diethylenetriamine	$C_4H_{13}N_3$	111-40-0	1	4.2		
Diethyl ether	$C_4H_{10}O$	60-29-7	400	1210	500	1520
Diethyl phthalate	$C_{12}H_{14}O_4$	84-66-2		5		
Diglycidyl ether	$C_6H_{10}O_3$	2238-07-5	0.1	0.53		
Diisopropylamine	$C_6H_{15}N$	108-18-9	5	21		
Diisopropyl ether	$C_6H_{14}O$	108-20-3	250	1040	310	1300
Dimethoxymethane	$C_3H_8O_2$	109-87-5	1000	3110		
N,N-Dimethylacetamide	C_4H_9NO	127-19-5	10	36		
Dimethylamine	C_2H_7N	124-40-3	5	9.2	15	27.6
N,N-Dimethylaniline	$C_8H_{11}N$	121-69-7	5	25	10	50
N,N-Dimethylformamide	C_3H_7NO	68-12-2	10	30		
2,6-Dimethyl-4-heptanone	$C_9H_{18}O$	108-83-8	25	145		
1,1-Dimethylhydrazine	$C_2H_8N_2$	57-14-7	0.01	0.025		
Dimethyl phthalate	$C_{10}H_{10}O_4$	131-11-3		5		

Substance	Molecular Formula	CAS Reg. No.	TWA		STEL	
			ppm	mg/m³	ppm	mg/m³
Dimethyl sulfate	$C_2H_6O_4S$	77-78-1	0.1	0.52		
Dinitrobenzene (all isomers)	$C_6H_4N_2O_4$	25154-54-5	0.15	1.0		
Dinitrotoluene (all isomers)	$C_7H_6N_2O_4$	25321-14-6		0.2		
1,4-Dioxane	$C_4H_8O_2$	123-91-1	25	90		
Diphenylamine	$C_{12}H_{11}N$	122-39-4		10		
Diphenyl ether	$C_{12}H_{10}O$	101-84-8	1	7	2	14
4,4'-Diphenylmethane diisocyanate	$C_{15}H_{10}N_2O_2$	101-68-8	0.005	0.051		
Divinyl benzene (all isomers)	$C_{10}H_{10}$	1321-74-0	10	53		
Epichlorohydrin	C_3H_5ClO	106-89-8	0.5	1.9		
1,2-Epoxy-4-(epoxyethyl)cyclohexane	$C_8H_{12}O_2$	106-87-6	0.1	0.57		
1,2-Ethanediamine	$C_2H_8N_2$	107-15-3	10	25		
Ethanethiol	C_2H_6S	75-08-1	0.5	1.3		
Ethanol	C_2H_6O	64-17-5	1000	1880		
Ethanolamine	C_2H_7NO	141-43-5	3	7.5	6	15
Ethyl acetate	$C_4H_8O_2$	141-78-6	400	1440		
Ethyl acrylate	$C_5H_8O_2$	140-88-5	5	20	15	61
Ethylamine	C_2H_7N	75-04-7	5	9.2	15	27.6
Ethylbenzene	C_8H_{10}	100-41-4	100	434	125	543
Ethylene glycol	$C_2H_6O_2$	107-21-1				100 C
Ethylene glycol dinitrate	$C_2H_4N_2O_6$	628-96-6	0.05	0.31		
Ethylene glycol monobutyl ether	$C_6H_{14}O_2$	111-76-2	25	121		
Ethylene glycol monoethyl ether	$C_4H_{10}O_2$	110-80-5	5	18		
Ethylene glycol monoethyl ether acetate	$C_6H_{12}O_3$	111-15-9	5	27		
Ethylene glycol monomethyl ether	$C_3H_8O_2$	109-86-4	5	16		
Ethylene glycol monomethyl ether acetate	$C_5H_{10}O_3$	110-49-6	5	24		
Ethyleneimine	C_2H_5N	151-56-4	0.5	0.88		
Ethylene oxide	C_2H_4O	75-21-8	1	1.8		
Ethyl formate	$C_3H_6O_2$	109-94-4	100	303		
N-Ethylmorpholine	$C_6H_{13}NO$	100-74-3	5	24		
Ethyl silicate	$C_8H_{20}O_4Si$	78-10-4	10	85		
Ferrocene	$C_{10}H_{10}Fe$	102-54-5		10		
Fluorine	F_2	7782-41-4	1	1.6	2	3.1
Fluorine monoxide	F_2O	7783-41-7			0.05 C	0.11 C
Formaldehyde	CH_2O	50-00-0			0.3 C	0.37 C
Formamide	CH_3NO	75-12-7	10	18		
Formic acid	CH_2O_2	64-18-6	5	9.4	10	19
Furfural	$C_5H_4O_2$	98-01-1	2	7.9		
Furfuryl alcohol	$C_5H_6O_2$	98-00-0	10	40	15	60
Germane	GeH_4	7782-65-2	0.2	0.63		
Glycerol	$C_3H_8O_3$	56-81-5		10		
Heptane	C_7H_{16}	142-82-5	400	1640	500	2050
2-Heptanone	$C_7H_{14}O$	110-43-0	50	233		
3-Heptanone	$C_7H_{14}O$	106-35-4	50	233		
4-Heptanone	$C_7H_{14}O$	123-19-3	50	233		
Hexachlorobenzene	C_6Cl_6	118-74-1		0.002		
Hexachloro-1,3-butadiene	C_4Cl_6	87-68-3	0.02	0.21		
Hexachloro-1,3-cyclopentadiene	C_5Cl_6	77-47-4	0.01	0.11		
Hexachloroethane	C_2Cl_6	67-72-1	1	9.7		
Hexachloronaphthalene (all isomers)	$C_{10}H_2Cl_6$	1335-87-1		0.2		
Hexamethylene diisocyanate	$C_8H_{12}N_2O_2$	822-06-0	0.005	0.034		
Hexane	C_6H_{14}	110-54-3	50	176		
1,6-Hexanediamine	$C_6H_{16}N_2$	124-09-4	0.5	2.3		
2-Hexanone	$C_6H_{12}O$	591-78-6	5	20		
sec-Hexyl acetate	$C_8H_{16}O_2$	108-84-9	50	295		
Hydrazine	H_4N_2	302-01-2	0.01	0.013		
Hydrazoic acid	HN_3	7782-79-8			0.11 C	
Hydrogen bromide	BrH	10035-10-6			3 C	9.9 C

Substance	Molecular Formula	CAS Reg. No.	TWA ppm	TWA mg/m³	STEL ppm	STEL mg/m³
Hydrogen chloride	ClH	7647-01-0			5 C	7.5 C
Hydrogen cyanide	CHN	74-90-8			4.7 C	5 C
Hydrogen fluoride	FH	7664-39-3			3 C	2.3 C
Hydrogen peroxide	H_2O_2	7722-84-1	1	1.4		
Hydrogen selenide	H_2Se	7783-07-5	0.05	0.16		
Hydrogen sulfide	H_2S	7783-06-4	10	14	15	21
p-Hydroquinone	$C_6H_6O_2$	123-31-9		2		
Indene	C_9H_8	95-13-6	10	48		
Iodine	I_2	7553-56-2			0.1 C	1.0 C
Iodomethane	CH_3I	74-88-4	2	12		
Iron(III) oxide	Fe_2O_3	1309-37-1		5		
Iron pentacarbonyl	C_5FeO_5	13463-40-6	0.1	0.23	0.2	0.45
Isobutyl acetate	$C_6H_{12}O_2$	110-19-0	150	713		
Isooctyl alcohol	$C_8H_{18}O$	26952-21-6	50	266		
Isopentyl acetate	$C_7H_{14}O_2$	123-92-2	100	532		
Isophorone	$C_9H_{14}O$	78-59-1			5 C	28 C
Isopropenylbenzene	C_9H_{10}	98-83-9	50	242	100	483
2-Isopropoxyethanol	$C_5H_{12}O_2$	109-59-1	25	106		
Isopropyl acetate	$C_5H_{10}O_2$	108-21-4	250	1040	310	1290
Isopropylamine	C_3H_9N	75-31-0	5	12	10	24
Isopropylbenzene	C_9H_{12}	98-82-8	50	246		
Ketene	C_2H_2O	463-51-4	0.5	0.86	1.5	2.6
Lead	Pb	7439-92-1		0.05		
Lead(II) arsenate	$As_2O_8Pb_3$	7784-40-9		0.15		
Lead(II) chromate	CrO_4Pb	7758-97-6		0.05		
Lithium hydride	HLi	7580-67-8		0.025		
Magnesium oxide	MgO	1309-48-4		10		
Maleic anhydride	$C_4H_2O_3$	108-31-6	0.25	1.0		
Mercury	Hg	7439-97-6		0.01		0.03
Mesityl oxide	$C_6H_{10}O$	141-79-7	15	60	25	100
Methacrylic acid	$C_4H_6O_2$	79-41-4	20	70		
Methanethiol	CH_4S	74-93-1	0.5	0.98		
Methanol	CH_4O	67-56-1	200	262	250	328
o-Methoxyaniline	C_7H_9NO	90-04-0	0.1	0.5		
p-Methoxyaniline	C_7H_9NO	104-94-9	0.1	0.5		
4-Methoxyphenol	$C_7H_8O_2$	150-76-5		5		
Methyl acetate	$C_3H_6O_2$	79-20-9	200	606	250	757
Methyl acrylate	$C_4H_6O_2$	96-33-3	2	7		
2-Methylacrylonitrile	C_4H_5N	126-98-7	1	2.7		
Methylamine	CH_5N	74-89-5	5	6.4	15	19
o-Methylaniline	C_7H_9N	95-53-4	2	8.8		
m-Methylaniline	C_7H_9N	108-44-1	2	8.8		
p-Methylaniline	C_7H_9N	106-49-0	2	8.8		
N-Methylaniline	C_7H_9N	100-61-8	0.5	2.2		
3-Methyl-1-butanol	$C_5H_{12}O$	123-51-3	100	361	125	452
3-Methyl-2-butanone	$C_5H_{10}O$	563-80-4	200	705		
Methyl tert-butyl ether	$C_5H_{12}O$	1634-04-4	40	144		
Methyl 2-cyanoacrylate	$C_5H_5NO_2$	137-05-3	2	9.1	4	18
Methylcyclohexane	C_7H_{14}	108-87-2	400	1610		
Methylcyclohexanol (all isomers)	$C_7H_{14}O$	25639-42-3	50	234		
2-Methylcyclohexanone	$C_7H_{12}O$	583-60-8	50	229	75	344
2-Methyl-3,5-dinitrobenzamide	$C_8H_7N_3O_5$	148-01-6		5		
2-Methyl-4,6-dinitrophenol	$C_7H_6N_2O_5$	534-52-1		0.2		
Methyl ethyl ketone peroxide	$C_8H_{18}O_2$	1338-23-4			0.2 C	1.5 C
Methyl formate	$C_2H_4O_2$	107-31-3	100	246	150	368
5-Methyl-3-heptanone	$C_8H_{16}O$	541-85-5	25	131		
5-Methyl-2-hexanone	$C_7H_{14}O$	110-12-3	50	234		
Methylhydrazine	CH_6N_2	60-34-4	0.01 C	0.019 C		

Substance	Molecular Formula	CAS Reg. No.	TWA ppm	TWA mg/m³	STEL ppm	STEL mg/m³
Methyl isocyanate	C_2H_3NO	624-83-9	0.02	0.047		
Methyl methacrylate	$C_5H_8O_2$	80-62-6	100	410		
Methyloxirane	C_3H_6O	75-56-9	20	48		
2-Methyl-2,4-pentanediol	$C_6H_{14}O_2$	107-41-5			25 C	121 C
4-Methyl-2-pentanol	$C_6H_{14}O$	108-11-2	25	104	40	167
4-Methyl-2-pentanone	$C_6H_{12}O$	108-10-1	50	205	75	307
2-Methyl-1-propanol	$C_4H_{10}O$	78-83-1	50	152		
Methylstyrene (all isomers)	C_9H_{10}	25013-15-4	50	242	100	483
N-Methyl-N,2,4,6-tetranitroaniline	$C_7H_5N_5O_8$	479-45-8		1.5		
Morpholine	C_4H_9NO	110-91-8	20	71		
Naphthalene	$C_{10}H_8$	91-20-3	10	52	15	79
Nickel	Ni	7440-02-0		1		
Nickel carbonyl	C_4NiO_4	13463-39-3	0.05	0.12		
Nicotine	$C_{10}H_{14}N_2$	54-11-5		0.5		
Nitric acid	HNO_3	7697-37-2	2	5.2	4	10
Nitric oxide	NO	10102-43-9	25	31		
p-Nitroaniline	$C_6H_6N_2O_2$	100-01-6		3		
Nitrobenzene	$C_6H_5NO_2$	98-95-3	1	5		
Nitroethane	$C_2H_5NO_2$	79-24-3	100	307		
Nitrogen dioxide	NO_2	10102-44-0	3	5.6	5	9.4
Nitrogen trifluoride	F_3N	7783-54-2	10	29		
Nitromethane	CH_3NO_2	75-52-5	20	50		
1-Nitropropane	$C_3H_7NO_2$	108-03-2	25	91		
2-Nitropropane	$C_3H_7NO_2$	79-46-9	10	36		
Nitrotoluene	$C_7H_7NO_2$	1321-12-6	2	11		
Nitrous oxide	N_2O	10024-97-2	50	90		
Nonane	C_9H_{20}	111-84-2	200	1050		
Octachloronaphthalene	$C_{10}Cl_8$	2234-13-1		0.1		0.3
Octane	C_8H_{18}	111-65-9	300	1400	375	1750
Osmium(VIII) oxide	O_4Os	20816-12-0	0.0002	0.0016	0.0006	0.0047
Oxalic acid	$C_2H_2O_4$	144-62-7		1		2
2-Oxetanone	$C_3H_4O_2$	57-57-8	0.5	1.5		
Oxiranemethanol	$C_3H_6O_2$	556-52-5	2	6.1		
Ozone	O_3	10028-15-6	0.1	0.2		
Pentaborane(9)	B_5H_9	19624-22-7	0.005	0.013	0.015	0.039
Pentachloronitrobenzene	$C_6Cl_5NO_2$	82-68-8		0.5		
Pentachlorophenol	C_6HCl_5O	87-86-5		0.5		
Pentaerythritol	$C_5H_{12}O_4$	115-77-5		10		
Pentanal	$C_5H_{10}O$	110-62-3	50	176		
Pentane	C_5H_{12}	109-66-0	600	1770	750	2210
Pentanedial	$C_5H_8O_2$	111-30-8			0.05 C	0.2 C
2-Pentanone	$C_5H_{10}O$	107-87-9	200	705	250	881
3-Pentanone	$C_5H_{10}O$	96-22-0	200	705		
Pentyl acetate	$C_7H_{14}O_2$	628-63-7	100	532		
sec-Pentyl acetate	$C_7H_{14}O_2$	626-38-0	125	665		
Perchloromethyl mercaptan	CCl_4S	594-42-3	0.1	0.76		
Perchloryl fluoride	$ClFO_3$	7616-94-6	3	13	6	25
Perfluoroacetone	C_3F_6O	684-16-2	0.1	0.68		
Perfluoroisobutene	C_4F_8	382-21-8			0.01 C	0.082 C
Phenol	C_6H_6O	108-95-2	5	19		
10H-Phenothiazine	$C_{12}H_9NS$	92-84-2		5		
Phenylenediamine (all isomers)	$C_6H_8N_2$	25265-76-3		0.1		
Phenyl glycidyl ether	$C_9H_{10}O_2$	122-60-1	0.1	0.6		
Phenylhydrazine	$C_6H_8N_2$	100-63-0	0.1	0.44		
Phosphine	H_3P	7803-51-2	0.3	0.42	1	1.4
Phosphoric acid	H_3O_4P	7664-38-2		1		3
Phosphorus (yellow)	P	7723-14-0	0.02	0.1		
Phosphorus(III) chloride	Cl_3P	7719-12-2	0.2	1.1	0.5	2.8

Substance	Molecular Formula	CAS Reg. No.	TWA ppm	TWA mg/m³	STEL ppm	STEL mg/m³
Phosphorus(V) chloride	Cl₅P	10026-13-8	0.1	0.85		
Phosphorus(V) oxychloride	Cl₃OP	10025-87-3	0.1	0.63		
Phthalic anhydride	C₈H₄O₃	85-44-9	1	6.1		
Propanoic acid	C₃H₆O₂	79-09-4	10	30		
1-Propanol	C₃H₈O	71-23-8	200	492	250	614
2-Propanol	C₃H₈O	67-63-0	400	983	500	1230
Propargyl alcohol	C₃H₄O	107-19-7	1	2.3		
Propyl acetate	C₅H₁₀O₂	109-60-4	200	835	250	1040
Propylene glycol monomethyl ether	C₄H₁₀O₂	107-98-2	100	369	150	553
Propyleneimine	C₃H₇N	75-55-8	2	4.7		
Propyl nitrate	C₃H₇NO₃	627-13-4	25	107	40	172
Propyne	C₃H₄	74-99-7	1000	1640		
2-Pyridinamine	C₅H₆N₂	504-29-0	0.5	1.9		
Pyridine	C₅H₅N	110-86-1	5	16		
Pyrocatechol	C₆H₆O₂	120-80-9	5	23		
Resorcinol	C₆H₆O₂	108-46-3	10	45	20	90
Selenium	Se	7782-49-2		0.2		
Selenium hexafluoride	F₆Se	7783-79-1	0.05	0.16		
Silane	H₄Si	7803-62-5	5	6.6		
Silicon dioxide (α-quartz)	O₂Si	14808-60-7		0.1		
Sodium bisulfite	HNaO₃S	7631-90-5		5		
Stibine	H₃Sb	7803-52-3	0.1	0.51		
Strychnine	C₂₁H₂₂N₂O₂	57-24-9		0.15		
Styrene	C₈H₈	100-42-5	20	85	40	170
Sulfur chloride	Cl₂S₂	10025-67-9			1 C	5.5 C
Sulfur dioxide	O₂S	7446-09-5	2	5.2	5	13
Sulfur hexafluoride	F₆S	2551-62-4	1000	5970		
Sulfuric acid	H₂O₄S	7664-93-9		1		3
Sulfur pentafluoride	F₁₀S₂	5714-22-7			0.01 C	0.10 C
Sulfur tetrafluoride	F₄S	7783-60-0			0.1 C	0.44 C
Sulfuryl fluoride	F₂O₂S	2699-79-8	5	21	10	42
Tantalum	Ta	7440-25-7		5		
Tellurium	Te	13494-80-9		0.1		
Tellurium hexafluoride	F₆Te	7783-80-4	0.02	0.10		
Terphenyl (all isomers)	C₁₈H₁₄	26140-60-3			0.53 C	5 C
1,1,2,2-Tetrabromoethane	C₂H₂Br₄	79-27-6	1	14		
Tetrabromomethane	CBr₄	558-13-4	0.1	1.4	0.3	4.1
1,1,1,2-Tetrachloro-2,2-difluoroethane	C₂Cl₄F₂	76-11-9	500	4170		
1,1,2,2-Tetrachloro-1,2-difluoroethane	C₂Cl₄F₂	76-12-0	500	4170		
1,1,2,2-Tetrachloroethane	C₂H₂Cl₄	79-34-5	1	6.9		
Tetrachloroethylene	C₂Cl₄	127-18-4	25	170	100	685
Tetrachloromethane	CCl₄	56-23-5	5	31	10	63
Tetrachloronaphthalene (all isomers)	C₁₀H₄Cl₄	1335-88-2		2		
Tetraethyl lead	C₈H₂₀Pb	78-00-2		0.1		
Tetrahydrofuran	C₄H₈O	109-99-9	200	590	250	737
Tetramethyl lead	C₄H₁₂Pb	75-74-1		0.15		
Tetramethyl silicate	C₄H₁₂O₄Si	681-84-5	1	6		
Tetranitromethane	CN₄O₈	509-14-8	0.005	0.04		
Thallium	Tl	7440-28-0		0.1		
Thioglycolic acid	C₂H₄O₂S	68-11-1	1	3.8		
Thionyl chloride	Cl₂OS	7719-09-7			1 C	4.9 C
Titanium(IV) oxide	O₂Ti	13463-67-7		10		
Toluene	C₇H₈	108-88-3	50	188		
Toluene-2,4-diisocyanate	C₉H₆N₂O₂	584-84-9	0.005	0.036	0.02	0.14
Tribromomethane	CHBr₃	75-25-2	0.5	5.2		
Trichloroacetic acid	C₂HCl₃O₂	76-03-9	1	6.7		
1,2,4-Trichlorobenzene	C₆H₃Cl₃	120-82-1			5 C	37 C
1,1,1-Trichloroethane	C₂H₃Cl₃	71-55-6	350	1910	450	2460

Substance	Molecular Formula	CAS Reg. No.	TWA		STEL	
			ppm	mg/m^3	ppm	mg/m^3
1,1,2-Trichloroethane	$C_2H_3Cl_3$	79-00-5	10	55		
Trichloroethylene	C_2HCl_3	79-01-6	50	269	100	537
Trichlorofluoromethane	CCl_3F	75-69-4			1000 C	5620 C
Trichloromethane	$CHCl_3$	67-66-3	10	49		
Trichloronaphthalene (all isomers)	$C_{10}H_5Cl_3$	1321-65-9		5		
Trichloronitromethane	CCl_3NO_2	76-06-2	0.1	0.67		
1,2,3-Trichloropropane	$C_3H_5Cl_3$	96-18-4	10	60		
1,1,2-Trichloro-1,2,2-trifluoroethane	$C_2Cl_3F_3$	76-13-1	1000	7670	1250	9590
Tri-*o*-cresyl phosphate	$C_{21}H_{21}O_4P$	78-30-8		0.1		
Triethanolamine	$C_6H_{15}NO_3$	102-71-6		5		
Triethylamine	$C_6H_{15}N$	121-44-8	1	4.1	3	12
Triiodomethane	CHI_3	75-47-8	0.6	10		
Trimellitic anhydride	$C_9H_4O_5$	552-30-7				0.04 C
Trimethylamine	C_3H_9N	75-50-3	5	12	15	36
Trimethylbenzene (all isomers)	C_9H_{12}	25551-13-7	25	123		
Trimethyl phosphite	$C_3H_9O_3P$	121-45-9	2	10		
Trinitroglycerol	$C_3H_5N_3O_9$	55-63-0	0.05	0.46		
2,4,6-Trinitrophenol	$C_6H_3N_3O_7$	88-89-1		0.1		
2,4,6-Trinitrotoluene (TNT)	$C_7H_5N_3O_6$	118-96-7		0.1		
Triphenylamine	$C_{18}H_{15}N$	603-34-9		5		
Triphenyl phosphate	$C_{18}H_{15}O_4P$	115-86-6		3		
Uranium	U	7440-61-1		0.2		0.6
Vanadium(V) oxide	O_5V_2	1314-62-1		0.05		
Vinyl acetate	$C_4H_6O_2$	108-05-4	10	35	15	53
4-Vinylcyclohexene	C_8H_{12}	100-40-3	0.1	0.44		
Xylene (all isomers)	C_8H_{10}	1330-20-7	100	434	150	651
Xylidine (all isomers)	$C_8H_{11}N$	1300-73-8	0.5	2.5		
Zinc chloride	Cl_2Zn	7646-85-7		1		2
Zinc oxide	OZn	1314-13-2		5		10

SYNONYM INDEX

Acetyl acetate: see Acetic anhydride

Acetylene tetrabromide: see 1,1,2,2-Tetrabromoethane

Acetylene tetrachloride: see 1,1,2,2-Tetrachloroethane

Acetylsalicylic acid: see 2-(Acetyloxy)benzoic acid

Allyl chloride: see 3-Chloropropene

Allyl trichloride: see 1,2,3-Trichloropropane

2-Aminoethanol: see Ethanolamine

2-Aminopyridine: see 2-Pyridinamine

Amyl acetate: see Pentyl acetate

sec-Amyl acetate: see *sec*-Pentyl acetate

o-Anisidine: see *o*-Methoxyaniline

p-Anisidine: see *p*-Methoxyaniline

Aziridine: see Ethyleneimine

Benzenamine: see Aniline

Benzenediamine: see Phenylenediamine

1,4-Benzenediol: see *p*-Hydroquinone

1,2,4-Benzenetricarboxylic anhydride: see Trimellitic anhydride

Benzyl chloride: see (Chloromethyl)benzene

Bis(2-amimoethyl)amine: see Diethylenetriamine

Bromoform: see Tribromomethane

1-Butanamine: see Butylamine

2-Butanol: see *sec*-Butyl alcohol

2-Butoxyethanol: see Ethylene glycol monobutyl ether

Butyl alcohol: see 1-Butanol

Butyl mercaptan: see 1-Butanethiol

tert-Butyl methyl ether: see Methyl *tert*-butyl ether

Butyl methyl ketone: see 2-Hexanone

Butyl-2-propenoate: see Butyl acrylate

Calcium silicate: see Calcium metasilicate

Carbon tetrabromide: see Tetrabromomethane

Carbon tetrachloride: see Tetrachloromethane

Catechol: see Pyrocatechol

Chlorine cyanide: see Cyanogen chloride

2-Chloro-1-ethanal: see Chloroacetaldehyde

Chloroform: see Trichloromethane

Chloromethyl ether: see Bis(chloromethyl) ether

(Chloromethyl)oxirane: see Epichlorohydrin

Chloropicrin: see Trichloronitromethane

Chloroprene: see 2-Chloro-1,3-butadiene

2-Chloropropionic acid: see 2-Chloropropanoic acid

Crotonaldehyde: see *trans*-2-Butenal

Cumene: see Isopropylbenzene

Cyclohexyl alcohol: see Cyclohexanol

2,6-Di-*tert*-butyl-*p*-cresol: see 2,6-Di-*tert*-butyl-4-methylphenol

2,2'-Dichlorethyl ether: see Bis(2-chloroethyl) ether

2,2-Dichloropropionic acid: see 2,2-Dichloropropanoic acid

Dicyclopentadienyl iron: see Ferrocene

Diethyl ketone: see 3-Pentanone

Diisobytyl ketone: see 2,6-Dimethyl-4-heptanone
Dimethylaniline: see Xylidine
Dimethylbenzene: see Xylene
N,N-Dimethylethanamide: see N,N-Dimethylacetamide
Dinitolmide: see 2-Methyl-3,5-dinitrobenzamide
Dinitro-o-cresol: see 2-Methyl-4,6-dinitrophenol
Di-sec-octyl phthalate: see Bis(2-ethylhexyl) phthalate
Diphenyl: see Biphenyl
Dipropyl ketone: see 4-Heptanone
Disulfur dichloride: see Sulfur chloride
Ethanal: see Acetaldehyde
1,2-Ethanediol: see Ethylene glycol
2-Ethoxyethanol: see Ethylene glycol monoethyl ether
2-Ethoxyethyl acetate: see Ethylene glycol monoethyl ether acetate
Ethyl alcohol: see Ethanol
Ethyl bromide: see Bromoethane
Ethyl butyl ketone: see 3-Heptanone
Ethyl chloride: see Chloroethane
Ethylene chlorohydrin: see 2-Chloroethanol
Ethylenediamine: see 1,2-Ethanediamine
Ethylene dichloride: see 1,2-Dichloroethane
Ethyl ether: see Diethyl ether
Ethylidene dichloride: see 1,1-Dichloroethane
Ethyl mercaptan: see Ethanethiol
Ethyl propenoate: see Ethyl acrylate
2-Furaldehyde: see Furfural
2-Furanmethanol: see Furfuryl alcohol
Germanium tetrahydride: see Germane
Glutaraldehyde: see Pentanedial
Glycidol: see Oxiranemethanol
Halon 1011: see Bromochloromethane
Halothane: see 2-Bromo-2-chloro-1,1,1-trifluoroethane
Hexafluoroacetone: see Perfluoroacetone
Hexahydro-2-azepinone: see Caprolactam
Hexamethylenediamine: see 1,6-Hexanediamine
Hexylene glycol: see 2-Methyl-2,4-pentanediol
Hydrobromic acid: see Hydrogen bromide
Hydrochloric acid: see Hydrogen chloride
Hydrocyanic acid: see Hydrogen cyanide
Hydrofluoric acid: see Hydrogen fluoride
4-Hydroxy-4-methyl-2-pentanone: see Diacetone alcohol
Iodoform: see Triiodomethane
Isoamyl acetate: see Isopentyl acetate
Isoamyl alcohol: see 3-Methyl-1-butanol
Isobutyl alcohol: see 2-Methyl-1-propanol
Isobutyl methyl ketone: see 4-Methyl-2-pentanone
Isopropyl alcohol: see 2-Propanol
Isopropyl ether: see Diisopropyl ether
Mesitylene: see Trimethylbenzene
Methanal: see Formaldehyde
Methanamide: see Formamide
Methanoic acid: see Formic acid
2-Methoxyethanol: see Ethylene glycol monomethyl ether
2-Methoxyethyl acetate: see Ethylene glycol monomethyl ether acetate
Methylacetylene: see Propyne
Methylal: see Dimethoxymethane
Methyl alcohol: see Methanol
Methylbenzene: see Toluene
Methyl bromide: see Bromomethane

Methyl chloride: see Chloromethane
Methyl chloroform: see 1,1,1-Trichloroethane
Methyl cyanide: see Acetonitrile
4,4-Methylene bis(2-chloroaniline): see Bis(4-amino-3chlorophenyl) methane
Methylene chloride: see Dichloromethane
4,4-Methylene dianiline: see 4,4'-Diaminodiphenylmethane
Methylene diphenyl diisocyanate: see 4,4'-Diphenylmethane diisocyanate
1-Methylethyl acetate: see Isopropyl acetate
Methyl ethyl ketone: see 2-Butanone
Methyl iodide: see Iodomethane
Methyl isobutyl carbinol: see 4-Methyl-2-pentanol
Methyl isopentyl ketone: see 5-Methyl-2-hexanone
Methyl isopropyl ketone: see 3-Methyl-2-butanone
Methyl mercaptan: see Methanethiol
Methyl 2-methylpropenoate: see Methyl methacrylate
Methylnitrobenzene: see Nitrotoluene
4-Methyl-3-penten-2-one: see Mesityl oxide
Methyl pentyl ketone: see 2-Heptanone
Methylphenol: see Cresol
2-Methyl-2-propanol: see tert-Butyl alcohol
2-Methylpropenenitrile: see 2-Methylacrylonitrile
Methyl propenoate: see Methyl acrylate
2-Methylpropenoic acid: see Methacrylic acid
1-Methylpropyl acetate: see sec-Butyl acetate
2-Methylpropyl acetate: see Isobutyl acetate
Methyl propyl ketone: see 2-Pentanone
Methyl silicate: see Tetramethyl silicate
α-Methyl styrene: see Isopropenylbenzene
Nitramine: see N-Methyl-N,2,4,6-tetranitroaniline
p-Nitrochlorobenzene: see 1-Chloro-4-nitrobenzene
Nitroglycerin: see Trinitroglycerol
Orthophosphoric acid: see Phosphoric acid
Osmium tetroxide: see Osmium(VIII) oxide
Oxirane: see Ethylene oxide
Oxolane: see Tetrahydrofuran
Oxygen difluoride: see Fluorine monoxide
Perchlorobutadiene: see Hexachloro-1,3-butadiene
Perchlorocyclopentadiene: see Hexachloro-1,3-cyclopentadiene
Perchloroethane: see Hexachloroethane
Perchloroethylene: see Tetrachloroethylene
Phenyl chloride: see Chlorobenzene
Phenyl ether: see Diphenyl ether
Phenyl mercaptan: see Benzenethiol
Phosgene: see Carbonyl chloride
Phosphorus pentachloride: see Phosphorus(V) chloride
Phosphorus trichloride: see Phosphorus(III) chloride
Phosphoryl chloride: see Phosphorus(V) oxychloride
m-Phthalodinitrile: see m-Dicyanobenzene
Picloram: see 4-Amino-3,5,6-trichloropyridinecarboxlic acid
Picric acid: see 2,4,6-Trinitrophenol
Pimelic ketone: see Cyclohexanone
1,2,3-Propanetriol: see Glycerol
2-Propanone: see Acetone
2-Propenal: see Acrolein
Propenenitrile: see Acrylonitrile
2-Propenoic acid: see Acrylic acid
2-Propen-1-ol: see Allyl alcohol

β-Propiolactone: see 2-Oxetanone
Propionic acid: see Propanoic acid
Propyl alcohol: see 1-Propanol
Propylene dichloride: see 1,2-Dichloropropane
1,2-Propylene oxide: see Methyloxirane
2-Propyn-1-ol: see Propargyl alcohol
Quartz: see Silicon dioxide (α-quartz)
Quinone: see p-Benzoquinone
Silicon tetrahydride: see Silane
Sulfinyl dichloride: see Thionyl chloride
Tetrachloro-1,1-difluoroethane: see 1,1,1,2-Tetrachloro-2,2-difluoroethane
Tetryl: see N-Methyl-N,2,4,6-tetranitroaniline
Titanium dioxide: see Titanium(IV) oxide

TNT: see 2,4,6-Trinitrotoluene (TNT)
Toluidine: see Methylaniline
Trichloroethene: see Trichloroethylene
Trifluorobromomethane: see Bromotrifluoromethane
3,5,5-Trimethyl-2-cyclohexen-1-one: see Isophorone
Valeraldehyde: see Pentanal
Vanadium pentoxide: see Vanadium(V) oxide
Vinylbenzene: see Styrene
Vinyl bromide: see Bromoethylene
Vinyl chloride: see Chloroethylene
Vinylcyclohexene dioxide: see 1,2-Epoxy-4-(epoxyethyl)cyclohexane
Vinylidene chloride: see 1,1-Dichloroethylene
Vinyltoluene: see Methylstyrene
m-Xylene diamine: see 1,3-Benzenedimethanamine

OCTANOL-WATER PARTITION COEFFICIENTS

The octanol-water partition coefficient, P, is a widely used parameter for correlating biological effects of organic substances. It is a property of the two-phase system in which water and 1-octanol are in equilibrium at a fixed temperature and the substance is distributed between the water-rich and octanol-rich phases. P is defined as the ratio of the equilibrium concentration of the substance in the octanol-rich phase to that in the water-rich phase, in the limit of zero concentration. In general, P tends to be large for compounds with extended non-polar structures (such as long chain or multi-ring hydrocarbons) and small for compounds with highly polar groups. Thus P (or, in its more common form of expression, log P) provides a measure of the lipophilic vs. hydrophilic nature of a compound, which is an important consideration in assessing the potential toxicity. A discussion of methods of measurement and accuracy considerations for log P may be found in Reference 1.

This table gives selected values of log P for about 450 organic compounds, including many of environmental importance. All values refer to a nominal temperature of 25°C. The source of each value is indicated in the last column. These references contain data on many more compounds than are included here.

Compounds are listed by molecular formula following the Hill convention. To locate a compound by name or CAS Registry Number when the molecular formula is not known, use the table "Physical Constants of Organic Compounds" in Section 3 and its indexes to determine the molecular formula.

REFERENCES

1. Sangster, J., *J. Phys. Chem. Ref. Data*, 18, 1111, 1989.
2. Mackay, D., Shiu, W.Y., and Ma, K.C., *Illustrated Handbook of Physical-Chemical Properties and Environmental Fate for Organic Chemicals*, Lewis Publishers/CRC Press, Boca Raton, FL, 1992.
3. Shiu, W.Y., and Mackay, D., *J. Phys. Chem. Ref. Data*, 15, 911, 1986.
4. Pinsuwan, S., Li, L., and Yalkowsky, S.H., *J. Chem. Eng. Data*, 40, 623, 1995.
5. *Solubility Data Series, International Union of Pure and Applied Chemistry, Vol. 20*, Pergamon Press, Oxford, 1985.
6. *Solubility Data Series, International Union of Pure and Applied Chemistry, Vol. 38*, Pergamon Press, Oxford, 1985.
7. Miller, M.M., Ghodbane, S., Wasik, S.P., Tewari, Y.B., and Martire, D.E., *J. Chem. Eng. Data*, 29, 184, 1984.

Mol. Form.	Name	log P	Ref.	Mol. Form.	Name	log P	Ref.
CCl_2F_2	Dichlorodifluoromethane	2.16	2	C_2H_4O	Acetaldehyde	0.45	1
CCl_3F	Trichlorofluoromethane	2.53	2	C_2H_4O	Ethylene oxide	-0.30	1
CCl_4	Tetrachloromethane	2.64	2	$C_2H_4O_2$	Acetic acid	-0.17	1
$CHBr_3$	Tribromomethane	2.38	2	C_2H_5Br	Bromoethane	1.6	2
$CHCl_3$	Trichloromethane	1.97	2	C_2H_5Cl	Chloroethane	1.43	2
CH_2BrCl	Bromochloromethane	1.41	2	C_2H_5I	Iodoethane	2	2
CH_2Br_2	Dibromomethane	2.3	2	C_2H_5NO	Acetamide	-1.26	1
CH_2Cl_2	Dichloromethane	1.25	2	$C_2H_5NO_2$	Nitroethane	0.18	1
CH_2F_2	Difluoromethane	0.20	1	C_2H_6O	Ethanol	-0.30	1
CH_2I_2	Diiodomethane	2.5	2	C_2H_6O	Dimethyl ether	0.10	1
CH_2O	Formaldehyde	0.35	1	C_2H_6OS	Dimethyl sulfoxide	-1.35	1
CH_2O_2	Formic acid	-0.54	1	$C_2H_6O_2S$	Dimethyl sulfone	-1.41	1
CH_3Br	Bromomethane	1.19	2	C_2H_7N	Ethylamine	-0.13	1
CH_3Cl	Chloromethane	0.91	2	C_2H_7N	Dimethylamine	-0.38	1
CH_3F	Fluoromethane	0.51	1	C_3H_3N	2-Propenenitrile	0.25	1
CH_3I	Iodomethane	1.5	2	$C_3H_4Cl_2$	*cis*-1,3-Dichloropropene	2.03	2
CH_3NO	Formamide	-1.51	1	C_3H_4O	Propargyl alcohol	-0.38	1
CH_3NO_2	Nitromethane	-0.33	1	C_3H_4O	Acrolein	-0.01	1
CH_4O	Methanol	-0.74	1	C_3H_5Br	3-Bromopropene	1.79	1
CH_5N	Methylamine	-0.57	1	C_3H_5ClO	Epichlorohydrin	0.30	2
$C_2Cl_3F_3$	1,1,2-Trichlorotrifluoroethane	3.16	2	$C_3H_5Cl_3$	1,2,3-Trichloropropane	2.63	2
C_2Cl_4	Tetrachloroethylene	2.88	2	C_3H_5N	Propanenitrile	0.16	1
C_2Cl_6	Hexachloroethane	4.00	4	C_3H_5NO	Acrylamide	-0.78	1
C_2HCl_3	Trichloroethylene	2.53	2	$C_3H_6Cl_2$	1,2-Dichloropropane	2.0	2
C_2HCl_5	Pentachloroethane	2.89	2	C_3H_6O	Allyl alcohol	0.17	1
$C_2H_2Cl_2$	1,1-Dichloroethylene	2.13	2	C_3H_6O	Propanal	0.59	1
$C_2H_2Cl_2$	*cis*-1,2-Dichloroethylene	1.86	2	C_3H_6O	Acetone	-0.24	1
$C_2H_2Cl_2$	*trans*-1,2-Dichloroethylene	1.93	2	C_3H_6O	Methyloxirane	0.03	1
$C_2H_2Cl_4$	1,1,2,2-Tetrachloroethane	2.39	2	$C_3H_6O_2$	Propanoic acid	0.33	1
C_2H_3Cl	Chloroethylene	1.38	2	$C_3H_6O_2$	Methyl acetate	0.18	1
$C_2H_3Cl_3$	1,1,1-Trichloroethane	2.49	2	C_3H_7Br	1-Bromopropane	2.1	2
$C_2H_3Cl_3$	1,1,2-Trichloroethane	2.38	2	C_3H_7Br	2-Bromopropane	1.9	2
C_2H_3N	Acetonitrile	-0.34	1	C_3H_7Cl	1-Chloropropane	2.04	1
$C_2H_4Cl_2$	1,1-Dichloroethane	1.79	2	C_3H_7Cl	2-Chloropropane	1.90	1
$C_2H_4Cl_2$	1,2-Dichloroethane	1.48	2	C_3H_7I	1-Iodopropane	2.5	2

Mol. Form.	Name	log P	Ref.	Mol. Form.	Name	log P	Ref.
C_3H_7N	Allylamine	0.03	1	C_5H_{10}	Cyclopentane	3.00	1
C_3H_7NO	N,N-Dimethylformamide	-1.01	1	$C_5H_{10}O$	2-Pentanone	0.84	1
C_3H_7NO	N-Methylacetamide	-1.05	1	$C_5H_{10}O$	3-Pentanone	0.82	1
$C_3H_7NO_2$	1-Nitropropane	0.87	1	$C_5H_{10}O$	3-Methyl-2-butanone	0.56	1
C_3H_8O	1-Propanol	0.25	1	$C_5H_{10}O$	Tetrahydropyran	0.82	1
C_3H_8O	2-Propanol	0.05	1	$C_5H_{10}O$	2-Methyltetrahydrofuran	1.85	2
C_3H_8S	1-Propanethiol	1.81	1	$C_5H_{10}O_2$	Pentanoic acid	1.39	1
C_3H_9N	Propylamine	0.48	1	$C_5H_{10}O_2$	Propyl acetate	1.24	1
C_3H_9N	Isopropylamine	0.26	1	$C_5H_{10}O_2$	Ethyl propanoate	1.21	1
C_3H_9N	Ethylmethylamine	0.15	1	$C_5H_{10}O_3$	Diethyl carbonate	1.21	1
C_3H_9N	Trimethylamine	0.16	1	$C_5H_{11}Br$	1-Bromopentane	3.37	1
C_4H_4O	Furan	1.34	1	$C_5H_{11}F$	1-Fluoropentane	2.33	1
C_4H_4S	Thiophene	1.81	1	$C_5H_{11}N$	Piperidine	0.84	1
C_4H_5N	Pyrrole	0.75	1	$C_5H_{11}NO_2$	1-Nitropentane	2.01	1
C_4H_6	1,3-Butadiene	1.99	1	C_5H_{12}	Pentane	3.45	1
C_4H_6	2-Butyne	1.46	1	C_5H_{12}	Neopentane	3.11	1
C_4H_6O	2,5-Dihydrofuran	0.46	1	$C_5H_{12}O$	1-Pentanol	1.51	1
$C_4H_6O_2$	Methacrylic acid	0.93	1	$C_5H_{12}O$	2-Pentanol	1.25	1
$C_4H_6O_2$	Vinyl acetate	0.73	1	$C_5H_{12}O$	3-Pentanol	1.21	1
$C_4H_6O_2$	Methyl acrylate	0.80	1	$C_5H_{12}O$	3-Methyl-1-butanol	1.28	1
C_4H_7N	Butanenitrile	0.60	1	$C_5H_{12}O$	2-Methyl-2-butanol	0.89	1
C_4H_8	cis-2-Butene	2.33	1	$C_5H_{12}O$	3-Methyl-2-butanol	1.28	1
C_4H_8	trans-2-Butene	2.31	1	$C_5H_{12}O$	2,2-Dimethyl-1-propanol	1.31	1
C_4H_8	Isobutene	2.35	1	$C_5H_{12}O$	Methyl tert-butyl ether	0.94	1
$C_4H_8Cl_2O$	Bis(2-chloroethyl) ether	1.12	2	$C_5H_{13}N$	Pentylamine	1.49	1
C_4H_8O	Ethyl vinyl ether	1.04	1	C_6Cl_6	Hexachlorobenzene	5.47	5
C_4H_8O	Butanal	0.88	1	C_6HCl_5	Pentachlorobenzene	5.03	5
C_4H_8O	2-Butanone	0.29	1	C_6HCl_5O	Pentachlorophenol	5.07	4
C_4H_8O	Tetrahydrofuran	0.46	1	$C_6H_2Cl_4$	1,2,3,4-Tetrachlorobenzene	4.55	5
$C_4H_8O_2$	Butanoic acid	0.79	1	$C_6H_2Cl_4$	1,2,3,5-Tetrachlorobenzene	4.65	5
$C_4H_8O_2$	Propyl formate	0.83	1	$C_6H_2Cl_4$	1,2,4,5-Tetrachlorobenzene	4.51	5
$C_4H_8O_2$	Ethyl acetate	0.73	1	$C_6H_3Cl_3$	1,2,3-Trichlorobenzene	4.04	5
C_4H_9Br	1-Bromobutane	2.75	1	$C_6H_3Cl_3$	1,2,4-Trichlorobenzene	3.98	5
C_4H_9Cl	1-Chlorobutane	2.64	2	$C_6H_3Cl_3$	1,3,5-Trichlorobenzene	4.02	5
C_4H_9F	1-Fluorobutane	2.58	1	$C_6H_4Cl_2$	o-Dichlorobenzene	3.38	5
C_4H_9I	1-Iodobutane	3	2	$C_6H_4Cl_2$	m-Dichlorobenzene	3.48	5
C_4H_9N	Pyrrolidine	0.46	1	$C_6H_4Cl_2$	p-Dichlorobenzene	3.38	5
C_4H_9NO	Butanamide	-0.21	1	$C_6H_4Cl_2O$	2,4-Dichlorophenol	3.23	4
C_4H_9NO	N,N-Dimethylacetamide	-0.77	1	C_6H_5Br	Bromobenzene	2.99	2
$C_4H_9NO_2$	1-Nitrobutane	1.47	1	C_6H_5Cl	Chlorobenzene	2.84	1
C_4H_{10}	Isobutane	2.8	2	C_6H_5F	Fluorobenzene	2.27	2
$C_4H_{10}O$	1-Butanol	0.84	1	C_6H_5I	Iodobenzene	3.28	2
$C_4H_{10}O$	2-Butanol	0.65	1	$C_6H_5NO_2$	Nitrobenzene	1.85	1
$C_4H_{10}O$	2-Methyl-1-propanol	0.76	1	C_6H_6	Benzene	2.13	1
$C_4H_{10}O$	2-Methyl-2-propanol	0.35	1	C_6H_6O	Phenol	1.48	4
$C_4H_{10}O$	Diethyl ether	0.89	1	C_6H_6S	Benzenethiol	2.52	1
$C_4H_{10}S$	1-Butanethiol	2.28	1	C_6H_7N	Aniline	0.90	1
$C_4H_{10}S$	Diethyl sulfide	1.95	1	C_6H_7N	2-Methylpyridine	1.11	1
$C_4H_{11}N$	Butylamine	0.86	1	C_6H_7N	3-Methylpyridine	1.20	1
$C_4H_{11}N$	tert-Butylamine	0.40	1	C_6H_7N	4-Methylpyridine	1.22	1
$C_4H_{11}N$	Diethylamine	0.58	1	C_6H_8	1,4-Cyclohexadiene	2.3	2
C_5H_5N	Pyridine	0.65	1	C_6H_8O	5-Hexyn-2-one	0.58	1
C_5H_6O	2-Methylfuran	1.85	1	C_6H_8O	2-Cyclohexen-1-one	0.61	1
C_5H_7N	1-Methylpyrrole	1.21	1	C_6H_8O	2-Ethylfuran	2.40	1
C_5H_8	1,4-Pentadiene	2.48	1	C_6H_{10}	1,5-Hexadiene	2.8	2
C_5H_8	1-Pentyne	1.98	1	C_6H_{10}	1-Hexyne	2.73	2
$C_5H_8O_2$	Methyl methacrylate	1.38	1	C_6H_{10}	Cyclohexene	2.86	1
$C_5H_8O_2$	Ethyl acrylate	1.32	1	$C_6H_{10}O$	5-Hexen-2-one	1.02	1
C_5H_9N	Pentanenitrile	0.94	1	$C_6H_{10}O$	Cyclohexanone	0.81	1
C_5H_{10}	1-Pentene	2.2	2	$C_6H_{10}O_2$	Ethyl methacrylate	1.94	1

Mol. Form.	Name	log P	Ref.	Mol. Form.	Name	log P	Ref.
$C_6H_{11}Br$	Bromocyclohexane	3.20	1	C_7H_{16}	Heptane	4.50	1
$C_6H_{11}N$	Hexanenitrile	1.66	1	$C_7H_{16}O$	1-Heptanol	2.62	1
C_6H_{12}	1-Hexene	3.40	1	$C_7H_{16}O$	2-Heptanol	2.31	1
C_6H_{12}	4-Methyl-1-pentene	2.5	2	$C_7H_{16}O$	3-Heptanol	2.24	1
C_6H_{12}	Cyclohexane	3.44	1	$C_7H_{16}O$	4-Heptanol	2.22	1
C_6H_{12}	Methylcyclopentane	3.37	2	$C_7H_{17}N$	Heptylamine	2.57	1
$C_6H_{12}O$	Cyclohexanol	1.23	1	C_8H_6	Phenylacetylene	2.40	1
$C_6H_{12}O$	Hexanal	1.78	1	C_8H_6O	Benzofuran	2.67	1
$C_6H_{12}O$	2-Hexanone	1.38	1	C_8H_6S	Benzo[b]thiophene	3.12	1
$C_6H_{12}O$	4-Methyl-2-pentanone	1.31	1	C_8H_7N	Benzeneacetonitrile	1.56	1
$C_6H_{12}O_2$	Hexanoic acid	1.92	1	C_8H_7N	Indole	2.14	1
$C_6H_{12}O_2$	Butyl acetate	1.82	1	C_8H_8	Styrene	3.05	1
$C_6H_{13}Br$	1-Bromohexane	3.80	1	C_8H_8O	Acetophenone	1.63	1
$C_6H_{13}N$	Cyclohexylamine	1.49	1	C_8H_8O	2-Methylbenzaldehyde	2.26	1
C_6H_{14}	Hexane	4.00	1	C_8H_8O	Benzeneacetaldehyde	1.78	1
C_6H_{14}	3-Methylpentane	3.60	2	C_8H_8O	2,3-Dihydrobenzofuran	2.14	1
C_6H_{14}	2,2-Dimethylbutane	3.82	1	C_8H_8O	Phenyloxirane	1.61	1
C_6H_{14}	2,3-Dimethylbutane	3.85	2	$C_8H_8O_2$	o-Toluic acid	2.32	4
$C_6H_{14}O$	1-Hexanol	2.03	1	$C_8H_8O_2$	m-Toluic acid	2.37	1
$C_6H_{14}O$	2-Hexanol	1.76	1	$C_8H_8O_2$	p-Toluic acid	2.34	1
$C_6H_{14}O$	3-Hexanol	1.65	1	$C_8H_8O_2$	Benzeneacetic acid	1.41	1
$C_6H_{14}O$	3,3-Dimethyl-2-butanol	1.48	1	$C_8H_8O_2$	Phenyl acetate	1.49	1
$C_6H_{14}O$	Dipropyl ether	2.03	1	$C_8H_8O_2$	Methyl benzoate	2.20	1
$C_6H_{14}O$	Diisopropyl ether	1.52	1	C_8H_{10}	Ethylbenzene	3.15	1
$C_6H_{15}N$	Hexylamine	2.06	1	C_8H_{10}	o-Xylene	3.12	1
$C_6H_{15}N$	Dipropylamine	1.67	1	C_8H_{10}	m-Xylene	3.20	1
$C_6H_{15}N$	Triethylamine	1.45	1	C_8H_{10}	p-Xylene	3.15	1
$C_7H_5BrO_2$	2-Bromobenzoic acid	2.20	4	$C_8H_{10}O$	o-Ethylphenol	2.47	1
$C_7H_5BrO_2$	3-Bromobenzoic acid	2.87	4	$C_8H_{10}O$	m-Ethylphenol	2.50	1
$C_7H_5BrO_2$	4-Bromobenzoic acid	2.86	4	$C_8H_{10}O$	p-Ethylphenol	2.50	1
C_7H_5N	Benzonitrile	1.56	1	$C_8H_{10}O$	2,4-Xylenol	2.35	1
C_7H_6O	Benzaldehyde	1.48	1	$C_8H_{10}O$	2,5-Xylenol	2.34	1
$C_7H_6O_2$	Benzoic acid	1.88	4	$C_8H_{10}O$	2,6-Xylenol	2.36	1
$C_7H_6O_2$	Phenyl formate	1.26	1	$C_8H_{10}O$	3,4-Xylenol	3.23	1
$C_7H_6O_3$	Salicylic acid	2.20	4	$C_8H_{10}O$	3,5-Xylenol	2.35	1
C_7H_7Br	(Bromomethyl)benzene	2.92	1	$C_8H_{10}O$	Benzeneethanol	1.36	1
C_7H_7Cl	o-Chlorotoluene	3.42	1	$C_8H_{10}O$	α-Methylbenzyl alcohol	1.42	1
C_7H_7Cl	m-Chlorotoluene	3.28	1	$C_8H_{10}O$	3-Methylbenzenemethanol	1.60	1
C_7H_7Cl	p-Chlorotoluene	3.33	1	$C_8H_{10}O$	4-Methylbenzenemethanol	1.58	1
C_7H_7Cl	(Chloromethyl)benzene	2.30	1	$C_8H_{10}O$	Phenetole	2.51	1
$C_7H_7NO_2$	p-Nitrotoluene	2.42	1	$C_8H_{10}O$	Benzyl methyl ether	1.35	1
C_7H_8	Toluene	2.73	1	$C_8H_{10}O$	2-Methylanisole	2.74	1
C_7H_8	1,3,5-Cycloheptatriene	2.63	2	$C_8H_{10}O$	3-Methylanisole	2.66	1
C_7H_8O	o-Cresol	1.98	1	$C_8H_{10}O$	4-Methylanisole	2.81	1
C_7H_8O	m-Cresol	1.98	1	$C_8H_{11}N$	p-Ethylaniline	1.96	1
C_7H_8O	p-Cresol	1.97	1	$C_8H_{11}N$	N,N-Dimethylaniline	2.31	1
C_7H_8O	Benzyl alcohol	1.05	1	$C_8H_{11}N$	Benzeneethanamine	1.41	1
C_7H_8O	Anisole	2.11	1	$C_8H_{14}O_2$	Butyl methacrylate	2.88	1
C_7H_9N	Benzylamine	1.09	1	$C_8H_{15}N$	Octanenitrile	2.75	1
C_7H_9N	o-Methylaniline	1.32	1	C_8H_{16}	1-Octene	4.57	1
C_7H_9N	m-Methylaniline	1.40	1	C_8H_{16}	Cyclooctane	4.45	2
C_7H_9N	p-Methylaniline	1.39	1	$C_8H_{16}O$	2-Octanone	2.37	1
C_7H_9N	N-Methylaniline	1.66	1	$C_8H_{16}O_2$	Octanoic acid	3.05	1
C_7H_{14}	1-Heptene	3.99	1	$C_8H_{17}Br$	1-Bromooctane	4.89	1
C_7H_{14}	Methylcyclohexane	3.88	1	C_8H_{18}	Octane	5.15	1
$C_7H_{14}O$	2-Heptanone	1.98	1	$C_8H_{18}O$	1-Octanol	3.07	1
$C_7H_{14}O$	5-Methyl-2-hexanone	1.88	1	$C_8H_{18}O$	2-Octanol	2.90	1
$C_7H_{15}Br$	1-Bromoheptane	4.36	1	$C_8H_{18}O$	4-Octanol	2.68	1
$C_7H_{15}Cl$	1-Chloroheptane	4.15	1	$C_8H_{18}O$	Dibutyl ether	3.21	1
$C_7H_{15}I$	1-Iodoheptane	4.70	1	C_9H_7N	Quinoline	2.03	1

Mol. Form.	Name	log P	Ref.	Mol. Form.	Name	log P	Ref.
C_9H_7N	Isoquinoline	2.08	1	$C_{12}H_2Cl_8$	2,2′,3,3′,5,5′,6,6′-Octachlorobiphenyl	7.10	3
C_9H_8	Indene	2.92	1	$C_{12}H_3Cl_7$	2,2′,3,3′,4,4′,6-Heptachlorobiphenyl	6.70	3
$C_9H_8O_2$	trans-Cinnamic acid	2.13	1	$C_{12}H_4Cl_6$	2,2′,3,3′,4,4′-Hexachlorobiphenyl	7.00	3
C_9H_9N	Benzenepropanenitrile	1.72	1	$C_{12}H_4Cl_6$	2,2′,4,4′,6,6′-Hexachlorobiphenyl	7.00	3
C_9H_{10}	Indan	3.33	1	$C_{12}H_4Cl_6$	2,2′,3,3′,6,6′-Hexachlorobiphenyl	6.70	3
$C_9H_{10}O$	1-Phenyl-1-propanone	2.19	1	$C_{12}H_5Cl_5$	2,3,4,5,6-Pentachlorobiphenyl	6.30	3
$C_9H_{10}O$	1-Phenyl-2-propanone	1.44	1	$C_{12}H_5Cl_5$	2,2′,4,5,5′-Pentachlorobiphenyl	6.40	3
$C_9H_{10}O$	4-Methylacetophenone	2.19	1	$C_{12}H_6Cl_4$	2,3,4,5-Tetrachlorobiphenyl	5.72	3
$C_9H_{10}O_2$	2-Phenylpropanoic acid	1.80	1	$C_{12}H_6Cl_4$	2,2′,4′,5-Tetrachlorobiphenyl	5.73	7
$C_9H_{10}O_2$	Benzyl acetate	1.96	1	$C_{12}H_7Cl_3$	2,4,5-Trichlorobiphenyl	5.60	3
$C_9H_{10}O_2$	4-Methylphenyl acetate	2.11	1	$C_{12}H_7Cl_3$	2,4,6-Trichlorobiphenyl	5.47	3
$C_9H_{10}O_2$	Ethyl benzoate	2.64	1	$C_{12}H_8Cl_2$	2,5-Dichlorobiphenyl	5.10	3
C_9H_{12}	Propylbenzene	3.69	1	$C_{12}H_8Cl_2$	2,6-Dichlorobiphenyl	5.00	3
C_9H_{12}	Isopropylbenzene	3.66	1	$C_{12}H_8O$	Dibenzofuran	4.12	1
C_9H_{12}	o-Ethyltoluene	3.53	1	$C_{12}H_9Cl$	2-Chlorobiphenyl	4.52	1
C_9H_{12}	p-Ethyltoluene	3.63	2	$C_{12}H_9Cl$	3-Chlorobiphenyl	4.58	1
C_9H_{12}	1,2,3-Trimethylbenzene	3.60	1	$C_{12}H_9Cl$	4-Chlorobiphenyl	4.61	1
C_9H_{12}	1,2,4-Trimethylbenzene	3.63	1	$C_{12}H_9N$	Carbazole	3.72	1
C_9H_{12}	1,3,5-Trimethylbenzene	3.42	1	$C_{12}H_{10}$	Acenaphthene	3.96	4
$C_9H_{12}O$	2-Propylphenol	2.93	1	$C_{12}H_{10}$	Biphenyl	3.76	6
$C_9H_{12}O$	4-Propylphenol	3.20	1	$C_{12}H_{10}N_2$	Azobenzene	3.82	1
$C_9H_{12}O$	2,3,6-Trimethylphenol	2.67	1	$C_{12}H_{10}O$	Diphenyl ether	4.21	1
$C_9H_{12}O$	2,4,6-Trimethylphenol	2.46	1	$C_{12}H_{10}S$	Diphenyl sulfide	4.45	1
$C_9H_{12}O$	Benzenepropanol	1.88	1	$C_{12}H_{11}N$	Diphenylamine	3.44	4
$C_9H_{13}N$	N,N-Dimethylbenzylamine	1.98	1	$C_{12}H_{12}$	1-Ethylnaphthalene	4.40	1
$C_9H_{13}N$	Amphetamine	1.76	1	$C_{12}H_{12}$	1,2-Dimethylnaphthalene	4.31	1
C_9H_{18}	1-Nonene	5.15	1	$C_{12}H_{12}$	1,4-Dimethylnaphthalene	4.37	1
$C_9H_{18}O$	2-Nonanone	3.16	1	$C_{12}H_{14}O$	4-Phenylcyclohexanone	2.45	1
$C_9H_{18}O$	5-Methyl-2-octanone	2.92	1	$C_{12}H_{18}$	Hexylbenzene	5.52	1
C_9H_{20}	Nonane	5.65	1	$C_{12}H_{18}$	Hexamethylbenzene	4.69	4
$C_9H_{20}O$	1-Nonanol	4.02	1	$C_{12}H_{22}O$	Cyclododecanone	4.10	1
$C_9H_{21}N$	Tripropylamine	2.79	1	$C_{12}H_{24}O_2$	Dodecanoic acid	4.6	1
$C_{10}H_7Cl$	1-Chloronaphthalene	3.90	1	$C_{12}H_{26}O$	1-Dodecanol	5.13	1
$C_{10}H_7Cl$	2-Chloronaphthalene	3.98	1	$C_{13}H_8O$	9H-Fluoren-9-one	3.58	1
$C_{10}H_8$	Naphthalene	3.34	4	$C_{13}H_9N$	Acridine	3.40	1
$C_{10}H_8$	Azulene	3.22	1	$C_{13}H_{10}$	9H-Fluorene	4.20	4
$C_{10}H_8O$	1-Naphthol	2.84	1	$C_{13}H_{10}O$	Benzophenone	3.18	1
$C_{10}H_8O$	2-Naphthol	2.70	1	$C_{13}H_{10}O_2$	Phenyl benzoate	3.59	1
$C_{10}H_{12}O_2$	Isopropyl benzoate	3.18	1	$C_{13}H_{11}NO$	N-Phenylbenzamide	2.62	1
$C_{10}H_{14}$	Butylbenzene	4.26	1	$C_{13}H_{12}$	Diphenylmethane	4.14	1
$C_{10}H_{14}$	tert-Butylbenzene	4.11	1	$C_{13}H_{12}$	4-Methylbiphenyl	4.63	1
$C_{10}H_{14}$	Isobutylbenzene	4.01	2	$C_{13}H_{12}O$	Diphenylmethanol	2.67	1
$C_{10}H_{14}$	p-Cymene	4.10	1	$C_{13}H_{12}O$	Benzyl phenyl ether	3.79	1
$C_{10}H_{14}$	1,2,4,5-Tetramethylbenzene	4.10	2	$C_{14}H_{10}$	Anthracene	4.56	4
$C_{10}H_{14}$	1,2,3,4-Tetramethylbenzene	4.00	1	$C_{14}H_{10}$	Phenanthrene	4.52	4
$C_{10}H_{14}$	1,2,3,5-Tetramethylbenzene	4.10	1	$C_{14}H_{12}$	trans-Stilbene	4.81	1
$C_{10}H_{14}O$	4-Butylphenol	3.65	1	$C_{14}H_{12}$	1-Methylfluorene	4.97	1
$C_{10}H_{20}O$	2-Decanone	3.77	1	$C_{14}H_{12}O$	2-Phenylacetophenone	3.18	1
$C_{10}H_{20}O_2$	Decanoic acid	4.09	1	$C_{14}H_{12}O_2$	Benzyl benzoate	3.97	1
$C_{10}H_{22}$	Decane	6.25	1	$C_{14}H_{14}$	1,2-Diphenylethane	4.70	1
$C_{10}H_{22}O$	1-Decanol	4.57	1	$C_{14}H_{14}$	4,4′-Dimethylbiphenyl	5.09	1
$C_{11}H_9N$	4-Phenylpyridine	2.59	1	$C_{14}H_{22}$	Octylbenzene	6.30	1
$C_{11}H_{10}$	1-Methylnaphthalene	3.87	1	$C_{14}H_{28}O_2$	Tetradecanoic acid	6.1	1
$C_{11}H_{10}$	2-Methylnaphthalene	4.00	1	$C_{15}H_{12}$	2-Methylanthracene	5.15	2
$C_{11}H_{16}$	Pentylbenzene	4.90	1	$C_{15}H_{12}$	9-Methylanthracene	5.07	1
$C_{11}H_{16}$	Pentamethylenzene	4.56	1	$C_{15}H_{12}$	1-Methylphenanthrene	5.14	2
$C_{11}H_{22}O$	2-Undecanone	4.09	1	$C_{16}H_{10}$	Fluoranthene	5.07	4
$C_{11}H_{22}O_2$	Methyl decanoate	4.41	1	$C_{16}H_{10}$	Pyrene	5.08	4
$C_{12}Cl_{10}$	Decachlorobiphenyl	8.26	3	$C_{16}H_{14}$	9,10-Dimethylanthracene	5.69	1
$C_{12}HCl_9$	2,2′,3,3′,4,5,5′,6,6′-Nonachlorobiphenyl	8.16	3	$C_{16}H_{32}O_2$	Hexadecanoic acid	7.17	1

Mol. Form.	Name	log P	Ref.	Mol. Form.	Name	log P	Ref.
$C_{17}H_{12}$	11H-Benzo[a]fluorene	5.40	1	$C_{18}H_{36}O_2$	Stearic acid	8.23	1
$C_{17}H_{12}$	11H-Benzo[b]fluorene	5.75	1	$C_{19}H_{16}O$	Triphenylmethanol	3.68	1
$C_{18}H_{12}$	Benz[a]anthracene	5.91	1	$C_{20}H_{12}$	Perylene	6.25	1
$C_{18}H_{12}$	Chrysene	5.73	4	$C_{20}H_{12}$	Benzo[a]pyrene	6.20	4
$C_{18}H_{12}$	Naphthacene	5.76	1	$C_{20}H_{32}O_2$	Arachidonic acid	6.98	1
$C_{18}H_{12}$	Triphenylene	5.49	4	$C_{20}H_{40}O_2$	Arachidic acid	9.29	1
$C_{18}H_{15}N$	Triphenylamine	5.74	1	$C_{21}H_{16}$	1,2-Dihydro-3-methylbenz[j]		
$C_{18}H_{30}O_2$	Linolenic acid	6.46	1		aceanthrylene	6.75	1
$C_{18}H_{32}O_2$	Linoleic acid	7.05	1	$C_{22}H_{12}$	Benzo[ghi]perylene	6.90	1
$C_{18}H_{34}O_2$	Oleic acid	7.64	1	$C_{24}H_{12}$	Coronene	6.05	4

PROTECTION AGAINST IONIZING RADIATION

The following data and rules of thumb are helpful in estimating the penetrating capability of and danger of exposure to various types of ionizing radiation. More precise data should be used for critical applications.

Alpha Particles

Alpha particles of at least 7.5 MeV are required to penetrate the epidermis, the protective layer of skin, 0.07 mm thick.

Electrons

Electrons of at least 70 keV are required to penetrate the epidermis, the protective layer of skin, 0.07 mm thick.

The range of electrons in g/cm^2 is approximately equal to the maximum energy (E) in MeV divided by 2.

The range of electrons in air is about 3.65 m per MeV; for example, a 3 MeV electron has a range of about 11 m in air.

A chamber wall thickness of 30 mg/cm^2 will transmit 70% of the initial fluence of 1 MeV electrons and 20% of that of 0.4 MeV electrons.

When electrons of 1 to 2 MeV pass through light materials such as water, aluminum, or glass, less than 1% of their energy is dissipated as bremsstrahlung.

The bremsstrahlung from 1 Ci of ^{32}P aqueous solution in a glass bottle is about 1 mR/h at 1 meter distance.

When electrons from a 1 Ci source of ^{90}Sr - ^{90}Y are absorbed, the bremsstrahlung hazard is approximately equal to that presented by the gamma radiation from 12 mg of radium. The average energy of the bremsstrahlung is about 300 keV.

Gamma Rays

The air-scattered radiation (sky-shine) from a 100 Ci ^{60}Co source placed 1 ft behind a 4 ft high shield is about 100 mrad/h at 6 ft from the outside of the shield.

Within ±20% for point source gamma emitters with energies between 0.07 and 4 MeV, the exposure rate (R/h) at 1 ft is $6C \cdot E \cdot n$ where C is the activity in curies, E is the energy in MeV, and n is the number of gammas per disintegration.

Neutrons

An approximate HVL (thickness of absorber for which the neutron flux falls to half its initial value) for 1 MeV neutrons is 3.2 cm of paraffin; that for 5 MeV neutrons is 6.9 cm of paraffin).

Miscellaneous

The activity of any radionuclide is reduced to less than 1% after 7 half-lives (i.e., $2^{-7} = 0.8\%$).

For nuclides with a half-life greater than 6 days, the change in activity in 24 hours will be less than 10%.

10 HVL (half-value layers) attenuates approximately by 10^{-3}.

There is 0.64 mm^3 of radon gas at STP in transient equilibrium with 1 Ci of radium.

The natural background from all sources in most parts of the world leads to an equivalent dose rate of about 0.04 to 4 mSv per year for the average person. About 84% of this comes from terrestrial sources, the remainder from cosmic rays. The U. S. average is about 3.6 mSv/yr but can range up to 50 mSv/yr in some areas. A passenger in a plane flying at 12,000 meters receives 5 µSv/hr from cosmic rays (as compared to about 0.03 µSv/hr at sea level).

The ICRP recommended exposure limit to man-made sources of ionizing radiation (Reference 2) is 20 mSv/yr averaged over 5 years, with the dose in any one year not to exceed 50 mSv.

A whole-body dose of about 3 Gy over a short time interval will typically lead to 50% mortality in 30 days assuming no medical treatment.

Units

The gray (Gy) is the SI unit of absorbed dose; it is a measure of the mean energy imparted to a sample of irradiated matter, divided by the mass of the sample. Gy is a special name for the SI unit J/kg.

The sievert (Sv) is the SI unit of equivalent dose, which is defined as the absorbed dose multiplied by a weighting factor that expresses the long-term biological risk from low-level chronic exposure to a specified type of radiation. The Sv is another special name for J/kg.

1 curie (Ci) = $3.7 \cdot 10^{10}$ becquerel (Bq); i.e., $3.7 \cdot 10^{10}$ disintegrations per second.

1 roentgen (R) = $2.58 \cdot 10^{-4}$ coulomb per kilogram (C/kg); a measure of the charge (positive or negative) liberated by x-ray or gamma radiation in air, divided by the mass of air.

1 rad = 0.01 Gy

1 rem = 0.01 Sv

REFERENCES

1. Padikal, T.N., and Fivozinsky, S.P., *Medical Physics Data Book, National Bureau of Standards Handbook 138*, U. S. Government Printing Office, Washington, D.C., 1981.
2. *1990 Recommendations of the International Commission on Radiological Protection,* ICRP Publication 60, *Annals of the ICRP,* Pergamon Press, Oxford, 1991.
3. *Radiation: Doses, Effects, Risks*, United Nations Sales No. E.86.III.D.4, 1985.
4. *Review of Particle Properties, Phys. Rev. D*, 50, 1173, 1994 (p. 1268).

ANNUAL LIMITS ON INTAKES OF RADIONUCLIDES

K. F. Eckerman

The following table lists, for workers, the annual limits on oral and inhalation intakes (ALI) for selected radionuclides based on the occupational radiation protection guidance of the International Commission on Radiological Protection (References 1 and 2). An intake of one ALI corresponds to an annual whole body dose of 0.02 Sv (2 rem).

The ALI is expressed in the SI unit of activity, the becquerel (Bq), and in the conventional unit, the microcurie (μCi); 1 μCi = $3.7 \cdot 10^4$ Bq. The chemical form of inhaled radionuclides is, in most instances, stated in terms of the rate of absorption to blood from the lungs and the fractional absorption from the small intestine. Type F, M, and S denote chemical forms which are absorbed from the lungs at rates characterized as fast, moderate, and slow, respectively. The time to absorb 90% of the deposited radionuclide, in the absence of radioactive decay, corresponds to about 10 minutes, 150 days, and 7000 days for Type F, M, and S compounds, respectively. Type F compounds can be considered to be more soluble than M or S, S being the most insoluble. Chemical form consideration for ingestion is specified by the fractional absorption from the small intestine, denoted as f_1. The f_1 values range from 10^{-5} to 1. Higher fractional absorption is associated with greater solubility of the compound.

REFERENCES

1. *1990 Recommendations of the International Commission on Radiological Protection, ICRP Publication 60, Annals of the ICRP 21, (1—3)*, Pergamon Press, Oxford, 1991.
2. *Dose Coefficients for Intakes of Radionuclides by Workers, ICRP Publication 68, Annals of the ICRP*, 24(4), Pergamon Press, Oxford, 1995.

	Physical half-life	Inhalation intakes			Oral intakes		
		Chemical form Type/f_1	ALI Bq	ALI μCi	Chemical form f_1	ALI Bq	ALI μCi
^{3}H	12.3 y	HT gas	1.1E+13	3.0E+08	1.000	1.1E+13	3.0E+08
		HTO vapor	1.1E+09	3.0E+04			
^{11}C	0.340 h	CO	1.7E+10	4.5E+05	1.000	8.3E+08	2.3E+04
		CO_2	9.1E+09	2.5E+05			
		Organic compounds	6.2E+09	1.7E+05			
^{14}C	5730 y	CO	2.5E+10	6.8E+05	1.000	3.4E+07	9.3E+02
		CO_2	3.1E+09	8.3E+04			
		Organic compounds	3.4E+07	9.3E+02			
^{18}F	1.83 h	F 1.000	3.7E+08	1.0E+04	1.000	4.1E+08	1.1E+04
		M 1.000	2.2E+08	6.1E+03			
		S 1.000	2.2E+08	5.8E+03			
^{22}Na	2.60 y	F 1.000	1.0E+07	2.7E+02	1.000	6.3E+06	1.7E+02
^{24}Na	15.0 h	F 1.000	3.8E+07	1.0E+03	1.000	4.7E+07	1.3E+03
^{32}P	14.3 d	F 0.800	1.8E+07	4.9E+02	0.800	8.3E+06	2.3E+02
		M 0.800	6.9E+06	1.9E+02			
^{35}S	87.4 d	Inorganic compounds					
		F 0.800	2.5E+08	6.8E+03	0.800	1.4E+08	3.9E+03
		M 0.800	1.8E+07	4.9E+02	0.100	1.1E+08	2.8E+03
		Vapor	1.7E+08	4.5E+03			
		Organic compounds			1.000	2.6E+07	7.0E+02
^{42}K	12.4 h	F 1.000	1.0E+08	2.7E+03	1.000	4.7E+07	1.3E+03
^{43}K	22.6 h	F 1.000	7.7E+07	2.1E+03	1.000	8.0E+07	2.2E+03
^{45}Ca	163 d	M 0.300	8.7E+06	2.4E+02	0.300	2.6E+07	7.1E+02
^{47}Ca	4.53 d	M 0.300	9.5E+06	2.6E+02	0.300	1.3E+07	3.4E+02
^{51}Cr	27.7 d	F 0.100	6.7E+08	1.8E+04	0.100	5.3E+08	1.4E+04
		M 0.100	5.9E+08	1.6E+04	0.010	5.4E+08	1.5E+04
		S 0.100	5.6E+08	1.5E+04			
^{54}Mn	312 d	F 0.100	1.8E+07	4.9E+02	0.100	2.8E+07	7.6E+02
		M 0.100	1.7E+07	4.5E+02			
^{52}Fe	8.28 h	F 0.100	2.9E+07	7.8E+02	0.100	1.4E+07	3.9E+02
		M 0.100	2.1E+07	5.7E+02			
^{55}Fe	2.70 y	F 0.100	2.2E+07	5.9E+02	0.100	6.1E+07	1.6E+03
		M 0.100	6.1E+07	1.6E+03			

	Physical half-life	Inhalation intakes			Oral intakes		
		Chemical form Type/f_1	ALI Bq	µCi	Chemical form f_1	ALI Bq	µCi
^{59}Fe	44.5 d	F 0.100	6.7E+06	1.8E+02	0.100	1.1E+07	3.0E+02
		M 0.100	6.3E+06	1.7E+02			
^{57}Co	271 d	M 0.100	5.1E+07	1.4E+03	0.100	9.5E+07	2.6E+03
		S 0.050	3.3E+07	9.0E+02	0.050	1.1E+08	2.8E+03
^{58}Co	70.8 d	M 0.100	1.4E+07	3.9E+02	0.100	2.7E+07	7.3E+02
		S 0.050	1.2E+07	3.2E+02	0.050	2.9E+07	7.7E+02
^{60}Co	5.27 y	M 0.100	2.8E+06	7.6E+01	0.100	5.9E+06	1.6E+02
		S 0.050	1.2E+06	3.2E+01	0.050	8.0E+06	2.2E+02
^{64}Cu	12.7 h	F 0.500	2.9E+08	7.9E+03	0.500	1.7E+08	4.5E+03
		M 0.500	1.3E+08	3.6E+03			
		S 0.500	1.3E+08	3.6E+03			
^{59}Ni	75000 y	F 0.050	9.1E+07	2.5E+03	0.050	3.2E+08	8.6E+03
		M 0.050	2.1E+08	5.8E+03			
		Vapor	2.4E+07	6.5E+02			
^{63}Ni	96.0 y	F 0.050	3.8E+07	1.0E+03	0.050	1.3E+08	3.6E+03
		M 0.050	6.5E+07	1.7E+03			
		Vapor	1.0E+07	2.7E+02			
^{65}Zn	244 d	S 0.500	7.1E+06	1.9E+02	0.500	5.1E+06	1.4E+02
^{67}Ga	3.26 d	F 0.001	1.8E+08	4.9E+03	0.001	1.1E+08	2.8E+03
		M 0.001	7.1E+07	1.9E+03			
^{68}Ga	1.13 h	F 0.001	4.1E+08	1.1E+04	0.001	2.0E+08	5.4E+03
		M 0.001	2.5E+08	6.7E+03			
^{68}Ge	288 d	F 1.000	2.4E+07	6.5E+02	1.000	1.5E+07	4.2E+02
		M 1.000	2.5E+06	6.8E+01			
^{75}Se	120 d	F 0.800	1.4E+07	3.9E+02	0.800	7.7E+06	2.1E+02
		M 0.800	1.2E+07	3.2E+02	0.050	4.9E+07	1.3E+03
^{79}Se	65000 y	F 0.800	1.3E+07	3.4E+02	0.800	6.9E+06	1.9E+02
		M 0.800	6.5E+06	1.7E+02	0.050	5.1E+07	1.4E+03
^{86}Rb	18.6 d	F 1.000	1.5E+07	4.2E+02	1.000	7.1E+06	1.9E+02
^{85}Sr	64.8 d	F 0.300	3.6E+07	9.7E+02	0.300	3.6E+07	9.7E+02
		S 0.010	3.1E+07	8.4E+02	0.010	6.1E+07	1.6E+03
^{87m}Sr	2.80 h	F 0.300	9.1E+08	2.5E+04	0.300	6.7E+08	1.8E+04
		S 0.010	5.7E+08	1.5E+04	0.010	6.1E+08	1.6E+04
^{89}Sr	50.5 d	F 0.300	1.4E+07	3.9E+02	0.300	7.7E+06	2.1E+02
		S 0.010	3.6E+06	9.7E+01	0.010	8.7E+06	2.4E+02
^{90}Sr	29.1 y	F 0.300	6.7E+05	1.8E+01	0.300	7.1E+05	1.9E+01
		S 0.010	2.6E+05	7.0E+00	0.010	7.4E+06	2.0E+02
^{99}Mo	2.75 d	F 0.800	5.6E+07	1.5E+03	0.800	2.7E+07	7.3E+02
		S 0.050	1.8E+07	4.9E+02	0.050	1.7E+07	4.5E+02
^{99m}Tc	6.02 h	F 0.800	1.0E+09	2.7E+04	0.800	9.1E+08	2.5E+04
		M 0.800	6.9E+08	1.9E+04			
^{99}Tc	213000 y	F 0.800	5.0E+07	1.4E+03	0.800	2.6E+07	6.9E+02
		M 0.800	6.3E+06	1.7E+02			
^{106}Ru	1.01 y	F 0.050	2.0E+06	5.5E+01	0.050	2.9E+06	7.7E+01
		M 0.050	1.2E+06	3.2E+01			
		S 0.050	5.7E+05	1.5E+01			
^{111}In	2.83 d	F 0.020	9.1E+07	2.5E+03	0.020	6.9E+07	1.9E+03
		M 0.020	6.5E+07	1.7E+03			
^{113m}In	1.66 h	F 0.020	1.1E+09	2.8E+04	0.020	7.1E+08	1.9E+04
		M 0.020	6.3E+08	1.7E+04			
^{113}Sn	115 d	F 0.020	2.5E+07	6.8E+02	0.020	2.7E+07	7.4E+02
		M 0.020	1.1E+07	2.8E+02			
^{123}I	13.2 h	F 1.000	1.8E+08	4.9E+03	1.000	9.5E+07	2.6E+03
		Vapor	9.5E+07	2.6E+03			
^{125}I	60.1 d	F 1.000	2.7E+06	7.4E+01	1.000	1.3E+06	3.6E+01
		Vapor	1.4E+06	3.9E+01			
^{129}I	$1.57 \cdot 10^7$ y	F 1.000	3.9E+05	1.1E+01	1.000	1.8E+05	4.9E+00

ANNUAL LIMITS ON INTAKES OF RADIONUCLIDES (continued)

		Inhalation intakes			Oral intakes		
	Physical half-life	Chemical form Type/f_1	ALI Bq	μCi	Chemical form f_1	ALI Bq	μCi
[131]I	8.04 d	Vapor	2.1E+05	5.6E+00			
		F 1.000	1.8E+06	4.9E+01	1.000	9.1E+05	2.5E+01
[129]Cs	1.34 d	Vapor	1.0E+06	2.7E+01			
		F 1.000	2.5E+08	6.7E+03	1.000	3.3E+08	9.0E+03
[134]Cs	2.06 y	F 1.000	2.1E+06	5.6E+01	1.000	1.1E+06	2.8E+01
[136]Cs	13.1 d	F 1.000	1.1E+07	2.8E+02	1.000	6.7E+06	1.8E+02
[137]Cs	30.0 y	F 1.000	3.0E+06	8.1E+01	1.000	1.5E+06	4.2E+01
[141]Ce	32.5 d	M 5.0E-04	7.4E+06	2.0E+02	5.0E-04	2.8E+07	7.6E+02
		S 5.0E-04	6.5E+06	1.7E+02			
[144]Ce	284 d	M 5.0E-04	8.7E+05	2.4E+01	5.0E-04	3.8E+06	1.0E+02
		S 5.0E-04	6.9E+05	1.9E+01			
[133]Ba	10.7 y	F 0.100	1.1E+07	3.0E+02	0.100	2.0E+07	5.4E+02
[140]Ba	12.7 d	F 0.100	1.3E+07	3.4E+02	0.100	8.0E+06	2.2E+02
[169]Yb	32.0 d	M 5.0E-04	9.5E+06	2.6E+02	5.0E-04	2.8E+07	7.6E+02
		S 5.0E-04	8.3E+06	2.3E+02			
[198]Au	2.69 d	F 0.100	5.1E+07	1.4E+03	0.100	2.0E+07	5.4E+02
		M 0.100	2.0E+07	5.5E+02			
		S 0.100	1.8E+07	4.9E+02			
[198m]Au	2.30 d	F 0.100	3.4E+07	9.2E+02	0.100	1.5E+07	4.2E+02
		M 0.100	1.0E+07	2.7E+02			
		S 0.100	1.1E+07	2.8E+02			
[197]Hg	2.67 d	Inorganic compounds					
		F 0.400	2.4E+08	6.4E+03	1.000	2.0E+08	5.5E+03
					0.400	1.2E+08	3.2E+03
		Vapor Organic compounds	4.5E+06	1.2E+02			
		F 0.020	2.0E+08	5.4E+03	0.020	8.7E+07	2.4E+03
		M 0.020	7.1E+07	1.9E+03			
[203]Hg	46.6 d	Inorganic compounds					
		F 0.400	2.7E+07	7.2E+02	1.000	1.1E+07	2.8E+02
					0.400	1.8E+07	4.9E+02
		Vapor Organic compounds	2.9E+06	7.7E+01			
		F 0.020	3.4E+07	9.2E+02	0.020	3.7E+07	1.0E+03
		M 0.020	1.1E+07	2.8E+02			
[201]Tl	3.04 d	F 1.000	2.6E+08	7.1E+03	1.000	2.1E+08	5.7E+03
[210]Pb	22.3 y	F 0.200	1.8E+04	4.9E-01	0.200	2.9E+04	7.9E-01
[207]Bi	38.0 y	F 0.050	2.4E+07	6.4E+02	0.050	1.5E+07	4.2E+02
		M 0.050	6.3E+06	1.7E+02			
[210]Po	138 d	F 0.100	2.8E+04	7.6E-01	0.100	8.3E+04	2.3E+00
		M 0.100	9.1E+03	2.5E-01			
[224]Ra	3.66 d	M 0.200	8.3E+03	2.3E-01	0.200	3.1E+05	8.3E+00
[226]Ra	1600 y	M 0.200	1.7E+03	4.5E-02	0.200	7.1E+04	1.9E+00
[228]Ra	5.75 y	M 0.200	1.2E+04	3.2E-01	0.200	3.0E+04	8.1E-01
[228]Th	1.91 y	M 5.0E-04	8.7E+02	2.4E-02	5.0E-04	2.9E+05	7.7E+00
		S 2.0E-04	6.3E+02	1.7E-02	2.0E-04	5.7E+05	1.5E+01
[230]Th	77000 y	M 5.0E-04	7.1E+02	1.9E-02	5.0E-04	9.5E+04	2.6E+00
		S 2.0E-04	2.8E+03	7.5E-02	2.0E-04	2.3E+05	6.2E+00
[232]Th	1.40·10^{10} y	M 5.0E-04	6.9E+02	1.9E-02	5.0E-04	9.1E+04	2.5E+00
		S 2.0E-04	1.7E+03	4.5E-02	2.0E-04	2.2E+05	5.9E+00
[234]U	2.44·10^5 y	F 0.020	3.1E+04	8.4E-01	0.020	4.1E+05	1.1E+01
		M 0.020	9.5E+03	2.6E-01	0.002	2.4E+06	6.5E+01
		S 0.002	2.9E+03	7.9E-02			

ANNUAL LIMITS ON INTAKES OF RADIONUCLIDES (continued)

	Physical half-life	Inhalation intakes			Oral intakes		
		Chemical form Type/f_1	ALI Bq	µCi	Chemical form f_1	ALI Bq	µCi
^{235}U	$7.04 \cdot 10^8$ y	F 0.020	3.3E+04	9.0E-01	0.020	4.3E+05	1.2E+01
		M 0.020	1.1E+04	3.0E-01	0.002	2.4E+06	6.5E+01
		S 0.002	3.3E+03	8.9E-02			
^{238}U	$4.47 \cdot 10^9$ y	F 0.020	3.4E+04	9.3E-01	0.020	4.5E+05	1.2E+01
		M 0.020	1.3E+04	3.4E-01	0.002	2.6E+06	7.1E+01
		S 0.002	3.5E+03	9.5E-02			
^{237}Np	$2.14 \cdot 10^6$ y	M 5.0E-04	1.3E+03	3.6E-02	5.0E-04	1.8E+05	4.9E+00
^{239}Np	2.36 d	M 5.0E-04	1.8E+07	4.9E+02	5.0E-04	2.5E+07	6.8E+02
^{238}Pu	87.7 y	M 5.0E-04	6.7E+02	1.8E-02	5.0E-04	8.7E+04	2.4E+00
		S 1.0E-05	1.8E+03	4.9E-02	1.0E-05	2.3E+06	6.1E+01
					1.0E-04	4.1E+05	1.1E+01
^{239}Pu	24100 y	M 5.0E-04	6.3E+02	1.7E-02	5.0E-04	8.0E+04	2.2E+00
		S 1.0E-05	2.4E+03	6.5E-02	1.0E-05	2.2E+06	6.0E+01
					1.0E-04	3.8E+05	1.0E+01
^{241}Pu	14.4 y	M 5.0E-04	3.4E+04	9.3E-01	5.0E-04	4.3E+06	1.2E+02
		S 1.0E-05	2.4E+05	6.4E+00	1.0E-05	1.8E+08	4.9E+03
					1.0E-04	2.1E+07	5.6E+02
^{241}Am	432 y	M 5.0E-04	7.4E+02	2.0E-02	5.0E-04	1.0E+05	2.7E+00
^{244}Cm	18.1 y	M 5.0E-04	1.2E+03	3.2E-02	5.0E-04	1.7E+05	4.5E+00
^{252}Cf	2.64 y	M 5.0E-04	1.5E+03	4.2E-02	5.0E-04	2.2E+05	6.0E+00

CHEMICAL CARCINOGENS

The following substances are listed in the *Seventh Annual Report on Carcinogens: Summary 1994*, released by the National Institute of Environmental Health Sciences (NIEHS) under the National Toxicology Program (NTP). Substances are grouped into three classes:

1. Known to be carcinogens from studies in humans
2. Reasonably anticipated to be carcinogens, based on limited human studies or studies on experimental animals
3. Associated with technological processes that are known to be carcinogenic

The table includes the name used by the NTP, the Chemical Abstracts Service Registry Number (CAS RN) as given in the NTP report, and, where possible, the systematic name of the substance (generally the CAS index name).

REFERENCE

Seventh Annual Report on Carcinogens: Summary 1994, Public Health Service, National Toxicology Program, Central Data Management, P.O. Box 12233, MD A0-01, Research Triangle Park, NC 27709.

NTP name	CAS RN	Systematic name
SUBSTANCES KNOWN TO BE CARCINOGENIC		
Aflatoxins	1402-68-2	
Aminobiphenyl	92-67-1	[1,1´-Biphenyl]-4-amine
Analgesic mixtures containing phenacetin		Acetamide, N-(4-ethoxyphenyl)- (in analgesic mixtures)
Arsenic	7440-38-2	Arsenic
Certain arsenic compounds		
Asbestos	1332-21-4	Asbestos
Azathioprine	446-86-6	Purine, 6-((1-methyl-4-nitroimidazol-5-yl)thio)
Benzene	71-43-2	Benzene
Benzidine	92-87-5	[1,1´-Biphenyl]-4,4´-diamine
Bis(chloromethyl) ether	542-88-1	Methane, oxybis[chloro-
Chloromethyl methyl ether (techical grade)	107-30-2	Methane, chloromethoxy-
1,4-Butanediol dimethyl-sulfonate (Myleran)	55-98-1	1,4-Butanediol, dimethanesulfonate
Chlorambucil	305-03-3	Benzenebutanoic acid, 4-[bis(2-chloroethyl)amino]-
(2-Chloroethyl)-3-(4-methylcyclohexyl)-1-nitrosourea (MeCCNU)	13909-09-6	Urea, 1-(2-chloroethyl)-3-(4-methylcyclohexyl)-1-nitroso-, (E)
Chromium	7440-47-3	Chromium
Certain chromium compounds		
Conjugated estrogens		
Cyclophosphamide	50-18-0	2H-1,3,2-Oxazaphosphorine, 2(bis(2-chloroethyl)amino)-tetrahydro, 2-oxide
Diethylstilbestrol	56-53-1	Phenol, 4,4´-(1,2-diethyl-1,2-ethenediyl)bis-, (E)-
Erionite	66733-21-9	Erionite (Calcium sodium aluminum silicate)
Melphalan	148-82-3	Alanine, 3-(p-(bis(2-chloroethyl)amino)phenyl)-, L
Methoxsalen with ultraviolet A therapy (PUVA)		
Mustard gas	505-60-2	Ethane, 1,1´-thiobis[2-chloro-
2-Naphthylamine	91-59-8	2-Naphthalenamine
Radon	10043-92-2	Radon
Thorium dioxide	1314-20-1	Thorium(IV) oxide
Vinyl chloride	75-01-4	Ethene, chloro-
SUBSTANCES ANTICIPATED TO BE CARCINOGENIC		
Acetaldehyde	75-07-0	Acetaldehyde
2-Acetylaminofluorene	53-96-3	Acetamide, N-9H-fluoren-2-yl-
Acrylamide	79-06-1	2-Propenamide
Acrylonitrile	107-13-1	2-Propenenitrile
2-Aminoanthraquinone	117-79-3	9,10-Anthracenedione, 2-amino-
o-Aminoazotoluene	97-56-3	Benzenamine, 2-methyl-4-[(2-methylphenyl)azo]-
1-Amino-2-methylanthraquinone	82-28-0	9,10-Anthracenedione, 1-amino-2-methyl-
Amitrole	61-82-5	1H-1,2,4-Triazol-3-amine
o-Anisidine hydrochloride	134-29-2	Benzenamine, 2-methoxy-, hydrochloride
Benzotrichloride	98-07-7	Benzene, (trichloromethyl)-

NTP name	CAS RN	Systematic name
Benzotrichloride	98-07-7	Benzene, (trichloromethyl)-
Beryllium	7440-41-7	Beryllium
Certain beryllium compounds		
Bischloroethyl nitrosourea	154-93-8	Urea, *N,N*´-bis(2-chloroethyl)-*N*-nitroso-
Bromodichloromethane	75-27-4	Methane, bromodichloro-
1,3-Butadiene	106-99-0	1,3-Butadiene
Butylated hydroxyanisole	25013-16-5	Phenol, (1,1-dimethylethyl)-4-methoxy-
Cadmium	7440-43-9	Cadmium
Certain cadmium compounds		
Carbon tetrachloride	56-23-5	Methane, tetrachloro-
Ceramic fibers (respirable size)		
Chlorendic acid	115-28-6	5-Norbornene-2,3-dicarboxylic acid, 1,4,5,6,7,7-hexachloro-
Chlorinated paraffins (C_{12}, 60% chlorine)	108171-26-2	
1-(2-Chloroethyl)-3-cyclohexyl-1-nitrosourea		
(CCNU)	13010-47-4	Urea, 1-(2-chloroethyl)-3-cyclohexyl-1-nitroso-
Chloroform	67-66-3	Methane, trichloro-
3-Chloro-2-methylpropene	563-47-3	1-Propene, 3-chloro-2-methyl-
4-Chloro-*o*-phenylenediamine	95-83-0	1,2-Benzenediamine, 4-chloro-
C.I. Basic Red 9 Monohydrochloride	569-61-9	Benzenamine, 4-[(4-aminophenyl)(4-imino-2,5-cyclohexadien-1-ylidene)methyl]-, monohydrochloride
Cisplatin	15663-27-1	Platinum(II), diamminedichloro-, *cis*
p-Cresidine	120-71-8	Benzenamine, 2-methoxy-5-methyl-
Cupferron	135-20-6	Benzenamine, *N*-hydroxy-*N*-nitroso-, ammonium salt
DDT	50-29-3	Benzene, 1,1´-(2,2,2-trichloroethylidene)bis[4-chloro-
2,4-Diaminoanisole sulfate	39156-41-7	1,3-Benzenediamine, 4-methoxy-, sulfate
2,4-Diaminotoluene	95-80-7	1,3-Benzenediamine, 4-methyl-
1,2-Dibromo-3-chloropropane	96-12-8	Propane, 1,2-dibromo-3-chloro-
1,2-Dibromoethane (Ethylene dibromide)	106-93-4	Ethane, 1,2-dibromo-
1,4-Dichlorobenzene	106-46-7	Benzene, 1,4-dichloro-
3,3´-Dichlorobenzidine	91-94-1	[1,1´-Biphenyl]-4,4´-diamine, 3,3´-dichloro-
3,3´-Dichlorobenzidine 2HCl	612-83-9	[1,1´-Biphenyl]-4,4´-diamine, 3,3´-dichloro-, dihydrochloride
1,2-Dichloroethane	107-06-2	Ethane, 1,2-dichloro-
Dichloromethane (Methylene chloride)	75-09-2	Methane, dichloro-
1,3-Dichloropropene (technical grade)	542-75-6	1-Propene, 1,3-dichloro- (mixed isomers)
Diepoxybutane	1464-53-5	2,2´-Bioxirane
Di(2-ethylhexyl) phthalate	117-81-7	1,2-Benzenedicarboxylic acid, bis(2-ethylhexyl) ester
Diethyl sulfate	64-67-5	Sulfuric acid, diethyl ester
Diglycidyl resorcinol ether	101-90-6	Oxirane, 2,2´-[1,3-phenylenebis(oxymethylene)]bis-
3,3´-Dimethoxybenzidine	119-90-4	[1,1´-Biphenyl]-4,4´-diamine, 3,3´-dimethoxy-
3,3´-Dimethoxybenzidine 2HCl	20325-40-0	[1,1´-Biphenyl]-4,4´-diamine, 3,3´-dimethoxy-, dihydrochloride
4-Dimethylamino-azobenzene	60-11-7	Benzenamine, *N,N*-dimethyl-4-(phenylazo)-
3,3´-Dimethylbenzidine	119-93-7	[1,1´-Biphenyl]-4,4´-diamine, 3,3´-dimethyl-
Dimethylcarbamoyl chloride	79-44-7	Carbamic chloride, dimethyl-
1,1-Dimethylhydrazine	57-14-7	Hydrazine, 1,1-dimethyl-
Dimethyl sulfate	77-78-1	Sulfuric acid, dimethyl ester
Dimethylvinyl chloride	513-37-1	1-Propene, 1-chloro-2-methyl-
1,4-Dioxane	123-91-1	1,4-Dioxane
Direct Black 38	1937-37-7	2,7-Naphthalenedisulfonic acid, 4-amino-3-((4´-((2,4-diaminophenyl)azo)(1,1´-biphenyl)-4-yl)azo)-5-hydroxy-6-(phenylazo)-, disodium salt
Direct Blue 6	2602-46-2	2,7-Naphthalenedisulfonic acid, 3,3´-((4,4´-biphenylylene)bis(azo))bis(5-amino-4-hydroxy-, tetrasodium salt
Epichlorohydrin	106-89-8	Propane, 1-chloro-2,3-epoxy-
Estrogens (not conjugated): Estradiol-17beta	50-28-2	Estra-1,3,5(10)-triene-3,17-diol (17ß)-
Estrogens (not conjugated): Estrone	53-16-7	Estra-1,3,5(10)-trien-17-one, 3-hydroxy-
Estrogens (not conjugated): Ethinylestradiol	57-63-6	19-Norpregna-1,3,5(10)-trien-20-yne-3,17-diol, (17α)-
Estrogens (not conjugated): Mestranol	72-33-3	17-α-19-Norpregna-1,3,5(10)-trien-20-yn-17-ol, 3-methoxy
Ethyl acrylate	140-88-5	2-Propenoic acid, ethyl ester
Ethylene oxide	75-21-8	Oxirane

NTP name	CAS RN	Systematic name
Ethylene thiourea	96-45-7	2-Imidazolidinethione
Ethyl methanesulfonate	62-50-0	Methanesulfonic acid, ethyl ester
Formaldehyde (gas)	50-00-0	Formaldehyde
Glasswool (respirable size)		
Glycidol	556-52-5	1-Propanol, 2,3-epoxy
Hexachlorobenzene	118-74-1	Benzene, hexachloro-
Hexachloroethane	67-72-1	Ethane, hexachloro-
Hexamethyl-phosphoramide	680-31-9	Phosphoric triamide, hexamethyl-
Hydrazine	302-01-2	Hydrazine
Hydrazine sulfate	10034-93-2	Hydrazine sulfate
Hydrazobenzene	122-66-7	Hydrazine, 1,2-diphenyl-
Iron dextran complex	9004-66-4	Imferon
Kepone (chlordecone)	143-50-0	1,3,4-Metheno-2H-cyclobuta[cd]pentalen-2-one, 1,1a,3,3a,4,5,5,5a,5b,6-decachlorooctahydro-
Lead acetate	301-04-2	Lead(II) acetate
Lead phosphate	7446-27-7	Lead(II) phosphate
Lindane and other hexachlorocyclohexane isomers		Cyclohexane, 1,2,3,4,5,6-hexachloro- (mixed isomers)
2-Methylaziridine (Propyleneimine)	75-55-8	Aziridine, 2-methyl-
4,4′-Methylenebis(2-chloroaniline) (MBOCA)	101-14-4	Benzenamine, 4,4′-methylenebis[2-chloro-
4,4′-Methylenebis(N,N-dimethylbenzenamine)	101-61-1	Benzenamine, 4,4′-methylenebis[N,N-dimethyl-
4,4′-Methylenedianiline	101-77-9	Benzenamine, 4,4′-methylenebis-
4,4′-Methylenedianiline 2HCl	13552-44-8	Aniline, 4,4′-methylenedi-, dihydrochloride
Methyl methanesulfonate	66-27-3	Methanesulfonic acid, methyl ester
N-Methyl-N-nitro-N-nitrosoguanidine	70-25-7	Guanidine, N-methyl-N′-nitro-N-nitroso-
Metronidazole	443-48-1	1H-Imidazole-1-ethanol, 2-methyl-5-nitro-
Michler′s ketone	90-94-8	Methanone, bis[4-(dimethylamino)phenyl]-
Mirex	2385-85-5	1,3,4-Metheno-1H-cyclobuta[cd]pentalene, 1,1a,2,2,3,3a,4,5,5,5a,5b,6-dodecachlorooctahydro-
Nickel	7440-02-0	Nickel
Certain nickel compounds		
Nitrilotriacetic acid	139-13-9	Glycine, N,N-bis(carboxymethyl)-
Nitrofen	1836-75-5	Benzene, 2,4-dichloro-1-(4-nitrophenoxy)-
Nitrogen mustard hydrochloride	55-86-7	Ethanamine, 2-chloro-N-(2-chloroethyl)-N-methyl-, hydrochloride
2-Nitropropane	79-46-9	Propane, 2-nitro-
N-Nitrosodi-n-butylamine	924-16-3	1-Butanamine, N-butyl-N-nitroso-
N-Nitrosodiethanolamine	1116-54-7	Ethanol, N-nitrosoimindi-
N-Nitrosodiethylamine	55-18-5	Ethanamine, N-ethyl-N-nitroso-
N-Nitrosodimethylamine	62-75-9	Methanamine, N-methyl-N-nitroso-
N-Nitrosodi-n-propylamine	621-64-7	1-Propanamine, N-nitroso-N-propyl-
N-Nitroso-n-ethylurea	759-73-9	Urea, N-ethyl-N-nitroso-
4-(N-Nitrosomethyl-amino)-1-(3-pyridyl)-1-butanone (NNK)	64091-91-4	Ketone, 3-pyridyl-3-(N-methyl-N-nitrosamino)propyl-
N-Nitroso-n-methylurea	684-93-5	Urea, N-methyl-N-nitroso-
N-Nitrosomethyl-vinylamine	4549-40-0	Ethenylamine, N-methyl-N-nitroso-
N-Nitrosomorpholine	59-89-2	Morpholine, 4-nitroso-
N-Nitrosonornicotine	16543-55-8	Pyridine, 3-(1-nitroso-2-pyrrolidinyl)-, (S)-
N-Nitrosopiperidine	100-75-4	Piperidine, 1-nitroso-
N-Nitrosopyrrolidine	930-55-2	Pyrrolidine, 1-nitroso-
N-Nitrososarcosine	13256-22-9	Glycine, N-methyl-N-nitroso-
Norethisterone	68-22-4	19-Nor-17-α-pregn-4-en-20-yn-3-one, 7-hydroxy
Ochratoxin A	303-47-9	Alanine, N-((5-chloro-8-hydroxy-3-methyl-1-oxo-7-isochromanyl)carbonyl)-3-phenyl-, (-)
4,4′-Oxydianiline	101-80-4	Benzenamine, 4,4′-oxybis-
Oxymetholone	434-07-1	5-α,17-ß-Androstan-3-one, 17-hydroxy-2-(hydroxymethylene)-17-methyl-
Phenacetin	62-44-2	Acetamide, N-(4-ethoxyphenyl)-
Phenazopyridine hydrochloride	136-40-3	Pyridine, 2,6-diamino-3-(phenylazo), monohydrochloride
Phenoxybenzamine hydrochloride	63-92-3	Benzenemethanamine, N-(2-chloroethyl)-N-(1-methyl-2-phenoxyethyl)-, hydrochloride

CHEMICAL CARCINOGENS (continued)

NTP name	CAS RN	Systematic name
Phenytoin	57-41-0	2,4-Imidazolidinedione, 5,5-diphenyl-
Polybrominated biphenyls		
Polychlorinated biphenyls	1336-36-3	
Polycyclic aromatic hydrocarbons, 15 listings		
Procarbazine hydrochloride	366-70-1	p-Toluamide, N-isopropyl-α-(2-methylhydrazino)-, monohydrochloride
Progesterone	57-83-0	Pregn-4-ene-3,20-dione
1,3-Propane sultone	1120-71-4	1,2-Oxathiolane, 2,2-dioxide
beta-Propiolactone	57-57-8	2-Oxetanone
Propylene oxide	75-56-9	Propane, 1,2-epoxy
Propylthiouracil	51-52-5	4(1H)-Pyrimidinone, 2,3-dihydro-6-propyl-2-thioxo-
Reserpine	50-55-5	Yohimban-16-carboxylic acid, 11,17-dimethoxy-18-[(3,4,5-trimethoxybenzoyl)oxy]-, methyl ester, (3ß,16ß,17α18ß,20α)-
Saccharin	128-44-9	1,2-Benzisothiazol-3(2H)-one, 1,1-dioxide, sodium salt
Safrole	94-59-7	1,3-Benzodioxole, 5-(2-propenyl)-
Selenium sulfide	7446-34-6	Selenium sulfide (SSe)
Silica, crystalline (respirable size)		Silicon dioxide
Streptozotocin	18883-66-4	Glucopyranose, 2-deoxy-2-(3-methyl-3-nitrosoureido)-, D
Sulfallate	95-06-7	Carbamodithioic acid, diethyl-, 2-chloro-2-propenyl ester
2,3,7,8-Tetrachlorodibenzo-p-dioxin (TCDD)	1746-01-6	Dibenzo-p-dioxin, 2,3,7,8-tetrachloro-
Tetrachloroethylene (perchloroethylene)	127-18-4	Ethene, tetrachloro-
Tetranitromethane	509-14-8	Methane, tetranitro-
Thioacetamide	62-55-5	Ethanethioamide
Thiourea	62-56-6	Thiourea
Toluene diisocyanate	26471-62-5	Benzene, 1,3-diisocyanato-methyl- (mixed isomers)
o-Toluidine	95-53-4	Benzenamine, 2-methyl-
o-Toluidine hydrochloride	636-21-5	Benzenamine, 2-methyl-, hydrochloride
Toxaphene	8001-35-2	Camphene, polychlorinated
2,4,6-Trichlorophenol	88-06-2	Phenol, 2,4,6-trichloro-
Tris(1-aziridinyl)phosphine sulfide (ThioTEPA)	52-24-4	Phosphine sulfide, tris(1-aziridinyl)
Tris(2,3-dibromopropyl) phosphate	126-72-7	1-Propanol, 2,3-dibromo-, phosphate (3:1)
Urethane	51-79-6	Carbamic acid, ethyl ester
4-Vinyl-1-cyclohexene diepoxide	106-87-6	7-Oxabicyclo[4.1.0]heptane, 3-oxiranyl-

SUBSTANCES ASSOCIATED WITH TECHNOLOGICAL PROCESSES KNOWN TO BE CARCINOGENIC

Coke oven emissions
Soots, tars, and mineral oils

Appendix A
Mathematical Tables

MISCELLANEOUS MATHEMATICAL CONSTANTS

π CONSTANTS

$$\pi = 3.14159\ 26535\ 89793\ 23846\ 26433\ 83279\ 50288\ 41971\ 69399\ 37511$$
$$1/\pi = 0.31830\ 98861\ 83790\ 67153\ 77675\ 26745\ 02872\ 40689\ 19291\ 48091$$
$$\pi^2 = 9.86960\ 44010\ 89358\ 61883\ 44909\ 99876\ 15113\ 53136\ 99407\ 24079$$
$$\log_e\pi = 1.14472\ 98858\ 49400\ 17414\ 34273\ 51353\ 05871\ 16472\ 94812\ 91531$$
$$\log_{10}\pi = 0.49714\ 98726\ 94133\ 85435\ 12682\ 88290\ 89887\ 36516\ 78324\ 38044$$
$$\log_{10}\sqrt{2\pi} = 0.39908\ 99341\ 79057\ 52478\ 25035\ 91507\ 69595\ 02099\ 34102\ 92128$$

CONSTANTS INVOLVING e

$$e = 2.71828\ 18284\ 59045\ 23536\ 02874\ 71352\ 66249\ 77572\ 47093\ 69996$$
$$1/e = 0.36787\ 94411\ 71442\ 32159\ 55237\ 70161\ 46086\ 74458\ 11131\ 03177$$
$$e^2 = 7.38905\ 60989\ 30650\ 22723\ 04274\ 60575\ 00781\ 31803\ 15570\ 55185$$
$$M = \log_{10}e = 0.43429\ 44819\ 03251\ 82765\ 11289\ 18916\ 60508\ 22943\ 97005\ 80367$$
$$1/M = \log_e 10 = 2.30258\ 50929\ 94045\ 68401\ 79914\ 54684\ 36420\ 76011\ 01488\ 62877$$
$$\log_{10}M = 9.63778\ 43113\ 00536\ 78912\ 29674\ 98645\ -10$$

π^e AND e^π CONSTANTS

$$\pi^e = 22.45915\ 77183\ 61045\ 47342\ 71522$$
$$e^\pi = 23.14069\ 26327\ 79269\ 00572\ 90864$$
$$e^{-\pi} = 0.04321\ 39182\ 63772\ 24977\ 44177$$
$$e^{1/2\pi} = 4.81047\ 73809\ 65351\ 65547\ 30357$$
$$i^i = e^{-1/2\pi} = 0.20787\ 95763\ 50761\ 90854\ 69556$$

NUMERICAL CONSTANTS

$$\sqrt{2} = 1.41421\ 35623\ 73095\ 04880\ 16887\ 24209\ 69807\ 85696\ 71875\ 37695$$
$$\sqrt[3]{2} = 1.25992\ 10498\ 94873\ 16476\ 72106\ 07278\ 22835\ 05702\ 51464\ 70151$$
$$\log_e 2 = 0.69314\ 71805\ 59945\ 30941\ 72321\ 21458\ 17656\ 80755\ 00134\ 36026$$
$$\log_{10}2 = 0.30102\ 99956\ 63981\ 19521\ 37388\ 94724\ 49302\ 67881\ 89881\ 46211$$
$$\sqrt{3} = 1.73205\ 08075\ 68877\ 29352\ 74463\ 41505\ 87236\ 69428\ 05253\ 81039$$
$$\sqrt[3]{3} = 1.44224\ 95703\ 07408\ 38232\ 16383\ 10780\ 10958\ 83918\ 69253\ 49935$$
$$\log_e 3 = 1.09861\ 22886\ 68109\ 69139\ 52452\ 36922\ 52570\ 46474\ 90557\ 82275$$
$$\log_{10}3 = 0.47712\ 12547\ 19662\ 43729\ 50279\ 03255\ 11530\ 92001\ 28864\ 19070$$

OTHER CONSTANTS

$$\text{Euler's Constant } \gamma = 0.57721\ 56649\ 01532\ 86061$$
$$\log_e \gamma = -0.54953\ 93129\ 81644\ 82234$$
$$\text{Golden Ratio } \phi = 1.61803\ 39887\ 49894\ 84820\ 45868\ 34365\ 63811\ 77203\ 09180$$

EXPONENTIAL AND HYPERBOLIC FUNCTIONS AND THEIR COMMON LOGARITHMS

	e^x		e^{-x}	sinh x		cosh x		tanh x
x	Value	log^{10}	(value)	Value	log^{10}	Value	log^{10}	(value)
0.00	1.0000	0.00000	1.00000	0.0000	— ∞	1.0000	0.00000	0.00000
0.01	1.0101	.00434	0.99005	.0100	2̄.00001	1.0001	.00002	.01000
0.02	1.0202	.00869	.98020	.0200	2̄.30106	1.0002	.00009	.02000
0.03	1.0305	.01303	.97045	.0300	2̄.47719	1.0005	.00020	.02999
0.04	1.0408	.01737	.96079	.0400	2̄.60218	1.0008	.00035	.03998
0.05	1.0513	.02171	.95123	.0500	2̄.69915	1.0013	.00054	.04996
0.06	1.0618	.02606	.94176	.0600	2̄.77841	1.0018	.00078	.05993
0.07	1.0725	.03040	.93239	.0701	2̄.84545	1.0025	.00106	.06989
0.08	1.0833	.03474	.92312	.0801	2̄.90355	1.0032	.00139	.07983
0.09	1.0942	.03909	.91393	.0901	2̄.95483	1.0041	.00176	.08976
0.10	1.1052	.04343	.90484	.1002	1̄.00072	1.0050	.00217	.09967
0.11	1.1163	.04777	.89583	.1102	1̄.04227	1.0061	.00262	.10956
0.12	1.1275	.05212	.88692	.1203	1̄.08022	1.0072	.00312	.11943
0.13	1.1388	.05646	.87809	.1304	1̄.11517	1.0085	.00366	.12927
0.14	1.1503	.06080	.86936	.1405	1̄.14755	1.0098	.00424	.13909
0.15	1.1618	.06514	.86071	.1506	1̄.17772	1.0113	.00487	.14889
0.16	1.1735	.06949	.85214	.1607	1̄.20597	1.0128	.00554	.15865
0.17	1.1853	.07383	.84366	.1708	1̄.23254	1.0145	.00625	.16838
0.18	1.1972	.07817	.83527	.1810	1̄.25762	1.0162	.00700	.17808
0.19	1.2092	.08252	.82696	.1911	1̄.28136	1.0181	.00779	.18775
0.20	1.2214	.08686	.81873	.2013	1̄.30392	1.0201	.00863	.19738
0.21	1.2337	.09120	.81058	.2115	1̄.32541	1.0221	.00951	.20697
0.22	1.2461	.09554	.80252	.2218	1̄.34592	1.0243	.01043	.21652
0.23	1.2586	.09989	.79453	.2320	1̄.36555	1.0266	.01139	.22603
0.24	1.2712	.10423	.78663	.2423	1̄.38437	1.0289	.01239	.23550
0.25	1.2840	.10857	.77880	.2526	1̄.40245	1.0314	.01343	.24492
0.26	1.2969	.11292	.77105	.2629	1̄.41986	1.0340	.01452	.25430
0.27	1.3100	.11726	.76338	.2733	1̄.43663	1.0367	.01564	.26362
0.28	1.3231	12160	.75578	.2837	1̄.45282	1.0395	.01681	.27291
0.29	1.3364	.12595	.74826	.2941	1̄.46847	1.0423	.01801	.28213
0.30	1.3499	.13029	.74082	.3045	1̄.48362	1.0453	.01926	.29131
0.31	1.3634	.13463	.73345	.3150	1̄.49830	1.0484	.02054	.30044
0.32	1.3771	.13897	.72615	.3255	1̄.51254	1.0516	.02187	.30951
0.33	1.3910	.14332	.71892	.3360	1̄.52637	1.0549	.02323	.31852
0.34	1.4049	.14766	.71177	.3466	1̄.53981	1.0584	.02463	.32748
0.35	1.4191	.15200	.70469	.3572	1̄.55290	1.0619	.02607	.33638
0.36	1.4333	.15635	.69768	.3678	1̄.56564	1.0655	.02755	.34521
0.37	1.4477	.16069	.69073	.3785	1̄.57807	1.0692	.02907	.35399
0.38	1.4623	.16503	.68386	.3892	1̄.59019	1.0731	.03063	.36271
0.39	1.4770	.16937	.67706	.4000	1̄.60202	1.0770	.03222	.37136
0.40	1.4918	.17372	.67032	.4108	1̄.61358	1.0811	.03385	.37995
0.41	1.5063	.17806	.66365	.4216	1̄.62488	1.0852	.03552	.33847
0.42	1.5220	.18240	.65705	.4325	1̄.63594	1.0895	.03723	.39693
0.43	1.5373	.18675	.65051	.4434	1̄.64677	1.0939	.03897	.40532
0.44	1.5527	.19109	.64404	.4543	1̄.65738	1.0984	.04075	.41364
0.45	1.5683	.19543	.63763	.4653	1̄.66777	1.1030	.04256	.42190
0.46	1.5841	.19978	.63128	.4764	1̄.67797	1.1077	.04441	.43008
0.47	1.6000	.20412	.62500	.4875	1̄.68797	1.1125	.04630	.43820
0.48	1.6161	.20846	.61878	.4986	1̄.69779	1.1174	.04822	.44624
0.49	1.6323	.21280	.61263	.5098	1̄.70744	1.1225	.05018	.45422
0.50	1.6487	.21715	.60653	.5211	1̄.71692	1.1276	.05217	.46212
0.51	1.6653	.22149	.60050	.5324	1̄.72624	1.1329	.05419	.46995
0.52	1.6820	.22583	.59452	.5438	1̄.73540	1.1383	.05625	.47770
0.53	1.6989	.23018	.58860	.5552	1̄.74442	1.1438	.05834	.48538
0.54	1.7160	.23452	.58275	.5666	1̄.75330	1.1494	.06046	.49299
0.55	1.7333	.23886	.57695	.5782	1̄.76204	1.1551	.06262	.50052
0.56	1.7507	.24320	.57121	.5897	1̄.77065	1.1609	.06481	.50798
0.57	1.7683	.24755	.56553	.6014	1̄.77914	1.1669	.06703	.51536
0.58	1.7860	.25189	.55990	.6131	1̄.78751	1.1730	.06929	.52267
0.59	1.8040	.25623	.55433	.6248	1̄.79576	1.1792	.07157	.52990
0.60	1.8221	.26058	.54881	.6367	1̄.80390	1.1855	.07389	.53705
0.61	1.8404	.26492	.54335	.6485	1̄.81194	1.1919	.07624	.54413
0.62	1.8589	.26926	.53794	.6605	1̄.81987	1.1984	.07861	.55113
0.63	1.8776	.27361	.53259	.6725	1̄.82770	1.2051	.08102	.55805
0.64	1.8965	.27795	.52729	.6846	1̄.83543	1.2119	.08346	.56490
0.65	1.9155	.28229	.52205	.6967	1̄.84308	1.2188	.08593	.57167
0.66	1.9348	.28664	.51685	.7090	1̄.85063	1.2258	.08843	.57836
0.67	1.9542	.29098	.51171	.7213	1̄.85809	1.2330	.09095	.58498
0.68	1.9739	.29532	.50662	.7336	1̄.86548	1.2402	.09351	.59152
0.69	1.9937	.29966	.50158	.7461	1̄.87278	1.2476	.09609	.59798
0.70	2.0138	.30401	.49659	.7586	1̄.88000	1.2552	.09870	.60437
0.71	2.0340	.30835	.49164	.7712	1̄.88715	1.2628	.10134	.61068
0.72	2.0544	.31269	.48675	.7838	1̄.89423	1.2706	.10401	.61691

x	e^x Value	e^x log^{10}	e^{-x} (value)	sinh x Value	sinh x log^{10}	cosh x Value	cosh x log^{10}	tanh x (value)
0.73	2.0751	.31703	.48191	.7966	$\bar{1}$.90123	1.2785	.10670	.62307
0.74	2.0959	.32138	.47711	.8094	$\bar{1}$.90817	1.2865	.10942	.62915
0.75	2.1170	.32572	.47237	.8223	$\bar{1}$.91504	1.2947	.11216	.63515
0.76	2.1383	.33006	.46767	.8353	$\bar{1}$.92185	1.3030	.11493	.64108
0.77	2.1598	.33441	.46301	.8484	$\bar{1}$.92859	1.3114	.11773	.64693
0.78	2.1815	.33875	.45841	.8615	$\bar{1}$.93527	1.3199	.12055	.65721
0.79	2.2034	.34309	.45384	.8748	$\bar{1}$.94190	1.3286	.12340	.65841
0.80	2.2255	.34744	.44933	.8881	$\bar{1}$.94846	1.3374	.12627	.66404
0.81	2.2479	.35178	.44486	.9015	$\bar{1}$.95498	1.3464	.12917	.66959
0.82	2.2705	.35612	.44043	.9150	$\bar{1}$.96144	1.3555	.13209	.67507
0.83	2.2933	.36046	.43605	.9286	$\bar{1}$.96784	1.3647	.13503	.68048
0.84	2.3164	.36481	.43171	.9423	$\bar{1}$.97420	1.3740	.13800	.68581
0.85	2.3396	.36915	.42741	.9561	$\bar{1}$.98051	1.3835	.14099	.69107
0.86	2.3632	.37349	.42316	.9700	$\bar{1}$.98677	1.3932	.14400	.69626
0.87	2.3869	.37784	.41895	.9840	$\bar{1}$.99299	1.4029	.14704	.70137
0.88	2.4100	.38218	.41478	.9981	$\bar{1}$.99916	1.4128	.15009	.70642
0.89	2.4351	.38652	.41066	1.0122	0.00528	1.4229	.15317	.71139
0.90	2.4596	.39087	.40657	1.0265	.01137	1.4331	.15627	.21630
0.91	2.4843	39521	.40242	1.0409	.01741	1.4434	.15939	.72113
0.92	2.5093	.39955	.39852	1.0554	.02341	1.4539	.16254	.72590
0.93	2.5345	.40389	.39455	1.0700	.02937	1.4645	.16570	.73059
0.94	2.5600	.40824	.39063	1.0847	.03530	1.4753	.16888	.73522
0.95	2.5857	.41258	.38674	1.0995	.04119	1.4862	.17208	.73978
0.96	2.6117	.41692	.38289	1.1144	.04704	1.4973	.17531	.74428
0.97	2.6379	.42127	.37908	1.1294	.05286	1.5085	.17855	.74870
0.98	2.6645	.42561	.37531	1.1446	.05864	1.5199	.18181	.75307
0.99	2.6912	.42995	.37158	1.1598	.06439	1.5314	.18509	.75736
1.00	2.7183	.43429	.36788	1.1752	.07011	1.5431	.18839	.76159
1.01	2.7456	.43864	.36422	1.1907	.07580	1.5549	.19171	.76576
1.02	2.7732	.44298	.36060	1.2063	.06146	1.5669	.19504	.76987
1.03	2.8011	.44732	.35701	1.2220	.08708	1.5790	.19839	.77391
1.04	2.8292	.45167	.35345	1.2379	.09268	1.5913	.20176	.77789
1.05	2.8577	.45601	.34994	1.2539	.09825	1.6038	.20515	.78181
1.06	2.8864	.46035	.34646	1.2700	.10379	1.6164	.20855	.78566
1.07	2.9154	.46470	.34301	1.2862	.10930	1.6292	.21197	.78946
1.08	2.9447	.46904	.33960	1.3025	.11479	1.6421	.21541	.79320
1.09	2.9743	.47338	.33622	1.3190	.12025	1.6552	.21886	.79688
1.10	3.0042	.47772	.33287	1.3356	.12569	1.6685	.22233	.80050
1.11	3.0344	.48207	.32956	1.3524	.13111	1.6820	.22582	.80406
1.12	3.0659	.48641	.32628	1.3693	.13649	1.6956	.22931	.80757
1.13	3.0957	.49075	.32303	1.3863	.14186	1.7083	.23283	.81102
1.14	3.1268	.49510	.31982	1.4035	.14720	1.7233	.23636	.81441
1.15	3.1582	.49944	.31644	1.4208	.15253	1.7374	.23990	.81775
1.16	3.1899	.50378	.31349	1.4382	.15783	1.7517	.24346	.82104
1.17	3.2220	.50812	.31037	1.4558	.16311	1.7662	.24703	.82427
1.18	3.2544	.51247	.30728	1.4735	.16836	1.7808	.25062	.82745
1.19	3.2871	.51681	.30422	1.4914	.17360	1.7957	.25422	.83058
1.20	3.3201	.52115	.30119	1.5095	.17882	1.8107	.25784	.83365
1.21	3.3535	.52550	.29820	1.5276	.18402	1.8258	.26146	.83668
1.22	3.3872	.52984	.29523	1.5460	.18920	1.8412	.26510	.83965
1.23	3.4212	.53418	.29229	1.5645	.19437	1.8568	.26876	.84258
1.24	3.4556	.53853	.28938	1.5831	.19951	1.8725	.27242	.83546
1.25	3.4903	.54287	.28650	1.6019	.20464	1.8884	.27610	.84828
1.26	3.5254	.54721	.28365	1.6209	.20975	1.9045	.27979	.85106
1.27	3.5609	.55155	.28083	1.6400	.21485	1.9208	.28349	.85380
1.28	3.5996	.55590	.27804	1.6593	.21993	1.9373	.28721	.85648
1.29	3.6328	.56024	.27527	1.6788	.22499	1.9540	.29093	.85913
1.30	3.6693	.56458	.27253	1.6984	.23004	1.9709	.29467	.86172
1.31	3.7062	.56893	.26982	1.7182	.23507	1.9880	.29842	.86428
1.32	3.7434	.57327	.26714	1.7381	.24009	2.0053	.30217	.86678
1.33	3.7810	.57761	.26448	1.7583	.24509	2.0228	.30594	.86925
1.34	3.8190	.58195	.26185	1.7786	.25008	2.0404	.30972	.87167
1.35	3.8574	.58630	.25924	1.7991	.25505	2.0583	.31352	.87405
1.36	3.8962	.59064	.25666	1.8198	.26002	2.0764	.31732	.87639
1.37	3.9354	.59498	.25411	1.8406	.26496	2.0947	.32113	.87869
1.38	3.9749	.59933	.25158	1.8617	.26990	2.1132	.32495	.88095
1.39	4.0149	.60367	.24908	1.8829	.27482	2.1320	.32878	.88317
1.40	4.0552	.60801	.24660	1.9043	.27974	2.1509	.33262	.88535
1.41	4.0960	.61236	.24414	1.9259	.28464	2.1700	.33647	.88749
1.42	4.1371	.61670	.24171	1.9477	.28952	2.1894	.34033	.88960
1.43	4.1787	.62104	.23931	1.9697	.29440	2.2090	.34420	.89167

	e^x		e^{-x}	sinh x		cosh x		tanh x
x	Value	log¹⁰	(value)	Value	log¹⁰	Value	log¹⁰	(value)
1.44	4.2207	.62538	.23693	1.9919	.29926	2.2288	.34807	.89370
1.45	4.2631	.62973	.23457	2.0143	.30412	2.2488	.35196	.89569
1.46	4.3060	.63407	.23224	2.0369	.30896	2.2691	.35585	.89765
1.47	4.3492	.63841	.22993	2.0597	.31379	2.2896	.35976	.89958
1.48	4.3929	.64276	.22764	2.0827	.31862	2.3103	.36367	.90147
1.49	4.4371	.64710	.22537	2.1059	.32343	2.3312	.36759	.90332
1.50	4.4817	.65144	.22313	2.1293	.32823	2.3524	.37151	.90515
1.51	4.5267	.65578	.22091	2.1529	.33303	2.3738	.37545	.90694
1.52	4.5722	.66013	.21871	2.1768	.33781	2.3955	.37939	.90870
1.53	4.6182	.66447	.21654	2.2008	.34258	2.4174	.38334	.91042
1.54	4.6646	.66881	.21438	2.2251	.34735	2.4395	.38730	.91212
1.55	4.7115	.67316	.21225	2.2496	.35211	2.4619	.39126	.91379
1.56	4.7588	.67750	.21014	2.2743	.35686	2.4845	.39524	.91542
1.57	4.8066	.68184	.20805	2.2993	.36160	2.5073	.39921	.91703
1.58	4.8550	.68619	.20598	2.3245	.36633	2.5305	.40320	.91860
1.59	4.9037	.69053	.20393	2.3499	.37105	2.5538	.40719	.92015
1.60	4.9530	.69487	.20190	2.3756	.37577	2.5775	.41119	.92167
1.61	5.0028	.69921	.19989	2.4015	.38048	2.6013	.41520	.92316
1.62	5.0531	.70356	.19790	2.4276	.38518	2.6255	.41921	.92462
1.63	5.1039	.70790	.19593	2.4540	.38987	2.6499	.42323	.92606
1.64	5.1552	.71224	.19398	2.4806	.39456	2.6746	.42725	.92747
1.65	5.2070	.71659	.19205	2.5075	.39923	2.6995	.43129	.92886
1.66	5.2593	.72093	.19014	2.5346	.40391	2.7247	.43532	.93022
1.67	5.3122	.72527	.18825	2.5620	.40857	2.7502	.43937	.93155
1.68	5.3656	.72961	.18637	2.5896	.41323	2.7760	.44341	.93286
1.69	5.4195	.73396	.18452	2.6175	.41788	2.8020	.44747	.93415
1.70	5.4739	.73830	.18268	2.6456	.42253	2.8283	.45153	.93541
1.71	5.5290	.74264	.18087	2.6740	.42717	2.8549	.45559	.93665
1.72	5.5845	.74699	.17907	2.7027	.43180	2.8818	.45966	.93786
1.73	5.6407	.75133	.17728	2.7317	.43643	2.9090	.46374	.93906
1.74	5.6973	.75567	.17552	2.7609	.44105	2.9364	.46782	.94023
1.75	5.7546	.76002	.17377	2.7904	.44567	2.9642	.47191	.94138
1.76	5.8124	.76436	.17204	2.8202	.45028	2.9922	.47600	.94250
1.77	5.8709	.76870	.17033	2.8503	.45488	3.0206	.48009	.94361
1.78	5.9299	.77304	.16864	2.8806	.45948	3.0492	.48419	.94470
1.79	5.9895	.77739	.16696	2.9112	.46408	3.0782	.48830	.94576
1.80	6.0496	.78173	.16530	2.9422	.46867	3.1075	.49241	.94681
1.81	6.1104	.78607	.16365	2.9734	.47325	3.1371	.49652	.94783
1.82	6.1719	.79042	.16203	3.0049	.47783	3.1669	.50064	.94884
1.83	6.2339	.79476	.16041	3.0367	.48241	3.1972	.50476	.94983
1.84	6.2965	.79910	.15882	3.0689	.48698	3.2277	.50889	.95080
1.85	6.3598	.80344	.15724	3.1013	.49154	3.2585	.51302	.95175
1.86	6.4237	.80779	.15567	3.1340	.49610	3.2897	.51716	.95268
1.87	6.4383	.81213	.15412	3.1671	.50066	3.3212	.52130	.95359
1.88	6.5535	.81647	.15259	3.2005	.50521	3.3530	.52544	.95449
1.89	6.6194	.82082	.15107	3.2341	.50976	3.3852	.52959	.95537
1.90	6.6859	.82516	.14957	3.2682	.51430	3.4177	.53374	.95624
1.91	6.7531	.82950	.14808	3.3025	.51884	3.4506	.53789	.95709
1.92	6.8210	.83385	.14661	3.3372	.52338	3.4838	.54205	.95792
1.93	6.8895	.83819	.14515	3.3722	.52791	3.5173	.54621	.95873
1.94	6.9588	.84253	.14370	3.4075	.53244	3.5512	.55038	.95953
1.95	7.0287	.84687	.14227	3.4432	.53696	3.5855	.55455	.96032
1.96	7.0993	.85122	.14086	3.4792	.54148	3.6201	.55872	.96109
1.97	7.1707	.85556	.13946	3.5156	.54600	3.6551	.56290	96185
1.98	7.2427	.85990	.13807	3.5923	.55051	3.6904	.56707	.96259
1.99	7.3155	.86425	.13670	3.5894	.55502	3.7261	.57126	.96331
2.00	7.3891	.86859	.13534	3.6269	.55953	3.7622	.57544	.96403
2.01	7.4633	.87293	.13399	3.6647	.56403	3.7987	.57963	.96473
2.02	7.5383	.87727	.13266	3.7028	.56853	3.8335	.58382	.96541
2.03	7.6141	.88162	.13134	3.7414	.57303	3.8727	.58802	.96609
2.04	7.6906	.88596	.13003	3.7803	.57753	3.9103	.59221	.96675
2.05	7.7679	.89030	.12873	3.8196	.58202	3.9483	.59641	.96740
2.06	7.8460	.89465	.12745	3.8593	.58650	3.9867	.60061	.96803
2.07	7.9248	.89899	.12619	3.8993	.59099	4.0255	.60482	.96865
2.08	8.0045	.90333	.12493	3.9398	.59547	4.0647	.60903	.96926
2.09	8.0849	.90768	.12369	3.9806	.59995	4.1043	.61324	.96986
2.10	8.1662	.91202	.12246	4.0219	.60443	4.1443	.61745	.97045
2.11	8.2482	.91636	.12124	4.0635	.60890	4.1847	.62167	.97103
2.12	8.3311	.92070	.12003	4.1056	.61337	4.2256	.62589	.97159
2.13	8.4149	.92505	.11884	4.1480	.61784	4.2669	.63011	.97215
2.14	8.4994	.92939	.11765	4.1909	.62231	4.3085	.63433	.97269
2.15	8.5849	.93373	.11648	4.2342	62677	4.3507	.63856	.97323

EXPONENTIAL AND HYPERBOLIC FUNCTIONS AND THEIR COMMON LOGARITHMS
(continued)

x	e^x Value	e^x log^{10}	(value)	sinh x Value	sinh x log^{10}	cosh x Value	cosh x log^{10}	tanh x (value)
2.16	8.6711	.93808	.11533	4.2779	.63123	4.3932	.64278	.97375
2.17	8.7583	.94242	.11418	4.3221	.63569	4.4362	.64701	.97426
2.18	8.8463	.94676	.11304	4.3666	.64015	4.4797	.65125	.97477
2.19	8.9352	.95110	.11192	4.4116	.64460	4.5236	.65548	.97526
2.20	9.0250	.95545	.11080	4.4571	.64905	4.5679	.65972	.97574
2.21	9.1157	.95979	.10970	4.5030	.65350	4.6127	.66396	.97622
2.22	9.2073	.96413	.10861	4.5494	.65795	4.6580	.66820	.97668
2.23	9.2999	.96848	.10753	4.5962	.66240	4.7037	.67244	.97714
2.24	9.3933	.97282	.10646	4.6434	.66684	4.7499	.67668	.97759
2.25	9.4877	.97716	.10540	4.6912	.67128	4.7966	.68093	.97803
2.26	9.5831	.98151	.10435	4.7394	.67572	4.8437	.68518	.97846
2.27	9.6794	.98585	.10331	4.7880	.68016	4.8914	.68943	.97888
2.28	9.7767	.99019	.10228	4.8372	.68459	4.9395	.69368	.97929
7.29	9.8749	.99453	.10127	4.8868	.68903	4.9881	.69794	97970
2.30	9.9742	.99888	.10026	4.9370	.69346	5.0372	.70219	.98010
2.31	10.074	1.00322	.09926	4.9876	.69789	5.0868	.70645	.98049
2.32	10.176	1.00756	.09827	5.0387	.70232	5.1370	.71071	.98087
2.33	10.278	1.01191	.09730	5.0903	.70675	5.1876	.71497	.98124
2.34	10.381	1.01625	.09633	5.1425	.71117	5.2388	.71923	.98161
2.35	10.486	1.02059	.09537	5.1951	.71559	5.2905	.72349	.98197
2.36	10.591	1.02493	.09442	5.2483	.72002	5.3427	.72776	.98233
2.37	10.697	1.02928	.09348	5.3020	.72444	5.3954	.73203	.98267
2.38	10.805	1.03362	.09255	5.3562	.72885	5.4487	.73630	.98301
2.39	10.913	1.03796	.09163	5.4109	.73327	5.5026	.74056	.98335
2.40	11.023	1.04231	.09072	5.4662	.73769	5.5569	.74484	.98367
2.41	11.134	1.04665	.08982	5.5221	.74210	5.6119	.74911	.98400
2.42	11.246	1.05099	.08892	5.5785	.74652	5.6674	.75338	.98431
2.43	11.359	1.05534	08804	5.6354	.75093	5.7235	.75766	.98462
2.44	11.473	1.05968	.08716	5.6929	.75534	5.7801	.76194	.98492
2.45	11.588	1.06402	.08629	5.7510	.75975	5.8373	.76621	.98522
2.46	11.705	1.06836	.08543	5.8097	.76415	5.8951	.77049	.98551
2.47	11.822	1.07271	.08458	5.8689	.76856	5.9535	.77477	.98579
2.48	11.941	1.07705	.08374	5.9288	.77296	6.0125	.77906	.98607
2.49	12.061	1.08139	.08291	5.9892	.77737	6.0721	.78334	.98635
2.50	12.182	1.08574	.08208	6.0502	.78177	6.1323	.78762	.98661
2.51	12.305	1.09008	.08127	6.1118	.78617	6.1931	.79191	.98688
2.52	12.429	1.09442	.08046	6.1741	.79057	6.2545	.79619	.98714
2.53	12.554	1.09877	.07966	6.2369	.79497	6.3166	.80048	.98739
2.54	12.680	1.10311	.07887	6.3004	.79937	6.3793	.80477	.98764
2.55	12.807	1.10745	.07808	6.3645	.80377	6.4426	.80906	.98788
2.56	12.936	1.11179	.07730	6.4293	.80816	6.5066	.81335	.98812
2.57	13.066	1.11614	.07654	6.4946	.81256	6.5712	.81764	.98835
2.58	13.197	1.12048	.07577	6.5607	.81695	6.6365	.82194	.98858
2.59	13.330	1.12482	.07502	6.6274	.82134	6.7024	.82623	.98881
2.60	13.464	1.12917	.07427	6.6947	.82573	6.7690	.83052	.98903
2.61	13.599	1.13351	.07353	6.7628	.83012	6.8363	.83482	.98924
2.62	13.736	1.13785	.07280	6.8315	.83451	6.9043	.83912	.98946
2.63	13.874	1.14219	.07208	6.9008	.83890	6.9729	.84341	.98966
2.64	14.013	1.14654	.07136	6.9709	.84329	7.0423	.84771	98987
2.65	14.154	1.15008	.07065	7.0417	.84768	7.1123	.85201	.99007
2.66	14.296	1.15522	.06995	7.1132	.85206	7.1831	.85631	.99026
2.67	14.440	1.15957	.06925	7.1854	.85645	7.2546	.86061	.99045
2.68	14.585	1.16391	.06856	7.2583	.86083	7.3268	.86492	.99064
2.69	14.732	1.16825	.06788	7.3319	.86522	7.3998	.86922	.99083
2.70	14.880	1.17260	.06721	7.4063	.86960	7.4735	.87352	.99101
2.71	15.029	1.17694	.06654	7.4814	.87398	7.5479	.87783	.99118
2.72	15.180	1.18128	.06587	7.5572	.87836	7.6231	.88213	.99136
2.73	15.333	1.18562	.06522	7.6338	.88274	7.6991	.89644	.99153
2.74	15.487	1.18997	.06457	7.7112	.88712	7.7758	.89074	.99170
2.75	15.643	1.19431	06393	7.7894	.89150	7.8533	.89505	.99186
2.76	15.800	1.19865	.06329	7.8683	.89588	7.9316	.89936	.99202
2.77	15.959	1.20300	.06266	7.9480	.90026	8.0106	.90367	.99218
2.78	16.119	1.20734	.06204	8.0285	.90463	8.0905	.90798	.99233
2.79	16.281	1.21168	.06142	8.1098	.90901	8.1712	.91229	.99248
2.80	16.445	1.21602	.06081	8.1919	.91339	8.2527	.91660	.99263
2.81	16.610	1.22037	.06020	8.2749	.91776	8.3351	92091	.99278
2.82	16.777	1.22471	.05961	8.3586	.92213	8.4182	.92522	.99292
2.83	16.945	1.22905	.05901	8.4432	.92651	8.5022	.92953	.99306
2.84	17.116	1.23340	.05843	8.5287	.93088	8.5871	.93385	.99320
2.85	17.288	1.23774	.05784	8.6150	.93525	8.6728	.93816	.99333
2.86	17.462	1.24208	.05727	8.7021	.93963	8.7594	.94247	.99346

	e^x		e^{-x}	$\sinh x$		$\cosh x$		$\tanh x$
x	Value	$\log^{10}$	(value)	Value	$\log^{10}$	Value	$\log^{10}$	(value)
2.87	17.637	1.24643	.05670	8.7902	.94400	8.8469	.94679	.99359
2.88	17.814	1.25077	.05613	8.8791	.94837	8.9352	95110	.99372
2.89	17.993	1.25511	.05558	8.9689	.95274	9.0244	.95542	.99384
2.90	18.174	1.25945	.05502	9.0596	.95711	9.1146	.95974	.99396
2.91	18.357	1.26380	.05448	9.1512	.96148	9.2056	.96405	.99408
2.92	18.541	1.26814	.05393	9.2437	.96584	9.2976	.96837	.99420
2.93	18.728	1.27248	.05340	9.3371	.97021	9.3905	.97269	.99431
2.94	18.916	1.27683	.05287	9.4315	.97458	9.4844	.97701	.99443
2.95	19.106	1.28117	.05234	9.5268	.97895	9.5791	.98133	.99454
2.96	19.298	1.28551	.05182	9.6231	.98331	9.6749	.98565	.99464
2.97	19.492	1.28985	.05130	9.7203	.98768	9.7716	.98997	.99475
2.98	19.688	1.29420	.05079	9.8185	.99205	9.8693	.99429	.99485
2.99	19.886	1.29854	.05029	9.9177	.99641	9.9680	.99861	.99496
3.00	20.086	1.30288	.04979	10.018	1.00078	10.068	1.00293	0.99505
3.05	21.115	1.32460	.04736	10.534	1.02259	10.581	1.02454	0.99552
3.10	22.198	1.34631	.04505	11.076	1.04440	11.122	1.04616	0.99595
3.15	23.336	1.36803	.04285	11.647	1.06620	11.690	1.06779	0.99633
3.20	24.533	1.38974	.04076	12.246	1.08799	12.287	1.08943	0.99668
3.25	25.790	1.41146	.03877	12.876	1.10977	12.915	1.11108	0.99700
3.30	27.113	1.43317	.03688	13.538	1.13155	13.575	1.13273	0.99728
3.35	28.503	1.45489	.03508	14.234	1.15332	14.269	1.15439	0.99754
3.40	29.964	1.47660	.03337	14.965	1.17509	14.999	1.17605	0.99777
3.45	31.500	1.49832	.03175	15.734	1.19685	15.766	1.19772	0.99799
3.50	33.115	1.52003	.03020	16.543	1.21860	16.573	1.21940	0.99818
3.55	34.813	1.54175	.02872	17.392	1.24036	17.421	1.24107	0.99835
3.60	36.598	1.56346	.02732	18.286	1.26211	18.313	1.26275	0.99851
3.65	38.475	1.58517	02599	19.224	1.28385	19.250	1.28444	9.99865
3.70	40.447	1.60689	.02472	20.211	1.30559	20.236	1.30612	0.99878
3.75	42.521	1.62860	.02352	21.249	1.32733	21.272	1.32781	0.99889
3.80	44.701	1.65032	.02237	22.339	1.34907	22.362	1.34951	0.99900
3.85	46.993	1.67203	.02128	23.486	1.37081	23.507	1.37120	0.99909
3.90	49.402	1.69375	.02024	24.691	1.39254	24.711	1.39290	0.99918
3.95	51.935	1.71546	.01925	25.958	1.41427	25.977	1.41459	0.99926
4.00	54.598	1.73718	.01832	27.290	1.43600	27.308	1.43629	0.99933
4.10	60.340	1.78061	.01657	30.162	1.47946	30.178	1.47970	0.99945
4.20	66.686	1.82404	.01500	33.336	1.52291	33.351	1.52310	0.99955
4.30	73.700	1.86747	.01357	36.843	1.56636	36.857	1.56652	0.99963
4.40	81.451	1.91090	.01227	40.719	1.60980	40.732	1.60993	0.99970
4.50	90.017	1.95433	.01111	45.003	1.65324	45.014	1.65335	0.99975
4.60	99.484	1.99775	.01005	49.737	1.69668	49.747	1.69677	0.99980
4.70	109.95	2.04118	.00910	54.969	1.74012	54.978	1.74019	0.99983
4.80	121.51	2.08461	.00823	60.751	1.78355	60.759	1.78361	0.99986
4.90	134.29	2.12804	.00745	67.141	1.82699	67.149	1.82704	0.99989
5.00	148.41	2.17147	.00674	74.203	1.87042	74.210	1.87046	0.99991
5.10	164.02	2.21490	.00610	82.008	1.91389	82.014	1.91389	0.99993
5.20	181.27	2.25833	.00552	90.633	1.95729	90.639	1.95731	0.99994
5.30	200.34	2.30176	.00499	100.17	2.00074	100.17	2.00074	0.99995
5.40	221.41	2.34519	.00452	110.70	2.04415	110.71	2.04417	0.99996
5.50	244.69	2.38862	.00409	122.34	2.08758	122.35	2.08760	0.99997
5.60	270.43	2.43205	.00370	135.21	2.13101	135.22	2.13103	0.99997
5.70	298.87	2.47548	.00335	149.43	2.17444	149.44	2.17445	0.99998
5.80	330.30	2.51891	.00303	165.15	2.21787	165.15	2.21788	0.99998
5.90	365.04	2.56234	.00274	182.52	2.26130	182.52	2.26131	0.99998
6.00	403.43	2.60577	.00248	201.71	2.30473	201.72	2.30474	0.99999
6.25	518.01	2.71434	.00193	259.01	2.41331	259.01	2.41331	0.99999
6.50	665.14	2.82291	.00150	332.57	2.52188	332.57	2.52189	1.00000
6.75	854.06	2.93149	.00117	427.03	2.63046	427.03	2.63046	1.00000
7.00	1096.6	3.04006	.00091	548.32	2.73904	548.32	2.73903	1.00000
7.50	1808.0	3.25721	.00055	904.02	2.95618	904.02	2.95618	1.00000
8.00	2981.0	3.47436	.00034	1490.5	3.17333	1490.5	3.17333	1.00000
8.50	4914.8	3.69150	.00020	2457.4	3.39047	2457.4	3.39047	1.00000
9.00	8103.1	3.90865	.00012	4051.5	3.60762	4051.5	3.60762	1.00000
9.50	13360.	4.12580	.00007	6679.9	3.82477	6679.9	3.82477	1.00000
10.00	22026.	4.34294	.00005	11013.	4.04191	11013.	4.04191	1.00000

X radians	X degrees	sin x	cos x	tan x	cot x	sec x	csc x		
.0000	0° 00′	.0000	1.0000	.0000	—	1.000	—	90° 00′	1.5708
.0029	10	.0029	1.0000	.0029	343.8	1.000	343.8	50	1.5679
.0058	20	.0058	1.0000	.0058	171.9	1.000	171.9	40	1.5650
.0087	30	.0087	1.0000	.0087	114.6	1.000	114.6	30	1.5621
.0116	40	.0116	.9999	.0116	85.94	1.000	85.95	20	1.5592
.0145	50	.0145	.9999	.0145	68.75	1.000	68.76	10	1.5563
.0175	1° 00′	.0175	.9998	.0175	57.29	1.000	57.30	89° 00′	1.5533
.0204	10	.0204	.9998	.0204	49.10	1.000	49.11	50	1.5504
.0233	20	.0233	.9997	.0233	42.96	1.000	42.98	40	1.5475
.0262	30	.0262	.9997	.0262	38.19	1.000	38.20	30	1.5446
.0291	40	.0291	.9996	.0291	34.37	1.000	34.38	20	1.5417
.0320	50	.0320	.9995	.0320	31.24	1.001	31.26	10	1.5388
.0349	2° 00′	.0349	.9994	.0349	28.64	1.001	28.65	88° 00′	1.5359
.0378	10	.0378	.9993	.0378	26.43	1.001	26.45	50	1.5330
.0407	20	.0407	.9992	.0407	24.54	1.001	24.56	40	1.5301
.0436	30	.0436	.9990	.0437	22.90	1.001	22.93	30	1.5272
.0465	40	.0465	.9989	.0466	21.47	1.001	21.49	20	1.5243
.0495	50	.0494	.9988	.0495	20.21	1.001	20.23	10	1.5213
.0524	3° 00′	.0523	.9986	.0524	19.08	1.001	19.11	87° 00′	1.5184
.0553	10	.0552	.9985	.0553	18.07	1.002	18.10	50	1.5155
.0582	20	.0581	.9983	.0582	17.17	1.002	17.20	40	1.5126
.0611	30	.0610	.9981	.0612	16.35	1.002	16.38	30	1.5097
.0640	40	.0640	.9980	.0641	15.60	1.002	15.64	20	1.5068
.0669	50	.0669	.9978	.0670	14.92	1.002	14.96	10	1.5039
.0698	4° 00′	.0698	.9976	.0699	14.30	1.002	14.34	86° 00′	1.5010
.0727	10	.0727	.9974	.0729	13.73	1.003	13.76	50	1.4981
.0756	20	.0756	.9971	.0758	13.20	1.003	13.23	40	1.4952
.0785	30	.0785	.9969	.0787	12.71	1.003	12.75	30	1.4923
.0814	40	.0814	.9967	.0816	12.25	1.003	12.29	20	1.4893
.0844	50	.0843	.9964	.0846	11.83	1.004	11.87	10	1.4864
.0873	5° 00′	.0872	.9962	.0875	11.43	1.004	11.47	85° 00′	1.4835
.0902	10	.0901	.9959	.0904	11.06	1.004	11.10	50	1.4806
.0931	20	.0929	.9957	.0934	10.71	1.004	10.76	40	1.4777
.0960	30	.0958	.9954	.0963	10.39	1.005	10.43	30	1.4748
.0989	40	.0987	.9951	.0992	10.08	1.005	10.13	20	1.4719
.1018	50	.1016	.9948	.1022	9.788	1.005	9.839	10	1.4690
.1047	6° 00′	.1045	.9945	.1051	9.514	1.006	9.567	84° 00′	1.4661
.1076	10	.1074	.9942	.1080	9.255	1.006	9.309	50	1.4632
.1105	20	.1103	.9939	.1110	9.010	1.006	9.065	40	1.4603
.1134	30	.1132	.9936	.1139	8.777	1.006	8.834	30	1.4573
.1164	40	.1161	.9932	.1169	8.556	1.007	8.614	20	1.4544
.1193	.50	.1190	.9929	.1198	8.345	1.007	8.405	10	1.4515
.1222	7° 00′	.1219	.9925	.1228	8.144	1.008	8.206	83° 00′	1.4486
.1251	10	.1248	.9922	.1257	7.953	1.008	8.016	50	1.4457
.1280	20	.1276	.9918	.1287	7.770	1.008	7.834	40	1.4428
.1309	30	.1305	.9914	.1317	7.596	1.009	7.661	30	1.4399
.1338	40	.1334	.9911	.1346	7.429	1.009	7.496	20	1.4370
.1367	50	.1363	.9907	.1376	7.269	1.009	7.337	10	1.4341
.1396	8° 00′	.1392	.9903	.1405	7.115	1.010	7.185	82° 00′	1.4312
.1425	10	.1421	.9899	.1435	6.968	1.010	7.040	50	1.4283
.1454	20	.1449	.9894	.1465	6.827	1.011	6.900	40	1.4254
.1484	30	.1478	.9890	.1495	6.691	1.011	6.765	30	1.4224
.1513	40	.1507	.9886	.1524	6.561	1.012	6.636	20	1.4195
.1542	50	.1536	.9881	.1554	6.435	1.012	6.512	10	1.4166
.1571	9° 00′	.1564	.9877	.1584	6.314	1.012	6.392	81° 00′	1.4137
.1600	10	.1593	.9872	.1614	6.197	1.013	6.277	50	1.4108
.1629	20	.1622	.9868	.1644	6.084	1.013	6.166	40	1.4079
.1658	30	.1650	.9863	.1673	5.976	1.014	6.059	30	1.4050
.1687	40	.1679	.9858	.1703	5.871	1.014	5.955	20	1.4021
.1716	50	.1708	.9853	.1733	5.769	1.015	5.855	10	1.3992
.1745	10° 00′	.1736	.9848	.1763	5.671	1.015	5.759	80° 00′	1.3963
.1774	10	.1765	.9843	.1793	5.576	1.016	5.665	50	1.3934
.1804	20	.1794	.9838	.1823	5.485	1.016	5.575	40	1.3904
.1833	30	.1822	.9833	.1853	5.396	1.017	5.487	30	1.3875
.1862	40	.1851	.9827	.1883	5.309	1.018	5.403	20	1.3846
.1891	50	.1880	.9822	.1914	5.226	1.018	5.320	10	1.3817
.1920	11° 00′	.1908	.9816	.1944	5.145	1.019	5.241	79° 00′	1.3788
.1949	10	.1937	.9811	.1974	5.066	1.019	5.164	50	1.3759
.1978	20	.1965	.9805	.2004	4.989	1.020	5.089	40	1.3730
.2007	30	.1994	.9799	.2035	4.915	1.020	5.016	30	1.3701
		cos x	sin x	cot x	tan x	csc x	sec x	x degrees	x radians

x radians	x degrees	sin x	cos x	tan x	cot x	sec x	csc x		
.2036	40	.2022	.9793	.2065	3.843	1.021	4.945	20	1.3672
.2065	50	.2051	.9787	.2095	4.773	1.022	4.876	10	1.3643
.2094	12° 00'	.2079	.9781	.2126	4.705	1.022	4.810	78° 00'	1.3614
.2123	10	.2108	.9775	.2156	4.638	1.023	4.745	50	1.3584
.2153	20	.2136	.9769	.2186	4.574	1.024	4.682	40	1.3555
.2182	30	.2164	.9763	.2217	4.511	1.025	4.620	30	1.3526
.2211	40	.2193	.9757	.2247	4.449	1.025	4.560	20	1.3497
.2240	50	.2221	.9750	.2278	4.390	1.026	4.502	10	1.3468
.2269	13° 00'	.2250	.9744	.2309	4.331	1.026	4.445	77° 00'	1.3439
.2298	10	.2278	.9737	.2339	4.275	1.027	4.390	50	1.3410
.2327	20	.2306	.9730	.2370	4.219	1.028	4.336	40	1.3381
.2356	30	.2334	.9724	.2401	4.165	1.028	4.284	30	1.3352
.2385	40	.2363	.9717	.2432	4.113	1.029	4.232	20	1.3323
.2414	50	.2391	.9710	.2462	4.061	1.030	4.182	10	1.3294
.2443	14° 00'	.2419	.9703	.2493	4.011	1.031	4.134	76° 00'	1.3265
.2473	10	.2447	.9696	.2524	3.962	1.031	4.086	50	1.3235
.2502	20	.2476	.9689	.2555	3.914	1.032	4.039	40	1.3206
.2531	30	.2404	.9681	.2586	3.867	1.033	3.994	30	1.3177
.2560	40	.2532	.9674	.2617	3.821	1.034	3.950	20	1.3148
.2589	50	.2560	.9667	.2648	3.776	1.034	3.906	10	1.3119
.2618	15° 00'	.2588	.9659	.2679	3.732	1.035	3.864	75° 00'	1.3090
.2647	10	.2616	.9652	.2711	3.689	1.036	3.822	50	1.3061
.2676	20	.2644	.9644	.2732	3.647	1.037	3.782	40	1.3032
.2705	30	.2672	.9636	.2773	3.606	1.038	3.742	30	1.3003
.2734	40	.2700	.9628	.2805	3.566	1.039	3.703	20	1.2974
.2763	50	.2728	.9621	2836	3.526	1.039	3.665	10	1.2945
.2793	16° 00'	.2756	.9613	.2867	3.487	1.040	3.628	74° 00'	1.2915
.2822	10	.2784	.9605	.2899	3.450	1.041	3.592	50	1.2886
.2851	20	.2812	.9596	.2931	3.412	1.042	3.556	40	1.2857
.2880	30	.2840	.9588	.2962	3.376	1.043	3.521	30	1.2828
.2909	40	.2868	.9580	.2994	3.340	1.044	3.487	20	1.2799
.2938	50	.2896	.9572	.3026	3.305	1.045	3.453	10	1.2770
.2967	17° 00'	.2924	.9563	.3057	3.271	1.046	3.420	73° 00'	1.2741
.2996	10	.2952	.9555	.3089	3.237	1.047	3.388	50	1.2712
.3025	20	.2979	.9546	.3121	4.204	1.048	3.356	40	1.2683
.3054	30	.3007	.9537	.3153	3.172	1.049	3.326	30	1.2654
.3083	40	.3035	.9528	.3185	3.140	1.049	3.295	20	1.2625
.3113	50	.3062	.9520	.3217	3.108	1.050	3.265	10	1.2595
.3142	18° 00'	.3090	.9511	.3249	3.078	1.051	3.236	72° 00'	1.2566
.3171	10	.3118	.9502	.3281	3.047	1.052	3.207	50	1.2537
.3200	20	.3145	.9492	.3314	3.018	1.053	3.179	40	1.2508
.3229	30	.3173	.9483	.3346	2.989	1.054	3.152	30	1.2479
.3258	40	.3201	.9474	.3378	2.960	1.056	3.124	20	1.2450
.3287	50	.3228	.9465	.3411	2.932	1.057	3.098	10	1.2421
.3316	19° 00'	.3256	.9455	.3443	2.904	1.058	3.072	71° 00'	1.2392
.3345	10	.3283	.9446	.3476	2.877	1.059	3.046	50	1.2363
.3374	20	.3311	.9436	.3508	2.850	1.060	3.021	40	1.2334
.3403	30	.3338	.9426	.3541	2.824	1.061	2.996	30	1.2305
.3432	40	.3365	.9417	.3574	2.798	1.062	2.971	20	1.2275
.3462	50	.3393	.9407	.3607	2.773	1.063	2.947	10	1.2246
.3491	20° 00'	.3420	.9397	.3640	2.747	1.064	2.924	70° 00'	1.2217
.3520	10	.3448	.9387	.3673	2.723	1.065	2.901	50	1.2188
.3599	20	.3475	.9377	.3706	2.699	1.066	2.878	40	1.2159
.3578	30	.3502	.9367	.3739	2.675	1.068	2.855	30	1.2130
.3607	40	.3529	.9356	.3772	2.651	1.069	2.833	20	1.2101
.3636	50	.3557	.9346	.3805	2.628	1.070	2.812	10	1.2072
.3665	21° 00'	.3584	.9336	.3839	2.605	1.071	2.790	69° 00'	1.2043
.3694	10	.3611	.9325	.3872	2.583	1.072	2.769	50	1.2014
.3723	20	.3638	.9315	.3906	2.560	1.074	2.749	40	1.1985
.3752	30	.3665	.9304	.3939	2.539	1.075	2.729	30	1.1956
.3782	40	.3692	.9293	.3973	2.517	1.076	2.709	20	1.1926
.3811	50	.3719	.9283	.4006	2.496	1.077	2.689	10	1.1897
.3840	22° 00'	.3746	.9272	.4040	2.475	1.079	2.669	68° 00'	1.1868
.3869	10	.3773	.9261	.4074	2.455	1.080	2.650	50	1.1839
.3898	20	.3800	.9250	.4108	2.434	1.081	2.632	40	1.1810
.3927	30	.3827	.9239	.4142	2.414	1.082	2.613	30	1.1781
.3956	40	.3854	.9228	.4176	2.394	1.084	2.595	20	1.1752
.3985	50	.3881	.9216	.4210	2.375	1.085	2.577	10	1.1723
.4014	23° 00'	.3907	.9205	.4245	2.356	1.086	2.559	67° 00'	1.1694
.4043	10	.3934	.9194	.4279	2.337	1.088	2.542	50	1.1665

		cos x	sin x	cot x	tan x	csc x	sec x	x degrees	x radians

x radians	x degrees	sin x	cos x	tan x	cot x	sec x	csc x		
.4072	20	.3961	.9182	.4314	2.318	1.089	2.525	40	1.1636
.4102	30	.3987	.9171	.4348	2.300	1.090	2.508	30	1.1606
4131	40	.4014	.9159	.4383	2.282	1.092	2.491	20	1.1577
.4160	50	4041	.9147	.4417	2.264	1.093	2.475	10	1.1548
.4189	24° 00′	.4067	.9135	.4452	2.246	1.095	2.459	66° 00′	1.1519
.4218	10	.4094	.9124	.4487	2.229	1.096	2.443	50	1.1490
.4247	20	.4120	.9112	.4522	2.211	1.097	2.427	40	1.1461
.4276	30	.4147	.9100	.4557	2.194	1.099	2.411	30	1.1432
.4305	40	.4173	.9088	.4592	2.177	1.100	2.396	20	1.1403
.4334	50	.4200	.9075	.4628	2.161	1.102	2.381	10	1.1374
.4363	25° 00′	.4226	.9063	.4663	2.145	1.103	2.366	65° 00′	1.1345
.4392	10	.4253	.9051	.4699	2.128	1.105	2.352	50	1.1316
.4422	20	.4279	.9038	.4734	2.112	1.106	2.337	40	1.1286
.4451	30	.4305	.9026	.4770	2.097	1.108	2.323	30	1.1257
.4480	40	.4331	.9013	.4806	2.081	1.109	2.309	20	1.1228
.4509	50	.4358	.9001	.4841	2.066	1.111	2.295	10	1.1199
.4538	26° 00′	.4384	.8988	.4877	2.050	1.113	2.281	64° 00′	1.1170
.4567	10	.4410	.8975	.4913	2.035	1.114	2.268	50	1.1141
.4596	20	.4436	.8962	.4950	2.020	1.116	2.254	40	1.1112
.4625	30	.4462	.8949	.4986	2.006	1.117	2.241	30	1.1083
.4654	40	.4488	.8936	.5022	1.991	1.119	2.228	20	1.1054
.4683	50	.4514	.8923	.5059	1.977	1.121	2.215	10	1.1025
.4712	27° 00′	.4540	.8910	.5095	1.963	1.122	2.203	63° 00′	1.0996
.4741	10	.4566	.8897	.5132	1.949	1.124	2.190	50	1.0966
.4771	20	.4592	.8884	.5169	1.935	1.126	2.178	40	1.0937
.4800	30	.4617	.8870	.5206	1.921	1.127	2.166	30	1.0908
.4829	40	.4643	.8857	.5243	1.907	1.129	2.154	20	1.0879
.4858	50	.4669	.8843	.5280	1.894	1.131	2.142	10	1.0850
.4887	28° 00′	.4695	.8829	.5317	1.881	1.133	2.130	62° 00′	1.0821
.4916	10	.4720	.8816	.5354	1.868	1.134	2.118	50	1.0792
.4945	20	.4746	.8802	.5392	1.855	1.136	2.107	40	1.0763
.4974	30	.4772	.8788	.5430	1.842	1.138	2.096	30	1.0734
.5003	40	.4797	.8774	.5467	1.829	1.140	2.085	20	1.0705
.5032	50	.4823	.8760	.5505	1.816	1.142	2.074	10	1.0676
.5061	29° 00′	.4848	.8746	.5543	1.804	1.143	2.063	61° 00′	1.0647
.5091	10	.4874	.8732	.5581	1.792	1.145	2.052	50	1.0617
.5120	20	.4899	.8718	.5619	1.780	1.147	2.041	40	1.0588
.5149	30	.4924	.8704	.5658	1.767	1.149	2.031	30	1.0559
.5178	40	.4950	.8689	.5696	1.756	1.151	2.020	20	1.0530
.5207	50	.4975	.8675	.5735	1.744	1.153	2.010	10	1.0501
.5236	30° 00′	.5000	.8660	.5774	1.732	1.155	2.000	60° 00′	1.0472
.5265	10	.5025	.8646	.5812	1.720	1.157	1.990	50	1.0443
.5294	20	.5050	.8631	.5851	1.709	1.159	1.980	40	1.0414
.5323	30	.5075	.8616	.5890	1.698	1.161	1.970	30	1.0385
.5352	40	.5100	.8601	.5930	1.686	1.163	1.961	20	1.0356
.5381	50	.5125	.8587	.5969	1.675	1.165	1.951	10	1.0327
.5411	31° 00′	.5150	.8572	.6009	1.664	1.167	1.942	59° 00′	1.0297
.5440	10	.5175	.8557	.6048	1.653	1.169	1.932	50	1.0268
.5469	20	.5200	.8542	.6088	1.643	1.171	1.923	40	1.0239
.5498	30	.5225	.8526	.6128	1.632	1.173	1.914	30	1.0210
.5527	40	.5250	.8511	.6168	1.621	1.175	1.905	20	1.0181
.5556	50	.5275	.8496	.6208	1.611	1.177	1.896	10	1.0152
.5585	32° 00′	.5299	.8480	.6249	1.600	1.179	1.887	58° 00′	1.0123
.5614	10	.5324	.8465	.6289	1.590	1.181	1.878	50	1.0094
.5643	20	.5348	.8450	.6330	1.580	1.184	1.870	40	1.0065
.5672	30	.5373	.8434	.6371	1.570	1.186	1.861	30	1.0036
.5701	40	.5398	.8418	.6412	1.560	1.188	1.853	20	1.0007
.5730	50	.5422	.8403	.6453	1.550	1.190	1.844	10	.9977
.5760	33° 00′	.5446	.8397	.6494	1.540	1.192	1.836	57° 00′	.9948
.5789	10	.5471	.8371	.6536	1.530	1.195	1.828	50	.9919
.5818	20	.5495	.8355	.6577	1.520	1.197	1.820	40	.9890
.5847	30	.5519	.8339	.6619	1.511	1.199	1.812	30	.9861
.5876	40	.5544	.8323	.6661	1.501	1.202	1.804	20	.9832
.5905	50	.5568	.8307	.6703	1.492	1.204	1.796	10	.9803
.5934	34° 00′	.5592	.8290	.6745	1.483	1.206	1.788	56° 00′	.9774
.5963	10	.5616	.8274	.6787	1.473	1.209	1.781	50	.9745
.5992	20	.5640	.8258	.6830	1.464	1.211	1.773	40	.9716
.6021	30	.5664	.8241	.6873	1.455	1.213	1.766	30	.9687
.6050	40	.5688	.8225	.6916	1.446	1.216	1.758	20	.9657
.6080	50	.5712	.8208	.6959	1.437	1.218	1.751	10	.9628
		cos x	sin x	cot x	tan x	csc x	sec x	x degrees	x radians

x radians	x degrees	sin x	cos x	tan x	cot x	sec x	csc x		
.6109	35° 00′	.5736	.8192	.7002	1.428	1.221	1.743	55° 00′	.9599
.6138	10	.5760	.8175	.7046	1.419	1.223	1.736	50	.9570
.6167	20	.5783	.8158	.7089	1.411	1.226	1.729	40	.9541
.6196	30	.5807	.8141	.7133	1.402	1.228	1.722	30	.9512
.6225	40	.5831	.8124	.7177	1.393	1.231	1.715	20	.9483
.6254	50	.5854	.8107	.7221	1.385	1.233	1.708	10	.9454
.6283	36° 00′	.5878	.8090	.7265	1.376	1.236	1.701	54° 00′	.9425
.6312	10	.5901	.8073	.7310	1.368	1.239	1.695	50	.9396
.6341	20	.5925	.8056	.7355	1.360	1.241	1.688	40	.9367
.6370	30	.5948	.8039	.7400	1.351	1.244	1.681	30	.9338
.6400	40	.5972	.8021	.7445	1.343	1.247	1.675	20	.9308
.6429	50	.5995	.8004	.7490	1.335	1.249	1.668	10	.9279
.6458	37° 00′	.6018	.7986	.7536	1.327	1.252	1.662	53° 00′	.9250
.6487	10	.6041	.7969	.7581	1.319	1.255	1.655	50	.9221
.6516	20	.6065	.7951	.7627	1.311	1.258	1.649	40	.9192
.6545	30	.6088	.7934	.7673	1.303	1.260	1.643	30	.9163
.6574	40	.6111	.7916	.7720	1.295	1.263	1.636	20	.9134
.6603	50	.6134	.7898	.7766	1.288	1.266	1.630	10	.9105
.6632	38° 00′	.6157	.7880	.7813	1.280	1.269	1.624	52° 00′	.9076
.6661	10	.6180	.7862	.7860	1.272	1.272	1.618	50	.9047
.6690	20	.6202	.7844	.7907	1.265	1.275	1.612	40	.9018
.6720	30	.6225	.7826	.7954	1.257	1.278	1.606	30	.8988
.6749	40	.6248	.7808	.8002	1.250	1.281	1.601	20	.8959
.6778	50	.6271	.7790	.8050	1.242	1.284	1.595	10	.8930
.6807	39° 00′	.6293	.7771	.8098	1.235	1.287	1.589	51° 00′	.8901
.6836	10	.6316	.7753	.8146	1.228	1.290	1.583	50	.8872
.6865	20	.6338	.7735	.8195	1.220	1.293	1.578	40	.8843
.6894	30	.6361	.7716	.8243	1.213	1.296	1.572	30	.8814
.6923	40	.6383	.7698	.8292	1.206	1.299	1.567	20	.8785
.6952	50	.6406	.7679	.8342	1.199	1.302	1.561	10	.8756
.6981	40° 00′	.6428	.7660	.8391	1.192	1.305	1.556	50° 00′	.8727
.7010	10	.6450	.7642	.8441	1.185	1.309	1.550	50	.8698
.7039	20	.6472	.7623	.8491	1.178	1.312	1.545	40	.8668
.7069	30	.6494	.7604	.8541	1.171	1.315	1.540	30	.8639
.7098	40	.6517	.7585	.8591	1.164	1.318	1.535	20	.8610
.7127	50	.6539	.7566	.8642	1.157	1.322	1.529	10	.8581
.7156	41° 00′	.6561	.7547	.8693	1.150	1.325	1.524	49° 00′	.8552
.7185	10	.6583	.7528	.8744	1.144	1.328	1.519	50	.8523
.7214	20	.6604	.7509	.8796	1.137	1.332	1.514	40	.8494
.7243	30	.6626	.7490	.8847	1.130	1.335	1.509	30	.8465
.7272	40	.6648	.7470	.8899	1.124	1.339	1.504	20	.8436
.7301	50	.6670	.7451	.8952	1.117	1.342	1.499	10	.8407
.7330	42° 00′	.6691	.7431	.9004	1.111	1.346	1.494	48° 00′	.8378
.7359	10	.6713	.7412	.9057	1.104	1.349	1.490	50	.8348
.7389	20	.6734	.7392	.9110	1.098	1.353	1.485	40	.8319
.7418	30	.6756	.7373	.9163	1.091	1.356	1.480	30	.8290
.7447	40	.6777	.7353	.9217	1.085	1.360	1.476	20	.8261
.7476	50	.6799	.7333	.9271	1.079	1.364	1.471	10	.8232
.7505	43° 00′	.6820	.7314	.9325	1.072	1.367	1.466	47° 00′	.8203
.7534	10	.6841	.7294	.9380	1.066	1.371	1.462	50	.8174
.7563	20	.6862	.7274	.9435	1.060	1.375	1.457	40	.8145
.7592	30	.6884	.7254	.9490	1.054	1.379	1.453	30	.8116
.7621	40	.6905	.7234	.9545	1.048	1.382	1.448	20	.8087
.7650	50	.6926	.7214	.9601	1.042	1.386	1.444	10	.8058
.7679	44° 00′	.6947	.7193	.9657	1.036	1.390	1.440	46° 00′	.8029
.7709	10	.6967	.7173	.9713	1.030	1.394	1.435	50	.7999
.7738	20	.6988	.7153	.9770	1.024	1.398	1.431	40	.7970
.7767	30	.7009	.7133	.9827	1.018	1.402	1.427	30	.7941
.7796	40	.7030	.7112	.9884	1.012	1.406	1.423	20	.7912
.7825	50	.7050	.7092	.9942	1.006	1.410	1.418	10	.7883
.7854	45° 00′	.7071	.7071	1.0000	1.0000	1.414	1.414	45° 00′	.7854

		cos x	sin x	cot x	tan x	csc x	sec x	x degrees	x radians

TRIGONOMETRIC FUNCTIONS

Function	$\sin\alpha$	$\cos\alpha$	$\tan\alpha$	$\cot\alpha$	$\sec\alpha$	$\csc\alpha$
$\sin\alpha$	$\sin\alpha$	$\pm\sqrt{1-\cos^2\alpha}$	$\dfrac{\tan\alpha}{\pm\sqrt{1+\tan^2\alpha}}$	$\dfrac{1}{\pm\sqrt{1+\cot^2\alpha}}$	$\dfrac{\pm\sqrt{\sec^2\alpha-1}}{\sec\alpha}$	$\dfrac{1}{\csc\alpha}$
$\cos\alpha$	$\pm\sqrt{1-\sin^2\alpha}$	$\cos\alpha$	$\dfrac{1}{\pm\sqrt{1+\tan^2\alpha}}$	$\dfrac{\cot\alpha}{\pm\sqrt{1+\cot^2\alpha}}$	$\dfrac{1}{\sec\alpha}$	$\dfrac{\pm\sqrt{\csc^2\alpha-1}}{\csc\alpha}$
$\tan\alpha$	$\dfrac{\sin\alpha}{\pm\sqrt{1-\sin^2\alpha}}$	$\dfrac{\pm\sqrt{1-\cos^2\alpha}}{\cos\alpha}$	$\tan\alpha$	$\dfrac{1}{\cot\alpha}$	$\pm\sqrt{\sec^2\alpha-1}$	$\dfrac{1}{\pm\sqrt{\csc^2\alpha-1}}$
$\cot\alpha$	$\dfrac{\pm\sqrt{1-\sin^2\alpha}}{\sin\alpha}$	$\dfrac{\cos\alpha}{\pm\sqrt{1-\cos^2\alpha}}$	$\dfrac{1}{\tan\alpha}$	$\cot\alpha$	$\dfrac{1}{\pm\sqrt{\sec^2\alpha-1}}$	$\pm\sqrt{\csc^2\alpha-1}$
$\sec\alpha$	$\dfrac{1}{\pm\sqrt{1-\sin^2\alpha}}$	$\dfrac{1}{\cos\alpha}$	$\pm\sqrt{1+\tan^2\alpha}$	$\dfrac{\pm\sqrt{1+\cot^2\alpha}}{\cot\alpha}$	$\sec\alpha$	$\dfrac{\csc\alpha}{\pm\sqrt{\csc^2\alpha-1}}$
$\csc\alpha$	$\dfrac{1}{\sin\alpha}$	$\dfrac{1}{\pm\sqrt{1-\cos^2\alpha}}$	$\dfrac{\pm\sqrt{1+\tan^2\alpha}}{\tan\alpha}$	$\pm\sqrt{1+\cot^2\alpha}$	$\dfrac{\sec\alpha}{\pm\sqrt{\sec^2\alpha-1}}$	$\csc\alpha$

Note: The choice of sign depends upon the quadrant in which the angle terminates.

HYPERBOLIC FUNCTIONS

Function	$\sinh x$	$\cosh x$	$\tanh x$
$\sinh x =$	$\sinh x$	$+\sqrt{\cosh^2 x-1}$	$\dfrac{\tanh x}{\sqrt{1-\tanh^2 x}}$
$\cosh x =$	$\sqrt{1+\sinh^2 x}$	$\cosh x$	$\dfrac{1}{\sqrt{1-\tanh^2 x}}$
$\tanh x =$	$\dfrac{\sinh x}{\sqrt{1+\sinh^2 x}}$	$\pm\dfrac{\sqrt{\cosh^2 x-1}}{\cosh x}$	$\tanh x$
$\operatorname{cosech} x =$	$\dfrac{1}{\sinh x}$	$\pm\dfrac{1}{\sqrt{\cosh^2 x-1}}$	$\dfrac{\sqrt{1-\tanh^2 x}}{\tanh x}$
$\operatorname{sech} x =$	$\dfrac{1}{\sqrt{1+\sinh^2 x}}$	$\dfrac{1}{\cosh x}$	$\sqrt{1-\tanh^2 x}$
$\coth x =$	$\dfrac{\sqrt{1+\sinh^2 x}}{\sinh x}$	$\pm\dfrac{\cosh x}{\sqrt{\cosh^2 x-1}}$	$\dfrac{1}{\tanh x}$

Function	$\operatorname{cosech} x$	$\operatorname{sech} x$	$\coth x$
$\sinh x =$	$\dfrac{1}{\operatorname{cosech} x}$	$\pm\dfrac{\sqrt{1-\operatorname{sech}^2 x}}{\operatorname{sech} x}$	$\dfrac{\pm 1}{\sqrt{\coth^2 x-1}}$
$\cosh x =$	$\pm\dfrac{\sqrt{\operatorname{cosech}^2 x+1}}{\operatorname{cosech} x}$	$\dfrac{1}{\operatorname{sech} x}$	$\pm\dfrac{\coth x}{\sqrt{\coth^2 x-1}}$
$\tanh x =$	$\dfrac{1}{\sqrt{\operatorname{cosech}^2 x+1}}$	$\pm\sqrt{1-\operatorname{sech}^2 x}$	$\dfrac{1}{\coth x}$
$\operatorname{cosech} x =$	$\operatorname{cosech} x$	$\pm\dfrac{\operatorname{sech} x}{\sqrt{1-\operatorname{sech}^2 zsx}}$	$\pm\dfrac{\sqrt{\coth^2 x-1}}{1}$
$\operatorname{sech} x =$	$\pm\dfrac{\operatorname{cosec} x}{\sqrt{\operatorname{cosech}^2 x+1}}$	$\operatorname{sech} x$	$\pm\dfrac{\sqrt{\coth^2 x-1}}{\coth x}$
$\coth x =$	$\sqrt{\operatorname{cosech}^2 x+1}$	$\pm\dfrac{1}{\sqrt{1-\operatorname{sech}^2 x}}$	$\coth x$

Whenever two signs are shown, choose + sign if x is positive, − sign if x is negative.

Derivatives*

In the following formulas u, v, w represent functions of x, while a, c, n represent fixed real numbers. All arguments in the trigonometric functions are measured in radians, and all inverse trigonometric and hyperbolic functions represent principal values.

1. $\dfrac{d}{dx}(a) = 0$

2. $\dfrac{d}{dx}(x) = 1$

3. $\dfrac{d}{dx}(au) = a\dfrac{du}{dx}$

4. $\dfrac{d}{dx}(u + v - w) = \dfrac{du}{dx} + \dfrac{dv}{dx} - \dfrac{dw}{dx}$

5. $\dfrac{d}{dx}(uv) = u\dfrac{dv}{dx} + v\dfrac{du}{dx}$

6. $\dfrac{d}{dx}(uvw) = uv\dfrac{dw}{dx} + vw\dfrac{du}{dx} + uw\dfrac{dv}{dx}$

7. $\dfrac{d}{dx}\left(\dfrac{u}{v}\right) = \dfrac{v\dfrac{du}{dx} - u\dfrac{dv}{dx}}{v^2} = \dfrac{1}{v}\dfrac{du}{dx} - \dfrac{u}{v^2}\dfrac{dv}{dx}$

8. $\dfrac{d}{dx}(u^n) = nu^{n-1}\dfrac{du}{dx}$

9. $\dfrac{d}{dx}(\sqrt{u}) = \dfrac{1}{2\sqrt{u}}\dfrac{du}{dx}$

10. $\dfrac{d}{dx}\left(\dfrac{1}{u}\right) = -\dfrac{1}{u^2}\dfrac{du}{dx}$

11. $\dfrac{d}{dx}\left(\dfrac{1}{u^n}\right) = -\dfrac{n}{u^{n+1}}\dfrac{du}{dx}$

12. $\dfrac{d}{dx}\left(\dfrac{u^n}{v^m}\right) = \dfrac{u^{n-1}}{v^{m+1}}\left(nv\dfrac{du}{dx} - mu\dfrac{dv}{dx}\right)$

13. $\dfrac{d}{dx}(u^n v^m) = u^{n-1}v^{m-1}\left(nv\dfrac{du}{dx} + mu\dfrac{dv}{dx}\right)$

14. $\dfrac{d}{dx}[f(u)] = \dfrac{d}{du}[f(u)] \cdot \dfrac{du}{dx}$

* Let $y = f(x)$ and $\dfrac{dy}{dx} = \dfrac{d[f(x)]}{dx} = f'(x)$ define respectively a function and its derivative for any value x in their common domain. The differential for the function at such a value x is accordingly defined as

$$dy = d[f(x)] = \dfrac{dy}{dx}dx = \dfrac{d[f(x)]}{dx}dx = f'(x)\,dx$$

Each derivative formula has an associated differential formula. For example, formula 6 above has the differential formula

$$d(uvw) = uv\,dw + vw\,du + uw\,dv$$

15. $\dfrac{d^2}{dx^2}[f(u)] = \dfrac{df(u)}{du} \cdot \dfrac{d^2u}{dx^2} + \dfrac{d^2f(u)}{du^2} \cdot \left(\dfrac{du}{dx}\right)^2$

16. $\dfrac{d^n}{dx^n}[uv] = \dbinom{n}{0}v\dfrac{d^nu}{dx^n} + \dbinom{n}{1}\dfrac{dv}{dx}\dfrac{d^{n-1}u}{dx^{n-1}} + \dbinom{n}{2}\dfrac{d^2v}{dx^2}\dfrac{d^{n-2}u}{dx^{n-2}}$

$$+ \cdots + \dbinom{n}{k}\dfrac{d^kv}{dx^k}\dfrac{d^{n-k}u}{dx^{n-k}} + \cdots + \dbinom{n}{n}u\dfrac{d^nv}{dx^n}$$

where $\dbinom{n}{r} = \dfrac{n!}{r'(n-r)'}$ the binomial coefficient, n non-negative integer and $\dbinom{n}{0} = 1$.

17. $\dfrac{du}{dx} = \dfrac{1}{\dfrac{dx}{du}} \qquad \text{if } \dfrac{dx}{du} \neq 0$

18. $\dfrac{d}{dx}(\log_a u) = (\log_a e)\dfrac{1}{u}\dfrac{du}{dx}$

19. $\dfrac{d}{dx}(\log_e u) = \dfrac{1}{u}\dfrac{du}{dx}$

20. $\dfrac{d}{dx}(a^u) = a^u(\log_e a)\dfrac{du}{dx}$

21. $\dfrac{d}{dx}(e^u) = e^u\dfrac{du}{dx}$

22. $\dfrac{d}{dx}(u^v) = vu^{v-1}\dfrac{du}{dx} + (\log_e u)u^v\dfrac{dv}{dx}$

23. $\dfrac{d}{dx}(\sin u) = \dfrac{du}{dx}(\cos u)$

24. $\dfrac{d}{dx}(\cos u) = -\dfrac{du}{dx}(\sin u)$

25. $\dfrac{d}{dx}(\tan u) = \dfrac{du}{dx}(\sec^2 u)$

26. $\dfrac{d}{dx}(\cot u) = -\dfrac{du}{dx}(\csc^2 u)$

27. $\dfrac{d}{dx}(\sec u) = \dfrac{du}{dx}\sec u \cdot \tan u$

28. $\dfrac{d}{dx}(\csc u) = -\dfrac{du}{dx}\csc u \cdot \cot u$

29. $\dfrac{d}{dx}(\text{vers } u) = \dfrac{du}{dx}\sin u$

30. $\dfrac{d}{dx}(\arcsin u) = \dfrac{1}{\sqrt{1-u^2}}\dfrac{du}{dx}, \qquad \left(-\dfrac{\pi}{2} \leq \arcsin u \leq \dfrac{\pi}{2}\right)$

31. $\dfrac{d}{dx}(\text{arc cos } u) = -\dfrac{1}{\sqrt{1 - u^2}}\dfrac{du}{dx},$ $(0 \le \text{arc cos } u \le \pi)$

32. $\dfrac{d}{dx}(\text{arc tan } u) = \dfrac{1}{1 + u^2}\dfrac{du}{dx},$ $\left(-\dfrac{\pi}{2} < \text{arc tan } u < \dfrac{\pi}{2}\right)$

33. $\dfrac{d}{dx}(\text{arc cot } u) = -\dfrac{1}{1 + u^2}\dfrac{du}{dx},$ $(0 \le \text{arc cot } u \le \pi)$

34. $\dfrac{d}{dx}(\text{arc sec } u) = \dfrac{1}{u\sqrt{u^2 - 1}}\dfrac{du}{dx},$ $\left(0 \le \text{arc sec } u < \dfrac{\pi}{2}, -\pi \le \text{arc sec } u < -\dfrac{\pi}{2}\right)$

35. $\dfrac{d}{dx}(\text{arc csc } u) = -\dfrac{1}{u\sqrt{u^2 - 1}}\dfrac{du}{dx},$ $\left(0 < \text{arc csc } u \le \dfrac{\pi}{2}, -\pi < \text{arc csc } u \le -\dfrac{\pi}{2}\right)$

36. $\dfrac{d}{dx}(\text{arc vers } u) = \dfrac{1}{\sqrt{2u - u^2}}\dfrac{du}{dx},$ $(0 \le \text{arc vers } u \le \pi)$

37. $\dfrac{d}{dx}(\sinh u) = \dfrac{du}{dx}(\cosh u)$

38. $\dfrac{d}{dx}(\cosh u) = \dfrac{du}{dx}(\sinh u)$

39. $\dfrac{d}{dx}(\tanh u) = \dfrac{du}{dx}(\text{sech}^2 u)$

40. $\dfrac{d}{dx}(\coth u) = -\dfrac{du}{dx}(\text{csch}^2 u)$

41. $\dfrac{d}{dx}(\text{sech } u) = -\dfrac{du}{dx}(\text{sech } u \cdot \tanh u)$

42. $\dfrac{d}{dx}(\text{csch } u) = -\dfrac{du}{dx}(\text{csch } u \cdot \coth u)$

43. $\dfrac{d}{dx}(\sinh^{-1} u) = \dfrac{d}{dx}[\log(u + \sqrt{u^2 + 1})] = \dfrac{1}{\sqrt{u^2 + 1}}\dfrac{du}{dx}$

44. $\dfrac{d}{dx}(\cosh^{-1} u) = \dfrac{d}{dx}[\log(u + \sqrt{u^2 - 1})] = \dfrac{1}{\sqrt{u^2 - 1}}\dfrac{du}{dx},$ $(u > 1, \cosh^{-1} u > 0)$

45. $\dfrac{d}{dx}(\tanh^{-1} u) = \dfrac{d}{dx}\left[\dfrac{1}{2}\log\dfrac{1 + u}{1 - u}\right] = \dfrac{1}{1 - u^2}\dfrac{du}{dx},$ $(u^2 < 1)$

46. $\dfrac{d}{dx}(\coth^{-1} u) = \dfrac{d}{dx}\left[\dfrac{1}{2}\log\dfrac{u + 1}{u - 1}\right] = \dfrac{1}{1 - u^2}\dfrac{du}{dx},$ $(u^2 > 1)$

47. $\dfrac{d}{dx}(\text{sech}^{-1} u) = \dfrac{d}{dx}\left[\log\dfrac{1 + \sqrt{1 - u^2}}{u}\right] = -\dfrac{1}{u\sqrt{1 - u^2}}\dfrac{du}{dx},$ $(0 < u < 1, \text{sech}^{-1} u > 0)$

48. $\dfrac{d}{dx}(\text{csch}^{-1} u) = \dfrac{d}{dx}\left[\log\dfrac{1 + \sqrt{1 + u^2}}{u}\right] = -\dfrac{1}{|u|\sqrt{1 + u^2}}\dfrac{du}{dx}$

49. $\dfrac{d}{dq} \displaystyle\int_p^q f(x)\,dx = f(q),$ [p constant]

50. $\dfrac{d}{dp} \displaystyle\int_p^q f(x)\,dx = -f(p),$ [q constant]

51. $\dfrac{d}{da} \displaystyle\int_p^q f(x,a)\,dx = \int_p^q \dfrac{\partial}{\partial a}[f(x,a)]\,dx + f(q,a)\dfrac{dq}{da} - f(p,a)\dfrac{dp}{da}$

INTEGRATION

The following is a brief discussion of some integration techniques. A more complete discussion can be found in a number of good text books. However, the purpose of this introduction is simply to discuss a few of the important techniques which may be used, in conjunction with the integral table which follows, to integrate particular functions.

No matter how extensive the integral table, it is a fairly uncommon occurrence to find in the table the exact integral desired. Usually some form of transformation will have to be made. The simplest type of transformation, and yet the most general, is substitution. Simple forms of substitution, such as $y = ax$, are employed almost unconsciously by experienced users of integral tables. Other substitutions may require more thought. In some sections of the tables, appropriate substitutions are suggested for integrals which are similar to, but not exactly like, integrals in the table. Finding the right substitution is largely a matter of intuition and experience.

Several precautions must be observed when using substitutions:

1. Be sure to make the substitution in the dx term, as well as everywhere else in the integral.
2. Be sure that the function substituted is one-to-one and continuous. If this is not the case, the integral must be restricted in such a way as to make it true. See the example following.
3. With definite integrals, the limits should also be expressed in terms of the new dependent variable. With indefinite integrals, it is necessary to perform the reverse substitution to obtain the answer in terms of the original independent variable. This may also be done for definite integrals, but it is usually easier to change the limits.

Example:

$$\int \frac{x^4}{\sqrt{a^2 - x^2}}\,dx$$

Here we make the substitution $x = |a| \sin \theta$. Then $dx = |a| \cos \theta\, d\theta$, and

$$\sqrt{a^2 - x^2} = \sqrt{a^2 - a^2 \sin^2 \theta} = |a|\sqrt{1 - \sin^2 \theta} = |a \cos \theta|$$

Notice the absolute value signs. It is very important to keep in mind that a square root radical always denotes the positive square root, and to assure the sign is always kept positive. Thus $\sqrt{x^2} = |x|$. Failure to observe this is a common cause of errors in integration.

Notice also that the indicated substitution is not a one-to-one function, that is, it does not have a unique inverse. Thus we must restrict the range of θ in such a way as to make the function one-to-one. Fortunately, this is easily done by solving for θ

$$\theta = \sin^{-1} \frac{x}{|a|}$$

and restricting the inverse sine to the principal values, $-\dfrac{\pi}{2} \le \theta \le \dfrac{\pi}{2}$.

Thus the integral becomes

$$\int \frac{a^4 \sin^4 \theta |a| \cos \theta \, d\theta}{|a| |\cos \theta|}$$

Now, however, in the range of values chosen for θ, $\cos \theta$ is always positive. Thus we may remove the absolute value signs from $\cos \theta$ in the denominator. (This is one of the reasons that the principal values of the inverse trigonometric functions are defined as they are.)

Then the $\cos \theta$ terms cancel, and the integral becomes

$$a^4 \int \sin^4 \theta \, d\theta$$

By application of integral formulas 299 and 296, we integrate this to

$$-a^4 \frac{\sin^3 \theta \cos \theta}{4} - \frac{3a^4}{8} \cos \theta \sin \theta + \frac{3a^4}{8} \theta + C$$

We now must perform the inverse substitution to get the result in terms of x. We have

$$\theta = \sin^{-1} \frac{x}{|a|}$$

$$\sin \theta = \frac{x}{|a|}$$

Then

$$\cos \theta = \pm\sqrt{1 - \sin^2 \theta} = \pm\sqrt{1 - \frac{x^2}{a^2}} = \pm\frac{\sqrt{a^2 - x^2}}{|a|}.$$

Because of the previously mentioned fact that $\cos \theta$ is positive, we may omit the $\pm$ sign. The reverse substitution then produces the final answer

$$\int \frac{x^4}{\sqrt{a^2 - x^2}} \, dx = -\tfrac{1}{4}x^3\sqrt{a^2 - x^2} - \tfrac{3}{8}a^2 x\sqrt{a^2 - x^2} + \frac{3a^4}{8} \sin^{-1} \frac{x}{|a|} + C.$$

Any rational function of x may be integrated, if the denominator is factored into linear and irreducible quadratic factors. The function may then be broken into partial fractions, and the individual partial fractions integrated by use of the appropriate formula from the integral table. See the section on partial fractions for further information.

Many integrals may be reduced to rational functions by proper substitutions. For example,

$$z = \tan \frac{x}{2}$$

will reduce any rational function of the six trigonometric functions of x to a rational function of z. (Frequently there are other substitutions which are simpler to use, but this one will always work. See integral formula number 484.)

Any rational function of x and $\sqrt{ax + b}$ may be reduced to a rational function of z by making the substitution

$$z = \sqrt{ax + b}.$$

Other likely substitutions will be suggested by looking at the form of the integrand.

The other main method of transforming integrals is integration by parts. This involves applying formula number 5 or 6 in the accompanying integral table. The critical factor in this method is the choice of the functions u and v. In order for the method to be successful, $v = \int dv$ and $\int v \, du$ must be easier to integrate than the original integral. Again, this choice is largely a matter of intuition and experience.

Example:

$$\int x \sin x \, dx$$

Two obvious choices are $u = x$, $dv = \sin x \, dx$, or $u = \sin x$, $dv = x \, dx$. Since a preliminary mental calculation indicates that $\int v \, du$ in the second choice would be more, rather than less, complicated than the original integral (it would contain x^2), we use the first choice.

$$u = x \qquad\qquad du = dx$$
$$dv = \sin x \, dx \qquad\qquad v = -\cos x$$

$$\int x \sin x \, dx = \int u \, dv = uv - \int v \, du = -x \cos x + \int \cos x \, dx$$
$$= \sin x - x \cos x$$

Of course, this result could have been obtained directly from the integral table, but it provides a simple example of the method. In more complicated examples the choice of u and v may not be so obvious, and several different choices may have to be tried. Of course, there is no guarantee that any of them will work.

Integration by parts may be applied more than once, or combined with substitution. A fairly common case is illustrated by the following example.

Example:

$$\int e^x \sin x \, dx$$

Let

$$u = e^x \qquad \text{Then} \quad du = e^x \, dx$$
$$dv = \sin x \, dx \qquad\qquad v = -\cos x$$

$$\int e^x \sin x \, dx = \int u \, dv = uv - \int v \, du = -e^x \cos x + \int e^x \cos x \, dx$$

In this latter integral,

$$\text{let} \quad u = e^x \qquad\qquad \text{Then} \quad du = e^x \, dx$$
$$dv = \cos x \, dx \qquad\qquad v = \sin x$$

$$\int e^x \sin x \, dx = -e^x \cos x + \int e^x \cos x \, dx = -e^x \cos x + \int u \, dv$$

$$= -e^x \cos x + uv - \int v \, du$$

$$= -e^x \cos x + e^x \sin x - \int e^x \sin x \, dx$$

This looks as if a circular transformation has taken place, since we are back at the same integral we started from. However, the above equation can be solved algebraically for the required integral:

$$\int e^x \sin x \, dx = \tfrac{1}{2}(e^x \sin x - e^x \cos x)$$

In the second integration by parts, if the parts had been chosen as $u = \cos x$, $dv = e^x \, dx$, we would indeed have made a circular transformation, and returned to the starting place.

In general, when doing repeated integration by parts, one should never choose the function u at any stage to be the same as the function v at the previous stage, or a constant times the previous v.

The following rule is called the extended rule for integration by parts. It is the result of $n + 1$ successive applications of integration by parts.

If

$$g_1(x) = \int g(x)\,dx, \qquad g_2(x) = \int g_1(x)\,dx,$$

$$g_3(x) = \int g_2(x)\,dx, \ldots, g_m(x) = \int g_{m-1}(x)\,dx, \ldots,$$

then

$$\int f(x) \cdot g(x)\,dx = f(x) \cdot g_1(x) - f'(x) \cdot g_2(x) + f''(x) \cdot g_3(x) - + \cdots$$

$$+ (-1)^n f^{(n)}(x)g_{n+1}(x) + (-1)^{n+1}\int f^{(n+1)}(x)g_{n+1}(x)\,dx.$$

A useful special case of the above rule is when $f(x)$ is a polynomial of degree n. Then $f^{(n+1)}(x) = 0$, and

$$\int f(x) \cdot g(x)\,dx = f(x) \cdot g_1(x) - f'(x) \cdot g_2(x) + f''(x) \cdot g_3(x) - + \cdots + (-1)^n f^{(n)}(x)g_{n+1}(x) + C$$

Example:

If $f(x) = x^2, g(x) = \sin x$

$$\int x^2 \sin x\,dx = -x^2 \cos x + 2x \sin x + 2 \cos x + C$$

Another application of this formula occurs if

$$f''(x) = af(x) \quad \text{and} \quad g''(x) = bg(x),$$

where a and b are unequal constants. In this case, by a process similar to that used in the above example for $\int e^x \sin x\,dx$, we get the formula

$$\int f(x)g(x)\,dx = \frac{f(x) \cdot g'(x) - f'(x) \cdot g(x)}{b - a} + C$$

This formula could have been used in the example mentioned. Here is another example.

Example:

If $f(x) = e^{2x}, g(x) = \sin 3x$, then $a = 4, b = -9$, and

$$\int e^{2x} \sin 3x\,dx = \frac{3\,e^{2x} \cos 3x - 2\,e^{2x} \sin 3x}{-9 - 4} + C = \frac{e^{2x}}{13}(2 \sin 3x - 3 \cos 3x) + C$$

The following additional points should be observed when using this table.

1. A constant of integration is to be supplied with the answers for indefinite integrals.
2. Logarithmic expressions are to base $e = 2.71828\ldots$, unless otherwise specified, and are to be evaluated for the absolute value of the arguments involved therein.
3. All angles are measured in radians, and inverse trigonometric and hyperbolic functions represent principal values, unless otherwise indicated.
4. If the application of a formula produces either a zero denominator or the square root of a negative number in the result, there is usually available another form of the answer which

avoids this difficulty. In many of the results, the excluded values are specified, but when such are omitted it is presumed that one can tell what these should be, especially when difficulties of the type herein mentioned are obtained.

5. When inverse trigonometric functions occur in the integrals, be sure that any replacements made for them are strictly in accordance with the rules for such functions. This causes little difficulty when the argument of the inverse trigonometric function is positive, since then all angles involved are in the first quadrant. However, if the argument is negative, special care must be used. Thus if $u > 0$,

$$\sin^{-1} u = \cos^{-1}\sqrt{1 - u^2} = \csc^{-1}\frac{1}{u}, \text{ etc.}$$

However, if $u < 0$,

$$\sin^{-1} u = -\cos^{-1}\sqrt{1 - u^2} = -\pi - \csc^{-1}\frac{1}{u}, \text{ etc.}$$

See the section on inverse trigonometric functions for a full treatment of the allowable substitutions.

6. In integrals 340–345 and some others, the right side includes expressions of the form

$$A \tan^{-1}[B + C \tan f(x)].$$

In these formulas, the $\tan^{-1}$ does not necessarily represent the principal value. Instead of always employing the principal branch of the inverse tangent function, one must instead use that branch of the inverse tangent function upon which $f(x)$ lies for any particular choice of x.

Example:

$$\int_0^{4\pi} \frac{dx}{2 + \sin x} = \frac{2}{\sqrt{3}}\tan^{-1}\frac{2\tan\frac{x}{2} + 1}{\sqrt{3}}\Bigg]_0^{4\pi}$$

$$= \frac{2}{\sqrt{3}}\left[\tan^{-1}\frac{2\tan 2\pi + 1}{\sqrt{3}} - \tan^{-1}\frac{2\tan 0 + 1}{\sqrt{3}}\right]$$

$$= \frac{2}{\sqrt{3}}\left[\frac{13\pi}{6} - \frac{\pi}{6}\right] = \frac{4\pi}{\sqrt{3}} = \frac{4\sqrt{3}\pi}{3}$$

Here

$$\tan^{-1}\frac{2\tan 2\pi + 1}{\sqrt{3}} = \tan^{-1}\frac{1}{\sqrt{3}} = \frac{13\pi}{6},$$

since $f(x) = 2\pi$; and

$$\tan^{-1}\frac{2\tan 0 + 1}{\sqrt{3}} = \tan^{-1}\frac{1}{\sqrt{3}} = \frac{\pi}{6},$$

since $f(x) = 0$.

7. B_n and E_n where used in Integrals represents the Bernoulli and Euler numbers as defined in tables of Bernoulli and Euler polynomials contained in certain mathematics reference and handbooks, as for example, Beyer, W. H., *Handbook of Mathematical Sciences,* 5th ed., CRC Press, Inc., West Palm Beach 1978, 577–583.

ELEMENTARY FORMS

1. $\displaystyle\int a\,dx = ax$

2. $\displaystyle\int a \cdot f(x)\,dx = a\int f(x)\,dx$

3. $\displaystyle\int \phi(y)\,dx = \int \frac{\phi(y)}{y'}\,dy,$ where $y' = \dfrac{dy}{dx}$

4. $\displaystyle\int (u + v)\,dx = \int u\,dx + \int v\,dx,$ where u and v are any functions of x

5. $\displaystyle\int u\,dv = u\int dv - \int v\,du = uv - \int v\,du$

6. $\displaystyle\int u\frac{dv}{dx}\,dx = uv - \int v\frac{du}{dx}\,dx$

7. $\displaystyle\int x^n\,dx = \frac{x^{n+1}}{n+1},$ except $n = -1$

8. $\displaystyle\int \frac{f'(x)\,dx}{f(x)} = \log f(x),$ $(df(x) = f'(x)\,dx)$

9. $\displaystyle\int \frac{dx}{x} = \log x$

10. $\displaystyle\int \frac{f'(x)\,dx}{2\sqrt{f(x)}} = \sqrt{f(x)},$ $(df(x) = f'(x)\,dx)$

11. $\displaystyle\int e^x\,dx = e^x$

12. $\displaystyle\int e^{ax}\,dx = e^{ax}/a$

13. $\displaystyle\int b^{ax}\,dx = \frac{b^{ax}}{a\log b},$ $(b > 0)$

14. $\displaystyle\int \log x\,dx = x\log x - x$

15. $\displaystyle\int a^x\log a\,dx = a^x,$ $(a > 0)$

16. $\displaystyle\int \frac{dx}{a^2 + x^2} = \frac{1}{a}\tan^{-1}\frac{x}{a}$

17. $\displaystyle\int \frac{dx}{a^2 - x^2} = \begin{cases} \dfrac{1}{a}\tanh^{-1}\dfrac{x}{a} \\ \text{or} \\ \dfrac{1}{2a}\log\dfrac{a+x}{a-x}, \quad (a^2 > x^2) \end{cases}$

18. $\displaystyle\int \frac{dx}{x^2 - a^2} = \begin{cases} -\dfrac{1}{a}\coth^{-1}\dfrac{x}{a} \\ \text{or} \\ \dfrac{1}{2a}\log\dfrac{x-a}{x+a}, \quad (x^2 > a^2) \end{cases}$

19. $\displaystyle\int \frac{dx}{\sqrt{a^2 - x^2}} = \begin{cases} \sin^{-1}\dfrac{x}{|a|} \\ \quad\text{or} \\ -\cos^{-1}\dfrac{x}{|a|}, \quad (a^2 > x^2) \end{cases}$

20. $\displaystyle\int \frac{dx}{\sqrt{x^2 \pm a^2}} = \log(x + \sqrt{x^2 \pm a^2})$

21. $\displaystyle\int \frac{dx}{x\sqrt{x^2 - a^2}} = \frac{1}{|a|}\sec^{-1}\frac{x}{a}$

22. $\displaystyle\int \frac{dx}{x\sqrt{a^2 \pm x^2}} = -\frac{1}{a}\log\left(\frac{a + \sqrt{a^2 \pm x^2}}{x}\right)$

FORMS CONTAINING $(a + bx)$

For forms containing $a + bx$, but not listed in the table, the substitution $u = \dfrac{a + bx}{x}$ may prove helpful.

23. $\displaystyle\int (a + bx)^n \, dx = \frac{(a + bx)^{n+1}}{(n + 1)b}, \quad (n \neq -1)$

24. $\displaystyle\int x(a + bx)^n \, dx$

$$= \frac{1}{b^2(n + 2)}(a + bx)^{n+2} - \frac{a}{b^2(n + 1)}(a + bx)^{n+1}, \quad (n \neq -1, -2)$$

25. $\displaystyle\int x^2(a + bx)^n \, dx = \frac{1}{b^3}\left[\frac{(a + bx)^{n+3}}{n + 3} - 2a\frac{(a + bx)^{n+2}}{n + 2} + a^2\frac{(a + bx)^{n+1}}{n + 1}\right]$

26. $\displaystyle\int x^m(a + bx)^n \, dx = \begin{cases} \dfrac{x^{m+1}(a + bx)^n}{m + n + 1} + \dfrac{an}{m + n + 1}\displaystyle\int x^m(a + bx)^{n-1} \, dx \\ \qquad\qquad\text{or} \\ \dfrac{1}{a(n + 1)}\bigg[-x^{m+1}(a + bx)^{n+1} \\ \qquad\qquad\qquad + (m + n + 2)\displaystyle\int x^m(a + bx)^{n+1} \, dx\bigg] \\ \qquad\qquad\text{or} \\ \dfrac{1}{b(m + n + 1)}\bigg[x^m(a + bx)^{n+1} - ma\displaystyle\int x^{m-1}(a + bx)^n \, dx\bigg] \end{cases}$

27. $\displaystyle\int \frac{dx}{a + bx} = \frac{1}{b}\log(a + bx)$

28. $\displaystyle\int \frac{dx}{(a + bx)^2} = -\frac{1}{b(a + bx)}$

29. $\displaystyle\int \frac{dx}{(a + bx)^3} = -\frac{1}{2b(a + bx)^2}$

30. $\displaystyle\int \frac{x \, dx}{a + bx} = \begin{cases} \dfrac{1}{b^2}[a + bx - a\log(a + bx)] \\ \qquad\text{or} \\ \dfrac{x}{b} - \dfrac{a}{b^2}\log(a + bx) \end{cases}$

31. $\displaystyle\int \frac{x\,dx}{(a+bx)^2} = \frac{1}{b^2}\left[\log(a+bx) + \frac{a}{a+bx}\right]$

32. $\displaystyle\int \frac{x\,dx}{(a+bx)^n} = \frac{1}{b^2}\left[\frac{-1}{(n-2)(a+bx)^{n-2}} + \frac{a}{(n-1)(a+bx)^{n-1}}\right], \qquad n \neq 1,2$

33. $\displaystyle\int \frac{x^2\,dx}{a+bx} = \frac{1}{b^3}\left[\frac{1}{2}(a+bx)^2 - 2a(a+bx) + a^2\log(a+bx)\right]$

34. $\displaystyle\int \frac{x^2\,dx}{(a+bx)^2} = \frac{1}{b^3}\left[a+bx - 2a\log(a+bx) - \frac{a^2}{a+bx}\right]$

35. $\displaystyle\int \frac{x^2\,dx}{(a+bx)^3} = \frac{1}{b^3}\left[\log(a+bx) + \frac{2a}{a+bx} - \frac{a^2}{2(a+bx)^2}\right]$

36. $\displaystyle\int \frac{x^2\,dx}{(a+bx)^n} = \frac{1}{b^3}\left[\frac{-1}{(n-3)(a+bx)^{n-3}}\right.$

$$\left. + \frac{2a}{(n-2)(a+bx)^{n-2}} - \frac{a^2}{(n-1)(a+bx)^{n-1}}\right], \qquad n \neq 1,2,3$$

37. $\displaystyle\int \frac{dx}{x(a+bx)} = -\frac{1}{a}\log\frac{a+bx}{x}$

38. $\displaystyle\int \frac{dx}{x(a+bx)^2} = \frac{1}{a(a+bx)} - \frac{1}{a^2}\log\frac{a+bx}{x}$

39. $\displaystyle\int \frac{dx}{x(a+bx)^3} = \frac{1}{a^3}\left[\frac{1}{2}\left(\frac{2a+bx}{a+bx}\right)^2 + \log\frac{x}{a+bx}\right]$

40. $\displaystyle\int \frac{dx}{x^2(a+bx)} = -\frac{1}{ax} + \frac{b}{a^2}\log\frac{a+bx}{x}$

41. $\displaystyle\int \frac{dx}{x^3(a+bx)} = \frac{2bx-a}{2a^2x^2} + \frac{b^2}{a^3}\log\frac{x}{a+bx}$

42. $\displaystyle\int \frac{dx}{x^2(a+bx)^2} = -\frac{a+2bx}{a^2x(a+bx)} + \frac{2b}{a^3}\log\frac{a+bx}{x}$

FORMS CONTAINING $c^2 \pm x^2$, $x^2 - c^2$

43. $\displaystyle\int \frac{dx}{c^2+x^2} = \frac{1}{c}\tan^{-1}\frac{x}{c}$

44. $\displaystyle\int \frac{dx}{c^2-x^2} = \frac{1}{2c}\log\frac{c+x}{c-x}, \qquad (c^2 > x^2)$

45. $\displaystyle\int \frac{dx}{x^2-c^2} = \frac{1}{2c}\log\frac{x-c}{x+c}, \qquad (x^2 > c^2)$

46. $\displaystyle\int \frac{x\,dx}{c^2 \pm x^2} = \pm\frac{1}{2}\log(c^2 \pm x^2)$

47. $\displaystyle\int \frac{x\,dx}{(c^2 \pm x^2)^{n+1}} = \mp\frac{1}{2n(c^2 \pm x^2)^n}$

48. $\displaystyle\int \frac{dx}{(c^2 \pm x^2)^n} = \frac{1}{2c^2(n-1)}\left[\frac{x}{(c^2 \pm x^2)^{n-1}} + (2n-3)\int \frac{dx}{(c^2 \pm x^2)^{n-1}}\right]$

49. $\displaystyle\int \frac{dx}{(x^2 - c^2)^n} = \frac{1}{2c^2(n-1)}\left[-\frac{x}{(x^2 - c^2)^{n-1}} - (2n-3)\int \frac{dx}{(x^2 - c^2)^{n-1}}\right]$

50. $\displaystyle\int \frac{x\,dx}{x^2-c^2} = \frac{1}{2}\log(x^2 - c^2)$

51. $\int \dfrac{x\,dx}{(x^2 - c^2)^{n+1}} = -\dfrac{1}{2n(x^2 - c^2)^n}$

FORMS CONTAINING $a + bx$ and $c + dx$

$$u = a + bx, \qquad v = c + dx, \qquad k = ad - bc$$

If $k = 0$, then $v = \dfrac{c}{a}u$

52. $\int \dfrac{dx}{u \cdot v} = \dfrac{1}{k} \cdot \log\left(\dfrac{v}{u}\right)$

53. $\int \dfrac{x\,dx}{u \cdot v} = \dfrac{1}{k}\left[\dfrac{a}{b}\log(u) - \dfrac{c}{d}\log(v)\right]$

54. $\int \dfrac{dx}{u^2 \cdot v} = \dfrac{1}{k}\left(\dfrac{1}{u} + \dfrac{d}{k}\log\dfrac{v}{u}\right)$

55. $\int \dfrac{x\,dx}{u^2 \cdot v} = \dfrac{-a}{bku} - \dfrac{c}{k^2}\log\dfrac{v}{u}$

56. $\int \dfrac{x^2\,dx}{u^2 \cdot v} = \dfrac{a^2}{b^2 ku} + \dfrac{1}{k^2}\left[\dfrac{c^2}{d}\log(v) + \dfrac{a(k - bc)}{b^2}\log(u)\right]$

57. $\int \dfrac{dx}{u^n \cdot v^m} = \dfrac{1}{k(m-1)}\left[\dfrac{-1}{u^{n-1} \cdot v^{m-1}} - (m + n - 2)b\int \dfrac{dx}{u^n \cdot v^{m-1}}\right]$

58. $\int \dfrac{u}{v}\,dx = \dfrac{bx}{d} + \dfrac{k}{d^2}\log(v)$

59. $\int \dfrac{u^m\,dx}{v^n} = \begin{cases} \dfrac{-1}{k(n-1)}\left[\dfrac{u^{m+1}}{v^{n-1}} + b(n - m - 2)\displaystyle\int \dfrac{u^m}{v^{n-1}}\,dx\right] \\[4pt] \text{or} \\[4pt] \dfrac{-1}{d(n - m - 1)}\left[\dfrac{u^m}{v^{n-1}} + mk\displaystyle\int \dfrac{u^{m-1}}{v^n}\,dx\right] \\[4pt] \text{or} \\[4pt] \dfrac{-1}{d(n-1)}\left[\dfrac{u^m}{v^{n-1}} - mb\displaystyle\int \dfrac{u^{m-1}}{v^{n-1}}\,dx\right] \end{cases}$

FORMS CONTAINING $(a + bx^n)$

60. $\int \dfrac{dx}{a + bx^2} = \dfrac{1}{\sqrt{ab}}\tan^{-1}\dfrac{x\sqrt{ab}}{a}, \qquad (ab > 0)$

61. $\int \dfrac{dx}{a + bx^2} = \begin{cases} \dfrac{1}{2\sqrt{-ab}}\log\dfrac{a + x\sqrt{-ab}}{a - x\sqrt{-ab}}, \qquad (ab < 0) \\[4pt] \text{or} \\[4pt] \dfrac{1}{\sqrt{-ab}}\tanh^{-1}\dfrac{x\sqrt{-ab}}{a}, \qquad (ab < 0) \end{cases}$

62. $\int \dfrac{dx}{a^2 + b^2 x^2} = \dfrac{1}{ab}\tan^{-1}\dfrac{bx}{a}$

63. $\int \dfrac{x\,dx}{a + bx^2} = \dfrac{1}{2b}\log(a + bx^2)$

64. $\int \dfrac{x^2\,dx}{a + bx^2} = \dfrac{x}{b} - \dfrac{a}{b}\int \dfrac{dx}{a + bx^2}$

65. $\int \dfrac{dx}{(a + bx^2)^2} = \dfrac{x}{2a(a + bx^2)} + \dfrac{1}{2a}\int \dfrac{dx}{a + bx^2}$

66. $\int \dfrac{dx}{a^2 - b^2 x^2} = \dfrac{1}{2ab} \log \dfrac{a + bx}{a - bx}$

67. $\int \dfrac{dx}{(a + bx^2)^{m+1}} = \begin{cases} \dfrac{1}{2ma}\dfrac{x}{(a + bx^2)^m} + \dfrac{2m-1}{2ma}\int \dfrac{dx}{(a + bx^2)^m} \\[4pt] \text{or} \\[4pt] \dfrac{(2m)!}{(m!)^2}\left[\dfrac{x}{2a}\sum_{r=1}^{m}\dfrac{r!(r-1)!}{(4a)^{m-r}(2r)!(a + bx^2)^r} + \dfrac{1}{(4a)^m}\int \dfrac{dx}{a + bx^2}\right] \end{cases}$

68. $\int \dfrac{x\,dx}{(a + bx^2)^{m+1}} = -\dfrac{1}{2bm(a + bx^2)^m}$

69. $\int \dfrac{x^2\,dx}{(a + bx^2)^{m+1}} = \dfrac{-x}{2mb(a + bx^2)^m} + \dfrac{1}{2mb}\int \dfrac{dx}{(a + bx^2)^m}$

70. $\int \dfrac{dx}{x(a + bx^2)} = \dfrac{1}{2a} \log \dfrac{x^2}{a + bx^2}$

71. $\int \dfrac{dx}{x^2(a + bx^2)} = -\dfrac{1}{ax} - \dfrac{b}{a}\int \dfrac{dx}{a + bx^2}$

72. $\int \dfrac{dx}{x(a + bx^2)^{m+1}} = \begin{cases} \dfrac{1}{2am(a + bx^2)^m} + \dfrac{1}{a}\int \dfrac{dx}{x(a + bx^2)^m} \\[4pt] \text{or} \\[4pt] \dfrac{1}{2a^{m+1}}\left[\sum_{r=1}^{m}\dfrac{a^r}{r(a + bx^2)^r} + \log \dfrac{x^2}{a + bx^2}\right] \end{cases}$

73. $\int \dfrac{dx}{x^2(a + bx^2)^{m+1}} = \dfrac{1}{a}\int \dfrac{dx}{x^2(a + bx^2)^m} - \dfrac{b}{a}\int \dfrac{dx}{(a + bx^2)^{m+1}}$

74. $\int \dfrac{dx}{a + bx^3} = \dfrac{k}{3a}\left[\dfrac{1}{2}\log \dfrac{(k + x)^3}{a + bx^3} + \sqrt{3}\tan^{-1}\dfrac{2x - k}{k\sqrt{3}}\right], \qquad \left(k = \sqrt[3]{\dfrac{a}{b}}\right)$

75. $\int \dfrac{x\,dx}{a + bx^3} = \dfrac{1}{3hk}\left[\dfrac{1}{2}\log \dfrac{a + bx^3}{(k + x)^3} + \sqrt{3}\tan^{-1}\dfrac{2x - k}{k\sqrt{3}}\right], \qquad \left(k = \sqrt[3]{\dfrac{a}{b}}\right)$

76. $\int \dfrac{x^2\,dx}{a + bx^3} = \dfrac{1}{3b}\log(a + bx^3)$

77. $\int \dfrac{dx}{a + bx^4} = \dfrac{k}{2a}\left[\dfrac{1}{2}\log \dfrac{x^2 + 2kx + 2k^2}{x^2 - 2kx + 2k^2} + \tan^{-1}\dfrac{2kx}{2k^2 - x^2}\right],$
$$\left(ab > 0, k = \sqrt[4]{\dfrac{a}{4b}}\right)$$

78. $\int \dfrac{dx}{a + bx^4} = \dfrac{k}{2a}\left[\dfrac{1}{2}\log \dfrac{x + k}{x - k} + \tan^{-1}\dfrac{x}{k}\right], \qquad \left(ab < 0, k = \sqrt[4]{-\dfrac{a}{b}}\right)$

79. $\int \dfrac{x\,dx}{a + bx^4} = \dfrac{1}{2bk}\tan^{-1}\dfrac{x^2}{k}, \qquad \left(ab > 0, k = \sqrt{\dfrac{a}{b}}\right)$

80. $\int \dfrac{x\,dx}{a + bx^4} = \dfrac{1}{4bk}\log \dfrac{x^2 - k}{x^2 + k}, \qquad \left(ab < 0, k = \sqrt{-\dfrac{a}{b}}\right)$

81. $\int \dfrac{x^2\,dx}{a + bx^4} = \dfrac{1}{4bk}\left[\dfrac{1}{2}\log \dfrac{x^2 - 2kx + 2k^2}{x^2 + 2kx + 2k^2} + \tan^{-1}\dfrac{2kx}{2k^2 - x^2}\right],$
$$\left(ab > 0, k = \sqrt[4]{\dfrac{a}{4b}}\right)$$

82. $\displaystyle\int \frac{x^2\, dx}{a + bx^4} = \frac{1}{4bk}\left[\log\frac{x - k}{x + k} + 2\tan^{-1}\frac{x}{k}\right], \qquad \left(ab < 0,\ k = \sqrt[4]{-\frac{a}{b}}\right)$

83. $\displaystyle\int \frac{x^3\, dx}{a + bx^4} = \frac{1}{4b}\log(a + bx^4)$

84. $\displaystyle\int \frac{dx}{x(a + bx^n)} = \frac{1}{an}\log\frac{x^n}{a + bx^n}$

85. $\displaystyle\int \frac{dx}{(a + bx^n)^{m+1}} = \frac{1}{a}\int \frac{dx}{(a + bx^n)^m} - \frac{b}{a}\int \frac{x^n\, dx}{(a + bx^n)^{m+1}}$

86. $\displaystyle\int \frac{x^m\, dx}{(a + bx^n)^{p+1}} = \frac{1}{b}\int \frac{x^{m-n}\, dx}{(a + bx^n)^p} - \frac{a}{b}\int \frac{x^{m-n}\, dx}{(a + bx^n)^{p+1}}$

87. $\displaystyle\int \frac{dx}{x^m(a + bx^n)^{p+1}} = \frac{1}{a}\int \frac{dx}{x^m(a + bx^n)^p} - \frac{b}{a}\int \frac{dx}{x^{m-n}(a + bx^n)^{p+1}}$

88. $\displaystyle\int x^m(a + bx^n)^p\, dx = \begin{cases} \dfrac{1}{b(np + m + 1)}\left[x^{m-n+1}(a + bx^n)^{p+1}\right. \\ \qquad\qquad \left. - a(m - n + 1)\displaystyle\int x^{m-n}(a + bx^n)^p\, dx\right] \\[2pt] \text{or} \\ \dfrac{1}{np + m + 1}\left[x^{m+1}(a + bx^n)^p\right. \\ \qquad\qquad \left. + anp\displaystyle\int x^m(a + bx^n)^{p-1}\, dx\right] \\[2pt] \text{or} \\ \dfrac{1}{a(m + 1)}\left[x^{m+1}(a + bx^n)^{p+1}\right. \\ \qquad\qquad \left. - (m + 1 + np + n)b\displaystyle\int x^{m+n}(a + bx^n)^p\, dx\right] \\[2pt] \text{or} \\ \dfrac{1}{an(p + 1)}\left[-x^{m+1}(a + bx^n)^{p+1}\right. \\ \qquad\qquad \left. + (m + 1 + np + n)\displaystyle\int x^m(a + bx^n)^{p+1}\, dx\right] \end{cases}$

FORMS CONTAINING $c^3 \pm x^3$

89. $\displaystyle\int \frac{dx}{c^3 \pm x^3} = \pm\frac{1}{6c^2}\log\frac{(c \pm x)^3}{c^3 \pm x^3} + \frac{1}{c^2\sqrt{3}}\tan^{-1}\frac{2x \mp c}{c\sqrt{3}}$

90. $\displaystyle\int \frac{dx}{(c^3 \pm x^3)^2} = \frac{x}{3c^3(c^3 \pm x^3)} + \frac{2}{3c^3}\int \frac{dx}{c^3 \pm x^3}$

91. $\displaystyle\int \frac{dx}{(c^3 \pm x^3)^{n+1}} = \frac{1}{3nc^3}\left[\frac{x}{(c^3 \pm x^3)^n} + (3n - 1)\int \frac{dx}{(c^3 \pm x^3)^n}\right]$

92. $\displaystyle\int \frac{x\, dx}{c^3 \pm x^3} = \frac{1}{6c}\log\frac{c^3 \pm x^3}{(c \pm x)^3} \pm \frac{1}{c\sqrt{3}}\tan^{-1}\frac{2x \mp c}{c\sqrt{3}}$

93. $\displaystyle\int \frac{x\, dx}{(c^3 \pm x^3)^2} = \frac{x^2}{3c^3(c^3 \pm x^3)} + \frac{1}{3c^3}\int \frac{x\, dx}{c^3 \pm x^3}$

94. $\displaystyle\int \frac{x\,dx}{(c^3 \pm x^3)^{n+1}} = \frac{1}{3nc^3}\left[\frac{x^2}{(c^3 \pm x^3)^n} + (3n-2)\int\frac{x\,dx}{(c^3 \pm x^3)^n}\right]$

95. $\displaystyle\int \frac{x^2\,dx}{c^3 \pm x^3} = \pm\frac{1}{3}\log(c^3 \pm x^3)$

96. $\displaystyle\int \frac{x^2\,dx}{(c^3 \pm x^3)^{n+1}} = \mp\frac{1}{3n(c^3 \pm x^3)^n}$

97. $\displaystyle\int \frac{dx}{x(c^3 \pm x^3)} = \frac{1}{3c^3}\log\frac{x^3}{c^3 \pm x^3}$

98. $\displaystyle\int \frac{dx}{x(c^3 \pm x^3)^2} = \frac{1}{3c^3(c^3 \pm x^3)} + \frac{1}{3c^6}\log\frac{x^3}{c^3 \pm x^3}$

99. $\displaystyle\int \frac{dx}{x(c^3 \pm x^3)^{n+1}} = \frac{1}{3nc^3(c^3 \pm x^3)^n} + \frac{1}{c^3}\int\frac{dx}{x(c^3 \pm x^3)^n}$

100. $\displaystyle\int \frac{dx}{x^2(c^3 \pm x^3)} = -\frac{1}{c^3 x} \mp \frac{1}{c^3}\int\frac{x\,dx}{c^3 \pm x^3}$

101. $\displaystyle\int \frac{dx}{x^2(c^3 \pm x^3)^{n+1}} = \frac{1}{c^3}\int\frac{dx}{x^2(c^3 \pm x^3)^n} \mp \frac{1}{c^3}\int\frac{x\,dx}{(c^3 \pm x^3)^{n+1}}$

FORMS CONTAINING $c^4 \pm x^4$

102. $\displaystyle\int \frac{dx}{c^4 + x^4} = \frac{1}{2c^3\sqrt{2}}\left[\frac{1}{2}\log\frac{x^2 + cx\sqrt{2} + c^2}{x^2 - cx\sqrt{2} + c^2} + \tan^{-1}\frac{cx\sqrt{2}}{c^2 - x^2}\right]$

103. $\displaystyle\int \frac{dx}{c^4 - x^4} = \frac{1}{2c^3}\left[\frac{1}{2}\log\frac{c + x}{c - x} + \tan^{-1}\frac{x}{c}\right]$

104. $\displaystyle\int \frac{x\,dx}{c^4 + x^4} = \frac{1}{2c^2}\tan^{-1}\frac{x^2}{c^2}$

105. $\displaystyle\int \frac{x\,dx}{c^4 - x^4} = \frac{1}{4c^2}\log\frac{c^2 + x^2}{c^2 - x^2}$

106. $\displaystyle\int \frac{x^2\,dx}{c^4 + x^4} = \frac{1}{2c\sqrt{2}}\left[\frac{1}{2}\log\frac{x^2 - cx\sqrt{2} + c^2}{x^2 + cx\sqrt{2} + c^2} + \tan^{-1}\frac{cx\sqrt{2}}{c^2 - x^2}\right]$

107. $\displaystyle\int \frac{x^2\,dx}{c^4 - x^4} = \frac{1}{2c}\left[\frac{1}{2}\log\frac{c + x}{c - x} - \tan^{-1}\frac{x}{c}\right]$

108. $\displaystyle\int \frac{x^3\,dx}{c^4 \pm x^4} = \pm\frac{1}{4}\log(c^4 \pm x^4)$

FORMS CONTAINING $(a + bx + cx^2)$

$$X = a + bx + cx^2 \text{ and } q = 4ac - b^2$$

If $q = 0$, then $X = c\left(x + \dfrac{b}{2c}\right)^2$, and formulas starting with 23 should be used in place of these.

109. $\displaystyle\int \frac{dx}{X} = \frac{2}{\sqrt{q}}\tan^{-1}\frac{2cx + b}{\sqrt{q}}, \qquad (q > 0)$

110. $\displaystyle\int \frac{dx}{X} = \begin{cases} \dfrac{-2}{\sqrt{-q}}\tanh^{-1}\dfrac{2cx + b}{\sqrt{-q}} \\[2mm] \qquad\text{or} \\[2mm] \dfrac{1}{\sqrt{-q}}\log\dfrac{2cx + b - \sqrt{-q}}{2cx + b + \sqrt{-q}}, \qquad (q < 0) \end{cases}$

111. $\displaystyle\int \frac{dx}{X^2} = \frac{2cx + b}{qX} + \frac{2c}{q}\int\frac{dx}{X}$

112. $\displaystyle \int \frac{dx}{X^3} = \frac{2cx + b}{q}\left(\frac{1}{2X^2} + \frac{3c}{qX}\right) + \frac{6c^2}{q^2}\int \frac{dx}{X}$

113. $\displaystyle \int \frac{dx}{X^{n+1}} = \begin{cases} \dfrac{2cx + b}{nqX^n} + \dfrac{2(2n-1)c}{qn}\displaystyle\int \frac{dx}{X^n} \\[2ex] \qquad\qquad \text{or} \\[2ex] \dfrac{(2n)!}{(n!)^2}\left(\dfrac{c}{q}\right)^n\left[\dfrac{2cx+b}{q}\displaystyle\sum_{r=1}^{n}\left(\dfrac{q}{cX}\right)^r\dfrac{\big((r-1)!r!\big)}{(2r)!} + \displaystyle\int \frac{dx}{X}\right] \end{cases}$

114. $\displaystyle \int \frac{x\,dx}{X} = \frac{1}{2c}\log X - \frac{b}{2c}\int \frac{dx}{X}$

115. $\displaystyle \int \frac{x\,dx}{X^2} = \frac{bx + 2a}{qX} - \frac{b}{q}\int \frac{dx}{X}$

116. $\displaystyle \int \frac{x\,dx}{X^{n+1}} = -\frac{2a + bx}{nqX^n} - \frac{b(2n-1)}{nq}\int \frac{dx}{X^n}$

117. $\displaystyle \int \frac{x^2}{X}\,dx = \frac{x}{c} - \frac{b}{2c^2}\log X + \frac{b^2 - 2ac}{2c^2}\int \frac{dx}{X}$

118. $\displaystyle \int \frac{x^2}{X^2}\,dx = \frac{(b^2 - 2ac)x + ab}{cqX} + \frac{2a}{q}\int \frac{dx}{X}$

119. $\displaystyle \int \frac{x^m\,dx}{X^{n+1}} = -\frac{x^{m-1}}{(2n-m+1)cX^n} - \frac{n-m+1}{2n-m+1}\cdot\frac{b}{c}\int \frac{x^{m-1}\,dx}{X^{n+1}}$
$$+ \frac{m-1}{2n-m+1}\cdot\frac{a}{c}\int \frac{x^{m-2}\,dx}{X^{n+1}}$$

120. $\displaystyle \int \frac{dx}{xX} = \frac{1}{2a}\log\frac{x^2}{X} - \frac{b}{2a}\int \frac{dx}{X}$

121. $\displaystyle \int \frac{dx}{x^2X} = \frac{b}{2a^2}\log\frac{X}{x^2} - \frac{1}{ax} + \left(\frac{b^2}{2a^2} - \frac{c}{a}\right)\int \frac{dx}{X}$

122. $\displaystyle \int \frac{dx}{xX^n} = \frac{1}{2a(n-1)X^{n-1}} - \frac{b}{2a}\int \frac{dx}{X^n} + \frac{1}{a}\int \frac{dx}{xX^{n-1}}$

123. $\displaystyle \int \frac{dx}{x^mX^{n+1}} = -\frac{1}{(m-1)ax^{m-1}X^n} - \frac{n+m-1}{m-1}\cdot\frac{b}{a}\int \frac{dx}{x^{m-1}X^{n+1}}$
$$- \frac{2n+m-1}{m-1}\cdot\frac{c}{a}\int \frac{dx}{x^{m-2}X^{n+1}}$$

FORMS CONTAINING $\sqrt{a + bx}$

124. $\displaystyle \int \sqrt{a + bx}\,dx = \frac{2}{3b}\sqrt{(a + bx)^3}$

125. $\displaystyle \int x\sqrt{a + bx}\,dx = -\frac{2(2a - 3bx)\sqrt{(a + bx)^3}}{15b^2}$

126. $\displaystyle \int x^2\sqrt{a + bx}\,dx = \frac{2(8a^2 - 12abx + 15b^2x^2)\sqrt{(a + bx)^3}}{105b^3}$

127. $\displaystyle \int x^m\sqrt{a + bx}\,dx = \begin{cases} \dfrac{2}{b(2m+3)}\left[x^m\sqrt{(a+bx)^3} - ma\displaystyle\int x^{m-1}\sqrt{a+bx}\,dx\right] \\[2ex] \qquad\qquad \text{or} \\[2ex] \dfrac{2}{b^{m+1}}\sqrt{a+bx}\displaystyle\sum_{r=0}^{m}\dfrac{m!(-a)^{m-r}}{r!(m-r)!(2r+3)}(a+bx)^{r+1} \end{cases}$

128. $\int \dfrac{\sqrt{a + bx}}{x} dx = 2\sqrt{a + bx} + a\int \dfrac{dx}{x\sqrt{a + bx}}$

129. $\int \dfrac{\sqrt{a + bx}}{x^2} dx = -\dfrac{\sqrt{a + bx}}{x} + \dfrac{b}{2}\int \dfrac{dx}{x\sqrt{a + bx}}$

130. $\int \dfrac{\sqrt{a + bx}}{x^m} dx = -\dfrac{1}{(m - 1)a}\left[\dfrac{\sqrt{(a + bx)^3}}{x^{m-1}} + \dfrac{(2m - 5)b}{2}\int \dfrac{\sqrt{a + bx}}{x^{m-1}} dx\right]$

131. $\int \dfrac{dx}{\sqrt{a + bx}} = \dfrac{2\sqrt{a + bx}}{b}$

132. $\int \dfrac{x\, dx}{\sqrt{a + bx}} = -\dfrac{2(2a - bx)}{3b^2}\sqrt{a + bx}$

133. $\int \dfrac{x^2\, dx}{\sqrt{a + bx}} = \dfrac{2(8a^2 - 4abx + 3b^2 x^2)}{15b^3}\sqrt{a + bx}$

134. $\int \dfrac{x^m\, dx}{\sqrt{a + bx}} = \begin{cases} \dfrac{2}{(2m + 1)b}\left[x^m\sqrt{a + bx} - ma\int \dfrac{x^{m-1}\, dx}{\sqrt{a + bx}}\right] \\ \text{or} \\ \dfrac{2(-a)^m\sqrt{a + bx}}{b^{m+1}}\sum_{r=0}^{m}\dfrac{(-1)^r m!(a + bx)^r}{(2r + 1)r!(m - r)!a^r} \end{cases}$

135. $\int \dfrac{dx}{x\sqrt{a + bx}} = \dfrac{1}{\sqrt{a}}\log\left(\dfrac{\sqrt{a + bx} - \sqrt{a}}{\sqrt{a + bx} + \sqrt{a}}\right), \quad (a > 0)$

136. $\int \dfrac{dx}{x\sqrt{a + bx}} = \dfrac{2}{\sqrt{-a}}\tan^{-1}\sqrt{\dfrac{a + bx}{-a}}, \quad (a < 0)$

137. $\int \dfrac{dx}{x^2\sqrt{a + bx}} = -\dfrac{\sqrt{a + bx}}{ax} - \dfrac{b}{2a}\int \dfrac{dx}{x\sqrt{a + bx}}$

138. $\int \dfrac{dx}{x^n\sqrt{a + bx}} = \begin{cases} -\dfrac{\sqrt{a + bx}}{(n - 1)ax^{n-1}} - \dfrac{(2n - 3)b}{(2n - 2)a}\int \dfrac{dx}{x^{n-1}\sqrt{a + bx}} \\ \text{or} \\ \dfrac{(2n - 2)!}{[(n - 1)!]^2}\left[-\dfrac{\sqrt{a + bx}}{a}\sum_{r=1}^{n-1}\dfrac{r!(r - 1)!}{x^r(2r)!}\left(-\dfrac{b}{4a}\right)^{n-r-1} \right. \\ \qquad\qquad\qquad \left. + \left(-\dfrac{b}{4a}\right)^{n-1}\int \dfrac{dx}{x\sqrt{a + bx}}\right] \end{cases}$

139. $\int (a + bx)^{\pm\frac{n}{2}}\, dx = \dfrac{2(a + bx)^{\frac{2\pm n}{2}}}{b(2 \pm n)}$

140. $\int x(a + bx)^{\pm\frac{n}{2}}\, dx = \dfrac{2}{b^2}\left[\dfrac{(a + bx)^{\frac{4\pm n}{2}}}{4 \pm n} - \dfrac{a(a + bx)^{\frac{2\pm n}{2}}}{2 \pm n}\right]$

141. $\int \dfrac{dx}{x(a + bx)^{\frac{m}{2}}} = \dfrac{1}{a}\int \dfrac{dx}{x(a + bx)^{\frac{m-2}{2}}} - \dfrac{b}{a}\int \dfrac{dx}{(a + bx)^{\frac{m}{2}}}$

142. $\displaystyle\int \frac{(a + bx)^{\frac{n}{2}} \, dx}{x} = b \int (a + bx)^{\frac{n-2}{2}} \, dx + a \int \frac{(a + bx)^{\frac{n-2}{2}}}{x} \, dx$

143. $\displaystyle\int f(x, \sqrt{a + bx}) \, dx = \frac{2}{b} \int f\left(\frac{z^2 - a}{b}, z\right) z \, dz, \qquad (z = \sqrt{a + bx})$

FORMS CONTAINING $\sqrt{a + bx}$ and $\sqrt{c + dx}$

$$u = a + bx \qquad v = c + dx \qquad k = ad - bc$$

If $k = 0$, then $v = \dfrac{c}{a}u$, and formulas starting with 124 should be used in place of these.

144. $\displaystyle\int \frac{dx}{\sqrt{uv}} = \begin{cases} \dfrac{2}{\sqrt{bd}} \tanh^{-1} \dfrac{\sqrt{bduv}}{bv}, \ bd > o, \ k < o \\[2ex] \text{or} \\[1ex] \dfrac{2}{\sqrt{bd}} \tanh^{-1} \dfrac{\sqrt{bduv}}{du}, \ bd > o, \ k > o. \\[2ex] \text{or} \\[1ex] \dfrac{1}{\sqrt{bd}} \log \dfrac{(bv + \sqrt{bduv})^2}{v}, \qquad (bd > 0) \end{cases}$

145. $\displaystyle\int \frac{dx}{\sqrt{uv}} = \begin{cases} \dfrac{2}{\sqrt{-bd}} \tan^{-1} \dfrac{\sqrt{-bduv}}{bv} \\[2ex] \text{or} \\[1ex] -\dfrac{1}{\sqrt{-bd}} \sin^{-1}\left(\dfrac{2bdx + ad + bc}{|k|}\right), \qquad (bd < 0) \end{cases}$

146. $\displaystyle\int \sqrt{uv} \, dx = \frac{k + 2bv}{4bd} \sqrt{uv} - \frac{k^2}{8bd} \int \frac{dx}{\sqrt{uv}}$

147. $\displaystyle\int \frac{dx}{v\sqrt{u}} = \begin{cases} \dfrac{1}{\sqrt{kd}} \log \dfrac{d\sqrt{u} - \sqrt{kd}}{d\sqrt{u} + \sqrt{kd}} \\[2ex] \text{or} \\[1ex] \dfrac{1}{\sqrt{kd}} \log \dfrac{(d\sqrt{u} - \sqrt{kd})^2}{v}, \qquad (kd > 0) \end{cases}$

148. $\displaystyle\int \frac{dx}{v\sqrt{u}} = \frac{2}{\sqrt{-kd}} \tan^{-1} \frac{d\sqrt{u}}{\sqrt{-kd}}, \qquad (kd < 0)$

149. $\displaystyle\int \frac{x \, dx}{\sqrt{uv}} = \frac{\sqrt{uv}}{bd} - \frac{ad + bc}{2bd} \int \frac{dx}{\sqrt{uv}}$

150. $\displaystyle\int \frac{dx}{v\sqrt{uv}} = \frac{-2\sqrt{uv}}{kv}$

151. $\displaystyle\int \frac{v \, dx}{\sqrt{uv}} = \frac{\sqrt{uv}}{b} - \frac{k}{2b} \int \frac{dx}{\sqrt{uv}}$

152. $\displaystyle\int \sqrt{\frac{v}{u}} \, dx = \frac{v}{|v|} \int \frac{v \, dx}{\sqrt{uv}}$

153. $\displaystyle\int v^m \sqrt{u} \, dx = \frac{1}{(2m + 3)d}\left(2v^{m+1}\sqrt{u} + k \int \frac{v^m \, dx}{\sqrt{u}}\right)$

154. $\int \dfrac{dx}{v^m \sqrt{u}} = -\dfrac{1}{(m-1)k}\left(\dfrac{\sqrt{u}}{v^{m-1}} + \left(m - \dfrac{3}{2}\right)b \int \dfrac{dx}{v^{m-1}\sqrt{u}}\right)$

155. $\int \dfrac{v^m\, dx}{\sqrt{u}} = \begin{cases} \dfrac{2}{b(2m+1)}\left[v^m\sqrt{u} - mk\int \dfrac{v^{m-1}}{\sqrt{u}}\, dx\right] \\ \text{or} \\ \dfrac{2(m!)^2\sqrt{u}}{b(2m+1)!} \displaystyle\sum_{r=0}^{m}\left(-\dfrac{4k}{b}\right)^{m-r}\dfrac{(2r)!}{(r!)^2}\, v^r \end{cases}$

FORMS CONTAINING $\sqrt{x^2 \pm a^2}$

156. $\int \sqrt{x^2 \pm a^2}\, dx = \frac{1}{2}\left[x\sqrt{x^2 \pm a^2} \pm a^2 \log\left(x + \sqrt{x^2 \pm a^2}\right)\right]$

157. $\int \dfrac{dx}{\sqrt{x^2 \pm a^2}} = \log\left(x + \sqrt{x^2 \pm a^2}\right)$

158. $\int \dfrac{dx}{x\sqrt{x^2 - a^2}} = \dfrac{1}{|a|} \sec^{-1}\dfrac{x}{a}$

159. $\int \dfrac{dx}{x\sqrt{x^2 + a^2}} = -\dfrac{1}{a}\log\left(\dfrac{a + \sqrt{x^2 + a^2}}{x}\right)$

160. $\int \dfrac{\sqrt{x^2 + a^2}}{x}\, dx = \sqrt{x^2 + a^2} - a\log\left(\dfrac{a + \sqrt{x^2 + a^2}}{x}\right)$

161. $\int \dfrac{\sqrt{x^2 - a^2}}{x}\, dx = \sqrt{x^2 - a^2} - |a|\sec^{-1}\dfrac{x}{a}$

162. $\int \dfrac{x\, dx}{\sqrt{x^2 \pm a^2}} = \sqrt{x^2 \pm a^2}$

163. $\int x\sqrt{x^2 \pm a^2}\, dx = \frac{1}{3}\sqrt{(x^2 \pm a^2)^3}$

164. $\int \sqrt{(x^2 \pm a^2)^3}\, dx = \dfrac{1}{4}\left[x\sqrt{(x^2 \pm a^2)^3} \pm \dfrac{3a^2 x}{2}\sqrt{x^2 \pm a^2}\right.$

$\left. + \dfrac{3a^4}{2}\log\left(x + \sqrt{x^2 \pm a^2}\right)\right]$

165. $\int \dfrac{dx}{\sqrt{(x^2 \pm a^2)^3}} = \dfrac{\pm x}{a^2 \sqrt{x^2 \pm a^2}}$

166. $\int \dfrac{x\, dx}{\sqrt{(x^2 \pm a^2)^3}} = \dfrac{-1}{\sqrt{x^2 \pm a^2}}$

167. $\int x\sqrt{(x^2 \pm a^2)^3}\, dx = \frac{1}{5}\sqrt{(x^2 \pm a^2)^5}$

168. $\int x^2\sqrt{x^2 \pm a^2}\, dx = \dfrac{x}{4}\sqrt{(x^2 \pm a^2)^3} \mp \dfrac{a^2}{8}x\sqrt{x^2 \pm a^2} - \dfrac{a^4}{8}\log\left(x + \sqrt{x^2 \pm a^2}\right)$

169. $\int x^3\sqrt{x^2 + a^2}\, dx = (\frac{1}{5}x^2 - \frac{2}{15}a^2)\sqrt{(a^2 + x^2)^3}$

170. $\int x^3\sqrt{x^2 - a^2}\, dx = \dfrac{1}{5}\sqrt{(x^2 - a^2)^5} + \dfrac{a^2}{3}\sqrt{(x^2 - a^2)^3}$

171. $\int \dfrac{x^2\, dx}{\sqrt{x^2 \pm a^2}} = \dfrac{x}{2}\sqrt{x^2 \pm a^2} \mp \dfrac{a^2}{2}\log\left(x + \sqrt{x^2 \pm a^2}\right)$

172. $\int \dfrac{x^3\,dx}{\sqrt{x^2 \pm a^2}} = \dfrac{1}{3}\sqrt{(x^2 \pm a^2)^3} \mp a^2\sqrt{x^2 \pm a^2}$

173. $\int \dfrac{dx}{x^2\sqrt{x^2 \pm a^2}} = \mp \dfrac{\sqrt{x^2 \pm a^2}}{a^2 x}$

174. $\int \dfrac{dx}{x^3\sqrt{x^2 + a^2}} = -\dfrac{\sqrt{x^2 + a^2}}{2a^2 x^2} + \dfrac{1}{2a^3}\log\dfrac{a + \sqrt{x^2 + a^2}}{x}$

175. $\int \dfrac{dx}{x^3\sqrt{x^2 - a^2}} = \dfrac{\sqrt{x^2 - a^2}}{2a^2 x^2} + \dfrac{1}{2|a^3|}\sec^{-1}\dfrac{x}{a}$

176. $\int x^2\sqrt{(x^2 \pm a^2)^3}\,dx = \dfrac{x}{6}\sqrt{(x^2 \pm a^2)^5} \mp \dfrac{a^2 x}{24}\sqrt{(x^2 \pm a^2)^3} - \dfrac{a^4 x}{16}\sqrt{x^2 \pm a^2}$

$$\mp \dfrac{a^6}{16}\log(x + \sqrt{x^2 \pm a^2})$$

177. $\int x^3\sqrt{(x^2 \pm a^2)^3}\,dx = \dfrac{1}{7}\sqrt{(x^2 \pm a^2)^7} \mp \dfrac{a^2}{5}\sqrt{(x^2 \pm a^2)^5}$

178. $\int \dfrac{\sqrt{x^2 \pm a^2}\,dx}{x^2} = -\dfrac{\sqrt{x^2 \pm a^2}}{x} + \log(x + \sqrt{x^2 \pm a^2})$

179. $\int \dfrac{\sqrt{x^2 + a^2}}{x^3}\,dx = -\dfrac{\sqrt{x^2 + a^2}}{2x^2} - \dfrac{1}{2a}\log\dfrac{a + \sqrt{x^2 + a^2}}{x}$

180. $\int \dfrac{\sqrt{x^2 - a^2}}{x^3}\,dx = -\dfrac{\sqrt{x^2 - a^2}}{2x^2} + \dfrac{1}{2|a|}\sec^{-1}\dfrac{x}{a}$

181. $\int \dfrac{\sqrt{x^2 \pm a^2}}{x^4}\,dx = \mp\dfrac{\sqrt{(x^2 \pm a^2)^3}}{3a^2 x^3}$

182. $\int \dfrac{x^2\,dx}{\sqrt{(x^2 \pm a^2)^3}} = \dfrac{-x}{\sqrt{x^2 \pm a^2}} + \log(x + \sqrt{x^2 \pm a^2})$

183. $\int \dfrac{x^3\,dx}{\sqrt{(x^2 \pm a^2)^3}} = \sqrt{x^2 \pm a^2} \pm \dfrac{a^2}{\sqrt{x^2 \pm a^2}}$

184. $\int \dfrac{dx}{x\sqrt{(x^2 + a^2)^3}} = \dfrac{1}{a^2\sqrt{x^2 + a^2}} - \dfrac{1}{a^3}\log\dfrac{a + \sqrt{x^2 + a^2}}{x}$

185. $\int \dfrac{dx}{x\sqrt{(x^2 - a^2)^3}} = -\dfrac{1}{a^2\sqrt{x^2 - a^2}} - \dfrac{1}{|a^3|}\sec^{-1}\dfrac{x}{a}$

186. $\int \dfrac{dx}{x^2\sqrt{(x^2 \pm a^2)^3}} = -\dfrac{1}{a^4}\left[\dfrac{\sqrt{x^2 \pm a^2}}{x} + \dfrac{x}{\sqrt{x^2 \pm a^2}}\right]$

187. $\int \dfrac{dx}{x^3\sqrt{(x^2 + a^2)^3}} = -\dfrac{1}{2a^2 x^2\sqrt{x^2 + a^2}} - \dfrac{3}{2a^4\sqrt{x^2 + a^2}}$

$$+ \dfrac{3}{2a^5}\log\dfrac{a + \sqrt{x^2 + a^2}}{x}$$

188. $\int \dfrac{dx}{x^3\sqrt{(x^2 - a^2)^3}} = \dfrac{1}{2a^2 x^2\sqrt{x^2 - a^2}} - \dfrac{3}{2a^4\sqrt{x^2 - a^2}} - \dfrac{3}{2|a^5|}\sec^{-1}\dfrac{x}{a}$

189. $\int \dfrac{x^m}{\sqrt{x^2 \pm a^2}}\,dx = \dfrac{1}{m}x^{m-1}\sqrt{x^2 \pm a^2} \mp \dfrac{m-1}{m}a^2\int\dfrac{x^{m-2}}{\sqrt{x^2 \pm a^2}}\,dx$

190. $\int \dfrac{x^{2m}}{\sqrt{x^2 \pm a^2}}\,dx = \dfrac{(2m)!}{2^{2m}(m!)^2}\left[\sqrt{x^2 \pm a^2}\sum_{r=1}^{m}\dfrac{r!(r-1)!}{(2r)!}(\mp a^2)^{m-r}(2x)^{2r-1}\right.$

$$\left. + (\mp a^2)^m \log(x + \sqrt{x^2 \pm a^2})\right]$$

191. $\int \dfrac{x^{2m+1}}{\sqrt{x^2 \pm a^2}}\,dx = \sqrt{x^2 \pm a^2}\sum_{r=0}^{m}\dfrac{(2r)!(m!)^2}{(2m+1)!(r!)^2}(\mp 4a^2)^{m-r}x^{2r}$

192. $\int \dfrac{dx}{x^m\sqrt{x^2 \pm a^2}} = \mp \dfrac{\sqrt{x^2 \pm a^2}}{(m-1)a^2 x^{m-1}} \mp \dfrac{(m-2)}{(m-1)a^2}\int\dfrac{dx}{x^{m-2}\sqrt{x^2 \pm a^2}}$

193. $\int \dfrac{dx}{x^{2m}\sqrt{x^2 \pm a^2}} = \sqrt{x^2 \pm a^2}\sum_{r=0}^{m-1}\dfrac{(m-1)!m!(2r)!2^{2m-2r-1}}{(r!)^2(2m)!(\mp a^2)^{m-r}x^{2r+1}}$

194. $\int \dfrac{dx}{x^{2m+1}\sqrt{x^2 + a^2}} = \dfrac{(2m)!}{(m!)^2}\left[\dfrac{\sqrt{x^2 + a^2}}{a^2}\sum_{r=1}^{m}(-1)^{m-r+1}\dfrac{r!(r-1)!}{2(2r)!(4a^2)^{m-r}x^{2r}}\right.$

$$\left. + \dfrac{(-1)^{m+1}}{2^{2m}a^{2m+1}}\log\dfrac{\sqrt{x^2 + a^2}+a}{x}\right]$$

195. $\int \dfrac{dx}{x^{2m+1}\sqrt{x^2 - a^2}} = \dfrac{(2m)!}{(m!)^2}\left[\dfrac{\sqrt{x^2 - a^2}}{a^2}\sum_{r=1}^{m}\dfrac{r!(r-1)!}{2(2r)!(4a^2)^{m-r}x^{2r}}\right.$

$$\left. + \dfrac{1}{2^{2m}|a|^{2m+1}}\sec^{-1}\dfrac{x}{a}\right]$$

196. $\int \dfrac{dx}{(x-a)\sqrt{x^2 - a^2}} = -\dfrac{\sqrt{x^2 - a^2}}{a(x-a)}$

197. $\int \dfrac{dx}{(x+a)\sqrt{x^2 - a^2}} = \dfrac{\sqrt{x^2 - a^2}}{a(x+a)}$

198. $\int f(x, \sqrt{x^2 + a^2})\,dx = a\int f(a\tan u, a\sec u)\sec^2 u\,du, \quad \left(u = \tan^{-1}\dfrac{x}{a}, a > 0\right)$

199. $\int f(x, \sqrt{x^2 - a^2})\,dx = a\int f(a\sec u, a\tan u)\sec u\tan u\,du, \quad \left(u = \sec^{-1}\dfrac{x}{a}\right.$

$$\left. a > 0\right)$$

FORMS CONTAINING $\sqrt{a^2 - x^2}$

200. $\int \sqrt{a^2 - x^2}\,dx = \dfrac{1}{2}\left[x\sqrt{a^2 - x^2} + a^2\sin^{-1}\dfrac{x}{|a|}\right]$

201. $\int \dfrac{dx}{\sqrt{a^2 - x^2}} = \begin{cases} \sin^{-1}\dfrac{x}{|a|} \\[2mm] \text{or} \\[2mm] -\cos^{-1}\dfrac{x}{|a|} \end{cases}$

202. $\int \dfrac{dx}{x\sqrt{a^2 - x^2}} = -\dfrac{1}{a}\log\left(\dfrac{a + \sqrt{a^2 - x^2}}{x}\right)$

203. $\int \dfrac{\sqrt{a^2 - x^2}}{x}\,dx = \sqrt{a^2 - x^2} - a\log\left(\dfrac{a + \sqrt{a^2 - x^2}}{x}\right)$

204. $\int \dfrac{x\,dx}{\sqrt{a^2 - x^2}} = -\sqrt{a^2 - x^2}$

205. $\int x\sqrt{a^2 - x^2}\,dx = -\dfrac{1}{3}\sqrt{(a^2 - x^2)^3}$

206. $\int \sqrt{(a^2 - x^2)^3}\, dx = \frac{1}{4}\left[x\sqrt{(a^2 - x^2)^3} + \frac{3a^2 x}{2}\sqrt{a^2 - x^2} + \frac{3a^4}{2}\sin^{-1}\frac{x}{|a|} \right]$

207. $\int \frac{dx}{\sqrt{(a^2 - x^2)^3}} = \frac{x}{a^2\sqrt{a^2 - x^2}}$

208. $\int \frac{x\, dx}{\sqrt{(a^2 - x^2)^3}} = \frac{1}{\sqrt{a^2 - x^2}}$

209. $\int x\sqrt{(a^2 - x^2)^3}\, dx = -\frac{1}{5}\sqrt{(a^2 - x^2)^5}$

210. $\int x^2\sqrt{a^2 - x^2}\, dx = -\frac{x}{4}\sqrt{(a^2 - x^2)^3} + \frac{a^2}{8}\left(x\sqrt{a^2 - x^2} + a^2\sin^{-1}\frac{x}{|a|} \right)$

211. $\int x^3\sqrt{a^2 - x^2}\, dx = \left(-\frac{1}{5}x^2 - \frac{2}{15}a^2 \right)\sqrt{(a^2 - x^2)^3}$

212. $\int x^2\sqrt{(a^2 - x^2)^3}\, dx = -\frac{1}{6}x\sqrt{(a^2 - x^2)^5} + \frac{a^2 x}{24}\sqrt{(a^2 - x^2)^3}$

$$+ \frac{a^4 x}{16}\sqrt{a^2 - x^2} + \frac{a^6}{16}\sin^{-1}\frac{x}{|a|}$$

213. $\int x^3\sqrt{(a^2 - x^2)^3}\, dx = \frac{1}{7}\sqrt{(a^2 - x^2)^7} - \frac{a^2}{5}\sqrt{(a^2 - x^2)^5}$

214. $\int \frac{x^2\, dx}{\sqrt{a^2 - x^2}} = -\frac{x}{2}\sqrt{a^2 - x^2} + \frac{a^2}{2}\sin^{-1}\frac{x}{|a|}$

215. $\int \frac{dx}{x^2\sqrt{a^2 - x^2}} = -\frac{\sqrt{a^2 - x^2}}{a^2 x}$

216. $\int \frac{\sqrt{a^2 - x^2}}{x^2}\, dx = -\frac{\sqrt{a^2 - x^2}}{x} - \sin^{-1}\frac{x}{|a|}$

217. $\int \frac{\sqrt{a^2 - x^2}}{x^3}\, dx = -\frac{\sqrt{a^2 - x^2}}{2x^2} + \frac{1}{2a}\log\frac{a + \sqrt{a^2 - x^2}}{x}$

218. $\int \frac{\sqrt{a^2 - x^2}}{x^4}\, dx = -\frac{\sqrt{(a^2 - x^2)^3}}{3a^2 x^3}$

219. $\int \frac{x^2\, dx}{\sqrt{(a^2 - x^2)^3}} = \frac{x}{\sqrt{a^2 - x^2}} - \sin^{-1}\frac{x}{|a|}$

220. $\int \frac{x^3\, dx}{\sqrt{a^2 - x^2}} = -\frac{2}{3}(a^2 - x^2)^{\frac{3}{2}} - x^2(a^2 - x^2)^{\frac{1}{2}} = -\frac{1}{3}\sqrt{a^2 - x^2}(x^2 + 2a^2)$

221. $\int \frac{x^3\, dx}{\sqrt{(a^2 - x^2)^3}} = 2(a^2 - x^2)^{\frac{1}{2}} + \frac{x^2}{(a^2 - x^2)^{\frac{1}{2}}} = -\frac{a^2}{\sqrt{a^2 - x^2}} + \sqrt{a^2 - x^2}$

222. $\int \frac{dx}{x^3\sqrt{a^2 - x^2}} = -\frac{\sqrt{a^2 - x^2}}{2a^2 x^2} - \frac{1}{2a^3}\log\frac{a + \sqrt{a^2 - x^2}}{x}$

223. $\int \frac{dx}{x\sqrt{(a^2 - x^2)^3}} = \frac{1}{a^2\sqrt{a^2 - x^2}} - \frac{1}{a^3}\log\frac{a + \sqrt{a^2 - x^2}}{x}$

224. $\int \frac{dx}{x^2\sqrt{(a^2 - x^2)^3}} = \frac{1}{a^4}\left[-\frac{\sqrt{a^2 - x^2}}{x} + \frac{x}{\sqrt{a^2 - x^2}} \right]$

225. $\int \frac{dx}{x^3\sqrt{(a^2 - x^2)^3}} = -\frac{1}{2a^2 x^2\sqrt{a^2 - x^2}} + \frac{3}{2a^4\sqrt{a^2 - x^2}}$

$$-\frac{3}{2a^5}\log\frac{a + \sqrt{a^2 - x^2}}{x}$$

226. $\displaystyle\int \frac{x^m}{\sqrt{a^2 - x^2}}\,dx = -\frac{x^{m-1}\sqrt{a^2 - x^2}}{m} + \frac{(m-1)a^2}{m}\int \frac{x^{m-2}}{\sqrt{a^2 - x^2}}\,dx$

227. $\displaystyle\int \frac{x^{2m}}{\sqrt{a^2 - x^2}}\,dx = \frac{(2m)!}{(m!)^2}\left[-\sqrt{a^2 - x^2}\sum_{r=1}^{m}\frac{r!(r-1)!}{2^{2m-2r+1}(2r)!}a^{2m-2r}x^{2r-1}\right.$

$$\left.+\frac{a^{2m}}{2^{2m}}\sin^{-1}\frac{x}{|a|}\right]$$

228. $\displaystyle\int \frac{x^{2m+1}}{\sqrt{a^2 - x^2}}\,dx = -\sqrt{a^2 - x^2}\sum_{r=0}^{m}\frac{(2r)!(m!)^2}{(2m+1)!(r!)^2}(4a^2)^{m-r}x^{2r}$

229. $\displaystyle\int \frac{dx}{x^m\sqrt{a^2 - x^2}} = -\frac{\sqrt{a^2 - x^2}}{(m-1)a^2 x^{m-1}} + \frac{m-2}{(m-1)a^2}\int \frac{dx}{x^{m-2}\sqrt{a^2 - x^2}}$

230. $\displaystyle\int \frac{ax}{x^{2m}\sqrt{a^2 - x^2}} = -\sqrt{a^2 - x^2}\sum_{r=0}^{m-1}\frac{(m-1)!m!(2r)!2^{2m-2r-1}}{(r!)^2(2m)!a^{2m-2r}x^{2r+1}}$

231. $\displaystyle\int \frac{dx}{x^{2m+1}\sqrt{a^2 - x^2}} = \frac{(2m)!}{(m!)^2}\left[-\frac{\sqrt{a^2 - x^2}}{a^2}\sum_{r=1}^{m}\frac{r!(r-1)!}{2(2r)!(4a^2)^{m-r}x^{2r}}\right.$

$$\left.+\frac{1}{2^{2m}a^{2m+1}}\log\frac{a - \sqrt{a^2 - x^2}}{x}\right]$$

232. $\displaystyle\int \frac{dx}{(b^2 - x^2)\sqrt{a^2 - x^2}} = \frac{1}{2b\sqrt{a^2 - b^2}}\log\frac{(b\sqrt{a^2 - x^2} + x\sqrt{a^2 - b^2})^2}{b^2 - x^2},$

$$(a^2 > b^2)$$

233. $\displaystyle\int \frac{dx}{(b^2 - x^2)\sqrt{a^2 - x^2}} = \frac{1}{b\sqrt{b^2 - a^2}}\tan^{-1}\frac{x\sqrt{b^2 - a^2}}{b\sqrt{a^2 - x^2}}, \quad (b^2 > a^2)$

234. $\displaystyle\int \frac{dx}{(b^2 + x^2)\sqrt{a^2 - x^2}} = \frac{1}{b\sqrt{a^2 + b^2}}\tan^{-1}\frac{x\sqrt{a^2 + b^2}}{b\sqrt{a^2 - x^2}}$

235. $\displaystyle\int \frac{\sqrt{a^2 - x^2}}{b^2 + x^2}\,dx = \frac{\sqrt{a^2 + b^2}}{|b|}\sin^{-1}\frac{x\sqrt{a^2 + b^2}}{|a|\sqrt{x^2 + b^2}} - \sin^{-1}\frac{x}{|a|}$

236. $\displaystyle\int f(x, \sqrt{a^2 - x^2})\,dx = a\int f(a\sin u, a\cos u)\cos u\,du, \quad \left(u = \sin^{-1}\frac{x}{a},\ a > 0\right)$

FORMS CONTAINING $\sqrt{a + bx + cx^2}$

$$X = a + bx + cx^2, q = 4ac - b^2, \text{ and } k = \frac{4c}{q}$$

If $q = 0$, then $\sqrt{X} = \sqrt{c}\left|x + \dfrac{b}{2c}\right|$

237. $\displaystyle\int \frac{dx}{\sqrt{X}} = \begin{cases} \dfrac{1}{\sqrt{c}}\log(2\sqrt{cX} + 2cx + b) \\[2mm] \text{or} \\[2mm] \dfrac{1}{\sqrt{c}}\sinh^{-1}\dfrac{2cx + b}{\sqrt{q}}, \quad (c > 0) \end{cases}$

238. $\displaystyle\int \frac{dx}{\sqrt{X}} = -\frac{1}{\sqrt{-c}}\sin^{-1}\frac{2cx + b}{\sqrt{-q}}, \quad (c < 0)$

239. $\displaystyle\int \frac{dx}{X\sqrt{X}} = \frac{2(2cx + b)}{q\sqrt{X}}$

240. $\displaystyle\int \frac{dx}{X^2\sqrt{X}} = \frac{2(2cx+b)}{3q\sqrt{X}}\left(\frac{1}{X}+2k\right)$

241. $\displaystyle\int \frac{dx}{X^n\sqrt{X}} = \begin{cases} \dfrac{2(2cx+b)\sqrt{X}}{(2n-1)qX^n} + \dfrac{2k(n-1)}{2n-1}\displaystyle\int \dfrac{dx}{X^{n-1}\sqrt{X}} \\[2ex] \text{or} \\[2ex] \dfrac{(2cx+b)(n!)(n-1)!4^n k^{n-1}}{q[(2n)!]\sqrt{X}} \displaystyle\sum_{r=0}^{n-1} \dfrac{(2r)!}{(4kX)^r(r!)^2} \end{cases}$

242. $\displaystyle\int \sqrt{X}\,dx = \frac{(2cx+b)\sqrt{X}}{4c} + \frac{1}{2k}\int \frac{dx}{\sqrt{X}}$

243. $\displaystyle\int X\sqrt{X}\,dx = \frac{(2cx+b)\sqrt{X}}{8c}\left(X+\frac{3}{2k}\right) + \frac{3}{8k^2}\int \frac{dx}{\sqrt{X}}$

244. $\displaystyle\int X^2\sqrt{X}\,dx = \frac{(2cx+b)\sqrt{X}}{12c}\left(X^2+\frac{5X}{4k}+\frac{15}{8k^2}\right) + \frac{5}{16k^3}\int \frac{dx}{\sqrt{X}}$

245. $\displaystyle\int X^n\sqrt{X}\,dx = \begin{cases} \dfrac{(2cx+b)X^n\sqrt{X}}{4(n+1)c} + \dfrac{2n+1}{2(n+1)k}\displaystyle\int X^{n-1}\sqrt{X}\,dx \\[2ex] \text{or} \\[2ex] \dfrac{(2n+2)!}{[(n+1)!]^2(4k)^{n+1}}\left[\dfrac{k(2cx+b)\sqrt{X}}{c}\displaystyle\sum_{r=0}^{n}\dfrac{r!(r+1)!(4kX)^r}{(2r+2)!}\right.\\[2ex] \qquad\qquad\qquad\qquad\qquad\qquad\qquad\left. + \displaystyle\int \dfrac{dx}{\sqrt{X}}\right] \end{cases}$

246. $\displaystyle\int \frac{x\,dx}{\sqrt{X}} = \frac{\sqrt{X}}{c} - \frac{b}{2c}\int \frac{dx}{\sqrt{X}}$

247. $\displaystyle\int \frac{x\,dx}{X\sqrt{X}} = -\frac{2(bx+2a)}{q\sqrt{X}}$

248. $\displaystyle\int \frac{x\,dx}{X^n\sqrt{X}} = -\frac{\sqrt{X}}{(2n-1)cX^n} - \frac{b}{2c}\int \frac{dx}{X^n\sqrt{X}}$

249. $\displaystyle\int \frac{x^2\,dx}{\sqrt{X}} = \left(\frac{x}{2c}-\frac{3b}{4c^2}\right)\sqrt{X} + \frac{3b^2-4ac}{8c^2}\int \frac{dx}{\sqrt{X}}$

250. $\displaystyle\int \frac{x^2\,dx}{X\sqrt{X}} = \frac{(2b^2-4ac)x+2ab}{cq\sqrt{X}} + \frac{1}{c}\int \frac{dx}{\sqrt{X}}$

251. $\displaystyle\int \frac{x^2\,dx}{X^n\sqrt{X}} = \frac{(2b^2-4ac)x+2ab}{(2n-1)cqX^{n-1}\sqrt{X}} + \frac{4ac+(2n-3)b^2}{(2n-1)cq}\int \frac{dx}{X^{n-1}\sqrt{X}}$

252. $\displaystyle\int \frac{x^3\,dx}{\sqrt{X}} = \left(\frac{x^2}{3c}-\frac{5bx}{12c^2}+\frac{5b^2}{8c^3}-\frac{2a}{3c^2}\right)\sqrt{X} + \left(\frac{3ab}{4c^2}-\frac{5b^3}{16c^3}\right)\int \frac{dx}{\sqrt{X}}$

253. $\displaystyle\int \frac{x^n\,dx}{\sqrt{X}} = \frac{1}{nc}x^{n-1}\sqrt{X} - \frac{(2n-1)b}{2nc}\int \frac{x^{n-1}\,dx}{\sqrt{X}} - \frac{(n-1)a}{nc}\int \frac{x^{n-2}\,dx}{\sqrt{X}}$

254. $\displaystyle\int x\sqrt{X}\,dx = \frac{X\sqrt{X}}{3c} - \frac{b(2cx+b)}{8c^2}\sqrt{X} - \frac{b}{4ck}\int \frac{dx}{\sqrt{X}}$

255. $\displaystyle\int xX\sqrt{X}\,dx = \frac{X^2\sqrt{X}}{5c} - \frac{b}{2c}\int X\sqrt{X}\,dx$

256. $\int x X^n \sqrt{X}\, dx = \dfrac{X^{n+1}\sqrt{X}}{(2n+3)c} - \dfrac{b}{2c}\int X^n \sqrt{X}\, dx$

257. $\int x^2 \sqrt{X}\, dx = \left(x - \dfrac{5b}{6c}\right)\dfrac{X\sqrt{X}}{4c} + \dfrac{5b^2 - 4ac}{16c^2}\int \sqrt{X}\, dx$

258. $\int \dfrac{dx}{x\sqrt{X}} = -\dfrac{1}{\sqrt{a}}\log\dfrac{2\sqrt{aX} + bx + 2a}{x}, \qquad (a > 0)$

259. $\int \dfrac{dx}{x\sqrt{X}} = \dfrac{1}{\sqrt{-a}}\sin^{-1}\left(\dfrac{bx + 2a}{|x|\sqrt{-q}}\right), \qquad (a < 0)$

260. $\int \dfrac{dx}{x\sqrt{X}} = -\dfrac{2\sqrt{X}}{bx}, \qquad (a = 0)$

261. $\int \dfrac{dx}{x^2 \sqrt{X}} = -\dfrac{\sqrt{X}}{ax} - \dfrac{b}{2a}\int \dfrac{dx}{x\sqrt{X}}$

262. $\int \dfrac{\sqrt{X}\, dx}{x} = \sqrt{X} + \dfrac{b}{2}\int \dfrac{dx}{\sqrt{X}} + a\int \dfrac{dx}{x\sqrt{X}}$

263. $\int \dfrac{\sqrt{X}\, dx}{x^2} = -\dfrac{\sqrt{X}}{x} + \dfrac{b}{2}\int \dfrac{dx}{x\sqrt{X}} + c\int \dfrac{dx}{\sqrt{X}}$

FORMS INVOLVING $\sqrt{2ax - x^2}$

264. $\int \sqrt{2ax - x^2}\, dx = \dfrac{1}{2}\left[(x - a)\sqrt{2ax - x^2} + a^2 \sin^{-1}\dfrac{x - a}{|a|}\right]$

265. $\int \dfrac{dx}{\sqrt{2ax - x^2}} = \begin{cases} \cos^{-1}\dfrac{a - x}{|a|} \\[2mm] \text{or} \\[2mm] \sin^{-1}\dfrac{x - a}{|a|} \end{cases}$

266. $\int x^n \sqrt{2ax - x^2}\, dx = \begin{cases} -\dfrac{x^{n-1}(2ax - x^2)^{\frac{3}{2}}}{n + 2} + \dfrac{(2n + 1)a}{n + 2}\int x^{n-1}\sqrt{2ax - x^2}\, dx \\[3mm] \text{or} \\[3mm] \sqrt{2ax - x^2}\left[\dfrac{x^{n+1}}{n + 2} - \displaystyle\sum_{r=0}^{n}\dfrac{(2n + 1)!(r!)^2 a^{n-r+1}}{2^{n-r}(2r + 1)!(n + 2)!n!}x^r\right] \\[4mm] \qquad + \dfrac{(2n + 1)!a^{n+2}}{2^n n!(n + 2)!}\sin^{-1}\dfrac{x - a}{|a|} \end{cases}$

267. $\int \dfrac{\sqrt{2ax - x^2}}{x^n}\, dx = \dfrac{(2ax - x^2)^{\frac{3}{2}}}{(3 - 2n)ax^n} + \dfrac{n - 3}{(2n - 3)a}\int \dfrac{\sqrt{2ax - x^2}}{x^{n-1}}\, dx$

268. $\int \dfrac{x^n\, dx}{\sqrt{2ax - x^2}} = \begin{cases} \dfrac{-x^{n-1}\sqrt{2ax - x^2}}{n} + \dfrac{a(2n - 1)}{n}\int \dfrac{x^{n-1}}{\sqrt{2ax - x^2}}\, dx \\[3mm] \text{or} \\[3mm] -\sqrt{2ax - x^2}\displaystyle\sum_{r=1}^{n}\dfrac{(2n)!r!(r - 1)!a^{n-r}}{2^{n-r}(2r)!(n!)^2}x^{r-1} \\[4mm] \qquad + \dfrac{(2n)!a^n}{2^n(n!)^2}\sin^{-1}\dfrac{x - a}{|a|} \end{cases}$

269. $\displaystyle\int \frac{dx}{x^n\sqrt{2ax - x^2}} = \begin{cases} \dfrac{\sqrt{2ax - x^2}}{a(1 - 2n)x^n} + \dfrac{n - 1}{(2n - 1)a}\displaystyle\int \dfrac{dx}{x^{n-1}\sqrt{2ax - x^2}} \\[2ex] \quad\text{or} \\[1ex] -\sqrt{2ax - x^2}\,\displaystyle\sum_{r=0}^{n-1}\dfrac{2^{n-r}(n - 1)!\,n!\,(2r)!}{(2n)!(r!)^2 a^{n-r}x^{r+1}} \end{cases}$

270. $\displaystyle\int \frac{dx}{(2ax - x^2)^{\frac{3}{2}}} = \frac{x - a}{a^2\sqrt{2ax - x^2}}$

271. $\displaystyle\int \frac{x\,dx}{(2ax - x^2)^{\frac{3}{2}}} = \frac{x\cdot}{a\sqrt{2ax - x^2}}$

MISCELLANEOUS ALGEBRAIC FORMS

272. $\displaystyle\int \frac{dx}{\sqrt{2ax + x^2}} = \log\left(x + a + \sqrt{2ax + x^2}\right)$

273. $\displaystyle\int \sqrt{ax^2 + c}\,dx = \frac{x}{2}\sqrt{ax^2 + c} + \frac{c}{2\sqrt{a}}\log\left(x\sqrt{a} + \sqrt{ax^2 + c}\right),$ $\quad(a > 0)$

274. $\displaystyle\int \sqrt{ax^2 + c}\,dx = \frac{x}{2}\sqrt{ax^2 + c} + \frac{c}{2\sqrt{-a}}\sin^{-1}\left(x\sqrt{-\frac{a}{c}}\right),$ $\quad(a < 0)$

275. $\displaystyle\int \sqrt{\frac{1 + x}{1 - x}}\,dx = \sin^{-1}x - \sqrt{1 - x^2}$

276. $\displaystyle\int \frac{dx}{x\sqrt{ax^n + c}} = \begin{cases} \dfrac{1}{n\sqrt{c}}\log\dfrac{\sqrt{ax^n + c} - \sqrt{c}}{\sqrt{ax^n + c} + \sqrt{c}} \\[2ex] \quad\text{or} \\[1ex] \dfrac{2}{n\sqrt{c}}\log\dfrac{\sqrt{ax^n + c} - \sqrt{c}}{\sqrt{x^n}}, \quad(c > 0) \end{cases}$

277. $\displaystyle\int \frac{dx}{x\sqrt{ax^n + c}} = \frac{2}{n\sqrt{-c}}\sec^{-1}\sqrt{-\frac{ax^n}{c}},$ $\quad(c < 0)$

278. $\displaystyle\int \frac{dx}{\sqrt{ax^2 + c}} = \frac{1}{\sqrt{a}}\log\left(x\sqrt{a} + \sqrt{ax^2 + c}\right),$ $\quad(a > 0)$

279. $\displaystyle\int \frac{dx}{\sqrt{ax^2 + c}} = \frac{1}{\sqrt{-a}}\sin^{-1}\left(x\sqrt{-\frac{a}{c}}\right),$ $\quad(a < 0)$

280. $\displaystyle\int (ax^2 + c)^{m+\frac{1}{2}}\,dx = \begin{cases} \dfrac{x(ax^2 + c)^{m+\frac{1}{2}}}{2(m + 1)} + \dfrac{(2m + 1)c}{2(m + 1)}\displaystyle\int(ax^2 + c)^{m-\frac{1}{2}}\,dx \\[2ex] \quad\text{or} \\[1ex] x\sqrt{ax^2 + c}\,\displaystyle\sum_{r=0}^{m}\dfrac{(2m + 1)!(r!)^2 c^{m-r}}{2^{2m-2r+1}m!(m + 1)!(2r + 1)!}(ax^2 + c)^r \\[2ex] \quad + \dfrac{(2m + 1)!c^{m+1}}{2^{2m+1}m!(m + 1)!}\displaystyle\int\dfrac{dx}{\sqrt{ax^2 + c}} \end{cases}$

281. $\displaystyle\int x(ax^2 + c)^{m+\frac{1}{2}}\,dx = \frac{(ax^2 + c)^{m+\frac{3}{2}}}{(2m + 3)a}$

282. $\displaystyle \int \frac{(ax^2 + c)^{m + \frac{1}{2}}}{x} \, dx = \begin{cases} \dfrac{(ax^2 + c)^{m + \frac{1}{2}}}{2m + 1} + c \displaystyle\int \dfrac{(ax^2 + c)^{m - \frac{1}{2}}}{x} \, dx \\ \text{or} \\ \sqrt{ax^2 + c} \displaystyle\sum_{r=0}^{m} \dfrac{c^{m-r}(ax^2 + c)^r}{2r + 1} + c^{m+1} \displaystyle\int \dfrac{dx}{x\sqrt{ax^2 + c}} \end{cases}$

283. $\displaystyle \int \frac{dx}{(ax^2 + c)^{m + \frac{1}{2}}} = \begin{cases} \dfrac{x}{(2m - 1)c(ax^2 + c)^{m - \frac{1}{2}}} + \dfrac{2m - 2}{(2m - 1)c} \displaystyle\int \dfrac{dx}{(ax^2 + c)^{m - \frac{1}{2}}} \\ \text{or} \\ \dfrac{x}{\sqrt{ax^2 + c}} \displaystyle\sum_{r=0}^{m-1} \dfrac{2^{2m - 2r - 1}(m - 1)!\,m!\,(2r)!}{(2m)!\,(r!)^2 c^{m - r}(ax^2 + c)^r} \end{cases}$

284. $\displaystyle \int \frac{dx}{x^m \sqrt{ax^2 + c}} = -\frac{\sqrt{ax^2 + c}}{(m - 1)cx^{m - 1}} - \frac{(m - 2)a}{(m - 1)c} \int \frac{dx}{x^{m - 2}\sqrt{ax^2 + c}}$

285. $\displaystyle \int \frac{1 + x^2}{(1 - x^2)\sqrt{1 + x^4}} \, dx = \frac{1}{\sqrt{2}} \log \frac{x\sqrt{2} + \sqrt{1 + x^4}}{1 - x^2}$

286. $\displaystyle \int \frac{1 - x^2}{(1 + x^2)\sqrt{1 + x^4}} \, dx = \frac{1}{\sqrt{2}} \tan^{-1} \frac{x\sqrt{2}}{\sqrt{1 + x^4}}$

287. $\displaystyle \int \frac{dx}{x\sqrt{x^n + a^2}} = -\frac{2}{na} \log \frac{a + \sqrt{x^n + a^2}}{\sqrt{x^n}}$

288. $\displaystyle \int \frac{dx}{x\sqrt{x^n - a^2}} = -\frac{2}{na} \sin^{-1} \frac{a}{\sqrt{x^n}}$

289. $\displaystyle \int \sqrt{\frac{x}{a^3 - x^3}} \, dx = \frac{2}{3} \sin^{-1} \left(\frac{x}{a}\right)^{\frac{3}{2}}$

FORMS INVOLVING TRIGONOMETRIC FUNCTIONS

290. $\displaystyle \int (\sin ax) \, dx = -\frac{1}{a} \cos ax$

291. $\displaystyle \int (\cos ax) \, dx = \frac{1}{a} \sin ax$

292. $\displaystyle \int (\tan ax) \, dx = -\frac{1}{a} \log \cos ax = \frac{1}{a} \log \sec ax$

293. $\displaystyle \int (\cot ax) \, dx = \frac{1}{a} \log \sin ax = -\frac{1}{a} \log \csc ax$

294. $\displaystyle \int (\sec ax) \, dx = \frac{1}{a} \log (\sec ax + \tan ax) = \frac{1}{a} \log \tan \left(\frac{\pi}{4} + \frac{ax}{2}\right)$

295. $\displaystyle \int (\csc ax) \, dx = \frac{1}{a} \log (\csc ax - \cot ax) = \frac{1}{a} \log \tan \frac{ax}{2}$

296. $\displaystyle \int (\sin^2 ax) \, dx = -\frac{1}{2a} \cos ax \sin ax + \frac{1}{2} x = \frac{1}{2} x - \frac{1}{4a} \sin 2ax$

297. $\displaystyle \int (\sin^3 ax) \, dx = -\frac{1}{3a}(\cos ax)(\sin^2 ax + 2)$

298. $\displaystyle \int (\sin^4 ax) \, dx = \frac{3x}{8} - \frac{\sin 2ax}{4a} + \frac{\sin 4ax}{32a}$

299. $\displaystyle \int (\sin^n ax) \, dx = -\frac{\sin^{n - 1} ax \cos ax}{na} + \frac{n - 1}{n} \int (\sin^{n - 2} ax) \, dx$

300. $\int (\sin^{2m} ax)\, dx = -\dfrac{\cos ax}{a} \sum\limits_{r=0}^{m-1} \dfrac{(2m)!(r!)^2}{2^{2m-2r}(2r+1)!(m!)^2} \sin^{2r+1} ax + \dfrac{(2m)!}{2^{2m}(m!)^2} x$

301. $\int (\sin^{2m+1} ax)\, dx = -\dfrac{\cos ax}{a} \sum\limits_{r=0}^{m} \dfrac{2^{2m-2r}(m!)^2(2r)!}{(2m+1)!(r!)^2} \sin^{2r} ax$

302. $\int (\cos^2 ax)\, dx = \dfrac{1}{2a} \sin ax \cos ax + \dfrac{1}{2} x = \dfrac{1}{2} x + \dfrac{1}{4a} \sin 2ax$

303. $\int (\cos^3 ax)\, dx = \dfrac{1}{3a}(\sin ax)(\cos^2 ax + 2)$

304. $\int (\cos^4 ax)\, dx = \dfrac{3x}{8} + \dfrac{\sin 2ax}{4a} + \dfrac{\sin 4ax}{32a}$

305. $\int (\cos^n ax)\, dx = \dfrac{1}{na} \cos^{n-1} ax \sin ax + \dfrac{n-1}{n} \int (\cos^{n-2} ax)\, dx$

306. $\int (\cos^{2m} ax)\, dx = \dfrac{\sin ax}{a} \sum\limits_{r=0}^{m-1} \dfrac{(2m)!(r!)^2}{2^{2m-2r}(2r+1)!(m!)^2} \cos^{2r+1} ax + \dfrac{(2m)!}{2^{2m}(m!)^2} x$

307. $\int (\cos^{2m+1} ax)\, dx = \dfrac{\sin ax}{a} \sum\limits_{r=0}^{m} \dfrac{2^{2m-2r}(m!)^2(2r)!}{(2m+1)!(r!)^2} \cos^{2r} ax$

308. $\int \dfrac{dx}{\sin^2 ax} = \int (\csc^2 ax)\, dx = -\dfrac{1}{a} \cot ax$

309. $\int \dfrac{dx}{\sin^m ax} = \int (\csc^m ax)\, dx = -\dfrac{1}{(m-1)a} \cdot \dfrac{\cos ax}{\sin^{m-1} ax} + \dfrac{m-2}{m-1} \int \dfrac{dx}{\sin^{m-2} ax}$

310. $\int \dfrac{dx}{\sin^{2m} ax} = \int (\csc^{2m} ax)\, dx = -\dfrac{1}{a} \cos ax \sum\limits_{r=0}^{m-1} \dfrac{2^{2m-2r-1}(m-1)!m!(2r)!}{(2m)!(r!)^2 \sin^{2r+1} ax}$

311. $\int \dfrac{dx}{\sin^{2m+1} ax} = \int (\csc^{2m+1} ax)\, dx =$

$\qquad -\dfrac{1}{a} \cos ax \sum\limits_{r=0}^{m-1} \dfrac{(2m)!(r!)^2}{2^{2m-2r}(m!)^2(2r+1)! \sin^{2r+2} ax} + \dfrac{1}{a} \cdot \dfrac{(2m)!}{2^{2m}(m!)^2} \log \tan \dfrac{ax}{2}$

312. $\int \dfrac{dx}{\cos^2 ax} = \int (\sec^2 ax)\, dx = \dfrac{1}{a} \tan ax$

313. $\int \dfrac{dx}{\cos^n ax} = \int (\sec^n ax)\, dx = \dfrac{1}{(n-1)a} \cdot \dfrac{\sin ax}{\cos^{n-1} ax} + \dfrac{n-2}{n-1} \int \dfrac{dx}{\cos^{n-2} ax}$

314. $\int \dfrac{dx}{\cos^{2m} ax} = \int (\sec^{2m} ax)\, dx = \dfrac{1}{a} \sin ax \sum\limits_{r=0}^{m-1} \dfrac{2^{2m-2r-1}(m-1)!m!(2r)!}{(2m)!(r!)^2 \cos^{2r+1} ax}$

315. $\int \dfrac{dx}{\cos^{2m+1} ax} = \int (\sec^{2m+1} ax)\, dx =$

$\qquad \dfrac{1}{a} \sin ax \sum\limits_{r=0}^{m-1} \dfrac{(2m)!(r!)^2}{2^{2m-2r}(m!)^2(2r+1)! \cos^{2r+2} ax}$

$\qquad\qquad\qquad + \dfrac{1}{a} \cdot \dfrac{(2m)!}{2^{2m}(m!)^2} \log (\sec ax + \tan ax)$

316. $\int (\sin mx)(\sin nx)\, dx = \dfrac{\sin(m-n)x}{2(m-n)} - \dfrac{\sin(m+n)x}{2(m+n)}, \qquad (m^2 \neq n^2)$

317. $\int (\cos mx)(\cos nx)\, dx = \dfrac{\sin(m-n)x}{2(m-n)} + \dfrac{\sin(m+n)x}{2(m+n)}, \qquad (m^2 \neq n^2)$

318. $\int (\sin ax)(\cos ax)\, dx = \dfrac{1}{2a} \sin^2 ax$

319. $\int (\sin mx)(\cos nx)\, dx = -\dfrac{\cos(m-n)x}{2(m-n)} - \dfrac{\cos(m+n)x}{2(m+n)}, \qquad (m^2 \neq n^2)$

320. $\int (\sin^2 ax)(\cos^2 ax)\, dx = -\dfrac{1}{32a}\sin 4ax + \dfrac{x}{8}$

321. $\int (\sin ax)(\cos^m ax)\, dx = -\dfrac{\cos^{m+1} ax}{(m+1)a}$

322. $\int (\sin^m ax)(\cos ax)\, dx = \dfrac{\sin^{m+1} ax}{(m+1)a}$

323. $\int (\cos^m ax)(\sin^n ax)\, dx = \begin{cases} \dfrac{\cos^{m-1} ax \sin^{n+1} ax}{(m+n)a} \\[2mm] \qquad + \dfrac{m-1}{m+n}\displaystyle\int (\cos^{m-2} ax)(\sin^n ax)\, dx \\[4mm] \text{or} \\[2mm] -\dfrac{\sin^{n-1} ax \cos^{m+1} ax}{(m+n)a} \\[2mm] \qquad + \dfrac{n-1}{m+n}\displaystyle\int (\cos^m ax)(\sin^{n-2} ax)\, dx \end{cases}$

324. $\int \dfrac{\cos^m ax}{\sin^n ax}\, dx = \begin{cases} -\dfrac{\cos^{m+1} ax}{(n-1)a \sin^{n-1} ax} - \dfrac{m-n+2}{n-1}\displaystyle\int \dfrac{\cos^m ax}{\sin^{n-2} ax}\, dx \\[4mm] \text{or} \\[2mm] \dfrac{\cos^{m-1} ax}{a(m-n)\sin^{n-1} ax} + \dfrac{m-1}{m-n}\displaystyle\int \dfrac{\cos^{m-2} ax}{\sin^n ax}\, dx \end{cases}$

325. $\int \dfrac{\sin^m ax}{\cos^n ax}\, dx = \begin{cases} \dfrac{\sin^{m+1} ax}{a(n-1)\cos^{n-1} ax} - \dfrac{m-n+2}{n-1}\displaystyle\int \dfrac{\sin^m ax}{\cos^{n-2} ax}\, dx \\[4mm] \text{or} \\[2mm] -\dfrac{\sin^{m-1} ax}{a(m-n)\cos^{n-1} ax} + \dfrac{m-1}{m-n}\displaystyle\int \dfrac{\sin^{m-2} ax}{\cos^n ax}\, dx \end{cases}$

326. $\int \dfrac{\sin ax}{\cos^2 ax}\, dx = \dfrac{1}{a\cos ax} = \dfrac{\sec ax}{a}$

327. $\int \dfrac{\sin^2 ax}{\cos ax}\, dx = -\dfrac{1}{a}\sin ax + \dfrac{1}{a}\log \tan\left(\dfrac{\pi}{4} + \dfrac{ax}{2}\right)$

328. $\int \dfrac{\cos ax}{\sin^2 ax}\, dx = -\dfrac{1}{a\sin ax} = -\dfrac{\csc ax}{a}$

329. $\int \dfrac{dx}{(\sin ax)(\cos ax)} = \dfrac{1}{a}\log \tan ax$

330. $\int \dfrac{dx}{(\sin ax)(\cos^2 ax)} = \dfrac{1}{a}\left(\sec ax + \log \tan \dfrac{ax}{2}\right)$

331. $\int \dfrac{dx}{(\sin ax)(\cos^n ax)} = \dfrac{1}{a(n-1)\cos^{n-1} ax} + \int \dfrac{dx}{(\sin ax)(\cos^{n-2} ax)}$

332. $\int \dfrac{dx}{(\sin^2 ax)(\cos ax)} = -\dfrac{1}{a}\csc ax + \dfrac{1}{a}\log \tan\left(\dfrac{\pi}{4} + \dfrac{ax}{2}\right)$

333. $\int \dfrac{dx}{(\sin^2 ax)(\cos^2 ax)} = -\dfrac{2}{a}\cot 2ax$

334. $\displaystyle\int \frac{dx}{\sin^m ax \cos^n ax} =$
$$\begin{cases} -\dfrac{1}{a(m-1)(\sin^{m-1} ax)(\cos^{n-1} ax)} \\ \qquad + \dfrac{m+n-2}{m-1}\displaystyle\int \dfrac{dx}{(\sin^{m-2} ax)(\cos^n ax)} \\ \text{or} \\ \dfrac{1}{a(n-1)\sin^{m-1} ax \cos^{n-1} ax} \\ \qquad - \dfrac{m+n-2}{n-1}\displaystyle\int \dfrac{dx}{\sin^m ax \cos^{n-2} ax} \end{cases}$$

335. $\displaystyle\int \sin(a+bx)\,dx = -\frac{1}{b}\cos(a+bx)$

336. $\displaystyle\int \cos(a+bx)\,dx = \frac{1}{b}\sin(a+bx)$

337. $\displaystyle\int \frac{dx}{1 \pm \sin ax} = \mp \frac{1}{a}\tan\left(\frac{\pi}{4} \mp \frac{ax}{2}\right)$

338. $\displaystyle\int \frac{dx}{1 + \cos ax} = \frac{1}{a}\tan\frac{ax}{2}$

339. $\displaystyle\int \frac{dx}{1 - \cos ax} = -\frac{1}{a}\cot\frac{ax}{2}$

*340. $\displaystyle\int \frac{dx}{a + b\sin x} =$
$$\begin{cases} \dfrac{2}{\sqrt{a^2-b^2}}\tan^{-1}\dfrac{a\tan\frac{x}{2}+b}{\sqrt{a^2-b^2}} \\ \text{or} \\ \dfrac{1}{\sqrt{b^2-a^2}}\log\dfrac{a\tan\frac{x}{2}+b-\sqrt{b^2-a^2}}{a\tan\frac{x}{2}+b+\sqrt{b^2-a^2}} \end{cases}$$

*341. $\displaystyle\int \frac{dx}{a + b\cos x} =$
$$\begin{cases} \dfrac{2}{\sqrt{a^2-b^2}}\tan^{-1}\dfrac{\sqrt{a^2-b^2}\tan\frac{x}{2}}{a+b} \\ \text{or} \\ \dfrac{1}{\sqrt{b^2-a^2}}\log\left(-\dfrac{\sqrt{b^2-a^2}\tan\frac{x}{2}+a+b}{\sqrt{b^2-a^2}\tan\frac{x}{2}-a-b}\right) \end{cases}$$

*342. $\displaystyle\int \frac{dx}{a + b\sin x + c\cos x}$
$$=\begin{cases} \dfrac{1}{\sqrt{b^2+c^2-a^2}}\log\dfrac{b-\sqrt{b^2+c^2-a^2}+(a-c)\tan\frac{x}{2}}{b+\sqrt{b^2+c^2-a^2}+(a-c)\tan\frac{x}{2}}, & \text{if } a^2 < b^2 + c^2, a \neq c \\ \text{or} \\ \dfrac{2}{\sqrt{a^2-b^2-c^2}}\tan^{-1}\dfrac{b+(a-c)\tan\frac{x}{2}}{\sqrt{a^2-b^2-c^2}}, & \text{if } a^2 > b^2 + c^2 \\ \text{or} \\ \dfrac{1}{a}\left[\dfrac{a-(b+c)\cos x-(b-c)\sin x}{a-(b-c)\cos x+(b+c)\sin x}\right], & \text{if } a^2 = b^2 + c^2, a \neq c. \end{cases}$$

*See note 6 on page A-19.

*343. $\int \dfrac{\sin^2 x\, dx}{a + b \cos^2 x} = \dfrac{1}{b}\sqrt{\dfrac{a+b}{a}}\, \tan^{-1}\!\left(\sqrt{\dfrac{a}{a+b}}\, \tan x\right) - \dfrac{x}{b}$, $(ab > 0,\ \text{or}\ |a| > |b|)$

*344. $\int \dfrac{dx}{a^2 \cos^2 x + b^2 \sin^2 x} = \dfrac{1}{ab}\, \tan^{-1}\!\left(\dfrac{b \tan x}{a}\right)$

*345. $\int \dfrac{\cos^2 cx}{a^2 + b^2 \sin^2 cx}\, dx = \dfrac{\sqrt{a^2 + b^2}}{ab^2 c}\, \tan^{-1}\dfrac{\sqrt{a^2 + b^2}\, \tan cx}{a} - \dfrac{x}{b^2}$

346. $\int \dfrac{\sin cx \cos cx}{a \cos^2 cx + b \sin^2 cx}\, dx = \dfrac{1}{2c(b-a)}\log(a \cos^2 cx + b \sin^2 cx)$

347. $\int \dfrac{\cos cx}{a \cos cx + b \sin cx}\, dx = \int \dfrac{dx}{a + b \tan cx} =$

$$\dfrac{1}{c(a^2 + b^2)}[acx + b \log(a \cos cx + b \,\text{si}$$

348. $\int \dfrac{\sin cx}{a \sin cx + b \cos cx}\, dx = \int \dfrac{dx}{a + b \cot cx} =$

$$\dfrac{1}{c(a^2 + b^2)}[acx - b \log(a \sin cx + b \cos x)]$$

*349. $\int \dfrac{dx}{a \cos^2 x + 2b \cos x \sin x + c \sin^2 x} = \begin{cases} \dfrac{1}{2\sqrt{b^2 - ac}}\log\dfrac{c \tan x + b - \sqrt{b^2 - ac}}{c \tan x + b + \sqrt{b^2 - ac}} \\ \qquad\qquad (b^2 > ac) \\ \text{or} \\ \dfrac{1}{\sqrt{ac - b^2}}\, \tan^{-1}\dfrac{c \tan x + b}{\sqrt{ac - b^2}},\quad (b_2 < ac) \\ \text{or} \\ -\dfrac{1}{c \tan x + b},\qquad (b^2 = ac) \end{cases}$

350. $\int \dfrac{\sin ax}{1 \pm \sin ax}\, dx = \pm x + \dfrac{1}{a}\tan\!\left(\dfrac{\pi}{4} \mp \dfrac{ax}{2}\right)$

351. $\int \dfrac{dx}{(\sin ax)(1 \pm \sin ax)} = \dfrac{1}{a}\tan\!\left(\dfrac{\pi}{4} \mp \dfrac{ax}{2}\right) + \dfrac{1}{a}\log \tan \dfrac{ax}{2}$

352. $\int \dfrac{dx}{(1 + \sin ax)^2} = -\dfrac{1}{2a}\tan\!\left(\dfrac{\pi}{4} - \dfrac{ax}{2}\right) - \dfrac{1}{6a}\tan^3\!\left(\dfrac{\pi}{4} - \dfrac{ax}{2}\right)$

353. $\int \dfrac{dx}{(1 - \sin ax)^2} = \dfrac{1}{2a}\cot\!\left(\dfrac{\pi}{4} - \dfrac{ax}{2}\right) + \dfrac{1}{6a}\cot^3\!\left(\dfrac{\pi}{4} - \dfrac{ax}{2}\right)$

354. $\int \dfrac{\sin ax}{(1 + \sin ax)^2}\, dx = -\dfrac{1}{2a}\tan\!\left(\dfrac{\pi}{4} - \dfrac{ax}{2}\right) + \dfrac{1}{6a}\tan^3\!\left(\dfrac{\pi}{4} - \dfrac{ax}{2}\right)$

355. $\int \dfrac{\sin ax}{(1 - \sin ax)^2}\, dx = -\dfrac{1}{2a}\cot\!\left(\dfrac{\pi}{4} - \dfrac{ax}{2}\right) + \dfrac{1}{6a}\cot^3\!\left(\dfrac{\pi}{4} - \dfrac{ax}{2}\right)$

356. $\int \dfrac{\sin x\, dx}{a + b \sin x} = \dfrac{x}{b} - \dfrac{a}{b}\int \dfrac{dx}{a + b \sin x}$

357. $\int \dfrac{dx}{(\sin x)(a + b \sin x)} = \dfrac{1}{a}\log \tan \dfrac{x}{2} - \dfrac{b}{a}\int \dfrac{dx}{a + b \sin x}$

358. $\int \dfrac{dx}{(a + b \sin x)^2} = \dfrac{b \cos x}{(a^2 - b^2)(a + b \sin x)} + \dfrac{a}{a^2 - b^2}\int \dfrac{dx}{a + b \sin x}$

*See note 6 on page A-19.

359. $\displaystyle\int \frac{\sin x\, dx}{(a + b \sin x)^2} = \frac{a \cos x}{(b^2 - a^2)(a + b \sin x)} + \frac{b}{b^2 - a^2} \int \frac{dx}{a + b \sin x}$

*360. $\displaystyle\int \frac{dx}{a^2 + b^2 \sin^2 cx} = \frac{1}{ac\sqrt{a^2 + b^2}} \tan^{-1} \frac{\sqrt{a^2 + b^2}\, \tan cx}{a}$

*361. $\displaystyle\int \frac{dx}{a^2 - b^2 \sin^2 cx} = \begin{cases} \dfrac{1}{ac\sqrt{a^2 - b^2}} \tan^{-1} \dfrac{\sqrt{a^2 - b^2}\, \tan cx}{a}, & (a^2 > b^2) \\[2mm] \text{or} \\[2mm] \dfrac{1}{2ac\sqrt{b^2 - a^2}} \log \dfrac{\sqrt{b^2 - a^2}\, \tan cx + a}{\sqrt{b^2 - a^2}\, \tan cx - a}, & (a^2 < b^2) \end{cases}$

362. $\displaystyle\int \frac{\cos ax}{1 + \cos ax} dx = x - \frac{1}{a} \tan \frac{ax}{2}$

363. $\displaystyle\int \frac{\cos ax}{1 - \cos ax} dx = -x - \frac{1}{a} \cot \frac{ax}{2}$

364. $\displaystyle\int \frac{dx}{(\cos ax)(1 + \cos ax)} = \frac{1}{a} \log \tan \left(\frac{\pi}{4} + \frac{ax}{2}\right) - \frac{1}{a} \tan \frac{ax}{2}$

365. $\displaystyle\int \frac{dx}{(\cos ax)(1 - \cos ax)} = \frac{1}{a} \log \tan \left(\frac{\pi}{4} + \frac{ax}{2}\right) - \frac{1}{a} \cot \frac{ax}{2}$

366. $\displaystyle\int \frac{dx}{(1 + \cos ax)^2} = \frac{1}{2a} \tan \frac{ax}{2} + \frac{1}{6a} \tan^3 \frac{ax}{2}$

367. $\displaystyle\int \frac{dx}{(1 - \cos ax)^2} = -\frac{1}{2a} \cot \frac{ax}{2} - \frac{1}{6a} \cot^3 \frac{ax}{2}$

368. $\displaystyle\int \frac{\cos ax}{(1 + \cos ax)^2} dx = \frac{1}{2a} \tan \frac{ax}{2} - \frac{1}{6a} \tan^3 \frac{ax}{2}$

369. $\displaystyle\int \frac{\cos ax}{(1 - \cos ax)^2} dx = \frac{1}{2a} \cot \frac{ax}{2} - \frac{1}{6a} \cot^3 \frac{ax}{2}$

370. $\displaystyle\int \frac{\cos x\, dx}{a + b \cos x} = \frac{x}{b} - \frac{a}{b} \int \frac{dx}{a + b \cos x}$

371. $\displaystyle\int \frac{dx}{(\cos x)(a + b \cos x)} = \frac{1}{a} \log \tan \left(\frac{x}{2} + \frac{\pi}{4}\right) - \frac{b}{a} \int \frac{dx}{a + b \cos x}$

372. $\displaystyle\int \frac{dx}{(a + b \cos x)^2} = \frac{b \sin x}{(b^2 - a^2)(a + b \cos x)} - \frac{a}{b^2 - a^2} \int \frac{dx}{a + b \cos x}$

373. $\displaystyle\int \frac{\cos x}{(a + b \cos x)^2} dx = \frac{a \sin x}{(a^2 - b^2)(a + b \cos x)} - \frac{b}{a^2 - b^2} \int \frac{dx}{a + b \cos x}$

*374. $\displaystyle\int \frac{dx}{a^2 + b^2 - 2ab \cos cx} = \frac{2}{c(a^2 - b^2)} \tan^{-1}\left(\frac{a + b}{a - b} \tan \frac{cx}{2}\right)$

*375. $\displaystyle\int \frac{dx}{a^2 + b^2 \cos^2 cx} = \frac{1}{ac\sqrt{a^2 + b^2}} \tan^{-1} \frac{a \tan cx}{\sqrt{a^2 + b^2}}$

*376. $\displaystyle\int \frac{dx}{a^2 - b^2 \cos^2 cx} = \begin{cases} \dfrac{1}{ac\sqrt{a^2 - b^2}} \tan^{-1} \dfrac{a \tan cx}{\sqrt{a^2 - b^2}}, & (a^2 > b^2) \\[2mm] \text{or} \\[2mm] \dfrac{1}{2ac\sqrt{b^2 - a^2}} \log \dfrac{a \tan cx - \sqrt{b^2 - a^2}}{a \tan cx + \sqrt{b^2 - a^2}}, & (b^2 > a^2) \end{cases}$

377. $\displaystyle\int \frac{\sin ax}{1 \pm \cos ax} dx = \mp \frac{1}{a} \log (1 \pm \cos ax)$

*See note 6 on page A-19.

378. $\int \dfrac{\cos ax}{1 \pm \sin ax}\, dx = \pm \dfrac{1}{a} \log\left(1 \pm \sin ax\right)$

379. $\int \dfrac{dx}{(\sin ax)(1 \pm \cos ax)} = \pm \dfrac{1}{2a(1 \pm \cos ax)} + \dfrac{1}{2a} \log \tan \dfrac{ax}{2}$

380. $\int \dfrac{dx}{(\cos ax)(1 \pm \sin ax)} = \mp \dfrac{1}{2a(1 \pm \sin ax)} + \dfrac{1}{2a} \log \tan\left(\dfrac{\pi}{4} + \dfrac{ax}{2}\right)$

381. $\int \dfrac{\sin ax}{(\cos ax)(1 \pm \cos ax)}\, dx = \dfrac{1}{a} \log\left(\sec ax \pm 1\right)$

382. $\int \dfrac{\cos ax}{(\sin ax)(1 \pm \sin ax)}\, dx = -\dfrac{1}{a} \log\left(\csc ax \pm 1\right)$

383. $\int \dfrac{\sin ax}{(\cos ax)(1 \pm \sin ax)}\, dx = \dfrac{1}{2a(1 \pm \sin ax)} \pm \dfrac{1}{2a} \log \tan\left(\dfrac{\pi}{4} + \dfrac{ax}{2}\right)$

384. $\int \dfrac{\cos ax}{(\sin ax)(1 \pm \cos ax)}\, dx = -\dfrac{1}{2a(1 \pm \cos ax)} \pm \dfrac{1}{2a} \log \tan \dfrac{ax}{2}$

385. $\int \dfrac{dx}{\sin ax \pm \cos ax} = \dfrac{1}{a\sqrt{2}} \log \tan\left(\dfrac{ax}{2} \pm \dfrac{\pi}{8}\right)$

386. $\int \dfrac{dx}{(\sin ax \pm \cos ax)^2} = \dfrac{1}{2a} \tan\left(ax \mp \dfrac{\pi}{4}\right)$

387. $\int \dfrac{dx}{1 + \cos ax \pm \sin ax} = \pm \dfrac{1}{a} \log\left(1 \pm \tan \dfrac{ax}{2}\right)$

388. $\int \dfrac{dx}{a^2 \cos^2 cx - b^2 \sin^2 cx} = \dfrac{1}{2abc} \log \dfrac{b \tan cx + a}{b \tan cx - a}$

389. $\int x(\sin ax)\, dx = \dfrac{1}{a^2} \sin ax - \dfrac{x}{a} \cos ax$

390. $\int x^2(\sin ax)\, dx = \dfrac{2x}{a^2} \sin ax - \dfrac{a^2 x^2 - 2}{a^3} \cos ax$

391. $\int x^3(\sin ax)\, dx = \dfrac{3a^2 x^2 - 6}{a^4} \sin ax - \dfrac{a^2 x^3 - 6x}{a^3} \cos ax$

392. $\int x^m \sin ax\, dx = \begin{cases} -\dfrac{1}{a} x^m \cos ax + \dfrac{m}{a} \int x^{m-1} \cos ax\, dx \\[2mm] \text{or} \\[2mm] \cos ax \displaystyle\sum_{r=0}^{\left[\frac{m}{2}\right]} (-1)^{r+1} \dfrac{m!}{(m-2r)!} \cdot \dfrac{x^{m-2r}}{a^{2r+1}} \\[4mm] + \sin ax \displaystyle\sum_{r=0}^{\left[\frac{m-1}{2}\right]} (-1)^{r} \dfrac{m!}{(m-2r-1)!} \cdot \dfrac{x^{m-2r-1}}{a^{2r+2}} \end{cases}$

Note: $[s]$ means greatest integer $\le s$; $[3\frac{1}{2}] = 3$, $[\frac{1}{2}] = 0$, etc.

393. $\int x(\cos ax)\, dx = \dfrac{1}{a^2} \cos ax + \dfrac{x}{a} \sin ax$

394. $\int x^2(\cos ax)\, dx = \dfrac{2x \cos ax}{a^2} + \dfrac{a^2 x^2 - 2}{a^3} \sin ax$

395. $\int x^3(\cos ax)\, dx = \dfrac{3a^2 x^2 - 6}{a^4} \cos ax + \dfrac{a^2 x^3 - 6x}{a^3} \sin ax$

396. $\int x^m(\cos ax)\,dx = \begin{cases} \dfrac{x^m \sin ax}{a} - \dfrac{m}{a}\int x^{m-1}\sin ax\,dx \\[4pt] \text{or} \\[4pt] \sin ax \sum\limits_{r=0}^{\left|\frac{m}{2}\right|}(-1)^r\,\dfrac{m!}{(m-2r)!}\cdot\dfrac{x^{m-2r}}{a^{2r+1}} \\[8pt] \quad + \cos ax \sum\limits_{r=0}^{\left|\frac{m-1}{2}\right|}(-1)^r\,\dfrac{m!}{(m-2r-1)!}\cdot\dfrac{x^{m-2r-1}}{a^{2r+2}} \end{cases}$

See note integral 392.

397. $\int \dfrac{\sin ax}{x}\,dx = \sum\limits_{n=0}^{r}(-1)^n\,\dfrac{(ax)^{2n+1}}{(2n+1)(2n+1)!}$

398. $\int \dfrac{\cos ax}{x}\,dx = \log x + \sum\limits_{n=1}^{r}(-1)^n\,\dfrac{(ax)^{2n}}{2n(2n)!}$

399. $\int x(\sin^2 ax)\,dx = \dfrac{x^2}{4} - \dfrac{x\sin 2ax}{4a} - \dfrac{\cos 2ax}{8a^2}$

400. $\int x^2(\sin^2 ax)\,dx = \dfrac{x^3}{6} - \left(\dfrac{x^2}{4a} - \dfrac{1}{8a^3}\right)\sin 2ax - \dfrac{x\cos 2ax}{4a^2}$

401. $\int x(\sin^3 ax)\,dx = \dfrac{x\cos 3ax}{12a} - \dfrac{\sin 3ax}{36a^2} - \dfrac{3x\cos ax}{4a} + \dfrac{3\sin ax}{4a^2}$

402. $\int x(\cos^2 ax)\,dx = \dfrac{x^2}{4} + \dfrac{x\sin 2ax}{4a} + \dfrac{\cos 2ax}{8a^2}$

403. $\int x^2(\cos^2 ax)\,dx = \dfrac{x^3}{6} + \left(\dfrac{x^2}{4a} - \dfrac{1}{8a^3}\right)\sin 2ax + \dfrac{x\cos 2ax}{4a^2}$

404. $\int x(\cos^3 ax)\,dx = \dfrac{x\sin 3ax}{12a} + \dfrac{\cos 3ax}{36a^2} + \dfrac{3x\sin ax}{4a} + \dfrac{3\cos ax}{4a^2}$

405. $\int \dfrac{\sin ax}{x^m}\,dx = -\dfrac{\sin ax}{(m-1)x^{m-1}} + \dfrac{a}{m-1}\int\dfrac{\cos ax}{x^{m-1}}\,dx$

406. $\int \dfrac{\cos ax}{x^m}\,dx = -\dfrac{\cos ax}{(m-1)x^{m-1}} - \dfrac{a}{m-1}\int\dfrac{\sin ax}{x^{m-1}}\,dx$

407. $\int \dfrac{x}{1\pm\sin ax}\,dx = \mp\dfrac{x\cos ax}{a(1\pm\sin ax)} + \dfrac{1}{a^2}\log(1\pm\sin ax)$

408. $\int \dfrac{x}{1+\cos ax}\,dx = \dfrac{x}{a}\tan\dfrac{ax}{2} + \dfrac{2}{a^2}\log\cos\dfrac{ax}{2}$

409. $\int \dfrac{x}{1-\cos ax}\,dx = -\dfrac{x}{a}\cot\dfrac{ax}{2} + \dfrac{2}{a^2}\log\sin\dfrac{ax}{2}$

410. $\int \dfrac{x+\sin x}{1+\cos x}\,dx = x\tan\dfrac{x}{2}$

411. $\int \dfrac{x-\sin x}{1-\cos x}\,dx = -x\cot\dfrac{x}{2}$

412. $\int \sqrt{1-\cos ax}\,dx = -\dfrac{2\sin ax}{a\sqrt{1-\cos ax}} = -\dfrac{2\sqrt{2}}{a}\cos\left(\dfrac{ax}{2}\right)$

413. $\int \sqrt{1+\cos ax}\,dx = \dfrac{2\sin ax}{a\sqrt{1+\cos ax}} = \dfrac{2\sqrt{2}}{a}\sin\left(\dfrac{ax}{2}\right)$

414. $\int \sqrt{1+\sin x}\,dx = \pm 2\left(\sin\dfrac{x}{2} - \cos\dfrac{x}{2}\right),$

[use $+$ if $(8k-1)\dfrac{\pi}{2} < x \le (8k+3)\dfrac{\pi}{2}$, otherwise $-$; k an integer]

415. $\int \sqrt{1 - \sin x}\, dx = \pm 2\left(\sin\dfrac{x}{2} + \cos\dfrac{x}{2}\right),$

[use $+$ if $(8k - 3)\dfrac{\pi}{2} < x \leq (8k + 1)\dfrac{\pi}{2}$, otherwise $-$; k an integer]

416. $\int \dfrac{dx}{\sqrt{1 - \cos x}} = \pm \sqrt{2} \log \tan \dfrac{x}{4},$

[use $+$ if $4k\pi < x < (4k + 2)\pi$, otherwise $-$; k an integer]

417. $\int \dfrac{dx}{\sqrt{1 + \cos x}} = \pm \sqrt{2} \log \tan\left(\dfrac{x + \pi}{4}\right),$

[use $+$ if $(4k - 1)\pi < x < (4k + 1)\pi$, otherwise $-$; k an integer]

418. $\int \dfrac{dx}{\sqrt{1 - \sin x}} = \pm \sqrt{2} \log \tan\left(\dfrac{x}{4} - \dfrac{\pi}{8}\right),$

[use $+$ if $(8k + 1)\dfrac{\pi}{2} < x < (8k + 5)\dfrac{\pi}{2}$, otherwise $-$; k an integer]

419. $\int \dfrac{dx}{\sqrt{1 + \sin x}} = \pm \sqrt{2} \log \tan\left(\dfrac{x}{4} + \dfrac{\pi}{8}\right),$

[use $+$ if $(8k - 1)\dfrac{\pi}{2} < x < (8k + 3)\dfrac{\pi}{2}$, otherwise $-$; k an integer]

420. $\int (\tan^2 ax)\, dx = \dfrac{1}{a} \tan ax - x$

421. $\int (\tan^3 ax)\, dx = \dfrac{1}{2a} \tan^2 ax + \dfrac{1}{a} \log \cos ax$

422. $\int (\tan^4 ax)\, dx = \dfrac{\tan^3 ax}{3a} - \dfrac{1}{a} \tan x + x$

423. $\int (\tan^n ax)\, dx = \dfrac{\tan^{n-1} ax}{a(n - 1)} - \int (\tan^{n-2} ax)\, dx$

424. $\int (\cot^2 ax)\, dx = -\dfrac{1}{a} \cot ax - x$

425. $\int (\cot^3 ax)\, dx = -\dfrac{1}{2a} \cot^2 ax - \dfrac{1}{a} \log \sin ax$

426. $\int (\cot^4 ax)\, dx = -\dfrac{1}{3a} \cot^3 ax + \dfrac{1}{a} \cot ax + x$

427. $\int (\cot^n ax)\, dx = -\dfrac{\cot^{n-1} ax}{a(n - 1)} - \int (\cot^{n-2} ax)\, dx$

428. $\int \dfrac{x}{\sin^2 ax}\, dx = \int x(\csc^2 ax)\, dx = -\dfrac{x \cot ax}{a} + \dfrac{1}{a^2} \log \sin ax$

429. $\int \dfrac{x}{\sin^n ax}\, dx = \int x(\csc^n ax)\, dx = -\dfrac{x \cos ax}{a(n - 1)\sin^{n-1} ax}$

$$-\dfrac{1}{a^2(n - 1)(n - 2)\sin^{n-2} ax} + \dfrac{(n - 2)}{(n - 1)} \int \dfrac{x}{\sin^{n-2} ax}\, dx$$

430. $\int \dfrac{x}{\cos^2 ax}\, dx = \int x(\sec^2 ax)\, dx = \dfrac{1}{a} x \tan ax + \dfrac{1}{a^2} \log \cos ax$

431. $\int \dfrac{x}{\cos^n ax}\, dx = \int x(\sec^n ax)\, dx = \dfrac{x \sin ax}{a(n - 1)\cos^{n-1} ax}$

$$-\dfrac{1}{a^2(n - 1)(n - 2)\cos^{n-2} ax} + \dfrac{n - 2}{n - 1} \int \dfrac{x}{\cos^{n-2} ax}\, dx$$

432. $\int \dfrac{\sin ax}{\sqrt{1 + b^2 \sin^2 ax}} dx = -\dfrac{1}{ab} \sin^{-1} \dfrac{b \cos ax}{\sqrt{1 + b^2}}$

433. $\int \dfrac{\sin ax}{\sqrt{1 - b^2 \sin^2 ax}} dx = -\dfrac{1}{ab} \log (b \cos ax + \sqrt{1 - b^2 \sin^2 ax})$

434. $\int (\sin ax) \sqrt{1 + b^2 \sin^2 ax}\, dx = -\dfrac{\cos ax}{2a} \sqrt{1 + b^2 \sin^2 ax}$

$$- \dfrac{1 + b^2}{2ab} \sin^{-1} \dfrac{b \cos ax}{\sqrt{1 + b^2}}$$

435. $\int (\sin ax) \sqrt{1 - b^2 \sin^2 ax}\, dx = -\dfrac{\cos ax}{2a} \sqrt{1 - b^2 \sin^2 ax}$

$$- \dfrac{1 - b^2}{2ab} \log (b \cos ax + \sqrt{1 - b^2 \sin^2 ax})$$

436. $\int \dfrac{\cos ax}{\sqrt{1 + b^2 \sin^2 ax}} dx = \dfrac{1}{ab} \log (b \sin ax + \sqrt{1 + b^2 \sin^2 ax})$

437. $\int \dfrac{\cos ax}{\sqrt{1 - b^2 \sin^2 ax}} dx = \dfrac{1}{ab} \sin^{-1} (b \sin ax)$

438. $\int (\cos ax) \sqrt{1 + b^2 \sin^2 ax}\, dx = \dfrac{\sin ax}{2a} \sqrt{1 + b^2 \sin^2 ax}$

$$+ \dfrac{1}{2ab} \log (b \sin ax + \sqrt{1 + b^2 \sin^2 ax})$$

439. $\int (\cos ax) \sqrt{1 - b^2 \sin^2 ax}\, dx = \dfrac{\sin ax}{2a} \sqrt{1 - b^2 \sin^2 ax} + \dfrac{1}{2ab} \sin^{-1} (b \sin ax)$

440. $\int \dfrac{dx}{\sqrt{a + b \tan^2 cx}} = \dfrac{\pm 1}{c\sqrt{a - b}} \sin^{-1} \left(\sqrt{\dfrac{a - b}{a}} \sin cx \right), \qquad (a > |b|)$

$$\left[\text{use } + \text{ if } (2k - 1)\dfrac{\pi}{2} < x \le (2k + 1)\dfrac{\pi}{2}, \text{ otherwise } - : k \text{ an integer} \right]$$

FORMS INVOLVING INVERSE TRIGONOMETRIC FUNCTIONS

441. $\int (\sin^{-1} ax)\, dx = x \sin^{-1} ax + \dfrac{\sqrt{1 - a^2 x^2}}{a}$

442. $\int (\cos^{-1} ax)\, dx = x \cos^{-1} ax - \dfrac{\sqrt{1 - a^2 x^2}}{a}$

443. $\int (\tan^{-1} ax)\, dx = x \tan^{-1} ax - \dfrac{1}{2a} \log (1 + a^2 x^2)$

444. $\int (\cot^{-1} ax)\, dx = x \cot^{-1} ax + \dfrac{1}{2a} \log (1 + a^2 x^2)$

445. $\int (\sec^{-1} ax)\, dx = x \sec^{-1} ax - \dfrac{1}{a} \log (ax + \sqrt{a^2 x^2 - 1})$

446. $\int (\csc^{-1} ax)\, dx = x \csc^{-1} ax + \dfrac{1}{a} \log (ax + \sqrt{a^2 x^2 - 1})$

447. $\int \left(\sin^{-1} \dfrac{x}{a} \right) dx = x \sin^{-1} \dfrac{x}{a} + \sqrt{a^2 - x^2}, \qquad (a > 0)$

448. $\int \left(\cos^{-1} \dfrac{x}{a} \right) dx = x \cos^{-1} \dfrac{x}{a} - \sqrt{a^2 - x^2}, \qquad (a > 0)$

449. $\int \left(\tan^{-1}\dfrac{x}{a}\right) dx = x \tan^{-1}\dfrac{x}{a} - \dfrac{a}{2}\log(a^2 + x^2)$

450. $\int \left(\cot^{-1}\dfrac{x}{a}\right) dx = x \cot^{-1}\dfrac{x}{a} + \dfrac{a}{2}\log(a^2 + x^2)$

451. $\int x[\sin^{-1}(ax)]\, dx = \dfrac{1}{4a^2}[(2a^2x^2 - 1)\sin^{-1}(ax) + ax\sqrt{1 - a^2x^2}]$

452. $\int x[\cos^{-1}(ax)]\, dx = \dfrac{1}{4a^2}[(2a^2x^2 - 1)\cos^{-1}(ax) - ax\sqrt{1 - a^2x^2}]$

453. $\int x^n[\sin^{-1}(ax)]\, dx = \dfrac{x^{n+1}}{n+1}\sin^{-1}(ax) - \dfrac{a}{n+1}\int \dfrac{x^{n+1}\, dx}{\sqrt{1 - a^2x^2}}, \qquad (n \neq -1)$

454. $\int x^n[\cos^{-1}(ax)]\, dx = \dfrac{x^{n+1}}{n+1}\cos^{-1}(ax) + \dfrac{a}{n+1}\int \dfrac{x^{n+1}\, dx}{\sqrt{1 - a^2x^2}}, \qquad (n \neq -1)$

455. $\int x(\tan^{-1} ax)\, dx = \dfrac{1 + a^2x^2}{2a^2}\tan^{-1} ax - \dfrac{x}{2a}$

456. $\int x^n(\tan^{-1} ax)\, dx = \dfrac{x^{n+1}}{n+1}\tan^{-1} ax - \dfrac{a}{n+1}\int \dfrac{x^{n+1}}{1 + a^2x^2}\, dx$

457. $\int x(\cot^{-1} ax)\, dx = \dfrac{1 + a^2x^2}{2a^2}\cot^{-1} ax + \dfrac{x}{2a}$

458. $\int x^n(\cot^{-1} ax)\, dx = \dfrac{x^{n+1}}{n+1}\cot^{-1} ax + \dfrac{a}{n+1}\int \dfrac{x^{n+1}}{1 + a^2x^2}\, dx$

459. $\int \dfrac{\sin^{-1}(ax)}{x^2}\, dx = a\log\left(\dfrac{1 - \sqrt{1 - a^2x^2}}{x}\right) - \dfrac{\sin^{-1}(ax)}{x}$

460. $\int \dfrac{\cos^{-1}(ax)\, dx}{x^2} = -\dfrac{1}{x}\cos^{-1}(ax) + a\log\dfrac{1 + \sqrt{1 - a^2x^2}}{x}$

461. $\int \dfrac{\tan^{-1}(ax)\, dx}{x^2} = -\dfrac{1}{x}\tan^{-1}(ax) - \dfrac{a}{2}\log\dfrac{1 + a^2x^2}{x^2}$

462. $\int \dfrac{\cot^{-1} ax}{x^2}\, dx = -\dfrac{1}{x}\cot^{-1} ax - \dfrac{a}{2}\log\dfrac{x^2}{a^2x^2 + 1}$

463. $\int (\sin^{-1} ax)^2\, dx = x(\sin^{-1} ax)^2 - 2x + \dfrac{2\sqrt{1 - a^2x^2}}{a}\sin^{-1} ax$

464. $\int (\cos^{-1} ax)^2\, dx = x(\cos^{-1} ax)^2 - 2x - \dfrac{2\sqrt{1 - a^2x^2}}{a}\cos^{-1} ax$

465. $\int (\sin^{-1} ax)^n\, dx = \begin{cases} x(\sin^{-1} ax)^n + \dfrac{n\sqrt{1 - a^2x^2}}{a}(\sin^{-1} ax)^{n-1} \\ \qquad\qquad -n(n - 1)\int (\sin^{-1} ax)^{n-2}\, dx \\ \qquad\qquad\text{or} \\ \displaystyle\sum_{r=0}^{\left[\frac{n}{2}\right]} (-1)^r \dfrac{n!}{(n - 2r)!} x(\sin^{-1} ax)^{n-2r} \\ \qquad + \displaystyle\sum_{r=0}^{\left[\frac{n-1}{2}\right]} (-1)^r \dfrac{n!\sqrt{1 - a^2x^2}}{(n - 2r - 1)!a}(\sin^{-1} ax)^{n-2r-1} \end{cases}$

Note: $[s]$ means greatest integer $\leq s$. Thus $[3.5]$ means 3: $[5] = 5$, $[\frac{1}{2}] = 0$.

466. $\displaystyle\int (\cos^{-1} ax)^n \, dx =$
$$\begin{cases} x(\cos^{-1} ax)^n - \dfrac{n\sqrt{1 - a^2x^2}}{a}(\cos^{-1} ax)^{n-1} \\ \qquad\qquad\qquad\qquad - n(n-1)\displaystyle\int (\cos^{-1} ax)^{n-2} \, dx \\ \qquad\qquad\qquad\text{or} \\ \displaystyle\sum_{r=0}^{\left[\frac{n}{2}\right]} (-1)^r \dfrac{n!}{(n-2r)!} x(\cos^{-1} ax)^{n-2r} \\ \qquad\qquad\qquad \displaystyle\sum_{r=0}^{\left[\frac{n-1}{2}\right]} (-1)^r \dfrac{n!\sqrt{1 - a^2x^2}}{(n-2r-1)!a}(\cos^{-1} ax)^{n-2r-1} \end{cases}$$

467. $\displaystyle\int \frac{1}{\sqrt{1 - a^2x^2}}(\sin^{-1} ax) \, dx = \frac{1}{2a}(\sin^{-1} ax)^2.$

468. $\displaystyle\int \frac{x^n}{\sqrt{1 - a^2x^2}}(\sin^{-1} ax) \, dx = -\frac{x^{n-1}}{na^2}\sqrt{1 - a^2x^2}\,\sin^{-1} ax + \frac{x^n}{n^2a}$
$$+ \frac{n-1}{na^2}\int \frac{x^{n-2}}{\sqrt{1 - a^2x^2}}\sin^{-1} ax \, dx$$

469. $\displaystyle\int \frac{1}{\sqrt{1 - a^2x^2}}(\cos^{-1} ax) \, dx = -\frac{1}{2a}(\cos^{-1} ax)^2$

470. $\displaystyle\int \frac{x^n}{\sqrt{1 - a^2x^2}}(\cos^{-1} ax) \, dx = -\frac{x^{n-1}}{na^2}\sqrt{1 - a^2x^2}\,\cos^{-1} ax - \frac{x^n}{n^2a}$
$$+ \frac{n-1}{na^2}\int \frac{x^{n-2}}{\sqrt{1 - a^2x^2}}\cos^{-1} ax \, dx$$

471. $\displaystyle\int \frac{\tan^{-1} ax}{a^2x^2 + 1} \, dx = \frac{1}{2a}(\tan^{-1} ax)^2$

472. $\displaystyle\int \frac{\cot^{-1} ax}{a^2x^2 + 1} \, dx = -\frac{1}{2a}(\cot^{-1} ax)^2$

473. $\displaystyle\int x\sec^{-1} ax \, dx = \frac{x^2}{2}\sec^{-1} ax - \frac{1}{2a^2}\sqrt{a^2x^2 - 1}$

474. $\displaystyle\int x^n\sec^{-1} ax \, dx = \frac{x^{n+1}}{n+1}\sec^{-1} ax - \frac{1}{n+1}\int \frac{x^n \, dx}{\sqrt{a^2x^2 - 1}}$

475. $\displaystyle\int \frac{\sec^{-1} ax}{x^2} \, dx = -\frac{\sec^{-1} ax}{x} + \frac{\sqrt{a^2x^2 - 1}}{x}$

476. $\displaystyle\int x\csc^{-1} ax \, dx = \frac{x^2}{2}\csc^{-1} ax + \frac{1}{2a^2}\sqrt{a^2x^2 - 1}$

477. $\displaystyle\int x^n\csc^{-1} ax \, dx = \frac{x^{n+1}}{n+1}\csc^{-1} ax + \frac{1}{n+1}\int \frac{x^n \, dx}{\sqrt{a^2x^2 - 1}}$

478. $\displaystyle\int \frac{\csc^{-1} ax}{x^2} \, dx = -\frac{\csc^{-1} ax}{x} - \frac{\sqrt{a^2x^2 - 1}}{x}$

FORMS INVOLVING TRIGONOMETRIC SUBSTITUTIONS

479. $\displaystyle\int f(\sin x) \, dx = 2\int f\left(\frac{2z}{1 + z^2}\right)\frac{dz}{1 + z^2}, \qquad \left(z = \tan\frac{x}{2}\right)$

480. $\displaystyle\int f(\cos x) \, dx = 2\int f\left(\frac{1 - z^2}{1 + z^2}\right)\frac{dz}{1 + z^2}, \qquad \left(z = \tan\frac{x}{2}\right)$

*481. $\displaystyle\int f(\sin x)\,dx = \int f(u)\,\frac{du}{\sqrt{1-u^2}},\qquad (u = \sin x)$

*482. $\displaystyle\int f(\cos x)\,dx = -\int f(u)\,\frac{du}{\sqrt{1-u^2}},\qquad (u = \cos x)$

*483. $\displaystyle\int f(\sin x, \cos x)\,dx = \int f(u, \sqrt{1-u^2})\,\frac{du}{\sqrt{1-u^2}},\qquad (u = \sin x)$

484. $\displaystyle\int f(\sin x, \cos x)\,dx = 2\int f\left(\frac{2z}{1+z^2}, \frac{1-z^2}{1+z^2}\right)\frac{dz}{1+z^2},\qquad \left(z = \tan\frac{x}{2}\right)$

LOGARITHMIC FORMS

485. $\displaystyle\int (\log x)\,dx = x\log x - x$

486. $\displaystyle\int x(\log x)\,dx = \frac{x^2}{2}\log x - \frac{x^2}{4}$

487. $\displaystyle\int x^2(\log x)\,dx = \frac{x^3}{3}\log x - \frac{x^3}{9}$

488. $\displaystyle\int x^n(\log ax)\,dx = \frac{x^{n+1}}{n+1}\log ax - \frac{x^{n+1}}{(n+1)^2}$

489. $\displaystyle\int (\log x)^2\,dx = x(\log x)^2 - 2x\log x + 2x$

490. $\displaystyle\int (\log x)^n\,dx = \begin{cases} x(\log x)^n - n\displaystyle\int (\log x)^{n-1}\,dx, & (n \neq -1) \\[2mm] \text{or} \\[2mm] (-1)^n n!\, x\displaystyle\sum_{r=0}^{n}\frac{(-\log x)^r}{r!} \end{cases}$

491. $\displaystyle\int \frac{(\log x)^n}{x}\,dx = \frac{1}{n+1}(\log x)^{n+1}$

492. $\displaystyle\int \frac{dx}{\log x} = \log(\log x) + \log x + \frac{(\log x)^2}{2\cdot 2!} + \frac{(\log x)^3}{3\cdot 3!} + \cdots$

493. $\displaystyle\int \frac{dx}{x\log x} = \log(\log x)$

494. $\displaystyle\int \frac{dx}{x(\log x)^n} = -\frac{1}{(n-1)(\log x)^{n-1}}$

495. $\displaystyle\int \frac{x^m\,dx}{(\log x)^n} = -\frac{x^{m+1}}{(n-1)(\log x)^{n-1}} + \frac{m+1}{n-1}\int \frac{x^m\,dx}{(\log x)^{n-1}}$

496. $\displaystyle\int x^m(\log x)^n\,dx = \begin{cases} \dfrac{x^{m+1}(\log x)^n}{m+1} - \dfrac{n}{m+1}\displaystyle\int x^m(\log x)^{n-1}\,dx \\[2mm] \text{or} \\[2mm] (-1)^n\dfrac{n!}{m+1}x^{m+1}\displaystyle\sum_{r=0}^{n}\frac{(-\log x)^r}{r!(m+1)^{n-r}} \end{cases}$

497. $\displaystyle\int x^p\cos(b\ln x)\,dx = \frac{x^{p+1}}{(p+1)^2 + b^2}\cdot[b\sin(b\ln x) + (p+1)\cos(b\ln x)] + c$

498. $\displaystyle\int x^p\sin(b\ln x)\,dx = \frac{x^{p+1}}{(p+1)^2 + b^2}\cdot[(p+1)\sin(b\ln x) - b\cos(b\ln x)] + c$

499. $\displaystyle\int [\log(ax + b)]\,dx = \frac{ax+b}{a}\log(ax+b) - x$

---

*The square roots appearing in these formulas may be plus or minus, depending on the quadrant of x. Care must be used to give them the proper sign.

500. $\int \dfrac{\log(ax + b)}{x^2}\, dx = \dfrac{a}{b}\log x - \dfrac{ax + b}{bx}\log(ax + b)$

501. $\int x^m[\log(ax + b)]\, dx = \dfrac{1}{m + 1}\left[x^{m+1} - \left(-\dfrac{b}{a}\right)^{m+1}\right]\log(ax + b)$

$$- \dfrac{1}{m + 1}\left(-\dfrac{b}{a}\right)^{m+1}\sum_{r=1}^{m+1}\dfrac{1}{r}\left(-\dfrac{ax}{b}\right)^r$$

502. $\int \dfrac{\log(ax + b)}{x^m}\, dx = -\dfrac{1}{m - 1}\dfrac{\log(ax + b)}{x^{m-1}} + \dfrac{1}{m - 1}\left(-\dfrac{a}{b}\right)^{m-1}\log\dfrac{ax + b}{x}$

$$+ \dfrac{1}{m - 1}\left(-\dfrac{a}{b}\right)^{m-1}\sum_{r=1}^{m-2}\dfrac{1}{r}\left(-\dfrac{b}{ax}\right)^r, (m > 2)$$

503. $\int \left[\log \dfrac{x + a}{x - a}\right] dx = (x + a)\log(x + a) - (x - a)\log(x - a)$

504. $\int x^m\left[\log \dfrac{x + a}{x - a}\right] dx = \dfrac{x^{m+1} - (-a)^{m+1}}{m + 1}\log(x + a) - \dfrac{x^{m+1} - a^{m+1}}{m + 1}\log(x - a)$

$$+ \dfrac{2a^{m+1}}{m + 1}\sum_{r=1}^{\left[\frac{m+1}{2}\right]}\dfrac{1}{m - 2r + 2}\left(\dfrac{x}{a}\right)^{m-2r+2}$$

See note integral 392.

505. $\int \dfrac{1}{x^2}\left[\log \dfrac{x + a}{x - a}\right] dx = \dfrac{1}{x}\log \dfrac{x - a}{x + a} - \dfrac{1}{a}\log \dfrac{x^2 - a^2}{x^2}$

506. $\int (\log X)\, dx = \begin{cases} \left(x + \dfrac{b}{2c}\right)\log X - 2x + \dfrac{\sqrt{4ac - b^2}}{c}\tan^{-1}\dfrac{2cx + b}{\sqrt{4ac - b^2}}, \\ \hspace{7cm}(b^2 - 4ac < 0) \\ \text{or} \\ \left(x + \dfrac{b}{2c}\right)\log X - 2x + \dfrac{\sqrt{b^2 - 4ac}}{c}\tanh^{-1}\dfrac{2cx + b}{\sqrt{b^2 - 4ac}}, \\ \hspace{7cm}(b^2 - 4ac > 0) \\ \text{where} \\ X = a + bx + cx^2 \end{cases}$

507. $\int x^n(\log X)\, dx = \dfrac{x^{n+1}}{n + 1}\log X - \dfrac{2c}{n + 1}\int \dfrac{x^{n+2}}{X}\, dx - \dfrac{b}{n + 1}\int \dfrac{x^{n+1}}{X}\, dx$

$$\text{where } X = a + bx + cx^2$$

508. $\int [\log(x^2 + a^2)]\, dx = x\log(x^2 + a^2) - 2x + 2a\tan^{-1}\dfrac{x}{a}$

509. $\int [\log(x^2 - a^2)]\, dx = x\log(x^2 - a^2) - 2x + a\log \dfrac{x + a}{x - a}$

510. $\int x[\log(x^2 \pm a^2)]\, dx = \tfrac{1}{2}(x^2 \pm a^2)\log(x^2 \pm a^2) - \tfrac{1}{2}x^2$

511. $\int [\log(x + \sqrt{x^2 \pm a^2})]\, dx = x\log(x + \sqrt{x^2 \pm a^2}) - \sqrt{x^2 \pm a^2}$

512. $\int x[\log(x + \sqrt{x^2 \pm a^2})]\, dx = \left(\dfrac{x^2}{2} \pm \dfrac{a^2}{4}\right)\log(x + \sqrt{x^2 \pm a^2}) - \dfrac{x\sqrt{x^2 \pm a^2}}{4}$

513. $\int x^m[\log(x + \sqrt{x^2 \pm a^2})]\, dx = \dfrac{x^{m+1}}{m + 1}\log(x + \sqrt{x^2 \pm a^2})$

$$- \dfrac{1}{m + 1}\int \dfrac{x^{m+1}}{\sqrt{x^2 \pm a^2}}\, dx$$

514. $\displaystyle\int \frac{\log (x + \sqrt{x^2 + a^2})}{x^2}\,dx = -\frac{\log (x + \sqrt{x^2 + a^2})}{x} - \frac{1}{a}\log \frac{a + \sqrt{x^2 + a^2}}{x}$

515. $\displaystyle\int \frac{\log (x + \sqrt{x^2 - a^2})}{x^2}\,dx = -\frac{\log (x + \sqrt{x^2 - a^2})}{x} + \frac{1}{|a|}\sec^{-1}\frac{x}{a}$

516. $\displaystyle\int x^n \log (x^2 - a^2)\,dx = \frac{1}{n+1}\Bigg[x^{n+1} \log (x^2 - a^2) - a^{n+1} \log (x - a)$

See note integral 392. $\qquad\qquad -(-a)^{n+1} \log (x + a) - 2 \sum_{r=0}^{\left[\frac{n}{2}\right]} \frac{a^{2r}x^{n-2r+1}}{n - 2r + 1}\Bigg]$

EXPONENTIAL FORMS

517. $\displaystyle\int e^x\,dx = e^x$

518. $\displaystyle\int e^{-x}\,dx = -e^{-x}$

519. $\displaystyle\int e^{ax}\,dx = \frac{e^{ax}}{a}$

520. $\displaystyle\int x\,e^{ax}\,dx = \frac{e^{ax}}{a^2}(ax - 1)$

521. $\displaystyle\int x^m e^{ax}\,dx = \begin{cases} \dfrac{x^m e^{ax}}{a} - \dfrac{m}{a}\displaystyle\int x^{m-1} e^{ax}\,dx \\[2mm] \qquad\text{or} \\[2mm] e^{ax}\displaystyle\sum_{r=0}^{m} (-1)^r \dfrac{m!\,x^{m-r}}{(m - r)!\,a^{r+1}} \end{cases}$

522. $\displaystyle\int \frac{e^{ax}\,dx}{x} = \log x + \frac{ax}{1!} + \frac{a^2x^2}{2\cdot 2!} + \frac{a^3x^3}{3\cdot 3!} + \cdots$

523. $\displaystyle\int \frac{e^{ax}}{x^m}\,dx = -\frac{1}{m-1}\frac{e^{ax}}{x^{m-1}} + \frac{a}{m-1}\int \frac{e^{ax}}{x^{m-1}}\,dx$

524. $\displaystyle\int e^{ax} \log x\,dx = \frac{e^{ax} \log x}{a} - \frac{1}{a}\int \frac{e^{ax}}{x}\,dx$

525. $\displaystyle\int \frac{dx}{1 + e^x} = x - \log (1 + e^x) = \log \frac{e^x}{1 + e^x}$

526. $\displaystyle\int \frac{dx}{a + be^{px}} = \frac{x}{a} - \frac{1}{ap}\log (a + be^{px})$

527. $\displaystyle\int \frac{dx}{ae^{mx} + be^{-mx}} = \frac{1}{m\sqrt{ab}}\tan^{-1}\left(e^{mx}\sqrt{\frac{a}{b}}\right), \qquad (a > 0, b > 0)$

528. $\displaystyle\int \frac{dx}{ae^{mx} - be^{-mx}} = \begin{cases} \dfrac{1}{2m\sqrt{ab}}\log \dfrac{\sqrt{a}\,e^{mx} - \sqrt{b}}{\sqrt{a}\,e^{mx} + \sqrt{b}} \\[2mm] \qquad\text{or} \\[2mm] \dfrac{-1}{m\sqrt{ab}}\tanh^{-1}\left(\sqrt{\dfrac{a}{b}}\,e^{mx}\right), \qquad (a > 0, b > 0) \end{cases}$

529. $\displaystyle\int (a^x - a^{-x})\,dx = \frac{a^x + a^{-x}}{\log a}$

530. $\displaystyle\int \frac{e^{ax}}{b + ce^{ax}}\,dx = \frac{1}{ac}\log (b + ce^{ax})$

531. $\displaystyle\int \frac{x\,e^{ax}}{(1 + ax)^2}\,dx = \frac{e^{ax}}{a^2(1 + ax)}$

532. $\int x\, e^{-x^2}\, dx = -\tfrac{1}{2}\, e^{-x^2}$

533. $\int e^{ax}\, [\sin (bx)]\, dx = \dfrac{e^{ax}[a \sin (bx) - b \cos (bx)]}{a^2 + b^2}$

534. $\int e^{ax}\, [\sin (bx)][\sin (cx)]\, dx = \dfrac{e^{ax}[(b - c)\sin (b - c)x + a \cos (b - c)x]}{2[a^2 + (b - c)^2]}$

$$-\dfrac{e^{ax}[(b + c)\sin (b + c)x + a \cos (b + c)x]}{2[a^2 + (b + c)^2]}$$

535. $\int e^{ax}[\sin (bx)][\cos (cx)]\, dx = \begin{cases} \dfrac{e^{ax}[a \sin (b - c)x - (b - c)\cos (b - c)x]}{2[a^2 + (b - c)^2]} \\[2ex] \quad + \dfrac{e^{ax}[a \sin (b + c)x - (b + c)\cos (b + c)x]}{2[a^2 + (b + c)^2]} \\[2ex] \text{or} \\[1ex] \dfrac{e^{ax}}{\rho}[(a \sin bx - b \cos bx)[\cos (cx - \alpha)] \\[1ex] \qquad\qquad\qquad -c(\sin bx)\sin (cx - \alpha)] \\[1ex] \text{where} \\[1ex] \rho = \sqrt{(a^2 + b^2 - c^2)^2 + 4a^2 c^2}, \\[1ex] \quad \rho \cos \alpha = a^2 + b^2 - c^2, \qquad \rho \sin \alpha = 2ac \end{cases}$

536. $\int e^{ax}[\sin (bx)][\sin (bx + c)]\, dx$

$$= \dfrac{e^{ax} \cos c}{2a} - \dfrac{e^{ax}[a \cos (2bx + c) + 2b \sin (2bx + c)]}{2(a^2 + 4b^2)}$$

537. $\int e^{ax}[\sin (bx)][\cos (bx + c)]\, dx$

$$= \dfrac{-e^{ax} \sin c}{2a} + \dfrac{e^{ax}[a \sin (2bx + c) - 2b \cos (2bx + c)]}{2(a^2 + 4b^2)}$$

538. $\int e^{ax}[\cos (bx)]\, dx = \dfrac{e^{ax}}{a^2 + b^2}[a \cos (bx) + b \sin (bx)]$

539. $\int e^{ax}[\cos (bx)][\cos (cx)]\, dx = \dfrac{e^{ax}[(b - c)\sin (b - c)x + a \cos (b - c)x]}{2[a^2 + (b - c)^2]}$

$$+ \dfrac{e^{ax}[(b + c)\sin (b + c)x + a \cos (b + c)x]}{2[a^2 + (b + c)^2]}$$

540. $\int e^{ax}[\cos (bx)][\cos (bx + c)]\, dx$

$$= \dfrac{e^{ax} \cos c}{2a} + \dfrac{e^{ax}[a \cos (2bx + c) + 2b \sin (2bx + c)]}{2(a^2 + 4b^2)}$$

541. $\int e^{ax}[\cos (bx)][\sin (bx + c)]\, dx$

$$= \dfrac{e^{ax} \sin c}{2a} + \dfrac{e^{ax}[a \sin (2bx + c) - 2b \cos (2bx + c)]}{2(a^2 + 4b^2)}$$

542. $\int e^{ax}[\sin^n bx]\, dx = \dfrac{1}{a^2 + n^2 b^2}\Bigg[(a \sin bx - nb \cos bx)\, e^{ax} \sin^{n-1} bx$

$$+ n(n - 1)b^2 \int e^{ax}[\sin^{n-2} bx]\, dx\Bigg]$$

543. $\int e^{ax}[\cos^n bx]\, dx = \dfrac{1}{a^2 + n^2 b^2}\Bigg[(a \cos bx + nb \sin bx)\, e^{ax} \cos^{n-1} bx$

$$+ n(n - 1)b^2 \int e^{ax}[\cos^{n-2} bx]\, dx\Bigg]$$

544. $\int x^m e^x \sin x \, dx = \frac{1}{2} x^m e^x(\sin x - \cos x) - \frac{m}{2} \int x^{m-1} e^x \sin x \, dx$

$$+ \frac{m}{2} \int x^{m-1} e^x \cos x \, dx$$

545. $\int x^m e^{ax}[\sin bx] \, dx = \begin{cases} x^m e^{ax} \dfrac{a \sin bx - b \cos bx}{a^2 + b^2} \\ \qquad - \dfrac{m}{a^2 + b^2} \int x^{m-1} e^{ax}(a \sin bx - b \cos bx) \, dx \\ \qquad\qquad \text{or} \\ e^{ax} \sum\limits_{r=0}^{m} \dfrac{(-1)^r m! x^{m-r}}{\rho^{r+1}(m-r)!} \sin\left[bx - (r+1)\alpha\right] \\ \qquad\qquad \text{where} \\ \rho = \sqrt{a^2 + b^2}, \qquad \rho \cos \alpha = a, \qquad \rho \sin \alpha = b \end{cases}$

546. $\int x^m e^x \cos x \, dx = \frac{1}{2} x^m e^x(\sin x + \cos x)$

$$- \frac{m}{2} \int x^{m-1} e^x \sin x \, dx - \frac{m}{2} \int x^{m-1} e^x \cos x \, dx$$

547. $\int x^m e^{ax} \cos bx \, dx = \begin{cases} x^m e^{ax} \dfrac{a \cos bx + b \sin bx}{a^2 + b^2} \\ \qquad - \dfrac{m}{a^2 + b^2} \int x^{m-1} e^{ax}(a \cos bx + b \sin bx) \, dx \\ \qquad\qquad \text{or} \\ e^{ax} \sum\limits_{r=0}^{m} \dfrac{(-1)^r m! x^{m-r}}{\rho^{r+1}(m-r)!} \cos\left[bx - (r+1)\alpha\right] \\ \qquad\qquad \text{where} \\ \rho = \sqrt{a^2 + b^2}, \qquad \rho \cos \alpha = a, \qquad \rho \sin \alpha = b \end{cases}$

548. $\displaystyle\int e^{ax}(\cos^m x)(\sin^n x)\,dx =$

$$\begin{cases}
\dfrac{e^{ax}\cos^{m-1} x \sin^n x[a\cos x + (m+n)\sin x]}{(m+n)^2 + a^2} \\[2ex]
\quad -\dfrac{na}{(m+n)^2 + a^2}\int e^{ax}(\cos^{m-1} x)(\sin^{n-1} x)\,dx \\[2ex]
\quad +\dfrac{(m-1)(m+n)}{(m+n)^2 + a^2}\int e^{ax}(\cos^{m-2} x)(\sin^n x)\,dx \\[2ex]
\qquad\qquad\text{or} \\[2ex]
\dfrac{e^{ax}\cos^m x \sin^{n-1} x[a\sin x - (m+n)\cos x]}{(m+n)^2 + a^2} \\[2ex]
\quad +\dfrac{ma}{(m+n)^2 + a^2}\int e^{ax}(\cos^{m-1} x)(\sin^{n-1} x)\,dx \\[2ex]
\quad +\dfrac{(n-1)(m+n)}{(m+n)^2 + a^2}\int e^{ax}(\cos^m x)(\sin^{n-2} x)\,dx \\[2ex]
\qquad\qquad\text{or} \\[2ex]
\dfrac{e^{ax}(\cos^{m-1} x)(\sin^{n-1} x)(a\sin x\cos x + m\sin^2 x - n\cos^2 x)}{(m+n)^2 + a^2} \\[2ex]
\quad +\dfrac{m(m-1)}{(m+n)^2 + a^2}\int e^{ax}(\cos^{m-2} x)(\sin^n x)\,dx \\[2ex]
\quad +\dfrac{n(n-1)}{(m+n)^2 + a^2}\int e^{ax}(\cos^m x)(\sin^{n-2} x)\,dx \\[2ex]
\qquad\qquad\text{or} \\[2ex]
\dfrac{e^{ax}(\cos^{m-1} x)(\sin^{n-1} x)(a\cos x\sin x + m\sin^2 x - n\cos^2 x)}{(m+n)^2 + a^2} \\[2ex]
\quad +\dfrac{m(m-1)}{(m+n)^2 + a^2}\int e^{ax}(\cos^{m-2} x)(\sin^{n-2} x)\,dx \\[2ex]
\quad +\dfrac{(n-m)(n+m-1)}{(m+n)^2 + a^2}\int e^{ax}(\cos^m x)(\sin^{n-2} x)\,dx
\end{cases}$$

549. $\displaystyle\int x\,e^{ax}(\sin bx)\,dx = \dfrac{x\,e^{ax}}{a^2 + b^2}(a\sin bx - b\cos bx)$

$$-\dfrac{e^{ax}}{(a^2 + b^2)^2}[(a^2 - b^2)\sin bx - 2ab\cos bx]$$

550. $\displaystyle\int x\,e^{ax}(\cos bx)\,dx = \dfrac{x\,e^{ax}}{a^2 + b^2}(a\cos bx + b\sin bx)$

$$-\dfrac{e^{ax}}{(a^2 + b^2)^2}[(a^2 - b^2)\cos bx + 2ab\sin bx]$$

551. $\displaystyle\int \dfrac{e^{ax}}{\sin^n x}\,dx = -\dfrac{e^{ax}[a\sin x + (n-2)\cos x]}{(n-1)(n-2)\sin^{n-1} x} + \dfrac{a^2 + (n-2)^2}{(n-1)(n-2)}\int \dfrac{e^{ax}}{\sin^{n-2} x}\,dx$

552. $\displaystyle\int \dfrac{e^{ax}}{\cos^n x}\,dx = -\dfrac{e^{ax}[a\cos x - (n-2)\sin x]}{(n-1)(n-2)\cos^{n-1} x} + \dfrac{a^2 + (n-2)^2}{(n-1)(n-2)}\int \dfrac{e^{ax}}{\cos^{n-2} x}\,dx$

553. $\displaystyle\int e^{ax}\tan^n x\,dx = e^{ax}\dfrac{\tan^{n-1} x}{n-1} - \dfrac{a}{n-1}\int e^{ax}\tan^{n-1} x\,dx - \int e^{ax}\tan^{n-2} x\,dx$

HYPERBOLIC FORMS

554. $\displaystyle\int (\sinh x)\,dx = \cosh x$

555. $\displaystyle\int (\cosh x)\,dx = \sinh x$

556. $\displaystyle\int (\tanh x)\,dx = \log \cosh x$

557. $\displaystyle\int (\coth x)\,dx = \log \sinh x$

558. $\displaystyle\int (\operatorname{sech} x)\,dx = \tan^{-1}(\sinh x)$

559. $\displaystyle\int \operatorname{csch} x\,dx = \log \tanh \left(\dfrac{x}{2}\right)$

560. $\displaystyle\int x(\sinh x)\,dx = x \cosh x - \sinh x$

561. $\displaystyle\int x^n(\sinh x)\,dx = x^n \cosh x - n \int x^{n-1}(\cosh x)\,dx$

562. $\displaystyle\int x(\cosh x)\,dx = x \sinh x - \cosh x$

563. $\displaystyle\int x^n(\cosh x)\,dx = x^n \sinh x - n \int x^{n-1}(\sinh x)\,dx$

564. $\displaystyle\int (\operatorname{sech} x)(\tanh x)\,dx = -\operatorname{sech} x$

565. $\displaystyle\int (\operatorname{csch} x)(\coth x)\,dx = -\operatorname{csch} x$

566. $\displaystyle\int (\sinh^2 x)\,dx = \dfrac{\sinh 2x}{4} - \dfrac{x}{2}$

567. $\displaystyle\int (\sinh^m x)(\cosh^n x)\,dx = \begin{cases} \dfrac{1}{m+n}(\sinh^{m+1} x)(\cosh^{n-1} x) \\[2mm] \qquad + \dfrac{n-1}{m+n}\displaystyle\int (\sinh^m x)(\cosh^{n-2} x)\,dx \\[2mm] \qquad\qquad \text{or} \\[2mm] \dfrac{1}{m+n}\sinh^{m-1} x \cosh^{n+1} x \\[2mm] \qquad - \dfrac{m-1}{m+n}\displaystyle\int (\sinh^{m-2} x)(\cosh^n x)\,dx, \quad (m+n \neq 0) \end{cases}$

568. $\displaystyle\int \dfrac{dx}{(\sinh^m x)(\cosh^n x)} = \begin{cases} -\dfrac{1}{(m-1)(\sinh^{m-1} x)(\cosh^{n-1} x)} \\[2mm] \qquad - \dfrac{m+n-2}{m-1}\displaystyle\int \dfrac{dx}{(\sinh^{m-2} x)(\cosh^n x)}, \quad (m \neq 1) \\[2mm] \qquad\qquad \text{or} \\[2mm] \dfrac{1}{(n-1)\sinh^{m-1} x \cosh^{n-1} x} \\[2mm] \qquad + \dfrac{m+n-2}{n-1}\displaystyle\int \dfrac{dx}{(\sinh^m x)(\cosh^{n-2} x)}, \quad (n \neq 1) \end{cases}$

569. $\displaystyle\int (\tanh^2 x)\,dx = x - \tanh x$

570. $\displaystyle\int (\tanh^n x)\,dx = -\dfrac{\tanh^{n-1} x}{n-1} + \int (\tanh^{n-2} x)\,dx, \quad (n \neq 1)$

571. $\displaystyle\int (\operatorname{sech}^2 x)\,dx = \tanh x$

572. $\displaystyle\int (\cosh^2 x)\,dx = \dfrac{\sinh 2x}{4} + \dfrac{x}{2}$

573. $\int (\coth^2 x)\, dx = x - \coth x$

574. $\int (\coth^n x)\, dx \doteq -\dfrac{\coth^{n-1} x}{n-1} + \int \coth^{n-2} x\, dx,\qquad (n \neq 1)$

575. $\int (\operatorname{csch}^2 x)\, dx = -\operatorname{ctnh} x$

576. $\int (\sinh mx)(\sinh nx)\, dx = \dfrac{\sinh(m+n)x}{2(m+n)} - \dfrac{\sinh(m-n)x}{2(m-n)},\qquad (m^2 \neq n^2)$

577. $\int (\cosh mx)(\cosh nx)\, dx = \dfrac{\sinh(m+n)x}{2(m+n)} + \dfrac{\sinh(m-n)x}{2(m-n)},\qquad (m^2 \neq n^2)$

578. $\int (\sinh mx)(\cosh nx)\, dx = \dfrac{\cosh(m+n)x}{2(m+n)} + \dfrac{\cosh(m-n)x}{2(m-n)},\qquad (m^2 \neq n^2)$

579. $\int \left(\sinh^{-1}\dfrac{x}{a}\right) dx = x\sinh^{-1}\dfrac{x}{a} - \sqrt{x^2 + a^2},\qquad (a > 0)$

580. $\int x\left(\sinh^{-1}\dfrac{x}{a}\right) dx = \left(\dfrac{x^2}{2} + \dfrac{a^2}{4}\right)\sinh^{-1}\dfrac{x}{a} - \dfrac{x}{4}\sqrt{x^2 + a^2},\qquad (a > 0)$

581. $\int x^n(\sinh^{-1} x)\, dx = \dfrac{x^{n+1}}{n+1}\sinh^{-1} x - \dfrac{1}{n+1}\int \dfrac{x^{n+1}}{(1+x^2)^{\frac{1}{2}}}\, dx,\qquad (n \neq -1)$

582. $\int \left(\cosh^{-1}\dfrac{x}{a}\right) dx = \begin{cases} x\cosh^{-1}\dfrac{x}{a} - \sqrt{x^2 - a^2}, & \left(\cosh^{-1}\dfrac{x}{a} > 0\right) \\[2mm] \qquad\qquad \text{or} \\[2mm] x\cosh^{-1}\dfrac{x}{a} + \sqrt{x^2 - a^2}, & \left(\cosh^{-1}\dfrac{x}{a} < 0\right), \quad (a > 0) \end{cases}$

583. $\int x\left(\cosh^{-1}\dfrac{x}{a}\right) dx = \dfrac{2x^2 - a^2}{4}\cosh^{-1}\dfrac{x}{a} - \dfrac{x}{4}(x^2 - a^2)^{\frac{1}{2}}$

584. $\int x^n(\cosh^{-1} x)\, dx = \dfrac{x^{n+1}}{n+1}\cosh^{-1} x - \dfrac{1}{n+1}\int \dfrac{x^{n+1}}{(x^2 - 1)^{\frac{1}{2}}}\, dx,\qquad (n \neq -1)$

585. $\int \left(\tanh^{-1}\dfrac{x}{a}\right) dx = x\tanh^{-1}\dfrac{x}{a} + \dfrac{a}{2}\log(a^2 - x^2),\qquad \left(\left|\dfrac{x}{a}\right| < 1\right)$

586. $\int \left(\coth^{-1}\dfrac{x}{a}\right) dx = x\coth^{-1}\dfrac{x}{a} + \dfrac{a}{2}\log(x^2 - a^2),\qquad \left(\left|\dfrac{x}{a}\right| > 1\right)$

587. $\int x\left(\tanh^{-1}\dfrac{x}{a}\right) dx = \dfrac{x^2 - a^2}{2}\tanh^{-1}\dfrac{x}{a} + \dfrac{ax}{2},\qquad \left(\left|\dfrac{x}{a}\right| < 1\right)$

588. $\int x^n\left(\tanh^{-1} x\right) dx = \dfrac{x^{n+1}}{n+1}\tanh^{-1} x - \dfrac{1}{n+1}\int \dfrac{x^{n+1}}{1 - x^2}\, dx,\qquad (n \neq -1)$

589. $\int x\left(\coth^{-1}\dfrac{x}{a}\right) dx = \dfrac{x^2 - a^2}{2}\coth^{-1}\dfrac{x}{a} + \dfrac{ax}{2},\qquad \left(\left|\dfrac{x}{a}\right| > 1\right)$

590. $\int x^n(\coth^{-1} x)\, dx = \dfrac{x^{n+1}}{n+1}\coth^{-1} x + \dfrac{1}{n+1}\int \dfrac{x^{n+1}}{x^2 - 1}\, dx,\qquad (n \neq -1)$

591. $\int (\operatorname{sech}^{-1} x)\, dx = x\operatorname{sech}^{-1} x + \sin^{-1} x$

592. $\int x\operatorname{sech}^{-1} x\, dx = \dfrac{x^2}{2}\operatorname{sech}^{-1} x - \dfrac{1}{2}\sqrt{1 - x^2}$

593. $\int x^n\operatorname{sech}^{-1} x\, dx = \dfrac{x^{n+1}}{n+1}\operatorname{sech}^{-1} x + \dfrac{1}{n+1}\int \dfrac{x^n}{(1 - x^2)^{\frac{1}{2}}}\, dx,\qquad (n \neq -1)$

594. $\int \operatorname{csch}^{-1} x\, dx = x\operatorname{csch}^{-1} x + \dfrac{x}{|x|}\sinh^{-1} x$

595. $\int x \operatorname{csch}^{-1} x\, dx = \dfrac{x^2}{2} \operatorname{csch}^{-1} x + \dfrac{1}{2} \dfrac{x}{|x|} \sqrt{1 + x^2}$

596. $\int x^n \operatorname{csch}^{-1} x\, dx = \dfrac{x^{n+1}}{n+1} \operatorname{csch}^{-1} x + \dfrac{1}{n+1} \dfrac{x}{|x|} \int \dfrac{x^n}{(x^2 + 1)^{\frac{1}{2}}}\, dx,\qquad (n \neq -1)$

DEFINITE INTEGRALS

597. $\displaystyle\int_0^\infty x^{n-1} e^{-x}\, dx = \int_0^1 \left(\log \dfrac{1}{x}\right)^{n-1} dx = \dfrac{1}{n} \prod_{m=1}^{\infty} \dfrac{\left(1 + \dfrac{1}{m}\right)^n}{1 + \dfrac{n}{m}}$

$$= \Gamma(n),\ n \neq 0,\ -1,\ -2,\ -3,\ldots \qquad \text{(Gamma Function)}$$

598. $\displaystyle\int_0^\infty t^n p^{-t}\, dt = \dfrac{n!}{(\log p)^{n+1}}, \qquad (n = 0, 1, 2, 3, \ldots \text{ and } p > 0)$

599. $\displaystyle\int_0^\infty t^{n-1} e^{-(a+1)t}\, dt = \dfrac{\Gamma(n)}{(a+1)^n}, \qquad (n > 0,\ a > -1)$

600. $\displaystyle\int_0^1 x^m \left(\log \dfrac{1}{x}\right)^n dx = \dfrac{\Gamma(n+1)}{(m+1)^{n+1}}, \qquad (m > -1,\ n > -1)$

601. $\Gamma(n)$ is finite if $n > 0$, $\Gamma(n+1) = n\Gamma(n)$

602. $\Gamma(n) \cdot \Gamma(1 - n) = \dfrac{\pi}{\sin n\pi}$

603. $\Gamma(n) = (n-1)!$ if $n = \text{integer} > 0$

604. $\Gamma(\frac{1}{2}) = 2 \displaystyle\int_0^\infty e^{-t^2}\, dt = \sqrt{\pi} = 1.7724538509 \cdots = (-\frac{1}{2})!$

605. $\Gamma(n + \frac{1}{2}) = \dfrac{1 \cdot 3 \cdot 5 \ldots (2n-1)}{2^n} \sqrt{\pi} \qquad n = 1, 2, 3, \ldots$

606. $\Gamma(-n + \frac{1}{2}) = \dfrac{(-1)^n 2^n \sqrt{\pi}}{1 \cdot 3 \cdot 5 \ldots (2n-1)} \qquad n = 1, 2, 3, \ldots$

607. $\displaystyle\int_0^1 x^{m-1}(1 - x)^{n-1}\, dx = \int_0^\infty \dfrac{x^{m-1}}{(1 + x)^{m+n}}\, dx = \dfrac{\Gamma(m)\Gamma(n)}{\Gamma(m+n)} = B(m, n)$

(Beta function)

608. $B(m, n) = B(n, m) = \dfrac{\Gamma(m)\Gamma(n)}{\Gamma(m+n)}$, where m and n are any positive real numbers.

609. $\displaystyle\int_a^b (x - a)^m (b - x)^n\, dx = (b - a)^{m+n+1} \dfrac{\Gamma(m+1) \cdot \Gamma(n+1)}{\Gamma(m+n+2)},$

$$(m > -1,\ n > -1,\ b > a)$$

610. $\displaystyle\int_1^\infty \dfrac{dx}{x^m} = \dfrac{1}{m-1}, \qquad [m > 1]$

611. $\displaystyle\int_0^\infty \dfrac{dx}{(1 + x)x^p} = \pi \csc p\pi, \qquad [p < 1]$

612. $\displaystyle\int_0^\infty \dfrac{dx}{(1 - x)x^p} = -\pi \cot p\pi, \qquad [p < 1]$

613. $\displaystyle\int_0^\infty \dfrac{x^{p-1}\, dx}{1 + x} = \dfrac{\pi}{\sin p\pi}$

$$= B(p, 1 - p) = \Gamma(p)\Gamma(1 - p), \qquad [0 < p < 1]$$

614. $\displaystyle\int_0^\infty \dfrac{x^{m-1}\, dx}{1 + x^n} = \dfrac{\pi}{n \sin \dfrac{m\pi}{n}}, \qquad [0 < m < n]$

615. $\int_0^\infty \dfrac{x^a\,dx}{(m+x^b)^c} = \dfrac{m^{\frac{a+1-bc}{b}}}{b}\left[\dfrac{\Gamma\left(\dfrac{a+1}{b}\right)\Gamma\left(c-\dfrac{a+1}{b}\right)}{\Gamma(c)}\right]$

$$\left(a>-1,\ b>0,\ m>0,\ c>\dfrac{a+1}{b}\right)$$

616. $\int_0^\infty \dfrac{dx}{(1+x)\sqrt{x}} = \pi$

617. $\int_0^\infty \dfrac{a\,dx}{a^2+x^2} = \dfrac{\pi}{2}$, if $a>0$; 0, if $a=0$; $-\dfrac{\pi}{2}$, if $a<0$

618. $\int_0^a (a^2-x^2)^{\frac{n}{2}}\,dx = \dfrac{1}{2}\int_{-a}^a (a^2-x^2)^{\frac{n}{2}}\,dx = \dfrac{1\cdot3\cdot5\ldots n}{2\cdot4\cdot6\ldots(n+1)}\cdot\dfrac{\pi}{2}\cdot a^{n+1}$ (n odd)

619. $\int_0^a x^m(a^2-x^2)^{\frac{n}{2}}\,dx = \begin{cases} \dfrac{1}{2}a^{m+n+1}B\left(\dfrac{m+1}{2},\dfrac{n+2}{2}\right) \\[1em] \text{or} \\[1em] \dfrac{1}{2}a^{m+n+1}\dfrac{\Gamma\left(\dfrac{m+1}{2}\right)\Gamma\left(\dfrac{n+2}{2}\right)}{\Gamma\left(\dfrac{m+n+3}{2}\right)} \end{cases}$

620. $\int_0^{\pi/2} (\sin^n x)\,dx = \begin{cases} \displaystyle\int_0^{\pi/2}(\cos^n x)\,dx \\[1em] \text{or} \\[1em] \dfrac{1\cdot3\cdot5\cdot7\ldots(n-1)}{2\cdot4\cdot6\cdot8\ldots(n)}\dfrac{\pi}{2}, \quad (n \text{ an even integer, } n\neq0) \\[1em] \text{or} \\[1em] \dfrac{2\cdot4\cdot6\cdot8\ldots(n-1)}{1\cdot3\cdot5\cdot7\ldots(n)}, \quad (n \text{ an odd integer, } n\neq1) \\[1em] \text{or} \\[1em] \dfrac{\sqrt{\pi}}{2}\dfrac{\Gamma\left(\dfrac{n+1}{2}\right)}{\Gamma\left(\dfrac{n}{2}+1\right)}, \quad (n>-1) \end{cases}$

621. $\int_0^\infty \dfrac{\sin mx\,dx}{x} = \dfrac{\pi}{2}$, if $m>0$; 0, if $m=0$; $-\dfrac{\pi}{2}$, if $m<0$

622. $\int_0^\infty \dfrac{\cos x\,dx}{x} = \infty$

623. $\int_0^\infty \dfrac{\tan x\,dx}{x} = \dfrac{\pi}{2}$

624. $\int_0^\pi \sin ax\cdot\sin bx\,dx = \int_0^\pi \cos ax\cdot\cos bx\,dx = 0$, $(a\neq b; a, b \text{ integers})$

625. $\int_0^{\pi/a} [\sin(ax)][\cos(ax)]\,dx = \int_0^\pi [\sin(ax)][\cos(ax)]\,dx = 0$

626. $\int_0^\pi [\sin(ax)][\cos(bx)]\,dx = \dfrac{2a}{a^2-b^2}$, if $a-b$ is odd, or 0 if $a-b$ is even

627. $\int_0^\infty \dfrac{\sin x\cos mx\,dx}{x}$

$$= 0, \text{ if } m<-1 \text{ or } m>1; \dfrac{\pi}{4}, \text{ if } m=\pm1; \dfrac{\pi}{2}, \text{ if } m^2<1$$

628. $\int_0^\infty \dfrac{\sin ax \sin bx}{x^2}\, dx = \dfrac{\pi a}{2}, \qquad (a \le b)$

629. $\int_0^\pi \sin^2 mx\, dx = \int_0^\pi \cos^2 mx\, dx = \dfrac{\pi}{2}$

630. $\int_0^\infty \dfrac{\sin^2 (px)}{x^2}\, dx = \dfrac{\pi p}{2}$

631. $\int_0^\infty \dfrac{\sin x}{x^p}\, dx = \dfrac{\pi}{2\Gamma(p) \sin(p\pi/2)}, \qquad 0 < p < 1$

632. $\int_0^\infty \dfrac{\cos x}{x^p}\, dx = \dfrac{\pi}{2\Gamma(p) \cos(p\pi/2)}, \qquad 0 < p < 1$

633. $\int_0^\infty \dfrac{1 - \cos px}{x^2}\, dx = \dfrac{\pi p}{2}$

634. $\int_0^\infty \dfrac{\sin px \cos qx}{x}\, dx = \left\{ 0, \; q > p > 0; \; \dfrac{\pi}{2}, \; p > q > 0; \; \dfrac{\pi}{4}, \; p = q > 0 \right\}$

635. $\int_0^\infty \dfrac{\cos (mx)}{x^2 + a^2}\, dx = \dfrac{\pi}{2|a|} e^{-|ma|}$

636. $\int_0^\infty \cos (x^2)\, dx = \int_0^\infty \sin (x^2)\, dx = \dfrac{1}{2}\sqrt{\dfrac{\pi}{2}}$

637. $\int_0^\infty \sin ax^n\, dx = \dfrac{1}{na^{1/n}} \Gamma(1/n) \sin \dfrac{\pi}{2n}, \qquad n > 1$

638. $\int_0^\infty \cos ax^n\, dx = \dfrac{1}{na^{1/n}} \Gamma(1/n) \cos \dfrac{\pi}{2n}, \qquad n > 1$

639. $\int_0^\infty \dfrac{\sin x}{\sqrt{x}}\, dx = \int_0^\infty \dfrac{\cos x}{\sqrt{x}}\, dx = \sqrt{\dfrac{\pi}{2}}$

640. (a) $\int_0^\infty \dfrac{\sin^3 x}{x}\, dx = \dfrac{\pi}{4}$ (b) $\int_0^\infty \dfrac{\sin^3 x}{x^2}\, dx \, \dfrac{3}{4} \log 3$

641. $\int_0^\infty \dfrac{\sin^3 x}{x^3}\, dx = \dfrac{3\pi}{8}$

642. $\int_0^\infty \dfrac{\sin^4 x}{x^4}\, dx = \dfrac{\pi}{3}$

643. $\int_0^{\pi/2} \dfrac{dx}{1 + a \cos x} = \dfrac{\cos^{-1} a}{\sqrt{1 - a^2}}, \qquad (a < 1)$

644. $\int_0^\pi \dfrac{dx}{a + b \cos x} = \dfrac{\pi}{\sqrt{a^2 - b^2}}, \qquad (a > b \ge 0)$

645. $\int_0^{2\pi} \dfrac{dx}{1 + a \cos x} = \dfrac{2\pi}{\sqrt{1 - a^2}}, \qquad (a^2 < 1)$

646. $\int_0^\infty \dfrac{\cos ax - \cos bx}{x}\, dx = \log \dfrac{b}{a}$

647. $\int_0^{\pi/2} \dfrac{dx}{a^2 \sin^2 x + b^2 \cos^2 x} = \dfrac{\pi}{2ab}$

648. $\int_0^{\pi/2} \dfrac{dx}{(a^2 \sin^2 x + b^2 \cos^2 x)^2} = \dfrac{\pi(a^2 + b^2)}{4a^3 b^3}, \qquad (a, b > 0)$

649. $\int_0^{\pi/2} \sin^{n-1} x \cos^{m-1} x\, dx = \dfrac{1}{2} B\left(\dfrac{n}{2}, \dfrac{m}{2}\right), \qquad m \text{ and } n \text{ positive integers}$

650. $\int_0^{\pi/2} (\sin^{2n+1} \theta)\, d\theta = \dfrac{2 \cdot 4 \cdot 6 \ldots (2n)}{1 \cdot 3 \cdot 5 \ldots (2n+1)}$, $(n = 1, 2, 3 \ldots)$

651. $\int_0^{\pi/2} (\sin^{2n} \theta)\, d\theta = \dfrac{1 \cdot 3 \cdot 5 \ldots (2n-1)}{2 \cdot 4 \ldots (2n)}\left(\dfrac{\pi}{2}\right)$, $(n = 1, 2, 3 \ldots)$

652. $\int_0^{\pi/2} \dfrac{x}{\sin x}\, dx = 2\left\{\dfrac{1}{1^2} - \dfrac{1}{3^2} + \dfrac{1}{5^2} - \dfrac{1}{7^2} + \cdots\right\}$

653. $\int_0^{\pi/2} \dfrac{dx}{1 + \tan^m x} = \dfrac{\pi}{4}$

654. $\int_0^{\pi/2} \sqrt{\cos \theta}\, d\theta = \dfrac{(2\pi)^{\frac{3}{2}}}{[\Gamma(\frac{1}{4})]^2}$

655. $\int_0^{\pi/2} (\tan^h \theta)\, d\theta = \dfrac{\pi}{2 \cos\left(\dfrac{h\pi}{2}\right)}$, $(0 < h < 1)$

656. $\int_0^{\infty} \dfrac{\tan^{-1}(ax) - \tan^{-1}(bx)}{x}\, dx = \dfrac{\pi}{2}\log\dfrac{a}{b}$, $(a, b > 0)$

657. The area enclosed by a curve defined through the equation $x^{\frac{b}{c}} + y^{\frac{b}{c}} = a^{\frac{b}{c}}$ where $a > 0$, c a positive odd integer and b a positive even integer is given by

$$\dfrac{\left[\Gamma\left(\dfrac{c}{b}\right)\right]^2}{\Gamma\left(\dfrac{2c}{b}\right)}\left(\dfrac{2ca^2}{b}\right)$$

658. $I = \iiint\limits_R x^{h-1} y^{m-1} z^{n-1}\, dv$, where R denotes the region of space bounded by

the co-ordinate planes and that portion of the surface $\left(\dfrac{x}{a}\right)^p + \left(\dfrac{y}{b}\right)^q + \left(\dfrac{z}{c}\right)^k = 1$,

which lies in the first octant and where $h, m, n, p, q, k, a, b, c$, denote positive real numbers is given by

$$\int_0^a x^{h-1}\, dx \int_0^{b\left[1 - \left(\frac{x}{a}\right)^p\right]^{\frac{1}{q}}} y^m\, dy \int_0^{c\left[1 - \left(\frac{x}{a}\right)^p - \left(\frac{y}{b}\right)^q\right]^{\frac{1}{k}}} z^{n-1}\, dz$$

$$= \dfrac{a^h b^m c^n}{pqk} \dfrac{\Gamma\left(\dfrac{h}{p}\right)\Gamma\left(\dfrac{m}{q}\right)\Gamma\left(\dfrac{n}{k}\right)}{\Gamma\left(\dfrac{h}{p} + \dfrac{m}{q} + \dfrac{n}{k} + 1\right)}$$

659. $\int_0^{\infty} e^{-ax}\, dx = \dfrac{1}{a}$, $(a > 0)$

660. $\int_0^{\infty} \dfrac{e^{-ax} - e^{-bx}}{x}\, dx = \log\dfrac{b}{a}$, $(a, b > 0)$

661. $\int_0^{\infty} x^n e^{-ax}\, dx = \begin{cases} \dfrac{\Gamma(n+1)}{a^{n+1}}, & (n > -1, a > 0) \\[2mm] \qquad \text{or} \\[2mm] \dfrac{n!}{a^{n+1}}, & (a > 0, n \text{ positive integer}) \end{cases}$

662. $\int_0^{\infty} x^n \exp(-ax^p)\, dx = \dfrac{\Gamma(k)}{pa^k}$, $\left(n > -1, p > 0, a > 0, k = \dfrac{n+1}{p}\right)$

663. $\int_0^{\infty} e^{-a^2 x^2}\, dx = \dfrac{1}{2a}\sqrt{\pi} = \dfrac{1}{2a}\Gamma\left(\dfrac{1}{2}\right)$, $(a > 0)$

664. $\int_0^{\infty} x e^{-x^2}\, dx = \tfrac{1}{2}$

665. $\displaystyle\int_0^\infty x^2 e^{-x^2}\,dx = \frac{\sqrt{\pi}}{4}$

666. $\displaystyle\int_0^\infty x^{2n} e^{-ax^2}\,dx = \frac{1\cdot 3\cdot 5\dots(2n-1)}{2^{n+1}a^n}\sqrt{\frac{\pi}{a}}$

667. $\displaystyle\int_0^\infty x^{2n+1} e^{-ax^2}\,dx = \frac{n!}{2a^{n+1}}, \qquad (a>0)$

668. $\displaystyle\int_0^1 x^m e^{-ax}\,dx = \frac{m!}{a^{m+1}}\left[1 - e^{-a}\sum_{r=0}^m \frac{a^r}{r!}\right]$

669. $\displaystyle\int_0^\infty e^{\left(-x^2 - \frac{a^2}{x^2}\right)}\,dx = \frac{e^{-2a}\sqrt{\pi}}{2}, \qquad (a \geq 0)$

670. $\displaystyle\int_0^\infty e^{-nx}\sqrt{x}\,dx = \frac{1}{2n}\sqrt{\frac{\pi}{n}}$

671. $\displaystyle\int_0^\infty \frac{e^{-nx}}{\sqrt{x}}\,dx = \sqrt{\frac{\pi}{n}}$

672. $\displaystyle\int_0^\infty e^{-ax}(\cos mx)\,dx = \frac{a}{a^2 + m^2}, \qquad (a>0)$

673. $\displaystyle\int_0^\infty e^{-ax}(\sin mx)\,dx = \frac{m}{a^2 + m^2}, \qquad (a>0)$

674. $\displaystyle\int_0^\infty x e^{-ax}[\sin(bx)]\,dx = \frac{2ab}{(a^2+b^2)^2}, \qquad (a>0)$

675. $\displaystyle\int_0^\infty x e^{-ax}[\cos(bx)]\,dx = \frac{a^2 - b^2}{(a^2+b^2)^2}, \qquad (a>0)$

676. $\displaystyle\int_0^\infty x^n e^{-ax}[\sin(bx)]\,dx = \frac{n![(a+ib)^{n+1} - (a-ib)^{n+1}]}{2i(a^2+b^2)^{n+1}}, \qquad (i^2 = -1, a>0)$

677. $\displaystyle\int_0^\infty x^n e^{-ax}[\cos(bx)]\,dx = \frac{n![(a-ib)^{n+1} + (a+ib)^{n+1}]}{2(a^2+b^2)^{n+1}}, \qquad (i^2 = -1, a>0)$

678. $\displaystyle\int_0^\infty \frac{e^{-ax}\sin x}{x}\,dx = \cot^{-1} a, \qquad (a>0)$

679. $\displaystyle\int_0^\infty e^{-a^2 x^2}\cos bx\,dx = \frac{\sqrt{\pi}}{2a}\exp\left(-\frac{b^2}{4a^2}\right), \qquad (ab \neq 0)$

680. $\displaystyle\int_0^\infty e^{-t\cos\phi}\, t^{b-1}\sin(t\sin\phi)\,dt = [\Gamma(b)]\sin(b\phi), \qquad \left(b>0, -\frac{\pi}{2} < \phi < \frac{\pi}{2}\right)$

681. $\displaystyle\int_0^\infty e^{-t\cos\phi}\, t^{b-1}[\cos(t\sin\phi)]\,dt = [\Gamma(b)]\cos(b\phi), \qquad \left(b>0, -\frac{\pi}{2} < \phi < \frac{\pi}{2}\right)$

682. $\displaystyle\int_0^\infty t^{b-1}\cos t\,dt = [\Gamma(b)]\cos\left(\frac{b\pi}{2}\right), \qquad (0<b<1)$

683. $\displaystyle\int_0^\infty t^{b-1}(\sin t)\,dt = [\Gamma(b)]\sin\left(\frac{b\pi}{2}\right), \qquad (0<b<1)$

684. $\displaystyle\int_0^1 (\log x)^n\,dx = (-1)^n \cdot n!$

685. $\displaystyle\int_0^1 \left(\log\frac{1}{x}\right)^{\frac{1}{2}}\,dx = \frac{\sqrt{\pi}}{2}$

686. $\displaystyle\int_0^1 \left(\log\frac{1}{x}\right)^{-\frac{1}{2}}\,dx = \sqrt{\pi}$

687. $\int_0^1 \left(\log \frac{1}{x}\right)^n dx = n!$

688. $\int_0^1 x \log (1 - x) \, dx = -\frac{3}{4}$

689. $\int_0^1 x \log (1 + x) \, dx = \frac{1}{4}$

690. $\int_0^1 x^m (\log x)^n \, dx = \dfrac{(-1)^n n!}{(m + 1)^{n+1}}, \qquad m > -1, n = 0, 1, 2, \ldots$

If $n \neq 0, 1, 2, \ldots$ replace $n!$ by $\Gamma(n + 1)$.

691. $\int_0^1 \dfrac{\log x}{1 + x} \, dx = -\dfrac{\pi^2}{12}$

692. $\int_0^1 \dfrac{\log x}{1 - x} \, dx = -\dfrac{\pi^2}{6}$

693. $\int_0^1 \dfrac{\log (1 + x)}{x} \, dx = \dfrac{\pi^2}{12}$

694. $\int_0^1 \dfrac{\log (1 - x)}{x} \, dx = -\dfrac{\pi^2}{6}$

695. $\int_0^1 (\log x)[\log (1 + x)] \, dx = 2 - 2 \log 2 - \dfrac{\pi^2}{12}$

696. $\int_0^1 (\log x)[\log (1 - x)] \, dx = 2 - \dfrac{\pi^2}{6}$

697. $\int_0^1 \dfrac{\log x}{1 - x^2} \, dx = -\dfrac{\pi^2}{8}$

698. $\int_0^1 \log \left(\dfrac{1 + x}{1 - x}\right) \cdot \dfrac{dx}{x} = \dfrac{\pi^2}{4}$

699. $\int_0^1 \dfrac{\log x \, dx}{\sqrt{1 - x^2}} = -\dfrac{\pi}{2} \log 2$

700. $\int_0^1 x^m \left[\log \left(\dfrac{1}{x}\right)\right]^n dx = \dfrac{\Gamma(n + 1)}{(m + 1)^{n+1}}, \qquad$ if $m + 1 > 0, n + 1 > 0$

701. $\int_0^1 \dfrac{(x^p - x^q) \, dx}{\log x} = \log \left(\dfrac{p + 1}{q + 1}\right), \qquad (p + 1 > 0, q + 1 > 0)$

702. $\int_0^1 \dfrac{dx}{\sqrt{\log \left(\dfrac{1}{x}\right)}} = \sqrt{\pi}$, (same as integral 686)

703. $\int_0^\infty \log \left(\dfrac{e^x + 1}{e^x - 1}\right) dx = \dfrac{\pi^2}{4}$

704. $\int_0^{\pi/2} (\log \sin x) \, dx = \int_0^{\pi/2} \log \cos x \, dx = -\dfrac{\pi}{2} \log 2$

705. $\int_0^{\pi/2} (\log \sec x) \, dx = \int_0^{\pi/2} \log \csc x \, dx = \dfrac{\pi}{2} \log 2$

706. $\int_0^\pi x(\log \sin x) \, dx = -\dfrac{\pi^2}{2} \log 2$

707. $\int_0^{\pi/2} (\sin x)(\log \sin x) \, dx = \log 2 - 1$

A-63

708. $\displaystyle\int_0^{\pi/2} (\log \tan x)\, dx = 0$

709. $\displaystyle\int_0^{\pi} \log (a \pm b \cos x)\, dx = \pi \log \left(\frac{a + \sqrt{a^2 - b^2}}{2} \right), \qquad (a \geq b)$

710. $\displaystyle\int_0^{\pi} \log (a^2 - 2ab \cos x + b^2)\, dx = \begin{cases} 2\pi \log a, & a \geq b > 0 \\ 2\pi \log b, & b \geq a > 0 \end{cases}$

711. $\displaystyle\int_0^{\infty} \frac{\sin ax}{\sinh bx}\, dx = \frac{\pi}{2b} \tanh \frac{a\pi}{2b}$

712. $\displaystyle\int_0^{\infty} \frac{\cos ax}{\cosh bx}\, dx = \frac{\pi}{2b} \operatorname{sech} \frac{a\pi}{2b}$

713. $\displaystyle\int_0^{\infty} \frac{dx}{\cosh ax} = \frac{\pi}{2a}$

714. $\displaystyle\int_0^{\infty} \frac{x\, dx}{\sinh ax} = \frac{\pi^2}{4a^2}$

715. $\displaystyle\int_0^{\infty} e^{-ax}(\cosh bx)\, dx = \frac{a}{a^2 - b^2}, \qquad (0 \leq |b| < a)$

716. $\displaystyle\int_0^{\infty} e^{-ax}(\sinh bx)\, dx = \frac{b}{a^2 - b^2}, \qquad (0 \leq |b| < a)$

717. $\displaystyle\int_0^{\infty} \frac{\sinh ax}{e^{bx} + 1}\, dx = \frac{\pi}{2b} \csc \frac{a\pi}{b} - \frac{1}{2a}$

718. $\displaystyle\int_0^{\infty} \frac{\sinh ax}{e^{bx} - 1}\, dx = \frac{1}{2a} - \frac{\pi}{2b} \cot \frac{a\pi}{b}$

719. $\displaystyle\int_0^{\pi/2} \frac{dx}{\sqrt{1 - k^2 \sin^2 x}} = \frac{\pi}{2}\left[1 + \left(\frac{1}{2}\right)^2 k^2 + \left(\frac{1 \cdot 3}{2 \cdot 4}\right)^2 k^4 \right.$
$$\left. + \left(\frac{1 \cdot 3 \cdot 5}{2 \cdot 4 \cdot 6}\right)^2 k^6 + \cdots \right], \text{ if } k^2 < 1$$

720. $\displaystyle\int_0^{\pi/2} \sqrt{1 - k^2 \sin^2 x}\, dx = \frac{\pi}{2}\left[1 - \left(\frac{1}{2}\right)^2 k^2 - \left(\frac{1 \cdot 3}{2 \cdot 4}\right)^2 \frac{k^4}{3} \right.$
$$\left. - \left(\frac{1 \cdot 3 \cdot 5}{2 \cdot 4 \cdot 6}\right)^2 \frac{k^6}{5} - \cdots \right], \text{ if } k^2 < 1$$

721. $\displaystyle\int_0^{\infty} e^{-x} \log x\, dx = -\gamma = -0.5772157 \ldots$

722. $\displaystyle\int_0^{\infty} e^{-x^2} \log x\, dx = -\frac{\sqrt{\pi}}{4}(\gamma + 2 \log 2)$

723. $\displaystyle\int_0^{\infty} \left(\frac{1}{1 - e^{-x}} - \frac{1}{x} \right) e^{-x}\, dx = \gamma = 0.5772157 \ldots \qquad \text{[Euler's Constant]}$

724. $\displaystyle\int_0^{\infty} \frac{1}{x}\left(\frac{1}{1 + x} - e^{-x} \right) dx = \gamma = 0.5772157 \ldots$

For n even:

725. $\displaystyle\int \cos^n x\, dx = \frac{1}{2^{n-1}} \sum_{k=0}^{\frac{n}{2} - 1} \binom{n}{k} \frac{\sin (n - 2k)\, x}{(n - 2k)} + \frac{1}{2^n} \binom{n}{\frac{n}{2}} x$

726. $\int \sin^n x\, dx \quad = \quad \dfrac{1}{2^{n-1}} \displaystyle\sum_{k=0}^{\frac{n}{2}-1} \binom{n}{k} \dfrac{\sin\left[(n-2k)\left(\frac{\pi}{2}-x\right)\right]}{2k-n} + \dfrac{1}{2^n} \binom{n}{\frac{n}{2}} x$

For n odd:

727. $\int \cos^n x\, dx \quad = \quad \dfrac{1}{2^{n-1}} \displaystyle\sum_{k=0}^{\frac{n-1}{2}} \binom{n}{k} \dfrac{\sin(n-2k)x}{(n-2k)}$

728. $\int \sin^n x\, dx \quad = \quad \dfrac{1}{2^{n-1}} \displaystyle\sum_{k=0}^{\frac{n-1}{2}} \binom{n}{k} \dfrac{\sin\left[n-2k)\left(\frac{\pi}{2}-x\right)\right]}{2k-n}$

DIFFERENTIAL EQUATIONS
SPECIAL FORMULAS

Certain types of differential equations occur sufficiently often to justify the use of formulas for the corresponding particular solutions. The following set of tables 1 to XIV covers all first, second, and nth order ordinary linear differential equations with constant coefficients for which the right members are of the form $P(x)e^{rx}\sin sx$ or $P(x)e^{rx}\cos sx$, where r and s are constants and $P(x)$, is a polynomial of degree n.

When the right member of a reducible linear partial differential equation with constant coefficients is not zero, particular solutions for certain types of right members are contained in tables XV to XXI. In these tables both F and P are used to denote polynomials, and it is assumed that no denominator is zero. In any formula the roles of x and y may be reversed throughout, changing a formula in which x dominates to one in which y dominates. Tables XIX, XX, XXI are applicable whether the equations are reducible or not.

The symbol $\binom{m}{n}$ stands for $\dfrac{m!}{(m-n)!n!}$ and is the $n+1$ st coefficient in the expansion of $(a+b)^m$. Also $0! = 1$ by definition.

The tables as herewith given are those contained in the text *Differential Equations* by Ginn and Company (1955) and are published with their kind permission and that of the author, Professor Frederick H. Steen.

Solution of Linear Differential Equations with Constant Coefficients

Any linear differential equation with constant coefficients may be written in the form

$$p(D)y = R(x)$$

where D is the differential operation

$$Dy = \frac{dy}{dx}$$

$p(D)$ is a polynomial in D,
y is the dependent variable,
x is the independent variable,
$R(x)$ is an arbitrary function of x.

A power of D represents repeated differentiation, that is

$$D^n y = \frac{d^n y}{dx^n}$$

For such an equation, the general solution may be written in the form

$$y = y_c + y_p$$

where y_p is any particular solution, and y_c is called the *complementary function*. This complementary function is defined as the general solution of the *homogeneous equation*, which is the original differential equation with the right side replaced by zero, i.e.

$$p(D)y = 0$$

The complementary function y_c may be determined as follows:

1. Factor the polynomial $p(D)$ into real and complex linear factors, just as if D were a variable instead of an operator.
2. For each nonrepeated linear factor of the form $(D - a)$, where a is real, write down a term of the form

$$ce^{ax}$$

where c is an arbitrary constant.
3. For each repeated real linear factor of the form $(D - a)^n$, write down n terms of the form

$$c_1 e^{ax} + c_2 x e^{ax} + c_3 x^2 e^{ax} + \cdots + c_n x^{n-1} e^{ax}$$

where the c_i's are arbitrary constants.

4. For each non-repeated conjugate complex pair of factors of the form $(D - a + ib)(D - a - ib)$, write down 2 terms of the form

$$c_1 e^{ax} \cos bx + c_2 e^{ax} \sin bx$$

5. For each repeated conjugate complex pair of factors of the form $(D - a + ib)^n (D - a - ib)^n$, write down $2n$ terms of the form

$$c_1 e^{ax} \cos bx + c_2 e^{ax} \sin bx + c_3 x e^{ax} \cos bx + c_4 x e^{ax} \sin bx + \cdots$$
$$+ c_{2n-1} x^{n-1} e^{ax} \cos bx + c_{2n} x^{n-1} e^{ax} \sin bx$$

6. The sum of all the terms thus written down is the complementary function y_c.

To find the particular solution y_p, use the following tables, as shown in the examples. For cases not shown in the tables, there are various methods of finding y_p. The most general method is called *variation of parameters*. The following example illustrates the method:

Find y_p for $(D^2 - 4) y = e^x$.

This example can be solved most easily by use of equation 63 in the tables following. However it is given here as an example of the method of variation of parameters.

The complementary function is

$$y_c = c_1 e^{2x} + c_2 e^{-2x}$$

To find y_p, replace the constants in the complementary function with unknown functions,

$$y_p = u e^{2x} + v e^{-2x}$$

We now prepare to substitute this assumed solution into the original equation. We begin by taking all the necessary dervatives:

$$y_p = u e^{2x} + v e^{-2x}$$
$$y_p' = 2u e^{2x} - 2v e^{-2x} + u' e^{2x} + v' e^{-2x}$$

For each derivative of y_p, except the highest, we set the sum of all the terms containing u' and v' to 0. Thus the above equation becomes

$$u' e^{2x} + v' e^{-2x} = 0 \quad \text{and} \quad y_p' = 2u e^{2x} - 2v e^{-2x}$$

Continuing to differentiate, we have

$$y_p'' = 4u e^{2x} + 4v e^{-2x} + 2u' e^{2x} - 2v' e^{-2x}$$

When we substitute into the original equation, all the terms not containing u' or v' cancel out. This is a consequence of the method by which y_p was set up.

Thus all that is necessary is to write down the terms containing u' or v' in the highest order derivative of y_p, multiply by the constant coefficient of the highest power of D in $p(D)$, and set it equal to $R(x)$. Together with the previous terms in u' and v' which were set equal to 0, this gives us as many linear equations in the first derivatives of the unknown functions as there are unknown functions. The first derivatives may then be solved for by algebra, and the unknown functions found by integration. In the present example, this becomes

$$u' e^{2x} + v' e^{-2x} = 0$$
$$2u' e^{2x} - 2v' e^{-2x} = e^x$$

We eliminate v' and u' separately, getting

$$4u' e^{2x} = e^x$$
$$4v' e^{-2x} = -e^x$$

Thus

$$u' = \tfrac{1}{4} e^{-x}$$
$$v' = -\tfrac{1}{4} e^{3x}$$

Therefore, by integrating

$$u = -\tfrac{1}{4} e^{-x}$$
$$v = -\tfrac{1}{12} e^{3x}$$

A constant of integration is not needed, since we need only one particular solution. Thus

$$y_p = u e^{2x} + v e^{-2x} = -\tfrac{1}{4} e^{-x} e^{2x} - \tfrac{1}{12} e^{3x} e^{-2x}$$
$$= -\tfrac{1}{4} e^x - \tfrac{1}{12} e^x = -\tfrac{1}{3} e^x$$

and the general solution is

$$y = y_c + y_p = c_1 e^{2x} + c_2 e^{-2x} - \tfrac{1}{3} e^x$$

The following examples illustrate the use of the tables.

Example 1. Solve $(D^2 - 4)y = \sin 3x$.

Substitution of $q = -4$, $s = 3$ in formula 24 gives

$$y_p = \frac{\sin 3x}{-9 - 4}$$

wherefore the general solution is

$$y = c_1 e^{2x} + c_2 e^{-2x} - \frac{\sin 3x}{13}$$

Example 2. Obtain a particular solution of $(D^2 - 4D + 5)y = x^2 e^{3x} \sin x$.

Applying formula 40 with $a = 2, b = 1, r = 3, s = 1$, $P(x) = x^2$, $s + b = 2$, $s - b = 0$, $a - r = -1$, $(a - r)^2 + (s + b)^2 = 5$, $(a - r)^2 + (s - b)^2 = 1$, we have

$$y_p = \frac{e^{3x} \sin x}{2} \left[\left(\frac{2}{5} - \frac{0}{1} \right) x^2 + \left(\frac{2(-1)2}{25} - \frac{2(-1)0}{1} \right) 2x \right.$$
$$\left. + \left(\frac{3 \cdot 1 \cdot 2 - 2^3}{125} - \frac{3 \cdot 1 \cdot 0 - 0}{1} \right) 2 \right]$$
$$- \frac{e^{3x} \cos x}{2} \left[\left(\frac{-1}{5} - \frac{-1}{1} \right) x^2 + \left(\frac{1 - 4}{25} - \frac{1 - 0}{1} \right) 2x \right.$$
$$\left. + \left(\frac{-1 - 3(-1)4}{125} - \frac{-1 - 3(-1)0}{1} \right) 2 \right]$$
$$= (\tfrac{1}{5} x^2 - \tfrac{4}{25} x - \tfrac{2}{125}) e^{3x} \sin x + (-\tfrac{2}{5} x^2 + \tfrac{28}{25} x - \tfrac{136}{125}) e^{3x} \cos x$$

The special formulas effect a very considerable saving of time in problems of this type.

Example 3. Obtain a particular solution of $(D^2 - 4D + 5)y = x^2 e^{2x} \cos x$. (Compare with Example 2.)

Formula 40 is not applicable here since for this equation $r = a$, $s = b$, wherefore the denominator $(a - r)^2 + (s - b)^2 = 0$. We turn instead to formula 44. Substituting $a = 2$, $b = 1$, $P(x) = x^2$ and replacing sin by cos, cos by $-\sin$, we obtain

$$y_p = \frac{e^{2x} \cos x}{4} (x^2 - \tfrac{2}{4}) + \frac{e^{2x} \sin x}{2} \int (x^2 - \tfrac{1}{2}) dx$$
$$= \left(\frac{x^2}{4} - \frac{1}{8} \right) e^{2x} \cos x + \left(\frac{x^3}{6} - \frac{x}{4} \right) e^{2x} \sin x$$

which is the required solution.

Example 4. Find z_p for $(D_x - 3 D_y) z = \ln (y + 3x)$.

Referring to Table XV we note that formula 69 (not 68) is applicable. This gives

$$z_p = x \ln (y + 3x)$$

It is easily seen that $- y/3 \ln (y + 3x)$ would serve equally well.

Example 5. Solve $(D_x + 2D_y - 4) z = y \cos (y - 2x)$.

Since R in formula 76 contains a polynomial in x, not y, we rewrite the given equation in the form

$$(D_y + \tfrac{1}{2} D_x - 2) z = \tfrac{1}{2} y \cos (y - 2x)$$

Then

$$z_c = e^{2y} F(x - \tfrac{1}{2} y) = e^{2y} f(2x - y)$$

and by the formula

$$z_p = -\tfrac{1}{2} \cos (y - 2x) \cdot \left(\frac{y}{2} + \frac{\tfrac{1}{2}}{2} \right)$$
$$= -\tfrac{1}{8} (2y + 1) \cos (y - 2x)$$

Example 6. Find z_p for $(D_x + 4D_y)^3 z = (2x - y)^2$.

Using formula 79, we obtain

$$z_p = \frac{\int \int \int u^2 du^3}{[2 + 4(-1)]^3} = \frac{u^5}{5 \cdot 4 \cdot 3 \cdot (-8)} = -\frac{(2x - y)^5}{480}$$

Example 7. Find z_p for $(D_x^3 + 5D_x^2 D_y - 7D_x + 4)z = e^{2x+3y}$.

By formula 87

$$z_p = \frac{e^{2x+3y}}{2^3 + 5 \cdot 2^2 \cdot 3 - 7 \cdot 2 + 4} = \frac{e^{2x+3y}}{58}$$

Example 8. Find z_p for

$$(D_x^4 + 6D_x^3 D_y + D_x D_y + D_y^2 + 9)z = \sin (3x + 4y)$$

A-67

Since every term in the left member is of even degree in the two operators D_x and D_y, formula 90 is applicable. It gives

$$z_p = \frac{\sin(3x + 4y)}{(-9)^2 + 6(-9)(-12) + (-12) + (-16) + 9}$$

$$= \frac{\sin(3x + 4y)}{710}$$

TABLE I: $(D - a)y = R$

R	y_p
1. e^{rx}	$\dfrac{e^{rx}}{r - a}$
2. $\sin sx$*	$-\dfrac{a \sin sx + s \cos sx}{a^2 + s^2} = -\dfrac{1}{\sqrt{a^2 + s^2}} \sin\left(sx + \tan^{-1}\dfrac{s}{a}\right)$
3. $P(x)$	$-\dfrac{1}{a}\left[P(x) + \dfrac{P'(x)}{a} + \dfrac{P''(x)}{a^2} + \cdots + \dfrac{P^{(n)}(x)}{a^n}\right]$
4. $e^{rx}\sin sx$*	Replace a by $a - r$ in formula 2 and multiply by e^{rx}.
5. $P(x)e^{rx}$	Replace a by $a - r$ in formula 3 and multiply by e^{rx}.
6. $P(x)\sin sx$*	$-\sin sx\left[\dfrac{a}{a^2 + s^2}P(x) + \dfrac{a^2 - s^2}{(a^2 + s^2)^2}P'(x) + \dfrac{a^3 - 3as^2}{(a^2 + s^2)^3}P''(x) + \cdots\right.$

$$+ \frac{a^k - \binom{k}{2}a^{k-2}s^2 + \binom{k}{4}a^{k-4}s^4 - \cdots}{(a^2 + s^2)^k}P^{(k-1)}(x) + \cdots\Bigg]$$

$$- \cos sx\left[\frac{s}{a^2 + s^2}P(x) + \frac{2as}{(a^2 + s^2)^2}P'(x) + \frac{3a^2s - s^3}{(a^2 + s^2)^3}P''(x) + \cdots\right.$$

$$+ \frac{\binom{k}{1}a^{k-1}s - \binom{k}{3}a^{k-3}s^3 + \cdots}{(a^2 + s^2)^k}P^{(k-1)}(x) + \cdots\Bigg]$$

7. $P(x)e^{rx}\sin sx$*	Replace a by $a - r$ in formula 6 and multiply by e^{rx}.
8. e^{ax}	xe^{ax}
9. $e^{ax}\sin sx$*	$-\dfrac{e^{ax}\cos sx}{s}$
10. $P(x)e^{ax}$	$e^{ax}\displaystyle\int P(x)\,dx$
11. $P(x)e^{ax}\sin sx$*	$\dfrac{e^{ax}\sin sx}{s}\left[\dfrac{P'(x)}{s^3} - \dfrac{P'''(x)}{s^3} + \dfrac{P^v(x)}{s^5} - \cdots\right] - \dfrac{e^{ax}\cos sx}{s}\left[P(x) - \dfrac{P''(x)}{s^2} + \dfrac{P^{iv}(x)}{s^4} - \cdots\right]$

* For $\cos sx$ in R replace "sin" by "cos" and "cos" by "$-$ sin" in y_p.

$$D^n = \frac{d^n}{dx^n} \qquad \binom{m}{n} = \frac{m!}{(m - n)!n!} \qquad 0! = 1$$

TABLE II: $(D - a)^2y = R$

R	y_p
12. e^{rx}	$\dfrac{e^{rx}}{(r - a)^2}$
13. $\sin sx$*	$\dfrac{1}{(a^2 + s^2)^2}[(a^2 - s^2)\sin sx + 2as \cos sx] = \dfrac{1}{a^2 + s^2}\sin\left(sx + \tan^{-1}\dfrac{2as}{a^2 - s^2}\right)$
14. $P(x)$	$\dfrac{1}{a^2}\left[P(x) + \dfrac{2P'(x)}{a} + \dfrac{3P''(x)}{a^2} + \cdots + \dfrac{(n + 1)P^{(n)}(x)}{a^n}\right]$
15. $e^{rx}\sin sx$*	Replace a by $a - r$ in formula 13 and multiply by e^{rx}.
16. $P(x)e^{rx}$	Replace a by $a - r$ in formula 14 and multiply by e^{rx}.
17. $P(x)\sin sx$*	$\sin sx\left[\dfrac{a^2 - s^2}{(a^2 + s^2)^2}P(x) + 2\dfrac{a^3 - 3as^2}{(a^2 + s^2)^3}P'(x) + 3\dfrac{a^4 - 6a^2s^2 + s^4}{(a^2 + s^2)^4}P''(x) + \cdots\right.$

$$+ (k - 1)\frac{a^k - \binom{k}{2}a^{k-2}s^2 + \binom{k}{4}a^{k-4}s^4 - \cdots}{(a^2 + s^2)^k}P^{(k-2)}(x) + \cdots\Bigg]$$

$$+ \cos sx\left[\frac{2as}{(a^2 + s^2)^2}P(x) + 2\frac{3a^2s - s^3}{(a^2 + s^2)^3}P'(x) + 3\frac{4a^3s - 4as^3}{(a^2 + s^2)^4}P''(x) + \cdots\right.$$

$$+ (k - 1)\frac{\binom{k}{1}a^{k-1}s - \binom{k}{3}a^{k-3}s^3 + \cdots}{(a^2 + s^2)^k}P^{(k-2)}(x) + \cdots\Bigg]$$

18. $P(x)e^{rx}\sin sx$*	Replace a by $a - r$ in formula 17 and multiply by e^{rx}.
19. e^{ax}	$\frac{1}{2}x^2e^{ax}$
20. $e^{ax}\sin sx$*	$-\dfrac{e^{ax}\sin sx}{s^2}$
21. $P(x)e^{ax}$	$e^{ax}\displaystyle\iint P(x)\,dx\,dx$
22. $P(x)e^{ax}\sin sx$*	$-\dfrac{e^{ax}\sin sx}{s^2}\left[P(x) - \dfrac{3P''(x)}{s^2} + \dfrac{5P^{iv}(x)}{s^4} - \dfrac{7P^{vi}(x)}{s^6} + \cdots\right]$

$$- \frac{e^{ax}\cos sx}{s^2}\left[\frac{2P'(x)}{s} - \frac{4P'''(x)}{s^3} + \frac{6P^v(x)}{s^5} - \cdots\right]$$

* For $\cos sx$ in R replace "sin" by "cos" and "cos" by "$-$ sin" in y_p.

TABLE III: $(D^2 + q)y = R$

R	y_p

23. e^{rx} $\dfrac{e^{rx}}{r^2 + q}$

24. $\sin sx^*$ $\dfrac{\sin sx}{-s^2 + q}$

25. $P(x)$ $\dfrac{1}{q}\left[P(x) - \dfrac{P''(x)}{q} + \dfrac{P^{iv}(x)}{q^2} - \cdots + (-1)^k \dfrac{P^{(2k)}(x)}{q^k} \cdots \right]$

26. $e^{rx} \sin sx^*$ $\dfrac{(r^2 - s^2 + q)e^{rx} \sin sx - 2rse^{rx}\cos sx}{(r^2 - s^2 + q)^2 + (2rs)^2} = \dfrac{e^{rx}}{\sqrt{(r^2 - s^2 + q)^2 + (2rs)^2}} \sin\left[sx - \tan^{-1}\dfrac{2rs}{r^2 - s^2 + q}\right]$

27. $P(x)e^{rx}$ $\dfrac{e^{rx}}{r^2 + q}\left[P(x) - \dfrac{2r}{r^2 + q}P'(x) + \dfrac{3r^2 - q}{(r^2 + q)^2}P''(x) - \dfrac{4r^3 - 4qr}{(r^2 + q)^3}P'''(x) \right.$

$$\left. + \cdots + (-1)^{k-1}\dfrac{\binom{k}{1}r^{k-1} - \binom{k}{3}r^{k-3}q + \binom{k}{5}r^{k-5}q^2 - \cdots}{(r^2 + q)^{k-1}}P^{(k-1)}(x) + \cdots \right]$$

28. $P(x)\sin sx^*$ $\dfrac{\sin sx}{(-s^2 + q)}\left[P(x) - \dfrac{3s^2 + q}{(-s^2 + q)^2}P''(x) + \dfrac{5s^4 + 10s^2q + q^2}{(-s^2 + q)^4}P^{iv}(x) + \cdots \right.$

$$\left. + (-1)^k \dfrac{\binom{2k+1}{1}s^{2k} + \binom{2k+1}{3}s^{2k-2}q + \binom{2k+1}{5}s^{2k-4}q^2 + \cdots}{(-s^2 + q)^{2k}}P^{(2k)}(x) + \cdots \right]$$

$$- \dfrac{s\cos sx}{(-s^2 + q)}\left[\dfrac{2P'(x)}{(-s^2 + q)} - \dfrac{4s^2 + 4q}{(-s^2 + q)^3}P'''(x) + \cdots \right.$$

$$\left. + (-1)^{k+1}\dfrac{\binom{2k}{1}s^{2k-2} + \binom{2k}{3}s^{2k-4}q + \cdots}{(-s^2 + q)^{2k-1}}P^{(2k-1)}(x) + \cdots \right]$$

TABLE IV: $(D^2 + b^2)y = R$

29. $\sin bx^*$ $-\dfrac{x\cos bx}{2b}$

30. $P(x)\sin bx^*$ $\dfrac{\sin bx}{(2b)^2}\left[P(x) - \dfrac{P''(x)}{(2b)^2} + \dfrac{P^{iv}(x)}{(2b)^4} - \cdots \right] - \dfrac{\cos bx}{2b}\int\left[P(x) - \dfrac{P''(x)}{(2b)^2} + \cdots \right]dx$

* For $\cos sx$ in R replace "sin" by "cos" and "cos" by "$-$ sin" in y_p.

TABLE V: $(D^2 + pD + q)y = R$

R	y_p

31. e^{rx} $\dfrac{e^{rx}}{r^2 + pr + q}$

32. $\sin sx^*$ $\dfrac{(q - s^2)\sin sx - ps\cos sx}{(q - s^2)^2 + (ps)^2} = \dfrac{1}{\sqrt{(q - s^2)^2 + (ps)^2}}\sin\left(sx - \tan^{-1}\dfrac{ps}{q - s^2}\right)$

33. $P(x)$ $\dfrac{1}{q}\left[P(x) - \dfrac{p}{q}P'(x) + \dfrac{p^2 - q}{q^2}P''(x) - \dfrac{p^3 - 2pq}{q^3}P'''(x) + \cdots \right.$

$$\left. + (-1)^n\dfrac{p^n - \binom{n-1}{1}p^{n-2}q + \binom{n-2}{2}p^{n-4}q^2 - \cdots}{q^n}P^{(n)}(x)\right]$$

34. $e^{rx}\sin sx^*$ Replace p by $p + 2r$, q by $q + pr + r^2$ in formula 32 and multiply by e^{rx}.

35. $P(x)e^{rx}$ Replace p by $p + 2r$, q by $q + pr + r^2$ in formula 33 and multiply by e^{rx}.

TABLE VI: $(D - b)(D - a)y = R$

36. $P(x)\sin sx^*$ $\dfrac{\sin sx}{b - a}\left[\left(\dfrac{a}{a^2 + s^2} - \dfrac{b}{b^2 + s^2}\right)P(x) + \left(\dfrac{a^2 - s^2}{(a^2 + s^2)^2} - \dfrac{b^2 - s^2}{(b^2 + s^2)^2}\right)P'(x) \right.$

$$\left. + \left(\dfrac{a^3 - 3as^2}{(a^2 + s^2)^3} - \dfrac{b^3 - 3bs^2}{(b^2 + s^2)^3}\right)P''(x) + \cdots \right]$$

$$+ \dfrac{\cos sx}{b - a}\left[\left(\dfrac{s}{a^2 + s^2} - \dfrac{s}{b^2 + s^2}\right)P(x) + \left(\dfrac{2as}{(a^2 + s^2)^2} - \dfrac{2bs}{(b^2 + s^2)^2}\right)P'(x) \right.$$

$$\left. + \left(\dfrac{3a^2s - s^3}{(a^2 + s^2)^3} - \dfrac{3b^2s - s^3}{(b^2 + s^2)^3}\right)P''(x) + \cdots \right]\dagger$$

37. $P(x)e^{rx}\sin sx^*$ Replace a by $a - r$, b by $b - r$ in formula 36 and multiply by e^{rx}.

38. $P(x)e^{ax}$ $\dfrac{e^{ax}}{a - b}\left[\int P(x)dx + \dfrac{P(x)}{(b - a)} + \dfrac{P'(x)}{(b - a)^2} + \dfrac{P''(x)}{(b - a)^3} + \cdots + \dfrac{P^{(n)}(x)}{(b - a)^{n+1}}\right]$

* For $\cos sx$ in R replace "sin" by "cos" and "cos" by "$-$ sin" in y_p.

† For additional terms, compare with formula 6.

TABLE VII: $(D^2 - 2aD + a^2 + b^2)y = R$

R y_p

39. $P(x) \sin sx$*

$$\frac{\sin sx}{2b}\left[\left(\frac{s+b}{a^2+(s+b)^2} - \frac{s-b}{a^2+(s-b)^2}\right)P(x) + \left(\frac{2a(s+b)}{[a^2+(s+b)^2]^2} - \frac{2a(s-b)}{[a^2+(s-b)^2]^2}\right)P'(x)\right.$$

$$\left. + \left(\frac{3a^2(s+b)-(s+b)^3}{[a^2+(s+b)^2]^3} - \frac{3a^2(s-b)-(s-b)^3}{[a^2+(s-b)^2]^3}\right)P''(x) + \cdots\right]$$

$$- \frac{\cos sx}{2b}\left[\left(\frac{a}{a^2+(s+b)^2} - \frac{a}{a^2+(s-b)^2}\right)P(x) + \left(\frac{a^2-(s+b)^2}{[a^2+(s+b)^2]^2} - \frac{a^2-(s-b)^2}{[a^2+(s-b)^2]^2}\right)P'(x)\right.$$

$$\left. + \left(\frac{a^3-3a(s+b)^2}{[a^2+(s+b)^2]^3} - \frac{a^3-3a(s-b)^2}{[a^2+(s-b)^2]^3}\right)P''(x) + \cdots\right]^\dagger$$

40. $P(x)e^{rx}\sin sx$*   Replace a by $a - r$ in formula 39 and multiply by e^{rx}.

41. $P(x)e^{ax}$ $\dfrac{e^{ax}}{b^2}\left[P(x) - \dfrac{P''(x)}{b^2} + \dfrac{P^{iv}(x)}{b^4} - \cdots\right]$

42. $e^{ax}\sin sx$* $\dfrac{e^{ax}\sin sx}{-s^2+b^2}$

43. $e^{ax}\sin bx$* $-\dfrac{xe^{ax}\cos bx}{2b}$

44. $P(x)e^{ax}\sin bx$* $\dfrac{e^{ax}\sin bx}{(2b)^2}\left[P(x) - \dfrac{P''(x)}{(2b)^2} + \dfrac{P^{iv}(x)}{(2b)^4} - \cdots\right] - \dfrac{e^{ax}\cos bx}{2b}\displaystyle\int\left[P(x) - \dfrac{P''(x)}{(2b)^2} + \dfrac{P^{iv}(x)}{(2b)^4} - \cdots\right]dx$

* For cos sx in R replace "sin" by "cos" and "cos" by "$-$ sin" in y_p.

† For additional terms, compare with formula 6.

TABLE VIII: $f(D)y = [D^n + a_{n-1}D^{n-1} + \cdots + a_1 D + a_0]y = R$

R y_p

45. e^{rx} $\dfrac{e^{rx}}{f(r)}$

46. $\sin sx$* $\dfrac{[a_0 - a_2 s^2 + a_4 s^4 - \cdots]\sin sx - [a_1 s - a_3 s^3 + a_5 s^5 + \cdots]\cos sx}{[a_0 - a_2 s^2 + a_4 s^4 - \cdots]^2 + [a_1 s - a_3 s^3 + a_5 s^5 - \cdots]^2}$

TABLE IX: $f(D^2)y = R$

47. $\sin sx$* $\dfrac{\sin sx}{f(-s^2)} = \dfrac{\sin sx}{a_0 - a_2 s^2 + \cdots \pm s^{2n}}$

TABLE X: $(D - a)^n y = R$

48. e^{rx} $\dfrac{e^{rx}}{(r-a)^n}$

49. $\sin sx$* $\dfrac{(-1)^n}{(a^2+s^2)^n}\{[a^n - \binom{n}{2}a^{n-2}s^2 + \binom{n}{4}a^{n-4}s^4 - \cdots]\sin sx + [\binom{n}{1}a^{n-1}s - \binom{n}{3}a^{n-3}s^3 + \cdots]\cos sx\}$

50. $P(x)$ $\dfrac{(-1)^n}{a^n}\left[P(x) + \binom{n}{1}\dfrac{P'(x)}{a} + \binom{n+1}{2}\dfrac{P''(x)}{a^2} + \binom{n+2}{3}\dfrac{P'''(x)}{a^3} + \cdots\right]$

51. $e^{rx}\sin sx$*   Replace a by $a - r$ in formula 49 and multiply by e^{rx}.

52. $e^{rx}P(x)$ Replace a by $a - r$ in formula 50 and multiply by e^{rx}.

53. $P(x)\sin sx$* $(-1)^n\sin sx[A_n P(x) + \binom{n}{1}A_{n+1}P'(x) + \binom{n+1}{2}A_{n+2}P''(x) + \binom{n+2}{3}A_{n+3}P'''(x) + \cdots]$

$$+ (-1)^n\cos sx[B_n P(x) + \binom{n}{1}B_{n+1}P'(x) + \binom{n+1}{2}B_{n+2}P''(x) + \binom{n+2}{3}B_{n+3}P'''(x) + \cdots]$$

$$A_1 = \frac{a}{a^2+s^2},\ A_2 = \frac{a^2-s^2}{(a^2+s^2)^2},\ \cdots,\ A_k = \frac{a^k - \binom{k}{2}a^{k-2}s^2 + \binom{k}{4}a^{k-4}s^4 - \cdots}{(a^2+s^2)^k}.$$

$$B_1 = \frac{s}{a^2+s^2},\ B_2 = \frac{2as}{(a^2+s^2)^2},\ \cdots,\ B_k = \frac{\binom{k}{1}a^{k-1}s - \binom{k}{3}a^{k-3}s^3 + \cdots}{(a^2+s^2)^k}.$$

54. $P(x)e^{rx}\sin sx$*   Replace a by $a - r$ in formula 53 and multiply by e^{rx}.

55. $e^{ax}P(x)$ $\qquad$ $e^{ax}\displaystyle\int\int\cdots\int P(x)dx^n$

56. $P(x)e^{ax}\sin sx*$ $\qquad$ $\dfrac{(-1)^{\frac{n-1}{2}}e^{ax}\sin sx}{s^n}\left[\binom{n}{n-1}\dfrac{P'(x)}{s}-\binom{n+2}{n-1}\dfrac{P'''(x)}{s^3}+\binom{n+4}{n-1}\dfrac{P^{v}(x)}{s^5}-\cdots\right]$

$$+\dfrac{(-1)^{\frac{n+1}{2}}e^{ax}\cos sx}{s^n}\left[\binom{n-1}{n-1}P(x)-\binom{n+1}{n-1}\dfrac{P''(x)}{s^2}+\binom{n+3}{n-1}\dfrac{P^{iv}(x)}{s^4}-\cdots\right]\ (n\ \mathbf{odd})$$

$$\dfrac{(-1)^{\frac{n}{2}}e^{ax}\sin sx}{s^n}\left[\binom{n-1}{n-1}P(x)-\binom{n+1}{n-1}\dfrac{P''(x)}{s^2}+\binom{n+3}{n-1}\dfrac{P^{iv}(x)}{s^4}-\cdots\right]$$

$$+\dfrac{(-1)^{\frac{n}{2}}e^{ax}\cos sx}{s^n}\left[\binom{n}{n-1}\dfrac{P'(x)}{s}-\binom{n+2}{n-1}\dfrac{P'''(x)}{s^3}+\binom{n+4}{n-1}\dfrac{P^{v}(x)}{s^5}-\cdots\right]\ (n\ \mathbf{even})$$

* For $\cos sx$ in R replace "sin" by "cos" and "cos" by "$-\sin$" in y_p.

TABLE XI: $(D-a)^n f(D)y = R$

57. e^{ax} $\qquad$ $\dfrac{x^n}{n!}\cdot\dfrac{e^{ax}}{f(a)}$

* For $\cos sx$ in R replace "sin" by "cos" and "cos" by "$-\sin$" in y_p.

TABLE XII: $(D^2+q)^n y = R$

R $\qquad\qquad$ y_p

58. e^{rx} $\qquad$ $e^{rx}/(r^2+q)^n$

59. $\sin sx*$ $\qquad$ $\sin sx/(q-s^2)^n$

60. $P(x)$ $\qquad$ $\dfrac{1}{q^n}\left[P(x)-\binom{n}{1}\dfrac{P''(x)}{q}+\binom{n+1}{2}\dfrac{P^{iv}(x)}{q^2}-\binom{n+2}{3}\dfrac{P^{vi}(x)}{q^3}+\cdots\right]$

61. $e^{rx}\sin sx*$ $\qquad$ $\dfrac{e^{rx}}{(A^2+B^2)^n}\{[A^n-\binom{n}{2}A^{n-2}B^2+\binom{n}{4}A^{n-4}B^4-\cdots]\sin sx-[\binom{n}{1}A^{n-1}B-\binom{n}{3}A^{n-3}B^3+\cdots]\cos sx\}$

$$A=r^2-s^2+q,\qquad B=2rs$$

TABLE XIII: $(D^2+b^2)^n y = R$

62. $\sin bx*$ $\qquad$ $(-1)^{\frac{n+1}{2}}\dfrac{x^n\cos bx}{n!(2b)^n}\ (n\ \mathbf{odd}),$ $\qquad\qquad$ $(-1)^{\frac{n}{2}}\dfrac{x^n\sin bx}{n!(2b)^n}\ (n\ \mathbf{even})$

TABLE XIV: $(D^n-q)y = R$

63. e^{rx} $\qquad$ $e^{rx}/(r^n-q)$

64. $P(x)$ $\qquad$ $-\dfrac{1}{q}\left[P(x)\dfrac{P^{(n)}(x)}{q}+\dfrac{P^{(2n)}(x)}{q^2}+\cdots\right]$

65. $\sin sx*$ $\qquad$ $-\dfrac{q\sin sx+(-1)^{\frac{n-1}{2}}s^n\cos sx}{q^2+s^{2n}}\ (n\ \mathbf{odd}),$ $\qquad$ $\dfrac{\sin sx}{(-s^2)^{n/2}-q}\ (n\ \mathbf{even})$

66. $e^{rx}\sin sx*$ $\qquad$ $\dfrac{Ae^{rx}\sin sr-Be^{rx}\cos sx}{A^2+B^2}=\dfrac{e^{rx}}{\sqrt{A^2+B^2}}\sin\left(sx-\tan^{-1}\dfrac{B}{A}\right)$

$$A=[r^n-\binom{n}{2}r^{n-2}s^2+\binom{n}{4}r^{n-4}s^4-\cdots]-q,\qquad B=[\binom{n}{1}r^{n-1}s-\binom{n}{3}r^{n-3}s^3+\cdots]$$

* For $\cos sx$ in R replace "sin" by "cos" and "cos" by "$-\sin$" in y_p.

TABLE XV: $(D_x+mD_y)z = R$

R $\qquad\qquad$ z_p

67. e^{ax+by} $\qquad$ $\dfrac{e^{ax+by}}{a+mb}$

68. $f(ax+by)$ $\qquad$ $\dfrac{\int f(u)du}{a+mb},\ u=ax+by$

69. $f(y-mx)$ $\qquad$ $xf(y-mx)$

70. $\phi(x,y)f(y-mx)$ $\qquad$ $f(y-mx)\int\phi(x,a+mx)dx\ \ (a=y-mx\ \text{after integration})$

TABLE XVI: $(D_z + mD_y - k)z = R$

71. e^{az+by} $\dfrac{e^{az+by}}{a + mb - k}$

72. $\sin(ax + by)$* $-\dfrac{(a + bm)\cos(ax + by) + k\sin(ax + by)}{(a + bm)^2 + k^2}$

73. $e^{\alpha x + \beta y}\sin(ax + by)$*   Replace k in 72 by $k - \alpha - m\beta$ and multiply by $e^{\alpha x + \beta y}$

74. $e^{xk}f(ax + by)$ $\dfrac{e^{kx}\int f(u)du}{a + mb}$, $u = ax + by$

75. $f(y - mx)$ $-\dfrac{f(y - mx)}{k}$

76. $P(x)f(y - mx)$ $-\dfrac{1}{k}f(y - mx)\left[P(x) + \dfrac{P'(x)}{k} + \dfrac{P''(x)}{k^2} + \cdots + \dfrac{P^{(n)}(x)}{k^n}\right]$

77. $e^{kx}f(y - mx)$ $xe^{kx}f(y - mx)$

* For $\cos(ax + by)$ replace "sin" by "cos," and "cos" by "$-$sin" in z_p.

$$D_z = \frac{\partial}{\partial x}; \qquad D_y = \frac{\partial}{\partial y}; \qquad D_z^k D_y^r = \frac{\partial^{k+r}}{\partial x^k \partial y^r}$$

TABLE XVII: $(D_z + mD_y)^n z = R$

R	z_p
78. e^{az+by}	$\dfrac{e^{az+by}}{(a + mb)^n}$
79. $f(ax + by)$	$\dfrac{\int\int \cdots \int f(u)du^n}{(a + mb)^n}$, $u = ax + by$
80. $f(y - mx)$	$\dfrac{x^n}{n!}f(y - mx)$
81. $\phi(x, y)f(y + mx)$	$f(y - mx)\int\int \cdots \int \phi(x, a + mx)dx^n$ ($a = y - mx$ after integration)

TABLE XVIII: $(D_z + mD_y - k)^n z = R$

82. e^{az+by} $\dfrac{e^{az+by}}{(a + mb - k)^n}$

83. $f(y - mx)$ $\dfrac{(-1)^n f(y - mx)}{k^n}$

84. $P(x)f(y - mx)$ $\dfrac{(-1)^n}{k^n}f(y - mx)\left[P(x) + \binom{n}{1}\dfrac{P'(x)}{k} + \binom{n+1}{2}\dfrac{P''(x)}{k^2} + \binom{n+2}{3}\dfrac{P'''(x)}{k^3} + \cdots\right]$

85. $e^{kx}f(ax + by)$ $\dfrac{e^{kx}\int\int \cdots \int f(u)du^n}{(a + mb)^n}$, $u = ax + by$

86. $e^{kx}f(y - mx)$ $\dfrac{x^n}{n!}e^{kx}f(y - mx)$

TABLE XIX: $[D_x^n + a_1 D_x^{n-1}D_y + a_2 D_x^{n-2}D_y^2 + \cdots + a^n D_y^n]z = R$

87. $e^{ax + by}$ $\dfrac{e^{ax + by}}{a + a_1 a^{n-1}b + a_2 a^{n-2}b^2 + \cdots + a_n b^n}$

88. $f(ax + by)$ $\dfrac{\int\int \cdots \int f(u)du^n}{a^n + a_1 a^{n-1}b + a_2 a^{n-2}b^2 + \cdots + a^n b^n}$, $(u = ax + by)$

TABLE XX: $F(D_x, D_y)z = R$

89. $e^{ax + by}$ $\dfrac{e^{ax + by}}{F(a, b)}$

TABLE XXI: $F(D_x^2, D_x D_y, D_y^2)z = R$

90. $\sin(ax + by)$* $\dfrac{\sin(ax + by)}{F(-a^2, -ab, -b^2)}$

* For $\cos(ax + by)$ replace "sin" by "cos", and "cos" by "$-$sin" in z_p.

DIFFERENTIAL EQUATIONS

Differential equation	Method of solution
Separation of variables $f_1(x)\,g_1(y)\,dx + f_2(x)\,g_2(y)\,dy = 0$	$$\int \frac{f_1(x)}{f_2(x)}\,dx + \int \frac{g_2(y)}{g_1(y)}\,dy = c$$
Exact equation $M(x,y)dx + N(x,y)dy = 0$ where $\partial M/\partial y = \partial N/\partial x$	$$\int M\,\partial x + \int \left(N - \frac{\partial}{\partial y}\int M\,\partial x \right) dy = c$$ where ∂x indicates that the integration is to be performed with respect to x keeping y constant.
Linear first order equation $\frac{dy}{dx} + P(x)y = Q(x)$	$$ye^{\int P dx} = \int Q e^{\int P dx}\,dx + c$$
Bernoulli's equation $\frac{dy}{dx} + P(x)y = Q(x)y^n$	$$ve^{(1-n)\int P dx} = (1-n)\int Q e^{(1-n)\int P dx}\,dx + c$$ where $v = y^{1-n}$. If $n = 1$, the solution is $$\ln y = \int (Q - P)\,dx + c$$
Homogeneous equation $\frac{dy}{dx} = F\left(\frac{y}{x}\right)$	$$\ln x = \int \frac{dv}{F(v) - v} + c$$ where $v = y/x$. If $F(v) = v$, the solution is $y = cx$
Reducible to homogeneous $(a_1 x + b_1 y + c_1)\,dx + (a_2 x + b_2 y + c_2)\,dy = 0$ $\frac{a_1}{a_2} \neq \frac{b_1}{b_2}$	Set $u = a_1 x + b_1 y + c_1$, $v = a_2 x + b_2 y + c_2$ Eliminate x and y and the equation becomes homogenous
Reducible to separable $(a_1 x + b_1 y + c_1)\,dx + (a_2 x + b_2 y + c_2)\,dy = 0$ $\frac{a_1}{a_2} = \frac{b_1}{b_2}$	Set $u = a_1 x + b_1 y$ Eliminate x or y and equation becomes separable

$y \, F(xy) \, dx + x \, G(xy)dy = 0$	$\ln x = \displaystyle\int \frac{G(v) \, dv}{v\{G(v) - F(v)\}} + c$ where $v = xy$. If $G(v) = F(v)$, the solution is $xy = c$.
Linear, homogeneous second order equation $\dfrac{d^2 y}{dx^2} + b \, \dfrac{dy}{dx} + cy = 0$ b, c are real constants	Let m_1, m_2 be the roots of $m^2 + bm + c = 0$. Then there are 3 cases: Case 1. m_1, m_2 real and distinct: $\qquad y = c_1 e^{m_1 x} + c_2 e^{m_2 x}$ Case 2. m_1, m_2 real and equal: $\qquad y = c_1 e^{m_1 x} + c_2 x e^{m_1 x}$ Case 3. $m_1 = p + qi, \; m_2 = p - qi$: $\qquad y = e^{px}(c_1 \, \cos qx + c_2 \, \sin qx)$ where $p = -b/2, \; q = \sqrt{4c - b^2}/2$
Linear, nonhomogeneous second order equation $\dfrac{d^2 y}{dx^2} + b \, \dfrac{dy}{dx} + cy = R(x)$ b, c are real constants	There are 3 cases corresponding to those immediately above: Case 1. $\qquad y = c_1 e^{m_1 x} + c_2 e^{m_2 x}$ $\qquad + \dfrac{e^{m_1 x}}{m_1 - m_2} \displaystyle\int e^{-m_1 x} R(x) \, dx$ $\qquad + \dfrac{e^{m_2 x}}{m_2 - m_1} \displaystyle\int e^{-m_2 x} R(x) \, dx$ Case 2. $\qquad y = c_1 e^{m_1 x} + c_2 x e^{m_1 x}$ $\qquad + x e^{m_1 x} \displaystyle\int e^{-m_1 x} R(x) \, dx$ $\qquad - e^{m_1 x} \displaystyle\int x e^{-m_1 x} R(x) \, dx$ Case 3. $\qquad y = e^{px} (c_1 \, \cos qx + c_2 \, \sin qx)$ $\qquad + \dfrac{e^{px} \sin qx}{q} \displaystyle\int e^{-px} R(x) \cos qx \, dx$ $\qquad - \dfrac{e^{px} \cos qx}{q} \displaystyle\int e^{-px} R(x) \sin qx \, dx$

DIFFERENTIAL EQUATIONS (Continued)

Euler or Cauchy equation $$x^2 \frac{d^2 y}{dx^2} + bx \frac{dy}{dx} + cy = S(x)$$	Putting $x = e^t$, the equation becomes $$\frac{d^2 y}{dt^2} + (b - 1) \frac{dy}{dt} + cy = S(e^t)$$ and can then be solved as a linear second order equation.
Bessel's equation $$x^2 \frac{d^2 y}{dx^2} + x \frac{dy}{dx} + (\lambda^2 x^2 - n^2)y = 0$$	$$y = c_1 J_n(\lambda x) + c_2 Y_n(\lambda x)$$
Transformed Bessel's equation $$x^2 \frac{d^2 y}{dx^2} + (2p + 1)x \frac{dy}{dx} + (\alpha^2 x^{2r} + \beta^2)y = 0$$	$$y = x^{-p} \left\{ c_1 J_{q/r}\left(\frac{\alpha}{r} x^r\right) + c_2 Y_{q/r}\left(\frac{\alpha}{r} x^r\right) \right\}$$ where $q = \sqrt{p^2 - \beta^2}$.
Legendre's equation $$(1 - x^2) \frac{d^2 y}{dx^2} - 2x \frac{dy}{dx} + n(n + 1)y = 0$$	$$y = c_1 P_n(x) + c_2 Q_n(x)$$

FOURIER SERIES

If $f(x)$ is a bounded periodic function of period 2L (i.e. $f x + 2L) = f(x)$, and satisfies the *Dirichlet conditions:*

A.   In any period $f(x)$ is continuous, except possibly for a finite number of jump discontinuities.

B.   In any period $f(x)$ has only a finite number of maxima and minima.

Then $f(x)$ may be represented by the *Fourier series*

$$\frac{a_0}{2} + \sum_{n=1}^{\infty} \left(a_n \cos \frac{n\pi x}{L} + b_n \sin \frac{n\pi x}{L} \right)$$

where a_n and b_n are as determined below. This series will converge to $f(x)$ at every point where $f(x)$ is continuous, and to

$$\frac{f(x^+) + f(x^-)}{2}$$

(i.e. the average of the left-hand and right-hand limits) at every point where $f(x)$ has a jump discontinuity.

$$a_n = \frac{1}{L} \int_{-L}^{L} f(x) \cos \frac{n\pi x}{L} \, dx, \quad n = 0, 1, 2, 3, \ldots,$$

$$b_n = \frac{1}{L} \int_{-L}^{L} f(x) \sin \frac{n\pi x}{L} \, dx, \quad n = 1, 2, 3, \ldots$$

We may also write

$$a_n = \frac{1}{L} \int_{\alpha}^{\alpha + 2L} f(x) \cos \frac{n\pi x}{L} \, dx \text{ and } b_n = \frac{1}{L} \int_{\alpha}^{\alpha + 2L} f(x) \sin \frac{n\pi x}{L} \, dx$$

where α is any real number. Thus if $\alpha = 0$,

$$a_n = \frac{1}{L} \int_{0}^{2L} f(x) \cos \frac{n\pi x}{L} \, dx, \quad n = 0, 1, 2, 3, \ldots,$$

$$b_n = \frac{1}{L} \int_{0}^{2L} f(x) \sin \frac{n\pi x}{L} \, dx, \quad n = 1, 2, 3, \ldots$$

2. If in addition to the above restrictions, $f(x)$), is even (i.e., $f(-x) = f(x)$ the Fourier series reduces to

$$\frac{a_0}{2} + \sum_{n=1}^{\infty} a_n \cos \frac{n\pi x}{L}$$

That is, $b_n = 0$. In this case, a simpler formula for a_n is

$$a_n = \frac{2}{L} \int_0^L f(x) \cos \frac{n\pi x}{L} dx, \quad n = 0, 1, 2, 3, \ldots$$

3. If in addition to the restrictions in (1), $f(x)$ is an odd function (i.e., $f(-x) = -f(x)$), then the Fourier series reduces to

$$\sum_{n=1}^{\infty} b_n \sin \frac{n\pi x}{L}$$

That is, $a_n = 0$. In this case, a simpler formula for the b_n is

$$b_n = \frac{2}{L} \int_0^L f(x) \sin \frac{n\pi x}{L} dx, \quad n = 1, 2, 3, \ldots$$

4. If in addition to the restrictions in (2) above, $f(x) = -f(L - x)$, then a_n will be 0 for all even values of n, including $n = 0$. Thus in this case, the expansion reduces to

$$\sum_{m=1}^{\infty} a_{2m-1} \cos \frac{(2m - 1)\pi x}{L}$$

5. If in addition to the restrictions in (3) above, $f(x), = f(L - x)$, then b_n will be 0 for all even values of n. Thus in this case, the expansion reduces to

$$\sum_{m=1}^{\infty} b_{2m-1} \sin \frac{(2m - 1)\pi x}{L}$$

(The series in (4) and (5) are known as *odd-harmonic series,* since only the odd harmonics appear. Similar rules may be stated for even-harmonic series, but when a series appears in the even-harmonic form, it means that $2L$ has not been taken as the smallest period of $f(x)$. Since any integral multiple of a period is also a period, series obtained in this way will also work, but in general computation is simplified if $2L$ is taken to be the smallest period.)

6. If we write the Euler definitions for $\cos \theta$ and $\sin \theta$, we obtain the complex form of the Fourier Series known either as the "Complex Fourier Series" or the "Exponential Fourier Series" of $f(x)$. It is represented as

$$f(x) = \frac{1}{2} \sum_{n=-\infty}^{n=+\infty} c_n e^{i\omega_n x}$$

where

$$c_n = \frac{1}{L} \int_{-L}^{L} f(x) e^{-i\omega_n x} dx, \quad n = 0, \pm 1, \pm 2, \pm 3, \ldots$$

with $\omega_n = \frac{n\pi}{L}, \quad n = 0, \pm 1, \pm 2, \ldots$

The set of coefficients $\{c_n\}$ is often referred to as the Fourier spectrum.

7. If both sine and cosine terms are present and if $f(x)$ is of period $2L$ and expandable by a Fourier series, it can be represented as

$$f(x) = \frac{a_0}{2} + \sum_{n=1}^{\infty} c_n \sin\left(\frac{n\pi x}{L} + \phi_n\right), \text{ where } a_n = c_n \sin \phi_n,$$

$$b_n = c_n \cos \phi_n, \quad c_n = \sqrt{a_n^2 + b_n^2}, \quad \phi_n = \arctan\left(\frac{a_n}{b_n}\right)$$

It can also be represented as

$$f(x) = \frac{a_0}{2} + \sum_{n=1}^{\infty} c_n \cos\left(\frac{n\pi x}{L} + \phi_n\right), \text{ where } a_n = c_n \cos \phi_n,$$

$$b_n = -c_n \sin \phi_n, \quad c_n = \sqrt{a_n^2 + b_n^2}, \quad \phi_n = \arctan\left(-\frac{b_n}{a_n}\right)$$

where ϕ_n is chosen so as to make a_n, b_n, and c_n hold.

8. The following table of trigonometric identities should be helpful for developing Fourier Series.

	n	n even	n odd	$n/2$ odd	$n/2$ even
$\sin n\pi$		0	0	0	0
$\cos n\pi$	$(-1)^n$	$+1$	-1	$+1$	$+1$
$*\sin \dfrac{n\pi}{2}$		0	$(-1)^{(n-1)/2}$	0	0
$*\cos \dfrac{n\pi}{2}$		$(-1)^{n/2}$	0	-1	$+1$
$\sin \dfrac{n\pi}{4}$			$\dfrac{\sqrt{2}}{2}(-1)^{(n^2+4n+11)/8}$	$(-1)^{(n-2)/4}$	0

*A useful formula for $\sin \frac{n\pi}{2}$ and $\cos \frac{n\pi}{2}$ is given by

$$\sin \frac{n\pi}{2} = \frac{(i)^{n+1}}{2}[(-1)^n - 1] \text{ and } \cos \frac{n\pi}{2} = \frac{(i)^n}{2}[(-1)^n + 1], \text{ where } i^2 = -1.$$

AUXILIARY FORMULAS FOR FOURIER SERIES

$$1 = \frac{4}{\pi}\left[\sin \frac{\pi x}{k} + \frac{1}{3}\sin \frac{3\pi x}{k} + \frac{1}{5}\sin \frac{5\pi x}{k} + \cdots \right] \qquad [0 < x < k]$$

$$x = \frac{2k}{\pi}\left[\sin \frac{\pi x}{k} - \frac{1}{2}\sin \frac{2\pi x}{k} + \frac{1}{3}\sin \frac{3\pi x}{k} - \cdots \right] \qquad [-k < x < k]$$

$$x = \frac{k}{2} - \frac{4k}{\pi^2}\left[\cos \frac{\pi x}{k} + \frac{1}{3^2}\cos \frac{3\pi x}{k} + \frac{1}{5^2}\cos \frac{5\pi x}{k} + \cdots \right] \qquad [0 < x < k]$$

$$x^2 = \frac{2k^2}{\pi^3}\left[\left(\frac{\pi^2}{1} - \frac{4}{1}\right)\sin \frac{\pi x}{k} - \frac{\pi^2}{2}\sin \frac{2\pi x}{k} + \left(\frac{\pi^2}{3} - \frac{4}{3^3}\right)\sin \frac{3\pi x}{k}\right.$$
$$\left. - \frac{\pi^2}{4}\sin \frac{4\pi x}{k} + \left(\frac{\pi^2}{5} - \frac{4}{5^3}\right)\sin \frac{5\pi x}{k} + \cdots \right] \quad [0 < x < k]$$

$$x^2 = \frac{k^2}{3} - \frac{4k^2}{\pi^2}\left[\cos \frac{\pi x}{k} - \frac{1}{2^2}\cos \frac{2\pi x}{k} + \frac{1}{3^2}\cos \frac{3\pi x}{k} - \frac{1}{4^2}\cos \frac{4\pi x}{k} + \cdots \right]$$
$$[-k < x < k]$$

$$1 - \frac{1}{3} + \frac{1}{5} - \frac{1}{7} + \cdots = \frac{\pi}{4}$$

$$1 + \frac{1}{2^2} + \frac{1}{3^2} + \frac{1}{4^2} + \cdots = \frac{\pi^2}{6}$$

$$1 - \frac{1}{2^2} + \frac{1}{3^2} - \frac{1}{4^2} + \cdots = \frac{\pi^2}{12}$$

$$1 + \frac{1}{3^2} + \frac{1}{5^2} + \frac{1}{7^2} + \cdots = \frac{\pi^2}{8}$$

$$\frac{1}{2^2} + \frac{1}{4^2} + \frac{1}{6^2} + \frac{1}{8^2} + \cdots = \frac{\pi^2}{24}$$

FOURIER EXPANSIONS FOR BASIC PERIODIC FUNCTIONS

$$f(x) = \frac{4}{\pi} \sum_{n=1,3,5\ldots} \frac{1}{n} \sin \frac{n\pi x}{L}$$

$$f(x) = \frac{2}{\pi} \sum_{n=1}^{\infty} \frac{(-1)^n}{n} \left(\cos \frac{n\pi c}{L} - 1 \right) \sin \frac{n\pi x}{L}$$

$$f(x) = \frac{c}{L} + \frac{2}{\pi} \sum_{n=1}^{\infty} \frac{(-1)^n}{n} \sin \frac{n\pi c}{L} \cos \frac{n\pi x}{L}$$

$$f(x) = \frac{2}{L} \sum_{n=1}^{\infty} \sin \frac{n\pi}{2} \frac{\sin \left(\frac{1}{2} n\pi c/L \right)}{\frac{1}{2} n\pi c/L} \sin \frac{n\pi x}{L}$$

$$f(x) = \frac{2}{\pi} \sum_{n=1}^{\infty} \frac{(-1)^{n+1}}{n} \sin \frac{n\pi x}{L}$$

$$f(x) = \frac{1}{2} - \frac{4}{\pi^2} \sum_{n=1,3,5\ldots} \frac{1}{n^2} \cos \frac{n\pi x}{L}$$

$$f(x) = \frac{8}{\pi^2} \sum_{n=1,3,5\ldots} \frac{(-1)^{(n-1)/2}}{n^2} \sin \frac{n\pi x}{L}$$

$$f(x) = \frac{1}{2} - \frac{1}{\pi} \sum_{n=1}^{\infty} \frac{1}{n} \sin \frac{n\pi x}{L}$$

$$f(x) = \tfrac{1}{2}(1 + a) + \frac{2}{\pi^2(1-a)} \sum_{n=1}^{\infty} \frac{1}{n^2} [(-1)^n \cos n\pi a - 1] \cos \frac{n\pi x}{L} \; ; \left(a = \frac{c}{2L}\right)$$

$$f(x) = \frac{2}{\pi} \sum_{n=1}^{\infty} \frac{(-1)^{n-1}}{n} \left[1 + \frac{\sin n\pi a}{n\pi(1-a)}\right] \sin \frac{n\pi x}{L} \; ; \left(a = \frac{c}{2L}\right)$$

$$f(x) = \tfrac{1}{2} - \frac{4}{\pi^2(1-2a)} \sum_{n=1,3,5\ldots} \frac{1}{n^2} \cos n\pi a \cos \frac{n\pi x}{L} \; ; \left(a = \frac{c}{2L}\right)$$

$$f(x) = \frac{2}{\pi} \sum_{n=1}^{\infty} \frac{(-1)^n}{n} \left[1 + \frac{1+(-1)^n}{n\pi(1-2a)} \sin n\pi a\right] \sin \frac{n\pi x}{L} \; ; \left(a = \frac{c}{2L}\right)$$

$$f(x) = \frac{4}{\pi} \sum_{n=1}^{\infty} \frac{1}{n} \sin \frac{n\pi}{4} \sin n\pi a \sin \frac{n\pi x}{L} \; ; \left(a = \frac{c}{2L}\right)$$

$$f(x) = \frac{9}{\pi^2} \sum_{n=1}^{\infty} \frac{1}{n^2} \sin \frac{n\pi}{3} \sin \frac{n\pi x}{L} \; ; \left(a = \frac{c}{2L}\right)$$

$$f(x) = \frac{32}{3\pi^2} \sum_{n=1}^{\infty} \frac{1}{n^2} \sin \frac{n\pi}{4} \sin \frac{n\pi x}{L} \; ; \left(a = \frac{c}{2L}\right)$$

$$f(x) = \frac{1}{\pi} + \frac{1}{2} \sin \omega t - \frac{2}{\pi} \sum_{n=2,4,6\ldots} \frac{1}{n^2-1} \cos n\omega t$$

Extracted from graphs and formulas, pages 372, 373, Differential Equations in Engineering Problems, Salvadori and Schwarz, published by Prentice-Hall, Inc., 1954.

THE FOURIER TRANSFORMS*

R. E. Gaskell

For a piecewise continuous function $F(x)$ over a finite interval $0 \leq x \leq \pi$, the *finite Fourier cosine transform* of $F(x)$ is

$$f_c(n) = \int_0^\pi F(x) \cos nx \, dx \quad (n = 0, 1, 2, \ldots) \tag{1}$$

If x ranges over the interval $0 \leq x \leq L$, the substitution $x' = \pi x/L$ allows the use of this definition, also. The inverse transform is written.

$$\bar{F}(x) = \frac{1}{\pi} f_c(0) - \frac{2}{\pi} \sum_{n=1}^\infty f_c(n) \cos nx \quad (0 < x < \pi) \tag{2}$$

where $F(x) = \dfrac{[F(x + o) + F(x - o)]}{2}$. We observe that $F(x) = F(x) =$ at points of continuity. The formula

$$f_c^{(2)}(n) = \int_0^\pi F''(x) \cos nx \, dx$$

$$= -n^2 f_c(n) - F'(0) + (-1)^n F'(\pi) \tag{3}$$

makes the finite Fourier cosine transform useful in certain boundary value problems.

Analogously, the *finite Fourier sine transform* of $F(x)$ is

$$f_s(n) = \int_0^\pi F(x) \sin nx \, dx \quad (n = 1, 2, 3, \ldots) \tag{4}$$

and

$$\bar{F}(x) = \frac{2}{\pi} \sum_{n=1}^\infty f_s(n) \sin nx \quad (0 < x < \pi) \tag{5}$$

Corresponding to (3) we have

$$f_s^{(2)}(n) = \int_0^\pi F''(x) \sin nx \, dx$$

$$= -n^2 f_s(n) - nF(0) - n(-1)^n F(\pi) \tag{6}$$

Fourier Transforms

If $F(x)$ is defined for $x \geq 0$ and is piecewise continuous over any finite interval, and if

$$\int_0^\infty F(x) \, dx$$

is absolutely convergent, then

$$f_c(\alpha) = \sqrt{\frac{2}{\pi}} \int_0^\infty F(x) \cos(\alpha x) \, dx \tag{7}$$

is the *Fourier cosine transform* of $F(x)$. Furthermore.

$$\bar{F}(x) = \sqrt{\frac{2}{\pi}} \int_0^\infty f_c(\alpha) \cos(\alpha x) \, d\alpha \tag{8}$$

If $\lim\limits_{x \to \infty} \dfrac{d^n F}{dx^n} = 0$, an important property of the Fourier cosine transform

$$f_c^{(2r)}(\alpha) = \sqrt{\frac{2}{\pi}} \int_0^\infty \left(\frac{d^{2r} F}{dx^{2r}} \right) \cos(\alpha x) \, dx$$

$$= -\sqrt{\frac{2}{\pi}} \sum_{n=0}^{r-1} (-1)^n a_{2r-2n-1} \alpha^{2n} + (-1)^r \alpha^{2r} f_c(\alpha) \tag{9}$$

where $\lim\limits_{x \to \infty} \dfrac{d^r F}{dx^r} = a_r$, makes it useful in the solution of many problems.

Under the same conditions.

$$f_s(\alpha) = \sqrt{\frac{2}{\pi}} \int_0^\infty F(x) \sin(\alpha x) \, dx \tag{10}$$

* From Beyer, W. H., Ed., *CRC Handbook of Mathematical Sciences*, 5th ed., CRC Press, Boca Raton, 1978, 592—598. With permission.

defines the *Fourier sine transform* of $F(x)$, and

$$\bar{F}(x) = \sqrt{\frac{2}{\pi}} \int_0^\infty f_s(\alpha) \sin(\alpha x)\, d\alpha \qquad (11)$$

Corresponding to (9) we have

$$f_s^{(2r)}(\alpha) = \sqrt{\frac{2}{\pi}} \int_0^\infty \frac{d^{2r}F}{dx^{2r}} \sin(\alpha x)\, dx$$

$$= -\sqrt{\frac{2}{\pi}} \sum_{n=1}^r (-1)^n \alpha^{2n-1} a_{2r-2n} + (-1)^{r-1} \alpha^{2r} f_s(\alpha) \qquad (12)$$

Similarly, if $F(x)$ is defined for $-\infty < x < \infty$, and if $\int_{-\infty}^\infty F(x)\, dx$ is absolutely convergent, then

$$f(\alpha) = \frac{1}{\sqrt{2\pi}} \int_{-\infty}^\infty F(x) e^{i\alpha x}\, dx \qquad (13)$$

is the *Fourier transform* of $F(x)$, and

$$\bar{F}(x) = \frac{1}{\sqrt{2\pi}} \int_{-\infty}^\infty f(\alpha) e^{-i\alpha x}\, d\alpha \qquad (14)$$

Also, if

$$\lim_{|x| \to \infty} \left| \frac{d^n F}{dx^n} \right| = 0 \quad (n = 1, 2, \ldots, r-1)$$

then

$$f^{(r)}(\alpha) = \frac{1}{\sqrt{2\pi}} \int_{-\infty}^\infty F^{(r)}(x) e^{i\alpha x}\, dx = (-i\alpha)^r f(\alpha) \qquad (15)$$

Finite Sine Transforms

	$f_s(n)$	$F(x)$
1	$f_s(n) = \displaystyle\int_0^\pi F(x) \sin nx\, dx \ (n = 1, 2, \cdots)$	$F(x)$
2	$(-1)^{n+1} f_s(n)$	$F(\pi - x)$
3	$\dfrac{1}{n}$	$\dfrac{\pi - x}{\pi}$
4	$\dfrac{(-1)^{n+1}}{n}$	$\dfrac{x}{\pi}$
5	$\dfrac{1 - (-1)^n}{n}$	1
6	$\dfrac{2}{n^2} \sin \dfrac{n\pi}{2}$	$\begin{cases} x & \text{when } 0 < x < \pi/2 \\ \pi - x & \text{when } \pi/2 < x < \pi \end{cases}$
7	$\dfrac{(-1)^{n+1}}{n^3}$	$\dfrac{x(\pi^2 - x^2)}{6\pi}$
8	$\dfrac{1 - (-1)^n}{n^3}$	$\dfrac{x(\pi - x)}{2}$
9	$\dfrac{\pi^2 (-1)^{n-1}}{n} - \dfrac{2[1 - (-1)^n]}{n^3}$	x^2
10	$\pi(-1)^n \left(\dfrac{6}{n^3} - \dfrac{\pi^2}{n} \right)$	x^3
11	$\dfrac{n}{n^2 + c^2} [1 - (-1)^n e^{c\pi}]$	e^{cx}
12	$\dfrac{n}{n^2 + c^2}$	$\dfrac{\sinh c(\pi - x)}{\sinh c\pi}$
13	$\dfrac{n}{n^2 - k^2} \ (k \neq 0, 1, 2, \cdots)$	$\dfrac{\sin k(\pi - x)}{\sin k\pi}$
14	$\begin{cases} \dfrac{\pi}{2} & \text{when } n = m \\ 0 & \text{when } n \neq m \end{cases} \quad (m = 1, 2, \cdots)$	$\sin mx$

	$f_s(n)$	$F(x)$
15	$\dfrac{n}{n^2 - k^2}[1 - (-1)^n \cos k\pi]$ $(k \neq 1, 2, \cdots)$	$\cos kx$
16	$\begin{cases} \dfrac{n}{n^2 - m^2}[1 - (-1)^{n+m}] \\ \quad \text{when } n \neq m = 1, 2, \cdots \\ 0 \quad \text{when } n = m \end{cases}$	$\cos mx$
17	$\dfrac{n}{(n^2 - k^2)^2}$ $(k \neq 0, 1, 2, \cdots)$	$\dfrac{\pi \sin kx}{2k \sin^2 k\pi} - \dfrac{x \cos k(\pi - x)}{2k \sin k\pi}$
18	$\dfrac{b^n}{n}$ $(|b| \leq 1)$	$\dfrac{2}{\pi} \arctan \dfrac{b \sin x}{1 - b \cos x}$
19	$\dfrac{1 - (-1)^n}{n} b^n$ $(|b| \leq 1)$	$\dfrac{2}{\pi} \arctan \dfrac{2b \sin x}{1 - b^2}$

Finite Cosine Transforms

	$f_c(n)$	$F(x)$
1	$f_c(n) = \displaystyle\int_0^\pi F(x) \cos nx\, dx$ $(n = 0, 1, 2, \cdots)$	$F(x)$
2	$(-1)^n f_c(n)$	$F(\pi - x)$
3	0 when $n = 1, 2, \cdots$; $f_c(0) = \pi$	1
4	$\dfrac{2}{n} \sin \dfrac{n\pi}{2}$; $f_c(0) = 0$	$\begin{cases} 1 \text{ when } 0 < x < \pi/2 \\ -1 \text{ when } \pi/2 < x < \pi \end{cases}$
5	$-\dfrac{1 - (-1)^n}{n^2}$; $f_c(0) = \dfrac{\pi^2}{2}$	x
6	$\dfrac{(-1)^n}{n^2}$; $f_c(0) = \dfrac{\pi^2}{6}$	$\dfrac{x^2}{2\pi}$
7	$\dfrac{1}{n^2}$; $f_c(0) = 0$	$\dfrac{(\pi - x)^2}{2\pi} - \dfrac{\pi}{6}$
8	$3\pi^2 \dfrac{(-1)^n}{n^2} - 6 \dfrac{1 - (-1)^n}{n^4}$; $f_c(0) = \dfrac{\pi^4}{4}$	x^3
9	$\dfrac{(-1)^n e^c \pi - 1}{n^2 + c^2}$	$\dfrac{1}{c} e^{cx}$
10	$\dfrac{1}{n^2 + c^2}$	$\dfrac{\cosh c(\pi - x)}{c \sinh c\pi}$
11	$\dfrac{k}{n^2 - k^2}[(-1)^n \cos \pi k - 1]$ $(k \neq 0, 1, 2, \cdots)$	$\sin kx$
12	$\dfrac{(-1)^{n+m} - 1}{n^2 - m^2}$; $f_c(m) = 0$ $(m = 1, 2, \cdots)$	$\dfrac{1}{m} \sin mx$
13	$\dfrac{1}{n^2 - k^2}$ $(k \neq 0, 1, 2, \cdots)$	$-\dfrac{\cos k(\pi - x)}{k \sin k\pi}$
14	0 when $n = 1, 2, \cdots$; $f_c(m) = \dfrac{\pi}{2}$ $(m = 1, 2, \cdots)$	$\cos mx$

Fourier Sine Transforms*

	$F(x)$	$f_s(\alpha)$
1	$\begin{cases} 1 & (0 < x < a) \\ 0 & (x > a) \end{cases}$	$\sqrt{\dfrac{2}{\pi}} \left[\dfrac{1 - \cos \alpha}{\alpha}\right]$
2	x^{p-1} $(0 < p < 1)$	$\sqrt{\dfrac{2}{\pi}} \dfrac{\Gamma(p)}{\alpha^p} \sin \dfrac{p\pi}{2}$
3	$\begin{cases} \sin x & (0 < x < a) \\ 0 & (x > a) \end{cases}$	$\dfrac{1}{\sqrt{2\pi}} \left[\dfrac{\sin[a(1 - \alpha)]}{1 - \alpha} - \dfrac{\sin[a(1 + \alpha)]}{1 + \alpha}\right]$
4	e^{-x}	$\sqrt{\dfrac{2}{\pi}} \left[\dfrac{\alpha}{1 + \alpha^2}\right]$
5	$xe^{-x^2/2}$	$\alpha e^{-\alpha^2/2}$
6	$\cos \dfrac{x^2}{2}$	$\sqrt{2} \left[\sin \dfrac{\alpha^2}{2} C\left(\dfrac{\alpha^2}{2}\right) - \cos \dfrac{\alpha^2}{2} S\left(\dfrac{\alpha^2}{2}\right)\right]^*$

	$F(x)$	$f_s(\alpha)$
7	$\sin \dfrac{x^2}{2}$	$\sqrt{2}\left[\cos \dfrac{\alpha^2}{2} C\left(\dfrac{\alpha^2}{2}\right) + \sin \dfrac{\alpha^2}{2} S\left(\dfrac{\alpha^2}{2}\right)\right]$ *

* $C(y)$ and $S(y)$ are the Fresnel integrals

$$C(y) = \frac{1}{\sqrt{2\pi}} \int_0^y \frac{1}{\sqrt{t}} \cos t \, dt,$$

$$S(y) = \frac{1}{\sqrt{2\pi}} \int_0^y \frac{1}{\sqrt{t}} \sin t \, dt$$

* More extensive tables of the Fourier sine and cosine transforms can be found in Fritz Oberhettinger, *Tabellen zur-Fourier Transformation*, Springer, 1957.

Fourier Cosine Transforms

	$F(x)$	$f_c(\alpha)$
1	$\begin{cases} 1 & (0 < x < a) \\ 0 & (x > a) \end{cases}$	$\sqrt{\dfrac{2}{\pi}} \dfrac{\sin a\alpha}{\alpha}$
2	$x^{p-1} \quad (0 < p < 1)$	$\sqrt{\dfrac{2}{\pi}} \dfrac{\Gamma(p)}{\alpha^p} \cos \dfrac{p\pi}{2}$
3	$\begin{cases} \cos x & (0 < x < a) \\ 0 & (x > a) \end{cases}$	$\dfrac{1}{\sqrt{2\pi}}\left[\dfrac{\sin[a(1-\alpha)]}{1-\alpha} + \dfrac{\sin[a(1+\alpha)]}{1+\alpha}\right]$
4	e^{-x}	$\sqrt{\dfrac{2}{\pi}}\left(\dfrac{1}{1+\alpha^2}\right)$
5	$e^{-x^2/2}$	$e^{-\alpha^2/2}$
6	$\cos \dfrac{x^2}{2}$	$\cos\left(\dfrac{\alpha^2}{2} - \dfrac{\pi}{4}\right)$
7	$\sin \dfrac{x^2}{2}$	$\cos\left(\dfrac{\alpha^2}{2} + \dfrac{\pi}{4}\right)$

Fourier Transforms*

	$F(x)$	$f(\alpha)$
1	$\dfrac{\sin ax}{x}$	$\begin{cases} \sqrt{\dfrac{\pi}{2}} & |\alpha| < a \\ 0 & |\alpha| > a \end{cases}$
2	$\begin{cases} e^{iwx} & (p < x < q) \\ 0 & (x < p, x > q) \end{cases}$	$\dfrac{i}{\sqrt{2\pi}} \dfrac{e^{ip(w+\alpha)} - e^{iq(w+\alpha)}}{(w+\alpha)}$
3	$\begin{cases} e^{-cx+iwx} & (x > 0) \\ 0 & (x < 0) \end{cases} \quad (c > 0)$	$\dfrac{i}{\sqrt{2\pi}(w+\alpha+ic)}$
4	$e^{-px^2} \quad R(p) > 0$	$\dfrac{1}{\sqrt{2p}} e^{-\alpha^2/4p}$
5	$\cos px^2$	$\dfrac{1}{\sqrt{2p}} \cos\left[\dfrac{\alpha^2}{4p} - \dfrac{\pi}{4}\right]$
6	$\sin px^2$	$\dfrac{1}{\sqrt{2p}} \cos\left[\dfrac{\alpha^2}{4p} + \dfrac{\pi}{4}\right]$
7	$|x|^{-p} \quad (0 < p < 1)$	$\sqrt{\dfrac{2}{\pi}} \dfrac{\Gamma(1-p) \sin \dfrac{p\pi}{2}}{|\alpha|^{(1-p)}}$
8	$\dfrac{e^{-a|x|}}{\sqrt{|x|}}$	$\dfrac{\sqrt{\sqrt{(a^2+\alpha^2)} + a}}{\sqrt{a^2+\alpha^2}}$
9	$\dfrac{\cosh ax}{\cosh \pi x} \quad (-\pi < a < \pi)$	$\sqrt{\dfrac{2}{\pi}} \dfrac{\cos \dfrac{a}{2} \cosh \dfrac{\alpha}{2}}{\cosh \alpha + \cos a}$
10	$\dfrac{\sinh ax}{\sinh \pi x} \quad (-\pi < a < \pi)$	$\dfrac{1}{\sqrt{2\pi}} \dfrac{\sin a}{\cosh \alpha + \cos a}$
11	$\begin{cases} \dfrac{1}{\sqrt{a^2-x^2}} & (|x| < a) \\ 0 & (|x| > a) \end{cases}$	$\sqrt{\dfrac{\pi}{2}} J_0(a\alpha)$

$F(x)$	$f(\alpha)$
12 $\dfrac{\sin[b\sqrt{a^2+x^2}]}{\sqrt{a^2+x^2}}$	$\begin{cases} 0 & (\vert\alpha\vert>b) \\ \sqrt{\dfrac{\pi}{2}}\,J_0(a\sqrt{b^2-\alpha^2}) & (\vert\alpha\vert<b) \end{cases}$
13 $\begin{cases} P_n(x) & (\vert x\vert<1) \\ 0 & (\vert x\vert>1) \end{cases}$	$\dfrac{i^n}{\sqrt{\alpha}}\,J_{n+1}(\alpha)$
14 $\begin{cases} \dfrac{\cos[b\sqrt{a^2-x^2}]}{\sqrt{a^2-x^2}} & (\vert x\vert<a) \\ 0 & (\vert x\vert>a) \end{cases}$	$\sqrt{\dfrac{\pi}{2}}\,J_0(a\sqrt{a^2+b^2})$
15 $\begin{cases} \dfrac{\cosh[b\sqrt{a^2-x^2}]}{\sqrt{a^2-x^2}} & (\vert x\vert<a) \\ 0 & (\vert x\vert>a) \end{cases}$	$\sqrt{\dfrac{\pi}{2}}\,J_0(a\sqrt{\alpha^2-b^2})$

* More extensive tables of Fourier transforms can be found in W. Magnus and F. Oberhettinger, *Formulas and Theorems of the Special Functions of Mathematical Physics.* Chelsea, 1949, 116—120.

The following functions appear among the entries of the tables on transforms.

Function	Definition	Name
$Ei(x)$	$\displaystyle\int_{-\infty}^{x}\frac{e^v}{v}\,dv$; or sometimes defined as $-Ei(-x)=\displaystyle\int_{x}^{\infty}\frac{e^{-v}}{v}\,dv$	Sine, Cosine, and Exponential Integral tables pages 548–556
$Si(x)$	$\displaystyle\int_{0}^{x}\frac{\sin v}{v}\,dv$	Sine, Cosine, and Exponential Integral tables pages 548–556
$Ci(x)$	$\displaystyle\int_{x}^{\infty}\frac{\cos v}{v}\,dv$; or sometimes defined as negative of this integral	Sine, Cosine, and Exponential Integral tables pages 548–546
$erf(x)$	$\dfrac{2}{\sqrt{\pi}}\displaystyle\int_{0}^{x}e^{-v^2}\,dv$	Error function
$erfc(x)$	$1-erf(x)=\dfrac{2}{\sqrt{\pi}}\displaystyle\int_{x}^{\infty}e^{-v^2}\,dv$	Complementary function to error function
$L_n(x)$	$\dfrac{e^x}{n!}\dfrac{d^n}{dx^n}(x^n e^{-x}),\ n=0,1,\cdots$	Laguerre polynomial of degree n

SERIES EXPANSION

The expression in parentheses following certain of the series indicates the region of convergence. If not otherwise indicated it is to be understood that the series converges for all finite values of x.

BINOMIAL

$$(x+y)^n = x^n + nx^{n-1}y + \frac{n(n-1)}{2!}x^{n-2}y^2$$
$$+ \frac{n(n-1)(n-2)}{3!}x^{n-3}y^3 + \cdots \quad (y^2 < x^2)$$

$$(1\pm x)^n = 1 \pm nx + \frac{n(n-1)x^2}{2!} \pm \frac{n(n-1)(n-2)x^3}{3!} + \cdots \quad (x^2<1)$$

$$(1\pm x)^{-n} = 1 \mp nx + \frac{n(n+1)x^2}{2!} \mp \frac{n(n+1)(n+2)x^3}{3!} + \cdots \quad (x^2<1)$$

$$(1\pm x)^{-1} = 1 \mp x + x^2 \mp x^3 + x^4 \mp x^5 + \cdots \qquad (x^2<1)$$
$$(1\pm x)^{-2} = 1 \mp 2x + 3x^2 \mp 4x^3 + 5x^4 \mp 6x^5 + \cdots \qquad (x^2<1)$$

REVERSION OF SERIES

Let a series be represented by

$$y = a_1 x + a_2 x^2 + a_3 x^3 + a_4 x^4 + a_5 x^5 + a_6 x^6 + \cdots \quad (a_1 \neq 0)$$

to find the coefficients of the series

$$x = A_1 y + A_2 y^2 + A_3 y^3 + A_4 y^4 + \cdots$$

$$A_1 = \frac{1}{a_1} \qquad A_2 = -\frac{a_2}{a_1^3} \qquad A_3 = \frac{1}{a_1^5}(2a_2^2 - a_1 a_3)$$

$$A_4 = \frac{1}{a_1^7}(5a_1 a_2 a_3 - a_1^2 a_4 - 5a_2^3)$$

$$A_5 = \frac{1}{a_1^9} \left(6a_1^2 a_2 a_4 + 3a_1^2 a_3^2 + 14a_2^4 - a_1^3 a_5 - 21a_1 a_2^2 a_3 \right)$$

$$A_6 = \frac{1}{a_1^{11}} \left(7a_1^3 a_2 a_5 + 7a_1^3 a_3 a_4 + 84a_1 a_2^3 a_3 - a_1^4 a_6 - 28a_1^2 a_2^2 a_4 - 28a_1^2 a_2 a_3^2 - 42a_2^5 \right)$$

$$A_7 = \frac{1}{a_1^{13}} \left(8a_1^4 a_2 a_6 + 8a_1^4 a_3 a_5 + 4a_1^4 a_4^2 + 120a_1^2 a_2^3 a_4 \right.$$
$$+ 180a_1^2 a_2^2 a_3^2 + 132a_2^6 - a_1^5 a_7$$
$$\left. - 36a_1^3 a_2^2 a_5 - 72a_1^3 a_2 a_3 a_4 - 12a_1^3 a_3^3 - 330a_1 a_2^4 a_3 \right)$$

TAYLOR

1. $f(x) = f(a) + (x-a)f'(a) + \dfrac{(x-a)^2}{2!} f''(a) + \dfrac{(x-a)^3}{3!} f'''(a)$
$$+ \cdots + \frac{(x-a)^n}{n!} f^{(n)}(a) + \cdots \quad \text{(Taylor's Series)}$$

(Increment form)

2. $f(x+h) = f(x) + hf'(x) + \dfrac{h^2}{2!} f''(x) + \dfrac{h^3}{3!} f'''(x) + \cdots$
$$= f(h) + xf'(h) + \frac{x^2}{2!} f''(h) + \frac{x^3}{3!} f'''(h) + \cdots$$

3. If $f(x)$ is a function possessing derivatives of all orders throughout the interval $a \leq x \leq b$, then there is a value X, with $a < X < b$, such that

$$f(b) = f(a) + (b-a)f'(a) + \frac{(b-a)^2}{2!} f''(a) + \cdots$$
$$+ \frac{(b-a)^{n-1}}{(n-1)!} f^{(n-1)}(a) + \frac{(b-a)^n}{n!} f^{(n)}(X)$$

$$f(a+h) = f(a) + hf'(a) + \frac{h^2}{2!} f''(a) + \cdots + \frac{h^{n-1}}{(n-1)!} f^{(n-1)}(a)$$
$$+ \frac{h^n}{n!} f^{(n)}(a + \theta h), \quad b = a + h, \, 0 < \theta < 1.$$

or

$$f(x) = f(a) + (x-a)f'(a) + \frac{(x-a)^2}{2!} f''(a) + \cdots + (x-a)^{n-1} \frac{f^{(n-1)}(a)}{(n-1)!} + R_n,$$

where

$$R_n = \frac{f^{(n)}[a + \theta \cdot (x-a)]}{n!} (x-a)^n, \quad 0 < \theta < 1.$$

The above forms are known as Taylor's series with the remainder term.

4. *Taylor's series for a function of two variables*

If $\left(h \dfrac{\partial}{\partial x} + k \dfrac{\partial}{\partial y} \right) f(x,y) = h \dfrac{\partial f(x,y)}{\partial x} + k \dfrac{\partial f(x,y)}{\partial y};$

$$\left(h \frac{\partial}{\partial x} + k \frac{\partial}{\partial y} \right)^2 f(x,y) = h^2 \frac{\partial^2 f(x,y)}{\partial x^2} + 2hk \frac{\partial^2 f(x,y)}{\partial x \partial y} + k^2 \frac{\partial^2 f(x,y)}{\partial y^2}$$

etc., and if $\left(h \dfrac{\partial}{\partial x} + k \dfrac{\partial}{\partial y} \right)^n f(x,y) \Big|_{\substack{x=a \\ y=b}}$ with the bar and subscripts means that after differentiation we are to replace x by a and y by b,

$$f(a+h, b+k) = f(a,b) + \left(h \frac{\partial}{\partial x} + k \frac{\partial}{\partial y} \right) f(x,y) \Big|_{\substack{x=a \\ y=b}} + \cdots$$
$$+ \frac{1}{n!} \left(h \frac{\partial}{\partial x} + k \frac{\partial}{\partial y} \right)^n f(x,y) \Big|_{\substack{x=a \\ y=b}} + \cdots$$

MACLAURIN

$$f(x) = f(0) + xf'(0) + \frac{x^2}{2!} f''(0) + \frac{x^3}{3!} f'''(0) + \cdots + x^{n-1} \frac{f^{(n-1)}(0)}{(n-1)!} + R_n,$$

where

$$R_n = \frac{x^n f^{(n)}(\theta x)}{n!}, \quad 0 < \theta < 1.$$

EXPONENTIAL

$$e = 1 + \frac{1}{1!} + \frac{1}{2!} + \frac{1}{3!} + \frac{1}{4!} + \cdots$$

$$e^x = 1 + x + \frac{x^2}{2!} + \frac{x^3}{3!} + \frac{x^4}{4!} + \cdots \qquad \text{(all real values of } x)$$

$$a^x = 1 + x \log_e a + \frac{(x \log_e a)^2}{2!} + \frac{(x \log_e a)^3}{3!} + \cdots$$

$$e^x = e^a \left[1 + (x - a) + \frac{(x - a)^2}{2!} + \frac{(x - a)^3}{3!} + \cdots \right]$$

LOGARITHMIC

$$\log_e x = \frac{x - 1}{x} + \frac{1}{2} \left(\frac{x - 1}{x} \right)^2 + \frac{1}{3} \left(\frac{x - 1}{x} \right)^3 + \cdots \qquad (x > \tfrac{1}{2})$$

$$\log_e x = (x - 1) - \tfrac{1}{2}(x - 1)^2 + \tfrac{1}{3}(x - 1)^3 - \cdots \qquad (2 \geq x > 0)$$

$$\log_e x = 2 \left[\frac{x - 1}{x + 1} + \frac{1}{3} \left(\frac{x - 1}{x + 1} \right)^3 + \frac{1}{5} \left(\frac{x - 1}{x + 1} \right)^5 + \cdots \right] \qquad (x > 0)$$

$$\log_e (1 + x) = x - \tfrac{1}{2} x^2 + \tfrac{1}{3} x^3 - \tfrac{1}{4} x^4 + \cdots \qquad (-1 < x \leq 1)$$

$$\log_e (n + 1) - \log_e (n - 1) = 2 \left[\frac{1}{n} + \frac{1}{3n^3} + \frac{1}{5n^5} + \cdots \right]$$

$$\log_e (a + x) = \log_e a + 2 \left[\frac{x}{2a + x} + \frac{1}{3} \left(\frac{x}{2a + x} \right)^3 + \frac{1}{5} \left(\frac{x}{2a + x} \right)^5 + \cdots \right]$$

$$(a > 0, \; -a < x < +\infty)$$

$$\log_e \frac{1 + x}{1 - x} = 2 \left[x + \frac{x^3}{3} + \frac{x^5}{5} + \cdots + \frac{x^{2n-1}}{2n - 1} + \cdots \right], \qquad -1 < x < 1$$

$$\log_e x = \log_e a + \frac{(x - a)}{a} - \frac{(x - a)^2}{2a^2} + \frac{(x - a)^3}{3a^3} - + \cdots, \qquad 0 < x \leq 2a$$

TRIGONOMETRIC

$$\sin x = x - \frac{x^3}{3!} + \frac{x^5}{5!} - \frac{x^7}{7!} + \cdots \qquad \text{(all real values of } x)$$

$$\cos x = 1 - \frac{x^2}{2!} + \frac{x^4}{4!} - \frac{x^6}{6!} + \cdots \qquad \text{(all real values of } x)$$

$$\tan x = x + \frac{x^3}{3} + \frac{2x^5}{15} + \frac{17x^7}{315} + \frac{62x^9}{2835} + \cdots + \frac{(-1)^{n-1} 2^{2n} (2^{2n} - 1) B_{2n}}{(2n)!} x^{2n-1} + \cdots,$$

$$\left[x^2 < \frac{\pi^2}{4}, \text{ and } B_n \text{ represents the } n \text{th Bernoulli number.} \right]$$

$$\cot x = \frac{1}{x} - \frac{x}{3} - \frac{x^3}{45} - \frac{2x^5}{945} - \frac{x^7}{4725} - \cdots - \frac{(-1)^{n+1} 2^{2n}}{(2n)!} B_{2n} x^{2n-1} - \cdots,$$

$$[x^2 < \pi^2, \text{ and } B_n \text{ represents the } n \text{th Bernoulli number.}]$$

$$\sec x = 1 + \frac{x^2}{2} + \frac{5}{24} x^4 + \frac{61}{720} x^6 + \frac{277}{8064} x^8 + \cdots + \frac{(-1)^n}{(2n)!} E_{2n} x^{2n} + \cdots,$$

$$\left[x^2 < \frac{\pi^2}{4}, \text{ and } E_n \text{ represents the } n \text{th Euler number.} \right]$$

$$\csc x = \frac{1}{x} + \frac{x}{6} + \frac{7}{360} x^3 + \frac{31}{15,120} x^5 + \frac{127}{604,800} x^7 + \cdots$$

$$+ \frac{(-1)^{n+1} 2 (2^{2n-1} - 1)}{(2n)!} B_{2n} x^{2n-1} + \cdots,$$

$$[x^2 < \pi^2, \text{ and } B_n \text{ represents } n \text{th Bernoulli number.}]$$

$$\sin x = x \left(1 - \frac{x^2}{\pi^2} \right) \left(1 - \frac{x^2}{2^2 \pi^2} \right) \left(1 - \frac{x^2}{3^2 \pi^2} \right) \cdots \qquad (x^2 < \infty)$$

$$\cos x = \left(1 - \frac{4x^2}{\pi^2} \right) \left(1 - \frac{4x^2}{3^2 \pi^2} \right) \left(1 - \frac{4x^2}{5^2 \pi^2} \right) \cdots \qquad (x^2 < \infty)$$

$$\sin^{-1} x = x + \frac{x^3}{2 \cdot 3} + \frac{1 \cdot 3}{2 \cdot 4 \cdot 5} x^5 + \frac{1 \cdot 3 \cdot 5}{2 \cdot 4 \cdot 6 \cdot 7} x^7 + \cdots \quad \left(x^2 < 1, \; -\frac{\pi}{2} < \sin^{-1} x < \frac{\pi}{2} \right)$$

$$\cos^{-1} x = \frac{\pi}{2} - \left(x + \frac{x^3}{2 \cdot 3} + \frac{1 \cdot 3}{2 \cdot 4 \cdot 5} x^5 + \frac{1 \cdot 3 \cdot 5 x^7}{2 \cdot 4 \cdot 6 \cdot 7} + \cdots \right) \quad (x^2 < 1, 0 < \cos^{-1} x < \pi)$$

$$\tan^{-1} x = x - \frac{x^3}{3} + \frac{x^5}{5} - \frac{x^7}{7} + \cdots \qquad (x^2 < 1)$$

$$\tan^{-1} x = \frac{\pi}{2} - \frac{1}{x} + \frac{1}{3x^3} - \frac{1}{5x^5} + \frac{1}{7x^7} - \cdots \qquad (x > 1)$$

$$\tan^{-1} x = -\frac{\pi}{2} - \frac{1}{x} + \frac{1}{3x^3} - \frac{1}{5x^5} + \frac{1}{7x^7} - \cdots \qquad (x < -1)$$

$$\cot^{-1} x = \frac{\pi}{2} - x + \frac{x^3}{3} - \frac{x^5}{5} + \frac{x^7}{7} - \cdots \qquad (x^2 < 1)$$

$$\log_e \sin x = \log_e x - \frac{x^2}{6} - \frac{x^4}{180} - \frac{x^6}{2835} - \cdots \qquad (x^2 < \pi^2)$$

$$\log_e \cos x = -\frac{x^2}{2} - \frac{x^4}{12} - \frac{x^6}{45} - \frac{17x^8}{2520} - \cdots \qquad \left(x^2 < \frac{\pi^2}{4}\right)$$

$$\log_e \tan x = \log_e x + \frac{x^2}{3} + \frac{7x^4}{90} + \frac{62x^6}{2835} + \cdots \qquad \left(x^2 < \frac{\pi^2}{4}\right)$$

$$e^{\sin x} = 1 + x + \frac{x^2}{2!} - \frac{3x^4}{4!} - \frac{8x^5}{5!} - \frac{3x^6}{6!} + \frac{56x^7}{7!} + \cdots$$

$$e^{\cos x} = e\left(1 - \frac{x^2}{2!} + \frac{4x^4}{4!} - \frac{31x^6}{6!} + \cdots\right)$$

$$e^{\tan x} = 1 + x + \frac{x^2}{2!} + \frac{3x^3}{3!} + \frac{9x^4}{4!} + \frac{37x^5}{5!} + \cdots \qquad \left(x^2 < \frac{\pi^2}{4}\right)$$

$$\sin x = \sin a + (x - a)\cos a - \frac{(x - a)^2}{2!}\sin a$$
$$- \frac{(x - a)^3}{3!}\cos a + \frac{(x - a)^4}{4!}\sin a + \cdots$$

VECTOR ANALYSIS

Definitions

Any quantity which is completely determined by its magnitude is called a *scalar*. Examples of such are mass, density, temperature, etc. Any quantity which is completely determined by its magnitude and direction is called a *vector*. Examples of such are velocity, acceleration, force, etc. A vector quantity is represented by a directed line segment, the length of which represents the magnitude of the vector. A vector quantity is usually represented by a boldfaced letter such as $\mathbf{V}$. Two vectors $\mathbf{V}_1$ and $\mathbf{V}_2$ are equal to one another if they have equal magnitudes and are acting in the same directions. A negative vector, written as $-\mathbf{V}$, is one which acts in the opposite direction to $\mathbf{V}$, but is of equal magnitude to it. If we represent the magnitude of $\mathbf{V}$ by v, we write $|\mathbf{V}| = v$. A vector parallel to $\mathbf{V}$, but equal to the reciprocal of its magnitude is written as $\mathbf{V}^{-1}$ or as $\frac{1}{\mathbf{V}}$.

The *unit vector* $\frac{\mathbf{V}}{v}$ $(v \neq 0)$ is that vector which has the same direction as $\mathbf{V}$, but has a magnitude of unity (sometimes represented as $\mathbf{V}_0$ or $\hat{v}$).

Vector Algebra

The vector sum of $\mathbf{V}_1$ and $\mathbf{V}_2$ is represented by $\mathbf{V}_1 + \mathbf{V}_2$. The vector sum of $\mathbf{V}_1$ and $-\mathbf{V}_2$, or the difference of the vector $\mathbf{V}_2$ from $\mathbf{V}_1$ is represented by $\mathbf{V}_1 - \mathbf{V}_2$.

If r is a scalar, then $r\mathbf{V} = \mathbf{V}r$, and represents a vector r times the magnitude of $\mathbf{V}$, in the same direction as $\mathbf{V}$ if r is positive, and in the opposite direction if r is negative. If r and s are scalars, $\mathbf{V}_1, \mathbf{V}_2, \mathbf{V}_3$, vectors, then the following rules of scalars and vectors hold:

$$\mathbf{V}_1 + \mathbf{V}_2 = \mathbf{V}_2 + \mathbf{V}_1$$

$$(r + s)\mathbf{V}_1 = r\mathbf{V}_1 + s\mathbf{V}_1; \qquad r(\mathbf{V}_1 + \mathbf{V}_2) = r\mathbf{V}_1 + r\mathbf{V}_2$$

$$\mathbf{V}_1 + (\mathbf{V}_2 + \mathbf{V}_3) = (\mathbf{V}_1 + \mathbf{V}_2) + \mathbf{V}_3 = \mathbf{V}_1 + \mathbf{V}_2 + \mathbf{V}_3$$

Vectors in Space

A plane is described by two distinct vectors $\mathbf{V}_1$ and $\mathbf{V}_2$. Should these vectors not intersect each other, then one is displaced parallel to itself until they do (fig. 1.) Any other vector $\mathbf{V}$ lying in this plane is given by

$$\mathbf{V} = r\mathbf{V}_1 + s\mathbf{V}_2$$

A *position vector* specifies the position in space of a point relative to a fixed origin. If therefore $\mathbf{V}_1$ and $\mathbf{V}_2$ are the position vectors of the points A and B, relative to the origin O, then any point P on the line AB has a position vector $\mathbf{V}$ given by

$$\mathbf{V} = r\mathbf{V}_1 + (1 - r)\mathbf{V}_2$$

The scalar "r" can be taken as the parametric representation of P since $r = 0$ implies $P = B$ and $r = 1$ implies $P = A$. (fig. 2). If P divides the line AB in the ratio $r:s$ then

$$\mathbf{V} = \left(\frac{r}{r + s}\right)\mathbf{V}_1 + \left(\frac{s}{r + s}\right)\mathbf{V}_2$$

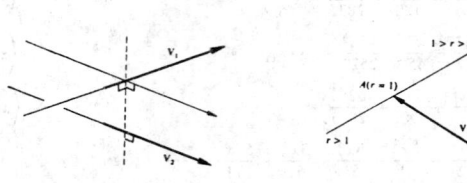

| Fig. 1. | Fig. 2. |

The vectors V_1, V_2, V_3, . . . , V_n are said to be *linearly dependent* if there exist scalars r_1, r_2, r_3, . . . , r_n, not all zero, such that

$$r_1V_1 + r_2V_2 + \cdots + r_nV_n = 0$$

A vector V is linearly dependent upon the set of vectors V_1, V_2, V_3, . . . , V_n if

$$V = r_1V_1 + r_2V_2 + r_3V_3 + \cdots + r_nV_n$$

Three vectors are linearly dependent if and only if they are co-planar.

All points in space can be uniquely determined by linear dependence upon three *base vectors* i.e. three vectors any one of which is linearly independent of the other two. The simplest set of base vectors are the unit vectors along the coordinate Ox, Oy and Oz axes. These are usually designated by i, j and k respectively.

If V is a vector in space, and a, b and c are the respective magnitudes of the projections of the vector along the axes then

$$V = ai + bj + ck$$

and

$$v = \sqrt{a^2 + b^2 + c^2}$$

and the direction cosines of V are

$$\cos \alpha = a/v, \quad \cos \beta = b/v, \quad \cos \gamma = c/v.$$

The law of addition yields

$$V_1 + V_2 = (a_1 + a_2)i + (b_1 + b_2)j + (c_1 + c_2)k$$

The Scalar, Dot, or Inner Product of Two Vectors V_1 and V_2

This product is represented as $V_1 \cdot V_2$ and is defined to be equal to $v_1v_2 \cos \theta$, where θ is the angle from V_1 to V_2, i.e.,

$$V_1 \cdot V_2 = v_1v_2 \cos \theta$$

The following rules apply for this product:

$$V_1 \cdot V_2 = a_1a_2 + b_1b_2 + c_1c_2 = V_2 \cdot V_1$$

It should be noted that this verifies that scalar multiplication is commutative.

$$(V_1 + V_2) \cdot V_3 = V_1 \cdot V_3 + V_2 \cdot V_3$$
$$V_1 \cdot (V_2 + V_3) = V_1 \cdot V_2 + V_1 \cdot V_3$$

If V_1 is perpendicular to V_2 then $V_1 \cdot V_2 = 0$, and if V_1 is parallel to V_2 then $V_1 \cdot V_2 = v_1v_2 = rw_1^2$

In particular

$$i \cdot i = j \cdot j = k \cdot k = 1,$$

and

$$i \cdot j = j \cdot k = k \cdot i = 0$$

The Vector or Cross Product of Vectors V_1 and V_2

This product is represented as $V_1 \times V_2$ and is defined to be equal to $v_1v_2(\sin \theta)1$, where θ is the angle from V_1 to V_2 and 1 is a unit vector perpendicular to the plane of V_1 and V_2 and so directed that a right-handed screw driven in the direction of 1 would carry V_1 into V_2, i.e.,

$$V_1 \times V_2 = v_1v_2(\sin \theta)1$$

and

$$\tan \theta = \frac{|V_1 \times V_2|}{V_1 \cdot V_2}$$

The following rules apply for vector products:

$$V_1 \times V_2 = -V_2 \times V_1$$
$$V_1 \times (V_2 + V_3) = V_1 \times V_2 + V_1 \times V_3$$
$$(V_1 + V_2) \times V_3 = V_1 \times V_3 + V_2 \times V_3$$
$$V_1 \times (V_2 \times V_3) = V_2(V_3 \cdot V_1) - V_3(V_1 \cdot V_2)$$
$$i \times i = j \times j = k \times k = 0.1 \text{ (zero vector)}$$
$$= 0$$

$$i \times j = k, \qquad j \times k = i, \qquad k \times i = j$$

If $V_1 = a_1i + b_1j + c_1k,$ $V_2 = a_2i + b_2j + c_2k,$ $V_3 = a_3i + b_3j + c_3k,$

then

$$V_1 \times V_2 = \begin{vmatrix} i & j & k \\ a_1 & b_1 & c_1 \\ a_2 & b_2 & c_2 \end{vmatrix} = (b_1c_2 - b_2c_1)i + (c_1a_2 - c_2a_1)j + (a_1b_2 - a_2b_1)k$$

It should be noted that, since $V_1 \times V_2 = -V_2 \times V_1$, the vector product is not commutative.

Scalar Triple Product

There is only one possible interpretation of the expression $V_1 \cdot V_2 \times V_3$ and that is $V_1 \cdot (V_2 \times V_3)$ which is obviously a scalar.

Further $V_1 \cdot (V_2 \times V_3) = (V_1 \times V_2) \cdot V_3 = V_2 \cdot (V_3 \times V_1)$

$$= \begin{vmatrix} a_1 & b_1 & c_1 \\ a_2 & b_2 & c_2 \\ a_3 & b_3 & c_3 \end{vmatrix}$$

$$= r_1 r_2 r_3 \cos \phi \sin \theta,$$

Where θ is the angle between V_2 and V_3 and ϕ is the angle between V_1 and the normal to the plane of V_2 and V_3.

This product is called the *scalar triple product* and is written as $[V_1V_2V_3]$.

The determinant indicates that it can be considered as the volume of the parallelepiped whose three determining edges are V_1, V_2 and V_3.

It also follows that cyclic permutation of the subscripts does not change the value of the scalar triple product so that

$$[V_1V_2V_3] = [V_2V_3V_1] = [V_3V_1V_2]$$

but $[V_1V_2V_3] = -[V_2V_1V_3]$ etc.   and $[V_1V_1V_2] \equiv 0$ etc.

Given three non-coplanar reference vectors V_1, V_2 and V_3, the *reciprocal system is* given by V_1^*, V_2^* and V_3^*, where

$$1 = r_1v_1^* = v_2v_2^* = r_3v_3^*$$

$$0 = v_1v_2^* = r_1r_3^* = v_2r_1^* \quad \text{etc.}$$

$$V_1^* = \frac{V_2 \times V_3}{[V_1V_2V_3]}, \qquad V_2^* = \frac{V_3 \times V_1}{[V_1V_2V_3]}, \qquad V_3^* = \frac{V_1 \times V_2}{[V_1V_2V_3]}$$

The system i, j, k is its own reciprocal.

Vector Triple Product

The product $V_1 \times (V_2 \times V_3)$ defines the *vector triple product*. Obviously, in this case, the brackets are vital to the definition.

$$V_1 \times (V_2 \times V_3) = (V_1 \cdot V_3)V_2 - (V_1 \cdot V_2)V_3$$

$$= \begin{vmatrix} i & j & k \\ a_1 & b_1 & c_1 \\ \begin{vmatrix} b_2 & c_2 \\ b_3 & c_3 \end{vmatrix} & \begin{vmatrix} c_2 & a_2 \\ c_3 & a_3 \end{vmatrix} & \begin{vmatrix} a_2 & b_2 \\ a_3 & b_3 \end{vmatrix} \end{vmatrix}$$

i.e. it is a vector, perpendicular to V_1, lying in the plane of V_2, V_3.

Similarly $(V_1 \times V_2) \times V_3 = \begin{vmatrix} i & j & k \\ \begin{vmatrix} b_1 & c_1 \\ b_2 & c_2 \end{vmatrix} & \begin{vmatrix} c_1 & a_1 \\ c_2 & a_2 \end{vmatrix} & \begin{vmatrix} a_1 & b_1 \\ a_2 & b_2 \end{vmatrix} \\ a_3 & b_3 & c_3 \end{vmatrix}$

$$V_1 \times (V_2 \times V_3) + V_2 \times (V_3 \times V_1) + V_3 \times (V_1 \times V_2) \equiv 0$$

If $V_1 \times (V_2 \times V_3) = (V_1 \times V_2) \times V_3$ then V_1, V_2, V_3 form an *orthogonal set*. Thus i, j, k form an orthogonal set.

Geometry of the Plane, Straight Line and Sphere

The position vectors of the fixed points A, B, C, D relative to O are V_1, V_2, V_3, V_4 and the position vector of the variable point P is V.

The vector form of the equation of the straight line through A parallel to V_2 is

$$V = V_1 + rV_2$$

or $(V - V_1) = rV_2$

or $(V - V_1) \times V_2 = 0$

while that of the plane through A perpendicular to V_2 is

$$(V - V_1) \cdot V_2 = 0$$

The equation of the line AB is

$$V = rV_1 + (1 - r)V_2$$

and those of the bisectors of the angles between $\mathbf{V}_1$ and $\mathbf{V}_2$ are

$$\mathbf{V} = r\left(\frac{\mathbf{V}_1}{v} \pm \frac{\mathbf{V}_2}{v_2}\right)$$

or $\qquad \mathbf{V} = r(\hat{\mathbf{v}}_1 \pm \hat{\mathbf{v}}_2)$

The perpendicular from C to the line through A parallel to $\mathbf{V}_2$ has as its equation

$$\mathbf{V} = \mathbf{V}_1 - \mathbf{V}_3 - \hat{\mathbf{v}}_2 \cdot (\mathbf{V}_1 - \mathbf{V}_3)\hat{\mathbf{v}}_2.$$

The condition for the intersection of the two lines,

$$\mathbf{V} = \mathbf{V}_1 + r\mathbf{V}_3$$

and $\qquad \mathbf{V} = \mathbf{V}_2 + s\mathbf{V}_4$

is $\qquad [(\mathbf{V}_1 - \mathbf{V}_2)\mathbf{V}_3\mathbf{V}_4] = 0.$

The common perpendicular to the above two lines is the line of intersection of the two planes

$$[(\mathbf{V} - \mathbf{V}_1)\mathbf{V}_3(\mathbf{V}_3 \times \mathbf{V}_4)] = 0$$

and $\qquad [(\mathbf{V} - \mathbf{V}_2)\mathbf{V}_4(\mathbf{V}_3 \times \mathbf{V}_4)] = 0$

and the length of this perpendicular is

$$\frac{[(\mathbf{V}_1 - \mathbf{V}_2)\mathbf{V}_3\mathbf{V}_4]}{|\mathbf{V}_3 \times \mathbf{V}_4|}.$$

The equation of the line perpendicular to the plane ABC is

$$\mathbf{V} = \mathbf{V}_1 \times \mathbf{V}_2 + \mathbf{V}_2 \times \mathbf{V}_3 + \mathbf{V}_3 \times \mathbf{V}_1$$

and the distance of the plane from the origin is

$$\frac{[\mathbf{V}_1\mathbf{V}_2\mathbf{V}_3]}{|(\mathbf{V}_2 - \mathbf{V}_1) \times (\mathbf{V}_3 - \mathbf{V}_1)|}.$$

In general the vector equation

$$\mathbf{V} \cdot \mathbf{V}_2 = r$$

defines the plane which is perpendicular to $\mathbf{V}_2$, and the perpendicular distance from A to this plane is

$$\frac{r - \mathbf{V}_1 \cdot \mathbf{V}_2}{v_2}$$

The distance from A, measured along a line parallel to $\mathbf{V}_3$, is

$$\frac{r - \mathbf{V}_1 \cdot \mathbf{V}_2}{\mathbf{V}_2 \cdot \hat{\mathbf{v}}_3} \quad \text{or} \quad \frac{r - \mathbf{V}_1 \cdot \mathbf{V}_2}{v_2 \cos \theta}$$

where θ is the angle beween $\mathbf{V}_2$ and $\mathbf{V}_3$.
(If this plane contains the point C then $r = \mathbf{V}_3 \cdot \mathbf{V}_2$ and if it passes through the origin then $r = 0$.)

Given two planes $\qquad \mathbf{V} \cdot \mathbf{V}_1 = r$

$$\mathbf{V} \cdot \mathbf{V}_2 = s$$

then any plane through the line of intersection of these two planes is given by

$$\mathbf{V} \cdot (\mathbf{V}_1 + \lambda\mathbf{V}_2) = r + \lambda s$$

where λ is a scalar parameter. In particular $\lambda = \pm r_1/r_2$ yields the equation of the two planes bisecting the angle between the given planes.

The plane through A parallel to the plane of $\mathbf{V}_2$, $\mathbf{V}_3$ is

$$\mathbf{V} = \mathbf{V}_1 + r\mathbf{V}_2 + s\mathbf{V}_3$$

or $\qquad (\mathbf{V} - \mathbf{V}_1) \cdot \mathbf{V}_2 \times \mathbf{V}_3 = 0$

or $\qquad [\mathbf{V}\mathbf{V}_2\mathbf{V}_3] - [\mathbf{V}_1\mathbf{V}_2\mathbf{V}_3] = 0$

so that the expansion in rectangular Cartesian coordinates yields

$$\begin{vmatrix} (x - a_1) & (y - b_1) & (z - c_1) \\ a_2 & b_2 & c_2 \\ a_3 & b_3 & c_3 \end{vmatrix} = 0 \qquad (\mathbf{V} \equiv x\mathbf{i} + y\mathbf{j} + z\mathbf{k})$$

which is obviously the usual linear equation in x, y and z.

The plane through AB parallel to $\mathbf{V}_3$ is given by

$$[(\mathbf{V} - \mathbf{V}_1)(\mathbf{V}_1 - \mathbf{V}_2)\mathbf{V}_3] = 0$$

or $\qquad [\mathbf{V}\mathbf{V}_2\mathbf{V}_3] - [\mathbf{V}\mathbf{V}_1\mathbf{V}_3] - [\mathbf{V}_1\mathbf{V}_2\mathbf{V}_3] = 0$

The plane through the three points A, B and C is

$$\mathbf{V} = \mathbf{V}_1 + s(\mathbf{V}_2 - \mathbf{V}_1) + t(\mathbf{V}_3 - \mathbf{V}_1)$$

or $\quad \mathbf{V} = r\mathbf{V}_1 + s\mathbf{V}_2 + t\mathbf{V}_3 \qquad (r + s + t \equiv 1)$

or $\quad [(\mathbf{V} - \mathbf{V}_1)(\mathbf{V}_1 - \mathbf{V}_2)(\mathbf{V}_2 - \mathbf{V}_3)] = 0$

or $\quad [\mathbf{V}\mathbf{V}_1\mathbf{V}_2] + [\mathbf{V}\mathbf{V}_2\mathbf{V}_3] + [\mathbf{V}\mathbf{V}_3\mathbf{V}_1] - [\mathbf{V}_1\mathbf{V}_2\mathbf{V}_3] = 0$

For four points A, B, C, D to be coplanar, then

$$r\mathbf{V}_1 + s\mathbf{V}_2 + t\mathbf{V}_3 + u\mathbf{V}_4 \equiv 0 \equiv r + s + t + u$$

The following formulae relate to a sphere when the vectors are taken to lie in three dimensional space and to a circle when the space is two dimensional. For a circle in three dimensions take the intersection of the sphere with a plane.

The equation of a sphere with center O and radius OA is

$$\mathbf{V} \cdot \mathbf{V} = v_1^2 \qquad (\text{not } \mathbf{V} = \mathbf{V}_1)$$

or $\quad (\mathbf{V} - \mathbf{V}_1) \cdot (\mathbf{V} + \mathbf{V}_1) = 0$

while that of a sphere with center B radius v_1 is

$$(\mathbf{V} - \mathbf{V}_2) \cdot (\mathbf{V} - \mathbf{V}_2) = v_1^2$$

or $\quad \mathbf{V} \cdot (\mathbf{V} - 2\mathbf{V}_2) = v_1^2 - v_2^2$

If the above sphere passes through the origin then

$$\mathbf{V} \cdot (\mathbf{V} - 2\mathbf{V}_2) = 0$$

(note that in two dimensional polar coordinates this is simply)

$$r = 2a \cdot \cos \theta$$

while in three dimensional Cartesian coordinates it is

$$x^2 + y^2 + z^2 - 2(a_2 x + b_2 y + c_2 x) = 0.$$

The equation of a sphere having the points A and B as the extremities of a diameter is

$$(\mathbf{V} - \mathbf{V}_1) \cdot (\mathbf{V} - \mathbf{V}_2) = 0.$$

The square of the length of the tangent from C to the sphere with center B and radius v_1 is given by

$$(\mathbf{V}_3 - \mathbf{V}_2) \cdot (\mathbf{V}_3 - \mathbf{V}_2) = v_1^2$$

The condition that the plane $\mathbf{V} \cdot \mathbf{V}_3 = s$ is tangential to the sphere $(\mathbf{V} - \mathbf{V}_2) \cdot (\mathbf{V} - \mathbf{V}_2) = v_1^2$ is

$$(s - \mathbf{V}_3 \cdot \mathbf{V}_2) \cdot (s - \mathbf{V}_3 \cdot \mathbf{V}_2) = v_1^2 v_3^2.$$

The equation of the tangent plane at D, on the surface of sphere $(\mathbf{V} - \mathbf{V}_2) \cdot (\mathbf{V} - \mathbf{V}_2) = v_1^2$, is

$$(\mathbf{V} - \mathbf{V}_4) \cdot (\mathbf{V}_4 - \mathbf{V}_2) = 0$$

or $\quad \mathbf{V} \cdot \mathbf{V}_4 - \mathbf{V}_2 \cdot (\mathbf{V} + \mathbf{V}_4) = v_1^2 - v_2^2$

The condition that the two circles $(\mathbf{V} - \mathbf{V}_2) \cdot (\mathbf{V} - \mathbf{V}_2) = v_1^2$ and $(\mathbf{V} - \mathbf{V}_4) \cdot (\mathbf{V} - \mathbf{V}_4) = v_3^2$ intersect orthogonally is clearly

$$(\mathbf{V}_2 - \mathbf{V}_4) \cdot (\mathbf{V}_2 - \mathbf{V}_4) = v_1^2 + v_3^2$$

The polar plane of D with respect to the circle

$$(\mathbf{V} - \mathbf{V}_2) \cdot (\mathbf{V} - \mathbf{V}_2) = v_1^2 \quad \text{is}$$

$$\mathbf{V} \cdot \mathbf{V}_4 - \mathbf{V}_2 \cdot (\mathbf{V} + \mathbf{V}_4) = v_1^2 - v_2^2$$

Any sphere through the intersection of the two spheres $(\mathbf{V} - \mathbf{V}_2) \cdot (\mathbf{V} - \mathbf{V}_2) = v_1^2$ and $(\mathbf{V} - \mathbf{V}_4) \cdot (\mathbf{V} - \mathbf{V}_4) = v_3^2$ is given by

$$(\mathbf{V} - \mathbf{V}_2) \cdot (\mathbf{V} - \mathbf{V}_2) + \lambda(\mathbf{V} - \mathbf{V}_4) \cdot (\mathbf{V} - \mathbf{V}_4) = v_1^2 + \lambda v_3^2$$

while the radical plane of two such spheres is

$$\mathbf{V} \cdot (\mathbf{V}_2 - \mathbf{V}_4) = -\tfrac{1}{2}(v_1^2 - v_2^2 - v_3^2 + v_4^2)$$

Differentiation of Vectors

If $\mathbf{V}_1 = a_1\mathbf{i} + b_1\mathbf{j} + c_1\mathbf{k}$, and $\mathbf{V}_2 = a_2\mathbf{i} + b_2\mathbf{j} + c_2\mathbf{k}$, and if $\mathbf{V}_1$ and $\mathbf{V}_2$ are functions of the scalar t, then

$$\frac{d}{dt}(\mathbf{V}_1 + \mathbf{V}_2 + \cdots) = \frac{d\mathbf{V}_1}{dt} + \frac{d\mathbf{V}_2}{dt} + \cdots.$$

where $\quad \dfrac{d\mathbf{V}_1}{dt} = \dfrac{da_1}{dt}\mathbf{i} + \dfrac{db_1}{dt}\mathbf{j} + \dfrac{dc_1}{dt}\mathbf{k}$, etc.

$$\frac{d}{dt}(\mathbf{V}_1 \cdot \mathbf{V}_2) = \frac{d\mathbf{V}_1}{dt} \cdot \mathbf{V}_2 + \mathbf{V}_1 \cdot \frac{d\mathbf{V}_2}{dt}$$

$$\frac{d}{dt}(\mathbf{V}_1 \times \mathbf{V}_2) = \frac{d\mathbf{V}_1}{dt} \times \mathbf{V}_2 + \mathbf{V}_1 \times \frac{d\mathbf{V}_2}{dt}$$

$$\mathbf{V} \cdot \frac{d\mathbf{V}}{dt} = v \cdot \frac{dv}{dt}$$

In particular, if **V** is a vector of constant length then the right hand side of the last equation is identically zero showing that **V** is perpendicular to its derivative.

The derivatives of the triple products are

$$\frac{d}{dt}[\mathbf{V}_1\mathbf{V}_2\mathbf{V}_3] = \left[\left(\frac{d\mathbf{V}_1}{dt}\right)\mathbf{V}_2\mathbf{V}_3\right] + \left[\mathbf{V}_1\left(\frac{d\mathbf{V}_2}{dt}\right)\mathbf{V}_3\right] + \left[\mathbf{V}_1\mathbf{V}_2\left(\frac{d\mathbf{V}_3}{dt}\right)\right]$$

and

$$\frac{d}{dt}\{\mathbf{V}_1 \times (\mathbf{V}_2 \times \mathbf{V}_3)\} = \left(\frac{d\mathbf{V}_1}{dt}\right) \times (\mathbf{V}_2 \times \mathbf{V}_3) + \mathbf{V}_1$$
$$\times \left(\left(\frac{d\mathbf{V}_2}{dt}\right) \times \mathbf{V}_3\right) + \mathbf{V}_1 \times \left(\mathbf{V}_2 \times \left(\frac{d\mathbf{V}_3}{dt}\right)\right)$$

Geometry of Curves in Space

s = the *length of arc*, measured from some fixed point on the curve (fig. 3).

$\mathbf{V}_1$ = the position vector of the point A on the curve

$\mathbf{V}_1 + \delta\mathbf{V}_1$ = the position vector of the point P in the neighborhood of A

$\hat{\mathbf{t}}$ = the *unit tangent* to the curve at the point A, measured in the direction of s increasing.

The *normal plane* is that plane which is perpendicular to the unit tangent. The principal normal is defined as the intersection of the normal plane with the plane defined by $\mathbf{V}_1$ and $\mathbf{V}_1 + \delta\mathbf{V}_1$ in the limit as $\delta\mathbf{V}_1 - 0$.

$\hat{\mathbf{n}}$ = the *unit normal* (principal) at the point A. The plane defined by $\hat{\mathbf{t}}$ and $\hat{\mathbf{n}}$ is called the *osculating plane* (alternatively plane of curvature or local plane).

ρ = the *radius of curvature* at A

$\delta\theta$ = the angle subtended at the origin by $\delta\mathbf{V}_1$.

$$\kappa = \frac{d\theta}{ds} = \frac{1}{\rho}$$

$\hat{\mathbf{b}}$ = the *unit binormal* i.e. the unit vector which is parallel to $\hat{\mathbf{t}} \times \hat{\mathbf{n}}$ at the point A:

λ = the *torsion* of the curve at A

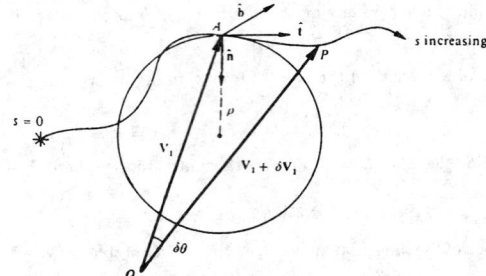

Figure 3.

Frenet's Formulae:

$$\frac{d\hat{\mathbf{t}}}{ds} = \kappa\hat{\mathbf{n}}$$
$$\frac{d\hat{\mathbf{n}}}{ds} = -\kappa\hat{\mathbf{t}} + \lambda\hat{\mathbf{b}}$$
$$\frac{d\hat{\mathbf{b}}}{ds} = -\lambda\hat{\mathbf{n}}$$

The following formulae are also applicable:

Unit tangent
$$\hat{\mathbf{t}} = \frac{d\mathbf{V}_1}{ds}$$

Equation of the tangent
$$(\mathbf{V} - \mathbf{V}_1) \times \hat{\mathbf{t}} = 0$$
$$\text{or} \quad \mathbf{V} = \mathbf{V}_1 + q\hat{\mathbf{t}}$$

Unit normal
$$\hat{\mathbf{n}} = \frac{1}{\kappa}\frac{d^2\mathbf{V}_1}{ds^2}$$

Equation of the normal plane
$$(\mathbf{V} - \mathbf{V}_1) \cdot \hat{\mathbf{t}} = 0$$

Equation of the normal
$$(\mathbf{V} - \mathbf{V}_1) \times \hat{\mathbf{n}} = 0$$
$$\text{or} \quad \mathbf{V} = \mathbf{V}_1 + r\hat{\mathbf{n}}$$

Unit binormal
$$\hat{\mathbf{b}} = \hat{\mathbf{t}} \times \hat{\mathbf{n}}$$

Equation of the binormal
$$(\mathbf{V} - \mathbf{V}_1) \times \hat{\mathbf{b}} = 0$$
$$\text{or} \quad \mathbf{V} = \mathbf{V}_1 + u\hat{\mathbf{b}}$$
$$\text{or} \quad \mathbf{V} = \mathbf{V}_1 + w\frac{d\mathbf{V}_1}{ds} \times \frac{d^2\mathbf{V}_1}{ds^2}$$

Equation of the osculating plane:
$$[(\mathbf{V} - \mathbf{V}_1)\hat{\mathbf{t}}\hat{\mathbf{n}}] = 0$$
$$\text{or} \quad \left[(\mathbf{V} - \mathbf{V}_1)\left(\frac{d\mathbf{V}_1}{ds}\right)\left(\frac{d^2\mathbf{V}_1}{ds^2}\right)\right] = 0$$

A *geodetic line* on a surface is a curve, the osculating plane of which is everywhere normal to the surface.

The differential equation of the geodetic is

$$[\hat{n}\,dV_1\,d^2V_1] = 0$$

Differential Operators—Rectangular Coordinates

$$dS = \frac{\partial S}{\partial x}\cdot dx + \frac{\partial S}{\partial y}\cdot dy + \frac{\partial S}{\partial z}\cdot dz$$

By definition

$$\nabla \equiv \text{del} \equiv \mathbf{i}\,\frac{\partial}{\partial x} + \mathbf{j}\,\frac{\partial}{\partial y} + \mathbf{k}\,\frac{\partial}{\partial z}$$

$$\nabla^2 \equiv \text{Laplacian} \equiv \frac{\partial^2}{\partial x^2} + \frac{\partial^2}{\partial y^2} + \frac{\partial^2}{\partial z^2}$$

If S is a scalar function, then

$$\nabla S \equiv \text{grad } S \equiv \frac{\partial S}{\partial x}\mathbf{i} + \frac{\partial S}{\partial y}\mathbf{j} + \frac{\partial S}{\partial z}\mathbf{k}$$

Grad S defines both the direction and magnitude of the maximum rate of increase of S at any point. Hence the name *gradient* and also its vectorial nature. ∇S is independent of the choice of rectangular coordinates.

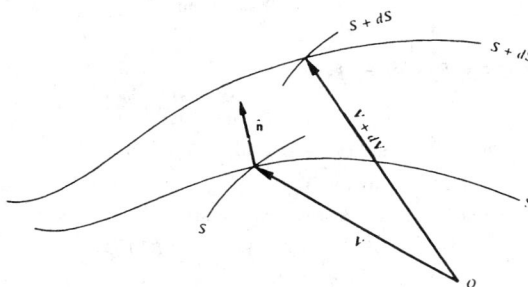

Figure 4

$$\nabla S = \frac{\partial S}{\partial n}\,\hat{n}$$

where $\hat{n}$ is the unit normal to the surface $S = $ constant, in the direction of S increasing. The total derivative of S at a point having the position vector $\mathbf{V}$ is given by (fig. 4)

$$dS = \frac{\partial S}{\partial n}\,\hat{n}\cdot d\mathbf{V}$$
$$= d\mathbf{V}\cdot \nabla S$$

and the directional derivative of S in the direction of $\mathbf{U}$ is

$$\mathbf{U}\cdot\nabla S = \mathbf{U}\cdot(\nabla S) = (\mathbf{U}\cdot\nabla)S$$

Similarly the directional derivative of the vector $\mathbf{V}$ in the direction of $\mathbf{U}$ is

$$(\mathbf{U}\cdot\nabla)\mathbf{V}$$

The *distributive* law holds for finding a gradient. Thus if S and T are scalar functions

$$\nabla(S + T) = \nabla S + \nabla T$$

The *associative* law becomes the rule for differentiating a product:

$$\nabla(ST) = S\nabla T + T\nabla S$$

If $\mathbf{V}$ is a vector function with the magnitudes of the components parallel to the three coordinate axes V_x, V_y, V_z, then

$$\nabla\cdot\mathbf{V} \equiv \text{div } \mathbf{V} \equiv \frac{\partial V_x}{\partial x} + \frac{\partial V_y}{\partial y} + \frac{\partial V_z}{\partial z}$$

The divergence obeys the distributive law. Thus, if $\mathbf{V}$ and $\mathbf{U}$ are vectors functions, then

$$\nabla\cdot(\mathbf{V} + \mathbf{U}) = \nabla\cdot\mathbf{V} + \nabla\cdot\mathbf{U}$$
$$\nabla\cdot(S\mathbf{V}) = (\nabla S)\cdot\mathbf{V} + S(\nabla\cdot\mathbf{V})$$
$$\nabla\cdot(\mathbf{U}\times\mathbf{V}) = \mathbf{V}\cdot(\nabla\times\mathbf{U}) - \mathbf{U}\cdot(\nabla\times\mathbf{V})$$

As with the gradient of a scalar, the divergence of a vector is invariant under a transformation from one set of rectangular coordinates to another.

$$\nabla\times\mathbf{V} \equiv \text{curl } \mathbf{V} \quad (\text{sometimes } \nabla\wedge\mathbf{V} \text{ or rot } \mathbf{V})$$

$$\equiv \left(\frac{\partial V_z}{\partial y} - \frac{\partial V_y}{\partial z}\right)\mathbf{i} + \left(\frac{\partial V_x}{\partial z} - \frac{\partial V_z}{\partial x}\right)\mathbf{j} + \left(\frac{\partial V_y}{\partial x} - \frac{\partial V_x}{\partial y}\right)\mathbf{k}$$

$$= \begin{vmatrix} \mathbf{i} & \mathbf{j} & \mathbf{k} \\ \dfrac{\partial}{\partial x} & \dfrac{\partial}{\partial y} & \dfrac{\partial}{\partial z} \\ V_x & V_y & V_z \end{vmatrix}$$

A-93

The *curl* (or *rotation*) of a vector is a vector which is invariant under a transformation from one set of rectangular coordinates to another.

$$\nabla \times (\mathbf{U} + \mathbf{V}) = \nabla \times \mathbf{U} + \nabla \times \mathbf{V}$$
$$\nabla \times (S\mathbf{V}) = (\nabla S) \times \mathbf{V} + S(\nabla \times \mathbf{V})$$
$$\nabla \times (\mathbf{U} \times \mathbf{V}) = (\mathbf{V} \cdot \nabla)\mathbf{U} - (\mathbf{U} \cdot \nabla)\mathbf{V} + \mathbf{U}(\nabla \cdot \mathbf{U}) - \mathbf{V}(\nabla \cdot \mathbf{U})$$

$$\text{grad } (\mathbf{U} \cdot \mathbf{V}) = \nabla(\mathbf{U} \cdot \mathbf{V})$$
$$= (\mathbf{V} \cdot \nabla)\mathbf{U} + (\mathbf{U} \cdot \nabla)\mathbf{V} + \mathbf{V} \times (\nabla \times \mathbf{U}) + \mathbf{U} \times (\nabla \times \mathbf{V})$$

If
$$\mathbf{V} = V_x\mathbf{i} + V_y\mathbf{j} + V_z\mathbf{k}$$
$$\nabla \cdot \mathbf{V} = \nabla V_x \cdot \mathbf{i} + \nabla V_y \cdot \mathbf{j} + \nabla V_z \cdot \mathbf{k}$$
and
$$\nabla \times \mathbf{V} = \nabla V_x \times \mathbf{i} + \nabla V_y \times \mathbf{j} + \nabla V_z \times \mathbf{k}$$

The operator ∇ can be used more than once. The number of possibilities where ∇ is used twice are

$$\nabla \cdot (\nabla \theta) \equiv \text{div grad } \theta$$
$$\nabla \times (\nabla \theta) \equiv \text{curl grad } \theta$$
$$\nabla(\nabla \cdot \mathbf{V}) \equiv \text{grad div } \mathbf{V}$$
$$\nabla \cdot (\nabla \times \mathbf{V}) \equiv \text{div curl } \mathbf{V}$$
$$\nabla \times (\nabla \times \mathbf{V}) \equiv \text{curl curl } \mathbf{V}$$

Thus: div grad $S \equiv \nabla \cdot (\nabla S) \equiv$ Laplacian $S \equiv \nabla^2 S$

$$\equiv \frac{\partial^2 S}{\partial x^2} + \frac{\partial^2 S}{\partial y^2} + \frac{\partial^2 S}{\partial z^2}$$

curl grad $S \equiv 0$; curl curl $\mathbf{V} \equiv$ grad div $\mathbf{V} - \nabla^2\mathbf{V}$;

div curl $\mathbf{V} \equiv$ 0

Taylor's expansion in three dimensions can be written

$$f(\mathbf{V} + \mathbf{\varepsilon}) = e^{\mathbf{\varepsilon} \cdot \nabla} f(\mathbf{V})$$

where $\mathbf{V} = x\mathbf{i} + y\mathbf{j} + z\mathbf{k}$

and $\mathbf{\varepsilon} = h\mathbf{i} + l\mathbf{j} + m\mathbf{k}$

(note the analogy with $f_p = e^{phD}f_0$ in finite difference methods).

Orthogonal Curvilinear Coordinates

If at a point P there exist three uniform point functions u, v and w so that the surfaces $u = $ const., $v = $ const., and $w = $ const., intersect in three distinct curves through P then the surfaces are called the *coordinate surfaces* through P. The three lines of intersection are referred to as the *coordinate lines* and their tangents a, b, and c as the *coordinate axes*. When the coordinate axes form an orthogonal set the system is said to define *orthogonal curvilinear coordinates* at P.

Consider an infinitesimal volume enclosed by the surfaces u, v, w, $u + du$, $v + dv$, and $w + dw$ (fig. 5).

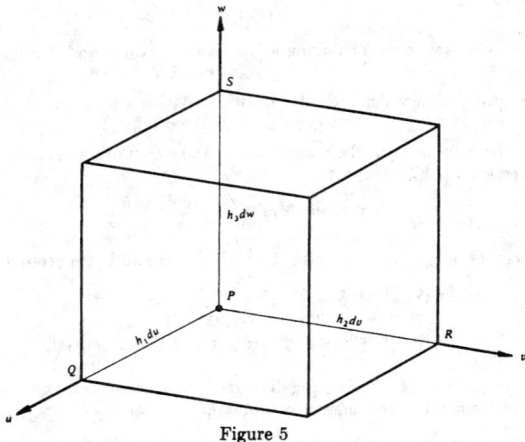

Figure 5

The surface $PRS \equiv u = $ const., and the face of the curvilinear figure immediately opposite this is $u + du = $ const. etc.

In terms of these surface constants

$$P = P(u,v,w)$$
$$Q = Q(u + du,v,w) \quad \text{and} \quad PQ = h_1 du$$

A-94

$$R = R(u, v + dv, w) \qquad PR = h_2 dv$$
$$S = S(u, v, w + dw) \qquad PS = h_3 dw$$

where h_1, h_2, and h_3 are functions of u, v, and w.

In rectangular Cartesians $\mathbf{i}$, $\mathbf{j}$, $\mathbf{k}$

$$h_1 = 1, \qquad h_2 = 1, \qquad h_3 = 1.$$

$$\frac{\hat{\mathbf{a}}}{h_1}\frac{\partial}{\partial u} = \mathbf{i}\frac{\partial}{\partial x}, \qquad \frac{\hat{\mathbf{b}}}{h_2}\frac{\partial}{\partial v} = \mathbf{j}\frac{\partial}{\partial y}, \qquad \frac{\hat{\mathbf{c}}}{h_3}\frac{\partial}{\partial w} = \mathbf{k}\frac{\partial}{\partial z}.$$

In cylindrical coordinates $\hat{\mathbf{r}}$, $\hat{\boldsymbol{\phi}}$, $\hat{\mathbf{k}}$

$$h_1 = 1, \qquad h_2 = r \qquad h_3 = 1.$$

$$\frac{\hat{\mathbf{a}}}{h_1}\frac{\partial}{\partial u} = \hat{\mathbf{r}}\frac{\partial}{\partial r}, \qquad \frac{\hat{\mathbf{b}}}{h_2}\frac{\partial}{\partial v} = \frac{\hat{\boldsymbol{\phi}}}{r}\frac{\partial}{\partial \phi}, \qquad \frac{\hat{\mathbf{c}}}{h_3}\frac{\partial}{\partial w} = \hat{\mathbf{k}}\frac{\partial}{\partial z}$$

In spherical coordinates $\hat{\mathbf{r}}$, $\hat{\boldsymbol{\theta}}$, $\hat{\boldsymbol{\phi}}$

$$h_1 = 1, \qquad h_2 = r, \qquad h_3 = r\sin\theta$$

$$\frac{\hat{\mathbf{a}}}{h_1}\frac{\partial}{\partial u} = \hat{\mathbf{r}}\frac{\partial}{\partial r}, \qquad \frac{\mathbf{b}}{h_2}\frac{\partial}{\partial v} = \frac{\hat{\boldsymbol{\phi}}}{r}\frac{\partial}{\partial \theta}, \qquad \frac{\hat{\mathbf{c}}}{h_3}\frac{\partial}{\partial w} = \frac{\hat{\boldsymbol{\phi}}}{r\sin\theta}\frac{\partial}{\partial \phi}$$

The general expressions for grad, div and curl together with those for ∇^2 and the directional derivative are, in orthogonal curvilinear coordinates, given by

$$\nabla S = \frac{\hat{\mathbf{a}}}{h_1}\frac{\partial S}{\partial u} + \frac{\hat{\mathbf{b}}}{h_2}\frac{\partial S}{\partial v} + \frac{\hat{\mathbf{c}}}{h_3}\frac{\partial S}{\partial w}$$

$$(\mathbf{V}\cdot\nabla)S = \frac{V_1}{h_1}\frac{\partial S}{\partial u} + \frac{V_2}{h_2}\frac{\partial S}{\partial v} + \frac{V_3}{h_3}\frac{\partial S}{\partial w}$$

$$\nabla\cdot\mathbf{V} = \frac{1}{h_1 h_2 h_3}\left\{\frac{\partial}{\partial u}(h_2 h_3 V_1) + \frac{\partial}{\partial v}(h_3 h_1 V_2) + \frac{\partial}{\partial w}(h_1 h_2 V_3)\right\}$$

$$\nabla\times\mathbf{V} = \frac{\hat{\mathbf{a}}}{h_2 h_3}\left\{\frac{\partial}{\partial v}(h_3 V_3) - \frac{\partial}{\partial w}(h_2 V_2)\right\} + \frac{\hat{\mathbf{b}}}{h_3 h_1}\left\{\frac{\partial}{\partial w}(h_1 V_1) - \frac{\partial}{\partial u}(h_3 V_3)\right\}$$
$$+ \frac{\hat{\mathbf{c}}}{h_1 h_2}\left\{\frac{\partial}{\partial u}(h_2 V_2) - \frac{\partial}{\partial v}(h_1 V_1)\right\}$$

$$\nabla^2 S = \frac{1}{h_1 h_2 h_3}\left\{\frac{\partial}{\partial u}\left(\frac{h_2 h_3}{h_1}\frac{\partial S}{\partial u}\right) + \frac{\partial}{\partial v}\left(\frac{h_3 h_1}{h_2}\frac{\partial S}{\partial v}\right) + \frac{\partial}{\partial w}\left(\frac{h_1 h_2}{h_3}\frac{\partial S}{\partial w}\right)\right\}$$

FORMULAS OF VECTOR ANALYSIS

	Rectangular coordinates	Cylindrical coordinates	Spherical coordinates
Conversion to rectangular coordinates		$x = r\cos\varphi \quad y = r\sin\varphi \quad z = z$	$x = r\cos\varphi\sin\theta \quad y = r\sin\varphi\sin\theta$ $z = r\cos\theta$
Gradient	$\nabla\phi = \frac{\partial\phi}{\partial x}\mathbf{i} + \frac{\partial\phi}{\partial y}\mathbf{j} + \frac{\partial\phi}{\partial z}\mathbf{k}$	$\nabla\phi = \frac{\partial\phi}{\partial r}\mathbf{r} + \frac{1}{r}\frac{\partial\phi}{\partial\varphi}\boldsymbol{\phi} + \frac{\partial\phi}{\partial z}\mathbf{k}$	$\nabla\phi = \frac{\partial\phi}{\partial r}\mathbf{r} + \frac{1}{r}\frac{\partial\phi}{\partial\theta}\boldsymbol{\theta} + \frac{1}{r\sin\theta}\frac{\partial\phi}{\partial\varphi}\boldsymbol{\phi}$
Divergence	$\nabla\cdot\mathbf{A} = \frac{\partial A_x}{\partial x} + \frac{\partial A_y}{\partial y} + \frac{\partial A_z}{\partial z}$	$\nabla\cdot\mathbf{A} = \frac{1}{r}\frac{\partial(rA_r)}{\partial r} + \frac{1}{r}\frac{\partial A_\varphi}{\partial\varphi}$ $+ \frac{\partial A_z}{\partial z}$	$\nabla\cdot\mathbf{A} = \frac{1}{r^2}\frac{\partial(r^2 A_r)}{\partial r} + \frac{1}{r\sin\theta}\frac{\partial(A_\theta\sin\theta)}{\partial\theta}$ $+ \frac{1}{r\sin\theta}\frac{\partial A_\varphi}{\partial\varphi}$
Curl	$\nabla\times\mathbf{A} = \begin{vmatrix} \mathbf{i} & \mathbf{j} & \mathbf{k} \\ \frac{\partial}{\partial x} & \frac{\partial}{\partial y} & \frac{\partial}{\partial z} \\ A_x & A_y & A_z \end{vmatrix}$	$\nabla\times\mathbf{A} = \begin{vmatrix} \frac{1}{r}\mathbf{r} & \boldsymbol{\phi} & \frac{1}{r}\mathbf{k} \\ \frac{\partial}{\partial r} & \frac{\partial}{\partial\varphi} & \frac{\partial}{\partial z} \\ A_r & rA_\varphi & A_z \end{vmatrix}$	$\nabla\times\mathbf{A} = \begin{vmatrix} \frac{\mathbf{r}}{r^2\sin\theta} & \frac{\boldsymbol{\theta}}{r\sin\theta} & \frac{\boldsymbol{\phi}}{r} \\ \frac{\partial}{\partial r} & \frac{\partial}{\partial\theta} & \frac{\partial}{\partial\varphi} \\ A_r & rA_\theta & rA_\varphi\sin\theta \end{vmatrix}$
Laplacian	$\nabla^2\phi = \frac{\partial^2\phi}{\partial x^2} + \frac{\partial^2\phi}{\partial y^2} + \frac{\partial^2\phi}{\partial z^2}$	$\nabla^2\phi = \frac{1}{r}\frac{\partial}{\partial r}\left(r\frac{\partial\phi}{\partial r}\right) + \frac{1}{r^2}\frac{\partial^2\phi}{\partial\varphi^2}$ $+ \frac{\partial^2\phi}{\partial z^2}$	$\nabla^2\phi = \frac{1}{r^2}\frac{\partial}{\partial r}\left(r^2\frac{\partial\phi}{\partial r}\right) + \frac{1}{r^2\sin\theta}\frac{\partial}{\partial\theta}\left(\sin\theta\frac{\partial\phi}{\partial\theta}\right)$ $+ \frac{1}{r^2\sin^2\theta}\frac{\partial^2\phi}{\partial\varphi^2}$

s = the distance along some curve "C" in space and is measured from some fixed point.

S = a surface area

V = a volume contained by a specified surface

$\hat{t}$ = the unit tangent to C at the point P

$\hat{n}$ = the unit outward pointing normal

F = some vector function

ds = the vector element of curve ($= \hat{t}\, ds$)

dS = the vector element of surface ($= \hat{n}\, dS$)

Then
$$\int_{(c)} \mathbf{F} \cdot \hat{t}\, ds = \int_{(c)} \mathbf{F} \cdot d\mathbf{s}$$

and when
$$\mathbf{F} = \nabla\phi$$

$$\int_{(c)} (\nabla\phi) \cdot \hat{t}\, ds = \int_{(c)} d\phi$$

Gauss' Theorem (Green's Theorem)

When S defines a closed region having a volume V

$$\iiint_{(v)} (\nabla \cdot \mathbf{F})\, dV = \iint_{(s)} (\mathbf{F} \cdot \hat{n})\, dS = \iint_{(s)} \mathbf{F} \cdot d\mathbf{S}$$

also
$$\iiint_{(v)} (\nabla\phi)\, dV = \iint_{(s)} \phi\hat{n}\, dS$$

and
$$\iiint_{(v)} (\nabla \times \mathbf{F})\, dV = \iint_{(s)} (\hat{n} \times \mathbf{F})\, dS$$

Stokes' Theorem

When C is closed and bounds the open surface S.

$$\iint_{(s)} \hat{n} \cdot (\nabla \times \mathbf{F})\, dS = \int_{(c)} \mathbf{F} \cdot d\mathbf{s}$$

also
$$\iint_{(s)} (\hat{n} \times \nabla\phi)\, dS = \int_{(c)} \phi\, d\mathbf{s}$$

Green's Theorem

$$\iint_{(s)} (\nabla\phi \cdot \nabla\theta)\, dS = \iint_{(s)} \phi\hat{n} \cdot (\nabla\theta)\, dS = \iiint_{(v)} \phi(\nabla^2\theta)\, dV$$

$$= \iint_{(s)} \theta \cdot \hat{n}(\nabla\phi)\, dS = \iiint_{(v)} \theta(\nabla^2\phi)\, dV$$

MOMENT OF INERTIA FOR VARIOUS BODIES OF MASS

The mass of the body is indicated by m

Body	Axis	Moment of inertia	Body	Axis	Moment of inertia
Uniform thin rod	Normal to the length, at one end	$m\frac{l^2}{3}$	Spherical shell, very thin, mean radius, r	Any diameter	$m\frac{2}{3}r^2$
Uniform thin rod	Normal to the length, at the center	$m\frac{l^2}{12}$	Right circular cylinder of radius r, length l	The longitudinal axis of the solid	$m\frac{r^2}{2}$
Thin rectangular sheet, sides a and b	Through the center parallel to b	$m\frac{a^2}{12}$	Right circular cylinder of radius r, length l	Transverse diameter	$m\left(\frac{r^2}{4} + \frac{l^2}{12}\right)$
Thin rectangular sheet, sides a and b	Through the center perpendicular to the sheet	$m\frac{a^2 + b^2}{12}$	Hollow circular cylinder, length l, radii r_1 and r_2	The longitudinal axis of the figure	$m\frac{(r_1^2 + r_2^2)}{2}$
Thin circular sheet of radius r	Normal to the plate through the center	$m\frac{r^2}{2}$	Thin cylindrical shell, length l, mean radius, r	The longitudinal axis of the figure	mr^2
Thin circular sheet of radius r	Along any diameter	$m\frac{r^2}{4}$	Hollow circular cylinder, length l, radii r_1 and r_2	Transverse diameter	$m\left[\frac{r_1^2 + r_2^2}{4} + \frac{l^2}{12}\right]$
Thin circular ring. Radii r_1 and r_2	Through center normal to plane of ring	$m\frac{r_1^2 + r_2^2}{2}$	Hollow circular cylinder, length l, very thin, mean radius	Transverse diameter	$m\left(\frac{r^2}{2} + \frac{l^2}{12}\right)$
Thin circular ring. Radii r_1 and r_2	Any diameter	$m\frac{r_1^2 + r_2^2}{4}$	Elliptic cylinder, length l, transverse semiaxes a and b	Longitudinal axis	$m\left(\frac{a^2 + b^2}{4}\right)$
Rectangular parallelopiped, edges a, b, and c	Through center perpendicular to face ab, (parallel to edge c)	$m\frac{a^2 + b^2}{12}$	Right cone, altitude h, radius of base r	Axis of the figure	$m\frac{3}{10}r^2$
Sphere, radius r	Any diameter	$m\frac{2}{5}r^2$	Spheroid of revolution, equatorial radius r	Polar axis	$m\frac{2r^2}{5}$
Spherical shell, external radius, r_1, internal radius r_2	Any diameter	$m\frac{2}{5}\frac{(r_1^5 - r_2^5)}{(r_1^3 - r_2^3)}$	Ellipsoid, axes $2a$, $2b$, $2c$	Axis $2a$	$m\frac{(b^2 + c^2)}{5}$

Bessel Functions*

1. Bessel's differential equation for a real variable x is

$$x^2 \frac{d^2y}{dx^2} + x \frac{dy}{dx} + (x^2 - n^2)y = 0$$

* From Beyer, W. H., Ed., *CRC Handbook of Mathematical Sciences*, 5th ed., CRC Press, Boca Raton, 1978, 500—503. With permission.

2. When n is not an integer, two independent solutions of the equation are $J_n(x)$, $J_{-n}(x)$, where

$$J_n(x) = \sum_{k=0}^{\infty} \frac{(-1)^k}{k!\,\Gamma(n+k+1)} \left(\frac{x}{2}\right)^{n+2k}$$

3. If n is an integer $J_{-n}(x) = (-1)^n J_n(x)$, where

$$J_n(x) = \frac{x^n}{2^n n!} \left\{ 1 - \frac{x^2}{2^2 \cdot 1!(n+1)} + \frac{x^4}{2^4 \cdot 2!(n+1)(n+2)} \right.$$
$$\left. - \frac{x^6}{2^6 \cdot 3!(n+1)(n+2)(n+3)} + \cdots \right\}$$

4. For $n = 0$ and $n = 1$, this formula becomes

$$J_0(x) = 1 - \frac{x^2}{2^2(1!)^2} + \frac{x^4}{2^4(2!)^2} - \frac{x^6}{2^6(3!)^2} + \frac{x^8}{2^8(4!)^2} - \cdots$$

$$J_1(x) = \frac{x}{2} - \frac{x^3}{2^3 \cdot 1!2!} + \frac{x^5}{2^5 \cdot 2!3!} - \frac{x^7}{2^7 \cdot 3!4!} + \frac{x^9}{2^9 \cdot 4!5!} - \cdots$$

5. When x is large and positive, the following asymptotic series may be used

$$J_0(x) = \left(\frac{2}{\pi x}\right)^{\frac{1}{2}} \left\{ P_0(x) \cos\left(x - \frac{\pi}{4}\right) - Q_0(x) \sin\left(x - \frac{\pi}{4}\right) \right\}$$

$$J_1(x) = \left(\frac{2}{\pi x}\right)^{\frac{1}{2}} \left\{ P_1(x) \cos\left(x - \frac{3\pi}{4}\right) - Q_1(x) \sin\left(x - \frac{3\pi}{4}\right) \right\}$$

where

$$P_0(x) \sim 1 - \frac{1^2 \cdot 3^2}{2!(8x)^2} + \frac{1^2 \cdot 3^2 \cdot 5^2 \cdot 7^2}{4!(8x)^4} - \frac{1^2 \cdot 3^2 \cdot 5^2 \cdot 7^2 \cdot 9^2 \cdot 11^2}{6!(8x)^6} + \cdots$$

$$Q_0(x) \sim - \frac{1^2}{1!8x} + \frac{1^2 \cdot 3^2 \cdot 5^2}{3!(8x)^3} - \frac{1^2 \cdot 3^2 \cdot 5^2 \cdot 7^2 \cdot 9^2}{5!(8x)^5} + - \cdots$$

$$P_1(x) \sim 1 + \frac{1^2 \cdot 3 \cdot 5}{2!(8x)^2} - \frac{1^2 \cdot 3^2 \cdot 5^2 \cdot 7 \cdot 9}{4!(8x)^4} + \frac{1^2 \cdot 3^2 \cdot 5^2 \cdot 7^2 \cdot 9^2 \cdot 11 \cdot 13}{6!(8x)^6} - + \cdots$$

$$Q_1(x) \sim \frac{1 \cdot 3}{1!8x} - \frac{1^2 \cdot 3^2 \cdot 5 \cdot 7}{3!(8x)^3} + \frac{1^2 \cdot 3^2 \cdot 5^2 \cdot 7^2 \cdot 9 \cdot 11}{5!(8x)^5} - \cdots$$

[In $P_1(x)$ the signs alternate from $+$ to $-$ after the first term]

6. If $x > 25$, it is convenient to use the formulas

$$J_0(x) = A_0(x) \sin x + B_0(x) \cos x$$
$$J_1(x) = B_1(x) \sin x - A_1(x) \cos x$$

where

$$A_0(x) = \frac{P_0(x) - Q_0(x)}{(\pi x)^{\frac{1}{2}}} \quad \text{and} \quad A_1(x) = \frac{P_1(x) - Q_1(x)}{(\pi x)^{\frac{1}{2}}}$$

$$B_0(x) = \frac{P_0(x) + Q_0(x)}{(\pi x)^{\frac{1}{2}}} \quad \text{and} \quad B_1(x) = \frac{P_1(x) + Q_1(x)}{(\pi x)^{\frac{1}{2}}}$$

7. The zeros of $J_0(x)$ and $J_1(x)$

If j_{0s} and j_{1s} are the sth zeros of $J_0(x)$ and $J_1(x)$ respectively, and if $a = 4_s - 1$, $b = 4_s + 1$

$$j_{0,s} \sim \frac{1}{4}\pi a\left\{ 1 + \frac{2}{\pi^2 a^2} - \frac{62}{3\pi^4 a^4} + \frac{15,116}{15\pi^6 a^6} - \frac{12,554,474}{105\pi^8 a^8} + \frac{8,368,654,292}{315\pi^{10}a^{10}} - + \cdots \right\}$$

$$j_{1,s} \sim \frac{1}{4}\pi b\left\{ 1 - \frac{6}{\pi^2 b^2} + \frac{6}{\pi^4 b^4} - \frac{4716}{5\pi^6 b^6} + \frac{3,902,418}{35\pi^8 b^8} - \frac{895,167,324}{35\pi^{10}b^{10}} + \cdots \right\}$$

$$J_1(j_{0,s}) \sim \frac{(-1)^{s+1}2^{\frac{1}{2}}}{\pi a^{\frac{1}{2}}}\left\{ 1 - \frac{56}{3\pi^4 a^4} + \frac{9664}{5\pi^6 a^6} - \frac{7,381,280}{21\pi^8 a^8} + \cdots \right\}$$

$$J_0(j_{1,s}) \sim \frac{(-1)^s 2^{\frac{1}{2}}}{\pi b^{\frac{1}{2}}}\left\{ 1 + \frac{24}{\pi^4 b^4} - \frac{19,584}{10\pi^6 b^6} + \frac{2,466,720}{7\pi^8 b^8} - \cdots \right\}$$

8. Table of zeros for $J_0(x)$ and $J_1(x)$

$$J_1(\alpha_n) = 0 \qquad J_0(\beta_n) = 0$$

Roots α_n	$J_1(\alpha_n)$	Roots β_n	$J_0(\beta_n)$
2.4048	0.5191	0.0000	1.0000
5.5201	−0.3403	3.8317	−0.4028
8.6537	0.2715	7.0156	0.3001
11.7915	−0.2325	10.1735	−0.2497
14.9309	0.2065	13.3237	0.2184
18.0711	−0.1877	16.4706	−0.1965
21.2116	0.1733	19.6159	0.1801

9. Recurrence formulas

$$J_{n-1}(x) + J_{n+1}(x) = \frac{2n}{x} J_n(x) \qquad\qquad nJ_n(x) + xJ_n'(x) = xJ_{n-1}(x)$$

$$J_{n-1}(x) - J_{n+1}(x) = 2J_n'(x) \qquad\qquad nJ_n(x) - xJ_n'(x) = xJ_{n+1}(x)$$

10. If J_n is written for $J_n(x)$ and $J_n^{(k)}$ is written for $\dfrac{d^k}{dx^k}\{J_n(x)\}$, then the following derivative relationships are important

$$J_0^{(r)} = -J_1^{(r-1)}$$

$$J_0^{(2)} = -J_0 + \frac{1}{x} J_1 = \frac{1}{2}(J_2 - J_0)$$

$$J_0^{(3)} = \frac{1}{x} J_0 + \left(1 - \frac{2}{x^2}\right)J_1 = \frac{1}{4}(-J_3 + 3J_1)$$

$$J_0^{(4)} = \left(1 - \frac{3}{x^2}\right)J_0 - \left(\frac{2}{x} - \frac{6}{x^3}\right)J_1 = \frac{1}{8}(J_4 - 4J_2 + 3J_0), \text{ etc.}$$

11. Half order Bessel functions

$$J_{\frac{1}{2}}(x) = \sqrt{\frac{2}{\pi x}} \sin x$$

$$J_{-\frac{1}{2}}(x) = \sqrt{\frac{2}{\pi x}} \cos x$$

$$J_{n+\frac{1}{2}}(x) = -x^{n+\frac{1}{2}} \frac{d}{dx}\{x^{-(n+\frac{1}{2})} J_{n+\frac{1}{2}}(x)\}$$

$$J_{n-\frac{1}{2}}(x) = x^{-(n+\frac{1}{2})} \frac{d}{dx}\{x^{n+\frac{1}{2}} J_{n+\frac{1}{2}}(x)\}$$

n	$\left(\dfrac{\pi x}{2}\right)^{\frac{1}{2}} J_{n+\frac{1}{2}}(x)$	$\left(\dfrac{\pi x}{2}\right)^{\frac{1}{2}} J_{-(n+\frac{1}{2})}(x)$
0	$\sin x$	$\cos x$
1	$\dfrac{\sin x}{x} - \cos x$	$-\dfrac{\cos x}{x} - \sin x$
2	$\left(\dfrac{3}{x^2} - 1\right)\sin x - \dfrac{3}{x}\cos x$	$\left(\dfrac{3}{x^2} - 1\right)\cos x + \dfrac{3}{x}\sin x$
3	$\left(\dfrac{15}{x^3} - \dfrac{6}{x}\right)\sin x - \left(\dfrac{15}{x^2} - 1\right)\cos x$	$-\left(\dfrac{15}{x^3} - \dfrac{6}{x}\right)\cos x - \left(\dfrac{15}{x^2} - 1\right)\sin x$

etc.

12. Additional solutions to Bessel's equation are

$Y_n(x)$ (also called Weber's function, and sometimes denoted by $N_n(x)$)

$H_n^{(1)}(x)$ and $H_n^{(2)}(x)$ (also called Hankel functions)

These solutions are defined as follows

$$Y_n(x) = \begin{cases} \dfrac{J_n(x)\cos(n\pi) - J_{-n}(x)}{\sin(n\pi)} & n \text{ not an integer} \\[2ex] \lim_{v \to n} \dfrac{J_v(x)\cos(v\pi) - J_{-v}(x)}{\sin(v\pi)} & n \text{ an integer} \end{cases}$$

$$H_n^{(1)}(x) = J_n(x) + iY_n(x)$$
$$H_n^{(2)}(x) = J_n(x) - iY_n(x)$$

The additional properties of these functions may all be derived from the above relations and the known properties of $J_n(x)$.

13. Complete solutions to Bessel's equation may be written as

$$c_1 J_n(x) + c_2 J_{-n}(x) \qquad \text{if } n \text{ is not an integer}$$

or

$$\left. \begin{array}{l} c_1 J_n(x) + c_2 Y_n(x) \\[2mm] c_1 H_n^{(1)}(x) + c_2 H_n^{(2)}(x) \end{array} \right\} \text{for any value of } n$$

or

14. The modified (or hyperbolic) Bessel's differential equation is

$$x^2 \frac{d^2 y}{dx^2} + x \frac{dy}{dx} - (x^2 + n^2)y = 0$$

15. When n is not an integer, two independent solutions of the equation are $I_n(x)$ and $I_{-n}(x)$, where

$$I_n(x) = \sum_{k=0}^{\infty} \frac{1}{k! \, \Gamma(n + k + 1)} \left(\frac{x}{2}\right)^{n+2k}$$

16. If n is an integer,

$$I_n(x) = I_{-n}(x) = \frac{x^n}{2^n n!} \left\{ 1 + \frac{x^2}{2^2 \cdot 1!(n+1)} + \frac{x^4}{2^4 \cdot 2!(n+1)(n+2)} \right.$$
$$\left. + \frac{x^6}{2^6 \cdot 3!(n+1)(n+2)(n+3)} + \cdots \right\}$$

17. For $N = 0$ and $n = 1$, this formula becomes

$$I_0(x) = 1 + \frac{x^2}{2^2(1!)^2} + \frac{x^4}{2^4(2!)^2} + \frac{x^6}{2^6(3!)^2} + \frac{x^8}{2^8(4!)^2} + \cdots$$

$$I_1(x) = \frac{x}{2} + \frac{x^3}{2^3 \cdot 1!2!} + \frac{x^5}{2^5 \cdot 2!3!} + \frac{x^7}{2^7 \cdot 3!4!} + \frac{x^9}{2^9 \cdot 4!5!} + \cdots$$

18. Another solution to the modified Bessel's equation is

$$K_n(x) = \begin{cases} \dfrac{1}{2} \pi \dfrac{I_{-n}(x) - I_n(x)}{\sin(n\pi)} & n \text{ not an integer} \\[4mm] \lim\limits_{v \to n} \dfrac{1}{2} \pi \dfrac{I_{-v}(x) - I_v(x)}{\sin(v\pi)} & n \text{ an integer} \end{cases}$$

This function is linearly independent of $I_n(x)$ for all values of n. Thus the complete solution to the modified Bessel's equation may be written as

$$c_1 I_n(x) + c_2 I_{-n}(x) \qquad n \text{ not an integer}$$

or

$$c_1 I_n(x) + c_2 K_n(x) \qquad \text{any } n$$

19. The following relations hold among the various Bessel functions:

$$I_n(z) = i^{-n} J_m(iz)$$

$$Y_n(iz) = (i)^{n+1} I_n(z) - \frac{2}{\pi} i^{-n} K_n(z)$$

Most of the properties of the modified Bessel function may be deduced from the known properties of $J_n(x)$ by use of these relations and those previously given.

20. Recurrence formulas

$$I_{n-1}(x) - I_{n+1}(x) = \frac{2n}{x} I_n(x) \qquad I_{n-1}(x) + I_{n+1}(x) = 2I_n'(x)$$

$$I_{n-1}(x) - \frac{n}{x} I_n(x) = I_n'(x) \qquad I_n'(x) = I_{n+1}(x) + \frac{n}{x} I_n(z)$$

The Gamma Function*

Definition: $\Gamma(n = \int_0^{\infty} t^{n-1} e^{-t} \, dt \; n > 0$

Recursion Formula: $\Gamma(n + 1) = n\Gamma(n)$
$\Gamma(n) + 1) = n!$ if $n = 0, 1, 2, \ldots$ where $0! = 1$
For $n < 0$ the gamma function can be defined by using

---

* From Beyer, W. H., Ed., *CRC Handbook of Mathematical Sciences*, 5th ed., CRC Press, Boca Raton, 1978, 484—485. With permission.

$$\Gamma(n) = \frac{\Gamma(n+1)}{n}$$

Graph:

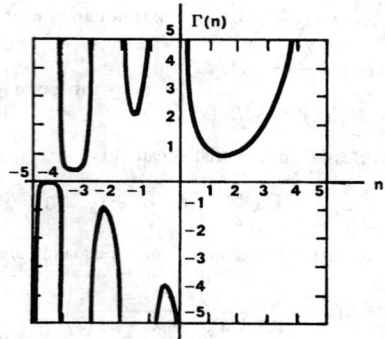

Special Values:

$$\Gamma(\tfrac{1}{2}) = \sqrt{\pi}$$

$$\Gamma(m + \tfrac{1}{2}) = \frac{1 \cdot 3 \cdot 5 \cdots (2m-1)}{2^m} \sqrt{\pi} \qquad m = 1,2,3,\ldots$$

$$\Gamma(-m + \tfrac{1}{2}) = \frac{(-1)^m 2^m \sqrt{\pi}}{1 \cdot 3 \cdot 5 \cdots (2m-1)} \qquad m = 1,2,3,\ldots$$

Definition:

$$\Gamma(x+1) = \lim_{k \to \infty} \frac{1 \cdot 2 \cdot 3 \cdots k}{(x+1)(x+2)\cdots(x+k)} k^x$$

$$\frac{1}{\Gamma(x)} = x e^{\gamma x} \prod_{m=1}^{\infty} \left\{ \left(1 + \frac{x}{m}\right) e^{-x/m} \right\}$$

This is an infinite product representation for the gamma function where γ is Euler's constant.

Properties:

$$\Gamma'(1) = \int_0^\infty e^{-x} \ln x \, dx = -\gamma$$

$$\frac{\Gamma'(x)}{\Gamma(x)} = -\gamma + \left(\frac{1}{1} - \frac{1}{x}\right) + \left(\frac{1}{2} - \frac{1}{x+1}\right) + \ldots + \left(\frac{1}{n} - \frac{1}{x+n-1}\right) + \ldots$$

$$\Gamma(x+1) = \sqrt{2\pi x}\, x^x e^{-x} \left\{ 1 + \frac{1}{12x} + \frac{1}{288x^2} - \frac{139}{51{,}840 x^3} + \ldots \right\}$$

This is called *Stirling's asymptotic series.*

If we let $x = n$ a positive integer, then a useful approximation for $n!$ where n is large (e.g., $n > 10$) is given by *Stirling's formula*

$$n! \approx \sqrt{2\pi n}\, n^n e^{-n}$$

The Gamma Function*

Values of $\Gamma(n) = \int_0^\infty e^{-x} x^{n-1}\, dx$; $\Gamma(n+1) = n\Gamma(n)$

n	$\Gamma(n)$	n	$\Gamma(n)$	n	$\Gamma(n)$	n	$\Gamma(n)$
1.00	1.00000	1.25	.90640	1.50	.88623	1.75	.91906
1.01	.99433	1.26	.90440	1.51	.88659	1.76	.92137
1.02	.98884	1.27	.90250	1.52	.88704	1.77	.92376
1.03	.98355	1.28	.90072	1.53	.88757	1.78	.92623
1.04	.97844	1.29	.89904	1.54	.88818	1.79	.92877
1.05	.97350	1.30	.89747	1.55	.88887	1.80	.93138
1.06	.96874	1.31	.89600	1.56	.88964	1.81	.93408
1.07	.96415	1.32	.89464	1.57	.89049	1.82	.93685
1.08	.95973	1.33	.89338	1.58	.89142	1.83	.93969
1.09	.95546	1.34	.89222	1.59	.89243	1.84	.94261
1.10	.95135	1.35	.89115	1.60	.89352	1.85	.94561
1.11	.94740	1.36	.89018	1.61	.89468	1.86	.94869
1.12	.94359	1.37	.88931	1.62	.89592	1.87	.95184
1.13	.93993	1.38	.88854	1.63	.89724	1.88	.95507
1.14	.93642	1.39	.88785	1.64	.89864	1.89	.95838
1.15	.93304	1.40	.88726	1.65	.90012	1.90	.96177
1.16	.92980	1.41	.88676	1.66	.90167	1.91	.96523
1.17	.92670	1.42	.88636	1.67	.90330	1.92	.96877
1.18	.92373	1.43	.88604	1.68	.90500	1.93	.97240
1.19	.92089	1.44	.88581	1.69	.90678	1.94	.97610
1.20	.91817	1.45	.88566	1.70	.90864	1.95	.97988
1.21	.91558	1.46	.88560	1.71	.91057	1.96	.98374
1.22	.91311	1.47	.88563	1.72	.91258	1.97	.98768
1.23	.91075	1.48	.88575	1.73	.91466	1.98	.99171
1.24	.90852	1.49	.88595	1.74	.91683	1.99	.99581
						2.00	1.00000

* For large positive values of x, $\Gamma(x)$ approximates Stirling's asymptotic series

$$x^x e^{-x} \sqrt{\frac{2\pi}{x}} \left[1 + \frac{1}{12x} + \frac{1}{288x^2} - \frac{139}{51840x^3} - \frac{571}{2488320x^4} + \cdots \right]$$

The Beta Function*

Definition: $B(m,n) = \int_0^1 t^{m-1} (1-t)^{m-1}\, dt \quad m > 0, n > 0$

Relationship with
Gamma Function: $B(m,n) = \dfrac{\Gamma(m)\Gamma(n)}{\Gamma(m+n)}$

Properties:

$$B(m,n) = B(n,m)$$

$$B(m,n) = 2\int_0^{\pi/2} \sin^{2m-1}\theta \, \cos^{2n-1}\theta \, d\theta$$

$$B(m,n) = \int_0^\infty \frac{t^{m-1}}{(1+t)^{m+n}}\, dt$$

$$B(m,n) = r^n (r+1)^m \int_0^1 \frac{t^{m-1}(1-t)^{n-1}}{(r+t)^{m+n}}\, dt$$

The Error Function

Definition: $erf\,x = \dfrac{2}{\sqrt{\pi}} \int_0^x e^{-t^2}\, dt$

Series: $erf\,x = \dfrac{2}{\sqrt{\pi}} \left(x - \dfrac{x^3}{3} + \dfrac{1}{2!}\dfrac{x^5}{5} - \dfrac{1}{3!}\dfrac{x^7}{7} + \dots \right)$

Property: $erf\,x = -erf\,(-x)$

Relationship with Normal Probability Function f(t): $\displaystyle\int_0^x f(t)\, dt = \dfrac{1}{2}\, erf\left(\dfrac{x}{\sqrt{2}}\right)$

To evaluate $erf\,(2.3)$, one proceeds as follows: Since $\dfrac{x}{\sqrt{2}} = 2.3$, one finds $x = (2.3)(\sqrt{2}) = 3.25$. In the normal probability function table (page A-), one finds the entry 0.4994 opposite the value 3.25. Thus $erf\,(2.3) = 2(0.4994) = 0.9988$.

* From Beyer, W. H., Ed., *CRC Handbook of Mathematical Sciences,* 5th ed., CRC Press, Boca Raton, 1978, 499. With permission.

$$erfc\ z = 1 - erf\ z = \frac{2}{\sqrt{\pi}} \int_z^\infty e^{-t^2}\ dt$$

is known as the complementary error function.

Orthogonal Polynomials*

I

Name: Legendre   *Symbol:* $P_n(x)$ *Interval:* $[-1,1]$

Differential Equation: $(1 - x^2)\,y'' - 2\,xy' + n(n + 1)\,y = 0$
$$y = P_n(x)$$

Explicit Expression: $P_n(x) = \dfrac{1}{2^n} \displaystyle\sum_{m=0}^{[n/2]} (-1)^m \binom{n}{m}\binom{2n - 2m}{n} x^{n-2m}$

Recurrence Relation: $(n + 1)\,P_{n+1}(x) = (2n + 1)\,x\,P_n(x) - nP_{n-1}(x)$

Weight: 1   *Standardization:* $P_n(1) = 1$

Norm: $\displaystyle\int_{-1}^{+1} [P_n(x)]^2\,dx = \dfrac{2}{2n + 1}$

Rodrigues' Formula: $P_n(x) = \dfrac{(-1)^n}{2^n n!} \dfrac{d^n}{dx^n} \{(1 - x^2)^n\}$

Generating Function: $R^{-1} = \displaystyle\sum_{n=0}^{\infty} P_n(x)\,z^n: \ -1 < x < 1,\ |z| < 1,$
$$R = \sqrt{1 - 2xz + z^2}$$

Inequality: $|P_n(x)| \leq 1,\ -1 \leq x \leq 1.$

II

Name: Tschebysheff, First Kind   *Symbol:* $T_n(x)$ *Interval:* $[-1,1]$

Differential Equation: $(1 - x^2)y - xy' + n^2 y = 0$
$$y = T_n(x)$$

Explicit Expression: $\dfrac{n}{2} \displaystyle\sum_{m=0}^{[n/2]} (-1)^m \dfrac{(n - m - 1)!}{m!(n - 2m)!} (2x)^{n-2m} = \cos(n \arccos x) = T_n(x)$

Recurrence Relation: $T_{n+1}(x) = 2xT_n(x) - T_{n-1}(x)$

Weight: $(1 - x^2)^{-1/2}$ *Standardization:* $T_n(1) = 1$

Norm: $\displaystyle\int_{-1}^{+1} (1 - x^2)^{-1/2}[T_n(x)]^2\,dx = \begin{cases} \pi/2, & n \neq 0 \\ \pi, & n = 0 \end{cases}$

Rodrigues' Formula: $\dfrac{(-1)^n(1 - x^2)^{1/2}\sqrt{\pi}}{2^{n+1}\Gamma(n + \frac{1}{2})} \dfrac{d^n}{dx^n} \{(1 - x^2)^{n-(1/2)}\} = T_n(x)$

Generating Function: $\dfrac{1 - xz}{1 - 2xz + z^2} = \displaystyle\sum_{n=0}^{\infty} T_n(x)\,z^n,\ -1 < x < 1,\ |z| < 1$

Inequality: $|T_n(x)| \leq 1,\ -1 \leq x \leq 1.$

III

Name: Tschebysheff, Second Kind   *Symbol:* $U_n(x)$ *Interval:* $[-1,1]$

Differential Equation: $(1 - x^2)\,y'' - 3\,xy' + n(n + 2)y = 0$
$$y = U_n(x)$$

Explicit Expression: $U_n(x) = \displaystyle\sum_{m=0}^{[n/2]} (-1)^m \dfrac{(m - n)!}{m!(n - 2m)!} (2x)^{n-2m}$

$$U_n(\cos\theta) = \dfrac{\sin[(n + 1)\theta]}{\sin\theta}$$

* From Beyer, W. H., Ed., *CRC Handbook of Mathematical Sciences*, 5th ed., CRC Press, Boca Raton, 1978, 557—560. With permission.

Recurrence Relation: $U_{n+1}(x) = 2xU_n(x) - U_{n-1}(x)$

Weight: $(1 - x^2)^{1/2}$ Standardization: $U_n(1) = n + 1$

Norm: $\displaystyle\int_{-1}^{+1} (1 - x^2)^{1/2}[U_n(x)]^2\, dx = \frac{\pi}{2}$

Rodrigues' Formula: $\displaystyle U_n(x) = \frac{(-1)^n(n + 1)\sqrt{\pi}}{(1 - x^2)^{1/2}2^{n+1}\Gamma(n + \frac{3}{2})}\frac{d^n}{dx^n}\{(1 - x^2)^{n+(1/2)}\}$

Generating Function: $\displaystyle\frac{1}{1 - 2xz + z^2} = \sum_{n=0}^{\infty} U_n(x)z^n,\ -1 < x < 1,\ |z| < 1$

Inequality: $|U_n(x)| \leqslant n + 1,\ -1 \leqslant x \leqslant 1.$

IV

Name: Jacobi   Symbol: $P_n^{(\alpha,\beta)}(x)$ Interval: $[-1,1]$

Differential Equation:

$$(1 - x^2)y'' + [\beta - \alpha - (\alpha + \beta + 2)x]y' + n(n + \alpha + \beta + 1)y = 0$$
$$y = P_n^{(\alpha,\beta)}(x)$$

Explicit Expression: $\displaystyle P_n^{(\alpha,\beta)}(x) = \frac{1}{2^n}\sum_{m=0}^{n}\binom{n + \alpha}{m}\binom{n + \beta}{n - m}(x - 1)^{n-m}(x + 1)^m$

Recurrence Relation: $2(n + 1)(n + \alpha + \beta + 1)(2n + \alpha + \beta)P_{n+1}^{(\alpha,\beta)}(x)$

$$= (2n + \alpha + \beta + 1)[(\alpha^2 - \beta^2) + (2n + \alpha + \beta + 2)$$
$$\times\, (2n + \alpha + \beta)x]\,P_n^{(\alpha,\beta)}(x)$$
$$- 2(n + \alpha)(n + \beta)(2n + \alpha + \beta + 2)P_{n-1}^{(\alpha,\beta)}(x)$$

Weight: $(1 - x)^\alpha(1 + x)^\beta;\ \alpha,\beta > 1$ Standardization: $P_n^{(\alpha,\beta)}(x) = \dbinom{n + \alpha}{n}$

Norm: $\displaystyle\int_{-1}^{+1}(1 - x)^\alpha(1 + x)^\beta[P_n^{(\alpha,\beta)}(x)]^2\, dx = \frac{2^{\alpha+\beta+1}\Gamma(n + \alpha + 1)\Gamma(n + \beta + 1)}{(2n + \alpha + \beta + 1)n!\,\Gamma(n + \alpha + \beta + 1)}$

Rodrigues' Formula: $\displaystyle P_n^{(\alpha,\beta)}(x) = \frac{(-1)^n}{2^n n!(1 - x)^\alpha(1 + x)^\beta}\frac{d^n}{dx^n}\{(1 - x)^{n+\alpha}(1 + x)^{n+\beta}\}$

Generating Function: $\displaystyle R^{-1}(1 - z + R)^{-\alpha}(1 + z + R)^{-\beta} = \sum_{n=0}^{\infty} 2^{-\alpha-\beta}P_n^{(\alpha,\beta)}(x)z^n,$

$$R = \sqrt{1 - 2xz + z^2},\ |z| < 1$$

Inequality: $\displaystyle\max_{-1\leqslant x\leqslant 1}|P_n^{(\alpha,\beta)}(x)| = \begin{cases} \dbinom{n + q}{n} \sim n^q\ \text{if}\ q = \max(\alpha,\beta) \geqslant -\frac{1}{2} \\[2mm] |P_n^{(\alpha,\beta)}(x')| \sim n^{-1/2}\ \text{if}\ q < -\frac{1}{2} \\ x'\ \text{is one of the two maximum points nearest} \\[2mm] \dfrac{\beta - \alpha}{\alpha + \beta + 1} \end{cases}$

V

Name: Generalized Laguerre   Symbol: $L_n^{(\alpha)}(x)$ Interval: $[0, \infty]$

Differential Equation: $xy'' + (\alpha + 1 - x)y' + ny = 0$
$$y = L_n^{(\alpha)}(x)$$

Explicit Expression: $\displaystyle L_n^{(\alpha)}(x) = \sum_{m=0}^{n}(-1)^m\binom{n + \alpha}{n - m}\frac{1}{m!}x^m$

Recurrence Relation: $(n + 1)L_{n+1}^{(\alpha)}(x) = [(2n + \alpha + 1) - x]L_n^{(\alpha)}(x) - (n + \alpha)L_{n-1}^{(\alpha)}(x)$

Weight: $x^\alpha e^{-x},\ \alpha > -1$ Standardization: $\displaystyle L_n^{(\alpha)}(x) = \frac{(-1)^n}{n!}x^n + \cdots$

Norm: $\displaystyle\int_0^\infty x^\alpha e^{-x}[L_n^{(\alpha)}(x)]^2\, dx = \frac{\Gamma(n + \alpha + 1)}{n!}$

Rodrigues' Formula: $\displaystyle L_n^{(\alpha)}(x) = \frac{1}{n!\,x^\alpha e^{-x}}\frac{d^n}{dx^n}\{x^{n+\alpha}e^{-x}\}$

Generating Function: $(1 - z)^{-\alpha-1} \exp\left(\dfrac{xz}{z-1}\right) = \displaystyle\sum_{n=0}^{\infty} L_n^{(\alpha)}(x)\, z^n$

Inequality: $|L_n^{(\alpha)}(x)| \leq \dfrac{\Gamma(n+\alpha+1)}{n!\,\Gamma(\alpha+1)}\, e^{x/2}; \quad \begin{array}{l} x \geq 0 \\ \alpha > 0 \end{array}$

$|L_n^{(\alpha)}(x)| \leq \left[2 - \dfrac{\Gamma(\alpha+n+1)}{n!\,\Gamma(\alpha+1)}\right] e^{x/2}; \quad \begin{array}{l} x \geq 0 \\ -1 < \alpha < 0 \end{array}$

Orthogonal Polynomials

Name: Hermite Symbol: $H_n(x)$ Interval: $[-\infty, \infty]$
Differential Equation: $y'' - 2xy' + 2ny = 0$

Explicit Expression: $H_n(x) = \displaystyle\sum_{m=0}^{[n/2]} \dfrac{(-1)^m\, n!\, (2x)^{n-2m}}{m!\,(n-2m)!}$

Recurrence Relation: $H_{n+1}(x) = 2x\, H_n(x) - 2n\, H_{n-1}(x)$
Weight: e^{-x^2} Standardization: $H_n(1) = 2^n x^n + \ldots$

Norm: $\displaystyle\int_{-\infty}^{\infty} e^{-x^2}\, [H_n(x)]^2\, dx = 2^n\, n!\, \sqrt{\pi}$

Rodriques' Formula: $H_n(x) = (-1)^n\, e^{x^2}\, \dfrac{d^n}{dx^n}\, (e^{-x^2})$

Generating Function: $e^{-z^2 + 2zx} = \displaystyle\sum_{n=0}^{\infty} H_n(x)\, \dfrac{z^n}{n!}$

Inequality: $|H_n(x)| < e^{\frac{x^2}{2}}\, k\, 2^{n/2}\, \sqrt{n!}\quad k \approx 1.086435$

NORMAL PROBABILITY FUNCTION

Areas under the Standard Normal Curve from 0 to z

z	0	1	2	3	4	5	6	7	8	9
0.0	.0000	.0040	.0080	.0120	.0160	.0199	.0239	.0279	.0319	.0359
0.1	.0398	.0438	.0478	.0517	.0557	.0596	.0636	.0675	.0714	.0754
0.2	.0793	.0832	.0871	.0910	.0948	.0987	.1026	.1064	.1103	.1141
0.3	.1179	.1217	.1255	.1293	.1331	.1368	.1406	.1443	.1480	.1517
0.4	.1554	.1591	.1628	.1664	.1700	.1736	.1772	.1808	.1844	.1879
0.5	.1915	.1950	.1985	.2019	.2054	.2088	.2123	.2157	.2190	.2224
0.6	.2258	.2291	.2324	.2357	.2389	.2422	.2454	.2486	.2518	.2549
0.7	.2580	.2612	.2652	.2673	.2704	.2734	.2764	.2794	.2823	.2852
0.8	.2881	.2910	.2939	.2967	.2996	.3023	.3051	.3078	.3106	.3133
0.9	.3159	.3186	.3212	.3238	.3264	.3289	.3315	.3340	.3365	.3389
1.0	.3413	.3438	.3461	.3485	.3508	.3531	.3554	.3577	.3599	.3621
1.1	.3643	.3665	.3686	.3708	.3729	.3749	.3770	.3790	.3810	.3830
1.2	.3849	.3869	.3888	.3907	.3925	.3944	.3962	.3980	.3997	.4015
1.3	.4032	.4049	.4066	.4082	.4099	.4115	.4131	.4147	.4162	.4177
1.4	.4192	.4207	.4222	.4236	.4251	.4265	.4279	.4292	.4306	.4319
1.5	.4332	.4345	.4357	.4370	.4382	.4394	.4406	.4418	.4429	.4441
1.6	.4452	.4463	.4474	.4484	.4495	.4505	.4515	.4525	.4535	.4545
1.7	.4554	.4564	.4573	.4582	.4591	.4599	.4608	.4616	.4625	.4633
1.8	.4641	.4649	.4656	.4664	.4671	.4678	.4686	.4693	.4699	.4706
1.9	.4713	.4719	.4726	.4732	.4738	.4744	.4750	.4756	.4761	.4767
2.0	.4772	.4778	.4783	.4788	.4793	.4798	.4803	.4808	.4812	.4817
2.1	.4821	.4826	.4830	.4834	.4838	.4842	.4846	.4850	.4854	.4857
2.2	.4861	.4864	.4868	.4871	.4875	.4878	.4881	.4884	.4887	.4890
2.3	.4893	.4896	.4898	.4901	.4904	.4906	.4909	.4911	.4913	.4916
2.4	.4918	.4920	.4922	.4925	.4927	.4929	.4931	.4932	.4934	.4936
2.5	.4938	.4940	.4941	.4943	.4945	.4946	.4948	.4949	.4951	.4952
2.6	.4953	.4955	.4956	.4957	.4959	.4960	.4961	.4962	.4963	.4964
2.7	.4965	.4966	.4967	.4968	.4969	.4970	.4971	.4972	.4973	.4974
2.8	.4974	.4975	.4976	.4977	.4977	.4978	.4979	.4979	.4980	.4981
2.9	.4981	.4982	.4982	.4983	.4984	.4984	.4985	.4985	.4986	.4986
3.0	.4987	.4987	.4987	.4988	.4988	.4989	.4989	.4989	.4990	.4990
3.1	.4990	.4991	.4991	.4991	.4992	.4992	.4992	.4992	.4993	.4993
3.2	.4993	.4993	.4994	.4994	.4994	.4994	.4994	.4995	.4995	.4995
3.3	.4995	.4995	.4995	.4996	.4996	.4996	.4996	.4996	.4996	.4997
3.4	.4997	.4997	.4997	.4997	.4997	.4997	.4997	.4997	.4997	.4998
3.5	.4998	.4998	.4998	.4998	.4998	.4998	.4998	.4998	.4998	.4998
3.6	.4998	.4998	.4999	.4999	.4999	.4999	.4999	.4999	.4999	.4999
3.7	.4999	.4999	.4999	.4999	.4999	.4999	.4999	.4999	.4999	.4999
3.8	.4999	.4999	.4999	.4999	.4999	.4999	.4999	.4999	.4999	.4999
3.9	.5000	.5000	.5000	.5000	.5000	.5000	.5000	.5000	.5000	.5000

F(z) below refers to area under Standard Normal Curve from $-\infty$ to z

z	1.282	1.645	1.960	2.326	2.576	3.090
F(z)	.90	.95	.975	.99	.995	.999
2[1 − F(z)]	.20	.10	.05	.02	.01	.002

This table gives values of t such that

$$F(t) = \int_{-\infty}^{t} \frac{\Gamma\left(\frac{n+1}{2}\right)}{\sqrt{n\pi}\,\Gamma\left(\frac{n}{2}\right)} \left(1 + \frac{x^2}{n}\right)^{-\frac{n+1}{2}} dx$$

for n, the number of degrees of freedom, equal to 1, 2, . . ., 30, 40, 60, 120, ∞; and for $F(t) = 0.60, 0.75, 0.90, 0.95, 0.975, 0.99, 0.995,$ and 0.9995. The t-distribution is symmetrical, so that $F(-t) = 1 - F(t)$

F n	.60	.75	.90	.95	.975	.99	.995	.9995
1	.325	1.000	3.078	6.314	12.706	31.821	63.657	636.619
2	.289	.816	1.886	2.920	4.303	6.965	9.925	31.598
3	.277	.765	1.638	2.353	3.182	4.541	5.841	12.924
4	.271	.741	1.533	2.132	2.776	3.747	4.604	8.610
5	.267	.727	1.476	2.015	2.571	3.365	4.032	6.869
6	.265	.718	1.440	1.943	2.447	3.143	3.707	5.959
7	.263	.711	1.415	1.895	2.365	2.998	3.499	5.408
8	.262	.706	1.397	1.860	2.306	2.896	3.355	5.041
9	.261	.703	1.383	1.833	2.262	2.821	3.250	4.781
10	.260	.700	1.372	1.812	2.228	2.764	3.169	4.587
11	.260	.697	1.363	1.796	2.201	2.718	3.106	4.437
12	.259	.695	1.356	1.782	2.179	2.681	3.055	4.318
13	.259	.694	1.350	1.771	2.160	2.650	3.012	4.221
14	.258	.692	1.345	1.761	2.145	2.624	2.977	4.140
15	.258	.691	1.341	1.753	2.131	2.602	2.947	4.073
16	.258	.690	1.337	1.746	2.120	2.583	2.921	4.015
17	.257	.689	1.333	1.740	2.110	2.567	2.898	3.965
18	.257	.688	1.330	1.734	2.101	2.552	2.878	3.922
19	.257	.688	1.328	1.729	2.093	2.539	2.861	3.883
20	.257	.687	1.325	1.725	2.086	2.528	2.845	3.850
21	.257	.686	1.323	1.721	2.080	2.518	2.831	3.819
22	.256	.686	1.321	1.717	2.074	2.508	2.819	3.792
23	.256	.685	1.319	1.714	2.069	2.500	2.807	3.767
24	.256	.685	1.318	1.711	2.064	2.492	2.797	3.745
25	.256	.684	1.316	1.708	2.060	2.485	2.787	3.725
26	.256	.684	1.315	1.706	2.056	2.479	2.779	3.707
27	.256	.684	1.314	1.703	2.052	2.473	2.771	3.690
28	.256	.683	1.313	1.701	2.048	2.467	2.763	3.674
29	.256	.683	1.311	1.699	2.045	2.462	2.756	3.659
30	.256	.683	1.310	1.697	2.042	2.457	2.750	3.646
40	.255	.681	1.303	1.684	2.021	2.423	2.704	3.551
60	.254	.679	1.296	1.671	2.000	2.390	2.660	3.460
120	.254	.677	1.289	1.658	1.980	2.358	2.617	3.373
∞	.253	.674	1.282	1.645	1.960	2.326	2.576	3.291

* This table is abridged from the "Statistical Tables" of R. A. Fisher and Frank Yates published by Oliver & Boyd. Ltd., Edinburgh and London, 1938. It is here published with the kind permission of the authors and their publishers.

PERCENTAGE POINTS, CHI-SQUARE DISTRIBUTION

This table gives values of χ^2 such that

$$F(\chi^2) = \int_{0}^{\chi^2} \frac{1}{2^{\frac{n}{2}}\,\Gamma\left(\frac{n}{2}\right)} x^{\frac{n-2}{2}} e^{-\frac{x}{2}} dx$$

for n, the number of degrees of freedom, equal to 1, 2, . . . , 30. For $n > 30$, a normal approximation is quite accurate. The expression $\sqrt{2\chi^2} - \sqrt{2n-1}$ is approximately normally distributed as the standard normal distribution. Thus χ_α^2, the α-point of the distribution, may be computed by the formula

$$\chi_\alpha^2 = \tfrac{1}{2}[x_\alpha + \sqrt{2n-1}]^2,$$

where x_α is the α-point of the cumulative normal distribution. For even values of n, $F(\chi^2)$ can be written as

$$1 - F(\chi^2) = \sum_{x=0}^{x'-1} \frac{e^{-\lambda}\lambda^x}{x!}$$

with $\lambda = \tfrac{1}{2}\chi^2$ and $x' = \tfrac{1}{2}n$. Thus the cumulative Chi-Square distribution is related to the cumulative Poisson distribution.

$$\chi_a^2 = n\left(1 - \frac{2}{9n} + z_a\sqrt{\frac{2}{9n}}\right)^3$$

n = degrees of freedom

z_a = the normal deviate, (the value of x for which $F(x)$ = the desired percentile).

x	1.282	1.645	1.960	2.326	2.576	3.090
$F(x)$	.90	.95	.975	.99	.995	.999

$\chi_{.99}^2 = 60[1 - 0.00370 + 2.326(0.06086)]^3 = 88.4$ is the 99th percentile for 60 degrees of freedom.

$$F(\chi^2) = \int_0^{\chi^2} \frac{1}{2^{\frac{n}{2}}\Gamma\left(\frac{n}{2}\right)} x^{\frac{n-2}{2}} e^{-\frac{x}{2}} dx$$

n \ F	.005	.010	.025	.050	.100	.250	.500	.750	.900	.950	.975	.990	.995
1	.0000393	.000157	.000982	.00393	.0158	.102	.455	1.32	2.71	3.84	5.02	6.63	7.88
2	.0100	.0201	.0506	.103	.211	.575	1.39	2.77	4.61	5.99	7.38	9.21	10.6
3	.0717	.115	.216	.352	.584	1.21	2.37	4.11	6.25	7.81	9.35	11.3	12.8
4	.207	.297	.484	.711	1.06	1.92	3.36	5.39	7.78	9.49	11.1	13.3	14.9
5	.412	.554	.831	1.15	1.61	2.67	4.35	6.63	9.24	11.1	12.8	15.1	16.7
6	.676	.872	1.24	1.64	2.20	3.45	5.35	7.84	10.6	12.6	14.4	16.8	18.5
7	.989	1.24	1.69	2.17	2.83	4.25	6.35	9.04	12.0	14.1	16.0	18.5	20.3
8	1.34	1.65	2.18	2.73	3.49	5.07	7.34	10.2	13.4	15.5	17.5	20.1	22.0
9	1.73	2.09	2.70	3.33	4.17	5.90	8.34	11.4	14.7	16.9	19.0	21.7	23.6
10	2.16	2.56	3.25	3.94	4.87	6.74	9.34	12.5	16.0	18.3	20.5	23.2	25.2
11	2.60	3.05	3.82	4.57	5.58	7.58	10.3	13.7	17.3	19.7	21.9	24.7	26.8
12	3.07	3.57	4.40	5.23	6.30	8.44	11.3	14.8	18.5	21.0	23.3	26.2	28.3
13	3.57	4.11	5.01	5.89	7.04	9.30	12.3	16.0	19.8	22.4	24.7	27.7	29.8
14	4.07	4.66	5.63	6.57	7.79	10.2	13.3	17.1	21.1	23.7	26.1	29.1	31.3
15	4.60	5.23	6.26	7.26	8.55	11.0	.4.3	18.2	22.3	25.0	27.5	30.6	32.8
16	5.14	5.81	6.91	7.96	9.31	11.9	15.3	19.4	23.5	26.3	28.8	32.0	34.3
17	5.70	6.41	7.56	8.67	10.1	12.8	16.3	20.5	24.8	27.6	30.2	33.4	35.7
18	6.26	7.01	8.23	9.39	10.9	13.7	17.3	21.6	26.0	28.9	31.5	34.8	37.2
19	6.84	7.63	8.91	10.1	11.7	14.6	18.3	22.7	27.2	30.1	32.9	36.2	38.6
20	7.43	8.26	9.59	10.9	12.4	15.5	19.3	23.8	28.4	31.4	34.2	37.6	40.0
21	8.03	8.90	10.3	11.6	13.2	16.3	20.3	24.9	29.6	32.7	35.5	38.9	41.4
22	8.64	9.54	11.0	12.3	14.0	17.2	21.3	26.0	30.8	33.9	36.8	40.3	42.8
23	9.26	10.2	11.7	13.1	14.8	18.1	22.3	27.1	32.0	35.2	38.1	41.6	44.2
24	9.89	10.9	12.4	13.8	15.7	19.0	23.3	28.2	33.2	36.4	39.4	43.0	45.6
25	10.5	11.5	13.1	14.6	16.5	19.9	24.3	29.3	34.4	37.7	40.6	44.3	46.9
26	11.2	12.2	13.8	15.4	17.3	20.8	25.3	30.4	35.6	38.9	41.9	45.6	48.3
27	11.8	12.9	14.6	16.2	18.1	21.7	26.3	31.5	36.7	40.1	43.2	47.0	49.6
28	12.5	13.6	15.3	16.9	18.9	22.7	27.3	32.6	37.9	41.3	44.5	48.3	51.0
29	13.1	14.3	16.0	17.7	19.8	23.6	28.3	33.7	39.1	42.6	45.7	49.6	52.3
30	13.8	15.0	16.8	18.5	20.6	24.5	29.3	34.8	40.3	43.8	47.0	50.9	53.7

PERCENTAGE POINTS, *F*-DISTRIBUTION

This table gives values of F such that

$$F(F) = \int_0^F \frac{\Gamma\left(\frac{m+n}{2}\right)}{\Gamma\left(\frac{m}{2}\right)\Gamma\left(\frac{n}{2}\right)} m^{\frac{m}{2}} n^{\frac{n}{2}} x^{\frac{m-2}{2}} (n + mx)^{-\frac{m+n}{2}} dx$$

for selected values of m, the number of degrees of freedom of the numerator of F; and for selected values of n, the number of degrees of freedom of the denominator of F. The table also provides values corresponding to $F(F)$ = .10, .05, .025, .01, .005, .001 since $F_{1-\alpha}$ for m and n degrees of freedom is the reciprocal of F_α for n and m degrees of freedom. Thus

$$F_{.05}(4, 7) = \frac{1}{F_{.95}(7, 4)} = \frac{1}{6.09} = .164 .$$

$$F(F) = \int_0^F \frac{\Gamma\left(\frac{m+n}{2}\right)}{\Gamma\left(\frac{m}{2}\right)\Gamma\left(\frac{n}{2}\right)} m^{\frac{m}{2}} n^{\frac{n}{2}} x^{\frac{m}{2}-1} (n+mx)^{-\frac{m+n}{2}} dx = .90$$

n \ m	1	2	3	4	5	6	7	8	9	10	12	15	20	24	30	40	60	120	∞
1	39.86	49.50	53.59	55.83	57.24	58.20	58.91	59.44	59.86	60.19	60.71	61.22	61.74	62.00	62.26	62.53	62.79	63.06	63.33
2	8.53	9.00	9.16	9.24	9.29	9.33	9.35	9.37	9.38	9.39	9.41	9.42	9.44	9.45	9.46	9.47	9.47	9.48	9.49
3	5.54	5.46	5.39	5.34	5.31	5.28	5.27	5.25	5.24	5.23	5.22	5.20	5.18	5.18	5.17	5.16	5.15	5.14	5.13
4	4.54	4.32	4.19	4.11	4.05	4.01	3.98	3.95	3.94	3.92	3.90	3.87	3.84	3.83	3.82	3.80	3.79	3.78	3.76
5	4.06	3.78	3.62	3.52	3.45	3.40	3.37	3.34	3.32	3.30	3.27	3.24	3.21	3.19	3.17	3.16	3.14	3.12	3.10
6	3.78	3.46	3.29	3.18	3.11	3.05	3.01	2.98	2.96	2.94	2.90	2.87	2.84	2.82	2.80	2.78	2.76	2.74	2.72
7	3.59	3.26	3.07	2.96	2.88	2.83	2.78	2.75	2.72	2.70	2.67	2.63	2.59	2.58	2.56	2.54	2.51	2.49	2.47
8	3.46	3.11	2.92	2.81	2.73	2.67	2.62	2.59	2.56	2.54	2.50	2.46	2.42	2.40	2.38	2.36	2.34	2.32	2.29
9	3.36	3.01	2.81	2.69	2.61	2.55	2.51	2.47	2.44	2.42	2.38	2.34	2.30	2.28	2.25	2.23	2.21	2.18	2.16
10	3.29	2.92	2.73	2.61	2.52	2.46	2.41	2.38	2.35	2.32	2.28	2.24	2.20	2.18	2.16	2.13	2.11	2.08	2.06
11	3.23	2.86	2.66	2.54	2.45	2.39	2.34	2.30	2.27	2.25	2.21	2.17	2.12	2.10	2.08	2.05	2.03	2.00	1.97
12	3.18	2.81	2.61	2.48	2.39	2.33	2.28	2.24	2.21	2.19	2.15	2.10	2.06	2.04	2.01	1.99	1.96	1.93	1.90
13	3.14	2.76	2.56	2.43	2.35	2.28	2.23	2.20	2.16	2.14	2.10	2.05	2.01	1.98	1.96	1.93	1.90	1.88	1.85
14	3.10	2.73	2.52	2.39	2.31	2.24	2.19	2.15	2.12	2.10	2.05	2.01	1.96	1.94	1.91	1.89	1.86	1.83	1.80
15	3.07	2.70	2.49	2.36	2.27	2.21	2.16	2.12	2.09	2.06	2.02	1.97	1.92	1.90	1.87	1.85	1.82	1.79	1.76
16	3.05	2.67	2.46	2.33	2.24	2.18	2.13	2.09	2.06	2.03	1.99	1.94	1.89	1.87	1.84	1.81	1.78	1.75	1.72
17	3.03	2.64	2.44	2.31	2.22	2.15	2.10	2.06	2.03	2.00	1.96	1.91	1.86	1.84	1.81	1.78	1.75	1.72	1.69
18	3.01	2.62	2.42	2.29	2.20	2.13	2.08	2.04	2.00	1.98	1.93	1.89	1.84	1.81	1.78	1.75	1.72	1.69	1.66
19	2.99	2.61	2.40	2.27	2.18	2.11	2.06	2.02	1.98	1.96	1.91	1.86	1.81	1.79	1.76	1.73	1.70	1.67	1.63
20	2.97	2.59	2.38	2.25	2.16	2.09	2.04	2.00	1.96	1.94	1.89	1.84	1.79	1.77	1.74	1.71	1.68	1.64	1.61
21	2.96	2.57	2.36	2.23	2.14	2.08	2.02	1.98	1.95	1.92	1.87	1.83	1.78	1.75	1.72	1.69	1.66	1.62	1.59
22	2.95	2.56	2.35	2.22	2.13	2.06	2.01	1.97	1.93	1.90	1.86	1.81	1.76	1.73	1.70	1.67	1.64	1.60	1.57
23	2.94	2.55	2.34	2.21	2.11	2.05	1.99	1.95	1.92	1.89	1.84	1.80	1.74	1.72	1.69	1.66	1.62	1.59	1.55
24	2.93	2.54	2.33	2.19	2.10	2.04	1.98	1.94	1.91	1.88	1.83	1.78	1.73	1.70	1.67	1.64	1.61	1.57	1.53
25	2.92	2.53	2.32	2.18	2.09	2.02	1.97	1.93	1.89	1.87	1.82	1.77	1.72	1.69	1.66	1.63	1.59	1.56	1.52
26	2.91	2.52	2.31	2.17	2.08	2.01	1.96	1.92	1.88	1.86	1.81	1.76	1.71	1.68	1.65	1.61	1.58	1.54	1.50
27	2.90	2.51	2.30	2.17	2.07	2.00	1.95	1.91	1.87	1.85	1.80	1.75	1.70	1.67	1.64	1.60	1.57	1.53	1.49
28	2.89	2.50	2.29	2.16	2.06	2.00	1.94	1.90	1.87	1.84	1.79	1.74	1.69	1.66	1.63	1.59	1.56	1.52	1.48
29	2.89	2.50	2.28	2.15	2.06	1.99	1.93	1.89	1.86	1.83	1.78	1.73	1.68	1.65	1.62	1.58	1.55	1.51	1.47
30	2.88	2.49	2.28	2.14	2.05	1.98	1.93	1.88	1.85	1.82	1.77	1.72	1.67	1.64	1.61	1.57	1.54	1.50	1.46
40	2.84	2.44	2.23	2.09	2.00	1.93	1.87	1.83	1.79	1.76	1.71	1.66	1.61	1.57	1.54	1.51	1.47	1.42	1.38
60	2.79	2.39	2.18	2.04	1.95	1.87	1.82	1.77	1.74	1.71	1.66	1.60	1.54	1.51	1.48	1.44	1.40	1.35	1.29
120	2.75	2.35	2.13	1.99	1.90	1.82	1.77	1.72	1.68	1.65	1.60	1.55	1.48	1.45	1.41	1.37	1.32	1.26	1.19
∞	2.71	2.30	2.08	1.94	1.85	1.77	1.72	1.67	1.63	1.60	1.55	1.49	1.42	1.38	1.34	1.30	1.24	1.17	1.00

$F = \dfrac{s_1^2}{s_2^2} = \dfrac{S_1}{m} \bigg/ \dfrac{S_2}{n}$, where $s_1^2 = S_1/m$ and $s_2^2 = S_2/n$ are independent mean squares estimating a common variance σ^2 and based on m and n degrees of freedom, respectively.

$$F(F) = \int_0^F \frac{\Gamma\left(\frac{m+n}{2}\right)}{\Gamma\left(\frac{m}{2}\right)\Gamma\left(\frac{n}{2}\right)} m^{\frac{m}{2}} n^{\frac{n}{2}} x^{\frac{m}{2}-1} (n+mx)^{-\frac{m+n}{2}} dx = .95$$

n \ m	1	2	3	4	5	6	7	8	9	10	12	15	20	24	30	40	60	120	∞
1	161.4	199.5	215.7	224.6	230.2	234.0	236.8	238.9	240.5	241.9	243.9	245.9	248.0	249.1	250.1	251.1	252.2	253.3	254.3
2	18.51	19.00	19.16	19.25	19.30	19.33	19.35	19.37	19.38	19.40	19.41	19.43	19.45	19.45	19.46	19.47	19.48	19.49	19.50
3	10.13	9.55	9.28	9.12	9.01	8.94	8.89	8.85	8.81	8.79	8.74	8.70	8.66	8.64	8.62	8.59	8.57	8.55	8.53
4	7.71	6.94	6.59	6.39	6.26	6.16	6.09	6.04	6.00	5.96	5.91	5.86	5.80	5.77	5.75	5.72	5.69	5.66	5.63
5	6.61	5.79	5.41	5.19	5.05	4.95	4.88	4.82	4.77	4.74	4.68	4.62	4.56	4.53	4.50	4.46	4.43	4.40	4.36
6	5.99	5.14	4.76	4.53	4.39	4.28	4.21	4.15	4.10	4.06	4.00	3.94	3.87	3.84	3.81	3.77	3.74	3.70	3.67
7	5.59	4.74	4.35	4.12	3.97	3.87	3.79	3.73	3.68	3.64	3.57	3.51	3.44	3.41	3.38	3.34	3.30	3.27	3.23
8	5.32	4.46	4.07	3.84	3.69	3.58	3.50	3.44	3.39	3.35	3.28	3.22	3.15	3.12	3.08	3.04	3.01	2.97	2.93
9	5.12	4.26	3.86	3.63	3.48	3.37	3.29	3.23	3.18	3.14	3.07	3.01	2.94	2.90	2.86	2.83	2.79	2.75	2.71
10	4.96	4.10	3.71	3.48	3.33	3.22	3.14	3.07	3.02	2.98	2.91	2.85	2.77	2.74	2.70	2.66	2.62	2.58	2.54
11	4.84	3.98	3.59	3.36	3.20	3.09	3.01	2.95	2.90	2.85	2.79	2.72	2.65	2.61	2.57	2.53	2.49	2.45	2.40
12	4.75	3.89	3.49	3.26	3.11	3.00	2.91	2.85	2.80	2.75	2.69	2.62	2.54	2.51	2.47	2.43	2.38	2.34	2.30
13	4.67	3.81	3.41	3.18	3.03	2.92	2.83	2.77	2.71	2.67	2.60	2.53	2.46	2.42	2.38	2.34	2.30	2.25	2.21
14	4.60	3.74	3.34	3.11	2.96	2.85	2.76	2.70	2.65	2.60	2.53	2.46	2.39	2.35	2.31	2.27	2.22	2.18	2.13
15	4.54	3.68	3.29	3.06	2.90	2.79	2.71	2.64	2.59	2.54	2.48	2.40	2.33	2.29	2.25	2.20	2.16	2.11	2.07
16	4.49	3.63	3.24	3.01	2.85	2.74	2.66	2.59	2.54	2.49	2.42	2.35	2.28	2.24	2.19	2.15	2.11	2.06	2.01
17	4.45	3.59	3.20	2.96	2.81	2.70	2.61	2.55	2.49	2.45	2.38	2.31	2.23	2.19	2.15	2.10	2.06	2.01	1.96
18	4.41	3.55	3.16	2.93	2.77	2.66	2.58	2.51	2.46	2.41	2.34	2.27	2.19	2.15	2.11	2.06	2.02	1.97	1.92
19	4.38	3.52	3.13	2.90	2.74	2.63	2.54	2.48	2.42	2.38	2.31	2.23	2.16	2.11	2.07	2.03	1.98	1.93	1.88
20	4.35	3.49	3.10	2.87	2.71	2.60	2.51	2.45	2.39	2.35	2.28	2.20	2.12	2.08	2.04	1.99	1.95	1.90	1.84
21	4.32	3.47	3.07	2.84	2.68	2.57	2.49	2.42	2.37	2.32	2.25	2.18	2.10	2.05	2.01	1.96	1.92	1.87	1.81
22	4.30	3.44	3.05	2.82	2.66	2.55	2.46	2.40	2.34	2.30	2.23	2.15	2.07	2.03	1.98	1.94	1.89	1.84	1.78
23	4.28	3.42	3.03	2.80	2.64	2.53	2.44	2.37	2.32	2.27	2.20	2.13	2.05	2.01	1.96	1.91	1.86	1.81	1.76
24	4.26	3.40	3.01	2.78	2.62	2.51	2.42	2.36	2.30	2.25	2.18	2.11	2.03	1.98	1.94	1.89	1.84	1.79	1.73
25	4.24	3.39	2.99	2.76	2.60	2.49	2.40	2.34	2.28	2.24	2.16	2.09	2.01	1.96	1.92	1.87	1.82	1.77	1.71
26	4.23	3.37	2.98	2.74	2.59	2.47	2.39	2.32	2.27	2.22	2.15	2.07	1.99	1.95	1.90	1.85	1.80	1.75	1.69
27	4.21	3.35	2.96	2.73	2.57	2.46	2.37	2.31	2.25	2.20	2.13	2.06	1.97	1.93	1.88	1.84	1.79	1.73	1.67
28	4.20	3.34	2.95	2.71	2.56	2.45	2.36	2.29	2.24	2.19	2.12	2.04	1.96	1.91	1.87	1.82	1.77	1.71	1.65
29	4.18	3.33	2.93	2.70	2.55	2.43	2.35	2.28	2.22	2.18	2.10	2.03	1.94	1.90	1.85	1.81	1.75	1.70	1.64
30	4.17	3.32	2.92	2.69	2.53	2.42	2.33	2.27	2.21	2.16	2.09	2.01	1.93	1.89	1.84	1.79	1.74	1.68	1.62
40	4.08	3.23	2.84	2.61	2.45	2.34	2.25	2.18	2.12	2.08	2.00	1.92	1.84	1.79	1.74	1.69	1.64	1.58	1.51
60	4.00	3.15	2.76	2.53	2.37	2.25	2.17	2.10	2.04	1.99	1.92	1.84	1.75	1.70	1.65	1.59	1.53	1.47	1.39
120	3.92	3.07	2.68	2.45	2.29	2.17	2.09	2.02	1.96	1.91	1.83	1.75	1.66	1.61	1.55	1.50	1.43	1.35	1.25
∞	3.84	3.00	2.60	2.37	2.21	2.10	2.01	1.94	1.88	1.83	1.75	1.67	1.57	1.52	1.46	1.39	1.32	1.22	1.00

$F = \dfrac{s_1^2}{s_2^2} = \dfrac{S_1}{m} \bigg/ \dfrac{S_2}{n}$, where $s_1^2 = S_1/m$ and $s_2^2 = S_2/n$ are independent mean squares estimating a common variance σ^2 and based on m and n degrees of freedom, respectively.

$$F(F) = \int_0^F \frac{\Gamma\left(\frac{m+n}{2}\right)}{\Gamma\left(\frac{m}{2}\right)\Gamma\left(\frac{n}{2}\right)} m^{\frac{m}{2}} n^{\frac{n}{2}} x^{\frac{m}{2}-1} (n+mx)^{-\frac{m+n}{2}} \, dx = .975$$

m \ n	1	2	3	4	5	6	7	8	9	10	12	15	20	24	30	40	60	120	∞
1	647.8	799.5	864.2	899.6	921.8	937.1	948.2	956.7	963.3	968.6	976.7	984.9	993.1	997.2	1001	1006	1010	1014	1018
2	38.51	39.00	39.17	39.25	39.30	39.33	39.36	39.37	39.39	39.40	39.41	39.43	39.45	39.46	39.46	39.47	39.48	39.49	39.50
3	17.44	16.04	15.44	15.10	14.88	14.73	14.62	14.54	14.47	14.42	14.34	14.25	14.17	14.12	14.08	14.04	13.99	13.95	13.90
4	12.22	10.65	9.98	9.60	9.36	9.20	9.07	8.98	8.90	8.84	8.75	8.66	8.56	8.51	8.46	8.41	8.36	8.31	8.26
5	10.01	8.43	7.76	7.39	7.15	6.98	6.85	6.76	6.68	6.62	6.52	6.43	6.33	6.28	6.23	6.18	6.12	6.07	6.02
6	8.81	7.26	6.60	6.23	5.99	5.82	5.70	5.60	5.52	5.46	5.37	5.27	5.17	5.12	5.07	5.01	4.96	4.90	4.85
7	8.07	6.54	5.89	5.52	5.29	5.12	4.99	4.90	4.82	4.76	4.67	4.57	4.47	4.42	4.36	4.31	4.25	4.20	4.14
8	7.57	6.06	5.42	5.05	4.82	4.65	4.53	4.43	4.36	4.30	4.20	4.10	4.00	3.95	3.89	3.84	3.78	3.73	3.67
9	7.21	5.71	5.08	4.72	4.48	4.32	4.20	4.10	4.03	3.96	3.87	3.77	3.67	3.61	3.56	3.51	3.45	3.39	3.33
10	6.94	5.46	4.83	4.47	4.24	4.07	3.95	3.85	3.78	3.72	3.62	3.52	3.42	3.37	3.31	3.26	3.20	3.14	3.08
11	6.72	5.26	4.63	4.28	4.04	3.88	3.76	3.66	3.59	3.53	3.43	3.33	3.23	3.17	3.12	3.06	3.00	2.94	2.88
12	6.55	5.10	4.47	4.12	3.89	3.73	3.61	3.51	3.44	3.37	3.28	3.18	3.07	3.02	2.96	2.91	2.85	2.79	2.72
13	6.41	4.97	4.35	4.00	3.77	3.60	3.48	3.39	3.31	3.25	3.15	3.05	2.95	2.89	2.84	2.78	2.72	2.66	2.60
14	6.30	4.86	4.24	3.89	3.66	3.50	3.38	3.29	3.21	3.15	3.05	2.95	2.84	2.79	2.73	2.67	2.61	2.55	2.49
15	6.20	4.77	4.15	3.80	3.58	3.41	3.29	3.20	3.12	3.06	2.96	2.86	2.76	2.70	2.64	2.59	2.52	2.46	2.40
16	6.12	4.69	4.08	3.73	3.50	3.34	3.22	3.12	3.05	2.99	2.89	2.79	2.68	2.63	2.57	2.51	2.45	2.38	2.32
17	6.04	4.62	4.01	3.66	3.44	3.28	3.16	3.06	2.98	2.92	2.82	2.72	2.62	2.56	2.50	2.44	2.38	2.32	2.25
18	5.98	4.56	3.95	3.61	3.38	3.22	3.10	3.01	2.93	2.87	2.77	2.67	2.56	2.50	2.44	2.38	2.32	2.26	2.19
19	5.92	4.51	3.90	3.56	3.33	3.17	3.05	2.96	2.88	2.82	2.72	2.62	2.51	2.45	2.39	2.33	2.27	2.20	2.13
20	5.87	4.46	3.86	3.51	3.29	3.13	3.01	2.91	2.84	2.77	2.68	2.57	2.46	2.41	2.35	2.29	2.22	2.16	2.09
21	5.83	4.42	3.82	3.48	3.25	3.09	2.97	2.87	2.80	2.73	2.64	2.53	2.42	2.37	2.31	2.25	2.18	2.11	2.04
22	5.79	4.38	3.78	3.44	3.22	3.05	2.93	2.84	2.76	2.70	2.60	2.50	2.39	2.33	2.27	2.21	2.14	2.08	2.00
23	5.75	4.35	3.75	3.41	3.18	3.02	2.90	2.81	2.73	2.67	2.57	2.47	2.36	2.30	2.24	2.18	2.11	2.04	1.97
24	5.72	4.32	3.72	3.38	3.15	2.99	2.87	2.78	2.70	2.64	2.54	2.44	2.33	2.27	2.21	2.15	2.08	2.01	1.94
25	5.69	4.29	3.69	3.35	3.13	2.97	2.85	2.75	2.68	2.61	2.51	2.41	2.30	2.24	2.18	2.12	2.05	1.98	1.91
26	5.66	4.27	3.67	3.33	3.10	2.94	2.82	2.73	2.65	2.59	2.49	2.39	2.28	2.22	2.16	2.09	2.03	1.95	1.88
27	5.63	4.24	3.65	3.31	3.08	2.92	2.80	2.71	2.63	2.57	2.47	2.36	2.25	2.19	2.13	2.07	2.00	1.93	1.85
28	5.61	4.22	3.63	3.29	3.06	2.90	2.78	2.69	2.61	2.55	2.45	2.34	2.23	2.17	2.11	2.05	1.98	1.91	1.83
29	5.59	4.20	3.61	3.27	3.04	2.88	2.76	2.67	2.59	2.53	2.43	2.32	2.21	2.15	2.09	2.03	1.96	1.89	1.81
30	5.57	4.18	3.59	3.25	3.03	2.87	2.75	2.65	2.57	2.51	2.41	2.31	2.20	2.14	2.07	2.01	1.94	1.87	1.79
40	5.42	4.05	3.46	3.13	2.90	2.74	2.62	2.53	2.45	2.39	2.29	2.18	2.07	2.01	1.94	1.88	1.80	1.72	1.64
60	5.29	3.93	3.34	3.01	2.79	2.63	2.51	2.41	2.33	2.27	2.17	2.06	1.94	1.88	1.82	1.74	1.67	1.58	1.48
120	5.15	3.80	3.23	2.89	2.67	2.52	2.39	2.30	2.22	2.16	2.05	1.94	1.82	1.76	1.69	1.61	1.53	1.43	1.31
∞	5.02	3.69	3.12	2.79	2.57	2.41	2.29	2.19	2.11	2.05	1.94	1.83	1.71	1.64	1.57	1.48	1.39	1.27	1.00

$F = \dfrac{s_1^2}{s_2^2} = \dfrac{S_1}{m} \Big/ \dfrac{S_2}{n}$, where $s_1^2 = S_1/m$ and $s_2^2 = S_2/n$ are independent mean squares estimating a common variance σ^2 and based on m and n degrees of freedom, respectively.

$$F(F) = \int_0^F \frac{\Gamma\left(\frac{m+n}{2}\right)}{\Gamma\left(\frac{m}{2}\right)\Gamma\left(\frac{n}{2}\right)} m^{\frac{m}{2}} n^{\frac{n}{2}} x^{\frac{m}{2}-1} (n+mx)^{-\frac{m+n}{2}} \, dx = .99$$

m \ n	1	2	3	4	5	6	7	8	9	10	12	15	20	24	30	40	60	120	∞
1	4052	4999.5	5403	5625	5764	5859	5928	5982	6022	6056	6106	6157	6209	6235	6261	6287	6313	6339	6366
2	98.50	99.00	99.17	99.25	99.30	99.33	99.36	99.37	99.39	99.40	99.42	99.43	99.45	99.46	99.47	99.47	99.48	99.49	99.50
3	34.12	30.82	29.46	28.71	28.24	27.91	27.67	27.49	27.35	27.23	27.05	26.87	26.69	26.60	26.50	26.41	26.32	26.22	26.13
4	21.20	18.00	16.69	15.98	15.52	15.21	14.98	14.80	14.66	14.55	14.37	14.20	14.02	13.93	13.84	13.75	13.65	13.56	13.46
5	16.26	13.27	12.06	11.39	10.97	10.67	10.46	10.29	10.16	10.05	9.89	9.72	9.55	9.47	9.38	9.29	9.20	9.11	9.02
6	13.75	10.92	9.78	9.15	8.75	8.47	8.26	8.10	7.98	7.87	7.72	7.56	7.40	7.31	7.23	7.14	7.06	6.97	6.88
7	12.25	9.55	8.45	7.85	7.46	7.19	6.99	6.84	6.72	6.62	6.47	6.31	6.16	6.07	5.99	5.91	5.82	5.74	5.65
8	11.26	8.65	7.59	7.01	6.63	6.37	6.18	6.03	5.91	5.81	5.67	5.52	5.36	5.28	5.20	5.12	5.03	4.95	4.86
9	10.56	8.02	6.99	6.42	6.06	5.80	5.61	5.47	5.35	5.26	5.11	4.96	4.81	4.73	4.65	4.57	4.48	4.40	4.31
10	10.04	7.56	6.55	5.99	5.64	5.39	5.20	5.06	4.94	4.85	4.71	4.56	4.41	4.33	4.25	4.17	4.08	4.00	3.91
11	9.65	7.21	6.22	5.67	5.32	5.07	4.89	4.74	4.63	4.54	4.40	4.25	4.10	4.02	3.94	3.86	3.78	3.69	3.60
12	9.33	6.93	5.95	5.41	5.06	4.82	4.64	4.50	4.39	4.30	4.16	4.01	3.86	3.78	3.70	3.62	3.54	3.45	3.36
13	9.07	6.70	5.74	5.21	4.86	4.62	4.44	4.30	4.19	4.10	3.96	3.82	3.66	3.59	3.51	3.43	3.34	3.25	3.17
14	8.86	6.51	5.56	5.04	4.69	4.46	4.28	4.14	4.03	3.94	3.80	3.66	3.51	3.43	3.35	3.27	3.18	3.09	3.00
15	8.68	6.36	5.42	4.89	4.56	4.32	4.14	4.00	3.89	3.80	3.67	3.52	3.37	3.29	3.21	3.13	3.05	2.96	2.87
16	8.53	6.23	5.29	4.77	4.44	4.20	4.03	3.89	3.78	3.69	3.55	3.41	3.26	3.18	3.10	3.02	2.93	2.84	2.75
17	8.40	6.11	5.18	4.67	4.34	4.10	3.93	3.79	3.68	3.59	3.46	3.31	3.16	3.08	3.00	2.92	2.83	2.75	2.65
18	8.29	6.01	5.09	4.58	4.25	4.01	3.84	3.71	3.60	3.51	3.37	3.23	3.08	3.00	2.92	2.84	2.75	2.66	2.57
19	8.18	5.93	5.01	4.50	4.17	3.94	3.77	3.63	3.52	3.43	3.30	3.15	3.00	2.92	2.84	2.76	2.67	2.58	2.49
20	8.10	5.85	4.94	4.43	4.10	3.87	3.70	3.56	3.46	3.37	3.23	3.09	2.94	2.86	2.78	2.69	2.61	2.52	2.42
21	8.02	5.78	4.87	4.37	4.04	3.81	3.64	3.51	3.40	3.31	3.17	3.03	2.88	2.80	2.72	2.64	2.55	2.46	2.36
22	7.95	5.72	4.82	4.31	3.99	3.76	3.59	3.45	3.35	3.26	3.12	2.98	2.83	2.75	2.67	2.58	2.50	2.40	2.31
23	7.88	5.66	4.76	4.26	3.94	3.71	3.54	3.41	3.30	3.21	3.07	2.93	2.78	2.70	2.62	2.54	2.45	2.35	2.26
24	7.82	5.61	4.72	4.22	3.90	3.67	3.50	3.36	3.26	3.17	3.03	2.89	2.74	2.66	2.58	2.49	2.40	2.31	2.21
25	7.77	5.57	4.68	4.18	3.85	3.63	3.46	3.32	3.22	3.13	2.99	2.85	2.70	2.62	2.54	2.45	2.36	2.27	2.17
26	7.72	5.53	4.64	4.14	3.82	3.59	3.42	3.29	3.18	3.09	2.96	2.81	2.66	2.58	2.50	2.42	2.33	2.23	2.13
27	7.68	5.49	4.60	4.11	3.78	3.56	3.39	3.26	3.15	3.06	2.93	2.78	2.63	2.55	2.47	2.38	2.29	2.20	2.10
28	7.64	5.45	4.57	4.07	3.75	3.53	3.36	3.23	3.12	3.03	2.90	2.75	2.60	2.52	2.44	2.35	2.26	2.17	2.06
29	7.60	5.42	4.54	4.04	3.73	3.50	3.33	3.20	3.09	3.00	2.87	2.73	2.57	2.49	2.41	2.33	2.23	2.14	2.03
30	7.56	5.39	4.51	4.02	3.70	3.47	3.30	3.17	3.07	2.98	2.84	2.70	2.55	2.47	2.39	2.30	2.21	2.11	2.01
40	7.31	5.18	4.31	3.83	3.51	3.29	3.12	2.99	2.89	2.80	2.66	2.52	2.37	2.29	2.20	2.11	2.02	1.92	1.80
60	7.08	4.98	4.13	3.65	3.34	3.12	2.95	2.82	2.72	2.63	2.50	2.35	2.20	2.12	2.03	1.94	1.84	1.73	1.60
120	6.85	4.79	3.95	3.48	3.17	2.96	2.79	2.66	2.56	2.47	2.34	2.19	2.03	1.95	1.86	1.76	1.66	1.53	1.38
∞	6.63	4.61	3.78	3.32	3.02	2.80	2.64	2.51	2.41	2.32	2.18	2.04	1.88	1.79	1.70	1.59	1.47	1.32	1.00

$F = \dfrac{s_1^2}{s_2^2} = \dfrac{S_1}{m} \Big/ \dfrac{S_2}{n}$, where $s_1^2 = S_1/m$ and $s_2^2 = S_2/n$ are independent mean squares estimating a common variance σ^2 and based on m and n degrees of freedom, respectively.

$$F(F) = \int_0^F \frac{\Gamma\left(\frac{m+n}{2}\right)}{\Gamma\left(\frac{m}{2}\right)\Gamma\left(\frac{n}{2}\right)} m^{\frac{m}{2}} n^{\frac{n}{2}} x^{\frac{m}{2}-1} (n+mx)^{-\frac{m+n}{2}}\, dx = .995$$

n \ m	1	2	3	4	5	6	7	8	9	10	12	15	20	24	30	40	60	120	∞
1	16211	20000	21615	22500	23056	23437	23715	23925	24091	24224	24426	24630	24836	24940	25044	25148	25253	25359	25465
2	198.5	199.0	199.2	199.2	199.3	199.3	199.4	199.4	199.4	199.4	199.4	199.4	199.4	199.5	199.5	199.5	199.5	199.5	199.5
3	55.55	49.80	47.47	46.19	45.39	44.84	44.43	44.13	43.88	43.69	43.39	43.08	42.78	42.62	42.47	42.31	42.15	41.99	41.83
4	31.33	26.28	24.26	23.15	22.46	21.97	21.62	21.35	21.14	20.97	20.70	20.44	20.17	20.03	19.89	19.75	19.61	19.47	19.32
5	22.78	18.31	16.53	15.56	14.94	14.51	14.20	13.96	13.77	13.62	13.38	13.15	12.90	12.78	12.66	12.53	12.40	12.27	12.14
6	18.63	14.54	12.92	12.03	11.46	11.07	10.79	10.57	10.39	10.25	10.03	9.81	9.59	9.47	9.36	9.24	9.12	9.00	8.88
7	16.24	12.40	10.88	10.05	9.52	9.16	8.89	8.68	8.51	8.38	8.18	7.97	7.75	7.65	7.53	7.42	7.31	7.19	7.08
8	14.69	11.04	9.60	8.81	8.30	7.95	7.69	7.50	7.34	7.21	7.01	6.81	6.61	6.50	6.40	6.29	6.18	6.06	5.95
9	13.61	10.11	8.72	7.96	7.47	7.13	6.88	6.69	6.54	6.42	6.23	6.03	5.83	5.73	5.62	5.52	5.41	5.30	5.19
10	12.83	9.43	8.08	7.34	6.87	6.54	6.30	6.12	5.97	5.85	5.66	5.47	5.27	5.17	5.07	4.97	4.86	4.75	4.64
11	12.23	8.91	7.60	6.88	6.42	6.10	5.86	5.68	5.54	5.42	5.24	5.05	4.86	4.76	4.65	4.55	4.44	4.34	4.23
12	11.75	8.51	7.23	6.52	6.07	5.76	5.52	5.35	5.20	5.09	4.91	4.72	4.53	4.43	4.33	4.23	4.12	4.01	3.90
13	11.37	8.19	6.93	6.23	5.79	5.48	5.25	5.08	4.94	4.82	4.64	4.46	4.27	4.17	4.07	3.97	3.87	3.76	3.65
14	11.06	7.92	6.68	6.00	5.56	5.26	5.03	4.86	4.72	4.60	4.43	4.25	4.06	3.96	3.86	3.76	3.66	3.55	3.44
15	10.80	7.70	6.48	5.80	5.37	5.07	4.85	4.67	4.54	4.42	4.25	4.07	3.88	3.79	3.69	3.58	3.48	3.37	3.26
16	10.58	7.51	6.30	5.64	5.21	4.91	4.69	4.52	4.38	4.27	4.10	3.92	3.73	3.64	3.54	3.44	3.33	3.22	3.11
17	10.38	7.35	6.16	5.50	5.07	4.78	4.56	4.39	4.25	4.14	3.97	3.79	3.61	3.51	3.41	3.31	3.21	3.10	2.98
18	10.22	7.21	6.03	5.37	4.96	4.66	4.44	4.28	4.14	4.03	3.86	3.68	3.50	3.40	3.30	3.20	3.10	2.99	2.87
19	10.07	7.09	5.92	5.27	4.85	4.56	4.34	4.18	4.04	3.93	3.76	3.59	3.40	3.31	3.21	3.11	3.00	2.89	2.78
20	9.94	6.99	5.82	5.17	4.76	4.47	4.26	4.09	3.96	3.85	3.68	3.50	3.32	3.22	3.12	3.02	2.92	2.81	2.69
21	9.83	6.89	5.73	5.09	4.68	4.39	4.18	4.01	3.88	3.77	3.60	3.43	3.24	3.15	3.05	2.95	2.84	2.73	2.61
22	9.73	6.81	5.65	5.02	4.61	4.32	4.11	3.94	3.81	3.70	3.54	3.36	3.18	3.08	2.98	2.88	2.77	2.66	2.55
23	9.63	6.73	5.58	4.95	4.54	4.26	4.05	3.88	3.75	3.64	3.47	3.30	3.12	3.02	2.92	2.82	2.71	2.60	2.48
24	9.55	6.66	5.52	4.89	4.49	4.20	3.99	3.83	3.69	3.59	3.42	3.25	3.06	2.97	2.87	2.77	2.66	2.55	2.43
25	9.48	6.60	5.46	4.84	4.43	4.15	3.94	3.78	3.64	3.54	3.37	3.20	3.01	2.92	2.82	2.72	2.61	2.50	2.38
26	9.41	6.54	5.41	4.79	4.38	4.10	3.89	3.73	3.60	3.49	3.33	3.15	2.97	2.87	2.77	2.67	2.56	2.45	2.33
27	9.34	6.49	5.36	4.74	4.34	4.06	3.85	3.69	3.56	3.45	3.28	3.11	2.93	2.83	2.73	2.63	2.52	2.41	2.25
28	9.28	6.44	5.32	4.70	4.30	4.02	3.81	3.65	3.52	3.41	3.25	3.07	2.89	2.79	2.69	2.59	2.48	2.37	2.29
29	9.23	6.40	5.28	4.66	4.26	3.98	3.77	3.61	3.48	3.38	3.21	3.04	2.86	2.76	2.66	2.56	2.45	2.33	2.24
30	9.18	6.35	5.24	4.62	4.23	3.95	3.74	3.58	3.45	3.34	3.18	3.01	2.82	2.73	2.63	2.52	2.42	2.30	2.18
40	8.83	6.07	4.98	4.37	3.99	3.71	3.51	3.35	3.22	3.12	2.95	2.78	2.60	2.50	2.40	2.30	2.18	2.06	1.93
60	8.49	5.79	4.73	4.14	3.76	3.49	3.29	3.13	3.01	2.90	2.74	2.57	2.39	2.29	2.19	2.08	1.96	1.83	1.69
120	8.18	5.54	4.50	3.92	3.55	3.28	3.09	2.93	2.81	2.71	2.54	2.37	2.19	2.09	1.98	1.87	1.75	1.61	1.43
∞	7.88	5.30	4.28	3.72	3.35	3.09	2.90	2.74	2.62	2.52	2.36	2.19	2.00	1.90	1.79	1.67	1.53	1.36	1.00

$F = \dfrac{s_1^2}{s_2^2} = \dfrac{S_1}{m} \Big/ \dfrac{S_2}{n}$, where $s_1^2 = S_1/m$ and $s_2^2 = S_2/n$ are independent mean squares estimating a common variance σ^2 and based on m and n degrees of freedom, respectively.

$$F(F) = \int_0^F \frac{\Gamma\left(\frac{m+n}{2}\right)}{\Gamma\left(\frac{m}{2}\right)\Gamma\left(\frac{n}{2}\right)} m^{\frac{m}{2}} n^{\frac{n}{2}} x^{\frac{m}{2}-1} (n+mx)^{-\frac{m+n}{2}}\, dx = .999$$

n \ m	1	2	3	4	5	6	7	8	9	10	12	15	20	24	30	40	60	120	∞
1	4053*	5000*	5404*	5625*	5764*	5859*	5929*	5981*	6023*	6056*	6107*	6158*	6209*	6235*	6261*	6287*	6313*	6340*	6366*
2	998.5	999.0	999.2	999.2	999.3	999.3	999.4	999.4	999.4	999.4	999.4	999.4	999.4	999.5	999.5	999.5	999.5	999.5	999.5
3	167.0	148.5	141.1	137.1	134.6	132.8	131.6	130.6	129.9	129.2	128.3	127.4	126.4	125.9	125.4	125.0	124.5	124.0	123.5
4	74.14	61.25	56.18	53.44	51.71	50.53	49.66	49.00	48.47	48.05	47.41	46.76	46.10	45.77	45.43	45.09	44.75	44.40	44.05
5	47.18	37.12	33.20	31.09	29.75	28.84	28.16	27.64	27.24	26.92	26.42	25.91	25.39	25.14	24.87	24.60	24.33	24.06	23.79
6	35.51	27.00	23.70	21.92	20.81	20.03	19.46	19.03	18.69	18.41	17.99	17.56	17.12	16.89	16.67	16.44	16.21	15.99	15.75
7	29.25	21.69	18.77	17.19	16.21	15.52	15.02	14.63	14.33	14.08	13.71	13.32	12.93	12.73	12.53	12.33	12.12	11.91	11.70
8	25.42	18.49	15.83	14.39	13.49	12.86	12.40	12.04	11.77	11.54	11.19	10.84	10.48	10.30	10.11	9.92	9.73	9.53	9.33
9	22.86	16.39	13.90	12.56	11.71	11.13	10.70	10.37	10.11	9.89	9.57	9.24	8.90	8.72	8.55	8.37	8.19	8.00	7.81
10	21.04	14.91	12.55	11.28	10.48	9.92	9.52	9.20	8.96	8.75	8.45	8.13	7.80	7.64	7.47	7.30	7.12	6.94	6.76
11	19.69	13.81	11.56	10.35	9.58	9.05	8.66	8.35	8.12	7.92	7.63	7.32	7.01	6.85	6.68	6.52	6.35	6.17	6.00
12	18.64	12.97	10.80	9.63	8.89	8.38	8.00	7.71	7.48	7.29	7.00	6.71	6.40	6.25	6.09	5.93	5.76	5.59	5.42
13	17.81	12.31	10.21	9.07	8.35	7.86	7.49	7.21	6.98	6.80	6.52	6.23	5.93	5.78	5.63	5.47	5.30	5.14	4.97
14	17.14	11.78	9.73	8.62	7.92	7.43	7.08	6.80	6.58	6.40	6.13	5.85	5.56	5.41	5.25	5.10	4.94	4.77	4.60
15	16.59	11.34	9.34	8.25	7.57	7.09	6.74	6.47	6.26	6.08	5.81	5.54	5.25	5.10	4.95	4.80	4.64	4.47	4.31
16	16.12	10.97	9.00	7.94	7.27	6.81	6.46	6.19	5.98	5.81	5.55	5.27	4.99	4.85	4.70	4.54	4.39	4.23	4.06
17	15.72	10.66	8.73	7.68	7.02	6.56	6.22	5.96	5.75	5.58	5.32	5.05	4.78	4.63	4.48	4.33	4.18	4.02	3.85
18	15.38	10.39	8.49	7.46	6.81	6.35	6.02	5.76	5.56	5.39	5.13	4.87	4.59	4.45	4.30	4.15	4.00	3.84	3.67
19	15.08	10.16	8.28	7.26	6.62	6.18	5.85	5.59	5.39	5.22	4.97	4.70	4.43	4.29	4.14	3.99	3.84	3.68	3.51
20	14.82	9.95	8.10	7.10	6.46	6.02	5.69	5.44	5.24	5.08	4.82	4.56	4.29	4.15	4.00	3.86	3.70	3.54	3.38
21	14.59	9.77	7.94	6.95	6.32	5.88	5.56	5.31	5.11	4.95	4.70	4.44	4.17	4.03	3.88	3.74	3.58	3.42	3.26
22	14.38	9.61	7.80	6.81	6.19	5.76	5.44	5.19	4.99	4.83	4.58	4.33	4.06	3.92	3.78	3.63	3.48	3.32	3.15
23	14.19	9.47	7.67	6.69	6.08	5.65	5.33	5.09	4.89	4.73	4.48	4.23	3.96	3.82	3.68	3.53	3.38	3.22	3.05
24	14.03	9.34	7.55	6.59	5.98	5.55	5.23	4.99	4.80	4.64	4.39	4.14	3.87	3.74	3.59	3.45	3.29	3.14	2.97
25	13.88	9.22	7.45	6.49	5.88	5.46	5.15	4.91	4.71	4.56	4.31	4.06	3.79	3.66	3.52	3.37	3.22	3.06	2.89
26	13.74	9.12	7.36	6.41	5.80	5.38	5.07	4.83	4.64	4.48	4.24	3.99	3.72	3.59	3.44	3.30	3.15	2.99	2.82
27	13.61	9.02	7.27	6.33	5.73	5.31	5.00	4.76	4.57	4.41	4.17	3.92	3.66	3.52	3.38	3.23	3.08	2.92	2.75
28	13.50	8.93	7.19	6.25	5.66	5.24	4.93	4.69	4.50	4.35	4.11	3.86	3.60	3.46	3.32	3.18	3.02	2.86	2.69
29	13.39	8.85	7.12	6.19	5.59	5.18	4.87	4.64	4.45	4.29	4.05	3.80	3.54	3.41	3.27	3.12	2.97	2.81	2.64
30	13.29	8.77	7.05	6.12	5.53	5.12	4.82	4.58	4.39	4.24	4.00	3.75	3.49	3.36	3.22	3.07	2.92	2.76	2.59
40	12.61	8.25	6.60	5.70	5.13	4.73	4.44	4.21	4.02	3.87	3.64	3.40	3.15	3.01	2.87	2.73	2.57	2.41	2.23
60	11.97	7.76	6.17	5.31	4.76	4.37	4.09	3.87	3.69	3.54	3.31	3.08	2.83	2.69	2.55	2.41	2.25	2.08	1.89
120	11.38	7.32	5.79	4.95	4.42	4.04	3.77	3.55	3.38	3.24	3.02	2.78	2.53	2.40	2.26	2.11	1.95	1.76	1.54
∞	10.83	6.91	5.42	4.62	4.10	3.74	3.47	3.27	3.10	2.96	2.74	2.51	2.27	2.13	1.99	1.84	1.66	1.45	1.00

* Multiply these entries by 100.

Index

The most efficient way to use this index is to look for the pertinent *property* (e.g., vapor pressure, entropy), *process* (e.g., disposal of chemicals, calibration), or *general concept* (e.g., units, radiation). Most primary entries are subdivided into several secondary entries, e.g., under heat capacity there are 17 secondary entries such as air, metals, water, etc. Primary entries will be found for certain *classes of substances*, such as alloys, elements, organic compounds, refrigerants, semiconductors, etc. Primary entries are also given for the individual chemical elements and for a few compounds such as water and carbon dioxide. However, only the most important tables are listed under these substances. Therefore, the user will find in most cases that it is best to look first for the property of interest, then examine the table or tables that are referenced.

The reference given for each index term is the inclusive pages of the pertinent table (e.g., **8**-45 to 55). The introduction to each table describes the method of ordering the substances within that table.

The Editor would be grateful for comments and suggestions on this index.

INDEX

STANDARD ATOMIC WEIGHTS (1995)

At. no.	Name	Symbol	Atomic weight	At. no.	Name	Symbol	Atomic weight
1	Hydrogen	H	1.00794(7)	55	Cesium	Cs	132.90545(2)
2	Helium	He	4.002602(2)	56	Barium	Ba	137.327(7)
3	Lithium	Li	6.941(2)	57	Lanthanum	La	138.9055(2)
4	Beryllium	Be	9.012182(3)	58	Cerium	Ce	140.116(1)
5	Boron	B	10.811(7)	59	Praseodymium	Pr	140.90765(2)
6	Carbon	C	12.0107(8)	60	Neodymium	Nd	144.24(3)
7	Nitrogen	N	14.00674(7)	61	Promethium	Pm	[145]
8	Oxygen	O	15.9994(3)	62	Samarium	Sm	150.36(3)
9	Fluorine	F	18.9984032(5)	63	Europium	Eu	151.964(1)
10	Neon	Ne	20.1797(6)	64	Gadolinium	Gd	157.25(3)
11	Sodium	Na	22.989770(2)	65	Terbium	Tb	158.92534(2)
12	Magnesium	Mg	24.3050(6)	66	Dysprosium	Dy	162.50(3)
13	Aluminum	Al	26.981538(2)	67	Holmium	Ho	164.93032(2)
14	Silicon	Si	28.0855(3)	68	Erbium	Er	167.26(3)
15	Phosphorus	P	30.973761(2)	69	Thulium	Tm	168.93421(2)
16	Sulfur	S	32.066(6)	70	Ytterbium	Yb	173.04(3)
17	Chlorine	Cl	35.4527(9)	71	Lutetium	Lu	174.967(1)
18	Argon	Ar	39.948(1)	72	Hafnium	Hf	178.49(2)
19	Potassium	K	39.0983(1)	73	Tantalum	Ta	180.9479(1)
20	Calcium	Ca	40.078(4)	74	Tungsten	W	183.84(1)
21	Scandium	Sc	44.955910(8)	75	Rhenium	Re	186.207(1)
22	Titanium	Ti	47.867(1)	76	Osmium	Os	190.23(3)
23	Vanadium	V	50.9415(1)	77	Iridium	Ir	192.217(3)
24	Chromium	Cr	51.9961(6)	78	Platinum	Pt	195.078(2)
25	Manganese	Mn	54.938049(9)	79	Gold	Au	196.96655(2)
26	Iron	Fe	55.845(2)	80	Mercury	Hg	200.59(2)
27	Cobalt	Co	58.933200(9)	81	Thallium	Tl	204.3833(2)
28	Nickel	Ni	58.6934(2)	82	Lead	Pb	207.2(1)
29	Copper	Cu	63.546(3)	83	Bismuth	Bi	208.98038(2)
30	Zinc	Zn	65.39(2)	84	Polonium	Po	[209]
31	Gallium	Ga	69.723(1)	85	Astatine	At	[210]
32	Germanium	Ge	72.61(2)	86	Radon	Rn	[222]
33	Arsenic	As	74.92160(2)	87	Francium	Fr	[223]
34	Selenium	Se	78.96(3)	88	Radium	Ra	[226]
35	Bromine	Br	79.904(1)	89	Actinium	Ac	[227]
36	Krypton	Kr	83.80(1)	90	Thorium	Th	232.0381(1)
37	Rubidium	Rb	85.4678(3)	91	Protactinium	Pa	231.03588(2)
38	Strontium	Sr	87.62(1)	92	Uranium	U	238.0289(1)
39	Yttrium	Y	88.90585(2)	93	Neptunium	Np	[237]
40	Zirconium	Zr	91.224(2)	94	Plutonium	Pu	[244]
41	Niobium	Nb	92.90638(2)	95	Americium	Am	[243]
42	Molybdenum	Mo	95.94(1)	96	Curium	Cm	[247]
43	Technetium	Tc	[98]	97	Berkelium	Bk	[247]
44	Ruthenium	Ru	101.07(2)	98	Californium	Cf	[251]
45	Rhodium	Rh	102.90550(2)	99	Einsteinium	Es	[252]
46	Palladium	Pd	106.42(1)	100	Fermium	Fm	[257]
47	Silver	Ag	107.8682(2)	101	Mendelevium	Md	[258]
48	Cadmium	Cd	112.411(8)	102	Nobelium	No	[259]
49	Indium	In	114.818(3)	103	Lawrencium	Lr	[262]
50	Tin	Sn	118.710(7)	104	Rutherfordium	Rf	[261]
51	Antimony	Sb	121.760(1)	105	Dubnium	Db	[262]
52	Tellurium	Te	127.60(3)	106	Seaborgium	Sg	[263]
53	Iodine	I	126.90447(3)	107	Bohrium	Bh	[262]
54	Xenon	Xe	131.29(2)	108	Hassium	Hs	[265]
				109	Meitnerium	Mt	[266]

Numbers in brackets are the mass numbers of the longest-lived isotope of elements for which a standard atomic weight cannot be defined.